2023 IEEE 50th Photovoltaic Specialists Conference (PVSC 2023)

San Juan, Puerto Rico, USA
11-16 June 2023

Pages 1-554

IEEE Catalog Number: CFP23PSC-POD
ISBN: 978-1-6654-6060-6

**Copyright © 2023 by the Institute of Electrical and Electronics Engineers, Inc.
All Rights Reserved**

Copyright and Reprint Permissions: Abstracting is permitted with credit to the source. Libraries are permitted to photocopy beyond the limit of U.S. copyright law for private use of patrons those articles in this volume that carry a code at the bottom of the first page, provided the per-copy fee indicated in the code is paid through Copyright Clearance Center, 222 Rosewood Drive, Danvers, MA 01923.

For other copying, reprint or republication permission, write to IEEE Copyrights Manager, IEEE Service Center, 445 Hoes Lane, Piscataway, NJ 08854. All rights reserved.

*** This is a print representation of what appears in the IEEE Digital Library. Some format issues inherent in the e-media version may also appear in this print version.

IEEE Catalog Number: CFP23PSC-POD
ISBN (Print-On-Demand):   978-1-6654-6060-6
ISBN (Online):   978-1-6654-6059-0

Additional Copies of This Publication Are Available From:

Curran Associates, Inc
57 Morehouse Lane
Red Hook, NY 12571 USA
Phone: (845) 758-0400
Fax: (845) 758-2633
E-mail: curran@proceedings.com
Web: www.proceedings.com

TABLE OF CONTENTS

3-Terminal Perovskite/Silicon Tandem Modules: A Dead End or a Bright Future of Tandem Based Photovoltaics 1
Miha Kikelj, Laurie-Lou Senaud, Jonas Geissbühler, Damien Lachenal, Derk Baetzner, Benjamin Lipovšek, Marko Topic, Christophe Ballif, Quentin Jeangros, Bertrand Paviet-Salomon

Temperature-Dependent Performance of Ultra-Thin Silicon Heterojunction Solar Cells for Space Applications 5
Anh Huy Tuan Le, Pradeep Balaji, André Augusto, Ziv Hameiri

Utilizing a Soft IZO Sputtering Process to Contact Buffer-Free Semitransparent Perovskite Pin Solar Cells 6
Roland Clausing, Annika Raugewitz, Benjamin Grimm, Marvin Diederich, Tobias Wietler, Felix Haase, Rolf Brendel, Robby Peibst

Effect of Surface Morphology on GaAs Solar Cells Grown on Planarized Spalled (100) GaAs Substrates 10
Anna K. Braun, Jacob T. Boyer, Kevin L. Schulte, John Simon, Steve W. Johnston, Myles A. Steiner, Corinne E. Packard, Aaron J. Ptak

Ultra-Light Environmental Protection for Solar Arrays in Space 11
Pilar Espinet-Gonzalez, Alexandra H. Teodor, Jann A. Grovogui, Yao Y. Lao

Highly Crystalline In2S3:V Thin Films Epitaxially Grown on Sapphire Substrates: A Potential Candidate for Intermediate Band Solar Cells 12
Tanja Jawinski, Chris Sturm, Roland Clausing, Heiko Kempa, Stefan Lange, Roland Scheer, Marius Grundmann, Holger Von Wenckstern

A Model-Free Approach for Estimating Service Transformer Capacity Using Residential Smart Meter Data 13
Joseph A. Azzolini, Matthew J. Reno, Jubair Yusuf

Bifacial PV Fed Electrolysis for Green Hydrogen Generation and Cofiring Hydrogen in an Aeroderivative Gas Turbine 14
Nicholas Pilot, Jim Harper

Reverse-Bias Testing of Perovskite Cells to Inform Bypass-Diode Design 15
Daniel Martinez, Dana Kern, Giles Eperon

Improving Deep Learning-Based Defect Classification in Solar Cells Using Conformal Prediction 16
Vitus B. Thomsen, Claire Mantel, Gisele A. Dos Reis Benatto, Allan P. Engsig-Karup, Søren Forchhammer

Off-The-Shelf Small Scale Photovoltaic Systems for Puerto Rico Sustainable Farms: Assisting Those Who Help Others 22
Catherine Lahood, Sarah Sirkisoon, Charles M. Fortmann

Flexible Photonic Cooler Based on Multi-Stacked Thin Films IR Filters with Anti-Dust Capability for PV-Desert Environment Applications 27
Brahim Aïssa, Mohammad I. Hossain

Loss Analysis and Performance Optimization Pathways of 729-MV V_{OC} Si Solar Cells with Poly-Si on Locally-Etched Dielectric Passivating Contacts 30
  Suchismita Mitra, Caroline Lima Anderson, Matthew Hartenstein, William Nemeth, Matthew Page, San Thiengi, David Young, Sumit Agarwal, Paul Stradins

Cadmium Telluride Accelerator Consortium (CTAC) 35
  Lorelle M. Mansfield, Matthew O. Reese, Michael J. Heben

Screen Printable Copper Pastes for Silicon Solar Cells 38
  Thad Druffel, Ruvini Dharmadasa, Apolo Nambo, Abafriseke Ebong, Sandra Huneycutt, Donald Ital

Optimization of Zinc Oxide Electron Transport Layers for Cs-Based Perovskite Solar Cells 39
  Zhuldyz Yelzhanova, Gaukhar Nigmetova, Gulzhan Bizhanova, Nurzhan Yermekov, Ulan Kashkimbayev, Annie Ng

UV Degradation of Formamidinium-Cesium Lead Halide Perovskite Solar Cells 42
  Kshitiz Dolia, Abasi Abudulimu, Sheng Fu, Tyler Brau, Karissa Jensen, Stephanie L. Moffitt, Randy J. Ellingson, Xiaohong Gu, Zhaoning Song, Yanfa Yan

Efficient Cd(Se, Te) Solar Cells with Cd(O, S, Se, Te) at the Front Interface 46
  Dengbing Li, Sabin Neupane, Sandip Singh Bista, Abasi Abudulimu, Kamala Khanal Subedi, Manoj K. Jamarkattel, Chuanxiao Xiao, Chun-Sheng Jiang, Jonathan D. Poplawsky, David A. Cullen, Adam B. Phillips, Michael J. Heben, Randy J. Ellingson, Yanfa Yan

Energy Yield and Economics of Single-Axis-Tracked Bifacial Photovoltaics with Artificial Ground Reflectors 47
  Mandy R. Lewis, Silvana Ovaitt, Byron McDanold, Chris Deline, Karin Hinzer

Advanced Encapsulants for Reduced Thermal Mechanical Stress in Photovoltaic Modules: A Quantitative Analysis Using FBGS 48
  Rik Van Dyck, Marta Casasola Paesa, Tine Engelen, Bin Luo, Tom Borgers, Jonathan Govaerts, Hariharsudan Radhakrishnan, Michael Daenen, Jef Poortmans, Aart Willem Van Vuure

TOPCon Solar Cell Degradation Via Pinhole Nucleation 51
  Andrew Diggs, Adam Goga, Zachary Crawford, Gergely T. Zimanyi

Soiling Model for PV Applications: Improved Parameterizations 54
  Vicente Lara-Fanego, Christian A. Gueymard, Leonardo Micheli

A Sustainable Energy Market Through Community-Based PV Systems 57
  Ramón Reyes-Colón, Efraín O'Neill-Carrillo

Radioisotope Thermoradiative Cell Power Generator 63
  Stephen J. Polly, Geoffrey A. Landis, Seth M. Hubbard

Robust Detection Method of Low-Voltage Islanding for Grid-Forming Inverters Operated in Conjunction with Existing PV Inverters 64
  Björn Oliver Winter, Julian Schwung, Bernd Engel

Improving Performance of III-V Solar Cells Grown on Spalled Germanium with Ex Situ Substrate Planarization 71
  John S. Mangum, Anna K. Braun, Allison Perna, John F. Geisz, Aaron J. Ptak, Corinne E. Packard, Ryan M. France

Fabrication and Characterization of III-V Photovoltaic Devices for Use as CO2 Reduction Photoelectrodes 75
Myles A. Steiner, Grace A. Rome, Ann L. Greenaway, Joel W. Ager, Emily L. Warren

Proton Degradation-Free Flexible Chalcopyrite Solar Cells Without Cover Glass and Adhesive 76
Hiroki Sugimoto, Tetsuya Nakamura, Mitsuru Imaizumi, Shin-Ichiro Sato, Takeshi Ohshima

Analysis of Hierarchical PV2PV Series Differential Power Processing Configuration for Photovoltaic Applications 80
Afshin Nazer, Patrizio Manganiello, Olindo Isabella

Plane of Array Irradiance Cleaning and Generation of Validated POA Readings for Plant Evaluation 81
Pramod N. Krishnani, Adonis E. Hajj, Clay Helms, Shreyas Nagarajan, Mark Mikofski

A Maximum Current Point Tracking Algorithm for Solar-To-Hydrogen Production 85
Kelvin Tan, Meng Tao

A Physics Based Approach for PV Lifetime and Degradation Signatures Prediction 86
Ismail Kaaya, Gofran Chowdhury, Arnaud Morlier

Epitaxial Growth and Testing of 1.1 eV Metamorphic InGaAs/GaAs Laser Power Converters 89
Katelynn E Fleming, Steve J Polly, Seth M Hubbard

Suitability of GaAsBi as a Candidate Junction in a III-V Multi-Junction Solar Cell 90
Thomas Wilson, Nicholas Ekins-Daukes

~20% Efficient Si PERC Solar Cell with Emitter Surface Passivated by H_2S Reaction 91
Tasnim K. Mouri, Ajay Upadhyaya, Ajeet Rohatgi, Youngwoo Ok, Amandee Hua, Dirk Hauschild, Lothar Weinhardt, Clemens Heske, Vijaykumar Upadhyaya, Brian Rounsaville, William N. Shafarman, Ujjwal K. Das

Investigations of Snail-Trail and Associated Microcrack Properities and Behavior in Brazil's Tropical Climate 94
Antonia Sonia A. C. Diniz, Neolmar De M. Filho, Cláudia K. B. Vasconcelos, Lawrence L. Kazmerski

A Study of POCl3 Deposition Reaction Rate with Residual Gas Analysis Method 97
Min Gu Kang, Sang Hee Lee, Kyung Taek Jung, Yunae Cho, Dohyung Kim, Sungeun Park, Munse Kim, Hee-Eun Song

Analysis of Photovoltaic Systems Penetration on Demand Curve and Locational Marginal Prices (LMPs) in PJM 98
Mesude Bayrakci Boz

Single Axis Tracker Performance Modeling on Sloped Terrain 101
Benjamin Pierce, Joshua Stein, Daniel Riley

Substrate Reuse of Hydride Vapor Phase Epitaxy Grown-GaAs Solar Cells for Low-Cost Photovoltaics 105
Yasushi Shoji, Ryuji Oshima, Kikuo Makita, Akinori Ubukata, Shuuichi Koseki, Takeyoshi Sugaya

Towards Polymer-Free, Fs Laser Welded Glass/Glass Modules 106
David L Young, Nick Bosco, Timothy J Silverman

Defects in RbF - Treated $Cu(In_xGa_{1-x})Se_2$ Solar Cells and Their Impact on V_{OC} 107
Michael F. Miller, Ana Kanevce, Alexandra M. Bothwell, Stefan Paetel, Darius Kuciauskas, Aaron R. Arehart

Improving the Space Silicon Solar Cell Efficiency by Adding the Layer Down-Converting UV Light to Visible 110
Alex Fedoseyev, Stan Herasimenka, Sergey Sarkisov

Near-Contactless Production I-V Testing of Silicon Solar Cells 113
Harrison Wilterdink, Ron Sinton, Adrienne Blum, Karoline Dapprich, Nick Degenhart, Wes Dobson

Self-Thermometry of PV Panels 119
Kaushal Chapaneri, Shahzada Pamir Aly, Jim Joseph John, Gerhard Mathiak, Vivian Alberts, Muhammad A. Alam

Inverter Clipping and Its Masking Effect on PV Soiling: Truth Or Myth? 122
Leonardo Micheli, Matthew Muller, Marios Theristis, Florencia Almonacid, Eduardo F. Fernandez

Effect of Arsenic Doping in Polycrystalline Thin Film CdTe Solar Cells 123
Mayank Mate, Akash Shah, Ramesh Pandey, Zachary Lustig, Walajabad Sampath, Amit H. Munshi

Thermal Modelling of a Renkube Panel 126
Deepika Gopal, Balaji Bangolae Lakshmikanth, Lakshmi Santhanam

Stable High Temperature Operation in Metal Halide Perovskite Solar Cells 132
Hadi Afshari, Shashi Sourabh, Sergio A. Chacon, Vincent R. Whiteside, Bibhudutta Rout, Giles E. Eperon, Joseph M. Luther, Ian R. Sellers

Characteristics of Detachable III-V Solar Cells Grown on Porous Germanium 133
Valentin Daniel, Thomas Bidaud, Jeremie Chretien, Abdelatif Jaouad, Jean-Francois Lerat, Nicolas Paupy, Bouraoui Ilahi, Jinyoun Cho, Kristof Dessein, Christian Dubuc, Gwenaelle Hamon, Abderaouf Boucherif, Maxime Darnon

Impact of Surface Roughness in Measuring Optoelectronic Characteristics of Thin-Film Solar Cells 134
David Magginetti, Seokmin Jeon, Yohan Yoon, Ashif Choudhury, Ashraful Mamun, Yang Qian, Jordan Gerton, Heayoung Yoon

Validation of Open-Source Distributed Energy Resources (OpenDER) Model with IEEE 1547-2018 Smart Inverter 138
Yiwei Ma, Charles Brewster, Aminul Huque

Upscaling of Perovskite Solar Cell Fabrication Via Slot-Die Coating: In Situ Tracking of the Drying and Crystallization Front During Gas Quenching 144
Kristina Geistert, Simon Ternes, David Benedikt Ritzer, Benjamin Hacene, Felix Laufer, Ulrich Wilhelm Paetzold

Inverted Metamorphic Photovoltaics Utilizing a Distributed Bragg Reflector Compatible with Epitaxial Lift-Off 145
Robert F McCarthy, David Rowell, Andree Wibowo, William Mohr, Chris Youtsey, Mark Osowski, Martin Drees, Roger E Welser, Noren Pan

Life-Cycle Analysis of Potentially Longer Life Expectancy CdTe PV Modules 146
Vasilis Fthenakis, Enrica Leccisi, Parikhit Sinha

Forecasting Day-Ahead Solar Irradiance for Puerto Rico Using the WRF Model and NSRDB 152
Manajit Sengupta, Jaemo Yang, Yu Xie

Coordinating the Frequency-Droop Controls of Inverter-Based Resources and Diesel Generators in an Isolated Microgrid 155
Mohan Du, Nayeem Ninad, Dave Turcotte

Probing Non-Equilibrium Hot Carrier Dynamics in Metal Halide Perovskite Solar Cells 161
Shashi Sourabh, Hadi Afshari, Vincent R. Whiteside, Giles E Eperon, Rebecca A. Scheidt, Varun N. Mapara, Madalina Furis, Matthew C Beard, Ian R. Sellers

On the Influence of Forced Convection in PV Energy Yield Models 162
Raed I. Bourisli, Bader S. Aldalali, Arttu Tuomiranta, Jef Poortmans

2-Terminal and 3-Terminal Subcell Characterization Platforms for Emerging Tandems 165
Jin Young Kim

Fast Cell Detection and Distortion Correction for Outdoor Electroluminescence Images 166
Evgenii Sovetkin, Bart E. Pieters, Andreas Gerber, Liviu Stoicescu, Pascal Koelblin

Limiting Factors on the Performenace of Luminescent Solar Concentrators for Building Integrated Photovoltaics 172
Bryce S. Richards, Ian A. Howard

3D Printed Transparent Sheet for Solar Panel Encapsulation and Thermal Management 173
Fahad Alam, Nazek El-Atab

Testing the Durability of Fluorine-Free Hydrophobic Coatings vs Porous Silica 174
Luke O. Jones, Adam M. Law, Gary Critchlow, John M. Law

Investigation of the Microstructure of Underdense Hydrogenated Amorphous Silicon Layers for Silicon Heterojunction Solar Cells by Raman Spectroscopy and Hydrogen Effusion 177
Benedikt Fischer, Maurice Nuys, Andreas Lambertz, Weiyuan Duan, Kaining Ding, Uwe Rau

Impact of Irradiation-Induced Filter Temperature Increase on Calibration of Reference Solar Cells with NIR-Longpass Filters 178
Tao Song, Larry Ottoson, Rafell Williams, John Geisz, Charles Mack, Jeremy Brewer, Nikos Kopidakis

Investigation of High Nitrogen Composition SiN_x for Textured Front Surface Passivation of n-Type Silicon Solar Cells in Terms of Light Stability of Injected Negative Charge and Cell Performance 179
Kwan Hong Min, Jeong-Mo Hwang, Christopher Chen, Wook-Jin Choi, Vijaykumar D Upadhyaya, Brian Rounsaville, Ajeet Rohatgi, Young-Woo Ok

Optimizing the Heat Sink for Concentrated Photovoltaic Systems for Different Heat Flux Conditions 182
Nemalipuri Surya Prathap, Harsh Chaurasia, K. S. Reddy

Comparison of Open-Source Photovoltaic Performance Models Against Multi-Year Field Data 188
Lelia Deville, Marios Theristis, Bruce H. King, Terrence L. Chambers, Joshua S. Stein

Q Cells Q.Antum Neo Technology with > 25% Conversion Efficiency Applying Mass-Production Processes 191

Matthias Junghänel, Ingmar Höger, Martin Schaper, Kai Petter, Enrico Jarzembowski, Christian Klenke, Anika Weihrauch, Michael Schley, Hans-Christoph Ploigt, Ohjin Kwon, Antje Schönmann, Osama Tobail, Axel Schwabedissen, Maximilian Kauert, Klaus Duncker, René Hönig, Janko Cieslak, Stefan Hörnlein, Florian Stenzel, Björn Faulwetter-Quandt, Jessica Scharf, Friederike Kersten, Cangming Ke, Sissel Tind Kristensen, Carsten Baer, Martina Queck, Gregor Zimmermann, Matthias Köhler, Nicole Lampa, Britta Pohl-Hampel, Lorenzo Burtone, Larissa Niebergall, Matthias Schütze, Susanne Schulz, Stefan Peters, Ansgar Mette, Fabian Fertig, Markus Fischer, Jörg W. Müller

Outdoor Study of Photovoltaic Mini-Modules with Different Perovskite Compositions 194

Vasiliki Paraskeva, Maria Hadjipanayi, Matthew Norton, Aranzazu Aguirre, Anurag Krishna, Rita Ebner, Tommasso Fontanot, Sabrina Pechmann, Silke Christiansen, George E. Georghiou

Achieving a New World Record Silicon Solar Cell Efficiency of 26.81% Using SHJ Device Structure 195

Xixiang Xu, Minghao Qu, Miao Yang, Xiaoning Ru, Shi Yin, Chengjian Hong, Fuguo Peng, Junxiong Lu, Liang Fang, Zhenguo Li, Yichun Wang, Tian Xie

Utilizing Particle Swarm Optimization for Autocalibration of LED Solar Simulators 198

Jann A Grovogui, John C Nocerino, Don Walker

Analysis of Cu(In, Ga)Se$_2$ Heterojunction Solar Cells in Terms of Their Balance of Thermodynamic Potentials 199

Uwe Rau, Felix Komoll, Tim Helder, Mario Zinßer, Ana Kanevce, Thomas Kirchartz, Theresa Magorian Friedlmeier

Epitaxy-Free, Thin-Film GaAs Solar Cells Fabricated with Diffusion Doping and Mechanical Spalling 202

Phillip R. Jahelka, Andrew W. Nyholm, Harry A. Atwater

Analysis of Impurity-Related Radiative Transitions in Silicon Materials Using Temperature-Dependent Photoluminescence 203

Tarek O. Abdul Fattah, Janet Jacobs, Vladimir P. Markevich, Nikolay V. Abrosimov, Matthew P. Halsall, Iain F. Crowe, Anthony R. Peaker

Reflector Candidates for a Vertical Bifacial Solar Canal 209

Jeremiah Reagan, Brandi McKuin, Sarah Kurtz

Evaluating Leafy Green Production in a Colorado Rooftop Agrivoltaic System 215

Armando Villa-Ignacio, Jennifer Bousselot

Isotropic Wet Etching of Acoustically-Spalled GaAs 218

Anica N Neumann, Myles A Steiner, Pablo G Coll, Mariana Bertoni, Emily L Warren

Degradation-Related Defect Level in Weathered Silicon Heterojunction Modules Characterized by Deep Level Transient Spectroscopy 219

Steve Johnston, Dirk C. Jordan, Dana B. Kern, Kristopher O. Davis, Helio R. Moutinho, George F. Kroeger

Interrogating Dominant Recombination Pathways in CdTe Solar Cells Using Wavelength-Dependent External Radiative Efficiency Measurements 220

Jared D. Friedl, Adam B. Phillips, Manoj K. Jamarkattel, Tyler Brau, Sabin Neupane, Scott L. Wenner, Abasi Abudulimu, Ebin Bastola, Yanfa Yan, Randy J. Ellingson, Michael J. Heben

Dynamic Calibration of Injection Dependent Carrier Lifetime from Time-Resolved Photoluminescence 227
Yan Zhu, Robert Lee Chin, Nursultan Mussakhanuly, Thorsten Trupke, Ziv Hameiri

Impacts of Dispatch Strategies and Forecast Errors on the Economics of Behind-The-Meter PV-Battery Systems 228
Brian T. Mirletz, Nicholas D. Laws

Advanced Characterization and Degradation Analysis of Perovskite Solar Cells Using Machine Learning and Bayesian Optimization 233
Joseph Chakar, Arthur Julien, Karim Medjoubi, Jorge Posada, Jean-François Guillemoles, Jean-Baptiste Puel, Yvan Bonnassieux

Statistical Analysis and Degradation Pathway Modeling of PERC PV Minimodules with Different Packaging Strategies in Indoor Accelerated Exposures 234
Sameera Nalin Venkat, Jiqi Liu, Xuanji Yu, William Oltjen, Xinjun Li, Jean-Nicolas Jaubert, Jennifer L. Braid, Roger H. French, Laura S. Bruckman

Hotspot Endurance of Pristine and Thermal Cycled Glass-Backsheet Photovoltaic Modules 237
Muhammad Afridi, Akash Kumar, Farrukh Ibne Mahmood, Govindasamy Tamizhmani

Integrated Large-Scale Data Management Platform for Photovoltaic Power Conversion Equipment (PCE) Reliability Data.. 240
Liwei Wang, Buck Brown, Shuan Dong, Tan Jin, Daniel Clemens, Joseph Hodges, Adam Reeves, Josh Ozbeytemur, Shuangshuang Jin, Zheyu Zhang

SnO2 Buffer Layers for High Efficiency CdSeTe/CdTe Devices.. 245
L. C. Infante-Ortega, Xiaolei Liu, Luksa Kujovic, Mustafa Togay, Luke O. Jones, Ali Abbas, Kieran Curson, R. C. Greenhalgh, Kurt L. Barth, Jake W. Bowers, John M. Walls, Ochai Oklobia, Stuart Irvine, Eric Colegrove, Brian Good, Matt Reese

Contact Interface Morphology of Screen-Printable Front-Side Contacts for Industrial N- TOPCon Crystalline Silicon Solar Cells... 248
Meijun Lu, Kurt R. Mikeska, Weilin Liao, Chaoying Ni, Yong Zhao, Jianming Wang, Kangping Zhang, Changgen Zhang, Yawen Xu, Baiqiang Liu

Light-Dark Cycling in Perovskite Solar Cells Studied by MPPT and Ion Migration Current Measurements.. 253
Takeshi Tayagaki, Kohei Yamamoto, Takurou N. Murakami, Masahiro Yoshita

Field Trial in Progress for Measuring Global, Direct, Diffuse, and Ground-Reflected Irradiance Using a Static Sensor Array... 254
Michael Gostein, Bruce H. King

The Proposition of a Public Policy to Stimulate Low-Income Communities' Assess to Distributed Energy Resources 256
Anna Carolina De Paula Sermarini, João Henrique Paulino Azevedo, Vanessa Cardoso De Albuquerque, Rodrigo Flora Calili, Felipe Gonçalves, Gilberto Jannuzzi

PV Plant Performance Review Methodology: Key Performance Indicators (KPI) Estimation........................ 259
Himanshu Gulati, Prashant Kumar Upadhyay, Yellasiri Bharath Kumar Reddy

Validation of Inverter Labeling with Plant Transfer Functions 263
Joseph Ranalli

Long Terms Stability and Metastable Behavior of Perovskite Solar Devices on Outdoor Conditions 268
Karim Medjoubi, Anne Migan Dubois, Jean Castillon, Thomas Guillemot, Johan Parra, Marion Provost, Jean-Baptiste Puel, Jean Rousset, Juan Pablo Medina Flechas, Camille Bainier, Dounya Barrit, Jordi Badosa, Jorge Posada

Ultrafast Dynamics of Photoexcited Carriers and Phonons in Tailored 1D Acoustic Phonon Potentials 271
Muhammad Hanif, Stephen Bremner, Michael P. Nielsen, Milos Dubajic, Gavin J. Conibeer

Improving PbS Colloidal Quantum Dot Solar Cell Performance Via Solution-Phase Engineering 272
Dhanvini Gudi, Arlene Chiu, Dana Kachman, Eric Rong, Serene Kamal, Yucheng Lan, Susanna M. Thon

Understanding and Advancing Bifacial Thin Film Solar Cells Under Dual Illumination 275
Adam B. Phillips, Jared D. Friedl, Zhaoning Song, Abasi Abudulimu, Ebin Bastola, Deng-Bing Li, Yanfa Yan, Randy J. Ellingson, Michael J. Heben

Spray-Assisted Passivation Strategy for Highly Efficient and Stable Perovskite Solar Cells 276
Rishabh Sahani, Neetesh Kumar, Cheng-Yu Lai, Daniela R. Radu

Approaching 19% Efficiency in (InxGa(1-X))2O3/CdSe/CdTe Solar Cells with Improved Front & Back Interfaces 279
Manoj K. Jamarkattel, Adam B. Phillips, Ebin Bastola, Sabin Neupane, Deng-Bing Li, Abasi Abudulimu, Jared D. Friedl, Tamanna Mariam, Yanfa Yan, Randy J. Ellingson, Michael J. Heben

Short-Circuit Current Density Chasing and Breakthroughs in High Efficiency Silicon Heterojunction Solar Cells 280
Weiyuan Duan, Karsten Bittkau, Andreas Lambertz, Kaining Ding

Benefits of Surface Engineered Silicon Quantum Dots in Formamidinium Lead Iodide Perovskite Solar Cells 281
Vladimir Svrcek, Calum McDonald, Dilli Babu Padmanaban, Ruairi McGlynn, Ankur Kambley, Bruno Alessi, Davide Mariotti, Takuya Matsui

Depth Profiling of Glass/POE/Transparent Backsheet Degradation for Bifacial Photovoltaics 282
Xiaohong Gu, Ashlee R. Aiello, Stefan Mitterhofer, Soshana Smith, Stephanie L. Moffitt, Lakesha N. Perry, Song-Syun Jhang, Stephanie S. Watson, Li-Piin Sung

Temporal Downscaling of GHI Clear-Sky Indices Using T-Copula 285
Jing Huang, Marc Perez

Extended FF-VOC Parameterization for Silicon Solar Cells 288
Karsten Bothe, David Hinken, Rolf Brendel

Towards Commercialisation with Lightweight, Flexible Perovskite Solar Cells for Residential Photovoltaics 289
Philippe Holzhey, Michael Prettl, Silvia Collavini, Nathan Chang, Michael Saliba

Thermal Stability of BiI3 Thin Films 290
Natália F. Coutinho, Thais Crestani, Otávio J. De Oliveira, M. M. M. Modesto Ana Paula De, Marcelo Villalva, Francisco C. Marques

In-Situ Smoothing of Facets on Spalled GaAs(100) Substrates During OMPVE Growth of III-V Solar Cells 291

William E. McMahon, Anna K. Braun, Allison N. Perna, Pablo G. Coll, Kevin L. Schulte, Jacob T. Boyer, Anica N. Neumann, John F. Geisz, Emily L. Warren, Aaron J. Ptak, Arno P. Merkle, Mariana I. Bertoni, I. Bertoni, Corinne E. Packard, Myles A. Steiner

Numerical Simulation Study for Analysis of Hydrogenated Amorphous Silicon/Crystalline Silicon Heterostructure by Reactive Molecular Dynamics Method 292

Kazuma Inoue, Naoya Uene, Kazuhiro Gotoh, Yasuyoshi Kurokawa, Takashi Tokumasu, Noritaka Usami

Planned Field Test of Soiling and Irradiance Measurement Uncertainties in Bifacial PV Systems Using In-Situ Reference Modules 295

Michael Gostein, Audrey Marquis, Marine Bila, Robert Campbell

Intermediate-Phase Engineering Via Dimethylammonium Cation Additive for Stable Perovskite Solar Cells 298

David P. McMeekin, Philippe Holzhey, Sebastian O. Fürer, Steven P. Harvey, Laura T. Schelhas, James M. Ball, Suhas Mahesh, Seongrok Seo, Nicholas Hawkins, Jianfeng Lu, Michael B. Johnston, Joseph J. Berry, Udo Bach, Henry J. Snaith

A Techno-Economic Analysis of Various Grid-Connected Photovoltaic System Configurations for Green Hydrogen Production 299

Rahul R Urs, Assia Chadly, Ahmad Mayyas

Multi-Layer Dense Antireflection Coatings 302

Yiyu Zeng, Martin Green, Jessica Yajie Jiang

Transparent Conductive Oxide Bi-Layer as Front Contact for Multijunction Thin-Film Silicon Solar Cells 305

Federica Saitta, Prashand Kalpoe, Govind Padmakumar, Paula Perez-Rodriguez, Gianluca Limodio, Rudi Santbergen, Arno H. M. Smets

12.3% Efficient Lifted-Off and Reconstructed As-Doped CdTe Thin Film Solar Cell 308

Ochai Oklobia, Deborah L. McGott, Giray Kartopu, Steve Jones, Stuart J. C. Irvine

Design, Fabrication, Test, and Flight Performance of the Parker Solar Probe Solar Array 311

Edward Gaddy, Andrew Gerger, Lew Roufberg, Richard Stall, Matthew J. Schurman

Investigating Electric Field and Light Induced Degradation in Perovskite Solar Cells Through Nanometer-Scale Potential Imaging 315

G. Paul, J. W. Schall, C.-S. Jiang, A. Louks, A. Palmstrom, N. S. Dutta, S. Johnston, H. Guthrey, A. Norman, M. M. Al-Jassim, D. B. Sulas-Kern

2D-MoS$_2$ Nano Structures to Enhance Silicon Solar Cells 321

Muntaser Abdelrahman Almansoori, Ayman Rezk, Ammar Nayfeh

Effect of Bidentate Ligand Additive in Tin Perovskite Solar Cells 324

Dhruba B. Khadka, Yasuhiro Shirai, Masatoshi Yanagida, Kenjiro Miyano

Optoelectronic Performance of Solution Processable MoS$_2$ for Application in Photovoltaic Devices 327

Dayanand Kumar, Ayman Rizk, Ammar Nayfeh, Nazek El-Atab

Mechanical Degradation of Perovskite Thin Films for Photovoltaics: In-Situ Microscopy & Digital Twin Modeling 330

Melissa A Davis, Mehul Tank, Michelena O'Rourke, Matthew Wadsworth, Zhibin Yu, Rebekah Sweat

Optimal Row Spacing for Monofacial and Bifacial Fixed-Tilt and Tracked Photovoltaic Systems Up to 75°N 331

Erin M. Tonita, Annie C. J. Russell, Christopher E. Valdivia, Karin Hinzer

Instability of Non-Fullerene Acceptors Used in Organic Solar Cells 332

Yongxi Li, Tonghui Wang, Aram Amassian, Stephen Forrest

Damp Heat Performance of Silicon Heterojunction Solar Cells with Reactive Silver Ink Metallization 333

Michael W Martinez-Szewczyk, Steven Digregorio, Subbarao Raikar, Owen Hildreth, Mariana Bertoni

Parametric Analysis of Photovoltaic Inverters Under Balanced and Unbalanced Voltage Phase Angle Jump Conditions 334

Rachid Darbali-Zamora, Jay Johnson, Matthew J. Reno

Chemical Reaction Kinetics of the Decomposition of Low Bandgap Tin-Lead Halide Perovskite Films and the Effect on the Ambipolar Diffusion Length 340

Yuhuan Meng, Preetham P. Sunkari, Marina Meila, Hugh W. Hillhouse

Demonstration of Thermoradiative Energy Conversion with InAs Cells 341

Eric J Tervo, Andrew J Ferguson, Jennifer Selvidge, Myles A Steiner, Ryan M France

InGaP/GaAs/In$_{0.35}$Ga$_{0.65}$As//In$_{0.53}$Ga$_{0.47}$As Four-Junction Solar Cells Integrated by Surface Activated Wafer Bonding 342

Kentaroh Watanabe, Takashi Shimasaki, Hassanet Sodabanlu, Yoshiaki Nakano, Masakazu Sugiyama

Efficiency Limits for Multi-Junction Coloured Photovoltaics 345

Phoebe M. Pearce, Janne Halme, Jessica Yajie Jiang, Farid Elsehrawy, Nicholas J. Ekins-Daukes

Microscale, High Aspect Ratio, Effectively Transparent Contacts (ETCs) Fabricated with String Printing 348

Mathis Van De Voorde, Janis A. Andersons, Rebecca Saive

Optimization of Bulk Heterojunction Layer Constituents in Organic Photovoltaic Device 351

Vilko Mandic, Dragana Vuk, Radovanovic-Peric Floren, Ivana Panzic

Towards Smart Integration of Cu-Plating for Silver-Free and Edge Passivated SHJ Shingle Modules 352

Samuel Harrison, Vincent Barth, Benoit Martel, Agata Lachowicz, Nicola Frasson, Marco Galiazzo

A Comparison of Emerging and Industry Benchmark Photovoltaic Backsheets Between Different Outdoor Locations 353

Elizabeth Palmiotti, Bruce King, Rachael Arnold, Sona Ulicná, Laura T. Schelhas, David C. Miller

Considering the Variability of Soiling in Long-Term PV Performance Forecasting 356

Matthew Muller, Faisal Rashed

Electron Paramagnetic Resonance Investigation of the Defect Responsible for Light- And Elevated-Temperature-Induced Degradation in Ga-Doped Czochralski Si 359

Chirag Mule, P. Craig Taylor, Abigail Meyer, William Nemeth, Vincenzo Lasalvia, Matthew Page, Sumit Agarwal, Pauls Strandins

Real-Time Regional PV Spinning Reserve Estimator with AGC Look-Ahead Windows 360
Mengmeng Cai, Govind Saraswat, Vahan Gevorgian

Investigation of Sputtered P-Type Electrical Contacts for Thin Film Cadmium Telluride-Based
Solar Cells 363
Blake Hill, Forrest Khulmann, Mayank Mate, Amit H. Munshi

Perovskite Bafacial Modules-Efficiency, Stability and Upscaling 366
Hangyu Gu, Jinsong Huang

Modular, Array-Mounted Photovoltaic Inspection Robot 367
Michael Y. Vazquez Nieves, Alanis M. Colón González, Jennifer L. Braid

ESSPI as a Fast Tool for Load Prioritization on Microgrids Design 370
Luis Colomba-Colon, Natanael Batista-Alvarez, Guillermo Lopez-Cardalda, Eduardo Ortiz-Rivera

Investigations on Absorber Type and Junction Position of GaAs Solar Cells 374
Gan Li, Hassanet Sodabanlu, Meita Asami, Kentaroh Watanabe, Masakazu Sugiyama, Yoshiaki Nakano

Transparent Tedlar® Frontsheet for Lightweight PV Module Designs 377
Hongjie Hu, Stela Chen, Oakland Fu, Michael Demko, Kaushik Roy Choudhury

Photovoltaic Design Projects Increase ECE Student Engagement 380
Devin C. Whalen, Peter Mark Jansson, Milton G. Newberry

Long-Term Degradation Rate of Photovoltaic Modules: A Meta-Analysis 385
Michael Straub-Mueck, Jerome Geyer-Klingeberg, Andreas Rathgeber

The Use of a Physics-Based DNI Model to Enhance the National Solar Radiation Database
(NSRDB) 386
Yu Xie, Jaemo Yang, Manajit Sengupta, Yangang Liu

Novel Module Architecture for Lower CapEx and Improved Recyclability for c-Si PV Modules 387
Ryan Ruhle, Larry Maple, Timothy Delazzer, Steve Johnston, Dana Kern, Walajabad Sampath

Snow Sensing for Photovoltaic Single Axis Tracker Systems 391
Ayush Chutani, Ana Dyreson, Laurie Burnham, Kyumin Lee

Operando Temperature Measurements of Photovoltaic Laser Power Converter Devices Under
Continuous High-Intensity Illumination 394
John F. Geisz, Daniel J. Friedman, Myles A. Steiner, Ryan M. France, Tao Song

Optimization of Back-Contact Diffusion Barrier for Solution-Processed CIGS Solar Cells: Case of
MoO_3 and MoN 395
Nada Benhaddou, Jacques Kenyon, Luke O. Jones, Liam M. Welch, Jake W. Bowers

Silicon Heterojunction Cell Metallization with Reactive Silver Inks: Printing Process, Ink Formula,
and Interconnection 400
Steven J. Digregorio, Michael W. Martinez-Szewczyk, Subbarao Raikar, Mariana I. Bertoni, Owen J. Hildreth

Optical Properties of (InxGal-X)2O3 Alloys and Evaluation as Emitter Layer in CST PV 401
Bishal Shrestha, Madan K. Mainali, Manoj K. Jamarkattel, Ebin Bastola, Adam B. Phillips, Michael J. Heben, Nikolas J. Podraza

Temperature Dependent Fill Factor in CdSe/CdTe PV Devices from -20°C to 60°C Under AM1.5G and AM0 Spectra 402

Nadeesha Katakumbura, Prabodika N. Kaluarachchi, Manoj Rajakaruna, Tyler Brau, Aesha P. Patel, Abudulimu Abasi, Ebin Bastola, David Raker, Adam B. Phillips, Michael J. Heben, Sorin Cioc, Randy J. Ellingson

Module Reliability in Winter: Field Analysis of Deflection and Cell Cracking Across Multiple Module Architectures 409

Laurie Burnham, Daniel Riley, Bruce King, William Snyder, Kevin Santistevan, Paul W. Dice

How Useful is a Field-Operable I-V Curve Tracer? 412

Alexander Cimaroli

Results of First Long Duration Space Flight of Hybrid Perovskite Thin Film 415

Lyndsey McMillon-Brown, William Delmas, Samuel Erickson, Jorge Arteaga, Mark Woodall, Michael Scheibner, Timothy Krause, Kyle Crowley, Kaitlyn Vansant, Joseph Luther, Jennifer Williams, Jeremiah McNatt, Sayantani Ghosh

Undergraduate Research Experience in the Design and Construction of a Photovoltaic Inspection Robot 416

Alanis M. Colón, Emmanuel J. González, Fernando J. Vargas, Samuel I. Hernandez, Michael Y. Vazquez, Eduarto I. Ortiz

Development of Gradient Layers to Improve the Efficiency of Transparent Passivating Contact Solar Cells 419

Alexander Eberst, Binbin Xu, Weiyuan Duan, Andreas Lambertz, Uwe Rau, Kaining Ding

A GIS-Based Approach for Prioritization of Photovoltaic Systems with Energy Storage Implementation for Vulnerable Community Resilience 420

Javier A. Moscoso-Cabrera, Edgar E. Cruz, Cristian R. Meléndez, Eduardo I. Ortiz

Position Dependence of the Performance Gain by Selective Ground Albedo Enhancement for Bifacial Installations 426

Nils-Peter Harder, Issam Smaine, Fadi Bourarach, Damien Cosme, Ines Arfaoui, Julien Chapon, Arttu Tuomiranta, Antonios Florakis

Microscopic Origins of Performance Losses in $(Ag,Cu)(In,Ga)Se_2$ Thin-Film Solar Cells 432

Sinju Thomas, Wolfram Witte, Dimitrios Hariskos, Rico Gutzler, Stefan Paetel, Chang-Yun Song, Heiko Kempa, Matthias Maiberg, Daniel Abou-Ras

Overview of Engineered Germanium Substrate Development for Affordable Large-Volume Multi-Junction Solar Cells 434

Jinyoun Cho, Valérie Depauw, Alexandre Chapotot, Waldemar Schreiber, Tadeáš Hanuš, Nicolas Paupy, Valentin Daniel, Guillaume Courtois, Bouraoui Ilahi, Abderraouf Boucherif, Clement Porret, Roger Loo, Jens Ohlmann, Stefan Janz, Kristof Dessein

X-RAYS Meet Neutrons Meet Ions Meet Electrons Meet Lasers Meet Magnets: Combined Access to Multiple Facilities Through EU Project "Remade@ARI" 435

Michael E. Stuckelberger, Christina Ossig, Barbara Schramm, Stefan Facsko

BIPV Market Potential Analysis with Building Shadow Simulation 436

Changyeol Yun, Myeongchan Oh, Boyoung Kim, Jehyun Lee, Hyungoo Kim, Deokoh Lim, Sangmin Jo

An Investigation on the Pollen-Induced Soiling Losses in Utility-Scale PV Plants 437

João Gabriel Bessal, Michael Valerino, Matthew Muller, Mike Bergin, Leonardo Micheli, Florencia Almonacid, Eduardo F. Fernández

The Role of PbS QDs on Strain and Optical Properties in Different Perovskite Matrix 441
Sofia Masi, Patricio Serafini, Iván Mora-Seró

Zr-Doped In2O3 Film for the Interlayer of Perovskite/Crystalline Silicon Tandem Solar Cells 442
Tappei Nishihara, Hyunju Lee, Ryuji Kaneko, Yoshio Ohshita, Atsushi Wakamiya, Atsushi Masuda, Atsushi Ogura

Design with Luminescent Solar Concentrator Photovoltaics in the Built Environment 443
Eli Shirazi, Wouter Eggink, Angele Reinders

Analysis of Solar Cell Electroluminescence Spectra for Daylight Inspection of c-Si PV Modules 448
Gisele A. Dos Reis Benatto, Alejandra A. Mayordomo, Rodrigo Del Prado Santamaria, Thøger Kari, Peter B. Poulsen, Sergiu V. Spataru

Artificial Neural Network and Peer-To-Peer Communications at the Grid-Edge to Mitigate Cyber Attacks on Distributed Photovoltaic Inverters 455
C. Birk Jones, Rachid Darbali-Zamora

Improvement of Radiation Tolerance in Solar Cells by Hetero P/N Junction Structure 463
Tetsuya Nakamura, Mitsuru Imaizumi, Meita Asami, Masakazu Sugiyama, Hidefumi Akiyama, Shin-Ichiro Sato, Takeshi Oshima, Yoshitaka Okada

Abnormal Responses of Residential Smart Photovoltaic Inverters to Cyberattacks 464
Thunchanok Kaewnukultorn, Sergio B. Sepúlveda-Mora, Steven Hegedus

Community Influence of Houses of Worship on Rooftop Solar Growth Rates 467
Ashley Degen, Laura Mogannam, Nisitaa Karen Clement Pradeep, Jillian Stern, Deborah A. Sunter

Decoupling Open-Circuit Voltage and Series Resistance in Electroluminescence Images Through Deep Learning 472
Gaia Maria N. Javier, Priya Dwivedi, Thorsten Trupke, Ziv Hameiri

Decomposition Mechanisms and Kinetics of Perovskite Semiconductors 473
Hugh Hillhouse, Yuhuan Meng, Spencer Cira, Preetham Sunkari

Development and Evaluation of Typical Plane of Array Year (TPY) for Solar Energy Systems Over the Americas 474
Aron Habte, Manajit Sengupta, Grant Buster, Yu Xie

10-Junction Edge-Illuminated Passivated-Contact Silicon Minimodules for Laser Power Conversion 475
Ryan M France, Matthew B Hartenstein, William Nemeth, San Theingi, Matthew Page, Sumit Agarwal, David Young, Paul Stradins

GaAs Betavoltaic Cell Modeling for Light to Medium Element Radiation Conversion into Electrical Power 476
Mathieu De Lafontaine, Gavin Forcade, Paige Wilson, Jayeshkumar Patel, Brian Ellis, Helmut Fritzsche, Sam Suppiah, John P. D. Cook, Christopher E. Valdivia, Karin Hinzer

Residential Electric Energy Storage System to Reduce Voltage and Thermal Violations in Distribution Lines and Increase PV Integration 477
Anny Huaman-Rivera, Agustin Irizarry-Rivera

Air-Bridge Cells for Higher Emission Temperatures 482
Bosun Roy-Layinde, Areefa Rahman, Jihun Lim, Sritoma Paul, Stephen R. Forrest, Andrej Lenert

On the Accuracy of Spectral Adjustment for Performance Measurements of Multijunction Solar Cells.. 485
    Nikos Kopidakis, Tao Song, John Geisz, Daniel Friedman

Spatially Resolved Degradation Analysis of Solar Modules After Combined Accelerated Aging 486
    Robert Heidrich, Anton Mordvinkin, Ralph Gottschalg

NiO as a P-Type TCO for Inorganic Thin-Film Photovoltaics 489
    Elline C. Hettiaratchy, Angus A. Rockett, Taylor D. Hill, Sachit Grover

Modeled Impacts of Solar Forecast Error on Utility Production Cost... 492
    William B. Hobbs, Jenner Tresan, Michael Kline, Mousumi Guha, Brent Duncan

Durability Testing of Porous SiO_2 Anti-Reflection Coatings for Solar Cover Glass.................................... 498
    Adam M Law, Luke O Jones, Michael Nasser, Ali Abbas, John M Walls

Performance Evaluation of Perovskite Solar Cell Modules with Tilt Angle Optimization in BIPV Application: A Case Study for Kazakhstan 502
    Yerassyl Olzhabay, Ikechi Ukaegbu, Annie Ng

CFD-Based Machine Learning Model for Agrivoltaic System Design 505
    Henry J. Williams, Emily Weed, Khaled Hashad, K. Max Zhang

Towards Highly Stable Metal-Halide Perovskite Materials for a Broad Range of Applications: Film Growth, Degradation Control, and Interfacial Engineering.. 508
    Lissette Rodriguez-Cabanas, Calvin Duong, Bradley Stanley, Andre Slonopas

First-Principles Study of Energy Band Alignment in Pristine $CdTe/TeO_2/Te$ Interfaces............................ 509
    Anthony P. Nicholson, James R. Sites, Walajabad S. Sampath

Horizon Profiling Methods for Photovoltaic Arrays.. 512
    Jennifer L. Braid, Benjamin G. Pierce

Field Effect Passivation Enables 2.2 V Open-Circuit Voltage All-Perovskite Tandems................................ 518
    Bin Chen, Ted Sargent

Distributed Generation Component Placement and Point of Common Coupling Allocation for Solar Rooftop Microgrid Sizing Costs Minimization 519
    Robert A. García Cooper, Marcel Castro Sitiriche, Agustín Irizarry Rivera, Fabio Andrade Rengifo

The Solar Boat: An Academic Research Experience... 525
    Guillermo Serrano, Erick Aponte, Eduardo I. Ortiz-Rivera

Evaluating the Use of Satellite Data and Machine Learning Models for PV Performance Monitoring... 528
    Daniel Fregosi, Rabin Dhakal, Devin Widrick

Drying Effects Upon Spin Coating of Solution-Processed Amine-Thiol Thin Film $Cu(In,Ga)(S,Se)_2$ Absorber Fabrication 534
    Jacques Kenyon, Nada Benhaddou, Liam Welch, Jake Bowers

Impact of Backsheet Versatility on Inverter Availability 540
    Claudia Buerhop-Lutz, Oleksandr Stroyuk, Jens Hauch, Ian Marius Peters

Evaluating the Weather Forecasting Models and the Impact to PV Generation Forecasting........................... 541
    Spyros Theocharides, Anastasios Koumis, George Makrides, George E. Georghiou

Plausibility Filtering of PV Outdoor Data 546

    T. S. Vaas, J. Körtgen, E. Sovetkin, U. Rau, B. E. Pieters

The European Solar Communication - Will it Strengthen the Photovoltaic Industry in the European
Union 550

    Arnulf Jäger-Waldau, Anatoli Chatzipanagi, Georgia Kakoulaki, Sandor Szábo

Measuring Sustainability of PV in Energy Transition: Mass, Energy, and Circularity 554

    *Heather M. Mirletz, Silvana Ovaitt, Macarena Mendez Ribo, Seetharaman Sridhar, Teresa M.
Barnes*

Photovoltaic Site Architecture Estimation Using Performance Data 555

    *Steven Koskey, Scott Sheppard, Corson Teasley, Christopher Perullo, Jared Kee, Daniel
Fregosi, Wayne Li*

Synthesis and Characterization of Bismuth Selenide and Copper Doped Bismuth Selenide Thin
Films by Chemical Bath Deposition 561

    Hamda A. Al-Thani, Shifaa M. Al-Baity, Falah S. Hasoon

CIGS Device Stability: A Comparison of Two Different Process Batches 566

    *Mohsen Jahandardoost, Curtis Walkons, Marco Nardone, Theresa M. Friedlmeier, Shubhra
Bansal*

Hierarchical Variance Analysis of Solar Cell Production Using Machine Learning and Numerical
Simulations 569

    Bernhard Klöter, Hannes Wagner-Mohnsen, Sven Wasmer

Identifying the Electrical Signature of Snow in Photovoltaic Inverter Data 572

    Emma C. Cooper, Jennifer L. Braid, Laurie M. Burnham

Band Tail Effects on Cd(Se,Te) Device Performance: A Numerical Simulation Approach 577

    Eva M Mulloy, Darius Kuciauskas, Craig L Perkins, Marco Nardone

Understanding the Dopability of as in Selenium-Alloyed Cadmium Telluride Solar Cells 578

    Xiaofeng Xiang, Aaron Gehrke, Scott Dunham

Modeling Transposition for Single-Axis Trackers Using Terrain-Aware Backtracking Strategies 581

    Kurt Rhee

What's New in the NSRDB 587

    Manajit Sengupta, Aron Habte, Grant Buster, Yu Xie, Brandon Benton, Michael Foster

Rapid, Contactless Measurements and Performance Predictions of Photovoltaic Materials 590

    *Brandon T Motes, Anthony T Troupe, Amy E Louks, Axel F Palmstrom, Joseph J Berry, Dane
W Dequilettes*

Optimizing the Packing Density of Building Integrated Concentrating Photovoltaic Systems for
Improved Performance and Reduced Embodied Carbon Through a Novel Polygonal Concentrator 591

    *Lewis Osikibo Tamuno-Ibuomi, Roberto Ramirez-Iniguez, A Sheila Holmes-Smith, Geraint
Bevan*

Human Health Risk Assessment for Improper Landfill Disposal of End-Of-Life CdTe PV 594

    Elaine Kupets, Garvin Heath

Accelerating Cycles of Learning for Silicon Heterojunction Architectures: Experimental Design and Data-Driven Degradation Pathway Prediction.. 597
Xuanji Yu, Diego Zubieta, Mirra Rasmussen, Chien-Hsuan Chen, Cécile Molto, Mariana Bertoni, Kristopher O. Davis, Laura S. Bruckman, Ina T. Martin

Modeling Reference Cell Performance Using Measured and Modeled Spectral Data.................................. 600
Josh Peterson, Frank Vignola, Afshin Andreas, Aron Habte, Manajit Sengupta

Silver Reflector-Driven Light Harvesting Enhancement in Large Area Dye Sensitized: Solar Cells 606
Navdeep Kaur, Faizan Syed, Cheng-Yu Lai, Daniela Radu

Survey of Snow Impacts on Bifacial Gain in Commercial Photovoltaic Arrays...................................... 607
Samantha S. Wilson, Stephen Lightfoote, Stephen Voss

New Theoretical Limits for Light Trapping in Solar Cells... 610
Stéphane Collin, Maxime Giteau

Supply Side Management with Agrivoltaics: Feasibility Study of Modeling Methodologies of Solar PV and Crop Response...611
Tadatoshi Takahashi

Effect of Solar Mounting Configurations on California Zero-Carbon Grid 617
Zabir Mahmud, Sarah Kurtz

Experimental Demonstration of Diffused Light Collimation in Free Space.. 622
Lisanne M. Einhaus, Geert C. Heres, Jelle Westerhof, Shweta Pal, Akshay Kumar, Anne Rikhof, Jian-Yao Zheng, Rebecca Saive

The Planet-Scale Performance Potential of Si-Perovskite Tandem Solar Farms................................. 626
Jabir Bin Jahangir, M. Tahir Patel, Reza Asadpour, M. Ryyan Khan, M. A. Alam

Investigating the Potential of Hydrogen Plasma Treated ALD-TiOx Films as Hole-Selective Passivating Contacts in Crystalline Silicon Solar Cells.. 629
Chien-Hsuan Chen, S. Novia Berriel, Taylor M. Currie, Jannatul Ferdous Mousumi, Titel Jurca, Parag Banerjee, Kristopher O. Davis

The Photovoltaic Exponential Model 630
Eduardo I. Ortiz Rivera

On the Unappreciated Impact of Se in as-Doped CdSexTe1-X.. 633
Deborah L McGott, Darius Kuciauskas, Craig L Perkins, Eric Colegrove, Matthew O. Reese

Leveraging High-Fidelity Sensor Data for Inverter Diagnostics: A Data-Driven Model Using High-Temperature Accelerated Life Testing Data 634
Sakir Karakaya, Murat Yildirim, Shijia Zhao, Feng Qiu, Jack David Flicker, Benjamin Peters, Zhaoyu Wang

Influence of Interfaces on Stability of Perovskite Solar Cells.. 641
Arkadi Akopian, Saba Sharikadze, Ranjith Kottokkaran, Vikram Dalal

A New Route to Facilitate Scaling of Lead-Tin Halide Perovskites: Thin Films Via Solvent Self-Volatilization 644
Jack R. Palmer, Apoorva Gupta, Sean P. Dunfield, David P. Fenning

Aerial Photoluminescence Imaging of PV Modules.. 645
Bernd Doll, Ernst Wittmann, Larry Lüer, Claudia Buerhop-Lutz, Jens A. Hauch, Christoph J. Brabec, Ian Marius Peters

Polarization Type Potential Induced Degradation Under Positive Bias in a Commercial PERC Module 648
> Farrukh Ibne Mahmood, Fang Li, Peter Hacke, Cécile Molto, Hubert Siegneur, Govindasamy Tamizhmani

Three-Dimensional and Multimodal X-Ray Microscopy Reveals the Impact of Voids in CIGS Solar Cells 651
> Giovanni Fevola, Christina Ossig, Mariana Verezhak, Jan Garrevoet, Martin Seyrich, Dennis Brueckner, Johannes Hagemann, Frank Seiboth, Andreas Schropp, Gerald Falkenberg, Peter Stanley Jorgensen, Christian Strelow, Tobias Kipp, Romain Carron, Jens Wenzel Andreasen, Michael E. Stuckelberger

Optical Characterization and Loss Simulation of Encapsulation Materials and Back Sheets for PERC+ Solar Modules 652
> Tim Lukas Brockmann, Henning Schulte-Huxel, Susanne Blankemeyer, Tobias Wietler

Integration of Lateral Power MOSFETs into IBC c-Si Solar Cells with Poly-Si Passivating Contacts 655
> David A. Van Nijen, Patrizio Manganiello, Yavuzhan Mercimek, Tristan Stevens, Guangtao Yang, René A. C. M. M. Van Swaaij, Miro Zeman, Olindo Isabella

The Formation of Dendrites in Overtreated CdSeTe/CdTe Solar Cells 656
> Vladislav Kornienko, Luke Jones, Kieran Curson, Ali Abbas, Rachael Greenhalgh, Yau Yau Tse, Michael Walls, Christian Drost, Bettina Spaeth, Bastian Siepchen

Carrier Dynamics in $Al_xGa_{1-x}As$/InAs-Based Photon Up-Conversion Solar Cells with a Doubled-Heterointerface 659
> Hambalee Mahamu, Shigeo Asahi, Takashi Kita

The Importance of Terrain-Shading Losses in PV Yield Assessment: The Case of Oahu 662
> Marc Perez, Upama Nakarmi, Philip Gruenhagen, Richard Perez

Utilizing PSO Technique for Locational-Dependent Feeder PV Hosting Capacity Evaluation 665
> Vinushika Panchalogaranjan, Paul Moses

Cryogenic Operation of GaAs Laser Power Converters 670
> Bora Kim, Mijung Kim, Brian D. Li, Ryan D. Hool, Minjoo L. Lee

Characterization of Field Exposed Photovoltaic Modules Featuring Signs of Contact Degradation 671
> Max Liggett, Dylan J. Colvin, Balaashwin Babu, William C. Oltjen, Xuanji Yu, Manjunath Matam, Hubert P. Seigneur, Mengjie Li, Andrew M. Gabor, Philip J. Knodle, Craig J. Neal, Sudipta Seal, Laura S. Bruckman, Roger H. French, Kristopher O. Davis

Energy Management in a Dynamic Microgrid Using Genetic Algorithms 674
> Ricardo Calloquispe-Huallpa, Rachid Darbali-Zamora, Erick E. Aponte-Bezares, Matthew S. Lave

Sn-Based Perovskite Thin Film Solar Cells with Enhanced Stability 682
> Wendy Reyes Ramos, Zeying Chen, Adam Thomas, Tara Dhakal

Cadmium Selenide (CdSe) as an Active Absorber Layer for Solar Cells with V_{OC} Approaching 750 mV 683
> Ebin Bastola, Adam B. Phillips, Abasi Abudulimu, Vlad Kornienko, Zulkifl Hussain, Manoj K. Jamarkattel, Tamanna Mariam, Prabodika N. Kaluarachchi, Jared Friedl, Dipendra Pokhrel, Kara B. Kile, Zhaoning Song, Yanfa Yan, Michael Walls, Randy J. Ellingson, Michael J. Heben

Mapping Stress in PV Modules: The Influence of Soldering, Tabbing, and Module Architecture.................. 689
Ian Slauch, Rico Meier, Kaushik Roy Choudhury, Mariana Bertoni

Elemental Vapor Transport Deposition of $CdSe_xTe_{1-X}$ Thin Films for n-Type CdTe Solar Cells 690
Wei Wang, Vasilios Palekis, Md Zahangir Alom, Sheikh Elahi Tawsif, Don Morel, Chris Ferekides

Intra-Grain Local Luminescence Properties of $CdSe_{0.1}Te_{0.9}$ Thin Films 695
Ganga R. Neupane, David S. Albin, Joel N. Duenow, Matthew O. Reese, Susanna Thon, Behrang H. Hamadani

Electroluminescence Imaging: A Study in the Impact of Microscopic Surface Defects 698
Meghan E. Bush, Timothy J. Peshek

Towards a Three-Terminal Perovskite/Silicon Tandem Solar Cell with Highest Efficiency 701
Michael Rienäcker, Somayeh Moghadamzadeh, Paul Fassl, Yevgeniya Larionova, Philipp Noack, Bianca Wattenberg, Ulrich Wilhelm Paetzold, Robby Peibst

Luminescence and Thermal Imaging Applied to Half-Cut-Cell and Emitter-Wrap-Through-Cell
Modules 702
Steve Johnston, Dana B. Kern, Kent Terwilliger, Ingrid L. Repins

Role of Solar Photovoltaics for a Sustainable Energy System in Puerto Rico in the Context of the
Entire Caribbean Featuring the Value of Offshore Floating Systems 708
Christian Breyer, Ayobami S Oyewo, Alejandro Kunkar, Rasul Satymov

Energy-Harvesting Efficiency Analysis for Solar Modules Using 2T and 4T Tandem Solar Cells............... 709
Robert Witteck, William E. McMahon, John F. Geisz, Qi Jiang, Emily L. Warren

A Reproducible Validation of Algorithms for Estimating Array Tilt and Azimuth from Photovoltaic
Power Time Series.. 710
Kirsten Perry, Bennet Meyers, Kevin Anderson, Matthew Muller

Investigation of P-Type Silicon Heterojunction Radiation Hardness 716
Romain Cariou, Adrien Danel, Nicolas Enjalbert, Frédéric Jay, Sébastien Dubois

Energy Injustice Metrics for Puerto Rico 717
Pablo Méndez-Curbelo, Carlos Peña, Willian Pacheco, Marcel Castro-Sitiriche

Inverse Design of Spectrally-Selective Films for PbS-CQD Tandem Solar Cells................................. 718
Sreyas Chintapalli, Tina Gao, Luna Singh, Serene Kamal, Arlene Chiu, Yijun Zhang, Susanna M. Thon

AM0 Optimized Dual Junction Quantum Well Solar Cells-Investigation of Radiation Tolerance
Designs and V_{OC} Retention at EOL... 721
Brandon M. Bogner, Stephen J. Polly, Seth M. Hubbard, Mitsuru Imaizumi, Roger E. Welser

Evaluation of Process Damage to Crystalline Silicon by Transparent Conductive Oxide Film
Deposition 724
Haruki Kojima, Tappei Nishihara, Yuta Ito, Hyunju Lee, Kazuhiro Gotoh, Noritaka Usami, Tomohiko Hara, Kyotaro Nakamura, Yoshio Ohshita, Atsushi Ogura

Brownfields to Brightfields: The Potential for Landfill Solar Redevelopment in New York State 727
Henry J. Williams, Hugh Peng, Olivia Goosay, K. Max Zhang

Carbon Footprint of Silicon Photovoltaics Manufacturing in North America................................. 730
Annick Anctil, Angela Farina, Luyao Yuan

The PV Efficiency Vs R&D Effort Learning Curve for Research-Stage Material Technologies 731
Phillip J. Dale, Michael A. Scarpulla

Perovskite/Silicon Tandem Solar Cells with Front Side Metallization Applied Prior to Top Cell
Fabrication Enabling High Curing Temperatures 733
Sara Baumann, Annika Raugewitz, Felix Haase, Tobias Wietler, Robby Peibst, Marc Köntges

Performance and Degradation in Silicon PV Systems Under Outdoor Conditions in Relation to
Reliability Aspects of Silicon PV Modules – Summary of Results of COST Action PEARL PV 737
*S. Lindig, J. Ascencio-Vásquez, J. Leloux, D. Moser, M. Aghaei, A. Fairbrother, A. Gok, S.
Ahmad, S. Kazim, K. Lobato, W. J. G. H. M. Van Sark, N. Pearsall, B. G. Burduhos, A.
Raghoebarsing, G. Oreski, J. Schmitz, M. Theelen, P. Yilmaz, J. Kettle, A. H. M. E. Reinders*

Progress Towards Scaling Perovskite/Silicon Tandem Modules 740
*Chris Eberspacher, Colin Bailie, Tim Gehan, Bryan Rosales, Tom Brenner, Mike Chen, Terry
Banks*

Identification of Module Replacements in US Utility-Scale Photovoltaic Installations 743
Chenyang Deng, Jacob T. Stid, Preeti Nain, Annick Anctil

Indoor and Outdoor Evaluation of Curved Modules for VIPV 746
*Rebeca Herrero, Ignacio Antón, Francisco Martín, Steve Askins, Javier Macías, Luis J. San
José, G. Vallerotto, R. Núñez, C. Domínguez*

Towards an Annual Terrawatt Photovoltaics Market - Comparison of the Social Acceptance in
Various IEA PVPS Countries 749
*Arnulf Jäger-Waldau, Georg Altenhöfer-Pflaum, Otto Bernsen, Christian Breyer, Jose
Donoso, Hubert Fechner, Kenn Henrik Bournonville, Jarand Hole, Izumi Kaizuka, Linda
Koschier, Johan Lindahl, Gaëtan Masson, Daniel Mugnier, Chinho Park, Lionel Perret,
Francesca Tilli*

Potential Induced Degradation Evaluation of Damp Heat Stressed PV Modules 755
Farrukh Ibne Mahmood, Akash Kumar, Muhammad Afridi, Govindasamy Tamizhmani

Data-Driven Photovoltaic Module Performance Analysis with FAIR Data 758
*Mengjie Li, Jarod Kaltenbaugh, Dylan J. Colvin, William C. Oltjen, Arafath Nihar,
Dominique Akissi Yao, Xuanji Yu, Alp Sehirlioglu, Roger H. French, Kristopher O. Davis*

Proposed Update of the Colombian Technical Standard NTC 4405 for Evaluating the Efficiency of
Photovoltaic Solar Systems and Their Components 761
Johann Hernández, Daniel H. Gamboa, Juan F. Beltrán

Aerosol-Deposited SnOx as an Electron Contact in Perovskite Solar Cells 767
David Quispe, David Matthews, Zhengshan J. Yu, Zachary C. Holman

The Temperature Dependence of Auger Recombination in Silicon 768
Jorge Ochoa, Simone Bernardini, Mariana Bertoni

Combining Perovskites and Quantum Dots: Application in Solar Cells 769
*Jose Raul Montes Bojorquez, Maria Fernanda Villa Bracamonte, Kevin J. Knebel, Arturo A.
Ayon*

Nanostructure Analysis of Parasitic Oxides and Contact Resisitivity Degradation During Annealing
of Silicon Heterojunction Solar Cells 770
*Stefan Lange, Angelika Hähnel, Christoph Luderer, David Adner, Martin Bivour, Christian
Hagendorf*

Impact of Selenium Doping in CdSeTe-Based Solar Cells at the Atomic-Scale 771
  Arashdeep S. Thind, Jack Farrell, Robert F. Klie

Performance of Vertical Bifacial 2T and 3T Perovskite/Silicon Tandem Solar Farms 772
  Syed Usama Bin Afzal, Hassan Imran, Suleman Sami Qazi, Muhammad Ashraful Alam,
  Nauman Zafar Butt

Influence of Insertion Position of a LiF Buffer Layer on Passivation Performance of Crystalline
Si/SiO$_y$/TiO$_x$/ Al Heterostructures 775
  Shohei Fukaya, Kazuhiro Gotoh, Takuya Matsui, Hitoshi Sai, Yasuyoshi Kurokawa, Noritaka
  Usami

Radiometric Standards and Best Practices: Recent Progress 778
  Aron Habte, Manajit Sengupta, Christian A. Gueymard

Compact and High Efficiency Micro-CPV Module with High Wafer Utilization Rate 781
  Corentin Jouanneau, Thomas Bidaud, Maxime Darnon, Gwenaelle Hamon

Selenium Diffusion During CdCl2 Treatment of CdSeTe Solar Cells 782
  Niranjana Mohan Kumar, Srisuda Rojsatien, Angel De La Rosa, Barry Lai, Arun K. M.
  Kanakkithodi, Maria Chan, Dan Mao, Mariana Bertoni

Analysis of Measured Operating Temperature of Perovskite Modules. 783
  D. Martinez Escobar, Aaron Wheeler, F. Brigham Pineda, Katty Kaydanik, Alan Murphy,
  Jing-Shun Huang, Mason Terry, Sarah R. Kurtz

A Broadband Anti-Reflection Coating for Thin Film CdSeTe/CdTe Solar Cells 789
  Adam M Law, Luksa Kujovic, Mustafa Togay, Xiaolei Liu, Kurt Barth, John M Walls

Can Solar+Storage Keep the Lights On? Assessing Solar+Storage for Backup Power During Long-
Duration Power Interruptions in the US 793
  Will Gorman, Galen Barbose, Juan Pablo Carvallo, Sunhee Baik, Chandler Miller

Close Roof Mounted System Temperature Estimation for Compliance to IEC TS 63126 794
  Michael D. Kempe, Silvana Ovaitt, Martin Springer, Matthew Brown, Dirk Jordan, William
  Sekulic, Colleen O'Brien, Jean-Nicolas Jaubert, Yuanjie Yu, Jaewon Oh, Govindasamy
  Tamizhmani, Bo Li

I-TOPCon Solar Cells Prepared by High Throughput Magnetron Sputtering of In-Situ Doped n-
Type Amorphous Silicon Layers 795
  Eric Schneiderloechner, Tina Dietsch, Jan Hoss, Jonathan Linke, Jana-Isabelle Polzin,
  Sebastian Mack, Henning Nagel, Volker Linss

Optimizing Demand Management to Enable Renewables: Why the Use of a Marginal Emissions
Signal is a Poor Choice 798
  Ronald A. Sinton

Charge Extraction and Recombination Dynamics of CdSe/CdTe Solar Cells Studied with Transient
Photovoltage/Photocurrent Techniques 804
  Abasi Abudulimu, Dengbing Li, Lei Chen, José Santos, Iwan Zimmermann, Nadeesha
  Katakumbura, Tyler Brau, Ebin Bastola, Adam Phillips, Zhaoning Song, Juan Cabanillas,
  Michael Heben, Mohammad K. Nazeeruddin, Nazario Martín, Yanfa Yan, Randy Ellingson

The Feasibility of Luminescent Solar Concentrators Overlays for Conventional Lens 809
  Xitong Zhu, Michael G. Debije, Angèle H. M. E. Reinders

Highly Efficient Bifacial Single Junction Perovskite Solar Cells.. 812
Qi Jiang, Zhaoning Song, Yanfa Yan, Kai Zhu

Efficiency Maps for Tandem Solar Cells Using High Resolution Spectral Data... 815
Rune Strandberg, Anne G. Imenes

Optimal Allocation of Voltage Regulations to Maximize the Hosting Capacity of Distribution
Systems.. 816
Bahman Ahmadi, Eli Shirazi

In-Situ Microscopy Characterization of Light-Induced Phase Segregation in Wide-Bandgap
Perovskite Materials 822
*Fangfang Cao, Liming Du, Zhiyu Gao, Minghui Li, Cong Chen, Dewei Zhao, Can Li, Zhen
Li, Jichun Ye, Chuanxiao Xiao*

Radiation Tolerance Studies of CdSe/CdTe Bilayer Solar Cells on Space-Qualified Cover Glass 823
*Aesha P. Patel, Adam B. Phillips, Ebin Bastola, Abasi Abudulimu, Zachary W. Zawisza,
Robert Snuggs, Manoj K. Jamarkattel, Deng-Bing Li, Richard Irving, Yanfa Yan, Michael J.
Heben, Randy J. Ellingson*

Advanced Production Line Monitoring with Time-Lag Sequential Analysis... 828
Gaia Maria N. Javier, Rhett Evans, Priya Dwivedi, Thorsten Trupke, Ziv Hameiri

NASA GRC Solar Cell Characterization: Facilities 829
Jeremiah D Sims

Application of Noise-Assisted Multivariate Data Analysis for Hour-Ahead GHI Forecasting 830
Priya Gupta, Rhythm Singh

Can Grid-Following DERs Operate in Parallel with Grid-Forming Resources Without
Compromising Microgrid Stability?... 835
Wenzong Wang, Aminul Huque

Developing Frequency Stability Constraint for Unit Commitment Problem Considering High
Penetration of Renewables 840
Ningchao Gao, Shuan Dong, Xin Fang, Andy Hoke, David Wenzhong Gao, Jin Tan

Precursor Ink Engineering for Scalable Slot-Die Coating of Perovskite Films for Photovoltaic
Mini-Module Production 844
*Manoj Rajakaruna, Jaehoon Chung, You Li, Tamanna Mariam, Muhammad Saeed Mohsin,
Prabodika N. Kaluarachchi, Amirhossein Rahimi, Zhaoning Song, Michael J. Heben, Yanfa
Yan, Randy J. Ellingson*

GaAs Solar Cells Grown on Acoustically-Spalled GaAs Substrates with 27% Efficiency 848
*Kevin L Schulte, Steve W Johnston, Anna K Braun, Jacob T Boyer, Anica N Neumann, William
E McMahon, Michelle Young, Pablo Guimerá Coll, Mariana I Bertoni, Emily L Warren,
Myles A Steiner*

Why Increased CdSeTe Charge Carrier Lifetimes and Radiative Efficiencies Did Not Result in
Voltage Boost for CdTe Solar Cells? 849
*Darius Kuciauskas, Alexandra Bothwell, Carey Reich, Chungho Lee, Eric Colegrove, Marco
Nardone*

Detection and Impact of Cracks Hidden Near Interconnect Wires in Silicon Solar Cells 850
Andrew M. Gabor, Hubert Seigneur, Philip J. Knodle, Dylan J. Colvin, Kristopher O. Davis

Net Zero Water Strategies and Impacts for PV Manufacturing 855
Parikhit Sinha, Sunil Sajja, Tzy Wei Ooi, Sreenivas Jayaraman, Sukhwant Raju

Characterization of Different Groups of Electricity Consumers and Measures Taken to Reduce
Energy Poverty 858
*Anna Carolina De Paula Sermarini, Lucas Aló Rodrigues Araujo Da Silva, Vanessa Cardoso
De Albuquerque, Rodrigo Flora Calili*

Using Neural Network Decomposition to Estimate Field Photovoltaic Performance Loss Rate 861
*Yangxin Fan, Raymond Wieser, Xuanji Yu, Jennifer Braid, Avishai Shaton, Adam Hoffman,
Thevenard Didier, Ben Spurgeon, Daniel Gibbons, Laura S. Bruckman, Yinghui Wu, Roger H.
French*

Capacitance Transients, Photoconductive Decay, and Impedance Spectroscopy on 19% to 22%
Efficient Silicon Solar Cells 865
Steve Johnston, Dana B. Kern

Development of Next Generation Solar Trackers Based on Shape Memory Alloy to Be Integrated in
CPV/PV Hybrid Modules 871
Alessandro Minuto, Edoardo Celi, Gianluca Timò

Quantifying Uncertainty Due to Climate Variability in Vehicle-Integrated Photovoltaic Yield
Predictions 874
Timofey Golubev

Phase Distributions and Local Bandgap Energies in Mixed-Halide Perovskite Nanoparticles 879
*Dan R. Wargulski, Tal Binyamin, Katrina Coogan, Benedikt Haas, Christoph Koch, Lioz
Etgar, Daniel Abou-Ras*

Third Generation Approaches for Low Cost, Radiation Tolerant, Efficienct Space Solar Cells 881
Gavin Conibeer, Santosh Shrestha

Effect of Novel Optimization Algorithm on the Performance of Photovoltaic Devices 885
Sheri F Michael

Towards Transfer Printing GaSb Membranes Using Selective Etchants 886
Margaret A Stevens, Jill A Nolde, Shawn Mack, Thomas C Mood, Kenneth J. Schmieder

Investigating the Role of Ag and Ga Content in the Stability of Wide-Gap (Ag,Cu)(In,Ga)Se2 Thin
Film Solar Cells 887
Patrick Pearson, Jan Keller, Charlotte Platzer-Björkman

A Horizontal Single-Axis Tracker Mock-Up to Quickly Assess the Influence of Geometrical
Factors on Bifacial Energy Gain.. 888
César Domínguez, Jesús Marcos, Sandra Ures, Steve Askins, Ignacio Antón

A New Method for the Evaluation of Majority and Minority Carrier Contact Resistivity of
Polysilicon on Oxide Contacts 891
Dirk Steyn, William Nemeth, David Young, Paul Stradins, Sumit Agarwal

Simultaneous Solar Power Generation and Bidirectional Data Transmission 894
*Emily Kessler-Lewis, Stephen J. Polly, Elijah Sacchitella, Seth M. Hubbard, Raymond
Hoheisel*

Localized Surface Plasmon Resonance of Quantum Dots in Two-Step Photon Up-Conversion Solar
Cell Structures 895
Yukihiro Harada, Mizuto Kawakami, Shigeo Asahi, Takashi Kita

Understanding the Degradation of Silicon Heterojunction Modules 896
Jorge Ochoa, Michael Martinez-Szewczyk, Nicholas Moser-Mancewicz, Dana Kern, Dirk Jordan, Mariana Bertoni

Enhancing Inverter Reliability: Current Status and Paths to Predictive Maintenance 897
Wayne Li, Rabin Dhakal, Daniel Fregosi, Curtis Fox, Michael Bolen

Improvements on Spectral Correction Predictive Modeling for CdTe Modules 903
Alan J. Curran, Boris Lin, Yuepeng Deng

Modification of PEDOT:PSS Hole Transporting Layer to Improve the Efficiency and Light Stability of Tin-Lead Perovskite Solar Cells 909
Lei Chen, Chongwen Li, Tyler R. Brau, Abasi Abudulimu, Randy J. Ellingson, Zhaoning Song, Yanfa Yan

Improving V_{OC} of CdSe/CdTe Solar Cells Via Incorporating Oxygenated CdS Between Front Buffer and Absorber 910
Abasi Abudulimu, Dengbing Li, Manoj Jamarkattel, Zachary Zawisza, Scott Wenner, Tyler Brau, Ebin Bastola, Adam Phillips, Michael Heben, Yanfa Yan, Randy Ellingson

From Accelerated Life Test to Accurate Degradation Prediction of CdTe PV Devices: A Modeling Approach 914
Da Guo, Jaliu Ma, Samuel Demtsu, Ryan Monnin, Igor Sankin, Jose A. Calderon, Markus Gloeckler

Quantifying Bulk and Surface Recombination in CdSeTe Absorbers by Modeling Terahertz and Photoluminescence Decays 915
Gregory A Manoukian, Calvin Fai, Deborah L McGott, Finley R Shapiro, Matthew O Reese, Charles J Hages, Jason B Baxter

Data Driven Energy Resilience for Low-To Middle-Income Communities in Puerto Rico 916
Christopher Gregory, Angel Echevarria, Yiamar Rivera-Matos, Daniel D. Campo-Ossa, Alex Routhier, Clark Miller, Richard King

Advances in GaAsP Top Cells for Use in GaAsP/Si Tandems 917
Tal Kasher, Lauren M. Kalizewski, Marzieh Baan, Chuqi Yi, Anastasia H. Soeriyadi, Atom Chang, Udo Römer, Gianluca Coletti, Stephen P. Bremner, Tyler J. Grassman, Steven A. Ringel

Key Areas of Due Diligence for Solar PV Project Financing 918
Eric R. Decristofaro, Matthew R. Thibodeau, Fitzgerald C. Okoli, Jake T. Silhavy, Evan S. Giacchino, Adam W. Loeding, Alexander E. Coologeorgen

Convergence of Efficiency, Stability, and Manufacturability in Perovskite Tandem Solar Cells 919
Rohit Prasanna, Tomas Leijtens, Jochen Titus, Laura E. Crowe, Hyunjong Lee, Annikki L. Santala, Maximilian T. Hoerantner, Giles E. Eperon

Validating View-Factor Approach and Spatial Albedo Models for Bifacial and AgriPV Modeling 920
Silvana Ovaitt, Matthew Boyd, Austin Kinzer, James Jones, Chris Deline

Field and Accelerated Aging of Cracked Solar Cells 921
Michael G. Deceglie, Timothy J Silverman, Ethan Young, William B. Hobbs, Cara Libby

Orange Button: Accelerating the Digital Transformation of Distributed Energy 922
Clifford W. Hansen, Jan Rippingale, Taos Transue, Philip Court, John Gorman

Analysis for Effects of Temperature Rise of Solar Cell Modules Upon the Driving Distance of Photovoltaics-Powered Vehicles.. 924
    Masafumi Yamaguchi, Taizo Masuda, Tsutomu Tanimoto, Yosuke Tomita, Yasuyuki Ota, Christian Thiel, Anastasios Tsakalids, Arnulf Jaeger-Waldau, Takashi Nakado, Kenji Araki, Kensuke Nishioka, Tatsuya Takamoto, Kyotaro Nakamura, Ryo Ozaki, Nobuaki Kojima, Yoshio Ohshita

Laser-Weld Qualification for a Reliable Aluminum Foil Interconnection of Copper-Metallized Back-Contact Silicon Solar Cells 925
    Barry Hartweg, Kathryn Fisher, Jason Ro, Zachary Holman

Conversion Efficiency Analysis of Tandem Solar Cells with Intermediate Band Tunnel Connection 926
    Shuhei Yagi, Hiroyuki Yaguchi

Effects of Period of Record Extension, Model Diversification, and DHI Measurements on Measure-Correlate-Predict Analyses for On-Site Solar Resource Assessments 927
    Lucila D. Tafur, Renn Darawali, Halley Darling

Solar Forecasting: The Value of Using Satellite Derived Irradiance Data in Machine Learning Based Forecasts 928
    Alex Kubiniec, Thomas Haley, Kyle Seymour, Richard Perez

Ageing Detection of Encapsulants and Backsheets in the Field Via NIR Spectroscopy 931
    Chiara Barretta, Sascha Lindig, Marton Bredács, Alexander Astigarraga, Eric Helfer, Gernot Oreski

In-Situ & Ex-Situ Study of Protons and Electrons Irradiations of Perovskite Solar Cells 934
    Carla Costa, Matthieu Manceau, Thierry Nuns, Sophie Duzellier, Romain Cariou

Numerical Investigation on Non-Radiative Recombination in InGaAs Front and Rear Hetero-Junction Solar Cell 935
    Depu Ma, Hassanet Sodabanlu, Gan Li, Meita Asami, Kentaroh Watanabe, Masakazu Sugiyama, Yoshiaki Nakano

Strategies to Improve the Mechanical Robustness of Metal Halide Perovskite Solar Cells............................ 938
    Muzhi Li, Siraj Sidhik, Lidon Gil-Escrig, Samuel Johnson, Aditya Mohite, Axel Palmstrom, Henk J. Bolink, Nicholas Rolston

Microstructure-Property Relationships in Epitaxial Cu(In, Ga)Se$_2$ Solar-Cell Absorbers.............................. 939
    Daniel Abou-Ras, Jiro Nishinaga, Takeyoshi Sugaya, Yukiko Kamikawa-Shimizu, Ulrike Bloeck, Henrik Prell, Sinju Thomas, Michael Tovar, Dan R. Wargulski, Harvey Guthrey, Pat Trimby, Aimo Winkelmann, Shogo Ishizuka

High-Performance Multi-Junction C-Band Photonic Power Converters: Calibrated Optoelectronic Model for Next Generation Designs... 941
    Gavin P Forcade, Meghan N Beattie, Christopher E Valdivia, Henning Helmers, Oliver H?hn, Paige Wilson, Louis-Philippe St-Arnaud, Robert Hunter, David Lackner, Jacob J Krich, Alexandre W Walker, Karin Hinzer

Strategies for High Fill Factor and Open-Circuit Voltage in Low-Doped c-Ge TPV Cells with Partially Contacted Surfaces Using 3D Simulations 942
    M. Gamel, D. Shojaei, J. M. López-González, G. López, M. Garín, I. Martín

Minimizing Sputter Damage-Induced Electrical Losses in Monolithic Perovskite/Silicon Tandem Solar Cells During Deposition of the Transparent Front-Electrode... 945
    Marlene Härtel, Bor Li, Silvia Mariotti, Philipp Wagner, Florian Ruske, Steve Albrecht, Bernd Szyszka

Nanometer-Scale Imaging on Electrical Potential in Absorber of As-Doped CdSeTe Solar Cells 946
Chun-Sheng Jiang, Eric Colegrove, Steve P. Harvey, Joel N. Duenow, Matthew O. Reese

Advanced Germanium TPV Cells for Latent Heat Thermal Batteries 947
A. Luque, P. Martin, R. Molinero, V. Orejuela, C. Sanchez-Perez, M. Zehender, I. García, I. Luque-Heredia, I. Rey-Stolle

Evaluation of Irradiance Variability Adjustments for Subhourly Clipping Correction 948
William B. Hobbs, Chloe L. Black, William F. Holmgren, Kevin S. Anderson

Evaluation of Beta-Phase Formation in the Failure of PVDF-Based Solar Module Backsheets 952
Stephanie L. Moffitt, Sona Ulicna, Song-Syun Jhang, Michael Owen-Bellini, Peter Hacke, Jared Tracy, Kaushik R. Choudhury, Laura T. Schelhas, Xiaohong Gu

Development of a Dynamic Photovoltaic Inverter Model with Grid-Support Capabilities for Power System Integration Analysis 953
Rachid Darbali-Zamora

PV Fleet Modeling Via Smooth Periodic Gaussian Copula 961
Mehmet G. Ogut, Bennet Meyers, Stephen P. Boyd

Unveiling the Structural Formation of Low Dimensional Layers Deposited on Lead Halide Perovskites by Thermal Evaporation 969
Carlo Andrea Riccardo Perini, Andres Felipe Castro Mendez, Tim Kodalle, Magdalena Ravello, Juanita Hidalgo, Juan-Pablo Correa-Baena

Ultrasonic Characterization of Ethylene Vinyl Acetate (EVA) Crosslinking for Quality Assurance and Lamination Process Control 970
Rico Meier, Ian M. Slauch, Mariana I. Bertoni

Detailed Raman Investigation on the Search for the Secondary Phases in the Chalcogenide Perovskite BaZrS3 975
Arif Yetkin Hasan, Ruiquan Yang, Charles Hages, Phillip J. Dale

A Direct Comparison of Thermoradiative and Thermophotovoltaic Operation of HgCdTe Photodiodes 976
Michael P. Nielsen, Muhammad H. Sazzad, Andreas Pusch, Phoebe M. Pearce, Peter J. Reece, Nicholas J. Ekins-Daukes

Solar Simulator Performance Metrics: Balloon Flown Calibration Standards Offer Real Time AM0 Solar Simulation Error Measurements 977
Scott J Ireton, Casey P Hare, Andrew J Schwab

Operation Efficiency Gains from Analyzing Minimal Solar Cells Production Data 978
Johnson Wong, Kissenger Chen, Dinica Li, Bryan Matthews, Jason Miller, Sal Sanci

Extracting Electrical Properties of CdTe, CdSeTe and CdSe Thin Films Using a Parallel Dipole Line Hall Effect System 981
Mustafa Togay, Rachael C. Greenhalgh, Kerrie Morris, Xiaolei Liu, Luksa Kujovic, Luis C. Infante-Ortega, Nicholas Hunwick, Adam M. Law, Tushar Shimpi, Sampath S. Walajabad, Eric Don, Gabor Parada, Kurt L. Barth, J. Michael Walls, Jake W. Bowers

Power Hardware-In-The-Loop Interface Method for Grid Forming Inverters Using a Voltage-Controlled Power Amplifier 984
Javier Hernandez-Alvidrez, Rachid Darbali-Zamora, Jack D. Flicker, Nicholas S. Gurule

Implications of Battery Storage for Solar Net-Metering Reforms 991
Galen Barbose, Sydney Forrester, Chandler Miller

III-V Solar Cells Grown Directly on V-Groove Si Substrates 994
Theresa E. Saenz, Jacob T. Boyer, John S. Mangum, Anica N. Neumann, Myles A. Steiner,
Ryan M. France, William E. McMahon, Jeramy D. Zimmerman, Emily L. Warren

Correlated Mapping of Raman Spectroscopy and Cathodoluminescence of Emerging Absorber
Bournonite (CuPbSbS3) 995
O M Rigby, C Hill, G Kusch, M Naylor, M Guennou, G Zoppi, M Szablewski, R A Oliver, L
Wirtz, P Dale, B G Mendis

Yb3+- Doped CsPbX3 Nanocrystals for Improving Free-Space Luminescent Solar Concentrators 996
Mathis Van De Voorde, Damien Hudry, Dmitry Busko, Bryce S. Richards, Rebecca Saive

Rear Heterojunction GaInP Solar Cells for Improved Performance at Elevated Temperatures 997
Mijung Kim, Yukun Sun, Ryan D. Hool, Minjoo L. Lee

Design and Fabrication of PERC-Like CdTe Solar Cells Using Micropatterned Al₂O₃ Layer 998
Etee Kawna Roy, Kaden Powell, Chungho Lee, Gang Xiong, Heayoung Yoon

Issues, Challenges, and Primary Factors in the Estimation of Floating Solar PV Performance 1002
Rick Meeker, Anna Brinck, Tom Lang, Jason Harrison

Partial Shading of Photovoltaic Modules: A Comparison Between Simulated and Measured IV
Characteristics 1005
Bianca Passarella, Maarten Verkou, Marco Leonardi, Fabrizio Coco, Youri Blom, Malte Vogt,
Rudi Santbergen, Agnese Di Stefano, Andrea Canino, Marina Foti, Antonino Ragonesi,
Marcello Sciuto, Francesco Rametta, Miro Zeman, Olindo Isabella, Cosimo Gerardi

Hierarchal Ti3C2Tx MXene and Aluminum Microgrid Back Contacts for Bifacial CdTe PV 1008
Benjamin E Sartor, Matthew O Reese, Chris Muzzillo, Chungho Lee, Andre D Taylor

Design of Electronic Control of PV Tracking Independent of Weather Forecast 1009
Sam Mil'Shtein, Dhawal Athana, Jeffrey Snell, William Brooks

Efficient and Stable All-Lead Perovskite Tandem Solar Cells Enabled by All-Inorganic CsPbI2Br
Top Cells.. 1019
Chongwen Li, Chuanxiao Xiao, Kamala Khanal Subedi, Bin Chen, Randy J. Ellingson, Song
Zhaoning, Yanfa Yan, Edward H. Sargent

(3-Aminopropyl)trimethoxysilane Surface Passivation Improves Perovskite Solar Cell Performance
by Reducing Surface Recombination Velocity 1020
Yangwei Shi, Esteban Rojas-Gatjens, Jian Wang, Justin Pothoof, Rajiv Giridharagopal, Kevin
Ho, Fangyuan Jiang, Margherita Taddei, Zhaoqing Yang, Carlos Silva-Acuña, David S.
Ginger

Machine Learning-Based Defect Identification Method at the c-Si/A-Si:H Interface 1021
Zitong Zhao, Gonglin Chen, Reza Vatan Meidanshahi, Gergely T. Zimányi

Recombination Analysis of Maxeon IBC Production Cells by Time-Resolved Photoluminescence 1024
David Jacob, Guillaume Von Gastrow, Nils-Peter Harder, Luis Buño, Gerly Reich, Maristel
Baldrias, Roderick J. Marstell, Arnold Castillo, David D. Smith, Michael J. Cudzinovic

Can Hierarchical Physics-Based Machine Learning De-Anonymize Solar Farm Locations? 1027
Jabir Bin Jahangir, Amandeep Singh Bhatia, Muhammad Alam

Investigation and Quantitative Understanding of Front Field Passivation in Rear Junction Selective Double-Side TOPCon Solar Cells 1030
  Wook-Jin Choi, Young-Woo Ok, Pradeep Padhamnath, Gabby De Luna, Kwan Hong Min, Ruohan Zhong, Sagnik Dasgupta, Vijaykumar D Upadhyaya, Ajay D Upadhyaya, Ajeet Rohatgi

Autonomous Control Strategies for Interconnected DC Microgrids with Geographical Separation 1034
  Emmanuel G. Robles-Rivera, Rachid Darbali-Zamora, Erick E. Aponte-Bezares, Jack D. Flicker, Andrew R. R. Dow, Felipe Palacios, Lee Rashkin

Enabling High Efficiency, Flexible, and Lightweight CdTe Solar Cells with a Cadmium Stannate Transparent Conducting Oxide 1041
  Manoj K. Jamarkattel, Adam B. Phillips, Ebin Bastola, Sabin Neupane, Deng-Bing Li, Yanfa Yan, Randy J. Ellingson, Michael J. Heben

Post-Mortem Failure Analysis of Metal Halide Perovskite Modules 1044
  Sona Ulicna, Nutifafa Y. Doumon, Michael Owen-Bellini, Jackson Schall, Dana B. Kern, Timothy J. Silverman, Lance M. Wheeler, Steven Hayden, Chengbin Fei, Md Aslam Uddin, Prem J. S. Rana, Jinsong Huang, Joseph J. Berry, Laura T. Schelhas

Tuning Device Interfaces for Improved Open Circuit Voltage in Wide-Bandgap Hybrid Perovskite Photovoltaics 1045
  Emily Smith, Sarah Brittman, David Scheiman, Woojun Yoon

Multiple-Reuse of Ge Substrates: Towards Cost-Effective and Sustainable III-V Solar Cells Fabrication 1051
  Alexandre Chapotot, Bouraoui Ilahi, Tadeáš Hanuš, Gwenaëlle Hamon, Jinyoun Cho, Kristof Dessein, Christian Dubuc, Maxime Darnon, Abderraouf Boucherif

Mitigation of Potential Induced Degradation in Perovskite Solar Cells 1052
  Laxmi Nakka, Shen Guibin, Armin G. Aberle, Fen Lin

A Robust Approach for Daily Solar Irradiance Clustering 1053
  Roshni Agrawal, Sivakumar Subramanian, Shashank Agarwal, Venkataramana Runkana

On the Role of Sn-Halide Post-Deposition Reactive Annealing for the Passivation of Defective Surfaces in Cu2ZnSnSe4 1059
  Alex Jimenez-Arguijo, Yuancai Gong, David Nowak, Devendrá Pareek, Levent Gütay, Lorenzo Calvo-Barrio, Zacharie Jehl Li-Kao, Sergio Giraldo, Edgardo Saucedo

Analysis of Thermal Behavior and Reliability of Bare Die Diodes Embedded Within PV Modules as Bypass Devices 1060
  Luis Eduardo Alanis, Julian Weber, Pascal Romer, Marc Andre Schüler, Louisa Winkler, Udo Steinebrunner, Martin Heinrich, Dirk Holger Neuhaus

Measurement and Analysis of Annual Solar Spectra at Different Installation Angles in Central Europe 1066
  Guillermo A. Farias-Basulto, Miguel Á. Sevillano-Bendezú, Maximillian Riedel, Mark Khenkin, Jan A. Töfflinger, Rutger Schlatmann, Reiner Klenk, Carolin Ulbrich

Photoluminescence Analysis of the Back Side of Cu(In,Ga)(S,Se)2 Absorbers 1067
  Aubin JC. M. Prot, Susanne Siebentritt, Anastasia Zelenina, Hossam Elanzeery, Alberto Lomuscio, Thomas Dalibor, Maxim Guc, Robert Fonoll-Rubio

Effects of Salt Spray on c-Si Photovoltaic Modules in the Brazilian Region 1068
  Mendelsson R. M. Neves, Allan Silveira, Hugo S. Alvarez, Rodrigo M. Garcia, Francisco C. Marques, Marcelo G. Villalva

Experimental Analysis of Distribution Network Voltage Regulation Using Smart Inverters 1074
Rasel Mahmud, Subhankar Ganguly, Jing Wang, Killian McKenna, Ning Li

Doped GaInAs/GaP Quantum Well Superlattice Solar Cells with 27.5% Efficiency 1081
Ryan M France, Myles A Steiner

Aggregated Three-Phase Photovoltaic Inverter Model with Sandia Frequency Shift Islanding
Detection 1082
Nelson E. Saavedra-Peña, Rachid Darbali-Zamora, Edgardo Desarden-Carrero, Erick
Aponte-Bezares

Time-Series Imputation Using Graph Neural Networks and Denoising Autoencoders 1088
Raymond Wieser, Yangxin Fan, Xuanji Yu, Jennifer Braid, Avishai Shaton, Adam Hoffman,
Ben Spurgeon, Daniel Gibbons, Laura S. Bruckman, Yinghui Wu, Roger H. French

Cradle to Cradle Recycling of Perovskite Solar Cells 1092
Zhenni Wu, Gülüsüm Babayeva, Zhang Jiyun, Mykhailo Sytnyk, Jens Hauch, Christoph J.
Brabec, Ian Marius Peters

Per- And Polyfluoroalkyl Substances (PFAS) Usage in Solar Photovoltaics 1095
Preeti Nain, Annick Anctil

Preliminary Gap Analysis of Existing IEEE 1547 and IEEE 2800 Standards Towards Grid-Forming
Technology 1096
Ganesh Marasini, Wenzong Wang, Wes Baker, Deepak Ramasubramanian, Aminul Huque,
Jens C. Boemer

Enhanced Bifaciality Factor with Sb_2Se_3 Devices Modeling Cu_2O Back Buffer 1102
Sanghyun Lee, Kent Price

Analysis of the Key Factors Influencing the Economic Competitiveness and Profitability of
Floating Photovoltaics 1105
Leonardo Micheli, Diego L. Talavera, Fredy A. Sepúlveda-Vélez

Towards Integrating Data Quality Assessments and Radiometer Uncertainty for Determining the
Expanded Uncertainty of Three-Component Solar Radiation Measurements 1106
Stephen Wilcox, Tom Stoffel, Aron Habte, Manajit Sengupta

Stable, High-Throughput Production of Robust Perovskites in Open-Air with Polymer Additives 1109
Nicholas Rolston, Carsen Cartledge, Vineeth Penukula, Muneeza Ahmad, Antonella Giuri,
Aurora Rizzo

Clear-Sky Detection Using Time-Averaged, Tilted-Plane Data 1110
Clifford W. Hansen, Dirk C. Jordan

Theoretical Performance Analysis for Thermo-Radiative Assisted Photovoltaic (TRAP™) Cell
Operating in Outer Space 1114
Jianjian Wang, Nathan Van Velson

A Study of Cell Cracks Formation During Freight Shipping : Monitoring Shock and Temperature in
Real-Time & Assessing Damages with Pre and Post-Transit Characterizations of PV Modules 1120
Cécile Molto, Dylan J. Colvin, Farrukh Ibne Mahmood, Fang Li, Ryan Smith, Govindasamy
Tamizhmani, Hubert Seigneur

Quantifying Real-World Sources of Error in Redundant GHI Measurements 1123
Josh Peterson, Julie Chard, Alex Bryan

Sputtering for the Formation of SI-Based Passivating Contacts..1129
   Christophe Allebé, Antoine Descoeudres, Patrick Wyss, Nicolas Pernès, Bertrand Paviet-Salomon, Christophe Ballif, Simon Hübner, Torsten Dippell

Copper Metallization for III-V Solar Cells..1130
   Theresa E. Saenz, Anica N. Neumann, Matthew R. Young, Jeramy D. Zimmerman, Emily L. Warren, Myles A. Steiner

An Enhanced Snow-Shedding Model: The Module Frame as a Key Variable 1131
   Daniel Riley, Laurie Burnham, Paul Dice, Jennifer Braid

The Effect of Tilt and Azimuth Angle Variations on Monthly and Annual Incident Solar Radiations
for Locations in Brazil...1134
   Eslam Mahmoudi, Tarcio Andre Dos Santos Barros, Hugo Alvarez, Rodrigo Garcia, Francisco C. Marques, Marcelo Gradella Villalva

Rear-Side Irradiance Simulation of Field PV Modules 1141
   Zelin Li, Raymond J. Wieser, Xuanji Yu, Stephanie L. Moffitt, Ruben Zabalza, Silvana Ayala, Matthew Brown, Xiaohong Gu, Liang Ji, Colleen O'Brien, Micheal D. Kempe, Laura S. Bruckman, Kenneth P. Boyce

Extreme Weather and PV Performance 1145
   Dirk Jordan, Kirsten Perry, Chris Deline

Viability of a Novel Methodology of Measure-Correlate-Predict for Albedo Estimation 1146
   Halley C. Darling, Renn Darawali, Lucila D. Tafur

Long Term Soiling Model Tuning for Enhanced PV Cleaning Schedule Optimization 1147
   Kyle Seymour, Akanksha Bhat, Thomas Haley

Coalescence of GaP on V-Groove Si for III-V/Si Solar Cells 1150
   Theresa E. Saenz, John S. Mangum, Olivia D. Schneble, Anica N. Neumann, Ryan M. France, William E. McMahon, Jeramy D. Zimmerman, Emily L. Warren

Mitigation of Phase Separation in High Ga Cu(In,Ga)S2 Absorbers to Achieve ~ 1 Volt 15.6%
Power Conversion Efficiency 1151
   Damilola Adeleye, Mohit Sood, Tobias Törndahl, Adam Hultqvist, Aline Vanderhaegen, Michele Melchiorre, Susanne Siebentritt

Germanium-Tin Diode for Thermophotovoltaic Energy Collection..1152
   Amanda Lemire, Kevin A. Grossklaus, Thomas E. Vandervelde

Rapid Thermal Annealing of Symmetric p-TOPCon Silicon Test Structures.................................1155
   Arpan Sinha, Sagnik Dasgupta, Ajeet Rohatgi, Mool C. Gupta

Temperature Dependent Carrier Extraction and the Effects of Excitons on Emission and
Photovoltaic Performance in Cs0.05FA0.79MA0.16Pb(I0.83Br0.17)3 Solar Cells...........................1156
   Hadi Afshari, Brandon K Durant, Ahmad R Kirmani, Sergio A Chacon, John Mahoney, Vincent R Whiteside, Rebecca A Scheidt, Matthew C Beard, Joseph M Luther, Ian R Sellers

Alba: Testing Emerging Photovoltaic Technologies in Low-Earth Orbit 1157
   Michael D. Kelzenberg, Phillip Jahelka, Richard G. Madonna, Harry A. Atwater

Optimizing the Design of 4-Terminal Perovskite/C-Si Tandem Photovoltaics 1158
   Paul F. Ndione, John F. Geisz, Qi Jiang, Gabriella D. Lahti, Rosemary C. Bramante, Kai Zhu, Axel F. Palmstrom, Emily L. Warren

Investigation of EMC Tests in Photovoltaic Inverter According to INMETRO Ordinance No. 140............1159
  Andrei C. Ribeiro, Geyciane P. De Lima, Guilherme C. S. Prym, Paulo R. D. R. Da Silva,
  Pedro O. C. M. Neto, Romullo R. M. Carvalho, Francisco V. E. Lemos, Mendelsson R. M.
  Neves, Tárcio A. S. Barros, Hugo Da S. Alvarez, Rodrigo M. Garcia, Francisco C. Marques,
  Marcelo G. Villalva

PV Inverter Testing for Momentary Cessation and Rate-Of-Change-Of-Frequency Events............1164
  Ramanathan Thiagarajan, Kumaraguru Prabakar, Li Yu, Ken Aramaki, Andy Hoke

Microgrid Design Toolkit Cost Optimization for a Rural Community in Puerto Rico............1170
  Rolando J. Tremont-Brito, Rachid Darbali-Zamora, Robert Broderick, Erick E. Aponte-
  Bezares, Efrain O'Neill-Carrillo, Matthew S. Lave

Co-Sputtered Sn-Doped ZnO Thin Film n-Type Layers for Incorporation into CdTe Based
Photovoltaics............1177
  Kerrie M. Morris, Mustafa Togay, Luke Jones, John M. Walls, Jake W. Bowers

Investigating the Impact of MACl Doping in FA-Based Perovskites by Multimodal Synchrotron X-
Ray Techniques............1183
  Yanqi Luo, Sanggyun Kim, Carlo A. R. Perini, Naveed Rahman, Luxi Li, Dina Sheyfer, Ross
  Harder, Barry Lai, Juan-Pablo Correa-Baena, Sarah Wieghold

Removing Barriers for Participation of Small PV Systems in Balancing Energy Markets by
Utilizing the Established Smart Meter Eco-System............1184
  Christoph Kondzialka, Manuela McCulloch, Rouven Taubmann, Jonas Dierenbach, Dietmar
  Graeber, Gerd Heilscher

A Comparative Study of the Reflectance of Commercial Photovoltaic Modules............1188
  Wei Wen Toh, Min Hsian Saw, Erik Birgersson, Mauro Pravettoni

The Microstructure of Thin Film CdSe Following Cadmium Chloride Activation Treatment............1193
  Rachael C. Greenhalgh, Vladislav Kornienko, Mustafa Togay, Ali Abbas, Ebin Bastola, Adam
  B. Phillips, Michael J. Heben, Jake Bowers, J. Michael Walls

High Open Circuit Voltage with Organic Hole Transport Layers in Group V Doped CdSeTe Solar
Cells............1197
  Sabin Neupane, Deng Bing Li, Sandip S Bista, Suman Rijal, Zhaoning Song, Alisha Adhikari,
  Lei Chen, You Li, Manoj K. Jamarkattel, Abasi Abudulimu, Dingyuan Lu, Xiaomeng Duan,
  Feng Yan, Michael Heben, Randy J. Ellingson, Gang Xiong, Yanfa Yan

Thin-Film Tandem Partners Based on Inline-Processed (Ag, Cu)(In,Ga)Se$_2$............1201
  Theresa Magorian Friedlmeier, Rico Gutzler, Tina Wahl, Dimitrios Hariskos, Stefan Paetel,
  Erik Ahlswede, Ana Kanevce, Jan-Philipp Becker

Development of 3D/2D Perovskite Solar Cells Using a Spray-Based Sequential Deposition............1206
  Neetesh Kumar, Rishabh Sahani, Daniel Muñoz, Cheng-Yu Lai, Daniela Radu

Numerical Modeling of Bifacial Thin Film Solar Cells............1210
  Briana Dokken, Sabin Neupane, Muhammad Mohsin Saeed, Jared D. Friedl, Adam B.
  Phillips, Michael J. Heben, Yanfa Yan, Zhaoning Song

Passivating Contacts with Engineered Pinhole Enabled Transport............1216
  Harvey L Guthrey, Caroline Lima Salles, William Nemeth, Sumit Agarwal, David Young,
  Pauls Stradins

InP-Based Tunnel Junctions for Micro-Concentrator Photovoltaics 1217
    Kenneth J Schmieder, Thomas C Mood, Eric A Armour, Mitchell F Bennett, Margaret A
    Stevens, Martin Diaz, Ziggy Pulwin, Matthew P Lumb

Residual Stress Limits Gridline Bridging in Cracked Solar Cells 1218
    Junki Joe, Timothy J Silverman, Michael Owen-Bellini, Nick Bosco

Optimizing the Laser Scribing Process to Achieve a Certified Efficiency of 25.9% for Over 240
cm^2 Four-Terminal Perovskite/Si Tandem Solar Cells 1221
    Yonglei Wang, Hongxu Zhang, Yang Liu, Yuan Qin, Jie Liu, Jiang Liu, Bo He, Xixiang Xu

Predicting Damp Heat Degradation in Heterojunction PV Modules Using Machine Learning 1224
    Zubair Abdullah-Vetter, Felix O'Kearney, Priya Dwivedi, Robert Lee Chin, Brendan Wright,
    Thorsten Trupke, Ziv Hameiri

Demand Following RE – a Demand Driven Approach for Rapid RE Capacity Addition in India 1227
    Prashant Kumar Upadhyay, Himanshu Gulati, Yellasiri Bharath Kumar Reddy

Perspectives on PV Adoption and Engaging Gen Z and Millennials in the Indian Scenario 1230
    Robins Anto, Rhythm Singh

An Analysis of the Current Status and Future Potential of Rooftop Solar Adoption in the United
States 1233
    AC Lemay, BP Rand

Investigation of Varying Se Vapor Pressure During Deposition of CdSeTe Thin Film PV Devices 1234
    Sushmakanth Myneni, Carey Reich, Daniel Shaw, Sampath Walajabad, Amit H. Munshi

Performance Optimization of the $CdSe_xTe_{1-x}$/CdTe Solar Cell 1237
    Md Zahangir Alom, Sheikh Tawsif Elahi, Vasilios Palekis, Wei Wang, Chris Ferekides

Analysis of Optoelectronic Characterization Data Via Bayesian Inference: A Desktop-Scale MCMC
Method 1241
    Calvin Fai, Gregory A. Manoukian, Jason B. Baxter, Anthony J. C. Ladd, Charles J. Hages

Generalizability of Neural Network-Based Identification of PV in Aerial Images 1242
    Joseph Ranalli, Matthias Zech

Modal Analysis of GaAs Nanowire Solar Cells for Optimal Device Design 1249
    Venkata S. A. Chaluvadi, Eduardo Camarillo Abad, Hannah J. Joyce, Louise C. Hirst

Benchmark Tests for IV Fitting Algorithms 1250
    Clifford W. Hansen, Abigail R. Jones, Taos Transue, Marios Theristis

Validation of Photovoltaic Plant Loss Estimation from Monitoring Data: String Faults, Shading and
Degradation 1253
    Karel De Brabandere, Maitheli Nikam, Julien Deckx, Gofran Chowdhury

Single-Junction Bifacial and Semitransparent $Sb_2(S,Se)_3$ Solar Cells 1256
    Chen Qian, Kaiwen Sun, Martin Green, Bram Hoex, Xiaojing Hao

The United States Renewable Energy Landscape: Siting, Management, and Potential Impacts 1259
    Jacob T. Stid, Anthony D. Kendall, Annick Anctil, Jeremy Rapp, David W. Hyndman

Interpreting Accelerated Tests on Perovskite Modules: Using Photooxidation of MAPbI3 as an Example 1260
Ingrid L. Repins, Michael Owen-Bellini, Michael D. Kempe, Michael G. Deceglie, Joseph J. Berry, Nutifafa Y. Doumon, Timothy J. Silverman, Laura T. Schelhas

Charge Carrier Diffusion and Recombination Near Misfit Dislocations in GaAsP/GaInP Heterostructures 1261
T. H. Gfroerer, A. Edmondson, Lilian Korir, Fan Zhang, Yong Zhang, M. W. Wanlass

Improving Operational Stability of High-Efficiency Inverted Perovskite Solar Cells 1265
Kai Zhu

Influence of Aluminum Co-Doping on Current-Induced Degradation and Regeneration Kinetics in Boron-Doped Cz PERC Solar Cells 1266
August Weber, Mahsa Mohammadi, Matthias Trempa, Thomas Buck, Johannes Heitmann, Matthias Müller

Reducing the Photovoltaic Operation and Maintenance Costs Through an Autonomous Control Operation Center 1270
Andreas Livera, Álvaro Fernández-Solas, Joao G. Bessa, Jesús Montes-Romero, Eduardo F. Fernández, Vassilis Papaeconomou, George E. Georghiou

Large-Area Uniformity Mapping of High-Speed Flexography-Printed Perovskite Solar Cells Via Scanning Photoluminescence 1276
Julia E. Huddy, William J. Scheideler

Measuring the Doping Concentration of Si and CdTe Absorbers Using Lock-In Amplified Quantitative QSSPL 1279
Mason P Mahaffey, Arthur Onno, Carey Reich, Adam Danielson, Walajabad Sampath, Zachary C Holman

Peer-To-Peer Energy Trading for PV Prosumers Using Fuzzy Logic Inference Systems 1280
Hector Lopez, Ali Zilouchian

Carbon Quantum Dots and Their Possible Application in Perovskites Passivation 1283
Maria Fernanda Villa-Bracamonte, Jose Raul Montes-Bojorquez, Alan Lopez-Becerra, Arturo Ayon

A New Combined Accelerated Stress Test Sequence for Rapid Reliability Screening of Photovoltaic Materials 1284
Yi Jiang, Xuanji Yu, Ben Huang, Ruirui Lv, Yuanjie Yu, Tao Xu, Guangchun Zhang

Improved Soiling Rate Estimation by Calculating PV Module Temperature Using a Distributed Thermal Model 1287
Shoubhik De, Yogeswara Rao Golive, Narendra Shiradkar, Anil Kottantharayil

Electrical and Electroluminescence Evaluation of 17 Year Old Monocrystalline Silicon Building Integrated Photovoltaic Modules 1292
Andrew M. Shore, Tali Schlenoff, Bakary Coulibaly, David Navon, Stephanie L. Moffitt, Brian Dougherty, Behrang H. Hamadani

Automated Photovoltaic Module Quality Assessment: Defect Identification and Classification from Luminescence Images Using Machine Learning 1295
Brendan Wright, James Petesic, Timothy Dawson, Ziv Hameiri

Influence of Alkali Iodide Fluxes on Cu2ZnSnS4 Monograin Powder Properties and Performance of Solar Cells 1298
  Kristi Timmo, Katri Muska, Maris Pilvet, Mare Altosaar, Valdek Mikli, Mati Danilson, Reelika Kaupmees, Jüri Krustok, Maarja Grossberg-Kuusk, Marit Kauk-Kuusik

Pyrolyzer Assisted Vapor Transport Deposition of Antimony-Doped Cadmium Telluride 1299
  Bin Du, Kevin Dobson, Brian McCandless, Aayush Nahar, Ujjwal Das, Shannon Fields, Aaron Arehart, William Shafarman

Early Degradation Trend Estimation of Bifacial PV, Investigating the Seasonality Effect 1304
  Gaetano Mannino, Giuseppe Marco Tina, Mario Cacciato, Lorenzo Todaro, Agnese Di Stefano, Fabrizio Bizzarri, Andrea Canino

The Rise and the Decay of the Photovoltage in Perovskite Solar Cells 1307
  Uwe Rau, Lisa Krückemeier, Sandheep Ravishankar, Thomas Kirchartz

Cadmium Zinc Telluride as an Electron Reflecting Back-Contact Layer for CdTe Solar Cells 1310
  Camden Kasik, Jennifer Drayton, James Sites

Innovative Layouts for Utility-Scale PV Modules: Module Characteristics, Shading Tolerance, and Electricity Costs 1313
  Li C. Rendler, Christian Reichel, Matthew F. Berwind, Anna Heimsath, Dirk H. Neuhaus

UV Absorption Utilizing a MoS2/Ge Nano-Junction for Solar Applications 1320
  Ayman Rezk, Ammar Nayfeh

Defect Signatures in Admittance Spectroscopy of Perovskite Solar Cells 1323
  Rasha A. Awni, Zhaoning Song, Jared D. Friedl, Abasi Abudulimu, Adam B. Phillips, Randall J. Ellingson, Michael J. Heben, Yanfa Yan

Modeling the Hardware Components of a Power Hardware-In-The-Loop Platform for Photovoltaic Applications 1326
  Edgardo Desarden-Carrero, Rachid Darbali-Zamora, Nelson E. Saavedra-Peña, Erick E. Aponte-Bezares

Effective and Equivalent Refractive Index Models for Patterned Solar Cell Films Via a Robust Homogenization Method 1334
  Ekin Gunes Ozaktas, Sreyas Chintapalli, Susanna M. Thon

Evaluation of Rear Contact Passivation Strategies Via Surface Photovoltage Spectroscopy 1337
  Nathan Rock, Chien-Hsuan Chen, Kristopher Davis, Amit Munshi, Michael Scarpulla

Machine Learning, Unmanned Vehicles, and Energy: A Review 1338
  Brian L. Reyes Santiago, Eduardo Ortiz-Rivera

III-V Epitaxy on Detachable Porous Germanium 4" Substrates 1341
  Waldemar Schreiber, Jens Ohlmann, Patrick Schygulla, Stefan Janz, Jinyoun Cho, Kristof Dessein

Effect of Soiling from Dust Particles on Solar Cell Efficiency in the United Arab Emirates (UAE) 1344
  Muntaser Abdelrahman Almansoori, Rawdah Almannaee, Mariam Aldhefairi, Meerh Alsuwaidi, Ayman Rezk, Ammar Nayfeh

Efficiency Limit of Transition Metal Dichalcogenide Solar Cells 1347
  Koosha Nassiri Nazif, Frederick U. Nitta, Alwin Daus, Krishna C. Saraswat, Eric Pop

Advanced Health-State Data Analytic Workflow for Utility-Scale Photovoltaic Power Plants 1348
Jesus Montes-Romero, Loucas Pikolos, Andreas Makrides, Nino Heinzle, George Makrides, Juergen Sutterlueti, Steve Ransome, George E. Georghiou

Monolithic Bifacial Perovskite-CdSeTe Tandem Solar Cells 1353
Zhaoning Song, Deng-Bing Li, Sabin Neupane, Kamala Khanal Subedi, Samuel Seibert, Randy J. Ellingson, Yanfa Yan

Improving the Performance and Yield of Colloidal Quantum Dot Solar Cells Through Electron Transport Layer Optimization 1356
Dana Kachman, Arlene Chiu, Dhanvini Gudi, Chengchangfeng Lu, Eric Rong, Sreyas Chintapalli, Yida Lin, Daniel Khurgin, Susanna M. Thon

Uncertainty Considerations in Bifacial Photovoltaic Systems with High Albedo Seasonality 1359
Javier Lopez-Lorente, Anja Neubert, Mike Hamer

India as an Emerging Solar Manufacturing Country 1366
Juzer Vasi, Mrunal Berad, Narendra Shiradkar, Anil Kottantharayil, Dinesh Kabra, Kedar Deshmukh, Aditi Chaubal, Rajeewa Arya, Probir Ghosh, Satyendra Kumar, Lawrence L. Kazmerski

IBC Technology Targeting Fast and Effective Silver Reduction Applying Advanced Screen: Printing 1372
Radovan Kopecek, Florian Buchholz, Valentin D. Mihailetchi, Joris Libal, Ning Chen, Haifeng Chu, Christoph Peter, Dominik Rudolph, Thomas Buck, Tudor Timofte, Andreas Halm, Yonggang Guo, Xiaoyong Qu, Xiang Wu, Jiaqing Gao, Peng Dong

Development of Machine Vision System for Detection of Wrap-Around in n-TOPCon Solar Cells 1373
Junhee Kim, Han-Jung Kim, Yohan Ko, Yoonkap Kim

Effect of CdS Annealing on the Performance of Antimony Selenosulfide Solar Cells 1376
Alisha Adhikari, Suman Rijal, Sabin Neupane, Manoj K. Jamarkattel, Deng-Bing Li, Tamanna Mariam, Michael J. Heben, Zhaoning Song, Yanfa Yan

20%-Efficient TOPCon Solar Cell with a Silicon Oxide Layer Deposited by Aerosol Impaction-Driven Assembly 1379
Maria Angelica M Garcia, William Weigand, Zachary C Holman

Understanding Practical Efficiency Limits for Tandem Solar Cells 1380
Emily L. Warren, Sirazul Haque, Qi Jiang, William E. McMahon, Lorelle M. Mansfield

Effect of Angle and Direction of Solar Panels in the Desert Climate of Abu Dhabi, United Arab Emirates 1381
Laith Nayfeh, Leia Nayfeh, Ammar Nayfeh

Per-Unit Dynamic Models for Grid-Following Photovoltaic Inverters 1384
Hyeonjung Tari Jung, D. Venkatramanan, Manish K. Singh, Sairaj Dhople

'There and Back Again': Reusable Germanium Wafers with Ge-On-Nothing Structures for Triple-Junction Solar Cells 1390
Valérie Depauw, Guillaume Courtois, Jinyoun Cho, Kristof Dessein, Clément Porret, Roger Loo

Characterizing Capacity Contribution of Renewable Resources Over Time in Renewable-Heavy Transmission System: MISO Case Study 1393
Hyeonjung Tari Jung, Megan Pamperin, Elspeth McGarvey, Eduardo Ibanez, James Okullo, Armando Figueroa Acevedo, Jordan Bakke

Device Modeling of HTL/BaSi$_2$ Heterojunction Solar Cells.. 1398
    Sho Aonuki, Carlos Mario Ruiz Tobon, Rudi Santbergen, Olindo Isabella, Takashi Suemasu

Modulating Efficiency and Stability of Methylammonium/Br-Free Perovskite Solar Cells Using
Fluoroarene Hydrazine 1401
    Dhruba B. Khadka, Yasuhiro Shirai, Masatoshi Yanagida, Kenjiro Miyano

In-Flight Validation of End-Of-Life Optimized Triple Junction Solar Cells Onboard ASTROBIO
Cubesat 1404
    *Luigi Schirone, Pierpaolo Granello, Matteo Ferrara, Matteo Avoli, Davide Imperatori, Nithin
    Maipan Davis, Lorenzo Iannascoli, Augusto Nascetti, Stefano Carletta, Claudio Paris,
    Erminio Greco, Roberta Campesato*

Evaluation of Module Mismatch Losses and Generation Impact in Utility Scale PV Systems....................... 1409
    Sara M. Macalpine, David A. Bowersox

SCAPS-1D Simulations of CdTe Based Solar Cells with an Amorphous Silicon-Based Back Buffer........... 1410
    *Abdul Quader, Venkanna Kanneboina, Prabin Dulal, Madan Mainali, Bishal Shrestha, Ebin
    Bastola, Adam B. Phillips, Ambalanath Shan, Nicholas J. Podraza, Michael J. Heben*

Impact of Current Collecting Grids on the Scalability of 3-Terminal Perovskite/Silicon Tandems
with Bipolar Transistor Architecture 1413
    Gemma Giliberti, Federica Cappelluti

Siting Optimization of PV Recycling Plants for Supply Chain Security and Critical Material
Recovery.. 1416
    Macarena Mendez Ribo, Silvana Ovaitt, Hope Wikoff, Heather Mirletz, Samantha Reese

Environmentally Controlled Electroluminescence/Photoluminescence Imaging System with
Current Density-Voltage Capabilities for Quantitative Degradation Analysis of Perovskite Thin
Film Solar Cells... 1417
    *Tamanna Mariam, Zahrah S. Almutawah, Adam B. Phillips, Sheng Fu, Jaehoon Chung, You
    Li, Manoj Rajakaruna, Kshitiz Dolia, Zhaoning Song, Randy J. Ellingson, Yanfa Yan, Michael
    J. Heben*

Method for Evaluating the Silicon Solar Cells Performances Under AM0 Thanks to AM1.5G
Spectrum.. 1423
    Philippe Voarino, Adem Dahi, Romain Cariou

Fatigue Debonding of EVA from Solar Glass at Elevated PV Service Temperatures.................................. 1428
    Gernot Wallner, Gabriel Riedl, Philipp Haselsteiner, Robert Pugstaller

Demonstration of Dual-Junction ELO Solar Cells with Strain-Balanced and Lattice-Matched
Quantum Well Absorbers.. 1429
    *Rao Tatavarti, Andree Wibowo, Mitsuru Imaizumi, Takeshi Ohshima, David Wilt, Roger
    Welser*

Cu2ZnSnS4 Monograin Layer Solar Cells for Flexible Photovoltaic Applications 1432
    *Marit Kauk-Kuusik, Kristi Timmo, Maris Pilvet, Katri Muska, Mati Danilson, Jüri Krustok,
    Raavo Josepson, Maarja Grossberg-Kuusk*

Advances in Flexible and Lightweight III-V Multijunction Solar Cells for High Power Density
Applications... 1433
    *Carlos Algora, Ivan Garcia, Clara Sanchez-Perez, Pablo Martin, Pablo Fernandez, Luis
    Cifuentes, Ivan Lombardero, Daniel Gomez, Mercedes Gabas, Ignacio Rey-Stolle*

Improving the Stability of Polycrystalline Silicon Passivated Contacts Using Titanium Dioxide 1434
Di Yan, Jesus Ibarra Michel, Yida Pan, Sieu Pheng Phang, Daniel Macdonald, Heping Shen, Leiping Duan, Kylie Catchpole, Jie Yang, Peiting Zheng, Xinyu Zhang, Hao Jin, James Bullock

Copper Oxide: A Potential Candidate for Hole Transport Material in Perovskite Solar Cells for Space 1437
Daniel Muñoz-Pinzon, Rishabh Sahani, Mateo Ferreira, Neetesh Kumar, Cheng-Yu Lai, Daniela Radu, Lyndsey McMillon-Brown

Innovative Methodology for an Advanced Characterization of Perovskite Systems to Reach Buried Interfaces: In-Depth Profile by Coupling GD-OES and XPS.. 1440
Mirella Al Katrib, Pia Dally, Armelle Yaiche, Jean Rousset, Muriel Bouttemy

Optimization of Optical and Electrical Properties of 2T Textured Perovskite/Silicon Tandem Solar Cell Structure... 1443
Chun-Hao Hsieh, Jun-Yu Huang, Yuh-Renn Wu

Development of an Ultra-Light Curvilinear Prismatic Window Which Mitigates Reflections and Glare for PV Modules and Other Surfaces 1446
Mark O'Neill, Chris Youtsey, Robert McCarthy

Evaluating Multi-Bias Modulation for Diagnostics of PV Modules in Daylight Electroluminescence Inspections.. 1452
Rodrigo Del Prado Santamaría, Gisele A. Dos Reis Benatto, Thøger Kari, Peter B. Poulsen, Sergiu V. Spataru

Recomibiation Center Defects Induced by TCO Reactive Plasma Deposition in Carrire Selective Contact Solar Cells.. 1453
Yoshio Ohshita, Tomohiko Hara, Taichi Tanaka, Keita Kimura, Yuto Ifuji

Predicting Site-Specific Adjustments to P50 Energy Production Estimates from Sub-Hourly Irradiance Data 1454
Faisal Rashed

Self-Assembled Monolayer Patterning for PolySi/SiO$_2$ Passivated Contacts 1459
B. Nemeth, D. L. Young, M. R. Page, V. Lasalvia, S. Theingi, P. Stradins

Non-Ionizing Radiation Effects on the Room Temperature Surface Recombination Velocity of Unintentionally Doped AlGaAs/GaAs Heterostructures 1462
Andrew Hudson, Daniele Monahan

Flexible Organic Solar Cells on Ti Foil Substrate 1466
Huiying Jia, Lei Kerr, Benjamin Leever

Novel Approach to Control Environmental Fatigue Tests on Glass/PV Encapsulant Laminates 1467
Gabriel Riedl, Gernot M. Wallner, Robert Pugstaller

Automated Analysis of Internal Quantum Efficiency Measurements of GaAs Solar Cells Using Machine Learning... 1468
Zubair Abdullah-Vetter, Priya Dwivedi, N. J. Ekins-Daukes, Thorsten Trupke, Ziv Hameiri

RF-Powered Sputtering of Iron Pyrite for Photovoltaic Applications 1471
Awais Zaka, Ayman Rezk, Sabina Abdul Hadi, Saeed Alhassan, Ammar Nayfeh

Analysis and Identification of Measurement Uncertainty Sources of a LED Sun Simulator with Double-Side Illumination for Bifacial PV Module Power Rating 1475
Sebastian Dittmann, Giuliano L. Martins, Ralph Gottschalg

The Effects of Global Damp Heat Ageing on Debonding of Polyolefin Glass Laminates 1481
Martin Tiefenthaler, Gernot M. Wallner, Robert Pugstaller

Nanographene (NG)-Based Hole Transporter with π- Interface Modifier for Thermally Stable Perovskite Solar Cells............ 1484
Seul-Gi Kim, Thybault De Monfreid, Jeong-Hyeon Kim, Fabrice Goubard, Joseph J. Berry, Kai Zhu, Thanh-Tuân Bui, Nam-Gyu Park

Automated Workflows for Machine Learning on Photovoltaic Timeseries and UV Fluorescence Image Datasets Using FAIR Principles 1485
William C. Oltjen, Xuanji Yu, Mengjie Li, Dylan J. Colvin, Yijia Sun, Hubert Seigneur, Philip Knodle, Andrew M. Gabor, Laura S. Bruckman, Kristopher O. Davis, Roger H. French

Post-Flight Analysis of Perovskite Solar Cells for NASA Materials International Space Station Experiment (MISSE)............ 1490
Kaitlyn Vansant, Ahmad Kirmani, Severin Habisreutinger, Steve Johnston, Brian Wieliczka, Joseph Luther, Timothy Peshek, Lyndsey McMillon-Brown

Cleaning Optimization for Photovoltaic Powerplants: A Novel Approach Combining Techno-Economic Modelling with Historic Rain and Soiling............ 1491
Thore Müller, Kostiantyn Pogorelov, Franco Clandestino

Characterization of Solar Cell Busbar Grid for Different Technologies by Time Domain Reflectometry Simulation: Transmission Line Approach............ 1492
A. M. C. Silveira, M. R. M. Neves, R. Garcia, H. Alvarez, M. G. Villalva, F. C. Marques, L. C. Kretly

19.5% Efficient CdSeTe/CdTe Solar Cells Using ZnO Buffer Layers 1497
Luksa Kujovic, Xiaolei Liu, Mustafa Togay, Luis C. Infante-Ortega, Kurt L. Barth, Jake W. Bowers, John M. Walls, Ochai Oklobia, Stuart J. C. Irvine, Wei Zhang, David W. Miller, Timothy Nagle, Rajni Mallick, Dingyuan Lu, Wyatt K. Metzger, Gang Xiong

Numerical Evaluation of Optimal Tilt Angle for Energy Production and Minimum Shadowing for Bifacial Solar Modules 1500
Roberto Corso, Fabio Matera, Salvatore A. Lombardo

Optimization of 1-Axis Tracking with N-S Rotating-Axis Orientation............ 1503
Jiahui Shi, Xitao Liu, Teliang Mu, Xiaotong Feng, Vasilis Fthenakis

A Crucial Role of Spin-Dry Cleaning on the Surface Passivation Quality of Crystalline Silicon............ 1509
Munan Gao, Vibhor Kumar, Ngwe Zin

Holistic Assessment of Monocrystalline Silicon (mono-Si) Solar Panels with Recycled Content Vs. Virgin-Grade Materials............ 1510
Christopher C. Bondoc, Ross Lee, Mary E. McRae, Pritpal Singh

Effect of Thickness of Electron Reflector Layer on the Efficiency of CdS/CdTe Heterojunction Thin-Film Solar Cell............ 1513
Chaitanya Santosh Rampalli, Hamid Fardi

Fabrication Au/TiO_x Nanoislands Systems by a Solid State Thermal Dewetting for Plasmonic Solar Cell Applications 1516
Brahim Aïssa, Mohammad I. Hossain, Adnan Ali

Detection and Analyze of Off-Maximum Power Points of PV Systems Based on PV-Pro Modelling........... 1519
Baojie Li, Xin Chen, Anubhav Jain

Laser Recycling of Silver from Waste Silicon Solar Cells... 1522
Mahantesh Khetri, Pawan K. Kanaujia, Mool C. Gupta

Study on Air Gap Effects on Photovoltaic Modules Operating Temperature on Typical Metal
Rooftop Appliation 1523
Quanzhi Wang, Yuanjie Yu, Tao Xu

Development of P-Type Silicon Heterojunction Solar Cells with 26.6% Efficiency 1526
Xiaoning Ru, Miao Yang, Yichun Wang, Jianqiang Wang, Chaowei Xue, Shi Yin, Chengjian
Hong, Fuguo Peng, Minghao Qu, Junxiong Lu, Liang Fang, Tian Xie, Zhenguo Li, Xixiang
Xu

Oxy-Fuel Combustion: A Threat Or an Opportunity for Solar? 1529
Mariela Colombo, Sarah Kurtz

High-Throughput In-Line Deposition of Silicon Oxide Passivation Layers in Silicon TOPCon Solar
Cells... 1533
Zachary B. Leuty, William J. Weigand, Jorge Ochoa, Joe V. Carpenter, Mariana I. Bertoni,
Zachary C. Holman

Measurement and Control of Mobile Ion Concentration in Halide Perovskites... 1534
Saivineeth Penukula, Nicholas Rolston

2D-GaSe/In_xSe_y Layer for Rapid ELO GaAs Technique.. 1535
Nobuaki Kojima, Yoshio Ohshita, Masafumi Yamaguchi

NSF Industry-University Cooperative Research Center (IUCRC) for a Solar Powered Future 2050
(SPF2050)... 1537
Amit Harenkumar Munshi, Walajabad S. Sampath, Brian Korgel

Indoor and Outdoor Characterization of III-V/Ge Solar Cells Assembled on Glass Substrate for
Concentrated Photovoltaic Applications... 1538
K. Kouame, J. Kinfack, D. Danovitch, P. Albert, T. Bidaud, A. Turala, M. Volatier, V. Aimez, A.
Jaouad, M. Darnon, G. Hamon

Survey of Module and System Quality in Brazil PV Deployments.. 1541
Lawrence L. Kazmerski, Denio Alves Cassini, Daniel Sena Braga, Suellen C. S. Costa,
Vinícius Camatta Santana, Antonia Sonia A. C. Diniz

690 WP N-Type i-TOPCon Modules in Mass Production with >25% Efficiency Solar Cells Based
on Large-Area 210 mm Wafers 1544
Yifeng Chen, Hong Chen, Shu Zhang, Le Wang, Chengfa Liu, Daming Chen, Jianmei Xu,
Pietro Altermatt, Zhiqiang Feng, Pierre Verlinden

Methylamine Post-Deposition Treatments of Vapor-Deposited Perovskite Thin Films 1545
Chaiwarut Santiwipharat, Austin G. Kuba, Kevin D. Dobson, Ujjwal K. Das, William N.
Shafarman

Optimization of Sb2Se3 Thin Films Prepared by Selenization of Sb Metallic Precursors for
Photovoltaic Application 1549
Woo Kyoung Kim, Vasudeva Reddy Minnam Reddy, Sreedevi Gedi, Salh Alhammadi, Ignatius
Andre Setiawan, Yujeong Ahn, Songhee Lee, Hyomin Kim

Narrow Bandgap Perovskite Solar Cell Degradation Monitoring by Spectroscopic Ellipsometry 1550
Marie Solange Tumusange, Madan K. Mainali, Lei Chen, Zhaoning Song, Yanfa Yan, Nikolas J. Podraza

Solution Processed N+ CdS/ n-CdTe/ Perovskite Heterojunction Thin-Film Solar Cells 1551
Isaiah Henry, Dakota Schwartz, Harry Larson, Shubhra Bansal

Effect of Double Cation Substitution on Nonradiative Recombination Losses in Cu2ZnSn(S,Se)4 Solar Cells 1556
Vijay Karade, Kiwhan Kim, Jae Ho Yun, Jin Hyeok Kim

Mapping Spatial Variations of Wide Band Gap Perovskite Thin Films.. 1557
Emily Miller, Kshitiz Dolia, Bailey Frye, Yanfa Yan, Zhaoning Song, Nikolas J. Podraza

Exploring Distributed PV Power Measurements for Real-Time Potential Power Estimation in Utility-Scale PV Plants.. 1558
Michael Gostein, William B. Hobbs

Controlling Photoexcited Carrier Relaxation Through Phonon Management in GaAs/AlAs Superlattices 1564
Muhammad Hanif, Milos Dubajic, Stephen P Bremner, Michael P Nielsen, Santosh Shrestha, Gavin J Conibeer

Controlling Residual Stresses for Scalable Open-Air Fabrication of Perovskite Solar Cells......................... 1565
Muneeza Ahmad, Carsen Cartledge, Nicholas Rolston

Silver Recovery Through a Fluoride Chemistry for Solar Module Recycling 1566
Theresa K Chen, Meng Tao

Thermal Models of Monofacial and Bifacial PV Modules: Machine Learning and Physical Estimation Models Comparison 1567
Marco Grisanti, Gaetano Mannino, Giuseppe Marco Tina, Alessandro Ortis, Mario Cacciato, Sebastiano Battiato, Fabrizio Bizzarri, Andrea Canino

In-Situ Photostability Analysis of Perovskite Solar Cells by Time-Evolving Photoluminescence Imaging.. 1570
Jackson W. Schall, Amy Louks, Goutam Paul, Nikita S. Dutta, Steve Johnston, Chun-Shen Jiang, Axel Palmstrom, Mowafak Al-Jassim, Dana B. Kern

Flexible Manufacturing of Colloidal Quantum Dot Solar Cells Via Spray-Casting Techniques.................... 1575
Lulin Li, Botong Qiu, Yida Lin, Laura Shimabukuro, Alex Ozbolt, Keyi Kang Yao, Stephen Farias, Samuel Rosenthal, Susanna M. Thon

Detection of PV Module Temperature Coefficient Using Machine Learning 1578
John M. Obrecht, Julián Ascencio-Vásquez

Widegap CdSe Solar Cells with V_{OC} >750mV 1582
Taylor Hill, Sachit Grover, James Sites

Eliminating the Need for Handling Individual Sub-Cells for Small Appliance PV Modules with Voltage Demands Above 12V.. 1588
Jan Paschen, Andreas Brand, Matheus Melati Menegassi, Oliver John, Jan Nekarda

Setting Priorities for Photovoltaic Reliability Research Using Criticality Analysis..................................... 1593
Ingrid L. Repins, Michael G. Deceglie, Timothy J. Silverman, David C. Miller, Dirk C. Jordan, Michael Woodhouse, Teresa M. Barnes

Uncertainties in PV Power Simulation Chain.. 1594
    Lubos Helienek, Jozef Rusnak, Branislav Schnierer, Martin Opatovsky, Lukas Dvonc, Vicente Lara Fanego, Artur Skoczek, Tomas Cebecauer

Characterizing TeO_2 Formation in CdTe Devices Using Transmission Electron Microscopy...................... 1600
    John Farrell, Ebin Bastola, Manoj Jamarkattel, Michael Heben, Walajabad S. Sampath, James Sites, Robert F. Klie

Dense Array TPV Modules with Alternating Polarity InGaAs Cells.. 1603
    Iván García, Aitana Cano, Víctor Orejuela, Pablo Martír, Ignacio Rey-Stolle

Influence of Spectral Albedo on the Performance of Lead-Free Perovskite Bifacial Tandem Solar
Cell 1606
    Atanu Purkayastha, Arun Tej Mallajosyula

Outdoor Characterization of a Bifacial Four-Terminal GaAs/Si Mini-Module Under Different
Albedo Conditions... 1609
    Roberto Corso, Fabio Matera, Andrea Scuto, Salvatore A. Lombardo

Evaluation of PV Snow Loss Models in the East Coast of Canada Using AI Computer Vision 1612
    Jessica Ma, Alexandre Khoury, Marianne Rodgers

Influence of NaF and KF Post-Deposition Treatment on the Sub-Band Gap Absorption of
$Cu(In,Ga)Se_2$ Absorber Layers 1617
    Sevan Gharabeiki, Michele Melchiorre, Susanne Siebentritt

The Importance of Data Quality for Reducing the Uncertainty of Site-Adapted Solar Resource
Datasets 1620
    Kristen Wagner, Alex Kubiniec, Tom McAlister, Richard Perez

Multifunctional Titanium Oxide Layers in Silicon Heterojunction Solar Cells by Selective
Anodization 1625
    Leonie Jakob, Leonard Tutsch, Thibaud Hatt, Johan Westraadt, Markus Glatthaar, Martin Bivour, Jonas Bartsch

Development of an Adaptive Droop Control Method for Interconnected Lunar DC Microgrids
Using Power Hardware-In-The-Loop.. 1626
    Andrew R. R. Dow, Rachid Darbali-Zamora, Felipe Palacios II, Jack D. Flicker, Marc A. Carbone, Jeffrey T. Csank

Assessing Degradation in Bifacial Photovoltaic by Sequential Stress and Outdoor Aging.......................... 1629
    Dennice M. Roberts, Sona Ulicna, Michael Owen-Bellini, Paul Ndione, Helio Moutinho, Kent Terwilliger, Steve Johnston, Laura T. Schelhas, Dana B. Kern

Damp Heat Exposure of Glass/Glass Coupons with Different Encapsulants 1630
    Chiara Barretta, Lisa Meinhart, Hannes Krebs, Andreas Brandstätter, Gernot Oreski

Trajectories to Reach 25% Efficiency CdTe Solar Cells with the Implementation of CdTe1-XSex
Band Gradient in SCAPs 1-D.. 1633
    Joel Saucedo, Hasitha Mahabaduge

How Do As-Local Structures in CdSexTe1-X Respond to Bias Conditions Under (X-Ray)
Illumination? 1634
    Srisuda Rojsatien, Niranjana Mohan Kumar, Barry Lai, Dan Mao, Arun Mannodi-Kanakkithodi, Maria K. Y. Chan, Mariana Bertoni

Trends in Solar PV Growth in Snowy Climates and Impact on Resource Adequacy 1635
Shelbie Wickett, Ana Dyreson

Modeling of Perovskite/Si Tandem Solar Cell 1645
Yegao Xiao, Michel Lestrade, Zhiqiang Li, Zhanming S. Li

Assessment of a DER Inverter Model for IEEE 1547 Ride-Through Requirements Using a Model
in the Loop Testbed 1648
Nayeem Ninad, Eugene Desjardins Couture

Methodology for the Analysis of Series Arc Fault Algorithms.. 1654
Paulo R. D. R. Da Silva, Guilherme C. S. Prym, Geyciane P. De Lima, Andrei C. Ribeiro,
Hugo Da S. Alvarez, Rodrigo M. Garcia, Francisco C. Marques, João A. F. G. Da Silva,
Mauricio Taconelli, Tárcio A. Dos S. Barros, Marcelo G. Villalva

Patterning the Front Polysilicon Contact for Silicon Solar Cells Using Laser Oxidation 1657
Sagnik Dasgupta, Pradeep Padhamnath, Vijaykumar Upadhyaya, Young-Woo Ok, Ruohan
Zhong, Wookjin Choi, Kyu-Hyeon Im, Ajeet Rohatgi

Evaluation of Motion-Induced Noise and Pixel-Bleeding in Electroluminescence Field Inspection
of PV Modules.. 1662
Thøger Kari, Rodrigo Del Prado Santamaria, Gisele A. Dos Reis Benatto, Pascal Koelblin,
Liviu Stoicescu, Sergiu V. Spataru

Interface Hydrogen and Passivation of Amorphous Silicon / Crystalline Silicon Heterojunction 1667
Ujjwal K. Das, Tasnim K. Mouri, Marissa Pina, Tyler Parke, Dhamelyz R. S. Quinones,
Andrew V. Teplyakov

Author Index

3-Terminal Perovskite/Silicon Tandem Modules: A Dead End or a Bright Future of Tandem Based Photovoltaics

Miha Kikelj[1], Laurie-Lou Senaud[2], Jonas Geissbühler[2], Damien Lachenal[3], Derk Baetzner[3], Benjamin Lipovšek[1], Marko Topič[1], Christophe Ballif[2], Quentin Jeangros[2] and Bertrand Paviet-Salomon[2]

[1] University of Ljubljana, Faculty of Electrical Engineering, Tržaška cesta 25, SI-1000 Ljubljana, Slovenia
[2] CSEM – Centre Suisse d'Electronique et de Microtechnique, Rue Jaquet-Droz 1, Neuchâtel, Switzerland
[3] Meyer Burger Research AG, Rouges-Terres 61, CH 2068 Hauterive

Abstract — We demonstrate an experimental 12-cell silicon-perovskite 3-terminal tandem string with an individual cell size of 24.5 cm² in a 2/1s voltage matched configuration. Based on the experimentally measured data we calibrate a string level SPCIE model, which we use to compare 3-terminal tandem strings to their 2-terminal equivalent.

I. INTRODUCTION

As silicon-based single-junction solar cells are slowly but steadily approaching their theoretical efficiency limit just shy of 30%, perovskite-silicon based tandem devices are one of the most promising solutions to drive the efficiency of commercial photovoltaic (PV) devices beyond the 30% mark [1]. Most of the research in the field has been focused on monolithic 2-terminal (2T) perovskite-silicon tandems which have already surpassed the 30% mark [2], [3] and are exhibiting a maximum efficiency of up to 33.7% at the laboratory scale [4].

Despite their high efficiency, 2T tandems exhibit an inherent limitation due to current matching between the two series-connected sub-cells, which can limit their performance under realistic in-field conditions if the device is not specifically adapted to its environmental conditions [5], [6]. As a solution, electrically decoupled, mechanically stacked 4-terminal (4T) perovskite-silicon tandems have been proposed [7]. Despite the insensitivity to current matching, they too come with their inherent problems of added optical losses, complex mechanical interconnections and module level electronics.

While the concept has been reported more than 20 years ago with other types of sub-cells [8], perovskite-silicon monolithic 3-terminal (3T) tandems based on IBC silicon heterojunction bottom cells have recently started to receive a lot more attention [1]. Trading in loose electrical coupling not prone to current mismatch losses for a decrease in optical losses thanks to a monolithic design, 3T perovskite-silicon tandems promise energy yield (EY) 3-9% relative higher than 2T or 4T tandems in an arbitrary climate [9]. In addition 3T voltage-matched (VM) strings have already been proposed [10] and studied [11], which should enable the production of strings with 2 terminals at the end of VM strings of 3T tandems, leading to the possibility to construct 2T modules interchangeable with the standard commercialized PV modules.

In our contribution we present and study (to the best of our knowledge) a first ever 3T perovskite-silicon tandem string in two different VM configurations and compare the results with an equivalent 2T string. Based on the string measurements, we calibrate and validate a 3T string model, which we use in further studies to evaluate the performance and robustness of 3T tandem strings.

II. 3T-TANDEM MINI-MODULE

We constructed a reconfigurable 12-cell 3T perovskite-silicon string, based on p-i-n perovskite cells on top of 5x5 cm² rear-textured n-type tunnel IBC bottom cells connected with a recombination junction. The average efficiency was around 26% when measured in a 2T configuration. The cells were interconnected using a custom chuck, which in addition to contacting the tandems, ensured temperature stability and the measurement of each individual sub-cell voltage. The contacting unit shown in Fig. 1 also allowed us to adaptively change the VM interconnection configuration from 2/1s to 3/2s and 2T equivalent (see [10] for interconnection details).

Fig. 1. 12-cell 3T tandem string constructed a custom designed reconfigurable contacting unit.

978-1-6654-6060-6/23 $31.00 © 2023 IEEE

We measured the string's integral *I-V* under STC as well as individual sub-cell voltages in each of the three configurations' maximum power points (MPPs). The measured integral *I-V* parameters of the 12-cell 2/1s configuration in comparison with their expected values simulated by using individually calibrated SPICE models (described in the following sections) are shown in Table I. Additionally, the comparison of the measured and the simulated *I-V* characteristics is shown in Fig. 2.

TABLE I
INTEGRAL VM STRING *I-V* PARAMETERS IN 2/1S CONFIGURATION

	Isc [mA]	Voc [mV]	FF [%]	Eff [%]
Experiment	998.8	6481	63.94	14.14
Simulation	1357.3	6256	69.02	20.02

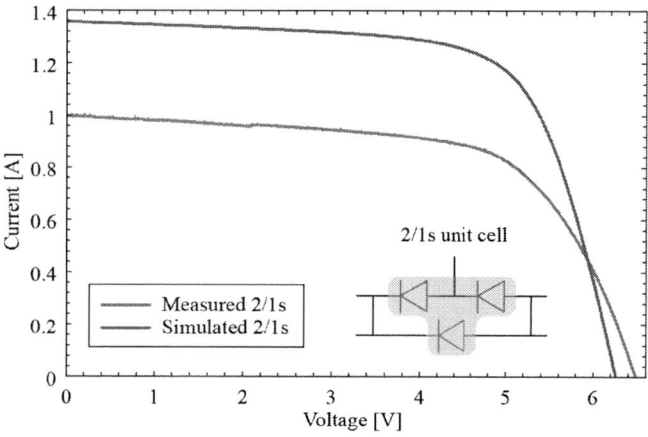

Fig. 2. Measured vs. simulated *I-V*s of the 12-cell 3-terminal tandem string in the 2/1s voltage-matched configuration.

One can observe that the simulated string results differ significantly from the measured *I-V* especially when it comes to short-circuit current (J_{SC}), which also appears to be the main contributor to the difference in efficiency between the measurements and the simulations. There is also a slight increase in the open-circuit voltage (V_{OC}) which we attributed to the additional series resistance introduced by the interconnecting wires. We therefore performed a detailed simulation-based analysis to refine our simulation model and figure out why such differences occur in the measured string.

A. Modelling Approach

The experimental results of each individual 3T tandem as well as the whole string measured in different VM configurations served as feedback to construct, calibrate and validate a 3TT VM string model. Each tandem in the string was measured individually in the IBC and the perovskite sub-cell configuration. Both sub-cells were then modeled as a SPICE two-diode model and calibrated to match the measured

characteristics. The two two-diode models were then combined appropriately to form an individually calibrated 3T tandem in the string. Those tandem models were then interconnected using an appropriate VM interconnection scheme to correctly represent the physical configuration.

B. Simulation Analysis

Using the individually calibrated SPICE 3T-tandem string model, we simulated the sub-cell voltages in the measured strings maximum power point (MPP) and compared them to the measured ones. The comparison is shown in Fig. 3. At first glance a good match between the simulated and the measured string sub-cell voltages at the string's MPP can be observed at the beginning and the end of the string. Towards the middle of the string experimental and simulated values diverge. The discrepancy is more significant for the perovskite sub-cells (shown in blue) which led us to believe that the main difference from the experiment has to do with perovskite sub-cells.

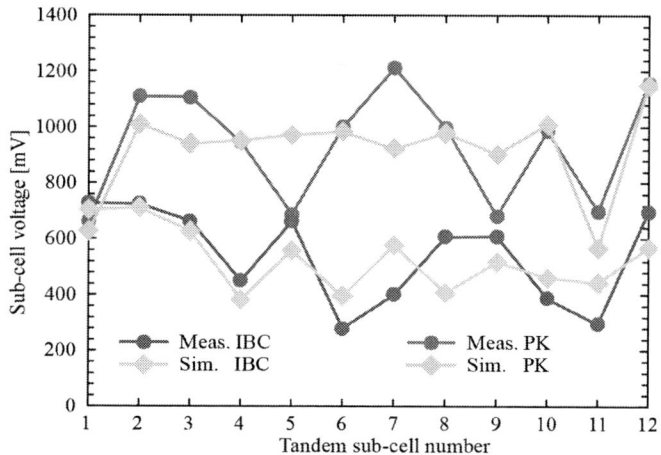

Fig. 3. Measured vs. simulated sub-cell voltages an the MPP of the 12-cell 3-terminal tandem string in the 2/1s voltage-matched configuration.

Based on the differences observed in Fig. 3, perovskite sub-cells #5, #7 and #9 within the string appeared to be modelled improperly. We, therefore, through simulations, studied two possible cases. In one of the cases, we shunted the perovskite sub-cells while in the other we increased top-cell's top electrode interconnection resistance in order to study the case of poor interconnection. We repeated the simulations for different combinations of the three crucial sub-cells and found that in our specific case the top contacts of perovskite sub-cells #5 and #9 were disconnected from the rest of the string. The *I-V* and sub-cell voltages at the string's MPP when taking this into account are shown in Fig. 4.a and 4.b. One can observe a perfect match in the sub-cell voltages at the string's MPP and at the MPP of the simulated *I-V* in Fig. 4.a. The discrepancy one can observe around the open-circuit as well as the short-circuit conditions are likely due to series resistance of the

978-1-6654-6060-6/23 $31.00 © 2023 IEEE

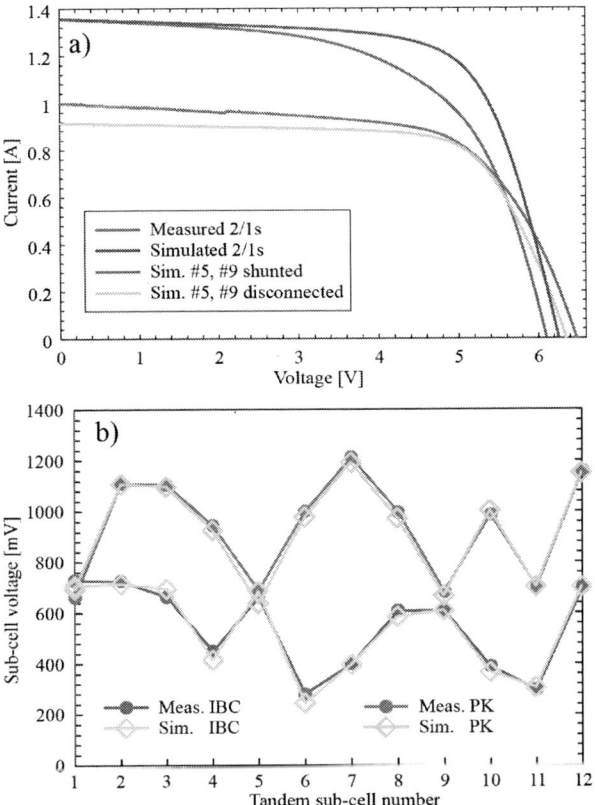

Fig. 4. Simulations vs. measurements of the 12-cell 3T tandem string in 2/1s VM configuration with disconnected top terminals of sub-cell #5 and #9.

interconnecting wires themselves, which were not accounted for.

Furthermore, the results reveal an interesting property of 3T tandem strings. Despite the fact that out of 24 sub-cells (12 tandems, 12 IBC sub-cells and 12 perovskite sub-cells) two perovskite sub-cell were contacted poorly, the string efficiency only dropped from around 20% to slightly above 14%. In a 2-terminal tandem string, a poor contact/disconnection would lead to a complete string malfunction. This result highlights the resilience of 3T tandem strings to interconnection interruptions or sub-cell degradation. This resilience originates from the fact that there are always more than one parallel current paths connecting the two ends of the string (3 in the case of 2/1, 5 in the case of 3/2…). Therefore, an interruption in one current path does not directly result in a complete string malfunction.

C. String Shading and Reverse Bias

We turned our attention to the shading resilience of 3T strings. The effect was partially experimentally demonstrated in 3T strings where the "perovskite" top-cell was replaced by a series connection of two silicon bottom cells [12]. We employed the previously discussed modelling approach to evaluate the effect of shading on perovskite-silicon 3T tandem solar cell strings. We constructed a 32-cell string with 16 cells

in one row as shown in Fig. 5.a. We postulated a worst-case shading scenario, where consecutive cells in the string are shaded. The most realistic scenario would be to shade the cells at the beginning or end of the string. As some of the sub-cell of tandems (#1, #2, #31 and #32) are partially disconnected because of the 3/2s VM interconnection scheme, we decided to shade the opposite end. We proceeded by consequently shading tandems #16, #15, #17 and #18 as shown in Fig. 5.a. The resulting string *I-V*s are shown in Fig. 5.b.

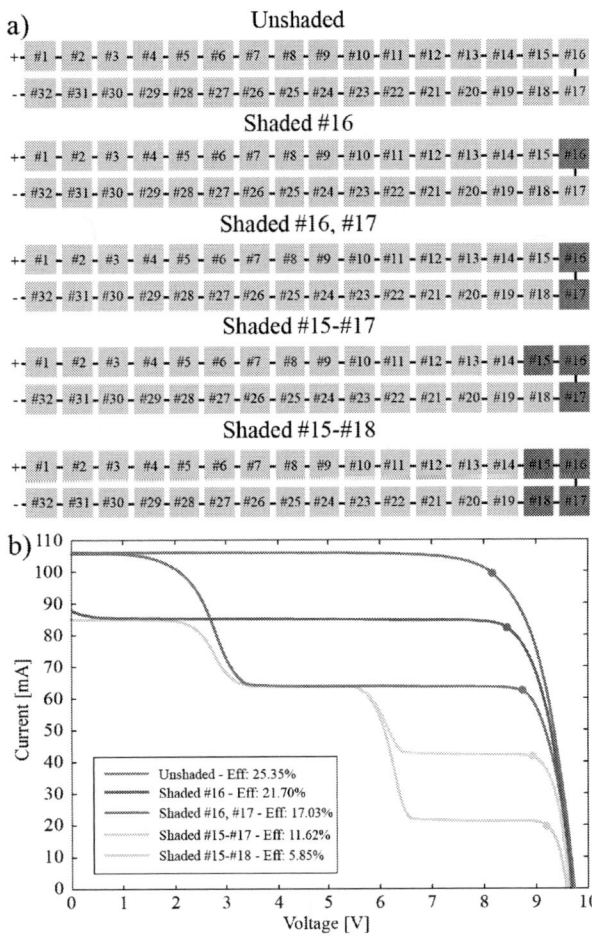

Fig. 5. Simulations of sequentially shaded 32-cell 3T tandem string in a 3/2s VM configuration. Note that the tandem active area was set to 1cm² but the conclusions also hold for other device dimensions.

One can observe that for a 3/2s configuration up to 4 neighboring cells can be shaded without reducing the strings efficiency to 0%. With 4 shaded cells the string still maintains an efficiency of nearly 6% in comparison with the initial 25%. Furthermore, for any of the presented cases there exists an MPP (denoted by a circle in Fig. 5.b) which ensures that neither the perovskite nor silicon sub-cells are reverse biased. This could prove beneficial as it is known that perovskite solar cells are relatively intolerant to reverse bias. In comparison, a 2T tandem string shaded in a similar way would exhibit an efficiency of

0% (assuming the cells are not passing any current in reverse bias). This further exhibits the robustness of 3T tandem strings.

IV. CONCLUSION

In conclusion, we demonstrated a physical 12-cell 3T tandems string in 2/1s voltage-matched configuration. We displayed the possibility to accurately model the current-voltage properties of the string using a SPICE modelling methodology. We showed that VM 3T tandem strings are relatively resistant to interconnection interruptions and cell degradation as there are multiple parallel conducting paths. In addition, we demonstrated that with lower VM ratios VM strings become more and more resistant to cell shading and therefore also reduce the possibility of reverse biasing the shaded sub-cells. The predicted 3-9% relative EY increase over 2T cells [9], combined with the demonstrated robustness of 3T tandem strings, VM 3T strings pose as an interesting alternative to conventional 2T series-connected strings.

REFERENCES

[1] P. Tockhorn et al., "Three-Terminal Perovskite/Silicon Tandem Solar Cells with Top and Interdigitated Rear Contacts," *ACS Appl. Energy Mater.*, vol. 3, no. 2, pp. 1381–1392, Feb. 2020, doi: 10.1021/acsaem.9b01800.

[2] "New world records: Perovskite-on-silicon-tandem solar cells." http://csem.ch/press/new-world-records-perovskite-on-silicon-tandem-solar (accessed Jan. 11, 2023).

[3] H.-Z. B. für M. und Energie, "World record back at HZB: Tandem solar cell achieves 32.5 percent efficiency," *HZB Website*. https://www.helmholtz-berlin.de/pubbin/news_seite?nid=24348;sprache=en (accessed Jan. 11, 2023).

[4] "KAUST team sets world record for tandem solar cell efficiency," *KAUST*. https://www.kaust.edu.sa/news/kaust-team-sets-world-record-for-tandem-solar-cell-efficiency (accessed May 03, 2023).

[5] O. Dupré, B. Niesen, S. De Wolf, and C. Ballif, "Field Performance versus Standard Test Condition Efficiency

of Tandem Solar Cells and the Singular Case of Perovskites/Silicon Devices," *J. Phys. Chem. Lett.*, vol. 9, no. 2, pp. 446–458, Jan. 2018, doi: 10.1021/acs.jpclett.7b02277.

[6] M. De Bastiani et al., "Efficient bifacial monolithic perovskite/silicon tandem solar cells via bandgap engineering," *Nat. Energy*, vol. 6, no. 2, Art. no. 2, Feb. 2021, doi: 10.1038/s41560-020-00756-8.

[7] T. Leijtens, K. A. Bush, R. Prasanna, and M. D. McGehee, "Opportunities and challenges for tandem solar cells using metal halide perovskite semiconductors," *Nat. Energy*, vol. 3, no. 10, Art. no. 10, Oct. 2018, doi: 10.1038/s41560-018-0190-4.

[8] T. Nagashima, K. Okumura, K. Murata, and Y. Kimura, "Three-terminal tandem solar cells with a back-contact type bottom cell," in *Conference Record of the Twenty-Eighth IEEE Photovoltaic Specialists Conference - 2000 (Cat. No.00CH37036)*, Sep. 2000, pp. 1193–1196. doi: 10.1109/PVSC.2000.916102.

[9] F. Gota, M. Langenhorst, R. Schmager, J. Lehr, and U. W. Paetzold, "Energy Yield Advantages of Three-Terminal Perovskite-Silicon Tandem Photovoltaics," *Joule*, vol. 4, no. 11, pp. 2387–2403, Nov. 2020, doi: 10.1016/j.joule.2020.08.021.

[10] W. E. McMahon et al., "Homogenous Voltage-Matched Strings Using Three-Terminal Tandem Solar Cells: Fundamentals and End Losses," *IEEE J. Photovolt.*, vol. 11, no. 4, pp. 1078–1086, Jul. 2021, doi: 10.1109/JPHOTOV.2021.3068325.

[11] H. Schulte-Huxel, R. Witteck, S. Blankemeyer, and M. Köntges, "Optimal interconnection of three-terminal tandem solar cells," *Prog. Photovolt. Res. Appl.*, vol. n/a, no. n/a, doi: 10.1002/pip.3643.

[12] R. Witteck, S. Blankemeyer, M. Siebert, M. Köntges, and H. Schulte-Huxel, "Partial shading of one solar cell in a photovoltaic module with 3-terminal cell interconnection," *Sol. Energy Mater. Sol. Cells*, vol. 219, p. 110811, Jan. 2021, doi: 10.1016/j.solmat.2020.110811.

Temperature-dependent performance of ultra-thin silicon heterojunction solar cells for space applications

Anh Huy Tuan Le, Pradeep Balaji, André Augusto, Ziv Hameiri

School of Photovoltaic and Renewable Energy Engineering, University of New South Wales, Sydney, Australia

School of Electrical, Computer and Energy Engineering, Arizona State University, Tempe, AZ, United States

Department of Information and Technology, Dalarna University, Falun, Sweden

Recently, ultra-thin silicon (Si) heterojunction (SHJ) solar cells with a base thickness of 40-60 μm and an efficiency above 20% have been successfully developed. These cells are a promising candidate to power small, light, and cost-effective satellites in the coming years. Therefore, insights into their performance under pertinent temperature and illumination operating conditions in space, i.e. a wide range of temperatures under the air mass zero (AM0) spectrum, are of significant interest. In this study, we investigate the temperature-dependent performance of ultra-thin SHJ solar cells under such conditions. Their temperature-dependent behaviour above 0 °C (under AM0) is then compared to that of thicker SHJ cells performing under the air mass 1.5 global (AM1.5G) spectrum. Below 0 °C, the cell electrical parameters remarkably decrease with temperature. This interesting behaviour of the open-circuit voltage and the fill factor might be attributed to the Schottky barrier height in this cell structure. Furthermore, we examine the temperature dependence of the specific power of the studied cells. At 25 °C, the obtained specific power is superior to those of most of the multi-junction cell structures, highlighting the advantage of using SHJ solar cells for space applications.

Utilizing a soft IZO sputtering process to contact buffer-free semitransparent perovskite *pin* solar cells

Roland Clausing[1], Annika Raugewitz[1], Benjamin Grimm[1], Marvin Diederich[1], Tobias Wietler[1], Felix Haase[1], Rolf Brendel[1,2], Robby Peibst[1]

[1]Institute for Solar Energy Research, Hamelin (ISFH), 31860 Emmerthal, Germany

[2]Institute for Solid State Physics, Leibniz University Hannover, 30167 Hannover, Germany

Abstract — Perovskite solar cells in *pin* structure are considered as advantageous for the implementation in perovskite silicon tandem solar cells. So far, one drawback of this approach is the requirement of buffer layers (e.g. SnO_x) between the front TCO and the carbon-based ETLs (e.g. C60 or PCBM). In this work, we show that a sufficiently soft process for indium zinc oxide sputtering allows a damage-free contacting of the *pin*-structure directly on top of the ETL (C60) while simultaneously enabling a leaner processing scheme by abandoning the need for a (ALD) buffer layer. Ray tracing simulations indicate a gain of 0.44 mA/cm² in photo generation in perovskite top cells by omitting the SnO_x buffer layer. For the optimized IZO sputter process, we obtain implied open circuit voltages in absolute photoluminescence measurements at least as high as for reference samples without IZO. The sputter process works for both vapor and wet chemically deposited perovskite solar cells. Under the constrains of an oxygen-free sputter process, our soft process leads to a 60 nm thick IZO layer with a high mobility of $\mu = 42.7 \pm 0.3$ cm²/Vs and low charge carrier density $N = 3.6 \pm 0.02 \times 10^{20}$ 1/cm³.

I. INTRODUCTION

Perovskite (Pk)-based absorbers are a promising candidate for top solar cells in terrestrial, non-concentrating tandem devices. Combined monolithically with Si bottom cells, a remarkable efficiency of 33.7 % has been achieved recently [1]. Regardless of the number of terminals (2-4) or the bottom cell material (Si, CIGS, low bandgap Pk, nanoclusters, ...), Pk top cells always feature a transparent and conductive oxide (TCO) as front electrode. Amorphous indium zinc oxide (IZO) is a common TCO for this application [2] due to its excellent optoelectronic properties achievable by room temperature deposition. However, radio frequency (RF) sputtering can damage sensitive substrates by high energetic particles. The most common Pk top cell configuration – *pin* – requires TCO sputtering on top of a few-nanometer thin electron transport layer (ETL) such as fullerenes (C60) or PCBM. Detrimental effects of direct IZO sputtering on the ETL on electrical cell characteristics have been reported [3,4]. Therefore, almost all recent perovskite top-cell structures in *pin* configuration implement a buffer layer on top of the ETL, protecting all layers and interfaces underneath from sputter damage. Atomic layer deposited (ALD) SnO_x is mostly used for this purpose.

However, regarding the industrialization of Pk/X tandem cell fabrication, it would be desirable to omit the ALD SnO_x buffer layer. As recently shown by M. Härtel et al. [5], Pk/Si tandem cells without SnO_x furthermore benefit from lower reflection losses. The authors furthermore point on thermal degradation of the perovskite, which might be implied by the elevated temperatures of the SnO_x deposition process and thus might be avoided by omitting the SnO_x layer.

Härtel [5] developed a « soft » IZO sputtering process and demonstrated its viability by an excellent efficiency of ~25.4 % for a SnO_x-free Pk/Si tandem solar cell – even slightly higher than for the reference device with ALD SnO_x buffer layer. However, they implemented an additional polymer polyethyleneimine ethoxylated (PEIE) layer on top of the C60, as well as a LiF_x passivation layer underneath. At least the former layer might have partially protected all layers and interfaces underneath during the IZO sputter process.

In this work, we present an optimized soft IZO sputter process, capable of contacting the ETL (C60) without damaging its interface to the perovskite or the perovskite itself – even without additional layers such as PEIE and LiF_x. This is done by lowering the power density of the RF-sputter process as much as possible and by increasing the process pressure to lower the free path length of particles in the plasma. Furthermore, we optimized the optical and electrical properties of the amorphous IZO layer while staying with a non-reactive sputter approach, without additional (and potentially detrimental for the perovskite) oxygen for an easier industrial process control.

Applying this process first on single junction (SJ) perovskite cell precursors, we evaluate the influence of the sputter parameters by monitoring the quasi-Fermi level splitting in the perovskite absorber by measuring absolute photoluminescence (aPL). We eventually fabricate perovskite single junction cells with and without sputtered IZO on the ETL (C60).

II. EXPERIMENTALS

A. Single junction perovskite solar cell precursors

We deposit two groups of SJ perovskite single junction cell precursors, as displayed in Fig. 1., on a 2.5 × 2.5 cm² glass substrate having an indium tin

978-1-6654-6060-6/23 $31.00 © 2023 IEEE

oxide (ITO) layer of 180 nm thickness on top possessing a sheet resistance of ~ 8 Ω/sq. The ITO on the substrate is structured by a laser and subsequently cleaned.

For the first group (evaluation of sputter damage by aPL) 10 nm of spiro-TTB (2,2',7,7'-Tetra(N,N-di-*p*-tolyl)amino-9,9-spirobifluorene) are thermally evaporated as hole transport layer (HTL). Afterwards the perovskite methylammonium lead iodide (MAPbI₃) is deposited via thermal co-evaporation.

For the second group (evaluation on cell level), we use a self-assembled monolayer (SAM, namely MeO-2PACz ([2-(3,6-Dimethoxy-9*H*-carbazol-9-yl)ethyl]phosphonic acid)) as HTL. Here, we spin-coat a $Cs_{0.05}(FA_{0.77}MA_{0.23})_{0.95}Pb(I_{0.77}Br_{0.23})_3$ perovskite absorber according to Ref. [6].

For both groups, 23 nm of C60 are evaporated as ETL. Up to here, processing is performed in inert atmosphere. For the subsequent IZO sputtering, the precursors are exposed to air for several minutes during the transfer into the sputter tool.

For opaque Pk single junction cells serving as reference for the second group, air exposure is not necessary. Here, 8 nm of Bathocuproine (BCP) are evaporated on top of C60, followed by an evaporation of 100 nm of Ag as for the whole second group. The perovskite thickness is always 500 nm.

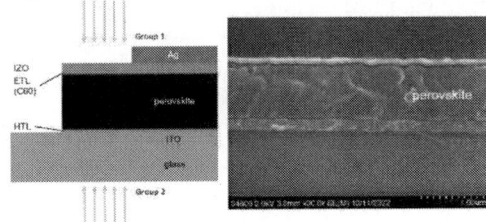

Fig. 1. Design of the Pk cell with indication of the side of illumination for both sample groups and SEM-image of a device without metal contacts (ETL/HTL not visible due to low thickness).

B. Soft IZO sputter Process

The IZO is RF-sputtered in a co-sputter deposition tool by Dreebit working with a frequency of 13.56 MHz. The used inert gas is pure Ar with a process pressure from 5×10^{-4} to 5×10^{-3} mbar, while the process is run with a power density between 0.66 and 4 W/cm². The target-substrate distance is about 20 cm. Henceforth, the term soft IZO refers to the film deposited at 0.66 W/cm². The thickness of the deposited IZO is usually set to 60 nm to have an optimum of optical transmittance and electrical conductivity at the front side of the targeted tandem device. For the second group a higher thickness of 100 nm was used for the IZO.

C. Characterization

The optical measurements are performed with a spectral photometer Varian Cary 5000, which allows the measurement of reflection and transmission of IZO-coated glass reference samples in a wavelength

region from 250 to 2500 nm. The intended IZO-thickness is confirmed with a Dektak profilometer and a J.A. Woollam M2000UI ellipsometer. We measure electrical parameters of the IZO-films, namely mobility, conductivity and charge carrier density, with a Linseis Hall measurement tool, using a 0.72 T magnetic field. The sheet resistance is measured with a Veeco Instruments FPP 5000 sheet resistance measurement tool.

aPL is measured with a PicoQuant FluoTime 300 fluorescence spectrometer with an integrating sphere, enabling the detection of all spherically emitted photoluminescence in the wavelength region from 300 to 900 nm. Aim of aPL is the determination of the photoluminescence quantum yield (PLQY, also known as photoluminescence quantum efficiency QE_{PL}), which is done according to the approach by de Mello [7]. Briefly, the method firstly measures the intensity $I_{Laser,blank}(\lambda)$ of the scattered laser light with a blank sample. Secondly, the intensity $I_{Laser,blank}(\lambda)$ of the PL emission peak is obtained from the sample. Eventually, the peak areas of both signals are compared to determine the PLQY following the equation

$$PLQY = \frac{\int I(PLpeak_{sample})d\lambda - \int I(background_{blank})d\lambda}{\int I(Laserpeak_{blank})d\lambda - \int I(Laserpeak_{sample})d\lambda}.$$

The resulting quasi-Fermi level splitting and thus the possible internal voltage iV_{oc} is determined by $\Delta E_F = e * iV_{oc,rad} + kT \ln(PLQY)$ with $iV_{oc,rad}$ being the theoretically possible maximal internal quasi-Fermi level splitting disregarding all non-radiative recombination losses. This value is calculated to be 1.33 eV for a MAPbI₃-PSC (E_G = 1.6 eV) at 1 sun-equal irradiation.

Raytracing is performed with the SunRays 3.1.3 Software by R. Brendel and with input data for a Perovskite-POLO-PERC (P3T) tandem device according to S. Mariotti et al. [8].

III. RESULTS

In Fig. 2, we show the optical properties of the soft IZO process at a process pressure of 5×10^{-3} mbar.

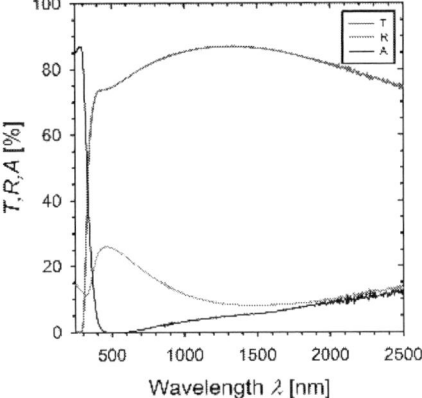

Fig. 2. Measured reflection and transmission spectra and calculated absorption spectrum of the 60 nm soft IZO film on a borofloat glass substrate.

978-1-6654-6060-6/23 $31.00 © 2023 IEEE

For wavelengths above 400 nm, transmission is constantly above 80 % (considering the mean transmission of the borofloat glass of over 90 %) and reflection only shows a peak at around 460 nm, influenced by the film thickness.

The sheet resistance of the soft IZO achieves values down to 70.6 Ω/sq at 60 nm thickness and 40.1 Ω/sq at 100 nm thickness. Hall-measurements result in values for mobility (μ = 42.7 ± 0.3 cm²/Vs), conductivity (σ = 2443 ±1/ Ωcm) and carrier density of $N = 3.6 \pm 0.02 \times 10^{20}$ 1/cm³, which are comparable to the values achieved by Härtel [5] for their soft sputtering process (48.6 cm²/Vs, 2418 1/ Ωcm). It is reported [2], that it is hardly possible to increase the mobility much further without adding oxygen to the sputter process, which we wanted to avoid due to the possible perovskite degradation.

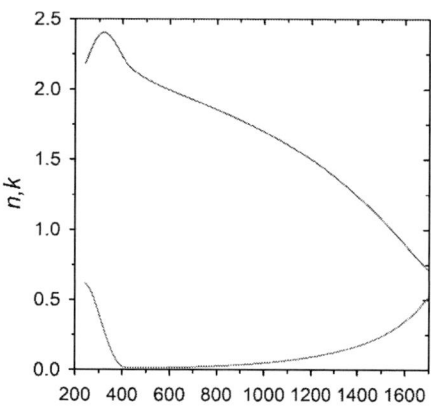

Fig. 3. Refractive index n (blue) and extinction coefficient k (red) of the 60 nm thick soft IZO.

Using the n- and k- values gained by ellipsometry measurements on our IZO films shown in Fig. 3, an optical simulation of a tandem solar cell containing a 60 nm thick IZO front layer was carried out. Following the cell model from [8] and inserting our own IZO data, the resulting photo generated current density in the perovskite top cell increases from 17.28 mA/cm² to 17.72 mA/cm² by omitting the 20 nm SnOₓ layer. Additionally, the photo generated current density in the bottom cell slightly increases from 18.51 mA/cm² to 18.60 mA/cm².

Evaluation of the internal voltage is possible with aPL measurements. The results are shown in Fig. 4. Every point represents the mean value of eight measured precursors. The absolute level of 1.11 eV for ΔE_F and therefore iV_{oc} for the reference sample without IZO is consistent to our so far best V_{oc} values measured with our co-evaporated MAPbI₃ absorbers (1.1 eV). With those we achieve up to 17 % efficiency, which is on par with other reports for this HTL/Pk/ETL stack on small area w./o. passivation of the MAPbI₃/C60 interface and antireflection coating. The IZO sputtering process shows a negative influence on the internal voltage iV_{oc} with increasing power density. For a power density of 1.33 W/cm² no degradation is resolvable.

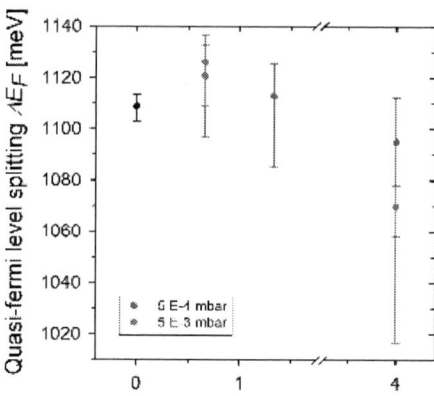

Fig. 4. Quasi-Fermi level splitting measured by aPL on the precursors from the first group after IZO sputtering. Green denotes low (5×10^{-4} mbar) and red high (5×10^{-3} mbar) process pressure. Error bars indicate the standard deviation.

In contrast, the lowest power density leads to an increase of 10 to 20 meV in the mean quasi Fermi level splitting while error bars of both groups are still overlapping. This shows that the soft IZO process does not yield any degradation. Whether this process even yields an increase of electronic quality of the perovskite and relevant interfaces still needs to be investigated with better statistics. The damage-mitigating influence of the higher chamber pressure is clearly visible for high power density. But for low power density the effect seems to be different, with slightly better internal voltages for the process with lower pressure.

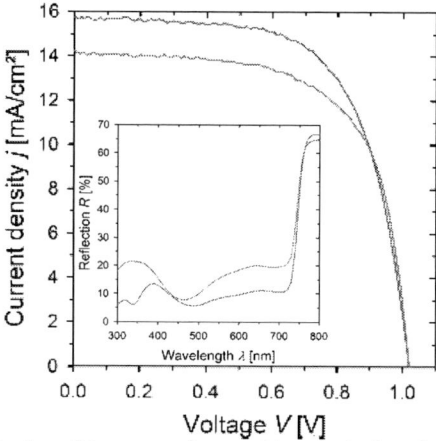

Fig. 5. JV-curves of perovskite single junction cells with 100 nm soft-sputtered IZO on top of the C60 (red) and reference sample (blue) without IZO, but direct metallization on top of an additional BCP layer. Inset: Reflection spectra of the same cells.

Figure 5 shows the best cells from the first successful implementation of the soft IZO with 100 nm thickness in the second group with wet-chemical perovskite single junction solar cells on 12 mm². This cell with an efficiency of 12 % is already comparable to the reference w./o. IZO but with 8 nm BCP, having almost identical V_{oc} (1016 mV with IZO, 1020 mV w/o). As confirmed by reflection

978-1-6654-6060-6/23 $31.00 © 2023 IEEE

measurements shown in the inset in Figure 5, the difference in current density (18.2 mA/cm² with IZO, 20.2 mA/cm² w/o) stems mainly from higher reflection losses of 1.6 mA/cm² for the Pk cells with IZO on the back side compared to a back side with 8 nm BCP.

This effect is not relevant for the targeted tandem application with IZO on the front. Fill factor values however are also comparable (65 % w. IZO and 64.5 % w./o. IZO), indicating that electron extraction at the perovskite / C_{60} interface is not compromised by the IZO sputtering process. Compared to best reports for this HTL/Pk/ETL stack, all *IV* parameters of our wet-chemical perovskite process still need to be improved. Nevertheless, our Pk cells are already sensitive enough (in particular with respect to V_{oc}) to verify the absence of sputter damage for out soft IZO process.

IV. CONCLUSION

The viability of the soft IZO process can be seen from the aPL measurements, showing no degradation in iV_{oc}. A buffer layer is not needed to mitigate sputter damage to the ETL, perovskite and interfaces, if the sputter power density is low enough. Compared to Härtel [5], who use 2.41 W/cm², we managed to lower the power density to 0.66 W/cm². Lower values are not achievable right now because the Ar plasma would not be stable. Despite the low power we measure similar optoelectronic properties of the IZO: a mobility of μ = 42.7 ± 0.3 cm²/Vs (Härtel: 48.6 cm²/Vs),

a conductivity of σ = 2443 ±1/ Ωcm (Härtel: 2418 1/ Ωcm) and carrier density of N = 3.6 ± 0.02 × 10^{20} 1/cm³. We obtain an even higher conductivity than [5] and only slightly lower mobility, which can be attributed to the absence of oxygen for reactive sputtering.

Compared to the damage mitigating effect at high power densities, at low power the higher process pressure does not show an advantage.

Using a more optimized perovskite cell process and its application to tandem solar cells illuminated through the IZO electrode, we also expect a clearer optical advantage of our soft IZO process as compared to top cells with SnO_x buffer layers. Raytracing simulations already indicate this, offering 0.44 mA/cm² gain in current density in a perovskite top cell from the used model [8].

REFERENCES

[1] NREL: Interactive best research-cell efficiency chart, https://www.nrel.gov/pv/interactive-cell-efficiency.html, last seen 25. June 2023

[2] M. Morales-Masis et.al., IEEE J. Photovoltaics, vol. 5, 1340-1347 (2015)

[3] J. Werner et.al., Sol. Energy Mater. Sol. Cells. 141, 407 (2015)

[4] T. Wahl et.al., Org. Electron physics, Mater. Appl. 54, 48–53 (2018)

[5] M. Härtel et.al., Sol. Energy Mater. Sol. Cells, 252, 112180 (2023)

[6] Al-Ashouri et al., Science 370, 1300–1309 (2020)

[7] J. C. de Mello et.al., Adv. Mater. 9, 230 (1997)

[8] S. Mariotti et.al., Solar RRL 6 (2022), 2101066/1-9

Effect of surface morphology on GaAs solar cells grown on planarized spalled (100) GaAs substrates

Anna K. Braun, Jacob T. Boyer, Kevin L. Schulte, John Simon, Steve W. Johnston, Myles A. Steiner, Corinnne E. Packard, Aaron J. Ptak

National Renewable Energy Laboratory, Golden, CO, United States

Colorado School of Mines, Golden, CO, United States

We analyze the effect of surface morphology created by planarizing spalled GaAs wafers on GaAs solar cells grown by HVPE. Controlled spalling of (100)-oriented GaAs has potential to reduce substrate costs for III-V photovoltaics (PV); however, it creates regularly faceted surfaces that complicate the growth of high quality III-V PV devices. We leverage the crystallographic-direction-dependent growth capability of hydride vapor phase epitaxy (HVPE) to planarize these faceted GaAs substrates, reducing the surface roughness and degree of faceting. We then demonstrate that GaAs solar cells grown on planarized surfaces are nominally identical to those grown on a planar, epi-ready GaAs surface. We observe slightly degraded device performance in cases where facets are not completely removed. We use device-scale imaging techniques combined with characterization of device cross sections to analyze this performance degradation. Lastly, we discuss the growth mechanisms that contribute to the planarization of faceted surfaces. These investigations into the mechanisms of both device degradation and planarizing growth will ultimately enable high-performing III-V PV with the cost reduction potential of controlled spalling. This advancement, when combined with low-cost epitaxy by HVPE, provides one of the most promising routes to low-cost III-V PV to date.

Ultra-light environmental protection for solar arrays in space

Pilar Espinet-Gonzalez, Alexandra H. Teodor, Jann A. Grovogui, Yao Y. Lao

The Aerospace Corporation, El Segundo, CA, United States

Every solar cell in space is protected with Ce-doped coverglass whose main function is to shield the solar cell against the radiation of high energy particles. The minimum thickness of coverglass is 2-3 mils, which ultimately limits the flexibility and weight of ultra-light solar arrays. Furthermore, the adhesion of coverglass on the solar cell is a low throughput step in the assembly process. In this work, the shielding required for different operating orbits, the lifetime of the mission, and the radiation hardness of the solar cell used is studied. Ultra-light and flexible alternatives to coverglass which allow low-cost solar array assembly integration will be evaluated. Their optical performance after different conditions emulating the space environment will be evaluated. A down-selection protocol which includes the characterization under UV, vacuum UV, outgassing, high energy particle radiation, and atomic oxygen will be performed to ultimately determine the potential and limitations of the different films and coatings studied.

Highly crystalline In2S3:V thin films epitaxially grown on sapphire substrates: A potential canditate for intermediate band solar cells

Tanja Jawinski, Chris Sturm, Roland Clausing, Heiko Kempa, Stefan Lange, Roland Scheer, Marius Grundmann, Holger von Wenckstern

Universität Leipzig, Leipzig, Germany

Martin-Luther-Universität Halle-Wittenberg, Halle, Germany

Fraunhofer-Center für Silizium Photovoltaik CSP, Halle, Germany

Indium sulfide hyperdoped with vanadium is a promising material to realize an intermediate band (IB) solar cell. We grew pure and vanadium-doped In2S3 using physical co-evaporation of the elements. Deposition parameters were varied to optimize the structural properties of undoped In2S3. Epitaxially grown samples on a-sapphire exhibit smoothest surfaces and highest crystallinity. We find a week and a strong absorption onset at 2.1 eV and 2.7 eV, respectively, which can be attributed to direct band-band transitions. All In2S3 thin films exhibit a strong persistent photoconductivity that we ascribe to deep defects within the bandgap. Using a combinatorial approach we grew In2S3:V with a wide range of doping concentrations. For vanadium doped In2S3 grown without seed layers on sapphire substrates we find phase-pure (103) oriented β-In2S3:V for low doping concentrations only. Doping concentrations above 3.2 at.% lead to secondary phases. Using a seed layer, we could stabilize the In2S3 β-phase and suppress the formation of secondary phases. We find a change of majority carrier type in temperature dependent transport measurements for doping concentrations above a critical concentration of 3.2 at.% vanadium, which might give evidence to the formation of an IB or is connected to the occurrence of a secondary phase.

978-1-6654-6060-6/23 $31.00 © 2023 IEEE

A Model-free Approach for Estimating Service Transformer Capacity Using Residential Smart Meter Data

Joseph A. Azzolini, Matthew J. Reno, Jubair Yusuf

Sandia National Laboratories, Albuquerque, NM, United States

Before residential photovoltaic (PV) systems are interconnected with the grid, various planning and impact studies are conducted on detailed models of the system to ensure safety and reliability are maintained. However, these model-based analyses can be time-consuming and error-prone, representing a potential bottleneck as the pace of PV installations accelerates. Data-driven tools and analyses provide an alternate pathway to supplement or replace their model-based counterparts. In this work, a data-driven algorithm is presented for assessing the thermal limitations of PV interconnections. Using input data from residential smart meters, and without any grid models or topology information, the algorithm can determine the nameplate capacity of the service transformer supplying those customers. The algorithm was tested on multiple datasets and predicted service transformer capacity with >98% accuracy, regardless of existing PV installations. This algorithm has various applications from model-free thermal impact analysis for hosting capacity studies to error detection and calibration of existing grid models.

Bifacial PV fed Electrolysis for Green Hydrogen Generation and Cofiring Hydrogen in an Aeroderivative Gas Turbine

Nicholas Pilot, Jim Harper

Electric Power Research Institute (EPRI), Palo Alto, CA, United States

Hybrid power systems that generate hydrogen through renewable energy sources can be used for dispatchable energy storage and carbon reduction in power generation, potentially improving the system efficiency and economics. However, determining the best overall plant design is complicated given the number of dependent variables included in the evaluation. A systems-design model was developed using the Modelon environment to perform a techno-economic analysis for a mixed generation facility including a co-located bifacial PV plant, lithium-ion battery, polymer electrolyte membrane (PEM) electrolyzer, hydrogen storage, desalination plant, and gas turbine which utilizes a blend of the green hydrogen and natural gas. NREL' System Advisor Model (SAM) was utilized to generate sub-hourly PV production profiles for various PV technologies, plant sizes, and orientations. These production profiles were imported into the Modelon environment along with specifications for each plant component within the facility to optimize both component & system design for metrics such as hydrogen generation and demand matching, optimal operational windows, systems sizes, levelized cost of electricity (LCOE), levelized cost of hydrogen (LCOH), capital cost, efficiency, curtailment, and performance. As a result, the bifacial PV system along with the PEM electrolyzer facility was sized to produce adequate green hydrogen to blend at 20% by Vol with natural gas for combustion in the gas turbine for nearly 6 hours per day, despite variable solar irradiance. Electricity generated by the gas turbine was transmitted to the grid via an existing interconnection point. This methodology can be utilized as an outline for future green hydrogen projects to optimize component and system-level size, technology, cost, performance, and operation.

978-1-6654-6060-6/23 $31.00 © 2023 IEEE

Reverse-Bias Testing of Perovskite Cells to Inform Bypass-Diode Design

Daniel Martinez, Dana Kern, Giles Eperon

Swift Solar, San Carlos, CA, United States

NREL, Golden, CO, United States

Perovskite-based PV devices promise the combination of low cost and high efficiency. While much effort has been spent characterizing perovskite stability at the cell level, less has been dedicated to module-level stability. Nonuniform illumination of modules can cause failure by forcing shaded cells to operate in reverse bias to pass current generated by illuminated cells. Bypass diodes are employed in conventional PV to re-route this current, sometimes shared across multiple cells to reduce cost. Furthermore, we envision a cell stack with thin-film diodes integrated into each cell. To understand how to best employ bypass diodes in perovskite modules, it is important to understand the cell-level electrical requirements to prevent damage. Here, we characterized the reverse bias behavior of spin-coated and thermally-evaporated single-junction and tandem perovskite-Si devices. We measured breakdown voltages and currents, and the amount of effective shading time cells could survive. The devices were also subjected to reverse-bias conditions and in-situ characterized via DLIT to identify local failure modes. We find that reverse-bias breakdown and irreversibly damaging conditions depend on both the device architecture and perovskite deposition. Solution-coating perovskite results in non-uniformities where breakdown occurs. Conformal evaporated perovskite layers have more robust and consistent reverse bias properties, and integration into perovskite-Si tandems results in higher breakdown voltages. We find that even in the conditions cells would experience when shaded and bypassed by a diode, some damage still occurs after hours of shading time. These findings have big implications for bypass diode integration into perovskite-based modules - setting limits on what conditions the cells can experience when considering stringing bypass diodes, providing design rules for integrating on-cell bypass diodes, and guiding use cases based on realistic shading scenarios.

Improving Deep Learning-Based Defect Classification in Solar Cells using Conformal Prediction

Vitus B. Thomsen, Claire Mantel, Gisele A. dos Reis Benatto, Allan P. Engsig-Karup, Søren Forchhammer

Technical University of Denmark, Kongens Lyngby, 2800, Denmark

Abstract — **Deep learning-based approaches have become popular for automatically detecting defects in electroluminescence images of solar cells. However, deep learning methods are those that require the most training data among machine learning approaches. Thus, the data available to train such models is currently a bottleneck for their performances due to expensive and possibly inaccurate labeling. To address this problem, we propose to use a model comprising a standard deep learning classifier to which we add conformal prediction. The model calculates a degree of confidence on new predictions and can send low-confidence predictions for human expert labeling in an uncertainty-aware active learning loop. In tests with a limited-size data set, using the conformal model to select and classify high-confidence samples yields significantly higher performance compared to the standard deep learning classifier, as the F1 score increases from 0.44 to 0.62 while only leaving out 9.4% of predictions as low-confidence that need human assessment for validation and model update, demonstrating the effectiveness of the framework.**

I. INTRODUCTION

Inspection of photovoltaic (PV) modules is crucial to ensure optimal power output. Electroluminescence (EL) imaging of PV cells is an effective way of performing inspection, as it allows for greater detail than when using thermal or visible imaging only, acting like an "X-ray" image of the PV device [1]. In recent years, there has been extensive research in using machine learning, especially convolutional neural networks (CNN), on EL images for detecting defective cells [2]-[4]. A major limitation of the performance of learning-based models is currently the data available. Indeed, training a model requires large datasets whose quality depends on the variety and distribution of defects. For supervised learning (the most common), models need to be trained on labeled datasets. These labels are annotations from human experts which make them both very time consuming to create and prone to containing potential errors (inaccuracies, inconsistencies, or omissions).

In machine learning, the idea of conformal prediction allows for uncertainty quantification of a model. Conformal prediction is a general framework for constructing prediction intervals for machine learning models with a guaranteed level of accuracy, regardless of the distribution of the underlying data, that can be implemented in a general way and comes with a wide array of applications [5][6].

Another important framework in machine learning is the concept of active learning, which can be used to define an algorithm driven training loop that relies on a machine learning model and some measure of information content that can be used to ask a human expert for input on new instances that are of most value to further improve the machine learning model. During active learning, the machine learning model is then conveniently retrained to be updated using newly labeled data, allowing for optimal model improvement over time while keeping the cost of labeling data to a minimum as only a minimal amount of new images is seen by an expert for labeling [7][8]. Active learning can be done in several ways, and in this work conformal prediction is used to define an uncertainty-aware active learning loop for determining the low-confidence data to be labeled by a human. This idea is also explored in other works, e.g. [9].

In this work, we develop and test an uncertainty-aware active learning framework for PV classification based on conformal prediction and apply it to a deep learning model. The objective is that this framework will make it easier to improve the accuracy of PV classification models as more data becomes

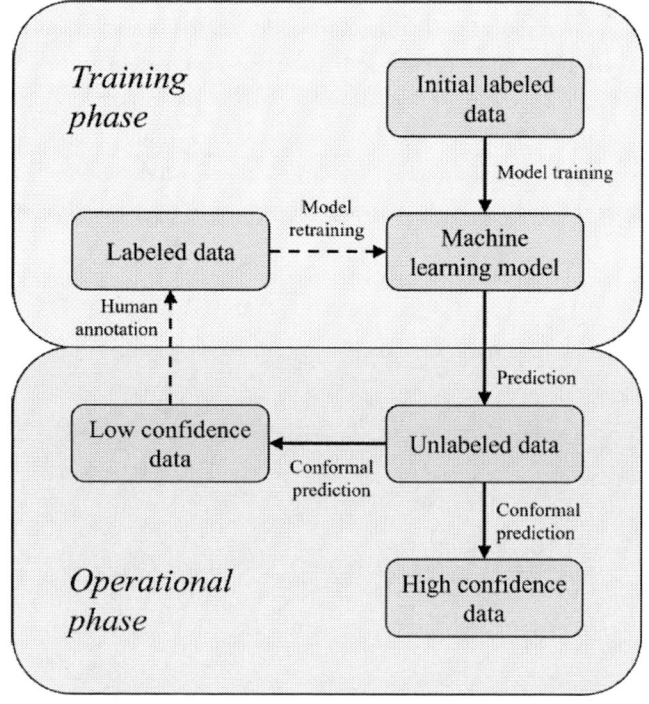

Fig. 1. Illustration of the proposed uncertainty-aware active learning framework based on conformal prediction. The dashed arrows indicate steps that are not directly considered in this work.

available, hence dealing with the challenges of doing manual annotations and improving model accuracy at the same time.

Fig. 1 illustrates the proposed active learning loop. Initially, a deep learning-based classification model is trained using existing labeled data (i.e., EL images of solar cells). Whenever new data is collected, the model is used to predict if each cell is defective or not. In this work, we also predict the type of defect, although the framework can be applied to simple binary classification as well. Using conformal prediction, each prediction is then identified as either high or low confidence. The high-confidence predictions are trusted, while the remaining samples are handed off to a human annotator (possibly along with the corresponding low-confidence predictions) who will manually go through these samples and correctly label the data. The newly labeled data is then used to retrain the model. The benefits of the active learning framework are thus twofold. The first benefit is that the predictions can be made more accurate and hence more trustworthy when restricted to the high confidence samples. The second benefit is that we can choose the examples for manual labeling that are most useful for retraining the model, and thus the model can be improved over time in an efficient manner. This work focuses on setting up the conformal prediction framework and the first of these benefits, i.e., optimizing the model accuracy on the high-confidence samples. We leave the annotation and retraining steps to be explored in more detail in future work.

We propose a simple way of splitting the predictions into high and low confidence, as well as a method of tuning the framework such that the high confidence predictions are accurate while having as few low confidence predictions as possible.

II. METHODOLOGY

A. Data filtering and preprocessing

The dataset consists of EL images collected from different solar farms using a high-resolution silicon-based detector camera at nighttime [10]. A perspective correction algorithm has been applied to the panel images, followed by local brightness normalization and algorithmic separation into individual cells [11]. Four different types of defects are considered: cracks of modes A, B, and C, as well as finger failures, as defined in [1]. Examples of defective cells are seen in Fig. 2. Each defective cell has been annotated with the type of defect and its location within the image as a binary mask.

Some cells in the dataset present potential induced degradation (PID), but this defect was not among the considered defects and those cells were excluded from the dataset by thresholding the mean pixel intensity. Furthermore, the cell images were checked for correct content, and those incorrectly cropped were discarded as well.

A small number of cells are annotated with more than one defect type. Since this work is focused on simple multiclass

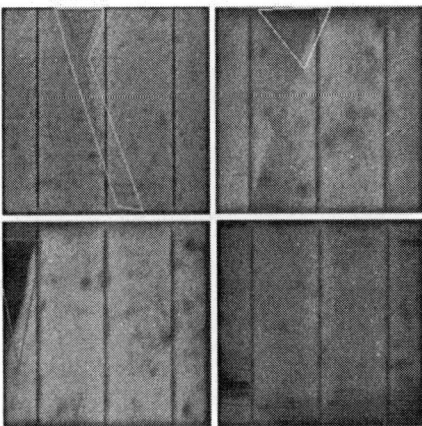

Fig. 2. Examples of cells with annotations from different defective classes. The defect types are crack A (top left), crack B (top right), crack C (bottom left), and finger failure (bottom right).

classification, where each cell belongs to only one of several classes (as opposed to multilabel classification, where a cell may belong to multiple classes at the same time), a single defect type is assigned to each of the multi-labeled cells. The assigned class is chosen using the following prioritized list based on how important or severe the four defect types are deemed, in order from highest to lowest priority: Crack C, crack B, finger failure, and crack A.

As the labeling was done on full modules, the marked locations of the defects were also split into cells. Special attention was made to avoid false positives from manual marking of a defect in one cell jutting out on neighboring cells.

After the data filtering, the dataset contains 35969 images of cells split into five classes – one class for the non-defective cells, and one class for each of the four defect types. The distribution of the five classes is seen in Table I. It is seen that the dataset is highly imbalanced, with only 993 cells in total labeled as defective.

TABLE I. DISTRIBUTION OF CLASSES

Class	Number of samples
No defect	34976
Crack A	130
Crack B	232
Crack C	124
Finger failure	507
Total	35969

B. Model selection

The dataset is split into a training, validation, and testing set in the proportions 60:20:20 with class stratification. The basic model is a classifier using a CNN, which is trained on the training set. Model selection is done using the macro-averaged F_1 score as a performance metric, evaluated on the validation set. The usual F_1 score for a single class is defined as

$$F_1 = 2 \cdot \frac{\text{Precision} \cdot \text{Recall}}{\text{Precision} + \text{Recall}}, \qquad (1)$$

where precision and recall are given by

$$\text{Precision} = \frac{\text{TP}}{\text{TP} + \text{FP}}, \qquad (2)$$

$$\text{Recall} = \frac{\text{TP}}{\text{TP} + \text{FN}}, \qquad (3)$$

where TP, FP, and FN denote true positives, false positives, and false negatives, respectively. The macro-averaged F_1 score is then given by the mean of the F_1 scores on the four defect classes (the no defect class is excluded).

After initial tests with different network architectures, a transfer learning model based on the VGG-13 architecture [12] with batch normalization is chosen. The first and last layers have been slightly modified, such that the network accepts grayscale images as input rather than RGB images, and such that the network has five output neurons, corresponding to the five classes. All layers of the network are trained simultaneously.

Model selection is done in several stages. A few different regularization techniques are tried to avoid overfitting, one at a time. Label smoothing is a technique used to avoid the model becoming overly confident in its predictions [13]. It is found that adding a small amount of label smoothing (smoothing level = 0.05) improves the F_1 score. Weight decay (L2 regularization) was also tried but was not found to improve the model. Hyperparameters (the learning rate and its decay rate) are also tuned by a small grid search. The networks are trained using the cross-entropy loss function. The Adam optimizer [14] is used with a learning rate of 0.001, which is set to exponentially decay by a factor of 0.995 after every epoch. Training is done for 300 epochs using a batch size of 64. Weighted random sampling is used during training, so the model sees roughly the same number of images from each class.

Due to the small number of images in the defective classes, data augmentation is used to introduce more variation to the training data. This is done by randomly applying transformations to each image on-the-fly during training, each with probability 1/2. The transformations that are found to improve the model are: horizontal/vertical flipping, 90-degree rotations, small rotations (±2°), brightness and contrast adjustments (±25%), gamma adjustments (γ chosen log-uniformly from [2/3, 3/2]), and random cropping, which is done in such a way that the bounding box of the defective region is kept (mostly) within the cropped image.

C. Conformal prediction

Conformal prediction is a general framework for constructing prediction intervals for machine learning models [5][6]. For a classifier, this allows for outputting a prediction set of classes rather than a single predicted class. A user-defined α allows for obtaining a set $C_{1-\alpha}$ that will contain the true class with probability at least 1 - α. Due to our limited amount of data, we here apply the cross-conformal prediction algorithm known as cross-validation+ [15]. In summary, the algorithm works by splitting the training set into K folds (we use $K = 5$), then training K different classifier networks, each using only K - 1 of the folds as a proper training set. For each of the samples in the remaining fold, a certain conformity score function is computed by comparing the raw model outputs to the true labels. These scores are referred to as the hold-out scores. When making predictions on new data, the prediction set is constructed as the set of classes such that the resulting conformity score of the new sample is smaller than $(1 - \alpha)(n + 1)$ corresponding hold-out scores, aggregated over all K classifiers. This method ensures that we meet the above-mentioned coverage guarantee of 1 - α. For a more detailed description of the algorithm, we refer to the original paper [15].

In the proposed active learning framework, the predictions are split into high and low confidence. The idea is that predictions for which the model has low confidence will be sent to a human expert for annotation and/or verification. In this study, we consider a prediction as high-confidence if its prediction set is a singleton, i.e., it contains exactly one class, and low-confidence otherwise (0 or ≥2 classes). Note that the prediction set can contain both the no defect class and any of the four defect classes.

We aim at maximizing jointly these two aspects: (i) The performance of the model when restricted to the high-confidence predictions, and (ii) The proportion of predictions that are considered high-confidence. With both these goals in mind, we define the following metric for evaluating the performance of the conformal model:

$$\tilde{F}_1 := \frac{N^{\text{conf}}}{N^{\text{val}}} F_1^{\text{conf}} \qquad (4)$$

where F_1^{conf} is the macro averaged F_1 score evaluated only on the trusted high-confidence predictions, N^{conf} is the number of high-confidence predictions, and N^{val} is the size of the validation set, i.e., the total number of predictions we make. We tune the conformal model by optimizing $\tilde{F}_1$ over the following two parameters: the significance level α (i.e., to what level of certainty should the prediction sets be constructed) and the number of epochs of training. The optimal set of values for these two parameters is determined by a simple grid search.

On a somewhat technical note, the cross-validation+ algorithm inherently involves randomness to construct the conformity scores and thus the prediction sets. To get more robust results, we generate 50 realizations of the prediction sets for each combination of parameters by using 50 different seeds for the random number generator, then average the relevant scores over these.

978-1-6654-6060-6/23 $31.00 © 2023 IEEE

Both the conformal model and a 'standard' classifier are then retrained using the union of the training set and the validation set as the new training set, then evaluated on the test set.

III. RESULTS

As a baseline model, a standard classifier is trained on the combined training/validation set and evaluated on the test set. The model is trained for 300 epochs. The macro-averaged F_1 score of the model during training is seen in Fig. 3. The final F_1 score obtained is 0.44, although the exact value fluctuates a lot between epochs. A notable observation is that after training for approximately 150 epochs, the F_1 score is not found to increase any further. This is explained by the fact that the precision is increasing with the epoch number, while the recall is decreasing, and these two effects seem to cancel each other out in the F_1 score. This suggests that the model tends to become

increasingly biased towards classifying cells as non-defective the longer it is trained.

When tuning the conformal model on the validation set, we find that the optimal $\tilde{F}_1$ value occurs at $\alpha = 0.03$ and after 125 epochs, with an estimated $\tilde{F}_1$ value of 0.54 (after some smoothing). The full grid search is seen in Fig. 4. In general, it is observed that lower values of α tend to give a greater $\tilde{F}_1$ value, although very small values such as $\alpha = 0.005$ gives very poor results. This is mainly because very small values of α result in prediction sets that are almost always of size ≥ 2 to achieve $1 - \alpha$ coverage. On the other hand, large values of α make it more likely for the model to give empty prediction sets.

Using the found optimal values, the conformal model is retrained on the combined training/validation set and applied to construct prediction sets for the test set. Once again, in order to get more robust results, we generate 50 realizations of the prediction sets and find the F_1^{conf} scores for each realization, however this time, we find the median of these F_1^{conf} scores and only consider the realization giving this median score. This is then considered representative of the typical outcome of the conformal model.

The confusion matrix for the standard classifier is shown in Table II, while the confusion matrix for the high-confidence predictions by the conformal model is shown in Table III (classes are abbreviated as ND = no defect, A/B/C = crack A/B/C, FF = finger failure). The key observation is that for the conformal model, the predictions are more concentrated on the diagonal of the confusion matrix. Table IV shows some key numbers summarizing the confusion matrices. We see

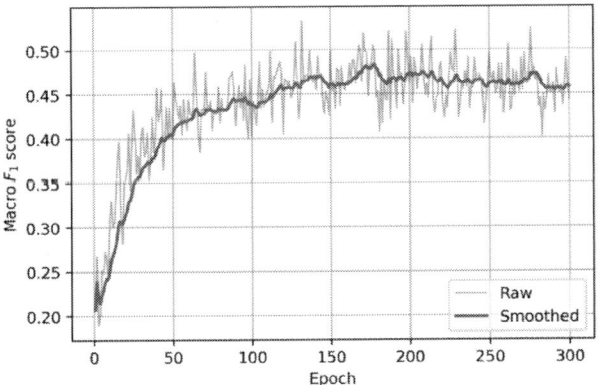

Fig. 3. Macro-averaged F_1 score of the final standard classifier, evaluated on the test set.

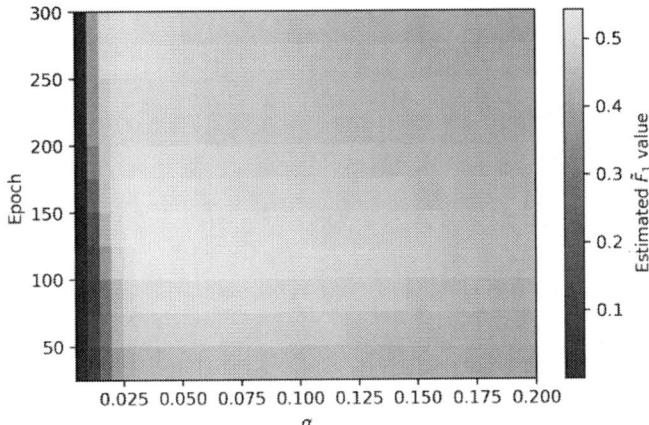

Fig. 4. Estimated $\tilde{F}_1$ values of the conformal model during the tuning process, evaluated at different α values and epoch numbers. A small amount of Gaussian smoothing has been applied in the epoch direction.

TABLE II. STANDARD CLASSIFIER

		Predicted					
		ND	A	B	C	FF	Total
Actual	ND	6907	6	17	4	61	6995
	A	18	5	2	0	1	26
	B	14	4	24	2	2	46
	C	4	1	5	13	2	25
	FF	46	1	1	2	52	102
	Total	6989	17	49	21	118	7194

TABLE III. CONFORMAL MODEL

		Predicted					
		ND	A	B	C	FF	Total
Actual	ND	6416	1	7	0	20	6444
	A	8	4	0	0	0	12
	B	3	1	10	0	0	14
	C	2	0	2	9	0	13
	FF	9	0	0	0	26	35
	Total	6438	6	19	9	46	6518

TABLE IV. METRICS FOR THE TWO CLASSIFIERS

	Standard	Conformal
Macro F1	0.4439	0.6277
Macro precision	0.4609	0.6895
Macro recall	0.4360	0.6207
Accuracy	0.9732	0.9919
Number of predictions	7194/7194	6518/7194

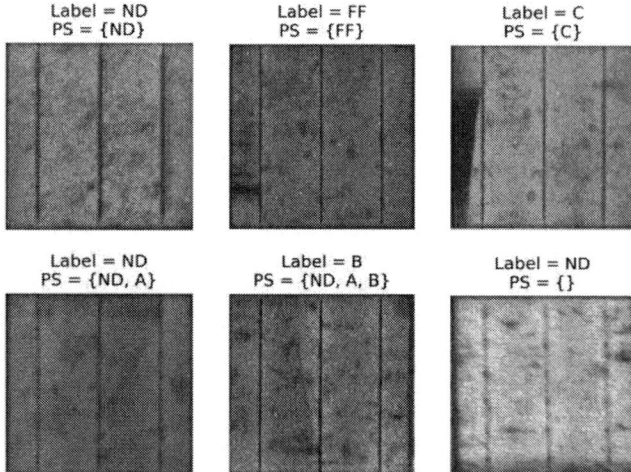

Fig. 5. Examples of test data with corresponding true labels and prediction sets (PS). The top three predictions are regarded as high-confidence since the prediction sets contain exactly one class, while the bottom three are regarded as low-confidence.

substantial improvements in F_1 score, precision, recall and accuracy for the conformal model. The price we pay for this is that 676 out of the 7194 test images (9.4%) are regarded as low-confidence, where a considerable proportion of these (18.5%) are from the defect classes.

Fig. 5 shows some examples of test data with their corresponding true labels and prediction sets, where the first three examples are high-confidence and the last three examples are low-confidence. In general, it is observed that the low-confidence samples indeed tend to be images where the true class is somewhat ambiguous, for example due to a blurry image or a crack where the type is not clear.

We remark that the conformal model has an empirical coverage of 97.9% across all prediction sets, exceeding the expected 97% (as $\alpha = 0.03$). We also remark that it is possible to use the conformal model as a 'regular' model, giving only a single prediction, by simply taking the predicted class to be the first class that would be included in the prediction set (thus using it as a sort of ensemble model over the five underlying classifiers). This method allows us to evaluate the conformal model on all test samples, not just the high-confidence ones. Doing this gives an F_1 score of 0.50, which is an improvement on the standard model but not on the conformal model restricted to high-confidence predictions. This further supports the finding that the separation into high- and low-confidence predictions is meaningful.

IV. DISCUSSIONS AND FUTURE WORK

The main factor limiting the performance of the models in this work is the 'real-life' quality of the dataset. Firstly, the dataset is heavily imbalanced with a lack of samples in the defective classes, which makes it difficult for a model to learn a general pattern, but also difficult to accurately assess the model performance due to a lack of test data. Secondly, the labeling has inconsistencies due to different people having labeled different parts of the dataset. Moreover, even though it was attempted to remove the factitious annotations, there may still have been some cells left with incorrect labels in the final dataset. It is certain that more data as well as relabeling the existing data can lead to further improvements of the models.

However, it was seen that the conformal prediction framework led to significant improvements when restricting to high-confidence predictions. This demonstrates that even with a dataset of non-ideal quality, the conformal prediction framework works as intended, in the sense that the high-confidence predictions by themselves are more accurate than when considering all predictions at once. The main benefit of the uncertainty-aware active learning framework is that it allows detecting the difficult samples for which the prediction can only be made with low confidence. Once these low-confidence predictions go through human evaluation to be labeled with certainty, the model is retrained to give better performance.

A significant downside of the cross-validation+ approach used in this work is its computational cost, as it requires training $K = 5$ distinct networks. This method was deemed necessary in this work due to limited data. There is also the limitation that the method cannot be easily applied to existing classifier systems without retraining. For future work, it would be relevant to explore cheaper methods such as simple split-conformal calibration [6][15] that are both computationally simpler and more readily applicable to existing systems.

Additionally, further work is required on streamlining the proposed active learning framework. The tuning process used in this work was necessary to give the desired performance improvements, and it may be revised. In a practical setting, one might want to choose the significance parameter α manually to gain control of the balance between the guaranteed level of confidence and the number of samples that must be manually labeled, depending on the application. Alternatively, one might want a fixed number of samples to be manually labeled in each iteration of the active learning loop. It is also important to stress that the parameters found in this work are by no means necessarily optimal in a general setting, as they may depend highly on the available data and the underlying deep learning model that is used.

Furthermore, while the conformal prediction framework is theoretically well-founded, there is no theoretical guarantee that the active learning framework used in this work chooses the optimal samples for learning. As such, the work done here should be seen as preliminary and a "proof-of-concept". In future work, more sophisticated ways of choosing samples for manual labeling may be considered, as has been done in other works [9]. Additionally, focus should be on incorporating tests that promote higher quality in the data. For example, it could be considered to modify the active learning loop such that we also predict on a small amount of data that was previously deemed high-confidence to ensure that the updated model is still confident in these samples, which can work as a quality assurance step.

REFERENCES

[1] M. Köntges, S. Kurtz, C.E. Packard, U. Jahn, K.A. Berger, K. Kato, T. Friesen, H. Liu, M. Van Iseghem, J. Wohlgemuth and D. Miller, "Review of failures of photovoltaic modules," IEA-PVPS T13-01:2014, 2014.

[2] A. Bartler, L. Mauch, B. Yang, M. Reuter and L. Stoicescu, "Automated detection of solar cell defects with deep learning," *2018 26th European Signal Processing Conference (EUSIPCO)*, Rome, Italy, 2018, pp. 2035-2039.

[3] S. Deitsch, V. Christlein, S. Berger, C. Buerhop-Lutz, A. Maier, F. Gallwitz and C. Riess, "Automatic classification of defective photovoltaic module cells in electroluminescence images," *Solar Energy*, 185, pp. 455-468, 2019.

[4] W. Tang, Q. Yang, K. Xiong and W. Yan, "Deep learning based automatic defect identification of photovoltaic module using electroluminescence images," *Solar Energy*, 201, pp. 453-460, 2020.

[5] V. Balasubramanian, S.S. Ho and V. Vovk, *Conformal prediction for reliable machine learning: theory, adaptations and applications*. Newnes, 2014.

[6] A.N. Angelopoulos and S. Bates, "A gentle introduction to conformal prediction and distribution-free uncertainty quantification", arXiv preprint, arXiv:2107.07511, 2021.

[7] B. Settles, "Active learning literature survey," Computer Sciences Technical Report 1648, University of Wisconsin-Madison, 2009.

[8] H.O. Ilhan and M.F. Amasyali, "Active learning as a way of increasing accuracy," *International Journal of Computer Theory and Engineering*, 6(6), p. 460, 2014.

[9] S. Matiz and K.E. Barner, "Inductive conformal predictor for convolutional neural networks: Applications to active learning for image classification," *Pattern Recognition*, 90, pp. 172-182, 2019.

[10] C. Mantel., F. Villebro, G.A. dos Reis Benatto, H.R. Parikh, S. Wendlandt, K. Hossain, P. Poulsen, S. Spataru, D. Sera and S. Forchhammer, "Machine learning prediction of defect types for electroluminescence images of photovoltaic panels", *Proc. SPIE*, vol. 11139, 2019, Art. no. 1113904.

[11] C. Mantel, F. Villebro, H.R. Parikh, S. Spataru, G.A. dos Reis Benatto, D. Sera, P.B. Poulsen and S. Forchhammer, "Method for estimation and correction of perspective distortion of electroluminescence images of photovoltaic panels," *IEEE Journal of Photovoltaics*, 10(6), pp. 1797-1802, 2020.

[12] K. Simonyan and A. Zisserman, "Very deep convolutional networks for large-scale image recognition," arXiv preprint, arXiv:1409.1556, 2014.

[13] C. Szegedy, V. Vanhoucke, S. Ioffe, J. Shlens and Z. Wojna, "Rethinking the inception architecture for computer vision," *2016 IEEE conference on computer vision and pattern recognition (CVPR)*, Las Vegas, NV, USA, 2016, pp. 2818-2826.

[14] D.P. Kingma and J. Ba, "Adam: A method for stochastic optimization," arXiv preprint, arXiv:1412.6980, 2014.

[15] Y. Romano, M. Sesia and E. Candes, "Classification with valid and adaptive coverage," *Advances in Neural Information Processing Systems*, 33, pp. 3581-3591, 2020.

Off-the-Shelf Small Scale Photovoltaic Systems for Puerto Rico Sustainable Farms: Assisting those Who Help Others

Catherine Lahood, Sarah Sirkisoon, and Charles M. Fortmann.

Department of Physics, St. John's University, 8000 Utopia Parkway, Queens NY 11439 USA

Abstract—**Puerto Rican small-scale sustainable farms struggle to provide wholesome food, preserve traditional methods and seeds, pass lessons on to the next generations, and are an oasis of shelter, food, water, and electric power in times of crisis. Since these efforts are certainly not fueled by greed or profiteering, appropriate technology must be cost-effective, long-lasting, on-site repairable, ecologically compatible with the environment, and in some cases, have long-term technological support. Small grids must be carefully sized to provide daily electrical usage during daylight hours, such as water pumping, water purification, and refrigeration, while not necessarily large enough to charge batteries for routine overnight and multiday usage. Since batteries have finite lifetimes that are discharge cycle dependent, costs can easily exceed $0.20/Watt; therefore, daylight use of solar power should be prioritized. However, battery and water storage must be large enough to cover prolonged sunless days. Agricultural water pumping is an ideal primary application of photovoltaic power as energy is converted to stored water at a minimal cost. Working together Non-Government Organizations, an Academic Research Team, and a Sustainable Farm have developed low-cost off-the-shelf engineering and enduring support structures that provide low-cost assistance that adapts to changing needs.**

Keywords— *sustainable, cost effective, mini-PV grid, water purification.*

I. INTRODUCTION

The demographic changes accompanying a society's industrialization challenge rural communities as people move to larger cities for employment and educational opportunities for children. These changes raise the standard of living and offer children an opportunity to utilize the greater resources of urban school systems. However, the resultant migration to urban centers dramatically reduces rural populations through the attrition of families with children and young adults. Industrial farms that take over rural lands, in turn, are as interdependent as urban centers on supply chains and transportation networks to maintain and export the extremely specialized livestock and crops produced. The recent storms and earthquakes in Puerto Rico reveal how vulnerable urban centers and industrial farms are with their specialized produce and how little support they provide to a populace in crisis.

The loss of rural populations also breaks the chain by which farming practices evolved over hundreds and thousands of years. These unfortunate events shine a beacon on an unexpected resource of the small sustainable farm based upon historical practice augmented and made viable by modern off-the-self photovoltaic technology and water purification methods.

This work examines small photovoltaic systems through the lens of rural farms and villages seeking resilience and sustainability in a fragile interconnected world.

II. EXPERIMENT

A. Micro photovoltaic panels for emergency preparedness

Our first intervention was an outreach to provide a measure of relief to a school for sustainable farming in the Orocovis region in the aftermath of Hurricane Maria. The request came through Luis Nicho, director of Nuestro Ideal, a non-governmental organization that had supplied mobile kitchens to Puerto Rico to assist in relief efforts. Orocovis had cell phone service but no means to recharge the phones due to the highly damaged electrical grid.

St. John's University mounted a multi-department effort to build 100 5W solar phone chargers. At the time, ready-made solar phone chargers cost about $30 each. The physics department tasked students with designing and testing phone chargers built using inexpensive off-the-shelf components. A design employing 5W mono-crystalline solar cells sourced from Alibaba at a cost of ~ $5.00 each. Each solar cell was fitted with a commercial diode at the cost of ~ $0.10, and importantly the standard automotive power plug and standard automotive type 12V to USB converter were again directly sourced from the manufacturer at the cost of ~ $1.00 each for a total cost of $7.00 per phone charger. During the construction, St. John's students and the recipient Orocovis students were in communication via Zoom.

B. Medium-scale photovoltaic arrays for farm water supply and emergency back-up.

A subsequent request was for a water pumping system for a sustainable farm near Lattes, Puerto Rico, compatible with an existing solar array. The pump would need to lift water from as much as 300 ft and need to pump as much as 2,000 gallons per day. An LZT2 ½ hp deep well pump[1] was chosen at a cost of ~ $700 from a U.S.-based supplier. A 300 ft. spool of two conductor 10-gauge PVC insulated wire was sourced in the U.S. at the cost of $150.00. Importantly, the two-wire 120 V pump was selected to be compatible with the existing solar array, which had a 120V 2-wire invertor and, at ~ 12A @ 115V, can pump 5 gallons per minute at a depth of 250ft. Note that these power ratings suggest a power draw of over 1,300W (~ 1.5 hp), significantly greater than ½ hp.

Upon installation, the inverter waveform was verified to ensure a pure-sine wave output, since a pump motor's operational lifetime can be negatively impacted by stepwise waveforms found in low-cost inverters. Another important consideration is that inverter capacity is reduced by half when driving inductive loads (e.g., pump and other motors).

Well casings (in this case 6 inch) mean pump motors must pack a lot of power into a narrow space. In turn, motors greater than ½ hp typically require 3-phase motors to achieve greater power density. Three-phase DC-AC power inverters are significantly more expensive than the common, more aggressively marketed 2-phase inverter. Here, the choice was made to keep the existing 2-phase inverter and allow the pump to run beyond sunset and on cloudy days using battery stored energy. However, on a sunny day, the requisite 2,000 gallons could easily be obtained in ~ 7 hours without storing energy. Of course, storing excess water obtained on a long, sunny day is more economical than battery energy storage.

C. Medium-scale photovoltaic arrays for farm water supply and emergency back-up.

A third system addressed a request for a backup power supply to run computers at the school in Orocovis. Two 100W mono-crystalline solar panels were sourced from Amazon. A 700W Renogy™ inverter and Renogy™ max power point tracking system (MPPT), and a size 27 marine deep cycle advanced glass mat battery was used for storage. Unlike water pumping applications, this system is used for emergency stand-by operations that need to withstand prolonged inactivity, and requires batteries to initiate the MPPT, which must have battery power on start-up. Lithium batteries would be superior but were too costly at the time.

The system was fitted into a waterproof plastic box (Pelican ™) modified with splash-resistant air-cooling vents and access holes for electrical connections. This system was first utilized during the power outages of Hurricane Fiona. An important aspect of these projects was the involvement of St. John's University undergraduate students in the project. All St. John's students are required to perform 6 hours of Academic Service-Learning activity during their studies. The project was one of the first STEM-oriented projects requiring students to apply knowledge learned for the greater good. In the case of the solar phone chargers, a two-tiered system was employed wherein more experienced physics students who had previously learned soldiering techniques mentored non-science majors during the building of the 100 solar phone chargers. Students gained genuine research experience as they worked alongside professors to establish the best ways to meet the project goals.

D. Water puurification

As in most locations, deep well water in Puerto Rico is typically clean, having few toxic elements or pathogens. However, this is not true of shallow wells, ponds, groundwater, and rainwater. Ground and pond water contain many pathogens from animal contamination. Even rainwater collected from rooftops contains pathogens chiefly from birds and the prey they feed upon, including frogs, frog eggs, and algae from local ponds. The most ecologically sound (no chemicals) and lowest environmental impact (little power draw) is Ultraviolet (UV) sanitation, wherein light is used to selectively kill pathogens without water heating. For UV sanitation to work, the water must be clarified. In the case of ponds, chemical methods for algae removal are available, and filtration is possible but requires periodic maintenance. The likely best solution involves pond aeration, shading with structures or water lilies, and filtration.

III. DISCUSSION

Current generation mono-crystalline silicon solar cells are both inexpensive and long-lasting, with many performing well beyond 20 years. However, batteries are relatively expensive and have a finite number of cycles. The initial outlay for batteries favors lithium-based batteries, with many LiFePO4 costing about $275 per kWh (e.g., see [2]). The price drops to $0.30 per kWh over a 1,000 deep cycle lifetime and some manufacturers claim as much as 4,000 cycles are possible. Note, old-fashioned lead acid batteries are limited to ~300 deep cycles and, unlike lithium-based batteries, have declining voltage and power as they age. Lithium batteries do not appear to age when not cycled. Therefore, in most locations battery stored energy may cost less than grid power (e.g., New York City is ~ $0.20/kWh). However, with the present Puerto Rico electrical rates, battery storage is a viable option.

Even when grid energy is inexpensive, although this does not appear to be the present case in Puerto Rico, it is still a significant expense for the small sustainable farm. In addition, the economics favors the immediate use of solar for water pumping with minimal routine energy storage but still a significant amount of reserved energy for emergency use during grid outages. Having enough stored power for emergency use is critical for maintaining precious local seed varieties and preserving produce. Importantly, the local, sustainable farm is a refuge for the local communities during disasters by providing food, shelter, and power for cell phones and critical medical devices. Therefore, suggesting batteries, even where grid energy is available, should be sized for multi-day backup power but not routinely fully cycled.

While the cost of installed solar electric generation is vastly more affordable than grid-based energy, battery storage is still expensive for independent, sustainable farms at current pricing. A strategic approach is required in Puerto Rico, as the present power grid cannot absorb excess solar power generation during daylight hours, since smart meters and other advanced grid adaptability are unavailable.

Therefore, running as much as possible directly from the installed solar arrays during daylight, is the most advantageous. Where excess solar power is installed, a novel approach to efficient, cost-effective energy use would be converting as much farm machinery and transportation to battery electric as possible.

Battery electric tractors, trucks, and cars, like the storage batteries discussed above, will in time, consume the useful life of the battery. However, there is no grid alternative to battery consumption in these applications. Rather, the comparison must be made to diesel and gasoline, which on a per-mile basis costs somewhat more than the electrical energy and battery consumption. Here, the cost of periodic battery replacement is favorable relative to fuel costs.

Students working with professors and NGO partners, in unison with the sustainable farming community, leaves a lasting impact on all involved. We found that the students and the professors learned far more from the sustainable farmers who live their philosophy, than the farmers learned from us. We also learned that our minor technological assistances are magnified a thousand-fold through the sustainable farm's assistance to its community. St. John's students learned about the power of technology, and how it can be utilized to improve the greater good. Students also gained a greater appreciation of science, teamwork, cooperation, and their ability to contribute to it. The ability for students to experience the application of science to helping others was extremely rewarding and heavily enhanced their undergraduate experience at St. John's University. The Puerto Rico sustainable farms and NGO partners we worked with live their lessons on sustainability and are an inspiration the world can learn from.

IV. FUTURE OPTOMIZATION

Lessons learned include the minimal use of battery cycles during routine operation when economical grid energy is available. Conversely, maximizing the immediate use of photovoltaic energy is the most economical application of this energy source. The savings to a sustainable farm can be extensive. Converting machinery and transportation to electric power where excess solar electric power is available could save even more energy, cost less in the long run, and help reduce the global CO_2 load.

V. CONCLUSIONS

Buying individual components from global marketplaces such as Alibaba and Amazon is typically much more economical than buying ready-to-use systems. Furthermore, such systems can be tailored to the needs and budget of the specific farm. However, as systems become cheaper, these should be taken advantage of when more economical. Understanding the needs of the sustainable farm and preserving and respecting their teachings remains the backbone of the project. We learned from and were inspired by the Puerto Rican sustainable farm community and our NGO partner.

ACKNOWLEDGMENT

We thank Gustavo Gonzales Ramos and Aixa Miranda of Proyecto Agroecológico Campesino for their discussion, partnership, friendship, and all the help and support they provide their community. And for teaching us and sharing their compassion for sustainability and community. We thank Luis Nicho of Nuestro Ideal for his partnership, networking, logistics, and support throughout this project, enabling our progress. We also thank the Society of Physics Students through which many of these activities was organized. Catherine Lahood and Sarah Sirkisoon thank The Clare Boothe Luce Program for support. We thank St. John's University's Vincentian Institute for Social Action, Lynn Stravino and Louis Saavedra of the Saint John's Academic Service-Learning Dept.

REFERENCES

[1] https://www.grainger.com/product/DAYTON-Submersible-Deep-Well-Pump-1LZT8

[2] https://www.litime.com/products/litime-12v-100ah-mini-lifepo4-lithium-battery-upgraded-100a-bms-max-1280wh-energy?gad=1&gclid=EAIaIQobChMI4ZaF4rfd_wIVEPzICh0TugBCEAAYASAAEgL9NfD_BwE

 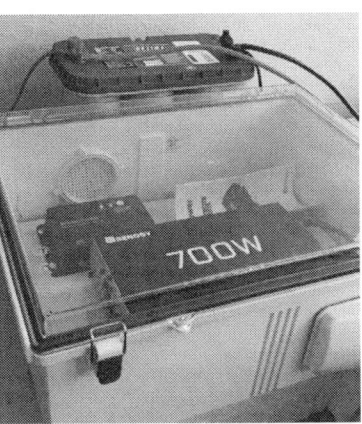

Figure 1: Solar-powered back-up power supply. Two 100W mono-crystalline solar panels sourced from Amazon, a 700W Renogy™ invertor, Renogy™ MPPT (max power point tracking system) and a 27M deep cycle marine battery was used for storage (left). The system was fitted into a Pelican™ water-proof plastic box that had been modified with splash-resistant air-cooling vents and access holes for electrical connections (right)

Figure 2: Algae growth in stored water. Without proper aeration and filtration algae growth is common in rainwater storage containers.

Figure 3: Animal Contamination. A horse is seen drinking from a pond, used as a source of crop irrigation for the sustainable farm.

Figure 4: Water pumping system. Water pump implemented in farm located in Vieques, Puerto Rico.

Flexible Photonic Cooler Based on Multi-Stacked Thin Films IR filters with Anti-Dust Capability for PV-Desert Environment Applications

Brahim Aïssa and Mohammad I. Hossain

Qatar Environment & Energy Research Institute (QEERI), Hamad bin Khalifa University (HBKU), P.O. Box 34110, Doha, Qatar

Abstract — **We report on the fabrication of metal-oxide/metal-oxide/metal thin films multi-stacked configuration used as a photonic cooler based on a near infrared light filter, with anti-dust and antireflective properties. The optimized structure considers TiOx as a top layer, NiO as the buffer layer, and Ag as the reflective hot-mirror layer of the near infrared (NIR) light spectrum. This configuration was developed by a thermal e-beam deposition process without breaking the vacuum, where all the oxides layers were grown reactively under oxygen atmosphere. TiOx layer has been found super-hydrophilic, thereby providing a self-cleaning adn anti-dust properties to the photonic cooler. Filtration (i.e. cutoff) of the IR spectrum has been improved with the stacking configuration of inorganic metal oxide (Low refractive index, TiOx)/metal oxide (High refractive index, NiO, MoOx)/ metal layer (Ag, Al). Among all the configurations, MoOx/TiOx and NiOx/TiOx layers have shown higher transmittance (T%) in the visible range compared with the single TiOx layer. Moreover, adding a metal layer of (Ag or Al) resulted in wavelength cut-off starting from 800 nm by reflecting (>70%) of light. Our results confirm the commercial potential of this multi-stack structure due to its multi-functionalities, such as IR filtering, anti-reflection coating in the visible range, and anti-soiling capability, in addition to its scalability.**

I. INTRODUCTION

The radiated solar spectrum on earth has three main sub-bodies, namely ultraviolet (UV), visible, and infrared (IR). The proportion of the IR spectrum is higher than visible and UV with 54% of incident radiation [1]. This part of solar spectrum is the main reason of heat by converting infrared wavelength into thermal energy, thus the energy consumption to cool a building increases significantly, especially in desert environment. Thus, developing thin film based technology to meet the purpose of reflecting IR spectrum becomes highly desirable [2]. An efficient mechanism that might be achieved consists on developing smart thin film layers that reflect back the IR spectrum starting from NIR while keeping a high transmissivity (T%) in the visible range. In photovoltaic (PV) field, a high T % in the visible range is essential to generate a descent PV power.

Multiple metal layers including Ag, Al, and Au have been already suggested to be used as filter for IR spectrum due to their capability of absorbing resonance frequency in the infrared range [3]. However, stability was found to be a major drawback to these metal layers [4]. Sandwich structures have then been proposed to keep such metal layer in between oxide layers [5]. However, oxygen diffusion of metal-oxide becomes obvious and in-situ stress of such films results in their cracking resulting in poor light T% in the visible spectrum [6]. A comprehensive literature survey (results of this study will be presented in the full version of this paper) shows that only four metal oxide films may meet the technical compatibility to develop such IR filters, namely TiO_x [7], SnO_x [8], MoO_x [9], and NiO_x [10] and this study served as a basis of our current work. Various physical vapor deposition techniques have been adapted to deposit these metal oxides films, including magnetron sputtering [11], vacuum evaporation 12 and atomic layer deposition (ALD) [13]. Among these techniques, the e-beam evaporation is the one of the most suitable technique due its degree of freedom, and scalability [14].

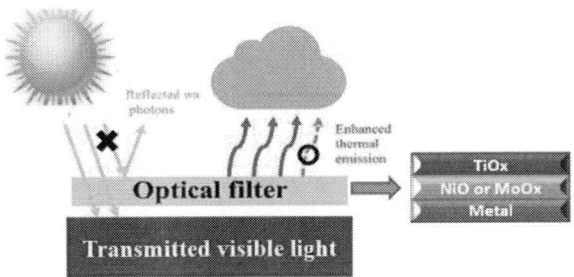

Fig. 1. Schematic structure of multilayered structures for infrared (IR) filter.

The goal of this work is to develop an efficient and cost-effective metal-oxide/metal-oxide/metal IR filter, with a high-T% in the visible range, and high-R% in the IR region. This filter is meant to operate in a hot and dusty desert environment, whose the need to integrate an anti-soiling property. Both numerical and experimental works have been considered to meet these requirements.

II. EQUIPMENT AND METHODS

Metal oxide thin films (NiO and TiO_x) were grown using e-beam evaporation at room temperature under a constant oxygen flow rate. Metal layers (Al, Ag) were also grown on glass and PET flexible substrates without any oxygen flow. Samples were optically measured using ellipsometry and UV–Vis spectroscopy. The wetting behavior was assessed by contact angle measurements. Surface topology was conducted by a three-dimensional (3D) stylus (Dektak) and the microstructure of the films was studied by FESEM and AFM. Structural characterization was measured using x-ray photoelectron spectroscopy (XPS, Fig. 2). At the beginning, thickness

978-1-6654-6060-6/23 $31.00 © 2023 IEEE

optimization was carried for each layer to maximize the optical properties. Comprehensive details about the different films deposition (i.e. growth parameters) and their characterization will be included in the full version of this paper.

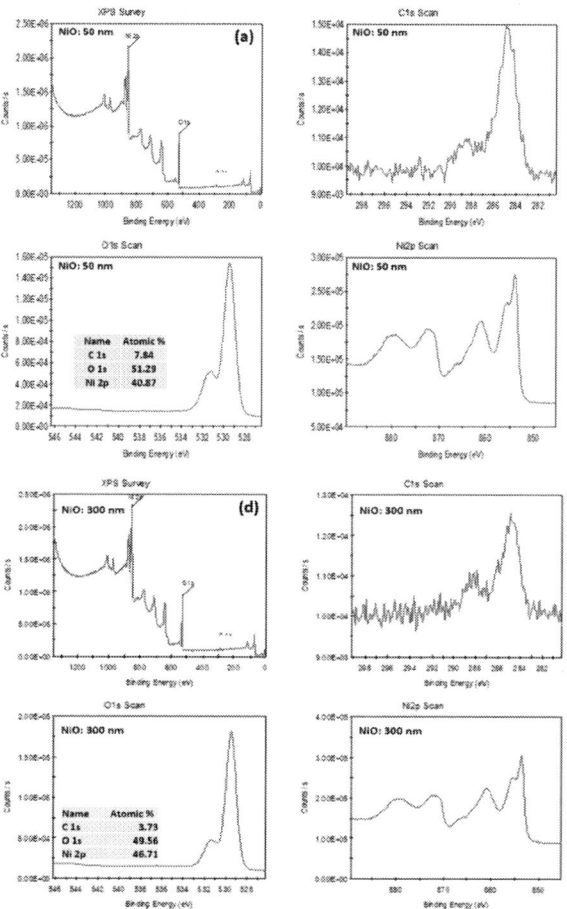

Fig. 1. XPS survey of NiO films grown at 2×10^{-4} Torr with four different thicknesses, examples shown here are for (a) 50 nm and (d) 300 nm.

III. RESULST AND DISCUSSIONS

Optical properties of Nickel oxide (NiO) thin films:

The optical properties of the films were investigated through the UV-Vis spectroscopy for the wavelength range from 200 to 2000 nm. Absorptance spectra was calculated using:

$$A(\%)=100-(T+R)$$

Where A is the absorptance, T is transmittance, and R is the reflectance. Transmission, reflectance and absorptance were found to vary with respect to the NiO thickness. As a matter of fact, while the transmittance was the highest (85%) for the wavelength (>500 nm) for 20 nm thick film, both absorptance and reflectance were higher for the films above 100 nm. Film thickness has been hence used to tune the optical properties. Same observation has been made for the metal layer. For instance, reflection was changing significantly with thickness between 10 nm and 15 nm of the metal layer. In our case, R%

should be lowest in visible and highest in IR region to cutoff the infrared spectrum. Among all samples, the lowest R% was found for all NiO/metal films except NiO 100 nm/Al 10nm, NiO 100 nm/ Ag 15nm starting from 700 nm (Fig. 3).

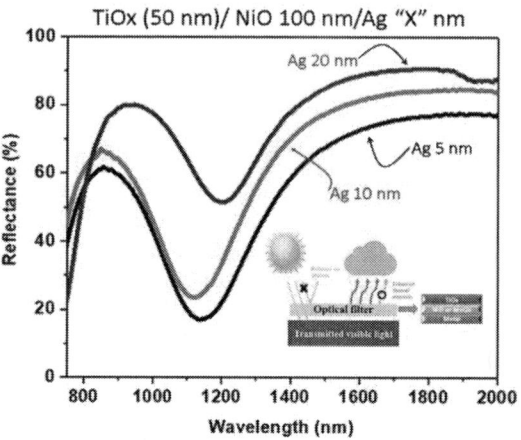

Fig. 2. Reflectance % as a function of wavelength (nm) for various Ag metal thicknesses.

Additional thickness optimization of both NiO and metal layers were carried out and the obtained results confirmed that the thickness of the metal films is the key to develop an optimize IR filter.

Morphological analysis:

As confirmed by a systematic SEM analysis, all the evaporated stacked layers were dense, homogenous, and without any pinholes or apparent cracks. Evaporated films were entirely covering the sample with uniformity, which is an essential characteristics to develop high-end optoelectronic devices such as infrared filters. Cracked surfaces with defects can suppress films properties of reflecting infrared solar spectrum. In our study, metal layers have been used as a template to grow metal oxide layers. Moreover, the average roughness has been found to change with respect to the thickness of the underneath metal layers. For instance, NiO /metal (Al, Ag=20 nm) stacked layers resulted in the highest average roughness (~17 nm) while NiO /metal (Al, Ag=10 nm) layers resulted in the lowest value (~10 nm). Such results can be correlated with the SEM results, where increasing smaller grains have been reported for thicker metal oxide layer. It is crucial to notice that metal layers play a vital role in tuning the roughness of top layers which might be used to tune the wettability of the films.

Optical properties of NIR filter grown on flexible substrate:

Later, TiO_x films were grown on NiO/Ag/PET flexible light weight structure (Fig. 4). Transmission, reflectance and absorptance spectra change accordingly after inserting the top layer TiO_x for the wavelength range of 200 to 2000 nm. For the films grown with TiOx (50nm)/NiO 100nm/Ag 20nm and TiOx (50nm)/NiO 100nm/Ag 25nm they show the highest reflection of around 20% starting from (>800 nm).

Contact angle properties:

Generally, the wetting behavior of thin films with a contact angle more than 60° can be used for anti-soiling applications [15]. Hydrophobic surface can be prepared to resist to dust accumulation, especially in humid climate. In this study, it is critical to be able to tune the hydrophilicity property. For instance, TiO_x (50 nm)/NiO (300 nm)/Ag (25 nm) showed the highest hydrophobicity with a CA of about 104°, whereas hydrophilic behavior has been systematically measured for the other stacked films. Previous study confirms that surface wettability changes significantly with roughness [16].

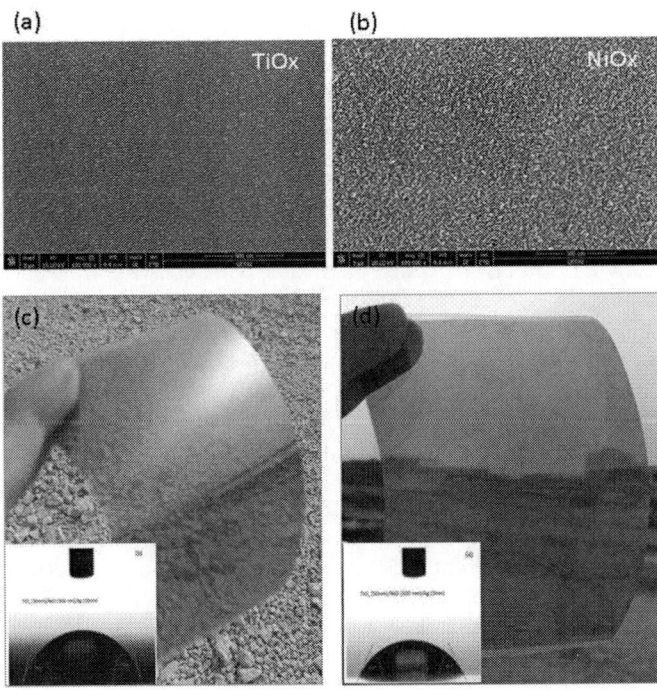

Fig. 4. SEM micrographs showing (a) the TiOx morphology (dense and pinhole-free films), (b) NiOx morphology (high quality films obtained by e-beam), (c-d) Optical photo of the flexible filters having two different silver thicknesses).

The lowest contact angle measurement of 61.3° has been found for TiO_x (50 nm)/NiO (300 nm)/Ag (5 nm) due to the roughness dependency. It is expected to develop super-hydrophobic surface with increasing roughness due to the reduced tension between surface and water molecule. The results of dust accumulations (which will be detailed in the full version of the paper) goes with the CA measurements and soiling was minimal (after two months of dust accumulation in outdoor conditions) for the most hydrophilic configuration.

IV. CONCLUSIONS

Infrared spectrum filters are useful to cut-off the infrared wavelengths responsible for heat generation. Our developed stacked layers can be used with three different functionalities such as infrared filtering, anti-reflection coating, and anti-

soiling coatings in a desert environment. In this work, we have developed a photonic cooler based on metal oxides and metal layers which were deposited as a stacking by a reactive e-beam evaporation process. Silver (Ag) with a thickness of 20 nm has been used as seed layer to grow NiO and TiOx layers for the stacking structure. The highest cut off was for NiO (300)/Ag (20 nm) films with a value of >75% above 750 nm. The highest T% was around 45% in the visible range for the metal thickness from 5 nm to 20 nm. Later, flexible substrates were used to develop such IR filter with other functionalities such as anti-reflection and anti-soiling coatings. TiO_x (50 nm)/NiO (300 nm)/Ag (25 nm) showed the highest hydrophobicity with a CA of about 104°. These results confirm the development of multi-stacked metal oxide/metal oxide/metal films using thermal e-beam evaporation to be used as near infrared light filter, anti-dust and antireflective coatings with large-scale fabrication feasibility. Mechanism of IR cut off and application to mini PV module will be discussed in the full version of this paper.

REFERENCES

1. Abundiz-Cisneros, N., R. Sanginés, R. Rodríguez-López, M. Peralta-Arriola, J. Cruz, and R. Machorro. Energy and Buildings 206 (2020): 109558.
2. Sahm, H., C. Charton, and R. Thielsch. Thin Solid Films 455 (2004): 819-823.
3. Hossain, M. I., A. Khandakar, M. E. H. Chowdhury, S. Ahmed et al. Journal of Electronic Materials (2021): 1-11.
4. Stamate, Eugen. Nanomaterials 10, no. 1 (2020): 14.
5. Tan, Wai Kian, Atsushi Yokoi, Go Kawamura, Atsunori Matsuda, and Hiroyuki Muto.
6. Liang, Chih-Hao, Sheng-Chau Chen, Xiaoding Qi, Chi-San Chen, and Chih-Chao Yang. Thin Solid Films 519, no. 1 (2010): 345-350.
7. J. T.-W. Wang, J. M. Ball, E. M. Barea, A. Abate, J. A. Alexander-Webber, J. Huang, et al., Nano letters, vol. 14 (2), pp. 724-730(2014).
8. P. Pinpithak, H.-W. Chen, A. Kulkarni, Y. Sanehira, M. Ikegami, and T. Miyasaka, Chemistry Letters, vol. 46 (3), pp. 382-384(2017).
9. C. Liu, W. Li, J. Chen, J. Fan, Y. Mai, and R. E. Schropp, Nano Energy, vol. 41 pp. 75-83(2017).
10. K. Cao, Z. Zuo, J. Cui, Y. Shen, T. Moehl, S. M. Zakeeruddin, et al., Nano Energy, vol. 17 pp. 171-179(2015).
11. P. Baroch, J. Musil, J. Vlcek, K. Nam, and J. Han, Surface and Coatings Technology, vol. 193 (1-3), pp. 107-111(2005)
12. J. Velevska and M. Ristova, Solar energy materials and solar cells, vol. 73 (2), pp. 131-139(2002)
13. X. Yang, P. Zheng, Q. Bi, and K. Weber, Solar Energy Materials and Solar Cells, vol. 150 pp. 32-38(2016)
14. I. S. Kim, E.-K. Jeong, D. Y. Kim, M. Kumar, and S.-Y. Choi, *Applied Surface Science,* vol. 255 (7), pp. 4011-4014(2009)
15. Feng, X. J., and Lei Jiang. Advanced Materials 18, no. 23 (2006): 3063-3078.
16. Hossain, Mohammad Istiaque, Brahim Aïssa, Ayman Samara, Said A. Mansour, Cédric A. Broussillou, and Veronica Bermudez Benito. *ACS omega* 6, no. 8 (2021): 5276-5286.

Loss Analysis and Performance Optimization Pathways of 729-mV V_{oc} Si Solar Cells with *Poly*-Si on Locally-Etched Dielectric Passivating Contacts

Suchismita Mitra[1*], Caroline Lima Anderson[1,2], Matthew Hartenstein[1,2], William Nemeth[1], Matthew Page[1], San Thiengi[1], David Young[1], Sumit Agarwal[1,2], Paul Stradins[1,2]

[1]National Renewable Energy Laboratory, Golden, Colorado, 80401, USA

[2]Chemical and Biological Engineering Department, Colorado School of Mines, Golden, Colorado, 80401, USA

Abstract — **In this article, the loss analysis of silicon solar cells with polysilicon on locally-etched dielectric passivating contacts with V_{oc}=729.0 mV and efficiency=22.6% has been presented. Experimentally, nano-pinholes were introduced in SiO_x (2.2 nm) and SiO_x/SiN_y (2.2 nm/8nm) stack using metal-assisted chemical etching (MACE). SunSolve and Quokka3 were used to simulate the experimental solar cell and investigate the optical and electrical power losses. Simulations suggest maximum power loss occurs due to recombination and resistive losses in the bulk (~0.76 mW/cm²) followed by power loss due to rear contact recombination (~0.35 mW/cm²). Recombination at the front surface also contributes to 0.24 mW/cm². The effect of improving the bulk lifetime and lowering the recombination current density at the rear side on V_{oc}, FF and hence, efficiency has been investigated. Further, advanced structures have been proposed to minimize recombination and parasitic absorption to achieve higher V_{oc} and J_{sc} of the solar cells with locally-etched dielectric passivating contacts.**

I. INTRODUCTION

Passivated contact solar cells using *poly*-Si/SiO_x interface has been attracting attention of PV researchers in the laboratories and industry for a few years now. In this article, we present the performance analysis of a passivated contact solar cell with **Poly-Si on Locally-Etched SiO_x (PLEO)** at the front side and **Poly-Si on Locally-Etched SiN_y/SiO_x (PLENO)** at the rear side. In this device, pinholes have been deliberately introduced in the SiO_x layer at the front and SiN_y/SiO_x at the rear side using metal-assisted chemical etching (MACE) technique creating a pathway for carriers to flow while maintaining excellent passivation, thereby leading to a V_{oc} of 729 mV and ~22.6% efficient solar cell. The loss analysis of the device was carried out by first performing optical simulation in SunSolve and then using the generation rate from SunSolve as an input for electrical simulation in Quokka3. Further, advanced structures have been proposed to reduce the recombination and current losses occurring in the present device and a pathway towards achieving higher V_{oc} and J_{sc} and hence, higher efficiency has been shown.

II. EXPERIMENTAL AND SIMULATION DETAILS

Fig. 1. shows the schematic diagram of passivated contact solar cell with poly-Si on locally-etched dielectrics (PLEO at front and PLENO at rear). The fabrication and characterization procedure have been discussed elsewhere [1-3].

In the device, a 122 μm thick, single side textured n-type Cz wafer is sandwiched between an ultra-thin (~2.1-2.2 nm) oxide layer. A second layer of SiN_y (~8 nm) was deposited at the rear side prevent boron accumulation in the SiO_x layer. After this, ~30 nm of a-Si:H films were grown on both sides of the wafer also via PECVD. Generally, passivated contact carrier transport occurs either predominantly through quantum tunnelling for SiO_x < 1.7 nm or pinholes for SiO_x > 2.1 nm. In this case, pinholes were deliberately introduced using the metal-assisted chemical etching (MACE) method using silver (Ag) nanoparticles at this stage. After pinhole formation, the Ag nanoparticles were chemically dissolved with a heated RCA 1 solution. This step was followed by RCA1-oxide removal, and subsequent etching of the a-Si:H using tetramethylammonium hydroxide (TMAH) to avoid undercuts [2]. Then, a 50 nm thick n^+ poly-Si at the front side and 50 nm thick p^+ poly-Si at the rear side was developed by first depositing ~30 nm of intrinsic a-Si:H films on both sides of the wafer by PECVD, followed by the depositions of ~20 nm of phosphorous- and boron-doped a-Si:H on the front and rear, respectively. For the rear boron doping, a 4cm² aperture mask was used to dope the rear side only locally, thus the surrounding poly-Si layer remained intrinsic This was followed by a hydrogenation step using Al_2O_3 and subsequent metallization on both sides. Metallization was carried out by evaporation of Al through specially designed masks at the front and Ti-Ag over the entire rear surface. Finally, a 60 nm SiN_y was deposited at the front side as an anti-reflection coating. Although the ARC covers front metal grids, the probes poke through SiN_y and makes probe-to-metal resistance low enough for proper JV measurement Please refer to the article by Anderson et al. [2], for the fabrication process in detail. The cell area was 4 cm² (2.5 cm x 1.6 cm) and a 2.5 cm x 1.6 cm aperture was used during the JV measurement independently certified by NREL.

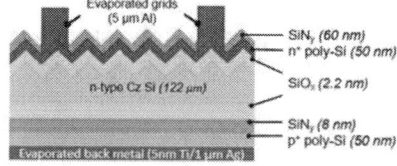

Fig. 1. Schematic diagram of the passivated contact solar cell with poly-Si on locally-etched dielectrics (pinholes omitted for clarity)

TABLE I
PARAMETERS FOR SIMULATION OF STRUCTURE

Structural Details		Metal Contacts		Passivation	
Wafer thickness	122 um	Width of fingers	20 um	$J_{0\,(no\ metal, front)}$	5.8 fA/cm^2
Base resistivity	1.6 ohm-cm	Pitch of Fingers	1.19 mm	$J_{0\,(metal, front)}$*	13 fA/cm^2
Cell Area	4 cm^2	Al for front contact	5um	$J_{0(no\ metal, rear)}$	0.6 fA/cm^2
n+ *Poly*-Si thickness	50 nm	TiAg for rear contact	0.005 um Ti + 1 um Ag	$J_{0(metal, rear)}$*	10 fA/cm^2
p+ *Poly*-Si thickness	50 nm			Bulk lifetime	2750 us
SiO$_x$ thickness (both sides)	2.2 nm				
Passivating SiN$_y$ in the rear side	8 nm				

*Note: The parameters have been derived from various characterizations such as J_0 measurements for $J_{0(no\ metal,\ front)}$ and $J_{0(no\ metal,\ rear)}$, minority carrier lifetime measurements using Sinton WCT for bulk lifetime, profilometer for width of metal contacts and PL imaging for observing degradation in surface recombination before and after metallization.*Values assumed to match simulation model with experimental IV curves.*

The structure shown in Fig. 1 was simulated in Quokka3 to understand the resistive and recombination losses incurred at different parts of the structure. The aim of the work is to find a way to reduce these losses and achieve higher efficiency.

Few parameters that were used in simulations have been found experimentally while some have been assumed to fit the simulation model with the experimental JV curve as shown in Table I. While $J_{0(no\ metal,\ front)}$=5.8 fA/cm^2 and $J_{0(no\ metal,\ rear)}$=0.6 fA/cm^2 have been experimentally determined [2], $J_{0(metal, front)}$=13 fA/cm^2 and $J_{0(metal, rear)}$=10 fA/cm^2 was assumed to fit the simulation model with the experimental JV curve, particularly V$_{oc}$. There is a possibility that $J_{0(metal, front)}$ and $J_{0(metal, rear)}$ have been overestimated. Experiments will be performed in future to determine the actual values.

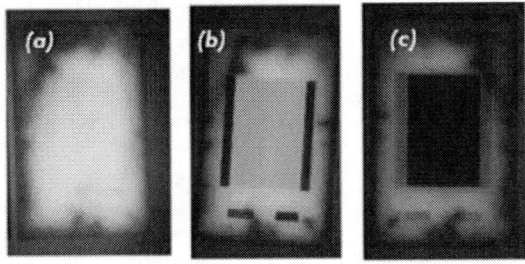

Fig. 2. Photoluminescence (PL) image of (a) before metallization (b) after metallization: front side (c) after metallization: rear side. 1/25s exposure time.

Fig. 2 shows the photoluminescence (PL) images of the device before and after metallization. Fig. 2(b) shows the PL image where the front side with PLEO contact was illuminated and in Fig. 2(c), rear side with PLENO contact was illuminated. It is to be noted that PLENO only exists in the cell area at the rear side. Around the cell area in the rear, it's intrinsic PLENO. The front gridlines in Fig. 2(b), appear black as they reflect the incoming light which is not absorbed by the detector in PL camera and hence, do not represent recombination. The remaining areas can be observed to understand the effects of PLENO/rear-metal interface. Similarly, the large rectangular area which appears dark in Fig. 2(c) represents the rear metal contact, while the two relatively lighter strips on the sides represent the recombination from the front busbars i.e., the PLEO/front-metal interface. Comparison of Fig. 2(b) and Fig. 2(c) shows that the recombination at PLENO/rear-metal interface is lower than PLEO/front-metal interface.

III. RESULTS AND DISCUSSIONS

It can be observed in Fig. 3, that the J-V curve from the simulation model in Quokka3 is in good agreement with the experimental results. The parameters in this model were varied to understand how the resistive and recombination losses can change and a higher efficiency can be achieved.

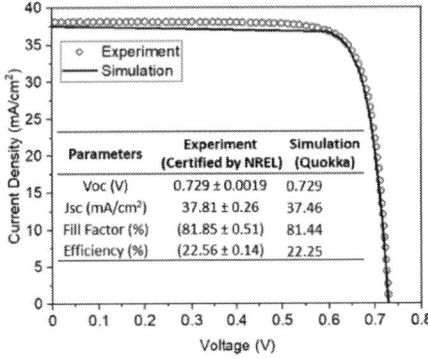

Fig. 3. Comparison of J-V for experimental result (certified at NREL) with simulation result using Quokka3.

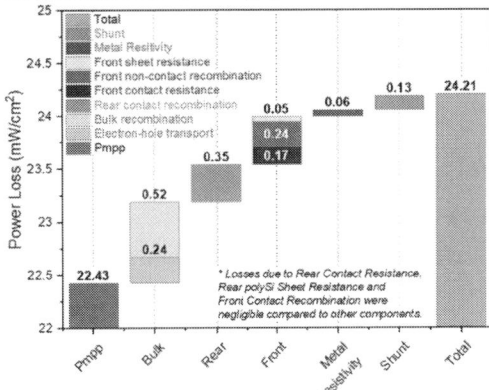

Fig. 4. Detailed power loss analysis using Quokka3 for the PLEO/PLENO passivated contact solar cell structure shown in Fig. 1

Fig. 4 shows the detailed power loss analysis which was obtained after simulating the device in Quokka3. The power at the maximum power point was found to be 22.43 mW/cm^2. If the electrical losses could be mitigated in the solar cell, the maximum power that could be extracted from the device is 24.21 mW/cm^2. The major losses are due to the recombination in the bulk and the resistance faced by the electron and hole transport through the bulk region (~0.76 mW/cm^2). The second largest source of power loss is due to rear contact recombination (0.35 mW/cm^2) followed by recombination at the front surface, particularly at the non-contact regions (0.24 mW/cm^2), i.e. where the metal is not in contact with the passivated layers. The contact resistance at the front side and shunt resistance contribute to 0.17 mW/cm^2 and 0.13 mW/cm^2 towards the total power loss. The metal resistivity of the front and rear contacts further contribute towards a power loss of 0.06 mW/cm^2. This is followed by the power loss due to resistance met by the lateral flow of carriers in the n$^+$ poly-Si (0.05 mW/cm^2). As the bulk recombination leads to maximum power loss in the device compared to other factors, the device performance was simulated as a function of minority carrier lifetime of the bulk substrate along with the rear recombination current density; J$_{0(metal, rear)}$ which is the second largest contributor towards the electrical power loss of the simulated device.

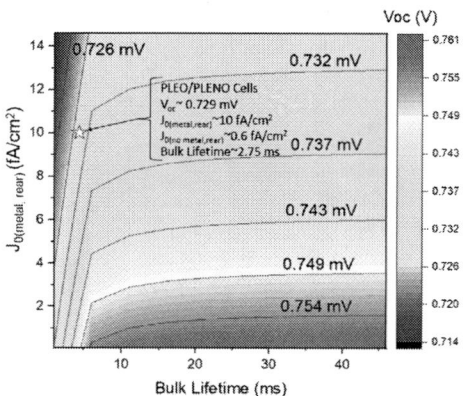

Fig. 5. Variation of V$_{oc}$ as the bulk lifetime and J$_{0(metal, rear)}$ for PLEO/PLENO passivated contact solar cells

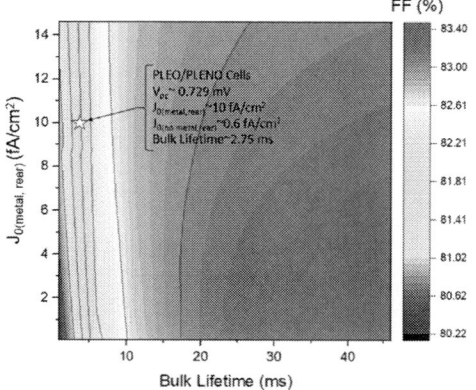

Fig. 6. Variation of FF with bulk lifetime and J$_{0(metal, rear)}$ for PLEO/PLENO passivated contact solar cells

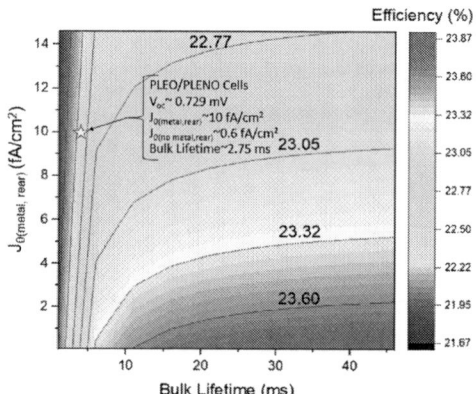

Fig. 7. Variation of efficiency with bulk lifetime and J$_{0(metal, rear)}$ for PLEO/PLENO passivated contact solar cells

Fig. 5, Fig. 6 and Fig. 7 show the variation of V$_{oc}$, FF and efficiency as the bulk lifetime and J$_{0(metal, rear)}$ is varied. It can be observed that for a bulk lifetime of 45 ms and J$_{0(metal, rear)}$~ 1 fA/cm^2, it is possible to achieve a V$_{oc}$~760 mV, FF~83.4% and efficiency ~23.7%.

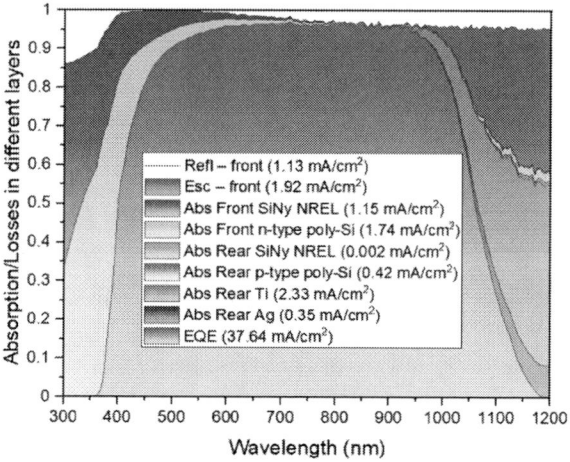

Fig. 8. Optical loss of current density simulated using SunSolve for the passivated contact solar cell with poly-Si on locally-etched dielectrics that has been fabricated experimentally (Shading loss ~1.56% has not been shown here but included in electrical simulation by Quokka 3).

Current loss analysis reveals that the absorption in front n$^+$ poly-Si, front SiN$_y$ and rear Ti layer results in a current loss of 1.74 mA/cm^2, 1.15 mA/cm^2 and 2.33 mA/cm^2, respectively. Based on the power and current loss analysis, we propose that by improving the bulk lifetime (~20 ms for practical reasons like availability) and better rear passivation (5fA/cm^2), a V$_{oc}$ = 0.745 V, J$_{sc}$ = 37.64 mA/cm^2, FF = 82.78 % and efficiency = 23.21 % can be achieved. Substituting Ti/Ag by Al reduces the absorption in the rear metal contact insignificantly. Ag can be used instead to reduce absorption in rear metal contacts, but it has been found during experiments in our laboratory that Ag layer peels off without underlying Ti layer. However, Ti results in higher parasitic absorption. The absorption in front n$^+$ poly-Si can be reduced by etching of n$^+$ poly-Si in non-metallized regions as shown in Fig. 9. Such devices have been demonstrated in the past by Chen et al. [4] using wet etching methods involving tetramethylammonium hydroxide (TMAH) to selectively etch *poly*-Si in the non-metallized regions of *poly*-Si/SiO$_x$ passivating contact cell to achieve a J$_{sc}$ of 39.8 mA/cm^2 in the champion cell. The removal of the n$^+$ *poly*-Si layer would force the majority carriers to flow through the tunneling oxide and the in-diffused n$^+$ c-Si region. Thus, an additional "skin layer" representing the in-diffused n$^+$ c-Si region has been included at the front side in the simulation structure as shown in Fig. 9.

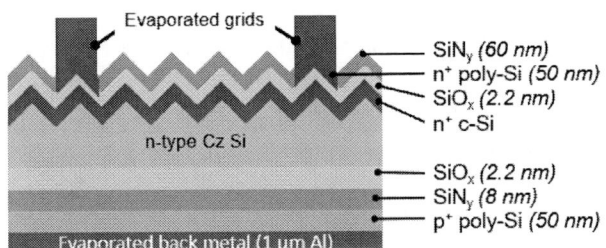

Fig. 9. Schematic diagram of the passivated contact solar cell with selectively etched front n$^+$ *poly*-Si on locally-etched SiO$_x$ at front and SiO$_x$/SiN$_y$ at rear

Simulations show that an improvement of 1.5 mA/cm^2 can be achieved leading to J$_{sc}$ = 39.14 mA/cm^2, V$_{oc}$ = 0.746 V, FF = 81.54 % and efficiency= 23.77 %. A loss in FF can be observed due to the loss of lateral conduction as front n$^+$ *poly*-Si has been removed. The V$_{oc}$ can be further improved to 760 mV in a bifacial solar cell with partial metallization at the rear side as shown in Fig. 10. In the simulations for bifacial structure, the rear anti-reflecting SiN$_y$ has been assumed to be non-absorbing in nature [5] and rear metal contacts are aligned to the front metal contacts.

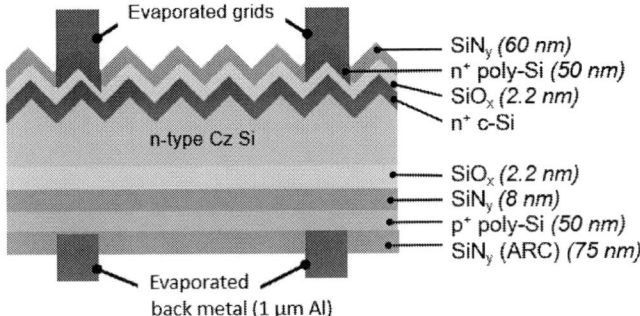

Fig. 10. Schematic diagram of the bifacial PLEO/PLENO cells with selectively etched front n$^+$ *poly*-Si

For front illumination of AM1.5 in bifacial PLEO/PLENO cells, a J$_{sc}$ = 39.53 mA/cm^2, FF = 77.39 % and efficiency = 23.21% can be achieved. The FF can be improved with optimization of the rear contacts which can lead to an efficiency >24% for FF>80%.

Fig. 11 shows the comparison of external quantum efficiency (EQE) for different structures discussed in this article. It can be observed that removal of front n$^+$ *poly*-Si improves the EQE at the lower wavelength region of 300 nm to 500 nm as parasitic absorption in n$^+$ *poly*-Si is reduced. In the bifacial structure, the EQE improves at higher wavelength region of 1000 nm to 1200 nm as rear recombination is reduced by lowering the metal fraction at the rear side.

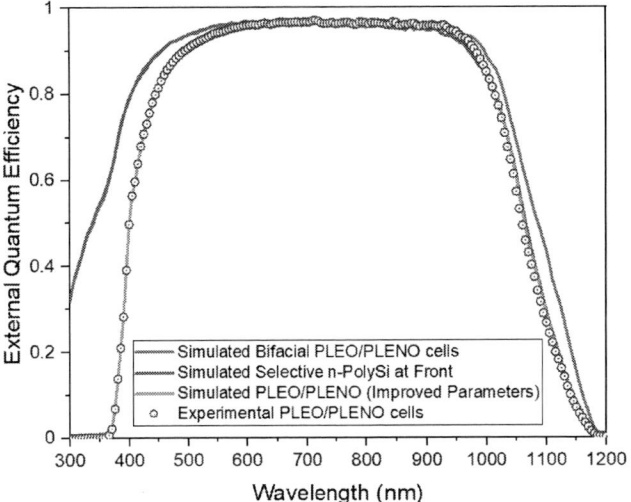

Fig. 11. Comparison of External Quantum Efficiency for different structures

IV. SUMMARY

In this article, electrical and optical loss analysis of a silicon solar cell experimentally realized with *poly*-Si on locally etched dielectric passivated contact using MACE method and having a V$_{oc}$=729.0mV and 22.56% efficiency has been attempted. Parameters like J$_{0(metal, front)}$ and J$_{0(metal, rear)}$ were deliberately speculated to allow the simulated JV curve to match the experimental JV curve. While simulated V$_{oc}$ and FF matched within the error margin of the certified device, simulated J$_{sc}$ was lower than experimental J$_{sc}$. Despite these limitations, we carried out the electrical power loss of the simulated cell. We show that electrical power loss can be reduced by improving the bulk lifetime and decreasing J$_{0(metal, rear)}$ of the simulated cell. On the other hand, current loss due to parasitic absorption in n$^+$ *poly*-Si can be reduced by removing the n$^+$ *poly*-Si layer in the non-metallized regions. Further, reducing the rear metal contact by adopting a bifacial structure with PLEO/PLENO contacts can increase the V$_{oc}$ to 760 mV. With optimization of rear contact design, an FF>80% can be achieved leading to an efficiency>24% for the bifacial PLEO/PLENO passivated contact solar cells..

ACKNOWLEDGEMENTS

This work was authored in part by the National Renewable Energy Laboratory, operated by Alliance for Sustainable Energy, LLC, for the U.S. Department of Energy (DOE) under Contract No. DE-AC36-08GO28308. Funding provided by U.S. Department of Energy Office of Energy Efficiency and Renewable Energy Solar Energy Technologies Office. The views expressed in the article do not necessarily represent the views of the DOE or the U.S. Government. The U.S. Government retains and the publisher, by accepting the article for publication, acknowledges that the U.S. Government retains a nonexclusive, paid-up, irrevocable, worldwide license to publish or reproduce the published form of this work, or allow others to do so, for U.S. Government purposes. Suchismita Mitra would like to thank the Fulbright Commission, the Institute of International Education (IIE) and the United States-India Educational Foundation (USIEF) for awarding the Fulbright Nehru Post-doctoral Fellowship (Award No. 2730 FNPDR/2021). The authors thank the High-Efficiency Silicon PV team at NREL and Colorado School of Mines for supporting the work. The authors thank Andreas Fell, for his valuable guidance when using Quokka3.

REFERENCES

[1] C. L. Salles, W. Nemeth, H. Guthrey, S. Agarwal, and P. Stradins, 2022, Controlling pinhole radius and areal density in a-Si/SiOx using metal-assisted chemical etching. In AIP Conference Proceedings (Vol. 2487, No. 1, p. 020011). AIP Publishing LLC.

[2] C. Lima Anderson, W. Nemeth, H. Guthrey, C.-S. Jiang, M. Page, S. Agarwal, and P. Stradins, " Nano-pinhole passivating contact Si solar cells fabricated with metal-assisted chemical etching," Advanced Energy Materials 13, no. 11 (2023): 2203579

[3] C.L. Anderson, H. L. Guthrey, W. Nemeth, C-S. Jiang, M. R. Page, P. Stradins, and S. Agarwal. "Pinhole electrical conductivity in polycrystalline Si on locally etched SiN_y/SiO_x passivating contacts for Si solar cells." Materials Science in Semiconductor Processing 165 (2023): 107655.

[4] Kejun Chen, Barry Hartweg, Michael Woodhouse, Harvey Guthrey, William Nemeth, San Theingi, Matthew Page et al. "Self-Aligned Selective Area Front Contacts on Poly-Si/SiOx Passivating Contact c-Si Solar Cells." IEEE Journal of Photovoltaics 12, no. 3 (2022): 678-689.

[5] Simeon C. Baker-Finch, and Keith R. McIntosh. "Reflection of normally incident light from silicon solar cells with pyramidal texture." Progress in Photovoltaics: Research and Applications 19, no. 4 (2011): 406-416.

978-1-6654-6060-6/23 $31.00 © 2023 IEEE

Cadmium Telluride Accelerator Consortium (CTAC)

Lorelle M. Mansfield,[1] Matthew O. Reese,[1] and Michael J. Heben[2]

[1]National Renewable Energy Laboratory, Golden, Colorado, 80401, USA

[2]University of Toledo, Toledo, Ohio, 43606, USA

Abstract — The United States is the leader in cadmium telluride (CdTe) photovoltaic (PV) manufacturing, and CdTe is the second most-deployed solar technology in the world. The CdTe Accelerator Consortium (CTAC) was formed to enhance U.S. technology leadership and competitiveness in CdTe photovoltaics (PV) including increasing cell efficiencies, decreasing module costs, and maintaining or increasing domestic CdTe PV material and module production. Leadership for the 3-year consortium was selected by an NREL team through a competitive solicitation. CTAC was formally announced on August 1st by U.S. Secretary of Energy Jennifer M. Granholm and U.S. Representative Marcy Kaptur at an event held at University of Toledo. We will give an overview of CTAC research directions along with their goals and activities. We can also discuss how CTAC relates to the National Renewable Energy Laboratory (NREL) and to the U.S. Manufacturing of Advanced Cadmium Telluride Photovoltaics Consortium (US-MAC). Visit our poster to find out who leads CTAC and how you can get involved.

I. INTRODUCTION

The United States is the leader in cadmium telluride (CdTe) photovoltaic (PV) manufacturing, and CdTe is the second most-deployed solar technology in the world. CdTe supplies 40% of the axis-based tracking market [1] and 5% of the world market. Among other PV technologies, CdTe also has the lowest environmental footprint [2], the lowest all-in cost structure, and the lowest degradation rate[3]. With CdTe in the manufacturing environment, rapid improvements have been made in the last 10 years. Projects deploying U.S. CdTe modules are also competing successfully on levelized cost of electricity (LCOE) with those using imported Si modules. In order to remain competitive, CdTe manufacturing is in need of readily adaptable advances at the device and module levels that simultaneously enable higher conversion efficiencies and manufacturing cost reductions without risking long-term module durability. Time is of the essence to ensure that this technology remains a valuable contributor to global renewable energy deployment efforts. The CdTe Accelerator Consortium (CTAC) was formed to enhance U.S. technology leadership and competitiveness in CdTe photovoltaics including increasing cell efficiencies, decreasing module costs, and maintaining or increasing domestic CdTe PV material and module production.

II. BACKGROUND AND ROLES

The U.S. Department of Energy (DOE) Solar Energy Technologies Office (SETO) works to accelerate the market competitiveness of solar energy technologies by targeting LCOE reductions and increased solar deployment. As such, DOE maintains a strong commitment to enhancing U.S. manufacturing competitiveness while advancing the nation's energy goals, boosting the economy, and contributing to energy security. To that end, the DOE strategically rallies experts from across the national laboratories, universities, nonprofit organizations, and private industry around opportunities for the United States to be the leader in solar energy manufacturing.

One such area of opportunity is enhancing U.S. competitiveness in cadmium telluride photovoltaics. In 2021, DOE asked NREL to competitively select a CdTe Consortium to:

- Accelerate the development of less expensive and more efficient cadmium telluride solar cells and modules.
- Enhance U.S. technology leadership and competitiveness in CdTe photovoltaics.
- Bring together CdTe supply chain companies and research centers to work on the most important challenges in U.S. CdTe production.

Specific goals include:

- Enabling cell efficiencies above 24% and sustainable module price below 20 ¢/W for domestic CdTe modules by 2025, and
- Enabling cell efficiencies above 26% and sustainable module price below 15 ¢/W for domestic CdTe modules by 2030.

CTAC is part of the $20 million that was provided for CdTe by congress. The successful proposal created a team (Fig. 1a) led by University of Toledo that includes Colorado State University, First Solar, Toledo Solar, and Sivananthan Laboratories. CTAC was officially announced on August 1, 2022 by U.S. Energy Secretary Jennifer Granholm and Ohio

Representative Marcy Kaptur at an event held at University of Toledo (Fig. 1b).

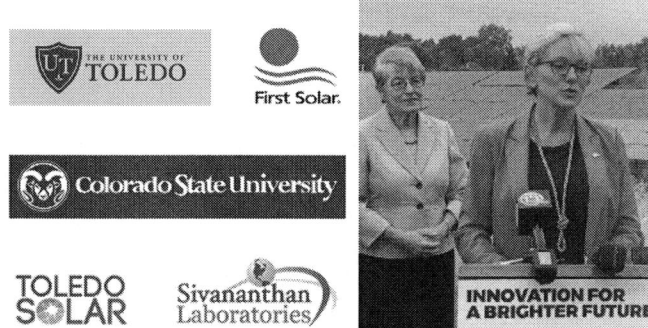

Fig. 1. Left – Leadership entities of the Cadmium Telluride Accelerator Consortium (CTAC). Right – Representative Marcy Kaptur and Secretary Jennifer Granholm (Photo by Toledo Blade).

A. CTAC Roles

CTAC leadership determines the direction of the CdTe work for the consortium. They will coordinate and conduct CdTe research and development projects (described in section III). They are tasked with developing a CdTe Technology Roadmap and maintaining it annually to reflect changes in research and industry directions. Input from stakeholders will be gathered through engagement activities such as workshops. Assessing the domestic CdTe supply chain on a regular basis is also a priority for CTAC. This will allow early identification of critical material or capacity constraints, technology transfer opportunities, and ways to expand and enhance U.S. manufacturing. Based on the roadmap and supply chain analysis, CTAC advises NREL on topic areas in which to solicit additional CdTe projects, and they can serve as non-voting members of the proposal selection committees. CTAC will seek opportunities to grow and diversify the market, supply chain, and research community.

B. NREL Roles

NREL has a unique role relating to the CdTe Accelerator Consortium. It includes acting as a resource, support network, program manager, and technical analysis center. NREL is expected to support the consortium efforts to develop a technology roadmap and to assist with stakeholder engagement for consortium meetings and events. NREL also supports the consortium as needed via applied research efforts including fabrication, measurements, and characterization. Initially, $1.5 million over 3 years has been allocated to do research at NREL in support of CTAC. NREL will also support the consortium in launching additional research efforts in order to meet the various targets set by the consortium's technology roadmap. NREL will administer solicitations, review, and award subcontracts for the selected technology development and demonstration projects. NREL is responsible for establishing all contracts and payments awarded under the solicitations. Project funding for all work efforts is sent directly from NREL to the corresponding partners.

C. Relationship to US-MAC

Although they share common goals and some common members, CTAC is not the same as the U.S. Manufacturing of Advanced Cadmium Telluride Photovoltaics Consortium (US-MAC). US-MAC is a larger organization consisting of industry, national labs, and universities dedicated to strengthening U.S. leadership in manufacturing cadmium telluride (CdTe). Membership is broader and not all organizations belonging to US-MAC have NREL or SETO funding. There is overlap in the leadership, with University of Toledo, First Solar, and Colorado State University holding leadership positions in both CTAC and US-MAC. NREL holds a leadership position in US-MAC, and a program management and supporting role in CTAC. The poster will have a visual representation to give more clarity. New members can be added to US-MAC by contacting the organization, presenting their case for membership to the Industrial Advisory and Executive Boards of US-MAC, and being voted in. To be formally admitted, new organizations are added to Memorandum of Understanding governing US-MAC. CTAC gathers new members as new projects are awarded by NREL.

III. PRIMARY RESEARCH ACTIVITIES

In addition to the roadmapping and supply chain activities, CTAC has several primary research areas. First Solar leads a task to investigate the next-generation of absorbers with improved Group-V doping for improving device and module efficiency. Studies will include dopant incorporation methods, profiles, and activation methods. Sivananthan Laboratories will contribute significantly to the doping and activation task as they have expertise in the growth of II-VI compounds and alloys via molecular beam epitaxy. Front-interface engineering, led by Colorado State University, will include exploring new emitter candidate materials and evaluating the interfaces with characterization and modelling. University of Toledo leads the passivated and selective back contacts task for efficiency improvements and a way to enable bifacial technology. Led by Toledo Solar's work, CTAC will explore new markets for CdTe PV such as rooftop and architectural solar. These tasks will help the U.S. CdTe community to reach more overarching goals such as:

- Developing a CdTe "toolkit"
- Expanding cell-design options
- Improving modules
- Diversifying applications

NREL research staff will assist CTAC with fabrication, measurents, and characterization. NREL can passivate surfaces, conduct studies of rapid thermal processing, and offer experience with As-doping. Scanning probe microscopy is used to investigate potential distributions, back contacts, and bifacial cells. Interface imaging with transmission electron microscopy (TEM) can detect lattice strain, dislocations, and defects. Surface science techniques such as X-ray photoelectron spectroscopy (XPS) will be use to characterize materials and interfaces to elucidate chemical structure, passivation, and band diagrams.

IV. SMALL PROJECTS

One way to get involved in CTAC is to propose small projects when NREL releases a request for proposals (RFP). Topic areas are recommend to NREL based on priorities in CTAC's CdTe Technology Roadmap. On September 19th, 2022, NREL released a follow-up RFP entitled *Small Projects to Accelerate Cadmium Telluride Technology Development*. Three topic areas were identified by CTAC as being the most important for additional funding. Topic 1 was high-efficiency devices, which inclued new approaches to ultra-high efficiency device configurations to achieve and exceed 26%. Topic 2 called for projects to increase tellurium production and yield in the U.S while lowering cost. Topic 3 was characterization, modelling, and simulation advances to guide technology development. Of the initial $3 million budgeted for external projects, approximately $2 million was awarded under this RFP (RFX-2022-10205). As of this evaluation abstract, the notification of successful offerors is imminent and subcontract negotiations will begin soon. At PVSC, a list of the successful small projects will be available. At least one additional RFP is planned which will fund more small CdTe projects. SETO has allocated an additional $1 million to the small project budget,

so there is approximately $2 million available. NREL expects to announce another RFP in 2023.

ACKNOWLEDGEMENT

This work was authored in part by the National Renewable Energy Laboratory, operated by Alliance for Sustainable Energy, LLC, for the U.S. Department of Energy (DOE) under Contract No. DE-AC36-08GO28308. Funding provided by U.S. Department of Energy Office of Energy Efficiency and Renewable Energy Solar Energy Technologies Office Award Number 37989. The views expressed in the article do not necessarily represent the views of the DOE or the U.S. Government. The U.S. Government retains and the publisher, by accepting the article for publication, acknowledges that the U.S. Government retains a nonexclusive, paid-up, irrevocable, worldwide license to publish or reproduce the published form of this work, or allow others to do so, for U.S. Government purposes.

REFERENCES

[1] "Electricity - Construction cost data for electric generators installed in 2020." https://www.eia.gov/electricity/generatorcosts/index.php (accessed Jan. 20, 2023).

[2] H. M. Wikoff, S. B. Reese, and M. O. Reese, "Embodied energy and carbon from the manufacture of cadmium telluride and silicon photovoltaics," *Joule*, vol. 6, no. 7, pp. 1710–1725, Jul. 2022, doi: 10.1016/j.joule.2022.06.006.

[3] D. C. Jordan *et al.*, "Photovoltaic fleet degradation insights," *Prog. Photovolt. Res. Appl.*, vol. 30, no. 10, pp. 1166–1175, 2022, doi: 10.1002/pip.3566.

Screen Printable Copper Pastes for Silicon Solar Cells

Thad Druffel, Ruvini Dharmadasa, Apolo Nambo, Abafriseke Ebong, Sandra Huneycutt, Donald Ital

Bert Thin Films, LLC, Louisville, KY, United States

University of North Caroplina Charlotte, Charlotte, NC, United States

The photovoltaic industry has been actively trying to replace silver with copper as the primary metallization material, but concerns around durability have limited deployment. To date, a screen printable copper-based paste that can fire through the SiN antireflective layer has yet to be commercialized. This work details the development of a screen printable copper paste that can be fired in atmospheric belt furnaces; thus, offering manufactures the ability to completely replace Ag without adding any processing costs. The performance and challenges of the copper-based paste are discussed, and accelerated aging tests are applied to copper metallized PERC cells to show the impact of processing conditions on long-term durability.

Optimization of Zinc Oxide Electron Transport Layers for Cs-based Perovskite Solar Cells

Zhuldyz Yelzhanova[1], Gaukhar Nigmetova[2], Gulzhan Bizhanova[1], Nurzhan Yermekov[1], Ulan Kashkimbayev[1], and Annie Ng[1,*]

[1]Department of Electrical and Computer Engineering, School of Engineering and Digital Sciences Nazarbayev University, Astana, Kazakhstan

[2]Department of Chemical and Materials Engineering, School of Engineering and Digital Sciences Nazarbayev University, Astana, Kazakhstan

e-mail:annie.ng@nu.edu.kz; phone:+7(7172)692684

Abstract

Zinc oxide (ZnO) possesses remarkable semiconductor properties. The compact ZnO (c-ZnO) and ZnO nanorod arrays (ZnO-NAs) are used as ETLs in $CsPbI_2Br$ perovskite solar cells (PSCs). It is found that c-ZnO doped with graphene oxide (GO) has an improvement in electron transport efficiency in PSCs. The devices based on c-ZnO:GO ETL exhibit an average power conversion efficiency (PCE) of 9.1%, which is an improvement of 7% compared to PSCs with the undoped ETL.

Keywords: electron transport layers, zinc oxide, graphene oxide, nanostructures, perovskite solar cells

Introduction

Since 2009, perovskite-based solar cells (PSCs) have attracted increasing attention in the photovoltaic community due to their outstanding optoelectronic and light-harvesting properties. PSCs beat efficiency records at a very rapid rate from 3.8% [1] to 25.7% recently [2]. Nowadays, the issues of stability, scalability and lead toxicity of PSCs remain challenging in the community. Intensive research efforts have been placed on material selections, optimization of device architectures, development of large-scale growth methods and implementation of interfacial engineering etc.

The electron-transport layer (ETL) plays a vital role in electron transport and hole blocking to suppress charge recombination in PSCs. Nowadays, the most common ETL material, titanium dioxide (TiO_2) can be replaced by other metal oxides such as SnO_2 [3] and ZnO [4]. TiO_2 ETLs are usually prepared at high processing temperatures (>500°C) [5], which is not feasible for low-cost and flexible devices. The electron mobility in mesoporous TiO_2 is relatively low, which intrinsically limits the electron transport efficiency in the PSCs [6]. The strong photocatalytic effect of TiO_2 is another concern. ZnO is a promising alternative to ETL, which has superior electron extraction and transport properties (electron mobility ~130-170 cm^2/Vs) [7]. ZnO can be prepared by low-temperature processing (100-200°C).

Organic-inorganic halide perovskites suffer from instability under long-term thermal and light illumination because the presence of organic cations (i.e. MA^+, FA^+), which are highly volatile and hydrophilic. In contrast, all-inorganic Cs based perovskites demonstrate good thermal stability, excellent carrier migration, and low exciton binding energy compared to the organic cation containing perovskites [8]. $CsPbI_3$ which has a bandgap of 1.73 eV, is easily converted to non-photoactive yellow phase [9]. $CsPbBr_3$ has better phase stability. However, the bandgap of $CsPbBr_3$ is too large (2.3 eV) to absorb low energetic wavelengths (>540 nm) [10]. The dual-halide system such as $CsPbI_2Br$ exhibits better phase stability than $CsPbI_3$ and smaller bandgap (1.91 eV) compared to $CsPbBr_3$. This work aims to study the performance of PSCs based on $CsPbI_2Br$ active layers. Since $CsPbI_2Br$ has a high conduction band, matching with the conduction band of ZnO [11], the investigations focus on ZnO/$CsPbI_2Br$ interface as well as the properties of different types of ZnO. It is believed that doped ZnO has enhancement in electron transport efficiency. One of the cost-effective materials, graphene, can be incorporated with carrier transport layers. Previous work of Zho *et al.* [12] reported the enhanced electron extraction driven by graphene interlayer between TiO_2 ETL and perovskite layer [12]. Balis *et al.* have demonstrated that graphene oxide doped TiO_2 can facilitate electron transport toward the anode in PSCs [13]. Several works also indicate the significant improvement in charge extraction and carrier conducting properties accomplished by incorporation of graphene derivatives in PSCs [14], [15]. In this work, the samples of compact ZnO layer (c-ZnO) were prepared by a low-temperature sol-gel method. In order to optimize the functionality of ZnO ETLs, the c-ZnO samples were doped with graphene oxide (GO), aiming to increase conductivity and improve the electron carrier transport efficiency across ETL/$CsPbI_2Br$ interface. Meanwhile, ZnO nanorod arrays (ZnO-NAs) were prepared on c-ZnO:GO samples via a hydrothermal growth method. Systematic characterizations were performed on the materials and devices.

Experimental Details

The FTO coated glasses were cleaned in detergent, deionized water (DI), acetone, and isopropanol (IPA) with ultrasonication. The c-ZnO layer was prepared by a sol-gel method. The precursor solution contained zinc acetate (40 mg), 2-methoxyethanol (1 mL), and ethanolamine (31 μL) was stirred at 80°C for 1 hour. The precursor solution was spin-coated on cleaned substrates at 4000 rpm for 30 s and then annealed at 150°C for 5 min and 400°C for 30 min. For the ZnO:GO samples, 4% GO (vol %), obtained from GO ammonia functionalized solution (1 mg/ml dispersed in

DI) was doped into the precursor solution. The doped solution was spin-coated at the identical conditions as mentioned above. ZnO nanorod arrays (NAs) were grown on c-ZnO via the hydrothermal method. The precursor solution was prepared by dissolving polyethylenimine (10 μL), zinc nitrate hexahydrate (0.2 g), and hexamethylenetetramine (0.2 g) in 30 mL DI water under stirring. The c-ZnO coated FTO substrates were immersed vertically in the prepared precursor solution and heated in the furnace under 90°C for 15 min.

Results and Discussions

In this work, three types of ZnO based ETL (c-ZnO, c-ZnO:GO and c-ZnO:GO/ZnO-NAs) were investigated for PSC applications. Their top and cross-sectional scanning electron microscope (SEM) images are shown in Figure 1. It is observed that the c-ZnO and c-ZnO:GO layers have a thickness of ~40 nm. The hydrothermal grown ZnO-NAs have a length varied from 45 nm-65 nm.

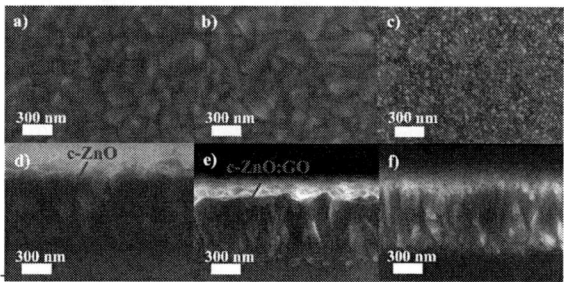

Fig.1: Top and cross-section SEM images of (a,d) c-ZnO, (b,e) c-ZnO:GO, (c,f) c-ZnO:GO/ZnO-NAs

The surface morphology and conductivity of three types of ETLs were investigated by atomic conductive force microscopy (c-AFM). The obtained images are shown in Figure 2. It is found that the surface roughness is reduced from 0.137 μm to 0.025 μm after incorporation of GO in c-ZnO. The smooth surface of ETL is desired as it can increase the interfacial quality of ETL/perovskite, leading to better charge transport and reduction in carrier recombination. It is found that the RMS value of ZnO:GO/ZnO-NAs sample has no significant increase compared to c-ZnO:GO, which is due to the uniform growth of short NAs. The technique of c-AFM characterizes the surface conductivity of three types of ZnO samples. The results show that ZnO:GO exhibits a higher surface conductivity compared to c-ZnO, indicating the advantages of doping GO in c-ZnO in conducting electrons. It is noteworthy that ZnO:GO/ZnO-NAs has similar surface conductivity compared to c-ZnO since GO was not introduced during the hydrothermal growth process. Therefore it is assumed that the impact of using ZnO:GO/ZnO-NAs on the performance of electron transport will be mainly caused by ZnO nanostructures instead of a combination effect of doping and increased interface area.

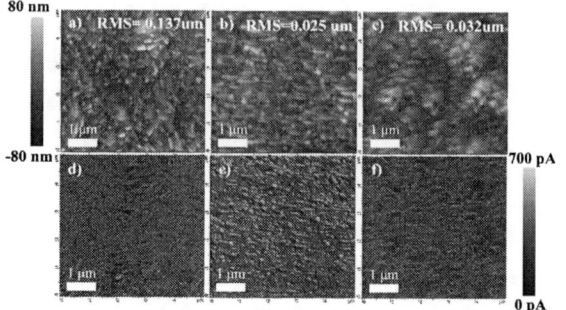

Fig.2: AFM images of (a) c-ZnO, (b) c-ZnO:GO, (c) c-ZnO:GO/ZnO-NAs; and c-AFM images of (d) c-ZnO, (e) c-ZnO:GO, (f) c-ZnO:GO/ZnO-NAs

Figure 3 demonstrates the photoluminescence (PL) spectra emitted from $CsPbI_2Br$ deposited on three different types of ZnO-based ETL. The PL signal obtained from the c-ZnO:GO sample exhibits a significant quenching effect compared to other samples, indicating the best electron extraction performance of c-ZnO:GO ETL among different ZnO ETLs. Meanwhile, the sample with c-ZnO:GO/ZnO-NAs ETL has a higher quenching effect compared to planar c-ZnO ETL.

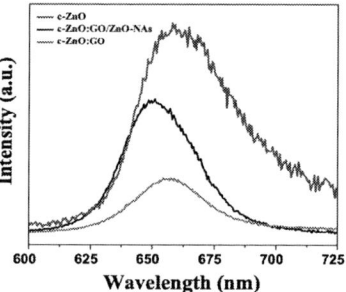

Fig.3: PL spectra of $CsPbI_2Br$ deposited on different ZnO-based ETL

Table 1 summarizes the photovoltaic parameters of PSCs using different types of ZnO ETL. It is found that the devices based on GO doped ZnO ETL have improved the short circuit current density (Jsc) and fill factor (FF), resulting in highest power conversion efficiency (PCE) among other types of PSCs. This result is consistent with the observation of the highest PL quenching effect from the c-ZnO:GO sample, confirming the positive effect of incorporating GO in c-ZnO ETL. It is noteworthy that the devices based on c-ZnO:GO/ZnO-NAs ETL exhibit lower performance. It is speculated that the introduction of an additional NA layer may increase recombination losses attributed to new generation of defects locating at increased ETL/perovskite interfaces. The PL results of the c-ZnO:GO/ZnO-NAs ETL containing sample show a higher quenching effect compared to c-ZnO sample. However, this measurement is not based on the device structure and the non-radiative recombination cannot directly reflect in the

PL spectra as shown in Figure 3. Therefore, the charge recombination mechanism and carrier transport dynamics at the interface of c-ZnO:GO/ZnO-NAs/perovskite should be further investigated. Nevertheless, the *IV* data for c-ZnO:GO ETL based devices shows the highest averaged PCE and smallest batch-to-batch variation, indicating the reproducible ETL processing condition. While, more efforts should be placed on optimization of c-ZnO:GO/ZnO-NAs ETL (e.g. NA density, length, diameter etc.). Furthermore, surface modification of ETL such as oxygen plasma treatment [3] and introduction of passivation layers [16] at the interface of ETL/perovskite can be considered during device fabrication.

Table 1: Photovoltaic parameters of ZnO-based PSCs

Conditions		V_{oc} (V)	J_{sc} (mA/cm)	FF (%)	PCE (%)
c-ZnO	Forward	1.1 ±0.07	14.9 ±2.0	56.1 ±6.0	8.5 ±2.0
	Reverse	0.8 ±0.15	14.1 ±2.0	40.6 ±8.8	5.0 ±2.3
c-ZnO: GO	Forward	1.1 ±0.04	15.1 ±0.9	57.2 ±2.6	9.1 ±0.8
	Reverse	0.9 ±0.07	14.1 ±1.6	36 ±5.7	4.4 ±1.3
c-ZnO: GO/ZnO-NAs	Forward	0.9 ±0.05	14.3 ±2.1	44.7 ±3.9	6.3 ±0.5
	Reverse	0.8 ±0.07	13.3 ±1.7	37.2 ±5.4	4.2 ±0.4

Conclusion

In conclusion, the ETL composed of c-ZnO, c-ZnO:GO or c-ZnO:GO/ZnO-NAs were investigated for application in CsPbI$_2$Br based PSCs. It is found that GO is an effective dopant for c-ZnO ETL, leading to a better electron transport and thus improving the PCE of the devices from an average of 8.5% to 9.1%. The sample of c-ZnO:GO/ZnO-NAs/CsPbI$_2$Br has a stronger PL quenching effect compared to c-ZnO/perovskite. However, c-ZnO:GO/ZnO-NAs based PSCs have lower device performance compared to c-ZnO based devices. It suggests that nanostructured ETLs can increase the interfacial areas for carrier transport, but it may also introduce more carrier recombination centers at the ETL/perovskite interface. Further investigation and optimization of c-ZnO:GO/ZnO-NAs/ CsPbI$_2$Br interface for PSC application is required.

Acknowledgements

A.N. acknowledges the financial support from the Science Committee of the Ministry of Education and Science of the Republic of Kazakhstan (Scientific Research Grant no. AP14869983) and the Nazarbayev University (Grant no. 021220CRP0422).

References

[1] A. Kojima, K. Teshima, Y. Shirai, and T. Miyasaka, "Organometal Halide Perovskites as Visible-Light γSensitizers for Photovoltaic Cells," *J. Am. Chem. Soc.*, vol. 131, no. 17, pp. 6050–6051, May 2009, doi: 10.1021/ja809598r.

[2] M. Kim *et al.*, "Conformal quantum dot–SnO$_2$ layers as electron transporters for efficient perovskite solar cells," *Science (80-.).*, vol. 375, no. 6578, pp. 302–306, Jan. 2022, doi: 10.1126/science.abh1885.

[3] Z. Yelzhanova *et al.*, "A Morphological Study of Solvothermally Grown SnO$_2$ Nanostructures for Application in Perovskite Solar Cells," *Nanomaterials*, vol. 12, no. 10. 2022, doi: 10.3390/nano12101686.

[4] J. Cao *et al.*, "Efficient, Hysteresis-Free, and Stable Perovskite Solar Cells with ZnO as Electron-Transport Layer: Effect of Surface Passivation," *Adv. Mater.*, vol. 30, no. 11, p. 1705596, Mar. 2018, doi: 10.1002/adma.201705596.

[5] Y. Bai, I. Mora-Seró, F. De Angelis, J. Bisquert, and P. Wang, "Titanium Dioxide Nanomaterials for Photovoltaic Applications," *Chem. Rev.*, vol. 114, no. 19, pp. 10095–10130, Oct. 2014, doi: 10.1021/cr400606n.

[6] M. Singh, C. W. Chu, and A. Ng, "Perspective on Predominant Metal Oxide Charge Transporting Materials for High-Performance Perovskite Solar Cells," *Frontiers in Materials*, vol. 8. 2021, [Online]. Available: https://www.frontiersin.org/articles/10.3389/fmats.2021.65520 7.

[7] N. A. Jayah, H. Yahaya, M. R. Mahmood, T. Terasako, K. Yasui, and A. M. Hashim, "High electron mobility and low carrier concentration of hydrothermally grown ZnO thin films on seeded a-plane sapphire at low temperature," *Nanoscale Res. Lett.*, vol. 10, no. 1, p. 7, 2015, doi: 10.1186/s11671-014-0715-0.

[8] J. Wang *et al.*, "Highly efficient all-inorganic perovskite solar cells with suppressed non-radiative recombination by a Lewis base," *Nat. Commun.*, vol. 11, no. 1, p. 177, 2020, doi: 10.1038/s41467-019-13909-5.

[9] B. Wang, N. Novendra, and A. Navrotsky, "Energetics, Structures, and Phase Transitions of Cubic and Orthorhombic Cesium Lead Iodide (CsPbI$_3$) Polymorphs," *J. Am. Chem. Soc.*, vol. 141, no. 37, pp. 14501–14504, Sep. 2019, doi: 10.1021/jacs.9b05924.

[10] J. Liang *et al.*, "All-Inorganic Perovskite Solar Cells," *J. Am. Chem. Soc.*, vol. 138, no. 49, pp. 15829–15832, Dec. 2016, doi: 10.1021/jacs.6b10227.

[11] J. Ma *et al.*, "Improve the oxide/perovskite heterojunction contact for low temperature high efficiency and stable all-inorganic CsPbI$_2$Br perovskite solar cells," *Nano Energy*, vol. 67, p. 104241, 2020, doi:10.1016/j.nanoen.2019.104241.

[12] Z. Zhu *et al.*, "Efficiency Enhancement of Perovskite Solar Cells through Fast Electron Extraction: The Role of Graphene Quantum Dots," *J. Am. Chem. Soc.*, vol. 136, no. 10, pp. 3760–3763, Mar. 2014, doi: 10.1021/ja4132246.

[13] N. Balis *et al.*, "Investigating the role of reduced graphene oxide as a universal additive in planar perovskite solar cells," *J. Photochem. Photobiol. A Chem.*, vol. 386, p. 112141, 2020, doi: 10.1016/j.jphotochem.2019.112141.

[14] M. M. Tavakoli, R. Tavakoli, Z. Nourbakhsh, A. Waleed, U. S. Virk, and Z. Fan, "High Efficiency and Stable Perovskite Solar Cell Using ZnO/rGO QDs as an Electron Transfer Layer," *Adv. Mater. Interfaces*, vol. 3, no. 11, p. 1500790, Jun. 2016, doi: 10.1002/admi.201500790.

[15] A. Saleem *et al.*, "Graphene Oxide–TiO$_2$ Nanocomposite Films for Electron Transport Applications," *J. Electron. Mater.*, vol. 47, no. 7, pp. 3749–3756, 2018, doi: 10.1007/s11664-018-6235-4.

[16] D. Aidarkhanov *et al.*, "Passivation engineering for hysteresis-free mixed perovskite solar cells," *Sol. Energy Mater. Sol. Cells*, vol. 215, p. 110648, 2020, doi: https://doi.org/10.1016/j.solmat.2020.110648.

UV Degradation of Formamidinium-Cesium Lead Halide Perovskite Solar Cells

Kshitiz Dolia,[1] Abasi Abudulimu,[1] Sheng Fu,[1] Tyler Brau,[1] Karissa Jensen,[2] Stephanie L. Moffitt,[2] Randy J. Ellingson,[1] Xiaohong Gu,[2] Zhaoning Song,[1] and Yanfa Yan[1]

1. Wright Center for Photovoltaics Innovation and Commercialization, Department of Physics and Astronomy, The University of Toledo, 2801 W. Bancroft St., Toledo, Ohio 43606 USA

2. Engineering Laboratory, National Institute of Standards and Technology, Gaithersburg, MD, 20899 USA

Abstract — **Perovskite solar cells show great promise for cost-effective and efficient solar energy production, but their stability remains a challenge. Understanding the degradation mechanisms of perovskite solar cells is essential for their successful implementation and widespread use. Here, we study the degradation mechanism of formamidinium-cesium lead halide perovskite solar cells under ultraviolet (UV) radiation. We measure the UV-induced device performance degradation and characterize the changes in the structural and optoelectronic properties of perovskite films after UV radiation. Further analysis reveals that UV-induced halide photo-redox reactions create iodine vacancies at the buried interface of perovskite films, leading to phase segregation and crystal reconstruction to form photoinactive CsI-rich defects, increasing non-radiative recombination and reducing the efficiency of the UV-exposed cells. Understanding the degradation mechanism will aid in directing attention toward developing encapsulants and the design of more stable perovskites.**

I. INTRODUCTION

Metal halide perovskite solar cells (PSCs) have been garnering substantial attention lately for their potential to revolutionize solar energy production. The promise of these cells lies in their low cost and high efficiency when converting sunlight into electricity. However, their stability under solar radiation is still a major hurdle to overcome if they are to be used as a viable energy source. To fully understand the stability issue of PSCs under operational conditions and develop corresponding mitigating strategies, it is crucial to develop proper accelerated aging tests to evaluate their stability and understand underlying degradation mechanisms.

The failure of perovskite solar cells (PSCs) under ultraviolet (UV) radiation presents a significant barrier to their durability under real operational conditions. The effects of UV light on the stability of PSCs have been extensively studied and found to be quite adverse compared to the effect of UV-free white LED light [1]. Ji et al. reported that the TiO_2 electron transport layer (ETL) used widely in n-i-p devices was a primary cause of photocatalytic device degradation under UV radiation [2]. Lee et al. studied the UV degradation of methylammonium lead iodide ($MAPbI_3$) PSCs and found a reversible UV degradation and recovery behavior [3]. Xu et al. investigated the impact of light illumination on MA-based mixed halide perovskites, revealing that photoexcited carriers induce localized trap states [4]. Datta et al. demonstrated that halide segregation in mixed halide perovskites under light illumination led to a significant loss in short-circuit current density [5]. However, most previous UV-degradation studies were performed on notoriously unstable MA-based perovskites and did not use calibrated UV light sources to correlate UV radiation to the standard solar spectrum.

Here, we investigate the UV photodegradation of MA-free, formamidinium-cesium (FA-Cs) based PSCs using a calibrated UV radiation source to simulate the UV component of terrestrial solar radiation. Combining the device performance analysis and material characterization, we find that UV radiation creates iodine vacancies on the surface of FA-Cs perovskite films through halide photo-redox reactions, providing sites for phase segregation and crystal reconstruction to form photoinactive CsI-rich defects, increasing non-radiative recombination and reducing the efficiency of PSCs. The phase segregation of perovskite and removal of organics and halides in various perovskite compositions under environmental stressors has been reported in numerous studies [6,7,8].

II. EXPERIMENTAL DETAILS

PSCs used in this study were fabricated in a structure of Glass/ indium tin oxide (ITO)/ carbazole phosphonic acid (MeO-2PACz)/ FA-Cs perovskite/ fullerene (C_{60})/ bathocuproine (BCP)/ Ag. The perovskite absorber layer has a composition of $FA_{1-x}Cs_xPb(I_{0.9}Br_{0.1})_3$ with Cs content ratio, x = 0.1, 0.2, and 0.3, which is hereafter referred to as Cs10Br10, Cs20Br10, and Cs30Br10, respectively. ITO glass substrates (15 Ω/sq) were rinsed with acetone and isopropyl alcohol (IPA) with sonication and treated with UV ozone for 30 min. MeO-2PACz was spin-coated using a solution of 0.5 mg/ml in ethanol and annealed at 100 °C for 10 min. A perovskite precursor was prepared using three inks, made by dissolving FAI and PbI_2, FABr and $PbBr_2$, and CsI and PbI_2 in a DMF and DMSO solvent mixture at a 3:1 ratio to make a 1.25 M solution. The three inks were then combined to get desired stoichiometric composition. The resulting precursor solution was spin-coated

978-1-6654-6060-6/23 $31.00 © 2023 IEEE

on the substrate and annealed at 50 °C for 2 min and 100 °C for 10 min. To complete a device, 25 nm C_{60} and 6 nm BCP layers were thermally evaporated. Finally, an 80 nm silver (Ag) electrode is thermally evaporated through a mask to make a device area of 0.16 cm^2. To encapsulate the device, UV epoxy was applied to a thin glass slide and placed on the exposed surface of the device, and then cured under UV illumination.

Three sets of PSCs were encapsulated and sent to the National Institute of Standards and Technology (NIST) for UV exposure using SPHERE (Simulated Photodegradation via High Energy Radiant Exposure) under 45 °C/0 % relative humidity (RH) for 22 h. Two UV intensities, ~22 and ~53 W/m^2, were achieved by adding the neutral density filters in front of the specimens to attenuate the irradiance. The corresponding UV doses were ~1.7 and ~ 4.2 MJ/m^2. A set of control samples traveled with UV-exposed devices. Current density-voltage (J-V) and external quantum efficiency (EQE) measurements were done before and after the exposure to gauge the effect of UV exposure on the PV parameters of the devices. For perovskite film characterization, samples were exposed from the glass side or film side to a monochromatic 395 nm UV light at 36 W/m^2 (M395L5, Thor Labs) in a nitrogen glovebox (25 °C/0 % RH) for 24 h, corresponding to a total UV dose of ~3.1 MJ/m^2. Scanning electron microscopy (SEM), Energy dispersive spectroscopy (EDS), X-ray diffraction (XRD), photoluminescence (PL), and UV-VIS absorbance measurements were done to characterize the UV degradation of perovskite films. XRD was carried out using Cu Kα radiation (1.54 Å) of a Rigaku X-ray diffractometer. Solar cell performances were measured using a Keithley 2400 source meter and a solar simulator (Sunbrick, G2V) calibrated to AM1.5G radiation. EQE measurements were done using a PV Instruments system (model IVQE8-C). PL measurements were done in a custom-built system with a Horiba Symphony-II CCD detector and iHR320 monochromator photoexciting the samples with a 633 nm continuous wave laser (6 mW/cm^2). Absorbance measurements were carried out using a UV-Vis spectrometer (Lambda 1050, Perkin Elmer).

III. RESULTS AND DISCUSSION

A. UV degradation of FA-Cs PSCs

In the first part of the study, we investigated the effect of UV exposure on device performance. Fig. 1 summarizes the PV performance statistics of PSCs with different compositions before and after UV exposure. The devices with all three perovskite compositions show a more severe performance drop with increasing UV exposure dose. All the PV performance parameters, including fill factor (FF), open-circuit voltage (V_{OC}), short-circuit current density (J_{SC}), and PCE, of these devices decreased after UV exposure.

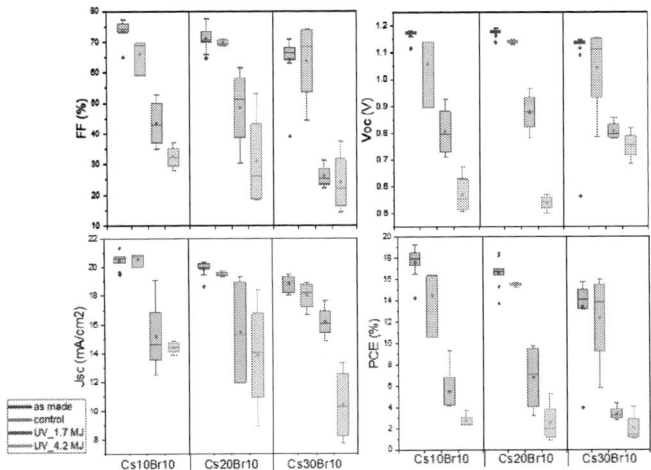

Fig. 1. Device performance parameters of PSCs with different perovskite compositions before and after UV exposure.

Fig. 2 compares the EQE spectra of representative cells of each composition before and after UV exposure. UV radiation led to significant degradation in all the devices. Cs30Br10 samples exhibit the largest drop in the integrated J_{SC}. The degradation is most rapid during the initial phase of UV exposure, which is in line with the observations made by other researchers on methylammonium (MA)-based perovskites [3].

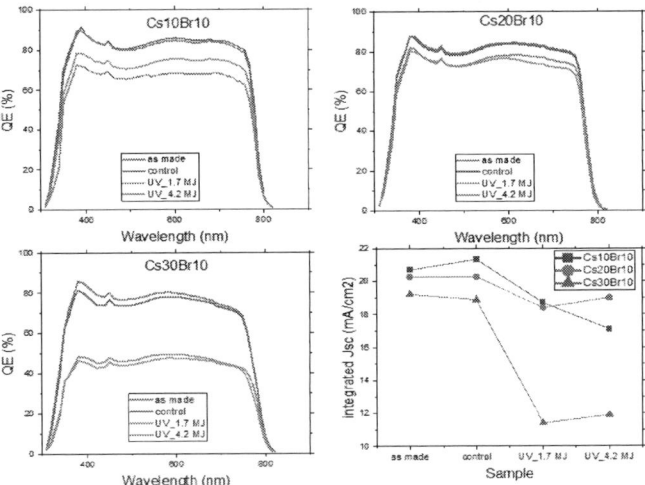

Fig. 2. EQE curves of PSCs with different perovskite compositions before and after UV exposure.

B. Characterization of UV-degraded perovskite films

To understand the mechanism of UV degradation of FA-Cs PSCs, we conducted a systematical characterization of UV-degraded perovskite films. Fig. 3 compares the SEM images of various perovskite films with and without UV exposure. Interestingly, the results show no change in the surface microstructures of the perovskite film and no significant physical defects (voids and deformation) formation due to UV exposure. Compared with MA-based perovskites, FA-Cs perovskite films show better tolerance against UV-induced decomposition.

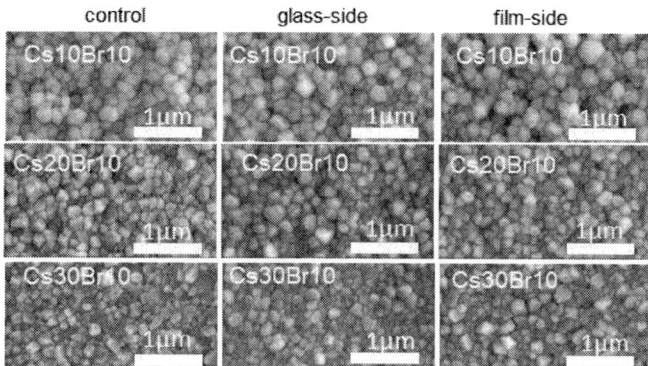

Fig. 3. SEM images of control and UV exposed perovskite films of different compositions.

Fig. 4a-c compares the XRD patterns of different perovskite films with and without UV exposure. In agreement with the SEM results (Fig. 3), the XRD analysis shows no change in the crystal structure of the perovskite films after UV exposure. No reduction in perovskite crystallinity or decomposition into PbI_2 was observed by XRD measurements. Additionally, no impurity phase was identified after UV exposure.

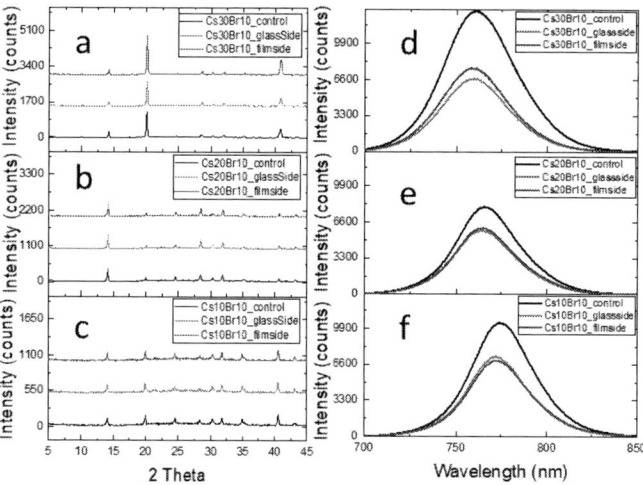

Fig. 4. XRD patterns and PL spectra of control and UV exposed perovskite films of different compositions.

Fig. 4d-f shows the PL spectra of control and UV exposed samples of different compositions. The PL emissions of all UV exposed samples are significantly suppressed compared with the control ones. The decreasing PL intensity indicates increased non-radiative recombination, likely caused by the formation of surface defects in the absorber. Additionally, PL peaks of all UV exposed samples slightly shift toward short

wavelengths, corresponding to an increased bandgap due to loss of iodine or FA. There is no significant difference between the glass side and film side exposed samples, indicating that UV attenuation by ITO glass and hole-transport layer has no significant impact on suppressing UV degradation. The combination of the SEM, XRD, and PL results suggests that UV radiation has no significant impact on the crystal structure and microstructure of the bulk perovskite film but creates electronic defects on the surface of the film.

To better understand how UV radiation creates defects on the surface of FA-Cs perovskite films, we measured the UV-degradation products. A small piece of a Cs10Br10 film was placed in toluene and exposed to UV radiation at ~330 W/m^2 for 6 days. Fig. 5 shows the absorbance spectra of the UV-exposed sample, Iodine, and toluene solution. The UV-exposed sample shows an iodine (I_2) absorption peak, which indicates the presence of iodine in the toluene solution [9,10]. The result reveals that FA-Cs perovskite films degrade under UV exposure through a halide photo-redox reaction. Under UV radiation, iodide anions in the perovskite films can be oxidized to form iodine, which can be dissolved in toluene. The photo-redox reaction creates halide vacancies on the surface of perovskite films, increasing surface recombination and reducing the PCE of the devices.

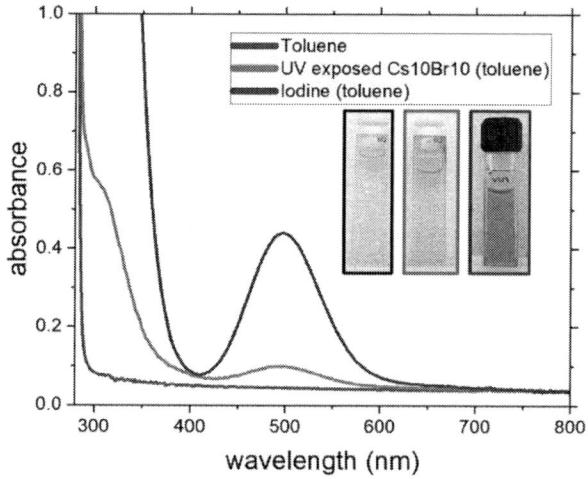

Fig. 5. Absorbance spectra of pure toluene solution, iodine dissolved in toluene solution and toluene immersed with a small film sample with UV exposure.

To find out how this iodine release impacts the device, we investigated the buried interface. The perovskite film sample was glued to a glass slide using epoxy and placed in a nitrogen glovebox overnight to set the epoxy. The sample was then exposed to a monochromatic 395 nm UV light at ~330 W/m^2 for 6 days. The perovskite film was peeled off, and we performed SEM on the buried interface of the control and UV exposed samples. Fig. 6a-c shows the SEM images of the

978-1-6654-6060-6/23 $31.00 © 2023 IEEE   44

control and UV-exposed samples. The significantly increased defect density is seen in the UV-exposed sample.

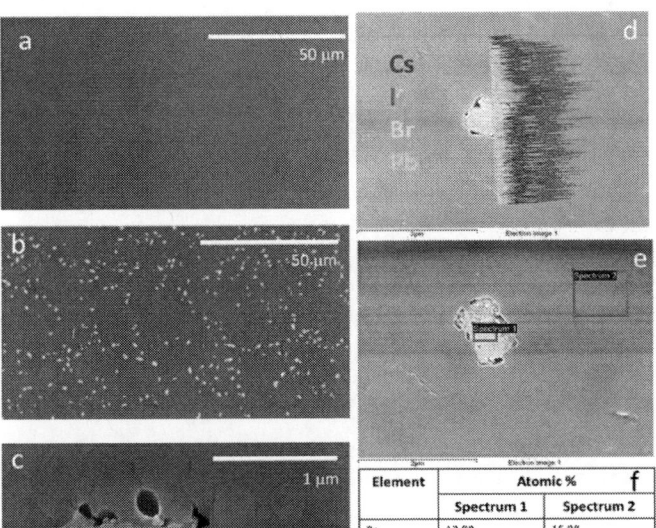

Fig. 6. SEM images of the buried interface of (a) control and (b, c) UV exposed sample. (d, e, f) EDS measurements on and around a defect center at multiple locations.

We performed EDS measurement (Fig. 6d-f) on the UV exposed sample at multiple locations, which suggests that these defective particles are rich in cesium iodide. This could be due to the irregular grain growth in the defective region left by iodine vacancies. The escape of volatile organic and iodine species leads to phase segregation and crystal reconstruction, resulting in these CsI-rich defects. Li et al. [5] has shown that the light generated carriers could provide thermodynamic force for phase segregation into Cs and FA rich phases and these Cs rich defects are current blocking and photoinactive.

IV. CONCLUSION

A UV degradation test was performed on different FA-Cs PSCs at film and device levels to identify the degradation mechanism occurring under UV exposure. Device measurements showed a clear performance drop in all PV parameters of PSCs after UV exposure. SEM, XRD, and PL measurements of perovskite films revealed that UV radiation created non-radiative recombination defects on the surface of the perovskite films. We further identified that UV-catalyzed halide redox reaction is the main cause of the surface degradation. The escape of volatile organic and iodine species leads to phase segregation and crystal reconstruction, resulting

in CsI-rich defects which are believed to be photoinactive and therefore result in device performance loss.

ACKNOWLEDGEMENT

This material is based upon work supported by the U.S. Department of Energy (DOE)'s Office of Energy Efficiency and Renewable Energy (EERE) under the Solar Energy Technology Office Award Numbers DE-EE0008753 and DE-EE0008970, under the Solar Energy Technologies Office Award Number DE-EE0008753. and by Air Force Research Laboratory (AFRL) under agreement FA9453-19-C-1002. The views expressed in the article do not necessarily represent the views of the DOE, AFRL, or the U.S. Government.

REFERENCES

[1] Domanski, K., Alharbi, E. A., Hagfeldt, A., Grätzel, M., & Tress, W. (2018). Systematic investigation of the impact of operation conditions on the degradation behaviour of perovskite solar cells. Nature Energy, 3(1), 61-67.

[2] Ji, J., Liu, X., Jiang, H., Duan, M., Liu, B., Huang, H., ... & Li, M. (2020). Two-stage ultraviolet degradation of perovskite solar cells induced by the oxygen vacancy-Ti4+ states. IScience, 23(4), 101013.

[3] Lee, S. W., Kim, S., Bae, S., Cho, K., Chung, T., Mundt, L. E., ... & Kim, D. (2016). UV degradation and recovery of perovskite solar cells. Scientific reports, 6(1), 1-10.

[4] Xu, R. P., Li, Y. Q., Jin, T. Y., Liu, Y. Q., Bao, Q. Y., O'Carroll, C., & Tang, J. X. (2018). In situ observation of light illumination-induced degradation in organometal mixed-halide perovskite films. ACS applied materials & interfaces, 10(7), 6737-6746.

[5] Datta, K., van Gorkom, B. T., Chen, Z., Dyson, M. J., van der Pol, T. P., Meskers, S. C., ... & Janssen, R. A. (2021). Effect of light-induced halide segregation on the performance of mixed-halide perovskite solar cells. ACS Applied Energy Materials, 4(7), 6650-6658.

[6] Li, N., Luo, Y., Chen, Z., Niu, X., Zhang, X., Lu, J., ... & Zhou, H. (2020). Microscopic degradation in formamidinium-cesium lead iodide perovskite solar cells under operational stressors. Joule, 4(8), 1743-1758.

[7] Ho, K., Wei, M., Sargent, E. H., & Walker, G. C. (2021). Grain transformation and degradation mechanism of formamidinium and cesium lead iodide perovskite under humidity and light. ACS Energy Letters, 6(3), 934-940.

[8] Song, Z., Wang, C., Phillips, A. B., Grice, C. R., Zhao, D., Yu, Y., ... & Yan, Y. (2018). Probing the origins of photodegradation in organic–inorganic metal halide perovskites with time-resolved mass spectrometry. Sustainable Energy & Fuels, 2(11), 2460-2467.

[9] Custer, J. J., & Natelson, S. (1949). Spectrophotometric determination of microquantities of iodine. Analytical Chemistry, 21(8), 1005-1.

[10] Kim, G. Y., Senocrate, A., Yang, T. Y., Gregori, G., Grätzel, M., & Maier, J. (2018). Large tunable photoeffect on ion conduction in halide perovskites and implications for photodecomposition. Nature materials, 17(5), 445-449.

Efficient Cd(Se,Te) Solar Cells with Cd(O,S,Se,Te) at the Front Interface

Dengbing Li, Sabin Neupane, Sandip Singh Bista, Abasi Abudulimu, Kamala Khanal Subedi, Manoj K. Jamarkattel, Chuanxiao Xiao, Chun-Sheng Jiang, Jonathan D. Poplawsky, David A. Cullen, Adam B. Phillips, Michael J. Heben, Randy J. Ellingson, Yanfa Yan

Department of Physics and Astronomy, and Wright Center for Photovoltaics Innovation and Commercialization, University of Toledo, Toledo, OH, United States

Materials Science Center, National Renewable Energy Laboratory, Golden, CO, United States

Center for Nanophase Materials Sciences, Oak Ridge National Laboratory, Oak Ridge, TN, United States

Cadmium sulfur selenide (Cd(S,Se)) at the front interface has been demonstrated to be a deleterious photo-inactive region in Cd(Se,Te) solar cells, which typically results in low open circuit voltages (VOCs) and short circuit current densities (JSCs). Thus, it is crucial to avoid the formation of Cd(S,Se) alloy to improve the device performance. In this work, we found that oxygen management during the device fabrication is a facile strategy to manipulate the atomic interdiffusion and avoid the Cd(S,Se) formation at the front interface. The key is to deposit both the CdS and CdSe layers and conduct the post CdCl2 treatment with oxygen incorporation, while oxygen should be absent during the CdTe deposition to avoid over-oxidization due to the high deposition temperature. Through oxygen management, CdS/CdSe/CdTe solar cells with efficiencies over 19% are demonstrated, which is among the champion CdTe solar cells.

Energy Yield and Economics of Single-Axis-Tracked Bifacial Photovoltaics with Artificial Ground Reflectors

Mandy R. Lewis, Silvana Ovaitt, Byron McDanold, Chris Deline, Karin Hinzer

University of Ottawa, Ottawa, ON, Canada

National Renewable Energy Laboratory, Golden, CO, United States

Artificial ground reflectors can potentially increase bifacial gain significantly, but their financial viability and ideal configuration are still unclear. We studied the performance of single-axis-tracked bifacial photovoltaic modules through ray-tracing modeling and site field measurements. Reflectors can increase rear irradiance by nearly 140%, front irradiance by 1.1%, and total irradiance by 6.8% over one year. Field measurements demonstrated that reflectors increased daily energy yield up to 6.2%. In both modeling and field tests, the ideal placement of the reflectors was directly under the module due to the significant resulting increase in rear irradiance. Levelized cost of energy calculations demonstrated that reflectors could be financially viable with costs of up to $2-4/m2.

Advanced Encapsulants for Reduced Thermal Mechanical Stress in Photovoltaic Modules: A Quantitative Analysis Using FBGS

Rik Van Dyck[1,2,3,4], Marta Casasola Paesa[2,3,4], Tine Engelen[2,3,4] Bin Luo[1,2,3,4], Tom Borgers[2,3,4], Jonathan Govaerts[2,3,4], Hariharsudan Radhakrishnan[2,3,4], Michael Daenen[2,3,4], Jef Poortmans[1,2,3,4], Aart Willem van Vuure[1]

[1]KU Leuven, Leuven, Belgium, [2]Hasselt University, imo-imomec, Hasselt, Belgium, [3]imec, Genk, Belgium, [4]EnergyVille, Genk, Belgium

Abstract — In this work, two mini-modules using a 3D multi-ribbon interconnection are fabricated. One with TPO and the other with glass fiber reinforced TPO (GF TPO) encapsulant. Using fiber Braggs grating sensors (FBGS) attached to the cell, in situ temperature and strain are quantified during reliability tests in the form of thermal cycling from -40°C to +85°C. It was found that the temperature of the cell surface reaches -36°C and +81°C at its minimum and maximum respectively. The measured cell strain followed the same cycling behavior between tension and compression. The strain in the GF TPO based module was found to have a lower peak-to-peak (difference between max tension and compression) value. Also, a consistent difference between strain in parallel and perpendicular directions relative to the busbars was observed, with the latter one being larger.

I. INTRODUCTION

The effect of the coefficient of thermal expansion (CTE) mismatch between module materials on the thermal stress induction has mainly been studied using finite element simulations [1]–[3]. The actual quantification of these stresses is limited to using external measurements methods such as spectroscopy [4]. The use of a Fiber Braggs Grating Sensor (FBGS) as an in-situ strain and temperature quantification method for photovoltaic (PV) modules was recently proposed [4], [5]. It was shown that temperature and strain could be measured, without affecting the module's thermal mechanical behavior. Indeed, the small FBGS diameter (~100 μm) allows effective integration without changing the PV module buildup, hence there is no significant impact on the thermal mass or mechanical stability. Also, optically there is no significant effect on the photon transmission due to the limited diameter and matching refracting indices . A brief explanation of the working principle of the FBGS is outside the scope of this abstract.

This work uses module-integrated FBGS to quantify temperature and thermally induced strain during reliability tests in the form of thermal cycling. The tested modules use a 3D multi-ribbon interconnection for back-contact cells. The concept [6] and reliability assessments [7] of this technology were previously published. Both a mini module with thermoplastic polyolefin (TPO) and glass fiber reinforced TPO (GF TPO) encapsulant will be fabricated and tested. The research objective is to find the effect of the encapsulant reinforcement on the thermally induced strain within the module.

II. MATERIALS AND METHODS

Encapsulant material selection

Two encapsulants are studied and compared for this work, a commercially available TPO, and the same TPO, reinforced with 10 w% randomly oriented short glass fibers: the GF TPO. The fibers have a diameter ranging from 10 to 12 μm. While previous publications explained the benefit of the GF addition on its processability [6], this work focuses on the effect on the CTE, and its associated thermal induced stresses. A thermal mechanical analysis was performed on the encapsulants to determine their CTE as shown in Fig. 1. This shows the CTE-reducing effect of the glass fiber reinforcement, being larger at higher temperatures.

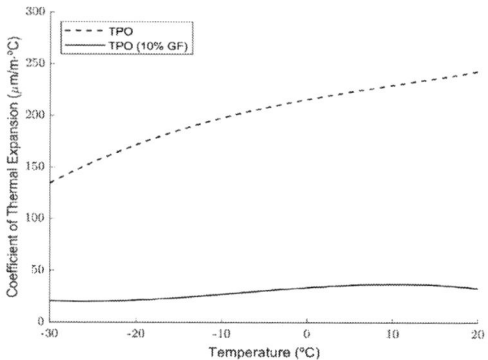

Fig. 1. Coefficient of thermal expansion (CTE) of a TPO and GF reinforced TPO as a function of temperature

Sample fabrication

For this work, two four-cell modules were studied. The architecture and fabrication procedure were identical, but one contains pure TPO encapsulant, while the other uses the GF TPO. Both modules were fabricated using a hand-made interconnection fabric of encapsulant with integrated metal ribbons. The metal ribbons have a copper core of 800x70 μm^2 and are coated in a 12 μm thick $Sn_{57}Bi_{42}Ag_1$ solder alloy, which has a melting temperature of 139 °C. An exploded view of the layup phase is shown in Fig. 2.

978-1-6654-6060-6/23 $31.00 © 2023 IEEE

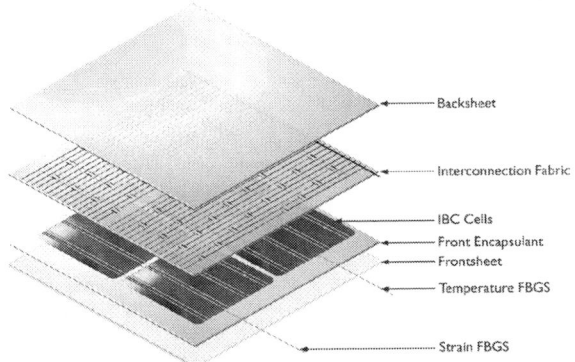

Fig. 2. Exploded view of the module layup

The established stack consists of a glass front sheet, encapsulant, four IBC cells, an interconnection fabric aligned to the cell metallization, and a glass backsheet. The front- and backsheet are both low Fe tempered 35x35 cm^2 glass plates with a thickness of 3 mm and without anti-reflection coating. During a lamination step, the stack is preheated to 165 °C in a vacuum for 10 minutes before 700 mbar of pressure is applied for 17 minutes using a membrane. An example picture of a module frontside is shown in Fig. 3(a).

Each module contains one strain and one temperature FBGS, which are attached to the front side of the cells using an adhesive before the lamination cycle. A schematic drawing of the configuration is shown in Fig. 3 (b), with the blue and red lines being the strain and temperature sensor respectively. The strain sensor has eight measuring points, two for each cell, allowing both X and Y direction strain quantification. The X direction corresponds to a measuring direction perpendicular to the busbars on the cell, while the Y direction corresponds to a measurement in parallel with the busbars

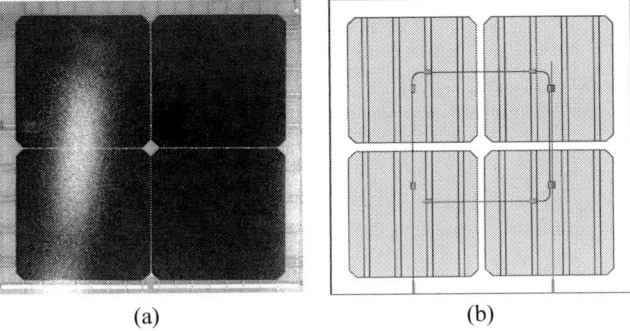

(a)   (b)

Fig. 3. (a) picture of the frontside of a laminated four-cell module, (b) schematic drawing of the module, with the red and blue lines indicating the strain and temperature FBGS respectively.

FBGS calibration and used formulas

After lamination, the modules underwent a step cycle of heating and cooling in order to calibrate the temperature sensor. The step cycle consists of 7 heating and cooling steps, where during each step, the module was kept at a constant temperature for 4 hours. The set step cycle and resulting output wavelengths are shown in Fig. 4 (a). The average wavelength on each plateau

was determined in order to find the wavelength-temperature combinations and fit the temperature as a function of the measured wavelength according to (1). The average wavelength and temperature data together with the fitted relationship are given in Fig. 4 (b).

$$ T(\lambda) = T_{ref} - \frac{S_1}{2S_2} + \frac{S_2}{|S_2|} \cdot \sqrt{\left(\frac{S_1}{2S_2}\right)^2 + \frac{1}{S_2} \cdot \ln\left(\frac{\lambda}{\lambda_{ref}}\right)} $$

(1)

With T temperature, λ the measured wavelength, S_1 and S_2 the linear and quadric temperature sensitivity coefficients, T_{ref} and λ_{ref} the temperature and wavelength at a known reference point.

Based on this model and the calibration data, S_1 and S_2 were found to be 5.98×10^{-6} and 8.14×10^{-9} respectively, with an R^2 value of 0.99995. These values can be filled in (2) in order to obtain a relationship between measured wavelength and strain.

$$ \varepsilon(\lambda) = \frac{1}{k}\left[\ln\left(\frac{\lambda}{\lambda_0}\right) - S_1(\Delta T_0) - S_2(\Delta T_0)^2\right] - (\alpha_{si} - \alpha_f)(\Delta T_0) $$

(2)

With

$$ \Delta T_0 = T - T_0 $$

and ε the actual strain, k the gage factor (7.77×10^{-7}), α_{si} and α_f the CTE of the Si cell (3.5×10^{-6} /°C) and FBGS (0.5×10^{-6} /°C) respectively.

This equation will be used in order to calculate the measured strain during thermal cycling according to the IEC 61215.2021 standard [8]. A total of 175 thermal cycles have been performed and monitored. Also, the temperature profile during the lamination cycle, showing the different stages has been measured.

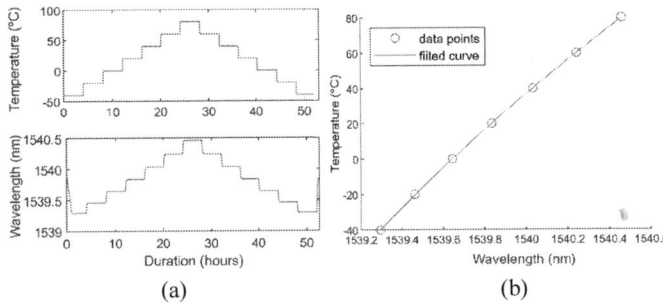

(a)   (b)

Fig. 4. (a) Set temperature profile during the calibration cycle (top) and measured wavelengths of the temperature sensor (bottom), (b) fitted temperature-wavelength relationship according to (1)

III. RESULTS AND DISCUSSION

Fig. 5. shows the measured (in module) and climate chamber (ambient) temperature and strain for one cell of both module during one thermal cycle. This is a zoom-in on one cycle during the first 24-hour thermal cycling test of 8 cycles. Similar results were observed for the 7 other cycles during this test. Both the in-module temperature and strain are calculated from the output wavelength using the equations discussed above, the strain being referenced to the module at 25 °C. The minimum and

978-1-6654-6060-6/23 $31.00 © 2023 IEEE

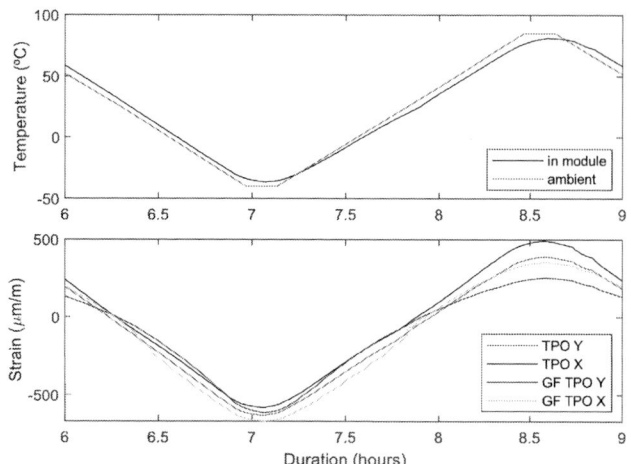

Fig. 5. Measured and set temperature (top) and strain (bottom) for one cell in X and Y direction during one thermal cycle, both for a module with TPO and GF TPO encapsulant

maximum in module temperatures are -36°C and 81°C degrees respectively, resulting in an offset with the set temperature profile of 4°C at these maxima.

When cooling down the module, the strain values go negative, which corresponds to compression. This is expected due to the larger CTE of the encapsulant compared to the Si cell. The encapsulant is attached to the cell, so during cooling, it pulls the cells into compression. The opposite happens during the heating cycle of the module. As mentioned above, the Y and X direction corresponds to the parallel and perpendicular directions compared to the busbars on the cell respectively.

TABLE 1 gives the maximum strain values in tension and compression. The peak-to-peak difference, in other words, the total strain variation to which the cell is exposed, is significantly smaller for the module with GF TPO encapsulant. The difference between the TPO and GF TPO module in compression is rather limited, with 36 and 31 µm/m for the X and Y direction respectively. A larger difference is observed in tension mode, with the difference being 140 and 139 for X and Y direction respectively. This might be attributed to the smaller CTE difference for the TPO and GF TPO at low temperatures compared to higher temperatures, as shown above in Fig. 1.

TABLE 1
MAXIMUM STRAIN DURING ONE THERMAL CYCLE IN
COMPRESSION AND TENSION

	Compression (µm/m)	Tension (µm/m)	Difference (µm/m)
TPO X	631	492	1123
TPO Y	580	391	971
GF TPO X	668	352	1020
GF TPO Y	611	252	863

Overall, the observed strain in compression is higher than in tension. This could again be attributed to the behavior of the encapsulant material behaving like a solid at lower temperatures and becoming more viscous at higher temperatures. This more viscous behavior can result in more internal slip in the polymer, resulting in stress relief. Dynamic mechanical analysis to show the viscous behavior is performed and results could be included in the final paper.

Longer thermal cycling tests (up to 100 cycles) are performed, and strain measurement results will be included in the final paper. Changes in strain during these tests can be an indication of degradation of the module. Also electrical characterization of the modules can be included.

IV. CONCLUSIONS

This work was able to utilize a previously proposed approach to incorporate FBGS in a PV mini-module in order to quantify temperature and strain during thermal cycling. Adding glass fiber reinforcement to a TPO encapsulant has been shown to reduce the strain on cell level, both in parallel and perpendicular directions to the busbars. This can be attributed to the reduction in CTE due to the glass fiber reinforcement, reducing the CTE mismatch in the module.

V. REFERENCES

[1] L. Yixian and A. A. O. Tay, "Finite element thermal stress analysis of a solar photovoltaic module," in *2011 37th IEEE Photovoltaic Specialists Conference*, 2011, pp. 3179–3184. doi: 10.1109/PVSC.2011.6186616.

[2] O. O. Ogbomo *et al.*, "Effect of Coefficient of Thermal Expansion (CTE) Mismatch of Solder Joint Materials in Photovoltaic (PV) Modules Operating in Elevated Temperature Climate on the Joint's Damage," *Procedia Manuf*, vol. 11, pp. 1145–1152, 2017, doi: 10.1016/j.promfg.2017.07.236.

[3] U. Eitner *et al.*, "Thermal Stress and Strain of Solar Cells in Photovoltaic Modules," in *Shell-like Structures: Non-classical Theories and Applications*, Ed: Springer Berlin Heidelberg, 2011, pp. 453–468.

[4] P. Nivelle *et al.*, "The Lamination Process Quantified Through In-Situ Optical Thermo-Mechanical Sensing," in *Proceedings in the WCPEC 8*, 2022.

[5] P. Nivelle *et al.*, "In situ quantification of temperature and strain within photovoltaic modules through optical sensing," *Prog. Photovolt.: Res. Appl.*, pp. 1–7, 2022, doi: 10.1002/pip.3622.

[6] R. van Dyck *et al.*, "Three-dimensional multi-ribbon interconnection for back-contact solar cells," *Prog. Photovolt.: Res. Appl.* vol. 29, no. 5, pp. 507–515, 2021, doi: https://doi.org/10.1002/pip.3390.

[7] R. van Dyck *et al.*, "Three-Dimensional Multi-Ribbon Back-Contact Interconnection: Latest Results on Reliability Testing," in *38th EUPVSEC*, 2021, pp. 728–731.

[8] "IEC 61215-1 Terrestrial photovoltaic modules – Design qualification and type approval," 2021.

TOPCon Solar Cell Degradation via Pinhole Nucleation

Andrew Diggs, Adam Goga, Zachary Crawford, and Gergely T. Zimanyi

Physics Department, University of California Davis, Davis, CA, 95616, USA

Abstract — **Reducing recombination at the interfaces is crucial to further improve the performance of silicon solar cells. One of the most promising cell designs, TOPCon cells suppress surface recombination by forming an ultra-thin SiOx passivating layer on the c-Si wafer. Efficient carrier extraction through the oxide layer via exponentially suppressed quantum tunneling is achieved by making the SiOx layer ultra-thin, ~1.5 nm. However, experiments indicate that at such extreme thinness, Si-rich pinholes start to pierce the SiOx layers that can act as regions of enhanced recombination. In this paper we demonstrate that over time the non-stoichiometric SiOx layer phase separates into Si-rich regions piercing a SiO2 layer. We believe that this is a probable mechanism of pinhole formation. We also report early indications that adding hydrogen increases this pinhole formation tendency.**

I. INTRODUCTION

One of the fundamental requirements needed to achieve high efficiencies in solar cells is a high minority carrier lifetime τ. High quality n-type c-Si wafers have demonstrated excellent bulk minority carrier lifetimes. This means that the lifetime and hence the cell performance is limited by recombination at the contacts and interfaces. One way to reduce such losses is via the use of passivating contacts. Currently the highest cell efficiencies approaching 27% are delivered by silicon heterojunction (HJ) cells, where a thin a-Si layer passivates the c-Si interface [1].

A leading alternative passivating contact combines a SiOx passivating layer with a doped polycrystalline silicon (poly-Si) contact layer. One of the most promising cell designs utilizing this structure is the Tunnel Oxide Passivating Contact, TOPCon cell, which demonstrated efficiencies in excess of 25% [2]. Due to their compatibility with current manufacturing infrastructure, TOPCon cells are rapidly becoming a leading replacement for the PERC/PERL/PERT cells currently dominating the market.

Charges are extracted from TOPCon cells via quantum tunneling through the SiOx layer [3,4]. Since the probability of quantum tunneling decreases exponentially with the barrier thickness, efficient carrier extraction is achieved by using an ultra-thin SiOx layer of 1.5 nm. While the thinness of the SiOx layer promotes charge transport, it also makes the TOPCon cells susceptible to pinhole formation. Pinholes are regions which create direct contact between the c-Si and the poly-Si layers, and thus enhance recombination. The formation of pinholes over long times was shown to be a major mechanism of cell degradation [5]. However, in spite of their importance, the mechanism of pinhole formation is not well understood. The present paper reports our project to model, analyze and understand pinhole formation in the SiOx passivating layer of TOPCon solar cells.

II. SIMULATIONS

All simulations were carried out with the LAMMPS Molecular Dynamics Simulator [6], using a ReaxFF force field [7], recently optimized for simulating Si/O/H at a c-Si/SiO2 interface [8]. ReaxFF potentials facilitate the analysis of the effects of intermolecular polarization and allow for charge transfer between atoms during simulations using a geometry-dependent electronegativity equalization method (EEM).

A. Creation of a-SiO2

We began by implementing and validating the creation of a-SiO2. To do this, we started with the β-crystobalite unit cell having a lattice parameter of 7.13 Å, which we replicated five times in all three dimensions. This produced a sample of c-SiO2, which had 1000 Si and 2000 O atoms with cubic simulation box dimensions of 35.65 Å. Then, in a canonical ensemble (using fixed volume "NVT" thermostating), the sample was heated up in a 5000 K bath for 200 ps, then cooled down in a bath starting at 5000 K and ending at 300 K over 200 ps. The sample was equilibrated in a 300 K bath for 1000 ps. Next, in an isothermal-isobaric "NPT" ensemble at 0 pressure, the sample was heated up by a 5,000 K bath for 200 ps, then gradually cooled down over 400 ps starting at 5000 K and ending at 300 K. The sample was left at 300 K for 200 ps. Finally, the energy of the sample was minimized using the

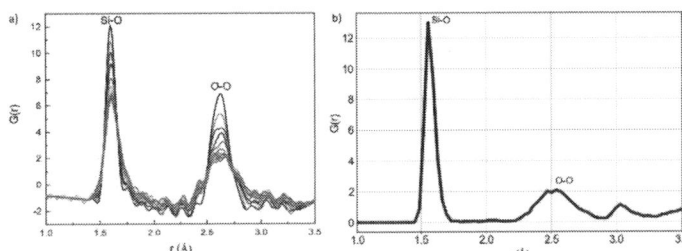

Fig. 1. a) Experimentally determined RDF plots of a-SiO2 samples over a temperature range from room temperature (black) to 950°C (red). The first Si-O and O-O peaks are approximately at 1.55Å and 2.6Å. [9]

b) RDF plot of our a-SiO2 sample after minimization. The first Si-O peak is located approximately at 1.57 Å, and the first O-O peak appears approximately at 2.53 Å.

Hessian-free truncated Newton minimization style, following previously validated protocols [8]. The NVT thermostating and the NPT barostating were performed by time integration of Nose-Hoover style equations of motion (EOM) and a velocity Verlet integration of Newtons EOM. **Fig. 1** shows that the radial correlation functions RDF in our simulated samples compared well with experimental data and previous simulations, thus validating the reliability of our methods.

B. *Phase separation of non-stoichiometric SiOx*

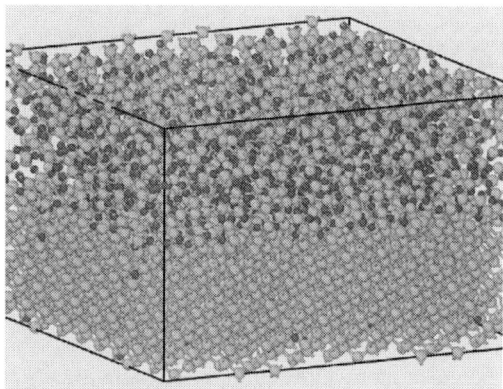

Fig 2. c-Si/SiOx slab, created with the described NVT-NPT-Newton minimization protocol.

After having validated our sample-creation protocols, we proceeded to investigate the effects of a spatially varying stoichiometry on the structural evolution of the SiOx layer during thermal treatment. We began by creating c-Si/SiOx cells of size of 5X5X6 c-Si unit cells. The cells were created by first creating a 5X5X3 c-Si base containing 650 Si atoms. Next, (319 + x) Si and (638 - 2x) O atoms were randomly inserted into the region above the c-Si. Atoms with a separation less than a pre-determined cutoff distance were deleted. The number of Si and O atoms and the overlap cutoff were chosen such that the mass density and stoichiometric ratios were consistent with previously reported values [10-12]. The SiOx region was then heated to 2000 K to ensure a thorough mixing of the Si and O, followed by a fixed volume NVT thermostating from 2000 K to 1000 K at a rate of 10^{10} K/s. The cells were then relaxed using a fixed pressure NPT barostating from 1000 K to 300 K at a rate of 10^9 K/s. Finally, the structural energy of the stacks was minimized using the conjugate gradient (CG) method.

To study the structural evolution of the SiOx layer under thermalization conditions we needed a large enough system to minimize the distorting effects of the periodic boundary conditions. Therefore, we constructed simulation "slabs" by replicating the above mentioned 1,400 atoms cells in a 2x2 pattern, resulting in a slab containing ~ 5,500 atoms. The slabs were then heated to and held at 1,100 K (~ 800 C). **Fig. 2** shows the slab, created by the here-described method.

To characterize the structural evolution of the c-Si/SiOx slabs, we determined the spatially dependent atomic fraction x of Si and O in the SiOx layer. Specifically, we computed the ratio of x=N(O)/N(Si) in 2.7 X 2.7 X 16.3 Å rectangular parallelepiped regions. The value of x for each region was then smoothed by computing the average of a 3x3 grid surrounding the region. **Fig. 3** shows that we constructed a 2D heatmap to present these x values. Our main result is that at temperatures consistent with the thermalization of TOPCon cells, the SiOx layer which initially was formed with a non-stoichiometric composition 0<x<2, manifested a profound phase separation from uniform disordered SiOx to into a well-defined Si-rich region piercing an SiO2 layer. Using another analogy: x~0

islands formed over an x~2 background "sea". The resulting Si-rich regions are ~2 nm in diameter which is consistent with reported values of the diameter of pinholes in TOPCon cells [4].

C. *Inducing Phase Separation with Hydrogen*

Next, we set out to investigate the possible role of hydrogen in inducing the formation of these pinholes in TOPCon cells.

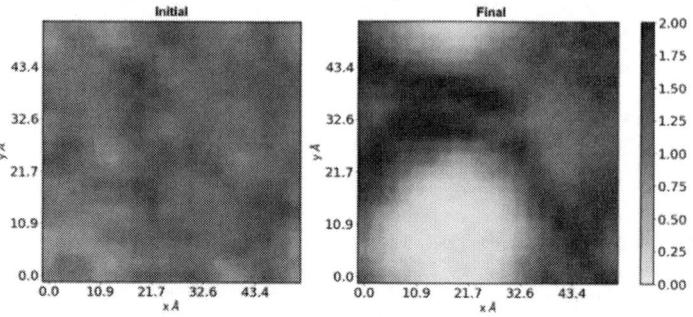

Fig 3. Spatially dependent heatmaps of the atomic fraction x of SiOx layers, in initial state, and after 1,000,000 timesteps at T=1,100 K.

For these trials, we created samples of SiOx with an atomic fraction of x=1.5. In a simulation cell of size 40.4x40.4x20.8 Å we randomly placed 720 Si and 1,440 O atoms with an overlap parameter of 0.4 Å over a max number of 100000 retries. Next, we annealed the sample with two heating-cooling cycles, first using an NVT protocol, then an NPT protocol. In the first NVT cycle, the sample was heated up in a 7000 K bath, and then cooled down in a 300 K bath, each over a time of 400 ps. Next, in the NPT cycle the sample was heated up in a 2500 K bath for 400 ps, while the z extent of the sample was allowed to vary. Then the sample was gradually cooled down over 400 ps from 2,500 K to 300 K. Next, the sample was equilibrated at 300 K for 200 ps. Finally, the energy was minimized using the Hessian-free truncated Newton minimization style, see above.

Three samples, created by this described protocol, were then hydrogenated by placing 0, 15, and 50 hydrogen atoms into random positions with an overlap parameter of 0.2 Å and a max number of 100,000 tries. Each sample was then put through the same thermal annealing protocol as earlier, with a mixed time-temperature protocol using a simple NVT loop, where the temperature was raised by 40 K every 200 ps.

Fig 4. shows the mixed time-temperature series of the O-coordination number of Si atoms in samples with the three hydrogen concentrations. In a-SiO2, Si atoms have 4 Oxygen atoms within a range of approximately 2 Å. Thus, tracking the total number of Si atoms that have 4 O atoms within a 2 Å radius indicates the size of the regions where a-SiO2 has been formed by oxygen in-drift, presumably leaving behind Si-rich regions. The simultaneous emergence of Si-rich and stoichiometric SiO2 regions are compelling indicators of phase-separation dynamics in the initial SiOx. Visibly, our initial study showed that the growth rate of the number of 4-fold O-coordinated Si atoms was increased by the introduction of H. However, longer and more comprehensive studies are planned to establish whether the number of 4-fold O-coordinated Si

978-1-6654-6060-6/23 $31.00 © 2023 IEEE

atoms is also boosted in the absolute sense by the introduction of H. We are even open to the possibility that the H-dependence will be non-monotonous, as reported in [20].

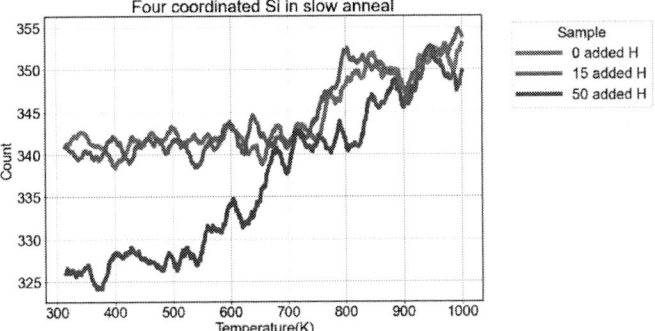

Fig. 4. The number of 4 coordinated silicon atoms during a slow perturbation of SiOx samples with hydrogen, versus samples without hydrogen. The x-axis indicates our mixed time-temperature variable, where the temperature was increased by 40 K every 200 ps.

III. CONCLUSIONS

In this project we investigated the formation of pinholes in non-stoichiometric SiOx using Molecular Dynamics (MD) simulations. We found compelling evidence that under conditions typical for the crystallization of the poly-Si contact, pinholes formed because the SiOx layer phase separated into a SiO_2 layer, pierced by regions of nearly pure Si, i.e. into an x~2 layer, pierced by x~0 pinholes. Building on this finding, we started to explore the effect of hydrogen on this phase separation mechanism.

This work was supported by the U.S. Department of Energy's Office of Energy Efficiency and Renewable Energy (EERE) under the Solar Energy Technologies Office Award Numbers DE-EE0008979 and DE-EE0009835. The views expressed herein do not necessarily represent the views of the U.S. Department of Energy or the United States Government.

REFERENCES

[1] K. Yoshikawa, H. Kawasaki, W. Yoshida, T. Irie, K. Konishi, K. Nakano, T. Uto, D. Adachi, M. Kanematsu, H. Uzu, and K. Yamamoto. "Silicon heterojunction solar cell with interdigitated back contacts for a photoconversion efficiency over 26%". *Nature Energy*, 2(5), 2017.

[2] A. Richter, J. Benick, F. Feldmann, A. Fell, M. Hermle, and S. W. Glunz. "n-type si solar cells with passivating electron contact: Identifying sources for efficiency limitations by wafer thickness and resistivity variation". *Solar Energy Materials and Solar Cells*, 173:96– 105, 2017. Proceedings of the 7th international conference on Crystalline Silicon Photovoltaics.

[3] F. Feldmann, M. Simon, M. Bivour, C. Reichel, M. Hermle, and S. W. Glunz. "Efficient carrierselective p- and n-contacts for si solar cells". *Solar Energy Materials and Solar Cells*, 131:100–104, 2014. SI: SiliconPV 2014.

[4] F. Feldmann, M. Bivour, C. Reichel, H. Steinkemper, M. Hermle, and S. W Glunz. "Tunnel oxide passivated contacts as an alternative to partial rear contacts". *Solar Energy Materials and Solar Cells*, 131:46–50, 2014. SI: SiliconPV 2014.

[5] Z. Zhang, Y. Zeng, C. Jiang, Y. Huang, M. Liao, T. Hui, M. Jassim, C. Shou, X. Zhou, B. Yan, and J. Ye. "Carrier transport through the ultrathin silicon-oxide layer in tunnel oxide passivated contact (topcon) c-si solar cells". *Solar Energy Materials and Solar Cells*, 187, 11 2018.

[6] A. P. Thompson, H. M. Aktulga, R. Berger, D. S. Bolintineanu, W. M. Brown, P. S. Crozier, P. J. in 't Veld, A. Kohlmeyer, S. G. Moore, T. D. Nguyen, R. Shan, M. J. Stevens, J. Tranchida, C. Trott, and S. J. Plimpton. "LAMMPS - a flexible simulation tool for particle-based materials modeling at the atomic, meso, and continuum scales". *Comp. Phys. Comm.*, 271:108171, 2022.

[7] J. C. Fogarty, H. M. Aktulga, Ananth Y. Grama, A. C. T. van Duin, and Sagar A. Pandit. "A reactive molecular dynamics simulation of the silica-water interface". *The Journal of Chemical Physics*, 132(17):174704, 2010.

[8] N. Nayir, A. C. T. van Duin, and S. Erkoc. "Development of the reaxff reactive force field for inherent point defects in the si/silica system". *The Journal of Physical Chemistry A*, 123(19):4303–4313, 2019. PMID: 31017438.

[9] Y. Shi, D. Ma, A. P. Song, B. Wheaton, M. Bauchy, and S. R. Elliott. "Structural evolution of fused silica below the glass-transition temperature revealed by in-situ neutron total scattering". *Journal of Non-Crystalline Solids*, 528:119760, 2020.

[10] S. Miyazaki, H. Nishimura, M. Fukuda, L. Ley, and J. Ristein. "Structure and electronic states of ultrathin sio2 thermally grown on si(100) and si(111) surfaces". *Applied Surface Science*, 113-114:585–589, 1997. Proceedings of the Eighth International Conference on Solid Films and Surfaces.

[11] Y. Sugita, S. Watanabe, N. Awaji, and S. Komiya. "Structural fluctuation of sio2 network at the interface with si". *Applied Surface Science*, 100-101:268–271, 1996.

[12] M. Jech, A. El-Sayed, S. Tyaginov, A. L. Shluger, and T. Grasser. "ab initio treatment of silicon-hydrogen bond rupture at si/sio2 interfaces". *Phys. Rev. B*, 100:195302, Nov 2019.

[13] R. R. King, R. A. Sinton, and R. M. Swanson. "Studies of diffused phosphorus emitters: saturation current, surface recombination velocity, and quantum efficiency." *IEEE Transactions on Electron Devices*, 37(2):365–371, 1990.

[14] A. S. Kale, W. Nemeth, S. U. Nanayakkara, H. Guthrey, M. Page, M. Al-Jassim, S. Agarwal, and P. Stradins. "Tunneling or pinholes: Understanding the transport mechanisms in siox based passivated contacts for highefficiency silicon solar cells". In *2018 IEEE 7th World Conference on Photovoltaic Energy Conversion (WCPEC) (A Joint Conference of 45th IEEE PVSC, 28th PVSEC 34th EU PVSEC)*, pages 3473–3476, 2018.

[15] Y. Hasnain, K. Muhammad Quddamah, C. Sanchari, P. D. Phong, K. Youngkuk, J. Minkyu, C. Younghyun, C. Eun-Chel, and Y. Junsin. "A review on topcon solar cell technology". *Current Photovoltaic Research*, 9(3):75–83.

[16] Y. Shi, D. Ma, A.P. Song, B. Wheaton, M. Bauchy, and S.R. Elliott. "Structural evolution of fused silica below the glass-transition temperature revealed by in-situ neutron total scattering". *Journal of Non-Crystalline Solids*, 528:119760, 2020.

[17] S. A. Pandit A. Y. Grama H. M. Aktulga, J. C. Fogarty. "Parallel reactive molecular dynamics: Numerical methods and algorithmic techniques". *Parallel Computing*, 38:245–259, 2012.

[18] D. Kang, H.-C. Sio, J. Stuckelberger, R. Liu, D. Yan, X. Zhang, and D. Macdonald. "Optimum Hydrogen Injection in Phosphorus Doped Polysilicon Passivating Contacts". *ACS App. Mat. Interfaces*, 55164-55171, 2021.

Soiling Model for PV Applications: Improved Parameterizations

Vicente Lara-Fanego[1], Christian A. Gueymard[2], and Leonardo Micheli[3]

[1] Solargis s.r.o., Bratislava (Slovakia), [2] Solar Consulting Services, Colebrook, NH (USA),
[3] Sapienza University of Rome (Italy)

Abstract — **Soiling of PV modules causes large financial losses worldwide. It is therefore essential to understand this process and model it precisely. The Coello & Boyle soiling model is widely used, but so far, its evaluations have shown widely variable performance depending on whether dry deposition velocities are considered constant or derived from a physics-based parameterization. This study shows how different possible parameterizations and optional adaptation of the parameters to the characteristics of PV plants explain this counterintuitive finding. The present results also highlight the importance of adjusting the most critical parameterizations to the realistic conditions of actual PV systems to improve the soiling model's performance and universal reliability.**

I. INTRODUCTION

Among the environmental factors that impact the actual power production of a PV plant, soiling (i.e., accumulation of airborne particles on optical surfaces) is a key one. It is unavoidable, though mitigable, and can produce significant direct losses, as well as contributing to the degradation of solar panels. It is thus critical for the solar industry to understand the soiling process, its impact on PV systems, and how to mitigate its effects. This requires models that can reliably quantify soiling losses and correctly assess their impact on solar projects.

The soiling issue is global, with a heterogeneous distribution over space and time. Thus, in general, it depends on local ambient conditions, time of year, and characteristics of each PV system. This complexity results from the multidimensional combination of different elements and processes that influence each other, acting simultaneously or independently on different spatial and time scales. The physico-chemical properties of the different particle species play a fundamental role. But so do meteorological variables such as temperature, relative humidity, wind, or the intensity and temporal distribution of precipitation. Actually, not all these factors increase soiling. Some of them might also induce the opposite effect, i.e., partial or total cleaning of the modules, through precipitation in particular. Indeed, rainwater can either increase soiling through wet deposition or remove some or all accumulated dirt, depending on various factors. Currently, however, the literature has only modelled the cleaning effect of rainwater, although no clear threshold for the activation of rainwater's cleaning action has been found yet [1].

Given the complexity of the soiling process, no model is currently able to integrate all the causal factors to consistently and reliably quantify soiling losses. Thus, soiling is one of the sources of uncertainty that is most difficult to estimate in a PV project. That uncertainty is the result of the contribution of several factors: the soiling model *per se*, its inputs, the cleaning events (e.g., rain or wind), and the quality of the observational data. In practice, it is difficult to isolate the model's uncertainty itself from that of the other sources just mentioned.

A few modeling approaches for soiling losses already exist in the literature, as reviewed in [2]. In particular, one of the most popular models is that of Coello & Boyle [3] (hereafter, C&B, but referred to as HSU in PVlib [4]). In that model, the mass accumulation, m, on a PV module over a period t is obtained by considering dry deposition velocities only:

$$m = \int \left(v_{2.5} \cdot C_{2.5} + v_{10-2.5} \cdot C_{10-2.5} \right) \cos(\beta) \cdot t dt \quad (1)$$

where β is the tilt angle of the module, C_i represents the mass concentration, and v_i is the dry deposition velocity for particles of diameter either less than 2.5 μm (PM2.5) or between 2.5 and 10 μm (PM10–2.5), respectively. (Wet deposition is beyond the scope of the present study.) Moreover, an empirical relationship provides an estimate of the transmission loss as a function of the accumulated mass. Ultimately, that transmission loss is directly related to the soiling ratio (SR) [3], a metric expressing the fraction of PV output left after soiling and calculated as 1-soiling loss:

$$\text{SR} = 1 - 0.3437 \text{erf} \left(0.17 m^{0.8473} \right). \quad (2)$$

C&B proposed default constant values for the deposition velocities. They also described an alternate physics-based formulation to derive variable deposition velocities. The reliability of C&B predictions has been evaluated in various studies, showing reasonably good performance at daily resolution [1][5][6]. Surprisingly, C&B obtained notably better results with their own model by using a *constant* deposition velocity, as opposed to the *variable* daily deposition velocities that are calculated dynamically by their physics-based approach. Similar counterintuitive results were found in [6]. A recent evaluation carried out by Solargis at 20 PV sites, whose results are partially shown here, also corroborates this finding. The constant values used by C&B were originally estimated for water surfaces. In contrast, [5] found better agreement with observations from their particular experimental setup by using dynamic deposition velocities based on [7], as implemented in PVlib, instead of C&B's default constant values.

The present work examines the discrepancy between the existing evaluation results by delving into the physical bases of the differing parameterizations in an attempt to find a better model for the dry deposition velocity. Intuitively, the ideal parameterization's *configuration* (i.e., the compendium of selected functions and coefficients of the various quantities that form the complete parameterization) should fit the particular conditions of each PV system. The ultimate goal here is to evaluate whether an updated parameterization for the estimation of the accumulated soiling on PV modules, using an "optimal" configuration adapted to their particular conditions, can effectively improve the reliability of this soiling model.

II. Methodology

An evaluation exercise is conducted here to compare different configurations for the calculation of deposition velocities. The values thus obtained result in differing estimates of the accumulated mass (Eq. 1), and ultimately in different predictions of the soiling ratio (Eq. 2).

This analysis is carried out on a daily basis, assuming a fixed tilt of 25° for the PV modules. To obtain the modeled SR predictions, daily mean values of all the required inputs are used. Temperature, atmospheric pressure, wind, and relative humidity are retrieved from the ERA5 reanalysis. The particulate matter concentrations (PM2.5 and PM10) are derived from the CAMS reanalysis. These datasets are recognized for their good quality, even though their spatial resolution is relatively coarse (tens of km).

Of all the analyzed configurations, only four are presented and discussed in this work. First, constant values of 0.9 mm/s and 4 mm/s for PM2.5 and PM10–2.5, respectively, are used for the deposition velocities, per the results in [3] and the PVlib library's recommendation. In the present work, this set (dubbed "Configuration 1", C1) is used as the reference against which other configurations are to be evaluated.

Configuration 2 (C2) relies on the results in [5], where the deposition velocities are calculated dynamically according to [7], as implemented in PVlib. This parameterization uses tabulated values for those parameters that are necessary in the calculation of aerodynamic and surface resistances as a function of preset land-use categories. Here, this is adjusted to reproduce the C&B configuration for dynamic deposition velocities

Configuration 3 (C3) is based on the previous one, but some of the methods and parameters are modified, based on various approaches from the literature. They are analyzed and tested against observational data (not described here for brevity). For instance, to estimate the collection efficiency from impaction, the Slinn method, as described in [8], is selected—instead of that discussed in [7] for smooth surfaces and used in PVlib. Moreover, a diffusion-advection method is used to calculate the deposition velocity, instead of the so-called electrical resistances method used in [7][8]. In parallel, proper values of certain parameters that best fit a PV plant's particular conditions have been selected. For instance, the α parameter needed to calculate the efficiency of the impaction process is assumed to be 100 here, which better fits the case of a PV module cover.

Finally, Configuration 4 (C4) is based on the previous one, but a fundamental physical parameter—the particle size—is modified in the calculation of the deposition velocities. Commonly, 2.5 μm and 10 μm are used as average diameters. Whereas 10 μm might be reasonable for large particles, 2.5 μm is likely too large for small particles, in general. Hence, in C4, a smaller average diameter (0.7 μm) is considered for PM2.5 particles. This emphasizes that it is important to scrutinize—and possibly reevaluate—all model inputs.

The daily modeled SR estimates are compared to SR observations obtained at 20 different PV plants. Their location spreads over different areas: the west-central and west coast of the USA, the Arabian Peninsula, and the east coast of Africa.

More specifically, 7 sites are in desert climates, 11 in arid temperate climates, and 2 in temperate climates, although the latter are affected by the presence of large bodies of water. Unfortunately, the soiling data being proprietary, it is not possible to reveal the exact locations for reasons of confidentiality. The periods vary from site to site, spanning from 2013 to 2020. All sites have ≈1–3 years of historical data.

The observed data have been pre-processed to detect and correct any quality issue, such as nonsensical values (e.g., SR>1). Scheduled cleaning events were not known a priori, so they had to be determined based on visual inspection of the data time series. All values considered erroneous, as well as days when a module cleaning action was detected, were eliminated. A reference value of 1 mm of accumulated rain per day was assumed to have a cleaning efficiency of 100% (hence yielding SR=1).

The results are presented in terms of the usual metrics, namely: standard deviation of the errors (STD), mean bias deviation (MBD), mean absolute deviation (MAD), and root mean-square deviation (RMSD).

It is stressed that the goal of this work is *not* to evaluate the performance of the C&B model itself. This would require the determination of each source of uncertainty (namely, observational data, model input data, as well as the cleaning events and their efficiency) separately from the model's uncertainty itself. Although desirable, such analysis is complex and beyond the scope of the present work. In practice, many sources of uncertainty exist and affect the present evaluation results. However, they do it equally for all configurations, thus the conclusions with respect to the objectives of this work remain valid.

III. Results

Fig. 1 shows the time series of observed and modeled SRs with each of the four configurations under scrutiny for a test site in the Arabian Peninsula, where soiling from dust particles can be intense. The daily totals of rainfall are also shown. (Only 4 days out of the 367-day test period had more than 1 mm of rainfall.) A pattern of regular jumps in the SR values is clearly observed, which normally corresponds to scheduled cleaning events. Nevertheless, some of these breaks also coincide with heavy rainfall events. C1, C4, C3, and C2 induce soiling ratios from higest to lowest (i.e. losses from lowest to highest).

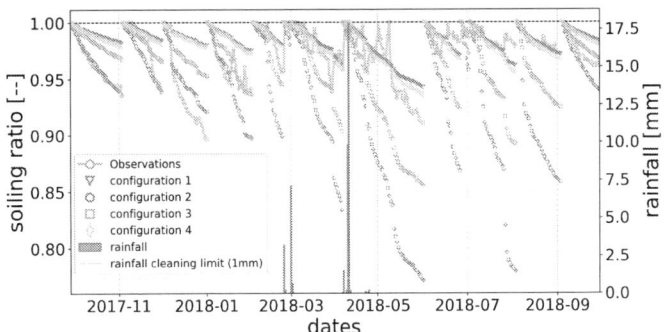

Fig. 1. Time-series of the modeled and observed SRs and of daily precipitation totals for a site in the Arabian Peninsula.

C2's SR predictions are strongly underestimated most of the time, with notable exceptions, such as at the end of 2017 and in early 2018. The overall performance statistics for the evaluation at the same site are shown in the upper plot of Fig. 2, confirming that C2 performs worst. Its underestimation of SR (and overestimation of the soiling mass) is indicated by the negative MBD. Moreover, the results for C2 are significantly worse (always overestimating the deposition velocities) than the other options at 11 of the 20 sites, while it is only slightly better at the other nine. An example of the latter case is shown at the bottom of Fig. 2 for a US site. Hence, in general, the deposition velocities estimated by C1, C3, and C4 fit the observed data much better than those estimated with C2.

The fact that C1 performs so well suggests that the deposition velocities are not that high for PV plants. Since C3 and C4 are dynamically calculated (just like C2), it follows that, to properly tune this type of model, it is necessary to fine tune the parameters so that they reflect the physical characteristics of each PV plant. This might explain the significant differences in SR results between the dynamically calculated deposition velocities and the recommended constant values used in the above-mentioned evaluations of C&B.

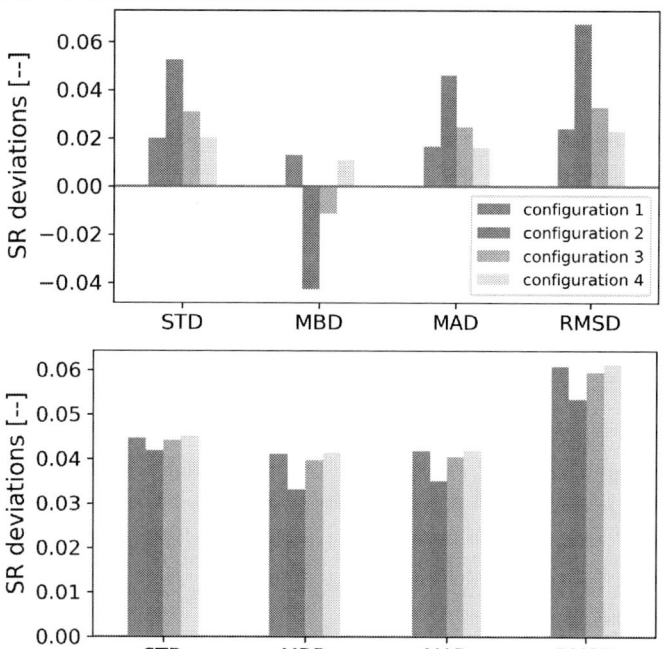

Fig. 2. Statistical results for the four configurations at a site in the Arabian Peninsula (top), and in the USA (bottom).

Table I summarizes the statistical results of the evaluation for all the 20 test sites combined. These results are in line with those reported in [6]. Overall, the present results confirm that the new parameterization, with proper adjustment for each PV plant's characteristics, performs quite well. The best results are obtained with C4, but the reference C1 is almost as good.

TABLE I.   AGGREGATED STATISTICAL RESULTS AT 20 TEST SITES

Configuration	STD	MBD	MAD	RMSD
1	0.018	0.011	0.015	0.022
2	0.030	-0.015	0.027	0.040
3	0.021	0.001	0.017	0.025
4	0.018	0.009	0.015	0.021

IV. CONCLUSIONS

An evaluation of daily soiling ratios has been carried out at 20 world sites. The Coello & Boyle (C&B) model has been used with four different parameterizations of the deposition velocity, using either a fixed value or dynamic physics-based functions. The results agree with those of other studies, and show that the soiling ratio predictions using constant values of deposition velocities fit the observational data remarkably better than when using the default setup of the parameterization in PVlib. Notwithstanding, the model results are substantially improved by updating the parameterization and adjusting the parameters to the particular conditions of the PV plant under scrutiny. The physically-based configuration that is found here to perform best overall does not appear to improve much over the satisfactory results of the reference C&B default configuration. Nevertheless, the latter only considers constant values for the deposition velocity, and therefore cannot adapt to the latter's natural spatiotemporal variability, which depends on atmospheric conditions. In contrast, the novel Configuration 4 does have the flexibility to realistically describe the behavior of this variable under any condition. Still, more improvements are certainly possible by better describing the input variables and fine-tuning the model parameters.

REFERENCES

[1] S. Pelland et al., "Testing Global Models of Photovoltaic Soiling Ratios Against Field Test Data Worldwide," in *IEEE-WCPEC Conf.*, 2018.

[2] C. Schill et al., "Impact on the Performance of Photovoltaic Power Plants", Rep. IEA-PVPS T13-21:2022, 2022.

[3] M. Coello and L. Boyle, "Simple Model for Predicting Time Series Soiling of Photovoltaic Panels," *IEEE J. Photovoltaics*, vol. 9, pp. 1382–1387, 2019.

[4] F. Holmgren et al., "Pvlib Python: a Python package for modeling solar energy systems," *J. of Open Source Software*, vol. 3, pp. 884, 2018.

[5] J. Polo et al., "Modeling soiling losses for rooftop PV systems in suburban areas with nearby forest in Madrid", *Renewable Energy*, vol. 178, pp. 420-428, 2021,

[6] L. Micheli et al., "Tracking Soiling Losses: Assessment, Uncertainty, and Challenges in Mapping," *IEEE J. of Photovoltaics*, vol. 12, pp. 114-118, 2022.

[7] L. Zhang et al., "A size-segregated particle dry deposition scheme for an atmospheric aerosol module," *Atmos. Environ.*, vol. 35, 549–560, 2001.

[8] A. De Visscher, "Air Dispersion Modeling: Foundations and Applications," Wiley, 2013.

A sustainable energy market through community-based PV systems

Ramón Reyes-Colón and Efraín O'Neill-Carrillo

Electrical and Computer Engineering Department, University of Puerto Rico, Mayagüez (UPRM), PR 00681

Abstract — **Conventional methods to optimize power systems may fall short in microgrids due to the penetration level and variety of resources; modern methods and techniques are needed to find efficient configurations. This paper presents the application of a new optimization algorithm for distributed energy resources in a microgrid while considering sustainability objectives (economic, social and environmental). The results show the sustainability potential of community-based microgrids that use local renewable resources. The results and the algorithm were used in applications for market scenarios in distribution systems.**

I. INTRODUCTION

Transactive energy (TES) markets are a promising approach to support energy transitions in power systems. These transactions between electric energy producers and consumers at distribution levels entail control techniques to balance electric energy generation and demand, for example among microgrids (MGs) [14]. Three stages have been proposed in order to achieve a market at the distribution level [15]. A very high integration of distributed energy resources (DERs) and a well-developed distribution platform are needed to achieve the third stage (a full market at the distribution level), as well as an optimal coordination of these resources due to their high penetration on the system. If one of the goals of a TES market is to be sustainable (i.e., balancing the economic, social and environmental dimensions) the DERs must rely heavily on renewable energy, efficiency and conservation measures.

Renewable resources are a key factor for energy transitions to cleaner alternatives, and to develop local socio-economic development. However, the integration of renewable energy to the conventional grid represents a challenge because the grid was designed to operate with controllable and constant-output generators. A significant amount of the research on renewable energy integration has dealt with these output power fluctuations [1]. Renewables are low inertia systems that can respond fast to disturbances on the system; unlike centralized generation with high inertia generators and limited ramp rates [2]. On one hand, this can be an advantage because they can respond faster to a disturbance, but on the other hand, the variabilities produced by renewables can affect the system's reliability and could cause power quality issues.

Renewable energy's intermittency will impact the utility's load factor; which is the ratio of the average energy in a determined period divided by the total peak energy that could have been used in that period [3], [4]. A high load factor can decrease the total energy production costs and can allow a higher integration of renewables in the system. Managing and coordinating storage and demand response (DR) with renewable-based, distributed energy resources (DERs), could minimize renewable-related fluctuations by compensating energy variations within a microgrid. This coordination of resources could be achieved through optimization techniques in energy management systems to enable energy transitions.

This paper presents the application of a new optimization approach to find efficient and stable configurations of DERs in community microgrid scenarios to maximize benefits, and to comply with the system constraints while considering sustainability objectives. The second part of the paper combines these optimized community microgrids to create a sustainable, transactive energy market. The same algorithm is applied to this market context to analyze sustainable energy scenarios in distribution systems in support of future transactive energy markets for community-based DERs.

II. COMMUNITY MICROGRID CASE STUDY

The Microgrids Optimal Power Flow (MOPF) developed by the authors [5], [6] is used in this paper to find efficient and stable configurations of DERs in community microgrid scenarios to maximize the economic, social and environmental benefits. This tool has the flexibility to adjust to a diversity of problems, system topologies, resource integration, system constraints and sustainability objectives. The MOPF is a combination of the General Pattern Search (GPS) optimization method and the Back/Forward Sweep algorithm and has better performance in this context than similar optimization methods [5]. GPS finds a sequence of points that approaches an optimal solution; the value of the objective function can either decrease or remain equal. At the time this work was completed, the GPS algorithm had not been used to study MGs with a very high integration of DERs, considering sustainability objectives or optimizing a dynamic system whose resources depend on time. This paper presents sample results obtained from 14 scenarios in a community-based microgrid (MG). The paper also shows for the first time the MOPF applied to an energy market.

An actual system with 200 houses was used; a subset of a residential community in Southern Puerto Rico [7]. The scenarios developed had different types and levels of DERs in a community microgrid. Twenty groups of 10 houses, all with photovoltaic (PV) systems, battery storage systems (BSS), and DR were modeled in a 45-bus system, 20 buses were residential loads. The utility delivered a contracted energy block to the 45-bus system.

There were 20 demand curves based on three profiles for buses 4 to 23 (10 residential clients on each bus) [7]. Each curve was created based on a percentage of one of the three profiles. The end-users are connected to single-phase distribution transformers. All loads have a 0.9 lagging power factor. The solar irradiation curves for southern Puerto Rico were used and discretized every 15 minutes [7]. Demand response (DR) was assumed, with different elasticities based on four categories of DR contribution: 0%, 8.33%, 16.67% and 25%. A 3-kW PV system was assumed for each end-user for a total of 30-kW aggregated PV system at each bus (10 clients per bus). The PV power output will follow the solar irradiation curves. Battery storage has a maximum capacity of 9.8-kWh and a maximum charge/discharge rate of 5 kW. The utility supplies a predefined energy amount through a 4.16 kV three phase distribution feeder and corresponding branches representing a constant demand contract for the MG for a constant demand of electric energy. The utility's rate was set at $0.20/kWh [8], PV was valued at $0.10/kWh, and storage value was set at $0.30/kWh.

III. COMMUNITY MICROGRID OPTIMIZATION RESULTS

Scenario #1 only had the utility to supply the demand (Fig. 1). The load factor obtained in this case was 43.14 %, which is a low. In scenario #4 all DERs are available in the community MG, with a sunny day and applying optimization. The algorithm found an efficient allocation of resources to supply the demand at the lowest cost possible while complying with the utility's power contract and obtaining a high load factor. Fig. 2 to 4 were obtained for scenario 4.

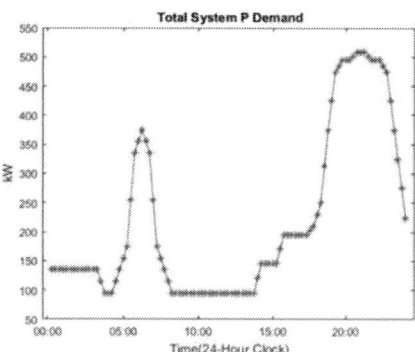

Fig. 1. Total real power demanded in each time step (Case #1).

Since all DERs are available and are being optimized within the MG, the utility only injects 90 kW (contracted) and no reactive power. Since the PVs are being optimized (managed and controlled), they do not inject their maximum energy available all the time, and the PVs real power curve does not necessarily follow the solar irradiance curve. Moreover, since the PVs can also provide and supply reactive power, the utility does not supply reactive power. The BSS curve have positive

(discharging) and negative (charging) values determined by the algorithm. The highest discharge values were obtained close to the demand peaks, and the highest charging values were obtained when the solar irradiation curve was reaching its maximum. The load factor obtained in this case was almost 100%, thus the optimization was successful.

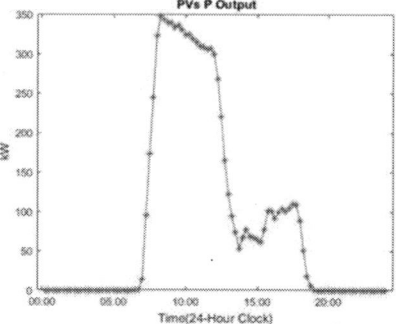

Fig. 2. PVs real power output in each time step (Case #4).

Scenario #5 is the same as #4, but for a cloudy day. The algorithm also found an efficient resource allocation and a load factor of 99%. However, even with optimization, a lower load factor could be obtained due to solar irradiation variations and less resources deployed (cases 8 and 9 showed a scenario with less resources). Nevertheless, scenarios 4 & 5 showed the potential of the MOPF because it achieved a high load factor for a sunny and a cloudy day with a high level of DERs; the algorithm managed to control the energy variations produced by the PVs using the BSS and DR resources (graphs for scenarios 4 and 5 show similar behavior).

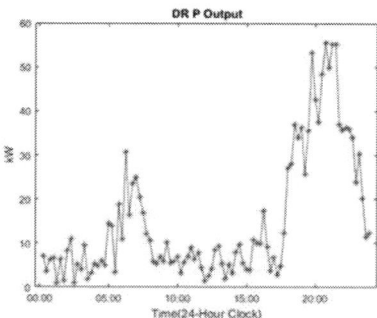

Fig. 3. Demand response applied in each time step (Case #4).

Scenario #8 was similar to scenario 4 but with less resources (not all customers had a PV system and a storage). The utility's real power injection is close to 90-kW except on the demand peaks where the injection was greater than 90-kW. The storage devices were not depleted all the time, but since there are less resources deployed, their contribution was less. Therefore, the utility was needed to cover the mismatch. This happens because the community MG does not have enough local energy resources to meet the demand, and even with optimization, a low load factor could result. For this particular scenario, the load factor was 39.97%, lower than the value obtained in case 4 (same scenario but with more energy resources). The highest power demanded from the utility was around 350 kW (approximately 72% of the energy demand at that time). If the utility cannot supply this power, or the ramp rate does not allow it, other actions are required (e.g., increased DR) to balance the generation with the energy demand. Another solution is to deploy more resources or increase the contracted power with the utility.

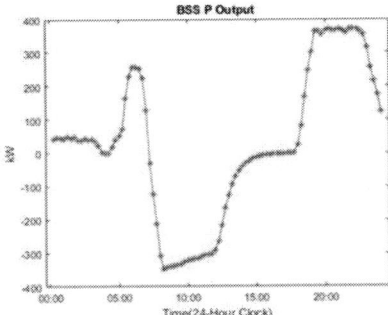

Fig. 4. BSS power output/input in each time step (Case #4).

In Fig. 6 for Case #8 almost all BSS resources were depleted (reached the minimum state of charge) when the energy demand was reaching the first peak (morning peak); therefore, the BSS contribution was less. The PVs were not injecting power at that time because the solar irradiation was zero; the same happened when the demand was reaching the second peak (afternoon peak). The only resource available at that time was DR and it could not supply the whole demand by itself (Fig. 5); thus, the utility had to supply the power mismatch. If the utility or a backup generator are not available, or the ramp rates limit their contribution, a possible solution to solve this could be to set a lower value to the minimum state of charge on the storage devices to allow them to deliver more energy and cover the demand on those times. Another option could be to set higher DR percentages and prioritize the use of this resource. Another option could be to install a higher capacity BSS.

The other scenarios are not shown due to space constraints. However, the results showed that a residential community MG could supply its energy demand, could work islanded from the utility, or can sell energy to the utility while satisfying its own energy demand if the MG has high integration of DERs, sufficient storage, and customers that are flexible in their energy use. The MOPF algorithm was able to find the best use of resources under different scenarios [16]. Even though each DER behaves differently, the MOPF managed these differences and also DER disturbances and intermittencies by balancing energy shortages with other local resources. The simulations showed that a community MG could operate in islanded mode, thus, they could be used as a resilience resource during and after natural disasters. For example, hurricane Maria destroyed or severely damaged almost half of Puerto Rico's transmission infrastructure, while destroying or

severely damaging over 75% of the distribution infrastructure. Many customers were in remote places and/or places with limited access. If MGs were available in those places, those customers would have had access to electricity. Even if the MGs receive damage during a hurricane, repairs could have lasted days or weeks, not months as was the case with the conventional grid [13]. Those MGs could have been used to supply energy to other neighborhoods or communities, while the bulk system was being repaired; a valuable social service provided by MGs.

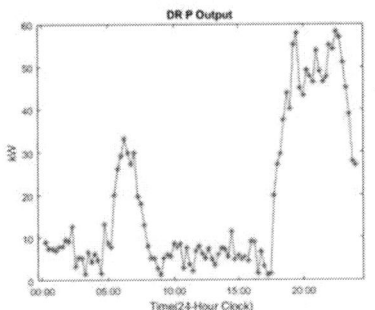

Fig. 5. Demand response applied in each time step (Case #8).

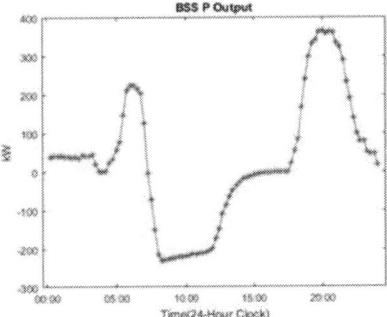

Fig. 6. Storage power output/input in each time step (Case #8).

IV. ENERGY MARKET

The MOPF was applied to optimize an energy market at the distribution level. The algorithm was used to study market scenarios with an adaptation of the community microgrid case study system previously described. The market had several community microgrids, energy buyers, energy sellers and the utility using market constraints (four market scenarios under different conditions). The utility only supplies energy if the community MGs and the sellers cannot supply the energy demand. The market scenarios had an 11-bus distribution system with five community MGs, two sellers, three buyers and the utility. The market optimization was performed for a 24-hour period with a 15-minute time step. The objective was to supply the demand using the community microgrids and the energy sellers with a low dispatch cost and comply with the system's constraints. The resources considered were fossil-fuel generators, large PV systems and community microgrids that could sell or buy electric energy.

978-1-6654-6060-6/23 $31.00 © 2023 IEEE

A. Energy market scenario #1: Seller #1 (Fossil-Gen), Seller #2 (Fossil-Gen), MGs with a sunny day.

Both sellers are 300 kVA fossil-fuel generators rated and the community MGs inject/demand energy with a sunny day irradiance curve (using the optimized results obtained in cases #4, #6, #8, #10 and #12). Seller #1 is located at 2 km and seller #2 at 0.5 km from the MGs. Both generators have a rate of $0.16 per kWh. The results showed that the resources available in the community MGs were not enough to cover the demand; the utility had to deliver energy at night. In Fig. 7 and 8 seller #2 injected more power than seller #1. This makes sense because seller #2 is closer to the MG. Seller #1 has to inject more power to supply the demand, which is more expensive and increases the dispatch cost in the market. For this scenario, the maximum power demanded from the utility was around 75 kW. Fig. 9 shows the total system's demand, including the buyers and the MGs demand/injection.

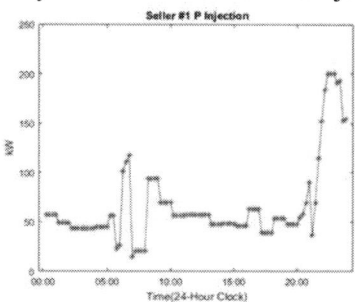

Fig. 7. Seller #1 real power injection (Case#1-energy market)

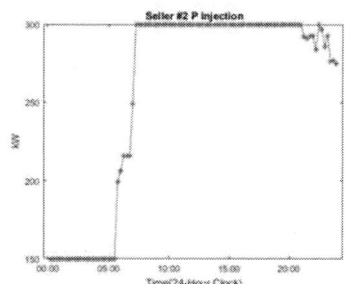

Fig. 8. Seller #2 real power injection (Case#1-energy market)

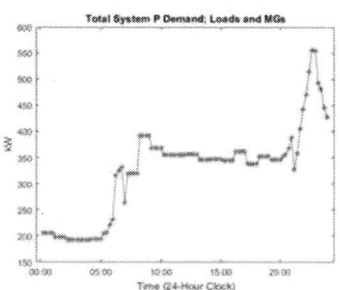

Fig. 9. Total system demand (Including loads and MGs) (Case#1- market)

B. Energy market scenario #2: Seller #1 (PV), Seller #2 (Fossil-Gen), MGs with a sunny day.

Seller #1 is a large PV system (can deliver a maximum of 300 kW and 300 kVAR but depends on the solar irradiation curve), seller #2 is a fossil-fuel generator (can deliver a maximum of 300 kW and 300 kVAR) and the MGs inject/demand energy with a sunny day irradiance curve.

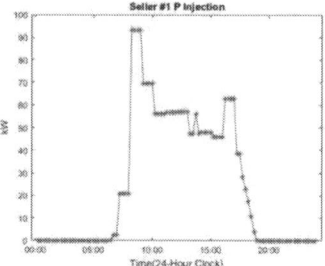

Fig. 10. Seller #1 real power injection (Case#2-energy market)

Seller #1 is located at 2 km and seller #2 at 0.5 km from the MG. The PV system has a rate of $0.10 per kWh, and the fossil generator has a rate of $0.16 per kWh. More power is being demanded from the utility because seller #1 cannot provide a constant power like seller #2. Since seller #2 is closer to the community MGs, it delivered more power that seller #1. For this case, the maximum power demanded from the utility was around 275 kW, significantly higher than in case #1 (3.67 times higher).

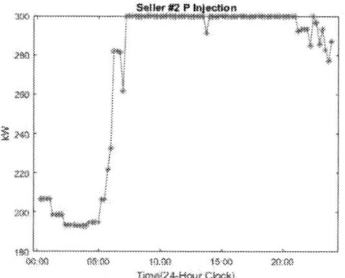

Fig. 11. Seller #2 real power injection (Case#2-energy market)

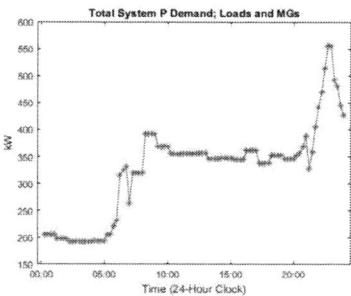

Fig. 12. Total system demand (Including loads and MGs) (Case#2-market)

C. Energy market scenarios #3 & 4 and additional discussion

Because of space limitations, this section presents a summary of the two remaining scenarios. Scenario #3 was the same as #2, but now seller #1 (Large PV system) is located at 10 km. The results are very similar to scenario #2, again, seller #2 dispatched more power because it was closer to the community MGs. Scenario #4 was the same as #3, but now MGs inject/demand energy during a cloudy day. On this case, more energy was demanded from the utility because the community MGs demand/injection and seller #1 could not produce the same amount of energy in a cloudy day, thus the utility delivered the energy mismatch to the market.

The four market scenarios were studied as initial tests to show the MOPF algorithm's flexibility to be modified and adapted to solve and optimize a particular problem [5]. The MOPF's output yields the efficient allocation of resources in each time step and thus was used to emulate energy bids in a distribution level market. Those bids already consider the community MG's constraints and sustainability objectives. If those bids are not optimized, a system constraint could be violated and cause harm to the system or the loads connected to it; an unfeasible configuration could also occur. Therefore, without optimization of bids (users with DERs) the market could fail. In these four scenarios, the market operator could use tools such as the MOPF to determine how much energy each seller would deliver, how much energy the MGs would demand or how much they would deliver, how much energy would be required from the utility and the dispatch costs of each resource. The tool could be modified to consider other objectives, penalties and constraints of the market.

V. SUSTAINABILITY IMPLICATIONS

The MOPF optimization considered the constraints and the limitations presented by the tree pillars of sustainability (economic, social and environmental aspects). Thus, a sustainable MG can be achieved through the efficient allocation of resources. Without optimization and a high integration of DERs, a sustainable MG cannot be achieved; or it can be very difficult and expensive. An important aspect the optimization was able to achieve was a reduction in the emissions produced by the utility. Fig. 13 shows the total utility emissions for each of the 14 cases studied.

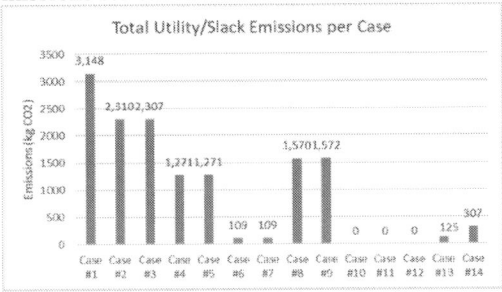

Fig. 13. Total kg CO2 Emissions per Case

The lowest emissions were obtained from the scenarios with optimization (4 to 14). The only scenarios with zero emissions were 10, 11 and 12 because they did not rely on the utility or any fossil-fuel generator. Those scenarios represent a community MG with 100% of renewable resources. DR was another important energy resource in this optimization that helped to balance the energy generation with the demand. This is why it is important to consider social aspects such as customers with different demand elasticities (different levels of willingness or capabilities to reduce their energy demand) because this will determine the amount of DR available in the system. If social considerations are not considered, DR cannot

be used properly as an energy resource. Fixing a determined DR percentage for the whole demand, as done in [7], will not be a very realistic or even feasible scenario, because every customer may not be able to contribute the same amount of DR. On the scenarios studied, DR was used when needed to balance the energy generation with the demand and it was mostly used during demand peaks.

Fig. 14 shows the proposed structure to study sustainable microgrids and markets that was used for this paper. It has three optimization blocks. The first one is for load optimization where the application of evolutionary algorithms for load forecasting can be applied. The output of this optimization block will be the maximum DR percentage that an end-user is willing to contribute; other social and environmental aspects could be added to this block. This data is one of the inputs for the DERs optimization block since DR is considered as an energy resource. The second block is for DERs optimization and its details, tests and results were already discussed in this paper. The third one is for market optimization where an auction energy system and a brokerage system can be implemented to support a market at the distribution level. Social and environmental aspects could also be considered in this block. The optimal allocation of the resources and the energy demanded by each end-customer will be the inputs used of this block since these are the energy bids issued by producers and consumers.

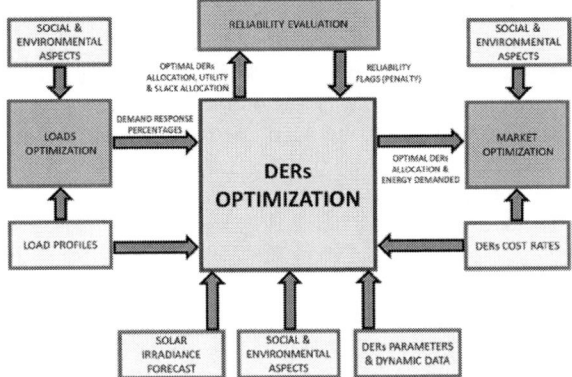

Fig. 14. Proposed structure for sustainable microgrids

There is also a reliability block, where the output obtained in the DERs optimization block can be used to evaluate reliability indices. If there is a reliability violation to the system, this block can raise a flag or sent a penalty signal to the DERs optimization block to let it know there is a reliability concern in the system and it needs to find another solution to the problem. In the future, other blocks with dynamic data such as the solar irradiance forecast, as well as other data that might affect the optimization blocks could be added.

VI. CONCLUSIONS

This paper presented the application of a new, microgrid optimal power flow (MOPF) approach to address and optimize problems related to microgrids and transactive energy markets.

The MOPF is a sustainable microgrid analysis and design framework that uses a multi-objective optimization approach. The MOPF was used to study energy allocation scenarios for the achievement of sustainability and self-reliance in a community microgrid case study with 200 houses, each with a PV system. The energy resources optimized also included the amount of non-renewable energy (from the utility), distributed storage, and consumption reduction strategies (demand response). To maximize the benefits provided by DERs in the microgrid, the best use of the community microgrid resources was found for each time step. The MOPF output was the efficient allocation of the available resources complying with the system's constraints, accounting for economic, social and environmental considerations, and with the best objective function value found in the search space (either local or global solution). The results showed that the MOFP can optimize resources for both sunny and cloudy days while obtaining a high load factor in a community MG with a high integration of DERs, increasing thusly the value of microgrids and their services. A high load factor ensures that the utility would not see major variations from the integration of renewable energy.

The GPS algorithm, used as the foundation for the MOPF, was able to efficiently prune unstable or unfeasible solutions out from the search space and produce good values in the process of balancing supply and demand in the MG and in the energy market. At the time of this work, this was the first instance of the use of the GPS algorithm in this context [16]. The simulations also showed that a community MG could operate in islanded mode, thus, they could be used as a resilience resource during and after natural disasters. This resilience feature should be accounted for when valuing and comparing community-based MGs with conventional power schemes. The potential to provide power to nearby neighborhoods or communities is another valuable social service that would improve the results of sustainability assessments. Furthermore, the MOPF has the flexibility to be modified and enhanced to better address user needs and future problems. For example, the framework could be extended to study the interaction of transactive energy markets with energy markets at the sub-transmission and transmission levels.

Stable and optimized community microgrid configurations found with the MOPF were used to study a transactive energy market. The MOPF was adapted to a market context to analyze sustainable energy scenarios. Since each microgrid was optimized as a sustainable microgrid, the market starts as a sustainable energy market and continues being sustainable since the market optimization has sustainability objectives as well. Four market scenarios were studied at the distribution level with sellers, buyers and community MGs. The MOPF's output was used to emulate energy bids that considered system constraints and sustainability objectives. An operator could use tools such as the MOPF to manage the market. The tool could also be modified to consider other scenarios, objectives, penalties and constraints of the market in support of existing or future community-based energy markets at the distribution level.

ACKNOWLEDGEMENT

This work was supported in part by the DOE under Grant DE-SC0020281.

REFERENCES

[1] J. Mossoba, M. Kromer, P. Faill, S. Katz, B. Borowy, S. Nichols, and L. Casey, "Analysis of Solar Irradiance Intermittency Mitigation Using Constant DC Voltage PV and EV Battery Storage," *IEEE Transp. Electrif. Conf. Expo*, 2012.

[2] N. Modi, "Low Inertia Power Systems: Frequency Response Challenges and a Possible Solution," *Australas. Univ. Power Eng. Conf.*, 2016.

[3] Energy Sentry., "Load Factor Calculations." [Online]. Available: http://energysentry.com/newsletters/load-factor-calculations.php

[4] "What Your Electrical Load Factor is Telling You," 2014. [Online]. Available at: https://efficiencyinnovascotia.wordpress.com/2014/02/14/what-your-electrical-load-factor-is-telling-you/

[5] R. A. Reyes Colón and E. O'Neill-Carrillo, "A General Pattern Search Algorithm to Optimize Microgrids," *9th Int. Symp. Power Electron. Distrib. Gener. Syst. (PEDG 2018)*, 2018.

[6] R. A. Reyes Colón and E. O'Neill Carrillo, "Optimal Use of Distributed Resources to Control Energy Variances in Microgrids," *45th IEEE PVSC*, Waikoloa, HI, 2018, pp. 1471-1476.

[7] I. Jordan, "Towards a Zero Net Energy Community Microgrid," 2017.

[8] AEE, "Estructura Tarifaria Autoridad de Energía Eléctrica." [Online]. Available: https://www.aeepr.com/Documentos/Ley57/Presentacion Tarifas para Pagina AEE Internet - 05-19-2015.pdf

[9] "PIKA ENERGY Harbor Plus datasheet." [Online]. https://www.pika-energy.com/files/datasheets/pika_harbor_plus_datasheet.pdf.

[10] Ultralife, "Li-Ion vs. Lead Acid," 2007. [Online]. http://www.beck-elektronik.de/uploads/media/lithium-ion-vs-lead-acid.pdf.

[11] G. Albright, J. Edie, and S. Al-Hallaj, "A Comparison of Lead Acid to Lithium-ion in Stationary Storage Applications," 2012.

[12] H. Ullah, S. Chalise, and R. Tonkoski, "Feasibility Study of Energy Storage Technologies for Remote Microgrid 's Energy Management Systems," *Symp. Power Electron. Electr. Drives, Autom. Motion*, 2016.

[13] FEMA, "Statistics Progress in Puerto Rico." [Online]. Available: https://www.fema.gov/hurricane-maria

[14] P. De Martini, R. Ambrosio, and E. Gunther, "Transactive Energy: GridWise Architecture Council Foundational Session," 2012.

[15] D. Sciano, "Distributed Resource Integration." [Online]. http://www.nyiso.com/public/webdocs/markets_operations/committees/environmental_a%09dvisory_council/meeting_materials/2016-05-06/Con_Ed_EAC_May_6_2016.pdf

[16] R. Reyes-Colón, "Optimal use of distributed energy resources in microgrids," Master's Thesis, Electrical and Computer Engineering Department, UPRM, 2018.

Radioisotope Thermoradiative Cell Power Generator

Stephen J. Polly, Geoffrey A. Landis, Seth M. Hubbard

Rochester Institute of Technology, Rochester, NY, United States

NASA John Glenn Research Center, Cleveland, OH, United States

The thermoradiative cell (TRC) is theoretically investigated, coupled with a radioisotope heat source, to provide primary power for a cubesat mission where photovoltaic power is impractical. Using a representative heat source, such as a single pellet of 238Pu from a General Purpose Heat Source (GPHS), provides 62.5 W of thermal energy. Using the 3K universe as a heat sink and a nominal power conversion efficiency of over 19%, a mass specific power of 30 W/kg is predicted. This represents a order-of-magnitude improvement over conventional radioisotope thermal generators. Realization of practical TRC devices by Metalorganic Vapor Phase Epitaxy (MOVPE), and development of identified low-bandgap materials such as $InAs_{0.91}Sb_{0.09}$ (0.28 eV) or $InP_{0.63}Sb_{0.37}$ (0.47 eV), is discussed.

Robust Detection Method of Low-Voltage Islanding for Grid-Forming Inverters Operated in Conjunction with Existing PV Inverters

Björn Oliver Winter*, Julian Schwung*, and Bernd Engel*

*elenia Institute for High Voltage Technology and Power Systems, TU Braunschweig, 38106 Germany
bjoern.winter@tu-braunschweig.de, ORCID: https://orcid.org/0000-0001-6860-3930

I. INTRODUCTION

In the distribution grid, raising the share of PV distributed energy resources (DER) increases the chance of unintended islanding to occur. This is a condition, where a subsection of the utility grid continues to operate after being disconnected from its point of coupling, such as during a grid fault, for an extended duration. It is suspected that the use of grid-forming inverters (GFI) for PV and BESS, that are discussed as future source of distributed grid inertia [1, 2], without appropriate countermeasures, further significantly increases the chance of stable unintended islanding [3]. The standard paradigm of islanding in low-voltage distribution grid sections would usually stipulate the cessation of operation of the island and the usual grid-supporing inverters (GSI) to be equipped with a set of mechanisms [4] to detect islanding and shutdown the island by ceasing operation to mitigate damage of equipment or accidental asynchronous reconnection of the island. Usually, this would encompass passive detection mechanisms, in which the inverter would trip, following measurements of abnormal voltage parameters, such as a high amount of harmonics, asymmetry, or a high rate of change of frequency (RoCoF), that would occur as a result from loss-of-mains connection. Furthermore, an inverter would possess an active islanding detection (AID) mechanism by which the inverter considered is able to actively destabilize an island, if passive islanding detection should fail, while minimizing disadvantageous effects during normal mains connected operation. The functionality of any new AID method has to be verified following a testing procedure [5, 6] with country-specific differences [7]. A common setup in all these procedures however is the operation of a GSI as device-under-test (DUT) on which the proposed AID procedure is installed in conjunction with a resonant circuit consisting of a resistive as well as an inductive and capacitive component (RLC-load) as depicted (green area only) in the following Fig. 1.

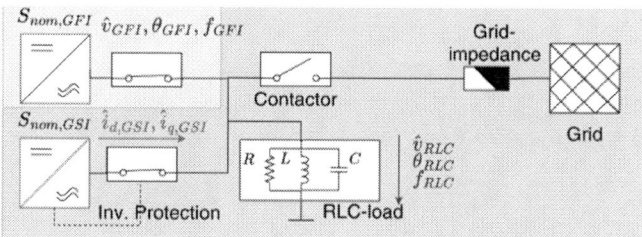

Fig. 1. Test setup for Islanding detection with a resonant load.

In this setup, the resonant circuit is adjusted so that it consumes the active power feed-in of the inverter up to a defined difference ΔP, which then flows through the grid connection. Furthermore, the inductive and capacitive components of the resonant circuit are adjusted in such a way that the reactive power consumed by them is offset, except for an adjustable reactive power difference ΔQ. This difference also flows via the grid connection, since the inverter initially does not feed in any reactive power. After settlement, the switch to the grid connection is opened, whereupon the DUT must destabilize the island and switch itself off, usually as a result of an actively induced exceedance of the voltage limits (frequency or amplitude) in the island. This represents a significant challenge for the inverter, as the GSI itself only controls its output current and the network voltage to be destabilized is consequently formed above the resonant circuit. The most difficult case for the inverter is the balanced combination $\Delta P, \Delta Q = 0$, since here, the network contactor is completely unloaded before opening and a islanding happens without changes in voltage properties that would usually go hand in hand with the sudden cessation of the power flow across the grid contactor.

One of the most common methods for active islanding detection is the Slip-Mode Frequency Shift (SMS) method, by means of which the GSI adjusts its power angle θ_{SMS} in dependence on the measured frequency f and the nominal frequency f_g according to the following equation [8]:

$$\theta_{SMS} = \frac{\pi}{180°} \cdot \theta_m \cdot \sin\left(\frac{\pi}{2} \cdot \frac{f-f_g}{f_m-f_g}\right) = \operatorname{atan}\left(\frac{Q_{SMS,inv}}{S}\right) \quad (1)$$

Where the parameter f_m is the frequency at which the maximum set phase shift θ_m occurs. The power balance in the islanded setup mandates the RLC-load with a given quality q_f and resonance frequency f_{res} to consume all the reactive power fed by the inverter, which shifts the system frequency f by [8]:

$$Q_{SMS,Inv} \stackrel{!}{=} Q_{RLC} = P_{RLC} \cdot q_f \cdot \left(\frac{f_{res}}{f} - \frac{f}{f_{res}}\right) \quad (2)$$

Successful islanding detection by destabilization is possible if the equations do not yield a common operation point for (Q,f) that is stable. This mutual dependency is plotted in Fig. 2:

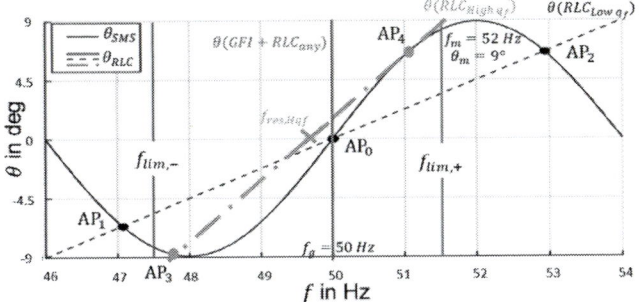

Fig. 2. Phase angle and frequency relationship in the islanded grid.

The plotted black sine-curve is the SMS curve of the inverter, in this case parameterized by $(f_m = 52\ Hz, \theta_m = 9°)$. The dotted black line is the frequency behavior of an RLC-load. As shown in equation (2), the load q_f has influence on the local gradient, its f_{res} on the intersection with the $\theta = 0$ axis. Possible operating points of the island are only feasible in the intersections between the curves. For any operating point OP to be stable (and thus islanding detection to possibly fail) however, the local gradient of the load curve must be higher than that of the islanding detection curve [8]:

$$\left.\left|\frac{d\theta_{RLC}}{df}\right|\right|_{f=f_{OP}} < \left.\left|\frac{d\theta_{SMS}}{df}\right|\right|_{f=f_{OP}} \qquad (3)$$

If equation (3) is fulfilled, small disturbances of the local θ/f lead the devices back to the original operating point which stabilizes the islanded grid. Here, OP_0 violates equation (3) and is unstable. For low q_f-loads, the curve is designed such that any stable OP (here: $OP_{1,2}$) can only be found outside of the allowed frequency limits for the grid, $f_{lim,\pm}$, which causes the system frequency to necessarily drift outside the allowed boundaries and thus the inverter's grid-and-plant protection (GPP) to trip and disconnect the device. Loads with very high quality q_f (orange dot-dashed line) can however stabilize (OP_3) or semi-stabilize (OP_4) operation within the allowed frequency bounds, which would cause islanding detection to fail. While a more aggressive parametrization of the SMS curve can mitigate this, this comes at the expense of undesired behavior in regular grid-connected operation, so a tradeoff between AID effectiveness and favorable mains connected operation will have to be made.

As shown, the load parameters quality factor q_f and resonance frequency f_{res} determine the success of the islanding detection and are thus important factors when the effectiveness of an active frequency shift AID procedure is measured.

II. Experimental/Theoretical Results

The setup from Fig. 1 was modeled in Simulink with the parameters from
Table I with varying q_f and f_{res} in a simulation series. The result is displayed in Fig. 3.

TABLE I
SUMMARY OF MODEL PARAMETERS

Parameter	Value
Voltage (V_{nom},f_{nom})	230 V$_{rms}$, 50 Hz
GFI (S_{nom})	10 kVA
GSI (S_{nom}, f_m θ_{SMS})	10 kVA, 52 Hz, 9 deg
Static grid support	Q(V),P(f), default parameters [9]
Fault-Ride-Through (FRT)	Deactivated, in acc. with [7]
RLC-load (q_f,f_{res})	(1e-1:10e1, 47:0.5:52 Hz)

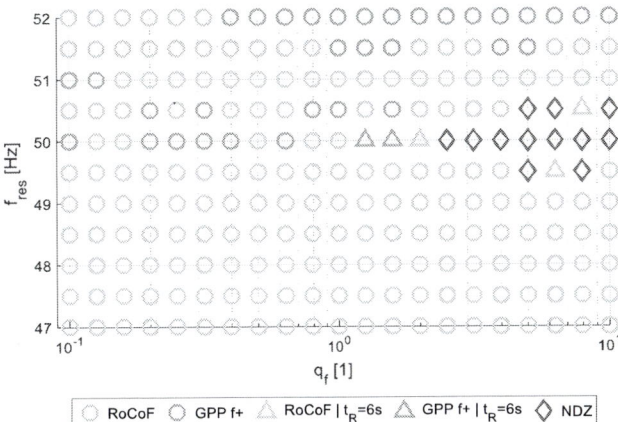

Fig. 3. Non-Detection Zone of the GFI test with SMS-Procedure.

Each one of the symbols in Fig. 3 corresponds to a simulation run, in which the RLC-load was tuned to exactly compensate the active power feed-in of the inverter while matching the desired load parameters. The color indicates the type of violation that occurred, as a result of which the inverter stops operating. A circle denotes the successful ID within 3 s after opening the contactor, a triangle shows successful destabilization within an extended time of 6 s. In most cases, the passive ID was able to trip the inverter due to an abnormal RoCoF detection. Other simulations were driven out of the upper frequency bound by the SMS AID. In the simulation cases with a red rhomb shape, for loads with higher q_f and $f_{res} \approx f_g$, the inverter did not cease operation within that allotted time. The sum of all failed detection runs is named the Non-Detection-Zone (NDZ). The general form and place of the NDZ correspond to the representations in literature [10].

When a GFI is added to the setup (yellow area in Fig. 1), the voltage amplitude and frequency of the islanded grid are now also actively controlled. The typical static control concept of a GFI is geared towards maintaining the nominal grid frequency and amplitude via use of an internal voltage phasor, and superposes these set points with controllers that cause a slight deviration in the set grid frequency when active power demand changes (P-f-dependency, active power control), and of the amplitude due to a shifted reactive power demand (Q-V-dependency, reactive power control) from other nearby units to ensure stable parallel operation of several GFI in the grid [11]. The most simple implementation is in form of droops, with cross-couplings induced between these to enhance stable power output in the low-voltage grid with high R-X grid impedance ratios. As a voltage source, however, the GFI actively stabilizes the voltage amplitude and frequency so that any attempt of the SMS-Proceduce to shift grid frequency by adjusting the GSI reactive power is largely compensated for by the GFI. To highlight this, the GSI in Fig. 1 can be replaced with a controlled current source to perform a sweep of θ. The result for the system frequency set by the GFI and load, quite independent from the load parameters, is the almost vertical red line in Fig. 2, which fulfils equation (3) at every intersection an thus stabilizes the island over a broad load parameter range despite the SMS procedure being active.

The result is a drastic increase of the NDZ in the simulation series when repeated with a GFI present, which is displayed in Fig. 4.

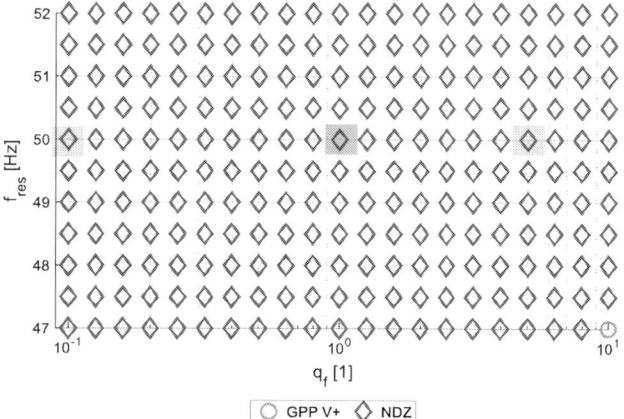

Fig. 4. Non-Detection Zone of the test with parallel GFI without ID

Grid simulation studies, that evaluated the effect of adding a GFI without ID next to several existing GSI with SMS in a low-voltage grid section, came to a similar result [12]. This motivates the development of own detection methods for GFI, since their use without these would most likely greatly increase the unwanted formation of islands in the grid. Generally, such methods must be oriented towards the characteristics of GFI, with its control being centered around the voltage parameters f, û as controlled output variables instead of the current.

The first thing to note is that, in the experimental setup above, the GFI is the only unit that actively controls the voltage, and so any number of different approaches can be defined that actively directly drive it outside the limits once the contactor is opened to pass the standard testing procedure [3]. However, this does not necessarily mean that this method would ensure reliable islanding detection in real grids, especially in parallel operation with other GFI. Thus, methods must be used that explicitly enable successful detection even in combined operation. In this paper, we propose the introduction of an artificial Q-f-dependency on the GFI inspired by the Japan Standard Islanding Detection (JSID) procedure [7], which is

implemented as a droop. This can in principle be done either by means of a -Q(f)-droop or an -f(Q)-droop, or a combination, which have different effects:

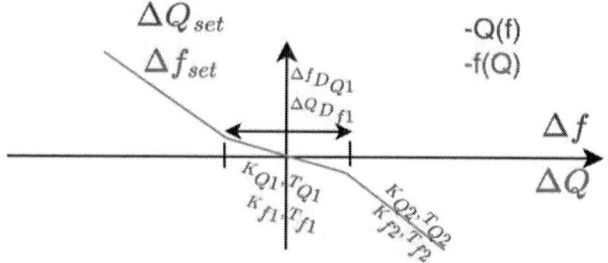

Fig. 5. -Q(f)/-f(Q)-Droop parameters for GFI islanding detection.

The proposed droops are implemented in form of positive feedback loops that serve to actively destabilize any islanded grid section where a notable q-f-dependency noted in formula (1) by a GSI with a corresponding procedure or (2) with a load are given. In our implementation, the -Q(f) droop influences the set point Q_{set} of the reactive power controller in reaction to a deviation of the internal frequency phasor $\Delta f = f_{meas} - f_g$ with a defined dynamics T_{f1} inside a narrow frequency band $\Delta f, D_{Q1}$ and optionally with a higher dynamics T_{Q2} in an outer frequency band, to enhance detection speed:

$$Q_{set,Qf} = Q_{set,0} - \left(\frac{1}{T_1 s+1} * K_{Q1} * \Delta f \right)_{|\Delta f| \leq \Delta f_{D_{Q1}}} \quad (4)$$

$$- \left(K_{Q1} * \Delta f_{D_{Q1}} + \frac{1}{T_2 s+1} * K_{Q2} * \left(\Delta f - \Delta f_{D_{Q1}} \right) \right)_{|\Delta f| > \Delta f_{D_{Q1}}}$$

In mains connected operation, the proposed -Q(f) droop has not shown destablizing behavior so far in simulation. Given the fact that GSI procedures such as the JSID or the SMS procedure work in a compatible fashion, this can be reasonably expected. Any inductive power generated as a result of frequency deviations would be transported to the medium voltage level. At very high penetrations of GFI with the droop enabled, the risk of voltage instability by a resulting cumulative reactive power flow can be mitigated by restricting K_1, K_2 to magnitudes similar to the SMS procedure of the GSI, and weighting effectiveness against stable mains operation. If however, the procedure would be used on higher voltage levels with larger inverters, further studies on the parametrization and adverse effects are required. In our implementation, $Q_{set,Qf}$ was limited to $\pm 0.8 * S_{nom}$, further tuning of this value might enhance the reliability at possible cost of decreased room for static grid support.

The -f(Q)-droop is a more direct droop, acting on the frequency controller in the lower level controls in GFI. As the name suggests, it induces a slight shift in internal target frequency in reaction to a reactive power draw from the grid. The negative operator illustrates the desired positive feedback loop, which is closed in an islanded grid only.

$$f_{set,Qf} = f_{set,0} - \left(\frac{1}{T_1 s+1} * K_{f1} * \Delta Q\right)_{|\Delta Q| \leq \Delta Q_{D_{f1}}} \quad (5)$$

$$- \left(K_{f1} * \Delta Q_{D_{f1}} + \frac{1}{T_2 s+1} * K_{f2} * \left(\Delta Q - \Delta Q_{D_{f1}}\right)\right)_{|\Delta Q| > \Delta Q_{D_{f1}}}$$

The -f(Q)-droop can act with more leverage, influencing directly the system frequency. However, the design of the droop parameters should be done carefully, since executing it in massive scales on distributed inverter fleets might have the potential of influencing frequency stability in mains connected operation as well. A feasible approach would be the implementation of a q-f-sensitivity of a magnitude similar to that of a load with low quality from equation (2)(3). This implementation would be intuitive since it would recreate the load behaviour on the GFI, thus re-enabling the requirements and leverages for the known frequency-shift procedures on the existing grid-supporting plants, the principle of which is qualitatively depicted on Fig. 6:

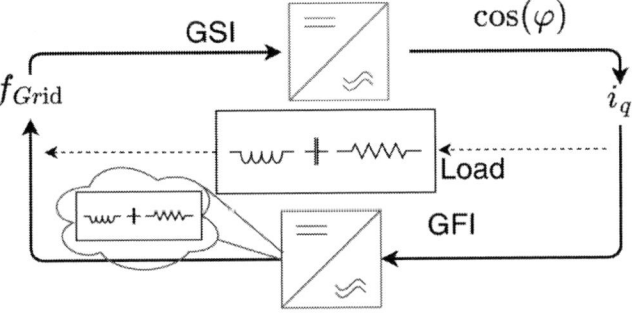

Fig. 6. Recreation of requirements for frequency-shift procedures by the -f(Q)-droop in the GFI.

However, the -f(Q)-droop and potential adverse effects on the grid are still under study. Due to this, we simulated the GFI with only the -Q(f)-droop activated in the following simulations with a first starting parametrization from Table II:

TABLE II
SUMMARY OF -Q(f)-DROOP PARAMETRIZATION

Parameter	Value
K_{Q1}, K_{Q2}	62 kvar/Hz
$Q_{set,Qf,limit}$	0.8* S_{nom}
T_1, T_2	0.33 s

Since the influence of a new inverter technology on the basic assumptions of established procedures is deemed the root cause of the issue of the observed deteoration of islanding effectiveness, we pay special attention to the interaction of different inverters with the proposed AID procedure. Here, the effect of the proposed droop is twofold. If the GFI with new AID is operated together with a state-of-the-art GSI as in Fig. 1, the GSI can now act on the grid parameters held by the grid former again by shifting its reactive power demand using the SMS method and thus destabilize the island. With favorable parameterization of the proposed droop, this even happens

more reliably in simulation than in the reference case, and an NDZ can again be almost completely avoided as shown in Fig. 7:

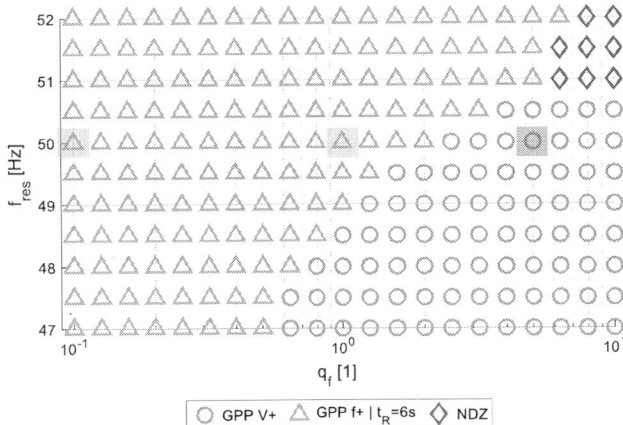

Fig. 7. Non-Detection Zone of the GSI and GFI with corresponding AID

As depicted, islanding detection can again be performed successfully in almost any region of the depicted load parameter space (q_f, f_{res}). For low quality loads or high resonance frequencies, the inverters are tripped by destabilization into over-frequency due to the feedback of the GSI SMS detection method. Islanding scenarios set with higher quality loads are tripped due to over-voltage destabilization, For combinations of very high quality loads with high resonance frequencies, the islanding cannot be ceased within the given 6 s. As a result, the NDZ of this setup is shifted in the higher frequency range. Notably, passive RoCoF detection does not take place anymore, since Grid-Forming inverters strongly act to dampen any highly dynamic voltage deviations that would happen during the islanding event. Overall, islanding detection in combined setups can again be performed as reliably as in GSI-only setups in the model, with the NDZ being diminished a little in direct comparison to Fig. 3 for the parametrization tested so far.

In future grid sections, especially for example in newly developed residential areas, the density of GFI might be much higher than of GSI. Therefore, a reasonable requirement for newly developed AID procedures should be that these should work reliably in GFI-only setups. Therefore, we repeated the simulation with the setup from Fig. 1, but replaced the GSI with a second GFI with the same AID procedure. The result is shown in Fig. 8:

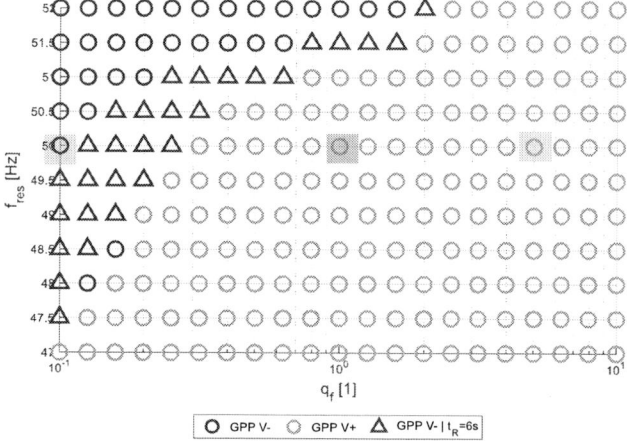

Fig. 8. Non-Detection zone of two parallel GFI with the proposed AID.

As seen here, the proposed -Q(f)-droop leads to a highly reliable islanding detection by destabilization of the voltage amplitude in parallel GFI setups in the simulation. The detection is possible over the whole load parameter space (LPS), and the under-/overvoltage-induced tripping is plausible, since the reactive power controller of GFIs is mainly acting on the inverter's set voltage amplitude.

III. LAB VALIDATION

We further verified the results of the simulation-based analysis in the laboratory. For this purpose, the inverter controllers were recreated in PHIL setups, each consisting of a real-time simulator with live measurements of the grid parameters and a corresponding amplifier, which, depending on the inverter type, is either operated in voltage- or in current controlled mode. The setup is further completed by a hardware RLC-load with physical load units that can be freely parameterized in the LPS. The setup is connected to an external grid via a contactor as shown in Fig. 9:

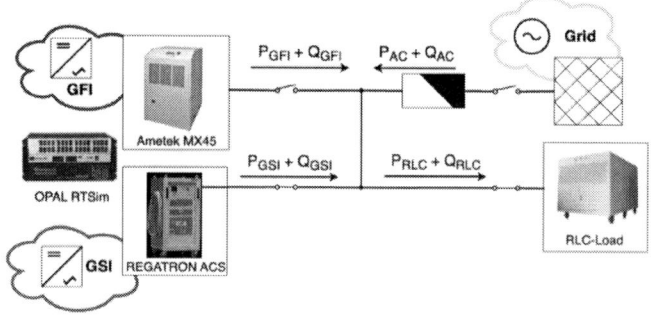

Fig. 9. Setup for lab-based verification of the islanding tests.

In this setup, we have validated individual runs of the simulation at the current time.

Firstly, the islanding by the GSI in the presence of a GFI without any AID method was run in the combinations highlighted in green in Fig. 4, and islanding could not be successfully detected by the GSI in any of the cases considered. The following Fig. 10 shows the run highlighted in blue in Fig. 4 in simulation, while Fig. 11 shows the corresponding laboratory trial.

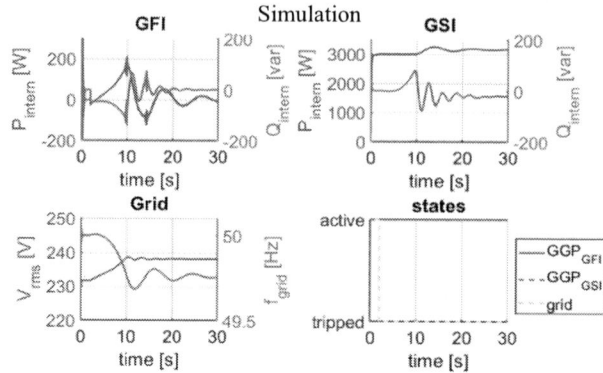

Fig. 10. Non-Detection of islanding in the presence of a GFI without AID, Simulation run, exemplification of the run in the blue area in Fig. 4.

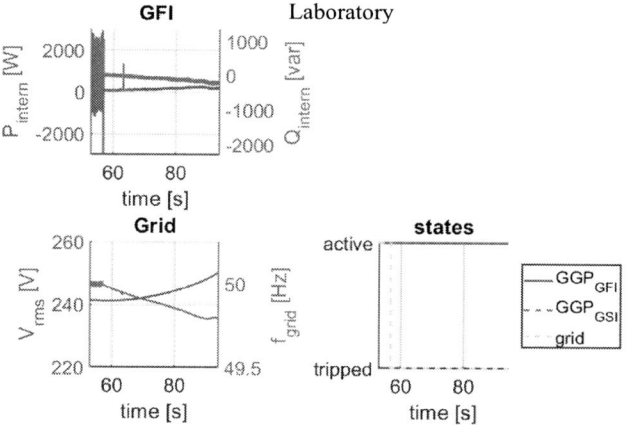

Fig. 11. Non-Detection of islanding in the presence of a GFI without AID, Lab trial run, exemplification of the run in the blue area in Fig. 4.

The upper graphs in the figures show the active and reactive power output of each inverter type in the setup. As can be noted, the power output of the GFI is largely deviating in opposite direction to the GSI output, the inverters are thus offsetting each other. The GSI power output in the lab is not available at the current time due to a recording error and will be recreated from raw voltage and current measurements. The lower right graph shows the grid and inverter states. The green line shows the moment at which the islanding is taking place. As can be seen from the lower left graph, the grid states frequency and amplitude do not deviate enough from the nominal values to induce inverter tripping, the islanding detection thus failed.

978-1-6654-6060-6/23 $31.00 © 2023 IEEE

If, on the other hand, the method described above is executed on the GFI, island grid detection in parallel operation with a GSI using the SMS method was successfully run in the lab. The parameter setups highlighted in green in Fig. 7 were validated with results corresponding to the simulation prediction. The following figures exemplify one of the verification runs in simulation (Fig. 12) and in the laboratory (Fig. 13).

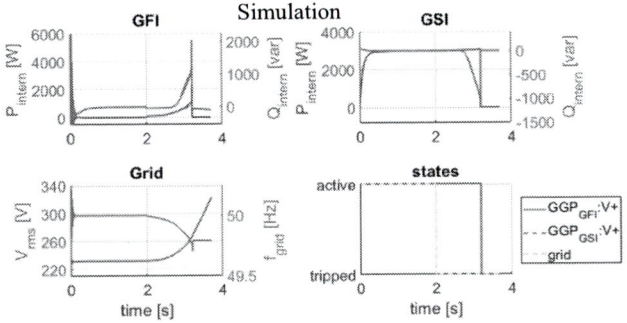

Fig. 12. Simulation run for blue area in Fig. 7, detection successful 1.2 s after islanding at t = 2 s.

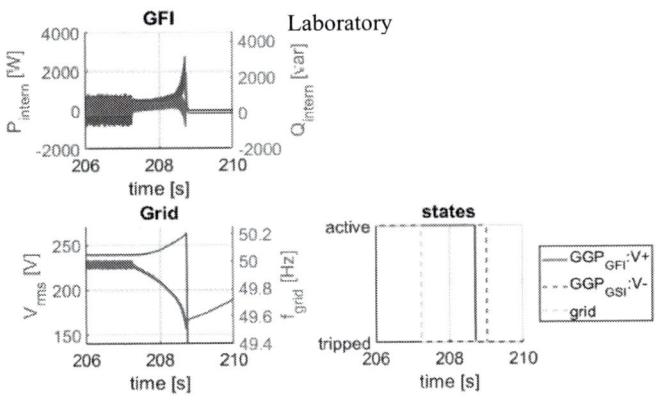

Fig. 13. Lab trial run for blue area in Fig. 7, detection successful 1.48 s after islanding at t = 207.24 s.

In the figures, the power feed-in from the inverters is qualitatively similar. As a result of the destabilization, reactive power is exchanged between the inverters, which causes the system voltage and frequency to be destabilized. It can be noted that at the time of measurement in the lab, the initial external grid voltage before islanding was a little over nominal levels, at about 238 V_{rms}, which might have accelerated the voltage induced tripping. The lower right graph shows the time and cause of inverter tripping, which ceased the islanded operation.

Lastly, the parallel operation of two GFI with AID was recreated in the lab. Here, the islanding could be detected and terminated within the available time in all verification runs highlighted in green in Fig. 8. The run highlighted in blue is

displayed in Fig. 14 in simulation and in Fig. 15 in the lab setup.

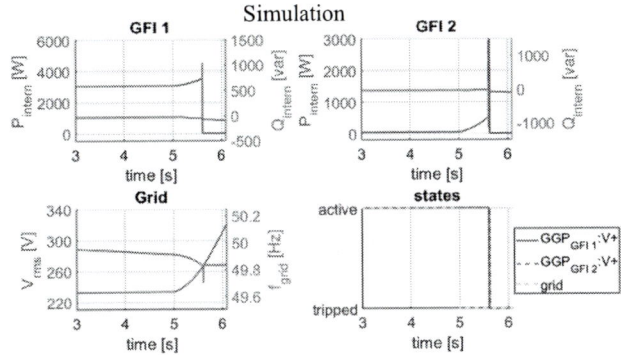

Fig. 14. Simulation run with two GFI, blue area in Fig. 8, detection successful 0.61 s after islanding at t=5 s.

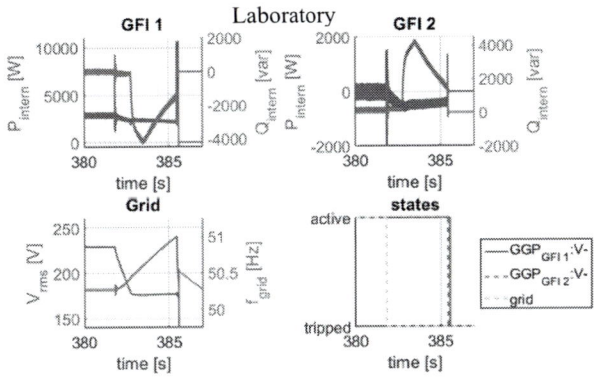

Fig. 15. Lab trial run with two GFI, blue area in Fig. 8 detection succesful 3.65 s after islanding at t=381.8 s

IV. DISCUSSION

The need for future inertia replacement mandates increased shares of DER to be operated with GFI in the near future [1]. However, as shown in this paper, if this inverter type would be utilized in the distribution grid without proper AID procedures, a stark increase of unintended islands due to existing procedures on installed inverters being rendered ineffective would need to be expected on the basis of the conducted considerations and simulations, which highlights the importance of developing backwards compatible islanding-detection methods for GFI.

The proposed detection method is an adaption of select JSID procedures [7] to the special characteristics of grid-forming control and seems to enable proper detection especially in combined operation cases with other inverters being present nearby in simulation. The lab trials conducted so far validate the simulation results.

Furthermore, the results show that the currently used islanding test procedures are lacking information value. Since the test is solemnly looking at a single inverter operated at a passive load, the AID procedures are optimized for this very setup. As seen in the simulations however, good performance of current and future inverters in this test might not accurately depict the effectiveness of its AID methods when connected to other inverters or even real and more complex grid infrastructure. The results therefore make a case for looking more closely at the effectiveness of a detection procedure in combined operation with other inverters, with grid-supporting and grid-forming control in future testing procedures.

V. SUMMARY

Based on a depiction of a state-of-the art active islanding (AID) detection procedure for a grid-supporting inverter (GSI) and its common test procedures, this paper looks at the effects of installing a grid-forming inverter (GFI) in its grid section. The results show that the considered existing GSI procedure largely looses effectiveness with the sheer presence of a GFI. This motivates the necessity for the exploration of underlying causes in the paper as well as developing adapted detection procedures for GFI that will on the one hand work reliably with existing infrastructure as well as in conjunction with other GFI. We introduce one such procedure with the -Q(f)-/-f(Q)-droop. We show that operating a GFI with that procedure in conjunction with the existing infrastructure enables for an even more robust islanding detection than before. The reliability further increases in setups where only GFI are used. The results are validated by lab trials in which the suggested procedure is also able to detect islanding in combined operation setups of several inverters.The results further show, that the current AID evaluation setup, consisting of a single inverter at an RLC-load might not accurately depict the issues faced by combined operation of several inverters in real grid infrastructure, and that combined operation setups should be more closely looked at when evaluating AID procedure effectiveness.

VI. ACKNOWLEDGEMENTS

We acknowledge the support of our work by the German Ministry for Economic Affairs and Climate Action and the Projektträger Jülich within the project "VN_2030plus - Safe and stable operation of the inverter-dominated distribution grid" (FKZ 03EI4067B). Only the authors are responsible for the content of this publication. This paper does not necessarily reflect the consolidated opinion of the project consortium "VN_2030plus".

VII. REFERENCES

[1] R. H. Lasseter, Z. Chen, and D. Pattabiraman, "Grid-Forming Inverters: A Critical Asset for the Power Grid," *IEEE J. Emerg. Sel. Topics Power Electron.*, vol. 8, no. 2, pp. 925–935, 2020, doi: 10.1109/JESTPE.2019.2959271.

[2] Y. Lin *et al.,* "Research Roadmap on Grid-Forming Inverters," National Renewable Energy Laboratory,

Golden, CO, 2020. [Online]. Available: https://t1p.de/nrel-GFRmap

[3] B. O Winter, F. Rauscher, and B. Engel, "Islanding Dependencies and Detection in low-Voltage grids with grid forming Inverters," in *The 9th Renewable Power Generation Conference (RPG Dublin Online 2021)*, 2021, pp. 25–30.

[4] S. Dutta, P. K. Sadhu, M. Jaya Bharata Reddy, and D. K. Mohanta, "Shifting of research trends in islanding detection method - a comprehensive survey," *Prot Control Mod Power Syst*, vol. 3, no. 1, 2018, doi: 10.1186/s41601-017-0075-8.

[5] *DIN EN 62116:2014-11. Utility-interconnected photovoltaic inverters - Test procedure of islanding prevention measures: Utility-interconnected photovoltaic inverters - Test procedure of islanding prevention measures*, IEC 62116:2014, German version EN 62116:2014.

[6] *JET GR 0003-4-1.0. Electrical Safety &Environment Technology Laboratories, Individual Test Method of Grid-connected Protective Equipment etc. for Multi-unit Grid-connected PV Power Generating Systems*, JET GR 0003-4-1.0 (2011).

[7] A. Ellis, S. Gonzalez, M. Ropp, D. Schutz, Y. Miyamoto, and T. Sato, "Comparative Analysis of Anti-Islanding Requirements and Test Procedures in the United States and Japan," 39th photovoltaic Specialists Conference (PVSC), pp. 3134–3140, 2013. [Online]. Available: https://t1p.de/P-AID

[8] L.A.C. Lopes and H. Sun, "Performance Assessment of Active Frequency Drifting Islanding Detection Methods," *IEEE Trans. On Energy Conversion*, vol. 21, no. 1, pp. 171–180, 2006, doi: 10.1109/TEC.2005.859981.

[9] *Generators connected to the low-voltage distribution network - Technical requirements for the interconnection to and parallel operation with low-voltage distribution networks*, VDE-AR-N 4105, VDE Verband der Elektrotechnik, Elektronik, Informationstechnik e.V., Berlin, Nov. 2018.

[10] M. Meyer, M. Dietmannsberger, and D. Schulz, *Ausgestaltung robuster und sicherer Regelungssysteme zur Vermeidung ungewollter Inselnetzbildung.* [Online]. Available: https://t1p.de/dena-isl

[11] P. Unruh, M. Nuschke, P. Strauß, and F. Welck, "Overview on Grid-Forming Inverter Control Methods," *Energies*, vol. 13, no. 10, p. 2589, 2020, doi: 10.3390/EN13102589.

[12] Grid Control 2.0. Research Project, *Grid Control 2.0 final report*, will be published before the conference.

Improving performance of III-V solar cells grown on spalled germanium with ex situ substrate planarization

John S. Mangum[1], Anna K. Braun[2], Allison Perna[2], John F. Geisz[1], Aaron J. Ptak[1], Corinne E. Packard[1,2] and Ryan M. France[1]

[1]National Renewable Energy Laboratory, Golden, CO, 80401, USA
[2]Colorado School of Mines, Golden, CO, 80401, USA

Abstract — Controlled spalling allows removal of devices and provides an opportunity for cost reduction through substrate reuse. However, the fracture-based process can leave behind morphological surface features, notably river lines, that can disrupt epitaxial growth and degrade device performance. We investigate the viability of various wet etch chemistries to planarize river lines to ensure high-quality device growth and performance without mechanical repolishing, and so maintain a route towards cost-effective reuse. Etching in a HF:HNO$_3$:CH$_3$COOH solution effectively planarizes river lines and produces a surface that yields devices with equivalent performance to those grown on epi-ready Ge wafer surfaces. Further studies will focus on optimizing etch composition, temperature, and time to minimize material removal while maintaining a suitable surface for high-quality epitaxy.

I. INTRODUCTION

Reducing the costs of III-V-based solar cells grown on Ge or GaAs substrates is a critical step for achieving production costs competitive enough for terrestrial adoption [1]. One major area of research for realizing this goal is focused on developing methods for reusing substrates multiple times due to their high price and opportunity for cost reduction. Only one method of substrate reuse, Epitaxial Lift-Off (ELO), is currently being utilized on an industrial scale [2], [3]. However, the ELO process requires long chemical etching times, on the order of hours, to release the devices from the substrate and eventually necessitates chemo-mechanical polishing (CMP) due to these prolonged harsh chemical etches [4],[5]. Techno-economic analysis shows that these CMP steps ultimately control the overall production cost of III-V solar cells, and should therefore be minimized or eliminated entirely [6]. Controlled spalling, a mechanical layer removal technique, is a potentially promising approach that has demonstrated device removal and regrowth without the need for CMP [7],[8].

Controlled spalling causes various localized morphological features that can be detrimental to device performance [8]. Our findings on the growth of III-V devices on spalled Ge surfaces indicated that one of these types of features, river lines, is the most harmful to overall solar cell performance [8]. Steep height non-uniformities (Δheight $\approx$ 300–500 nm over 1–2 μm lateral distance) characteristic of river lines disrupt continuous growth of the devices layers, resulting in the formation of shunt pathways and crystalline defects in epitaxial overgrowth. Therefore, the primary focus of our current efforts is the mitigation or elimination of river line features on the spalled Ge surfaces to achieve a greater yield of high-quality devices grown on spalled Ge wafers without the need for CMP.

Here, we investigate various wet etch chemistries that reduce the height and steepness/slope of river lines. The majority of existing Ge etchant studies focus on preparing the Ge surface for epitaxial growth with etchants that offer relatively low etch rates (<20 nm/min). Faster etch rates and/or highly crystallographically-selective etch chemistries are necessary for reducing or eliminating river lines on spalled Ge surfaces. An optimal etch chemistry planarizes the spalled Ge surface while minimizing the total etch time and volume removed. In this work, we investigate common Ge and Si wet etch chemistries used in the semiconductor industry (e.g., NH$_4$OH:H$_2$O$_2$:H$_2$O, HCl:H$_2$O$_2$:H$_2$O, H$_3$PO$_4$.H$_2$O$_2$:H$_2$O and HF:HNO$_3$:CH$_3$COOH). These etchants include an oxidizing agent (e.g. H$_2$O$_2$, HNO$_3$) for generating GeO$_X$ at the surface along with an etchant to remove the GeO$_X$ (e.g. HCl, HF, NH$_4$OH). We evaluate each etchant's effectiveness in reducing river line steepness as a proof of efficacy of ex situ etching for surface preparation of spalled Ge.

II. EXPERIMENTAL DETAILS

All controlled spalling work utilized 2" diameter p-type Ge (001) substrates with a 6° offcut toward <111>. Electrodeposition of a Ni stressor layer across the entire wafer was performed in a 0.6 M NiCl$_2$·6H$_2$O and 5.0 mM H$_3$PO$_3$ solution at 60 °C using a current density of 40 mA/cm^2 and deposition time from 7–8 min. Spalling was initiated by applying an external force to an adhesive film applied to the Ni stressor layer. The spalled films that were removed from the wafer were not used in this work. The resulting spalled wafers were used for the etching and regrowth experiments described below. All spalled Ge wafers were immersed in Transene TFG etchant to remove any residual nickel from the wafer edge, followed by an acetone/isopropanol solvent rinse.

Etch solutions were created by mixing an oxidizing agent with a germanium oxide etching agent and diluted with either deionized H$_2$O or acetic acid. Oxidizing agents were either H$_2$O$_2$ or HNO$_3$. Germanium oxide etching agents were either NH$_4$OH, HCl, H$_3$PO$_4$, or HF. Etch solutions were stirred with a magnetic stir bar at $\geq$500 rpm and no external heating was applied. Etching times ranged from 5 to 20 minutes.

Approximate etch rates were calculated by measuring mass loss of wafers after etching relative to their areal dimensions.

Solar cell devices were grown on both 2" diameter commercial, epi-ready Ge wafers and the wafers that resulted from the controlled spalls (with and without etching). Upright, front-junction III-V solar cell devices were grown in an organometallic vapor phase epitaxy (OMVPE) reactor. The device structure consisted of an approximately 400 nm Ge buffer layer followed by a nominally 2 μm-thick (spalled surface morphology can introduce growth rate anisotropy), Zn-doped $Ga_{.99}In_{.01}As$ lateral conduction layer, a $Ga_{0.5}In_{0.5}P$ back surface field (BSF) layer, a p-type GaInAs base, n-type GaInAs emitter, a two-layer GaInP/AlInP window and a GaInAs top contact layer. Following growth, solar cell devices with size 5 mm x 5 mm were defined using standard photolithography processing.

Nomarski optical microscopy and laser scanning confocal microscopy (Keyence 6000) were used to image and measure height profiles across the river line features, perpendicular to the river line ridges. *J-V* performance was measured on an adjustable continuous solar simulator, calibrated to the AM1.5G spectrum at 1000 W/m^2. Electron channeling contrast imaging (ECCI) was performed using a vCD backscatter detector inserted underneath the polepiece on a FEI Nova NanoSEM 630 operating at 25 kV accelerating voltage and 3.2 nA beam current.

III. RESULTS AND DISCUSSION

A. Wet Etching of Spalled Ge Surfaces

Preliminary results from our etching studies using standard Ge etches showed that the more dilute solutions containing H_2O (2 NH_4OH:1 H_2O_2:10 H_2O and 2 HCl:1 H_2O_2:10 H_2O) do not etch the Ge surface enough to have a significant impact on river line morphology. The basic solution containing NH_4OH actually showed steepening of the river lines, resulting in worse morphology for epitaxial growth. The acidic HCl-containing solution appeared to etch the flatter regions between river lines faster than the river lines, effectively increasing their height, but also showed some lateral etching of the river lines which could lead to their eventual removal if etching is conducted for long enough. So far, attempts at increasing the etch rate of these solutions by decreasing the H_2O content have resulted in rough surfaces due to aggressive bubbling and/or non-uniform etch rate across the wafer surface. A similar acidic etch chemistry where HCl is replaced by H_3PO_4 was also studied because H_3PO_4 can be mixed with H_2O_2 in more concentrated solutions (i.e., with less H_2O) without aggressive bubbling, which allowed for greater river line etch rate without compromising the rest of the spalled Ge surface that is free from morphological defects. Etching in various mixtures/ratios of H_3PO_4:H_2O_2:H_2O solution for 5-15 minutes exhibited no significant surface roughening in initially flat regions, but resulted in river line height reductions up to 100's of nm. Further experiments are being conducted to assess the etching behavior and resulting river line morphology when different ratios of H_3PO_4 and H_2O_2 are used to etch the spalled Ge surface.

One etch chemistry has stood out thus far for effectively etching river lines – HF:HNO_3:CH_3COOH (hereafter referred to as HNA). This etch chemistry is much more aggressive (etch rate ≈ 2 μm/min) compared to those discussed previously and maintains a Ge surface suitable for epitaxial growth. Height profiles of river lines acquired by laser profilometry before and after etching in HNA for 20 min (Figure 1) show a dramatic reduction in both height and steepness of the river lines. Of those measured and presented here, the as-spalled river lines had an average step height of 321 nm over 1.5 μm lateral distance (slope = 0.214), which are values representative of the vast majority of as-spalled river lines we have encountered in this work. The HNA etched river lines showed dramatic morphological planarization, averaging 34 nm step height over 2.9 μm lateral distance (slope = 0.012). Further studies will focus on varying the constituent ratios and/or temperature to search for an optimal HNA etch composition that minimizes etch time and volume removal while maintaining the planarization effects demonstrated in this preliminary investigation.

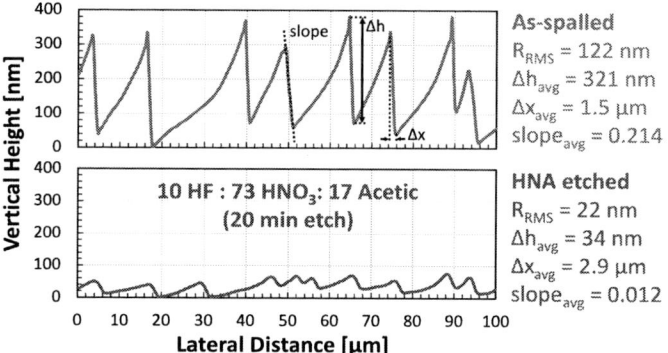

Fig 1. Representative laser profilometry height profiles of spalled Ge river lines as spalled and after 20 min of etching in HNA solution showing significant reduction of river line height and slope. These profiles are not from the exact same area before and after etching.

B. Solar Cell Device Performance

Solar cell devices were fabricated over river lines on spalled Ge wafers etched in this HNA solution and device performance was characterized (Figure 2). Other morphological spalling defects that affect device performance, such as major arrest lines, were present on the wafer presented in Figure 2 but the following analysis is solely focused on regions containing only river line defects. Comparing devices grown over river lines with HNA etching treatment to those without any etching showed a dramatic increase in cell efficiency. Significant increases in all J-V performance metrics relative to devices grown on as-spalled substrates were observed when river lines were etched in the HNA solution prior to device growth. Additionally, both ECCI characterization and the J-V performance of the cells grown over HNA-etched river lines indicated no increase in crystalline defect density compared to cells grown on epi-ready Ge. No significant shunting was observed in the J-V curves corresponding to these cells as well, further indicating the effective mitigation of river line morphology through HNA etching. The J-V metrics of the device grown over HNA-etched river lines shown in Figure 2d were comparable to control devices grown on epi-ready Ge that exhibited V_{OC} = 1.0 V, J_{SC} = 22.0 mA/cm^2, FF = 82.4 %, and efficiency = 18.2 %. These results indicate that the river line morphology has been significantly ameliorated to the point of minimally impacting device performance. It is also important to note that while the surface roughness of the etched river lines was still much greater than a typical epi-ready surface (R_{RMS} of 22 nm vs <1 nm, respectively), device growth and performance were not significantly impacted by these rougher surfaces. Based on our previous study, the smoothing of the river line features with HNA etching likely decreases the opportunity for discontinuous growth of thin device layers and subsequently reduces the formation of shunt pathways and crystalline defects [8].

IV. Summary and Outlook

We investigated a variety of potential wet etch chemistries for effectively planarizing morphological spalling defects, notably river lines, on the surface of spalled Ge wafers. We found that the more dilute standard industrial chemistries used for etching Ge (H_2O_2 with either HCl or NH_4OH, diluted with H_2O) were either ineffective for planarizing river lines or too slow to reduce the height of river lines within a useful time frame for low-cost mass production. A mixture of H_3PO_4 and H_2O_2 with minor dilution with H_2O etched river lines considerably more than the more dilute HCl- and NH4OH-based chemistries, which shows promise but requires further study. Etching in HF:HNO$_3$:CH$_3$COOH for 20 min reduced river line heights and planarized them to such a degree that solar cell devices grown over the etched river lines showed no discernable difference in performance compared to devices grown on pristine epi-ready Ge wafer surfaces. Further studies on spalled Ge surface planarization by wet etching will focus on altering the ratios of etch constituents and/or temperature to find more efficient planarization conditions. However, this initial success shows the promise of wet chemical etchants to enable substrate reuse after spalling without chemo-mechanical repolishing.

Acknowledgements

John Goldsmith provided III-V growth support. This work was authored in part by Alliance for Sustainable Energy, LLC, the Manager and Operator of the National Renewable Energy Laboratory for the U.S. Department of Energy (DOE) under Contract No. DE-AC36-08GO28308. Funding was provided by the U.S. Department of Energy Office of Energy Efficiency and

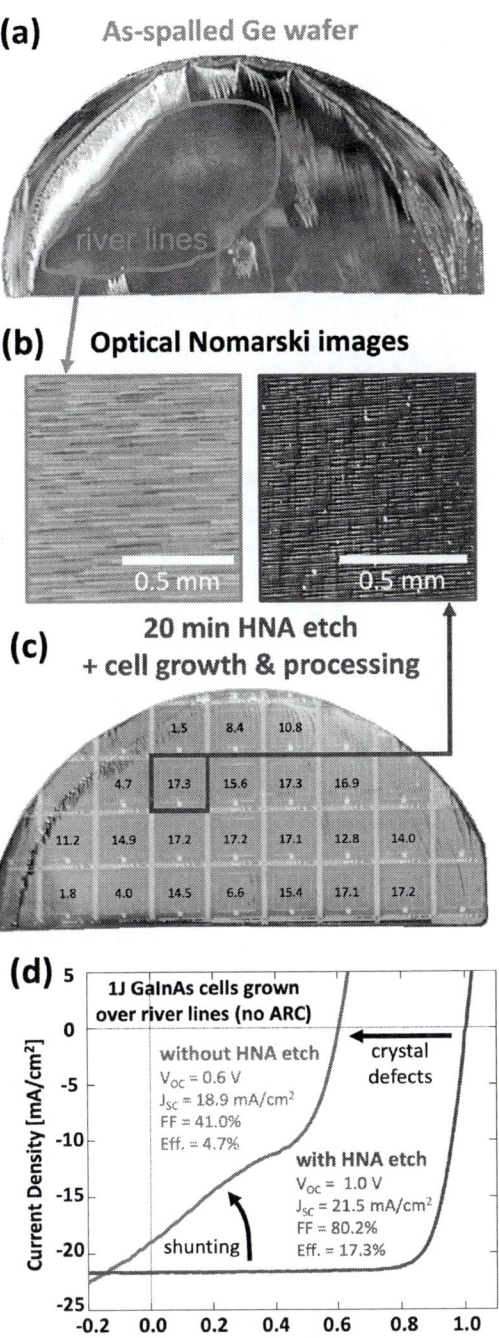

Fig 2. (a) Half of a 2" spalled Ge wafer with large region of dense river lines. (b) Zoomed-in representative optical Nomarski micrographs showing the river line morphology on the as-spalled surface and the surface of a cell grown over HNA-etched river lines. (c) Same wafer in (a) after a 20 min etch in HNA followed by OMVPE solar cell growth and processing. Cell efficiencies (in %) are indicated for each cell. (d) J-V metrics from representative cells grown over river lines with and without HNA etching treatment showing the dramatic cell performance improvement when river lines are etched in HNA prior to cell growth.

Renewable Energy Solar Energy Technologies Office under Award No. 38261. Anna K. Braun's contributions to this material are based upon work supported by the National Science Foundation Graduate Research Fellowship Program under Grant No. DGE-1646713. Allison N. Perna acknowledges support of the U.S. Department of Education Graduate Assistance in Areas of National Need fellowship. Any opinions, findings, and conclusions or recommendations expressed in this material are those of the author(s) and do not necessarily reflect the views of the National Science Foundation. The views expressed in the article do not necessarily represent the views of the DOE or the U.S. Government. The U.S. Government retains and the publisher, by accepting the article for publication, acknowledges that the U.S. Government retains a nonexclusive, paid-up, irrevocable, worldwide license to publish or reproduce the published form of this work, or allow others to do so, for U.S. Government purposes.

REFERENCES

[1] K. A. Horowitz, *et al.*, "Techno-economic analysis and cost reduction roadmap for III-V solar cells." Golden, CO: National Renewable Energy Laboratory. *NREL/TP-6A20-72103.* 2018.

[2] A. P. Kirk, *et al.*, "Recent progress in epitaxial lift-off solar cells." *2018 IEEE 7th World Conference on Photovoltaic Energy Conversion*, 0032-0035, 2018.

[3] J. Adams, *et al.*, "Demonstration of multiple substrate reuses for inverted metamorphic solar cells." *IEEE Journal of Photovoltaics*, 899-903, 2013.

[4] J. J. Schermer, *et al.*, "High rate epitaxial lift off of GaInP films from GaAs substrates." *Applied Physics Letters,* 76: 2131–2133, 2000.

[5] G. J. Bauhuis, *et al.*, "Wafer reuse for repeated growth of III-V solar cells." *Prog. Photovolt Res. Appl.* 18: 155–159, 2010.

[6] J. S. Ward, *et al.*, "Techno-economic analysis of three different substrate removal and reuse strategies for III-V solar cells." *Prog. Photovolt: Res. Appl.*, 24: 1284– 1292, 2016.

[7] J. Chen and C. E. Packard, "Controlled spalling-based mechanical substrate exfoliation for III-V solar cells: A review," *Solar Energy Materials and Solar Cells*, vol. 225, p. 111018, Jun. 2021.

[8] J. S. Mangum *et al.*, "High-Efficiency Solar Cells Grown on Spalled Germanium for Substrate Reuse without Polishing," *Advanced Energy Materials*, p. 2201332, Jun. 2022.

Fabrication And Characterization Of III-V Photovoltaic Devices For Use As CO2 Reduction Photoelectrodes

Myles A. Steiner, Grace A. Rome, Ann L. Greenaway, Joel W. Ager, Emily L. Warren

National Renewable Energy Laboratory, Golden, CO, United States

Colorado School of Mines, Golden, CO, United States

Lawrence Berkeley National Laboratory, Berkeley, CA, United States

In addition to their demonstrated high efficiency as photovoltaic devices, III-V materials are promising photoelectrodes for the production of solar fuels. While there are many demonstrations of multijunction III-V devices driving the production of H_2 from water, there is growing interest in the photoelectrochemical reduction of CO_2 to carbon-based fuels. We have begun fabricating superstrate p-on-n GaInP/GaAs photoelectrodes for cascade CO_2 reduction. Following deposition, the target devices are removed from the growth substrate and bonded to a glass handle from the top side. Processing the photoelectrodes as three-terminal devices allows for separate metal catalysts on the solution contacts of the GaInP and GaAs cells, enabling two reaction steps that together form a cascade for CO_2 reduction to multi-carbon products.

Proton Degradation-free Flexible Chalcopyrite Solar Cells without Cover Glass and Adhesive

Hiroki Sugimoto[1], Tetsuya Nakamura[2], Mitsuru Imaizumi[2], Shin-ichiro Sato[3], and Takeshi Ohshima[3]

1 PXP Corporation, Sagamihara, Kanagawa 252-0131, Japan
2 Japan Aerospace Exploration Agency, Tsukuba, Ibaraki 305-8505, Japan
3 National Institutes for Quantum Science and Technology, Takasaki, Gunma 370-1292, Japan

Abstract— **Strong radiation hardness against low-energy and high-energy proton irradiation on newly developed flexible chalcopyrite Cu(In, Ga)(Se, S)$_2$ solar cells without cover glass and adhesive was confirmed. Degradation of efficiency after 300 keV and 3 MeV proton irradiation (fluence varied from 10^{11} to 10^{13} cm^{-2}) was fully recovered by heat light soaking. Capacitance-voltage measurement indicated recovery of the net carrier concentration after the heat light soaking. Compositional depth profile measurement suggested the full-recovery characteristics should be delivered from Ga-less and Cu-poor composition in the depletion layer in Cu(In, Ga)(Se, S)$_2$ absorber layer. We believe these results are one of big steps for achieving light-weight and low-cost space solar cells with long lifetime.**

Keywords— *flexible, chalcopyrite, Cu(In, Ga)(Se, S)$_2$ solar cell, radiation hardness, proton irradiation*

I. INTRODUCTION

In order to endure space missions for more than 10 years in comparably severe irradiation environments, such as the geostationary orbit, the space solar cells need to survive until the fluence of 10^{16} cm^{-2} for the electron irradiation and that of 10^{13} cm^{-2} for the proton irradiation [1,2]. Chalcopyrite Cu(In, Ga)(Se, S)$_2$ (CIGSS) solar cells have been known to have strong radiation hardness. For the electron irradiation, they show almost no degradation until the fluence of 10^{16} cm^{-2} [3,5,8,9][14]-[17], however, there is no reports for proton

degradation-free solar cells due to the severe defect introduction by the proton irradiation [2,4,6,7,10,11]. For the low-energy proton irradiation less than 200 keV, the penetration depth of proton corresponds to the depth of the transparent conductive oxide (TCO) and electron transport layer (ETL), fortunately they hardly show any degradation thanks to their strong crystalline structures [4]. However, the low-energy proton irradiation ranging from 200 to 400 keV introduces a large number of defects in the CIGSS absorber layer because the penetration depth of proton just corresponds to the inside of the CIGSS absorber layer, which results in the large degradation of electrical performances of the CIGSS solar cells [2,4,6,7]. Therefore, all space solar cells need cover glass and adhesive to prevent the damages of the low-energy proton irradiation. Regarding the high-energy proton irradiation, the cover glass and adhesive no longer prevent the proton damage, all space solar cells are damaged more or less.

So far, self-recovery characteristics for the radiation damages of the CIGSS solar cells by light soaking (LS) [18,19], heat treatment [25,26] and heat light soaking (HLS) [19,27] have been reported. Among them, the HLS shows drastic recovery for the degradation of the short circuit current (J_{sc}) and the fill factor (FF). Regarding the degradation of open circuit voltage (V_{oc}), complete recovery has been difficult because the Ga-related antisite (Ga$_{Cu}$) defects are considered to be hardly passivated, while the In-related antisite (In$_{Cu}$) defects and metastable donor-

TABLE I. SUMMARY OF PROTON IRRADIATION EFFECTS ON CHALCOPYRITE SOLAR CELLS

Irradiation			Introduced defects (as irrad.)					LS recovery				HLS recovery			
Dose [1,2]	Energy [2,4]	Damage [2,4,6,7]	Defects [10,11]	Type [12]-[13]	Impact on cells [6,7,10,11,16,17]			Modification [12]-[13]	Impact on cells [19]			Modification [12]-[13] [20]-[24]	Impact on cells [19], [25]-[27]		
					V_{oc}	J_{sc}	FF		V_{oc}	J_{sc}	FF		V_{oc}	J_{sc}	FF
Proton ~10^{13} cm^{-2}	Low 200~400 keV	Extra large	so many $V_{Se,S}$-V_{Cu}	Donor (meta-stable)	much down		much down	Acceptor (meta-stable)	recover		recover	Acceptor (meta-stable)	recover		recover
			so many In$_{Cu}$	Deep level	much down	down	much down	Weak deep level	down	slightly down	down	Donor (passivated)	recover	recover	recover
			so many Ga$_{Cu}$	Deep level	much down	down	much down	Weak deep level	down	slightly down	down	Weak deep level (passivated)	recover (this work) down	recover	recover
	High >1 MeV	Large	many $V_{Se,S}$-V_{Cu}	Donor (meta-stable)	down		down	Acceptor (meta-stable)	recover		recover	Acceptor (meta-stable)	recover		recover
			many In$_{Cu}$	Deep level	down	slightly down	down	Weak deep level	down	recover	recover	Donor (passivated)	recover	recover	recover
			many Ga$_{Cu}$	Deep level	down	slightly down	down	Weak deep level	down	recover	recover	Weak deep level (passivated)	recover (this work) down	recover	recover

978-1-6654-6060-6/23 $31.00 © 2023 IEEE

like defects related to the defect pairs of Se,S-vacancy ($V_{Se,S}$) and Cu-vacancy (V_{Cu}) could be easily passivated [12,13]. These proton irradiation effects on the chalcopyrite solar cells are summarized in Table I. The purpose of this paper is to realize the proton degradation-free CIGSS solar cells by restricting the effect of the antisite defects.

II. EXPERIMENTAL DETAIL

In order to investigate the proton irradiation hardness, newly developed flexible CIGSS solar cells with initial air-mass (AM) 1.5 efficiency (Eff) of about 15.8% (17.0% with anti-reflective coating) were prepared. The CIGSS solar cells used for this study were specially developed for severe space environments. Not only the radiation hardness but also high temperature tolerance, thermal cycle tolerance and mechanical vibration tolerance were much improved than conventional CIGSS solar cells. The CIGSS solar cells were fabricated by highly productive sputtering-based process. The basic structure was Ag grid electrode/In-based TCO/Zn-based ETL/CIGSS absorber layer/Mo back electrode on Ti film without cover glass and adhesive. The thickness of the CIGSS device layer and the Ti substrate was about 3 um and 50 um, respectively. The weight of the CIGSS solar cells were about 250 g/m^2.

After the cell fabrication, all samples were stabilized by the HLS (30 min at 200C in N$_2$ box) and initial electrical parameters were measured before the proton irradiation in order to eliminate over estimation of recovery effects after the proton irradiation. We performed 300 keV and 3 MeV proton irradiation tests on the CIGSS solar cells at the National Institutes for Quantum Science and Technology, Takasaki. Then, the electrical parameters were checked after 1 day stock in the dry-air box. Then, the all samples were put on the temperature-controlled plate and the LS (3 hours at 25C in air) were applied. After the electrical parameter measurements, the HLS (1 hour at 150C in N$_2$ box) were conducted. Finally, we checked the electrical parameters again.

The current–voltage characteristics were measured by a class A solar simulator (XI-05A1V2-L, SERIC Ltd., Japan) with AM 1.5 and 100 mW/cm^2 illumination at 25C. Capacitance–voltage (C–V) profiles were recorded at the frequency of 10 kHz using a 4294A Precision Impedance Analyzer (Hewlett-Packard Company, USA) in the dark at room temperature. Compositional depth profiles in the CIGSS absorber layer were measured by Scanning Electron Microscope with Energy Dispersive X-ray Spectroscopy (JSM-7001F, JEOL Ltd., Japan) on the cross section polished by the ion milling.

III. RESULTS AND DISCUSSION

Figure 1 shows the remaining factors of electrical parameters of the CIGSS solar cells after the 300 keV and 3 MeV proton irradiation as a function of different fluences. The results of as irradiated samples showed the 300 keV proton irradiation introduced larger degradation than the 3 MeV proton irradiation. Large degradation of both V_{oc} and FF should be delivered from the decrease in the acceptor concentration in the CIGSS absorber layer, as reported in the other papers [11,19]. Increase in recombination centers could cause additional V_{oc} and J_{sc} degradation. After the LS, the FF showed big recovery while the degradation of V_{oc} still remained, which suggested the acceptor

concentration should be recovered while the recombination centers were not passivated yet. After the HLS, surprisingly all parameters were much recovered. Actually, the 3 MeV proton irradiated samples showed rather improvement than the initial performance. Judging from the improvement of FF, the acceptor concentration seemed to be increased. One possibility for the increasement of acceptor concentration was considered to be due to the irradiation-induced acceptor-like V_{Cu} defects, the other reason could be due to the irradiation-induced metastable $V_{Se,S}$-V_{Cu} defect pairs. Even in the severe 300 keV proton irradiation, all parameters were fully recovered, which indicated the recombination centers should be almost passivated as well as the acceptor concentration seemed to be recovered.

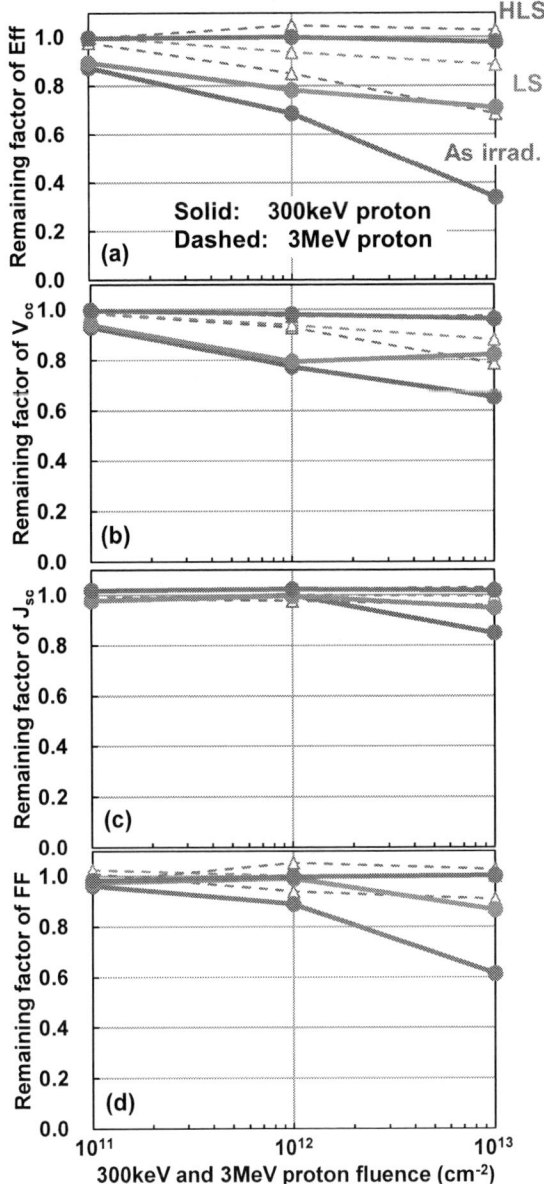

Fig. 1. Remaining factors of (a) Eff, (b) V_{oc}, (c) J_{sc} and (d) FF of CIGSS solar cells after 300 keV (solid line) and 3 MeV (dashed line) proton irradiation as a function of different fluences. Gray, green and red lines represent as irradiated, after LS and HLS, respectively.

Net carrier concentration (N_{CV}) profiles investigated by the C-V measurement on the CIGSS solar cells before and after 300 keV proton irradiation with fluences of 1×10^{11} and 1×10^{13} cm^{-2} are shown in Fig. 2. So far, the decrease in the N_{CV} of the CIGSS solar cells by the proton irradiation and the recovery of N_{CV} by the HLS have been reported [11,19]. As shown in Fig 2, we confirmed the N_{CV} of the proton-irradiated CIGSS solar cells were also recovered to almost comparable level of their initial value by the HLS. These results indicated the metastable donor-like defects and deep-donor antisite defects should be passivated by the HLS.

Figure 3 shows compositional depth profiles of Ga/(In+Ga) and Cu/(In+Ga) ratio on the newly developed CIGSS solar cells. Compared with conventional chalcopyrite solar cells [28,29], notable structures of Ga-less and Cu-poor compositions in the depletion layer in the CIGSS absorber layer were observed. The activation energy of self-recovery by the HLS has been reported as low as around 0.8~1.0 eV [18,27], which corresponds to the migration energy of V_{Cu} [23,24]. From these results, these full-recovery characteristics should be delivered from the passivation of the antisite defects, such as In_{Cu} and Ga_{Cu} defects, by the thermally migrated V_{Cu} as well as the restriction of Ga_{Cu} defects by the Ga-less composition. We also found that there was the strong Ga grading at front side of the CIGSS absorber layer, which should also help the full-recovery characteristics by enhancing the generated electron collection.

Calculated from the activation energy of the HLS recovery, the current HLS condition of 1 hour at 150C corresponds to around 16~69 days at operation temperature of 60C, which indicates sufficient self-recovery could be conducted also in actual space environments. Further investigations about the actual activation energy for the newly developed CIGS solar cells are required, however, these findings allow us to eliminate the expensive cover glass and adhesive for space solar cells.

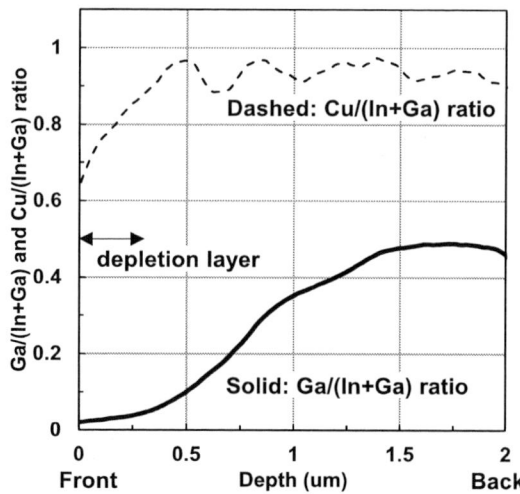

Fig. 3. Compositional depth profiles of Ga/(In+Ga) (solid line) and Cu/(In+Ga) (dashed line) ratio on newly developed CIGSS solar cells.

IV. SUMMARY

Strong radiation hardness against low-energy and high-energy proton irradiation on newly developed flexible chalcopyrite CIGSS solar cells without cover glass and adhesive was confirmed. Degradation of Eff after the proton irradiation was fully recovered by HLS. The full-recovery characteristics should be delivered from Ga-less and Cu-poor composition in the depletion layer in the CIGSS absorber layer. We believe these results are one of big steps for achieving light-weight and low-cost space solar cells with long lifetime.

ACKNOWLEDGMENT

We would like to thank Mr. Y. Mori and Mr. M. Sugai of Advanced Engineering Services Co., Ltd. for their assistance in conducting the irradiation tests, and Dr. Y. Okuno of RIKEN Center for Advanced Photonics for helpful discussion.

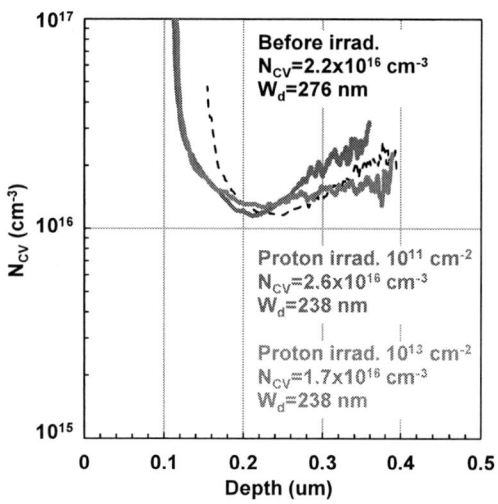

Fig. 2. N_{CV} profiles of CIGSS solar cells before (black dashed line) and after 300 keV proton irradiation with fluences of (blue line) 1×10^{11} cm^2 and (red line) 1×10^{13} cm^{-2}. W_d denotes width of depletion layer. All samples were treated by HLS before measurements.

REFERENCES

[1] C. Inguimbert and S. Messenger, "Equivalent displacement damage dose for on-orbit space applications," *IEEE Transactions on Nuclear Science,* vol. 59, pp. 3117-3125, 2012.

[2] J. R. Woodyard, "Investigation of proton radiation resistance of CIGS solar cells," in *31st IEEE Photovoltaic Specialist Conference*, 2005, p. 834.

[3] S. Kawakita, M. Imaizumi, S. Ishizuka, S. Niki, S. Okuda, and H. Kusawake, "Influence of electrical performance on Cu-related defects generated by 250 keV electron irradiation in Cu (In, Ga) Se$_2$ thin-film solar cells," *Thin Solid Films,* vol. 535, pp. 353-356, 2013.

[4] Y. Hirose, M. Warasawa, I. Tsunoda, K. Takakura, and M. Sugiyama, "Effects of proton irradiation on optical and electrical properties of Cu(In,Ga)Se$_2$ solar cells," *Japanese Journal of Applied Physics,* vol. 51, pp. 111802, 2012.

[5] A. Jasenek and U. Rau, "Defect generation in CuInGaSe$_2$ heterojunction solar cells by high-energy electron and proton irradiation," *Journal of Applied Physics,* vol. 90, pp. 650-658, 2001.

[6] S. Kawakita, M. Imaizumi, T. Sumita, K. Kushiya, T. Ohshima, M. Yamaguchi, S. Matsuda, S. Yoda, and T. Kamiya, "Super radiation tolerance of CIGS solar cells demonstrated in space by MDS-1 satellite," in *3rd World Conference on Photovoltaic Energy Conversion*, 2003, p. 693.

[7] S. Kawakita, M. Imaizumi1, K. Kibe, S. Yoda, T. Ohshima, H. Itoh, and M. Yamaguchi, "Analysis of proton induced defects in Cu(In,Ga)Se$_2$ thin-film solar cells," *Materials Research Society Symposia Proceedings,* vol. 865, pp. F5.17.1-6, 2005.

[8] S. Kawakita, M. Imaizumi1, S. Ishizuka, H. Shibata, S. Niki, S. Okuda, and H. Kusawake, "Influence of electron irradiation on electroluminescence of Cu(In,Ga)Se$_2$ solar cells," *Japanese Journal of Applied Physics,* vol. 53, pp. 05FW08, 2014.

[9] I. Khatri, T.-Y. Lin, T. Nakada, and M. Sugiyama, "The effect of electron irradiation on cesium fluoride-free and cesium fluoride-treated Cu(In$_{1-x}$,Ga$_x$)Se$_2$ Solar Cells," *Physica Status Solidi RRL,* vol. 13, pp. 1900415, 2019.

[10] H. Afshari, B. K. Durant, K. Hossain, D. Poplavskyy, B. Rout, and I. R. Sellers, "CIGS solar cells for outer planetary space applications: the effect of proton irradiation," in *47th IEEE Photovoltaic Specialist Conference*, 2020, p. 2635.

[11] S. Kawakita, M. Imaizumi, K. Kibe, T. Ohshima, H. Itoh, S. Yoda, and O. Odawara, "Analysis of anomalous degradation of Cu(In,Ga)Se$_2$ thin-film solar cells irradiated with protons," *Japanese Journal of Applied Physics,* vol. 46, pp. L670-L672, 2007.

[12] B. Huang, S. Chen, H.-X. Deng, L.-W. Wang, M. A. Contreras, R. Noufi, and S.-H. Wei, "Origin of reduced efficiency in Cu(In,Ga)Se$_2$ solar cells with high Ga concentration: alloy solubility versus intrinsic defects," *IEEE Journal of Photovoltaics,* vol. 4, pp. 477-482, 2014.

[13] S. Siebentritt, M. Igalson, C. Persson, and S. Lany, "The electronic structure of chalcopyrites—bands, point defects and grain boundaries," *Progress in Photovoltaics: Research and Applications,* vol. 18, pp. 390-410, 2010.

[14] A. Jasenek, U. Rau, T. Hahn, G. Hanna, M. Schmidt, M. Hartmann, H.W. Schock, J.H. Werner, B. Schattat, S. Kraft, K.-H. Schmid, and W. Bolse, "Defect generation in polycrystalline Cu(In, Ga)Se$_2$ by high-energy electron irradiation," *Applied Physics A,* vol. 70, pp. 677-680, 2000.

[15] Y. Hirose, M. Warasawa, K. Takakura, S. Kimura, S.F. Chichibu, H. Ohyama, M. Sugiyama, "Optical and electrical properties of electron-irradiated Cu(In,Ga)Se$_2$ solar cells," *Thin Solid Films,* vol. 519, pp. 7321-7323, 2011.

[16] M. Imaizumi, Y. Okuno, S. Sato, and T. Ohshima, "Displacement damage dose analysis on alfa-ray degradation of output of a CIGS solar cell," in *48th IEEE Photovoltaic Specialist Conference*, 2021, p. 1876.

[17] M. Imaizumi, Y. Okuno, T. Takamoto, S. Sato, and T. Ohshima, "Displacement damage dose analysis of the output characteristics of In$_{0.5}$Ga$_{0.5}$P and Cu(In,Ga)(S,Se)$_2$ solar cells irradiated with alpha ray simulated helium ions," *Japanese Journal of Applied Physics,* vol. 61, pp. 044002, 2022.

[18] A. Jasenek, U. Rau, K. Weinert, H. W. Schock, and J. H. Werner, "Illumination-induced recovery of Cu(In,Ga) Se$_2$ solar cells after high-energy electron irradiation," *Applied Physics Letters,* vol. 82, pp. 1410-1412, 2003.

[19] I. Khatri, T.-Y. Lin, T. Nakada, and M. Sugiyama, "Proton irradiation on cesium-fluoride-free and cesium fluoride-treated Cu(In,Ga)Se$_2$ solar cells and annealing effects under Illumination," *Physica Status Solidi RRL,* vol. 13, pp. 1900519, 2019.

[20] K. Yoshida, M. Tajima, S. Kawakita, K. Sakurai, S. Niki, and K. Hirose, "Photoluminescence analysis of proton irradiation effects in Cu(In,Ga)Se$_2$ solar cells," *Japanese Journal of Applied Physics,* vol. 47, pp. 857-861, 2008.

[21] I. Khatri, T.-Y. Lin, T. Nakada, and M. Sugiyama, " Temperature-dependent current–voltage and admittance spectroscopy analysis on cesium-treated Cu(In$_{1-x}$,Ga$_x$)Se$_2$ solar cell before and after heat-light soaking and subsequent heat-soaking treatments," *Progress in Photovoltaics: Research and Applications,* vol. 28, pp. 1158-1166, 2020.

[22] C. Walkons, M. Jahandardoost, T. M. Friedlmeier, W. Hempel, S. Paetel, M. Nardone, B. Ursprung, E. S. Barnard, K. E. Kweon, V. Lordi, and S. Bansal, "Behavior of Na and RbF-treated CdS/Cu(In,Ga)Se$_2$ solar cells with stress testing under heat, light, and junction bias," *Physica Status Solidi RRL,* vol. 15, pp. 2000530, 2021.

[23] S. Nakamura, T. Maeda, and T. Wada, "First-principles study of diffusion of Cu and In atoms in CuInSe$_2$," *Japanese Journal of Applied Physics,* vol. 52, pp. 04CR01, 2013.

[24] T. Maeda, A. Kawabata, and T. Wada, "First-principles study on alkali-metal effect of Li, Na, and K in CuInSe$_2$ and CuGaSe$_2$," *Japanese Journal of Applied Physics,* vol. 54, pp. 08KC20, 2015.

[25] A. Jasenek, H. W. Schock, J. H. Werner, and U. Rau, "Defect annealing in Cu(In,Ga)Se$_2$ heterojunction solar cells after high-energy electron irradiation," *Applied Physics Letters,* vol. 79, pp. 2922-2924, 2001.

[26] C. R. Brown, V. R. Whiteside, D. Poplavskyy, K. Hossain, M. S. Dhoubhadel, and I. R. Sellers, "Flexible Cu(In,Ga)Se$_2$ solar cells for outer planetary missions: investigation under low-intensity low-temperature conditions," *IEEE Journal of Photovoltaics,* vol. 9, pp. 552-558, 2019.

[27] S. Kawakita, M. Imaizumi, M. Yamaguchi, K. Kushiya, T. Ohshima, H. Itoh, and S. Matsuda, "Annealing enhancement effect by light illumination on proton irradiated Cu(In,Ga)Se$_2$ thin-film solar cells," *Japanese Journal of Applied Physics,* vol. 41, pp. L797-L799, 2002.

[28] C. Frisk, C. Platzer-Björkman, J. Olsson, P. Szaniawski, J. T. Wätjen, V. Fjällström, P. Salomé, and M. Edoff, "Optimizing Ga-profiles for highly efficient Cu(In, Ga)Se$_2$ thin film solar cells in simple and complex defect models," *Journal of Physics D: Applied Physics,* vol. 47, pp. 485104, 2014.

[29] H. Liang, U. Avachat, W. Liu, J. V. Duren, and M. Le, "CIGS formation by high temperature selenization of metal precursors in H$_2$Se atmosphere," *Solid-State Electronics,* vol. 76, pp. 95-100, 2012.

Analysis of Hierarchical PV2PV Series Differential Power Processing Configuration for Photovoltaic Applications

Afshin Nazer, Patrizio Manganiello, Olindo Isabella

Delft University of Technology, Delft, Netherlands

Mismatches among PV (sub)modules reduce energy yield. One solution is PV2PV Series Differential Power Processing (SDPP), but this configuration suffers from the so-called accumulation effect leading to higher total processed power and component power rating. A hierarchical PV2PV SDPP configuration mitigates the accumulation effect. This paper describes the accumulation effect and details hierarchical PV2PV SDPP performance and equations for power flow analysis. Based on this mathematical model, a comparison between the conventional and hierarchical PV2PV SDPP is done in terms of total power processing and converter rating. Finally, Matlab/Simulink simulation results verify the ability of the hierarchical configuration to mitigate the accumulation effect.

978-1-6654-6060-6/23 $31.00 © 2023 IEEE

Plane of Array Irradiance cleaning and generation of Validated POA readings for plant evaluation

Pramod N. Krishnani[1], Adonis E.Hajj[2], Clay Helms[3], Shreyas Nagarajan[3], Mark Mikofski[4]

[1]DNV Singapore Pte.Ltd., Singapore 118227, Singapore
[2]DNV UK Ltd., Vienna, Austria
[3]Silicon Ranch Corporation, Nashville, TN 37211, United States
[4]DNV USA, Oakland, CA 94612, United States

Abstract— Every solar power plant of any size will have multiple installed irradiance measurement sensors, which are supposed to measure the quantity of irradiance received on the asset location throughout the day. These irradiance sensor measurements are key constituents to the evaluation of the true performance of solar power plants using solar analytics. As the foundation to nearly all understanding of plant performance, high-quality irradiance data is critical to the long-term ownership of healthy solar power plants. The most common issues with irradiance sensors which are related to incorrect readings are: missing data for a particular time series, readings out of range, missing sensor recordings and dead sensor readings. Apart from the above mentioned issues, which are clearly defined in the IEC 61724-3 standards, sensors with misaligned orientation and lost calibration contribute to more than ~5% of the identified bad data recorded from irradiance sensors. In this technical paper, the methodology of cleaning and scientifically correcting the aggregation of bulk irradiance sensors i.e., more than three irradiance sensors, is defined with an example solar power plant asset. The asset which has eight irradiance sensors is aggregated after removing all erroneous data and cleaned validated and representative data is produced using this methodology.

Keywords— irradiance, cleansing, insolation, solar pv, photovoltaic, radiation

I. INTRODUCTION

Solar radiation is the input for all solar energy generation systems. In photovoltaics, the measurement of solar irradiance is essential for research, quality control, feasibility studies, investment and operational decisions, plant monitoring, field operations and maintenance (O&M) optimization, shareholder reporting, site comparison, and as an input for short/long term irradiance and energy forecasting which is generally referred to as annual operational plans by the independent power providers (IPP).

Solar power plants measure solar irradiance using at least one, but usually several pyranometers installed in either the plane of array of the PV modules (Plane of Array (POA) sensors) or horizontal to the ground (Global Horizontal Irradiance (GHI) sensors). These sensors can encounter several issues that can cause erroneous measurement recordings due to reasons either related to the data loggers recording the measurements or due to sensor malfunctions. Therefore, it is

essential for the stakeholders of PV power plants to be able to work with aggregated cleaned irradiance data.

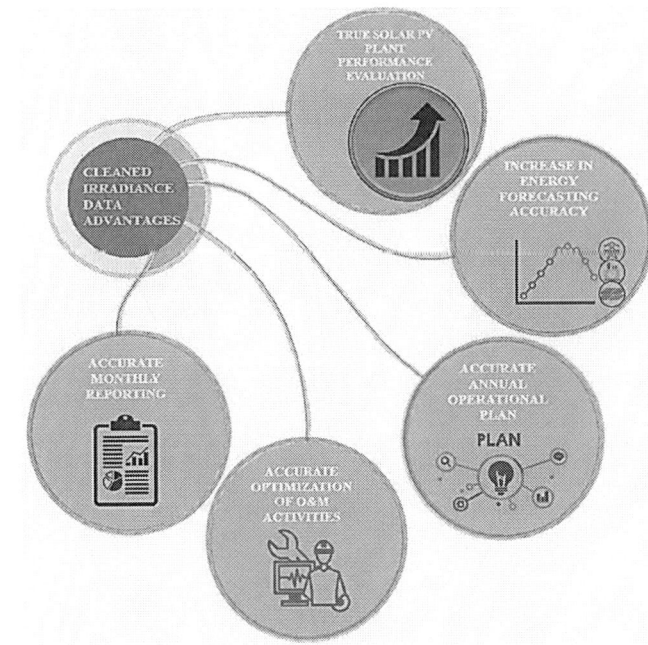

Fig.1 Importance of cleaning aggregated irradiance data

As illustrated in Fig.1, cleaned irradiance data provides the following advantages to the stakeholders:

1. True representative irradiance readings will form a good basis to calculate the correct expected energy of a power plant based on its PV model. Thereby, stakeholders such as power plant owners, operators, equity partners and off takers are informed about its true expected performance. This enables the stakeholders to make crucial decisions related to the asset management and performance management of a power plant.

2. It is essential to have cleaned historical irradiance data from a PV power plant in order to provide accurate short term and long-term energy forecast of the power plants. If the forecast is accurate, it will empower the operators and scheduling coordinators to minimize grid dispatch

losses and increase reliability/predictability of plant operations.

3. Cleaned irradiance data will help the owners of PV power plants to accurately evaluate and prepare their yearly budgets, working from a reliable understanding of historical performance.

4. The Operations and Maintenance team of a PV power plant can use the cleaned historical irradiance data and accurate forecast of energy production, also based on cleaned irradiance data, to improve their planning of the maintenance and minimize lost production of the power plants.

5. Lastly, cleaned irradiance data will provide stakeholder confidence in the accuracy of performance reports of power plants and portfolios of plants. The gain in confidence of key stakeholders will open the doors for introducing more investments in the field of solar PV assets.

II. GENERAL ISSUES WITH IRRADIANCE SENSORS

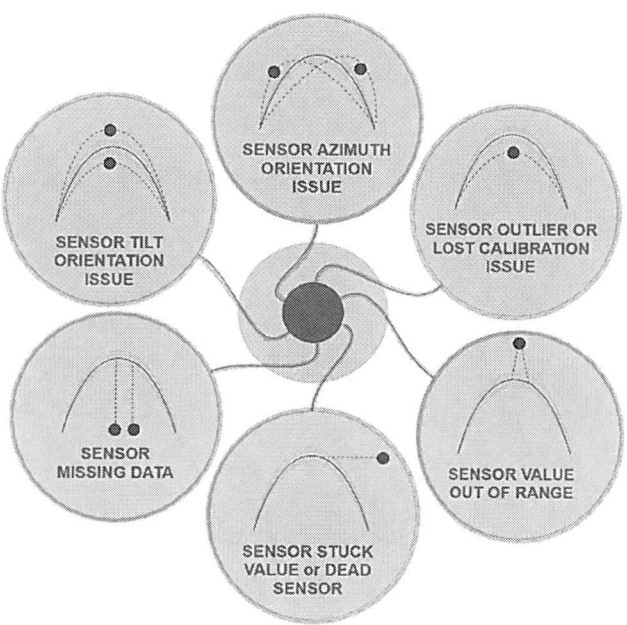

Fig. 2 General irradiance sensor measurement issues

Fig. 2 illustrates the most common causes behind erroneous irradiance data.

1. Tilt Orientation: The sensor has a different tilt angle orientation compared to that of the angle at which the solar modules are tilted.

2. Azimuth Orientation: The sensor has a different azimuth angle orientation compared to that of the site's azimuth angle.

3. Sensor Drift/Lost Calibration: The irradiance data starts drifting during a perfect clear sky period due to the sensor experiencing a loss of calibration.

4. Missing Data: The sensor fails to collect measurements and thus registers missing values for a specific period, either due to issues with its communication protocols

or associated equipment. This is aligned with industry standards as illustrated in IEC TS 61724-3 standards [1] and by Kurtz et al [2].

5. Dead or Stalled Data: The sensor reports measurements with either readings of zero or constant stalled non-zero readings during the time frame when the sun elevation angle is above zero degrees. This is aligned with industry standards as illustrated in IEC 61724-3 standards [1] and by Kurtz et al [2].

6. Data Out of Range: Sensors report irradiance measurements outside the range of -5 and 1400 W/m² as specified by IEC 61724-3 standards[1].

III. METHODOLOGY

The methodology described in this section illustrates how multiple irradiance sensors are cleaned and aggregated to a single validated irradiance reading, which is representative of the power plant in consideration.

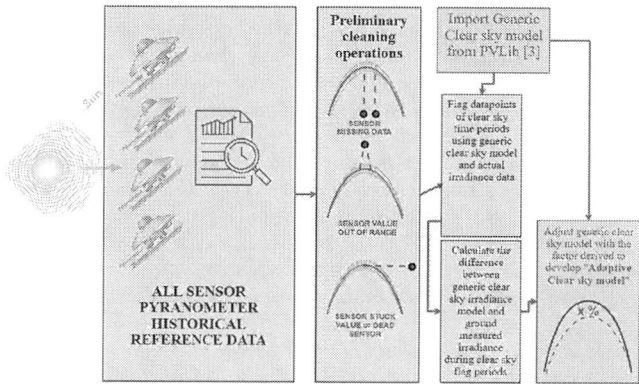

Fig. 3 Derivation of Adaptive Clear Sky model

In order to identify time periods when the irradiance sensors have lost their orientation or calibration, it is necessary to have a baseline site adaptive clear sky model for comparison during the clear sky periods. As illustrated in Fig. 3, known historical time period data is collected and processed through general cleaning algorithms, where the irradiance data is cleaned for missing data, out of range data and dead/stuck sensors using IEC 61724-3 standards [1, 2]. Once the sensors are cleaned, the clean data is parsed using the PVLib Python package [3] to generate a generic clear sky model. Next, using the cleaned preliminary irradiance and generic clear sky model, clear sky periods are flagged. A correction factor is then derived to generate the Adaptive Clear Sky model. This Adaptive clear sky model is representative of the power plant in consideration and used in the following cleaning operations.

As shown in Fig. 4, all raw irradiance measurements are compared to the adaptive clear sky model during flagged clear sky periods. If any sensor drifts from the clear sky model by more than a defined +/- threshold, the sensor is flagged as an outlier. It is kept in flagged status until the algorithm detects that the sensor has been corrected in a later clear sky period. The irradiance sensors are then passed through the next logic flow, where the irradiance is checked for missing data, whether it is within the bounds specified in section II, and whether the sensors are reporting null/stuck data. All such datapoints which

978-1-6654-6060-6/23 $31.00 © 2023 IEEE

do not satisfy the parameters of the logic outlined previously are then removed from the aggregation.

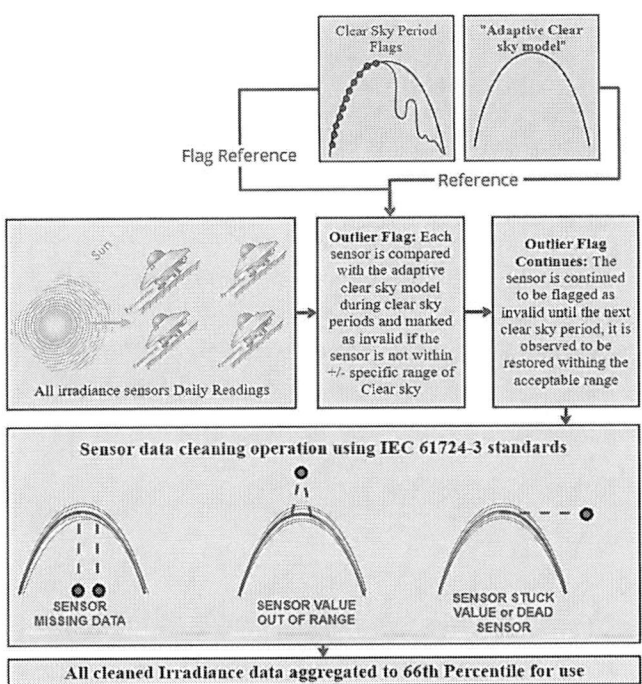

Fig. 4 Irradiance cleansing operations methodology

The data from the remaining valid sensors are aggregated to a single site representative irradiance measurement. The 66th percentile is taken as the aggregation method due to a better corelation with the power measurement of the plant.

IV. CASE STUDY

The irradiance cleansing methodology described in was tested for the complete year of 2022 on a single-axis tracking site in the United States. The site has nine irradiance POA sensors, some of which were reporting invalid data due to sensor and alignment issues. Manual correction of the data can be susceptible to human errors, which could have misrepresented the true performance of the power plant. All the irradiance POA sensor readings are then fed into the python program of the above-mentioned algorithm, and a final validated POA irradiance is generated.

As shown in Fig. 5, due to the algorithm, one irradiance sensor (sensor S6) was detected as an outlier due to a tracker malfunction during the early morning. Additionally, sensors S1, S4 & S7 were detected to have dead values throughout the day. After the flagged invalid datapoints from each sensor were removed, the remaining data is aggregated to a single validated cleaned POA data called Validated Irradiance (Cleaned).

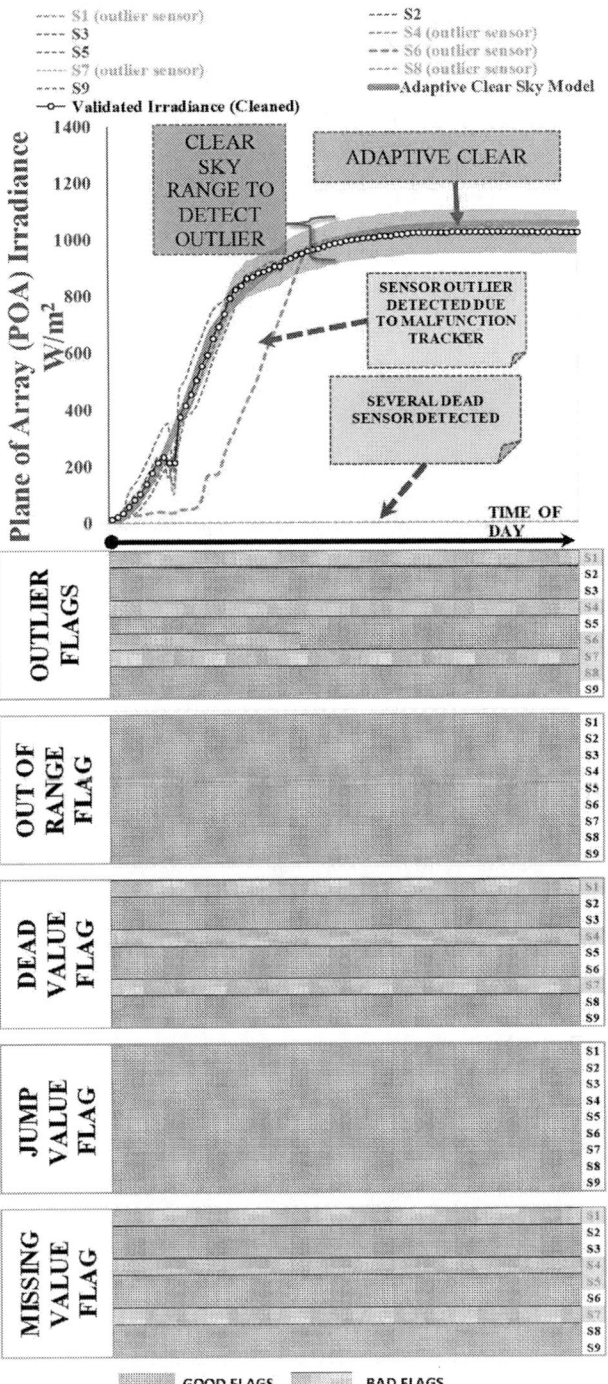

Fig. 5 Case study of irradiance data cleansing operation

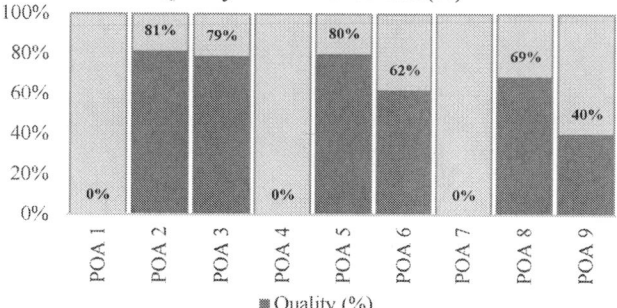

Fig. 6 Results of Case Study

Fig. 6 highlights the results of processing actual irradiance data for 2022 through the cleaning model. Three of the nine sensors were detected as malfunctioning, and several other issues were successfully identified from the remaining sensors. The cleansing operations allowed the identification and removal of the invalid irradiance data, which ranged from approx. 20% to 60% of the data from each sensor.

V. CONCLUSION

It is essential for any power plant operator to ensure that ground station irradiance sensor measurements are cleaned as per IEC standards [1, 2]. The approach described above enables automation of cleaning irradiance data by automating several otherwise-manual processes for identifying outliers, null/stale data, data drift due to calibration issues etc. Only once invalid data has been removed should it be further used in reporting and analytics of PV systems and operational losses.

REFERENCES

[1] International Electrotechnical Commission [IEC], "IEC/TS 61724-3 Ed. 1.0 en cor.1:2018, Corigendum 1 - Photovoltaic System Performance - Part 3: Energy Evaluation Method", Jan 2018.

[2] Sarah Kurtz, Pramod Krishnani, Janine Freeman, Robert Flottemesch, Evan Riley, Tim Dierauf, Jeff Newmiller, Lauren Ngan, Dirk Jordan and Adrianne Kimber, "PV system energy test" in 2014 IEEE 40th Photovoltaic Specialist Conference (PVSC), 8-13 June 2014.

[3] William F. Holmgren, Clifford W. Hansen, and Mark A. Mikofski. "pvlib python: a python package for modeling solar energy systems." Journal of Open Source Software, 3(29), 884, (2018). https://doi.org/10.21105/joss.00884

A Maximum Current Point Tracking Algorithm for Solar-to-Hydrogen Production

Kelvin Tan, Meng Tao

Arizona State University, Tempe, AZ, United States

The integration of electrolyzers with photovoltaic (PV) systems allows green hydrogen production. Conventional solar electrolyzers require a power converter for maximum power point tracking (MPPT) to maximize the power output of the PV array. However, the production of hydrogen by water electrolysis is proportional to the electrical current, not the power. In this paper, we propose and demonstrate a maximum current point tracking (MCPT) algorithm that maximizes the current output of the PV array. The algorithm is implemented in a load-matching PV system and also incorporates voltage regulation. It is demonstrated that the MCPT algorithm is able to produce more hydrogen than the MPPT algorithm, as well as increase the time the electrolytic loads spend in a specified voltage range. The MCPT system does not require a power converter making it an inexpensive, scalable, and efficient system for green hydrogen.

A physics based approach for PV lifetime and degradation signatures prediction

Ismail Kaaya[1], Gofran Chowdhury[2] and Arnaud Morlier[1]

[1]Imec, Imo-Imomec, Thor Park, Genk, 3001, Flanders, Belgium
[2]3E, Quai à la Chaux 6, 1000 Brussels, Belgium

Abstract—We present a physics-based approach to model the degradation rate/lifetime of Photovoltaic (PV) module considering various aspects that might influence the PV reliability. Our approach considers the optical and thermal properties of a PV module. The electrical model is modelled as time and stress factors dependent based on the degradation rate and reliability models of the PV electrical circuit parameters; series and shunt resistances, transmittance, and the saturation current. Since our model is mostly physics-based, it allows to perform 'what-if' simulations and is hence capable of simulating different PV components, technologies and designs. Moreover, it allows to extract information about the degradation signatures and how they vary with different technologies and climate conditions.

Index Terms—Degradation rate, degradation signatures, Modelling, PV modules

I. INTRODUCTION

PV technology and bill of materials have been been rapidly evolving in the last 3 decades. Additionally, PV modules are being installed in a wide range of locations with different climatic conditions. These technological, materials and climate variations impact the PV operational performance differently and in particular the degradation rates and mechanisms [1]–[3]. Moreover, the emerging PV installation methods, designs and applications (e.g., Building integrated PV, Vehicle integrated PV, floating etc.) might have implications—positive or negative—on PV reliability and lifetime.

However, the developments are too frequent compared to the life expectancy of PV modules which makes it almost impossible to track the reliability issues associated with them. A quick option to evaluate the reliability issues that might be associated with these new developments is by using physics-based models that consider all necessary degradation aspects/parameters. There is still a knowledge gap for such a modelling approach.

To bridge this knowledge gap, we proposed a starting point for a generalized physics-based PV degradation modelling approach, which was presented at the 8th PV world conference last year [4]. Our approach is based on multi-aging models that can allow to simulate the aging behavior of different PV modules technologies, designs, and packaging to cater for the ever-changing PV industry. In this work we present our new developments for this physics-based modelling approach and we demonstrate how it can be applied to extract information

Part of this work has received funding from the the Horizon 2020 Research and invention Programme , under Grant Agreement No 952957, Trust-PV project.

about the degradation signatures and how they vary with different technologies and climate conditions.

II. METHODOLOGY

Our approach involves several steps as shown in the schematic diagram Fig. 1. Highlights of each step are presented in this abstract and in the final manuscript, more details will be presented.

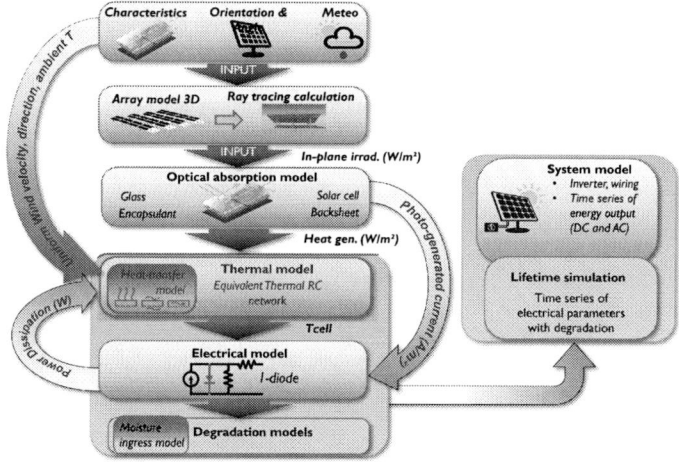

Fig. 1: Schematic of the modelling approach. Energy yield to degradation.

Step 1: Starts with our bottom-up physics-based modelling approach [5]. The illumination model uses the weather data to calculate the plane-of-array irradiation ($Gpoa$) on all PV elements. The electrical model and thermal model [4] uses the cell, module, and array characteristics along with their thermal properties to derive the IV characteristics of the system. This is an important step to simulate the effect of temperature profiles of novel bill of materials and effects of PV configurations and plant design on PV lifetime.

Step 2: The degradation rate (k) models of the equivalent circuit parameters; series resistance (R_s), shunt resistance (R_{sh}), transmittance (τ) and saturation current (I_o) are proposed. The degradation rate models relate the climatic/stress factors (SF) to material properties. The models will be presented in the final manuscript.

Step 3: The reliability models of R_s, R_{sh}, τ and I_o are proposed. The reliability models describe the nature of degradation evolution over time. Several reliability models for each parameter are proposed to simulate a wide range of degradation curves observed in the field (to be presented in the final manuscript). The choice of which degradation curve to use might depend on expert knowledge, data from accelerated aging experiments, or field data if available.

Step 4: Model the electrical equivalent circuit as a function of time (t) and climatic/stress factors (SF) using the degradation rate and reliability models. Hence the generated photocurrent (I_{PV}) is expressed as:

$$I(t, k[SF]) = I_{PV}(t, k[SF]) - I_0(t, k[SF]) \bigg[$$
$$\exp\left(\frac{V + R_s(t, k[SF]) \cdot I}{n \cdot V_t}\right) - 1\bigg] - \left(\frac{V + R_s(t, k[SF]) \cdot I}{R_{sh}(t, k[SF])}\right) \quad (1)$$

Given;

$$I_{PV}(t, k[SF]) = I_{PV_{ref}}\left[1 + K_I(T - T_{ref})\frac{G}{G_{ref}} \cdot \tau(t, k[SF])\right] \quad (2)$$

$$V_{OC}(t, k[SF]) = \frac{n \cdot k \cdot T}{q} ln\left[\frac{I_{PV}(t, k[SF])}{I_0(t, k[SF])} + 1\right] \quad (3)$$

where I, and I_o are the generated solar cell current and reverse saturation current respectively. V is the solar cell voltage, n is the ideality factor, and V_t is the thermal voltage.

Step 6: Evaluate the degradation of I-V curve parameters (i.e., V_{OC}, I_{SC}, FF, P_{max}), performance ratio and the lifetime energy yield with degradation effect. At this step the degradation signatures are also assessed.

Steps 2 -6 are specific for degradation rate estimations using inputs from step 1.

III. PRELIMINARY RESULTS

Historical meteo data of over 18 years (hourly timeseries from 2004 to 2022) provided by 3e is used. Three locations; Adrar, Accra and Brussels to represent hot and dry, hot and humid and moderate climates respectively are considered. A 317 Wp module with $I_{SC} = 6.0$ A, $V_{OC} = 64.9$ V, $I_{max} = 5.7$ A and $V_{max} = 55.6$ A is simulated. To evaluate the impact of PV technology a traditional Al-BSF and PERC modules are simulated by changing their sensitivity to specific degradation mechanisms (detailed will be discussed in the final manuscript).

A. Simulation of climate and technological impacts

Fig. 2 shows the simulated evolution of series resistance (a) and transmittance (b) in the three locations. It is visible that there is more series resistance increase in Adrar (Hot and dry) compared to Accra (hot and humid) and Brussels. On-contrary, in hot and humid climate, higher transmittance loss is predicted compared to hot and dry climate. The higher transmittance loss in Accra can be linked to the browning/yellowing effect of PV encapsulate materials caused by the combination of high UV dose, higher temperatures and relative humidity.

Fig. 3 shows the simulated degradation rates of the current-voltage (I-V) parameters in the three locations. The degradation rates are evaluated using the year-on-year (YOY) method based on NREL Rdtools [4] as shown in Fig. 3 (a,b). Both temperature and irradiance corrections are considered in the evaluations. The irradiance filter of 200 W/m^2 -1200 W/m^2 is used to remove night time data and non-uniform irradiance scenarios.

Fig. 3 (c) and Fig. 3 (d) show the simulated degradation rates considering scenarios of Al-BSF and PERC modules. The main difference between the two technologies is that Al-BSF module show higher short-circuit current (I_{SC}) and fill factor (FF) degradation and very small (V_{OC}) loss. On-contrary, the PERC module shows relatively high V_{OC} loss on comparison to Al-BSF module. Such degradation trends have been reported by some authors [3]. According to the simulations, the increased V_{OC} degradation in PERC modules showed a negative impact on P_{max} degradation in hot and dry climate on comparison to other climates. We hope to carry out more extensive literature review and data analysis from operational PV modules/systems to validate these degradation trends.

Additionally, it is visible that P_{max} degrades more in hot and dry climates compared to other climates. This is consistent to what is reported in the extensive degradation rates surveys [1], [2]. The higher series resistance increase in hot and dry climates leads to higher fill factor degradation.

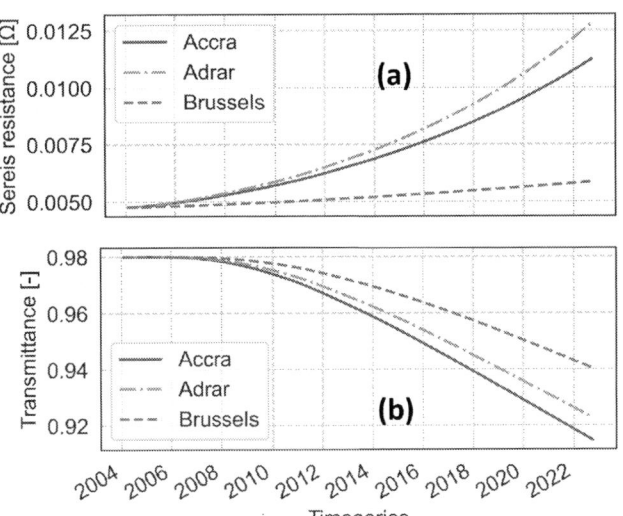

Fig. 2: Simulated degradation of series resistance (a) and transmittance (b) in different locations.

B. Validation

The validation of degradation rate/lifetime models for PV is indeed, a very challenging task due to the scarcity of the validation data. However, different aspects can be considered to reduce the uncertainties of the degradation models, such as: (i) improving accuracy of input parameters, (ii) good

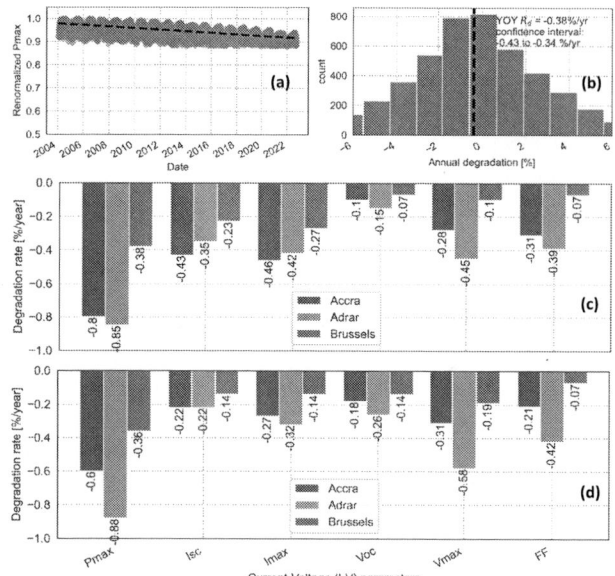

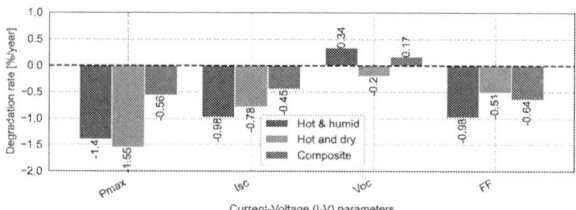

Fig. 4: Measured degradation rates in different climate zone. Figure plotted using data from [2]. Positive values mean increased perfomance.

Fig. 3: Evolution of daily renormalized P_{max} (a), and the corresponding histogram of the annual degradation rate (Rd) based on the YOY analysis (b) (dahed line shows the median). Simulated I-V parameters degradation rates for Al-BSF module (c) and a PERC module (d).

understanding of specific degradation mechanisms and the factors influencing them, and (iii) correlating the simulated degradation rates trends in different climates to what is observed in the field. We intend to deploy all these aspects to improve the reliability and accuracy of our models. For example, in this work the simulated climate and technology impacts where correlated with the degradation rates trends reported in the literature [1]–[3] (also Fig. 4). It should be noted that since the bill of materials, installation conditions and other specific factors impacting PV reliability are varying, the focus is to validate the degradation trends rather than absolute values. First results show that our simulated degradation trends in different climates relatively agree with those reported in different studies (i.e more degradation of P_{max} and I_{SC} in hot and dry followed by hot and humid climates). Further understanding of degradation mechanisms using results from accelerated ageing experiments and field degradation data analysis are underway to help us improve our model development, calibration and validation to make it more reliable.

IV. SIGNIFICANCE OF THE WORK FOR THE FIELD

There are numerous aspects where physics-based models are useful in the ever-growing PV industry, for example: (a) to correlate indoor accelerated aging measurements to outdoor conditions, which is a yet unsolved question, (b) to model and predict the degradation behavior of new PV applications and installation methods (VIPV, BIPV, floating PV, etc.), which all may have unique degradation patterns, (c) to model the degradation rates of new PV designs, technologies, and

materials (glass-glass vs glass-backsheet, bifacial vs monofacial etc.), (d) to model the degradation rates in different environmental conditions. A generalized model/approach that considers all these aspects is required to simulate and predict the degradation rates and lifetime of PV modules. Such an approach is still missing in the PV industry but could save time and money to evaluate the long-term benefits of some new PV aspects in terms of reliability. We are working towards developing this simulation approach [4] and it is good to see that other researchers agree that a PV reliability modelling framework is still a knowledge gap in the PV industry [7].

V. SUMMARY OF THE WORK

We propose a physics-based approach to model the degradation/lifetime of PV modules considering different bill of materials, technology, installation, design, and the different climate conditions. The first results show the potential of the proposed approach to evaluate the climate and technology impacts on PV module degradation rates. We also demonstrate the possibility to simulate the dominating or influencing degradation mechanisms on the lifetime of a PV module.

REFERENCES

[1] Dirk C. Jordan, John H. Wohlgemuth, and Sarah R. Kurtz, "Technology and Climate Trends in PV Module Degradation", Presented at the 27th EUPVSEC, 2012.

[2] R. Dubey et al., "Performance degradation in field-aged crystalline silicon PV modules in different indian climatic conditions," 2014 IEEE 40th Photovoltaic Specialist Conference (PVSC), Denver, CO, USA, 2014, pp. 3182-3187, doi: 10.1109/PVSC.2014.6925612.

[3] D.C. Jordan et al. "High Efficiency Silicon Module Degradation - From Atoms to Systems, presented at the 37th EUPVSEC, 2020. DOI: 10.4229/EUPVSEC20202020-4BO.14.2

[4] I. Kaaya, et al., "Lifetime Prediction of Photovoltaic Modules: Towards a Generalized Physics-Based Approach", 8th World Conference on Photovoltaic Energy Conversion, Milan, Italy, 2022.

[5] I.T. Horvath, H. Goverde, P. Manganiello, A. Schils, A.S.H. Van der Heide, J. Govaerts, E. Voroshazi, G.H. Yordanov, J. Moschner, I. Oroutzoglou, L.A. Radkar, N.-P. Harder, T. Mueller, A. Lambert, S. Scheerlinck, B. Aldalali, D. Soudris, A.H.M.E. Reinders, F. Catthoor, J. Poortmans, "Next Generation Tools for Accurate Energy Yield Estimation of Bifacial PV Systems – Best Practices, Improvements and Challenges" 36th EUPVSEC, 2019

[6] M. G. Deceglie, A. Nag, A. Shinn, G. Kimball, D. Ruth, D. Jordan, J. Yan, K. Anderson, K. Perry, M. Mikofski, M. Muller, W. Vining, and C.s Deline "RdTools", version: 2.2.0-beta.1, Compuer Software, https://github.com/NREL/rdtools. DOI:10.5281/zenodo.7411201

[7] Springer, M, Jordan, DC, Barnes, TM. Future-proofing photovoltaics module reliability through a unifying predictive modeling framework. Prog Photovolt Res Appl. 2022; 1- 8. doi:10.1002/pip.3645

978-1-6654-6060-6/23 $31.00 © 2023 IEEE

Epitaxial Growth and Testing of 1.1 eV Metamorphic InGaAs/GaAs Laser Power Converters

Katelynn E Fleming, Steve J Polly, Seth M Hubbard

Rochester Institute of Technology, Rochester, NY, United States

Laser power converters (LPCs) are photovoltaic devices used in laser power beaming (LPB) systems to transmit power from one location to another through free space or an optical fiber via laser light. This application is well suited for extraterrestrial exploration, such as on the moon or Mars, for accessing regions hostile to a heavy or solar generator. Current challenges include increasing the voltage of LPCs and the energy conversion efficiency of the full system. This work concerns the epitaxial growth and fabrication of a LPC and tunnel junction (TJ) for a high voltage multijunction device using metamorphically graded $In_{0.20}Ga_{0.80}As$ with a bandgap of approximately 1.1 eV on GaAs. The bandgap was chosen to effectively convert 1064 nm laser light, due to the availability of high power, high coherence lasers at this wavelength. The devices were grown via metal organic chemical vapor deposition at RIT. The LPCs and TJs have been grown. The pin TJ structure attained a current of $37 A/cm^2$. The nip TJ structure did not show a discernable negative differential resistance region. It is hypothesized that zinc diffused across the junction during the higher temperature growth of the upper layers of the diode. Development of the LPC and TJs continues in an effort to create a multijunction device.

Suitability of GaAsBi as a candidate junction in a III-V multi-junction solar cell

Thomas Wilson, Nicholas Ekins-Daukes

Imperial College, London, United Kingdom

UNSW, Sydney, Australia

The introduction of bismuth into GaAs results in a large band-gap bowing and a dramatic reduction in band-gap energy. This enables GaAs alloys containing dilute fractions of Bi to offer technologically useful absorption thresholds in multi-junction solar cells. Here we investigate the opportunity that GaAsBi holds as a component in an upright metamorphic solar cell architecture which, in principle, could offer a better match to the AM0 spectrum over the conventional InGaAs based UMM cell. We show that the strong and inherent alloy disorder associated with the dilute Bi fraction results in a degradation in the GaAsBi sub-cell by as much as 202 mV with 5.5% Bi incorporation. Despite this degradation in voltage we find that a potential GaAsBi based design that accounts for finite diffusion length and alloy disorder matches the conventional InGaAs UMM architecture at AM0, requiring only 2.8% Bi incorporation and 0.25% compressive strain.

~20% efficient Si PERC solar cell with emitter surface passivated by H$_2$S reaction

Tasnim K. Mouri[1,2], Ajay Upadhyaya[4], Ajeet Rohatgi[4], YoungWoo Ok[4], Amandee Hua[5], Dirk Hauschild[5,6,7], Lothar Weinhardt[5,6,7], Clemens Heske[5,6,7], Vijaykumar Upadhyaya[4], Brian Rounsaville[4], William N. Shafarman[1,2], Ujjwal K. Das[1,2,3]

[1]Institute of Energy Conversion, University of Delaware, Newark, DE 19716, USA
[2]Materials Science & Engineering, University of Delaware, Newark, DE 19716, USA
[3]Department of Electrical and Computer Engineering, University of Delaware, Newark, DE 19716, USA
[4]School of Electrical and Computer Engineering, Georgia Institute of Technology, Atlanta, USA
[5]Department of Chemistry and Biochemistry, University of Nevada Las Vegas, Las Vegas, USA
[6]Institute for Photon Science and Synchrotron Radiation, Karlsruhe Institute of Technology, Karlsruhe, Germany
[7]Institute for Chemical Technology and Polymer Chemistry, Karlsruhe Institute of Technology, Karlsruhe, Germany

Abstract — **Phosphorous (n$^+$) diffused emitter surface was passivated by hydrogen sulfide (H$_2$S) gas reaction and fabricated to small area (4 cm^2) p-type passivated emitter and rear contact (PERC) Si solar cells using an industry standard screen printed metallization process. A promising implied open circuit voltage (iV$_{OC}$) of 686 mV was achieved with the emitter (n$^+$) passivated by sulfur. A completed cell V$_{OC}$ of $\approx$ 645 − 650 mV was recorded in light JV curves, and an efficiency of 19.93% was demonstrated. Degradation during the metal contact firing step was found, limiting cell performance. X-ray photoelectron spectroscopy (XPS) studies reveal an increase in sulfur at the surface after high temperature exposure, suggesting the diffusion of sulfur and/or thinning of the SiN$_x$ surface passivation layer.**

Keywords: p-PERC cell, n$^+$ diffused emitter, surface passivation, hydrogen sulfide reaction passivation, screen printed metal contact firing, x-ray photoelectron spectroscopy

I. INTRODUCTION

The passivated emitter and rear contact (PERC) has emerged as the mainstream commercial Si solar cell in the past decade with global production capacity of ~120 GW in 2022. The success of this technology comes from the relentless improvements in cell and module efficiency by reducing the back surface recombination loss with improved rear optics. The current record solar cell efficiency, in laboratory scale, is 24.5% for a p-type PERC solar cell demonstrated by Trina Solar [1]. The most important feature of PERC cells is the introduction of an aluminum oxide (Al$_2$O$_3$) rear surface contact passivation structure, which provides excellent surface passivation of p-type Si surfaces, less shunt parasitic current and enhanced light absorption by improved rear reflection. The localized laser removal of Al$_2$O$_3$ provides low resistance contacts [2]. The front emitter junction surface is passivated by old but established methodology and materials, i.e., by hydrogenated amorphous Si nitride (a-SiN$_x$:H) and/or a stack of SiO$_2$/a-SiN$_x$:H. In this work, we have explored an alternative front emitter passivation method using a hydrogen sulfide (H$_2$S) gas reaction. Over the years, many materials have been developed for surface passivation to improve the cell performance. Most commonly used passivating materials include a-SiN$_x$:H, Al$_2$O$_3$, SiO$_2$ and hydrogenated amorphous silicon (a-Si:H). All these

materials can effectively reduce Si surface recombination (surface recombination velocity < 5 cm/s) on specific wafer surfaces and in different device structures. But they have their drawbacks as well, such as the fact that amorphous silicon (a-Si:H) passivation degrades at high temperatures, limiting the downstream cell processing temperature to <300°C, and suffers from parasitic light absorption loss [3]. Silicon dioxide (SiO$_2$) is another commonly used passivation layer that can be grown either by dry oxidation or wet steam oxidation [4,5] at temperatures >850°C, which introduces a challenge to maintain the Si bulk quality. Having a substantial negative fixed charge density makes aluminum oxide (Al$_2$O$_3$) passivation more suitable for p-type doped surfaces [6]. Therefore, the search for alternative passivation layers has been a subject of extensive research and a detail review of them can be found in the literature [7].

In the quest for an alternative passivation material, it was observed that hydrogen sulfide (H$_2$S) interacts with Si (100) by sitting in the bridge position (Si-S-Si) and passivating the dangling bonds, analogous to H$_2$O [8]. For exposure of Si(100) surfaces to H$_2$S gas in an ultra-high vacuum chamber (base pressure ~4×10^{-11} Torr), dissociative adsorption (H$_2$S → H+ HS) was found at low temperatures ranging from −145 to 425°C [9]. A desorption of hydrogen, as well as S diffusion into the Si crystal with formation of Si–S–Si bonds by breaking the Si dimer over the temperature range 525–625°C, was observed by temperature-programmed desorption (TPD) and Auger electron spectroscopy (AES) measurements [9]. A minority carrier lifetime >2000 μs for n-type [1] and >250 μs for p-type Si(100) planar wafers was reported after H$_2$S gas phase reaction [10].

In this work, we have applied the S-passivation approach for the n$^+$ diffused front emitter surface in PERC solar cell structures and compared the results with and without S-passivation. An implied open circuit voltage (iV$_{OC}$) of $\approx$ 680 mV (highest 686 mV) is observed after emitter passivation by S, and a cell V$_{OC}$ $\approx$ 650 mV and efficiency $\approx$19.93% is recorded with S-passivation. Chemical changes at the surface of S-passivated Si with a SiN$_x$ capping layer after high temperature

978-1-6654-6060-6/23 $31.00 © 2023 IEEE

exposure are identified using x-ray photoelectron spectroscopy (XPS).

II. EXPERIMENTAL

The PERC solar cells were fabricated on ~2 Ωcm boron-doped p-type Cz Si wafers. The fabrication process of the PERC cell involved saw damage removal in a heated potassium hydroxide (KOH) solution, followed by alkaline texturing of both sides of 6" pseudo-square Si wafers. A conventional Centrotherm tube furnace was used to form the front phosphorus oxychloride (POCl$_3$) homogeneous emitter with ~75 Ω/□ sheet resistance. The rear side was then planarized by KOH solution with a front SiN$_x$ mask, followed by removal of the front mask in hydrofluoric acid (HF) and cleaning of the wafers. This was followed by annealing at 700°C in nitrogen (N$_2$) ambient. All wafers had an identical rear planar surface, passivated by atomic layer deposition (ALD) of Al$_2$O$_3$, and a SiN$_x$:H layer stack was prepared by plasma enhanced chemical vapor deposition (PECVD). The wafers were then cut into 75 x 37 mm^2 large pieces and further cleaned in 10% HF solution for 1 min to remove any native oxide on the exposed n-type diffused (n$^+$) surface. The wafers were then immediately loaded into the H$_2$S reactor, which was pumped down to < 1x10^{-6} Torr. The S-passivation of the emitter was performed by reacting in a 3.4% H$_2$S – Ar gas mixture at 550°C for 60 mins. Since only the n$^+$ diffused emitter of the sample is required to be passivated by S, S-passivation was performed simultaneously on two samples by loading the samples vertically back-to-back in some runs. The S-passivated surfaces were capped with a low temperature PECVD deposited nitride (LT-SiN$_x$) at 300°C with a thickness of ~30 nm immediately after taking them out of the H$_2$S reactor. This was followed by a high-temperature nitride (HT-SiN$_x$) deposition at 450°C (thickness of ~70 nm) to achieve anti-reflection property. The rear surface was laser ablated to open the ontact area and the front Ag and back Al metals were screen printed. The samples were then co-fired at a temperature of ~760°C for ~ 3 sec in a commercial belt furnace to form the contacts. Two PERC cells of areas 2 x 2 cm^2 were defined by laser scribing at the rear surface in each 75 x 37 mm^2 wafers. The schematic structure of a fabricated solar cell is shown in Fig. 1. The iV$_{OC}$ values reported here are measured from a Sinton WCT-100 tool which uses the quasi-steady-state photoconductance (QSSPC) method to evaluate the passivation quality [11].

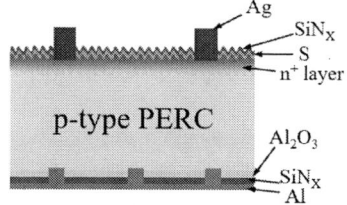

Figure 1: S-passivated PERC cell structure developed during this work.

To study the surfaces after high temperature exposure, a separate sample set was prepared consisting of two phosphorous-doped n-type diffused (n$^+$) S-passivated Si wafers with 30 nm LT-SiN$_x$ capping layers, using the same procedure as discussed above. One of the wafers was exposed to rapid thermal processing (RTP) at 700°C for 2 minutes in Ar atmosphere to simulate the firing process. After preparation, samples were sealed in inert atmosphere and shipped to UNLV for XPS which was performed with a SPECS PHOIBOS 150 MCD electron analyzer and a SPECS XR 50 Mg K$_\alpha$ x-ray source. The analyzer was calibrated according to [12].

III. RESULTS & DISCUSSION

Fig. 2 shows the measured iV$_{oc}$ values before metallization and the cell V$_{oc}$ obtained from illuminated J-V curves after metallization on the finished small area cells (4 cm^2). The black squares, red circles, and white squares represent iVoc for no S (LT nitride + HT nitride), with S (S + LT nitride + HT nitride), and cell V$_{oc}$, respectively. An iV$_{OC}$ ≈ 680 mV (highest 686 mV) with emitter (n$^+$) passivated by S was recorded. For wafers with only SiN$_x$ passivation of the emitter, iV$_{oc}$ values were in the range of 625-627 mV. This indicates the effectiveness of the S-passivation, as the wafers had higher iV$_{oc}$ than the wafers with only silicon nitride passivation. The best PERC cell with S-passivation had an efficiency of 19.93%, FF of 76.79%, J$_{sc}$ of 40.03 mA/cm^2, but with a V$_{oc}$ of only ≈649 mV. The cell Voc measured from light JV curves (white squares in Fig. 2) are, however, similar for both, with and without S-passivation, and in the range of 640 – 650 mV. This suggests that degradation occurs during the cell fabrication steps after the S-passivation (e.g., during application of the PERC pattern and metal firing). One possible reason for the low performance could be degradation occuring after metal firing, as changes at the surface are observed after high temperature exposure (see below).

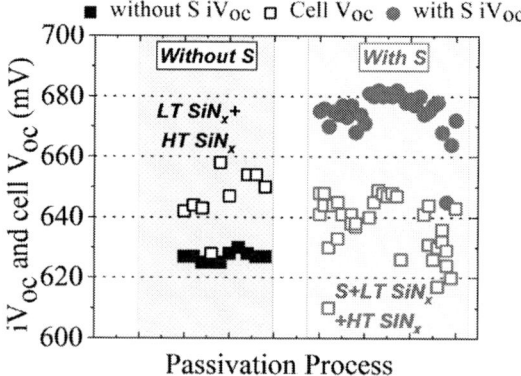

Figure 2: Implied open circuit voltage (iV$_{oc}$) of cells with n$^+$ emitter without (black squares) and with S-passivation (red circles). Corresponding cell V$_{oc}$ values after device fabrication are shown in open symbols.

Fig. 3 shows the XPS survey spectra of the S-passivated n-n$^+$ silicon wafers with and without RTP. Expected signals of nitrogen and silicon are visible, in addition to weak sulfur-related intensities. A significant amount of oxygen is present on the sample without RTP. After RTP, the oxygen signal is even further increased. A (small) presence of sodium is likely due to residues in the RTP chamber, and a small zinc

978-1-6654-6060-6/23 $31.00 © 2023 IEEE

signal is ascribed to the fact that Zn is a common trace metal in silicon. Additionally, the presence of a small fluorine signal is likely due to the use of HF.

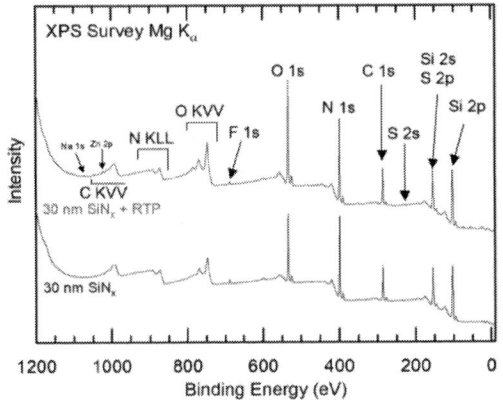

Figure 3: Mg K_α XPS survey spectra of the S-passivated n-n$^+$ silicon wafers with a 30 nm SiN$_x$ layer, before and after RTP. Prominent photoemission and Auger peaks are labeled [12,13].

Details of the S 2s region are shown in Fig. 4. Prior to RTP, a weak sulfur signal is visible and found to be in a sulfite/sulfate-like chemical environment. This is not unexpected, as oxygen is strongly present at the sample surface.

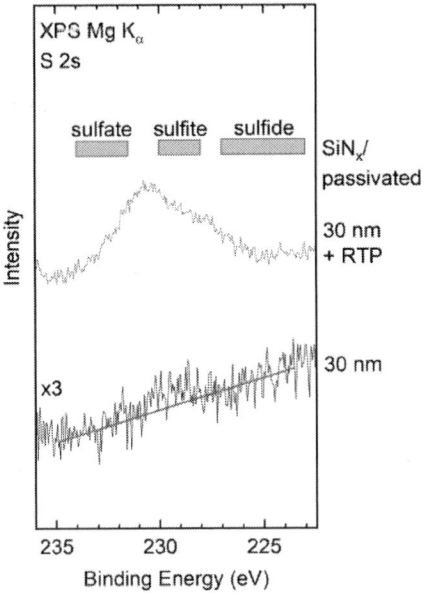

Figure 4: XPS detail region of the S 2s core level. Gray boxes represent typical literature-based binding energies for various chemical environments of sulfur [12]. A linear background (in red) is drawn for the "30 nm" spectrum to emphasize the presence of sulfur.

After RTP, a significant increase in the sulfur signal is visible, in addition to a broadening of the sulfur peak to lower binding energies (i.e., a less oxidized environment). The increased sulfur signal after RTP suggests a diffusion of S towards the SiN$_x$ surface (RTP-induced diffusion) and/or a

modification of the SiN$_x$ layer (e.g., a thinning or the formation of islands or pinholes). In both scenarios, sulfur subsequently oxidizes, as also seen in the increased presence of oxygen at the surface. Nevertheless, additional, less oxidized species, are also observed.

IV. CONCLUSION

In this work, a novel passivation approach with hydrogen sulfide was integrated into a p-type PERC cell, which exhibited an effective emitter surface passivation with an $iV_{OC} \approx 686$ mV in a PERC cell structure. However, a low cell V_{OC} of only 649 mV suggests degradation during cell and contact firing processes. XPS analysis suggest the diffusion of sulfur and/or modification of the SiN$_x$ layer after high temperature exposure.

Acknowledgment

This work was supported by the US Department of Energy's Office of Energy Efficiency and Renewable Energy (EERE) under Solar Energy Technologies Office (SETO) Agreement Number DE-EE0008554.

REFERENCES

[1] https://www.pv-magazine.com/2022/07/13/trina-solar-achieves-24-5-efficiency-for-210-mm-p-type-perc-solar-cell/

[2] V. I. Kuznetsov, M. A. Ernst and E. H. A. Granneman, "Al2O3 surface passivation of silicon solar cells by low cost ald technology," *2014 IEEE 40th Photovoltaic Specialist Conference (PVSC)*, pp. 0608-0611 2014

[3] Z. C. Holman, A. Descoeudres, L. Barraud, F. Z. Fernandez, J. P. Self J P, S. De Wolf, C.Ballif, "Current losses at the front of silicon heterojunction solar cells," *IEEE J. Photovolt.* 2 7 2012.

[4] A. G. Aberle, S. W. Glunz, A. W. Stephens, and M. A. Green, "High-efficiency Si Solar Cell: Si/SiO2, interface parameters and their impact on device performance," *Prog. Photovolt: Res. Appl.* 2 265 1994.

[5] J. Benick, K. Zimmermann, J. Spiegelman, M. Hermle and S W Glunz, "Rear side passivation of PERC-type solar cells by wet oxides grown from purified steam," *Prog. Photovolt: Res. Appl.*, 19, 361, 2011.

[6] J. Schmidt, A. Merkle, R. Brendel, B. Hoex, M. C. M. van de Sanden, and W. M. M. Kessels, "Surface passivation of high-efficiency silicon solar cells," *Prog. Photovolt Res. Appl.*, vol. 16, pp. 461–466, 2008.

[7] L E Black, B W H van de Loo, B Macco, J Melskens, W J H Berghuis and W M M Kessels, "Explorative studies of novel silicon surface passivation materials: Considerations and lessons learned," *Sol. EnergyMater. Sol. Cells,* 188, 182, 2018.

[8] V. Barone, "The cluster approach in the study of atomic and molecular chemisorption on silicon", *Surface Science*, 189-190, 106–113, 1987.

[9] M. Han, Y. Luo, N. Camillone and R. M. Osgood, "Reaction of H2S with Si (100)", *J. Phys. Chem* B 104 6576 2000.

[10] H-Y. Liu, U. K. Das and R. W. Birkmire, "Surface Defect Passivation and Reaction of c-Si in H2S", *Langmuir* 33 14580 2017.

[11] R. A. Sinton and A. Cuevas, "Contactless determination of current–voltage characteristics and minority-carrier lifetimes in semiconductors from quasi-steady-state photoconductance data", *Appl. Phys. Lett.*, 69, 2510, 1996.

[12] J.F. Moulder, W.F. Stickle, P.E. Sobol, and K.D. Bomben, "Handbook of x-ray photoelectron spectroscopy," Physical Electronics Division, Perkin-Elmer: Eden Prairie, MN, USA, pp. 15, 1992.

Investigations of Snail-Trail and Associated Microcrack Properities and Behavior in Brazil's Tropical Climate

Antonia Sonia A.C. Diniz[1], Neolmar de M. Filho[1], Cláudia K.B. Vasconcelos[1,2], and Lawrence L. Kazmerski[3]

[1]Pontifícia Universidade Católica de Minas Gerais (PUC Minas), Belo Horizonte, Brazil; [2]Universidade Federal de Minas Gerais (UFMG), Belo Horizonte, Brazil; [3]Renewable and Sustainable Energy Institute (RASEI), University of Colorado Boulder, Boulder CO 80309 USA

Abstract — **PV technology reliability and operating lifetime are priority interests for financial investors, developers, ensuring consumer confidence, and research investments. Performance degradation of the PV module is a major concern for technology viability. Cell and module defects and damage can be generated during the manufacturing process, transportation, handling, installation, maintenance, and operating environmental and climate conditions. The relationship between the creation of cell microcracks and the appearance of snail trails or visible discolored regions has been previously studied and reported. This study focuses on installations in the tropical region of Brazil in which high densities of snail trails are observed. This investigation examines further the formation of the snail trails, their microchemistry and origins, and the conditions under which the underlying microcracks can affect the cell and module performances. Specifically, the formation of Ag nanoparticles that react with the presence of oxygen, CO_2, and the organic encapsulant is shown to form compounds (Ag_2CO_3 and AgO_2) accounting for the snail-trail discoloration. Over time, bubble formation is observed, starting at the intersection of the microcrack with the Ag fingers then evolving along the snail trail itself. The reversible changes in P_m of the cells and modules is reported under thermal cycling. Direct evidence is presented to show this is due to the opening of the microcracks as the temperature rises above a certain point, and closing or healing of the regions as the temperature returns to near its original state.**

I. INTRODUCTION

With worldwide PV cumulative installed capacity now exceeding 1-TW, reliability is a priority to ensurse continued market growth and consumer confidence. Certainly, the understanding of potential failure mechanisms and the ability to predict and avoid performance limitations for specific climate conditions remains a prominent research venture. This case study arose from the observation of one aspect of PV durability that has occurred in several sites in tropical Minas Gerais, Brazil, for modules installed about a decade ago. Snail trail or narrow discoloration regions were recorded with high densities during visual (spreadsheet) inspections of several sites. One site in particular ememplified significant occurences of these visual defects and is the subject of this case study.

The appearance, origins, nature, and possible effects of snail trails have been investigated and published [1-7]. This case study reports on the relatively large population of snail trails observed in crystalline-Si modules operating the tropical conditions (high humidity, elevated UV-solar radiation

component, high temperature) in Minas Gerais, Brazil, over the past 5-8 years. These conditions have been reported to enhance the discoloration of EVA encapsulation and associated with the formation of snail trails [7].

The associated discoloration process was confirmed to be associated with Ag nanoparticles migrating from the grid contacts, forming the compound Ag_2CO_3 [17,20-23]. The cause was associated with the reaction of the Ag ions with CO_2 and acetic acid from the degradation of the ethylene-vinal acetate (EVA) encapsulant. The origins of the atmospheric CO_2 and moisture is primarily due to diffusion through the PV-module backsheet [5-7]—and can be significantly enhanced if these protective layers are damaged.

This paper reports updated and new results on the processes involved and especially the idenication of the effects on cell and module electrical performance. Specifically, the occurrence of the Ag nanoparticles and reactions is confirmed by several microanalysis and imaging techniques. The change in power output these modules is further explained due to the development and widening of the microcrack defects with increased operating temperature. The reversibility of opening and healing of these microcracks is shown in cell and module characterizations.

II. METHODOLOGY

The case study methodology is summarized in the research flow chart in Fig. 1. The investigations started with the speadsheet inspections of the sites followed by the identification of modules that were of particular interest to the more-detailed snail trail investigations. As noted before, one site was of particular interest—and is the focus of this paper. Modules were chosen for the distributive and non-destructive characterization procedures noted in the flow chart. The spreadsheets provided information on the considtion of the modules—particularly damage to the backsheets.

III. RESULTS AND DISCUSSION

The presentation & publication will provide complementary imaging by (a) normal photography, (b) scanning UV fluorescence imaging, and (c) EL luminescence imaging for modules that were operating at this paticular site over ~9–year period. These are compared to reference modules that were

978-1-6654-6060-6/23 $31.00 © 2023 IEEE

stored for periodically checking the performance at the power plant. The major focus of this submission is the chemical and compositional analysis of the snail trail discolorations and the electrical performance effects of the associated microcracks.

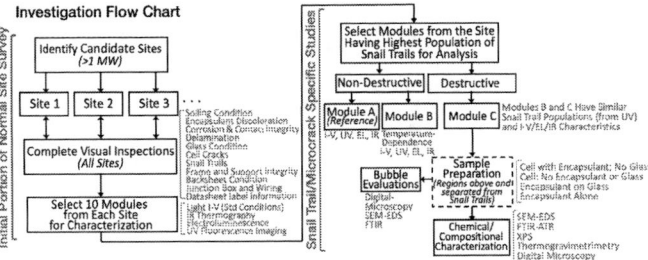

Fig. 1. Flow chart summarizing methodology [7].

A. Chemical and Compositional Characterization

Prevsiously, we provided complementary FTIR, EDS, and micro-thermogravmetric analysis of the discolored materials above the microcrack regions giving initial indiations of the chemistry and compositional changes of these regions compared to those separated from the snail trails. In this paper we report higer-rolution scanning electron microscopy of the regions around the snail trails and confirmed the co-existence of corresponding microcracks in the Si solar cell. Complementary energy dispersive spectroscopy (EDS) has gives strong indication of the existence of low levels of Ag in the discolored regions exposed in EVA samples removed from the module. These regions have been further analyzed using X-ray photoelectron spectroscopy to identify the chemical nature of the distince sample regions.

EVA samples from regions immediately above (discolored) and regions separated (clear) from the snail trails were analyzed by XPS to compare any differences in composition and chemical bonding between the areas. Fig. 2 shows 3-representative spectra for the samples. The bottom (blue} spectrum (Fig. 5a) is for the clear EVA (away from the snail trail region) and is typical for EVA that has not degraded. Two scans are shown for the region above the microcrack/snail trail (Fig. 2b,c) and show the presence of small amounts of Ag (compounds).

The Ag-3d peaks at 373.9 eV [$3d_{3/2}$] and 367.8 eV [$3d_{5/2}$] (Fig. 5b,c) are characteristic Ag_2CO_3 [8] with possible small concentrations of Ag_2O. This follows the chemical reaction cited in the previous section that accounts for the discoloration associated with the snail trails. These results support the formation of the silver carbonate in the reaction in the presence of the Ag-nanoparticles. The CO_2 ingresses through the damaged backsheet, although the cited chemical breakdown of the EVA could also release this component. The data also show that the Ag_2CO_3 nanoparticles are not uniformly distributed—and are highest in concentration near the center of the underlying microcrack. These XPS results are the strongest direct confirmation for the reaction responsible for the snail trail discolorations.

The presentation will provide complementary direct imaging of the nanoparticles (compounds) along the snail trail/

microcrack regions for the first time using high-resolution electron microsopy and specialized spectroscopic scanning probe microscopy tcchniques. While under negative polarization conditions, the water (vapor) is reduced on the surface of the Ag, generating hydrogen and OH⁻:

$$2e^- + 2H_2O \rightarrow H_2 + 2OH^-, \qquad (1)$$

causing the bubbling above the Ag fingers and eventually the Ag contaning regions above the microcracks [7].

Imaging shows initial reactions are concentrated at the intersections where the silver concentration is higher, leaving the region and the Ag finger with a brownish discoloration. The imaging shows the reaction proceeding with the evolution of the bubbles along the snail-trail itself.

B. Module Performance Characterization

The snail trails present themselves as relatively inconsequential optical impact on short-circuit current, the parameter expected to be most affected due to the snail-trail discoloration. However, the microcrack associated with the appearance of the snail trails are more of a potential issue over longer module operating periods and operating conditions, with the concern that they could develop into cracks and associated loss of power generation in portions of the cells. In order to better examine the possible effects of the microcracks associated with the snail trails, the differences in module P_{max} at low and higher temperatures is investigated, giving evidence

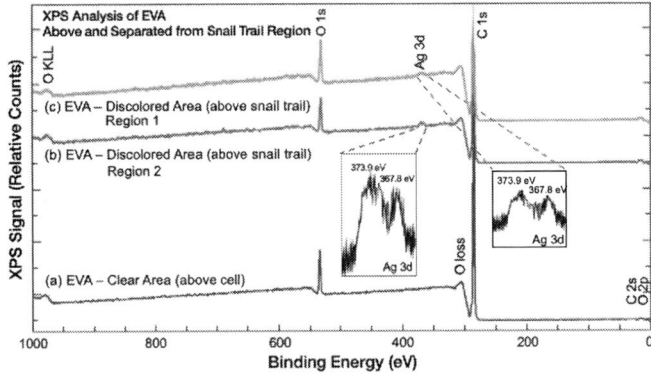

Fig. 2. XPS spectra of EVA-exposed sample areas with and without snail trails indicating presence of Ag_2CO_3 and AgO_2.in the snail trail regions. (a) Blue scan is over clear EVA; (b) Red scan is at center of snail-trail discoloration. Insets show the Ag-$3d_{3/2}$ and Ag-$3d_{5/2}$ peaks corresponding to Ag_2CO_3. (c) Green scan is for location with less discoloration.

for differences in losses in these different temperature regimes. The measured changed in P_m under these different temperature conditions can account for the previous reports of why these defects can have some effects on the module performance operating in the field.

Fig. 3 illustrates the change in module P_m for a selected PV module. This shows the expected (following the maximum power temperature coefficient, $T_k(P_m)$) characteristics compared to the experimental data. The module with very few

978-1-6654-6060-6/23 $31.00 © 2023 IEEE

snail trails (used as a reference) followed the expected temperature dependence (blue line in Fig. 3). This P_m behavior would correspond to the report of the opening of the microcracks, degrading the module performance. Upon cooling, the module IV characteristics return to their original (25 °C) condition. The measurement after the module was cooled is indicated by the red data point in Fig. 3.

The presentation will provide images of the opening of the microcrack regions on laboratory cells that complement and confirm this process. (This includes EL characterizations.) The microcracks are shown to open above a certain temperature. Subsequesntly, these regions are shown to close or heal as the sample is cooled to its original temperature condition.

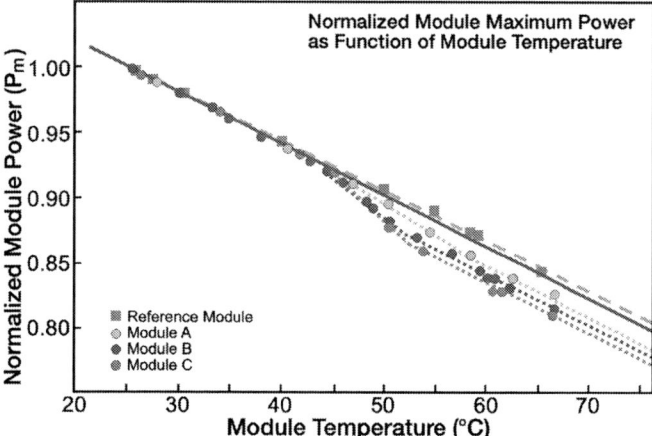

Fig. 3. Module P_m as function of operating temperature indicating effect of microcrack effects for 4 modules. Reference module (green) has no observed snail trails. Temperature coefficient is shown by green dashed line. The deviation from the expected temperature change for the other 3 modules (yellow, blue, red) with high snail trail (microcrack) populations is attributed to opening of microcracks at elevated temperatures.

Fig. 4 compares the resulting EL measurements for a 4-cell module selection corresponding to the I-V shown in Fig. 3. On increasing operating T, the microcracks are observed to open (between 40 °C and 50 °C), causing loss in regions. On cooling, most of the microcracks heal with corresponding observation of the EL signals. Not all microcracks heal, but the majority are observed to follow this behavior for the modules of this study.

IV. SUMMARY

This study centers on the formation, composition, and evolution of related chemical artifacts for snail trails in Si module. The case study is for modules operating in the tropical climate in Brazil, where higher populations of snail trails are observed for modules commissioned in the period 2010-2015. The major contributions include the direct confirmation of the discoloration process involving Ag nanoparticles and the first time direct imaging of the microcrack behavior as a function of thermal cycling.

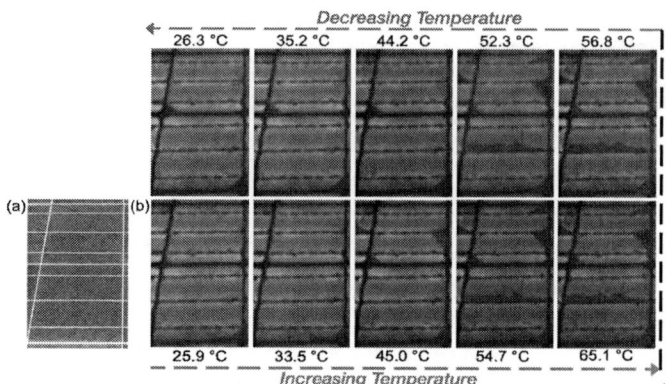

Fig. 4. Comparison of (a) snail trail and underlying microcrack locations with the electrical response of affected (in-module) cell areas—shown for two selected cells. (b) Temperature-dependent EL data indicating the opening of the microcracks on increasing operating temperature and healing of most of these defects on return to lower operating temperatures. Observations correspond to I-V and P_m data dependences on module temperature (Fig. 3).

Acknowledgements: The authors would like to thank the collaborators of the GREEN PUC Minas Laboratory (Energy Study Group) and the Graduate Program in Mechanical Engineering of the Pontifical Catholic University of Minas Gerais (PUC Minas) for their support, technical assistance, facilities, and constructive inputs. This work was partially carried out with the financial support of the Federal Center for Technological Education of Minas Gerais (CEFET-MG). We also acknowledge the assistance and cooperation of CEMIG in working with the module technologies. The authors gratefully acknowledge the Fulbright Foundation which supported this study through a 2022 Fulbright Scholar project, as well as CNPq, and CAPES-MEC.

REFERENCES

[1] S. Meyer, S.Timmel, S. Richter, M. Werner, et al. (2014). "Silver nanoparticles cause snail trails in photovoltaic modules," Solar Energy Materials and Solar Cells 121, 171-175.

[2] S. Meyer, S. Richter, S. Timmel, S. M. Gläser, et al. (2015), "Snail Trails: Root Cause Analysis and Test Procedures," Energy Procedia 38, 498-505.

[3] A. Dolara, S. Leva, G, Manzolini, & E. Ogliari, (2014). "Investigation on performance decay on photovoltaic modules: Snail trails and cell microcracks." IEEE J, of Photovolt. 4, 1204-11.

[4] P. Peng, A. Hu, W. Zheng, W., et al. (2012). "Microscopy study of snail trail phenomenon on photovoltaic modules," RSC Advances, Vol. 2, pp. 11359-11365.

[5] H-C. Liu, C-T. Huang, W-K. Lee et al. (2015). "A defect formation as snail trails in photovoltaic modules." Energy and Power Engineering 07, 348-353.

[6] M.W. Akram, G. Li, Y. Jin, & X, Chen, X. (2022). "Failures of photovoltaic module and their detection: A Review." Applied Energy 313, 118822.

[7] N. de M. Filho, A.S.A.C. Diniz, et al. "Snail trails on PV modules in Brazil's tropical climate: Detection, chemical properies, bubble formation, and performance effects." Sust. Energy Technol. Assessments 54, 102808.

[8] C. Yu, L. Wei, W. Zhou, et al. (2014). "Enhancement of the visible light activity and stability of Ag_2CO_3 by formation of AgI/Ag_2CO_3 heterojunction," Appl. Surf. Science 319, 312-318.

A Study of POCl3 Deposition Reaction Rate with Residual Gas Analysis Method

Min Gu Kang, Sang Hee Lee, Kyung Taek Jung, Yunae Cho, Dohyung Kim, Sungeun Park, Munse Kim, Hee-eun Song

Korea Institute of Energy Research, Daejeon, Korea

Real-time process monitoring technology is needed to detect and diagnose abnormal changes in manufacturing process equipment and products. This study proposes the possibility of real-time process monitoring by in-situ monitoring the change of residual gas emitted during the thin film deposition process through chemical vapor deposition technique through residual gas analyzer (RGA). RGA monitoring allows immediate identification of abnormal conditions in the thin film deposition process. To this end, PSG thin films were deposited using industrial CVD equipment. According to the change in gas inflow, the composition ratio of exhaust gas was confirmed by RGA. The reaction rate according to the gas amount was obtained from the composition ratio of the exhaust gas, and the thin film deposition rate was obtained from this.

Analysis of Photovoltaic Systems Penetration on Demand Curve and Locational Marginal Prices (LMPs) in PJM

Mesude Bayrakci Boz
Penn State Hazleton
Hazleton, PA
mzb187@psu.edu

Abstract—It is important to analyze the effect of PV penetration on the demand curve and Locational Marginal Prices (LMPs) to establish better control of energy dispatches, economic optimization and a better plan for future PV penetration. This study first analyzes how the demand curve changes with PV penetration in the PECO zone in the PJM region. Next, it predicts LMP prices in all PJM regions according to PV penetration in the PECO zone using the Seemingly Unrelated Regression (SUR) model monthly base. The power production is calculated using suitable rooftops with possible minimum and maximum solar irradiation in Philadelphia, PA region. Comparative analyses are then performed between LMP prices with and without PV penetration; hourly percentage differences are also calculated and the ripple effect is studied. The findings of this study suggest that PV penetration in the PECO zone affects the PECO demand curve, and in particular, that it causes ramps during morning and afternoons. The findings also indicate that PV penetration decreases the LMPs in the PECO zone around 10% and the LMPs in the neighboring zones along the descent in the spring, summer, and fall. It decreases in some zones in the winter.

Index Terms—Photovoltaic, Demand Curve, Locational Marginal Price (LMP), SUR model, PJM.

I. INTRODUCTION

Over the past decade, energy demand has increased while concerns about environmental issues and the security of energy suppliers have grown. Therefore, renewable energy resources, especially photovoltaic systems (PV), have received much attention and become important parts of the energy resource mix. PV systems produce power during the day, thereby affecting the demand curve. This is especially the case between 9:00 am to 3:00 pm during the summer and 11:00 am to 2:00 pm during the winter when the solar peak power is high. Therefore, PV systems have the potential to meet energy demands while alleviating congestion within the grid infrastructure and decreasing locational marginal prices (LMPs). However, solar energy is a variable generation source, and it causes a more volatile net load than more conventional energy sources; it might effect the grid structure and causes the cost. Therefore, it is important to analyze the effect of PV penetration on the net load curve and LMPs to establish better control of energy dispatches, achieve economic optimization, and better plan for future PV penetration.

The demand curve is a graph of variations in demand versus time. PV systems effects the demand curve. When the power

generated by PV systems is less than or equal to the demand, the power can be considered to be depleted. When the power generated by PV systems is more than the energy consumed at the point of use, the surplus electricity flows back to the grid. If the generated power surplus is high and there are enough PV systems, the demand curve changes dramatically within a 3-4 hour period since the peak demand for electricity is between 3:00 pm to 7:00 pm in summer, and 6:00 am to 8:00 am and 5:00 pm to 7:00 pm in the winter, thereby causing problems for the grid such as short, steep ramps and an increased risk of over-generation.

Even though there are number of studies offers methods to analyze how renewable energy effect demand curve, there is, however, a lack of models able to predict LMP prices in a RTO level with PV integration and quantify the magnitude of the associated impact. This study analyzes the impact of PV integration on the demand curve and LMPs in the PJM region (RTO level) when PV systems are incorporated into the PECO zone, and provides a method that allows predicting LMP prices with PV penetration using different scenarios, using the seemingly unrelated regression (SUR) approach.

A. Seemingly Unrelated Regression Model

In general, there is more than one equation that can be used to describe models. The disturbance terms of these equations may be correlated since they may be influenced by the same unconsidered factors. In order to arrive at efficient estimates, such covariance should be considered. This procedure is called Seemingly Unrelated Regression (SUR) [1]. In this model, consider a system of G equations, where the ith equation is of the form:

$$y_i = X_i * \beta_i + u_i, i = 1, 2, ..., G, \qquad (1)$$

where y_i is a vector of the dependent variable, x_i is a matrix of the exogenous variables, i is the coefficient vector and u_i is a vector of the disturbance terms of the ith equation [2]. There are two assumptions in the SUR model. First, the model assumes that there is no correlation of the disturbance terms across observation. On the other hand, the model assumes there is contemporaneous correlation.

978-1-6654-6060-6/23 $31.00 © 2023 IEEE

II. DATA AND METHODOLOGY

A. Load Data and Locational Marginal Price (LMP) Data

There are various zones in the PJM region. The website includes 17 zones with subzones for load data and 20 zones for LMP data. 3 years metered load data (Megawatt-hour) and LMPs ($/MWh) were obtained from PJM for these zones.

B. Solar Power Simulation

PV power production was calculated on an hourly basis for the city of Philadelphia, PA using TRNSYS (TRaNsient SYstems Simulation) for all suitable rooftops [3]. As an initial step, an automated ArcGIS model was used to obtain tilt, azimuth, and area information for the suitable rooftops in Philadelphia, PA. Secondly, hourly measurements of solar irradiation for Philadelphia, PA, ten year data, were used. The data was obtained from the Solar Anywhere website [4]. Solar irradiation can change year to year due to atmospheric conditions. Two scenarios were created in order to analyze the effects of possible minimum and maximum solar irradiation on power production and the demand curve. In Scenario I, the maximum value of solar irradiation of one hour over 10 years was used. In Scenario II, the minimum value of solar irradiation of one hour over 10 years was used. Finally, these minimum and maximum solar irradiation data file and outputs of the ArcGIS model was used as the input, while power production is calculated on an hourly basis in TRNSYS.

C. The Demand Curve and SUR Analyses

In order to analyze how the demand curve would change with PV penetration in the PECO zone, power production in Philadelphia, PA region was calculated and the net load curves were created for each month.

LMPs can be expressed as a function of demand in the zone and the demand in rest of the PJM, Equation 2.

$$LMP = b_0 + b_1 \times Demand_{PJM} + b_2 \times Demand_{PJM}^2 \quad (2)$$
$$+ b_3 \times Demand_{zone} + b_4 \times Demand_{zone}^2 + u_1$$

where b_0, b_1, b_2, and b_4 are coefficients and u is the residuals (disturbance term).

A mathematical relationship was posed and it holds for all zones, except coefficients is different for each zone. SUR technique was used to estimate the parameters of this model. Therefore, Equation 2 was written for all 20 zones in the PJM region and can be used to predict LMPs. All process is applied in R and all of the analyses were performed on a monthly basis. The SUR was applied, and all the coefficients and residuals for each zone were obtained. Then, using the power production results, a new load was calculated for the PECO zone and PJM. In all of the equations, the PJM load was changed; however, with the exception of the PECO zone, the zone loads were not changed. Next, all coefficients and residuals were substituted into the equations with the new loads, and possible LMPs were calculated.

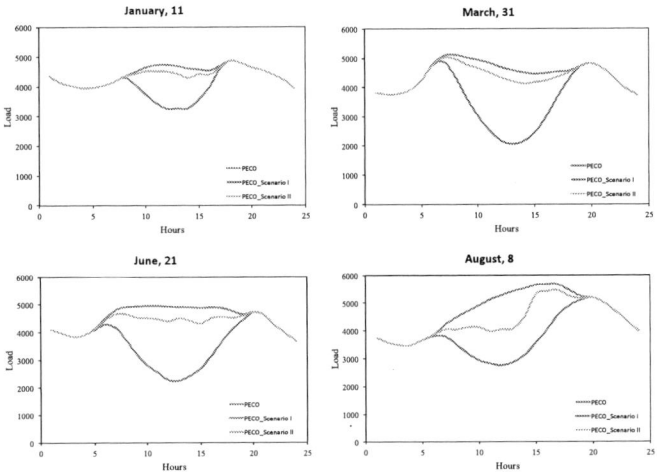

Fig. 1. The demand curves for PECO zone

D. Hourly LMPs' Difference Evaluation

The LMP results from the SUR model and the real LMP values were compiled on a comparative basis in two ways. First, the hourly percentage difference for each month based on the real LMP value was determined. Equation 3 was used for this calculation.

$$\text{Hourly Percent Difference}(\%) = \left(\left(\frac{\text{Estimated LMP}}{\text{Real LMP}} \right) - 1 \right) \times 100\% \quad (3)$$

III. RESULTS AND DISCUSSION

A. The Demand Curve

The effects of solar power production on hourly ramp-rate in net load were investigated for each month, and January, March, June, and August are shown in Figure 1 as an example. It can be seen that the effect of PV on the load curve is obvious, even with possible minimum power production (Scenario II); moreover, PV creates ramps in the mornings and afternoons. In order to compare CA (Duck Curve) [5], HI (Nessie Curve) [6], and PECO's net demand curves, the steepest ramp was considered (3:00 pm to 6:00 pm for PECO); then, the percentage of increase in the net load was calculated and the time period was found. The results are shown in Table I. The increase in net load in PECO was not higher than those of CA during January and March. Even when the power production was at its maximum (Scenario I) due to PV systems, the increase was not as great as those of CA on March 31 and January 11. The time periods used were 5:00 pm to 8:00 pm for CA and 3:00 pm and 6:00 pm for PECO. On the other hand, August and June might be problematic for PECO: in Scenario I, the increase in net load was around 60%, but it was still smaller than the rise in CA in March. These ramping have to be addressed with some other matching hybridization power.

978-1-6654-6060-6/23 $31.00 © 2023 IEEE

TABLE I
Percent Increase in Net Loads and Time Periods for CA, HI and Philadelphia, PA

	Date	Time Period	Percent Increase in Net Load
California	January 11	4:00 pm to 7:00 pm	47
	March 31	5:00 pm to 8:00 pm	108
Hawaii	August 11	2:00 pm to 6:00 pm	600
PECO (Scenario I)	January 11	3:00 pm to 6:00 pm	35
	March 31	3:00 pm to 6:00 pm	20
	June 21	3:00 pm to 6:00 pm	55
	August 8	2:00 pm to 6:00 pm	60
PECO (Scenario II)	January 11	–	–
	March 31	4:00 pm to 6:00 pm	6
	June 21	3:00 pm to 6:00 pm	5
	August 8	1:00 pm to 4:00 pm	33

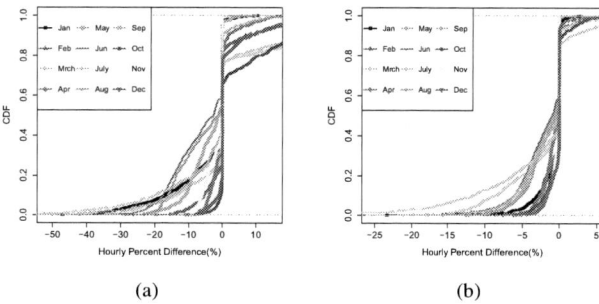

(a)   (b)

Fig. 2. CDF of Hourly Percent Differences of LMP in PECO zone a)Scenario I b) Scenario II

B. SUR Results

Eq. 2 for all 20 zones in the PJM region were solved and the coefficients for all zones and the correlation between residuals for each month was ontained. All SUR results for all months will be included in the full paper. The results shows the SUR approach is good way to estimate LMP prices.

C. Hourly Percent Differences

Figure 2 and 3 shows the cumulative distribution function (cdf) plots of the PECO and APS zones for all months. As seen from the plots, 70-80% of the hourly percentage differences indicate a decrease in LMPs in the PECO zone. LMPs decreased to about 60% with Scenario I and to about 20% with Scenario II in July. In the APS zone, 80% of the hourly percentage differences indicate a decrease, which drop to about 30% during the winter months. In the PPL zone, around 90% of the data indicate a decrease in LMP. On the other hand, 10-30% of the hourly percent differences in the zones indicate an increase in LMP. The reason for that are disturbance terms.The other plots of all of the zones also be created and similar pattern were observed.

IV. Summary and Conclusion

This study analyzes the changes in demand curve in the PECO zone and uses the SUR model to predict LMPs for all the zones in the PJM given PV penetration in the Philadelphia region. Two scenarios were considered for power production. In Scenario I, the maximum possible irradiation value based on ten-year data was used, whereas in Scenario II, the minimum possible irradiation value based on ten-year data was

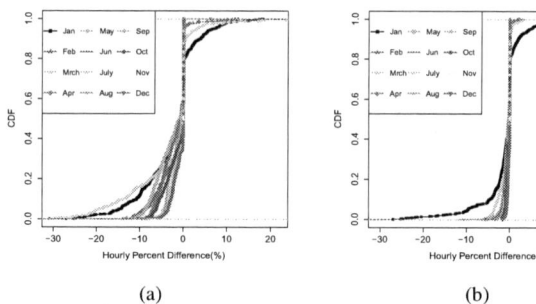

(a)   (b)

Fig. 3. CDF of Hourly Percent Differences of LMP in APS zone a)Scenario I b) Scenario II

employed. The net load was then calculated using metered load data from the PJM and power production and compared with California's duck curve and Hawaii's Nessie curve. The SUR model was then applied using real LMPs and metered load data for all zones in the PJM. Coefficients and residuals were calculated using 20 equations, one equation for each zone in the SUR model. These calculated coefficients and residuals were used to predict LMPs in PJM zones. Finally, comparative analyses were conducted. It was found that LMPs decrease in the PECO zone during the year; hourly percentage differences decrease as much as 50% percent in Scenario I and as much as 20% in Scenario II. Moreover, the LMPs in PECO's neighboring zones decrease during the spring, summer, and fall along the descent and a ripple effect can be observed. This study provides electric operators and system planners plots and maps that can allow for better control of energy dispatches and better plans for PV penetration. Knowing the impact of PV on LMPs will also help to inform planners and policymakers.

V. Acknowledgment

The author acknowledges the review and feedback by Jeffrey Brownson and Seth Blumsack of The Pennsylavaia State University.

References

[1] Arnold Zellner and Henri Theil. Three-stage least squares: simultaneous estimation of simultaneous equations. Econometrica: Journal of the Econometric Society, pages 54–78, 1962.
[2] Arne Henningsen and Jeff D. Hamann. A package for estimating systems of simultaneous equations in r, 2017.
[3] S.A. et al. Klein. Trnsys 18: A transient system simulation program. Solar Energy Laboratory, University of Wisconsin, Madison, USA., 2017. http://sel.me.wisc.edu/trnsys.
[4] Clean Power Research. Solar anywhere, 2017.
[5] California ISO. What the duck curve tells us about managing a green grid, 2016.
[6] Jeff John GTM. Hawaii's solar-grid landscape and the 'nessie curve', 2014.

Single Axis Tracker Performance Modeling on Sloped Terrain

Benjamin Pierce, Joshua Stein, Daniel Riley
Sandia National Laboratories, Albuquerque, NM, 87123, USA

Abstract—As solar photovoltaics (PV) become more widespread, installations have become more nuanced to model due to complex terrain, such as hills or mountains. Single axis trackers (SATs) provide the best economic performance on the market, and as such, modeling their performance is quite relavent. However, SATs are more difficult to model then traditional fixed-tilt systems due to a vareity of factors. This work focuses on modeling slope-aware backtracking, a modification to the current backtracking algorithm that takes terrain slope into account to avoid row-on-row shading. We present a methodology that uses digital elevation models (DEMs) to automatically calculate the terrain slope and its orientation, which can then be used to calculate tracker positions. Furthermore, we present a comparison of modeled tracker performance with and without awareness of the terrain, and show a 1% performance difference between the two. The magnitude of this effect is then useful to know when evaluating the performance of installed and proposed SAT installations.

Index Terms—single axis tracker, digital elevation model, solar resource

I. INTRODUCTION

Ideally, SAT installations would be on un-sloped level ground; this scenario is well modeled and studied. However, as SATs have become more prevalent outside of the desert Southwest, sub-optimal terrains have become more common, to the point where industry has begun to adopt customized terrain-aware single axis tracking algorithms [1]. Thus, it is necessary to model the effect of the terrain on SAT performance, which can be useful to evaluate potential sites or explain underperformance in existing plants. This modeling process generally requires two things: a way to programatically retrieve local elevation and geography information to calculate ground slope, and a modified tracker algorithm to use this information. This work focuses on the first step, and uses the method described by Anderson *et. al.* in [2] as the terrain-aware tracking algorithm. Anderson's approach is already part of the open source pvlib-python tool-chain, but as input requires the slope of the ground and its orientation to function, which is the contribution of this work.

In order to calculate the slope of the land, we use Digital Elevation Models (DEMs) provided by NASA's Shuttle Radar Topography Mission (SRTM). Elevation maps for much of

Sandia National Laboratories is a multimission laboratory managed and operated by National Technology & Engineering Solutions of Sandia, LLC, a wholly owned subsidiary of Honeywell International Inc., for the U.S. Department of Energy's National Nuclear Security Administration under contract DE-NA0003525. This work was supported by the U.S. Department of Energy's Office of Energy Efficiency and Renewable Energy (EERE) under the Solar Energy Technologies Office under Award 34367.

the world are available in 30m and 90m resolution, and are distributed in 1x1 degree tiles, as shown in Figure 1. DEMs

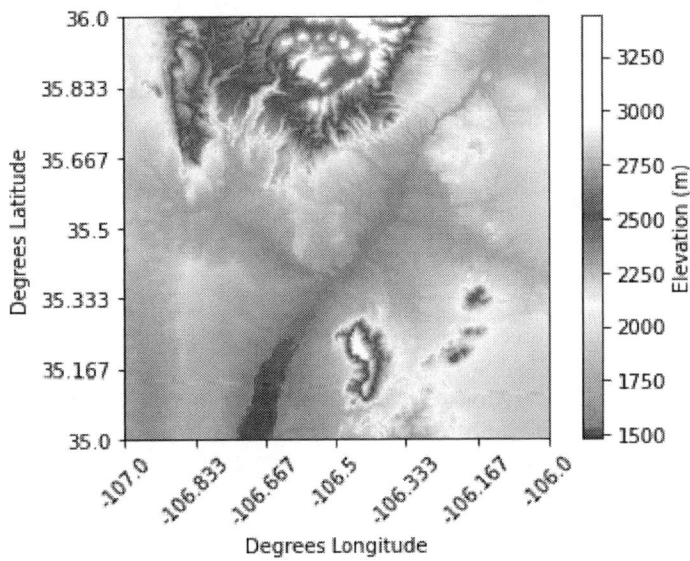

Fig. 1. A DEM of Albuquerque, NM and surrounding area.

can be manually downloaded from USGS or other sources; we developed a convenience function to do this programatically, which will be included in an upcoming release of pvlib-python. Once acquired, the metadata such as the corresponding map projection and resolution are used to calculate latitudes and longitudes for each point on the raster. The data is then ready for processing.

II. METHODOLOGY

Once imported the following the land slope and the direction it slopes in (called the "aspect" in geological literature) must be calculated. To calculate the slope, we use the method delineated in Horn, 1981 [3]. Horn's method calculates the slope with respect to a point on the DEM by fitting a plane to the point and its neighbours (as shown in Figure 2, and then using a central difference estimation to select the maximum slope. The direction of this slope is the corresponding aspect. Then, for each point on the DEM, the following information is known:

1) Geographic latitude and longitude
2) Elevation in meters
3) Slope of best fit plane of local neighbourhood

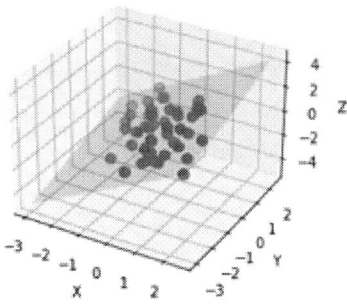

Fig. 2. A plane fitted to a collection of points

4) Aspect of the slope

This information is sufficient for simulating a SAT with Anderson's method.

However, in practical use, there are a few problems with this procedure. One such issue is speed; the DEM tiles are distributed as 3600x3600 tiles, meaning that there is a very large quantity of data to be simulated. Although calculating the slope and aspect are quite fast, the simulation itself can be quite time consuming; for survey purposes where several prospective sites are modeled, the performance can be lackluster. Additionally, the DEM data is 30m/90m accurate, which is a very fine resolution for these purposes. Without any preprocessing, small-scale variation can cause the results to be very dependent on relatively minor changes in the terrain, especially on less steep slopes. Furthermore, the results of a SAT simulated in very steep slopes (where it would be impossible/impractical to even install a SAT) are nonsensical. To summarize, the following issues must be mitigated:

1) Time-consuming computation and simulation
2) Outsized effects of small-scale terrain differences
3) Un-physical results on areas where SATs would not be installed.

The first strategy to fix these issues is to use calculated slopes to select areas where SATs would be viable. Particularly, slopes are classified as Viable, Borderline, and Nonviable. This classification is used by computing the percentiles of the distribution of slopes for the entire map, and using these

as guidelines for a heuristic approach. First, slopes greater then 15 degrees are said to be Nonviable in this study, which can be adjusted to the installer's capabilities and preference. Slopes less then this threshold and greater then the 75th percentile are classified as Borderline. Finally, slopes between zero degrees and the 75th percentile are said to be Viable. This selection process was applied to Albuquerque, NM and surrounding areas in Figure 3. Note that this selection process is only

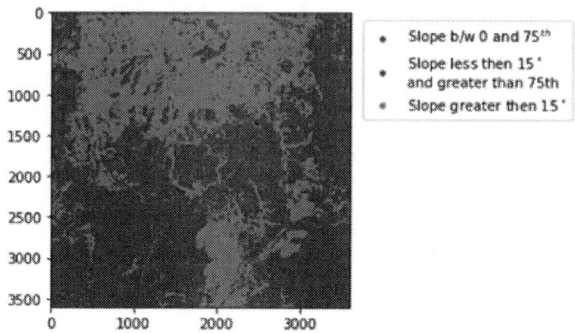

Fig. 4. Viability map for ABQ, NM and surroundings. Highly mountainous areas (red) are correctly classified as Nonviable, where as the flatter desert terrain (blue) is viable.

concerned with slopes; in practical application, we would also be concerned with topics like proximity to transmission lines. However, by using this simple heuristic process, we can eliminate much of the map that would be unrealistic to consider in the first place, thereby improving performance.

To address the outsized impacts of small changes in the DEM, we propose down-scaling the DEM by taking the average over small regions. Naturally, information is lost when down-scaling, and the idea is to downscale enough to smooth the data and make computation feasible, but not so much as to lose a large quantity of information. In areas with vector data, Shannon's entropy [4] is commonly used to quantify information loss. In the case of matrix or image data, the Structural Similarity Index Measure (SSIM) [5] can be used. However, these metrics can be somewhat misleading when applied to *downscaling* rather then *interpolation* or *smoothing*. That is, our DEMs actually decrease in resolution; obviously, a larger object will contain much more information. This

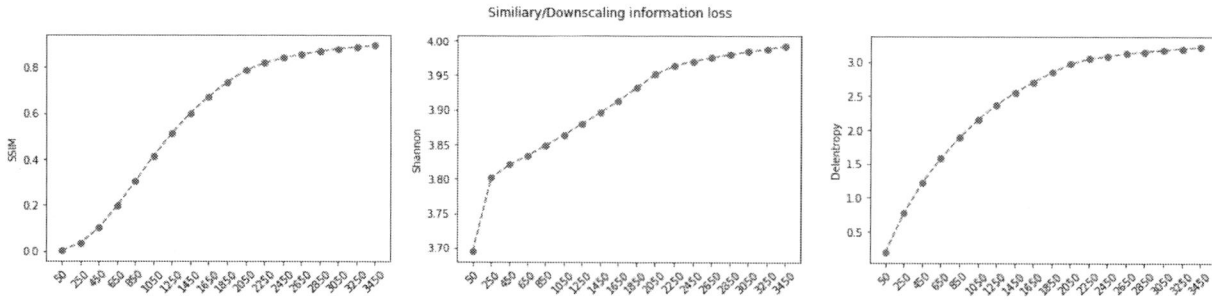

Fig. 3. Comparison of information loss quantification methods for a 2D DEM or image. Note the odd shape of Shannon's entropy. For Shannon's entropy, the curve should by definition be logarithmic but is instead somewhat piece-wise linear.

978-1-6654-6060-6/23 $31.00 © 2023 IEEE

causes problems for SSIM, and as Shannon's entropy does not preserve spacial variation (as the matrix must be flattened into a vector), neither are quite appropriate for this situation.

Instead, we employ Larkin's modification to entropy [6], termed "delentropy," which *does* consider spacial information by defining a joint "deldensity" distribution over the image, which can then be interpreted as a two dimensional probability density.

Using delentropy as a guide, we can select a point where the balance of information content to size is optimal. Note that the delentopy function is convex; we can then find the maximum point in a similar manner to the maximum power point (P_{mp}) on a diode curve. For this example, a good compromise between image size and information content is located at a map of roughly 800x800 pixels. Note that a result of the downscaling process effectively makes each pixel cover a larger area; at a ratio of the original 3600:800, each pixel is corresponds to a 4.5x larger area.

Once the DEM is appropriately preprocessed and the slope and aspect for all points on it are calculated, we simulate the performance of a single axis tracker for each point's corresponding slope, and compare to the same system where slope is assumed to be zero. This step is handed by pvlib-python which implements Anderson *et. al.*'s method. Clearsky irradiance is modeled using the Haurwitz [7] and Erbs [8] models, which do *not* incorporate a turbidity term, as we found including turbidity with our modeled data to be problematic. Plane-of-array (POA) irradiance was then calculated for each point in time over a simulation period of one year. POA is then used as a proxy for power to avoid complications due to different types of modules. For example, if half-cut modules with two junction boxes are employed, then the effects of row-on-row shading can be less drastic then a traditional module as each half of the module effectively functions separately.

III. RESULTS

Simulation is conducted for each of the three categories of land (Viable, Borderline, Nonviable). Note that in Figure 5, the Borderline and Nonviable cases experience a very large degree of spread, which supports our methodology of separating the categories to avoid large errors contributed by regions that would not have single axis trackers installed in any case.

To determine the projected gain in power, we perform a Welch $t-$test for the mean on the pixel difference (between slope-aware and slope-unaware simulations) for the regions classified as Viable. The $t-$test results in an estimation of a gain in received POA of $1.35 \pm 0.92\%$. The t statistic is very negative and the corresponding p-value is consequently very close to zero. Thus, by considering the slope of the terrain in a tracking algorithm, even on relatively level ground, the trackers will receive more POA irradiance and correspondingly produce more power.

IV. CONCLUSION

In this work, a method to model the projected POA/power gains by considering the topography of the land was demonstrated to result in a positive change in production. By using

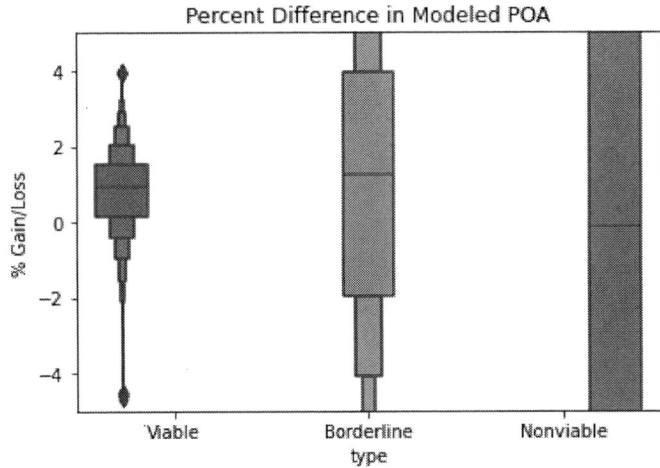

Fig. 5. Shown is the difference in modeled POA irradiance between a tracker that considers the actual topography of the terrain and a tracker that assumes completely flat terrain. Note the large variance in the Nonviable and Borderline types; steep slopes have outsized, nonphyiscal impact on the results as there would never be trackers there. For the Viable type of terrain, 75% of all data points see a positive impact on POA, with some outliers observing a loss in modeled POA. This is likely a consequence of using the same parameters, such as GCR and angle of maximum rotation for all sites; in some sites, the default values would be suboptimal.

Digital Elevation Models (DEMs), the slope and aspect of arbitrary locations were calculated. To lessen the effects of local errors in DEMs, generalize performance over a large installation, and improve computation time, a downscaling approach was proposed and the effects were quantified using delentropy. Trackers that incorporated land sloped into their control strategy were found to receive $1.35 \pm 0.92\%$ more POA irradiance, which was calculated using a clearsky simulation. When modeling single axis trackers, this work shows that slope aware trackers produces a non-negligible difference in received irradiance and thus projected power output.

REFERENCES

[1] "Nextracker's All-terrain Solar Tracker, NX Horizon-XTR." [Online]. Available: https://www.nextracker.com/horizon-xtr/

[2] K. Anderson and M. Mikofski, "Slope-aware backtracking for single-axis trackers," no. NREL/TP-5K00-76626. [Online]. Available: https://www.osti.gov/biblio/1660126

[3] B. Horn, "Hill shading and the reflectance map," *Proceedings of the IEEE*, vol. 69, no. 1, pp. 14–47.

[4] C. E. Shannon, "A mathematical theory of communication," *Bell System Technical Journal*, vol. 27, no. 3, pp. 379–423, bell System Technical Journal. [Online]. Available: https://people.math.harvard.edu/~ctm/home/text/others/shannon/entropy/entropy.pdf

[5] Z. Wang, A. Bovik, H. Sheikh, and E. Simoncelli, "Image quality assessment: from error visibility to structural similarity," *IEEE Transactions on Image Processing*, vol. 13, no. 4, pp. 600–612.

[6] K. G. Larkin, "Reflections on shannon information: In search of a natural information-entropy for images." [Online]. Available: http://arxiv.org/abs/1609.01117

[7] B. Haurwitz, "Insolation in relation to cloudiness and cloud density," *Journal of the Atmospheric Sciences*, vol. 2, no. 3, pp. 154–166. [Online]. Available: https://journals.ametsoc.org/view/journals/atsc/2/3/1520-0469_1945_002_0154_iirtca_2_0_co_2.xml

978-1-6654-6060-6/23 $31.00 © 2023 IEEE

[8] D. G. Erbs, S. A. Klein, and J. A. Duffie, "Estimation of the diffuse radiation fraction for hourly, daily and monthly-average global radiation," *Solar Energy*, vol. 28, no. 4, pp. 293–302. [Online]. Available: https://www.sciencedirect.com/science/article/pii/0038092X82903024

Substrate reuse of hydride vapor phase epitaxy grown-GaAs solar cells for low-cost photovoltaics

Yasushi Shoji, Ryuji Oshima, Kikuo Makita, Akinori Ubukata, Shuuichi Koseki, Takeyoshi Sugaya

National Institute of Advanced Industrial Science and Technology (AIST), Tsukuba, Ibaraki, Japan

Taiyo Nippon Sanso Corporation, Tsukuba, Ibaraki, Japan

High production costs are a significant obstacle to expanding applications for III-V solar cells. Therefore, we studied to combine the substrate reuse technique and hydride vapor phase epitaxy (HVPE) toward realizing low-cost III-V solar cells. For HVPE, a low epitaxial growth cost can be realized because of the use of group-III precursors through in situ high-efficiency reactions involving less expensive pure metals and HCl gas. The substrate reuse was performed through the epitaxial lift-off (ELO). All samples were grown via HVPE. For the ELO process, a 100 nm-thick AlAs release layer was grown on GaAs substrate. The AlAs was grown using $AlCl_3$ precursor. The GaAs epitaxial layer grown on reused substrate showed good surface flatness comparable to that grown on new epi-ready substrates. The arithmetic mean roughness values were approximately 0.5 nm for GaAs layers grown on new epi-ready and reused substrates. Further, the GaAs single-junction solar cells grown on reused substrate exhibited similar performances with samples grown on new substrates. For the GaAs cell grown on the new substrate, the short-circuit current density (J_{SC}), open-circuit voltage (V_{OC}), fill factor (FF), and conversion efficiency (η) were 19.6 mA cm-2, 0.991 V, 0.817, and 15.8%, respectively. Whereas, for the GaAs cells grown on the reused substrate, J_{SC}, V_{OC}, FF, η were 19.4 mA cm-2, 0.984 V, 0.817, and 15.6%, respectively. The combination of substrate reuse and HVPE can reduce solar cell fabrication costs. This study is a major achievement in reducing the production cost of III-V solar cells.

Towards Polymer-Free, fs Laser Welded Glass/Glass Modules

David L Young, Nick Bosco, Timothy J Silverman

National Renewable Energy Laboratory, Golden, CO, United States

This project explores the use of femto-second (fs) lasers to form glass-to-glass welds for hermetically sealed, polymer-free modules. Solite glass coupons were welded together without the use of fillER using a fs laser with dedicated optics to elongate the focal plane parallel to the incident beam. The resulting welds were then stress tested to failure to reveal the critical stress intensity factor, KIc. These values were then used in a Finite element analysis (FEA) model of a 1 m by 2 m glass/glass module under a simulated static load test. The results show that the fs laser welds are strong enough for a suitably framed module to pass the IEC 61215 static load test of 5400 Pa. Key to this finding is that the module must be framed and braced, and the glass must be ribbed to allow pockets for the cells and welds inside the border of the module. The result is a module design that is completely polymer-free, hermetically sealed, and easily recycled.

Defects in RbF-Treated Cu(In$_x$Ga$_{1-x}$)Se$_2$ Solar Cells and Their Impact on V$_{OC}$

Michael F. Miller[1], Ana Kanevce[2], Alexandra M. Bothwell[3], Stefan Paetel[2], Darius Kuciauskas[3], and Aaron R. Arehart[1]

[1]Department of Electrical & Computer Engineering, The Ohio State University, Columbus, OH 43210, USA
[2]Zentrum fuer Sonnenenergie-und Wasserstoff-Forschung (ZSW), 70563 Stuttgart, Germany
[3]National Renewable Energy Laboratory, Golden, CO 80401, USA

Abstract— **Cu(In,Ga)Se$_2$ solar cell efficiency is limited by V$_{OC}$ due in large part to bulk defects limiting lifetime, but alkali treatments such as RbF recover some of the V$_{OC}$ loss. In this work, defects in RbF-treated and untreated CIGS were quantitatively characterized using DLTS and DLOS, and three main defects were identified in each sample. The RbF-PDT resulted in a large decrease in the mid-gap trap concentration, which was accompanied by a large improvement in minority carrier lifetime. This lifetime improvement combined with a change in doping accounted for a significant portion of the V$_{OC}$ improvement in the RbF CIGS.**

I. INTRODUCTION

Cu(In,Ga)Se$_2$ (CIGS) solar cells are optimal for terrestrial solar applications due to the low LCOE (levelized cost of energy) and the tunable bandgap. However, while efficiencies are above 20% and increasing, they are limited by reduced open-circuit voltage (V$_{OC}$), largely due to semiconductor defects (traps) within the CIGS absorber, which increase the nonradiative recombination rate and reduce the minority carrier lifetime [1]. A promising method to reduce these defects is to treat the CIGS with alkali-halides, such as rubidium fluoride (RbF), which has resulted in larger V$_{OC}$ [2]. While the large Rb atoms are proposed to passivate grain boundary defects [3], the specific impacts on the trap states are unknown. Here we investigate the impacts of a RbF treatment on defects in the CIGS absorber and determine the impacts of these defects on the solar cell performance.

In this study, commercial CIGS solar cells from ZSW production line were characterized, which provide insight into expected RbF effects on modules. Two sample sets were compared, each with an untreated sample (no-PDT) and a sample with RbF post-deposition treatment (RbF-PDT). Sample Set 1 was treated with RbF with a source temperature of 520 ºC and had a Cu-content of ~0.84, whereas Sample Set 2 was treated with RbF at 540 ºC and had a Cu-content of ~0.79. All samples were full solar cells as described in [4] with a soda-lime glass substrate, sputtered Mo back contact, 2-2.5 μm CIGS absorber (grown using multi-stage co-evaporation to a GGI of ~0.20 in the depletion region), CdS buffer, and i-ZnO and Al-ZnO window layers. The solar cells were then physically scribed to isolate devices with an area of ~2 mm^2.

Light current- voltage (LIV) measurements were performed with a WACOM Class A (IEC-60904-9) solar simulator at one sun illumination and temperature of 25 ºC. The CIGS defects were measured using deep level transient and optical spectroscopies (DLTS/DLOS) that probe defects in the lower and upper half of the bandgap, respectively. The DLTS was measured using fill/empty voltages of -0.3 V and -1 V with a fill pulse of 20 ms, and DLOS was performed using fill/empty voltages of 0 V and -1 V with a fill and measure time of 10 s and 300 s, respectively. Further details of the measurement setup and equipment are provided in [1], [2], [5]. To measure the carrier lifetime for each sample, time-resolved photoluminescence (TRPL) measurements were performed using excitation at 640 nm and time-correlated single photon counting with a fluence range of $1 \times 10^{11} - 1 \times 10^{14}$ photons/(cm^{2}pulse).

II. RESULTS AND DISCUSSION

LIV measurements using AM 1.5 illumination were performed, and the results are shown in Table I. The RbF-PDT resulted in higher efficiencies in both samples sets, largely due to improved V$_{OC}$ and fill factor.

The DLTS spectra, shown in Fig. 1a, show two peaks in all four samples, and the Arrhenius plot (Fig. 1b) indicates the same two traps are likely present independent of treatment although with different concentrations. The higher temperature trap overlaps in Arrhenius space with the "mid-gap" trap previously reported in CIGS with similar bandgaps indicating this is likely the same trap [1], [2], [5]. While the measured trap energy of E$_V$+0.71 eV is slightly higher than in previous studies, this is

TABLE I: SOLAR CELL PARAMETERS

		J$_{SC}$ (mA/cm^2)		V$_{OC}$ (mV)		Fill Factor (%)		Efficiency (%)	
		Value	Relative Increase	Value	Relative Increase	Value	Relative Increase	Value	Relative Increase
Sample Set 1	No PDT	31.5	0.0%	637	11.8%	69.6	10.2%	14.1	22.9%
	RbF PDT	31.5		712		76.7		17.2	
Sample Set 2	No PDT	31.4	0.6%	683	4.0%	71.4	2.0%	15.3	4.6%
	RbF PDT	31.6		710		72.8		16.0	

978-1-6654-6060-6/23 $31.00 © 2023 IEEE

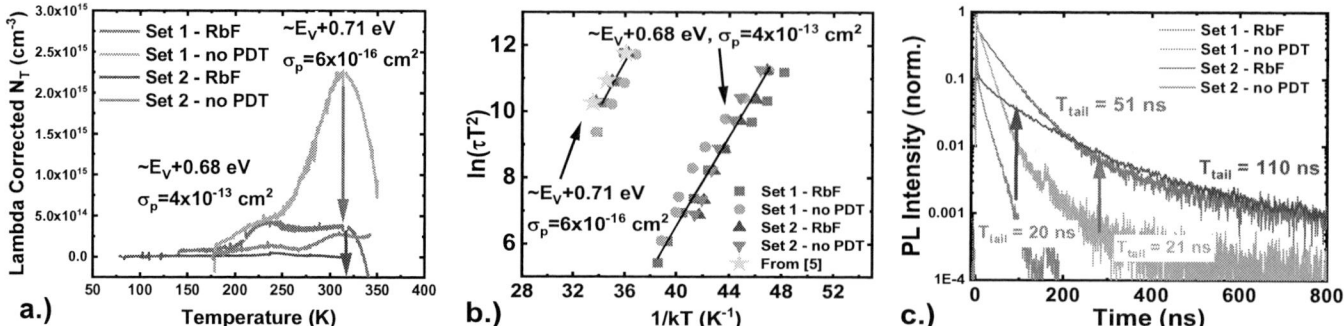

Fig. 1. a) DLTS spectra showing two traps, at ~E_V+0.68 and ~E_V+0.71 eV. The trap concentration is calculated from the DLTS peak height. We accounted for the volume where the traps changed occupation (i.e. lambda correction) and did not assume trap modulation throughout the depletion region [8]. N_T is reduced in the RbF-PDT CIGS for both sample sets, especially for the ~E_V+0.71 eV trap. (b) Arrhenius plot shows good agreement for the traps in each sample, and the ~E_V+0.71 eV trap appears near mid-gap in similar bandgap CIGS [5]. (c) TRPL data for device (no applied bias) show improved lifetimes after PDT. In both sample sets, the lifetime is improved in the RbF-PDT CIGS, consistent with the decreased trap concentration observed from DLTS.

likely due to a small temperature difference caused by the 2.5 mm thick ZSW glass substrates with low thermal conductivity. Previous scanning-DLTS measurements have shown this mid-gap level is localized to specific grain boundaries, which is consistent with previous observations that alkali treatments preferentially passivate traps near grain boundaries [3], [6]. It has been linked with a Cu$_{In/Ga}$ antisite defect, and has also been shown to have a negative impact on the minority carrier lifetime [1], [7]. In both sample sets, there was a large reduction in N_T in the RbF-PDT CIGS indicating that RbF has a positive impact on this trap concentration. A second, lower temperature trap is also observed with DLTS at E_V+0.68 eV, which is likely a new trap that has not been observed in previous CIGS samples. Again, the measured trap energy is likely higher than the actual energy due to the poor thermal resistance, and work is underway to improve the Arrhenius plots. For this trap, there is a slight N_T decrease in the RbF-treated CIGS, indicating that the RbF is slightly beneficial for this trap as well.

To investigate the relation to the carrier lifetime, TRPL measurements were performed and are shown in Fig. 1c. In each sample set, the minority carrier lifetime showed a large improvement in the RbF-treated CIGS, with an over 5X improvement in Sample Set 2. The front interface recombination velocity, S_{front}, was determined by fitting the initial TRPL decay (τ_1) according to $S_{front}=(\alpha\tau_1)^{-1}$ where α is the absorption coefficient. S_{front} was ~1×10^3 cm/s for each sample, indicating that improvement in the front interface is likely not responsible for the observed TRPL lifetime improvement. The radiative recombination rate, estimated from the sample doping ($\tau_{Rad}^{-1} = \beta N_A$ where β is the radiative recombination coefficient), decreased in Sample Set 1 (141 to 83 ns) and was much larger than the TRPL lifetime in Sample Set 2 (313 to 100 ns), indicating that the improved TRPL lifetime was also not due to improved radiative lifetime. The improved TRPL lifetime was likely due to improved bulk lifetime; trap concentration is linear with the bulk recombination rate ($\tau_B^{-1} = \sigma v_{th} N_T$), so improved bulk lifetime is expected from reduced mid-gap trap concentration, which is consistent with previous results [1]. Indeed, the RbF-treated CIGS from Sample set 2 had both the highest TRPL lifetime (110 ns) and the lowest mid-gap trap concentration (2.0×10^{13} cm^{-3}). Voltage-biased TRPL results are

underway to directly measured the bulk lifetime, which will highlight the impact of the reduced trap concentration.

DLOS was used to measure deep traps in the CIGS, and the resulting SSPC is shown in Fig. 2. A single trap was observed, which was fit to the Lucovsky photoionization model (Fig. 2 inset) resulting in a trap energy of E_V+1.02 eV with negligible lattice relaxation energy. This near-conduction band defect has been observed in previous CIGS samples, and is likely due to the V$_{Se}$-V$_{Cu}$ predicted by DFT [10]. The SSPC peak heights shown in Fig. 2 show a slight increase in E_V+1.02 eV trap concentration for both sample sets indicating that RbF-PDT causes an increased near-conduction band trap concentration. While this trap has previously been shown to cause light-induced J_{SC} reduction [11], the doping in these samples (Table II) is much higher than the trap concentration, and combined with a long diffusion length due to the high carrier lifetime, there likely will be little light-induced J_{SC} reduction for both the RbF-PDT and untreated CIGS.

Through the lens of these measurements, we can begin to make estimates of the expected voltage improvement, which is ultimately the area of improvement for these solar cells. As shown in Table I, Sample Set 1 had a 75 mV V_{OC} improvement, and Sample Set 2 had a 27 mV improvement. Sample doping,

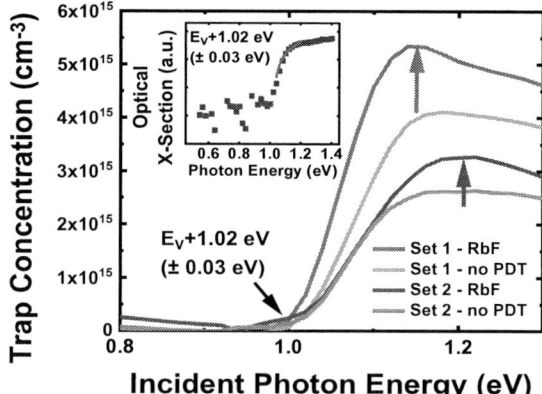

Fig. 2. Steady-state photocapacitance from DLOS shows a single deep trap. The trap energy, fit from the optical cross section using the Lucovsky model [9] (inset) was found to be E_V+1.02 eV (±0.03 eV due to wavelength spread in measurement optics). The trap concentration, found from the SSPC peak height, was increased in both sample sets in the RbF-PDT CIGS.

TABLE II: MEASURED DEVICE PARAMETERS

		E_V+0.68 eV Trap Conc. (cm⁻³)	E_V+0.71 eV Trap Conc. (cm⁻³)	E_V+1.02 eV Trap Conc. (cm⁻³)	Net Acceptor conc. (cm⁻³)	Measured lifetime (ns)
Sample Set 1	RbF	4.3×10^{14}	3.8×10^{14}	5.5×10^{15}	1.2×10^{17}	51
	No PDT	4.6×10^{14}	2.2×10^{15}	4.3×10^{15}	7.1×10^{16}	21
Sample Set 2	RbF	4.9×10^{13}	2.0×10^{13}	3.3×10^{15}	1.0×10^{16}	110
	No PDT	1.4×10^{14}	2.8×10^{14}	2.9×10^{15}	3.2×10^{16}	20

shown in Table II, can impact the cell V_{OC} but the different growths resulted in opposite doping trends where doping increased in Sample Set 1, but decreased in Sample Set 2 (possibly due to the different RbF source temperatures). Lifetime also has an impact on V_{OC}, and both devices demonstrated an increase in carrier lifetime, shown in Table II. The expected change in V_{OC} is predicted by [12] .

$$\Delta V_{OC} \leq kTln\left(\frac{N_{A,RbF}}{N_{A,no-PDT}}\right) \tag{1}$$

$$\Delta V_{OC} \leq kTln\left(\frac{\tau_{RbF}}{\tau_{no-PDT}}\right) \tag{2}$$

where N_A is the acceptor concentration, τ is the minority carrier lifetime, k is Boltzmann's constant, and T is temperature. Through these equations, we can add the expected voltage changes from the untreated to the RbF-PDT CIGS and compare to the observed voltage improvements shown in Fig. 3. In each sample set, around half of the observed voltage improvement is predicted by Eqns. 1 and 2, which indicates RbF has other impacts that lead to larger improvements than are predicted by this simple model. Several other mechanisms are being considered to construct a more accurate model of the V_{OC} improvement in the RbF-PDT CIGS.

III. CONCLUSIONS

In this study, two sample sets of RbF-PDT and untreated commercial CIGS solar cells were measured to study their defect characteristics and understand the observed V_{OC} improvements. Three traps were observed using DLTS and DLOS, with the RbF-PDT effectively reducing the mid-gap trap concentration, which resulted in improved minority carrier lifetime. This

lifetime improvement, along with a doping change, helped account for a significant portion of the observed V_{OC} improvement in the RbF-PDT CIGS. The next step is to continue investigating sources of the observed V_{OC} improvement in efforts to construct a more accurate model.

ACKNOWLEDGMENT

This material is based upon work supported by the U.S. Department of Energy's Office of Energy Efficiency and Renewable Energy (EERE) under the Solar Energy Technology Office (SETO) Award Number DE-EE0008755.

REFERENCES

[1] A. M. Bothwell, S. Li, R. Farshchi, M. F. Miller, J. Wands, C. L. Perkins, A. Rockett, A. R. Arehart, and D. Kuciauskas, "Large-Area (Ag,Cu)(In,Ga)Se2 Thin-Film Solar Cells with Increased Bandgap and Reduced Voltage Losses Realized with Bulk Defect Reduction and Front-Grading of the Absorber Bandgap," *Solar RRL*, vol. 6, no. 8, p. 2200230, 2022.

[2] S. Karki, P. Paul, G. Rajan, B. Belfore, D. Poudel, A. Rockett, E. Danilov, F. Castellano, A. Arehart, and S. Marsillac, "Analysis of Recombination Mechanisms in RbF-Treated CIGS Solar Cells," *IEEE Journal of Photovoltaics*, vol. 9, no. 1, pp. 313–318, Jan. 2019.

[3] M. Chugh, T. D. Kühne, and H. Mirhosseini, "Diffusion of Alkali Metals in Polycrystalline CuInSe₂ and Their Role in the Passivation of Grain Boundaries," *ACS Appl. Mater. Interfaces*, vol. 11, no. 16, pp. 14821–14829, Apr. 2019.

[4] A. Kanevce, S. Paetel, D. Hariskos, and T. M. Friedlmeier, "Impact of RbF-PDT on Cu(In,Ga)Se2 solar cells with CdS and Zn(O,S) buffer layers," *EPJ Photovolt.*, vol. 11, p. 8, 2020.

[5] S. Karki, P. K. Paul, G. Rajan, T. Ashrafee, K. Aryal, P. Pradhan, R. W. Collins, A. Rockett, T. J. Grassman, S. A. Ringel, A. R. Arehart, and S. Marsillac, "In Situ and Ex Situ Investigations of KF Postdeposition Treatment Effects on CIGS Solar Cells," *IEEE Journal of Photovoltaics*, vol. 7, no. 2, pp. 665–669, Mar. 2017.

[6] P. K. Paul, D. W. Cardwell, C. M. Jackson, K. Galiano, K. Aryal, J. P. Pelz, S. Marsillac, S. A. Ringel, T. J. Grassman, and A. R. Arehart, "Direct nm-Scale Spatial Mapping of Traps in CIGS," *IEEE Journal of Photovoltaics*, vol. 5, no. 5, pp. 1482–1486, Sep. 2015.

[7] J. I. Deitz, P. K. Paul, R. Farshchi, D. Poplavskyy, J. Bailey, A. R. Arehart, D. W. McComb, and T. J. Grassman, "Direct Nanoscale Characterization of Deep Levels in AgCuInGaSe2 Using Electron Energy-Loss Spectroscopy in the Scanning Transmission Electron Microscope," *Advanced Energy Materials*, vol. 9, no. 35, p. 1901612, 2019.

[8] P. Blood and J. W. Orton, *The Electrical characterization of semiconductors: Majority carriers and electron states*. London: Academic Press, 1992.

[9] G. Lucovsky, "On the photoionization of deep impurity centers in semiconductors," *Solid State Communications*, vol. 3, no. 9, pp. 299–302, Sep. 1965.

[10] S. Lany and A. Zunger, "Light- and bias-induced metastabilities in Cu„In, Ga...Se2 based solar cells caused by the „VSe-VCu... vacancy complex," *J. Appl. Phys.*, p. 15.

[11] P. K. Paul, T. Jarmar, L. Stolt, A. Rockett, and A. R. Arehart, "Role of Ev+0.98 Ev trap in light soaking-induced short circuit current instability in CIGS solar cells," in *2017 IEEE 44th Photovoltaic Specialist Conference (PVSC)*, 2017, pp. 30–32.

[12] Arno Smets, *Solar Energy*, 1st ed. UIT Cambridge, 2015.

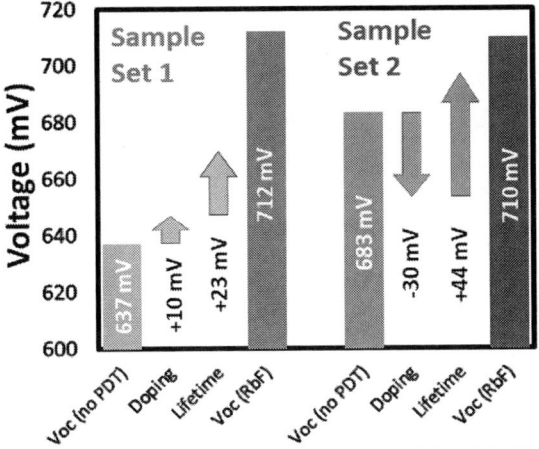

Fig. 3. Open circuit voltage improvements observed in RbF-PDT CIGS combined with estimates of voltage improvements from sample doping and carrier lifetime (arrows). In each sample set, the changes in doping and lifetime accounted for around half of the observed V_{OC} improvement.

Improving the Space Silicon Solar Cell Efficiency by Adding the Layer Down-converting UV Light to Visible

Alex Fedoseyev[1], Stan Herasimenka[1], Sergey Sarkisov[2]

[1]Solestial Inc, Tempe, Arizona, USA, [2]SSS Optical Technologies LLC, Huntsville, Alabama, USA

Abstract — We present the approach to the efficiency improvement of the space silicon solar cells that involves a layer of down-converting material on the top of the cell. The layer consist of polymer nanocomposite coating impregnated with the nanoparticles of inorganic compounds (fluorides or oxides) doped with rare earth ions, down-converting solar UV light to visible matching the spectrum of the responsivity of the cell. We discuss the choice of the polymer host for the nanoparticles to withstand both UV and space radiation and the nanoparticles with maximal down-conversion capabilities. Experimental results demonstrate a significant increase of the generated current and photoelectric efficiency.

I. INTRODUCTION

When solar cells are used in outer space or in Lunar environment, they are subject to bombardment by high-energy particles, which induce a degradation referred to as radiation damage. Additional damaging phenomenon in space is the UV radiation. Cover glass is a typical choice to minimize solar cell degradation.

Alternative to cover glass is a polymer layer that provides similar protection of the solar cell from UV and space radiation. We propose to use such layer to increase solar cell photoelectric efficiency by down-converting the UV part of solar spectrum to visible-NIR light that matches the spectrum of responsivity of a silicon cell. The layer in this case can be designed as a polymer nanocomposite impregnated with the nanoparticles of the inorganic compounds (fluorides or oxides) doped with the ions of rare earth (RE) that down-convert solar UV to visible-NIR light.

II. DOWN-CONVERSION PRINCIPLE AND IMPLEMENTATION

The principle of operation of a nanocomposite layer improving PV power conversion efficiency is illustrated in Fig. 1. Nanocomposite layer made of RE nanoparticles embedded in a transparent polyimide is deposited on the top of the polyimide film that protects PV cell from UV radiation. Both UV and visible-NIR components of solar radiation reach the layer. UV component is partially down-converted into visible-NIR radiation. All the remaining UV radiation is blocked by the protective polyimide film. The down-converted radiation along with the major visible-NIR component of solar radiation pass unabsorbed through the protective polyimide film to the PV cell and get transformed into electricity. The down-converted radiation contributes to the production of extra electricity thus increasing PV efficiency.

Fluorides doped with the ions of rare earth (RE) elements $NaYF_4$:Yb^{3+}, Er^{3+} [1, 2] have been used as a spectrum down-converting matter.

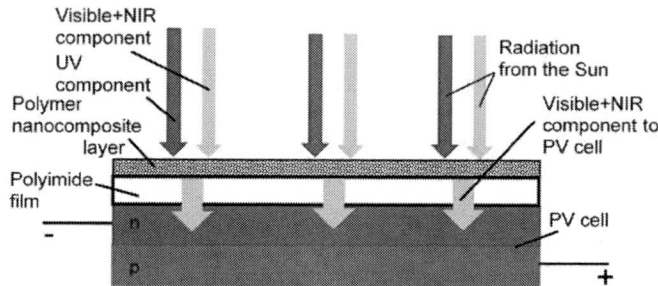

Fig. 1. Schematic illustrating the principle of operation UV down converting layer resulting in PV conversion efficiency improving.

The selection of fluoride $NaYF_4$ as a host for RE ions is justified by its low phonon energy (~ 300 cm^{-1}) that makes multi-phonon assisted nonradiative relaxation of excited ions negligible [3]. The nanoparticles (NPs) made of such compound are environmentally stable and not toxic. The energy level diagram of Er^{3+} and Yb^{3+} ions in the compound is presented in Fig. 2a.

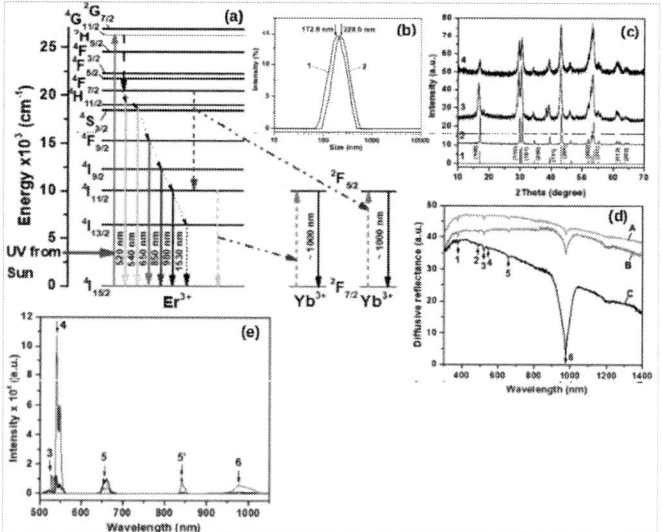

Fig. 2. a) Energy level diagram of coupled ions Er^{3+} - Yb^{3+}. (b) Size distributions of two exemplary colloids of ball-milled NPs of phosphor $NaY_{0.83}F_4$: $Yb^{3+}_{0.14}Er^{3+}_{0.03}$ in 1-propanol obtained with dynamic light scattering (DLS) measurement. Vertical arrows down mark the average size of the QDs (assuming their spherical shape) on size scale (c) X-ray diffraction (XRD) spectra. Plot 1- XRD spectrum

978-1-6654-6060-6/23 $31.00 © 2023 IEEE

(computed, JCPDS card No. 28-1192) of $NaYF_4$ beta–phase (hexagonal). 2- synthesized $NaY_{0.83}F_4$: $Yb^{3+}_{0.14}Er^{3+}_{0.03}$ powder baked at 500°C for 1h before ball milling; 3, 4 – two exemplary ball milled nanopowders of $NaY_{0.83}F_4$: $Yb^{3+}_{0.14}Er^{3+}_{0.03}$. (d) Diffuse reflectance spectra of the powder samples of RE-doped fluoride down-conversion phosphors. Curve A corresponds to $NaY_{0.94}F_4$: $Yb^{3+}_{0.03}Er^{3+}_{0.03}$; B - $NaY_{0.78}F_4$: $Yb^{3+}_{0.2}Er^{3+}_{0.02}$; C - $NaY_{0.69}F_4$: $Yb^{3+}_{0.3}Er^{3+}_{0.01}$. Absorption peak 1 corresponds to transition in ion of Er^{3+}: $^4I_{15/2} -> {}^4G_{11/2}$ (380 nm); 2 - Er^{3+}: $^4I_{15/2} -> {}^4F_{7/2}$ (480 nm); 3 - Er^{3+}: $^4I_{15/2} -> {}^2H_{11/2}$ (520 nm); 4 - Er^{3+}: $^4I_{15/2} -> {}^4S_{3/2}$ (540 nm); 5 - Er^{3+}: $^4I_{15/2} -> {}^4F_{9/2}$ (650 nm); 6 — Er^{3+}: $^4I_{15/2} -> {}^4I_{11/2}$ and Yb^{3+}: $^2F_{7/2} -> {}^2F_{5/2}$ (~ 980 - 1000 nm) as presented in the energy level diagram (a). All the spectra are normalized to peak 3: its depth is (-1). (e) Visible and NIR emission spectrum of the powder of phosphor $NaY_{0.83}F_4$: $Yb^{3+}_{0.14}Er^{3+}_{0.03}$ excited by a 372-nm UV laser (black curve) and by 488-nm line of an Ar-ion laser (blue). The peaks correspond to the absorption bands with the same numbers as in plot (d). New peak 5' corresponds to the band at 850 nm. This band does not exhibit prominent absorption in plot (d).

The erbium ion gets excited with a solar UV photon and emits down-shifted visible and NIR radiation (single photon) in one of the spectral bands around 520, 540, 650, 850, or 980 nm (marked with vertical arrows down). It can also perform energy exchange with the ytterbium ion through two quantum-cutting mechanisms (marked by slanted dash-dotted arrows down) when the energy of the UV solar photon is cut into the energies of two NIR photons emitted by two ytterbium ions.

We synthesized the RE-doped fluoride phosphor using the wet method followed by baking the obtained micro-powder in ambient air and reducing it with ball-milling to nano-powder [1]. Average size of the QDs as measured by the dynamic light scatterometry (DLS) with a Zetasizer instrument from Malvern Instruments was estimated as 172 to 228 nm (Fig. 2b). The X-ray diffraction (XRD) spectrum of the ball-milled nanopowder (Fig. 2c) showed that the NPs were the nanocrystals of hexagonal beta-phase of $NaYF_4$. Photoluminescence (PL) spectral bands of the phosphor excited with UV radiation included downshifting and quantum cutting components (Fig. 2e) that corresponded to the optical absorption bands (Fig. 2d). Photoluminescence quantum yield (PLQY) determined as the ratio of the emitted visible-NIR photons to the absorbed UV photons was < 10%.

We used polymers CP-1 and CORIN as hosts for the NPs. They are space qualified polymers.

We called the polymer nanocomposite layer made in such way the Layer of UV CONverting coating or LUVCON.

III. EXPERIMENTAL RESULTS

UV stability of LUVCON coating and its ability to protect PV cells was tested using Flood Lamp System Model 38125 from Dymax (Fig. 3). UV irradiance on the top of the atmosphere is ~ 8% of the total Solar irradiance [Brune 2001]. UV irradiance in Dymax is equivalent to 37 mW/cm²/0.08 = 425 mW/cm2 or ~ 3 total Solar irradiances. One hour of exposure in Dymax is thus ~ 3 ESH.

Fig. 3. Dymax UV-A+visible light source (metal halide lamp) used in the experiments on UV stability.

We have identified the material that satisfy all the key requirements: stable coating under UV light for 2500+ hours, high optical transmittance for visible light, minimal optical darkening under UV light, good suppression of the UV light.

III. EXPERIMENTAL RESULTS AT AM0 CONDITIONS

As the main product of Solestial Inc is silicon solar cells (ultrathin and radiation hard UT-Si) [4], we have performed all the experiments with silicon solar cells. A solar simulator with a Xenon arc lamp and AM0 filter has been used to measure the AM0 I-V characteristics of our solar cells coated with LUVCON.

For comparison, we also provide the results for the cells coated with pure polymer coating. We present the recent results for solar cells from Solestial made on 5" silicon wafers: the original bare silicon PV cells are shown in Fig. 4(a), the CORIN coated PV cells - in Fig. 4(b), and CORIN + LUVCON (CORIN + NPs) coated cells illuminated with a UV LED source (365 nm peak) - in Fig. 4(c).

The UV light is down converted to NIR light (invisible) and red light seen in Fig. 4(c).

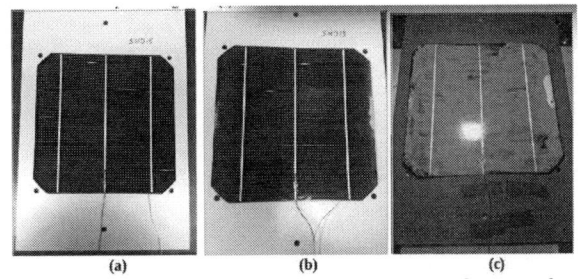

Fig. 4. Large Solestial silicon CICs made on 5" silicon wafer with contacts: (a) bare PV cells, (b) CORIN coated cells, and (c) CORIN+LUVCON (CORIN+Red NPs) cells illuminated with a UV LED source (365 nm peak). The incident UV light is down converted to NIR and red light. The latter one is seen in (c).

Solar cell improvement with CORIN +LUVCON coating (CORIN + Red NPs) is demonstrated in Figure 13. Silicon solar cell coated with LUVCON increased the current by

22.0% compared to the cells coated with the polymer only. That will be the absolute 4.3% efficiency increase for a Solestial UT-Si call with 22% efficiency (FF=0.80), resulting in 26.3% BOL efficiency. As UT-Si solar cells degrade very little under space radiation, the EOL efficiency is expected to be 26.0% (10 years in space).

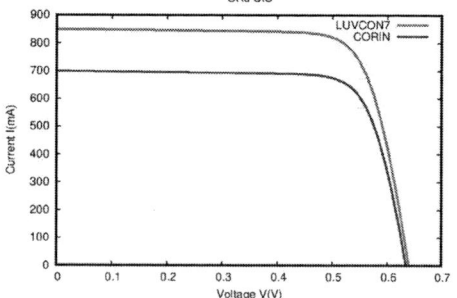

Fig. 5. Improving PV power conversion efficiency of large Solestial silicon CICs made on a 5" silicon wafer for AM0 conditions: measured IV curves for CORIN coating only (blue curve) and LUVCON coating (red curve). Silicon solar cell coated with LUVCON increased the current by 22.0% compared to the cells coated with CORIN polymer only. That translates into absolute 4.3% efficiency increase for 22% efficient UT-Si silicon solar cells, resulting in 26.3% BOL efficiency. As UT-Si solar cells degrade very little under space radiation, the EOL efficiency is expected to be 26.0% (FF=0.80).

IV. DISCUSSION

UT-Si are radiation hard, low cost and mass-produced, and the EOL efficiency of UT-Si cells drops minimally when exposed to space radiation making them more attractive for the use in space (EOL = 21%, "Upgraded Si" in Fig. 6).

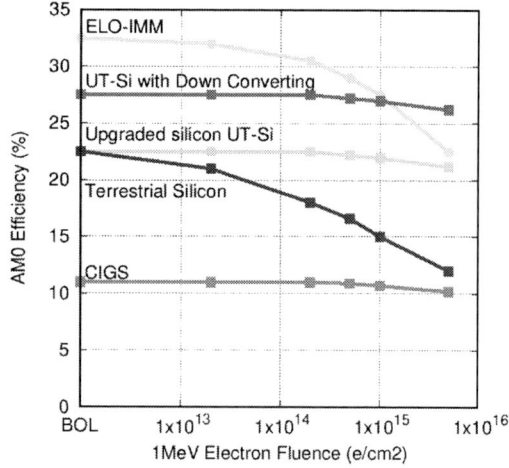

Fig. 6. Targeted improvement of the EOL conversion efficiency of UT-Si solar cells with down converting.

With the introduction of down converting into the UT-Si solar cells their EOL efficiency becomes the same or higher than the efficiency of ELO-IMM solar cells making even them more attractive for the use in space (EOL 26%).

V. CONCLUSION

The experimental results have proved the feasibility of the proposed down converting concept, a significant improvement of Silicon solar cell efficiency by LUVCON coating, down-converting UV light to near infra-red and therefore increasing the current generated. UT-Si solar cell coated with LUVCON increased the current by 22.0% compared to cells coated with CORIN polymer only. That translates into absolute 4.3% efficiency increase for 22% efficient UT-Si silicon solar cells, resulting in 26.3% efficiency. As UT-Si solar cell degrades very little under space radiation, the EOL efficiency expected to be 26.0%. EOL efficiency of such solar cells becomes the same or higher than the efficiency of ELO-IMM solar cells making even them attractive for the use in space applications

REFERENCES

[1] Sarkisov S.S., Patel D.N., et al. "Polymer nanocomposite luminescent films for solar energy harvesting made by concurrent multi-beam multi-target pulsed laser deposition". Proc. of SPIE 2018;10755,1075502..

[2] Sarkisov S.S., Patel D.N., et al. Polymer nanocomposite sunlight spectrum down-converters made by open-air PLD. Nanotech Rev 2020;9,1044–58.

[3] Baker C.C., Fontana J., et al. Nanoparticle doping for high power fiber lasers at eye-safer wavelengths. Opt Express 2017;25(12):3903-13915.

[4] S. Y. Herasimenka, A. Fedoseyev et al., "Recovery of electron and proton damage in Ultrathin Silicon Solar Cells at 65C and 80C Under Light: Wafer Thickness, Wafer Type and Cell Technology Dependence," pres. at the 27th Space Photovoltaic Research and Technology Conference, Cleveland, OH, 2022

Near-Contactless Production I-V Testing of Silicon Solar Cells

Harrison Wilterdink, Ron Sinton, Adrienne Blum, Karoline Dapprich, Nick Degenhart, Wes Dobson

Sinton Instruments, Boulder, CO, 80301, USA

Abstract — We demonstrate a new tool capable of performing nearly contactless current-voltage (I-V) and efficiency measurements for binning in silicon solar cell production lines. We validate the technique against conventional test methods for over 400 cells representing a range of technologies including 5-busbar passivated emitter rear contact (PERC), 5-busbar heterojunction technology (HJT), and both 9-busbar and 16-busbar tunnel oxide passivated contact (TOPCon). The tool leverages a wide variety of metrology techniques including Suns-Voc, light I-V, dark I-V, and a series resistance (R_s) analysis based on a single photoluminescence (PL) image. The R_s analysis reports multiple key resistive loss components including grid/finger loss (R_{grid}) as well as the combined effect of emitter and contact loss ($R_{contact}$). The technique is designed especially for present and future cell concepts (e.g. "multi-busbar" and "busbarless" cells) that use minimal silver and are thus difficult to electrically probe for standard I-V characterization.

I. INTRODUCTION

Modern silicon solar cell designs have long been driven by silver costs, leading to industry-wide reductions in silver consumption per watt over time [1]. Simply looking at metallization designs shows the trend clearly—over the last several decades, busbars have simultaneously become much narrower and more numerous (see Fig. 1). As a result, it is increasingly difficult to reliably contact modern cells, often having 9 or more busbars, using conventional probe bars and pogo pins. Besides decreasing the reliability of production line I-V testing and binning, this also introduces several other conflicting demands into the I-V test station: namely, probe bars are often made thinner (and mechanically less rigid) in order to reduce shading and increase visibility of the cell for electroluminescence (EL) imaging. In a somewhat extreme example, the authors have personally seen at least one production line remove up to half the probe bars from their test station to address shading and visibility issues while, unfortunately, greatly exacerbating their test reliability issues.

Arguably, the ultimate response to silver cost pressures is the busbarless cell design, which some predict will account for almost one-third of cell production in the next decade [1], and will be accompanied by a phase out of designs having ≤ 10 busbars. Although at least one commercial solution for contacting busbarless cells has been available for some time [2], it is unclear to us that such systems have been widely adopted, perhaps simply due to their greater complexity than conventional pogo pin solutions. At least one group has recently demonstrated proof-of-concept for fully contactless I-V binning based largely on PL measurements [3], but its transferability to production lines is yet to be seen.

Given the industry's ongoing need to accurately characterize these difficult-to-measure multi-busbar and busbarless cell designs, this paper demonstrates a new tool capable of performing *nearly* contactless I-V binning using a single conventional probe bar (see Fig. 2, right). The tool leverages a range of techniques including Suns-Voc, light I-V, dark I-V, and a physics-based R_s analysis based on a single PL image. The reported R_s value is additionally broken down into multiple key resistive loss components including R_{grid} and $R_{contact}$. This is complemented by a detailed lifetime analysis highlighting losses due to surface, bulk, and shunt components [4]. We show good correlations in efficiency ($R^2 > 0.9$) between the new tool and a standard I-V test tool for over 400 cells encompassing PERC, TOPCon, and HJT concepts, thereby proving its applicability to production line efficiency binning. We also highlight an additional use of the same hardware for an on-the-fly PL imaging system (a technique previously demonstrated with modules [5]) that is capable of generating high-resolution images for further defect recognition and binning.

Fig. 2. Comparison of probing designs for conventional contacting with 16 probe bars (left) and proposed contacting in this work with a single probe bar (right).

Fig. 1. Comparison of a 3-busbar cell from the early 2000s (left) to a 12-busbar cell circa 2019 (right), illustrating the trend towards narrower busbars of increasing number.

978-1-6654-6060-6/23 $31.00 © 2023 IEEE

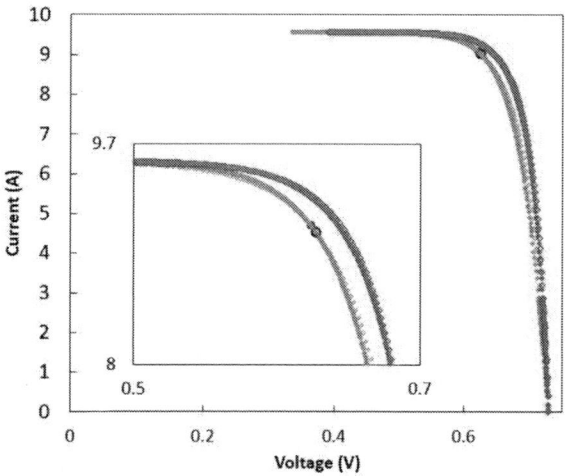

Fig. 4. Measured data for a single sample, including the Suns-Voc curve (red) and I-V curve (blue). The open symbols are for a standard I-V test tool, while the closed symbols are for the new tool. The inset graph highlights the maximum power region of the curves.

II. METHODS

At its core, our nearly contactless tool reports I-V curves by measuring the cell's R_s-free I-V curve (the Suns-Voc curve [6]), measuring the cell's R_s, and then combining the two to report an implied I-V curve (see Fig. 4). The cell's Suns-Voc curve is measured under open-circuit conditions (no current flow) with a Xe flash light source, so the electrical probing requirements are minimal. The cell's R_s is measured using the same concepts we previously applied to diagnosing shunted solar cells [7]. Specifically, we acquire a single PL image of the cell while it is non-uniformly illuminated by a single line of light emitting diodes (LEDs). In analogy to EL, the LEDs act as a "virtual busbar" where carriers are injected and then flow laterally to the rest of the cell. A single voltage probe, placed far from the virtual busbar in a region of the cell with low current injection, provides a voltage reference so the PL image carrier profiles can be converted to voltage using well-established formulas [8] after the image has been corrected for typical distortions due to vignetting and camera perspective. Finally, we fit the voltage profiles to an electrical transport model having a grid loss component (R_{grid}) and a component representing the combination of both emitter and contact losses ($R_{contact}$, for

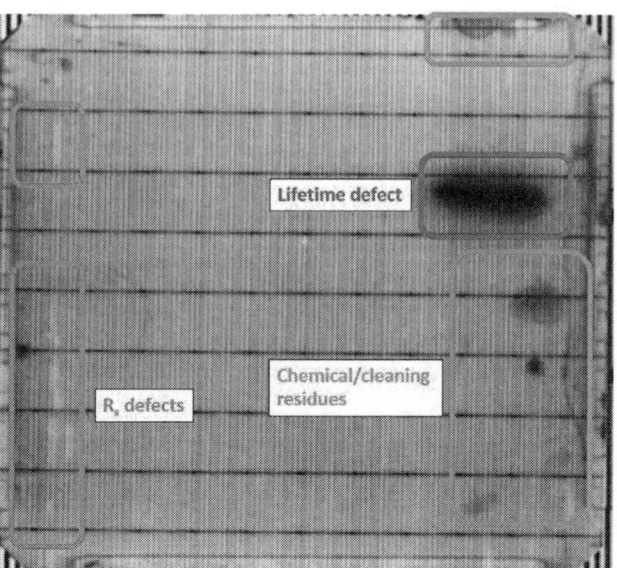

Fig. 3. On-the-fly PL image of a 9-busbar TOPCon cell showing a mix of several defects. Brightness and contrast are enhanced to aid visibility.

simplicity). The sum of these two components gives the cell's total series resistance (R_s).

Several additional features complement the core I-V curve measurement. One feature is a dark I-V sweep that diagnoses issues with shunt resistance and reverse-bias current leakage. Another feature is a patented substrate doping measurement [9] that is designed for the finished cell stage and therefore reports the substrate doping after thermal donors have been annihilated by upstream high-temperature process steps. Knowing the substrate doping at cell test allows us to convert the Suns-Voc curve into minority carrier lifetime data, which, depending on the injection levels reached, can be used to extract bulk lifetime (τ_{bulk}) and emitter saturation current density (J_{0s}) values on either a per-cell basis [10] or an aggregated per-cell-group basis [11]. We also leverage these key lifetime parameters to report a comprehensive physics-based power loss analysis [4] that can highlight losses due to bulk recombination, surface recombination, shunt resistance, and series resistance on a per-cell basis at line speed. The reported power loss parameters can give line operators a detailed view of production line quality at any given moment, and can also enable big data studies of incremental optimizations to the production line. Additionally,

Table 1. Summary of cell samples measured for this paper. The right-most column shows the designation code and color used throughout the paper to differentiate sample groups.

Description	Approximate quantity	Group designation code (and color)
5-busbar production PERC	100	PERC (blue)
5-busbar production HJT	100	HJT (yellow)
9-busbar research & development (R&D) TOPCon	100	9BB TOPCon (gray)
16-busbar production TOPCon	100	16BB TOPCon (green)

the product of substrate doping and nominal wafer thickness yields an approximation of the base/substrate resistive loss (R_{base}) [12], which can be used to further complement the electrical transport model described earlier. The final feature is an on-the-fly PL image of the cell as it moves in or out of the contacting station, which is acquired using an additional camera and virtual busbar. As mentioned before, the virtual busbar generates lateral carrier flow across the cell, which means R_s defects (like broken fingers) that impede carrier flow appear bright in the image [5], whereas they would normally appear dark in standard EL or not at all in standard open-circuit PL. This makes it easier to visually identify R_s defects within the mix of dark features caused by scratches, cracks, or low-lifetime regions (see Fig. 3).

III. RESULTS AND DISCUSSION

We measured over 400 cells (see Table 1) from four different manufacturers on the new tool, a standard I-V test tool (Sinton Instruments' FCT-650), and the on-the-fly PL imaging system. Additionally, to validate the new tool's breakdown of resistive loss components, we compared its reported R_{grid} to standard 4-wire busbar-to-busbar resistance (BBR) measurements. We collected all data and images using measurement and analysis times that are compatible with production line speeds ($\geq$ 3600 wafers per hour), again proving the tool's applicability to production line binning.

Fig. 5 compares the I-V parameters reported by the new tool and the standard I-V tool using standard linear regression analysis. The new tool shows excellent agreement on short-circuit current, open-circuit voltage, and efficiency (I_{sc}, V_{oc}, and

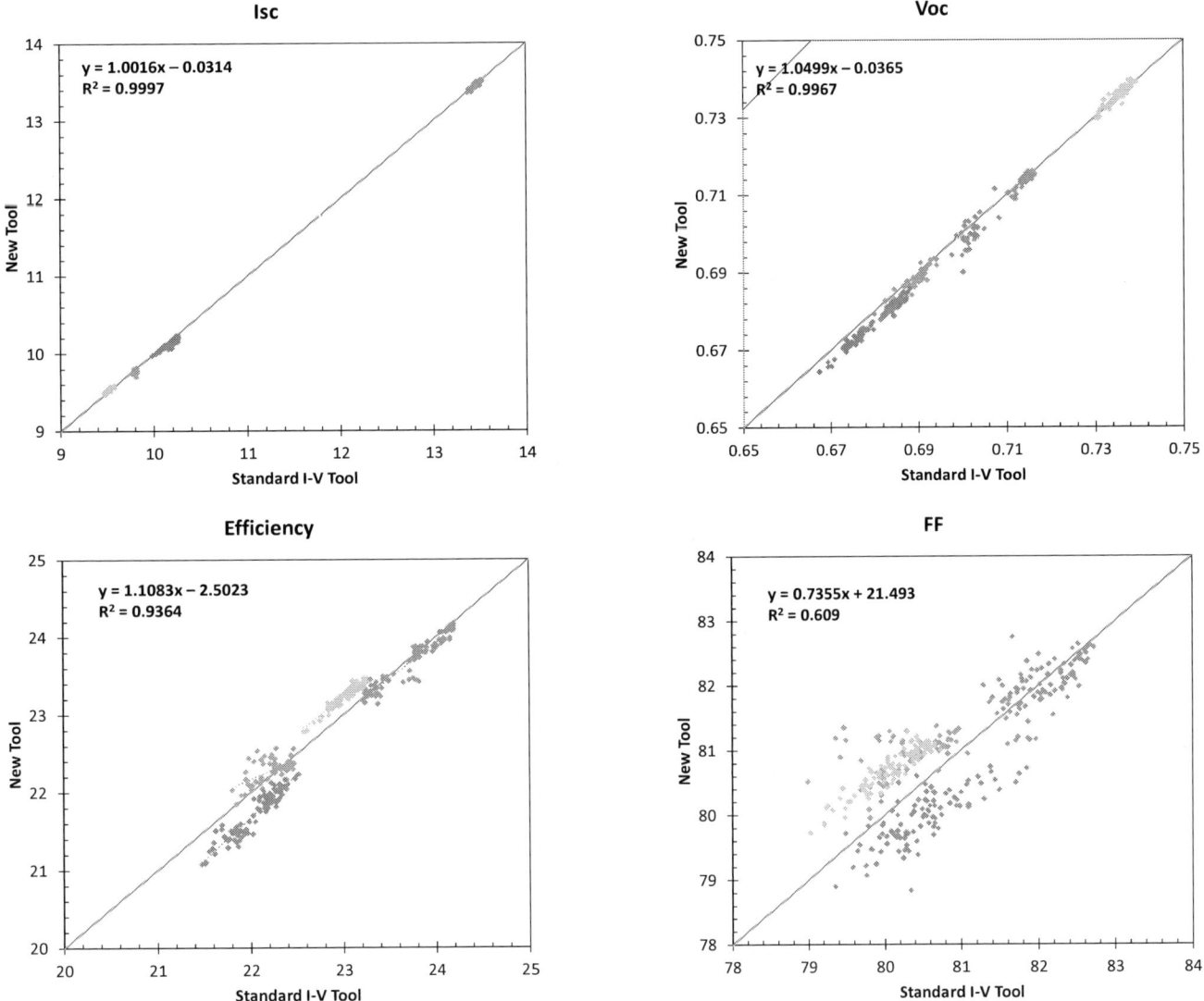

Fig. 5. Comparisons of I-V parameters reported by new tool vs. standard I-V tool (I_{sc} in amps, V_{oc} in volts, η and FF in percent). Solid orange lines represent equality ($y = x$), and different cell groups are color coded according to Table 1. The linear regression summary is reported for the entire sample set.

η, respectively), as evidenced by high values of the linear correlation coefficient ($R^2 > 0.9$). The root-mean-square relative error (RMSRE) is quite low for I_{sc} and V_{oc} ($< 0.4 \%_{rel}$) and slightly higher for η ($< 1.2 \%_{rel}$), though the latter value would decrease if we offset the reported η value on a per-cell-group basis, as is commonly done in production environments when tool matching different test systems. Although we have not done so here for clarity, such an offset would especially bring the PERC and HJT cell groups closer to the $y = x$ line in Fig. 5 (bottom-left), reducing the overall RMSRE for η and

bringing it closer to the levels seen for I_{sc} and V_{oc}. The new tool shows moderate agreement on fill factor (FF, $R^2 = 0.6$, RMSRE = 0.8 $\%_{rel}$), since this is a derived parameter whose uncertainty is a combination of the individual uncertainties in I_{sc}, V_{oc}, and η. The same comment as before regarding tool matching offsets also applies to the RMSRE for FF.

Approximately 30 cells from the 9BB TOPCon group were excluded from our analysis due to large-scale defects seen in their on-the-fly PL image. Table 2 shows examples of these excluded cells, whose defects would almost certainly lead them

Table 2. Example on-the-fly PL images of cells in the 9BB TOPCon group that were excluded from comparison with the standard I-V tool. Brightness and contrast are enhanced to aid visibility.

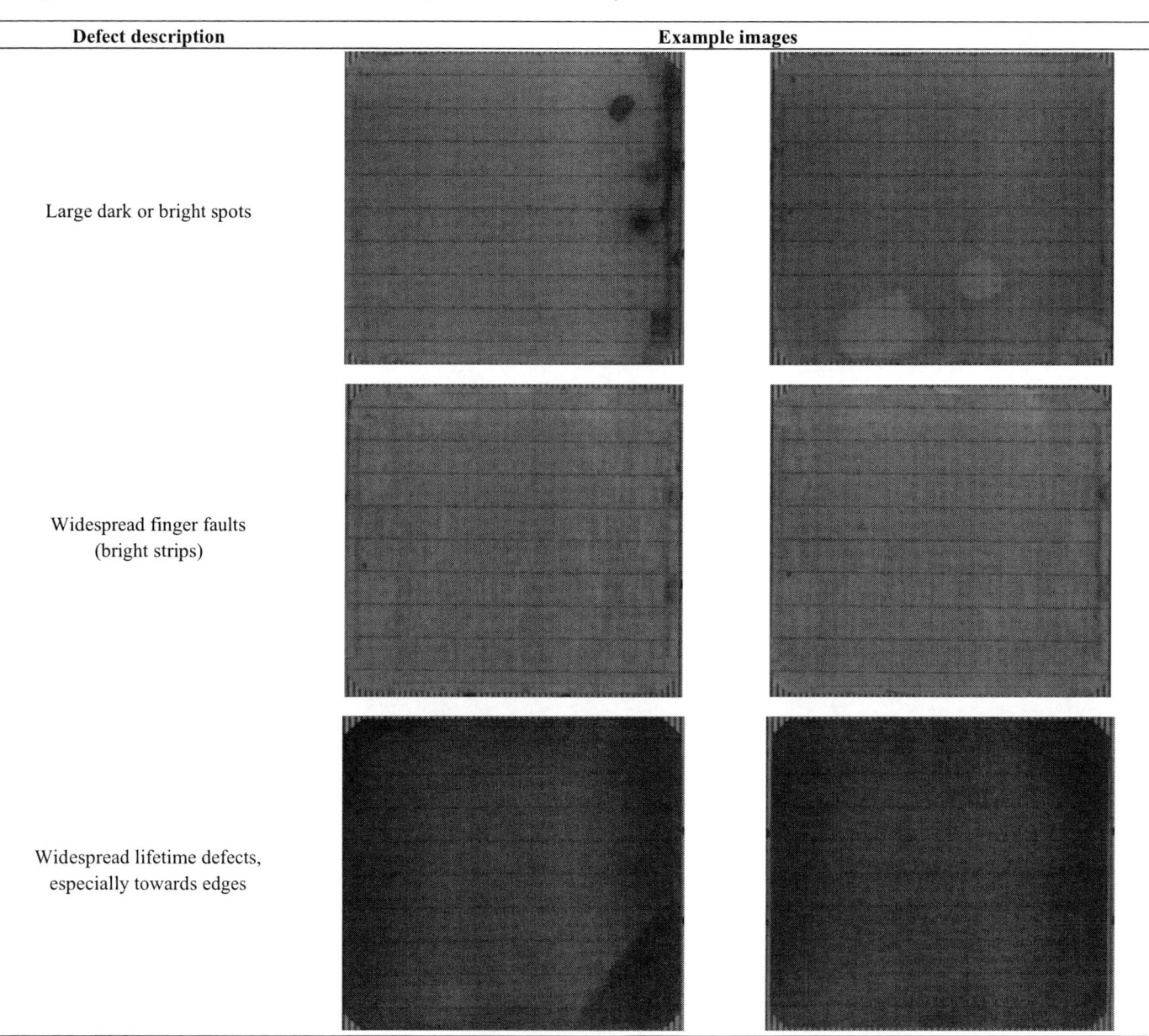

Defect description	Example images
Large dark or bright spots	
Widespread finger faults (bright strips)	
Widespread lifetime defects, especially towards edges	

to be rejected in a production line scenario. Interestingly, this entire cell group showed a wide variety of defects that were significantly more severe and more numerous compared to other groups, which is likely due to the R&D nature of this group's production processes (all other groups consisted of production line cells). We believe the prevalence of such defects in the 9BB TOPCon group also explains why it had the worst agreement on η (Fig. 5, bottom-left), as these widespread defects violate an assumption of uniformity made by the electrical transport model we use to report R_s, which then increases uncertainty in the reported R_s, η, and FF.

Fig. 6 compares the R_{grid} value reported by the new tool and BBR measurements. The agreement between methods is strong overall ($R^2 > 0.9$), but varies considerably by cell group. To our knowledge, this is the first demonstration of an alternative R_{grid} measurement method apart from the standard 4-wire BBR approach. Besides the novelty of our method (being primarily image-based), it also has the benefit of being executed with surprisingly simple hardware and minimal requirements for electrical contacting. Except for the PERC group, we generally observed tightly distributed R_{grid} values on the order of 0.1 $\Omega\,cm^2$, which is only about 20 % of the total R_s (approximately 0.5 $\Omega\,cm^2$) for most of the cells we measured. This suggests that R_s variability within cell groups predominantly comes from the $R_{contact}$ parameter, which in our model encompasses the combined effects of emitter transport loss and metal-semiconductor contact loss. Although our electrical transport model currently cannot separate these two distinct losses, it could be supplemented with additional outside information. For example, using the emitter sheet resistance ($\Omega/\square$), we could compute the emitter transport loss using either a geometrical formula [12, Eqn. (4)] or the modeling capabilities of any number of commercially- or freely-available solar cell physics toolkits. The emitter sheet resistance input into these calculations could simply be the process target value, or it could be the individual cell value determined by in-line wafer sheet resistance measurements taken upstream before and after emitter formation (e.g. using the Sinton Instruments IL-800 tool [13]).

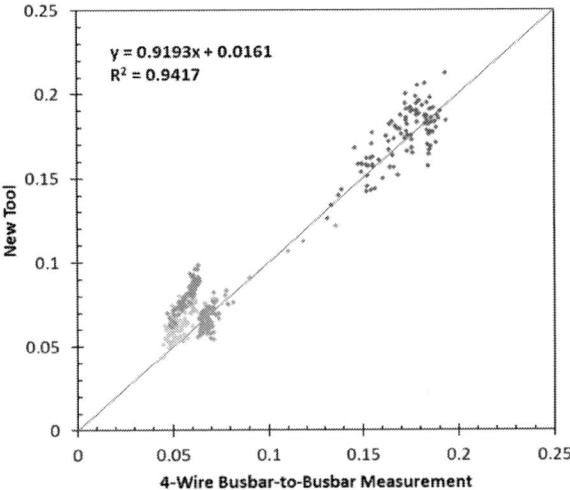

Fig. 6. Comparisons of R_{grid} (in $\Omega\,cm^2$) reported by the new tool vs. BBR measurements. The solid orange line represents equality ($y = x$), and different cell groups are color coded according to Table 1. The linear regression summary is reported for the entire sample set.

IV. CONCLUSIONS

We demonstrated a new and nearly contactless tool for production line I-V binning of solar cells that shows good agreement to a standard I-V tool in a benchmark of over 400 cells representing a production-relevant range of manufacturers and cell technologies. The new tool leverages a variety of measurement techniques to enable accurate measurement of multi-busbar and busbarless cells that are becoming more prevalent in the industry due to silver cost pressures, but are increasingly difficult to measure with conventional pogo-pin-based contacting techniques. The new tool not only provides standard I-V binning functionality, but can also enhance normal process control capabilities by reporting a breakdown of multiple series resistance components, as well as power losses due to bulk recombination, surface recombination, and shunt resistance on a per-cell basis at line speed.

ACKNOWLEDGMENTS

This material is based upon work supported by the U.S. Department of Energy's Office of Energy Efficiency and Renewable Energy (EERE) under the Solar Energy Technologies Office (SETO) award number DE-EE0009335. This report was prepared as an account of work sponsored by an agency of the United States Government. Neither the United States Government nor any agency thereof, nor any of its employees, makes any warranty, express or implied, or assumes any legal liability or responsibility for the accuracy, completeness, or usefulness of any information, apparatus, product, or process disclosed, or represents that its use would not infringe privately owned rights. Reference herein to any specific commercial product, process, or service by trade name, trademark, manufacturer, or otherwise does not necessarily constitute or imply its endorsement, recommendation, or favoring by the United States Government or any agency thereof. The views and opinions of authors expressed herein do not necessarily state or reflect those of the United States Government or any agency thereof.

REFERENCES

[1] VDMA "International technology roadmap for photovoltaics," 14th Edition, 2023. https://www.vdma.org/international-technology-roadmap-photovoltaic

[2] N. Bassi, C. Clerc, Y. Pelet, J. Hiller, V. Fakhfouri, C. Droz, M. Despeisse, J. Levrat, A. Faes, D. Bätzner, and P. Papet, "GridTOUCH: innovative solution for accurate IV measurement of busbarless cells in production and laboratory environments," in *29th European Photovoltaic Solar Energy Conference and Exhibition (EU PVSEC)*, 2014.

[3] J. Greulich, W. Wirtz, H. Höffler, N. Wöhrle, M. Juhl, O. Kunz, A. Paduthol, S. Rein, and A. Bett, "Contactless measurement of current-voltage characteristics for silicon solar cells," in *12th SiliconPV Conference*, 2022.

[4] H. Wilterdink, R. Sinton, A. Blum, K. Dapprich, "Power loss analysis for silicon PV cells and modules using the Richter recombination limit," in *NREL PV Reliability Workshop (PVRW)*, 2021.

[5] I. Zafirovska, M. Juhl, J. Weber, O. Kunz, and T. Trupke, "Module inspection using line scanning photoluminescence imaging," in *32nd European Photovoltaic Solar Energy Conference and Exhibition (EU PVSEC)*, 2016.

[6] R. Sinton and A. Cuevas, "A quasi-steady-state open-circuit voltage method for solar cell characterization," in *16th European Photovoltaic Solar Energy Conference (EU PVSEC)*, 2000.

[7] R. Sinton, "Contactless electroluminescence for shunt-value measurement in solar cells," in *23rd European Photovoltaic Solar Energy Conference and Exhibition (EU PVSEC)*, 2008.

[8] M. Glatthaar, J. Haunschild, R. Zeidler, M. Demant, J. Greulich, B. Michl, W. Warta, S. Rein, and R. Preu, "Evaluating luminescence based voltage images of silicon solar cells," in *Journal of Applied Physics* **108**, 014501 (2010); http://doi.org/10.1063/1.3443438

[9] A. Blum, R. Sinton, W. Dobson, H. Wilterdink, and J. Dinger, "Lifetime and substrate doping measurements of solar cells and application to in-line process control," in *IEEE 43rd Photovoltaic Specialists Conference (PVSC)*, 2016.

[10] D. Kane and R. Swanson, "Measurement of the emitter saturation current by a contactless photoconductivity decay method," in *IEEE 18th Photovoltaic Specialists Conference (PVSC)*, 1985.

[11] K. Dapprich, R. Sinton, H. Wilterdink, W. Dobson, C. Sainsbury, and J. Dinger, "Using doping variation to determine J_{0s} at I-V test," *11th SiliconPV Conference*, 2021.

[12] G. Micard and G. Hahn, "Discussion and simulation about the evaluation of the emitter series resistance," in *28th European Photovoltaic Solar Energy Conference and Exhibition (EU PVSEC)*, 2013.

[13] Sinton Instruments, IL-800 Product Page; https://www.sintoninstruments.com/products/il800/

Self-Thermometry of PV Panels

Kaushal Chapaneri[1], Shahzada Pamir Aly[1], Jim Joseph John[1], Gerhard Mathiak[1], Vivian Alberts[1]
and Muhammad A. Alam[2]

[1]*Research and Development Center, Dubai Electricity and Water Authority, Dubai, United Arab Emirates*
[2]*Electrical and Computer Engineering Department, Purdue University, West Lafayette, IN, USA*

Abstract—During field operations, it is an unfortunate reality that over time, the temperature sensors of the PV panels get detached, damaged, or miscalibrated. Since the performance and reliability of a panel are determined by panel temperature (T_{mod}), a complementary/alternate approach is needed. In this regard, the Sandia model can be used to predict panel temperature, if the model parameters are calibrated by site-measured historical data. However; this is often impossible because historical data may not be available. In this study, we show that by merely incorporating the panel's electrical efficiency (degrading over time) into the equation, the panel temperature can be predicted with precision, obviating the need for calibration. Based on an analysis of more than 100 different PV panels installed at the outdoor test facility of DEWA R&D (Dubai, UAE), we show that the generalized model anticipates the actual temperature within 3.1 °C (worst case). The approach offers a new paradigm where the panel serves as its own thermometer and flags when the temperature sensor malfunctions.

Keywords—PV panel, Outdoor testing, Thermal model, Empirical model, Temperature Sensors.

I. INTRODUCTION

Time-dependent power production is the most crucial metric for evaluating the performance and reliability of a solar farm. Its evaluation requires several parametric inputs, one of which is the operating temperature (T_{mod}) of the PV panel, which may be determined either experimentally or via a thermal model. Indeed, the power (and ultimately the efficiency) decreases linearly and the lifetime decreases exponentially with T_{mod}. Therefore, it is critically important to monitor T_{mod} of a PV panel.

The T_{mod} can be measured by mounting various types of temperature sensors (i.e. thermocouples) at the back of the panel (IEC 67724-1). Typical sensors have a broad range (from 40 to 130 °C), high accuracy (~0.5 °C), and excellent stability (less than 0.1 °C per year). Unfortunately, despite careful installation, it is common to find the sensors damaged, degraded, or miscalibrated over time. In this regard, we suggest that the thermal model of the panel can be used as an alternative "thermometer" of the panel.

In the literature, various thermal models have been reported. Some models rely on empirical correlations [1,2], and deduced analytical equations using the heat equation [3], while others use comprehensive models developed by means of numerical techniques [4,5]. Based on the geometry of the PV panels and the boundary conditions they are typically subjected to, numerical models are better fitted for accurately estimating the PV panel temperature, especially under rapidly varying ambient conditions in the field [6–8]. Fortunately, in the literature, many empirical equations have been shown to provide adequate estimates of the PV panel temperature, such as the Faiman [2] and the King's (or Sandia) model [1].

In this study, we will determine PV panel temperature for 100 plus PV panels of different types and configurations installed at the outdoor test facility of DEWA R&D in Dubai, UAE (shown in Fig. 1). We will compare two approaches for model-based self-thermometry. In the traditional approach, we will derive the empirical constants of the Sandia model by fitting the local field data. For the proposed approach, we will generalize the Sandia Model (with default parameters) by accounting for the panel's electrical efficiency but obviating the empirical fitting [9].

Fig. 1. More than 100 different types of PV panels were installed at the outdoor test facility (OTF) of DEWA R&D, Dubai, UAE.

II. MODEL IMPROVEMENT

The following equation is used for estimating the PV panel temperature using the Sandia model [1]:

$$T_{mod} = \text{POA} \times e^{a+b \times ws} + T_{amb} \qquad (1)$$

Here, POA is the incident plane of array irradiance, ws is the wind speed and T_{amb} is the ambient temperature. As shown in TABLE I, the empirical constants a and b depend on the panel type and its configuration.

TABLE I.   SANDIA MODEL DEFAULT PARAMETERS [1]

Panel Type	Mount	a	b
Glass/cell/glass	Open rack	-3.47	-.0594
	Close roof mount	-2.98	-.0471
Glass/cell/polymer sheet	Open rack	-3.56	-.0750
	Close roof mount	-2.81	-.0455
Polymer/thin-film/steel	Open rack	-3.58	-.113
22X Linear Concentrator	Tracker	-3.23	-.130

978-1-6654-6060-6/23 $31.00 © 2023 IEEE

The default parameters of the Sandia model are obtained from at V_{oc} condition, and depend on local conditions. To determine T_{mod} of a panel operating in the field, one must account for its electrical efficiency (η). This is because, for a PV panel under operation, only part of the incident irradiance (equal to the efficiency - η) is converted to electrical power. The remaining is converted to heat before finally being dissipated to the surroundings via convection and radiation.

$$T_{mod} = \text{POA} \times [1 - \eta(t)] \times e^{a+b \times ws} + T_{amb} \qquad (2)$$

Here, $\eta(t)$ is the electrical efficiency of the panel degrading over time.

III. FIELD DATA

The 100-plus PV panels shown in Fig. 1 consist of various mono-crystalline, multi-crystalline, thin-film, and bifacial PV technologies. Each of the PV panels is installed in a stand-alone configuration, with a dedicated electronic load measuring its complete I-V characteristics every 10 mins and maximum power point currents-voltages every 30 secs. Also, for each PV panel, there is a dedicated temperature sensor attached in the middle of the rear side, which also records data every 30 secs. Most of these PV panels have field data from March 2015 to the present day (January 2023 at the moment). Similarly, the POA, ws and T_{amb}, which are needed by the Sandia model as input were also measured every 30 secs.

IV. MODEL IMPLEMENTATION AND VALIDATION

A. Estimating panel temperature with fitted parameters

For all the 100-plus panels, parameters a and b in Eq. (1) were one by one fitted individually using the field measured T_{mod}, POA, ws and T_{amb}. For one of the installed multi-crystalline panels, the estimated temperature using fitted parameters vs actual measured temperature is shown in Fig. 2. The scatter plot in Fig. 2(a) gives an impression that there is a wide spread in the correlation between the actual and estimated temperatures. However, the 2-D histogram in Fig. 2(b) shows that there is a very strong correlation between the actual and estimated temperatures, with less than 1% of data points as outliers, which is remarkable considering there are almost 4.5 million data entries from March 2015 to January 2023, with 30 seconds recording interval. Indeed, the field data used for fitting the Sandia parameters in Eq. (1) was not pre-processed, except for basic filtering. Cleaning the field data would further improve fitting and eliminate the outliers. Moreover, the Sandia model is a steady-state model, i.e., it does not account for the thermal mass of the PV panel, therefore it does not perform well under rapidly changing weather conditions. Apart from the strong correlation observed in Fig. 2(b), it can also be observed that the Sandia temperature model slightly underestimates the actual temperature, but not more than 3-4 °C on average.

Likewise, RMSE for estimated temperature using fitted parameters vs actual measured temperature of all the 100-plus panels was calculated and as shown in Fig. 3, the RMSE does not exceed 4.01 °C on the worst case (left y-axis and blue color). The lower RMSE values from panel 90 onwards are for panels installed in vertical positions. A key conclusion of this paper is that a well-calibrated traditional Sandia model can serve as a virtual thermometer of a panel. The challenge is that this rich dataset is often unavailable. The approach discussed below addresses the concern and should be more broadly applicable.

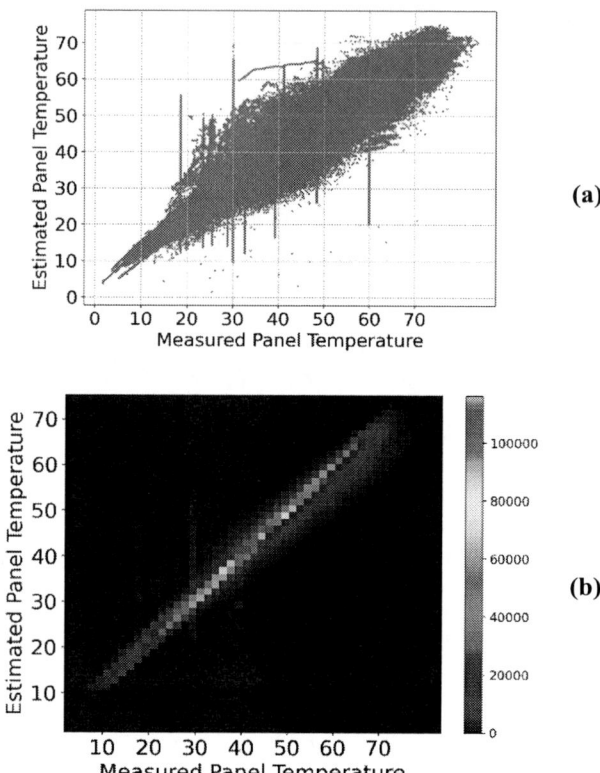

Fig. 2. Estimation of panel temperature using Sandia model with fitted 'a' and 'b' parameters: (a) scatter plot and (b) 2-D histogram, of actual measured temperature vs. estimated temperature.

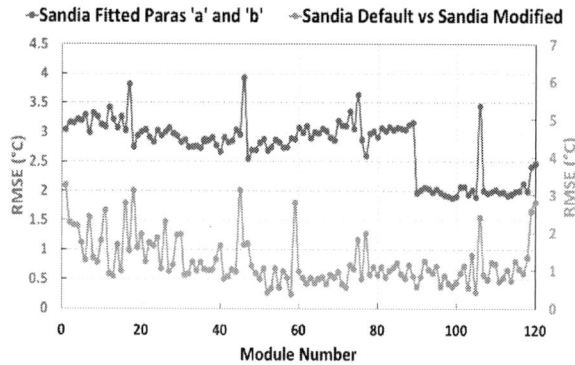

Fig. 3. RMSE calculation for estimated temperature using fitted parameters vs actual measured temperature (left y-axis in blue color) and with default Sandia parameters and linearly degrading efficiency (right y-axis in orange color), of all the 100-plus panels.

B. Panel temperature by the generalized Sandia Model

The multi-crystalline panel discussed above has $\eta = 16.19\%$, and a measured efficiency loss of 0.5%/year, see Fig. 4. With these parameters and with default a and b values from Table 1, Eq. (2) can be used to determine T_{mod}.

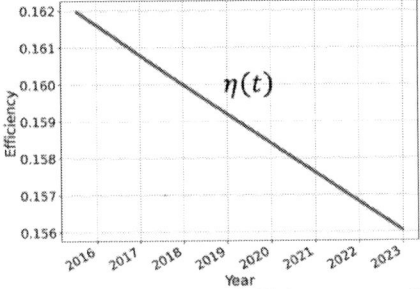

Fig. 4. Considering a 0.5% linear loss of efficiency per year for a PV panel, the efficiency drops from 16.19% to 15.61% from Mar 2015 to Jan 2023.

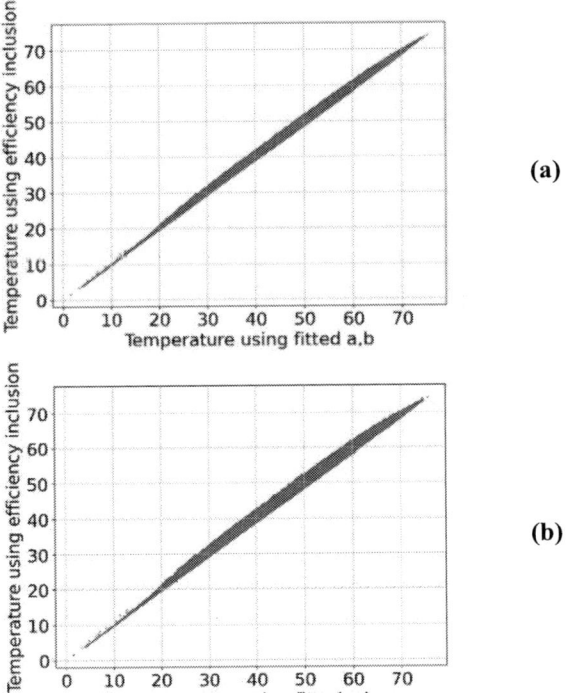

Fig. 5. Estimated temperature of the panel using fitted parameters and the modified Sandia model in Eq. (2) with default parameters and linearly degrading efficiency. (a) For the best correlation, the RMSE was 0.38 °C and (b) for the worst correlation, the RMSE was 3.27 °C.

Fig. 5 shows that the generalized model (with default parameters) yields temperatures very close to that estimated by fitted parameters using the field data through the traditional model. Remarkably, these results show that in cases where the historical field data is not available for fitting, Eq. (2) can be used to estimate the panel temperature with high confidence. Similarly, the RMSE for temperature estimation using the default Sandia parameters and linearly degrading efficiency (Eq. (2)) vs temperature estimation

using fitted parameters, for all the 100-plus panels is shown in Fig. 3 (right y-axis and orange color).

Moreover, real-time estimates of T_{mod} using these virtual thermometers can also help monitor the health of the sensor (in cases of detachment, damage, or calibration drift) and the health of the PV panel itself, by raising alarms or flags in cases the measured temperature exceeds a certain range from the estimated temperature value.

V. CONCLUSIONS

It is well known that tracking panel temperature is important for monitoring the energy output and degradation rate of the PV panel. Based on the work and analyses presented, the following are the takeaways from this study:

1. With historical data, the Sandia model (with empirical a, b parameters) can serve as an excellent proxy thermometer of a PV panel, with best and worst case RSME of 1.88 °C and 4.01 °C.
2. By including time-dependent efficiency, the generalized Sandia model panel obviates the need for historical data, and yet serves as an equally effective thermometer, with best/worst RMSE of 0.38 °C/3.27 °C.
3. The temperature difference and loss of correlation between self-thermometry (solar panel as its own thermometer) and the physical temperature sensor indicates that one of the sensors (e.g., POA, wind speed, or temperature) is damaged/miscalibrated and must be replaced. This would address a major challenge of corrupted/missing historical field data by flagging the issue in real-time.

REFERENCES

[1] D.L. King, J.A. Kratochvil, W.E. Boyson, Photovoltaic array performance model, (2004).

[2] F. David, Assessing the outdoor operating temperature of photovoltaic modules, Progress in Photovoltaics: Research and Applications. 16 (2008) 307–315. https://doi.org/10.1002/pip.813.

[3] D.V. Widder, The heat equation, Academic Press, 1976.

[4] S.P. Aly, N. Barth, S. Ahzi, A numerical study for approximating cells temperature inside a PV module, in: Proceeding of 3rd Thermal and Fluids Engineering Conference (TFEC), Begellhouse, Connecticut, 2018: pp. 1575–1584. https://doi.org/10.1615/TFEC2018.rce.022023.

[5] S.P. Aly, N. Barth, S. Ahzi, A Three-Dimensional Finite Element Based Dynamic Thermal Model of PV Modules with an Improved Thermal Network, in: 2017 International Renewable and Sustainable Energy Conference (IRSEC), IEEE, 2017: pp. 1–6. https://doi.org/10.1109/IRSEC.2017.8477421.

[6] S.P. Aly, S. Ahzi, N. Barth, A. Abdallah, Using energy balance method to study the thermal behavior of PV panels under time-varying field conditions, Energy Convers Manag. 175 (2018) 246–262. https://doi.org/10.1016/j.enconman.2018.09.007.

[7] S.P. Aly, S. Ahzi, N. Barth, B.W. Figgis, Two-dimensional finite difference-based model for coupled irradiation and heat transfer in photovoltaic modules, Solar Energy Materials and Solar Cells. 180 (2018) 289–302. https://doi.org/10.1016/j.solmat.2017.06.055.

[8] S.P. Aly, J.J. John, G. Mathiak, O. Albadwawi, L. Pomares, V. Alberts, A thermal model for bifacial PV panels, in: 2022 IEEE 49th Photovoltaics Specialists Conference (PVSC), IEEE, 2022: pp. 457–459. https://doi.org/10.1109/PVSC48317.2022.9938549.

[9] Alam, Muhammad A., and M. Ryyan Khan. *PRINCIPLES OF SOLAR CELLS: Connecting Perspectives on Device, System, Reliability, and Data Science.* 2022.

Inverter clipping and its masking effect on PV soiling: truth or myth?

Leonardo Micheli, Matthew Muller, Marios Theristis, Florencia Almonacid, Eduardo F. Fernandez

DIAEE, Sapienza University of Rome, Rome, Italy

NREL, Golden, CO, United States

Sandia National Laboratories, Albuquerque, NM, United States

University of Jaen, Jaen, Spain

Clipping is caused by the saturation of the inverter in a PV plant. Indeed, in utility-scale systems, the inverter is commonly undersized compared to the total DC capacity of the modules. Because of this, in some conditions, the DC power might exceed the capacity of the inverter and as a result, part of the photo-generated DC energy cannot be converted into AC. This means that, if the magnitude of the DC losses is lower than the amount of "clipped" energy, clipping functions as a mask. This work attempts to quantify the impact of clipping on the U.S. PV energy generation with a particular attention on soiling losses. These affect the DC energy generation, and are often expected to be fully masked by clipping. The preliminary results of this investigation show that, despite the common belief, under the typical configurations, clipping has limited impact on soiling losses.

978-1-6654-6060-6/23 $31.00 © 2023 IEEE

Effect of Arsenic Doping in Polycrystalline Thin Film CdTe Solar Cells

Mayank Mate, Akash Shah, Ramesh Pandey, Zachary Lustig, Walajabad Sampath and Amit H. Munshi

Colorado State University, Fort Collins, Colorado, 80538, Department of Mechanical Engineering, Department of Physics, USA

Abstract — Previous work has shown the positive effects of Arsenic doping in polycrystalline CdTe solar cells, particularly on charge carrier lifetimes, implied voltage due to the quasi Fermi level splitting, and improved device performance. In this study, we investigate the effect of varying arsenic concentrations, optimizing Cd vapor overpressure on the device performance. The devices were fabricated using a novel closed space sublimation method, in which, CdSeTe:As and CdTe:As would be sublimated under excess Cd vapor. Since CdSeTe is a material with a potential to achieve efficiency >25% cell efficiency, devices with the structure CdSeTe/CdTe:As were fabricated and characterized.

Index Terms —CdTe, Arsenic doping, cadmium overpressure, carrier lifetime, device efficiency.

I. INTRODUCTION

CdTe based solar cells is an inexpensive, easy to manufacture technology that is imperative to drive the mission of renewable energy generation. Currently, small area research devices have shown a record 22.1% efficiency as compared to commercially fabricated modules which show efficiencies up to 19.0% [1]. At Colorado State University (CSU), we have fabricated a champion $CdSe_xTe_{1-x}$/CdTe bilayer absorber device with an efficiency of 20.14% [2]. Further efficiency improvements have not yet been achieved limited by lower open-circuit voltage (V_{OC}). The Shockley-Queisser efficiency limit for CdTe-based semiconductor with a bandgap of ~1.4eV is about 33% with a V_{OC} >1.1V [3]. However, for polycrystalline absorbers, V_{OC} greater than ~885 mV, and therefore higher device efficiencies, have not yet been achieved. Research is ongoing to overcome the challenge of achieving V_{OC}>900mV in polycrystalline CdTe to further device performance.

This paper reviews the effects of optimizing Cd overpressure for doped polycrystalline CdTe devices to improve carrier concentration and device efficiency. This can be achieved by doping the absorber with p-type dopants using group V elements in the periodic table. Previously, P, Sb and As have been utilized to investigate these effects and increase hole concentration to >10^{15} cc^{-1}.

Recently, As as a dopant has attracted the attention of researchers. Burst et al showed that an open-circuit voltage (V_{OC}) > 900mV can be achieved with a single crystal device using As doping [4]. A study by Yang et al showed that As substitutes Te sites under Cd rich growth conditions and is expected to yield higher carrier concentrations than conventional Cu doping [5]. Kanevce et al performed numerical models that suggest device efficiencies over 24% could be achieved with surface recombination velocity S < 100cm/s,

doping density ρ >10^{16} cc^{-1}, and bulk recombination lifetime of τ > 10 ns [6].

Atomistic models using density functional theory (DFT) have been used to understand CdTe surface and interface properties. Studies performed by Shah et al have suggested that higher As concentrations at CdTe surface would be required to bend the valence band maximum (VBM) upwards to aid hole charge carrier transport at the back electrode [7].

Various experimental pathways will also be explored to optimize As activation. This will aid in understanding As diffusion as a monoatomic species which acts as a p-type dopant, instead of As clusters which create defects, lowering device performance. Previous work by Burton et al presented with similar findings of As clusters in MBE grown films which prevented dopant activation [8]. Previous work explored and compared deposition of CdTe:As with diffusion of the As by annealing during the CdCl$_2$ treatment. It shows that the diffused process has a much higher carrier concentration than the direct deposition of CdTe:As films[9].

II. EXPERIMENTAL

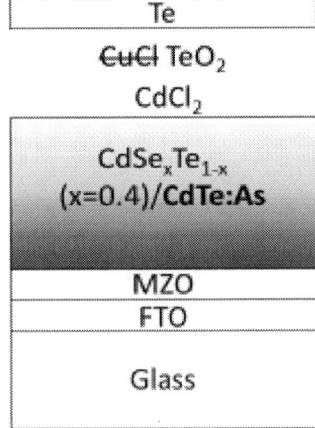

Fig. 1. Device structure of the fabricated 19% efficiency device

As shown in figure 1, thin film $CdSe_xTe_{1-x}$ /CdTe:As devices were fabricated using closed space sublimation (CSS). CSU's proprietary advanced co-sublimation system was used to deposit CdTe:As film under over-pressure of Cd vapor so as to grow a film with higher V_{Te} sites to enable greater dopant activation [10]. A control device was fabricated with no intentional doping to compare against the doped devices. Another control device was also fabricated with traditional Cu doping. Arsenic doped devices were fabricated by sublimating undoped $CdSe_xTe_{1-x}$ (CST), with a film thickness of 700 nm followed by co-sublimating CdTe:As with Cd overpressure. The Cd overpressure was applied to create V_{Te} sites that would be occupied by As leading to activated doping. The CdTe:As source material was fabricated using High Pressure Bridgeman method provided by Washington State University was used.

Following the deposition of CdTe:As, the device was treated with $CdCl_2$ for defect passivation and also acts as an annealing treatment. Further, TeO_x was sputtered at the back surface followed thermal evaporation of Te to create an ohmic contact. Conductive electrode contact was made by carbon and nickel conductive paint and delineating each substrate into 25 small area devices with an area of ~0.65 cm². Performance characterization of the devices was done by using current density vs voltage (JV) measurements. Further, carrier concentration vs distance from junction (N_A vs X) was deduced from capacitance-voltage (CV) measurement recombination lifetime using time resolved photoluminescence (TRPL) was also measured.

III. RESULTS

A. Electrical Characterization

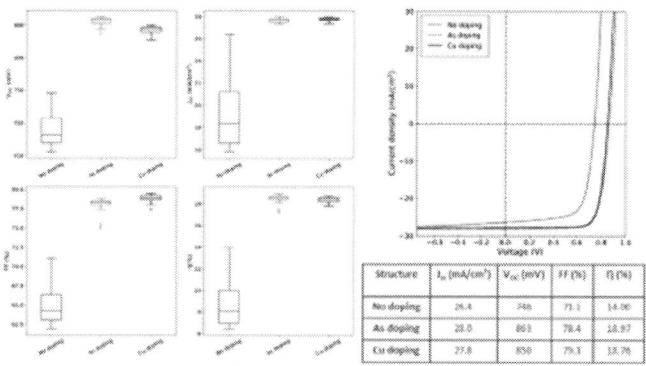

Fig. 2. J-V measurement comparison for undoped, As and Cu doped devices

From Figure 2, it can be observed that As doping leads to device performance comparable to Cu doped devices with the similar structure, with a slight increase in performance. Cu doping increases the cell efficiency from 9% to 18.75% as compared to undoped devices. However, As doping increases the efficiency to ~19% compared to undoped devices. The V_{oc} of Cu doped devices was increased to 850 mV from 746 mV and with As doping, the V_{oc} rises to >863 mV. The JV characteristics from the figure are from the best performing cell from each substrate.

B. Effect of Cd Overpressure on CST/CdTe:As devices with a different structure.

Figure 3 shows the device structure that was fabricated to study the effect of Cd overpressure on device performance. As seen from the boxplots in Figure 4, Cd overpressure during CdTe:As film growth is vital to achieving >750 mV V_{OC}, >25 mA/cm² current density and efficiencies of >15%.

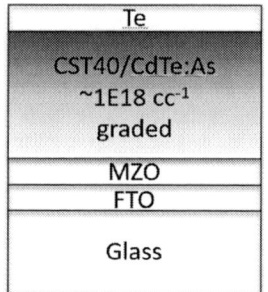

Fig. 3. Device structure for optimizing Cd over pressure co-sublimation temperature.

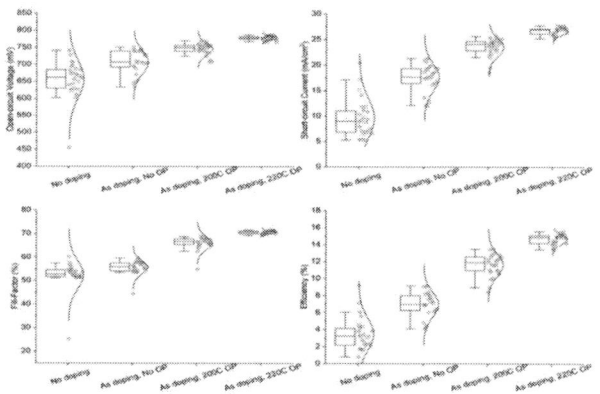

Fig. 4. Effect of optimizing and varying Cd overpressure on the performance of As doped devices.

C. Spectroscopic Characterization

Figure 5 shows how Cd overpressure affects carrier concentration. It can be observed that 210°C, carrier concentration was highest and measured to be 2.3E16 cc⁻¹ at zero bias.

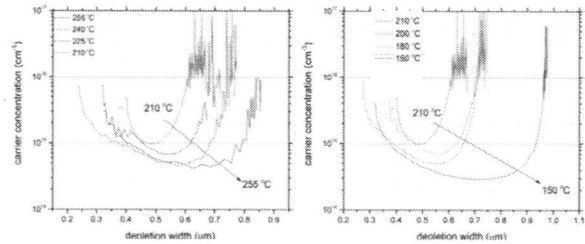

Fig. 5. Carrier concentration vs depletion width as a function of Cd overpressure

TRPL characterization presented in figure 6 shows high recombination lifetimes.

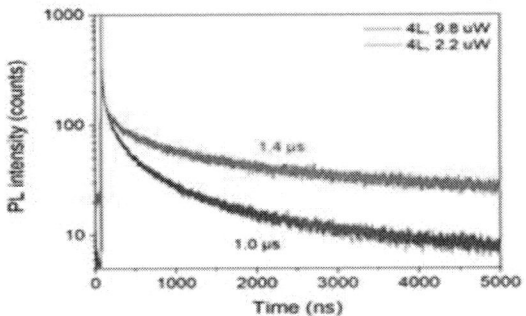

Fig. 6. TRPL of As doped devices.

IV. SUMMARY

Arsenic has been investigated to improve carrier concentration and efficiency of CdTe thin film photovoltaic devices. The best performing As doped device fabricated at CSU had an efficiency of ~19%. Investigation of the effects of process parameters such as absorber thickness, As concentration and Cd overpressure would need to be optimized to further improve device performance.

ACKNOWLEDGEMENT

The authors would also like to acknowledge the support from the National Science foundation under the Industry/University Cooperative Research Center (I/UCRC) for Solar Powered Future SPF 2050 (Award Numbers – 1821526 and 2052735), and the I/UCRC Industrial Advisory board.

REFERENCES

[1] Green et al., "Solar Cell Efficiency Tables, Progress in Photovoltaics", doi.org/10.1002/pip.3371.

[2] T. Shimpi *et al.*, "Influence of Process Parameters and Absorber Thickness on Efficiency of Polycrystalline CdSeTe/CdTe Thin Film Solar Cells," *2020 47th IEEE Photovoltaic Specialists Conference (PVSC)*, Calgary, AB, Canada, 2020, pp. 1933-1935, doi: 10.1109/PVSC45281.2020.9300840.

[3] Sven Rühle, "Tabulated values of the Shockley–Queisser limit for single junction solar cells", *Solar Energy*, Volume 130, 2016, Pages 139-147, ISSN 0038-092X.

[4] J. M. Burst *et al.*, "CdTe solar cells with open-circuit voltage breaking the 1V barrier," *Nat. Energy*, vol. 1, no. 4, 2016, doi: 10.1038/NENERGY.2016.15.

[5] J.H. Yang, W.J. Yin, J.S. Park, J. Burst, W.K. Metzger, T. Gessert, T.Barnes, S.H. Wei. Enhanced p-type dopability of P and As in CdTe using non-equilibrium thermal processing. *J. Appl. Phys.* 118 (2015) 025102.

[6] A. Kanevce, M. O. Reese, T. M. Barnes, S. A. Jensen, and W. K. Metzger, "The roles of carrier concentration and interface, bulk, and grain-boundary recombination for 25% efficient CdTe solar cells," *J. Appl. Phys.*, vol. 121, no. 21, 2017, doi: 10.1063/1.4984320.

[7] A. Shah et al., "First principles guided device fabrication of arsenic doped CdTe photovoltaics", *2021 IEEE 48th Photovoltaic Specialists Conference (PVSC)*, Fort Lauderdale, FL, USA, 2021, pp. 1527-1529, doi: 10.1109/PVSC43889.2021.9518988.

[8] G. L. Burton *et al.*, "Understanding arsenic incorporation in CdTe with atom probe tomography," *Sol. Energy Mater. Sol. Cells*, vol. 182, no. February, pp. 68-75, 2018, doi: 10.1016/j.solmat.2018.02.023.

[9] A. Danielson et al., "Electro-optical characterization of arsenic-doped CdSeTe and CdTe solar cell absorbers doped in-situ during close space sublimation", *Solar Energy Materials and Solar Cells* Volume 251, March 2023, 112110.

[10] A. H. Munshi et al., "Arsenic Doping of Polycrystalline CdSeTe Devices for Microsecond Life-times with High Carrier Concentrations", *2020 47th IEEE Photovoltaic Specialists Conference (PVSC)*, Calgary, AB, Canada, 2020, pp. 1824-1828, doi: 10.1109/PVSC45281.2020.9301003.

Thermal Modelling of a Renkube Panel

Deepika Gopal, Balaji Bangolae Lakshmikanth and Lakshmi Santhanam

Renkube Private Limited, Bangalore, Karnataka, India - 560072

Abstract—**Renkube's new Solar PV panel which uses Motion Free Optical Tracking (MFOT) technology gives a 20% increase in energy yield over Fixed tilt solar panels due to increased redirected light on the cell [1]. This is achieved by using fixed glass light redirecting reflectors. The 20% increase in yield is in the same improvement range as Single Axis Trackers but there are no moving parts and no maintenance costs. Since it is expected that the increased light on the cell will lead to a slight increase in cell temperature, thermal modeling needs to be carried out to predict the temperature of the cell and to ensure that it is well below the "Solar Panel Maximum Temperature" which is on average around 85°C. Furthermore, we show that all cells of the panel receive equal illumination. We validate our findings with our prototype set up on the roof of our office in Bangalore, India. We conduct thermal modelling simulations for warmer locations all over the world such as India, US, Australia and UAE to understand the maximum cell temperatures obtained.**

Keywords—Solar, Photovoltaics, Motion Free Optical Tracking (MFOT), Heat Modelling, Uniform Illumination, Simulation

I. INTRODUCTION

Global warming and increased natural disasters around the world have made it imperative that we take immediate and urgent action towards solving the climate change crisis. Solar PV panels can play a huge role in reducing the usage of fossil fuels for the generation of electricity. Yield improving technologies can help drive down the cost and will help increase the adoption of solar around the world. Renkube's Motion Free Optical Tracking (MFOT) is a new variant of a Fixed tilt PV solar panel which increases the energy yield of a panel by 20% by using light redirecting reflectors to increase the light incident on the cell during periods of lower solar insolation [1]. The design is such that it will work anywhere in the world, and it is cell technology agnostic.

To test our design against a non-MFOT panel to find the increase in energy yield, we used the Ray Tracing Software Tracepro and observed a 22% increase in yield over the course of the year. But this is not realistic as the effect of temperature on the efficiency of the cell must be considered. The electrical efficiency of a cell decreases with increase in cell temperature dependent on electrical properties, thermal properties, meteorological conditions, and PV installation characteristics. If we consider the efficiency of PV cells to be 20% at STC, most of the remaining light is converted to heat which contributes to the rise of temperature of the cell. The subsequent increase in temperature leads to reduction of cell efficiency proportional to the temperature coefficient of the cell. It is

expected that the temperature of an MFOT panel would be slightly higher when compared to a similarly situated cell of a non-MFOT panel due to the extra redirected light. We carry out Heat Modelling using COMSOL Multiphysics software to calculate the temperature and updated efficiency at each point in time. COMSOL simulates the effects of conduction, convection, and radiation to calculate the temperature of the cell – we explain our Methodology in Section 3.

Most solar panels have a rated "solar panel max temperature" of around 85°C. Damp heat tests also use this as the maximum temperature to test panels. With our simulations, we show that the temperature of Renkube panels is slightly higher for a given location but is well below 85°C. In the following section, we run simulations for locations which have large solar installations - Rajasthan (India), California (US), Sydney (Australia) and Dubai (UAE) to understand the maximum cell temperatures obtained. To validate our results, we have set up multiple panels on the roof of our office in Bangalore. Data loggers are attached to the modules which measure voltage, current and temperature every minute and the data is sent to a server.

The design of Light Redirecting Reflectors has been done to ensure that there is increased light incident on the cell during times of lower sun elevation angle and reduces as the sun moves to a directly overhead position. This ensures that the intensity of light on the cell never crosses the one-sun limit where "one sun" is equivalent to the irradiance of one solar constant - 1000 W/m². Crossing of this one sun limit can lead to cell deterioration in case of PV cells and is strictly avoided.

The effects of shading even a small part of a single cell of a panel are well known – the power of the entire panel and string gets affected [9]. In Renkube panels, the redirecting structures are arranged such that all cells (and thus all panels) receive the exact same illumination and we show this in the proceeding sections with various tests and simulations.

This paper studies the effects of increased incident light on the temperature of the cell as well as demonstrates that all cells receive the same illumination with no concentrated light on the cell. We validate our findings using measurements from our prototype.

978-1-6654-6060-6/23 $31.00 © 2023 IEEE

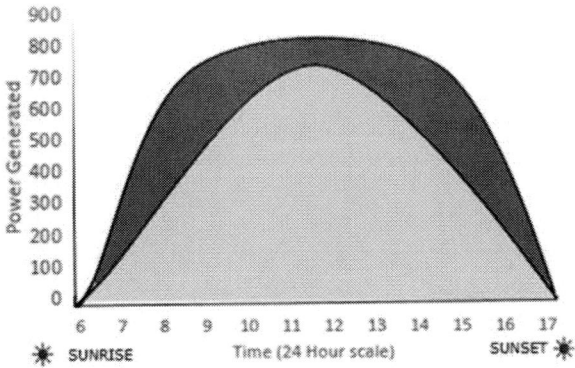

Figure 1: Performance of Renkube vs Baseline Panels

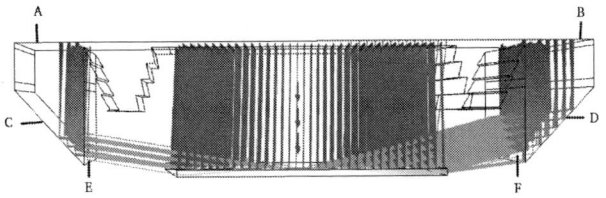

Figure 2: Equinox rays for Renkube Panel

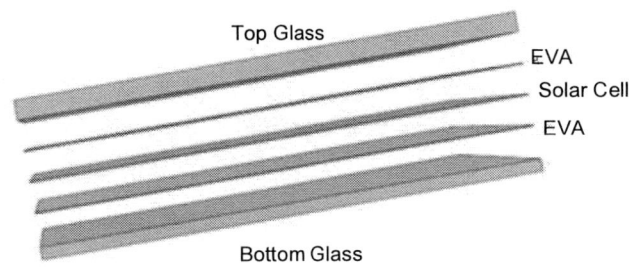

Figure 3: Base module for Fixed tilt and Renkube panels

II. DESCRIPTION OF THE MFOT PANEL

Renkube's MFOT panel is a Fixed tilt panel with absolutely no moving parts. It uses a light harvesting glass redirecting reflector which is designed based on ray optics principle and co-joined with the solar glass and it increases the illumination on the solar cell. It redirects extra sunlight on to the solar cell and boosts the panel's energy generation by 20% [1]. No changes need to be made to the cell - MFOT is cell technology agnostic.

When the sun is at its peak, a Fixed tilt panel performs close to its Rated Power and hardly any performance gain is possible. It is in the mornings and evenings that the panel would not be performing at its best. For example, under ideal conditions, a 100W panel would give an instantaneous power value of 100W with the sun directly overhead. But at 8 A.M for instance, when the sun's elevation angle is 30°, the panel may produce only about 50W – half of its maximum (conditional on the azimuth angle).

MFOT focuses on improving the performance for lower elevation angles to compensate for the cosine losses and bring the output up to peak performance. In Fig. 1, we show the power generated on December 21st in Bangalore, India when the elevation angle does not cross 70°. We see an improvement all through the day.

Fig. 2 shows how the MFOT panel handles the equinox sunshine with the outer light redirecting reflectors A and B. They redirect the equinox sunshine towards the solar cell and thereby increase the yield of the panel. An equinox sunlight incident on the top surface of A undergoes TIR at the surface C and leaves via the output surface E and gets redirected towards the solar cell. In this way, different parts of the light redirecting reflectors work year-round to give a 20% increase in yield.

III. METHODOLOGY

The Renkube panel is a glass-on-glass panel. Typically, the material used on the underside of a panel is glass or Tedlar. In the case of Renkube panels, glass has been used for its lower cost and better protection of the cell against moisture. To obtain a fair comparison of the effects of temperature on the cell, the non-MFOT panel considered is also a glass-on-glass panel.

The structure of the base module is the same for both types of panels - as shown in Fig. 3. The solar cell is sandwiched between a top and bottom glass cover using Ethylene Vinyl Acetate (EVA) as the encapsulation layer [5]. In the case of Renkube panels, redirecting structures on either side are made of glass as shown in Fig. 2.

A 160mm x 80mm cell is considered for our simulations which are carried out in a two-step process. Tracepro, which is the industry's leading Ray tracing tool for simulating sun rays on 3D objects is used to obtain the power of incident light on the cell in 15-minute intervals. These values are then fed to COMSOL's Heat Transfer module to simulate changes of temperature on the cell and the corresponding cell efficiency is calculated over the course of a day. COMSOL models the ambient temperature based on the location and time. The thermal model analyzed in this study is similar to that modeled by Jones and Underwood [3].

Of the total incident light on the cell(q_{rad}), only a certain percentage is converted to electricity decided by the rated efficiency of the cell(η_{ref}) at a temperature of 25°C. But when the temperature is greater than 25°C, the cell efficiency reduces as per [3] –

$$\eta_{pv} = \eta_{ref} \left[1 - \beta \left(T_{cell} - T_{ref}\right)\right] \quad (1)$$

The value for the temperature coefficient β is dependent on the material. For crystalline silicon, the value is taken to be 0.0045°C^{-1} which means for every degree above 25°C, the efficiency of the panel decreases by 0.45%. η_{ref} is the PV efficiency of the solar cell at the reference temperature T_{ref}. where η_{ref} = 20% with T_{ref}=25°C in our case. T_{cell} is the temperature of the cell and η_{pv} is the corresponding efficiency at that temperature. Material properties have been used as per Table 1 [2, 3, 4, 5]. Most of the remaining light is converted to heat q_{heat} as per [3] –

$$q_{heat} = q_{rad} \left(1 - \eta_{pv}\right) \quad (2)$$

The following effects of heat have been modelled in COMSOL as shown in Fig. 4 –
1. Conduction - In each of the topologies, conduction effects between solar cell-top glass and the solar cell-back glass is modelled.
2. Convection - In each of the topologies, heat loss due to convection is modelled on the top and bottom surfaces of the solar panel. The heat transfer coefficient used on the top of the panel is 5.8 W/(m^2.K) and on the bottom is 2.9 W/(m^2.K).
3. Radiation - Heat loss due to radiation from the top glass and bottom glass have been modelled.

As part of the simulation, temperature and updated efficiency at each unit time is calculated.

IV. THERMAL MODELLING RESULTS

A. Simulation in Software

Simulation experiments were first run using Tracepro with location set to Bangalore, India. Output power over the course of the day on the 21st of every month for the whole year has been measured. As the next step, this power is fed to the COMSOL Heat Transfer Module to find the variation of temperature of the cell over the course of the day and this is carried out for every month. Fig. 5 plots the maximum temperature of the MFOT panel for each month with the corresponding non-MFOT and ambient temperatures. We see that MFOT panels reach a maximum temperature of 65°C in April when the ambient temperature is the highest at 31°C and non-MFOT panel temperature is 60°C.

To confirm that this holds true for popular warm solar installation locations around the globe, we have run similar simulations for Rajasthan (India), California (US), Dubai (UAE) and Sydney (Australia) as shown in Fig. 7. We see that the

	Glass	Tedlar	Solar Cell
Thermal Conductivity W/(m.K)	2	0.15	130
Density kg/m^3	3000	1200	2330
Heat Capacity J/(kg.K)	500	1250	677
Refractive Index	1.54	1.46	3.72
Emissivity	0.85	0.9	0.9

Table 1: Material Properties

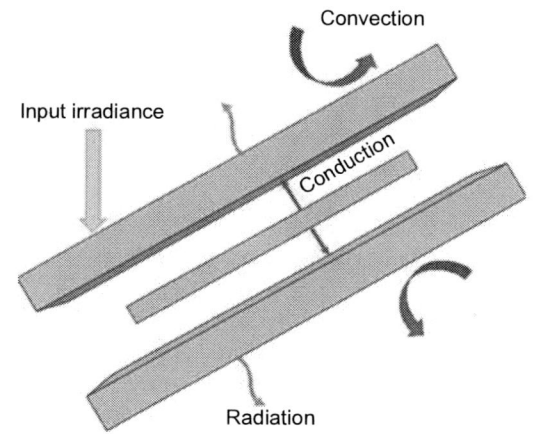

Figure 4: Heat Modelling

maximum temperature reached by the MFOT panel is in Dubai with 70°C in August when the ambient temperature is 40°C and non-MFOT panel temperature is 65.3°C.

B. Prototype Results in Bangalore, India

Our Experimental Setup consists of MFOT and non-MFOT panels set up on the roof of our office building in Bangalore, India. Both panels are placed next to each other with the same tilt and there are no objects near them which cause shading. A data logger is connected at the back of both panels to measure the temperature. The setup is as shown in Fig. 6. We have started collecting the data and the results are encouraging. In Fig. 8, we plot the temperature variation on June 12th and July 6th between the Baseline and MFOT Panel. We observe the maximum

978-1-6654-6060-6/23 $31.00 © 2023 IEEE

temperature is 50°C and 46°C respectively for MFOT panels.

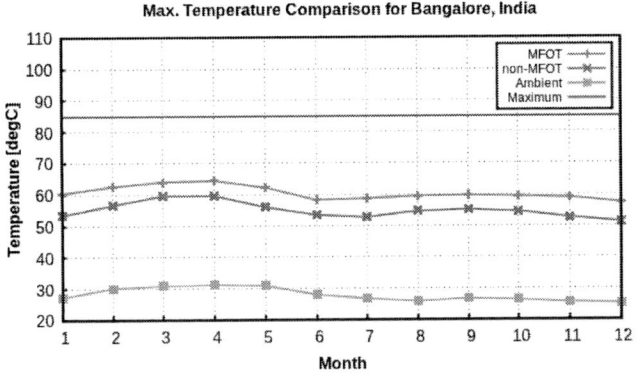

Figure 5: Comparison of maximum temperatures for Bangalore

Figure 6: Prototype on the roof of our office in Bangalore

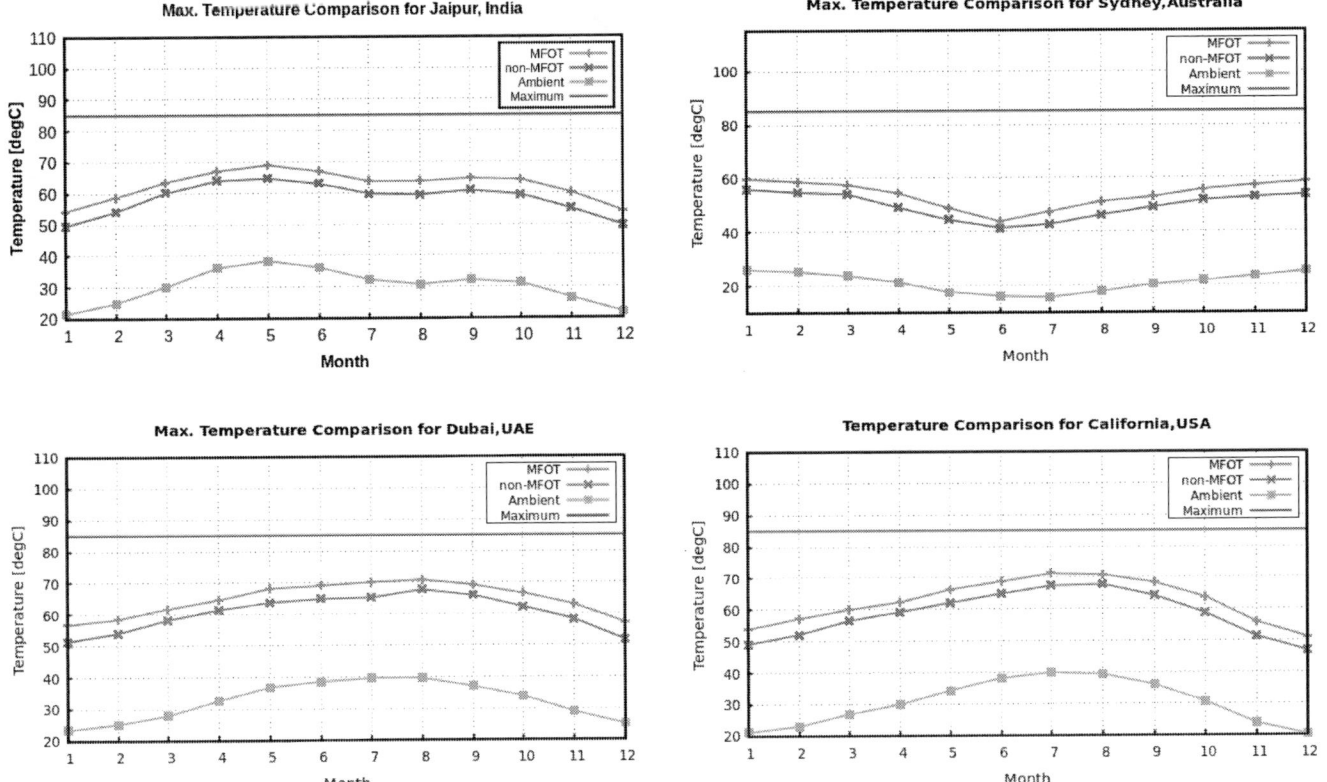

Figure 7: Maximum Temperature comparison for different locations

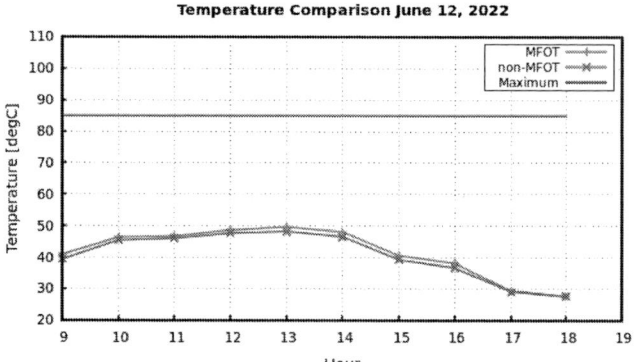
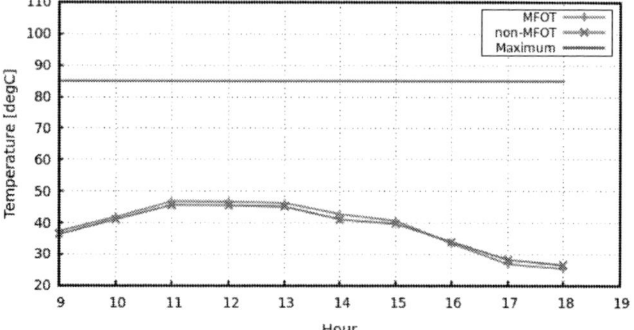

Figure 8: Prototype measurements in Bangalore

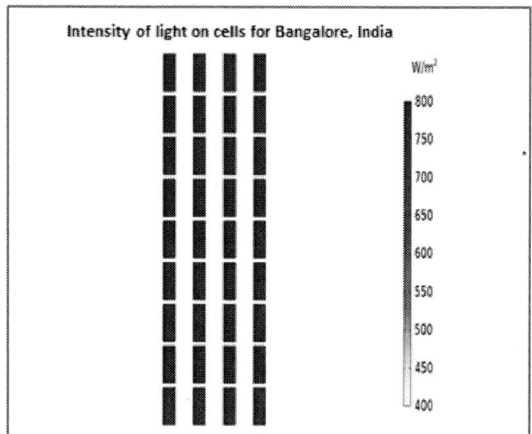

Figure 9: Uniform intensity of light on all cells

V. UNIFORM ILLUMINATION ACROSS CELLS/PANELS

Issues faced by panels if even a single cell is partially shaded is well known - the yield of the module and the entire string falls [9]. Since Renkube panel's improvement in yield is due to redirected light, our solution will not work unless every single cell always receives the exact same illumination. Suppose the cells at the ends do not receive the redirected light from the light redirecting reflectors, the yield of the entire panel will be reduced and it may even lead to hot spots. Thus, the design has been carefully constructed to ensure that there is uniform illumination across all cells of a panel.

We have conducted simulations for a single cell as well as for a panel with 36 cells and the power output has been measured. Fig. 9 shows the results for a complete panel with 9 cells in a row and 4 such rows. We can see all cells receive uniform illumination on January 21st at 11am. Fig. 10 displays the difference in power between the rows and Fig. 11 shows the difference of power between the cells of a single row. The difference goes up to a maximum of 1.4% – but this is because of CPU and memory constraints due to which the number of rays were not increased beyond a certain number. This

difference becomes lower as the number of rays in the simulation increases and accuracy increases.

VI. ABSENCE OF CONCENTRATED LIGHT

PV panels are tested using solar simulators under standard test conditions of cell temperature 25°C and an irradiance of 1000 W/m^2 also known as one-sun. This value is important because it is the irradiance of sunlight when the sun is directly overhead. When the intensity of light on the cell is greater than this one-sun value, it is considered as concentrated light and PV cells may undergo damage with such concentration. In Renkube panels, PV cells are used and hence intensity of light should not cross 1000 W/m^2. Design of light redirecting reflectors has been done to ensure that there is increased light incident on the cell during lower elevation angles and the intensity of light on the cell never crosses the one-sun limit at any point in time. Since we don't cross the one sun limit there is no risk of damage to the PV cell used. In Fig. 12, we plot the maximum intensity of light on the cell for the MFOT panel for different locations. In every case we observe the intensity does not cross 1000 W/m^2.

VII. SIGNIFICANCE

Since Renkube panels can give a 20% increased yield, this will play a major role in reducing LCOE making solar more affordable. This in turn will encourage more people to switch from non-renewable sources of energy like coal to solar for their electricity needs. Before it is commercially available to customers, we need to conduct simulations and tests to ensure that the glass redirectors do not adversely affect the cell or module in any way. To this end, we present here the results of our simulations and experiments with respect to checking the effect of the light redirectors on the following aspects of the cells – Temperature, Uniform light across cells and non-concentration of light. With the findings as presented in this paper, we can state with certainty that our new technology does not cause any problems to the cell or module.

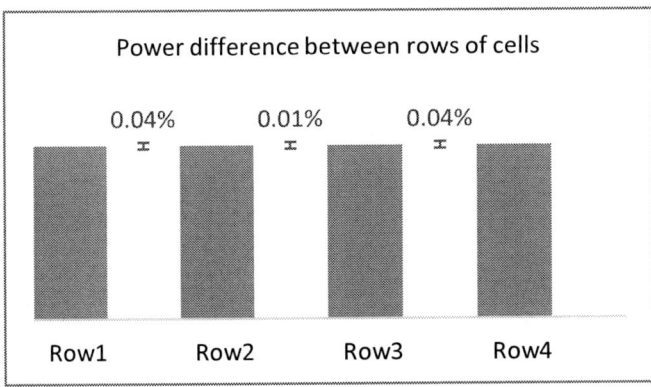

Figure 10: Difference in power between rows of cells

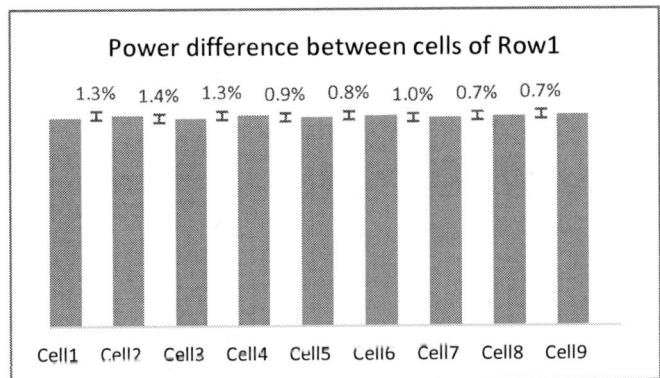

Figure 11: Difference in power between cells of the same row

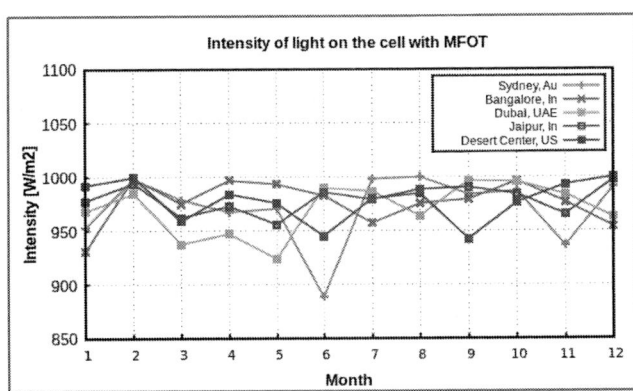

Figure 12: Maximum intensity of light on a cell in MFOT panel

conduct experiments with our test setup in Bangalore and will publish experimental results for the whole year in future work.

VIII. Conclusion

In this paper, Ray tracing and Heat modelling have been carried out using Tracepro and COMSOL Multiphysics Software for MFOT and non-MFOT panels. Measurements have been recorded on the roof of our office in Bangalore, India and we study the results. Firstly, in Simulation and Prototype cases for Bangalore we observe that temperature of the panel does not cross 65°C. To ensure that temperature does not cross the Solar Panel Maximum temperature of 85°C for other locations as well, we conducted simulations in various places that have a large number of PV installations around the world and confirmed that the temperature does not cross 70°C with our model. Secondly, our simulations show that illumination across all the cells is equal, and we show this with a panel with 36 cells. Thirdly, we show that intensity of light on the cell with an MFOT panel does not cross the one-sun value and the maximum intensity of light on the panel is 980 W/m² on March 21st in Bangalore, India. As future work, we will be setting up our prototype in Rajasthan and Dubai and conduct similar experiments in these locations as well. We will continue to

References

[1] Lakshmi Santhanam, Balaji Lakshmikanth Bangolae and Deepika Gopal, "Performance Analysis of BDRF based Reflectors, MFOT and Fixed Tilt PV" in 47th IEEE Photovoltaic Specialists Conference (PVSC), 2020.

[2] Yixian Lee and Andrew A. O. Tay, "Finite Element Thermal Analysis of a Solar Photovoltaic Module" in International Conference on Materials for Advanced Technologies 2011, Symposium O

[3] A.D.Jones and C.P.Underwood, "A thermal model for photovoltaic systems" in Solar EnergyVol. 70, No. 4, pp. 349–359, 2001

[4] Manel Hammami, Simone Torretti, Francesco Grimaccia and Gabriele Grandi, "Thermal and Performance Analysis of a PhotovoltaicModule with an Integrated Energy Storage System" in Appl. Sci. 2017, 7(11), 1107

[5] G.Notton, C.Cristofari, M.Mattei and P.Poggi,"Modelling of a double-glass photovoltaic module using finite differences" in Applied Thermal Engineering Volume 25, Issues 17–18, December 2005, Pages 2854-2

[6] R. K. Koech, H. O Ondieki, J. K. Tonui and S. K Rotich, "A Steady State Thermal Model For Photovoltaic/Thermal (PV/T) System Under Various Conditions" in International Journal of Scientific & Technology Research, December 2012

[7] Tony Kerzmann and Laura Schaefer, "System simulation of a linear concentrating photovoltaic system with an active cooling system" in Renewable Energy 41, May 2012, Pages 254-261

[8] I. Santiago, D. Trillo-Montero, I.M. Moreno-Garcia, V. Pallarés-López and J.J. Luna-Rodríguez, "Modeling of photovoltaic cell temperature losses: A review and a practice case in South Spain" in Renewable and Sustainable Energy Reviews Volume 90, July 2018, Pages 70-89

[9] J.C.Teo, Rodney H. G. Tan, V. H. Mok, Vigna K. Ramachandaramurthy and ChiaKwang Tan, "Impact of Partial Shading on the P-V Characteristics and the Maximum Power of a Photovoltaic String" in Energies, MDPI, vol 11(7), pages 1-22, July

[10] Marek Jaszczur, Qusay Hassan, Mateusz Szubel and Ewelina Majewska, "Fluid flow and heat transfer analysis of a photovoltaic module under varying environmental conditions" in Journal of Physics Conference Series 1101(1):012009

[11] S. Kaplanis, "Determination of the electrical characteristics and thermal behaviour of a c-Si cell under transient conditions for various concentration ratios" in International Journal of Sustainable Energy Volume 35, 2016 - Issue 9

Stable High Temperature Operation in Metal Halide Perovskite Solar Cells

Hadi Afshari, Shashi Sourabh, Sergio A. Chacon, Vincent R. Whiteside, Bibhudutta Rout, Giles E. Eperon, Joseph M. Luther, Ian R. Sellers

University of Oklahoma, Norman, OK, United States

University of North Texas, Denton, TX, United States

Swift Solar, San Carlos, CA, United States

National Renewable Energy Laboratory, Golden, CO, United States

$FA_{0.8}Cs_{0.2}PbI_{2.4}Br_{0.6}Cl_{0.02}$ triple halide perovskite solar cells have been shown to be amongst the most stable metal halide perovskites to date, and while their band gap is too high for use in high efficiency single gap solar cells, these systems have shown considerable promise as the high energy junction in perovskite-perovskite, silicon-perovskite, and CIGS-perovskite tandem solar cells. While in early work we have demonstrated the high radiation tolerance of these systems for potential space power applications, here high temperature data is presented which systematically and statistically demonstrates the high thermal stability of this system at temperatures in excess of 200 °C. Here, device measurements between 300 K and 490 K show that while some loss of performance is evident at higher temperature, this is driven by reversible halide segregation with no evidence of any structural phase transitions over the measurement range probed. Moreover, when the temperature of the device is reduced, then much of the operation conditions are recovered over multiple cycles. The observed stability is attributed to a combination of the high structural and crystallographic purity of this system as well as, the application of an encapsulation layer that inhibits the loss of constituent elements from the system at higher temperatures.

Characteristics of detachable III-V solar cells grown on porous germanium

Valentin Daniel, Thomas Bidaud, Jeremie Chretien, Abdelatif Jaouad, Jean-francois Lerat, Nicolas Paupy, Bouraoui Ilahi, Jinyoun Cho, Kristof Dessein, Christian Dubuc, Gwenaelle Hamon, Abderaouf Boucherif, Maxime Darnon

Institut Interdisciplinaire d'Innovation Technologique (3IT), Universite de Sherbrooke, Sherbrooke, QC, Canada

Laboratoire Nanotechnologies Nanosystemes (LN2) - CNRS IRL-3463 Institut Interdisciplinaire d'Innovation Technologique (3IT), Universite de Sherbrooke, Sherbrooke, QC, Canada

Umicore Electro-Optic Materials, Olen, Belgium

Saint-Augustin Canada Electric Inc. , Saint-Augustin, QC, Canada

III-V photovoltaic cells typically use germanium (Ge) as a substrate for the epitaxial growth, however, this material contributes significantly to the overall price of the multijunction solar cells. In order to reduce the environmental and economic cost of the solar cells, we have developed a porosification technique using bipolar electrochemical etching (BBE) to create a weak layer between the Ge substrate and the epitaxial layers. This approach allows the easy separation of the grown layers and the subsequent reuse of germanium. As evidence of the potential of this method, we have compared the performances of non-detached single-junction III-V solar cells grown and fabricated (without anti-reflection coating-ARC) on porosified Ge, and on bulk Ge as a reference. All the final cells show mirror-like monocrystalline III-V layers with comparable characteristics notably concerning the Voc (VocGePorous=0.862V vs VocGeBulk=0.882V) and good efficiencies (Eff.GePorous=13.03% vs Eff.GeBulk=15.96%) in comparison with current literature values on similar substrates (Vocliterature= 0.75V and FFliterature=61%). These promising results open the path toward thin III-V solar cells and multiple Ge substrates reuse.

Impact of Surface Roughness in Measuring Optoelectronic Characteristics of Thin-Film Solar Cells

David Magginetti[1], Seokmin Jeon[2], Yohan Yoon[2,3], Ashif Choudhury[1], Ashraful Mamun[1], Yang Qian[1], Jordan Gerton[4], and Heayoung Yoon[1]

[1] Department of Electrical and Computer Engineering, University of Utah, UT 84112, USA
[2] US Naval Research Laboratory, Washington, DC 20375, USA
[3] Department of Materials Engineering, Korea Aerospace University, Goyang, 10540 South Korea
[4] Department of Physics and Astronomy, University of Utah, UT 84112, USA

Abstract — **Microstructural properties of thin-film absorber layers play a vital role in developing high-performance solar cells. Scanning probe microscopy is frequently used for measuring spatially inhomogeneous properties of thin-film solar cells. While powerful, the nanoscale probe can be sensitive to the roughness of samples, introducing convoluted signals and unintended artifacts into the measurement. Here, we apply a glancing-angle focused ion beam (FIB) technique to reduce the surface roughness of CdTe while preserving the subsurface optoelectronic properties of the solar cells. We compare the nanoscale optoelectronic properties "before" and "after" the FIB polishing. Simultaneously collected Kelvin-probe force microscopy (KPFM) and atomic force microscopy (AFM) images show that the contact potential difference (CPD) of CdTe pristine (peak-to-valley roughness > 600 nm) follows the topography. In contrast, the CPD map of polished CdTe (< 20 nm) is independent of the surface roughness. We demonstrate the smooth CdTe surface also enables high-resolution photoluminescence (PL) imaging at a resolution much smaller than individual grains (< 1 μm). Our finite-difference time-domain (FDTD) simulations illustrate how the local light excitation interacts with CdTe surfaces. Our work supports low-angle FIB polishing can be beneficial in studying buried sub-microstructural properties of thin-film solar cells with care for possible ion-beam damage near the surface.**

I. INTRODUCTION

Thin-film CdTe solar cells are a leading photovoltaic (PV) technology owing to cost-effective manufacturing and reliable power production for 20+ years [1, 2]. Optimized close-space sublimation (CSS) and vertical traveling deposition (VTD) enable the rapid production of high-quality CdTe thin films. Such absorber layers consist of inhomogeneous microstructures with a typical grain size of approximately 1 μm after post-deposition processing. Previous studies suggested that electrically-active point defects and structural defects of microstructures (i.e., grain bulk, grain boundaries) can significantly impact the cell efficiencies and long-term stability and reliability of CdTe solar cells [3, 4].

Scanning probe microscopy (SPM) has been extensively used for characterizing inhomogeneous semiconductor materials and devices. Examples include atomic force microscopy, electron-beam microscopy, and confocal optical microscopy, where a local excitation source is raster-scanned on the area of interest to image spatially resolved properties of solar cells [5-9]. While powerful, the nanoscale probe used in SPM can be sensitive to the surface roughness of the sample, introducing convoluted signals and unintended artifacts into the measurement. Various polishing techniques have been proposed and applied to reduce the sample roughness, including focused ion beam (FIB), gas-assisted etching, wet etching, and low-temperature milling [10-12]. Among them, previous efforts demonstrated an optimized argon ion (Ar⁺) beam could efficiently produce a smooth surface with minimizing possible beam damage [13, 14].

In this work, we use glancing-angle FIB milling to reduce the topographical variation of CdTe while retaining the quality of the sample. AFM/KPFM is used to measure the surface potential of CdTe "before" and "after" the polishing. Analysis of KPFM shows the different CPD profiles of pristine and polished devices. We demonstrate PL imaging that resolves the luminescence characteristics of individual grains and grain boundaries. We perform FDTD simulations to visualize the light absorption profile of CdTe microstructures, illustrating the non-uniform interactions of the local light source to grains and grain boundaries in nanoscale PL imaging.

II. EXPERIMENTAL

This study used conventional CdS/CdTe solar cells extracted from a solar panel [6]. Figure 1(a) shows a representative topographic image of the pristine CdTe obtained by atomic force microscopy (AFM). Irregular grain sizes range from < 1 μm to a few μm with a peak-to-valley variation as high as 0.6 μm in this sampling area (10 μm × 10 μm). To reduce the roughness, we performed Ar⁺ polishing at an incident beam

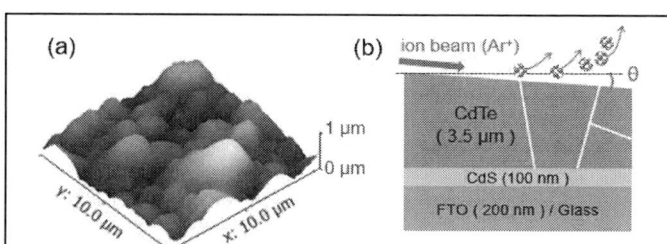

Figure 1. (a) Topography of CdTe obtained using atomic force microscopy. (b) Schematic of ion beam milling.

energy of 3 keV for one hour. The ion beam was irradiated at a glancing angle of 1°, only removing a thin layer of CdTe while preserving the CdTe bulk unchanged (Fischione 1060). A schematic in Figure 1(b) illustrates the ion-beam milling process. KPFM (CPD) and AFM (topography) images were simultaneously collected on the samples of "before (pristine)" and "after (polished)" the milling (Bruker AFM probes; OSCM-PT-R3). The diameter of a cantilever tip was 20 nm, and the resonance frequency of the tip was set to ≈ 71 kHz with a spring constant of 2 N/m. PL imaging at 405 nm was performed using a customized confocal microscopy system. The PL maps were collected via immersion objective lens (100x) in an oil media (Cargille type B). Finite-difference time-domain (FDTD) simulations were conducted to study the absorption profiles of the laser beam in CdTe for both smooth and rough sample surfaces (Ansys Lumerical). FDTD methods are often used to solve electromagnetic interaction in a sample of interest. Maxwell equations are solved numerically on a discrete grid (mesh) in space and time defined in the model. We used a standard material library (J. A. Wollam Ellipsometry Solutions) to formulate the refractive index and extinction coefficient of CdTe at different wavelengths. The mesh size was set to 0.25 nm for the simulations.

III. RESULTS AND DISCUSSION

Nanoscale SPM measurements were performed on two CdTe solar cells extracted from the same solar panel. To polish one of the samples, we used a 3 kV Ar^+ beam that is irradiated at an angle of 1° parallel to the CdTe surface (Figure 1b). Our milling conditions intend to remove the prominent structures on CdTe (≈ 100 nm) rather than to produce an atomically smooth surface, maintaining a low-level surface roughness after the polishing. Figures 2(a, b) show the topography of the "before (pristine)" and the "after (polished)" CdTe surface. The maximum peak-to-valley roughness of ≈ 600 nm of pristine CdTe was noticeably reduced to ≈ 80 nm after the milling.

The KPFM maps in Figures 2(c, d) show the potential difference of CdTe microstructures. It is apparent that the potential distribution (i.e., CPD) of the pristine CdTe closely follows the topography. Low CPDs are seen on the hills of grains (≈ 1.020 V), while grain boundaries show higher CPDs (≈ 1.035 V) compared to their adjacent grain interiors. We observe the overall CPDs decreases to approximately 30 mV (Figure 2d) after the Ar^+ beam milling. This reduction could be attributed to the eliminated TeO_x-rich CdTe surface, which exposes the bare CdTe subsurface. Unlike the pristine, the CPD variation of the polished CdTe is independent of the remaining surface roughness, marked with the red boxes in the area in Figure 2. The CPD difference between grain interior and grain boundary of the polished CdTe (≈ 20 mV) is similar to the pristine. The CPD distribution near grain boundaries observed in our KPFM shows a good agreement with the previous work

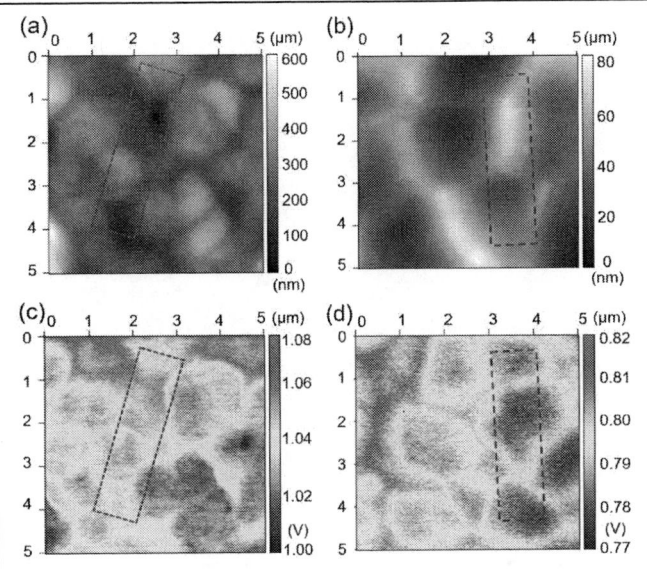

Figure 2. Topography and surface potential map of pristine (a, c) and polished (c, d) CdTe solar cells.

by Jiang *et al.*, where a CdTe sample was polished with a 4 kV Ar^+ beam followed by 250 °C annealing [5].

The polished CdTe surface enables high-resolution PL imaging with a spatial resolution much smaller than individual grains (< 1 μm). In our confocal PL system, a 100x objective (numerical aperture [NA] = 1.4) focused the collimated laser beam (λ = 405 nm) on CdTe, which was facing down on a cover glass. We used an immersion lens oil (n = 1.5) to match the refractive indices in the PL setup (immersion objective lens / oil / cover glass / CdTe), improving the resolution.

Figure 3 compares the representative PL images obtained on pristine and polished CdTe solar cells. Overall, the PL emission near grain boundaries is lower than grain interiors for both samples. This observation is consistent with previous studies, where the defects near grain boundaries serve as non-radiative

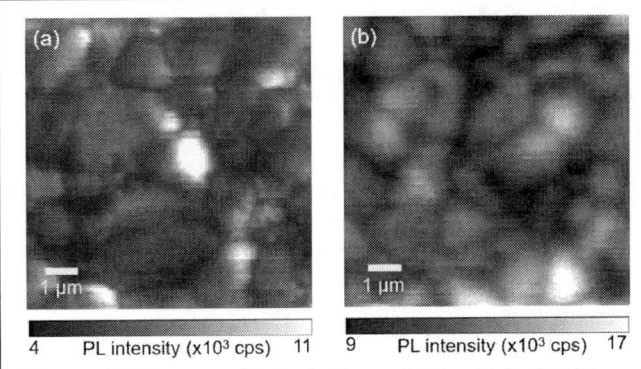

Figure 3. PL maps of (a) pristine and (b) polished CdTe solar cells excited with a 405 nm laser beam. Each pixel of the image (128 × 128 pixels) represents the integrated PL intensities near the band-gap (≈ 820 nm).

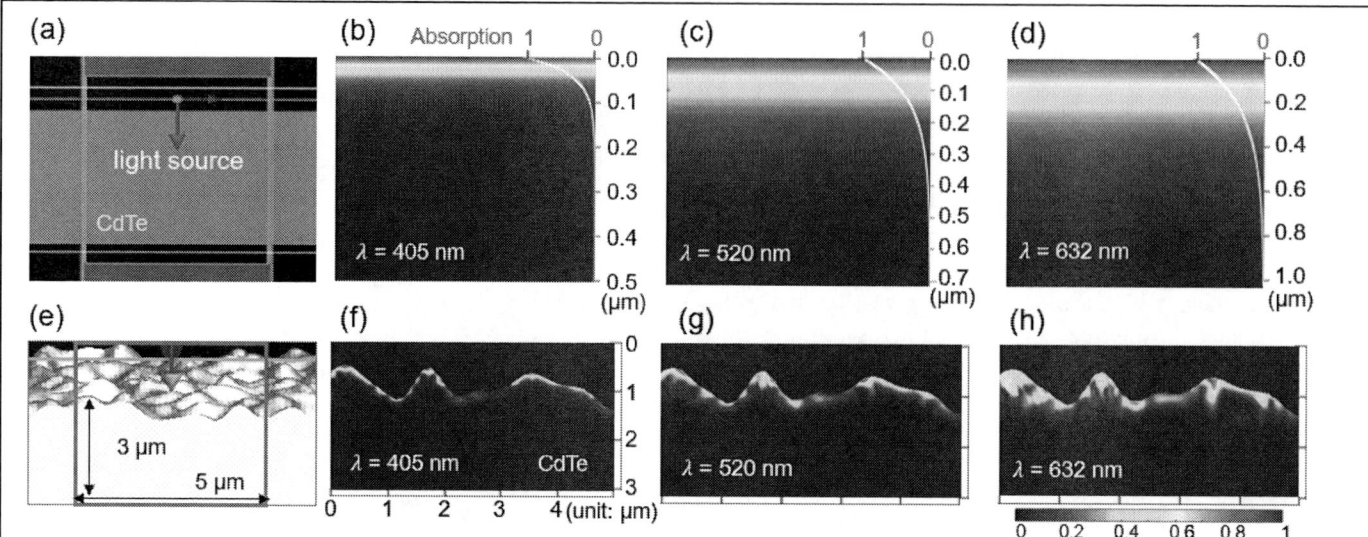

Figure 4. Absorption profiles of planar and rough surface CdTe at 405 nm laser beam illuminations. FDTD simulation regions (orange color box in a, e) include the perfect match layer (PMLs) and periodic boundary conditions.

recombination centers reducing the PL emission near the bandgap [9, 15]. The brightness contrast of the pristine PL map appears to follow the topography of CdTe microstructures, with random bright spots on the edge of some grains. In contrast, the PL emission of the polished sample shows a relatively uniform contrast between the grain bulk and grain boundary of CdTe. The bright spots are mainly observed in the center of the grain bulk, decaying proportionally to the adjacent grain boundaries. The dark areas in the PL map are likely associated with the concentrated defects rather than unintended artifacts introduced by the topography.

We developed FDTD models of the smooth and rough surface CdTe to understand how the light source (i.e., a laser beam) interacts with the topography, coupled with inhomogeneous microstructures, affects the high-resolution PL imaging. Schematics of the models are shown in Figures 4 (a, e). Our model defined a rough surface CdTe based on the AFM surface roughness seen in Figure 1(a). In addition to the light source of 405 nm used in the PL imaging (Figure 3), we also simulated the absorption profile for 520 nm and 632 nm, frequently used in PL measurements. The absorption profiles of the rough surface are compared to their planar counterparts in Figure 4. As expected, the light absorption decreases exponentially in the planar structures, where the estimated absorption depth of approximately 39 nm, 113 nm, and 215 nm at an illumination of 405 nm, 520 nm, and 632 nm, respectively. Figure 4 (f ~ h) compares the representative absorption landscapes simulated for the rough CdTe surface. It is apparent that the extruded structures of the CdTe film could absorb more photons from an irradiating laser beam. The light absorption near grain boundaries can be influenced by the topography of adjacent grains due to light scattering. Our simulations support that the random bright spots observed in the PL image of the rough CdTe can be attributed to its topography rather than its intrinsic properties.

IV. CONCLUSIONS

In summary, we have shown the impact of surface roughness in measuring nanoscale optoelectronic characteristics of CdTe solar cells. By reducing the surface roughness using shallow-angle ion beams, the KPFM probe can measure the subsurface CPD distribution of CdTe solar cells. The polished surface enables PL imaging at a spatial resolution much smaller than individual grains (< 1 μm). Our FDTD simulations show inhomogeneous light absorption on rough CdTe, indicating the impact of surface roughness in nanoscale optoelectronic characterizations. Our work supports low-angle FIB polishing can be beneficial in studying buried sub-microstructural properties of thin-film solar cells with care for possible ion-beam damage near the surface.

ACKNOWLEDGEMENT

This research was supported by the U.S. Department of Energy's Office of Energy Efficiency and Renewable Energy (EERE) under the DE-FOA-0002064 program award number DE-EE0008983.We acknowledge support in part by the National Science Foundation (NSF) CAREER Award No. 2048152. This work was support by the USTAR shared facilities at the University of Utah, in part, by the MRSEC Program of NSF under Award No. DMR-1121252. We thank P. Perez, M. Wang, and Z. Liu for valuable inputs and training of the measurement systems.

References

[1] G. M. Wilson *et al.*, "The 2020 photovoltaic technologies roadmap," (in English), *Journal of Physics D-Applied Physics,* vol. 53, no. 49, Dec 2 2020.

[2] S. G. Kumar and K. S. R. K. Rao, "Physics and chemistry of CdTe/CdS thin film heterojunction photovoltaic devices: fundamental and critical aspects," (in English), *Energy & Environmental Science,* vol. 7, no. 1, pp. 45-102, Jan 2014.

[3] J. D. Major, "Grain boundaries in CdTe thin film solar cells: a review," (in English), *Semiconductor Science and Technology,* vol. 31, no. 9, Sep 2016.

[4] I. Visoly-Fisher, S. R. Cohen, K. Gartsman, A. Ruzin, and D. Cahen, "Understanding the beneficial role of grain boundaries in polycrystalline solar cells from single-grain-boundary scanning probe microscopy," (in English), *Advanced Functional Materials,* vol. 16, no. 5, pp. 649-660, Mar 20 2006.

[5] C. S. Jiang *et al.*, "Imaging hole-density inhomogeneity in arsenic-doped CdTe thin films by scanning capacitance microscopy," (in English), *Solar Energy Materials and Solar Cells,* vol. 209, Jun 1 2020.

[6] H. P. Yoon *et al.*, "Local electrical characterization of cadmium telluride solar cells using low-energy electron beam," (in English), *Solar Energy Materials and Solar Cells,* vol. 117, pp. 499-504, Oct 2013.

[7] C. X. Xiao *et al.*, "Microscopy Visualization of Carrier Transport in CdSeTe/CdTe Solar Cells," (in English), *Acs Applied Materials & Interfaces,* vol. 14, no. 35, pp. 39976-39984, Sep 7 2022.

[8] Y. Yoon *et al.*, "Nanoscale imaging and spectroscopy of band gap and defects in polycrystalline photovoltaic devices," (in English), *Nanoscale,* vol. 9, no. 23, pp. 7771-7780, Jun 21 2017.

[9] A. Abudulimu *et al.*, "Photophysical Properties of CdSe/CdTe Bilayer Solar Cells: A Confocal Raman and Photoluminescence Microscopy Study," in *2022 IEEE 49th Photovoltaics Specialists Conference (PVSC)*, 2022, pp. 1088-1090.

[10] L. A. Giannuzzi and F. A. Stevie, *Introduction to focused ion beams : instrumentation, theory, techniques, and practice.* New York: Springer, 2005.

[11] N. I. Kato, "Reducing focused ion beam damage to transmission electron microscopy samples," *Journal of Electron Microscopy,* vol. 53, no. 5, pp. 451-458, 2004.

[12] E. M. Turner, K. R. Sapkota, C. Hatem, P. Lu, G. T. Wang, and K. S. Jones, "Wet-chemical etching of FIB lift-out TEM lamellae for damage-free analysis of 3-D nanostructures," *Ultramicroscopy,* vol. 216, p. 113049, 2020/09/01/ 2020.

[13] M. J. Campin, C. S. Bonifacio, P. Nowakowski, P. E. Fischione, and L. A. Giannuzzi, "Narrow-Beam Argon Ion Milling of Ex Situ Lift-Out FIB Specimens Mounted on Various Carbon-Supported Grids," 2018, pp. 339-344: ASM International.

[14] Y.-L. Hsu *et al.*, "Subsurface Characteristics of Metal-Halide Perovskites Polished by Argon Ion Beam," p. arXiv:2212.13694Accessed on: December 01, 2022Available: https://ui.adsabs.harvard.edu/abs/2022arXiv221213694H

[15] D. Kuciauskas *et al.*, "Spectrally and time resolved photoluminescence analysis of the CdS/CdTe interface in thin-film photovoltaic solar cells," (in English), *Applied Physics Letters,* vol. 102, no. 17, Apr 29 2013.

Validation of Open-source Distributed Energy Resources (OpenDER) Model with IEEE 1547-2018 Smart Inverter

Yiwei Ma, Charles (Chuck) Brewster, Aminul Huque

EPRI, Knoxville, TN, 37932, USA

Abstract — **With the latest edition of IEEE Standard 1547™ published in 2018 and amended in 2020, and IEEE 1547.1-2020 tested / UL 1741 SB certified inverters entering the market, there are strong interests to have accurate distributed energy resource (DER) models to support the accurate assessment of DER impacts on power systems. EPRI has developed an open-source distributed energy resource (OpenDER) model, which aims to represent the functional definitions and requirements of IEEE Standard 1547-2018. This paper presents the OpenDER model validation results. The model outputs are compared against the lab test results of a commercial energy storage smart inverter that has completed UL1741 SB / IEEE 1547.1-2020 certification tests. In general, the OpenDER model is able to represent the inverter with proper parameterization. The mismatches are also identified for future possible model improvements.**

I. INTRODUCTION

IEEE Standard 1547 [1] has been developed to define the technical minimum interconnection requirements and criteria for distributed energy resources (DERs), including PV and energy storage. With the growing penetration of DERs, there are strong interests to have uniform and standardized DER behavior, such that utilities can accurately model, predict and understand DER behavior, perform analysis to quantify grid impacts, and determine how to optimize the utilization of DER grid support functions and configure control settings to address potential issues and increase feeder hosting capacity.

In 2018, the latest edition of the IEEE Standard 1547 was published with a significant number of additions and modifications from the original 2003 version and 2014 amendment. The new additions include multiple grid support functions such as voltage and reactive power (volt-var) control, and requirements such as low voltage ride through, that the DER must be able to provide or comply with. The edition from 2018 was subsequently amended in 2020 to provide greater flexibility to grid planners in terms of utilizing the voltage ride-through capability for Category III DERs [2]. Many U.S. state public utility commissions (PUCs) and utilities are considering requiring 1547-2018 inverter conformity by IEEE 1547.1 testing and UL 1741 SB certification from 2023 onwards [3].

Many DER manufacturers, integrators, and certification laboratories are designing, developing, and testing new products following the latest standard requirements. Developers of power system analysis tools are updating the DER models in their product libraries to represent the performance that matches these products. Utilities and researchers can benefit from an accurate and configurable

generic DER model that becomes available in simulation software when conducting system planning and impact studies.

EPRI has developed an open-source DER (OpenDER) model, with the objective to harmonize accurate interpretations of the IEEE Std 1547-2018 among applicable stakeholders and help the industry to properly model DERs for interconnection or planning analyses, as shown in Fig. 1 [4]. It is developed following the functional definition and requirements of the IEEE Std 1547-2018, and reviewed and improved by an industry-wide DER model user's group (DERMUG).

EPRI intends to maintain, improve, and expand the OpenDER model going forward, and plans to include additional DER control functions that are required by other standards and country codes. The ambiguities and gaps in the IEEE Std 1547-2018 identified during the model development process are documented and will be provided as input to the revision of the standard started in 2023.

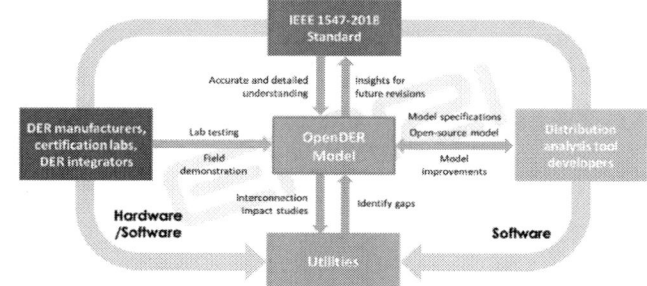

Fig. 1.   OpenDER development objective: Harmonizing understanding of the DER behavior among all stakeholders

In this paper, the model validation results using a single phase inverter, which has completed IEEE 1547.1-2020 / UL 1741 SB certification tests, are presented.

III. OPEN-SOURCE DER (OPENDER) MODEL

The IEEE Std 1547-2018 is technology agnostic and applicable to all types of DERs. The OpenDER model aims to accurately represent steady-state and dynamic (fundamental frequency phasor component) behavior of inverter-based DERs, including PV and battery energy storage systems. This model can be used to run snapshot, Quasi-Static Time Series (QSTS), and a variety of dynamic analyses to study the impacts of DERs on a distribution system.

The OpenDER model is maintained in two formats: A publicly available model specification document presenting the DER model in terms of equations and block/flow diagrams [5],

978-1-6654-6060-6/23 $31.00 © 2023 IEEE

and an open-source DER model software implemented in Python [6]. Both can be used as a reference by stakeholders who develop their own DER models, such as power system analysis tool developers, R&D organizations, consultants, and academia. In addition, the software code can be used to benchmark and validate existing and future DER models, and interface with circuit simulation tools for system studies. The OpenDER model development process and potential usage is shown in Fig. 2.

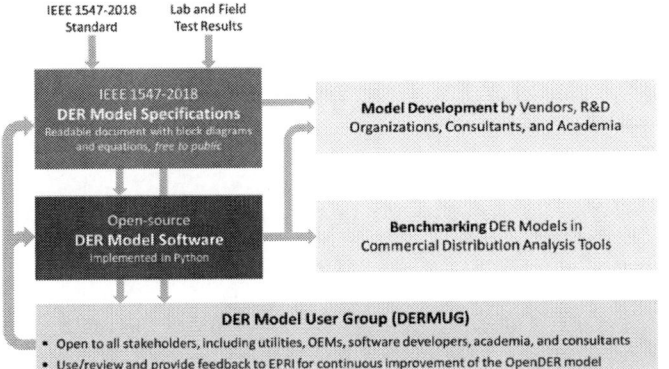

Fig. 2.   OpenDER model formats, potential uses, and DERMUG

The OpenDER model captures the function definitions and performance requirements defined in IEEE Std 1547-2018, including grid support functions, DER enter service performance, and abnormal voltage and frequency trip and ride-

through requirements. As shown in Fig. 3, the model is developed and documented in a modular fashion. Each module represents a specific standard requirement or expected performance behavior from DER. The OpenDER model as a whole, or each individual module can be used as a reference by stakeholders to develop their own models.

The basic inputs of the OpenDER model include nameplate information, function setting variables, and operating condition variables. To the extent possible, the model input variable names and labels are consistent with those in IEEE Std 1547.1-2020 [7] and the common file format developed by EPRI [8]. Additional variables are included in this model to better represent the DER behaviors that are not clearly defined in the standard, following a similar naming convention. The default control settings specified in the IEEE Std 1547-2018 are used as the default model parameter values unless the user specifies different values. Any input values that are outside of the defined or allowed ranges in the IEEE Std 1547-2018 are flagged by the model for user awareness.

Regarding the DER model outputs, different system analyses require different levels of detail. For example, steady-state power flow simulation may only require active and reactive power values from the DER model, whereas fault current contribution analysis may require current injection from each individual phase from the DER model. With the intention to cover most of the simulation types and use cases, the OpenDER model provides multiple options to interface with the distribution feeder model, including power source, current source, and voltage source behind impedance.

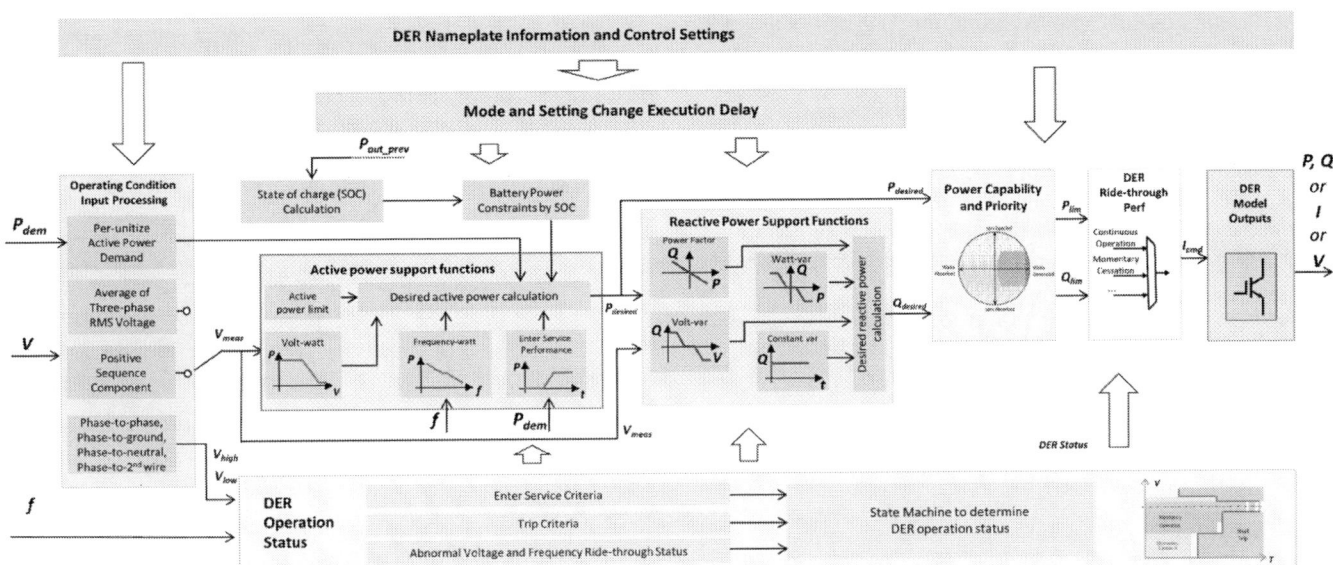

Fig. 3.   Block Diagram of OpenDER model

III. MODEL VALIDATION WITH IEEE 1547-2018 SMART INVERTER

To make sure the OpenDER model can represent actual commercial DERs in distribution analysis, model validations are being conducted to compare the OpenDER model outputs with IEEE 1547-2018 smart inverters, by applying the same test conditions (i.e., voltage, output power level/command, and control settings). In this paper, model validation results using a single-phase 7.68kW energy storage inverter are presented.

The following sections present the model validation results for DER power capability, constant power factor, volt-var, frequency-droop, fault current contribution, and momentary cessation performance. The relevant standard requirements are presented. Mismatches in the results are also identified as future opportunities for model improvement.

A. Power Capabilities

IEEE Std 1547-2018 Section 5.2 requires Category B DER to have its reactive power capability to be at least 44% of its nameplate apparent power rating, if the DER is generating greater than 20% of its rated active power. The reactive power capability of the inverter under test is 100% of the apparent power rating.

The capability requirement on reactive power by the standard is for the full extent of ANSI C84.1 voltage range A. The standard does not provide guidance on the apparent power and active power capability if a non-nominal voltage is provided to the DER. Currently, the OpenDER model considers both current limitations and power limitations: When voltage is above 1pu, the model output is limited by the apparent, active, and reactive power nameplate ratings. And when voltage is below 1pu, the model output is limited by the current ratings.

When the inverter terminal voltage is at 1pu, the OpenDER results match the inverter output with reasonable accuracy, as shown in Fig. 4. However, when the inverter terminal voltage is reduced to 0.89, there are some mismatches, as shown in Fig. 5. As highlighted in red, the actual inverter reduces more active power than the OpenDER model outputs. This indicates the inverter may have additional output limitations than the ones currently modeled in OpenDER.

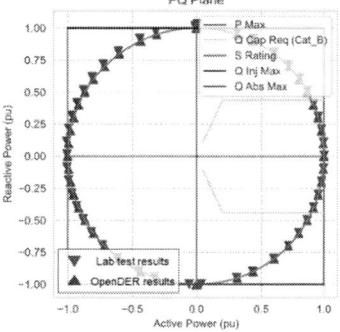

Fig. 4.   Model validation of active and reactive power capability when voltage is at 1pu

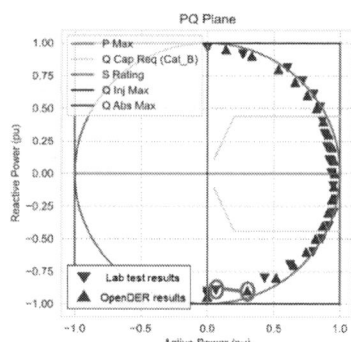

Fig. 5.   Model validation of active and reactive power capability when voltage is at 0.89pu

B. Constant Power Factor Function

IEEE Std 1547-2018 Section 5.3.2 requires the DER to have the capability to operate at a constant power factor. In addition, IEEE Std 1547-2018 Section 5.2 requires the DER not to constrain the delivery of reactive power up to the minimum reactive power capability range, which is also known as "reactive power priority". For constant power factor mode, the DER is thus required to maintain the constant power factor setting, and give priority to reactive power over active power when the apparent power reaches the nameplate rating. The OpenDER model is developed following the standard requirements.

The model validation results are shown in Fig. 6. The inverter is set to 0.95 constant power factor absorption. The active power demand is set to -1 to 1 pu, with 0.1pu increments. As can be seen, OpenDER model outputs match with the inverter test results, following the reactive power priority as required by the standard. Also, when charging active power, both the model and the inverter still maintain the reactive power priority, which the standard does not have requirements for. The reactive power output errors between model outputs and lab results are less than 0.016 pu.

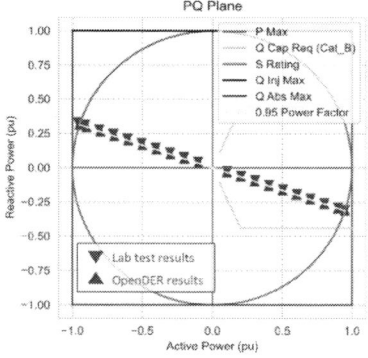

Fig. 6.   Model validation of constant power factor function

C. Volt-var Function

IEEE Std 1547-2018 Section 5.3.3 requires the DER to have the capability to autonomously control its reactive power in relation to voltage at its reference point of applicability (RPA). This function is known as voltage-reactive power (volt-var) mode. The DER is also subjected to the reactive power priority requirement in this mode, in which the operation of active power not to constrain the ability of the inverter to absorb/inject minimum reactive power specified by the standard. The standard also includes an "open loop response time (OLRT)" requirement for the volt-var function. The OLRT is defined as the duration from a step change in control signal input until the output changes by 90% of its final value, before any overshoot. However, the characteristics of how the DER output reaches the final value is not specified in the standard. In OpenDER model, the OLRT behavior is currently modeled as a first-order lag followed by a time delay block that emulates the reaction time of the control system.

Fig. 7 shows the model validation results when the volt-var is configured with IEEE Std 1547-2018 Category B default settings. The OpenDER model is exposed to the same voltage profile as the inverter, and the model generated results are shown in the same graph as the measured lab test results. The OpenDER output reactive power matches well with the lab test results, indicating the modeling of OLRT behavior represents the actual inverter. On the other hand, the active power performance of the OpenDER model depends on the nameplate current rating (*NP_CURRENT_PU*) parameter, which impacts the inverter's capability to maintain the output apparent power level if the terminal voltage is less than 1 pu. As can be seen, when the nameplate current rating is set to 1.07 pu, the model output matches the test results better than other values.

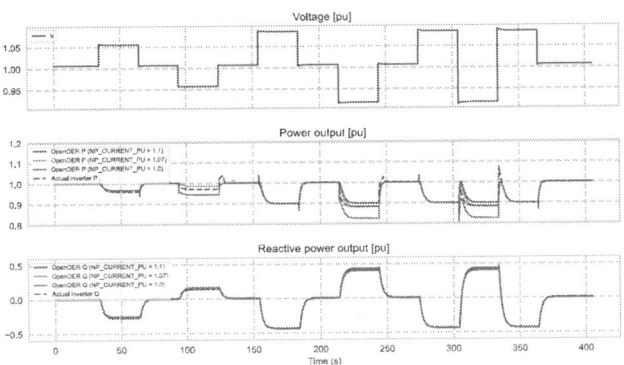

Fig. 7.   Model validation of volt-var function

D. Frequency-droop Function

IEEE Std 1547-2018 Section 6.5.2.7 requires the DER to adjust its active power output from the pre-disturbance levels if the system frequency is outside of the adjustable dead-band. The standard specifies a slope formula for this function, which was corrected in the errata [9]. This function is enabled by default with IEEE Std 1547-2018 default settings.

Fig. 8 shows the model validation results by comparing the OpenDER model outputs with lab test results. Various frequencies are applied to the inverter and the OpenDER model, with 1.0, 0.8, and 0.5 pu pre-disturbance active power generation. The lab test results and the active power output errors between model and lab results are less than 0.006 pu.

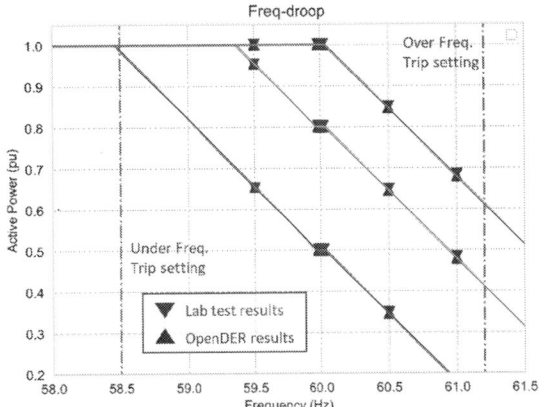

Fig. 8.   Model validation of frequency-droop function

D. Fault Current Contribution

IEEE Std 1547-2018 Section 6.4.2 requires the DER to ride through abnormal voltage disturbance, and contribute at least 80% of the pre-disturbance current to the area electric power system in Mandatory Operation region. As the standard does not have further guidance on the fault current contribution, OpenDER model currently assumes the inverter to continue to maintain the output power level by injecting more current, within its capability constraint, when the voltage is low.

Fig. 9 shows the model validation results when the voltage reduces to 0.6 pu for 9.8 second. The steady-state current calculated by OpenDER model matches well with the actual inverter's output, for both magnitude and angle. However, there are some mismatches when the fault starts and ends. The actual inverter has a slower response than the OpenDER model. This is likely due to the dynamic performance introduced by inverter current closed-loop control, which is not modeled in detail by OpenDER model. Fig. 10 shows another model validation result when volt-var function is enabled, with the same voltage drop. As can be seen, OpenDER calculated fault current matches well with the actual inverter, with minor inaccuracy at the beginning and the end of the fault.

E. Momentary Cessation Performance

IEEE Std 1547-2018 Section 6.4.2 also requires Category III DERs to enter momentary cessation within 0.083s after voltage drops to less than 0.5 pu. During momentary cessation, the DER shall not deliver active power, and reactive power change shall exclusively result from passive devices, such as inverter filter. When the system voltage returns, the DER is also required to restore 80% of its pre-disturbance active power level within

0.4s. The OpenDER model is developed following such requirements. It also offers the capability to parameterize the susceptance of the filter capacitance, which determines the reactive current contribution during the momentary cessation stage.

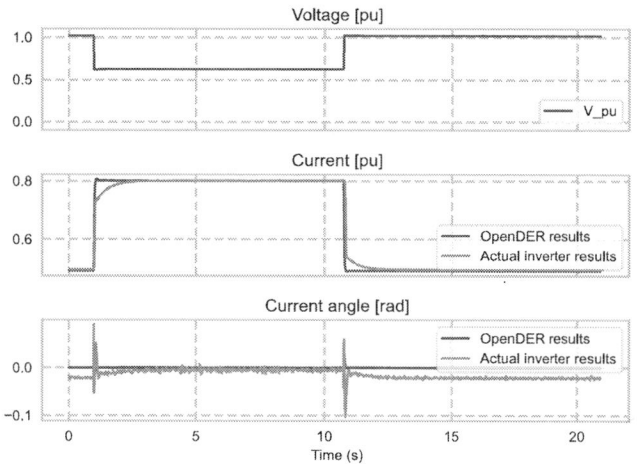

Fig. 9. Model validation of fault current contribution when voltage drops to 0.6pu.

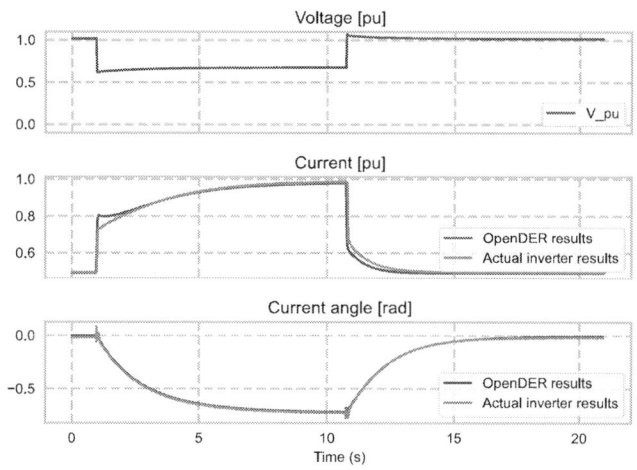

Fig. 10. Model validation of fault current contribution when voltage drops to 0.6pu, and volt-var function is enabled.

Fig. 11 and Fig. 12 show the model validation results when the inverter terminal voltage drops to 0 and 0.2pu, respectively. As can be seen, the current magnitude and angle calculated from OpenDER follow the actual inverter model with reasonable accuracy. There are some mismatches on the current overshoot when the fault begins, and some mismatches on the delay time when the inverter starts to restore outputs after the fault recovers.

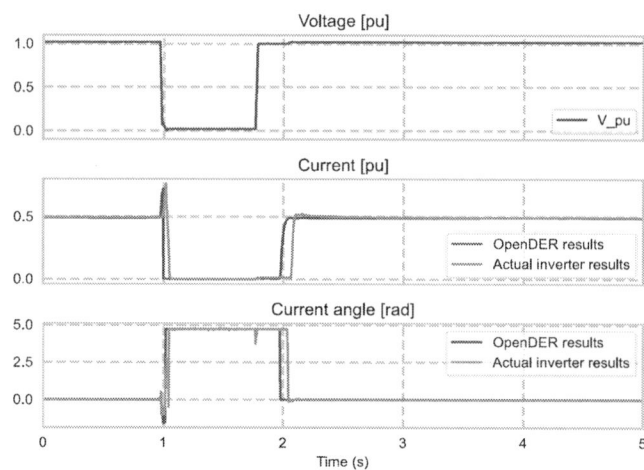

Fig. 11. Model validation of momentary cessation performance when voltage drops to 0pu.

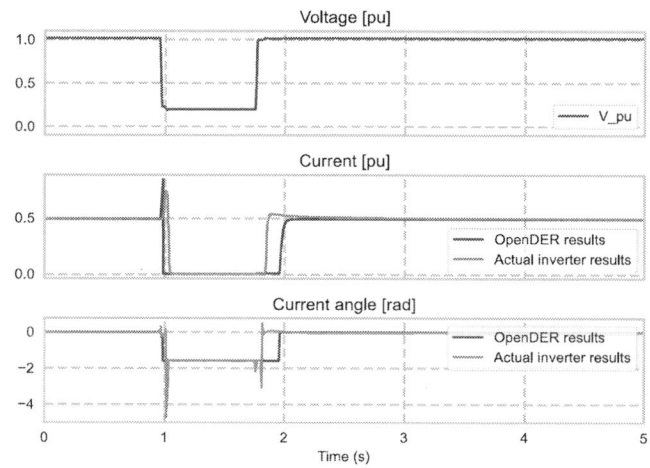

Fig. 12. Model validation of momentary cessation performance when voltage drops to 0.2pu.

IV. CONCLUSIONS AND FUTURE WORK

EPRI has developed an Open-source DER (OpenDER) model, which aims to capture all performance requirements specified by the IEEE Std 1547-2018 in sufficient detail. This paper presents the model validation results against a single phase energy storage smart inverter that has completed UL 1741SB certification tests.

The OpenDER output results match the actual inverter output with reasonable accuracy, both in steady-state and dynamically. Some mismatches are identified, likely due to the measurement inaccuracy of actual inverter and additional capability constraints or dynamic behavior introduced by inverter closed-loop control, which are not yet captured by the OpenDER model.

EPRI intends to maintain, improve, and expand the OpenDER model going forward, and to conduct model validation utilizing other IEEE 1547-2018 certified smart inverters.

REFERENCES

[1] IEEE Std 1547-2018 *Standard for Interconnection and Interoperability of Distributed Energy Resources with Associated Electric Power Systems Interfaces.* https://standards.ieee.org/ieee/1547/5915/

[2] IEEE Std 1547a-2020, *Amendment to IEEE Std 1547-2018 to provide more flexibility for adoption of abnormal operating performance Category III.* https://standards.ieee.org/content/ieee-standards/en/standard/1547a-2020.html

[3] IEEE Std 1547-2018 (Revision of IEEE Std 1547-2003), IEEE SCC21 sponsor website, https://sagroups.ieee.org/scc21/standards/1547rev/

[4] Open-Source Distributed Energy Resource (OpenDER) Model, https://www.epri.com/opender

[5] *IEEE 1547-2018 OpenDER Model: Version 2.0*, EPRI, Palo Alto, CA: 2022. 3002025583. https://www.epri.com/research/products/3002025583

[6] Open-Source Distributed Energy Resource (OpenDER) Model Software Repository, https://github.com/epri-dev/opender

[7] IEEE Std 1547.1-2020. *IEEE Standard Conformance Test Procedures for Equipment Interconnecting Distributed Energy Resources with Electric Power Systems and Associated Interfaces* https://standards.ieee.org/standard/1547_1-2020.html.

[8] *Common File Format for Distributed Energy Resources Settings Exchange and Storage: Version 2.0*, EPRI, Palo Alto, CA: 2022. 3002025445. https://www.epri.com/research/products/3002025445

[9] *Errata to IEEE Standard for Interconnection and Interoperability of Distributed Energy Resources with Associated Electric Power Systems Interfaces.* https://standards.ieee.org/wp-content/uploads/import/documents/erratas/1547-2018_errata.pdf

978-1-6654-6060-6/23 $31.00 © 2023 IEEE

Upscaling of Perovskite Solar Cell Fabrication via Slot-Die Coating: In Situ Tracking of the Drying and Crystallization Front during Gas Quenching

Kristina Geistert, Simon Ternes, David Benedikt Ritzer, Benjamin Hacene, Felix Laufer, Ulrich Wilhelm Paetzold

Light Technology Institute, Karlsruhe, Germany

Institute of Microstructure Technology, Eggenstein-Leopoldshafen, Germany

Remarkable progress in efficiency and stability has been demonstrated for the next generation photovoltaic (PV) technology of thin-film perovskite solar cells (PSCs) on the laboratory scale. Small-area PSCs already exceed 25 % of power conversion efficiency (PCE) since they were first investigated about one decade ago [1,2]. However, besides the unquestionable potential of perovskite-absorber films for efficient thin-film PV, there are significant challenges when scaling the technology. The reason is that established, industrial deposition techniques such as slot-die coating are difficult to control during the entirety of the complex perovskite film formation - causing low operational stability when fabricating larger areas [3]. In response, this work leverages prior studies of our group on a quantitative model of the drying dynamics based on local heat transfer measurements [9], for controlling the crystallization of slot-die coated perovskite absorber layers dried via gas-quenching. More specifically, we apply and validate and this knowledge by systematically implementing and evaluating the upscaling strategy for fabricating slot-die coated modules on an area of 100 cm2 with an elongated slot nozzle. In situ monitoring with a CCD camera is used to critically evaluate the position of the crystallization front on the sample. This position depends directly on the drying dynamics and thus the applied process parameters. Remarkably, we succeed in demonstrating the accurate correlation of the position of the drying front with the evolving thin-film morphology as predicted by the drying models. We further show that these morphological differences have a direct impact on the achievable PCE of perovskite test devices. These methological findings denote a corner stone toward scaling perovskite fabrication preventing expensive brute force optimization. Thus, the transition from perovskite PV to large scale commercially viable manufacturing plants can be significantly facilitated.

Inverted Metamorphic Photovoltaics Utilizing a Distributed Bragg Reflector Compatible with Epitaxial Lift-Off

Robert F McCarthy, David Rowell, Andree Wibowo, William Mohr, Chris Youtsey, Mark Osowski, Martin Drees, Roger E Welser, Noren Pan

MicroLink Devices, Inc., Niles, IL, United States

Consultant, Providence, RI, United States

MicroLink Devices fabricates state-of-the art triple-junction inverted metamorphic (IMM) solar cells on GaAs wafers with a release layer that is selectively etched in an epitaxial lift-off (ELO) process to create ultra-thin foils of semiconductor with metal backing. They achieve >2000W/kg specific power with efficiencies over 30% in production. These cells have been utilized to great affect on high-altitude long-endurance (HALE) planes, and are now being adapted for space applications. To improve radiation tolerance, each absorber layer was thinned with lower doping, and the bottom subcell used a reflective back structure to maintain current. The middle subcell required a distributed Bragg reflector (DBR) to increase its absorption path length and maintain high current. MicroLink utilized a non-conventional $In_x(Al_yGa_{1-y})_{1-x}P$/GaAs DBR that was compatible with ELO. The DBR had a bandwidth of ~110nm and a peak reflectance in air near 95%. Average cell efficiencies were 28% with power retention of 84% after 1E15cm-2 electron dosing (1MeV energy). Large area cells and coupons underwent internal stress testing and showed no damage or degradation. Pathways to achieve 29% and 30% efficiency with >85% power retention are discussed.

Life-cycle analysis of potentially longer life expectancy CdTe PV modules

Vasilis Fthenakis[1], Enrica Leccisi[1] and Parikhit Sinha[2]

[1]Center for Life Cycle Analysis, Columbia University, 918 Mudd, 500W 120th street, New York, NY 10027

[2]First Solar, 350 West Washington Street, Tempe, AZ 85281

Abstract — **Research on PV module longevity shows that with lower degradation rates and better encapsulation, PV modules can significantly outlast their current life expectancy of 25-30 years. The focus of this study is on alternative CdTe PV module back-contact and encapsulation materials that show the potential of increasing module operation lives from 30 to 40 and 50 years. Life-cycle-analyis shows that some alternative materials may slightly increase cradle-to-gate environmental impacts per m² of module, but increased operational life more than counterbalances such impacts. The life-cycle carbn footprint of systems operating under average US insolation conditions of 1800 kWh/m2/yr is reduced from 10 gCO2eq/kWh to 6 gCO2eq/kWh when the operation life increases from 30 to 50 yrs. Similar reductions are noted for all life-cycle impact indicators, whereas the EROI of the considered system increases from 67 to 177 as life increases from 30 to 50 yrs.**

I. INTRODUCTION

The U.S.-based CdTe PV solar technology has achieved the lowest levelized costs of electricity (LCOE) [1] and the highest Energy Return On Investment (EROI) among all PV commercial technologies [2]. Yet, there is still headroom to improve if, concurrently with projected module efficiency enhancements, module stability is maintained by minimizing degradation rates, and protection against moisture ingress is enhanced. In an effort to increase their PV module life from 25-30 years to 40-50 years, First Solar (FS), the leading US manufacturer of CdTe PV, considers alternative back-contact dopants, aiming to decrease the degradation rate of their next generation modules. With a target for decreasing degradation from the current ~0.3%/yr to ≤ 0.2%/yr, longer service times become more meaningful. Thus, a parallel improvement of module encapsulation to enhance moisture ingress prevention can extend module lifetime, enabling LCOE at or below 2 cents/kWh for utility installations at sunny locations [3].

To this end, several back-contact and edge isolation compounds were investigated for their stability, life-cycle environmental impact and recyclability potential.

II. BACKGROUND

Fundamentally, PV modules have two primary mechanisms of power loss during the course of time: cell degradation and module corrosion.

Degradation with time manifests itself through different mechanisms. For CdTe/CdS(Se) degradation may occur at the interface of the back-contact and the semiconductor. Hall et al. [4] gives a review of several back-contact materials suitable for CdTe PV modules. Experiments by Metzger et al. [5] showed that Group V-doping of the back-contact significantly improved the stability of CdTe thin-film solar cells.

The second major reason for limiting lifetime is corrosion from moisture ingress into the module. In most cases, encapsulant (typically EVA) and/or backsheet film degradation seem to play a major role in allowing moisture ingress and corrosion [6]. EVA has been the industry standard for solar cell encapsulation but degradation of EVA can form acetic acid which can cause corrosion. However, FS has transitioned to a polyolefin encapsulant for their Series 6 modules and encapsulant degradation is now less of a problem. Since FS modules have glass sheets both at the front and the back, the remaining potential pathway for moisture ingress then becomes the edge sealant [7]. Long-lasting encapsulation and edge-sealing of the modules can minimize the potential for corrosion.

II. METHODOLOGY

System boundaries and assumptions

The analyzed PV systems are composed of First Solar PV Series 6 modules and BOS (mounting and supporting structures, inverters, transformers, and cables), as well as system operation and maintenance).

The LCA starts by normalizing life-cycle indicators in terms of m2 of PV module, which is useful to capture directly the material and energy utilization improvements over the years, it proceeds with using rated power (kWp) of modules as the functional unit, to capture the variation in module efficiencies; and subsequently are expressed in terms of kWh, at the location of deployment, taking into the account, the solar irradiation conditions, the performance ratio (PR), and the PV system operational lifetime.

The analysis was performed using the LCA software package SimaPro 9 (Pré Consultants, Amersfoort, The Netherlands), and impact assessment was performed by means of the Centrum voor Milieukunde Leiden (CML) method [36] developed by Leiden University in the Netherlands [37], with the exception of the human toxicity and ecotoxicity impact indicators for which the USEtox method was used.

Life cycle inventories and data sources
In this analysis, the Ecoinvent V3 Database (Ecoinvent, Zurich, Switzerland) was the main background LCI data source, and the data were adapted to the US average electricity grid. The foreground, module manufacturing data were provided by First Solar and LCI for the considered material alternatives were developed as the first part of this study.

III. RESULTS

Life cycle inventory compilation

The project started by compiling a life-cycle-inventory (LCI) database representing a global production of FS Series 6 modules which are double-glass, 2.47 m² large, framed modules. Then LCI data of alternative materials were compiled. Several alternatives were examined under the prism of environmental sustainability; this paper shows some of the considered alternatives.

A. Alternative back contact materials

Of interest for their applicability to tandem and bifacial cells, are back contact materials with good visible and near-infrared transparency. LCIs of the following materials were used from the Ecoinvent database

a. In_2O_5Sn (ITO), ZnO
b. ITO or ZnO plus ZnTe and doping element (D)
c. Aluminum (not transparent, reference case)

B. Alternative edge sealant materials

a. Polyisobutylene (PIB) with dessicant filler
b. Silicone + Polyisobutylene
c. Aclar™ [poly-chloro-tri-fluoro-ethylene (PCTFE)]
The reference case uses "synthetic rubber" LCI data from ecoinvent. We compiled new LCI inventories of polyisobutylene (PIB) with desiccant and Aclar™.

Cradle to Grave Life Cycle Analysis (LCA) and Net Energy Analysis (NEA)

This LCA started with investigating the differences in each of these indicators per unit mass (kg) of compound, proceeded with comparisons at the module level (m²) of the highest impact combinations of alternatives with the reference case Series 6 module, and by adding the balance of system (BOS) components we included installation and operation in average US and southwest (US-SW) regions. The task entails the assessment of the following sustainability indicators and metrics for PV modules employing the compounds considered in the previous section. a) Cumulative energy demand (CED),

b) Global Warming Potential (GWP) ; c) Acidification Potential; d) Human Toxicity Potential (HTP); e) Eco-toxicity potential (ETP); f) Abiotic Resource Depletion (ADP), g) Ozone Depletion Potential (ODP), g) Eutrophication Potential (EP, and h) Photochemical Oxidation [8, 9].
The NEA uses the CED results of the LCA and estimates of solar electricity production by the considered systems in specific regions to determine energy payback times (EPBT) and energy return of energy investment (EROI); the later is calculated as the ratio of the energy delivered to society to the sum of energy carriers diverted from other societal uses [10].

Material focus

As shown in Figures 1 and 2, and Table 1, the CED, GWP and AP per m2 of alternative back contact materials do not differ significantly. However large differences are noted for some other impact indicators (Table 1). The very large increase in ADP is due to the high ADP of Indium which is a minor byproduct in the production of zinc.

Table 1. Back-contact comparisons per m2

Environmental Impact Indicator Comparisons of Back-Contact Alternatives per m2				
Impact Indicator	Units	Al	ITO+ZnTe+D	% difference
Cumulative Energy Demand (CED)	MJ/m2	2.39E-01	3.72E-01	56
Global Warming Potential (GWP)	kgCO2eq/m2	2.05E-02	1.97E-02	-4
Acidification Potential (AP)	kgSO2eq/m2	1.14E-04	1.23E-04	7
Human toxicity potential (HTP) cancer	CTU/m2	1.43E-09	4.62E-09	223
Human toxicity potential (HTP) non cancer	CTU/m2	3.58E-09	1.97E-08	450
Eco-toxicity potential (ETP)	CTU/m2	5.73E+00	1.29E+01	125
Ozone depletion potential (ODP)	CFC-11 eq/m2	6.57E-10	1.60E-09	143
Abiotic resource depletion potential (ADP)	kg Sb eq/m2	4.20E-08	7.83E-06	18,559
Eutrophication potential (EP)	kg PO4-eq/m2	3.04E-05	8.12E-05	167
Photochemical Oxidation	kg C2H4 eq/m2	7.02E-06	1.21E-05	72

Large increases in all environmental indicators are projected when Aclar substitutes synthetic rubber or polyisobutylene as an edge sealant. (Figures 3 and 4 and Table 2). Contribution analyses of these compounds showed that the production of Aclar involves more materials and processes than the production of the other materials.

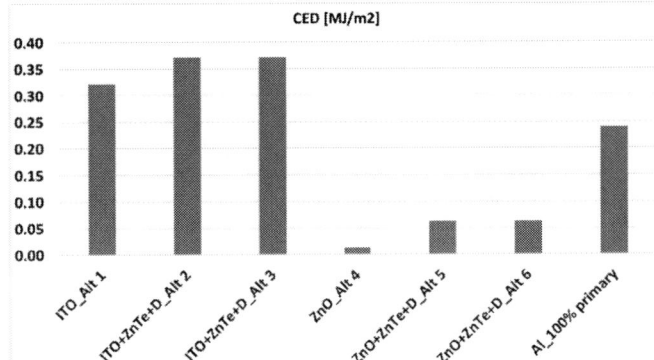

Fig. 1. CED (MJ/m2) of Back-contact material alternatives

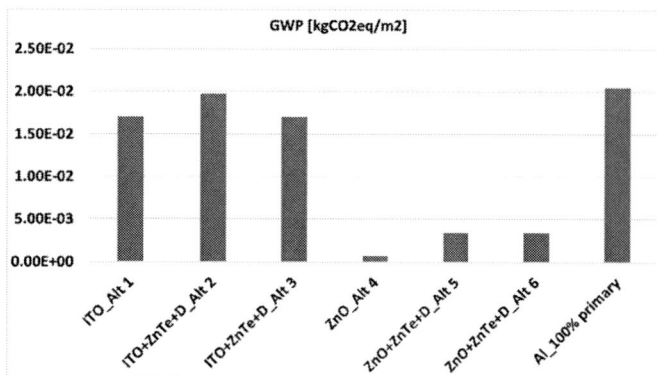

Fig. 2. GWP (kgCO2eq/m2) of Back-contact material alternatives

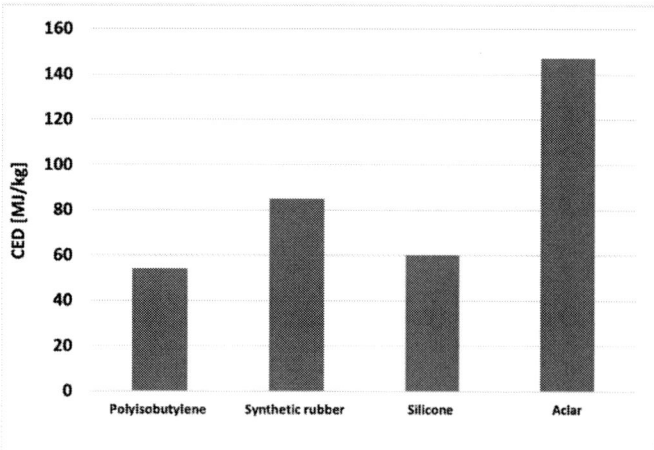

Fig. 3. CED (MJ/kg) of edge sealant materials

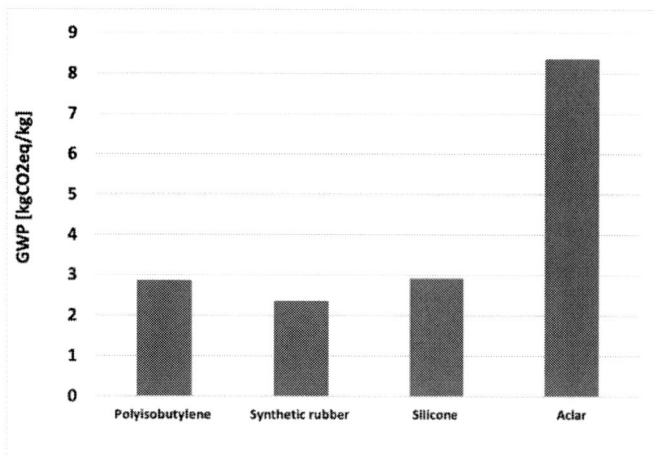

Fig. 4. GWP (kgCO2eq/kg) of edge sealant materials

Table 2. Edge sealant material comparisons per kg

Environmental Impact Indicator Comparisons of Encapsulation Alternatives per kg

Impact Indicator	Units	Synthetic rubber	Aclar	% difference
Cumulative Energy Demand (CED)	MJ/kg	8.51E+01	1.47E+02	73
Global Warming Potential (GWP)	kgCO2eq/kg	2.35E+00	8.36E+00	255
Acidification Potential (AP)	kgSO2eq/kg	1.10E-02	6.31E-02	474
Human toxicity potential (HTP) cancer	CTU/kg	6.29E-08	2.46E-07	290
Human toxicity potential (HTP) non cancer	CTU/kg	2.15E-07	1.80E-06	734
Eco-toxicity potential (ETP)	CTU/kg	4.99E+02	2.73E+03	447
Ozone depletion potential (ODP)	CFC-11 eq/kg	5.09E-07	3.65E-06	618
Abiotic resource depletion potential (ADP)	kg Sb eq/kg	5.03E-05	2.34E-04	366
Eutrophication potential (EP)	kg PO4-eq/kg	3.67E-03	1.66E-02	352
Photochemical Oxidation	kg C2H4 eq	6.39E-04	2.64E-03	313

Module focus

Although the per kg impact indicator values differ significantly, the differences are essentially diluted in the context of module impacts as these compounds are used in relatively small quantities. As shown by Figures 5 and 6, the CED difference between a reference and an alternative module case is only 1 MJ/m^2 or 0.15%. The module materials that are substituted by alternatives are red-highlighted.

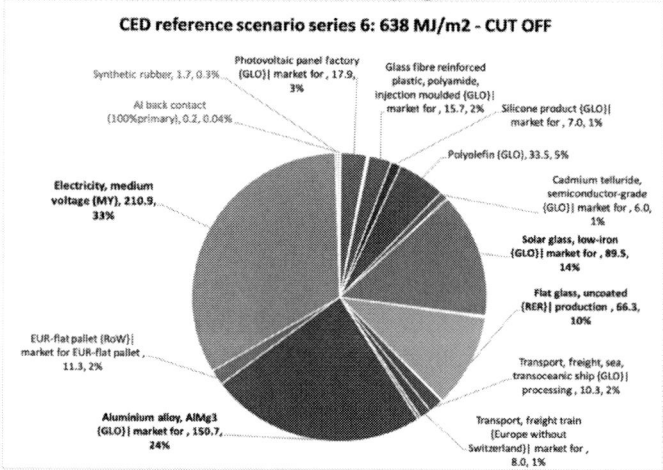

Fig. 5. Cradle to gate CED of reference Series 6 CdTe modules

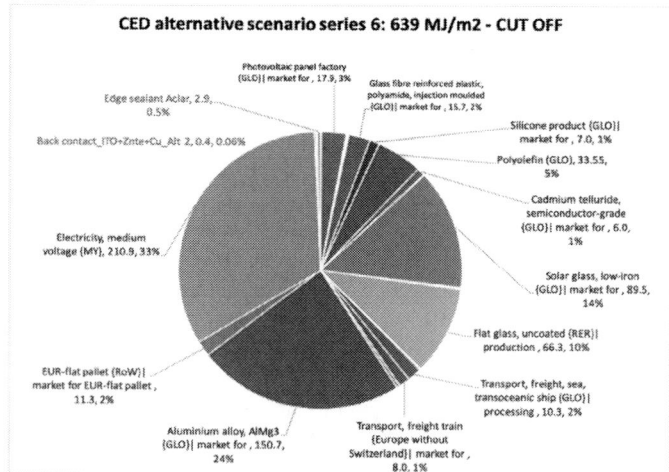

Fig. 6. Cradle to gate CED of alternative Series 6 CdTe modules

Also a small difference (0.26%) was found by comparing the GWP of the two module designs (Figures 7 and 8).

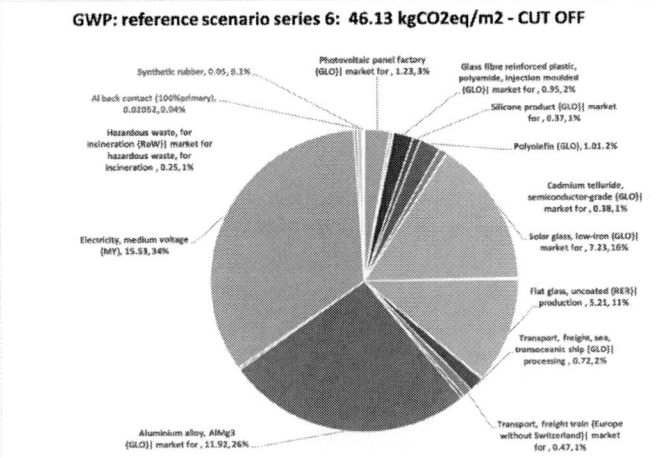

Fig. 7. Cradle to gate GWP of reference Series 6 CdTe modules

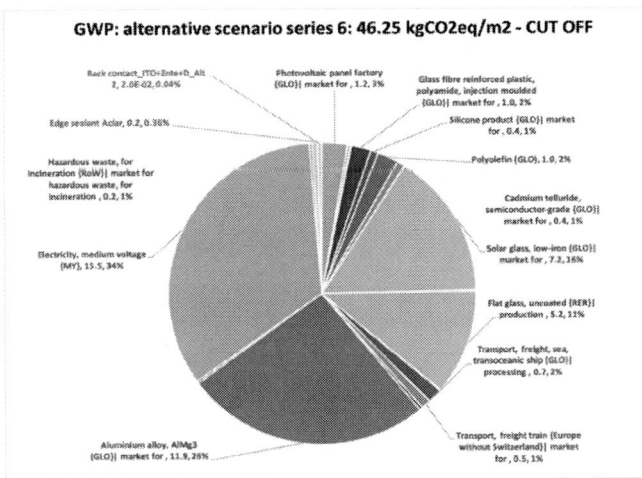

Fig. 8. Cradle to gate GWP of alternative Series 6 CdTe modules

Comparisons of the reference and the longer life modules based on other environmental impact indicators, is shown in Table 3. As shown, the difference is less than 1% with the exception of the ADP and ODP where a difference of 2.6% and 2.2 % is primarily due to the impact of Aclar.

Table 3. Series 6 reference and alternative module comparisons

Impact Indicator	Units	Reference	Alternative	% difference
Cumulative Energy Demand (CED)	MJ/m2	638	639	0.16
Global Warming Potential (GWP)	kgCO2eq/m2	46.13	46.25	0.26
Acidification Potential (AP)	kgSO2eq/m2	2.64E-01	2.65E-01	0.38
Human toxicity potential (HTP) non cancer	CTU/m2	4.61E-06	4.66E-06	1.04
Eco-toxicity potential (ETP)	CTU/m2	8.15E+03	8.20E+03	0.61
Ozone depletion potential (ODP)	CFC-11 eq/m2	2.71E-06	2.77E-06	2.21
Abiotic resource depletion potential (ADP)	kg Sb eq/m2	4.52E-04	4.64E-04	2.65
Eutrophication potential (EP)	kg PO4-eq/m2	6.59E-02	6.62E-02	0.46
Photochemical Oxidation	kg C2H4 eq	1.13E-02	1.13E-02	0.44

Thus, although the differences per kg or per m2 of the environmental impact indicators for the materials considered in the "alternative" module are in the range of 50% to

18000%, those differences attribute only 0.2% to 2.6% impact increases at the module m2 level, as the considered alternative materials are only a very small fraction of the module mass.

Module Recyclability
The considered alternative module architectures were also evaluated for their recyclability. The evaluation considers whether the proposed design change impacts the current recycling technology and whether any mitigation measures are needed. As part of First Solar's management change system, proposed changes to PV module design were evaluated by a cohort of First Solar subject matter experts (SME). Potential impacts on the module recycling process were evaluated with experimental leaching tests. Coupon-sized modules were tested in the recycling R&D V4 leaching reactor and full-sized modules were tested in the commercial-scale recycling facility. It was found that changes in encapsulant and back-contact materials did not affect the recycling process.

System Installation and Operability focus
The BOS components for latitude-tilt installations were added to the module LCI data and the performance of the full system is evaluated under the following assumptions:
Irradiation on latitude-tilt plane of array: 1800 kWh/m2/yr
US-SW: 2300 kWh/m2/yr
Performance Ratio Utility Ground-mount installations: 0.85
Module Degradation Rates/Operation Life: 0.3%/yr -30 yrs; 0.2%/yr -40 yrs; 0.1%/yr -50 yrs.
Inverter Replacement: Every 15 yrs.

Life Cycle Assessment Results

Tables 4 and 5 summarize the LCA results for systems operating in average US conditions and US-SW conditions correspondingly. As shown, the system GWP can be reduced from 10 gCO_{2eq}/kWh to 6 gCO_{2eq}/kWh when the system operational life under average US conditions increases from 30 yrs to 50 yrs. In US-SW operation the GWP can be decreased to less than 5 gCO_{2eq}/kWh. Simarly proportional decreases are estimated for all environmental impact indicators.

Table 4. LCA results for CdTe PV operating under average US insolation

		Operation Life (years) under	1800	kWh/m2/yr
Environmental Impact Indicator	Units	30	40	50
Global Warming Potential (GWP)	gCO_{2eq}	10.3	7.6	5.9
Acidification Potential (AP)	$gSO2eq$	1.5E-01	1.1E-01	8.4E-02
Human toxicity potential (HTP)	CTU	3.8E-09	2.8E-09	2.2E-09
Eco-toxicity potential (ETP)	CTU	11.7	8.7	6.7
Ozone depletion potential (ODP)	g CFC-11 eq	8.5E-07	6.3E-07	4.9E-07
Abiotic resource depletion potential (ADP)	g Sb eq	2.0E-03	1.5E-03	1.1E-03
Eutrophication potential (EP)	g PO4--- eq	2.0E-02	1.5E-02	1.1E-02
Photochemical Oxidation	kg C2H4 eq	6.4E-03	4.8E-03	3.7E-03

Table 5. LCA results for CdTe PV operating under in the US-SW

Environmental Impact Indicator	Units	Operation Life (years) under	2300	kWh/m2/yr
		30	40	50
Global Warming Potential (GWP)	gCO$_{2eq}$	8.1	6.0	4.6
Acidification Potential (AP)	gSO2eq	1.1E-01	8.5E-02	6.6E-02
Human toxicity potential (HTP)	CTU	3.0E-09	2.2E-09	1.7E-09
Eco-toxicity potential (ETP)	CTU	9.2	6.8	5.3
Ozone depletion potential (ODP)	g CFC-11 eq	6.7E-07	5.0E-07	3.8E-07
Abiotic resource depletion potential (ADP)	g Sb eq	1.6E-03	1.2E-03	9.0E-04
Eutrophication potential (EP)	g PO4--- eq	1.5E-02	1.1E-02	8.9E-03
Photochemical Oxidation	kg C2H4 eq	5.0E-03	3.7E-03	2.9E-03

Net Energy Analysis Results

The Energy Payback Times (EPBT$_{PE}$) in terms of primary energy of ground-mount, fixed latitude-tilt systems incoporating Series 6 modules is 0.4 years under US-average solar irradiation of 1800 kWh/m2/yr when solar electricity displaces electricity in a fossil-fuel grid of a 30% conversion efficiency. This system's EPBT is increased to 1.4 years in a future scenario of 100% grid conversion efficiency (i.e., a grid fully comprised of solar, wind and hydro). We denote it as EPBT$_{el.}$

For 1-axis tracking systems operating under the high irradiation of the US-SW, the EPBT$_{PE}$ is 0.3 and the EPBT$_{el}$ is 1.1 years. The Energy Return on Energy Investment (EROI) is greatly affected by the degradation rate and associated life expectancy. As shown in Figure 8, EROI$_{PE}$ under US average insolation ranges from 67.6 to 117.3 and the EROI$_{el}$ ranges from 20.3 to 35.2 as the operation life ranges from 30 to 50 years.

For 1-axis tracking systems operating in the US-SW under 2300 kWh/yr , the EROI$_{PE}$ ranges from 86.4 to 149.9 and the EROI$_{el}$ ranges from 25.9 to 45.0 as the operation life ranges from 30 to 50 years, under the PR and degradation rate assumptions listed above.

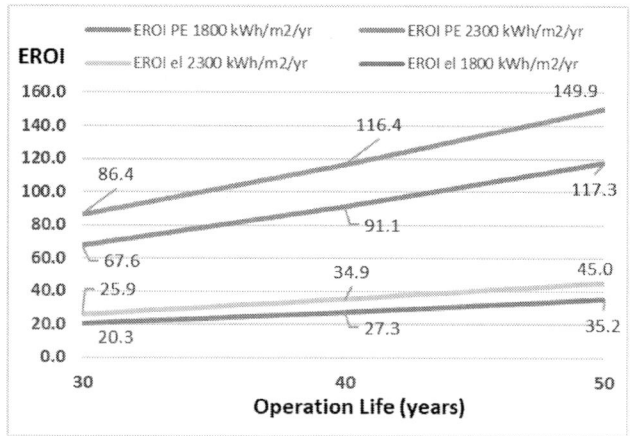

Fig. 9. EROI as a function of insolation, operation life and grid composition

It is noted that all the estimates presented herein assume that the balance of system (BOS) will not be affected by the extension of the system life from 30 to 40 and 50 years. This is a valid assumption for the galvanized steel and aluminum support and mounting parts and the coper and aluminum cables but it is likely than with current vintage inverters and transformers lasting 15-years an additional inverter and a third inverter/transformer may be needed in extended lives. This will add 3-4% on the BOS, thus about 1-2% on the whole system environmental impacts.

IV. SUMMARY

Sustainability investigations were conducted of back-contact doping for reduction of degradation rate, and edge sealant alternatives for improving encapsulation of CdTe PV modules, along two dimensions: Environmental and energy impacts and recyclability. Degradation and encapsulation affect the lifetime of a PV module, and the lifetime affects the rate of replacement.

The environmental impact assessment was informed by LCA using LCI of promising back-contact and edge sealant materials. It was found that the alternatives only slightly increased the environmental impact of Series 6 modules. Longer operation lives result in significant reductions in all environmental impact indicators and increases energy returns on energy investment. Longer lives also postpone the decommissioning costs and reduce waste.

Also the recyclability of alternative module architectures was examined and found that the considered changes had a negligible effect on the recyclability of the new architectures.

The current investigation is based on developments towards lower module degradation rates established by First Solar R&D and on reasonable expectations that edge sealing may be further improved in modules designed for longer life. However, it is noted that current PV module qualification tests are not intended to establish long term reliability and there is a a need for developing appropriate testing protocols and measurement methods. The International Electrotechnical Commission (IEC) TS 63209-1:2021[11] is a starting point, intended to provide a standardized method for extended reliability testing of PV.

Levelized Cost of Electricity (LCOE)

Starting from a LCOE of $0.04/kWh which reflects utility scale PV plants in the US-SW, and a reference degradation rate of 0.3%/yr, as degradation deacreses to 0.1%/yr and operational life increases to 50 years, the LCOE deceases to $0.023kWh (Fig. 10) This LCOE reduction is similar to LCOE reduction that would result from a module efficiency increase of FS series 6 from 18% to ~25%. A combination of increased efficiency and operational life will enable drastic LCOE reductions.

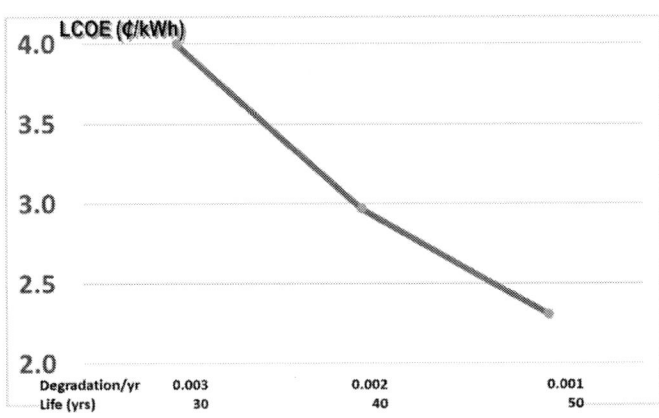

Fig. 10. LCOE as a function operation life and degradation

ACKNOWLEDGEMENT

This material is based upon work supported by the U.S. Department of Energy's Office of Energy Efficiency and Renewable Energy (EERE) under the Solar Energy Technologies Office Award Number DE-EE0009830. **Disclaimer**: This report was prepared as an account of work sponsored by an agency of the United States Government. Neither the United States Government nor any agency thereof, nor any of their employees, makes any warranty, express or implied, or assumes any legal liability or responsibility for the accuracy, completeness, or usefulness of any information, apparatus, product, or process disclosed, or represents that its use would not infringe privately owned rights. Reference herein to any specific commercial product, process, or service by trade name, trademark, manufacturer, or otherwise does not necessarily constitute or imply its endorsement, recommendation, or favoring by the United States Government or any agency thereof. The views and opinions of authors expressed herein do not necessarily state or reflect those of the United States Government or any agency thereof.

REFERENCES

[1] Lazard, 2021. Levelized Cost of Energy Analysis. Available at: https://www.lazard.com/research-insights/levelized-cost-of-energy-levelized-cost-of-storage-and-levelized-cost-of-hydrogen-2021/

[2] E. Leccisi, M. Raugei, and V. Fthenakis, 2016. The energy and environmental performance of ground-mounted photovoltaic systems—a timely update. Energies, 9(8), p.622.

[3] I. M. Peters, P. Sinha, Value of stability in photovoltaic life cycles, in *48th IEEE Photovoltaic Specialist Conference*, 2021, pp. 0416-0419.

[4] R.S. Hall, D. Lamb, S. J. C. Irvine, Back contacts materials used in thin film CdTe solar cells—A Review, Energy Sci Eng. vol. 9, 606–632, 2021.

[5] W. K. Metzger 1, S. Grover, D. Lu, E. Colegrove, J. Moseley, C. L. Perkins, X. Li, R. Mallick, W. Zhang, R. Malik, J. Kephart, C.-S. Jiang, D. Kuciauskas, D. S. Albin, M. M. Al-Jassim, G. Xiong and M. Gloeckler, Exceeding 20% efficiency with in situ group V doping in polycrystalline CdTe solar cells, *Nature Energy*, vol. 4, 837-845, 2019.

[6] IEA PVPS Task 13, 2014. Performance and Reliability of Photovoltaic Systems. Report IEA-PVPS T13-01:2014. Available at: https://iea-pvps.org/wp-content/uploads/2020/01/IEA-PVPS_T13-01_2014_Review_of_Failures_of_Photovoltaic_Modules_Final.pdf

[7] M. D. Kempe, D. L. Nobles, L. Postak, J. A., Calderon, "Moisture ingress prediction in polyisobutylene-based edge seal with molecular sice desiccant", *Prog Photovolt Res Appl.*, vol 26, 93-101, 2018.

[8] Universiteit Leiden. CML-IA Characterisation Factors. 2016. Available at: at: https://www.universiteitleiden.nl/en/research/research-output/science/cml-ia-characterisation-factors

[9] P. Fantke, M. Bijster, C. Guignard, M. Hauschild, M. Huijbregts, O. Jolliet, A. Kounina, V. Magaud, M. Margni, T. McKone,. and L. Posthuma, 2017. USEtox 2.0: Documentation (Version 1).

[10] M. Raugei, R. Frischknecht, C. Olson, P. Sinha, G. Heath. Methodological guidelines on net energy analysis of photovoltaic electricity, International Energy Agency PVPS, 2016, IEA Report T12-07:2016.

[11] IEC TS 63209-1:2021, Photovoltaic modules - Extended-stress testing - Part 1: Modules, March 2023. https://webstore.iec.ch/publication/63120

Forecasting Day-Ahead Solar Irradiance for Puerto Rico using the WRF Model and NSRDB

Manajit Sengupta[1], Jaemo Yang[1], Yu Xie[1]

[1]National Renewable Energy Laboratory, Golden, CO, 80401, USA

Abstract — Accurately predicting solar energy resources is a major challenge in integrating photovoltaics generation on the electric grid. Numerical weather prediction has been recognized by the solar energy community as a major approach to provide solar resource forecasts at various locations and for a variety of timescales. In this study, as a part of the Puerto Rico Grid Resilience and Transitions to 100% Renewable Energy Study (PR100), we develop day-head solar irradiance forecast data using the Weather Research and Forecasting (WRF) model at 3 km and hourly/5-minute. The global horizontal irradiance (GHI) and direct normal irradiance (DNI) forecasts simulated from the WRF model are postprocessed by a simple optimization method using satellite-derived gridded observations from the National Solar Radiation Data Base (NSRDB) to reduce error and bias of the solar irradiance forecasts covering 2018-2020. The NSRDB contributes to improving the GHI and DNI forecasts and also offers the opportunity for an in-depth analysis to evaluate their accuracy over a wide range of Puerto Rico regions. Preliminary results show overall improvements of GHI forecasts up to 37% (DNI: 15%) for mean absolute error and 97% (DNI: 76%) for mean bias error by applying a postprocessing technique to WRF model output.

I. INTRODUCTION

The Weather Research and Forecasting (WRF) model is a publicly available numerical weather prediction (NWP) model that is used worldwide by renewable energy researchers to forecast renewable resources for various energy applications (e.g., solar and wind) [1]–[2]. The WRF model is capable of dynamically downscaling global-scale weather data to a finer grid and predicting meteorological variables. Recently, the U.S. Department of Energy and six national laboratories commenced the Puerto Rico Grid Resilience and Transitions to 100% Renewable Energy Study (PR100). Through PR100, a broad analysis of possible pathways is provided for Puerto Rico's energy future, with a goal to actualize 100% renewable energy by 2050 [3]. As a part of renewable energy potential assessment for the project, day-ahead solar irradiance forecast using the WRF model is developed to support analyses for potential solar power applications for Puerto Rico. To produce accurate data, the WRF-based solar irradiance forecasts are postprocessed using the National Solar Radiation Database (NSRDB) [4] to improve their prediction accuracy. The raw and postprocessed global horizontal irradiance (GHI) and direct normal irradiance (DNI) are compared and evaluated using the NSRDB in different regions of Puerto Rico.

II. METHODOLOGY

A. WRF Model Configuration

The atmospheric model used in this study is the WRF model, a publicly available NWP model, that is developed and distributed by the National Center for Atmospheric Research. For Puerto Rico, we selected the WRF Version 4.3 [5]. The WRF model is configured with two nests with horizonal grid spacing of 9 km and 3 km (Fig. 1) respectively. The inner domain covering Puerto Rico and the U.S. Virgin Islands is the focus area of this work.

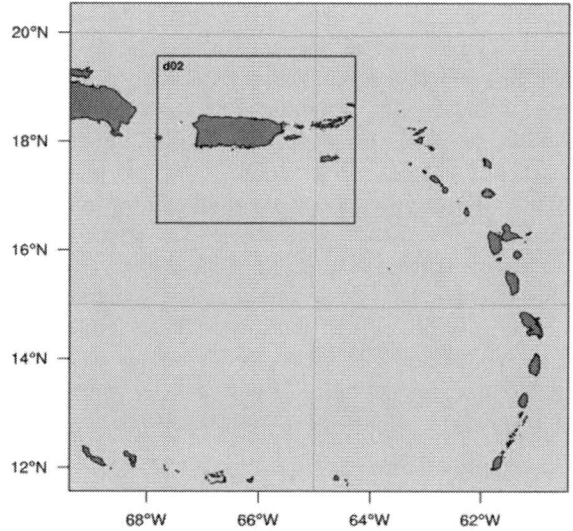

Fig. 1. WRF domain with two nests (outer: 9 km and inner: 3 km) for Puerto Rico.

The National Centers for Environmental Prediction (NCEP) Global Forecast System (0.25°×0.25°, 3-hourly intervals) forecast was used for initial and boundary conditions of the WRF model. The forecast horizon of one day-ahead forecast was 64 hours. Each WRF run was initialized at 12 UTC and then the first 16 hours were regarded as the model spinup time and discarded from the analysis so as to only cover midnight-to-midnight forecasts (for the second day) over Puerto Rico regions in local time. The WRF output of the 3-km domain was stored at every 5 minute interval.

We performed five WRF experiments focused on using different cumulus, shallow cumulus, cloud fraction, and PBL schemes for 2019. The experiments, did not come up with a configuration that provides a bias-free solar irradiance forecast (not shown). Thus, two WRF configurations that show different bias characteristics (i.e., positive and negative) were selected and combined to improve the quality of the forecasts. A summary of the configuration including the main physics schemes for the two WRF experiments (E01 and E02) are summarized in Table 1. Data sets of day-ahead forecasts covering 2017-2020 were simulated from the E01 and E02. Additionally, we processed the 5-minute WRF outputs to generate two data sets which have hourly and 5-minute resolutions.

TABLE I
SUMMARY OF WRF PHYSICS CONFIGURATIONS

WRF Scheme	Experiment #	
	E01	E02
Microphysics	Thompson-aerosol-aware	Thompson-aerosol-aware
Planetary boundary layer	Shin-Hong	MYJ
Surface layer	Revised MM5 Monin-Obukhov	Monin-Obukhov (Janjic) scheme
Land surface model (LSM)	Noah LSM	Noah LSM
Cumulus	Grell-Freitas	Grell-Freitas
Shallow cumulus (9-km domain)	-	Deng
Cloud fraction; cloud effect to the optical depth in radiation	CLD3	Xu-Randall

B. Satellite-derived Gridded Observation: NSRDB

The NSRDB is a publicly available data set that provides solar irradiance variables (e.g., GHI, DNI, and diffuse horizontal irradiance (DHI)) and meteorological data globally and includes North and Central America (https://nsrdb.nrel.gov/). The NSRDB, which has 4-km and 30-minute spatiotemporal resolution was regridded to the 3-km WRF grid. The averaged GHI and DNI over 2019 is shown in Figure 2.

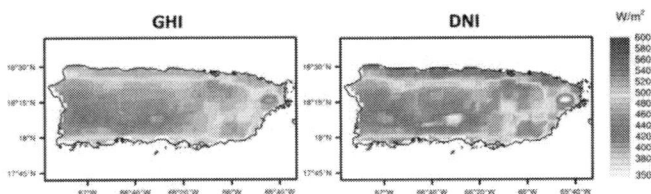

Fig. 2.   2019 Mean GHI and DNI from NSRDB (3-km grid).

C. Postprocessing of Solar Forecast from WRF

A simple optimization method is applied as the postprocessing scheme to improve day-ahead prediction of solar irradiance simulated from E01 and E02 (Figure 3).

The basic concept behind the approach is to limit the weaknesses related to systematic bias in the numerical prediction of physical processes for clouds. The postprocessing method is based on the linear combination of two predictors defined by

$$Y_{i,j,t} = a_{i,j,t} X_{E01,i,j,t} + b_{i,j,t} X_{E02,i,j,t} \qquad (1)$$

where $X_{E01,i,j,t}$ and $X_{E02,i,j,t}$ are the predictors (i.e., GHI) from E01 and E02 for grid point i,j and time t, $a_{i,j,t}$ and $b_{i,j,t}$ indicate the weights for the two predictors, and $Y_{i,j}$ is the adjusted GHI for grid point i,j and time t where t ranges from 0-48 hours with a 1 hour time step. The optimization problem to estimate $a_{i,j,t}$ and $b_{i,j,t}$ is defined as follows

$$minimize\ MAE_{i,j}$$
$$subject\ to\ a_{i,j,t} + b_{i,j,t} = 1 \qquad (2)$$

where $MAE_{i,j}$ is the mean absolute error (MAE) of adjusted GHI forecasts calculated at each NSRDB grid point and for 365 sets of modeled-observed data at a given hour. The optimization is implemented given the weight constraint, $a_{i,j,t} + b_{i,j,t} = 1$. One year of GHI forecasts is used as the training data set to estimate the weights, and the next year is used for the out-of-sample application (i.e., validation of estimated weights from the previous year). In this work, the approach is chronologically applied and validated for 2018-2020. For 5-minute forecast data, the linear interpolation is applied to generate 5-minute scale of weights from the hourly estimated $a_{i,j,t}$ and $b_{i,j,t}$.

The optimization method was not applied to the DNI forecasts because the relationship between GHI, DNI, and DHI (i.e., GHI = DNI × cosθ + DHI) cannot be physically satisfied if the DNI is also separately adjusted by the optimization. Thus, the empirical Direct Insolation Simulation Code (DISC) [6] was used as a postprocessing tool for generating the DNI forecast. In this case the adjusted GHI forecast using the estimated weights was used as input of the DISC model to generate the postprocessed DNI forecast.

III. RESULTS

The postprocessed GHI and DNI forecasts (hourly data sets) are evaluated against NSRDB. Figure 3 displays the yearly MAE maps for raw GHI forecasts from WRF (E01 and E02) and the postprocessing results.

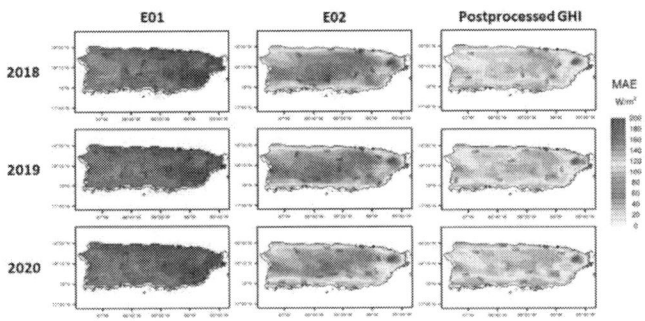

Fig. 3.   MAE maps of raw and postprocessed GHI forecasts for 2018-2020.

E01 shows high MAEs which greater than 140 W/m² for most of the Puerto Rico regions. Compared to E01, E02 performs better in predicting GHI, but still exhibits high MAE (>140 W/m²) for the island with the exception of southern and eastern coasts of the Puerto Rico. It is evident that a large reduction in MAE is achieved by the postprocessing approach across all regions.

Figure 4 shows the MAE maps for DNI prediction. Consistent with results shown in Figure 3, improvements from the postprocessed DNI forecasts are indicated by the overall reductions in MAE across most of the regions in Puerto Rico.

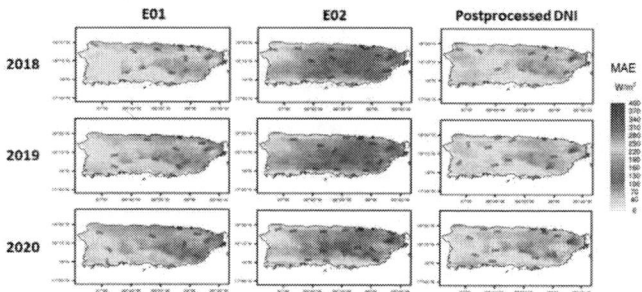

Fig. 4.   MAE maps of raw and postprocessed DNI forecasts for 2018-2020.

Figure 5 shows yearly calculated MAE and and mean bias error (MBE) for E01, E02, and postprocessed irradiance foreacsts. The postprocessing method reduces the MAE in E01 and E02 up to 37% and 14%, respectively for 2018-2020. The GHI MBE is reduced by 97% and 91% for E01 and E02. For DNI MAE, improvements of up to 10% and 15% for E01 and E02 are attained by the postprocessing approach. The DNI MBE is reduced by 76% and 72% for E01 and E02, respectively.

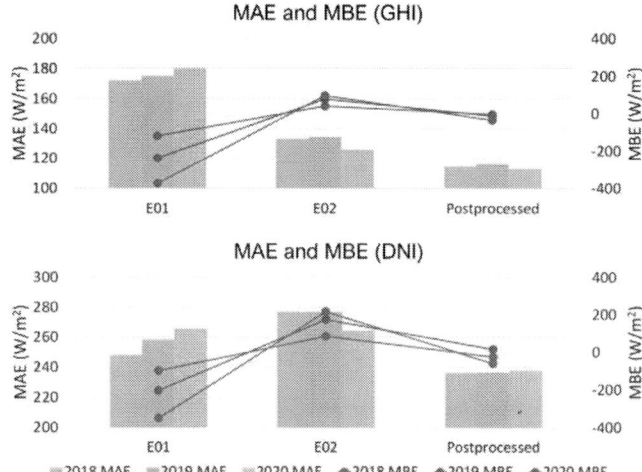

Fig. 5.   MAE (bar) and MBE (line with markers) of GHI and DNI forecasts for 2018-2020.

IV. CONCLUSION

Day-ahead solar irradiance forecasts were developed for Puerto Rico and postprocessed to reduce their error and bias. The WRF model using two different configurations was used to simulate GHI and DNI forecasts covering 2017-2020. The NSRDB data sets which are satellite-derived gridded observations were used to improve and validate the solar irradiance forecasts for 2018-2020 across Puerto Rico regions. A simple optimization technique that minimizes MAE of GHI prediction and the DISC model were applied to postprocessing of GHI and DNI forecasts. For GHI forecasts, MAE and MBE reductions of up to 37% (DNI: 15%) and 97% (DNI: 76%) were achieved by the postprocecssing method for Puerto Rico. An analog technique using the NSRDB will be considered to further improve the day-ahead solar irradiance foreacsts and produce probabilistic predictions of the ranges of GHI and DNI in a future study.

ACKNOWLEDGMENT

This work was authored by the National Renewable Energy Laboratory, operated by Alliance for Sustainable Energy, LLC, for the U.S. Department of Energy (DOE) under Contract No. DE-AC36-08GO28308. Funding provided by the U.S. Department of Energy Office of Electricity. The views expressed in the article do not necessarily represent the views of the DOE or the U.S. Government. The U.S. Government retains and the publisher, by accepting the article for publication, acknowledges that the U.S. Government retains a nonexclusive, paid-up, irrevocable, worldwide license to publish or reproduce the published form of this work, or allow others to do so, for U.S. Government purposes.

REFERENCES

[1] D. Yang, W. Wang, C.A. Gueymard, T. Hong, J. Kleissl, J. Huang, M.J. Perez, R. Perez, J.M. Bright, X.A. Xia, and D. van der Meer, "A review of solar forecasting, its dependence on atmospheric sciences and implications for grid integration: Towards carbon neutrality," *Renewable and Sustainable Energy Reviews*, vol. 161, p.112348, 2022.

[2] P. Duffy, G.R. Zuckerman, T. Williams, A. Key, L.A. Martínez-Tossas, O. Roberts, N. Choquette, J. Yang, H. Sky, and N. Blair, "Wind energy costs in Puerto Rico through 2035," (No. NREL/TP-5000-83434), National Renewable Energy Lab.(NREL), Golden, CO (United States), 2022.

[3] R. Burton, M. Baggu, J. Rhodes, N. Blair, T. Harris, M. Sengupta, C. Barrows, H. Sky, J. Yang, J. Elsworth, and P. Das, "Analysis and technology needs for getting to 100% renewable energy: Puerto Rico Grid Resilience and Transitions to 100% Renewable Energy Study (PR100) example," (No. NREL/PR-5C00-84318), National Renewable Energy Lab.(NREL), Golden, CO (United States), 2022.

[4] M. Sengupta, Y. Xie, A. Lopez, A. Habte, G. Maclaurin, and J. Shelby, "The national solar radiation data base (NSRDB)," *Renewable and Sustainable Energy Reviews*, vol. 89, pp.51-60, 2018.

[5] W.C. Skamarock, J.B. Klemp, J. Dudhia, D.O. Gill, Z. Liu, J. Berner, W. Wang, J.G. Powers, M.G. Duda, D.M. Barker, and X.Y. Huang, "A description of the advanced research WRF model version 4," National Center for Atmospheric Research: Boulder, CO, USA, 145, p.145, 2019.

[6] E.L. Maxwell, "A quasi-physical model for converting hourly global horizontal to direct normal insolation," (No. SERI/TR-215-3087). Solar Energy Research Inst., Golden, CO (USA), 1987.

Coordinating the Frequency-Droop Controls of Inverter-Based Resources and Diesel Generators in an Isolated Microgrid

Mohan Du, Nayeem Ninad and Dave Turcotte
CanmetENERGY, Natural Resources Canada (NRCan), Varennes, J3X 1P7, Canada

Abstract—With the increasing penetration level of renewable energy sources (RESs), microgrids (MGs) implemented with inverter-based resources (IBRs) and diesel generators (DGs) are considered to be new solutions to supply power in remote areas and islands. Despite the introduction of power and voltage balancing services for distributed energy resources (DERs) and DGs in recent standards and codes, the coordination between the services provided by DERs and DG is not thoroughly investigated. This paper assesses the frequency-droop control defined in IEEE 1547-2018 std. for a DER, investigates the coordination between the DER's and DGs' droop controls, and reveals a risk of malfunction—exaggerating frequency deviation and power waste—if their droop controls do not coordinate well. Furthermore, this paper presents a frequency-restoration function to resolve the observed malfunction. The coordination of DER's and DG's control parameters is also examined. Moreover, the power sharing between multiple DGs and DERs during significant load changes is investigated. Simulation results show that the presented frequency-restoration function, suggested control parameters, and developed power sharing method can significantly benefit the coordination of multiple DERs and DGs in an isolated MG.

Index Terms—Frequency-droop, inverter, frequency restoration, diesel generator, IEEE 1457, Solar PV, DER.

I. INTRODUCTION

The penetration level of renewable energy sources (RESs) reported by the International Energy Agency (IEA) has significantly increased from 19.5% in 2010 to 26.0% in 2019 [1] and is expected to continue its astonishing growth to occupy 60% of global electricity generation by 2035 [2]. The Government of Canada has committed to achieve a net-zero electricity grid (i.e., no or through offset of direct GHG emissions from the electricity sector) by 2035 [3] and additionally, has committed to achieving net-zero emissions by 2050 [4]. Meanwhile, with more and more inverter-based RESs (IBRs) integrated into the power system, renewable microgrids (MGs)—distribution grids that can operate autonomously with the support of renewable distributed energy resources (DERs)—are attracting more and more research attention as the traditional operation concept is changing rapidly.

Small remote isolated MG typically operates with a diesel power plant which generally includes one or more droop-controlled diesel generators (DGs) operating in parallel [5], [6]. Therefore, the frequency and voltage of the DG vary based on the active and reactive power demand of the MG load. The studied MG system does not include automatic generation control (AGC) for frequency regulation. Recent standards have included the frequency-droop function based on pre-disturbance power for the DERs so that they can contribute to power-sharing with other generation sources [7], [8]. This frequency droop function and corresponding control parameters are widely assessed by researchers [9]–[13]. [9] points out that most currently manufactured solar PV inverters do not implement the over-frequency droop control defined in IEEE 1547-2018 and CSA C22.3 No. 9 standards [7], [8] and the paper analyzes the percentage of curtailed energy, including comparison with real-life data, given the over-/under-frequency droop control is implemented in PV inverters. [10] assesses the frequency-droop control with the Oahu MG in the Hawaiian island and recommends a range of droop slopes and points out that the deadband for the frequency-droop control should not be too small to activate the control frequently. [11] conducts a detailed sensitivity analysis on the parameters of the frequency-droop control. [12] assesses the capability of the PV inverters implemented with the frequency-droop control to provide contingency reserve and frequency regulation. [13] proposed a rapid active power control method coupled with a maximum power point estimation method that allows PV inverters to operate in under-frequency droop operation. This can mitigate grid frequency contingency events, including fast power-frequency droop, inertia emulation, and fast frequency response, without the need for energy storage.

This paper examines the challenges in applying the current frequency-droop control in the latest IEEE 1547-2018 to small isolated MGs. It highlights that the lack of frequency restoration services can interfere with the DERs' frequency-droop control. To address this, a frequency-restoration method for DGs is proposed, enabling the consistent contribution of DERs to frequency stabilization. The paper also explores the coordination of control parameters between DERs and DGs, and investigates power-sharing strategies and the impact of DER's power limit values.

978-1-6654-6060-6/23 $31.00 © 2023 IEEE

II. METHODOLOGY

A. Test System

For analyzing the droop control of a DER device in parallel operation with droop-controlled DGs, a one-bus isolated MG as shown in Fig. 1 is considered. The DG is rated at 90 kW and the DER has a rated power of 30 kW.

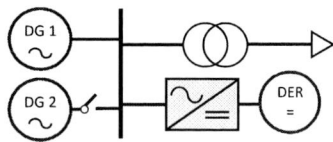

Fig. 1: One-bus test MG.

B. Modeling of DG

The 90-kW DG is adapted from the Matlab-Simulink model presented in [14]. The governor, which receives the measured and reference frequency as input and produces the reference active power as output, employs droop control. To constrain the rate of increase for the reference power, a rate limiter with a 50 kW/s limit is incorporated within the governor.

C. Modeling of DER

The 30-kW DER is modeled using the smart inverter toolbox developed by [15] in the Matlab-Simulink environment, and is compatible with the OPAL-RT eMEGAsim platform. The inverter-based DER model comprises phase locked loop (PLL), wave reference, primary control, secondary control, and protection system blocks. The frequency-droop control, which is the focus of this paper, resides within the secondary control block.

The synchronous reference frame PLL allows the DER to maintain synchronization with the DG. The wave reference block generates the modulating signal for the semiconductor bridge based on the reference real and reactive current components from the primary control block. The current control is achieved in dq-frame with appropriate feed-forward decoupling. The primary control block produces reference real and reactive current components through the DC voltage controller and a closed AC reactive power loop. The input reference power values are obtained from the secondary control block. The secondary control block implements grid support functions defined in the IEEE 1547-2018 Standard, where the frequency-droop function takes the power limit command, measured frequency, and active power setpoint as inputs, and outputs the reference active power. The protection system facilitates the inverter's bridge, voltage, and frequency protection as well as islanding detection.

In the simulation, the secondary control loop time constant is set to 0.5 s, and the power limit values range from 50% to 100%.

D. Droop Control for the DER and DG

This paper employs the revised frequency-droop function proposed by [9] for the DER inverter, which is,

$$p_{ref}^{der} = \min_{f<60-db_{uf}^{der}} \left(p_{pre}^{der} + \frac{(60 - db_{uf}^{der}) - f}{60 \cdot k_{uf}^{der}}, p_{avl}^{der} \right) \quad (1a)$$

$$p_{ref}^{der} = \max_{f>60+db_{of}^{der}} \left(\min \left(p_{pre}^{der} - \frac{f - (60 - db_{of}^{der})}{60 \cdot k_{of}^{der}}, p_{avl}^{der} \right), p_{min}^{der} \right) \quad (1b)$$

where p_{ref}^{der} is the reference active power, p_{pre}^{der} is the pre-disturbance active output power (before the frequency deviates out of the deadband), p_{avl}^{der} is the available active power from the primary source, p_{min}^{der} is the minimum active output power, f is the system frequency measured at the DER inverter terminal, db_{uf}^{der} and db_{of}^{der} are the single-sided deadbands for the under- and over-frequency scenarios, respectively, and k_{uf}^{der} and k_{of}^{der} are the per-unit values of frequency change due to one unit change of output power for the under- and over-frequency scenarios, respectively [7], [9]. The characteristic curve of the droop control for the DER inverter is presented in Fig. 2(a). In this work, the DER inverter is considered to be Category III definition for the frequency-droop function of IEEE 1547-2018 std. The values of these parameters are presented in Table I.

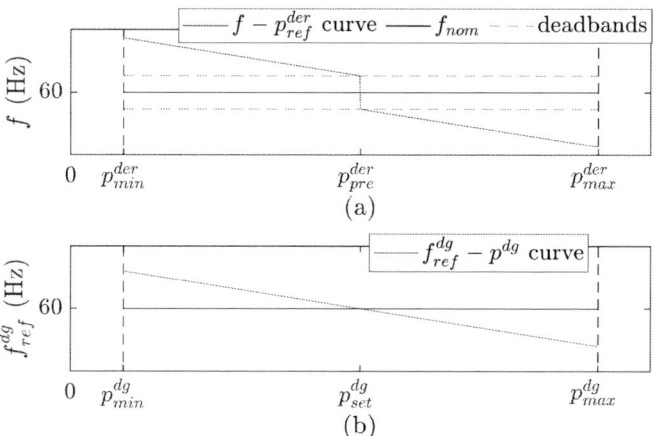

Fig. 2: Frequency-droop curves for (a) DER, and (b) DG.

The frequency-droop function for DG is expressed as:

$$f_{ref}^{dg} = f_{nom} - k^{dg}(p^{dg} - p_{set}^{dg}) \quad (2)$$

where f_{ref}^{dg} is the reference output frequency at the DG terminal, f_{nom} is the nominal frequency, p^{dg} is the DG's active output power, p_{set}^{dg} is the setpoint of the DG's output power given at output frequency of f_{nom}, and k^{dg} is the per-unit value of frequency change due to one unit change of DG's output power. The characteristic curve for DG's droop control is presented in Fig. 2(b). According to ISO 8528-5 std. [16], the DG is of type G2 for operation in the microgrid and some relevant parameters are also presented in Table II.

TABLE I: Parameters for DER Frequency-Droop Control [7]

Parameter	Default value	Allowable setting range
Dead-band (Hz)	0.036	$0.017 - 1.0$
Droop factor (%)	5	$2 - 5$
Time response (s)	5	$0.2 - 10$

978-1-6654-6060-6/23 $31.00 © 2023 IEEE

E. Restoration of System Frequency

The droop control enables the DER and DG to respond to dynamics and guarantee local stability, however, the systemwide service is still required for global frequency stability. Otherwise, the local droop control may malfunction and contribute to the frequency diversion, which will be presented in Fig. 3. As a result, the isolated MG urgently requires the service of frequency restoration.

TABLE II: Operating Limit Values of DG in G2 setting [16]

Parameter	Unit	Limit Values
Frequency droop	%	$\leqslant 5$
Steady-state frequency band	%	$\leqslant 1.5$
Rate of change of frequency setting	%/s	0.2 to 1
Frequency change for 100% power decrease	%	$\leqslant +12$
Frequency change for sudden power increase	%	$\leqslant -(10 + k^{dg})$
Frequency recovery time	s	$\leqslant 5$
Related frequency tolerance band	%	2

A frequency-restoration function based on an example concept from ISO 8528-5 std. [16] is implemented in the DG. A load change results in frequency deviation from the nominal value. Following a load change in the MG, the main idea of the frequency-restoration function for the DG is changing the DG's setpoint p^{dg}_{set} (power setpoint for nominal frequency) to its output power p^{dg} when the system enters a steady state. The frequency restoration process is defined as follows:

Step 1: When the fluctuation of the real-time measured system frequency exceeds a pre-defined threshold or the deadband defined as $db^{dg}_{uf/of}$, the modified droop control process to change the DG's setpoint p^{dg}_{set} is activated.

Step 2: When the system frequency enters a new steady state, i.e., the frequency oscillation stays within 0.1 Hz for ten seconds, the setpoint p^{dg}_{set} is set to the output power p^{dg}, therefore, this new setpoint corresponds to nominal frequency operation. As such, the system frequency will be restored to the rated value.

It should be noted that certain coordination of the restoration deadband parameters with the DER droop parameters are required for proper frequency restoration of the MG, e.g., $db^{dg}_{uf/of}$ should be equal to or smaller than $db^{der}_{uf/of}$. Otherwise, the DER will still undertake the risk of being operated out of the deadband for a long time, e.g., it will continue operating based on pre-disturbance power even though the available power could have increased in the meantime. Also, typically small remote MG are subjected to frequent frequency variation, therefore, the default values of DER droop with the default deadband of 0.036 Hz from IEEE 1514 [7] may cause rapid frequency support and may also cause undesirable consequences. Therefore, the deadband needs to be wide as well for the DER. In this study, the deadband for DER and frequency restoration is considered to be 0.1 Hz [12].

F. Parallel Operation for Two DGs

In the case that the load in the MG increase significantly over the capability of the DG and DERs, a second diesel generator (DG 2) should be connected to the MG to share the power generation with the primary one (DG 1). To facilitate the power-sharing between two DGs in the MG, a two-DG cooperation method is developed for an MG, which is defined as follows:

Step 1: DG 1 is always connected to the MG and is under operation to respond to the load changing and frequency variation.

Step 2: When the load is increasing, the output power of DG 1 (p^{dg1}) increases correspondingly. When p^{dg1} exceeds 95% of DG 1's rated power, the cooperation process is started and the cooperation operator is activated.

Step 3: The cooperation operator measures the output power of DG 1 and DG 2, p^{dg1} and p^{dg2}, and generates the setpoints for these two DGs, p^{dg1}_{set} and p^{dg2}_{set}. Assume DG 1 and DG 2 are of the same rated power, the two setpoints are determined as per the following step.

Step 4: When the frequency exceeds the deadband, then for the transient the setpoints are calculated based on the pre-disturbance power of the DGs $p^{dg1}_{set} = p^{dg2}_{set} = \frac{p^{dg1}_{pre} + p^{dg2}_{pre}}{2}$, where p^{dg1}_{pre} and p^{dg2}_{pre} are the output power of DG 1 and DG 2 when the frequency exceeds the deadbands, respectively.

Step 5: When the system frequency enters a new steady state, i.e., the frequency oscillation stays within 0.1 Hz for ten seconds, the final new setpoints are set based on the total loading of the DGs as $p^{dg1}_{set} = p^{dg2}_{set} = \frac{p^{dg1} + p^{dg1}}{2}$, which will restore the system frequency to the nominal value. Thus in the steady state, the two DGs will share the load equally.

III. SIMULATION RESULTS

The MG systems including the DER inverter and DGs are modeled in the Matlab-Simulink environment. Section III-A discusses the dynamic responses of the MG both with and without the frequency restoration function when one DG operates in parallel with the DER. In contrast, Section III-B presents the cooperation between two DGs alone, as well as their collaboration with the DER when different power limit values (for power reserve) are considered.

A. One DG with a DER

This section will show the importance of restoration of the DG, as presented in Fig. 3 and 4. Until 5 s, the input renewable energy resource power to the DER is zero, so the DG supplies the load ($p^{dg} = p_{load} = 45$ kW), and the system frequency is stable at 60 Hz. A load change in the MG will result in a frequency change. When this frequency change exceeds the deadband, the pre-disturbance power of the DER inverter will be zero. So, even though the resource available power could have increased to a value of rated power, the DER inverter will not be able to increase the output power at all. To show such misoperation, a programmable freeze time for the frequency droop function of DER was adopted in the model that helps during the start-up transients by keeping 60 Hz as input for the

978-1-6654-6060-6/23 $31.00 © 2023 IEEE

function and switches to measured frequency at the end of the freeze time. In this analysis, a freeze time of 20 s is considered for the following process. At 5 s, the load increases to 90 kW, and the input renewable energy increases to the rated value, which leads to a rated output power of 30 kW. During this process, the droop control of DER is still frozen/deactivated. Since the load change is higher than p^{der}, p^{dg} changes to 60 kW, and f drops to 59.5 Hz. At 20 s, the freeze time of DER droop control ends and the DER follows regular droop function based on the measured frequency. Instead of at least maintaining the maximum rated power, the droop control brings p^{der} down to 10.81 kW, which exaggerates the power unbalance, forces the DG to increase its output power, and further brings down the system frequency. The reason for this behavior of DER's droop control is that p^{der}_{pre} is constantly zero when f goes out of the deadband at 5 s. Since f never returns to the deadband from 5 s to 20 s, p^{der}_{pre} is never reset to $p^{der} = 30$ kW at this period. As a result, when the droop control is activated, p^{der}_{ref} is determined by (1) with $p^{der}_{pre} = 0$.

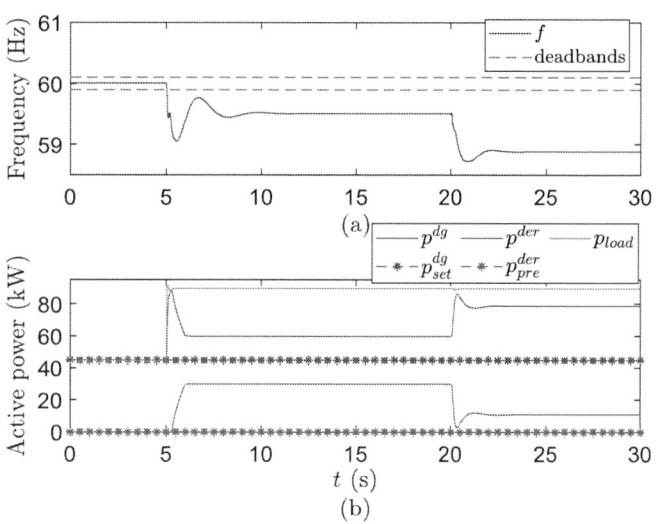

Fig. 3: Dynamic response without restoration.

Fig. 4 presents a more complex process showing how the DER's droop control can co-operate with DG's droop control and restoration function and cope with the load change. The same load and irradiance change are applied until 20 s. The frequency restoration method gets initiated after the load change and, it changes the setpoint of DG to operate the system at nominal frequency after ten seconds when the system frequency oscillates within 0.1 Hz. Thus it was restored around $t = 20$ s. For illustration purposes, the DER droop control is activated when the frequency restores within the deadband. Thus, the DER output power does not change. Then, the load decreases to 65 kW at $t = 25$ s. As a result of the load decrease, f increases beyond 60.1 Hz. Both p^{der} and p^{dg} decrease according to the droop control. After 10 seconds when the system frequency is stable, DG restores the system frequency to 60 Hz. Thus, after 30 s, p^{der} is restored to the maximum value again, and p^{dg} becomes 35 kW.

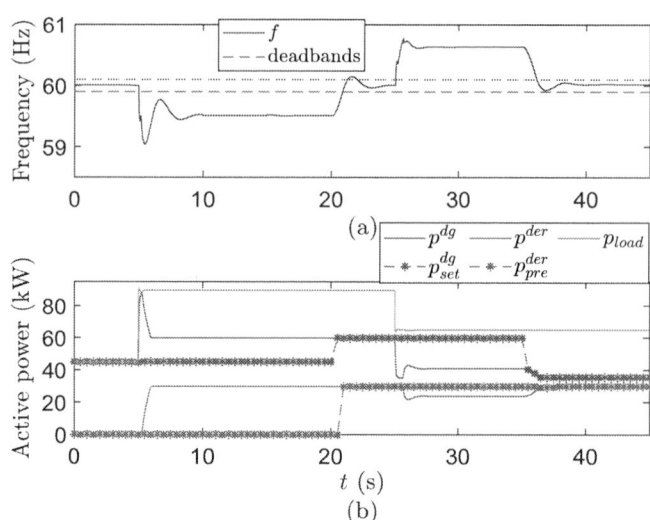

Fig. 4: Dynamic response with restoration.

B. Two DGs with a DER

Fig. 5 illustrates the cooperation (power sharing) of two DGs during significant load changes when there is no participation of DER. Initially, the load is 45 kW supplied by DG 1 ($p^{dg1} = p_{load} = 45$ kW), and the system frequency is stable at 60 Hz. At $t = 5$ s, the load increases from 45 kW to 140 kW, bringing the system frequency down out of the deadband of 60 ± 0.1 Hz and resulting in an increase of p^{dg1}. Detecting p^{dg1} exceeding 95% of DG 1's rating, the cooperation operator is activated and DG 2 is activated shortly after $t = 5$ s. Following to the cooperation algorithm, the setpoints p^{dg1}_{set} and p^{dg2}_{set} are set to half of the pre-cooperation p^{dg1}, which are $p^{dg1}_{set} = p^{dg2}_{set} = p^{dg1}_{pre}/2 = 45$ kW. From $t = 10$ s to 20 s, DG 1 and 2 share the same setpoint of 45 kW, output power of 70 kW, and output frequency of $f^{dg1}_{ref} = f^{dg2}_{ref} = f_{nom} - k^{dg}(p^{dg1} - p^{dg1}_{set}) = 59.18$ Hz. After 10 seconds when the system frequency is stable, p^{dg1}_{set} and p^{dg2}_{set} are set to 70 kW to bring the system frequency back to 60 Hz, as shown in Fig. 5(a) from $t = 20$ s to 30 s.

At $t = 30$ s, the load decreases from 140 kW to 110 kW, resulting in decreases of p^{dg1} and p^{dg2} to 55 kW and a frequency increase to 60.5 Hz. After 15 seconds of the load decrease, p^{dg1}_{set} and p^{dg2}_{set} are set to 55 kW, and the system frequency is restored to 60 Hz. During the process of load increase and decrease, DG 1 and DG 2 share the same setpoint, output power, and reference frequency. Besides, the frequency variation during the transients meets the requirement of ISO 8528 std. that the frequency variation during large transients should be within $\pm$ 6 Hz. As such, Fig. 5 justified the feasibility of the presented DG-cooperation algorithm.

IEEE 1547-2018 std. requires DER to include power limit capability. Fig. 6 illustrates the co-operated operation of two DGs and the DER inverter when the DER operates with a default power limit value of 1 pu (without the curtailment). In this scenario, the load is initially 75 kW and is supplied by DG 1 ($p^{dg1} = 45$ kW) and DER ($p^{der} = 30$ kW), and

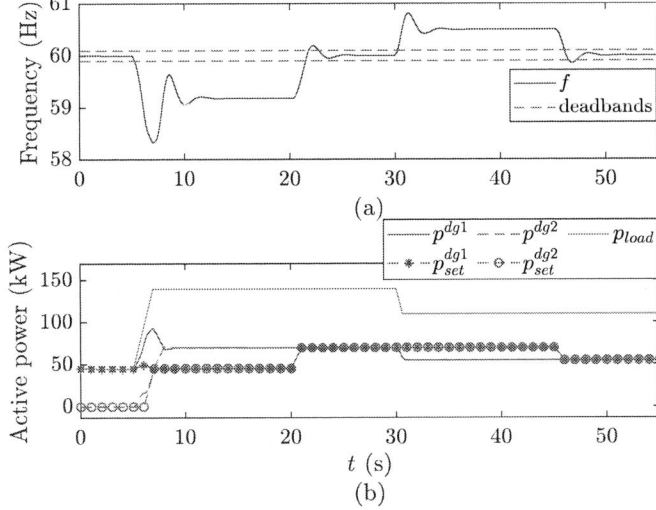

(a)

(b)

Fig. 5: Power-sharing between two DGs with no DER.

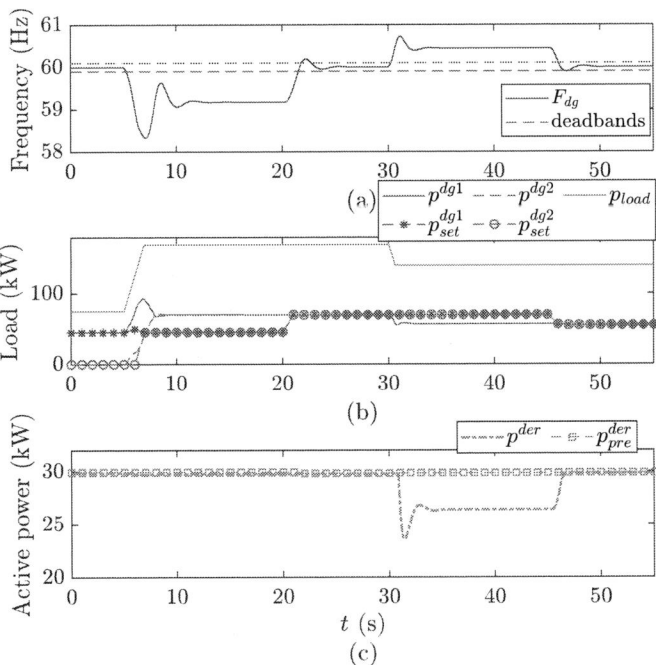

(a)

(b)

(c)

Fig. 6: Power-sharing between two DGs and a DER with no curtailment.

the system frequency is stable at 60 Hz. At $t = 5$ s, the load increases from 75 kW to 170 kW, bring the system frequency down and resulting in an increase of p^{dg1}. Then, DG 2 is activated to equally share the power with DG 1. Since the DER is operating at its maximal power, the DER does not contribute to the power sharing during the load increase. At the steady state from $t = 10$ s to 20 s, DG 1 and 2 share the output power of 70 kW, and the system frequency is 59.18 Hz before being restored to 60 Hz at $t = 20$ s.

At $t = 30$ s, the load decreases from 170 kW to 140 kW, resulting in decreases of p_{set}^{dg1} and p_{set}^{dg2} to 56.63 kW and p^{der}

to 26.74 kW. Since the DER participates in the power sharing, the power decrease of p_{set}^{dg1} and p_{set}^{dg2} is smaller than that in the preceding scenario, resulting in a smaller frequency increase to 60.44 Hz. By conclusion, the participation of an un-curtailed DER can help to reduce the frequency variation during a load decrease but not a load increase.

Fig. 7 and 8 illustrate the cooperation of two DGs and the DER inverter operating at capabilities (with power limit values) of 90% and 50%, respectively. In these two scenarios, the same load increase of 95 kW and load decrease of 30 kW occurs at $t = 5$ s and 30 s, respectively. For the system with a DER operating at 90% capacity, the steady state p^{der} is 30 kW and the system frequency is 59.23 Hz from $t = 10$ s to 20 s. From $t = 35$ s to 45 s, p^{der} and system frequency are 23.79 kW and 60.44 Hz, respectively. For the system with a DER operating at 50% capacity, the steady state p^{der} is 21.12 kW and the system frequency is 59.28 Hz from $t = 10$ s to 20 s. From $t = 35$ s to 45 s, p^{der} and system frequency are 11.55 kW and 60.44 Hz, respectively.

For the system where the DER has a smaller power limit value, e.g., operating at 50% capacity, p^{der} can participate more in the power sharing when the load increase and lead to a smaller frequency decrease. During the load decrease, since no DER reaches its minimal output power, the system frequency stays the same in the two scenarios, Fig. 7 and 8.

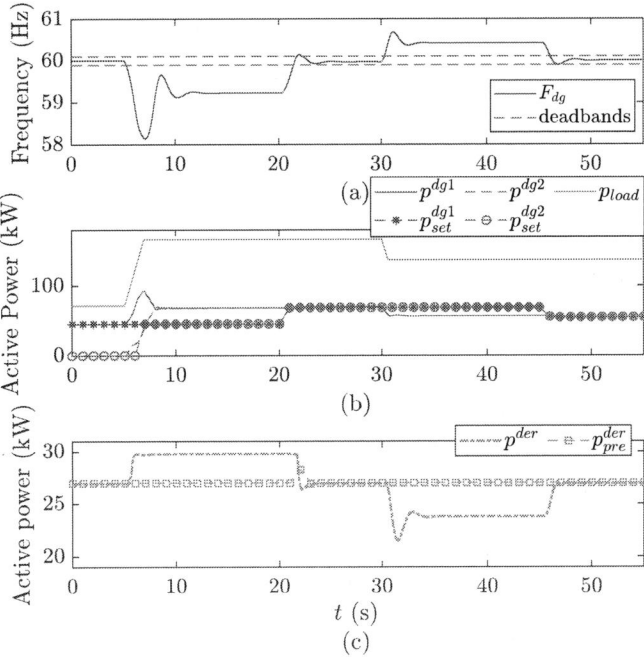

(a)

(b)

(c)

Fig. 7: Power-sharing between two DGs and a DER with a power limit of 0.9 pu (power limit value of 90%).

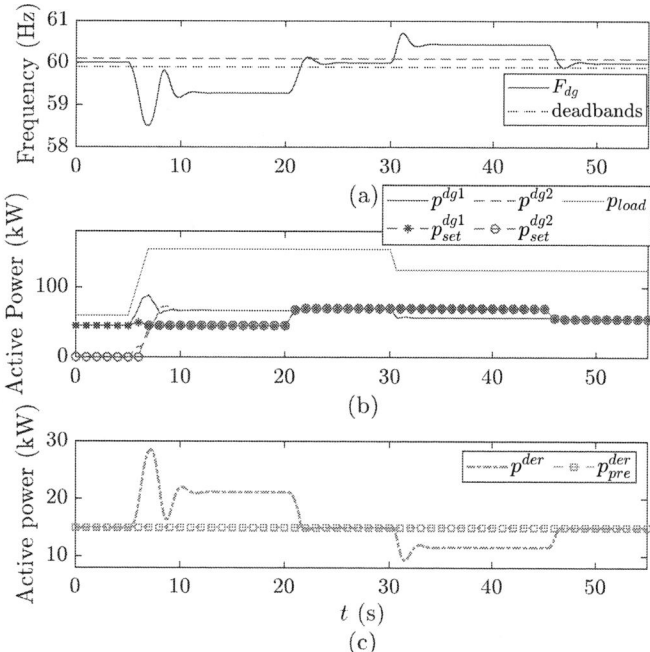

Fig. 8: Power-sharing between two DGs and a DER with a power limit of 0.5 pu (power limit value of 50%).

IV. CONCLUSION AND FUTURE WORK

This work assessed the frequency-droop control for DG and inverter-based DER in an isolated MG, the restoration of frequency in the MG, and the cooperation between DGs' and DER's droop control. For the assessment of the frequency-droop control for DG and DER, this work revealed the control's risks of malfunction during certain scenarios. If the system frequency of the MG is not restored to the rated value frequently, the droop control of the DER may malfunction and may present the worst case by decreasing active power during under-frequency events or vice versa. Moreover, the malfunctioning droop control may result in significant renewable energy waste. The work presented a frequency-restoration method to address the droop control's malfunction and the solution enhanced the power-sharing between the DER and DG. Simulation results show that the restoration method can restore the system frequency to the nominal value automatically in tens of seconds after a frequency transient. This work also investigated the coordination of the DG's and DER's droop parameters. The deadbands of DG's droop control are suggested to be equal to or smaller than DER's for higher dependability of frequency restoration. Furthermore, this work investigated the power sharing of multiple DGs and DERs in an MG and the effect of DERs' power limit value on power sharing. Simulation results show that the proposed power-sharing method between DGs can respond to transients of load changing quickly and stably and share the load equally. The investigation of DERs' power limit values shows that a smaller power limit can enhance the DER's ability to assist load shedding.

V. ACKNOWLEDGEMENT

Financial support for this research work was provided by the Government of Canada through the Program on Energy Research and Development (PERD) in the framework of the REN-2 Smart Grid and Microgrid Control for Resilient Power Systems Project.

REFERENCES

[1] IEA, "Renewables share of electricity generation in selected countries and regions, 2000-2020," 2020. [Online]. Available: https://www.iea.org/data-and-statistics/charts/renewables-share-of-electricity-generation-in-selected-countries-and-regions-2000-2020

[2] McKinsey, "Global Energy Perspective 2022 — McKinsey," apr 2022. [Online]. Available: https://www.mckinsey.com/industries/oil-and-gas/our-insights/global-energy-perspective-2022

[3] Government of Canada, "A clean electricity standard in support of a net-zero electricity sector: discussion paper - Canada.ca," mar 2022. [Online]. Available: https://www.canada.ca/en/environment-climate-change/services/canadian-environmental-protection-act-registry/achieving-net-zero-emissions-electricity-generation-discussion-paper.html

[4] ——, "Net-Zero Emissions by 2050 - Canada.ca," nov 2022. [Online]. Available: https://www.canada.ca/en/services/environment/weather/climatechange/climate-plan/net-zero-emissions-2050.html

[5] R. Tonkoski, L. Lopes, and D. Turcotte, "Active power curtailment of pv inverters in diesel hybrid mini-grids," in *2009 IEEE Electrical Power & Energy Conference (EPEC)*, 2009, pp. 1–6.

[6] N. A. Ninad and L. A. C. Lopes, "A bess control system for reducing fuel-consumption and maintenance costs of diesel-hybrid mini-grids with high penetration of renewables," in *2013 IEEE ECCE Asia Downunder*, 2013, pp. 409–415.

[7] IEEE, *IEEE Standard for Interconnection and Interoperability of Distributed Energy Resources with Associated Electric Power Systems Interfaces*, 1547th ed., apr 2018. [Online]. Available: https://standards.ieee.org/ieee/1547/5915/

[8] CSA, *Interconnection of distributed energy resources and electricity supply systems*, 2020. [Online]. Available: https://www.csagroup.org/store/product/CSAC22.3NO.9:20/

[9] N. Ninad, D. Turcotte, and M. Bui, "Assessment of the IEEE 1547-2018 Frequency-Droop Function for PV Inverter Operation," *Conference Record of the IEEE Photovoltaic Specialists Conference*, pp. 360–366, jun 2021.

[10] A. Hoke, M. Elkhatib, A. Nelson, J. Johnson, J. Tan, R. Mahmud, V. Gevorgian, J. Neely, C. Antonio, D. Arakawa, and K. Fong, "The Frequency-Watt Function: Simulation and Testing for the Hawaiian Electric Companies," 2017.

[11] E. Mohamed, N. Jason, and J. Jay, "Evaluation of Fast-Frequency Support Functions in High Penetration Isolated Power Systems — IEEE Conference Publication — IEEE Xplore," in *2017 IEEE 44th Photovoltaic Specialist Conference (PVSC)*, 2017. [Online]. Available: https://ieeexplore.ieee.org/document/8366506

[12] J. Johnson, J. C. Neely, J. J. Delhotal, and M. Lave, "Photovoltaic Frequency-Watt Curve Design for Frequency Regulation and Fast Contingency Reserves," *IEEE Journal of Photovoltaics*, vol. 6, no. 6, pp. 1611–1618, nov 2016.

[13] A. F. Hoke, M. Shirazi, S. Chakraborty, E. Muljadi, and D. Maksimovic, "Rapid active power control of photovoltaic systems for grid frequency support," *IEEE Journal of Emerging and Selected Topics in Power Electronics*, vol. 5, no. 3, pp. 1154–1163, 2017.

[14] F. Katiraei, D. Turcotte, A. Swingler, and J. Ayoub, "Modeling and dynamic analysis of a medium penetration pv-diesel mini-grid system," in *4th European Conference on PV-Hybrid and Mini-Grid*, Athens, Greece, May 2008.

[15] N. Ninad, J.-P. Bérard, and S. Q. Ali, "Smart inverter modeling toolbox for emt simulation studies of power systems," in *2021 IEEE Canadian Conference on Electrical and Computer Engineering (CCECE)*, 2021, pp. 1–6.

[16] ISO, "ISO - ISO 8528-5:2005 - Reciprocating internal combustion engine driven alternating current generating sets — Part 5: Generating sets," jul 2005. [Online]. Available: https://www.iso.org/standard/39047.html

978-1-6654-6060-6/23 $31.00 © 2023 IEEE

Probing Non-Equilibrium Hot Carrier Dynamics in Metal Halide Perovskite Solar Cells

Shashi Sourabh, Hadi Afshari, Vincent R. Whiteside, Giles E Eperon, Rebecca A. Scheidt, Varun N. Mapara, Madalina Furis, Matthew C Beard, Ian R. Sellers

University of Oklahoma, Norman, OK, United States

Swift Solar, San Carlos, CA, United States

NREL, Golden, CO, United States

The presence of hot carriers is presented in the operational properties of (FA,Cs)Pb(I, Br, Cl)3 solar cells at various temperatures and under practical solar concentration. At 100 K, clear evidence of hot carriers is observed in both the high energy tail of the photoluminescence spectra and from the appearance of a non-equilibrium like photocurrent at higher fluence in light J-V measurements. At room temperature, the presence of hot carriers in the emission is screened at high laser fluence as photo induced halide segregation begins to occur at higher lattice temperature. However, simultaneous J-V measurements indicate the continued presence of non-equilibrium carriers at 300 K. Finally, the presence of hot carriers is also presented in these devices in transient absorption measurements and the dynamics of the non-equilibrium carriers assessed at Voc, Jsc, and at the maximum power point.

On the influence of forced convection in PV energy yield models

Raed I. Bourisli[1], Bader S. Aldalali[2], Arttu Tuomiranta[3] and Jef Poortmans[3]

[1] Mechanical Engineering Department, Kuwait University, PO Box 5969, Safat 13060, Kuwait.
[2] Electrical Engineering Department, Kuwait University, PO Box 5969, Safat 13060, Kuwait.
[3] IMEC, Kapeldreef 75, B-3001 Leuven, Belgium.

Abstract — **The efficiency of PV cells is greatly affected by their operating temperatures. Depending on the locale, this change in efficiency can be significant. However, most energy yield frameworks only consider cooling by natural convection due to ambient conditions. What many of these frameworks neglect is the effect of forced convection on the temperature of the cells. In this study, PV modules are modeled numerically to quantify the effect of forced convection on the temperature and thus efficiency. The initial results show that for a typical summer day, the reduction in temperature due to forced convection—as opposed to natural convection alone—can exceed 12°C. The module efficiency increases from 15.25% to 16.53%. This 1.3% increase in efficiency is significant, and warrants reconsideration of forced convection in energy yield modeling.**

I. INTRODUCTION

The use of photovoltaic (PV) cells has become central in the recent drive to move away from fossil fuels and towards clean energy. A major form of clean, renewable energy is solar energy, where the design, operation and maintenance of the photovoltaic solar cells is at the heart of the discussion/problem. At the core of their efficacy lies their reported/assumed *efficiency*. The efficiency of any given PV cell, in turn, is usually dependent on its operating temperature. Reported efficiencies of PV cells are obtained by testing them at standard testing conditions (STC), that is, at an ambient temperature of 25°C. The rule-of-thumb is that for every 1 degree increase in cell temperature above STC, the efficiency drops by something between 0.375%/°C to 0.52%/°C of the reported value [1]. Such rules-of-thumb, in addition to the nature of heat dissipation from PV cells, are the heart of most energy yield frameworks in use today.

PV cells' temperature constantly increase due to solar irradiation. The main mechanism of heat dissipation from the cells, assumed in most energy yield models, is *natural convection*, often to stagnant atmospheric air at that STC. The resulting increase in temperature is factored in when computing the actual energy yield of modules. The assumed reduction in efficiency due to the elevated temperature of the modules rarely considers the effect of additional heat removal by forced convection due to atmospheric wind effects. In this study, we investigate the change in the assumed heat removal capacity of the surrounding air due to forced convection. The enhanced heat removal reduces the maximal temperatures of the cells in a PV module and thus increases its overall efficiency.

To quantify the influence of wind on PV energy yield, a windy location that experiences high ambient temperatures is ideal. The desert of Kuwait is used since it meets both criteria.

II. BASE CASE DESCRIPTION

The base case is a single crystalline silicon, bifacial PV module composed of six 0.16-by-0.16-m cells with 22.5° inclination, on top of each other, due south. The electrical properties of each cell are shown in Table 1.

TABLE 1
ELECTRICAL PROPERTIES OF PHOTOVOLTAIC CELLS

Parameter	Value
V_{oc}	0.675 V
I_{sc}	11.354 A
V_{max}	0.571 V
I_{max}	10.799 A

The base case considers fluid flow around the module in stagnant air—that is, only natural convection is considered. The setup mimics the one used in the energy yield framework developed by IMEC, Belgium, where only natural convection is assumed as the heat loss mechanism from modules. The energy yield framework has already been validated for desert climates using onsite experimental data. More information on IMEC's energy yield framework can be found in ref. [2].

The problem is modeled using the finite element model and validated with the results of the framework for the sample case conditions. Numerically, the flow in the fluid is governed by the Navier-Stokes and energy equations,

$$\frac{\partial \rho}{\partial t} + \nabla \cdot (\rho \boldsymbol{u}) = 0 \tag{1}$$

$$\frac{\partial (\rho \boldsymbol{u})}{\partial t} + \nabla \cdot (\rho \boldsymbol{u} \times \boldsymbol{u}) = -\nabla p + \mu \nabla \cdot (\nabla \boldsymbol{u} + (\nabla \boldsymbol{u})^T) + \rho \boldsymbol{f} \tag{2}$$

$$\frac{\partial (\rho c_p T)}{\partial t} + \boldsymbol{u} \cdot \nabla (c_p T) = \frac{\partial p}{\partial t} + \boldsymbol{u} \cdot \nabla p - \nabla \cdot \boldsymbol{q} \tag{3}$$

where viscous dissipation is neglected in eq. (3). Within the cells, the flow of energy is governed by the 3D heat equation in stationary media.

For the hydrodynamic and thermal boundary conditions, the 90th-percentiles of wind speed and ambient air temperature are used, which are approximately 42.63°C and 3.82 m/s, respectively. The heat flux applied at the upper surface of the cells was 682.7 W/m², which is balance of the irradiance minus the power generated per unit area. Due to reflection from the ground, about 10% of this value is applied at the bottom surface as well. The hydrodynamic velocity only applies to the forced convection case. The dry bulb temperature at that hour was 42.63°C. Only natural convection is used for this part of the base case calculations. The complementary part of the base case is the forced convection problem.

III. PRELIMINARY RESULTS

A. Validation of the Natural Convection Case

The natural convection case takes about 5 machine hours to converge to a steady-state solution. Due to the turbulence incurred by the eddies accumulating and ejecting from the ends of the module, the surface temperature converges to a cyclic stead state. This average temperature at the surface comes out to be 72.52°C, which is very close to the 72.67°C that comes from the energy yield framework for the parallel case.

B. Natural Convection Flow and Energy Contours

A temperature field at some instant in the 15 minutes simulated, with only natural convection, is shown in Figure 1 below. It is seen that there is a significant build-up of heat. The vortices accumulate gradually and are shed periodically, either off the top or the bottom of the module.

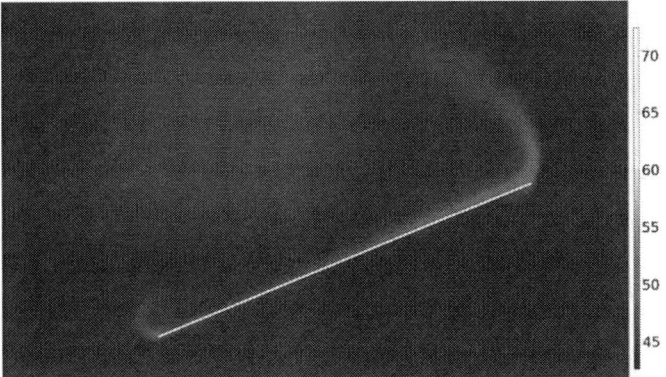

Fig. 1. Temperature field halfway through the 15 minutes simulated time span of the natural convection case. (Numbers in °C.)

C. Forced Convection Flow and Energy Contours

As expected, the forced convection case results in significantly different fields. On one hand, the temperature of the module is seen to be reduced significantly; the average surface temperature dropped from 72.52 to 59.71°C. On the other hand, a large recirculation zone behind the module is present. The heat buildup causes a slight increase in temperature of the module, but it does not seem to cancel the effect of the overall enhanced convection. Figures 2 and 3 show the energy and flow fields, respectively.

IV. IMPLICATIONS AND FUTURE WORK

The above results confirm the significance of forced convection when it comes to improving the efficiency of PV modules [3]. This importance cannot be disregarded in thermal analyses of solar systems, much less in energy yield frameworks used the world over. Forced convection will yield lower temperatures, which result in higher efficiencies and larger energy yields. For example, for the simulated conditions above, the reduction in module temperature is 12.81 degrees Celsius. A reduction of 0.32%/°C is assumed in the used framework, causing a 20%-efficient cell to reduce in efficiency to 16.96% when no forced convection is considered. However, if the simulated drop in temperature is considered, the efficiency is 17.78%. This 0.82 increase in efficiency is not insignificant.

The other significant takeaway from the above is the existence of a threshold, in terms of ambient temperature and ambient wind velocity, above which forced convection must be considered to render more realistic estimates of the total energy yield of modules.

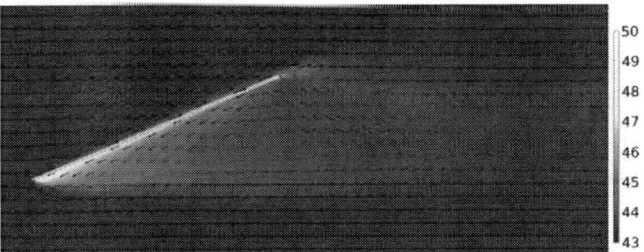

Fig. 2. Temperature field at the end of the forced convection case simulation. Note that the color range was limited to 50°C to show more demarcation. (Numbers in °C.)

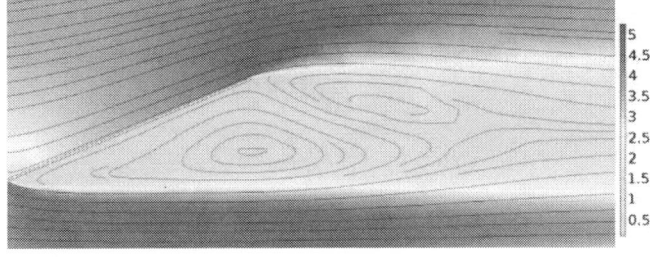

Fig. 3. Velocity flow field at the end of the forced convection case simulation. Note that the color range was limited to 50°C to show more demarcation. (Numbers in m/s.)

The goal of this study is to quantify the limits of such thresholds. The ultimate goal, however, is to come up with an empirical correlation, to be embedded in energy yield frameworks, which takes into account these modes of heat loss and their effect on the temperature and thereby the efficiency of the modules. The aim is to provide a formula that designers can consult, given the location, wind, and ambient temperature data, to "correct" for forced convection. A better estimate of the yearly energy yield will result.

REFERENCES

[1] M. Akhsassi *et al.*, "Experimental investigation and modeling of the thermal behavior of a solar PV module," *Solar Energy Materials and Solar Cells*, vol. 180, pp. 271–279, Jun. 2018, doi: 10.1016/j.solmat.2017.06.052.

[2] I. T. Horváth *et al.*, "Photovoltaic energy yield modelling under desert and moderate climates: What-if exploration of different cell technologies," *Solar Energy*, vol. 173, no. July, pp. 728–739, 2018, doi: 10.1016/j.solener.2018.07.079.

[3] J. K. Kaldellis *et al.*, "Temperature and wind speed impact on the efficiency of PV installations. Experience obtained from outdoor measurements in Greece," *Renewable Energy*, vol. 66, June, pp. 612-624, 2014, doi.org/10.1016/j.renene.2013.12.041.

2-Terminal and 3-Terminal Subcell Characterization Platforms for Emerging Tandems

Jin Young Kim

Department of Materials Science and Engineering, Research Institute of Advanced Materials, Seoul National University, Seoul, South Korea

Significant improvements in performance has been reported for emerging tandem solar cells like the monolithic perovskite/Si tandem solar cell, the lack of simple electrical characterization methods impedes further development and successful commercialization of them. In this study, we report comprehensive electrical characterization methods based on either 2-terminal or 3-terminal platform that can accurately and easily analyze the subcells of monolithic emerging tandems. In addition to the current-voltage and external quantum efficiency measurements, advanced characterization methods for subcells such as impedance spectroscopy and thermal admittance spectroscopy will be demonstrated.

Fast cell detection and distortion correction for outdoor electroluminescence images

Evgenii Sovetkin[1†], Bart E. Pieters[1], Andreas Gerber[1], Liviu Stoicescu[2] and Pascal Koelblin[3]

[1]IEK-5 Photovoltaics, Forschungszentrum Jülich, 52425 Jülich, Germany
[2]Solarzentrum Stuttgart GmbH, 70197 Stuttgart, Germany
[3]Institute for Photovoltaics and Research Center SCoPE, University of Stuttgart, 70569 Stuttgart, Germany

Abstract—**This paper proposes a fast and robust method for electroluminescence image preprocessing, where lens and perspective distortions are corrected, and individual cells in the module are detected. Our approach works with low-resolution (640 × 512 pixels) images, uses an image-to-image translation neural network, and leverages the geometric properties of a photovoltaic module. The fast computational speed of the neural network allows us to complete image analysis in under 0.5 seconds, which is ten times faster than currently published methods. In addition, the geometry-based postprocessing makes our approach robust to small misdetections in the neural network output.**

Index Terms—**electroluminescence, pix2pix neural network, distortion correction**

I. INTRODUCTION

The International Energy Agency (IEA) forecasts that solar energy will have at least 27% of the global share of energy production by 2050 [1]. With the increasing photovoltaic (PV) production capacity, there is a growing demand for monitoring and early-fault detection to minimize PV plant losses.

Time series data of inverters and strings may indicate a potential fault; however, for a thorough investigation, spatially-resolved measurements (imaging) provide a deeper understanding of the module's health and allows for diagnosing faults and defects, which is important for warranty or insurance claims.

With the increase in the installed PV capacity and the emergence of automated drone-based measurements, fast and real-time image processing algorithms for module and cell detection will become crucial for automated plant inspections.

In particular, camera and perspective distortion corrections and module and cell detection are essential routines for processing in-field-collected images. These algorithms are important as these processing steps are always required and applied before other analysis methods are used.

Previous studies of such algorithms are based on a series of image-processing filters. First, the original image undergoes a series of transformations, amplifying PV module characteristics such as grid lines and busbars. Those characteristics are then used to correct for lens and perspective distortions and detect individual cells. Sovetkin et al. [2] use Hough Transform to find short pieces of straight lines, followed by robust linear regression predicting the slope change of those lines hence estimating the parameters of the perspective distortion. Deitsch et al. [3] use tensor voting to amplify the line features that

point in a similar direction and subpixel-accuracy routine to compute the location of cell corners. Kölblin et al. [4] detect cell gaps using an iterative linear Hough transform applied on an adaptive locally thresholded image.

This paper replaces image-processing filters with a trained image-to-image translation neural network, significantly reducing the runtime. Furthermore, we leverage the typical PV module's geometrical properties to overcome misdetections in the neural network and improve the algorithm's robustness.

II. METHODOLOGY

Our image correction approach consists of four steps. In the first step, we apply the trained image-to-image neural network that takes as input the EL image and outputs a binary image of a cell grid (see Section II-B and Figure 2). This grid hints at the location of individual cells, and the neural network is trained on the data obtained from the previously published method (Section II-A). In the second step, we apply several routines that rank cell corners and the corresponding cell locations within the module (Section II-C). The ranking of the cells corresponds to the quality of detection. In the third step, the accuracy of the cell corner coordinates is improved with a subpixel-accuracy method (Section II-D). Lastly, we use the best cell corners to either estimate the lens distortion parameters or correct the perspective distortion (Section II-E).

The flowchart in Figure 1 illustrates our approach pipeline. The original image can be used either to accumulate information to estimate the camera distortion matrix or to estimate homography and extract individual cell images.

A. Data

We develop our methods using the daylight electroluminescence (EL) images collected with the DaySy system of Solarzentrum Stuttgart, GmbH [5]. We use a low-resolution InGaAs sensor with a 640 × 512 pixels resolution. Our study uses data from several commercial PV fields in Germany containing several types of c-Si modules. Figure 2 (top) shows an example of an EL image.

We utilize the methods proposed in [4] to generate training data, which works well with our low-resolution images. First, we generate a binary grid image for every EL image using the computed cell data (see Figure 2, bottom). Then, we manually clear the training dataset, removing faulty images. For this

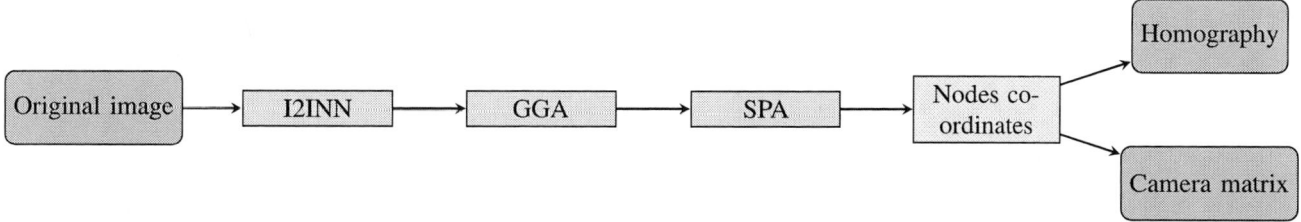

Fig. 1. A flowchart describing our approach pipeline. The original image is processed by an image-to-image neural network (I2INN, Section II-B). The graph-geometric analysis (GGA) is applied to rank the cell corner quality (Section II-C). The accuracy of the node coordinates is improved with the subpixel-accuracy method (Section II-D). The nodes resulting node coordinates are then used to estimate homography or a camera matrix (Section II-E)

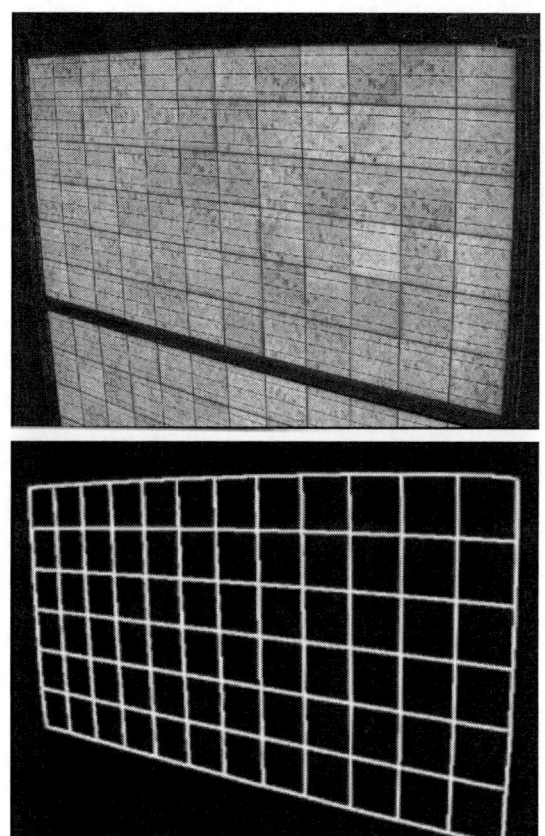

Fig. 2. Above: an EL image with 640×512 pixels resolution. Below: output of the neural network

paper, we use a sample 1800 number of modules to train the neural network.

B. Image-to-image neural network (I2INN)

We use the Pix2Pix [6] neural network for the image-to-image neural network (I2INN). First, we prepared a training dataset discussed in Section II-A. Then, the training images are augmented using a random perspective transformation, shift, symmetry reflection, contrast scaling, and random noise. Lastly, the resulting image is resized to have a 256×256 pixels resolution. From 1800 module images, we generate a training dataset with 126000 images. The training uses NVIDIA

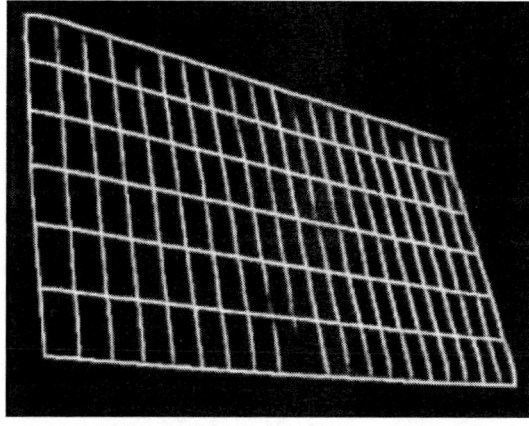

Fig. 3. An example of a faulty output of the neural network. The method is applied to an out-of-sample image of a half-cell PV module

GeForce GTX 970 graphic card for 20 epochs. In the Pix2Pix architecture, we experimented with several different generators and discriminators, with all networks yielding similar results.

The output of an I2INN is prone to small misidentification especially applied to new module types or modules with severe defects. For example, some grid lines may not be detected, or several cells may be merged to one area into the grid image (see Figure 3). Therefore, postprocessing is required to improve the robustness of the method.

C. Graph geometric analysis (GGA)

This paper utilizes the geometrical properties of the PV module structure to discover any potential misidentification from the neural network. Geometrically speaking, we expect a binary grid to consist of a grid of distorted rectangles, where each inner rectangle has precisely eight neighbors sharing a vertex, every rectangle on the module border has five neighbors, and every corner cell has three neighbors.

To analyze the structure of the binary grid, we focus on its graph-theoretic properties. To this end, we first apply the image thinning operator. Then, the resulting image can be treated as a pixel graph P, where nodes of the graph are non-zero pixels, and there is an edge between nodes if two pixels are neighbors. We compute pixel graph using the `pixel_graph` routine in skimage [7].

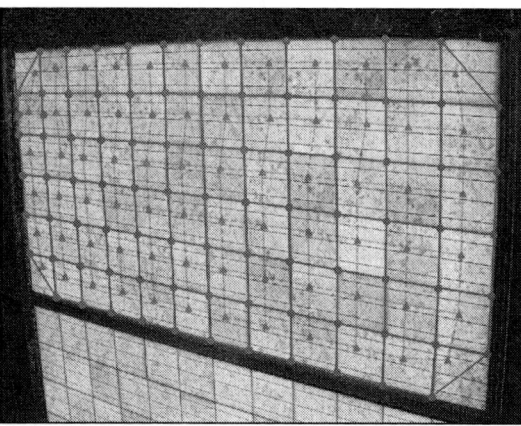

Fig. 4. An EL image with graph S (green nodes and edges) and graph $S^\star$ (red colour triangles with dashed edges)

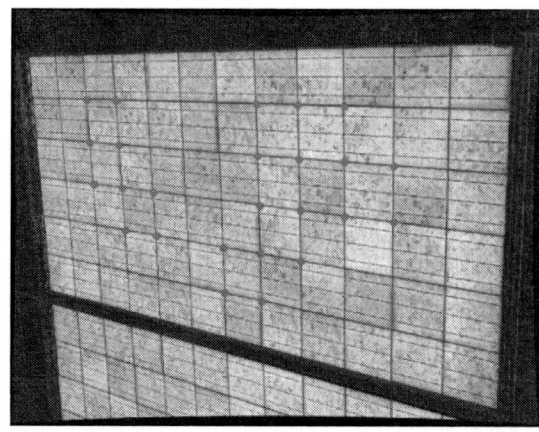

Fig. 5. Best points (red) are used to either the estimate camera matrix or correct homography (need at least 4 points)

The number of nodes in graph P can be further reduced, as most nodes have degree two with both edges oriented in the same direction. By applying a version of the breath-first-search algorithm, we obtain a weighted graph S, where every node has a degree of at least 3, and the weight of each edge corresponds to the length of the path connecting two nodes in the original graph P. We remark that the nodes of graph S are not a subset of graph P, as we average the coordinates of closely located nodes in P with a degree of at least 3. Such averaging guarantees that all our edges do not intersect one another, and graph S remains planar (with pixel coordinates being an embedding).

Further, since graph S is planar, the dual graph $S^\star$ is well-defined. Recall that the dual graph's nodes are the faces of the original graph and an edge in $S^\star$ connects two faces if they have a common edge in S. The dual graph is beneficial to us as it captures all the necessary geometric structures about the grid. Figure 4 depicts a module image and the corresponding graph S (green dots connected with lines) and dual graph $S^\star$ (red triangles connected with dashed lines).

We use a version of the depth-first-search algorithm to compute the dual graph $S^\star$. During the computation, we calculate various geometric quantities, such as the area of each face, circumference, minimal enclosing parallelogram, its edge ratio and area. We compute minimal enclosing parallelograms using the algorithm from [8].

The depth-first search algorithm identifies a face in graph S by walking around this face in a counter-clockwise direction. Apart from individual cell faces, there are outer faces when we walk around the module boundary in the clockwise direction. We define the area of the outer face to be a negative value, which equals to the module's area. We select the largest component when multiple modules are in a single image.

To avoid problems with the I2INN output inaccuracies (such as shown in Figure 3), we introduce a ranking of the nodes of the dual graph. This ranking aims to select the best faces with respect to some heuristic rule. We use geometric characteristics of the graph S faces and compare those values with the values of its neighbors. Our heuristic is based on the fact that in the PV module all cells are identical and similar in a distorted image.

To this end, every face in graph $S^\star$ comes with a set of qualitative characteristics $v_i \in \mathbb{R}^d$. Then, for each face $i \in S^\star$, we calculate the Mahalanobis distance d_M to the neighboring faces $N(i) \subset S^\star$ sharing a vertex with i:

$$w_i := \{d_M(v_i, v_j)\}_{j \in N(i)}, \quad i \in S^\star,$$

where

$$d_M(x, y) := \sqrt{(x - y)^T \Sigma^{-1}(x - y)}.$$

and the covariance matrix Σ is estimated using a set of training images.

In our current implementation, our vector v_i is three-dimensional and consists of the area of the face (normalized by the dimension of the image), the area ratio between the face and the area of the minimal enclosing parallelogram, and the ratio of the sides of the parallelogram.

A q-quantile, $w_i^{(q)} \in \mathbb{R}, i \in S^\star$ ($q > 50\%$), is computed on a sample of those distances. This quantile demonstrates how much face i differs from *most* of its neighbors. Note, here we consider two faces are neighbors when they share at least one node (i.e., each face inside a module has eight neighbors). We expect $w_i^{(q)}$ to be small if most neighbors are close in its characteristic to the face $i \in S^\star$. The ranking of faces is then performed according to the ordering of $w_i^{(q)}, i \in S^\star$.

We select the first half of the ordered list of faces. Then, the selected faces are reordered so that the first face has the highest rank; the second face is furthest away from the first, and so on. Such a strategy allows the distribution of the selected points throughout the image, improving homography estimation. Figure 5 demonstrates an example of selected faces.

Additionally, the dual graph determines the module *square lattice coordinates*. For that, we choose any node $i \in S^\star$ with degree 4 and assign to this face a square lattice co-ordinate $(0, 0)$. The nodes below and above the initial face get coordinates $(0, −1)$ and $(0, 1)$, respectively, nodes on the left and the right get coordinates $(−1, 0)$ and $(1, 0)$. Walking over the complete grid allows us to determine a square lattice coordinate for each node in $S^\star$ with degree 4. Furthermore, the square lattice coordinates' range determines the detected module's size.

D. Subpixel accuracy correction (SPA)

The coordinate of the discovered nodes is accurate up to a few pixels. Improving the accuracy is beneficial for the camera distortion matrix and homography estimation. Therefore in the last step, we apply a subpixel-accuracy coordinate correction method, a procedure similar to the one used in [3].

To this end, for each node, we sample the original EL image in the direction of the edge. The image is then reduced to a vector, where each coordinate equals the sum of a column in the sampled image. Figure 6 shows the resulting vector with a red line. Since the image is sampled towards the direction of the graph S edge, the resulting vector is expected to contain a single peak corresponding to the cell border in the module. The x-axis is given relative to the original node coordinate.

To estimate the location of the shift, we use the following parametric model:

$$m(x; A, B, \mu, \sigma) = A + B \exp\left(\frac{(x − \mu)^2}{\sigma^2}\right). \quad (1)$$

where parameter μ describes the desired shift value.

Unlike the subpixel-accuracy routine in [3], model (1) is a shifted Gaussian bell; hence fast Gaussian model fits like [9] are inapplicable in our scenario. Therefore, we rely on a generic curve-fitting method based on the trust-region optimization method [10]. We use the implementation from `curve_fit` in the scipy library [11]. The blue line in Figure 6 shows an example of the model fit.

Among the parameters of the model (1), the σ and its variance of the estimator can be utilized to identify problematic fits. Namely, we ignore any nodes with $\sigma^2 > 0$ and $\mathrm{Var}(\hat{\sigma}^2) > 10$. Such thresholding allows ignoring curves where peaks are not prominent or, in the scenario when a dark cell makes the bell asymmetric.

Optimization is performed twice for each node: in vertical and horizontal directions. The resulting parameters μ and the direction of the edges define the correction shift vector, with which the node's coordinate is adjusted.

E. Camera and homography corrections

The resulting cell corner and square lattice coordinates are used to estimate the camera matrix and perform a lens distortion correction or image homography. In addition, the square lattice coordinates allow us to determine the number of rows and columns in the module and extract those cells according to the square lattice grid.

For camera calibration and homography estimation, we use routines implemented in the OpenCV library [12]. Namely, `calibrateCamera` [13] and `findHomography`. The

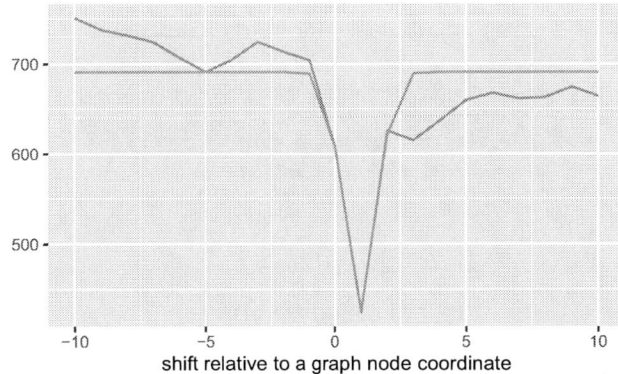

Fig. 6. Subpixel-accuracy coordinate correction method to improve point locations. Sample average pixel intensity of the original image in the direction on graph edges (red) and fit a parametric model (blue)

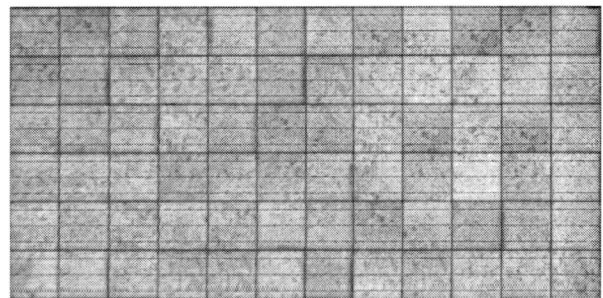

Fig. 7. Corrected image

camera matrix is estimated from identified coordinates from several images.

Lastly, since we use only a subset of good nodes for estimating homography, we can verify the detection with the remaining nodes. Namely, we apply the corresponding homography transformation and compute the root-mean-square error (RMSE) with the excepted square lattice coordinates. If the RMSE is above a chosen value, we report an error for the given image.

III. RESULTS

Figure 7 depicts the desired corrected image for the module in Figure 2. Here we selected each cell to have a dimension of 100×100 pixels as those modules have square cells. The resulting image dimension equals 1200×600. We can easily extract individual cell images by shifting a window within an image.

Several factors contribute to the robustness of our method. Firstly, the ability for neural networks to generalize to new data, our I2INN step can be applied to new types of modules. Furthermore, graph-based geometric property analysis allows filtering potential misdetections in the grid images. Again, our approach can handle images with multiple modules, as all our routines are applied for each component. Moreover, the thresholding of covariances of the estimator in the SPA routines allows us to ignore nodes where cells are defective.

978-1-6654-6060-6/23 $31.00 © 2023 IEEE

TABLE I

I2INN: TIME REQUIRED DEPENDS ON THE DEVICE

device	time
gpu:jetson	70 ms
cpu:jetson	400 ms
gpu:nvidia gv 102	5 ms
cpu:intel xeon w-2123	70 ms

TABLE II

OTHER STEP: RUNS ON 1-CPU PER IMAGE, SUBPIXEL CORRECTION PER NODE (MEASURED ON INTEL XEON W-2123). IO STANDS FOR INPUT/OUTPUT IMAGE OPERATION

procedure	time
IO	60 ms
GGA	89 ms
SPA	7 ms
lens correction	5 ms
homography estimation	4 ms

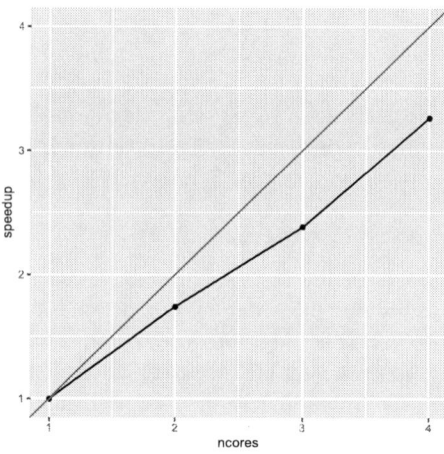

Fig. 8. Comparison of pipeline speedup with the theoretical one (red line). Benchmark using Intel Xeon W-2123

Lastly, even after all those steps, the module is not accurately detected; computation RMSE with all the nodes allows for identifying such failures.

We benchmark our pipeline on several devices to demonstrate our approach's computational performance. First, we use the NVIDIA Jetson machine, a lightweight computer with a GPU module for potential in-field applications. Further, we use the Intel Xeon W-2123-based machine for large-scale processing with the NVIDIA GV 102 graphic card. For all results, we process images in maximum batches, device memory permits.

Table I compares the performance of the I2INN step for different devices. Here, we have an option of running the routine on CPU or GPU devices for each of the machines. On the Jetson, we used batches of size 16; on a larger machine, we used batches of size 256.

Table II shows the running time of the GGA step. This step has no parallel implementation; hence each image is processed on a single CPU. Here the subpixel correction routine timing is given per node. The homography requires calling routine at least four times, whereas, for the camera distortion matrix computation, we need 20–30 nodes.

The complete pipeline can be computed on multiple CPUs for processing many images in batches. Figure 8 demonstrates how our parallel implementation of batch processing deviates from the ideal linear scaling measured on the Intel Xeon W-2123 machine. Table III provides timings for the complete pipeline on small and large machines. The memory requirement of the pipeline depends on the batch size. For instance, a batch size 256 requires up to 5.6 GB of memory on an Intel Xeon W-2123 machine.

TABLE III

COMPLETE PIPELINE USING 4 CPU

machine	time/image
intel xeon + gpu	125 ms/image
nvidia jetson	527 ms/image

These performance benchmark values demonstrate a significant improvement upon the previously published model, for which the running time ranges from 6 s [4] to 360 s [3, 2] per image. The performance of our method can be further improved by implementing graph algorithms in a fast-compiled language (the current implementation is in Python), and the input/output can be further optimized. However, we expect a running time to be at most 60 ms per image.

To estimate the accuracy of the proposed approach, we compute the RMSE values for the camera calibration and homography correction routines. First, we estimate the camera distortion matrix with the iterative aruco-based calibration method suitable for low-resolution images [14]. Then we calculate the camera distortion matrix with our approach based on the 320 EL images. Then we project every pixel coordinate of the image using those two camera matrices and compute an RMSE of 2.9 pixels between the two projections.

For homography accuracy, we compare the cell coordinates between our method and the results of the method published in [4]. Based on the data from 1500 images, we obtained an RMSE of 3.2 pixels. We remark that the method in [4] does not perform subpixel-accuracy coordinate correction, hence neither of the methods can be considered as ground truth. Figure 9 shows that visually, the SPA routine gives a more accurate position of the cell coordinate (red points) compared to the green points obtained from [4].

We remark that the proposed approach is not restricted to EL images, for example, can be applied to drone-based infrared images or EL for complete module identification. Figure 10 depicts individual module identification in low-resolution IR images, where the camera distortion matrix is estimated from a sample of 13 images.

IV. SUMMARY

Image data volumes will grow with the emergence of drone-based automatic PV plant inspection. Therefore, automatizing module inspection and defect detection require fast and reliable image processing algorithms.

Fig. 9. Comparison of cell corners between [4] (green points) and our method (red points). The SPA in our approach visually improves the node estimation

Fig. 10. The approach can be applied to IR/EL drone-based images. Image source: Aerial PV Inspection GmbH

This paper proposed a method for camera and homography distortion corrections and cell extraction in EL images. By leveraging the computational speed of neural networks and the geometric properties of a PV module, we improved the running times of such preprocessing algorithms by a factor of 10. Generally, our approach applies to any images containing regular patterns.

ACKNOWLEDGMENT

This work is supported by the "PARK3" project (Förderkennzeichen: 03EE1104) and "ReliaREN-Pro" project (Förderkennzeichen: 03EI4052B). The authors would like to thank Solarzentrum Stuttgart GmbH for providing the data.

REFERENCES

[1] IEA. *Net Zero by 2050.* 2021. URL: https://www.iea.org/reports/net-zero-by-2050.

[2] E. Sovetkin and A. Steland. "Automatic processing and solar cell detection in photovoltaic electroluminescence images". In: *Integrated Computer-Aided Engineering* 26.2 (2019), pp. 123–137. DOI: 10.3233/ICA-180588.

[3] S. Deitsch et al. "Segmentation of photovoltaic module cells in uncalibrated electroluminescence images". In: *Machine Vision and Applications* 32.4 (2021), p. 84. DOI: 10.1007/s00138-021-01191-9.

[4] P. Kölblin, A. Bartler, and M. Füller. "Image Preprocessing for Outdoor Luminescence Inspection of Large Photovoltaic Parks". In: *Energies* 14.9 (2021), p. 2508. DOI: 10.3390/en14092508.

[5] L. Stoicescu, M. Reuter, and J. Werner. "DaySy: luminescence imaging of PV modules in daylight". In: *29th European Photovoltaic Solar Energy Conference and Exhibition.* 2014, pp. 2553–2554. DOI: 10.4229/EUPVSEC20142014-5DO.16.2.

[6] P. Isola, J.-Y. Zhu, T. Zhou, and A. A. Efros. "Image-to-image translation with conditional adversarial networks". In: *Proceedings of the IEEE Conference on Computer Vision and Pattern Recognition (CVPR).* 2017, pp. 1125–1134.

[7] S. van der Walt et al. "scikit-image: image processing in Python". In: *PeerJ* 2 (June 2014), e453. DOI: 10.7717/peerj.453.

[8] C. Schwarz, J. Teich, E. Welzl, and B. Evans. "On finding a minimal enclosing parallelogram". In: *International Computer Science Institute, Berkeley, CA, Tech. Rep. tr-94-036* (1994).

[9] E. Pastuchová and M. Zákopčan. "Comparison of algorithms for fitting a gaussian function used in testing smart sensors". In: *Journal of Electrical Engineering* 66.3 (2015), pp. 178–181. DOI: 10.2478/jee-2015-0029.

[10] M. A. Branch, T. F. Coleman, and Y. Li. "A subspace, interior, and conjugate gradient method for large-scale bound-constrained minimization problems". In: *SIAM Journal on Scientific Computing* 21.1 (1999), pp. 1–23. DOI: 10.1137/S1064827595289108.

[11] P. Virtanen et al. "SciPy 1.0: Fundamental Algorithms for Scientific Computing in Python". In: *Nature Methods* 17 (2020), pp. 261–272. DOI: 10.1038/s41592-019-0686-2.

[12] Itseez. *Open Source Computer Vision Library.* https://github.com/itseez/opencv. 2015.

[13] Z. Zhang. "A flexible new technique for camera calibration". In: *IEEE Transactions on pattern analysis and machine intelligence* 22.11 (2000), pp. 1330–1334.

[14] S. Schramm, J. Ebert, J. Rangel, R. Schmoll, and A. Kroll. "Iterative feature detection of a coded checkerboard target for the geometric calibration of infrared cameras". In: *Journal of Sensors and Sensor Systems* 10.2 (2021), pp. 207–218. DOI: 10.5194/jsss-10-207-2021.

Limiting Factors on the Performenace of Luminescent Solar Concentrators for Building Integrated Photovoltaics

Bryce S. Richards, Ian A. Howard

Institute of Microstructure Technology, Karlsruhe Institute of Technology, Karlsruhe, Germany

Light Technology Institute Technology, Karlsruhe Institute of Technology, Karlsruhe, Germany

This paper examines the application of luminescent solar concentrators (LSCs) for building integrated photovoltaics (BPIV) with two possible implementations being considered: i) opaque façade elements and ii) semi-transparent window elements with ~50% visible light transmission. Here a large number of ray-tracing simulations are conducted to illustrate the technical challenges to maintaining efficiency when scaling LSCs from the typical lab-scale devices (< 25 cm2) to pilot-scale (1000 cm2) and ultimatey commercial-scale (10 m2) BPIV elements. The key variables investigated via ray-tracing relate to i) the LSC waveguide itself (area, total host attenuation) and ii) the chosen luminophores (absorbance, emission peak, quantum yield, degree of overlap between absorption and emission spectra). Which of these factors restricts energy conversion efficiency the most will be discussed in detail. Based on the results, the target efficiencies for opaque and semi-transparent LSCs based are proposed: i) 11.0% and 5.5% opaque and semi-transparent for 25 cm2 devices; ii) 9.0% and 4.5% for 1000 cm2 pilot-scale modules; and iii) 8.0% and 4.0% for 10 m2 modules. It is difficult to see opaque LSCs successfully competing against standard flat-plate PV modules in the BIPV market. However, the application of semi-transparent LSCs as power-generating window elements could have potential for BIPV and an economic analysis of the inclusion of LSCs into commercial glazing elements is also presented.

978-1-6654-6060-6/23 $31.00 © 2023 IEEE

3D printed Transparent Sheet for Solar Panel Encapsulation and Thermal Management

Fahad Alam, Nazek El-Atab

King Abdullah University of Science and Technology, Thuwal, Saudi Arabia

The current work demonstrates the potential of 3D-printed polymer sheets for the encapsulation of solar panels for the purpose of prevention and thermal management. The sheets were prepared via a vat-photopolymerization-based 3D printing technique utilizing a photocurable polymer. The optimized sheet showed ˇ92% transparency with a UV wavelength blocking capability from the range of 200-400 nm. A reduction in around 5 °C in the temperature from the surface of the solar panel was observed when the solar panel was exposed to a solar simulator for 5 min as compared to the surface temperature from the bare panel. The 3D printed sheet is expected to act as the encapsulant and aid in the thermal management of the solar panel.

Testing the Durability of Fluorine-Free Hydrophobic Coatings vs Porous Silica

Luke O. Jones[1,2], Adam M. Law[1], Gary Critchlow[2], and John M. Law[1]

[1]Centre for Renewable Energy Systems Technology (CREST), Wolfson School of Mechanical, Electrical and Manufacturing Engineering, [2]Department of Materials, Loughborough University, Loughborough, Leicestershire LE11 3TU, UK

Abstract — Current porous SiO_2 anti-reflective coatings used on solar modules have recently been found to be susceptible to mechanical abrasion and environmental degradation. A fluorine-free polymeric anti-soiling coating is tested as an alternative coating to porous SiO_2. Two hydrophobic coatings, porous SiO_2, and a glass control were subjected to a linear abrasion test under 5N of force for 50 cycles. The porous SiO_2 demonstrated higher initial optical transmittance and reflectance but degraded by -0.7 T% and +1.6 R% after 50 cycles of abrasion. The hydrophobic coating with SiO_2 nanoparticles demonstrated resistant to abrasion with 0 T% and +0.1 R% increase, outperforming the control glass. The nanoparticle filled hydrophobic coating also retained its hydrophobic properties after abrasion with a WCA of 96° which is beneficial in their role as an anti-soiling coating.

I. INTRODUCTION

A major issue faced with solar production is the issue of soiling on the solar cover glass. Soiling is the accumulation of dirt, dust, debris and biological matter on the surface of a solar panel cover glass which attenuates the light into the solar cell, overall decreasing efficiency over the panel's lifetime. The reduction of transmitted light into the solar cell will reduce the current density and therefore power output [1]. Soiling losses depend on the geolocation of the solar utility but can decrease by 0.01% a day in Europe and North America and reach as high as 1-2% a day in arid and / or dusty areas such as India, East Asia, and the Middle East and North Africa (MENA) [2].

The glass cover sheet provides important functions for the solar module such as protecting the solar cell from environmental degradation, physical damage, water ingress, and chemical interactions. The glass sheet must also yield high optical transmittance and low reflectance to allow light to pass through to the cells underneath. For this reason, solar panels in the field typically feature a porous SiO_2 (n = 1.27 at 550 nm) anti-reflective (AR) coating on the top side to decrease the optical reflectance of the cover glass due to lower refractive index compared to soda-lime glass (n = 1.52 at 550 nm) whilst retaining mechanical strength and chemical inertness [3]. However, these AR coatings, typically applied via sol-gel process, have been found to be vulnerable to degradation in high humidity environments due to water ingress that fills the air voids in the porous structure [4], increasing the refractive index due to water replacing air in the voids. Additionally, porous SiO_2 is vulnerable to mechanical abrasion that occurs during regular site cleaning which removes the porous layer; observed on modules in the field [5, 6]. The high

surface energy of porous SiO_2 creates a hydrophilic surface that is likely to attract and adhere dust to the surface which reduces the power output of the module over its operational lifespan. One proposed method to mitigate this issue is to implement an anti-soiling coating, such as a hydrophobic coating.

Hydrophobic coatings are a simple and effective way to reduce the effects of soiling. Optically clear, hydrophobic coatings decrease the wettability of the cover glass. This combined with high water contact angles (WCA), greater than 90°, and low roll-off angles (RoA), ideally less than 30°, create a 'self-cleaning' surface [7]. Cohesive forces within the liquid cause the drop to move freely on the coating surface collecting, dislodging, and removing soiling particles on the surface of the coating, maintaining high optical transmittance during operation. Different chemistries can be used to achieve this effect [8], however, typical chemistries used for hydrophobic coatings, which are optically clear, durable, chemically resistant, and environmentally stable are fluoropolymer based; and with recent climate goals proposed by COP27 targeting Polyfluoroalkyl Substances (PFAS) [9], industries are looking to move towards fluorine free alternatives.

In this work, the durability of a multi-layer fluorine free polymer-based hydrophobic coating was tested against a porous SiO_2 AR coating that is used on commercial solar panels. Abrasion tests were conducted to test the mechanical durability of each type of coating. Optical properties of the coatings, contact angle, and surface chemistry were observed during the tests to understand the benefits and weaknesses of using a fluorine-free coatings over a porous SiO_2 AR coating. Further durability testing is planned that will incorporate dew-dust-drying soiling tests, UV accelerated ageing, damp heat accelerated ageing, salt-spray, and thermal cycling to test the resistance to each type of environmental degradation.

II. EXPERIMENTAL

A. Sample Preparation

Two polymeric hydrophobic coatings were prepared: a Polydimethylsiloxane-based (PDMS) multilayer compound (coating A), and a second PDMS-based coating with 2.5 wt% of functionalized SiO_2 nanoparticles to increase hardness and add surface nano-texture (coating B). Coatings were applied using a wipe-on wipe-off method. After application of each layer, the coating was left to partial-cure for 30 minutes. A total of 4 layers were applied for each coating. After the final layer was added, the coatings were left to cure for 24 hours. The samples were then split for characterisation and performance testing.

5cm×10cm samples of commercially available porous SiO_2 AR-coated glass were obtained for testing. Such coatings typically have a thickness between 100 nm and 150 nm. Commercially available soda-lime glass slides were used as a control.

B. Sample Testing

Surface chemical analysis was conducted using X-ray photoelectron spectroscopy (XPS). All spectra were recorded using a Thermo Scientific K-Alpha XPS with a monochromated Aluminium Kα source. The sample surface was characterised using survey scans and high-resolution scans with the following parameters: Surveys were measured using 100 eV pass energy, 1 eV step size, 10 ms dwell time averaged over 10 Scans. High-resolution scans were measured using 50 eV pass energy, 0.1 eV step size, 50 ms dwell time averaged over 5 scans for each element of interest. All results were charge corrected to carbon C1s at 284.6 eV (PDMS peak).

Sample optical performance was measure using transmission and reflection measurements taken across a wavelength range of 200-1200 nm using a Varian Cary5000 UV-VIS spectrophotometer with an integrating sphere attachment. Contact angle measurements were made using the static sessile drop method with a Dataphysics OCA50 ESr-N contact angle measurement system. Roll of angle (RoA) was measured using a custom built RoA rig.

Abrasion testing was performed using a Elcometer 1720 abrasion tester. Samples were subject to both a 20 and 50 dry abrasion cycle at a rate of 37 strokes per minute, where 1 cycle constitutes a complete forward and backward stroke of the machine. The abrasive material used was denoted as an "abrasive pad" and follows the ISO 11998 standard. A complete stroke length was performed across the entire breadth of the sample, with a weight of 5N applied. Abraded samples were cleaned with compressed air following abrasion to remove any loose debris on the surface.

III. RESULTS & DISCUSSION

A. Sample Characterisation

After coatings A and B had fully cured, each coating had a non-uniform haze in some areas specifically around the sample edges, however, visible sample clarity was quite high. Coating B did contain small amounts of unadhered, agglomerated SiO_2 nanoparticles on the surface. XPS was used to identify the surface elements. Initial observation using survey scans revealed the coatings contained C, Si, and O which is typical of PDMS chemistry. Additionally, small amounts Na, Mg and F was found as contaminants, with Na and Mg being common surface contaminants, whilst F is likely to have come from the polytetrafluoroethylene (PTFE) trays used to hold the samples. Fluorine contamination was minimal and detected in limited areas of the sample.

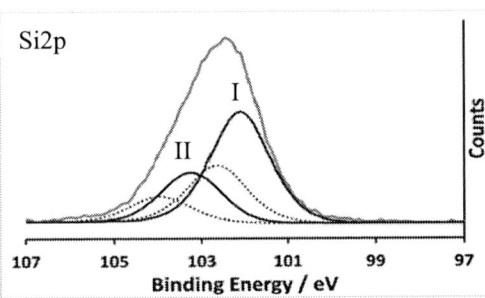

Fig. 1: Si2p spectra of B after deposition. Due to the spin-split orbit of the p-shell electrons, solid peaks represent the Si2p3 peaks, whilst the dotted peaks represent the Si2p1 electrons.

Analysing the C1s and Si2p high resolution (HR) spectra we can see both coating A and B have a peak at 284.6 eV for C1s and 102.0 eV (I) for Si2p [10], which are typical peak binding energy values for PDMS due to the influence of Si on the C-C bonds, as opposed to 284.8 eV for C-C / C-H bonds in pure hydrocarbon-based polymers. The difference between coating A and B can be seen in the O1s and Si2p spectra where there is a higher presence of SiO_2 At% at 103.1 eV (II) in coating B due to the presence of SiO_2 nanoparticles (Fig. 1).

B. Abrasion Testing

The transmittance and reflection for each sample was measured across a wavelength range of 200-1200 nm. The transmission and reflection measurements can be seen in (Fig 2). At higher wavelengths (>800 nm) Coating A and B have higher transmittance than the soda-lime glass substrate but decreases at lower wavelengths. Additionally, the hydrophobic coatings demonstrate lower reflectance across all wavelengths compares to the control glass sample.

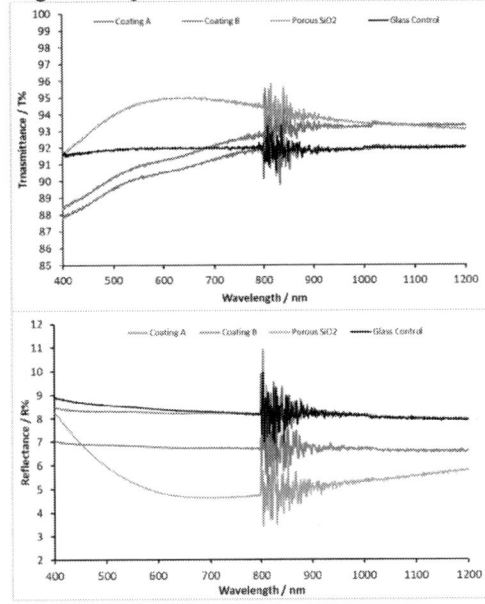

Fig. 2: (top) Transmittance and (bottom) reflectance of as deposited hydrophobic coatings A and B compared to porous SiO_2 Ar coating and soda-lime glass control. The feature between 800 nm and 900 nm is due to detector crossover from the Cary5000 UV-VIS spectrophotometer.

The transmittance of the hydrophobic coatings and control glass are similar; however, the reflectance is lower due to the lower refractive index of PDMS (n = 1.43 at 600 nm), this exhibits an anti-reflective property when the polymer is applied to a soda-lime substrate.

The weight average transmittance and reflectance values (WAT & WAR) for 0, 20 and 50 cycles are shown in Table 1 for all coatings. Both Coating A and porous SiO_2 reduce in WAT after 50 cycles of abrasion, suggesting mechanical abrasion caused by the abrasive pad. However, the porous SiO_2 coating reduces from 94.0 T% to 93.3 T% which is still greater than control glass, indicating at this level of abrasion it can still function as an AR coating, whereas Coating A reduced from 92.8 T% to 91.5 T% which is similar to the glass control, indicating the coating may have been removed. Coating B starts with a WAT lower than coating A, likely due to the incorporation of nanoparticles of silica which would increase the refractive index compared to the virgin polymer, however, coating B shows lno change in WAT even after 50 cycles of abrasion, suggesting the nanoparticles of silica improve the abrasion resistance of the coating.

The WAR values follow a similar trend to the WAT values, with Coating A and porous SiO_2 increasing in reflectance from 6.8 and 5.4 R% to 8.3 and 7.0 R% respectively. Coating B demonstrates negligible change in reflectance at 8.2 R% which suggests the nano-SiO_2 increased mechanical resistance compared to Coating A. With the porous SiO_2 showing signs of degradation after 50 cycles, further cycles could suggest complete removal of the coating, therefore losing the benefit provided by the AR properties of the porous layer.

Table 1: WAT and WAR for each sample over 0, 20, and 50 cycles of abrasion.

	Transmittance / T%			Reflectance / R%		
Abrasion Cycle	0	20	50	0	20	50
Coating A	92.8	91.8	91.5	6.8	8.1	8.3
Coating B	91.9	91.6	91.9	8.2	8.2	8.3
Porous SiO_2	94.0	93.3	93.3	5.4	6.3	7.0
Control Glass	91.9	91.4	91.9	8.3	8.6	8.6

C. Surface Wettability

Coating A and B achieved a WCA of 112.4° and 113.1° respectively. The presence of the SiO_2 nanoparticles in coating B likely increased the WCA due to the surface nano-texturing. After 50 cycles of abrasion, coating A had neutral wettability, whilst coating B maintained its hydrophobic properties, indicating both that the polymeric coating remained on the sample and the SiO_2 nanoparticles increased abrasion resistance. Porous SiO_2 and glass started with a hydrophilic surface but increased WCA after 20 cycles with the control glass achieving a WCA of 90° after 20 cycles, likely due to the abrader creating nanotexturing on the surface, creating a Wenzel surface, as supported by the high RoA.

The RoA for all samples increased to over 30° by 50 cycles of abrasion which limits their ability to function as a passive self-cleaning coating, however it should be noted that coating A and B performed well after 20 cycles and can be reapplied in the field by cleaners, whilst the porous SiO_2 cannot be repaired once damaged.

Table 2: WCA and RoA for each sample over 0, 20, and 50 cycles of abrasion.

	WCA / °			RoA / °		
Abrasion Cycle	0	20	50	0	20	50
Coating A	112.4	102.6	83.0	25.4	30.2	36.9
Coating B	113.1	105.5	96.0	30.7	18.5	55.6
Porous SiO_2	10.4	56.4	53.0	19.6	36.2	40.0
Control Glass	14.2	90.2	61.5	23.8	83.8	54.1

IV. Conclusion

Two hydrophobic coatings were prepared on Soda-lime glass substrates and compared to porous SiO_2 and a glass control sample. Coating A and B demonstrated lower optical transmittance and higher reflectance to porous SiO_2 however, after 50 cycles of abrasion testing coating A and porous SiO_2 degraded by ~-0.5 T% and +1.5 R% respectively. Coating B demonstrated resistance to abrasion testing, showing no change in transmittance and reflectance, whilst also retaining hydrophobic properties with a WCA over 90°. Further degradation testing is required to compare the coatings durability to environmental degradation including UV exposure, damp heat exposure, salt spray, dust, and thermal cycling.

References

[1] K. K. Ilse, B. W. Figgis, V. Naumann, C. Hagendorf and J. Bagdahn, "Fundamentals of soiling processes on photovoltaic modules," Renewable and Sustainable Energy Reviews, vol. 98, pp. 239-254, 2018.

[2] M. Mahamudul Hasan Mithhu, T. Ahmed Rima and M. Ryyan Khan, "Global analysis of optimal cleaning cycle and profit of soiling affected solar panels," Applied Energy, vol. 285, p. 116436, 2021.

[3] A. M. Law, F. Bukhari, L. O. Jones, A. Abbas and J. M. Walls, "Testing the Abrasion Resistance of Porous SiO2 Anti-reflection Coatings for Solar Cover Glass," 2022 IEEE 49th Photovoltaics Specialists Conference (PVSC), pp. 0786-0791, 2022.

[4] G. Womack, K. Isbilir, F. Lisco, G. Durand, A. Taylor and J. M. Walls, "The performance and durability of single-layer sol-gel anti-reflection coatings applied to solar module cover glass," Surface and Coatings Technology, vol. 358, pp. 76-83, 2019.

[5] A. M. Law, F. Bukhari, L. O. Jones, P. J. M. Isherwood and J. M. Walls, "Multilayer Antireflection Coatings for Cover Glass on Silicon Solar Modules," IEEE Journal of Photovoltaics, vol. 12, no. 5, pp. 1205-1210, 2022.

[6] D. C. Miller, et al. "The Abrasion of Photovoltaic Glass: A Comparison of the Effects of Natural and Artificial Aging," IEEE Journal of Photovoltaics, vol. 10, no. 1, pp. 173-180, 2020.

[7] G. Hassan, B. Yilbas, A. Al-Sharafi and H. Al-Qahtani, "Self-cleaning of a Hydrophobic Surface by a Rolling Water Droplet," Scientific Reports, vol. 1, no. 9, pp. 1-14, 2019.

[8] L. O. Jones, A. M. Law, G. Critchlow and J. M. Walls, "Comparing Fluorinated and Non-Fluorinated Anti-Soiling Coatings for Solar Panel Cover Glass," 2022 IEEE 49th Photovoltaics Specialists Conference (PVSC), p. 0683, 2022.

[9] L. Moosmann, et al. "The COP27 Climate Change Conference: Status of Climate Negotiations and Issues at Stake," ENVI Committee, Luxembourg, 2022.

[10] NIST X-ray Photoelectron Spectroscopy Database, "NIST Standard Reference Database Number 20," National Institute of Standards and Technology, 2000. [Online]. [Accessed 15 January 202

Investigation of the Microstructure of Underdense Hydrogenated Amorphous Silicon Layers for Silicon Heterojunction Solar Cells by Raman Spectroscopy and Hydrogen Effusion

Benedikt Fischer, Maurice Nuys, Andreas Lambertz, Weiyuan Duan, Kaining Ding, Uwe Rau

Forschungszentrum Jülich GmbH, Jülich, Germany

The application of thin underdense hydrogenated amorphous silicon (a-Si:H) films for passivation of crystalline Si (c-Si) and for avoiding epitaxy in silicon heterojunction (SHJ) solar cell technology has recently been proposed and successfully applied. Here, we investigate the microstructure of such underdense a-Si:H films, as used in Jülich solar cell technology. From the refractive index and hydrogen content, the density of our films is estimated to 2.1-2.2 g/cm3. In H effusion experiments (besides effusion at higher temperature) a low temperature H effusion peak near 400 °C shows up which has been attributed to the diffusion of molecular H2 through a void network. The dependence of the H effusion peaks on film thickness is similar as observed previously for low substrate temperature a-Si:H material. By applying different plasma power (0.08 W/cm2 - 0.16 W/cm2) the Si-H microstructure parameter measured by Raman decreases from about 0.5 to 0.25 for about 20 nm film thickness. The nucleation zone is estimated from the Raman results to < 10 nm. The substrate type (HF etched c-Si, c-Si with native oxide, glass) shows no influence on the Raman microstructure parameter. The fact that with such material good passivation of c-Si solar cells was achieved suggests that in the c-Si passivation process molecular hydrogen plays an important role. The second layer of dense a-Si:H as usually required for good c-Si passivation may be primarily a H2 out-diffusion barrier.

978-1-6654-6060-6/23 $31.00 © 2023 IEEE

Impact of Irradiation-induced Filter Temperature Increase on Calibration of Reference Solar Cells with NIR-Longpass Filters

Tao Song, Larry Ottoson, Rafell Williams, John Geisz, Charles Mack, Jeremy Brewer, Nikos Kopidakis

National Renewable Energy Laboratory, Golden, CO, United States

Reference solar cells are essential to the performance rating of photovoltaic devices. Si, KG-filtered Si, and GaAs reference cells are the common reference cell types used in the calibration chain of mainstream PV technologies. For emerging tandem PV technologies such as, GaAs/Si, CdTe/Si, and perovskite/Si, the spectral responses of their bottom junctions are only in near-infrared (NIR) region. Hence, these conventional reference cells are no longer suitable for their performance rating as their poor spectral matching can introduce large measurement errors. Si reference cells with colored glass NIR-longpass filters are preferrable as they have a much better spectral matching. However, this type of reference cells shows ~4-8 times larger Isc variation than other conventional reference cells in our primary calibration process. It is found that the cut-on wavelength of NIR-longpass filters (e.g., RG715 and 850 nm Schott filters) assembled on these reference cells is temperature-sensitive. In a 5-min calibration period, the longpass filters can be gradually heated up by over 5 °C due to the absorption of visible light and accordingly their cut-on wavelength increases by ~1-3 nm. As a result, their measured Isc gradually decreases by ~0.5-1.0%, causing additional large measurement uncertainty in the primary calibration. We also show that actively cooling the filters and/or choosing thinner filters are effective approaches to reduce the Isc variation to a ~0.2% level as they enable the NIR filters to reach a stabilized temperature faster.

978-1-6654-6060-6/23 $31.00 © 2023 IEEE

Investigation of high nitrogen composition SiN$_x$ for textured front surface passivation of n-type silicon solar cells in terms of light stability of injected negative charge and cell performance

Kwan Hong Min[1], Jeong-Mo Hwang[2], Christopher Chen[2], Wook-Jin Choi[1], Vijaykumar D Upadhyaya[1], Brian Rounsaville[1], Ajeet Rohatgi[1] and Young-Woo Ok[1]

[1]Georgia Institute of Technology, Atlanta, GA 30332, USA

[2]Inert Plasma Charging LLC, Chandler, AZ 85286, USA

Abstract— In this paper, we explored the potential of passivating the B-doped emitter with SiN$_x$ film(s) with an externally injected negative charge by using our plasma charge injection technology to replace the more costly Al$_2$O$_3$ passivation technology, by addressing some issues and challenges in the light stability for light exposure. At first, we investigated the stability of injected negative charge and passivation quality of single layer SiN$_x$ films with low and high-x (x~1.01 and 1.30) as well as dual-x (x~1.30/1.01) SiN$_x$ films on B doped emitter. About ~1E13 cm^{-2} negative charge was injected into single- and dual-x SiN$_x$ layers by a plasma charge injection system. It was found that a single SiN$_x$ film with x~1.01 could not hold injected negative charges after several hours of exposure to simulated visible sunlight. However, excellent charge stability was confirmed in an 80 nm thick single SiN$_x$ layer with high x~1.30, which was able to retain more than ~ 2E12 cm^{-2} negative charge density after the extrapolated 50K hours (~25 years) exposure to sunlight. For the dual-x SiN$_x$ stacks with a total thickness of ~80 nm, the charge stability got even better with the increasing thickness of the high-x SiN$_x$. In parallel, we investigated the passivation quality of the single layer SiN$_x$ and dual-x SiN$_x$ stacks on the B emitter before charge injection. The uncharged low-x single SiN$_x$ layer showed very good passivation both before and after metallization with iV$_{oc}$ ~680mV. However, the uncharged single layer SiN$_x$ with high-x showed inferior passivation quality before metallization and significant degradation after metallization with iV$_{oc}$ ~590mV. In contrast, the dual-x SiN$_x$ stacks showed comparable passivation quality to the single layer low-x SiN$_x$ (x~1.01). Passivation properties of the charged SiN$_x$ films are being studied now and will be reported in the paper with cell performance.

Keywords—Passivation, SiN$_x$, negative charge injection, Boron doped emitter, Al$_2$O$_3$ passivation

I. INTRODUCTION

The fixed charge in a dielectric layer on silicon surface is known to induce field-effect passivation by creating band bending in Si surface. Typically, the field-effect passivation of p-type silicon surface or boron doped emitter in Si solar cells is accomplished by depositing aluminum oxide (Al$_2$O$_3$) which contains a high fixed negative charge density [1]. However, Trimethyl-aluminum (TMA) material, which is the precursor to depositing Al$_2$O$_3$ film, has high operational cost and safety concerns. Likewise, silicon nitride (SiN$_x$) films deposited by plasma-enhanced chemical vapor deposition (PECVD) are widely used for passivating n-type Si surfaces due to their fixed positive charge. However, the low-cost PECVD SiN$_x$ is not effective in passivating the p-type Si because high positive charge density in SiN$_x$ can induce an inversion layer and cause parasitic shunting, resulting in degradation of cell performance [2]. Therefore, if we can introduce built-in a negative charge into SiN$_x$ by electron charge injection, it can be used for passivating p-type wafers and B emitters, which can overcome the cost and safety drawbacks of Al$_2$O$_3$. In a previous study, Hwang et al. demonstrated the high negative charge injection into SiN$_x$ films using plasma charge injection [3]. In addition, we demonstrated performance enhancement by introducing negative charge injection into the rear SiN$_x$ layer of bifacial p-PERC and in the B emitter of n-PERT solar cells [4-7]. Recently, Chen et al. reported that the negative charge injected thicker SiN$_x$ film (~150 nm) with x-value (x=N/Si≥1.29) has excellent charging stability after the exposure AM1.5 simulation visible light and full-spectrum light at temperatures ranging from 55 to 78 °C for up to 300 h. [8].

In this work, we investigated the injected negative charge stability and challenges for the single thin SiN$_x$ (~80 nm) and dual-x SiN$_x$ (~80 nm, x~1.30/1.01) and explored the effective lifetime degradation during after high temperature firing process with and without metallization on the boron emitter. Passivation properties of negatively charged thin SiN$_x$ films are under investigation. Further optimized SiN$_x$ with cell performance will be reported in the final paper.

EXPERIMENTAL DETAIL

First, test samples with an oxide/nitride/oxide (ONO) stack on the planar surface were prepared to investigate the injected negative charge stability. Single layer SiN$_x$ films (80nm, x~1.30 and 1.01) and dual-x SiN$_x$ (x~1.30/1.01) stack films with a total thickness of ~80 nm were grown on top of thermally grown SiO$_2$ (~10 nm) followed by PECVD-grown SiO$_x$ as a top layer to form an ONO stack as shown in figure 1(a). The rear side received the phosphorus diffusion process with screen-printed Ag metal contact fired through SiO$_2$/SiN$_x$ stack passivation. Rear metal ohmic contact can improve the uniformity of negative charge injection on the front ONO stacks. The negative charge injection into these samples was performed using a plasma charging system with optimized charging conditions described in a previous publication [3]. The charge density in the ONO stack was determined by Capacitance-Voltage (C-V) using a mercury probe. For the testing of injected charge stability, the test samples were

978-1-6654-6060-6/23 $31.00 ©2023 IEEE

exposed to the simulated visible sunlight followed by C-V measurement.

To investigate the passivation properties of single- and dual-x SiN$_x$ films before charge injection on the B emitter, a test cell structure was prepared with the same ONO stacks on the B emitter with a sheet resistance of 120$\Omega/\square$ as shown in figure 1(b). The passivation quality was measured by the quasi-steady state photoconductance (QSSPC) technique after SiN$_x$ deposition and simulated contact firing with no metal. At the same time, some samples were screen-printed Ag on only B emitter followed by the same firing process to investigate the effect of metal contact on the cell performance. This was done after removing the metal bulk by HCl/H$_2$O$_2$/DI mixed solution but leaving behind embedded metal crystallites in Si.

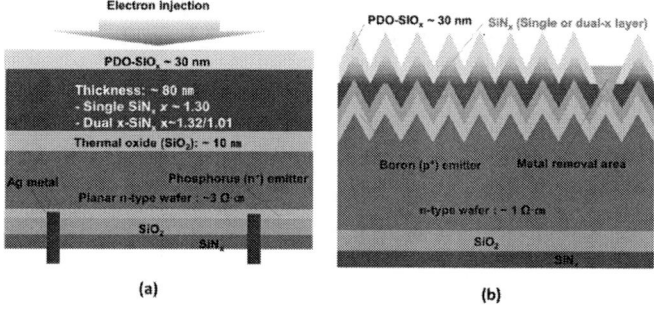

(a)   (b)

Figure 1. Schematic sample structure; (a) ONO stack structure for the injected negative charge stability (b) Cell structure with single or dual-x SiN$_x$. The left side is the sample structure of after SiN$_x$ deposition and simulated contact firing with no metal, and the right side is the structure of simulated contact firing with metal electrode and access metal removal.

II. RESULTS AND DISCUSSION

A. Charge stability of Single Layer SiN$_x$ with high-x value (~1.30) and a dual-x SiNx stack (x ~1.30/1.01)

Figure 2 shows C-V curves before and after negative charge injection from the test sample (figure 1(a)). As expected, the negative flat band voltage (V_{fb}), prior to charge injection shifts to the positive voltage regime after the negative charge injection. This confirms that electrons are successfully injected into the SiN$_x$ thin film, resulting in the charge polarity change from positive to negative.

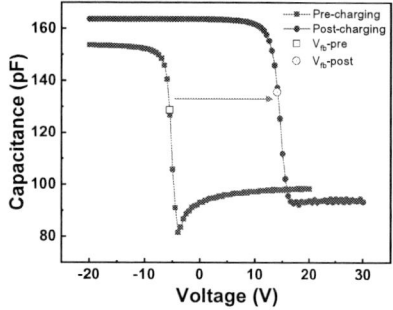

Figure 2. C-V curves before and after negative charge injection

Figure 3(a) shows V_{fb} decay curves for the ONO films with single layer SiN$_x$ films ($x \sim 1.01$ and ~ 1.30) as a function of simulated visible light exposer time. We analyzed the measured

V_{fb} data with a power-law fit to assess the loss rate of the injected negative charge.

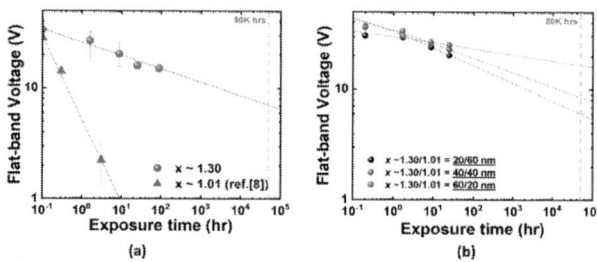

(a)   (b)

Figure 3. Flat-band voltage (V_{fb}) curves due to charge loss by visible light source at room temperature for (a) two different x-value (x ~1.30 and 1.01) and (b) three different dual x-value (x ~1.30/1.01). Note that SiN$_x$ with x ~ 1.01 data referred to literature [8]

Note that the extrapolated flat-band voltage for single layer SiN$_x$ ($x \sim 1.30$) sample remains above 0V at time=50,000 hours illumination, assuming that the power law time dependence holds for positive V_{fb}. However, the single layer SiN$_x$ film ($x \sim 1.01$) showed a rapid drop in V_{fb} only after several hours, which means that injected charge in not stable in this film. Chen et al. explained the mechanism of this charge loss in single layer SiN$_x$ layer with high- and low-x value on basic of dispersive hydrogen transport [8]. Table 1 shows the calculated negative charge density before and after 50K hours light exposure from the extrapolated flat band voltage. The injected negative charge density decreased from 1.1E13 cm^{-2} to 2.2E12 cm^{-2} after visible light exposure for 50K hours for the single SiN$_x$ layer with $x \sim 1.30$. This remaining charge density after 50K hours exposure is very similar to our reference sample passivated with Al$_2$O$_3$(3nm)/SiN$_x$, as shown in Table 1.

Next, we investigated the charge stability of dual-x SiN$_x$ stack to examine the effect of high-x SiN$_x$ thickness on the charge stability. We adjusted the thickness of high-x ($x \sim 1.30$) and low-x ($x \sim 1.01$) SiNx while keeping the total thickness at ~ 80 nm. Figure 3(b) shows the measured V_{fb} decay as a function of light exposure time with the power law time fitted curve using the last three measured points. Table 3 shows the calculated negative charge density after 0 and 50K hours light exposure from the fitted V_{fb} curve. The V_{fb} decay slows with increased high-x SiN$_x$ thickness in the 80 nm dual-x SiN$_x$ stack. It means that the charge stability becomes better with the increase in the thickness of the high-x SiN$_x$. It is found that if the thickness of high-x SiN$_x$ is above 40 nm, the retained negative charge density is very similar to Al$_2$O$_3$ passivated samples.

Table 1. Calculated negative charge density for the single- and dual-x SiNx layers with Al$_2$O$_3$

x-value of SiN$_x$	Thickness (nm)	Q$_{neg}$ (10^{12} cm^{-2})	
		@ 0 hours	@ 50K hours
~1.30	80	10.5	2.19
~1.30/~1.01	20/60	6.22	1.22
	40/40	7.71	1.76
	60/20	7.51	3.40

978-1-6654-6060-6/23 $31.00 © 2023 IEEE

Al$_2$O$_3$(3nm)/SiN$_x$ (ref.)	2.5E12 cm^{-2}

B. Passivation properties of SiN$_x$ with single and dual-x layer on boron emitter surface Prior to Negative Charge Injection

Figure 4 shows the measured implied V_{oc} (iV_{oc}) of the test cell structures with B emitter passivated with the different ONO stacks (figure 1(b)). For the reference sample with single layer SiN$_x$ (x~1.01), iV_{oc} increased from 655 to 690 mV after simulated firing without metallization. This is attributed to the release of hydrogen from the SiN$_x$ which can improve surface passivation, resulting in iV_{oc} increase. After front metallization on B emitter, iV_{oc} showed ~15mV drop (~675 mV) due to increased metal-induced recombination. For the single layer high-x SiN$_x$ sample, starting iV_{oc} was lower (~660 mV) than the low-x SiN$_x$, which did not change much after simulated firing process probably due to the less hydrogen in the high-x SiN$_x$ layer [9]. It is also noted that iV_{oc} decreased rapidly from 660 to 610 mV with the front-side metallization through the single layer high-x SiN$_x$. The reason for this is unclear at this stage, but our preliminary investigation reveals that the wafer bulk lifetime also degraded after the firing process for the front metalized sample with high-x SiN$_x$. We will investigate this further. For the dual-x SiN$_x$ (50/30 nm) film on the B emitter, the trend is very similar to single layer low-x SiN$_x$ layers, although measured iV_{oc} is slightly lower. After the simulated firing process, iV_{oc} increased from 670 to 685 mV and decreased to 670 mV with the metallized B emitter.

Based on the charge stability and passivation behavior and quality of uncharged films, we think negatively charged dual-x SiN$_x$ stack may provide the best combination of charge stability and passivation quality of B emitter. Work is in progress on passivation quality of charged films and will be reported at the conference.

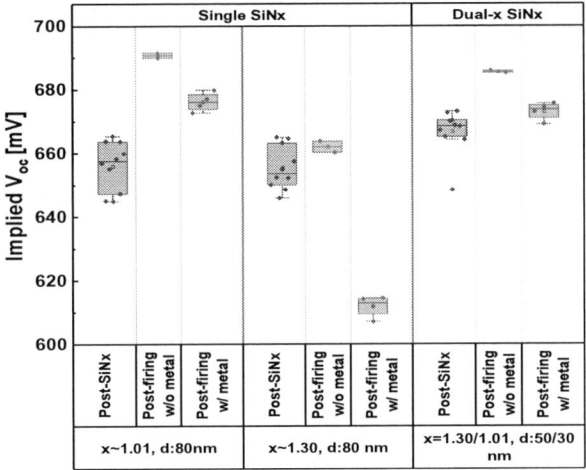

Figure 4. Implied V_{oc} of cell structure with single SiN$_x$ and dual-x SiN$_x$.

III. Conclusions

In this work, we report on the stability of injected negative charge in single layer SiN$_x$ films with x~1.01 and 1.30 as well as and dual-x SiN$_x$ stack (x~1.30/1.01). Injected negative charge was very stable in the single high-x and dual-x SiN$_x$ layers after simulated visible light exposure. We demonstrated that the inject charge density was retained at ~2E12 cm^{-2} after the 50K hours exposure, which is comparable to Al$_2$O$_3$ layers. We also found that the charge stability improves with the increasing thickness of the high-x SiN$_x$ in the dual-x SiN$_x$ stack. In addition, we investigated the passivation quality of the uncharged single and dual-x SiN$_x$ on the B emitter. The single layer high-x SiN$_x$ showed inferior passivation quality and significant iV_{oc} degradation after screen printed metallization. However, the dual-x SiN$_x$ layer showed comparable passivation to a single layer low-x SiN$_x$. Based on the charge stability and passivation quality, we will apply the dual-x SiN$_x$ layer on our B emitter of the n-type cells, followed by negative charge injection in the final n-type cells and compare the result with Al$_2$O$_3$ passivation.

Acknowledgment

This material was based upon work supported by the U.S. Department of Energy's Office of Energy Efficiency and Renewable Energy (EERE) under the Solar Energy Technologies Office Award No. DE-EE0008566. The views expressed herein do not necessarily represent the views of the U.S. Department of Energy or the United States Government.

References

[1] B. Hoex, S.B.S Heil, E. Langereis, M.C.M. van de sanden, and W.M.M. Kessels, Appl. Phys. Lett. 89, 042112 (2006)

[2] S. Dauew, L. Middestadt, A. Metz, and R. Hezel, Prog. Photovolt.: Res. Appl. 10, 271 (2002)

[3] J.-M. Hwang, J. Appl. Phys. 125, 173301 (2019)

[4] E. Cho, Y.-W. Ok, J. Hwang, A. D. Upadhyaya, J. K. Tate, F. Zimbardi, and A. Rohatgi, in Proceedings of 43rd IEEE Photovoltaic Specialist Conference (IEEE, New York, 2016), p. 2874.

[5] E. Cho, Y.-W. Ok, J. Hwang, A. D. Upadhyaya, J. K. Tate, F. Zimbardi, and A. Rohatgi, in Proceedings of 44th IEEE Photovoltaic Specialist Conference (IEEE, New York, 2017), p. 333.

[6] J. -M. Hwang, C, Chen, Y. -W. Ok, W. Choi, A. Upadhyaya, V. Upadhyaya, B. Rounsaville, and A. Rohatgi, in proceedings of 48th IEEE Photovoltaic Specialist Conference (IEEE, New York, 2021), p. 2119.

[7] K.H. Min, J.-M Hwang, E. Cho, H. Song, S. Park, A. Rohatgi, D. Kim, H.-S. Lee, Y. Kang, Y.-W. Ok, and M.G. Kang, Prog. Photovolt. **29**, 54 (2021).

[8] C. Chen, J. -M. Hwang, Y. -W. Ok, W. Choi, V. Upadhyaya, B. Rounsaville, and A. Rohatgi, J. Appl. Phys. 132, 213302 (2022)

[9] J.-F. Lelièvre, E. Fourmond, A. Kaminski, O. Palais, D. Ballutaud, M. Lemiti, Sol. Energy Mater. Sol. Cells 93, 1281-1298 (2009)

Optimizing the Heat Sink for Concentrated Photovoltaic Systems for Different Heat Flux Conditions

Nemalipuri Surya Prathap, Harsh Chaurasia, K. S. Reddy[*]

Heat Transfer and Thermal Power Laboratory, Department of Mechanical Engineering, Indian Institute of Technology Madras, Chennai 600036, India

Abstract — Concentrated photovoltaic (CPV) technology has gained a lot of attention in the past few decades as researchers look to overcome the problems caused by conventional sources of energy. CPV technology uses low-cost optical concentrators to increase the solar radiation intensity on the photovoltaic cell, which then converts it to electricity due to the photovoltaic effect. The increased solar intensity causes the temperature of the cell to increase considerably, and hence heat sink becomes a must. In the present work, the central aim is to optimize the heat sink to be employed in the CPV system for various heat flux conditions. Firstly, micro-finned heat sinks of two different structures made of undoped silicon wafer are designed and numerically studied for Concentrator Standard Test Conditions (CSTCs) and Worst-Case Conditions (WCCs). The results obtained are compared with a square cross-section flat and finless base plate. Further, the study aims at finding the optimum arrangement for two CPV cells on a single heat sink under 2000,3000, and 4000 concentration ratios without comprising efficiency. Subsequently, the variation of heat transfer coefficient (HTC), maximum operating temperature of CPV cell, and mass-specific HTC with variation in height (H) of the fin to the space between the fins (S) are numerically analyzed in ANSYS 2022 R2. The mass-specific power of the circular-shaped heat sink is investigated to be highest, with a value of 327.41 W/kg. For the second case, the optimum distance between the two CPV cells arranged on a single heat sink was found to be 40 mm. The results also clearly suggest that the heat transfer coefficient significantly changes with the H/S ratio without significant changes in the mass-specific HTC.

Keywords: mass-specific power, passive cooling, concentrated photovoltaics

I. INTRODUCTION

Renewable energy has gained a lot of attention in the past few decades as researchers look to overcome the problems caused by conventional sources of energy. Conventional sources of energy which primarily include coal, oil, and natural gas, cause higher carbon emissions, greenhouse gases, and other environmental impacts. Solar energy has been recognized as the leading solution to overcome the problems caused by conventional sources of energy and produce electricity. Electrical power from solar energy can be utilized in two ways, direct and indirect method. The indirect method involves conversion into thermal energy which is later converted to electrical power. By utilizing photovoltaic (PV) technology, solar energy is directly converted into electricity. As compared to the indirect method, direct conversion using PV is considered more efficient and superior [1]. The high cost of PV technology limits its widespread application. Thus, to achieve higher

application, it is necessary to find ways to reduce the cost associated with them. PV cost can be primarily reduced in two ways, increasing the efficiency of solar cell or reducing the area of the cell by using concentrated photovoltaic (CPV) technology. Recently, CPV has received a lot of attention because of its ability to reduce the cost considerably by employing low-cost optical concentrators. Concentrators increase the solar flux incident on the PV cell surface, leading to reduced requirement of cell surface area, hence, less PV material is required. Less requirement of PV material supports the employment of high cost much efficient multi- junction cells which have an efficiency of around 40% [2].

CPV technology addresses many of the problems of traditional PV technology, but the increased solar intensity causes the temperature of the cell to increase considerably. With higher temperature of operation, the PV cell efficiency drops, and in long term irreversible decay of the cell structure occur [3]. Therefore, large attention has been paid in recent years to cooling of CPV systems. Various active and passive cooling methods have been analyzed by authors for application in the CPV system. Active cooling methods require external power to create necessary cooling and include forced air convection, microchannel cooling, jet impingement, and thermoelectric cooling. Active cooling has the potential to maintain the cell at optimal temperature, given the cost of the system is not an issue. The increased complexity of active cooling methods increases the cost of the system. Passive cooling methods overcome this cost constraint since it needs no external energy to run. Passive cooling method includes finned-based heat sinks, PCM-based cooling, heat pipe, and liquid immersion cooling.

The finned-based cooling is the simplest technique for maintaining the cell at the optimum temperature as the application of fins increases the thermal exchanging surface[4]. Natarajan [5] numerically investigated the temperature of solar cells for CPV system under a concentration ratio (CR) of 10 for both finned and finless aluminum base plates. The investigation concluded that better temperature regulation of cells was achieved in the case of aluminum base plate with fins compared to an aluminum base plate with no fins. Bar-Cohen et al.[6] proposed optimization and modeling techniques for maximizing the heat transfer rate and minimizing the weight of

978-1-6654-6060-6/23 $31.00 © 2023 IEEE

heat sink. Least material approach (LMA) was adopted for optimizing the thermal design, which involved determination of the fin aspect ratio having maximum rate of heat transfer for a particular fin mass and volume. Micheli et al.[7] adopted LMA for analyzing vertically oriented rectangular plate fins under Worst Case Conditions (WCCs) and Concentrator Standard Test Conditions (CSTCs). In WCCs, the PV cell is considered to produce no electricity, and all incident radiation is converted to heat. The cell is considered to operate at its highest efficiency at CSTCs. A temperature of 63.3 °C was reported for CSTCs and 91.5 °C for WCCs.

Leonardo Micheli et. al.[8] studied opportunities in cooling CPV cells using Mass Specific power approach. Mass specific power (MSP) approach involves increasing the electrical power output per unit mass of the heat sink. Copper Diamond composite heat sink consisting of 63% copper was found to be the best compromise between pure Copper and pure diamond-based heat sink. Consequently, the convective HTC increased with an increase in fin spacing and decreased fin height.

The above review of the literature shows that most of the work for passive cooling has been done for low-concentrating photovoltaics. Furthermore, there hardly exists studies that analyze the comparative effectiveness of micro and macro finned heat sinks in keeping CPV cells at optimum temperature. Therefore, the central aim of the present work is to analyze and optimize the different configurations of micro-finned heat sinks for application in high concentrating photovoltaic systems

having concentration ratio more than 500. Moreover, the study also focuses on finding the optimum arrangement for two CPV cells on a single heat sink. This analysis becomes necessary since, in most practical scenarios, the weight of the heat sink is too high, causing the system to be bulky and, at the same time, increasing the cost due to an increase in the material of the heat sink. To further optimize the heat sink, the H/S ratio effect on convective HTC is investigated. For all the considered heat sinks, the convective HTC for air is obtained using an iterative procedure and is implemented using an in-house C++ code.

II. SYSTEM DESCRIPTION

The concentrated photovoltaic system consists of optical elements called reflectors for focusing the sun rays on the PV cell. The system under consideration has a Parabolic Trough Concentrator (PTC) as a primary reflector which focuses the rays than to Compound Parabolic Concentrator (CPC) which acts as a secondary reflector. Furthermore, to homogenize the highly nonuniform irradiance falling on PV cell, a homogenizer is used. Accordingly, to maintain the cell temperature, a heat sink is fixed to the cell with a solder having a thermal conductivity of 48 W/mK and 0.125 mm thickness. The cell under consideration is a 3 mm x 3 mm germanium cell developed by Azur Space Solar Power [9]. The cell has a thermal conductivity of 60 W/mK and thickness of 0.19 mm. The entire system is represented in Fig.1.

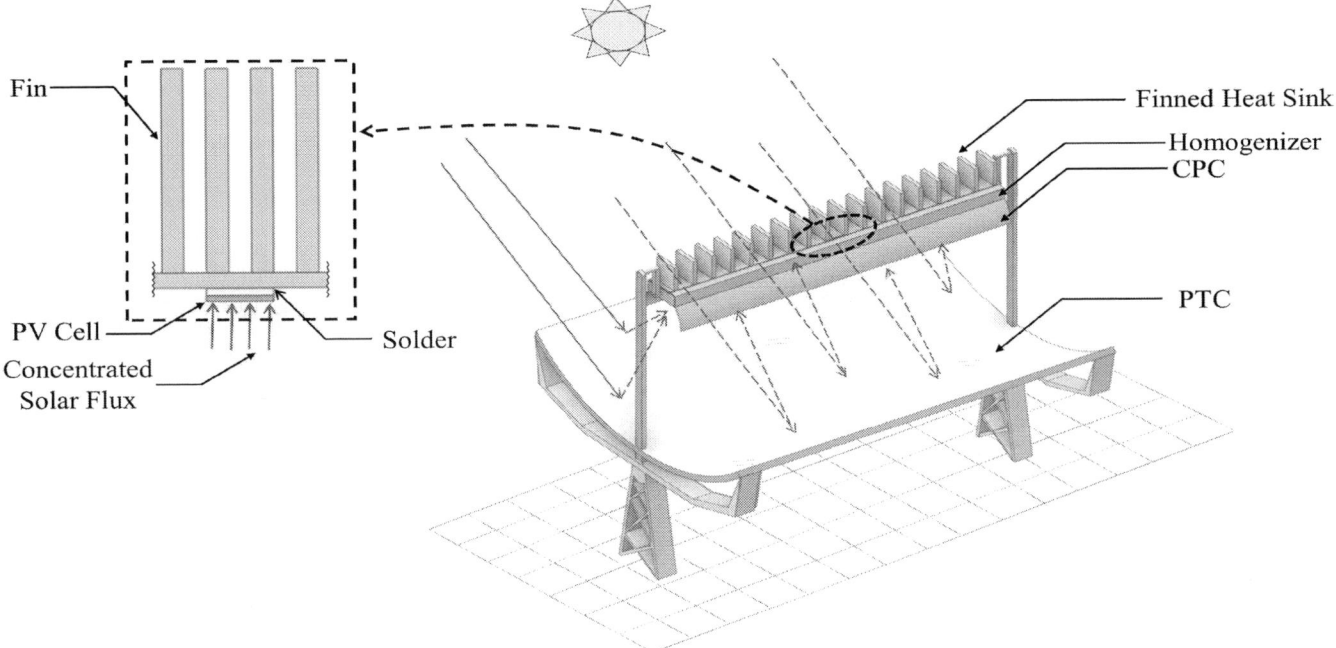

Fig. 1.   Schematic of CPV System

III. NUMERICAL MODELING

The main aim of the study is to optimize the heat sink to be employed in the CPV system for various heat flux conditions using a numerical model. To achieve this, firstly, micro-finned heat sinks of two different structures made of undoped silicon wafer are numerically studied for CSTCs and WCCs. Secondly, the optimum distance between two CPV cells on a single heat sink under 2000,3000, and 4000 concentration ratios is determined without compromising efficiency. Finally, the variation of heat transfer coefficient (HTC), maximum operating temperature of CPV cell, and mass-specific HTC with variation in height (H) of the fin to the space between the fins (S) are analyzed.

A. Governing Equations and Boundary Conditions

The energy of rays falling on the solar cell is partly converted to electricity, and the rest is converted to heat. Accordingly, in the present work, the volumetric heat generated (q_{Ge}) is modeled as a boundary condition and is calculated using the direct normal irradiation (DNI), concentration ratio, thickness (t), concentrator's optical efficiency (η_{opt}), and cell efficiency (η_{cell}) as per (1).

$$q_{Ge} = DNI * CR * \frac{1}{t} * (1 - \eta_{cell}) * \eta_{opt} \qquad (1)$$

For all the investigated cases, the DNI is 900 W/m² and η_{opt} is taken as 0.85. The heat sink surface is subjected to convective heat transfer ($q_{conv,s-amb}$) with ambient as per (2).

$$q_{conv,s-amb} = h(T_s - T_{amb}) \qquad (2)$$

Where T_s is heat sink temperature and T_{amb} is ambient temperature. Accordingly, to obtain the value of the convective HTC (*h*), an iterative procedure based on the work of Roncati [10] has been adopted and is implemented using an in-house C++ code. The converged value of HTC obtained is then used as input during the analysis. For all the considered layers, 3D conduction equation has been solved given by (3).

$$\nabla.(k_i \nabla T_i) + q_i = 0 \qquad (3)$$

where k_i is the isotropic thermal conductivity of the material and q_i is the volumetric heat generation. It must be noted that for heat sink and solder, the volumetric heat generation is zero. The heat transfer through different surfaces and interfaces is governed by (4-6) [11]. For the top surface of the heat sink, the heat transfer is governed by (4):

$$-k_s \left(\frac{\partial T_s}{\partial y}\right) = q_{conv,s-amb} \qquad (4)$$

The interface between germanium and solder is modeled as thermally coupled, as given in (5):

$$k_{Ge} \nabla T_{Ge} = k_{sol} \nabla T_{sol} \text{ and } T_{Ge} = T_{sol} \qquad (5)$$

A similar boundary condition is applied to the interface between solder and the heat sink as per (6):

$$k_{sol} \nabla T_{sol} = k_s \nabla T_s \text{ and } T_{sol} = T_s \qquad (6)$$

B. Micro-finned Heat Sinks

Two different micro-finned heat sinks made of silicon have been studied and compared with a fin-less base plate attached to PV cell under a concentration ratio of 500. The finless base plate has length of 50 mm, width 50 mm and thickness of 0.92 mm. The two micro-finned heat sink configuration includes a square-shaped base plate (50 mm*50 mm*0.8 mm) with micro-fins and a circular-shaped base plate (50 mm diameter) with micro-fins. For both the investigated micro-finned heat sinks, the fins have a thickness of 0.2 mm and height of 0.6 mm with the spacing between the fins to be 0.8 mm. Accordingly, the number of fins were found to be 50. Thermal analysis is then performed under both CSTCs and WCCs to obtain the average temperature of the cell. The CSTC corresponds to condition that cell produces electrical power at its peak efficiency which in this case is 41% [4]. Subsequently for WCCs, the cell is modelled to produce no electrical power [4]. Mass specific

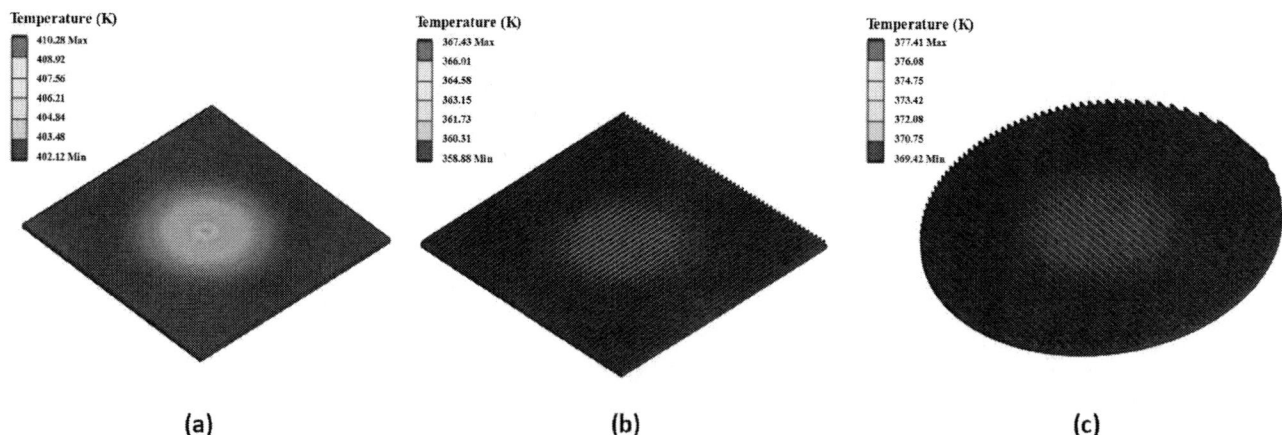

Fig. 2. Temperature Contour (a) Fin-less base plate, (b) micro-finned heat sink with square base plate, (c) micro-finned heat sink with circular base plate

power, the ratio of electrical power output to the weight of the heat sink, is then evaluated.

MSP analysis is necessary for tracked CPV systems because the tracking system utilizes part of the output energy in moving the module, and thus the reduction in weight can significantly enhance the electrical efficiency. The effect of operating temperature is then incorporated by refining the electrical power by considering a drop of 2.1 mW in power output per degree temperature increase [4,9].

C. Optimizing the Distance Between the Cells on the Same Heat Sink

In this model, the distance between the two CPV cells mounted on a heat sink has been varied from 3 mm to 70 mm. The dimensions of the heat sink are given in Table I. Three concentration ratios have been considered for the same: 2000,3000 and 4000. Accordingly, the optimum distance is determined by the lowest maximum temperature of the cells obtained.

TABLE I
Dimensions of the heat sink for two cells on the same heat sink

Length (mm)	Width (mm)	Fin Thickness (mm)	Fin Spacing (mm)	No. of fins	Fin Height (mm)	Base Thickness (mm)
79	79	5.2	6	7	54.4	3

D. H/S Ratio Variation

For the final case, the effect of variation in height (H) of the fin to the space between the fins (S) on HTC, maximum operating temperature of CPV cell, and mass-specific HTC are analyzed. The length (80 mm), width (80 mm), and thickness (3 mm) of the base plate are kept constant, along with fin thickness (4 mm) and fin spacing (4 mm). The number of fins on the base plate is 10. Only fin height is varied from 4 mm to 40 mm with a step size of 4 mm. The concentration ratio is considered to be 2000

IV. RESULTS AND DISCUSSIONS

The results from analysis of micro-finned heat sinks are presented in Table II. For CSTCs, the lowest maximum temperature is obtained for squared shaped plate with micro-fins as represented Fig. 2. Accordingly, the refined maximum electrical power is obtained for the all considered heat sinks. It is noteworthy to mention that mass specific power is maximum for cell mounted on circular shaped base plate with micro-fins due to the low weight of the heat sink. Under WCCs, the square shaped flat plate without micro-fins has the highest heat transfer coefficient when compared with the micro-fin heat sinks. The maximum temperature of the square-shaped flat plate without micro-fins reaches beyond 150 °C, indicating the unreliability of the heat sink.

For, the case of two cells on a single heat sink, the maximum temperature of the cell is lowest when the distance between the cell is 40 mm for all the considered concentration ratios. Fig. 3 indicates the variation in maximum, minimum and average temperature for the cells under 2000 CR with variation in the distance between them. Similar trend is observed for the cells under 3000 CR and 4000 CR as represented by Fig. 4 and Fig. 5 respectively.

For the heat sink models with H/S variation, the heat transfer coefficient, maximum temperature of the cell, and the mass-specific HTC decrease with an increase in H/S ratio as depicted in Fig. 6. HTC and maximum temperature decrease as a cubic function of H/S ratio according to trendlines given in Table III.

TABLE II
Comparison between micro-finned heat sinks with fin-less base plate

S. No.	Type of Heat Sink	Test Condition	Average Temperature (K)	Maximum Temperature (K)	Heat Transfer Coefficient (W/m²K)	Refined maximum electrical power (W)	Weight of the heat Sink (kg)	Mass Specific Power (W/kg)
1	Square shaped base plate without micro-fins	CSTCs	402.97	410.28	8.93	1.317	0.005358	245.80
		WCCs	450.88	463.14	9.85	0	0.005358	0
2	Square shaped base plate with micro-fins	CSTCs	359.82	367.42	7.66	1.394	0.005358	260.20
		WCCs	385.03	397.79	8.47	0	0.005358	0
3	Circular shaped base plate with micro-fins	CSTCs	370.19	377.41	8.03	1.376	0.004202	327.41
		WCCs	399.53	412.94	8.88	0	0.004202	0

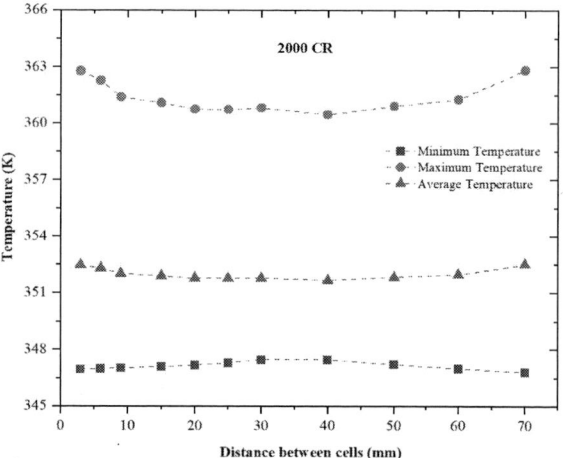

Fig. 3. Variation in temperature with distance between the cell for cells under 2000 CR.

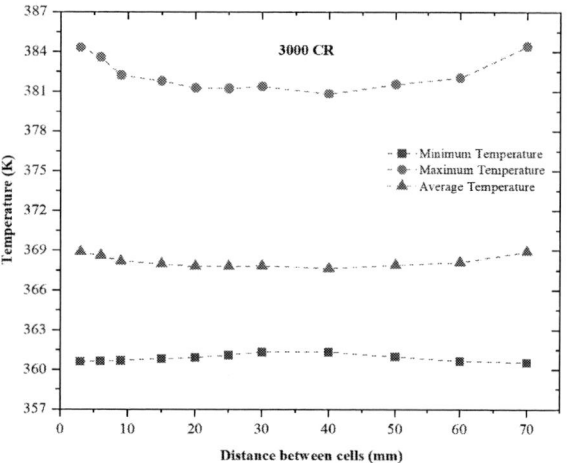

Fig.4. Variation in temperature with distance between the cell for cells under 3000 CR.

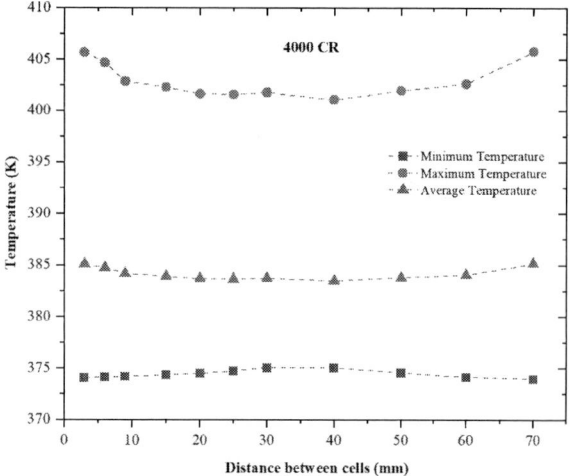

Fig.5. Variation in temperature with distance between the cell for cells under 4000 CR.

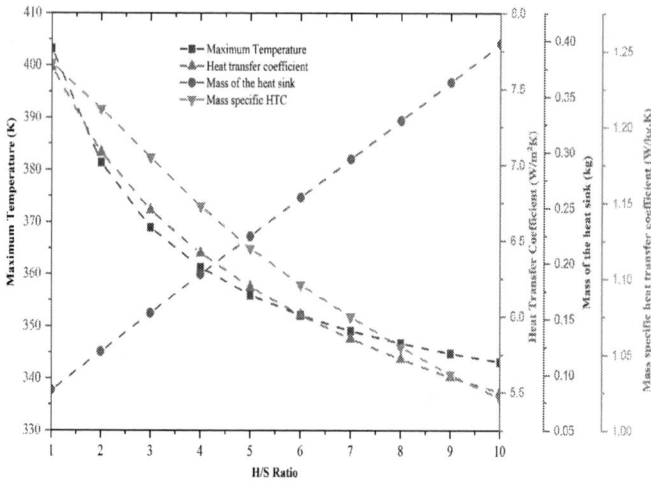

Fig. 6. Variation of different parameters with H/S ratio.

TABLE III
Trendlines for various parameters (y) with H/S ratio (x)

Parameter	Trendline	R^2
Heat transfer Coefficient	$y = -0.00299x^3 + 0.0727x^2 - 0.706x + 8.26$	0.999
Maximum temperature	$y = -0.157x^3 + 3.54x^2 - 28.20x + 426$	0.996
Mass of the heat sink	$y = 0.0346x + 0.0518$	1
Mass specific HTC	$y = 0.00117x^2 - 0.0375x + 1.28$	1

V. CONCLUSIONS

In light of the current study, some significant conclusions can be drawn. The circular-shaped base plate gives the highest mass-specific power. In addition, the optimum distance between two cells mounted on the same heat sink for minimum temperature is independent of the concentration ratio. The results also clearly suggest that the HTC significantly changes with the H/S ratio without significant changes in mass-specific HTC. The maximum value of HTC obtained was equal to 7.645 W/m²K at H/S ratio 1.

REFERENCES

[1] M. S. Yousef, A. K. Abdel Rahman, and S. Ookawara, "Performance investigation of low – Concentration photovoltaic systems under hot and arid conditions: Experimental and numerical results," *Energy Convers. Manag.*, vol. 128, pp. 82–94, 2016, doi: 10.1016/j.enconman.2016.09.061.

[2] A. Radwan, M. Ahmed, and S. Ookawara, "Performance enhancement of concentrated photovoltaic systems using a microchannel heat sink with nanofluids," *Energy Convers. Manag.*, vol. 119, pp. 289–303, 2016, doi: 10.1016/j.enconman.2016.04.045.

[3] A. Radwan and M. Ahmed, "The influence of microchannel heat sink configurations on the performance of low concentrator photovoltaic systems," *Appl. Energy*, vol. 206, no. August, pp. 594–611, 2017, doi: 10.1016/j.apenergy.2017.08.202.

[4] L. Micheli, K. S. Reddy, and T. K. Mallick, "Plate micro-fins in

natural convection: An opportunity for passive concentrating photovoltaic cooling," *Energy Procedia*, vol. 82, pp. 301–308, 2015, doi: 10.1016/j.egypro.2015.12.037.

[5] S. K. Natarajan, T. K. Mallick, M. Katz, and S. Weingaertner, "Numerical investigations of solar cell temperature for photovoltaic concentrator system with and without passive cooling arrangements," *Int. J. Therm. Sci.*, vol. 50, no. 12, pp. 2514–2521, 2011, doi: 10.1016/j.ijthermalsci.2011.06.014.

[6] A. Bar-Cohen, M. Iyengar, and A. D. Kraus, "Design of optimum plate-fin natural convective heat sinks," *J. Electron. Packag. Trans. ASME*, vol. 125, no. 2 SPEC., pp. 208–216, 2003, doi: 10.1115/1.1568361.

[7] L. Micheli, E. F. Fernandez, F. Almonacid, K. S. Reddy, and T. K. Mallick, "Enhancing ultra-high CPV passive cooling using least-material finned heat sinks," *AIP Conf. Proc.*, vol. 1679, no. September 2015, 2015, doi: 10.1063/1.4931563.

[8] L. Micheli, N. Sarmah, X. Luo, K. S. Reddy, and T. K. Mallick, "Opportunities and challenges in micro- and nano-technologies for concentrating photovoltaic cooling: A review," *Renew. Sustain. Energy Rev.*, vol. 20, pp. 595–610, 2013, doi: 10.1016/j.rser.2012.11.051.

[9] AZUR SPACE Solar Power GmbH, "Concentrator Triple Junction Solar Cell Cell Type : 3C44C- 3 × 3 mm2 Azur Space." https://www.azurspace.com/images/products/0004357-00-01_3C44_AzurDesign_3x3.pdf (accessed Aug. 19, 2022).

[10] D. Roncati, "Iterative calculation of the heat transfer coefficient," no. 2, 2015, [Online]. Available: www.progettazioneottica.it.

[11] O. Rejeb *et al.*, "Numerical analysis of passive cooled ultra-high concentrator photovoltaic cell using optimal heat spreader design," *Case Stud. Therm. Eng.*, vol. 22, no. September, p. 100757, 2020, doi: 10.1016/j.csite.2020.100757.

Comparison of Open-Source Photovoltaic Performance Models Against Multi-Year Field Data

Lelia Deville[1,2], Marios Theristis[1], Bruce H. King[1], Terrence L. Chambers[2], Joshua S. Stein[1]

[1]Sandia National Laboratories, Albuquerque, New Mexico
[2]University of Louisiana at Lafayette, Lafayette, Louisiana

Abstract — All freely available plane-of-array (POA) transposition models, photovoltaic (PV) module/cell temperature models, and PV performance models were examined against multi-year field data from Albuquerque, New Mexico. The data include different PV systems comprised of c-Si modules that vary in cell type, module construction, and materials. These systems have been characterized via IEC 61853 testing and the input data for each model were sourced from these test results. Six POA transposition models, seven temperature models, and twelve performance models are included in this comparative analysis. These freely available models were proven effective across many different types of c-Si technologies. Overall, it was observed that model complexity and/or availability of module-specific characterization data does not guarantee greater accuracy, at least in Albuquerque. The POA transposition and PV performance models exhibited average normalized mean bias errors (NMBE) within ±3%; the mean and median residuals of the PV temperature models were within ±5°C.

I. INTRODUCTION

One of the most powerful tools for the planning and performance analysis of a photovoltaic (PV) system is a performance model. Performance models take as inputs irradiance and weather time series data, parameters that describe the performance characteristics of the PV modules, as well as system design specifications and output time series of simulated power or efficiencies produced by the system. Modeling can serve as both a simulation and optimization tool and can be used at various stages of development in a PV system (e.g., site assessment, design evaluation, technology comparisons, etc.).

Many comparisons of freely available models exist (e.g. Marion et al. [1]), but these comparisons usually only consider two to three models at a time or models that require the exact same inputs [2] and/or benchmarked on a limited number of systems [3] over a one-year period.

The purpose of this study is to compare twelve open-source PV performance models with measured data from a variety of different PV systems comprised of PV modules that vary in terms of PV cell type, module construction, and materials. The study includes a comparison of other types of models that are necessary to early steps in PV performance modeling. Six plane-of-array (POA) transposition models and seven module and cell temperature models are compared against data measured in Albuquerque, New Mexico (NM).

II. FIELD AND MODULE CHARACTERIZATION DATA

Seven PV systems of varying c-Si technologies located at Sandia's Photovoltaic Systems Evaluation Laboratory were considered. These systems have been characterized by both Sandia National Laboratories and an external laboratory using various methods to obtain different types of test data, like PAN files and IEC 61853-1 matrix data. Using these test data, the systems' power and efficiency were analyzed against PV performance model predictions. Measured POA, back-of-module, and string voltage and current data at 1-min intervals were used to validate temperature and transposition models. Figure 1 shows the different modeling pipelines that were developed in order to predict a system's output.

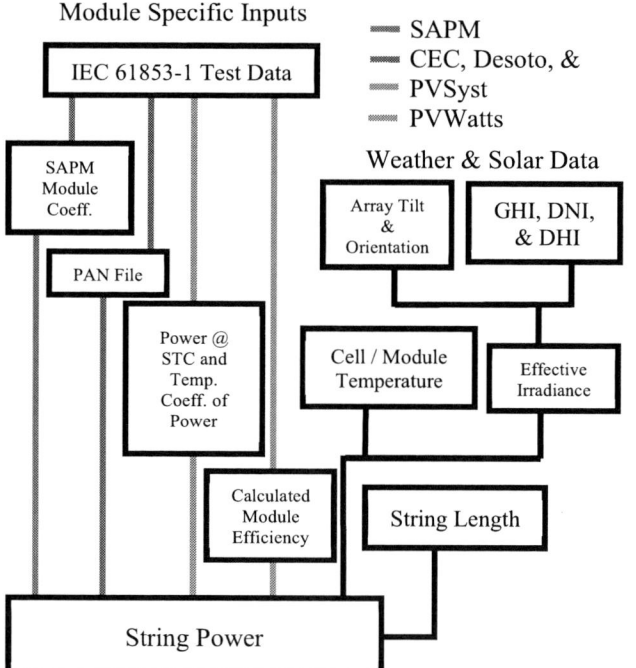

Figure 1: Flowchart describing the method of calculation for the models including weather data and module specific inputs.

III. OVERVIEW OF MODELS

A. POA Transposition Models

The irradiance transposition models compared in this study were run using *pvlib-python* [4]. Overall, six models were included: Isotropic [5], Haydavies [6], Klucher [7], Reindl [8], King [9], and Perez [10] models. The Perez model has multiple versions based on data from specific geographical locations; as such, all eleven Perez models were tested.

B. PV Module Temperature Models

All temperature models examined in this study were run and compared using *pvlib-python* [4] using steady- and transient-state assumptions. The steady-state analysis includes two module temperature models (Sandia Array Performance Model [SAPM] [9] and Faiman [11]), and five cell temperature models (Ross [12], PVSyst [13] , SAM NOCT [14], SAPM Cell [9], and Fuentes [15]). For a fair comparison, the results of the cell temperature models were converted to module temperature using the SAPM method. To examine the influence of transient-state assumptions on the temperature predictions, the same models were re-run by incorporating the Prilliman's transient temperature model [16], which acts as a moving average filter.

C. PV Performance Models

The performance models compared in this study were taken from two different python packages. SAPM [9], PVWatts (*PVW*) [17], CEC [18], Desoto (*DES*) [19], and PVSyst (*PVS*) were run using *pvlib-python* [4], and ADR [2], Heydenreich (*HEY*) [20], MotherPV (*MOT*) [21], PVGIS (*PVG*) [22], MPM5 [23], MPM6 [23], and Bilinear Interpolation (*BIL*) were run using *pvpltools-python* [24].

IV. RESULTS

Overall, most of the transposition models performed similarly with a NMBE of ±3%. The Isotropic model under-estimated irradiance by 1.86% whereas the King model over-estimated by up to 2.74%. The NMBE and RMSE values for all models are summarized in Table 1. The remaining models showed minimal to zero bias, whereas the RMSE values of Perez and Klucher were lower (< 40 W/m^2) as compared to the rest (> 40 W/m^2).

Table 1: Summary of NMBE and RMSE for all POA transposition models considered in this study.

Model	NMBE (%)	RMSE (W/m^2)
Isotropic	-1.86	41.25
Perez – abq1988	-0.83	39.59
Haydavies	-0.73	46.24
Klucher	0.51	38.61
Reindl	-0.52	46.39
King	2.74	44.20

The transposition models' performance varied at different irradiance levels, as shown in Figure 2. Isotropic, Klucher, Reindl, and Haydavies were the best performing models at very low irradiance (< 150 W/m^2). The King model performed much worse than other models at low irradiance and it consistently over-estimated irradiance until it reached similar levels of NMBE at around > 650 W/m^2. The Perez – abq1998 model

exhibited the most consistent performance at all irradiance levels.

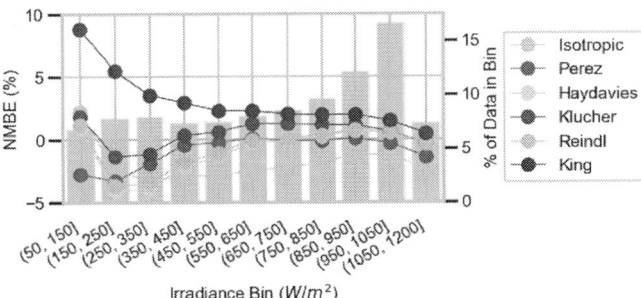

Figure 2: NMBE (%) of POA transposition models at different irradiance levels.

For all cell and module temperature models, the mean and median residuals were within ±5°C when all systems were considered. Not all models performed similarly on a given system; this shows that model performance was more dependent on the model parameters and not the specific PV technology. Figure 3 shows the model performance per system, in which all residuals were within ±6.5°C. The most accurate model was the Fuentes temperature (cell temperature was converted to module temperature) model, which had the lowest mean residual of -0.9°C and lowest RMSE of 3.6°C.

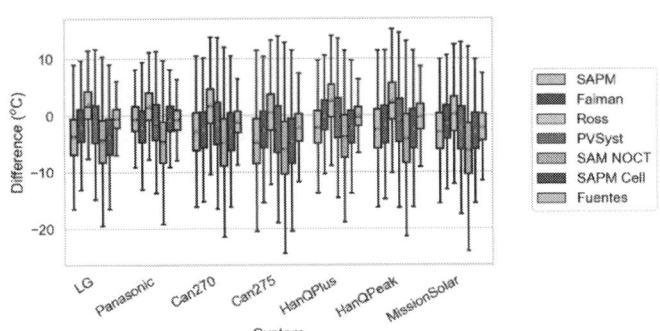

Figure 3: Residuals (°C) of module and cell temperature models by system.

Table 2 shows the changes in RMSE before (i.e. steady-state assumptions) and after applying the transient temperature model.

Table 2: RMSE (°C) of module and cell temperature models without (i.e., steady-state) and with the transient model applied.

Model	RMSE of Steady-State	RMSE of Transient
SAPM	5.6	5.2
Faiman	**5.0**	**4.4**
Ross	5.3	4.8
PVSyst	5.4	5.2
SAM NOCT	7.4	7.1
SAPM Cell	5.6	5.2
Fuentes	**3.6**	**3.7**

These results indicate that considering transient behavior reduces the spread, even in Albuquerque, NM where the sky conditions are relatively constant all year round. It is speculated that locations with more dynamic conditions would show larger improvements when applying the transient temperature model. In this case, the model with the greatest reduction in RMSE was the Faiman model whereas Fuentes showed no improvement.

Although the 12 PV performance models varied widely in their inputs and calculations, the performance of models for a given system were very similar. Figure 3 shows the NMBE for all models and all systems after a flat 2% derate was applied to account for soiling, wiring losses, etc.

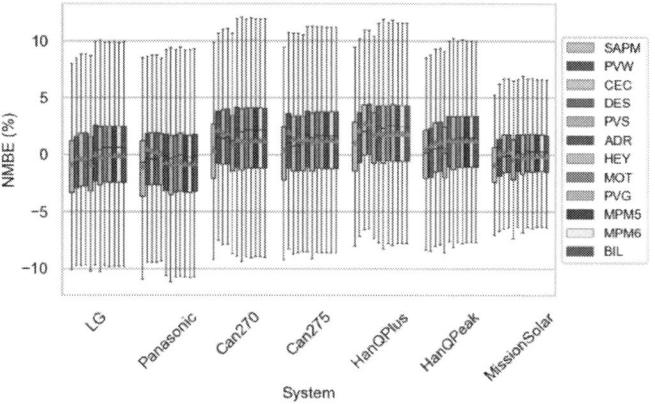

Figure 4: NMBE (%) of all PV performance models across all systems.

All models exhibited a first and third quartile NMBE within ±4.5%. The average NMBE for all models was within ±3% of the measured values. The simplest model, PVW considered only two module specific inputs: the STC power and the temperature coefficient of power. Even so, this model performed on par with and sometimes exceeded the performance of more detailed models, like PVS.

V. CONCLUSIONS AND LIMITATIONS

The results of this study can aide in the selection of an appropriate POA transposition, module temperature, or PV performance model. These results indicate that availability of module specific data should not be a hinderance in modeling of a system. They also indicate that model complexity does not guarantee any greater accuracy.

It should be noted that all of the temperature and performance models were compared with seven c-Si systems; no thin-film modules were available. The systems that were considered are small-scale laboratory systems that are monitored closely, and therefore, typical derate assumptions do not apply.

VI. ACKNOWLEDGEMENTS

This work was supported by the U.S. Department of Energy's Office of Energy Efficiency and Renewable Energy (EERE) under the Solar Energy Technologies Office Award Number 38267. Sandia National Laboratories is a multimission laboratory managed and operated by National Technology &

Engineering Solutions of Sandia, LLC, a wholly owned subsidiary of Honeywell International Inc., for the U.S. Department of Energy's National Nuclear Security Administration under contract DE-NA0003525. This paper describes objective technical results and analysis. Any subjective views or opinions that might be expressed in the paper do not necessarily represent the views of the U.S. Department of Energy or the United States Government.

REFERENCES

[1] B. Marion. Comparison of Predictive Models for Photovoltaic Module Performance.

[2] A. Driesse, et al. "A new photovoltaic module efficiency model for energy prediction and rating," *IEEE Journal of Photovoltaics,* vol. 11, no. 2, pp. 527-534, 2021.

[3] S. Kichou, et al, "Comparison of two PV array models for the simulation of PV systems using five different algorithms for the parameters identification," *Renewable Energy,* vol. 99, pp. 270-279, 2016.

[4] J. S. Stein, et al, "PVLIB: Open source photovoltaic performance modeling functions for Matlab and Python," in *2016 IEEE 43rd photovoltaic specialists conference (PVSC),* 2016: IEEE, pp. 3425-3430.

[5] H. Hottel et al, "Evaluation of flat-plate solar collector performance," in *Trans. Conf. Use of Solar Energy;(),* 1955, vol. 3.

[6] J. E. Hay, "Calculating solar radiation for inclined surfaces: Practical approaches," *Renewable energy,* vol. 3, no. 4-5, pp. 373-380, 1993.

[7] T. M. Klucher, "Evaluation of models to predict insolation on tilted surfaces," *Solar energy,* vol. 23, no. 2, pp. 111-114, 1979.

[8] D. Reindl, et al, "Evaluation of hourly tilted surface radiation models," *Solar energy,* vol. 45, no. 1, pp. 9-17, 1990.

[9] D. L. King, et al, "Photovoltaic array performance model," *Sandia Report No. SAND 2004-3535,* 2004. [Online]. Available: http://www.osti.gov/scitech//servlets/purl/919131-sca5ep/.

[10] R. Perez, et al, "The development and verification of the Perez diffuse radiation model," Sandia National Lab.(SNL-NM), Albuquerque, NM (United States); State Univ …, 1988.

[11] D. Faiman, "Assessing the outdoor operating temperature of photovoltaic modules," *Progress in Photovoltaics: Research and Applications,* vol. 16, no. 4, pp. 307-315, 2008.

[12] R. Ross, "Design techniques for flat-plate photovoltaic arrays," in *Proceedings of the 15th Photovoltaic Specialists Conference, Orlando, FL, USA,* 1981, pp. 12-15.

[13] A. Mermoud, et al., "PVSYST user's manual," *Switzerland, January,* 2014.

[14] P. Gilman, et al, "SAM photovoltaic model technical reference update," *NREL: Golden, CO, USA,* 2018.

[15] M. K. Fuentes, "A simplified thermal model for flat-plate photovoltaic arrays," Sandia National Labs., Albuquerque, NM (USA), 1987.

[16] M. Prilliman, et ali, "Transient Weighted Moving-Average Model of Photovoltaic Module Back-Surface Temperature," *IEEE Journal of Photovoltaics,* vol. 10, no. 4, pp. 1053-1060, 2020, doi: 10.1109/jphotov.2020.2992351.

[17] A. P. Dobos, "PVWatts version 5 manual," National Renewable Energy Lab.(NREL), Golden, CO (United States), 2014.

[18] A. P. Dobos, "An improved coefficient calculator for the California energy commission 6 parameter photovoltaic module model," *Journal of solar energy engineering,* vol. 134, no. 2, 2012.

[19] W. De Soto, et al., "Improvement and validation of a model for photovoltaic array performance," *Solar energy,* vol. 80, no. 1, pp. 78-88, 2006.

[20] W. Heydenreich, et al, "Describing the world with three parameters: a new approach to PV module power modelling," in *23rd European PV Solar Energy Conference and Exhibition (EU PVSEC),* 2008, pp. 2786-2789.

[21] A. G. de Montgareuil, et al, "A new tool for the MotherPV method: modelling of the irradiance coefficient of photovoltaic modules," in *24th European Photovoltaic Solar Energy Conference (EU PVSEC),* 2009, pp. 21-25.

[22] T. Huld *et al.,* "A power-rating model for crystalline silicon PV modules," *Solar Energy Materials and Solar Cells,* vol. 95, no. 12, pp. 3359-3369, 2011.

[23] S. Ransome and J. Sutterlueti, "How to Choose the Best Empirical Model for Optimum Energy Yield Predictions," in *44th IEEE Photovoltaic Specialist Conference (PVSC),* 2017: IEEE, pp. 652-657.

[24] A. Driesse, "PV Performance Labs Tools for Python," *GitHub repository at* https://github.com/adriesse/pvpltools-python, 2020.

Q CELLS Q.ANTUM NEO technology with > 25% conversion efficiency applying mass-production processes

Matthias Junghänel, Ingmar Höger, Martin Schaper, Kai Petter, Enrico Jarzembowski, Christian Klenke, Anika Weihrauch, Michael Schley, Hans-Christoph Ploigt, Ohjin Kwon, Antje Schönmann, Osama Tobail, Axel Schwabedissen, Maximilian Kauert, Klaus Duncker, René Hönig, Janko Cieslak, Stefan Hörnlein, Florian Stenzel, Björn Faulwetter-Quandt, Jessica Scharf, Friederike Kersten, Cangming Ke, Sissel Tind Kristensen, Carsten Baer, Martina Queck, Gregor Zimmermann, Matthias Köhler, Nicole Lampa, Britta Pohl-Hampel, Lorenzo Burtone, Larissa Niebergall, Matthias Schütze, Susanne Schulz, Stefan Peters, Ansgar Mette, Fabian Fertig, Markus Fischer, Jörg W. Müller

Hanwha Q CELLS GmbH, Sonnenallee 17-21, 06766 Bitterfeld-Wolfen, Germany

Abstract — **Within the last 3 years Q CELLS has developed its Q.ANTUM NEO technology based on a passivating contact solar cell and has transferred this technology to mass production. With a very lean and cost-effective process flow a conversion efficiency of 25.2 % has been demonstrated with this solar cell concept while average pilot line efficiency is currently at 24.7 % with a steep learning curve. This enables a high module power output of 444 W (132 half cells M6) corresponding to a module efficiency of 23.9 % on aperture area. The cell structure features a solar cell with passivating contacts on the rear side, double-sided screen-printed metal contacts, a module-optimized anti-reflective coating (ARC) and a homogeneous front-side emitter on a *n*-type silicon substrate.**

I. INTRODUCTION

In 2022, solar cell manufacturing was dominated by the passivated emitter and rear solar cell (PERC) structure [1] which features dielectric rear-side passivation with local contacts [2]. An example of this success story is Q CELLS' Q.ANTUM technology [3-6], which comprises a PERC-like structure with additional features. So-called passivating contacts to further reduce charge carrier recombination are an alternative to dielectric surface passivation. These contacts electrically separate the highly recombination-active metal contacts from the silicon absorber while simultaneously allowing for efficient charge extraction [7]. Examples of passivating contacts are amorphous silicon [8,9], metal oxides [10] and polycrystalline silicon in combination with an interfacial oxide [11-13].

While PERC-based solar cell structures are close to reach their inherent limits mainly due to high saturation current densities on the rear side [14], passivating contacts are the most promising solar cell scheme to keep the annual learning rate. Recently, significant production capacity expansion has been ramped up or been announced. ITRPV predicts that the market share of passivating contacts will increase from 11 % in 2022 to 28 % in 2026 [15]. Though the efficiency potential of passivating contact solar cells has been evident for a long time and champion cells of up to 26.4% have been demonstrated [16], the main challenge remains to develop a low-cost manufacturing process.

This work summarizes the development of Q CELLS' innovative Q.ANTUM NEO technology based on a passivating contact solar cell approach. It discusses the main requirements to enable a cell efficiency of up to 25.2 % and average efficiencies of 24.7 % with a lean and cost-effective process flow which has been successfully transferred to mass production.

II. APPROACH

Cell structure and fabrication

Fig. 1 shows a sketch of the solar cell structure based on our new Q.ANTUM NEO technology. The solar cell features a passivating contact on the rear side, double-sided screen-printed metal contacts, a module-optimized ARC, a homogeneous diffused front-side emitter and a *n*-type silicon substrate. For our Q.TRON module integration, state-of-the-art technology such as half-cells, multi-wire interconnection, standard encapsulants and zero-gap technology can be applied.

Starting in 2019, the pilot line of Q CELLS' R&D Center in Thalheim, Germany, has been upgraded with additional process tools to evaluate different process flows and to identify a lean and cost-effective sequence that can be transferred to mass production. The main focus has been on an evolutionary development design based on the PERC-like Q.ANTUM technology with only two additional process steps that can be easily integrated into existing production lines. Standard commercial-grade *n*-type Czochralski-grown silicon (Cz-Si) with a surface area up to 330.2 cm² ("M10-format") has been utilized and only mass-production compatible processes have been applied. Core of the Q.ANTUM NEO technology is the development of a full-area passivation on the rear side based on polycrystalline silicon [17] which enables an extremely low saturation current density of $J_{o,rear} < 3$ fAcm^{-2}. This technology has been recently transferred to mass production. Applying state-of-the-art machine learning algorithms [18] with Q CELLS single wafer tracking system [19], has been key to a very short ramp-up time and process integration from pilot line to continuous production mode.

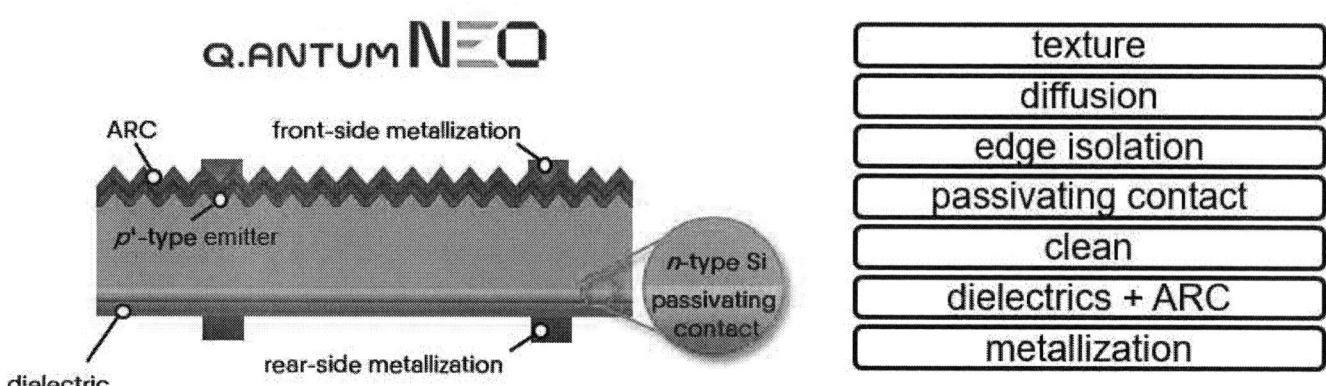

Fig. 1.   Solar cell structure and processing sequence of Q CELLS' Q.ANTUM NEO technology.

III. RESULTS AND DISCUSSION

A. Cell performance

Fig. 2 shows the average efficiency of cells with Q CELLS' Q.ANTUM NEO technology as a function of time in pilot line production. Within the first two years a learning rate of >1 %abs/year has been observed. This is much higher than the average +0.5 %abs/year for Q.ANTUM solar cells that has been achieved from 2013 to today [20]. All reported values are total-area efficiencies measured with an industrial cell tester and the solar cells do not comprise an ARC matched to air for highest cell efficiency but for integration into a module for highest module efficiency.

The crucial milestone in the development process has been the improvement of the rear- and front-side saturation current densities based on mass production compatible processes. Besides process optimization and stabilization, the interaction between individual process steps has proved to be essential for the steep learning curve.

After the preparation of the rollout in 2021, the learning curve has resumed to ca. 0.5%abs annual increase. This indicates that Q CELLS' Q.ANTUM NEO technology has reached a certain level of maturity and improvements will be rather incremental.

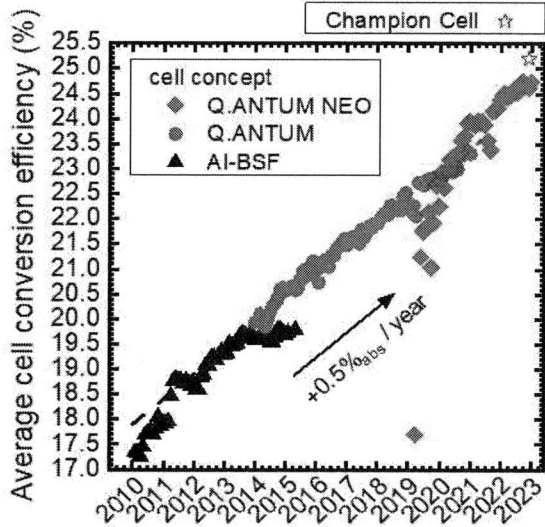

Fig. 2.   Solar cell efficiency of pilot line production as a function of time at Q CELLS. All reported values are total-area efficiencies.

To date, the champion cell has an efficiency of 25.2% (independently certified by ISFH). It has been fabricated in a regular cell batch applying the processing sequence shown in Fig. 1. The high open circuit potential of 728 mV has been mainly achieved by reduction of the front metal contact recombination in combination with optimization of the emitter doping. The *I-V* parameters are summarized in Table 1.

V_{oc} [mV]	J_{sc} [mAcm^{-2}]	FF [%]	η [%]
728	41.3	83.8	25.2

Table 1. *I-V* parameters of the Q.ANTUM NEO champion cell externally certified at ISFH

Converting Q.ANTUM NEO solar cells using state-of-the-art technology such as half-cells, multi-wire interconnection, standard encapsulants and zero-gap technology yields power output of up to 444 W (132 half cells with M6 format) as shown in Table. 2. This corresponds to a module conversion efficiency of 23.9 % on aperture area.

V_{oc} [V]	I_{sc} [A]	FF [%]	P_{MPP} [W]
47.8	11.31	82.1	444

Table 2. *I-V* parameters of the Q.TRON champion module (pending independent certification)

B. Loss analysis and efficiency potential

An energy loss channel analysis has been done aided by Quokka2 simulation to estimate the contribution by bulk, front and rear side of the solar cell as shown in Fig. 3.

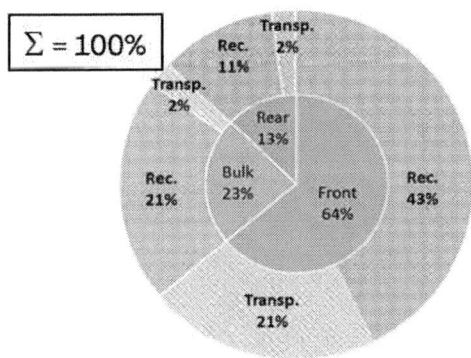

Fig. 3 Different loss channels for Q CELLS' Q.ANTUM NEO technology based on Quokka2 simulation

Whereas in typical PERC-like cell structures the rear side contribution to the total loss is in the range of one third, the Q.ANTUM NEO technology has an almost ideally passivated rear side and the main energy loss channel is on the front side of the solar cell. This indicates the importance to further improve the surface and metal contact passivation and to minimize Auger recombination within the emitter. Procedures that are currently being evaluated on R&D level suggest that the emitter saturation current density can be decreased to < 8 fAcm^{-2} which will enable a cell efficiency of > 25.5 % in the near future. Midterm improvements on both front and rear side indicate an efficiency potential of above 26 %. It is predictable that the Q.ANTUM NEO technology will maintain the annual efficiency increase of ca. 0.5%$_{abs}$ for the coming years in contrast to PERC-like cell concepts.

IV. SUMMARY, CONCLUSION AND OUTLOOK

This work summarizes the results of 3 years development of Q CELLS' Q.ANTUM NEO technology on a solar cell structure with passivated rear-side contacts. By applying a lean and cost-effective process sequence, an average energy conversion efficiency of 24.7 % in pilot line production has been demonstrated while the champion solar cell features a conversion efficiency of 25.2 % as independently confirmed by ISFH. This results in a high module power output of 444 W (132 M6 half cells) corresponding to a module conversion efficiency of 23.9 % on aperture area. The developed solar cell structure can be incorporated into modules applying state-of-the-art standard module interconnection and encapsulation technology. Hence, existing multi-GW scale-passivated emitter and rear cell and module fabrication capacities are currently being upgraded for the new technology and will be used, accordingly, for mass manufacturing of this structure at very competitive manufacturing costs. Midterm improvements on both front and rear sides imply an efficiency potential of above 26 % which will support Q CELLS efficiency roadmap for the coming years.

REFERENCES

[1] ITRPV: 2020 Res., Ed. 12, October 2020. www.itrpv.vdma.org.
[2] A. W. Blakers *et al.*, Appl. Phys. Lett. **55**, no. 13, p. 1363, 1989.
[3] P. Engelhart *et al.*, in Proc. 26th EUPVSEC, 2011, pp. 821–826.
[4] A. Mohr *et al.*, in Proc. 26th EUPVSEC, 2011, pp. 2150–2153.
[5] F. Kersten *et al.*, Solar Energy Materials and Solar Cells **142**, pp. 83–86, 2015.
[6] F. Fertig *et al.*, Energy Procedia **124**, pp. 338–345, 2017.
[7] R. Brendel *et al.*, IEEE Journal of PV **6** (6), pp. 1413–20, 2016.
[8] K. Masuko *et al.*, IEEE Journal of PV **4** (6), pp. 1433–5, 2014.
[9] S. de Wolf *et al.*, Green **2** (1), pp. 7-24, 2012.
[10] J. Bullock *et al.*, Nat. Energy **1** (3), p. 63016, 2016.
[11] F. Feldmann *et al.*, Solar Energy Materials and Solar Cells **159**, p. 265-71, 2017.
[12] R. Peibst *et al.*, Prog. Photovolt. Res. Appl. **28** (6), p. 503-16, 2020.
[13] N. Nandakumar *et al.*, in Proc. 46th IEEE PVSC, 2019, pp. 1463-5.
[14] B.G. Lee *et al.*, "Development and mass production of bifacial Q. ANTUM p-Cz PERC cells," Proceedings of the 46th IEEE PVSC, Chicago, USA, 2019, pp. 1460-1462.
[15] ITRPV: 2022 Res., Ed. 13, November 2022 www.itrpv.vdma.org.
[16] https://ir.jinkosolar.com/news-releases/news-release-details/jinkosolars-high-efficiency-n-type-monocrystalline-silicon-2
[17] F. Fertig *et al.*, IEEE Journal of PV **12** (1), pp. 22-25, 2021
[18] S. Wasmer *et al.*, in Proc. 38th EUPVSEC, 2021, p. 107-110
[19] S. Wanka *et al.*, in Proc. 26th EUPVSEC, 2011, p. 1104-1106
[20] F. Kersten *et al.*, AIP Conference Proceedings 2487, 130007 (2022)

Outdoor study of Photovoltaic Mini-Modules with different perovskite compositions

Vasiliki Paraskeva, Maria Hadjipanayi, Matthew Norton, Aranzazu Aguirre, Anurag Krishna, Rita Ebner, Tommasso Fontanot, Sabrina Pechmann, Silke Christiansen, George E. Georghiou

University of Cyprus, Nicosia, Cyprus

Imec, imo-imomec,Thin Film PV Technology , Genk, Belgium

EnergyVille, imo-imomec, Genk, Belgium

Hasselt University,imo-imomec, Hasselt, Belgium

AIT Austrian Institute of Technology, Vienna, Austria

Fraunhofer Institute for Ceramic Technologies and Systems IKTS, Forchheim, Germany

Max Planck Institute for the Science of Light, Erlangen, Germany

A long-term outdoor study of photovoltaic mini-modules with different perovskite compositions was undertaken to detect differences in their long-term performance that could be attributed to their composition. Diurnal efficiency degradation and overnight recovery was observed over the outdoor testing period. In addition, the performance recovery of the mini-modules was investigated after their removal from the field to detect differences in the reversible process mechanisms between the different module compositions. Light current-voltage (IV) scans, Resonance Raman spectroscopy techniques and spatially-resolved Electroluminescence (EL) measurements were utilized for this purpose.

Achieving a new world record silicon solar cell efficiency of 26.81% using SHJ device structure

Xixiang Xu, Minghao Qu, Miao Yang, Xiaoning Ru, Shi Yin, Chengjian Hong, Fuguo Peng, Junxiong Lu, Liang Fang, Zhenguo Li

Central R&D Institute, Longi Green Energy Technology Co., Ltd. Xi'an, Shaanxi, China

Yichun Wang, Tian Xie

R&D Centre-Wafer Business Unit, Longi Green Energy Technology Co., Ltd. Xi'an, Shaanxi, China

Abstract — **As the cornerstone of photovoltaitics industry, silicon solar cell draws extensive interests and its progress on conversion efficiency concerns the implementation of carbon neutrality promise. In order to achieve high efficiency, good surface passivation, low contact resistance and transparent front skin are the indispensable requirements, bringing about variety of technologies and strategies to balance out for maximum efficiency. Recently, our group achieved a new world record, 26.81%, in silicon solar cells using improved heterojunction technology (SHJ). Thanks to the successful integration of nanocrystalline doped hydrogenated silicon (nc-Si:H), surface passivation quality is further enhanced by field effect and the contact resistance is highly restrained. On the other hand, oxygen doping at front nc-Si:H broadens the band gap and the transparency to short-wavelength sunlight is increased for higher short-circuit current density. Combined with new transparent conductive oxide and advanced metallization, intrinsic properties of silicon emerge from intricate power loss mechanism, causing unprecedented fill factor and record conversion efficiency in silicon solar cells. Moreover, our record solar cell is based on front and rear contacted architecture with full-size commercial Czochralski silicon wafer and total-area certification. With the simplicity and compatibility of our SHJ technology to mass production, it is easy to transfer the result into industrial manufacture with high mass production cell conversion efficiency.**

I. INTRODUCTION

Silicon is the most fundamental and decisive material in photovoltaics (PV) industry, which has boosted up dramatic expansion of silicon solar cells and provides a feasible technology solution for carbon neutruality promise in the globe. As a strong area-correlated product, power conversion efficiency (PCE) is the eternal pursuit in the community of silicon solar cells and, to a large extent, determines the competitiveness of various high-efficiency solar cell technologies. At present, passivated emitter and rear cell (PERC) solar cells dominate the PV market due to its relatively high PCE and excellent adaptability for mass production, long-term reliability and economic efficiency. Thus, last decade witnesses the rapid development of PERC solar cells over the others. However, limited by carrier recombination at Al/Si contact and graded doping homojunction, further improvement of conversion efficiency on PERC solar cells is hardly maintained. Thus, the so-called passivating contact technologies, like tunnel oxide passivating contact (TOPCon) and silicon heterojunction (SHJ) technology, were developed in order to surpress the recombination loss and thus keep improving the conversion efficiency.

SHJ technology employs the stack of intrinsic hydrogenated amorphous silicon (i-a-Si:H) layer and n-/p-type doped hydrogenated silicon layer, which is either amorphous structure (a-Si:H) or nanocrystalline structure (nc-Si:H), to passivate and selectively contact the bulk silicon. With high-quality chemical passivation by i-a-Si:H and field-effect passivation by doped silicon layer, this technology helps to minimize the surface recombination of the bulk silicon and thus enhance the open-circuit voltage (V_{OC}). The passivation quality of SHJ technology surpasses that of TOPCon technology which designs an ultra-thin SiOx film superposed by a few-hundred-nm layer of heavily doped poly-silicon for passivating contact. However, conventional SHJ technology using a-Si:H layer as selective contact suffers from low short-circuit current density (J_{SC}) and small fill factor (FF), due to its heavy absorption at short wavelength and high resistivity, thus leading to an ordinary PCE and less competive to TOPCon. To avoid the above shortcoming, adoption of nc-Si:H as the selective contact in SHJ solar cell has been experimentally and theoretically proved as an effective method.

In this work, n-type oxygen-alloyed nc-Si:H (n-nc-SiOx:H) layer is used at the front as the band gap is widened for better transparency at the short-wavelength range. At the rear, p-type nc-Si:H (p-nc-Si:H) layer is technically realized and the contact resistance at the emitter is largely reduced. Finally, combined with new transparent conductive oxide (TCO), advanced metalliztion and iterative optimization, our SHJ champion cell reaches 26.81% in PCE, a new world record in silicon solar cell [1], which demonstrates a feasible way to high conversion efficiency.

II. EXPERIMENTAL DETAILS

A. Manufacture process

978-1-6654-6060-6/23 $31.00 © 2023 IEEE

SHJ solar cells were fabricated by LONGi R&D line, which is based on LONGi Cz n-type monocrystalline silicon (n-cSi) wafer with M6 format and (100) crystalline orientation. The bulk resistivity of the wafer is 1.2-1.5 Ω·cm and the final thickness with ingoring grid electrode is around 130 μm. The n-cSi wafer was first cleaned and textured by wet chemical process. Then, i-a-Si:H layers on both sides were deposited by radio-frequency plasma-enhanced chemical vapor deposition (RF-PECVD) at 13.56 MHz and covered by nc-Si:H which is realized by very-high-frequency PECVD (VHF-PECVD) at 40 MHz. More details can be found in our previous work [2]. In order to acquire p-nc-Si:H layer with limited thickness, CO_2 plasma treatment was carried out on i-a-Si:H and high flow rate H_2 was introduced during deposition. TCO layers were coated by reactive plasma deposition (RPD) facility using a cylindrical target of ceria-doped indium oxide (ICO). Silver grid electrodes were printed by advanced metallization process with the finger width around 25 μm, followed by a 190°C annealing process for 30 min. In order to further increase J_{SC}, a 150-nm-thick MgF_2 layer was evaporated on the front TCO layer as the second anti-reflective coating. Finally, light soaking under 60 suns was performed for 90 sec at the temperature of 190°C.

B. Characterization

Light current-voltage (IV) curves of our SHJ champion solar cells were tested and certified by ISFH CalTec system under standard AM1.5G illumination. Raman spectra were collected by Horiba LabRAM Odyssey Raman spectrometer with a 523 nm excitation laser. The test layers were deposited on planar glass substrate, using the same deposition processes of the solar cells. The cross section on the stack of ICO/n-nc-SiOX:H/i-a-Si:H/n-cSi at the front and ICO/p-nc-Si:H/i-a-Si:H/n-cSi at the rear was observed by transmission electron microscopy (TEM), and the test samples were cut from the finished solar cell by focused ion beam (FIB).

C. Simulation

Power loss analysis (PLA) was performed using Quokka3 software with the measured parameters published where else. Here, Richter model [3] for Auger recombination is employed with considering photon recycling effect. Optical calculation is simplified by scaling the optical generation of Green $4n^2$ model in order to obtain the measured J_{SC}. The simplication avoids the intractable measurement on the optical parameters of the solar cell since they are excluded in PLA.

III. RESULTS AND DISCUSSION

Our first piece of SHJ solar cell was delivered after R&D facilities move-in on Feb. 2021 and its conversion efficiency reaches 21.39% measured by our own IV tester. The first certified SHJ solar cell using similar manufacture process as described in Ref. 2 has the improved conversion efficiency at 25.26%, which adopted n-nc-SiOX:H at the front and p-a-Si:H

as the rear emitter. Here, new TCO target with 1wt%-doped indium oxide was employed for its higher transparency and lower sheet resistance than conventional indium oxide with 10wt% tin oxide (ITO). A giant improvement occuring at 25.82% and 26.30% SHJ solar cell introduces RPD-deposited ICO and highly crystallized p-nc-Si:H at the rear, respectively. The introduction of p-nc-Si:H as the rear emitter largely reduces the contact resistance, where direct band-to-band tunneling dominates the carrier transport through the heterojunction and proved to be the most efficient mannar. It's worth to note that the record FF of 86.59% in silicon solar cell was measured by 26.30%-PCE SHJ solar cell, although its successors hold higher PCE. Continual increase of the conversion efficiency to 26.50% and 26.81% includes the employment of advanced metallization with high-aspect-ratio grid fingers, elaborate optimization on the window layers and enhanced back reflection. Finally, our SHJ champion cell based on n-cSi wafer with front and rear contacted device structure, rear emitter and 18 busbars was certified with a PCE of 26.81% which is the new world record in silicon solar cells.

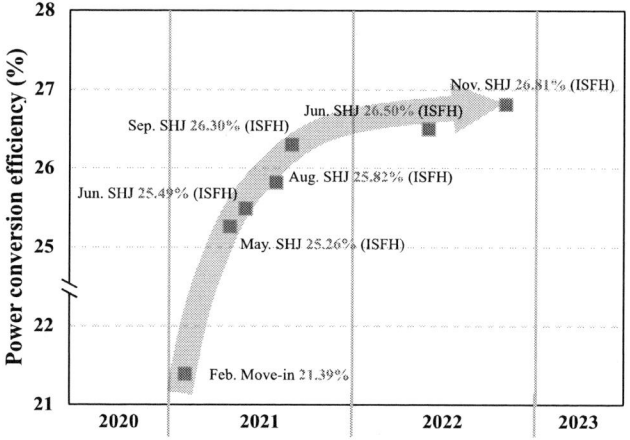

Fig. 1. Improvement of conversion efficiency by LONGi SHJ technology since facility move-in on Feb. 2021.

In Fig. 2, ISFH-certifed light IV characteristics of the SHJ champion cell is shown with the parameters in the inset. Compared to our first certified SHJ solar cell with 25.26% PCE and similar technologies in Ref. 2, V_{OC} gains by 2.9 mV to 751.4 mV, FF gains by 0.61% to 86.07%, and J_{SC} gains 1.98 mA/cm^2 to 41.16 mA/cm^2, raising the efficiency by 1.55%. Here, the gain of V_{OC} is caused by the application of nc-Si:H at both sides which leads to enhanced field-effect passivation due to its higher doping efficiency. The improvement of FF is mainly derived from two aspects, i.e. the gain of V_{OC} and the reduction of series resistance (R_S). In the former, recombination loss is reduced and only intrinsic recombination dominates in the loss mechanism. The latter is mainly attributed to the application of ICO layers with low sheet resistance and the smaller contact resistivity caused by doped nc-Si:H structures. In addition to the electrical improvment, the optical aspect is largely enhanced as the results of further optimization of n-nc-SiO$_X$:H, adoption of high-transparency ICO layer and high-aspect-ratio grid fingers at the front. This certification was performed on total area illumination.

PLA analysis was performed by Quokka 3 software based on the successful simulation on the certified IV curve. As the result, the intrinsic recombination in bulk silicon dominates the power loss by ~60% and limits further increase in conversion efficiency. A considerably low series resistance (ρ_C) value of 0.20 mΩ·cm^2 has been achieved. As the light-IV measurement by ISFH excludes the resistive contribution from the front busbars and rear grid, the resistance here implies contact resistance of ICO/n-nc-SiO$_X$:H(or p-nc-Si:H)/i-a-Si:H/n-cSi stack. The recombination current density (J_0) on both sides reaches a low level of 0.5 fA/cm^2, while the front and rear contact resistivity are improved to 30 mΩ·cm^2 and 10 mΩ·cm^2, respectively. These improvements have led to extrinsic recombination loss of ~0.2% and resistive loss of ~0.4%. According to Schmidt's formula [4] for predicting the conversion efficiency with the passivating contacts by various technologies, our SHJ champion cell could result in a theoretical limiting effcieincy at 29.0%.

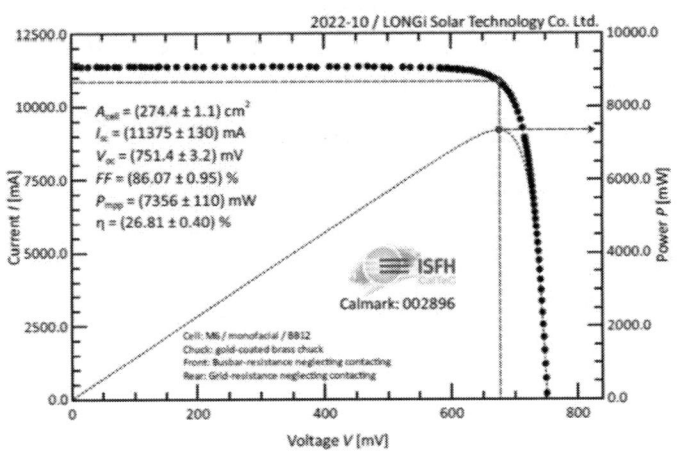

Fig. 2.   ISFH-certificated SHJ solar cell with a PCE of 26.81%.

IV. CONCLUSION

A new world record (26.81%) of silicon solar cell was achieved by our SHJ technology using industry-compatible silicon wafer and mass production facilities. With the application of i-a-Si:H, n-nc-SiO$_X$:H and p-nc-Si:H, transparent passivating contact is implemented on both sides of silicon wafer where intrinsic properties of bulk silicon dominates the loss mechanism and the power loss relating to bulk silicon is limited. Combined with related technologies, such as new TCO layer and advanced metallization, electrical and optical performance were highly ameliorated.

Due to the simplicity and compatibility of our SHJ technologies to mass production facilities, it is easy to tranfer our achievement to industrial applications with high throughout. Based on our pilot line, the batch average conversion efficiency has reached 25.41% in PCE.

REFERENCES

[1]   K. Yoshikawa, H. Kawasaki, W. Yoshida, T. Irie, K. Konishi, K. Nakano, T. Uto, D. Adachi, M. Kanematsu, H. Uzu, and K. Yamamoto, "Silicon heterojunction solar cell with interdigitated back contacts for a photoconversion efficiency over 26%," *Nature Energy*, vol. 2, pp. 17032, 2017.

[2]   X. Ru, M. Qu, J. Wang, T. Ruan, and X. Xu, "25.11% efficiency silicon heterojunction solar cell with low deposition rate intrinsic amorphous silicon buffer layers," *Solar Energy Materials and Solar Cells*, vol. 215, pp. 110643, 2020.

[3]   A. Richter, M. Hermle, and S. W. Glunz, "Reassessment of the Limiting Efficiency for Crystalline Silicon Solar Cells," *IEEE Journal of Photovoltaics*, vol. 3, pp. 1184-1191, 2013.

[4]   J. Schmidt, R. Peibst, and R. Brendel, "Surface passivation of crystalline silicon solar cells: Present and future," *Solar Energy Materials and Solar Cells*, vol. 187, pp. 39-54, 2018.

Utilizing Particle Swarm Optimization for Autocalibration of LED Solar Simulators

Jann A Grovogui, John C Nocerino, Don Walker

The Aerospace Corporation, El Segundo, CA, United States

LED solar simulators offer a low-cost, scalable, and versatile way to reproduce the solar spectrum. Increasing the number of unique LEDs increases the ability of users to accurately recreate the solar spectrum, but also increases the time to achieve an accurate calibration due to the increasing complexity of the system. This study introduces the Particle Swarm Optimization (PSO) algorithm as a method to assist users in quickly achieving an accurate calibration. The PSO algorithm provides a unique method to solar simulator calibration because it not only attempts to reduce spectral mismatch, but also accounts for Jsc that would be produced by an illuminated cell. The algorithm is flexible in that it can currently be applied to any single junction solar cell technology (given that cell' spectral response) and any solar spectrum (AM0, AM1.5, etc.). This investigation shows that a PSO autocalibration algorithm can generate a theoretical calibration for an LED solar simulator with 36 channels in ~14 seconds. Furthermore, this theoretical calibration for a silicon cell was calculated to have a Jsc that deviates less than 0.5% from the ideal Jsc under AM0 illumination. Although the real spectrum collected from the simulator after calibration usually differs by a few percent error from the theoretical spectrum, this can be due to a variety of factors such as the temperature of the system and the method in which channel power settings are derived from the spectrum irradiance. This presentation seeks to open the door to conversations about the use of optimization algorithms (not just the PSO algorithm) for auto-calibration and to further the discussion of best practices for calibration and use of LED solar simulators.

Analysis of Cu(In,Ga)Se$_2$ heterojunction solar cells in terms of their balance of thermodynamic potentials

Uwe Rau[1,2], Felix Komoll[1,2], Tim Helder[3,4], Mario Zinßer[3,4], Ana Kanevce[3], Thomas Kirchartz[1,5], Theresa Magorian Friedlmeier[3]

[1]IEK-5 Photovoltaik, Forschungszentrum Jülich GmbH, Wilhelm-Johnen Straße, 52425 Jülich, Germany
[2]Jülich Aachen Research Alliance (JARA-Energy) and Faculty of Electrical Engineering and Information Technology, RWTH Aachen University, Schinkelstr. 2, 52062 Aachen, Germany
[3]Center for Solar Energy and Hydrogen Research, Stuttgart, Germany
[4]Light Technology Institute (LTI), Karlsruhe Institute of Technology (KIT), Karlsruhe, Germany
[5]Faculty of Engineering and CENIDE, University of Duisburg-Essen, Carl-Benz-Str. 199, 47057 Duisburg, Germany

Abstract— **This paper analyses theoretically electronic loss mechanisms in Cu(In,Ga)Se₂ heterojunction solar cells. The analysis is based on the balance of thermodynamic potentials (BoTPot), a novel approach that considers the complete set of thermodynamic potentials: electrical, chemical, and electrochemical potentials for electrons and holes. Here we concentrate on loss mechanisms that are specifically important for ZnO/CdS/Cu(In,Ga)Se₂ heterojunction solar cells, like interface recombination and carrier transport across barriers at heterointerfaces. We attribute the recombination losses at interfaces to losses in chemical potential leading to a reduction of the open-circuit voltage. In contrast, carrier transport across energy barriers may result either in losses of the electrostatic potential (for majority carriers) or in losses of the chemical potential (for minority carriers). This different attribution of minority and majority carrier losses sheds new light on the question of how to describe the selectivity of contacts in solar cells.**

Keywords— *thermodynamic loss analysis, Cu(In,Ga)Se₂ solar cells, numerical simulation, resistive losses, recombination losses*

I. INTRODUCTION

A solar cell transforms radiation power from the sun into electrical power at the cell's terminals. The investigation of this transformation is usually done in two stages: (i) the transformation of the solar light into free charge carriers in the photovoltaic absorber material (optical analysis) and (ii) the collection of these charge carriers at the terminals of the device in order to create electrical energy (electrical analysis). Thermodynamic descriptions are available for both the optical [1-3] and the electronic stage [4]. Recently, the description of the charge carrier dynamics in terms of free energy balance [4] was extended towards an analysis of the full balance of all thermodynamic potentials (BoTPot) [5], treating chemical, electrical, and electrochemical potentials separately.

The present paper presents the application of the BoTPot method to ZnO/CdS/Cu(In,Ga)Se₂ heterojunction solar cells.

Starting from numerical device simulations (SCAPS [6]) that yield the device's band diagrams in thermal equilibrium and in non-equilibrium, i.e. under illumination and with a given voltage bias.

While the final contribution will present a complete thermodynamic loss analysis of optimum and realistic Cu(In,Ga)Se₂ devices, the current extended abstract concentrates on the description and the illustration of the method (Section II) and the investigation of losses at the interfaces between the absorber and the back contact as well as between the absorber and the front contact (Section III). These two types of losses are closely related to the concept of contact selectivity. Here, two distinct approaches are available, namely the resistive model by Brendel and Peibst [7], and the kinetic model by Roe et al. [8]. As discussed in Ref. [9], these models are not compatible. Rather they describe specific situations where the losses show up as losses either in the electrostatic potential (resistive) or in the chemical potential (kinetic), exclusively.

The present results show that a back-contact barrier affects the collection of majority carriers (holes). In this situation, the power loss across the contact results from a loss in the electrostatic potential, i.e. the loss is resistive. In contrast, losses at the front contact are related to the extraction of electrons (minority carriers) through the space charge region of the junction and across the CdS buffer layer. Here, it turns out that the loss of output power follows from a loss in the chemical potential (kinetic loss).

II. THEORY AND ILLUSTRATION

With the excess electrostatic potential of electrons defined via $\delta\varphi_n = -q\delta\Phi$ (in units of electron volts) the excess free energy of electrons is given by

$$\delta\eta_n = \delta\mu_n - q\delta\Phi = \delta\mu_n + \delta\varphi_n. \tag{1}$$

Likewise, we have

$$\delta\eta_p = \delta\mu_p + q\delta\Phi = \delta\mu_p + \delta\varphi_p \qquad (2)$$

for the excess free energy of holes, with the excess free energy potentials $\delta\eta_{n/p}$, the excess electrostatic potentials $\delta\varphi_{n/p}$ and the excess chemical potentials $\delta\mu_{n/p}$ for electrons and holes, respectively. Figure 1a shows a non-equilibrium band diagram of a simple ZnO/CdS/Cu(In,Ga)Se$_2$ device obtained from numerical device simulations (SCAPS [6]).

Note that the settings for these calculations are kept simple, in order to illustrate how the thermodynamic potentials $\delta\eta_{n/p}$, $\delta\varphi_{n/p}$, and $\delta\mu_{n/p}$ (shown in Fig. 1b) are obtained technically from the band diagram. Since we have $E_{Fn}^{eq} = E_{Fn}^{eq} = 0$ for the equilibrium quasi-Fermi levels of electrons and holes, the non-equilibrium electrochemical potentials are obtained from $\delta\eta_n = E_{Fn}$ and $\delta\eta_p = -E_{Fp}$. The excess electrostatic potential $\delta\varphi_n$ for electrons is defined as the change of the energy E_C of the conduction band with respect to its equilibrium value, i.e., $\delta\varphi_n = E_C - E_C^{eq}$. Because of the opposite charge sign, the electrostatic potential of holes is $\delta\varphi_p = -\delta\varphi_n$. Finally, the chemical potentials are obtained from Eqs. (1) and (2) via $\delta\mu_{n/p} = \delta\eta_{n/p} - \delta\varphi_{n/p}$.

Figure 1b shows the potentials $\delta\eta_{n/p}$, $\delta\varphi_{n/p}$, and $\delta\mu_{n/p}$ as obtained from the band diagram in Fig. 1a. In the neutral bulk of the absorber (x < 1 μm) we have $\delta\eta_n = \delta\mu_n$, $\delta\eta_p = \delta\mu_p$, and $\delta\varphi_n = -\delta\varphi_p = 0$, i.e., the free energy (electrochemical potential) of the charge carriers is entirely represented by their chemical potentials. This situation is maintained towards the back contact because of the idealized assumptions (flat band, high collection velocity for holes, zero recombination velocity for electrons). Within the space charge region, the chemical potential $\delta\mu_n$ goes to zero whereas the electrostatic potential $\delta\varphi_n$ increases up to $\delta\varphi_n = \delta\eta_n = qV$. This transformation of the chemical potential of photogenerated charge carriers in the neutral absorber into the electrostatic potential and finally to a photovoltage V at the terminals of the device is the electric stage of the photovoltaic effect, as mentioned in the introduction.

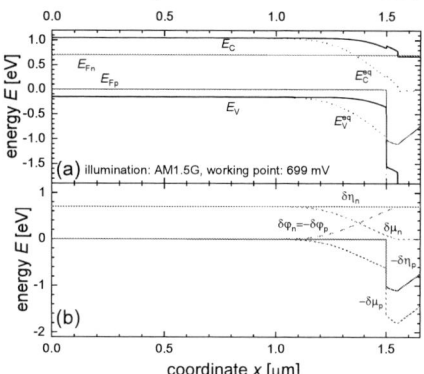

Fig. 1. (a) Non-equilibrium band diagram of a simplified ZnO/CdS/Cu(In,Ga)Se$_2$ solar cell under AM1.5G illumination and at the voltage V_{mpp}= 699 mV. At the back contact (x=0), flat band conditions are assumed. The CdS buffer layer (x=1.50 to 1.55 μm) has a thickness of 50 nm and a band offset of 0.1 eV with respect to the absorber. The equilibrium conduction band and valence band energies are shown as gray dashed lines.

(b) Non-equilibrium thermodynamic potentials $\delta\eta_n$, $\delta\mu_n$, $\delta\varphi_n$ for electrons and $\delta\eta_p$, $\delta\mu_p$, $\delta\varphi_p$ for holes. Note that the potentials for holes are shown with a minus sign for better visibility.

The chemical and the electrochemical potentials $\delta\mu_p$, $\delta\eta_p$ are relatively high in the CdS buffer layer (notice the minus sign for the hole potentials in Fig.1b). This is because holes photogenerated in the high band gap CdS are minority carriers and contain more free energy than holes generated in the lower band gap Cu(In,Ga)Se$_2$, where they are majority carriers.

III. ANALYSIS OF FRONT- AND BACK-CONTACT LOSSES

A. Back contact

Figure 2a shows the band diagram of the same Cu(In,Ga)Se$_2$ solar cell as in Fig.1 with the only difference that at the back contact we have introduced a barrier of 250 mV resulting in a downward band bending towards the back contact. This band bending affects the collection of holes, the majority carriers, and significantly influences the electronic properties of the device [10]. In the potential diagram (Fig. 2b) we see that, as expected, changes with respect to the reference system (Fig. 1b) only occur at the back contact. These changes are especially significant for the potentials of the holes (see inset).

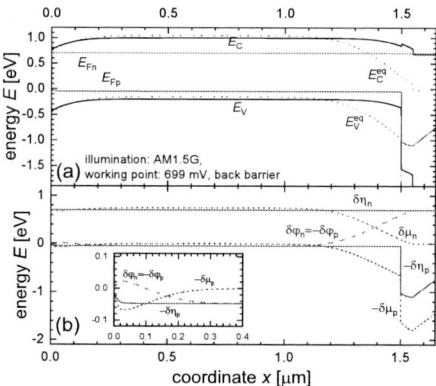

Fig. 2. (a) Non-equilibrium band diagram of a simplified ZnO/CdS/Cu(In,Ga)Se$_2$ solar cell under AM1.5G illumination and at the voltage V_{mpp}= 699 mV. The cell has a barrier of 250 mV at the back contact leading to a downward bending of the bands towards the back contact. (b) Non-equilibrium thermodynamic potentials $\delta\eta_n$, $\delta\mu_n$, and $\delta\varphi_n$ for electrons. The inset highlights the resistive losses of 48 meV at the back contact.

The free energy $\delta\eta_p$ sharply drops from 48 mV to zero close to the interface. The chemical potential $\delta\mu_p$ is fixed to zero in the bulk of the absorber (x > 0.4μm) by the doping and to zero at the back contact by definition. The accumulation of holes due to the transport restriction through the barrier leads to an increase of $\delta\mu_p$ within the space charge region (0<x<0.4μm). The electrostatic potential $\delta\varphi_p$ is influenced by both the accumulated holes and negative space charges of the space charge region. At the contact, the electrostatic potential $\delta\varphi_p$ is fixed to zero, whereas in the bulk $\delta\varphi_p$ approaches $\delta\eta_p \approx 48$ meV. Considering the loss of free energy $\Delta\eta_p^{back}$ we find

$$\Delta\eta_p^{back} = \Delta\varphi_p^{back} \approx 48\text{ meV}, \qquad (3)$$

i.e., the potential loss at the back contact (majority carrier contact) is a loss in electrostatic potential, hence a resistive loss. This finding corresponds to our earlier result on silicon solar cells [5], where losses at the majority carrier contact have been found to be of resistive nature as well.

B. Front contact

Due to the band offset between the CdS buffer layer and the Cu(In,Ga)Se$_2$ absorber, transport of electrons towards the ZnO windows can be inhibited [11]. In the present simulations, we assumed a band offset of 0.1 eV. A significant reduction of the mobility of electrons in the CdS buffer layer leads to a similar effect as increasing the band offset.

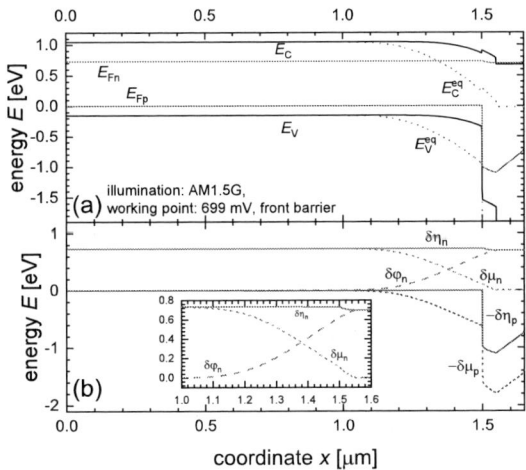

Fig. 3. (a) Non-equilibrium band diagram of a simplified ZnO/CdS/Cu(In,Ga)Se$_2$ solar cell under AM1.5G illumination and at the voltage V_{mpp}= 699 mV. The cell has a significant reduction of the electron mobility in the CdS buffer layer. (b) Non-equilibrium thermodynamic potentials $\delta\eta_n$, $\delta\mu_n$, and $\delta\varphi_n$ for electrons and (c) $\delta\eta_p$, $\delta\mu_p$, and $\delta\varphi_p$ for holes. (d) Fluxes p_F, p_{elec}, and p_{chem} of the electrochemical, the electrostatic, and the chemical potentials.

As shown in Fig. 3a, there is now a drop of the electron quasi-Fermi level through the buffer layer by about 34 meV. Accordingly, the free energy $\delta\eta_n$ of electrons shown in Fig. 3b (see also inset) decreases from 733 meV inside the absorber (x < 1 μm) towards 699 meV at the front contact (corresponding to the voltage V = 699 mV). It is also seen from Fig. 1b that in the absorber the chemical potential equals the electrochemical potential (free energy) $\delta\mu_n = \delta\eta_n = 733$ meV, whereas $\delta\mu_n$ becomes zero at the front contact. We now identify the portion of chemical energy that is *not* translated into the electrostatic potential of 699 meV as the loss in chemical energy. With this we have

$$\Delta\mu_n^{front} = \Delta\eta_n^{front} \approx 34\text{ meV}. \qquad (4)$$

Thus, the loss of free energy at the front (minority carrier) contact can be entirely attributed to a loss in chemical energy, which is the definition of a loss according to the kinetic model. In contrast, for the minority carrier contact we found the opposite: The free energy loss directly corresponds to a loss of the electrostatic potential (resistive loss) as shown by Eq. (3).

IV. SUMMARY

The present contribution has demonstrated the application of the balance of thermodynamic potentials (BoTPot) analysis [5] to Cu(In,Ga)Se$_2$ solar cells. Concentrating on losses at the back and at the front contacts, we found that losses at the back contact (majority carrier contact) are resistive, i.e., losses of electrostatic potential. In contrast, losses at the front contact (minority carrier contact) are kinetic, i.e., these losses result from losses in the chemical potential. We argue that the attribution of resistive losses to the majority carrier contact and of kinetic losses to the majority carrier contact is quite general for solar cells.

ACKNOWLEDGEMENT

The authors acknowledge financial support by the German Federal Ministry of Economic Affairs and Climate Action (BMBK) under contract # 0324353B ('CIGSTheoMax').

REFERENCES

[1] T. Markvart, Thermodynamics of losses in photovoltaic conversion, Appl. Phys. Lett. 91, 064102, 2007.

[2] U. Rau, U. W. Paetzold, and T. Kirchartz, Thermodynamics of light management in photovoltaic devices, Phys. Rev. B 90, 035211, 2014.

[3] T. Markvart, Shockley-Queisser detailed balance limit after 60 years, Wiley Interdisciplinary Reviews: Energy and Environment 11, e430, 2022.

[4] R. Brendel, S. Dreissigacker, N. P. Harder, and P. Altermatt, Theory of analyzing free energy losses in solar cells, Appl. Phys. Lett. 93, 173503, 2008.

[5] F. Komoll and U. Rau, The balance of thermodynamic potentials in solar cells investigated by numerical device simulations, IEEE J. Photov. 12, 1463-1468, 2022.

[6] M. Burgelman, P. Nollet, and S. Degrave, Modelling polycrystalline semiconductor solar cells, Thin Solid Films 361-362, 527, 2000.

[7] R. Brendel and R. Peibst, Contact selectivity and efficiency in crystalline silicon photovoltaics, IEEE J. Photov. 6, 1413, 2016.

[8] E. T. Roe, K. E. Egelhofer, and M. C. Lonergan, Limits of contact selectivity/recombination on the open-circuit voltage of a photovoltaic, ACS Appl. Energy Mater. 1, 1037, 2018.

[9] U. Rau and T. Kirchartz, Charge carrier collection and contact selectivity in solar cells, Adv. Mat. Interf. 6, 1900252, 2019.

[10] T. Eisenbarth, T. Unold, R. Caballero, C. A. Kaufmann, H.-W. Schock, Interpretation of admittance, capacitance-voltage, and current-voltage signatures in Cu(In,Ga)Se$_2$ thin film solar cells, J. Appl. Phys. 107, 034509, 2010.

[11] A. Niemeegers, M. Burgelman, and A. deVos, On the CdS/CuInSe$_2$ conduction band discontinuity, Appl. Phys. Lett. 67, 843, 1995.

978-1-6654-6060-6/23 $31.00 © 2023 IEEE

Epitaxy-Free, Thin-Film GaAs Solar Cells Fabricated with Diffusion Doping and Mechanical Spalling

Phillip R. Jahelka, Andrew W. Nyholm, Harry A. Atwater

California Institute of Technology, Pasadena, CA, United States

We report a process for the fabrication of ultralight thin-film GaAs solar cells without epitaxial growth. To fabricate the cell, we first create a pn junction by zinc diffusion to realize p-type doping in bulk n-type crystals. We then spall a thin film of GaAs that includes the pn junction on an electroplated nickel backing layer directly deposited onto the diffusion doped p-GaAs. The electroplated nickel serves as a mechanical handle layer and rear electrical contact. A Pd/Ge/Au top contact becomes ohmic after a 180 C anneal, a process which is compatible with the nickel and tape handling layers. Under AM 1.5G illumination the cell achieves a Voc of 856 mV, and including the metal and GaAs, we reach a specific power of 500 W/kg at an areal mass density of 32 g/m2.

Analysis of Impurity-related Radiative Transitions in Silicon Materials using Temperature-dependent Photoluminescence

Tarek O. Abdul Fattah[1], Janet Jacobs[1], Vladimir P. Markevich[1], Nikolay V. Abrosimov[2], Matthew P. Halsall[1], Iain F. Crowe[1], and Anthony R. Peaker[1]

[1]Photon Science Institute and Department of EEE, University of Manchester, Manchester, M13 9PL, United Kingdom

[2]Leibniz-Institut für Kristallzüchtung (IKZ), Max-Born-Straße 2, 12489 Berlin, Germany

Abstract — **The contradictory reports in the literature about the stability of Ga-doped silicon (Si) material for photovoltaic applications, in comparison to those doped with B, necessitate a more detailed understanding of the characteristics of this material before solid conclusions about degradation mechanisms can be made. In this work, high-resolution low-temperature photoluminescence (PL) has been used to investigate and analyze the luminescence from Ga-doped and P+Ga co-doped Czochralski-grown silicon (Cz-Si) materials. Comparison of thermally induced changes in luminescence features for these materials are compared to those occurring in B-doped and P+B co-doped Si materials. It has been found that the Ga bound exciton (BE) exhibits a triplet luminescence structure which is preserved in the co-doped material, explained by the splitting of the exciton ground state. A similar effect does not occur for the BE-related PL signal in B-doped silicon material. A low temperature (10-20 K) range was then used to investigate the temperature-induced changes in impurity related photon emission lines in the PL spectra of the studied materials. The effect of thermal energy on the PL intensity of different radiative recombination channels is elucidated. It has been argued that the presence of compensating impurities causes enhanced radiative recombination of some excitonic emissions while others behave in a similar way as in a single-doped material. The possible relationship of the observed effects on electron-hole recombination at room temperature is discussed.**

I. Introduction

With silicon-based solar photovoltaic (PV) technology anticipated to play a major role in global warming mitigation plans, i.e. terawatt (TW) scale PV deployment, concerns about the supply issues of high purity Si feedstock and carrier-induced degradation mechanisms in Si-based solar cells are raised. To address these concerns, deeper understanding of the fundamental properties of recently used dopants in Si for PV applications, e.g. gallium (Ga) doping, and effects of compensation on electronic properties of dopants is needed.

Ga-doped Si solar cells have rapidly penetrated in the PV market in the last few years, moving towards phasing out boron (B)-doped ones completely [1], based on arguments that Ga-doped Si solar cells are more stable and overcomes the issues of light and/or temperature induced efficiency loss occurring in B-doped ones. This statement was recently contradicted by experimental observation of degradation in Ga-doped material upon illumination at room or elevated temperatures [2, 3].

Before such contrary observations can be explained, a detailed understanding of the characteristics of Ga-doped Si material and how Ga as a dopant behaves in Si compared to B is paramount. Even the studies reporting degradation-free Ga-based solar cells have not proposed mechanisms for the observed results. Further importance to this understanding is given by the need to find alternative less pure Si feedstock, such as compensated solar grade silicon (SOG-Si), to overcome the anticipated shortage in the supply of high purity Si in the future.

High-resolution low-temperature Photoluminescence (PL) is one of the best techniques to study impurities in Si materials. Many previous reports demonstrated the use of PL to identify impurity species, their properties and their respective concentrations in Si [4], but very little has been published on Ga-doped Si material. PL also offers the advantage of investigating Si materials containing compensating impurities, thus overcoming the limitations of many available characterization techniques. Compared to relatively old PL studies on Si, advances in luminescence detection equipment allow the achievement of higher resolution and the access to important radiative features in the material. These advantages are important for studies of impurity-related emissions in narrow energy ranges and to resolve overlapping PL features, especially in the presence of compensating impurities.

Impurity-related PL features in Si were mainly used to quantitatively estimate the impurity concentration by the analysis of excitonic luminescence. For solar PV applications, the calibration curves for B- and P-doped Si materials have been obtained based on the intensity ratio of principal bound exciton (BE) relative to free exciton (FE) [5, 6]. Further, the influence of doping concentration in both single-doped and co-doped Si materials on different radiative recombination features in PL has been studied in details [4]. Surprisingly however, despite the known importance of temperature in PL, little work has been done on analyzing thermally-induced effects on impurity-related PL features.

In this study, the low-temperature PL spectra of single (B, Ga and P) doped and co-doped (P+B and P+Ga) Cz-Si materials have been recorded and analyzed. The main spectral features dominating the radiative recombination processes have been identified and compared. We demonstrate that the use of temperature-dependent PL allows understanding the changes in

978-1-6654-6060-6/23 $31.00 © 2023 IEEE

excitonic emissions and deriving important characteristics of the dopants in Si materials.

II. EXPERIMENTAL DETAILS

The co-doped samples used for the present study were cut from Si wafers grown by the Czochralski technique (Cz) at the Leibniz-Institut für Kristallzüchtung (IKZ) in Berlin, intentionally doped with phosphorus (P) and either boron (B) or gallium (Ga). Test growth of single doped ingots was carried out to first to plot profiles of single dopant concentration vs ingot height. The plotted profiles were then used to calibrate and determine dopants concentrations in co-doped materials grown under similar conditions as single doped test ones. The determined concentrations in the co-doped samples reported here are [P]~3×10^{16} cm^{-3}, co-doped with either [Ga]~1.5×10^{16} cm^{-3} or [B]~2×10^{16} cm^{-3}.

Ga-doped, B-doped and P-doped Cz-grown Si samples of resistivities ρ~1-2 Ω·cm were used as control samples. All samples received an etch in a solution of HF:HNO$_3$ to reduce the surface damage and minimize surface recombination.

Samples were fixed on 1-inch copper disk using thermally conducting silver DAG which sits at the bottom of Leybold re-cycling helium cryostat on the cold finger. The samples inside the cryostat were kept at a vacuum level of ~10^{-6} mbar for 24 hours before cooling. At the intended temperature, mirrors were used to direct the exciting 780 nm laser spot to the sample surface. Two Si diodes, one mounted to the rear of the cold finger and second sitting in the front next to the samples, were used to monitor the actual samples temperature. The measured laser power at the cryostat entrance window was ~5.5 mW and the focused spot size on the sample was ~2mm. Luminescence from samples was collected and dispersed with an iHR550 Horiba spectrometer using a 1200 l/mm grating. Dispersed spectra were detected by cooled InGaAs array, giving an overall spectral resolution of 60 μeV. Calibration and correction for system response were applied before the analysis of data.

III. RESULTS AND DISCUSSION

A. PL spectra of single doped Si materials

The PL spectra of B-doped and Ga-doped Cz-Si samples with acceptor concentrations of ~1×10^{16} cm^{-3} recorded at 9 K are shown in Fig. 1a and Fig. 1b. The recorded PL spectra mainly show luminescence lines in three different energy ranges: i) no-phonon (NP) region centered at ~1.15 eV, ii) transverse acoustic phonon (TA) range centered at ~1.13 eV and iii) transverse optical (TO) phonon range centered at ~1.09 eV. The two samples show dominant peaks in the TO region where the recombination of excitons bound to acceptor atoms (BE) is assisted by the TO phonons. A noticeable difference occurs in the NP region (recombination of BE without the involvement of phonons) at which B-doped sample shows only very weak peaks whereas the peaks are powerful and comparable to those

in the TO region in the case of Ga. Importantly, a closer look at the NP region of the Ga-doped sample shows the existence of three peaks labelled as Ga_{NP}^{α}, Ga_{NP}^{β} and Ga_{NP}^{γ} respectively in order of increasing photon energy, in agreement with a previous study [7]. The three peaks are also present in the TO region but are a little bit broader due to the involvement of phonons in the recombination process. This is not the case for the B-doped sample where only bound multi-exciton complexes (BMEC) peaks are observed at the low energy side of the principal BE (B_{TO}) [7]. The existence of this triplet in Ga-BE structure can be explained by the splitting of the Ga bound exciton ground state into three levels as governed by the j-j coupling scheme [8]. An interesting observation is the change in the most intense transition among the triplet structure, from Ga_{NP}^{α} in the NP region to Ga_{TO}^{β} in the TO region. All of these differences suggest that the structure of the acceptor impurity has a role to play in the dominant radiative transitions. It seems for the case of heavier atoms like Ga and P, radiative recombination is efficient with and without the involvement of TO phonons whereas only phonon-assisted recombination is efficient for lighter atoms like B. This could be due to the local vibrational mode of these atoms being more delocalized than that of Boron and/or a stronger polarization field generated by the lighter atom. A similar effect has been observed in the P-doped PL spectrum (not shown here) in which the NP BE transitions are at least as efficient as the TO-assisted ones.

Weak peaks can also be seen in Fig. 1a and Fig. 1b at ~1.13 eV which is ~19 meV from NP region peaks. This corresponds to recombination of BEs with the involvement of the transverse acoustic (TA) phonons. A weak shoulder that can be observed at ~1.098 eV is related to the recombination of free excitons (FE) which have a very small population at this temperature due to the efficient trapping of excitons by dopant atoms.

B. PL spectra of co-doped Si materials

Fig. 1c and Fig. 1d show the PL spectra of P+B and P+Ga co-doped Cz-Si samples recorded at 9 K. The concentrations of the dopants are shown in the plots. It is clearly noticeable that the simultaneous presence of two impurities at these concentration levels added many luminescence features. At least two additional recombination channels in the PL spectra can be identified: 1) impurity cluster bound excitons (ICBE) channel where excitons are bound by a pair or cluster of dopant atoms and 2) donor acceptor pair (DAP) recombination where an electron bound to a donor atom recombines radiatively with a hole localized at an acceptor atom. ICBE is responsible for the observed structures at ~1.143 eV (NP region), 1.085 eV (TO-sideband) and less clearly at 1.123 eV (TA-sideband).

The DAP emission is more complicated and is responsible for broad bands at the low energy side (two and three peaks are apparent in the P+B and P+Ga spectra, respectively) and a discrete fine structure in the region 1.10-1.13 eV, which is observable in the P+B case but more evident in the P+Ga spectrum. Note that even in the P+B case, a third broad band exists but is not very efficient under the current experimental conditions and/or is overlapped by more efficient emissions.

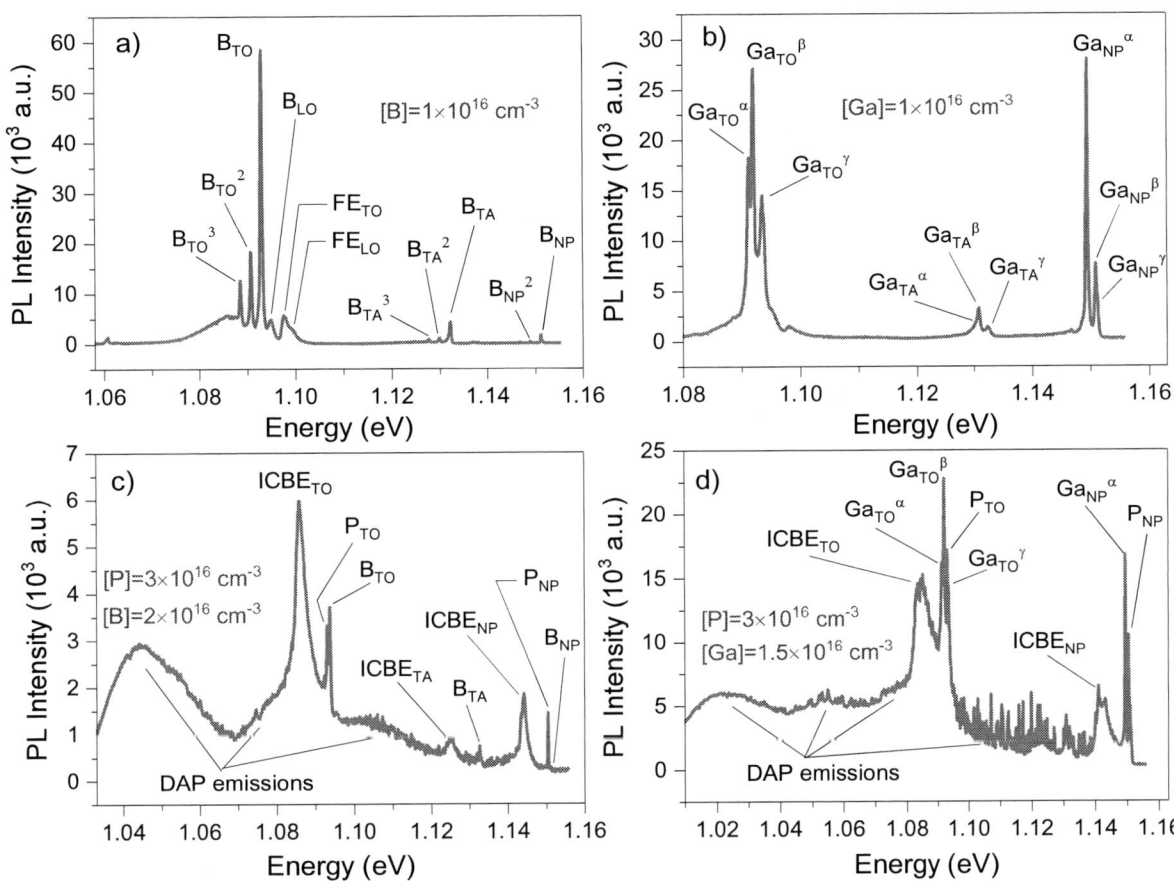

Fig. 1. PL spectra of a) B-doped, b) Ga-doped, c) P+B co-doped and d) P+Ga co-doped Cz-grown Si samples recorded at 9 K. Dominant luminescence lines are named based on respective recombination channels with and without the involvement of phonons in the same notation used in [4, 7]. Concentrations of impurities in each of the studied samples are also shown in the plots.

The comparison between single-doped and co-doped spectra (Fig. 1a vs Fig. 1c and Fig. 1b vs Fig. 1d) shows that, at the given impurity concentrations, the single acceptor-related luminescence peaks still exist in the co-doped materials spectra with the addition of P-related peaks (NP and TO sidebands).

B. Temperature-dependent PL in the range 10-20 K

One way of interpreting the large amount of information in these PL spectra is the investigation of the effect of thermal energy on the observed luminescence features. The PL spectra of P+Ga co-doped Cz-Si sample recorded in the temperature range 10-20 K is presented in Fig. 2. Similar measurements have been also carried out for the P+B co-doped and single doped Si samples. The first clear observation in the figure is the increase in the PL intensity of FE with increasing temperature. This is expected as more excitons can acquire enough energy to escape from the dopant atoms.

For ICBE bands, a continuous reduction of intensity has been observed over all the studied temperature range. This is clearly visible in Fig. 2 for ICBE (TO) and ICBE (NP) at ~1.085 and ~1.142 eV, respectively.

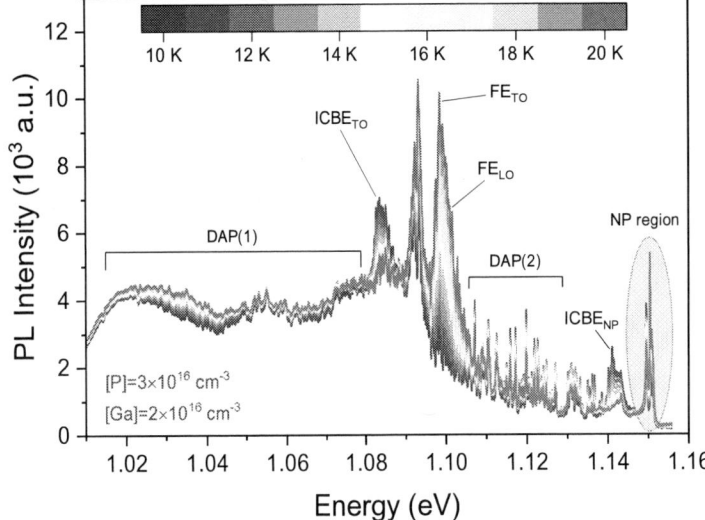

Fig. 2. Temperature-dependent PL spectra of the same P+Ga co-doped sample in the temperature range 10-20 K.

The case is different for the DAP recombination channel where two different structures that can be recognized in Fig. 2

and labelled as DAP1 and DAP2. DAP1 entitles three broad peaks centered at about 1.02, 1.055 and 1.075 eV and are related to the recombination of electrons and holes bound to distant P and Ga atoms.

They can be labelled as P+Ga(TO), P+Ga(TA) and P+Ga(NP), respectively [9]. We will not discuss the details of this assignment here due to limited space. DAP2 refers to the discrete luminescence structure with many narrow sharp peaks which correspond to the DAP recombination at close P and Ga atoms. The two structures behave differently as the temperature increases. The lowest energy peak in the DAP1 structure, P+Ga(TO), at 1.02 eV shows a small increase in intensity. The other two broad peaks almost preserve the same PL intensity in the studied temperature range. The sharp peaks in DAP2, however, constantly show a reduction in the PL intensity with increased temperature. This different behavior is attributed to the difference in the recombination probability between these two structures. As the donor and acceptor impurities get closer, the bound electron and hole wave functions overlap, and their respective recombination probability increases. With the supply of some thermal energy, it seems like that the electron and hole acquire enough energy to escape from the close pairs but not enough to overcome the barrier of more distance DA pairs. It is important to mention that the intensity of the three peaks in the DAP1 structure is preserved up to ~50 K after which thermal quenching starts to occur and non-radiative recombination is enhanced.

C. Temperature-dependent PL of single vs co-doped Si materials in the NP region

A closer look at the NP region (emitted photon energy in the range 1.148-1.152 eV, highlighted by the circle in Fig. 2) offers some useful information about the isolated impurity-related luminescence features. Fig. 3 shows the temperature-dependent PL spectra recorded in the range 10-20 K for B- and P-doped Si samples and the P+B co-doped sample.

For the B_{NP} peak, a slight increase in the PL intensity followed by a more rapid drop can be observed. In the case of a single P-doped Si, an increase in the temperature produces a constant decrease in all bound excitonic features including P_{NP} peak (Fig. 3b). Interestingly, the presence of both impurities in the same material leads to a change in the response of both B_{NP} and P_{NP} peaks to temperature. An increasing PL intensity has been recorded from both peaks in the temperature range 10-17 K, as shown in Fig. 3c. Beyond 17 K, the intensity of B_{NP} peak is preserved whereas that of the P_{NP} peak starts to decrease. Note that at all temperatures, the P_{NP} peak dominates the NP region. The same is observed in p-type co-doped Si material (B concentration is higher than that of P), thus confirming the conclusion made in section III.A that this is function of the impurity structure in Si.

The temperature evolution of the PL lines in the NP region of the Ga-doped and P+Ga co-doped samples is presented in Fig. 4. It is clear that the triplet structure in Ga-doped spectra behaves differently as function of temperature. The dominating peak at 9 K in the single-doped spectra, Ga_{NP}^{α}, constantly loses

intensity with increasing temperature whereas the other two peaks, Ga_{NP}^{β} and Ga_{NP}^{γ}, experience only slight changes.

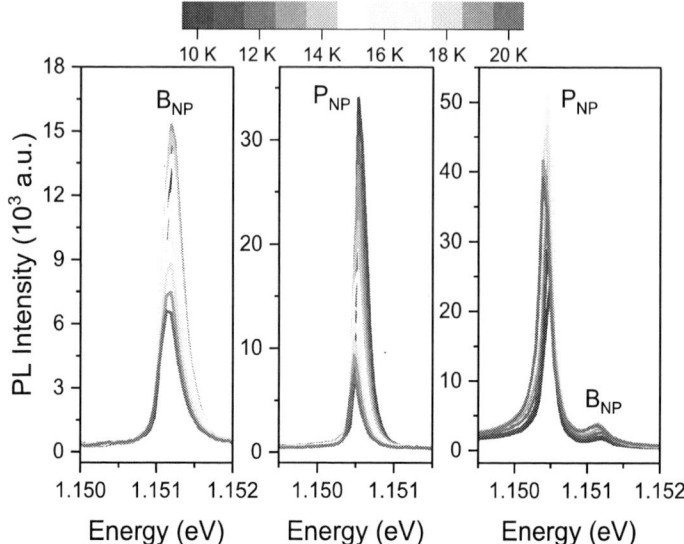

Fig. 3. Temperature-dependent PL intensity for a) B-doped, b) P-doped and c) P+B co-doped samples in the NP region.

In the P+Ga co-doped spectra in Fig. 4b, the P_{NP} peak shows an overall increase in intensity with increasing temperature. This is similar to some extent to the change in the P+B sample and completely opposite to that in the single P-doped PL spectra. The rate of increase starts fast at low temperatures and decreases as the temperature approaches T=20 K where it saturates. The case is very similar for Ga_{NP}^{β} and Ga_{NP}^{γ} peaks where an intensity increase followed by saturation has been observed in the P+Ga co-doped material, unlike the case in the Ga-doped material (Fig. 4a). In contrast, the intensity of Ga_{NP}^{α} peak showed a continuous decrease with increasing temperature, similar to the case in the single-doped material. Note that the dominant peak in the P+Ga co-doped spectra is changed from Ga_{NP}^{α} at 9 K to the P_{NP} peak for T > 12K. The further increase in temperature above 20 K causes the reduction in PL intensity magnitudes of all the BE related peaks.

The efficiency of the different radiative recombination channels in the temperature range 10-20 K are seen to be changing in the co-doped materials, where the released excitons from one channel are consumed by other channels, thus different recombination channels should be considered simultaneously. A possible source of excitons responsible for the increased intensity of single dopant luminescence, in addition to ICBE, is the DAP2 channel where charge carriers at close DA pairs have a high recombination probability and thus their respective intensity continuously decreases with temperature as shown in Fig. 2. With increasing temperature above 20 K, it appears that excitons start to acquire enough energy to overcome most of the impurity-related radiative recombination channel barriers where they recombine radiatively via the FE channel, and/or the non-radiative recombination starts to compete with the occurring radiative transitions. The behavior of the Ga BE, particularly the Ga_{NP}^{α} peak, is not fully understood at the moment. Several questions remain open, e.g. why the Ga_{NP}^{α} peaks dominates the NP region of n-type P+Ga material at 9 K, even though [P] > [Ga].

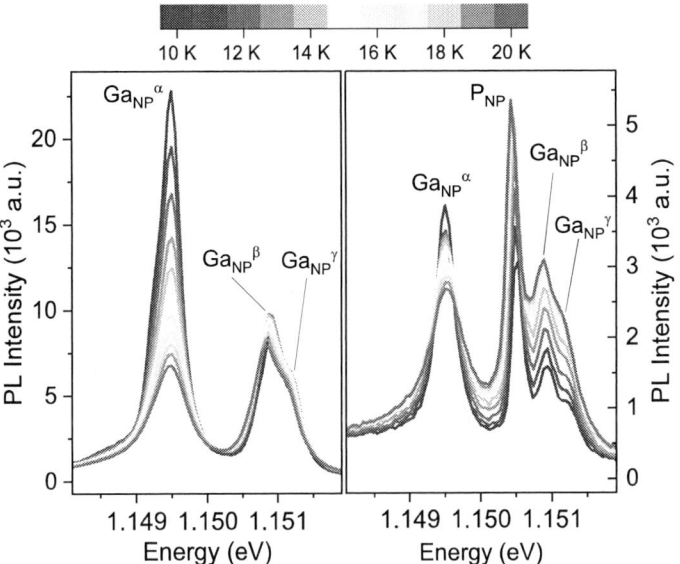

Fig. 4. Temperature-dependent PL intensity for a) Ga-doped and b) P+Ga co-doped Si samples in the NP region.

These results reveal an important difference in the behavior of quasiparticles in differently doped Si materials. The significant difference in the thermally-induced changes to radiative recombination between single and co-doped Si materials suggests a non-trivial effect of the presence of compensating impurities on the behavior of quasiparticles in the Si material which could affect the electronic and optical properties of it. The occurring transitions, radiative and non-radiative, are substantially affected by the structure and behavior of the dopant atoms which can be controlled by different factors, e.g. presence of compensating impurities, temperature, etc.. This complication potentially explains the yet-to-come penetration of low-cost compensated Si into the PV market, which can be accelerated by studies underlying the fundamental behavior of impurities in this material.

The literature on thermally and/or light-induced degradation of Si solar cells, including theoretical models and experimental observations, points to the crucial role played by electrons and/or holes in the occurrence of these phenomena. In addition, it has been argued that the degradation behavior is substantially affected by the acceptor species [2, 10]. However, very little is known about the details of electron/hole interactions with different acceptor-related complexes, possible recombination-enhanced reactions and how this turns into differences in the observable degradation of different materials and under different conditions. Comparisons like the one we present in this study are a prerequisite for understanding fundamental differences between solar cell materials and would pave the way for a detailed understanding of degradation behavior and ways of mitigating it.

IV. SUMMARY

Both, addressing light and/or thermally induced degradation problems of Si-based solar cells in the field, and finding alternative low-cost Si materials to achieve TW capacity of PV makes it essential to learn more about fundamental characteristics of dopants in different Si materials. Understanding these properties is the first step in expanding the knowledge on improving the Si PV technology technically and sustainably. In the current study, we have conducted a detailed analysis of the excitonic features in the PL spectra of single doped (B, Ga and P) and co-doped (P+Ga and P+B) Cz-Si materials. The full energy range PL spectra are first presented and discussed followed by a more focused look at changes in the isolated impurity-related luminescence features in the no-phonon (NP) region. We have found fundamental differences in the dominating radiative transitions function of the doping species in the single doped materials. It has been suggested that the size of the dopant atom determines the dominating radiative recombination channel, where excitons in smaller atoms (like B) recombine most effectively via a TO phonon assisted process unlike bigger atoms in which the most efficient recombination route is the one with no phonon assistance. Also, the Ga BE has been found to exhibit a triplet luminescence structure that is not present in the case of boron and is preserved in the co-doped material. Temperature dependent PL spectra in the range 10-20 K were recorded for all studied Si materials. The results showed important differences in the thermally induced changes in PL intensity of NP region peaks in the co-doped materials and the single doped ones. Even in the single doped case, the response of the principal NP bound exciton is different in differently doped materials. These findings are important in the context of electron-hole recombination mechanisms and compensated Si materials.

REFERENCES

[1] VDMA, *International Technology Roadmap for Photovoltaic*, 2023.

[2] S. Jafari, M. Figg, and Z. Hameiri, "Investigation of light-induced degradation in gallium- and indium-doped Czochralski silicon," *Solar Energy Materials and Solar Cells*, vol. 251, pp. 112121, 2023.

[3] N. E. Grant, J. R. Scowcroft, A. I. Pointon, M. Al-amin, P. P. Altermatt, and J. D. Murphy, "Lifetime instabilities in gallium doped monocrystalline PERC silicon solar cells," *Solar Energy Materials and Solar Cells*, vol. 206, pp. 110299, 2020.

[4] M. Tajima, H. Toyota, and A. Ogura, "Systematic variation of photoluminescence spectra with donor and acceptor concentrations ranging from 1×10^{10} to 1×10^{20} cm^{-3} in Si," *Japanese Journal of Applied Physics*, vol. 61, no. 8, pp. 080101, 2022.

[5] M. Tajima, "Determination of boron and phosphorus concentration in silicon by photoluminescence analysis," *Applied Physics Letters*, vol. 32, no. 11, pp. 719-721, 1978.

[6] T. Iwai, M. Tajima, and A. Ogura, "Quantitative analysis of impurities in solar-grade Si by photoluminescence spectroscopy around 20 K," *Physica Status Solidi C,* vol. 8, no. 3, pp. 792-795, 2011.

[7] M. L. W. Thewalt, "Fine Structure of the Luminescence from Excitons and Multiexciton Complexes Bound to Acceptors in Si," *Physical Review Letters,* vol. 38, no. 9, pp. 521-524, 1977.

[8] M. L. W. Thewalt, "Details of the structure of bound excitons and bound multiexciton complexes in Si," *Canadian Journal of Physics,* vol. 55, no. 17, pp. 1463-1480, 1977.

[9] M. Tajima, K. Tanaka, M. Forster, H. Toyota, and A. Ogura, "Donor-acceptor pair luminescence in B and P compensated Si co-doped with Ga," *Journal of Applied Physics,* vol. 113, no. 24, pp. 243701, 2013.

[10] W. Kwapil, J. Dalke, R. Post, and T. Niewelt, "Influence of Dopant Elements on Degradation Phenomena in B- and Ga-Doped Czochralski-Grown Silicon," *Solar RRL,* vol. 5, no. 5, pp. 2100147-2100147, 2021.

Reflector Candidates for a Vertical Bifacial Solar Canal

Jeremiah Reagan, Brandi McKuin, and Sarah Kurtz

University of California Merced, Merced, California, 95340, USA

Abstract—Five candidate materials were assessed for suitability of fabricating a horizontal reflector as part of a hypothetical vertical bifacial solar canal. Albedo values of candidate materials were measured under sunlight and used to model the impact on production and cost of energy of such a system. Despite the diamond-patterned aluminized mylar material by Vivosun leading by a wide margin in both total reflectivity and lowest material cost, its plastic deformation under stress makes it unsuitable for fabricating a tarp-like structure on its own. However, attaching it as a top layer to one of the other materials as a base layer remains viable, with such hybrid reflectors projected to be more cost effective than single-material reflectors. A hybrid reflector can be optimized by making the top layer width shorter than that of the base layer until it reaches a width to height ratio with the panels of 1.25 − 1.9. This metric can further be used to assess suitability of potential sites for such vertical systems based on canal width. While cost of energy improved significantly with height of the system, more work is needed to assess the lifetimes of the materials and the structural requirements of taller systems. While polyvinyl-chloride-coated polyester has much lower cost, it is also substantially weaker, while the preferred choice for implementation between CoverMax, CoverTuff, and polyester-canvas materials may come down to logistical issues of production and contracts struck with individual vendors.

I. INTRODUCTION

The topic of Solar Canals, shading waterways with solar panels to conserve water while producing electricity, has received increased attention in recent months, particularly in light of California's recent steps to begin a pilot project [1]. A prior paper by Reagan and Kurtz [2] modeled the output of a hypothetical variant approach to solar canals using a wall of vertical bifacial panels and a reflective cover stretched across the water. This paper builds on that approach, exploring the suitability of several candidate materials for the construction of the reflector. The albedo values of samples were recorded with a homemade albedometer designed to mimic photovoltaic response and put into the same model as before to simulate the annual production of the vertical panel + reflector system. The weather data, along with several parameters of system size, orientation, and bifaciality, were altered to better match a prospective site location in Ceres, California for a future test system. Details on these changes are listed below in the Experimental Design and Methodology section. The associated increase in system production, along with varying cost and expected lifetimes of the material candidates were used to calculate their impact on the levelized cost of energy of the whole system.

II. EXPERIMENT DESIGN AND METHODOLOGY

Five sample swatches of candidate materials were acquired for reflectivity testing. The first two, CoverMax and CoverTuff, were provided as part of a free sample kit of custom materials from company Covers and All, which produces custom tarps and covers. Both are polyvinyl chloride (PVC) coated polyester with additional weatherproofing treatment, the details of which were not provided. While CoverMax has a glossy laminate coating on both sides, CoverTuff has a single-sided finish with the bare PVC coating on the back. It was this more reflective PVC-coated side that was used for testing. A third sample of PVC-coated polyester (PVC Poly) was obtained from OnlineFabricStore.com, and a fourth sample of polyester canvas (Poly Canvas), without any coatings, was obtained from MyTarp.com. Finally, a roll of 6-mil (0.15-mm) diamond-patterned aluminized Mylar film, produced by Vivosun was ordered from Amazon. This material is designed to line walls of indoor grow rooms and greenhouses and claims 92% - 97% total reflectivity. Rather than attempting to determine specific spectral reflectance by wavelength, measurements were taken using silicon solar cells under real sunlight conditions to best mimic intended use in the field. Albedo was determined via the ratio of the outputs of downward- and upward-facing solar cells.

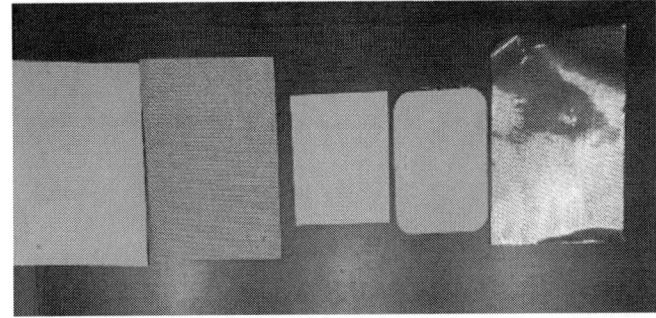

Fig. 1. Material samples. Left to right: Covermax, CoverTuff, PVC Poly, Poly Canvas, Vivosun aluminized Mylar film.

Two identical miniature solar cells were connected to a datalogger and held horizontally under clear skies at solar noon to calibrate against each other. One was then inverted to face downward and the ratio of the downward cell vs the upward cell used to determine the albedo of the surface below them (Fig. 2). Sample materials were then placed underneath and measurements recorded under midday and evening conditions (17:30 − 18:15).

978-1-6654-6060-6/23 $31.00 © 2023 IEEE

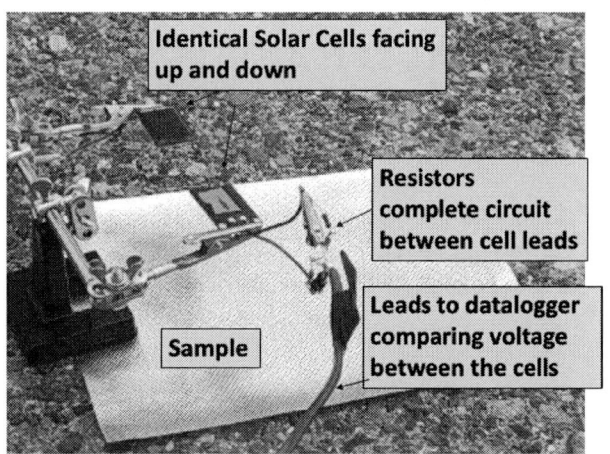

Fig. 2. Albedometer made from two identical solar cells. Leads of each cell are bridged with three 1-Ω resistors so wires running to datalogger read operating voltage rather than open-circuit voltage. The ratio of the two voltages gives the albedo of the sample material.

Once albedo values were recorded, *pvlib* was used to model the first year of energy production of hypothetical vertical bifacial systems (facing 245° azimuth with bifaciality of 0.8) along a canal in the Turlock area using 2019 weather data. System sizes of 3.6 kW and 5.4 kW were chosen to represent arrays of 8 and 12 bifacial modules (approximately 1 m by 2 m) in landscape, 4 wide by 2 high and 4 wide by 3 high, respectively and will be referred to as "short system" and "tall system" from this point on. Bifacial panels were modeled by adding the DC output of two opposite facing monofacial panels before sending to the virtual inverter, and the presence of a 25' by 30' (7.6 m by 9.1 m) reflective cover was simulated by changing the value of the ground albedo on the canal side to that of one of the measured materials, while leaving the other side at a default of 0.3. For the condition of no reflector being present, canal side albedo was left at 0.3 for empty canals and set to $0.05 - 0.10$ for full canals during the months of March through October, which represent irrigation season when water is typically flowing.

While the other samples retained stable albedo values at lower-light (*e.g.* evening) conditions or when shaded from direct irradiance, the Vivosun sample showed a high degree of variability and strong sensitivity to sample geometry such as creases and changes in orientation. Measurement was repeated with two Apogee pyranometers placed 1' (30 cm) over an 8' by 8' (2.4 m by 2.4 m) sheet of material weighted down with bricks (Fig. 3), recording continuously and averaging over five-minute intervals for several days (5/11/2023 - 5/15/2023). Datapoints were grouped by hour and averaged to create hour-specific albedo values (Fig. 4) that the *pvlib* model was revised to accept.

Limited strength testing was conducted using 8.5" by 6" (22 cm by 15 cm) samples in an Instron tensile-strength tester. Though tear strength does not translate directly to predicted lifetime, it could be used as a comparative metric for ranking samples against one another in terms of overall reliability. However, mechanical failures caused slipping of the CoverMax, CoverTuff, and Poly Canvas materials, preventing establishment of an upper strength limit for these materials.

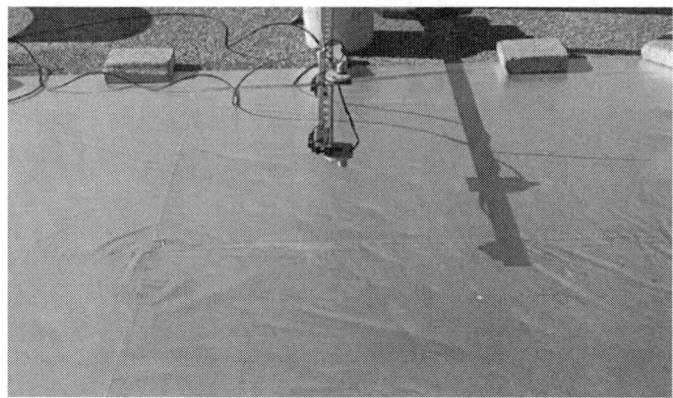

Fig. 3. Measurement of Vivosun material. Upward- and downward-facing pyranometers connected to a datalogger take measurements every minute and record averages of five-minute intervals.

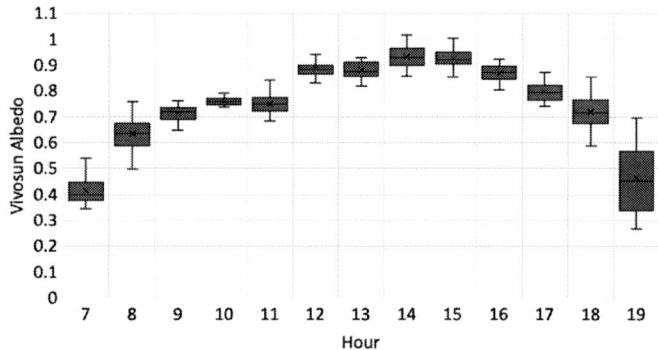

Fig. 4. Hourly albedo values for Vivosun material. Individual measurements were recorded every 5 minutes over 5 days and grouped by hour. Values for each hour were averaged and fed to the model, with hours <7 set to the same average albedo value associated with 7, while hours > 19 were set to the value associated with 19.

III. HYBRID REFLECTOR COST OF ENERGY CALCULATIONS

Due to the low material cost of Vivosun material (Table 1), we considered the possibility of adhering a surface layer of highly reflective material to a stronger base layer. From this point on, the list of reflector candidates is expanded to include hybrids of Vivosun material on top of the other candidate materials, with the hybrid cost being the full fabrication cost of the base layer and the additional material cost of the top layer (we neglected the additional labor cost).

As the contributions of the far end of the reflector to the panels provide diminishing returns with increasing width, the cost of energy can be used to optimize the hybrid reflectors with respect to the amount of top-layer coverage. While the base layer needs to span the entire canal to shade the water, the top layer is not under any such restriction.

The marginal cost of energy provided by the hybrid reflector, $COE_{reflector}$ ($/MWh), is given in Eq. 1:

TABLE 1

MEASURED ALBEDO AND ESTIMATED COSTS OF CANDIDATE MATERIALS

Material	Albedo	Material Cost $/ft^2	Full fabricated Cost $/ft^2	Cost for 25'X30' Test Reflector ($)	% of Panel Cost (short)	% of Panel Cost (tall)
CoverMax	0.46 - 0.48	*0.65*	**0.91**	680	21 - 27%	14 - 18%
CoverTuff	0.38 - 0.40	*0.76*	**1.07**	800	25 - 32%	17 - 21%
PVC Poly	0.44 - 0.45	**0.45**	*0.63*	470	15 - 19%	10 - 13%
Poly Canvas	0.42 - 0.45	**0.86**	**1.2**	900	28 - 36%	19 - 24%
Vivosun	0.41 - 0.93	**0.29**	*0.41*	310	9 - 12%	6 - 8%

Bold numbers are direct quotes, while italics are inferred values from the ratio between Poly Canvas values.

$$COE_{reflector} = \frac{C_{Total}}{E_{m,Total}} \tag{1}$$

where C_{Total} ($) is the total cost of a hybrid reflector, and $E_{m,Total}$ (MWh) is the total marginal energy production of the reflector defined as the difference in system generation with and without the reflector. The cost of the hybrid reflector, C_{Total} ($) is given by Eq. 2:

$$C_{Total} = C_{Base} + x \cdot y \cdot C_{Top} \tag{2}$$

where C_{Base} ($) is the cost of the 25' X 30' (7.6 m X 9.1 m) base reflector (given in Table 1), x is the width of the reflector across the canal (25 ft. (7.6 m) by default), y is the shared length of the panels and reflector top layer (26.25 ft (8 m); 4 landscape panels at 2 m each), and C_{Top} ($/ft^2) is the material cost of the reflector top layer per unit area (given in Table 1).

The total marginal energy provided by the hybrid reflector, $E_{m,Total}$ ($/MWh) is given by Eq. 3:

$$E_{m,Total} = R(x) \cdot E_{m,Top} + [1 - R(x)] \cdot E_{m,Bottom} \tag{3}$$

where $R(x)$ is the ratio (see below) of reflected light of the cropped reflector to an uncropped reflector, $E_{m,top}$ (MWh) is the marginal energy produced by a full span reflector of the top material and is defined as the difference between the energy produced with and without the reflector (using the albedo of the uncovered canal for the case without the reflector), $E_{mBottom}$ is the marginal energy produced by the a full span reflector of the base material and is defined as the difference between the energy produced with and without the reflector, and the quantity $[1 - R(x)]$ corresponds to the ratio of the production of the uncovered far end of the base material vs a full span of the base material.

The numerator of the ratio of reflected light sent to the panels, $R(x)$, is a function of the reflector width and the denominator is evaluated at the maximum reflector width (25 ft). Because in this case, the reflector length is constant and because the material of cropped and uncropped material is the same, the albedo is also constant, simplifying the expression for $R(x)$ as shown in Eq. 4:

$$R(x) = \frac{x \cdot y \cdot F(x) \cdot \alpha}{x_{max} \cdot y \cdot F(x_{max}) \cdot \alpha} = \frac{x \cdot F(x)}{25 \cdot F(25)} \tag{4}$$

where α is the albedo, and $F(x)$ is the view factor between the horizontal reflector and the vertical panels as a case of perpendicular rectangular plates [3].

IV. RESULTS

Table 1 shows recorded albedo values of each sample, as well as cost estimates in terms of $/ft^2 for base material cost, total fabrication cost for a 25' by 30' (7.6 m X 9.1 m) reflector, and the cost of said reflector as a percentage of the base cost of the modules in each system, using $0.7/W - $0.9/W purchase price for bifacial modules (does not include cost of installation), which is a retail price for the modules, reflecting the relatively small size of the anticipated vertical systems. Bold numbers under material and fabrication costs represent direct quotes for the material from a relevant sales website. As most only listed prices for a roll of material or a finished tarp and not both, numbers in italics represent inferred values using the ratio of the listed price for Poly Canvas tarps on MyTarps.com and the listed price on Amazon for a roll of the same material.

Fig. 5 shows the effect of each reflector as a percentage of total gain in annual production over the scenario of no reflector being present. Percentage gain was identical between short and tall system sizes.

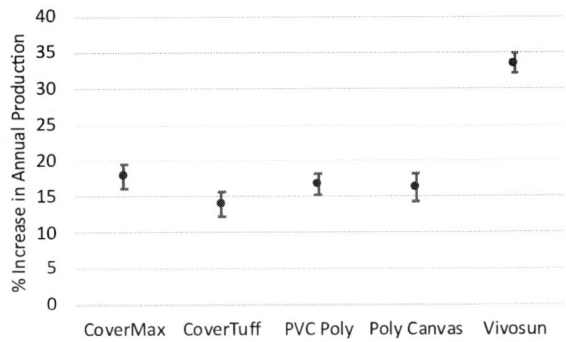

Fig. 5. Percentage increase in annual production due to presence of reflector compared to scenario of no reflector. Range of values stems from uncertainty in albedo values for reflectors shown in Table 1 and for that of water in the canal (albedo = 0.05 − 0.1) during irrigation season in the no-reflector scenario.

Though the Vivosun material has superior reflectivity, it was discovered over the course of handling and subsequent tensile strength testing that it is not suitable for fabricating a cover on its own. This is due to the ease at which it plastically deforms under stress making the retention of tautness infeasible. This motivated the addition of Vivosun as a top layer to stronger base layers creating hybrid reflectors.

Table 2 lists the optimized top layer widths calculated through the method outlined in Section III for both the short and tall system, as well as the ratio of marginal cost of energy between the optimized and non-optimized (full cover) hybrid reflectors, representing the savings of trimming unnecessary material.

TABLE 2

OPTIMUM TOP LAYER WIDTHS OF HYBRID REFLECTORS AND COMPARISON TO NON-OPTIMIZED HYBRIDS.

Base Material	Short (6.56')		Tall (9.85')	
	Optimum Top Width	Ratio of Energy Cost to non-optimized hybrid	Optimum Top Width	Ratio of Energy Cost to non-optimized hybrid
CoverMax	10'	0.92	12.5'	0.94
CoverTuff	12.5'	0.95	15'	0.97
PVC Poly	10'	0.88	12.5'	0.91
Poly Canvas	12.5'	0.95	15'	0.97

Fig. 6 shows the reflector cost of energy for single-material and optimized hybrid reflectors over a 20-year period, assuming an average reflector lifetime of 5 years. The average reflector lifetime was chosen as a reference point because it is the minimum warranty associated with the CoverTuff material. For additional reference, horizontal lines mark the bounds of 20-year levelized cost of energy for utility, commercial, and residential scale PV as of 2022 according to NREL's Annual Technology Baseline [4].

production in the no reflector scenario, we obtain an estimated range of total 20-year cost for the PV array by itself. We can then add the base cost of the reflector and any replacements and divide by the total production of the PV + reflector scenario to see how LCOE of the whole system is affected by the addition of a reflector. Note that this approach assumes that the cost of the PV is independent of mounting configuration, *e.g.* that the cost of the short and tall systems will be the same per installed watt, neglecting any cost increase associated with higher wind loading for the tall system. The following four figures show this for each of the optimized hybrid reflectors as a percentage shift in the total LCOE with respect to the PV system alone. Fig. 9 and Fig. 10 assume utility-scale cost for the construction of the PV array while Fig. 11 and Fig. 12 assume commercial cost.

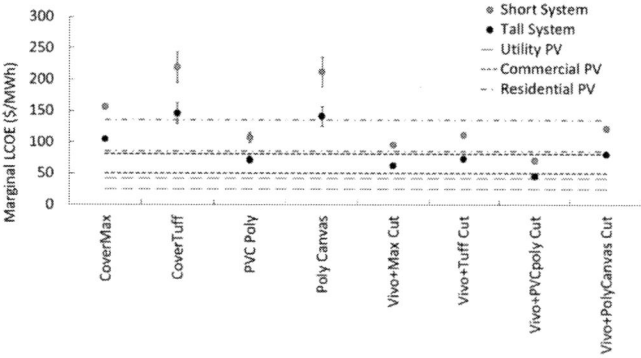

Fig. 6. Marginal levelized cost of energy (LCOE) generated by the reflector over 20 years. Based on fabrication cost of 25' by 30' (7.6 m X 9.1 m) reflector divided by the difference in generation between system with and without reflector. Assumes complete reflector replacement every 5 years at same initial cost. Dashed lines are LCOE ranges for utility, commercial, and residential PV as of 2022, added for reference.

As the LCOE depends strongly on the assumed lifetime of the reflector, Fig. 7 and Fig. 8 repeat this calculation for a range of possible reflector lifetimes for short and tall systems, respectively.

By taking the estimates for PV LCOE provided by the Annual Technology Baseline and multiplying by the modeled

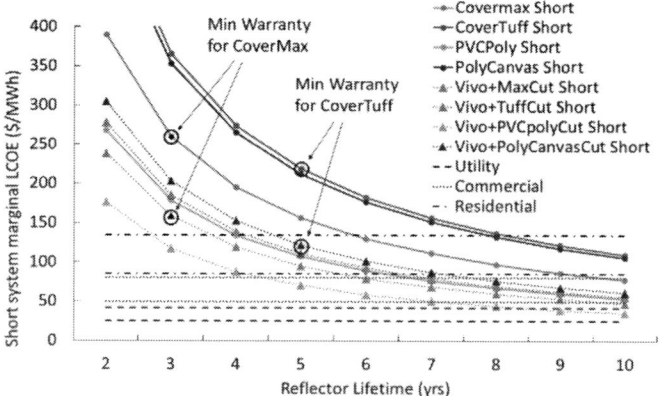

Fig. 7. Marginal levelized cost of electricity (LCOE) generated by the short system reflector over 20 years. Solid lines with dots are single-material reflectors. Dotted lines with triangles are hybrid reflectors with optimized Vivosun top layers. Circled points are warranties provided by Covers and All for CoverMax and CoverTuff products. Dashed lines are LCOE ranges for utility, commercial, and residential PV as of 2022.

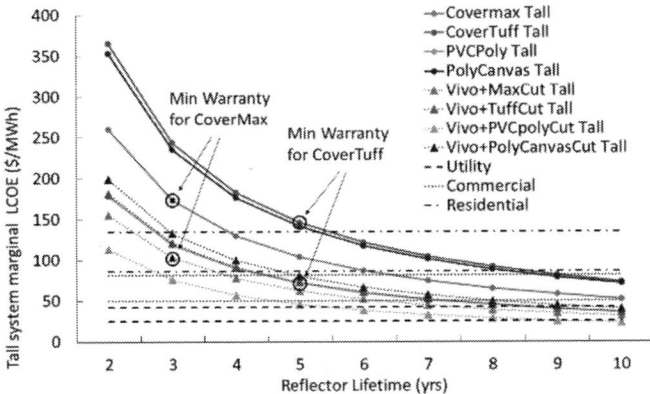

Fig. 8. Marginal levelized cost of electricity (LCOE) generated by the tall system reflector over 20 years. Dotted lines with triangles are hybrid reflectors with optimized Vivosun top layers. Circled points are warranties provided by Covers and All for CoverMax and CoverTuff products. Dashed lines are LCOE ranges for utility, commercial, and residential PV as of 2022.

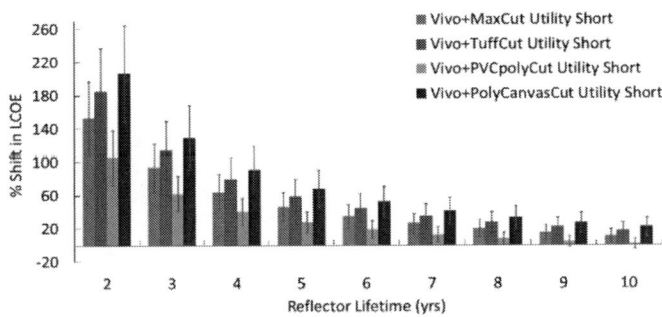

Fig. 9. Percentage shift in 20-year LCOE of short system with addition of reflector vs no reflector, assuming utility-scale cost for PV. Assumes same cost for each replacement reflector.

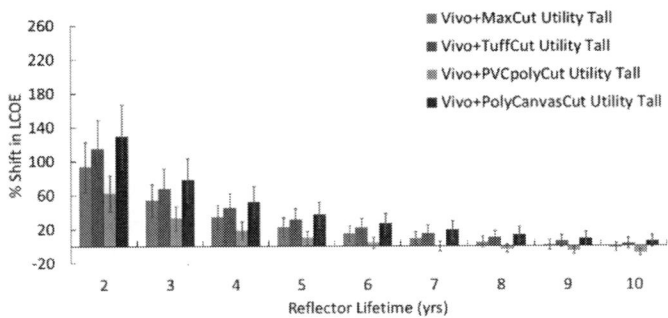

Fig. 10. Percentage shift in 20-year LCOE of tall system with addition of reflector vs no reflector, assuming utility-scale cost for PV. Assumes same cost for each replacement reflector.

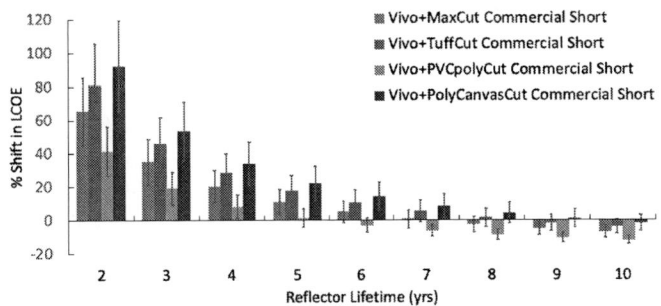

Fig. 11. Percentage shift in 20-year LCOE of short system with addition of reflector vs no reflector, assuming commercial cost for PV. Assumes same cost for each replacement reflector.

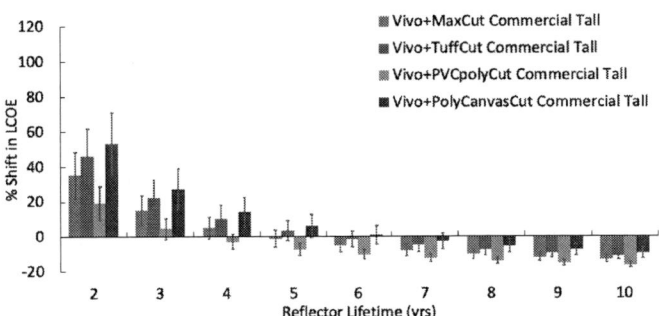

Fig. 12. Percentage shift in 20-year LCOE of tall system with addition of reflector vs no reflector, assuming commercial cost for PV. Assumes same cost for each replacement reflector.

V. ANALYSIS OF RESULTS AND SIGNIFICANCE OF FINDINGS

Though Figs. 9 and 10 show little benefit to total utility-scale system LCOE for materials with lifetimes under a decade (in large part due to just how low utility-scale PV costs have dropped in recent years), it is important to keep in mind that energy production is a secondary objective of the reflector. The primary objective remains the shading of the water, and thus a more comprehensive cost analysis of such a system should additionally consider the value of evaporation reduction and mitigation of aquatic weeds, both identified as major benefits of solar canal systems by McKuin et al [4] and the further analysis of which is part of California's planned pilot project.

Furthermore, the difference between the short and tall systems under both utility and commercial conditions underscores the strong dependence on the ratio of vertical panel height to canal width. This suggests a "go tall or go home" approach with wind loading on the vertical panels being the primary limiting factor for the height. Similarly, while this paper is concerned with 25'-wide (7.6-m-wide) canals, narrower canals would see increased benefit due to reduced reflector cost. These savings should continue down to the point at which the width to height ratio reaches those similar to the top layer of the optimized hybrid reflectors shown in Table 2, roughly $1.25 - 1.9$ depending on material, though the albedo of the far shore of the canal may need to be considered for narrower canals. This approach could prove useful for assessing suitability of prospective canal sites for such vertical systems.

Figures 7 and 8 provide a useful means of comparing between reflector candidates, assuming an accurate assumption of lifetime for each can be made. For example, a hybrid PVC Poly reflector that lasts X years is generally more cost effective than a hybrid CoverMax reflector that lasts X+1 years if X ≥ 3, and better than one that lasts X+2 years if X ≥ 7. Unfortunately, characterizing lifetimes of each material is challenging.

Tensile strength testing was attempted to assess comparative reliability between samples. However, mechanical failure in the grips resulted in only the PVC Poly material being tested to full failure at 728 N, while the other samples slipped from the grips at the 1100 N − 1400 N range, making the PVC Poly the weakest of the four by a substantial margin. CoverMax and CoverTuff come with 3-year and 5-year warranties, respectively, which at least serve as a reasonable floor to expected lifetime. Using the comparative weights of each material for reference (10.5 oz/yd^2 or 356 g/m^2, 12 oz/yd^2 or 407 g/m^2, 14.5 oz/yd^2 or 492 g/m^2, and 18 oz/yd^2 or 610 g/m^2 for PVC Poly, CoverMax, Poly Canvas, and CoverTuff, respectively), the Poly Canvas is likely to fall somewhere between the two in terms of lifetime.

Another factor to consider is the possible effect of weathering on each material. Though they are all intended for outdoor use, with some degree of treatment against UV degradation, some amount of discoloration may occur and affect long term albedo values. An exposure and weathering study of sample swatches of each material is currently underway to assess this impact.

Finally, the actual cost of production for the reflectors will likely depend on the fabrication process. For example, Covers and All does not sell their CoverMax and CoverTuff material, but only fabricates tarps and covers to order. Thus, in the absence of any special contract arrangements, a hybrid reflector would require fabrication of the base and modification by a second party, increasing costs and making the use of materials that can be sourced independently, such as the Poly Canvas, more attractive by comparison. Similarly, additional economies of scale may come into play when scaling up depending on what deals can be arranged with various vendors.

VI. SUMMARY OF WORK

Experimental testing under sunlight conditions yielded effective albedo values for a variety of sample materials. This included a dynamic hour-dependent albedo value for the Vivosun material. These values were put into a model to simulate the impact of a 25' by 30' (7.6 m by 9.1 m) reflector on the production of a vertical bifacial array in a 4 panel long by 2 panel high arrangement and a 4 panel long by 3 panel high arrangement.

Modeled production values were used to determine the cost of energy of the additional production associated with the reflector as well as the impact on the overall system-wide cost of energy over a 20-year period for variable assumptions of reflector lifetime. These cost of energy values were also used to optimize the design of hybrid reflectors, where a top layer of Vivosun material is attached to a stronger base layer. Additionally, this optimization method provides a guideline for determining suitability of canal sites for vertical PV + reflector systems based on the ratio of canal width to panel height.

More accurate comparison between reflector candidate materials and associated cost of energy modeling depends on more accurate characterization of lifetimes for each material, which necessitates more extensive field testing under a range of conditions. In addition, a study of the possible impact of UV and general weather exposure on long-term albedo values of each material is currently ongoing.

ACKNOWLEDGEMENTS

We gratefully acknowledge the California Department of Water Resources for funding, administered through Turlock Irrigation District.

REFERENCES

[1] https://www.tid.org/about-tid/current-projects/project-nexus/

[2] J. Reagan and S. Kurtz, "Vertical Bifacial Solar Panels as a Candidate for Solar Canal Design," in 49th IEEE Photovoltaic Specialist Conference, 2022

[3] Howell, J.R., "A catalog of radiation configuration factors", McGraw-Hill, 1982. http://www.thermalradiation.net/tablecon.html

[4] NREL, 2022 Electricity ATB Technologies and Data Overview https://atb.nrel.gov/electricity/2022/index

[5] B. McKuin, et al, "Energy and water co-benefits from covering canals with solar panels". Nat Sustain 4, 609–617 (2021). https://doi.org/10.1038/s41893-021-00693-8 .

Evaluating Leafy Green Production in a Colorado Rooftop Agrivoltaic System

Armando Villa-Ignacio and Jennifer Bousselot, Ph.D.

Department of Horticulture and Landscape Architecture, Colorado State University, Fort Collins, CO 80523, USA

Abstract — By 2050, it is estimated that more than 2/3 of the world's populations will live in urban areas. With land management, food security, and energy production becoming increasingly challenging, mitigating the limiting variables has becoming progressively important. Rooftop agrivoltaics combines effective land management, food security, and energy production. This study evaluates the growth of high value leafy greens under opaque CdTe framed solar panels, opaque CdTe frameless panels, 40% semi-transparent CdTe frameless panels, and in full sun as a control treatment. A rooftop agrivoltaic system is simulated at grade in Fort Collins, Colorado, at the Foothills Campus of Colorado State University. Biomass accumulation, growth rate, stomatal conductance, and environmental conditions were evaluated through multiple growth cycles for five leafy greens. Upon preliminary analysis, leafy greens grown under semi-transparent panels accumulated the largest fresh weights comparative to the other panel types and full sun treatments. It was also found that the plants have a relatively similar stomatal conductance between the traditional opaque panel and the frameless opaque panel, and the shade from the opaque and frameless panels resulted in lower stomatal conductance across all species, except kale. There is greater biomass accumulation under solar panel treatments than in full sun. Understanding the growth characteristics and growing environment of high value crops under these treatments will increase understanding of how these crops will grow in a green roof system under solar panels. These systems can be used to ease food, energy, and land needs within urban centers and supplement the population's basic needs.

I. INTRODUCTION

As the world continues to move towards urbanization, access to food and energy becomes increasingly difficult. By 2050, 68% of the world's population will reside within urban areas [1]. It is also projected that the overall population would grow by 2 billion, or by 20%. Food production would need to more than double that, about 50% more, to feed the growing population [2]. Finding new ways to feed a population without ready access to land for farming would be necessary.

Agrivoltaics systems have been studied for over a decade, with the first prototype of this system in Montpellier, France, in the Spring of 2010 [3]. It was found that the combination of plants and solar panels may increase crop productivity and had a 60%-70% increase of overall land productivity [3]. Plants growing in an agrivoltaic setting under solar panels receive less light, but this has now been shown to be associated with positive trade-offs in terms of reduced evaporative loss of soil moisture in a dryland area [4]. Rooftop agrivoltaics combines photovoltaic arrays with agriculture and green roof systems. This system is an effective land use system in urban areas where space is limited due to the increase in buildings and infrastructure to serve our growing population.

The primary objectives of our rooftop agrivoltaics research are to:

1. Explore the growth and yield of five leafy green crops (lettuce, arugula, spinach, kale, and Swiss chard) in rooftop agrivoltaics systems.
2. Study the reciprocal interactions between vegetation and solar panels. Specifically, test to what extent does the shade of solar panels alter crop growth, and reciprocally, whether the presence of crops increases solar panel performance.
3. Evaluate the stomatal conductance (rate of gas exchange through the stomata) of the vegetables, under the different treatment of solar panels.

II. EXPERIMENTAL DESIGN

A simulated green roof system was installed at the Foothills Campus of Colorado State University [5]. This system contains a 20-mil root barrier and Extenduct drainage/water retention layer, and was supplied by Green Roof Solutions (Glenview, Illinois, USA). The substrate is comprised of 60% expanded shale aggregate, 20% compost, 10% vermiculite, and 10% peat moss, by volume. Four treatments were done, one under full sun conditions, one in deep shade under opaque CdTe framed solar panels, one under opaque CdTe frameless, and one under 40% semitransparent CdTe frameless solar panels, The panels are mounted to a standard ground-mounted racking system angled at approximately 35 degrees due south. The front edge of the panels is 35 cm (14 in) above the substrate and the back edge is 122 cm (48 in) above the substrate. Irrigation was supplied by 1.5 lph (0.4 gph) Netafim drip emitters spaced at 15 cm (6

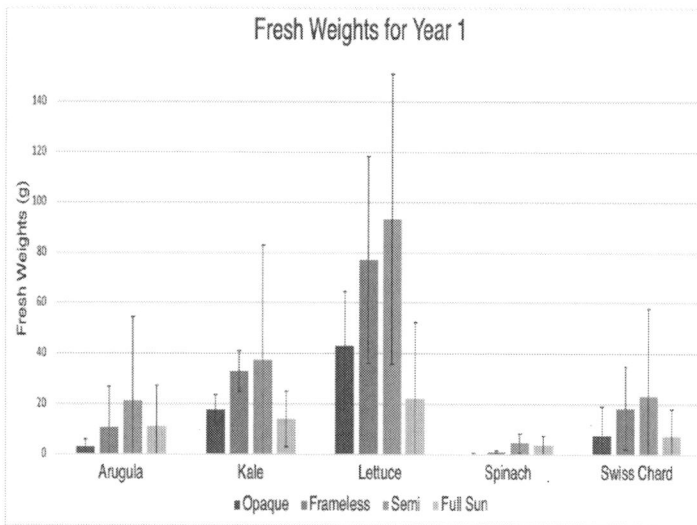

Figure 1. Fresh weight of leafy greens after 1 year

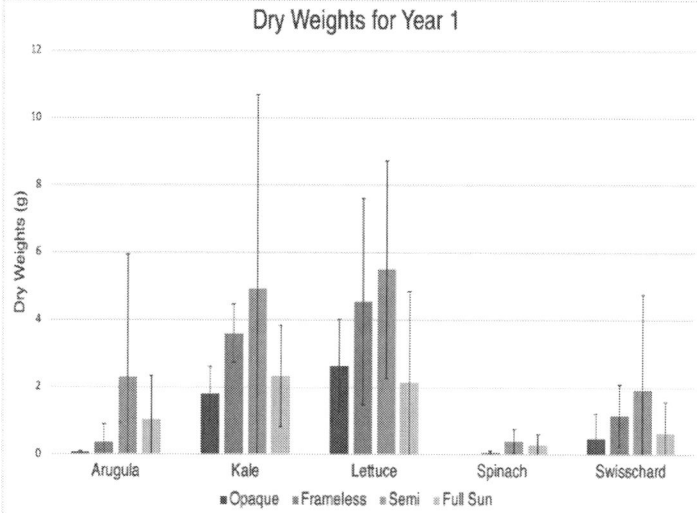

Figure 2 Dry weights of leafy greens after 1 year

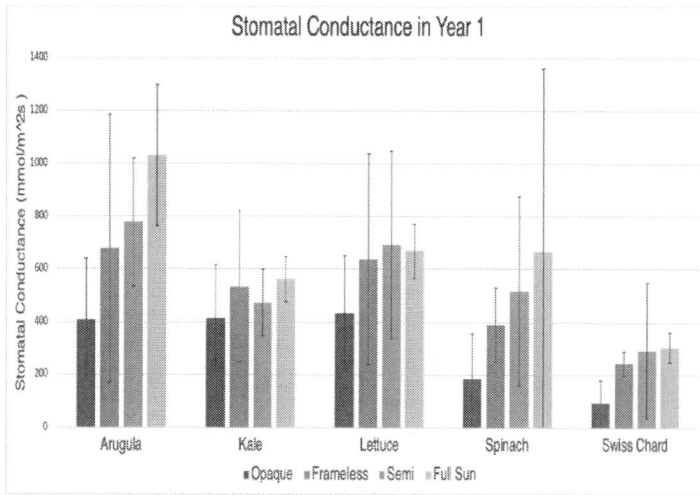

Figure 3 Stomatal Conductance of leafy greens after 1 year

in) intervals and lines were spaced 30 cm (12 in) apart. Arugula (Standard Arugula, *Eruca sativa*), kale ('Winterbore', *Brassica oleracea*), lettuce ('Salvius', *Lactuca sativa*), spinach ('Space', *Spinacia oleracea*), and Swiss chard ('Charbell', *Beta vulgaris*) was grown and obtained from Johnny's Selected Seeds (Winslow, Maine, USA).

Growing conditions are continuously monitored using HOBO H21-USB micro station data loggers (Onset Computer Corporation, Bourne, MA, USA) every 15 minutes. Solar panel surface and air temperatures (measured at 30 cm [12 in] above the surface with solar shield) are measured using HOBO 12-bit temperature smart sensors and are measured every 30 minutes. Stomatal conductance was collected using an SC-1 Leaf Porometer System (Meter Group, Pullman, WA, USA). Height and width were measured. Height was measured from soil level to the tallest point. Width was measured twice, measuring once at the widest part of the canopy of the plant, and directly perpendicular. Chlorophyll content was collected using the atLEAF CHL BLUE (atLEAF, Wilmington, DE, USA). Stomatal conductance, height and width, and chlorophyll content were collected at every harvest.

III. RESULTS

Preliminary results for the first year of data collection show that the 40% semi-transparent CdTe panel resulted in the greatest fresh weight compared to the other treatments (Fig 1). This trend is also seen in Fig 2, where the 40% semi-transparent CdTe panel had a higher dry weight compared to the other treatments. Lower water use measured by stomatal conductance was found under the solar panels compared to full sun. Full sun exposure had a higher stomatal conductance than the solar panel treatments, except for lettuce, where the 40% semi-transparent treatment showed a higher stomatal conductance (Fig 3.).

IV. DISCUSSION

The 40% semi-transparent panel is currently showing greater plant growth compared to the other treatments. The increased light availability under the semi-transparent panels compared to both opaque types allowed the plants to acquire the necessary quantity of sunlight for adequate growth while also using less water. We are investigating the pattern between overall plant size and stomatal conductance over the next growing season. Overall, leafy greens grown under solar panels are showing higher productivity, both in fresh weight and dry weight accumulation.

During the 2023 growing season, a second site will be used at the Colorado State University Spur Campus in Denver, Colorado, USA. This site will include silicon opaque panels and silicon bifacial panels on the Hydro building of Spur. The same leafy greens will be grown and evaluated at both locations.

REFERENCES

[1] Ritchie, H., and M. Roser. 2018. "Urbanization." Our World in Data.org

[2] Deutsch, L., Dyball, R., Steffen, W. (2013). Feeding Cities: Food Security and Ecosystem Support in an Urbanizing World. In:, *et al.* Urbanization, Biodiversity and Ecosystem Services: Challenges and Opportunities. Springer, Dordrecht. https://doi.org/10.1007/978-94-007-7088-1_26

[3] C. Dupraz, H. Marrou, G. Talbot, L. Dufour, A. Nogier, Y . Ferard, Combining solar photovoltaic panels and food crops for optimising land use: towards new agrivoltaic schemes Renewable Energy, 36 (10) (2010), pp. 2725-2732

[4] Barron-Gafford, G., M. Pavao-Zuckerman, R. Minor, L. Sutter, I. Barnett-Moreno, D. Blackett, M. Thompson, Y. Dimond, A. Gerlak, G. Nabhan, and J. Macknick. 2019."Agrivoltaics provide mutual benefits across the food–energy–water nexus in drylands." Nature Sustainability, 2: 1-8. https://doi.org/10.1038/s41893-019-0364-5

[5] Uchanski, M., Hickey, T., Bousselot, J., & Barth, K. L. (2023, March 25). *Characterization of agrivoltaic*

[6] *cropenvironment conditions using opaque and thin-film semi-transparent modules*. MDPI. Retrieved April 6, 2023, from https://www.mdpi.com/1996-1073/16/7/3012

Isotropic Wet Etching of Acoustically-Spalled GaAs

Anica N Neumann, Myles A Steiner, Pablo G Coll, Mariana Bertoni, Emily L Warren

National Renewable Energy Laboratory, Golden, CO, United States

Colorado School of Mines, Golden, CO, United States

Crystal Sonic, Inc., Tempe, AZ, United States

Arizona State University, Tempe, AZ, United States

While III-V photovoltaics are among the most efficient, they suffer from a high manufacturing cost. One of the methods of reducing this cost is through GaAs substrate reuse, with a promising technology being that of acoustic spalling. Acoustic spalling, however, is a developing technology that leaves a rough, faceted GaAs surface. Chemo-mechanical polishing is a method for preparing the acoustically-spalled surface for photovoltaic growth, processing and subsequent respalling of the substrate, but at substantial cost. A less expensive method of substrate preparation is the application of a smoothing etch. Six etching systems were chosen to explore for this application: $HCl:H_2O_2:H_2O$ (40:4:1 and 80:4:1), $H_3PO_4:H_2O_2:H_2O$ (1:1:3), and $H_2SO_4:H_2O_2:H_2O$ (3:1:1, 5:1:1, and 8:1:1). Through testing at room temperature and a moderate stir rate of 600 rpm, the $HCl:H_2O_2:H_2O$ (40:4:1 and 80:4:1), $H_3PO_4:H_2O_2:H_2O$ (1:1:3) etchants proved to have a negative post-etch surface morphology for subsequent growth, processing, and/or re-spalling. The $H_2SO_4:H_2O_2:H_2O$ (8:1:1) etchant at 30°C and 600 rpm stirring showed to be the best etchant at a high percent roughness reduction to mass loss. This etchant was then applied to a full 2-inch acoustically-spalled GaAs wafer prior to single-junction GaAs cell growth. The processed cells will be compared to an unetched, spalled substrate, and an epi-ready substrate with the same growth/processing to compare cell performance metrics (external quantum efficiency, open circuit voltage, short circuit current) and yield. The growth and comparison is ongoing, but the substrate with the $H_2SO_4:H_2O_2:H_2O$ (8:1:1) etchant applied is expected to increase yield and show improvements in average cell performance metrics as compared to the unetched substrate.

Degradation-related Defect Level in Weathered Silicon Heterojunction Modules Characterized by Deep Level Transient Spectroscopy

Steve Johnston, Dirk C. Jordan, Dana B. Kern, Kristopher O. Davis, Helio R. Moutinho, George F. Kroeger

National Renewable Energy Laboratory, Golden, CO, United States

University of Central Florida, Orlando, FL, United States

Kroeger, Inc., Phoenix, AZ, United States

Commercial silicon heterojunction modules known as heterojunction with intrinsic (amorphous silicon) thin-film layer (HIT) modules show average degradation after 10 years in the field. HIT modules weathered outdoors in Colorado and Florida have reduced photoluminescence intensity compared to a control module and have degradation dominated by voltage loss. Deep level transient spectroscopy (DLTS) detects three electron-trap defect states in all modules with activation energies of 0.07, 0.16, and 0.50eV. DLTS on the weathered modules shows an additional deep-level, electron-trap defect state with an activation energy of 0.52eV.

Interrogating Dominant Recombination Pathways in CdTe Solar Cells Using Wavelength-Dependent External Radiative Efficiency Measurements

Jared D. Friedl, Adam B. Phillips, Manoj K. Jamarkattel, Tyler Brau, Sabin Neupane, Scott L. Wenner, Abasi Abudulimu, Ebin Bastola, Yanfa Yan, Randy J. Ellingson, and Michael J. Heben

Wright Center for Photovoltaics Innovation and Commercialization, Department of Physics and Astronomy, University of Toledo, Toledo, Ohio, 43606, USA

Abstract — **External radiative efficiency (ERE) measurements are increasingly being used by the CdTe community to identify sources of loss in these devices. So far, ERE has narrowly been applied to determining the implied open-circuit voltage (iV$_{OC}$) throughout processing and determining the selectivity of contacts by comparing iV$_{OC}$ to measured open-circuit voltage (V$_{OC}$). These applications fail to address more specific questions about the relative importance of the primary recombination channels in CdTe devices. Here, we simulate the V$_{OC}$ of CdTe devices under different monochromatic illumination and find that ERE should be able to distinguish differences in iV$_{OC}$ that correspond to differences in the location of the limiting recombination channel. The intensity dependence of the difference in iV$_{OC}$ between two wavelengths further refines the possible conclusions, as do back-illuminated ERE simulations. These principles are applied to wavelength- and intensity-dependent ERE measurements of a set of CdTe solar cells with purposefully varied front and back interface qualities. Experimental data shows agreement with simulation in several ways, however a more thorough treatment will be necessary to identify the most effective implementation.**

I. INTRODUCTION

CdTe-based photovoltaics have achieved significant improvements in the past several years [1]. The introduction and advancement of CdSe intermixing [2] has greatly improved the bulk, shifting attention to the interfaces. Various front emitters and back buffers have so far been investigated but performance has stagnated, mainly due to persistently low open-circuit voltage (V$_{OC}$). While minority carrier lifetime in the bulk is readily quantified by time-resolved photoluminescence (TRPL) measurements, a similar diagnostic of interface quality has yet to be established. Systematic identification of the limiting recombination channel remains difficult, hindering the ability of researchers to accurately target improvements where most necessary and possibly obscuring improvements made elsewhere.

One technique that has gained popularity in the CdTe community is the measurement of external radiative efficiency (ERE), which quantifies the fraction of radiative recombination to the total recombination at open circuit and provides an implied V$_{OC}$ (iV$_{OC}$) which signifies the maximum electrical potential extractible from a device. Eq. (1) shows the relationship between iV$_{OC}$, ERE, and the radiative V$_{OC}$ limit V$_{OC,ideal}$.

$$iV_{OC} = V_{OC,ideal} + (k_B T/q)\ln(ERE) \qquad (1)$$

where k$_B$ is Boltzmann's constant, T is the temperature, and q the elementary charge. As with other optical measurements, ERE is expected to be dependent on both the magnitude and spatial profile of electron-hole pair generation throughout a device. While various studies have shown the utility of intensity-dependent ERE measurements [3,4], the literature lacks significant examination of the degree of wavelength dependence of ERE and its implications. By manipulating the importance of recombination either at the front interface, in the bulk, or at the back interface through differences in generation brought on by different monochromatic wavelength excitations, wavelength-dependent ERE measurements may provide insight into the relative significance of recombination occurring at each of these locations.

Here, we use numerical simulation to investigate the conditions under which wavelength-dependent ERE is expected to provide insight into the location of limiting recombination in CdTe solar cells. The difference in iV$_{OC}$ measured by different wavelengths at equivalent photon fluxes ($\Delta iV_{OC}(\Phi) = iV_{OC}(\lambda_1, \Phi) - iV_{OC,\lambda_2}(\lambda_2, \Phi)$) removes the need to separately determine V$_{OC,ideal}$ altogether, as separate ERE measurements occur on the same device. This difference is readily determined by the ratio of separate ERE measurements in the following way:

$$\Delta iV_{OC} = (k_B T/q)\ln(ERE_1/ERE_2) \qquad (2)$$

The intensity dependence of this value, in some situations, provides clarifying information with respect to the dominant recombination channel. After simulating both front- and back-illuminated ΔiV_{OC} values over large ranges of front interface, bulk, and back interface quality, we provide

978-1-6654-6060-6/23 $31.00 © 2023 IEEE

experimental data testing the validity of the simulated results. The wavelength dependence of ERE at low intensity seemingly confirms the relative significance of front interface recombination in the prepared devices, motivating further investigation of the viability of this characterization technique.

II. SIMULATION

A. Numerical Model

Simulations were carried out using SCAPS-1D [5]. The material parameters used in this model have been used previously [6]. Properties relevant to the following analysis include a CdTe thickness of 4 μm and a CdTe bulk lifetime of 100 ns. Additionally, flat bands are assumed at the front and back electrodes. By imposing flat bands at the contacts, no band bending is introduced as a result of fixed metal work functions and it is expected that the simulated V_{OC} represents the iV_{OC} that would be measured by ERE.

Unless otherwise noted, the conduction band offset (CBO) at the front interface and initial Fermi level offset (IFLO) at the back interface [7] are set to 0 eV. The CBO and IFLO are set using the emitter electron affinity and the back buffer shallow acceptor density, respectively. Fixed band bending at the interfaces leads the surface recombination velocity (SRV) at each interface to more completely determine the interface recombination current density, J_{rec}^{int}, as J_{rec}^{int} is the product of the minority carrier density and an effective SRV determined by band bending and SRV.

The front surface recombination velocity (FSRV) at the emitter/CdTe interface and the back surface recombination velocity (BSRV) at the CdTe/back buffer interface are used to control the quality of the front and back interfaces, respectively. The FSRV and BSRV are varied by changing the total defect density of their corresponding interface, in accordance with $SRV = v_{th}\sigma N_t$, with v_{th} the thermal velocity of carriers, σ the capture cross section, and N_t the total defect density.

Simulation of monochromatic excitation is possible using spectrum files provided with the SCAPS software. 600 and 800 nm are chosen due to the relative availability and low cost of laser sources of comparable wavelengths (633 nm and 785 nm). The $1/\alpha$ absorption depth of 600 and 800 nm light in this model are 133 nm and 453 nm, respectively. This value is 149 and 357 nm for 633 and 785 nm light, respectively, so the simulated difference in penetration depth is somewhat higher than the expected experimental value. The irradiance of monochromatic excitation is normalized to a 1-sun condition based on the AM1.5G photon flux above the absorber bandgap, as is common in experimental measurements [3]. The integer wavelength nearest to the bandgap of CdTe is 827 nm; the cumulative photon flux in the AM1.5G spectrum at or below this

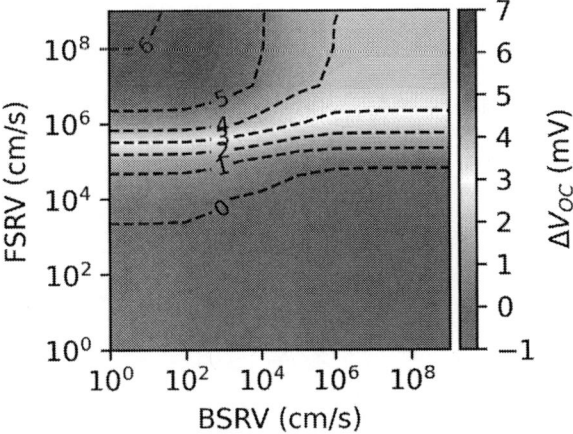

Fig. 1. Difference in simulated open-circuit voltage between 800 and 600 nm monochromatic illumination (ΔV_{OC}) at 1-sun from the front as a function of front and back surface recombination velocity.

wavelength is 1.806x10²¹ photons/m²s. This value can be converted to irradiance by multiplying by the Joules per photon for a given wavelength. The 1-sun irradiance values for 600 and 800 nm excitations are therefore 598 and 448 W/m², respectively. In this investigation, the difference in V_{OC} between 800 and 600 nm excitation is modeled under various conditions. This difference will be denoted using ΔV_{OC}, equal to $V_{OC}(800\ nm) - V_{OC}(600\ nm)$. As a larger absorption depth generates more carriers away from the illuminated interface, greater ΔV_{OC} should in general correspond to high recombination in the vicinity of that interface, and vice versa.

B. Front-Illuminated Simulations

To understand the impact of the relative quality of each interface on the wavelength dependence of ERE, ΔV_{OC} was simulated as a function of a range of FSRV and BSRV values. Fig. 1 shows the front-illuminated $\Delta V_{OC} = V_{OC}(800nm) - V_{OC}(600nm)$ as a function of FSRV and BSRV at 1 sun. The value of ΔV_{OC} shifts from more negative to more positive as the color shifts from red to blue.

Over much of the range of simulated interface conditions, ΔV_{OC} is between -1 and 0 mV. A negative ΔV_{OC} indicates that the deeper absorption depth of the 800 nm light causes a higher share of non-radiative recombination. This occurs at FSRVs low enough that front interface recombination no longer limits V_{OC} for either wavelength. Limiting recombination instead occurs in the bulk when BSRV is low and transitions to the back interface as BSRV increases. In either case, increased penetration shifts carrier generation away from the non-limiting front, slightly accelerating the bias-dependent accumulation of minority carriers in the limiting channel (bulk or back) that leads to open-circuit. A $\Delta V_{OC} < 0$ when illuminating from the front indicates that the front interface recombination is essentially negligible

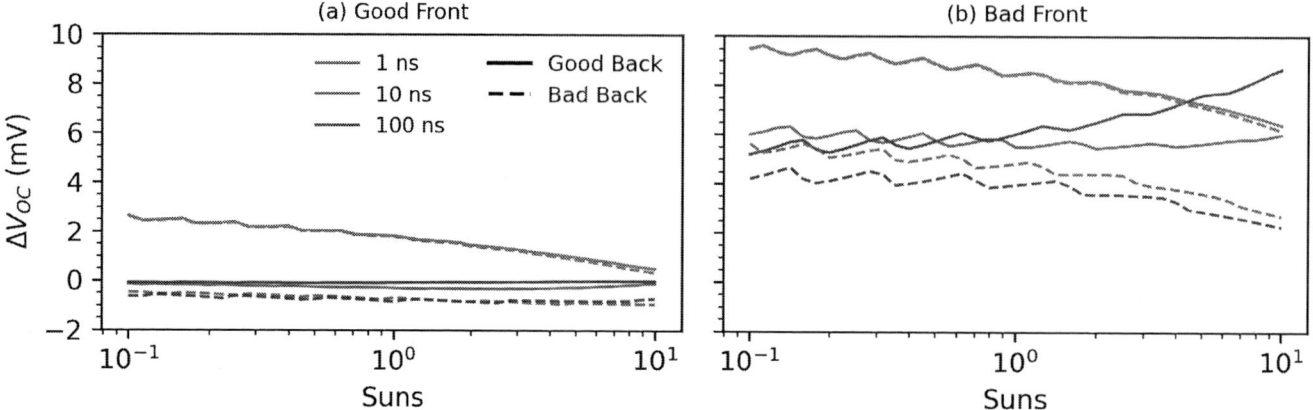

Fig. 2.   Intensity, lifetime, and back interface quality dependence of difference in simulated open-circuit voltage between 800 and 600 nm monochromatic illumination (ΔV_{OC}) in the case of a (a) Good or (b) Bad Front as defined in the text.

relative to that of the bulk and/or back. According to (2), measuring a $\Delta i V_{OC}$ on the order of 1 mV requires distinguishing a relative ERE difference of ~4% at room temperature, requiring high precision when dealing with the low photoluminescence of CdTe.

To the contrary, a positive ΔV_{OC} suggests that front interface recombination plays a significant role in limiting V_{OC}. Because differences in generation are greatest nearest to the front interface, ΔV_{OC} exhibits a much greater capacity for distinguishing differences in the degree of front interface recombination. Simulated ΔV_{OC} greater than 6 mV occurs when FSRV is highest, but this requires the BSRV to simultaneously be very low. Detecting $\Delta i V_{OC}$ of this magnitude requires a relative ERE difference of ~26%, a difference that can be distinguished with some confidence for reasonably luminescent CdTe stacks. The influence of BSRV is itself dependent on the value of FSRV, such that ΔV_{OC} from the front alone is not necessarily able to distinguish the degree to which each interface contributes to limiting recombination.

The absorber lifetime determines the bulk contribution to the recombination current density, which may further affect the functionality of ΔV_{OC}. To examine the roles of lifetime and intensity in determining ΔV_{OC}, the intensity dependence of ΔV_{OC} was simulated for discrete series of FSRV, BSRV, and CdTe lifetime. The FSRV and BSRV are each separately set to 1 cm/s (a "Good" interface) and 10^9 cm/s (a "Bad" interface), corresponding to data points at each of the four corners of Fig. 1. Previously fixed at 100 ns, 3 CdTe bulk lifetimes are now examined ranging from 1 to 100 ns. Lifetime is modulated using the density of the CdTe bulk defect. The intensity is varied logarithmically between 0.1 and 10 suns. Fig. 2 shows the intensity dependence of ΔV_{OC} with these combinations of interface and bulk properties.

As expected from the discussion of Fig. 1, a Good Front (Fig. 2(a)) produces low or even negative values of ΔV_{OC} as the deeper absorption depth of the longer wavelength excitation more strongly emphasizes the limiting channel (either the bulk or back interface). However, positive values of ΔV_{OC} do still occur with a Good Front when the lifetime is lowest, 1 ns in Fig 2(a). With such a limiting bulk, the lower absorption coefficient of longer wavelength light better avoids high concentrations of minority carriers that facilitate bulk recombination. The degree to which the bulk is limiting here is evident in the inability to distinguish a Good or Bad Back with ΔV_{OC}. Increasing the lifetime decreases the value of ΔV_{OC} and increases the distinction possible between a Good and Bad Back as carriers are more effectively transported to the back in greater numbers.

Fig. 2(b) shows that a Bad Front not only causes larger ΔV_{OC}, but more variation compared to a Good Front as the bulk, back, and intensity are changed. As with a Good Front (Fig. 2(a)), a 1 ns lifetime produces highly positive ΔV_{OC} that decreases with intensity and varies little as a function of the back interface. With increasing lifetime, distinct behaviors arise as a function of the back interface due to the increased transport of carriers across the bulk. At a lifetime of 10 ns or above, ΔV_{OC} trends either upward or downward with increased intensity for a Good or Bad Back, respectively. This intensity dependence partly clarifies the relative quality of the front and back interface: increasing ΔV_{OC} indicates that the front is more strongly limiting, while decreasing ΔV_{OC} indicates that the back is at least not significantly better than the front.

C. Back-Illuminated Simulations

Comparisons of back-illuminated PL measurements are often used to diagnose the relative quality of the back interfaces of different samples. Photoluminescence of CdTe films is often much lower when excited from the back, however these measurements remain useful given the availability of measurement systems with sufficient sensitivity. Front-illuminated wavelength-dependent ERE measurements are unable to unambiguously distinguish the

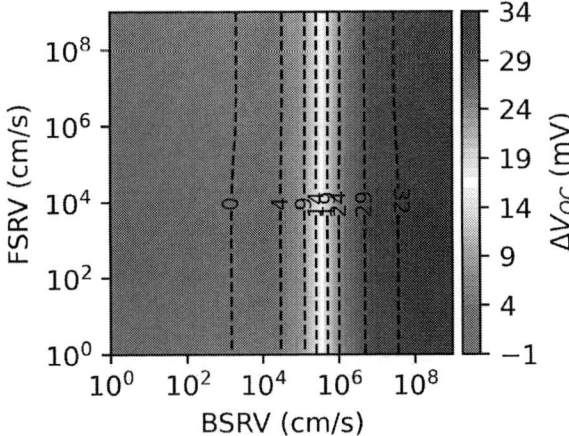

Fig. 3. Difference in simulated open-circuit voltage between 800 and 600 nm monochromatic illumination (ΔV_{OC}) at 1-sun from the back as a function of front and back surface recombination velocity.

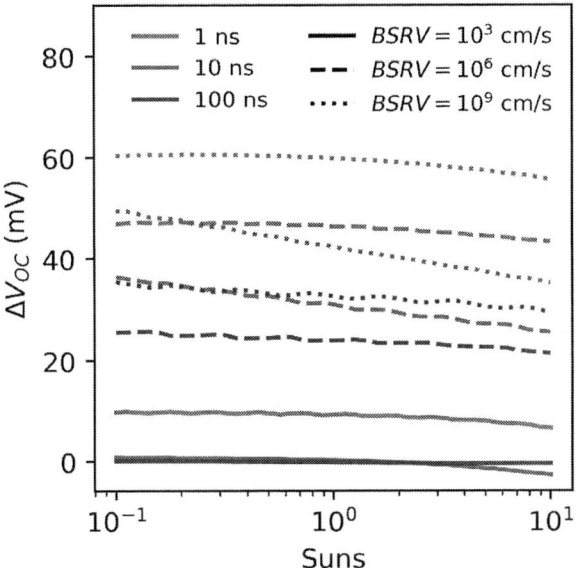

Fig. 4. Intensity dependence of the difference in simulated open-circuit voltage between 800 and 600 nm monochromatic illumination (ΔV_{OC}) from the back for combinations of lifetime and back surface recombination velocity.

limiting interface on their own; companion measurements taken from the back side may provide a fuller picture of the limiting recombination channel.

As was done for front illumination, the impact of interface quality on the wavelength dependence of ERE was simulated with light incident from the back. Fig. 3 shows the back-illuminated $\Delta V_{OC} = V_{OC}(800nm) - V_{OC}(600nm)$ as a function of FSRV and BSRV at 1 sun. It is evident not only that the expected magnitude of ΔV_{OC} in a similar range of FSRV and BSRV values is potentially much higher during back illumination, but also that ΔV_{OC} is much more exclusively a product of BSRV than it was of either recombination velocity for front illumination. This is sensible do the lack of band bending at the back interface; there is no large built-in field to repel minority carriers, so the assigned BSRV more completely determines the effective BSRV. A very abrupt transition occurs between BSRVs of 10^4 cm/s and 10^7 cm/s from low (or slightly negative) ΔV_{OC} to highly positive ΔV_{OC}.

Because back-illuminated ΔV_{OC} is so dominated by the BSRV, ΔV_{OC} dependence on the FSRV for back illumination is not seen on the scale that BSRV dependence was seen observed for front illumination. Instead, it is more informative to observe the interplay of lifetime and BSRV for a greater number of BSRV values, rather than just the two extremes of "Good" and "Bad." Figure 4 shows the intensity dependence of back-illuminated ΔV_{OC} for identical lifetimes as in Fig. 2, now for specific BSRV values of 10^3, 10^6, and 10^9 cm/s. The FSRV is fixed at 1 cm/s.

As expected, Fig. 4 shows that back-illuminated ΔV_{OC} is potentially much larger than that from the front for similar values of lifetime and the SRV of the illuminated interface. With no band bending at the back, there is no field to effectively separate generated carriers within its depletion width. Electrons must now diffuse away from the back

interface and across the bulk before being collected at the front, making the lifetime much more influential to back-illuminated ΔV_{OC} as well. The presence of a field at the back interface, contributed by a positive IFLO, would likely decrease ΔV_{OC} similar to the strong front interface band bending.

III. EXPERIMENT

A. Experimental Details

In order to apply the principles revealed through numerical simulation in the previous section, CdTe samples were prepared using combinations of methods that are known to create varying interface and bulk qualities. In a previous work, co-sputtered In_2O_3 and Ga_2O_3 were used to produce $(In_xGa_{1-x})_2O_3$ (IGO) alloys as front emitter layers for CdSe/CdTe solar cells, whose wide, tunable bandgaps were used to alter the front interface recombination via changes in band bending and the band alignment with the absorber [8]. Separately, copper thiocyanate (CuSCN) has been demonstrated as an effective back buffer to CdTe whose short, low-temperature application confines copper to the back of the device and improves hole transport there [9]. Using each of these methods either separately or together allows us to control the relative quality of each interfaces.

Film stacks of the structure TEC12/(In_xGa_{1-x})$_2O_3$/CdSe/CdTe were deposited and CdCl$_2$-treated as described in our previous work [8]. The IGO for one set of samples used x = 0.71 which is expected to lead to a poor

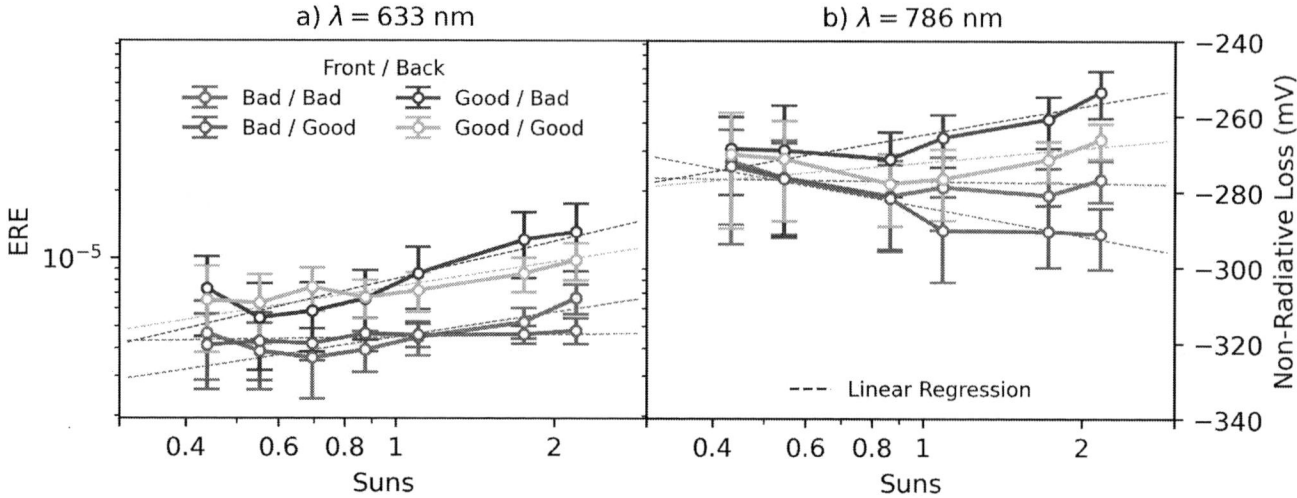

Fig. 5. Intensity-dependent external radiative efficiency of the four prepared samples at excitation wavelength of a) 633 nm and b) 786 nm.

fronter interface, while another set of samples used the best-performing IGO composition x = 0.36. Samples with IGO compositions of x = 0.71 and x = 0.36 are referred to as having a Bad and Good front, respectively. These sets of samples were further divided by their copper treatments: one sample from each set underwent a standard $CuCl_2$ doping procedure [10] while the other received CuSCN as a back buffer. Samples treated with $CuCl_2$ and CuSCN are referred to as having a Bad and Good back, respectively. Complete devices are referred to by their corresponding front and back interface qualities as summarized in Table I. This simple set of devices aims to provide a variety of mixed front and back interface qualities. Differences in the bulk, and even the front to a lesser extent, are also expected due to the differences in copper incorporation.

B. Front-Illuminated Results

Glass-side intensity-dependent ERE measurements were taken at two wavelengths, 633 nm and 786 nm, for each of the four samples. Accounting for the variable power and beam shapes of each light source, an overlapping range of suns was achieved from approximately 0.4 to 2. Figure 5 shows the ERE measurements for each sample as a function of sun-equivalent intensity for excitation wavelength of a) 633 nm and b) 786 nm, with the corresponding non-radiative voltage loss on the right axis as determined using the rightmost term in (1). The ERE values measured using 633

nm excitation (Fig. 5a) lie between $2x10^{-6}$ and $2x10^{-4}$ for all intensities, corresponding to non-radiative potential loss between 340 and 280 mV. These values are quite typical for CdTe. It is evident that the ERE measured using 786 nm excitation (Fig. 5b) is higher for all samples than when excited with the shorter wavelength light. This is reasonable, given that simulated ΔV_{OC} for front illumination was often positive, indicating a higher expected ERE for longer wavelengths. Log-linear regressions of non-radiative loss vs. intensity are imposed over the data for each sample. All samples show an increase in ERE with intensity, except for the two samples with a Bad front interface measured using 786 nm excitation. This is contrary to expectations, and may

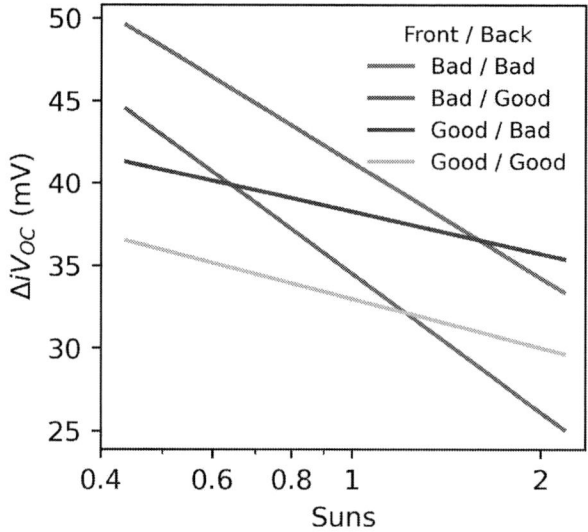

Fig. 6. Linear fits of the intensity dependence of the difference in implied open-circuit voltage measured between 786 nm and 633 nm excitation for the four CdTe samples.

TABLE I
SAMPLE LABELLING CONVENTIONS

Emitter	Cu Source/ Buffer	Label
$(In_{0.71}Ga_{0.29})O_3$	$CuCl_2$	Bad / Bad
	CuSCN	Bad / Good
$(In_{0.36}Ga_{0.64})O_3$	$CuCl_2$	Good / Bad
	CuSCN	Good / Good

indicate large error in the low-intensity data for those samples at that wavelength.

The measurement of wavelength- and intensity-dependent ERE makes possible the construction of a ΔiV_{OC} plot, shown in Fig. 6. As in the simulations, ΔiV_{OC} represents the iVoc of the longer wavelength minus that of the shorter wavelength, $iV_{OC}(786nm) - iV_{OC}(633nm)$. Despite the indication in Fig. 2 that ΔV_{OC} may not necessarily evolve linearly with intensity, the limited range of intensities possible in the measurements precludes the use of higher-order fits. The regression curves shown in Fig. 5 are used to plot ΔiV_{OC} over the measured intensity range. This simple fit reveals several features consistent with expectations developed in the previous section.

A large and downward-sloping ΔV_{OC} was found to indicate either an extremely low bulk lifetime, or jointly poor front and back interfaces. Both samples with a Bad front show higher ΔiV_{OC} than those with a Good front at the lowest intensity, showing a strong imperative to avoid the front interface using longer wavelengths. Furthermore, smaller distinctions in the slope of ΔiV_{OC} correspond to differences in time-resolved photoluminescence (TRPL) tail lifetimes. The ΔiV_{OC} of the Bad/Good device, whose lifetime is 3.6 ns, decreases somewhat more rapidly with intensity than the Bad/Bad device, whose lifetime is 16.4 ns. A corresponding distinction occurs for the Good/Good and Good/Bad devices, whose lifetimes are 76.4 and 112 ns, respectively. A more rapid decrease in ΔiV_{OC} with intensity is then consistent with both lower front quality and lower lifetime, as expected, though the data show that this is much more clearly a function of the front. The effect of the front interface on the TRPL lifetimes themselves should also be accounted for.

Nearly independent of intensity, there is a split in ΔiV_{OC} between samples with similar front interface quality. This is somewhat inconsistent with the measured lifetimes as increased lifetime was shown to consistently decrease ΔV_{OC}, and here the opposite is observed. CuSCN is meant to improve hole transport at the back, as well, which was also associated with increased ΔV_{OC}. However, the front could have also been further modified as a result of the prevention of Cu migration afforded by the CuSCN treatment [9], and the sensitivity to the front discussed previously may remain dominant.

Further interpretation of the data is made difficult by its limited scope and the resulting simplicity of the extrapolated points. The intersection of the fit lines of each Cu dopant between 1 and 2 suns, for example, which was not observed in our simulations, cannot be assigned significance without additional certainty.

IV. SUMMARY

It was shown using numerical simulation that differences in Voc, and therefore iVoc, are expected to be measureable using different wavelengths of monochromatic excitation. When front-illuminated, it is expected that differences in wavelength- and intensity-dependent iVoc will have mixed dependence on the share of recombination occurring at the front interface, in the bulk, and at the back interface. When back-illuminated, this difference was much more uniquely indicative of BSRV, especially when lifetimes are not too low. CdTe devices were fabricated with various combinations of front and back interface qualities as indicated by previous studies. Front-illuminated experimental wavelength- and intensity-dependent ERE results were able to distinguish the intentional differences between the front and back interfaces, but bulk recombination also played an unclear role. Increased intensity overlap between wavelengths, back-illuminated measurements, and correspondence with additional characterization should increase the confidence with which this method is able to identify the limiting channel of recombination in CdTe solar cells.

ACKNOWLEDGMENTS

This report is based on research sponsored by the U.S. DOE's Office of Energy Efficiency and Renewable Energy (EERE) under Solar Energy Technologies Office (SETO) Agreement DE-EE0008974, through the Alliance for Sustainable Energy, LLC, Managing and Operating Contractor for the National Renewable Energy Laboratory for the U.S. Department of Energy, under Award Number 37989.

REFERENCES

[1] Green, Martin A., et al. "Solar cell efficiency tables (Version 61)." *Progress in Photovoltaics: Research and Applications* (2023).

[2] Paudel, Naba R., and Yanfa Yan. "Enhancing the photo-currents of CdTe thin-film solar cells in both short and long wavelength regions." *Applied Physics Letters* 105.18 (2014): 183510.

[3] Kuciauskas, Darius, et al. "Radiative efficiency and charge-carrier lifetimes and diffusion length in polycrystalline CdSeTe heterostructures." *physica status solidi (RRL)–Rapid Research Letters* 14.3 (2020): 1900606.

[4] Caprioglio, Pietro, et al. "On the relation between the open-circuit voltage and quasi-fermi level splitting in efficient perovskite solar cells." *Advanced Energy Materials* 9.33 (2019): 1901631.

[5] Burgelman, Marc, Peter Nollet, and Stefaan Degrave. "Modelling polycrystalline semiconductor solar cells." *Thin solid films* 361 (2000): 527-532.

[6] Liyanage, Geethika K., et al. "The role of back buffer layers and absorber properties for> 25% efficient CdTe solar cells." *ACS Applied Energy Materials* 2.8 (2019): 5419-5426.

[7] Phillips, Adam B., et al. "Understanding cdte performance with engineered front and back interfaces." *2019 IEEE 46th Photovoltaic Specialists Conference (PVSC)*. IEEE, 2019.

[8] Jamarkattel, Manoj K., et al. "Indium Gallium Oxide Emitters for High-Efficiency CdTe-Based Solar Cells." *ACS Applied Energy Materials* 5.5 (2022): 5484-5489.

[9] Li, Deng-Bing, et al. "CuSCN as the back contact for efficient ZMO/CdTe solar cells." *Materials* 13.8 (2020): 1991.

[10] Jamarkattel, Manoj K., et al. "Reduced recombination and improved performance of CdSe/CdTe solar cells due to cu migration induced by light soaking." *ACS Applied Materials & Interfaces* 14.17 (2022): 19644-19651.

Dynamic Calibration of Injection Dependent Carrier Lifetime from Time-Resolved Photoluminescence

Yan Zhu, Robert Lee Chin, Nursultan Mussakhanuly, Thorsten Trupke, Ziv Hameiri

University of New South Wales, Sydney, Australia

Time-resolved photoluminescence is widely used to measure the charge carrier lifetime of thin film semiconductor materials. Nevertheless, the essential injection dependency of the carrier lifetime, which is hidden in these measurements, is often neglected. In this study, a novel dynamic calibration method is proposed to extract injection-dependent carrier lifetime from time-resolved photoluminescence measurements. The proposed method is based on the combination of transient and steady-state measurements. The measured relative photoluminescence signal is calibrated into excess carrier concentration, hence, the injection dependency of the carrier lifetime can be extracted. The method is demonstrated experimentally using a perovskite thin film. The obtained injection-dependent lifetime is then fitted to investigate the recombination mechanisms within this sample. This proposed method significantly leverages the capability of time-resolved photoluminescence and provides many potential applications for a wide range of emerging photovoltaic materials.

Impacts of Dispatch Strategies and Forecast Errors on the Economics of Behind-the-Meter PV-Battery Systems

Brian T. Mirletz
National Renewable Energy Laboratory
Golden, Colorado, USA
Email: brian.mirletz@nrel.gov

Nicholas D. Laws
Walker Department of Mechanical Engineering
The University of Texas at Austin
Austin, TX, USA Email: nlaws@utexas.edu

Abstract—To assess the economic value of batteries in hybrid PV-battery systems, one must create a dispatch profile for the battery. Many analyses of battery value assume perfect forecasts of PV generation and load, determining an upper limit on the value of the battery. Prior work that accounts for forecast uncertainty often does so in the context of a single dispatch algorithm, which does not provide a baseline for comparison. Furthermore, when multiple dispatch algorithms are assessed with uncertainty, the benefits considered are for diesel generation in a microgrid, not retail rate savings. This work addresses the gaps in the literature by comparing the performance of both heuristic and optimal dispatch algorithms for retail rate savings under forecast uncertainty, and provides comparisons of the robustness of these algorithms and their associated estimates of economic value. We find that using a perfect forecast can overestimate the value of hybrid PV-battery systems between 1% and 8% compared to the reality of using a day-ahead forecast, depending on the dispatch algorithm used. Thus, accounting for forecast uncertainty in system design and analysis will significantly improve the accuracy of modeled system values.

Keywords—solar plus storage, batteries, battery dispatch, System Advisor Model, SAM, behind-the-meter

I. INTRODUCTION

Pairing behind-the-meter (BTM) batteries with solar photovoltaic systems (PV) can provide numerous benefits to the customer, including bill reductions, increased PV self-consumption, and backup power [1]. However, the value of these benefits is highly dependent on the local conditions, such as the customer's utility rate, the battery control method, and forecast errors [2, 3, 4, 5]. Therefore, accurate estimation of system value is important for driving adoption.

A key component of system value is the choice of dispatch algorithm. Mirletz and Guittet 2021 explores the trade-offs of these dispatch strategies when varying utility rates, with estimated net present value (NPV) varying by factors of 2.3 or more for the same system design [4]. Dispatch strategies can vary by the use case, and both heuristic and optimal algorithms have been evaluated in the literature [6, 7]. However, these assume perfect forecasts; day-ahead weather forecasts include PV production errors ranging from 3% to 14% [8, 9], and forecasts of peak load have average error rates of 4 to 7% [10]. These forecast errors can cause reductions in system value. For example, Vedullapalli and Hadidi observed 1.3 to 7.8 percentage point reductions in system profit in demand charge

reduction with optimal dispatch depending on the forecast used [11]. Mazzola et al. studied the effect of forecast errors on dispatch strategies, including load following, cycle charging, and optimization with a rolling forecast, finding forecast errors could increase the cost of diesel generation in a microgrid case between 1.2% and 6.6% [5]. Sahoo et al. accounted for forecast uncertainty for residential batteries with a sliding window method and a swarm intelligence-based optimization algorithm, but only compared their proposed algorithm to a manual schedule, and no heuristic algorithms that used a forecast were evaluated [12]. This paper provides the additional analysis of the financial impact of forecast uncertainty on grid connected systems, including comparisons between heuristic and optimal dispatch algorithms that utilize forecasts.

II. METHODOLOGY

This work combines two state-of-the-art techno-economic decision support tools developed at the National Renewable Energy Laboratory (NREL). The System Advisor Model (SAM) contains technology performance models, including PV+battery systems, and allows for techno-economic analysis of systems in multiple financial contexts. The Renewable Energy Integration and Optimization (REopt) platform is used to evaluate optimal dispatch strategies, as well as determine the cost-optimal PV and battery capacities.

The SAM battery model utilizes subcomponent models, including voltage, lifetime, and thermal models for multiple battery chemistries [13]. SAM includes several choices for heuristic dispatch algorithms, including time-based manually specified schedules [14], automated dispatch for peak shaving based on forecasted generation and load (peak shaving dispatch) [15], and price signals dispatch, which considers utility rates in addition to those factors [4].

The peak shaving dispatch algorithm in SAM uses forecasts of generation and load to set a target for power drawn from the grid at each step, referred to as the grid power target [15]. If the power from the grid at a step is forecast to be greater than the target, the battery will discharge. Similarly, the battery will charge if grid use is forecast to be less than the target. Prior to this work, the peak shaving and price signals dispatch algorithms in SAM used the forecast both for the grid power target and to determine charge and discharge power at each step. For this paper, the battery is able to observe the actual load and PV generation at each time step and charge or discharge based on the difference between the grid usage and the target in real time. This does not change the performance of perfect forecasts, but improved the financial performance of the look-behind method by a factor of 6 for peak shaving[1].

To evaluate the performance of optimal dispatch, we used REopt's model predictive control module (MPC). REopt uses mixed integer linear programming to determine cost-optimal sizing and dispatch for distributed energy resources, including behind-the-meter batteries with PV [16]. The MPC module re-uses some of the core functionality from the REopt sizing and dispatch optimization to determine optimal dispatch strategies for use in real-time controls. It should be noted that the stability of model predictive control relies on only implementing the first interval of the optimal control plan. However, for this work only day-ahead forecasts were available so the entire 24 hour optimal dispatch from the REopt MPC module was used for each day in the simulated year [17]. For this work, the MPC module re-planned dispatch each hour, with the actual load and PV generation in the first step of each forecast.

We conducted a case study using actual day-ahead forecasts of weather from locations in Puerto Rico for 2018 from the Puerto Rico Grid Resilience and Transitions to 100% Renewable Energy study. Code and data used in the case study are available on GitHub[2]. Load profiles were generated for a hospital using EnergyPlus, using the day-ahead weather forecast to generate a day-ahead load forecast in addition to the actual weather and load [18]. The resulting load profile uses 6,800 MWh of electricity annually, with a peak load of 1,267 kW. The utility rate is the time-of-use at primary distribution voltage rate[3], which includes both time-of-use rates and demand charges. We assumed excess net metering credits would be compensated at $0.075/kWh at the end of June each year [19]. Costs for the system is based on the 2022 NREL benchmark [20], which are default parameters in SAM version 2022.11.21r2. The analysis period for the system is 25 years, with a 6.4% real discount rate. We assume that the project can receive a 30% ITC through direct pay and that the battery can charge from the grid. As of writing, direct pay requirements are still not published, so if the system was not eligible, it would uniformly decrease the net present value of the systems (NPVs). Batteries are replaced when they reach 80% of their nameplate capacity. Additional defaults (such as AC to DC and DC to AC efficiencies of 96%) are also used.

III. RESULTS

A. System Sizing

System sizing was determined by REopt. Optimal sizing without considering resilience led to 3.63 MW-ac of PV, 151 kW of battery power, and 280.8 kWh of battery energy. However, given the building model is a hospital, resilience is a key consideration, so we re-sized the system with 50% critical load for a 24-hour outage. This resulted in 3.66 MW-ac of PV, 514 kW of battery power, and 4,884 kWh of battery energy. When testing with this size, the SAM battery dispatch algorithms were unable to meet the critical load during the

outage, since the SAM grid outage dispatch algorithm does not currently have foresight of the outage. In order to meet the critical load, we restricted the economic dispatch of the battery to a minimum state of charge of 70%. With this constraint, the optimal system size was 3.66 MW-ac of PV, 506 kW of battery power, and 6,574 kWh of battery energy. We assumed that batteries would be available in 50-kW increments, so rounded up to 550 kW of battery power and 6,500 kWh of battery energy for the remainder of the analysis. Note that the rated battery power is based on the nominal voltage, the actual maximum power is higher at timesteps with a high state of charge, and therefore a high voltage. This allows up to 600 kW of discharge power.

B. Forecast Error

The root mean squared error (RMSE) in direct normal irradiance between the actual weather data and the forecast weather data is 140 W/m^2, and the mean absolute error (MAE) is 75 W/m^2, or around 13.6% of the maximum DNI. This is in line with typical errors in day-ahead forecasts [9]. The RMSE for the PV production data was 493 kW, with MAE of 249 kW, corresponding to 13.6% of the maximum generation. The RMSE for the load data was 33.4 kW, with MAE of 25.2 kW, or around 2.6% of the peak load.

C. Dispatch Results

For this system, we tested the battery with both perfect forecasts and actual day-ahead forecasts using the following dispatch methods:

- Peak shaving dispatch with a rolling 24-hour forecast (peak shaving).

- Price signals dispatch with a rolling 24-hour forecast (price signals).

- Optimal dispatch with a rolling 24-hour forecast (MPC 24 hour).

- Optimal dispatch with a full-year forecast horizon (Full-year horizon).

- A time-based schedule with no forecast (Manual dispatch).

The SAM dispatch models and the REopt dispatch models have different underlying models for the underlying battery and utility rate structures, and the battery power and utility rate calculations presented here were performed in SAM. Price signals dispatch is able to account for surplus or consumed kWh by time of use period during each month, whereas the MPC algorithm does not track net metering credits over the course of the month. The full-year horizon forecast does not have information about the real-time PV and load values, and is not able to re-plan given the forecast errors. Therefore, we will not focus on specific NPV values, rather the differences within each algorithm's performance when run with and without forecast errors.

The algorithms respond differently to forecast errors. Peak shaving will perform charging and discharging according to its grid power target, but forecast errors may introduce differences in the target. For example, on the day when the peak occurs in

---

[1] https://github.com/NREL/ssc/pull/683, https://github.com/NREL/ssc/pull/1031
[2] https://github.com/NREL/SAM-analyses/tree/main/2023/mirletz_laws_pvsc
[3] https://apps.openei.org/IURDB/rate/view/5bfdc7925457a33744146c53

TABLE I. NPV REDUCTIONS FROM USING DAY-AHEAD FORECASTS.

Dispatch	Peak shaving	Price Signals	MPC 24 hour	Full-year horizon	Manual Dispatch
Perfect Forecast NPV	$8,119k	$8,840k	$8,433k	$8,519k	$7,639k
Day-ahead Forecast NPV	$8,045k	$8,732k	$7,739k	$7,605k	N/A
NPV Reduction %	0.91%	1.23%	8.23%	10.7%	N/A

TABLE II. CHANGES IN ELECTRICITY BILL COMPONENTS FROM FORECAST ERRORS.

Dispatch	Peak shaving	Price Signals	MPC 24 hour	Full-year horizon
Increase in demand charge	$8.0k	$5.3k	$49.7k	$102.1k
Percent change versus perfect forecast	9.4%	9.4%	52.4%	117.8%
Increase in energy charge %	$4.0k	$7.1k	$32.3k	$5.1k
Percent change versus perfect forecast %	0.3%	0.5%	2.3%	0.5%

June, the actual weather and load data produces a grid power target of 722.4 kW, whereas the grid power target based on the day-ahead forecast data was 649.5 kW. This tighter grid power target meant that the battery using the day-ahead forecast data fully discharged before the peak, resulting in a higher peak demand charge. The other dispatch algorithms specify power to or from the battery in each time step, meaning that if there is not as much PV production as expected, the batteries will charge from the grid instead. Batteries are allowed to discharge when load does not exceed PV output, which could allow the discharge to push PV power to the grid, but the batteries are not allowed to discharge to the grid by SAM's powerflow code.

The differences in NPV for each dispatch algorithm are reported in Table I. Peak shaving dispatch shows the smallest reduction in system value, followed by price signals dispatch, MPC, and then full year horizon. Manual dispatch results are provided for reference, but would be the same regardless of forecast.

While demand charges only make up 9.6% of the bill without system, in all cases they make up the majority of the difference between the bill savings with the perfect forecast and the day-ahead forecast. Table II summarizes these changes. Changes in energy savings are smaller, given that all systems have the same PV production.

Figure 1 shows how the forecast errors reduce peak shaving. The peak demand for June occurs at 11 am, and with the actual forecast every algorithm except manual dispatch is discharging at that time. However, the forecast expects the highest PV production of the day at that time, so the algorithms using the forecast schedule a charge. The algorithms with real-time information (peak shaving, price signals, and MPC) stop charging, while the full-year forecast charges anyway. This increases the peak demand charge for the hour.

For additional metrics, Table III shows battery throughput (energy discharged in year 1), round trip efficiency, and percent of energy charged from PV. The main trend in these data is

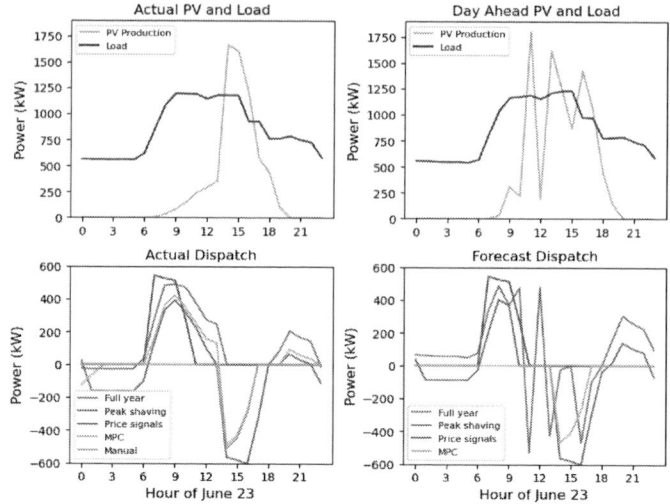

Fig. 1. A comparison of dispatch using a day-ahead forecast versus perfect forecast on June 23, the day of peak grid demand. Negative numbers for battery power indicate charging, while positive numbers indicate discharging. The perfect forecast allows the algorithms to plan to discharge at 11 am and reduce the peak, whereas the expected PV production at that time in the day-ahead forecast discourages discharging.

TABLE III. BATTERY METRICS. MOST ALGORITHMS DISCHARGE LESS WITH FORECAST ERRORS. ALL ALGORITHMS CHARGE FROM THE GRID MORE WITH FORECAST ERRORS.

Dispatch and forecast	Energy Discharged Year 1	Round Trip Efficiency	Charge Percent from PV
Peak shaving - perfect	398 MWh	90.28%	89.6%
Peak shaving - day ahead	311 MWh	90.36%	83.7%
Price signals - perfect	546 MWh	89.62%	99.6%
Price signals - day ahead	546 MWh	89.52%	99.4%
MPC - perfect	715 MWh	90.08%	99.8%
MPC - day ahead	588 MWh	89.60%	99.9%
Full-year - perfect	802 MWh	90.72%	99.2%
Full-year - day ahead	471 MWh	90.99%	97.3%
Manual	470 MWh	90.50%	100.0%

that the cases using the day-ahead forecast for dispatch tend to have a lower percentage of energy charged from PV. This is another potential source of the cost increases in Table II, as energy from the grid is more expensive.

Due to the 70% state of charge restriction for resilience, the depth of discharge is low for all cases, and SAM's battery degradation models predict that the battery will last for the entire 25-year lifetime of the system. This is likely to be an optimistic assumption, but it is consistent between all dispatch algorithms so we did not examine the costs of battery replacement in this analysis.

IV. CONCLUSIONS

We find that using a perfect forecast can overestimate the value of hybrid PV-battery systems between 1% and 8% compared to the reality of using a day-ahead forecast, depending on the dispatch algorithm used. For large systems such as those studied here, this can be $674,000 of value. Given the impact to project value, accounting for forecast uncertainty in system design and analysis will significantly improve the accuracy of modeled system values.

Improving the performance of dispatch while using a forecast could come from real-time controls, improved forecasts, or algorithmic improvements. Batteries in the field would run controls sub-hourly, allowing for real-time correction decisions while facing forecast errors and sub-hourly variability. When we added real-time information to the price signals dispatch forecast processing, the NPV increased by 2.9%. Hour-ahead forecasts of PV generation and load may be more accurate than day-ahead forecasts.

The most robust algorithm to forecast errors was also the simplest: peak shaving dispatch. This algorithm's use of grid power targets allows for flexible adaptation to changes in actual load and generation. However, it does not account for load shifting in its dispatch, which is a major source of value for the other algorithms. Therefore, future work could develop different grid power targets for different time of use periods, allowing for load shifting using this method. Future work could also examine additional utility rate structures and locations, ensuring the robustness of the results.

ACKNOWLEDGMENT

The authors would like to acknowledge Janine Keith and Nate Blair for comments and suggestions on drafts, and Liz Breazeale for editing support. This work was authored in part by the National Renewable Energy Laboratory, operated by Alliance for Sustainable Energy, LLC, for the U.S. Department of Energy (DOE) under Contract No. DE-AC36-08GO28308. Funding provided by the U.S. Department of Energy's Office of Energy Efficiency and Renewable Energy (EERE) under the Solar Energy Technologies Office Award Number 38407. The views expressed in the article do not necessarily represent the views of the DOE or the U.S. Government. The U.S. Government retains and the publisher, by accepting the article for publication, acknowledges that the U.S. Government retains a nonexclusive, paid-up, irrevocable, worldwide license to publish or reproduce the published form of this work, or allow others to do so, for U.S. Government purposes.

REFERENCES

[1] G. Fitzgerald, J. Mandel, and J. Morris. (). The economics of battery energy storage, RMI, [Online]. Available: https://rmi.org/insight/economics-battery-energy-storage/ (visited on 12/28/2022).

[2] J. McLaren, N. Laws, K. Anderson, N. DiOrio, and H. Miller, "Solar-plus-storage economics: What works where, and why?" *The Electricity Journal*, vol. 32, no. 1, pp. 28–46, Jan. 1, 2019, ISSN: 1040-6190. DOI: 10.1016/j.tej.2019.01.006. [Online]. Available: https://www.sciencedirect.com/science/article/pii/S1040619018302744 (visited on 12/28/2022).

[3] A. Prasanna, K. McCabe, B. Sigrin, and N. Blair, "Storage futures study: Distributed solar and storage outlook: Methodology and scenarios," NREL/TP-7A40-79790, 1811650, MainId:37010, Jul. 27, 2021, NREL/TP–7A40–79 790, 1811650, MainId:37010. DOI: 10.2172/1811650. [Online]. Available: https://www.osti.gov/servlets/purl/1811650/ (visited on 12/28/2022).

[4] B. T. Mirletz and D. L. Guittet, "Heuristic dispatch based on price signals for behind-the-meter PV-battery systems in the system advisor model," in *2021 IEEE 48th Photovoltaic Specialists Conference (PVSC)*, ISSN: 0160-8371, Jun. 2021, pp. 1393–1400. DOI: 10.1109/PVSC43889.2021.9519013.

[5] S. Mazzola, C. Vergara, M. Astolfi, V. Li, I. Perez-Arriaga, and E. Macchi, "Assessing the value of forecast-based dispatch in the operation of off-grid rural microgrids," *Renewable Energy*, vol. 108, pp. 116–125, Aug. 1, 2017, ISSN: 0960-1481. DOI: 10.1016/j.renene.2017.02.040. [Online]. Available: https://www.sciencedirect.com/science/article/pii/S096014811730126X (visited on 06/21/2021).

[6] P. P. Mishra, A. Latif, M. Emmanuel, Y. Shi, K. McKenna, K. Smith, and A. Nagarajan, "Analysis of degradation in residential battery energy storage systems for rate-based use-cases," *Applied Energy*, vol. 264, p. 114632, Apr. 15, 2020, ISSN: 0306-2619. DOI: 10.1016/j.apenergy.2020.114632. [Online]. Available: https://www.sciencedirect.com/science/article/pii/S0306261920301446 (visited on 01/13/2023).

[7] J. Cai, H. Zhang, and X. Jin, "Aging-aware predictive control of PV-battery assets in buildings," *Applied Energy*, vol. 236, pp. 478–488, Feb. 15, 2019, ISSN: 0306-2619. DOI: 10.1016/j.apenergy.2018.12.003. [Online]. Available: https://www.sciencedirect.com/science/article/pii/S0306261918318208 (visited on 01/13/2023).

[8] S. Theocharides, G. Tziolis, J. Lopez-Lorente, G. Makrides, and G. E. Georghiou, "Impact of data quality on day-ahead photovoltaic power production forecasting," in *2021 IEEE 48th Photovoltaic Specialists Conference (PVSC)*, ISSN: 0160-8371, Jun. 2021, pp. 0918–0922. DOI: 10.1109/PVSC43889.2021.9518471.

[9] Y. Wang, D. Millstein, A. D. Mills, S. Jeong, and A. Ancell, "The cost of day-ahead solar forecasting errors in the united states," *Solar Energy*, vol. 231, pp. 846–856, Jan. 1, 2022, ISSN: 0038-092X. DOI: 10.1016/j.solener.2021.12.012. [Online]. Available: https://www.sciencedirect.com/science/article/pii/S0038092X21010616 (visited on 04/24/2023).

[10] B. Yildiz, J. Bilbao, and A. Sproul, "A review and analysis of regression and machine learning models on commercial building electricity load forecasting," *Renewable and Sustainable Energy Reviews*, vol. 73, pp. 1104–1122, 2017. DOI: 10.1016/j.rser.2017.02.023.

[11] D. T. Vedullapalli and R. Hadidi, "Effect of forecaster performance on peak shaving in a university building by battery scheduling," in *2020 IEEE/IAS 56th Industrial and Commercial Power Systems Technical Conference (I CPS)*, ISSN: 2158-4907, Jun. 2020, pp. 1–6. DOI: 10.1109/ICPS48389.2020.9176774. [Online]. Available: https://ieeexplore.ieee.org/document/9176774.

[12] E. Oh and S.-Y. Son, "Theoretical energy storage system sizing method and performance analysis for wind power

forecast uncertainty management," *Renewable Energy*, vol. 155, pp. 1060–1069, Aug. 1, 2020, ISSN: 0960-1481. DOI: 10.1016/j.renene.2020.03.170. [Online]. Available: https://www.sciencedirect.com/science/article/pii/S0960148120305036 (visited on 12/27/2022).

[13] N. DiOrio, A. Dobos, S. Janzou, A. Nelson, and B. Lundstrom, "Technoeconomic modeling of battery energy storage in SAM," National Renewable Energy Lab. (NREL), Golden, CO (United States), NREL/TP-6A20-64641, Sep. 1, 2015. DOI: https://doi.org/10.2172/1225314. [Online]. Available: https://www.osti.gov/biblio/1225314.

[14] N. DiOrio, A. Dobos, and S. Janzou, "Economic analysis case studies of battery energy storage with SAM," National Renewable Energy Lab. (NREL), Golden, CO (United States), NREL/TP–6A20-64987, 1226239, Nov. 1, 2015, NREL/TP–6A20–64987, 1226239. DOI: 10.2172/1226239. [Online]. Available: http://www.osti.gov/servlets/purl/1226239/.

[15] N. A. DiOrio, "An overview of the automated dispatch controller algorithms in the system advisor model (SAM)," National Renewable Energy Lab. (NREL), Golden, CO (United States), NREL/TP-6A20-68614, Nov. 22, 2017. DOI: https://doi.org/10.2172/1410499. [Online]. Available: https://www.osti.gov/biblio/1410499.

[16] S. Mishra, J. Pohl, N. Laws, D. Cutler, T. Kwasnik, W. Becker, A. Zolan, K. Anderson, D. Olis, and E. Elgqvist, "Computational framework for behind-the-meter DER techno-economic modeling and optimization: REopt lite," *Energy Systems*, vol. 13, no. 2, pp. 509–537, May 1, 2022, ISSN: 1868-3975. DOI: 10.1007/s12667-021-00446-8. [Online]. Available: https://doi.org/10.1007/s12667-021-00446-8 (visited on 12/27/2022).

[17] J. Keith, B. Mirletz, M. Prilliman, N. Blair, D. Guittet, S. Janzou, and P. Gilman, "FY19-FY21 final technical report: Foundational open source solar system modeling through improvement and validation of the system advisor model and PVWatts," NREL/TP-7A40-82478, 1870820, MainId:83251, May 26, 2022, NREL/TP–7A40–82478, 1870820, MainId:83251. DOI: 10.2172/1870820. [Online]. Available: https://www.osti.gov/servlets/purl/1870820/ (visited on 12/28/2022).

[18] "EnergyPlus^{TM}," National Renewable Energy Lab. (NREL), Golden, CO (United States); Lawrence Berkeley National Lab. (LBNL), Berkeley, CA (United States), EnergyPlus^{TM} (e+); 005462MLTPL00, Sep. 30, 2017. [Online]. Available: https://www.osti.gov/biblio/1395882 (visited on 01/12/2023).

[19] (). Puerto rico - net metering, Database of State Incentives for Renewables and Efficiency, [Online]. Available: https://programs.dsireusa.org/system/program/detail/2846/puerto-rico-net-metering (visited on 05/17/2023).

[20] V. Ramasamy, J. Zuboy, E. O'Shaughnessy, D. Feldman, J. Desai, M. Woodhouse, P. Basore, and R. Margolis, "U.s. solar photovoltaic system and energy storage cost benchmarks, with minimum sustainable price analysis: Q1 2022," NREL/TP-7A40-83586, 1891204, MainId:84359, Sep. 28, 2022, NREL/TP–7A40–83586, 1891204, MainId:84359. DOI: 10.2172/1891204. [Online]. Available: https://www.osti.gov/servlets/purl/1891204/ (visited on 01/12/2023).

Advanced Characterization and Degradation Analysis of Perovskite Solar Cells using Machine Learning and Bayesian Optimization

Joseph Chakar, Arthur Julien, Karim Medjoubi, Jorge Posada, Jean-François Guillemoles, Jean-Baptiste Puel, Yvan Bonnassieux

Laboratoire de Physique des Interfaces et des Couches Minces (LPICM), Centre National de la Recherche Scientifique (CNRS UMR 7647), École Polytechnique, Institut Polytechnique de Paris (IP Paris), Palaiseau, France

Institut Photovoltaïque d'Île-de-France (IPVF), Palaiseau, France

EDF R&D, Palaiseau, France

IPVF, UMR 9006, Air Liquide, Chimie ParisTech, CNRS, EDF, IPVF SAS, Total, Ecole Polytechnique, IP Paris, Palaiseau, France

In the race to design the next generation of solar cells, perovskites are widely seen as the likely candidate to push the performance of conventional silicon solar cells. Despite recent breakthroughs, this technology still has to overcome several challenges before it can be commercially deployed. With a vast array of material combinations and fabrication processes to choose from, researchers often spend years trying to optimize their solar cell design and understand performance bottlenecks. Herein, we combine physics modeling, machine learning, and experimentation to better understand the complex relationship between device performance and the underlying material parameters. Specifically, we use Bayesian inference and traditional machine learning techniques to extract material properties from simulated and experimental current-voltage curves of perovskite solar cells measured under indoor and outdoor conditions. This approach, which can be generalized to other semiconductor devices, allows us to gain valuable insight into photovoltaic performance and degradation without resorting to time-consuming and laborious characterization techniques, thus accelerating the pace of experimental research.

Statistical Analysis and Degradation Pathway Modeling of PERC PV Minimodules with Different Packaging Strategies in Indoor Accelerated Exposures

Sameera Nalin Venkat*, Jiqi Liu*, Xuanji Yu*, William Oltjen*, Xinjun Li†,
Jean-Nicolas Jaubert‡, Jennifer L. Braid§*, Roger H. French*, Laura S. Bruckman*
*SDLE Research Center, Department of Materials Science and Engineering, Case Western Reserve University (CWRU),
Cleveland, OH, 44106, USA
†Cybrid Technologies Inc., Suzhou, Jiangsu, China
‡CSI Solar Co. Ltd., 199 Lushan Road, SND, Suzhou, Jiangsu, China
§Sandia National Laboratories, Albuquerque, NM, USA

Abstract—To prolong the PV system lifetime, it is important to understand the impact of various packaging strategies on the degradation of PV modules. Statistical analysis and data-driven network structural equation modeling has been used to assess degradation patterns in 4-cell PV minimodule variants fabricated at two different units. The packaging strategies implemented in the fabrication of minimodules are in terms of architecture (glass/backsheet versus double glass) along with various types of encapsulants and backsheets. Stepwise electrical characterization (I-V and $Suns$-V_{oc}) were utilized to collect data which was used for analysis.

Index Terms—degradation modeling, netSEM, indoor accelerated exposures, electrical characterization, packaging strategies

I. INTRODUCTION

As the field of photovoltaics (PV) is evolving in terms of installation capacity, size, and market share in the renewable energy sector, there is a greater need to prolong the lifetime of PV modules to minimize the levelized cost of electricity. In this respect, it is of utmost importance to understand the causes of degradation and understand the impact of packaging strategies on the overall PV module performance. Even though novel polymeric materials have entered the PV market, the long-term durability and behavior under a variety of stressors are not well-understood.

The existing models in literature do not take the complexity of PV module degradation into account and are often difficult to generalize. In this study, statistical and degradation pathway

This material is based upon work supported by the U.S. Department of Energy's Office of Energy Efficiency and Renewable Energy (EERE) under Solar Energy Technologies Office (SETO) Agreement Number DE-EE-0008550. The views expressed herein do not necessarily represent the views of the U.S. Department of Energy or the United States Government. Sandia National Laboratories is a multimission laboratory managed and operated by National Technology & Engineering Solutions of Sandia, LLC, a wholly owned subsidiary of Honeywell International Inc., for the U.S. Department of Energy's National Nuclear Security Administration under contract DE-NA0003525.

modeling are implemented to study how 4-cell PV minimodules with different packaging strategies experience power loss. Statistical analysis is performed using 83.4% and 95% confidence intervals (CIs) to compare between minimodule variants and identify if they are durable/degrading, respectively. Network structural equation modeling (netSEM) is employed as a data-driven modeling technique to identify power loss patterns in minimodule variants as well as analyze pairwise relationships to identify dominant degradation pathways [1], [2].

II. EXPERIMENTAL SECTION

This section is divided into three parts. Section II-A provides an overview of the components in the study protocol. Having a clearly defined study protocol is crucial for effective data collection, processing, and modeling. Section II-B deals with 4-cell PV minimodules specifications and quantities. Section II-C involves the details of data processing and R packages used in the study. The details of these parts are discussed in our previous studies [2]–[4].

A. Study Protocol

The study protocol for this work has four main components: fabrication of 4-cell PV minimodules, exposure of minimodules in indoor accelerated conditions, stepwise electrical characterization, and data-driven modeling.

Fabrication of PV minimodules involved soldering of 4 PERC cells in a series manner and lamination of the soldered cells, polymeric packaging materials, and glass panes. The as-fabricated PV minimodules were pre-conditioned to ensure their stability. The minimodules were exposed in two different accelerated conditions: modified damp heat (mDH), with or without full spectrum light (FSL). mDH refers to 80°C and 85% relative humidity (RH). FSL involves the usage of 420 Wm^{-2} intensity light from a discharge lamp.

After every 504 hours (21 days), stepwise electrical measurements (I-V and $Suns$-V_{oc}) were performed for each of the minimodules. Subsequently, after the data collection was done, additional processing steps were performed (highlighted in Section II-C). By implementing data-driven modeling and statistical analysis, durable/degrading minimodule variants were identified using netSEM and confidence intervals.

B. Minimodule Specifications

Five sets of minimodules were fabricated at two manufacturing units, A and B. Each minimodule variant includes two minimodules, to increase the stastistical significance of the results. Table I refers to the minimodule specifications. The encapsulants used in the minimodule sets are ethylene vinyl acetate (EVA), polyolefin elastomer (POE), and co-extruded EVA-POE-EVA (EPE). The backsheet materials used are polyvinylidene fluoride/polyethylene terephthalate/fluorocoating (commonly referred to as KPf), and transparent backsheets. The transparent backsheets used in the study are of two types: with and without grid pattern. The different module architectures are glass/backsheet (GB) and double glass (DG).

Please note that in fabricating set #5 minimodules at unit A, only DG minimodules were made; after pre-conditioning, they were subsequently exposed in mDH+FSL.

TABLE I
SPECIFICATIONS OF SETS #1-#5 4-CELL PV MINIMODULES.

Set #	PERC Cell Type	Encapsulant	Rear Encapsulant	Architecture
1	Monofacial	EVA	UV-cutoff	DG
				GB
		POE		DG
				GB
2	Bifacial	EVA	Opaque	DG
				GB
		POE		DG
				GB
3	Monofacial	EVA	Opaque	DG
				GB
		POE		DG
				GB
4	Bifacial	EVA	Transparent UV-cutoff	DG
				GB
		POE	Transparent UV-cutoff	DG
				GB
5	Bifacial	EPE	Transparent UV-cutoff	DG
				GB
		POE	Transparent UV-cutoff	DG
				GB

C. Data Methods

For constructing data-driven degradation pathway models, netSEM R package (version 0.7.0) was utilized. For extracting I-V features, $ddiv$ R package (version 0.1.1) was used. The data points were normalized based on baseline measurements from exposure step 0. For calculating the CIs, $Rmisc$ package (version 1.5.1) was used.

III. RESULTS

This section consists of stastical analysis using 83.4% and 95% confidence intervals and degradation pathway modeling results using netSEM. The results are included for minimodule variants fabricated by manufacturers A and B. Majority of the

results will focus on set #5; however, the minimodule sets are compared based on a particular variant.

A. Statistical Analysis Using Confidence Intervals

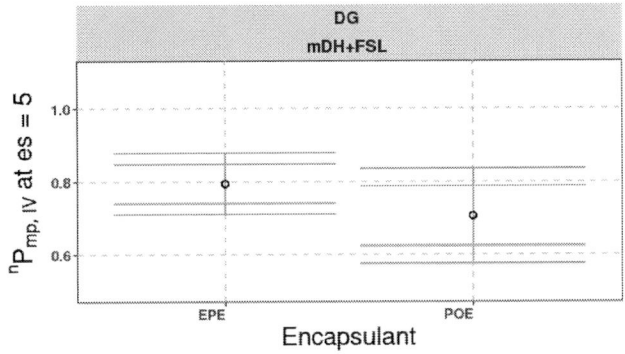

Fig. 1. 83.4% and 95% CIs of $^{n}P_{mp,IV}$ at final exposure step (i.e. step 5) for minimodule variants manufactured at unit A.

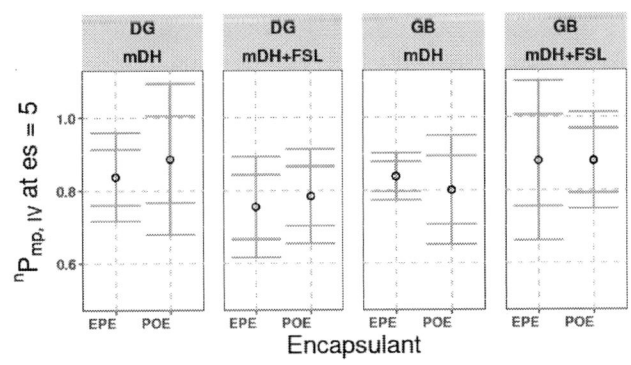

Fig. 2. 83.4% (orange) and 95% (blue) CIs of $^{n}P_{mp,IV}$ at final exposure step (i.e. step 5) for minimodule variants manufactured at unit B.

83.4% (orange) and 95% (blue) CIs were constructed for various minimodule variants. Minimodules with similar/different behavior at the end of exposure cycle were identified using 83.4% CIs. Using 95% CIs, durable/degrading minimodule variants were identified.

From Fig. 1 which consists of DG minimodules exposed in mDH+FSL conditions manufactured at unit A, the average power loss for DG with POE is more than the one with EPE. Due to overlap in the 83.4% and 95% CIs, the minimodules may be behaving similarly as per inference by eye [5].

Fig. 2 shows the results for minimodule variants manufactured at unit B. The average power loss for DG with EPE exposed in mDH+FSL is greater than the rest of the variants. There is significant CI overlap between minimodule variants. In some cases, there is widening of CIs (especially for DG with POE in mDH and GB with EPE in mDH+FSL) suggesting that the data points may have high variability and do not provide a precise population mean estimate.

B. Degradation Pathway Modeling Using Network Structural Equation Modeling: <Stressor|Response> Results

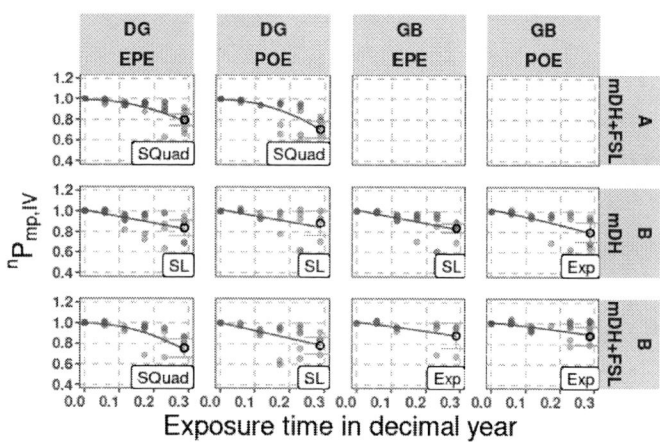

Fig. 3. Variation of $^nP_{mp,IV}$ with dy (<S|R>) manufactured at units A and B. The best model equation line and name in text, data points and 83.4% CIs (orange) at the end of exposure cycle are shown..

Fig. 3 shows how $^nP_{mp,IV}$ changes with dy. This is referred to as the <S|R> pathway in the <S|M|R> model. It can be observed that all of the minimodule variants, irrespective of whether they were fabricated at unit A or B, experience some extent of power loss. On average, the power loss is at least 10%. Among the DG minimodules fabricated at manufacturing unit A and exposed in mDH+FSL, the one with EPE has greater stability than the one with POE. Under the influence of mDH+FSL, GB minimodules fabricated at unit B seem to have similar power loss patterns. In mDH exposure, DG with POE fabricated by unit B has the least amount of power loss among the other variants.

To compare between minimodule sets, the variant DG with POE fabricated by manufacturer A is chosen as an example (this will be referred to as minimodule variant from here on). Fig 4 shows how the variant changes with each set. As highlighted in Table I, it is to be noted that even though the variant name is the same across all the sets, the cell type and the rear encapsulant are different. Set #5 minimodule variant seems to undergo higher power loss over the exposure cycle compared to other sets. Set #1 minimodule variant seems to have the least power loss. Sets #3 and #4 minimodule variant have similar variation in $^nP_{mp,IV}$ with exposure time. Further investigation into the properties of the materials is to be done to validate the observations.

IV. DISCUSSION

EPE is a relatively new encapsulant material that has been introduced to the PV market recently. Not many studies have been done to explore its degradation behavior and how it impacts the overall module performance. All of the set #5 minimodule variants undergo power loss but the reason is not

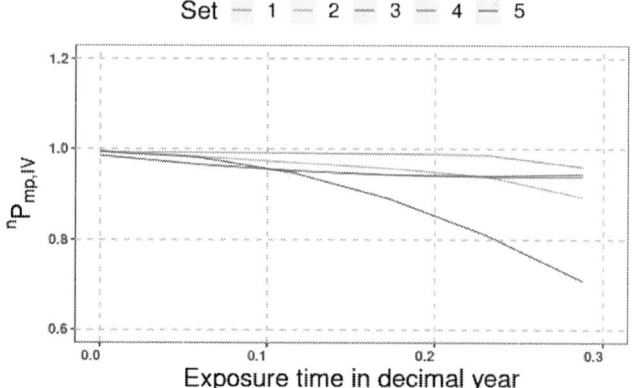

Fig. 4. Variation of $^nP_{mp,IV}$ with dy (<S|R>) for minimodule variant DG with POE in mDH+FSL exposure manufactured at unit A for all sets. The best model equation line is shown for each set.

apparent from looking at CIs and <S|R> results. Further exploration into the mechanistic variables need to be explored in order to identify the reason for degradation. We plan on using netSEM Principle 2, in which power loss is simultaneously impacted by stressor and mechanistic variables, to understand the root cause of this behavior.

V. CONCLUSION

Using statistical and data-driven netSEM results, we were able to see the power loss patterns in minimodule variants with novel packaging strategies. We observe that set #5 minimodule variants undergo a power loss of at least 10%. There seems to be some similarity in the overall power loss behavior of DG minimodules fabricated at units A and B and exposed in mDH+FSL exposure but it cannot be concluded if EPE is better or POE is better without further investigation into other components of analysis (mechanistic pathways and Principle 2). Comparing DG with POE fabricated by manufacturer A across different minimodule sets, set #5 undergoes greater power loss over the exposure cycle of mDH+FSL.

REFERENCES

[1] L. S. Bruckman, N. R. Wheeler, J. Ma, E. Wang, C. K. Wang, I. Chou, J. Sun, and R. H. French, "Statistical and Domain Analytics Applied to PV Module Lifetime and Degradation Science," *IEEE Access*, vol. 1, pp. 384–403, 2013.

[2] S. Nalin Venkat, "Network Structural Equation Modeling of PV Minimodule Variants Under Indoor Accelerated Exposures," MS Thesis, Case Western Reserve University.

[3] S. N. Venkat, J. Liu, J. Wegmueller, B. Yu, B. Gould, X. Li, J.-N. Jaubert, J. L. Braid, L. S. Bruckman, and R. H. French, "Degradation Pathway Modeling of PV Minimodule Variants with Different Packaging Materials Under Indoor Accelerated Exposures," in *2021 IEEE 48th Photovoltaic Specialists Conference (PVSC)*, Jun. 2021, pp. 1725–1731.

[4] S. N. Venkat, J. Liu, J. Yu, Xuanji Wegmueller, K. Rath, X. Li, J.-N. Jaubert, J. L. Braid, L. S. Bruckman, and R. H. French, "Evaluation of PV Module Packaging Strategies of Monofacial and Bifacial PERC Using Degradation Pathway Network Modeling," in *2022 IEEE 49th Photovoltaic Specialists Conference (PVSC)*, Jun. 2022, pp. 1020–1027.

[5] G. Cumming and S. Finch, "Inference by Eye: Confidence Intervals and How to Read Pictures of Data," *American Psychologist*, vol. 60, no. 2, pp. 170–180, 2005.

Hotspot Endurance of Pristine and Thermal Cycled Glass-Backsheet Photovoltaic Modules

Muhammad Afridi, Akash Kumar, Farrukh ibne Mahmood, GovindaSamy TamizhMani

Photovoltaic Reliability Laboratory, Arizona State University, Mesa, Arizona, USA

Abstract — Hotspots can be a detrimental reliability issue that can adversely affect the long-term performance of (PV) modules in the field. The endurance of hotspot stress is usually evaluated on pristine modules, which do not represent the hotspot issues related to long-term field-aged modules. PV modules can become vulnerable to hotspots when exposed to long-term field operating conditions. Hence, in this study, we evaluated the endurance of hotspot stress of a fresh half-cell glass-backsheet (GB) module along with a prestressed half-cell GB module that was subjected to 600 accelerated thermal cycling stress. The maximum power (Pmax) of both GB modules (fresh and prestressed) experienced a negligible decrease of approximately 1% after the sequential thermal cycling followed by the hotspot stress test. This indicates that GB modules having half-cell GB configuration can potentially endure the hotspot stress even after being exposed to the field for over 30 years. However, these findings were derived from a sample size lacking statistical significance and, therefore, should not be generalized to encompass all half-cell GB modules.

I. INTRODUCTION

In field conditions, PV modules experience various stresses concurrently, potentially resulting in unanticipated degradation and failure mechanisms. The majority of these problems occur because traditional testing methods are applied individually only one stress at a time, or the stress duration is too short to detect actual field failures [1]. Moreover, the stressors used in conventional testing might not be applied in the correct order to accurately replicate the effects of the actual environment [1].

Hotspot stress is a significant problem in photovoltaic (PV) modules that can lead to permanent damage or even module failure [2]. The phenomenon arises when a portion of the solar cell in a PV module becomes shaded or damaged, causing it to generate less current than the surrounding cells [2]. This creates a localized "hotspot" region that experiences increased current density, which leads to further heating and can exacerbate the damage, creating a positive feedback loop.

Hotspots can be triggered by a variety of factors, such as shading from nearby objects or structures, manufacturing defects such as microcracks or scratches on the cell surface, and environmental factors such as soiling or weather events [2]. Moreover, temperature gradients within the module can also exacerbate the formation of hotspots, as different areas of the module may experience uneven thermal conditions.

In this study, two commercial GB modules were used to investigate the impact of extended thermal cycling on the hotspot endurance of these modules. A GB module was prestressed for 600 accelerated thermal cycles (TC600) to simulate over 30 years of field operation. After performing thermal cycling, both GB modules (fresh and prestressed with TC600) were tested for hotspot endurance. Several pre- and post-characterization techniques were performed in this study to evaluate the change in performance characteristics for each GB module, including outdoor light current-voltage (IV) measurements to determine short-circuit current (Isc), open-circuit voltage (Voc), peak-current (Ipeak), peak-voltage (Vpeak) and fill factor (FF). The IV data results were translated into standard test conditions (STC) at 1000 W/m^2 and 25°C. Electroluminescence (EL) imaging was conducted at Isc with a 60-second exposure for EL images to determine the mean grayscale value, and dark IV was performed to obtain shunt resistance (Rsh) and series resistance (R$_S$).

II. EXPERIMENTAL METHODOLOGY

This investigation utilized two identical commercial GB modules. Each module features a glass-encapsulant-cell-encapsulant-backsheet-frame construction and weights 23 kg, with dimensions measuring 2004 mm x 996 mm x 35 mm. The module consists of a 144-cell half-cut mono-facial design with a power rating of 380 W. Utilizing the passivated emitter and rear contact (PERC) cell technology, the cells measure 78 mm x 156 mm in size. Featuring a 3.2mm coated tempered glass front, the module's frame is constructed from anodized aluminum alloy, and EVA serves as the encapsulant.

The endurance of hotspot stress was investigated by comparing a fresh GB module with a prestressed GB module that represented a simulated field-aged module. This GB module was stressed under thermal cycling for 600 cycles in an indoor environmental chamber where the ambient temperature cycled between -40°C and 85°C as per IEC 61215-2 [3]. After the TC600 stress, the modules underwent the hotspot stress endurance test outdoors according to the procedure defined according to IEC 61215-2 [3]. The hotspot stress test was performed on one high-shunt resistance cell and three low-shunt resistance cells that were selected based on the light IV [3] of the GB modules, and these cells were partially shaded under worst-case shadowing conditions individually.

To determine the worst-case shadowing condition for the GB modules and obtain the highest temperature due to hotspot heating, IV curves were collected for the test cells while they were shaded at different ratios individually [3]. After the identification of the worst-case shadowing scenario, the test cells were individually shaded with that shadowing level, and the module was short-circuited for a minimum of one hour while recording the operating temperature of the shaded cell.

978-1-6654-6060-6/23 $31.00 © 2023 IEEE

III. RESULTS AND DISCUSSION

A. GB Modules Temperature During the Hotspot Test

Fig. 1 and 2 exhibits the infrared (IR) images demonstrating the dissipation of extreme level of heat by the stressed cells in the fresh and prestressed GB modules throughout the hotspot experiment. Based on these IR images, it is evident that during the hotspot experiment, the solar cells in the GB modules endured a considerable amount of thermal stress after they were partially shadowed. The considerable thermal stress is caused by the localized heating resulting from the reverse current flow when the module experiences partial shading under the worst-case shadow condition. This condition was established when each test cell in the GB modules was shaded by 25%. In this situation, the electrical current at the maximum power point of the shadowed module is nearly the same as the peak current indicated on the modules' nameplate. In the worst-case shadow condition, the shadowed cells function as an electrical load, consuming power produced by the module. They dissipate a significant amount of thermal energy and operate under reverse bias condition.

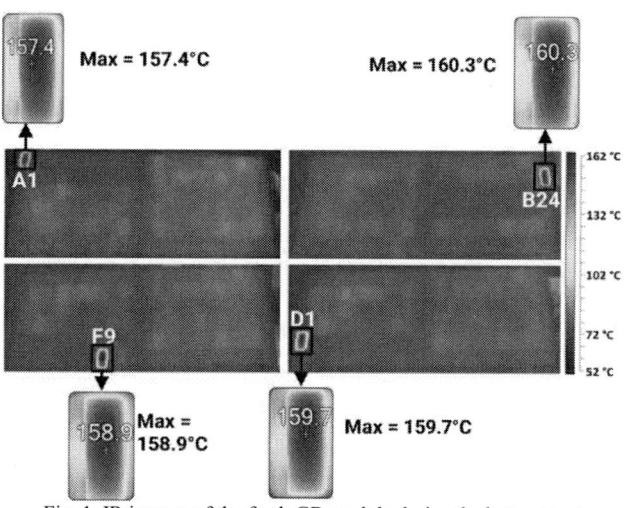

Fig. 1. IR images of the fresh GB module during the hotspot test

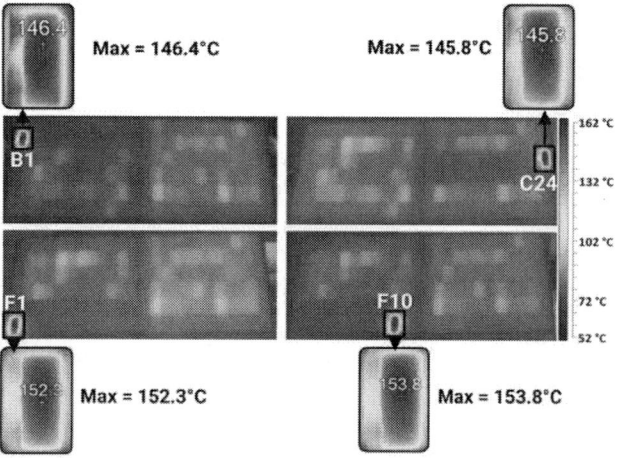

Fig. 2. IR images of the prestressed GB module during the hotspot test

B. EL Imaging Analysis of GB Modules Post-Hotspot Experiment

The performance of the GB modules was evaluated by analyzing the EL images. The EL images pre- and post-hotspot test for the fresh GB module are illustrated in Fig. 3, whereas the EL images pre- and post-hotspot test for the prestressed GB module are illustrated in Fig. 4. The EL images of the prestressed GB module (Fig. 4) reveal a minor crack, which resulted from improper handling of the module prior to initiating the sequential testing. After subjecting this module to TC600, no significant variation in the brightness of its solar cells was evident in the EL image of this module shown in Fig. 4 (left), indicating a uniform flow of electrical current. This signifies that the thermal cycling test had no significant impact on the prestressed GB module.

Similarly, no cell darkening or module defects were evident in the post-hotspot EL image of the fresh GB module shown in Fig. 3 (right). Furthermore, the EL image in Fig. 4 (right) of the prestressed GB module also did not show any significant defects, demonstrating the ability of this module to endure degradation subsequent to the hotspot experiment. The hotspot test was performed on a low shunt resistance cell highlighted with red and three high shunt resistance cells highlighted with yellow, as shown in Fig. 3 and 4. The post-hotspot EL images for both GB modules doesn't indicate any major defects or cell-level damage. No substantial damage was evident for both modules, indicating the reliability of GB modules to withstand degradation and endure hotspot stress after being exposed to decades of field-use operation.

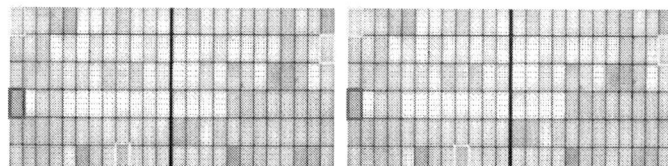

Fig. 3. EL images for fresh GB module pre-hotspot (left) and post-hotspot (right) collected at an exposure of 60 seconds at Isc

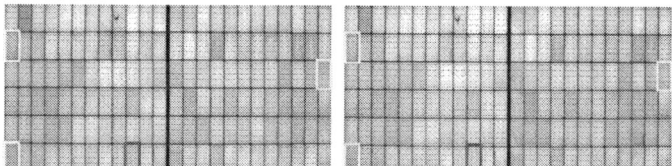

Fig. 4. EL images for prestressed GB module pre-hotspot (left) and post-hotspot (right) collected at an exposure of 60 seconds at Isc

In order to evaluate the change in the EL image pixel intensity pre- and post-hotspot experiment, a grayscale analysis was conducted for the GB modules EL images, as shown in Fig. 5 and 6. The grayscale analysis is a technique employed to assess the uniformity of solar cells within a photovoltaic module, offering a non-destructive and non-invasive approach. It is clearly evident from Fig. 5 and 6 that both the fresh as well as prestressed GB modules did not experience any significant reduction in the mean grayscale value post-hotspot experiment.

This signifies that negligible change occurred in the pixel intensity of the EL images post-hotspot for GB modules.

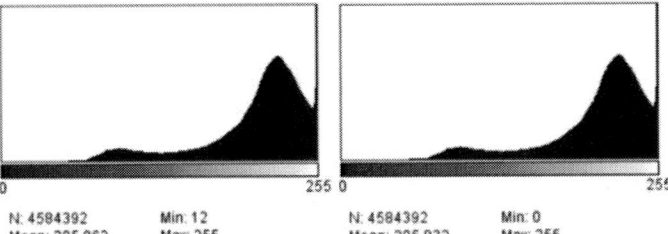

N: 4584392   Min: 12
Mean: 205.862   Max: 255

N: 4584392   Min: 0
Mean: 205.832   Max: 255

Fig. 5. Grayscale histogram of fresh GB module pre-hotspot (left) and post-hotspot test (right)

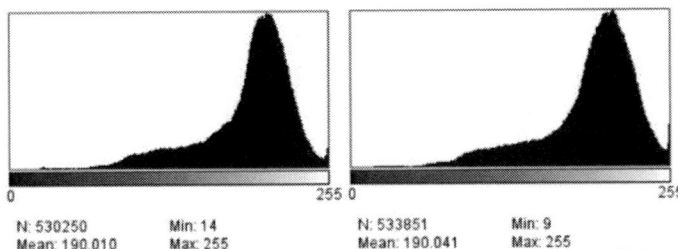

N: 530250   Min: 14
Mean: 190.010   Max: 255

N: 533851   Min: 9
Mean: 190.041   Max: 255

Fig. 6. Grayscale histogram of prestressed GB module pre-hotspot (left) and post-hotspot test (right)

C. Deterioration of Electrical Parameters for GB Modules Following the Hotspot Experiment

The impact of the hotspot stress on the GB modules was also analyzed by the change observed in the electrical parameters displayed in the bar plot in Fig. 7, which was quantified by measuring the IV curves. After undergoing 600 thermal cycles, the Pmax of the prestressed GB module dropped trivially by 2.6 W from 377.2 W to 374.6 W. Additionally, following the hotspot experiment, this Pmax value further decreased by nearly 2.3 W to 372.4 W, resulting in a cumulative 1.3% drop of almost 5 W. A trivial decline of less than a percent was observed in the electrical performance parameters of Isc, Voc, Ipeak, Vpeak, and FF of this module, as displayed in Fig. 7, and no change was evident in the Rs or the Rsh of this module. This validates that the performance of this module was not affected by the sequential thermal cycling and hotspot stress test. Similarly, the Pmax of the fresh GB module was initially 374.7 W which decreased to 371.6 W, indicating a 0.8% loss of Pmax following the hotspot experiment. A negligible decrease of less than a percent was also observed in the electrical performance parameters of Isc, Voc, Ipeak, Vpeak, and FF of this module, as displayed in Fig. 7. Post-hotspot analysis indicated no considerable change in Rsh or Rs.

The robustness of the GB modules featuring a half-cell design can be ascribed to its superior cell-to-module power ratio, coupled with reduced consumption of materials [4]. The performance of half-cell GB modules under partially shaded conditions is also enhanced due to the decreased current flow, leading to reduced loss of power and reduced thermal stress experienced by the module during operation [4]. Moreover, half-cell GB modules exhibit a decrease in interconnection fatigue, which can result from thermal cycling, and display a 50% reduction in cell displacement [4].

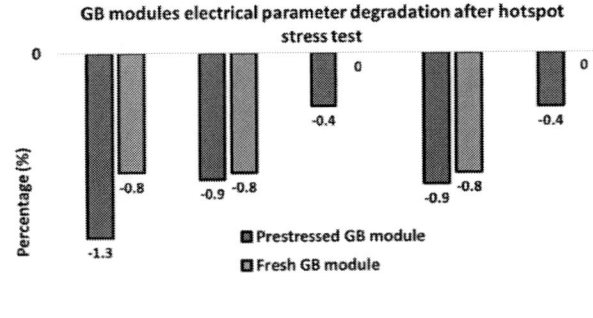

Fig. 7. GB modules electrical parameter degradation after hotspot experiment

IV. CONCLUSIONS

In this study, we investigated the endurance of hotspot stress in fresh and thermally cycled (600 cycles) GB modules and compared their performance. The fresh as well as the prestressed GB modules did not indicate any significant cell-level damage or decrease in its electrical parameter's performance after the hotspot experiment. This potentially implies that the GB modules having a half-cell and series-parallel configuration can successfully endure the hotspot stress with only minimal deterioration in their electrical parameters even after being in field operating conditions for over 30 years. Given that subjecting the module to extended thermal cycling resulted in a mere 0.5% increase in Pmax degradation, the impact can be considered negligibly small. Nonetheless, these findings were derived from a sample size lacking statistical significance and, therefore, should not be generalized to encompass all half-cell GB modules.

REFERENCES

[1] F. ibne Mahmood, A. Kumar, M. Afridi, and G. TamizhMani, "Potential induced degradation in c-Si glass-glass modules after extended damp heat stress," *Solar Energy*, vol. 254, pp. 102–111, 2023.

[2] M. Afridi, A. Kumar, F. ibne Mahmood, and G. Tamizhmani, "Hotspot testing of glass/backsheet and glass/glass PV modules pre-stressed in extended thermal cycling," *Solar Energy*, vol. 249, pp. 467–475, 2023.

[3] *IEC 61215-2:2016 Terrestrial photovoltaic (PV) modules - Design qualification and type approval - Part 2: Test procedures.*

[4] H. Hanifi, B. Jaeckel, M. Pander, D. Dassler, S. Kumar, and J. Schneider, "Techno-Economic Assessment of Half-Cell Modules for Desert Climates: An Overview on Power, Performance, Durability and Costs," *Energies (Basel)*, vol. 15, no. 9, 2022.

Integrated Large-Scale Data Management Platform for Photovoltaic Power Conversion Equipment (PCE) Reliability Data

Liwei Wang[1], Buck Brown[1], Shuan Dong[5], Tan Jin[5], Daniel Clemens[2], Joseph Hodges[3], Adam Reeves[4], Josh Ozbeytemur[4], Shuangshuang Jin[1], and Zheyu Zhang[1]

[1]Clemson University, North Charleston, SC, 29405, USA

[2]SMA Solar Technology AG, Niestetal, Kassel, 34266, Germany

[3]Dominion Energy, Cayce, SC, 29033, USA

[4]Hannah Solar Government Services, Summerville, SC, 29483, USA

[5]National Renewable Energy Laboratory, Golden, CO, 80401, USA

Abstract — To meet the demand for accuracy and real-time capability of PV system degradation evaluation, massive volume data is needed to run high-fidelity and high-efficiency simulations and perform advanced data analysis. However, PV farm operators have a series of difficulties with PV inverter data, such as data collection from multiple channels, massive data storage, data management and massive data analysis. To address these challenges, we developed an integrated data management platform capable of data acquisition, processing, storage, query, and performing big data analysis utilizing AI algorithms. The platform can also achieve data correctness verification and provide an effective distributed data management solution to retrieve massive data and establish a connection to distributed computational frameworks.

Keywords— data management platform, distributed computing, field reliability data, PV inverter

I. INTRODUCTION

The installation of PV systems has grown across the globe in the first half of 2022 [1]. Influenced by the enactment of the Inflation Reduction Act (IRA), the increase of photovoltaic (PV) penetration into the U.S. power market can be estimated to be further incentivized in the future. As more PV systems come into service rapidly, component operation and maintenance (O&M) costs need to be noticed. It is evidenced by field data from PV power plant operators that power electronic converters contribute most to O&M events, responsible for between 43% and 70% of the service calls [2][4]. Field reliability data is an important indicator to help operators to monitor the operational status under measured environmental stressors. Based on the historical performance data, operators can better evaluate the current state of health condition and predict the lifetime of components. Meaningful field reliability data may include parameters of the local environment (temperature, humidity, irradiance), PV farm (configuration considering grounding), grid data (grid command, power quality, grid disturbance), and inverter data (DC-link voltage, P/Q reference). The collection and analysis of field reliability data may help operators to better understand the degradation pattern of inverters. In addition, operators can schedule necessary maintenance in advance, reducing the likelihood of PV system failure.

The traditional field data collecting and managing approaches meet several challenges:

1. Some field reliability data is measured by the built-in measurement tools inside the component. It is only accessible from the designated data portal provided by the manufacturers. Environmental data and inverter operational data are usually collected by distinct channels. There are still several difficulties with integrating data from different sources.
2. The correctness of collected field data is not always dependable. For example, if the solar irradiance sensors are blocked by other objects, the irradiance recorded at that time is inaccurate. It is difficult to verify the correctness of collected field data.
3. More advanced measurement tools with higher accuracy and higher time resolution are applied widely. Massive data with higher precision is accumulated as time grows. The way to efficiently store, retrieve, and analyze mass historical data is also a challenge.

In this paper, we propose an integrated field data management platform that can address the challenges mentioned above. Field data collection is introduced in Section II. The data management platform capability is introduced in Section III. Section IV discusses conclusions and future work.

II. FIELD RELIABILITY DATA COLLECTION

The importance of renewable energy cannot be overstated as the world transitions to net zero carbon emissions. PVs are a type of renewable energy source that have vast potential yet currently face reliability issues. In particular, PV inverters are one of the most unreliable subsystems within the larger PV system due to their complexity. Some of the components within PV inverters that have the worst reliability are capacitors, cooling fans, metal oxide varistors, printed circuit

978-1-6654-6060-6/23 $31.00 © 2023 IEEE

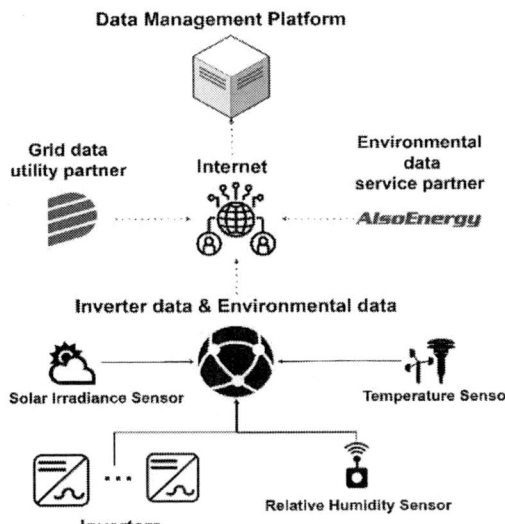

Fig. 1. Field reliability data acquisition dataflow.

30000TL-US inverters (30000TL-US-10) are used at the Otarre solar farm in Cayce, South Carolina, USA. Data sources provided by our utility and service partner, their

TABLE I. DATA SOURCE REFERENCE

Data Source	Utility & Service Partner (Online Portal)	Time Resolution
Grid Data Source	Dominion Energy	10 minutes
Inverter Data Source	SMA (Sunny Portal Powered by ennexOS)	5 minutes
Environmental Data Source A	SMA (Sunny Portal Powered by ennexOS)	1 minute
Environmental Data Source B	Also Energy (PowerTrack)	5 minutes

corresponding web portal, and their time resolution can be found in TABLE I.

III. DATA MANAGEMENT PLATFORM

To overcome the challenge of managing the field reliability data efficiently, we established a data management platform to integrate the data and perform the preliminary analysis based on the collected historical data. It 1) integrates field data from different channels, 2) applies big data solutions to optimize data retrieval and analysis capability, and 3) performs cross-data validation to increase the accuracy and reliability of collected field data. Related technical details of these features will be introduced in the rest of this section.

A. Data Integrations

In our data acquisition system, there are online data portals for grid data, environmental data, and inverter data. Each online data portal provides its own user interface to help operators monitor the operating condition and environmental information as well as track historical data. However, monitoring and tracking historical data across different channels is not supported by any existing online portals. As a result, we developed a third-party, python-based data management system to integrate multi-channel data and provide an efficient data retrieval solution to monitor and track the inverter operating condition and surrounding environmental information.

The workflow of the data management is shown in Fig. 2. At first, dedicated web scraping scripts extract field data from the three online portals separately. The web scrapings simulate the web browser to send requests for target data and gather the target data by fetching the response returned from the web portal server. Request, BeautifulSoup, and Selenium libraries are utilized in this application to perform the data scrap and parse the data.

If some data source is missing, this data is marked with an error flag to be filled with the proper value later. Then, data is inserted into the corresponding MySQL databases and awaits further processing.

B. Data Correlation and Fusion

boards, power modules, and relays/contactors. Collecting data on the stressors that cause these components to fail, and thus, the entire PV system to fail, could provide invaluable insight. As such, a field reliability data collection technique that captures stressors such as temperature, humidity, power, voltage, current, etc. is paramount.

For the purpose of understanding the comprehensive performance of PV inverters, a field data acquisition system is established. The structure of the data acquisition system is shown in Fig. 1. Two categories of field data are collected: operating condition data and environmental data. Operating condition data is comprised of grid data and inverter data. Environmental data provides historical environmental conditions around PV inverters.

Grid data is provided by the grid data source. It can be acquired via the online portal provided by the grid data utility partner. The grid data includes three-phase current, voltage, and power. It also includes transient data records to indicate the operating condition data when faults or dynamic events happen on the grid. **Inverter data** includes AC and DC power, voltage, and current. It is recorded by the measurement tool designed in the PV inverters. The data is transmitted to the user interface that can help operators to conduct energy monitoring, managing, and grid-compliant power control. **Environmental data** includes the ambient temperature, relative humidity in the PV farm, Global Horizontal Irradiance (GHI) received by the PV farm, and module temperature around critical electrical components in the PV inverter. These types of environmental data have two data resources. One is from the sensors deployed around the PV inverters. This data is captured, uploaded to the web portal, and known as environmental data source A. The other is from the sensors deployed around the PV farm. This data is captured, uploaded to the web portal, and known as environmental data source B.

To demonstrate the validation of our design, the following case study is conducted. **SMA SUNNY-TRIPOWER-**

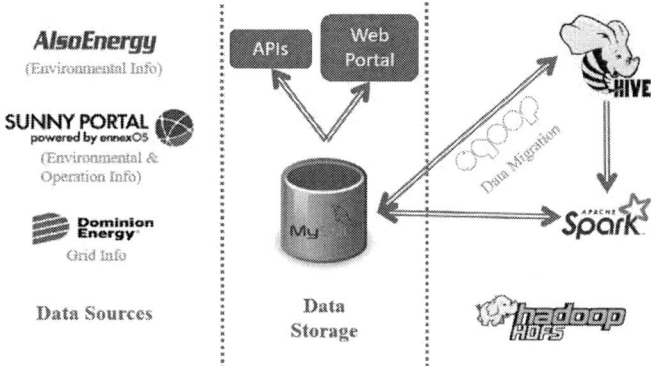

Fig. 2. Data management platform workflow.

TABLE II. DATA FIELD AND SOURCE

Information Type	Data Field	Data Source
Environmental	Ambient Temperature (°F)	Environmental Data Sources A and B
	Solar Irradiance (W/m²)	Environmental Data Sources A and B
Operational	AC Power (W)	Inverter and Grid Data Sources
	AC Voltage (V)	Inverter and Grid Data Sources
	AC Current (A)	Inverter and Grid Data Sources

The same data field may have multiple data sources, therefore, the measurements from multiple channels can be merged into a more accurate dataset. The list of data that have multiple channels is shown in TABLE II.

Data correlation: To avoid the failure of data fusion caused by invalid source data, several data validation actions are taken to verify the consistency and reasonability of the data. The Module Temperature Check [6] is performed to reduce the measurement noise. Besides, the isolation Forest algorithm [5] is performed to detect outliers. Isolation Forest constructs iTree based on the features of the dataset. The node near the root node of iTree is detected as an outlier. The outlier data is filled with the reasonable value derived based on the nearest data point.

Data fusion: After assuring the validation of the data source, data fusion of multiple data channels can be performed. However, due to discrepancies in the location of sensors, the measurement scope, and the measurement accuracy, the same type of datasets from different channels are not always identical. The field data from Sunny Portal is collected by the dedicated sensors deployed in the PV inverter closure to evaluate the operational status of PV inverters and the environmental conditions around inverters, and hence, the data from Sunny Portal is more reliable. Other data channels are regarded as complementary sources to increase the fidelity and accuracy of the data. As for operational information, like AC power, the power data from Sunny Portal provides the operating condition of each inverter, and the power data from the grid data source provides the operating condition of the grid. When dynamic events are detected on the grid, the data collected by individual inverters is not as accurate as the data from the grid because of higher resolution measurement instrumentation equipped at the grid side by the utility company.

The metadata from different channels is organized in different formats and time resolutions. Therefore, the data from the channel with a lower time resolution can augment to a higher resolution. For example, the temperature from environmental data source A is updated every minute, and the temperature from environmental data source B is updated every five minutes. Temperature data from Sunny Portal is compensated by referring to the data from the other channel.

Compared to the traditional interpolation method, the data from the other channel provides a more accurate estimation. After finishing the data correlation and data fusion, the merged dataset is integrated into the database.

C. Data Management

With the field data collected over time, proper data management solutions are needed. To satisfy the demand for data storage, query, and further analysis, the relational database, distributed data framework, Apache Hadoop, and other techniques are utilized. These utilization details are discussed in the rest of this section.

Metadata management: Due to the need for the prompt query for a large number of metadata, the field metadata is stored in the relational database MySQL. The field data can be updated and retrieved via SQL commands efficiently. For the convenience of queries, an online demo portal is developed to acquire specific data as shown in

Fig. 3. APIs are also provided for querying the field data from the database with the specific data field in the designated time range. The metadata can be easily acquired for future R&D needs.

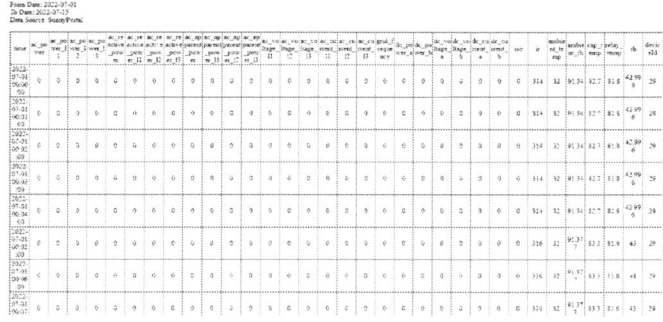

Fig. 3. Metadata query online portal.

Distributed data management: As the metadata accumulates, performing data processing and analysis on massive data becomes more expensive. It takes a long time to read data, perform computation, and write data back to the database when the data volume becomes large. Therefore, distributed data management solutions are needed. The Hadoop Distributed File System (HDFS) is a distributed data storage system used in the Hadoop ecosystem [9]. HDFS stores the data across multiple machines or nodes in a cluster. It is the foundation of other Apache applications. Hadoop applications follow the master-slave architecture to achieve storing and processing of massive amounts of data effectively. A portion of data is assigned to each node and the data on each node is replicated across the node for failure recovery. Considering the data update and query efficiency, the metadata is stored in MySQL, but for the convenience of performing distributed data processing and analysis, metadata needs to be migrated to HDFS. Apache Sqoop is used as a translator to migrate the metadata from structured data in MySQL to unstructured data in Apache Hive called Apache Spark. Apache Spark is designed for fast computation in performing data science and machine learning on single machines or clusters. This workflow can be deployed on cloud service platforms, like AWS and AZURE.

IV. CASE STUDY

A case study is performed to demonstrate the efficiency and feasibility of the workflow we designed. We selected the operating condition data for the same inverter with a time resolution of 5 minutes and environmental condition data with a time resolution of 1 minute as a use case to introduce the workflow of the data management platform. All the data is measured at Ottare PV farm located in Cayce, South Carolina. To be specific, we select the data for a whole day on May 15th, 2022, to introduce the data correlation and fusion workflow. And we select the data for a whole year from May 2022 May 2023 to demonstrate the distributed data management.

Both operating condition data and environmental data are collected by web scraper scripts from three data resources daily. The collected data will do the data correlation. Module Temperature Check will be used to evaluate the validation of the temperature and irradiance information. The completeness of data will be evaluated. Empty, missing data and outlier values will be filled with the value measured at the closest moment if the completeness is over 99% (The ratio of Nan value to total values). Or else, also including other situations for data fusion, it will be reported as the missing data period. The data from the second data channel will be used and filled into the missing data period. Because the data from two different channels vary caused of location differences, the similarity between the data from the two channels will be evaluated. If the cosine similarity between data from two channels is lower than 99.5%, the second channel data will be converted based on the gap between the data from the two channels before being filled. And the data from the second channel will be filled directly if the data from the two channels are similar. The comparison of ambient temperature data before and after data correlation and fusion is shown in Fig. 4. After the data correction and fusion, metadata will be stored and managed by the relational database MySQL for prompt data queries.

When the data accumulates to a large volume, the data manipulation and advanced data analysis are very time-consuming. In order to increase the operation efficiency, the metadata will be migrated from MySQL via the Sqoop to HIVE for later use by Apache SPARK. Advanced statistics can be performed via PySpark more efficiently compared to traditional data management platforms.

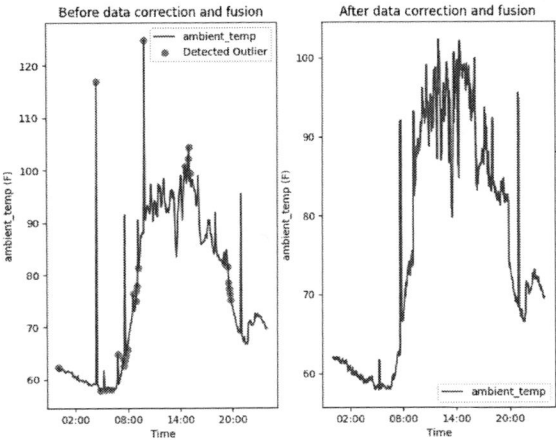

Fig. 4. Comparison of results before and after data correction and fusion.

Besides, we also can perform machine learning algorithms using Mllib [8] for fast advanced analysis when handling large-scale data. For example, cluster algorithms can be used to categorize environmental condition data of different days into different weather patterns. To better visualize the clustering result, the data has been decomposed into two-dimensional data, which is shown in Fig. 5. The red points are the locations of the cluster centers. The blue points are locations of the one-day environmental conditions mission profile in two-dimension. The clustering result can help to merge similar mission profiles and decrease the simulation workload by skipping using similar mission profiles as input.

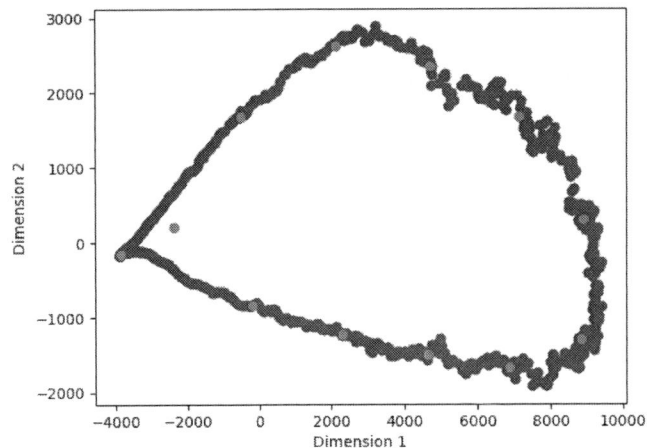

Fig. 5. Clustering results for environmental conditions of one year.

V. CONCLUSION AND FUTURE WORK

To meet the demand for accuracy and real-time capability of PV system degradation evaluation [10], massive volume data is needed to run high-fidelity and high-efficiency simulations [11] and perform advanced data analysis. In this paper, the implementation details of field data management are introduced. And a demo example use case is described to show the feasibility. **our main contribution is developing an integrated data management platform capable of data acquisition, processing, storage, query, and performing big data analysis utilizing AI algorithms.** The platform can also achieve data correctness and provide an effective distributed data management solution to retrieve massive data and establish a connection to distributed computational frameworks. This data management platform provides a large-scale computation capability for PV-related R&D needs. Existing analysis tools and algorithms which are subject to computation capability and large-scale data manipulation efficiency can also be moved to this platform. For the current stage, we focus more on the environmental condition data process and analysis. In the future, we will add some support targets to the data process and analysis on the operating condition data process.

ACKNOWLEDGMENT

This work is supported by the Solar Energy Technologies Office, Office of Energy Efficiency and Renewable Energy, Department of Energy, USA under the award number DE-EE0009348.

REFERENCES

[1] V. Ramasamy, J. Zuboy, E. O' Shaughnessy, D. Feldman, J. Desai, M. Woodhouse, P. Basore, R. Margolis, "U.S. Solar Photovoltaic System and Energy Storage Cost Benchmarks, With Minimum Sustainable Price Analysis: Q1 2022", No. NREL/TP-7A40-83586, National Renewable Energy Lab. (NREL), Golden, CO (United States), 2022.

[2] F. David, K. Dummit, J. Zuboy, and R. Margolis, "Spring 2022 Solar Industry Update", National Renewable Energy Lab.(NREL), Golden, CO (United States), 2022.

[3] J. Flicker "Reliability of power conversion systems in photovoltaic applications," Sandia, NW, 2015J. Clerk Maxwell, A Treatise on Electricity and Magnetism, 3rd ed., vol. 2. Oxford: Clarendon, 1892, pp.68–73.

[4] P. Hacke, et al. "A status review of photovoltaic power conversion equipment reliability, safety, and quality assurance protocols." Renewable and Sustainable Energy Reviews 82 (2018)

[5] F. Liu, K. Ting, and Z. Zhou. "Isolation forest." 2008 eighth IEEE international conference on data mining. IEEE, 2008.

[6] Perry, Kirsten, et al. PVAnalytics: A Python Package for Automated Processing of Solar Time Series Data. No. NREL/PR-5K00-83824. National Renewable Energy Lab.(NREL), Golden, CO (United States), 2022.

[7] Aravinth, S. S., et al. "An efficient HADOOP frameworks SQOOP and ambari for big data processing." International Journal for Innovative Research in Science and Technology 1.10 (2015): 252-255.

[8] Meng, Xiangrui, et al. "Mllib: Machine learning in apache spark." The Journal of Machine Learning Research 17.1 (2016): 1235-1241.

[9] K. Shvachko, H. Kuang, S. Radia and R. Chansler, "The Hadoop Distributed File System," *2010 IEEE 26th Symposium on Mass Storage Systems and Technologies (MSST)*, 2010, pp. 1-10, doi: 10.1109/MSST.2010.5496972.

[10] I. Vernica, H. Wang, F. Blaabjerg, "Design for reliability and robustness tool platform for power electronic systems—Study case on motor drive applications," in 2018 IEEE Applied Power Electronics Conference and Exposition (APEC), 2018.

[11] L. Wang, R. Thiagarajan, S. Jin, and Z. Zhang, "Accelerating Simulation for High-Fidelity PV Inverter System Reliability Assessment with High-Performance Computing". In 2022 IEEE 49th Photovoltaics Specialists Conference (PVSC) (pp. 0178-0182). IEEE.

SnO2 buffer layers for high efficiency CdSeTe/CdTe devices.

L. C. Infante-Ortega, Xiaolei Liu, Luksa Kujovic, Mustafa Togay, Luke O. Jones, Ali Abbas, Kieran Curson,
R.C. Greenhalgh, Kurt L. Barth, Jake W. Bowers and John M. Walls
CREST, Loughborough University, Loughborough, LE11 3TU, UK

Ochai Oklobia and Stuart Irvine
CSER, Swansea University, Swansea, SA2 8PP, UK

Eric Colegrove, Brian Good, Matt Reese
NREL, 15013 Denver West Parkway, Golden, CO 80401, USA

Abstract — **SnO2 buffer layers of different thickness were deposited onto TEC 15 Fluorine doped tin oxide coated glass substrates using rf magnetron sputtering. The buffer layers were then incorporated into Cu-doped CdSeTe/CdTe devices using a range of CdCl₂ activation treatments and CuCl₂ annealing temperatures to determine the effects of buffer layer thickness on device performance. Results show that all devices fabricated with thinner buffer layers resulted in much better $J-V$ characteristics than their thicker counterparts. This was mainly due to a reduced open-circuit voltage (V_{oc}) when using thicker buffer layers. The best device produced a conversion efficiency of 16.59%, fill factor of 71.62%, Jsc of 28.44 mA/cm² and V_{oc} of 814.23 mV.**

I. INTRODUCTION

Cadmium Telluride (CdTe) solar cells are a thin film solar technology which use a graded CdSeTe/CdTe absorber layer which absorbs incident light, with energy above its bandgap, within 1 μm from its surface. The main advantages of CdTe include its low cost, high optical absorption and direct band gap of 1.45eV which is near optimal for photoconversion as it corresponds to the maximum of the solar spectrum. [1] Recent improvements have been made by incorporating Se as a CdSeTe alloy at the front of the absorber and by using a transparent buffer layer above the Transparent Conducting contact. [2] The current efficiency record for CdTe cells is 22.1%, achieved by First Solar. [3]

Research is ongoing to improve layer and material properties to increase Voltage, diode quality, and ways to passivate interface defects causing recombination. Efforts are also being made to improve material recycling to reduce waste and mitigate material scarcity. [4]

'Buffer' layers are used in CdTe photovoltaics to improve device performance. [5] This investigation examines the use of SnO2 thin films of different thickness as buffer layers for CdSeTe/CdTe solar cells. SnO_2 is a transparent n-type semiconductor with a band gap of $3.6-4$ eV. It is popular due to its low cost, chemical stability, and ease of manufacture. SnO_2 films are made with a variety of processes such as chemical vapor deposition (VCD) and spray pyrolysis, but in this investigation we opted to use rf magnetron sputtering which enables uniform thin films to be deposited on a large area

with excellent thickness control. The optical and electrical properties of the thin film can be controlled via the sputtering process parameters. ([6],[7])

II. METHODOLOGY

A. Thin Film Deposition

The buffer layers were deposited using rf magnetron sputtering using an AJA ATC 2200 system, using a SnO2 compound target measuring 4 inches in diameter. The substrates were TEC 15 glass (NSG-Pilkington UK) which had been subjected to 5 minutes of UV cleaning TEC 15 is 3.8mm thick glass which is coated with a fluorine doped tin oxide (FTO) [8] transparent conductor with a rated sheet resistance of 12-14 Ω/sq [9]. The buffer thin films were also deposited on to Eagle glass (Abrisa Technologies) to characterise their optical and electrical properties. The Eagle glass had been cleaned in a sonicator beforehand using soapy water, acetone and Isopropyl alcohol.

The depositions used a target power of 120W and a substrate temperature of 500°C. The sputtering was carried out in an Argon atmosphere with 25% oxygen partial pressure. The substrates were rotated at 10rpm during sputtering to assist uniformity. These conditions yielded a deposition rate of 107nm/hr, which was used to calculate the time needed to prepare 50nm and 100nm thick films.

The electrical properties of the SnO2 films were determined with a Biorad Hall Effect system. The 100nm film had a carrier concentration of -3.57E17cm⁻³, a carrier mobility of 2.14cm²/Vs and a resistivity of 8.179Ωcm. The 50nm film was found to be too resistive to provide an accurate Hall measurement.

B. Device Fabrication

The SnO2 buffer layers on TEC15 were used to fabricate CdSeTe/CdTe devices using VTD. These devices are made from a series of layers as shown schematically in figure 1. A cadmium selenium telluride (CdSeTe) alloy layer was deposited on the buffer layer, and this was followed by the deposition of a CdTe layer. The CdSeTe then diffused into the CdTe during the cadmium chloride activation step to produce

978-1-6654-6060-6/23 $31.00 © 2023 IEEE

band gap grading. The lower band gap at the front of the device improves device efficiency by increasing the photocurrent without reducing the open-circuit voltage (V_{oc}) ([2],[10]). Devices were also fabricated by depositing the CdSeTe/CdTe absorber directly on to TEC15.

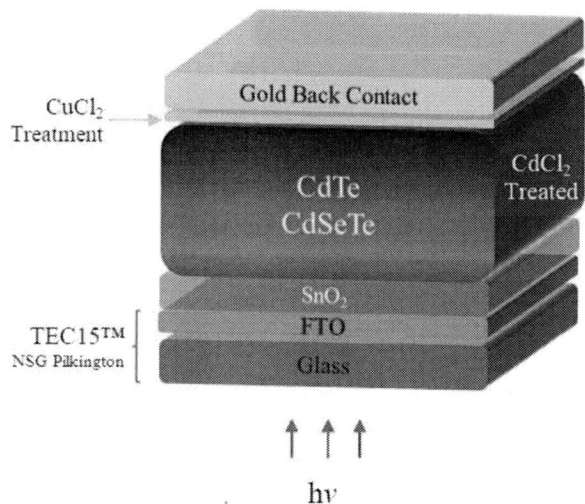

Fig. 1.   Structure of CdSe/CdTe devices made with sputtered buffer layers of 50nm and 100nm thickness.

Substrates with SnO2 buffer layers were loaded into a thermal evaporator (Angstrom Engineering). 500nm of a 30% CdSeTe alloy was deposited followed by 3um of CdTe. The substrate was kept at 400°C and rotated at 15 rpm to ensure film uniformity. The substrates were then removed from the chamber, rinsed with IPA and then CdCl2 treated using a CSS system at 450°C. They were then treated in a diluted CuCl2 solution for 3 minutes and annealed immediately after in a tube furnace for 25 minutes at 200°C. The devices were then placed in a metal evaporator and gold contacts were deposited onto the CdTe layer surface. Finally, the device area around the gold contacts was mechanically scribed using a razor and Indium contacts were deposited onto the now exposed TCO layer using an Ultrasonic solder.

Using the process described, four devices were made using the 50nm and 100nm SnO2 buffer layers on TEC 15. Two of these, named A1 and A2, were CdCl2 treated for 10 minutes and used one buffer layer of each thickness. The other two, named B1 and B2, were identical except that the CdCl2 treatment, was extended to 20 minutes. A device was also made using TEC 15 without a buffer layer, using the same CdCl2 treatment as A1 and A2, to examine how the absence of this layer would affect performance. The effects of the thickness of the buffer layers on the resulting J − V curves of the devices were then analysed.

III. RESULTS AND ANALYSIS

A finished device consists of four electrically isolated solar cells. Two devices were made with each buffer layer, making a total of eight cells for each buffer layer thickness and process

combination. Each cell had a 0.5cm x 0.5cm gold contact on top of the CdSeTe/CdTe layer stack.

J-V curves were measured in a superstrate solar simulator. The system was calibrated using a GaAs reference cell which was connected to a Keithley digital multimeter, ensuring that all devices placed at a certain position and height from the lamp were under the same 1 sun intensity. Devices were measured at this position with probes connected to the gold and indium contacts. Light and dark I-V measurements were then obtained from a device area of $0.25cm^2$.

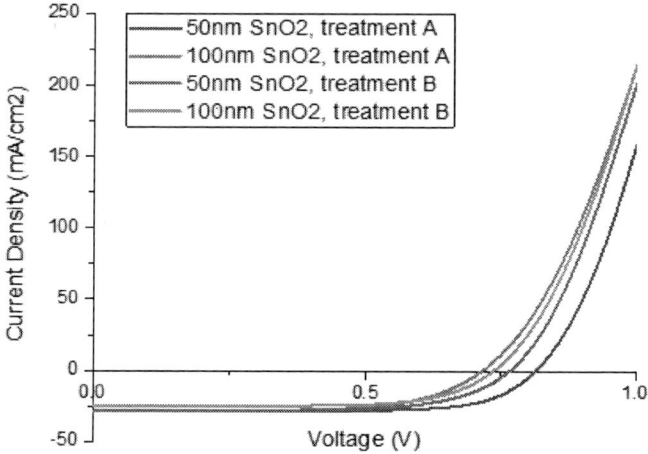

Fig. 2.   J − V curves of devices made with 50nm and 100nm SnO2 buffer layers, using two different CdCl2 treatments.

Figure 2 shows the J-V curves of the highest efficiency cell of each device with sputtered SnO2 buffer layers of 50nm and 100nm thickness. Table 2 contains the J − V parameters.

The best performing cell (cell A1) used a 50nm SnO2 buffer layer and was CdCl2 treated for 10 minutes It produced an efficiency of 16.59% fill factor (FF) of 71.62%, open-circuit voltage (V_{oc}) of 814.23mV and a short-circuit current density of 28.44 mA/cm^2.

Cell A2, identical to A1 except for a 100nm thick buffer layer, had a near-identical J_{sc} to A1 but the V_{oc} was almost 100mV lower. Comparing devices B1 and B2 also showed that the 50nm buffer layer performed better than its thicker counterpart. Furthermore, both devices with a 50nm buffer layer outperformed all 100nm devices when using PCE and V_{oc} as metrics.

The J − V measurements of the device B3 shown in Fig 3 is interesting since a working device has been fabricated even with the absence of a buffer layer. Note that the J_{sc} remains relatively unaffected although the Voc is significantly degraded.

978-1-6654-6060-6/23 $31.00 © 2023 IEEE   246

TABLE II
J-V CHARACTERISTICS OF DEVICES USING SPUTTERED SNO2 BUFFER LAYERS

Device name	Buffer layer	Process used	One sun efficiency (%)	Measured FF (%)	V_{oc} (mV)	Jsc (mA/cm^2)
A1	50nm SnO2	A	16.59	71.62	814.23	28.44
A2	100nm SnO2	A	12.88	63.41	715.08	28.41
A3	None	A	5.34	57.56	341.95	27.13
B1	50nm SnO2	B	14.11	64.81	765.60	28.43
B2	100nm SnO2	B	12.36	66.70	735.86	25.19

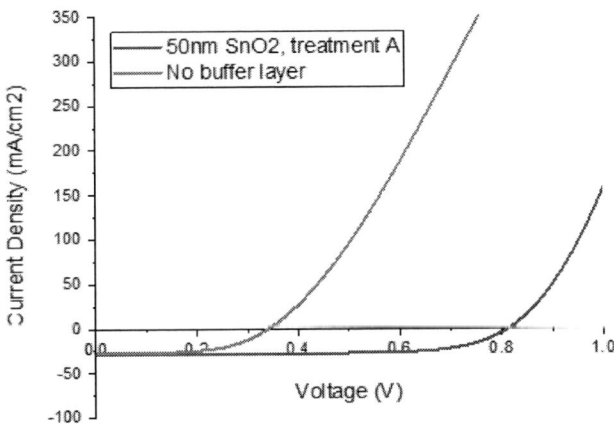

Fig. 3. J − V curves of CdSeTe/CdTe devices made with and without a buffer layer. The cell without the buffer layer has a significantly reduced V_{oc}.

IV. CONCLUSION

SnO2 buffer layers with two different thicknesses were deposited on TEC15 using rf magnetron sputtering. These were then used to fabricate CdSeTe/CdTe solar cells. Two alternative CdCl2 treatments were used.

Devices using 50nm buffer layers significantly outperformed those with 100nm buffer layers, based on their J − V characteristics. Results show that the use of thicker buffer layers results in a lower V_{oc}, while the J_{sc} remains relatively unaffected. Surprisingly, a working device was fabricated using the same conditions without a buffer layer Although the open-circuit voltage Voc was substantially reduced the open circuit current density Jsc was comparable.

The best performing buffer layers were more resistive than their thicker counterparts. This suggest that the buffer layer is acting as a high-resistance transparent (HRT) layer indicating that an important relationship exists between the transparent conductor and the buffer layer.

In the full paper we will present results on a more complete range of thickness 40nm to 200nm. This will help to optimise the thickness and the associated device efficiency. It may also help to explain how these devices function and the precise role played by the buffer layer. This work is part of a study that will also include devices made with new buffer layers made from alloys of SnO2 and ZnO.

This will be supported by materials characterisation with cross-sectional TEM and EBSD analysis of the devices. This will reveal any diffusion occurring between layers in the device caused by the changes made during the CdCl$_2$ process.

REFERENCES

[1] A. Bosio, S. Pasini and N. Romeo, "The History of Photovoltaics with Emphasis on CdTe Solar Cells and Modules," *Coatings 2020 Special Issue: "Advances in Thin Films for Photovoltaic Applications."*

[2] X. Zheng, D. Kuciauskas, J. Moseley, et al., "Recombination and bandgap engineering in CdSeTe/CdTe solar cells," *APL Materials*, vol. 7, 2019.

[3] First Solar Press Release: "First Solar Achieves Yet another Cell Conversion Efficiency World Record"; First Solar Press Release: Tempe, AZ, USA, 2016.

[4] Solar Energy Technologies Office, "Cadmium Telluride," available at: https://www.energy.gov/eere/solar/cadmium-telluride#:~:text=CdTe%20solar%20cells%20are%20the,to%20 conventional%20silicon%2Dbased%20technologies.

[5] J. M. Kephart, J. W. McCamy, Z. Ma, A. Ganjoo, F. M. Alamgir, W. S. Sampath, "Band alignment of front contact layers for high-efficiency CdTe solar cells," *Solar Energy Materials and Solar Cells*, vol. 157, p. 266-275, December 2019.

[6] A. F. Khan, M. Mehmood, A. M. Rana, M. T. Bhatti, "Effect of annealing on electrical resistivity of rf-magnetron sputtered nanostructured SnO2 thin films," *Applied Surface Science*, vol. 255, p. 8562–8565, 2019.

[7] D. Leng, L. Wu, H. Jiang, Y. Zhao, J. Zhang, W. Li, and L Feng, "Preparation and Properties of SnO2 Film Deposited by Magnetron Sputtering," *International Journal of Photoenergy*, vol. 2012.

[8] Pilkington NSG TEC™ overview, available at: https://www.pilkington.com/en/global/digital-signage/products/applications/coated-cover-glass-for-touch-screens-and-displays/nsg-tec#

[9] NSG-Pilkington UK, NSG TEC™ technical data, available at: https://www.pilkington.com/en/global/digital-signage/products/applications/coated-cover-glass-for-touch-screens-and-displays/nsg-tec

[10] A. H. Munshi , J. Kephart, A. Abbas, J. Raguse , J. N. Beaudry , K. Barth, J. Sites , J. Walls, and W. Sampath, "Polycrystalline CdSeTe/CdTe Absorber Cells With 28 mA/cm2 Short-Circuit Current," *IEEE journal of photovoltaics*, vol. 8, p. 310-314, January 2018.

Contact Interface Morphology of Screen-Printable Front-Side Contacts for Industrial N-TOPCon Crystalline Silicon Solar Cells

Meijun Lu[a], Kurt R. Mikeska[a], Weilin Liao[a], Chaoying Ni[b], Yong Zhao[b], Jianming Wang[c], Kangping Zhang[c], Changgen Zhang[a], Yawen Xu[a] and Baiqiang Liu[a]

[a]Jiangxi Jiayin Science and Technology Co. LTD, No.2266 Yingxiong Street, Nanchang 330013, China

[b]University of Delaware, Department of Materials Science and Engineering, Newark, DE 19716, USA

[c]DAS Solar Co. LTD, Research and Development Department, Quzhou, 32400, China

Abstract — **High-resolution TEM/STEM was used to characterize the microstructure and chemistry of the front-side contact interface for optimally prepared screen-printed industrial n-TOPCon crystalline silicon solar cells. A front-side screen-printable paste comprising silver, metallic aluminum, and an inorganic lead-silicate frit was designed to contact boron-diffused p^+ emitter surfaces with SiN_x:H-Al_2O_3 antireflection-passivation layers. The final front-side bulk silver metal region microstructure shows isolated metallic aluminum particles surrounded by solidified liquid phase within the bulk sintered silver conductor line. TEM/STEM characterization of the front-side silver metal-p^+ boron-diffused emitter contact region shows continuous, amorphous interfacial films decorated with silver colloids located between the bulk conductor metal and emitter surface. STEM/EDS analysis shows the amorphous interfacial films were metal oxide-based ions comprised of lead, aluminum, silicon and silver. The interfacial films were relatively thick compared to interfacial films observed in front-side screen-printed contacts for p-type c-Si cells. Free energy calculations indicate aluminum ions dissolved in the interfacial film do not participate in SiN_x:H etch-through. Microcopy observations and free energy calculations suggest the function of aluminum metal additions in the paste is to increase the number density of the Ag colloids in the IF film which is necessary to reduce contact resistance in the relatively thick interfacial films.**

I. INTRODUCTION

Tunnel oxide passivated contacts (TOPCon) is the leading next-generation crystalline silicon cell technology choice after mono PERC (passivated emitter and rear cell). Compared to technologies such as HJT (heterojunction) and IBC (interdigitated back contact), TOPCon architectures can be achieved by upgrading current industrial PERC or PERT (passivated emitter rear totally diffused) manufacturing lines which means a lower capital expenditure is needed for existing PERC or PERT manufacturers who want to upgrade their existing production lines. We investigated n-type TOPCon since n-type silicon wafers have inherent advantages over p-type silicon such as higher minority carrier diffusion length and less sensitivity to metal impurities, and n-type cells don't have boron-oxygen (BO) related light induced degradation (LID) [1-3]. A TOPCon structure offers a gain in cell efficiently of ~1% absolute compared to previous generation technologies [4,5].

Industrial solar cell manufactures utilize screen-printable pastes as a cost-effective metallization solution because of their low cost, high throughput, and relatively high performance. Among the challenges for introducing a TOPCon architecture to industrial solar cell manufacturing is utilizing cost-effective screen-printable pastes that effectively contact both the front-side (FS) and rear-side (RS) surfaces of the solar cell in a single metallization firing step while maintaining the benefits of the TOPCon architecture [6].

We are exploring both FS and RS state-of-the-art screen-printable pastes on an industrial n-type TOPCon bifacial architecture shown in Fig. 1. It has a FS p^+ boron diffused emitter and a RS n^+ phosphorus diffused passivating contact, with screen-printed conductor lines on both the front and rear sides to obtain bifaciality. The challenge for screen-printable pastes is to contact the underlying semiconductor surface while minimizing damage to the TOPCon passivation layers and underlaying emitter surface.

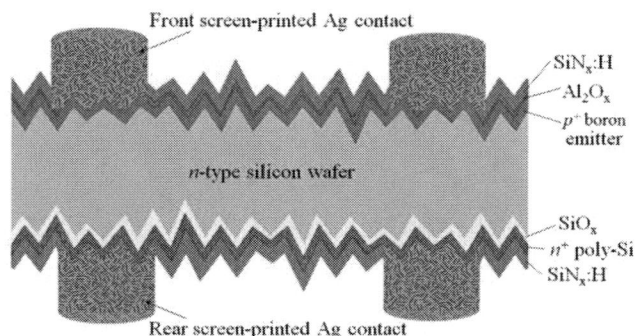

Fig. 1. Industrial n-TOPCon c-Si bifacial solar cell with front and rear-side screen-printed contacts.

Previously, we reported on the FS and RS electrical properties and contact regions microstructure using SEM/FIB characterization for screen-printed industrial bifacial n-type TOPCon solar cells [7,8]. In this work, high-resolution TEM/STEM techniques were used to further characterize the microstructure and chemistry of the FS screen-printed contact region. TEM/STEM characterization is essential for elucidating the details of the electrical contact mechanism which, in our opinion, has not yet been conclusively established for the FS screen-printed contact of an n-type cell.

978-1-6654-6060-6/23 $31.00 © 2023 IEEE

II. Experimental

RS screen-printable pastes were prepared by mixing silver (Ag) metal powder (~90 wt%), inorganic frit powder (~2 wt%) and organic media (~8 wt%) in a planetary mixer, while FS pastes were prepared by mixing Ag metal powder (~85 wt%), inorganic frit powder (~5 wt%), aluminum (Al) metal additive (~2 wt%) and organic media (~8 wt%) in a planetary mixer, both pastes were then 3-roll milled and viscosity adjusted (measured by a Brookfield DV2T-HB). Pastes were roll milled to a fineness of grind (FOG) of 3~5 μm in accordance with ASTM Test Method D 1210-05. Pastes were adjusted to a final viscosity between 200 and 450 Pa-s. at 10 rpm.

Pastes were evaluated on commercially available (DAS Solar) industrial processed bifacial n-type TOPCon mono c-Si pseudo-square (182 mm × 182 mm) solar cells with front surface random pyramids and flat back surface. The cell fabrication process included alkali texturing, p^+ boron diffusion (BBr₃), rear p^+ removal (single side HF/HNO₃ etching), thermally deposited SiO_x and intrinsic poly-Si films (LPCVD), n^+ phosphorus diffusion (POCl₃), front n^+ removal (single side etching), and double sided SiN_x:H (PECVD) deposition. For the metallization, an industrial Baccini screen printer was used to print front and rear surface five busbar H-patterned conductor lines. The fired FS conductor line mean width was ~30 μm and height was ~11 μm. The fired RS line mean width was ~40 μm and height was ~9μm. An industrial 9-zone Despatch furnace was used to fire the screen-printed wafers. The final screen-printed TOPCon cell configuration is shown in Fig. 1.

An industrial Berger I-V tester was used to measure solar cell efficiency (Eff), fill factor (FF), open circuit voltage (V_{OC}) and short circuit current (I_{SC}). Electrical data values are median and standard deviation values for about ten cells. Gibbs free energy (ΔG) values were calculated using FactSage™ [9].

A thin section sample for high resolution TEM/STEM characterization was prepared in cross-section using a focused ion beam (FIB) Auriga 60 CrossBeam™ (FIB/FE-SEM). The monocrystalline Si solar cell sample was FIBed along the <101> plane and through the front surface silver conductor line. This geometry allows for the sample to be viewed orthogonal (edge-on) to the surface of the Si etch pyramid facets. The thin section was nominally estimated to be 80 − 100 nm thick. The sample was analyzed with a JOEL JEM-2010F field-emission transmission electron microscope equipped with an Ametek TEAM™ X-ray energy dispersive spectroscopy (EDS) system.

III. Results and Discussion

A critical challenge for industrial scale TOPCon solar cell manufacturing is the implementation of cost-effective screen-printable pastes on both the front and rear sides of the cell. Each surface presents unique paste development challenges. For a n-type cell, the front surface requires a paste optimized for contact to a p^+ doped Si surface; the rear surface requires a paste optimized for contact to a n^+ doped Si surface. This report focuses on the FS contact (i.e., contact to a boron doped p^+ surface). A screen-printable paste for the FS contact typically comprises an organic phase, Ag metal particles, metallic Al additions, and an inorganic frit. The paste is screen printed on the wafer surface and rapidly fired at relatively high temperature (~720°C - 780°C). During the firing process the inorganic frit forms a liquid phase flux that helps sinter the Ag particles to high bulk density and, ideally, etches-through the SiN_x:H antireflective coating (ARC), which is electrically insulating, and then stops; leaving the underlaying layers undamaged. In practice, as previously discussed, precise control of the etch-through process is quite difficult [10].

Screen-printable pastes for contact to p-type wafers with phosphorous doped n^+ emitter surfaces currently contain telluride-based frits [11] and are well established. Pastes designed to contact n-type wafers with boron doped p^+ emitter surfaces are less well established and usually contain Al metal additions to accomplish contact, but which suffer high emitter damage and relatively high contact resistance and conductor line resistance compared to pastes for p-type wafers. TOPCon surface passivation layers (on both front and rear sides) further complicate contact. Several studies have speculated on an n-type electrical contact mechanism and the role metallic Al plays in establishing contact [12, 13], but the precise role of Al additions and a definitive contact model have yet to be fully established. We use high resolution TEM/STEM to characterize the microstructure and chemical composition of the FS screen-printed contact for an n-type cell to elucidate the contact mechanism and clarify the role metallic Al additions play in establishing contact.

The FS paste in this study contained Ag metal, a lead-silicate based frit, and Al metal particles. Metallic Al is required for contact to p^+ surfaces. The Al metal is thought to promote electrical contact by reacting with and penetrating the p^+ doped Si surface during the firing process [12]. The paste must contain both frit and Al metal for good contact [8].

TABLE I

Screen-Printed *N*-TOPCon Cell JV Data

Eff (%)	FF (%)	V_{OC} (mV)	J_{SC} (mA/cm²)
24.12 ±0.03	82.92 ±0.25	697.03 ±1.40	41.79 ±0.03

Table I lists JV data for optimally prepared n-TOPCon bifacial cells printed with front and rear-side screen printable pastes. Samples for microscopy characterization were randomly selected from the group of cells used to measure JV in Table 1.

Fig. 2 is an SEM image of the FS contact region. The FS has a boron doped p+ emitter coated with an Al_2O_x layer and PECVD SiN_x:H ARC. The light contrast area (top) is the sintered bulk Ag conductor and the dark contrast area (bottom) is the bulk silicon emitter. An interfacial (IF) film is observed at the interface between the bulk Ag conductor metal and silicon emitter. During the paste firing process, the inorganic frit in the paste forms a low viscosity liquid phase flux that migrates to the silver-silicon interface region where it enables the oxidation, dissolution and subsequent removal of the SiN_x:H layer. The final IF film is decorated with Ag colloids from precipitation of silver ions dissolved in the IF liquid phase

[10]. The SEM image shows the IF film to be continuous along the interface and relatively thick with no directed contact between the bulk Ag conductor and Si emitter layer. Voids, pockets of solidified liquid, and an isolated metallic Al particle are observed in the bulk silver. The Al particle is surround by solidified liquid phase decorated with Ag colloids [8]. The isolated Al metal particle is not in contact with the surrounding bulk Ag; therefore, it does not contribute to electrical conduction and increases conductor line resistance since Al metal is inherently less conductive than Ag metal.

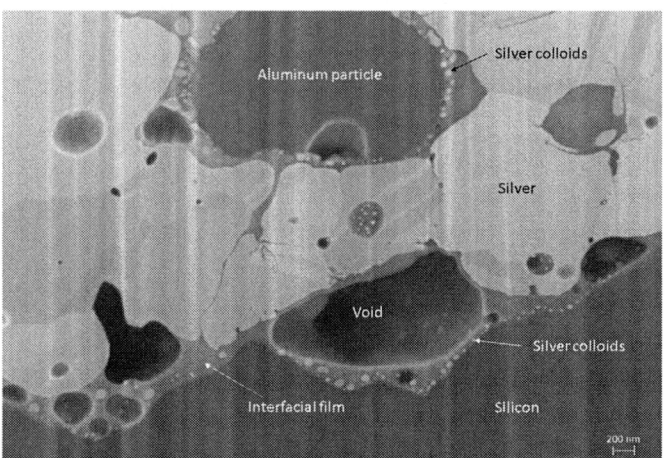

Fig. 2. SEM cross section micrograph of the final front-side fire-through metallization interface region.

Fig 3 is a collage of bright field STEM images of the IF region. The dark contrast region (top) is the bulk Ag conductor; the light contrast region (bottom) is the Si emitter. The images have several notable features. The IF film is continuous along the interface between the bulk Ag conductor and Si emitter with no direct contact between the Ag conductor and Si emitter; the IF film is relatively thick and decorated with Ag colloids; numerous voids are observed and solidified liquid phase decorated with Ag colloids extends into the bulk Ag conductor. Note that the thin section sample varied in thickness which results in contrast nonuniformity in the microcopy images. The large number of voids along the IF also creates thickness and contrast nonuniformity where a void may be "behind" the image surface plane.

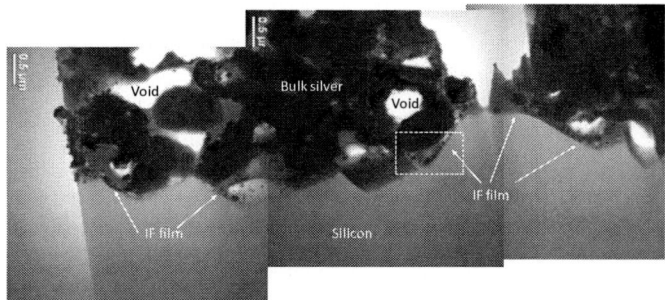

Fig 3. Collage of low-resolution bright field STEM images of front-side contact region.

Figures 4 is a high-resolution bright field images of the IF film (box inset shown in Fig. 3). Again, Ag colloids decorate the IF film. The diffraction image of the solidified IF liquid region has a characteristic amorphous diffraction pattern. The diffraction image of the Ag colloid has a crystalline diffraction pattern. The formation mechanism of Ag colloids in c-Si solar cell IF films has previously been discussed [10].

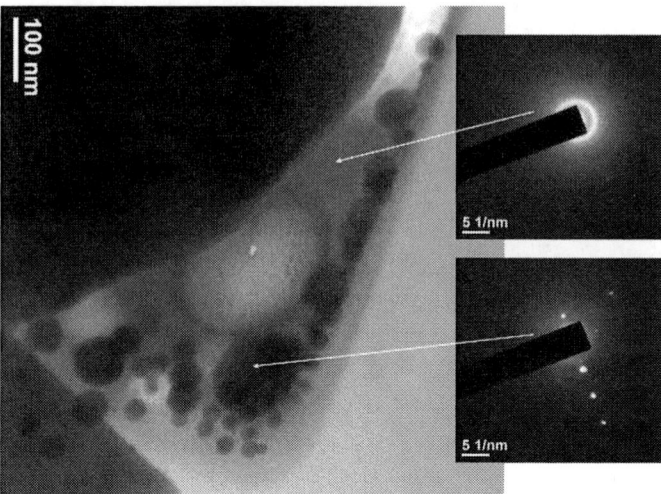

Fig 4. High-resolution bright field STEM image of the IF film (inset shown in Fig. 3).

Fig. 5 are TEM/EDS elemental spot maps of the IF film shown in Fig. 4. The EDS maps have several notable features. The maps confirm that the spherical colloids decorating the IF film are silver; the IF film is an inorganic oxide as indicated by the oxygen map; the IF film contains Pb and Al ions. The Pb ions are from Pb in the starting frit. The presence of Al ions in the IF film is noteworthy. The source of Al ions in the IF film is from the metallic Al additions in the starting paste. Al ions dissolve into the liquid phase flux during the firing process. Al ions are ubiquitous in the IF film along the interface.

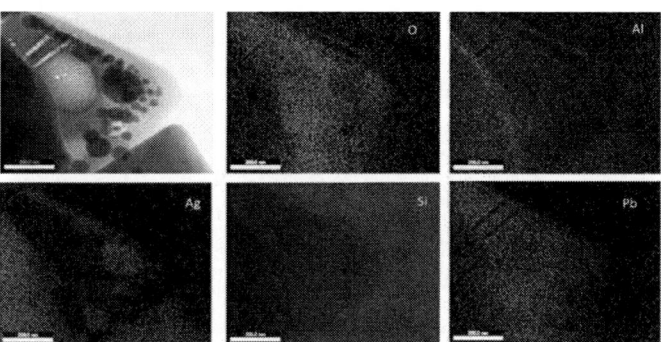

Fig 5. TEM/EDS elemental spot maps of IF film.

Previously, we discussed the redox chemistry for the SiN_x:H etch-through process in establishing contact during the firing process, and the role soluble metal ions dissolved in the liquid phase flux play in the oxidation and subsequent dissolution and removal (etch-through) of the SiN_x:H ARC layer [10]. ΔG calculations show metal ions such as Pb dissolved in the liquid phase flux have a high thermodynamic driving force (large negative ΔG) for the oxidation of SiN_x:H and subsequent dissolution of the SiN_x:H ARC layer [10].

Al ions dissolved in the liquid phase flux, however, have a low driving force for the oxidation and dissolution of the SiN_x:H layer. The ΔG value is positive as shown in equation 1; therefore, this reaction will not occur indicating Al ions in the flux do not participate in the removal of the SiN_xH ARC layer.

$$Si_3N_{(4-x)}H_{x \text{ (solid)}} + 2Al_2O_{3 \text{ (liquid flux)}} = 3SiO_{2 \text{ (liquid)}} + 4Al_{\text{(solid)}}$$
$$+ (2 - 0.5x)N_{2 \text{ (gas)}} + (0.5x)H_{2 \text{ (gas)}}$$
$$\Delta G = 215 \text{ kcal.} \quad (1)$$

Similarly, Al ions in the flux have a positive ΔG and low driving force for the reaction of Al ions with the Si emitter (once the SiN_x:H layer has been removed). Therefore this reaction will not occur.

$$3Si_{\text{(solid)}} + 2Al_2O_{3 \text{ (liquid flux)}} = 3SiO_{2 \text{ (liquid)}} + 4Al_{\text{(solid)}}$$
$$\Delta G = 125 \text{ kcal.} \quad (2)$$

The observation that Al ions are present in the liquid phase flux is not surprising since the paste contains metallic Al additions. But, the low driving force for reaction with both SiN_x:H and Si leaves Al's function in establishing contact uncertain. Al metal does, however, favor the formation of Ag metal colloids in the liquid phase flux as indicated by the large negative ΔG for this reaction as shown in equation 3, and the observation in Fig.2 of Ag colloids surrounding a residual Al metal particle.

$$4Al_{\text{(solid)}} + 6Ag_2O_{\text{(liquid flux)}} = 2Al_2O_{3 \text{ (solid)}} + 12Ag_{\text{(solid)}}$$
$$\Delta G = -695 \text{ kcal.} \quad (3)$$

This suggests that Al metal additions increase the number density of Ag colloids in the liquid phase flux and IF film. The microscopy images show many Ag colloids in the solidified liquid phase flux and IF film regions. The relatively thick and continuous IF films decorated with Ag colloids suggests electrical contact occurs by a tunneling mechanism through the Ag decorated IF film. An increase in Ag colloid density in the IF film supports an Ag colloid assisted tunnel contact mechanism [14]. The microcopy observations and ΔG calculations suggests the function of the Al metal additions in the paste is to increase the number density of the Ag colloids in the IF film which is necessary to reduce contact resistance in the relatively thick IF films.

Ag-Al and/or Ag-Si eutectic reaction products such as micro alloy spikes or alloy crystallites [12, 13] were not observed directly along the interface or in direct contact with the Si emitter surface. If these occur, they are sporadic and outside the cross-sectional viewing area, or they occur at firing temperatures different from this study or in non-optimally prepared solar cells. Ag crystallites were also not observed along the interface. Alloy spikes and crystallites along the interface increase Jo recombination and are, therefore, undesirable in optimally prepared solar cells.

The function of Al metal additions is to increase the number density of Ag colloids in the IF film in optimally prepared solar cells. We continue to develop and study both front and rear side screen-printed contacts on n-TOPCon solar cells.

IV. CONCLUSIONS

High resolutions STEM/TEM characterization of the FS contact of an industrial n-type TOPCon solar cell shows the Ag conductor metal-p^+ boron-diffused Si emitter contact region has relatively thick, continuous, amorphous IF films decorated with Ag colloids located between the bulk Ag conductor metal and silicon emitter surface. STEM/EDS analysis shows the IF films are oxide-based comprised of lead, aluminum, silicon and silver ions. ΔG calculations indicate Al ions dissolved in the IF film do not participate in the SiN_x:H etch-through process. Al metal, does, however, favor formation of Ag colloids in the liquid phase flux. The presence of relatively thick and continuous IF films decorated with Ag colloids suggests electrical contact occurs by a tunnel mechanism through the decorated IF film. The function of Al metal additions in the screen-printable paste is to increase the number density of the Ag colloids in the IF film which is necessary to reduce contact resistance in the relatively thick IF films of optimally prepared n-type TOPCon solar cells.

REFERENCES

[1] J. Benick, B. Hoex, M. C. M. Van De Sanden, W. M. M. Kessels, O. Schultz, and S. W. Glunz, "High efficiency n-type Si solar cells on Al_2O_3-passivated boron emitters," *Appl. Phys. Lett*, vol. 92, no. 25, 253504, 2008.

[2] A. Edler, "Development of bifacial n-type solar cells for industrial application," PhD dissertation, Univ. of Konstanz, Dec.3, 2014.

[3] J. Schmidt and K. Bothe, "Structure and transformation of the metastable boron- and oxygen-related defect center in crystalline silicon," *Phys. Rev. B*, vol.69, no. 2, 024107, Jan. 2004.

[4] F. Feldmann, M. Bivour, C. Reichel, H. Steinkemper, M. Hermle, S. W. Glunz, "Tunnel oxide passivated contacts as an alternative to partial rear contacts," *Sol. Energy Mater. Sol. Cells*, vol. 131, pp. 46–50, Dec. 2014.

[5] F. Feldmannn, M. Simon, M. Bivour, C. Reichel, M. Hermle, and S.W. Glunz, "Efficient carrier-selective p- and n-contacts for Si solar cells," *Sol. Energy Mater. and Sol. Cells*. vol. 131, pp. 100–104. Dec. 2014.

[6] D. Chen, Y. Chen, Z. Wang, et al., "24.58% total area efficiency of screen-printed, large area industrial silicon solar cells with the tunnel oxide passivated contacts (i-TOPCon) design", *Sol. Energy Mater. and Sol. Cells*, vol. 206, 110258, Mar. 2020.

[7] M. Lu, K.R. Mikeska, C. Ni, Y. Zhao., F. Chen, X. Xie, Y. Xu and C. Zhang, "Screen-Printable Conductor Metallizations for Industrial n-TOPCon Crystalline Silicon Solar Cells," in *Proc. 48th IEEE Photovoltaic Specialists (PVSC) Conference*, Miami, FL, USA, June 20-25, 2021.

[8] M. Lu, K.R. Mikeska, C. Ni, Y. Zhao., F. Chen, X. Xie, Y. Xu and C. Zhang, "Screen-Printable Contacts for Industrial *N*-TOPCon Crystalline Silicon Solar Cells." *IEEE Journal of Photovoltaics*, vol. 12, no. 2, pp. 469-473, March 2022.

[9] C.W. Bale, E. Belisle, P. Chartrand et.al., "FactSage Thermochemical Software and Databases, 2010-2016," *Calphad*, vol. 5, pp. 35-53, Sept. 2016.

[10] K.R. Mikeska, M. Lu, and W. Liao, "Tellurium-based screen-printable conductor metallizations for crystalline silicon solar cells", *Prog. Photovolt. Res. Appl.*, vol. 27, no. 12, pp. 1071- 1080, Aug. 30, 2019.

[11] A.F. Carrol, K.W. Hang, B.J. Laughlin, K.R. Mikeska, C. Torardi, and P.D. VerNooy, "Thick-film pastes containing lead- and tellurium-oxides, and their use in the manufacture of semiconductor devices," US Patent 8497420 B2, July 30, 2013.

[12] W. Wua, K.E. Roelofs, S. Subramoney, K. Lloyd, and L. Zhang, "Role of aluminum in silver paste contact to boron-doped silicon emitters", *AIP Advances*, vol. 7, 015306, Jan. 2017.

[13] L. Liang, Z.G. Li, L.K. Cheng, N. Takeda, and A.F. Carroll, "Microstructural characterization and current conduction mechanisms of front-side contact of n-type crystalline Si solar cells with Ag/Al pastes", *J. Appl. Phys.*, vol. 117, no. 21, pp 215102, March 2015.

[14] Z.R. Li, L. Liang, and L.K. Cheng, "Electron microscopy study of front-side Ag contact in crystalline Si solar cells", *J. Appl. Phys.,* vol. 105, no. 6, pp. 66102, March 2009.

Light-dark cycling in perovskite solar cells studied by MPPT and ion migration current measurements

Takeshi Tayagaki, Kohei Yamamoto, Takurou N. Murakami, Masahiro Yoshita

National Institute of Advanced Industrial Science and Technology (AIST), Tsukuba, Japan

Light-dark cycling is an important test protocol for understanding the degradation and metastability of perovskite solar cells. We first investigated light-dark cycling in a triple-cation mixed-halide lead perovskite solar cell using maximum power point tracking (MPPT) technique and ion migration current measurements and showed that the energy yield during 8 hours of illumination was reduced by light-dark cycles at 45 °C. In this case, the concentration of mobile ions, especially slow mobile ions, increased. Next, we investigated light-dark cycling with varying temperatures between 15 to 45 °C. The rate of power decrease under illumination increased with increasing temperature, where slow mobile ions dominated. Moreover, when the power recovered similar values to the initial ones after storage in the dark, slow and fast mobile ion concentrations decreased and increased, respectively, and the total concentrations remained almost unchanged. This indicates the conversion between the fast and slow mobile ions under illumination and dark storage. Our results indicate that the increase in the slow mobile ions, possibly due to deep trap centers, is the primary cause of power reduction for reversible change and irreversible degradation, which can address the long-term reliability of perovskite solar cells.

978-1-6654-6060-6/23 $31.00 © 2023 IEEE

Field Trial In Progress for Measuring Global, Direct, Diffuse, and Ground-Reflected Irradiance Using a Static Sensor Array

Michael Gostein[1], Bruce H. King[2]

[1]Atonometrics, Austin, USA; [2]Sandia National Laboratories, Albuquerque, USA

Abstract **We report on plans for a field trial now in progress to test measurement of global horizontal irradiance (GHI), direct normal irradiance (DNI), diffuse horizontal irradiance (DHI), and reflected horizontal irradiance (RHI) using an array of static sensors with no moving parts. As in our recent work in this area, the collection of static sensors includes reference cells in multiple orientations. In addition, our current system under test includes a modified reference cell with a collimation tube to admit only diffuse light contributions from a limited region of the sky. We are developing an analysis model to determine GHI, DNI, DHI, and RHI from the combined sensor data. Field trials have recently begun. Results will be published at a later date.**

Index Terms — **photovoltaic systems, resource assessment, pyranometer, reference cell**

I. INTRODUCTION

Key steps in financing, commissioning, and monitoring commercial solar photovoltaic (PV) power plants require measurements of solar irradiance to predict and assess plant performance. Solar irradiance has multiple components, including direct, diffuse, and ground-reflected light. As PV systems become more sophisticated, especially for bifacial and single-axis tracker systems, there is an increasing need for more sophisticated irradiance measurements that resolve all irradiance components.

The most complete assessments of solar irradiance include determination of beam, diffuse, and ground-reflected components – i.e., direct normal irradiance (DNI), diffuse horizontal irradiance (DHI), and reflected horizontal irradiance (RHI) – which allows determination of irradiance on any plane via transposition. However, due to the cost and maintenance requirements of specialized equipment for DNI and DHI measurement, which usually involves moving parts for sun tracking, these measurements are typically omitted from power plant meteorological stations.

Here we discuss a novel method to provide a complete irradiance assessment at lower cost. The method uses an array of static sensors to determine GHI, DNI, DHI, and RHI, with no moving parts. The method is similar to that used in previous work on multi-pyranometer arrays [1][2][3]. However, we use PV reference cells instead of pyranometers, we introduce a correction for incidence angle dependence of the sensors, and we explicitly measure and correct for ground-reflected irradiance instead of blocking it with a horizon blocking ring. Since our analysis method determines the beam component of irradiance, it allows the effect of the reference cells' incidence

angle response [4][5][6] to be automatically compensated, enabling accurate measurements of GHI with reference cells.

In previous work [7] we reported on results obtained with identical prototype reference cell arrays at two stations, in New Mexico and California. Outputs from the reference cell array analysis for DNI and DHI were compared with data from a tracking pyrheliometer and tracking diffusometer, respectively. Results showed good correlation between the reference cell array and tracking instruments over a wide range of DHI. However, for the most challenging conditions, corresponding to mid-day in clear sky with low diffuse irradiance, results from the reference cell array showed greater systematic error than desired.

In the current work we are making several modifications to the prototype and test environment to reduce systematic error.

II. EXPERIMENT

For this work the updated sensor array concept includes five sky-facing reference cells (Atonometrics RC22), with four cells facing nominally north, east, south, and west on a 35-degree tilt,

Fig. 1. Reference cell array prototype design with five sky-facing reference cells (center, north, east, south, west) and one ground-facing cell (on extension arm, not shown). The north-facing cell has a diffuse isolator tube that blocks direct and ground-reflected light.

978-1-6654-6060-6/23 $31.00 © 2023 IEEE

and a central cell at a 5-degree south-facing tilt, as shown in Fig. 1. The central cell is intended to be nominally horizontal, with a 5-degree tilt for water roll-off.

To provide better discrimination of direct and diffuse light during clear-sky conditions, the north-facing reference cell is covered with a diffuse isolator tube that blocks all contributions from direct and ground-reflected light. This cell's internal shunt resistance value was modified to provide proper sensitivity at the very low signal levels resulting from the reduced angular acceptance defined by the isolator tube.

The prototype also includes a sixth ground-facing reference cell (not shown) mounted on an extension arm for measuring reflected horizontal irradiance (RHI) and albedo (RHI / GHI). In our previous study, the ground-facing reference cell was located somewhat remotely from the sky-facing cells and viewed a different environment, which may have contributed to systematic error. In the current work, we have mounted the ground-facing reference cell near the sky-facing cells and have positioned the entire assembly in an area of uniform albedo to improve consistency of the readings.

We deployed the prototype recently, in May 2023 at Sandia National Laboratories in Albuquerque, New Mexico.

The site includes a research-grade two-axis sun-tracking system (Kipp & Zonen, SOLYS) with a pyrheliometer and a shaded pyranometer for DNI and DHI measurement, as well as separate pyranometers for GHI measurement. In addition, the site includes multiple PV arrays with plane-of-array (POA) irradiance measurements. These reference instruments will provide comparisons for assessing the prototype sensor array.

IV. ASSESSMENT METHODS

Our ultimate objective is to provide a low-cost instrument that can be used for improved performance analysis of monofacial and bifacial PV systems. In the absence of DHI measurement instrumentation at typical PV power plants, performance analysis based on GHI measurements typically relies on GHI decomposition models [8] to determine DNI and DHI for estimating POA irradiance. We wish to determine whether the low-cost static sensor array can provide measurements of DNI and DHI of sufficient quality to provide improved performance analysis of PV systems versus what can be achieved with GHI measurements and decomposition alone.

Therefore, as part of our assessment of the prototype sensor array, we plan to use the DNI, DHI, and RHI measurements from the sensor array in performance models of monofacial and bifacial PV arrays at the site, including fixed and tracking systems, and compare these results with those obtained from GHI decomposition, to determine if the sensor array provides practical utility for PV system performance modeling.

IV. RESULTS

Field measurements have been underway for only a few weeks. We will present some preliminary results in our conference poster. Full results will be presented elsewhere.

ACKNOWLEDGEMENT

This material is based upon work supported by the U.S. Department of Energy's Solar Energy Technologies Office under Award Number DE-SC0020831.

Sandia National Laboratories is a multimission laboratory managed and operated by National Technology and Engineering Solutions of Sandia, LLC, a wholly owned subsidiary of Honeywell International Inc., for the U.S. Department of Energy's National Nuclear Security Administration under contract DE-NA0003525.

REFERENCES

[1] D. Faiman, D. Feuermann, and A. Zemel, "Site-independent algorithm for obtaining the direct beam insolation from a multipyranometer instrument," *Sol. Energy*, vol. 50, no. 1, pp. 53–57, Jan. 1993, doi: 10.1016/0038-092X(93)90007-B.

[2] B. Marion, "Multi-Pyranometer Array Design and Performance Summary," *Proc. 1998 Am. Sol. Energy Annu. Conf.*, 1998.

[3] J. C. Baltazar, Y. Sun, and J. Haberl, "Improved methodology to evaluate clear-sky direct normal irradiance with a multi-pyranometer array," *Sol. Energy*, vol. 121, pp. 123–130, Nov. 2015, doi: 10.1016/j.solener.2015.07.015.

[4] N. Martin and J. M. Ruiz, "Calculation of the PV modules angular losses under field conditions by means of an analytical model," *Sol. Energy Mater. Sol. Cells*, vol. 70, no. 1, pp. 25–38, Dec. 2001, doi: 10.1016/S0927-0248(00)00408-6.

[5] B. H. King and C. D. Robinson, "Differential Analysis of the Angle of Incidence Response of Utility-Grade PV Modules," in *2019 IEEE 46th Photovoltaic Specialists Conference (PVSC)*, Jun. 2019, pp. 77–81, doi: 10.1109/PVSC40753.2019.8981355.

[6] B. H. King, D. Riley, C. D. Robinson, and L. Pratt, "Recent advancements in outdoor measurement techniques for angle of incidence effects," *2015 IEEE 42nd Photovolt. Spec. Conf. PVSC 2015*, Dec. 2015, doi: 10.1109/PVSC.2015.7355849.

[7] M. Gostein, A. Hoffman, B. H. King, and A. Marquis, "Measuring Global, Direct, Diffuse, and Ground-Reflected Irradiance Using a Reference Cell Array," in *2022 IEEE 49th Photovoltaics Specialists Conference (PVSC)*, Nov. 2022, pp. 0285–0290, doi: 10.1109/PVSC48317.2022.9938489.

[8] M. Lave, W. Hayes, A. Pohl, and C. W. Hansen, "Evaluation of global horizontal irradiance to plane-of-array irradiance models at locations across the United States," *IEEE J. Photovoltaics*, vol. 5, no. 2, pp. 597–606, Mar. 2015, doi: 10.1109/JPHOTOV.2015.2392938.

978-1-6654-6060-6/23 $31.00 © 2023 IEEE

The proposition of a public policy to stimulate low-income communities' assess to distributed energy resources

Anna Carolina de Paula Sermarini[2], João Henrique Paulino Azevedo[1,2], Vanessa Cardoso de Albuquerque[2], Rodrigo Flora Calili[2], Felipe Gonçalves[1], Gilberto Jannuzzi[3]

[1]Getúlio Vargas Foundation, Rio de Janeiro/RJ, 22250-040, [2]Pontifical Catholic University of Rio de Janeiro, Rio de Janeiro/RJ, 22451-900, [3]University of Campinas, Campinas/SP, 13083-970

Abstract — The Solar Photovoltaic Distributed Generation (PVDG) has proven to be an essential strategy for achieving the goals of the 2030 Agenda, especially SDG 7 and SDG 11. However, we note that most projects involving this technology are restricted to the most favored society classes, not yet reaching the low-income population, which contradicts the theories of energy justice. Thus, the objective of this study is to assess the economic feasibility of a policy of replacing the social electricity tariff (the current policy aimed to subsidize electricity tariff for low-income population in Brazil) by PVDG projects. In this sense, a specific methodology was developed, carrying out an analysis in the Brazilian context based on the proposal to replace the Social Electricity Tariff (TSEE) by PVDG, providing this way an alternative exit strategy to the subsidy scheme for low-income households.

I. INTRODUCTION

In September 2015, 193 United Nations (UN) Member of States established 17 Sustainable Development Goals (SDGs), organized in the 2030 Agenda, with the mission of improving people's quality of life, promoting sustainable, inclusive and fair development. To achieve the SDGs, in particular, SDGs 7 - Clean and affordable energy - and SDGs 11 - Sustainable Cities and Communities, the advent of renewable energy sources, such as photovoltaic solar energy, is an essential contribution to meeting the Agenda [1].

However, the low-income population is even less likely to use photovoltaic solar systems than families with higher purchasing power [2][3][4]. According to [1][5], the adoption of these systems by low-income families is an important political objective for achieving energy justice, in line with SDG 10 - Reducing inequalities. Although, for this objective to be achieved, incentives are needed and a regulation supports it.

Thus, to keep up with the progress of renewable energy sources, regulation of the Brazilian electricity sector has undergone significant changes in recent years, and together with theories of economic and social regulation, such as energy justice, energy democracy [6], and distributed economy [7], there has been a pursuit of equity in the adoption of solar energy and the mitigation of energy poverty [1][5][8].

Currently, the so-called Social Electricity Tariff (TSEE) is regulated in Brazil, which, in general terms, aims to guarantee access to electricity to the low-income residential population through regressive discounts on the energy tariff according to consumption. However, it is believed that the simple granting of such a benefit may not be the most appropriate and it is possible to reconcile it with the application of technologies in full expansion, such as distributed energy resources (DER), more precisely with photovoltaic solar energy, in line with the referred SDGs. In this context, Brazilian Law 14.300/2022 created the legal framework for distributed micro-generation and mini-generation, the Electricity Compensation System (SCEE) and the Social Renewable Energy Program (PERS), the latter focusing on the democratization of DER.

Since there is a gap in methodologies that demonstrate the benefit of using DER to expand social justice, the primary purpose of this study is to assess the economic feasibility of the policy of replacing the social electricity tariff by Photovoltaic Solar Distributed Generation (PVDG). To this end, a specific methodology was developed, and an analysis was carried out in the Brazilian context, based on the proposal for the replacement of the TSEE, the current policy aimed at the low-income population, by PVDG.

II. REGULATORY ASPECTS

Law 10.438/2002 instituted the Social Electricity Tariff program in Brazil, and currently, the TSEE is regulated by Law 12.212/2010 and by Decree 7.583/2011. Through it, discount ranges are implemented in energy tariffs that allow servicing families in the low-income residential subclass registered in the Unified Registration for Social Programs of the Federal Government (CadÚnico), a database used by several Brazilian social programs and policies. TSEE beneficiaries are exempt from the Energy Development Account (CDE) cost and from the cost of the Incentive Program for Alternative Electricity Sources (PROINFA), which are passed on to other regulated consumers.

DERs are defined as electric energy generation and/or storage technologies located within the limits of the area of a particular distribution concessionaire, usually next to consumer units and include Distributed Generation (DG). In Brazil, DG is regulated by the Normative Resolutions of the Regulatory Agency (ANEEL) 482/2012, 687/2015, and Law 14.300/2022. According to them, the energy injected into the grid must be used to fully deduct the energy consumed, with all tariff components.

Despite the potential and wide dissemination of DG in the country, especially with the photovoltaic solar source, it is observed that most small-scale projects are restricted to the most favored classes due to the high costs of implementing the systems, as clarified by [4]. From this perspective, this work sought to formulate guidelines for a policy that makes it possible to better meet the energy demand of the low-income population through sustainable business models. This policy is based on replacing the TSEE subsidy with photovoltaic solar DG, which was analyzed in the various Brazilian municipalities.

III. METHODOLOGY

This section presents the methodology used. Initially, information was collected regarding the diagnosis of electricity use by low-income families in the municipalities, divided into energy consumption ranges, an indicator related to the benefit of the TSEE. After mapping the target audience, some stakeholders were consulted to identify the variables that would be modeled in the economic-financial analysis of a possible policy of replacing the TSEE subsidy by installing photovoltaic DG, which are presented below.

A. *Dimension Regulatory*
- Compensation System Rules
- TSEE rules

B. *Socio-environmental*
- Irradiation by location
- Selected Beneficiaries

C. *Technical (or technological)*
- DG modality
- System components
- Project load

D. *Economic/financial*
- Tariff adjustment rate
- Taxes
- Capital Remuneration
- CAPEX
- Costs - O&M

To carry out the economic-financial analysis of photovoltaic solar energy projects, was applied simulations based on the NREL methodology [9] and the RETScreen software [10].

IV. RESULTS AND DISCUSSIONS

In this section, the results of the simulations of the economic analyses applied to the proposed study will be presented. The results will be presented for the two possible PVDG's modalities in the country: shared power plant and consumers behind the meter. The municipalities are represented by five different colors, according to the economic feasibility of the proposed policy. They are based on the difference between two kinds of NPV (net present value), NPV_{TSEE} (corresponding to the adoption of the TSEE by the user) and the NPV_{DG} (corresponding to the application of the benefits of the DG to the user). The explanation for each color is given below: (i) The green color represents the municipalities in which TSEE beneficiaries would benefit from the proposed policy, also

resulting in a subsidy reduction for other consumers connected to the grid (consumer benefit, lower subsidy); (ii) The blue color indicates the municipalities in which the TSEE beneficiaries would benefit, but with an increase in the subsidy to other consumers (consumer benefit, more significant subsidy); (iii) The yellow color corresponds to the municipalities in which the TSEE beneficiaries would have a lower benefit from the policy, but there would be a reduction in the subsidy for other consumers (no consumer benefit, lower subsidy); (iv) The red color represents the municipalities in which the TSEE beneficiaries would not benefit from the policy, and the other consumers would not receive a subsidy reduction (no consumer benefit, more significant subsidy); (v) The municipalities in black are those that do not have TSEE beneficiaries (Municipality without TSEE consumers).

Beneficiaries of the TSEE are considered to be benefiting when the value of the electricity bill paid by them is, on average, lower with the proposed policy than with the subsidy provided by the social electricity tariff.

Based on assumptions adopted in the simulations for each variable in Table 1 and considering the information about NPV, the following maps were prepared.

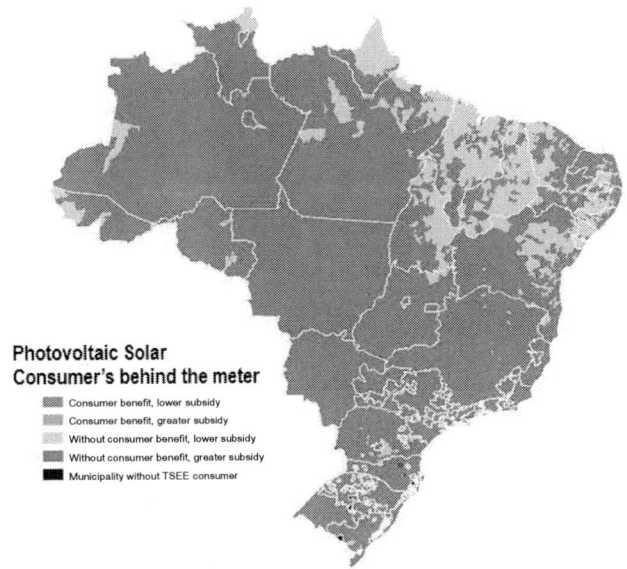

Fig. 1. Result of the economic feasibility analysis for the consumer's behind the meter modality

In the consumer's behind the meter modality, most municipalities are represented in green, as shown in Figure 1. The proposed policy presents a more significant benefit to TSEE consumers and a lower subsidy charge for other consumers connected to the grid. It is also verified that some municipalities are represented in yellow, in which the proposed policy does not offer benefits to the TSEE consumer. However, there is a lower subsidy for the other consumers.

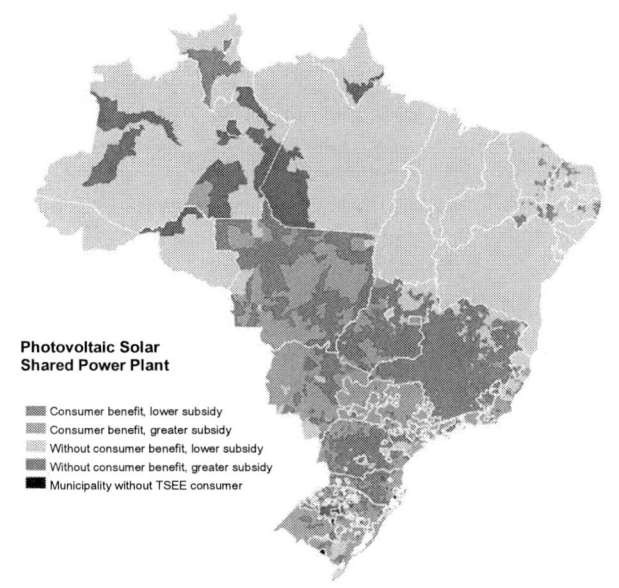

**Photovoltaic Solar
Shared Power Plant**

▨ Consumer benefit, lower subsidy
▨ Consumer benefit, greater subsidy
▨ Without consumer benefit, lower subsidy
▨ Without consumer benefit, greater subsidy
■ Municipality without TSEE consumer

Fig. 2.   Result of the economic feasibility analysis for the shared plant

Regarding the shared power plant modality (Figure 2), it is observed that, in most municipalities, consumers benefiting from the TSEE are not benefited by the policy, expressed by the yellow color. However, it is possible to verify that TSEE beneficiaries have more significant benefit from the policy (green color), mainly in the Midwest, Southeast, and South regions of the country.

Comparing the maps in Figures 1 and 2, it is verified that the consumer's behind the meter modality presents a more significant number of municipalities in green color, that is, with a benefit for consumers of TSEE and lower subsidy burden for other consumers connected to the grid, so that this modality has proven to be more adequate.

Based on the results obtained, an attempt was made to analyze which parameters could impact and even make the proposed policy unfeasible. Consumption was one of these observed parameters, and the greater the consumption of electricity by low-income families who opt for the proposed policy, the greater the cost for other consumers connected to the network. Another parameter was the impact of the cost of using the distribution network (TUSD, acronym in Portuguese), related to the distance between the distribution networks and the consumer units. Thus, economic analyzes on the use of DER by the low-income population show that there is a distinction between the impacts that the proposed policy can cause to each municipality. When considering a large country like Brazil, public policies must consider the differences of each location to be efficient in mitigating energy inequality.

V. FINAL CONSIDERATIONS

In the quest for equity in the use of distributed generation systems, economic feasibility analyses were carried out with a focus on demonstrating the effectiveness of a policy in which resources from the social electricity tariff, in particular the Social Electricity Tariff (TSEE, acronym in Portuguese), are applied in a Solar Photovoltaic Distributed Generation project aimed at the low-income population. The study found that, in some regions, the maintenance of the TSEE presents more economic benefits than its replacement by PVDG, although the proposed policy presents many opportunities. The possibility of regionalizing a subsidy/charge can be an opportunity to carry out specific energy planning, according to local potential.

ACKNOWLEDGMENT

The authors acknowledge support from the UK government under the Brazil Energy Programme (2019-2022), from FGV Energia.

REFERENCES

[1] MILČIUVIENĖ, S.; KIRŠIENĖ, J.; DOHEIJO, E.; URBONAS, R.; MILCIUS, D. The Role of Renewable Energy Prosumers in Implementing Energy Justice Theory. **Sustainability**, 11, 5286, 2019.

[2] O'SHAUGHNESSY, E., BARBOSE, G., WISER, R. ET AL. The impact of policies and business models on income equity in rooftop solar adoption. **Nat Energy**, v. 6, p. 84–91, 2021.

[3] MACIEL, L. S. B.; BONATTO, B. D.; ARANGO, H.; ARANGO, L. G. Evaluating Public Policies for Fair Social Tariffs of Electricity in Brazil by Using an Economic Market Model. **Energies**, v. 13, article number 4811, 2020.

[4] MONTUENGA, E. C.; WEISS, M.; CELAYA, R.; RAVILLARD, P.; TOLMASQUIM, M.; HALLACK, M. **Shedding light on the unequal distribution of residential sola PV adoption in LAC.** Cataloging-in-Publication data provided by the Inter-American Development Bank. Available at: < https://bityli.com/Jjisl>. Accessed on:March 30, 2022.

[5] SIGRIN, B.; SEKAR, A.; TOME, E. The solar influencer next door: Predicting low income solar referrals and leads. **Energy Research & Social Science**, v. 86, 102417, ISSN 2214-6296, 2022.

[6] BURKE, M.J; STEPHENS, J.C. Energy democracy: Goals and policy instruments for sociotechnical transitions. **Energy Research & Social Science**, v. 33, p. 35-48, 2214-6296, 2017.

[7] JOHANSSON, A.; KISCH, P.; MIRATA, M. Distributed economies – A new engine for innovation. **Journal of Cleaner Production**, v. 13, Issues 10–11, p. 971-979, 0959-6526, 2005.

[8] DAY, R.; WALKER, G.; SIMCOCK, N. Conceptualising energy use and energy poverty using a capabilities framework. **Energy Policy**, v. 93, p. 255-264, 0301-4215, 2016.

[9] SHORT, W.; PACKEY, D.; HOLT, T. A manual for the economic evaluation of energy efficiency and renewable energy technologies. **Renewable Energy**, v. 95, n. March, p. 73–81, 1995.

[10] NATURAL RESOURCES CANADA. Clean Energy Project Analysis. [s.l: s.n.].

PV Plant Performance Review Methodology: Key Performance Indicators (KPI) Estimation

Himanshu Gulati, Prashant Kumar Upadhyay, and Yellasiri Bharath Kumar Reddy

Solar Energy Corporation of India Limited, New Delhi, India

Abstract— **The detailed procedure to estimate two key performance indicators (KPIs) of Solar PV power plant i.e., Performance Ratio (PR) & Capacity Utilization Factor (CUF) using statistical methods has been presented. Calculation of PR and CUF by simply using the standard formulas results in wide inaccuracies in the results. Comprehensive methodology to deal with large data sets has been set down to estimate these key performance indicators with more accuracy.**

Keywords—performance ratio, confidence interval, data profiling, capacity utilization factor, radiation correction factor

I. INTRODUCTION

Ever increasing demand for energy, dwindling fossil fuels and environmental concerns have led to deployment of renewable energy projects throughout the globe for meeting the growing energy demands. International Energy Alliance (IEA), estimates the supply of renewable energy is expected to grow at a rate of 13% per year from 2022 to 2030 to meet the net zero requirements of India. [1]. This rapid growth is the result of several factors viz. competitiveness in the market, better policy support by the government, and investors' confidence on the projects.

The return on investment of the solar projects directly relies on the performance of the projects in line with the estimated values before construction. This requires a comprehensive study on the performance monitoring methodology of the PV plant. As presented in the report by SERENDI-PV [2], there are several factors which affect the PV energy production viz., location of the plant, technology of the PV module, solar radiation, weather, system architecture, system availability & its operation & maintenance. To assess & quantify the above factors, the following parameters of the projects are to be continuously monitored:

1. Performance Ratio (PR)

2. Capacity Utilization Factor (CUF)

Performance Ratio (PR) is the quality indicator of the plant, which assesses how effectively the plant has converted solar radiation into electrical energy & Capacity Utilization Factor (CUF) is the quantity indicator, which assesses the targeted quantum of the energy produced.

Continuous measurement and monitoring of the plants is key to ensure the intended generation from the projects. However various uncertainties in measurement of data needs to be accounted for to arrive at the actual performance of the plant.

A detailed methodology to estimate these two parameters has been laid down in the present study.

II. ESTIMATION OF PERFORMANCE RATIO (PR)

A. Definition

To assess the reliability & the quality of the PV plant, Performance Ratio (PR) is considered as the most prominent indicator. As per IEC 61724-1, it is defined as the ratio between the final energy yield (Y_f) & the reference yield (Y_r) [3]. To estimate the PR in the practical conditions, where, temperature variations are there, it is calculated for a particular reporting period as follows:

$$PR_{Temp.\ Corrected} = \frac{Y_f}{\sum Y_r * C_k * t_k} \qquad (1)$$

$$Y_f = \frac{E_{out}}{P_0}$$

$$Y_r = \frac{G_{i,k}}{G_{ref,k}}$$

Where,

E_{out} is the cumulative AC energy measured at the Plant End (ABT meter) over the duration of reporting period (kWh)

P_0 is the installed nominal peak power of PV modules at STC

$G_{i,k}$ is the average irradiance measured at the Plane of Array (POA) at the commencement of time interval t_k (kW/m^2) of the reporting period (average of all Pyranometers installed)

t_k is the duration of k_{th} recording interval of reporting period

$G_{ref,k}$ is the reference irradiance at which P_0 is determined

$C_k = 1 + \gamma\ (T_{avg,k} - T_{ref})$, where, γ is the temperature coefficient, $T_{avg,k}$ is the average module temperature measured the commencement of time interval t_k & T_{ref} is the reference module temperature at which P_0 is determined

B. Methodolgy

A Performance Ratio (PR) test is commenced within 60 days of the commissioning of plant facilities to demonstrate that the plant has achieved the guaranteed Performance Ratio after completion of minor activities in the plant. This is one of the pre-conditions for operational acceptance of the plant.

978-1-6654-6060-6/23 $31.00 © 2023 IEEE

The PR test is carried out for a period of 30 days. The pyranometers & tilt angle are verified before the test commences and then visually inspected at regular intervals for cleanliness during the tests.

PR Test Procedure:

i. Data Collection:

  a. Plant irradiance data is collected from the installed pyranometers with a temporal resolution of 1 minute.

  b. Weather data is collected from the weather monitoring station (WMS) installed at plant facility with a temporal resolution of 1 minute

  c. Energy injected into the grid is cumulated at the plant end meter with a temporal resolution of 15 minutes (Energy reporting time).

In spite of detailed procedure established for data collection, the data received from the plant is having various data errors due to factors such as latency in communication, signal loss etc.

ii. Data Profiling:

  a. Missing GHI/POA data is removed. If there are more than five 1-minute time bocks missing in a 15-minunte block, the block is discarded.

  b. Data is filtered from nuisance or bad data which has high degree of error due to either calibration of the sensor or bad weather conditions.

  c. Grid interruptions/shutdowns are removed from the data.

  d. Data is filtered as per minimum POA radiation of $200 \ W/m^2$ to eradicate low radiation recording errors.

Minimum 24 Nos of 15-minute time blocks are considered to account in a day for PR measurement.

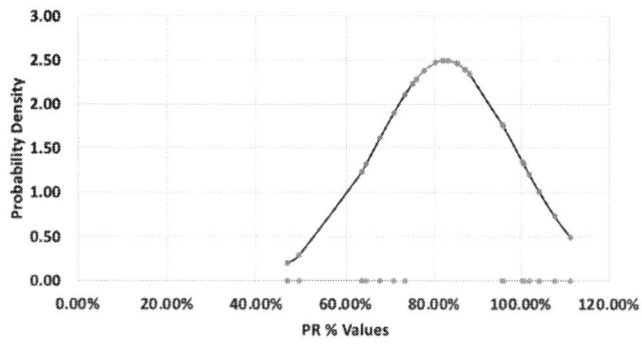

Fig. 1. Normal Distribution of PR values with confidence interval of 99%

iii. PR Estimation:

  a. Since, generation data (plant end meter) has a temporal resolution of 15 minutes, valid weather & radiation data are integrated for 15-minute time blocks.

  b. PR is calculated for each reporting period of 15 minutes using the equation (1).

  c. Considering the calculated PR values of 30 days (minimum 24 15-minute time blocks each day) as random variables derived from the radiation data, a confidence interval of 99% is created on the normal distribution as shown in Fig.1.

  d. Finally, PR is estimated using the average value in the 99% confidence interval. Depiction of a particular sample day PR is shown in Fig.2.

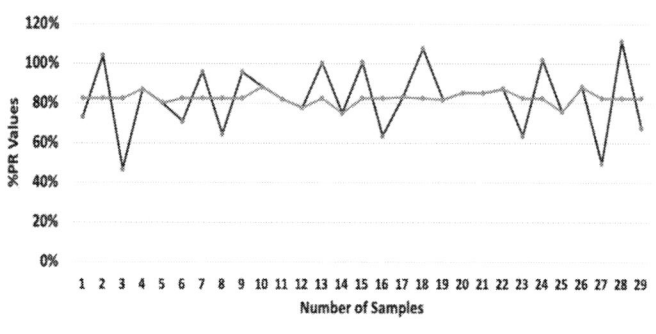

Fig. 2. Sample day PR values with confidence interval of 99%

III. ASSESSMENT OF ENERGY YIELD: CAPACITY UTILISATION FACTOR (CUF)

A. Definition

To ascertain the economic viability of the project, the quantum of energy produced is the most important parameter. Capacity utilization factor (CUF) is the ratio between the annual produced energy & the energy that would be produced by the plant operating the whole day at its maximum power output. [2] In the practice, factors such as plant outage hours, plant degradation & radiation correction factor (RCF) are also included as follows:

$$CUF_{RCF \ Corrected} = \frac{E_{ac} + E_{Outage}}{8760 * P_{ac} * \left(1 - D_f * (N-1)\right) * RCF} \quad (2)$$

Where,

E_{ac} is the cumulative annual AC energy measured at the Plant End ABT meter (kWh)

E_{Outage} is the total energy estimated during grid outage hours (kWh)

P_{ac} is the plant AC capacity (kW)

D_f is the module degradation factor

N is the number of years of operation after the operation acceptance

$RCF = \frac{Radiation \ measured \ at \ site}{Reference \ radiation \ during \ plant \ design}$, is the radiation correction factor to account radiation changes as per plant design

Since, plant estimated CUF (at the time of design) differs from the calculated CUF due to variation in radiation, CUF is corrected using radiation correction factor (*RCF*).

B. Methodology

CUF is estimated on the annual basis from the date of operational acceptance.

CUF Calculation Procedure:

i. Data Collection

    a. Plant irradiance data (GHI & POA) is collected from the pyranometers installed at the site with a temporal resolution of 1 minute.

    b. Reference irradiance data is taken from the standard data sources such as Meteonorm & SolarGIS [4].

    c. Energy injected into the grid is cumulated at the plant end ABT meter with a temporal resolution of 15 minutes (Energy reporting time).

The radiation data received from the plants have various missing values and is found to have high variations with respect to the reference values (radiation databases Meteonorm, Solar GIS).

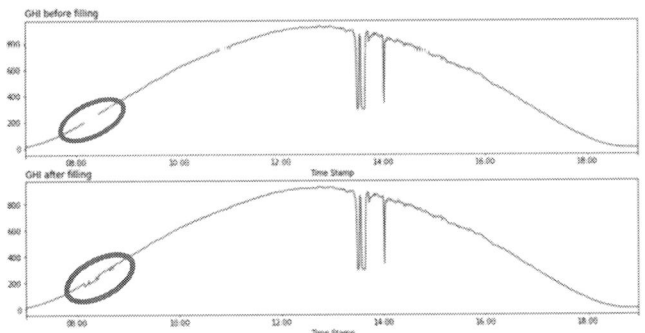

Fig. 3. Interpolated GHI values for a particular sample day

ii. Data Profiling

    a. Missing GHI data is interpolated with average values of preceding & succeeding GHI values as shown in Fig.3.

    b. Grid outages are rectified by estimating the energy output during the outage hours. Estimated E_{Outage} for a 15 minutes time block is the average value of preceding & succeeding energy values.

    c. Interpolated GHI values are compared with reference GHI source. SolarGIS database is found to be have better correlation with the measured data and is taken as the reference GHI for calculation of RCF. To validate the same, GHI data of the nearest projects are also referred.

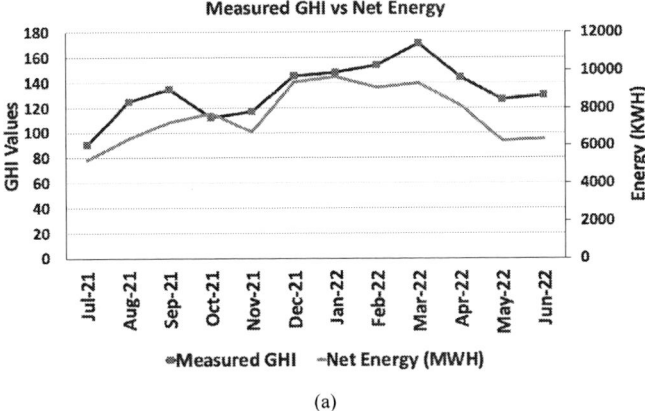

(a)

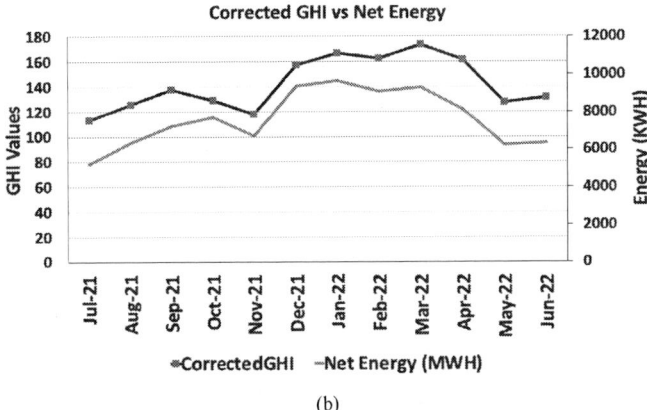

(b)

Fig. 4.   (a) Measured GHI at site Vs Net energy measured (b) Corrected GHI Vs Net energy measured

iii. CUF Estimation

    a. A correlation check is carried between GHI & energy values on a monthly basis. Correlation coefficient, $r \geq 0.9$ is considered as satisfactory.

    b. Monthly values of a sample measured GHI & corrected GHI vs Net energy curve is shown in Fig.4. The gap between the corrected GHI & energy can be explained by the high temperature during summer season which subsides during winter season. This needs to be further investigated.

    c. The degradation factor from the module datasheet, shall also be accounted for CUF estimation using the equation (2).

IV. Conclusion

The uncertainties in data collection for estimating PR and CUF of the solar plants are addressed using statistical methods. Missing data and low radiation data are not considered for PR measurement. Normal distribution of the calculated PR values data points is plotted and the confidence interval data of 99% is considered for estimation of final PR. Also, radiation correction factor (RCF) is computed for the deviation of the CUF from the estimated values.

REFERENCES

[1] IEA (2022), Renewables, IEA, Paris https://www.iea.org/reports/renewables, License: CC BY 4.0

[2] SERENDIPV, "Key Performance Indicators on State of the Art PV Reliability, Perfromance, Profitability and Grid Intigration," July 15, 2022. [Online]. Available: https://serendipv.eu/outputs/

[3] *Photovoltaic system performance-Part 1:Monitoring*, IEEE 61724-1,2021

[4] *Solar Resource Maps & GIS Data*, SolarGIS. [Online]. Available: https://solargis.com/maps-and-gis-dat

Validation of Inverter Labeling with Plant Transfer Functions

Joseph Ranalli

Penn State Hazleton, Hazleton, PA, 18202, USA

Abstract — The large quantity of data sources found within a utility scale photovoltaic plant presents data quality control challenges. One potential issue is mislabeling of the plant's component outputs (e.g. production measurements made at the combiner or inverter level). If a component's output is incorrectly labeled, it presents an obstacle to plant monitoring and maintenance, as operators will not know where fixes are needed. This study aims to demonstrate the possibility of utilizing the Cloud Advection Model to perform quality checks on the labeling of production outputs at a plant component level based on information about the plant's spatial layout. Results utilizing simulated data showed that the plant transfer function predicted by the CAM could provide discrimination between plant segments that are separated in the cloud motion direction. The discrimination occurred primarily through the phase of the transfer function, but in cases where the spatial dispersion of the plant varied significantly in the cloud motion direction, changes to the transfer function bandwidth were also observable. This methodology shows promise using the simulated plant data in this study, which warrants further study and practical validation of this method utilizing real plant data.

I. INTRODUCTION

Efficient operation of utility scale photovoltaic (PV) plants requires the collection and handling of extremely large quantities of generation data that can be used to monitor the performance of the plant. These data can be used to assess plant component failures, damage, or other contingencies that impact the plant's overall generation, and thus, the economic return on the investment. Due to its importance, plant operators are regularly concerned about quality control for this data. One potential data quality issue is verification of correct labeling for all plant data sources. Specifically, the possibility exists that mistakes could be made during initial construction when labeling specific strings/combiners/inverters within the plant. Due to the large size of utility scale plants and the large number of labeled entities, performing audits would be a time consuming and cost intensive process that would require manual inspection of each component and its connections to the plant's data acquisition system. The present study simulates the use of variability models for performing a validation of data source location within an overall plant in an effort to provide an analytical method that would avoid this expense. The approach relies on the individual time series of plant sub-segment generation measurements and knowledge of the plant's layout.

II. METHODOLOGY

Since this paper only aims to demonstrate the feasibility of the concept, a distributed irradiance dataset was used to simulate the output of an actual plant, rather than utilizing real plant generation data. Different groupings of the individual sensors from the irradiance measurement network were used to simulate the various inverter segments of the plant. The previously described Cloud Advection Model (CAM) [1] was used to predict the output for each of the plant subsections based on its transfer function relative to a reference irradiance measurement. Comparing these predictions with the spatially aggregated irradiance measurement data allows interpretation to be made as to whether the segments were correctly labeled.

A. Cloud Advection Model

The Cloud Advection Model (CAM) was first proposed as a method to represent spatial aggregation of irradiance by a spatially distributed plant [1], [2] and served a similar purpose to the well-known Wavelet Variability Model (WVM) [3]. Where the WVM was derived to represent the effects of the aggregation process matched to long-term trends in variability, the CAM was shown to better represent aggregation in detail on short timescales when cloud advection dominates the variability [1].

The CAM models frozen advection of clouds over a hypothetical distributed plant by representing the plant as a transfer function with a low-pass filter characteristic. This transfer function has the effect of smoothing the irradiance time series. The transfer function defined by the CAM relates the frequency domain representation of a single reference point's irradiance time series, $G_{ref}(f)$, to the plant's aggregate irradiance, $P(f)$. Due to the convolutional nature of the frozen advection phenomenon, the form of this transfer function can be analytically derived and is written as the Fourier transform of the plant's 1-D spatial distribution, d^*, as in (1). For plants that are distributed over a two-dimensional area, the plant's spatial distribution can be projected into a 1-D form along the cloud motion vector, albeit with some degradation to the model's effectiveness [1].

$$ TF(f) = \frac{P(f)}{G_{ref}(f)} = \mathcal{F}\left[d^*\left(\frac{x}{V_c}\right) \right] \qquad (1) $$

B. Irradiance Data

The irradiance dataset used was that from the HOPE-Melpitz campaign [4]. The campaign utilized 50 distributed sensors that measured irradiance with a 1 second temporal resolution. Sensors were arranged into various groups in order to simulate the effects of the inverter or combiner arrangements within a hypothetical plant. Cloud advection speeds, V_c, are required as an input by the CAM, and were identified from the irradiance

978-1-6654-6060-6/23 $31.00 © 2023 IEEE

measurements using the method of Jamaly and Kleissl [5] on the entire field. The spatial distributions of the plant segments were computed by projecting the individual sensor positions onto the cloud motion vector as described previously.

III. RESULTS AND DISCUSSION

The example calculation conducted in this study shows how this method might be applied for the purpose of verifying plant segment positions. Plant segments were defined within the irradiance measurement network, and transfer function predictions were computed using the CAM, relative to a common reference point. An example of selected plant component segments is shown in Fig. 1. The cloud motion vector for this plant was in the south-to-north direction, shown by the arrow in the figure. The aggregate outputs from the plant segments (shown on colored lines in Fig. 2) are similar and would be difficult to distinguish in the time domain. However, when computing transfer functions, differences arise that can be used to discriminate which segment is which. As seen in Fig. 3, the blue segment is observed to exhibit a rising phase, while green and blue both show falling phase. This results from the fact that blue is to the south of the reference point, causing its phase to lead the input. Conversely, the red and green signals are located north of the reference point and consequently lag the reference in time. The phase trends predicted by the CAM predictions match those found in the real data reasonably while coherence remains high, but fail to agree at higher frequencies.

Unlike the blue segment, the red and green segments in Fig. 1 are co-located with respect to the south-to-north cloud motion direction, and thus, have the same predicted phase behavior in Fig. 3. This prevents them from being differentiated on this basis. Attempts to fully map the plant therefore require investigation of multiple cloud motion directions that would induce spatial separation between the segments along the cloud motion vector. For example, using the same plant segments described in Fig. 1, but instead focusing on a time period where the clouds move from west-to-east, different transfer functions behavior can be obtained, shown in Fig. 4. In this case the red segment exhibits leading delay in the phase (due to its position west of the reference), while blue and green have indistinguishable lagging phase due to their similar eastward position.

Unlike the phase, the magnitude does not provide a clear discriminator between the segments for the previous examples. In part, this is due to the fact that when segments have the same projected size and shape in the cloud motion direction, the CAM's physical basis would lead to prediction of identical transfer function magnitude shapes. In this case, all plant segments described in Fig. 1 have similar spatial distributions and extents relative to the cloud motion directions tested, so it is unsurprising that their predicted and measured transfer function magnitudes are similar. On the other hand, for plant segments with different spatial extents in the cloud motion

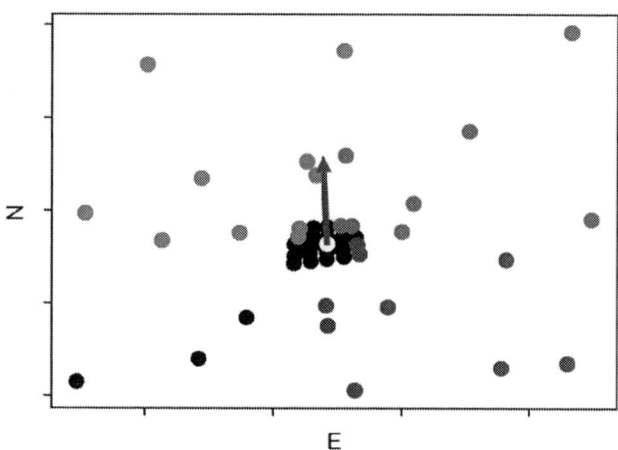

Fig. 1. Example plant layout. Yellow dot shows reference irradiance station. Red, Blue and Green sets of dots represent the two plant segments. Purple arrow shows south-to-north cloud motion direction.

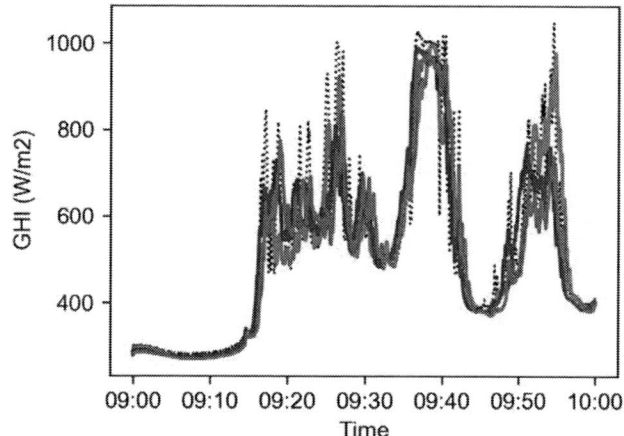

Fig. 2. Sample time series for the three plant subsets shown in Fig. 1. Dashed line is the reference irradiance measurement.

direction (e.g. in Fig. 5), the bandwidth does provide some discrimination capability. As seen in Fig. 6, the blue segment exhibits a higher bandwidth, resulting from its more compact size relative to the cloud motion and its reduced effect at smoothing the variability. This result is also predicted by the output of the CAM model. As before, the phase characteristics also provide discrimination between these two segments due to their spatial separation along the cloud motion direction.

Some discussion is warranted on the impact of the temporal resolution of the data. Though the low-frequency phase provides the clearest discrimination of the phase delay difference between the plant segments, it is necessary that the sampling rate for the data is high enough to actually capture the temporal delay between the signals. Delay can be imagined as proportional to the segments' spatial separation distance along the cloud advection direction and inversely proportional to the cloud motion speed. In the case of the south-to-north cloud

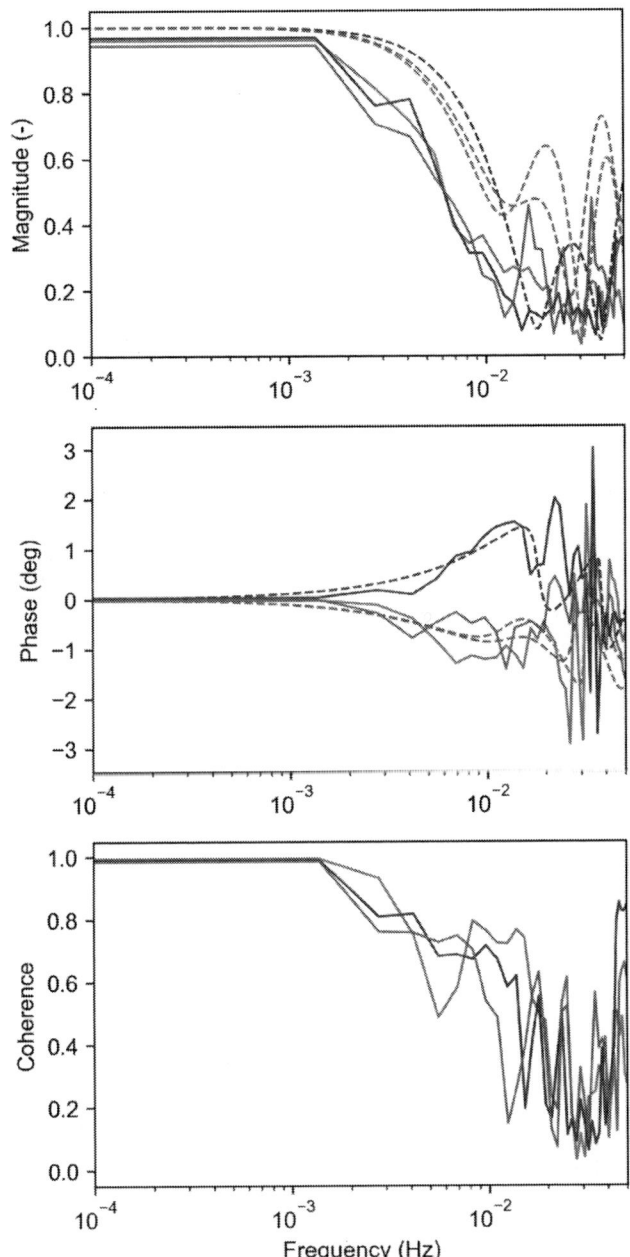

Fig. 3. Transfer functions for the plant subsets shown in Fig. 1. Dashed line is the CAM model prediction for each segment.

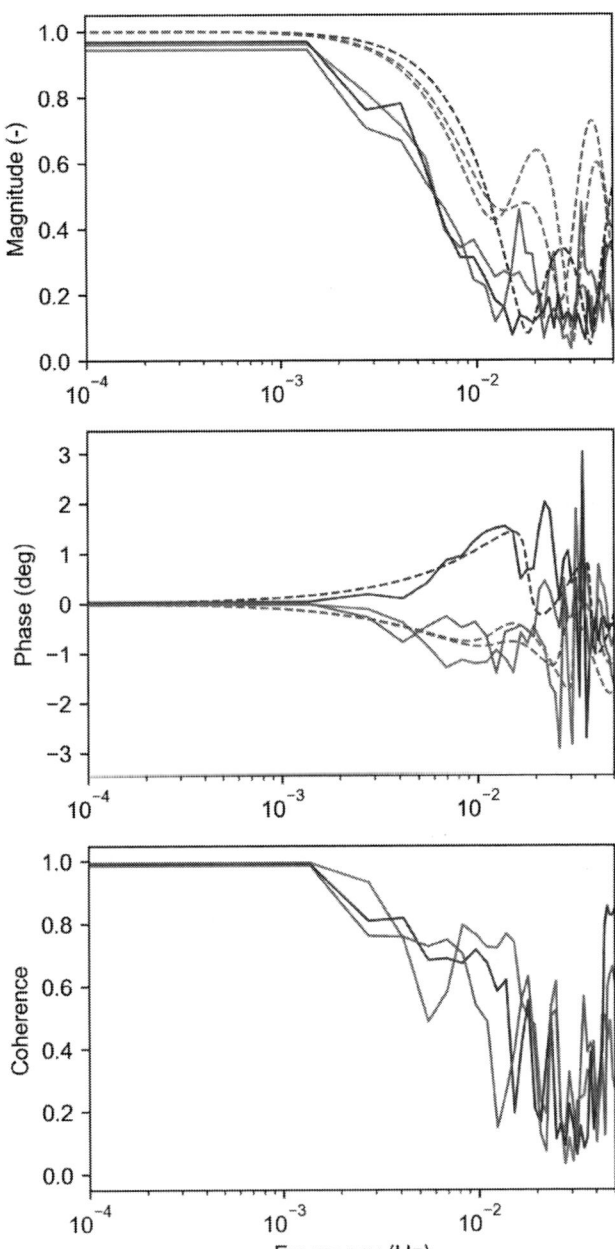

Fig. 4. Transfer functions for the same plant subsets shown in Fig. 1, but for a different time period with a west-to-east cloud motion. Dashed line is the CAM model prediction for the segments.

motion, the midpoint separation between the red and green segments in Fig. 1 was approximately 1 km, with cloud motion speeds of around 20 m/s, corresponding to an expected delay of approximately 50 seconds between the two segments. The 1 second resolution data used in this study was therefore sufficient to resolve this delay, but 1 minute or 5 minute resolution data might be sampled too slowly to meaningfully observe the difference.

III. CONCLUSION

This study demonstrated the feasibility of using the previously reported Cloud Advection Model (CAM) for matching of the transfer function between a plant segment and a reference irradiance measurement. This approach is proposed as a method to cross-check the accurate labeling of plant

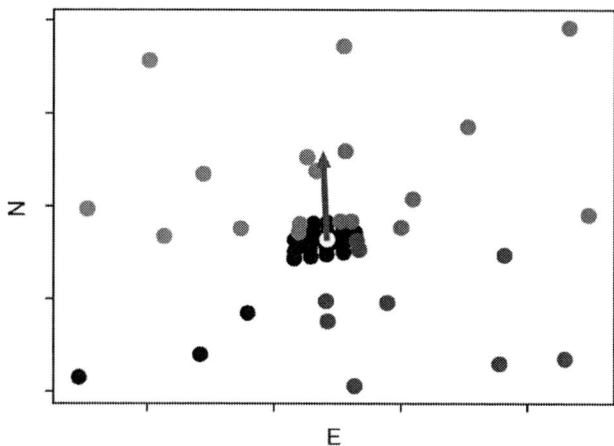

Fig. 5. Layout of plant with different spatial dispersion in the cloud motion direction. Yellow dot shows reference irradiance station. Red and Blue dots represent the two plant segments. Purple arrow shows cloud motion direction.

segments (e.g. inverter- or combiner-level outputs) within a larger PV plant.

The results showed that the transfer function appears to provide some discrimination of the position of sub-plant segments. The transfer function phase provides a more consistent representation of the location of the segment, as compared to the transfer function magnitude. This results from the fact that the phase characteristics are dominated by the group delay, which was apparent when considering segments with a spatial separation in the cloud motion direction. The delay between these segments was matched reasonably well by the CAM model at low frequencies, where the coherence between the signals remained relatively high. Due to the advective physics dominating these relationships, it was observed that discrimination of segment position within the field is difficult when working with segments that are co-located with respect to the direction of the cloud motion. However, the ability to discriminate those segments was regained by considering additional time periods with perpendicular cloud motion directions (i.e. aligned with the spatial separation between the plant segments).

The bandwidth inferred from the transfer function magnitude was observed to provide some discriminatory capability when considering two plant segments with different aspect ratios or orientations with respect to the cloud motion direction. However, detailed dynamical characteristics of the transfer function magnitude were too noisy to be matched effectively by the model, due to the loss of coherence. Thus, the transfer function magnitude seems unlikely to provide a good target for identifying plant segments, unless a high degree of non-uniformity is present within the plant segment configurations.

Some limitations of the method can be inferred from the present work. Implementing this analysis requires temporally coincident data from the various site segments along with some indication of the cloud motion vector. Such time periods require

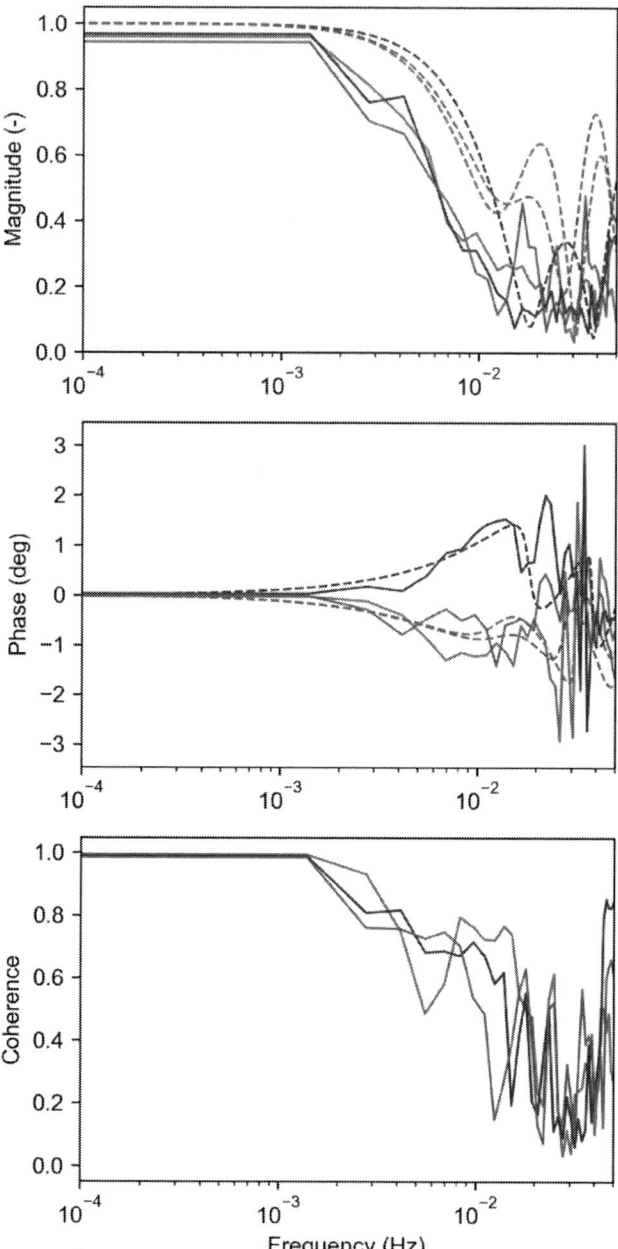

Fig. 6. Transfer functions for the two plant subsets shown in Fig. 5. Dashed line is the CAM model prediction for the segment.

data with moderate variability induced by consistent cloud motion, such that the frozen cloud advection assumption of the CAM is satisfied. In order to map a full two-dimensional plant, it is necessary to investigate multiple such time periods with sufficiently perpendicular cloud motion vector components. It is also necessary for the generation data to be sampled at a sufficiently high rate to resolve the temporal delay between segments.

While the present results indicate that this method shows promise for plant quality control applications, this demonstration was based on distributed irradiance sensor data, rather than actual outputs from a plant. Future work would be needed to test this method on real plant data to determine its effectiveness for practical applications and its suitability for quality control in a real setting. Additionally, automation of this process would be desirable to facilitate its adoption across an entire utility scale PV plant. difference.

ACKNOWLEDGEMENT

The author would like to acknowledge financial support from Penn State Hazleton and the Penn State School of Engineering Design and Innovation.

REFERENCES

[1] J. Ranalli and E. E. M. Peerlings, "Cloud advection model of solar irradiance smoothing by spatial aggregation," *J. Renew. Sustain. Energy*, vol. 13, no. 3, p. 033704, May 2021, doi: 10.1063/5.0050428.

[2] J. Ranalli, E. E. M. Peerlings, and T. Schmidt, "Cloud Advection and Spatial Variability of Solar Irradiance," in *47th IEEE Photovoltaic Specialist Conference (PVSC)*, Virtual: IEEE, Jun. 2020, p. 8.

[3] M. Lave and J. Kleissl, "Cloud speed impact on solar variability scaling – Application to the wavelet variability model," *Sol. Energy*, vol. 91, pp. 11–21, May 2013, doi: 10.1016/j.solener.2013.01.023.

[4] A. Macke *et al.*, "The HD(CP)^2 Observational Prototype Experiment (HOPE) – an overview," *Atmospheric Chem. Phys.*, vol. 17, no. 7, pp. 4887–4914, Apr. 2017, doi: https://doi.org/10.5194/acp-17-4887-2017.

[5] M. Jamaly and J. Kleissl, "Robust cloud motion estimation by spatio-temporal correlation analysis of irradiance data," *Sol. Energy*, vol. 159, pp. 306–317, Jan. 2018, doi: 10.1016/j.solener.2017.10.075.

Long Terms Stability and Metastable Behavior of Perovskite Solar Devices on Outdoor Conditions

Karim Medjoubi[1], Anne Migan Dubois[2], Jean Castillon[1], Thomas Guillemot[1], Johan Parra[3], Marion Provost[1]
Jean-Baptiste Puel[1,4], Jean Rousset[1,4], Juan Pablo Medina Flechas[1,5], Camille Bainier[1,5], Dounya Barrit[1,5],
Jordi Badosa[3], Jorge Posada[1,4]

[1]Institut Photovoltaïque d'Île-de-France (IPVF), 18 Boulevard Thomas Gobert, 91120, Palaiseau, France
[2]GeePs, CNRS, SUPELEC, UPMC, University Paris-Sud, 91192 Gif-sur-Yvette Cedex, France
[3]LMD, IPSL, École Polytechnique, Université Paris-Saclay, ENS, CNRS, 91128 Palaiseau, France
[4]EDF R&D, IPVF, 18 boulevard Thomas Gobert, 91120 Palaiseau, France
[5]TotalEnergies S.E. Renewables, 2 Place Jean Millier, 92078 Paris La Défense Cedex, France

Abstract — **Major Research and development in the field of photovoltaics is currently focused on the commercialization of perovskite technology for the solar energy market. Although recent advances in performance enhancement, this technology must overcome long-terms stability issues before its commercialization. Currently, research are focused on the reliability under continuous illumination and damp heat conditions. Perovskite materials, on the other hand, show different behavior under real operating conditions. In this work, we are interested in the behavior of perovskite solar cells and modules under outdoor conditions for more than 9 months. In the long term, a strong degradation is observed at the beginning of the test followed by a strong recovery and stabilizes at relatively initial values for 5000 hours. At daily variation level, perovskite shows a metastable behavior. The temperature coefficients were calculated for the different electrical parameters and compared to that of silicon.**

I. INTRODUCTION

The emerging solar cell technology based on metal halide perovskite materials has unlocked a new potential for revolutionizing the PV field [1]. Indeed, extensive research has been carried out on the improvement of the photovoltaic conversion efficiency by optimizing the different layers of the stack and/or the manufacturing process [2]. Currently, the certified conversion records are set at 25.7% and 32.5%[1] single junction and monolithic tandem PK/Si cells respectively [3].

However, other characteristics must be met in order to reach the commercial level such as long-term stability [4]. Even though several research studies [5], [6] have been conducted on the behavior of perovskite cells and modules in external conditions, it still lacks quantitative accurate data. Multiples studies have shown that the degradation mechanisms observed between aging under indoor and outdoor conditions can be different depending on the perovskite technology [7]. In certain cases the degradation observed is more pronounced for outdoor cycling than for indoor conditions. However, in other case study a non negligeable recovery is observed after dark cycle.

Hence the interest to study the behavior of this technology in realistic working conditions.

In this work, we were interested in evaluating the metastable behavior of solar cells and modules based on perovskite materials under real operating conditions. A long-term analysis as well as a focus on the behavior under diurnal variation will be explored. The temperature coefficients of different electrical parameters are calculated and compared to that of silicon.

II. METHODOLOGY

Outdoor testing were conducted with the aim to assess the behavior and the long-terms stability of perovskite cells and module in real operating conditions complementary to indoor testing. Indeed, reversible and irreversible degradation mechanisms can take place simultaneously due to the complexity of outdoor environment (relative humidity variation, light and temperature cycling ...).

A. Used perovskite solar devices

The baseline NIP architecture perovskite solar devices developed at IPVF was fabricated as depicted in Fig. 1-left and correspond to the following stack: glass substrate, fluorine-doped tin oxide (FTO), titanium dioxide (TiO2), mesoporous titanium dioxide (mp-TiO2), triple-cation perovskite absorber, poly(triarylamine) (PTAA) hole transport layer (HTL), gold back electrode.

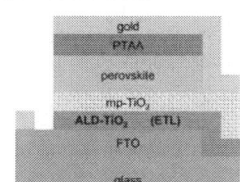

Fig. 1. Schematic representation of baseline perovskite solar cells and modules developed at IPVF (right) photographic view of the samples placed on the outdoor platform for ISOS-O2 aging test.

---

[1] NREL : https://www.nrel.gov/pv/cell-efficiency.html

Comparative study between baseline architecture and passivated interface using 2D layer [8], or passivated bulk using Pb(SCN)$_2$ [9] was conducted in this study. For this purpose, 4x4 mm² cells and 2x2 cm² modules were used. The samples were encapsulated in glass-glass configuration with polyisobutylene (PIB) and polyolefin (PO). For seek of simplicity only 2D sample behavior will be shown here.

B. Outdoor platform and aging protocol

Outdoor performances were measured in the city of Palaiseau on the SIRTA platform[2] (see Fig. 1-right) where the climate zone classification according to Köppen-Geiger is a Cfb — temperate oceanic climate. The samples were placed facing South at an angle of 27°. The irradiance profile was measured using a calibrated Si photodiode on the Plane-Of-Arrays (POA) for each point of an I-V curve. Additional detectors such as Pt100 placed at the back surface of the modules and an encapsulated thermocouple were placed for temperature recording during daily variation. These measurements allow precise calculation of temperature coefficients for the different electrical parameters. I-V measurements were performed every 10 min in both direction: forward and reverse for an irradiance higher than 50 W/m². In addition, tracking at different conditions V_{OC}, fixed resistance, MPPT were conducted. This experimental set-up corresponds to the ISOS-O2 protocol defined in the consensus statement for PSCs reliability testing [4].

Through the SIRTA platform, different other environmental parameters were measured during the test phase, i.e., relative humidity, wind speed, spectral distribution, ratio DNI/GHI. The above parameters will assist in understanding the sensitivity of the perovskite devices for each parameter using AI models.

III. RESULTS AND DISCUSSION

The evolution of different electrical parameters, i.e., normalized J_{SC} by irradiance, V_{OC}, FF, and efficiency, of a baseline perovskite solar cell with 2D interface passivation as well as irradiance and temperature profile are presented as a function of exposure time in Fig. 2. The experiment is currently in progress with an exposure of more than 9 months.

If we consider the behavior of the baseline solar cell with 2D passivation layer on a macroscopic scale, four phases can be distinguished over the whole test duration:

- **Phase 1:** during the first 500 hours, a continuous degradation is observed for all electrical parameters. However, not all parameters are impacted in the same way with a higher degradation for the normalized J_{SC} compared to V_{OC} and FF where the degradation reaches 80%, 20% and 30% respectively. As the efficiency is the product of these parameters, the degradation reaches 90%.

- **Phase 2:** All electrical parameters stabilize at their low level (mentioned above) for a period of 1000 hours.

- **Phase 3:** A significant recovery is observed for all electrical parameters during a period of 500 hours where they reach their initial values.

- **Phase 4:** The different electrical parameters remain relatively stable at a level close to the initial values for more than 4700 hours.

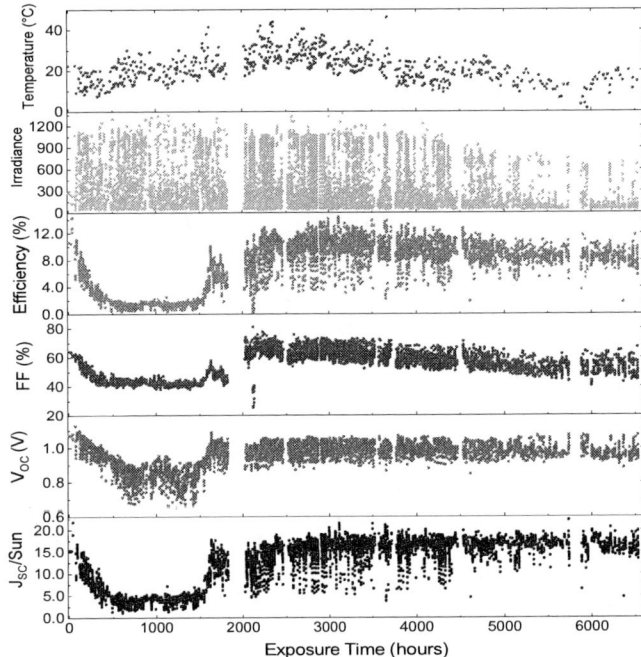

Fig. 2. Evolution of electrical parameters and irradiance/temperature profile as a function of exposure time for Baseline + 2D passivation layer sample placed at SIRTA for outdoor measurements.

In the following, we consider a microscopic analysis of the solar cell behavior during a daily variation. Fig. 3 shows the evolution of the efficiency as well as the irradiance and temperature profile as a function of the local time. The data reports the behavior on the two stabilized levels, i.e., 500-1500 hours in blue and 2200-6600 hours in red.

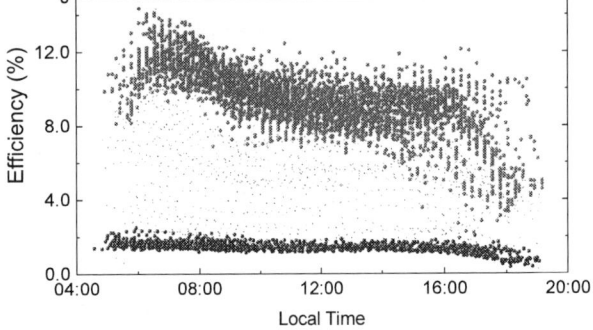

Fig. 3. Evolution of efficiency profile as a function of local time for the tow stabilized phases 500-1500 hours and 2200-4700 hours.

---

[2] SIRTA: https://sirta.ipsl.fr/

A very interesting behavior is observed with an increase of the efficiency at the beginning of the day until 08h then degrades until reaching a stabilization plateau between 10h-15h then degrades until reaching the lowest level at the end of the day. The same trend is observed for the next day and so on. This behavior is driven by that of the normalized J_{SC} (not shown here). By comparing the efficiency profile with that of irradiance and temperature, it is clear that they are not correlated with each other. This effect is more related to the metastability of the perovskite solar devices under light cycling environment, where significant recovery is observed after a dark exposition, i.e., during the night. These behavior can be attributed to the formation of metastable defect and/or reversible ion redistribution under illumination that can explain the increase of efficiency at the beginning of the day [10] [11].

TABLE I
ELECTRICAL PARAMETERS TEMPERATURE
COEFFICIENT FOR FW AND RV IV CURVES

Electrical parameters	Temperature Coefficient	
	Forward IV	Reverse IV
Normalized J_{SC} (mA/cm²/sun/°C)	-0.0298	-0.0294
V_{OC} (mV/°C)	0.748	0.576
Fill Factor (%/°C)	0.566	0.328
Efficiency (%/°C)	0.0823	0.0389

One key indicator of solar devices is the electrical parameter temperature coefficient (see Table I). The calculation of the temperature coefficient was performed taking into account: (i) exposure time from 2200 to 4700 hours (ii) a local time from 10:00 to 15:00. The purpose is to avoid the degradation, recovery and metastable behavior of perovskite as mentioned above. A positive and negative temperature coefficient is calculated for V_{OC} and normalized J_{SC} respectively, which is opposite to that of Si taken as a reference. This can be explained by an increase of the bandgap with temperature [12]–[14]. In the case of the fill factor, a positive temperature coefficient is calculated with higher values, which results in a positive temperature coefficient for the efficiency. It is also important to mention that the electrical parameters in forward direction are more sensitive to temperature compared to the reverse ones.

IV. CONCLUSION

The reliability in real operating conditions of perovskite solar devices was addressed in this study. The first results show a strong recovery of all electrical parameters after a stabilization at very low values for a period of 1000 hours. The recovery process brings the electrical parameters to values similar to those initially measured where they stabilized at this plateau for more than 5000 hours. The efficiency behavior is driven by that of the J_{SC}. However, by analyzing the efficiency during daily variation, a metastable behavior is observed with an increase of efficiency at the beginning of the day that can be related to light soaking phenomenon then it degrades and stabilize for more than 5 hours then degrades at the end of the day. This can be related to reversible degradation on the perovskite devices.

Temperature coefficients were calculated at both stabilized region, and found that the perovskite devices show a positive efficiency coefficient that is driven by that of the fill factor. High sensitivity is observed for FW than RV curves Furthermore, the effect of seasonality is also studied on the behavior of perovskite solar devices through experimental results and using AI models.

REFERENCES

[1] N. K. Elumalai, M. A. Mahmud, D. Wang, et A. Uddin, « Perovskite Solar Cells: Progress and Advancements », *Energies*, vol. 9, n° 11, Art. n° 11, nov. 2016, doi: 10.3390/en9110861.

[2] R. Wang, M. Mujahid, Y. Duan, Z. Wang, J. Xue, et Y. Yang, « A Review of Perovskites Solar Cell Stability », *Adv. Funct. Mater.*, vol. 29, n° 47, p. 1808843, nov. 2019, doi: 10.1002/adfm.201808843.

[3] A. Al-Ashouri *et al.*, « Monolithic perovskite/silicon tandem solar cell with >29% efficiency by enhanced hole extraction », *Science*, déc. 2020, doi: 10.1126/science.abd4016.

[4] M. V. Khenkin *et al.*, « Consensus statement for stability assessment and reporting for perovskite photovoltaics based on ISOS procedures », *Nat. Energy*, vol. 5, n° 1, p. 35-49, janv. 2020, doi: 10.1038/s41560-019-0529-5.

[5] E. Velilla, D. Ramirez, J.-I. Uribe, J. F. Montoya, et F. Jaramillo, « Outdoor performance of perovskite solar technology: Silicon comparison and competitive advantages at different irradiances », *Sol. Energy Mater. Sol. Cells*, vol. 191, p. 15-20, mars 2019, doi: 10.1016/j.solmat.2018.10.018.

[6] M. Jošt *et al.*, « Perovskite Solar Cells go Outdoors: Field Testing and Temperature Effects on Energy Yield », *Adv. Energy Mater.*, vol. 10, n° 25, p. 2000454, juill. 2020, doi: 10.1002/aenm.202000454.

[7] H. Köbler *et al.*, « The challenge of designing accelerated indoor tests to predict the outdoor lifetime of perovskite solar cells ». 2 février 2022. doi: 10.21203/rs.3.rs-777413/v1.

[8] J. Xia *et al.*, « Deep surface passivation for efficient and hydrophobic perovskite solar cells », *J. Mater. Chem. A*, vol. 9, n° 5, p. 2919-2927, févr. 2021, doi: 10.1039/D0TA10535J.

[9] A. Mahapatra, D. Prochowicz, M. M. Tavakoli, S. Trivedi, P. Kumar, et P. Yadav, « A review of aspects of additive engineering in perovskite solar cells », *J. Mater. Chem. A*, vol. 8, n° 1, p. 27-54, déc. 2019, doi: 10.1039/C9TA07657C.

[10] M. V. Khenkin *et al.*, « Dynamics of Photoinduced Degradation of Perovskite Photovoltaics: From Reversible to Irreversible Processes », *ACS Appl. Energy Mater.*, vol. 1, n° 2, p. 799-806, févr. 2018, doi: 10.1021/acsaem.7b00256.

[11] K. Domanski, E. A. Alharbi, A. Hagfeldt, M. Grätzel, et W. Tress, « Systematic investigation of the impact of operation conditions on the degradation behaviour of perovskite solar cells », *Nat. Energy*, vol. 3, n° 1, Art. n° 1, janv. 2018, doi: 10.1038/s41560-017-0060-5.

[12] O. Dupré, R. Vaillon, et M. A. Green, « Specificities of the Thermal Behavior of Current and Emerging Photovoltaic Technologies », in *Thermal Behavior of Photovoltaic Devices: Physics and Engineering*, O. Dupré, R. Vaillon, et M. A. Green, Éd. Cham: Springer International Publishing, 2017, p. 105-128. doi: 10.1007/978-3-319-49457-9_4.

[13] W. L. Leong *et al.*, « Identifying Fundamental Limitations in Halide Perovskite Solar Cells », *Adv. Mater.*, vol. 28, n° 12, p. 2439-2445, 2016, doi: 10.1002/adma.201505480.

[14] L. Lin et N. M. Ravindra, « Temperature dependence of CIGS and perovskite solar cell performance: an overview », *SN Appl. Sci.*, vol. 2, n° 8, p. 1361, juill. 2020, doi: 10.1007/s42452-020-3169-2.

Ultrafast Dynamics of Photoexcited Carriers and Phonons in Tailored 1D Acoustic Phonon Potentials

Muhammad Hanif, Stephen Bremner, Michael P. Nielsen, Milos Dubajic, Gavin J. Conibeer

School of Photovoltaics and Renewable Energy Engineering, UNSW Sydney, Sydney, Australia

Department of Physics, University of Cambridge, Cambridge, United Kingdom

Above bandgap photoexcitation of a semiconductor absorber results in kinetically energetic electrons and holes, collectively called hot carriers. Hot carriers lose their excess energy through a plethora of inelastic scatterings among themselves and through interactions with lattice vibrations-phonons. Frustrating one or more carrier energy loss mechanisms is a key challenge for realizing novel devices like hot carrier solar cells, which promise efficiencies beyond the Shockley-Quiesser limit. Despite evidence of hot carrier solar cell operation, experimental demonstration of performance requires control of the electron-phonon interactions, necessitating developing a deeper understanding of these processes. Controlling phonon properties directly through nanostructures should allow insights into electron-phonon interactions and phonon dynamic processes such as phonon bottleneck. Acoustic cavities spatially confine acoustic phonons, analogous to optical cavities for photons. Since the wavelengths of visible light and low-frequency (GHz) phonon modes are of the same order, the confinement of both photons and phonons is possible in III-V semiconductor superlattices. Superlattices realized by stacking two dissimilar materials give rise to phonon minibands at the zone centre resulting in phonon distributed Bragg Reflector (DBR). Recently, a more compact cavity design has been proposed to achieve arbitrary phononic potential by smoothly varying the thicknesses of the unit cell's constituent layers, keeping the unit cell's length constant in the superlattice. We have demonstrated phonon cavities operating at 96GHz with tailored 1D phonon potentials realized using GaAs/AlAs multilayers. Room temperature ultrafast vibrational spectroscopy showed long-lived coherent acoustic phonon modes compared to DBRs. By also performing time-resolved photoluminescence, the impact of the phonon confinement on photogenerated carriers' energy loss in the phonon potential and DBR has been studied. The carriers' excess energy, estimated using a full spectrum photoluminescence fitting, reveals a significantly slower carrier energy loss rate in the phononic cavity.

Improving PbS colloidal quantum dot solar cell performance via solution-phase engineering

Dhanvini Gudi[1], Arlene Chiu[1], Dana Kachman[1], Eric Rong[1], Serene Kamal[1], Yucheng Lan[2] Susanna M. Thon[1]

[1]Department of Electrical and Computer Engineering, Johns Hopkins University, Baltimore, Maryland, 21218, USA

[2]Department of Physics, Morgan State University, Baltimore, Maryland, 21218, USA

Abstract — **Lead Sulfide (PbS) colloidal quantum dots (CQDs) are promising materials for flexible and wearable photovoltaic devices and technologies due to their low cost, solution processibility and bandgap tunability with quantum dot size. However, PbS CQD solar cells have limitations on performance efficiency due to charge transport losses in the CQD layers and hole transport layer (HTL). This study pursues two promising techniques in parallel to address these challenges. Solution-phase annealing of the absorbing PbS-PbX$_2$ (X = Br, I) layer can reduce charge transport losses by removing oleic acid and parasitic hydroxyl ligands. Additionally, optoelectronic simulations are used to show that HTL performance can be improved by the addition of a 2D transition metal dichalcogenide (TMD) layer to the PbS CQD-based HTL. We use solution-phase exfoliation to produce and incorporate 2D WSe$_2$ nanoflakes into the HTL. We report a power conversion efficiency (PCE) increase of up to 3.4% for the solution-phase-annealed devices and up to 1% for the 2D WSe$_2$ HTL augmented devices. A combination of these two techniques should result in high-performing PbS CQD solar cells, paving the way for further advancements in flexible photovoltaics.**

I. Introduction

Third generation photovoltaic research has lead to advancements in the development of novel materials that allow for higher efficiency while reducing processing costs. Lead Sulfide (PbS) colloidal quantum dots (CQDs) are one class of materials being investigated due to their properties such as low-cost solution processibility, size – bandgap tunability and infrared absorption [1], [2]. Owing to these beneficial properties, PbS CQDs are being explored for flexible and wearable photovoltaic device applications. Due to their solution processability, these materials can also be used with scalable manufacturing technologies reducing the processing costs further. The current standard PbS CQD solar cell device architecture is shown in Fig. 1a, where the illumination is through a transparent fluorine-doped tin oxide (FTO) coated glass substrate, on top of which a zinc oxide (ZnO) based electron transport layer (ETL) is deposited, followed by a solution-phase ligand exchanged PbS-PbX$_2$ (X = Br, I) absorbing layer, a PbS CQD hole transport layer (HTL) with ethanedithiol (EDT) ligands introduced during a solid-state exchange and top gold contacts [3].

While the advantages of PbS CQDs make them a promising candidate for flexible and wearable photovoltaic devices, PbS CQD solar cells have not been able to achieve efficiencies close to the theoretical limits, with a current record power conversion efficiency (PCE) for PbS CQD solar cells of 15.45%, demonstrated by Ding et. al. in 2022 [4]. The limited efficiencies have largely been attributed to non-radiative

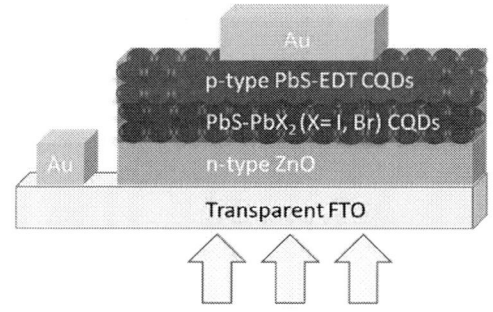

a.

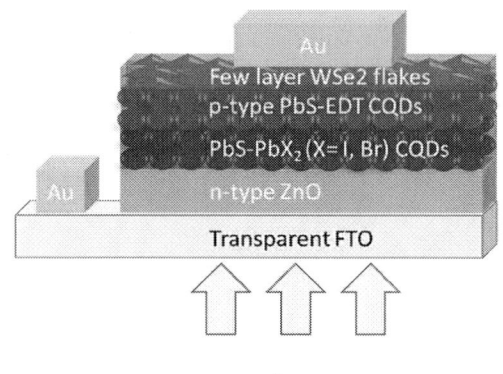

b.

recombination losses caused by poor charge transport within the CQD layer and carrier losses in the HTL [5], [6]. Therefore, improving the properties of PbS CQD films is an important problem to solve to push efficiencies forward.

Fig. 1. a. A standard PbS CQD solar cell device architecture. b. The device architecture with a WSe$_2$ HTL layer incorporated on the PbS-EDT HTL layer.

This study takes a two-pronged approach to this problem. The first approach is improvement of the charge transport properties of the PbS-PbX$_2$ absorbing layer by solution-phase annealing of the ligand-exchanged absorbing layer material. This technique has been shown to improve charge transport within the absorbing layer by the thermal desorption of bulky oleic acid (OA$^-$) and parasitic OH$^-$ ligands [7]. The second approach is augmentation of the existing PbS-EDT HTL layer by the addition of 2D transition metal dichalcogenide (TMD) nanoflakes to improve HTL charge mobility and carrier transport. Our previous compuational studies have predicted improved carrier mobility in the HTL with the addition of a few-layer-thick WSe$_2$ film onto the PbS–EDT layer[8]. This study involves experimental fabrication of solar cells with solution-exfoliated few-layer 2D WSe$_2$ on top of the PbS-EDT HTL as shown in Fig 1.b. We test the two approaches separately and measure the current density – voltage (J-V) profiles of the resulting solar cells to extract figures of merit and compare performances. We also optically characterize the few-layer 2D WSe$_2$ to study its properties. We hypothesize that combining these two approaches would greatly improve the performance of PbS CQD solar cells by addressing some of their challenges and paving the way for further advancements to reach theoretical efficiency limits.

II. EXPERIMENTS AND RESULTS

A. Solution phase anneal of ligand-exchanged absorbing layer

PbS-PbX$_2$ (X = Br, I) is synthesized by a solution-phase ligand exchange process that involves replacing oleic acid ligands on PbS CQDs with PbX$_2$. There is a need to remove OA$^-$ and OH$^-$ ligands from the ligand-exchanged solution as they can cause an electronic barrier to charge transport and unwanted carrier recombination, respectively. The solution-

TABLE I
SUMMARY OF DEVICE PERFORMANCE FOR DIFFERENT SOLUTION-ANNEAL TEMPERATURES

Device	PCE [%] best, (average)	Voc [V] best, (average)	Jsc [mA/cm^2] best, (average)	Fill Factor best, (average)
Control	6.9, (5.9±0.7)	0.61, (0.59±0.01)	22.5, (19.4±1.8)	0.47, (0.42±0.04)
50°C solution annealed	8.2, (7.1±0.7)	0.58, (0.57±0.01)	24.4, (21.9±1.8)	0.47, (0.47±0.02)
60°C solution annealed	8.9, (7.5±2.1)	0.57, (0.57±0.01)	28.3, (24.2±2.4)	0.45, (0.48±0.04)
70°C solution annealed	10.3, (9.1±0.7)	0.56, (0.56±0.01)	28.4, (27.2±1.9)	0.54, (0.49±0.03)

phase ligand exchange is performed by following standard procedures in which oleic acid capped PbS CQDs are added to PbX$_2$ in dimethylformamide and vortexed until solvent separation is observed[9]. Next, the solution is washed with

octane three times and annealed for 1 hour, after which it is dispersed in a solution composed of butylamine, pentalyamine and hexylamine. In this work, we optimize a solution-phase annealing procedure to maximize device PCE.

We chose three anneal temperatures – 50°C, 60°C and 70°C based on literature and experimental observations [7], [10], [11]. We varied the anneal time periods from 30 minutes to 1.5 hours, and found 1 hour to be the optimized period. Shorter anneal times did not show much change from the control performance whereas longer anneal times caused degradation of the absorbing layer material. The temperature dependent studies are summarized in Table 1. From the table, it can be seen that compared to the control device with no solution-phase anneal, the device with a 70°C solution-phase anneal exhibited the highest increase in PCE by 3.4%. These findings demonstrate that this technique significantly improves device performance with a noted increase in the Jsc, indicating that there is a high likelihood of oleic acid ligands being desorbed from the PbS-PbX$_2$, thereby improving charge transport in the active layer.

B. 2D WSe$_2$ as a HTL additive for PbS CQD solar cells

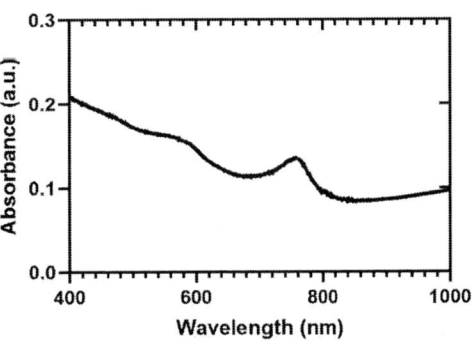

Fig. 2. Absorption spectrum of few-layer solution-phase exfoliated WSe$_2$ nanoflakes .

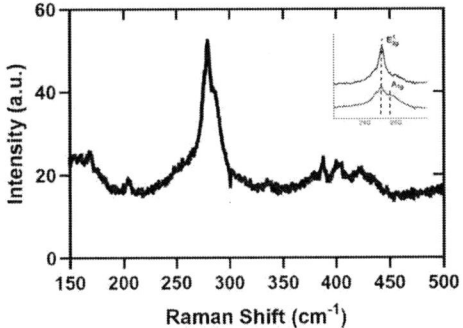

Fig. 3. Raman spectrum of few-layer solution-phase exfoliated WSe$_2$ nanoflakes with a prominent E$_{2G}$ peak at 250 cm^{-1} . Inset shows the E$_{2G}$ peak at 250 cm^{-1} and a shoulder A$_{1G}$ peak at 257 cm^{-1} for 2D nanoflake WSe$_2$ (blue) compared to bulk WSe$_2$ (black) with no A$_{1G}$ peak.

Following previous investigations from our group using 1-D SCAPS [12] to model the effect of WSe_2 as a HTL in PbS CQD solar cells, solution-processed WSe_2 nanoflakes were produced, characterized and incorporated into PbS CQD solar cells to study their impact on device performance [8]. We performed solution-phase exfoliation of commercial bulk WSe_2 powder in ethanol using probe sonication to obtain 2D WSe_2 nanoflakes. These flakes were spin-cast onto the PbS-EDT layer of the solar cell at different spin speeds to study coverage and device performance. Optical characterization of the-few layer 2D WSe_2 was performed using UV-Visible-Near-Infrared spectrophotometry (UV-Vis-NIR) to study absorption as shown in Fig. 2, and Raman spectroscopy was performed as shown in Fig. 3 to qualitatively determine if the films were bulk-like or a few layers thick. From Fig. 2 we can observe an onset absorption peak at 750 nm, consistent with literature [13]. From the Raman spectroscopy (Fig. 3 inset), we observed that bulk WSe_2 powder (black) only showed the dominant 250 cm^{-1} characteristic E_{2G} peak whereas the solution-phase exfoliated film (blue) also showed a shoulder A_{1G} peak at 257 cm^{-1}, characteristic of a few-monolayer-thick film [13]–[15].

The solution-phase exfoliated 2D WSe_2 nanoflakes were spin-cast after the HTL on standard devices at different speeds as listed in Table II. It can be seen that WSe_2 spin-cast at 1000

TABLE II
SUMMARY OF DEVICE PERFORMANCE FOR DIFFERENT SPIN SPEEDS OF SOLUTION-PROCESSED WSE$_2$ FILMS

Device	PCE [%] best, (average)	Voc [V] best, (average)	Jsc [mA/cm^2] best, (average)	FF best, (average)
Control	8.7, (7.3±0.7)	0.65, (0.61±0.02)	24.0, (23.5±1.5)	0.56, (0.50±0.04)
WSe$_2$ spun at 2000 rpm	9.5, (7.9±0.8)	0.63, (0.60±0.03)	26.0, (23.6±1.7)	0.58, (0.55±0.03)
WSe$_2$ spun at 1000 rpm	9.7, (8.4±1.2)	0.58, (0.62±0.04)	27.8, (23.7±2.7)	0.60, (0.54±0.04)
WSe$_2$ spun at 500 rpm	9.4, (8.0±1.0)	0.62, (0.62±0.03)	25.9, (23.8±2.2)	0.59, (0.50±0.03)

rpm offered the highest increase in PCE of 1% compared to the control. Scanning electron microscopy (SEM) measurements suggest that this was likely due to slower spin speeds (500 rpm) causing aggregation of flakes and non uniformity of the HTL layer, whereas higher spin speeds enabled more uniform coverage. It is important to note that the control devices reported in Tables I and II differed slightly in performance due to slight differences in processing conditions and materials.

III. CONCLUSIONS AND SUMMARY

This study experimentally verified two promising techniques to improve PbS CQD solar cell device performance, with a PCE increase of up to 3.4% obtained by solution-phase annealing of the PbS-PbX$_2$ layer and a PCE improvement of up to 1% obtained by integration of a few-layer 2D WSe_2 HTL. Future steps include combining these two techniques to test and validate the performance improvement of devices and further characterization of material and optical properties. This study aimed to improve the performance of PbS CQD solar cells and allow for further innovation in next generation solar cell technologies.

IV. ACKNOWLEDGEMENTS

National Science Foundation (DMR-1807342) and US Department of Defense (W911NF2120213) for funding. Prof. Natalia Drichko for Raman Spectroscopy.

REFERENCES

[1] S. Malhotra, L. Gupta, R. Pandey, and R. Sharma, "A critical review on the recent progress in the area of PbS CQDs based solar cell technology," 2022, p. 020016. doi: 10.1063/5.0110868.

[2] M. Yuan et al., "Phase-Transfer Exchange Lead Chalcogenide Colloidal Quantum Dots: Ink Preparation, Film Assembly, and Solar Cell Construction," Small, vol. 18, no. 2, p. 2102340, Jan. 2022, doi: 10.1002/smll.202102340.

[3] A. Chiu, C. Bambini, E. Rong, Y. Lin, and S. M. Thon, "New Hole Transport Materials via Stoichiometry-Tuning for Colloidal Quantum Dot Photovoltaics," in 2020 47th IEEE Photovoltaic Specialists Conference (PVSC), Jun. 2020, pp. 1096–1097. doi: 10.1109/PVSC45281.2020.9300891.

[4] C. Ding et al., "Over 15% Efficiency PbS Quantum-Dot Solar Cells by Synergistic Effects of Three Interface Engineering: Reducing Nonradiative Recombination and Balancing Charge Carrier Extraction," Adv Energy Mater, vol. 12, no. 35, p. 2201676, Sep. 2022, doi: 10.1002/aenm.202201676.

[5] S. Liu et al., "Enhancing the Efficiency and Stability of PbS Quantum Dot Solar Cells through Engineering an Ultrathin NiO Nanocrystalline Interlayer," ACS Appl Mater Interfaces, vol. 12, no. 41, pp. 46239–46246, Oct. 2020, doi: 10.1021/acsami.0c14332.

[6] L. Hu et al., "Optimizing Surface Chemistry of PbS Colloidal Quantum Dot for Highly Efficient and Stable Solar Cells via Chemical Binding," Advanced Science, vol. 8, no. 2, p. 2003138, Jan. 2021, doi: 10.1002/advs.202003138.

[7] X. Liu et al., "Solution Annealing Induces Surface Chemical Reconstruction for High-Efficiency PbS Quantum Dot Solar Cells," ACS Appl Mater Interfaces, vol. 14, no. 12, pp. 14274–14283, Mar. 2022, doi: 10.1021/acsami.2c01196.

[8] E. Rong, A. Chiu, C. Bambini, Y. Lin, C. Lu, and S. M. Thon, "New Chalcogenide-Based Hole Transport Materials for Colloidal Quantum Dot Photovoltaics," in 2021 IEEE 48th Photovoltaic Specialists Conference (PVSC), Jun. 2021, pp. 0750–0753. doi: 10.1109/PVSC43889.2021.9518695.

[9] A. Chiu, E. Rong, C. Bambini, Y. Lin, C. Lu, and S. M. Thon, "Sulfur-Infused Hole Transport Materials to Overcome Performance-Limiting Transport in Colloidal Quantum Dot Solar Cells," ACS Energy Lett, vol. 5, no. 9, pp. 2897–2904, Sep. 2020, doi: 10.1021/acsenergylett.0c01586.

[10] Y. Shi et al., "Influence of the Post-Synthesis Annealing on Device Performance of PbS Quantum Dot Photoconductive Detectors," physica status solidi (a), Jul. 2018, doi: 10.1002/pssa.201800408.

[11] B. Ding, Y. Wang, P.-S. Huang, D. H. Waldeck, and J.-K. Lee, "Depleted Bulk Heterojunctions in Thermally Annealed PbS Quantum Dot Solar Cells," The Journal of Physical Chemistry C, vol. 118, no. 27, pp. 14749–14758, Jul. 2014, doi: 10.1021/jp503256d.

[12] M. Burgelman, P. Nollet, and S. Degrave, "Modelling polycrystalline semiconductor solar cells," Thin Solid Films, vol. 361–362, pp. 527–532, Feb. 2000, doi: 10.1016/S0040-6090(99)00825-1.

[13] W. Zhao et al., "Evolution of Electronic Structure in Atomically Thin Sheets of WS2 and WSe2," ACS Nano, vol. 7, no. 1, pp. 791–797, Jan. 2013, doi: 10.1021/nn305275h.

[14] A. Arora, M. Koperski, K. Nogajewski, J. Marcus, C. Faugeras, and M. Potemski, "Excitonic resonances in thin films of WSe2 : from monolayer to bulk material," Nanoscale, vol. 7, no. 23, pp. 10421–10429, 2015, doi: 10.1039/C5NR01536G.

[15] Q. Cui, F. Ceballos, N. Kumar, and H. Zhao, "Transient Absorption Microscopy of Monolayer and Bulk WSe2," ACS Nano, vol. 8, no. 3, pp. 2970–2976, Mar. 2014, doi: 10.1021/nn500277y.

Understanding and Advancing Bifacial Thin Film Solar Cells Under Dual Illumination

Adam B. Phillips, Jared D. Friedl, Zhaoning Song, Abasi Abudulimu, Ebin Bastola, Deng-Bing Li, Yanfa Yan, Randy J. Ellingson, Michael J. Heben

University of Toledo, Toledo, OH, United States

Bifacial thin film (BTF) solar cells are starting to receive more attention in the literature. However, deleterious band bending and high interface recombination velocities are often a problem for thin film devices. These issues may affect how BTF devices operate under bifacial illumination compared to monofacial illumination. Using numerical modeling we investigate a BTF device with bifaciality of 0.76 for varying illumination conditions. We show that the power generated from a BTF device is not simply the sum of powers generated by a front illuminated and back illuminated device and use the band diagram to explain the difference. In addition we investigate how the angle of the light incident on the backside of the device affects the device performance.

Spray-assisted Passivation Strategy for Highly Efficient and Stable Perovskite Solar Cells

Rishabh Sahani[1], Neetesh Kumar[1], Cheng-Yu Lai[1,*], Daniela R. Radu[1,*]

[1]Department of Mechanical and Materials Engineering, Florida International University, Miami, FL, 33174

*Corresponding Author

Abstract — **Despite the wide attention and remarkable power conversion efficiencies (~25%) for 3D-perovskite solar cells (PSCs), their commercialization is inhibited by poor stability and moisture ingress. This can be attributed to the irregular distribution of the defects in the crystal lattice and grain boundaries that hinders the charge transport and results in non-radiative recombination. This acts as a bottleneck towards achieving higher power conversion efficiency (PCE) and stability. Herein, we report an unprecedented technique to form a 2D layer atop the 3D perovskite via spray-assisted passivation methodology using organic salt, n-octylammonium iodide as a bulky organic cation (BOC) forming a 3D-2D multidimensional perovskite. This defect passivation strategy is a promising route to further boost the PCE and stability thereby suppressing the defect states at the grain boundaries, resulting in a decrease in ion migration and recombination. The deposition of the thin 2D capping layer is further confirmed by advanced characterization techniques such as X-ray diffraction, scanning electron microscopy, atomic force microscopy, and optical absorbance spectroscopy.**

Keywords—Bulky organic cation, spray deposition, passivation, defects, 3D-2D perovskite.

I. INTRODUCTION

Archetypal 3D-perovskites have gained immense research interest and attention due to their remarkable power conversion efficiencies (PCE ~ 25% for single junction and ~30% for tandem devices) [1] and exceptional optoelectronic properties such as long diffusion length, high absorption coefficient, high defect tolerance, low exciton binding energy, higher mobility, and tunable bandgap. Other than its comparable PCE with crystalline silicon solar cells it possesses properties such as scalability, solution-processability, is lightweight, and has a simplified fabrication route. However, the technological exploitation and real-life application of 3D-perovskites are inhibited by their poor long-term stability and rapid degradation upon exposure to air, moisture, and intense UV irradiation. Moreover, upon exposure to the ambient atmosphere PSCs results in non-radiative recombination, the evolution of trap states, and bulk and surface defects. These defects states render the PSCs with decreased efficiency and stability [2]. This fragility presents a great challenge for the achievement of their full practical potential in photovoltaic technologies. Therefore, we aim to achieve a balanced trade-off between the performance and stability of a hybrid perovskite system by tailoring its dimensionality and structural configuration. Low-dimensional, structurally-versatile perovskite particularly two-dimensional, (2D) hybrid perovskite systems with superior heat/moisture stability have been developed as an alternate to replace 3D-perovskites. 2D-perovskites or Ruddlesden-Popper phase perovskite, have the general formula $R_2A_{n-1}B_nX_{3n+1}$, where R represents bulky organic cations with variable inorganic layers 'n' (n = 1, 2, 3…) held-together [3]. The hydrophobic nature and ion-migration suppression ability of these bulky organic cations the have blessed 2D-perovskite system with superior stability. In consideration of this, it is desirable to design a 3D-2D multi-dimensional perovskite that simultaneously features the exceptional optoelectronic properties of 3D-perovskite and the excellent stability of 2D-perovskite.

The core idea of this project lies in expanding the diversity of 3D-2D multi-dimensional perovskites and exploring how the different varieties influence the performance of optoelectronic devices. Following this idea, herein we report the unprecedented strategy involving spray-assisted deposition of a thin 2D-capping layer of, n-octylammonium iodide resulting in a 3D-2D multidimensional structure. The spray deposition technique here allows a room temperature deposition of the thin capping layer resulting in lesser degradation of the 3D perovskite with the solvent compared to spin-coating deposition counterparts [4] [5]. The thin 2D-capping layer encapsulates the 3D-perovskite from the moisture boosting the stability thereby passivating and suppressing the defects in the grain boundaries. The insitu 2D growth can be varied depending on the number of cycles/solution sprayed. Further, the 2D formation is confirmed by advanced characterization techniques.

II. EXPERIMENTAL DETAILS

PSC Device Fabrication and characterization: In this study, *n-i-p* type planar PSCs with a device structure of glass/indium tin oxide (ITO)/SnO$_2$/perovskite/perovskite/Spiro-OMeTAD/Au were fabricated. First, ITO-coated glass substrates were sequentially sonicated in acetone, ethanol, and 2-propanol, and UV-ozone (UVO) treated for 15 min. The SnO$_2$

978-1-6654-6060-6/23 $31.00 © 2023 IEEE

solution (2.7 wt%.) was deposited onto the ITO substrates at 3000 rpm for 30 seconds to produce ~30 nm thick film. The SnO₂ films were post-annealed at 150 °C for 30 min in air. The perovskite precursor was prepared by blending PbI₂ (530 mg), FAI (189 mg), MABr (18.5 mg), MACl (4.5 mg) and PbBr₂ (75 mg) in 4:1 ratio of DMF/DMSO to form a 1 ml solution.[6]

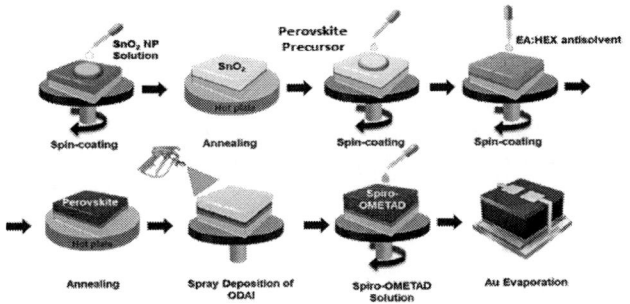

Figure 1. Device Fabrication Procedure.

Solutions was spin-coated onto SnO₂-coated substrates using a two-step spin-coating, *i.e.* 1000 rpm for 10 seconds, followed by 5000 rpm for 30seconds. 100 μL of ethyl acetate-hexane (7:3 vol/vol) [6] [7] mixed antisolvent was cast onto the perovskite film at the 10th-second interval during the second step of spin-coating. The perovskite films were post-annealed at 150 °C for 20 min. For passivation, 10 mg of ODAI in 10 ml of IPA was prepared and sprayed using a homemade spray setup with a solution flow rate of 3 ml/min. The number of spray cycles was increased to attain an optimized thickness of the 2D layer Further, Spiro-MeOTAD HTL solution was prepared by spin-coated onto the perovskite layer. Finally, ~90 nm thick Au electrodes were sputtered onto the active layers using metal masks (aperture area ~0.09 cm²). X-ray diffractometer (XRD, Rigaku, Miniflex) with CuKα radiation (λ = 1.5405 Å) was used to characterize the perovskite and the passivated films for crystallographic properties. The topography measurements were conducted using atomic force microscopy from Anton Paar instruments. Additionally, scanning electron microscopy images were obtained using the FESEM. Jeol JSM-F100 at Advanced Materials Engineering Research Insitute (AMERI) at Florida International University. Optoelectronic properties were characterized using Shimadzu UV-3600 and FS5 spectrofluorometer.

III. RESULTS AND DISCUSSION

Structural Characterization

OAI is a BOC and serves as a passivator at the perovskite/hole transport layer interface which could be attributed to their hydrophobic nature rendering the PSC devices with greater stability. Further, the long chain alkylammonium cations passivate the surface of the perovskite chemically resulting in making up for the undercoordinated bonds on the surface. This further increases the charge transfer and enhances the PCE of the devices [8] [9]. Furthermore, BOCs help in suppressing the undercoordinated Pb²⁺ions at the surface by passivating those defects. This can be further confirmed by XRD spectra. The spectra indicated in **Figure. 2** consist of three spectra namely, ODAI, reference and OAI passivated.

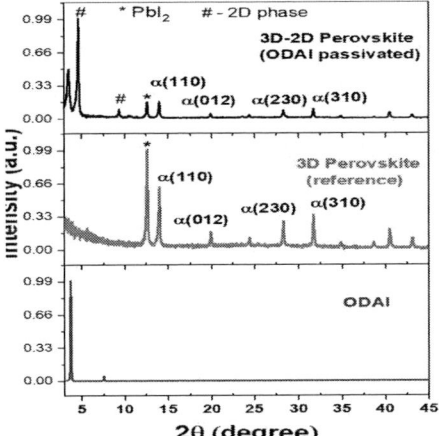

Figure 2. XRD patterns of reference and passivated perovskite films.

OAI spectra indicate the formation of pure 2D phase without any perovskite phase at 2θ values ~3.7 and 7.5 degrees respectively. Reference spectra contain only perovskite without any passivation show peaks at 2θ values 14°, 19.8°, 28.35°, 31.71° corresponding to perovskite phase with α(110), α(012), α(230) and α(310). On the contrary, the spectra depicted by ODAI passivated 3D perovskite result in the quenching of the PbI₂ (2θ ~ 12.65°) and a slight increase in the crystallinity of the perovskite phase. This is a desirable change and hence suggests the reduction in the recombination passivation of the undercoordinations ions at the surface resulting in greater stability. Apart from this, there is also a 2D peak at 2θ value ~9.3°. To further confirm this 2D peak more cycles were sprayed and it was observed that the intensity of the peak further increased indicating an increase in 2D growth.

Morphological and Topographical Characterization

The SEM micrographs for the reference film shows some remnant PbI₂ at the surface depicted by irregular white grains in between the grain boundaries (**Figure. 3. a-b**). However, the passivated film shows more compact morphology with 2D flakes growth throughout the film. This further indicates the vertical growth of 2D sheets. Moreover, the passivation helps in tailoring the grain boundaries by stitching them and leaving no or fewer trap state sites compared to the reference counterparts. AFM micrographs (**Figure 3. c-d**) also depicts the less root mean square roughness (r.m.s) for the passivated film (~ 37 nm) compared to reference (~ 43 nm) counterparts.

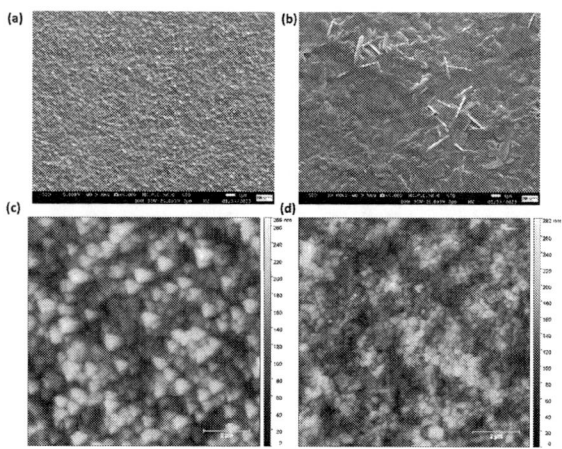

Figure 3 SEM micrographs for (a) reference (b) ODAI passivated. AFM micrographs for (c) reference (d) ODAI passivated perovskite film.

Optoelectronic Characterization

In the optical absorbance spectra, it is clearly visible that a bump ~ 500 nm indicates the formation of 2D perovskite. There is also a blue shift (towards the left: **Figure 4a**) indicating an increase in the bandgap for the OAI passivated film. The bandgap increase correlates well with the formation of 3D-2D multidimensional perovskite. Further, the 2D growth atop the 3D perovskite is confirmed by the photoluminescence (PL) measurements. The PL intensity (**Figure 4b**) increases for the OAI passivated film indicating a decrease in the trap states and non-radiative recombination compared to the reference counterparts.

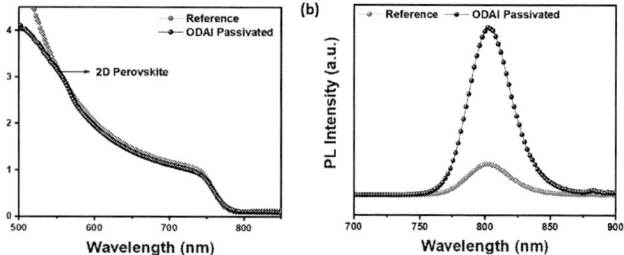

Figure 4. (a) Optical absorbance spectra **(b)** PL spectra.

IV. SUMMARY AND FUTURE WORK

The depicted results suggest that the unprecedented spray-assisted technique could help in preserving the perovskite film for a prolonged period of time and thereby reducing the trap states, ion migration, and non-radiative recombination. The thin 2D capping layer could boost efficiency and stability by acting as an encapsulating layer. XRD results vividly describe the formation of the 2D phase and quenching of the PbI_2 phase intensity that shows the passivation of the surface defects. Further by optimization of the spray cycles the 2D growth can be controlled and device performance can be tuned rendering the PSC devices with a greater lifetime.

ACKNOWLEDGMENTS

R.S: Conceptualization, writing, methodology, data curing, and experimental work. N.K: Data analysis and experimental work. C.Y. and D.R.: Supervision, validation, project administration, funding acquisition.

REFERENCES

[1] E. Aydin, M. D. Bastiani and S. D. Wolf, "Defect and Contact Passivation for Perovskite Solar Cells," 07 5 2019.

[2] F. Gao, Y. Zhao, X. Zhang and J. You, "Recent Progresses on Defect Passivation toward Efficient Perovskite Solar Cells," 14 10 2019.

[3] H. B. Lee, N. Kumar, B. Tyagi, S. He, R. Sahani and J.-W. Kang, "Bulky organic cations engineered lead-halide perovskites: a review on dimensionality and optoelectronic applications," 16 4 2021.

[4] E. Jokar, P.-Y. Cheng, C.-Y. Lin, . Narra, S. Shahbazi and . W.-G. Diau, "Enhanced Performance and Stability of 3D/2D Tin Perovskite Solar Cells Fabricated with a Sequential Solution Deposition".

[5] J. J. Yoo, . S. Wieghold and . M. G. Bawendi, "An interface stabilized perovskite solar cell with high stabilized efficiency and low voltage loss".

[6] H. B. Lee, . R. Sahani, . V. Devaraj, N. Kumar and J.-W. Kang, "Complex Additive-Assisted Crystal Growth and Phase Stabilization of α-FAPbI3 Film for Highly Efficient, Air-Stable Perovskite Photovoltaics".

[7] H. B. Lee, .-K. Jeon, N. Kumar and J.-W. Kang, "Boosting the Efficiency of SnO2-Triple Cation Perovskite System Beyond 20% Using Nonhalogenated Antisolvent".

[8] N. Mozaffari, T. Duong and . M. Shehata, "Above 23% Efficiency by Binary Surface Passivation of Perovskite Solar Cells Using Guanidinium and Octylammonium Spacer Cations".

[9] S. Gharibzadeh, B. A. Nejand and U. W. Paetzold, "Record Open-Circuit Voltage Wide-Bandgap Perovskite Solar Cells Utilizing 2D/3D Perovskite Heterostructure".

Approaching 19% Efficiency in (InxGa(1-x))2O3/CdSe/CdTe Solar Cells with Improved Front & Back Interfaces

Manoj K. Jamarkattel, Adam B. Phillips, Ebin Bastola, Sabin Neupane, Deng-Bing Li, Abasi Abudilimu, Jared D. Friedl, Tamanna Mariam, Yanfa Yan, Randy J. Ellingson, Michael J. Heben

Wright Center for Photovoltaic Innovation and Commercialization, University of Toledo, Department of Physics and Astronomy, Toledo, OH, United States

Emitter layers play an important role in limiting front surface recombination in thin film solar cells. Recently, we presented experimental results for a new class of emitter layers based on (InxGa(1-x))2O3 (IGO). By varying x, the mole fraction of indium (In) in the alloy, we were able to tune the bandgap and conduction band offset (CBO) at the interface between the emitter and the CdSexTe(1-x) (CST) absorber layer. A 16.1% efficient device was fabricated when the (InxGa(1-x))2O3 bandgap was 4.02 eV, corresponding to a composition of x=0.36. Here we present results for the same (In0.36Ga0.64)2O3 emitter composition while the composition of the CdSexTe(1-x) (CST) layer was varied. Additionally, the back contact was formed using copper thiocyanide (CuSCN) as a source of copper (Cu) doping as well to introduce as a hole transport layer to minimize back surface recombination. A champion cell had an 18.8% efficiency with a Voc of 856 mV, a FF of 76.4 % FF, and a Jsc of 28.7 mAcm-2 when the device was fabricated with a CST layer that exhibited a bandgap of 1.41 eV after processing. The results demonstrate the potentiality for IGO as an emitter layer for high-efficiency CdTe-based solar cells.

Short-circuit Current Density Chasing and Breakthroughs in High Efficiency Silicon Heterojunction Solar Cells

Weiyuan Duan, Karsten Bittkau, Andreas Lambertz, Kaining Ding

IEK5-Photovoltaik, Forschungszentrum Jülich GmbH, Jülich, Germany

Silicon heterojunction (SHJ) solar cell shows high efficiency and open-circuit voltage due to its double sides passivated contacts. However, the hydrogenated amorphous silicon (a-Si:H) layers and transparent conductive oxide (TCO) layers could cause parasitic absorption, leading to lower short-circuit current density (Jsc) compared to other crystalline record cells. This abstract reviews the eminent work which has been done in Forschungszentrum Jülich for chasing up Jsc in high efficiency SHJ solar cells. Three different parts of this work have been presented and discussed, including implementation of nanocrystalline silicon carbide based transparent passivating contact (TPC), front-side TCO-free design and Catalytic doping (Cat-doping) in SHJ solar cells. We have achieved certified efficiency of 23.99% with a Jsc of 40.87 mA/cm2 for TPC solar cells. The front-side TCO-free SHJ solar cells have been demonstrated by effective modulation between the metal contact and a-Si:H layers. Finally, we showed the present status and potential of Cat-doping in SHJ solar cells. The challenges and future research directions for these three methods were also discussed in between. We would like to show more detailed information and better results in the conference.

978-1-6654-6060-6/23 $31.00 © 2023 IEEE

Benefits of surface engineered silicon quantum dots in formamidinium lead iodide perovskite solar cells

Vladimir Svrcek, Calum McDonald, Dilli Babu Padmanaban, Ruairi McGlynn, Ankur Kambley, Bruno Alessi, Davide Mariotti, Takuya Matsui

National Institute of Advanced Industrial Science and Technology (AIST), Tsukuba, Japan

Ulster University, Ulster, United Kingdom

Hybrid quantum dot (QD) solar cells based on solution-processed blends of perovskites with Si QDs might be a potential candidate toward practical use of its outstanding optoelectronic properties. On the other hand, to overcome the stability bottleneck of perovskites we avoid using methylammonium (MA) ions and instead favor the more thermally stable formamidinium (FA) cation. In this contribution we present details on the synthesis and wide characterizations (e.g. energy bands evaluation, elemental content analysis, etc.) of FA lead iodide (FAPbI3) stable thin films and single junction FAPbI3 solar cells with efficiencies exceeding > 20%. We show that FAPbI3 with the energy gap 1.54 eV considerably enhances the short-circuit current density (JSC) and does not compromise the VOC, whereby VOC exceeds >1.1 eV. We furthermore show that the integration of surface engineered (SE) Si quantum dots into the FAPbI3 absorber does not significantly change the material properties such as energy band gap, while exhibiting superior solar cell properties over long time durations. We observed superior properties of solar cells made from SE Si-QDs in FAPbI3 over long time durations (5 months stored in a dry box between I-V measurements) compared to FAPbI3 without SE Si-QDs. Typical performance for the reference FAPbI3 solar cell (JSC = 20.6 mA/cm2, VOC =0.98 V, FF =67%, η=13.7 %) and that with SE Si-QDs (JSC = 22.5 mA/cm2, VOC =1.05 V, FF =65%, η=15.4 %) after 5 months. The external quantum efficiency (EQE) of SE Si-QDs in FAPbI3 cells remains superior after 5 months and the evaluated JSC roughly correspond to that from the I-V curves.

Depth Profiling of Glass/POE/Transparent Backsheet Degradation for Bifacial Photovoltaics

Xiaohong Gu, Ashlee R. Aiello, Stefan Mitterhofer, Soshana Smith, Stephanie L. Moffitt,
LaKesha N. Perry, Song-Syun Jhang, Stephanie S. Watson, Li-Piin Sung

Engineering Laboratory, National Institute of Standards and Technology, Gaithersburg, MD 20899, USA

Abstract—Decarbonatization drives rapid growth of the global bifacial solar market. However, the long-term durability of bifacial technology has not yet been clearly demonstrated. Here, the durability of three types of glass/polyolefin elastomer (POE)/transparent backsheet (G/CB) coupons was investigated under UV/65 °C/50 % RH, followed by thermal cycling. Spatially resolved depth-dependent techniques were used to characterize the degradation of transparent backsheets, POEs, and their interfaces in G/CB coupons. Non-destructive depth profiling of optical and chemical degradation of G/CB was developed by confocal-based Raman microscopy and fluorescence microscopy. Cross-sectional mechanical analyses of transparent backsheets and encapsulants were performed by nanoindentation. The results indicate that aged G/CB coupons show a higher yellowness index than the glass/glass construction, and their values depend on the type of backsheet. Depth-dependent confocal Raman spectra and fluorescence mapping show that the highest emission comes from the POE region closest to the glass, and the yellowness of the glass side-exposed G/CB comes from both encapsulant and backsheet yellowing, which could be related to the migration and interaction of additives in these components. Furthermore, the results from nanoindentation clearly reveal the depth-dependent mechanical changes of individual layers of G/CB coupons after aging.

Keywords - glass/transparent backsheet; bifacial modules; polyolefin elastomer; depth profiling; degradation; yellowing

I. INTRODUCTION

The goal of reaching net-zero carbon emissions by 2050 drives rapid growth of the global bifacial solar market. Bifacial modules, which are capable of absorbing light from both front and rear sides, can gain as high as 30 % - 40 % more power than traditional monofacial modules, depending on the installation, location, and bifaciality of the solar cells [1]. However, due to the short history of deployment and the lack of field data, the long-term durability of this technology has not been clearly demonstrated. Recent results [2] indicate that bifacial cells and modules can increase the prevalence of failure modes such as potential induced degradation (PID), therefore, new reliability challenges should be addressed.

Glass/glass (G/G) and glass/clear backsheet (G/CB) are two common configurations for bifacial modules. G/G is currently predominant in the market due to its resistance to moisture, fire, scratching, and sand abrasion. Polyolefin elastomers (POE) are usually used as encapsulants for this structure, replacing ethylene-vinyl acetate copolymers (EVA), which can produce acetic acid and lead to corrosion and potential induced degradation (PID) [3]. G/CB offers lighter weight, smaller heat capacity, higher stain resistance, and lower risk for hail damage. Results from the PID tests under damp heat indicate that transparent backsheets can accelerate encapsulant degradation [4] and electrochemical degradation [5] due to increased moisture ingress than G/G. However, recent work based on the combined-accelerated stress testing [2] suggests that G/G construction would be more susceptible to PID in the field than G/CB, and UV light is important to the accelerated laboratory testing for bifacial modules including PID tests.

Compared to opaque backsheets, transparent backsheets can be more vulnerable to photo-degradation because UV light would penetrate the transparent outer layer into the backsheet core and inner layers when UV protection of the outer layer is not sufficient. Our previous studies [6] showed that significant degradation was observed not only in a fluoropolymeric outer layer, but also in the PET core layer of a transparent backsheet exposed to UV light from the backsheet side, leading to severe backsheet cracking. Due to the complexity of the multilayer, multicomponent structure, it is crucial to develop depth-dependent characterization capabilities to better understand the degradation of the encapsulant, transparent backsheet, and their adjacent interfaces inside a module or a laminated coupon.

In this study, we performed the laboratory accelerated testing of G/CB coupons under UV/65 °C/50 % RH for up to 3600 h, followed by thermal cycling. Three types of clear backsheets - polyvinyl fluoride (PVF)-based, polyvinylidene fluoride (PVDF)-based, and fluoroethylene vinyl ether (FEVE)-based- were used for the construction of G/CB coupons. Spatially resolved depth-profiling techniques were developed to characterize the optical, chemical and mechanical degradation across the thickness of aged and unaged G/CB coupons. Examples of non-destructive optical and chemical depth profiling of aged G/CB coupons by confocal-based Raman microscopy and fluorescence microscopy, and the cross-sectional mechanical analysis of individual layers of aged transparent backsheets by nanoindentation will be presented.

II. EXPERIMENTAL[*]

A. Materials

Three types of Glass/POE/transparent backsheet coupons (i.e., G/CB1, G/CB2, G/CB3) were prepared using a 3.2 mm-thick polished fused silica wafer. The materials and structures of

---

*Certain commercial products or equipment are described in this paper to specify adequate experimental procedures. In no case does such identification imply recommendation or endorsement by the NIST, nor does it imply that it is necessarily the best available for the purpose.

978-1-6654-6060-6/23 $31.00 © 2023 IEEE

these backsheets and coupons can be found in our previous work [6]. CB1, CB2 and CB3 represent three types of bacheets, i.e., PVF/polyethylene terephthalate (PET)/FEVE, PVDF/PET/FEVE, and FEVE/PET/EVA, respectively. A UV-transparent POE and a UV cut-off POE were used as top and bottom encapsulants, respectively, for lamination.

B. Accelerated laboratory weathering

Laboratory accelerated weathering was performed using the NIST Simulated Photodegradation via High Energy Radiant Exposure (SPHERE) at 65 °C/50 % RH for coupons under approximately 140 W/m^2 of UV irradiance (295 nm to 400 nm). G/POE/CB coupons of 75 mm x 100 mm were mounted on top of a ≈ 2 mm thick white Teflon sheet in a black anodized sample holder with either the backsheet side or the glass side facing the UV light. Thermal cycling tests (TC) were performed using an ESPEC BTX-475 environmental chamber between − 40 °C and 85 °C after UV exposure. Samples were subjected to two rounds of 100 cycles of TC tests, one was at ≈1000 MJ/m^2 UV dose and the other at ≈1800 MJ/m^2.

C. Depth-dependent degradation characterization

Confocal Raman spectroscopy (Senterra II, Bruker) of G/CB coupons was performed using a 785 nm laser at 50mW with a 50 μm pinhole aperture, 4 cm^{-1} resolution, 15-s integration time, 5 acquisitions, and 5 μm step size. A 10x objective (NA = 0.3) was used for glass-side measurements to maximize working distance (11 mm). A 100x objective (NA = 0.8) was used for backsheet side measurements for individual layer identification.

Confocal fluorescence mapping of G/CB coupons was conducted using a Zeiss model LSM800 (Carl Zeiss Microscopy) laser scanning confocal microscope (LSCM) with a 405 nm laser. The sample surface was first brought into focus using reflection mode and then switched to fluorescence mode, collecting from the glass surface to the bottom backsheet with a step size of 5 μm, a 5x objective, and a 90 μm pinhole size.

Young's moduli of the individual layers of the epoxy-mounted aged and unaged G/CB cross-sections were measured with a G200 nano-indenter system (KLA) using a 1 μm radius, 60° diamond cone indenter for backsheets, and 25 μm radius, 90° diamond flat indenter for POE encapsulants, respectively. Fives lines of nanoindentations at an angle of 45°, with a spacing of (15-30) μm between each indent, were performed to evaluate the depth-dependency of the mechanical properties of backsheets and encapsulants. Additional measurements were taken in the thin layers of the backsheets. The indent marks were then imaged in the LSCM in reflection mode.

III. RESULTS AND DISCUSSION

A. Yellowing of G/CB and G/G coupons during exposure

Fig. 1 illustrates the changes in yellowness index (YI) of G/CB and G/G coupons as a function of UV dose. Results from both glass side (a) and backsheet side exposures (b) are displayed. G/POE/G shows the overall least yellowing after exposure, while the backsheet side exposed G/CB3 shows the most yellowing at the same condition (Fig. 1b). Our previous work [6] indicated that the yellowing of G/CB3 was due to the substantial photodegradation of CB3 (FEVE/PET/EVA), implying that the FEVE outer layer was insufficient to protect

the lower PET core layer from UV degradation [7]. CB1 and CB2 display better optical stability than CB3. For glass-side exposure (Fig. 1a), all aged G/CB coupons show an increase in YI with increasing UV dose, and their rates seem to depend on the type of transparent backsheet. The yellowing seen on the glass front is believed to be associated with UV absorbers or additives migrating from the bottom UV cut-off POE encapsulant layer during lamination [2]. Since G/G shows little yellowing after the same exposure, the effect of the transparent backsheet on interactions and migration of additives in G/CB will be studied.

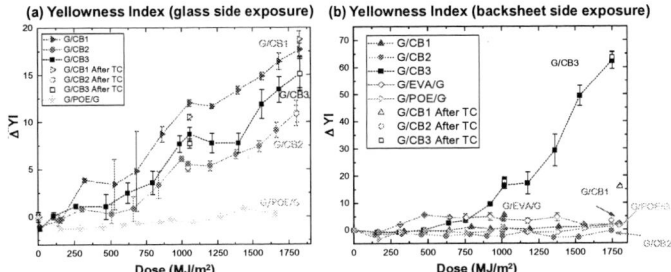

Fig. 1. Yellowness index of glass-side exposed (a) and backsheet-side exposed (b) G/CB and G/G coupons as a function of UV dose.

B. Non-destructive optical and chemical depth profiling

To better understand the degradation of encapsulant and backsheet inside a module or a laminated coupon, it is important to develop non-destructive depth profiling techniques. Raman spectra obtained from the glass side (Fig. 2a) and the backsheet side (Fig. 2b) of the unexposed G/CB2 clearly demonstrate that chemical structures of POE encapsulant and individual layers of CB2 backsheet (i.e., PVDF outer, adhesive, PET core) can be identified, and the depth-dependent chemical information across the thickness of G/CB2 can be effectively revealed by confocal Raman depth profiling.

The raw Raman spectra of the G/CB2 coupon exposed to UV/65 °C/50 % RH from the glass side show a strong background fluorescence (Fig. 3). The color scale in Fig. 3a represents the distance from the glass surface. The closer to the glass, the higher the background emission intensity. This depth-dependency is consistent with our previous studies on G/EVA/PET [8, 9], indicating the UV intensity, which decays as a function of the penetration depth, plays an important role in the amount of chromophore formation. The aged backsheet also exhibits emissions, particularly near the outer layer (Fig. 3b). Note that the backsheet is not only under direct light from the glass side, but also under indirect reflected/scattering lights from the backside because the white Teflon sheet was used as a backing for G/CB coupons during the glass side exposure. Therefore, the yellowness of this coupon is a result of both encapsulant and backsheet yellowing, with a major contribution from the POE region closest to the glass. These results are further confirmed by the confocal fluorescence mapping of the same specimen (Fig. 3c). The top POE and the bottom CB2 show a higher fluorescence than the interior POE. The background-subtracted Raman spectra will be further analyzed.

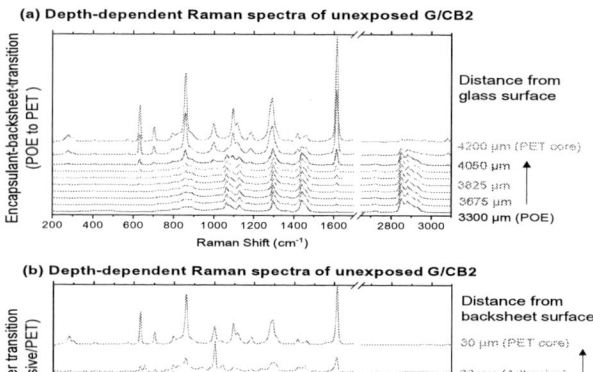

(a) Depth-dependent Raman spectra of unexposed G/CB2

(b) Depth-dependent Raman spectra of unexposed G/CB2

Fig. 2. Depth-dependent Raman spectra of unexposed G/CB2, scanning from a) glass side and b) backsheet side.

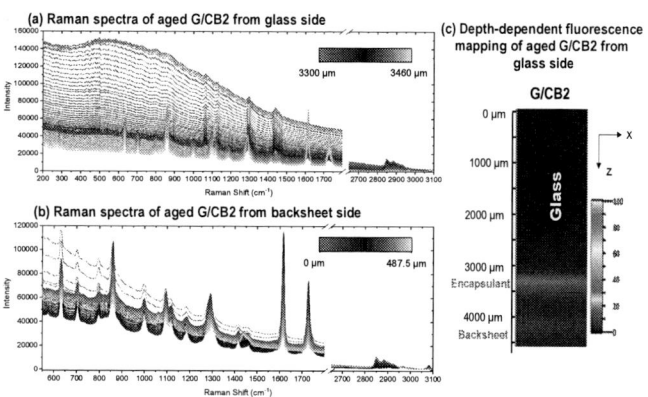

(a) Raman spectra of aged G/CB2 from glass side

(b) Raman spectra of aged G/CB2 from backsheet side

(c) Depth-dependent fluorescence mapping of aged G/CB2 from glass side

Fig. 3. Depth-dependent Raman spectra (a-from glass side, b-from backsheet side) and fluorescence mapping (c) of G/CB2 after glass side exposure to 1800 MJ/m^2 UV at 65 °C/ 50 % RH and 200 TC.

C. Cross-sectional characterization of mechanical degradation

Cross-sectional optical, chemical and mechanical analyses were also carried out on the aged and unaged G/CB coupons. Fig. 4 shows an example of nanoindentation performed on a cross-sectional G/CB1 sample mounted in epoxy resin. The residual indents across the thickness of CB1 (PVF/PET/FEVE) are seen in the LSCM images (Fig. 4a). Fig. 4b compares Young's moduli of individual layers of CB1 between glass side exposure and backsheet side exposure. The more pronounced changes observed for the backsheet side exposure demonstrate that the effect of UV lights (intensity and wavelength) on the degradation of the transparent backsheet is not limited to the outer layer, but also the lower layers such as core or adhesive layer. The cross-sectional studies of the POE encapsulants also indicate that the strong depth-dependent yellowing observed from the glass side exposure may not be directly linked to the chemical or mechanical degradation of POE copolymers; the migration and interaction of the additives (UV absorbers, antioxidant, etc.) in encapsulants and backsheets can play an important role in their optical changes. The distribution of the additives along the thickness of G/CB coupons will be further studied.

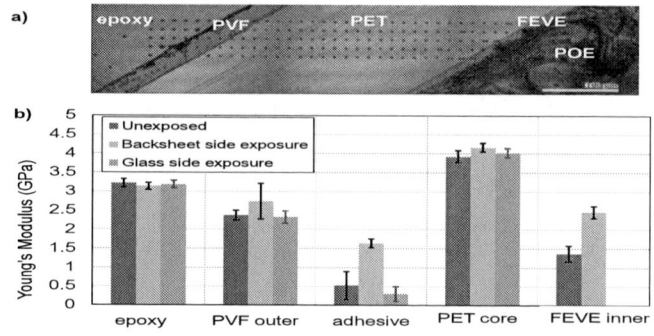

Fig. 4. (a) LSCM images of the nanoindentations in the cross-section of G/CB1. (b) Young's moduli of individual layers of the backsheet in G/CB1after exposure to 1800 MJ/m^2 UV dose and 200 TC.

IV. SUMMARY

The durability of three types of G/CB coupons was investigated using NIST SPHERE under UV/65 °C/50 % RH, followed by thermal cycling. Non-destructive depth profiling of optical and chemical degradation of G/CB was developed by confocal-based Raman microscopy and fluorescence microscopy. Cross-sectional mechanical analyses of transparent backsheets and encapsulants were performed by nanoindentation. The results indicate that aged G/CB coupons have a higher yellowness index than G/G, and their values depend on the type of the backsheet. The depth-dependent confocal Raman spectra and fluorescence mapping of the glass side-exposed G/CB show the highest emission comes from the POE region closest to the glass, and the yellowness is a combined result of both encapsulant and backsheet yellowing. Furthermore, the nanoindentation results clearly reveal the depth-dependent mechanical changes of individual layers of G/CB coupons after aging.

REFERENCES

[1] G. W. Wilson, et al., "The 2020 Photovoltaic technologies roadmap," J. Phys. D: Appl. Phys, 53 (2020) 493001.

[2] P. Hacke, et al., "Evaluation of bifacial module technologies with combined-accelerated stress testing," Prog Photovolt: Res Appl. 2022. doi: 10.1002/pip.3636

[3] C. Barretta, et al., "Comparison of degradation behavior of newly developed encapsulation materials for photovoltaic applications under different artificial ageing tests," Polymer, vol. 13, 271, 2021.

[4] L. Spinellal, et al., "Chemical and mechanical interfacial degradation in bifacial glass/glass and glass/transparent backsheet photovoltaic modules," Prog Photovolt: Res Appl. 2022. doi: 10.1002/pip.3602

[5] D. Sulas-Kern, et al., "Electrochemical degradation modes in bifacial silicon photovoltaic modules," Prog Photovolt: Res Appl. 2021. doi: 10.1002/pip.3530

[6] S. Smith, et al., "Transparent backsheets for bifacial photovoltaic (PV) modules: Material characterization and accelerated laboratory testing," Prog Photovolt: Res Appl. 2021. doi: 10.1002/pip.3494

[7] X. Gu, et al., "Investigating Long-term UV-Durability of Glass/Transparent Backsheet Laminates for Bifacial Photovoltaics," Proceedings of 8th World Conference on Photovoltaic Energy Conversion, Milan, Italy, Setpember 26-30, 2022.

[8] C.C. Lin, et al., "Depth profiling of degradation of multilayer photovoltaic backsheets after accelerated laboratory weathering: Cross-sectional Raman imaging," Sol. Energy Mater Sol. Cells, 144, pp.289–299, 2016.

[9] Y. Lyu, et al., "Fluorescence imaging analysis of depth-dependent degradation in photovoltaic laminates: insights to the failure," Prog Photovolt: Res Appl. 2019. doi: 10.1002/pip.3212

Temporal downscaling of GHI clear-sky indices using T-Copula

Jing Huang and Marc Perez

Clean Power Research, Bellevue, WA 98005, USA

Abstract — **Access to high quality granular solar irradiance data is important for many solar applications. However, they are not always available due to practical and technical challenges. Copula is able to synthetically generate scenarios reproducing observed correlation structures. In this paper, we built on our previous study to use a type of elliptical copula, T-Copula, to temporally downscale clear-sky indices of global horizon irradiance. One of the promising features of our downscaling method is its ability to simulate binary events caused by the passage of thick and broken clouds with adjustable frequency.**

I. INTRODUCTION

Modeling solar variabilities at various scales has many implications for solar power generation and its integration into central grids. For example, it is known that using hourly solar irradiance data tends to overestimate the average yield of solar photovoltaic (PV) farms, in particular when the AC capacity of power inverters is much lower than the DC capacity of solar PV modules [1]. Hoff and Perez [2] studied the output variability of PV fleet systems at local scale and established the relationship between the minimal variability of system power generation and cloud transit speed. It has also been shown that the incorporation of correlation structures improves the forecasting accuracy of aggregated power output of scattered solar PV farms at regional scale [3].

Huang et al. [4] proposed a novel algorithm to temporally downscale the clear-sky index (CSI) of global horizontal irradiance (GHI) from its hourly averages to minute level. They used Gaussian copula to produce 60 fold of synthetic time series data representing the observed probability distribution and correlation structures. It has also been shown that the error of clipping loss caused by limited AC inverter capacity is significantly reduced by using the synthetically downscaled minute-level data for power modeling. Although the use of Gaussian copula achieved promising results, there is still no systematic guidance on what copula family suits solar applications the best due to lack of research on the topic [5]. In this study, we built on our previous research to explore the use of another elliptical copula, student-t copula or simply t-copula, in modeling solar variabilities.

II. METHODOLOGY

We first introduce the ground station data used in this study and then t-copula.

A. Ground station data

Figure 1 provides information on the reference ground stations used in this study. The measurement GHI data span the entire year of 2020 with a temporal resolution of 1 minute (same as in [4]). In addition, we obtain the corresponding clear-sky GHI at all the reference stations from SolarAnywhere® V3.5 [6]. The CSI, or kt, is then calculated as the ratio of GHI to the clear-sky GHI. We cap kt values at 1.3.

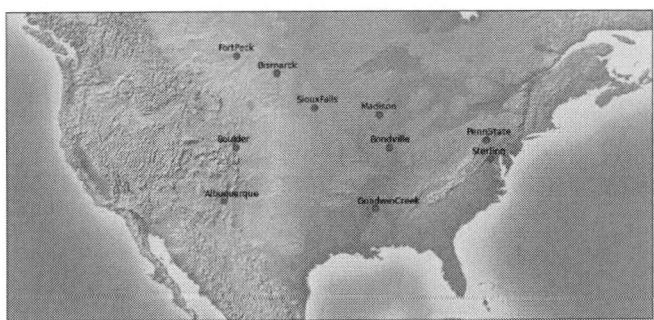

Figure 1. 10 reference ground stations in the eastern United States are used in this study.

B. From Gaussian- to T-copula

A copula is a joint multivariate cumulative distribution function (CDF) where all marginal probability distributions are uniform within [0, 1]. Copula is normally categorized by and named after its joint CDF. For example, Gaussian copula implies that the distribution of its joint CDF is Gaussian [4]. Student's t-distribution (or simply t-distribution) was invented to estimate the mean of normal distributions in situations where the sample size is small, and the standard deviation is unknown. The probability density function of a multivariate t-distributed random variable is:

$$f(\mathbf{x}) = \frac{\Gamma(\nu+p)/2}{\Gamma(\nu/2)\nu^{p/2}\pi^{p/2}|\Sigma|^{1/2}}\left[1+\frac{1}{\nu}\mathbf{x}^{\mathrm{T}}\Sigma^{-1}\mathbf{x}\right]^{-(\nu+p)/2}, \quad (1)$$

where Γ is the gamma function, p is the dimension of $\mathbf{x}$, Σ is the $p \times p$ dimensional shape matrix, and ν is the degrees of freedom.

Figure 2 illustrates the difference between Gaussian copula and T-copula. It can be discerned that T-copula mimics the pattern of kt better than Gaussian copula for at least two reasons: (1) with the same ρ, the scatter plot of the observed kt is much more centralized than of Gaussian copula. This is well captured by T-copula; (2) Gaussian copula is unable to simulate strong anti-correlated events, which can be captured by T-copula as well. In addition, degree of freedom ν can adjust the frequency of those anti-correlated events. These anti-correlated

978-1-6654-6060-6/23 $31.00 © 2023 IEEE

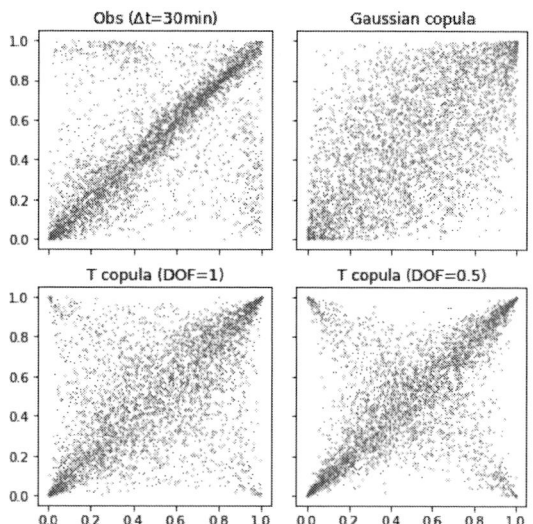

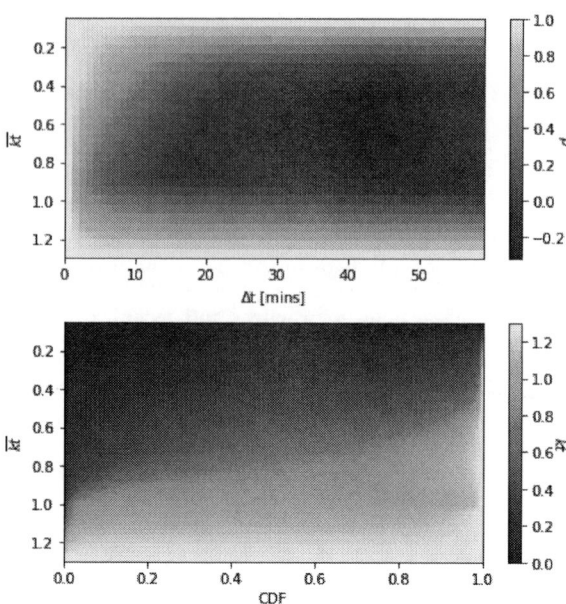

Figure 2. Comparison of scatter plots between (top left) rank normalized *kt* observation with 30min lag (ρ=0.61); (top right) 2D Gaussian copula with the same ρ; (bottom left) T-copula with the same ρ and ν=1; (bottom right) T-copula with the same ρ and ν=0.5.

Figure 3. Parameters required for kt downscaling using copula: (top) $\rho(\Delta t)$ and (bottom) CDF color plots for $\overline{kt} = 0.05, 0.1, \ldots, 1.3$.

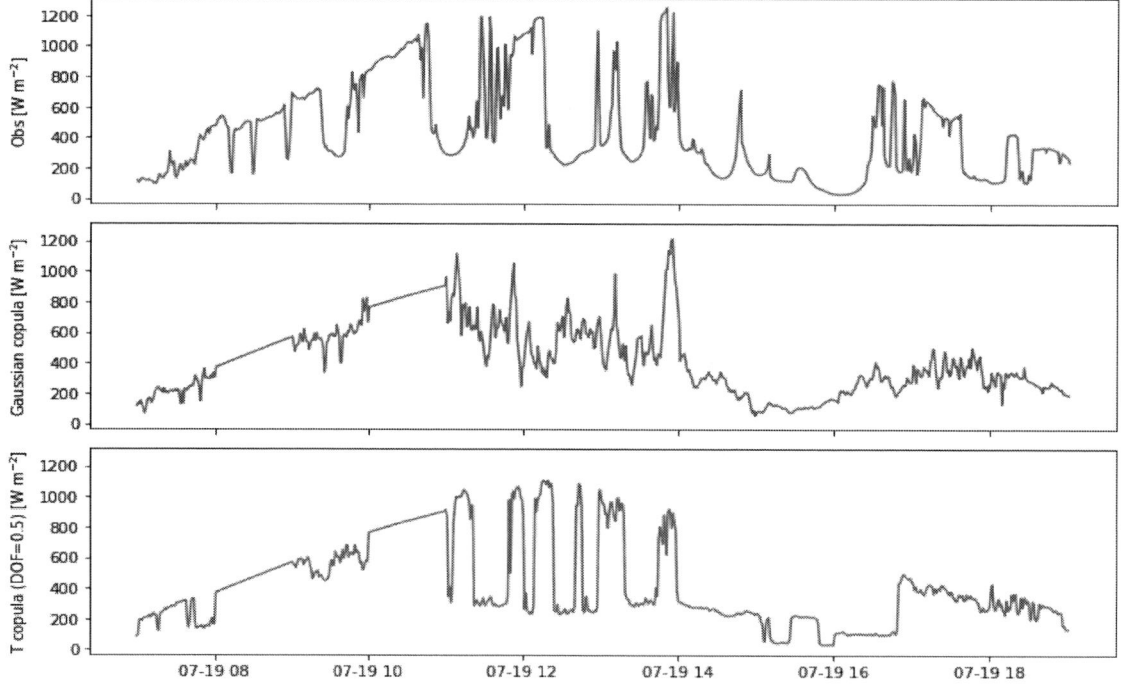

Figure 4. One example day at Boulder: (top) 1-min GHI measurement; (middle) 1-min downscaled GHI from hourly averages using Gaussian copula; and (bottom) 1-min downscaled GHI from hourly averages using T-copula with $\nu = 0.5$.

events normally occur during the passage of thick and broken clouds.

C. Temporal downscaling with T-Copula

Let's briefly review the procedures of temporally downscaling kt using copula as described in [4]. First, we calculate $\overline{kt}$ dependent correlations and distributions from observed 1-min time series data for all the sites we have. Second, the universal parameters are obtained by taking the medians of correlations and distributions across all sites. These parameters are directly taken from [4] and shown in Figure 3. Third, given a $\overline{kt}$ to downscale, we interpolate from Figure 3 to obtain a correlation structure and a distribution. Fourth, we use the chosen copula to generate multiple correlated scenarios and select one out based on the proximity of the averaged value of that scenario to $\overline{kt}$.

Figure 4 demonstrates the downscaled 1-min GHIs using Gaussian copula and T-copula, respectively. It can be seen from observation that there are many large swing events of GHI on that day due to fast movement of thick clouds. Gaussian copula is unable to simulate those events because its kernel of distribution is relatively more homogeneous than kt as shown in Figure 2. On the other side, T-copula is able to reproduce those large swing events because its kernel of distribution allows for more extreme anticorrelated events, in particular with small degree of freedom.

D. Validation

After demonstrating the downscaled GHI time series using T-copula, we validate them by comparing kt spectrum and how much loss would a GHI truncation result in. Figure 5 shows that T-copula is able to lift up the spectrum of kt with decreasing v. As shown for Goodwin Creek, the spectrum of T-copula downscaled kt fits that of observed kt much better. However, there are also deviations from observation. For example, with $v = 0.5$, the spectrum of T-copula downscaled kt is lower than observation at medium scales but higher at small scales. Finally, the downscaled results are benchmarked by the relative loss error of truncated GHI at certain thresholds. This is to mimic the scenarios of power clipping loss when the DC to AC ratio is high. However, no power modeling is involved here for simplicity. It can be observed that the simulation of Gaussian copula still underestimates the clipping loss although its error is significantly smaller than just using the hourly averaged GHI time series. The use of T-copula can further reduce the underestimation error with the extent depending on the clipping threshold and v.

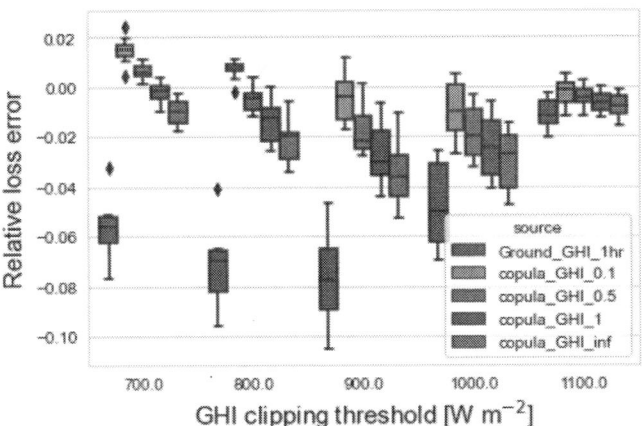

Figure 6. Box plots of relative error of curtailed irradiance using interpolated hourly-averaged observation and 1-min copula-downscaled time series ($v = 0.1, 0.5, 1$ for T-copula and $v = $ inf for Gaussian copula) with the clipping threshold of GHI ranging from 700 W m^{-2} to 1100 W m^{-2}.

III. CONCLUSIONS

Copula is a useful tool for modeling solar variabilities. Built on our previous study, we investigated the use of T-copula to temporally downscale clear-sky indices from their hourly averages to minute level. We demonstrated that T-copula is more effective to model intra-hourly solar variabilities than Gaussian copula, in particular in terms of simulating binary events caused by the passage of thick clouds.

REFERENCES

[1] K. Bradford, R. Walker, D. Moon and M. Ibanez, "A regression model to correct for intra-hourly irradiance variability bias in solar energy models", 2020 47th IEEE Photovoltaic Specialists Conference (PVSC), 2020, pp. 2679-2682.

[2] T. E. Hoff and R. Perez (2010), "Quantifying PV power Output Variability", Solar Energy, vol 84: 1782-1793.

[3] J. Huang (2019), "Time-space dependency of utility-scale solar photovoltaic power generation in the National Electricity Market", 23rd International Congress on Modelling and Simulation, Canberra, ACT, Australia.

[4] J. Huang, M. Perez, R. Perez, D. Yang, P. Keelin and T. E. Hoff (2022), "Nonparametric Temporal Downscaling of GHI Clear-sky Indices using Gaussian Copula", 2022 IEEE 49th Photovoltaic Specialists Conference (PVSC), Philadelphia, PA, USA.

[5] F. Golestaneh, H. B. Gooi and P. Pinson (2016), "Generation and evaluation of space-time trajectories of photovoltaic power", Applied Energy 176:80-91.

[6] P. Keelin, A. Kubiniec, A. Bhat, M. Perez, J. Dise, R. Perez and J. Schlemmer, (2021) "Quantifying the solar impacts of wildfire smoke in western North America," 2021 IEEE 48th Photovoltaic Specialists Conference (PVSC), 2021, pp. 1401-1404

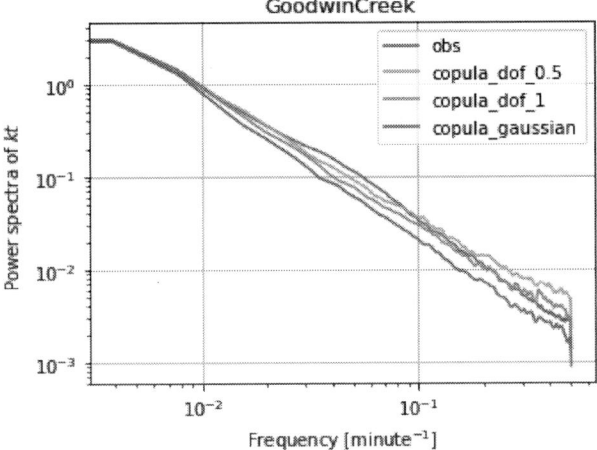

Figure 5. Power spectra of 1-min kt time series at Goodwin Creek for observation, T-copula downscaled with $v = 0.5$ and $v = 1$, and Gaussian copula downscaled, respectively.

Extended FF-Voc parameterization for silicon solar cells

Karsten Bothe, David Hinken, Rolf Brendel

Institut für Solarenergieforschung (ISFH), Emmerthal, Germany

Leibniz Universität Hannover (LUH), Hannover, Germany

This work is concerned with maximal and currently obtained fill factors of crystalline silicon solar cells. Recent research activities has led to a drastically decreased recombination in the volume and at the surfaces of crystalline silicon solar cells. As a result, the reported open-circuit voltages (Voc) and fill factor (FF) values increased significantly. In order to classify how good the achieved improvements are, it is necessary to know the maximum fill factor (FFmax) achievable for a certain open circuit voltage. Thus, in this work we calculate FFmax-Voc-pairs for an ideal resistance free single junction silicon solar cell limited by intrinsic recombination only as a function of dopant concentration and thickness by using state-of-the art analytical models. The obtained curves are compared to recently published record FF-Voc-pairs showing that all currently published record FF-Voc-pairs stay below this limit. For calculating FFmax-Voc-pairs we provide a simple extension of Green' well known FF0-equation.

Towards commercialisation with lightweight, flexible perovskite solar cells for residential photovoltaics

Philippe Holzhey, Michael Prettl, Silvia Collavini, Nathan Chang, Michael Saliba

Adolphe Merkle Institute, Fribourg, Switzerland

University of Oxford, Oxford, United Kingdom

Department of Physics and Astronomy, Vrije Universiteit Amsterdam, Amsterdam, Netherlands

POLYMAT, University of the Basque Country UPV/EHU, Donostia-San Sebastián, Spain

The Australian Centre for Advanced Photovoltaics (ACAP), School of Photovoltaic and Renewable Energy Engineering, University of New South Wales, Sydney, Australia

Institute for Photovoltaics (ipv), University of Stuttgart, Stuttgart, Germany

Helmholtz Young Investigator Group FRONTRUNNER, IEK-5 Photovoltaik, Forschungszentrum Jülich GmbH, Juelich, Germany

Metal-halide perovskites have emerged as a promising class of next-generation solar cells. Here, we assess what lifetimes and efficiencies perovskite solar cells (PSCs) have to reach to lower the price of commercial residential photovoltaic (PV) further. We find that using light and flexible substrates, as opposed to heavy and rigid ones, reduces the total installed system cost of PSCs. The flexibility and lighter weight culminate in a lower balance of systems (BOS) cost, as it is possible to use different mounting methods. Concretely, we analyse the scenario when the modules are directly stuck onto a roof without requiring a racking. That reduces both labour and material costs. This effectively lowers the necessary efficiency or lifetime of PSCs (T80 value) required to achieve the same electricity cost as commercialised silicon. For 2021, we find that a rigid perovskite module with 17% efficiency would need at least 24 years to be competitive with residential-installed silicon. In comparison, a light, flexible module with the same efficiency would only need to last 19 years. In 2030, with projected BOS costs, a 23% efficient perovskite module would need to last 24 years if rigid but only 17 years if flexible. Finally, we extended our analysis towards tandem structures with silicon or perovskite-junction. We find that flexible PSCs present a most promising commercialisation route because they can enable low manufacturing and BOS deployment costs, which opens up commercial viability at lower efficiencies or lifetimes.

Thermal Stability of BiI3 Thin Films

Natália F. Coutinho, Thais Crestani, Otávio J. de Oliveira, Ana Paula de M. M. M. Modesto, Marcelo Villalva, Francisco C. Marques

IFGW-Unicamp, Campinas, Brazil

FEEC-Unicamp, Campinas, Brazil

BYD, Campinas, Brazil

Bismuth triiodide (BiI3) is a possible candidate to replace lead in perovskite-like materials for photovoltaic applications. Besides this applicability as an active layer in solar cells, BiI3 can also be converted, using methylammonium iodide, to the perovskite-like material MA3Bi2I9, that is also suitable for photovoltaic applications. Here we investigate the thermal stability of thermally evaporated BiI3 thin films annealead up to 150 °C in ambient atmosphere (in a closed oven in the absence of illumination). BiI3 films show similar X-ray diffratrograms with indexed peaks related to an R-3 rhombohedral crystal structure. The bandgaps of BiI3 thin films were determined by Tauc plots of transmittance data in UV-vis range and indicate a bandgap of 1.72 eV regardless of annealing temperature and suitable for photovoltaic applications. We verified that its morphological properties, observed through SEM images, are also not changed with respect to its annealing at temperatures up to 150 °C. Thereby, the structural, morphological and optical properties of annealing on thermally evaporated BiI3 are not considerably altered, the films retaining their properties when heated up to 150 °C, temperature higher than the operation temperature of solar cells and below the melting point of BiI3.

In-situ smoothing of facets on spalled GaAs(100) substrates during OMPVE growth of III-V solar cells

William E. McMahon, Anna K. Braun, Allison N. Perna, Pablo G. Coll, Kevin L. Schulte, Jacob T. Boyer, Anica N. Neumann, John F. Geisz, Emily L. Warren, Aaron J. Ptak, Arno P. Merkle, Mariana I. Bertoni, Corinne E. Packard, Myles A. Steiner

National Renewable Energy Laboratory, Golden, CO, United States

Colorado School of Mines, Golden, CO, United States

Crystal Sonic Inc., Phoenix, AZ, United States

Arizona State University, Tempe, AZ, United States

This presentation describes how faceted GaAs(100) surfaces can be flattened during organometallic vapor-phase epitaxy (OMVPE) growth, by systematically examining results for different material/dopant combinations. This work is motivated by the use of spalling for substrate removal and reuse, as a pathway toward reducing the cost of III-V solar cells. Here we consider the growth of III-V cells on spalled GaAs(100) substrates, which typically have facetted surfaces after spalling. To facilitate the growth of high-quality cells, these facetted surfaces should be smoothed prior to cell growth. This study presents results on the OMVPE growth of surface-smoothing buffer layers prior to cell growth, and finds that the material/dopant combination used for smoothing must be carefully chosen for a good outcome. Some material/dopant combinations (i.e. C:GaAs) smooth the surface quite quickly, but others (i.e. Zn:GaAs) can significantly roughen it. Here we provide a systematic study based upon representative cases, and discuss the impact of these material/dopant combinations on surface smoothing and subsequent cell growth. Representative examples will be presented, along with a discussion of the underlying growth processes.

Numerical simulation study for analysis of hydrogenated amorphous silicon/crystalline silicon heterostructure by Reactive Molecular Dynamics Method

Kazuma Inoue[1], Naoya Uene[2,3], Kazuhiro Gotoh[1], Yasuyoshi Kurokawa[1],
Takashi Tokumasu[3], Noritaka Usami[1,4]

[1] Graduate School of Engineering, Nagoya University, Nagoya, Aichi, 464-8603, Japan
[2] Graduate School of Engineering, Tohoku University, Sendai, Miyagi, 980-8577, Japan
[3] Institute of Fluid Sciences, Tohoku University, Sendai, Miyagi, 980-8577, Japan
[4] Institutes of Innovation for Future Society, Nagoya University, Nagoya, Aichi, 464-8601, Japan

Abstract — Crystalline silicon heterojunction solar cells (SHJ) use intrinsic hydrogenated amorphous silicon (a-Si:H(i)) as a high-performance passivation film for crystalline silicon (c-Si) surface. The high passivation performance is considered to be achieved by hydrogen atoms in the a-Si:H(i) film terminating dangling bonds at the a-Si:H(i)/c-Si heterointerface. It has been reported that the crystallization of the a-Si:H(i) film during the deposition decreases passivation performance. In this study, reactive force-field molecular dynamics (ReaxFF MD) simulations were performed to investigate the effects of the crystallization and the hydrogen concentrations in a-Si:H(i) films on the passivation performance. We prepared a-Si:H(i)/c-Si heterostructure and simulated diffusion of the hydrogen atoms and the crystallization in the a-Si:H(i) films. A simulation system was constructed for the structure of a-Si:H(i) close to the actual structure, and an increase in the crystallinity of the a-Si:H(i) during annealing treatment was obtained. The crystallization was observed to progress with localized hydrogen diffusion close to the a-Si:H(i)/c-Si heterointerface. This suggests that the diffusion of the hydrogen atoms affects the crystallization at a-Si:H(i)/c-Si heterointerface.

I. Introduction

Crystalline silicon heterojunction solar cells (SHJ) employing hydrogenated amorphous silicon (a-Si:H) are known for their high energy conversion efficiency [1]. The high energy conversion efficiency is attributed to the high passivation performance of intrinsic a-Si:H (a-Si:H(i)) thin film against crystalline silicon (c-Si), which leads to long photogenerated carrier lifetime owing to suppression of the carrier recombination at the a-Si:H(i)/c-Si interface. It is known that carrier recombination depends on the number of dangling bonds in materials [2]. For the SHJ, the dangling bonds are terminated by the hydrogen atoms in a-Si:H(i); thus, recombination sites are reduced. Therefore, the hydrogen atoms are the key to the passivation mechanism of the SHJ.

In general, the a-Si:H(i) films are fabricated from silane and hydrogen gases by plasma-enhanced chemical vapor deposition (PECVD) method. Recently, Zhang *et al.* reported that a-Si:H(i) dual layer exhibited a high passivation level compared with the a-Si:H(i) single layer [3]. The dual layer was prepared by depositing the a-Si:H(i) overlayer using diluting the silane gas with hydrogen gas on the a-Si:H(i) layer without diluting the silane gas. We have reproduced the improved passivation performance using the dual layer. The transmission electron microscope (TEM) observation shows that the a-Si:H(i)/c-Si interfaces are partially crystallized. Furthermore, a-Si:H(i) crystallizes with too much high hydrogen concentrations during the deposition process, and passivation performance decreases [4]. It is commonly accepted that the crystallization of a-Si:H(i) degrades its passivation performance. Therefore, the local structure needs to be clarified to further improve the passivation performance. However, the structure of the a-Si:H(i)/c-Si heterointerface is not fully understood since conventional experimental methods are insufficient to reveal local structures such as the interface.

In this research, we studied the process of the crystallization of a-Si:H(i) films and the diffusion of hydrogen atoms in a-Si:H(i) films using reactive force-field molecular dynamics (ReaxFF MD) [5]. ReaxFF MD can investigate the crystallization of a-Si:H(i) and diffusion of hydrogen atoms in a thin film. We prepared the a-Si:H(i)/c-Si heterostructure and calculated the process of a-Si:H(i) crystallization and hydrogen atoms diffusion during annealing treatment.

II. Calculation methods

The melt-quenching (MQ) method [6] was used to prepare the a-Si:H(i)/c-Si heterointerface. The initial system consists of bulk c-Si(100) in the size of 30.7225 Å × 26.8822 Å × 70.0 Å in the x, y and z directions with a vacuum layer of 20 Å in the z direction, which enables to observe hydrogen desorption from the a-Si:H(i) layer. A c-Si(100) has oriented in the $[\bar{1}10]$ and $[110]$ directions in the x and y direction, respectively. The c-Si is composed of 51 atomic layers in the z direction. The bottom 26 layers of c-Si were fixed, and the other 27 to 51 layers were melted by raising the temperature from 1000 K to 4000 K by Berendsen thermostat and then amorphized c-Si by quenching. Then the hydrogen atoms were randomly placed, and the structure was relaxed to prepare a-Si:H(i)/c-Si heterostructure with various hydrogen concentrations (C_H). Periodic boundary conditions were imposed in the x and y direction. Fixed boundary conditions were imposed in the z direction. Timestep was set at 0.25 fs, and the velocity Verlet algorithm was employed for time integration methods. The temperature of the prepared a-Si:H(i)/c-Si heterostructure was heated up from 300 K to 2000 K and then cooled down to 300 K. The crystallization of a-Si:H(i) and the behavior of the hydrogen atoms in a-Si:H(i) were analyzed by open visualization tool (OVITO) [7]. The progression of crystallization was evaluated by quantifying the time evolution of crystallinity of the a-Si:H(i) films.

III. RESULTS AND DISCUSSIONS

Fig. 1 shows the radial distribution function (RDF) of amorphous silicon (a-Si) prepared by the MQ method. The obtained RDF agreed well with the RDF of a-Si in the previous study [8]. Thus, this simulation appropriately reproduced the structure of a-Si. Fig. 2 shows the structure of the a-Si:H(i) obtained by the MQ method.

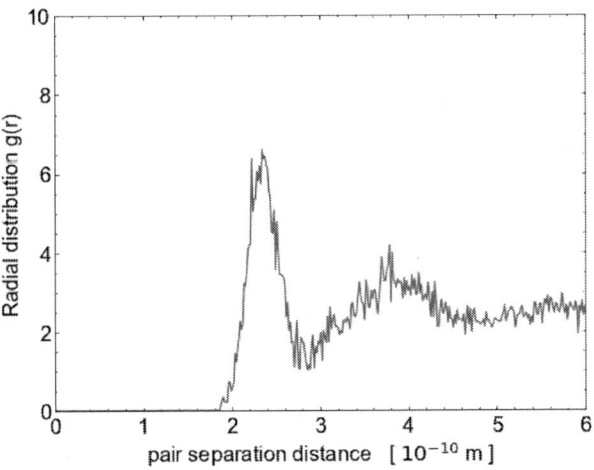

Fig. 1.   Radial distribution function of the a-Si obtained by the MQ method.

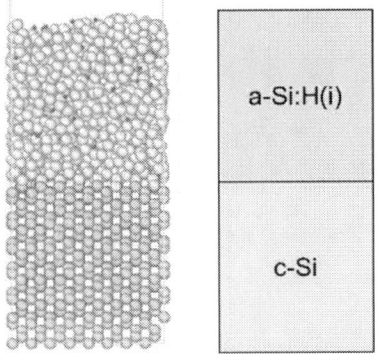

Fig. 2.   Structure of the a-Si:H(i) obtained by the MQ method. Crystalline Si and amorphous Si are colored in light blue and gray, respectively. Pink spheres are hydrogen atoms and others are silicon atoms.

Fig. 3 shows the time development of the crystallinity of a-Si:H during the annealing process. Crystallinity was defined as the below equation [9], [10].

$$(\text{Crystallinity}) = \frac{\text{Cubic} + \text{1st neighbor} + \text{2nd neighbor}}{\text{number of Si atoms in a-Si:H(i) layer}}$$

Where Cubic is a number of Si atoms having all of its first and second nearest neighbors positioned on cubic diamond lattice sites, 1st neighbor is a number of Si atoms being a first neighbor of an atom which is classified as cubic diamond, and 2nd neighbor is a number of Si atoms being a second nearest neighbor of an atom which is classified as cubic diamond. Significant changes were not observed until 2.6 ns and drastically increased after 2.6 ns. Fig. 4 shows the simulated a-Si:H/c-Si heterostructure at an annealing duration of (a) 0.00, (b) 2.60, (c) 3.72, and (d) 4.50 ns. Blue spheres are cubic diamond Si atoms, light blue are 1st neighbor Si atoms, green are 2nd neighbor Si atoms, white are other Si atoms, and pink is hydrogen atoms. The upper figure shows the structure as it originally was, and for simplicity, only Si atoms, which are similar to the diamond structure, are drawn in the lower figure. As shown in Fig. 4 (a), the diamond-like crystal structure was observed at the heterointerface before annealing. Comparing Fig. 4 (a) and (b), the a-Si:H(i) layers have similar structures. However, Fig. 4 (c) and (d) show that crystallization of a-Si:H(i) is more advanced as cooling progresses. These results indicate that crystalline growth was observed at the a-Si:H(i)/c-Si heterointerface from the beginning of cooling, thus improving the crystallinity. In addition, local hydrogen diffusion was observed at the a-Si:H(i)/c-Si heterointerface with crystallization of a-Si:H(i) This suggests that the crystallization at a-Si:H(i)/c-Si heterointerface is strongly affected by the diffusion of the hydrogen atoms. From Fig. 4. (c), the intermittent increase in crystallinity suggests the formation of metastable nanocrystals in the a-Si:H(i) film.

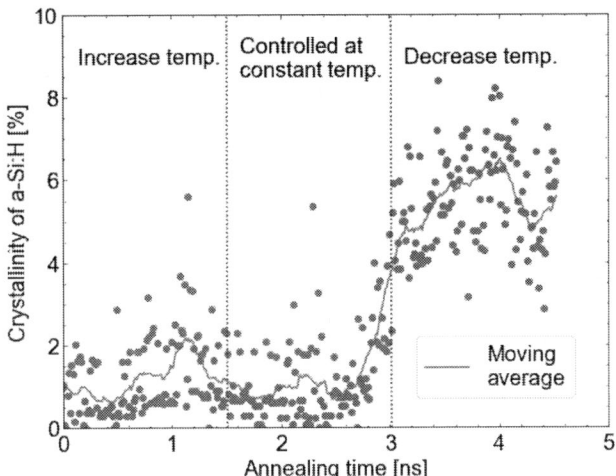

Fig. 3.   Time evolution of crystallinity of a-Si:H during the annealing process. The red line represents a moving average.

978-1-6654-6060-6/23 $31.00 © 2023 IEEE

ACKNOWLEDGMENT

This work was supported by New Energy and Industrial Technology Development Organization (NEDO) and MEXT, Grants-in-Aid for Scientific Research on Innovative Areas "Hydrogenomics," JP18H05514. Numerical simulations were performed on the Supercomputer system "AFI-NITY" at the Advanced Fluid Information Research Center, Institute of Fluid Science, Tohoku University.

REFERENCES

[1] P. Wagner *et al.*, "Interdigitated back contact silicon heterojunction solar cells: Towards an industrially applicable structuring method," *AIP Conf Proc*, vol. 1999, 2018.

[2] S. Olibet, E. Vallat-Sauvain, and C. Ballif, "Model for a-Si:H/c-Si interface recombination based on the amphoteric nature of silicon dangling bonds," *Phys Rev B Condens Matter Mater Phys*, vol. 76, no. 3, 2007.

[3] Y. Zhang *et al.*, "Significant Improvement of Passivation Performance by Two-Step Preparation of Amorphous Silicon Passivation Layers in Silicon Heterojunction Solar Cells," *Chinese Physics Letters*, vol. 34, no. 3, 2017.

[4] S. de Wolf and M. Kondo, "Abruptness of a-Si:H/c-Si interface revealed by carrier lifetime measurements," *Appl Phys Lett*, vol. 90, no. 4, pp. 1–4, 2007.

[5] A. C. T. van Duin, S. Dasgupta, F. Lorant, and W. A. Goddard, "ReaxFF: A reactive force field for hydrocarbons," *Journal of Physical Chemistry A*, vol. 105, no. 41, pp. 9396–9409, 2001.

[6] M. D. Kluge, J. R. Ray, and A. Rahman, "Amorphous-silicon formation by rapid quenching: A molecular-dynamics study," *Phys Rev B*, vol. 36, no. 8, pp. 4234–4237, 1987.

[7] A. Stukowski, "Visualization and analysis of atomistic simulation data with OVITO-the Open Visualization Tool," *Model Simul Mat Sci Eng*, vol. 18, no. 1, 2010.

[8] S. Hara, T. Kumagai, S. Izumi, and S. Sakai, "Evaluation of Structural and Mechanical Properties of Amorphous Silicon Surfaces by a Combination Approach of Ab-initio and Classical Molecular Dynamics," *J. Soc. Mat. Sci., Japan*, vol. 54, no. 1, pp. 45–50, 2005.

[9] N. Uene, T. Mabuchi, M. Zaitsu, S. Yasuhara, and T. Tokumasu, "Reactive Force-Field Molecular Dynamics Study of SiGe Thin Film Growth in Plasma Enhanced Chemical Vapor Deposition Processes," *ECS Trans*, vol. 98, no. 5, pp. 177–184, 2020.

[10] N. Uene, T. Mabuchi, M. Zaitsu, S. Yasuhara, and T. Tokumasu, "Reactive Force-Field Molecular Dynamics Study of the Effect of Gaseous Species on Silicon Germanium Alloy Growth by PECVD Techniques," *2021 International Conference on Simulation of Semiconductor Processes and Devices (SISPAD)*, pp. 238–241, 2021.

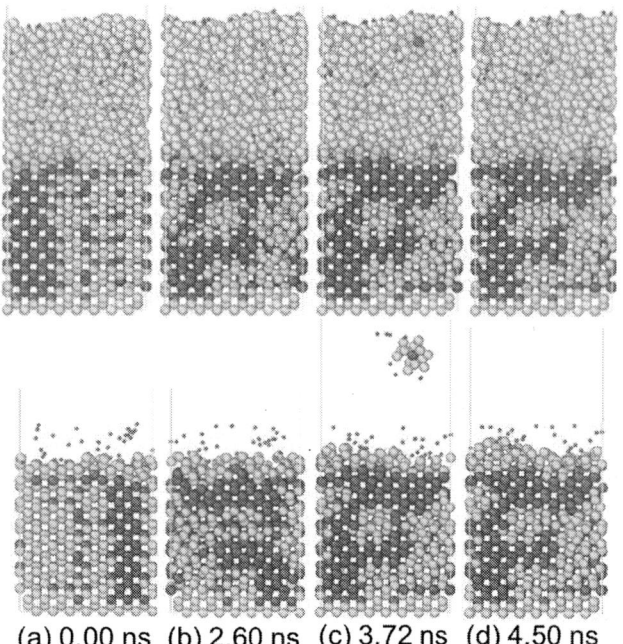

(a) 0.00 ns (b) 2.60 ns (c) 3.72 ns (d) 4.50 ns

Fig. 4.   Time evolution of crystallized part of the a-Si:H(i)/c-Si heterostructure during the annealing process. The upper figure shows the structure as it originally was, and for simplicity, hydrogen atoms, which are close to the heterointerface, and Si atoms, which are similar to the diamond structure, are drawn in the lower figure.

IV. CONCLUSIONS

We constructed the model of a-Si:H(i)/c-Si heterostructure in MD simulation and studied the processes of the crystallization of a-Si:H(i) and the diffusion of the hydrogen atoms in a-Si:H(i) films. The constructed model showed a good reproduction of the actual a-Si. We demonstrated that the ReaxFF MD simulation could be used to elucidate the processes in a-Si:H(i) films from an atomic scale. The computational system constructed in this study is applicable to a-Si:H thin films with various hydrogen concentration distributions, and is expected to contribute to process proposals for improvement in the passivation performance of the a-Si:H(i) films.

Planned Field Test of Soiling and Irradiance Measurement Uncertainties in Bifacial PV Systems Using In-Situ Reference Modules

Michael Gostein[1], Audrey Marquis[1], Marine Bila[2], Robert Campbell[2]

[1]Atonometrics, Austin, TX, USA; [2]EDF Renewables, San Diego, CA, USA

Abstract — We present a study plan to investigate uncertainties in soiling and irradiance measurement in a bifacial photovoltaic (PV) power plant using in-situ I-V measurements of module power. Soiling ratio is the ratio of actual module power output to expected power under clean conditions. However, precise determination of expected module power output for clean conditions is challenging for bifacial systems because of the need to account not only for front-side but also for rear-side irradiance contributions which have greater sources of variability. We present plans for a recently initiated field test to assess two methods of determining soiling ratio. In both methods we will use in-situ module I-V measurement to directly measure power output of a soiled reference module within the plant, yielding the numerator of soiling ratio. In method one, for the denominator of soiling ratio, the expected module power, we will use power measured from another reference module that is routinely cleaned. In method two, for the denominator we will use module power estimated from front and rear-side cleaned reference cells. Our aim is to determine the precision of soiling ratio measurements in a bifacial system given the challenge of normalizing for rear-side irradiance. For both methods we plan to quantify drift in the soiling ratio baseline corresponding to zero soiling loss and use this as a measure of soiling ratio precision.

Index Terms — photovoltaic (PV) systems, soiling, bifacial PV

I. INTRODUCTION

The accumulation of dust and other contaminants on photovoltaic (PV) modules, known as soiling, is one of the principal energy loss factors in PV systems. Soiling is quantified by the soiling ratio (*SR*), the ratio of actual PV power output to expected output under clean conditions [1]. Early work on direct measurement of soiling ratio compared the output of soiled modules to regularly cleaned modules [2], using short-circuit current as a proxy for module power. This concept has also been extended to comparing short-circuit current of soiled and clean mini-modules or reference cells. In recent years soiling sensors employing all-optical detection of accumulated dust have been developed [3][4][5]. These approaches work well when soiling accumulations are uniformly distributed on module surfaces. However, soiling accumulations are often non-uniform, with concentrated bands of soiling at module edges, especially bottom edges. For non-uniform soiling, more accurate measurements of soiling ratio are achieved by measuring module power and calculating the ratio of measured soiled power to expected power [6][7][8] for clean conditions. Expected power, the denominator of *SR*, can be determined either by using clean modules identical to the soiled modules [7] ("module-module") or by using periodically cleaned reference cells [8] ("module-cell").

For bifacial PV systems, measurement of soiling ratio based on measured soiled module power becomes more challenging. To compute the denominator of *SR*, clean module expected power must be determined considering the impact of total irradiance including contributions from both front and rear sides. In the module-module approach, a clean module identical to the soiled module should be used, but, in addition, for bifacial systems care should be taken that factors affecting rear-side irradiance contributions are identical between soiled and clean modules. These factors include position within a module row, albedo variations, and variations in structural shading. In the module-cell approach, reference cells should be positioned on both the front and rear sides of the module, and rear-side reference cells must be placed in optimal locations for measuring average rear-side irradiance across the module area, as in [9].

Equipment for measuring soiled module power output is also evolving. In-situ module-level I-V measurement systems [10][11] now provide the ability to directly measure maximum power of modules while they are still connected to a PV array. This eliminates the need for standalone reference modules for soiling measurement and other applications and ensures that soiling reference modules are always identical to standard modules within an array.

In this work we present plans for our recently initiated field study of soiling and irradiance measurement in a bifacial PV system using in-situ I-V measurement to determine soiled module power. We will be evaluating both module-module and module-cell methods for normalizing soiled module power to clean expectations based on total front plus rear-side irradiance contributions. Our primary goals are to demonstrate effective normalization methods for total irradiance contributions and to quantify expected precision levels for soiling ratio measurements in bifacial systems.

II. EXPERIMENT PLAN

Measurements will be performed at a utility-scale single-axis tracking bifacial PV power plant operated by EDF Renewables in North America. Atonometrics in-situ I-V measurement units (RDE300i) and bifacial-optimized reference cells (RC22) are in process of installation at the site as of May 2023. Fig. 1 illustrates the site and Fig. 2 illustrates the equipment layout. Module I-V data including short-circuit current and maximum power will be collected once per minute on two reference modules, one of which will be designated a "clean" module and washed weekly and the other of which will be designed as a

Fig. 1. Left: In-situ I-V unit for measuring soiled reference module. Right: rear plane of array; rear-side of reference cells will be mounted on either side of the torque tube.

"soiled" module. Three reference cells will be used for determination of total effective irradiance, one on the module front side and two on the rear. Rear-side reference cells will be positioned about mid-way between the tracker torque tube and the module edge, so that the average of the two reference cell readings closely represents the overall average rear-side irradiance as suggested in [9]. Reference cells will also be washed weekly.

Using this equipment set we will calculate daily average values for SR using two different methods to normalize for clean expectations. In both methods the numerator for SR will

be the temperature-corrected measured soiled module power. In the module-module method, the denominator will be the temperature-corrected clean module power multiplied by a calibration factor determined upon initial setup. In the module-cell method, the denominator will be an expected module power calculated by using the reference cells for front and rear irradiance, taking into account module bifaciality factors, temperature correction, and initial calibration.

Our aim is to determine precision of soiling measurements and any uncertainty due to the normalization methods. Therefore, our primary focus is on drifts in the baseline value of SR, the value corresponding to zero soiling. Ideally, this value is identically 1. However, any uncertainty in normalizing for total irradiance would cause SR to deviate from 1 even if the soiled module is clean.

We plan to quantify SR baseline drift in several ways. By cleaning the designated soiled module several times during the study, we can recheck the baseline value of SR to quantify any deviation over time. In addition, by comparing daily SR values in between the weekly cleanings of the clean module and clean reference cells, when SR should remain approximately constant due to all devices soiling simultaneously, we can further assess baseline drift, provided that soiling rates of clean and dirty devices are nearly the same. We also plan to calculate hourly averaged soiling ratio to elucidate systematic within-day variations in normalization.

In addition, we will highlight the use of in-situ I-V measurement for determining soiled module power and the

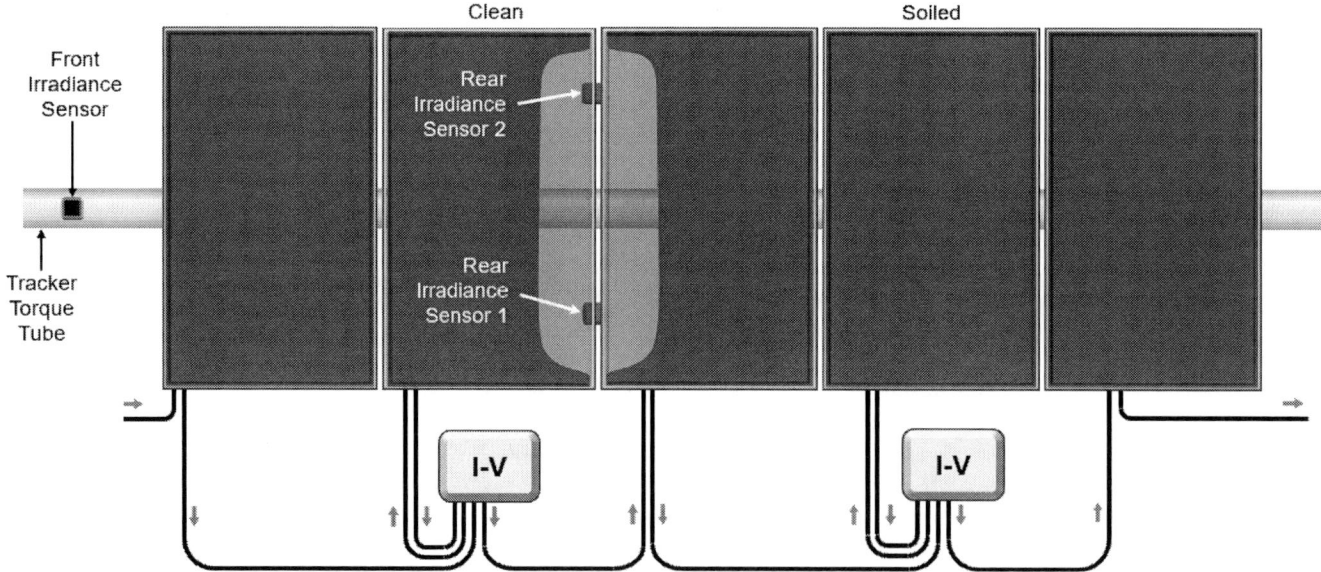

Fig. 2. Layout of the experimental setup, showing string and reference module wiring to in-situ I-V units and front and rear reference cells.

978-1-6654-6060-6/23 $31.00 © 2023 IEEE

compatibility of this measurement method with the PV power plant inverter.

III. RESULTS

Measurements will begin in June 2023. Some preliminary results may be presented on our conference poster. Complete results will be presented later in another forum.

ACKNOWLEDGEMENT

This material is partially based upon work supported by the U.S. Department of Energy's Solar Energy Technologies Office under Award Number DE-SC0020831.

REFERENCES

[1] "IEC 61724-1 Ed. 1.0 en:2017 - Photovoltaic system performance - Part 1: Monitoring."

[2] M. Gostein, J. R. Caron, and B. Littmann, "Measuring soiling losses at utility-scale PV power plants," in *2014 IEEE 40th Photovoltaic Specialist Conference, PVSC 2014*, 2014, pp. 885–890, doi: 10.1109/PVSC.2014.6925056.

[3] M. Gostein, S. Faullin, K. Miller, J. Schneider, and B. Stueve, "Mars Soiling Sensor™," in *2018 IEEE 7th World Conference on Photovoltaic Energy Conversion, WCPEC 2018 - A Joint Conference of 45th IEEE PVSC, 28th PVSEC and 34th EU PVSEC*, 2018, pp. 3417–3420, doi: 10.1109/PVSC.2018.8547767.

[4] M. Gostein, B. Bourne, F. Farina, and B. Stueve, "Field Testing of Mars™ Soiling Sensor," in *Conference Record of the IEEE Photovoltaic Specialists Conference*, Jun. 2020, vol. 2020-June, pp. 0524–0527, doi:

[5] M. Korevaar, J. Mes, P. Nepal, G. Snijders, and X. van Mechelen, "Novel soiling detection system for solar panels," in *33rd European Photovoltaic Solar Energy Conference*, 2017, pp. 2349–2351.

[6] M. Gostein, B. Littmann, J. R. Caron, and L. Dunn, "Comparing PV power plant soiling measurements extracted from PV module irradiance and power measurements," *Conf. Rec. IEEE Photovolt. Spec. Conf.*, pp. 3004–3009, 2013, doi: 10.1109/PVSC.2013.6745094.

[7] M. Gostein, T. Duster, and C. Thuman, "Accurately measuring PV soiling losses with soiling station employing module power measurements," *2015 IEEE 42nd Photovolt. Spec. Conf. PVSC 2015*, Dec. 2015, doi: 10.1109/PVSC.2015.7355993.

[8] S. Kagan *et al.*, "Impact of Non-Uniform Soiling on PV System Performance and Soiling Measurement," *2018 IEEE 7th World Conf. Photovolt. Energy Conversion, WCPEC 2018 - A Jt. Conf. 45th IEEE PVSC, 28th PVSEC 34th EU PVSEC*, pp. 3432–3435, Nov. 2018, doi: 10.1109/PVSC.2018.8547728.

[9] N. Riedel-Lyngskar, M. Bartholomaus, J. Vedde, P. B. Poulsen, and S. Spataru, "Measuring Irradiance With Bifacial Reference Panels," *IEEE J. Photovoltaics*, Nov. 2022, doi: 10.1109/JPHOTOV.2022.3201468.

[10] J. E. Quiroz, J. S. Stein, C. K. Carmignani, and K. Gillispie, "In-situ module-level I-V tracers for novel PV monitoring," *2015 IEEE 42nd Photovolt. Spec. Conf. PVSC 2015*, Dec. 2015, doi: 10.1109/PVSC.2015.7355608.

[11] A. Marquis, M. Gostein, and B. H. King, "Validation of In-Situ I-V Measurement Unit for PV System Monitoring Applications," in *2022 IEEE 49th Photovoltaics Specialists Conference (PVSC)*, Nov. 2022, pp. 0291–0294, doi: 10.1109/PVSC48317.2022.9938898.

10.1109/PVSC45281.2020.9300975.

Intermediate-Phase Engineering via Dimethylammonium Cation Additive for Stable Perovskite Solar Cells

David P. McMeekin, Philippe Holzhey, Sebastian O. Fürer, Steven P. Harvey, Laura T. Schelhas, James M. Ball, Suhas Mahesh, Seongrok Seo, Nicholas Hawkins, Jianfeng Lu, Michael B. Johnston, Joseph J. Berry, Udo Bach, Henry J. Snaith

Clarendon Laboratory, Department of Physics, University of Oxford, Oxford, United Kingdom

Department of Chemical Engineering, Monash University, Melbourne, Australia

ARC Centre of Excellence for Exciton Science, Monash University, Melbourne, Australia

Material Science Center, National Renewable Energy Laboratory, Golden, CO, United States

Applied Energy Programs, SLAC National Accelerator Laboratory, Menlo Park, CA, United States

Chemistry and Nanoscience Center, National Renewable Energy Laboratory, Golden, CO, United States

Department of Zoology, University of Oxford, Oxford, United Kingdom

Achieving the long-term stability of perovskite solar cells is arguably the most critical challenge to enabling their widespread commercialization. Understanding the perovskite crystallization process and its direct impact on device stability is essential to achieve this goal. Surprisingly, we find that intermediate phases that occur during the crystallization process strongly influence the long-term perovskite device stability. The commonly employed dimethyl formamide/dimethyl sulfoxide (DMF/DMSO) solvent system preparation method results in poor crystal quality and microstructure of the polycrystalline perovskite films. In this work, we introduce a high-temperature DMSO-free processing method that utilizes dimethylammonium chloride (DMACl) as an additive to control the perovskite intermediate precursor phases accurately. By precisely controlling the 2H to 3C perovskite phase crystallization sequence, we tune the grain size, texturing, orientation (corner-up vs face-up), and crystallinity of the formamidinium $(FA)_yCs_{1-y}Pb(I_xBr_{1-x})_3$ perovskite system. Encapsulated devices show significantly improved operational stability, with a champion device showing a T80 of 490 hours under simulated sunlight at 85 °C in air, under open circuit conditions. Our work introduces a new processing method that allows higher overall perovskite device stability by controlling the intermediate phase domains during the perovskite formation. This work highlights the importance of material quality to achieve long-term operational stability of perovskite optoelectronic devices.

978-1-6654-6060-6/23 $31.00 © 2023 IEEE

A Techno-Economic Analysis of Various Grid-Connected Photovoltaic System Configurations for Green Hydrogen Production

Rahul R Urs, Assia Chadly, Ahmad Mayyas

Department of Industrial and Systems Engineering, Khalifa University, Abu Dhabi 127788

I. INTRODUCTION

Annual solar photovoltaic (PV) installations will reach 162 GW by 2022, over 50% more than the pre-pandemic level of 2019 [1]. This tremendous increase in PV system integration is mainly due to the reduced system component costs, attractive incentives, subsidies, and feed-in-tariff (FiT) rates offered by governments worldwide. However, although historical trends are not always indicative of future situations, the analysis from Gilmore J. et al., [2] shows the solar industry's long-term concerns regarding costs are not declining swiftly enough to ensure sustained reliability. Green Hydrogen has been gaining wider interest throughout the years due to its numerous qualities and unlimited potential namely as a clean and efficient energy carrier. However, producing Hydrogen can be a costly process both technically and economically. As such, its production relies majorly on fossil fuels. This led to the introduction of green Hydrogen production via renewable energy resources, such as photovoltaics (PV), by electrolyzing the Hydrogen [3]. The electrolysis process produces green Hydrogen with zero carbon footprint, making it the most efficient method of production [4]. From the economic perspective, producing Hydrogen is still a relatively expensive process, as it depends on different factors related to the PV and the electrolyzer such as the PV module cost, the electrolyzer stack cost, the balance of plant (BOP), the balance of system (BOS), in addition to the cost of operation and maintenance (O&M), among others [5]. Therefore, the type of PV has a major impact on the LCOH. In fact, the efficiency of the PV is one of the most influential factors on the LCOH as well. Gallardo et al. [6] concluded that some country-specific characteristics such as capital expenditures and electricity prices significantly impact the LCOH. Therefore, the LCOH is lower in areas with higher levels of solar radiation and vice-versa. Sevik [7] evaluated the economic impact of a hybrid PV-trigeneration-hydrogen on electricity and Hydrogen production in a university campus in Turkey, and concluded that a higher efficiency eventually resulted in lower production costs, with the lowest LCOE and LCOH values recorded when 100% of the grid power is used. Consequently, and depending on the operating hours, the resulting LCOE and LCOH ranged between $0.068/kWh and $0.073/kWh, and between $1.78/kg and $3.4/kg, respectively.

The location influences the amount of solar irradiation

TABLE I
PV SYSTEM SPECIFICATION

Parameter	Value
Global Horizontal Irradiance	2015 kWh/m^2
Optimum tilt in degrees	30
Weather database	Meteonorm 8.0
Location	Abu Dhabi
Latitude	24.24 N
Longitude	54.65 E
Type/ efficiency of inverter	String/ 94%
Type/ efficiency of monofacial module	Monocrystalline/ 15.36%
Type/ efficiency of bifacial module	Monocrystalline/ 17.69%
Ohmic loss (DC circuit)	1.50%
Module quality loss	0.80%
Module power loss at MPP	2.00%
String power loss at MPP	0.10%

received by the solar modules, but it also influences the expenses of operation of the PV systems. While the literature provides research on the performance of PV systems with different mounting/tracking structures and solar cell types, this work analyses the techno-economic performance of PV systems with different module technology (monofacial and bifacial) and mounting and single-axis tracking structures. Upto our current knowledge, there is no literature discussing the impact of different PV configurations on LCOH. Hence in this study we consider different configurations PV systems and analyse its impacts on LCOH.

II. SYSTEM DESCRIPTION

The PV system in this study is assumed to be capacity of 30kWp. The system location is Khalifa University Campus, Abu Dhabi, UAE. The PV system specification and components are described in Table I and Table II, respectively. The PV modules are either monofacial or bifacial, and either a mounting structure or a single-axis tracking system were adopted in each PV system. As a result, each PV system is tested in four different configurations as tabulated in Table III. The electrolyzer size was assumed to be 50% of PV size, that is 15kW with 67% efficiency or specific consumption of 58kWh/kg.

TABLE II
PV SYSTEM DESCRIPTION

Parameter	System
Peak Power in kW	30
PV module power in W (Monofacial)	250
PV module power in W (Bifacial)	300
Inverter power in kW	30
No of PV modules (Monofacial)	120
No of PV modules (Bifacial)	100
No of Inverters	1

TABLE III
CONFIGURATIONS OF PV SYSTEMS

Parameter	Module Type	Structure	Total Cost in USD
Configuration 1 (C1)	Monofacial	Mounting	36,435
Configuration 2 (C2)	Bifacial	Mounting	36,885
Configuration 3 (C3)	Monofacial	Single axis	39,435
Configuration 4 (C4)	Bifacial	Single axis	39,885

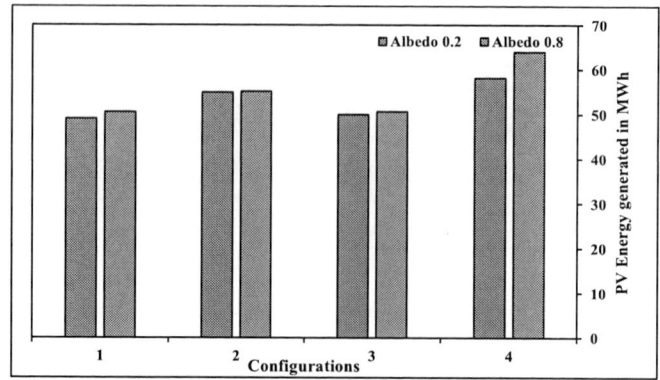

Fig. 1. Annual energy generation of PV system configurations

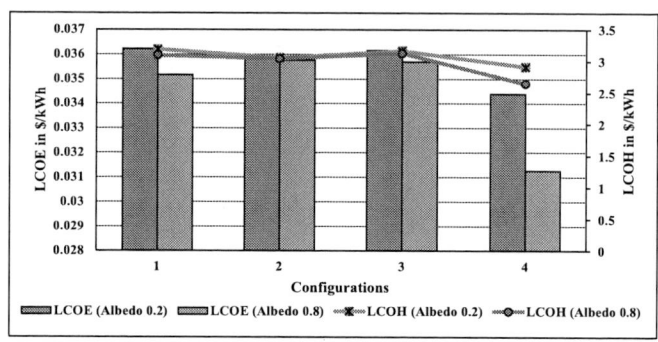

Fig. 2. LCOE and LCOH of PV system configurations

III. METHODOLOGY

The key objective of analyzing a PV system before installation is to determine the system's feasibility by estimating the total energy produced and transmitted to the grid. Furthermore, this assessment allows for determining the system's levelized cost of energy and hydrogen. PVsyst is a techno-economic platform mainly used to assess the performance of photovoltaic systems. Using PVsyst software, the PV system and its configuration described in the system description section were simulated.

All four configurations were assessed using generic PV system components from software. As shown in Table 3, the PV system was designed in a wide range of configurations and its related cost including system cost, Balance of system costs, installation cost, and total O&M cost for 25 years was estimated. For example, the monofacial PV module was chosen for configuration 1, and the bifacial PV module was chosen for configuration 2, with mounting structures in the first two configurations. Similarly, for the last two configurations with single-axis tracking, the monofacial PV module was chosen for configuration 3 and the bifacial PV module for configuration 4.

The LCOEpv and LCOH are calculated for each case using Eq. (1) and Eq. (2), respectively. The term LCOEpv2ez in Eq. (2) is the total cost of assessnergy consumed by the electrolyzer for hydrogen generation. A sensitivity analysis was also performed to examine the uncertainty of PV energy yield (5% and 10%) on LCOH values.

$$LCOE_{PV} = \frac{\text{Annualised PV Cost}}{\text{Yearly PV energy generated}} \quad (1)$$

$$LCOH = \frac{\text{Total EZ and HT Cost} + \text{Total } LCOE_{PV2EZ}}{\text{Total Hydrogen produced}} \quad (2)$$

IV. RESULTS AND DISCUSSION

The annual energy generation pattern for each configuration was different throughout the PVsyst simulation as shown in Figure 1. The difference in energy yield between monofacial and bifacial PV modules and changes in mounting systems induced this variation (fixed and single-axis tracking structures). In addition, the fluctuation in energy generation was influenced by the albedo of the surface. PV modules were installed on fixed mounting frames in the first two configurations. Even though the PV module in the second configuration (C2) was bifacial, there was no substantial increase in PV generation compared to the first configuration (C1). In the last two configurations, PV modules were installed on single-axis tracking mechanisms. With surface albedo 0.2 and surface albedo 0.8, the fourth configuration (C4) with bifacial module generated 5.8% and 16% more than the third configuration (C3). This significant gain over C4 was attributed to the increased energy yield and the single-axis tracking structure of the bifacial module.

The levelized cost of energy (in USD/kWh) and levelized cost of hydrogen were computed for each configuration and are shown in Figure 2. The LCOE and LCOH of C4 were the lowest for each system, with a surface albedo of 0.2 and 0.8, whereas the LCOE and LCOH of C1 were the highest. Although the cost of system components was higher for C4

978-1-6654-6060-6/23 $31.00 © 2023 IEEE   300

TABLE IV
P95 AND P90 COMPUTATION FOR LCOH BASED ON PV GENERATION
UNCERTAINTY

5% change in PV generation					
		C1	C2	C3	C4
Albedo 0.2	P95	3.159	3.136	3.027	2.885
	P90	3.165	3.142	3.033	2.891
Albedo 0.8	P95	3.065	3.014	3.096	2.620
	P90	3.071	3.102	3.020	2.626
10% change in PV generation					
Albedo 0.2	P95	3.137	3.113	3.006	2.859
	P90	3.149	3.125	3.018	2.871
Albedo 0.8	P95	3.042	3.071	2.991	2.592
	P90	3.054	3.083	3.003	2.604

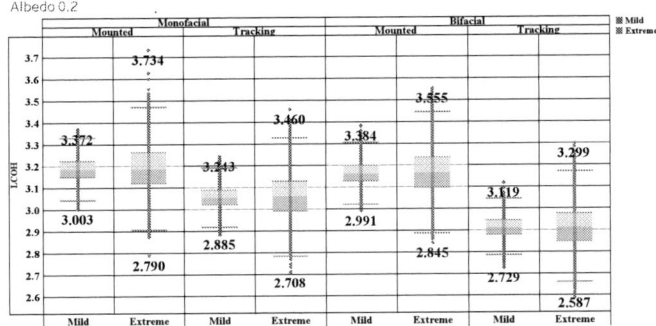

Fig. 3. Impact of PV generation uncertainty with albedo 0.2

and lower for C1, the difference in LCOE and LCOH was also due to the difference in lifetime energy generation.

The results show that under uncertain PV generation conditions, the LCOH values remain within acceptable ranges between \$2.6/kg and \$3.7/kg, slightly varying from the base case values. To give a better presentation of the results, 95% and 90% exceedance probabilities were determined for each case as illustrated in Table IV.

V. CONCLUSION

Based on the type of module and mounting structure, four possible configurations were defined. PVsyst software was used to model these configurations for surface albedo 0.2 and 0.8 for a university in Abu Dhabi, UAE. From the technical

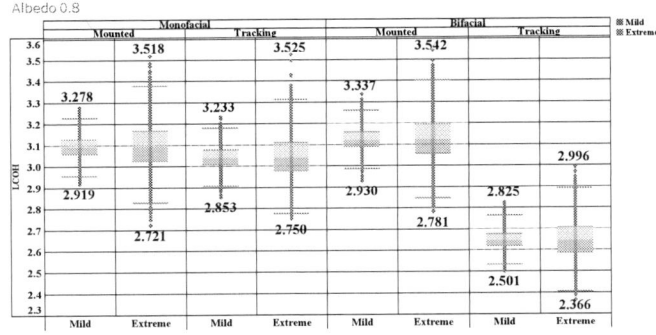

Fig. 4. Impact of PV generation uncertainty with albedo 0.8

assessment, the fourth configuration of each system, which included a combination of bifacial modules and a single-axis tracking structure, generated more energy than any other configuration. However, compared to the first configuration, the second configuration with a bifacial module and mounting structure did not observe a substantial increase in energy generation. Specifically, configurations with bifacial modules performed well with a surface albedo of 0.8, boosting the performance ratio of PV systems. Furthermore, with albedo 0.2 and 0.8, the fourth configuration generated more energy than the first configuration, generating more than 18.32% and 26.3%, respectively. Furthermore, the fourth configuration in each system with the maximum annual energy generation reduced the levelized cost of energy and hydrogen, despite the higher capital cost of these systems. This reduction is due to the prevalence of increased energy generation over additional costs. For example, for albedo 0.2 and 0.8, the levelized cost of energy of the fourth configuration was 8.3% and 15.32% lower, respectively and the levelised cost of hydrogen was 9.42% and 16.85% lower respectively than the first configuration. The uncertainty analysis also ensured that C4 is the best configuration as it results in the lowest LCOH despite uncertainty. As a result of these observations, hydrogen production from a PV system with a bifacial module and single-axis axis tracking structure is more feasible than a PV system with a monofacial module combined with the mounting structure for the UAE location.

REFERENCES

[1] "Renewable Energy Market Update 2021 – Analysis - IEA." https://www.iea.org/reports/renewable-energy-market-update-2021 (accessed Oct. 14, 2021).

[2] J. Gilmore, B. Vanderwaal, I. Rose, and J. Riesz, "Integration of solar generation into electricity markets: an Australian National Electricity Market case study," IET Renewable Power Generation, vol. 9, no. 1, pp. 46–56, Jan. 2015, doi: 10.1049/IET-RPG.2014.0108.

[3] J. A. Azzolini, M. Tao, K. Ayers, and J. Vacek, "A Load-Managing Photovoltaic System for Driving Hydrogen Production," Conf. Rec. IEEE Photovolt. Spec. Conf., vol. 2020-June, pp. 1927–1932, Jun. 2020, doi: 10.1109/PVSC45281.2020.9300922.

[4] B. Mali, D. Niraula, R. Kafle, and A. Bhusal, "Green Hydrogen: Production Methodology, Applications and Challenges in Nepal," 2021 7th Int. Conf. Eng. Appl. Sci. Technol. ICEAST 2021 - Proc., pp. 68–76, Apr. 2021, doi: 10.1109/ICEAST52143.2021.9426300.

[5] M. A. Khan, I. Al-Shankiti, A. Ziani, and H. Idriss, "Demonstration of green hydrogen production using solar energy at 28% efficiency and evaluation of its economic viability," Sustain. Energy Fuels, vol. 5, no. 4, pp. 1085–1094, Feb. 2021, doi: 10.1039/D0SE01761B.

[6] F. I. Gallardo, A. Monforti Ferrario, M. Lamagna, E. Bocci, D. Astiaso Garcia, and T. E. Baeza-Jeria, "A Techno-Economic Analysis of solar hydrogen production by electrolysis in the north of Chile and the case of exportation from Atacama Desert to Japan," Int. J. Hydrogen Energy, vol. 46, no. 26, pp. 13709–13728, Apr. 2021, doi: 10.1016/J.IJHYDENE.2020.07.050.

[7] S. Şevik, "Techno-economic evaluation of a grid-connected PV-trigeneration-hydrogen production hybrid system on a university campus," Int. J. Hydrogen Energy, vol. 47, no. 57, pp. 23935–23956, Jul. 2022, doi: 10.1016/J.IJHYDENE.2022.05.193.

Multi-layer dense antireflection coatings

Yiyu Zeng, Martin Green, Jessica Yajie Jiang

University of New South Wales, Sydney, NSW, 2033, Australia

Abstract — **To match the lifespan of installed modules, ARCs are expected to resist the harsh abrasion from brush cleaning and withstand severe environments. It was found that dense ARCs present much better durability compared to commercial inner-pore and open-pore ARCs in the market, enabling a much longer lifespan although optical benefits were reduced. However, the optical benefit of single-layer dense silica is limited. In this work, Five-layer dense ARCs were investigated for boosting durability as well as maintaining a reasonable optical enhancement compared to single-layer ARC. These findings provide substantial guidance on the optimization of sustainable future commercial ARCs.**

I. INTRODUCTION

PV glass is an effective barrier that protects the fragile PV cells against external damage such as abrasion, water, dirt, and soiling. According to the Fresnel reflection formula, there is about 4.26% sunlight reflected at front of the panel glass (refractive index, RI, n_{air}=1 and n_{glass}=1.52), which does not contribute to photon generation.[1,2] The most common approach to suppress such reflection loss is introducing a single-layer antireflection coating with the RI value intermediate between the n_{air} and n_{glass}, ideally fulfilling the relationship of $n_{ARC} = \sqrt{n_{air}n_{glass}}$ with an optical thickness equal to a quarter of the desired wavelength. Under this condition, the destructive interference between two reflected beams (from the front and rear surface) occurs, which leads to the cancellation of these beams and thereby minimizes the unwanted reflection.[3] Typical single-layer ARCs enhance the total solar transmittance by around 2%, especially over the 500-600 nm wavelength range, increasing PV module power generation. Such small increases in solar transmittance through the PV glass can impart a significant commercial advantage.

The theoretical optimum RI of a single layer ARC (nARC) is around 1.23. However, it is difficult to source such a naturally occurring low-RI material. Approaches to creating porous silica have been widely investigated to reduce the density of the film and then reduce the RI. Sol-gel coating by roll-coater is currently the most common way to deposit silica ARC as it is low-cost and suitable for mass production in the solar industry. 1968, Stöber et al. fabricated ARCs comprise homogeneous spherical silica nanoparticles of uniform shape and size by means of base-catalyst method, the RI of fabricated silica film is 1.22, considered as the ideal RI for SLARC. Since 1990s, this technology has been widely reported for the silica ARCs. However, the is so called open-pore film has poor mechanical strength. To increase the mechanical strength of open-pore silica ARC while maintaining the optical enhancement, an inner-pore structure where hollow silica spheres are embedded in the dense silica coating has been widely used, with the size of the spheres and hence pores usually within the 30–100 nm range. This technology was first reported by Boilot et al., in 2010 and then introduced at a commercial scale by DSM in 2012 and it still represents the state-of-the-art technology. Recently, we collected the dominant ARCs from the solar industry and did a systematic optical and durability characterization by comparing with the lab-fabricated dense silica film, it was found that the dense SLARC gives the least pronounced optical enhancement but shows the most robust wear-resistance.[4]

In order to take the durability advantage of the dense silica and to beat the optical performance of the commercial inner-pore SLARCs, a more sophisticated approach to increasing the solar transmittance of PV glass is to introduce a multilayer ARC designs that broaden the anti-reflection spectral bandwidth. SiO_2 is commonly used as a low-index material. Among the high-index coatings, TiO_2 is a cost-effective, non-toxic photocatalyst with self-cleaning function that can be used in the multi-layer ARCs.

Besides, the multi-layer ARCs has the potential to reduce the operating temperature of solar modules by designing the RI and thickness of each layer of multi-layer ARCs, It is well know commercial PERC cells respond only to wavelengths < 1200 nm, in the standard AM1.5G spectrum, there is 836 W/m² below this wavelength, with 164 W/m² above it, unusable by the cell that will be transferred to heat. The certain multi-layer ARCs can improve the transmittance in the usable solar spectrum range and reject some portion of light in the unusable spectrum range.

In this abstract, a 5-layer ARC made of various mixture TiO_2/SiO_2 layers have been investigated. The fabricated 5-layer ARC presents comparable transmittance compared to the commercial ARCs but with much better mechanical strength.

II. DETAILED EXPERIMENT

A. Optical performance of Commercial ARCs

To understand the optical performance of the mainstream product in the market, we collected and characterized 6 SLARC PV glass samples provided by various companies. All SLARC are coated on low-iron super-white tempered glass. Such PV glasses are roller patterned with shallow texturing on the backside. The coatings are deposited on the smooth side, and the textured side will be attached to encapsulant in a laminated

978-1-6654-6060-6/23 $31.00 © 2023 IEEE

Table 1 The optical parameters of various commercial SLARCs

Specimen	RI	Thickness	Weighted transmittance
A	1.45	31 nm	92.9%
B	1.37	80 nm	94.3%
C	1.34	123 nm	94.5%
D	1.35	126 nm	94.1%
E	1.37	106 nm	94.2%
F	1.32	130 nm	94.9%
Glass	1.51	2 mm	92.1%

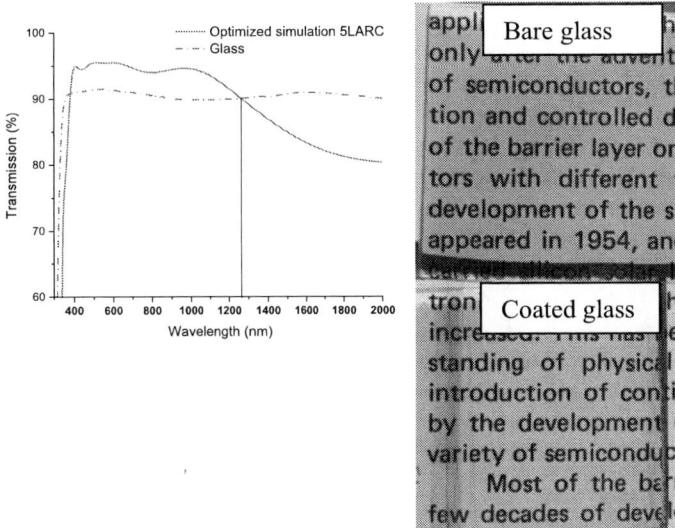

Figure 1 The optimized transmittance curve of the 5-layer ARC

module. The RI and thickness of each ARC were determined by ellipsometry. The RI of a typical dense silica coating is around 1.45 at 550 nm. As shown in Table 1, five companies use the inner-pore silica ARC with a RI of about 1.35, while only one company uses dense silica with a RI of 1.45. As shown in Table 1, the transmittance in the visible range is remarkably improved with all ARCs compared to bare glass. For practical PV application, the solar spectral irradiance at air mass (AM) 1.5 and the cell external quantum efficiency (EQE) should be considered. Therefore, we introduce a weighted PV transmittance (TW) for performance evaluation:

$$T_W = \frac{\int_{350\,nm}^{1200\,nm} E_{AM1.5G}(\lambda)\, EQE(\lambda)\, T(\lambda)\, \lambda\, d\lambda}{\int_{350\,nm}^{1200\,nm} E_{AM1.5G}(\lambda)\, EQE(\lambda)\, \lambda\, d\lambda}$$

B. Optical performance of Multi-layer ARCs

Herein, we have run a simulation of the reflection loss (R_EQE, the EQE of a PERC cell is considered) by minimizing the reflection loss within the usable solar spectrum and by maximizing the energy loss within the unusable spectrum as mentioned in the section A. As plotted on the Figure 1, the reflection loss is gradually reduced by increasing the number of layers from 1 to 10 while the heat energy loss is gradually

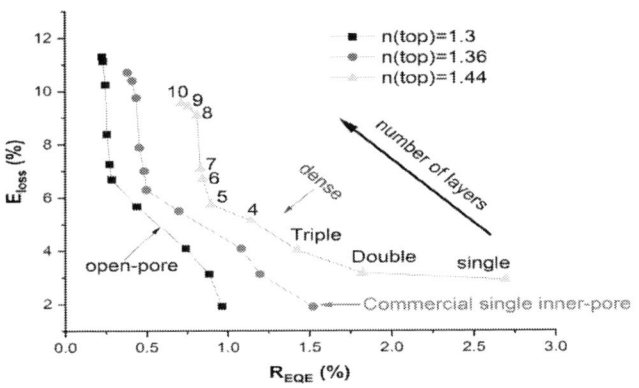

Figure 2 The simulation results of optimum R_EQE and E_Loss of the ARCs with various layers.

increased. It can be found the reflection reduction is much more significant when adding the layers from 1 to 5 compared to 5 to 10 layers. Therefore, the optical benefit is very limited when the number of the coating layers is above 5 layers. Furthermore, the open pore structures present better performance than the inner-pore and dense structures. As we aim to fabricate the dense structures, based on the simulation results, the 5-layer ARC is a compromised choice with good optical property and moderate cost. It is clear seen in the Figure 1, the R_EQE can be lowered to 1% by using the 5-layer dense structure, which is much better compared to the commercial inner-pore structure (1.5%), while the E_Loss is about 3 times higher than the commercial coatings. The optimized ARC of the 5-layer structure is illustrated in Figure 2, the weighted transmittance of the 5-layer ARC is 94.58%, which is comparable to the textured commercial single-layer coating. It should be noted the transmittance can be further improved by replacing the flat glass substrate with the textured commercial glass. The transmittance at above 1200 nm of such 5-layer coating is reduced below the 80% while the transmittance is above 92% during the whole spectrum for the commercial SLARC. The rejection of those light in infrared range can lead to a operating temperature reduction of the module.

C. ABRASION TESTING

It is important to investigate the abrasion resistance of different ARCs to evaluate the ability to withstand the typical cleaning process for long-term PV application. A standard artificial linear machine abrasion test (IEC 62788-7-3) designed for PV applications was performed on all the coatings.[5] The dry brush (without abrasive medium) and wet slurry (with abrasive medium mixed in DI water) abrasion testing were employed in this section following the procedures reported in IEC 62788-7-

Table 2 The layer thickness, refractive index, surface roughness, water contact angel, abrasion scratch width and depth of all ARC samples

Specimen	Material	Coating thickness/Index	RMS (nm)	Contact Angle (degree)	Scratch width (dry/wet) (μm)	Scratch depth (dry/wet) (nm)
O1	Open-pore Silica	131 nm/1.24	2.7	27	21/40	110/90
I1	Inner-pore Silica	80 nm/1.36	10	46	0.24/4.1	68/80
D1	Dense silica	110 nm/1.44	0.3	0	0.12/0.25	11/30

3. To quantitively compare the abrasion damage, AFM scanning was conducted to probe the scratch depth of these coatings. The scanning size is 10 μm × 10 μm, and only the scratched area was selected for analysis. The scratch width and depth after the dry-brush and the wet-slurry test are listed in Table 2. The depth of the scratch damage became deeper after the slurry test, indicating the damage from the slurry abrasion test is more detrimental than the dry-brush abrasion testing. Nevertheless, much shallower scratch depth and narrower scratch width were found on the dense ARCs, indicating both the dense SLARC have much better abrasion resistance than the commercial inner-pore ARCs.

IV. CONCLUSION

Five-layer dense ARC have been investigated and fabricated. The dense coatings exhibited much better wear-resistance compared to the commercial coatings while it presented comparable optical improvements compared to the commercial single-layer coatings. The fabricated 5-layer coating can improve the transmittance at selected usable wavelength range and reflect some energy at unusable infrared range to reduce the operating temperature.

REFERENCES

1 D. B. Judd, *J. Res. Natl. Bur. Stand. (1934).*, 1942, **29**, 329.

2 C. Ballif, J. Dicker, D. Borchert and T. Hofmann, 2004, **82**, 331–344.

3 B. İkizler, *Res. Eng. Struct. Mat*, 2020, **6**, 1–21.

4 Y. Zeng, N. Song, S. Lim, M. Keevers, Y. Wu, Z. Yang, S. Pillai, J. Y. Jiang and M. Green, *Sol. Energy Mater. Sol. Cells*, 2023, **251**, 112122.

5 *IEC 62788-7-3 Measurement procedures for materials used in photovoltaic modules – Part 7-3: Accelerated stress tests – Methods of abrasion of PV module external surfaces*, Geneva, 1st edn., 2022.

Transparent Conductive Oxide Bi-layer as Front Contact for Multijunction Thin-film Silicon Solar Cells

Federica Saitta, Prashand Kalpoe, Govind Padmakumar, Paula Perez-Rodriguez, Gianluca Limodio, Rudi Santbergen and Arno H.M. Smets

Photovoltaic Materials and Devices (PVMD) group, Delft University of Technology. Building 36, Mekelweg 4, 2628CD, Delft, The Netherlands

Abstract — Transparent conductive oxides (TCOs) are used as front electrode of thin film silicon (TF-Si) solar cells to increase power conversion efficiency. Metal oxides doped with different materials can be deployed as TCO. The preferred TCO is usually selected using a trade-off between transparency and conductivity. This work proposes a bi-layer front contact to address the limitation of this trade-off. IOH and i-ZnO are chosen as the best candidates for such architecture due to their good opto-electrical properties. A thin layer of IOH ensures good lateral conductivity and high transparency in the visible part of the solar spectrum. An additional i-ZnO layer provides minimized parasitic absorption losses along with low transverse resistivity. The best opto-electrical properties are achieved when deposition temperature and power density are set at 25 °C and 1.5 W/cm², 200 °C and 2 W/cm² for IOH and i-ZnO respectively.

I. INTRODUCTION

The optical and electrical properties of metal oxides rely on the oxidation state of the metal, and on the amount/nature of impurities embedded in the films [1]. The simultaneous optimization of both transparency and conductivity in a single TCO layer may lead to a compromise solution. Therefore, this investigation is directed towards a front electrode with two film-stack to maximize the opto-electrical trade-off. According to H. Tan *et al.* [2], the implementation of a TCO bi-layer front contact diminishes parasitic absorption losses and contributes to increasing the efficiency of multijunction thin-film silicon (TF-Si) solar cells.

The TCOs contribute to parasitic absorption across the optical bandgap: i) in the ultraviolet (UV) region, and ii) in the near infrared (NIR) region of the solar spectrum due to free carrier absorption (FCA). The first issue can be tackled by selecting a material with high energy bandgap to reduce the absorption in the short wavelength range and meet the opto-electrical trade-off. The second issue is more challenging and thus addressed in this work through a comprehensive analysis.

This paper characterizes the optical and electrical properties of multiple TCO materials, and then focuses on the bi-layer architecture of two of these TCO films embedded in the front electrode.

II. EXPERIMENTAL PROCEDURE

A. Fabrication

Four types of TCO materials are investigated in order to select the best candidates for a bi-layer TCO structure. They are hydrogenated indium-, indium tin-, intrinsic zinc- and aluminum doped zinc- oxides (IOH, ITO, i-ZnO, AZO). Corning glass (10 cm × 10 cm) is used as substrate for TCO film depositions. Each glass substrate is cleaned in acetone and isopropyl alcohol ultrasonic baths for 10 min respectively. IOH, ITO, i-ZnO and AZO layers are grown by using Radio Frequency (RF) magnetron sputtering technique. Time, power, and temperature are tuned to explore multiple deposition conditions for TCO fabrication. In the case of IOH, partial H_2O pressure is used as additional setting.

To improve film quality, TCO samples are subjected to a post-deposition annealing (PDA) treatment in an atmospheric environment. The saturated annealing time is 20 min and the annealing temperature (T_a) is varied between 130 and 250 °C.

B. Characterization

Given the complex interplay between optical and electrical properties, two main metrics are analyzed.

Electrically, the TCO is characterized by the conductivity (σ), which is directly proportional to carrier mobility (μ), free carrier concentration (N) and elementary charge (e). The Hall measurement allows to determine resistivity (ρ), measured by the so-called van der Pauw method, and then derive N and μ combined with the Hall effect [3]. Rather than conductivity, resistivity is also utilized as a device-relevant parameter to evaluate TCO electrical properties.

Optically, the sample is characterized by the absorption coefficient spectra (α). It is determined by a spectroscopic ellipsometry (SE) M-2000DI system (J.A. Woollam Co., Inc.) for single TCO layers. SE provides additional parameters such as bulk thickness (t_b), refractive index (n), extinction coefficient (k) and optical bandgap (E_g) from Tauc plot according to SE-fitted α curve. In SE analysis, the dielectric function of TCOs sample is considered homogenous in depth and modelled by combining two oscillators: Cody-Lorentz and Drude [4]. For the bi-layers, a spectrophotometric PerkinElmer Lambda 1050 system is utilized to measure transmittance (T) and reflectance (R), and α is derived by applying the Lambert-Beer law. This second method is employed due to the complexity of the SE fitting using dedicated oscillator theories on bi-layers.

III. RESULTS AND DISCUSSION

The front contact architecture is designed by two layers (TCO_1 and TCO_2) in order to optimize the trade-off between the opto-electrical properties.

978-1-6654-6060-6/23 $31.00 © 2023 IEEE

The conductivity can be improved by increasing N and/or μ. However, there is an upper limit for σ due to the limitation of dopant scattering and doping efficiency. Mobility and carrier concentration are related to each other by a rule of $\mu \propto N^{-2/3}$. Therefore, a trade-off between N and μ imposes that σ cannot continuously increase [5].

Additionally, intra-band transitions within conduction band lead to parasitic free carrier absorption in the NIR region. In metals and semiconductors, FCA is modelled by the Drude oscillator theory which defines a direct dependency between α_{NIR} and N. The optimized trade-off results in TCO films with high mobility and limited free carrier concentration from both optical and electrical perspectives.

The PDA process is utilized to further enhance the opto-electrical properties of TCOs and determine a temperature optimum, where N and α_{NIR} are minimized and, μ is simultaneously maximized.

The overall thickness should not exceed 700 nm. The TCO layer deposited as the first layer of the stack on glass (TCO$_1$) should be sufficiently thin (100-200 nm) to guarantee good lateral conductivity for the collection of photo-generated carriers at the front contact. The layer deposited as second (TCO$_2$) should ensure low absorptance in the NIR region, along with good transverse electrical properties. TCO$_2$ should be thick enough (500-600 nm) to make it conductive.

First, each TCO material is characterized individually on glass at different deposition conditions as deposited (no PDA). Fig. 1 shows α_{NIR} (cm^{-1}) versus ρ (Ω cm) for ITO, IOH, i-ZnO and AZO. The values of the absorption coefficient at wavelength $\lambda = 1100$ nm are taken as a good indicator for FCA in NIR. The bottom-left corner of this plot (yellow triangle) represents the ideal situation for a TCO, in which α_{NIR} and ρ are below 10^2 cm^{-1} and 10^{-2} Ω cm respectively. Among the TCOs investigated, AZO and ITO are highly absorptive, between $3 \cdot 10^3$ and $1 \cdot 10^4$ cm^{-1}, and ρ values are smaller than $1 \cdot 10^{-2}$ Ω cm. IOH samples reveal the lowest ρ values between $9 \cdot 10^{-3}$ and $7 \cdot 10^{-3}$ Ω cm. Even if the characteristics of i-ZnO are widely spread, it underlines the poor conductivity of such intrinsic material, with the lowest α_{NIR} values.

Fig. 2 illustrates the relation between μ and N. The top-left corner (yellow triangle) represents the ideal electrical trade-off: μ above 30 cm^2/Vs and N below 10^{19} cm^{-3}. The order of magnitude of N is between 10^{20} and 10^{21} cm^{-3} for IOH, ITO and AZO samples, whereas it is between 10^{18} and 10^{19} cm^{-3} for i-ZnO films. In terms of μ, IOH stands out with values above 30 cm^2/Vs in as-deposited condition.

Characterized mono-film TCOs samples do not meet the optimal trade-off indicated in the yellow region in Fig. 1 and Fig. 2. Hence, high transparency and high conductivity are suggested to be optimized separately in a bi-layer front contact

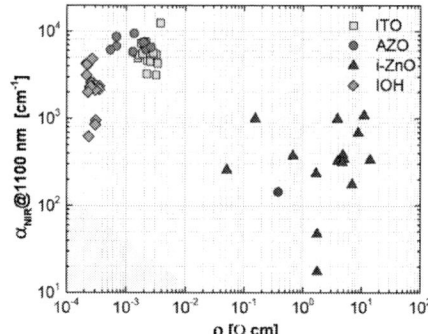

Fig. 1.   Absorption coefficient @ $\lambda = 1100$ nm vs resistivity of processed TCOs. Absorption coefficient is calculated through SE.

design. Intrinsic ZnO appears to be the most suitable TCO in terms of low FCA losses, while IOH clearly achieves the lowest ρ values of all TCOs examined, by preserving high free carrier mobility. Therefore, IOH is selected as TCO$_1$ and i-ZnO as TCO$_2$ in the bi-layer structure.

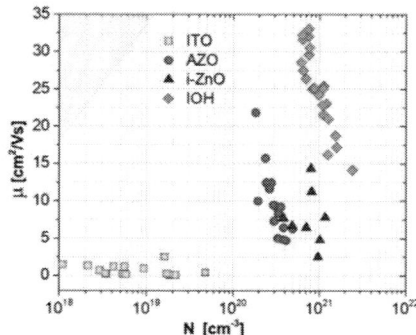

Fig. 2.   Mobility vs free carrier concentration of processed TCOs.

A. TCO$_1$: IOH deposition conditions and PDA

Once the IOH material has been selected as the best candidate for TCO$_1$ in the bi-layer structure, the deposition conditions are refined using a power density window between 1.3 and 1.8 W/cm^2, and the effect of PDA on α_{NIR}, μ and N is explored. Deposition temperature, H$_2$O partial pressure and thickness are set at 25 °C, $3 \cdot 10^{-5}$ mbar and 200 nm respectively. Deposited samples are then annealed at 130, 180, 200 and 250 °C.

Fig. 3 shows how the progressive increase of annealing temperature rapidly decreases α_{NIR} and N values, and displays a simultaneous μ gain. A plateau occurs at 200 °C, representing the optimized PDA temperature for IOH. Higher T_a might cause hydrogen atoms effusion during the growth of the IOH film, leading to a deterioration of the film quality. Power density is chosen to be 1.5 W/cm^2 on account of the best opto-electrical trade-off seen in Fig. 3.

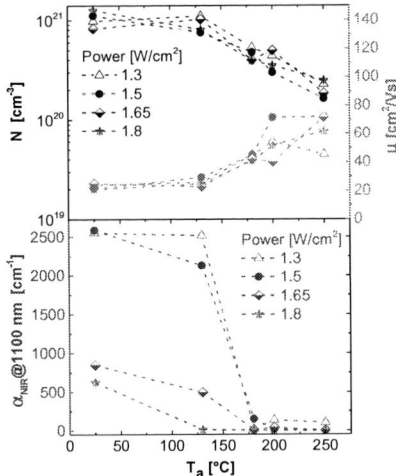

Fig. 3.   Processed IOHs at different power densities (W/cm²): mobility and carrier concentration vs annealing temperature (plot above);   absorption coefficient @ λ = 1100   vs annealing temperature (plot below).

B.   TCO₂: i-ZnO deposition conditions

The deposition conditions for i-ZnO are dependent on TCO_1. For instance, high power may damage the IOH surface due to ion bombardment. Additionally, i-ZnO cannot be deposited at higher temperature than the optimal T_a, since it would have negative effect on TCO_1's opto-electrical properties. Deposition temperature, thickness and power density are set at 200 °C, 500 nm and 2 W/cm² respectively.

C.   Towards bi-layer architecture

To investigate the opto-electrical metrics of the bi-layer design, 8 samples are processed: 6 bi-layers and 2 IOH films as reference. The IOH thicknesses are 180 and 100 nm. Each IOH thickness is then combined with i-ZnO processed with varying thicknesses of 600, 500 and 427 nm.

Fig. 4 shows the shift of opto-electrical properties from IOH references (orange) to bi-layer samples (blue). Bi-layer samples further minimize the FCA losses (α_{NIR} ~ 10^2 cm⁻¹) compared to IOH films, as shown in Figure 4 (a). However, the IOH single

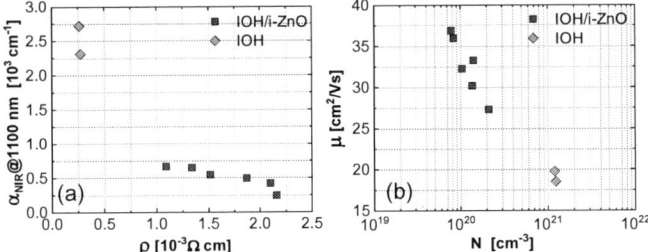

Fig. 4.   (a) Absorption coefficient @ λ = 1100 nm vs resistivity of IOH/i-ZnO samples and IOH references. Absorption coefficient is calculated through R,T measurements; (b) mobility against free carrier concentration of IOH/i-ZnO samples and IOH references.

layer is more conductive than the bi-layer design, whose ρ values are above $1 \cdot 10^{-3}$ Ω cm. Figure 4 (b) points out that high μ (above 30 cm²/Vs) and low N (10^{19}- 10^{20} cm⁻³) are preserved in IOH/i-ZnO structure without any PDA treatment.

IV. CONCLUSIONS

The complex interplay between opto-electrical properties makes the optimization of single TCOs challenging. A bi-layer architecture can be an alternative solution to single TCO layer as front electrode in TF-Si application. Hence, this work designs a bi-layer structure in which IOH and i-ZnO are the best TCO candidates. This stack is able to combine good lateral conductivity and low absorption in the NIR.

The investigation is then focused on IOH and i-ZnO deposition conditions. The effect of PDA is studied on IOH. The annealing temperature (T_a) equal to 200 °C shows optimized N, μ and α_{NIR} when power density is 1.5 W/cm². T_a also determines a threshold for the deposition temperature of the i-ZnO material. Additionally, the power density of the i-ZnO deposition is kept under control and equal to 2 W/cm² due to potential damage on the IOH surface.

A bi-layer structure is fabricated using these deposition conditions, combining several thicknesses for each of the layers. This makes it possible to range the overall thickness and investigate opto-electrical metrics on multiple samples.

Results exhibit α_{NIR} between 2.5 and $7 \cdot 10^2$ cm⁻¹, and ρ between $1 \cdot 10^{-3}$ Ω cm and $2.5 \cdot 10^{-3}$ Ω cm. The majority of μ values are above 30 cm²/Vs and N is in the range of 10^{19} - 10^{20} cm⁻³ without any post deposited annealing (PDA) treatment. Further improvements on the opto-electrical trade-off can be then achieved by including PDA in the fabrication of the bi-layer structure. Preliminary findings show μ up to 103 cm²/Vs for annealed bilayers.

REFERENCES

[1]   Bel Hadj Tahar, R., Ban, T., Ohya, Y., & Takahashi, Y., "Tin doped indium oxide thin films: Electrical properties". *Journal of Applied Physics*, 83(5), 2631-2645, 1998.

[2]   Tan, H., Moulin, E., Si, F. T., Schüttauf, J. W., Stuckelberger, M., Isabella, O., ... & Smets, A. H., "Highly transparent modulated surface textured front electrodes for high-efficiency multijunction thin-film silicon solar cells". *Progress in Photovoltaics: Research and Applications*, 23(8), 949-963, 2015.

[3]   Philips'Gloeilampenfabrieken, O. "A method of measuring specific resistivity and Hall effect of discs of arbitrary shape." *Philips Res. Rep* 13.1: 1-9, 1958.

[4]   Ferlauto, A. S., Ferreira, G. M., Pearce, J. M., Wronski, C. R., Collins, R. W., Deng, X., & Ganguly, G. "Analytical model for the optical functions of amorphous semiconductors from the near-infrared to ultraviolet: Applications in thin film photovoltaics". *Journal of Applied Physics*, 92(5), 2424-2436, 2002.

[5]   Han, C. "High-Mobility TCO-Based Contacting Schemes for c-Si Solar Cells.", PhD thesis, 2022, https://doi.org/10.4233/uuid:d6f35adf-486e-453a-9ae9-679a81105bed.

12.3% efficient lifted-off and reconstructed As-doped CdTe thin film solar cell

Ochai Oklobia[1], Deborah L. McGott[2], Giray Kartopu[3], Steve Jones[1], Stuart J. C. Irvine[1]

[1]Centre for Solar Energy Research, Faculty of Science & Engineering, Bay Campus, Swansea University, SA1 8EN, UK

[2]National Renewable Energy Laboratory (NREL), 16253 Denver West Parkway, Golden, CO 80401, USA

[3]Department of Mathematics, Physics and Electrical Engineering, Northumbria University, Newcastle upon Tyne, NE1 8ST, UK

Abstract — **CdTe:As devices grown by MOCVD in the superstrate configuration have been reconfigured to a substrate structure, by employing a cleaving and reconstruction technique. Indium and aluminium doped Zinc oxide (IZO and AZO) layers were used as the transparent front interface without a metal grid. A substrate efficiency of 12.3% was realized with AZO front contact, retaining 80% of the superstrate device efficiency and an impressive ~99% of the open-circuit voltage. The IZO–based substrate device on the other hand performed at ~69% retained efficiency. The crucial role of front interfaces is highlighted by the two contacts used here. The results in this work show promise for the development of highly efficient substrate CdTe:As devices.**

I. INTRODUCTION

Conventionally, thin film CdTe solar cells are fabricated according to superstrate configuration [1, 2], with a transparent conductive oxide (TCO)/n-type emitter /p-CdTe/back contact stack on a transparent sheet. Superstrate CdTe thin film solar cells have achieved 22.1% record cell efficiency and ~19% module efficiency [3]. So far, the superstrate configuration is the best option in terms of efficiency. With the substrate configuration, the stack fabrication sequence is reversed and the substrate can be opaque [2, 3]. It has an appealing advantage of possibly using lightweight transparent or opaque flexible substrates, for example polyimide or metal foils [4]. Achieving high efficiency devices in the substrate configuration, however, has been challenging [2, 4]. Ensuring the formation and retention of a low resistance ohmic back contact through subsequent stack fabrication and post growth processing conditions is a steep requirement of the substrate device design, because the buried back contact can be subjected to degradation during the preceeding high temperature steps [2]. In 2013, Gretner *et al.* [3] demonstrated a substrate CdTe thin film solar cell with 11.3% efficiency, where a combination of MoO_3, Te, and Cu were used as the back contact buffer layer. Mo sputtered onto a glass substrate is widely used as the back contact [3, 5, 6]. Reported efficiencies from earlier studies were below 10% [5, 6]. Lingg *et al.*, [7] demonstrated a 12.2% efficient cell in substrate configuration using a CdTe absorber graded with $CdTe_{1-x}Se_x$ towards the CdS layer. Using a water-assited liftoff process, Bista and co-workers were able to demonstrate a flexible substrate CdTe cell with 12.6% efficiency [8], and before that Kranz *et al.*, [9] reported an impressive 13.6% efficient CdTe on flexible metal foil in the substrate configuration. A record efficiency of 15.1% was recently achieved by McGott *et al.*, [10], where a superstrate $CdSe_xTe_{1-x}$ solar cell was mechanically reconfigured into a substrate structure, by cleaving the device stack at the emitter/absorber interface and depositing MgZnO/ZnO:Al bilayer as a front contact onto the exposed CdSeTe absorber. Devices here were Cu-doped. >80% of initial superstrate cell efficiency was retained. As – doped reconstructed substrate CdSeTe, in comparison, performed much worse, retaining <50% of the superstrate cell efficiency [10]. This was attributed to increased sensitivity to front interface quality, which may be worse in reconstructed devices from changes in chemical state / sputter damage and/or poorly matched emitter properties (i.e., low electron density in MgZnO), as a result of higher absorber hole density.

In this work, MOCVD grown superstrate As-doped CdTe (Se free) devices are cleaved at the TCO/emitter interface and reconstructed as substrate devices, using either IZO or AZO for the front interface. This allowed the emitter/absorber interface to remain intact and greatly improved substrate device efficiency from <50% retention to 80%. A 12.3% efficiency was measured with AZO – based substrate-type device, where the majority of loss was in current density as the AZO used here was designed to be used with a metal grid, which was not used here. This demonstrates promise for high efficiency substrate As – doped CdTe thin film solar cells.

II. EXPERIMENTAL DETAILS

CdTe:As devices in superstrate configuration were fabricated at Swansea University on fluorine-doped tin oxide (FTO)/glass substrates by conventional metalorganic chemical vapor deposition (MOCVD) in a horizontal-tube reactor at atmospheric pressure. CdZnS (emitter layer)/CdTe:As device structure was grown at 350°C under Cd-saturated conditions; details can be found elsewhere [11]. Dimethylcadmium (DMCd), diisopropyltellurium (DiPTe) and tris-dimethylaminoarsine (tDMAAs) organometallic precursors were used as sources for Cd, Te and As respectively, with H_2 as the carrier gas. The partial pressure ratio DMCd/DiPTe of 4 was used. By this method, the absorber carrier density is usually 10^{16} cm^{-3}. For growing the CdZnS emitter layer, Diethylzinc (DEZn) and ditertiarybutylsulphide (DtBS) were used as the metalorganic precursor for Zn and S, respectively. A heavily doped As cap layer on the back of the absorber was followed by CdCl$_2$ heat treatment (CHT), comprising a thick CdCl$_2$ layer deposited at 200°C and a 10 minutes anneal above 420°C (both in H_2 ambient), where tertiarybutylchloride was used as the precursor for Cl. Prior to completing the device by evaporating Au contacts, excess surface CdCl$_2$ was rinsed off with deionised water and the superstrate structure was annealed in air ambient at 170°C, for 90 minutes for further activation of the absorber. Substrate-type CdTe:As solar cells were constructed at NREL using a novel stack lift-off/reconstruction technique developed earlier, as described in [12]. AZO was sputtered in pure argon to ~120 nm thick, resulting in a sheet resistance of ~70 ohm/sq. IZO was sputtered in an oxygen/argon ambient to ~340 nm thick, resulting in a sheet resistance of ~10 ohm/sq.

Current density – voltage (J-V) characteristics (under standard AM 1.5G testing conditions) and external quantum efficiency (EQE) spectra responses of CdTe:As solar cells in the superstrate configuration for "As-grown" devices and in the substrate configuration for reconstructed devices were all measured at NREL. Capacitance – voltage (C-V) were measured to determine device acceptor concentration and depletion width profiles respectively. Data was collected by characterizing superstrate devices, cleaving and reconstructing, then characterizing the resulting substrate device to get a direct comparison.

III. RESULTS

Fig. 1 shows the illuminated J-V curves of devices before and after reconstruction (with either IZO or AZO as front interface contacts). The J-V parameters are summarized with boxplots in Fig. 2. J-V curves for reconstructed devices show good and working cells (Fig. 1). Device efficiency drops following reconstruction, mainly due to J_{SC} loss, particularly with the IZO-based sample (see Fig. 2). While the V_{OC} of the IZO-based device dropped by ~50 mV, the AZO-based device only dropped by 12 mV. V_{OC} loss has been an issue for previous reconstructed CdTeSe:As devices which were cleaved at the emitter/absorber interface and paired with a low-doped MgZnO reconstructed emitter [10]. That is not the case here since the CdZnS emitter is removed with the stack during cleave.

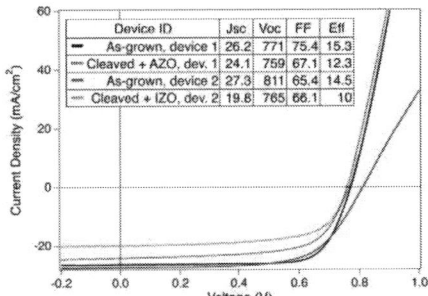

Fig. 1: Illuminated J-V curve of as-grown (superstrate) cells compared to cleaved and reconstructed (substrate) cells with AZO and IZO front contact layers.

FF remained relatively similar in reconstructed cases, with what appears to be even a slight improvement in the IZO-based device relative to its superstrate parent device. With all reconstructed devices, a loss in conversion efficiency relative to the superstrate (as-grown) sample, was noted. 10% efficiency was measured in substrate-type device reconstructed with IZO front interface, whilst in the case of AZO front interface, 12.3% was measured meaning 80% efficiency retention with respect to parent device.

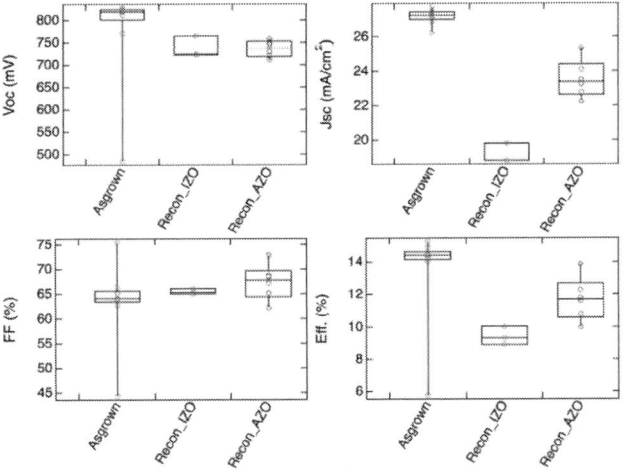

Fig. 2: Boxplots of J-V parameters for as-grown (superstrate) and reconstructed (substrate) solar cells shown in Fig. 1.

Spectral response (EQE) of these devices was measured to observe the changes in the J_{sc}. EQE loss can be seen in most of the visible region for especially the IZO-based substrate-type sample, which agrees with the loss in J_{sc} (Fig. 3). This suggests how the IZO front contact, which has a lower sheet resistance but is also much thicker than the AZO, may result in lower J_{SC}. On a different note, some improvement in the near IR (750 – 900 nm) region is also observable for substrate devices, resulting in more square profiles, which hints at a reduction in bulk recombination through a decrease in the absorber acceptor concentration [13].

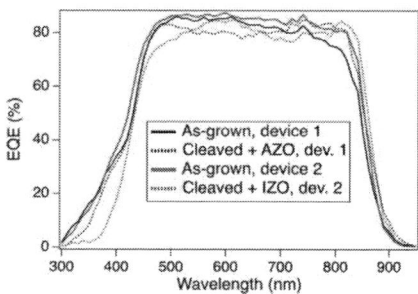

Fig. 3: EQE spectra comparison between as-grown (superstrate) cells and cleaved and reconstructed (substrate) cells with AZO and IZO front interface.

As noted it is encouraging to point out that there was not a significant loss in the V_{OC} with the substrate-type devices, following reconstruction with different front interface layers. C-V measurements were used to determine the acceptor concentration; the drop in this parameter was greater for the IZO-based device (see Fig. 4). AZO based device incurred a relatively small penalty on the acceptor concentration. Further work is required to derive the full correlation between the V_{oc} and acceptor concentration parameters and their dependence on the choice of front contact material.

Some of the collection and FF losses can be ascribed to the absence of a metal collector grid on the TCO layers. Thus, it can be expected to achieve higher performance CdTe:As substrate devices in the future with the utilization and optimization of the TCO + metal grid contact pair.

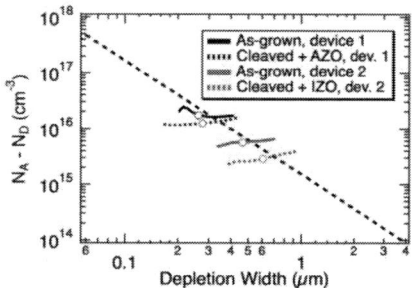

Fig. 4: Acceptor concentration vs depletion width from As grown (superstrate) cells, and reconstructed substrate cells with AZO and IZO front interface.

IV. CONCLUSION AND OUTLOOK

CdTe:As substrate devices are presented with a 12.3% conversion efficiency, using a cleave and recontruction technique. Two different front interfaces were explored in the reconstructed devices. J_{SC} loss was much more significant with the IZO front interface. This work shows that there is need for front interface optimization in order to realise the potential of higher efficiency substrate CdTe:As cells. This will help to further improve the current collection. Assessment of junction quality will also be explored and discussed.

ACKNOWLEDGMENT

Funding by the Engineering and Physical Sciences Research Council (EPSRC), United Kingdom via the grant EP/W000555/1 and from the European Regional Development Fund (ERDF) and the Welsh European Funding Office (WEFO) for funding the 2nd Solar Photovoltaic Academic Research Consortium (SPARC II) which supported this research. Funding was also provided by the U.S. Department of Energy Office of Energy Efficiency and Renewable Energy Solar Energy Technologies Office under agreement #38257.

REFERENCES

[1] D. Bonnet and P. Meyers, "Cadmium-telluride – Material for thin film solar cells" *J. Mater. Res.*, Vol. 13, 10, 1998.
[2] B. M. Başol and B. McCandless, "Brief review of cadmium telluride-based photovoltaic technologies", Journal of Photonics for Energy, Vol. 4, Issue 1, 040996, June 2014.
[3] M.A. Green, K. Emery, Y. Hishikawa, W. Warta, E.D. Dunlop, "Solar cell efficiency tables (version 50)", *Prog. Photovolt: Res. Appl.* vol. 25, no. 7, pp. 668-676, 2017.
[4] C. Gretener, J. Perrenoud, L. Kranz, L. Kneer, R. Schmitt, S. Buecheler and A. N. Tiwari, "CdTe/CdS thin film solar cells grown in substrate configuration", *Prog. Photovolt: Res. Appl.*; 21: pp. 1580–1586, 2013.
[5] N. Romeo, A. Bosio, V. Canevari, "A new method to prepare efficient CdTe/CdS thin film backwall solar cells." *Proceedings of 11th E.C. Photovoltaic Solar Energy Conference*; 972–974, 1992.
[6] R. G. Dhere et al.,, "Analysis of the Junction Properties of CdS/CdTe Devices in Substrate and Superstrate Configurations." *Proceedings of 26th European Photovoltaic Solar Energy Conference and Exhibition*; 2456–2459, 2011.
[7] M. Lingg et al., "Structural and electronic properties of CdTe$_{1-x}$Se$_x$ films and their application in solar cells." *Science and Technology of Advanced Materials*, 19:1, pp. 683-692, 2018.
[8] S. S. Bista et al., "Water-Assisted Lift-Off Process for Flexible CdTe Solar Cells. " *ACS Appl. Energy Mater.* 6, 2, pp. 885–891, 2023.
[9] L. Kranz et al., "Doping of polycrystalline CdTe for high-efficiency solar cells on flexible metal foil." *Nat Commun* 4, 2306, 2013.
[10] D. L. McGott, E. Colegrove, J. N. Duenow, C. A. Wolden, and M. O. Reese, "Revealing the Importance of Front Interface Quality in Highly Doped CdSe$_x$Te$_{1-x}$ Solar Cells", *ACS Energy Lett.*, 6, pp. 4203–4208, 2021.
[11] O. Oklobia et al., "Impact of In-Situ Cd Saturation MOCVD Grown CdTe Solar Cells on As Doping and VOC," in *IEEE Journal of Photovoltaics*, vol. 12, no. 6, pp. 1296-1302, 2022.
[12] D. L. McGott et al., "3D/2D passivation as a secret to success for polycrystalline thin-film solar cells." *Joule*, 5, Issue 5, pp. 1057-1073, 2021.
[13] G. Kartopu et al., "Study of thin film poly-crystalline CdTe solar cells presenting high acceptor concentrations achieved by in-situ arsenic doping". *Solar Energy Materials and Solar Cells*, 194, pp. 259-267, 2019.

Design, Fabrication, Test, and Flight Performance of the Parker Solar Probe Solar Array

Edward Gaddy
Space Exploration Sector
Johns Hopkins University
Applied Physics Laboratory
Laurel, MD, USA
Edward.Gaddy@jhuapl.edu

Andrew Gerger
Space Exploration Sector
Johns Hopkins University
Applied Physics Laboratory
Laurel, MD, USA
Andrew.Gerger@jhuapl.edu

Lew Roufberg
Space Exploration Sector
Johns Hopkins University
Applied Physics Laboratory
Laurel, MD, USA
Lew.Roufberg@jhuapl.edu

Richard Stall
newForge Technologies
Wrightstown, PA, USA
rick@newforgetechnologies.com

Matthew J. Schurman
newForge Technologies
Wrightstown, PA, USA
matt@newforgetechnologies.com

Abstract—This paper describes the design, fabrication, test, and flight performance of the Parker Solar Probe's solar array. PSP travels closer to the sun than any other spacecraft ever has. This presented a number of unusual challenges to its array: keeping the cells at an acceptable temperature, preventing inter-cell grouting from deteriorating, testing at irradiances up to 70 suns, and predicting optical degradation of the coverglass to cell adhesive. The array produced less power than predicted early in its life and now produces slightly more than the predicted power. After more than four and a half years in orbit, it appears the array will exceed its seven year end of life power requirements.

Keywords—spacecraft solar array, spacecraft solar cell, high temperature solar array

I. INTRODUCTION

The PSP mission, under NASA's Living with a Star Program, will fly as close as 8.86 solar radii from the Sun's surface. The mission duration is seven years. The nominal mission requires 24 orbits about the Sun. It was launched on August 12, 2018.

The Applied Physics Laboratory purchased the photovoltaics for the array from SolAero. newForge Technologies performed extensive development testing. Collins Aerospace Systems provided the water cooled substrates.

II. DESIGN

The array configuration is shown in Fig. 1. The spacecraft varies the wing flap angles between those depicted for near Earth and near Sun depending on distance from the Sun. The array wings are divided into two sections. Near perihelion the secondary section becomes the only source of power; in other segments of the trajectory both sections provide power. To keep the array wings below 150 °C near perihelia, the wings operate in the penumbra of the thermal protection shield and at a high angle.

The authors gratefully acknowledge funding by the Heliophysics Division of NASA's Science Mission Directorate.

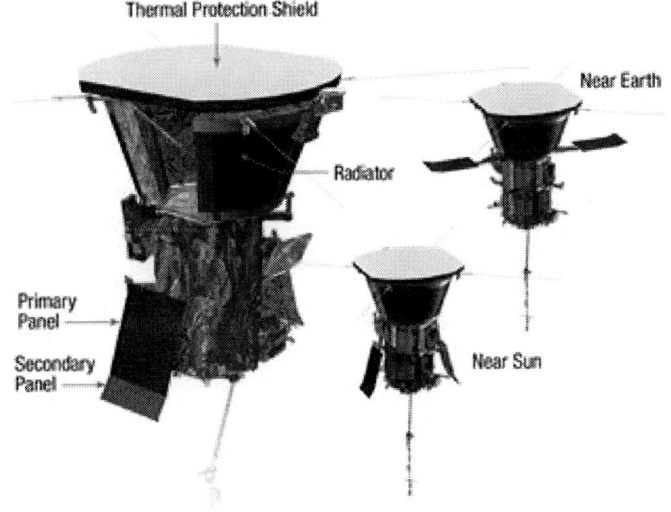

Fig. 1. The PSP solar array wings shown near earth and near sun.

To further cool the wings, high pressure cooling water circulates through the wings' titanium substrates and to the spacecraft's radiators. If the wings are completely shadowed by the heat shield, the water can freeze even at perihelion, so safeties are present to be certain the wings cannot be totally shadowed for a significant interval.

During nominal operation, the irradiance on the wings will exceed 16 Suns during the last few perihelia, which are closer to the Sun than the earlier perihelia. This greater than 16 suns occurs after considering the angle of the wings to the Sun and the partial shading of the wings by the spacecraft thermal protection shield. With this irradiance, the usual method of using Kapton as the cell to substrate electrical insulator would cause the cells to reach unacceptably high temperatures. So ceramic pieces, approximately the same size as the cells, were used instead. This insured electrical isolation while providing good thermal conductance to the water cooled substrate. To reduce the probability

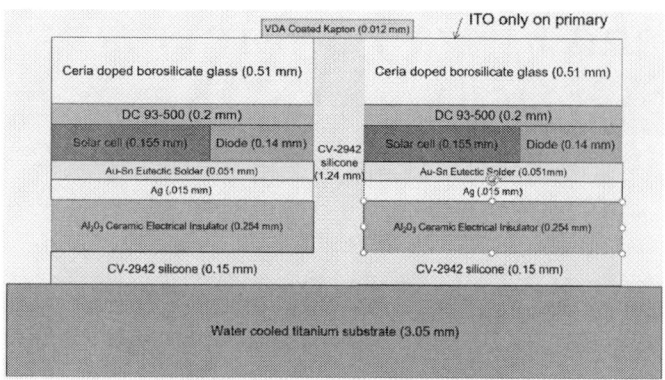

Fig. 2. The solar array's cell stack

of substrate penetration by dust, silicone adhesive grouting was placed between the cells. Unless protected from UV, this grouting would deteriorate and become powdery through the mission. So, the grouting was covered by Kapton tape coated with vacuum deposited aluminum (VDA). The tape itself would burn up unless it was thermally coupled to the grout. So care was taken to ensure this adherence. These features are displayed in Fig. 2.

Because of the high current generated by the high irradiance, heritage welds between the interconnects and the solar cell contacts would overheat. SolAero overcame this issue by using a eutectic gold tin solder. For the joints on the rear of the cell, the soldering was performed in vacuum. Regular tin lead solder could not be used because it would deteriorate at the sometimes high operating temperature of the array.

PSP is a scientific spacecraft measuring, among other things, fields and particles near the Sun's surface. This required that the array not form electric fields that would unduly disturb the natural electric field near the spacecraft. To this end, covers with a conductive transparent indium tin oxide (ITO) coating were used on the primary section of each wing. The VDA Kapton tape was placed between and made contact with the edges of the coverglasses. The tape's adhesive had embedded conductive adhesive that electrically connected each cover's ITO to adjacent covers

and finally to the spacecraft structure. This was not done on the secondary sections of the wings because these sections are always illuminated through the trajectory; the ultraviolet light from this illumination prevents accumulation of electrons on the covers. As it happens, spacecraft tends to charge negatively accumulating excess electrons rather than excess protons.

The wing output at perihelion follows from the wing's view of the Sun, which is the limb of the Sun. This is shown in Fig. 3. The limb is both redder and dimmer than the Sun as a whole; of course, this must be considered in computing the wing power output. Formulas for determining the spectrum and irradiance as a function of distance from the Sun have been derived from empirical evidence gathered by Neckel et al. [1], [2] and others [3].

Of course, the configuration shown near Sun in Fig. 1 and in Fig. 3 meant that the irradiance on each wing decreased going toward the inboard regions. This led to the requirement that every whole string had to go from one long edge of a wing to the other. In other words the strings had to be on a single row of the wing.

III. TEST

The project ran a number of development tests to qualify the array against degradation due to high irradiance and high temperature. Included was a test on a full size secondary water cooled coupon exposed to a non-nominal exposure that went as high as 70 suns. The irradiance source for this test, and many other tests, was a heliostat field that newForge manufactured. The heliostat directed light into a trailer with a portion of its side cut out. Inside the trailer was a vacuum chamber with a large window facing the cut out, Fig. 4 and Fig. 5. The heliostat was required as solar simulators, at the time the project started, were not able to attain the required irradiance over the size needed. However, several years into the project, newForge developed LED simulators that did have this capability to a useful extent if not the full extent required for some tests. The heliostat field proved useful early on

Fig. 4. The heliostat field at SolAero that provided irradiance to test coupons, qualification panels and flight wings for the PSP solar array build. The field could generate irradiance in excess of 70 suns at the test plane and vary the irradiance across a PSP wing.

Fig. 3. View of the sun from the secondary section of a wing near perihelion.

978-1-6654-6060-6/23 $31.00 © 2023 IEEE

Fig. 5. Vacuum chamber being installed in test facility trailer.

and later for very high irradiance tests.

The LED simulators were used to measure array output and to illuminate the outer regions of the qualification panel's secondary section at 22 suns with smoothly decreasing irradiance further inboard. This was done for a test to prove the power system electronics could perform adequately with a graded irradiance on the secondary. The LED simulators were also used to condition the cover to cell adhesive.

The extensive testing with LED simulators showed that exposing solar cell assemblies with UV light from a LED simulator degraded the transmission of the DC 93-500 cover to cell adhesive. It also showed that this exposure, if done at particular temperatures and irradiances, then slowed later degradation. This led to the process of preconditioning the flight arrays with LED light to lessen the overall degradation.

The qualification wing was tested largely in accordance with AIAA S-112A-2013 [4]. Functional tests were specified by S-112 as were thermal cycling and exposure to an acoustic environment. Special tests included measurements from the primary at 5.3 suns, output at air mass zero versus temperature, output under an irradiance graded from 0 suns on the most outboard primary string to 22 suns on the most outboard string with a most outboard string temperature of 141 °C, thermal vacuum bake and exposure to ultra-violet light to pre-condition the cell to coverglass adhesive, a 30 hour high irradiance high temperature test with the irradiance graded from 0 suns on the most outboard primary string to 24.2 suns on the most outboard string at a temperature of 151 °C, an exposure of the entire wing to 5.8 suns for 40 hours at 140 °C, arc sensitivity tests, impedance characterization, and exposure to non-nominal high irradiance.

IV. Effects of Ultra-Violet Light

Every array that the primary author knows about that went close to the Sun and stayed there suffered substantial anomalous degradation, which we now believe is explained by the work described below. An example of this is the degradation of the MESSENGER array [5], which degraded approximately 50% over its life of 370,000 hours with an exposure of 195,000 sun hours. All of the arrays that went near to the Sun and degraded produced adequate power for their spacecraft as their array had to be large enough to operate the spacecraft near Earth where the irradiance was much lower. This would not work for PSP because PSP orbits the Sun with aphelia at Sun distances much larger than the perihelia. The PSP array has to have sufficient power to operate far from the Sun.

newForge technologies ran extensive and long duration tests, in some cases 16,000 hours, to determine the cause and predict the amount of this degradation, which was initially guessed to be either from out gassing depositing on the arrays or ultra- violet light degradation of the cover to cell adhesive.

Testing determined that the cause was darkening of the coverglass to cell adhesive, and these tests; also predicted the amount of degradation due to this cause [6], [7]. This was used in the development of a model that included degradation from other sources, see Fig. 6 on the next page. The UV degradation predicted for close to air mass zero is significantly greater than the frequently used 2%, in spite of taking steps to minimize the UV degradation through keeping the solar panel at a low temperature and UV preconditioning.

V. Flight Performance

Fig. 6 shows array power output for the nominal life of the spacecraft. Initially the spacecraft power output was below predictions and now is reasonably close. The reasons for the inaccuracies in the prediction are related to significant challenges in performing long term ultra-violet exposures and making ground measurements on the output of solar cells that cover a wide range of temperature, irradiance, and charged particle radiation levels. The solar array peak power data can be measured in flight because the power electronics is designed to temporarily sweep SA voltage upon command to obtain current versus voltage characterization.

Acknowledgment

The authors thank Brian Abel for development of the method of applying VDA Kapton with connection to ITO, Ed Abrams for contracts, Andreea Boca for early design and testing, Larry Cisneros for directing the flight build, Kevin Crist for contracts and schedule, Karen deZetter for testing coupons and the flight wings, Paul Johnson for testing coupons, Faith Kahler for schedule, Rick Mitchell for sensor cell testing, Shawn Morrow for fabricating the flight array, Rick Pfisterer for quality control, Paul Sharps for oversight, Charles Sarver and Cory Tourino for developing the gold-tin solder attachment, and Christopher Sulyma for UV degradation tests with continuous spectrum xenon lamps.

References

[1] H. Neckel, "Analytical reference functions F(λ) for the

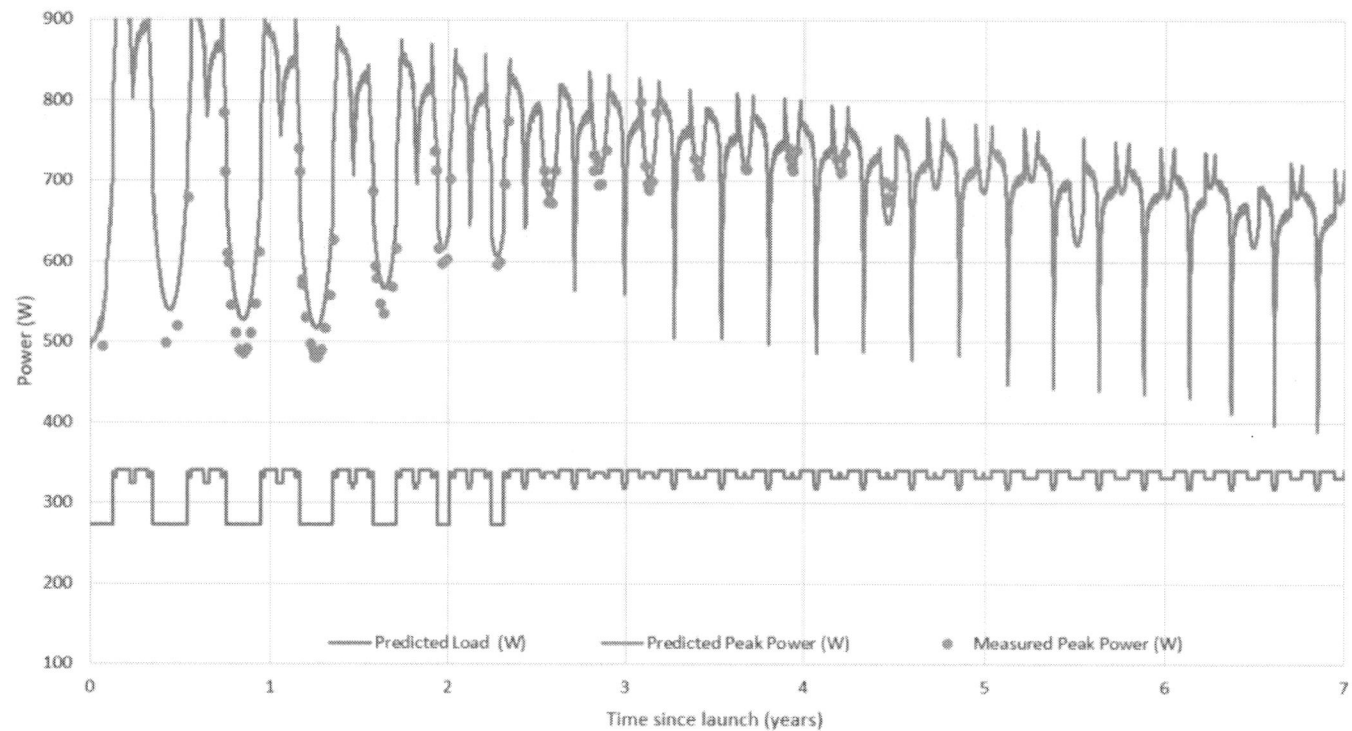

Fig. 6. PSP array pre-launch predicted peak power (W), measured peak power, and predicted load.

Sun's limb darkening and its absolute continuum intensities," Solar Phys., 229, 2005, pp. 13-33.

[2] H. Neckel, "On the Sun's absolute disk-center and mean disk intensities, its limb darkening, and its 'limb temperature' (λλ330 to 1099nm)" *Solar Physics*, 212, pp. 239-250, 2003.

[3] A. K. Pierce., C. D. Slaughter, and D. Weinberger, D., "Solar Limb Darkening in the Interval 7407 - 24018Å, II," *Solar Phys.*, 52, 1977, pp. 179-189.

[4] AIAA S-112A-2013, Qualification and Quality Requirements for Electrical Components on Space Solar Panels.

[5] D. Gallagher, "MESSENGER Solar Array Operations and Flight Data," *Space Power Workshop Proceedings*, 2014.

[6] R. Stall, M. Schurman, Chris Sulyma, A. Gerger, E. Gaddy, "UV degradation of space solar cell assemblies under high temperature and irradiance," *Proceedings of the Seventh World Photovoltaic Energy Conference* (28th PVSC), Waikoloa, Hawaii, June, 2018

[7] J. Schurman, A. Gerger, R. Stall, P. Sharps, "Path Dependent Power Degradation via UV Exposure for the Parker Solar Probe, *Space Power Workshop Proceedings*, 2018.

Investigating Electric Field and Light Induced Degradation in Perovskite Solar Cells through Nanometer-Scale Potential Imaging

G. Paul[1], J. W. Schall[1,2], C.-S. Jiang[1], A. Louks[1,2], A. Palmstrom[1], N. S. Dutta[1], S. Johnston[1], H. Guthrey[1], A. Norman[1], M. M. Al-Jassim[1], D. B. Sulas-Kern[1]

[1]National Renewable Energy Laboratory, Golden, Colorado, USA
[2]Colorado School of Mines, Golden, Colorado, USA

Abstract — Electric field and light induced degradations in perovskite solar cells were evaluated through nanometer-scale potential imaging across the device by using *in-situ* Kelvin probe force microscopy (KPFM). We derived the electric field profile from potential profiles at different bias voltages to evaluate the locations and quality of junctions across the device. We found relative changes in electric field peak intensity at the HTL/perovskite and perovskite/ETL interfaces upon stressing devices separately under voltage or light. KPFM results during 12-hour stress/rest cycling under electrical bias show both reversible and irreversible changes in the device's interfacial fields. We also observed change in the electric field profile between control and degraded devices after 100 hours of stress/rest cycling under light. Our results demonstrate how nanometer-scale potential imaging can be used to understand the impacts of external electric fields and light soaking on both irreversible degradation and reversible metastability in perovskite solar cells.

I. INTRODUCTION

Perovskite solar cells (PSCs) are a leading candidate for future photovoltaic technology, with a record power conversion efficiency of 25.5% [1]. Recently, stability and reliability studies of PSCs have become crucial for industrial-scale commercialization. The characteristics of PSCs are very sensitive to humidity, UV and sunlight, temperature, and electric fields [2]. The ionic nature of perovskite materials makes them very sensitive to optically- or electrically- induced fields [3,4], and PSCs may exhibit different (but simultaneous) mechanisms for reversible metastability and irreversible degradation under applied voltage and light soaking [5,6]. For achieving long-term reliability and commercialization, it is a prerequisite to separately understand reversible changes and irreversible degradation.

Fundamental understanding of such reversible changes and irreversible degradation through nanoscale characterization can inform the path forward for perovskite reliability and process optimization. Recently, Kelvin probe force microscopy (KPFM) has been used to understand the operational mechanisms of solar cells at the nanometer scale [7]. Herein, we employ *in-situ* KPFM during stress/rest cycling to obtain the nanometer-scale potential profiles of PSCs under varied conditions. By deriving the electric field profile across the device and comparing it under different stress/rest conditions, we observe both reversible and irreversible evolution of interfacial electric field peaks over time. Our work contributes an important step toward understanding the reliability of PSCs.

II. EXPERIMENTAL

KPFM is an extension of atomic force microscopy, where a conducting tip is used to probe the local surface potential of a sample. Here, we used a home-made KPFM inside an Ar-filled glove box with ~30 nm spatial and ~10 mV voltage resolutions. We utilize the second harmonic resonant oscillation frequency of the probe cantilever at 300–400 kHz for the KPFM surface potential measurement, while the first harmonic resonance frequency at ~60 kHz is used for the AFM topography imaging. A Pt/Ir-coated silicon probe with the tip apex radius less than 25 nm is used. In order to prepare a cross-sectional sample, devices were mechanically cleaved inside the glove box. As perovskites are sensitive materials, we didn't perform any further cleaning or polishing. We performed KPFM on the cleaved cross-section in clean and smooth regions in topography.

We used *p-i-n* perovskite solar cells with a mixed-halide absorbers in construction of glass / indium tin oxide (ITO) / NiO_x /N 4,N 4'-Di (naphthalen-1-yl)-N 4,N 4'-bis (4-vinylphenyl)biphenyl 1-4,4'-diamine(VNPB)/ $FA_{0.77}Cs_{0.13}Pb(I_{0.95}Br_{0.05})_3$ / C_{60} / bathocuproine (BCP) /Ag. To study the electric field-induced changes, the cleaved devices were stressed for 12 hours with 6 stress/rest cycles inside an Ar-filled glove box. Each cycle consisted of 1 hour stress under 1V forward bias followed by 1 hour rest at 0 V while potential profiles were recorded after each stressing and resting period. During KPFM measurements, the transparent conductive oxide (TCO) at the hole-transport interface was grounded, and a bias voltage was applied to the Ag at the electron-transport side. For in-situ KPFM with optical bias, we used a halogen lamp with intensity greater than 1 Sun. In addition, we also measured KPFM on control and light-soaked devices which were stressed for 100 hours under 450 nm LED at 5 Sun intensity.

III. RESULTS AND DISCUSSION

A. *Baseline KPFM of Perovskite Solar Cells*

We record the current-voltage (I-V) characteristics of the devices before KPFM measurement for screening. Fig. 1 represents the I-V characteristics of six different devices under 1 Sun illumination. All devices show almost similar I-V characteristics. Table 1. represents photovoltaic parameters of all devices along with their average values. The perovskite solar cells show an average efficiency of 19.6%.

Fig. 2(a) shows the cross-sectional topography of a cleaved perovskite solar cell, along with surface potential images at 0V, 1V, and -0.5V. We obtain the electric field profile over the device cross-section first by taking the difference between the biased scan and the 0V scan to remove impacts of static surface charge, and then taking the derivative of the potential change. This procedure is because localized charges trapped at the cross-sectional surface do not move significantly under a small bias voltage, so the measured surface potential change by applying a bias voltage can be used to approximately assess the potential change in the device bulk.

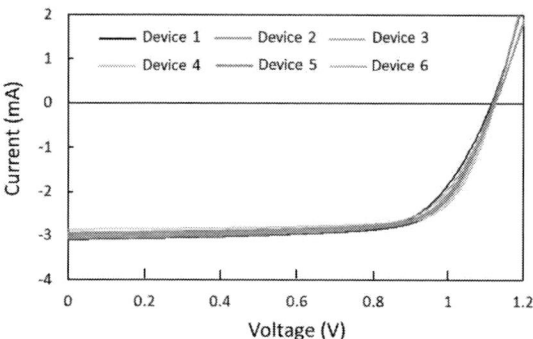

Fig. 1. I-V characteristics of different pixels of the perovskite solar cell used for KPFM.

Table 1. I-V parameters of different pixels of the perovskite solar cell used for KPFM.

Device	Scan direction	V_{oc} (V)	J_{sc} (mA/cm²)	FF	Efficiency (%)
1	Forward	1.113	25.22	0.688	19.30
	Reverse	1.117	25.33	0.709	20.07
2	Forward	1.113	24.72	0.709	19.50
	Reverse	1.117	25.24	0.705	19.86
3	Forward	1.117	24.18	0.705	19.04
	Reverse	1.126	24.90	0.697	19.55
4	Forward	1.122	23.43	0.747	19.62
	Reverse	1.126	24.05	0.735	19.90
5	Forward	1.117	24.11	0.722	19.45
	Reverse	1.122	24.39	0.719	19.69
6	Forward	1.122	24.30	0.713	19.43
	Reverse	1.126	24.78	0.708	19.75
Average		1.120	24.55	0.713	19.597

Fig. 2(b) shows the surface potential profiles averaged over each image along with potential change and electric field. Although the potential on the Ag/ETL (electron transport layer) side is higher than that on the TCO/HTL (hole transport layer) side at 0V external bias, the amplitude is smaller than the built-in potential across the device and the potential distribution is different from a typical linear potential across a p-i-n structure, because of charges trapped on the irregular surface defects. In the electric field profile, we observe two different peaks corresponding to the electrical junctions at the ETL/perovskite and perovskite/HTL interfaces. The peak intensity is higher at the ETL/perovskite side compared to the HTL side indicating greater drop in the external voltage at ETL/perovskite interface (i.e., greater relative resistance to charge movement).

We note that the potential drop across the device is determined by the equivalent resistances of different layers and interfaces of the device. This is because constant current flows through the whole device stack under the applied bias voltage. For this reason, the electric field profile does not represent the bulk built-in field of the perovskite solar cells. Rather, it represents relative resistance to current flow across the various layers, which can change due to variations in properties such as injection barriers caused by energy band misalignments, different densities of trap states, defects causing recombination, etc [8-11].

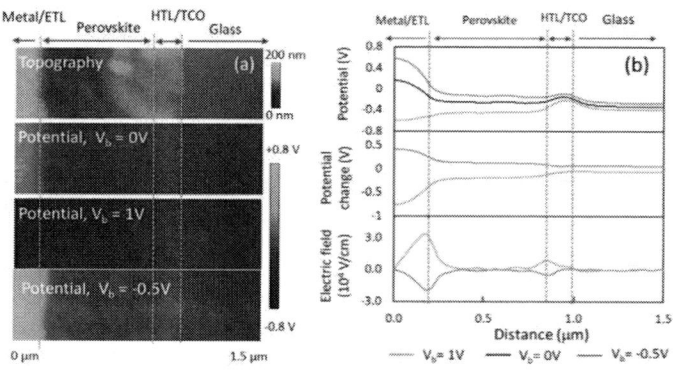

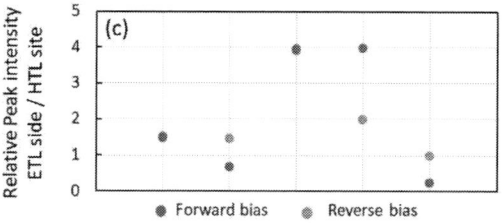

Fig. 2. (a) Topography and surface potential at different V_b and (b) potential profiles, the potential changes from that at $V_b=0$, and the corresponding electric field profiles. (c) Relative peak intensities of five different devices under forward and reverse bias.

B. Comparison Between Electrical and Optical Bias

We further measured the cross-sectional KPFM on the device at 0 V, +0.5 V, +1 V, -0.5 V, -1 V and at open circuit (OC) under light to present a comparison between the electrical and optical bias conditions on the same device cross-sectional area. For comparison with optical bias, we choose a range of electrical bias and try to find the similarity between potential and electric field profiles under electrical biases and optical bias.

Fig. 3(a) shows the surface potential images at different electrical biases and OC conditions under light (optical bias). We observe similar surface potential images under +1V electrical bias and optical bias (OC+light). The potential profiles, potential changes from $V_b = 0$ V, and the derived electric field profiles at different electrical and optical bias conditions are presented in figure 3(b). We find similar potential and electric field profiles under optical bias and +1V electrical bias. This is because the perovskite solar cells used in this investigation have a V_{OC} of around +1V. Therefore, it is expected that at OC under illumination, a similar potential difference of around +1V is generated between the two counter electrodes of the device. Based on these results, we propose that ion migration or charging occurs in the perovskite devices in a similar fashion during electrical and optical bias and causes analogous changes in surface potential across the device cross section.

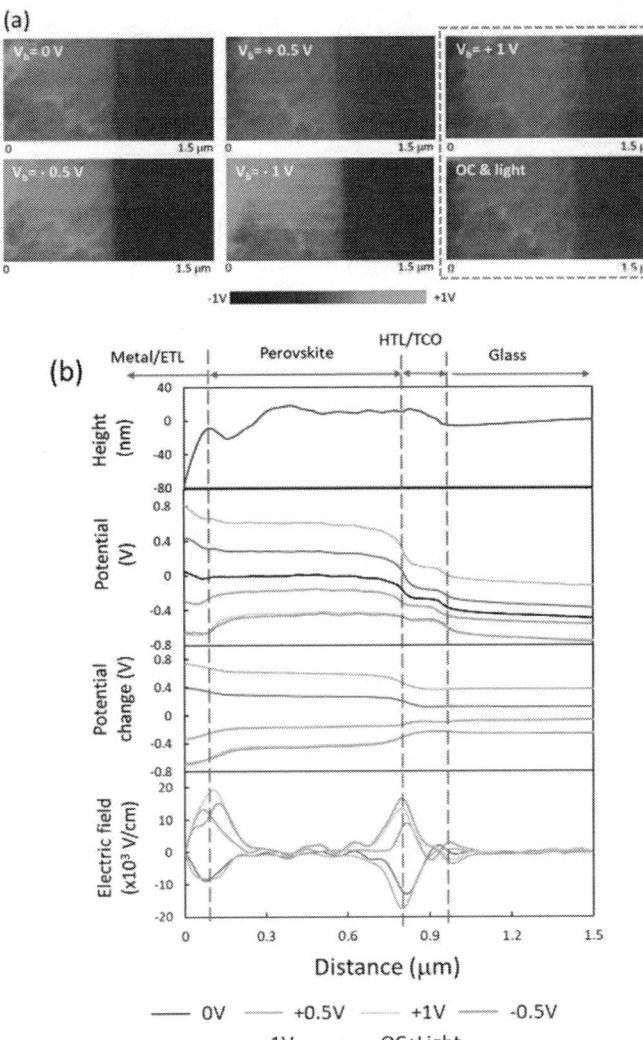

Fig. 3. (a) Surface potential images and (b) potential profiles, the potential changes from that at $V_b=0$, and the corresponding electric field profiles under different electrical bias and optical bias

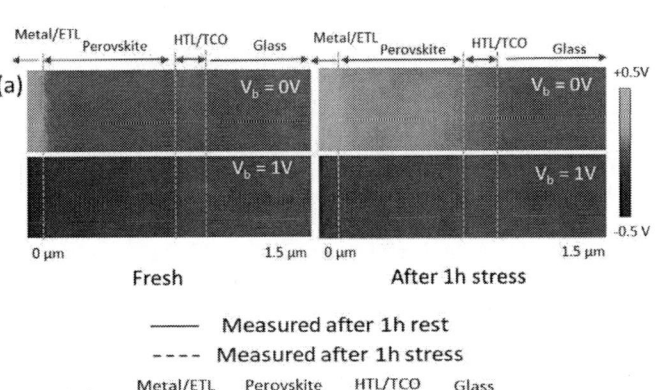

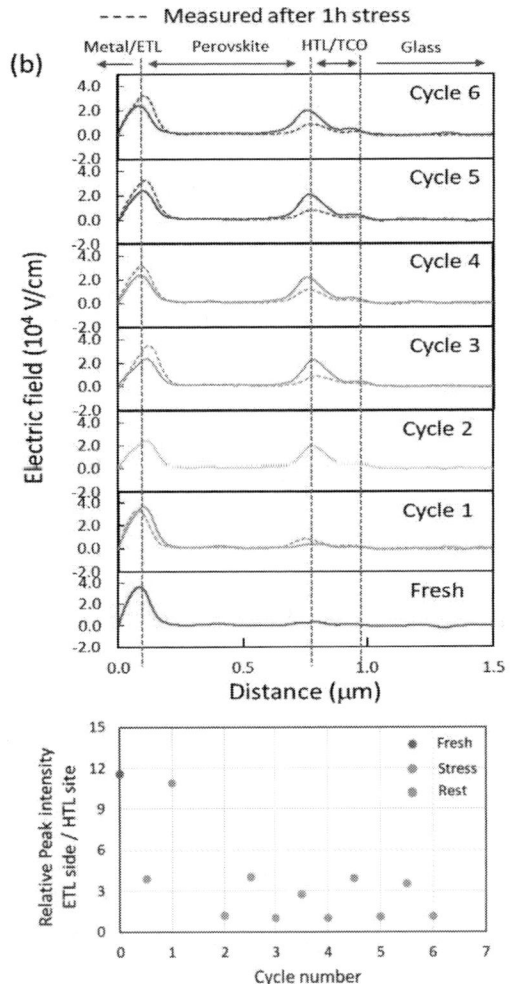

Fig. 4. (a) Surface potential at different V_b of fresh and stressed devices (b) evolution of electric field profiles during a complete stress/rest cycle, $V_b = 1$V and (c) relative peak intensity after each complete cycle.

C. Reversible Interface Changes under Electrical Bias Stress

To explore reversible changes in the perovskite solar cells, we measure cross-sectional KPFM on devices during constant voltage stress/rest cycling over 12 hours. Each cycle consists of 1-hour stress and 1-hour rest intervals, repeated 6 times. Fig. 4(a) represents surface potential images of a device in its fresh state and after 1 hour stressing at 1V forward bias voltage (V_b).

We observe a significant change in the surface potential and electric field profiles of the device upon stressing.

Fig. 4(b) summarizes the changes in the electric field profiles due to the constant voltage stress/rest cycles. We observe a similar electric field profile at the fresh condition and after the first stress/rest cycle. Interestingly, after the second cycle, we observe a different electric field profile compared to the fresh condition, which indicates an irreversible change in the device. Despite this irreversible change, cycles 2-6 continue to show reversible behavior.

The irreversible change at cycle 2 results in decreased electric field intensity at the ETL side, and greater peak intensity at the HTL side. This result suggests either a greater equivalent resistance at the HTL side, or a greater conductance at the ETL side. One possibility is a sudden change in local shunting near the measurement area (e.g., shunt formation on the ETL side or shunt burn-out on the HTL side). Other origins of this irreversible change may also be possible, such as the junction at the ETL side becoming leaky through creation of defects.

In Fig. 4(c), we represent the reversible nature of the changes in interfacial electric fields by plotting the ratio of peak intensity at the ETL/perovskite side to that at the perovskite/ HTL side. Initially, this ratio is very high for a fresh device, where peak intensity at the ETL side is much higher than at the HTL side. The ratios are comparable for the device at fresh condition and after one stress/rest cycle. For the rest of the cycles, the ratio differs drastically from that of the fresh condition and after the first cycle. However, for cycles 2–6, the peak ratio appears to be reversible between stress and rest conditions. Such a result indicates reversible metastabilities in the perovskite devices that are often attributed to ion migration or charging due to an external electric field [12-14].

D. Characterization of Reversible Interface Changes under Optical Bias Stress

Optical excitation is another way to induce electric fields in perovskite devices that may also cause reversible changes such as ion migration or charging effects [15-17], in addition to possible irreversible degradation pathways. To explore reversible changes in the perovskite solar cells under optical bias, we measure cross-sectional KPFM on devices during one stress/rest cycle over 4 hours. The cycle consists of 2-hour stress and 2-hour rest intervals. For optical bias, we used a halogen lamp having an intensity greater than 1 Sun.

Fig. 5(a) shows surface potential images of a device in its fresh state and after 2 hours stressing at open circuit (OC) under illumination. We observe a significant change in the surface potential of the device upon stressing under light. Fig. 5(b) represents the electric field profiles of the device stack during one complete optical bias stress/rest cycle. In fresh condition, the electric field peak intensity is larger at the ETL/perovskite interface compared to the perovskite/HTL interface, and it becomes opposite after 2-hour stress under optical bias. Such a

change in the electric field profile under optical bias may be related to ion migration [15,16]. Interestingly, we observe similar electric field profiles at the fresh condition and after the complete cycle (2 hour stress + 2 hour rest), which indicates a reversible change of the device under an optical bias cycle. This change in electric field ratio between the ETL and HTL interfaces has the same trend with the change during cycle 1 under electrical bias. The ETL/HTL field ratio decreased after the optical/electrical stressing and increased after rest, indicating that the electrical and optical bias may have similar effect on the perovskite solar cell.

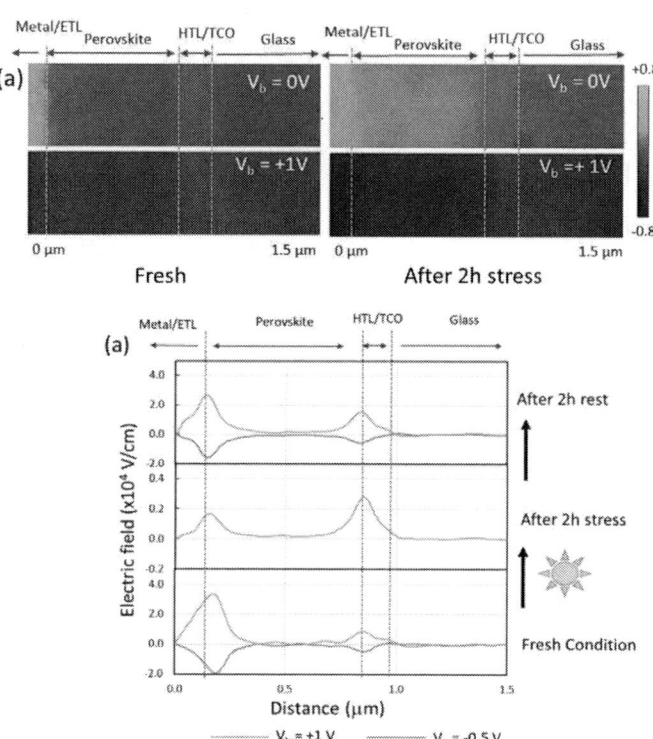

Fig. 5. (a) Surface potential at different V_b of fresh and stressed devices and (b) Evolution of electric field profiles of a device during a complete stress/rest cycle under optical bias

E. Characterization of Irreversible Interface Changes

Optical excitation also induces irreversible degradation in perovskite solar cells. To understand the effects of light soaking, we measured a control cell and a cell that has gone through a 100-hour stress/rest experiment under 450 nm LED illumination at 5 sun intensity. Figure 6(a) shows the I-V characteristics of control and test devices before and after stressing. Before stress, the control cell and the test cell have similar I-V curves. Upon light soaking, the I-V curve degrades mainly in fill factor and short-circuit current, with a smaller extent of V_{oc} loss. The stress process and in-situ characterization of this device are further discussed in a separate PVSC submission [18].

Fig. 6(b) shows the electric field profile derived from

978-1-6654-6060-6/23 $31.00 © 2023 IEEE

potential changes in the control and stressed devices. We observe a relative change in the peak intensities of the electric field profile for the device upon stressing it under light. For the control device, the peak intensity at the ETL side is higher than the HTL side, whereas in a stressed device, the electric field profile shows almost similar peak intensities at the ETL and HTL sides under +1 V. This represents an irreversible change in the electrode interfaces. As we observe a device-to-device variation of the relative intensity of the peaks at ETL and HTL sides, we derive the electric field profiles of several control and stressed devices and obtain the average of the ratio of peak intensity at ETL and HTL sides. Fig. 6(c) represents the average value (along with the variation) of the ratio of peak intensity of control and stressed devices under forward and reversed bias. Under forward bias, the control device shows a higher ratio compared to the stressed device. But under reverse bias, the control and stressed devices show almost similar peak ratios. Maybe the small reverse bias (-0.5 V) is not enough to distinguish the control and stressed devices.

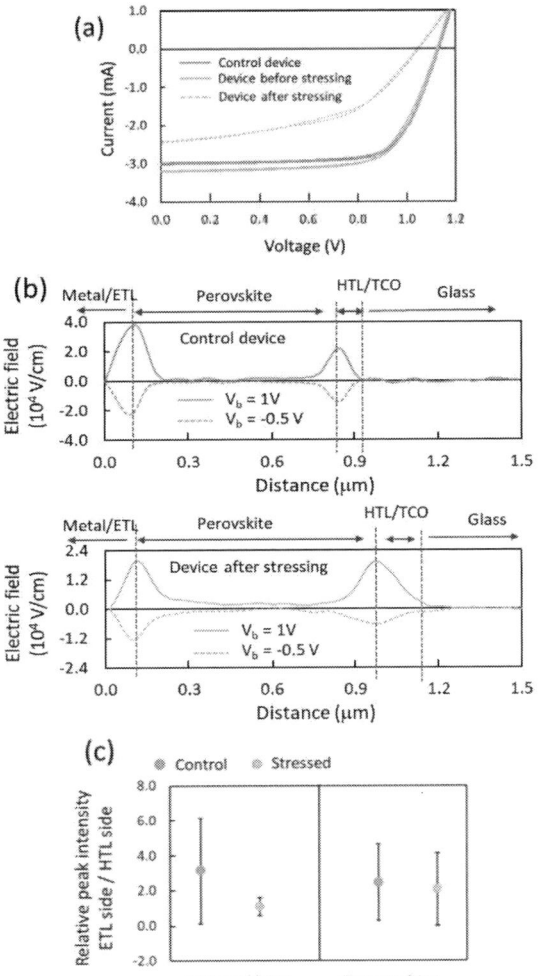

Fig.6. (a) I-V characteristics and (b) Electric field profiles of control device and device stressed under 450 nm LED at 5 Sun intensity (c) average value of the peak ratio (along with variation) of several control and stressed devices.

IV. Conclusion

We have investigated the electric field and light-induced degradation in perovskite solar cells from the perspective of the electric field across the device using cross-sectional KPFM measurements. To understand (meta)stability of perovskite devices, we employed a unique method of stress/rest cycling and compared the relative intensities of the peaks in the electric field profile at the charge-transport layer interfaces. We have observed both metastability and irreversible degradation in the device during constant voltage stress/rest cycling. A reversible metastability has also been observed during the stress/rest cycle under optical bias. The KPFM results under electrical and optical biases indicate that the profiles of steady potential/field changes are similar under these two bias conditions. Furthermore, we observed a significant change in the electric field profile in a perovskite device after 100 hours of stress/rest cycling under 450 nm illumination. The change in electric field profile indicates irreversible degradation of the device under light soaking, which is also supported by the I-V characteristics. Our results demonstrate that nanometer-scale potential imaging can serve as an important tool for assessing reliability concerns at various interfaces across perovskite device stacks.

Acknowledgement

This work was authored by the National Renewable Energy Laboratory, operated by Alliance for Sustainable Energy, LLC, for the U.S. Department of Energy (DOE) under Contract No. DE-AC36-08GO28308. Funding provided by the U.S. Department of Energy's Office of Energy Efficiency and Renewable Energy (EERE) Solar Energy Technologies Office (SETO) Agreement with agreement number 38044. The views expressed in the article do not necessarily represent the views of the DOE or the U.S. Government. The U.S. Government retains and the publisher, by accepting the article for publication, acknowledges that the U.S. Government retains a nonexclusive, paid-up, irrevocable, worldwide license to publish or reproduce the published form of this work, or allow others to do so, for U.S. Government purposes.

References

[1] NREL. Best Research-Cell Efficiencies. https://www.nrel.gov/pv/assets/images/efficiency-chart.png (accessed Dec 2021).

[2] D. Zhang et al., "Degradation pathways in perovskite solar cells and how to meet international standards", Commun. Mater., vol 3, pp 00281, August 2022

[3] H. J. Hovel Bae et al., "Electric-Field-Induced Degradation of Methylammonium Lead Iodide Perovskite Solar Cells", J. Phys. Chem. Lett., vol 7, pp 3091– 3096, July 2016.

[4] Y. Patikirige et al., "Linking Transient Voltage to Spatially-Resolved Luminescence Imaging to Understand Reliability of Perovskite Photovoltaics", 2021 IEEE 48th Photovoltaic Specialists Conference (PVSC), Fort Lauderdale, FL, USA, pp. 0660-0663, 2021.

[5] H. J. S. Ruan et al., "Light induced degradation in mixed-halide perovskites", J. Mater. Chem. C, vol 7, pp 9326–9334, July 2019.

[6] Z. Xu et al., "Iodine Electrochemistry Dictates Voltage-Induced Halide Segregation Thresholds in Mixed-Halide Perovskite Devices", Adv. Funct. Mater., vol 32, pp 2203432, June 2022.

[7] C. S. Jiang et al., "Electrical potential investigation of reversible metastability and irreversible degradation of CdTe solar cells", Sol. Energy Mater. Sol. Cell., vol 238, pp 111610, February 2022.

[8] C. Chen et al., "Achieving a high open-circuit voltage in inverted wide-bandgap perovskite solar cells with a graded perovskite homojunction", Nano Energy, vol 61, pp. 141-147, July 2019.

[9] C. Xiao et al., "Perovskite quantum dot solar cells: Mapping interfacial energetics for improving charge separation", Nano Energy, vol 78, pp. 105319, December 2020.

[10] C. Xiao et al., "Junction Quality of SnO2-Based Perovskite Solar Cells Investigated by Nanometer-Scale Electrical Potential Profiling.", ACS Appl. Mater. Interfaces, vol 9, pp. 38373–38380, October 2017.

[11] C. S. Jiang et al., "Real-space distributions of electrical potential in planar and porous perovskite solar cells: Carrier separation and transport," 2015 IEEE 42nd Photovoltaic Specialist Conference (PVSC), New Orleans, LA, USA, 2015, pp. 1-5, December 2015.

[12] Z. Li et al., "Extrinsic ion migration in perovskite solar cells.", Energy Environ. Sci., vol 10, pp. 1234-1242, April 2017.

[13] H. Lee et al., "Effect of Halide Ion Migration on the Electrical Properties of Methylammonium Lead Tri-Iodide Perovskite Solar Cells", J. Phys. Chem. C, vol 123, pp. 17728–17734, July 2019.

[14] X. Deng et al., "Electric field induced reversible and irreversible photoluminescence responses in methylammonium lead iodide perovskite", J. Mater. Chem. C, vol 4, pp. 9060-9068, September 2019.

[15] J. Xing et al., "Ultrafast ion migration in hybrid perovskite polycrystalline thin films under light and suppression in single crystals", Phys. Chem. Chem. Phys., vol 18, pp. 30484-30490 October 2016.

[16] V. W. Bergmann et al., "Local Time-Dependent Charging in a Perovskite Solar Cell", ACS Appl. Mater. Interfaces, vol 8, pp. 19402–19409, July 2016.

[17] B. Roose et al., "Ion migration drives self-passivation in perovskite solar cells and is enhanced by light soaking", RSC Adv., 2021, vol 11, pp. 12095-12101, March 2021.

[18] J. W. Schall et al., "In-Situ Photostability Analysis of Perovskite Solar Cells by Time-Evolving Photoluminescence Imaging", 2023 IEEE 50th Photovoltaic Specialist Conference (PVSC), San Juan, PR, USA, June 2023.

2D-MoS$_2$ Nano Structures to Enhance Silicon Solar Cells

Muntaser Abdelrahman Almansoori, Ayman Rezk, and Ammar Nayfeh

Khalifa University, Abu Dhabi, 127788, UAE *Corresponding Author: ammar.nayfeh@ku.ac.ae*

Abstract — MoS$_2$ is a promising 2D material for solar energy harvesting applications due to its size-dependent tunable bandgap and attractive magnetic, optical, and electrical properties. This study shows an easy way to deposit a 2D layer of MoS$_2$ nanoparticles on top of AZO/Quartz, AZO/Si, and bare Si. We investigate their UV-Vis spectral responses and potential for application in optoelectronic systems. Furthermore, compare the IV characteristics of solar structure with and without the MoS$_2$ 2D nanoparticles. Initial results indicate modification of the absorption spectrum over the UV-Vis range with an evident increase in the higher visible and lower UV range.

I. INTRODUCTION

Silicon-based solar cells make the majority of photovoltaic (PV) cells used in solar renewable energy generation around the globe. In 2014, 92% of PV panels were Si-based technology and predicted to still hold ~70% of the solar cells market share by 2030[1]. This is driven by industries' investments in the Si fabrication sector. In addition, Si is one of the most abundant materials in the earth's crust [2], [3]. This indeed pushed the cost of energy production per dollar, in the Middle East, to as low as 1.35 cents/kWh [4]. However, Si solar cells still face challenges regarding photons and bandgap mismatching. This leads to thermalization losses and losses from non-absorbed lower energy photons [5], among other losses due to reflection or photons scattering away from the cell. Ways to overcome these limitations include down-conversion/-shifting, up-conversion, and using plasmonic light trapping and scattering [6].

Molybdenum disulfide (MoS$_2$) is one of the promising 2D materials that caught the interest of many research fields [7], [8] due to their size-dependent tunable bandgap and attractive magnetic, optical, and electrical properties. Furthermore, recently there has been a growing interest in utilizing MoS$_2$ for solar cell applications that demonstrated measurable device enhancements [9], [10]. Hence, there is a great interest in understanding its potential for solar energy harvesting. Moreover, 2D and quantum dots structures of MoS$_2$ have shown an increase in the absorbance spectrum near the UV region. Also, it exhibits photon upconversion and downconversion photoluminescence, respectively [11], [12]. This can be utilized in light management techniques to overcome some of the limitations of the Si-based cell.

In this study, we show a straightforward way to deposit a 2D layer of MoS$_2$ nanoparticles on top of 1) an aluminum-doped zinc oxide (AZO) layer on quartz, 2) AZO on Si, and 3) bare Si. We investigate the spectral response of these different stacks and their potential for application in optoelectronic systems. Also, we examine the I-V characteristics of single junction MoS$_2$/Si solar cells under 1Sun and dark conditions using a solar simulator.

II. METHODOLOGY

An 80 nm AZO layer thin film is grown on a 3x3 cm^2 quartz substrate and 3x3 cm^2 p-Si EPI substrate using ALD with a 1:19 ratio. The AZO layer has shown good electrical and optical qualities for solar cell applications [13]. The 2D-MoS$_2$ nanoparticles solution is from Sigma Aldrich (SKU: 902012-25ML), which was deposited after sonicating by spin-coating on our AZO/Quartz, AZO/Si and Si substrates for 60 sec at 1000 rpm. Using a precise pipet, 500 μL of MoS$_2$ was spin-coated three times, covering a total volume of 1500 μL. Characterization measurements were conducted for each spin-coated layer.

Additional samples were prepared for I-V characterization using a 2.5x2.5 cm^2 p-Si substrate. We spin-coated 2000 μL of MoS$_2$ on the Si substrate and sputtered gold contacts onto MoS$_2$ through a finger pattern shadow mask (1x1 cm^2).

III. CHARACTERIZATION

A. UV-Vis Measurements

The samples were characterized using a UV-Vis-NIR spectrometer (Perkin Lambda 1050) to cover the UV-Vis

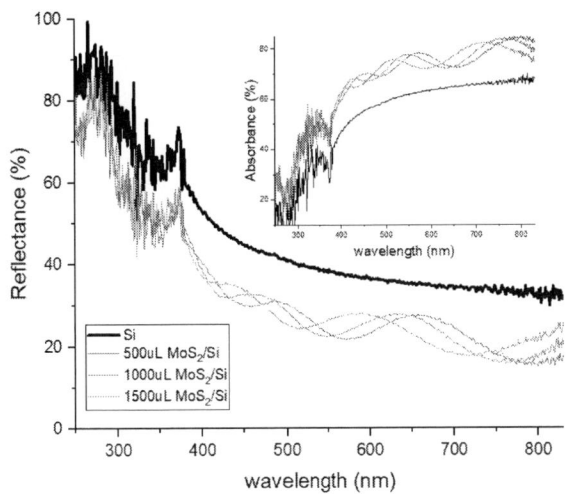

Fig. 1. Reflectance spectrum of MoS$_2$/Si, and inset show the absorbance spectrum.

978-1-6654-6060-6/23 $31.00 © 2023 IEEE

wavelength range (250-830 nm). Furthermore, the base AZO/Quartz, AZO/Si, and Si background spectra signal are measured before and after spin-coating.

B. IV Measurements

I-V measurements of the sample stack of Au(150 nm)/MoS₂(2000 µL)/Si were conducted using a solar simulator (Newport Sol3A Solar Simulator). We did a voltage sweep from -1.5 V to 1.5 V while monitoring the current response across the device. A black mask with a 1x1 cm^2 opening is used to cover the area of the cell not coated by the sputtered gold fingers.

IV. RESULTS AND DISCUSSION

A. Bare Si

Fig. 1 shows a decrease in reflectance across the UV-Vis range shifted relative to each other depending on the amount of MoS₂ deposited. In MoS₂ samples spectra, there are oscillations of two types caused by wave interference in the MoS₂ thin film and the thick substrate. No appreciable change in transmittance was found (fluctuating around 0%). Revealing that such a decrease in reflectance is attributed to an increase in absorbance caused by the MoS₂. This is either due to absorbance by MoS₂ or reflectance reduction due to the increase of photons scattering into the Si. Also, we note that the instrument detects much noise at the region close to the detector edge for high reflectance samples such as Si.

B. AZO/Si

AZO is a promising alternative transparent conductive oxide (TCO) that is abundant, environmentally friendly and has good anti-reflectance qualities[14]. Hence, it is valuable to understand how the addition of MoS₂ to such a stack behaves to investigate its potential to enhance solar cell designs. The effect of growing an AZO antireflective (ARC) layer on the Si substrate before spin-coating the MoS₂ is also investigated. Fig. depicts the UV-Vis data of AZO/Si stack before and after MoS₂ coatings, where first we notice an expected reduction in reflection of AZO/Si compared to bare Si due to the anti-reflection properties of AZO film. Additionally, the MoS₂ coating here led to a wavelength-dependent trend, where we see an increase of reflectance from wavelengths above 450 nm and a decrease in the range of ~300-450 nm. Similar relation among different quantities of MoS₂ is witnessed; however, they behave differently compared to bare Si, probably due to the extra interaction with the AZO layer.

C. AZO/Quartz

The absorption of the MoS₂/AZO system is measured on quartz substrates. The obtained spectrometer data plotted in Fig. shows a high absorbance effect due to the addition of 2D-MoS2

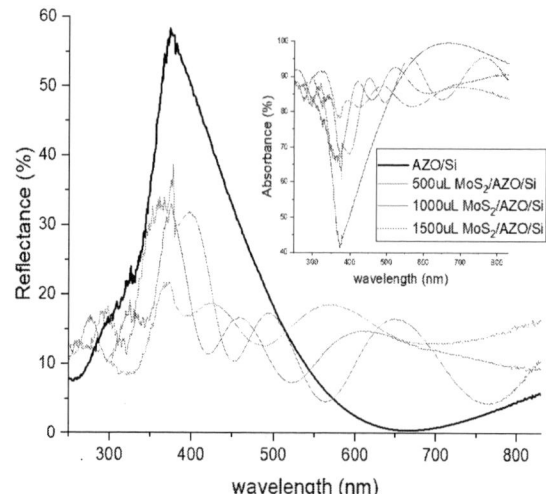

Fig. 2. Reflectance spectrum of MoS2/AZO/Si and inset show the absorbance spectrum.

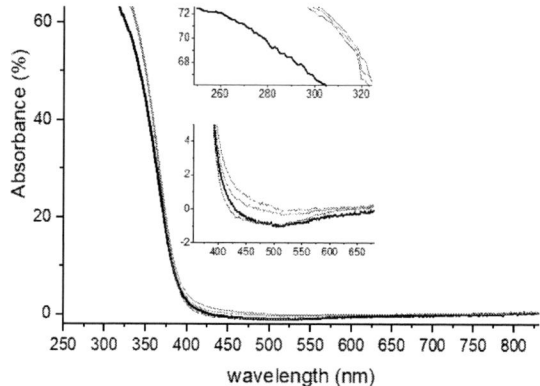

Fig. 3. Absorbance spectrum of MoS₂/AZO/Quartz, while insets show zoomed plot regions of interest.

nanoparticles at the higher visible and lower UV spectrum range. It peaks around 340 nm with an approximate absorbance increase of ~6.7%. Utilizing s transparent substrate such as quartz enables us to measure the absorbance by the 2D-MoS₂ layers.

Upon further examination, we see that the behavior is not linear across the whole spectrum and is a function of 1) wavelength and 2) MoS₂ quantity. This could be due to the quantum confinement effect of several layers of stacked 3D MoS₂ nanoparticles[15]. This phenomenon could open the possibility of utilizing this material for low-wavelength filters or UV sensing applications. Also, it can potentially be used for down-conversion[16] of high-energy photons. If engineered to re-emit photons at lower energies, it can enhance solar cells' efficiencies and reduce thermal burden; however, further investigation is needed.

D. Au/2000 µL MoS₂/Si

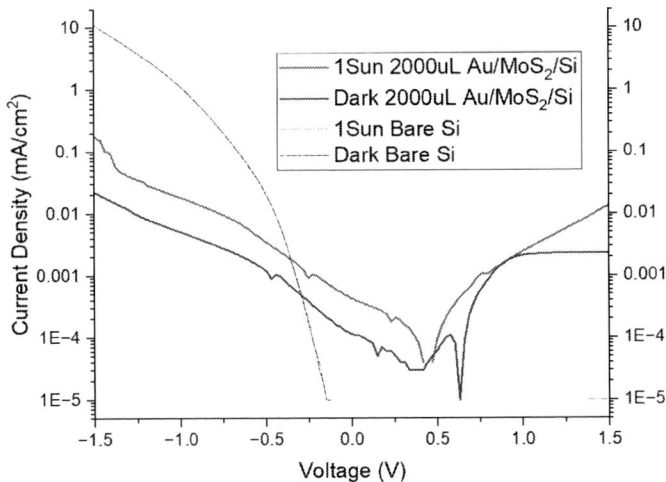

Fig. 4. Current density versus voltage for Au/MoS₂/Si sample.

We extracted current density values from the IV curve measurements for the sample, as shown in Fig. 4. We see a slight increase of current under one sun compared to dark measurements. This indicates an increased current generation under light conditions. However, this performance can result from the non-uniform deposition of MoS₂ layers. It is challenging to have a thin uniform film by spin-coating 2D MoS₂ nanoparticles, where it was shown that MoS₂ flakes tend to aggregate[17].

Finally, MoS₂ has the potential to enhance current Si-based solar cells. While further studies need to be conducted, it is apparent that controlling growth mechanisms is essential. It allows the synthesis of nanostructures with suitable dimensions and shapes to enhance solar cells as absorption layers, scattering centers, or down-conversion/-shifting layers.

REFERENCES

[1] S. Weckend, A. Wade, and G. Heath, "End of Life Management: Solar Photovoltaic Panels," Aug. 2016. doi: 10.2172/1561525.

[2] "Mineral commodity summaries 2020," U.S. Geological Survey, Reston, VA, USGS Unnumbered Series, 2020. doi: 10.3133/mcs2020.

[3] V. E. Ferry, "Light Trapping in Plasmonic Solar Cells," phd, California Institute of Technology, 2011. doi: 10.7907/AMD4-Q845.

[4] H. Apostoleris, A. Al Ghaferi, and M. Chiesa, "What is going on with Middle Eastern solar prices, and what does it mean for the rest of us?," *Prog Photovolt Res Appl*, vol. 29, no. 6, pp. 638–648, Jun. 2021, doi: 10.1002/pip.3414.

[5] J. Day, S. Senthilarasu, and T. K. Mallick, "Improving spectral modification for applications in solar cells: A review," *Renewable Energy*, vol. 132, pp. 186–205, Mar. 2019, doi: 10.1016/j.renene.2018.07.101.

[6] H. A. Atwater and A. Polman, "Plasmonics for improved photovoltaic devices," *Nature Mater*, vol. 9, no. 3, pp. 205–213, Mar. 2010, doi: 10.1038/nmat2629.

[7] P. Zhou, C. Chen, X. Wang, B. Hu, and H. San, "2-Dimentional photoconductive MoS2 nanosheets using in surface acoustic wave resonators for ultraviolet light sensing," *Sensors and Actuators A: Physical*, vol. 271, pp. 389–397, Mar. 2018, doi: 10.1016/j.sna.2017.12.007.

[8] H. Dong *et al.*, "Fluorescent MoS ₂ Quantum Dots: Ultrasonic Preparation, Up-Conversion and Down-Conversion Bioimaging, and Photodynamic Therapy," *ACS Appl. Mater. Interfaces*, vol. 8, no. 5, pp. 3107–3114, Feb. 2016, doi: 10.1021/acsami.5b10459.

[9] Y. Tsuboi *et al.*, "Enhanced photovoltaic performances of graphene/Si solar cells by insertion of a MoS ₂ thin film," *Nanoscale*, vol. 7, no. 34, pp. 14476–14482, 2015, doi: 10.1039/C5NR03046C.

[10] Y.-J. Huang, H.-C. Chen, H.-K. Lin, and K.-H. Wei, "Doping ZnO Electron Transport Layers with MoS ₂ Nanosheets Enhances the Efficiency of Polymer Solar Cells," *ACS Appl. Mater. Interfaces*, vol. 10, no. 23, pp. 20196–20204, Jun. 2018, doi: 10.1021/acsami.8b06413.

[11] W. Xing *et al.*, "MoS ₂ Quantum Dots with a Tunable Work Function for High-Performance Organic Solar Cells," *ACS Appl. Mater. Interfaces*, vol. 8, no. 40, pp. 26916–26923, Oct. 2016, doi: 10.1021/acsami.6b06081.

[12] A. P. Sunitha, P. Hajara, M. Shaji, M. K. Jayaraj, and K. J. Saji, "Luminescent MoS2 quantum dots with reverse saturable absorption prepared by pulsed laser ablation," *Journal of Luminescence*, vol. 203, pp. 313–321, Nov. 2018, doi: 10.1016/j.jlumin.2018.06.004.

[13] S. Abdul Hadi, G. Dushaq, and A. Nayfeh, "Effect of atomic layer deposited Al ₂ O ₃ :ZnO alloys on thin-film silicon photovoltaic devices," *Journal of Applied Physics*, vol. 122, no. 24, p. 245103, Dec. 2017, doi: 10.1063/1.4990871.

[14] K. Ellmer, A. Klein, and B. Rech, *Transparent conductive zinc oxide basics and applications in thin film solar cells*. Berlin: Springer, 2008. Accessed: Apr. 25, 2014. [Online]. Available: http://public.eblib.com/EBLPublic/PublicView.do?ptiID=373032

[15] T. Li and G. Galli, "Electronic Properties of MoS ₂ Nanoparticles," *J. Phys. Chem. C*, vol. 111, no. 44, pp. 16192–16196, Nov. 2007, doi: 10.1021/jp075424v.

[16] A. P. Sunitha, P. Praveen, M. K. Jayaraj, and K. J. Saji, "Upconversion and downconversion photoluminescence and optical limiting in colloidal MoS2 nanostructures prepared by ultrasonication," *Optical Materials*, vol. 85, pp. 61–70, Nov. 2018, doi: 10.1016/j.optmat.2018.08.038.

[17] G. G. Politano *et al.*, "Physical Investigation of Spin-Coated MoS2 Films," in *The 2nd International Online-Conference on Nanomaterials*, Nov. 2020, p. 3. doi: 10.3390/IOCN2020-08005.

Effect of Bidentate Ligand Additive in Tin Perovskite Solar Cells

Dhruba B. Khadka[1], Yasuhiro Shirai[1], Masatoshi Yanagida[1] and Kenjiro Miyano[1]

[1]Photovoltaics Materials Group, Global Research Center for Environment and Energy based on Nanomaterials Science (GREEN), National Institute for Materials Science (NIMS), 1-1 Namiki, Tsukuba, Ibaraki 305-0044, Japan

Abstract — Tin- perovskite solar cell (Sn-PSC) is limited by their poor stability arising from facile tin oxidation and uncontrolled film growth. Here, we introduced a multifunctional bidentate ligand, formohydrazide (FHZ) as an additive into FASnI$_3$ perovskite film. This additive is found to be effective for the control of Sn-oxidation and forms pin-hole free film morphology. The Sn-PSCs with FHZ additive enhanced the device efficiency from 9.93% (for control Sn-HaP) to 13.14 % (for device with FHZ additive) with higher reproducibility and superior device stability. The device analysis suggests that the hydrazide additive significantly passivates the bulk and surface defect in the Sn-PSCs. This report gives insights into the film growth properties, device photo-physics, and defect analysis correlating with device performance and device stability.

I. INTRODUCTION

Lead perovskite solar cells (Pb-PSCs) have scaled up their power conversion efficiency (PCE) >25% in the past decade.[1], [2] But the toxic Pb has imposed a hurdle in commercialization due to health hazards. Therefore, lead-free perovskite candidates have centred colossal attention on replacing Pb.[3]–[6] Among the various alternative candidates, tin perovskite (Sn-HaP) derivates are excellent candidates with ideal bandgaps of 1.2- 14 eV.[7] It is reported that the intrinsic instability is due to the facile oxidation of Sn^{2+}, poor film growth, and shorter carrier lifetime.[8]

Several reports have been documented on reducing Sn^{2+} oxidation, surface passivation, structural regulation, and bulk or interface engineering, for the improvement in PCE as well as device stability in the Sn-PSCs.[9]–[12] Ning and co-workers have documented well-controlled crystal orientation using precursors SnI$_2$ adduct comprised with PEABr.[13] A PCE of 14.81% has been reported by modulating the 2D/3D microstructures using fluorine functionalized F-PEABr.[11] Indeed, functional additives engineering is effective in attuning the growth of the highly orientated crystalline film, defect attenuation, and carrier transport.

In this report, we used a bidentate ligand, formohydrazide (FHZ) as an additive in the Sn-HaP precursor solution to control the oxidation and film growth. The PCE of FHZ added device enhanced from 9.93 to 13.14 % with superior device stability. It is found that the FHZ additive results in high-quality Sn-HaP film with highly oriented crystalline growth, well-compact film texture, and control of the extent of Sn^{2+} oxidation. This work underscores a detailed insight into the effect of bidentate ligand in Sn-PSCs.

II. EXPERIMENTAL

A. Device fabrication

For the fabrication of FASnI$_3$; Sn-based HaP precursor solution (0.85 M) was prepared by dissolving a 0.92:0.08:10:1 stoichiometric ratio of FAI, RbCl, SnI$_2$, and SnF2 adding EDAI-0.01 M and PEABr-0.05 M in dimethyl sulfoxide (DMSO) solvent. For FHZ additive precursor, FHZ at a molar ratio of 2, 4, 8, 12, 16, 20, and 30% to SnI$_2$ was added to the above solution. The Sn-HaP precursor was deposited on the PEDOT:PSS (30 nm)/ITO substrate. These films were annealed on the hot plate at 60 ℃ for 1min and 850 ℃ for 10 min. Then, the device is completed depositing PCBM /BCP thin films were spun-coated on top of the Sn-HaP films. Finally, Ag (100 nm) was thermally evaporated and get device. The detailed fabrication can be found in our earlier report.[14]–[16]

B. Materials and device characterizations

The XRD results were collected using Rigaku Smart Lab, CuKα radiation, λ=1.5405Å. The SEM images were obtained by a high-resolution scanning electron microscope (SEM) at 5 kV accelerating voltage (Hitachi, S-4800). XPS spectra were obtained using a Versa Probe II (ULVAC-PHI, Japan). The current density–voltage (J-V) curves were measured under 1 sun with an AM1.5G spectral filter (100 mW/cm^2) coupled with an MPPT system (Systemhouse Sunrise Corp.). capacitance spectra (C–f) were collected using an LCR meter (IM3536, Hioki) under dark.

III. RESULTS AND DISCUSSION

The Sn-HaP films were fabricated by spin-coating precursor solution with FHZ, bidentate ligand for the improvement in the crystallization and passivating defect chemistry in the film.

To evaluate the effect of the FHZ additive on the photovoltaic performance, we fabricated Sn-PSCs with the inverted device architectures of ITO/PEDOT:PSS/Sn-HaP/ICBA/BCP/Ag as shown in the cross-sectional image (Fig. 1). The density-voltage (J-V) characteristics with varying FHZ contents are given in Fig. 1b. The statistical data (Fig. 1c) show theb trend of device efficiency with varying FHZ additive contentration. It demonstrated the best device with FHZ additive (12 mol%). The control device achieved a PCE of 9.93% (with J$_{SC}$ ~19.62

mA/cm², V_{OC} ~0.734 V, and FF ~69.01%). The device with FHZ additive (≤12 mol%) in Sn-HaP (hereafter, Sn-PSC with FHZ) improved the device PCE of ~13.14% with a significant increase in V_{OC} ~0.892 V and FF ~75.2%.

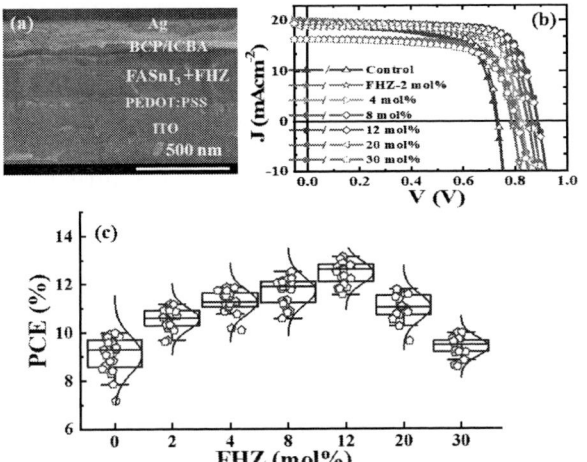

Fig. 1. Cross-sectional image of the device (a). The J-V curves of FAPSnI₃+FHZ additive (for x=0 – 30 mol%) (b). Device efficiency trend with content of FHZ additive.

X-ray diffraction (XRD) patterns of Sn-HaP films with FHZ additive (selective mol%) are shown in Fig. 2a. XRD patterns reveal the characteristic diffraction assigned to highly oriented crystallographic planes of (100) and (200) which can be assigned to the orthorhombic phase of FASnI₃.[9,35] There is no shifting of characteristics XRD peak for the control or FHZ-added Sn-HaP films suggesting not incorporated in the Sn-HaP crystal lattice. The Sn-HaP film with FHZ additive (20 mol%) showed comparatively intensified XRD characteristic peaks indicating improved crystallinity. This is parallel to the trend of device performance.

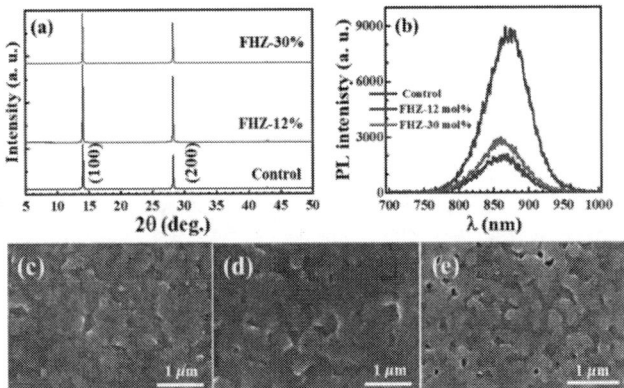

Fig. 2. Effect of FHZ additiveon; XRD patterns (a), PL spectra (b), and SEM images of Sn-HaP film FHZ additives.

Figure 2b depicts the PL spectra of corresponding films. The Sn-HaP film with the FHZ additive shows a higher peak intensity compared to the control film indicating ameliorated film quality with the FHZ additive. This observation is also consistent with the XRD peak intensity. The characteristic PL peaks are found to be centred at 1.430 ± 0.02 eV for the control and 1.416 ± 0.02 eV for 12 mol% FHZ.

Figure 2c-e depicts scanning electron microscopic images of the control and FHZ additive films. The films with FHZ additive show compact and better film coverage by suppressing the pinhole's densities. Small granular features in the control film disappear in the FHZ-added Sn-HaP film. However, the Sn-HaP with a higher FHZ additive (30 mol%) was found to grow with pinholes and uneven morphology.

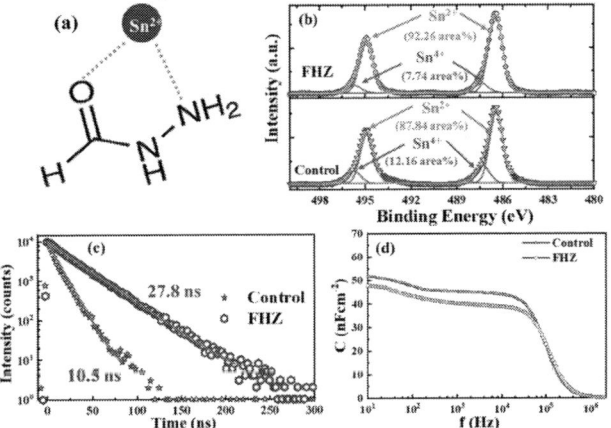

Fig. 3. XPS spectra (S-2p, Sn-3d) of the respective film (a,b). TRPL results of films (c). C-f spectra at room temperature (d).

Figure 3a shows the shematics of bonding interation of FHZ additive to the Sn-sites in Sn-perovskite. The FHZ additive strongly binds the Sn²⁺ molecules via -NH₂ and -O=C bidentate ligand which control the extent of Sn²⁺ oxidation.

Figure 3b presents the two characteristic peaks deconvoluted into the Sn 3d (3d5/2 (3d3/2)) at ~486.7 (495.2) eV and 487.3 (495.7) eV which are attributed to the Sn²⁺ and Sn⁴⁺ species, respectively. The Sn-HaP film shows suppression of the ionic percentage of Sn⁴⁺ from 12.16 to 7.74%. It corporates that FHZ can act as a bidentate ligand with metal. It forms a coordination complex with Sn²⁺ which control the formation of deleterious chemical derivatives such as Sn⁴⁺.

Figure 3c shows the time-resolved photoluminescence (TRPL). The FHZ-treated HaP film demonstrated a longer carrier lifetime (τ-1~27.8 ns) compared to the control film (τ-~10.5 ns) suggesting suppression of the recombination pathways in the Sn-HaP film with FHZ additive. Hence, it leads to improvement in device performance.

Moreover, Fig. 3d shows the capacitance-frequency (C-f) spectra of the control and FHZ additive devices. Note that the capacitance at a lower frequency accounts for the ionic motion or charge accumulation.[17] The device with the FHZ additive has a lower value indicating the reduction in ion or charge accumulation at the interfacial layer or electrode. While the control device reveals a slightly larger value in the frequency range of 1kHz < f < 50 kHz, which indicates a higher defect density in the control Sn-HaP film. It is concurrent with the device results.

IV. SUMMARY AND CONCLUSIONs

We have achieved the Sn-PSC of efficiency ~13.14 % using formohydrazide additive (9.93 % for the control device) with superior stability. The FASnI$_3$ film with the FHZ additive not only improved the film morphology and highly oriented crystal growth but also inhibited Sn^{2+}/Sn^{4+} oxidation benefiting from bidentate legend. The FASnI$_3$ film with FHZ additive increases the carrier lifetime suggesting the suppression of defect densities in the bulk and at the interface as supported capacitance spectra. This report substantiates that the antioxidative functional additive is expedient for the control of oxidation and defect passivation which scales up for the performance and stability of Sn-PSCs.

ACKNOWLEDGMENT

This work was supported by JST-Mirai Program Grant Number JPMJMI21E6, Japan.

REFERENCES

[1] O. Almora et al., "Device Performance of Emerging Photovoltaic Materials (Version 3)," Adv. Energy Mater., vol. 13, no. 1, p. 2203313, Jan. 2023, doi: 10.1002/aenm.202203313.

[2] D. B. Khadka, Y. Shirai, M. Yanagida, T. Tadano, and K. Miyano, "Interfacial Embedding for High-Efficiency and Stable Methylammonium-Free Perovskite Solar Cells with Fluoroarene Hydrazine," Adv. Energy Mater., vol. 12, no. 38, p. 2202029, Oct. 2022, doi: 10.1002/aenm.202202029.

[3] M. G. M. Pandian et al., "Effect of solvent vapour annealing on bismuth triiodide film for photovoltaic applications and its optoelectronic properties," J. Mater. Chem. C, vol. 8, no. 35, pp. 12173–12180, 2020, doi: 10.1039/D0TC02455D.

[4] G. Nasti and A. Abate, "Tin Halide Perovskite (ASnX3) Solar Cells: A Comprehensive Guide toward the Highest Power Conversion Efficiency," Adv. Energy Mater., vol. 10, no. 13, p. 1902467, Apr. 2020, doi: 10.1002/aenm.201902467.

[5] S. T. Umedov, D. B. Khadka, M. Yanagida, A. Grigorieva, and Y. Shirai, "A-site tailoring in the vacancy-ordered double perovskite semiconductor Cs2SnI6 for photovoltaic application," Sol. Energy Mater. Sol. Cells, vol. 230, p. 111180, Sep. 2021, doi: 10.1016/j.solmat.2021.111180.

[6] D. B. Khadka, Y. Shirai, M. Yanagida, and K. Miyano, "Tailoring

the film morphology and interface band offset of caesium bismuth iodide-based Pb-free perovskite solar cells," J. Mater. Chem. C, vol. 7, no. 27, pp. 8335–8343, 2019, doi: 10.1039/C9TC02181G.

[7] M. Baranowski and P. Plochocka, "Excitons in Metal-Halide Perovskites," Adv. Energy Mater., vol. 10, no. 26, p. 1903659, Jul. 2020, doi: 10.1002/aenm.201903659.

[8] X. Liu et al., "Templated growth of FASnI 3 crystals for efficient tin perovskite solar cells," Energy Environ. Sci., vol. 13, no. 9, pp. 2896–2902, 2020, doi: 10.1039/D0EE01845G.

[9] M. A. Kamarudin et al., "Suppression of Charge Carrier Recombination in Lead-Free Tin Halide Perovskite via Lewis Base Post-treatment," J. Phys. Chem. Lett., vol. 10, no. 17, pp. 5277–5283, Sep. 2019, doi: 10.1021/acs.jpclett.9b02024.

[10] D. B. Khadka, Y. Shirai, M. Yanagida, and K. Miyano, "Attenuating the defect activities with a rubidium additive for efficient and stable Sn-based halide perovskite solar cells," J. Mater. Chem. C, vol. 8, no. 7, pp. 2307–2313, 2020, doi: 10.1039/C9TC06206H.

[11] B. Bin Yu et al., "Heterogeneous 2D/3D Tin-Halides Perovskite Solar Cells with Certified Conversion Efficiency Breaking 14%," Adv. Mater., vol. 33, no. 36, p. 2102055, Sep. 2021, doi: 10.1002/adma.202102055.

[12] D. B. Khadka, Y. Shirai, M. Yanagida, and K. Miyano, "Passivation of the Recombination Activities with Rubidium incorporation for Efficient and Stable Sn- HaP Solar Cells," in 2020 47th IEEE Photovoltaic Specialists Conference (PVSC), 2020, pp. 0113–0116, doi: 10.1109/PVSC45281.2020.9300783.

[13] X. Jiang et al., "One-Step Synthesis of SnI2·(DMSO)x Adducts for High-Performance Tin Perovskite Solar Cells," J. Am. Chem. Soc., vol. 143, no. 29, pp. 10970–10976, Jul. 2021, doi: 10.1021/jacs.1c03032.

[14] D. B. Khadka, Y. Shirai, M. Yanagida, and K. Miyano, "Pseudohalide Functional Additives in Tin Halide Perovskite for Efficient and Stable Pb-Free Perovskite Solar Cells," ACS Appl. Energy Mater., vol. 4, no. 11, pp. 12819–12826, Nov. 2021, doi: 10.1021/acsaem.1c02496.

[15] D. B. Khadka, Y. Shirai, M. Yanagida, and K. Miyano, "Effect of Phenethylammonium Thiocyanate Additive in Tin Perovskite for Efficient and Stable Pb-free Perovskite Solar Cells," in 2022 IEEE 49th Photovoltaics Specialists Conference (PVSC), 2022, pp. 0004–0006, doi: 10.1109/PVSC48317.2022.9938869.

[16] D. B. Khadka, Y. Shirai, M. Yanagida, T. Masuda, and K. Miyano, "Enhancement in efficiency and optoelectronic quality of perovskite thin films annealed in MACl vapor," Sustain. Energy Fuels, vol. 1, no. 4, pp. 755–766, 2017, doi: 10.1039/C7SE00033B.

[17] D. B. Khadka, Y. Shirai, M. Yanagida, and K. Miyano, "Insights into Accelerated Degradation of Perovskite Solar Cells under Continuous Illumination Driven by Thermal Stress and Interfacial Junction," ACS Appl. Energy Mater., vol. 4, no. 10, pp. 11121–11132, Oct. 2021, doi: 10.1021/acsaem.1c02037.

Optoelectronic Performance of Solution Processable MoS₂ for Application in Photovoltaic Devices

Dayanand Kumar[1], Ayman Rizk[2], Ammar Nayfeh[2], and Nazek El-Atab[1*]

[1]Smart, Advanced Memory Devices and Applications (SAMA) Laboratory, Electrical and Computer Engineering, King Abdullah University of Science and Technology (KAUST), 23955, Saudi Arabia, [2]Khalifa University, Abu Dhabi, 127788, UAE.
[*]Corresponding Author: nazek.elatab@kaust.edu.sa

Abstract — Currently, the transition metal dichalcogenides (TMDCs) such as MoS₂ have open up the way to the fulfilment of ultrathin and ultralight optoelectronic devices. With the possible applications of TMDCs, the evolvement of ultrathin solar cells is appealing due to the possible reduction in cost and suitability of these materials for flexible and ultralight photovoltaics. In this work, a 70-nm thick MoS₂ based MOS device is demonstrated. The results confirm that electron-hole pairs are generated when optical light with different wavelengths was incuced on it. The optical results confirm that the MoS₂ has excellent potential to be used in solar cell applications.

I. INTRODUCTION

Currently, the renewal energy's demand is swiftly rising due to the limited energy resources worldwide. Appropriately, new materials and new technologies have been examined for energy production, including the search for new semiconducting materials for harvesting clean energy. The transition metal dichalcogenides (TMDCs) have shown great attraction for solar cell applications because of their electrical and optical features. In particular, their exellent optical light absorption and semiconducting band gap have allowed exhibitions of photovoltaic response from heterostructures composed of TMDCs and other organic or inorganic materials [1-3].

The TMDs materials are possible to be grown *via* the chemical vapor deposition (CVD) technique, or to be transferred *via* physical and chemical exfoliation methods. More specifically, the mechanical exfoliation method has enabled high quality TMDs like MoS₂, WS₂, ZrS₂, MoSe₂, NbSe₂, TiS₂, TaS₂, and WSe₂ for various applications [4]. Le *et al.,* reported the development of single and multilayer MoS₂ and WSe₂ based on mechanical exfoliation [5]. Jin An *et al.* reported that TMDs based nanosheets such as MoSe₂, MoS₂, WS₂, and WSe₂ could be achieved through the exfoliation method using high-power laser [6, 7]. The main disadvantage of this approach is that large sized bulk material is needed for obtaining the desired nanosheets as well as challenges in terms of getting large-sized monolayers. While CVD allows for a larger coverage of the 2D material, its main disadvantage is the high thermal budget that is required and which makes it incompatible with back-end-of-line processing or for wearables and flexible electronics. On the other hand, spin coating of 2D materials allows for obtaining a large density of the material at low temperature, with the main disadvantage being the lack of control and uniformity. Nevertheless, Matsuba *et al.,* have shown the ability to achieve full coverage of a monolayer using the spin coating technique which makes solution processing very promising [8].

In this work, in an attempt to explore the peroformace of MoS₂ in converting photons with different energies to electron-hole pairs, the solution processable MoS₂ based Al/Al₂O₃/MoS₂/Al₂O₃/P⁺-Si device which was fabricated on P⁺-Si substrate. The device shows electrons storage at different wavelengths of visible lights which confirms that photons are being absorbed and electron-hole pairs being generated. The results show that MoS₂ is highly suitable for solar cell applications in near future.

II. METHODOLOGY

Al/Al₂O₃/MoS₂/Al₂O₃/p⁺-Si Device Fabrication:

The device was fabricated on P⁺-Si substrate. First, the Si substrate was wet etched in buffered oxide etchant (BOE) for 3 minutes and then cleaned with deionized water (DI) water and nitrogen gun to remove the native oxide from the wafer. After

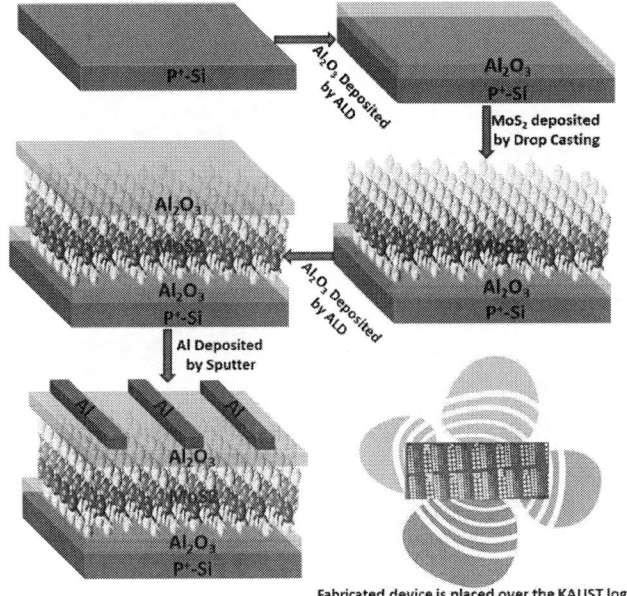

Figure 1. Schametic and fabricated MoS₂ based device

978-1-6654-6060-6/23 $31.00 © 2023 IEEE

that, a 3 nm thick Al_2O_3 as a tunnel oxide layer was deposited by plasma enhanced atomic layer deposition (PE-ALD) at 250^0 C using Al (CH_3), trimethylaluminium (TMA) and O_2 plasma. Next, a 70-nm thick Molybdenum Disulfide (MoS_2) film was deposited using the drop casting method. Thereafter, a 7 nm thick Al_2O_3 layer was again deposited by PE-ALD as a blocking oxide layer at 250^0 C. Finally, 50 nm thick Al film was deposited as a top electrode by DC sputtering using a metal shadow mask.

III. CHARACTERIZATIONS

The surface mrhology of drop casted MoS_2 sample was analyzed, which is depicted in Fig. 2. The scanning electron microscope (SEM) image shows that the MoS_2 sample has large number of flakes with high density. We believe that the large number of flakes and high density of MoS_2 absorbed the more light, which is highly suitable for solar cell applications.

To check the crystal structure of the MoS_2, the X-ray diffraction (XRD) spectrum was analyzed, which is depicted in Fig. 3. XRD diffraction confirms a sharp and narrow peak at about 14.5°, which is recognized to be the (002) plane of MoS_2 [9]. The XRD spectra of the MoS_2 was matched with the already reported data of Hexagonal MoS_2 thin film (JCPDS: 37-1492). After matching the XRD spectrum of the sample with JCPDS, it is confirmed that there is no shift in the 2θ value for the MoS_2 thin film. The XRD spectra of the sample also suggests that the (002) plane is highly oriented.

To explain the surface chemical states and coordination geometry of the drop casted MoS_2 flakes, XPS analysis was carried out to determine the binding energies of MoS_2, which is

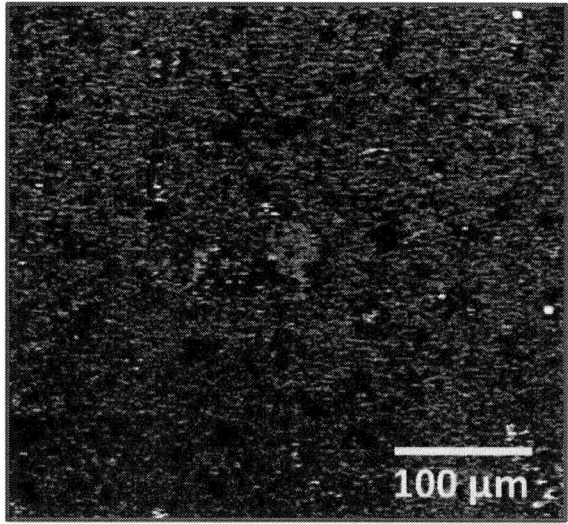

Figure 2. Surface morphology of drop casted MoS_2 sample

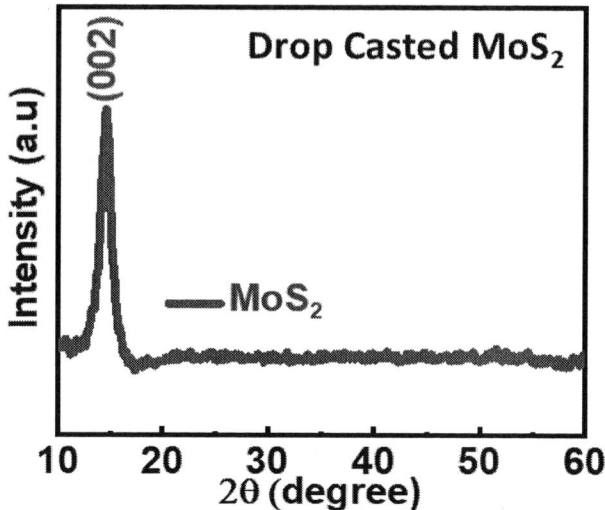

Figure 3. XRD spectra of drop casted MoS_2 sample.

shown in Fig. 4(a) and 4(b), respectively. From the Fig. 4(a), the Mo 3d depicts the three binding energy peaks which are located at about 226.8, 229.4 and 232.8 eV. The binding energy peaks of 229.4 eV and 232.8 eV are ascribed to the doublet of Mo $3d_{5/2}$ and Mo $3d_{3/2}$, respectively [10]. The S 2 s peak (weak sulfur) which is located at the binding energy of 226.8 eV (Fig. 4(a)) and the divalent sulfide ion (S^{2-}) peaks with the binding energy of around 162.4 eV and 163.7 eV are allocated to $2p_{1/2}$ and $2p_{3/2}$, as shown in Fig. 4(b). In addition, from the XPS analysis, we found a small peak which is located at a higher binding energy of 235.6 eV which is attributed to Mo^{6+}, signifying that the Mo edges in drop casted MoS_2 are oxidized during the deposition of the MoS_2 film. It is suggested that the interfacial oxidation between MoS_2 and Al_2O_3 layer is due to the deposition of MoS_2. Conclusively, the XPS analysis confirmed that the drop casted MoS_2 layer shows the hexagonal structure [10].

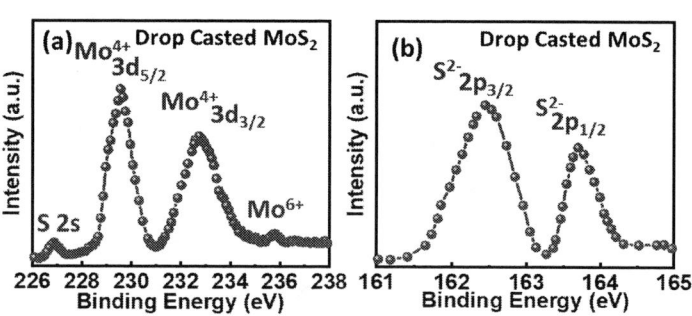

Figure 4. XPS analysis of drop casted MoS_2 sample. (a) The Mo 3d and S 2s and (b) S 2p XPS peaks.

978-1-6654-6060-6/23 $31.00 © 2023 IEEE

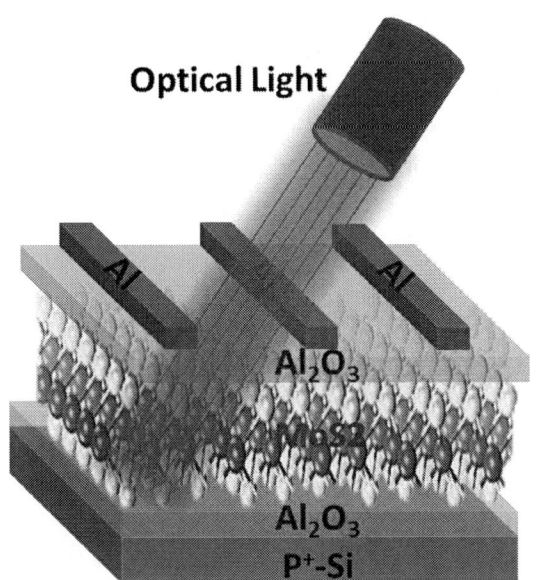

Figure. 5. Schematic diagram of the device with optical light illumination.

IV. RESULTS AND DISCUSSION

The device was measured using the irradiation of different wavelengths of light. The schematic structure of the device is shown in Fig. 5. Fig. 6 (a) shows the C-V characteristics of the device when exposed to light with different wavelengths. The optical characteristics of the device were measured using the different wavelength of light from 600 nm to 400 nm. When the device was in dark conditions, the threshold voltage of the device was observed to be about 2.5 V. When the light was turned on for 1 s, onto the device, with 600 nm wavelength (Intensity: 2 mW/cm^2) during the programming condition of +6/-6 V, the threshold voltage of the device was increased from 2.8 V to 3.3V, which confirms that photons absorption is leading to electrons generation and storage.

Thus, this means that, due to the illumination of optical light, the MoS$_2$ absorbed the photons. Therefore, threshold voltage is shifting higher voltage side. We further checked the effect of different wavelengths of light from 550 nm to 400 nm with the interval of 50 nm using the same programming voltage, same intensity, and same illumination time. We got the tremendous shift in threshold voltage form 3.3V to more than 6 V. The

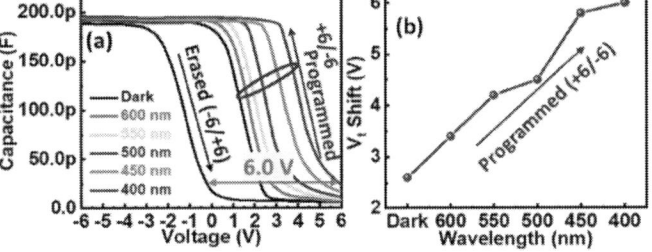

Figure. 6. (a) C-V curves of the device using different wavelengths of optical light from 600 nm to 400 nm with interval of 50 nm. (b) Wavelength dependent threshold voltage of the device.

maximum shift in threshold voltage was observed for 400 nm wavelength. Thus, it is confirmed that MoS$_2$ layer absorb more light for 400 nm wavelength. Figure 6 (b) depicts the threshold voltage shift with the optical light wavelength. The device shows that the threshold voltage increases with decreasing the wavelength of light.

IV. CONCLUSION

To summarize, The SEM images shows a good coverage of the MoS$_2$ on the substrate. The XRD and XPS confirm the crystallinity and crystal structure of the MoS$_2$. The excellent optical properties of MoS$_2$ based device and its ability to convert photons to electron-hole pairs confirms the potential of MoS$_2$ in solar cell applications in near future.

REFERENCES

[1] T. A. Shastry, I. Balla, H. Bergeron, S. H. Amsterdam, T. J. Marks, and M. C. Hersam, " Mutual Photoluminescence Quenching and Photovoltaic Effect in Large-Area Single-Layer MoS$_2$−Polymer Heterojunctions," *ACS Nano*, vol. 10, pp. 10573−10579, 2016.

[2] M. Bernardi, M. Palummo, J. C. Grossman, "Extraordinary Sunlight Absorption and One Nanometer Thick Photovoltaics Using Two-Dimensional Monolayer Materials. *Nano Lett.*, vol. 13, pp. 3664− 3670, 2013.

[3] M. M. Furchi, A. Pospischil, F. Libisch, J. Burgdorfer, T. Mueller, "Photovoltaic Effect in an Electrically Tunable van der Waals Heterojunction," *Nano Lett.* 2014, vol. 14, pp. 4785−4791, 2014.

[4] S. Lee1, R. Peng, C. Wu, and M. Li, "Programmable black phosphorus image sensor for broadband optoelectronic edge computing," *Nature communications*, vol. 13, 1485, 2022.

[5] V. Shanmugam, R. A. Mensah, K. Babu, S. Gawusu, A. Chanda, Y. Tu, R. E. Neisiany, M. Forsth, G. Sas, O. Das, "A Review of the Synthesis, Properties, and Applications of 2D Materials. *Particle*, vol. 39, 2200031, 2022.

[6] H. Li, J. Wu, Z. Yin, and H. Zhang, "Preparation and Applications of Mechanically Exfoliated Single-Layer and Multilayer MoS$_2$ and WSe$_2$ Nanosheets," *Acc. Chem. Res.*, vol. 47, pp. 1067–1075, 2014.

[7] S. J. An, Y. H. Kim, C. Lee, D. Y. Park, and M. S. Jeong Exfoliation of transition metal dichalcogenides by a high-laser power femtosecond laser," *Sci. Rep.*, vol 8, pp. 12957, 2018.

[8] K. Matsuba, C. Wang, K. Saruwatari, Y. Uesusuki, K. Akatsuka, M. Osada, Y. Ebina, R. Ma, T. Sasaki, "Neat monolayer tiling of molecularly thin two-dimensional materials in 1 min," *Sci. Adv.*, vol. 3, 1700414, 2017.

[9] B. Lei, G. R. Li, and X. P. Gao, "Morphology dependence of molybdenum disulfide transparent counter electrode in dye-sensitized solar cells," *J. Mater. Chem. A.* vol. 2, pp. 3919-3925, 2014.

[10] F. Wu, S. Si, P. Cao, W. Wei, X. Zhao, T. Shi, X. Zhang, J. Ma, R. Cao, L. Liao, T. Y. Tseng, and Q. Liu, "Interface engineering via MoS2 insertion layer for improving resistive switching of conductive-bridging random access memory,". *Adv. Electron. Mater.*, vol. 5, 1800747, 2019.

Mechanical Degradation of Perovskite Thin Films for Photovoltaics: In-Situ Microscopy & Digital Twin Modeling

Melissa A Davis, Mehul Tank, Michelena O'Rourke, Matthew Wadsworth, Zhibin Yu, Rebekah Sweat

High-Performance Materials Institute, Tallahassee, FL, United States

National Renewable Energy Laboratory, Golden, CO, United States

Notre Dame College of Engineering, Notre Dame, IN, United States

Flexible perovskite solar cells introduce opportunities for high throughput, high specific weight, and low energy payback time photovoltaics. However, they require additional investigation into their mechanical resiliency. This work investigates the mechanical properties and behaviors of perovskite thin films and builds a robust model for future research. A two-pronged approach was utilized. Perovskite thin films were flexed in a three-point bend mode with in-situ SEM. Novel insights into the perovskite mechanical behaviors with varying substrate layers were gained. Modeling and validation, the second prong, was completed with finite element analysis. Model coupons of the imaged perovskite architectures were built, with sensitivity analysis completed to provide mechanical property estimates. Results demonstrate mechanical degradation of perovskite thin films on polyethylene terephthalate, or PET, primarily crack in the grain boundaries between crystals. Perovskite thin films on Indium Tin Oxide, or ITO, and PET primarily crack in a periodic pattern regardless of the placement of perovskite crystals.

Optimal row spacing for monofacial and bifacial fixed-tilt and tracked photovoltaic systems up to 75°N

Erin M. Tonita, Annie C. J. Russell, Christopher E. Valdivia, Karin Hinzer

SUNLAB, University of Ottawa, Ottawa, ON, Canada

The inter-row spacing of photovoltaic arrays is an influential design parameter that impacts both a system' energy yield and land-use. Optimization of PV arrays within a constrained area is required, and rule-of-thumb approaches to row spacing which focus solely on eliminating shading for conventional monofacial fixed-tilt PV arrays may not be appropriate. Here, we quantify how variations in ground coverage ratio (GCR) between 0-1 for fixed-tilt and horizontal single-axis tracked (HSAT) monofacial and bifacial PV arrays affect the amount of energy yield lost due to inter-row shading between latitudes of 17-75°N. We additionally optimize the tilt of fixed-tilt systems for these latitudes and GCRs. We demonstrate that fixed-tilt and HSAT arrays located >55°N require similar land-use, while for low-to-moderate latitudes marginal changes in GCR result in significant changes to shading loss for HSAT arrays compared to fixed-tilt arrays. For example, a shift in GCR from 0.3 to 0.4 in Tuxtla Gutierrez at 17°N increases the percent of module energy yield lost to inter-row shading effects by 0.5% abs. for a fixed-tilt array compared to 2.4% abs. for a HSAT array. We additionally calculate that bifacial PV arrays require GCRs lower by 0.03 on average than monofacial arrays to achieve the same shading loss, regardless of tracking type.

Instability of Non-fullerene Acceptors Used in Organic Solar Cells

Yongxi Li, Tonghui Wang, Aram Amassian, Stephen Forrest

University of Michigan, Ann Arbor, MI, United States

North Carolina State University, Raleigh, NC, United States

Inspired by the recent progress of acceptor-donor-acceptor (A-D-A) type non-fullerene acceptors (NFAs) with easily tunable molecular structures and strong near-infrared (NIR) absorption, the Organic photovoltaic (OPV) cells achieved great progress in power conversion efficiencies (PCEs), especially ternary OPVs. The ternary OPVs with two different NFAs and a polymer donor have resulted in a record efficiency for a single-junction opaque OPV cell with PCE of 19.6% and a semitransparent cell with PCE of 11.4% and 47% visible transparency. Unfortunately, their ability to withstand use in adverse environments over long periods is, as yet largely unproven, and the causes of instability are not well understood. In addition, high reproducibility of ternary solar cells has yet to be achieved in the ternary NFA-based cells. In this work, we find that almost all the non-fullerene acceptors (NFAs) used in high efficiency ternary organic photovoltaic cells undergo reactions during blending that create a plethora of reaction products. Briefly, if two NFAs are blended in a solution at elevated temperature, the acceptor end groups exchange between the molecules, creating at times up to 6 different NFA molecules within the bulk heterojunction mixture. These reaction products are dipolar, leading to a decrease in charge extraction efficiency, changes in film morphology and in the photogeneration dynamics. This unanticipated result has significant impacts on ternary OPV performance, reliability and reproducibility.

Damp Heat Performance of Silicon Heterojunction Solar Cells with Reactive Silver Ink Metallization

Michael W Martinez-Szewczyk, Steven DiGregorio, Subbarao Raikar, Owen Hildreth, Mariana Bertoni

Arizona State University, Tempe, AZ, United States

Colorado School of Mines, Golden, CO, United States

The use of reactive silver ink (RSI) metallization for silicon heterojunction (SHJ) solar cells has been shown to outperform standard low-temperature silver paste (LT-SP) with a dramatically different form factor due to its markedly improved electrical properties, which results in 88% lower finger silver usage. As is necessary with any new technology, it must pass accelerated testing in order to ensure viability compared to the current standard. Damp heat testing is vital as it evaluates various modes of degradation: such as, moisture ingress, temperature driven diffusion processes, encapsulant decomposition, and particularly corrosion of interconnects and electrodes. Degradation of contacts are especially crucial for SHJ solar cells due to the required increase in silver content of the silver paste arising from processing temperatures below 200 °C. In order to gain an understanding of these mechanisms and their role, SHJ mini-modules fabricated with two different metallization types: LT-SP and RSI, as well as two common encapsulants: ethylene vinyl acetate (EVA) and polyolefin elastomer (POE) were damp heat tested. These modules fabricated with a glass-backsheet configuration were subsequently exposed to 85 °C/ 85% relative humidity for 1000 hours total. The EVA samples generally demonstrated higher rates of degradation versus POE as has been previously found, and the RSI is observed to have similar degradation rates when compared with these LT-SP samples. Out of the four sample sets, the RSI/POE cells demonstrated the lowest degradation of 1.5% absolute efficiency over the course of the test, which emphasizes the impact of interconnect corrosion due to acetic acid in EVA. The results suggest that degradation is dominated by the cell itself rather than any issues related to the novel silver ink technology, which showcases for the first time high performance of SHJ with 88% less finger silver usage.

978-1-6654-6060-6/23 $31.00 © 2023 IEEE

Parametric Analysis of Photovoltaic Inverters Under Balanced and Unbalanced Voltage Phase Angle Jump Conditions

Rachid Darbali-Zamora, Jay Johnson, and Matthew J. Reno

Renewable Energy and Distributed Systems Integration, Sandia National Laboratories,
Albuquerque, New Mexico, 87185, USA

Abstract – **With growing interest in renewable energy, more photovoltaic (PV) inverters and other distributed energy resources (DERs) are being connected to the grid. Grid interconnection standards such as IEEE Std. 1547 have been developed to define acceptable DER electrical behavior, that are evaluated with stringent compliance test protocols like IEEE Std. 1547.1. Executing IEEE Std. 1574.1 compliance tests for a given DER can be challenging because it requires adjusting the testing platform and verifying measurement data. The System Validation Platform (SVP) is a highly versatile platform that can automate interconnection standard certification experiments. This paper illustrates the ability of the SVP to automate phase-angle change ride-through (PCRT) variation 1 tests. To demonstrate this capability, the IEEE Std. 1547.1-2020 voltage PCRT test protocol was performed for an unbalanced phase jump event. A parametric analysis was performed to summarize the recovery time of the RMS current and phase angle alignment of the commercial PV inverter under test. To understand DER behaviors for other phase jump events, the PV inverter was subjected to a wider range of phase jump angles from $10°$ to $120°$ at increments of $10°$. Trip times were calculated for the PV inverter.**

Keywords – *photovoltaic inverter, automation, system validation platform, short-circuit analysis, phase jump, ride-through.*

I. INTRODUCTION

With increasing interest in clean energy, interactions between renewable energy sources and the grid are more critical [1], [2]. Distributed energy resources (DER), such as photovoltaic (PV) inverters help reduce the dependency on conventional fossil fuel generation and decrease harmful carbon dioxide (CO_2) emissions. Despite their benefits, these systems can have unforeseen impacts on the electrical grid [3]. Understanding how these devices operate under abnormal conditions provides insights into how to prevent adverse effects on the electric grid. Characterization of DER response to abnormal scenarios such as frequency changes, phase jumps, faults, and voltage deviations has become a critical subject [4], [5], [6]. Standardized certification to verify the performance of DERs plays an important role, ensuring DER equipment will operate and communicate as anticipated in the field by inspecting "corner cases" that could lead to unexpected device behavior. Grid interconnection standards such as IEEE Std. 1547 have been developed to define acceptable DER electrical behavior [7]. In 2018, IEEE Std. 1547, was updated with new grid-support functionality and interoperability requirements [8], [9]. Compliance test protocols such as IEEE Std. 1547.1 have also been developed to ensure that DERs are compliant [10], [11]. IEEE Std. 1547.1 has been updated to include the conformance testing requirements for IEEE Std.1547-2018. This includes hundreds of electrical type test cases for each DER along with interoperability compliance tests [12], [13]. This not only includes certifying a device, but also requires verifying a variety of measurement parameters as well as adjusting the grid simulator settings and DER operating modes. The System Validation Platform (SVP) allows for automating certification experiments designed to accelerate and automate IEEE Std.1547.1 [14], [15], [16], [17].

This paper demonstrates how the SVP can automate grid interconnection standard testing by providing a parametric analysis for the IEEE Std. 1547.1-2020 "test for voltage phase-angle change ride-though." The PV inverter's ride-through capabilities were studied to help provide insight into PV inverter performance when subjected to phase jump changes. This work expands on previous phase jump experiments performed for balanced and unbalanced phase jumps by considering a wider range of phase jump angles [18]. Experimental results are presented for the PV inverter's RMS current and phase angle alignment recovery times at incrementing phase angles when put through three different phase jump conditions. The commercial PV inverter is subjected to an unbalanced phase jump on phase A, unbalanced phase jump on phases A and B and a balanced phase jump on phases A, B and C. The motivation for this work is twofold: (a) to demonstrate IEEE Std. 1547.1-2020 experiments with a fully autonomous test platform, and (b) to estimate the ride-through characteristics of DER equipment for a range of phase jumps to better characterize fault behavior of PV inverters.

We found that the SVP could automate the phase jump interconnection compliance testing for PV inverters as specified by IEEE Std. 1547.1. In addition, the commercial PV inverter used for these interconnection compliance experiments was able to perform the voltage ride-through when subjected to unbalanced phase jump on phase A, unbalanced phase jump on phases A and B and a balanced phase jump on phases A, B and C for a range of phase jump angles.

II. SYSTEM VALIDATION PLATFORM TESTBED

The SVP is a versatile platform developed for certification experiments designed to accelerate and automate IEEE Std. 1547.1 interoperability compliance tests. It has been used for a variety of IEEE Std. 1547.1 test procedures by interacting with a device under test (DUT) and laboratory equipment [19], [20]. Fig. 1 illustrates the diagram of the SVP used for PV inverter interconnection compliance tests. The SVP can communicate with programmable AC power supplies (grid simulators), DC power supplies (PV simulators), data acquisition systems, controllable loads, and switches.

978-1-6654-6060-6/23 $31.00 © 2023 IEEE

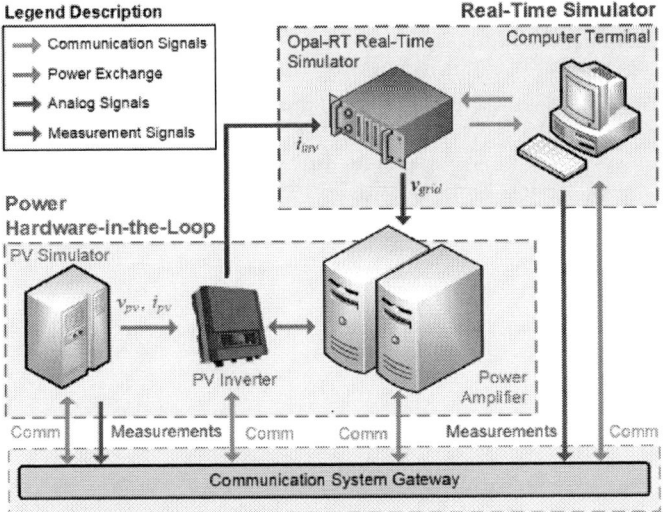

Fig. 1. Diagram of the System Validation Platform (SVP) used for testing three-phase photovoltaic inverters under different phase jump conditions.

In Fig. 1, three-phase voltage signals are generated utilizing the *MATLAB/Simulink* simulation model executed in an Opal-RT real-time simulator [21]. These voltage signals are amplified by the AC power supply via three analog signals, which replicates the testing environment conditions for the PV inverter. A DC power supply, that emulates PV array dynamics, is used to provide power to the three-phase PV inverter. Three-phase voltages and currents from the PV inverter are measured by the Opal-RT real-time simulator. The SVP automatically modifies conditions for each phase jump test, providing flexibility during each experiment.

III. EXPERIMENTAL PHASE JUMP TEST SETUP

The IEEE Std. 1547.1 phase jump angle test was developed to help determine if a PV inverter can perform a ride-through when the voltage phase angle changes. The experimental tests are composed of a series of phase angle change sequences. For this test, there is a total of five sequence combinations that must be performed. Three of these are unbalanced phase jumps and consider only changing one of the phase angles. The remaining two are balanced phase jumps that consider changing all three-phase angles at the same time. Table I summarizes the IEEE Std. 1547.1 phase jump test procedure. Notice from Table I, that for the balanced phase jump tests, (*A-E-A* and *A-F-A*), the duration of the event is significantly longer than for unbalanced phase jumps (*A-B-A*, *A-C-A*, and *A-D-A*).

TABLE I.
IEEE STD. 1547.1 THREE-PHASE JUMP TEST PROCEDURE

Test	Phase A	Phase B	Phase C	Time (s)
A	0°	-120°	120°	30 - 40
B	60°	-120°	120°	0.32 - 0.50
C	0°	-120°	180°	0.32 - 0.50
D	0°	- 60°	120°	0.32 - 0.50
E	20°	-100°	140°	55-65
F	-20°	-140°	100°	55-65

For these tests, the DUT is a three-phase commercial PV inverter with a power rating of 24 kVA, operating at a voltage of 277 V. The DUT is designed to be compliant with IEEE Std. 1547-2018. Three different tests consisting of three different phase jump events are presented.

To comply with IEEE Std. 1547.1, the PV inverter performance must adhere to criteria in Section 5.5.6.4, which specifies the minimum time that the DUT must respond and the RMS current percentage. IEEE Std. 1547.1 criteria 5.5.6.4 specifies that the RMS current must reach 80% of the rated RMS current in less than 0.5 s after the phase jump is removed.

IV. EXPERIMENTAL RESULTS

The SVP is used to automate the IEEE Std. 1547.1-2020 voltage phase jump tests. Experimental results are obtained using a commercial three-phase PV inverter rated at 24 kVA, an operational line-line voltage of 480 V and a current rating of 86.64 A. This PV inverter is subjected three different phase jumps, including balanced and unbalanced phase jumps. The PV inverter's RMS current measurements are used to calculate the RMS current recovery time. The PV inverter RMS current recovery time is calculated from the moment the phase jump event is removed until the time the RMS current reaches 80% of its rated value. The PV inverter's phase angles are used to calculate the phase angle recovery time from the moment the phase jump event is removed until the PV inverter phase angle alignment is below a minimum threshold of 6°. Fig. 2 shows an example of the instantaneous voltage and current waveforms when a voltage phase jump is applied to phase A. Fig. 3 shows the phase angle difference between the current and voltage phase angles for the phase jump shown in Fig. 2. In addition, the PV inverter is also subjected to a wider range of phase jump angles from 10° to 120° at increments of 10°. Each test was performed 5 times per IEEE Std. 1547.1 requirements and the data was used summarize the results in box plots. Five samples are considered a minimum box plot sample size [22].

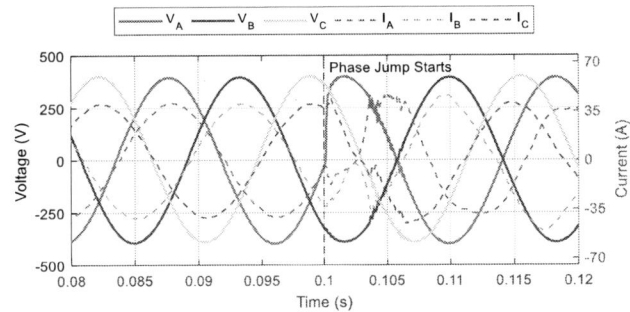

Fig. 2. Example of the PV inverter per phase voltages and currents.

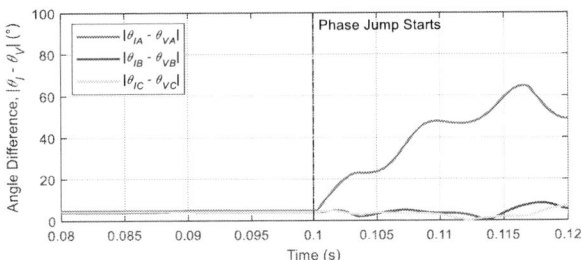

Fig. 3. Example of the PV inverter phase angle difference.

978-1-6654-6060-6/23 $31.00 © 2023 IEEE

A. Unbalanced Phase Jump Test on Phase A

For the unbalanced phase jump on phase A, the phase jump event is introduced at 0.1 s and lasts 0.5 s until it is removed at 0.6 s. Fig. 4 shows the results for the PV inverter RMS current when subjected to an unbalanced phase shift on phase A. Again, note that only a phase jump of 60° is required by IEEE 1547.1. Results show that when the unbalanced phase jump on phase A is removed, the RMS current can return to its nominal value. The greater the phase shift, the longer the recovery time for the RMS current to reach 80% of the rated current. Fig. 5 shows the results for the difference between the current and voltage phase angles for phase A. As the phase shift angle increases, so does the recovery time for the angle difference. This is partially due to the phase-locked loop (PLL) resynchronizing.

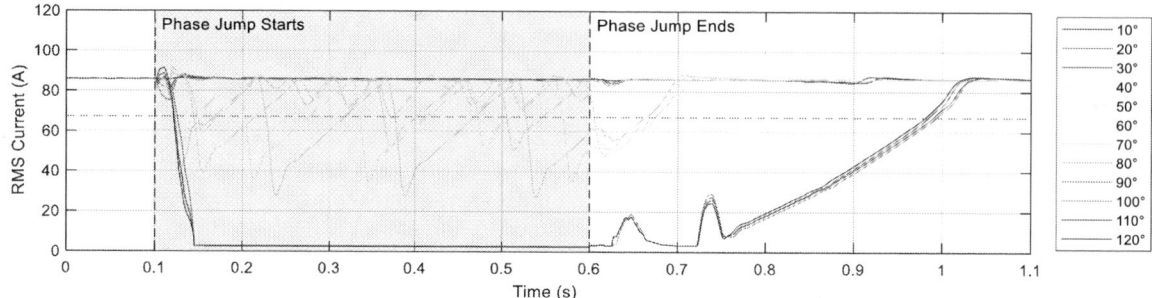

Fig. 4. Experimental time series results for the PV inverters RMS current when subjected to an unbalanced phase shift on phase A under different phase shifts.

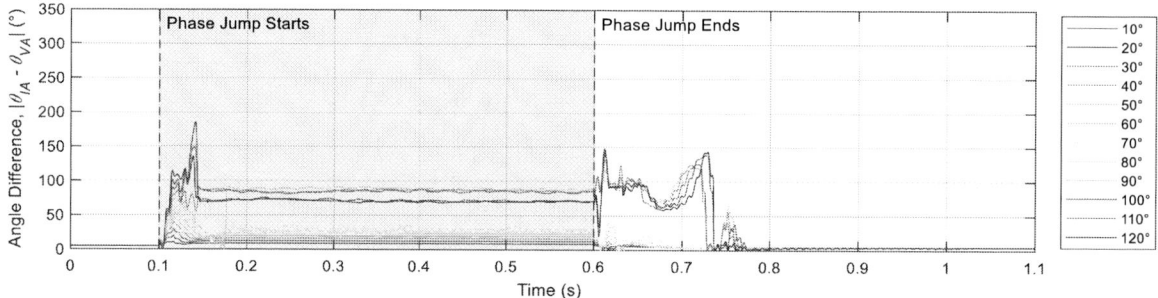

Fig.5. Experimental time series results for the PV inverter's phase A current angle and phase A voltage angle difference when subjected to an unbalanced phase shift on phase A under varying phase shifts.

Fig. 6 illustrates a box plot that summarizes the experimental results obtained for the RMS current and phase angle realignment recovery times when subjected to an unbalanced phase shift on phase A under different phase shift increments.

Initially, for phase angles below 50°, the RMS current reaches 80% of the rated RMS current as soon as the phase jump event is removed. For phase jump between 50° and 70° there is a slight increase in the RMS currents recovery time, reaching up to 58 ms. When subjected to an 80° phase jump, the recovery time of the RMS current varies between 50 ms and 390 ms, with an average recovery time of 190 ms. RMS currents subjected to phase jumps greater than 80° on average lasted 400 ms before reaching 80% nominal current. A similar behavior is observed for the phase angle alignment recovery times. These results demonstrate the RMS current of the PV inverter is able recover before the required 0.5 s.

The recovery time of the phase angle alignment follows a similar pattern. When a phase jump of 10° is introduced, the phase angle alignment recovery time is 10 ms. While phase jumps between 20° and 70° on average have a phase angle alignment recovery time of 45 ms. For experimental results obtained at a phase angle change of 80°, alignment recovery time varied between 60 ms and 170 ms, with an average of 100 ms. Finally, the phase angle alignment recovery time for phase jumps between 90° and 120° lasted on average 170 ms.

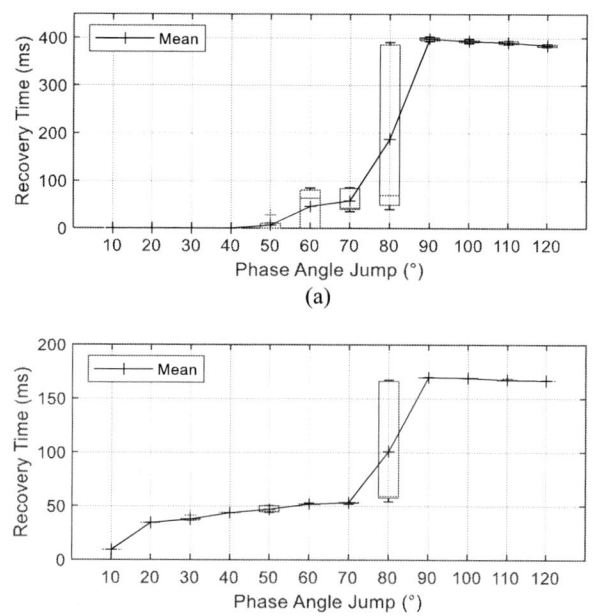

Fig. 6. Box plot recovery time results for an unbalanced phase shift on phase A under varying phase angle shifts. (a) RMS current. (b) Phase angle alignment.

978-1-6654-6060-6/23 $31.00 © 2023 IEEE

B. Unbalanced Phase Jump Test on Phases A and B

Fig. 7 shows the experimental results for the PV inverter RMS current when subjected to an unbalanced phase shift on phases A and B. These results show that when the unbalanced phase jump on phases A and B is removed, the RMS current is able to return to its nominal value. Results show that for phase jump angles higher than 70°, the RMS current recovery time takes longer to return to 80% of its nominal value. Fig. 8 shows the results for the difference between the current and voltage phase angles for phase A. These experimental results show that when the phase jump is removed, there is an increase in the phase angle difference. As the phase shift is increased higher than 70°, the recovery time for the PV inverters angle difference to return to 0° is longer.

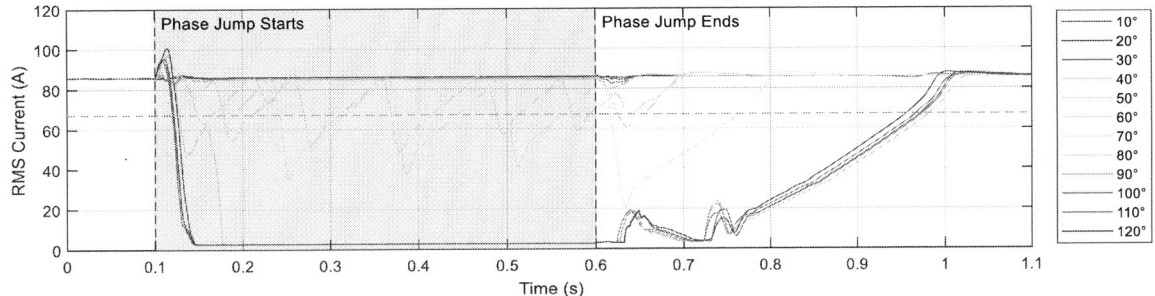

Fig. 7. Experimental time series results for the PV inverters RMS current recovery time when subjected to an unbalanced phase shift on phase A and B under varying phase shifts.

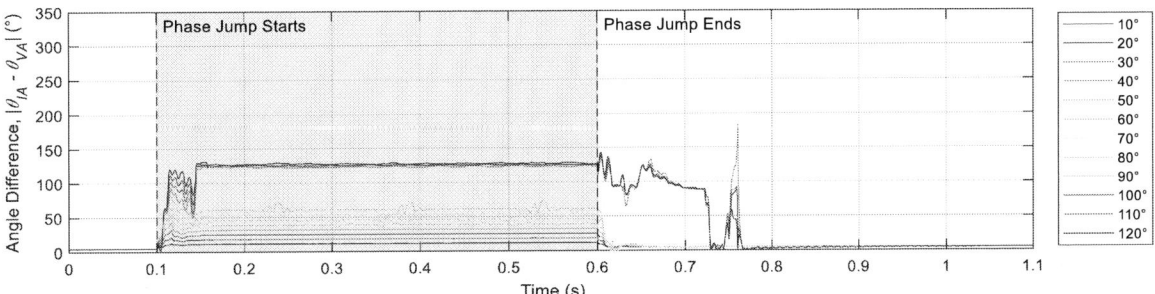

Fig. 8. Experimental time series results for the PV inverter's phase A current angle and phase A voltage angle difference when subjected to an unbalanced phase shift on phase A and B under varying phase shifts.

Fig. 9 illustrates a box plot that summarizes the experimental results obtained for the RMS current and phase angle realignment recovery times when subjected to an unbalanced phase shift on phases A and B under different phase shift increments.

Results illustrate that for phase angles below 40°, the RMS current reaches 80% of the rated RMS current as soon as the phase jump event is removed. Phase angles jump angles that range between 50° and 70° have an average RMS current recovery time of 100 ms. As the phase jump angles is increased to 80°, there is a significant increase in RMS current recovery time of approximately 390 ms. The results for RMS current recovery time for an unbalanced phase jump on phase A and B are similar to the results observed for the unbalanced phase jump on phase A. This is partially due to both phase jumps being unbalanced and having similar duration times.

A similar pattern is observed for the phase angle alignment, as what is observed with the RMS current recovery times. Notice from the results for the phase alignment that there is a steady increase in phase angle alignment recovery time between 10° and 70°, ranging from 40 ms to 84 ms. At 80°, there is an increase in phase angle alignment recovery time of 82 ms. Results show that for phase jumps ranging between 80° to 120°, the phase angle alignment recovery time is constant, averaging 170 ms.

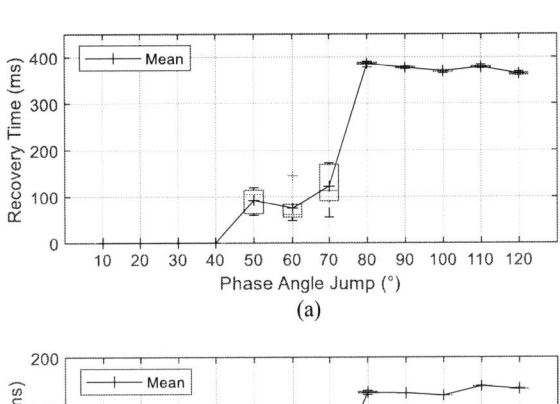

(a)

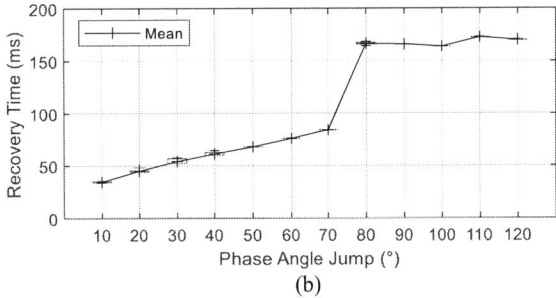

(b)

Fig. 9. Box plot recovery time results for an unbalanced phase shift on phases A and B under varying phase angle shifts. (a) RMS current. (b) Phase angle alignment.

C. Balanced Phase Jump Test on Phases A, B and C

For the balanced phase jump on phases A, B and C, the phase jump event is introduced at 0.1 s and lasts 60 s until it is removed at 60.1 s. Fig. 10 shows the experimental results for the current recovery time when subjected to a balanced phase shift on phase A, B and C. Results show that for phase jump angles higher than 30°, the RMS current recovery time takes longer to return to 80% of its nominal value. Fig. 11 shows the results for the difference between the current and voltage phase angles for phase A.

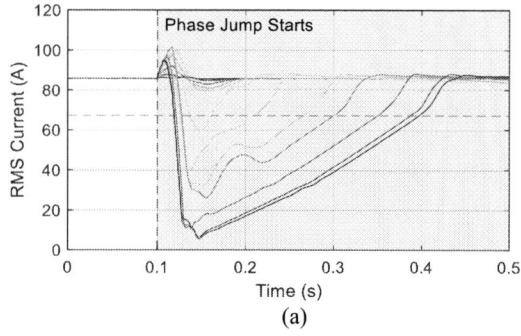
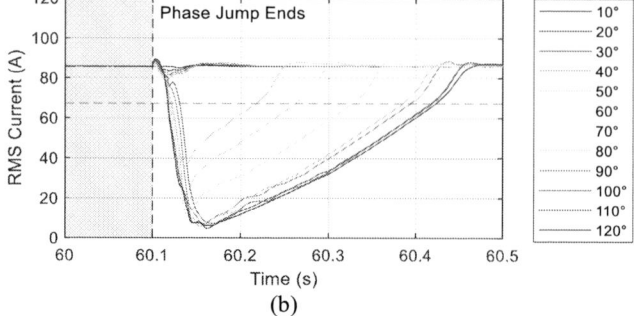

Fig. 10. Experimental time series results for the PV inverters RMS current recovery time when subjected to a balanced phase shift on all phases under varying phase shifts. (a) Event start. (b) Event end.

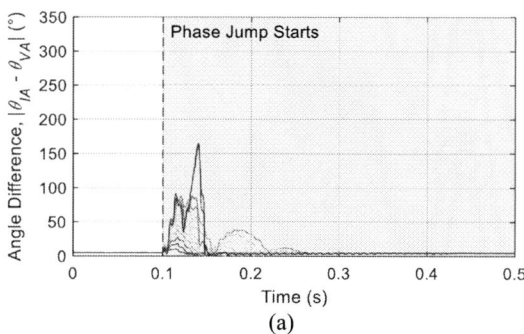
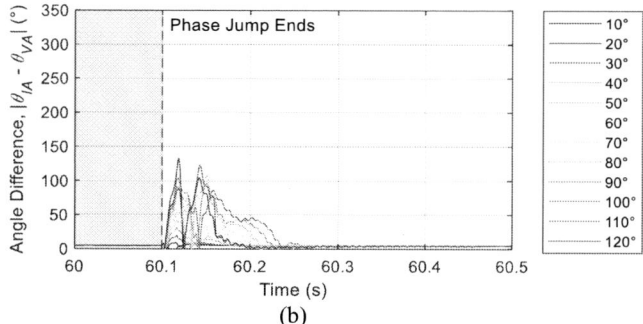

Fig. 11. Experimental time series results for the PV inverter's phase A current angle and phase A voltage angle difference when subjected to a balanced phase shift on phase A, B and C under varying phase shifts. (a) Phase jump start. (b) Phase jump end.

When performing balanced phase jumps the PV inverter had difficulties under certain phase jump angles. Moreover, this test had to be performed multiple times due to the severity of the 120° phase jump which caused the PV inverter to trip. Fig. 12 illustrates a box plot that summarizes the experimental results obtained for the RMS current and phase angle realignment recovery times when subjected to a balanced phase shift on phases A, B and C under different phase shift increments.

Results illustrate that for phase angles below 40°, the RMS current reaches 80% of the rated RMS current as soon as the phase jump event is removed. For phase angles above 40°, has a steady increase until reaching approximately 310 ms. In comparison with the experimental result obtained for an unbalanced phase jump, notice that for a balanced phase jump the recovery time for the RMS current is less.

For the phase angles between 10° and 90° there is a steady increase in phase angle alignment recovery time ranging from 40 ms to 125 ms. At a phase angle of 100°, the phase angle alignment recovery time increases to approximately 158 ms. Notice that between the phase angles of 110° and 120°, the phase angle alignment recovery time decreases to approximately 90 ms. For both the RMS current and phase angle alignment recovery times, results were consistent for each time the testes were performed.

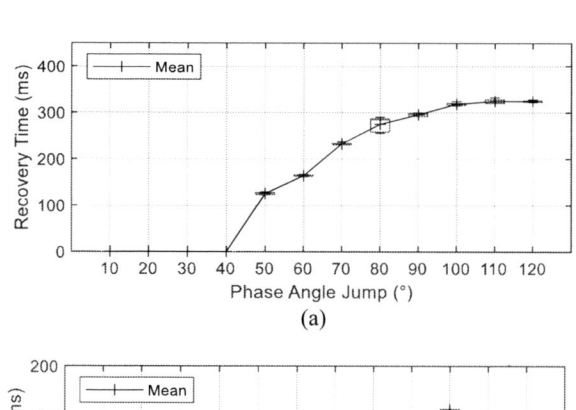
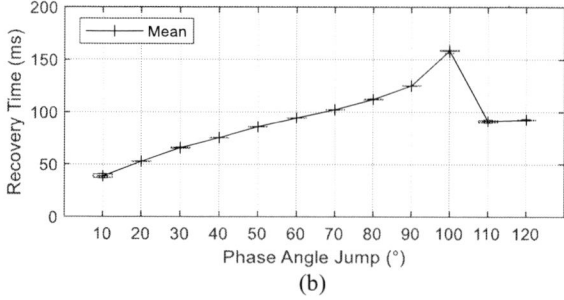

Fig. 12. Box plot recovery time results for a balanced phase shift on phases A, B and C under varying phase angle shifts. (a) RMS current. (b) Phase angle alignment.

V. CONCLUSION

This paper presents a parametric analysis based on experimental results obtained from automated IEEE Std. 1547 phase jump tests conducted on a commercial PV inverter. The commercial PV inverter is subjected to an unbalanced phase jump on phase A, unbalanced phase jump on phases A and B and a balanced phase jump on phases A, B and C. The presented parametric results demonstrate that the SVP can automate the IEEE Std. 1547.1 phase jump tests. In addition, the SVP is able to automate phase jump tests outside of IEEE Std. 1547.1 by performing phase jumps ranging from 10° to 120° degrees, with increments of 10°. These ride-through characteristics were summarized in box plots to better characterize fault behavior of PV inverters. Moreover, these results demonstrated that the commercial PV inverter was able to perform a voltage phase change ride-through as specified by IEEE Std. 1547.1.

ACKNOWLEDGEMENT

This article has been authored by an employee of National Technology & Engineering Solutions of Sandia, LLC under Contract No. DE-NA0003525 with the US Department of Energy (DOE). The employee owns all right, title and interest in and to the article and is solely responsible for its contents. The US government retains and the publisher, by accepting the article for publication, acknowledges that the US government retains a non-exclusive, paid-up, irrevocable, world-wide license to publish or reproduce the published form of this article or allow others to do so, for US government purposes. The DOE will provide public access to these results in accordance with the DOE Public Access Plan.

REFERENCES

[1] B. Mather, O. Aworo, R. Bravo and P. E. David Piper, "Laboratory Testing of a Utility-Scale PV Inverter's Operational Response to Grid Disturbances", *2018 IEEE Power & Energy Society General Meeting (PESGM)*, 2018, pp. 1-5.

[2] Q. Zhang, L. Zhou, M. Mao, B. Xie, and C. Zheng, "Power quality and stability analysis of large-scale grid-connected photovoltaic system considering non-linear effects", *IET Power Electronics*, vol. 11, no. 11, pp. 1739-1747, 18 9 2018.

[3] North American Electric Reliability Corporation (NERC), "1,200 MW Fault Induced Solar Photovoltaic Resource Interruption Disturbance Report: Southern California 8/16/2016 Event", *Report*, Version 1, June 8, 2017.

[4] C. Li and R. Reinmuller, "Fault Responses of Inverter-based Renewable Generation: On Fault Ride-Through and Momentary Cessation", *2018 IEEE Power & Energy Society General Meeting (PESGM)*, 2018, pp. 1-5.

[5] G. Kou, L. Chen, P. VanSant, F. Velez-Cedeno and Y. Liu, "Fault Characteristics of Distributed Solar Generation", *IEEE Transactions on Power Delivery*, vol. 35, no. 2, pp. 1062-1064, Apr. 2020.

[6] R. Darbali-Zamora *et al.*, "Distribution Feeder Fault Comparison Utilizing a Real-Time Power Hardware-in-the-Loop Approach for Photovoltaic System Applications", *2019 IEEE 46th Photovoltaic Specialists Conference (PVSC)*, 2019, pp. 2916-2922.

[7] "IEEE Standard for Interconnection and Interoperability of Distributed Energy Resources with Associated Electric Power Systems Interfaces", *IEEE Std 1547-2018 (Revision of IEEE Std 1547-2003)*, pp.1-138, 6 Apr. 2018.

[8] E. Desarden-Carrero, R. Darbali-Zamora and E. E. Aponte-Bezares, "Analysis of Grid Support Functionality Dynamics under Ride-Through Requirements Using Power-Hardware-in-the-Loop Implementation", *2021 IEEE 48th Photovoltaic Specialists Conference (PVSC)*, 2021, pp. 1795-1802.

[9] I. Stefani, V. Stokic, S. Dzaleta and B. Brbaklic, "Grid optimization using DER grid support functions", *2022 IEEE PES Innovative Smart Grid Technologies Conference Europe (ISGT-Europe)*, 10-12 Oct. 2022, pp. 1-5.

[10] "IEEE Standard Conformance Test Procedures for Equipment Interconnecting Distributed Energy Resources with Electric Power Systems and Associated Interfaces", in IEEE P1547.1/D9.9, January 2020, pp.1-283, 6 Jan. 2020.

[11] E. Desarden-Carrero, R. Darbali-Zamora, N. S. Gurule, E. Aponte-Bezares and S. Gonzalez, "Evaluation of the IEEE Std 1547.1-2020 Unintentional Islanding Test Using Power Hardware-in-the-Loop", *2020 47th IEEE Photovoltaic Specialists Conference (PVSC)*, 2020, pp. 2262-2269.

[12] N. Ninad *et al.*, "Development and Evaluation of Open-Source IEEE 1547.1 Test Scripts for Improved Solar Integration", *EU PVSEC*, 9-13 Sept. 2019.

[13] N. Ninad *et al.*, "PV Inverter Grid Support Function Assessment using Open-Source IEEE P1547.1 Test Package", *2020 47th IEEE Photovoltaic Specialists Conference (PVSC)*, 2020, pp. 1138-1144.

[14] J. Johnson, B. Fox, K. Kaur, and J. Anandan, "Evaluation of Interoperable Distributed Energy Resources to IEEE 1547.1 Using SunSpec Modbus, IEEE 1815, and IEEE 2030.5", *IEEE Access*, vol. 9, pp. 142129-142146, 2021.

[15] J. Johnson, R. Ablinger, R. Bruendlinger, B. Fox and J. Flicker, "Interconnection Standard Grid-Support Function Evaluations Using an Automated Hardware-in-the-Loop Testbed", *IEEE Journal of Photovoltaics,* vol. 8, no. 2, pp. 565-571, Mar. 2018.

[16] J. Johnson *et al.*, "International Development of a Distributed Energy Resource Test Platform for Electrical and Interoperability Certification", *7th World Conference on Photovoltaic Energy Conversion (WCPEC-7)*, 10-15 Jun. 2018.

[17] J. Johnson, R. Ablinger, R. Bruendlinger, B. Fox, J. Flicker, "Design and Evaluation of SunSpec-Compliant Smart Grid Controller with an Automated Hardware-in-the-Loop Testbed," *Technology and Economics of Smart Grids and Sustainable Energy*, vol. 2, no. 16, Dec. 2017.

[18] R. Darbali-Zamora *et al.*, "Evaluation of Photovoltaic Inverters Under Balanced and Unbalanced Voltage Phase Angle Jump Conditions", *2020 47th IEEE Photovoltaic Specialists Conference (PVSC)*, 2020, pp. 1562-1569.

[19] R. Darbali-Zamora, N. S. Gurule, J. Hernandez-Alvidrez, S. Gonzalez and M. J. Reno, "Performance of a Grid-Forming Inverter Under Balanced and Unbalanced Voltage Phase Angle Jump Conditions", *2021 IEEE 48th Photovoltaic Specialists Conference (PVSC)*, 2021, pp. 1409-1416.

[20] C. Hansen, J. Johnson, R. Darbali-Zamora, and N. Gurule, "Modeling Efficiency of Inverters with Multiple Inputs", *2022 IEEE 49th Photovoltaics Specialists Conference (PVSC)*, 2022, pp. 1335-1337.

[21] R. Darbali-Zamora, J. E. Quiroz, J. Hernández-Alvidrez, J. Johnson and E. I. Ortiz-Rivera, "Validation of a Real-Time Power Hardware-in-the-Loop Distribution Circuit Simulation with Renewable Energy Sources", *2018 IEEE 7th World Conference on Photovoltaic Energy Conversion (WCPEC) (A Joint Conference of 45th IEEE PVSC, 28th PVSEC & 34th EU PVSEC)*, 2018, pp. 1380-1385.

[22] M. Krzywinski, N. Altman, "Visualizing Samples with Box Plots", *Nature Methods* 11, 119–120, Feb. 2014.

Chemical Reaction Kinetics of the Decomposition of Low Bandgap Tin-Lead Halide Perovskite Films and the Effect on the Ambipolar Diffusion Length

Yuhuan Meng, Preetham P. Sunkari, Marina Meila, Hugh W. Hillhouse

University of Washington, Seattle, WA, United States

Understanding the degradation mechanism and quantifying the degradation rate of hybrid organic-inorganic halide perovskite (HP) materials are essential for achieving the commercialization of HP solar cells. Here, we focus on evaluating the reaction products and kinetics of the degradation of mixed Sn-Pb halide perovskite ($FA_{0.75}Cs_{0.25}Pb_{0.5}Sn_{0.5}I_3$) thin films in response to oxygen, moisture, and illumination using in-situ measurements of optical transmittance, photoluminescence, X-ray diffraction, and UV-Vis-NIR spectroscopy. We determined that the decomposition occurs by a dry oxidation pathway (1x10-9 mol/(m2s) at 25 °C in air) and a water-accelerated oxidation pathway (3x10-9 mol/(m2s) at 25 °C in 50% RH air). In 50% RH air, the water-accelerated oxidation dominates at low temperatures, but the dry oxidation pathway (which has a higher activation energy) dominates at temperatures greater than 40 °C. A kinetic rate expression for the decomposition is determined using 32 degradations at different environmental conditions. The subsequent collection of an additional 33 degradation runs was used to evaluate the accuracy and robustness of the model with a mean test error of 18%. Further, we developed a predictive model of the decay of the diffusion length (predicting the time required for the diffusion length to decrease to 80% of its initial value) with a prediction test error of 24%. Out of 59 possible features, the chemical decomposition rate expression was selected as the most dominant feature in all 4-feature models with low prediction test error. The results highlight the importance and utility of quantitative measurements of perovskite degradation.

Demonstration of Thermoradiative Energy Conversion with InAs Cells

Eric J Tervo, Andrew J Ferguson, Jennifer Selvidge, Myles A Steiner, Ryan M France

University of Wisconsin-Madison, Madison, WI, United States

National Renewable Energy Laboratory, Golden, CO, United States

Thermoradiative diodes are devices that convert thermal energy to electricity while emitting light as waste heat to their surroundings. These are similar to solar- and thermo-photovoltaics in many ways, but they operate with a net emission of light instead of a net absorption. Despite only being recently proposed, thermoradiative converters have gained substantial interest for their potential to achieve high efficiency energy conversion with low heat source temperatures, but there have been very few experimental investigations. Here, we design, fabricate, and demonstrate a large-area (0.8 cm2) InAsP / InAs / InAsP double heterojunction device for thermoradiative energy conversion at elevated temperatures of several hundred degrees Celsius. The cell achieves a record-low reverse saturation current for InAs below 15 mA/cm2 at 25 °C via minority carrier extraction, which minimizes nonradiative Auger losses. The device generates electricity with power densities >0.4 mW/m2 and short-circuit currents on the order of milliAmps, resulting in the first demonstration of measurable power from a thermoradiative device at elevated temperatures. Our results show that pathways exist to practical thermoradiative converters through careful device design.

978-1-6654-6060-6/23 $31.00 © 2023 IEEE

InGaP/GaAs/In$_{0.35}$Ga$_{0.65}$As//In$_{0.53}$Ga$_{0.47}$As Four-Junction Solar Cells Integrated by Surface Activated Wafer Bonding

Kentaroh Watanabe[1], Takashi Shimasaki[2], Hassanet Sodabanlu[1], Yoshiaki Nakano[2], and Masakazu Sugiyama[1,2]

[1]Research Center for Advanced Science and Technology, University of Tokyo, Tokyo, 1538904 Japan
[2]Department of Electrical Engineering, University of Tokyo, Tokyo, 1138656 Japan

Abstract — The four-junction solar cell was fabricated by surface activated wafer bonding (SAB) method with the inverted metamorphic (IMM) triple-junction InGaP/GaAs/In$_{0.35}$Ga$_{0.65}$As integrated on the In$_{0.53}$Ga$_{0.47}$As single junction cell. Applying the direct integrated p$^+$-AlGaAs/n$^+$-InGaAs tunnel junction at the bonding interface, electrically connected 4-junction solar cell improved the open-circuit voltage Voc = 3.24 V was realized. The current density at the maximum power point was restricted by photocurrent in the In$_{0.53}$Ga$_{0.47}$As 4th subcell compared with ~11 mA/cm^2 of short-circuit current density (Jsc) expected for the IMM-3J under the AM1.5G illumination.

I. INTRODUCTION

The III-V compound semiconductor is frequently used for composing a multi-junction solar cells owing to their availability of high quality single crystalline layers with epitaxial growth. In recent years, development of the multi-junction cells increased the number of sub cells over 4 and represented over 45% efficiency. In order to realize the high-efficiency multi-junction cell, it is required to reduce the photo-current mismatch. Therefore, the selection of materials with suitable bandgap is necessary. However, the lattice matching for epitaxial growth is the limited by the availability of the wafers. To break this limitation, the metamorphic epitaxial growth with graded buffer layer or direct integration by mechanical staking is a hopeful alternative technique for an ideal selection of the materials. For example, the IMM grown 6-junction cell recoeded 47.1% and the wafer-bonded 4-junction cell showed 46.0% under sunlight concentration, respectively [1,2]. In this study, we tried to fabricate the 4-junction cell integrating a IMM-3J cell on an InGaAs single junction (SJ) cell by SAB. The IMM triple-junction is well studied and possible to prepare the samples with high quality of device performance. We developed both the InGaAs single junction solar cell with high open-circuit voltage (Voc) and the method of direct integration of GaAs/InGaAs tunnel diode (TD) by SAB [3]. In this study, the results of the directly integrated

4-juntion solar cell composed by the IMM-3J and the InGaAs-SJ are reported.

II. EXPERIMENTAL

Fig.1 shows the schematic structure of 4-junctoin solar cell devices fabricated in this study. The InGaP/GaAs/In$_{0.35}$Ga$_{0.65}$As IMM-3J and the In$_{0.53}$Ga$_{0.47}$As solar cell structure were grown by metalorganic vapor phase epitaxy (MOVPE) on the GaAs (001) 5° off to <111>A and the p-InP (001) just substrates, respectively. The top surface layer of IMM-3J is heavily p-type doped AlGaAs with rough surface due to the accumulated strain causes a so called cross hatched pattern. The roughness of the surface was reduced by chemical-mechanical polishing (CMP) process to be less than 0.5 nm in RMS. The surface of the epitaxially grown lattice-matched In$_{0.53}$Ga$_{0.47}$As SJ to InP substrate was kept mirror-like with roughness less than 0.5 nm.

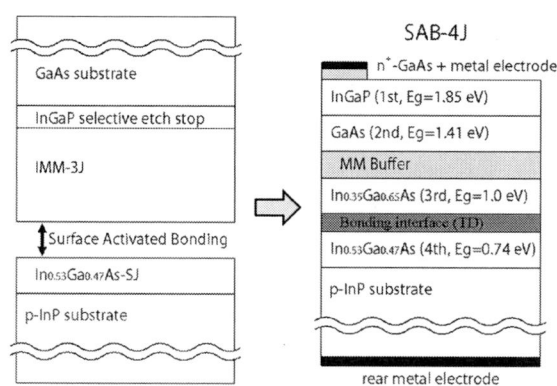

Fig. 1. Schematic structure of the SAB-4J solar cell and process flow of the SAB integration.

After cleaning the both surfaces of the specimen, SAB process was performed in the vacuum chamber with the background pressure less than 1×10^{-5} Pa. The surface activation

978-1-6654-6060-6/23 $31.00 © 2023 IEEE

process was conducted with fast atom beam (FAB) of neutralized Ar ion with 100 mA of current injection and 1 kV of applied acceleration electric field. And then interconnection was formed with enough mechanical strength between both samples by attaching both surfaces under the pressure of 10 kN. Here, p^+-AlGaAs/n^+-InGaAs tunnel junction was formed at the SAB interface to connect the top side IMM-3J and the bottom side InGaAs-SJ electrically. After the SAB process, the GaAs substrate of the IMM-3J was removed by wet-chemical etching in aqueous ammonia solution. The etching process was automatically stopped at the InGaP selective etching stop layer. And then the InGaP layer was etched by HCl solution to expose the n^+-GaAs topside contact layer. The backside planner metal electrode was formed by Ti/Au deposition by the thermal evaporation. And the 5×5 mm^2 size isolated PV devices were fabricated by forming the topside grid-type metal electrode with ~10% of shadow loss.

The fabricated solar cells were evaluated their performance by ligh-IV measurement under the AM1.5G standard sunlight illumination (100 mW/cm^2) and spectral external quantum efficiency (EQE) measurement by variable monochromatic light illumination of constant power density of 200 µJ/cm^2. The reflectance and transmittance of the IMM-3J were also measured. For this measurement, the sample was attached to the glass by the transparent epoxy resin without top- and back-electrode. To analyse the performance of the SAB-4J, IMM-3J and InGaAs-SJ devices with the same structures were also fabricated and the performances were evaluated, individually. There was no anti-reflection coatings (ARCs) applied to the PV devices in this study. And the effective area excluding the shadow loss was taken into account to evaluate the performance to simplify the analysis.

III. RESULTS AND DISCUSSION

The Fig.2 is the measurement results for IV characteristics under the AM1.5G illumination for IMM-3J and SAB-4J. The IV under the dark condition was also measured and plot with the dash lines. The typical performance parameters for solar cell: the short-circuit current density (Jsc), the open-circuit voltage (Voc), the fill factor (FF) and the conversion efficiency (η) are also summarized in this figure. According to this result, the SAB-4J showed clear improvement in Voc from that of the IMM-3J. However, it can be seen two problems are caused in IV curve for SAB-4J that prevents the improvement in η. The first issue is a non-ohmic behaviour of light-IV around the point of V=Voc. Because the both top- and backside metal electrode showed sufficiently low contact resistivity less than 1×10^2 Ω/cm^2 confirmed by the transfer length method (TLM) measurement, this non-ohmic behaviour is due to the TD at bonding interface. The other issue of light-IV is the photocurrent changing between V = 1 to 2 V range. In spite of the Jsc of SAB-4J which shows the value close to that of the

IMM-3J, the current density was decreased to ~6 mA/cm^2 in the voltage range of V = 2 to 3 V. This phenomenon is possibly caused by the breakdown of 4th InGaAs subcell.

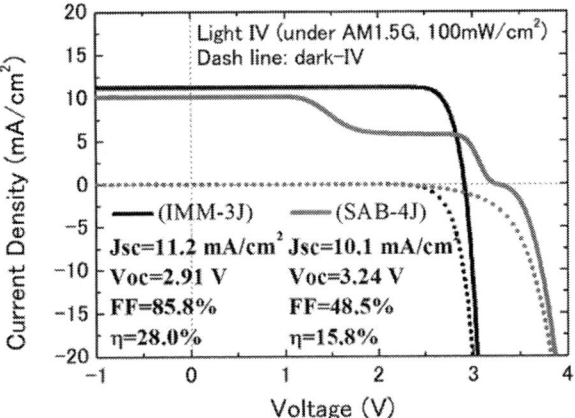

Fig. 2. IV characteristics of the SAB-4J and the IMM-3J solar cells under the AM1.5G illumination.

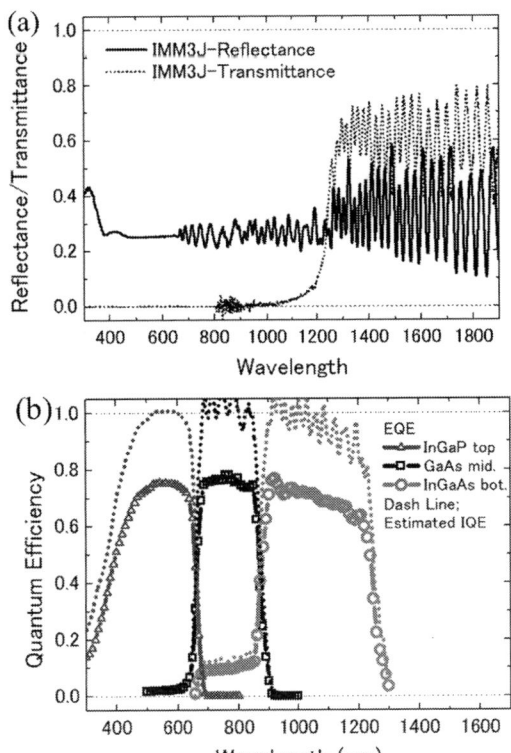

Fig. 3. Measurement results of the transmittance and reflectance for IMM-3J attached to the glass substrate (a) and EQE spectra for each subcells (b).

The measurement results of individual IMM-3J and $In_{0.53}Ga_{0.47}As$-SJ are summarized in Fig. 3 and Fig. 4. From the reflectance and transmittance spectrum of the IMM-3J shown in Fig. 3 (a), there is enough transparency in the wavelength range longer than the cutoff by the 3rd $In_{0.35}Ga_{0.65}As$ subcell ($\lambda > 1250$ nm). The observed EQE shown in Fig. 3 (b) indicates that the photocurrent in the light-IV for IMM-3J cell is limited by 1st InGaP subcell to be 11.2 mA/cm^2. Regarding the measured EQE and reflectance (R), the simplified internal quantum efficiency (IQE, neglecting the internal reflection) is estimated by IQE = EQE/(1-R) and plotted in Fig. 3 (b). This IQE indicates the possible photocurrent density in each subcell under the desirable ARCs condition with R = 0. Similar analysis for the $In_{0.53}Ga_{0.47}As$ 4th subcell is also carried out and the IQE for the InGaAs-SJ is shown in Fig. 4 (b) as well as the measured EQE. From the estimated IQE and standard sunlight spectrum, the possible current density in each 1st InGaP, 2nd GaAs, 3rd InGaAs and 4th InGaAs is summarized in Table. I. In this calculation, the current density in 4th subcell is derived under the assumption of filtered by IMM-3J from the transmittance profile. From this result, the current density around 6 mA/cm^2 at V = 2 to 3 V in the IV characteristics of SAB-4J agrees well with the estimated current density of 4th subcell under the IMM-3J without ARCs. And the breakdown tendency of the IV characteristics for the InGaAs-SJ caused around at V = -1.0 V to reverse bias voltage also matched to the drop in the photocurrent for the SAB-4J. Therefore, it was confirmed that the IV characteristics under AM1.5G illumination for the SAB-4J is limited by the photocurrent density in the $In_{0.53}Ga_{0.47}As$ 4th subcell. Compared to the photocurrent estimation in each subcell for the case of under AM1.5G and AM0 shown in the Table. I, it was revealed the SAB-4J should be limited by the $In_{0.53}Ga_{0.47}As$ 4th subcell for both cases even if there were ideal ARCs were applied. Further structural or optical balance optimization for current-balance in the SAB-4J is required to improve its conversion efficiency.

TABLE I
SUMMARY OF POSSIBLE PHOTOCURRENT IN SUBCELLS

Sunlight condition	Estimated photocurrent density in each Subcell from obtained IQE (mA/cm^2)			
	1st InGaP	2nd GaAs	3rd $In_{0.35}Ga_{0.65}As$	4th $In_{0.53}Ga_{0.47}As$
AM1.5G	16.48	15.04	17.97	7.18
AM0	20.72	16.67	22.48	13.52

IV. CONCLUSION

In this study, we tried to fabricate the four-junction solar cells integrating the IMM-3J on the InGaAs-SJ by SAB method. The obtained SAB-4J showed improved Voc to 3.24 V compared with the IMM-3J owing to the 4-junction operation. However, the SAB-4J showed still unsatisfactory conversion efficiency due to insufficient performance of the direct integrated TD and

the limitation of photocurrent in the $In_{0.53}Ga_{0.47}As$ 4th subcell. To achieve the enhanced conversion efficiency of the SAB-4J, the TD performance and subcell structure design should be improved.

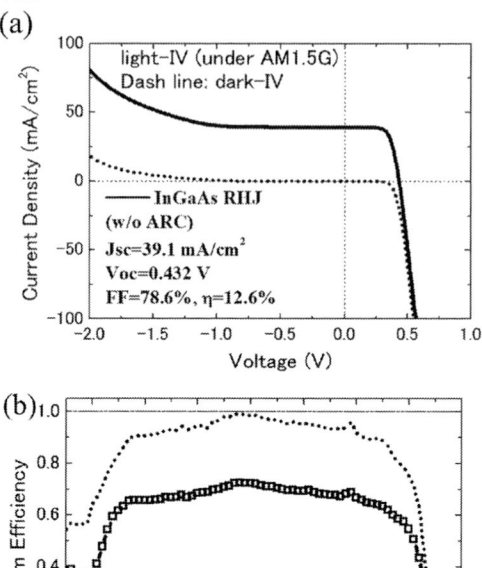

Fig. 4. IV characteristics and QE measurement result for the InGaAs SJ soalr cell.

ACKNOWLEDGEMENT

A part of this study is based on results obtained from a project, JPNP20015, subsidized by the New Energy and Industrial Technology Development Organization (NEDO).

REFERENCES

[1] J. F. Geisz, R. M. France, K. L. schulte, M. A. Steiner, A.G. Norman, Hl. L. Guthrey, M. R. Young, T. Song, and T. Moriarty, "Six-junction III-V solar cells with 47.1% conversion efficiency under 143 suns concentration," *Nat. Energy*, vol. 5, pp. 326-335, 2020.

[2] F. Dimroth, T. N. D. Tibbits, M. Niemeyer, F. Predan, P. Beutel, C. Karcher, E. Oliva, G. Siefer, D. Lackner, P. Guiot, J. Wasselin, A. Tuzin, and T. Signamarcheix, "Four-junction waferbonded concentrator solar cells," *IEEE Photovoltaics*, vol. 6, p.p. 343-349, 2016.

[3] R. Yokota, K. Wataname, H. Sodabanlu, H. Xu, M. Asami, Y. Nakano, M. Sugiyama, "High Efficiency InGaAs Cell with Enhanced External Radiative Efficiency" " *30th International Photovoltaic Science and Engineering conference (PVSEC-30)*, 2020.

Efficiency limits for multi-junction coloured photovoltaics

Phoebe M. Pearce[1], Janne Halme[2, 1], Jessica Yajie Jiang[1], Farid Elsehrawy[2], Nicholas J. Ekins-Daukes[1]

[1] School of Photovoltaic and Renewable Energy Engineering, UNSW Sydney, Kensington, NSW 2052, Australia

[2] Department of Applied Physics, Aalto University School of Science, P.O. Box 15100, 00076 Aalto, Finland

Abstract — **Coloured photovoltaic cells are of interest for product or building-integrated photovoltaics. Limiting efficiencies have previously been reported for single-junction semi-transparent and opaque cells; in this work, we expand this analysis to multi-junction devices with up to six junctions, with opaque colours produced through reflection of the incident Sunlight. We introduce a multi-objective optimization framework in Python which uses differential evolution to find parameters (e.g. central wavelength and width of reflection peaks, sub-cell bandgaps) which maximize both efficiency and colour accuracy. We find that for the 18 standard chromatic ColorChecker colours, the limiting efficiencies for cells up to six junctions are within 18% (relative) of those of a black cell with the same number of junctions, while achromatic white or light grey cells with high luminance (lightness) are much less efficient. We find that, as reported previously for single-junction cells, only two box-shaped reflection peaks are sufficient for reaching the highest efficiencies and producing any target colour which can be formed by reflecting Sunlight. Generally, colours with higher luminance cause a redshift in the optimal bandgap of the top junction to compensate for the loss of photons (and thus current) required to produce colour. However, as the number of junctions increases to five or more, one or more of the optimal bandgaps must be placed in the visible wavelength range (380-750 nm) near the reflection peaks, which affects the placement of the optimal bandgap beyond the expected redshift. Finally, we will consider technologically relevant material combinations such as perovskite or III-V materials on silicon tandems, and III-V multi-junction cells, and assess their suitability for use as a platform for coloured PV.**

I. Introduction

Coloured photovoltaic cells are of interest for building and product integration, for example to provide aesthetically pleasing building materials or for use as a decorative feature [1]. In order to produce colour which can be perceived by the human eye, the cell must reflect or otherwise emit photons within the visible wavelength range (approximately 380-750 nm); thus, the limiting efficiency of a coloured cell under a given incident spectrum will be lower than that of a black cell, since high-energy photons which could otherwise be used to generate current in photovoltaic operation are instead used to produce colour. Efficiency limits for coloured single-junction cells under an AM1.5G spectrum across a range of colours have been calculated previously for semi-transparent [2] and opaque cells [3]; in this work, we will investigate limits for opaque, coloured multi-junction cells with up to six junctions.

In order to produce these results, we developed a model in Python which uses differential evolution (DE) to

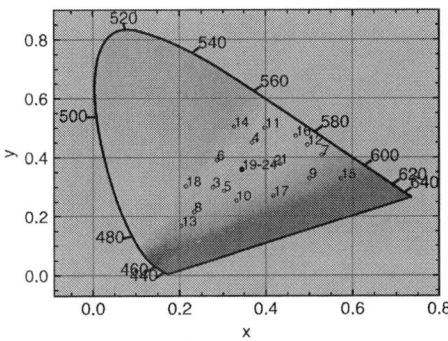

Fig. 1. The 24 MacBeth ColorChecker colours in the CIE 1931 colour space, with the colours produced by monochromatic wavelengths (in nm) along the edge. The solid point labelled 19-24 represents achromatic grey points (ColorChecker colours 19-24). Note that this plot is shown at fixed luminance ($Y = 0.5$), while the ColorChecker colours have varying luminance values.

simultaneously optimize for colour accuracy for one or more target colours chosen by the user, and cell efficiency. This can be done for arbitrary reflection spectra, although we focus here on limiting efficiencies with optimal reflection spectra consisting of narrow box-shaped peaks. This flexible approach has the advantage of being suitable for use with more realistic peak shapes or cell models, such as peak shapes measured or simulated for e.g. Bragg reflectors, and can be used with any number of junctions, depending on the user's preferences.

II. Methods

A. Colour formation

It was assumed that the colour of the cells is produced by reflecting part of the incident solar spectrum (as opposed to e.g. emission from a quantum dot or similar). Thus the observer sees an incident spectrum equal to $R(\lambda)\Phi(\lambda)$, where $\Phi(\lambda)$ is the incident spectrum (e.g. AM1.5G). The reflectivity $R(\lambda)$ was assumed to be $R = 1$ inside reflection bands and $R = 0$ elsewhere to calculate limiting efficiencies. Colour perception by the human eye is calculated using the CIE 1931 standards [4]. The colour matching functions of the CIE standard observer are shown in Fig. 4. Colours are represented by three coordinates, usually (x, y, Y), or (X, Y, Z); see further details of how these coordinates are calculated from the reflection spectra in [3]. The 24 colours of the MacBeth ColorChecker chart form the basis of the optimizations; the (x, y) coordinates of these colours are shown in Fig. 1.

978-1-6654-6060-6/23 $31.00 © 2023 IEEE

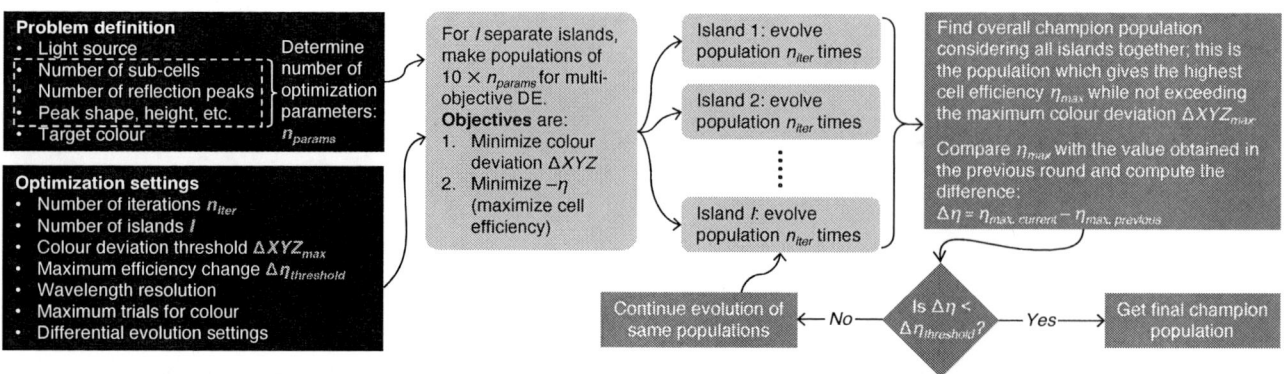

Fig. 2. Schematic of the optimization algorithm implemented in Python. Setup steps are shown in black boxes, the differential evolution steps are shown in yellow, and post-processing of the data is shown in orange. The island 1 to I can proceed in parallel to decrease computation time.

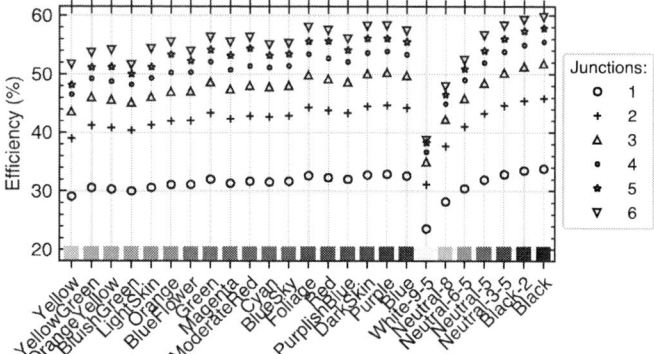

Fig. 3. Efficiency limits of coloured cells with 1-6 junctions. The first 18 (non-neutral) ColorChecker colours are ordered by decreasing luminance (i.e. from light to dark), as are the neutral colours from white to black.

B. Differential evolution

Designing a high-efficiency coloured solar cell has two key objectives: (1) minimizing the deviation from the target colour coordinates (X_t, Y_t, Z_t) and (2) maximizing efficiency. The parameters which can be varied are those which determine the reflectivity spectrum $R(\lambda)$ (in the case of simple rectangular reflection peaks, the centre and width of the reflection peaks) and the bandgaps of the sub-cells. The two objectives will not be optimized by the same set of parameters (except for a black cell), since producing colour requires some efficiency loss. The multi-objective different evolution capabilities of the pygmo2 library [5] are used to define the two-objective optimization problem; a flowchart for the optimization process is shown in Fig. 2. The results presented are generated by allowing a maximum colour deviation of $\Delta XYZ = 0.004$ (a difference not perceptible to the human eye) with a wavelength resolution of 0.1 nm in sets of $n_{iter} = 100$ iterations on $I = 10$ separate islands. The threshold efficiency change between sets of DE iterations for terminating the optimization is 0.0001 percentage points. ΔXYZ is defined as:

$$\Delta XYZ = \left(\left| \frac{X_{produced} - X_t}{X_t} \right|, \left| \frac{Y_{produced} - Y_t}{Y_t} \right|, \left| \frac{X_{produced} - X_t}{X_t} \right| \right)$$

where none of the elements of the vector may exceed the colour deviation threshold (= 0.004).

C. Detailed balance model

The analytical form for the maximum power point for multi-junction cells in the detailed-balance limit presented in [6] is used (as this is faster than a numerical calculation). The cells are assumed to be current-matched, two-terminal devices. The input spectrum for the cell calculation is $(1 - R(\lambda))\Phi(\lambda)$, i.e. all photons which are not used to produce colour are absorbed.

Both an AM1.5G and a 5778K black body incident spectrum were investigated for 1-6 junctions. In addition, some specific cases with relevant technological applications will be considered, including a tandem cell with an Si bottom junction ($E_g = 1.12$ eV) and a top cell with flexible bandgap, relevant to e.g. perovskite on Si or III-V on Si applications, and a standard InGaP/(In)GaAs/Ge (1.9/1.44/0.67 eV) triple-junction cell.

III. RESULTS

Fig. 3 shows the efficiency limits for cells with 1 to 6 junctions obtained using the DE optimization algorithm for the 24 standard ColorChecker colours. The single-junction results are in good agreement with the results presented in [3]. As expected, we see an increase in efficiency with number of junctions, with diminishing returns as more junctions are added. The white cell is severely limited in efficiency ($< 40\%$ efficient for up to 6 junctions compared to 60% for a black cell) due to the wide reflection band required to produce the target colour severely limiting the current available to the higher-bandgap junctions (see Fig. 4); however, the 18 chromatic colours all have limiting efficiencies within 18% of the detailed balance efficiency for a black cell with the same number of junctions. The effect of the luminance on the limiting efficiency can also be seen in Fig. 5, which further demonstrates that green cells have the highest limiting efficiencies, due to the high sensitivity of the human eye around 550 nm (see the $\bar{x}$ and $\bar{y}$ functions in Fig. 4). It is not possible to form any arbitrary colour (hue and luminance) by reflecting unconcentrated Sunlight, since the required number of photons may simply not be available, as is very clear for blue and red cells in Fig. 5.

978-1-6654-6060-6/23 $31.00 © 2023 IEEE

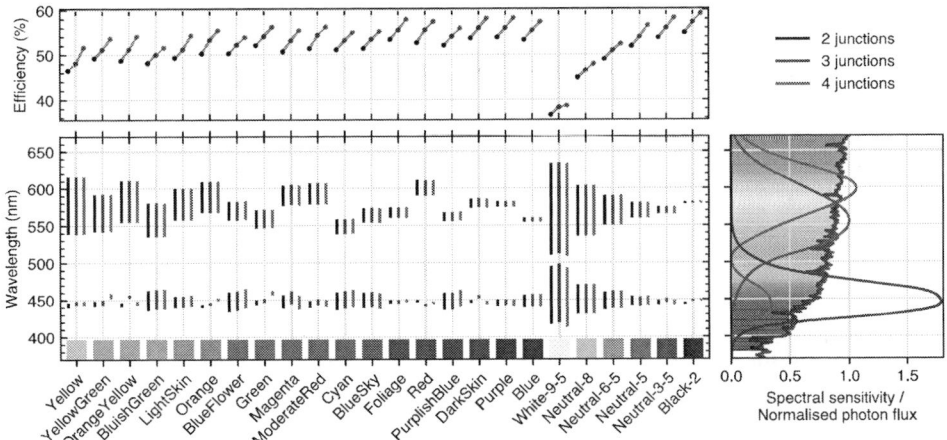

Fig. 4. Top: limiting efficiency of 2, 3 and 4 junction cells for the 24 ColorChecker colours. Bottom left: Positions of the reflection peaks required to obtain maximum efficiency for a given colour for each number of junctions. Bottom right: the CIE 1931 colour matching functions $\bar{x}$ (red), $\bar{y}$ (green) and $\bar{z}$ (blue), and the normalized AM1.5G photon flux in the visible wavelength range (shaded area).

Fig. 4 shows the position of the optimal reflection peaks to produce each ColorChecker colour using an AM1.5G spectrum. We see that the reflection peaks are centred around the peaks of the color matching functions, which describe the chromatic response of an average human observer; as the solar photon flux varies relatively smoothly in the visible wavelength region, this is expected to be the most efficient way to produce any colour. All the colours have a peak around 450 nm (blue), although for colours with very little blue this peak is of negligible importance, with a second peak centred around 550-600 nm, depending on whether the colour is 'more green' or 'more red'. In fact, even if the reflection spectrum is allowed to have more than two peaks, it was found that only two rectangular peaks were necessary to produce maximal efficieny. As shown in Fig. 4, adding additional junctions also does not change the optimal position of the reflection peaks.

Up to and including four junctions, we see a general trend of the optimal bandgap of the top junction increasing as the luminance of the colours decreases (data not shown here), which is expected; brighter colours require more photons to be reflected, so to compensate for this loss of current we expect the bandgap of the top junction to shift to lower energies to allow additional photons to be absorbed and current-match with the lower-bandgap junctions, which are not affected by the colour formation. For 5 and 6 junctions, one or two of the optimal bandgaps will be placed in the visible wavelength range, and will thus be affected by the location of the reflection peaks. The effect of the AM1.5G spectrum vs. a black body spectrum on these trends will also be investigated.

IV. Conclusions

We present limiting efficiency results for coloured multi-junction solar cells with up to six junctions, showing that coloured cells can be very efficient if optimal colour formation can be achieved. Achromatic (i.e. white, grey) cells with high luminance are much more limited in their efficiency and are thus not expect to be good candidates for multi-junction applications. For cells with a high number of junction (≥ 5), the interplay between reflection band positioning and optimal bandgap must be considered. In addition to the results presented, the Python code will be published under an open-source license.

References

[1] A. Reinders, *Designing with Photovoltaics*, 1st ed. CRC Press, 2020. doi: 10.1201/9781315097923.

[2] R. R. Lunt, 'Theoretical limits for visibly transparent photovoltaics', *Appl. Phys. Lett.*, vol. 101, no. 4, p. 043902, Jul. 2012, doi: 10.1063/1.4738896.

[3] J. Halme and P. Mäkinen, 'Theoretical efficiency limits of ideal coloured opaque photovoltaics', *Energy Environ. Sci.*, vol. 12, no. 4, pp. 1274–1285, Apr. 2019, doi: 10.1039/c8ee03161d.

[4] T. Smith and J. Guild, 'The C.I.E. colorimetric standards and their use', *Trans. Opt. Soc.*, vol. 33, no. 3, pp. 73–134, Jan. 1931, doi: 10.1088/1475-4878/33/3/301.

[5] F. Biscani and D. Izzo, 'A parallel global multiobjective framework for optimization: pagmo', *J. Open Source Softw.*, vol. 5, no. 53, p. 2338, Sep. 2020, doi: 10.21105/joss.02338.

[6] A. Pusch, P. Pearce, and N. J. Ekins-Daukes, 'Analytical Expressions for the Efficiency Limits of Radiatively Coupled Tandem Solar Cells', *IEEE J. Photovolt.*, vol. 9, no. 3, pp. 679–687, May 2019, doi: 10.1109/JPHOTOV.2019.2903180.

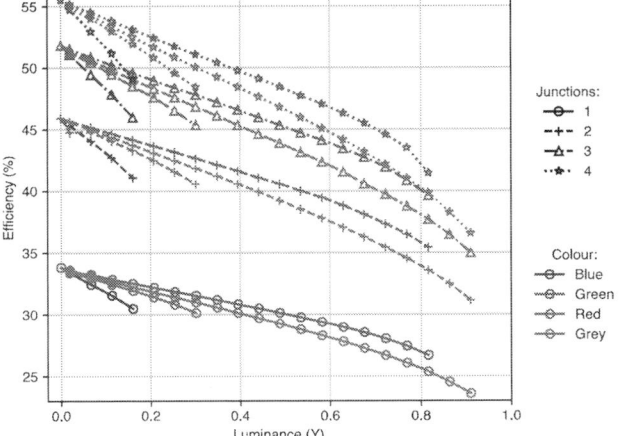

Fig. 5. Limiting efficiency as a function of luminance (lightness) for blue, green, red and grey cells with 1-4 junctions.

Microscale, high aspect ratio, effectively transparent contacts (ETCs) fabricated with string printing

Mathis Van de Voorde, Janis A. Andersons, Rebecca Saive

University of Twente, Enschede, 7522NB, the Netherlands

Abstract — **We present the fabrication of microscale, high aspect ratio front contacts for solar cells using our recently developed string printing method. String printing allows fabrication of contact lines from inks or pastes and offers high flexibility with respect to contact dimension, aspect ratio, and shape. The process can be optimized to yield triangular cross-section effectively transparent contacts (ETCs). We will present the fabrication method, scanning electron microscopy images, as well as properties of silicon heterojunction solar cells with string-printed contacts.**

I. INTRODUCTION

Front contacts are an important topic for the photovoltaic industry as their ability to conduct generated charge carriers while simultaneously accepting photons has a major influence on the performance of solar cells. Furthermore, metallization is a significant consumer of raw material (i.e. silver). Currently, the most prevalent front contact fabrication technique is screen printing which focuses on high throughput, but not necessarily on the best performances, nor on the reduction of raw material consumption. Screen printing, while being highly scalable, produces thick and low aspect ratio (i.e. height to width ratio) contacts, resulting in a decreased cell performance due to optical shading. As the terawatt future draws near, new or optimized front contact manufacturing methods which improve performance, throughput, and materials consumption become inevitable.

As one example, approaches have been investigated to reduce the shading effect of printed contacts. Kik et al. [1] showed that shaped catoptric electrodes significantly increase transmission of incident light with large metal coverages up to 50% of the wafer area. The shaped contacts redirect incident light and trap it in the structure with total internal reflection (TIR). Sachs et al. [2] also proposed patterned interconnection ribbons allowing for light trapping with TIR. Up to 81% of the light incident on the wires was recaptured in the cell, leading to a power gain of 2% of the module. Heavy texturing of the upper glass layer in solar cells however reduce the efficiency of the light trapping by 30%. Recent works by Saive et al. [3–5] on effectively transparent contacts (ETCs) have shown excellent performance metallization with transparencies up to 99% due to the triangular geometry and high aspect ratio (A/R) of the contacts. However, the fabrication of these contacts is more involved than traditional screen-printing, making an entry into the solar cell mass production market difficult. Here, we propose a fabrication process to tackle performance, throughput, and materials consumption simultaneously which we call string printing. In this process, silver paste is coated on ultra-thin (25-100 µm) metal strings. The paste is then transferred onto a silicon substrate or solar cell by approach and withdrawal to and from the substrate and the released silver then forms the contact. We believe this method could compete with the throughput of screen printing, while also providing performances similar to ETCs. In this report, we present an experimental demonstration to fabricate high A/R triangular contacts on Si substrates using string printing. We investigated the role of solid paste content and thread diameter to establish a route to reliably fabricating high performance structures. The scanning electron microscopy (SEM) images of the fabricated samples show that an aspect ratio of up to 3 was achieved with a single print. The aspect ratio can be tuned and the results are reproducible. First tests on solar cells show a redirection efficiency or effective transparency of the individual contacts of up to 70%. A detailed description of the method, first results and supplemental materials was recently published by Progress in Photovoltaics [1].

II. STRING PRINTING METHOD

In this section, we describe the string printing method. In Fig. 1, a schematic of the screen printing process is presented. A string made of tungsten alloy is loaded into our coating device. A thick layer of paste, which is resting on the tip of a syringe needle, is brought into contact with the string. The needle is moved along the string such that some of the ink is transferred from the syringe needle to the string, on the desired printing length. The interfacial energy allows for the paste to wet the thread ideally homogeneously. To obtain full control of the applied ink amount, we fed the string through a die. Once the paste is applied, the string is lowered on to the substrate until contact is made. There is a short adjustable delay time which precedes the withdrawal of the thread. The withdrawal speed can be modulated through an Arduino code. Once the process is done the sample is placed on a hot plate at 120°C to evaporate the remaining solvent.

Figure 2 presents a render of a silicon substrate in the string printing device. The string for paste deposition can be seen right above the substrate and in this render, several lines were already printed. Prior to the printing, a two-step cleaning process was performed on the substrates using ethanol

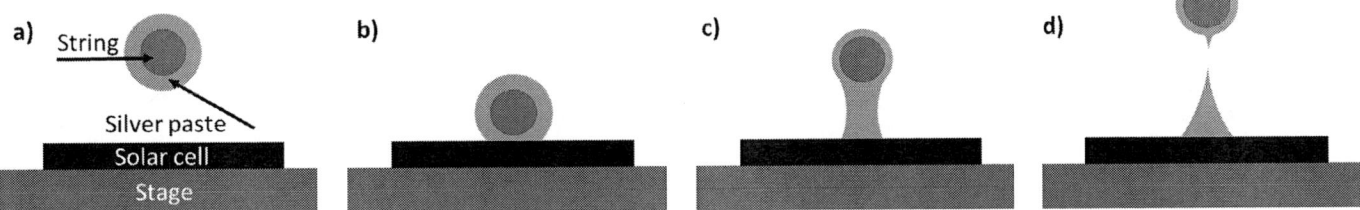

Fig. 1 Schematic of the string printing process. a) A string is coated with silver paste and the solar cell is loaded on the stage. b) The stage is lifted such that contact with solar cell and ink is established. c) The stage is slowly withdrawn. d) At a certain distance, the paste snaps off and a high aspect ratio, triangular cross-section contact is left if parameters are adjusted accordingly.

followed by water in an ultrasonic bath. The sample is then dried with a nitrogen gun and placed on the sample holder of the coating device. A render of the string-printing device is shown in Fig. 3.

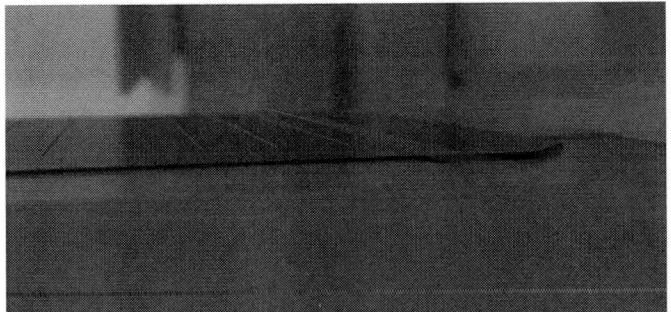

Fig. 2. Render of a silicon substrate within the string printing device. The string is visible just above the substrate.

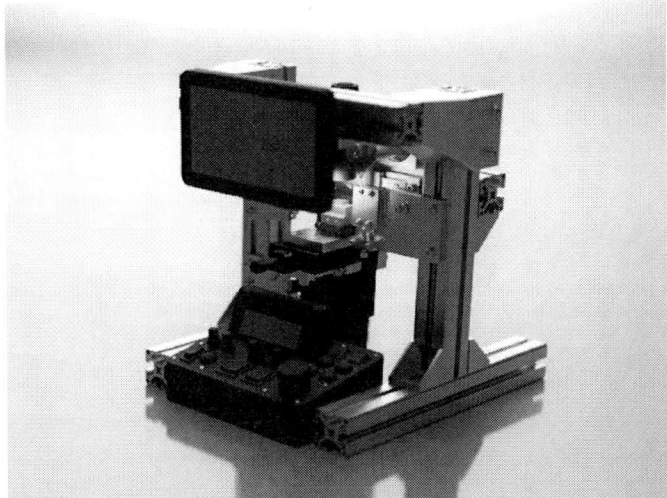

Fig. 3. Render of the string-printing prototype.

III. RESULTS

We performed prints with different parameters and on different substates such as polished silicon wafers and silicon heterojunction solar cells. Figure 4 shows scanning electron microscopy images (SEM) of several printed lines. It can be seen that by carefully tuning the process parameters, we can obtain a variety of aspect ratios and sizes within one print. Furthermore, the results are reproducible and we can print lines with similar properties parallel to each other. Furthermore, it is possible to deposit lines perpendicular to the fingers with string-printing. In [1] images are shown that employ such perpendicular silver prints on top of a silicon heterojunction solar cell. These perpendicular silver lines served as busbars in the subsequent solar cell characterization. Furthermore, it can be seen that these contacts feature triangular-cross section [1], thereby offering the possibility to redirect incoming light towards the solar cell [3-5].

To characterize the contact lines, we performed current density-voltage (j-V) measurements on silicon heterojunction solar cells with string-printed contacts. We define effective transparency or redirection efficiency as the fraction of the light that falls onto the contact and is redirected towards the solar cell thereby contributing to the photocurrent. The difficulty in the experiments was to ensure for a good reference to neither over nor underestimate the performance. We employed different size apertures to compare areas with and without contacts on the same solar cell and furthermore, we compared different solar cells with different contact layouts with each other.

Fig. 4. Scanning electron microscopy images of different string-printed silver lines.

Effective transparency of different contacts is shown in Fig. 5. For the best print, we achieved an effective transparency of 70%. We noticed a surprisingly low effective transparency for 3 of our prints which we later related to degradation from oxidation as these contacts were cured in air. For better results, we will cure contacts under nitrogen atmosphere in the future.

The relative error in determining the aperture size can get large for decreasing aperture size, therefore, we are reporting a large error margin in our results.

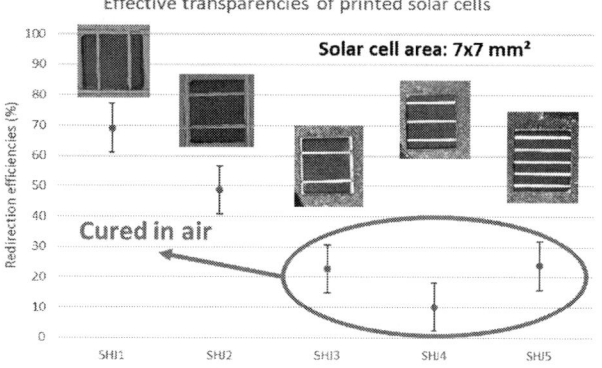

Fig. 5.   Effective transparency of different contact lines.

IV. CONCLUSIONS

We have presented a novel method for the fabrication of high-aspect ratio, micro-scale and triangular cross-section front contacts for solar cells. This method, called string printing, can be tuned to yield contact lines of different dimensions, aspect ratios and shapes. We have demonstrated aspect ratio of up to 3 and widths ranging between 8μm and 150 μm. Measurements of silicon heterojunction solar cells employing string-printed contact lines show promising results with effective transparency of up to 70%. Further results and information can be found in [1].

REFERENCES

[1] M. Van de Voorde, J. Andersons, R. Saive, "High aspect ratio triangular front contacts for solar cells fabricated by string-printing." Prog Photovolt Res Appl. 2023

[2] P. G. Kik, "Catoptric electrodes: transparent metal electrodes using shaped surfaces," Optics letters, vol. 39, pp. 5114-5117, 2014.

[3] E. M. Sachs, J. Serdy, A. Gabor, F. Van Mierlo, and T. Booz, "Light-Capturing Interconnect Wire for 2% Module Power Gain," in Proceedings of the 24th European Photovoltaic Solar Energy Conference and Exhibition, 4CO, 2009..

[4] Saive R, Borsuk AM, Emmer HS, et al. Effectively Transparent Front Contacts for Optoelectronic Devices. *Advanced Optical Materials.* 2016;4(10):1470-1474.

[5] Saive R, Bukowsky CR, Yalamanchili S, et al. Effectively transparent contacts (ETCs) for solar cells. In: *Conference Record of the IEEE Photovoltaic Specialists Conference.* Vol 2016-November. Institute of Electrical and Electronics Engineers Inc.; 2016:3612-3615.

[6] Saive R, Boccard M, Saenz T, et al. Silicon heterojunction solar cells with effectively transparent front contacts. Published online 2017.

[7] E. Jewell, S. Hamblyn, T. Claypole, and D. Gethin, "Deposition of high conductivity low silver content materials by screen printing," *Coatings*, vol. 5, no. 2, pp. 172–185, Jun. 2015.

[8] S. Tepner, N. Wengenmeyr, M. Linse, A. Lorenz, M. Pospischil, and F. Clement, "The Link between Ag-Paste Rheology and Screen-Printed Solar Cell Metallization," *Advanced Materials Technologies*, vol. 5, no. 10, Oct. 2020.

Optimization of bulk heterojunction layer constituents in organic photovoltaic device

Vilko Mandic, Dragana Vuk, Floren Radovanovic-Peric, Ivana Panzic

Faculty of Chemical Engineering and Technology University of Zagreb, ZAgreb, Croatia

Organic photovoltaic devices (OPVs) are a viable alternative to traditional solar cells for conversion of solar into electricity. Recently, numerous variants of OPVs have been developed as promising systems due to their simple, flexible and relatively facile production. Organic absorbers of the bulk heterojunction absorbing layers (BHJs) stand out as the most promising price-efficient systems. Among these, aminosquaraine constituents are investigated for their high absorptivity and emission extending to IR part of the visible light spectrum. In addition to implementation of symmetric aminosquaraines, we show that some extending to asymmetric squaraines may further facilitate alignment of electron donor behavior with optical properties. Here we report broad range of activities related with: design of the molecules, conducting of the organic synthesis, blending of the derived electron donors and acceptors into BHJs, BHJ deposition and processing, and OPV cell assembly and characterization. The laboratory synthesis efforts are material extensive, making the course of the optimization challenging. Proportionally, this work dedicated considerable attention to describe the bottlenecks encountered during the optimization and routes proposed to overcome issues. This thorough approach aims in correlating aminosquaraine phenomenology with performance of the functional OPV cells. Finally, selected candidates proved fully viable for BHJ OPV systems, considering chemical and morphological aspects of the derived films.

Towards smart integration of Cu-plating for silver-free and edge passivated SHJ Shingle modules

Samuel Harrison, Vincent Barth, Benoit Martel, Agata Lachowicz, Nicola Frasson, Marco Galiazzo

CEA-INES, Bourget Du Lac, France

CSEM, Neuchatel, Switzerland

AMAT, Olmi Di S.Biaggio, Italy

Moving towards silver free cell/modules seems unavoidable for PV in a near future for both cost & sustainability reasons. This is even more true for shingle heterojunction, where large amount of silver is needed to insure good module final performances. First integration tests conducted with copper plating has thus been successfully initiated, with very promising first outcomes: strong gain in cell/module FF in particular, related to a significant improvement of metal grid resistance. Combined with the extremely low ECA deposit needed, we thus show that a simple and efficient (almost) &"ilver free" path can be considered for SHJ shingle. Furthermore, the self-aligned plating process chosen integrates insulating layers, which also enables reduction of TCO thickness, optical improvement or serve as humidity barrier or for cell-edge management, thus improving even further the potential of the technology.

A Comparison of Emerging and Industry Benchmark Photovoltaic Backsheets Between Different Outdoor Locations

Elizabeth Palmiotti[1], Bruce King[1], Rachael Arnold[2], Soňa Uličná[2], Laura T. Schelhas[2], David C. Miller[2]

[1] Sandia National Laboratories, Albuquerque, NM, 87185, USA

[2] National Renewable Energy Laboratory, Golden, CO, 80401, USA

Abstract — Recent interest within the photovoltaic (PV) module industry is largely directed toward enhanced lifetimes in the field, balanced with improved recyclability. Traditionally, fluoropolymer-based backsheets have been used, however, are difficult to recycle. Emerging polyolefin (PO)-based backsheets are more recyclable and can be formulated to be robust. Properties of different fluoropolymer- and non-fluoropolymer-based backsheet coupons and in encapsulated silicon mini modules that have been fielded in Albuquerque, NM and Cocoa, FL are reported here. Seven backsheets were examined: two novel PO's, TPT, APO, PPE, AAA, and KPf. Methods of examination include module electrical performance (I-V flash test), surface morphology (optical microscope and gloss), polymer chemical structure (FTIR), EL imaging, mechanical tensile testing, DC breakdown voltage, DSC (phase transitions), and optical performance (reflectance spectra).

I. Introduction

The rear surface of photovoltaic (PV) modules is typically protected by a polymeric backsheet. PV backsheets are traditionally multi-layer composites where the internal layer aids adhesion to the encapsulant, the core provides electrical insulation and mechanical strength, and the outer layer is environmentally durable [1]–[3]. These composite products often consist of polyethylene terephthalate -based (PET) cores and polyethylene inner and fluoropolymer outer layers. Such backsheets have demonstrated some reliability risks as well as poor recyclability. Polyamide (PA) layers have since been explored towards PV backsheets. Novel co-extruded polyolefin-based backsheets are a more recent promising alternative. Polyolefins (PO's) have demonstrated both durability and enhanced recyclability compared to traditional backsheet materials [4].

Commercial and novel backsheets have been compared through accelerated stress testing of backsheet coupons and mini-modules (MiMo's) [5], [6]. For the same study, coupons and MiMo's were also fielded in Albuquerque, NM and Cocoa, FL and characterized every six months. This work provides a summary of the fielded results and a comparison to accelerated stress test results.

II. Experimental

A. Samples and Aging

Seven backsheets were compared in this work: three commercially relevant products (TPT, PPE, and KPf), two novel PO's (PO1 and PO2), APO – which uses both PA and PO, and one known-bad (AAA). The backsheets were indexed with a number; details are summarized in Table I. Per backsheet type, three types of sample were prepared: backsheet coupons, backsheet coupons laminated to EVA encapsulant, and fully-functional MiMo's. MiMo's were completed with four, 156 mm, Si-Cz, p-PERC cells connected with ribbon to an edge-mounted junction box. Coupons were run through the same laminator and procedure as the MiMo's to ensure the same thermal history.

TABLE I. BACKSHEET SAMPLE IDENTIFICATION, MATERIAL, AND CONSTRUCTION.

Backsheet ID	Material	Construction
BS-1	PO1	Co-extruded
BS-2	PO2	Co-extruded
BS-3	TPT	Laminate
BS-4	APO	Co-extruded
BS-5	PPE	Laminate
BS-6	AAA	Co-extruded
BS-7	KPf	Laminate

Three MiMo's of each backsheet type were fielded in Albuquerque and Cocoa, respectively. Four backsheet coupons of each type were fielded in both locations; every six months one coupon was retrieved from the field for characterization. Two backsheets laminated with EVA were fielded in both locations; every twelve months one backsheet was retrieved from the field for characterization. The replicate MiMos were characterized every six months and returned the field for continued aging.

Details about accelerated aging can be found in [5], [6]. There, steady-state accelerated aging was conducted via three hygrometric and four photolytic conditions with read points at 1000 hours for up to 4000 hours.

978-1-6654-6060-6/23 $31.00 © 2023 IEEE

B. Characterization

Electrical characterization of the MiMo's was conducted by I-V flash testing using a Spire 4600SLP Class AAA solar simulator at STC conditions (25°C, 1000 W·m² and AM1.5). All coupons and MiMo backsheet surfaces were characterized optically and for gloss and color (L, a*, and b*). Surface images were collected using a Keyence VHX 5000 Digital Microscope. Gloss was collected by a Konica Minolta Multi Gloss 268A. Gloss measures the specular reflectance of a surface by measuring the intensity of a light beam of known intensity, for the separate incidence angles of 20, 65, and 80°. Color data was collected using a Konica Minolta CM-600d Spectrophotometer; the instrument was allowed to warm up prior to calibration using a white standard before each measurement session. Tensile testing was conducted on 1 cm wide strips cut from coupons following the method in IEC TS 62788-2. Fourier transform infrared (FTIR) spectroscopy was conducted using an Agilent 4300 Handheld FTIR Spectrometer. Multiple spectra were collected per sample to ensure valid results.

III. RESULTS AND DISCUSSION

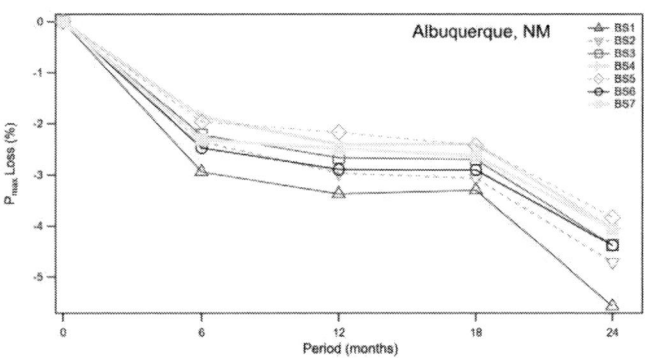

Fig. 1. Electrical performance data from MiMo's with different backsheet materials fielded in Albuquerque, NM up to 24 months.

Fig. 1 shows the percent maximum power (P_{max}) loss for each type of backsheet over 24 months of field exposure in Albuquerque. All backsheets showed similar, moderate degradation with the greatest decrease in P_{max} between 18 and 24 months. No substantive features that might explain the loss of P_{max} at 24 m were observed in EL imaging.

Fig. 2 shows surface images of the coupons fielded in Cocoa on the sun-facing sides. Images were also collected on the

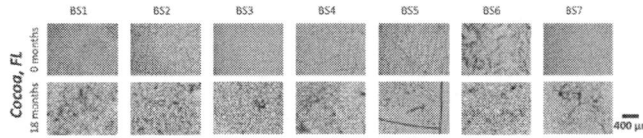

Fig. 2. Images of the sun-side of backsheet coupons after 0 and 18 months of fielding in Cocoa, FL.

opposite, air-facing side, and in Albuquerque, however, are not shown as they showed similar, though less drastic, trends relative to Fig. 2. All backsheets fielded in Cocoa saw the formation and growth of organic matter on both sides; this did not occur on samples fielded in Albuquerque. Different morphology was observed for each backsheet, suggesting the surface or material enabled growth of different species of organic matter. While different organic matters secrete acidic substances that could catalyze degradation, the long term effects of biological species on PV are not well understood. BS-5 coupons show the development of cracks (mm-scale) in both climates starting after 12 months of exposure. Note that backsheets integrated into MiMo's did not crack. BS-6 shows the onset of micro-scale cracking after 24 months of fielding in Albuquerque. Accelerated testing incorporating UV weathering also resulted in cracking of the sun-side of BS-5.

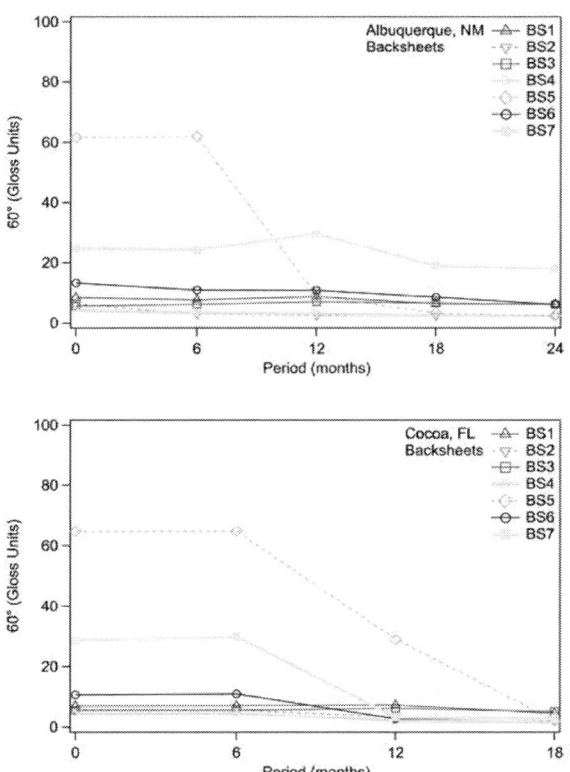

Fig. 3. Gloss measurements for the sun-side of backsheet coupons fielded in Albuquerque, NM and Cocoa, FL.

Gloss measurements for the sun-facing side of the backsheet coupons are shown in Fig. 3. The Albuquerque data is representative of the gloss measurements collected for the MiMo's. The reported values are in gloss units (GU) which range from 0, for a perfectly matte surface, to 100, for a polished black glass reference. Decreases in gloss are noted for most backsheets, though most drastically for BS-5 and BS-7 in both Albuquerque and Cocoa. Decreased gloss may indicate surface roughening from erosion or surface cracking. The

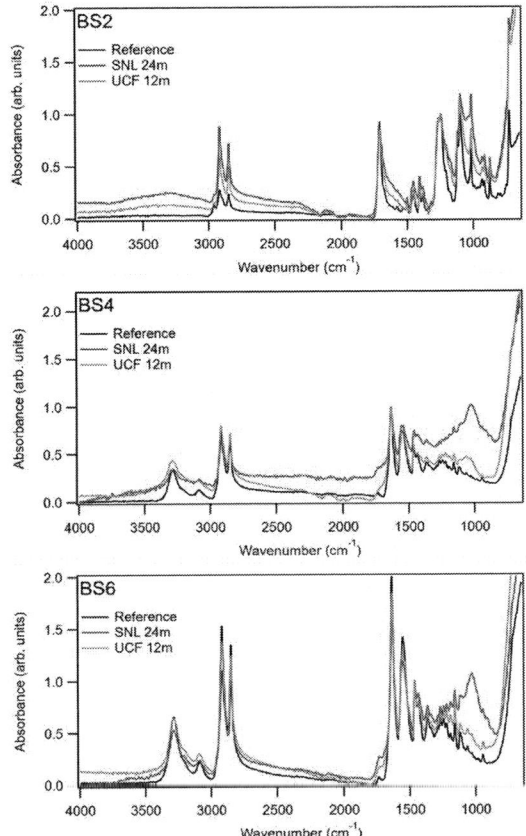

Fig. 4. FTIR spectra collected on the sun-facing side of backsheet samples fielded in Albuquerque, NM and Cocoa, FL.

significant reduction in gloss for BS-5 may be explained from the changes in surface morphology seen in Fig. 2. Similar decreases in gloss were observed for BS-5 samples that underwent UV weathering.

FTIR spectra collected for reference (black) and fielded (Albuquerque in red and Cocoa in blue) BS-2, BS-4, and BS-6 are shown in Fig. 4. Relative to unaged references, greater changes were seen for the coupon samples (plotted) compared to the MiMo's, likely because the backsheets were facing the sun during fielding reducing the effect of UV radiaiton. Samples BS-1, BS-3, BS-5, and BS-7 demonstrated minimal changes with fielding, thus were not shown here. Enhanced intensity and peak broadening are noted for BS-2 samples. Enhancement of the peak around 1000 cm⁻¹ is noted, especially in Albuquerque and was also observed in accelerated UV weathering. BS-4 and BS-6 show the emergence of a new peak around 1000 cm⁻¹ particularly in Albuquerque also observed in Cocoa. A PA material is used for the air side surface of both BS-4 and BS-6. Peak formation suggests molecular structure changes in the outer backsheet layer, attributed to UV degradation. In comparison, the most substantive FTIR results were observed in accelerated UV weathering rather than hygrometric aging [7]. Additional results will be presented at the conference and included in the final manuscript. This will include the all read points, from 0 to 24 months, of all fielded data from Albuquerque and Cocoa. Additional data and analysis includes MiMo electrical performance data, relevant electroluminescence images, Keyence images, gloss, color, tensile, breakdown voltage, FTIR, and differential scanning calorimetry.

IV. CONCLUSION

Three commercially relevant (TPT, PPE, and KPf), two novel POs, one APO, and one known bad (AAA) backsheets were fielded in Albuquerque, NM and Cocoa, FL for 24 months and characterized in 6 month increments. The outdoor weathering was compared to accelerated aging by hygrometric and photolytic methods for up to 4000 hours. Fielded and accelerated results showed the formation of cracks on BS-5 (PPE). Minimal changes to gloss were seen with fielding, except for BS-5, which may be attributed to enhanced texturing or roughness from crack formation. Notably, cracking was not observed in MiMo's, suggesting a latent degradation is occurring. FTIR revealed changes in the polymer structure of BS-2 (PO), BS-4 (APO), and BS-6 (AAA) in Albuquerque, NM, which is corroborated in accelerated UV weathering tests. This suggests material changes to the outer layer from UV photolysis, where the net effect on backsheets performance, including surface or bulk damage, may be compared between the many characteristics examined in this study.

REFERENCES

[1] K. J. Geretschläger et al., "Structure and basic properties of photovoltaic module backsheet films," Solar Energy Materials, vol. 144, pp. 451–456, 2016, doi: 10.1016/j.solmat.2015.09.060.
[2] C.-C. Lin et al., "Depth profiling of degradation of multilayer photovoltaic backsheets after accelerated laboratory weathering: Cross-sectional Raman imaging," Solar Energy Materials, vol. 144, pp. 289–299, 2016, doi: 10.1016/j.solmat.2015.09.021.
[3] M. Aghaei et al., "Review of degradation and failure phenomena in photovoltaic modules," Renewable and Sustainable Energy Reviews, vol. 159, p. 112160, 2022, doi: 10.1016/j.rser.2022.112160.
[4] A. Omazic et al., "Increased reliability of modified polyolefin backsheet over commonly used polyester backsheets for crystalline PV modules," Journal of Applied Polymer Science, vol. 137, no. 30, p. 48899, 2020, doi: 10.1002/app.48899.
[5] M. Thuis et al., "A Comparison of Emerging Nonfluoropolymer-Based Coextruded PV Backsheets to Industry-Benchmark Technologies," IEEE Journal of Photovoltaics, vol. 12, no. 1, pp. 88–96, 2022, doi: 10.1109/JPHOTOV.2021.3117915.
[6] N. Al Hasan et al., "Arrhenius Analysis of the Degradation Modes in BackFLIP Study of Emerging Photovoltaic Backsheets," in 49th IEEE Photovoltaic Specialists Conference (PVSC 49), Philadelphia, PA, 2022. doi: NREL/CP-5K00-8263
[7] S. Uličná et al., "BACKFLIP: Identification of Materials and Changes Upon Aging of Emerging Fluoropolymer-Free and Industry-Benchmark PV Backsheets." IEEE 48th Photovoltaic Specialists Conference (PVSC 48), Fort Lauderdale, FL, 2021. doi: 10.1109/PVSC43889.2021.9518781

Considering the Variability of Soiling in Long-term PV Performance Forecasting

Matthew Muller[1], Faisal Rashed[2]

[1]National Renwewable Energy Laboratory, Golden, CO, USA
[2]Leidos, Denver, CO, USA

Abstract—This study presents the development of a methodology for evaluating the variability associated with soiling on long-term PV forecasting. Independent engineering firms typically build P50 forecasts for large PV plants through the use of the PVsyst software, where monthly soiling losses are one of many inputs to the P50 model. Subsequently, long-term performance distributions, or Pvalues, are constructed through a Monte Carlo analysis that includes various factors such as: satellite irradiance modeling uncertainty, uncertainty in the PVsyst model, and long-term irradiance variability. Often the PVsyst model uncertainty is increased to account for sites with significant soiling concerns but no systematic method has been presented in the literature to specifically include soiling variability within Pvalues. In this work soiling information from 16 sites in the U.S. Southwest are combined with 20 years of rainfall data to generate 20 years of energy production with soiling losses and then subsequently generate Pvalues. The results show that the spread of Pvalues (P1-P99) can increase from 0-13% when interannual soiling variability is included.

Keywords—photovoltaic soiling, performance forecasting, uncertainty, Pvalues, interannual variability

I. INTRODUCTION

Photovoltaic (PV) soiling loss is the well-known phenomenon where dust or other airborne particulates accumulate on the surface of PV modules causing light blockage and therefore power loss to the PV system. Soiling losses depend on local climate, geography, nearby pollution sources, module orientation and various other factors [1]. Annualized soiling losses can be as low as 0.5%/year in temperate climates with frequent rainfall and as high 30%/year in deserts such as the middle east [2, 3]. Revenue losses due to soiling losses depend on the specific PV system but can easily reach millions of dollars per year for large utility scale systems [4]. Independent Engineers (IEs) typically model utility scale PV system P50 performance (annual energy yield that expected to be exceeded 50% of the time) using PVsyst or other software where key inputs are satellite site irradiance, temperature, and wind speed, losses due to irradiance transposition to plane of array, PV module electrical parameters, various other electrical losses, and monthly soiling losses. These monthly soiling losses, specifically consideration for their interannual variability and a method to propagate this variability into the plant probabilistic performance (Pvalues) is the focus of this work. Monte Carlo simulations of annual plant energy yield are used to generate a P1 (1% of all observations are estimated to exceed this energy yield) and a P99 (99% of all observations are estimated to

exceed this energy yield) among other Pvalues that might be desired. While there are various methods to generate PV system Pvalues, all methods generally include uncertainty of the satellite derived global horizontal irradiance (GHI), interannual variability of the weather (i.e. irradiance and temperature), and uncertainty in the PV power production model (i.e. uncertainties associated with irradiance transposition, electrical, availability, soiling, and other losses [5]. IEs typically have internal proprietary methods that have been developed through years of experience to assign an uncertainty distribution to each site-specific energy model. While soiling has traditionally been included as part of this overall model uncertainty, there has been sufficient progress in soiling research in recent years to consider an approach for separately accounting for soiling interannual variability similarly to the handling of weather. In this work we describe a transparent method for calculating soiling interannual variability and incorporating the results directly into Pvalue calculations. We first present a methodology section and then we provide results from applying the approach to 16 sites in the Southwest U.S. with well-established data on soiling rates.

II. METHODOLOGY

The Kimber soiling model [6] is commonly used to estimate soiling losses, where the basic assumption is that soiling occurs linearly during dry periods followed by cleaning or recovery through rainfall events above a minimum threshold. PVlib currently provides a free implementation of the Kimber model using Python [7]. The two primary inputs to the model are daily rainfall (available through PRISM for the continental U.S. [8]) and soiling rates for the site under investigation. Proposed utility scale sites are often subjected to an irradiance and soiling measurement campaign in order to capture data for reducing irradiance and soiling model uncertainty. For similar reasons NREL has been working to build a soiling data map through the extraction of soiling information from PV time series data [9, 10]. To examine the interannual variability of soiling losses we have selected 16 Southwest U.S. sites from the NREL soiling map that have soiling losses greater than 1% and therefore also report data on monthly soiling rates (see Fig.1). We simulate soiling losses for 20 years using the basic Kimber model with the following assumptions: cleaning to 99.5% occurs for daily PRISM rainfall totals greater than 2.5 mm, no grace period is included, the monthly median soiling rates are input from the NREL soiling map, and in the event that data isn't available for a specific month then the lowest median rate from all other months is used for that month. 20 years of simulation is chosen because 1999-2018 is currently available through both PRISM

978-1-6654-6060-6/23 $31.00 © 2023 IEEE

and NREL's free PSM3 satellite based solar irradiance data through the NSRDB [11]. As an NSRDB update is currently in progress it is expected that 1998-2021 can be run for the full paper.

It is expected that to best account for soiling variability that daily soiling losses be weighted by daily insolation totals or optimally as a loss within the appropriate step in the PV performance model. For example, a 5% raw soiling loss results in significantly lower energy loss during short sunny winter day as compared to long sunny summer day. If the PV system DC/AC ratio is significantly greater than 1 it is critical to apply the soiling losses within the PV model as system clipping can mitigate soiling losses and reduce interannual variability. Similarly, if the PV system contract mandates cleanings, these cleanings should be included in the soiling model to correctly capture the impact on interannual soiling variability. In this work we use the algorithms within PVlib's ModelChain class to model hourly PV energy output for each of the 16 sites in Fig. 1. The baseline model is 100 megawatt single-axis tracking system (±60°) with a 0.33 ground coverage ratio, and DC/AC ratio of 1 (no clipping). Each of the 20 years the monthly soiling losses resulting from the Kimber model are input into the PV model with the given years PSM3 irradiance to generate an annual energy production for that year. Specifically, the effective irradiance profile within the PVlib model is multiplied by the monthly soiling loss factor. The simulation is performed with and without soiling to estimate the impact soiling has on interannual performance variability.

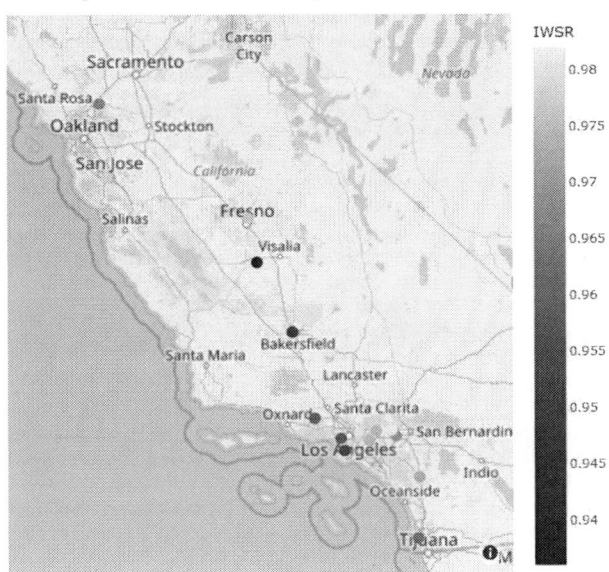

Fig. 1. 16 sites selected from the NREL soiling map.

Pvalues are generated with and without soiling using a Monte Carlo analysis. 20,000 samplings are made from the 20 years of modeled performance data in conjunction with a sampling from a normal distribution for the PV models uncertainty, which is defined by a mean of zero with a ± 2.5% uncertainty at one standard deviation. This uncertainty distribution is intended to represent all the uncertainties that go into the P50 PV model (for example, uncertainties associated with the soiling and irradiance models are considered here). This uncertainty is typically calculated by IEs through various

evaluations of the plant specific model, including factors such ground tuning of satellite irradiance and site soiling measurement campaigns. Here we are not evaluating individual PV plants and therefore we apply a general model uncertainty to all sites. In the full paper we intend to examine a separate accounting for soiling uncertainty based on the range of soiling rates provided for each month on the NREL soiling map.

III. SOILING VARIABILITY RESULTS

The box and whisker plot in Fig. 2 provides the 20 years of energy-weighted soiling losses for each of the 16 sites. The average soiling losses range from 4.3-15.5% while the full range of soiling losses varies significantly depending on the site. For example, sites 5219 and 5286 are examples of low to moderate soiling with only about a 5% spread in losses over the 20 years. Alternatively, site 7052 has a median loss of about 15% and the spread of values over the 20 years is about 20%. It is also clear that for 20 years the data is not necessarily normally distributed.

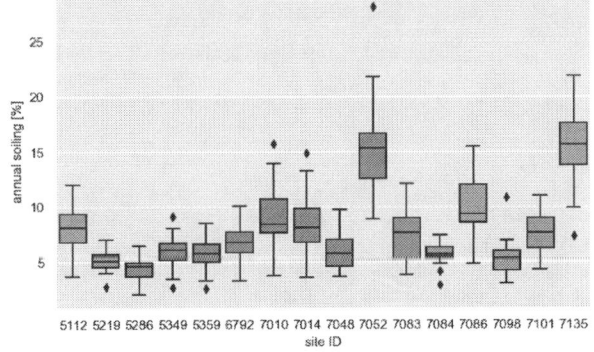

Fig. 2. Box and whisker plot of annual energy-weighted soiling losses for each site (the box represents the interquartile range, whiskers represent the maximum and minimum boundaries if an outlier/diamond is not plotted).

Fig. 3 provides a box and whisker plot comparing the interannual variability of energy production for models with and without soiling (given as percentage change from the P50 for each model). It is important to note that for some sites interannual variability over 20 years is significantly different between the models with and without soiling. This demonstrates that the relationship between irradiance, rain, and soiling each year can be important to accurately representing interannual variability.

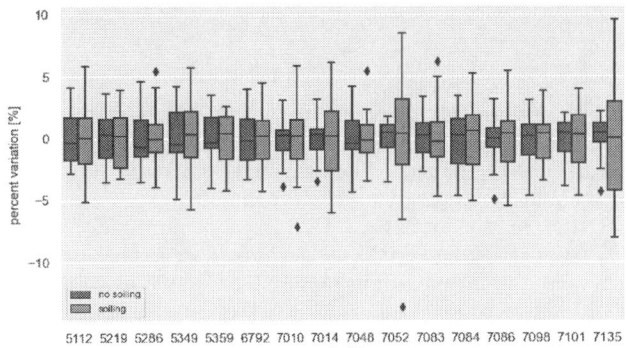

Fig. 3. Interannual variation of PV energy generation with and without soiling (the box represents the interquartile range, whiskers represent the maximum and minimum boundaries if an outlier/diamond is not plotted).

IV. PVALUE RESULTS

Table 1 provides the spread of Pvalues (P1-P99) generated from the Monte Carlo simulation considering PV interannual performance variability and ± 2.5% model uncertainty at one standard deviation. Mean and standard deviation of soiling for each site are provided for context. The final column in Table 1 provides the increase in the spread of Pvalues when soiling is included in the interannual performance calculations. For three of the sites the Pvalue spread is nearly the same (± 0.1%). The other thirteen sites show the Pvalue spread increase anywhere from 0.4-12.7% showing that including soiling within interannual variability can be especially important for PV plants with higher soiling rates. While these results will vary with system design, especially DC/AC overbuild or the inclusion of contracted cleaning schedules, they do point to the importance of accounting for interannual soiling variability within Pvalue calculations.

TABLE 1. Pvalues:16 sites with and without soiling

Site	Mean soiling [%]	Soiling stdev [%]	P1-P99 no soil [%]	P1-P99 soiling change [%]
5112	8.0	1.9	14.9	3.0
5219	5.1	1.0	14.7	0.4
5286	4.3	1.1	15.4	0.8
5349	5.9	1.5	15.5	1.8
5359	5.7	1.4	14.8	-0.1
6792	6.8	1.7	14.9	0.8
7010	9.0	2.6	14.1	4.6
7014	8.7	2.8	13.7	4.9
7048	6.1	1.7	14.9	0.1
7052	15.6	4.1	13.4	12.7
7083	7.5	2.4	13.9	2.9
7084	5.8	1.1	15.2	1.2
7086	10.0	2.6	14.5	2.9
7098	5.4	1.7	15.1	-0.1
7101	7.6	1.9	13.7	1.9
7135	15.5	3.7	13.6	9.8

V. CONCLUSIONS

This work has demonstrated a methodology that can be applied through existing tools (the PVlib Kimber model, the NREL soiling map, PRISM, and NSRDB) to estimate interannual soiling variability. The methodology was applied to 16 sites in the Southwest U.S. as a demonstration of what interannual soiling variability can look like. In the case of these 16 sites, the mean soiling ranged from 4.3% to 15.5%. The inclusion of soiling calculations within 20 years of annual PV performance calculations resulted in the Pvalue spread (P1-P99) increasing from -0.1% to 12.7%. These results are exemplary as

actual results with depend on specific PV system design parameters like DC/AC overbuild and contract features around operations and maintenance (specifically cleaning schedules). Additionally, the specific calculated PV model uncertainty propagated into the Monte Carlo simulation will impact final results.

ACKNOWLEDGMENT

This work was authored in part by Alliance for Sustainable Energy, LLC, the manager and operator of the National Renewable Energy Laboratory for the U.S. Department of Energy (DOE) under Contract No. DE-AC36-08GO28308. Funding provided by the U.S. Department of Energy's Office of Energy Efficiency and Renewable Energy (EERE) under Solar Energy Technologies Office (SETO) Agreement Numbers 38258. The views expressed in the article do not necessarily represent the views of the DOE or the U.S. Government. The U.S. Government retains and the publisher, by accepting the article for publication, acknowledges that the U.S. Government retains a nonexclusive, paid-up, irrevocable, worldwide license to publish or reproduce the published form of this work, or allow others to do so, for U.S. Government purposes.

REFERENCES

[1] L. Micheli, M. Muller, An investigation of the key parameters for predicting PV soiling losses. Prog Photovolt Res Appl. 2017;25 (4):291 - 307. https://doi.org/10.1002/pip.2860S.

[2] S. Costa, A. Diniz, L. Kazmerski, Solar energy dust and soiling R&D progress: Literature review update for 2016. Renew. Sustain. Energy Rev. https://doi.org/10.1016/j.rser.2017.09.015

[3] S. Costa, A. Diniz, L. Kazmerski, Dust and soiling issues and impacts relating to solar energy systems: Literature review update for 2012–2015. Renew. Sustain. Energy Rev. 63, 33–61. https://doi.org/10.1016/j.rser.2016.04.059G. Eason, B. Noble, and I. N. Sneddon, "On certain integrals of Lipschitz-Hankel type involving products of Bessel functions," Phil. Trans. Roy. Soc. London, vol. A247, pp. 529–551, April 1955. *(references)*

[4] K. Ilse, L. Micheli, B.W. Figgis, K. Lange, D. Daßler, H. Hanifi, F. Wolfertstetter, V. Naumann, C. Hagendorf, R. Gottschalg, J. Bagdahn, Techno-Economic Assessment of Soiling Losses and Mitigation Strategies for Solar Power Generation. Joule 2303–2321. 2019. https://doi.org/10.1016/j.joule.2019.08.019

[5] https://solargis.com/blog/best-practices/how-to-calculate-p90-or-other-pxx-pv-energy-yield-estimates

[6] A. Kimber, L. Mitchell, S. Nogradi, and H. Wenger, "The Effect of Soiling on Large Grid-Connected Photovoltaic Systems in California and the Southwest Region of the United States," in Photovoltaic Energy Conversion, Conference Record of the 2006 IEEE 4th World Conference on, 2006, pp. 2391–2395.PVlib

[7] W. F. Holmgren, C. W. Hansen, and M. A. Mikofski. "pvlib python: a python package for modeling solar energy systems." Journal of Open Source Software, 3(29), 884, (2018). https://doi.org/10.21105/joss.00884ib

[8] PRISM Climate Group - Oregon State University, PRISM Gridded Climate Data (AN81d), http://www.prism.oregonstate.edu/explorer/

[9] National Renewable Energy Laboratory, "Photovoltaic modules soiling map," https://www.nrel.gov/pv/soiling.html

[10] M. Deceglie, L. Micheli, M. Muller, Quantifying Soiling Loss Directly from PV Yield. IEEE J. Photovoltaics 8, 547–551. 2018.

[11] M. Sengupta, Y. Xie, A. Lopez, A. Habte, G. Maclaurin, and J. Shelby. 2018. "The National Solar Radiation Data Base (NSRDB)." Renewable and Sustainable Energy Reviews 89 (June): 51-60.

978-1-6654-6060-6/23 $31.00 © 2023 IEEE

Electron paramagnetic resonance investigation of the defect responsible for light- and elevated-temperature-induced degradation in Ga-doped Czochralski Si

Chirag Mule, P. Craig Taylor, Abigail Meyer, William Nemeth, Vincenzo LaSalvia, Matthew Page, Sumit Agarwal, Pauls Strandins

Colorado School of Mines, Golden, CO, United States

National Renewable Energy Laboratory, Golden, CO, United States

In this work, we report electron paramagnetic resonance (EPR) spectroscopic study of the defect that causes light- and elevated-temperature-induced degradation (LeTID) in p-type Ga-doped Cz Si wafers. The EPR signal contains two features - a silicon-dangling-bond signal and hydrogen hyperfine-doublet. We report the light dependence of the hydrogen-doublet signal which gives information about the spin-activity of the defect centre. We also report the ratio of defect densities of the two features to be ~1, based on the EPR signal intensities of the degraded sample. This new information would help in understanding the atomic structure of the defect.

Real-time Regional PV Spinning Reserve Estimator with AGC Look-ahead Windows

Mengmeng Cai, Govind Saraswat, and Vahan Gevorgian

National Renewable Energy Laboratory (NREL), Golden, CO, USA.

Abstract—Curtailed PV generation is a zero-marginal cost spinning reserve that can be used for a number of active power control services. However, unlike the traditional spinning reserve providers, i.e., fossil-fueled generators, who have well-defined operating characteristics, e.g., available headroom or potential high limit (PHL), PV plants have by nature variable and uncertain operating characteristics. To ensure the effective coordination between PV plants and the system operator during an active power control event, accurate forecasts of the PV PHL are essential. A novel reference-control grouping based scaling method has been proposed by NREL to estimate the PV PHL in real-time. This work further enhances the methodology by: 1) improving the model accuracy through machine learning; 2) considering look-ahead windows introduced by the computation and communication latencies; 3) applying the method to regional spinning reserve estimation. A significant performance improvement, over 99% of estimation error reduction, has been observed based on real-world data collected by CAISO and PV plant operators.

Keywords—*PV, Spinning Reserves, Reliability Services, Potential High Limit*

I. INTRODUCTION

The integration of variable energy resources (VERs) is gaining momentum in the U.S., adding great variability and uncertainty to the bulk power system. More system flexibilities from various energy resources are therefore needed to maintain the system reliability, including flexibility coming from the wind and solar plants themselves [1]. To resolve the emerging need, California independent system operator (CAISO) recommended the active power control of VERs, by which the VERs operate at curtailed generation points following automated dispatch instructions in response to a grid service need. Several demonstration projects conducted in Texas, Puerto Rico, and California under the collaboration between the U.S. Department of Energy (DOE) and industry have proven the capability of utility-scale PV plants in providing a full spectrum of reliability services, i.e., primary frequency response, automatic generation control, inertial response and ramp control, etc., as opposed to being just a source of variable bulk energy production [2], [3].

However, unlike the traditional operating reserve providers - fossil-fueled generators - who have well-defined operating

This work was authored in part by NREL, operated by Alliance for Sustainable Energy, LLC, for the U.S. Department of Energy (DOE) under Contract No. DE-AC36-08GO28308. The views expressed in the article do not necessarily represent the views of the DOE or the U.S. Government. The U.S. Government retains and the publisher, by accepting the article for publication, acknowledges that the U.S. Government retains a nonexclusive, paid-up, irrevocable, worldwide license to publish or reproduce the published form of this work, or allow others to do so, for U.S. Government purposes.

characteristics, e.g., available operation headroom, PV plants by nature have variable and uncertain operating characteristics. To ensure an optimal and feasible coordination between PV plants and system operator during an active power control event, it is important to accurately estimate the potential maximum available power output, or potential high limit (PHL), of the plant at any moment even when the plant is curtailed to a lower production level. The PHL is driven by the solar irradiation and the plant condition [4] and therefore varies across time and cannot be measured directly. Estimating the PV plant PHL in curtailed mode is not a trivial task, especially for large PV plants spanning large geographical spaces. Both model-based and data-driven methods have been proposed for the PV plant PHL estimation [5]–[8]. However, all of them are highly dependent on accurate knowledge of plant-level and device-level parameters as well as the weather conditions. A simpler and potentially more accurate method has been developed by NREL that uses only a subset of inverters (control group) to achieve desired levels of curtailment and uses the other uncurtailed subset of inverters (reference group) to estimate the plant-level PHL in real-time. This method demonstrates superior performance compared to the ones found in the literature given its robustness, simplicity, and independence of PV modules, inverter types, array topology, solar irradiance variation, cloud movements, panel temperatures, and panel soiling, etc. Its real-world application has been shown in [3] under clear-sky days for a single 300 MW PV plant in California. Full detail of the NREL method can be found in [9].

In this work, NREL further enhances the methodology by:

1) Improving the estimation accuracy through machine learning (ML).
2) Accounting for look-ahead windows to reflect the computation and communication latencies occurring during the implement of the active power control.
3) Applying it for the regional reserve estimation.

II. METHODOLOGY

The NREL PV PHL estimation approach is enabled by splitting inverters in a PV plant into a control group and a reference group. While inverters in the control group are curtailed to track with the dispatch instructions, inverters in the reference group are reserved to operate at their maximum operating limits. Fig. 1 illustrates how the group splitting and PHL estimator fit into the PV plant active power control application, taking the automatic generation control (AGC) as the example.

As depicted in Fig. 1, the plant PHL estimator first collects inverter-level generation data from the reference group and

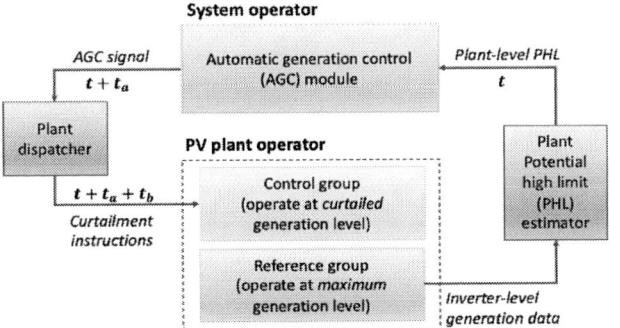

Fig. 1. Illustration of the automatic generation control enabled by the control-reference grouping and PHL estimation.

estimates the plant-level PHL to be sent to the system operator at time t. Then, the system operator calculates the optimal AGC signals for all participating generators, including the PV plant, based on their available headroom and response speed, which are communicated back to the plant dispatcher at time $t + t_a$. Finally, after receiving the AGC signal, the plant dispatcher determines the PV curtailment instructions at the inverter-level for the control group and executes the controls at time $t + t_a + t_b$.

Note that t_a and t_b reflect the computation and communication delays occurring in the AGC control loop. Since PV is a variable generation, the actual available power of the plant at time interval $t + t_a + t_b$ could differ from the estimated available power at t if no look-ahead window is considered, which will increase the estimation error. In what follows, we compare two mathematical models for estimating the PHL: 1) the scaling model, originally proposed in [9], that is incapable of considering the look-ahead window, and 2) the proposed ML model, introduced in this study, with configurable look-ahead windows.

A. Scaling Method

The scaling method simply estimates the plant-level PHL at time $t + t_a + t_b$ as a scaled value of the reference-group PV generation at time t, based on the ratio between numbers of inverters in the reference group versus in the PV plant, as given in Eq. (1):

$$\hat{P}_{PHL}^{t+t_a+t_b} = \frac{N}{N_{ref}} \sum_{k=1}^{N_{ref}} P_{ref}^{t,k}, \tag{1}$$

where N and N_{ref} indicate the numbers of inverters in the plant and the reference group. $P_{ref}^{t,k}$ represents the measured power output of reference inverter k at time t.

B. Machine Learning Method

Despite the fact that the scaling method has been successfully demonstrated and used in a number of projects, it has two drawbacks: First, same weights are assumed for inverters in the reference group when estimating the plant-level PHL, whereas varying weights may actually apply given different modules, capacities, and locations of the reference inverters. Second, the scaling method does not have any foresight, given that it

by default assumes an equal relationship between generations at two separate time steps. To address the above-mentioned gaps, a linear regression based PHL estimation model, $f(\cdot)$, is introduced. As stated in Eq. (2), given a linear model $f(\cdot)$ whose coefficients are trained based on historical data, the estimated plant-level PHL at time $t + t_a + t_b$, $\hat{P}_{PHL}^{t+t_a+t_b}$, can be computed as a weighted linear combination of reference inverter generations at time t, $\mathbf{P_{ref}^t}$, and two time indexes corresponding to the forecasting time t, TI^t, and execution time $t + t_a + t_b$, $TI^{t+t_a+t_b}$.

$$\hat{P}_{PHL}^{t+t_a+t_b} = f(\mathbf{P_{ref}^t}, TI^t, TI^{t+t_a+t_b}) \tag{2}$$

Note that the time index, TI, is innovatively introduced in our study to capture the variation of the clear-sky generation across time, which is unique to each plant given its geographical location and tracking technique being applied. The intuition behind is that the ratio between all-sky generation and clear-sky generation (caused by the cloud cover) should keep relatively constant in near term as the cloud condition won't change drastically for a matter of seconds. Such that, by taking the two time indexes into account, the model can capture the trend of generation growth/drop between two time steps. It is designed as a normalized upper envelope of PV generation profiles collected from recent historical days, as exampled in Fig. 2.

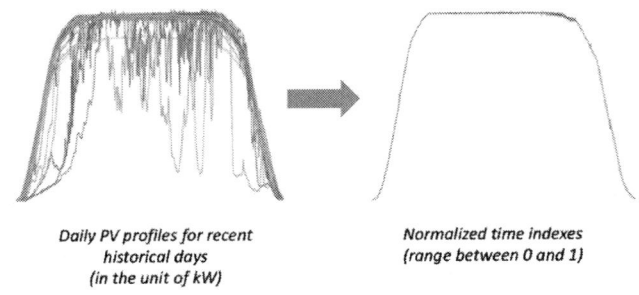

Fig. 2. Illustration of the time indexing.

C. Regional Estimation

Once the estimated plant-level PHLs for PV plants located in the same balancing authority are obtained, the regional PHL can be estimated by simply summing the plant-level values up, as given in Eq. (3), assuming M PV plants are in the region.

$$\hat{P}_{PHL,regional}^{t+t_a+t_b} = \sum_{m=1}^{M} \hat{P}_{PHL,m}^{t+t_a+t_b} \tag{3}$$

III. CASE STUDY

A. Dataset

A case study is performed applying a month-length high-bandwidth inverter-level dataset, collected by 4 utility-scale PV plants in California, to evaluate the performance of the proposed PHL estimator. All data were cleaned and interpolated to one-second resolution. In addition, CAISO provided plant-level production and curtailment set points data covering the same month at 15-min interval to help identify periods with curtailments. 25 out of the 31 days' data are picked out after a

TABLE I. REGIONAL MEAN ABSOLUTE ERRORS UNDER VARYING
LOOK-AHEAD WINDOWS FOR BOTH APPROACHES

MAE (MW)	ML-based approach	Scaling-based approach
0 AGC step	0.88	262.72
1 AGC step	0.90	262.00
2 AGC steps	0.96	262.60
3 AGC steps	1.05	263.78

quality check. And we further split the 25 days' data into the training and testing datasets based on a 7:3 ratio. One linear regression model is created for each PV plant for a particular length of look-ahead window.

B. Performance Evaluation

Fig. 3 depicts the histograms of percentage errors (with respect to the plant rated power) for four plants with 0-3 AGC steps (0-12s) look-ahead windows, obtained by the proposed ML-based method. It is noted that the percentage errors are symmetrically centered around zero. For over 96% of time, the estimation errors are below 1% of the rated power for all four plants. Moreover, trivial performance drops are observed as the look-ahead window grows.

Table. I compares the regional estimation performance between the scaling-based and the ML-based methods, using mean absolute errors (MAE) as the performance metric. It is shown that the ML-based approach provides a significant improvement in regional spinning reserve estimation compared to the scaling-based method at all three AGC intervals (MAE is reduced to essentially zero from 250 MW level).

IV. CONCLUSION

A ML-based model is proposed to further enhance the control-reference grouping based PHL estimation approach developed at NREL. It demonstrates superior performance compared with the original scaling-based model given real-world data collected by CAISO and PV plant owners.

REFERENCES

[1] S. Dahlke, M. Morjaria, V. Gevorgian, and B. Mather, "The economics of flexible solar for electricity markets in transition," Tech. Rep., May, Tech. Rep., 2020.

[2] V. Gevorgian and B. O'Neill, "Advanced grid-friendly controls demonstration project for utility-scale pv power plants," National Renewable Energy Lab.(NREL), Golden, CO (United States), Tech. Rep., 2016.

[3] C. Loutan, P. Klauer, S. Chowdhury, S. Hall, M. Morjaria, V. Chadliev, N. Milam, C. Milan, and V. Gevorgian, "Demonstration of essential reliability services by a 300-mw solar photovoltaic power plant," National Renewable Energy Lab.(NREL), Golden, CO (United States), Tech. Rep., 2017.

[4] W. Hobbs, D. Ault, V. Gevorgian, and G. Saraswat, "Accuracy of potential high limit estimation for solar plants in the southeast us," IEEE Photovoltaic Specialists Conference (PVSC), 2022.

[5] A. Hoke, E. Muljadi, and D. Maksimovic, "Real-time photovoltaic plant maximum power point estimation for use in grid frequency stabilization," in 2015 IEEE 16th workshop on Control and Modeling for Power Electronics (COMPEL). IEEE, 2015, pp. 1–7.

[6] E. I. Batzelis, G. E. Kampitsis, and S. A. Papathanassiou, "Power reserves control for pv systems with real-time mpp estimation via curve fitting," IEEE Transactions on Sustainable Energy, vol. 8, no. 3, pp. 1269–1280, 2017.

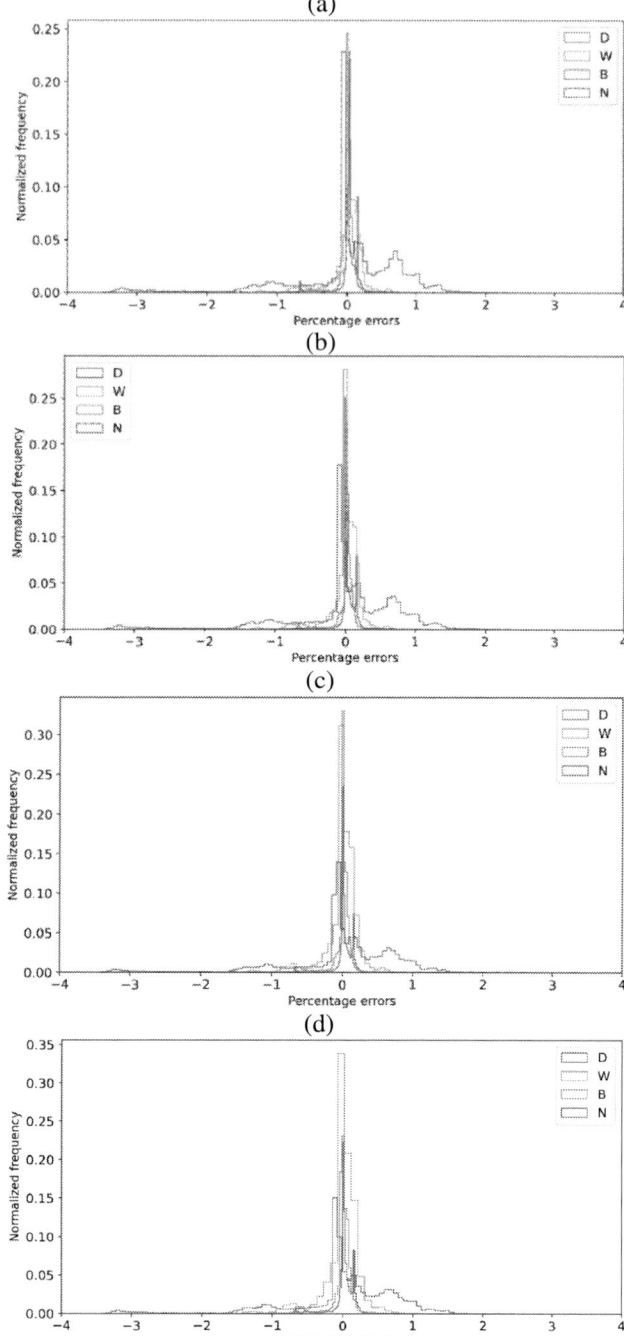

Fig. 3. Histograms of plant-level percentage errors for (a) 0, (b) 1, (c) 2, and (d) 3 AGC steps ahead (D, W, B, N represents IDs of the four plants)

[7] G. Kumar and A. K. Panchal, "Geometrical prediction of maximum power point for photovoltaics," Applied energy, vol. 119, pp. 237–245, 2014.

[8] Y. Liu, L. Chen, L. Chen, H. Xin, and D. Gan, "A newton quadratic interpolation based control strategy for photovoltaic system," in International Conference on Sustainable Power Generation and Supply (SUPERGEN 2012). IET, 2012, pp. 1–6.

[9] V. Gevorgian, "Highly accurate method for real-time active power reserve estimation for utility-scale pv power plants," National Renewable Energy Lab.(NREL), Golden, CO (United States), Tech. Rep., 2019.

Investigation of Sputtered P-Type Electrical Contacts for Thin Film Cadmium Telluride-Based Solar Cells

Blake Hill, Forrest Khulmann, Mayank Mate, and Amit H. Munshi

Colorado State University, Fort Collins, Colorado, 80523, United States of America

Abstract —The PV device structure has evolved significantly over the past few years and efficiencies have improved from ~11% to >20%. While the semiconductor layers and thin film growth processes at CSU are well established and documented, the p-type electrical contact has been fabricated by painting carbon (C) and nickel (Ni) conductive paint. This introduces significant variability due to many reasons, including human factors. The presented study aims to replace this traditional electrode deposition method with a sputtered method. The effect of ambient exposure as well as variation with different thicknesses and deposition rates are also being investigated.

I. INTRODUCTION

Cadmium telluride (CdTe) thin film solar photovoltaics (PV) technology has grown increasingly popular to meet global clean energy demands. Its advantages include high manufacturing throughput, low material requirement and higher material utilization, ease of recyclability, operational reliability, low degradation rate, and one of the lowest levelized cost of energy (LCOE) [1]. While the technology is young in relation to the c-Si PV, it has already acquired about 5% of global PV market owing to it numerous advantages [2]. It continues to grow in populatiry within the energy industry. Improving the performance and reliability of this technology using scalable methods would allow greater adoption of scientific advances leading to expanding the penetration into the commercial energy market.

A PV device includes a semiconductor material in which absorption of light creates charged carriers (electron-hole pairs). The charge carriers travel through the circuit to generate electric current. In practice nearly all photovoltaic energy conversion uses semiconductor materials in the form of a p-n junction [2]. The primary aim of this research is to improve upon the p-type electrode in thin film CdTe PV device. The p-type junction has significant impact on the fill factor and is one of the main focuses throughout this study. The fill factor (FF) is a parameter that determines the maximum power from a solar cell in conjunction with open circuit voltage (Voc) and short circuit current (Jsc) [2].

$$FF = P_{MP}/(V_{OC} \times J_{SC})$$

A PV device is a semiconductor diode and FF is measured as the "squareness" of the JV diode curve. Higher value of FF indicates a higher maximum power point (P_{MP}) which is a point on the JV curve where the product of J_{SC} and V_{OC} is the greatest [2]. The goal of this project is to maximize FF more reliably and repeatably by minimizing the series and shunt resistance losses through optimization of sputtered metal contacts. Preliminary Research has been done at the NGPV to replace carbon and nickel conductive paint with sputtered nickel.

Metals tend to form an electrical barrier to conduction of holes in the p-type cadmium telluride (CdTe). Such a barrier, called the Schottky barrier, that can affect the Voc and FF [3]. Materials such as nickel, gold and platinum are highly suitable, but gold and platinum are cost prohibitive. [3]. To be able to utilize other metals there are a few ways to avoid the formation of a Schottky barrier. At the NGPV Munshi et al and Song et al, shows that depositing 30-50nm of elemental tellurium forms an ohmic contact [4,5]. This significantly mitigates the Schottky barrier but Te being a metalloid with more semiconductor-like properties and therefore is not optimum for use as an electrical conductor. Therefore, a metallic conducting film is required to form an electrode. Furthermore, this study focuses on investigating single and multilayer metal electrodes to find an optimal combination.

II. EXPERIMENTAL

A. Device Fabrication

Devices were fabricated in a superstrate configuration on NSG Tec10 soda lime glass that had fluorine-doped tin oxide layer deposited by the manufacturer as the transparent conducting oxide (TCO). The superstrates were cleaned and prepared for deposition of a buffer layer. A 100nm thick layer of $Mg_xZn_{1-x}O$ (MZO) high resistivity transparent buffer (HRT) was sputter deposited via RF magnetron with no substrate heating [6]. After the deposition of MZO buffer the substrates were introduced in a single vacuum chamber with multiple sublimation sources [7]. The superstrates were preheated to ~540°C followed by an in-situ growth of an ~600nm $CdSe_xTe_{1-x}$ and a ~3.5μm CdTe thin film without exposure to ambient between these layers. Following this step an in-situ $CdCl_2$ passivation treatment was performed without breaking vacuum. Thereafter the substrates were removed from the vacuum chamber and excess $CdCl_2$ was rinsed using deionized water. Then substrate was treated with CuCl and ~30nm of evaporated Te was deposed on top of the absorber layer. This process formed the baseline device structure and different electrodes were investigated to understand their

978-1-6654-6060-6/23 $31.00 © 2023 IEEE

effects on device performance. Figure 1 shows the schematic device configuration used in the NGPV (not to scale). Each superstrate after deposition of all the active layers were then delineated into 25 small area devices with an area of ~0.65cm² each.

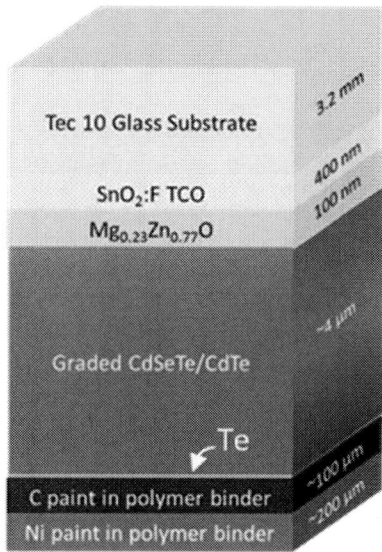

Fig.1. Superstrate device structure for baseline device. (Not to Scale)

B. Characterization

Current-voltage (J-V) measurements were performed on all devices, to analyze their performance. Surface and cross-sectional scanning electron microscope (SEM) and energy dispersive x-ray spectroscopy (EDS) are being performed to investigate crystallographic properties, film morphology and elemental distribution within the film stacks. Transmission electron microscope (TEM) is being performed to characterize the interfacial features and film morphology.

III. PRELIMINARY WORK

Preliminary work was focused on exploring if sputtered metals would be able to replace carbon and nickel conductive paint. Initially single layer sputtered electrode was investigated. This is due to current limitations with fabrication equipment at NGPV. The DC sputter system is being upgraded which will enable for multiple metal layers to be sputtered without breaking vacuum. For these experiments, Ni has been the focus because the metal does not oxidize spontaneously when exposed to air. Aluminum has been studied as well but because of spontaneous oxidation it forms an alumina that is highly resistive and unsuitable for a PV device. However, Al is an excellent electrical conductor and the limitation can be overcome by deposition of a thin layer of Chromium (Cr) depostited in-situ to ensure Al does not oxidize since Cr does not readily tarnish.

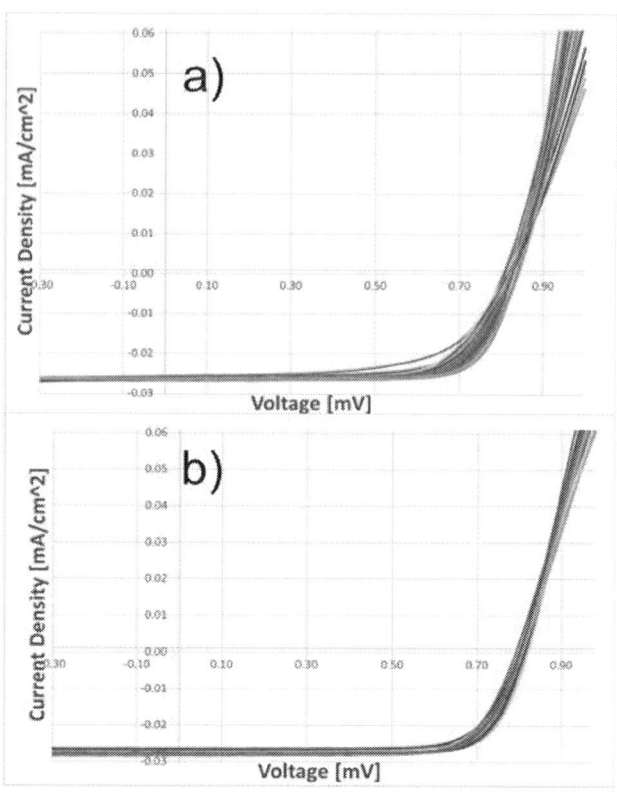

Fig.2. a) Representative JV cures of C+Ni paint b) Representative JV curves of Ni sputtered

TABLE 1
BEST DEVICE DATA FROM CARBON-NICKEL PAINT VS. NICKEL SPUTTERED

Best Devices from JV Curves	J_{SC} [mA/cm²]	V_{OC} [mV]	Fill Factor [%]	Efficiency [%]
Paint Contact	27.4	838	77.3	17.75
Ni Sputter Contact	29.1	843	76	18.63

In Figure 2, J-V measurements for 50 devices from 2 substrates are shown for two electrodes investigated. The variability throughout the superstrate with painted C/Ni electrode is noticeable. Furthermore, this JV data shows an increase in series and shunt resistance in the substrate which needs to be avoided. In Figure 2, the J-V curve of the Ni sputtered electrode shows better uniformity than the C/Ni substrate. It also shows preliminary data that the sputtered Ni minimize the series and shunt resistance losses. In Table 1, sputtered Ni showed a noticeable increase in efficiency in best performing device. The devices with sputtered Ni electrode gave the highest efficiency of 18.63% within preliminary experiments. Further investigation to optimize the Ni electrode, along with other bilayer electrodes such as Al/Cr are being investigated and will be reported.

Based on the data described previously an experiment was planned and executed to compare the statistical spread between sputtered Ni vs the painted C/Ni contacts. Four substrates were fabricated with carbon and nickel paint and four with sputtered Ni.

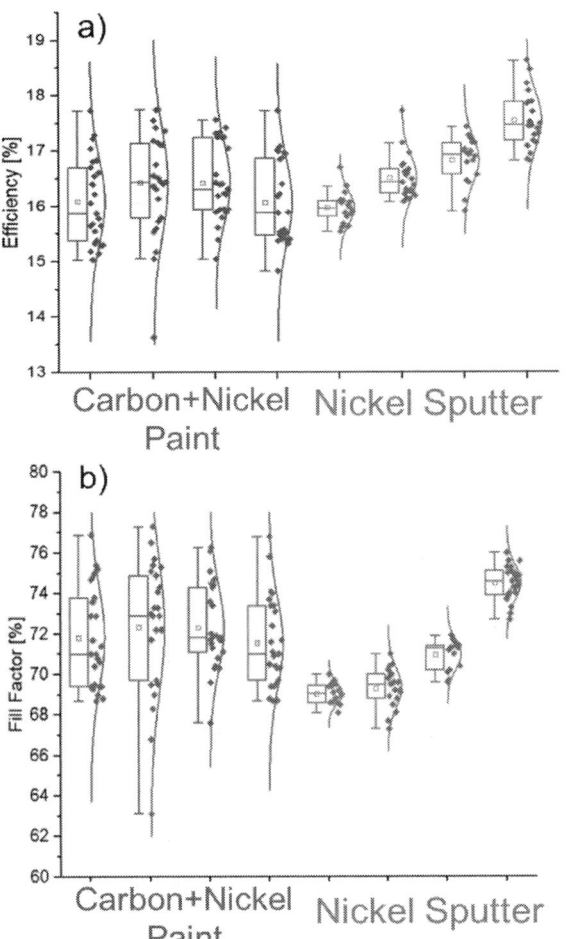

Fig.3. a) Box plot of efficiency for carbon and nickel paint vs. nickel sputter. B) Box plot of Fill Factor for carbon and nickel paint vs. nickel sputter.

In Figure 3, the efficiency between the C/Ni paint was consistent throughout each of the four substrates but the variance throughout superstrate was significant. It can also be seen in Figure 3 that the efficiencies for nickel sputter show lower of variance throughout each of the substrates with desirable uniformity. The increasing trend in efficiency between the four nickel sputtered is due to the changing layer thickness of $CdSe_xTe_{1-x}$. Once properly calibrated the nickel sputtered substrates show high efficiencies. Figure 3 shows how the variance in the substrates decreases using sputtered Ni vs. the C/Ni painted electrodes. Although the fill factor

showed higher values for carbon and nickel paint the variance was too high. Therefore, this experiment has provided the foundation needed to continue to pursue this study to optimize the electrode as well as understand the effect of varying thickness and deposition rates on device performance.

IV. CONCLUSION

Preliminary work shows that the sputtered Ni gives better efficiency and uniformity. Optimization of sputtered metal electrode should lead to better efficiency and uniformity as well as process repeatability. Based on these preliminary results, other metals will be studied which include aluminum, molybdenum, chromium, etc. Further characterization will be performed to understand the mechanism that leads to improvement in device performance.

ACKNOWLEDGEMENTS

This abstract was developed in part based upon funding from the Alliance for Sustainable Energy, LLC, Managinf and Operating contractor for the National Renewable Energy Laboratory for the U.S. Department of Energy. Devices fabricated for this study used CdTe, CdSeTe and $CdCl_2$ provided by 5N Plus Inc. CSU authors also acknowledge the National Science Foundation Industry/University Cooperative Research Center (IUCRC) for Solar Powered Future 2050 and the Industrial Advisory Board support under award number 2052735

REFERENCES

[1] D. pal Singh, "Levelized cost of energy and of storage 2020," Lazard.com.[Online].Available: https://www.lazard.com/perspective/levelized-cost-of-energy-levelized-cost-of-storage-and-levelized-cost-of-hydrogen-2020/. [Accessed: 22-Jan-2023].

[2] PVEducation.[Online]. Available: https://www.pveducation.org/. [Accessed: 03-Oct-2022].

[3] R. S. Hall, D. Lamb, and S. J. Irvine, "Back contacts materials used in thin film CdTe solar cells—a review," Energy Science & Engineering, vol. 9, no. 5, pp. 606–632, 2021

[4] Munshi, Amit H., et al. "Polycrystalline CdTe photovoltaics with efficiency over 18% through improved absorber passivation and current collection." Solar Energy Materials and Solar Cells 176 (2018): 9-18.

[5] T. Song, A. Moore and J. R. Sites, "Te Layer to Reduce the CdTe Back-Contact Barrier," in IEEE Journal of Photovoltaics, vol. 8, no. 1, pp. 293-298, Jan. 2018, doi: 10.1109/JPHOTOV.2017.2768965.

[6] J. M. Kephart, "Opimization of the Front Contact to Minimize Short-Circuit Current Losses in CdTe Thin-Film Solar Cells," dissertation, 2015.

[7] D. E. Swanson, J. M. Kephart, P. S. Kobyakov, K. Walters, K. C. Cameron, K. L. Barth, W. S. Sampath, J. Drayton, and J. R. Sites, "Single vacuum chamber with multiple close space sublimation sources to fabricate CdTe solar cells," Journal of Vacuum Science & Technology A: Vacuum, Surfaces, and Films, vol. 34, no. 2, p. 021202, 2016.

Perovskite Bafacial Modules-Efficiency, Stability and Upscaling

Hangyu Gu, Jinsong Huang

University of North Carolina at Chapel Hill, Chapel Hill, NC, United States

Bifacial solar modules can produce 5% to over 30% more energy than monofacial ones, which can further reduce the levelized cost of electricity from photovoltaic devices. The market share of bifacial silicon modules is rising. To compete with silicon photovoltaics, perovskite solar cells also need to go bifacial structure for an increased energy yield. In addition, the bifacial perovskite modules are essentially semitransparent modules which are needed to realize efficient perovskite-silicon 4-terminal tandem modules. However, the efficiency of bifacial perovskite modules was far below that of monofacial ones. Here I will report bifacial perovskite minimodules with a high certified efficiency by addressing several unique challenges in bifacial module design and fabrication. The front efficiency without albedo light already reached that of best opaque monofacial minimodules, which is the highest efficiency a bifacial module can reach, while these bifacial modules gain additional power from albedo light. The additional of hydrophobic additive in hole transport layer surprisingly protects the perovskite films from moisture damage during atomic layer deposition. Integrating silica nanoparticles with proper size and spacing in perovskite films recovers the absorption loss induced by the absence of reflective metal electrodes while maintains the charge collection properties of perovskites. The small area single junction bifacial cells have an equivalent stabilized efficiency of 26.4% at an albedo of 0.2, which is already higher than any reported single junction perovskite solar cells. The bifacial solar minimodules show front and rear aperture efficiencies of 19.2% and 14.1%, respectively, certified by National Renewable Energy Laboratory, which yield an equivalent aperture efficiency of 22% at albedo of 0.2. We also show that the bifacial minimodules are extremely stable, with 97% of its initial efficiency retained after light soaking under one simulated sun for over 6000 hours at 60 ± 5 °C, representing the most stable perovskite module reported so far. We believe these efficiency and stability of the perovskite modules represent a significant advance, while most other reported work still focus on small area devices.

Modular, Array-Mounted Photovoltaic Inspection Robot

Michael Y. Vazquez Nieves, Alanis M. Colón González, and Jennifer L. Braid*

University of Puerto Rico – Mayagüez Campus, Mayagüez, Puerto Rico, 00681, United States

*Sandia National Laboratories, Albuquerque, New Mexico, 87123, United States

Abstract — Due to the exponential deployment of new photovoltaic systems in recent years, there is a pressing need for efficient array inspection methods. While drone-based methods can detect hotspots and large outages in an array, diagnostic and prognostic measurements of PV modules are more suited for stable, proximal use. To enable these measurements, students were tasked with the design and creation of a robotic platform which would autonomously traverse PV arrays, fit to various array sizes and configurations, and be able to deploy various characterization tools including imaging and spectroscopic techniques. This document details the development and demonstration of the robot.

I. INTRODUCTION

With the continued acceleration of photovoltaic (PV) system deployment around the globe, efficient and accurate inspection of fielded modules is paramount to ensuring these systems are safe and continue to operate at expected performance. PV modules are prone to various degradation modes based on regular fielding and climatic conditions, and can even be damaged during transportation, installation, or extreme weather. As a result, there are various points throughout the life of a PV array when it would be advantageous to conduct detailed inspection and valuation. These may include, but are not limited to, commissioning/installation, at time of sale, after a major weather event, or as part of regular operations and maintenance (O&M) schedules.

Understanding the health status of an array has many advantages for the PV system owner/operator. First, knowing the types and extend of module damage and degradation present in the field helps to more accurately predict the amount of power the system should be generating now, as well as the expected power loss in future years. Secondly, the calculated revenue loss based on expected power loss can inform financial decisions such as the cost-benefit of replacing modules. Third, if a system is underperforming either after commissioning or after a weather event, understanding the nature of underperformance allows the owner to file a claim (guarantee, legal, or insurance as appropriate) and have the issue financially resolved by the responsible party.

In the past decade, drone-based PV inspection methods have become popular for identifying degradation and outages in utility-scale PV power plants. Typically these methods use infrared cameras to detect hotspots in modules, or entire strings/arrays that are underperforming or disconnected. While this is a highly efficient inspection method for current operating state of the PV array, it has a few limitations. First, the imaging types and resolutions for drones only useful for identifying severe and immediate problems in the field. They do not indicate modules that will become problematic in the future, nor do they reveal the underlying cause of the outage. A second limitation is that there is little opportunity for deployment of additional measurement types by drone. There have been efforts to deploy electroluminescence and ultraviolet fluorescence imaging techniques by drone, but coordination of the various requirements of these methods may be too complicated or expensive to be practical. Deployment of other inspection methods such as surface-contacting and spectroscopic methods are completely impractical for drone deployment.

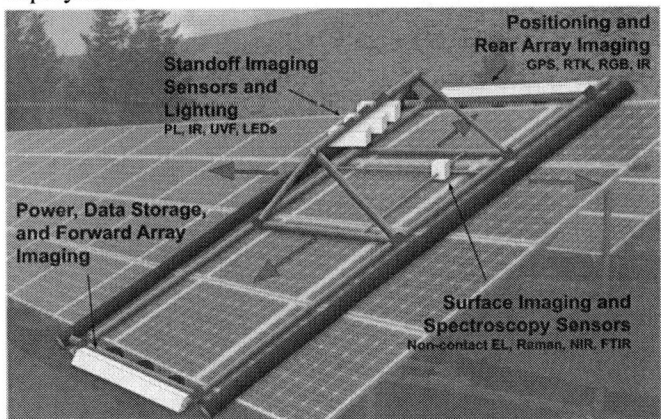

Fig. 1.   PV robot concept showing a movable platform, with a carriage to deploy characterization cameras and sensors.

In order to address the above limitations, an array-mounted robotic solution (Fig. 1) was conceptualized to automate detailed characterization of large PV systems. In addition to autonomous movement and measurements, the robot was planned to be self-locating, adaptable to various array sizes, and modular to accommodate a range of characterization tools to be selected as appropriate for the array and conditions.

II. MECHANICAL DESIGN

Based on the concept previously shown it can be seen that this robot will be placed on top of a photovoltaic array where the entire robot will move horizontally across the array and the area where the cameras and sensors are placed will move vertically in the robot's structure. After the functionality of the robot was established some contraints were identified such as the need to safely displace the weight of the robot in the surface of the PV array so that it wouldn't cause any damage to the modules. Another constraint identified was the adjustability in length of said robot since not all arrays are built to a standard measurement, even within our test site. Additionally, the placement of the wheels for the horizontal movement also had to be adjustable because the thickness of PV panels vary depending the model and manufacturer. Considering these constraints, the design process started using 80/20 aluminum pieces and some other pieces to integrate the moving mechanisms. The CAD models of the pieces provided online were used to create an assembly of over 75 pieces. After the general structure was designed, the parts for the moving mechanisms such as the chains and gears were also incorporated. The final design is shown in Fig. 2.

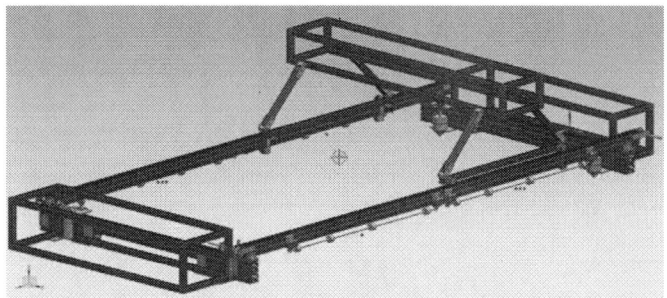

Fig. 2.   Final CAD design of the robot.

III. Electrical Design

To develop the first prototype of this robot the following electrical components were selected: batteries, voltage regulators, power distribution block, microcontrollers, motor controller, port expansion, breadboard, ultrasonic sensor, DC motors, encoders, fans, toggle switch, push buttons, LED lights, infrared camera, and HD camera. For the battery calculation, the robot's expected operation time was used together with the power demanded by all the electrical components. Due to lack of compatibility for a peak current of 4 DC motors in stall, a voltage regulator of 24 V at 40 A was chosen. Therefore, two 12 V @ 20 Ah/each batteries were selected, which were connected in series to boost the voltage to 24 V. This allowed compatibility with the commercially available stepdown voltage regulator, and the power demand. On the other hand, for the selection of the DC motors, the maximum force that each DC motor could drive was calculated based on an estimate of the load. Using the calculated force, torque was calculated according to the allowable distance of the load. This series of calculations resulted in 12 V @ 20 A DC motors with a torque of 57 Nm and 21 Nm for the horizontal and vertical movement respectively. The DC motors contained an encoder attached that counted the shaft revolutions. The microcontroller selected was the Raspberry Pi 4 which works with Thonny IDLE, which is an interface that uses Python as a programming language. This has as a limitation, the maximum input of 40 pins or GPIO. However, an expansion was used to be able to connect more

electrical components. Multiple motor controllers are connected to the microcontroller's pin expansion, to achieve power supply and control of the DC motors (circuit example in Fig. 3). Safety switches were used to turn the entire system on and off in case of emergency, and buttons with programmed functions to start, stop or restart the programming and a series of LEDs to indicate the status of the robot during autonomous use. Fans controlled by voltage regulators were installed to prevent the DC motors from overheating during operation. Finally, HD and infrared cameras were installed to collect images of the photovoltaic cells for future study.

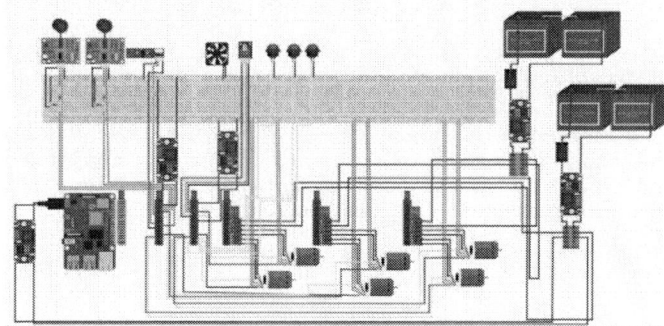

Fig. 3.   Illustration of the electrical circuit used for one of the movement actuators.

IV. Software Approach

The microcontroller used was the Raspberry Pi Model 4. It contains a Thonny IDLE interface which uses Python as a programming language. Multiple libraries and open source python scripts were explored for the configuration of each of the electrical components. Multiple tests were performed on each device to verify proper operation as well as amending the code of the devices that failed such tests. Once each electrical component was functioning individually, a flowchart was developed to define the steps of operation (Fig. 4). The operational schema consists of turning on the robot, detecting the vertical limits of the scanning zone using ultrasonic sensors, returning to home, capturing and storing HD and infrared photos, and moving vertically in predefined steps until reaching the upper limit of the scanning zone. Once the upper position limit is reached, the carriage returns to home (lowest position), and the robot begins to move horizontally by a predetermined distance, defined using the values read by the encoders of each DC motor. When the robot detects the boundary of the PV array by ultrasonic sensor, it halts operation.

V. Results and Future Work

As a work in progress what was accomplished in this first stage was the construction of the robot and its first series of functionality test that involved the electromechanical area with the software development. The first prototype of the PV Array Inspection Robot can be seen in Fig. 5.

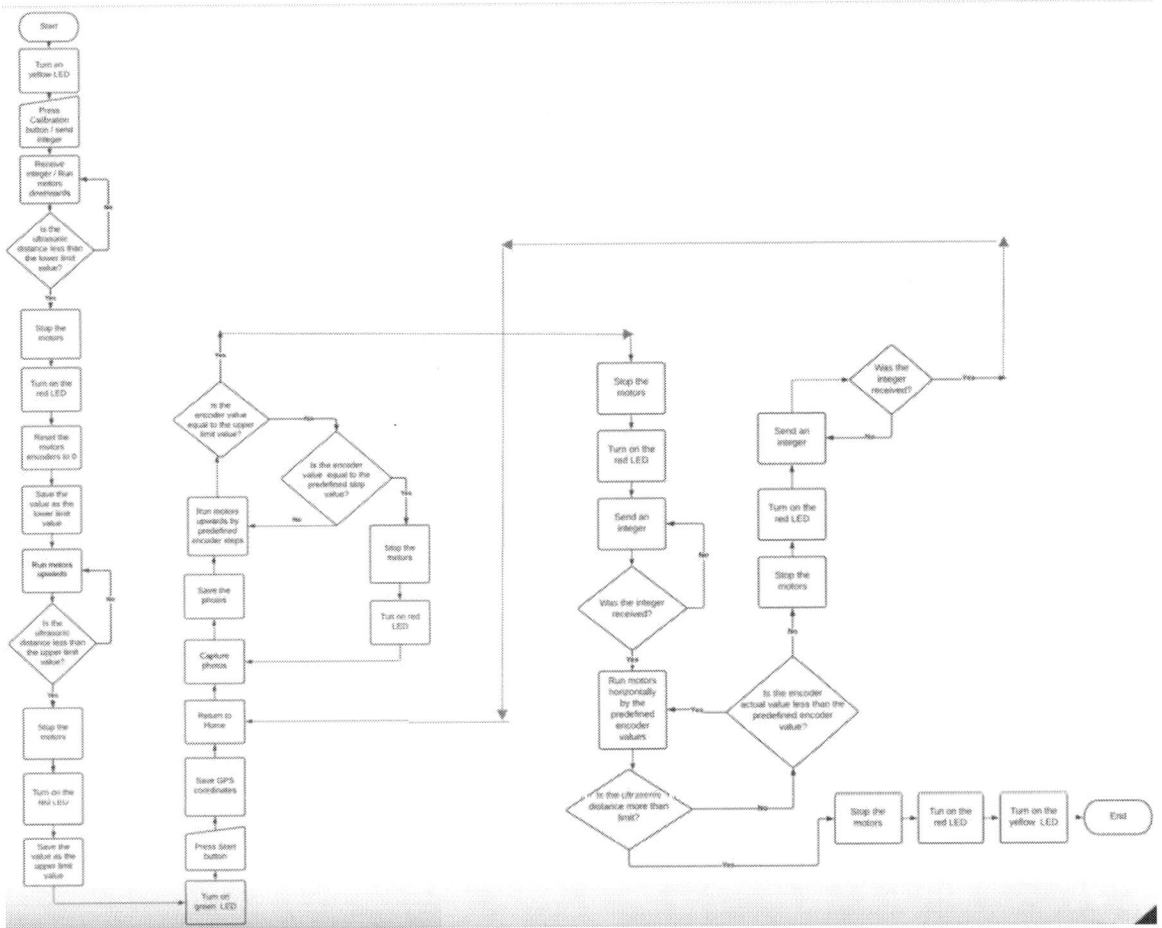

Fig. 4. Flowchart of the software logic of the robot including self-location, vertical and horizontal movement, and measurement steps.

To achieve a more complete robot in the next stage of this project some additional implementations should be made. First of some mechanical modification would be needed to allow cleaning solar modules or melting snow prior scanning. A very useful development that would automate the data analysis would be the application of computer vision to the images gathered for recognition of solar module degradation. Additionally, advanced imaging technologies such as Photoluminescence (PL) Imaging, Ultraviolet Fluorescence (UVF) Imaging, Non-contact Electroluminescent (EL) Imaging, and spectroscopic techniques including Raman and Fourier-transform infrared spectroscopy (FTIR) could be implemented to collect more specialized data. The addition of real-time data transfer to remote storage would also enable a more effective workflow and status monitoring of the robot.

VII. ACKNOWLEDGMENTS

MYVN and AMCG contributed equally to this work. Sandia National Laboratories is a multimission laboratory managed and operated by National Technology and Engineering Solutions of Sandia, LLC, a wholly owned subsidiary of

Fig. 5. Image of the first prototype of the PV Array Inspection Robot placed on a PV array.

Honeywell International Inc., for the U.S. Department of Energy's National Nuclear Security Administration under contract DE-NA0003525. William Snyder, Kevin Santistevan, and Charles Robinson of Sandia are acknowledged for their valuable contributions.

ESSPI as a Fast Tool for Load Prioritization on Microgrids Design

Luis Colomba-Colon, Natanael Batista-Alvarez, Guillermo Lopez-Cardalda and Eduardo Ortiz-Rivera

Department of Electrical and Computer Engineering, University of Puerto Rico, Mayaguez, Puerto Rico

Abstract—**The Energy Storage System Priority Index (ESSPI) was originally presented in the past as a methodology to prioritize load. In this paper, the ESSPI is going to be used as a fast tool for load prioritization in microgrid design. The tool consists of the user inputting the load information based on the classification of the load. The inputs needed are the maximum critical recovery time and the energy consumption of the loads. Loads should be classified into different categories: critical load, essential load, discretionary load, non-essential load, and expendable load. The tool outputs would be the loads arranged based on their priority, the size, and the cost of the microgrid system depending on the user's desired investment for short-, mid-, and long-term investments.**

Keywords—energy storage, ESSPI, load classification

I. INTRODUCTION

This paper uses the Energy Priority Index presented in [1] to create a fast tool for load prioritization on microgrid design. The idea of the tool is to identify and arrange the order of priority of the loads using the index (ESSPI). By doing this now we can also consider those loads for a microgrid design based on the necessity of those loads. The ranking system outlined in this section separates loads into five categories ranging from least important to most important (note: examples presented here are based on the reality of Puerto Rico, it is suggested that the loads be ranked according to the experience for each region/country): Expendable (EXL), Non-Essential (NEL), Discretionary (DL), Essential (EL) and Critical (CL)

The proposed load ranking system is related to the projected time recovery horizon of each load. The Reliability, Resiliency & Recovery Energy Systems (RE3) concept presented in this paper is treated as three kinds/types of investment:

- Short-Term Investment – guarantees a reliable system that includes Critical Loads (CL) and Essential Loads (EL).

- Mid-Term Investment – guarantees a resilient system that includes Discretionary Loads (EL) and Non-Essential Loads (DL).

- Long-Term Investment – guarantees a fully recovered system that includes Expendable Loads (EXL).

The RE3 Segmentation Curve presented in Figure 1 is the expected energy restoration time curve for all loads after a blackout. The idea of the RE3 concept is to minimize the impact of a blackout to society's wellbeing and economy as mentioned before.

For more information on black start and BESS integration guidelines see "System Restoration from Blackstart Resources" and "Modeling, and Simulations of BPS Connected Battery Energy Storage Systems and Hybrid Power Plants" from the North American Electric Reliability Corporation (NERC), and Order No. 841-845 of the Federal Energy Regulatory Commission (FERC) [2], [3], [4],[5]. per and style the text. All margins, column widths, line spaces, and text fonts are prescribed; please do not alter them. You may note peculiarities. For example, the head margin in this template measures proportionally more than is customary. This measurement and others are deliberate, using specifications that anticipate your paper as one part of the entire proceedings, and not as an independent document. Please do not revise any of the current designations.

II. ESSPI EQUATIONS

The equation used for the App presented in this paper comes from the following [1] and is shown below. *Note in this section is not our intention to define or explain each of the meanings of all the variables or how to use the equations. If you desire to know more information read [1].*

1. $t_{(n)}$ – Maximum Critical Recovery Time (*day*).
2. $T_{(n)}$ – Complete Recovery Time (*day*)
3. $CLE_{(n)}$– Critical Load Energy: Total energy needed from loads classified as critical in section 1. Units are in *kilowatt-hours (kWh)*.
4. $ELE_{(n)}$– Essential Load Energy: Total energy needed from loads classified as essential in section 1. Units are in *kilowatt-hours (kWh)*.
5. $DLE_{(n)}$– Discretionary Load Energy: Total energy needed from loads classified as discretionary in section 1. Units are in *kilowatt-hours (kWh)*.
6. $NELE_{(n)}$ – Non-Essential Load Energy: Total energy needed from loads classified as non-essential in section 1. Units are in *kilowatt-hours (kWh)*.
7. $EXLE_{(n)}$– Expendable Load Energy: Total energy needed from loads classified as expendable in section 1. Units are in *kilowatt-hours (kWh)*.
8. VLE – Vital Load Energy: Units are in *kilowatt-hours (kWh)*.

$$VLE = \sum_{n=1}^{i} CLE_{(n)} + \sum_{n=1}^{j} ELE_{(n)} \qquad (1)$$

9. *SLE–* Supplementary Load Energy: Units are in *kilowatt-hours (kWh)*.

$$SLE = \sum_{n=1}^{m} DLE_{(n)} + \sum_{n=1}^{k} NELE_{(n)} \quad (2)$$

10. $STESSPI_{(n)}$ – Short-Term Energy Storage Selection Prioritization Index:

$$STESSPI_{(n)} = \frac{CLE_{(n)}}{VLE}\left(\frac{T_{(n)}}{t_{(n)}}\right) \quad (4)$$

$$STESSPI_{(n)} = \frac{ELE_{(n)}}{VLE}\left(\frac{T_{(n)}}{t_{(n)}}\right) \quad (5)$$

11. $MTESSPI_{(n)}$ – Mid-Term Energy Storage Selection Prioritization Index:

$$MTESSPI_{(n)} = \frac{DLE_{(n)}}{VLE + SLE}\left(\frac{T_{(n)}}{t_{(n)}}\right) \quad (6)$$

$$MTESSPI_{(n)} = \frac{NELE_{(n)}}{VLE + SLE}\left(\frac{T_{(n)}}{t_{(n)}}\right) \quad (7)$$

III. ESSPI Test Case Scenario Culebra, PR

The following ESSPI test case scenario was used [1] and it is going to compare the results obtained from the application in the next section. The data was obtained from NREL report (Contract No. DE-AC36-08GO28308) [6]. The report provides information related to critical and essential loads in this municipality. From the NREL report the following buildings were considered as critical (Table 1) and essential (Table 2). These buildings are the Health Clinic, Wastewater Treatment Plant (WWTP), Municipal Building, Police Station, and Fire Station. For the demonstration of the case study the municipal building will be considered as discretionary (Table 2). The blue color indicates the buildings considered as CL, the green color indicates the buildings considered as EL and the yellow color indicates the buildings considered as DL. The t for the hospital is based on research on how time a hospital can withstand without energy [7]. The other scenarios are estimations based on community feedback and other scenarios are modified for proof of concept of ESSPI [8]. T is presented as a hypothetical case where the service is down for any reason just to prove with different cases for the ESSPI analysis. Table 3 and Table 4 present the inputs and results for the ESSPI analysis.

TABLE 1: BUILDINGS THAT ARE CONSIDERED AS CL.

Infrastructure Type (Culebra, PR)	Health Clinic	WWTP
Annual Energy Use (KWh)	689.622	680

TABLE 2: BUILDINGS CONSIDERED EL AND DL.

Infrastructure Type (Culebra, PR)	Police Station	Fire Station	Municipal Building
Annual Energy Use (KWh)	112.564	74.838	527.493

TABLE 3: ESSPI ANALYSIS FOR SHORT-TERM.

Infrastructure Type (Culebra, PR)	Health Clinic	WWTP	Police Station	Fire Station
t (days)	3	0.25	1	1
T (days)	2	2	1	3
STESSPI	0.2953	3.4938	0.0723	0.1442

TABLE 4: ESSPI ANALYSIS FOR MID-TERM

Infrastructure Type (Culebra, PR)	Municipal Building
t (days)	1
T (days)	2
MTESSPI	0.5265

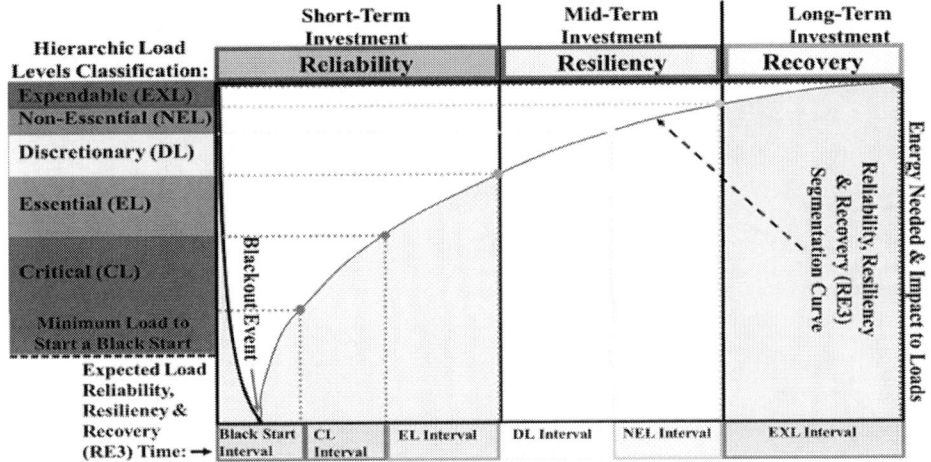

Fig. 1: Proposed Load Reliability, Resiliency & Recovery Load (RE3) Curve. This figure illustrates how should be the RE3 after a blackout.

978-1-6654-6060-6/23 $31.00 © 2023 IEEE

IV. ESSPI APP

The program was made with the Java programming language because Java already has its own methods to make a simple and efficient GUI. The program's main objective is to calculate the Short-Term and Mid-Term Energy Storage Selection Prioritization Index. The user can input the ESSPI variables: the maximum critical recovery time, the complete recovery time, the load and that the user specifies the type of energy load. We decided that applying an Object-Oriented approach would be the most efficient design for the program. The program has a public class "Load" that will oversee creating the Load Energy objects with the user inputs. It will also have methods to calculate the individual STESSPI and MTESSPI of each energy load. and the class implements a comparator to be able to compare each load with each other and determine which one is greater. There is also a public class "Calculator" that initializes all the variables needed to update the GUI. In a frame we add buttons, labels, and text fields to make the interface intuitive for the user to enter data. Following equations 1 and 2, there are two methods that calculate the SLE and VLE based on user input. With these methods, the STESSPI and MTESSPI of the energy loads are calculated following the equations 4, 5, 6, 7. Depending on whether it is short or midterm, it is placed in an Array List where it is organized from highest to lowest. To receive data from the user the program uses text fields and buttons. The buttons are also used to display the output and to change the power load class. By pressing the "results" button, we can see the STESSPI and MTESSPI results of the sorted energy loads in order of priority. And then the user can press enter to start again. For cost estimation was used [8] and [9].

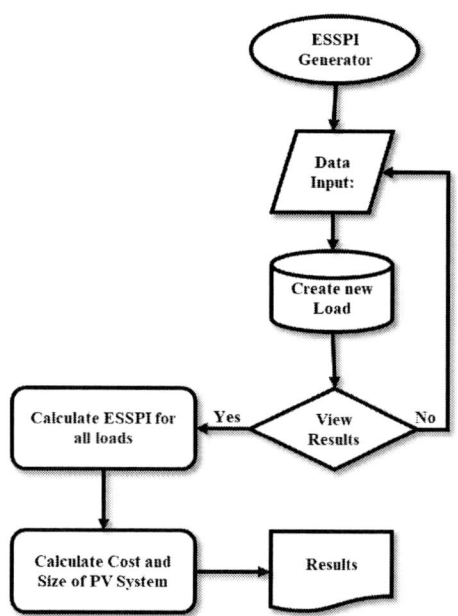

Fig 2. Flow Chart for the app functions. We can see how the user input is used to create loads and then display the ESSPI calculations.

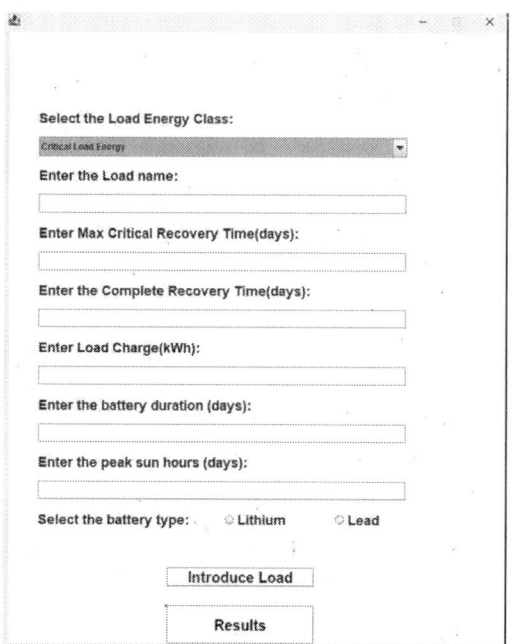

Fig. 3. The interface for the Calculator, here the user creates and saves the load energies.

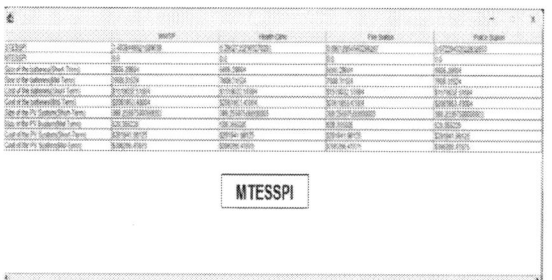

Fig. 4. Results, showing the STESSPI and MTESSPI for all the loads from the Culebra Example. The Results are ordered from largest to smallest.

V. USING THE TEMPLATE

In this Section is presented the different results obtained with the ESSPI application. In Figure 5 are the obtained values of ESSPI at different $T_{(n)}$. In Table 5 are presented the different results obtained for Short- and Mid-term Investment of the: *Size of the batteries, Cost of the batteries , Size of the PV System and Cost of the PV System.* The graph presented in Figures 6 and 7 are the values obtained from Table 5.

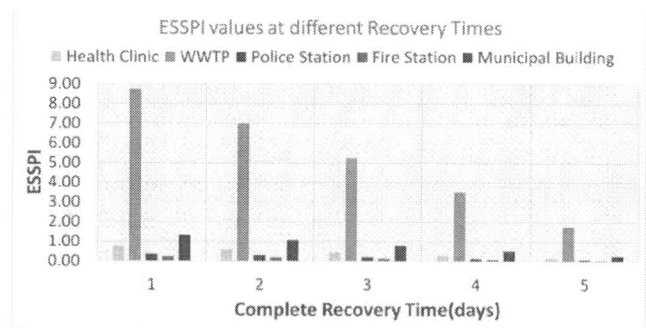

Fig. 5. ESSPI Values at different Recovery Times

978-1-6654-6060-6/23 $31.00 © 2023 IEEE

TABLE 5: SIZE OF BATTERIES AND PV SYSTEM SIZE WITH COST

Investment	Size of the batteries(kW)	Cost of the batteries	Size of the PV System(kW)	Cost of the PV System
Short-Term	5605.28	$1,519,032.50	389.25	$291,940
Mid-Term	7608.32	$2,061,853.40	528.4	$396,300

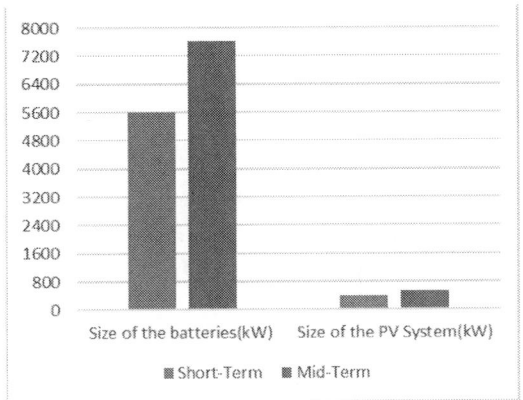

Fig. 6. Comparison graph of system sizes by Term.

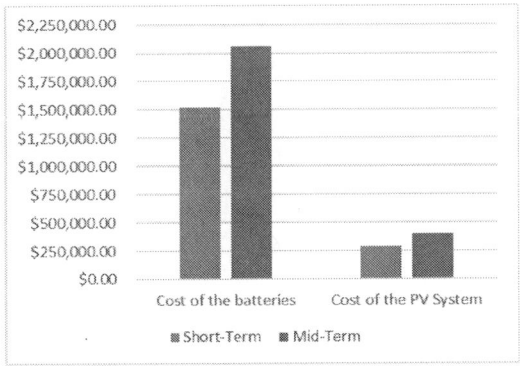

Fig. 7. Comparison of system cost by Term.

VI. CONCLUSION

The idea of the tool is to facilitate the identification and prioritization of loads for the user for a microgrid. The user can also obtain other useful information such as the size of the batteries needed to maintain the system and the estimated cost. As well as the size of the PV system and its cost. Then based on that information, the next objective of the tool is to suggest to the user, in a fast way, the size of the system needed to satisfy the energy and power requirements of those loads. Also, sizes and systems cost may vary depending on if the user wants the analysis for short-, mid-, and long-term investment.

ACKNOWLEDGMENT

This work was sponsored in part by the Consortium for Hybrid Resilient Energy Systems (CHRES) under grant number DE-NA0003982 from the National Nuclear Security Administration part of the U.S. Department of Energy. Also, is sponsored by universal interoperability for grid-forming inverters (unifi) Consortium is a U. S. Department of Energy funded effort to advance grid-forming (GFM) inverter technology.

REFERENCES

[1] G. Lopez-Cardalda and E. I. Ortiz-Rivera, "Proposed Methodology Using the Energy Storage System Prioritization Index," 2021 IEEE 48th Photovoltaic Specialists Conference (PVSC), Fort Lauderdale, FL, USA, 2021, pp. 2529-2533, doi: 10.1109/PVSC43889.2021.9518618.

[2] NERC, "EOP-005-3 – System Restoration from Blackstart Resources", Nerc.com. [Online]. Available: https://www.nerc.com/pa/Stand/Reliability%20Standards/EOP-005-3.pdf.

[3] NERC, "Reliability Guideline Improvements to Interconnection Requirements for BPS-Connected Inverter-Based Resources", Nerc.com, 2019. [Online]. Available: https://www.nerc.com/comm/PC_Reliability_Guidelines_DL/Reliability _Guideline_IBR_Interconnection_Requirements_Improvements.pdf.

[4] NERC, "Reliability Guideline Performance, Modeling, and Simulations of BPSConnected Battery Energy Storage Systems and Hybrid Power Plants", Nerc.com, 2021. [Online]. Available: https://www.nerc.com/comm/RSTC_Reliability_Guidelines/Reliability_ Guideline_BESS_Hybrid_Performance_Modeling_Studies_.pdf#search =blackstart.

[5] FERC, "Electric Storage Participation in Markets Operated by Regional Transmission Organizations and Independent System Operators", Federal Energy Regulatory Commission, 2018. [Online]. Available: https://ferc.gov/sites/default/files/2020-06/Order-841.pdf.

[6] Salasovich, J. and Gail M. (2019). "Energy Resilience Assessment for Culebra, Puerto Rico". Golden, CO: National Renewable Energy Laboratory.NREL/TP-7A40-73885.

[7] FEMA, Healthcare Facilities and Power Outages: Guidance for State, Local, Tribal, Territorial, and Private Sector Partners. 2020.

[8] "U.S. Census Bureau QuickFacts: Culebra Municipio, Puerto Rico", Census Bureau QuickFacts, [Online]. Available: https://www.census.gov/quickfacts/culebramunicipiopuertorico.

[9] V. Ramasamy, J. Zuboy, E. O'Shaughnessy, D. Feldman, J. Desai, M. Woodhouse, P. Basore, and R. Margolis, "U.S. Solar Photovoltaic System and Energy Storage Cost Benchmarks, With Minimum Sustainable Price Analysis: Q1 2022," National Renewable Energy Laboratory, Golden, CO, USA, Tech. Rep. NREL/TP-7A40-83586, 2022. https://www.nrel.gov/docs/fy22osti/83586.pdf.

[10] K. Mongird, V. Fotedar, V. Viswanathan, V. Koritarov, P. Balducci, and B. Hadjerioua, "Energy Storage Technology and Cost Characterization Report," July 2019. Available: https://energystorage.pnnl.gov/pdf/PNNL-28866.pdf..

Investigations on Absorber Type and Junction Position of GaAs Solar Cells

Gan Li[1], Hassanet Sodabanlu[2], Meita Asami[1],
Kentaroh Watanabe[2], Masakazu Sugiyama[1,2] and Yoshiaki Nakano[1]

[1]School of Engineering, The University of Tokyo, Bunkyo-ku, Tokyo, 113-8656, Japan
[2]Research Center for Advanced Science and Technology, Meguro-ku, Tokyo, 153-8904, Japan

Abstract — **This paper presents a study comparing GaAs solar cells with different structures grown by metal-organic vapor phase epitaxy (MOVPE). The impacts of absorber type and junction position are individually analyzed regarding the solar cell performance. When using the same structures, the n-type absorber samples had enhanced open-circuit voltage (Voc) and radiative efficiency than the p-type absorber ones, indicating superior crystal quality of n-GaAs. Besides, when using the same type of absorber, front-junction (FJ) samples showed an all-around boosted performance than rear-junction (RJ) ones, refreshing our understanding of the role of RHJ structure. Based on these observations, a p-on-n FJ structure using an n-GaAs absorber is recommended for high-performance solar cells.**

I. INTRODUCTION

Recently, some of the best InGaP [1], GaAs [2], and InGaAs [3] cells have been fabricated by placing the pn junction at the back side of the cell, forming a heterojunction with a high bandgap material layer, as known as rear-heterojunction (RHJ) structure. RHJ structure is considered able to boost the open circuit voltage (V_{oc}) of a device by suppressing non-radiative Sah-Noyce-Shockley (SNS) recombination in the space-charge region (SCR) compared with conventional front-junction (FJ) one. Though this gain in the V_{oc} is at a cost of short-circuit current density (J_{sc}) due to insufficient carrier extraction, the degraded current can be recovered when a backside reflector is equipped. Further studies evaluated FJ and RHJ structures of metamorphic materials [4], [5] and came to similar conclusions. However, all these researches compared the FJ with a p-type absorber and the RHJ with an n-type absorber, ignoring the different properties of the absorber material brought about by the two dopant types [6]. These differences include the type and

density of native point defects that are induced in metal-organic vapor phase epitaxy (MOVPE) growth. The disregard for the difference in absorber may be owing to the prejudice of an uppermost n-type layer since tandem cells grown on Ge substrate requires such a polarity of n-on-p.

In this paper, we systematically analyzed the GaAs solar cells' performance regarding, not only the junction positions, but also their absorber types, see Fig. 1. Mutual comparisons were carried out for cells with either the same active layers or the same structures to isolate the impacts of junction position and absorber type, respectively. It is found that the enhanced V_{oc} of RHJ may be mainly attributed to the crystal quality in n-GaAs rather than junction engineering.

II. EXPERIMENTAL DETAILS

Seven different structures of GaAs solar cells, with either n-type or p-type base absorber layers, were prepared for this study. All these samples have an identical thickness of 2 μm for the base layers, whereas the position and material of the emitter layers differ from each other, forming either FJ, front-heterojunction (FHJ), rear-junction (RJ) or RHJ, see Table I. Details of the layer sequences of these structures are given in Table II. When InGaP was used as the front hetero-emitter, an extra InAl(Ga)As layer was hired as the window layer; while when InGaP was used for the rear hetero-emitter, the BSF was omitted. The epitaxial growth in this

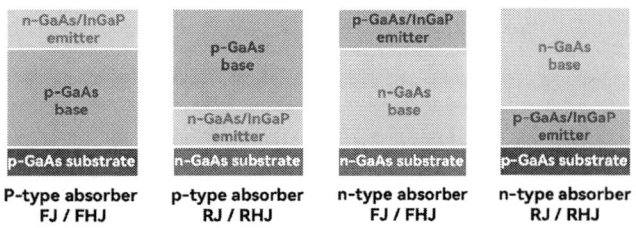

Fig. 1. Schematic diagrams of 4 types of structures compared in this work regarding their absorber types and junction positions.

TABLE III
SUMMARY OF SOLAR CELL MEASUREMENTS

Sample ref.	Absorber type	Emitter type	Substrate
FJ-p	p-GaAs	n⁺-GaAs	p
FJ-n	n-GaAs	p⁺-GaAs	n
FHJ-p	p-GaAs	n⁺-InGaP	p
FHJ-n	n-GaAs	p⁺-InGaP	n
RHJ-p	p-GaAs	n⁺-InGaP	n
RHJ-n	n-GaAs	p⁺-InGaP	p
RJ-n	n-GaAs	p⁺-GaAs	p

TABLE II
DETAILS OF THE SOLAR CELL STRUCTURES

Layer type	Materials (FJ/FHJ)	Material (RJ/RHJ)	Thickness (nm)	Doping Concentration (cm-3)
Contact	GaAs	GaAs	100	1×10^{19}
Window	InGaP / Al(Ga)InP	InGaP	25	$5 / 5 \times 10^{18}$
Front emitter	GaAs / InGaP	-	$100 \sim 150 / 50 \sim 100$	$2 \sim 3 / 3 \sim 5 \times 10^{18}$
Base	GaAs	GaAs	2000	1×10^{17}
Rear emitter	-	GaAs / InGaP	$100 / 30 \sim 100$	$1 / 1 \times 10^{18}$
BSF	InGaP	InGaP / -	30 / -	$3 \sim 4 \times 10^{18} / -$

paper was carried out in a planer MOVPE reactor (Aixtron, AIX2000HT). Standard metal-organic precursors were utilized for III-group sources and dopants including trimethylgallium (TMGa), trimethlindium (TMIn), trimethlaluminum (TMAl), diethyltellurium (DETe), and dimethylzinc (DMZn). Low-toxic tertiarybutylarsine (TBAs) and tertiarybutylphosphine (TBP) were used for V-group sources. Both Si-doped n-type and Zn-doped p-type GaAs (0 0 1) substrates were used in the growth process regarding the polarity of the cell devices. The solar cells were processed into a 0.5×0.5 cm² mesa area after standard metal evaporation, photolithography, and wet chemical etching steps. After removing the GaAs contact layer, the current-voltage (J-V) characteristics under AM1.5G (100mW/cm²) illumination and external quantum efficiency (EQE) without anti-reflection coating (ARC) were examined for each cell. An internal quantum efficiency (IQE) spectrum was then calculated from EQE by

$$IQE(\lambda) = EQE(\lambda) / (1 - R(\lambda)) \qquad (1)$$

where $R(\lambda)$ is the surface reflectance. A bandgap (E_g) and V_{oc} at the radiative limit (V_{oc}^{rad}) are extracted from the EQE spectrum for each sample. Based on them, we can further calculate external radiative efficiency (ERE) and W_{oc}:

$$ERE = e^{(V_{oc} - V_{oc}^{rad}) / V_{th}} \qquad (2)$$

$$W_{oc} = E_g / q - V_{oc} \qquad (3)$$

where q is the element charge and V_{th} is the thermal voltage.

W_{oc} and ERE are good figures of merit to compare solar cells representing their space of improvement, especially when E_g differs in samples.

III. RESULTS AND DISCUSSION

The solar cells' performance parameters of these GaAs devices are summarized in Table III. All seven samples in this study have very close E_g of 1.428 eV. Focusing on the overall efficiency performance, sample FHJ-p with n⁺-InGaP stacking on p-GaAs was found the champion compared with other designs. On the other hand, sample RHJ-p with a p-GaAs stacking on n⁺-InGaP turned out to be a poor harvesting device. It could be also noted that these two samples also possess the highest and lowest fill factor (FF), respectively. It is surprising that the RHJ structure with a p-type absorber has the worst performance. In expectation, its prolonged electron diffusion length compared with one with the n-type absorber was supposed to provide even higher carrier extraction while keeping the high V_{oc} benefitted from the RHJ structure.

A. Absorber Type

Among these samples, FJ-p & FJ-n, FHJ-p & FHJ-n, and RHJ-p & RHJ-n have the same structure in pairs, respectively. The polarities of samples in a pair are opposite: FJ-p, FHJ-p, and RHJ-p have a p-GaAs layer as the absorber, while FJ-n, FHJ-n, and RHJ-n use n-GaAs as the absorber.

TABLE III
SUMMARY OF SOLAR CELL MEASUREMENTS

Sample ref.	Absorber type	V_{oc} (V)	J_{sc} (mA/cm²)	FF (%)	Eff (%)	E_g (eV)	W_{oc} (V)	V_{oc}^{rad} (V)	ERE (%)
FJ-p	p-GaAs	1.021	19.6	80.9	15.979	1.427	0.406	1.150	0.70
FJ-n	n-GaAs	1.036	20.0	80.2	16.631	1.427	0.391	1.152	1.19
FHJ-p	p-GaAs	1.023	21.4	82.2	18.000	1.428	0.406	1.154	0.64
FHJ-n	n-GaAs	1.048	21.2	80.5	17.906	1.428	0.379	1.152	1.89
RHJ-p	p-GaAs	0.948	19.6	74.0	13.730	1.428	0.481	1.154	0.04
RHJ-n	n-GaAs	1.036	19.1	80.0	15.824	1.427	0.391	1.148	1.07
RJ-n	n-GaAs	1.009	19.0	81.7	15.647	1.428	0.418	1.152	0.75

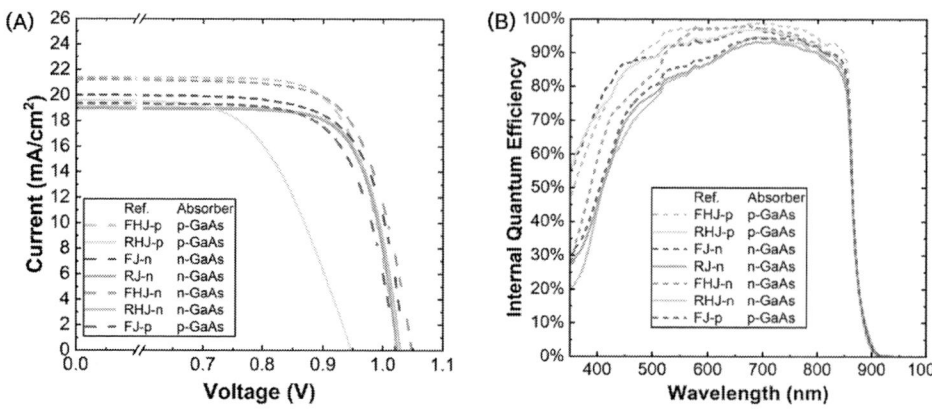

Fig. 2. (A) J-V characteristics under AM1.5G (100 mW/cm²) illumination and (B) IQE spectrum of the GaAs cell samples. To better distinguish, FJ/FHJ samples and RJ/RHJ samples are plotted in dash and straight line, respectively, and the samples with identical active layers are plotted in the same color tone. Note that the display order of samples is different to that in the previous tables.

Such comparison in pairs allows us to understand the isolated effect of absorber type for different solar cell structures. As given in Table III, The samples with an n-GaAs absorber all show a superior V_{oc} than those with a p-GaAs one. Increments of V_{oc} of 0.015 V, 0.026 V, and 0.089 V are observed for FJ, FHJ, and RHJ structures, respectively, after the absorber layer exchanged from p-type to n-type. The explanation is most likely the fewer As_{Ge} antisite defects that act as non-radiative recombination centers. This result is also in line with the ERE measurement. The samples with an n-type absorber show exhibit higher ERE above 1%, indicating a superior crystal quality with suppressed non-radiative recombination. As for current performance, samples using an n-type absorber show modest J_{sc} and degraded FF, which is plausibly caused by the inferior diffusion length of the minority hole in the n-type absorber.

B. Junction Position

Similarly, the effect of junction position is investigated by comparison in pairs between samples with identical active layers. Sample FHJ-p & RHJ-p, FJ-n & RJ-n, and FHJ-n & RHJ-n have the same combination of the emitter and base layers in pairs whereas the polarity is inversed. The J-V and IQE spectra of these samples are plotted in Fig. 2 (A) and (B), respectively. Looking at the J-V curve in Fig. 2 (A), one can soon figure out that both J_{sc} and V_{oc} can be significantly enhanced when shifting the junction from the rear to the front. An increased IQE in FJ/FHJ samples can be also observed from the RJ/RHJ baselines in Fig (B), possibly due to the strong drift force applied by the built-in electric field.

These comprehensive improvements by using front junction address the fact that RJ and RHJ structures are impeding the solar cell performance. Combined with the observation in the previous section, it is plausible that the advantage of the n-on-p RHJ devices reported so far is attributed to the prominent crystal quality of n-GaAs rather than to the structure adjustment, while RHJ structure actually weakens carrier extraction and radiative efficiency compared with its FHJ variant. This way, we can explain the substandard performance of the p-on-n RHJ-p sample, as it suffers from both poor p-GaAs quality and deficient carrier extraction in the RHJ structure.

IV. SUMMARY

Individual effects of absorber types and structures on the performance of GaAs solar cells are investigated for the first time. The n-GaAs samples express superior V_{oc} and ERE than p-GaAs, which is likely owing to a lower density of point defects that suppress the non-radiative recombination. On the other hand, the FJ and FHJ structures show enhanced carrier extraction and decreased recombination current than RJ and RHJ ones that have the same active layers. These observations reveal that the improved V_{oc} of the n-on-p RHJ solar cells in reports is on account of the absorber quality rather than engineering the structure. Further study should focus on the p-on-n FJ GaAs cells using an n-GaAs absorber for their outstanding J_{sc} and V_{oc}.

REFERENCES

[1] J. F. Geisz, M. A. Steiner, I. García, S. R. Kurtz, and D. J. Friedman, Appl. Phys. Lett., vol. **103**, no. 4, p. 041118, 2013

[2] M. A. Steiner, R. France, J. Buencuerpo, J. Geisz, M. Nielsen, A. Pusch, W. Olavarria, M. Young, N. Ekins - Daukes, Adv. Energy Mater., vol. **11**, no. 4, pp. 2002874, 2021

[3] R. Yokota, K, Watanabe, H. Sodabanlu, M. Asami, H. Xu, Y. Nakano, M. Sugiyama, in *68h Spring meeting of Japan Society of Applied Physics*, 2021, 16p-Z02-6

[4] B. D. Li, P. Dhingra, R. D. Hool, S. Fan, and M. L. Lee, in *48th IEEE Photovoltaic Specialists Conference*, 2021, pp. 2614–2615

[5] M. Kim, Y. Sun, R. D. Hool, and M. L. Lee, in *48th IEEE Photovoltaic Specialists Conference*, 2021, pp. 1762-1762

[6] H. Sodabanlu, A. Ubukata, K. Watanabe, T. Sugaya, Y. Nakano, and M. Sugiyama, in *46th IEEE Photovoltaic Specialists Conference*, 2020, pp. 0152–0155

Transparent Tedlar® Frontsheet for Lightweight PV Module Designs

Hongjie Hu[1], Stela Chen[1], Oakland Fu[1], Michael Demko[2], and Kaushik Roy Choudhury[2]

[1]DuPont (China) Research & Development and Management Co. Ltd., Shanghai, P.R.C.
[2]E. I. du Pont de Nemours and Company, 200 Powder Mill Road, Wilmington, DE, USA

Abstract — **The weight of conventional photovoltaic (PV) solar panels is a significant challenge in applications that can benefit from the integration of this renewable energy source, but are sensitive to increased weight loads. Numerous commercial buildings and auxiliary structures are designed with little to no spare structural capacity. Vehicle fuel efficiency is directly tied to the total vehicle weight. Lightweight modules can address this challenge by reducing weight, enabling PV elements to be installed or retrofitted at low cost for parking roofs, in building structures (Building Integrated PV or BIPV), and in vehicles (Vehicle Integrated PV VIPV). The main challenge in acheiving lightweight PV modules is replacing the glass frontsheet while maintaining transparency, mechanical stability and weatherability over the lifetime of the module. Transparent Tedlar® PVF films have been used for decades in protecting graphics and signage and photovoltaic backsheets. Building on years of successful field performance, a new transparent Tedlar® TFS15BM3 film has been designed to provide a high level of outdoor UV stability and protection, offering a significantly lighter technologically advanced alternative to traditional glass. In this paper, we present the performance of this film as a protective frontsheet of solar modules. The unique balance of durability, UV resistance, high level of light transmittance, lasting UV protection, mechanical toughness, chemical resistance, good adhesion to encapsulant, easy cleaning, light weight and flexibility makes this an attractive alternative to traditional glass in applications necessitating lightweight.**

I. INTRODUCTION

In designing PV modules to perform reliably over decades of sustained operation in terrestrial environments, it is critical that the performance of the constituent components are ensured through careful design and selection of materials, the durability is is characterized, degradation mechanisms identified, and the collective impact on module performance understood. This is especially true for newer application areas that are witnessing renewed growth. Integration of PV into existing and new building structures and into mobile applications (eg. vehicles) is highly attractive. However, for these segments the weight of conventional front-glass or double-glass solar panels is a significant challenge. Oftentimes, commercial buildings and auxiliary structures (carports, awnings) are designed with little to no spare structural load-bearing capacity. For vehicles, fuel efficiency is heavily impacted by total vehicle weight. Lightweight modules can address this challenge by reducing the weight of panels, enabling PV elements to be installed or retrofitted at low cost for parking roofs, in building structures, and in vehicles.

The main challenge in lightweighting PV modules is replacing the glass frontsheet while maintaining performance, aesthetics, mechanical stability, weatherability and safety. For a typical 72 cell glass-backsheet panel that weighs about 22 kg, the 3.2mm glass frontsheet can weigh more than 10 kg. Therefore, reducing the contribution of the frontsheet to the overall weight of the panel can significantly impact the integration of PV modules in these new applications. In the past, for crystalline silicon cells, polymer frontsheets have been implemented to provide protection for supporting lattices like glass-fiber reinforced polymer (GFRP) structures. For thin film technologies like copper indium gallium selenide (CIGS), flexible substrates and polymer frontsheets have been applied as durable insulation. However, widespread adoption of these lighter components has not happened due to performance shortcomings compared to glass, and due to reliability concerns arising out of a lack of field data.

Transparent Tedlar® films have been used for decades in protecting graphics and signage in the outdoor environment and have been successfully used for several years in photovoltaic backsheets for bifacial modules. Building on years of successful performance, a new transparent Tedlar® TFS15BM3 film has been designed to provide the highest level of outdoor stability and protection, offering a significantly lighter alternative to traditional glass frontglass. In this paper, we present the performance of this film and laminates as a protective frontsheet of solar modules.

II. PERFORMANCE AND DURABILITY OF FRONTSHEET FILM

Transparent Tedlar® films have been used for decades in graphics and signage applications to protect substrates from harsh environmental conditions. Free-standing transparent films have been exposed for ten years in south Florida and showed no degradation in transmission, haze, or color over ten years of exposure. This transparent Tedlar® film was reformulated to substantially increase its ability to survive in harsh environments outdoors for the front side of PV modules. The new film represents the highest level of performance in Tedlar® transparent films. Table 1 lists typical properties of this Tedlar® TFS15BM3 film, with high optical transparency and robust mechanical properties.

978-1-6654-6060-6/23 $31.00 © 2023 IEEE

Property	Typical Value	Method
Thickness	38 µm	Micrometer
Optical Transmission	93%	ASTM D1003
Haze	44	ASTM D1003
Gloss (60°)	38	ASTM D2457
MD Elongation at Break	240%	ASTM D882
TD Elongation at Break	140%	ASTM D882

Table 1. Typical thickness, optical and mechanical properties of the Tedlar® film for frontsheet application.

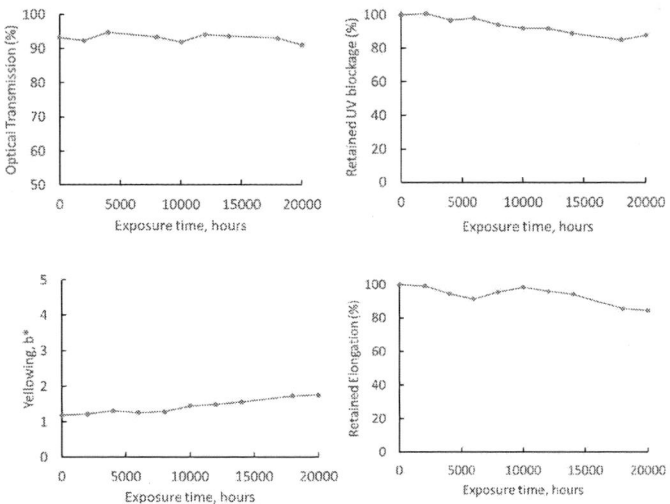

Figure 1. a) Retention of transmission, b) Retention of UV blocking, c) Discoloration, and d) Retention of elongation in ASTM 7869.

To evaluate the performance and durability of the film in outdoor environment, a sequential accelerated test incorporating broad-spectrum Xenon Arc light, thermal and dark cycles, and water spray (described by ASTM D7869) was used to simulate the full range of environmental conditions on the front side of a photovoltaic module. Fig. 1 shows the relevant optical and mechanical properties from this extended accelerated testing. The film survived for 20,000 hours (911 kWh/m² of UV light from 290 to 400 nm) with minimal change in transmission and retained more than 90% of UV blocking ability in the range of 290 to 370 nm. The yellowness index of the film, indicated by the b* value, was well below 2 even after the 20,000 hours of exposure. As a result of this excellent UV screening, the film retained more than 85% of elongation through this test, confirming its suitability for use in a solar application.

III. ENHANCED LONGEVITY OF COMPOSITE FRONTSHEETS

The Tedlar® film is designed to absorb UV light and provide protection to underlying substrates, enabling composite frontsheets with a combination of outstanding weathering properties, robust mechanical strength and required electrical insulation at reasonable cost. Previously, transparent Tedlar® film has been successfully laminated to fiberglass panels for architecture applications, where it has substantially improved panel life in humid and corrosive environments with high ultraviolet light.

For the PV application, composite frontsheets were constructed with an outer layer of Tedlar® and an inner layer of PET. The composite laminates were then subjected to accelerated UV weathering to asses their durability. In Fig. 2, the composite Tedlar® frontsheet (TFS15BM3) laminated with UV-stabilized transparent PET sheet retains its transmission and mechanical properties after 14,000 hours of an accelerated weathering test in continuous Xenon Arc light at elevated temperatures (IEC 62788-7-2, Method A3).

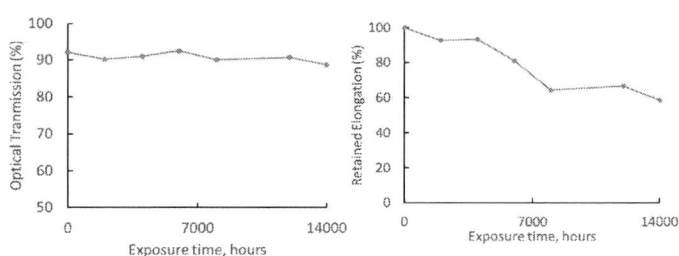

Figure 2. a) Retention of transmission, b) Retention of mechanical property of Tedlar® frontsheet (TFS15BM3) laminate with UV-stabilized transparent PET sheet after 14,000 hours (1120 kWh/m2 of UV light from 295-400 nm) of Xenon Arc exposure following IEC 62788-7-2, Method A3 (Xenon Arc, 0.8 W/m2-nm @ 340 nm, 90 °C BPT, 20% R.H.).

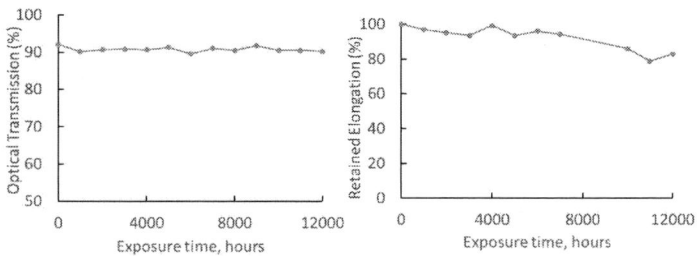

Figure 3. a) Retention of transmission, b) Retention of mechanical property of Tedlar® frontsheet (TFS15BM3) laminate with UV-stabilized transparent PET sheet after 12,000 hours of UVA exposure per ASTM G154.

In Fig. 3, the same composite Tedlar® frontsheet is tested using a UVA fluorescent light source (ASTM G154, UVA-340 fluorescent bulb, 1.2 W/m²-nm @ 340 nm, 70 °C BPT) for 12,000 hours. Additionally, the frontsheet was also tested in an accelerated test protocol (data not shown here) that combines exposure of UV light, dark cycle, water spray, and condensation

(ASTM D7869). In both tests, the optical and mechanical properties of the composite frontsheet are preserved. These extreme levels of testing demonstrate the suitability of this frontsheet for use on the front side of the PV module where intense light and weather provide high stress to materials.

III. ADHESION AND CHEMICAL RESISTANCE

A potential source of failure of polymer backsheet and frontsheets in field operation is delamination from the encapsulant layer. To test the robustness of the frontsheet/EVA encapsulant interface test samples were made with EVA laminated to the frontsheet and were subjected to UV Xenon exposure per ASTM J1960 and 85°C/85%RH DH. For comparison and ETFE frontsheet was also tested. The results outlined in Table 2 clearly show that while 3000 hours of DH exposure does not affect the adhesion of any frontsheet, after 12000 hours of UV Xenon exposure the adhesion of ETFE frontsheet degraded significantly, white the Tedlar® based frontsheet maintained good adhesion.

90° peel test(lbf/in)	PVF 1	PVF 2	ETFE
XENON UV*	**lbf/in****	**lbf/in**	**lbf/in**
kJ/m²	**EVA**	**EVA**	**EVA**
0	15.9	15.9	7.4
2400	11.4	11.4	3.2
4800	7.8	7.8	2.4
7200	*	*	2.1
9600	19.3	19.3	1.5
12000	16.2	16.2	1.9
14400			
DAMP HEAT	**lbf/in**	**lbf/in**	**lbf/in**
Hours @ 85°C/85%RH	**EVA**	**EVA**	**EVA**
0	15.9	15.9	7.4
500	23.1	23.1	7.8
1000	23.5	23.5	7.4
1500	14.5	14.5	6.9
2000	10.5	10.5	8.1
2500	11.3	11.3	8
3000	19.1	19.1	7.5

Table 2. Adhesion of two Tedlar® based frontsheets to EVA in UV Xenon DH accelerated testing.

Chemical (1 week immersion)	Appearance	Color, ΔE*	UV Spectrum
10% HCl in water	Normal	0.06	No change
10% H₂SO₄ in water	Normal	0.07	No change
10% NaOH in water	Normal	0.08	No change
Isopropanol	Normal	0.1	No change
Window cleaner	Normal	--	No change

Table 3. Resistance of the Tedlar® TFS15BM3 film to acidic chemicals, alcohols and commercial cleaners.

The fluorocarbon nature of Tedlar® film is the basis for the film's outstanding durability and resistance to a variety of solvents and harsh chemicals (Table 3). The Tedlar® film stands up well to atmospheric pollutants and resists acid rain attack and mildew. Most airborne dirt does not adhere to Tedlar® film; if soiling does occur, rainwater or a commercial cleaner can be used to restore the surface to its original appearance without any loss in light transmittance.

IV. ADVANTAGE IN SYSTEM COST AND MAINTENANCE

A Tedlar® film frontsheet on a typical 1m*2m module weighs less than 130 grams, while a glass frontsheet would weigh more than 15 kg. This significant difference can help reduce the weight of a full size c-Si module by as much as 70%, as outlined in Fig. 4. The lightweight frontsheet is also flexible and resistant to chipping, making it easier to install and safer to handle than glass. All of these attributes not only enable it to be used in applications where a lightweight module is necessary, but they also help lower the total system cost. The lightweight modules reduce BOS costs and also enable 3x-5x faster installation.

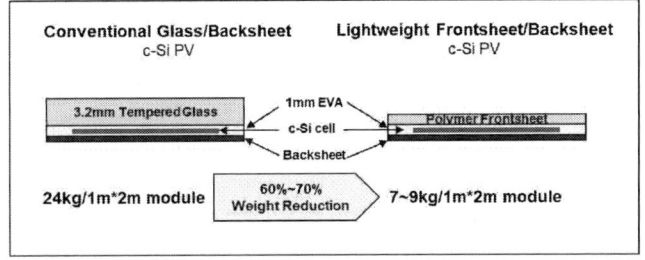

Figure 4. Structure of a lightweight PV module compared to an conventional one, highlighting the difference in weight.

The Tedlar® film also improves the cleanability of the frontsheet as it is a fluoropolymer and contains no plasticizers. Additives in uncoated glass frontsheet can migrate to the surface of the module and collect dirt after field aging, making cleaning difficult. This can add to the overall maintenance cost of a system.

V. SUMMARY

In this paper, we presented a new Tedlar® film based frontsheet for PV modules. The performance of this film and laminates as a protective frontsheet of solar modules is presented in detail. The unique balance of durability, UV resistance, high level of light transmittance, lasting UV protection, mechanical toughness, chemical resistance, good adhesion to encapsulant, easy cleaning, light weight and flexibility makes this an attractive alternative to traditional glass in applications necessitating lightweight.

Photovoltaic Design Projects Increase ECE Student Engagement

Devin C. Whalen, Peter Mark Jansson, and Milton G. Newberry III

Bucknell University, Lewisburg, Pennsylvania, 17837, United States

Abstract — **In this paper, we share the pedagogical change of converting half the semester [previously focused on laboratories reinforcing basic concepts and application of circuit theory] to projects based on designing novel applications of photovoltaic technologies at the end-use level in a residential home. In this electrical and computer engineering required course [ECEG 210], we scaffolded students in the design and prototyping of novel PV systems while increasing the project-based learning and engineering design aspects of the course. This novel approach introduces second year ECE students to some of the pressing challenges and opportunities presented by the decreasing costs of PV hardware. In this work, we examine the project innovations made to the ECEG 210 course to develop students' skills and knowledge of PV materials and aid in them learning about the newest applications of PV already available in the industry literature and marketplace.**

I. INTRODUCTION

The normal semester at our university consists of 15 weeks, and the ECEG 210 Applications of Circuit Theory course offered through the recent COVID-19 pandemic included approximately five or six laboratories spaced throughout the autumn semester. In our pedagogical redesign following the COVID-19 pandemic, we devised a comprehensive approach to enhance student learning. We implemented a series of nine weekly labs that aim to reinforce the fundamental concepts of the course.

Lab 1 served as a foundation, focusing on familiarizing students with the usage of the benchtop multimeter and power supply. Lab 2 delved deeper into the fundamental principles of electrical circuits, including Ohm's law, Kirchhoff's current and voltage laws, and nodal analysis. Building upon this knowledge, Lab 3 centered around diode fundamentals and introduced students to the design of a simple digital-to-analog converter. Lab 4 continued the exploration of diode behavior and expanded the students' toolkit with the desktop oscilloscope and function generator. In Lab 5, students applied Thevenin and Norton equivalent techniques to analyze and simplify complex circuits. Lab 6 introduced the concept of operational amplifiers, demonstrating their role in forming basic linear equations. The subsequent labs, Labs 7 and 8, focused on inductors and capacitors, respectively, providing a comprehensive understanding of these important circuit elements. Finally, Lab 9 delved into complex power analysis, emphasizing the distinction between leading and lagging phases in electricity.

After establishing fundamental concepts, we introduced a 6-week photovoltaic (PV) project, carefully divided into six segments, which provided students with an outlet to apply the skills they learned during the former part of the semester. Although service learning, extracurricular activities, and project-based learning experiences involving renewable energy technologies are familiar to engineering educators [1, 2], incorporating these experiences into the core curriculum of an Electrical and Computer Engineering (ECE) program poses a greater difficulty and challenge [3-7].

The PV project design exercise was broken down into parts A through F, where we led the students through all critical aspects of the project. Each of these parts can be shared upon request. In general, the project teams comprised three ECE students. It was their charge to analyze several potential end-use applications of PV in a residential home and then create a minimum viable product [MVP] to demonstrate the proof of concept and assess its potential success in the energy market. Students designed, built, and evaluated a PV-powered electrical appliance of some type. Teams first broadly explored the challenge of taking typical residential end-use devices off-grid by researching three different end-use devices of their choice, ultimately focusing on one of these ideas. In our final project list, we saw porch lights, laptop and tool chargers, PV room air conditioners, well pumps, smart lawnmowers, and a high efficiency washer/dryer system.

The teams first calculated the benefits of their idea [including environmental (carbon) savings] and the economic and energy savings given the potential customer base across the United States. It was up to the teams to propose their preferred final project topic based on the value proposition they believed their potential innovation could offer.

All teams identified metrics upon which the performance of their end-use MVP powered by PV could be assessed. Each product, before construction, had to be clearly represented via a physical block diagram, which showed any required data flow. The representations conveyed not only the functionality of the device but also clearly showed what the construction of their end-use PV systems would entail. Further, these diagrams illustrate how each subsystem and component contribute to the finished, full-scale product.

A generous donation from the William Corrington Renewable Energy Fund enabled students to purchase all required components for their projects. After the design and specification work were completed and a Bill of Materials was generated for both the MVP and a full-scale version of their design, the teams constructed and evaluated their MVP to demonstrate the viability of their innovation and collect data. The tests that each team performed were based on the performance metrics they had established for their innovation of the PV end-use. The goal was to demonstrate with specific data that the MVP met all their established design criteria. To document their MVPs, each team also developed two videos: one was functionality and performance-focused directed at a technical audience, and the other was a marketing video directed toward potential customers (a non-technical audience).

These projects allowed students to challenge themselves and engage with concepts through their own research, resulting in innovative designs that have the potential to save a lot of energy. The MVPs included designs across many applications in the residential space and are detailed in the section following.

II. RESULTS: MVP DESCRIPTIONS

Descriptions of student-generated MVPs employing photovoltaics follow:

A. Solar-Powered Porch Light

By replacing a single grid-powered porch light with an alternative PV-powered system, this team calculated an annual savings of 30.7 kWh of electricity and 11.87kg of CO_2, equivalent to driving a vehicle 34 miles. If 5% of U.S. households replaced two grid-powered porchlights, it would result in an annual savings of over 152 Megatons of CO_2. The team's design uses an 84Wh battery to guarantee at least two hours of light operation per night as activated by a motion sensor. The device can be powered with a single Evergreen solar panel even during the winter when less sunlight is available. The total cost of their minimum viable product was $107.

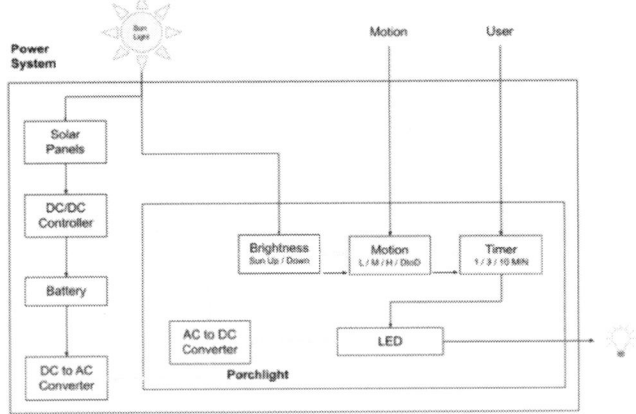

Fig 1. Solar-Powered Porch Light Level 2 Block Diagram

B. Solar-Powered Laptop Charger

An average laptop uses 0.055 kWh per day, or 1.7 kWh per month. For each laptop charged using their PV-powered system, this team calculated an annual energy savings of 146kWh resulting in a projected reduction of over 328,000 tons of CO_2. The total cost of the MVP was $220.

C. Solar-Powered Air Conditioner

Air conditioning's high energy use in the United States generates an estimated 140 million tons of carbon dioxide each year. If 5% of grid-powered AC units were replaced with PV alternatives, it would result in a CO_2 reduction of over 1.5 million tons annually. AC usage is the highest between the months of May and August. Four CS3U-345 PB panels would satisfy the 183-273kWh necessary to power a solar alternative AC between 5 and 7 hours a day during the summer. The total cost of the MVP was $106.

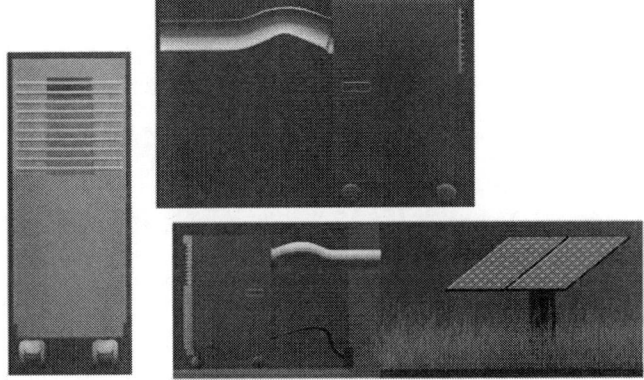

Fig 2. CAD Drawings for Solar AC Implementation

D. Solar-Powered Washer/Dryer Combo

One 2kW solar panel array can power a high-efficiency washer and dryer machine for a single 2-hour load in a day, amounting to approximately 1.5kWh of power consumption. The team implemented a battery with enough capacity for two to three loads to provide more flexibility to the user and utility during darker day1 's. According to the team's calculations, the adoption of this solution by 5% of households in the U.S. instead of using a traditional washer and dryer would lead to an annual reduction of more than 840,000 tons of CO_2 emissions. These calculations assume that an average American household performs five loads of laundry per week. The total cost of the MVP was $107.

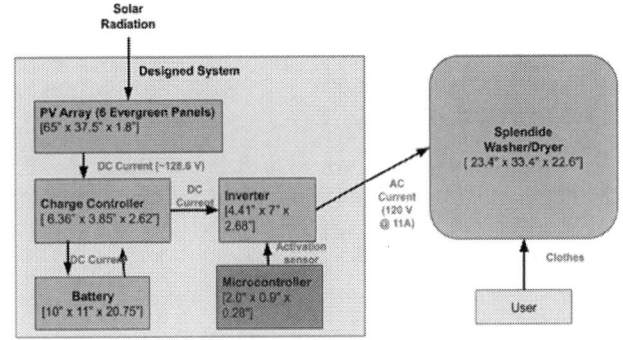

Fig 3. Solar-Powered Washer and Dryer Level 2 Block Diagram

E. Solar-Powered Phone Charging Station

This group designed a solar-powered phone charging station equipped with a 200W solar panel and a 12V battery. Their MVP can fully charge a phone or portable charging brick in 2 ½ to 5 hours. Based on their calculations, a typical household would save 44 kWh resulting in an alternative carbon reduction of over 110,000 tons of CO_2 assuming a 5% market penetration in the U.S. A single cell can produce the required power, up to 63W on the darkest day of the year. The total cost of the MVP was $90.

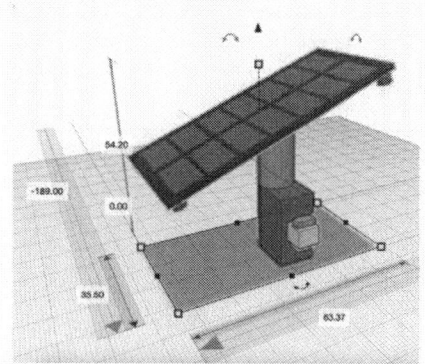

Fig 4. Solar Charging Station CAD Render

F. Solar-Powered Lawnmower Charging

Around 53,000 Americans own a robotic lawnmower, producing 3,155 metric tons of CO_2 yearly. If 5% of these individuals charged their mowers with renewable energy, it would reduce yearly CO_2 production by 158 metric tons. To achieve the desired reduction, this group developed a solar-powered electric lawnmower charging station. The total cost of the MVP was $148.

G. Solar-Powered Universal Tool Battery Charger

This group developed a solar-powered charger for electric tool batteries. From their calculations, if 5% of electric power tool batteries in the U.S. were charged using renewable energy, it would result in a yearly reduction of 9,175 metric tons of carbon emissions. They estimated that their product could save 2.9 kWh per year per battery. The MVP requires 1920Wh to charge over a 24-hour period, though realistically it would only need to run for a few hours at a time. The group calculated that almost 1kWh could be produced with the PV device used, even on a dark day in December. The total cost of the MVP is $242.

H. Solar-Powered Well Pump

This group developed a well pump that can be powered by solar energy when the grid is not available. According to their calculations, if 5% of Americans who use an electric well pump switch to a PV-powered option, it would result in a reduction of over 400,000 metric tons of CO_2 annually. Individual consumers would save up to 256 kWh of energy per year. The total cost of the MVP was $51.

I. Solar-Powered Toolbox Charger

This group developed a toolbox with a PV-powered charger built in to charge tools and batteries on the go. The product is designed to supply power to leaf blowers, drills, lawn mowers, and chainsaws, among others. According to their research, 170 million Americans own and operate gas or electric powered tools, which use over 76.5TWh of electricity per year. This energy usage equates to 79.5 billion pounds of carbon emissions. A product like this encourages and facilitates the transition to all-electric tools. The solar-powered toolbox eliminates anxiety about running out of battery during a job. The device requires 30kWh monthly, which can be supplied by a single Canadian Solar 345W PV module. The total cost of the MVP was $169.

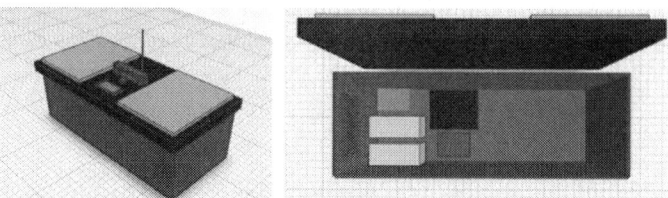

Fig 5. Solar Toolbox Charger CAD Render

III. ANALYSIS OF RESULTS

To evaluate the effectiveness of the design project, instructors required students to provide feedback on the benefits of this design project to the course and rate the value they felt the project phase provided. Students described methods that they found useful as well as changes they would like to see to benefit those taking the course in the future. To avoid survey fatigue, students provided feedback only once after the conclusion of the course. Bonus points were offered as an incentive and 20 out of the 28 students enrolled in the class participated in the evaluation. In future iterations of the course, it may be beneficial to poll students for feedback at least once before the end of the semester to allow for time to implement suggested changes into the current course iteration.

A. Course Evaluations

Feedback revealed students were enthusiastic about hands-on learning and experience in practical designing and testing. We asked students to rate how effective they found various areas at enhancing their engineering and ECE education and contributing to the overall project.

The three most highly rated areas were "Creating my MVP of the design we developed," "Testing/Evaluating the MVP we created," and "Creating representations including block and data-flow diagrams," see Table 1. Seventy percent of students who completed the course evaluation felt that the project allowed them to apply core ECE concepts that they learned through the labs and coursework. Among the 20 respondents, 70% also expressed that the project effectively imparted valuable knowledge beyond the scope of their regular labs and coursework. Commonly cited areas of learning included the significant impact of solar power, calculation techniques for carbon reduction, and a comprehensive understanding of PV cells and charge controllers.

The project and course effectively engaged students with PV concepts that they would not have explored in a

978-1-6654-6060-6/23 $31.00 © 2023 IEEE

traditional lecture setting. By immersing students in practical design related to solar power, the project provided a unique opportunity to bridge theory and real-world applications. This engagement with PV concepts not only enhanced their technical knowledge but also cultivated enthusiasm and motivation for the PV industry.

On a Scale from 1 to 10, How Effective Did You Find the Following Areas at Enhancing Your Overall Engineering Education (20 students)	Average
Evaluating multiple potential designs before deciding on a final topic	6.9
Creating my MVP of the design we developed	8.5
Testing/Evaluating the MVP we created	7.4
Creating representations including block and data-flow diagrams	7.5
Evaluating carbon emission savings of my design	6.6
Estimating economic benefits of my novel product / design using PV to take a particular end-use off-grid	7.0
Market analysis (evaluating similar available products and estimating potential markets)	6.5
Evaluating PV cell / module capabilities	7.1
Creating a technical video for my project	6.7
Creating a marketing video for my project	6.4

Table 1. Feedback on Final Project Components

By integrating PV-related curricula into course projects and activities, we can help solidify pro-renewable energy and sustainability attitudes in college students. Several social psychology theories such as the Theory of Planned Behavior [8] and the Value-Belief-Norm theory [9] state that attitudes are strong predictors of behaviors and contribute to their saliency. Thus, we posit that student engagement with our structure of an engineering course can lead to lifelong, positive attitudes towards renewable energy and the adoption of renewable energy technologies (e.g., residential PV). This adoption can, in turn, help make measurable changes to contributing factors to climate change (e.g., CO_2 emissions). At the same time, our course structure serves as a model for other engineering and STEM-focused courses in facilitating an experiential learning experience using PV.

How Strongly Do You Agree with the Following Statements on a Scale From 1 to 5 (20 students)	Average
Laboratories were supportive of course learning outcomes	4.4
Learning meetings (lecture period) was very helpful to my learning	3.8
The textbook (online available) was very helpful to my learning	3.5
The homework assignments were very helpful to my learning	4.0
The project experience was very helpful to my learning & experience	3.4
I believe I have learned many ECE / circuit fundamentals in this course	4.7
I can effectively analyze electric circuits using Ohm's law KVL and KCL	4.7
I can calculate the voltage, current and power across capacitors and inductors	4.2
I can effectively apply nodal analysis to electric circuits	4.7
I am able to solve first-order RC and RL circuits	4.0
I am able to solve simple second order RLC circuits	3.5
I am able to use complex algebra to perform phasor domain analysis	3.7
I am able to work effectively in teams to solve problems and deliver reports	4.5
I can effectively use ECE tools (multimeter, signal gen, scope, AD2, MultiSim)	4.6
I can effectively test and measure parameters in electric circuits	4.4

Table 2. Feedback on Laboratory and Coursework Instruction

IV. CONCLUSION

Encouraging future engineers to participate in the PV industry has far-reaching benefits. It not only contributes to a sustainable future but also leads to a workforce that is passionate, motivated, educated, and engaged. By exposing students to the potential and possibilities of solar power through hands-on projects, they are inspired to pursue careers in the PV industry. This, in turn, will foster a workforce that is dedicated to advancing renewable energy solutions and addressing the challenges of climate change. By incorporating PV concepts and providing opportunities to work on PV-related projects, the course instills a sense of purpose and relevance in students. They develop a deeper understanding of the practical applications of their education and are equipped with the skills necessary to make a positive impact in the field of solar energy. This engagement not only benefits the students individually but also contributes to a future workforce that is well-prepared, enthusiastic, and dedicated to advancing sustainable energy solutions. Going forward, we will continue to analyze student feedback and project evaluations for future iterations of the ECEG 210 curriculum.

REFERENCES

[1] M.G. Newberry, S.M. Myers and P.M. Jansson, "Sustainability Experiential Learning Laboratory – A Method for SELLing Photovoltaic Technology and Education to College Students," American Solar Energy Society – SOLAR 2021: Empowering a Sustainable Future (*50th Annual National Solar Conference*), University of Boulder, Boulder, Colorado, 3-6 August 2021.

[2] P.M. Jansson, J. Murphy, J. Stewart, R. Molner, P. Tomkiewicz and W. Heston, "Undergraduate Service Learning: Photovoltaic System Design and Construction," *ASEE 2005 Annual Conference, Seattle*, WA, June 12-15, 2005.

[3] P.M. Jansson, K. Whitten, C. Delia, M. Angelow, B. Ferraro, M. Giordano, M. Colosa, "EE Students Complete Photovoltaic R&D for Industry in Electrical Engineering Curriculum," *Proceedings of the 118th ASEE Annual Conference*, Vancouver, BC, Canada, 26-29 June 2011.

[4] W.T. Riddell, E. Constans, J. Courtney, K. Dahm, R. Harvey, P.M. Jansson, M. Simone, P. von Lockette, B. Wolff, "Lessons Learned from Teaching Project Based Learning Communication and Design Courses," *2007 ASEE Middle Atlantic Section Fall 2007 Conference*, Philadelphia, PA, November 3, 2007.

[5] P.M. Jansson and R. Elwell, "Design of Photovoltaic Systems for Municipal and School Buildings in Ocean City, New Jersey," *ASEE 2007 Annual Conference Proceedings*, Honolulu, HI, June 24-27, 2007.

[6] S. Hazel and P.M. Jansson, "Photovoltaic System Feasibility Assessments: Engineering Clinics Transforming Renewable Markets," *ASEE 2006 Annual Conference Proceedings*, Chicago, IL, June 18-21, 2006.

[7] P.M. Jansson, S. A. Mandayam and J.L Schmalzel, "Green Power Engineering: Pedagogy for the Next Generation of Electrical Engineers," *Proceedings of the 2004 IEEE Annual Power Engineering Society Conference*, Denver, CO, June 6 10, 2004, IEEE Xplore Digital Object Identifier: 10.1109/PES.2004.1372755, Vol. 1, pp. 65-70.

[8] A.H. Seyal and M.N. Abd Rahman MN, eds. "Theory of Planned Behavior: New Research". New York: Nova Science; 2017. http://search.ebscohost.com/login.aspx?direct=true&scope=site&db=nlebk&db=nlabk&AN=1530922. Accessed February 13, 2023.

[9] L. Zhang et al., "Predicting Climate Change Mitigation and Adaptation Behaviors in Agricultural Production: A Comparison of the Theory of Planned Behavior and the Value-Belief-Norm Theory", 68 Journal of Environmental Psychology (2020).

Long-term Degradation Rate of Photovoltaic Modules: A Meta-analysis

Michael Straub-Mueck, Jerome Geyer-Klingeberg, Andreas Rathgeber

University of Augsburg, Augsburg, Germany

A critical factor in determining the ecological and economic benefits of photovoltaic (PV) investments is the projected lifespan of the installed PV modules. A well-founded estimate of the decline in power output over an extended period of time is essential in assessing whether an installation under a specific set of conditions can meet performance expectations. This power decline, commonly referred to as degradation rate (DR), is dependent on multiple conditions that result in a large heterogeneity of contingency factors determining the DR. To derive the summarized effect of all reported DRs of outdoor exposed PV installations across the entire literature and explain the large heterogeneity, we conduct a meta-regression analysis (MRA) using a wide set of moderator variables, including publication characteristics, installation differences, methodological differences, and geographical characteristics. Our analysis of 99 primary studies comprising 837 DR estimates reveals a median DR of 1 %/year, which is higher than those reported in previous reviews, with the technology of PV modules and the climatic conditions being the main drivers of the differences in reported DR estimates.

978-1-6654-6060-6/23 $31.00 © 2023 IEEE

The Use of a Physics-based DNI Model to Enhance the National Solar Radiation Database (NSRDB)

Yu Xie, Jaemo Yang, Manajit Sengupta, Yangang Liu

National renewable Energy Laboratory, Golden, CO, United States

Brookhaven National Laboratory, Upton, NY, United States

Direct normal irradiance (DNI) is often interpreted differently in ground measurements and forecasts using numerical weather prediction (NWP) models, leading to substantial bias in DNI computation and forecasting especially under cloudy-sky conditions. To mitigate the bias, we use the Fast All-sky Radiation Model for Solar applications with DNI (FARMS-DNI) to conduct physics-based simulations of solar radiation in the circumsolar region. The DNI is quantified using the sum of the radiation along the sun direction and the scattered radiation within the circumsolar region. FARMS-DNI is implemented in the National Solar Radiation Database (NSRDB) to assess 5-minute DNI in 2-km pixels over the contiguous United States. The DNI computation for 2019-2021 are validated using surface-based observations at the Atmospheric Radiation Measurement (ARM), Surface Radiation Budget Network (SURFRAD), National Renewable Energy Laboratory (NREL), Solar Radiation (SOLRAD), and University of Oregon (UO). The results show that the cloudy-sky DNIs from the NSRDB are improved as evaluated using absolute percentage error and mean absolute error.

Novel Module Architecture for Lower CapEx and Improved Recyclability for c-Si PV Modules

Ryan Ruhle,[1] Larry Maple,[1] Timothy DeLazzer,[1] Steve Johnston,[2] Dana Kern,[2] and Walajabad Sampath,[1]

[1] Colorado State University (CSU), Fort Collins, Colorado 80523, US

[2] National Renewable Energy Laboratory (NREL), Golden, Colorado 80401, US

Abstract — Photovoltaic (PV) energy production is currently increasing at a rate at which recycling is becoming necessary. A novel module architecture has been demonstrated that has potential for high value recycling for c-Si PV. This architecture eliminates the vacuum lamination process and cross-linked encapsulants. Functioning prototypes of c-Si have been fabricated for stress testing in collaboration with NREL. These modules are being tested against traditionally manufactured modules. Based on preliminary results, this module architecture is a potentially viable solution for improving the manufacturing cost and recyclability of PV modules while retaining module performance.

I. INTRODUCTION

This architecture uses a glass-glass packaging and perimeter seal. This structure eliminates typical cross-linked encapsulants and is therefore suitable for high value recycling (near 100% recovery of silicon, metals, and glass) [1]. Traditional module architecture requires costly vacuum lamination which significantly increases the factory floor space required, capital expense (CapEx), labor, and production time of PV modules [1]. By eliminating this step in the process, it would eliminate a significant amount of the initial manufacturing cost.

As well as improving the cost of manufacturing, this technology represents a significant step in the sustainability for end-of-life treatment of PV modules. Studies indicate that, currently, end-of-life treatment of PV costs $5.00-$9.00 to dispose of each module in a hazardous waste treatment facility, while the cost to recycle a module can reach $45.00[2]-[3]. A major consequence of this cost difference is that there exists a strong economic incentive to throw away modules containing hazardous materials instead of recycling. This represents a potential issue given the rapidly increasing scale of module production.

The major cost associated with recycling originates from the cost of crushing the module and chemical separation processes which are necessary using traditional module architecture [4]. With the novel module architecture being developed by CSU, the simplified process reduces the cost of module disassembly by enabling the PV device to be easily separated into constituent parts rather than crushed and chemically separated.

Not only can this fabrication process improve recyclability, but it is also important to address current degradation mechanisms for PV modules. In order to meet the goals of the DOE and achieve a 50-year module lifetime, this type of technology may also be necessary for improved robustness against moisture ingress and resistance to other degradation mechanisms such as degradation of the encapsulant and Potential Induced Degradation (PID) [1],[5].

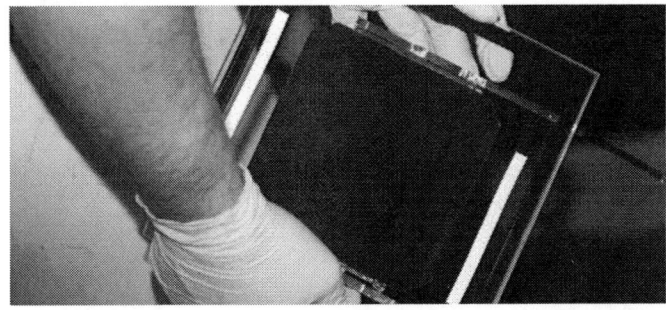

Figure 1: c-Si mini-module fabricated by the advanced encapsulation research laboratory at CSU. This sample utilizes acrylic nanostructure on the rear of the glass and top of the cell to improve transmission of light through various interfaces. More testing is needed to improve the application the c-Si cell.

One major hurdle which has been identified is internal reflection losses. This is due to a mismatch of indices of refraction between the glass, inert gas, and finally the cell itself. While some research has been conducted to demonstrate the possibility of utilizing anti-reflective (AR) coatings on the interior of the module, which may represent a significant barrier

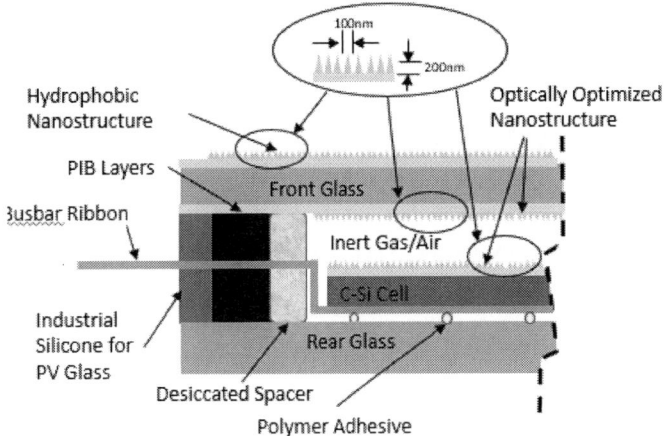

Figure 2: Cross-sectional view of proposed module architecture. The issue created by the optical mismatch between the layers of the material is solved utilizing a textured nanostructure which has been shown to reduce reflectance to 0.2% at glass interfaces.

to adopting this style of module. This is because in order to reduce internal reflection losses utilizing AR, this design would require multiple AR layers. These layers would need to be resistant to UV, have similar thermal expansion rates, and not outgas chemicals which may be detrimental to PV cells. Since these requirements are somewhat stringent, this multi-layer AR coating is somewhat prohibitive. CSU's module architecture addresses this issue via internal textures which reduce reflectivity more uniformly across the spectrum than AR coatings.

Further, optical modeling of the reflectance at each interface shows that the reflectance can be sufficiently lowered to match the power of traditionally encapsulated modules. This may be further improved by the addition of a textured surface on the top of the front glass of the PV module shown in figure 2.

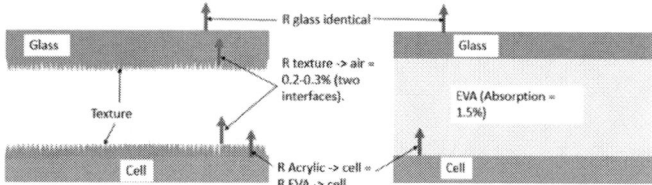

Figure 3: Optical model which demonstrates transmission of light based on reflectance modeling can be matched to traditional module architecture. This model assumes that internal losses for the interface from the texture to the cell are equivalent to the optical losses from the EVA to the cell and similarly for the texture to glass and EVA to glass.

It is the intent of this ongoing research to demonstrate the viability of this module architecture through experimentation and stress testing for c-Si. This architecture is based on manufacturing techniques developed at CSU and modules fielded by Abound Solar for Cadmium Telluride modules in 2009.

II. EXPERIMENTAL

The work relies on past optimization experiments done by CSU in order to rapidly produce non-vacuum non-laminated modules [4]. This work has included process time development, uniform deposition of materials for improved edge seal, reliability experiments for moisture ingress, and strength testing. The tests shown below were completed on 30cm x 30cm mini-modules with moisture indicating strips and no circuit plates.

- Damp Heat (DH) at 85 C and 85% RH (IEC 61215 10.1)
- Thermal cycles from -40C to 85 C (IEC 61216 10.2)
- 2400 Pa mechanical loading cycles (IEC 61217 10.6)

Under thermal and mechanical stress, no sign of delamination, moisture ingress, or discoloration of imbedded moisture indicator was seen. Modules were also fabricated of commercial size measuring 120cm x 60cm and subjected to various stress tests including mechanical loading, sample creep testing, and humidity freeze testing outlined in IEC 61251

10.12. At 40 cycles (4 x the standard test), and again, no sign of delamination, moisture ingress, or indicator color changes were seen.

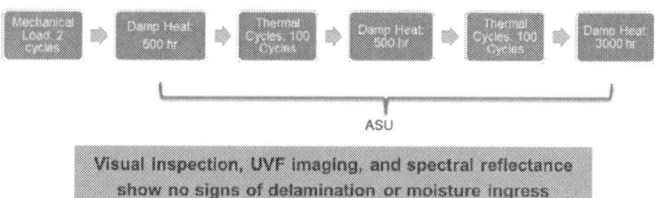

Figure 4: Accelerated stress test cycle for 120 x 60cm edge seal modules with moisture indicator.

Modules of size 20cm x 20cm, with c-Si cells fabricated by CSU shown in figure 1, are going through advanced stress testing in collaboration with NREL. Device fabrication is managed by CSU which utilizes an automated encapsulation tool and pass-through pneumatic press for streamlined production and programmability for varying module architectures as shown in figure 5[4]. Modules were also subject to 2,000 hours in damp heat. All modules undergo testing alongside traditionally manufactured modules.

This technology incorporates tested processes in a novel manner and is capable of manufacturing prototype module devices 120cm x 60cm maximum size. This process has been adapted for unique sizes to match testing capabilities at NREL. Utilizing equipment shown in figure 6, CSU has begun the fabrication of unique prototype modules which are currently undergoing testing.

Another significant consideration for c-Si modules which utilize a layer of gas between the front glass and device is the lack of optical coupling. Initially, CSU has begun testing acrylic nanostructure. Measurements of reflectance were taken as a function of wavelength and transmission is approximated as 1-R where R is reflectance and absorption is minimal in glass. This data combined with industrially provided data sheets on testing materials yields a potentially achievable improvement of 5.1% relative transmission through each side of the glass it is applied to. These tests and application optimization are

Figure 5: CSU's edge seal manufacturing equipment which has been developed initially for CdTe. It has now been adapted to produce greater variability in size of PV modules as well as c-Si modules for testing and prototyping.

currently in progress.

A prototype of the edge sealed module has undergone DH testing. This architecture utilizes a prototype side-junction box in order to lower manufacturing costs and further simplify the design. This test has been conducted for 6000 hours with no junction box, and 5000 hours with wires mounted internally and

Figure 6: Proof of concept for high-value recycling. Internal components are revered and separated without complex chemical processes.

led through to the exterior using a prototype side-junction box. The moisture indicators have not changed color during the test when the design is fully realized.

Finally, preliminary experiments to demonstrate high value recycling of this module architecture for c-Si modules have been conducted. This includes module disassembly and removal of device from the glass-glass structure in figure 6.

III. RESULTS AND DISCUSSION

Previous results attained by CSU have shown no signs of delamination or change in color of moisture indicator for 5,000 in DH testing, as well as standard module testing for both 30cm x 30cm modules and 120cm x 60cm glass-only modules. While these results are indicative of the potential for this module architecture, it is necessary to fabricate and test functioning c-Si modules. This has led to collaboration with NREL in order to accommodate testing for functioning c-Si modules produced by CSU.

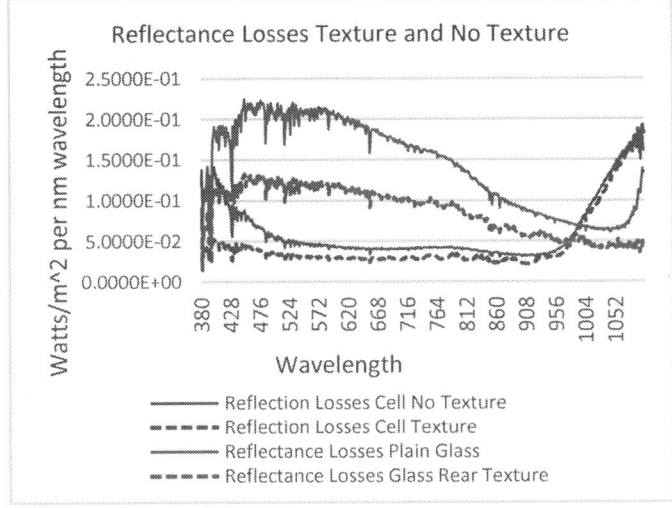

Figure 7: Reflectance losses measured comparing glass and c-Si cells reflectance with and without addition of acrylic nanostructure.

Results of testing indicate the viability of this module architecture. Under LED I-V testing CSU's novel architecture shows a power reduction of 8-9% relative compared to standard lamination. This is being addressed by utilizing a variety of anti-reflective textures and layers which can be applied at low cost. Preliminary results suggest a potential improvement of 5.1% relative improvement to transmission of the front glass, approximately 1.5% improvement from application of texture on the cell. Combined with a fully optimized glass package which increases transmission another 1.8% on average, experimental measurements indicate that this module architecture could potentially match the power output of traditionally manufactured PV modules.

Modules produced by CSU have been subject to initial measurements, DH200 (200 hours at 85 C, 85% relative humidity), A3 (Ultraviolet exposure at 85.8 W/m^2 irradiance at 300-400nm and 65 degrees Celsius), TC50 (50 thermal cycles), and are currently undergoing HF10 testing (10 Humidity Freeze Cycles). EL and PL imaging also shows no significant degradation at the current stage of testing. This is likely due to lack of micro crack formation during the manufacturing process in the c-Si cells which can mitigate micro crack growth during stress testing. EL and PL imaging of

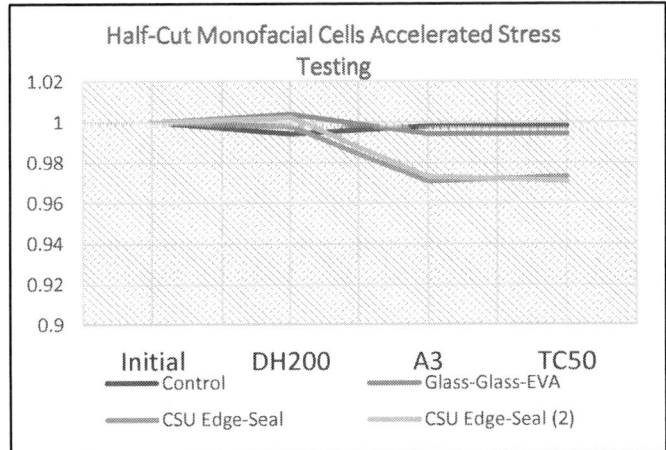

Figure 8: Relative power retained from initial testing after 2000 hours of PID exposure at 1,000 volts. CSU's edge seal architecture is compared to traditionally encapsulated modules.

standard lamination indicates some micro-crack formation and growth, while this has not been indicated through imaging in current testing with CSU's novel architecture yet.

This, in combination with better retention of performance indicates that utilizing improvements from textured nanostructure in future samples will be a cost-effective method to retain most of the initial power relative to traditional module architecture and perform even better under real-world module degradation.

CSU has also demonstrated that the utilization of a side-junction box can be included in this module architecture. This is a key factor in improving the technology as semi-transparent modules become more prevalent for agrivoltaics and windows which may require wires which do not interfere with the light transmission.

PID testing results (figure 9) also demonstrate robustness

against PID. This is likely due to the lack of encapsulant which

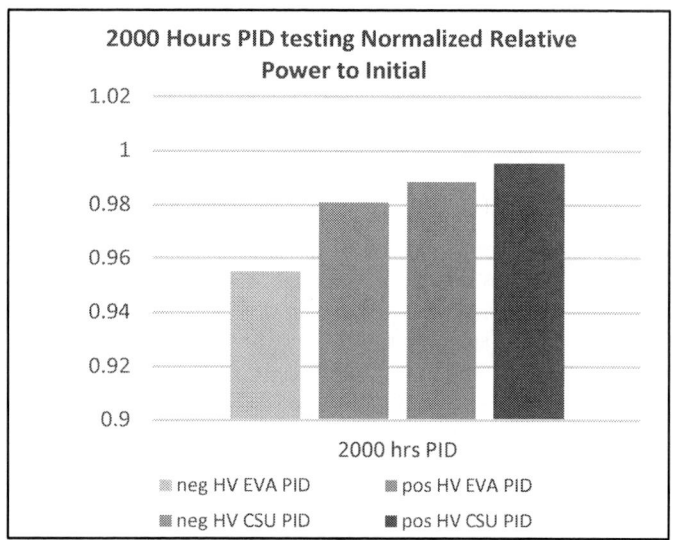

Figure 9: Relative power compared to initial power. CSU's edge seal module architecture is compared to traditionally encapsulated PV modules.

prevents sodium ions from transporting from the glass to the cell under high voltages. Test results are compared normalized to the initial performance of each cell. Each is exposed to a voltage difference either positive or negative of 1,000 volts for a total of 2,000 hours. Edge-sealed modules developed by CSU saw a significant improvement over traditionally encapsulated modules during this stress test.

During accelerated stress testing, CSU's edge-sealed modules performed well during all tests with the exclusion of exposure to UV light during A3 testing. This is likely because the cells utilize boron as a dopant. Modern and better performing cells will typically rely on utilizing gallium dopants which do not have the same issues as boron dopants. Best use cells for this technology should be explored in future iterations of experimentation.

IV. CONCLUSIONS

A novel module architecture has been developed that eliminates typical cross-linked encapsulants and is therefore suitable for high value recycling (near 100% recovery of silicon, metals, and glass) [1]. The structure promises extreme robustness to moisture ingress (5000 hrs. vs 1000 hrs. DH), improved mechanical robustness, significant reduction in Potential Induced Degradation (PID), and small manufacturing footprint (90% less) while improving module reliability [1],[4]. Challenges do exist in terms of the application of the texture onto the c-Si cell. Proper selection of test cells and continued improvements to the manufacturing process will be key points moving forward. These tasks in future research will improve the adoptability of this module architecture, however initial results are promising.

REFERENCES

[1] K.L. Barth, J. Morgante, W. S. Sampath, T. Shimpi and L. Maple, "Advanced encapsulation technology for reduced costs, high durability and significantly improved manufacturability," *2018 IEEE 7th World Conference on Photovoltaic Energy Conversion (WWCPEC) (A Joint Conference of 45th IEEE PVSC, 28th PVSEC & 34th EU PVSEC),* 2018.

[2] "Solar energy technologies office photovoltaics end-of -life action plan", *DOE,* 2022.

[3] Deng, R., Chang, N. L., Ouyang, Z., amp; Chong, C. M. "A techno-economic review of silicon photovoltaic module recycling," *Renewable and Sustainable energy Reviews,* vol. 109, pp. 532=550, 2019.

[4] Ellis, Samuel, T. Shimpi, A. Pavgi, G. Tamizhamani, W. Sampath, and K. Barth, "Reliability and manufacturing demonstrations of a new photovoltaic module architecture and streamlined approach to encapsulation," *2021 IEEE 48th Photovoltaic Specialists Conference* (PVSC). IEEE, 2021.

[5] Horowitz, Kelsey, T.W.Remo, B. Smith, and A.J. Ptak, "a techno-economic Analysis and cost Reduction roadmap for III-V," 2018.

ACKNOWLEDGEMENT:

The authors would also like to acknowledge the support from the National Science foundation under the Industry/University Cooperative Research Center (I/UCRC) for Solar Powered Future SPF 2050 (Award Numbers − 1821526 and 2052735), and the I/UCRC Industrial Advisory board. Support under the NSF INTERN program is also acknowledged

Snow sensing for photovoltaic single axis tracker systems

Ayush Chutani[a], Ana Dyreson[a], Laurie Burnham[b], and Kyumin Lee[c]

[a] Michigan Technological University, Houghton, Michigan, 49931, USA; [b] Sandia National Laboratories, Albuquerque, New Mexico, 87185, USA; [c] Array Technologies, Inc., Albuquerque, New Mexico, 87109, USA

Abstract — With the rapid deployment of photovoltaic (PV) systems around the world, PV is expanding in northern climate regions where snowfall is a major challenge for solar PV farms. Single axis trackers (SATs) increase the generation of solar plant at a competitive price and using bifacial panels is most effective in snow clad white ground due to high albedo. The objective of this project is to design and validate a snow measurement system that can measure snow on the top of moving panels by using laser-based sensors and digital image capture and to develop advanced tracker controls for improved snow shedding and higher efficiency in winters. The study is complemented by power generation measurement at panel and string levels as well as plane of array solar irradiation; meteorological data at the location is recorded for the correct weather assessment and ultrasonic snow depth measurement at ground level provides a comparison for snow measurements. The target panels are observed by various cameras to provide a holistic image of snowfall and shedding events. Considering the need for correct measurement of snow on the surface of PV panels, a snow measurement system enables the development of an empirical classification that can help in correlation of the angles needed to shed the snow. Snow shedding is an important factor in northern climates as it can increase the period for which the PV panels are generating power. This paper focuses on the novel snow measurement system design and prototype that measures snow on the top of moving solar PV panels.

I. INTRODUCTION

Photovoltaic (PV) systems are considered one of the most affordable and scalable renewable energy methods and more regions of the world are exploring the option of viable solar plants. Traditionally it was believed that solar PV is more beneficial in only dry and desert areas due to high solar irradiation, but the northern latitudes have become viable also. System designs that include bifacial solar PV modules especially benefit from the high solar reflectance or albedo of ground snow in snowy climates.

Despite significant research on wind loading and stow algorithms for hurricane-type events, there is a lack of studies on tracker behavior in snow. Optimizing for snow can improve system efficiency, increase energy availability, and reduce stress on modules and motors. Thus, there is a pressing need for research on snow monitoring and advanced controls for solar tracker systems. Even though snow has high albedo it comes with a basic problem of covering the top surface of solar PV and rendering the generation from top part essentially zero. Even after considering the number of overcast days installing a solar PV system or a plant in northern latitudes and regions like the northern United States, Canada, and northern Europe, the economic benefits and climate change awareness to offset carbon footprint makes solar PV more attractive than conventionally powered electricity generation. There has been a lot of interest in making solar PV systems more reliable and robust in extreme weather conditions [1] and snow is one part of the problem. This project focuses on single axis solar trackers (or SAT), which provide increased overall generation by following the elevation of the sun, and are increasing in market share across the U.S., including in Northern climates [2].

The current project focuses on snow on SAT solar panels. There have been studies on the fixed tilt solar PV systems in areas with high snowfall and models for the optimum tilt angle of the structure [3] but this work is the first known attempt to understand the parameters that contribute to snow shedding from SATs. Single axis tracking system can be used for snow removal due to its controlled drive system so that we do not have to invest in external snow removal process. In theory, we can simply change the angle of tilt to induce snow shedding based on the amount of snow on panels as was proven for fixed tilt systems.

In this study, the objective is to develop a system to measure and quantify the amount of snow that accumulates on SAT solar panels and then quantify the relationship between back of module temperature, energy generation, tilt angles, and snow shedding. To that end, we have partnered with Array Technologies on the design and installation of an experimental SAT system at the Michigan Regional Test Center (MI RTC) for Emerging Solar Technologies in Calumet, MI, at a latitude of $47°\ 07'\ 02''$ N and longitude of $88°\ 33'\ 45''$ W. The site routinely sees some of the highest annual snowfall in the US, 325.6 inches in 2021-2022 season due to lake effect snow [4].

II. SENSOR SELECTION

The design for the sensor system was based on the objective to have quantified and comprehensive data for snow depth on the project site as well as on the SAT using snow depth sensors. The idea behind the design approach is that at any time an observer should be able to tell how much snow is on the snow panels under observation as well as the weather conditions affecting it. The MI RTC already has a meteorological station but previously lacked snow instrumentation. Our first step was to identify and compare different snow depth measurement technologies to design a comprehensive snow-monitoring system for the SAT.

The different snow measurement technologies and sensors we evaluated are listed in Table 1.

S no.	Model	Type	Functions
1	Cambell Scientific SnowVUE 10	Ultrasonic	Snow depth
2	Cambell Scientific SR50 ATH	Ultrasonic	Snow depth
3	Cambell Scientific SDM S40	Multi-point laser	Snow depth
4	**Sommer USH9**	Ultrasonic	Snow depth
5	**Lufft SHM31**	Multi-point laser	Snow depth
6	Snowscale SSC230	Load	snow load
7	Kipp and Zonen Dust IQ	Soiling monitor	Soiling losses
8	Cambel scientific CCFC-R2	Image analysis	snow cover
9	Luft RS45	Image analysis	snow cover
10	**Reolink RLC-810A**	Image analysis	snow cover
11	Snowowl	Digital	Precipitation
12	**Lufft WS100**	Radar	Precipitation type
13	**Campbell Scientific CS241**	Temperature	Module back temperature
14	Tmodul	Temperature	Module back temperature
15	**Cambell Scientific CR1000X**	Data Logger	onsite data logging
16	Sutron XLink500	Data Logger	onsite data logging
17	**AcuDC 240**	Generation Data	DC energy meter
18	**SUNNY TRIPOWER CORE1 33-US**	Generation Data	Invertor
19	Kipp and Zonen RT1	POA	Irradiation sensor
20	**Kipp and Zonen SMP12**	POA	Pyranometer
21	Lufft WS600	MET	All-Weather sensor
22	Flow cap 4 Isaw	Snow drift snesor	Particle flux
23	Sommer SND V1.41	Snow drift snesor	Particle flux

Table 1. Commercially available snow sensors researched for the snow measurement system.

The sensor system is designed for the Array Technologies single axis tracker DuraTrack HZ v3 with a SmartTrack control algorithm and Trina bifacial solar panels. The layout of the PV plant is 10 rows of panels of 14 solar modules in each row.

Parallel 1 configuration and four dummy modules on the edges to offset the edge effects of snow and wind drifting are used. Snow can be measured by both weight and volume. Using the weight as a quantifying parameter provides data on loads and strain; volume provides data on snow density. With our project, the latter one has been chosen. From the sensors in Table 1, we selected SHM31 laser sensor since it can be used at an angle to measure snow which is needed to avoid shadows on panels. The laser sensor also has higher accuracy than ultrasonic sensor for the snow depth due to motion of SAT. We selected security night mode 4k cameras to work in conjunction with the monitoring cameras at site. We selected SMP12 pyranometers that offer inbuilt heating option which is required in cold temperatures and costs less than other counterparts in market.

III. SNOW SENSING SYSTEM

A. Measurement Scheme

The measurement system design in Fig. 1 shows the sensors and communications for the study. We designed the system to include a laser-based snow depth sensor and two plane of array (POA) pyranometers, one facing the top and one facing downwards to measure solar irradiation in the plane of the solar panels under observation. The site has a research-quality meteorological station that measures ambient temperature, humidity, dew, solar irradiation, wind speed and direction as well as precipitation amount. In addition, we installed a precipitation radar that identifies the exact time and type of precipitation. Ultrasonic snow depth measurement is also installed to provide the accurate measurement of snow depth on the ground level at site. Our monitoring system has three major data streams:1) measurements from sensors on the moving solar PV panels, 2) measurements from sensors installed on the ground level and 3) current and voltage measurements from inverters and DC energy meters.

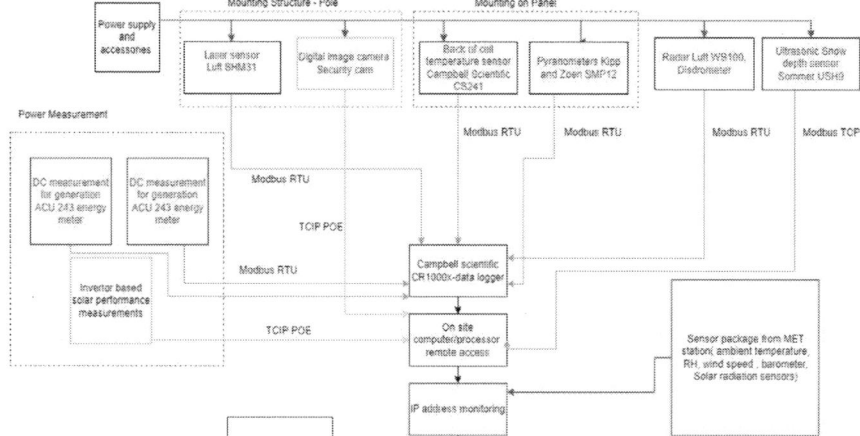

Fig. 1. Line diagram of snow depth measurement scheme and communications. The sensor system communicates via Modbus RTU and TCP connections as well as SDI12 protocols for Laser sensor. The data logger at the center is chosen to be Campbell Scientific 1000x for its utility and usage in extreme cold temperatures. The site can be accessed remotely for the real time data acquisition as well as monitoring.

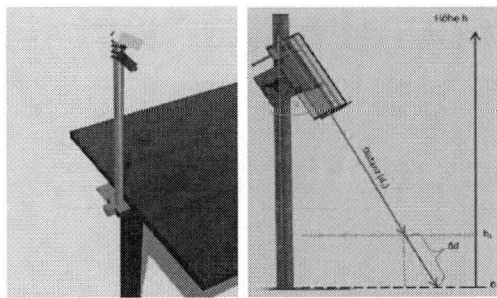

Fig. 2. a) 3D design of mounting on the rotating panels, b) Measurement diagram for laser sensor

The panel-based sensors include the laser sensor and a digital camera mounted on a perpendicular fixture to the SAT system The camera rotates along with solar panels so as the relative motion between the sensor and the panels is zero as in Fig. 2. There are also POA pyranometers and back-of-module temperature sensors for the modules installed on panels under observation. The ground-based snow depth sensors and meteorological station account for ambient conditions. The energy generation data from the inverter is used as well as DC monitoring system and the images from cameras installed at different angles on the site to run the image processing models for snow coverage.

B. Conceptual Design of Moving Sensors

The measurement of snow over a rotating platform is challenging due to the constant motion of the panels and because the sensors available commercially, use a stationary horizontal surface as target. The sensor should be aligned to the observation surface with a mounting angle α which should lie in the range of 10° and 30° [6].On one hand where the smaller angles below 10° can lead to a false reading and inaccurate measurements due to falling snow from the sensor housing, increasing angles greater than 30° results in increased noise in the measurements. The snow depth is calculated by following formula where h_r is the mounting height of the laser and mean(d) is the average result of distance measurements by laser sensor from the snow surface.

$$h_1 = h_r - h = h_r - mean(d).\cos{(\alpha)} \qquad (1)$$

Given the specified installation constraints, our design is to mount the laser sensor on the northern side of the array to eliminate any potential shading on the panels. The laser will be mounted on a custom-built rig that will be affixed to the torque tube of the entire array. This will ensure the alignment between the measurement surface and the laser sensor is always maintained. The rig is built from extruded aluminum which has slots for different positioning of sensors as well as equipment for future studies. The digital camera which collects time-series images is installed along with the laser sensor to capture the same surface the laser sensor is capturing.

IV. LONG-TERM MONITORING PLANS

The project will be conducted over two winters with the first winter season focused on creating a monitoring system for snow on SATs and deploying the system to establish a baseline for energy generation and snow measurement. This will include the energy generation for the whole year, the impact of snow on generation losses, a comparison of snow measurements from the different sensors, and data on the amount of snow on ground and panel surface as well as solar irradiation in plane of array. The second winter will be focused on developing smart control algorithm for the SAT drive system to initiate snow shedding or optimizing the tilt angles to minimize the losses due to snow. If our study shows increased performance in winter because of having better data, the control algorithm could become a part of ongoing tracker control architecture and in turn help in increasing the energy generation.

V. CONCLUSION

Snow depth measurement on a rotating panel is a novel problem and with our concept of measurement, the results can be used to create empirical relationships between generation data, images and ground-based snow depth and onsite weather data. Even though there are some unknown parameters like snow drift and frosting, the current studies nevertheless provide new information about snow behavior on SAT systems which are important as SAT increase in market share. The research we are conducting represents an important step toward increasing the efficiency and availability of SATs in northern latitudes and therefore is intended to accelerate the transition to a solar-dominant energy economy.

ACKNOWLEDGMENTS

This work is funded in part or whole by the U.S. Department of Energy Solar Energy Technologies Office, under Award Number 38527. We would like to acknowledge Array Technologies, Inc.

REFERENCES

[1] Bolinger, M., Seel J., Warner, C., Robson, D., *Utility-Scale Solar*, 2021 Edition, Lawrence Berkeley National Laboratory, October 2021, https://emp.lbl.gov/utility-scale-solar

[2] Jackson, N. D., & Gunda, T. (2021). Evaluation of extreme weather impacts on utility-scale photovoltaic plant performance in the United States. *Applied Energy*, *302*, 117508. https://doi.org/10.1016/j.apenergy.2021.117508

[3] Marion, B., Schaefer, R., Caine, H., & Sanchez, G. (2013). *Measured and modeled photovoltaic system energy losses from snow for Colorado and Wisconsin locations*. Solar Energy, 97, 112–121. https://doi.org/10.1016/j.solener.2013.07.029

[4] (n.d.). *Keweenaw County Snowfall*. Keweenaw County Online. Retrieved January 13, 2023, from https://www.keweenawcountyonline.org/snowfall-full.php

[5] Burnham, L., Riley, D., King, B. H., Braid, J., Dice, P., Dyreson, A., Snyder, W., & Pike, C. (2022). Dedicated cold-climate field laboratory for photovoltaic system and component studies: the Michigan Regional Test Center as a case study. *2022 IEEE 49th Photovoltaics Specialists Conference (PVSC)*, 0333–0335. https://doi.org/10.1109/PVSC48317.2022.9938861

[6] *User Manual SHM 31 Snow depth sensor*. (2018 Rev 1.9). Fellbach, Germany: Lufft.

Operando Temperature Measurements of Photovoltaic Laser Power Converter Devices under Continuous High-Intensity Illumination

John F. Geisz, Daniel J. Friedman, Myles A. Steiner, Ryan M. France, Tao Song

National Renewable Energy Laboratory, Golden, CO, United States

Photovoltaic devices that operate under extremely high irradiances, such as laser power converters, may also operate at elevated temperatures as a result of large temperature gradients. We demonstrate the operation of an inverted GaAs laser power converter device under irradiances over six orders of magnitude in a full-featured monochromatic laser simulator with an active cooling stage. The VOC as a function of irradiance is shown to droop at high irradiance as a result of junction heating, but the junction temperature can be difficult to measure by conventional methods. Fast measurements of the open-circuit voltage under these extreme operating conditions are used to determine the actual junction temperature. Empirical voltage temperature coefficients of the device at each irradiance of interest are determined and used as an integral part of this technique.

Optimization of Back-Contact Diffusion Barrier for Solution-Processed CIGS Solar Cells: Case of MoO_3 and MoN

Nada Benhaddou, Jacques Kenyon, Luke O. Jones, Liam M. Welch, and Jake W. Bowers.

Centre for Renewable Energy Systems Technologies, CREST, Wolfson School of Mechanical, Electrical and Manufacturing Engineering, Loughborough University, Loughborough, LE11 3TU, UK

Abstract — This study aims to investigate the effect of MoN and MoO_3 diffusion barriers on the back contact properties of solution-processed CIGS solar cells. The deposition conditions were varied to assess the optimal conditions inhibiting the degradation of Mo to $MoSe_2$ during CIGS selenization. The results show that a proper adjustment of the sputtering gas mix leads to better control over the subsequent amount of $MoSe_2$ formation. 5% efficiency can be obtained by using MoN for an amine-thiol solution-produced CIGS absorber using a Na-free glass substrate.

I. INTRODUCTION

Chalcopyrite Cu(In,Ga)Se$_2$ (CIGS) solar cells are recognized as a promising technology arousing the interest of both academic researchers and industry. This absorber material is characterized by a direct band gap and a high absorption coefficient ($10^7 cm^{-1}$), therefore allowing the use of thinner films compared to other photovoltaic materials while maintaining good yields [1].

In the case of CIGS-based solar cells, the fabrication starts with back contact deposition. This step has to follow many requirements, e.g., high conductivity for efficient charge transfer, chemical stability to prevent reaction with the absorbing layer, and strong adhesion with both the glass substrate and the absorber [2]. Molybdenum (Mo) is the most commonly used back-contact material for substrate-configuration chalcogenide thin film solar cells.

Although Mo withstands high temperatures during CIGS thermal treatment, a small portion of it reacts with selenium (Se) vapor in the furnace to form a layer of $MoSe_2$. This layer can be beneficial within a certain thickness to provide good ohmic contact, and good adhesion at the Mo/CIGS interface. Nevertheless, it can be deleterious for carrier collection not only due to its high resistivity but also because it affects the band alignment at the back interface [3].

It is worth mentioning that the porous microstructure of the solution-processed absorber leads to more exposure of Mo to Se vapor, hence, its optimization is of utmost importance to obtain good device efficiencies.

In an attempt to prevent the overselenization of Mo, many research groups have suggested the introduction of intermediate layers at the Mo/absorber interface as diffusion barriers. For instance, transition metals and their nitrides (e.g., TiN, Ta-N, W-N, and Mo-N) have been used as effective barriers against excessive diffusion of Se towards Mo [4]-[5].

On the other hand, Mo oxides have proved their efficacy in controlling the formation of $MoSe_2$, as is the case of a thermally evaporated MoO_2 layer in a tri-layer Mo configuration that has led to a drastic increase in Voc and more than a 2% absolute increase in efficiency [6].

The present work sheds light on the Mo back contact, with an emphasis on the effect of the sputtering gas mix on the ability of the MoN and MoO_3 interfacial layer to cope with rear contact degradation during selenization for solution-based CIGS solar cells.

Overall, it is widely known that the formation of the $MoSe_2$ layer is influenced by not only the selenization but also by the deposition conditions of the Mo film. Unlike what has been published for Mo oxide barrier layer in the case of chalcopyrite solar cells, this work proposes an in-situ introduction of O_2 without breaking the vacuum during Mo deposition [7,8]

Preliminary results showed that tuning the N_2/Ar gas mixture ratio in the Mo-N deposition can improve the overall cell efficiency by reducing the series resistance from 4.8 to 1.3 Ωcm^2, leading to the production of an amine-thiol solution-based CIGS solar cell with an efficiency of 5%.

II. EXPERIMENTAL

A. Molybdenum multilayer deposition

The multi-stack Mo back contact used in this work follows a multilayer approach as depicted in Fig. 1 The Mo layer deposited at a high sputtering pressure (HPr Mo) is used to enhance the adhesion to the substrate whereas the Mo layer deposited at a low sputtering pressure (LPr Mo) ensures a good conductivity.

MoOx and MoN barrier layers have been deposited by introducing O_2 or N_2 as reactive gases to the sputtering chamber for Mo nitridation/ oxidation along with Ar. A thin sacrificial layer of Mo (Mo-sac) is deposited as a top layer (~25 nm) to be intentionally selenized to improve the contact with the absorber.

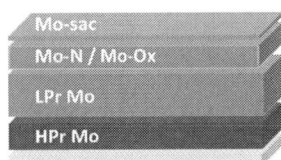

Fig. 1. Schematic representing the Mo multilayers used in this work.

The different layers of the back contact were deposited onto Na-free EAGLE XG (Corning) substrates. The deposition was carried out using a D.C. magnetron sputtering system (AJA International ATC 2200-V). The Mo target was pre-cleaned by sputtering for 30 min with shutters closed before use and for 30s prior to each layer deposition.

B. CIGS solar cell fabrication

CIGS precursor solution of 0.2 M is prepared in a 3-vial method, by dissolving individual metal chalcogenide precursors indium sulfide (In_2S_3), copper sulfide (Cu_2S), and elemental gallium with selenium powder (Ga + Se) in 1,2-ethylenediamine (EDA)/1,2- ethanedithiol (EDT) solvent mixture with 10/1 v/v ratio. The final goal ratios are of Cu/(Ga+In)= 0.9, and Ga/(Ga+In)= 0.3. Once dissolved, the 3 vials are combined and filtered before use.

The precursor layer is subsequently deposited by spin coating at 1500rpm for 60s onto Mo-coated Eagle glass and annealed on an N_2- connected hotplate (310°C, 60s). This operation is repeated 15 times to obtain the desired thickness of CIGS layer (~1.5µm).

The selenization of the CIGS precursor occurs in an RTP furnace by putting the as-deposited CIGS in a closed graphite box with Se pellets (~600mg). The details of the thermal treatment conditions are illustrated in Fig. 2.

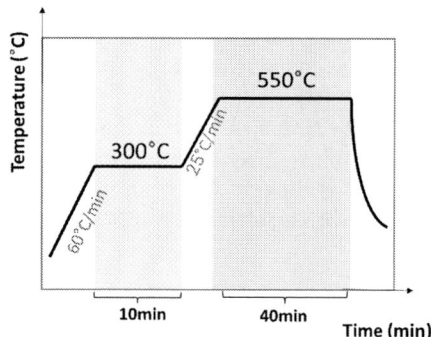

Fig. 2. Selenization regime of CIGS absorber in the RTP.

CIGS solar cells were completed by depositing CdS via a chemical bath, followed by the RF sputtering of i-ZnO and AZO transparent conductive oxide layers. Finally, Ag front contacts were deposited by thermal evaporation.

III. RESULTS

A. Effect of gas mixture ratio on intrinsic Mo-N and MoOx barrier layers

For the nitridation or oxidation of Mo, N_2 or O_2 reactive gases were introduced into the sputtering chamber with different proportions to Ar (33%, 50%, 66%), while other parameters were kept constant (sputtering power, pressure, total gas flow). The deposition rate varies depending on the gas mix ratio and the reactive gas. To compare the behavior of both barrier layers,

the deposition rate was determined to compare the same film thickness for all films. Fig. 3 elucidates the effect of the gas mix ratio on the deposition rate. The introduction of more N_2 into the chamber lowers the deposition rate of Mo-N, in contrary to O_2 showing thicker films with high O_2 in the gas mixture.

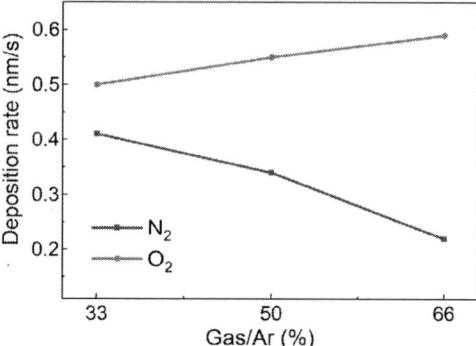

Fig. 3. Variation of the deposition rate of Mo-N and MoOx barriers layers with gas/Ar ratios

To assess how the composition of the gas mix used during Mo deposition as well as variations in film thickness affect the electrical resistivity of Mo-O_x and Mo-N films, the sheet resistance of each individual barrier layer was measured by four-point probe, and the thickness was determined using SEM cross-section.

Fig. 4 illustrates the resistivity behavior of Mo-O_x and Mo-N films with approximately the same film thickness deposited on glass. it is noticeable that the electrical resistance increases rapidly with more incorporation of N in the film during the formation of molybdenum nitride, while the trend of increase is less pronounced in the case of Mo-O_x.

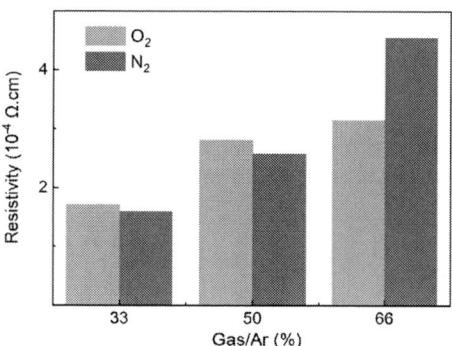

Fig. 4. dependence of resistivity to the gas mix ratios in MoN and MoO_3 barrier layers

Table I summarizes the correlation between the gas mix ratio of single Mo-O_x and Mo-N layers, as well as the thickness with their corresponding resistivities.

It was evidenced that the sheet resistance increases with the increasing the reactive gas content in Mo regardless of its nature and increasing the layer's thickness helps in reducing the resistance. The lowest values of sheet resistance are obtained with 33% O_2 in Mo-O_x.

TABLE I: SHEET RESISTANCE OF DIFFERENT THICKNESSES OF MoOx AND Mo-N LAYERS ON GLASS USING VARIED O_2 AND N_2 TO Ar GAS RATIOS

	Gas %	Thickness (nm)	Sheet resistance (Ωsq)	Resistivity (10^{-4}Ωcm)
MoO₃	33%	110	15.6	1.7
		150	4.8	0.7
		300	2.4	0.7
	50%	114	24.7	2.8
		170	13.5	2.3
		330	6.5	2.1
	66%	94	33.5	3.1
		190	18.7	3.5
		335	12.5	4.2
MoN	33%	95	28.4	2.7
		114	14.1	1.6
		247	6.7	1.6
	50%	104	44.7	4.6
		120	21.5	2.6
		205	10	2
	66%	60	150	9
		85	74.5	6.3
		130	35	4.5

The chemical analysis of Mo-O_x and Mo-N barrier layers was conducted using X-ray photoelectron spectroscopy (XPS). All spectra were recorded using a Thermo Scientific K-Alpha XPS with a monochromated Aluminium Kα source. The sample surface was characterized using survey scans and high-resolution scans with the following parameters: Surveys were measured using 100 eV pass energy, 1 eV step size, 10 ms dwell time averaged over 10 Scans. High-resolution scans were measured using 25 eV pass energy, 0.1 eV step size, 250 ms dwell time averaged over 1 scan for each element of interest. All results were charge corrected to carbon C1s at 284.8 eV. A 30s depth profile using Argon ions was used on the MoN samples to analyze the chemistry of the bulk material. The effect of gas ratios on both barrier layers is illustrated in the XPS scan of Mo 3d doublet core levels (Figure 5).

In the case of Mo-N barrier layer, the Mo3d5 peak at 228.1eV indicates the presence of Mo (Metal and/or MoN). The presence of MoN is supported by N1s peak at 397 eV (not represented here). It is worth mentioning that the amount of metallic Mo decreases linearly with increasing the nitrogen content in the sputtering chamber (Metallic Mo 61% with 33% N_2/Ar to 30% with 66%), leading the way to the formation of MoN phase. The slight shift to higher binding energies noticed in the sample with 66% of N_2 emphasizes the incorporation of more N atoms into Mo and the formation of more MoN

compared to Mo metal. Along with MoN phase, some other oxides (MoO_x with Mo: O =1:0.5) were also detected. This is due to the reduction of Mo Oxides due to Ar sputtering and X-ray exposure.

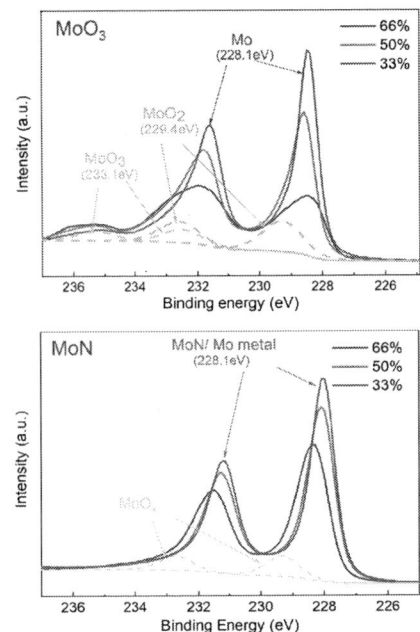

Fig. 5. XPS spectra of Mo 3d region for both barrier layers, MoN (bottom), and MoO₃ (top), with different O_2 and N_2 ratios to Ar.

XPS spectra of Mo oxide films indicated the presence of metallic Mo with different proportions in all samples quantified from Mo3d$_5$ main peak at a binding energy of 228 eV. Metallic Mo decreases from 29% to 9.6% as the amount of O_2/Ar increases to mainly form MoO₃ phase at 232.3 eV. MoO₂ phase is present in all samples (229.1eV). As aforementioned, the analysis of MoOx samples was superficial, thus, without excluding the native oxides on the surface. According to the literature [9], this phase is prone to form by exposure to air because MoO₃ is not influenced much by storage.

B. Resilience of MoO₃ and MoN barrier layers against selenization

Mo back contacts with 100 nm barrier layers of each gas mix ratio were selenized with and without the CIGS absorber layer to investigate the effect of the N_2 or O_2 content on the MoSe₂ formation. The back contacts using MoO₃ and MoN barrier layers were selenized under the same conditions as CIGS absorbers and are shown in Fig. 6.

Higher O_2 concentrations than 33% in MoOx films have led to severe delamination of the whole Mo layer, while samples containing Mo-N survived the selenization only when N_2/Ar exceeded 50%.

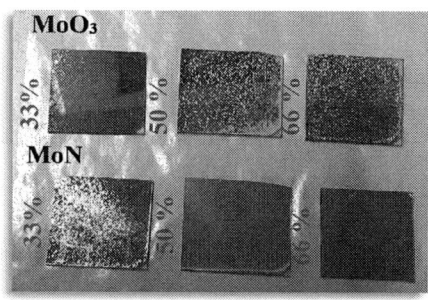

Fig. 6. photograph of Mo back contact with MoO₃ (top), MoN(bottom) with different gas ratios.

The thickness of the MoN diffusion barrier is expected to affect its ability to control the amount of Mo converted into $MoSe_2$. Therefore, we have varied the deposition time of this layer and probed its structural properties before and after the annealing by XRD analysis.

XRD data were collected using a Bruker D2 Phaser diffractometer equipped with Lynxeye™ detector and Cu Kα source.

As illustrated in Fig. 7.a, XRD spectra show a distinct peak at ~ 36.9°, attributed to the hexagonal MoN phase (200) for the sample with a thickness of ~100nm. The largest peak observed at 40.5° is assigned to the (110) plane of Mo with a body-centered cubic (BCC) crystal structure.

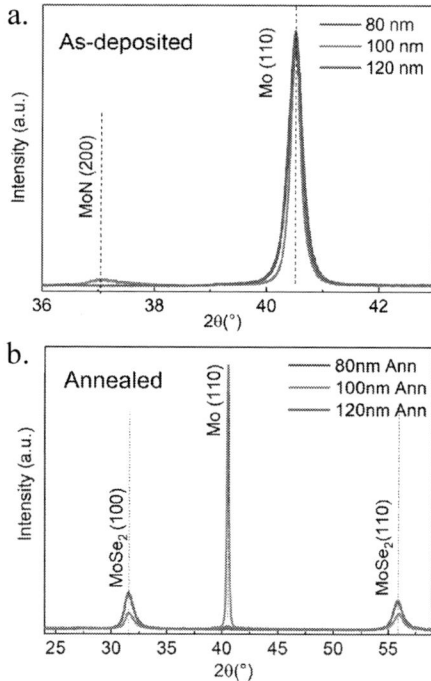

Fig. 8. XRD spectra of Mo back contact with different MoN thicknesses, as-deposited (a), and annealed (b).

The barrier effectiveness was clearly noticed for the annealed samples, where an obvious reduction in $MoSe_2$ amount using this thickness was observed whilst Mo was completely consumed in the case of the other Mo_2N film thicknesses (Fig. 7.b).

Scanning electron microscope (SEM) cross-section images of Mo back contact with different barrier layers were acquired using JEOL JSM-7800F FE-SEM equipped with the Oxford Instruments energy. dispersive Xray spectroscopy (EDX) detector used for elemental composition

Fig. 8. shows that $MoSe_2$ thickness increases proportionally with decreasing the amount of N_2 in the mixture. It was noticed also, that the MoSe2 thickness remained almost similar with and without a CIGS layer on top of Mo, indicating that the absorber layer does not stop Se diffusion and the barrier effect stems only from the MoN layer.

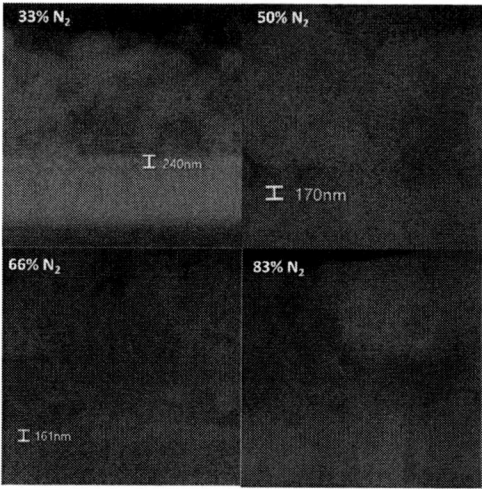

Fig. 7: EDX-SEM cross-sections of annealed CIGS-based solar cells with different N₂/Ar ratios in MoN layer of the back contact.

C. Effect of barrier layers on optoelectrical properties

Completed solar cells were fabricated with the different Mo samples. Table I regroups the electrical parameters extracted from the J-V curve. The results show an N_2/Ar ratio of 66% has led to the least series resistance, thus producing a better FF and overall efficiency.

TABLE II

ELECTRICAL PARAMETERS OF COMPLETED SOLAR CELLS WITH DIFFERENT N_2/AR RATIOS IN MO BACK CONTACT

N_2/Ar	33%	50%	66%	83%
J_{sc} (mA/cm²)	18	18	18	18
V_{oc} (mV)	392	366	383	368
FF (%)	48	52	58	52
Efficiency (%)	3.4	3.4	4.1	3.6
R_s(Ω.cm²)	4.8	1.9	1.3	2

Using the optimal parameters obtained from this study, completed devices were fabricated. Fig. 9-a shows current density – voltage (J-V) curves, and Fig. 9-b, the external quantum efficiency (EQE) spectrum of the best device obtained from absorbers prepared on the optimized Mo back contact.

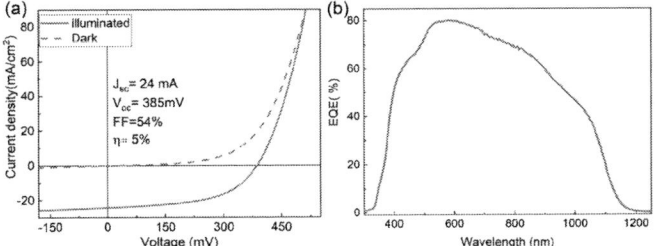

Fig. 9: J-V curve (a) and EQE spectrum (b) of CIGS solar cells using optimized Mo back contact parameters

IV. CONCLUSIONS

It has been demonstrated through this work that a fine-tuning of barrier layer deposition parameters can have a drastic effect on the selenization of Mo during the absorber thermal treatment. The use of 66% of N_2/Ar sputtering gas mixture for Mo-N barrier layer can lead to reduced series resistance in the final device (PCE= 5%, FF=54%, Voc=385 mV, Jsc=24 mAcm^{-2}). Further work is still ongoing to allow a better understanding of both barrier layers, and more precisely to investigate the effect of thickness variation in both barrier layers on solar cell performance. Additionally, the effect on the absorber crystallinity at the bottom interface will be studied.

REFERENCES

[1] B.Barman, and P. K. Kalita. "Influence of back surface field layer on enhancing the efficiency of CIGS solar cell." *Solar Energy 216* 329-337. (2021).

[2] N. Bansal, K. Pandey, K. Singh, and B. C. Mohanty. "Growth control of molybdenum thin films with simultaneously improved adhesion and conductivity via sputtering for thin film solar cell application." *Vacuum* 161 347-352. (2019).

[3] K.H. Ong, et al. "Review on substrate and molybdenum back contact in CIGS thin film solar cell." *International Journal of Photoenergy* (2018).

[4] C.W. Jeon, T. Cheon, H. Kim, M.S. Kwon, and S.H. Kim. "Controlled formation of MoSe2 by MoNx thin film as a diffusion barrier against Se during selenization annealing for CIGS solar cell." *Journal of Alloys and Compounds 644* 317-323. (2015).

[5] S. Uličná, A. Panagiota, A. Abbas, M. Togay, L. M. Welch, M. Bliss, A. V. Malkov, J. M. Walls, and J. W. Bowers. "Deposition and application of a Mo–N back contact diffusion barrier yielding a 12.0% efficiency solution-processed CIGS solar cell using an amine–thiol solvent system." *Journal of Materials Chemistry A* 7, 12 7042-705.(2019)

[6] S. L. Marino, M. E. Rodriguez, Y. Sanchez, X. Alcobe, F. Oliva, H. Xie,M. Neuschitzer, S. Giraldo, M. Placidi, R. Caballero, et al., "The importance of back contact modification in Cu2ZnSnSe4 solar cells: the role of a thin moo2 layer,"*Nano Energy*, vol. 26, pp. 708–721, (2016).

[7] Gretener, C., J. Perrenoud, L. Kranz, C. Baechler, S. Yoon, Y. E. Romanyuk, S. Buecheler, and A. N. Tiwari. "Development of MoOx thin films as back contact buffer for CdTe solar cells in substrate configuration." Thin Solid Films 535 (2013): 193-197.

[8] Chen, Wenjian, Teoman Taskesen, David Nowak, Ulf Mikolajczak, Mohamed H. Sayed, Devendra Pareek, Jörg Ohland et al. "Modifications of the CZTSe/Mo back-contact interface by plasma treatments." RSC advances 9, no. 46 (2019): 26850-26855.

[9] S. Lin, L. Wei, Z. Yunxiang, C. Shiqing, F. Yu, Z. Zhiqiang, H. Qing, Z. Yi, and S. Yun. "Adjustment of alkali element incorporations in Cu (In, Ga) Se2 thin films with wet chemistry Mo oxide as a hosting reservoir." Solar Energy Materials and Solar Cells 174 (2018): 16-24.

Silicon Heterojunction Cell Metallization with Reactive Silver Inks: Printing Process, Ink Formula, and Interconnection

Steven J. DiGregorio, Michael W. Martinez-Szewczyk, Subbarao Raikar, Mariana I. Bertoni, Owen J. Hildreth

Colorado School of Mines, Golden, CO, United States

Arizona State University, Tempe, AZ, United States

Silver is the most expensive non-silicon component in photovoltaic cells. This is particularly salient for Silicon Heterojunction (SHJ) cells which can consume between 200-400 mg of silver in low-temperature silver pastes. Replacing low-temperature silver pastes with reactive silver ink can reduce silver consumption by producing more conductive fingers and lower contact resistances while avoiding high temperatures that can damage SHJ cells. This work furthers the research on SHJ metallization with RSI by discussing the high throughput printing scheme developed for this work. The process uses arrays of flexible needles that contact the substrate while printing. This unique approach solves many of the issues with drop-on-demand printing and can reach competitive cell throughputs. Next, we investigate reactive ink formulas that enable printing temperatures above 150 °C to achieve low resistivities and low finger widths. Lastly, we explore how the minuscule silver amount in RSI metallization impacts solderability for interconnection. We found that the solder can dissolve enough silver during soldering to delaminate the busbars, but solder already saturated with silver can enable soldering.

Optical properties of (InxGa1-x)2O3 alloys and evaluation as emitter layer in CST PV

Bishal Shrestha, Madan K. Mainali, Manoj K. Jamarkattel, Ebin Bastola, Adam B. Phillips, Michael J. Heben, Nikolas J. Podraza

The University of Toledo, Toledo, OH, United States

Indium gallium oxides (IGO) are wide bandgap, highly transparent materials with remarkable opto-electronic properties used in wide varieties of semiconducting devices. Ease of fabrication and tunability of its intrinsic properties controlled by In-to-Ga content, has lately leveraged its use in the field of cadmium selenide telluride (CST) based photovoltaics. We have evaluated the THz and infrared to ultraviolet complex dielectric function (ε) spectra and carrier transport properties of (InxGa(1-x))2O3 with x = 0.71, 0.55, 0.45, 0.36, and 0.28 using spectroscopic ellipsometry. With decreasing In content, bandgap energies increase from 3.62 to 4.14 eV, carrier mobilities increase from 397 to 1031 cm2/Vs, and carrier concentrations decrease from 7.7×10^{18} to 1.4×10^{17} cm^{-3}. Near infrared to ultraviolet spectra in ε spectra obtained for IGO is used to simulate external quantum efficiencies (EQE) for CST device and compared to reported experimental EQE. For all IGO compositions, EQE simulations incorporating incomplete carrier collection near the heterojunction and at the back contact best match experimental EQE to provide sources of optical and electronic losses in these devices.

Temperature Dependent Fill Factor in CdSe/CdTe PV Devices from -20 °C to 60 °C Under AM1.5G and AM0 Spectra

Nadeesha Katakumbura, Prabodika N. Kaluarachchi, Manoj Rajakaruna, Tyler Brau, Aesha P. Patel, Abudulimu Abasi, Ebin Bastola, David Raker, Adam B. Phillips, Michael J. Heben, Sorin Cioc, Randy J. Ellingson

Wright Center for Photovoltaic Innovation and Commercialization, Department of Physics and Astronomy, University of Toledo, Toledo, OH, 43606, USA

Abstract—In this work, a comparison between theoretical and experimental calculations, is presented on the temperature dependency of Fill Factor (FF) of a CdSe/CdTe PV layer stack under both irradiances, AM1.5G and AM0. An inhouse built Irradiance and Temperature Dependent JV Measurement Chamber is used to control the temperature and irradiance variation and thermal cycling on the PV modules and FF is obtained via measuring the JV parameters at different temperatures from -20°C to 60°C under AM1.5G and AM0. Well defined theoretical model is used to obtain the theoretical values for FF at each temperature. Slight increment in current and a respectively large drop of voltage were observed as expected with increasing temperature. Theoretically calculated FF was larger than the experimentally obtained FF and this difference highly dependent on r_s and r_{sh}. Both theoretical and experimental temperature coefficients are approximately equal and behave same at temperatures higher than 20 °C for AM1.5 and 40 °C for AM0. Fast recovery of cell performances with a little degradation was observed, based on FF, during thermal cycling. Future work is necessary on studying and understanding the overall temperature dependency of CdSe/CdTe layer stacks.

Keywords—fill factor, irradiance, temperature, solar cell

I. INTRODUCTION

CdTe thin film solar cells consist of direct band gap of 1.45 eV and high absorption coefficient as high as 10^5 cm^{-1}. And the capability of photovoltaic energy conversion is respectively high in these cells due to its band structure matches perfectly with the available solar spectrum [1], [2]. Hence, in the present, CdTe based thin film solar cells has been demonstrated to be a promising candidate and successfully commercialized in the photovoltaic market. Even though, unlike on Si and GaAs solar cells, the studies have been carried out on material properties and performances of CdTe based solar cells are very low. It is important to focus on these studies also as the CdTe solar cell structure is much different from that of Si and GaAs [3].

Therefore, CdTe solar cells may have its own characteristic, electronic and performance variations when it is exposed to general use and harsh environment conditions. The temperature these cells get exposed to is one of the major factors which varies over time, geography or the application they are being employed. Hence, it is important to study the behavior of CdTe solar cells with temperature as, in general use they get exposed to temperature variations from 15 °C (288 K) to 50 °C (323 K) [4]. When the extra-terrestrial applications are considered, this range can be even higher [5]. With the help of previous work [6]–[8], generally, it is known that increase in the operational temperature of a solar cell tend to lower its performances. These performances are mainly determined by the short circuit current (I_{SC}), Short Circuit Voltage (V_{OC}), Fill Factor (FF) and Efficiency (Eff). Here the FF is defined as a measure of quality of the solar cell. It compares the maximum power to the theoretical power that would be output at both the V_{OC} and I_{SC} together. Higher this value is the better for a solar cell. Theoretically this value is calculated using above mentioned V_{OC}, I_{SC}, and Maximum Operating Current (I_{MAX}) and Maximum Operating Voltage (V_{MAX}).

It is reported in previous studies that with increasing T, V_{OC} tend to decrease but slight increment happens in I_{SC}. Hence both FF and Eff decrease as the temperature increases. It is also shown that this Eff degradation is mainly due to the drop of V_{OC} [6]–[8]. Furthermore, studies have shown the significant effect of solar spectrum, light intensity, curve factor as well as the slight effect of series resistance (Rs), parallel resistance (Rsh) on the variation of FF along with the temperature [9]. And also, earlier work can be found out on FF experiment of a mc-Si solar cell within a temperature range of 25 °C to 60 °C at different but constant light intensities (215-515 Wm^{-2}) where, the results show that the FF temperature coefficient is negative [10]. And studies on the relationship between FF and temperature for Si solar cells at temperatures lower than 40 °C can also be found where FF varies significantly [11]. However, the conclusions are not consistent and almost nothing under the relationship between temperature and FF of CdTe based PV devices.

The work presented in this paper investigates the variation of performance parameters within the temperature range of -20 °C and 60 °C, hence discuss the temperature dependency of the FF and Eff of CdSe/CdTe PV devices. Here an inhouse built Irradiance and Temperature Dependent JV Measurement Chamber is used to achieve the precise temperature and irradiance variations on PV modules and V_{OC}, I_{SC}, V_{MAX}, I_{MAX} are measured at different temperatures under both AM1.5 and

978-1-6654-6060-6/23 $31.00 © 2023 IEEE

AM0 irradiance spectra [12]. And also, a comparison is carried out between the theoretical expectation and the experimental observation of the temperature dependency of the FF of CdSe/CdTe PV devices. A temperature cycling has also been carried out in order to understand the overtime FF behavior. The work presented here will be useful in predicting the CdTe based PV module performances between the temperature range of -20 °C and 60 °C, both for the terrestrial and extra-terrestrial use.

II. THEORETICAL BASIS

A. General Fomula

Fill factor is defined as the ratio of the maximum power output (P_{MAX}) at the maximum power point to the product of the V_{OC} and I_{SC} density and can be expressed as;

$$FF = \frac{V_{MAX} \cdot I_{MAX}}{V_{OC} \cdot I_{SC}} \qquad (1)$$

As per the characteristic curve in Figure 1, if the ideal conditions are considered, $V_{MAX}.I_{MAX} = V_{OC}.I_{SC}$. But when the general use of PV modules is considered, the shape of the characteristic curve measured tend to deviate from its ideal rectangular shape and FF is used as a numerical representation of this deviation between actual maximum power and the ideal maximum power.

In general, the most affecting factor for this deviation of FF with temperature is the variation of open circuit voltage (V_{OC}). Typically, as in Figure 2, the V_{OC} decreases as the cell temperature increases. This is due to an increase of thermally generated electron-hole pairs, which normally result in an increase of reverse saturation current of the cell. This also results in a decrease in the voltage at maximum power point of the cell, hence FF drops. When the short circuit current (I_{SC}) is considered, it typically increases with increasing temperature due to an increase in the number of available charge carriers. However, this increase may not always be linear and can saturate at higher temperatures due to increased recombination losses in the semiconductor material which again may result in a decrease of FF [13], [14].

Solar cells generally possess a parasitic series and shunt resistance and various physical mechanisms in the cells are responsible for these resistances. Major contribution for the series resistance, R_s, is given by bulk resistance of the semiconductor materials and the contacts resistance. The shunt resistance, R_{SH}, is generally caused by leakage across the p-n junction around the edge of the cell and due to the presence of defects and impurities at the junction region [5], [14]–[16]. Both of these resistances can cause a reduction in FF, as indicated in Figure 3. The magnitude of this effect of R_s and R_{SH} on the FF can be identified by comparing them with the Characteristic Resistance (R_{CH}) which can be calculated from Equation 3.

In this work, a comparison between the experimental and the theoretical FF is expected to carried out. Hence, following equations were found out from previous work, which claimed to be expressions for the theoretical calculation of FF to a higher accuracy [17]. Therefore, these equations have been used to derive temperature dependency of theoretical FF based on the measured V_{OC} and I_{SC} at each temperature.

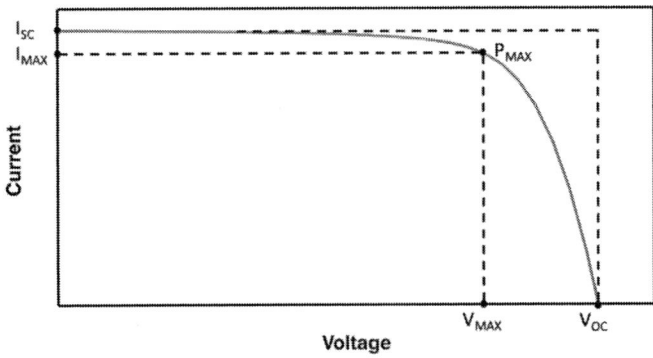

Figure 1. Load characteristic curve of a solar cell.

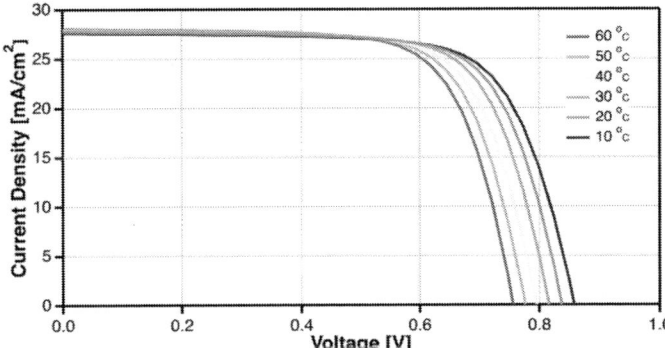

Figure 2. I-V curves at different temperatures [18].

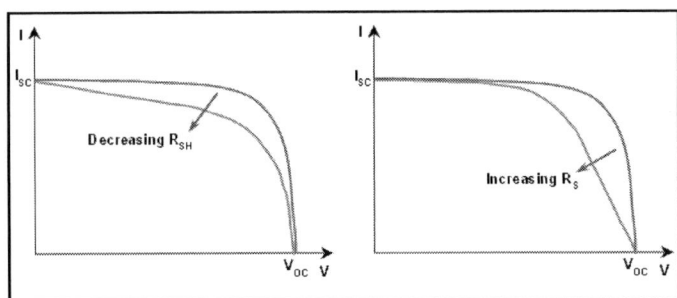

Figure 3. Effect of parasitic resistances on the output characteristics of solar cells. Effect of shunt resistance, R_{SH} (Left) and Effect of series resistance, R_S (Right).

$$FF_O = \frac{v_{OC} - \ln(v_{OC} + 0.72)}{v_{OC} + 1} \qquad (2)$$

$$R_{CH} = \frac{V_{OC}}{I_{Sc}} \qquad (3)$$

$$v_{OC} = \frac{V_{oc}}{V_{th}} \qquad (4)$$

978-1-6654-6060-6/23 $31.00 © 2023 IEEE   403

If $R_{CH} \gg R_S$ and $R_{CH} \ll R_{SH}$,

$$FF_{s,sh} = (1 - r_s)\left\{1 - \frac{(v_{OC} + 0.7)}{v_{OC}}\frac{FF_O}{r_{sh}}\right\} \quad (5)$$

Where,

$$r_s = \frac{R_S}{R_{CH}}, \quad r_{sh} = \frac{R_{SH}}{R_{CH}}$$

B. Solar Spectra

The performance measurements of PV devices mainly depend on the spectral distribution they get exposed to. Hence, it is the norm to follow the standard spectra distributions defined by the American Society for Testing and Materials (ASTM). For AM1.5, they define two standard terrestrial spectral distributions, the Direct Normal (AM1.5D) where the spectrum of sun incident on earth directly without any contribution from diffused rays and the other is the Global (AM1.5G) which the spectrum consists of both direct and diffused rays from the sun. In this work, solar simulator is calibrated, using Global AM1.5 (1000 Wm^{-2}, AM1.5G) where the light is considered to be contributed from both direct and diffused rays from the sun. The solar spectral-irradiance spectra AM0 (1353 Wm^{-2}, AM0) is used to calibrate the simulator to perform outer space situations. All the spectra data in this paper has been taken from Ref. [19], [20].

III. EXPERIMENT

A. Experimental Setup

This is a custom-built chamber with two sub-chambers: Test Chamber (to the left) and Power Chamber (to the right). In the test chamber, the thermal stage is placed at the bottom and the light source is mounted on the top [12]. A mirror assembly is precisely mounted onto a custom-built small chamber inside the test chamber maintaining 90° angle between each mirror and normal to the thermal stage to concentrate the spectrum. All the outer walls covered and sealed to make the test chamber nominally airtight, and dry airflow inlet and outlet are designed at the bottom and top of the test chamber respectively to purge the system of with dry air to avoid condensation and water frosting on the sample stage when low temperature measurements are carried out. Schematic flow chart of the system is shown in Figure 5.

A multi-array hybrid light source with 9 similar sections (3x3) is used in the system, manufactured by Alfartec (La Chaux de-Fonds, Neuchâtel, Switzerland). Each section consists of 9 different colors of LEDs and one halogen lamp. A LABVIEW program came with the light source has modified as per the experiment requirements and is used to control this light source [12].

B. Light Source

When the light source calibration is considered, three main categories are addressed under IEC standards: Spectral Match, Homogeneity, and Temporal Stability. And this system managed to accomplish a Class A classification in each section.

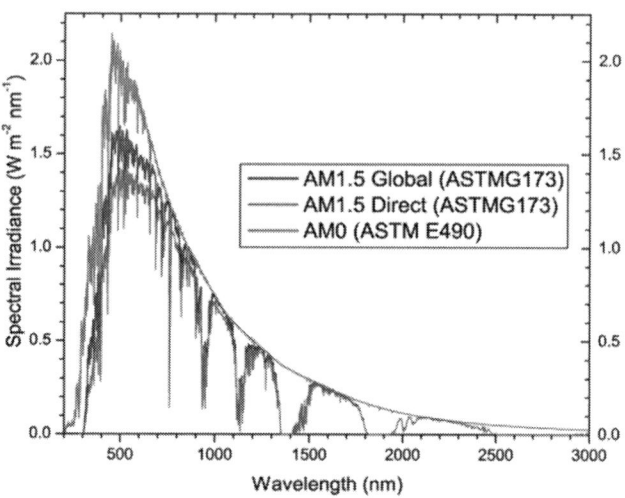

Figure 4. Standard Solar Spectra for space and terrestrial use [19].

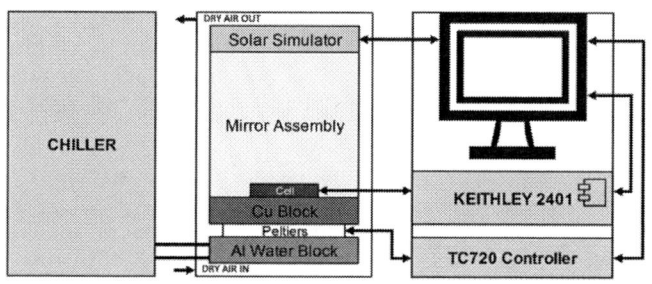

Figure 5. Schematic flow chart of the Irradiance and Temperature Control Chamber.

Note that, although each specific LED type's spectral peaks remain constant with varying intensity, the halogen bulb's spectrum varies considerably with operating current. A fiber-optic spectrometer has been used to identify these variation of individual spectra at different intensity level and multi-start global minimization algorithm was used to match the spectral output of the light source to both AM1.5G and AM0. Hence, we managed to produce and control a wide range of spectra with intensity ranges from 0.1 sun up to 1.5 suns [12].

C. Thermal Stage

This stage having an area of 10 cm x 10 cm and mainly consists of aluminum water block (connected to an external chiller), a Peltier/thermoelectric coolers (TECs) configuration, copper block, electrically insulated thermal conductive silicon sheet, pogo pin contacts, KEITHLEY2400 instrument and sample holding clips. Here the aluminum water block is used to achieve the initial cooling/heating process of the sample stage and to serve as a heat sink for the Peltier coolers when they are in use. This is of course with the help of the external chiller unit which uses water and propylene glycol mixture as the heat transferring medium.

In between this water block and top copper block, the Peltierapplying thermally conductive paste on all Peltier surfaces. The test cell is directly connected to the KEITHLEY2400 instrument via pogo pin configuration, using a 4-probe design to reduce the resistance loss of the measurements and enhance accuracy. Good and leveled contact of the testing solar cell with the stage is maintained all the time with sample clips

A TC-720 temperature controller (TE Technology, Inc.) is used to control the temperature of this thermal stage between -20 °C and 100 °C by controlling the power availability to the Peltiers. Two thermistors (MP-3193, MP-2444) are used as sensors, one to measure the copper block temperature and the other to read the cell temperature. These are used to maintain the sample stage temperature at a set point during an experiment [12].

D. Experimental Procedures

The experiment consists of measuring IV characteristics of thin film solar cell samples between the temperature range of -20 °C and 60 °C and a comparison between experimentally and theoretically obtained FF, under both spectrums, AM1.5G and AM0. Here, during the temperature ramping process, all measurements were measured about 10 minutes after stage reaching the set temperature. This is to give enough time to the cell to reach and settle at that temperature.

Separately, a thermal cycling was done on the same cell to get an understanding on the FF behavior when it exposed to a continues temperature variation between a low and a high temperature. Here, each cycle started from 20 °C and went up to a maximum of 50 °C and all the way back down to -20 °C and then back to 20 °C, temperature increase and decrease done in 10 °C steps and made sure to keep the device for about 10 mins at each temperature to settle at that temperature, prior taking the measurements. Each of these cycles took about 120 mins (single cycle) to complete. 20+ such cycles were performed over a total time period of 3000 minutes. Only AM1.5G spectra was used in this process.

A solar cell sample with CdSe/CdTe layer stack is used in this experiment. Here, 0.1mM $CuCl_2$ solution in deionized water is used to submerge the original samples for 2 min followed by rising in DI water and drying with N_2, followed by 250 °C anneal for 15 min to achieve the back contact Cu activation as reported previously [14]. Finally, 60 nm back contact is deposited with thermally evaporated gold using a tungsten boat at a base pressure of $\sim 4 \times 10^{-6}$ Torr. A cell area of 0.09 cm^2 was defined using laser scribing.

IV. RESULTS AND DISCUSSION

IV characteristic curves at different temperatures obtained for both spectra, AM1.5G and AM0. Temperature dependency of each parameter is discussed separately in the coming sections.

A. Temperature dependency of V_{OC}, I_{SC}

As it is shown in Figure 6, V_{OC} decreases with temperature and a slight increment in I_{SC} can be observed, under both spectrums. As the V_{OC} is mainly determined by the band gap of the device which has a strong temperature dependency, a reasonable drop of the V_{OC} with increase of temperature is expected [21]. Here, the temperature coefficients of V_{OC} under AM1.5G and AM0 are -2.07 ± 0.015 mV and -2.11 ± 0.054 mV, respectively.

Under both spectrums, the slight increment in I_{SC} can be identified as the response of the layer stack to longer wavelength regions, as a result of bandgap drop occurs with temperature increase and correspondingly increase in band-to-band absorption coefficient across the spectrum as reported in previous studies [21]. Hence, both temperature coefficients of I_{SC} are positive, which are 0.031 ± 0.001 mAcm^{-2} and 0.047 ± 0.001 mAcm^{-2}, respectively under AM1.5 and AM0.

B. Temperature Dependancy of FF

In Figure 7 (top), temperature dependency of FF is compared between the experimentally obtained and theoretically calculated results with respective to AM1.5. The same, respective to AM0 is shown in Figure 7 (bottom). In the process of obtaining the experimental FF, Eqs. (1) is used and Voc, Ioc, Vmax, Imax measured at each temperature are plugged in. Eqs (4) is used in all the theoretical calculations to identify the temperature dependency of FF theoretically.

When the change of FF with the temperature is considered under AM1.5, throughout the temperature range, almost a constant negative coefficient of -0.013 ± 0.0012 1/C° can be identified in theoretically obtained results. But experimentally, a negative temperature coefficient can only be seen at temperatures higher than 30 °C and it is -0.017 ± 0.0063 1/C° which is close to the theoretically calculated value. At temperatures lower than 30 °C, FF of this CdSe/CdTe layer stack follows a positive temperature coefficient, but the magnitude decreases with increase in temperature.

Under AM0 spectrum also, similar behaviors can be observed for theoretical and experimental temperature dependency of FF. Here, the obtained theoretical temperature coefficient is -0.020 ± 0.0014 1/C° and the experimental is -0.041 ± 0.0001 1/C° which is almost the double. As in the AM1.5 case, here also the positive temperature coefficient of FF is in agreement with the same range of temperature, from -20 °C to 30 °C.

It is also clear that, for any given temperature between -20 °C and 60 °C, the experimentally obtained FF is lower than the theoretically calculated value. This difference can be a result of omitting the temperature effect on saturation current, series and shunt resistance, defects etc, when calculations are carried out for theoretical FF. Furthermore, it is observed that r_s and r_{sh} values used in the theoretical calculations has a high impact on this difference but not on the temperature coefficient of the theoretical results. Hence, even though pervious work has been reported that r_s and r_{sh} decrease exponentially with increasing temperature [22], it is reasonable to assume that the experimental FF behavior shown at lower temperatures (-20 °C to 20 °C) may have an impact of some other temperature dependent physical and electronic parameters of CdSe/CdTe layer stack which overrides the effect of r_s and r_{sh} temperature dependency.

C. Effect of Temperature Cycling on FF

As it can be seen in Figure 8, the FF trend observed in Figure 7 (bottom) repeats throughout all the cycles where the minimums at extreme temperatures and maximums within the range between 20 C° and 40 C°. This signifies the fast recovery

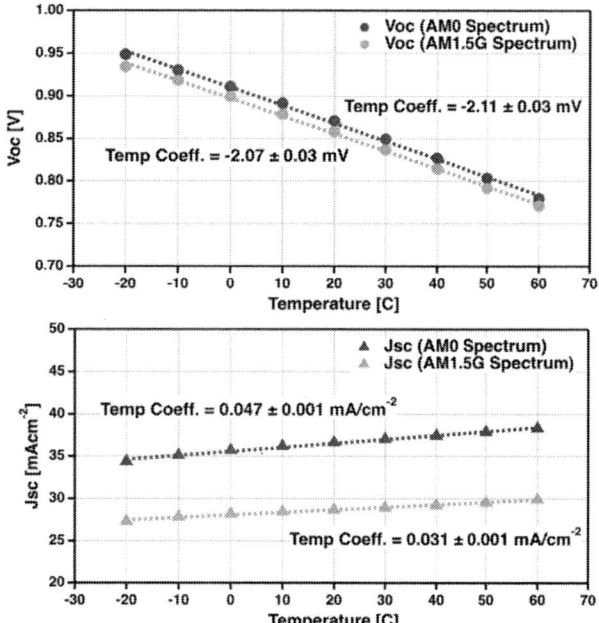

Figure 6. Theoretical and experimental temperature dependencies of V_{OC} (top) and J_{SC} (bottom)

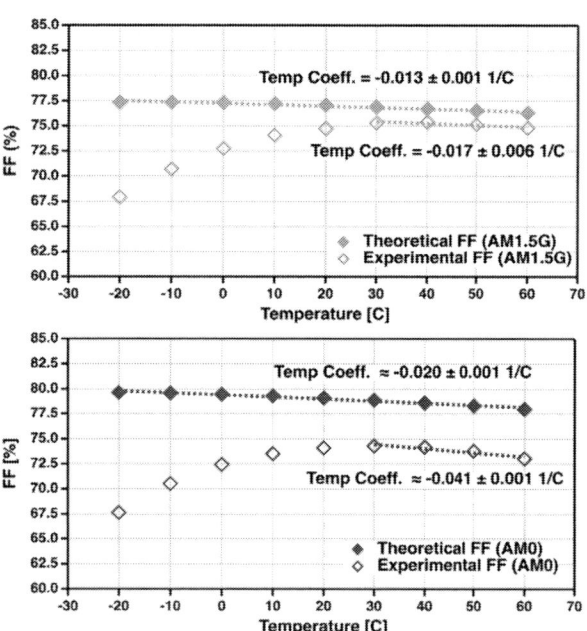

Figure 7. Theoretical and experimental temperature dependencies of FF under AM1.5G (top) and AM0 (bottom).

of cell performances without any major drawback within the temperature range of -20 C° and 50 C°. However, within the first few cycles, a very little overall decrement of FF can be scene and overtime the difference between the maximum and minimum of FF in a single cycle tend to increase as the number of cycles increases. This may be due to the permanent structure changes and cell degradation which need further experimental analysis and studies to clarify.

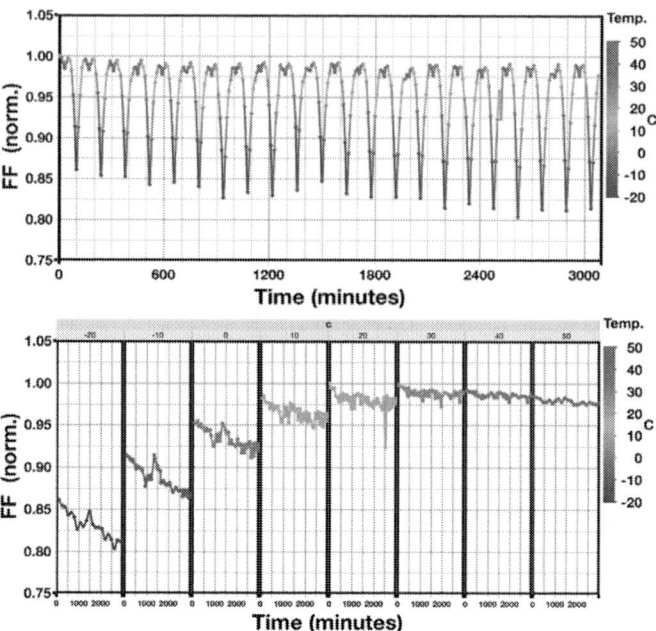

Figure 8. Normalized FF behavior of thermal cycling under AM1.5G.

V. CONCLUSION

It is a known fact that, increasing solar cell temperature results in narrow band gaps. Hence more carriers can move from valance band to conduction band which positively take part in increasing the current. In addition, carrier recombination probability between the conduction band and valance band tend to increase, which causes a drop in operating voltage. This slight increment in current and the respectively large drop of voltage with the temperature can be identified here.

Even though the theoretically calculated FF is larger than the experimentally obtained FF, both temperature coefficients are approximately equal and behave same at temperatures higher than 20 °C for AM1.5 and 40 °C for AM0, which is in agreement with most of the previous work. But at lower temperatures (between -20 °C and 20 °C), this specific CdSe/Cdte layer stack seems to be showing interesting characteristics in terms of FF dependency. It is also observed that r_s and r_{sh} values used in the theoretical calculations has a high impact on this difference between the theoretical and experimental results but not on the overall temperature coefficient of the theoretical results. Even though a little drop of overall FF factor is observed during first few cycles, a fast recovery of cell performances at each cycle is observed from thermal cycling, without any major degradation. Overall, it is reasonable to assume that the experimental FF behavior observed here may have an impact of some other temperature dependent physical and electronic parameters of CdSe/CdTe layer stack which are not discussed here, specially at lower temperatures. Hence, more future work is focused on studying and understanding this dependency.

ACKNOWLEDGMENTS

The authors would like to acknowledge First Solar Inc. for providing CdSeTe-based film stack samples for devices. This work is based on research sponsored by Air Force Research

978-1-6654-6060-6/23 $31.00 © 2023 IEEE

Laboratory under agreement number FA9453-21-C0056, and by the U.S. DOE's Office of Energy Efficiency and Renewable Energy (EERE) under the Solar Energy Technologies Office (SETO), through Agreement DE-EE0008974 and through the Alliance for Sustainable Energy, LLC, Managing and Operating Contractor for the National Renewable Energy Laboratory for the U.S. Department of Energy, under Award Number 37989. The U.S. Government is authorized to reproduce and distribute reprints for Governmental purposes notwithstanding any copyright notation thereon. The views expressed are those of the authors and do not reflect the official guidance or position of the United States Government, the Department of Defense or of the United States Air Force. The appearance of external hyperlinks does not constitute endorsement by the United States Department of Defense (DoD) of the linked websites, or the information, products, or services contained therein. The DoD does not exercise any editorial, security, or other control over the information you may find at these locations. Approved for public release; distribution is unlimited. Public Affairs release approval # AFRL-2023-019.

REFERENCES

[1] X. Wu, "High-efficiency polycrystalline CdTe thin-film solar cells," *Solar Energy*, vol. 77, no. 6, pp. 803–814, Dec. 2004, doi: 10.1016/j.solener.2004.06.006.

[2] C. S. Ferekides *et al.*, "High efficiency CSS CdTe solar cells," *Thin Solid Films*, vol. 361–362, pp. 520–526, Feb. 2000, doi: 10.1016/S0040-6090(99)00824-X.

[3] A. MORALESACEVEDO, "Physical basis for the design of CdS/CdTe thin film solar cells," *Solar Energy Materials and Solar Cells*, vol. 90, no. 6, pp. 678–685, Apr. 2006, doi: 10.1016/j.solmat.2005.04.004.

[4] S. M. Sze and K. K. Ng, *Physics of Semiconductor Devices*. Hoboken, NJ, USA: John Wiley & Sons, Inc., 2006. doi: 10.1002/0470068329.

[5] P. Singh and N. M. Ravindra, "Temperature dependence of solar cell performance—an analysis," *Solar Energy Materials and Solar Cells*, vol. 101, pp. 36–45, Jun. 2012, doi: 10.1016/j.solmat.2012.02.019.

[6] J. J. Wysocki and P. Rappaport, "Effect of Temperature on Photovoltaic Solar Energy Conversion," *J Appl Phys*, vol. 31, no. 3, pp. 571–578, Mar. 1960, doi: 10.1063/1.1735630.

[7] J. C. C. Fan, "Theoretical temperature dependence of solar cell parameters," *Solar Cells*, vol. 17, no. 2–3, pp. 309–315, Apr. 1986, doi: 10.1016/0379-6787(86)90020-7.

[8] P. SINGH, S. SINGH, M. LAL, and M. HUSAIN, "Temperature dependence of I–V characteristics and performance parameters of silicon solar cell," *Solar Energy Materials and Solar Cells*, vol. 92, no. 12, pp. 1611–1616, Dec. 2008, doi: 10.1016/j.solmat.2008.07.010.

[9] H. Qu and X. Li, "Temperature dependency of the fill factor in PV modules between 6 and 40 °C," *Journal of Mechanical Science and Technology*, vol. 33, no. 4, pp. 1981–1986, Apr. 2019, doi: 10.1007/s12206-019-0348-4.

[10] S. Chander, A. Purohit, A. Sharma, Arvind, S. P. Nehra, and M. S. Dhaka, "A study on photovoltaic parameters of mono-crystalline silicon solar cell with cell temperature," *Energy Reports*, vol. 1, pp. 104–109, Nov. 2015, doi: 10.1016/j.egyr.2015.03.004.

[11] W. He, A. Guo, F. Meng, and S. Feng, "Effect of temperature on performance parameters of multicrystalline silicon solar cells and modules," *Taiyangneng Xuebao/Acta Energiae Solaris Sinica*, vol. 31, pp. 454–457, 2010.

[12] N. Katakumbura *et al.*, "Irradiance and Temperature Control Chamber for Testing Solar Cell Performance," in *2021 IEEE 48th Photovoltaic Specialists Conference (PVSC)*, IEEE, Jun. 2021, pp. 1813–1820. doi: 10.1109/PVSC43889.2021.9518734.

[13] Priyanka, M. Lal, and S. N. Singh, "A new method of determination of series and shunt resistances of silicon solar cells," *Solar Energy Materials and Solar Cells*, vol. 91, no. 2–3, pp. 137–142, Jan. 2007, doi: 10.1016/j.solmat.2006.07.008.

[14] P. Singh and N. M. Ravindra, "Temperature dependence of solar cell performance—an analysis," *Solar Energy Materials and Solar Cells*, vol. 101, pp. 36–45, Jun. 2012, doi: 10.1016/j.solmat.2012.02.019.

[15] F. Khan, S. N. Singh, and M. Husain, "Determination of diode parameters of a silicon solar cell from variation of slopes of the $I - V$ curve at open circuit and short circuit conditions with the intensity of illumination," *Semicond Sci Technol*, vol. 25, no. 1, p. 015002, Jan. 2010, doi: 10.1088/0268-1242/25/1/015002.

[16] Q. Li *et al.*, "Comparative study of GaAs and CdTe solar cell performance under low-intensity light irradiance," *Solar Energy*, vol. 157, pp. 216–226, Nov. 2017, doi: 10.1016/j.solener.2017.08.023.

[17] M. A. Green, *Solar cells: Operating principles, technology, and system applications*. Prentice-Hall, Englewood Cliffs, NJ, 1982.

[18] R. Hossain *et al.*, "New Design of Solar Photovoltaic and Thermal Hybrid System for Performance Improvement of Solar Photovoltaic," *International Journal of Photoenergy*, vol. 2020, pp. 1–6, Jul. 2020, doi: 10.1155/2020/8825489.

[19] "American Society for Testing and Materials (ASTM). Reference solar spectral irradiance: Air mass 1.5. Available:"

[20] C. Riordan and R. Hulstron, "What is an air mass 1.5 spectrum? (solar cell performance calculations)," in *IEEE Conference on Photovoltaic Specialists*, IEEE, 1990, pp. 1085–1088. doi: 10.1109/PVSC.1990.111784.

[21] W. Zhou, H. Yang, and Z. Fang, "A novel model for photovoltaic array performance prediction," *Appl*

Energy, vol. 84, no. 12, pp. 1187–1198, Dec. 2007, doi: 10.1016/j.apenergy.2007.04.006.

[22] D. M. Fébba, R. M. Rubinger, A. F. Oliveira, and E. C. Bortoni, "Impacts of temperature and irradiance on polycrystalline silicon solar cells parameters," *Solar Energy*, vol. 174, pp. 628–639, Nov. 2018, doi: 10.1016/j.solener.2018.09.051.

Module Reliability in Winter: Field analysis of deflection and cell cracking across multiple module architectures

Laurie Burnham[1], Daniel Riley[1], Bruce King[1], William Snyder[1], Kevin Santistevan[1],

Paul W.Dice[2]

Sandia National Laboratories[1], Albuquerque, NM, 87185 USA
Michigan Technological University[2], Houghton, MI, 49931 USA

Abstract — **Snow, which affects vast areas of the northern hemisphere, can be problematic for photovoltaic (PV) systems for three reasons; one, the shading of solar cells by snow stifles power generation[1]; two, the broad geographic sweep of most snowstorms challenges the availability of solar to meet regional energy needs; and three, the weight of snow, combined with wind and cold loading, can physically damage modules and cells. Not only are load stressors in winter poorly understood but almost nothing is known about the robustness of different module technologies exposed to those stressors. This paper will present data from a field study of multiple module types, including half-cut cells, bifacial and large-form factor architectures, installed at the Michigan Regional Test Center in Calumet, Michigan. Modules were installed on single-module racks instrumented with load cells, deflection sensors, and anemometers. Each module was subjected to electro-luminescent imaging at the beginning and end of winter to document the presence or absence of cell damage; in addition, time-series data were collected to quantify snow, wind and temperature loads and images collected to compare rates of snow-shedding across module technologies.**

I. INTRODUCTION

The increase in frequency and severity of hurricanes, typhoons and other high-wind weather events around the world has raised concerns about the long-term impact of extreme wind-loading on photovoltaic (PV) panels, including those that show no visible damage. Far less attention has been paid to the phenomenon of extreme snow-loading on PV panels and racking hardware, yet in regions that see frequent snow in winter, snow loading can be persistent, repetitive, and non-uniform. When heavy snow loads are combined with high winds and low temperatures, as are typical of blizzard conditions, the risks to module integrity increase [2].

This paper, which aims to quantify the meteorological conditions in winter under which PV modules deflect to the point of cell breakage and frame distortion, is important for two reasons. One, single winter-storms, especially in the US, cover large geographic areas: as much as 60% of the country, putting gigawatts of solar capacity at risk of both catastrophic damage, as well as longer-term degradation, which can include moisture ingress and cell cracking. And two, having empirical evidence regarding the resilience of different module architectures under winter field conditions and data documenting how well—or poorly—specific components and design parameters perform under loading—is essential to the build-out of a reliable solar sector in regions that see heavy snow. While resource availability and power losses remain serious challenges, they are not the focus of this paper. The focus of this paper is instead on the comparative reliability of multiple module technologies and their performance relative to each other under the same field conditions.

It is important to emphasize here that the rationale for this work is to collect critical information not captured in accelerated tests and other simulated environments, notably the long-term unpredictable, uneven, persistent and repetitive patterns of wind and snow loading in the field, combined with low temperatures. Investigations of cell cracking under thermomechanical load have established a general principle, but little is known about cell or polymer behavior in response to actual and uneven snow load or to the prolonged low-temperature-exposure typical of a fielded installation. In addition, virtually nothing is known about how different module technologies handle snow and cold loading. Parameters such as cell size and thickness, which determine surface to volume ratio, and the different thermal expansion coefficients of the modules' components, influence loading sensitivity but the phenomenon has been poorly quantified outside of the laboratory environiment.

International certification standards (e.g IEC 61215) set baseline performance parameters for the snow loading of PV modules and most utility-scale modules meet those specifications. For many years, the standards called for load testing at a horizontal tilt, which does not capture the dynamics of shedding or weight of snow that piles up on a module's lower edge. Recognizing those limitations, IEC 62938 was published in 2020, specifically to describe a method for measuring non-uniform snow loading on inclined PV modules. Those standards now specify that modules must withstand snow loads as heavy as 5.4kPa but the specified method has a duration of one hour and is conducted at above-freezing temperatures. Standards for calculating snow load for rooftop solar arrays, as described by the ASCE 7-10 building code, also have their limitations., as they assume the roof to be a uniform pitch and free of obstructions. But roofs are rarely simple planar surfaces and have unbalanced snow loads caused by drifting and shedding from one plane to another. More important, none of the certification methods capture real-world operating conditions that include temperature-related brittleness and long-term snow loading, which can include freeze-thaw cycles, frequent load shifts, and persistent concentrated loading on the lower half of modules.

What is known, however, is that extreme weather, including heavy seasonal snowstorms, which can drop three feet of more of snow in a single event, is increasingly factored into the risk calculations of insurers and investors: insurance rates for renewable energy projects are seeing steep price increases (as much as 400% in the last two years) and some insurers are refusing to cover cell cracking, at any price.

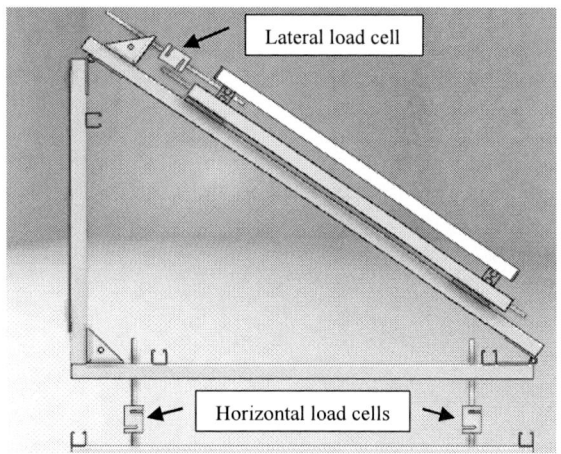

Figure 1. Side view of the top portion of the snow-load measurement station, depicting location of the load cells, four of which are normal to the ground; two of which are installed parallel to the plane of the module to capture shear forces created by snow loading.

TABLE I
MODULE TECHNOLOGIES DEPLOYED FOR SNOW-LOAD STUDY, WINTER OF 2022-2023

Stn #	Module Type	# Cells	Module Specifications	Module Format
1	Mono-facial	60	Glass/backsheet (bs), 270W	960 mm x 1994 mm
2	Mono-facial	72	3.2mm glass/bs, 440W	1016 mm x 1999 mm x 40 mm
3	Mono-facial	144	3.2mm glass/bs, 400W	1002 mm x 2008 mm x 39.8"
4	Bifacial	144	M6 cells, 3.2mm glass/bs, 450W	1048 mm x 2108 mm x 40 mm
5	Bifacial	144	M10 cells 3.2mm glass/bs, 530W	1134 mm x 2279 mm x 40 mm

II. METHODOLOGY

Sandia designed a single-module racking station to measure snow and wind loading, and also the module's deflection under load (see Figure 1). Five of these snow-load stations were installed at the Michigan Regional Test Center for Emerging Solar Technologies [3] in Calumet, Michigan, in the fall of 2022, prior to the onset of snow. Each PV module is affixed to a rigid structural frame and four are set at an angle of 35 degrees relative to horizontal; one module is horizontal. Each station has six load cells, which are integrated into the rack frame, and measure both downward and lateral loading once a second. At 15-second intervals, the minimum, maximum, and average values of those one-second load measurements are recorded. In

addition, an anemometer is affixed to the frame in the plane-of-array and measures wind speed and direction at one-second frequencies. Measuring direction is important: wind normal to the front of the module adds load to the existing snow load, whereas wind normal to the rear of the module negates the downward pressure from snow-loading. A camera facing the snow-load stations captures time-series image data every 10-minutes, providing visual evidence of the accumulation and loss of snow from the modules, including patterns of non-uniform distribution. Onsite meteorological instrumentation collects data on wind speed and direction, ambient temperature, humidity, and snow depth.

Prior to installation, each module was subjected to baseline electroluminescent imaging and each image carefully examined and confirmed to have intact, i.e., crack-free, solar cells.

TABLE II
MONITORING EQUIPMENT FOR SNOW-LOAD STUDY 2022-2023

Instrument	Function	Make/Model
Load cells	Downward & lateral snow load	AmCells STL Series S-Type, 1000 lb rated.
Ultrasonic proximity sensor	Deflection of module under load	XX9 cylindrical 18A3C2M18 (default res. 0.55 mm; calibrated res.approx. 0.1 mm)
Anemometer	Wind speed & direction	
Rel.humidity and ambient air		Vaisala HMP60-L
Snow sensor	Snow depth	Lufft laser sensor SHM41
Pyranometer	Plane-of-array irradiance	Hukseflux SR30 D1
Camera	Images of snow coverage	Cambell Scientific CC5MPXWD with defroster

III. RESULTS (TO BE UPDATED IN APRIL 2023)

Time-series data collected as of January 13, 2023 indicates that the snow loading stations are accurately capturing both the total downward force and shear forces that act on PV modules. By the end of winter, we will have collected the data measuring the simultaneous loading of the PV modules and the deflection of the center those modules under the load of snow.

The extent to which module deflection translates into cell damage will be determined at the end of winter when the modules are re-imaged under electroluminescence. By comparing these images with baseline images taken at the beginning of winter, we will be able to identify cracks that formed as a result of winter snow loading. Utilizing existing methods for crack analysis, we will compare the length and number of cracks per cell and per module, and tabulate the results as percentage of damage per module.

Although the precise point in time at which cell cracking occurred cannot be determined from our data nor can it be determined that the cell cracking resulted from one extreme loading event or from repetitive events, we will plot time-series deflection data for each module, along with snow-load and wind data (see Figure 2).

A. Module Deflection Under Load

We expect to show that module deflection under heavy snow load varies across the module technologies as a result of their varied construction factors and architectures. We will demonstrate maximum module deflection from an unloaded state for each module type and correlate this deflection with the loads measured as a result of snow.

We will also have data from this single winter showing the variation and cyclic nature of snow loading, under both uniform and non-uniform conditions, demonstrating that snow loading, as a discreet variable, fluctuates throughout the winter season and also from year to year. In addition, we will show changes in module deflection that occur as snow begins to shed and under non-uniform loading patterns. Maximum load measurements for this winter (pounds/sq ft) for the horizontal and tilted modules will be presented. Table to be added, including data on lateral loading. In contrast, maximum loading for non-uniform snow-loading with weight concentrated on the lower half of a module was XX, a situation that represented partially thawing followed by [TBD: the formation of an ice dam and addition of new snow.]

Wind loading data will also be provided and a graph will be inserted showing time-series temperature, snow-load and wind load data for each module type, identifying peak snow and wind loads, defined as the combination of temperatures below XX and winds in excess of XX.

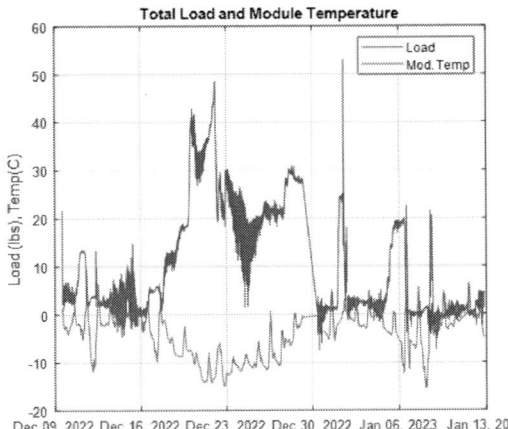

Figure. 2. Example time-series temperature and snow-load data, from horizontal sensors, showing total downward loading for module #2. Load increases indicate new snow.

B. Rates and Patterns of Snow Shedding

In addition to varying in their resilience to snow and wind loading, our time-series images reveal different rates and patterns of shedding across the different module types. Although single modules are not fully representative of a string, a row or an entire array, we saw consistent shedding patterns that translate into significant performance differences, with power data provided by micro-inverters. Calculations of power loss based on the number of cells shaded and their electrical design [3] reveal that module had the greatest power loss (X% more than the highest performing module. [See Table X].

IV. CONCLUSIONS AND FUTURE WORK

One winter of data does not reveal the full range of stochastic weather events nor does it provide conditions that are fully representative of different northern climates. In addition, the multiple stressors in winter that can damage modules and solar cells are challenging to isolate as they are unpredictable and interact in complex and synergistic, ways. Not yet known, for example, are the environmental conditions that contribute most to cell fragility, notably the contributions of a single loading event versus persistent or repetitive exposure.

Even so, the data collected and presented here provides a strong rationale for expanded field studies of different module architectures, with a focus not on establishing the precise boundary conditions in winter under which cell integrity is compromised (although we believe our data will inform a new generation of accelerated testing protocols) but in providing a high-fidelity technical platform for evaluating the comparative performance of different module technologies in winter. In addition to showing differences in deflection behavior [to be described], our image data shows that module design contributes to different rates and patterns of snow shedding, differences that translate into different amounts of power loss. The intent is to expand this study by installing a greater diversity of modules in 2023-24 and to continue monitoring the modules for a minimum of ten years.

REFERENCES

[1] E. Andenæs, B. P. Jelle, K. Ramlo, et al., The influence of snow and ice coverage on the energy generation from photovoltaic solar cells, Solar Energy,159, 2018, pp. 318-328,ISSN 0038-092X.

[2] E. J. Schneller, H. Seigneur, J. Lincoln and A. M. Gabor, The impact of cold emperature exposure in mechanical durability testing of PV modules," *2019 IEEE 46th PVSC Conf*, Chicago, IL, pp. 1521-1524, doi: 10.1109/PVSC40753.2019.8980533.

[3] L. Burnham, D. Riley, B. King, J. Braid, et al, "Dedicated cold-climate field laboratory for photovoltaic system and component studies: the Michigan Regional Test Center," IEEE 49th PVSC Conf, Philadelphia, 2022.

[4] 2020. J. Braid, D. Riley, J. Pearce, and L.Burnham, Image analysis method for quantifying snow losses on PV systems, IEEE 47th PVSC, June 15-August 21 virtual meeting; 7pp.

How Useful is a Field-Operable I-V Curve Tracer?

Alexander Cimaroli

Fluke, Everett, WA, 98203, USA

Abstract — **Solar power is continuing its upward trend of megawatts installed per year across the board in the residential, commercial, and utility sectors. To assess the performance of PV modules onsite, field-capable I-V curve tracers have proven to be an essential tool in identifying and troubleshooting PV performance issues. Coupled with solar irradiance meters, surface temperature probes, and a PV module database, field-operable I-V curve tracers can calculate whether a particular PV module, or string of modules, is performing to manufacturer specifications. The key parameter that technicians typically want to know is the maximum power output of the PV system. However, the electrical, irradiance, and temperature measurements have measurement uncertainty associated with them, as well as technical issues that limit how accurate they can be for a given PV setup. This work looks at several different I-V curve tracers, including the Fluke SMFT-1000, to understand the differences between different technologies and how accurate an I-V curve tracer can actually be when measuring the maximum power output of different PV systems.**

I. INTRODUCTION

By the end of 2022, the US installed 142.3 gigawatts GW of solar power [1], and in 2021, 36% of all new electricity generation capacity in the country came from solar [2]. Over the past decade, there has been regular annual increases in the total installed solar capacity worldwide. Larger and larger solar farms are either planned or have already begun development, in recent years, where millions of solar panels will be used to generate hundreds of megawatts to gigawatts [3]-[4]. Installation, commissioning, and operation of large solar farms, or even small residential installations, often require quick and accurate evaluation of the performance of solar panels. A simple multimeter can be used to measure the short-circuit (Isc) current and open-circuit voltage (Voc) of a solar module. While this method does not give a measurement at the operating maximum power point of a solar module, it can provide some insight as to whether there is something wrong.

An emerging requirement in the United States for commissioning new solar panel installations that are grid connected is ISO/IEC 62446-1 [5]. Within standards like ISO/IEC 62446-1, there are provisions for simple Isc and Voc measurements, however, there is guidance for the use of so-called I-V curve tracers to document and evaluate solar panel, or solar panel string, performance. I-V curve tracers have been used for decades to measure the performance of solar modules, and solar cells alike [6].

Coupled with an irradiance meter, surface temperature probe, and a solar module database, I-V curve tracers can perform I-V curve shifting to standard test conditions (STC) or nominal operating cell temperature (NOCT) and compare to the manufacturer specified performance parameters at STC and/or NOCT. However, in order for an I-V curve tracer to be useful, it must be accurate enough to adequately determine the performance of a solar system. Irradiance, temperature, voltage, and electrical current all have measurement uncertainty associated with them. Additionally, measurement errors are commonly introduced, and are highly dependent on the type of I-V curve tracer and solar module being measured. This work will focus on dissecting two major aspects of field-capable I-V curve tracer measurements, and how these aspects affect the accuracy of the measured I-V curve:

1. The two main styles of field-capable I-V curve tracer, resistive and capacitive
2. How the capacitance of individual solar panels, and strings of solar panels, affect the accuracy of field-capable I-V curve tracers

With the advent of Perovskite solar modules, or even high efficiency Silicon architectures, the capacitance of the solar module under test is going to play a central role in the overall efficacy of field capable I-V curve tracers [7].

II. TYPES OF FIELD-CAPABLE I-V CURVE TRACERS

Today, field-capable I-V curve tracers must be able to measure single solar panels as well as strings of solar panels to be competitive in the commercial and utility installation market. At short-circuit currents as high as 20 A and voltages as high as 1500 V, great care is taken in the design of a field-capable I-V curve tracer so as to handle the potentially high amounts of power delivered to them. There are two main designs of field-capable I-V curve tracers commercially available today: resistive (common) and capacitive (less common). Both types have benefits and deficits to their approach. Since solar panels and strings of solar panels

A. Resistive Type

By far, the simplest implementation of a field-capable I-V curve tracer, the resistive type uses a variable resistance as the load presented to the solar module(s) under test. When the resistance is set, the I-V curve tracer simultaneously measures the voltage across the solar panel and the electrical current flowing through the resistor to complete one I-V curve test point.

To prevent overheating, this type of I-V curve tracer remains in the open-circuit state for most of the measurement, and only briefly presents the needed resistance for the I-V curve test point. This pulsed method is relatively slow, taking around 15 to 20 seconds to complete one I-V curve measurement [7].

978-1-6654-6060-6/23 $31.00 © 2023 IEEE

Since the resistance jumps to open-circuit between each test point, the resistive type of I-V curve tracer typically has a settling period for the solar panel output to settle at the current test point before making a measurement.

Lastly, with the resistive type of I-V curve tracer, it is difficult to provide characteristic I-V curve sweep parameters, such as those described in with ISO/IEC 60904-1 [8]. Parameters such as sweep direction and sweep rate being the ones of primary concern. Neither of these can truly be defined for this type of method due to the module under test going to open-circuit between each test point in the I-V curve.

B. Capacitive Type

While not entirely new technology [9], capacitive field-capable I-V curve tracers use a capacitor instead of a load resistor. Due to the significantly lower power transfer, this reduces the size and weight of the I-V curve tracer since large heat sinks are not required [9]. Additionally, the capacitive I-V curve tracer approach can finish one I-V curve measurement in less than 1 second [7].

The rapid acquisition of I-V curve points for the capacitive method, understandably, brings into question the overall accuracy of such a measurement. This will be discussed further in the next section.

III. EFFECT OF SOLAR MODULE CAPACITANCE

As mentioned previously, the capacitance of the solar module under test will affect the overall accuracy of I-V curve measurements in the field, both resistive and capacitive alike. Figure 1 shows a simplified equivalent circuit diagram of a solar panel with parallel capacitance.

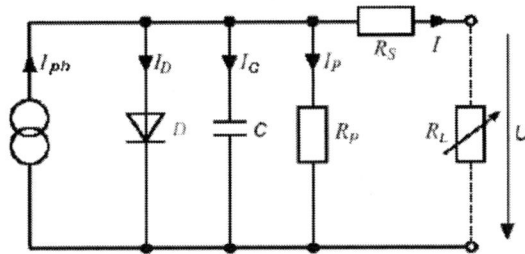

Fig. 1. Equivalent circuit diagram representation of a solar module. The load presented by the I-V curve tracer (R_L) can be seen on the far right, whereas the inherent capacitance of the solar module can be seen as a capacitor with capacitance C.

While the resistive type of I-V curve tracer has a settling period at each I-V curve test point to allow the transients to die off, this settling time is usually less than 1 millisecond, due to power dissipation restrictions. With more advanced Silicon solar module architectures, such as TOPCon and PERC, and Perovskite solar modules, 1 millisecond may not be enough time for capacitive transients to sufficiently dissipate to consider the measurement to be in the stead-state of the solar module under test.

Detailed in the Solmetric application note [7], there is an issue with measuring "high efficiency" solar modules. A large in-rush current, higher than the short-circuit current of the solar module under test, can cause the Solmetric I-V curve tracer to issue an "overcurrent pulse" warning. The in-rush current is partly due to the capacitance of the solar module under test. With rapid changes in voltage across the inherent capacitance of the solar module comes large injections of current from this capacitor, thus, pushing the system from a steady state to a dynamic state. The purpose of I-V curve tracers is to measure the steady-state performance of the solar module under test. The capacitive method may not allow sufficient time for the solar module capacitive transients to sufficiently dissipate either.

IV. IRRADIANCE, TEMPERATURE, AND I-V CURVE SHIFTING

Modern-day, commercially available, field-capable I-V curve tracers may come with the ability to provide I-V curve shifting and solar module database comparison. Shifting an I-V curve to STC or NOCT allows the shifted performance parameters, such as maximum power point voltage (Vmpp) and maximum power point current (Impp), to be compared directly to manufacturer specifications. The Fluke SMFT-1000 as well as other high-end I-V curve tracers provide this functionality. Neglecting the uncertainty in I-V curve shifting due to the choice of mathematical model and assumptions within the model, measurement uncertainty from solar irradiance and surface temperature measurements plays a key role in the accuracy of I-V curve shifting. Electrical measurement uncertainty, such as voltage and electrical current, will serve as the basis of comparison for the effect of irradiance and temperature measurements.

To help illustrate measurement uncertainty in I-V curve shifting, a Silicon solar module with a Voc of 45 V and an Isc of 10.5 A, at STC, will be used. With these parameters, a common maximum power point operating voltage and current is roughly 38.9 V and 9.8 A, at STC.

Table I shows the effect of irradiance measurement uncertainty at various levels of solar irradiance.

TABLE I

EFFECT OF SOLAR IRRADIANCE MEASUREMENT UNCERTAINTY

Irradiance [W/m^2]	Irradiance Uncertainty [W/m^2]	Shifted Impp Uncertainty [A]	Shifted Vmpp Uncertainty [V]
200	7.5	0.07 (0.22)	0.09 (0.37)
200	15	0.15 (0.22)	0.19 (0.37)
200	30	0.29 (0.22)	0.36 (0.37)
600	17.5	0.17 (0.26)	0.09 (0.39)
600	35	0.34 (0.26)	0.17 (0.39)
600	70	0.69 (0.26)	0.34 (0.39)
1200	32.5	0.32 (0.32)	0.09 (0.4)
1200	65	0.64 (0.32)	0.18 (0.4)
1200	130	1.27 (0.32)	0.35 (0.4)

Table I. Values in parentheses are the tolerance limits of the respective electrical measurements of industry-standard I-V curve tracers.

Table II shows the effect of solar panel temperature measurement uncertainty at various panel temperatures.

TABLE II
EFFECT OF SOLAR MODULE TEMPERATURE MEASUREMENT UNCERTAINTY

Module Temperature [°C]	Temperature Uncertainty [°C]	Shifted Impp Uncertainty [A]	Shifted Vmpp Uncertainty [V]
15	1	0.00 (0.30)	0.11 (0.40)
15	3	0.01 (0.30)	0.34 (0.40)
15	8	0.04 (0.30)	0.90 (0.40)
30	1	0.00 (0.30)	0.11 (0.39)
30	3	0.01 (0.30)	0.31 (0.39)
30	8	0.04 (0.30)	0.83 (0.39)
75	1	0.00 (0.30)	0.08 (0.37)
75	3	0.01 (0.30)	0.25 (0.37)
75	8	0.04 (0.30)	0.66 (0.37)

Table II. Values in parentheses are the tolerance limits of the respective electrical measurements of industry-standard I-V curve tracers.

For the most part, solar irradiance measurement uncertainty has a significant effect on the accuracy of shifting the electrical current, whereas solar panel temperature measurement uncertainty mostly affects voltage shifting. Comparing to the electrical specifications of the Fluke SMFT-1000, which are typical in the industry, the appropriate temperature and irradiance measurement tools must be chosen to interfere as little as possible when shifting the I-V curve.

V. CONCLUDING REMARKS

The accuracy of field-capable I-V curve tracers greatly depends on three factors: the type of I-V curve tracer used, the amount of capacitance inherent to the solar module (string) under test, and the measurement uncertainty of solar irradiance and surface temperature. Understanding the source and propagation of measurement uncertainty/error is critical to successfully evaluating the performance of solar modules (strings) in the field. The usefulness of a field-capable I-V curve tracer is limited by the measurement uncertainty of the solar irradiance and solar module temperature measurements, as well as, errors introduced by capacitive effects of the solar module, or string, under test.

REFERENCES

[1] "Solar Industry Research Data," Solar Energy Industries Association, 2023.
[2] IRENA (2022), Renewable Energy Statistics 2022, The International Renewable Energy Agency, Abu Dhabi
[3] https://gegrenewables.com/mammoth-solar/
[4] https://www.rosendin.com/project/athos-i-ii/
[5] ISO/IEC International Standard 62446-1, "Photovoltaic (PV) systems – Requirements for testing, documentation and inspection," Edition 1.1, 2018.
[6] K. A. Emery, "Solar Simulators and I-V Measurement Methods," *Solar Cells*, vol. 18, pp. 3-4, 1986.
[7] Solmetric PV Analyzer Application Note, "I-V Curve Tracing of High Efficiency PV Modules," December 2022.
[8] ISO/IEC International Standard 60904-1, "Photovoltaic devices – Part 1: Measurement of Photovoltaic Current-voltage Characteristics," Edition 3.0, 2020.
[9] C. H. Cox and T. H. Warner, "Photovoltaic I-V Curve Measurement Techniques," *Electrical Power Systems Engineering Laboratory Report to DOE*, 1982.

978-1-6654-6060-6/23 $31.00 © 2023 IEEE

Results of First Long Duration Space Flight of Hybrid Perovskite Thin Film

Lyndsey McMillon-Brown, William Delmas, Samuel Erickson, Jorge Arteaga, Mark Woodall, Michael Scheibner, Timothy Krause, Kyle Crowley, Kaitlyn Vansant, Joseph Luther, Jennifer Williams, Jeremiah McNatt, Sayantani Ghosh

NASA Glenn Research Center, Cleveland, OH, United States

University of California, Merced, CA, United States

National Renewable Energy Laboratory, Golden, CO, United States

Wilberforce University, Wilberforce, OH, United States

In support of NASA's Artemis program with the goal of a sustained human-lunar presence, there is a need for very large (>100kW) and high-voltage-capable solar arrays, estimated to cost over $150M. Perovskite-based thin-film photovoltaics offer substantial advantages over state of the art solar arrays from the perspective of manufacturing large arrays. Perovskites have also demonstrated some of the lowest temperature coefficients and highest defect tolerance, which make them excellent candidates for aerospace applications. However, metal halide perovskites (MHP) must demonstrate durability in space which presents different challenges than terrestrial operating environments. To decisively test the viability of perovskites being used in space, a perovskite thin film is positioned in low earth orbit for 10 months on the International Space Station, which was the first long-duration study of an MHP in space. Postflight high-resolution ultrafast spectroscopic characterization and comparison with control samples reveal that the flight sample exhibits superior photo-stability, no irreversible radiation damage, and a suppressed structural phase transition temperature by nearly 65 K, broadening the photovoltaic operational range. Further, significant photo-annealing of surface defects is shown following prolonged light-soaking postflight. These results emphasize that methylammonium lead iodide can be packaged adequately for space missions, affirming that space stressors can be managed as theorized.

978-1-6654-6060-6/23 $31.00 © 2023 IEEE

Undergraduate Research Experience in the Design and Construction of a Photovoltaic Inspection Robot

Alanis M. Colón, Emmanuel J. González, Fernando J. Vargas, Samuel I. Hernandez, Michael Y. Vazquez, Eduarto I. Ortiz

University of Puerto Rico at Mayagüez (UPRM), Mayagüez, Puerto Rico, 00681, United States

Abstract — **Renewable energy has become an area of intense development and innovation as the world moves towards net zero emission goals. Thus, creating demand for tools to perform specialized processes such as gathering scientific data, inspection, and maintenance of photovoltaic (PV) arrays. Said processes are needed to ensure that existing and future systems are working efficiently and safely to meet the energy industry's critical standards. The approach to address this need was to develop a robot to perform said specialized processes. The prototype design and construction of the PV Inspection Robot used power electronics, control systems, and algorithms as part of a complete undergraduate research experience with a project-based learning (PBL) approach.**

I. INTRODUCTION

Renewable energy has become an area of intense development and innovation as the world moves towards net zero emission goals. Subsequently, many industries, such as the photovoltaic industry, have been exponentially growing for the past decade. As a result of this fast growth in the industry, the need for specialized tools or machines to achieve specific tasks emerged. Some of these needs are in PV systems' maintenance, inspection, and scientific research. Over the years, innovators have become aware of the deficits in said areas and started to develop machines and robots to perform specialized tasks that otherwise would have a significant amount of manual labor that sometimes is dangerous and very time-consuming.

Some examples of the issues that directly impact the efficiency of PV systems are the maintenance of green areas at solar farm installations, snow covering of PV arrays in the winter, dust covering of PV surfaces, and natural disaster damage such as hail. Nevertheless, panel degradation due to time and climate are not equally evident but still impacts the efficiency of the PV arrays. In a PV module, a hot spot describes an over-proportional heating of a single solar cell or a cell part compared to the surrounding cells. It is a typical degradation mode in PV modules. [1] As the output power of a single silicon solar cell is not enough to meet the actual needs, many silicon solar cells usually make up the PV module with series and parallel connections. Hot spots may occur in a PV module when the solar cells are mismatched, have certain defects, or are partially shaded. [2] Furthermore, micro-fractures, also known as micro-cracks, represent another form of solar cell degradation and can affect both energy output and the system lifetime of a solar photovoltaic (PV) system. [3]

Micro-cracks result from manufacturing defects and mechanical stress due to the fragile composition of the silicon used to manufacture the PV modules. Overall, there are many degradation modes for PV modules, and as a result, researchers need tools or methods to study and find ways to improve certain aspects of PV modules.

Over the years, researchers have developed specialized imaging techniques and tests to study all kinds of phenomena that happen to PV modules. Some of these techniques are Photoluminescence imaging, Ultraviolet Fluorescence Photography (UVF), Near Infrared Spectroscopy and Imaging (NIRS), and electroluminescence crack detection (ELCD) testing, among many other techniques used. Additionally, PV manufacturers use some of these techniques to perform quality control. The problem is that most of these processes occur in controlled environments; therefore, in the real world, once the module is out of the controlled and automated environment of factories, the process of taking data to study the behavior of the PV modules becomes very time-consuming and complicated. Unfortunately, the most common method to collect specialized images is by unmounting the PV modules taking the pictures in laboratories, and then mounting them again to continue collecting data.

In this case, the topics of the specialized tasks of development in this research were a combination of all the ones mentioned before (maintenance, inspection, and scientific research). The goal was to create a robot to clean the panel's surface and then take regular and specialized images of the PV array. The main idea is that the robot simultaneously cleans the modules and automates the process of gathering data to keep studying the behavior of PV modules in a more detailed way as they encounter all nature elements. Finally, the UPRM's Industry Affiliates Program (IAP) and CHRES Consortium sponsored this project.

II. PURPOSE

The Photovoltaic Inspection Robot project intends to help undergraduate students improve their skills in software, hardware, and control systems using tools and devices commonly used in the industry. As a result, this prepares them for future jobs while solving actual problems and enhancing their undergraduate learning experience. This project aimed to design a robot (Photovoltaic Inspection Robot) that efficiently

978-1-6654-6060-6/23 $31.00 © 2023 IEEE

inspects solar panel arrays without having to dismount them from the systems. The Photovoltaic Inspection Robot is fully autonomous. This project aims to stimulate the students to design, program, and build a small robot to inspect solar panels most efficiently and cost-effectively. As a result, students apply project-based learning (PBL) to better understand theoretical concepts discussed in class.

III. DESIGN APPROACH

Undergraduate Electrical, Computer, Software, and Mechanical Engineering students designed and developed the PV Array Inspection Robot as an interdisciplinary project using their abilities and skills. The multidisciplinary team of students used tools such as 3D CAD software, Fritzing, and Python to design and develop each area of the robot's functionality.

A. Mechanical Engineering Approach

The structure of the Photovoltaic Inspection Robot was carefully thought through because the robot had to move across a solar panel array to collect data, be firmly attached to the solar panel, and move in an X-Y axis direction without disrupting any other components. Additionally, not all PV arrays have a unique standard structure; therefore, some features had to be adjustable.

3D-printed parts were designed and printed to construct a relatively complex prototype to create the robot's structure. First, the linear actuators used to convert rotational movement to linear movement were systems composed of a ball screw and a ball nut as its primary mechanism. Then, rotational and linear ball bearings were used to ensure smooth rotational and linear movements. The last mechanical component integrated into the design was a timing belt to distribute the rotational motion of some of the motors used.

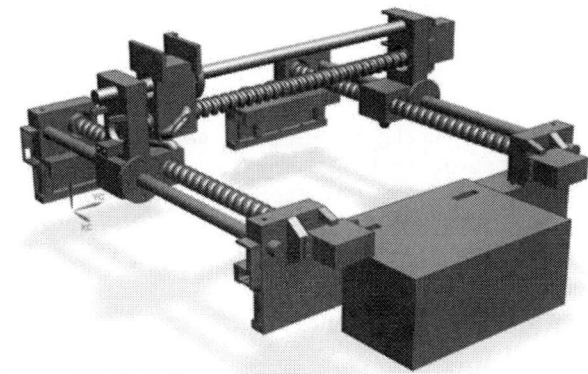

Fig. 1.   3D CAD Structural Design

B. Electrical Engineering Approach

After the mechanical design was done, the electrical components included ultrasonic sensors to measure the limits of the PV array and the distance to determine the position of the vertical linear actuators. Limit switches were also implemented to stop the vertical linear actuators in case the ultrasonics weren't working and to calibrate the limits of the robot. A power distribution block connected to the battery was used to distribute energy to several voltage regulators to provide the appropriate voltage to all the components. An HD camera was also integrated camera to take the images. Lastly, the electromechanical components, such as the brushless DC motors, were incorporated into the circuit.

The use of the appropriate motor is essential and must be made, considering several factors, such as the weight of the robot and the energy source power capacity. Brushless DC Motors, Brushed DC Motors, and Stepper motors, amongst several others, will be tested and selected for this project based on their advantages and disadvantages. The servos will oversee the movement of the scanner that will be moving in the x and y axis to scan every individual location of the solar panel thoroughly. Two DC motors move the complete system in the x-axis since these motors need a much bigger value of torque in contrast to the servos. Another critical aspect of the correct functionality of the system and the individual electric components is the battery sizing. Each component's stall current must be added to calculate the power the battery must supply for efficient system functioning. The stall current is the current drained by each piece when first powered on; this current is typical of a much greater value than that of components at rest. After summing each stall current, the result was added 25% of itself to ensure that the battery would be sufficient for the system. Various skills were needed to develop the electrical system, such as soldering and electrical schematic designing. Soldering was essential to correctly connect components and pins to the raspberry pi and the battery.

C. Computer and Software Engineering Approach

The programming paradigm established was the Imperative programming style. The reason for using this paradigm is that it is straightforward to understand. The codebase needed to be more significant to consider another paradigm; the imperative style with python is easy to manage, maintain and upgrade. The structure of the code base consists of a directory where all the image data is organized and a source file where the code base is stored. In the source folder, there is the main file where the main logic is taken place; inside this folder, we can find a component folder where the code of all the different devices takes place. The component code base consists of four python scripts (these are the camera controllers), a limit switch controller, a motor controller, and an ultrasound controller. The Pimorini stack hat was used to extend the GPIO of the raspberry pi to add more modules and sensors.

The routine consisted of initially resetting the robot's position to one of the edges of the solar panels when one of the limit switch sensors was triggered. Then, the robot moved horizontally towards the other edge of the solar panels and took pictures after traveling certain distances. After the limit sensor at the opposite side of the reset position was triggered, the robot moved vertically to repeat the same process but across the center row. Finally, after repeating the same pattern across the

978-1-6654-6060-6/23 $31.00 © 2023 IEEE

center and upper row, the robot resets its initial position. As a result, the robot now has pictures of every cell in the solar panel. A machine-learning algorithm will eventually analyze to detect whether the panels are in proper condition.

IV. Current Prototype and Future Work

After the prototype of the robot was completed, it effectively was able to scan the cells in the solar panels and saved the pictures to analyze them later. In Fig. 2 can be seen the designed prototype. However, the Machine Vision algorithm needed to detect defects in the solar panels still needs to be adequately improved. Because of this, the main priority for the future will be to create an effective algorithm with deep learning and computer vision that can detect various types of defects in the cells. Besides this, the robot's movement can be optimized by adding additional sensors to the motors. Some of the sensors can be Encoders and Gyroscopes since it allows for monitoring the motor's speed. Thus, it would be possible to correct the speed of motors working in parallel when needed by applying control algorithms. Finally, after perfecting the prototype currently being worked on, a real-sized version will be created with the ability to scan industrial layouts of solar panels autonomously.

Fig. 2. First Prototype of the PV Inspection Robot.

Additionally, to everything previously mentioned, another future goal is to take the purpose of the robot further in such a manner that it can be applied in different fields. In other words, it increases the robot's adaptability in aerospace applications. The robot could also be adapted to inspect solar panels, but different types of surfaces and applications could be countless.

V. Students Educational and Professional Benefits

Some of the vital skills gained by the students as part of this project include the opportunity to apply concepts learned in class, build soft skills, obtain experience in scientific research, and add value to their resume by learning how to use different tools used in the industry, learning how to develop patents, and how to commercialize products. This strategy helps the students develop their scientific research skills and prepares them to excel in the engineering industry.

A. Educational Benefits

In developing the robot, the students could apply concepts learned in class and better understand them. Additionally, they expanded their experience in scientific research, which fosters critical thinking and analytical skills through hands-on learning; helps them define academic, career, and personal interests; and expands their knowledge and understanding of a chosen field outside the classroom. [4] Equally important in any chosen career path, soft skills such as teamwork, communication, problem-solving, time management, critical thinking, decision-making, leadership, creativity, and resourcefulness also flourished in this project. Consequently, another area developed is short- and long-term academic development. More specifically, the short term is the preparation to create the final compulsory project of the bachelor's degree, in which everything learned must be applied. The long-term objective is the development of graduate studies, where all these skills and more would be used.

B. Professional Benefits

When undergraduate students search for internships or coops to obtain experience in different fields before and after graduating, recruiters observe their academic performance and the student's extracurricular activities. These extracurricular activities vary from student associations for professional development and technical projects to research projects. These are very important because students obtain leadership roles and learn new tools that add value to their resumes and help them learn new skills. Some of these skills are the dominance of specific programs, patent development, and even product commercialization which are vital skills in the industry since it's a skill set that the employer doesn't have to invest time and money. Not only do these skills help students find jobs in recognized companies, but they also nourish the opportunity for them to create their own.

VI. Conclusion

After working on this project, the students could work as a team, reinforce their engineering skills, and learn some new vital skills along the way. The project's outcome was a total success, and the undergraduate students will continue working on this project to improve and implement more essential functionalities so that the project objective is wholly achieved.

References

[2] S. Deng, Z. Zhang, C. Ju, J. Dong, Z. Xia, X. Yan, T. Xu, and G. Xing, "Research on hot spot risk for high-efficiency solar module," Energy Procedia, vol. 130, pp. 77–86, 2017.

[3] "Micro-Fractures in Solar Modules: Causes, Detection and Prevention," Ajg.com, 2023. https://www.ajg.com/us/news-and-insights/2020/jan/micro-fractures-in-solar-modules-causes-detection-and-prevention/#:~:text=(PV)%20system.

Development of gradient layers to improve the efficiency of transparent passivating contact solar cells

Alexander Eberst, Binbin Xu, Weiyuan Duan, Andreas Lambertz, Uwe Rau, Kaining Ding

Forschungszentrum Jülich GmbH, IEK-5 Photovoltaik, Jülich, Germany

RWTH Aachen University, Aachen, Germany

One limitation to currently applied passivating contacts like hydrogenated amorphous silicon or polycrystalline silicon is the strong absorption of light in the passivating and contacting layers. To increase the amount of light reaching the absorber material of the solar cell, a more transparent material is needed. Hydrogenated nanocrystalline silicon carbide (nc-SiC:H) is highly transparent, as well as highly conductive and has excellent passivating properties, showing very high generated currents in combination with a good fill factor (FF) and high open-circuit voltage (VOC) when applied in a solar cell. This approach is called transparent passivating contact (TPC). However, the FF is still lower as compared to other state-of-the-art approaches. In this work, a closer look on this reduced FF is taken. It is found that there is a direct trade-off between the passivation and the FF depending on the thickness of the SiC layer. In previous works, the SiC layer consists of two SiC layers, whereas the first deposited SiC layer was grown at a soft deposition condition to not harm the SiO passivation layer and therefore referred as the passivating SiC layer and the subsequent SiC layer was more crystalline and conductive and therefore referred as the conducting SiC layer. We found out that an even thinner passivating SiC layer in combination with a SiC layer with continuous transition of its material properties along the growth direction from more passivation SiC layer-like to conduction SiC layer-like, realized by a slow transition of the temperature of the catalytic filament leads to higher iVOCs in combination with an increased FF, leading to solar cell efficiencies exceeding the previous double-layer nc-SiC:H stack and even the amorphous silicon reference cell

A GIS-based approach for prioritization of photovoltaic systems with energy storage implementation for vulnerable community resilience

Javier A. Moscoso-Cabrera, Edgar E. Cruz, Cristian R. Meléndez, and Eduardo I. Ortiz

University of Puerto Rico-Mayagüez Campus, Mayagüez, Puerto Rico, PO Box 9000-00681, USA

javier.moscoso1@upr.edu, edgar.cruz2@upr.edu, cristian.melendez@upr.edu, eduardo.ortiz7@upr.edu

Abstract — More than 40.5% of the local Puerto Rican population live under the US federal poverty level. For context, there are 36 of 78 municipalities (46.15% of the total) with economic trends that identify these as low income-areas and the median income per household of three members is around $22,000. There is sufficient evidence indicating that hurricane response and critical service restoration in Puerto Rico can last more than a year for many communities, particularly the most vulnerable. On top of this, losing electricity service also means losing water service for many because when power outages occur, most of the water pump plants also lose service and potable water service is interrupted. There are more than 400 community centers all around the Puerto Rican archipelago that have been established throughout the years. However, many seem to be abandoned or not in continuous use. Recent events have led to the news of millions of dollars in United States federal funds for the installation of solar PV systems. By combining the use of Geographic Information Systems (GIS) and Multi-Criteria Decision-Making methodologies (MCDM) like Analytic Hierarchy Process (AHP) and Grey Relational Analysis (GRA), the authors propose a prioritization model for the implementation of emergency PV systems equipped with energy storage as a means to support community resilience, address critical needs and revitalize the sense of community all over Puerto Rico. This framework can serve as an example in many countries all around the world as it can be replicated and customized to the needs of the particular use case or administrator preferences.

I. INTRODUCTION

The United Nations Office for Disaster Risk Reduction defines vulnerability as "the conditions determined by physical, social, economic and environmental factors or processes which increase the susceptibility of an individual, a community, assets or systems to the impacts of hazards" [1]. Puerto Rico has a long history of severe hurricane activity, and in recent years, these events have proven to test the energy infrastructure integrity and resilience as well as the mental health, among other issues of those who face these challenges continuosly [2, 3]. With climate change and global warming, storms are increasing in both frequency and intensity. Community vulnerability becomes even more serious when considering that nearly half (40.5%) of the entire population lives under the federal poverty level, which is considered to be $24,860 for a 3-member family in 2023 [4, 5]. Energy generation is heavily dependent on imported fossil fuels, which are also subject to global markets and supply. Moreover, a centralized power generation infrastructure with outdated transmission and distribution grids lead to serious problems particularly in times of need. Considering these challenges, people all over Puerto Rico have set ambitious goals and ar opting to increase the adoption of renewable energy technologies, with a particular focus on rooftop solar energy systems. In order to achieve optimal photovoltaic (PV) system integration, it is essential to identify the suitable locations considering multiple factors related to geo-location as well as socioeconomic needs.

This paper presents a GIS-based approach for ranking potential PV system locations in Puerto Rico, based on user priorities using numerical data tables from the latest census as input. The aim was to formulate a customizable model for identifying areas that could enable the most impact from PV system installations, this was done by evaluating community centers and the surrounding community profiles. There are several efforts targeting solar PV systems development, such as local organization Fundación Comunitaria is working in the implementation of solar PV systems for community health clinics and main waterways [6]. Other organizations, such as Casa Pueblo and AMANESER 2025, are working on empowering communities through self-management in order to provide access to solar energy for residential critical needs [7].

The main focus for this study are Community Centers as they represent centric points of contact, for most sub-counties, and could enable widespread benefits from renewable energy systems to serve the collective. Additionally, this research considers the vulnerability of communities to ensure that the proposed locations for PV systems have significant potential in terms of benefits and feasibility. According to The Caribbean Handbook on Risk Management, there are many aspects of vulnerability, arising from various physical, social, economic, and environmental factors some of which will be mentioned in the methodology section [8]. After processing multiple vulnerability factors and the available data, the results should provide an objective roadmap for decision-makers, organizations, authorities and academics interested in the deployment of PV systems in Puerto Rico, specifically in the context of increasing energy independence, reliability and resilience in the face of natural disasters for vulnerable populations.

II. METHODOLOGY

Previous relevant studies discuss that an ideal plan for increasing sustainable energy adoption is through a community-led and decentralized approach considering rooftop solar as a means to address energy security and justice. Although this study favors that approach, it focuses on the potential of community centers as a pathway to empower the collective and foster a sense of community, solidarity and empathy among neighbors in everyday life and emergency scenarios.

To grasp the status of vulnerability in the 902 sub-counties ("barrios") accounted for, 2020 U.S. Census and the American Community Survey (ACS) 5-year estimates were considered. Survey data was assessed as well as available solar irradiation information for Puerto Rico [9]. The considered vulnerability factors in this study include median household income ($), demographics (total population as well as percent of population by age), flood zones, geographic location, and irradiation for solar PV generation potential. Solar performance simulations and analyses were performed using Excel, first comparing daily generation potential using the softwares PV Watts Calculator and the PV GIS Online Tool. Community centers were manually identified in 409 locations. The Quantum Geographical Information System (QGIS) software tool enabled a combination of these factors in order to visualize trends in a geographical setting. The datasets were aggregated within the QGIS tool for a visual interpretation on top of the landscape of Puerto Rico. The aggregated Census datasets were then extracted and subjected to MCDM methodologies, such as AHP and GRA methodologies to be able to both subjectively and objectively consider potential locations based on the community profiles, while taking into account qualitative or unquantifiable reasoning. The application of AHP and GRA concluded in justifiable ranking systems to prioritize the implementation of emergency solar PV systems with storage.

Both methods resulted in two ranked lists of all the community centers based solely on demographic criteria correlations. The AHP method provides a subjective framework for MCDM through pairwise comparisons and arbitrary preference [10,11]. The comparison scores are then validated in terms of consistency to ultimately rank alternatives based on benefit potential. The GRA method accounts for various input parameters to objectively determine the alternative that statistically would allow for the most benefit [11]. To provide a combined representative result, an average rank between the two could be considered. With the proper dataset collection, both methods could be applied in future efforts to optimize their combined results and consider additional factors such as average cost of transportation, availability of labor and proximity to critical service suppliers and infrastructure, health related matters, among other considerations.

A. Vulnerability Scoring Framework

The datasets used were comprised of median household income, demographics (total population by age, sex, disability) per sub-county, flood zones, and the amount of neighboring community centers. Table 1 displays the subjective pairwise comparison scoring that was applied to the datasets for processing using the AHP method. In the case of this study, 332 out of the total number of community centers were considered due to their location outside of floodzones. Mitigation efforts and alternative measures could be suggested for those specific centers and communities, which have that increased vulnerability.

Table 1. AHP Method: Pairwise Comparison Matrix with subjective scores and resulting weights to define the hierarchy.

	Total Population	Population Under 18	Population Over 64	Total Disabled	Nearby Community Centers	Nearby Pump Stations	Median Household Income	Weights
Total Population	1.000	3.000	0.333	0.500	3.000	3.000	2.000	15.83%
Population Under 18	0.333	1.000	0.250	0.500	2.000	2.000	2.000	9.88%
Population Over 64	3.000	4.000	1.000	0.500	3.000	4.000	2.000	23.52%
Total Disabled	2.000	2.000	2.000	1.000	4.000	8.000	5.000	32.13%
Nearby Community Centers	0.333	0.500	0.333	0.250	1.000	0.500	0.500	5.15%
Nearby Pump Stations	0.333	0.500	0.250	0.125	2.000	1.000	3.000	7.05%
Median Household Income	0.500	0.500	0.500	0.200	2.000	0.333	1.000	6.44%

Table 2. Top ranking community Centers using the AHP methodology, featuring only 15 out of 332 community centers candidates to prioritize for PV system installation.

Community Center	Municipality	Rank
Centro Comunitario Tras Talleres	San Juan	1
Centro Comunal Barrio Obrero	San Juan	2
Centro Comunal Campo Alegre	San Juan	3
Centro Comunal Cantera	San Juan	4
Centro Comunal de Sabana Seca	Toa Baja	5
Centro Comunal Estancias de Rv≠o Hondo	Bayamon	6
Centro Comunal Habra Estrecha	Bayamon	7
Centro Comunal De Villa Panteon	Bayamon	8
Centro Comunal Santa Teresita	Bayamon	9
Centro Comunal Monte Carlo	Bayamon	10
Centro Comunal de Ayuda Social	San Juan	11
Centro Comunal Summit Hills	San Juan	12
Centro Comunal La Marina	San Juan	13
Centro Comunal Venus Gardens Norte	San Juan	14
Centro Comunal Villa Capri	San Juan	15

This method conventionally makes use of the Saaty scale which assigns values to represent the degree of importance or preference between two elements in a pairwise comparison. The scale ranges from 1 to 9 and the yellow cells in Table 1 represent the input values. Note that the values that are less than 1 are the inverse of the comparison, meaning that the paired counterpart has a greater degree of importance.

After calculating the weights per factor, normalizing the values, calculating the weighted sum a consistency ratio is determined to assess the consistency of judgments made during pairwise comparison. The consistency ratio was then divided by the Random index which is acquired from the Random Index Table developed by Satyy (1980) based the amount of factors

[12,13], ultimately validating that the judgements made are acceptable.

For an objective approach, the GRA method incorporates a set of formulas depending on the relevance of the data which can be interpreted in two forms, the "Lower the better" or the "Higher the better". In decision-making and performance evaluation scenarios, these approaches refer to two different perspectives on how to assess and prioritize outcomes or performance metrics. These equations are stated are stated below, respectively:

Lower The Better - Approach

$$x_i = \frac{\max(y_i) - y_j}{\max(y_i) - \min(y_i)} \quad (1)$$

Higher The Better - Approach

$$x_i = \frac{y_j - \min(y_i)}{\max(y_i) - \min(y_i)} \quad (2)$$

Almost all factors are analized using the "Higher The Better" approach which means a higher value of the metric is considered more desirable or favorable. The "Lower The Better" approach is only used for the Median Household Income as the PV system location should favor those with less economic resources.

After subjecting the data to the equation, deviations and Grey Relational Coefficients (GRC) per factor, the Grey Relational Grade (GRG) was determined. The GRG provides a measure of the closeness or similarity between the target sequence, which in this case compares the reference sequence per community center to one with maximum values for each factor.

Table 3. Top ranking community Centers using the GRA methodology, featuring only 15 out of 332 community centers candidates to prioritize for PV system installation.

Centro	Pueblo	Rank
Centro Comunal Campeche	Carolina	1
Centro Comunal Santa Cruz	Carolina	2
Centro Comunal de Boqueron Juan de Leon Cruz	Las Piedras	3
Centro Comunal Gonzalez	Las Piedras	4
Centro Comunal Parcelas Nuevas	Las Piedras	5
Centro Comunal Loma del Viento	Guayama	6
Centro Comunal Rivieras de Cupey	San Juan	7
Centro Comunal San Gerardo	San Juan	8
Centro Comunal Las Curias	San Juan	9
Centro Comunal Playita	Yabucoa	10
Centro Comunal Barrio Quebradillas	Yabucoa	11
Centro Comunal de Calabazas	Yabucoa	12
Centro Comunal La Hormiga	Juncos	13
Centro Comunal La Esperanza	Vega Alta	14
Centro Comunal Villa Josco	Toa Alta	15

B. Community Center Ranking

AHP and GRA methodologies were used to rank the community centers in order of preference for PV systems implementation, considering demographic data and proximity to other centers as a means to be a center of collective service. Table 1, displays subjective weight criteria defined for the elements of interest expressed, while Tables 2 and 3 include the top ranking community centers after the two multi-criteria decision making methodologies were implemented.

III. RESULTS AND DISCUSSION

In Figure 1, all the community centers can be observed in their sub-county setting. The color scale in the map goes accordingly to the amount of median household income in said barrio. It was defined that the darker colors relate to higher income. Figure 2 presents a representation of poverty level between lighter green for families below the poverty level and darker green for families above said level. A trend is observed of higher income in the Metropolitan Area. This helps to visually determine in which barrios to start focusing, ideally prioritizing those with lower income.

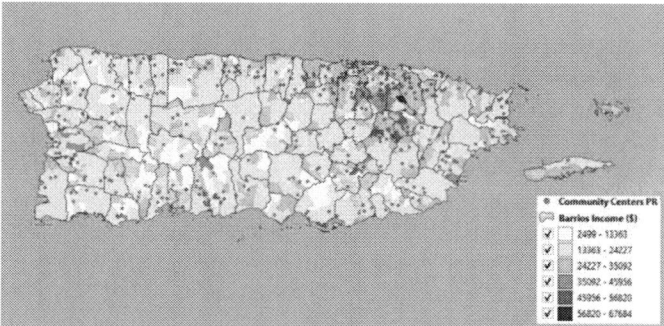

Fig. 1. Community Centers and Median Household Income per county sub divisions.

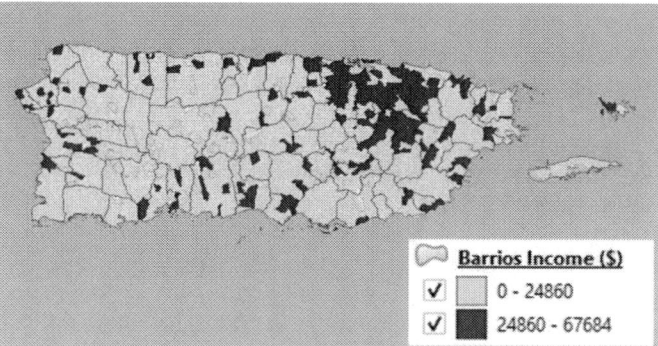

Fig. 2. Poverty Level per county sub divisions.

In Figure 3, community centers are observed along with the percent of vulnerable population concerning people 64 years or older. The gradient of colors represents the proportion of this population as a percent of the total population of the subcounties. A trend is observed of higher concentration of over 64 year-old population in the West area of Puerto Rico.

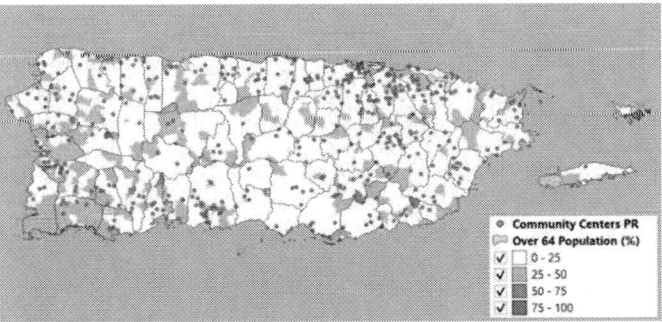

Fig. 3. Over 64 year-old Population Distribution by Percent.

In Figure 4, community centers are observed along with the percent of vulnerable population of minors, 18 years or younger older. The gradient of colors represents the proportion of this population as a percent of the total population of their subcounties. A general impression is observed as to notice that there is not a great amount of young population, but particularly in the center and south there seems to be more younger generations.

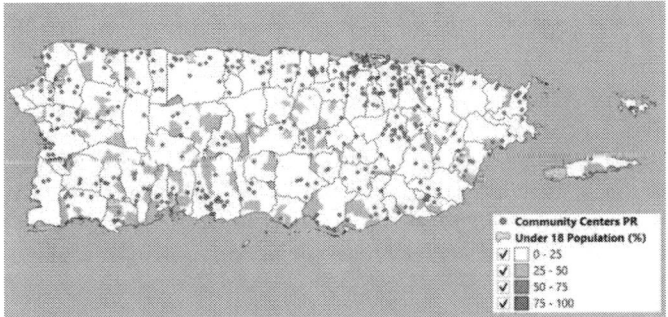

Fig. 4. Under 18 year-old Population Distribution by Percent.

In Figure 5, we can appreciate the visualization of community centers distribution along with flood zones both for 100 and 500 years of occurrence. This helps the analysis of locations by serving as a penalty consideration factor for the centers that could be more affected by floods.

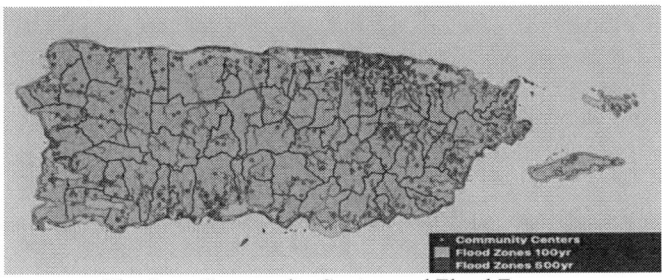

Fig. 5. Community Centers and Flood Zones.

In Figure 6, community centers can be observed distributed around Puerto Rico in relation to solar irradiance. The colors represent the energy output thorught the day for a 1 kWp solar PV system. Around Puerto Rico, we find that it ranges from 3.61 kWh to 5.01 kWh. These results can depend on

geographical location, elevation, shading, among other factors. The colorscale was defined as to consider the color blue to be where lowest solar output is generated and red the areas where there is the most solar PV potential. This helps to visually determine which systems would benefit form the available solar resource with the least amount of area of panels installed.

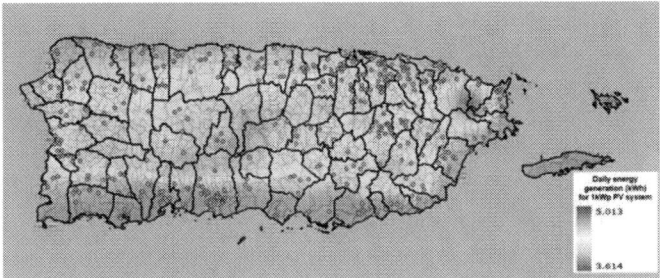

Fig. 6. Community Centers and Solar PV Irradiation.

As a general proposed solution, we consider **2.58 kWp** in solar PV panels, along with **10.24 kWh** in lithium iron phosphate battery storage and a **5 kW** inverter to be an alternative for community centers. This follows considerations based of a low, medium and high daily electricity consumption scenarios of 6.69, 7.78 and 9.39 kWh, respectively. These consider LED lights, fans, a 3.9 cubic feet washing machine, a 9.82 cubic feet refrigerator, medical equipment needs and small electronic devices charging, such as phones for communications. In Figure 7, daily consumption scenarios are considered to take into account for the customized solar PV design. Figure 8 presents a daily average PV generation simulation (comparing results between the tools PV Watts and PV GIS) for the system and in Figure 9 we include a schematic diagram of connections for the proposed solution.

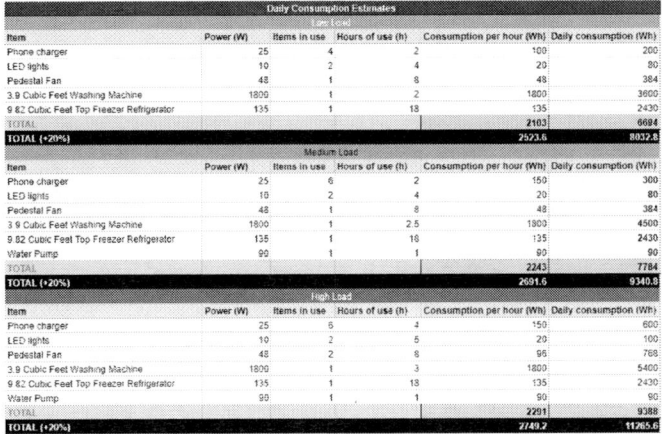

Fig. 7. Daily consumption scenarios considered for the design of the PV system.

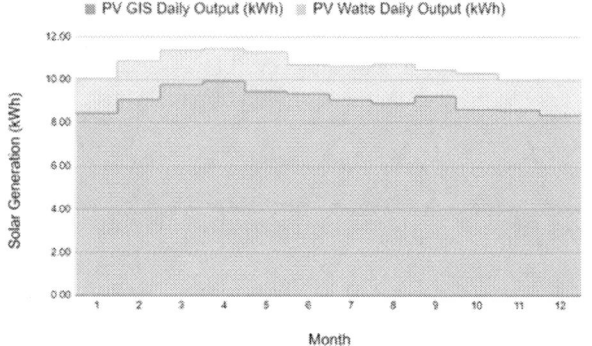

Fig. 8. Solar PV daily average simulations for the 2.58 kWp PV.

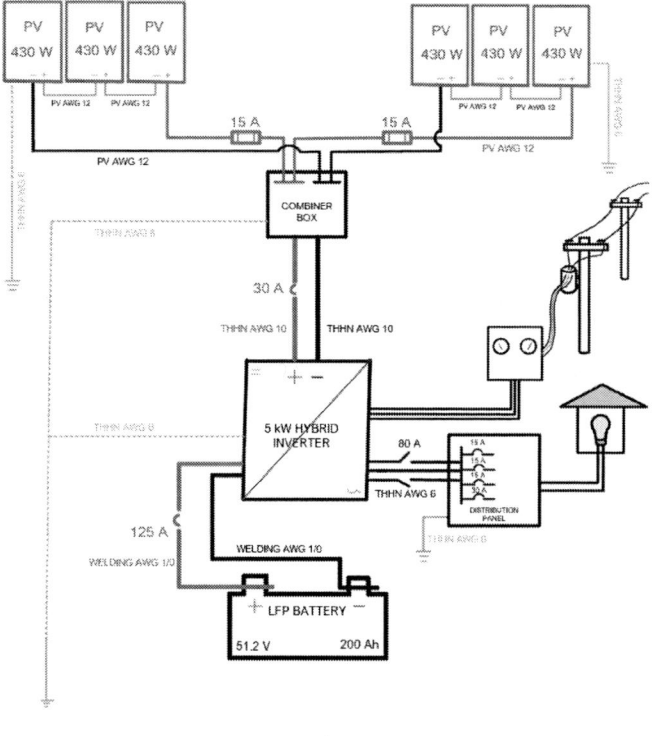

system.

Fig. 9. Proposed 2.58 kW solar PV system schematic diagram.

In terms of the technology and its features, with hybrid inverters three types of load working operation modes can be considered (PV priority, utility priority or inverter mode) as well as four different battery charging modalities. A hybrid inverter can be capable of operating in a grid-tie mode, backup mode, hybrid mode and off-grid mode. It provides flexibility for operation when the grid is available as well as in an emergency when power outages occur. Charging processes can be configured with a solar charging, electric utility, solar consumption and hybrid charging. For future considerations

there is possibility for a grid forming inverter implementation for a resiliency improvement.

IV. CONCLUSIONS

The meaning of power may be contained in the words we decide to use to communicate a message. An energy transition is an effort that considers shifting away from the norm of fossil fuels and heading into the use of renewables, while continuing to use electricity in the usual manner. An example would mean installing a solar PV system but continuing to leave all the lights on at the same time. On the other hand, an energy transformation includes this shift but considers a change in awareness, behaviors and patterns of consumption. This can be motivated through education, awareness efforts, assertive action, among other means. In summary, to transform a reconsideration of how people, engage with energy as a resource to sustain life and empower decisions, since it is the people, ultimate end users of electric service, the ones which should benefit from technology implementation.

This effort presents a GIS-based approach for prioritizing potential PV system locations in Puerto Rico using a combination of publicly available census data and geo-location data such as demographics, income, solar irradiation, and physical characteristics of the terrain, using open data sources. It also considers a combination of MCDM methodologies that can be further developed in future studies to assess and improve its accuracy.

The proposed model and the input data can be adjusted to account for any criteria that addresses a similar use-case or customized with other preferences. Authors expect this work to begin a conversation and provide a guideline basis for the implementation of solar PV energy systems equipped with energy storage all around Puerto Rico based on different vulnerabilities identified of the population. This work is highlights the importance and role of community centers as a means to revitalize the sense of community and provide such centers with tools to serve critical needs in their respective areas. Although the focus on residential systems must be crucial and the priority, with this approach the authors provide an alternative to pave the way for community development and gathering through access to energy in a way it can prioritize the most vulnerable. Focusing energizing efforts on community centers can serve as an example in Puerto Rico and for the rest of the world. Moreover, prioritizing vulnerable communities with various needs promotes energy justice.

Having community, particularly in moments of emergency and need, can be crucial in order to survive. It is important to emphasize that this paper does not intend to provide a sole solution given communities face different challenges and have different realities. As part of the future development for this framework, a more in-depth data analysis and processing will be performed. Multiple data points from Census and geological data could be added to improve the current study as well as more community centers, which represent additional layers of data from population, disabilities, landslides, among other aspects. Also, proximity to critical infrastructure is to be added to the analysis, as most of the emergency services are provided from these locations in the aftermath of severe atmospheric events. A more comprehensive approach can be worked with collaboration of diverse disciplines in social and political sciences as well as humanities to consider a more holistic understanding of energy transformation matters, considering having access to electricity service can be a matter of life and death for many, and goes beyond just technology and engineers.

ACKNOWLEDGEMENTS

The team would like to acknowledge the Consortium for Hybrid Resilient Energy Systems (CHRES), Minds2Create Research Team, the Universal Interoperability for Grid Forming Inverters (UNIFI) Consortium and Augusto Gandía Ojeda for their support throughout this research effort.

REFERENCES

[1] United Nations Office For Disaster Risk Reduction, "Vulnerability", undrr.org. [Online] Available: https://www.undrr.org/terminology/vulnerability#:~:text=The%20conditions%20determined%20by%20physical,to%20the%20impacts%20of%20hazards.

[2] M. Castro, Y. Cintrón and J. Gómez, "The Longest Blackout in History and Energy Poverty", ICAT, pp. 36-48, 2018.

[3] A. Kwasinski, F. Andrade, M. Castro-Sitiriche and E. O'Neill-Carrillo, "Hurricane Maria Effects on Puerto Rico Electric Power Infrastructure", in IEEE Power and Energy Technology, 2019.

[4] B. Glassman, "Puerto Rico outmigration increases poverty declines," Census.gov, 28-Oct-2021. [Online]. Available: https://www.census.gov/library/stories/2019/09/puerto-rico-outmigration-increases-poverty-declines.html.

[5] "En pobreza el 50% o más de la población en 36 Municipios de Puerto Rico | State Data Center," Estadisticas.pr, Dec. 19, 2019.

https://censo.estadisticas.pr/Comunicado-de-prensa/2019-12-19t145558#:~:text=A%20nivel%20de%20Puerto%20Rico.

[6] R. V. del Sur, "Instalarán Sistemas Solares en acueductos comunitarios del sur," Voces del Sur, 19-Apr-2021. [Online]. Available: https://www.vocesdelsurpr.com/2021/04/instalaran-sistemas-solares-en-acueductos-comunitarios-del-sur/.

[7] "Creating Sustainable Communities in Puerto Rico (Amaneser 2025)," Global Ministries, 28-Oct-2022. [Online]. Available: https://www.globalministries.org/project/creating_sustainable_communities_in_puerto_rico/.

[8] "5.3 Vulnerability | CHARIM," www.charim.net. http://www.charim.net/methodology/53

[9] Census.gov, 2020. https://www2.census.gov/geo/tiger/TIGER_DP/2020ACS/ACS_2020_5YR_COUSUB_72.gdb.zip

[10] H. -C. Liu, M. Yang, M. Zhou and G. Tian, "An Integrated Multi-Criteria Decision Making Approach to Location Planning of Electric Vehicle Charging Stations," in IEEE Transactions on Intelligent Transportation Systems, vol. 20, no. 1, pp. 362-373, Jan. 2019, doi: 10.1109/TITS.2018.2815680.

[11] M. Gerus-Gościewska and D. Gościewski, "Grey Relational Analysis (GRA) as an Effective Method of Research into Social Preferences in Urban Space Planning," Land, vol. 11, no. 1, p. 102, Jan. 2022, doi: 10.3390/land11010102. [Online]. Available: http://dx.doi.org/10.3390/land11010102

[12] X. Zhang, J. Lu and Y. Peng, "Hybrid MCDM Model for Location of Logistics Hub: A Case in China Under the Belt and Road Initiative," in IEEE Access, vol. 9, pp. 41227-41245, 2021, doi: 10.1109/ACCESS.2021.3065100.

[13] E. Isagba and N. Ihimekpen, "The Use of AHP (Analytical Hierarchy Process) as Multi Criteria Decision Tool for the Selection of Best Water Supply Source for Benin City". [Online]. Available: https://www.researchgate.net/publication/318532544_The_Use_of_AHP_Analytical_Hierarchy_Process_as_Multi_Criteria_Decision_Tool_for_the_Selection_of_Best_Water_Supply_Source_for_Benin_City

Position dependence of the performance gain by selective ground albedo enhancement for bifacial installations

Nils-Peter Harder[1], Issam Smaine[1], Fadi Bourarach[1], Damien Cosme[2], Ines Arfaoui[1], Julien Chapon[1], Arttu Tuomiranta[1], and Antonios Florakis[1,3]

[1]TotalEnergies, 91120 Palaiseau, France

[2]TotalEnergies Research Center Qatar, Qatar Science and Technology Park, P.O. Box 9803 – Doha – Qatar

[3]Advanced Solar Energy Technologies Consulting BV, 3000 Leuven, Belgium

Abstract — We explore how positioning high-albedo material on the ground impacts the performance gain obtainable by these selective ground albedo enhancements. The ground albedo can be improved, for example, by placing geosynthetic materials, white stones, or paint under or between PV panels. Specifically, we determine how the most effective positioning of albedo enhancement material (AEM) is influenced by geographic latitude, by the diffuse-light content in the total irradiance, by the choice between tracking versus fixed-tilt mounting, and by module mounting height. We find for fixed-tilt systems that albedo improvement strips with a width similar to the module table or less, should be under the tables, but shifted towards the sun-facing leading (lower) edge of the modules. This preference for placement towards the lower edge is slightly more pronounced the closer the location is to the equator. For typical / high-mounted (1.5m torque-tube height) tracked systems, for all albedo strip widths, the optimum placement is centered at the center of the module table. However, for low-mounted tracked systems and a ground area coverage by the AEM of about equal or less the module table width, we find that it is favorable to split the material into two off-centric strips rather than one strip centered to the trackers. Interestingly, we find for both fixed-tilt and tracking systems that the optimum placement of such albedo enhancement strips is not notably influenced by the diffuse content of the irradiation. However, the magnitude of the gain is influenced by the diffuse content and we find higher relative gain from albedo improvement in case of a high diffuse light content. Particularly for tracked systems the relative production gain by ground albedo enhancement is larger for higher diffuse content. Of course, the overall performance of tracked systems prefers low diffuse light content and dominance of DNI.

I. INTRODUCTION

Bifaciality of photovoltaic (PV) modules has notably contributed to driving down PV electricity generation cost, and large utility power plants (UPPs) are these days increasingly designed as bifacial installations [1]. Compared to monofacial installations, a bifacial system favors higher module racking (and wider row-to-row spacing), to allow more light under the modules and more efficient back-reflection from the ground onto the rear sides of the modules. Another option for improving the bifacial gain of a PV system is to change the properties of the ground. This can involve placing white gravel, geotextiles, or paint on the ground under or between the modules [2], [3]. Such ground albedo improvement is cheaper

per area than the module cost per area, yet the cost of such ground improvement can be a notable factor in the total system cost. Because different positions on the ground are differently efficient for reflecting sunlight back onto the rear sides of the panels, it can be economically beneficial to apply the albedo improvement only on parts of the ground. Albedo is very often to a good degree of approximation Lambertian (that is: angle-independent radiant reflection intensity for all directions). Generally, ground areas closer to the modules reflect incoming light more efficiently back onto the modules (unless view factor effects minimize the radiation transfer, such as for ground areas underneath vertically mounted panels). At the same time, the ground areas closest to the modules, for example directly under the modules of a fixed-tilt installation, may receive less light to reflect at all, because these regions will be more frequently shaded by the modules (see Fig. 1). It is therefore a non-trivial optimization task to define the best placements for ground albedo improvement.

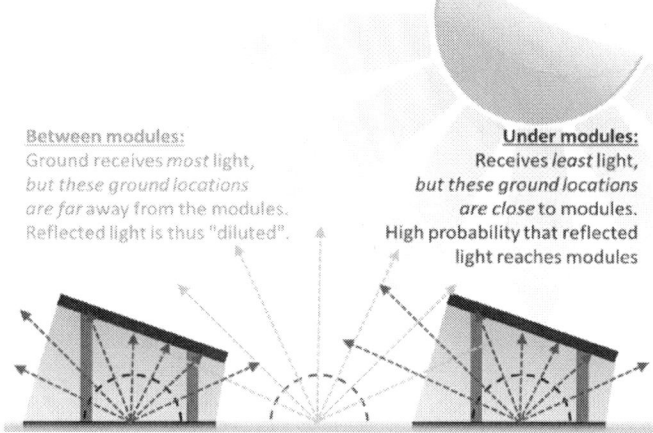

Fig. 1. Optimization problem for placement of albedo enhancement material that covers less than 100% of the ground area. Ground areas that receive most sunlight are far from modules, and vice versa.

Jaubert *et al.* [2] and Rhazi *et al.* [3] already explored several configurations of albedo enhancement materials (AEM) under PV modules in bifacial solar systems. In our study, we provide a systematic investigation on the dependence of the obtainable

978-1-6654-6060-6/23 $31.00 © 2023 IEEE

gain on the position of the ground albedo improvement on flat ground. Mandy Lewis presented at this conference, where our work was presented, too, highly related results [4]. We highlight general effects of several parameters on this position dependence for partial (or "selective") ground albedo improvement. In particular, we investigate the effect of latitude, diffuse-light content in the solar irradiance, and module mounting height. We explore these effects for single-axis tracking systems as well as for fixed-tilt systems.

While any project will profit from dedicated simulations of the best placement for albedo improvement, our study allows us to estimate the main effects for a wide range of situations.

II. SIMULATION MODEL

We use SolarOPS PV system performance simulation software [5], developed by TotalEnergies. SolarOPS incorporates both ray-tracing (RT) and view-factor (VF) optical engines, which are based on various open-source modules from PV Performance Modeling Collaborative [6]-[8] and Radiance [8]. We radically re-designed parts of these modules to add several additional features that enable us to model specific complex scenarios. One of these additional features relevant for this work is the ability to set zones of differentiated albedo also in the VF model, instead of having only a uniform albedo distribution. Strips of material with increased albedo can be placed on various positions and with various widths between or under the two module rows, allowing for the optimization of the strip placement. (For the RT version, this functionality is intrinsically available as RT allows direct assignment of the optical properties of the different zones of the scene.) Via RT, it is possible to include the impact of the racking and torque tube geometry and its shading, and thus it is more suitable for detailed simulation studies and often necessary for simulating small R&D test installations, while the typical implementation of VF models only represent systems with long rows of modules where edge effects can be neglected [9]. On the other hand, the current implementation of RT in SolarOPS is computationally demanding and, therefore, not as practical for large-scale optimization studies. For this purpose, we often opt in such cases for the much faster VF-based modeling. The main limitation of VF is the neglection of said edge effects [10] due to the implicit assumption of semi-infinite module table rows. We show in our validation example in section III that this neglection/idealization of VF diminishes with increased length of the module rows and that VF is therefore suitable for simulating UPP scenarios.

We use astronomic tracking with backtracking in the simulations presented in this paper.

III. MODEL VALIDATION

Our R&D team, based in Palaiseau, France and Doha, Qatar, studies module and PV system performance at various installation locations across the world. By comparing simulation results to production data from various R&D PV test installations and UPPs and also by doing "peer-to-peer" comparison with e.g., PVsyst, we validate and improve our system simulation software SolarOPS continuously. This section of our paper reports on a comparative study for two R&D systems in Doha, thanks to our collaboration with the Qatar Environment and Energy Research Institute (QEERI). Both systems are installations of 2x3 half-cell modules, and one of them features a geomembrane for improved albedo on a section of the ground (Fig. 2).

Fig. 2. Photo of experimental PV test-stand, used in this study as an example-validation case, representing regular other validations of our model. In this study, we compare the energy yield of this set up against an identical system without artificial ground albedo enhancement.

Fig. 3 shows data from these systems, comparing the experimental observations to RT and VF modelling. RT modelling closely reproduces the experimentally observed energy gain from improving the ground albedo with a geomembrane. VF-based modelling is not precise for small systems [10], but converges with more accurate RT modelling for longer rows of modules (Fig. 3); the latter represents the case of UPPs that we address in the main section of this paper.

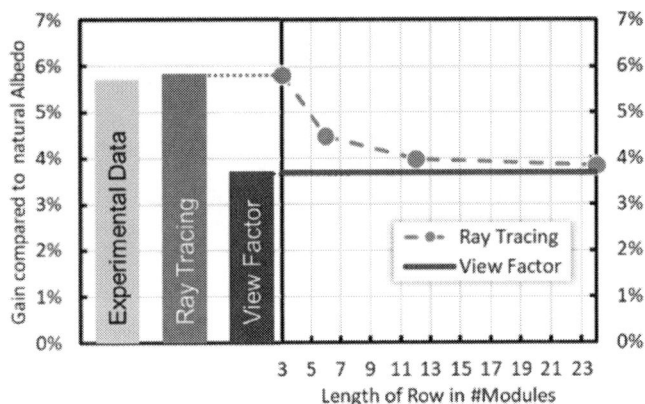

Fig. 3. Left: Measured and simulated energy yield gains from artificial ground albedo improvement. Right: View-Factor-based simulation converges with more accurate Ray-Tracing for long module rows typical of UPP installations.

978-1-6654-6060-6/23 $31.00 © 2023 IEEE

IV. DEFINITION OF SYSTEMS FOR ALBEDO STUDY

Using VF modeling, we simulate systems in Doha, Qatar, and Toulouse, France, with 6.5m row-to-row pitch and a ground-coverage ratio (GCR) of 32.6% with 2.12m-long modules in portrait orientation. The modules in these simulations are Jolywood D72N 410 Monocyrstalline Silicon, with a bifaciality factor of 80%. The corresponding PAN file was exported from PVsyst [11, 12]. The inverter in this study was represented by an OND file for the inverter UEP-4700 of Gamesa.

Our study has the aim to determine the best positions on the ground where to place high-albedo material, to achieve the highest production gain for a given amount of AEM placed on the ground. We approach this purpose by first presenting simulation results that are free of clipping effects. Specifically, we used a DC/AC ratio of 0.84, that is 73 383 kWp DC, with a total AC capacity of the inverters of 87 106 kWp. After determining clipping-free ("real") ground position sensitivity to albedo enhancement, we discuss the effect of clipping on the shape of such ground sensitivity curves.

Figure 4 illustrates how these sensitivity curves are constructed, comparing the gain achieved by strips of enhanced ground albedo to the operation of the same system with only natural albedo.

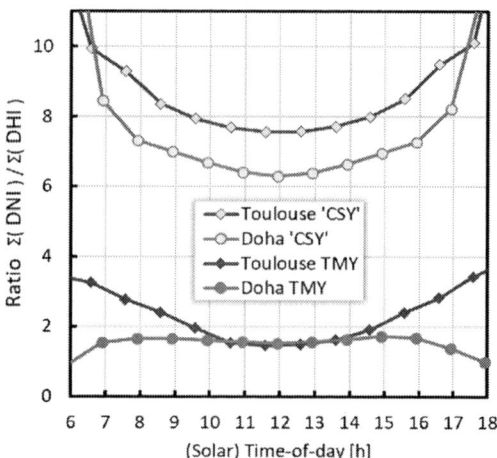

Fig. 5. Ratio of the annual sum of DNI, and sum of the annual DHI plotted for each hour of the day. Shown are results for the four meteo files used in the simulation study: TMY data for Toulouse and Doha, and "clear-sky-years" (CSY) in the same locations, where the CSY is an artificially created meteo file consisting of only clear-sky days.

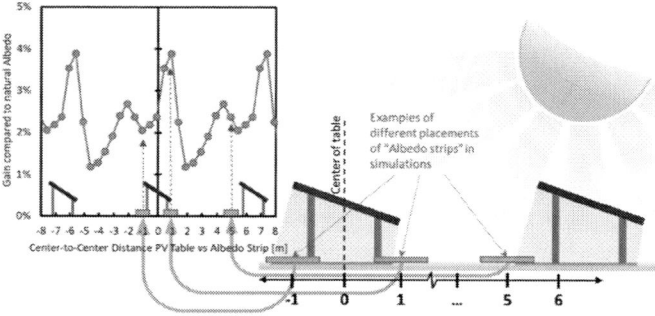

Fig. 4. Schematic representation of a bifacial PV system on the right, with examples of different placement options of albedo enhancement material on the ground. The arrows point to the corresponding x-axis values of a graph that plots the gain created by the AEM on the ground as a function of the placement position, and compared to operating the system with only natural ground albedo.

In the fixed-tilt cases, we use an inclination angle of 22° for Doha and 32° for Toulouse. Other than the inclination angles, latitudes, and meteorological data, we use the same system definitions in Doha and in Toulouse, with a natural ground albedo of 20% and for the areas with artificial ground albedo enhancement we use an albedo value of 75%.

The tracking systems follow an astronomical tracking algorithm, with a limiting angle of 60° and backtracking.

The center heights of the module tables in the tracked cases and in the fixed-tilt cases are 1.1m and 1.5m. For tracked installations this corresponds essentially to the torque-tube height. For fixed-tilt cases, one typically specifies the "ground clearance" i.e., the distance between the lower ("leading") edge

of the modules and the ground. Since we are using (according to the different latitudes) different mounting angles in Doha and in Toulouse, the same center heights imply different ground clearances. We chose to keep the same center height, as to provide a more direct comparison of the optical configuration between the fixed-tilt cases in Doha and Toulouse, and a more direct comparison to the tracked cases.

Please also note that a "one-portrait" configuration is unusual for fixed-tilt installations. Again, this choice was made to have a more direct comparison between the tracked and the fixed-tilt cases. The fixed-tilt results remain applicable to other cases (such as "two-portrait") by noting that performance of the "one-portrait" case is identical to the performance of a "two-portrait" case, if the center height and row-to-row spacing is scaled accordingly i.e., by a factor of "2".

In order to explore the effect of the content of the diffuse light and the direct light on the ground sensitivity curves that were sketched in Fig. 4, we generated for both locations, Doha and Toulouse, "clear-sky years" (CSY). To this end, we used a function in PVsyst [12] that allows for a given location to construct meteorological data where each day of the year is a clear-sky day. Figure 5 provides an impression on the relative amount of direct light compared to diffuse light, as a function of the hour of the day, shown for regular typical meteorological year (TMY) files for Toulouse and Doha and for the corresponding CSY artificial meteo data. Also seen in Fig. 5 are slight asymmetries for times before and after "solar noon", leading in our simulation results to corresponding slight asymmetries for the albedo ground sensitivity curves. Note that otherwise for north-south-oriented single-axis tracker systems the ground sensitivity curves would be perfectly symmetric in a climate where morning irradiance matches the afternoon irradiance.

978-1-6654-6060-6/23 $31.00 © 2023 IEEE

V. RESULTS

Figure 6 shows our simulation results for the fixed-tilt systems in the two different locations, Doha and Toulouse. Each data point on the curves represents a simulation of a full year, either a TMY as indicated by the closed symbols, or a year with only clear-sky days (calculated from the TMY), indicated as open symbols in the graph. The x-axis value of each data point refers to the position of the strip of improved albedo that was used in the simulation (compare to Fig. 4). We can see that for fixed tilt systems, the best position for narrow strips (equal or less than the module table width) is to center the strip near the front (sun-facing lower edge) side of the module row. For albedo strips that are wider than the module table, the optimum position shifts to the rear side (higher edge) of the module row. This observation holds true for both module mounting heights: Indicated in the graph are the heights of the center of the module table above ground.

Interestingly, the use of a CSY as meteo input did not change the qualitative shapes of these ground sensitivity curves. However, as visible by the slightly different scale for the right axis for the open symbols, a more direct light content (CSY) creates less performance gain by ground albedo enhancement than under TMY conditions.

While some details of the graphs for Toulouse and Doha are different, the main conclusion about the design rules for placing ground enhancement materials is the same: For fixed-tilt systems, the best position for strips of albedo enhancement is

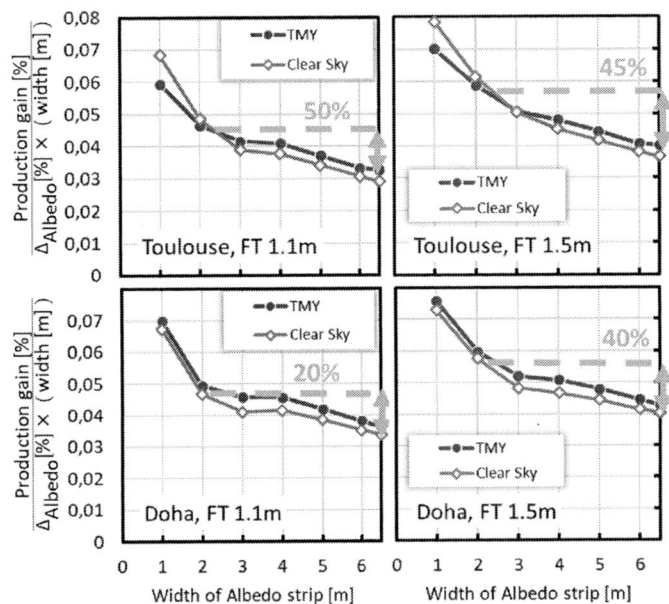

Fig. 7. The efficiency or "value" per area of the albedo enhancement material placed on the ground, plotted as a function of the width of optimally placed strips of AEM. The module bifaciality in these fixed-tilt (FT) simulations was 80%, and the curves scale accordingly for modules with a different bifaciality factor. The "value" per material for strip widths similar to the module table (2.12m here) can be up to 50% higher than for full-area ground coverage.

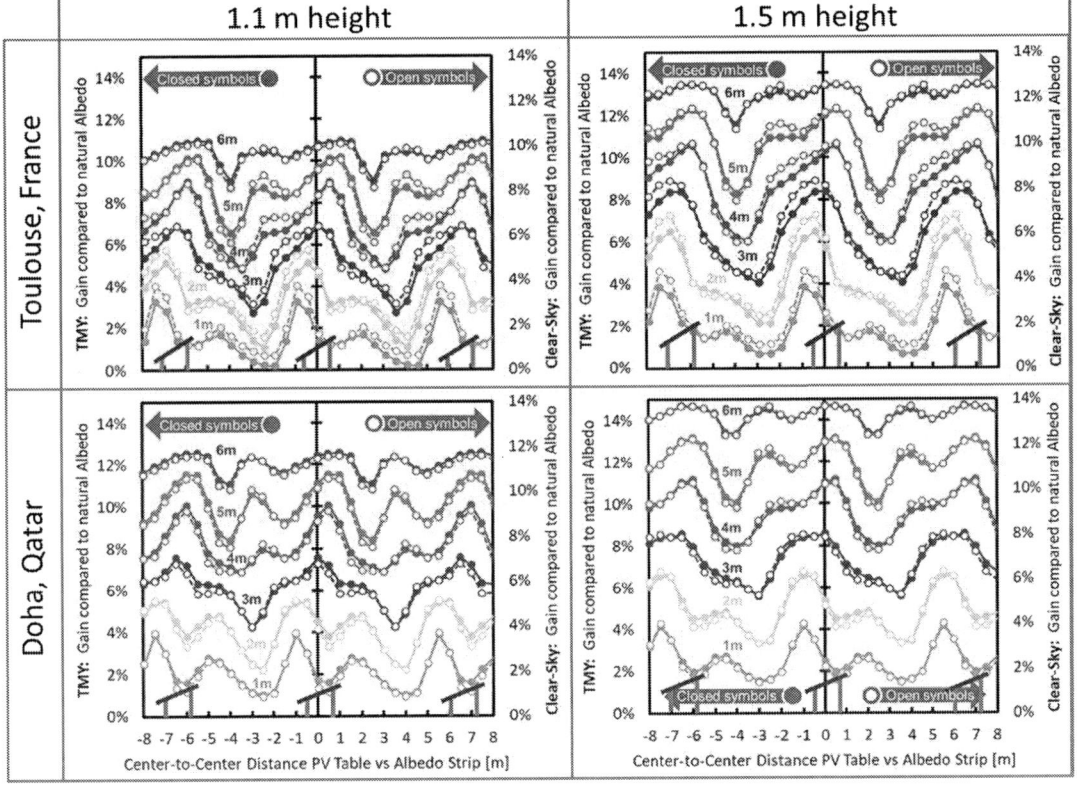

Fig. 6. Fixed-tilt energy production gain by arranging strips of enhanced albedo (75%) on the ground (20% albedo). Module bifaciality: 80%. Each graph shows results for six strip widths (1m to 6m), indicated next to curves. Horizontal axis indicates the position of albedo strips on the ground. Module racking schematically sketched (to scale) at bottom of each figure. Upper graphs: Toulouse. Lower graphs: Doha. Left graphs: center height 1.1m, right graphs: 1.5m. Left axes (all same scale) are for TMY (closed symbols). Right axes (all same scale, but different to left axis) are for the "clear-sky years" i.e., simulation results with artificial meteo files that have only clear-sky days.

978-1-6654-6060-6/23 $31.00 © 2023 IEEE

near the lower edge of the module table in case the strip widths are comparable to the module table width or smaller. For wider strips, a position centered under the module table is preferred.

Figure 7 plots for all four fixed-tilt systems the obtainable performance gain versus the witdth of the albedo enhancement strip (red and blue curves). For each width of the strip, the optimum position was assumed. Note the quantity and units of the vertical y-axis: Gain versus "only natural albedo", divided by the difference (ΔAlbedo) between the natural albedo and the enhanced albedo value, and divided by the width of the albedo strip. Except for bifacility of the modules (here: 80%) this quantity divides the gain by those variables that can be expected to cause a proportional effect on the gain, and it therefore expresses in a generalized form how much gain can be expected "per meter (width) of invested AEM". The graphs show, that for an albedo strip of the width of the module table (here 2.12m) an optimized placement produces 20-50% more production gain "per invested meter width" than a full-area coverage with AEM.

Figures 8 and 9 show the same fashion as Figs. 6 and 7 the corresponding simulation results for tracked systems. Generally, we find for the SAT tracking systems ground sensitivity curves that are symmetrical to the center of the tracker torque tube. As a general rule-of-thumb, one can see that for SAT tracking systems, the most efficient placement of AEM is right under the tracker, centered to the torque tube. Again, the qualitative difference of the shapes of these curves is not much

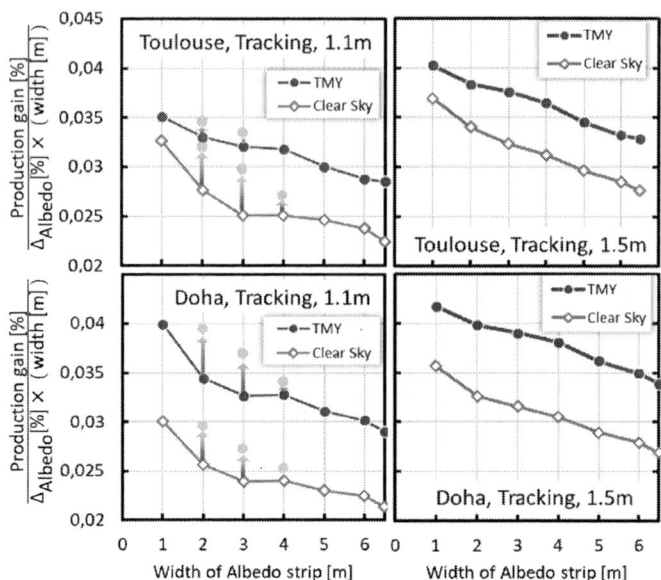

Fig. 9. Efficiency or "value" per area of the albedo enhancement material placed on the ground, plotted as a function of the width of optimally-placed strips of AEM. The module bifaciality in these SAT tracking simulations was 80%, and the curves scale accordingly for modules with a different bifaciality factor. The "value" per material can be increased (yellow points) if the material of the total strip width is divided into two narrower strips, placed off-centric to the tracker.

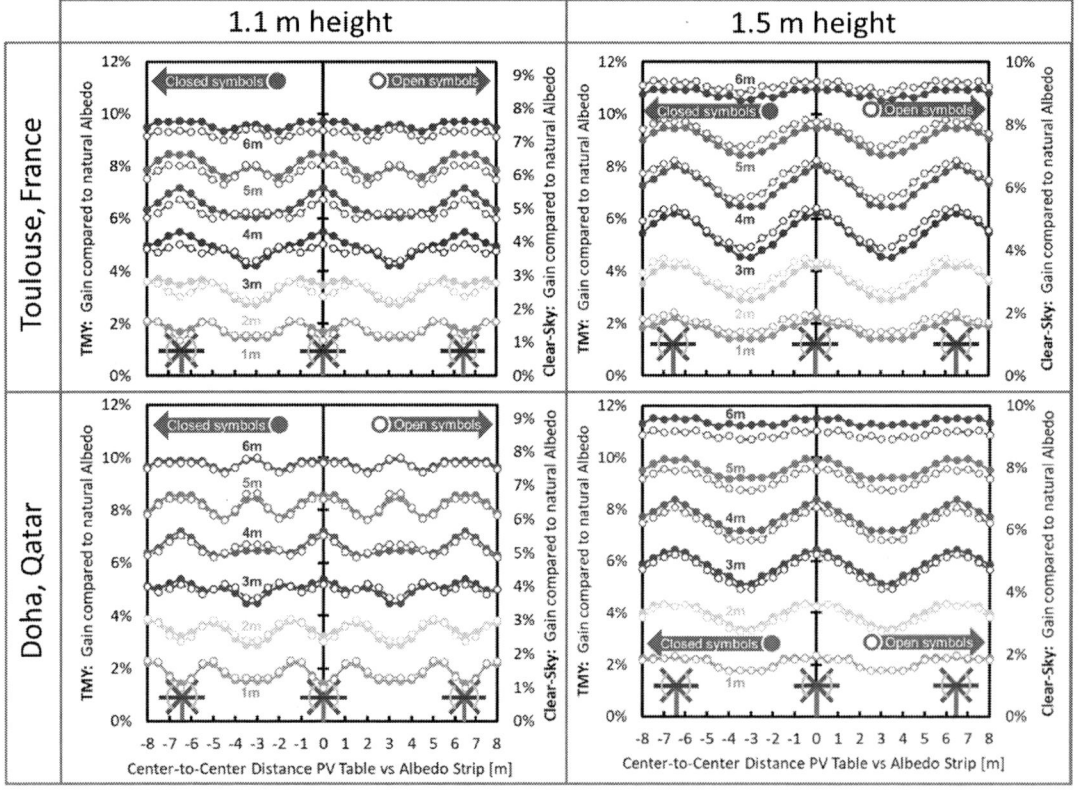

Fig. 8. Tracking system (SAT) energy production gain by arranging strips of enhanced albedo (75%) on the ground (20% albedo). Module bifaciality: 80%. Each graph shows results for six strip widths (1m to 6m), indicated next to curves. Horizontal axis indicates the position of albedo strips on the ground. Module racking schematically sketched (to scale) at bottom of each figure. Upper graphs: Toulouse. Lower graphs: Doha. Left graphs: center height 1.1m, right graphs: 1.5m. Left axes (all same scale) are for TMY meteo (closed symbols). Right axes (slightly different for left and right graphs) are for the "clear-sky years", i.e. simulation results with artificial meteo files that have only clear-sky days.

different between the more diffuse TMY meteo situation and the clear-sky DNI-dominated CSY meteo data. We do see, however, that for the tracked case, the relative gain obtainable for albedo enhancement is more pronounced for more diffuse light (compare y-axis units for the solid and the open symbols).

The other notable difference is the development of two peaks to the east and west of the torque tube position (with a minimum at the center) for 1m and 2m wide albedo strips for lowly-mounted tracked systems. This has a notable effect for the optimization of the placement of AEM: The yellow symbols in Fig. 9 indicate that placing material in the form of two narrower strips (instead of one wide strip) can use the invested material notably more efficiently for lowly-mounted tracking systems as one may encounter in more windy regions.

All of the results shown up to this point were generated in scenarios without clipping. While clipping does not change the general trends, it does reduce the relative gain obtainable by albedo enhancements, and also some of the finer details and optimization options found in the system. Figure 10 shows the effect of clipping for two example cases of the tracking scenarios: 1.5m center height with 3m AEM strip width, and 1.1m center height with 2m AEM strip width. Apart from a general lowering of the relative gain obtainable, we see a reduction of the finer features in the ground sensitivity curves for high degree of clipping. In particular, the two separate double-peaks for the lowly-mounted tracking systems vanish, so that the general rule that centering ground AEM centered under the tracker also re-emerges again for this case.

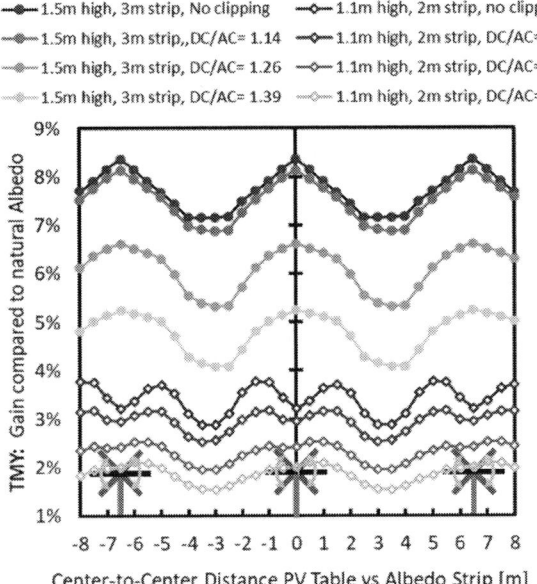

Fig. 10. Selected SAT tracking configurations from Fig. 8 (see graph legend, and compare to the non-clipped cases in Fig. 8) SAT Doha 1.1m height with 2m wide strip and Doha, 1.5m with 3m wide strip, to show the effect of clipping on the ground sensitivity curves for AEM placement.

VI. CONCLUSIONS

This paper presents the main effects of the positioning of albedo enhancement material (AEM) on the ground. In particular we have explored how the PV production gain by these AEM depends on where on the ground it is placed. In some cases up to 50% more gain per invested material on the ground can be achieved, when placing the AEM in the best positions only, as compared to a full-area ground coverage. We show these effects for tracking or fixed-tilt bifacial solar systems in different latitudes and climates. Our results allow deriving the general rule of thumb, that a positioning of AEM centered under the modules is optimal for AEM widths notably wider than the module table. For AEM strips of about the module table width (or less), fixed-tilt systems prefer centering the AEM near the lower (front) edge of the module table. For tracked systems with a low mounting height, it can be beneficial to split the AEM into two off-center strips to east and west of the SAT tracker.

REFERENCES

[1] R. Kopecek, J. Libal, "Bifacial Photovoltaics 2021: Status, Opportunities and Challenge", *Energies* vol. 14, p. 2076, (2021).

[2] J.-N. Jaubert, Yuanjie Yu, Baohua He, Gang Yan, Zhigen Zhang, Ray Zhao, "Layout optimization of Albedo enhancer materials used in bifacial PV systems," *6th BiFi PV Workshop*, 2019.

[3] O. Rhazi, M. Chiodetti, J. Dupuis, S. Benyakhlef, K. Radouane, P. Dupeyrat, "Optimizing the utilization of reflective materials for bifacial PV plants," *37th EU-PVSEC*, 2020, p.1286.

[4] "Energy Yield and Economics of Single-Axis-Tracked Bifacial Photovoltaics with Artificial Ground Reflectors", M.R. Lewis, S. Ovaitt, B. McDanold, C. Deline, K. Hinzer. Presented at this same conference: 50th IEEE-PVSC (2023)

[5] A. Florakis, F. Bourarach, E. Houzay, N.-P. Harder, I. Smaine, A. Poquet, A. Buzy-Debat, S. Ait-Tilat, J. Chapon, P. Biver, T. Chugunova, A. Tuomiranta, and G. Poulain (2023, May 9). SolarOPS: A versatile PV system simulation software. 16th PV Performance Modeling Workshop, Salt Lake City, Utah, USA. https://www.sandia.gov/app/uploads/sites/243/2023/06/5-6_PVPMC23-SolarOPS-Gen2-final-1.pdf

[6] W. F.. Holmgren, C. W. Hansen, and M. A. Mikofski. "pvlib python: a python package for modeling solar energy systems," *Journal of Open Source Software*, 3(29), P. 884, (2018).

[7] B. Marion, S. MacAlpine, C. Deline, A. Asgharzadeh, F. Toor, D. Riley, J. Stein, and C. Hansen, "A Practical Irradiance Model for Bifacial PV Modules," *44th IEEE Photovoltaic Specialists Conference*, pp. 1537-1542, (2017).

[8] S. A. Pelaez, and C. Deline, "bifacial_radiance: a python package for modeling bifacial solar photovoltaic systems," *Journal of Open Source Software*, 5(50), p. 1865, (2020)

[9] G. W. Larson, and R. A. Shakespeare, "Rendering with Radiance: The Art and Science of Lighting Visualization", San Francisco: Morgan Kaufmann Publishers, Inc. (1998)

[10] S. A. Pelaez, C. Deline, S. M. MacAlpine, B. Marion, J. S. Stein, and R. K. Kostuk, "Comparison of Bifacial Solar Irradiance Model Predictions With Field Validation," in *IEEE Journal of Photovoltaics*, vol.9(1), pp. 82-88, Jan. 2019.

[11] A. Mermoud, "PVSYST: a user-friendly software for PV-systems simulation", Proceedings of 12th EU-PVSEC 1994, p. 1703

[12] https://www.pvsyst.com/

Microscopic origins of performance losses in (Ag,Cu)(In,Ga)Se$_2$ thin-film solar cells

Sinju Thomas[1], Wolfram Witte[2], Dimitrios Hariskos[2], Rico Gutzler[2], Stefan Paetel[2], Chang-Yun Song[3], Heiko Kempa[3], Matthias Maiberg[3], Daniel Abou-Ras[1]

1. Helmholtz-Zentrum Berlin für Materialien und Energie GmbH, Hahn-Meitner-Platz 1, 14109 Berlin, Germany
2. Zentrum für Sonnenenergie- und Wasserstoff-Forschung Baden-Württemberg (ZSW), Meitnerstr. 1, 70563 Stuttgart, Germany
3. Martin-Luther-Universität Halle-Wittenberg, Institut für Physik, Fachgruppe Photovoltaik, von-Danckelmann-Platz 3, 06120 Halle (Saale)

Abstract— Ag alloying of CIGSe absorbers in thin film solar cells provides the means to fabricate absorber layers with slightly wider band-gap energies at lower substrate temperatures than for CIGSe layers without Ag. In the present study, three solar cells with ACIGSe photoabsorber of same band-gap energies of about 1.2 eV exhibit different open-circuit voltages V_{OC}. The influence of the microscopic properties of the ACIGSe absorber on the V_{OC} of the solar cell was investigated. Several characterization techniques in scanning electron microscopy were applied in a correlative manner on the identical specimen positions in addition to time-resolved photoluminescence and external quantum efficiency measurements. Differences in microstructural and optoelectronic properties such as average grain size, effective electron lifetime, absorption edge broadening by compositional gradients, and fluctuations in the spatial luminescence distribution were identified as origins of radiative and nonradiative loss mechanisms.

I. INTRODUCTION

Wide gap Cu(In,Ga)Se$_2$ (CIGSe) thin-film solar cells finds applications as top cells in tandem devices [1]. Fabrication of CIGSe absorbers with large band-gap energies (E_g) to increase the open-circuit voltage (V_{OC}) of corresponding solar cells is possible by increasing the [Ga]/[Ga]+[In] (GGI) ratio. However, increasing this ratio imposes changes in the absorber, such as decrease in grain size, increase in the density of point defects and alterations in the deep defect states which are considered as origins of V_{OC} deficit [2][3][4].

Ag alloying of CIGSe layers has also been reported as an approach to slightly increase the E_g of the photoabsorbers [5]. Achieved record efficiencies are close to 23% (in-house measurement). In order to improve the device performance, it is important to investigate the origins of performance losses more in detail.

In the present work, we make an approach down from the microscopic scale, by studying the possible origins of V_{OC} deficit in three (Ag,Cu)(In,Ga)Se$_2$ (ACIGSe) solar cells with identical E_g. Strong compositional gradients within individual grains were identified as substantial contributors to radiative V_{OC} deficits. Relatively larger V_{OC} deficit in the solar cells of absorbers with smaller grains indicates that device performance is also limited by non-radiative recombination at grain boundaries (GBs) in the ACIGSe absorbers.

II. EXPERIMENTAL DETAILS

ACIGSe thin-film solar cells were fabricated at ZSW with an in-line multi-stage co-evaporation process on Mo-coated soda-lime glass. Solution-grown CdS and sputtered i-ZnO were used as buffer system with a ZnO:Al as front contact. Current-voltage (I-V) curves were recorded on the as-grown ACIGSe solar cells with a WACOM solar simulator at standard testing conditions with a simulated AM1.5G spectrum and a Si referenc.

Integral chemical composition was determined via X-ray fluorescence (XRF), performed with a Fisherscope X-ray XDV-SDD on ACIGSe absorbers on Mo-coated soda-lime glass substrates.

Time-resolved photoluminescence (TRPL) was acquired on bare absorber layers at room temperature. External quantum efficiency (EQE) measurements were performed under light bias over a wavelength range of 300 to 1400 nm using a grating monochromator.

The electron backscatter diffraction (EBSD) maps maps were acquired using Zeiss UltraPlus scanning electron microscopes in combination with Oxford Instruments NordlysNano EBSD detector. Cathodoluminescence (CL) spectroscopy was

TABLE I

MATERIAL PROPERTIES OF THE INVESTIGATED ACIGSE AS WELL AS PHOTOVOLTAIC PARAMETERS OF THE CORRESPONDING SOLAR CELLS. HERE, AAC IS [AG]/([AG]+[CU]) AND ACGI = ([AG]+[CU])/([GA]+[IN]).

Solar cell	AAC	GGI	ACGI	η (%)	FF (%)	E_g (eV)	j_{sc} (mA)	V_{OC} (mV)	ΔV_{OC} (mV)	d_{grain} (μm)	S_{GB} (cm/s)	τ_{eff} (ns)	σ_{total} (meV)
#1	0.14	0.29	0.81	17.5	79.2	1.19	31.0	714	476	4.7	200	200	46
#2	0.04	0.34	0.72	16.5	77.4	1.20	31.4	677	523	0.8	80	180	57
#3	0.14	0.34	0.68	15.6	77.6	1.22	30.3	664	556	0.8	200	54	49

performed using Zeiss Merlin scanning electron microscope equipped with DELMIC SPARC CL system.

III. RESULTS AND DISCUSSIONS

Figure 1 shows an example of the various microscopic properties investigated via correlative scanning electon microscopy techniques applied on identical positions of the solar cell #2 with E_g=1.20 eV as extracted from *EQE*. The EBSD map shows that most of the grains in the ACIGSE absorber extend across the entire layer thickness. Average grain sizes (d_{grain}) were calculated from the etched ACIGSE surfaces containing more than 1000 grains.

Decrease in the CL intensity (Figure 1) at the GBs is an indication of enhanced nonradiative recombination at the GBs of the absorber. We quantified the recombination velocities at more than 20 GBs in each of the studied ACIGSE layers. The CL emission-peak energy increases from the ACIGSE/buffer to the ACIGSE/Mo interface within individual grains, confirming the presence of strong compositional gradient perpendicular to the substrate. Luminescence energy distribution of the absorber fluctuates both parallel ($\sigma_{lat.CL}$) and perpendicular ($\sigma_{ver.CL}$) to the substrate. The range of these fluctuations were quantified from the standard deviation of the CL peak energy over 50 pixels. Local compositional changes, defects and (micro)strain are the features in the ACIGSE absorbers that contribute to the $\sigma_{lat.CL}$ and $\sigma_{ver.CL}$.

Table 1 shows the average grain sizes, the GB recombination velocities, the electron lifetimes, and the absorption edge broadening for the three solar cells. Cell #1 exhibits the largest grain size of about 4.7 μm and an electron lifetime of 200 ns. Therefore, the V_{oc} of this cell is not limited mainly by enhanced nonradiative recombination at the GBs. Indeed, the strong compositional gradient in this ACIGSe film leads to substantial broadening of the EQE onset and thus, to considerable radiative V_{OC} losses.

In the presentation, we will give a complete account of the microscopic differences between the three solar cells and their possible impact on the device performance.

IV. CONCLUSION

Increased electron lifetimes of ~ 200 ns and GB recombination velocities on the order of few 100 cm/s were achieved via increasing the grain size upon Ag addition. However, solar cells with identical E_g exhibit different V_{OC} owing to the differences in the d_{grain}, τ_{eff} and σ_{total}. Enhanced nonradiative recombination at the GBs, small electron lifetime and local variations in chemical composition were identified as substantial contributors to radiative and nonradiative V_{OC} losses in ACIGSE thin-film solar cells.

REFERENCES

[1] Jošt M, et al. "Perovskite/CIGS Tandem Solar Cells: From Certified 24.2% toward 30% and beyond," *ACS Energy Letters*, 7(4), pp.1298–1307, 2022.

[2] Abou-Ras D, et al., "Impact of the Ga concentration on the microstructure of $CuIn_{1-x}Ga_x Se_2$," *Physica Status Solidi—Rapid Research. Letters*, vol. 2, pp. 135–137, 2008

[3] Hanna, G., et al. "Influence of the Ga-content on the bulk defect densities of Cu(In,Ga)Se₂," *Thin Solid Films*, 387(1–2), pp.71–73, 2001.

[4] Spindler C, Babbe F, Wolter MH, Ehré F, Santhosh K, Hilgert P, et al. "Electronic defects in Cu(In,Ga)Se₂: Towards a comprehensive model," *Physical Review Materials*, 3(9), pp.1-20, 2019.

[5] Edoff M, Jarmar T, Nilsson NS, et al. "High V_{oc} in (Cu,Ag)(In,Ga)Se₂ solar cells," *IEEE Journal of Photovoltaics*, 7(6), pp.1789-1794, 2017.

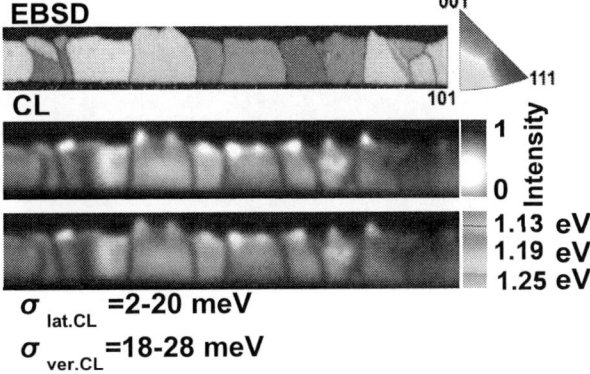

Fig. 1.EBSD map showing the microstructure and preferred grain orientation. CL map with the luminescence intensity and peak energy distribution on the same ACIGSE absorber of cell #2 with E_g=1.20 eV.

Overview of Engineered Germanium Substrate Development for Affordable Large-Volume Multi-Junction Solar Cells

Jinyoun CHO, Valérie Depauw, Alexandre Chapotot, Waldemar Schreiber, Tadeáš Hanuš, Nicolas Paupy, Valentin Daniel, Guillaume Courtois, Bouraoui Ilahi, Abderraouf Boucherif, Clement Porret, Roger Loo, Jens Ohlmann, Stefan Janz, Kristof Dessein

Umicore, Olen, Belgium

imec, Leuven, Belgium

Université de Sherbrooke & 3iT, Sherbrooke, QC, Canada

Fraunhofer ISE, Feriburg, Germany

New massive markets for space multi-junction solar cells are being discussed globally. For such an explosive increase in demand to materialize, a more sustainable and affordable Ge substrate technology is required. To this end, lithography-based Ge-on-Nothing and electrochemical process-based porous Ge wafers were developed. Both approaches yield uniform and smooth monocrystalline Ge-on-Ge engineered substrates after annealing, of which the top layer is weakly attached to the mother substrate. High-quality space solar cells were grown on them, followed by successful foil detachment and surface reconditioning. These results clearly demonstrate the feasibility of the reusable Ge substrate concept.

X-RAYS meet NEUTRONS meet IONS meet ELECTRONS meet LASERS meet MAGNETS: COMBINED ACCESS TO MULTIPLE FACILITIES THROUGH EU PROJECT “REMADE@ARI”

Michael E. Stuckelberger, Christina Ossig, Barbara Schramm, Stefan Facsko

Deutsches Elektronen-Synchrotron DESY, Hamburg, Germany

Helmholtz-Zentrum Dresden-Rossendorf e.V., Dresden, Germany

A radical shift to the Circular Economy is urgently needed to cope with the challenge of finite resources decreasing at a frightening pace while the quantity of waste increases alarmingly. The European Commission' Circular Economy Action Plan (CEAP) adopted in March 2020 has identified seven key product value chains that must rapidly become circular, given their environmental impacts and circularity potentials. This requires substantial research on materials with a very high recycling capability while exhibiting competitive functionalities. In ReMade@ARI, the most significant European analytical research infrastructures join forces to pioneer a support hub for materials research facilitating a step change to the Circular Economy. ReMade@ARI offers coordinated access to more than 50 European analytical research infrastructures, comprising the majority of the facilities that constitute the Analytical Research Infrastructures in Europe network. ReMade@ARI offers comprehensive services suiting any research focusing on the development of new materials for the Circular Economy in the key areas highlighted in the CEAP and plays an important role in the preparation of the common technology roadmap for circular industries. Senior scientist, facility experts and highly trained young researchers contribute scientific knowledge and extensive support to realize a user service of unprecedented quality, making each promising idea a success. Particular attention is attributed to the implementation of attractive formats to support researchers and developers from industry. The comprehensive service catalog is complemented by an extensive training program.

978-1-6654-6060-6/23 $31.00 © 2023 IEEE

BIPV Market Potential Analysis with Building Shadow Simulation

Changyeol Yun, Myeongchan Oh, Boyoung Kim, Jehyun Lee, Hyungoo Kim, Deokoh Lim, Sangmin Jo

Korea Institute of Energy Research, Daejeon, Korea

Korea Energy Economics Institute, Ulsan, Korea

Photovoltaic potential is directly affected by solar radiation. In particular, in the case of densely populated areas, the shadow effect has a fatal effect on the reduction of insolation on the building surface. In this study, in order to calculate the building-type photovoltaic market potential in Korea, the solar radiation reduction statistics due to shadow was extracted by region and applied to the calculation. We created building models with 1m spatial resolution and extracted solar radiation data for individual points of each building surface using Rhinoceros software and plugins Grasshopper and Ladybug tools. To verify the performance of the applied model, the results were compared with the other simulation model output. In order to extract the standard coefficient to be applied to the whole country, the sensitivity according to the data distribution such as building density, building orientation, building height in the target area was analyzed. The market potential is affected by the solar power supply curve in connection with LCOE(Levelized Cost of Electricity) and site condition for the energy facility. For the simulation, the entire Korean peninsula was divided into 1km X 1km grids and exclusion area values were input into the each grid through GIS(Geographical Information System) calculation. The market potential is affected by the SMP (System Marginal Price) and REC(Renewable Energy Certificate) conditions at the base time point. If this calculation model is used, it is possible to simulate the change in market potential due to changes in regulatory and support policies. The results were created in the form of a distribution map so that quantitative figures by region can be revealed. Using the supply curve derived in this study, the market potential that can actually be supplied under the current market conditions was analyzed.

An investigation on the pollen-induced soiling losses in utility-scale PV plants

João Gabriel Bessa1[1], Michael Valerino[2], Matthew Muller[3], Mike Bergin[4], Leonardo Micheli[5], Florencia Almonacid[1] and Eduardo F. Fernández[1]

[1]Advances in Photovoltaic Technology Research Group (AdPVTech), University of Jaén, Las Lagunillas Campus, Jaén 23071, Spain

[2]Solar Unsoiled, Inc. 303 White Pine Drive, Durham Nc 27705, USA

[3]National Renewable Energy Laboratory, Golden, CO, 80401, USA

[4]Civil and Environmental Engineering Department, Duke University, Durham, NC 27705, USA

[5] Dept. of Astronautical, Electrical and Energy Eng. (DIAEE), Sapienza University of Rome, Rome, Italy

Abstract — Soiling, the accumulation of dust and other contaminants on the surface of photovoltaic (PV) modules, is a common factor that can negatively impact the performance of PV systems. In this study, the authors aim to analyze the impact of pollen on soiling losses in PV systems located in North Carolina, USA, particularly during the spring season. The performance data of two utility-scale PV plants was collected and analyzed using the two soiling extraction methods. Environmental data, including croplands and vegetation was also collected and analyzed to identify correlations with soiling losses. The results of the study may help improve understanding of necessary operation and maintenance activities for PV plants and provide new insights into the phenomenon of pollen deposition on PV systems.

I. INTRODUCTION

Environmental factors can negatively impact the performance of photovoltaic(PV) systems by reducing the amount of energy they generate. Soiling, the process in which dust, dirt and others contaminants accumulate on the surface of PV modules, is a common but often overlooked and underestimated factor that affects PV performance. Despite its widespread effects, soiling is a local phenomenon that varies over time and space, making it necessary to address it on a case-by-case basis.

Alternative methods to ground based soiling sensors, that use performance data to quantify the soiling losses are available and widely functional. At least two extraction methods are present in the literature: the Fixed Rate Precipitation (FRP) [1] and the Stochastic Rate and Recovery (SRR) [2].

This work is focused on the application of methods for quantifying soiling losses in PV systems that are affected by pollen. By doing so, this work aims to improve the understanding of PV plant owners about the necessary operation and maintenance activities and to provide new insights for the scientific community on the phenomenon of pollen deposition in PV systems.

Estimating pollen soiling impacts from ambient concentrations first requires modelling of particle deposition velocities. A wide range of particle sizes with diameters from 10 to 100 μm, along with a diversity of physical morphologies both impact the dry deposition velocity of pollen particles[3], [4]. Modelling of pollen deposition is complicated by changing physical properties while suspended. Particle hygroscopicity has been shown to change the size, shape, and density of pollen particles while suspended[10], [11], all properties which impact deposition velocity and considerations for dispersion. For solar energy, the distance to the pollen source will also impact the particle deposition velocity[12].

There is only a few studies that analyze the influence of pollen in PV systems. Partial cleanings have been reported in Belgium [13] where some local spots with resin particles (probably pine) remained in the PV module surface. In Evora, Portugal, the soiling losses during spring reached ~4%[14]. Non-uniform soiling patterns where observed in Madrid, Spain after a light rain during the pollen season. In addition, an indoor experiment found a linear correlation between transmittance losses and the density of artificially deposited pollen[10].

The present study aims to analyze the soiling losses in PV systems located in North Carolina,USA that are impacted by pollen, particularly during spring season. By investigating the unexpected behavior of the soiling processes, such as accumulation and removal, the study aims to interpret the challenges faced by PV systems in this region and bring up the discussion for improvments to the current tools and methods used for soiling extraction.

II. METHODOLOGY

A. Performance metrics and soiling extraction

---

[1] Correspondence: jbessa@ujaen.es

978-1-6654-6060-6/23 $31.00 © 2023 IEEE

TABLE I
PV PLANTS IN THE STUDY

Site	Period	Count	Land Cover	Crop land	Climate
A	Feb 2020 – Aug 2022	Cumberland,NC	Cultivate crops, Evergreen Forest, Scrub, a few herbaceous and hay/pasture.	Soybeans, Corn	Humid subtropical (Cfa)
B	April 2020 – Sept 2022	Catawba,NC	Cultivated crops, Hay/pasture, Deciduous Forest, Herbaceous and Evergreen Forest.	Soybeans, Corn	Humid subtropical (Cfa)

Table 1 summarizes the sites under evaluation in this study that are all located in North Carolina or the southeastern USA. They correspond to two utility scale PV plants with an installed power ranging from 4 to 5 MW. The performance data was extracted at the inverter level and it includes irradiance and module temperature data measured on site. 15-minute data was collected and then averaged to obtain daily values. Filtering criteria were applied to remove any unreliable data, which might have been caused by cloudy days, incorrect instrument readings, malfunctions or inverter clipping. Irradiance, temperature and normalized energy filters were applied according to the method present in the RdTools Python package[15].

Two soiling extraction algorithms were applied to extract soiling losses. The FRP method is present in the Pvlib Python package[16] and the SRR is present in RdTools[15].

In the FRP model, a fixed soiling rate, determined from the reduction in performance during the longest dry period, is applied to any dry period to generate a sawtooth-shaped soiling profile. The dry periods are defined according to a predetermined cleaning threshold(i.e. amount of rainfall that restores the PV performance to a cleaned condition). This model requires as input performance, irradiance and rainfall data.

The SRR model uses a Monte Carlo simulation to estimate stochastically the soiling rates between cleanings and generate potential soiling profiles. This method does not require any weather data, as cleanings are identified based on the changes in performance, identified as statistically significant positive shits in the rolling median of the performance index. If the soiling process behaves as expected by this method, the knowledge of the cleaning events, natural or artificial, are not necessary. Several cleaning thresholds[17], partial cleanings assumptions and supporting evidence[13] has been reported in the literature.

B. Environmental data

The agriculture activities in this region are mostly covered by pastureland and croplands of corn, soybeans, wheat,cotton and tobacco [18]. The vegetation in the area is covered mostly by deciduous forest(e.g. maple and oak), woody wetlands and evergreen forest(e.g. pine)[19]. The agricultural descriptions for each site were extracted from the National Agricultural Statistic Service, where the changes overtime were observed (e.g. rotation of the land usage in the crop land) [19]. The temporal and spatial changes in the croplands or any changes in the land were additional analyzed using satellite images sourced from PlanetScope by Planet Labs Inc., that provides satellite observation data[20]. This data allowed detection of the rotation of the croplands, the harvesting periods and the changes in the vegetation.

The data on rainfall used in this analysis were obtained from the PRISM database.[21] The period of analysis comprises at least 2 spring seasons, which are the periods of higher pollen airbone concentration. The evidence of high pollen deposition were reported by the O&M personal of each PV site.

III. RESULTS AND DISCUSSION

This section presents the results of using two extraction algorithms to determine the pollen-induced soiling losses at the investigated site. Fig. 1 shows the daily performance metrics, as well as the results of the SRR and FRP methods. It is noticeable that there is a drop in performance during the spring seasons, where losses reach up to 10% in both sites (see Fig. 2 for both sites). Site A experienced the most unexpected soiling behaviour when looking at the performance and rainfall data. After the period of high particle deposition (end of spring), the PV system does not fully recover to its "cleaned" level even though it rains often. Instead, it takes around 4 months for the performance to gradually increase to a maximum over the summer. Furthermore, the manual cleaning in the summer of 2022 demonstrates that this maximum was 5% less than the fully cleaned state. The mentioned satellite data and the croplands maps were utilized to identify the possible pollen sources from vegetation, as well the changes in the environment. During the spring season, pollen sources may include maple, oak and pine trees, as well cultivated crops such as corn and soybeans. Further investigations using daily county levelpollen data will be examined in the full paper.

Examining the rainfall data, it difficult to claim that recoveries are occurring to rains above a certain threshold. It was expected that cleaning thresholds within 1-10mm/day would produce positive shifts indicating recovery in performance while rains as high as 50 mm/day occur and still no clear positive shifts are seen.

The expected behavior of the soiling profile can be seen in the bottom graph of Fig. 1, which shows the profiles generated by the FRP model. The soiling rate for the longest dry period identified by SRR (0.0015%/day) for site A was used for this model. The FRP method was applied using cleaning thresholds of 1, 5, and 10 mm/day (5 mm/day shown) but none of the sawtooth profiles matched the grey data in the upper part of Fig. 1.

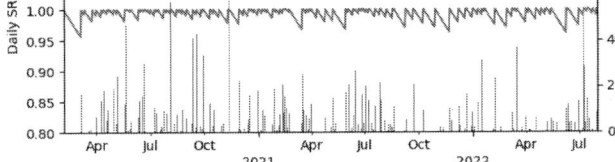

Fig. 1. Upper graph : Performance metrics in gray and soiling periods in orange identified by SRR for site A. The red dashed line represents the manual cleaning. The black dashed lines represents the increment in performance due to the manual cleaning for site A. Lower graph: Modeled soiling profile using 5mm/day as cleaning threshold by FRP for site A.

For SRR, the median soiling ratio of all possible values generated by Monte Carlo simulations is shown in Fig. 1. The SRR method looks specifically for profiles with gradual accumulation and a sharp recovery, which is nearly the opposite of the soiling profile in this case. It was expected that this method would not perform well with this type of signal because SRR is designed to reject periods with positive slopes.

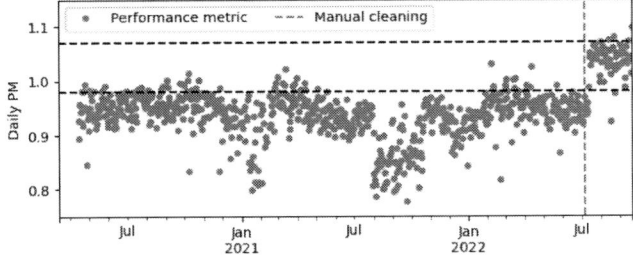

Fig. 2.Improved performance due to manual cleaning performed in Site B.

There is currently insufficient data to prove if the periods with positive slopes are gradual removal of sticky pollen from the module surface or if the trend is residual seasonality in the PM due to modeling errors.

In June and July 2022, a manual cleaning was carried out at site A and site B respectively, Fig.2. The dashed lines represent the soiling impact up to the cleaning date. Site A experienced an increase in performance of around 5% after the manual cleaning. This value was around 10% for Site B. One can expect that for this period, no rainfall event was capable to produce such increments and restore the PV systems to a 'cleaned state'.

IV. PRELIMINARY CONCLUSIONS

This study analyzes the unexpected behavior of soiling accumulation and removal in two sites affected by seasonal pollen events in the southeast region of USA.

The PV systems experienced a slow recovery after the periods of high particle deposition. The rainfall pattern had a low correlation with the soiling accumulation periods, recovery and the expected natural cleaning process. Manual cleaning was necessary to remove remaining soiling on the module surface not removed by rainfall events.Neither the SRR or FRP methods are appropriate for use with these sites as they both assume abrupt cleaning events should occur due to rainfall.

From the performance data, the environmental characteristcs of the site and local observations from the O&M team, it it is clear that soiling which is persistent to rainfall is occurring on these PV systems located in the southeast region of the USA. O&M reporting indicates this unexpected behavior is at least partially explained by pollen deposition. It is recommended that further studies be conducted to increase knowledge of the relationship between pollen characteristics, its' morphology and physical properties, and persistent PV soiling losses.

Pollen is a distinct form of soiling that is site dependent and behaves differently, requiring alternative methods or improvments to existing methods for proper identification. In order to achieve a comprehensive understanding of site specific pollen concentrations, addtitional studies using precise pollen data will be conducted.

REFERENCES

[1] A. Kimber, L. Mitchell, S. Nogradi, and H. Wenger, "The effect of soiling on large grid-connected photovoltaic systems in California and the Southwest Region of the United States," *Conf. Rec. 2006 IEEE 4th World Conf. Photovolt. Energy Conversion, WCPEC-4*, vol. 2, pp. 2391–2395, 2007.

[2] M. G. Deceglie, L. Micheli, and M. Muller, "Quantifying Soiling Loss Directly from PV Yield," *IEEE J. Photovoltaics*, vol. 8, no. 2, pp. 547–551, 2018.

[3] K. Dong, C. Woo, and N. Yamamoto, "Plant assemblages in atmospheric deposition," *Atmos. Chem. Phys.*, vol. 19, no. 18, pp. 11969–11983, 2019.

[4] F. Di-Giovanni, P. G. Kevan, and M. E. Nasr, "The variability in settling velocities of some pollen and spores," *Grana*, vol. 34, no. 1, pp. 39–44, 1995.

[5] J. Wang, M. Qi, H. Huang, R. Ye, X. Li, and C. Neal Stewart, "Atmospheric pollen dispersion from herbicide-resistant horseweed (Conyza canadensis L.)," *Aerobiologia (Bologna).*, vol. 33, no. 3, pp. 393–406, 2017.

[6] F. Hofmann *et al.*, "Accumulation and variability of maize pollen deposition on leaves of European Lepidoptera host plants and relation to release rates and deposition determined by standardised technical sampling," *Environ.*

Sci. Eur., vol. 28, no. 1, 2016.

[7] S. Dupont, Y. Brunet, and N. Jarosz, "Eulerian modelling of pollen dispersal over heterogeneous vegetation canopies," *Agric. For. Meteorol.*, vol. 141, no. 2–4, pp. 82–104, 2006.

[8] R. W. Arritt, C. A. Clark, A. S. Goggi, H. Lopez Sanchez, M. E. Westgate, and J. M. Riese, "Lagrangian numerical simulations of canopy air flow effects on maize pollen dispersal," *F. Crop. Res.*, vol. 102, no. 2, pp. 151–162, 2007.

[9] O. Souhar, A. Marceau, and B. Loubet, "Modelling and inference of maize pollen emission rate with a Lagrangian dispersal model using Monte Carlo method," *J. Agric. Sci.*, vol. 158, no. 5, pp. 383–395, 2020.

[10] C. S. Saiz, J. P. Martínez, and N. M. Chivelet, "Influence of pollen on solar photovoltaic energy: Literature review and experimental testing with pollen," *Applied Sciences (Switzerland)*, vol. 10, no. 14. MDPI AG, 01-Jul-2020.

[11] D. E. Aylor, M. T. Boehm, and E. J. Shields, "Quantifying aerial concentrations of maize pollen in the atmospheric surface layer using remote-piloted airplanes and Lagrangian stochastic modeling," *J. Appl. Meteorol. Climatol.*, vol. 45, no. 7, pp. 1003–1015, 2006.

[12] and J. V. H. Raynor, Gilbert S., Eugene C. Ogden, "Dispersion and deposition of ragweed pollen from experimental sources," *J. Appl. Meteorol. Climatol.*, vol. 9.6, pp. 885–895, 1970.

[13] R. Appels *et al.*, "Effect of soiling on photovoltaic modules," *Sol. Energy*, vol. 96, pp. 283–291, 2013.

[14] R. Conceição, H. G. Silva, J. Mirão, and M. Collares-Pereira, "Organic soiling: The role of pollen in PV module performance degradation," *Energies*, vol. 11, no. 2, pp. 1–13, 2018.

[15] M. G. Deceglie *et al.*, "RdTools." .

[16] W. F. Holmgren, C. W. Hansen, and M. A. Mikofski, "Pvlib Python: a Python Package for Modeling Solar Energy Systems," *J. Open Source Softw.*, vol. 3, no. 29, p. 884, 2018.

[17] J. G. Bessa, L. Micheli, F. Almonacid, and E. F. Fernández, "Monitoring Photovoltaic Soiling: Assessment, Challenges and Perspectives of Current and Potential Strategies," *iScience*, vol. 24, no. 3, p. 102165, 2021.

[18] N. A. S. Service, "Census of Agriculture." [Online]. Available: https://www.nass.usda.gov/Publications/AgCensus. [Accessed: 18-Dec-2022].

[19] National Agricultural Statistic Service, "CroplandCROS," *https://croplandcros.scinet.usda.gov/.* .

[20] A. E. Frazier and B. L. Hemingway, "A technical review of planet smallsat data: Practical considerations for processing and using planetscope imagery," *Remote Sens.*, vol. 13, no. 19, 2021.

[21] P. C. G.-O. S. University, "PRISM Gridded Climate Data." [Online]. Available: http://prism.oregonstate.edu. [Accessed: 18-Sep-2022].

The role of PbS QDs on strain and optical properties in different perovskite matrix

Sofía Masi, Patricio Serafini, Iván Mora-Seró

INAM, Universidad Jaume I, Spain, Castellón de la Plana, Spain

The mixture of inorganic quantum dots and hybrid perovskite in one nanocomposite is considered a favorable approach to overcome restrictions of metastable perovskite. However, to date only few examples of improved opto-electronic perovskites have been realized with such materials. Here, we show a systematic approach to characterize the standard methylammonium lead iodide (MAPbI3) perovskite system by: (i) the substitution of some MA by guanidinium (Gu); (ii) the incorporation of PbS quantum dot (QD) additives and (iii) addition of both Gu and PbS at the same time. We studied the effect of the incorporations of the big cation on the film strain and crystal cell unit volume, and on the solar cell device efficiency and stability. With the control of Gu and PbS QD content, higher performance and longer solar cell stability are obtained. PbS QDs aid Gu incorporation, resulting in an expected more stable material and devices with high amount of guanidnium. From the optical point view, the tuning of the energy levels of perovskite nanocomposite with PbS with different ligands will be also exemplified, to show that a slight shift of the energy level leads to improved efficiencies especially in the case of formamidinium/PbI2 capping ligand. With this study, we demonstrated the reason why the solar cells performances are improved up to 20%, even when the optical properties are detrimental.

Zr-doped In2O3 film for the interlayer of perovskite/crystalline silicon tandem solar cells

Tappei Nishihara, Hyunju Lee, Ryuji Kaneko, Yoshio Ohshita, Atsushi Wakamiya, Atsushi Masuda, Atsushi Ogura

School of Science and Technology, Meiji University, Kawasaki, Kanagawa, 214-8571, Japan

JSPS Reserch Fello DC2, Chiyoda, Tokyo, 102-0083, Japan

Meiji Renewable Energy Laboratory, Meiji University, Kawasaki, Kanagawa, 214-8571, Japan

Toyota Technological Institute, Nagoya, Aichi, 468-8511, Japan

Institute for Chemical Research, Kyoto University, Uji, Kyoto, 611-0011, Japan

Graduate School of Science and Technology, Niigata University, Nishi, Niigata, 950-2181, Japan

The potential of Zr-doped In2O3 (IZrO) as a transparent conductive oxide film for the interlayer in perovskite/Si tandem solar cells, one of the structure expected to exceed the theoretical conversion efficiency of crystalline Si solar cells, is investigated. From optical simulations, the optimal film thickness for the intermediate layer was calculated to be 20 nm. The fabricated IZrO successfully achieved a transmittance over 95% and a carrier mobility over 20 cm2/Vs. Furthermore, the formation of oxygen vacancies in IZrO by annealing was suppressed with Zr doping.

Design with Luminescent Solar Concentrator Photovoltaics in the Built Environment

Eli Shirazi[1], Wouter Eggink[1], Angele Reinders[1,2]

1) University of Twente, Enschede, 7500 AE, The Netherlands, 2) Eindhoven University of Technology, Eindhoven, 5612 AZ, The Netherlands.

Abstract — This study explores the possible designs of Luminescent Solar Concentrator PV technologies which can be integrated in various environments by optimally using design features of this technology, such as colour, form giving, and transparency. These properties give a lot of freedom for designing an aesthetically pleasing product to be integrated into the built environments. The presented designs covering a wide range of applications from a landmark to a bus stop, are conceptual designs created by student teams of University of Twente in 2022. These conceptual designs illustrate how user experiences with sustainable energy technologies could be improved, especially in terms of aesthetics.

Keywords - Luminescent Solar Concentrator, LSC PV, Design, Building Integrated PV, Product Integrated PV.

I. INTRODUCTION

Growing concerns about energy crisis and climate change, push countries to take appropriate measures to meet ever-increasing energy demand through sustainable approaches. Luminescent Solar Concentrators Photovoltaics (LSC PV) are among technologies that can address these issues by harvesting solar energy. LSC are usually made from transparent polymer materials doped with luminescent dyes which act as a spectrum-converting lightguide. Also, they can be made from glass, polycarbonate, and other transparent plastics, where luminophores are added to coatings or in intermediate layers.

When solar irradiance enters the LSC, it is absorbed by luminescent dyes and then re-emitted at a longer wavelength. This light is trapped in the lightguide, by total internal reflections, then absorbed and converted to electricity through the photovoltaic (PV) cells at the edges of the lightguide [1]. LSCs can be produced in different colours, shapes, and transparencies. These features together with low cost of production give increased design freedom and widens the possibilities of applying LSC PV.

Regardless of all abovementioned properties, there have been little experience with LSC PV applications. Therefore, we would like to explore possible LSC PV applications in the built environment through design driven projects. This study builds on prior design studies with sustainable energy technologies which were executed and reported since 2008 within the scope of the master course "Sources of Innovation" at University of Twente [2], [3].

II. DESIGN CONTEXT

A design study was conducted on possible applications for LSC PV technologies that can be applied in the built environment within the scope of the course "Sources of Innovation". The goal of this course is to develop novel designs for innovative technologies. This approach has been successfully applied to designing with PV and LSC PV technologies as well as for design projects with other renewable energy technologies from 2006 onwards [3]–[10].

To come up with a concept using LSC PV technologies an ideation phase is needed, during which the design brief is translated into an initial concept description. As the LSC PV technology offers a lot of possibilities, implementing LSC PV in the built environment, therefore, leaves a pile of possible design directions. In this step, the design goal will be torn down to the essence by describing it on a high level of abstraction. For example "design an LSC sunshield" can be rephrased in "design a system to regulate the sunlight radiance on a terrace, powered by the sunlight itself". To visualize this abstract design vision, disruptive images will be made, where two or more images that are related to the vision, but not related to each other are combined to present novel and uncommon concepts. This will help to explore possibilities which are not directly obvious and to stimulate creative 'outside the box' design [11].

Innovative design and styling (IDS) is a helpful tool in this phase to derive positive and negative associations of each design in terms of the communication function of design, the balance between novelty and typicality, and the proper application of metaphors [12]. The outcome will be requirements for the design such as function, performance requirements, user expectations, materials, forms, and components. To achieve the final design, the concepts need to be further iterated by applying various methods such as TRIZ, technology road mapping, constructive technology assessment (CTA), risk diagnosing methodology, Delft Innovation Model (DIM) and platform-driven product development [6].

III. CONCEPTUAL DESIGNS

A total of 20 designs were created by students covering a wide range of applications including a pier, a micro-algae production system, lanterns, bus stops, floating docks, playgrounds, a highway median barrier, trash cans, a water filtration system, modular green house, shades, blinds, public toilet, and landmarks. A selection of designs will be explained and presented below.

A. Funky Fungi

The Funky Fungi is a city landmark as a point of interest for both tourists and locals to visit and meet at. The exotic look of the structures would create an aesthetically pleasing cityscape that stands out without disrupting the viewer. The idea is that many different structures would be placed in the same city square, creating an environment of shade, and promoting the idea of walkable cities. The rotating mushroom caps have been designed with three layers of colour gradients which creates a dynamic show of colour changing shadows. At the same time the LSC PVs get powered by the constant sunlight which allows the structure to power itself since enough energy gets produced to rotate the layers. The main objectives of this project were to provide a point of interest and to both promote walkable cities and the value of LSC PVs. This is to show that sustainable energy technology can also have a positive contribution to the city landscape by banking on one of the strengths of the LSC technology: colourfulness.

Fig. 1. Funky Fungi concept courtesy of Nick Jelle Buhling, Pablo Jimenez Chillon, Gonzalo Vidal Prieto

B. Solar Pier

The main part of this design is the pier itself. The floor of this pier is constructed out of LSC PV panels which are held together by a metal frame. This metal frame is supported by specially designed beams, secured into the waterbed. The beams elevate the pier above the water. The room beneath the panels should allow the water to move freely without causing stresses on the pier-flooring.

It is chosen to have panels in a 3- or 9- panel grid, the panels will be encapsulated in a metal frame which is supported by pillars in the water. Taking the results of the load-calculations into account, it might be required to support the outer edges of each structure with pillars to prevent bending stresses in the

frame. The pillars are made from steel which performs well underwater and is resistant to corrosion. They are secured at the bottom of the lake with a concrete counterweight.

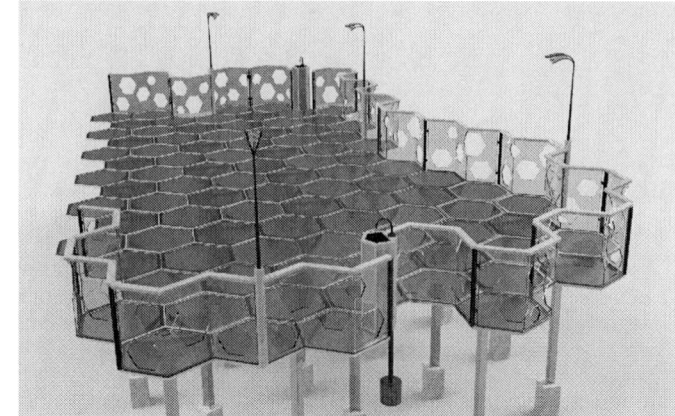

Fig. 2. Solar Pier, concept courtesy of Joris van Den Brekel, Maria Luisa Puga Jardim, Ilse Akkermans

As the solar pier is located near water facilities, the panels are expected to get wet. Due to the inherent characteristics of the LSC panels, their surface is naturally less adhesive. The combined effect of both aspects increases the risk of slipping or falling. To prevent this risk adding a protective later with higher surface roughness is needed. The protective layer can increase the lifetime of LSC panels, however, adding an extra layer on top of LSC panel, can decrease its efficiency. Considering all the circumstances, in addition to the cost, durability and maintenance, adding a polymer layer of polycarbonate can meet all the requirements. It is amorphous and therefore transparent, transmittance of over 90% of the incoming light, highly resistant to abrasion and fracture, lightweight, and good electrical insulation properties.

C. Shade and blind

There are two designs within the context of shades and blinds. In the designed blind system, the blind slats are replaced by LSC PV panels. The blinds are designed similar to regular blinds, with the only major difference being the energy production. Embedded in the top of the blinds is a battery with a control system. This battery is used to control the electric components of the blinds and to provide power for a LED light, a USB hub, or an air purifier.

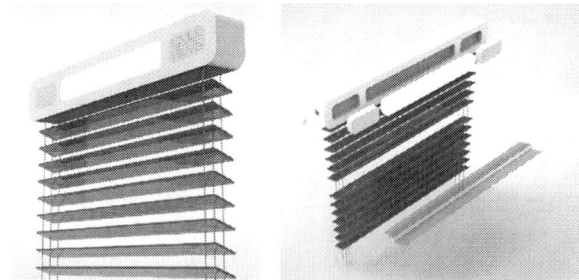

Fig. 3. LSC PV blinds, concept courtesy of Merijn Kendziorra, Sanne Meijer, Pim de Smit

LSTREE is an overhanging sunshield that provides a comfortable place on public terraces. The overhang is fitted with LSC PV leaves that change opacity to simulate a tree canopy. The sun shining through the leaves takes away its harshness, making the space underneath lit but shady, just like underneath a canopy. It can also function as a weather shield on a terrace.

Each of these leaves are 5x5cm, with the efficiency around 3 percent. Taking the total area of the leaves combined into account, the LSTree can generate up to 31.1 Wp. With the inclusion of the second LSC layer, power generation is significantly increased. The system can also use the generated power to change the opacity of the leaves and turning them from opaque to see-through.

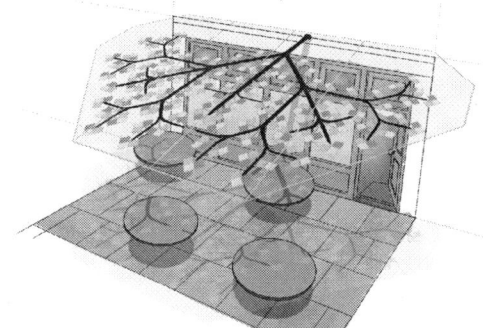

Fig. 4 LSTree, concept courtesy of Tirsa van der Ouw, Huub Rijnders, Robin Venhuizen

D. Lumient - Lantern

Lumient is a modular LSC PV lantern for the built environment, with the goal of permanent placement in the city context. It has

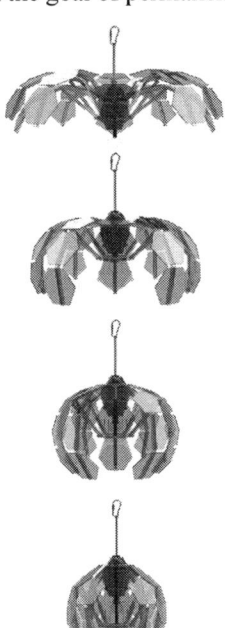

Fig. 5. Lumient concept courtesy of Anne Roos de Gooijer, Julia Scheeper, Roel Spierenburg

different functions during the day and night. During the day, the product acts as a decoration. The sides are folded upwards toward the sun, just like a flower. When it is dark, the arms are folded downwards, just like a lantern. The generated energy is used for the lamp and the light shines through the transparent LSC PV panels, aiming to create a feeling of safety for the citizens. The shape of the product is already common in the street scene. Important parts of the products are the LSC PV panels, dynamic mechanism, attachment system, and lighting. The Lumient consists of 20 hexagon and 10 pentagons shaped LSC PV panels in addition to battery, gears, micro servo motor, hooks, and arms. The size of the panels was based on calculations for the required energy to light an LED for one night, and the energy needed for the servo motor.

Beams made of folded sheet metal facilitate a safe space for wiring. The closed compartment in the Lumient is for all electronics. Inside is one servo motor which spins anti clockwise to put the Lumient into flower-mode. This is done by pulling on the ropes, which lift the pentagon base, while simultaneously moving the hooks outside. A spur gear on the same axis as the servo gear moves a rack gear down, which forces the hooks to move outward. The pentagon base can now rest on the hooks, making sure the servo only must function twice a day. Moving the servo motor clockwise does the opposite: It moves the hooks inside, which makes the arms drop down and puts the Lumient into lantern-mode. The battery is also stored in this chamber, which is connected to the LED at the bottom.

E. Micro Algae Production

As LSC PV panels absorb some wavelengths, they can be used to filter specific wavelengths for more efficient and sustainable micro-algae production systems. The concept of a vertical photobioreactor with LSC PV panels to cultivate micro-algae is displayed in 6. The designed photobioreactor incorporates wavelength isolating LSC panels into a steel mounting structure. This structure can pivot around top and bottom hinges using a set of linear bearings. Using three photovoltaic cells along the corners of the panels, the design collects energy to either fully drive or supplement the driving of a top mounted stepper motor. The stepper motor drives a 1:1 gear system to pivot the structure, allowing for directional shading and in-creased energy harvesting and algae production by following the direction of the sun. While it would have been possible to directly drive the structure without a gear system, such a system negates the need to fully dismount the stepper motor in the event of cleaning or replacing of the panels. A metal frame surrounds the system and tubes on the bottom allow for the inlet and outlet of the liquid and gas. Reactors can be added or taken away, making the system modular.

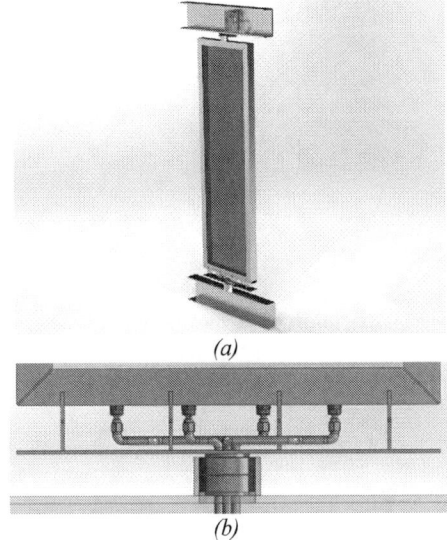

Fig. 6. Micro-algae production (a) photobioreactor, (b) piping system, concept courtesy of Berk Yılmaz, Giacomo serra, Femke Dijksterhuis

Furthermore, it has great potential to be integrated in the built environment, such as on roofs, walls, or facades of buildings. LSC PV panels rotate to control the light that dissipates in the environment, using the generated energy. Technical Medical Center at University of Twente is considered as a case study for this design (see fig.7).

Fig. 7. Photobioreactors on Technical Medical Center (conceptual design), concept courtesy of Berk Yılmaz, Giacomo serra, Femke Dijksterhuis

F. Neo-Holland

The main goal of the Neo-Holland project is to raise more awareness regarding potentials of the LSC technology. The idea is to create a new architectural movement, which wants to make a statement by including LSC into typical Dutch infrastructure and buildings. Think of mills, bridges, noise barriers, stations or public transport in a cyber punk, neon, futuristic style throughout different spots in the Netherlands. Solar mill is one design in the Neo-Holland project, where LSC panels are integrated into the blades of a traditional windmill. It is meant as a demonstrator that proves the feasibility of the design. The mill will be a large building that generates its own power with both wind and solar energy through LSC's. Next to a proof of concept the mill will also serve as publicity stunt and full prototype of the NEO-HOLLAND architecture.

Fig. 8. Solar mill, concept courtesy of Délisa Jutstra, Nathan van Emous, Sietse Oosterhout

G. Green Box

Green Box is a modular LSC powered green house for domestic use. The box element size is 50 cm in width, 50 cm in height, and 40 cm in depth. The measurements determine the inside volume of the box, meaning that the distance from outer edge to outer edge would be 4 cm larger, since the outer frames are 2 cm in thickness. Based on the research, the bottom plate of the cabinet has been designed hollow and extruded slightly downwards. Advantages of this are that water and dirt resulting from redundant sand can now be collected separately. The bottom floor can be cleaned when needed. Additionally, two tubes can be seen at the bottom: one for water transportation and the other for air ventilation (fig. 9). The individual plant boxes also feature a system to connect them both horizontal and vertically..

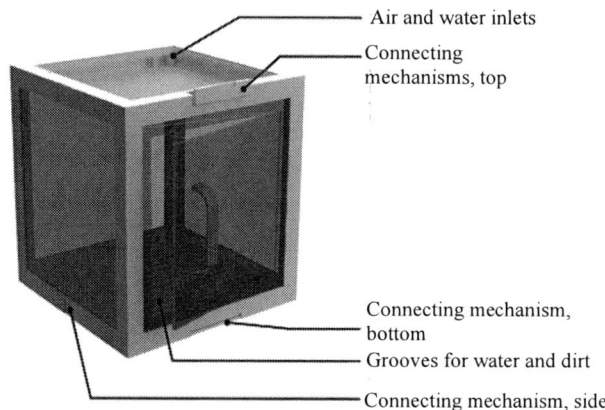

Fig. 9. Green Box, concept courtesy of Rinke Hiel, Nayim Manik, Rick Schildkamp

An example of a complete shed can be seen in Fig. 10. The idea is that the client can order separate roofs as well. These roofs are also modular and should be placed on each vertical bar of boxes. The transparent windows are LSC panels as well and can be stacked onto the plant boxes. For the ventilation and water, there will be channels integrated in the LSC windows.

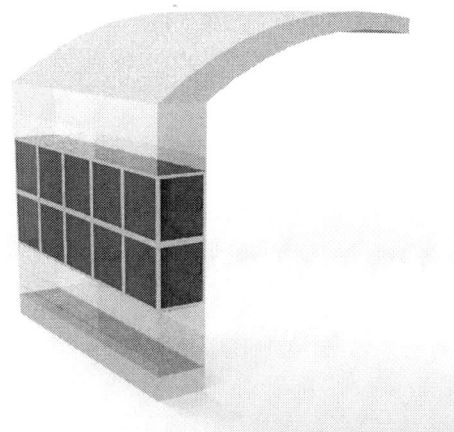

Fig. 10. Integration of the Green Boxes into a shed, concept courtesy of Rinke Hiel, Nayim Manik, Rick Schildkamp

  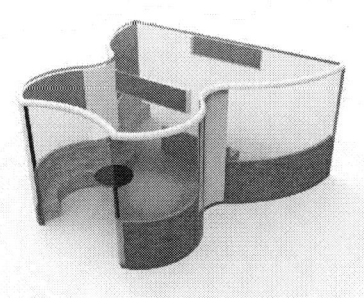

| (a) | (b) | (c) |

Fig. 11. Bus stops, concept courtesy of (a) Mark van der Heijden, Niene Keizer, Jasmijn Mennink, (b) Lisanne Piers, Nadia Holstege, Alin Nassri, (c) Fons Ellermeijer, Miguel Garcia, Charlotte de Graaf

H. Bus stop

There are three different designs for bus stops. The goal is to create a nicely looking shelter or bus stop, which is well integrated into its surrounding and improves the aesthetics in cities, while generating electricity to be used on site.

Although these designs share the same concept, they have different forms and colours. All the designed bus stops use the aesthetic features of LSC PV panels to create a comfortable place to wait for the bus, while one of the designs uses the bending feature of the LSC PV to create a novel type of design (Fig.11). There are screens in all designs communicating travel information which are (partially) powered by LSC PV.

IV. DISCUSSION AND CONCLUSION

This project aimed to design products that fully takes advantage of the properties of LSC PV technology including color, transparency, bendability, and low cost. In total 20 designs were created, out of which a selection of 10 concepts was presented in this study. Some of the designs were conventional products but with a different materials and look (bus stops), some integrated the novel technology into an existing structure (Neo-Holland), some of the designs stretched the idea out of the box (micro-algae production) and some designed a functional product which can be used in everyday life (blinds). As expected, the aesthetics of LSC PV namely colour and transparency played a key role in designing most of the products specially landmarks (Funky fungi) and furniture (Lumient). Colourfulness made designs vibrant while transparency helped with feeling safe and connected to the environment in almost any of the designs. The selected examples show how the design possibility of the technology with the functional and emotional requirements of the chosen applications lead to aesthetically pleasing products, which can help to improve user experience and consequently adopting the LSC PV technology.

ACKNOWLEDGEMENT

The authors would like to thank all students who participated in the 2022 course *Sources of Innovation* at *University of Twente* which generated the results presented in this paper, as well as two guest lecturers, Michael Debije, who introduced LSC technology, and Valerie Souchkov who taught TRIZ and Technology Road mapping.

REFERENCES

[1] M. G. Debije and P. P. C. Verbunt, "Thirty Years of Luminescent Solar Concentrator Research: Solar Energy for the Built Environment," *Adv Energy Mater*, vol. 2, no. 1, pp. 12–35, Jan. 2012.

[2] W. Eggink and A. Reinders, "Product Integrated PV: The Future is Design and Styling," *32nd European Photovoltaic Solar Energy Conference and Exhibition*, pp. 2823–2826, Jul. 2016.

[3] E. Shirazi, W. Eggink, X. Zhu, and A. Reinders, "Design with Integrated PV Technologies in Various Products and Environments," pp. 0731–0731, Nov. 2022.

[4] A. Reinders, "Designing with Photovoltaics," *Designing with Photovoltaics*, Apr. 2020.

[5] A. Reinders and W. Eggink, "A Short History of Photovoltaic-Powered Products," *Designing with Photovoltaics*, pp. 27–60, Apr. 2020.

[6] A. Reinders, J. C. Diehl, and H. Brezet, *The Power of Design: Product Innovation in Sustainable Energy Technologies*. London, UK: John Wiley and Sons, 2012.

[7] A. Reinders, R. Kishore, L. Slooff, and W. Eggink, "Luminescent solar concentrator photovoltaic designs," *Jpn J Appl Phys*, vol. 57, no. 8, p. 08RD10, Aug. 2018.

[8] A. Reinders, A. de Boer, A. de Winter, and M. Haverlag, "Designing PV powered led products - Sensing new opportunities for advanced technologies," *Conference Record of the IEEE Photovoltaic Specialists Conference*, pp. 000415–000420, 2009.

[9] U. Obinna, A. Reinders, P. Joore, and L. Wauben, "A design-driven approach for developing new products for smart grid households," *IEEE PES Innovative Smart Grid Technologies Conference Europe*, vol. 2015-January, no. January, Jan. 2015.

[10] W. Eggink and A. Reindere, "Design it with LSCs; an exploration of applications for Luminescent Solar Concentrator PV technologies," in *IEEE 44th Photovoltaic Specialist Conference (PVSC)*, Washington, 2017.

[11] W. Eggink, "Disruptive Images: Stimulating Creative Solutions by Visualizing the Design Vision / The Design Society," in *The 13th International Conference on Engineering and Product Design Education*, 2011, pp. 97–102.

[12] W. Eggink and A. Reinders, "Explaining The Design & Styling of Future Products," in *International Conference on Engineering and Product Design Education*, Dublin, 2013.

Analysis of Solar Cell Electroluminescence Spectra for Daylight Inspection of c-Si PV Modules

Gisele A. dos Reis Benatto, Alejandra A. Mayordomo, Rodrigo Del Prado Santamaria, Thøger Kari, Peter B. Poulsen and Sergiu V. Spataru

Department of Electrical and Photonics Engineering, DTU Electro, Technical University of Denmark, Frederiksborgvej 399 4000 Roskilde, Denmark.

Abstract — As outdoor electroluminescence (EL) PV inspections are becoming more commercially available, this work has the objective to address the missing knowledge of what can be expected from the PV modules EL signal in the field, especially for daylight conditions, when the signal to be detected is very low in relation to the noise. Here we study the details of the EL emission spectral data statistics from each cell or representative selected cells in a PV module and verify that the crystalline silicon material characteristics of the solar cells are stable, even when faults were detected in EL image. Additionally, when comparing EL peak luminescence intensity of modules with different solar cell technologies, PERC modules where 9.7 and 14.5 times more intense than Al-BSF modules and the IBC module were 73.2% more intense, matching recent qualitative observations during outdoor luminescence imaging. The quantified results in absolute irradiance presented here will lead to the development of models for the prediction of the daylight EL imaging performance at different sunlight irradiance conditions, and for suggestions of optical filtering requirements during image acquisition.

Keywords: Electroluminescence, luminescence spectra, Al-BSF, PERC, IBC.

I. INTRODUCTION

Electroluminescence (EL) imaging was introduced in 2005 for crystalline silicon (c-Si) solar cells [1] and fast evolved to be largely used for PV diagnostic in laboratory and industry for PV cells and full modules. The reason for this fast evolution is the practicality of performing it, even for full size modules. More recently, PV inspection companies have been performing EL images in the field, where preventing the removal of the modules from the racks meant a great advantage for cost and module integrity. In addition to that, there are scenarios where warranty claims require module level and sublevel imaging and EL measurements are necessary to accurately identify cell defects and module faults.

EL imaging of PV modules during nighttime and indoor conditions is relatively straightforward. However imaging in daylight, requires more complex methods, such as lock-in EL and Photoluminescence (PL) imaging using the sunlight as excitation source, which both exhibit very low luminescence signal in comparison with sunlight [2], [3]. The PL and EL emission spectra are known to be equivalent when normalized, and the difference for silicon solar cells will be on pathway of carriers injection [3].

In the past, it was frequently reported that multi-crystalline silicon (multi-Si) wafers and cells presented one or many defect related bands in the PL spectrum [4]–[7], where the area fraction of defective regions observed in PL images presented a correlation with the final cell electrical parameters. In a module scale, a connection between the details of the spectral luminescence emission of the c-Si solar cells is lacking, while many new PV module technologies are widely available commercially nowadays, such as aluminum back surface field cells (Al-BSF), passivated emitter and rear cells (PERC), interdigitated back contact (IBC), among others.

The aim of this paper is to study and quantify the luminescence emission spectra distribution of cells in commercial c-Si PV modules, in terms of in-cell and module luminescence spectra uniformity, healthy and defective cell luminescence, as well as quantify differences in the EL emission between cell technologies and correlate with their electrical parameters. This is motivated by recent works that have observed a clear difference in the luminescence emission intensity of modules with more recent cell technologies but not quantified it [8], [9]. With these EL emission quantification results, this work should lead to modeling for the prediction of the daylight EL imaging performance during daytime at different sunlight irradiance conditions, and prediction of exposure time and optical filtering requirements during image acquisition.

II. METHODOLOGY

A. Experimental Details

Table I shows the sample specifications of the five PV modules studied in this work. They include PV modules manufactured in different years from 2011 to 2023, with different cells sizes, and cells technologies such as Al-BSF cells, IBC and PERC cells for mono-Si, and one Al-BSF for multi-Si. The electrical parameters in Table I are from measurements taken at standard test conditions (STC) and include a percentage of power loss compared with the manufacturer datasheet specifications.

The EL image of the modules included in Table II. From the characterization and EL images, the PV with <2% power loss are considered to be in initial stages of degradation, for instance Al-BSF Mono-Si module has few cracked cells, and PERC 1 and PERC 2 have evidence of initial potential induced

TABLE I

DETAILS AND MEASURED ELECTRICAL PARAMETERS OF THE PV MODULES STUDIED

Module Type	Module Manuf. Year	Measured cells	Cell size (mm)	Cell #	Cell Voc (V)	Jsc (mA/cm^2)	Module FF (%)	Module MPP (W)	Cell Eff. (%)	Fielded Time	Power loss
Al-BSF Mono-Si	2011	All cells	125x125	72	0.620	35.84	76.0	189.95	16.57	10 years	-3%
Al-BSF Multi-Si	2015	All cells	156x156	36	0.627	35.90	75.1	148.00	17.40	3 years	-1%
IBC	2023	All cells	40x125	16	0.771	34.40	79.2	16.80	21.00	None	0%
PERC 1	2018	All cells	156x156	60	0.657	39.48	76.7	290.35	20.64	None	-2%
PERC 2	2020	20 selected cells	105x210	120	0.672	40.43	78.6	564.75	21.34	2 years	-5%

TABLE II

EL IMAGES AND HEATMAPS OF EL SPECTRA PARAMETERS

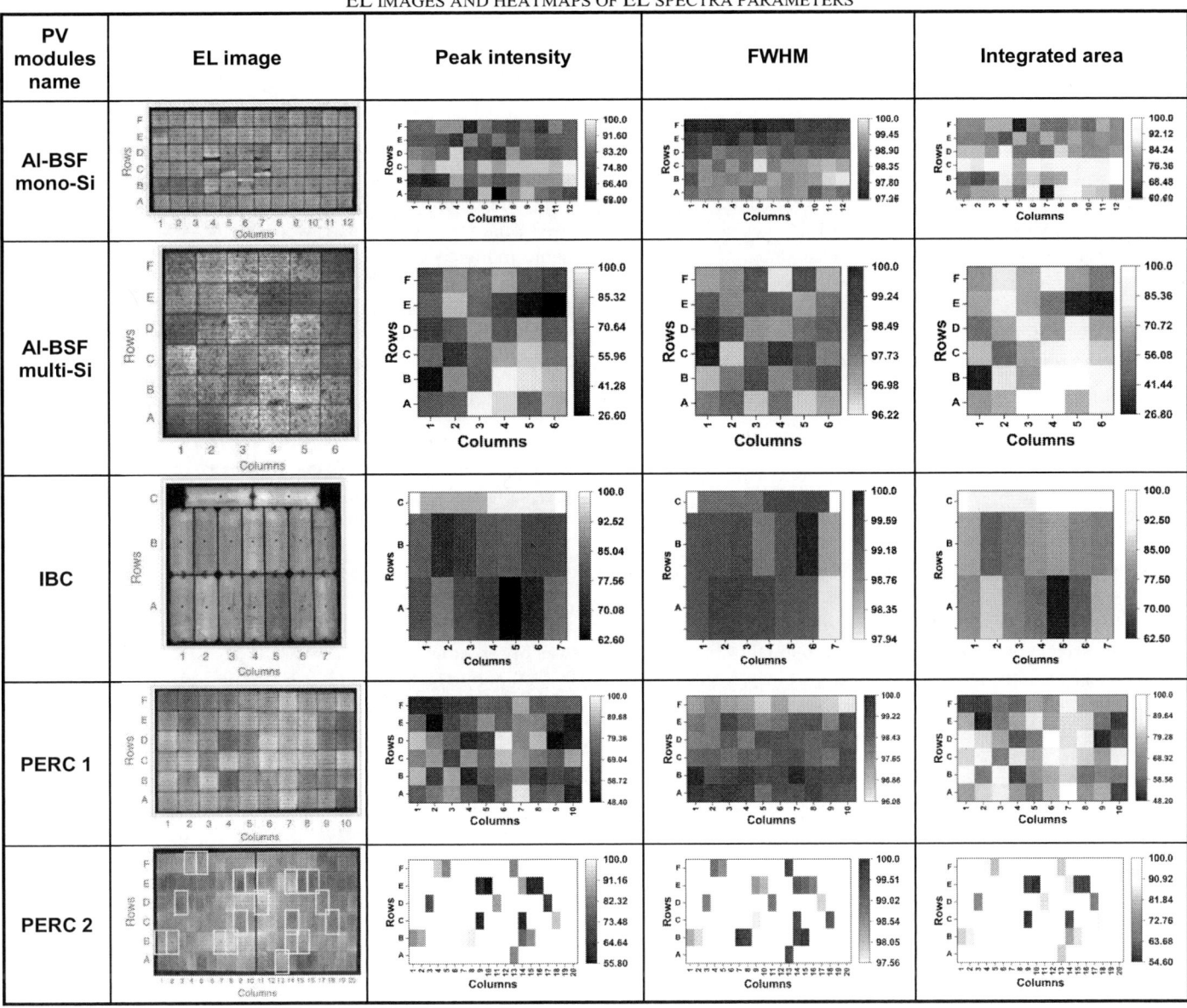

978-1-6654-6060-6/23 $31.00 © 2023 IEEE

degradation (PID). The cells of the modules are labeled as suggested in the IEC technical specification on PV testing requirements [10]. Spectral measurements were all acquired at the center of the cell, unless a busbar prevented it and then the measurement was performed slightly beside it. In some images, it can be seen as a small circle in the center of the cells showing a sticker placed to mark the measurement position. This is a non-defective spot on the cell and the sticker was removed when the spectrum acquisition was performed. For module PERC 2, the selected cells evaluated are marked in yellow, and are representative of the entire module luminescence characteristics. The other modules had the spectrum of all their cells measured in this study. The spectral data was acquired using an InGaAs spectrometer B&W Tek spectrometer Sol 1.7 (900-1700 nm range) with five averaged acquisitions. Each module was biased at 100% of its respective datasheet short-circuit current and the exposure time was adjusted to be maximum possible without saturation. The settled exposure times were 50 ms (IBC), 250 ms (PERC 2), 400 ms (PERC 1) and 1 sec (both Al-BSF modules). To assure a steady state emission, the PV module was current biased for 15 min before the spectra measurements started. The laboratory ambient temperature was controlled and set to 25 ± 2 °C. The EL images in Table I were acquired in a dark room using an InGaAs camera with 40 ms exposure time and were contrast corrected for better visualization. The outdoor EL images were acquired with the same camera with 2 ms exposure time.

B. Data Analysis

The spectral data acquired was dark-subtracted and the absolute irradiance values were defined via calibration using an incandescent lamp. Spectrum parameter calculations were performed in the 950-1300 nm range.

III. Results and Discussion

A. Module Statistics

From the luminescence spectra of c-Si, four main parameters can be extracted:

- Peak intensity, which has a correlation with the general electrical parameters for that location in the cell [1], [11].
- Peak position, related to temperature dependency of the material bandgap and red shifts with the higher temperature [11], [12].
- Full width at half maximum (FWHM), correlated with bulk carrier lifetime, where the narrower FWHM, the longer the lifetime [11].
- Integrated area, which embodies both peak intensity and FWHM into the curve area value.

The first observation of the EL spectra for the five modules studied is that all the spectra matched remarkably well within the same module when normalized. For the peak position, the variations were 0.28% for all the modules. When keeping their

original intensities, in Table II, the module heatmaps show the distribution of the spectrum parameters per solar cell expressed in percentage of the highest value extracted from the EL spectrum. The distribution of peak intensity and integrated area are app. 40% for the Al-BSF mono-Si; 73% for the Al-BSF multi-Si; 37% for the IBC; 52% for the PERC 1; and 60% for the PERC 2 (in this last case, for the 20 selected cells). Their heatmaps resemble the intensity distribution observed with the camera in the EL image. PERC 2 that has significant power loss, does present a higher variation than PERC 1, but at the same time, Al-BSF Mono-Si that also has power loss, presents a variation comparable to the IBC module which is brand new. By observing the EL intensity-mismatch, some trends could be found for PID affected modules, but in principle a point measurement in the center of the cell is not a reliable method to evaluate the health of the module, especially when cracks are involved (Al-BSF Mono-Si). Al-BSF Multi-Si clearly shows more variation due to multi-Si cell properties. More sampling comparing new and older modules should be addressed to evaluate the expected cell EL intensity variation within the same module and if trends can be drawn towards what is acceptable to assure a good quality PV module. Al-BSF multi-Si and PERC 1 modules had the highest variation of the FWHM of about 4%. Contrary to what is observable in the EL images, cells that seem to be degraded such as F1-4 in module PERC 1 are in the range of the narrower FWHM, while cell B1 of the same module is one of the broadest, indicating that different degradation mechanisms can be taking place; one that is keeping luminescence intensity but decreasing carrier lifetime.

B. Module Comparison

Fig. 1 shows the average spectrum curve of five modules studied in linear (Fig. 1a) and semilogarithmic scale (Fig. 1b). The peak at approximately 1145 nm corresponds to indirect band-to-band emission of the c-Si material and a shoulder in a app. 240 nm region is attributed to a phonon sideband [1]. A defect-band reported in literature [4], [5] is not observed at longer wavelengths in any of the modules on the non-defective positions (see EL images in Table II) analyzed. Looking closer at the Al-BSF multi-Si, the spectrum was acquired in 10 points on the area of three cells B4, D2 and F5 and still no defect band was observed.

Comparing the average curves in Fig. 1, the EL peak luminescence irradiance of the IBC cells had the highest emission intensity. PERC 1 peak intensity reached 14% of the IBC emission, PERC 2 reached 21%, and the Al-BSF had only 2 and 1% for the Mono- and Multi-Si respectively of the correspondent IBC peak luminescence intensity. This translates to the indicative number at the peak of the curves, corresponding to the number of times that the average curve intensity of the module was higher than the Al-BSF Multi-Si module, which had the lowest average EL emission peak. It is valid to remember that these number and percentages are subjected to the variation of peak intensity within the same PV

978-1-6654-6060-6/23 $31.00 © 2023 IEEE

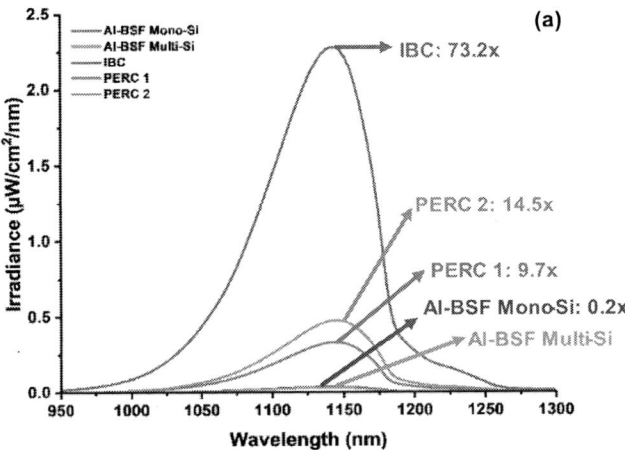

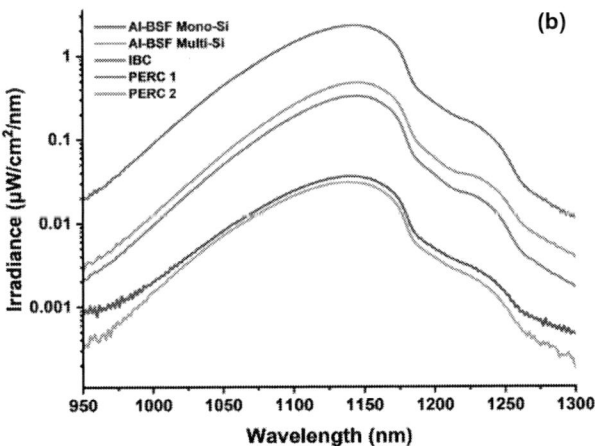

Fig. 1. Average curves of EL spectra from the PV modules in linear scale plot (top) and semilogarithmic scale plot (bottom).

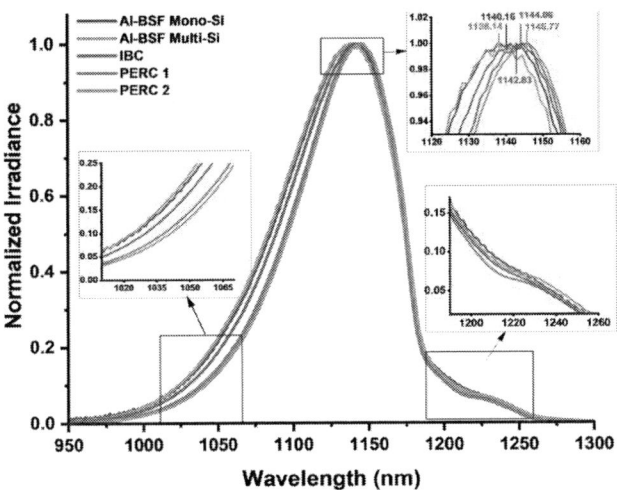

Fig. 2. Normalized average curve of EL spectra from the PV modules with insets emphasizing details.

module, as discussed in the previous section. Such differences between modules are usually associated with cell current density, open circuit voltage, rear surface reflectance, among other cell properties [12]–[14] and correlation with the electrical parameters found in Table I will be made in the next section.

Fig. 2 shows the normalized averaged curves presented previously in Fig. 1a in absolute irradiance units. Between technologies it is observed a shift of 7.63 nm in the main peak position (see inset in the top right of Fig. 2). As for semiconductors, the bandgap shrinks as the temperature increases from room temperature [7], [11], this can be due to the stabilization temperature difference of the c-Si cells which have different short-circuit current and temperature coefficients [15]. For instance, this temperature was 35.9°C for the Al-BSF mono-Si and 39.7°C for the PERC 1 module.

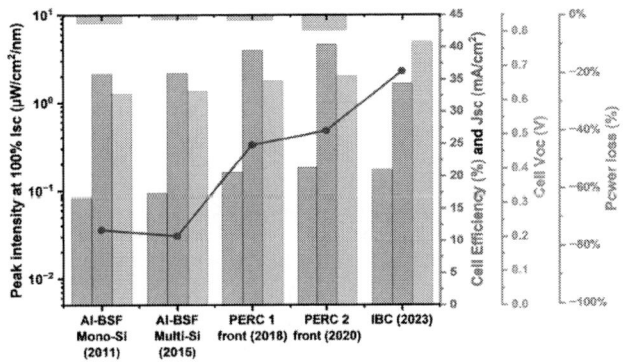

Fig. 3. Peak luminescence intensity of averaged EL spectra from the PV modules cells compared with their electrical parameters.

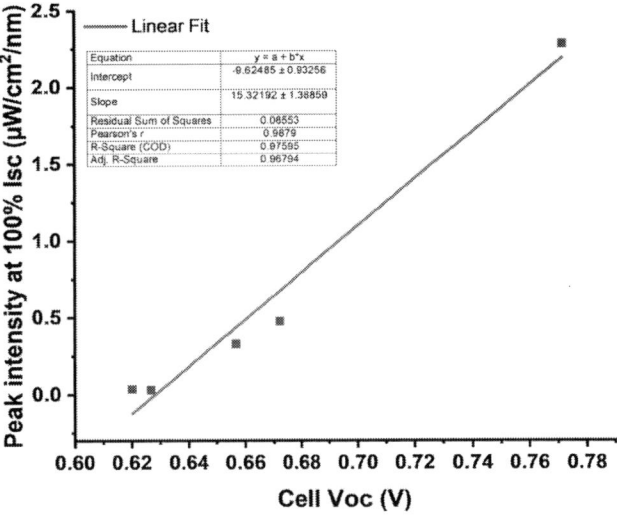

Fig. 4. Peak luminescence intensity of averaged EL spectra correlation with cell Voc.

The insets on the bottom left and right of Fig. 2 show how the broadness and shoulder of the modules differ, with the older technologies being broader, with a less defined shoulder.

C. Electrical Parameters and Peak Luminescence Intensity

Fig. 3 shows the comparison of EL peak intensity with the electrical parameters measured at the module level at STC. The electrical parameters are also shown in Table I. Even though the peak intensity can be correlated with all the electrical parameters in the solar cells, in this module level approach, a clear correlation can be observed with the cell open circuit voltage only. For cell efficiency, PERC 1, PERC 2 and IBC have similar values (20.64%, 21.34% and 21.00% respectively), with PERC 2 with even higher efficiency than IBC, however IBC have a much higher luminescence intensity. Fig. 4 shows the correlation of average curve peak intensity with cell V_{OC}, indicating a close to linear relationship. The linear correlation between luminescence intensity and V_{OC} has been reported previously for PL in implied V_{OC} measurements [16], [17]. More statistics on different modules with same cell technologies studied here is required for this correlation with the EL emission to be established with more certainty and that other uncertainty factors are not interfering or hiding other possible correlations.

D. Implications for Outdoor Luminescence Imaging

In this work we quantified the EL emission of solar cells in PV modules in terms of absolute irradiance. In Fig. 5 we show the same spectra from Fig. 1a with the specular reflection of sunlight on the IBC module during daylight EL image acquisition with the EL signal (I_{SC} current bias) ON and OFF. The IBC EL intensity shows a clear difference on the spectrum measurement from the PV module under sunlight on the 1100 to 1180 nm range. This is an indication that even without optical filters and signal enhancement via the acquisition of several

images [18] using an InGaAs camera detector, the EL signal should be visible. It has been shown previously in the literature that usually is not the case, as daylight luminescence imaging requires signal enhancement [2], [19], [20], and by the magnitude of the signal of the Al-BSF, this is probably the case.

Fig. 6 shows two images from the IBC module with and without EL signal taken with an InGaAs camera without optical filters. Their profiles marked with horizontal line of the images

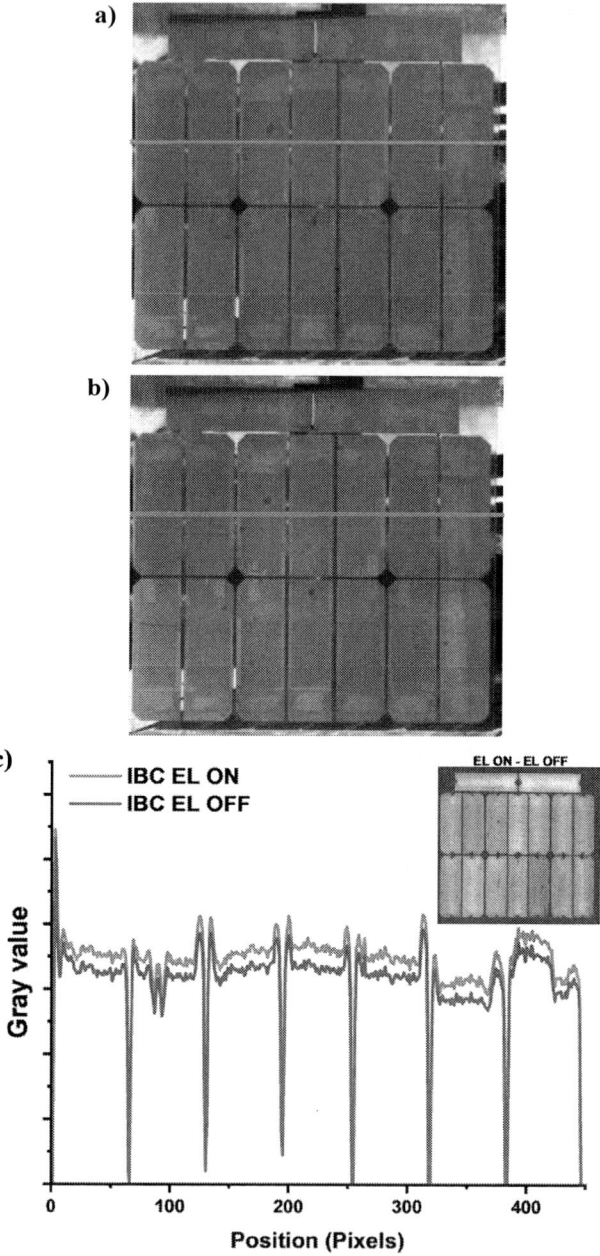

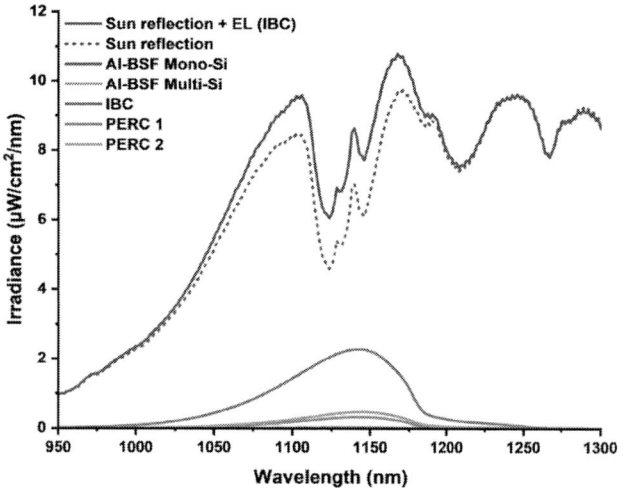

Fig. 5. Sunlight specular reflection on the IBC module at 611 W/m² irradiance at the POA compared with the EL spectra average curves from the PV modules.

Fig. 6. Outdoor EL acquisition of the IBC module at 611 W/m² sun irradiance at the POA using an InGaAs camera with 2ms exposure time without optical filter. (a) EL signal OFF; (b) EL signal ON; (c) Profiles of the images from the line marked with the blue and green line respectively and background subtracted image in the inset.

are shown in Fig. 6c. It is clear to see that the EL signal from this module in the image profile difference and the background subtraction is sufficient to generate a high-quality daylight EL image (inset of profile plot). This example shows that quantifying the EL signal does provide an indication of successful daylight EL image acquisition. In the future, we will use indoor spectral data from a module type and simulate daylight EL acquisition under different daylight conditions, highlighting connection sunlight irradiance noise on the need for optical filters, design of ideal optical filters, requirements for camera settings, among other relevant parameters for EL imaging. The same modeling may be applicable for outdoor daylight PL imaging.

IV. SUMMARY

In this study, we have shown the statistics of the EL emission spectral parameters of five different PV module technologies and compared them in terms of EL spectral data and electrical parameters. The results point out the variation in percentages of what can be expected in modules with no or little power loss. Cell V_{OC} is shown to be the electrical parameter that correlates the most with the intensity of the EL signal.

Additionally, in the comparison between the average EL emission from modules with different solar cell technologies, IBC cells had the highest emission intensity, being 73.2 times higher than former preeminent Al-BSF module technology, at the same time, PERC modules presented 9.7 and 14.5 times more intense EL emission than Al-BSF modules. These relative values do match qualitative observations during outdoor luminescence imaging found in recent literature. The quantified EL emission results in absolute irradiance should lead to modeling for the prediction of the daylight EL imaging performance at different sunlight irradiance conditions, possibility of performing low current bias EL outdoors, prediction of exposure time requirements and optical filtering during image acquisition. Further studies and bigger sampling for modeling are required to establish the correlation between EL luminescence with datasheet or STC electrical parameters.

ACKNOWLEDGMENT

The authors would like to acknowledge the discussions with Lukas Koester about the stabilization temperature difference between PV modules when forward current bias is applied during EL. This research has been supported by the Eurostars project "Automated Daylight Electroluminescence Inspection of Large Photovoltaic Systems" (E115687 ADELI).

REFERENCES

[1] T. Fuyuki, H. Kondo, T. Yamazaki, Y. Takahashi, and Y. Uraoka, "Photographic surveying of minority carrier diffusion length in polycrystalline silicon solar cells by electroluminescence," *Appl. Phys. Lett.*, vol. 86, no. 26, pp. 1–3, 2005, doi: 10.1063/1.1978979.

[2] L. Stoicescu, M. Reuter, and J. H. Werner, "DaySy: Luminescence Imaging of PV Modules in Daylight," *29th Eur. Photovolt. Sol. Energy Conf. Exhib.*, pp. 2553–2554, 2014, doi: 10.4229/EUPVSEC20142014-5DO.16.2.

[3] R. Bhoopathy, O. Kunz, M. Juhl, T. Trupke, and Z. Hameiri, "Outdoor photoluminescence imaging of photovoltaic modules with sunlight excitation," *Prog. Photovoltaics Res. Appl.*, vol. 26, no. 1, pp. 69–73, Jan. 2018, doi: 10.1002/pip.2946.

[4] I. Burud, T. Mehl, A. Flo, D. Lausch, and E. Olsen, "Hyperspectral photoluminescence imaging of defects in solar cells," *J. Spectr. Imaging*, vol. 5, no. August, pp. 1–5, 2016, doi: 10.1255/jsi.2016.a8.

[5] F. Yan, S. Johnston, K. Zaunbrecher, M. Al-Jassim, O. Sidelkheir, and K. Ounadjela, "Defect-band photoluminescence imaging on multi-crystalline silicon wafers," *Phys. Status Solidi - Rapid Res. Lett.*, vol. 6, no. 5, pp. 190–192, 2012, doi: 10.1002/pssr.201206068.

[6] T. Fuyuki and A. Kitiyanan, "Photographic diagnosis of crystalline silicon solar cells utilizing electroluminescence," *Appl. Phys. A Mater. Sci. Process.*, vol. 96, no. 1, pp. 189–196, 2009, doi: 10.1007/s00339-008-4986-0.

[7] T. Trupke *et al.*, "Temperature dependence of the radiative recombination coefficient of intrinsic crystalline silicon," *J. Appl. Phys.*, vol. 94, no. 8, pp. 4930–4937, 2003, doi: 10.1063/1.1610231.

[8] M. Vuković, M. Jakovljević, A. S. Flø, E. Olsen, and I. Burud, "Noninvasive photoluminescence imaging of silicon PV modules in daylight," *Appl. Phys. Lett.*, vol. 120, no. 24, p. 244102, Jun. 2022, doi: 10.1063/5.0097576.

[9] B. Doll *et al.*, "High-throughput, outdoor characterization of photovoltaic modules by moving electroluminescence measurements," *Opt. Eng.*, vol. 58, no. 08, p. 1, 2019, doi: 10.1117/1.oe.58.8.083105.

[10] *IEC TS 62446-3:2017 - Photovoltaic (PV) systems — Requirements for testing, documentation and maintenance – Part 3: Photovoltaic modules and plants – Outdoor infrared thermography.* 2017.

[11] W. S. Yoo, K. Kang, G. Murai, and M. Yoshimoto, "Temperature Dependence of Photoluminescence Spectra from Crystalline Silicon," *ECS J. Solid State Sci. Technol.*, vol. 4, no. 12, pp. P456–P461, 2015, doi: 10.1149/2.0251512jss.

[12] T. Fuyuki, H. Kondo, Y. Kaji, A. Ogane, and Y. Takahashi, "Analytic findings in the electroluminescence characterization of crystalline silicon solar cells," *J. Appl. Phys.*, vol. 101, no. 2, 2007, doi: 10.1063/1.2431075.

[13] K. Bothe and D. Hinken, *Quantitative luminescence characterization of crystalline silicon solar cells*, 1st ed., vol. 89. Elsevier Inc., 2013. doi: 10.1016/B978-0-12-381343-5.00005-7.

[14] H. T. Nguyen, S. C. Baker-Finch, and D. MacDonald, "Temperature dependence of the radiative recombination coefficient in crystalline silicon from spectral photoluminescence," *Appl. Phys. Lett.*, vol. 104, no. 11,

pp. 1–4, 2014, doi: 10.1063/1.4869295.

[15] L. Koester, E. Vallarella, A. Louwen, S. Lindig, and D. Moser, "Technical Considerations Resulting from Photovoltaic Module Heating During Electroluminescence Inspection," in *8th World Conference on Photovoltaic Energy Conversion*, 2022, pp. 1112–1118. doi: 10.4229/WCPEC-82022-4DO.4.4.

[16] B. Hallam, Y. Augarten, B. Tjahjono, T. Trupke, and S. Wenham, "Photoluminescence imaging for determining the spatially resolved implied open circuit voltage of silicon solar cells," *J. Appl. Phys.*, vol. 115, no. 4, 2014, doi: 10.1063/1.4862957.

[17] Y. Augarten, A. Wrigley, U. Rau, and B. E. Pieters, "Calculation of the TCO sheet resistance in thin film modules using electroluminescence imaging," *Conf. Rec. IEEE Photovolt. Spec. Conf.*, vol. 2016-Novem, no. 4, pp.

1527–1531, 2016, doi: 10.1109/PVSC.2016.7749874.

[18] G. A. dos Reis Benatto *et al.*, "Daylight Electroluminescence of PV Modules in Field Installations: When Electrical Signal Modulation is Required?," in *8th World Conference on Photovoltaic Energy Conversion*, 2022, pp. 735–739. doi: 10.4229/WCPEC-82022-3BV.3.44.

[19] M. Guada *et al.*, "Daylight luminescence system for silicon solar panels based on a bias switching method," *Energy Sci. Eng.*, no. June, pp. 1–15, 2020, doi: 10.1002/ese3.781.

[20] G. A. dos Reis Benatto *et al.*, "Development of outdoor luminescence imaging for drone-based PV array inspection," *44th IEEE Photovolt. Spec. Conf.*, pp. 2682–2687, 2017, doi: 10.1109/PVSC.2017.8366602.

Artificial Neural Network and Peer-to-Peer Communications at the Grid-Edge to Mitigate Cyber Attacks on Distributed Photovoltaic Inverters

C. Birk Jones & Rachid Darbali-Zamora

Abstract—Resilient control of photovoltaic (PV) inverters using a local Artificial Neural Network (ANN) and peer-to-peer communications can maintain grid services during a cyberattack. High penetrations of PV systems presents grid performance challenges, and alterations to connected systems can introduce additional problems. To tackle these issues, this paper introduces a methodology for controlling PV inverters that are under attack using the Laterally Primed Adaptive Resonance Theory (LAPART) ANN to predict the best reactive power control input when communications between the central command center are down or cannot be trusted. This work tested the approach using a 6-bus feeder model with a high penetration of PV. The experiment found that the algorithm can predict the appropriate reactive power setting with high accuracy, and when embedded inside the grid model, the algorithm can predict a reactive power that improved system voltages.

Index Terms—aggregators, cyber-attacks, cybersecurity, distributed energy resources, PV inverters, resilience

I. Introduction

A distributed Artificial Neural Network (ANN)-based control that uses peer-to-peer (P2P) communications will maintain photovoltaic (PV) services during a cyber-attack event. The approach uses an embedded ANN that learns the PV inverters behaviors (i.e. reactive power (VAR) injection or absorption) by associating its performance with nearby peers. It then reacts to abnormal behavior by setting the correct VAR for maintaining the appropriate level of grid services.

PV inverters perform three key functions, which are to shape the current into a waveform, invert the current into AC, and boost the voltage to match the grid [1]. These power electronic devices support grid voltage and frequency by implementing voltage ride-through operations, VAR injection or absorption, and others [2]. Implementation of the services, in some cases, include centralized controllers that send signals through the open internet. The high penetration of these internet connected devices presents challenges associated with cybersecurity vulnerabilities within centralized control system architectures. To provide grid services, PV inverter control settings and firmware are updated using the open internet, and therefore subject to manipulation or obstruction.

Cybersecurity of PV inverters is a recognized risk for the electric grid. A roadmap for PV cybersecurity reviews the communications infrastructure, protection considerations, and industry "best" practices [3]. Key challenges include divided administration, privacy, control functions, and increased cyber-physical inter-dependencies [4]. Past cybersecurity risk assessments, that used MATLAB/Simulink models, of PV inverters capable of reactive power injection found that the spoofing of sensors could result in grid voltage violations [5]. Another experiment found that denial of service (DOS), packet replay, and man-in-the-middle (MitM) attacks affect the integrity and confidentiality of actual PV inverters [6]. Clearly intrusion detection will provide a necessary level of defense.

Detecting malicious activity directed at PV systems has been explored in past work. One paper leveraged analysis deployed on a microcomputer to provide an Intrusion Detection System (IDS) for a PV inverter [7]. Another paper built on the IDS work to create an unsupervised learning approach for detecting malicious behaviors [8]. Another research effort performed both detection and mitigation of the attack [9].

This supplementary monitoring and control, that is initiated during a successful attack, includes P2P communications, embedded analysis, and local control of the PV inverter's VAR injection or absorption. The P2P communications allows the inverters to share sensor values with its neighbors. The shared data, which includes bus voltage (V_b), real power (P), and VAR injection/absorption, is analyzed using the Laterally Primed Adaptive Resonance Theory (LAPART) ANN to predict the VAR control value that overrides the unwanted setting. The predicted VAR is sent to the PV inverter control board to override incorrect control commands or settings caused by a cyber-attack.

This work assumes that the performance of neighboring PV inverters is shared using one of many P2P communication platforms [10] [11], but does not implement the actual communication standard. This paper introduces a distributed monitoring and analysis platform for resilient control of distributed PV inverters that reacts to an attack by maintaining appropriate control functions using an on-board ANN algorithm.

II. Background

PV inverters, connected to centralized aggregators based on IEC 61850, are subject to intervention from an intruder through MitM attacks that alter the control functionality and cause a loss in service [12]. Alterations to smart PV inverter parameters and sensor values can impact grid performance. For example, Teymouri *et al.* found that changes to the reactive power reference and other sensor measurements impact the grid voltage [5].

C. Birk Jones is with Camus Energy in San Francisco, CA USA

R. Darbali-Zamora is with the Renewable and Distributed Systems Integration Group at Sandia National Laboratories in Albuquerque, NM USA

978-1-6654-6060-6/23 $31.00 © 2023 IEEE

PV inverter control schemes leverage both onboard software and centralized aggregators to define operations [13], [14]. Often PV inverters operate autonomously to support grid voltage and frequency [15]. Aggregators can also intervene, and control PV inverters from a central location. [16] demonstrated, through a power hardware-in-the-loop experiment, the potential for an extremum seeking control algorithm to modulate the reactive power for voltage deviation minimization. A less intrusive approach involves the definition of specific Volt-Var Curves (VVC) that vary based on its aggressiveness [17]. Without proper oversight adversaries can exploit the various control methodologies and cause harm to the electric grid.

A decentralized control architecture provides resilient control during times of distress on the distribution grid. Unlike Shelar et al., that describes a distributed control strategy for non-compromised distributed energy resources [18], this work introduces a method for controlling PV inverters operating in normal operating conditions or under-attack. The approach deployed learning algorithms at the grid-edge to monitor and control the inverters based on local and nearby PV inverter and grid performance.

The LAPART ANN, used in this work to determine the best course of action, was previously used for other PV monitoring and analytics applications. The algorithm successfully detected faults in PV system performance [19] and accurately forecasted irradiance for PV smoothing [20]. Also, there is evidence that the algorithm can work well on a single-board computer out in the field. The Adaptive Resonance Theory (ART) algorithm, which is used by LAPART, was deployed on Raspberry Pi computer to detect network irregularities in building automation systems [21], [22]. The algorithms ability to work on a small computer allows it to work at the edge of the grid. Data collection and analysis at the edge helps reduce the latency in control response, saves on network bandwidth, eliminates cloud based storage costs, and supports the deployment of robust and intelligent analysis and control methodologies.

III. METHODOLOGY

To mitigate the risk associated with cyber-threats on centralized distributed PV inverter control system, the present work introduces a resilient methodology for controlling the PV inverter's VARs in the event of an attack. This resilient control reacts to alarms produced by an IDS system, depicted in Fig. 1. Fig. 1 shows that the IDS system was not in the scope of this work. Instead, the "Resilient Control" assumed that an alarm was created and then mitigated any issues associated with non-optimized reactive power support.

The approach leverages P2P communications among PV inverters. A local learning algorithm then analyzes the shared data to determine the correct VAR injection or absorption value. The paper documents an initial attempt at the implement and used a 6-bus electric feeder model that includes 3 loads and 9 PV systems to test the distributed, learning-based methodology.

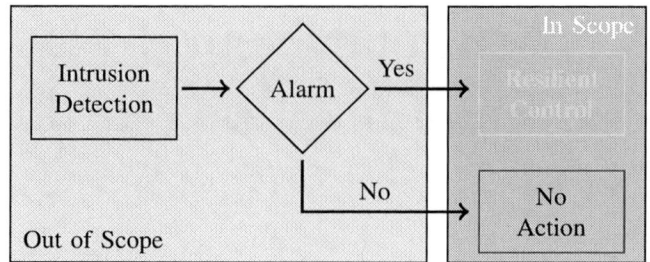

Fig. 1. Diagram depicts the scope of this paper. The detection of malicious activity is not included but assumed to initiate the resilient control described in this paper.

A. Communication Infrastructure

Current standards require that all PV inverters have the capability to communicate with the respective utility using SunSpec Modbus, DNP3, or IEEE 2030.5. This enables utilities or aggregators to perform read and write requests with individual inverters from a central location. These connections are represented in Fig. 2, where the three aggregators (A, B, and C) are connected directly to a total of nine (9) distributed PV nodes to support centralized control operations. The dashed lines in-between the distributed PV nodes represent the P2P communications that support the "resilient" control functions.

The P2P interconnections provides each PV inverter with some awareness of the surrounding PV inverters. The connections with neighbors allows for the sharing of real power, reactive power, and voltage. The interactions support on-board analysis under the following scenarios:

1) PV inverter learns using its own performance data.
2) The on-board algorithm learns using data from neighbors on the same bus.
3) The local analysis uses data from neighbors on different buses.

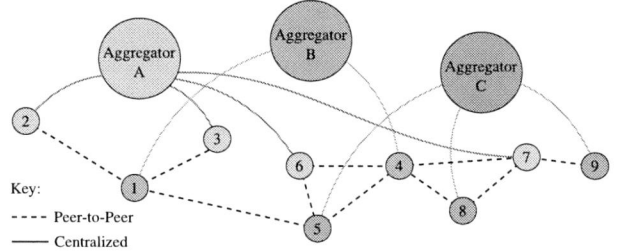

Fig. 2. Image depicts the assumed communications for the PV systems. Aggregators communicate with each devices from a central location. The PV systems can communicate with their neighbors through peer-to-peer connections.

B. On-Board Analysis & Control

The resilient control of PV inverters requires the intelligence of an ANN to define the control actions in the event of a cyber-attack. A successful attack could include the Denial-of-Service (DOS) of the aggregator control signals or an intrusion that altered the Volt-Var Curve (VVC) settings. During an attack, the on-board analysis will work with its peers to determine the correct VARs for maintaining grid services. In this case, LAPART ANN algorithm determines the amount of VARs.

978-1-6654-6060-6/23 $31.00 © 2023 IEEE

1) Artificial Neural Network: The LAPART algorithm was introduced by Healy and Caudell for logical inference and supervised learning [23], [24]. The algorithm offers light-weight solution in comparison to others, such as those that use multi-layer perceptron methods. It can also train and test on a small single-board computer in the field without any computational issues. The algorithm's architecture couples two Fuzzy Adaptive Resonance Theory (ART) [25] algorithms to create a mechanism for making predictions based on learned associations.

The LAPART algorithm is the coupling of two Fuzzy ARTs. The A and B Fuzzy ARTs are connected through the L matrix that associates the A and B templates (ANN memory). Each Fuzzy ART has its respective vigilance parameters ρA and ρB, and during the learning process inputs are presented to the A and B side simultaneously. The A and B sides create and update templates while at the same time producing links between one another by developing the L matrix. After training finishes, testing inputs are only applied to the A side and allowed to resonate with the previously learned templates. Then the associations in the L matrix are used to connect with the B side and provide the prediction (or testing) outputs.

2) Accuracy Evaluation: The ability of the LAPART algorithm to predict the PV inverters VARs at different operating conditions was assessed by varying the free parameters within a K-Fold cross validation process. The K-Fold process began by randomizing a data set of 6.9 million points, that represent well performing operations, created in the simulation (over an 80 day period) described in Section III-C; the data was then split into K equal parts or folds. This division of data for K = 4 divided the data up into four 1,728,000 sets. For each fold K $\in$ {1,2,..,K} the model was trained on the data that was located in all of the folds except for the Kth, and the Kth fold was used for testing [26]. This process was conducted in a round-robin manner until each of the folds was used for training and testing.

The output from the testing process was an R^2 value; the R^2 evaluated the amount of explained variation versus the total variation between the actual and predicted values. The R^2 metric can be between -1 and 1, where a value close to 1 indicates that there is a strong linear relationship between the two dependent variables. In this case, an R^2 for multiple ρA and ρB settings was calculated to determine the overall accuracy of the algorithm at different free parameter conditions. The final outcome from the K-Folds iterations was the optimal free parameter settings that produce the most accurate results.

The evaluation concluded by examining the performance of the algorithm at the free parameters that generated the best R^2 values. The review used the training data from the 4th fold in the K-Fold process and tests the algorithms performance on data left out of the K-Fold iterations. The test data included about 500,000 data points for a 6-day period. The results for this evaluation are described in Section IV-B2.

C. Experiment - Simulation Test

Testing of the proposed communications, and learning algorithm control involved the implementation of a simple simula-tion electric power system. The environment included a feeder model that operated under normal and adverse conditions. The normal operations required the PV inverters to perform Volt-Var services to maintain the appropriate bus voltages. It also had the ability to operate in modes that were detrimental to the system and required intervention from the proposed learning-based control to rectify inappropriate system control settings. This inimical situation represented a successful cyber-attack that alters the ability of the inverters to provide the appropriate service necessary for optimizing the voltage on the feeder. Under these conditions the LAPART algorithm stepped in and controlled the inverter.

1) Feeder Model: To demonstrate the capabilities of the approach, a MATLAB/Simulink secondary distribution feeder with a high penetration of PV was modeled [27]. The model simulates a power distribution feeder, depicted in Fig. 3, consisting of a 7200/250 V transformer (rated at 40 kVA), 6 buses, 4 transmission lines, 3 loads (16 kW), and nine PV inverters with a total rating of 29 kVA. The assumed line distance, resistance, and inductance values for the feeder are defined in Table I. And, the predefined capacity for each PV inverter and load are described in Table II and III, respectively. The capacity provided an overall limit for the load and PV systems, and the profiles varied over the entire simulation effort that equated to about 86 days.

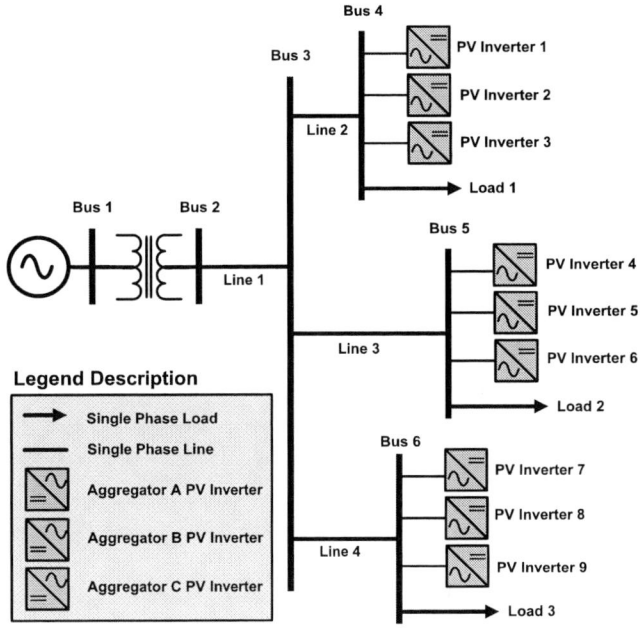

Fig. 3. Distribution feeder model that includes 6 buses, 3 loads, and 9 PV inverters.

TABLE I
DISTRIBUTION FEEDER LINE PARAMETERS

Line #	Resistance (Ω)	Inductance (mH)	Distance (km)
1	0.1470	0.0834	0.122
2	0.0735	0.0417	0.061
3	0.0073	0.0042	0.006
4	0.0110	0.0063	0.009

978-1-6654-6060-6/23 $31.00 © 2023 IEEE

TABLE II
GENERATION CAPACITIES

PV Inverter #	Bus	Capacity (kVA)
1	4	5.0
2	4	3.0
3	4	4.0
4	5	8.0
5	5	3.0
6	5	2.0
7	6	1.8
8	6	1.2
9	6	1.0

TABLE III
LOAD CAPACITIES

Load #	Bus	Capacity (kW)
1	4	2.0
2	5	9.0
3	6	5.0

The load and PV generation systems followed realistic profiles. The load profiles emulated typical residential behavior that changed throughout the day. The peak occurred during the day and the low point was during the middle of the night [28], as shown in Fig. 4. In this case, the load profiles were the same except that each were scaled depending on their capacities. The PV systems, on the other hand, were based on a mathematical model that accepts irradiance (shown in Fig. 4) and the ambient temperature to simulate both active and reactive power [29].

The simulated PV systems were able to perform Volt-Var control based on a pre-programmed characteristic curve. When PV generation was higher than the load demand the inverter absorbed the VARs to decrease the V_{bus}. Inversely, when the load demand exceeded the available PV power, the PV inverter controls injected VARs and as a result increased the V_{bus}. This grid service was not limited to day-time operations; the PV inverters in the simulation environment provide VARs during the night to maintain the appropriate voltage on the bus [30].

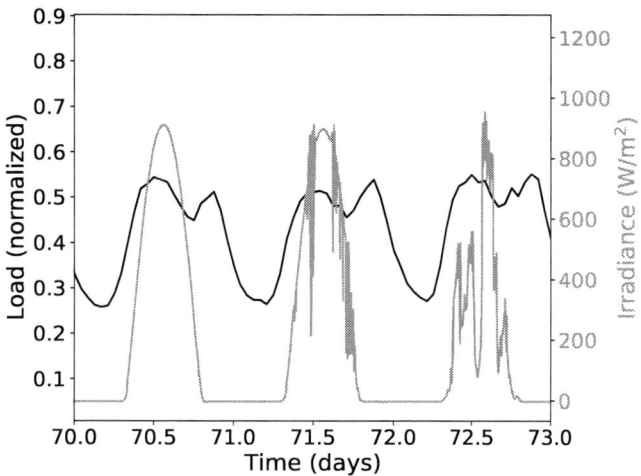

Fig. 4. The simulation effort includes realistic load and PV operations. The loads throughout the day are based on a typical load profile. The PV power generation is based on a mathematical model that considers actual irradiance.

2) Cyber-Attack & Resilient Control: A cyber-attack that prevents aggregator control signals from reaching PV inverters, using various methods including a MitM attack, (as depicted in Figs. 5 and 6) can result in non-optimal VVC settings. The unwanted situation, where the characteristics of the VVC settings have incorrect values can cause undesired operations. In this experiment the aggressiveness of the VVC to provide VARs in each PV inverter was set to be 60% of the rated power so that the voltage was maintained within the ANSI standard on each of the buses. However, when subjected to a cyber-attack that altered or stopped the communications between the aggregator and its PV inverters, the aggressiveness cannot be corrected or updated. In this case, the VVC was set to an undesirable 1% of rated power during the cyber-attack.

The on-board learning algorithm learned from its own performance data and data from its peers to define the correct VARs that support the grid until an operator or network technicians can rectify the issue. The experiment tested the ability of the LAPART algorithm to operate under the communications scenarios depicted in Figs. 5 and 6. The evaluation also reviewed the algorithms ability to control its VARs without help from its peers, which is not depicted in a diagram.

The first scenario, tested in the current work, predicted the VARs based on local data only. In this case, the two PV inverters (1 and 4) lost communications with the aggregator and the nearby neighbors. The analysis, performed by LAPART, used the following inputs to predict the VARs:

$\Rightarrow$ Scenario 1A: $\quad Q_1 = f(P_1, Q_1, V_4)$

$\Rightarrow$ Scenario 4A: $\quad Q_4 = f(P_4, Q_4, V_5)$

where Q_x is the amount of VARs, P_x is the real power, and V_y is the bus voltage for the respective PV inverters x and bus voltage y.

The second scenario, shown in Fig. 5, subjected PV in-

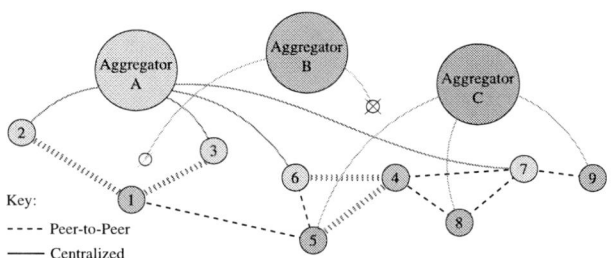

Fig. 5. Depiction of the system communications and the loss in communications from a centralized location to inverters 1 and 4. The two inverters used data from neighbors on the same bus.

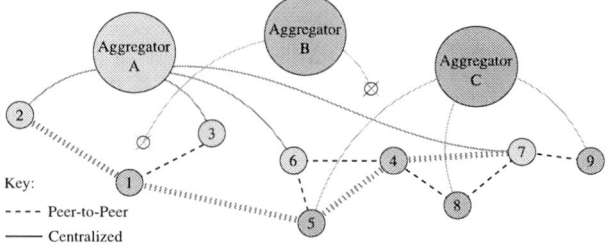

Fig. 6. Similar to loss in communications to Fig. 5, but this time the two inverters used data from neighbors on different buses to inform their control actions.

978-1-6654-6060-6/23 $31.00 © 2023 IEEE

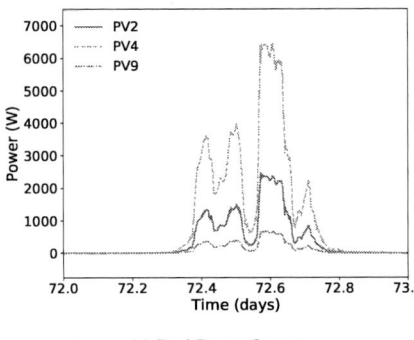
(a) Real Power Output

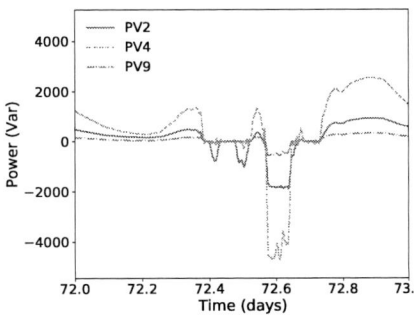
(b) Reactive Power Output

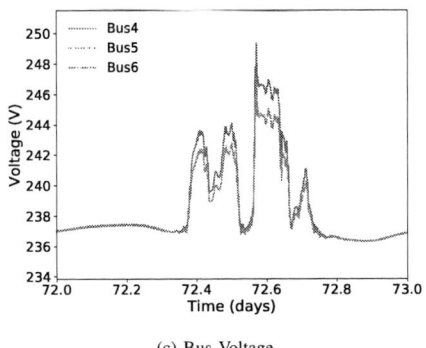
(c) Bus Voltage

Fig. 7. The model provides realistic power generation values for each of the inverters. This plot shows the power output of inverters located on the three different buses on a cloudy day. The simulation environment accurately modeled the reactive power injection or absorption for the 9 PV inverters. The amount of Q depends on the system's capacity and the amount of PV generation. The bus voltage fluctuated with the amount of solar generation and remained below the 250 V with the help of the reactive power control.

verters 1 and 4 to an attack that disabled Aggregator B's communication signals so that it could not update or change the VVC characteristics. Without centralized oversight, the on-board control used what it could from its peers to determine the best VARs for the PV inverter. In this case, PV inverter 1 learned from PV inverters 2 and 3, which were controlled by Aggregator A and resided on the same bus. Similarly, PV inverter 4 learned from PV inverter 6, controlled by Aggregator A, and PV inverter 5, connected to Aggregator C. The three PV inverters (4, 5, and 6) were all on the same bus. This learning scenario is represented by the following functions:

⇒ Scenario 1B: $Q_1 = f(Q_2, Q_3, V_4)$

⇒ Scenario 4B: $Q_4 = f(Q_5, Q_6, V_6)$

where Q_1 was the amount of VARs for the PV inverter under attack, Q_2 and Q_3 are its peer's VARs, and V_4 was the bus voltage for each of the PV inverters (1, 2, and 3). The prediction of PV inverter 4 VARs (Q_4) was based on inputs Q_5, Q_6, V_6 that represent the VARs from PV inverters 5 and 6 and the bus voltage respectively.

Similar to the second scenario, the third case depicted in Fig. 6) responded to an attack that disableed communications from Aggregator B to PV inverters 1 and 4. In this situation, the algorithm learned from PV inverters that reside on the same and different buses. PV inverter 1, for instance, was connected to PV inverter 5, which was on a different bus, while PV inverter 2 was on the same bus. The same was true for PV inverter 4 that communicated with 5 and 7. PV inverter 5 was on the same bus as 4 but PV inverter 7 was not. The inputs used by the algorithm to predict the individual PV inverters VARs were as follows:

⇒ Scenario 1C: $Q_1 = f(Q_2, Q_5, V_4, V_5)$

⇒ Scenario 4C: $Q_4 = f(Q_5, Q_7, V_5, V_6)$

where Q_1 was the amount of VARs for the PV inverter under attack, Q_2 and Q_5 are its peers, and V_4 and V_5 were the bus voltage for each of the peers. PV inverter 4's VARs (Q_4) was based on inputs $Q_5, Q_7, V_5,$ and V_6 that represent the VARs from PV inverters 5 and 6 and the bus voltages for each respectively.

IV. RESULTS

The results section summarizes the distribution system simulation outputs, reviews the dependence of Volt-VAR control for the simulated feeder, evaluates the accuracy of the LAPART algorithm, and discusses the impact of the proposed resilient control methodology on the grid's performance.

A. Simulation Feeder Performance

The simulation of the electric feeder in MATLAB/Simulink successful emulated the P, VARs, and V for each PV inverter and bus. For example, three PV inverters (PV2, PV4, and PV9) that reside on three different buses had realistic power generation profiles that followed the sun as shown over a one day period in Fig. 7(a). The injected or absorbed VARs provided by each of the PV inverters on the three different buses resembled realistic operations as shown in Fig. 7(b). Fig. 7(b) showed that the inverters provide VARs support during day-light and night-time hours. The amount of VARs depended on the system's capacity and the amount of PV generation. The bus voltage, shown for the same single day period in Fig. 7(c), varied based on the amount of P generated by the PV systems. The PV inverters maintained a voltage below 252 V using the VARs support functions.

B. LAPART Accuracy

The assessment of the LAPART algorithm's accuracy for this application included two review methods. The first used the K-Folds process to evaluate the R^2 of the predictions versus the actual values at different free parameter settings. The second tested the algorithm on previously unseen data using the best free parameter combination.

1) Find Parameters using K-Folds: The accuracy of the algorithm varied based on the value of the pre-set free parameters, ρA and ρB. The K-Folds method subjects the algorithm to multiple variations of the free parameters to identify the best combination and evaluate its overall performance. In this experiment, the R^2 value provided an overall score for each of the K-Fold iterations. The test subjected the algorithm to the six input scenarios, described in Section III-C2, used to compute the VARs for PV inverters 1 and 4. Fig. 8 provides an

978-1-6654-6060-6/23 $31.00 © 2023 IEEE

example three dimensional mesh of all the average R^2 scores of the LAPART algorithm outputs for each combination of the free parameters when processing Scenario 4B inputs to predict the appropriate VARs.

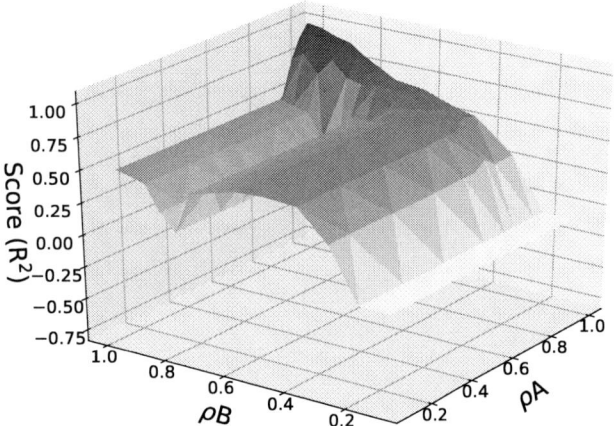

Fig. 8. The reactive power predication for PV inverter 4 using an PV inverter on the same bus (PV inverter 5) and on a different bus (PV inverter 7) resulted in high R^2 values near 1 when the free parameters where close to 1. The score got worse as the ρB parameter decreased to zero.

The R^2 score varied between -0.5 and 0.99 for each of the scenarios. A high R^2 value greater than or equal to 0.99 was achieved using ρA and ρB equal to 0.99 and 0.99 for Scenarios 1B, 4B, and 4C, as described in Table IV. Scenarios 1A and 4A, that used the PV inverter's internal data only, as expected, produced the lowest R^2 values at 0.94 and 0.86, respectively. Based on these results, the final testing of the LAPART algorithm's accuracy for each scenario used ρA equal to 0.99 and ρB equal to 0.99.

TABLE IV
GOODNESS OF FIT FOR EACH SCENARIO

PV Inverter #	Scenario	ρA	ρB	R^2
	1A	0.99	1.00	0.940
1	1B	0.99	0.99	0.998
	1C	0.99	1.00	0.995
	4A	0.95	0.90	0.860
4	4B	0.99	0.99	0.995
	4C	0.99	0.99	0.991

2) One-Off Evaluation: The final step in the accuracy evaluation considered the algorithms R^2 and absolute error on previously unseen data. The LAPART algorithm used previous created A, B, and L matrices from the 4th fold in the K-Folds training that used ρA = 0.99 and ρB = 0.99 and tested on 6 days of data. As a result the predicted VARs for Scenarios 1A, 1B, and 1C generated R^2 values equal to 0.896, 0.997, and 0.983 respectively. Similarly, Scenarios 4A, 4B, and 4C tests resulted in R^2 values of 0.867, 0.992, and 0.986 respectively.

The absolute error results followed similar patterns with the R^2 evaluation. The algorithm performed the worst with Scenario A inputs that used data from internal monitoring sources only. PV inverter 1 had an absolute error that was around 273 and PV inverter 4's error was at 484 for Scenario

A. However, the Scenario B inputs, where external data from peers on the same bus, produced better results with errors at 58 and 119 for PV inverters 1 and 4 respectively. Scenario C, that used external data from peers located on the same and different buses, had error slightly higher than Scenario B at 83 and 170 for PV inverters 1 and 4, respectively.

C. Volt-Var Control Impacts

The high penetration of PV on the simulation feeder required Volt-Var control to maintain bus voltages. Often, the PV inverters had to absorb VARs to lower the bus voltage because of the high level of generation in comparison to the load on the feeder. This is evident in Fig. 9, where the voltage exceeded the ANSI standard of 252 V without the Volt-Var control during each of the three days plotted. Over the course of the entire 80 day simulation, the no control case average voltage was about 249 V as shown in Fig. 10. In contrast, the

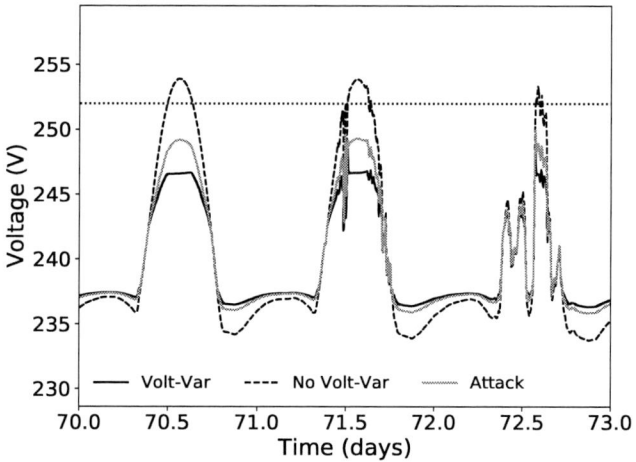

Fig. 9. The voltage exceeded the ANSI standard of 252 V in the uncontrolled case. The simulation with Volt-Var control limited the voltage to 246 V. An attack caused two inverters to not provide Volt-Var support.

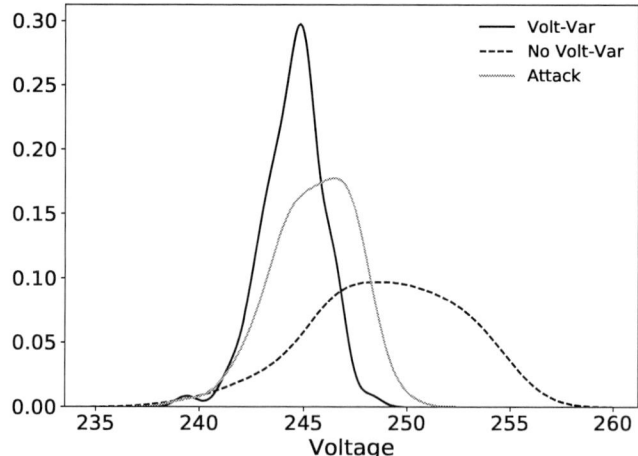

Fig. 10. The simulation that includes Volt-Var control maintains a bus voltage far below the ANSI limit of 252 V. However, the simulation that uses no control the voltage exceeded the ANSI standard in many instances. The attack scenario, that rendered two PV inverters useless, caused the bus voltage to rise but did not exceed the 252 volt limit.

simulation with the Volt-Var control enabled in each of the PV inverters the voltage never exceeded 247 V over the three day period in Fig. 9. Also, the average voltage over the 80 days was around 245 V, a standard deviation that was much lower than the no control case, and it never exceeded 250 V, as shown in Fig. 10.

The simulated cyber-attack changed the VVC for two of the nine PV inverters on different buses, which resulted in an undesirable increase in voltage. For example, the maximum voltage went from about 246 V in the complete control case to 249 V in the attacked scenarios over the three day period in Fig. 9. The distribution of the bus voltages for the cyber-attack scenario over the 80-day period had an average value of about 246 V and a standard deviation that was slightly larger than the control case but much smaller than the no control simulation results (Fig. 10).

The LAPART algorithm provided resilient control of PV inverters 1 and 4. The algorithm was deployed inside the two inverters within the MATLAB/Simulink environment. The

nine PV inverters in the simulation received a signal that enabled the VVC each day from their respective aggregators. Fig. 11 shows results for the 79th day of the simulation. On this day Aggregator B could not send the signal to PV inverters 1 and 4. As a result, the PV inverters were not able to provide Volt-Var support. The attacked caused the voltage to increase above the baseline simulation, that included VVC control, by about 1.5 V at the peak as shown in Fig. 12. The implementation of the LAPART algorithm to provide resilient control improved operations, however.

The LAPART algorithm countered the cyber-attack by defining the optimal control for the PV inverter. Fig. 11 shows the scenario where the algorithm controlled the VARs in PV inverters 1 and 4 based on data from peers on the same bus. Fig. 12 shows that the proposed mitigation controls resulted in a voltage that was closer to nominal then the baseline case where the typical VVC was used. The successful deployment of the algorithm into the simulation environment provided evidence that it can accurately predict a PV inverters VARs during a cyber-attack to maintain desired grid operations.

V. CONCLUSION

The analysis of the LAPART-based PV inverter control system involved a review of its accuracy and a realistic evaluation of its impact on the feeders performance. To evaluate the algorithm's accuracy at different free parameters, the full paper will describe results from a K-Folds process to define the R^2 value for various training and testing scenarios and free parameter conditions. This analysis showed that the LAPART algorithm can accurately represent the system and produced R^2 values well above 0.9.

Then, to provide evidence that the approach can positively impact the grid's performance, the algorithm was deployed into inverters within the simulation environment that emulated normal and attack conditions. During an attack the algorithm controlled PV inverters 1 and 4 using three different data availability situations.

The LAPART algorithm could determine the appropriate VARs with high accuracy given data from itself and its peers on the same and different buses. This is evident in the K-Folds assessment process that evaluate the algorithms ability to predict VARs at different free parameters and at the best free parameters. Additionally, the integration of the algorithm into the simulation showed a slight improvement in the bus voltage over a six day period. However, the work does not end here.

This paper describes an initial and successful implementation of the algorithm in a simulation environment. The next step is to deploy the methodology into simulations that include more buses and nodes and get a more realistic evaluation of the approach. Then, if successful within larger systems, the plan is to test the approach using actual hardware and perform extensive testing and development within a hardware-in-the-loop experiment. This future work will test its ability work with IDS, learn actual PV inverter operations in the field, operations in small computers at the edge of the grid, test its ability to perform on-line learning, and others.

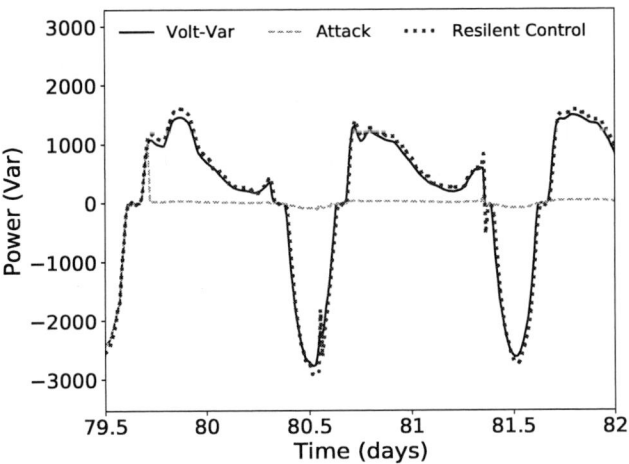

Fig. 11. The cyber-attack causes the impacted PV inverters to not provide Volt-Var support. The LAPART algorithm and the peer-to-peer data sharing provided a reactive power output that matched with the actual.

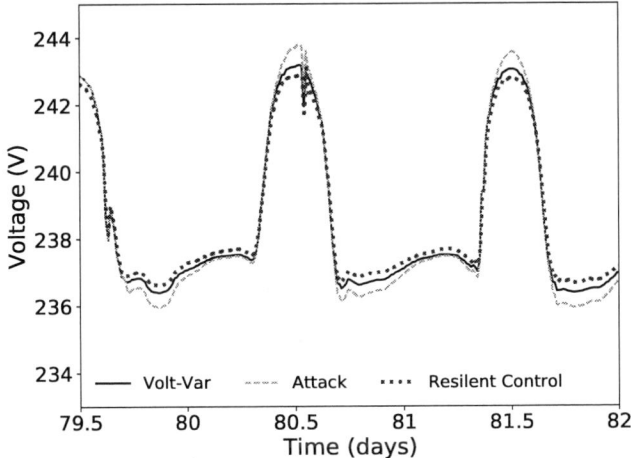

Fig. 12. The attack on the two PV inverters controlled by Aggregator B caused the voltage to deviate from the controlled case. The LAPART predicted a reactive power that resulted in a voltage closer to normal.

ACKNOWLEDGMENT

This article has been authored by an employee of National Technology & Engineering Solutions of Sandia, LLC under Contract No. DE-NA0003525 with the U.S. DOE. The employee owns all right, title and interest in and to the article and is solely responsible for its contents. The U.S. Government retains and the publisher, by accepting the article for publication, acknowledges that the U.S. Government retains a non-exclusive, paid-up, irrevocable, world-wide license to publish or reproduce the published form of this article or allow others to do so, for U.S. Government purposes. The DOE will provide public access to these results of federally sponsored research in accordance with the DOE Public Access Plan.

This material is based upon work supported by the U.S. Department of Energy's Office of Energy Efficiency and Renewable Energy (EERE) under the Solar Energy Technology Office (SETO), Award Number DE-EE00034234.

REFERENCES

[1] M. Calais, J. Myrzik, T. Spooner, and V. G. Agelidis, "Inverters for single-phase grid connected photovoltaic systems-an overview," in *2002 IEEE 33rd Annual IEEE Power Electronics Specialists Conference. Proceedings (Cat. No.02CH37289)*, vol. 4, Jun. 2002, pp. 1995–2000 vol.4.

[2] B. Crăciun, T. Kerekes, D. Séra, and R. Teodorescu, "Overview of recent Grid Codes for PV power integration," in *2012 13th International Conference on Optimization of Electrical and Electronic Equipment (OPTIM)*, May 2012, pp. 959–965.

[3] J. Johnson, "Johnson J. Roadmap for photovoltaic cyber security," Sandia National Laboratories, Sandia Technical Report SAND2017-13262, 2017.

[4] J. Qi, A. Hahn, X. Lu, J. Wang, and C.-C. Liu, "Cybersecurity for distributed energy resources and smart inverters," *IET Cyber-Physical Systems: Theory & Applications*, vol. 1, no. 1, pp. 28–39, Dec. 2016, tex.bdsk-url-2: https://doi.org/10.1049/iet-cps.2016.0018 tex.date-modified: 2023-01-13 16:42:11 -0700.

[5] A. Teymouri, A. Mehrizi-Sani, and C. Liu, "Cyber Security Risk Assessment of Solar PV Units with Reactive Power Capability," in *IECON 2018 - 44th Annual Conference of the IEEE Industrial Electronics Society*, Oct. 2018, pp. 2872–2877.

[6] C. Carter, I. Onunkwo, P. Cordeiro, and J. Johnson, "Cyber Security Assessment of Distributed Energy Resources," in *2017 IEEE 44th Photovoltaic Specialist Conference (PVSC)*, Jun. 2017, pp. 2135–2140.

[7] C. B. Jones, A. R. Chavez, R. Darbali-Zamora, and S. Hossain-McKenzie, "Implementation of Intrusion Detection Methods for Distributed Photovoltaic Inverters at the Grid-Edge," in *2020 IEEE Power & Energy Society Innovative Smart Grid Technologies Conference (ISGT)*, Feb. 2020, pp. 1–5, iSSN: 2472-8152.

[8] C. B. Jones, A. Chavez, S. Hossain-McKenzie, N. Jacobs, A. Summers, and B. Wright, "Unsupervised Online Anomaly Detection to Identify Cyber-Attacks on Internet Connected Photovoltaic System Inverters," in *2021 IEEE Power and Energy Conference at Illinois (PECI)*, Apr. 2021, pp. 1–7.

[9] J. Johnson, C. B. Jones, A. Chavez, and S. Hossain-McKenzie, "SOAR4DER: Security Orchestration, Automation, and Response for Distributed Energy Resources," in *Power Systems Cybersecurity: Methods, Concepts, and Best Practices*, ser. Power Systems, H. Haes Alhelou, N. Hatziargyriou, and Z. Y. Dong, Eds. Cham: Springer International Publishing, 2023, pp. 387–411.

[10] M. Conoscenti, A. Vetrò, and J. C. D. Martin, "Blockchain for the Internet of Things: A systematic literature review," in *2016 IEEE/ACS 13th International Conference of Computer Systems and Applications (AICCSA)*, Nov. 2016, pp. 1–6.

[11] P. Bellavista and A. Zanni, "Towards better scalability for IoT-cloud interactions via combined exploitation of MQTT and CoAP," in *2016 IEEE 2nd International Forum on Research and Technologies for Society and Industry Leveraging a better tomorrow (RTSI)*, Sep. 2016, pp. 1–6.

[12] B. Kang, P. Maynard, K. McLaughlin, S. Sezer, F. Andrén, C. Seitl, F. Kupzog, and T. Strasser, "Investigating cyber-physical attacks against IEC 61850 photovoltaic inverter installations," in *2015 IEEE 20th Conference on Emerging Technologies Factory Automation (ETFA)*, Sep. 2015, pp. 1–8.

[13] N. Karthikeyan, B. R. Pokhrel, J. R. Pillai, and B. Bak-Jensen, "Coordinated voltage control of distributed PV inverters for voltage regulation in low voltage distribution networks," in *2017 IEEE PES Innovative Smart Grid Technologies Conference Europe (ISGT-Europe)*, Sep. 2017, pp. 1–6.

[14] R. A. Jabr, "Robust Volt/VAr Control With Photovoltaics," *IEEE Transactions on Power Systems*, vol. 34, no. 3, pp. 2401–2408, May 2019.

[15] W. Peng, C. Hicks, O. Gonzalez, B. Blackstone, and Y. Baghzouz, "Experimental test on some autonomous functions of advanced PV inverters," in *2016 IEEE Power and Energy Society General Meeting (PESGM)*, Jul. 2016, pp. 1–5.

[16] J. Johnson, A. Summers, R. Darbali-Zamora, J. Hernandez-Alvidrez, J. Quiroz, D. Arnold, and J. Anandan, "Distribution Voltage Regulation Using Extremum Seeking Control With Power Hardware-in-the-Loop," *IEEE Journal of Photovoltaics*, vol. 8, no. 6, pp. 1824–1832, Nov. 2018.

[17] M. Jafari, T. O. Olowu, and A. I. Sarwat, "Optimal Smart Inverters Volt-VAR Curve Selection with a Multi-Objective Volt-VAR Optimization using Evolutionary Algorithm Approach," in *2018 North American Power Symposium (NAPS)*, Sep. 2018, pp. 1–6.

[18] D. Shelar, J. Giraldo, and S. Amin, "A distributed strategy for electricity distribution network control in the face of DER compromises," in *2015 54th IEEE Conference on Decision and Control (CDC)*, Dec. 2015, pp. 6934–6941.

[19] C. B. Jones, J. S. Stein, S. Gonzalez, and B. H. King, "Photovoltaic system fault detection and diagnostics using Laterally Primed Adaptive Resonance Theory neural network," in *2015 IEEE 42nd Photovoltaic Specialist Conference (PVSC)*, Jun. 2015, pp. 1–6.

[20] A. Mammoli, A. Ellis, A. Menicucci, S. Willard, T. Caudell, and J. Simmins, "Low-cost solar micro-forecasts for PV smoothing," in *2013 1st IEEE Conference on Technologies for Sustainability (SusTech)*, Aug. 2013, pp. 238–243.

[21] C. B. Jones and C. Carter, "Trusted Interconnections Between a Centralized Controller and Commercial Building HVAC Systems for Reliable Demand Response," *IEEE access : practical innovations, open solutions*, vol. 5, pp. 11 063–11 073, 2017.

[22] C. B. Jones, C. Carter, and Z. Thomas, "Intrusion Detection Response using an Unsupervised Artificial Neural Network on a Single Board Computer for Building Control Resilience," in *2018 Resilience Week (RWS)*, Aug. 2018, pp. 31–37.

[23] M. J. Healy, T. P. Caudell, and S. D. G. Smith, "A neural architecture for pattern sequence verification through inferencing," *IEEE Transactions on Neural Networks*, vol. 4, no. 1, pp. 9–20, Jan. 1993.

[24] M. J. Healy and T. P. Caudell, "Acquiring rule sets as a product of learning in a logical neural architecture," *IEEE Transactions on Neural Networks*, vol. 8, no. 3, pp. 461–474, May 1997.

[25] G. A. Carpenter, S. Grossberg, and D. B. Rosen, "Fuzzy ART: Fast stable learning and categorization of analog patterns by an adaptive resonance system," *Neural Networks*, vol. 4, no. 6, pp. 759–771, Jan. 1991, tex.bdsk-url-1: https://doi.org/10.1016/0893-6080(91)90056-B tex.date-modified: 2023-01-13 16:41:38 -0700.

[26] K. P. Murphy, *Machine Learning: A Probabilisitc Perspective.* Cambridge, MA: MIT Press, 2012.

[27] R. Darbali-Zamora, J. E. Quiroz, J. Hernández-Alvidrez, J. Johnson, and E. I. Ortiz-Rivera, "Validation of a Real-Time Power Hardware-in-the-Loop Distribution Circuit Simulation with Renewable Energy Sources," in *2018 IEEE 7th World Conference on Photovoltaic Energy Conversion (WCPEC) (A Joint Conference of 45th IEEE PVSC, 28th PVSEC 34th EU PVSEC)*, Jun. 2018, pp. 1380–1385.

[28] A. Safdarian, M. Fotuhi-Firuzabad, and M. Lehtonen, "Optimal Residential Load Management in Smart Grids: A Decentralized Framework," *IEEE Transactions on Smart Grid*, vol. 7, no. 4, pp. 1836–1845, Jul. 2016.

[29] R. Darbali-Zamora and E. I. Ortiz-Rivera, "Optimal duty ratio maximum power point tracking technique using the SEPIC topology for photovoltaic systems applications," in *2016 IEEE ANDESCON*, Oct. 2016, pp. 1–4.

[30] A. Maknouninejad, N. Kutkut, I. Batarseh, and Zhihua Qu, "Analysis and control of PV inverters operating in VAR mode at night," in *ISGT 2011*, Jan. 2011, pp. 1–5.

978-1-6654-6060-6/23 $31.00 © 2023 IEEE

Improvement of radiation tolerance in solar cells by hetero p/n junction structure

Tetsuya Nakamura, Mitsuru Imaizumi, Meita Asami, Masakazu Sugiyama, Hidefumi Akiyama, Shin-ichiro Sato, Takeshi Oshima, Yoshitaka Okada

Japan Aerospace Exploration Agency, Tsukuba, Japan

Department of Advanced Interdisciplinary Studies, Graduate School of Engineering, The University of Tokyo, Tokyo, Japan

Research Center for Advanced Science and Technology, The University of Tokyo, Tokyo, Japan

Institute for Solid State Physics, The University of Tokyo, Kashiwa, Japan

AIST-UTokyo OPERANDO-OIL, The University of Tokyo, Kashiwa, Japan

National Institutes for Quantum Science and Technology, Takasaki, Japan

This paper describes the radiation effect of hetero p/n junction solar cells that can control the influence of native defects on the electrical properties of solar cells. We obtained the radiation damage coefficients for the carrier lifetime in the depletion region of GaAs p/n homojunction and n-InGaP/p-GaAs heterojunction solar cells by simulations and experiments. Our simulations and experiments reveal that the hetero p/n junction structure can also control the effects of radiation defects in the depletion region on the effective minority-carrier lifetime. In other words, by designing an optimal hetero p/n junction structure, hetero p/n junction solar cells can be expected to improve not only the initial efficiency but also the radiation tolerance.

978-1-6654-6060-6/23 $31.00 © 2023 IEEE

Abnormal Responses of Residential Smart Photovoltaic Inverters to Cyberattacks

Thunchanok Kaewnukultorn[1], Sergio B. Sepúlveda-Mora[2], and Steven Hegedus[1]

[1]Institute of Energy Conversion & Department of Electrical and Computer Engineering, University of Delaware, Newark, DE, 19716, USA

[2]Departamento de Electricidad & Electrónica, Universidad Francisco de Paula Santander, Cúcuta, Norte de Santander, 540006, Colombia

Abstract— **Smart PV inverters will play a significant role in supporting the electric grid and mitigating the effect of high solar energy penetration. Since smart inverters provide communication capabilities with real–time grid control and monitoring, they could be vulnerable to cyberattacks. Understanding the responses of smart inverters to cyberattacks is essential for system operators and PV owners to protect the system from malicious attacks. In this work, we designed and demonstrated two intermittent cyberattacks on two smart PV inverters from different manufacturers to observe their responses in terms of real and reactive power output. The results show that different inverters have their unique response during and after the cyberattacks which need to be recovered using different methods.**

Keywords—smart inverters, grid instability, inverter responses, cyberattacks, photovoltaic

I. INTRODUCTION

Renewable energy systems have been increasingly integrated into the centralized electric grid during the last decade. Solar energy has shown the fastest growth and is expected to account for 35% of global electricity by 2050 [1]. Smart photovoltaic (PV) inverters have been introduced to alleviate the effect of high solar energy penetration and to address grid instability issues by controlling voltage and frequency and serving residential loads during power outages [2]. Network communications for smart grid devices provide higher system flexibility and reliability with real–time monitoring, grid abnormality detection, and centralized control. However, the synchronous controls and embedded intelligence could lead to vulnerability in smart inverters from exchanging information with other devices under the same communication network [3]. Consequently, it is important to study the impact of cyberattacks on smart inverters and analyze their responses to malicious controls which will be useful for developing cyberthreat detection models to protect and secure the grid.

To the best of our knowledge, there have been very few studies conducted to observe the behaviors of residential PV inverters from different manufacturers under cyberattacks using actual hardware equipment. Our work focuses on identifying the signatures of smart PV inverter's responses using a power hardware–in–the–loop (P–HIL) system that includes residential PV inverters comparable to the ones installed in the field. The goal is to identify unexpected reactions of the PV inverters to provide guidance for the grid operator and system owners when the inverters were subjected to various hostile attacks.

II. EXPERIMENTAL SETUP

A. P–HIL laboratory configuration

A P–HIL laboratory has been installed at the Institute of Energy Conversion, whose configuration is shown in Fig. 1 [4]. Two commercially available PV inverters from different brands are tied to the grid simulator supplied by the actual electric grid in the building. Each inverter is connected to its own PV power supply. As we focus on evaluating the responses of PV inverters, we excluded from this analysis any cyberattacks on the power meter, automatic backup unit, battery inverter, and Li–ion battery. Grid supporting functionalities and inverter efficiency tests were performed and reported in [4] which also described the power and communication diagram of P–HIL testbed in elaborated details. The grid supporting functions of the two inverters have already been tested based on IEEE 1547–2018 standard [5]. Two inverter brands will be labeled as Inverter A and B to avoid sensitive cyberattack information disclosure.

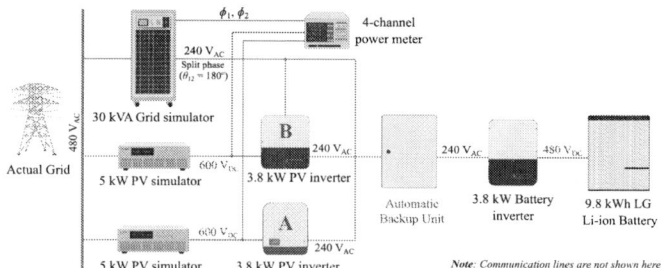

Fig. 1. A power diagram of the P–HIL laboratory with voltage levels annotated on power line.

B. Cyberattack simulation and data acquisition

The two types of cyberattacks, namely intermittent active (P) and reactive (Q) power were designed and simulated based on

978-1-6654-6060-6/23 $31.00 © 2023 IEEE

their possibilities of either causing grid instability or sabotaging inverter operations. A description of each cyberthreat is shown in Table I. Two residential inverter controls were attacked simultaneously via Modbus communication. We used threading in Python to execute the dual attacks and obtain real time measurements. We set the DC power of the PV simulator at 3800 W_{DC} for all experiments which is expected to result in approximately 3650 W_{AC} output from the inverter during normal operation. We reset the power factor (PF) to 1 after cyberattacks so each inverter will be expected to operate at the full power without outputting any reactive power. Note that inverter B was not involved in the intermittent P attack due to a restriction of cyclical changes by the manufacturer as discussed below.

TABLE I

DESCRIPTION OF THE CYBERATTACKS AND PROCESS OF DATA ACQUISITION

Type of attack	Cyberattack description
Intermittent P	Changing P from 100% to 5% of P_{MAX} back and forth every 3 seconds (Fig. 2) and every second (Fig. 3) for 120 seconds. Grid voltage is always operating at 120 V during the attack.
Intermittent Q	Changing PF from 0.85 underexcited (negative or absorbing Q) to overexcited (positive or injecting Q) back and forth every second for 90 seconds. Grid voltage is always operating at 120 V during the attack.

III. RESULTS AND DISCUSSION

A. Intermittent P attack

In this experiment, the intermittent P attack has been launched to inverter A for 120 seconds. When the inverter control was attacked and forced to oscillate from 3600 W to 180 W every 3 seconds, the inverter responded by changing the active power output cyclically every 3 seconds. However, as shown in Fig. 2, the time for stabilization as the designated power level was greater than 3 seconds causing the inverter active power to bounce back and forth between $1500 - 2500$ W. Thus, it failed to reach either the maximum or minimum power during the attack. There was 12.5% of the total attack duration where the inverter was able to reach the full power.

We repeated this type of intermittent P attack but with a shorter attack interval; i.e. changing the control every second from 100–5–100% of the inverter maximum power. As demonstrated in Fig. 3, during 120 seconds of attack, the inverter was never able to reach the desired active power level and the pattern of response was exposed more clearly compared to the 3–second changes attack. Note that both 3–second and 1–second test results have been validated in repeated experiments, with nearly identical results. Consequently, this cyberattack could lead to a drastic instability in active power output of the inverter, causing the PV owners to lose their revenue due to the lower total PV power being exported and/or supplied to the AC local loads. Additionally, this severe fluctuation of active power

could result in grid instability when multiple inverters that are tied to the same point of common coupling (PCC) are simultaneously attacked. Finally, we found that this type of attack will cause controls of inverter B to lock–up at some point and become stuck at a random output condition.

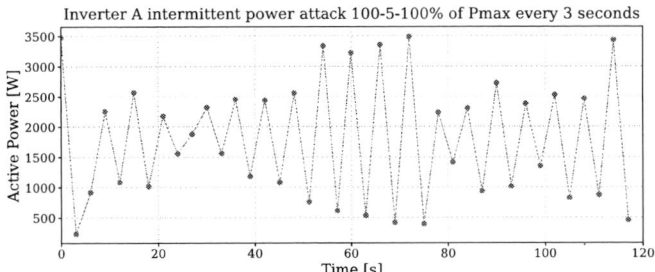

Fig. 2. Intermittent P attack on inverter A. The control was being changed every 3 seconds for 120 seconds.

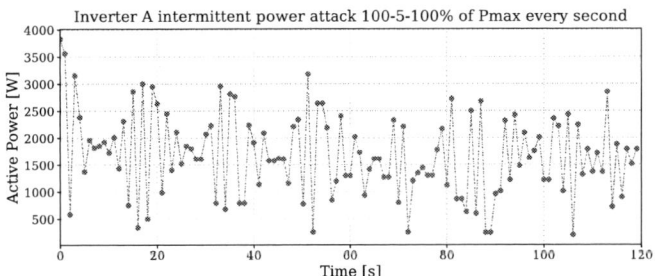

Fig. 3. Intermittent P attack on inverter A. The control was being changed every second for 120 seconds.

B. Intermittent Q attack

In addition to studying an attack on the real power, a reactive power attack was also evaluated. The intermittent Q attack was tested on inverter A and B simultaneously for 120 seconds. As described in Table I, we changed the constant PF of both inverters from –0.85 (–2300 VAR) to +0.85 (+2300 VAR) and vice versa every second. Different tests were created to end at different values. Fig. 4 shows the responses of inverter A (blue) and B (red) regarding the intermittent Q attack where the last instruction reset PF = 1. During the attack, inverter A was able to respond quickly to 1–second aggressive changes of PF by outputting high amount of reactive power periodically while inverter B reactive power only reached the designated amount of ±2300 VAR few times and the response is unpredictable. However, inverter B successfully recovered from the malicious control and operated at the unity PF after we cleared the attack, whereas inverter A was not able to control the reactive power back to zero and kept operating at the same setpoint of –0.85 even after the cyberattack had been properly removed.

Fig. 5 represents results from the second experiment of intermittent Q attack where we repeated the test but intentionally set the last point of the attack to be PF = +0.85. The result confirmed that inverter A has a potential to block its communication after several seconds of recurring changes to the PF controls. This post–attack response of inverter A could

978-1-6654-6060-6/23 $31.00 © 2023 IEEE

raise a concern on system operators not being able to clear the attack remotely or manually at the front panel, resulting in grid voltage violations and loss of PV revenue as the inverter will be forced to curtail active power to output high amounts of reactive power. We found the only solution is to disconnect inverter A from the grid and perform a factory–reset.

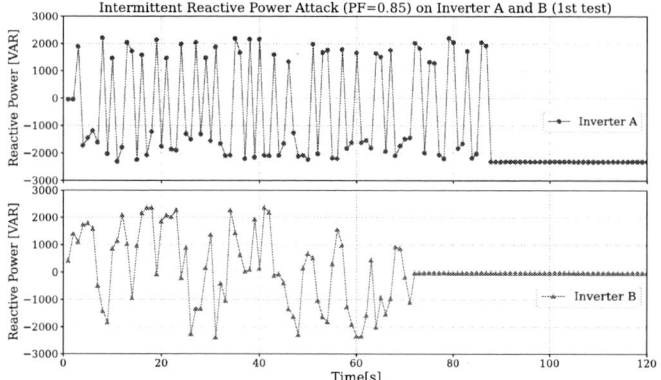

Fig. 4. Intermittent Q attack on inverter A (blue) and B (red) inverters. The control was being changed every second for 90 seconds using threading. The data collection was performed for 120 seconds.

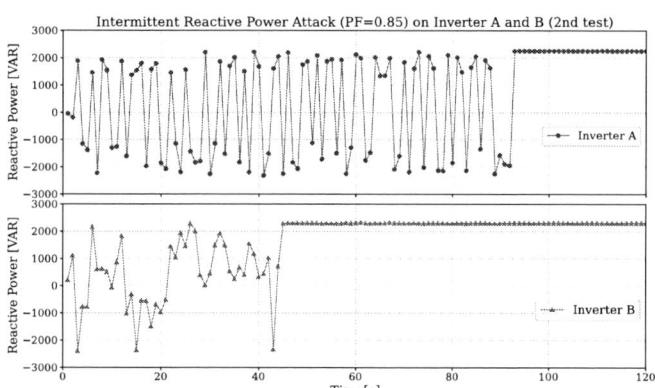

Fig. 5. Intermittent Q attack on inverter A (blue) and B (red) inverters. The control was being changed every second for 90 seconds using threading. The data collection was performed for 120 seconds.

Regarding inverter B, the second test in Fig. 5 showed that inverter B also has a post–attack response which is similar to inverter A, yet the reason leading to the fixed and loss of communication control is different. After being attacked for about 70 cycles (i.e. at 70th second), inverter B stopped listening to any parameter changes and remained operating at the last point before the control got locked. Thus, one similarlity between both PV inverters is that their grid controls prevent further malicious attacks by freezing the inverter setpoints and control modes to the last point before the remote communication is disconnected. According to the information provided by inverter B technical support, cyclical changes of the inverter controls could lead to a damage of an inverter chipset. Once the number of changes have reached that limit, the inverter will not receive any further commands remotely or manually. Lock–out on the inverter A does not have the same

cause but rather occurs as an after–attack symptom which is only specific to the intermittent reactive power attack. All experiments reported in this section are repeatable and the results were almost identical.

These findings indicate that cyclical cyberthreats on smart inverters could have a more significant impact than just their behavior during the attack. Their vulnerability extends to post–attack symptoms, requiring them to be reset from the grid operator or the inverter manufacturer manually or remotely. If multiple inverters tied to the same PCC are attacked, it could lead to inverter failure, grid instibility, and economic loss.

IV. CONCLUSIONS

Two types of intermittent attacks were tested on inverter A and B residential PV inverters to study the response during and after 90 second attack periods. Our key finding shows that despite their having met the same IEEE 1547–2018 standard, they can still have different responses during and after the attack, level of vulnerability and alternatives to recover from some simple cyberthreats. We believe that it will be crucial for the grid operator and PV owners to have appropriate detection, protection and recovery strategies to the intermittent cyberattacks. These results were obtained under standard stable grid conditions. Our future work will include inverter's responses in larger scale with multiple inverters via software simulation and a study of inverter controls and interactions under unstable electric grid conditions including other inverter brands.

ACKNOWLEDGMENT

This work was supported by the U.S. Department of Energy's Office Energy Efficiency and Renewable Energy (EERE) under agreement number DE–EE0008768. Any opinions, findings, and conclusions or recommendations expressed in this material are those of the authors.

REFERENCES

[1] R. K. Varma, Smart solar PV inverters with advanced grid supporting fucntionalities, 1st ed., Wiley, 2022.

[2] B. Mirafzal and A. Adib, On grid-interactive smart inverters: features and advancements, IEEE Access, September 2020. DOI: 10.1109/ACCESS.2020.3020965

[3] S. Zuo, O. A. Beg, F. L. Lewis, A. Davoudi, "Resilient networked AC microgrids under unbounded cyber attacks", IEEE Transactions on Smart Grid, 2020. DOI: 10.1109/TSG.2020.2984266

[4] T. Kaewnukultorn, S. B. Sepúlveda-Mora, and S. Hegedus, "Characterization of voltage stabilization functions of residential PV inverters in a power Hardware-in-the-loop environment", IEEE Access, November 2022. DOI: 10.1109/ACCESS.2022.3217472

[5] "IEEE Standard for Interconnection and Interoperability of Distributed Energy Resources with Associated Electric Power Systems Interfaces," *IEEE Std 1547-2018 (Revision of IEEE Std 1547-2003)*, 2018, DOI: 10.1109/IEEESTD.2018.8332112

Community Influence of Houses of Worship on Rooftop Solar Growth Rates

Ashley Degen, Laura Mogannam, Nisitaa Karen Clement Pradeep, Jillian Stern, and Deborah A. Sunter

Tufts University, Medford, MA, 02155, USA

Abstract — With a high upfront cost and only eventual return on investment of the technology, wealthier communities have adopted more rooftop solar than low- and moderate-income communities (LMI). Even after correcting for differences in income, there are observed racial and ethnic differences in rooftop solar installations. A better understanding of solar adoption influences is needed to reduce the observed disparities in rooftop solar across demographic lines. While studies suggest peer effects influence solar adoption, there is a gap in the literature investigating the influence of community institutions on solar installation rates. Of particular interest are houses of worship since they are trusted, community-level organizations that may be able to engage with a broader audience. In this study, the influence of houses of worship adopting rooftop solar on residential installation growth rates in the surrounding community were assessed. Shapiro-Wilk tests, Mood's median tests, and LOWESS models were used to draw conclusions about the differences in growth rates between census tracts in the United States with and without rooftop solar on a house of worship. The data was segmented by race and ethnicity and studied across demographic features including median household income, home ownership, and education level. Overall, the census tracts with a house of worship installation had higher growth rates for residential installations. However, results varied by racial and ethnic group when additional demographic features were included. Diverse, Hispanic, and White majority census tracts showed consistent, statistically significant (at a significance level of 0.05) higher solar adoption growth rates in communities with house of worship installs. However, Asian, and Black majority communities showed inconsistent patterns.

I. INTRODUCTION

While the economic benefits of rooftop solar may be transformative, these benefits have not been reaped evenly across demographic lines. Many methods of spreading information about solar panels are targeted toward a high-income population [1]. Expensive initial installation costs may be daunting for a lower-income community despite long-term economic benefits [1]. Even after correcting for income differences, Black- and Hispanic-majority census tracts have installed less rooftop PV compared to White-majority census tracts [2].

To better understand the disparities in solar adoption, it is important to study how information is spread. Past studies show that rooftop solar can be spread through neighbor-to-neighbor adoption, called peer effects. If low- and moderate-income (LMI) communities already have a local rooftop solar installation, other members of the community are more likely to follow suit [3-4]. For example, an installation in the same zip code increases the probability of more installations by 0.78%

[5] while an installation within 0.5 miles increases the probability of installations in the radius by 0.44% in the following six months [6].

Although studies suggest peer effects influence solar adoption, there is a gap in the literature investigating the influence of community institutions on solar installation rates. Houses of worship (HoW) are community-level organizations that may be able to engage with a broader audience. Members at the same HoW may already have shared values that can build trust in other aspects of their lives. Previous research has been conducted regarding the effect of faith-based organizations promoting HIV testing in African American and Latino congregations, as churches are seen as trusted sources within these communities [7]. Another study examined how HoWs can use the established trust within the community to promote behavior changes surrounding physical activity and improved nutrition [8].

To better understand the role that HoWs may play in both increasing rooftop solar adoption and reducing the observed demographic disparities in adoption rates, this paper aims to answer: *i)* do communities with solar installations on HoW have faster growth rates in rooftop solar adoption in the surrounding community, and *ii)* does the difference in growth rates between communities with and without HoW installations differ across demographic groups.

II. METHODOLOGY

The "Tracking the Sun" report from Lawrence Berkeley National Laboratory [9] contains data on the location and installation date of roughly 2.5 million grid-connected, distributed solar systems in the United States. To identify installations on HoWs, the LBNL dataset was merged geospatially with the Homeland Infrastructure Foundation Level Data (HIFLD) Open Data [10] on places of worship (e.g., churches, temples, and mosques). Any HIFLD address that was coded as a residential building by Melissa Data [11] was not categorized as a house of worship. Rooftop solar installations were grouped by census tract and each census tract was labeled as either having an installation on a HoW or not. The Google Project Sunroof data [12] on the total number of buildings that could support rooftop solar, using criteria based on space availability and shading, was used as the rooftop solar carrying capacity in each census tract.

The demographic features of household income, education level, homeownership, race, and ethnicity were considered due

978-1-6654-6060-6/23 $31.00 © 2023 IEEE

to their presence within the literature as features impacting rooftop solar adoption rates [2,13-15]. This data was collected from the 2019 5-year American Community Survey using data from tables B03002, B19013, S1501, and S2502 [16]. If 50% or more of a census tract self-identified as the same race or ethnicity, the tract was classified as being that majority race/ethnicity; otherwise, the tract was categorized as having no majority.

Rooftop solar growth rates for each census tract are calculated using the sigmoid curve formula:

$$y = \frac{c}{1 + ae^{-bt}} \qquad (1)$$

where y is the number of rooftop solar installations at time t, b is the uncorrected growth rate, and c is the carrying capacity from Google Project Sunroof. The constant a can be calculated as c divided by the number of rooftop solar installations at $t = 0$, which is the time of the first influential seed installation. The first influential seed installation occurs when the next installation in the census tract occurs within 2 years. Examples of growth curves can be seen in Fig. 1.

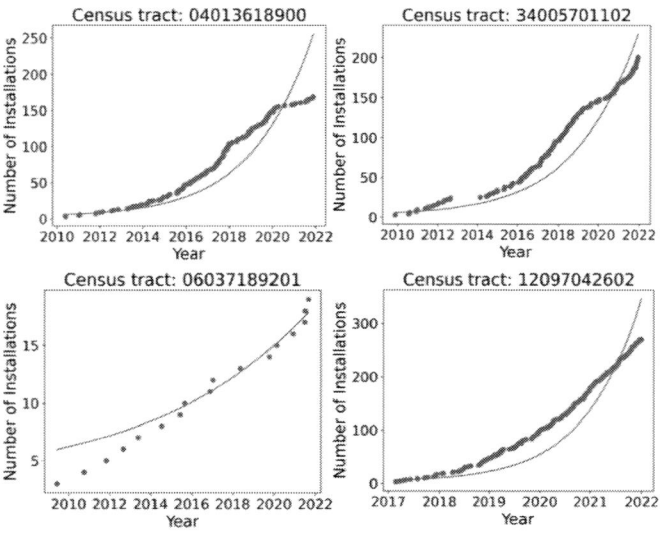

Fig. 1. Growth curves, shown in green, fitted to rooftop solar installations, shown in blue, for a random set of census tracts. These growth curves were calculated using Eqn. 1 to find the uncorrected growth rates, b.

Uncorrected growth rates were calculated using Python's SciPy optimizer curve.fit [17] and the accuracy of curve fits were measured by sMAPE. Census tracts with *i)* fewer than 10 installations, *ii)* sMAPE greater than 75, *iii)* Project Sunroof data coverage of less than 95% of the buildings, *iv)* invalid data entries (i.e., entries before 1980, entries in future dates, or dates that do not exist) or *v)* the median annual household income below the 2019 poverty threshold of $26,172 for a 4-person household [18] were excluded from the study.

There are confounding factors that must be considered when comparing growth rates nationally. Each state has different policies and incentives for solar energy that affects rooftop solar adoption rates [19]. Additionally, the costs of solar has dropped significantly in recent years [20]. To correct for these factors, a linear regression of the uncorrected growth rates (calculated in Eqn. 1) on the date of the first influential seed was fit for each state separately. The linear regression predicts the expected growth rate based on the date of the influential seed installation and the state. Then, each uncorrected growth rate was divided by the predicted growth rate to get the corrected growth rate to be used in subsequent analyses.

After segmenting the census tracts by majority race and ethnicity and grouping by census tracts with and without installations on HoWs, Shapiro-Wilk tests were used to determine if the distributions were normal. If so, Mood's median tests were performed, and p-values calculated.

LOWESS models were run for each of the three selected features — median income, homeownership, and high school education — against each census tract's corrected growth rates segmented by census tracts with and without at least one solar installation on a HoW for each racial/ethnic majority group. The optimal span for each model was found using 5-fold cross-validation based on the one-standard error rule of the lowest mean squared error. Ten spans from 0.1 to 1 in 0.1 increments were considered. The 95% confidence intervals of each model were found based on 100 bootstrap replicas.

III. RESULTS AND DISCUSSION

All census tracts studied, as well as each of the groups of census tracts separated by racial and ethnic majority, were found to have a non-normal distribution using the Shapiro-Wilk test at a significance level of 0.05. Therefore, we used the Mood's median test, continuing to apply a significance level of 0.05. The results of these tests can be seen in Fig. 2. When considering all census tracts collectively, we found statistically significant evidence that census tracts with rooftop solar installations on HoWs had higher median corrected rooftop solar growth rates in their surrounding community. When separated by racial and ethnic demographics, White majority, Hispanic majority, and No majority (diverse) census tracts had significantly higher median corrected solar growth rates, whereas Black and Asian majority census tracts did not. This suggests that proximity to a HoW with a solar installation may be correlated with increased residential solar adoption in White, Hispanic, and diverse communities.

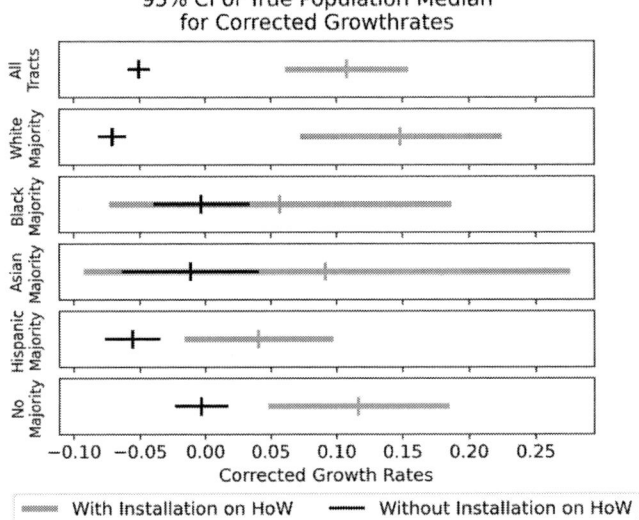

Fig. 2. The median corrected growth rates (vertical lines) and 95% confidence intervals (horizontal lines) in census tracts with (shown in red) and without (shown in black) at least one rooftop solar installation on a house of worship for all census tracts, as well as census tracts segmented by racial/ethnic majority.

The LOWESS models for corrected growth rates as a function of median income can be seen in Fig. 3. The optimal span was found to be 1.0. When all census tracts, No-Majority (diverse), Hispanic-majority, and White-majority tracts are considered (Figs. 3a,b,e,f), the 95% confidence intervals for census tracts with and without at least one solar installation on a HoW generally do not overlap with corrected growth rates being higher in census tracts with at least one solar installation on a HoW. There are some exceptions in regions with exceptionally low and high income; however, a limited number of census tracts can be found in those regions and the overlap may be a result of tail effects. This indicates that proximity to a HoW with solar installations may be correlated to higher adoption rates on overall solar rooftop adoption within the same census tract across most household incomes. In contrast, Black majority census tracts (Fig. 3d) only show a significant difference at household incomes below $40,000 and Asian majority census tracts do not show a significant difference at any household income. It is important to note that were substantially fewer Asian majority census tracts in this study (439 census tracts), when compared to the other races and ethnicities explored (11,341 White, 1,075 Black, 3,583 Hispanic and 4,038 No-Majority census tracts).

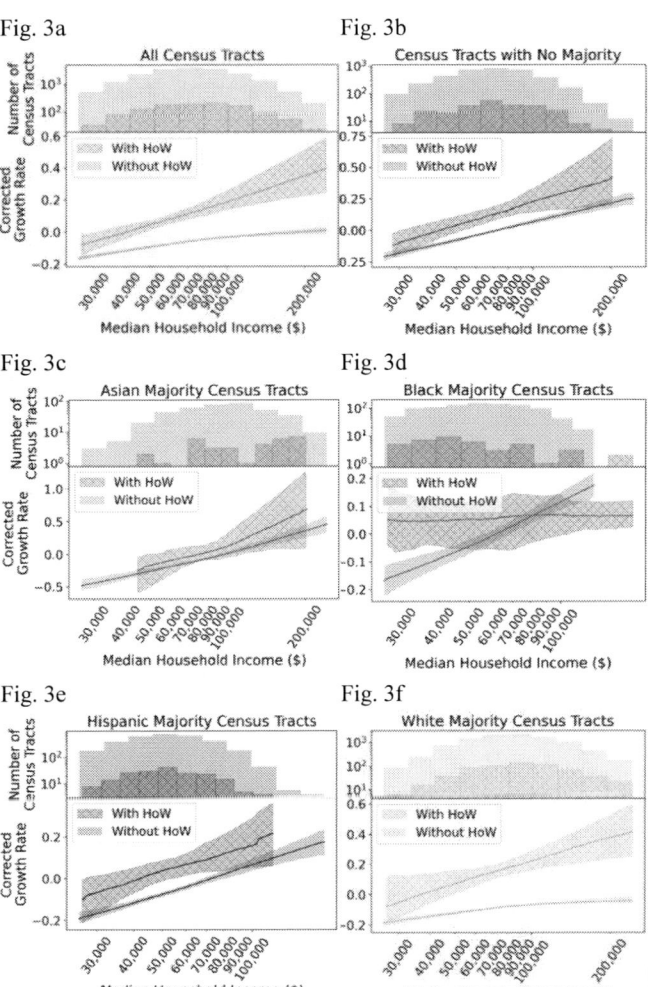

Fig. 3. Number of census tracts with and without houses of worship are plotted as a histogram above LOWESS models of corrected growth rates of solar installations as a function of median household income for (a) all census tracts studied, (b) tracts with no racial/ethnic majority, (c) Asian majority tracts (d) Black majority tracts, (e) Hispanic majority tracts, and (f) White majority census tracts. Dark continuous curves represent the results of the LOWESS method applied to all data in each racial and ethnic majority group. Lighter shading represents the 95% CIs based on 100 bootstrap replications of each racial and ethnic majority group. Hatched shading indicates a model for census tracts with at least one solar installation on a HoW. Note that the x axes and the y axes of the histograms are plotted on a base 10 logarithmic scale.

The LOWESS models for corrected growth rates as a function of percent renter occupied households can be seen in Fig. 4. The optimal span was found to be 1.0. When all census tracts, No-Majority (diverse), Hispanic-majority, and White-majority tracts are considered (Figs. 4a,b,e,f), the 95% confidence intervals for census tracts with and without at least one solar installation on a HoW do not overlap, with corrected growth rates being higher in census tracts with at least one solar installation on a HoW. This indicates that proximity to a HoW with solar installations may be correlated to higher adoption rates on overall solar rooftop adoption within the same census tract, regardless of home ownership status. In contrast, the 95% confidence intervals for Asian and Black majority census tracts (Figs. 4c-d) overlap between census tracts with and without installations on HoWs when accounting for household ownership.

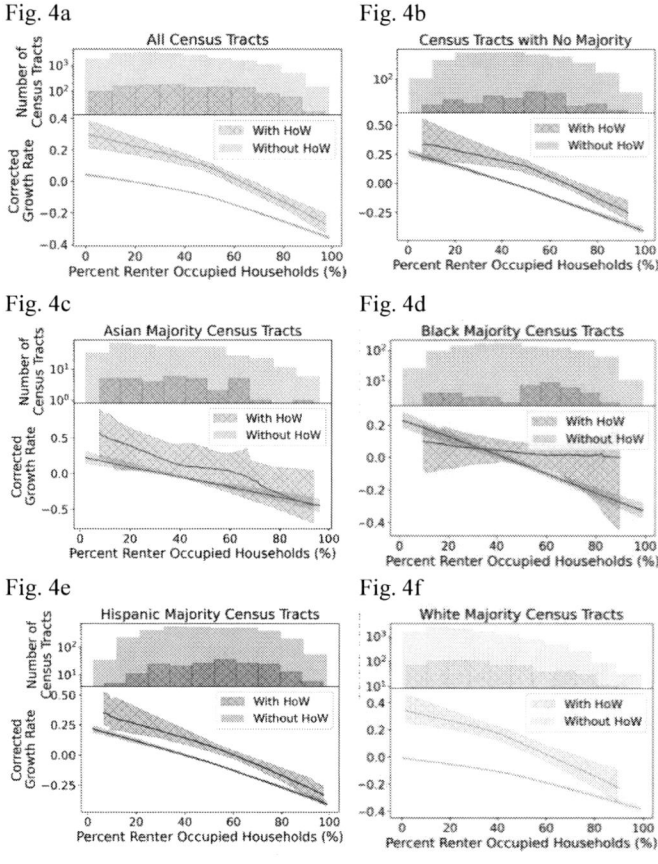

Fig. 4. Number of census tracts with and without HoWs are plotted as a histogram above corrected growth rates of solar installations as a function of percentage of renter occupation for (a) all census tracts studied, (b) tracts with no racial/ethnic majority, (c) Asian majority tracts (d) Black majority tracts, (e) Hispanic majority tracts, and (f) White majority census tracts. Dark continuous curves represent the results of the LOWESS method applied to all data in each racial and ethnic majority group. Lighter shading represents the 95% CIs based on 100 bootstrap replications of each racial and ethnic majority group. Hatched shading indicates a model for census tracts with at least one solar installation on a HoW.

The LOWESS models for corrected growth rates as a function of percentage of high-school degree completion can be seen in Fig. 5. The optimal span was found to be 1.0. When all census tracts, No-Majority (diverse), Hispanic-majority, and White-majority tracts are considered (Figs. 5a,b,e,f), the 95% confidence intervals for census tracts with and without at least one solar installation on a HoW generally do not overlap, with

corrected growth rates being higher in census tracts with at least one solar installation on a HoW. This indicates that proximity to a HoW with solar installations may be correlated to higher adoption rates on overall solar rooftop adoption within the same census tract, regardless of education levels. The confidence intervals at the tails of the all tracts, No-Majority (diverse), and Hispanic-majority models overlap, where there were fewer census tracts and potential tail effects. In contrast, Asian majority census tracts (Fig. 5c) only show a significant difference with census tracts with a percentage of high school degree holders above 85% and Black majority census tracts do not show a significant difference at any percentage of high school educational attainment with the census tracts.

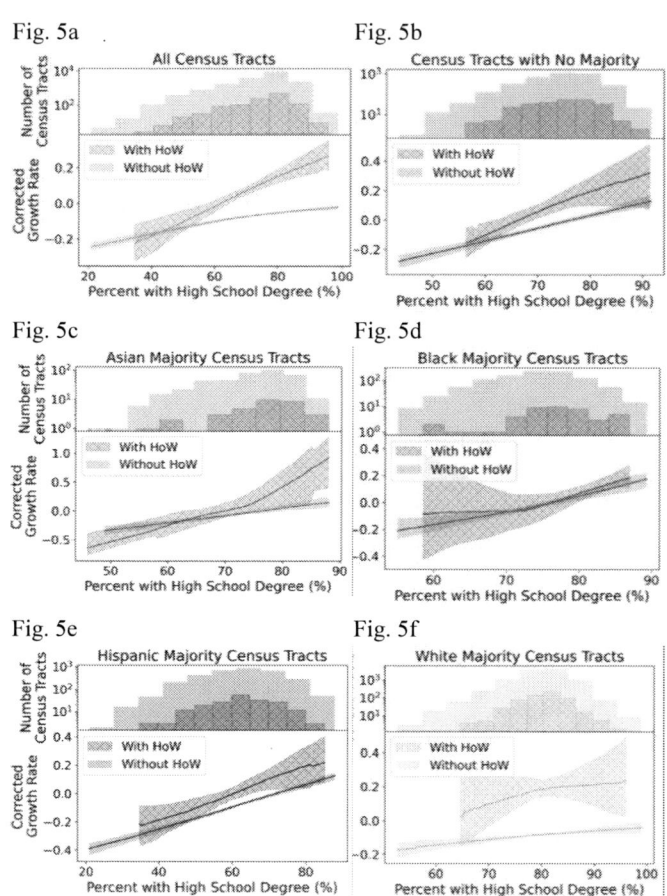

Fig. 5. Number of census tracts with and without HoWs are plotted as a histogram above corrected growth rates of solar installations as a function of percentage of high school degree completion for (a) all census tracts studied, (b) tracts with no racial/ethnic majority, (c) Asian majority tracts (d) Black majority tracts, (e) Hispanic majority tracts, and (f) White majority census tracts. Dark continuous curves represent the results of the LOWESS method applied to all data in each racial and ethnic majority group. Lighter shading represents the 95% CIs based on 100 bootstrap replications of each racial and ethnic majority group. Hatched shading indicates a model for census tracts with at least one solar installation on a HoW.

IV. CONCLUSIONS

Communities with solar installations on HoWs have higher median growth rates in rooftop solar adoption in the surrounding community when considering all census tracts studied. When segmented by race and ethnicity, diverse, Hispanic, and White majority census tracts showed significantly higher median rooftop solar installation growth rates in census tracts with at least one HoW with rooftop solar than those without, whereas Black and Asian majority census tracts do not show a significant difference. When additional demographic features such as household income, home ownership, and education levels are included in the analysis, the result also differ across racial and ethnic majority census tracts. While diverse, Hispanic, and White majority communities did show a difference at all income, home ownership and education levels at a significance level of 0.05, Asian and Black majority communities did not. Asian majority census tracts only show a significant difference in census tracts with a percentage of high school degree holders above 85%. It is important to note that there is a substantially lower number of Asian majority census tracts compared to all other racial and ethnic groups. In Black majority census tracts, we find that there are significantly higher growth rates when the median income is below $40,000. This suggests that some demographic communities may be influenced by installations on HoWs differently.

While this paper illuminates higher solar adoption rates in census tracts with installations on HoWs for some communities, there are limitations to the study. We have only identified correlations, not causations. For example, it is possible that aggressive marketing strategies were deployed causing rapid growth in local rooftop solar adoptions for both residential homes and HoWs. Further research would require causation analysis. Other considerations include additional community infrastructure with solar installations (e.g., libraries, schools) and community-owned renewable energy projects that are not locally deployed.

V. DATA AVAILABILITY

All data and computer codes used for this study are available online at https://github.com/DeborahSunter/ Houses-of-Worship-and-PV-Growth-Rates.

V. ACKNOWLEDGEMENTS

The authors thank NREL and LBNL for providing valuable data, and Galen Barbose and Isa Ferrall for their insightful discussions.

REFERENCES

[1] "Solar Industry Research Data." *SEIA*, 2022.
[2] Sunter, D.A., Castellanos, S. & Kammen, D.M. Disparities in rooftop photovoltaics deployment in the United States by race and ethnicity. *Nat Sustain* **2**, 71–76 (2019).
[3] O'Shaughnessy, Eric, Rooftop solar incentives remain effective for low- and moderate-income adoption, *Energy Policy*, 163 (2022).
[4] Wolske, K.S., Gillingham, K.T. & Schultz, P.W. Peer influence on household energy behaviours. *Nat Energy*, **5** (2020).
[5] Bollinger, Bryan, et al. "Peer Effects in the Diffusion of Solar Photovoltaic Panels." Marketing Science (2012).
[6] Graziano, M., Gillingham, K., Spatial patterns of solar photovoltaic system adoption: The influence of neighbors and the built environment, *Journal of Economic Geography* (2015).
[7] Flórez, K.R, et al, Process Evaluation of a Peer-Driven, HIV Stigma Reduction and HIV Testing Intervention in Latino and African American Churches, *Health Equity*, (2017).
[8] Evans K.R., Hudson S.V., Engaging the community to improve nutrition and physical activity among houses of worship, *Prev Chronic Dis*, (2014).
[9] LBNL, Tracking the Sun: Pricing and Design Trends for Distributed Photovoltaic Systems in the United States, (2022).
[10] Homeland Infrastructure Foundation-Level Data, https://hifld-geoplatform.opendata.arcgis.com/.
[11] Melissa Data, https://www.melissa.com/.
[12] Project Sunroof Data Explorer: Description of Methodology and Inputs Technical Report (Google Project Sunroof, 2017).
[13] Schelly, C. Testing Residential Solar Thermal Adoption. *Environment and Behavior*, 42(2), 151–170. (2010).
[14] Borenstein, S. Private net benefits of residential solar PV: the role of electricity tariffs, tax incentives, and rebates. *J. Assoc. Environ. Resour. Econ.* 4, S85–S122. (2017).
[15] Elgar, Edward, Handbook of sustainable development. (2007).
[16] 2019 5-year American Community Survey (US Census Bureau, 2019).
[17] Pauli Virtanen, et al, SciPy 1.0: Fundamental Algorithms for Scientific Computing in Python, *Nat Methods*, 17(3), (2020).
[18] Poverty Thresholds (US Census Bureau, 2019).
[19] Michaud, G. Perspectives on community solar policy adoption across the United States. *Renewable Energy Focus* (2020).
[20] Branker, K., et al. A review of solar photovoltaic levelized cost of electricity, *Renewable & sustainable energy reviews*, (2011).

Decoupling open-circuit voltage and series resistance in electroluminescence images through deep learning

Gaia Maria N. Javier, Priya Dwivedi, Thorsten Trupke, Ziv Hameiri

The University of New South Wales, Sydney, Australia

Electroluminescence (EL) imaging is one of the most common characterisation techniques for photovoltaic cells and modules. EL images contain information on both the open-circuit voltage (Voc) and series resistance (Rs) of the device. However, separating the two effects and identifying features related to each parameter can be challenging. In this study, a novel approach for decomposing EL images into Voc and Rs maps using a convolutional neural network architecture is presented. A deep learning model was first trained on paired EL and photoluminescence images that were generated using a simulation tool. Results obtained using the validation set show that the trained model is able to accurately differentiate between features related to Voc and Rs in EL images, thus replacing the need for multiple types of measurements. The proposed method presents a unique approach to analyse EL images, unlocking new capabilities that have the potential to advance solar cell characterisation.

Decomposition Mechanisms and Kinetics of Perovskite Semiconductors

Hugh Hillhouse, Yuhuan Meng, Spencer Cira, Preetham Sunkari

University of Washington, Seattle, WA, United States

Understanding the chemical reactions that halide perovskite semiconductors undergo in the presence of moisture, oxygen, and light are essential to the commercial development of perovskite solar cells. The presentation will summarize our results and conclusions from the in-situ measurement optical transmittance, light scattering, quantitative spatially resolved photoluminescence, and photoconductivity (from which an ambipolar diffusion length is calculated) during the decomposition of MAPbI3 and two representative low and high bandgap perovskite compositions: $(FA_{0.75},Cs_{0.25})(Pb_{0.50},Sn_{0.50})I_3$ and $(FA_{0.80},Cs_{0.20})Pb(I_{0.83},Br_{0.17})_3$ in the presence of moisture, oxygen, and light (individually and combinations thereof). We report the quantitative chemical decomposition rate constants for based on optical absorbance above and below the perovskite bandgap as a function of the environmental species present. These data are used to derive a rate law for each material. These first-of-their-kind measurements reveal the dominant degradation pathways in each of these three archetypal perovskites. Briefly, for MAPbI3 we discovered a new chemical reaction pathway referred to as a water-accelerated photo-oxidation (WPO) pathway, which is the dominant chemical degradation pathway when light, oxygen, and humidity are present (and much faster than other dry photooxidation pathways) [1]. For the low bandgap $(FA_{0.75},Cs_{0.25})(Pb_{0.50},Sn_{0.50})I_3$ kinetic measurements that a simple oxidation pathway is the most dominant [2]. More recent results from monitoring the decomposition dynamics of $(FA_{0.80},Cs_{0.20})Pb(I_{0.83},Br_{0.17})_3$ reveal there are two dominant pathways, depending on the presence of oxygen and moisture, which lead to distinct degradation products. These latest results are from a forthcoming manuscript on this composition. [1] Siegler, Dunlap-Shohl, Meng, Yang, Kau, Sunkari, Tsai, Armstrong, Chen, Beck, Meila & Hillhouse, "Water-Accelerated Photooxidation of CH3NH3PbI3 Perovskite," J. Am. Chem. Soc. 2022, 144, 12, 5552-5561. [2] Meng, Sunkari, Meila & Hillhouse, "Chemical Reaction Kinetics of the Decomposition of Low Bandgap Tin-Lead Halide Perovskite Films and the Effect on the Ambipolar Diffusion Length," ACS Energy Lett. 2023, Accepted.

Development and Evaluation of Typical Plane of Array Year (TPY) for Solar Energy Systems Over the Americas

Aron Habte, Manajit Sengupta, Grant Buster, Yu Xie

NREL, Golden, CO, United States

A Typical Meteorological Year (TMY) dataset consists of a concatenation of 12 months of data extracted from a long-term time series of meteorological and radiation data, assuming a horizontal surface. However, the deployment of solar energy systems uses various orientations such as fixed, single- ,or dual-axis tracking systems. This paper proposes a new method to develop and generate a typical dataset for the specific solar energy technologies (such as photovoltaics) that rely, for example, on fixed planar collectors. This new approach considers the plane-of-array (POA) irradiance as the main driving variable for the selection of 12 typical months that make up the new entity, referred to as Typical POA Year (TPY). The justification for this new method is that the current application of TMY datasets is prone to bias when converted to POA and can be significantly different from a TPY. This study uses the NSRDB dataset (1998-2018) version 3 to produce both TPYs and TMYs for solar energy and various other applications. Hypothetical solar systems information and POA irradiance data for both fixed-tilt and single-axis tracking are used to generate TPYs and associated generation capacity profiles for both TMYs and TPYs. A comparison between the new TPY and the existing TMY demonstrates that the latter mis-predicts the energy yield of typical PV systems by about ±5 % for the interquartile range as compared to the more appropriate TPY.

10-junction Edge-Illuminated Passivated-Contact Silicon Minimodules for Laser Power Conversion

Ryan M France, Matthew B Hartenstein, William Nemeth, San Theingi, Matthew Page, Sumit Agarwal, David Young, Paul Stradins

National Renewable Energy Laboratory, Golden, CO, United States

Colorado School of Mines, Golden, CO, United States

Laser power conversion efficiency is greatest when the laser emission wavelength is matched to the absorption edge of the photovoltaic, resulting in minimized thermalization loss. Thus, a wide variety of III-V photovoltaics have been bandgap-tuned to the various laser wavelengths, demonstrating efficiencies over 60%. However, high power lasers with emission around 1064 nm have been developed, which nicely align with the bandgap of Silicon, potentially providing a low-cost, high performance laser power conversion option. Silicon has been investigated for laser power conversion previously, using top-illuminated planar single junction cells, as well as edge-illuminated vertical multijunction structures. However, single junction Silicon cells become limited by Auger-Meitner recombination at high irradiance, and vertical multijunction devices can have low efficiency and need careful fabrication, which can impact cost. Here, we investigate Si solar cells with passivated contacts for high efficiency, fabricated into vertical multijunction minimodules using laser micromachining. The results step towards a higher efficiency vertical multijunction structure fabricated using a scalable, cost-effective technique. Using a 1000-nm filtered flash simulator, over 40% efficiency is demonstrated in these initial devices at current densities from 1 to 10 A/cm2, showing the potential for high power laser conversion.

GaAs Betavoltaic Cell Modeling for Light to Medium Element Radiation Conversion into Electrical Power

Mathieu de Lafontaine, Gavin Forcade, Paige Wilson, Jayeshkumar Patel, Brian Ellis, Helmut Fritzsche, Sam Suppiah, John P.D. Cook, Christopher E. Valdivia, Karin Hinzer

SUNLAB, University of Ottawa, Ottawa, ON, Canada

Canadian Nuclear Laboratories Ltd., Chalk River, ON, Canada

GaAs betavoltaic cells are modeled using Monte Carlo simulations to study electron-matter interactions and drift-diffusion calculations to assess device performance under radioactive sources. Simulations show that most electron-hole pair generation occurs in the first micrometer of the heterostructure. Since most of the carriers are generated near the surface, the maximum power density drops by 20 % as the emitter thickness is increased from ultra-thin to thin designs. The base thickness must also be thinner than a traditional first-generation GaAs solar cell design.

Residential Electric Energy Storage System to Reduce Voltage and Thermal Violations in Distribution Lines and Increase PV Integration

Anny Huaman-Rivera[1], and Agustin Irizarry-Rivera[1]

[1]University of Puerto Rico-Mayagüez, Mayagüez, Puerto Rico 00682, USA

Abstract—Electrical power systems are in constant transformation, more noticeably power distribution systems since the penetration of distributed energy resources (DERs) has increased in recent years. DER penetration may lead to increased reverse power flows and overvoltages in low voltage (LV) networks causing deterioration of power quality and limiting the increase of DERs in distribution systems. Energy storage systems (ESSs) are useful to decrease reverse flows that cause thermal violations in distribution transformers and conductors. This study employs residential energy storage systems (RESS) to mitigate voltage and thermal violations, thereby enhancing the integration potential of rooftop photovoltaic (PV) systems on urban distribution feeders. In addition, the methodology used, which combines the OpenDSS distribution system simulation tool with Matlab to process the results in different PV penetration scenarios, is presented. Through this methodology, the feasibility of achieving a higher level of PV penetration through the implementation of RESS will be demonstrated.

Index Terms—Distribution system, Energy storage systems, Thermal violations, Voltage violations.

I. INTRODUCTION

The need to reduce greenhouse gas emissions and address the impacts of climate change has become increasingly urgent in recent years. As a result, many countries, including Puerto Rico, have implemented policies aimed at promoting more environmentally friendly means of electricity generation. Puerto Rico, in particular, has experienced a significant increase in the deployment of distributed PV systems on rooftops.

In addition to the environmental benefits, there are other reasons driving the adoption of DERs in Puerto Rico, such as cost savings and improved electric service reliability. Furthermore, the use of DERs aligns with the targets outlined in the *"Public Policy on Energy Diversification by Means of Sustainable and Alternative Renewable Energy in Puerto Rico Act"* 82-2010 Second Amendment, which aims to achieve 100% renewable energy generation on the island by 2050 [1].

To ensure the resilience of the electricity grid, especially in the face of severe storms like Hurricane Maria in 2017, the proposal involves the installation of autonomous residential PV systems with batteries. These systems would provide continuous electricity service even after a hurricane, which is crucial considering that over 35% of the population remained without power for five months after Hurricane Maria [2].

However, as the number of residential PV systems increases, challenges may arise in the distribution systems due to variable electric generation and the lack of generation control. Improper integration of this technology at high penetration levels can also have negative impacts on power quality [3]. The severity of these impacts depends on three key parameters: penetration level, generation distribution, and distribution circuit conditions.

To mitigate the negative impacts of excess PV power production, the use of storage systems has been proposed [4]–[6]. These storage systems can be centralized, distributed throughout the grid, or installed at the residential level. They offer multiple benefits, such as reducing the inflow of reverse flows into the power system, lowering energy costs for consumers, enhancing grid stability and power quality, and providing increased security in the event of power system failures.

This study focuses on evaluating the impact of increasing PV penetration and the effectiveness of RESS in mitigating challenges related to voltage and thermal violations. Our study analyzed two case studies: Case 1, PV deployment without distributed RESS, and Case 2, PV deployment with distributed RESS. We developed these case studies using the Open-Source Distribution System Simulator (OpenDSS) interfaced with MATLAB and the GridPV toolbox.

II. THEORETICAL BACKGROUND

A. Limiting Factors

An extensive adoption of solar power systems can lead to power quality issues within the distribution system. As a result, certain factors restrict the incorporation of DERs into distribution networks, as per established standards or norms, in order to uphold power quality. The factors examined in this study encompass voltage and thermal breaches, as they are the initial limitations encountered when the level of photovoltaic (PV) penetration increases in a feeder. These violations are classified as steady-state problems, so they will be analyzed at each hour of the day, running for 24 hours (1 day).

1) Voltage Violations: The voltage analysis considers overvoltages and under voltages (on the primary and secondary sides of the system). All voltages on a power distribution systems are compared to ANSI C84.1 A-range [7]. For steady state simulations, the acceptable voltage range is 0.975-1.05 pu on the primary and 0.95-1.05 pu on the secondary. Any value above this range can be considered an overvoltage violation, and any below, an under-voltage violation.

2) Thermal Violations: On the other hand, excess power generated by DER units during times of high production and low consumption flows back into the distribution substation through distribution lines and distribution transformers. Both the lines and transformers have a rated current limit that they can withstand. Exceeding this limit causes these components to overheat, resulting thermal violations [8].

B. PV Penetration Level

PV systems can be installed at the residential scale, on LV networks (120V, 240V), or directly to medium voltage (MV) networks (4.16kV, 13kV). The PV penetration level in low voltage networks (PVLV) is defined here as the percentage of LV customers of every secondary substation, within the analyzed distribution network, with a rooftop PV unit installation [9]. In this work, PV systems were deployed on the load side, that is, the LV. No PV systems were considered on the MV side.

C. Residential Energy Storage Systems

RESSs generally operate in conjunction with PV systems in Puerto Rico. These are used in small-scale power systems and are mostly customer owned. Therefore, the electric utility has no control over them. The technologies commonly implemented for RESS are solid state lead-acid, lead-carbon and lithium ion batteries.

III. METHODOLOGY

A. Algorithm Description

The methodology used to determine the impact of increased photovoltaic penetration is shown in Figure 1. The impact of increased photovoltaic penetration analysis is performed for 24 hours with one-hour steps. The simulation is quasi-static, since for each hour, a static power flow analysis is performed. Figure 1 shows that the algorithm is divided into two blocks. The first block refers to the simulation actions performed in the OpenDSS program. OpenDSS has a COM Server interface that allows it to interact with MATLAB. In the OpenDSS program, the distribution system is modeled, collecting information on the technical characteristics of the distribution system under study. Next, files for the different PV penetration scenarios are created in OpenDSS. The penetration range considered is between 0 and 150 % of feeder total demand (in steps of 10 %). At this stage, no RESS is considered. Then, the power flow is solved in OpenDSS for the different PV penetration scenarios per hour. The data is collected in MATLAB for a 0 % PV penetration and the first hour of the morning at 1:00 a.m.(t=1). After that, voltage and current levels at each node of the system are compared to the limits and constraints set. Finally, a summary of the constraint violations found for each PV penetration level is displayed.

1) RESS implementation: Once Case 1 has been evaluated, RESSs are added to the model, as shown in Figure 2 (i.e., if a household has a PV system, it also has a RESS) and applies to all PV penetration scenarios.

For Case 2, the algorithm described in Figure 1 is used, including a charge and discharge algorithm to account for the RESS. The resulting algorithm is shown in Figure 3, where PV refers to the power delivered by the PV system, PL the power of the load, and PB the power in the battery.

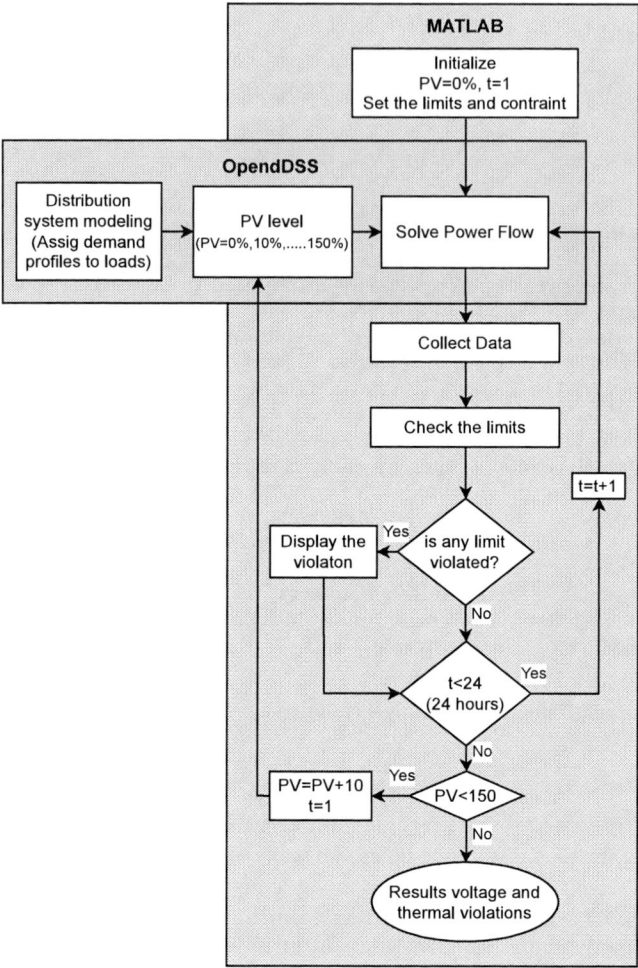

Fig. 1. Algorithm to evaluate the impact of increased photovoltaic penetration.

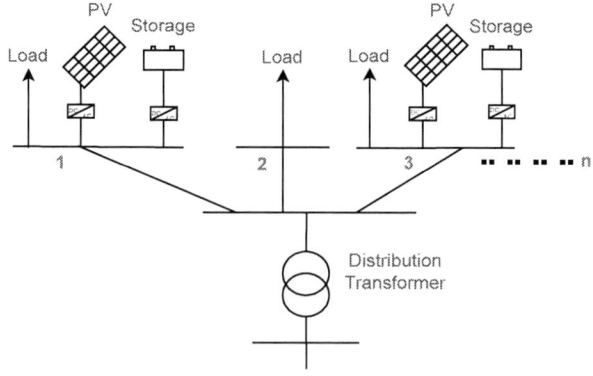

Fig. 2. Single-line diagram of the LV network.

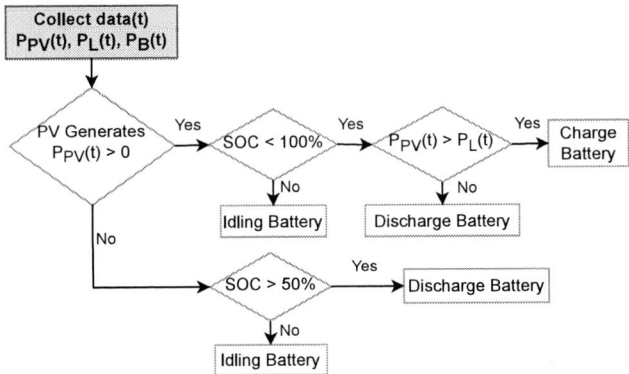

Fig. 3. Flowchart of the battery charging and discharging algorithm.

IV. CASE STUDY

A. Distribution Network

The feeder modeled in the simulations is an actual LV feeder in Puerto Rico, shown in Figure 4. This feeder, with a typical radial structure, was modeled in OpenDSS by collecting technical data from geographic information system (GIS) layers, obtained from the public database of the Puerto Rico Electric Power Authority. The main characteristics of the distribution network are summarized in Table I.

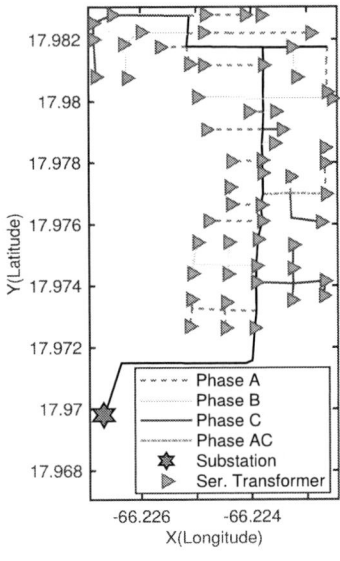

Fig. 4. Network topology (different color per phase).

TABLE I
CHARACTERISTICS OF THE DISTRIBUTION NETWORK

Substation	Voltage	38kV - 4.16kV
	Capacity	2MVA
Service Transformer	Voltage	4.16kV - 120/240kV
	Capacity	25kVA - 1 unit
		50kVA - 30 units
		75kVA - 25 units
Feeder	Peak current	66.08A
	Peak load	638.4kVA
	Length of feeder	3.7km

B. Load Profiles

To perform an analysis of the daily PV demand increase impact, curves were developed based on actual demand data. The distribution system analyzed supplies 809 households. When designing the daily demand profiles, the number of inhabitants per household was considered to develop five representative curves to model all households. The average daily demand profiles, shown in Figure 5, were developed from residential electric service bills. The curves represent typical daily behavior [10].

Table II shows a summary of the number of residents and households per demand profile.

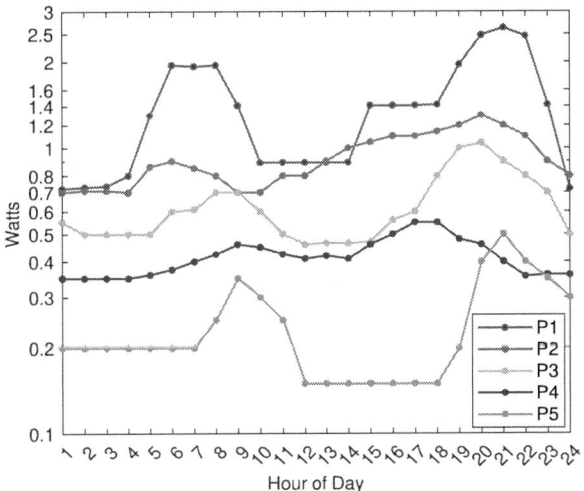

Fig. 5. Typical load profiles.

TABLE II
DEMAND PROFILES DISTRIBUTED

Demand Profile	Family Members	House Demand (kWh/day)	Household Distributed Percent(%)	Number of Households
P1	6	33	8.40%	68
P2	5	22	29.40%	238
P3	4	15	28.20%	228
P4	2	10	10.10%	82
P5	1	5.75	23.90%	193
			100.00%	809

C. Sizing PV Array

The irradiance and temperature profiles used to model the PV systems were obtained from NASA's POWER database, which provides historical temperature and irradiance data for the entire planet. Figure 6 shows the temperature and irradiance averages for the year 2020 for one day. The area where the distribution system under study is located has an average annual irradiance of 5.3 kWh/m2/day [11]. The PV array in a house, for penetration levels below 100 % is composed of 12 PV panels of 330W delivering 19.8 kWh. For penetration levels above 100 %, the number of PV panels in a house is increased. Some houses have up to 22 PV panels to reach a PV penetration level of 150 %.

978-1-6654-6060-6/23 $31.00 © 2023 IEEE

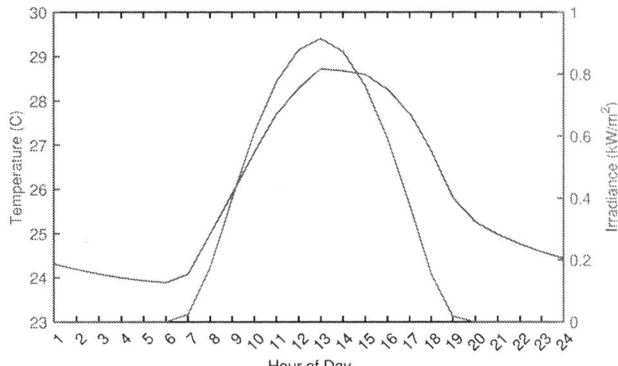

Fig. 6. Temperature and irradiance averages for 2020.

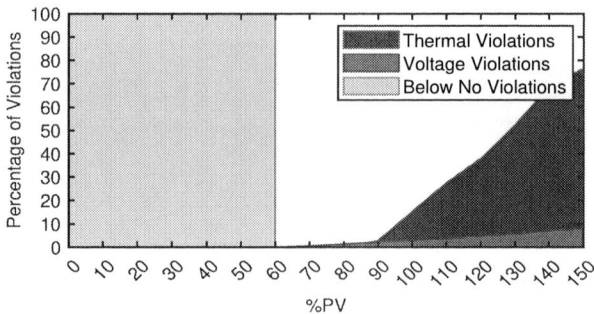

Fig. 7. Results for Case 1

D. RESS Sizing

RESS was sized for resiliency, to supply the estimated critical loads shown in Table III. A minimum critical energy demand of 5kWh per day was to be supplied. In addition, two days of autonomy and a depth of discharge of 50 % were imposed as further restrictions. This results in a storage system composed of 16 (12 Vdc nominal) batteries, for a nominal ESS voltage of 48 V and system energy capacity of about 20 kWh.

TABLE III
APPLIANCES TO BE SUPPLIED BY STORAGE

Loads	Unit	Power (W)	Hours Used(hrs)	Total power (kWh/day)
Refrigerator	1	140	24	3360
Pedestal Fan	2	100	6	600
Ceiling Fan	1	75	4	300
LCD TV	1	105	4	420
Radio	1	7	12	84
Smart Phone	2	6	4	24
Lights	4	20	8	160
			Total	5

V. RESULTS

A. Case 1

In Case 1, distributed PV systems were gradually increased, reaching up to 60 % PV penetration before the first voltage violations occurred. When PV penetration reached 90 %, thermal violations were also observed, as depicted in Figure 7. The color scheme in Figure 7 indicates the levels of PV penetration associated with different types of violations, with green representing no violations, red indicating voltage violations, and blue denoting thermal violations. The vertical axis in the figure shows the percentage violations on the allowed voltage and thermal limits. Figure 8 (a) shows the distribution system nodes where voltage violations occurred once the system reached 70 % PV penetration. As PV penetration increased to 90 % (b) the node where the transformer with the highest load (24 households) begins showing thermal violations.

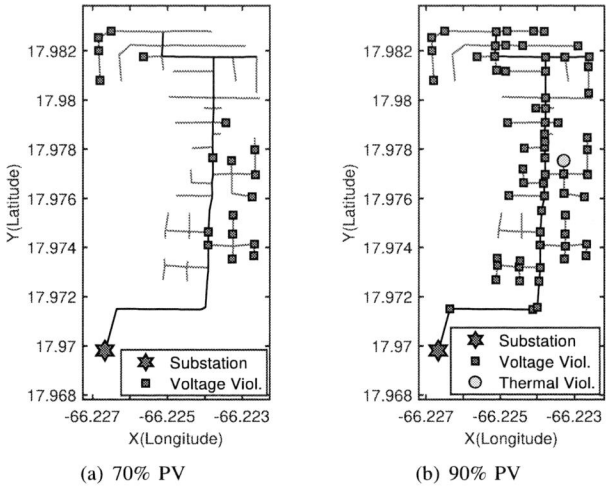

(a) 70% PV   (b) 90% PV

Fig. 8. Network topology where violations occur in Case 1.

B. Case 2

The results of Case 2 demonstrate a noteworthy enhancement in the distribution feeder's photovoltaic penetration level, as illustrated in Figure 9. With the integration of RESS, the feeder can accommodate up to 110 % of its demand from PV. However, voltage violations start to emerge after 120 % PV penetration and thermal violations occur after 125 % PV penetration. Figure 10(a) shows the location of the first voltage violation for 120 % PV penetration. Figure 10(b) shows the locations of thermal violations for 140 % PV penetration. These are consistent with the base case violation locations.

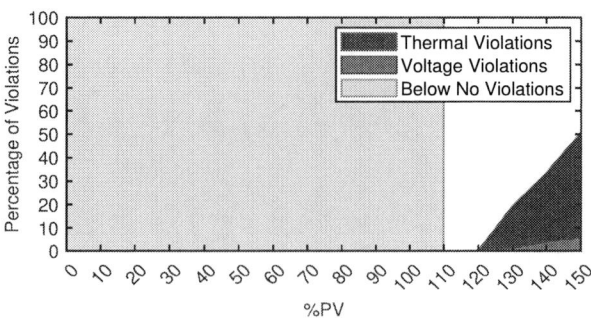

Fig. 9. Results for Case 2

978-1-6654-6060-6/23 $31.00 © 2023 IEEE

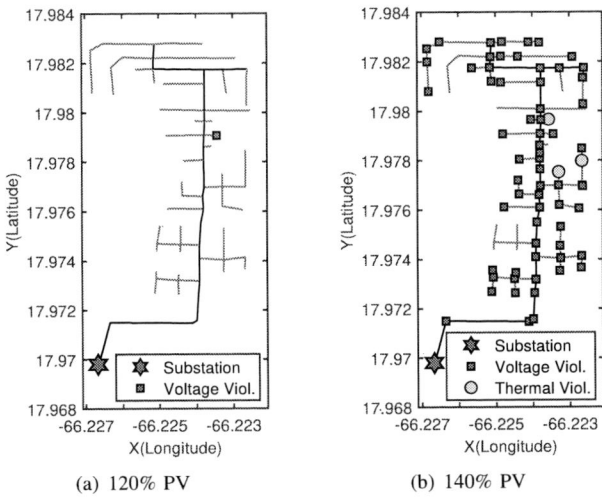

(a) 120% PV (b) 140% PV

Fig. 10. Network topology where violations occur in Case 2.

Table IV compares the results between the Case 1 (PV only), and Case 2 (adding RESS). From this comparison, it is observed that RESS systems contribute to a 50 % increase in PV penetration.

TABLE IV
COMPARISON OF RESULTS

	Only PV		PV and Storage	
PV%	Voltage Violation(%)	Thermal Violation(%)	Voltage Violation(%)	Thermal Violation(%)
<60	0	0	0	0
70	0.5	0	0	0
80	1.4	0	0	0
90	2.2	2.6	0	0
100	3.0	14.9	0	0
110	3.7	27.2	0	0
120	4.8	37.6	0.05	0
130	5.7	51.9	1.07	19.30
140	7.0	69.5	3.98	33.26
150	7.9	78.6	5.54	50.64

VI. CONCLUSION

We have developed a practical algorithm using OpenDSS and MATLAB for testing different solar energy generation and storage systems, providing different configuration options for communities. Using these tools, we analyzed the impact of increasing distributed PV systems in an actual LV distribution feeder in Puerto Rico and show that increasing the level of PV penetration i.e., increased hosting capacity, is achievable through the deployment of RESS.

In this simulation, we used realistic irradiance, temperature, and load profiles for a daily cycle with one-hour steps. Voltage violations occur first, at the nodes where the loads are connected (transformers). This suggests that an economical increase in feeder photovoltaic penetration can be achieved by implementing solutions on the load side rather than on the feeder side, thus avoiding major modifications in the LV network structure, equipment, and conductors.

The use of RESS increases both resiliency of the electric supply for residential customers and the photovoltaic penetration of the distribution feeder. This type of analysis may

also assist the electric utility in identifying the nodes of the system where more attention is required when increasing the penetration of PV systems by identifying which system nodes are prone to voltage and thermal violations.

REFERENCES

[1] A. Legislativa and P. Rico, "Law no. 82-law for diversification through sustainable renewable energy and alternative."

[2] G. Maria-Arroyo, "Reflexiones para una mayor resiliencia tras el paso del huracán María por Puerto Rico: Introducción al volumen especial," *Revista de Administración Pública*, vol. 49, pp. 7–9, 2018.

[3] J. Smith, "Distributed photovoltaic feeder analysis preliminary findings from hosting capacity analysis of 18 distribution feeders," 12 2013. [Online]. Available: www.epri.com

[4] M. Bartecka, G. Barchi, and J. Paska, "Time-series PV hosting capacity assessment with storage deployment," *Energies*, vol. 13, no. 10, p. 2524, 2020.

[5] N. Ertugrul and F. Castillo, "Maximizing PV hosting capacity and community level battery storage," in *2019 29th Australasian Universities Power Engineering Conference (AUPEC)*. IEEE, 2019, pp. 1–6.

[6] P. H. Divshali and L. Söder, "Improvement of RES hosting capacity using a central energy storage system," in *2017 IEEE PES Innovative Smart Grid Technologies Conference Europe (ISGT-Europe)*. IEEE, 2017, pp. 1–6.

[7] ANSI, "Standard c84.1-2011 american national standard for electric power systems and equipment - voltage ratings (60 hz)," 2011.

[8] M. Zain ul Abideen, O. Ellabban, and L. Al-Fagih, "A review of the tools and methods for distribution networks' hosting capacity calculation," *Energies*, vol. 13, no. 11, p. 2758, 2020.

[9] A. Ballanti and L. F. Ochoa, "On the integrated PV hosting capacity of MV and LV distribution networks," in *2015 IEEE PES Innovative Smart Grid Technologies Latin America (ISGT LATAM)*. IEEE, 2015, pp. 366–370.

[10] G. Carrión, R. Cintrón, M. Rodríguez, W. Sanabria, R. Reyes, and E. O'Neill-Carrillo, "Community microgrids to increase local resiliency," in *2018 IEEE International Symposium on Technology and Society (ISTAS)*. IEEE, 2018, pp. 1–7.

[11] E. O'Neill-Carrillo and M. A. Rivera-Quinones, "Energy policies in Puerto Rico and their impact on the likelihood of a resilient and sustainable electric power infrastructure." *Centro Journal*, vol. 30, no. 3, 2018.

AIR-BRIDGE CELLS FOR HIGHER EMISSION TEMPERATURES

Bosun Roy-Layinde, Areefa Rahman, Jihun Lim, Sritoma Paul, Stephen R. Forrest and Andrej Lenert

University of Michigan, Ann Arbor, Michigan, USA

Abstract — Interest in thermal batteries for inexpensive grid-scale storage of renewable energy motivates the development of photovoltaics that efficiently convert very high temperature thermal emission to electrical energy. We have previously shown that InGaAs air-bridge cells can increase TPV efficiency by ~30% compared to cells with more conventional back surface reflectors. In this study, we design and experimentally characterize airbridge cells with wider bandgaps for applications at higher emission temperatures. Parametric studies with varying bandgap and emitter temperature identify high performance regimes. At temperatures up to 2000K, predicted device efficiencies of single-junction air-bridge cells match that of record-holding multi-junction cells. Furthermore, a novel platform for device testing using porous graphite emitters is designed and experimentally demonstrated.

I. INTRODUCTION

Thermophotovoltaic (TPV) conversion of thermal radiation to electrical power is a readily scalable process for on-demand electricity generation. A TPV system consists of two major parts, a thermal emitter and a photovoltaic cell in close proximity, as shown in Fig. 1a. Heat supplied by an upstream primary energy source maintains the emitter at elevated temperatures (>1000 K) which drives emission of infrared radiation. Photons with energy greater than its bandgap (in-band) excite electron-hole pairs. Meanwhile, absorption of emitted photons with energy less than the bandgap (out-of-band) generates undesirable low-grade waste heat.

TPVs can be used for a broad range of applications, which include waste heat recovery (800-1300 K), direct solar thermal energy conversion (1000-1300 K), space exploration (1200-1400 K), cogeneration of heat and power (1300-1500 K), and grid scale thermal storage (1500-2400 K) [1]–[4]. However, current approaches to TPV generation are typically optimized for a specific temperature range and do not translate easily to other temperatures and applications.

In our prior work, we demonstrated a record high 98.5% out-of-band reflectance R_{out} enabled by an air cavity situated between the absorber and reflector [5]. This airbridge cell improved the TPV efficiency by 25-30% compared to a cell without the air cavity.

The objective of this work is to identify a set of cell designs, all within one material system, that enable highly efficient conversion across a very wide range of blackbody emission temperatures and that are specifically suitable for energy storage applications. To this end, we design airbridge cells (ABCs) within the $In_{1-x}Ga_xAs_{1-y}P_y$ (InGaAsP) material system, which is lattice-matched to InP substrates. The binary materials for this quaternary system include GaP, GaAs, InAs and InP. These four materials play a role in determining the material properties for the InGaAsP quaternary cell by compositional

mixing. Fig. 1b shows the bandgap energy as a function of lattice constants where the red dashed line indicates the bandgap range when lattice matched to InP. The bandgap range is wide, from 0.74 eV to 1.35 eV, which makes it possible to match the bandgap to different emission temperatures.

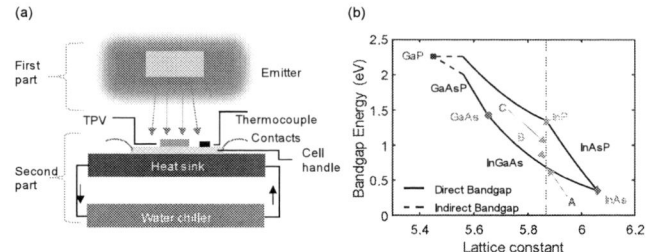

Fig. 1 (a) Thermophotovoltaic system consisting of the heat source (emitter) and the cold source (photovoltaic cell). **(b)** Bandgap diagram as a function of lattice constant for binary and ternary III-V semiconductors.

II. THEORY AND DESIGN

The conversion efficiency η of a TPV system is defined as the ratio of the electrical power output P_{mpp} to the heat absorbed by the cell Q_{abs}. Absorption of out-of-band photons ($E<E_g$) represents one of the greatest losses. Therefore, TPV systems must realize wavelength-selective radiative transfer between the emitter and cell to achieve high efficiencies. One method for achieving selectivity is photon recuperation, a process by which low energy photons are reflected back to the emitter are reabsorbed [5]–[7]. The other common approach involves the use of selective emitters that preferentially emit in-band photons [8]. The TPV conversion efficiency also depends on the cell's ability to collect photoexcited charge carriers. Aligning the bandgap to the emission spectrum to ensure proper photon utilization is another major design consideration in improving device performance. Other major losses in the device include resistive (ohmic) losses and non-radiative recombination.

To understand the different factors that contribute to efficiency, we can decouple the conversion efficiency into the product of two meaningful performance metrics [8], namely spectral management (*SE*)(*IQE*) and carrier management (*VF*)(*FF*) efficiencies, as given by:

$$\eta = \left(\frac{P_{mpp}}{Q_{abs}}\right) = (SE)(IQE)(VF)(FF) \qquad (1)$$

In this work, we model the electronic and optical properties of the quaternary InGaAsP/InP heterojunction for the three

978-1-6654-6060-6/23 $31.00 © 2023 IEEE

different systems. To achieve bandgaps of 0.74 eV, 1.0 eV and 1.2 eV, we use alloy compositions of $In_{0.53}Ga_{0.47}As_1P_0$, $In_{0.76}Ga_{0.24}As_{0.52}P_{0.48}$ and $In_{0.9}Ga_{0.1}As_{0.21}P_{0.79}$ respectively, as determined by equation 2 [9].

$$M = x \cdot y \cdot B_{GaP} + x \cdot (1-y) \cdot B_{GaAs} + (1-x) \cdot y \cdot B_{InP} + (1-x) \cdot (1-y) \cdot B_{InAs} \quad (2)$$

The absorption coefficients (α) for the cells were modeled using a piece-wise function (from [10]), given by:

$$\alpha = \begin{cases} \alpha_0 \exp\left(\dfrac{E - E_g}{E_0}\right), E \leq E_g \\ \alpha_0 \exp\left(1 + \dfrac{E - E_g}{E'}\right), E \geq E_g \end{cases} \quad (3)$$

These coefficients were fed into a transfer matrix model that resolves the electric field distribution inside the cell.

The electronic properties (current-voltage characteristics) are modeled as in our prior work using a two-diode model [11] [12], which describes different rates of recombination in the depleted and quasi-neutral regions. This model assumes the cell has a series resistance of 30 mΩ.cm^2 and is being cooled to a constant cell temperature of 298 K.

III. RESULTS AND DISCUSSION

To understand how charge collection differs for the three systems, we plot $(VF)(FF)$ as a function of increasing temperature as shown in Fig. 2a. Larger bandgaps produce better $(VF)(FF)$ at higher emitter temperatures. At relatively low temperatures ($<$ 1000 K), the three different systems have comparable $(VF)(FF)$. As the temperature increases, we see a common trend where $(VF)(FF)$ increases until a maximum point where further increase in heat source temperature leads to a reduction in performance. The eventual decline in $(VF)(FF)$ is due to resistive losses which scale quadratically with the increasing photocurrent density.

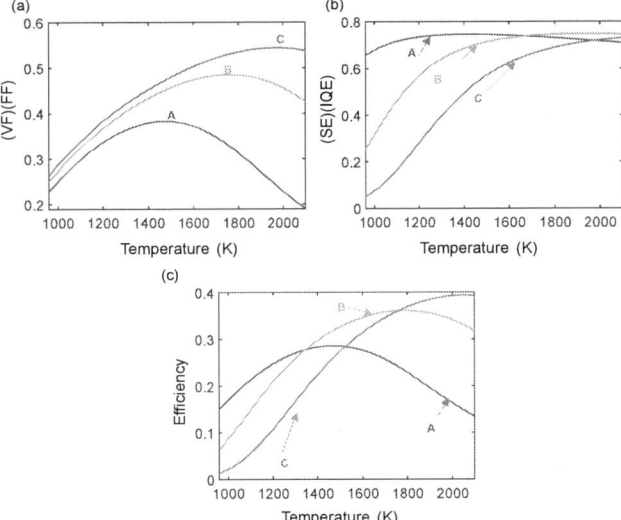

(grey), B (orange), and C (blue) have bandgaps of 0.74 eV, 1 eV and 1.2 eV respectively.

With regards to spectral management $(SE)(IQE)$, which depends primarily on R_{out}, the three systems exhibit relatively high and temperature-insensitive performance, as shown in Fig. 2b. We find R_{out} (weighted to a 1700K blackbody emitter) are 98.8%, 99.1% and 99.2% for the 0.74 eV, 1.0 eV and 1.2 eV bandgap cells, respectively. The slight improvement in reflectance with increasing bandgap results since at a constant temperature, there is less power lost to band-edge absorption. We also see that $(SE)(IQE)$ decreases at lower temperatures, but the onset of this decrease shifts to higher temperatures with increasing bandgap.

Overall, there is a tradeoff between spectral and carrier management for the three different systems, which gives rise to a distinct optimal temperature range for each device. The 0.74 eV device (A) has improved efficiency in the range of 1000 K to 1300 K, making it suitable for applications such as waste heat recovery and concentrated solar thermal. In the moderate temperature range of 1300 K to 1700 K, device B produces the best efficiency, with potential applications in space exploration and distributed cogeneration. Lastly, at >1700 K, device C has the highest conversion efficiency, suitable for applications in thermal energy grid storage. Of note, with these three devices, all with the same architecture and from the same material system, one can achieve near-optimal performance over wide range of emitter temperatures, spanning 1000 to 2200 K.

IV. EXPERIMENT

Fig. 3a shows a scanning electron microscopy (SEM) image featuring the airbridge architecture. The emitter used for the testing these device is a Joule-heated graphite felt suspended over the cell. This setup allows us to achieve high emitter temperatures and large view factors using a single configuration. We have demonstrated emitter temperatures up to 2273 K as shown in Fig. 3b. The heat absorbed, Q_{abs}, by these cells can be quantified using a calorimeter consisting of a copper pedestal with embedded temperature measurements that determine the heat flux. Detailed results from these experimental studies will be presented.

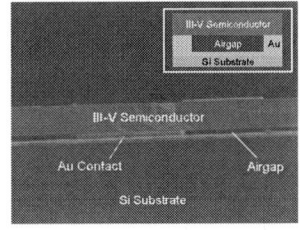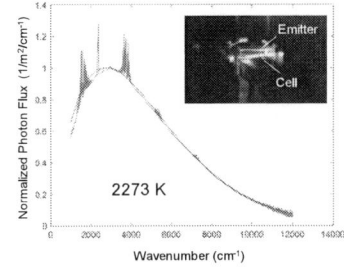

Fig. 2 Carrier management (**a**), spectral management (**b**), and overall TPV efficiency (**c**) as a function of emitter temperature. Device A

Fig. 3 (a) Cross section SEM of the airbridge cell. **(b)** Measured and simulated emission spectra from the graphite felt heater.

V. CONCLUSION

This work provides insights into how efficiency can be maximized across a wide range of emission temperatures by designing airbridge cells with varying bandgaps within a single lattice-matched material system. Experimental devices and a novel characterization platform are presented to support the theoretical results.

REFERENCES

[1] A. Lenert, D. M. Bierman, Y. Nam, W. R. Chan, I. Celanovic, and E. N. Wang, "A nanophotonic solar thermophotovoltaic device," *Nat. Nanotechnol.*, vol. 9, no. 2, pp. 126–130, 2014, doi: 10.1038/nnano.2013.286.

[2] A. Datas, A. Ramos, A. Martí, C. del Cañizo, and A. Luque, "Ultra high temperature latent heat energy storage and thermophotovoltaic energy conversion," *Energy*, vol. 107, pp. 542–549, 2016, doi: 10.1016/j.energy.2016.04.048.

[3] A. LaPotin *et al.*, "Thermophotovoltaic efficiency of 40%," *Nature*, vol. 604, no. 7905, pp. 287–291, 2022, doi: 10.1038/s41586-022-04473-y.

[4] L. M. Fraas, J. E. Avery, and H. X. Huang, "Thermophotovoltaic furnace-generator for the home using low bandgap GaSb cells," *Semicond. Sci. Technol.*, vol. 18, no. 5, 2003, doi: 10.1088/0268-1242/18/5/316.

[5] D. Fan, T. Burger, S. Mcsherry, B. Lee, A. Lenert, and S. R. Forrest, "Near-perfect photon utilization in an air-bridge thermophotovoltaic cell," *Nature*, no. February, 2020, doi: 10.1038/s41586-020-2717-7.

[6] Z. Omair, L. M. Pazon-Outon, and E. Yablonovitch, "Practical challenges towards 50% efficient thermophotovoltaic energy conversion," vol. 1112002, no. September 2019, p. 1, 2019, doi: 10.1117/12.2534511.

[7] T. Burger, B. Roy-Layinde, R. Lentz, Z. J. Berquist, S. R. Forrest, and A. Lenert, "Semitransparent thermophotovoltaics for efficient utilization of moderate temperature thermal radiation," *Proc. Natl. Acad. Sci.*, vol. 119, no. 48, p. e2215977119, 2022, doi: 10.1073/pnas.2215977119.

[8] T. Burger, C. Sempere, B. Roy-Layinde, and A. Lenert, "Present Efficiencies and Future Opportunities in Thermophotovoltaics," *Joule*, vol. 4, no. 8, pp. 1660–1680, 2020, doi: 10.1016/j.joule.2020.06.021.

[9] R. L. Moon, G. A. Antypas, and L. W. James, "Bandgap and lattice constant of GaInAsP as a function of alloy composition," *J. Electron. Mater.*, vol. 3, no. 3, pp. 635–644, 1974, doi: 10.1007/BF02655291.

[10] P. Jurczak, A. Onno, K. Sablon, and H. Liu, "Efficiency of GaInAs thermophotovoltaic cells: the effects of incident radiation, light trapping and recombinations," *Opt. Express*, vol. 23, no. 19, p. A1208, Sep. 2015, doi: 10.1364/oe.23.0a1208.

[11] B. Roy-Layinde *et al.*, "Sustaining efficiency at elevated power densities in InGaAs airbridge thermophotovoltaic cells," *Sol. Energy Mater. Sol. Cells*, vol. 236, no. December 2021, p. 111523, 2022, doi: 10.1016/j.solmat.2021.111523.

[12] T. Burger, D. Fan, K. Lee, S. R. Forrest, and A. Lenert, "Thin-Film Architectures with High Spectral Selectivity for Thermophotovoltaic Cells," *ACS Photonics*, vol. 5, no. 7, 2018, doi: 10.1021/acsphotonics.8b00508.

On the Accuracy of Spectral Adjustment for Performance Measurements of Multijunction Solar Cells

Nikos Kopidakis, Tao Song, John Geisz, Daniel Friedman

National Renewable energy Laboratory, Golden, CO, United States

Accurate measurement of the performance of multijunction solar cells under Standard Test Conditions (STC) requires adjustment of the spectral output of the solar simulator to achieve a spectral irradiance matching condition under which the current in each junction under the solar simulator is the same as the current that the particular junction would carry under the reference spectrum. To quantify if this condition has been achieved, a "current balance" is calculated for each junction, typically labeled Zi for the i-th junction and it is generally accepted that 0.99

Spatially Resolved Degradation Analysis of Solar Modules After Combined Accelerated Aging

Robert Heidrich[12], Anton Mordvinkin[1], and Ralph Gottschalg[12]

[1]Fraunhofer Center for Silicon Photovoltaics CSP, Halle (Saale), 06120, Germany
[2]Anhalt University of Applied Sciences, Koethen, 06366, Germany

Abstract— The degradation behavior of solar modules was investigated under combined damp heat and UV weathering conditions. A full electrical characterization, including electro luminescence measurements, shows a severe module degradation after 2000h, with power losses exceeding 60%. A chemical analysis of the encapsulation material on the basis of ethylene-vinyl acetate copolymer, extracted directly from the weathered modules, reveals a clear dependence of the encapsulant degradation on the position in the module and the weathering time. Remarkably, the consumption of the UV stabilizer in the encapsulant correlates to the degradation of the encapsulation material and finally to the degradation of the solar module. This finding emphasizes the relevance of the quality control of the encapsulant formulation for the module reliability.

Keywords—accelerated aging, I-V, EL, GCMS, FTIR

I. INTRODUCTION

The degradation behavior of solar modules and encapsulation materials is mainly analyzed with help of the accelerated weathering. However, one mostly relies on individual stressors, e.g., damp heat (DH) and UV weathering, thus separating the water ingress and UV-driven degradation. A more realistic approach requires a combination of the UV irradiation with the moisture ingress. Furthermore, a solar module is a complex object for physical investigations consisting of several multi-component materials, including encapsulants and backsheets interacting at their interfaces. Thus, to address this complexity, the degradation of encapsulation materials must be studied enabling these interactions rather than weathering individual polymer films.

Considering the aforementioned points, the given work shows the analysis of one-cell mini modules, built using an EVA encapsulant and a PET-based backsheet, which were aged under combined DH and UV weathering conditions. The macroscopic analysis such as the electrical characterization and electro luminescence (EL) measurements were correlated with space-resolved polymer investigations both on the chain and additive level.

II. RESULTS AND DISCUSSION

A. Electrical Characterization

Mini modules were weathered under combined damp heat (DH) and UV irradiation (UV) conditions for 2000h. The following parameters were used: 80°C chamber temperature,

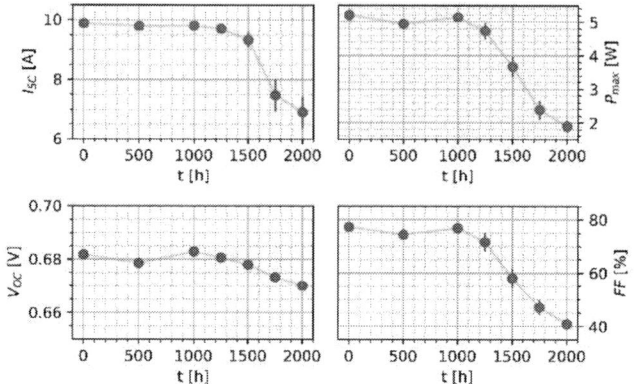

Figure 1: Electrical performance data of the one-cell mini modules weathered under combined UV and DH conditions.

60% relative humidity (r.h.) and 227.5 W/m² integrated UV intensity between 280nm and 380nm.

Figure 1 shows the electrical characterization. An exemplary mini module after 2000h of the UV+DH weathering is visualized in Figure 2. The colored rectangles mark the positions for the EVA-sample extraction (Green: Backside EVA – middle of the module (m_m_b). Purple: Frontside EVA – top right corner of the module, on top of the electrical contact (t_r_f).). After approximately 1000h of weathering significant performance losses were detected. Within the next 1000h of weathering, the mini modules fully degraded with a total power loss of over 65%. While the series resistance was increasing, I-V curves suggest more complex degradation mechanism such as non-radiative recombination and surface short circuits. Furthermore, the mini modules showed prominent browning originating from the formation of chromophores [1]. Degradation effects regarding the encapsulation materials were investigated by several characterization methods on the polymer and additive levels.

Figure 2 demonstrates the EL images of a single solar module every 500h of weathering. The EL images show a significant decrease in the radiative recombination confirming the assumptions made due to the I-V curves. Non-radiative recombination effects are assumed to occur mainly due to the humidity-based degradation. Thus, the path of moisture ingress is visualized by the EL images, starting with the diffusion

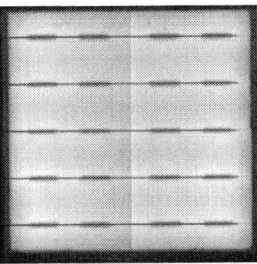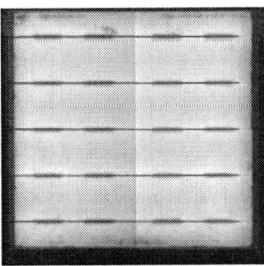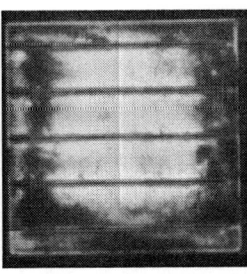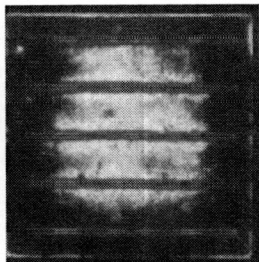

Figure 2: Left: Exemplary mini module after 2000h of weathering. The colored rectangles mark the position for the EVA-sample preparation. Middle left to right: Electro luminescence (EL) characterization every 500h of weathering.

through the backsheet, followed by the diffusion around the solar cell edges to the top of the cell. The moisture is known to interact with the EVA encapsulant leading to the formation of acetic acid and subsequently causing the decrease in EL [2].

The chemical analysis of the mini modules was performed destructively. Therefore, the data points for every weathering step were evaluated for different modules and include effects of inhomogeneities and local differences between samples.

B. GCMS measurements

The degradation behavior of additives within the encapsulant was investigated quantitatively with pyrolysis-gas chromatography-mass spectrometry (PY-GCMS) in dependence of the weathering time [3]. In particular, the content of the UV absorber Cyasorb UV 531 and the UV stabilizer Tinuvin 770 was of interest, serving as UV protectors. Figure 3 visualizes the PY-GCMS quantification in dependence of the weathering time and sample position in the module. The large difference of the initial UV absorber concentration can be caused by the local spatial inhomogeneities, introduced by the polymer film production. Furthermore, the degradation behavior

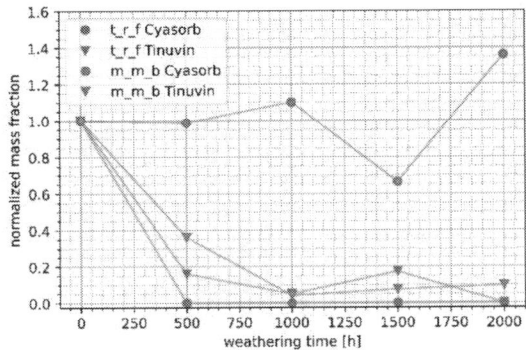

Figure 3: Consumption of the UV absorber Cyasorb UV 531 and the UV stabilizer Tinuvin 770 in dependence of the sample position and weathering (values normalized to the initial mass fraction).

strongly diverges in dependence of the sample position. While the concentration in the top EVA above the cell is already consumed after 500h of the UV irradiation, the concentration in the bottom EVA below the cell remains unchanged. Therefore, Cyasorb UV 531 becomes consumed in the course of photo degradation.

In contrast to the UV absorber, the UV stabilizer behaves nearly similar for the frontside and backside EVA. Thus, the UV

irradiation seems to be not relevant for the degradation of the UV stabilizer. Tinuvin 770 as a hindered amine light stabilizer (HALS) reacts according to the Denisov cycle [4]. It degrades by interacting with free radicals, formed during the degradation, which results in the concentration-dependent reaction rate. Therefore, the data suggests that the radical formation is mainly driven by humidity rather than UV irradiation.

The solar modules showed a strong decrease in the power output after approximately 1000h of weathering, after the UV stabilizer was already consumed. Thus, it suggests that the degradation of the UV stabilizer can be considered as an early signature for the subsequent module degradation. After the free radicals cannot be neutralized by the UV stabilizer, the degradation of the mini modules intensifies leading to a significant power loss.

C. ATR -FTIR- measurements

The differences in the encapsulant degradation on the polymer level depending on the position and weathering time were investigated by ATR-FTIR measurements. Figure 4 shows the FTIR spectrum of EVA samples at the t_r_f and m_m_b positions. Especially the peaks at 1370 cm^{-1} (symmetric deformation of CH3), 1465 cm^{-1} (asymmetric deformation vibration of CH2), 1715 cm^{-1} (C=O stretching vibration of ketones), between 1800 cm^{-1} and 1680 cm^{-1} (C=O stretching vibration) and between 3700 cm^{-1} and 3100 cm^{-1} (hydroxyl groups vibrations) will be discussed [5] [6] [7]. As a measure of the chain oxidation, carbonyl indices were calculated referring the C=O stretching vibration peak to the area of the 1465 cm^{-1} CH2 peak, thus normalizing away the effects of the variable ATR-crystal contact with the EVA sample. These results are shown in Figure 5.

Except for the largely growing hydroxyl peak after 2000h of weathering, the spectrum of the front side EVA shows minor changes. The large hydroxyl peak can be caused by hydroperoxides forming due to high energy irradiance. Because these peaks also occur by aging EVA in a hot-air fan oven, the presence of moisture is probably not required for the hydroxyl formation [5].

The degradation of the backside EVA differs strongly from the degradation of the frontside EVA, as it is affected only by the moisture ingress. The carbonyl index strongly increases after 1500h of weathering. Furthermore, the 1715 cm-1 peak becomes more pronounced. This appearance in combination with the emerging peak at approximately 1175 cm^{-1} indicates an

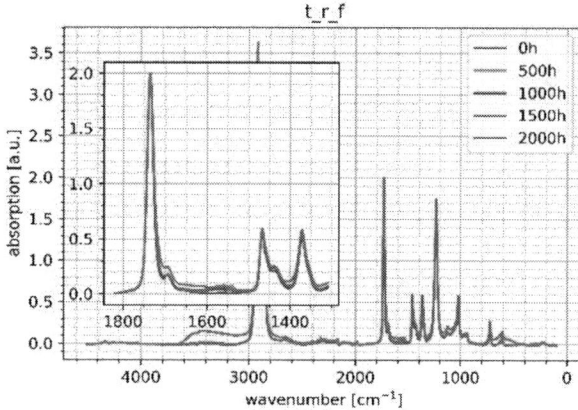 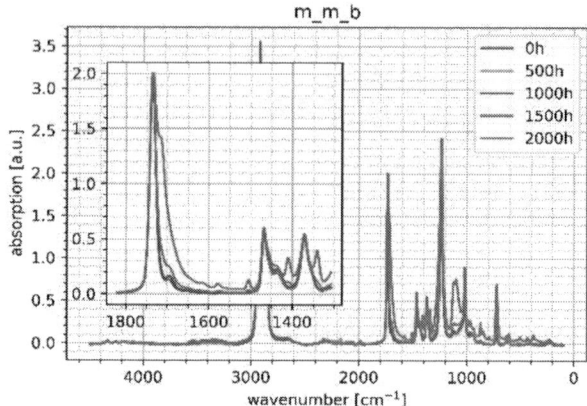

Figure 4: FTIR-ATR measurements in dependence of sample localization and weathering time.

increased presence of ketone carbonyl groups in the polymer chains, acting as chromophores [8] [9]. Considering that the 1715 cm-1 peak significantly increases after the UV stabilizer is consumed, it is conceivable that these reactions were prevented in the beginning. HALS amines trapped the formed radicals stopping the ketone formation.

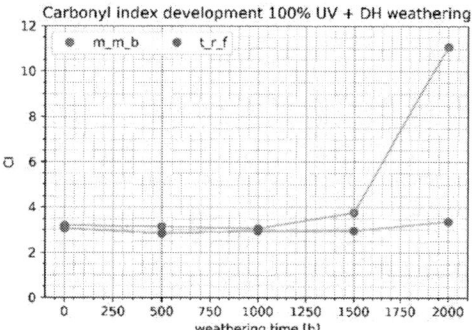

Figure 5: Change of the carbonyl index in dependence of sample localization weathering time.

III. SUMMARY

Mini modules have been weathered under combined damp heat and UV conditions. The electrical characterization showed a performance decrease after approximately 1000h of combined weathering. A total power loss of approximately 65% occurred after 2000h. The I-V curve and EL images suggested the presence of non-radiative recombination centers. Furthermore, the EL images showed a moisture path which starts 2 dimensionally through the backsheet and crawling around the edges of the solar cell afterwards.

The degradation analysis of the encapsulant and the imbedded additives shows diverging results in dependence of the sample position. The UV absorber in the frontside EVA was already consumed after 500h of weathering while it stayed nearly unchanged in the backside EVA. The UV stabilizer degraded qualitatively similarly in the frontside and backside EVA and was completely consumed after approximately 1000h. The frontside EVA shows the formation of hydroperoxides due to the photo degradation which was not the case for the backside EVA. In the latter case, after the UV stabilizer was consumed,

the moisture-driven formation of ketone carbonyl groups was observed because of radical forced polymer chain degradation.

These findings show that the occurring degradation reactions in the encapsulant are highly dependent on the position in the module. In addition, they occur after the UV stabilizer is fully consumed. Further studies should investigate the dependence of the UV stabilizer concentration as well as the dependence on weathering parameters on the encapsulant and module degradation. This finding emphasizes that the UV stabilizer can serve as an early-stage degradation marker for the module degradation. Further, it stresses that the quality control of the EVA formulation, especially the UV-stabilizer content, plays a key role for the solar module reliability.

[1] Pern, F. J. "Ethylene-vinyl acetate (EVA) encapsulants for photovoltaic modules: Degradation and discoloration mechanisms and formulation modifications for improved photostability." Die Angewandte Makromolekulare Chemie: Applied Macromolecular Chemistry and Physics 252.1 (1997): 195-216.

[2] Jankovec, Marko, et al. "In-situ monitoring of moisture ingress in PV modules using digital humidity sensors." IEEE journal of photovoltaics 6.5 (2016): 1152-1159.

[3] Heidrich, Robert, Anton Mordvinkin, and Ralph Gottschalg. "Quantification of UV protecting additives in ethylene-vinyl acetate copolymer encapsulants for photovoltaic modules with pyrolysis-gas chromatography-mass spectrometry." *Polymer Testing* (2022): 107913.

[4] Hodgson, Jennifer L., and Michelle L. Coote. "Clarifying the mechanism of the Denisov cycle: how do hindered amine light stabilizers protect polymer coatings from photo-oxidative degradation?." Macromolecules 43.10 (2010): 4573-4583.

[5] Allen, Norman S., et al. "Aspects of the thermal oxidation, yellowing and stabilisation of ethylene vinyl acetate copolymer." Polymer degradation and stability 71.1 (2000): 1-14.

[6] Sharma, Bhuwanesh K., et al. "Effect of vinyl acetate content on the photovoltaic-encapsulation performance of ethylene vinyl acetate under accelerated ultra-violet aging." Journal of Applied Polymer Science 137.2 (2020): 48268.

[7] Barretta, Chiara, et al. "Comparison of degradation behavior of newly developed encapsulation materials for photovoltaic applications under different artificial ageing tests." Polymers 13.2 (2021): 271.

[8] Jin, Jing, Shuangjun Chen, and Jun Zhang. "UV aging behaviour of ethylene-vinyl acetate copolymers (EVA) with different vinyl acetate contents." Polymer degradation and stability 95.5 (2010): 725-732.

[9] Çopuroğlu, Mehmet, and Murat Şen. "A comparative study of thermal ageing characteristics of poly (ethylene-co-vinyl acetate) and poly (ethylene-co-vinyl acetate)/carbon black mixture." Polymers for Advanced Technologies 15.7 (2004): 393-399

NiO as a p-type TCO for inorganic thin-film photovoltaics

Elline C. Hettiaratchy[1], Angus A. Rockett[2], Taylor D. Hill[1,3], Sachit Grover[1]

[1] First Solar Inc, Santa Clara, California, 95050, USA

[2] Colarado School of Mines, Golden, Colorado, 80401, USA

[3] Colorado State University, Fort Collins, Colorado 80523 USA,

A p-type TCO for wider applicability in inorganic absorbers is urgently needed to enable bifacial and tandem thin-film architectures. NiO is a p-type transparent conductive oxide (TCO) that is successfully used in low-temperature processed organic and perovskite photovoltaics. Thin-film absorbers such as chalcogenides often employ temperatures that are higher than those used in organic or perovskite devices. In this paper we study the feasibility of using sputtered NiO as a hole transport layer in chalcogenide PV. NiO thin films deposited using RF magnetron sputtering were studied at varying substrate-temperature, gas pressure and composition. The antiferromagnetic property of NiO prevented the determination of carrier properties using Hall effect, therefore a hot-point-probe employing the Seebeck effect was used to determine the carrier type. This was shown to change from p-type to n-type as the resistivity decreases. NiO films with resistivity less than 0.3 Ω-cm were n-type, whereas those with resistivities larger than 1 Ω-cm are p-type. Even though n-type in transport behavior, the workfunction of the highly conductive films remained close to the valence band-position suggesting that the n-type behavior is likely due electron transport in a acceptor-like defect-band. The conductivity degraded when the films were subjected to post-deposition temperatures > 200C. The Haacke figure-of-merit was calculated for the as-deposited films with a median value of $1.1 \times 10^{-9}\ \Omega^{-1}$ which suggests NiO is a promising p+ layer as long as it is maintained at lower post-deposition temperatures.

I. INTRODUCTION

NiO is a transparent conductive oxide (TCO) that is being studied as a candidate for a hole transporting material in photovoltaic cells. NiO can either be n- or p-type, but the majority of NiO films tend to be p-type because they are deposited under oxygen rich conditions resulting in Ni vacancies [1]. NiO is pursued here as a promising p-type TCO to make a transparent heterojunction with chalcogenide absorbers. NiO has a band gap between 3.6-4.0 eV and has been used in conventional, UV and perovskite photovoltaics [2]. NiO films can be created using spin-coating and wet chemical processes, but for optoelectronic applications it is typically deposited using sputtering [2]. Recently, NiO has been doped with N by introducing reactive N_2 mixed with nonreactive Ar as the background gas in the sputter system [3].

In this study, we deposit NiO thin films at various substrate temperatures, and working gas pressures and compositions, to determine the film properties as a function of deposition conditions.

II. METHODS AND RESULTS

NiO thin films were deposited on soda lime glass substrates using an AJA Orion RF magnetron sputter system with a 99.9% pure NiO target. Various sputtering conditions were studied with varying gases, pressures and substrate temperatures. All the depositions were done with identical rotations per minute (RPM), operating power, and at a fixed working distance. The deposition time was kept fixed at 1 hour.

The glass substrates are pre-cleaned before introduction into the sputter chamber. The NiO target is pre-cleaned for ~3 minutes with the shutter closed before the deposition is carried out at the desired conditions.

After the films are deposited, the samples undergo 4-point probe, XRD, ellipsometry and UV-VIS measurements to investigate their resistivity, crystallinity, thicknesses and transmission behavior.

Figure 1 shows the XRD spectra for some of the samples in this study. The NiO film deposited with the O_2 has a resistivity of 106 Ω-cm and the XRD data shows the presence of the (200)

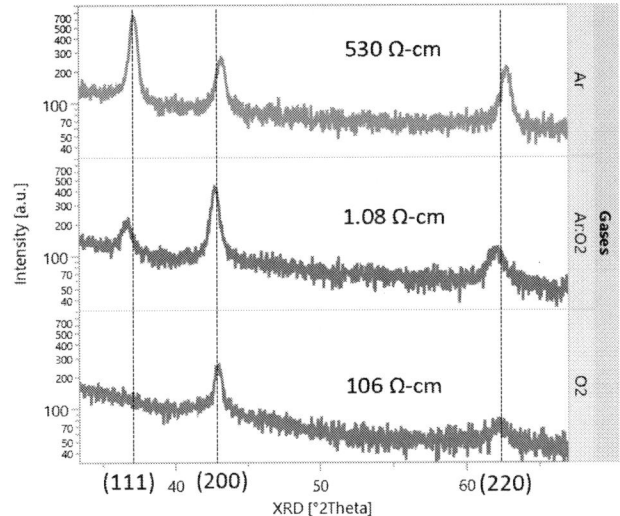

Fig. 1. XRD spectra of NiO films sputtered with different background gases. The resistivity of the films are also shown. Dashed lines show the (111), (200) and (220) diffraction conditions for NiO.

and (220) NiO diffraction peaks. The XRD spectra for a NiO thin film deposited with Ar shows an additional sharp peak corresponding to the (111) diffraction condition and this film has a higher resistivity of 530 Ω-cm. A NiO thin film that was deposited with both Ar and O_2 gas shows the presence of the (200), (220) and (111) diffraction peaks and has a significantly lower resistivity of 1.08 Ω-cm. The XRD results confirm the presence of NiO thin films on soda lime glass substrates.

Figure 2 shows UV-VIS spectra for NiO thin films deposited in pure O_2. With increasing sputtering gas pressure, the transmission of the NiO films increases from 20% to 64%. It can also be observed that with increasing pressure, the resisitivity of the films also increases from 0.004 Ω-cm to 106 Ω-cm. This provides an avenue to tune the transmission and resisitivity of the NiO films by selecting the working gas pressure during deposition. The median Haacke figure-of-merit (FOM) [4] for these NiO films is $1.1 \times 10^{-9} \Omega^{-1}$.

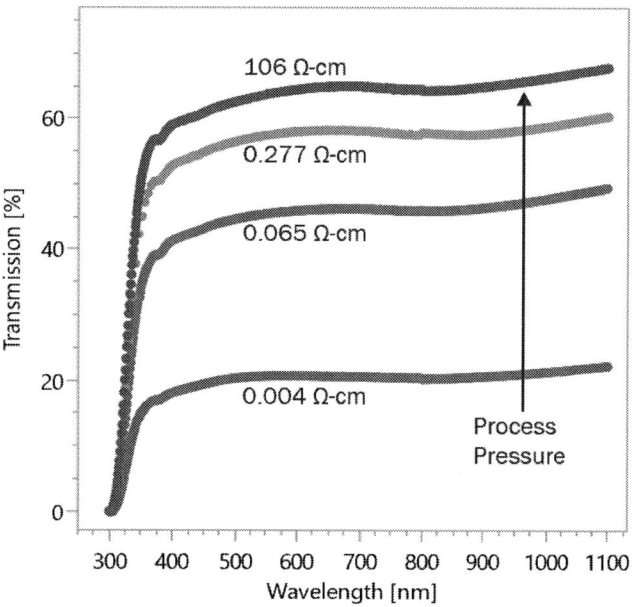

Fig. 2. Transmission spectra of NiO films sputtered at various background pressures in a pure O2 environment and at the same substrate temperature.

Hall effect measurements of the NiO films (carrier density and carrier type) are difficult due to its antiferromagnetic behavior,[5] which interacts with the background magnetic field in the Hall effect measurement system. We were unable to obtain reproducible carrier concentrations and carrier type with standard Hall effect measurements. Instead, we determined the carrier type using hot-point probe measurements. Figure 3 shows the resistivity of NiO thin films as a function of transmission at 730 nm. The thin films with the largest transmission at 730 nm tend to be p-type with resistivities >1

Ω-cm. A significant number of the thin films deposited in this study are p-type, with only 7 thin films showing n-type behavior. Ours results show that the NiO films with resistivity <0.3 Ω-cm are n-type, and those with larger resistivities are p-type. Note that the data in Fig 3 shows the classic behavior for a transparent conductor with an optimal performance near the lower right hand corner where the transmission is greatest and resistivity lowest.

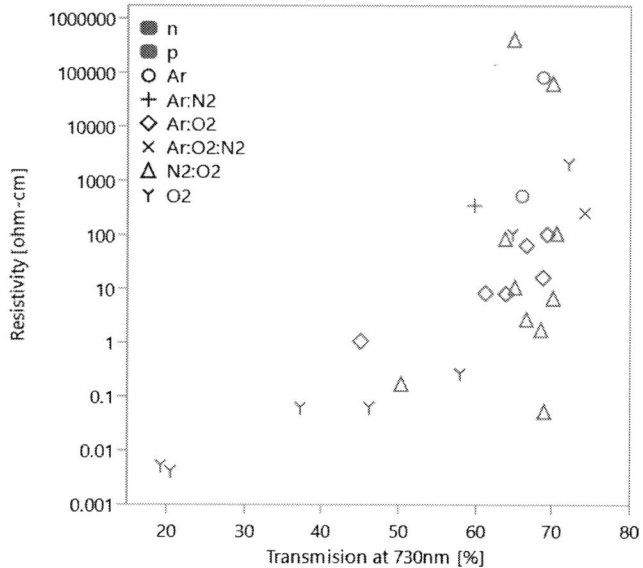

Fig. 3. Resistivity vs Transmission at 730nm for NiO films sputtered with various gas mixtures at the same substrate temperature. The different background gas mixtures are shown as different symbols. p-type NiO films are shown in red and n-type films are shown in blue.

The observation of this carrier type flip has been documented in literature as the oxygen concentration in the gas mixture during RF sputtering is decreased to <30% O_2 [1]. It was reported that n-type layers are more nickel-rich and p-type layers are more oxygen-rich [1]. However, in this study, the carrier type flip is observed regardless of the O_2 concentration during sputtering and is strongly dependent on the conductivity of the films, as determined primarily by the working gas pressure. The mechanism of this phenomenon observed in this study is unknown. Using Kelvin probe measurement, we are able to confirm that despite the apparent carrier type change measured by hot-point-probe, the position of the Fermi-level remained close to the valence band as it would for a p-type material. Fig. 4 shows the work function measurements as a function of resistivity and is > 5.4eV for these films. These results suggest that there is a change in carier type without a

978-1-6654-6060-6/23 $31.00 © 2023 IEEE   490

large change in the work function of the NiO films. We theorize that when the conductivity is low, the dominating carriers are holes with high mobility in the valence band which results in a p-type film. However, when the conductivity increases, this could be attributed to an increase in the mobility of electrons or number of electrons in the defect band. At lower working gas pressures, we suspect that a composition change could increase the acceptor states. This causes the electrons conducting through the defect band to merge with the valence band edge. Defect band electrons at the valence band edge could explain the result in n-type films as measured through hot-point probe.

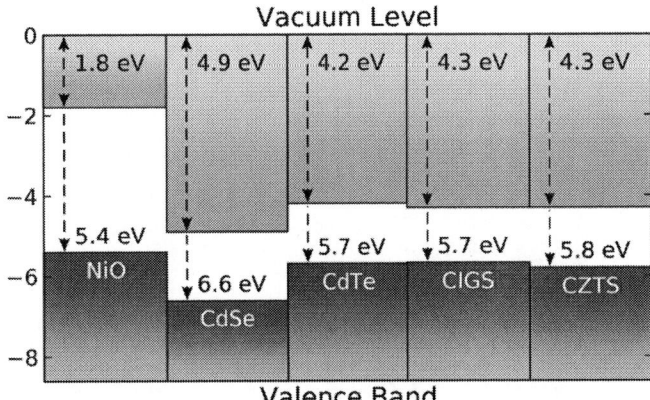

Fig. 5   Band alignment of NiO, CdSe, CdTe, CIGS and CZTS showing the conduction band and valence band relative to the vacuum level.

gas mixtures to determine the film properties as a function of deposition conditions. The carrier type of the NiO thin films changes from p-type to n-type as the resistivity decreases to <1 Ω-cm. The work function of the highly doped films remained close to the valence band-position suggesting that the n-type behavior observed in the NiO films with low resistivities is likely due to polaron conduction. The median Haacke figure-of-merit of $1.1 \times 10^{-9} \ \Omega^{-1}$ suggests that NiO is a promising p+ layer at lower post-deposition temperatures.

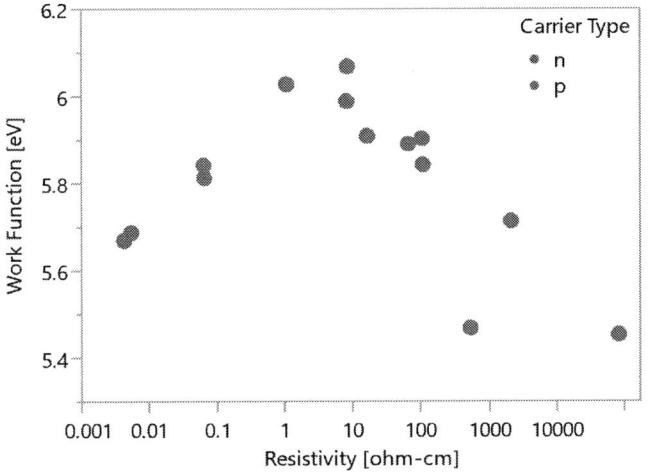

Fig. 4.   Resisitivity vs Work Function for NiO films. p-type NiO films are shown in red and n-type films are shown in blue.

III. DISCUSSION

In Fig. 5 we survey the band-alignment of NiO (Eg= 3.6 eV) with chalcogenide and chalcopyrite thin-film absorbers. NiO forms a type-II heterojunction (staggered gap) with these absorbers and is a promising candidate to form a transparent hole-contact especially with CdTe, CIGS and CZTS thin films. A study by Pintor-Monroy et. al. (2019) [1] investigated NiO thin films that were deposited with varying O_2/Ar gas backgrounds at a fixed substrate temperature of 200C. This study extends this work by characterizing the properties of NiO thin films deposited at colder substrate temperatures and the impact of heat treatments. By annealing conductive NiO films even at 200C in a tube furnace, we observed a degradation in the conductivity. This change in properties would limit the applicability of NiO as p+ layers in processes where high temperature treatment is required in the presence of NiO.

IV. SUMMARY

NiO thin films were deposited with RF magnetron sputtering at various substrate temperatures, background pressures and

REFERENCES

[1] Pintor-Monroy, Maria Isabel and Murillo-Borjas, Bayron L and Catalano, Massimo and Quevedo-Lopez, Manuel A. 2019. "Controlling carrier type and concentration in NiO films to enable in situ PN homojunctions." ACS applied materials and interfaces (ACS Publications) 11: 27048-27056.

[2] Aivalioti, Chrysa and Papadakis, Alexandros and Manidakis, Emmanouil and Kayambaki, Maria and Androulidaki, Maria and Tsagaraki, Katerina and Pelekanos, Nikolaos T and Stoumpos, Constantinos and Modreanu, Mircea and Craciun, Gabriel and others. 2022. "An assessment of sputtered nitrogen-doped nickel oxide for all-oxide transparent optoelectronic applications: The case of hybrid NiO: N/TiO2 heterostructure." In Recent Trends in Chemical and Material Sciences Vol. 6, 86 - 111

[3] Tian, Yuan and Gong, Lianguo and Qi, Xueqian and Yang, Yibiao and Zhao, Xiaodan. 2019. "Effect of substrate temperature on the optical and electrical properties of nitrogen-doped NiO thin films." Coatings (MDPI) 9: 634.

[4] Haacke, G. 1976. "New figure of merit for transparent conductors." *Journal of Applied Physics* (American Institute of Physics) 47: 4086-4089.

[5] Nachman, M and Popescu, FG and Rutter, J. 1965. "Hall Effect in NiO." Physica status solidi (b) (Wiley Online Library) 10: 519-524.

Modeled Impacts of Solar Forecast Error on Utility Production Cost

William B. Hobbs[1], Jenner Tresan[2], Michael Kline[2], Mousumi Guha[2], and Brent Duncan[1]

[1]Southern Company, Birmingham, AL, USA
[2]Guidehouse, Washington, D.C., USA

Abstract—Solar forecast error can be represented in electric utility production cost modeling to assess impacts of uncertainty in different scenarios. We use reforecasts (a.k.a., historical forecasts, hindcasts) that were developed to be concurrent with synthetic solar generation profiles (matching each hour in the same year), along with asynchronous forecasts (i.e., with errors remapped from different hours and/or years), to explore impacts on model results. We demonstrate that production cost increases non-linearly with forecast error, and that our method for remapping asynchronous forecast errors to a solar generation profile produces results that are similar to the concurrent forecasts, but with about 20% lower cost impacts due to forecast error.

Index Terms—production cost modeling, solar forecasting, grid integration

I. INTRODUCTION

Production cost modeling is commonly used by researchers and electric utility planners to assess the operational costs of different generation portfolio scenarios. These models simulate an attempt to optimally dispatch a fleet of assets to meet electric load over a range of time, e.g., every hour in a year. Some modeling tools can mimic real-world uncertainties, e.g., where future solar generation is not known, and imperfect forecasts are used in the model for scheduling assets for dispatch. Forecast data can be archived versions of real forecasts from prior operation, or reforecasts (a.k.a., historical forecasts, hindcasts) can be developed using realistic forecasting methods. In cases where developing reforecasts is not feasible and where archived real forecasts are not concurrent with the time period being modeled (i.e., matching each hour in the same year), an asynchronous reforecast can be used, where forecast errors have been remapped to different hours within the year.

In this research work we explore the impacts of concurrent and asynchronous reforecasts on an approximation of Southern Company's generation fleet with a hypothetical 11 GW portfolio of solar PV (approximately 33% of the 32 GW peak load by nameplate capacity, and 15% of the 176 TWh annual energy). We look at the impact of solar forecast error by linearly scaling the concurrent reforecast errors in different model runs, and we evaluate the suitability of an asynchronous forecast error remapping method by comparing results with the concurrent reforecast run.

TABLE I
GENERATION CAPACITY BY UNIT TYPE USED FOR MODELING.

Unit Type	Installed Capacity (MW)
Combined Cycle (CC)	13,618
Coal (C)	8,036
Energy Storage (ES)	65
Internal Combustion (IC), Gas Turbine (GT)	9,249
Nuclear (N)	8,018
Solar PV (PV)	10,690
Steam Turbine (ST)	2,919
Wind (W)	735

II. METHODS

A. Production Cost Modeling

Modeling was performed using PSO by Polaris via the ENELYTIX platform [1]. First, a capacity expansion model was run for Southern Company's system, using a target of net zero carbon by 2050 [2]. Capital and operating costs of new generation were based on NREL ATB Advanced Technology Innovation Scenarios [3]. Load and weather data were based on 2019, with load data coming from FERC Form 714 data [4], but modified to account for increased building electrification and electric vehicle infrastructure due to the Inflation Reduction Act. Unit retirement assumptions were based on published planned retirement dates and typical operating lifetimes by generation type.

For production cost modeling runs, one year in the mid-2030s was run. A year with high levels of solar penetration but low levels of batteries and other ultra-flexible resources was chosen to best capture the impact of forecast error on production costs. In the forecast, the Southern Company system still relies heavily on gas and coal to meet demand. Table I includes capacity by generation type in the modeled portfolio.

To reflect Southern Company's dispatch, a multi-cycle modeling approach was used. In a multi-cycle production cost simulation, the model makes unit commitment and dispatch decisions at different points in time. Some of the early decisions can be altered at later chronological points in the simulation as the model gains access to new information. Other decisions cannot. For example, for this simulation the decision to commit coal plants is made in the earliest cycle and cannot be changed in later cycles. By contrast, the model has the ability to alter peaking unit dispatch in all cycles. This setup

978-1-6654-6060-6/23 $31.00 © 2023 IEEE

TABLE II
COMMITMENT CYCLE PARAMETERS.

Cycle	Decision Time (hrs)	Horizon Length (hrs)	Look Ahead (hrs)	Resource Types Committed
Week Ahead	25	24	168	C, N, H
Day Ahead	15	24	48	Large CC, ST
4-Hours Ahead	2	1	4	Small CC
Real Time	0	1	0	PV, W, ES, CT

mimics the way real world system operators make decisions to dispatch a portfolio of resources that have different levels of operational flexibility.

Key attributes of cycles are the decision time (the number of hours before the cycle begins that decisions must be made), the look ahead (the number of hours after the cycle ends that the model considers when making decisions for the cycle), and the horizon length (the number of hours decided in the cycle). Cycles with different look aheads were used to dictate which resources were committed and dispatched. A total of 4 cycles were modeled each with different decision times, look ahead, and horizon lengths: a week ahead, a day ahead, four-hour ahead, and real time dispatch. Table II describes the durations, commitment and dispatch decisions for the resource types within Southern Company's System. These cycles nest such that decisions made in the week ahead cycle are passed down to the day ahead, and again into the more granular cycles.

To quantify the impact of solar forecast uncertainly, each cycle was given a solar forecast representative of the expected solar generation at the decision time for each cycle. A description of the solar generation profiles supplied is provided in Section II-B. As the hour of actual dispatch becomes closer, the forecasted solar generation generally becomes closer to the actual observed value, up to the point that the real time dispatch cycle occurs, which uses the actual solar generation. Because commitment and, in some cases, dispatch for less flexible resources is decided in earlier cycles, decreased solar forecast error tends to reduce total production cost and CO2 emissions. This occurs because with access to more accurate information, the model is able to dispatch the thermal portfolio more efficiently.

This process was repeated with a forecast with no solar forecast error (all cycles had the same solar generation profile), with half error (non-real time cycles had solar forecasts halfway between the real time and the originally forecasted values for the cycle length), and with synthetic error (days with similar weather conditions had their forecasts swapped, see below for more detail). The added costs of solar forecast error can be calculated by subtracting the system production costs of the no error case from the production cost from the error cases.

B. Solar Generation Profiles

Solar generation profiles were produced using commercial satellite-based weather data to represent multiple sites across Southern Company's Southeastern US territory. The PV performance model was similar to PVWatts [5] and was run at one-hour intervals, which is the resolution of the weather data used. A generic plant design was assumed, which included single axis tracking, a 1.5 DC to AC ratio, and 10% DC losses. The distribution of the 18 sites across Southern Company's territory approximately matches the peak generating capacity of the three service territories included.

C. Reforecast Data

Reforecast, or hindcast, data was produced through the same commercial service as the satellite-based weather data used for generation profiles. Reforecasts were produced in each hour for each hour beginning at one hour ahead through 168 hours, or one week. The reforecast is intended to be representative of real operational forecasts, except it does not include satellite-based cloud vectoring that might be used for short-term forecasts in the zero to 6 hours ahead range.

Because the solar generation profiles are synthetic, and because they are produced by the same vendor as the reforecast data, the reforecast may not include realistic errors and biases that could occur in real forecasts. However, we found that the error statistics, when looking at single sites and subsets of the 18 sites, were representative of real observed forecast errors.

D. No Error and Reduced Error Reforecasts

To explore the sensitivity of production cost modeling results to forecast error, two additional forecast error datasets were produced: one with no error (perfect forecast in every hour and horizon), and one where the error was reduced by one half. To reduce error by half, we calculated the error in all hours and horizons by subtracting the forecast from the actual generation, multiplied the error value by 0.5, and then added the new reduced error to the actual generation to get the new forecast for each hour and horizon.

E. Remapping Method for Asynchronous Reforecasts

We previously developed a method to remap normalized forecast errors from a pair of concurrent actual and forecasted power profiles (reference profiles) to a new actual power profile. This can be used to take real forecasts from some time range and create a realistic synthetic forecast profile from a different weather time range, resulting in asynchronous forecasts. To make the new forecast more realistic, despite asynchronous weather between reference and new profiles, forecast errors are assigned by matching days based on similar performance (total daily energy) relative to approximate clear-sky generation for the month of year, referred to as clearness index (CI).

Forecast error is normalized to total generation capacity (max observed hourly value for each month). To prevent the remapped new forecast from going below zero or above clear-sky generation, the forecasts are limited to be between those values in each hour. This "clipping" of forecast error does result in some small overall reduction in overall forecast error.

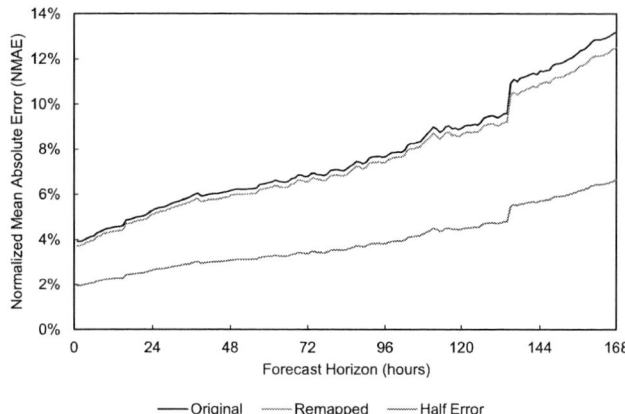

Fig. 1. Annual Normalized Mean Absolute Error values for the three forecast datasets for all horizon hours.

Methods to redistribute these clipped errors are expected to be feasible, but were not attempted in this work.

For this work, the method was modified slightly so that the same weather data could be used for the reference and new (remapped) sets of profiles. We chose not to use forecasts from another year to avoid the possibility of different forecast errors from one year to the next influencing our results. The existing method of matching days based on clearness index would have resulted in duplicating the reference forecasts exactly, so days were shuffled by alternating every-other day, starting with the lowest CI to the highest CI, and for months with odd numbers of days, the highest 3 CI days were shuffled. Other methods were tested, including shifting uniformly up or down by one day and repeating the first or last day, but those methods had three issues: 1) not all days of error were used (one day per month was dropped), 2) one day of error was used twice, and 3) one day of error was not remapped at all.

Normalized Mean Absolute Error (NMAE) values for the forecast datasets are shown in Fig. 1. For the half error case, NMAE is exactly one half of the original forecast's NMAE in all horizona. In the remapped error case, NMAE is very close to the original forecast, but a small reduction in error due to previously discussed "clipping" is visible.

III. RESULTS

Original forecast error ranged from about 4% NMAE at one hour ahead up to about 1% NMAE at 168 hours ahead. The reduced error forecast resulted in an exact 50% (relative) reduction in errors. The remapped forecast errors were about 5-8% (relative) lower than the original forecast error, dependent on the horizon. This slight reduction is due to the "clipping" of errors in some hours.

Annual production costs were about $2,666M for the no forecast error cases, and increased by about $3.4M, $10.1M, and $8.3M for the half error, full error, and remapped error cases, respectively. These values, with 24-hour ahead forecast errors for reference, are included in Table III and Fig. 2. Key features include 1) a non-linear increase in cost with an increase in forecast error (where cost increases at a higher

TABLE III
RESULTS SUMMARY FOR EACH FORECAST ERROR CASE.

Case	24-Hour Ahead Error (NMAE)	Annual Production Cost ($M)	Increased cost due to Forecast Error ($M)
No Error	0.00%	$2,666	–
Half Error	2.65%	$2,669	$3.4
Original	5.29%	$2,675	$10.1
Remapped	5.11%	$2,674	$8.3

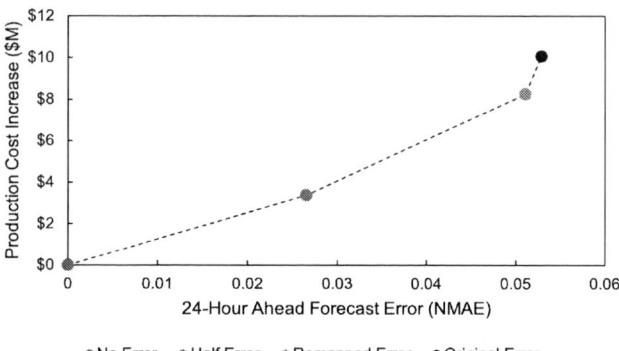

Fig. 2. Increase in production cost, relative to no error, and 24-hour ahead forecast error for the range of cases.

rate with higher error); and 2) while remapped (asynchronous) forecasts result in about 5-8% less error and about 18% less cost increase than original ("full", synchronous) error, the remapped forecast is much more similar to the original forecast than it is the half-error forecast.

To demonstrate cost impacts of forecast error at a more granular level than annual costs, Fig. 3 illustrates cost differences on a daily basis for each case relative to the no error case. Approximately 70-80% of days result in an increase in cost when forecast error is introduced, with the highest cost days in the original and remapped error cases costing over $500,000 more than the no error case. The 20-30% of days with cost savings relative to the no error case is counter-intuitive, but may be partially explained by sub-optimal scheduling of units with minimum up- or down-times that span multiple days. We ran a similar analyses with five-day rolling average costs, and the fraction of intervals where forecast error resulted in unexpected savings was reduced by about 10 percentage points, to approximately 10-20% of days. This indicates that forecast error can result in savings on individual days but increased costs over longer time spans. The remaining fraction of intervals with savings due to error may be due to sub-optimal unit scheduling around outages, but more investigation is needed.

Solar curtailment and total emissions are shown in Table IV. Solar curtailment has a clear trend of increasing with forecast error in each case, illustrated in Fig. 4, while CO_2 emissions are less strongly correlated. Table V includes production cost by unit type and forecast error case, and Table VI shows annual energy generated by unit type and forecast error case.

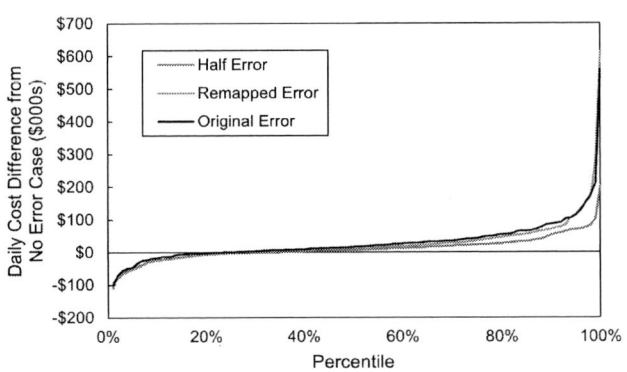

Fig. 3. Daily cost difference from the no error case in thousands of dollars, where positive values indicate an increase in cost.

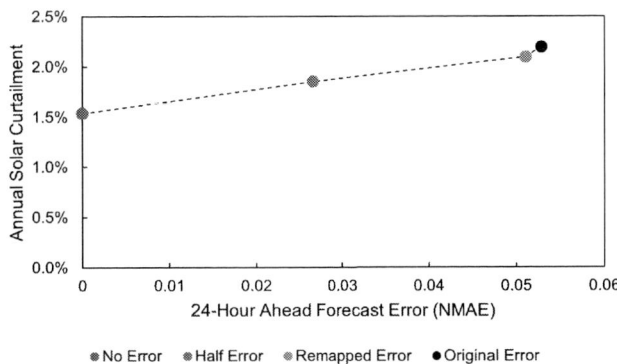

● No Error ● Half Error ● Remapped Error ● Original Error

Fig. 4. Annual solar curtailment, as a percentage of potential solar energy generation, and 24-hour ahead forecast error for the range of cases.

TABLE IV
SOLAR CURTAILMENT AND EMISSIONS FOR EACH FORECAST ERROR CASE.

Case	Solar Curtailment	Emissions (000s Tons)		
	(MWh, %)	CO_2	NOX	SO2
No Error	423,466, 1.5%	49,822	13	2
Half Error	509,908, 1.9%	49,914	13	2
Original	604,537, 2.2%	50,044	13	2
Remapped	577,067, 2.1%	49,822	13	2

TABLE V
ANNUAL COSTS BY UNIT TYPE FOR EACH FORECAST ERROR CASE.

Unit Type	Cost by Case (2020 000$)			
	Original	Half	Remapped	No Error
PV	$0	$0	$0	$0
ES	$0	$0	$0	$0
IC/GT	$104,363	$95,087	$99,430	$94,369
ST Gas	$14,361	$15,241	$15,198	$15,292
Nuclear	$420,892	$420,893	$420,892	$420,893
CC	$1,675,667	$1,680,408	$1,687,087	$1,680,417
Wind	$0	$0	$0	$0
Coal	$460,291	$457,259	$451,159	$454,540

TABLE VI
ANNUAL GENERATION BY UNIT TYPE FOR EACH FORECAST ERROR CASE.

Unit Type	Generation by Case (GWh)			
	Original	Half	Remapped	No Error
PV	26,947	27,042	26,975	27,128
ES	-23	-22	-24	-23
IC/GT	2,613	2,394	2,499	2,376
ST Gas	389	412	409	412
Nuclear	62,528	62,528	62,528	62,528
CC	59,825	59,995	60,196	59,992
Wind	1,835	1,838	1,836	1,838
Coal	22,147	22,074	21,842	22,010

To illustrate changes in dispatched generation type in different error cases, two sample days are shown in Fig. 5, representing an overforecast, where solar produced less than was forecasted, and an underforecast, where solar produced more than was forecasted. Cost savings for under forecasted solar is primarily seen through reduced curtailment. If the forecasted solar is below actual solar, the system will commit more resources than necessary, resulting in higher CC generation and lower IC/GT generation which is a more efficient dispatch, but results in more solar curtailment causing a higher level of thermal generation to be online increasing costs and emissions. A system with no forecast error will commit the necessary amount of CCs and will use IC/GTs to fill in the ramp periods around the full solar generation. In contrast, over forecasted days result in excessive IC/GT generation as an insufficient amount of the more efficient CC units are committed and peakers have to fill in for non-peak conditions. At the studied level of penetration baseload generation does not significantly change as solar only makes up 33% of peak load. However, at higher penetrations, a much more dramatic separation in costs could be observed if a coal unit were to be turned off or on unnecessarily. The above pattern described would be the same, but the cost impacts would be more severe.

IV. DISCUSSION

We have demonstrated the impacts of solar forecast error on annual utility operations and costs, for a single solar penetration case of approximately 11 GW, or 33% of peak load by nameplate capacity and 15% of annual energy.

A. Remapping Results

The remapping process demonstrated here explores how asynchronous forecasts might compare with synchronous forecasts in solar integration studies using production cost models. Relative to no error or a 50% relative reduction in error, the remapping process produces production cost results that close to results using the original synchronous error.

Because the method resulted in slight decreases in error, and error reduction has a disproportionate reducing effect to production cost, it can be assumed that the at least some of

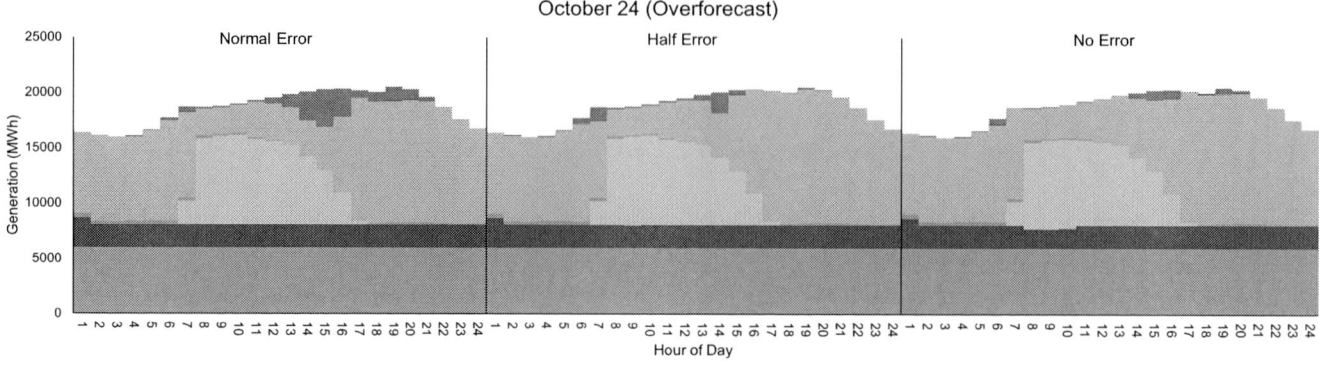

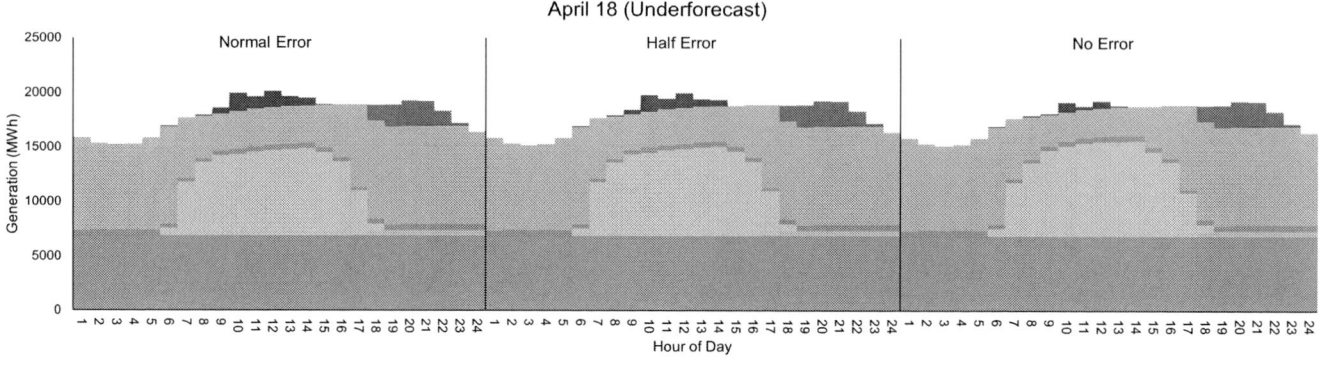

Fig. 5. Generation by source for two sample days with an overforecast (top, solar produced less than was forecasted) and underforecast (bottom, solar produced more than was forecasted) and , showing the original error (left), error reduced by half (middle), and no error (right) cases.

the production cost differences are due to the error reduction. An improved remapping method that does not clip errors, and instead reassigns them in a more complex way, may address this.

B. General Results

The overall impacts of forecast error in this work can be compared with previous work by Martinez-Anido, et al. [6]. Their work focused on ISO-NE in the Northeast US, looking at solar penetrations of 0, 4.5, 9, 13.5, and 18% by energy, and uniform forecast error improvements of 25, 50, 75, and 100%. Their annual production costs were similar to our work (approximately $2.5-3B/yr for solar penetrations of 9-18% by energy). Interpolating their results to a solar penetration of about 15%, their 100% error reduction results in about $23M in annual savings compared to our $10.1M, and their 50% error reduction results in about $15M in savings compared to our $6.7M in savings when going from full to half error. In both cases, our savings are about 2.2-2.3 times lower than in [6]. See Fig. 6 for an illustration of these values. Reasons for this difference could include differences in regional climate, generation mix and fuel cost, production cost modeling limitations, and solar forecast models used, with the expectation that a commercial forecast in 2022/2023 should be better than forecasts available in 2016. Considering

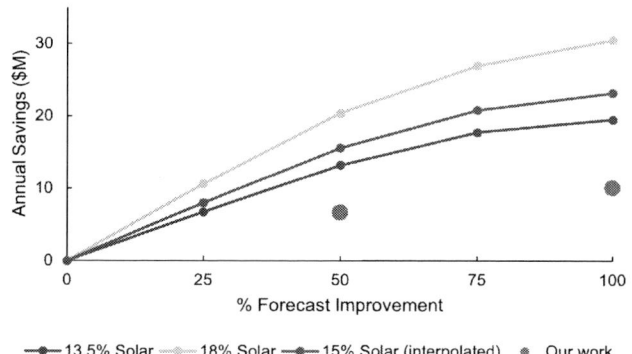

Fig. 6. Annual cost savings for ISO-NE from improved solar forecasts at 13.5% (orange) and 18% (purple) solar penetration, adapted from [6], along with linearly interpolated values for 15% penetration (gray), which can be compared to our results (red).

potential forecasts improvements alone, we think that our results compare well those from [6].

REFERENCES

[1] Polaris Systems Optimization, Inc., "Power System Optimizer (PSO)." [Online]. Available: https://www.enelytix.com/home/pso

[2] "Implementation and action toward net zero," Southern Company, Atlanta, GA, Report, September 2020. [Online]. Available: https://www.southerncompany.com/content/dam/southern-company/pdf/public/Net-zero-report.pdf

[3] "2022 Annual Technology Baseline," NREL (National Renewable Energy Laboratory),Golden, CO, Data Set, 2022. [Online]. Available: https://atb.nrel.gov/

[4] Federal Energy Regulatory Commission, "Form No. 714 - Annual Electric Balancing Authority Area and Planning Area Report," December 2021. [Online]. Available: https://www.ferc.gov/industries-data/electric/general-information/electric-industry-forms/form-no-714-annual-electric/data

[5] A. P. Dobos, "PVWatts version 5 manual," National Renewable Energy Laboratory, Golden, CO, Tech. Rep. NREL/TP-6A20-62641, 2014. [Online]. Available: https://doi.org/10.2172/1158421

[6] C. Brancucci Martinez-Anido, B. Botor, A. Florita, C. Draxl, S. Lu, H. Hamann, and B.-M. Hodge, "The value of day-ahead solar power forecasting improvement," *Solar Energy*, vol. 129, pp. 192–203, 05 2016.

Durability Testing of Porous SiO$_2$ Anti-reflection Coatings for Solar Cover Glass

Adam M Law, Luke O Jones, Michael Nasser, Ali Abbas, and John M Walls

CREST, Wolfson School of Mechanical and Manufacturing Engineering,
Loughborough University, Loughborough, LE11 3TU, United Kingdom

Abstract—Solar photovoltaic (PV) modules experience an optical loss of just over 4% at the front cover glass surface, as a result of the difference in refractive index between glass and air. This loss can be reduced by applying an anti-reflection (AR) coating, and currently over 90% of commercial modules contain a single layer of porous silica (SiO$_2$). These AR coating are effective at reducing reflection losses, but have been shown to exhibit poor durability, especially to abrasion caused by module cleaning. In this work, porous SiO$_2$ coated glass samples have been subject to dry linear abrasion testing, using an Elcometer abrasion tester with a brush and test fixture adapted from a SunBrush, a brush used in industrial module cleaning operations. The coating shows significant degradation in AR performance after abrasion, with almost complete removal after 1000 cycles, with weighted average reflectance (WAR) values increasing from 5.42% to 7.70%. A multilayer AR coating with all-dielectric metal oxide layers has been tested alongside the porous SiO$_2$ to provide a comparison, and shows significantly higher abrasion resistance with very little change in reflectance after 1000 abrasion cycles, offering a durable alternative to the current industry standard.

Index Terms—solar, photovoltaics (PV), anti-reflection (AR), coatings, durability, abrasion

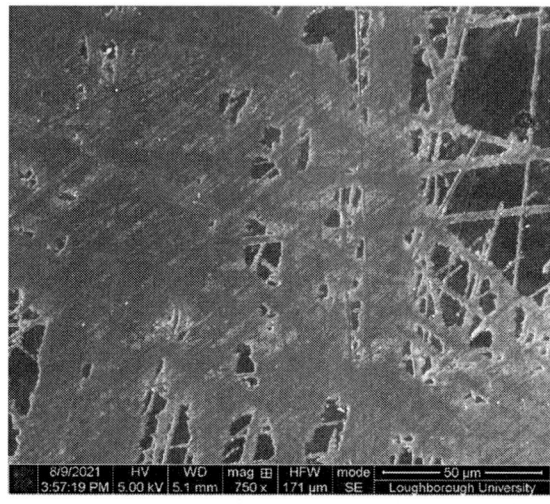

Fig. 1. SEM image showing severe abrasion damage to the AR coating on module cover glass after 6 years in operation

I. INTRODUCTION

Almost all solar photovoltaic (PV) modules include a sheet of cover glass at the front surface, to provide mechanical and chemical protection as well as high optical transmission into the absorber. PV modules experience a loss of over 4% from reflection of incident light at the front glass surface, as a result of the refractive index mismatch between glass and air. This loss can be reduced by depositing an anti-reflection (AR) coating on the glass surface. The most common type of AR coating is a single layer coating with a refractive index between that of glass and air, and the ideal value for this is ~1.22. To obtain a refractive index this low, porous structures are required, introducing voids into a material to lower its refractive index. SiO$_2$ is often used, with a refractive index of 1.47. Adding porosity to the coating can reduce the refractive index to the low values required by AR coatings, and porous SiO$_2$ AR coatings are now the industry standard, covering over 90% of commercial modules [1].

Porous SiO$_2$ AR coatings are effective at reducing reflection losses by up to 3%, but the use of porous structures compromises the durability of the coating. These coatings are especially vulnerable to abrasion which occurs when the modules are cleaned, and multiple studies have highlighted the poor abrasion resistance of these coatings [2]–[6]. PV

modules are regularly cleaned, with varying frequency depending on location, to remove the build-up of dust and dirt on the surface, known as soiling. Soiling is a significant problem for PV modules and can result in power losses as high as 50% in desert areas [7], so module cleaning is necessary. The poor durability of current AR coatings, however, has raised questions over their suitability. Recently, Lange et al. [8] has suggested that in areas with high soiling, and subsequently a higher frequency of cleaning, it may not be worth using a porous SiO$_2$ AR coating at all, given how quickly it would be removed. Fig. 1 shows an SEM image of a piece of module cover glass taken directly from a UK PV utility after ~6 years. Modules in the UK are typically cleaned twice a year, meaning this module has been subject to ~12 cleaning cycles. The abrasion damage to the AR coating as a result of cleaning is clearly visible, and large parts of the coating have been removed entirely.

In this work, porous SiO$_2$ samples have been subject to linear abrasion testing using a brush adapted from a SunBrush, a type of brush used in PV module cleaning operations. Samples have undergone up to 1000 abrasion cycles, with reflectance measurements taken after each set. Multilayer all-dielectric AR coatings described in previous work [9] have

978-1-6654-6060-6/23 $31.00 © 2023 IEEE

been included as a direct comparison, as these coatings have been developed for superior durability.

II. EXPERIMENTAL DETAILS

Samples of commercial porous SiO_2 AR coated glass samples, 5cm x 10cm x 3mm thick, have been obtained for abrasion testing. Multilayer antireflection (MAR) coatings of 6 layers of alternating SiO_2 and ZrO_2 are used for a direct comparison, and details of the deposition process for these coatings can be found in [9]. Linear abrasion testing is performed using an Elcometer 1720 Washability Tester (Fig. 2), for varying numbers of cycles up to 1000, at a speed of 37 cycles per minute. The brush used is a bespoke design adapted from the Elcometer brush fixtures, using bristles taken from a Sunbrush, a type of brush used directly in solar utilities for cleaning of PV modules. The existing brush fixtures compatible with the abrasion tester have been combined with bristles from a Sunbrush, to provide a realistic test. The bristle length has been shortened to fit into the existing test fixture, with the characteristic splayed ends of the bristle retained. A weight of 3.5N has been applied. After testing, samples are washed with DI water and dried with compressed air to remove any loose debris prior to characterisation.

Reflection measurements are taken before and after abrasion testing using a Varian Cary5000 UV-VIS-NIR spectrophotometer with an integrating sphere attachment, across a wavelength range of 300-1200nm. Weighted average reflectance (WAR) values, which account for the AM1.5g solar spectrum, are calculated using Equation (1). Where Φ is the photon flux and R is the percent reflectance at each given wavelength, λ. The calculations were performed using a custom Matlab script.

$$WAR(\lambda_{min}, \lambda_{max}) = \int_{\lambda_{min}}^{\lambda_{max}} \frac{\Phi \cdot R}{R} d\lambda \qquad (1)$$

III. RESULTS & DISCUSSION

A. Reflectance Measurements

The measured reflectance of porous SiO_2-coated glass, MAR-coated glass, and uncoated glass between 300-1200nm, prior to any abrasion testing, is shown in Fig. 4. The porous SiO_2 has an initial WAR of 5.42%, with 5.83% for the MAR-coated glass. The uncoated glass has a WAR of 7.82% including back-surface reflectance. All measurements in this work include the uncoated back surface of the glass. Noise in the data caused by a detector changeover in the spectrophotometer at 800nm has been smoothed, using a 15-point adjacent averaging method, to reduce the impact on data presentation.

The measured reflectance for the porous SiO_2 samples after each set of abrasion cycles is shown in Fig. 5. The porous SiO_2 samples show a significant amount of degradation following

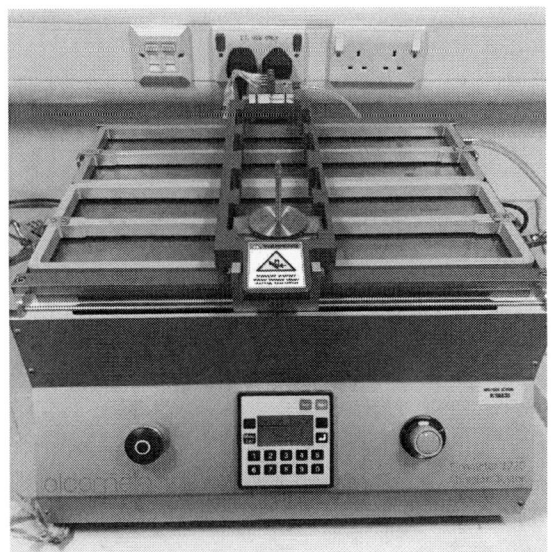

Fig. 2. The Elcometer 1720 washability tester used for abrasion testing

Fig. 3. The bespoke brush fixture used for abrasion testing alongside a section of bristle

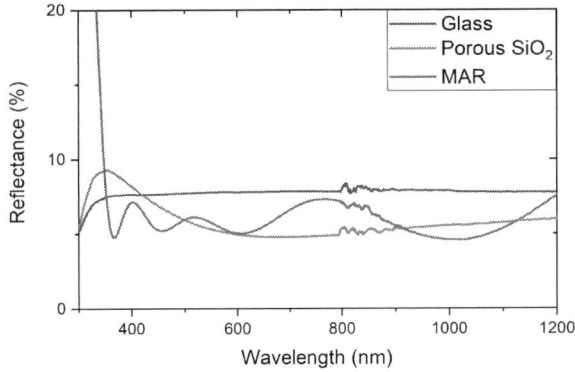

Fig. 4. Measured reflectance of as-received porous SiO_2-coated glass and MAR-coated glass, compared to uncoated glass, prior to abrasion testing

the abrasion testing, with reflectance increasing after each set of cycles. The majority of the damage can be seen in the first 100 cycles, with the WAR increasing from 5.53% to 6.21% and 6.83% after 50 and 100 cycles, respectively, suggesting that a significant amount of the coating has been removed. The observed increase in reflectance magnitude and the shift in the reflectance minimum are indicative of both coating thinning and removal occurring simultaneously [3]. Subsequent abrasion cycles result in further damage, increasing WAR to 7.70% after 1000 cycles, close to that of uncoated glass.

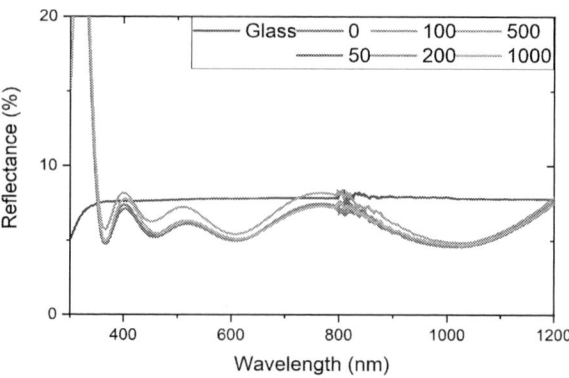

Fig. 6. Measured reflectance of MAR-coated glass samples after each set of abrasion cycles, with a comparison to uncoated glass

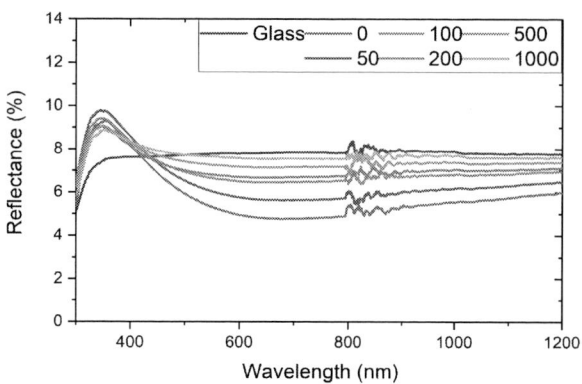

Fig. 5. Measured reflectance of porous SiO_2-coated glass samples after each set of abrasion cycles, with a comparison to uncoated glass

In contrast to the porous SiO_2, the MAR-coated sample show very little damage after 1000 cycles (Fig. 6), with WAR increasing very slightly from 5.83% to 5.98%. The WAR increases in small increments with increasing cycles, suggesting a small, but insignificant, amount of damage is occurring to the coating. After 500 cycles, however, the reflectance increases by 0.7%, suggesting a larger amount of damage has occurred, but for the sample undergoing 1000 cycles the reflectance is much closer to the as-deposited coating. This anomaly at 500 cycles may be caused by additional damage resulting from a piece of debris e.g., from the brush itself, or a piece of dirt, glass shard etc. being dragged across the surface by the brush. In general, the MAR coating exhibits much greater resistance to abrasion than the porous SiO_2 coating, suggesting a longer lifetime in the field if it were to be applied to module cover glass. A longer lifetime given by the AR coating would improve power output over the lifetime of the PV module.

The WAR of both coatings after each set of abrasion cycles is shown in Fig. 7. A red dashed reference line is included to show the WAR of uncoated glass, highlighting the increase in reflectance of the porous SiO_2 coated samples, with most of the damage caused in the first 100 cycles. The MAR coating initially has a higher WAR than the porous SiO_2, although after 50 cycles of abrasion this is no longer the case.

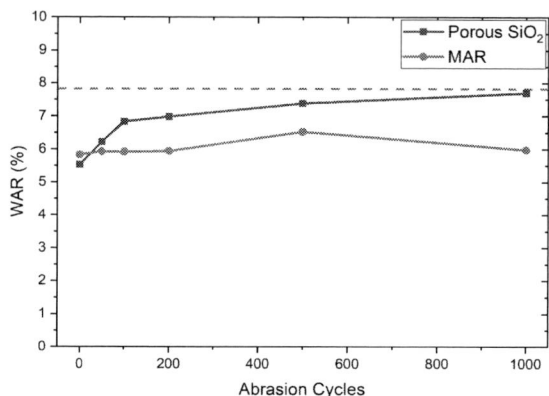

Fig. 7. Weighted average reflectance (WAR) for each set of abrasion cycles for both the porous SiO_2 and MAR coating, with a reference line for uncoated glass

IV. SUMMARY & CONCLUSIONS

Single layer porous SiO_2 AR coatings are applied over 90% of PV modules and are effective at reducing front surface reflection losses by up to 3%. However, the durability of these coatings is poor, and they are vulnerable to abrasion damage from the cleaning processes used to remove soiling. Abrasion testing has been carried out on both porous SiO_2 and MAR coated glass samples using an Elcometer washability tester with a brush fixture adapted from a SunBrush, which is used in commercial PV utility maintenance operations. The adaptation of a commercially-used brush is intended to bring accelerated testing closer to real-world results. The porous SiO_2 shows considerable damage as a result of abrasion, with 1000 abrasion cycles resulting in an increase in WAR from 5.42% to 7.70%, with reflectance curves indicative of both coating removal and thinning. The MAR coating on the other hand, shows excellent abrasion resistance with very little change after 1000 cycles, as WAR increases slightly from 5.83% to 5.98%. This work highlights the vulnerabilities of porous SiO_2 AR coatings to abrasion, and shows that MAR coatings with all-dielectric layers offer a durable alternative.

978-1-6654-6060-6/23 $31.00 © 2023 IEEE

ACKNOWLEDGMENT

The authors are grateful to Everblue Solar for providing test materials.

REFERENCES

[1] VDMA. International Technology Roadmap for Photovoltaic, 2021.

[2] Klemens Ilse, Charlotte Pfau, Paul-T Miclea, Stephan Krause, and Christian Hagendorf. Quantification of abrasion-induced arc transmission losses from reflection spectroscopy. In *2019 IEEE 46th Photovoltaic Specialists Conference (PVSC)*, pages 2883–2888. IEEE, 2019.

[3] Muhammad Zahid Khan, Charlotte Pfau, Matthias Schak, Paul-Tiberiu Miclea, Volker Naumann, Ahmed Debess, Christian Hagendorf, and Klemens Ilse. Resilience of industrial pv module glass coatings to cleaning processes. *Journal of Renewable and Sustainable Energy*, 12(5):053504, 2020.

[4] B. Figgis and V. Bermudez. Pv coating abrasion by cleaning machines in desert environments–measurement techniques and test conditions. *Solar Energy*, 225:252–258, 2021.

[5] Jimmy M Newkirk, Illya Nayshevsky, Archana Sinha, Adam M Law, QianFeng Xu, Bobby To, Paul F Ndione, Laura T Schelhas, John M Walls, Alan M Lyons, et al. Artificial linear brush abrasion of coatings for photovoltaic module first-surfaces. *Solar Energy Materials and Solar Cells*, 219:110757, 2021.

[6] Adam M Law, Farwa Bukhari, Luke O Jones, Ali Abbas, and John Michael Walls. Testing the abrasion resistance of porous sio 2 anti-reflection coatings for solar cover glass. In *2022 IEEE 49th Photovoltaics Specialists Conference (PVSC)*, pages 0786–0791. IEEE, 2022.

[7] Muhammed J Adinoyi and Syed AM Said. Effect of dust accumulation on the power outputs of solar photovoltaic modules. *Renewable energy*, 60:633–636, 2013.

[8] Katja Lange, Charlotte Pfau, Erik Grunwald, Matthias Schak, Eric Matthes, Stefan Grob, Marko Turek, Christian Hagendorf, and Klemens Ilse. Abrasion testing of anti-reflective coatings under various conditions. *Solar Energy Materials and Solar Cells*, 240:111732, 2022.

[9] Adam M Law, Farwah Bukhari, Luke O Jones, Patrick JM Isherwood, and John M Walls. Multilayer antireflection coatings for cover glass on silicon solar modules. *IEEE Journal of Photovoltaics*, 12(5):1205–1210, 2022.

Performance Evaluation of Perovskite Solar Cell Modules with Tilt Angle Optimization in BIPV application: A Case Study for Kazakhstan

Yerassyl Olzhabay, Ikechi Ukaegbu*, Annie Ng*

Department of Electrical and Computer Engineering, School of Engineering and Digital Sciences,
Nazarbayev University, Kabanbay Batyr Ave. 53, Astana 010000 Kazakhstan
Corresponding emails: ikechi.ukaegbu@nu.edu.kz, annie.ng@nu.edu.kz

Abstract — Perovskite solar cells (PSCs) have gained increasing attention in the fields of Internet-of-things (IoT) and building integrated photovoltaics (BIPV) applications. PSCs have promising indoor performance for powering indoor IoT devices and potentially low module costs for applications in BIPVs. This study focuses on the performance evaluation of perovskite solar cell modules (PSCMs) integrated with the roof of buildings located in Astana, the capital of Kazakhstan. The results obtained in this work aim to show the possibility of deploying perovskite PV modules in this region for applications in BIPVs. In this work, the energy harvesting and generation of tilted angles of PSCM installation were investigated for a full cycle of one year. Based on the results, the energy generation from the roof PSCMs increased by 16.1 %, 17.3 %, and 19.5 % when 32-degree, seasonal, and monthly tilt angles were applied for installation. Combining 32-degree and seasonal tilt angles can result in a 20 % energy generation increase.

I. INTRODUCTION

PSCs are one of the emerging photovoltaic (PV) technologies that use a perovskite-structured material as the active layer for converting sunlight into electricity. These cells have attracted significant attention in recent years because of their high efficiency, low cost, and versatility. Perovskite materials possess promising optoelectronic properties such as strong absorption of light and high charge carrier mobility. The efficiency of PSCs has been rapidly increasing in recent years, with laboratory cells reaching efficiencies of 25.7% [1]. While this is still below the efficiencies of traditional silicon-based solar cells, PSCs have the potential to be significantly less expensive to manufacture in the future.

BIPV integrates photovoltaic cells into the building envelope, such as the roof or walls, to provide structural and electrical functionality. BIPV systems can be used for both new construction and existing buildings. PSCs have been considered a suitable technology for BIPV applications due to their high efficiency, mechanical flexibility, and ease of processing. PSCs can be easily printed, coated, or deposited onto various building materials, like glass and metal, making them suitable for the BIPV system. The building structures such as windows [2], facades [3], [4], roofs [5], and bus stop shelters [6], [7] are typical examples of integrating with PVs.

The tilt angle for a solar panel on the roof should be considered since it determines the desirable light incident angle, maximizing the sunlight the solar panels can harvest throughout the year [8]. For PSCMs, the optimal tilt angle is one of the critical installation criteria because it directly affects the electrical power generation from the solar module. Hence, to show the possibility of deploying perovskite PV in buildings, it is imperative to find out the desirable angles for installing solar panels and then compare the difference in the energy generation from perovskite-powered energy harvesting systems with and without tilting. In this work, for the Central Asian region, Astana in Kazakhstan is taken as a case study.

II. MAIN EXPERIMENTAL AND/OR THEORETICAL RESULTS

This study considers a roof of the building with nontransparent PSCMs. The energy harvesting system consists of PSCMs, maximum power point tracker (MPPT), batteries, and loads.

A. Perovskite solar cell modules

The PSCMs consist of multiple PSCs connected in series and parallel to yield a higher voltage and current than the individual PSC. This study considers a nontransparent PSCM with a power conversion efficiency (PCE) of 12.6 % and an active area of 354 cm^2 [9]. This PSCM is placed on the roof of the building. The PSCM installation on the roof has a 1.5 by 4.5 meters dimension and contains 100 PSCMs. According to the report of Jacobson and Jadhav, the desirable tilting angle for photovoltaic cells in Kazakhstan is 32° [10].

B. Maximum power point tracking

MPPT is a technique used to optimize the performance of a PV system by extracting the maximum power that the PV module can generate under varying environmental conditions. The Hill Climbing Algorithm is one of the MPPT techniques used to maximize the power that can be extracted from a PV module by adjusting the operating point of the module to the point at which the power output is maximized. The algorithm works by monitoring the power output of the PV module and then adjusting the operating point of the module in small increments in the direction that increases the power until the maximum power point is found until the maximum power is reached. The algorithm starts with an initial operating point and then repeatedly takes small steps

C. Energy harvesting system

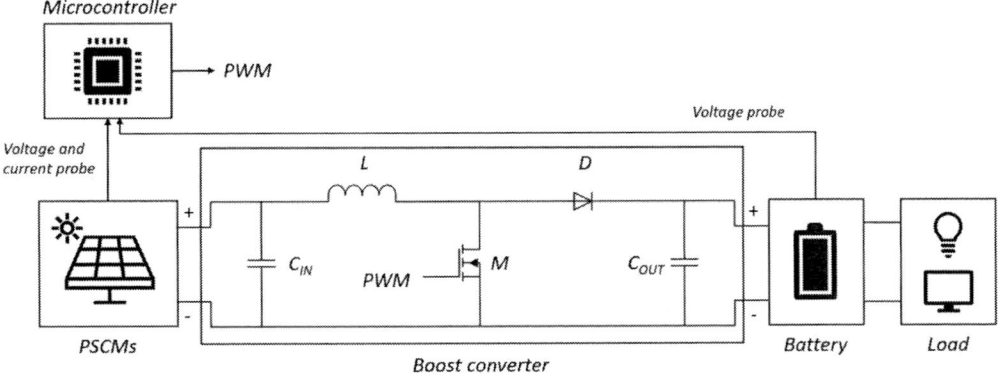

Fig. 1. Energy harvesting circuit block diagram.

The harvesting system uses PSCMs to convert sunlight into electrical energy. Fig. 1 illustrates the circuit schematic of the proposed energy harvesting system used in this work. PSCM is connected to the boost converter to match the battery voltage. The boost converter consists of the inductor L, input C_{IN} and output C_{out} capacitors, MOSFET M, and diode D.

A microcontroller is used to run the MPPT algorithm. Voltage and current probes of PSCMs are required for MPPT operation. The voltage probe from the battery is used for safe battery charging.

The electrical energy generated by the PSCMs is then stored in batteries for use when the PSCM energy generation is insufficient. This allows the system to continue providing power to the load, even during periods of low sunlight or at night.

III. RESULTS ANALYSIS

Simulations were conducted in Matlab Simulink to evaluate the performance of the energy harvesting system. Irradiation data used in simulations was retrieved from PVsyst software. Non-tilted PSCMs have no tilt angle between the PSCM and horizon plane, i.e., no tilt angle between PSCMs and flat roof. The energy generation of 15.7, 28.1, 52.9, 65.6, 88.7, 85.7, 79.5, 68.5, 47.7, 29.9, 15.5, and 11.7 kWh from January to December, respectively. Fig. 2 illustrates the energy generation from the roof with tilting PSCMs and non-titling PSCMs. The investigation was performed on the case of a tilting angle of 32-degree, seasonal, and monthly tilting. Astana's seasonal tilt angle is 51.2 degrees for Spring and Fall due to the vernal and autumnal equinox. For Astana's latitude of 51 degrees, the seasonal tilt angles for the winter and summer periods are 74.6 and 27.7 degrees, respectively. Tilt angles for every month are calculated according to Karafil *et al.* [8]. Monthly tilt angles are 71.7, 64.0, 53.0, 41.1, 32.0, 27.9, 30.1, 38.0, 49.4, 61.2, 70.3, and 74.1 degrees from January to December.

According to Fig. 2, the PSCM tilted at 32 degrees yearly generated more energy than without a tilt angle for most of the year. Seasonally tilted PSCMs improve the energy generation

TABLE I
COMPARISON BETWEEN ENERGY GENERATIONS IN ASTANA

	Increase in roof PSCM energy generation compared to the non-tilted scenario [%]		
	32 degree tilt	Seasonal tilt	Monthly tilt
January	82.2	107.1	108.7
February	64.0	73.5	79.8
March	35.7	38.8	38.4
April	12.1	5.2	10.1
May	-0.3	-11.7	0.7
June	-5.7	-1.9	-2.0
July	-4.3	-0.9	-1.9
August	5.4	7.0	4.3
September	21.3	18.6	19.3
October	39.2	45.1	42.7
November	53.7	66.6	63.7
December	99.7	136.9	137.4
Year	16.1	17.3	19.5

of the non-tilted PSCMs; however, such methods decrease energy generation between May and June. Table I summarize the energy difference generated between non-tilted and tilted PSCMs (32 degrees, seasonal tilt, and monthly tilt). Non-tilted

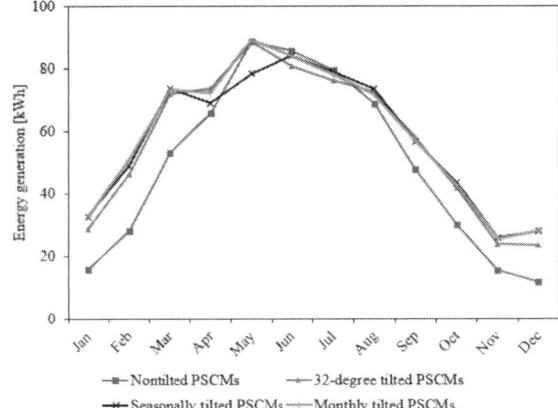

Fig. 2. Energy generation from non-tilting and tilting PSCMs on the roof in Astana.

PSCMs generate 589 kWh annually. Notably, the most significant difference in energy generation (for 32-degree tilt) is in winter, where the tilted PSCMs have an increase of 82.2 %, 64.0 %, and 99.7 % for January, February, and December, respectively. For seasonal tilt, the energy generation increases are 107.1 %, 73.5 %, and 136.9 % for the same winter months. Monthly tilt angles demonstrate 108.7 %, 79.8 %, and 137.4 % increases in energy generation for January, February, and December.

Although energy generation from 32-degree tilted PSCMs decreased in summer. However, the total energy generation of 32-degree tilted PSCMs is 16.1 % more compared to the non-tilted scenario. Monthly tilted PSCMs generate the highest amount of energy among all scenarios present. The energy generation difference between seasonal and monthly tilted PSCMs is small and negligible throughout the year except April, May, and September. It is possible to generate energy by combining 32-degree and seasonal tilt angles instead of monthly tilted angles. In such a combined scenario, the energy generation is increased by 20 % compared to the non-tilted scenario.

IV. DISCUSSION OF THE SIGNIFICANCE OF THE WORK FOR THE FIELD

Having the proper tilt angle can significantly affect the overall performance and energy yield of the PSCMs. For example, a module installed at the optimal tilt angle will receive more sunlight and generate more electricity than one installed at a less optimal tilt angle. Such a study can demonstrate the importance of the optimal tilted angle of PSCMs in BIPV systems, leading to better performance and economic benefits for the installations. Every situation has an optimal tilt angle to generate maximum energy. According to Fig. 2, non-tilted PSCMs have the lowest energy generation compared to tilted PSCMs. The yearly 32-degree tilt angle is lower than seasonal and monthly tilt angles, especially in winter and summer.

Moreover, according to Table 1, total annual PSCM energy generation with seasonal tilt resulted in an increase of 17.3 % than non-tilted PSCM energy generation. The monthly tilted angle of PSCMs resulted in an increase of 19.5 % than the non-tilted scenario. Notably, achieving a 20 % energy generation increase is possible by combining 32-degree and seasonal tilt angles to achieve higher value. Combining all tilt angles to achieve a higher energy generation increase to achieve 20.3 % is possible by adjusting PSCM tilt more frequently than using only 32-degree and seasonal tilt angles.

V. SUMMARY OF THE WORK

In summary, the tilt angle is critical for PSCMs, as it directly affects the amount of sunlight they receive and, thus, the amount of electricity they can generate. The optimal tilt angle varies depending on the location and design of the module, and it is vital to consider the installation site's specific conditions when determining the optimal tilt angle for PSCMs. According to the results, there is an increase in roof PSCM energy generation for buildings in Astana by 16.1 %, 17.3 %, and 19.5 % for 32-degree, seasonal, and monthly tilt angles. Moreover, a 20 % energy generation increase can be achieved by combining only 32-degree and seasonal tilted angles. Based on the achieved results, the PSCM can be deployed in this region for energy generation for BIPV applications.

ACKNOWLEDGMENT

A. N. thanks the grant from Ministry of Education and Science of the Republic of Kazakhstan (No. AP14869983) and Nazarbayev University (Grant no. 021220CRP0422).

REFERENCES

[1] "Best Research-Cell Efficiencies, NREL." https://www.nrel.gov/pv/assets/pdfs/best-research-cell-efficiencies-rev220630.pdf (accessed Nov. 10, 2022).

[2] H. Ravula and S. Bollapragada, "Solar Window as an Energy Source : A Patent Study," *2020 IEEE International Power and Renewable Energy Conference, IPRECON 2020*, Oct. 2020, doi: 10.1109/IPRECON49514.2020.9315230.

[3] H. Alrashidi, W. Issa, N. Sellami, S. Sundaram, and T. Mallick, "Thermal performance evaluation and energy saving potential of semi-transparent CdTe in Façade BIPV," *Solar Energy*, vol. 232, pp. 84–91, Jan. 2022, doi: 10.1016/J.SOLENER.2021.12.037.

[4] N. Martín-Chivelet, J. Polo, C. Sanz-Saiz, L. T. Núñez Benítez, M. Alonso-Abella, and J. Cuenca, "Assessment of PV Module Temperature Models for Building-Integrated Photovoltaics (BIPV)," *Sustainability 2022, Vol. 14, Page 1500*, vol. 14, no. 3, p. 1500, Jan. 2022, doi: 10.3390/SU14031500.

[5] A. Kumar Singh, V. R. Prasath Kumar, and L. Krishnaraj, "Emerging technology trends in the C&I rooftop solar market in India: Case study on datacentre – Retrofit with BIPV by U-Solar," *Solar Energy*, vol. 238, pp. 203–215, May 2022, doi: 10.1016/J.SOLENER.2022.04.033.

[6] T. Santos, K. Lobato, J. Rocha, and J. A. Tenedório, "Modeling photovoltaic potential for bus shelters on a city-scale: A case study in Lisbon," *Applied Sciences (Switzerland)*, vol. 10, no. 14, Jul. 2020, doi: 10.3390/APP10144801.

[7] M. Sánchez-Aparicio, S. Lagüela, J. Martín-Jiménez, S. del Pozo, E. González-González, and P. Andrés-Anaya, "Smart Mobility in Cities: GIS Analysis of Solar PV Potential for Lighting in Bus Shelters in the City of Ávila," *Communications in Computer and Information Science*, vol. 1359, pp. 154–166, 2021, doi: 10.1007/978-3-030-69136-3_11.

[8] A. Karafil, H. Ozbay, M. Kesler, and H. Parmaksiz, "Calculation of optimum fixed tilt angle of PV panels depending on solar angles and comparison of the results with experimental study conducted in summer in Bilecik, Turkey," *ELECO 2015 - 9th International Conference on Electrical and Electronics Engineering*, pp. 971–976, Jan. 2016, doi: 10.1109/ELECO.2015.7394517.

[9] H. Higuchi and T. Negami, "Largest highly efficient 203 × 203 mm2 CH3NH3PbI3 perovskite solar modules," *Jpn J Appl Phys*, vol. 57, no. 8, p. 08RE11, Aug. 2018, doi: 10.7567/JJAP.57.08RE11/XML.

[10] M. Z. Jacobson and V. Jadhav, "World estimates of PV optimal tilt angles and ratios of sunlight incident upon tilted and tracked PV panels relative to horizontal panels," *Solar Energy*, vol. 169, pp. 55–66, Jul. 2018, doi: 10.1016/J.SOLENER.2018.04.030.

CFD-Based Machine Learning Model for Agrivoltaic System Design

Henry J. Williams, Emily Weed, Khaled Hashad, K. Max Zhang

Sibley School of Mechanical and Aerospace Engineering, Cornell University, Ithaca, NY, 14853, USA

Abstract — **Agrivoltaics, the co-location of solar and food production, is a promising solution to land-use conflict between solar photovoltaics (PV) and agriculture. Microclimate studies indicate that agrivoltaic systems enhance solar farm cooling, leading to increased panel efficiency, but stakeholders lack efficient design tools to quickly evaluate the consequences of various agrivoltaic designs. Here we present a computational fluid dynamics (CFD)-based machine learning (ML) model which is utilized to develop an early version of an agrivoltaic design tool, where solar cell operating temperature is optimized based on panel height and ground cover type selection. Results indicate that Random Forest Regression (RFR) and Gradient Boosted Trees (GBT) perform better than Support Vector Regression (SVR) and Linear Regression (LR) in this case, with RFR and GBT achieving under 2 °C RMSE compared to SVR and LR over 3 °C. Using dual annealing optimization with the GBT model, findings indicate that panel heights up to 3.2m and ground albedo up to 64% are preferred in warm months to maximize panel cooling.**

I. INTRODUCTION

The solar energy industry is rapidly expanding with the help of technological innovations and aggressive climate policies [1], but with increasing land requirements for solar PV leading to land-use competition with agriculture, many stakeholders are looking to agrivoltaics as a solution to co-locate agricultural activities and solar PV [2]. Agrivoltaic systems may experience different microclimates compared to traditional solar farms due to altered ground cover types and panel management strategies required for food production [3]. For example, studies have shown reduced module surface temperatures when crops are planted beneath solar arrays due to increased cooling from evapotranspiration (ET) and high ground albedo [3-7]. This is particularly attractive to solar developers, as decreasing operating temperatures can increase efficiency and improve the longevity of a solar cell [8-9]. However, there are no existing design tools which provide recommendations for ground cover properties and solar farm design to achieve optimal module surface temperatures. Furthermore, there is a gap in understanding how different seasons effect optimal site design for passive solar panel temperature regulation.

Computational fluid dynamics (CFD) simulations have been developed to simulate module surface temperatures based on the thermal-fluid environment of a solar farm [3], but these are computationally expensive and might rely on commercial software. Machine learning (ML) is a promising tool to improve modeling efficiency, and utilizing training data from thermal fluid CFD simulations ensures a physics-based approach is maintained throughout the model development. Numerous studies have presented CFD-based ML models [10-11], but not in the context of agrivoltaic site design. In this study, we evaluate several CFD-based ML algorithms and select the best

model to optimize agrivoltaic site design targeting a solar module operating temperature of 25 °C, which is the standard test conditions (STC) temperature for solar cells.

II. METHODS

A. CFD model development

This study utilizes a CFD model which was developed to characterize the solar farm microclimate based on weather conditions, ground cover, and site design. The first version of the CFD model was presented in 2023 by Williams et al. to investigate the potential for evapotranspiration, ground albedo, and panel height to provide passive cooling for solar modules, and was evaluated against module surface temperature data obtained from a solar facility in Ontario, Canada [3]. This paper presents an updated version which implements the Food and Agriculture Organization (FAO) Penman-Monteith equation to calculate evapotranspiration as a function of weather conditions (Table I). Here, we use ML to learn from CFD data and predict module surface temperature, thus creating a computationally efficient alternative to the CFD model.

TABLE I

EVAPOTRANSPIRATION HEAT FLUX CALCULATION

FAO P-M equation: $\lambda ET_o = c$	$\dfrac{0.408\Delta(R_n - G) + \gamma\left(\frac{37}{T_{hr}+273}\right)u(e^o(T_{hr}) - e_a)}{\Delta + \gamma(1 + 0.34u)}$	
$\Delta = \dfrac{4098\left[0.6108\exp\left(\frac{17.27 T_{hr}}{T_{hr} + 237.3}\right)\right]}{(T_{hr} + 137.3)^2}$		$G = \pm\begin{cases} 0.1R_n, \ daytime \\ 0.5R_n, \ nighttime \end{cases}$
$e^o(T_{hr}) = 0.6108\exp\left[\dfrac{17.27 T_{hr}}{T_{hr} + 237.3}\right]$		$e_a = e^o(T_{hr})\dfrac{RH}{100}$
λET_o	evapotranspiration heat flux (W/m^2)	
c	277.8; unit conversion from MJ/m2-hr to W/m^2	
Δ	saturation slope vapor pressure curve (kPa/°C)	
R_n	net radiation (MJ/m^2-hr)	
G	soil heat flux density (+ warming, - cooling) (MJ/m^2-hr)	
γ	psychrometric constant (kPa/°C)	
T_{hr}	mean hourly air temperature (°C)	
u	mean hourly wind speed (m/s)	
$e^o(T_{hr})$	saturation vapor pressure (kPa) at T_{hr}	
e_a	mean hourly actual vapor pressure (kPa)	
RH	relative humidity (%)	

B. Feature space

A total of 20 days throughout all four seasons were simulated to develop the ML model. Table II displays the mean, standard deviation (std), minimum (min) and maximum (max) of the feature space which was standardized prior to modeling. Input weather conditions were measured in Ontario, Canada nearby the solar farm originally used to evaluate the CFD model accuracy [3].

978-1-6654-6060-6/23 $31.00 © 2023 IEEE

TABLE II
FEATURE SPACE

Type	Feature	Mean	Std	Min	Max
Constant	Panel height (m)	2.1	1.3	0.50	4.5
	Albedo (%)	48	24	20	80
Variable (30min interval)	λET (W/m^2)	26	32	0	120
	Wind (m/s)	2.6	1.7	0	9.5
	Relative humidity (%)	47	32	4.8	100
	Radiation (W/m^2)	160	230	0	750

C. ML model evaluation

A total of 2340 data points were randomly split into training and testing data using an 80-20% split. Five-fold cross validation was utilized on the training set to tune hyperparameters, and a held-out test set was used for final model evaluation. Models were evaluated using R^2 and RMSE.

D. ML algorithms and feature selection

Using Python's Scikit-learn package, this study explored four models: Linear Regression (LR), Support Vector Regression (SVR), Random Forest Regression (RFR) and Gradient Boosted Trees (GBT) for regression. LR assumes there to be a linear relationship between the features and the output. SVR is adapted from Support Vector Machines, a classification model which finds the optimal hyperplane maximizing the margin of data. RFR is an ensemble model that combines a number of decision trees that are trained on a subset of the data. GBT is also an ensemble method, fitting weak decision trees to the residual of the previous tree. Each model was used to measure performance with forward stepwise selection, backward stepwise selection, and best subsets selection. Table III shows feature importance for each model.

TABLE III
MODEL PERFORMANCE ON TEST SET
FEATURE IMPORTANCE FROM 1 (MOST) TO 6 (LEAST)

ML Model:	LR	SVR	RFR	GBT
Ambient temp.	1	1	1	1
Panel height	2	3	2	2
Relative humidity	4	2	3	3
Solar irradiation	3	5	6	5
Evapotranspiration	5	4	4	4
Ground albedo	6	6	5	6

E. Parameter tuning and learning curves

Hyperparameters were tuned and potential for overfitting was assessed using grid search with five-fold cross validation. Validation R^2 and RMSE were evaluated to select optimal parameters for each model. The GBT model achieved the highest validation R^2 value using the best parameters found during grid search. For each algorithm, convergence is achieved with the available training set.

F. Optimization framework

Utilizing the CFD-based ML model, an agrivoltaic system design tool was developed to optimize module surface temperature by changing panel height and ground cover type

(i.e., crop albedo). This represents the preliminary version of a tool which aims to support design and management decisions for agrivoltaic systems.

The optimization framework presented in this study was developed using dual annealing optimization with the Scipy Optimization Library, which provides support for global minimization of an objective function subject to parameter bounds. The objective function was defined as the absolute difference between predicted value and targeted value of 25 °C, which is selected as a realistic target for passive panel cooling from agrivoltaic design in conditions representative of summertime in Ontario, Canada. Utilizing the fully trained models, optimal solar farm design parameters are obtained for panel height and ground albedo based on minimizing the objective function. This optimization problem utilizes the GBT model.

Multiple wrapper functions were developed around the base optimization problem, creating a function which accepts the weather conditions as parameters and optimizes over the panel height and albedo value. This shows how site design and ground cover can be altered to maximize solar cell efficiency in a given set of weather conditions. This study evaluates weather conditions throughout warm months (June to September) in which solar panel cooling is particularly relevant.

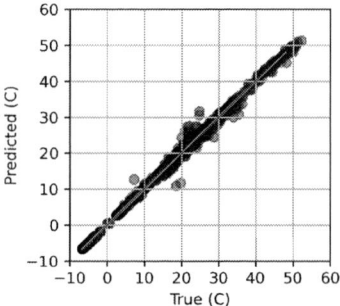

Fig. 1. Predicted vs true data for GBT

III. RESULTS AND DISCUSSION

A. Performance on test sets

Table IV shows test R^2 and RMSE values obtained by LR, SVR, RFR, and GBT on the test set which was held out from the beginning, with RFR and GBT performing the best. The error is relatively low, indicating that the training data covered a large range of possible parameters within the given weather conditions. Nevertheless, model performance may be specific to a given location, as training and test data in this study both come from the input dataset.

TABLE IV
MODEL PERFORMANCE ON TEST SET

ML Model	LR	SVR	RFR	GBT
Test R^2	91	90	98	98
Test RMSE (°C)	3.7	3.9	1.7	1.8

B. Panel temperature optimization

The optimization tool was used to determine which values of panel height and albedo would lead to the target module surface temperature for a given set of weather conditions from four months in Ontario, Canada. This concept is representative of a stakeholder making decisions on site design and ground cover type to maximize solar cell efficiency. To determine which crop to select, it is important to check crop trials in similar weather conditions and at various stages of development, as albedo can vary with season and crop maturity [12-13].

Figure 2 shows the optimization results for panel height and ground albedo between the months of June and September using weather data from Ontario, Canada. In terms of generalizability, this model can perform well with previously unseen data drawn from similar distributions as the training data, but might be less applicable to data from different distributions. Nevertheless, the overall implementation process presented is repeatable for different locations and shows promising overall results.

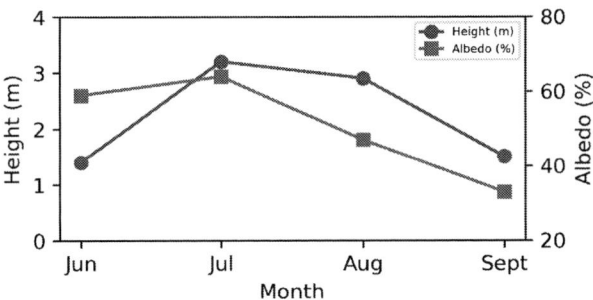

Fig. 2. Monthly optimization results between June and September

IV. CONCLUSION AND FUTURE WORK

With the rapid increase in large-scale solar energy projects globally, site design must evolve to incorporate agriculture within solar PV facilities. In this study, we utilize inputs and outputs from a thermal-fluid CFD model to develop ML models including Linear Regression (LR), Random Forest Regression (RFR), Support Vector Regression (SVR), and Gradient Boosted Trees (GBT). The best models, RFR and GBT, obtained an RMSE below 2 °C for the test dataset. The GBT model was utilized in an optimization problem to design a solar farm around a desired module surface temperature of 25 °C, the results of which indicate that optimal panel height and albedo values vary from month to month, with relatively taller panels and higher albedo desired in warm months to provide passive cooling for panels. This shows the potential for an efficient design tool which can be used to maximize power output by adjusting solar site design parameters for a given facility location and vegetative cover, but current results may be dependent on locational weather conditions. Additionally, selecting an appropriate crop type based on albedo is dependent on satellite data evaluated with sufficient ground truth data. Future work will focus on expanding the training data to include weather conditions from more locations, and incorporating more generalizable crop type selection metrics. Furthermore, the design tool will be developed to support design and management recommendations to maximize agricultural production based on microclimate conditions achieved by various solar site designs.

REFERENCES

[1] N. Kannan and D. Vakeesan, "Solar energy for future world: - A review," Renewable and Sustainable Energy Reviews, vol. 62. Elsevier Ltd, pp. 1092–1105, Sep. 01, 2016. doi: 10.1016/j.rser.2016.05.022.

[2] Dupraz, C., Marrou, H., Talbot, G., Dufour, L., Nogier, A., & Ferard, Y. (2011). Combining solar photovoltaic panels and food crops for optimising land use: Towards new agrivoltaic schemes. Renewable Energy, 36(10), 2725–2732. https://doi.org/10.1016/j.renene.2011.03.005.

[3] Henry J. Williams, Khaled Hashad, Haomiao Wang, K. Max Zhang, The potential for agrivoltaics to enhance solar farm cooling, Applied Energy, Volume 332, 2023, 120478, ISSN 0306-2619, https://doi.org/10.1016/j.apenergy.2022.120478.

[4] Adeh, E. H., Good, S. P., Calaf, M., & Higgins, C. W. (2019). Solar PV Power Potential is Greatest Over Croplands. Scientific Reports, 9(1). https://doi.org/10.1038/s41598-019-47803-3

[5] Skoplaki, E., & Palyvos, J. A. (2009). Operating temperature of photovoltaic modules: A survey of pertinent correlations. Renewable Energy, 34(1), 23–29. https://doi.org/10.1016/j.renene.2008.04.009.

[6] Macknick, J., Beatty, B., & Hill, G. (2013). Overview of Opportunities for Co-Location of Solar Energy Technologies and Vegetation. www.nrel.gov/publications.

[7] Bloom, D. E., Choi, C. S., Li, Y., Macknick, J., McCall, J., and Ravi, S., "Interactions between solar photovoltaic arrays and the underlying soil-vegetation system", vol. 2019, 2019.

[8] Skoplaki, E., & Palyvos, J. A. (2009). On the temperature dependence of photovoltaic module electrical performance: A review of efficiency/power correlations. Solar Energy, 83(5), 614–624. https://doi.org/10.1016/j.solener.2008.10.008.

[9] Pan, R., Kuitche, J., & Tamizhmani, G. (2011). Degradation analysis of solar photovoltaic modules: Influence of environmental factor. Proceedings - Annual Reliability and Maintainability Symposium. https://doi.org/10.1109/RAMS.2011.5754514.

[10] Khaled Hashad, Jiajun Gu, Bo Yang, Morena Rong, Edric Chen, Xiaoxin Ma, K. Max Zhang, Designing roadside green infrastructure to mitigate traffic-related air pollution using machine learning, Science of The Total Environment, Volume 773, 2021, 144760, ISSN 0048-9697, https://doi.org/10.1016/j.scitotenv.2020.144760.

[11] D. Xiao, C.E. Heaney, L. Mottet, F. Fang, W. Lin, I.M. Navon, Y. Guo, O.K. Matar, A.G. Robins, C.C. Pain, A reduced order model for turbulent flows in the urban environment using machine learning, Building and Environment, Volume 148, 2019, Pages 323-337, ISSN 0360-1323, https://doi.org/10.1016/j.buildenv.2018.10.035.

[12] Michael Abraha et al (2021) Albedo-induced global warming impact of Conservation Reserve Program grasslands converted to annual and perennial bioenergy crops. Environ. Res. Lett. 16 084059, DOI: 10.1088/1748-9326/ac1815.

[13] Allen, Richard G. et al (1998) Crop evapotranspiration - Guidelines for computing crop water requirements. Chapter 5 – Introduction to crop evapotranspiration. ISBN 92-5-104219-5.

ACKNOWLEDGEMENTS

This work was supported by the New York State Energy and Research Development Authority (NYSERDA).

Towards highly stable metal-halide perovskite materials for a broad range of applications: film growth, degradation control, and interfacial engineering.

Lissette Rodriguez-Cabanas, Calvin Duong, Bradley Stanley, Andre Slonopas

1U.S. Army Futures Command, APG, MD, United States

Metal-halide perovskites remain one of the most discussed and promising materials for optoelectronic applications. Tunable optoelectrical properties of the materials have inherent qualities that find applications in both military and civilian purposes. One of the main applications for the materials is large-scale low-cost solar energy generation. Perovskite Solar Cells (PSCs) have continuously demonstrated Power Conversion Efficiencies (PCEs) above 24%, cost reductions above 30%, direct band-gap structures, and low-temperature malleability for utility-scale applications. Industry is actively exploring potential applications for the perovskite materials and pursuing efforts to leverage these materials in CO_2 footprint reduction. Current challenges of the materials lie in the acquisition of high-quality stable, and ruggedized films with lifetimes above 5 years. Encapsulation can provide stability from film oxidation, but hydrophilicity, thermal and light stability of the materials remain to be solved. This work demonstrates progress in obtaining metal-halide perovskite stabilities and describes the possibility of integrating these materials for a broad range of uses.

First-Principles Study of Energy Band Alignment in Pristine CdTe/TeO₂/Te Interfaces

Anthony P. Nicholson, James R. Sites, and Walajabad S. Sampath

Colorado State University, Fort Collins, CO, 80521, USA

Abstract — **A first-principles atomistic modeling approach is utilized to investigate the critical factors responsible for electronic structure and charge transport phenomena within back heterointerfacial regions of cadmium telluride thin-film photovoltaic devices. Within CdTe PV, the absorber/back contact interface is a major limiting factor to solar cell device performance and energy conversion efficiencies, even with a tellurium back contact layer present. It is speculated that a natively-forming tellurium dioxide layer could passivate the CdTe back interface prior to Te deposition and contribute to achieving higher device efficiencies. However, there is not a clear fundamental understanding of how atomic-scale properties at the CdTe/TeO₂ and TeO₂/Te heterointerfaces based on plane orientation modify the charge transport behaviors in the PV device as a whole. This study addresses the gap in knowledge of CdTe-based absorber/back contact interfacial effects by computationally describing electronic structures of pristine CdTe/TeO₂/Te heterointerfaces and providing insight on efficient pathways for interface engineering in CdTe-based thin-film PV applications.**

I. Introduction

Noteworthy advances in solar cell efficiency for CdTe-based thin-film photovoltaic devices have mainly been attributed to the well-known cadmium chloride ($CdCl_2$) passivation treatment. However, other avenues of improvements in relation to interfacial characteristics within CdTe PV are being explored as well. One such effort is the formation of native oxides that may act as passivating layers, especially for the back interface where recombination velocities remain a major limiting factor in CdTe PV device performance.

Oxygen interactions on II-VI compound surfaces have been a topic of study for several decades regarding their role in II-VI semiconductor device performance. [1,2] Oxide layers have proven to be indispensable for stability and passivation in Si-based technologies, leading to great interest for oxide layer formation in II-VI semiconductor applications such as CdTe-based thin-film solar cells. [3]

Although a series of possible native oxide compounds is expected to form on CdTe surfaces [4], one oxide compound known as tellurium dioxide (TeO_2) is known to develop on CdTe-based semiconductors under ambient conditions. Three crystalline polymorphs exist for tellurium dioxide as tetragonal α-TeO_2 (paratellurite), orthorhombic β-TeO_2 (tellurite) [5], and orthorhombic γ-TeO_2 [5,6]. TeO_2 formation has been observed along multiple plane orientations including the CdTe{111} polar surface [7] and the CdTe{110} non-polar surface. [1] Although a native tellurium dioxide layer formation at the back interface region of CdTe PV is possible, it is not clear how

morphological factors such as plane orientation affect the electronic structure of CdTe/TeO₂ interfaces. A first-principles atomic-scale approach capable of determining the energy band alignment of the the back region of pristine CdTe/TeO₂ interfaces offers an avenue of understanding on the role of native oxides for various plane orientations.

II. Computational Details

The first-principles-based calculations via density functional theory (DFT) were performed using the QuantumATK *ab initio* software package under version T-2022.03. All DFT computations used the linear combination of atomic orbitals method for representing the calculated wavefunctions of the supercell model. The model consists of three distinct layers that make up the expected back interface regions of the CdTe PV device: 1) CdTe, 2) β-TeO_2 (tellurite), and 3) Te. The bulk CdTe, TeO₂, and Te primitive cells were geometrically optimized using the HSE06 hybrid functional along with the SG15 pseudopotential and medium basis set. A self-consistent field (SCF), force, and stress convergence threshold of 1×10^{-6} eV, 0.01 eV/Å, and 1×10^{-4} eV/Å³, respectively, were implemented during the geometric optimization of the bulk primitive cells of each material system. Slab models of the CdTe/TeO₂/Te interfaces were relaxed using the GGA-PBE exchange-correlation functional with an SCF and force convergence threshold of 1×10^{-4} eV and 0.05 eV/Å, respectively, prior to constructing the main interface models. At least two unit cells of the CdTe and Te layers had their atoms respectively fixed and rigidly held together during the relaxation process. Each interface model, also called a two-probe device model, [8] consists of a central region and two semi-infinite bulk electrode regions that act as boundary conditions. All two-probe device models implemented the HSE06 functional to obtain the electronic structures of the

Table I
Pertinent Details of DFT Models for Pristine Interfaces

Plane Orientation		CdTe/TeO₂/Te Details		CdTe/ TeO₂	TeO₂/ Te	
CdTe	TeO₂	# of total atoms	In-plane area (Å²)	ε_{av} (%)	ε_{av} (%)	$\varepsilon_{ps,Te}$ (%)
(2$\bar{2}$0)	($\bar{1}$3$\bar{2}$)	640	242.2	4.93	4.60	-4.09
(220)	(110)	516	181.63	1.86	0.73	-0.85
(111)	(110)	636	185.37	1.2	1.42	-1.66

CdTe/TeO$_2$/Te interfaces, after which the potential alignment was calculated similar to Weston *et al.* [9] The TeO$_2$ layer was kept less than 2 nm to mitigate the influence of TeO$_2$ length on the calculated interfacial properties while ensuring that the differing plane orientations were the main focus of the DFT-based computational study.

The construction of both the CdTe/TeO$_2$ and TeO$_2$/Te interfaces utilized a lattice-matching method [10] to calculate the resulting in-plane strain. Only the TeO$_2$ and Te layers underwent in-plane strain during the lattice matching step. Ideally, an in-plane strain below 2% should be selected for each investigated overlayer that forms an interface to reduce strain effects on the band gaps of each layer. However, additional in-plane strain on Te based on the in-plane strain of the pre-selected CdTe/TeO$_2$ interface was not known *a priori* during the construction of the TeO$_2$/Te interface. Thus, no specific threshold of in-plane strains was used for either interface. Instead, the in-plane strain was chosen to be as low as possible without exhausting the computational resources available to simulate the large modeling system size. Table 1 provides the modeling details for each interface. The average in-plane strain ε_{av} is calculated by averaging all in-plane lattice strains (i.e., ε_{11}, ε_{22}, and ε_{12}) that appear for the TeO$_2$ and Te overlayers when they are fitted to the lattice vector of the CdTe substrate layer. The Poisson strain $\varepsilon_{ps,Te}$ was calculated for the Te(0001) overlayer to determine how the Te helical chains respond to the in-plane strain caused by the preceeding CdTe and TeO$_2$ lattice constants. Negative (positive) values denote a compressive (tensile) strain that occurs on the Te overlayer.

III. PRELIMINARY RESULTS AND DISCUSSION

The energy band alignments are provided in Figure 1 for various CdTe and TeO$_2$ plane orientations while the Te plane is maintained as (0001). The selected CdTe and TeO$_2$ facets were chosen to evaluated the effect of non-polar vs polar facets on the electronic structure of the CdTe/TeO$_2$ interface. The band gaps of each layer besides the Te overlayer are well represented under the HSE06 hybrid functional using the default Fock-PBE exchange energy mixing parameter $\alpha = 0.25$. However, the Te bandgap is significantly greater than the experimental band gap due to multiple reasons including the absence of spin-orbit coupling in the model as well as larger mixing parameter than usual for an appropriate Te band gap magnitude (i.e., $\alpha = 0.125$ [11]). Admittedly, the discussion on energy band alignment and subsequent charge transport through the TeO$_2$/Te interface is difficult to make with the currently inaccurate Te band gap, which will be immediately addressed in updated DFT models. Presently, the results will be solely focused on the CdTe/TeO$_2$ interface and what the energy band alignments suggest in terms of charge transport through the back interface.

All CdTe/TeO$_2$ interfaces presented in Figure 1 show notably large valence and conduction band offsets regardless of plane orientation that lead to a Type I straddling gap band

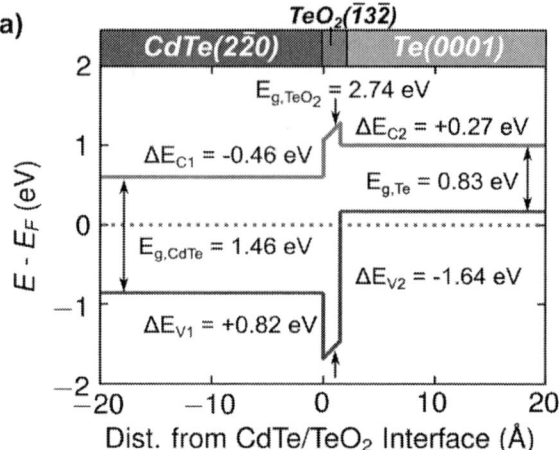

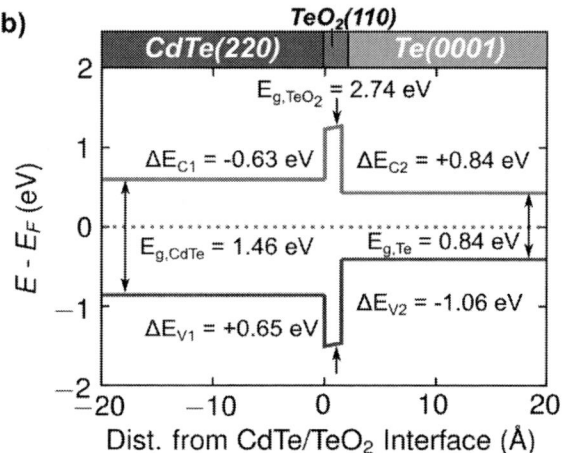

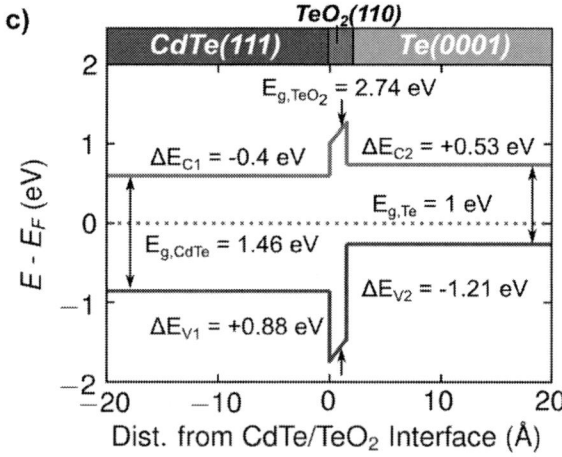

Figure 1. Calculated CdTe/TeO$_2$/Te energy band alignments for various (a,b) non-polar CdTe and (c) polar CdTe facets coupled to differing TeO$_2$ plane orientations. Band gaps and band offsets are provided within each band diagram with the subscript values "1" and "2" denoting the band alignment magnitudes associated with the CdTe/TeO$_2$ and TeO$_2$/Te, respectively.

978-1-6654-6060-6/23 $31.00 © 2023 IEEE

alignment at the CdTe/TeO$_2$ interface. Under classical charge transport conditions, the downward VBO would act as a major limiting barrier for hole majority carriers generated within the p-type CdTe absorber layer. As a result, the predicted VBO would be highly unfavorable at the back interface of CdTe-based PV devices. Experimental data reports thin native-oxide thicknesses on metal-chalcogenide single crystals [12] comparable to the currently presented models, but in some cases form much thicker layers at tens of nanometers [1]. Considering TeO$_2$ thicknesses of less than 2 nm as modeled in the current DFT study, there is a possibility for specialized charge transport modes such as quantum tunneling or interface state assisted carrier transport to effectively allow hole carriers through the CdTe/TeO$_2$ interface.

Figure 1a shows the how the non-polar CdTe($2\bar{2}0$) facet combined with a unique TeO$_2$($\bar{1}3\bar{2}$) plane orientation leads to VBO and CBO magnitudes of 0.82 and 0.46 eV, respectively. However, simply changing to the CdTe(220)/TeO$_2$(110) interface causes the VBO to decrease to 0.65 eV while the CBO increases to 0.63 eV in response. The atomic arrangement of the TeO$_2$ layer in Figure 1b creates a band shift at the interface that is likely due to an induced interfacial dipole forming at the CdTe/TeO$_2$ interface region. A quantified value of the induced dipole will be determined to verify this claim and further characterize the differences between each interface case. Lastly, Figure 1c shows how a highly polar facet such as the CdTe(111) plane coupled with the TeO$_2$(110) plane creates the largest (smallest) VBO (CBO) magnitude at 0.88 eV (0.4 eV) out of the evaluated planar facets. The progression of VBO/CBO magnitudes for various planar combinations of CdTe/TeO$_2$ interfaces highlights the intrinsic role that interface morphology plays in energy band alignment across the back interface region of CdTe-based thin-film PV devices.

IV. NEXT STEPS

Besides further corrections to the Te band gap discrepancy, the projected local density of states (PLDOS) for each interface will be used to analyze the electronic states across the entire supercell length and provide further insight on the strain-dependent effects within the thin TeO$_2$ layer. The band alignment results from the PLDOS images are expected to differ from the potential alignment method since the direct measurement of the density of states is provided for each atomic layer. Furthermore, doping in the form of external electrostatic potentials will be incorporated in the model to evaluate, at least from a first-order perspective, how higher p-type concentrations in the CdTe absorber layer, TeO$_2$ native oxide layer, or Te overlayer affect the predicted band alignments. Additionally, interfacial dipole calculations will be performed to quantify their effects on band alignment and whether any trends develop due to the electronic structures of non-polar vs polar CdTe facets.

ACKNOWLEDGEMENTS

This work utilized the Summit supercomputer, which is supported by the National Science Foundation (awards ACI-1532235 and ACI-1532236), the University of Colorado Boulder, and Colorado State University. The research was also performed using computational resources sponsored by the Department of Energy's Office of Energy Efficiency and Renewable Energy and located at the National Renewable Energy Laboratory. A.P.N. received funding from the National Science Foundation MPS-Ascend Postdoctoral Research Fellowship under Grant No. 2138081. Any opinions, findings, and conclusions or recommendations expressed in this material are those of the author(s) and do not necessarily reflect the views of the National Science Foundation. This material is based in part upon work supported by the U.S. Department of Energy's Office of Energy Efficiency and Renewable Energy (EERE) under the Solar Energy Technologies Office Award No. DE-EE0008974.

REFERENCES

[1] F. A. Ponce, R. Sinclair, and R. H. Bube, "Native tellurium dioxide layer on cadmium telluride: A high–resolution electron microscopy study," *Appl. Phys. Lett.* **39** (12), pp. 951–953, 1981.

[2] T. L. Chu, S. S. Chu, and S. T. Ang, "Surface passivation and oxidation of cadmium telluride and properties of metal–oxide–CdTe structures," *J. Appl. Phys.* **58** (8), pp. 3206–3210, 1985.

[3] R. S. Hall, D. Lamb, S. J. C. Irvine, "Back contact materials used in CdTe solar cells – A review," *Energy Sci. Eng.* **9**, pp. 606 – 632, 2021.

[4] E. Menéndez-Proupin, G. Gutiérrez, E. Palmero, and J. L. Peña, "Electronic structure of crystalline binary and ternary Cd-Te-O compounds," *Phys. Rev. B* **70**, p. 035112, 2004.

[5] Y. Li, W. Fan, H. Sun, X. Cheng, P. Li, and X. Zhao, "Structural, electronic, and optical properties of α,β, and γ-TeO$_2$," *J. Appl. Phys.* **107**, p. 093506, 2010.

[6] N. Dewan, K. Sreenivas, and V. Gupta, "Properties of crystalline γ-TeO$_2$ thin film," *J. Cryst. Growth* **305**, pp.237–241, 2007.

[7] B. Kowalski, B. Orłowski, and J. Ghijsen, "Oxide formation on the CdTe(111)A (1×1) surface," *Appl. Surf. Sci.* **166** (1), pp. 237–241, 2000.

[8] D. Stradi, U. Martinez, A. Blom, M. Brandbyge, and K. Stokbro, "General atomistic approach for modeling metal-semiconductor interfaces using density functional theory and nonequilibrium Green's function," *Phys. Rev. B* **93**, p. 155302, 2016.

[9] L. Weston, H. Tailor, K. Krishnaswamy, L. Bjaalie, and C. G. Van de Walle, "Accurate and efficient band-offset calculations from density functional theory," *Comp. Mater. Sci.* **151**, pp. 174–180, 2018.

[10] D. Stradi, L. Jelver, S. Smidstrup, and K. Stokbro, "Method for determining optimal supercell representation of interfaces," *J. Phys. Condens. Matter* **29**, p. 185901, 2017.

[11] S. Yi, Z. Zhu, X. Cai, Y. Jia, and J.-H. Cho, "The nature of bonding in bulk tellurium composed of one-dimensional helical chains," *Inorg. Chem.* **57**, pp. 5083–5088, 2018.

[12] S. Babar, P. J. Sellin, J. F. Watts, and M. A. Baker, "An XPS study of bromine in methanol etching and hydrogen peroxide passivation treatments for cadmium zin telluride radiation detectors," *Appl. Surf. Sci.* **264**, pp. 681 – 686, 2013.

Horizon Profiling Methods for Photovoltaic Arrays

Jennifer L. Braid and Benjamin G. Pierce

Sandia National Laboratories, Albuquerque, NM 87113 USA

Abstract — **In this work, we introduce and compare the results of several methods for determining the horizon profile at a PV site, and compare their use cases and limitations. The methods in this paper include horizon detection from time-series irradiance or performance data, modeling from GIS topology data, manual theodolite measurements, and camera-based horizon detection. We compare various combinations of these methods using data from 4 Regional Test Center sites in the US, and 3 World Bank sites in Nepal. The results show many differences between these methods, and we recommend the most practical solutions for various use-cases.**

I. INTRODUCTION

Nearly all photovoltaic (PV) arrays experience some degree of shading due to objects on the horizon such as geographical features, trees, buildings, or utility poles. These objects affect the performance of the array to varying degrees depending on their size and proximity to the array [1]. The imposition of these objects on the incident sunlight for a specific location are summarized as a horizon profile: pairs of solar azimuth and elevation corresponding to the upper edges of the effective horizon. A horizon profile can be used in various applications to estimate times or percentage of direct solar irradiance that will be regularly blocked. This can be useful for accurately estimating the total solar resource availability for siting or performance modeling and analysis.

A 2017 study on the sensitivity of solar shading to horizon uncertainty showed that $0.5°$ uncertainty in the relative elevation of the horizon resulted in up to 3% error in the estimated direct and diffuse irradiance [2]. We show here that horizons generated by different profiling methods can vary by much more than $0.5°$ in azimuth and elevation.

Because a horizon profile is typically measured or calculated at a single point, the accuracy of this horizon profile varies within the PV array. For example, the edge of a large array may be near a forest or building, causing a high horizon profile elevation in that direction. However, to the opposite edge of the array, the same object(s) will appear to be at a much lower elevation. This is an issue for any PV array and horizon profiling method, though some methods may be used at multiple points within the array.

Various methods exist to measure the horizon profile at an actual or potential PV site. These methods vary greatly in input datatype, processing approach, and measurement time. There is also high variation in ease of use and accuracy. Here we compare collection and results from different horizon profiling methods, and discuss ideal use cases for each.

II. METHODS AND PRELIMINARY RESULTS

We will explore the horizon profiling methods summarized in Table I. These methods use different input datatypes and have different accuracies for various horizon objects and proximities.

A. Theodolite

A theodolite (Fig. 1) is a device which can be used to manually map the horizon profile by measuring the relative distances between reference points in vertical and horizontal axes. The device is oriented using a built-in compass and level, then the user locates the azimuth and elevation angles corresponding to points on the apparent horizon. This process can lead to accurate results if performed correctly, but requires many measurements and careful setup.

Fig. 1. Theodolite used to manually measure the horizon profile in Albuquerque.

TABLE I
SUMMARY OF HORIZON DETECTION METHODS

Method	Operation	Datatype	Use Cases	Accuracy Dependence	Limitations
Theodolite	In-field, manual	Manual horizon measurements	Siting	Placement accuracy and number of measurements	Time-consuming
Cameras	In-field, automated or manual	Manually collected mage	Siting, performance analysis	Image processing and photography accuracy	Results may not be reproducible
GIS	Remote, automated	GIS topography	Siting, modeling, performance analaysis	DEM accuracy and computationally expense	Only captures GIS topography features
Time-Series	Remote, automated	Irradiance or performance	Performance analysis	Data density and length	Requires long timeseries

978-1-6654-6060-6/23 $31.00 © 2023 IEEE

B. Camera-based methods

Various camera-based methods have been used to measure horizon profiles, with the earliest published method dating to 2000 [3] showing that an analog or digital camera mounted on a rotating platform could be used to take a series of images at various azimuths, which could then be manually or automatically analyzed to create a horizon profile. Generally all modern camera-based methods for horizon detection follow the same general image processing steps for determining the horizon. These steps are demonstrated in Figure 2 on a panoramic photo.

Original

HSV

Saturation Channel

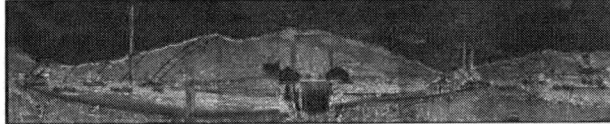

Canny Edge Filter

Fig. 2. Image processing method for horizon detection. The image is first converted to HSV space, and then binarized by selecting the saturation channel. The Canny edge filter is applied to calculate lines in the image, and the top-most line is selected to be the horizon.

Wide-angle lens camera methods: An example of a panoramic photo from a wide-angle lens for horizon detection is shown in Fig. 2. Additionally, smartphone applications exist to capture the horizon profile with the built-in camera, either with the user tracing the horizon manually, or with automated image processing.

In either case, lens distortion and camera stitching of images can cause inaccuracies in the shading profile. These methods use continuity of features to match consecutive image frames and find transition points between various views. While smartphones can determine their orientation, the camera does not typically use this information to construct an image, and therefore geographic features may not be proportionally or locationally correct in a final image.

Fish-eye lens camera methods: A fish-eye lens captures an image of a full or near hemispherical view. One device that uses a fish-eye lens to determine the horizon profile is the SolMetric SunEye [4]. The SunEye is placed at the location and orientation of a planned or existing array, then captures the hemispherical view from the location. The internal (proprietary) software then analyzes the image to create a horizon profile. The SunEye can be very accurate, but user error in placement/orientation can result in inaccuracies.

Another horizon profiling approach using a fish-eye lens is to use images from a sky camera, which uses a fish-eye lens to record clouds or other weather for a variety of analyses and applications. The resulting hemispherical image is then divided into sky/cloud and non-sky/cloud (horizon objects). This approach is similarly susceptible to camera distortion, especially because the features to be measured occur exclusively at the edges of a fish-eye image, and because this analysis requires manual calibration of the camera orientation.

D. GIS digital elevation map methods

Horizon profiles are often created through geographic information system (GIS) data like digital elevation maps (DEMs), which provide rasterized elevation data in resolutions up to 30m per pixel. To calculate horizon maps from a DEM, lines are drawn from the observer point out to a fixed radius of interest, and the maximum elevation angle on each line is calculated using the arctangent of the horizontal distance from the observer and vertical height of each point on the line. This approach is used in common modeling software like PVGIS.

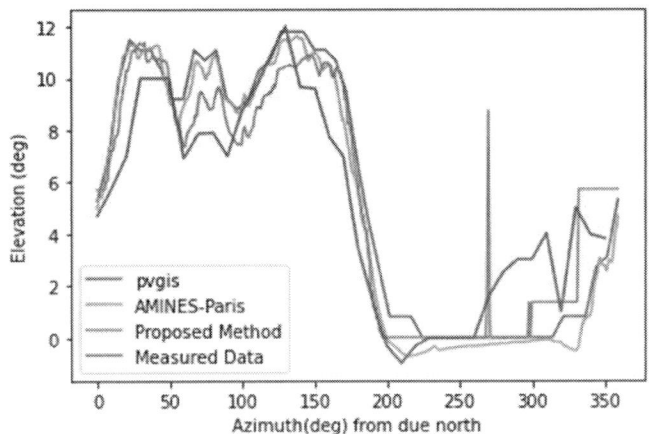

Fig. 3. Horizon profiles for a site near Albuquerque. Measured data was collected via theodolite, other methods used DEMs.

Note that the performance of this method is dependent on the accuracy of the DEM as well as the chosen sampling rate of points on the circular region of interest.

Because DEMs capture only the underlying terrain, these horizon profiling methods are limited to geographical features and do not capture other objects on the apparent horizon such as trees and buildings. Therefore these methods are typically only used for large-scale studies or preliminary siting.

As shown in Fig. 3, there are numerous implementations of the horizon profiling using DEMs. Two online sources include PVGIS and a tool hosted by AMINES-Paris. The algorithm behind AMINES-Paris' tool is unknown, but PVGIS uses GRASS-GIS as its backend, a GIS-focused software package written in C. Furthermore, we have implemented our own method for horizon profiling, which appears to give similar performance. However, for practical use, we recommend using a tool like PVGIS or AMINES-Paris. DEMs are large files, and require numerous steps to download and import into a useful format, whereas PVGIS, for example, is a simple HTML GET request. To support this, we have opened a pull request for pvlib-python to add horizon reterival from PVGIS to the library, slated for an upcoming release. To this end, we use PVGIS as a representative of the GIS DEM methods for further comparison in this work.

E. Time-series data-driven method

The proposed time-series method takes long-term irradiance or irradiance-like data with a DNI component as input. The irradiance data can be plane-of-array (POA), global horizontal (GHI), or direct normal (DNI). In the case of PV performance data, a model is used to convert the power or current to effective measured POA irradiance. A modeled irradiance is also generated using pvlib-python, and measured irradiance is divided by the modeled irradiance to create a time-series of normalized irradiance.

The next step is to bin the normalized irradiance by solar position. Solar position is calculated using location and timestamp data as input to the irradiance model, so we recycle those measurements here. Then a bin size is determined (usually 1 degree in elevation and azimuth, but can be larger for sparse datasets), and data is sorted into elevation/azimuth pairs.

To determine the shading profile from the binned normalized irradiance values, the values must first be summarized for each bin. The appropriate method may be different for different sites. Generally taking a percentile (~75th) of each bin is necessary to avoid effects of weather patterns (Fig. 4). In some cases it may be helpful to take an average or maximum value for each bin. The bin values are thresholded to create a 2-D mask of azimuth/elevation. Finally, the horizon profile is extracted as the lower edge of this mask. Depending on the density and quality of irradiance data, some additional image processing methods such as smoothing or closing may be used prior to the binarization step.

This method is limited in detecting horizon where objects interrupt the sun's path, so objects under or outside the sun's arc will not be detected.

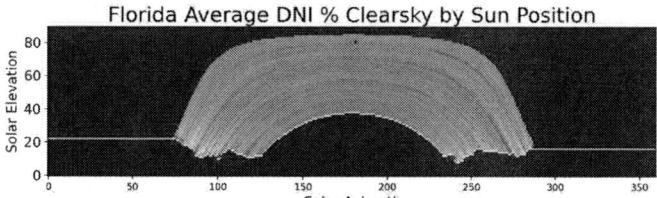

Fig. 4. Visualization of normalized DNI measured in Cocoa, FL, binned and averaged by solar azimuth and elevation, along with the detected horizon (yellow line).

III. SITES AND DATASETS

The latter two methods described above, both new in this publication, were developed and evaluated against other methods on various PV sites in the US and Nepal, summarized in Table II. Each of these datasets include independent GHI, DNI, and DHI measurements collected at 1 minute intervals. For the results presented here, we use 1-minute data DNI data filtered to >1° solar elevation.

TABLE II
SUMMARY OF DATASETS

Maintainer	Country	City	Latitude (°)	Longitude (°)	Elevation (m)	Days of Data
RTC	USA	Albuquerque, NM	35.055	-106.540	1657	3000+
		Cocoa, FL	28.387	-80.758	9	2500+
		Livermore, CA	37.680	-121.700	199	1556
		Williston, VT	44.467	-73.102	108	1551
World Bank	Nepal	Jumla	29.272	82.194	2383	732
		Lumle	28.297	83.818	1740	794
		Nepalgunj	28.113	81.589	150	732

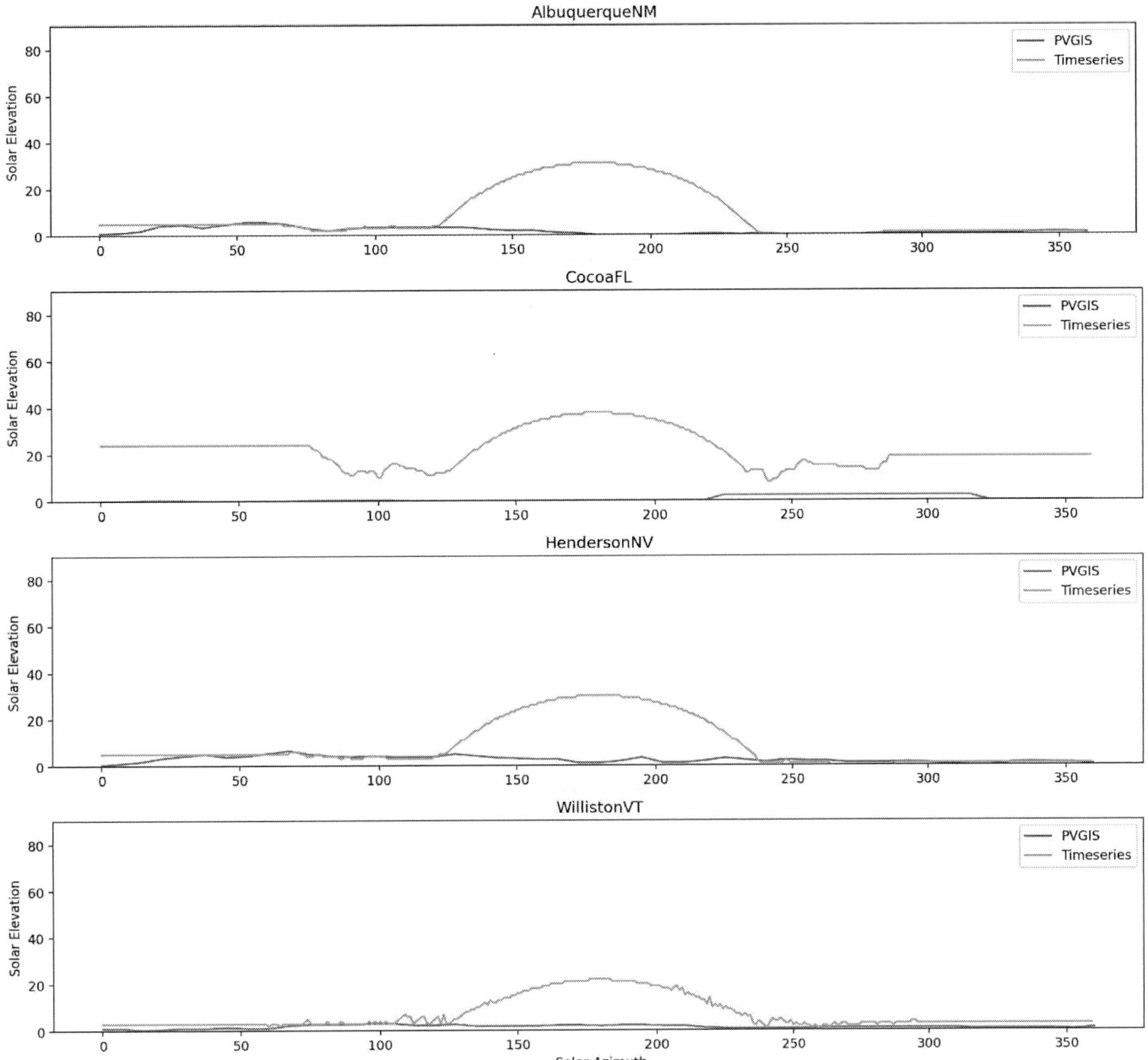

Fig. 5.   Horizon profiles detected via the PVGIS DEM method and the time-series data-driven method are shown for the 4 RTC sites listed in Table II. The time-series method utilized all available data for these calculations.

IV. RESULTS

Horizon profiless for the sites listed above each include at minimum DEM and time-series data-driven results. Some also include image-based results.

A. RTC Sites

DEM and time-series data methods are compared for the RTC sites in Fig.5. These figures demonstrate how these methods agree well for systems with scarce foliage or buildings (Albuquerque and Henderson), and poorly where foliage and/or manmade structures are plentiful (other sites).

Of additional interest is comparing horizon profiles extracted with the time-series method for different years. Fig. 6 shows this comparison for the Cocoa, FL site from 2015-2022. This figure shows the growth of foliage on the left side of the profile, which indicates that changes in horizon profile can be captured dynamically by this method.

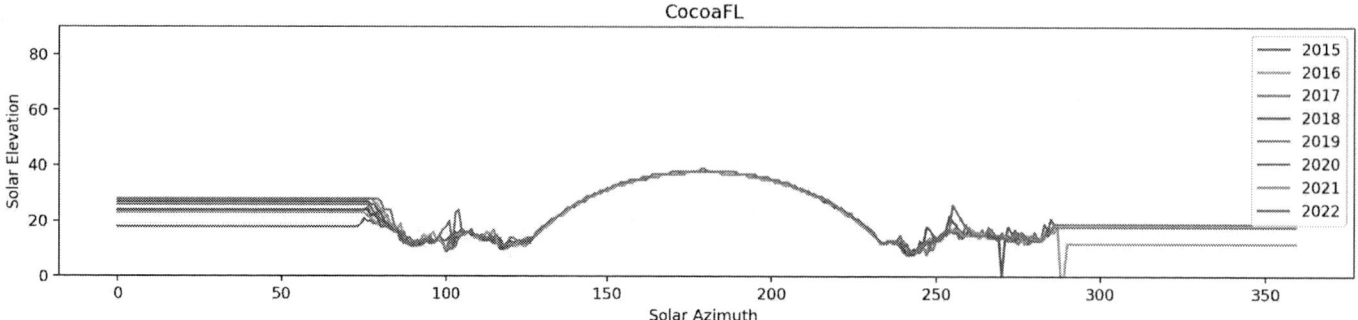

Fig. 6.   Horizon profiles generated with the time-series data-driven method at the Florida RTC for each year 2015 through 2022. The figure shows foliage growing on the left side of the graph and the increase in the amount of data filtered for horizon shading over time.

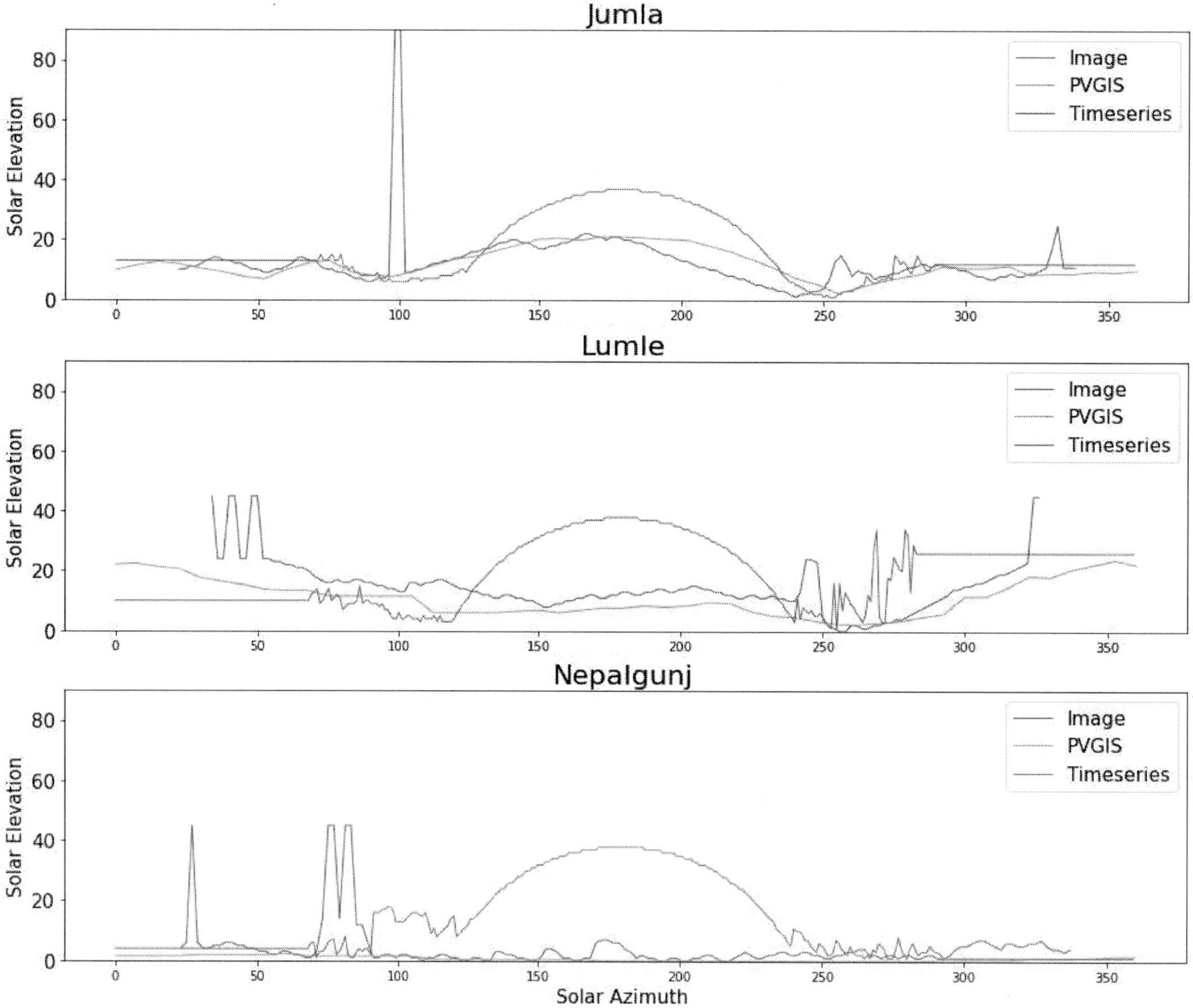

Fig. 7.   Horizon profiles detected via the PVGIS DEM method, the time-series data-driven method, and panoramic imaging are shown for the 3 Nepal sites listed in Table II. The time-series method utilized all available data for these calculations.

B. Nepal Sites

978-1-6654-6060-6/23 $31.00 © 2023 IEEE   516

The sites in Nepal show the differences between DEM, time-series data-driven, and image methods. These comparisons are shown in Fig. 7. While some agreement can be seen between all 3 methods, the PVGIS method's lack of sight for trees is quite apparent. Also apparent is the sun arc consistently seen in the time-series method results. However what is most striking to the authors is the perspective shift in the image-based horizon profiles as compared with the DEM and time-series data-driven results. Tall features in these images become distorted, both in azimuthal location as well as in apparent elevation, relative to the time-series result. This distortion can result from either or both a) the location of the camera relative to the irradiance sensor, and b) lens or image processing distortion in capturing and creating a panoramic image.

V. DISCUSSION

The horizon profiling methods presented here each have limitations, advantages and disadvantages corresponding to the data used. These methods broadly fall into categories of manual and automated data collection and processing.

We classify manual horizon profiling methods to include theodolite and image-based methods. These methods require physical access to a site, and manual collection of the data used to generate the horizon profile. The accuracy and applicability of horizon profiles generated by these methods are very dependent on the location at which they were taken relative to the array, which is also a major source of discrepancy between these and the automated methods. Furthermore, the manual methods do not inherently capture the orientation of a measured horizon profile; that is, they do not automatically capture reference points, and so the location at which they were taken and the relative positions of landmarks or reference points must be carefully documented. The risk in these methods, particularly image-based methods, is the perspective dysmorphia caused by camera lenses. The inherent deformation in the creation of a panoramic image with a wide-angle lens, or in the capture of a fish-eye hemispherical image, warps the relative size of features, particularly at the edges of the image. This was shown to create large differences in what is seen by an irradiance sensor vs. a camera. However, these methods are effective in capturing a complete horizon profile including foliage and man-made objects. They also evaluate areas beyond the sun path, which if captured correctly are useful in estimating diffuse contributions, as well as direct shading.

Automated methods explored in this work include the DEM-based (PVGIS) and time-series data-driven methods. DEM-based methods have distinct advantages for siting, including the ability to evaluate many sites with low effort, lack of need for access to the site, and also provides a complete 360° horizon profile. Its main disadvantage is the lack of awareness of trees, poles, and buildings. DEM methods can also be computationally expensive for high resolution/accuracy.

The time-series data-driven method for horizon profiling offers other advantages and disadvantages compared to the DEM-based method. First, this method requires long and high-density time-series (at least 6 months at the beginning or end of the year) to capture the full sun path and at high spatial resolution. However, if available, this data offers high spatial sensitivity to objects including man-made and foliage, even as these change through time. This makes the time-series method ideal for dynamic control of single-axis trackers, and filtering of historic time-series data for which horizon profiles were not previously measured. Additionally, the data-driven method can be applied to system power or current data, allowing the array to act as its own irradiance sensor and determine the effective horizon for the array, rather than a weather station.

VI. CONCLUSIONS

This work explored manual and novel automated methods for measuring the horizon profile at a PV site, including the introduction and methodology for time-series data-driven horizon profiling. Evaluations of the different methods for different application spaces were discussed, revealing clear frontrunning methods for various application spaces, including siting (digital elevation mapping methods), single-axis tracking (time-series data-driven method), diffuse irradiance estimation (manual and image methods), and historical data analysis and filtering (time-series data driven method).

VII. ACKNOWLEDGMENTS

This material is based upon work supported by the U.S. Department of Energy's Office of Energy Efficiency and Renewable Energy (EERE) under the Solar Energy Technologies Office Award Number 38530, Single Axis Tracker Reliability and Performance Improvement, led by Daniel Riley of Sandia. Sandia National Laboratories is a multimission laboratory managed and operated by National Technology and Engineering Solutions of Sandia, LLC, a wholly owned subsidiary of Honeywell International Inc., for the U.S. Department of Energy's National Nuclear Security Administration under contract DE-NA0003525. The authors acknowledge data contributions from the World Bank dataset managed by CSP Services and hosted at energydata.info.

REFERENCES

[1] G. Damian, K. Dariusz, and T. Grzegorz, "Analysis of the influence of shading by horizon of PV cells on the operational parameters of a photovoltaic system," Przeglad Elektrotechniczny, no. 04/2014, pp. 78–80, 2014, doi: 10.12915/pe.2014.04.17.

[2] J. Ranalli, R. Vitagliano, M. Notaro, and D. J. Starling, "Sensitivity of shading calculations to horizon uncertainty," Solar Energy, vol. 144, pp. 399–410, Mar. 2017, doi: 10.1016/j.solener.2017.01.017.

[3] R. Frei, C. Meier, and P. Eichenberger, "A fast, efficient and reliable way to determine the PV-shading horizon," in Sixteenth European Photovoltaic Solar Energy Conference, 2000.

[4] "SunEye 210 Shade Tool." www.solmetric.com

Field effect passivation enables 2.2 V open-circuit voltage all-perovskite tandems

Bin Chen, Ted Sargent

Northwestern University, Evanston, IL, United States

University of Toronto, Toronto, ON, Canada

All-perovskite tandem solar cells are promising to break the efficiency limits of single-junction cells at low cost. However, such promise has been limited by the open circuit voltage (VOC) deficit in tandems. The VOC loss is greater in wide bandgap (>1.7 eV) top cells than in ~1.5 eV perovskites. Quasi-Fermi level splitting (QFLS) measurements reveal VOC-limiting recombination at the electron transport layer (ETL) contact. Our study shows that treating the perovskite surface with 1,3-propane diammonium (PDA) increases QFLS by 90 mV. We found that the suppressed interfacial recombination is due to reduced minority carrier population at the perovskite/C60 interface, and improved surface homogeneity. PDA treated 1.78 eV perovskite cells achieve a certified 1.33 V VOC, and > 19% power conversion efficiency (PCE). Incorporating this layer into a monolithic all-perovskite tandem, we report a tandem VOC of 2.2 V and > 28% PCE. Encapsulated tandems retain more than 85% of their initial PCE after 500 hrs operation under 1 sun illumination in ambient conditions.

978-1-6654-6060-6/23 $31.00 © 2023 IEEE

DISTRIBUTED GENERATION COMPONENT PLACEMENT AND POINT OF COMMON COUPLING ALLOCATION FOR SOLAR ROOFTOP MICROGRID SIZING COSTS MINIMIZATION

Robert A. García Cooper
Student Member, IEEE

Marcel Castro Sitiriche
Member, IEEE

Agustín Irizarry Rivera
Senior Member IEEE

Fabio Andrade Rengifo
Senior Member IEEE

Mayagüez Campus - Dept. of ECE,
University of Puerto Rico,
Mayagüez, PR, 00681

Abstract — One of the pitfalls in renewable energy system (RES) component sizing for solar rooftop microgrid planning is disregarding the interconnectedness of the aggregate installations that would ultimately form a communal unit. Each establishment within the collective has specific constraints that need to be met and will ultimately impact all neighboring installations linked within and without the microgrid. To focus on supplying the needs of the disaggregated establishments, without considering their interconnectedness, or the aggregate without contemplating customer specific restrictions will eventually lead to higher capital and levelized costs from component oversizing or higher internal distribution losses due to improper component allocation. To overcome this challenge, the design process must be singularly and holistically approached in a phased manner, keeping in mind that interconnection decisions will require the addition or subtraction of RES components that will impact the operational conditions of other elements. This paper proposes a methodological framework aimed at attaining time efficient and cost-effective sizing in microgrid design through the optimal allocation of RES components, that maximize the use and availability of the solar resource to both meet aggregate demand and be accessible for power purchase agreements (PPA) beyond the microgrid's points of common coupling (PCC). The proposed methodology has been developed and applied to the pre-design of renewable energy system based secondary voltage microgrids for the remote Castañer and Maricao communities within the Puerto Rican Midwestern zone and can be applied to minimize projected microgrid sizing costs in any part of the world.

Keywords— Rooftop Microgrid Sizing, LCOE Minimization, Point-of-common-coupling, Power Purchase Agreements

I. INTRODUCTION

As most of modern civilization has grown dependent on electric power to fuel their livelihoods, interests, and sustain their communal infrastructures, ensuring affordable access to electricity has become of key importance for many communities, particularly those who are susceptible to frequent or extended electric service interruptions due to power grid vulnerability to atmospheric or telluric phenomena or poor utility maintenance. Electric service interruption costs tend to be much greater than levelized replacement costs of emergency generators [1], while renewable energy system (RES) levelized costs gravitate towards even greater cost-effectiveness when properly sized and financed [2].

The fact that RESs impose a lesser burden over individuals and communities, as they eliminate indirect electric service interruption costs related to lost personal leisure time [3] spent purchasing fuel and providing maintenance to emergency generators, a rising number of consumers to shift their attention towards RES acquisition. Yet, for low-to-medium income households, current RES costs can border on the inaccessible, because of improper design to lacking awareness of the costs that they incur to replace services when utility sourced electric power is unavailable [4]. Many consumers are now considering collective RES electric power generation to lower capital and levelized costs by interconnecting systems and forming the aggregate that we presently call a microgrid. While the ordinary dangers of potential sizing error with their consequential higher costs remain present, RES component allocation now becomes a concave cost-effectiveness problem that aggravates as interconnectivity constraints become ever more present, such as the point of common coupling (PCC) allocation.

Power sector modeling solutions have been instrumental in estimating single and multiscale RES sizing designs [5]-[6]-[7]. But additional attention must be given due to interconnection restrictions. In the past years, teams of researchers have provided methods designed to optimally size off-grid and grid-connected microgrids mostly focusing on photovoltaic (PV) and battery energy storage system (BESS) optimality [8]. Since the least levelized costs of an RES is obtained through a convex function, researchers recommended methods that employed different types of optimization approaches, such as Self-adaptive Bee Swarm Optimization aimed at finding optimal net present value [9], Particle Swarm Optimization to find Net Present Cost, a probabilistic method based on the loss of power supply probability [10], a three objective model optimization algorithm for hybrid generation sources [11], and even optimization focused on considering demand response [12]. The common thread is that they highlight how to optimize multi-scale RES sizing yet fall short when addressing a method on approaching microgrid interconnectivity constraints.

The authors employed a power sector sizing optimization solution they had developed, the "Renewable Energy Algorithm Tool for Rural Electrification and Appropriate technology" or REAL-TREAT optimization model, to estimate

RES sizing with the lowest levelized cost of energy (LCOE) while yielding year-long hourly potential energy exchange data from hybrid power and storage sources between prosumers and their respective communities and/or utility, for power purchase agreements [2]. REAL-TREAT was applied to the Puerto Rican community microgrid studies of the Castañer and Maricao town centers, where the development of new constraints, otherwise absent in single site RES designs, were required to achieve RES sizing LCOE minimization.

II. EXPERIMENTAL PROCEDURE

Microgrid design ultimately depends on the decisions that communities and other stakeholders agree upon when interconnecting. Since these agreements require decades long commitments of both passive and active shared capital, communities need to have a broad set of options available before committing. Developing a step-by step framework that would make the potential microgrid's design process more time-efficient and as cost-effective as possible became an objective. The following is a summary:

1) Acquire load measurements or estimates of individual microgrid members.
   - Aggregate load data.
2) Assess PCC allocation with minimal distribution power flow loss of utility supplied load.
3) Generation resource and installation site data acquisition of individual microgrid members.
4) Estimate optimal isolated RES Sizing LCOE for each participating microgrid member.
5) Estimate RES Sizing LCOE Minimization for aggregate microgrid loads without power flow losses and distribute RES components amongst participating prosumers.

   - Reduce inverter sizing and capital expenditures by distributing PV modules amongst prosumers.
6) Estimate net hourly power dynamics for each participating member with component sizing yielded in Step 5.
7) Assess PCC allocation with minimal distribution loss of net metered consumption and supplied power for each participating member.
8) Estimate RES Sizing LCOE Minimization of aggregate microgrid including both consumption and net power losses, of steps 2 and 7 respectively.
9) Repeat steps 6-8 for all the lowest possible PV module combinations that do not violate the roof-top area limits of participating customer installation sites.
10) Present findings to stakeholders.

III. ANALYSIS OF RESULTS

The optimized costs of each customer financing an isolated RES for their residence or establishment must first be estimated, as shown in TABLE I. Differences in consumption patterns within each establishment will suggest a particular RES sizing that yields the minimum LCOE. In this case study, the local residential utility rate is $0.28/kWh, and the commercial rate is $0.30/kWh presuming that the electric utility's power grid is reliable and no outage costs are present. Commerce 6 lacks the area necessary for RES installation, thus it must forgo an RES installation within its premises or enter into an agreement to interconnect with other customers and/or use their rooftop area.

A. Microgrid Solar Resource Placement Problem

The following step is to obtain the minimum RES sizing costs that will meet the collective load, simulating the aggregate if it

TABLE I

OPTIMAL RES COMPONENT SIZING FOR ISOLATED ESTABLISHMENTS WITH ISLANDED RES'S AND THE UTILITY AS A BACK-UP

Customer	% of Total Load	PV Modules	Inverters No. (Size)	DESS (kWh)	Charging Power (kW)	RES Supplied Load	CAPEX	RES LCOE	Net Met. LCOE
Residence 1	1%	6	1 (3 kW)	5.12	3.84	87%	$ 12,400	$ 0.63	$ 0.59
Residence 2	2%	6	1 (3 kW)	3.84	3.84	53%	$11,800	$ 0.42	$ 0.39
Residence 3	1%	3	1 (2 kW)	2.56	3.84	73%	$ 9,900	$ 0.74	$ 0.74
Commerce 1	9%	27	2 (6 kW)	7.68	3.84	45%	$26,700	$ 0.32	$ 0.25
Commerce 2	3%	12	1 (6 kW)	7.68	3.84	82%	$ 16,700	$ 0.39	$ 0.33
Commerce 3	2%	6	1 (3 kW)	2.56	3.84	83%	$11,500	$ 0.50	$ 0.47
Commerce 4	13%	42	2 (10 kW)	23.04	3.84	77%	$39,500	$ 0.28	$ 0.22
Commerce 5	9%	36	2 (8 kW)	30.72	7.68	89%	$39,000	$ 0.31	$ 0.27
Commerce 6	37%	none	none	none	none	none	none	none	none
Commerce 7	24%	105	5 (10 kW)	30.72	7.68	83%	$80,800	$ 0.25	$ 0.19
Totals		243 (104 kW)		114 kWh	42.24 kW	48%	$ 248,300	$ 0.31	

were a single site RES installation and distributing the components amongst the different customers to further reduce costs. If customers decide to not curtail PV generation, then the different inverter sizes will vary depending on the PV's installed capacity. Since inverter sizing will ultimately depend on the PV sizing uncurtailed generation, combining PV modules within each participating establishment would ultimately reflect in the lossless levelized costs (LCOE).

TABLE II shows up to four different combinations of PV modules allocated over the different establishments, manually iterated to comply with available roof-top area limits. Lossless LCOE estimations for each iterated combination of PV modules were first obtained through the initial microgrid aggregate sizing LCOE minimization process described in Step 5. In all four cases, installed storage capacity was sized as 92.16 kWh and installed charging power capacity was sized at 23.04 kW.

Three different LCOE scenarios, depending on the supply or generation sources are presented:

- The microgrid RES combined with the utility as a back-up for demand unmet by the RES (MG + Utility),
- The microgrid RES combined with an auxiliary generator as a back-up for demand and DESS charging, (MG + AG)

TABLE II
LEVELIZED COSTS OF ENERGY PER COMBINATIONS OF ALLOCATED PV

Number of PV Panel Allocation per Building				
Customer	1st	2nd	3rd	4th
Residence 1	0	12	12	21
Residence 2	21	21	21	21
Residence 3	21	9	21	21
Commerce 1	42	42	42	42
Commerce 2	21	21	21	21
Commerce 3	21	21	21	21
Commerce 4	63	63	63	63
Commerce 5	21	21	30	21
Commerce 6	0	0	0	0
Commerce 7	105	105	84	84
Totals	315	315	315	315
Preliminary Lossless Interconnected LCOE ($/kWh)				
Microgrid + Utility	$ 0.2712	$ 0.2742	$ 0.2748	$ 0.2724
Microgrid + Aux. Gen	$ 0.5670	$ 0.5700	$ 0.5700	$ 0.5682
Net Metering	$ 0.1896	$ 0.1926	$ 0.1926	$ 0.1908
Preliminary LCOE with Power Distribution Losses ($/kWh)				
Microgrid + Utility	$ 0.2838	$ 0.2862	$ 0.2868	$ 0.2844
Microgrid + Aux. Gen	$ 0.6072	$ 0.6096	$ 0.6102	$ 0.6078
Net Metering	$ 0.1914	$ 0.1938	$ 0.1944	$ 0.1920

- Net metering PPA between the microgrid and utility.

It is important to note that in this case study, although the first iteration of PV module allocation yields the minimum LCOE compared to the other combinations, Residence 1 would not have access to electric power should the microgrid's internal distribution circuit become compromised or unavailable. To ensure that all participants have at least some degree of resiliency, with PV generation as a minimum, the other combinations should be evaluated, with the fourth yielding the least LCOE among the remaining options.

Even when the difference of lossless LCOE between each iterated combination is less than a cent, we must bear in mind that for this case study, a difference of half a cent per kWh could represent up to $1000 a year that could be saved or diverted towards operation and maintenance. This may not seem a considerable amount for some commercial, industrial, and high-income customers, but it may have a larger impact on low to medium income communities. Another thing to keep in mind is how the different allocations of PV modules will generate different internal distribution losses within the microgrid's side of the PCC. Since the microgrid will be responsible for supplying and/or paying for said distribution losses, factoring them within the microgrid's RES sizing costs minimization will yield a considerate difference, depending on the location of generation sources, plus the distance and sizing of the radial distribution wiring. Adding said distribution losses to Microgrid RES sizing is our next problem.

B. Microgrid Optimal PCC Problem

Typically, the PCC between the single site of an RES and an electric utility can be found in the utility's electricity meter. The same happens for an interconnected microgrid, except that the utility and the microgrid's operator must agree on how and where they will exchange energy. Since this exchange will usually be held at a single location and the community will be responsible for all distribution losses that are generated on the microgrid's side of the PCC, they must find the point of least distribution loss. But before the point of exchange or the point of common coupling (PCC) is defined, the community must choose the type of service they want their microgrid to provide and what will be the contractual relationship that the community will have with the electric utility.

If the community wants their microgrid to be their primary supplier of electric power, where they will only be interconnected with the electric utility so that the later can serve as a back-up in the moment that their renewable energy systems cannot supply demand. The utility will then charge the community for the loads that the utility supplies as a back-up, but they will also charge for any internal distribution losses generated when supplying said loads. For our case study, TABLE III illustrates power distribution losses depending on the location of the potential PCC when the utility is solely supplying customer consumption. In this example, it was estimated that the lowest distribution losses would be obtained

TABLE III
POWER DISTRIBUTION LOSSES (KWH) PER POTENTIAL PCC FOR MICROGRID SUPPLIED ONLY BY THE GRID POWER

Potential PCC	2ndary Dist. 120 V	1ary Dist. 4.16 kV	Sub-transmission 38kV to 38kV	Transmission 38kV to 115kV	Transmission 115kV to 115kV	Generation 230kV to MPP	Total
Residence 1	492	6.32	1.59	0.82	0.0096	0.0279	501
Residence 2	322	5.72	1.44	0.74	0.0087	0.0252	330
Residence 3	781	7.53	1.890	0.980	0.011	0.033	792
Commerce 1	675	7.10	1.78	0.92	0.01	0.03	685
Commerce 2	413	6.09	1.53	0.79	0.01	0.03	422
Commerce 3	305	5.70	1.43	0.74	0.0087	0.0251	313
Commerce 4	254	5.51	1.38	0.72	0.0084	0.0243	262
Commerce 5	222	5.40	1.36	0.70	0.0082	0.0238	229
Commerce 6	262	5.52	1.39	0.72	0.0084	0.0244	270
Commerce 7	397	5.96	1.50	0.78	0.0091	0.0263	405

when the PCC was allocated in Commerce 5. Hence, if the community wishes to have a microgrid that only uses the utility as a back-up, then the PCC should be allocated in Commerce 5.

If the community instead wished to not curtail PV generation and instead sell electric power to the electric utility via Net Metering or another PPA, then the PCC should be where the least amount of net supplied and consumed energy is obtained after factoring in distribution losses. In a microgrid net metering PPA, the most net supply after deducting distribution losses is pursued. In cases where the net difference of all potential PCC's is negative, as is shown in TABLE IV, then the least net consumption is sought.

Although said net difference will depend on the aggregate RES sizing, the potential PCC allocation must also not violate

TABLE IV
PROJECTED POWER CONSUMPTION VS SUPPLY (KWH) PER POTENTIAL PCC FOR INTERCONNECTED MICROGRID SAMPLE CASE STUDY WITH NET METERING PPA

Potential PCC	Total Consumed	Total Supplied	Net Diff.	Wiring CAPEX
Residence 1	71436	67122	(4314)	$ (22,499)
Residence 2	71786	68137	(3649)	$ (22,524)
Residence 3	71324	68072	(3252)	$ (27,117)
Commerce 1	72797	66407	(6390)	$ (25,479)
Commerce 2	71391	67768	(3623)	$ (22,705)
Commerce 3	70735	69020	(1716)	$ (21,747)
Commerce 4	70465	69613	(852)	$ (21,507)
Commerce 5	72353	66449	(5904)	$ (21,689)
Commerce 6	70657	63982	(6675)	$ (24,040)
Commerce 7	70426	69796	(630)	$ (23,246)

the distribution wire ampacity, limits of each of the radial distribution circuit's segments, for more than a specific amount of time. In this case study, the maximum distribution system's total load was not to surpass the neutral wiring's ampacity limit by more than 1.1 per unit for more than 250 hours of the year. In TABLE IV, regardless of whether Commerce 7 forecasted a more favorable net consumption before and after including distribution losses, Commerce 4 was chosen as the potential PCC.

To place the PCC in Commerce 7 meant that neutral wiring ampacity limits would probably be violated by over 250 hours in at least four segments of the proposed radial distribution circuit, with one of said segment being over 700 hours. To allocate the PCC in Commerce 4 violated none of the segment's ampacity limits for more than 220 hours, requiring smaller distribution wiring sizes and ultimately translating to smaller distribution wiring capital costs (CAPEX).

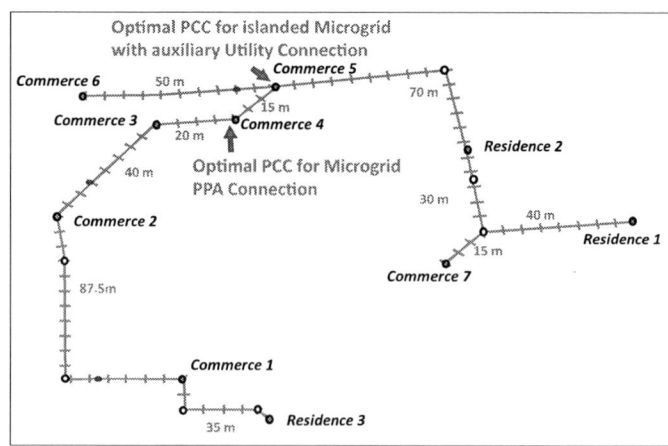

Fig. 1. Optimal Point of Common Coupling Assessment for sample Community Microgrid Case Study.

TABLE V
Minimized RES Component Sizing Costs for Interconnected Establishments within Islanded Microgrid with the Utility as a Back-Up and the PCC in Commerce 4

Customer	% of Total Load	PV Modules	Inverters No. (Size)	DESS kWh	Charge Controllers	Aggregate Supplied Load	CAPEX	Microgrid LCOE	NET Met. LCOE
Residence 1	1%	21	1 (10 kW)	-	-	67.9%	$ 3,500	$ 0.28	$ 0.19
Residence 2	2%	21	1 (10 kW)	-	-	67.9%	$ 6,000	$ 0.28	$ 0.19
Residence 3	1%	21	1 (10 kW)	-	-	67.9%	$ 2,700	$ 0.28	$ 0.19
Commerce 1	9%	42	2 (10 kW)	15.36	3.84	67.9%	$ 25,600	$ 0.28	$ 0.19
Commerce 2	3%	21	1 (10 kW)	-	-	67.9%	$ 9,600	$ 0.28	$ 0.19
Commerce 3	2%	21	1 (10 kW)	-	-	67.9%	$ 4,500	$ 0.28	$ 0.19
Commerce 4	13%	63	3 (10 kW)	30.72	7.68	67.9%	$ 37,300	$ 0.28	$ 0.19
Commerce 5	9%	21	1 (10 kW)	30.72	7.68	67.9%	$ 25,400	$ 0.28	$ 0.19
Commerce 6	37%	none	none	none	none	67.9%	$106,300	$ 0.28	$ 0.19
Commerce 7	24%	90	4 (10 kW)	30.72	7.68	67.9%	$ 69,900	$ 0.28	$ 0.19
Totals		321 (138 kW)		107.52 kWh	26.88 kW	67.9%	$ 290,800	$ 0.28	$ 0.19

Once the Microgrid's potential islanded PCC and interconnected PCC are identified, the projected yearlong hourly power distribution losses must be added to the microgrid's aggregate load for it to undergo RES Sizing LCOE minimization. The islanded RES LCOE should be employed when the microgrid is to supply most of the aggregate load and will only use the utility grid as a backup through the PCC in commerce 5, as shown in Fig. 1. When the microgrid is to have a Net Metering or other PPA with the electric utility, the Net LCOE calculations can be factored by again adding the hourly distribution losses to the microgrid's aggregate load in order to size the aggregate RESs and then add the yearlong hourly net supply vs consumption distribution losses to the microgrid's projected yearlong hourly net supply vs consumption data. It is important to ensure that both consumption and the net losses are obtained with the same PCC, in this case commerce 4 will be used.

TABLE V yields the results of the microgrid's sizing costs minimization process, which clearly differs from TABLE I since interconnecting requires a greater amount of PV resource, as it must now supply for commerce 6, even when Residence 1 has no RES component but instead pays to supplied by the microgrid. Storage capacity is smaller when interconnected and it is arbitrarily distributed amongst or nearest to the larger consumers in order tax the DESS with the least amount of distribution losses. Capital expenditures are greater in TABLE V as interconnection capital expenditures are included, without labor costs and contractor fees. It is expected that the Microgrid's distributed RESs would supply approximately 67.9% of the aggregate yearlong load while also supplying projected interconnection losses. This means the utility would have to supply up to 32.1% of the yearlong load, with costs levelized to grid parity at $0.28 and an even smaller net

metering levelized cost of $0.19, assuming that the electric utility's power grid is reliable.

The results of TABLE V also differ from those of TABLE II, because once the distribution losses were added to the aggregate load, the REAL-TREAT model suggested a slight increase of installed PV, storage and charging power capacity. The result, for an islanded operated microgrid using the utility as a back-up, was an aggregate LCOE of $0.2832/kWh instead of the $0.2838/kWh obtained from the first combination of TABLE II. Although the slight increase in storage and charging capacity raises the net metering LCOE to 0.192/kWh instead of the $0.1914 and hikes annual payments by $131, should it not be possible to supply energy to the utility, the RES can now supply an additional 1.9% of the year-long consumption. This small added resiliency translates into being able to supply an additional 4,220 kWh of annual consumption that would otherwise have to be supplied by the utility or lost, in case of an electric service interruption.

The added resiliency was also enough to increase the Average Time of Expected Resiliency (AToER), from nine to ten hours per day. The AToER represents the average amount of daily time that the microgrid's aggregate RESs will supply demand throughout the whole year. This is particularly useful when one wishes to estimate the level of RES robustness that stakeholders are willing to pay for, as it gives a clear idea of the average time per day that they would have power during the year, should the utility grid or any other source of power not be available. This parameter carries a lot of weight when would-be prosumers are serviced by an unreliable electric power grid or when they are exposed to high probabilities of electric service interruptions due to atmospheric and/or telluric phenomena.

TABLE VI
COSTS AND AVERAGE TIME OF EXPECTED RESILIENCE PER ELECTRICITY SOURCE

Electricity Source(s)	Annual Costs	LCOE	ROI	AToER
Utility	$ 49,219	$ 0.295	-	-
Emergency Generator	$240,400	$ 1.442	-	-
PV no PPA	$ 49,200	$ 0.295	6	-
PV w/ Net Metering	$ 30,200	$ 0.181	5	-
RES w/ Utility	$ 47,200	$ 0.283	6	10
RES w/ Aux. Gen.	$88,900	$ 0.533	31	5
RES w/ Net Metering	$ 32,000	$ 0.192	5	10

Presenting findings to stakeholders is an important step of microgrid design. TABLE VI summarizes various scenarios, while including ancillary information that may interest potential microgrid stakeholders. The first scenario is the annual costs of purchasing the utility supplied power. The second scenario represents the annual costs related to supplying the year-long load solely with an emergency generator, which does not include the time spent obtaining fuel nor operation and maintenance costs. It is against these first two scenarios that the annual costs and the levelized costs of energy yielded by the different or hybrid combinations of power sources can be compared to. Return of investment calculations can then be estimated by comparing scenario based annual costs with what is already being paid to the utility, particularly when the utility exclusively supplies the yearlong load.

$$ROI = \begin{cases} \dfrac{CAPEX}{EAC}, & \dfrac{CAPEX}{EAC} > 0 \\ 0, & otherwise \end{cases} \quad (1)$$

Where,

$$EAC = UAC + \Delta AC \quad (2)$$

UAC = Average electric utility service annual costs

ΔAC = Average electric utility service annual costs minus the projected annual costs of a PV system or one of the hybrid combinations of power sources described in TABLE VI.

IV. SUMMARY

This paper provides a brief general overview of the methodology employed to achieve minimum RES sizing costs for potential Microgrid pre-designs. The REAL-TREAT optimization model for single site installations was adapted to employ the proposed microgrid RES sizing cost minimization procedure that can be applied to any locality, presuming a reliable electric utility power grid. Factoring electric power grid unreliability in single site or microgrid aggregate RES sizing costs has been developed and will be presented in future work.

The proposed method solves two problems that are exclusive to microgrid or multiple interconnected RES design, being PV component distribution amongst participating stakeholders and PCC allocation that reduces power distribution losses and overall levelized cost. Future work must be undertaken to explore cost differences related to distribution losses resulting from DESS and charge control unit allocation. The authors hope that their research will contribute to the discussion on how to provide low-to-medium income communities with electrical resiliency, via access to renewable energy system sourced electric power.

REFERENCES

[1] London Economics International LLC, "Estimating the Value of Lost Load," Electric Reliability Council of Texas, Inc. (ERCOT), 2013.

[2] R. García Cooper and M. Castro Sitiriche, "Renewable Energy ALgorithm Tool for Rural Electrification and Appropriate Technology," in *International Conference on Appropriate Technology (ICAT)*, Khartoum, Sudan, 2022.

[3] T. Schröder and W. Kuckshinrichs, "Value of Lost Load - An Efficient Economic Indicator for Power Supply Security A Literature Review," *Frontiers in Energy Research*, vol. 3, no. 55, 2015.

[4] Sandia National Laboratories, "Analysis of Microgrid Locations Benefitting Community Resilience for Puerto Rico," 2018. [Online]. Available: https://www.osti.gov/biblio/1481633-analysis-microgrid-locations-benefitting-community-resilience-puerto-rico. [Accessed 7 January 2022].

[5] NREL – U.S. National Renewable Energy Laboratory, "System Advisor Model (SAM)," 2018. [Online]. Available: https://sam.nrel.gov/. [Accessed 6 January 2022].

[6] NREL – U.S. National Renewable Energy Laboratory, " Engage: Open Access Energy System Planning," [Online]. Available: https://engage.nrel.gov/en/login/?next=/en/. [Accessed 6 January 2022].

[7] HOMER Energy, "Homer Software.," [Online]. Available: https://www.homerenergy.com/index.html. [Accessed 6 January 2022].

[8] L. N. An, T. Quoc-Tuan, B. Seddik and N. Van-Linh, "Optimal sizing of a Grid-connected Microgrid," in *International Conference on Industrial Technology (ICIT)*, Seville, Spain, 2015.

[9] B. Mozafari and S. Mohammadi, "Optimal sizing of energy storage system for microgrids," *Sādhanā*, vol. 39, no. 4, pp. 819-841, 2014.

[10] F. Tooryan, E. Collins, A. Ahmadi and S. Rangarajan, "Distributed generators optimal sizing and placement in a microgrid using PSO," in *International Conference on Renewable Energy Research and Applications (ICRERA)*, Instanbul, Turkey, 2017.

[11] G. Zhao, T. Cao, Y. Wang, H. Zhou and C. Zhang, "Optimal Sizing of Isolated Microgrid Containing Photovoltaic/Photothermal/Wind/Diesel/Battery," *International Journal of Photoenergy*, vol. 10, pp. 1-19, 2021.

[12] X. Shao, X. Ren, Y. Li, Z. Song, Y. Ye and X. Xu, "Capacity Allocation Optimization of PV-and-storage Microgrid Considering Demand Response," in *Power System and Green Energy Conference (PSGEC)*, Shanghai, China, 2022.

The Solar Boat: An Academic Research Experience

Guillermo Serrano, *IEEE Member*, Erick Aponte, *IEEE Member*, Eduardo I. Ortiz-Rivera, *IEEE Senior Member*
University of Puerto Rico, Mayagüez
Electrical and Computer Engineering Department, Mechanical Engineering Department
gserrano@ece.uprm.edu, eaponte@ece.uprm.edu, eduardo.ortiz7@upr.edu

Abstract — This paper aims to explain the unique learning process students from the UPRM's Solar Boat Team passed through while working on the design, simulation, and construction of a solar boat. The research is interdisciplinary, combining students from mechanical, electrical, and industrial engineering majors. The ultimate objective for UPRM is to compete at the Solar Splash Competition. In 2022, UPRM received second place overall at the Solar Splash.

Keywords: photovoltaic, interdisciplinary, boat, solar panels.

I. INTRODUCTION

Academic research can be defined as an investigation based on the idea of scientific inquiry. It is based on a multi-step cyclical process that doesn't always go in a straight line, leading the researcher to return to previous stages as the project gets perfected. Academic research allows students to expand their knowledge in new areas of interest and develop skills that cannot be attained in regular classes. Students who perform investigative work get exposed to incredible opportunities for technical knowledge and abilities.

The University of Puerto Rico-Mayagüez (UPRM) has been involved in power systems, power electronics, and renewable energy projects for over forty years. One of the UPRM's insignia projects is the UPRM's Solar Boat. The primary purpose of the UPRM's Solar Boat is to compete at the Solar Splash Competition. On 2022, UPRM received the second place overall at the Solar Splash.

The Solar Splash objective is for collegiate student teams to design and construct a human-crewed photovoltaic solar-powered boat within a school year at a reasonable cost, then compete against other college teams in a 5-day series of events. The competition is designed to provide practical engineering experiences, development of project and program management skills, and exposure to many technical disciplines while learning the efficient use of energy and systems to create a thriving craft and competitive team.

The Solar Boat Research Team expanded the scope of skills accessible to a student working in their lab and provided the opportunity to grow, not only in a technical capacity but also in a professional and academic area. This paper aims to illustrate how this research team developed and experienced an educational research process focused on developing skills that only sometimes worked in other research groups. The following section will present a quick overview of the development process of the research project.

Fig. 1: Solar Splash Competition [1]

II. UNDERGRADUATE RESEARCH DEVELOPING PROCESS

The Solar Boat Research Team is an interdisciplinary group which integrate engineers from electrical, mechanical and industrial department. Worked by undergraduate student from the University of Puerto Rico, Mayaguez Campus (UPRM). The objective of their research is to design and construct a solar boat that competes at the Solar Splash Competition [1]. Also, the team is working towards creating a cost effective solar boat. Figure 2 shows an example of the 3D CAD structure and components used for a solar boat.

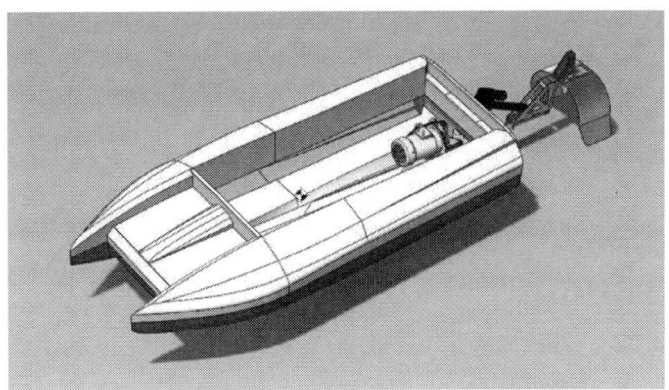

Fig. 2: Main components of a simulator using 3D CAD design.

A. Student Selection

Before starting the main research work, the project relevance and objectives had to be defined. Students must consider the work and research already done on the topic selected. This experience gives students the opportunity and knowledge to identify areas yet to be explored. This phase can also prove crucial for the student's understanding of the type of work and expected responsibilities. Students with no previous research experience may believe that they will be working the same way as for a class project, an assignment, or some side extracurricular project. This experience could lead the students to underestimate the magnitude of work and the level of analysis required, which might lead to underperformance and/or lack of commitment.

B. Literature Review

The main focus for this phase was understanding the different components of the solar boat and how to simulate each piece and a system as a whole. To fill in the knowledge gaps, consulting papers, books, and class presentation material professors and colleagues were necessary. The databases used contain many documentations related to sun simulation, photovoltaic panels, dc/dc converters, batteries, structural boat designing process, load measurements, Maximum Power Point Tracker technology, photovoltaic systems, solar energy, and irradiation, etc. was revised and discussed. Through the literature review process, the students were exposed to analyzing and gathering information through the secondary research methodology, interacted with professors in different departments, and developed critical review thinking in the understanding of technicalities in professional papers. The industry and market were also analyzed, including different types and models of solar simulators. From all the revised structural simulations and their various functions, it was concluded that a solar boat with hydrofoils was the optimal option due to cost benefits, the life span, and the possibility of reducing the water's drag (friction).

C. Goals and Objective

The main goals identified were:
- The creation of a low-cost and cost-efficient solar boat using hydrofoils
- Maximize the solar energy conversion of the PV modules using MPPT technology.
- Transportable and easy to move

D. Team Organization

These areas were selected to optimize the work inside the research, in fields concentrated on the capabilities and abilities of the students, taking advantage of the interdisciplinary research as part of the experimental learning process in Fig. 2.
- The construction of a test platform for testing the PV panels, DC/DC converters, batteries, and motors.
- The creation of an electrical system can control different power extraction depending on the solar boat motor speeds.
- The development of different testing procedures could produce the data required to validate the solar boat for final construction and implementation.

Fig. 3: UPRM's Solar Boat Research Team for the Solar Splash..

E. Work Methodology

For the solar boat, the most critical components are the PV panels, DC/DC converter, batteries, motors, boat structure, and propels. The core of the process was analyzing, designing, and developing a solar boat that could meet the minimum Solar Splash Standards.

1. Electric Team

The electrical team has to work with the specifications established before, optimizing the PV array, battery selection, and dc/dc converter design that should be made to achieve the optimal power to move the solar boat. An MPPT algorithm must be implemented in the circuit that delivers the energy needed to the motor. Different PV array configurations were tested, but the electrical team decided to use the square shape (i.e. m x m PV array).

2. Mechanic Team

The mechanical team designed the testing platform using NX software, as shown in Figure 4. The specification provided by the different marine standards for the parameters and requirements for the solar boat propels some of the principal specifications the students identified for the testing platform. The propel had been adapted to the motor load used, the heating that could be generated, and the materials used that could affect testing.

Fig. 4: Propel design in NX program as potential prototype

3. Sensors and Test Team

The sensor and testing team has to work with the methodology of the testing and how to analyze the data produced by the solar boat, and the sensors needed for each testing. The tests were divided in single components (e.g. only dc/dc converter) and combination of multiples subsystems from two and more (e.g. PV array and dc/dc converter).

III. ACADEMIC BACKGROUND

It is important to mention that for a successful research experience, it is critical to have an excellent academic background. The courses mentioned in Table I are offered by the University of Puerto Rico Mayagüez Campus [2]. For the students that are working in this research or have an interest in the field, the following courses presented are strongly recommended as preparation for this research project.

Field	Course	Description	About
MECHANICAL ENGINEERING	INME 4001	THERMODINAMICS	A study of the fundamental laws of thermodynamics as applied to closed and open systems
	INGE 4011	INTRODUCTION TO MECHANICS OF MATERIALS	Stresses and strains due to axial, torsional, and flexural loads; shear and moment diagrams
	INME 3809	CREATIVE DESIGN1	Introduction to the underlying principles and methodologies of engineering graphic
PHYSICS	FISI 3172	PHYSICS TWO	Principles of electricity, optics, and modern physics for engineering
INDUSTRIAL ENGINEERING	ININ 4015	ECONOMIC ANALYSIS	Criteria and techniques of economic analysis
ELECTRICAL ENGINEERING	INEL 4405	ELECTRIC MOTORS AND DRIVES	Analysis of electric motors
	INEL 4201	ELECTRONICS ANALYSIS	Semiconductors characteristics and amplifiers
	INGE 3016	COMPUTER PROGRAMMING	Algorithm and high level language
	INEL 4416	POWER ELECTRONICS	Circuit Rectification, inversion, direct current and alternate current control

Fig. 5: UPRM's Solar Boat

IV. SKILLS DEVELOPMENT

The students are suggested to have previous knowledge related to the work done to perform the given tasks. If not, they were expected to develop the vital skills needed during the work and activities involved. Being a team composed of only undergraduate students, they were allowed to work outside of their majors. Mechanical engineering students learned how to design, understand, and weld circuits, while electrical engineering students learned how to create, understand and mount a mechanical platform. Some of the skills the student acquires and develops are:

A. Vital skills:

- Ability to develop organization skills, a working plan, communications skills, group decisions, and critical thinking.
- Ability to give oral and poster presentations to other members of the team and professionals and company recruitments.
- Ability to identify learning gaps in their colleagues and support them.

B. Management skills:

- Ability to make a budget plan and adjust it to the limitations.
- Create good technical writing habits. Students could write technical reports of the work done with personal growth to maintain feedback on the learning options.

C. Technical skills:

- Learning to design and do simulations in NX, PSpice and Simulink in order to create possible design for the mechanical and electrical systems.
- Learn how to weld in circuits, assemble mechanical components, and integrate mechanical an electrical systems.
- Make a detail manual for people outside the research to understand. The manual includes specific details for the solar boat.

Fig. 6: UPRM's Solar Boat Testing Facilities at lake Cerrillos, Ponce, PR.

V. CONCLUSION

The solar boat is an excellent opportunity for students to grow and develop abilities that will help them become better professionals. The research does not limit the knowledge gained by the students to a single field; it motivates them to look for more. Interdisciplinary research helps students expand their ability to other areas outside their major. It prepares the student to perform better in any job in the future. They are giving the student a complete experience in approaching a work challenge and making better decisions that books or classes do not provide or lack. Also, it allows the students to know professors mentoring them, advising students, and the critical thinking they need to develop to resolve problems and help shape them to be a professional.

REFERENCE

[1] The Solar Splash Competition, http://www.solarsplash.com

[2] UPRM Catalogue. Undergraduate Catalogue. [Online]. Avaliable: http://www.uprm.edu/catalog

Evaluating the use of Satellite Data and Machine Learning Models for PV Performance Monitoring

Daniel Fregosi, Rabin Dhakal & Devin Widrick

Electric Power Research Institute (EPRI), Pala Alto, CA, 94304, USA

Abstract — Understanding and quantifying the performance of PV plants is crucial for predicting energy production, optimizing maintenance activities to cost-effectively increase the energy production, and improving future designs and construction. PV plant performance data and weather inputs along with models to generate expected power are the basis of performance analysis. This work evaluates the accuracy of expected power models across sources of weather input data and model types. Variations include free and commercial satellite-based and ground-based weather data. For PV plant models machine learning models are compared with physics-based and simple regression models. Additionally, the sensitivity of the resulting performance loss rate (PLR) calculation to different weather inputs is studied.

Keywords—Photovoltaics (PV), performance modelling, performance loss rate, PLR, irradiance sensor, satellite, machine learning, data-driven

I. INTRODUCTION

Over the past 20 years, the cumulative annual growth rate of PV capacity has exceeded 20 % globally, and it is anticipated that this trend will continue [1]. With this rate of ongoing growth, there are far more new plants than older ones, and only a tiny proportion have amassed a considerable amount of operational data. More than 90% of plants are under ten years old [2]. Since plants are anticipated to operate for more than 25 years, there is a knowledge gap between anticipated and actual performance. To inform decisions about future fleet and improve operation and maintenance of current fleet, lessons are being learned and documented from genesis PV plants.

Performance loss rate (PLR) refers to the pace at which a system's nameplate capacity degrades over time due to both reversible and irreversible events, like module PLR, increasing soiling, compounded unmitigated maintenance failures and so forth [3]. Knowledge of PLR is critical for many decisions that are made throughout the service life of a PV plant, such as predicting energy, informing economic evaluation of project during development stage, identifies opportunities for operation and maintenance and so on.

In the recent years, there is growing number of PV plant installations which brings opportunity to analyze field collected performance data to determine fleet-wide PLR. The important of this analysis is indicated by "PV Fleet Performance Data Initiative" launched by U.S. Department of Energy [4]. EPRI is also taking an initiative to launch a performance and PLR benchmarking database to collect data, analyze it, and then publish the results for large scale PV plants [5]. EPRI's PLR method is building upon a method developed by the National Renewable Energy Laboratory (NREL) which is publicly available in the RdTools Python package [6-7]. Prior to calculating the PLR, which is the median of the distribution, the analysis method in RdTools first normalizes, filters, and aggregates the performance data to build a loss rate distribution. This three-step procedure is designed to normalize meteorologically related performance variations for each measured time interval and allow a comparison of year-on-year performance changes over time.

One of the important steps in a PLR analysis is to compare how much power a PV plant produces with the amount one would expect with given the weather conditions. Plane of Array (POA) irradiance, ambient temperature (T_{amb}) and wind speed are basic weather inputs for calculating expected power. The weather inputs are measured by meteorological station which can be ground based on the same PV plant location or a satellite based. Ground-based sensors have issues such as sensor drift, data shifts, soiling, calibration, maintenance, etc. [8]. Satellite based measurements overcome these issues; however, the time resolution of the data is typically 10-15 minutes or more and this may impact the accuracy of the analysis which would limit its usefulness.

In this study, we first to evaluate the PLR of 24 utility-scale plants using ground and satellite-based weather data sources and compare the results. Next, we investigate deeper on the accuracy of different satellite data sources and machine learning models used to normalize PV plant performance.

II. PLR ANALYSIS: DATA, METHODS AND SENSITIVITY RESULTS

A. Data Overview

PLR analysis combines PV plant performance data with weather data to isolate the overall trend in a plant's performance from the variations in weather. The PV plant performance data used in this study was collected from 24 utility-scale plants with total nameplate capacity of 1860 MW AC. The 24 plants consist of a total of 1132 inverters. The performance data consists of data collected over a period of 4 years, on average.

978-1-6654-6060-6/23 $31.00 © 2023 IEEE

In addition to the performance data, weather input is also required to calculated expected power for the normalization step (Figure 1). The weather data is collected from two sources: ground-based sensors and satellite-based sensors. The data from the ground-based sensors is provided by the utility operating the plant from their local ground mounted sensors. The satellite data is collected programmatically using an application programming interface (API) through the National Solar Radiation Database (NSRDB) [9]. The NSRDB is a serially complete collection of meteorological data, at hourly and sub-hourly time resolution, and includes the most common measurements for solar irradiance: global horizontal, direct normal, and diffuse horizontal irradiance. The data set consists of locations all over the US and a growing subset of international locations.

B. Analysis Methodology

The PLR analysis conducted by EPRI consists of three analysis steps: normalization, filtering, and trend analysis.

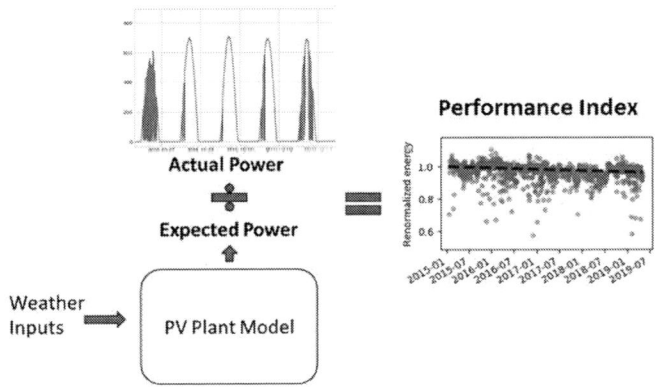

Fig 1. Visualization of the normalization step in the PLR analysis.

Normalization: The goal of the normalization step is to account for weather and other local conditions by calculating the expected plant production. The model used in the RdTools package is a linear regression model defined by: $P = f(POA, T_{amb}, Wind, AOI)$. This model does not account for the non-linearity associated with variables of model. . EPRI adopted a second order multiple regression model. A simple example of second order combination of two independent variables would be $[1, a, b, ab, a^2, b^2]$.

Irregular Performance Filter: Irregularities can obscure the normal performance of the PV plant. Hence, in the second step, filtering is employed on the normalized data to exclude the impact of erroneous measurement and noise. The filters employed in this work include irradiance, outlier, clipping, etc. The filters flag performance and other issues in the data.

Trend Analysis: The remaining normalized data left over after the filtering is then used to determine the average PLR of the

plant. The year-on-year approach is used to compute this average PLR and is calculated by comparing the percent change in normalized power between the same calendar days for all the years in the dataset. It isolates trend from seasonality and has a reduced sensitivity to outliers, snow, and soiling events [6]. The result is multiple PLR rates that have been calculated for each calendar day and the final PLR rate is calculated as the median from all these values.

In this study, we have also incorporated the clear sky irradiance for the trend analysis. The clear-sky approach lessens the impact of sensor-related problems on the PLR rate calculation such as sensor drifts, data shifts, soiling, calibration, maintenance of sensor, etc. The irradiance values during clear-sky conditions can be estimated using lookup tables [10].

C. Overall PLR Analysis Results

The PLR analysis is conducted using three different weather/irradiance sources: ground, satellite, and clear sky-based irradiance. Two different models were used for calculating expected power in the normalization steps: the PVWatts model from RdTools and the second order multiple regression model from EPRI. Moreover, there is also a variation in the filtering steps (i.e., with or without a performance filter).

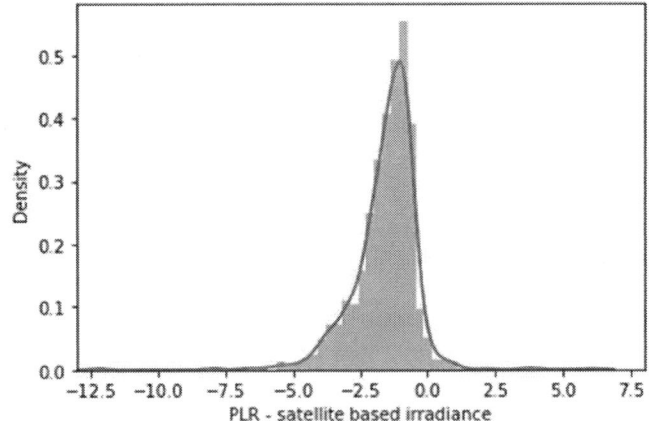

Fig 2. Distribution of performance loss rate of all inverters

The performance loss rate obtained on each inverter and each plant from the different analyses have a wide distribution (~-1--2%). A sample distribution of performance loss rate on the inverters using satellite-based weather data is show in Figure 2. Plant-by-plant breakdown of performance loss rate is shown in Figure 3. Global module performance loss rate, climate, soiling and maintenance level might impact the calculated PLR on a plant-by-plant level whereas Balance of System (BOS) faults, individual module performance loss rate, inverter localized soiling/shading, etc., might impact the calculated performance loss rate on an array-to-array level.

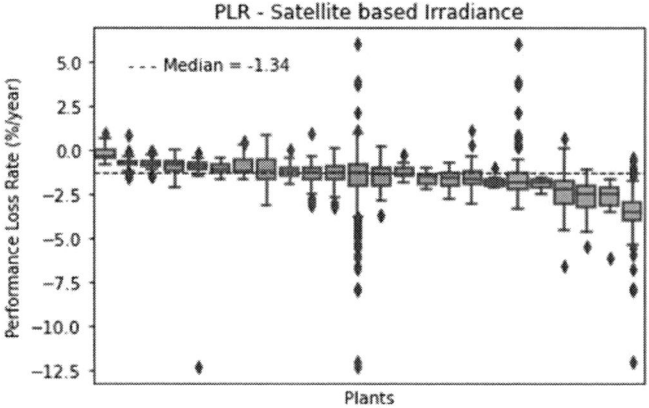

Fig 3.   Plant-by-plant breakdown of performance loss rate.

D. Comparison of Ground vs. Satellite

The performance loss rate obtained from analyzing 1132 inverter data using ground, satellite and clear sky-based irradiance data are -1.18, -1.34 and -1.11 respectively as shown in figure 4. All these results are slightly lower than other industry estimates which are around -0.75% or -1% per year. This is likely due to differences in model structure, normalization approach, and filtering choices [8].

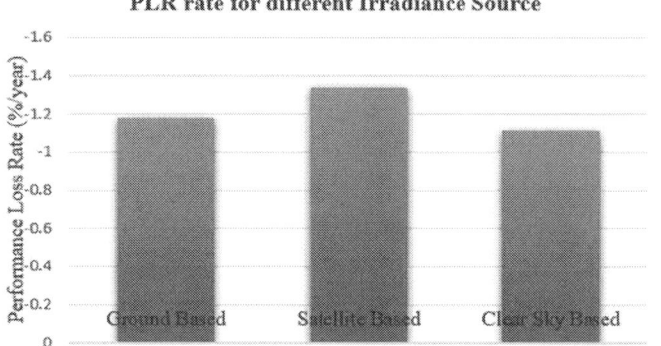

Fig 4.   Performance loss rate for different irradiance sources.

The performance loss rate obtained using satellite-based irradiance is slightly lower than ground based sensor as shown in Figure 5. It is unclear why this bias exists. It may be due to a combination of drift in the ground-based sensors and resolution of weather data used for normalization introducing modelling error, such as the average-then-clip bias.

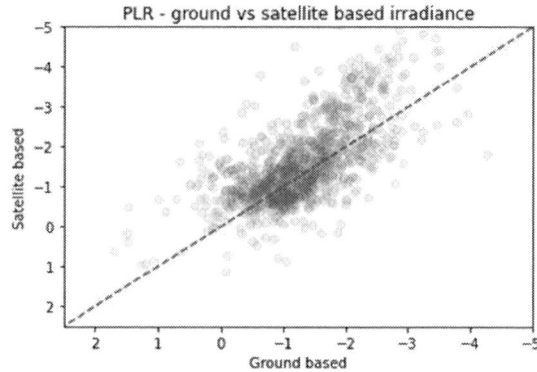

Fig 5.   Comparison of performance loss rate using ground and satellite-based irradiance

III. EVALUATION OF ADDITIONAL DATA SOURCES AND MODELS

A. Experimental Setup for Improving Normalization Accuracy

For performance monitoring applications, including PLR analysis, normalizing for weather is the critical initial step, as described earlier in Section IIB. Normalization removes the impacts of weather and predictable or repeatable behavior. What remains in the resulting performance index are impacts of soiling, faults, and degradation. Improvements in normalization will reduce the amount of noise or scatter in the performance index. We seek to improve the accuracy of expected power calculations by examining the two components of that calculation, the weather input, and the PV plant model.

For the weather input, we seek to compare the performance of satellite-derived weather to ground-based sensors installed at the PV plants. Ground-based sensors offer the highest potential accuracy since they measure the actual conditions from the perspective of the PV array. However, ground sensors require regular maintenance and calibration [11], which is expensive and time-consuming, therefore it is not always done properly. Poorly maintained sensors may significant drift [11]. In addition, error arises with improper installation. Lastly, communication systems can introduce issues like missing data, improper scaling, interpolation, etc. While satellite data is not as accurate as a perfectly maintained ground sensor, it offers the benefits of completeness and consistency. There is no missing data and no drift over time due to measurement error.

Typical ground-based weather inputs are plane-of-array (POA) irradiance, ambient temperature, and wind speed. These three inputs are used in this study. Some sites may have additional inputs such as global horizontal irradiance (GHI), module temperature, wind direction, precipitation, soiling ratio, and albedo, but these are less common. The satellite data has all these channels except POA irradiance since that is specific to the PV system design. In addition, satellite sources have direct normal irradiance (DNI), diffuse horizontal irradiance (DHI), precipitable water vapor, linke-turbidity factor, snowfall, and clear sky indication. Ground sensors are typically recorded at 1- or 5-minute averaged intervals. Satellite data is slightly less frequent, ranging from 5 to 15 minutes in frequency. The special resolution of satellite data is around 4-16 square kilometers. Lastly, several derived inputs are used in both the sensor and satellite cases to improve model performance. These inputs are angle of incidence (AOI), cosine of AOI, log of POA, and the zenith angle.

The objective of the model is to emulate the performance of the healthy plant in order to calculate the expected plant production given the weather inputs. The model is a digital twin of the healthy pv plant. Examining differences between the modeled expected power and the actual power allows operators to detect changes in performance, such as soiling and degradation. We have chosen to use purely data-driven models since the setup is automated. An automated setup allows deployment at scale (inverter or combiner level), whereas manually configured physics-based models require significant labor cost to create, tune, and maintain. Theoretically, the data-driven model can implicitly learn the plant specs like tilt angle, azimuth, wire loss, inverter efficiency, etc. In this study, the models are trained and tested on data from the first year of operation, to minimize the impacts of PLR. Lastly, the model is time-sequence independent. Each input/output pair is treated independently, with no impact from the neighboring pairs. This is because we do not wish the model to learn time-dependent impacts like soiling and degradation. Those should show up as deviations between the model and the actual power [12].

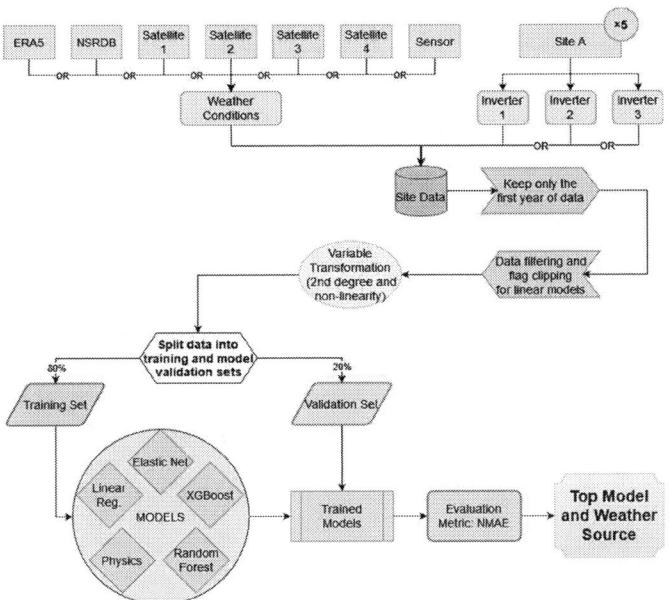

Fig 6. Experimental setup for evaluating the accuracy of weather inputs and models.

The experimental setup is pictured in Figure 6. Four commercial satellite data sources, the free NSRDB data source, and the ground-based sensor comprise the 6 weather inputs. For the ground-based sensor, the irradiance value is the average of 3 or more pyranometers with erroneous values removed. Erroneous values were found by comparing the sensors to each other and removing points that fall outside 5% of the average. Three inverters are taken from each of five PV plants for a total of 15 PV performance datasets. There are four regression model types, linear regression, elastic net, random forest, and XGBoost. In addition, a physics-based model is considered. The physics-based model is built using pvlib and plant-specific information including the module, inverter, number of modules per string, number of strings per inverter, and location. The models are evaluated on daily normalized mean absolute error of the energy prediction. The error is the difference in expected and actual energy divided by the average daily energy for that plant.

B. Normalization Results

The performance by model type is given in Figure 7. 1st order linear regression is the lowest performer by a wide margin. With second order inputs, linear regression performs similar to elastic net and other models. In subsequent plots, 1st order linear regression is removed. The two machine learning models, random forest and XGBoost, and the physics model perform best. Adding second order inputs does not significantly increase performance of XGBoost. The runtime of random forest with second order inputs was considered too long for inclusion in the study, however preliminary results showed little improvement in using second order inputs for the random forest. XGBoost has a runtime of 5 seconds, whereas random forest has a runtime of over 2 minutes, on a standard laptop.

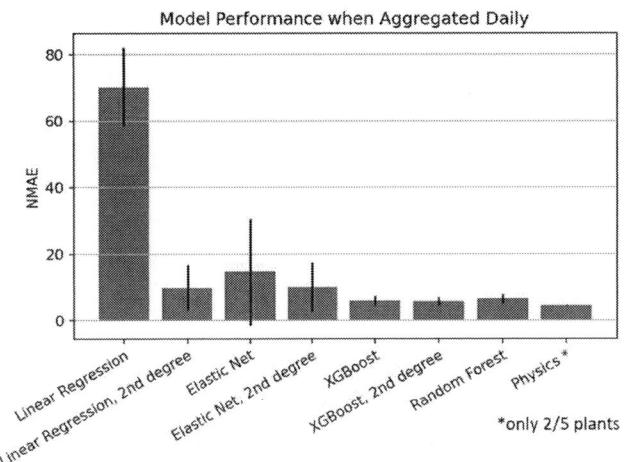

Fig 7. Comparison of model performance across all sensor inputs

For a closer look data-driven versus physics model performance, linear regression was removed to create a better scale on the plot for Figure 8. For the physics model, only 2 of the 5 were modeled due to inaccuracies in metadata for the others. Figure 8a. shows the data-driven results on all 5 plants, whereas Figure 8b. shows the results for just the two plants that were modelled by physics. The results are similar for the advanced models. For the simple regression models, the error is higher with the full set of 5 plants. This demonstrates the consistency and adaptability of the advanced models. From Figure 8b., the physics model slightly outperforms the advanced data-driven models. However, when the results are broken down by input data source, the data-driven models are better in some cases, as described next.

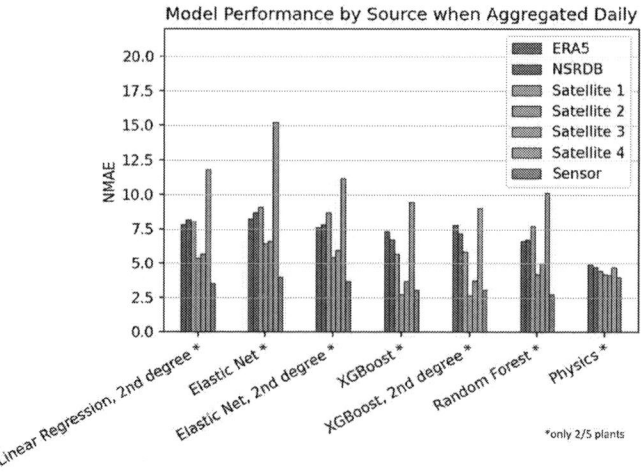

Fig 9. Comparison of model performance (without 1st order linear) for A) all plants and B) 2 plants that have accurate physics models.

In evaluating the input data sources, the ground sensor data performs at least as well as the satellite data all models. However, for XGBoost, a similar level of performance to the ground sensor is achieved with two of the 4 commercial data sources. Lastly, the free sources perform relatively well, especially with the physics model.

The highest performance of 2.5 NMAE is achieved with XGBoost. To visualize the level of error, the performance index of an inverter is plotted using a model with an NMAE 2.5 and 7 in Figure 10. and Figure 11.

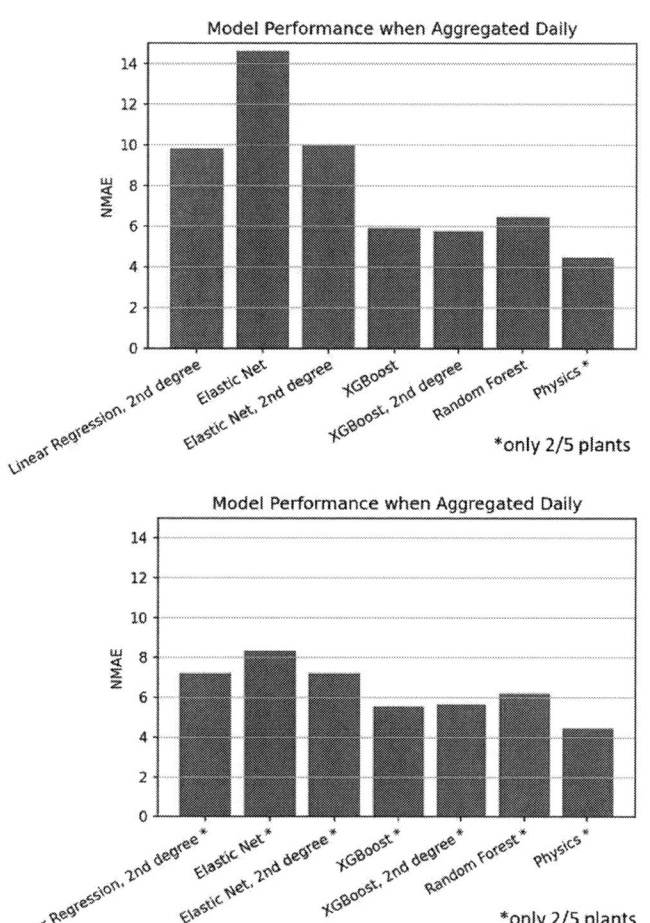

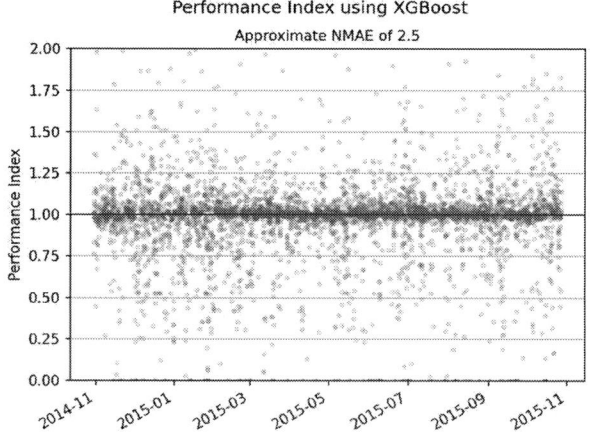

Fig 10. Performance index using XGBoost model with NMAE of 2.5

Fig 8. Comparison of model performance (without 1st order linear) for A) all plants and B) 2 plants that have accurate physics models.

In Figure 9. the model accuracy is shown for each input weather source for each model type. The blue bars are the free satellite data sources, while the yellow ones are the commercial ones. Lastly, the green bars represent the ground-based sensor inputs. Regarding model evaluation, the physics model is more uniform across the input data sources, however, the physics model is the lowest performer for the ground sensor input among all models. XGBoost outperforms the physics model with the ground sensor and two of the satellite sources.

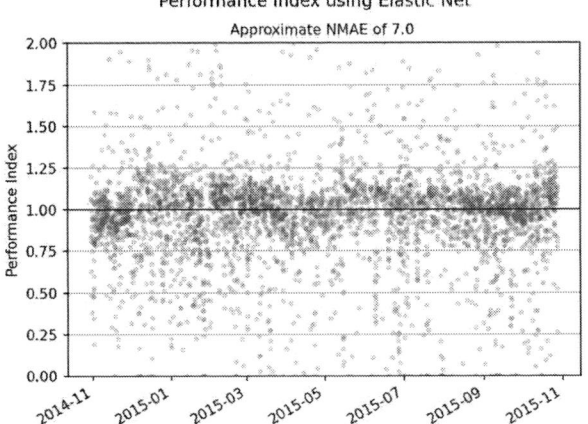

Fig 11. Performance index using Elastic Net model with NMAE of 7

IV. CONCLUSION AND FUTURE WORKS

The calculated PLR depends on the source of the weather/irradiance data. Moreover, different modeling and analysis choices yield different results. For an informed decision on which weather/irradiance data source is best, an exploration of the use of different paid and freely available satellite-based irradiance data is suggested to see the accuracy and cost of PLR analysis.

For the highest accuracy in normalization modeling, it was found that two of satellite sources as well as the ground sensor can achieve a NMAE of 2.5 when using the XGBoost model. The ground-based sensor performed at least as well as the satellite sources for all models. Finally, the physics-based model performed reasonably well on all input data sources but did not achieve the high accuracy of the XGBoost model, nor did it outperform any of the data driven models when using the ground-based sensor.

Future work includes evaluating the conditions in which models perform best and worst, such as in variable cloudiness, low or high irradiance, and across daily and seasonal patterns. This will help identify additional derived inputs that can be added such that the model can learn behaviors previously overlooked by the existing inputs.

REFERENCES

[1] New Energy Outlook 2019. Bloomberg NEF. 2019
[2] Analyzing Performance Loss Rates in PV Plants using Operational (SCADA) Data, Palo Alto, CA: 2022. 3002021060
[3] D. Fregosi, M. Bolen and B. Paudyal, "Analysis of Variability in Calculated Performance Loss Rates of Large-Scale PV Plants," 2020 47th IEEE Photovoltaic Specialists Conference (PVSC), 2020, pp. 1742-1748, 2020
[4] T. Golnas, I. Kozinsky, "PV Fleet Performance Data Initiative: Long-Term Photovoltaic (PV) System Performance Benchmark,"energy.gov.eere/solar. DOE/EE-1967, April 2019
[5] D. Fregosi, M. Bolen, "Large-Scale Solar Photovoltaic Plant Performance and PLR Benchmarking," EPRI, Palo Alto, CA: 2020. 3002019581
[6] D. C. Jordan, C. Deline, S. R. Kurtz, G. M. Kimball, and M. Anderson,"Robust PV degradation Methodology and Application," IEEE Journal of PV, vol. 8, no. 2, p. 525, 2018
[7] NREL, "RdTools," NREL, 2018. [Online]. Available: https://github.com/NREL/rdtools
[8] B. Paudyal, M. Bolen, and D. Fregosi. "PV plant performance loss rate assessment: Significance of data filtering and aggregation." 2019 IEEE 46th Photovoltaic Specialists Conference (PVSC), pp. 0866-0869. IEEE, 2019.
[9] NSRDB: National Solar Radiation Database [Online]. Available: https://nsrdb.nrel.gov/
[10] Reno, M.J. and C.W. Hansen, "Identification of periods of clear sky irradiance in time series of GHI measurements" Renewable Energy, 2016
[11] D. Fregosi, M. Bolen and B. Paudyal, "An Assessment of In-Field Irradiance Sensor Accuracy and Error Mitigation Techniques," 2021 IEEE 48th Photovoltaic Specialists Conference (PVSC), Fort Lauderdale, FL, USA, 2021, pp. 1430-1436, doi: 10.1109/PVSC43889.2021.9518813.
[12] D. Fregosi and M. Bolen, "An Evaluation of Empirical Models for use in Normalizing PV Plant Performance Data," 2022 IEEE 49th Photovoltaics Specialists Conference (PVSC), Philadelphia, PA, USA, 2022, pp. 0116-0120, doi: 10.1109/PVSC48317.2022.9938822.

Drying Effects Upon Spin Coating of Solution-Processed Amine-Thiol Thin Film Cu(In,Ga)(S,Se)$_2$ Absorber Fabrication

Jacques Kenyon, Nada Benhaddou, Liam Welch, Jake Bowers*

Loughborough University, Loughborough, Leicestershire, LE11 3TT, England, United Kingdom

Abstract—**This paper utilises solution-processed spin coating of CIGS precursor solutions in order to fabricate thin film PV absorbers (1-2μm thick) in air. Opto-electronic properties of the device have been shown to improve under post-spin air annealing of the precursor film, due to removing carbon in the bulk of the film, left by improperly dried amine-thiol-based solvents used for the absorber. A champion device was reached for a 20min post-spin anneal time in air, yielding a best cell J$_{sc}$ of 26.4mA/cm^2, a V$_{oc}$ of 587mV and an efficiency of 10.1%.**

Index Terms—**Cu(In,Ga)(S,Se)$_2$, Solution-Processed, Air Annealing of Precursor, Ga Effect on Band Gap, Ga Loss, Na Doping, Rapid Thermal Annealing.**

I. INTRODUCTION

Spin coating of amine-thiol solution-processed CIGS [1] has been recorded to produce 15%+ efficiency solar cells [2] and can be translated into industrial deposition methods such as inkjet/slot-die printing [3] at a fraction of the cost of vacuum-based procedures [4]. However, the deposition mechanics vary between fabrication types, in which case the final precursor film needs to be as translatable as possible to one which could be made in industry using the same solution conditions and treatment.

One key element is uniformity, which requires optimisation of spin coating parameters such as rotation speed and time, in order produce a homogeneous film identical to other non-rotating fabrication methods. The other is stoichiometric control, as chemical bonding, site-exchanging and thermal treatment of the device throughout fabrication can introduce material loss and segregation, affecting the true final absorber stoichiometry. Therefore, the stoichiometry of the absorber must be controlled between each stage of the fabrication in order to be reproducible.

Also, it is crucial that good crystallinity of CIGS must be obtained to achieve high-efficiency devices [5], which can be influenced by initial stoichiometry (e.g. copper rich/poor CIGS [6]), spin coating deposition parameters [7], doping mechanisms ([8]- [9]), and selenisation conditions/profile [10]. Specifically, NaCl doping will be implemented to counteract the lack of alkali metals present in the glass used (Eagle) to fabricate the PV device.

In the fabrication of amine-thiol-based solution-processed CIGS, when drying spin-coated films in air, it has been revealed from Energy-Dispersive X-Ray (EDX) Spectroscopy that over 50% of the film consists of carbon (At%). The source of the carbon is most likely residual carbon-based solvents not being removed from the film during the drying process. In order to remove carbon from the film, different drying conditions are to be implemented to remove as much carbon as possible from the CIGS precursor before selenisation.

II. RESULTS & ANALYSIS

A. Preliminary Condition Testing

Initially, two different annealing techniques (post-spin) were adopted in an attempt to remove the carbon from the bulk of the precursor:

- Drying/Annealing in Air for 10mins at 310°C
- Drying/Annealing in N$_2$ (flowing, 2.4Torr) for 30mins at 500°C

These heating parameters were chosen according to previous research, following a potential precursor mass reduction in Thermogravimetric Analysis for N$_2$ annealing, and up to a maximum annealing temperature for air annealing, so as to not oxidise Mo.

Firstly, for indirect characterisation of the gallium to gallium-indium (GGI) ratio, an empirical equation exists to provide a quantitative estimation of the stoichiometry, relating to band gap (in eV) [11]:

$$E_g = 1.65x + 1.01(1 - x) - 0.151(1 - x)x \qquad (1)$$

where x is referred to as GGI. Therefore, rearranging this equation and solving for the real root, we can prove that:

$$GGI_{implied} = 2.573\left(\sqrt{E_g - 0.614}\right) - 1.619 \qquad (2)$$

Table I shows the bandgap values from Photoluminescence data in Fig. 1 obtained by a Gaussian fit peak of the PL wavelengths (calculated using $E_g = hc/\lambda$ [12]), and the GGI values calculated from equation (2).

TABLE I
PHOTOLUMINESCENCE DATA OVERVIEW

Drying Condition	PL Data	
	Bandgap (eV)	*GGI (Implied)*
None	1.197	0.345
Air	1.188	0.330
N$_2$	1.201	0.353

Figures 2 & 3 represent Current Density-Voltage (JV) (seen in Table II) and External Quantum Efficiency (EQE) characteristics (seen in Table III), extracted at 1000W/m^2 for a AM1.5

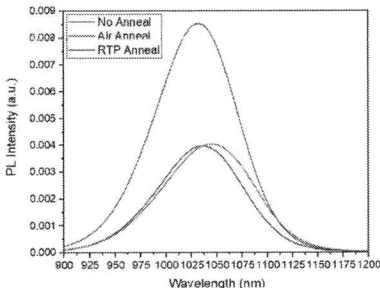

Fig. 1. Single-spot Photoluminescence measurement data of CIGS with different drying conditions

spectrum, using box and peak cell plots. The bandgap values were obtained from EQE in fig. 2b), using an extrapolation method via calculating the wavelength associated with the largest negative derivative of EQE with respect to wavelength (see [13]). Similarly to the PL data in Table I, the GGI values were also calculated using equation (2):

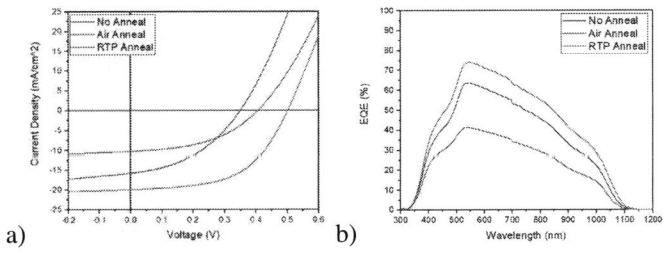

Fig. 2. a) JV and b) EQE data of CIGS with various drying conditions

TABLE II
JV DATA OVERVIEW

Drying Condition	JV Characteristics					
	J_{sc} (mA/cm^2)	V_{oc} (mV)	FF (%)	R_s (Ω)	R_{sh} (Ω)	η (%)
None	15.7	348	41.5	8.9	526	2.28
Air	19.9	501	54.4	19.6	557	5.42
N$_2$	10.3	405	45.9	13.1	1685	1.91

TABLE III
EQE DATA OVERVIEW

Drying Condition	EQE Data	
	Bandgap (eV)	GGI (Implied)
None	1.198	0.347
Air	1.187	0.328
N$_2$	1.198	0.347

Note: The EQE data can also be used to verify the J_{sc} via the area bounded by the EQE function with respect to wavelength [14]:

$$J_{sc} = q \int \Phi(\lambda) EQE(\lambda) d\lambda \qquad (3)$$

where Φ is the wavelength-dependent AM1.5 spectral irradiance and q is the charge of an electron.

Below (in Fig. 4) are Scanning Electron Microscopy (SEM) Images with structure, stoichiometry, and thickness analysis of the absorber and completed PV device:

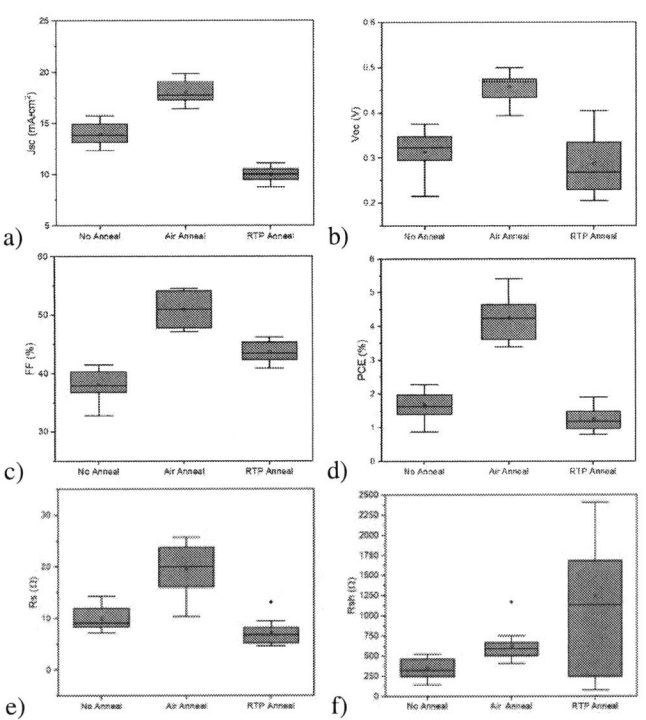

Fig. 3. Box Plot Evaluation of a) J_{sc}, b) V_{oc}, c) Fill Factor, d) Efficiency, e) Series Resistance, and f) Shunt Resistance characteristics for CIGS with various drying conditions:

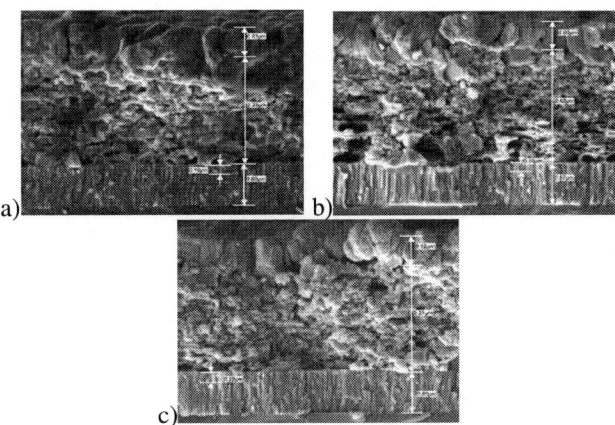

Fig. 4. Cross-sectional SEM data for CIGS with a) No annealing, b) Air annealing and c) N$_2$ annealing

TABLE IV
SEM DATA OVERVIEW

Drying Condition	Material Thicknesses (μm)		
	Mo	MoSe$_2$	CIGS
None	0.69	0.19	2.28
Air	0.68	0.19	2.42
N$_2$	0.64	0.22	2.27

Below (in Fig. 5) is Energy-Dispersive X-Ray Spectroscopy (referred to as EDX) data mapped over surface SEM regions at 20kV accelerating voltage (seen in Fig. 6):

978-1-6654-6060-6/23 $31.00 © 2023 IEEE

No Anneal: $Cu_{0.90}(In_{0.69},Ga_{0.31})(S_{0.08},Se_{1.83})_{1.91} + 4.95C + 0.14O$
Air Anneal: $Cu_{0.93}(In_{0.70},Ga_{0.30})(S_{0.00},Se_{1.75})_{1.75} + 4.54C + 0.12O$
RTP Anneal: $Cu_{0.96}(In_{0.69},Ga_{0.31})(S_{0.06},Se_{1.77})_{1.83} + 5.07C + 0.20O$

Fig. 5. EDX data of CIGS for varying drying conditions, presented as stoichiometric ratios relative to a Ga+In ratio of 1 (Key: Red-Excess, Yellow-Small Excess, Green-No Change, Blue-Loss)

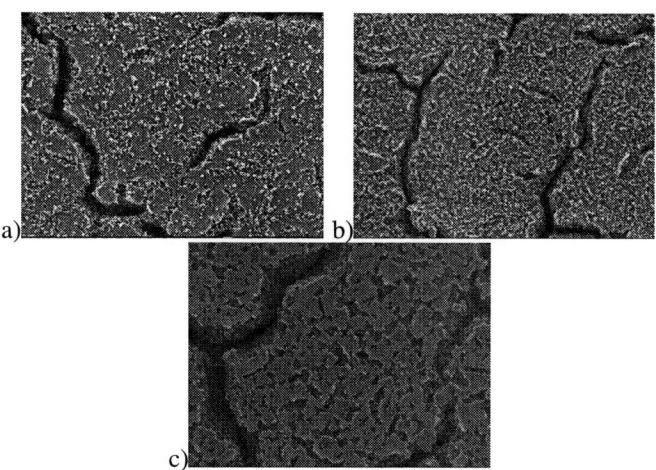

Fig. 6. Surface SEM data for CIGS with a) No annealing, b) Air annealing and c) N$_2$ annealing, used for EDX

B. NaCl Doping

It has been evident that in order to aid crystalline CIGS growth upon selenisation, alkali-metals must be present to improve and preserve the grain structure of large-grain, pure CIGSe [15]. In this fabrication, Eagle (GX) Corning glass is used as the reliable substrate for the device, as it consists of a higher surface roughness, higher melting point, and smaller heat expansion coefficient, as compared to industrially used SLG, reducing the possibility of delamination during selenisation or chemical bath processes [8].

However, Eagle is characterised as low-alkali metal glass (primarily Na), in which case, Na doping has been implemented through thermal evaporation of NaCl, and has produced the following (relevant) results for different NaCl thicknesses (Note: Thicker Mo was used, and only 80% of precursor was deposited for this):

TABLE V
TRPL DATA OVERVIEW

NaCl Thickness	Normalised TRPL Curve-Fit Coefficients			
	A_1	τ_1 (ns)	A_2	τ_2 (ns)
0nm (control)	1.413	0.221	0	?
10nm	1.196	0.294	0.021	?
20nm	1.158	0.351	0.071	?
30nm	0.785	0.518	0.318	1.590

Note: The Time-Resolved Photoluminescence (TRPL) data, shown in Table V, has been extracted by best-fit estimators of a two-term exponential fitting function with the following equation:

$$I/I_0 = A_1 e^{t/\tau_1} + A_2 e^{t/\tau_2} \qquad (4)$$

where I/I_0 corresponds to a normalised count intensity function relative to the maximum number of decay counts.

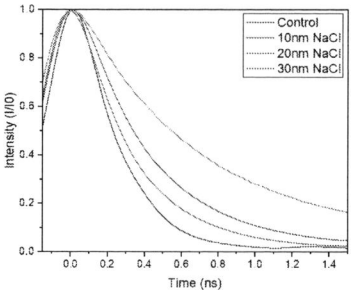

Fig. 7. TRPL data for CIGS with different thicknesses of NaCl evaporated on precursor

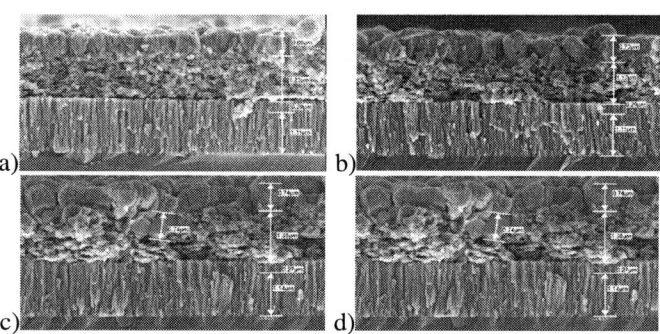

Fig. 8. Cross-sectional SEM data for CIGS with a) 0nm, b) 10nm c) 20nm and d) 30nm thick NaCl evaporated onto precursor

TABLE VI
SEM DATA OVERVIEW

NaCl Thickness	Material Thicknesses (μm)		
	Mo	MoSe$_2$	CIGS
0nm (control)	1.11	0.29	1.22
10nm	1.11	0.26	1.12
20nm	1.14	0.27	1.38
30nm	1.12	0.29	1.28

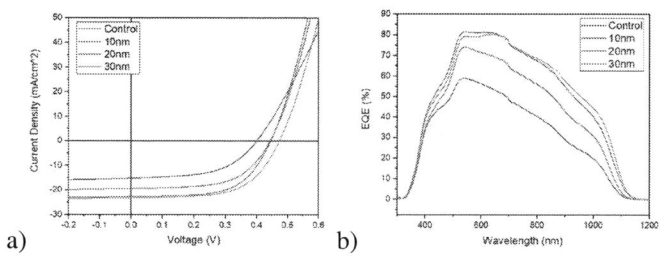

Fig. 9. a) JV and b) EQE data of CIGS with different thicknesses of NaCl evaporated on precursor

TABLE VII
JV DATA OVERVIEW

NaCl Thickness	JV Characteristics					
	J_{sc} (mA/cm²)	V_{oc} (mV)	FF (%)	R_s (Ω)	R_{sh} (Ω)	η (%)
0nm (control)	15.2	400	53.7	10.7	1253	3.26
10nm	19.4	443	58.5	4.2	1559	5.02
20nm	22.6	445	60.3	4.3	1283	6.06
30nm	23.1	473	59.8	5.1	1316	6.55

TABLE VIII
EQE DATA OVERVIEW

NaCl Thickness	EQE Data	
	Bandgap (eV)	GGI (Implied)
0nm (control)	1.187	0.328
10nm	1.176	0.309
20nm	1.164	0.290
30nm	1.159	0.280

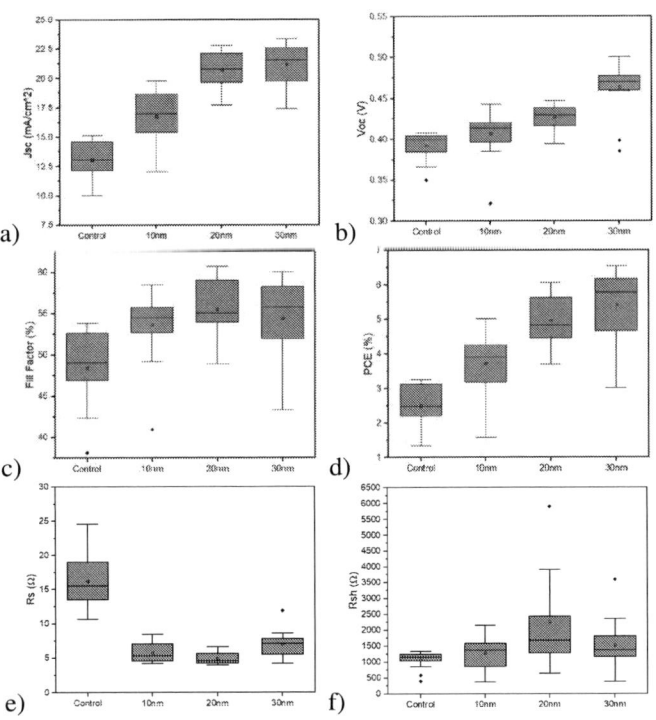

Fig. 10. Box Plot Evaluation of a) J_{sc}, b) V_{oc}, c) Fill Factor, d) Efficiency, e) Series Resistance, and f) Shunt Resistance characteristics for CIGS with different thicknesses of NaCl evaporated on precursor:

C. Precursor Air Annealing

From the preliminary condition testing (section 2A), there is evidence to suggest a small reduction in carbon when air-annealing the precursor post-spin coating (Fig. 5) which also improves all opto-electronic properties (except R_s). In addition to doping the precursor with 30nm of evaporated NaCl which also improved all opto-electronic properties and crystallinity, post-spin air-annealing times will be tested incrementally for 0mins, 10mins, 15mins, and 20mins. This produced the following collection of results:

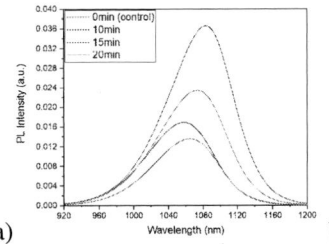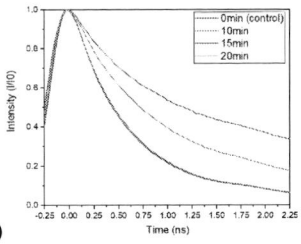

Fig. 11. a) PL and b) TRPL data of CIGS for varying precursor air-annealing times

TABLE IX
PHOTOLUMINESCENCE DATA OVERVIEW

Air Anneal Time	PL Data	
	Bandgap (eV)	GGI (Implied)
0min (control)	1.172	0.303
10min	1.165	0.291
15min	1.145	0.256
20min	1.155	0.274

TABLE X
TRPL DATA OVERVIEW

Air Anneal Time	Normalised TRPL Curve-Fit Coefficients			
	A_1	τ_1 (ns)	A_2	τ_2 (ns)
0min (control)	0.853	0.448	0.232	1.622
10min	0.939	0.494	0.179	1.948
15min	0.370	0.472	0.692	3.098
20min	0.612	0.588	0.434	2.351

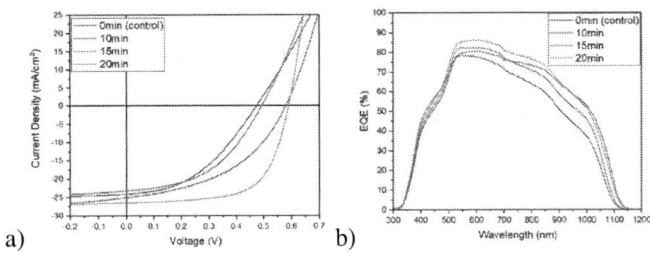

Fig. 12. a) JV and b) EQE data of CIGS with different annealing times in air post-spin

TABLE XI
JV DATA OVERVIEW

Air Anneal Time	JV Characteristics					
	J_{sc} (mA/cm²)	V_{oc} (mV)	FF (%)	R_s (Ω)	R_{sh} (Ω)	η (%)
0min (control)	24.1	473	42.1	22.4	1242	4.79
10min	23.2	492	46.4	13.2	906	5.29
15min	25.0	577	45.2	5.6	395	6.54
20min	26.4	587	65.1	3.1	1194	10.09

Below (in Fig. 13) is EDX data mapped over surface SEM regions at 20kV accelerating voltage (seen in Fig. 13):

TABLE XII
EQE DATA OVERVIEW

Air Anneal Time	EQE Data	
	Bandgap (eV)	GGI (Implied)
0min (control)	1.175	0.309
10min	1.170	0.299
15min	1.148	0.261
20min	1.159	0.280

0min (control): $Cu_{1.07}(In_{0.64},Ga_{0.36})(S_{0.06},Se_{2.00})_{2.06} + 4.81C + 0.230$
10min: $Cu_{1.08}(In_{0.63},Ga_{0.37})(S_{0.06},Se_{1.96})_{2.02} + 4.43C + 0.210$
15min: $Cu_{1.10}(In_{0.63},Ga_{0.37})(S_{0.04},Se_{1.92})_{1.96} + 4.53C + 0.120$
20min: $Cu_{1.07}(In_{0.63},Ga_{0.37})(S_{0.05},Se_{1.96})_{2.00} + 4.65C + 0.290$

Fig. 13. EDX data of CIGS for varying precursor air-annealing times, presented as stoichiometric ratios relative to a Ga+In ratio of 1 (Key: Red-Excess, Yellow-Small Excess, Green-No Change, Blue-Loss)

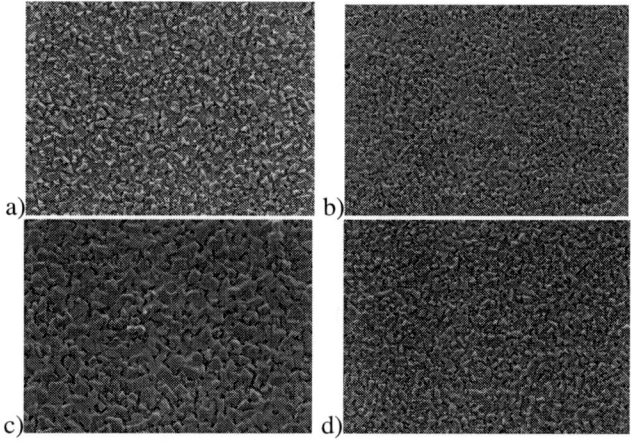

Fig. 14. Surface SEM data for CIGS with post-spin air annealing with a) 0mins, b) 10min, c) 15mins and d) 20mins anneal timed, used for EDX

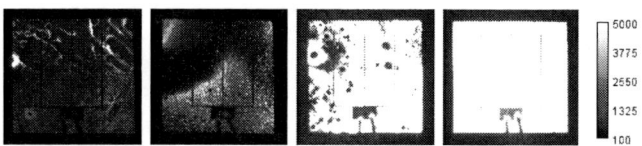

Fig. 15. Contrast-Relative Electroluminescence (EL) imaging for CIGS with post-spin air annealing with (left to right) 0mins, 10min, 15mins and 20mins anneal time

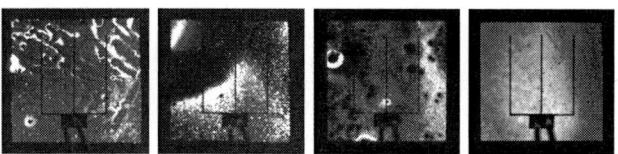

Fig. 16. Qualitative EL imaging for CIGS with post-spin air annealing with (left to right) 0mins, 10min, 15mins and 20mins anneal time

III. DISCUSSION OF RESULTS

From the preliminary condition testing, it can be seen that all bandgap values from PL (Table I) and EQE (Table III)

show a blue shift, indicating an implied GGI greater than the expected value of 0.3, which contradicts the true stoichiometric GGI seen in EDX (Fig. 5). This initial bandgap shift by all devices without any stoichiometric losses, could implicate no material losses in the bulk, however, there exists some mechanism which removes some of the indium from the photo-active region of the absorber.

One factor which could theoretically influence the bandgap data is solvent ageing, where it may seem that the increased blue shifting of the bandgap values occurs over a longer period of time, when the primary ethylene-diamine solvent has not been replaced. In this case, the solvent has aged over a six month period, which may have affected the bandgap, compared with results which used a newer solvent, only consisting of bandgap values lower than that of the expected bandgap (redshift only).

However, we can also see from the PL and EQE data (Tables I and III) that an air-annealed device consists of a bandgap that is more red-shifted (also seen in Tables IX and XII), in comparison to not being annealed or annealed in N_2. This effect is also observed in the more in-depth precursor air annealing (in part 2C) and has been seen in other experiments when drying the precursor in air, which sees an effect upon some of the gallium being removed from the photoactive region of the absorber. The primary difference in all cases is that there is (more) oxygen present during drying, which could imply that the gallium is being oxidised.

The JV data in Table II and Fig. 3 suggest when annealing in N_2, the fill factor and shunt resistance seem to increase, however, it reduces in Jsc reducing the overall efficiency slightly. However, all properties except series resistance improve when annealing the precursor in air for 10 minutes post-spin, which supported by EDX (Fig. 5) correlates to a successful reduction in carbon content in the bulk. With NaCl doping (and a newer solvent), this pattern seems to alter, as the series resistance decreases in Rs with increasing post-spin anneal time in air.

From the EDX data in Fig. 13 (supported by data in Fig. 5), it can be seen that there is an initial carbon reduction between 0 and 10mins of air-annealing time, however does not seem to decrease the carbon content further (this may need a more spacially-averaged set of EDX data across a whole film to determine the true carbon content). There also exists an indium loss from the bulk, which increases the GGI and CGI (copper to gallium-indium) ratios, which may need cross sectional EDX mapping analysis to verify any elemental diffusion from the bulk (where it may not be detected through surface EDX).

The surface crystalline CIGS grain sizes (seen in Fig. 14) seem to increase in density and size in correlation with a few quantitative data values, such as increases in minority carrier lifetime (estimated as τ_2 in Table X) and reduction of oxygen content (seen in Fig. 13). The latter could be cross-referenced with the images in Fig. 5, where the surface crystalline density seems to have a similar trend with oxygen content. However, for both possibilities, additional/repeated data sets would have to be carried out to verify any correlation.

From EL imaging, it can be seen in contrast-relative EL (Fig. 15) that the overall illumination intensity of the devices correlate to the Voc quite distinctively, in addition to consisting of potential links with series resistance (R_s). The density of dead space and pinholing, seen in the qualitative EL imaging for 10 and 15min annealed devices, correlate quite significantly with shunt resistance values (R_{sh}). It is also apparent that there are multiple sources of non-uniformity which may require optimisation of fabrication procedures, as the 0min anneal EL image in Fig. 16 indicates flow trailing of a chemical procedure such as the CdS layer, in addition to a circular edge pattern apparent on the 20min anneal cell, linking to spin-coating non-uniformity.

IV. SUMMARY

From the results, it can be seen that there is an initial speculation of gallium loss/segregation, supported by the shift in PL and EQE bandgap data, which may occur due to oxidisation. In order to support this claim (and identify its source), multiple characterisation techniques will be carried out alongside repeated experiments, such as X-Ray Diffraction (XRD), X-Ray Flourescence (XRF), and absorber surface or Transmission Electron Microscopy (TEM) Energy-Dispersive X-ray (EDX) Spectroscopy. These techniques will measure the optical properties and stoichiometry to determine whether the gallium is oxidising or is being lost/segregating through other mechanisms.

Alongside this, it is evident that drying/annealing the films in air, introduces a reduction in carbon content, and improves the opto-electronic properties of the device. Further experimentation and analysis will therefore be carried out to find any limitations to this annealing, but also to gain a more accurate understanding into the effects of precursor drying.

V. METHOD OF FABRICATION

Initial dissolution of Cu, In, Ga, and Se metals in Ethylenediamine and 1,2 Ethane-dithiol, at a ratio of 10:1 respectively, with a GGI of 0.3, a CGI of 0.9 and a SeGI (selenium to gallium-indium) of 3, at a concentration of 0.98M. The solution is kept in a pre-purged Nitrogen atmosphere and stirred under heating between 50 and 60°C until no residue is visible and the solution is a dark-brown transparent consistency. This solution can then be spin coated for 1min/layer at 1100rpm for 15 layers per device, onto a Mo-deposited Na-free glass substrate, where each layer is dried under localised N_2 flow (open system) at 310°C for 1min then cooled. The precursor is then selenised in a Rapid Thermal Processing (RTP) furnace, in a closed system, at an initial pressure of 350Torr (47kPa) using 560mg of Se, with the selenisation profile shown in Fig. 6.

A 50nm CdS layer is then deposited via Chemical Bath Deposition (CBD), along with 50nm ZnO and 500nm Al:ZnO

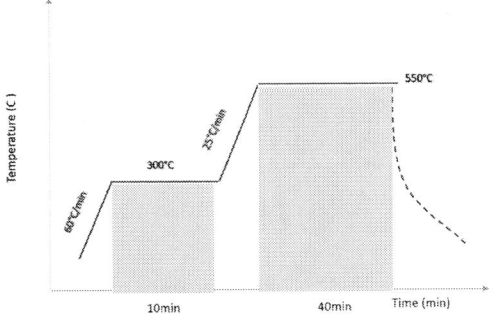

Fig. 17. RTP Selenisation Profile

TCO layers deposited using RF magnetron sputtering. Finally, Ag top grids are evaporated and isolated to measure different cells of the device for JV, EQE and other opto-electronic properties of the device.

VI. ACKNOWLEDGEMENTS

I would like to express my deepest gratitude to Loughborough Materials Characterisation Centre for training and usage of their Electron Microscopy equipment, in addition to various academics and researchers at CREST, for the construction and training of equipment for PV characterisation and fabrication associated with the project.

REFERENCES

[1] N. Sahu et al. Fundamental understanding and modeling of spin coating process : A review. *Indian J. Phys*, 83:493–502, 2009.

[2] Y. Zhao et al. Controllable formation of ordered vacancy compound for high efficiency solution processed cu(in,ga)se2 solar cells. *Advanced Functional Materials*, 31(10):2007928, 2019.

[3] R. Patidar et al. Slot-die coating of perovskite solar cells: An overview. *Applied Materials Today*, 22(100808), 2020.

[4] K. Yong D. Lee. Non-vacuum deposition of cigs absorber films for low-cost thin film solar cells. *Korean Journal of Chemical Engineering*, 30:1347–1358, 2013.

[5] H. Hiroi et al. 960mv open circuit voltage chalcopyrite solar cell. *2015 IEEE 42nd Photovoltaic Specialist Conference (PVSC), New Orleans, LA, USA*, pages 1–4, 2015.

[6] J. Liu et al. Preparation and characterization of cu(in,ga)se2 thin films by selenization of cu0.8ga0.2 and in2se3 precursor films. *nternational Journal of Photoenergy*, (149210):7, 2012.

[7] M. Pichumani et al. Dynamics, crystallization and structures in colloid spin coating. *Soft Matter*, 9(12):3220–3229, 2013.

[8] S. Ulicna. *Solution-processing of Cu(In,Ga)(S,Se)2 solar cells from metal chalcogenides: aspects of absorber crystallisation and interface formation*. PhD thesis, Loughborough University, 2019.

[9] L. Welch. *Extrinsic Doping of Amine-Thiol Solution-Processed Cu(In,Ga)(S,Se)2 Thin Film Photovoltaics*. PhD thesis, Loughborough University, 2019.

[10] Y. Cui et al. Dmf-based large-grain spanning cu2znsn(sx,se1-x)4 device with a pce of 11.76%. *Advanced Science*, 9(20):2201241, 2022.

[11] B. J. Stanbery. Copper indium selenides and related materials for photovoltaic devices. *Critical Reviews in Solid State and Materials Sciences*, 27(2):73–117, 2002.

[12] K. A. Connors. The phenomenological theory of solvent effects in mixed solvent systems. *Handbook of Solvents*, 1(2):467–490, 2014.

[13] H. Fujiwara et al. Analysis of optical and recombination losses in solar cells. *Spectroscopic Ellipsometry for Photovoltaics*, 214:29–82, 2019.

[14] H. Choi. High performance of pbse/pbs core/shell quantum dot heterojunction solar cells: short circuit current enhancement without the loss of open circuit voltage by shell thickness control. *Nanoscale*, 7(4):17473–17481, 2015.

[15] D. Columbara et al. The fox and the hound: in-depth and in-grain na doping and ga grading in cu(in,ga)se2 solar cells. *Journal of Material Chemistry A*, 8:6471–6479, 2020.

Impact of backsheet versatility on inverter availability

Claudia Buerhop-Lutz, Oleksandr Stroyuk, Jens Hauch, Ian Marius Peters

Forschungszentrum Juelich GmbH, HI ERN, Erlangen, Germany

Studies point out that the polymer materials can be crucial for inverter availability, operation and PV-system performance. All modules of inverters need our attention because the versatility of backsheets (BSs) can be high with 73% of inverters having modules with differing, mixed BSs. We evaluated time-series of ground impedances measured by the inverters. For correlating the electrical data with the BS-material, we identified BSs of 29,118 modules using near-infrared spectroscopy and visual inspection of a 6.3 MWp PV power station. Most critical for inverter operation are modules with fluorinated coatings (FC) followed by polyamide, according to instances and distribution of low ground impedance values. The risk for inverter tripping is high for mixed strings including at least one module with FC-BS. In year nine, 41% of the days are affected by critical low GI-values, trend increasing.

978-1-6654-6060-6/23 $31.00 © 2023 IEEE

Evaluating the Weather Forecasting Models and the Impact to PV Generation Forecasting

Spyros Theocharides, Anastasios Koumis, George Makrides, George E. Georghiou

PV Technology Laboratory, FOSS Research Centre for Sustainable Energy, Department of Electrical and Computer Engineering, University of Cyprus, Nicosia, 1678, Cyprus

Abstract — Accurate photovoltaic (PV) generation forecasting is an important feature that can assist utilities and plant operators in the direction of energy management and dispatchability planning. However, the accuracy of the PV forecasting is directly related to the quality of the weather forecasts. In this work, an evaluation of the various weather forecasting models is performed. Additionally, an investigation of the effect of the quality of the weather forecasts on PV generation forecasting will be examined. Finally, this study focused to propose an approach to improve the quality of the weather forecasts and eventually the PV production forecasting accuracy by employing machine learning and linear regression models that could record the behaviour of the local weather.

I. INTRODUCTION

Photovoltaic (PV) generation forecasting can mitigate the power quality effects posed by large shares of distributed systems through active grid management and is, therefore, an important feature that can assist utilities and plant operators in the direction of energy management and dispatchability planning. More specifically, short-term PV production forecasts (intra-hour) are necessary for power ramp and voltage flicker prediction as well as control operations and dispatch management. On the other hand, mid-term PV production forecasting (intra-day and day-ahead) is used for load consumption and production monitoring to control voltage and frequency levels and reduce secondary reserve [1]. However, the accuracy of the PV generation forecasting is largely dependent on the accuracy of the underlying weather forecasting model used to acquire the respective weather forecasts. Specifically, weather forecasting models are mathematical models used to predict the weather.

The most common models include the Global Forecast System (GFS) [2], the European Centre for Medium-Range Weather Forecasts (ECMWF) model [3] and the Weather and Research Forecasting (WRF) model [4]. These models use a range of data such as solar irradiance, temperature, pressure, wind direction, precipitation and cloud cover to create a forecast of the weather over a given area.

In this aspect, this work demonstrates a performance evaluation between the three weather forecasting models with respect to the accuracy of the global horizontal irradiance (GHI) and ambient temperature (T_{amb}) forecasts. Additionally, to examine the actual impact of the quality of the weather forecasts on the PV generation forecasting, a machine learning

model was trained using data sets from a reference poly-crystalline silicon (poly-c-Si) PV system located in the University of Cyprus used and tested using data sets from the three weather forecasting models. Finally, this work will propose a methodology to improve the accuracy and the quality of the weather forecasts by employing a combination of an ensemble method and machine learning models.

II. OUTDOOR TESTING FACILITY

A. Location Characteristics

Table I describes the location characteristics (latitude, longitude and altitude) as well as the description of the pyranometer used for the measurements.

TABLE I
LOCATION AND PYRANOMETER CHARACTERISTICS

Latitude	35.142927°
Longitude	33.405993°
Altitude	139.0m
Pyranometer Brand	Kipp and Zonen
Pyranometer Model	CM21-CV 2
Pyranometer Uncertainty	± 2%

B. Experimental Setup

The OTF of the PV Technology Laboratory at the UCY in Nicosia, Cyprus is a flexible and scalable testing, demonstration and R&D facility for smart grid and other advanced energy technologies. The infrastructure includes a test-bench grid-connected PV system used for the forecasting analysis commenced in this study. The test-bench PV system comprises 5 poly-crystalline Silicon (poly-c-Si) PV modules of rated power 235 Wp that are installed in an open-field arrangement at the optimal yearly energy yield inclination angle of 30°.

Furthermore, the PV system was connected to a data acquisition (DAQ) platform used to monitor and store meteorological and PV operational data. The platform comprises meteorological and PV operational measuring sensors connected to a central data acquisition system. The performance of the system and the prevailing meteorological conditions were recorded according to the requirements set by IEC 61724 [5]. In particular, the meteorological measurements include the global plane-of-array irradiance (G_{POA}), relative humidity (RH), wind direction (W_a) and speed (W_s), as well as ambient temperature (T_{amb}). The PV system operational measurements include maximum power current (I_{mp}), voltage (V_{mp}) and power (P_{mp}), as measured at the DC side of the system [6].

III. METHODOLOGY

In this study, several steps were undertaken to assess and enhance the accuracy of weather forecasting models and their application in photovoltaic generation forecasting. Firstly, the performance of three prominent weather forecasting models, namely ECMWF, GFS, and WRF, was meticulously evaluated by comparing their outputs with ground measurements. This initial step provided a comprehensive understanding of the strengths and weaknesses of each model. Subsequently, the collected data was ingested into a GBM (Gradient Boosting Machine) model, enabling the development of a robust photovoltaic generation forecasting system. Additionally, a methodology was proposed to enhance the existing weather forecasting models, taking into account the findings from the previous evaluations. Once the weather forecasting models were improved, their outputs were utilized for photovoltaic generation forecasting. Finally, the two photovoltaic generation forecasts, one based on the original weather forecasting models and the other utilizing the enhanced models, were compared to assess the impact of the proposed methodology on the accuracy of the photovoltaic generation predictions. This comprehensive approach aimed to refine weather forecasting models and optimize their application in predicting photovoltaic generation, ultimately contributing to more reliable renewable energy management. Figure 1 summarises the steps to be followed.

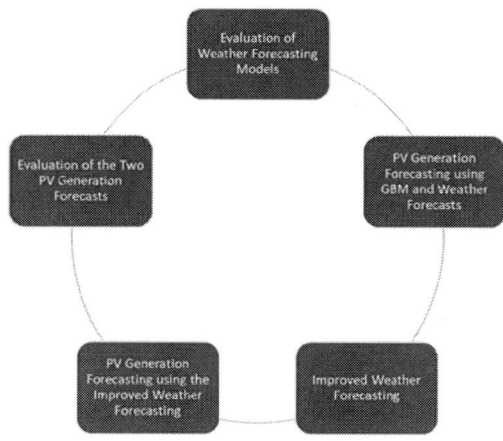

Figure 1 Summary of the methodology.

A. Performance Evaluation Methods

To compare the three weather forecasting methods against weather measurements both in terms of accuracy and quality, several methods were employed and will be further expanded during the final manuscript. Specifically, the first approach is correlation analysis. This evaluation measures the similarity between two-time series data sets by calculating the correlation coefficient, which ranges from -1 to 1. A coefficient close to 1 indicates a strong positive correlation, a coefficient close to -1 indicates a strong negative correlation, and a coefficient close to 0 indicates no correlation. Additionally, dynamic time wrapping (DTW) will be utilized to estimate the local stretch and/or compression to apply to the time axes of two timeseries in order to optimally map one set onto the other. Moreover, the comparison of the frequency components of two time series data sets (spectral analysis) will be performed by utilising the power spectral density (PSD) of the time series data Finally, statistical tests and measures were used to determine if there is a significant difference between two time series data sets. Specifically, the measures used are the normalized RMSE (nRMSE) to the capacity of the investigated system and the mean absolute percentage error (MAPE). The performance metrics used in this study are given:

$$nRMSE = \frac{100}{Y_{\text{nominal}}} \sqrt{\frac{\sum_1^n (e_i)^2}{n}} \quad (1)$$

$$MAPE = \frac{100}{n} \cdot \sum_{i=1}^{n} \left| \frac{e_i}{y_{observed,i}} \right| \quad (2)$$

$$e_i = y_{\text{observed},i} - y_{\text{forecasted},i} \quad (3)$$

where n is the amount of data points, $y_{i,observed}$ and $y_{i,forecasted}$ is the observed and forecasted data, respectively. Y_{nominal} is the nominal capacity of the investigated system.

B. PV Generation Forecasting

The methodology followed to develop an optimal machine learning forecasting model for day-ahead PV power forecasts included a training phase (to effectively apply a learning technique to a performance function), a validation phase (to identify the important features and architectural parameters of each model) and a testing phase (to assess the forecasting accuracy).

For the development optimisation and benchmarking of the machine learning model, the data-sets of the PV system installed at the UCY test facility and the data-sets from the weather forecasting models were used. The data-set was separated into the train, validation and test set.

The training, validation and testing set, comprised of the model inputs, which included the measurements of GHI and Tamb. Finally, the dc Pmp was the output feature of the developed model.

The model used to develop the PV generation forecasting was a Gradient Boosting Machine (GBM) [7]. GBM utilise preservative regression models by consecutively fitting a function (base learner) to current pseudo-residuals with the least squares method [8]. In the function estimation, there is a random output variable y and a set of random inputs $X = \{x_1 \ldots x_n\}$. Subsequently, a training sample $\{y_i, x_i\}_N$ of known (y, x) is used in order to obtain an estimation for the function $F^*(x)$. This minimises the expected value of a specified loss function $L(y, F(x))$ over the joint distribution of all (y, x)-values and is given by [10]:

$$F^* = \arg\min_F E_{y,x} L(y, F(x)) = \arg\min_F E_x[E_y(L(y, F(x)))|x] \quad (1)$$

Frequently employed loss functions $L(y, F)$ include squared errors $(y - F)^2$ for $y^{TM} R^1$ (regression) and negative binomial log-likelihood, $\log(1 + e^{-2yF})$ when $y^{TM} \{-1, 1\}$ (classification) [10]. A common procedure is to restrict $F(x)$ to be a member of the parametrised class of functions $F(x; P)$, where $P = \{P_1, P_2...\}$ is a finite set of parameters whose joint values identify individual class members [10]:

$$F(x; \{\beta_m, \alpha_m\}_1^M = \sum_{m=1}^M \beta_m h(x; \alpha_m) \quad (2)$$

where $h(x, a)$ is a simple parametrised function of the input variables x, characterised by parameters $\alpha = \{\alpha_1, \alpha_2...\}$.

C. Improved Weather Forecasting Method

To enhance the quality and precision of weather forecasts, a combination of machine learning models and an ensemble methodology will be employed. The objective is to develop a more accurate weather forecasting system by incorporating advanced techniques.

Firstly, a Bayesian regularised neural network (BRNN) [9] will be created for each of the three primary weather forecasting models: ECMWF, GFS, WRF. These BRNN models were trained using weather measurements obtained from the testing location. The BRNN models are chosen due to their ability to handle complex patterns and relationships in the data.

Next, a voting system will be employed to determine the two best-performing BRNN weather forecasting models among the three. This voting process will evaluate the performance of each BRNN model based on various criteria such as accuracy, reliability, and consistency. The two models with the highest scores will be selected to form an ensemble.

The ensemble model combined the outputs of the two chosen BRNN models to generate a more robust and reliable forecast. By aggregating the predictions from multiple models, the ensemble can capture a wider range of possible outcomes and improve the overall accuracy of the forecast.

To evaluate the effectiveness of the ensemble model, its results compared against the forecasts provided directly by the three original weather forecasting models. This evaluation assess the performance of the ensemble in terms of its ability to outperform or match the existing forecasting models.

Finally, the specific outputs of the improved weather forecasts will be utilized to evaluate the impact on PV generation forecasting. By integrating the enhanced weather forecasts into PV generation forecasting models, it becomes possible to assess the influence of accurate weather predictions on estimating the amount of electricity generated by PV systems. This evaluation will help understand how improved weather forecasts can contribute to more precise predictions of

renewable energy generation, aiding in planning and optimizing PV systems.

IV. RESULTS

A. Performance Evaluation of the Weather Forecasts

A typical example of the behaviour of the three weather forecasting models against the measured GHI located at the University of Cyprus can be found in Figure 1. As can be seen, the data from the three weather forecasting models follow the behaviour of the weather measurements from the testing location, however, specific patterns were not captured from the models (e.g. high ramping rates).

Additionally, Table II summarises the initial results of the three weather forecasting models (GHI evaluation only). The initial results demonstrated that the best-performing model was the WRF. Specifically, the model demonstrated an absolute difference of 2% - 4% and 2.5% - 3% for the nRMSE and MAPE respectively.

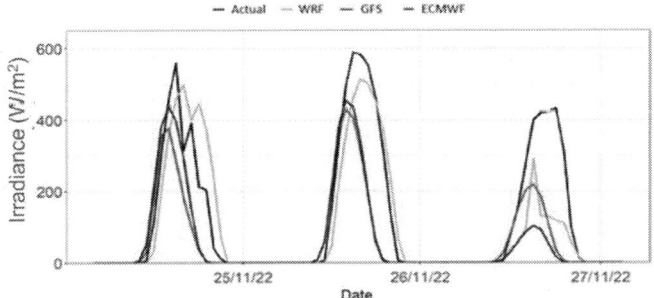

Figure 2 Typical behavior of the 3 weather forecasting models against the weather measurements acquired from the testing location.

TABLE II
SUMMARY OF THE PERFORMANCE EVALUATION RESULTS FOR THE
WEATHER FORECASTING MODELS

Weather Forecasting Model	Performance Evaluation	
	nRMSE (%)	MAPE (%)
GFS	14.23	11.98
ECMWF	16.77	13.76
WRF	12.18	9.41

B. PV Generation Forecasting

Furthermore, weather forecasts were employed to evaluate their impact on PV generation forecasting. As can be seen from Table III, as expected the best-performing model was the PV generation forecasting model fed with the WRF datasets. It is important to mention that the error demonstrated for the PV generation forecasting was increased compared to the error of the GHI forecasts.

TABLE III
SUMMARY OF THE PERFORMANCE EVALUATION RESULTS FOR THE PV GENERATION FORECASTING

Weather Forecasting Model Used	Performance Evaluation	
	nRMSE (%)	MAPE (%)
GFS	14.99	12.73
ECMWF	18.83	14.61
WRF	13.41	9.56

C. Improved Weather Forecasting Method

Upon applying the improved weather forecasting models (ECMWF, GFS, and WRF), notable enhancements in the accuracy and reliability of PV generation forecasting were observed compared to the forecasts obtained in the. The refined models exhibited improved precision in predicting weather conditions relevant to photovoltaic generation, resulting in more accurate forecasts of the energy output from photovoltaic systems. Table IV demonstrates the summary of the performance evaluation of the improved weather forecasting. As can be observed the improved weather forecasting outperforms the three models by 5.50%-3.90% and 6.50% - 3.30% for the nRMSE and MAPE respectively.

The enhanced forecasts provided valuable insights into the expected photovoltaic generation, allowing for better planning, optimization, and management of renewable energy resources. The utilization of improved weather forecasting models led to increased confidence in predicting photovoltaic generation levels, thereby enabling more efficient utilization of solar energy resources and facilitating the integration of renewable energy into the power grid.

The comparison between the two photovoltaic generation forecasts (one based on the original models and the other incorporating the improved models) demonstrated the tangible benefits of the proposed methodology. The improved forecasts exhibited reduced errors and a closer alignment with ground measurements, indicating the positive impact of the methodology in enhancing the accuracy and reliability of photovoltaic generation predictions.

Figure 3 illustrates the performance of the photovoltaic (PV) generation forecasting in terms of daily normalized root mean square error (nRMSE) (Figure 3a) and mean absolute percentage error (MAPE) (Figure 3b) over a test set period of 300 days. The results indicate that the majority of the days exhibited errors below 10% for nRMSE and 5% for MAPE. Furthermore, the specific PV generation forecasting utilizing the improved weather forecasts showcased notable improvements. The nRMSE decreased by approximately 5.60% to 3.40%, while the MAPE reduced from 6.10% to 3.15%. These enhancements reflect the positive impact of the improved weather forecasts on the accuracy and reliability of PV generation predictions.

TABLE IV
SUMMARY OF THE PERFORMANCE EVALUATION RESULTS FOR THE IMPROVED WEATHER FORECASITNG

Weather Forecasting Model Used	Performance Evaluation	
	nRMSE (%)	MAPE (%)
Improved Weather Forecasting	9.13	6.28

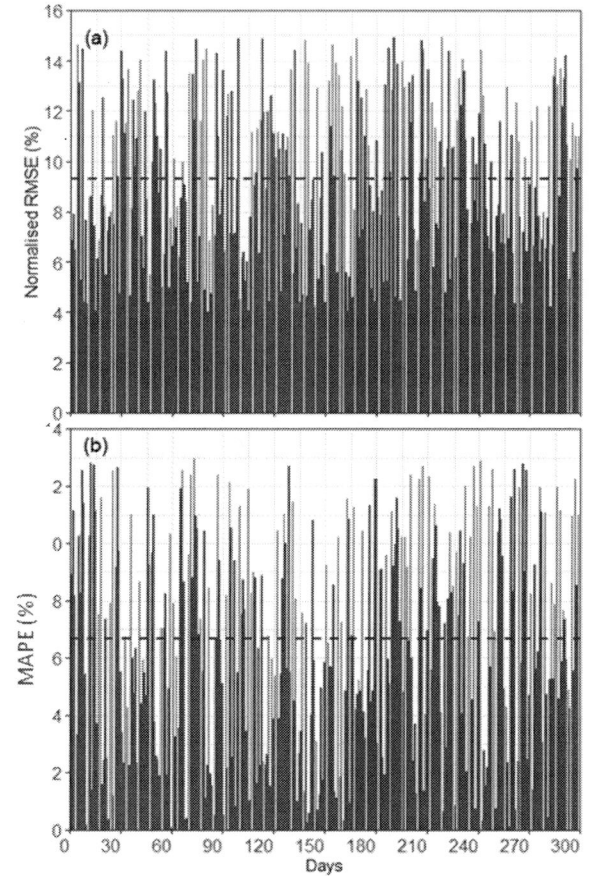

Figure 3 Performance evaluation of the PV generation forecasting: (a) Daily nRMSE evaluation and (b) daily MAPE evaluation. The blue dashed line demonstrates the value of the total test set period.

IV. CONCLUDING REMARKS

The accuracy of the PV generation forecasting is largely dependent on the accuracy of the underlying weather forecasting model used to acquire the respective weather forecasts. The scope of this work was to demonstrate a performance evaluation method to compare the three weather forecasting models (WRF, GFS, ECMWF) and to investigate the impact of the quality and accuracy of weather forecasting to PV generation forecasting.

Additionally, this work will provide a combination of machine learning and ensemble structure in order to improve the quality and accuracy of the weather forecasts. Initial results from this work demonstrated that the best-performing weather forecasting model for the specific testing location and data sets was the WRF model.

Furthermore, the improved weather forecasting modes significantly enhanced accuracy in both of weather and PV generation forecasting. Evaluation showed 5.50%-3.90% and 6.50%-3.30% improvements in nRMSE and MAPE respectively. This enabled better planning, optimization, and integration of renewable energy. These findings highlight the positive impact of improved weather forecasts on PV generation forecasting.

Overall, the outcome of the implementation of the methodology showcased the potential to advance weather forecasting models for PV generation forecasting, contributing to the effective utilization and management of renewable energy resources in a more sustainable manner.

ACKNOWLEDGEMENT

This work has received funding from the European Union's Horizon 2020 research and innovation programme under Grant Agreement no 864537, project title Flexible Energy Production, Demand and Storage-based Virtual Power Plants for Electricity Markets and Resilient DSO Operation (FEVER).

REFERENCES

[1] M. Abdel-Nasser and K. Mahmoud, "Accurate photovoltaic power forecasting models using deep LSTM-RNN," *Neural Comput. Appl.*, vol. 31, no. 7, pp. 2727–2740, Jul. 2019, doi: 10.1007/s00521-017-3225-z.

[2] NOAA/NCEP, "GFS analysis data.," 2022. https://www.ncep.noaa.gov/data/global-forecast-system-gfs/ (accessed Jan. 15, 2023).

[3] ECMWF, "ERA5 reanalysis data," 2019. https://www.ecmwf.int/en/data/data-catalogue/reanalysis-datasets/era5 (accessed Jan. 15, 2023).

[4] J. G. Powers *et al.*, "The weather research and forecasting model," *Bull. Am. Meteorol. Soc.*, 2017, doi: 10.1175/BAMS-D-15-00308.1.

[5] A. Drews, H. G. Beyer, and U. Rindelhardt, "Quality of performance assessment of PV plants based on irradiation maps," *Sol. Energy*, vol. 82, no. 11, pp. 1067–1075, Nov. 2008, doi: 10.1016/j.solener.2008.04.009.

[6] G. Makrides, B. Zinsser, M. Norton, G. E. Georghiou, M. Schubert, and J. H. Werner, "Potential of photovoltaic systems in countries with high solar irradiation," *Renewable and Sustainable Energy Reviews*, vol. 14, no. 2. pp. 754–762, 2010, doi: 10.1016/j.rser.2009.07.021.

[7] J. H. Friedman, "Greedy function approximation: A gradient boosting machine," *Ann. Stat.*, vol. 29, no. 5, 2001, doi: 10.1214/aos/1013203451.

[8] J. H. Friedman, "Stochastic gradient boosting," *Comput. Stat. Data Anal.*, vol. 38, no. 4, 2002, doi: 10.1016/S0167-9473(01)00065-2.

[9] R. Neal, "Bayesian Learning for Neural Networks," *Lect. NOTES Stat. -NEW YORK- SPRINGER VERLAG-*, 1996.

Plausibility Filtering of PV Outdoor Data

T. S. Vaas*[†], J. Körtgen*, E. Sovetkin*, U. Rau*[†] and B. E. Pieters*

*IEK5-Photovoltaik, Forschungszentrum Jülich, 52425 Jülich, Germany
[†]Faculty of Electrical Engineering and Information Technology, RWTH Aachen University, 52074 Aachen, Germany

Abstract—High-quality input data is necessary for calculating performance loss and predicting energy output and the lifetime of photovoltaic (PV) modules. Therefore, filtering PV outdoor data for unplausable measurements is crucial for analyzing PV outdoor performance. Photovoltaic outdoor data consists of electrical measurements (e.g. complete current-voltage (IV-) characteristics or single performance parameters), and meteorological data (e.g. irradiation and temperature). Here, the various electrical and meteorological measurement all constitute different dimensions in the measured data. Currently there is no standard for outdoor PV data filtering. However, commonly data filtering is based on simple thresholds in single dimension of the data. Since thresholding usually results in information loss, we propose using a plausibility filter, which simultaneously considers the various dimensions in the data. To this end we use well known correlations between the various dimensions in the data to compute a measure of plausibility by means of the Mahalanobis distance. In this work we demonstrate this concept by combining the solar cell IV parameters, module temperature, and various irradiance measurements (plane-of-array, global-horizontal, and diffuse horizontal irradiance).

I. INTRODUCTION

The first step in measurement data analysis should always be validation and filtering of outliers. Especially when the measurements are performed under conditions which are hard to control, as is generally the case for long-term outdoor experiments. Disruptions in normal operating conditions may considerably affect the result of an analysis. Since there is usually little information about the exact effects of outliers on the measurement, a statistical analysis of the measurements is impeded. In general, the superposition of the natural uncertainty on measured values with outlier deviations leads to a complex distribution.

For the evaluation of PV Outdoor data, the PV community lack standardizations of outdoor data filtering [1]. Common filter approaches focus on threshold filtering of, i.e. imposing thresholds on values of, e.g. point of array irradiance (POA), nominal output power, ambient, or module temperature. Thresholding has the advantage of simplicity, but generally leads to information loss as valid data may be removed, and, at the same time, invalid datapoints are not always reliably removed. Furthermore, Jordan and Kurtz showed that for PV degradation rate estimation, different filter approaches lead to different results [2]. Lindig et al. present a comprehensive overview of common filter approaches and their advantages and disadvantages [1].

An interesting approach is presented by Hansen, who proposes filtering while fitting a diode model to IV-characteristics, and remove data for which the diode model parameters are unreasonable [3]. This method may be somewhat hampered by the diode model parametrization, which may be challenging on its own. However, rather than simply imposing a threshold on a single measured value, it filters based on the shape of the complete IV-characteristics, i.e. it imposes restrictions on how the measured values correlate.

In this paper we seek to utilize the many correlations between the various measured dimensions in the data to evaluate its plausability. To this end we simultaneously consider all Solar Cell Parameters (SCP) (short circuit current, I_{SC}, open circuit voltage, V_{OC}, the maximum power point current, I_{MPP}, and the maximum power point voltage, V_{MPP}), and meteorological data. Using the Mahalanobis distance in combination with well known correlations between the various dimensions, we combine all this data into a plausability measure. We demonstrate the presented method on publicly available data published by the National Renewable Energy Laboratory (NREL) [4].

II. THEORY AND EXPERIMENTAL DATA

A. Condition correction of IV characteristics

To correct IV-characteristics regarding their temperature T and irradiation G dependency we use procedure two from the IEC60891:2021 norm [5]. The correction procedure implements the two equations:

$$I_2 = I_1 \cdot \frac{G_2}{G_1} \cdot \frac{1 + \alpha \cdot (T_2 - 25°C)}{1 + \alpha \cdot (T_1 - 25°C)}, \tag{1}$$

$$\begin{aligned}
V_2 = &V_1 - R_{S1} \cdot (I_2 - I_1) - \kappa \cdot I_2 \cdot (T_2 - T_1) \\
&+ V_{OC,STC} \cdot \beta \cdot [f(G_2)(T_2 - 25°C) - f(G_1)(T_1 - 25°C)] \\
&+ V_{OC,STC} \cdot \beta \cdot [\frac{1}{f(G_2)} - \frac{1}{f(G_1)}],
\end{aligned} \tag{2}$$

where

$$f(G) = B_2 \cdot \ln^2\left(\frac{1000\mathrm{Wm}^{-2}}{G}\right) + B_1 \cdot \ln\left(\frac{1000\mathrm{Wm}^{-2}}{G}\right) + 1, \tag{3}$$

$$R_{S1} = R_S + \kappa \cdot (T_1 - 25°C). \tag{4}$$

Such the procedure translates an IV-characteristics, point by point, from one condition (T_1, G_1), to another (T_2, G_2). The procedure uses 6 coefficients:

- α temperature coefficient of current [K^{-1}]
- β temperature coefficient of the voltage [K^{-1}]
- B_1 lin. irradiation correction factor of the voltage [-]
- B_2 quad. irradiation correction factor of the voltage [-]
- R_S series resistance coefficient of the device [Ω]

978-1-6654-6060-6/23 $31.00 © 2023 IEEE

- κ temperature coefficient of R_S [$\Omega \cdot \mathrm{K}^{-1}$]

In this work we apply the IEC60891 norm to translate the SCP from the measured conditions to some standard conditions, i.e., $T_S = 25°C$ and $G_S = 500\mathrm{Wm}^{-2}$.

B. Simple Sky Dome Projector

To model the POA irradiance we use the open-source SSDP [6] library. The Library implements the Perez All-Weather sky model [7] and accounts for diffusive and direct irradiation contributions. As input, the model uses the location (latitude, longitude and elevation), orientation and tilt of the PV module as well as the time and local effective albedo of the surrounding and measured global horizontal direct and diffusive irradiation GHI and DHI. Optionally a local topography may be provided to simulate shading from the surrounding structures and terrain. More information on SSDP can be found in [8].

C. Mahalanobis distance

The Mahalanobis distance of a point x to a mean of the distribution $\mathbb{Q}$ on $\mathbb{R}^n$ is given by

$$d_M(x, \mu) = \sqrt{(x - \mu)^T \Sigma^{-1} (x - \mu)}, \qquad (5)$$

where Σ is the non negative definite covariance matrix of $\mathbb{Q}$. Assuming $\mathbb{Q}$ as a normal distribution in $\mathbb{R}^n$, d_M^2 follows the $\mathcal{X}^2$-distribution with n degrees of freedom. Therefore, equating the cumulative distribution function (CDF) of the $\mathcal{X}^2$-distribution with a given quantile q gives

$$F_{\mathcal{X}^2}(n, d_{M,\mathrm{threshold}}) = q, \qquad (6)$$

where $F_{\mathcal{X}^2}(n, x)$ is the CDF of the $\mathcal{X}^2$-distribution with n degrees of freedom, and $d_{M,\mathrm{threshold}}$ is the Mahalanobis distance threshold for which only $1 - q$ of the datapoints have a higher Mahalanobis distance than $d_{M,\mathrm{threshold}}$.

D. Used data

We use several datasets provided by the National Renewable Energy Laboratory (NREL) [4]. From these datasets we used data on 11 different modules, operated for 13 months in Cocoa, Florida, and another 13 months in Eugene, Oregon. The different modules cover various thin-film as well as wafer based technologies (mono-crystalline Silicon, multi-crystalline Silicon, Cadmium Telluride, Copper Indium Gallium Selenide, amorphous/crystalline heterojunction, amorphous/microcrystalline tandem, amorphous silicon tandem and amorphous silicon triple junction).

III. MAHALANOBIS DISTANCE FILTER

From the datasets introduced in the previous section we used the time series of Solar Cell Parameters (SCP), I_{SC}, V_{OC}, I_{MPP} and V_{MPP}, the module temperature, T_{Mod}, and the in-plane irradiance, G_{POA}, as well as GHI and DHI. To filter the datasets we compare these parameters with modeled values.

This way we can account for the expected relations between the SCP, T_{Mod}, G_{POA}, GHI, and DHI. More formally we write

$$\Delta X_i = X_i - X_{\mathrm{expected},i}$$
$$dX_i = \frac{\Delta X_i}{\sigma_{\Delta X}}, \qquad (7)$$

where X_i is the i-th measurement of a particular parameter, $X_{\mathrm{expected},i}$ is the corresponding expected value, and $\sigma_{\Delta X}$ is the standard deviation of ΔX over all measurements.

For the parameter G_{POA} we use Eq. (7) and substitute $X = G_{POA}$. The expected G_{POA} (X_{expected}) is computed from the measured GHI and DHI using SSDP and the given time stamp and coordinate (in Eugene and Cocoa). For the albedo we use a value of 20% for the vegetation near the test locations in Eugene and Cocoa[9]. Note that occasionally there are missing values. For simplicity we remove incomplete datapoints.

In a similar fashion we treat the SCP. Here we use the correction procedure 2 of the IEC60891 norm to correct each measured SCP to a set of standard conditions. In this work the standard conditions are defined as $T_{Mod,S} = 25°C$ and $G_{POA,S} = 500\mathrm{Wm}^{-2}$. Note that we do not use the more commonly used $G_{POA,STC} = 1000\mathrm{Wm}^{-2}$ as this high irradiance is somewhat untypical for actual operation conditions. Thus, we substitute a corrected SCP for X in Eq. (7). The expected value, X_{expected}, is set to the mean of the corrected SCP. The correction procedure 2 of the IEC60891 norm requires 6 correction coefficients. These coefficients were obtained from regressions of the SCP to the measured T_{Mod} and G_{POA}.

The Mahalanobis distances, $d_{M,i}$, are computed in 5 dimensions (dI_{SC}, dV_{OC}, dI_{MPP}, dV_{MPP}, dG_{POA}) using Eq. 5. To filter the dataset we use a threshold distance of $d_M = d_{M,\mathrm{threshold}} \approx 3.884$, representing the distance where the CDF of a Mahalanobis distance distribution of a 5-dimensional normal distributed vector, reaches the quantile of $q = 99\%$.

IV. RESULTS

The Mahalanobis distance filter is applied to in total $\approx 880,000$ datapoints devided in 22 dataset (11 modules at the two locations). Table I comprehensively shows the size of the used data sets (after measurement points with missing GHI and DHI are filtered out) and the share (percentage) of the data our Mahalanobis filter removes. In total $\approx 13.9\%$ of the datapoints are filtered out. For the individual datasets the share ranges from 11.7% to 16.1%. The threshold distance was set to filter 1% of the data in case the variables are normal distributed. Therefore the result of 13.9% filtered data indicates a variable distribution with significant deviation from a normal distribution, i.e., a heavy tailed distribution of the Mahalanobis distance.

Figure 1 shows exemplary the scatter density plots for the modeled against measured G_{POA} for the Eugene mSi0166 dataset. The color of the points in the scatter is the estimated local density of scatter points. From the unfiltered dataset in Fig. 1 we see the SSDP estimate of the G_{POA} generally matches the merasured data quite closely. Note that the density color scale is logarithmic and that the highest density of points

978-1-6654-6060-6/23 $31.00 © 2023 IEEE

	Eugene		Cocoa	
module type	data set size	filtered	data set size	filtered
mSi0166	42908	12.8%	35669	13.7%
mSi0188	42773	13.8%	38012	13.7%
mSi460A8	42755	15.4%	37864	15.3%
xSi12922	42829	13.7%	37905	14.1%
HIT05667	42912	13.2%	37313	12.8%
aSiMicro03036	42969	15.7%	37949	16.0%
aSiTandem72-46	42905	16.1%	38109	15.3%
aSiTriple28324	42353	14.8%	37407	14.2%
CdTe75638	41959	11.7%	37993	13.0%
CIGS8-001	42791	13.3%	37860	11.7%
CIGS39017	42312	13.2%	33791	12.0%
total	879338	datapoints	13.9%	filtered

TABLE I

OVERVIEW OF DATASET SIZE AND SHARE OF FILTERED DATA

is closly matching the identity line. Comparing the unfiltered and filtered dataset in Fig. 1 (12.8% filtered out), we see the Mahalanobis filter reliably removes measurements, where the with SSDP modeled G_{POA} deviates from the measured G_{POA}. Note, that this two-dimensional representation is a projection of our multi-dimensional data, i.e., the applied filter removes data points according to the multidimensional Mahalanobis distance filter and not only due to deviations of the depicted dimension.

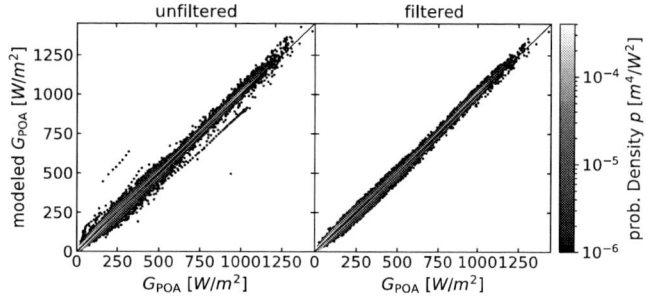

Fig. 1. Scatter density plots of the modeled vs. measured G_{POA} for the unfiltered and filtered Eugene mSi0166 dataset. The straight black line is the identity line, where the modeled and measured G_{POA} are equal.

Figure 2 shows scatter density plots for the corrected I_{SC} against measured G_{POA} for the unfiltered and the filtered Eugene mSi0166 dataset. For irradiance values of $G_{POA} > 500 \mathrm{Wm}^{-2}$, the current correction is generally reliable. However, below this value the corrected values of the unfiltered dataset in Fig. 2 exhibit a considerable scatter. The highest densities of scatter points are around the same corrected current values as for $G_{POA} > 500 \mathrm{Wm}^{-2}$, i.e. in most cases the correction also works at low irradiance values. However, there is clearly some structure in the scatter density plot where the corrected currents split over two branches, where one branch exhibits considerably lower corrected currents. This branching appears to be the effect of partial shading of the module, occuring primarily in the mornings and evenings when the sun is low. Note that the irradiance sensors are, judging from the pictures of the setups in [4], positioned higher than the modules. Thus the modules may be partially shaded when the sensors are not. The corrected I_{SC} of the filtered dataset

in Fig. 2, shows the large scatter for $G_{POA} < 500 \mathrm{Wm}^{-2}$ is effectively trimmed off, effectively filtering partial shading from the dataset. Depending on the purpose of the data filtering this may or may not be desiarable. In case this is not desirable, a more elaborate model could be used to correct or predict the I_{SC}, which includes the effects of partial shading. This, however, is beyond the scope of this paper.

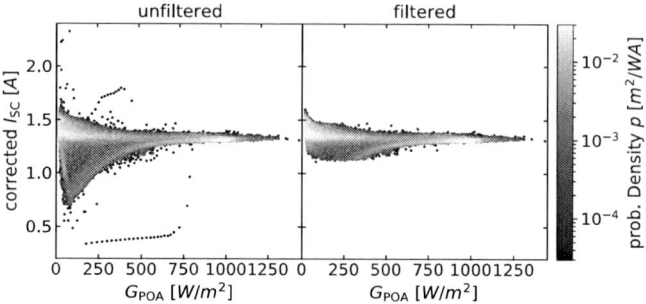

Fig. 2. Scatter density plot of corrected I_{SC} vs. measured G_{POA} for the unfiltered and filtered Eugene mSi0166 dataset

Finally, in Fig. 3, we show the scatterdensity plots for the corrected V_{OC} as a function of the measured T_{Mod} for the Eugene mSi0166 dataset. Here it is quite notable that for temperatures above 310 K (37 °C) there is considerably less scatter in the corrected V_{OC} values. Also here, it apears the partial shading events lead to an increased scatter. We observe the filtered dataset exhibits considerably less scatter.

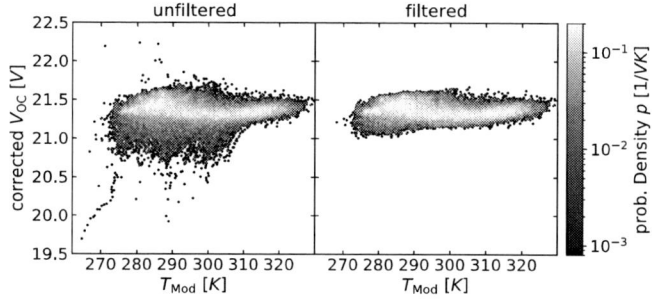

Fig. 3. Scatter density plot of corrected V_{OC} vs. T_{Mod} for the unfiltered and filtered Eugene mSi0166 dataset

V. SUMMARY AND DISCUSSION OF SIGNIFICANCE

A variaty of data filtering methods exist for the analysis of datasets for PV applications. However, most these filtering methods are based on simple thresholding. In this work we presented a plausibility filter concept, utilizing known correlations between the various measured quantities to define a measure of plausibility for each datapoint. To obtain such a measure of plausability, we used the Mahalanobis distance combined with models to describe the correlations between several measured dimensions in a dataset. We demonstrated this concept using the SCP, T_{Mod}, G_{POA}, GHI and DHI. It should be noted that this concept may be extended to use more dimensions in the data, for example, provided wind speed, ambient temperature, G_{POA}, and T_{Mod} are avialable, the consistency between these dimensions may be validated with the Faiman model [10]. More general, this approach may be used whenever a model is available which allows predicting one measured dimension from one or more other dimensions available in the data.

REFERENCES

[1] S. Lindig, A. Louwen, D. Moser, and M. Topic, "Outdoor pv system monitoring—input data quality, data imputation and filtering approaches," *Energies*, vol. 13, no. 19, p. 5099, 2020.

[2] D. C. Jordan and S. R. Kurtz, "The dark horse of evaluating long-term field performance—data filtering," *IEEE Journal of Photovoltaics*, vol. 4, no. 1, pp. 317–323, 2013.

[3] C. Hansen, "Parameter estimation for single diode models of photovoltaic modules," Sandia National Lab.(SNL-NM), Albuquerque, NM (United States), Tech. Rep., 2015.

[4] W. Marion, A. Anderberg, C. Deline, S. Glick, M. Muller, G. Perrin, J. Rodriguez, S. Rummel, K. Terwilliger, and T. Silverman, "User's manual for data for validating models for pv module performance," National Renewable Energy Lab.(NREL), Golden, CO (United States), Tech. Rep., 2014.

[5] "Photovoltaic devices – procedures for temperature and irradiance corrections to measured i-v characteristics," International Electrotechnical Commission, Geneva, CH, Standard IEC 60891, 2021.

[6] "SSDP: Simple Sky Dome Projector," github.com/IEK-5/SSDP, accessed: Oct, 2022.

[7] R. Perez, R. Seals, and J. Michalsky, "All-weather model for sky luminance distribution—preliminary configuration and validation," *Solar energy*, vol. 50, no. 3, pp. 235–245, 1993.

[8] E. Sovetkin, J. Noll, N. Patel, A. Gerber, and B. E. Pieters, "Vehicle-integrated photovoltaics irradiation modeling using aerial-based lidar data and validation with trip measurements," *Solar RRL*, p. 2200593, 2022.

[9] B. Marion, M. G. Deceglie, and T. J. Silverman, "Analysis of measured photovoltaic module performance for florida, oregon, and colorado locations," *Solar energy*, vol. 110, pp. 736–744, 2014.

[10] D. Faiman, "Assessing the outdoor operating temperature of photovoltaic modules," *Progress in Photovoltaics: Research and Applications*, vol. 16, no. 4, pp. 307–315, 2008.

The European Solar Communication

—

Will it strengthen the Photovoltaic Industry in the European Union

Arnulf Jäger-Waldau
European Commission
Joint Research Centre (JRC),
Ispra, Italy
arnulf.jaeger-
waldau@ec.europa.eu

Anatoli Chatzipanagi
European Commission
Joint Research Centre (JRC),
Ispra, Italy
Anatoli.Chatzipanagi
@ec.europa.eu

Georgia Kakoulaki
European Commission
Joint Research Centre (JRC),
Ispra, Italy
Georgia.Kakoulaki@ec.europa.eu

Sandor Szábo
European Commission
Joint Research Centre (JRC),
Ispra, Italy
sandor.szabo@ec.europa.eu

Abstract—Since the introduction of the first European Renewable Energy Directive in 2009, PV installations have significantly increased to reach more than 211 GWp in the European Union at the end of 2022. The European Solar Communication and it's Solar Strategy can help to accelerate not only the deployment of urgently needed new renewable energy and photovoltaic power capacity, but also help to revamp a competitive European solar value chain, which will not only provide local jobs and wealth creation, but also hedge against the risk of global supply chain disruptions in the future.

Keywords— European Renewable Energy Directive, European Solar Communication Green Deal, Solar Strategy, greenhouse gas emission, solar photovoltaics, PV deployment, competitive European solar value chain

I. INTRODUCTION

The publication of the 6th IPCC Assessment Report in April 2022 [1] and the geopolitical developments in the first half of 2022 have highlighted the urgency of the clean energy transition. The European Commission had reacted with the REPowerEU Communication and the Solar Strategy Communication in March and May 2022 respectively [2, 3]. As an intermediate step towards climate neutrality (European Green Deal) by 2050, in December 2020, the European leaders endorsed the Commission's proposed target to reduce net emissions by at least 55% by 2030. To achieve this, the European Commission proposed the "Fit for 55" package of EU legislative measures, which amongst other actions sets a target of 40% participation of renewable energies in EU's energy mix [4].

In December 2022, the European Council reached a provisional deal with the European Parliament on the revision of the ETS. This deal includes an increase of the overall ambition of emissions reductions by 2030 in the sectors covered by the EU ETS to 62%, compared to the 61% target proposed by the European Commission.

Since the end of 2010, the capacity of grid-connected solar photovoltaic (PV) systems in the European Union has increased from 34.2 GWp to about 211 GWp at the end of 2022 (Fig. 1) [5].

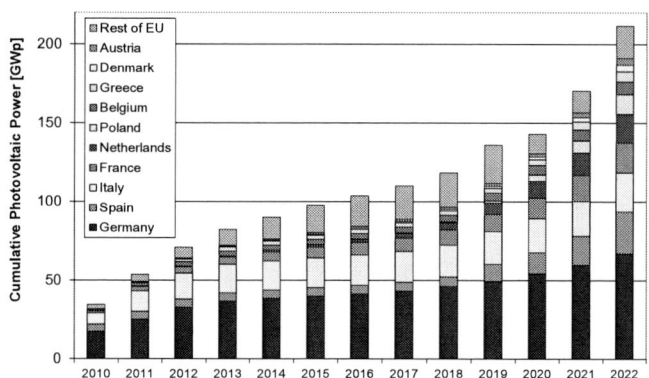

Fig. 1.   Grid-connected PV capacity in EU [5]

II. CURRENT SITUATION

In the European Union, solar energy and photovoltaics in particular were identified as one of the cornerstones of a rapid and more ambitious deployment of renewable energy technologies in order to meet the climate-neutrality objective in 2050 and a significant reduction of the EU's dependence on imported fossil fuels. Although, the currently proposed measures include a strong component to diversify the source of fossil fuel imports, away from Russia, a path on how to phase out entirely their use is not yet so clear. The full implementation of the "Fit for 55" proposals would lower the Union's gas consumption by 30%, still requiring over 200 bcm, by 2030.

Each EU Member State had to prepare a national recovery and resilience plan, which outlines their individual reform and

978-1-6654-6060-6/23 $31.00 © 2023 IEEE

investment agendas for the years 2021-2023 in order to be eligible for the Recovery and Resilience Facility. A minimum of 37% of expenditure are earmarked for actions to combat climate change.

During the first nine months of 2022, more than 70 GWp of modules were imported into the European Union, but due to shortages of inverters (chip shortage) and the labour market, the annual market for new PV installations grew just a little more than 33% to about 41 GWp in 2022 (Fig. 2) [5]. Twelve countries installed more than 1GWp, namely Germany and Spain with (7.4 to 7.6 GWp) followed by Poland (4.8 to 5 GWp), the Netherlands (3.9 to 4.1 GWp), France (2.6 to 2.8 GWp), Italy and Portugal (2.4 to 2.6 GWp), Denmark (1.4 to 1.6 GWp), Greece (1.3 to 1.5 GWp), Austria (1.2 to 1.4 GWp), Sweden and Belgium (1.0 to 1.2 GWp).

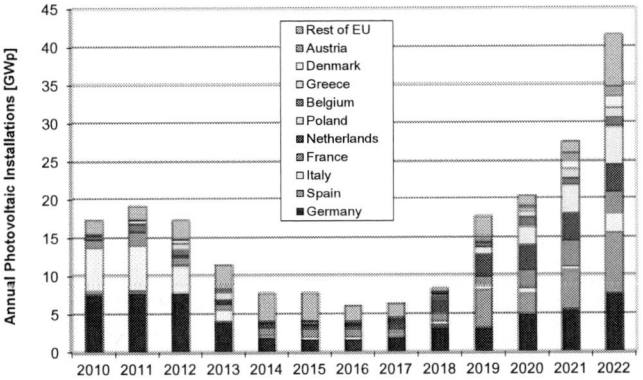

Fig. 2.   Annual installation of grid-connected PV capacity in the EU [5].

The Netherlands are leading in terms of installed capacity per capita in the EU with 1051 W_p, second only to Australia with 1168 W_p (Fig. 3). Six EU countries have more than the European Union average, namely, Germany (800 W_p), Denmark (760 W_p), Malta (750 W_p), Belgium (690 W_p), Greece (614 W_p), and Spain (558 W_p). So far, only five countries have installed less than the world average of 148 Wp per capita. Based on previous analysis for 2019, Germany was surpassed by the Netherlands, and Denmark entered the top five EU countries, while Italy lost its fourth place position and has now less than the EU average [6]. Furthermore, according to the above-mentioned analysis and our calculations, the EU average installed PV capacity per capita has increased by only 220 W_p between 2019 and 2022.

The Solar Strategy calls for an additional photovoltaic capacity 450 GWp between 2021 and 2030, which would mean a roughly fourfold increase of the nominal capacity to over 720 GWp by 2030. Compared to 2022, this would require an annual market volume increase to over 100 GWp annually by 2030, which is achievable if the current market trend can be maintained.

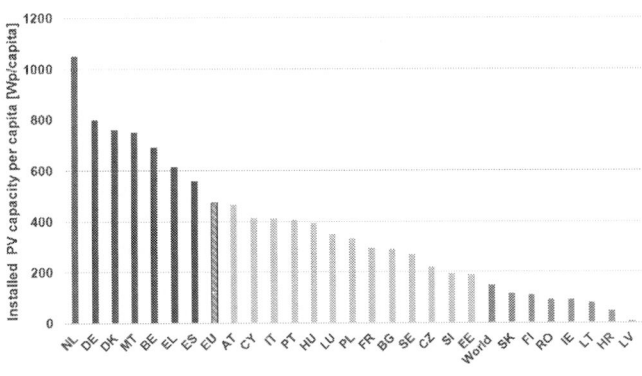

Fig. 3.   Grid-connected PV capacity/capita in the EU [5].
(colour code: red – No 1; dark blue – above EU average; hatched – EU average; yellow, green – world average and light blue – below EU average; light blue – below 100 Wp/capita)

With such a market increase, the issue of the security of supply becomes more important. In order to hedge against supply chain disruptions, which could be seen in the last two years during the COVID pandemic, the European Union has to rebuild its local manufacturing capabilities along the full value chain.

Therefore, the European Solar Communication includes a number of building blocks to achieve this in a timely manner. The following initiatives, are aimed to deliver the expected outcome:

1) European Solar Rooftops Initiative
2) Utility scale deployment including multi use of land (e.g. agri-PV, floating-PV, PV on noise barriers, etc.)
3) Solar value for buildings, districts and cities
4) Preparing the energy network for the efficient distribution of solar energy
5) Establishment of a resilient supply chain
6) Supporting investments regarding EU PV manufacturing (de-risking, funding)

In October 2022, the European Commission endorsed the creation of a new European Solar Industry Alliance, similar to the already existing Battery Alliance, which was launched in 2017. The Solar Industry Alliance will support the objectives of the EU's Solar Energy Strategy, an essential component of the REPowerEU plan, which set out how to massively scale-up and speed-up the production of renewable energy in Europe to regain our independence from Russian fossil fuels, and make our energy system more resilient.

The Solar Industry Alliance was formally launched on 9 December 2022. It is aimed to revamp an EU solar PV industrial ecosystem to help secure and diversify supplies of solar photovoltaic components. This should be done by scaling up EU manufacturing capacities for competitive, innovative, and sustainable solar PV products as well as diversification of the international PV value chain components and supply of raw materials.

The alliance will elaborate and implement a strategic action plan based by

- identifying manufacturing scale-up bottlenecks and providing recommendations
- facilitating access to finance, including by establishing commercialisation pathways for solar PV manufacturing
- providing a framework for cooperation on the development and uptake
- maintaining international partnerships and resilient global supply chains
- supporting solar PV research and innovation
- facilitating and increasing communication on circularity and sustainability
- exploring and developing a skilled workforce for the PV manufacturing sector

III. OUTLOOK

The introduction of green hydrogen and the European Commission's hydrogen strategy will play an important role for the future need of solar PV capacity as current hydrogen production is still 97% based on fossil fuels [7]. The Communication related to the hydrogen strategy for a climate-neutral Europe mentions a first phase of at least 6 GW of green hydrogen electrolysers by 2024 and at least 40 GW of RES-powered electrolysers by 2030 [8]. The full-time operation of this level of electrolyser capacity would require the equivalent electricity output of 256 TWh. Under the assumption that PV generated electricity would need to supply about half of this electricity and this capacity is fairly distributed close to the hydrogen demand this would require approximately 115 GWp of PV capacity.

The European Green Deal also targets 90% reduction in transport-related greenhouse gas emissions by 2050 [9]. The goals set together with the hydrogen strategy in the communication of the Sustainable and Smart Mobility Strategy could only be reached if the future electricity based mobility part will rely on a fossil fuel free electricity portfolio, which implies also huge increase of the PV shares in the electricity mix.

Additional demand for green electricity and rooftop PV systems will come from the foreseen Renovation Wave and the transformation of the Energy Performance of Building Directive with the concept of Nearly Zero-Energy Buildings (NZEBs) in national legislation [10] At municipal and regional level in the EU there is an ongoing discussion to make the installation of renewable energy systems in new buildings mandatory. According to the European Central Bank, approximately 1.6 million new residential buildings were constructed in 2016 in the EU [11]. If on average a 4 kW PV system would be needed on each building this would add more than 6 GWp per year or another 60 GWp until 2030. In case the annual number of housing completions rebounds to that of the 2000-2009 (2.2 – 2.8 million annually), new buildings could host up 11 GWp of solar PV, annually.

An invigorated EU PV market needs a development strategy for the full PV value chain, supported by research and innovation. This should include new cell and module manufacturing in the EU, with significant job creation potential. This is acknowledged by the launch of the new European Solar Industry Alliance, which should accelerate this process.

The Green Deal's circular economy action plan is of high importance, with its aim of promoting sustainability across the whole value chain and encouraging businesses to offer reusable, durable and repairable products. In addition the proposed 'renovation wave' of public and private buildings can also be an important stimulus for using PV products to achieve near-zero energy buildings. The European Solar Communication is an important signal that Europe is strongly committed to rapid PV deployment scale up and wants to avoid the "Stop-and-Go" policy obstacles of the past (like in case of setting the Feed-in tariff or Premiums in the past policies) and by this it can secure a more business friendly environment for the European industry to grow.

The recently proposed Net-Zero Industry Act [12] aims to set the required environment to scale up manufacturing of net-zero industry in the EU. One of the identified strategic net-zero technologies is PV. A simplification of the regulatory framework (permitting) for the PV manufacturing and a skills development support are among the actions included in the Act that will help increase the EU PV competitiveness. The acceleration of fossil fuel phase out, a precondition for the 1.5°C goal, will require an even faster deployment of renewable energy sources and photovoltaics. Together with the uptake of electromobility and green hydrogen for industrial processes, the demand for photovoltaics in the European Union could exceed 1 TWp by 2030.

The cumulative PV power capacity additions under the EU solar energy strategy requires a compound annual growth rate (CAGR) of 17.5%, which is slightly lower than current market trend of 18% [13]. To reach 1 TWp by 2030, CAGR would have to increase to 22%, which is feasible according to industry analysts.

Last but not least, the development of the PV markets need a dimension for a just transition by ensuring a significant share of decentralised PV to provide local jobs and ensure citizen participation.

Disclaimer: The scientific output expressed is based on the current information available to the authors, and does not imply a policy position of the European Commission.

IV. REFERENCES

[1] IPCC Sixth Assessment Report, WG III, 2022, https://www.ipcc.ch/report/ar6/wg3/

[2] European Commission, REPowerEU Communication, 08.03.2022, COM(2022) 108 final

[3] European Commission, Solar Strategy Communication, 18.05.2022, COM(2022) 221 final

[4] European Council, Fit for 55%, https://www.consilium.europa.eu/en/policies/green-deal/fit-for-55-the-eu-plan-for-a-green-transition/

[5] A. Jäger-Waldau, Snapshot of Photovoltaics – May 2023, EPJ Photovoltaics

[6] Wolniak, R., Skotnicka-Zasadzień, B. Development of Photovoltaic Energy in EU Countries as an Alternative to Fossil Fuels. *Energies* **2022**, *15*, 2. https://doi.org/10.3390/en15020662

[7] Kakoulaki G, Kougias I, Taylor N, Dolci F, Moya J, Jäger-Waldau A. Green Hydrogen in Europe - a regional assessment: Substituting Existing Production with Electrolysis Powered by Renewables. Energy Convers Manag 2020;113649

[8] European Commission. A hydrogen strategy for a climate-neutral Europe. vol. 53. Brussels, Belgium: 2020. https://doi.org/10.1017/CBO9781107415324.004

[9] European Commission, Communication on Sustainable and Smart Mobility Strategy – putting European transport on track for the future 9.12.2020 COM (2020) 789 final

[10] Official Journal of the European Union. Directive (EU) 2018/844 amending Directive 2010/31/EU on the energy performance of buildings and Directive 2012/27/EU on energy efficiency. Off J Eur Union 2018;L 156/75. [4] Official Journal of the European Union. Directive (EU) 2018/844 amending Directive 2010/31/EU on the energy performance of buildings and Directive 2012/27/EU on energy efficiency. Off J Eur Union 2018;L 156/75.

[11] European Central Bank. Statistical Data Warehouse. SHI Struct Hous Indic Stat 2020. https://sdw.ecb.europa.eu/browse.do?node=70499

[12] European Commission, Proposal for a REGULATION OF THE EUROPEAN PARLIAMENT AND OF THE COUNCIL on establishing a framework of measures for strengthening Europe's net-zero technology products manufacturing ecosystem (Net Zero Industry Act), 12.05.2023, COM (2023) 161 final.

[13] Chatzipanagi A. and Jäger-Waldau A., The European Solar Communication – Will it pave the road to achieve 1 TW of Photovoltaic System Capacity in the European Union by 2030?, *Sustainability* 2023, *15*(8), 6531; doi: 10.3390/su15086531

Measuring Sustainability of PV in Energy Transition: Mass, Energy, and Circularity

Heather M. Mirletz, Silvana Ovaitt, Macarena Mendez Ribo, Seetharaman Sridhar, Teresa M. Barnes

National Renewable Energy Lab, Golden, CO, United States

Advanced Energy Systems Graduate Program, Colorado School of Mines, Golden, CO, United States

Ira A. Fulton Schools of Engineering, Arizona State University, Tempe, AZ, United States

Transition to a carbon-free energy system is crucial for global decarbonization and underpins Circular Economy (CE) goals. Photovoltaic (PV) technology is required for Energy Transition, but manufacturing and circular pathways can be material, energy, and carbon-intensive. Therefore, we need a prioritization of sustainability strategies for PV evolution and lifecycle management in the context of Energy Transition. This study employs a suite of quantitative metrics to compare different proposed sustainability strategies for PV modules on their ability to achieve Energy Transition. Proposals for sustainable PV range from high-yield, high-efficiency paradigms, to short-lived and fully recyclable, to long-lasting, indestructible modules. We leverage a global decarbonization deployment schedule through 2100 with the open-source PV in Circular Economy (PV ICE) tool to quantify the impacts of different evolving module design scenarios covering the range of proposed sustainability strategies. First, modules are compared on effective capacity and required replacements to meet and maintain decarbonization capacity targets through 2100. We demonstrate the effects of lifetime, degradation, and reliability on effective capacity. Next, we quantify and compare virgin material demands and lifecycle wastes, examining the impacts of lifetime and recycling rates. Finally, and critically for renewable energy technologies, we quantify the energy demands required to achieve the decarbonization capacity targets and calculate energy balance metrics (net energy, energy return on investment). These results are then summarized into a metric matrix, demonstrating tradeoffs and the importance of longevity. Our suite of mass and energy metrics provides stakeholders and decision-makers with quantitative data on circular economy choices for PV in the energy transition, enabling informed evaluation of tradeoffs of different PV module designs and CE pathways.

AUTHOR INDEX

Abad, Eduardo Camarillo 1249
Abasi, Abudulimu 402
Abbas, Ali 245, 498, 656, 1193
Abdullah-Vetter, Zubair 1224, 1468
Aberle, Armin G. 1052
Abou-Ras, Daniel 432, 879, 939
Abrosimov, Nikolay V. 203
Abudulimu, Abasi 42, 46, 220, 275, 279, 683, 804,
.... 823, 909, 910, 1197, 1323
Acevedo, Armando Figueroa 1393
Adeleye, Damilola 1151
Adhikari, Alisha 1197, 1376
Adner, David 770
Afridi, Muhammad 237, 755
Afshari, Hadi 132, 161, 1156
Afzal, Syed Usama Bin 772
Agarwal, Shashank 1053
Agarwal, Sumit 30, 359, 475, 891, 1216
Ager, Joel W. 75
Aghaei, M. 737
Agrawal, Roshni 1053
Aguirre, Aranzazu 194
Ahlswede, Erik 1201
Ahmad, Muneeza 1109, 1565
Ahmad, S. 737
Ahmadi, Bahman 816
Ahn, Yujeong 1549
Aiello, Ashlee R. 282
Aimez, V. 1538
Aïssa, Brahim 27, 1516
Akiyama, Hidefumi 463
Akopian, Arkadi 641
Al Katrib, Mirella 1440
Alam, Fahad 173
Alam, M. A. 626
Alam, Muhammad A. 119
Alam, Muhammad Ashraful 772
Alam, Muhammad 1027
Alanis, Luis Eduardo 1060
Al-Baity, Shifaa M. 561
Albert, P. 1538
Alberts, Vivian 119
Albin, David S. 695
Albrecht, Steve 945
Aldalali, Bader S. 162
Aldhefairi, Mariam 1344
Alessi, Bruno 281
Algora, Carlos 1433

Alhammadi, Salh 1549
Alhassan, Saeed 1471
Ali, Adnan 1516
Al-Jassim, M. M. 315
Al-Jassim, Mowafak 1570
Allebé, Christophe 1129
Almannaee, Rawdah 1344
Almansoori, Muntaser Abdelrahman 321, 1344
Almonacid, Florencia 122, 437
Almutawah, Zahrah S. 1417
Alom, Md Zahangir 690, 1237
Alsuwaidi, Meerh 1344
Altenhöfer-Pflaum, Georg 749
Altermatt, Pietro 1544
Al-Thani, Hamda A. 561
Altosaar, Mare 1298
Alvarez, H. 1492
Alvarez, Hugo Da S. 1159, 1654
Alvarez, Hugo S. 1068
Alvarez, Hugo 1134
Aly, Shahzada Pamir 119
Amassian, Aram 332
Anctil, Annick 730, 743, 1095, 1259
Anderson, Caroline Lima 30
Anderson, Kevin S. 948
Anderson, Kevin 710
Andersons, Janis A. 348
Andreas, Afshin 600
Andreasen, Jens Wenzel 651
Anto, Robins 1230
Antón, Ignacio 746, 888
Aonuki, Sho 1398
Aponte, Erick 525
Aponte-Bezares, Erick E. 674, 1034, 1170, 1326
Aponte-Bezares, Erick 1082
Araki, Kenji 924
Aramaki, Ken 1164
Arehart, Aaron R. 107
Arehart, Aaron 1299
Arfaoui, Ines 426
Armour, Eric A. 1217
Arnold, Rachael 353
Arteaga, Jorge 415
Arya, Rajeewa 1366
Asadpour, Reza 626
Asahi, Shigeo 659, 895
Asami, Meita 374, 463, 935
Ascencio-Vásquez, J. 737

Ascencio-Vásquez, Julián 1578
Askins, Steve 746, 888
Astigarraga, Alexander 931
Athana, Dhawal 1009
Atwater, Harry A. 202, 1157
Augusto, André 5
Avoli, Matteo 1404
Awni, Rasha A. 1323
Ayala, Silvana 1141
Ayon, Arturo A. 769
Ayon, Arturo 1283
Azevedo, João Henrique Paulino 256
Azzolini, Joseph A. 13
Baan, Marzieh 917
Babayeva, Gülüsüm 1092
Babu, Balaashwin 671
Bach, Udo 298
Badosa, Jordi 268
Baer, Carsten 191
Baetzner, Derk 1
Baik, Sunhee 793
Bailie, Colin 740
Bainier, Camille 268
Baker, Wes 1096
Bakke, Jordan 1393
Balaji, Pradeep 5
Baldrias, Maristel 1024
Ball, James M. 298
Ballif, Christophe 1, 1129
Banerjee, Parag 629
Banks, Terry 740
Bansal, Shubhra 566, 1551
Barbose, Galen 793, 991
Barnes, Teresa M. 554, 1593
Barretta, Chiara 931, 1630
Barrit, Dounya 268
Barros, Tárcio A. Dos S. 1654
Barros, Tárcio A. S. 1159
Barros, Tarcio Andre Dos Santos 1134
Barth, Kurt L. 245, 981, 1497
Barth, Kurt 789
Barth, Vincent 352
Bartsch, Jonas 1625
Bastola, Ebin 220, 275, 279, 401, 402, 683, 804,
.... 823, 910, 1041, 1193, 1410, 1600
Batista-Alvarez, Natanael 370
Battiato, Sebastiano 1567
Baumann, Sara 733
Baxter, Jason B. 915, 1241
Beard, Matthew C 161, 1156
Beattie, Meghan N 941
Becker, Jan-Philipp 1201

Beltrán, Juan F. 761
Benatto, Gisele A. Dos Reis 16, 448, 1452, 1662
Benhaddou, Nada 395, 534
Bennett, Mitchell F. 1217
Benton, Brandon 587
Berad, Mrunal 1366
Bergin, Mike 437
Bernardini, Simone 768
Bernsen, Otto 749
Berriel, S. Novia 629
Berry, Joseph J. 298, 590, 1044, 1260, 1484
Bertoni, I. 291
Bertoni, Mariana I. 291, 400, 848, 970, 1533
Bertoni, Mariana 218, 333, 597, 689, 768, 782, 896, 1634
Berwind, Matthew F. 1313
Bessa, Joao G. 1270
Bessal, João Gabriel 437
Bevan, Geraint 591
Bhat, Akanksha 1147
Bhatia, Amandeep Singh 1027
Bidaud, T. 1538
Bidaud, Thomas 133, 781
Bila, Marine 295
Binyamin, Tal 879
Birgersson, Erik 1188
Bista, Sandip S 1197
Bista, Sandip Singh 46
Bittkau, Karsten 280
Bivour, Martin 770, 1625
Bizhanova, Gulzhan 39
Bizzarri, Fabrizio 1304, 1567
Black, Chloe L. 948
Blankemeyer, Susanne 652
Bloeck, Ulrike 939
Blom, Youri 1005
Blum, Adrienne 113
Boemer, Jens C. 1096
Bogner, Brandon M. 721
Bojorquez, Jose Raul Montes 769
Bolen, Michael 897
Bolink, Henk J. 938
Bondoc, Christopher C. 1510
Bonnassieux, Yvan 233
Borgers, Tom 48
Bosco, Nick 106, 1218
Bothe, Karsten 288
Bothwell, Alexandra M. 107
Bothwell, Alexandra 849
Boucherif, Abderaouf 133
Boucherif, Abderraouf 434, 1051
Bourarach, Fadi 426
Bourisli, Raed I. 162

Bournonville, Kenn Henrik....................749
Bousselot, Jennifer....................215
Bouttemy, Muriel....................1440
Bowers, Jake W....................245, 395, 981, 1177, 1497
Bowers, Jake....................534, 1193
Bowersox, David A....................1409
Boyce, Kenneth P....................1141
Boyd, Matthew....................920
Boyd, Stephen P....................961
Boyer, Jacob T....................10, 291, 848, 994
Boz, Mesude Bayrakci....................98
Brabec, Christoph J....................645, 1092
Bracamonte, Maria Fernanda Villa....................769
Braga, Daniel Sena....................1541
Braid, Jennifer L....................234, 367, 512, 572
Braid, Jennifer....................861, 1088, 1131
Bramante, Rosemary C....................1158
Brand, Andreas....................1588
Brandstätter, Andreas....................1630
Brau, Tyler R....................909
Brau, Tyler....................42, 220, 402, 804, 910
Braun, Anna K....................10, 71, 291, 848
Bredács, Marton....................931
Bremner, Stephen P....................917, 1564
Bremner, Stephen....................271
Brendel, Rolf....................6, 288
Brenner, Tom....................740
Brewer, Jeremy....................178
Brewster, Charles....................138
Breyer, Christian....................708, 749
Brinck, Anna....................1002
Brittman, Sarah....................1045
Brockmann, Tim Lukas....................652
Broderick, Robert....................1170
Brooks, William....................1009
Brown, Buck....................240
Brown, Matthew....................794, 1141
Bruckman, Laura S....................234, 597, 671, 861, 1088, 1141, 1485
Brueckner, Dennis....................651
Bryan, Alex....................1123
Buchholz, Florian....................1372
Buck, Thomas....................1266, 1372
Buerhop-Lutz, Claudia....................540, 645
Bui, Thanh-Tuân....................1484
Bullock, James....................1434
Buño, Luis....................1024
Burduhos, B. G....................737
Burnham, Laurie M....................572
Burnham, Laurie....................391, 409, 1131
Burtone, Lorenzo....................191
Bush, Meghan E....................698
Busko, Dmitry....................996

Buster, Grant....................474, 587
Butt, Nauman Zafar....................772
Cabanillas, Juan....................804
Cacciato, Mario....................1304, 1567
Cai, Mengmeng....................360
Calderon, Jose A....................914
Calili, Rodrigo Flora....................256, 858
Calloquispe-Huallpa, Ricardo....................674
Calvo-Barrio, Lorenzo....................1059
Campbell, Robert....................295
Campesato, Roberta....................1404
Campo-Ossa, Daniel D....................916
Canino, Andrea....................1005, 1304, 1567
Cano, Aitana....................1603
Cao, Fangfang....................822
Cappelluti, Federica....................1413
Carbone, Marc A....................1626
Cariou, Romain....................716, 934, 1423
Carletta, Stefano....................1404
Carpenter, Joe V....................1533
Carron, Romain....................651
Cartledge, Carsen....................1109, 1565
Carvalho, Romullo R. M....................1159
Carvallo, Juan Pablo....................793
Cassini, Denio Alves....................1541
Castillo, Arnold....................1024
Castillon, Jean....................268
Castro-Sitiriche, Marcel....................717
Catchpole, Kylie....................1434
Cebecauer, Tomas....................1594
Celi, Edoardo....................871
Chacon, Sergio A....................132, 1156
Chadly, Assia....................299
Chakar, Joseph....................233
Chaluvadi, Venkata S. A....................1249
Chambers, Terrence L....................188
Chan, Maria K. Y....................1634
Chan, Maria....................782
Chang, Atom....................917
Chang, Nathan....................289
Chapaneri, Kaushal....................119
Chapon, Julien....................426
Chapotot, Alexandre....................434, 1051
Chard, Julie....................1123
Chatzipanagi, Anatoli....................550
Chaubal, Aditi....................1366
Chaurasia, Harsh....................182
Chen, Bin....................518, 1019
Chen, Chien-Hsuan....................597, 629, 1337
Chen, Christopher....................179
Chen, Cong....................822
Chen, Daming....................1544

Chen, Gonglin..1021
Chen, Hong...1544
Chen, Kissenger..978
Chen, Lei...................804, 909, 1197, 1550
Chen, Mike..740
Chen, Ning..1372
Chen, Stela...377
Chen, Theresa K..1566
Chen, Xin...1519
Chen, Yifeng..1544
Chen, Zeying..682
Chin, Robert Lee.............................227, 1224
Chintapalli, Sreyas................718, 1334, 1356
Chiu, Arlene......................272, 718, 1356
Cho, Jinyoun..........133, 434, 1051, 1341, 1390
Cho, Yunae...97
Choi, Wookjin...1657
Choi, Wook-Jin.................................179, 1030
Choudhury, Ashif..134
Choudhury, Kaushik R...................................952
Choudhury, Kaushik Roy.........................377, 689
Chowdhury, Gofran............................86, 1253
Chretien, Jeremie...133
Christiansen, Silke.......................................194
Chu, Haifeng..1372
Chung, Jaehoon.............................844, 1417
Chutani, Ayush..391
Cieslak, Janko..191
Cifuentes, Luis...1433
Cimaroli, Alexander.......................................412
Cioc, Sorin...402
Cira, Spencer...473
Clandestino, Franco.......................................1491
Clausing, Roland..6, 12
Clemens, Daniel...240
Coco, Fabrizio..1005
Colegrove, Eric.................245, 633, 849, 946
Coletti, Gianluca...917
Coll, Pablo G..................................218, 291
Coll, Pablo Guimerá.......................................848
Collavini, Silvia...289
Collin, Stéphane..610
Colomba-Colon, Luis.......................................370
Colombo, Mariela..1529
Colón, Alanis M...416
Colvin, Dylan J...........671, 758, 850, 1120, 1485
Conibeer, Gavin J..............................271, 1564
Conibeer, Gavin...881
Coogan, Katrina...879
Cook, John P. D...476
Coologeorgen, Alexander E.................................918
Cooper, Emma C..572

Cooper, Robert A. García..................................519
Correa-Baena, Juan-Pablo.......................969, 1183
Corso, Roberto..................................1500, 1609
Cosme, Damien...426
Costa, Carla..934
Costa, Suellen C. S.......................................1541
Coulibaly, Bakary...1292
Court, Philip...922
Courtois, Guillaume............................434, 1390
Coutinho, Natália F.......................................290
Couture, Eugene Desjardins................................1648
Crawford, Zachary..51
Crestani, Thais...290
Critchlow, Gary...174
Crowe, Iain F...203
Crowe, Laura E..919
Crowley, Kyle...415
Cruz, Edgar E...420
Csank, Jeffrey T..1626
Cudzinovic, Michael J.....................................1024
Cullen, David A..46
Curran, Alan J..903
Currie, Taylor M..629
Curson, Kieran..................................245, 656
Da Silva, João A. F. G....................................1654
Da Silva, Lucas Aló Rodrigues Araujo......................858
Da Silva, Paulo R. D. R.........................1159, 1654
Daenen, Michael..48
Dahi, Adem..1423
Dalal, Vikram...641
Dale, P...995
Dale, Phillip J.................................731, 975
Dalibor, Thomas...1067
Dally, Pia..1440
Danel, Adrien...716
Daniel, Valentin...............................133, 434
Danielson, Adam...1279
Danilson, Mati.................................1298, 1432
Danovitch, D..1538
Dapprich, Karoline..113
Darawali, Renn.................................927, 1146
Darbali-Zamora, Rachid..........334, 455, 674, 953, 984,
...........................1034, 1082, 1170, 1326, 1626
Darling, Halley C...1146
Darling, Halley...927
Darnon, M...1538
Darnon, Maxime.................133, 781, 1051
Das, Ujjwal K...................91, 1545, 1667
Das, Ujjwal...1299
Dasgupta, Sagnik...............1030, 1155, 1657
Daus, Alwin...1347
Davis, Kristopher O...........219, 597, 629, 671, 758, 850, 1485

Davis, Kristopher 1337
Davis, Melissa A 330
Davis, Nithin Maipan 1404
Dawson, Timothy 1295
De Albuquerque, Vanessa Cardoso 256, 858
De Brabandere, Karel 1253
De La Rosa, Angel 782
De Lafontaine, Mathieu 476
De Lima, Geyciane P. 1159, 1654
De Luna, Gabby 1030
De Monfreid, Thybault 1484
De Oliveira, Otávio J. 290
De, M. M. M. Modesto Ana Paula 290
De, Shoubhik 1287
Debije, Michael G. 809
Deceglie, Michael G. 921, 1260, 1593
Deckx, Julien 1253
Decristofaro, Eric R. 918
Degen, Ashley 467
Degenhart, Nick 113
Delazzer, Timothy 387
Deline, Chris 47, 920, 1145
Delmas, William 415
Demko, Michael 377
Demtsu, Samuel 914
Deng, Chenyang 743
Deng, Yuepeng 903
Depauw, Valérie 434, 1390
Dequilettes, Dane W 590
Desarden-Carrero, Edgardo 1082, 1326
Descoeudres, Antoine 1129
Deshmukh, Kedar 1366
Dessein, Kristof 133, 434, 1051, 1341, 1390
Deville, Lelia 188
Dhakal, Rabin 528, 897
Dhakal, Tara 682
Dharmadasa, Ruvini 38
Dhople, Sairaj 1384
Di Stefano, Agnese 1005, 1304
Diaz, Martin 1217
Dice, Paul W. 409
Dice, Paul 1131
Didier, Thevenard 861
Diederich, Marvin 6
Dierenbach, Jonas 1184
Dietsch, Tina 795
Diggs, Andrew 51
Digregorio, Steven J. 400
Digregorio, Steven 333
Ding, Kaining 177, 280, 419
Diniz, Antonia Sonia A. C. 94, 1541
Dippell, Torsten 1129

Dittmann, Sebastian 1475
Dobson, Kevin D. 1545
Dobson, Kevin 1299
Dobson, Wes 113
Dokken, Briana 1210
Dolia, Kshitiz 42, 1417, 1557
Doll, Bernd 645
Domínguez, C. 746
Domínguez, César 888
Don, Eric 981
Dong, Peng 1372
Dong, Shuan 240, 840
Donoso, Jose 749
Dougherty, Brian 1292
Doumon, Nutifafa Y. 1044, 1260
Dow, Andrew R. R. 1034, 1626
Drayton, Jennifer 1310
Drees, Martin 145
Drost, Christian 656
Druffel, Thad 38
Du, Bin 1299
Du, Liming 822
Du, Mohan 155
Duan, Leiping 1434
Duan, Weiyuan 177, 280, 419
Duan, Xiaomeng 1197
Dubajic, Milos 271, 1564
Dubois, Anne Migan 268
Dubois, Sébastien 716
Dubuc, Christian 133, 1051
Duenow, Joel N. 695, 946
Dulal, Prabin 1410
Duncan, Brent 492
Duncker, Klaus 191
Dunfield, Sean P. 644
Dunham, Scott 578
Duong, Calvin 508
Durant, Brandon K 1156
Dutta, N. S. 315
Dutta, Nikita S. 1570
Duzellier, Sophie 934
Dvonc, Lukas 1594
Dwivedi, Priya 472, 828, 1224, 1468
Dyreson, Ana 391, 1635
Eberspacher, Chris 740
Eberst, Alexander 419
Ebner, Rita 194
Ebong, Abafriseke 38
Echevarria, Angel 916
Edmondson, A. 1261
Eggink, Wouter 443
Einhaus, Lisanne M. 622

Ekins-Daukes, N. J. ... 1468
Ekins-Daukes, Nicholas J. ... 345, 976
Ekins-Daukes, Nicholas ... 90
Elahi, Sheikh Tawsif ... 1237
Elanzeery, Hossam ... 1067
El-Atab, Nazek ... 173, 327
Ellingson, Randall J. ... 1323
Ellingson, Randy J. ... 42, 46, 220, 275, 279, 402, 683, 823, 844, 909, 1019, 1041, 1197, 1353, 1417
Ellingson, Randy ... 804, 910
Ellis, Brian ... 476
Elsehrawy, Farid ... 345
Engel, Bernd ... 64
Engelen, Tine ... 48
Engsig-Karup, Allan P. ... 16
Enjalbert, Nicolas ... 716
Eperon, Giles E. ... 132, 161, 919
Eperon, Giles ... 15
Erickson, Samuel ... 415
Escobar, D. Martinez ... 783
Espinet-Gonzalez, Pilar ... 11
Etgar, Lioz ... 879
Evans, Rhett ... 828
Facsko, Stefan ... 435
Fai, Calvin ... 915, 1241
Fairbrother, A. ... 737
Falkenberg, Gerald ... 651
Fan, Yangxin ... 861, 1088
Fanego, Vicente Lara ... 1594
Fang, Liang ... 195, 1526
Fang, Xin ... 840
Fardi, Hamid ... 1513
Farias, Stephen ... 1575
Farias-Basulto, Guillermo A. ... 1066
Farina, Angela ... 730
Farrell, Jack ... 771
Farrell, John ... 1600
Fassl, Paul ... 701
Fattah, Tarek O. Abdul ... 203
Faulwetter-Quandt, Björn ... 191
Fechner, Hubert ... 749
Fedoseyev, Alex ... 110
Fei, Chengbin ... 1044
Feng, Xiaotong ... 1503
Feng, Zhiqiang ... 1544
Fenning, David P. ... 644
Ferekides, Chris ... 690, 1237
Ferguson, Andrew J ... 341
Fernández, Eduardo F. ... 122, 437, 1270
Fernandez, Pablo ... 1433
Fernández-Solas, Álvaro ... 1270
Ferrara, Matteo ... 1404

Ferreira, Mateo ... 1437
Fertig, Fabian ... 191
Fevola, Giovanni ... 651
Fields, Shannon ... 1299
Filho, Neolmar De M. ... 94
Fischer, Benedikt ... 177
Fischer, Markus ... 191
Fisher, Kathryn ... 925
Flechas, Juan Pablo Medina ... 268
Fleming, Katelynn E ... 89
Flicker, Jack D. ... 984, 1034, 1626
Flicker, Jack David ... 634
Florakis, Antonios ... 426
Floren, Radovanovic-Peric ... 351
Fonoll-Rubio, Robert ... 1067
Fontanot, Tommasso ... 194
Forcade, Gavin P ... 941
Forcade, Gavin ... 476
Forchhammer, Søren ... 16
Forrest, Stephen R. ... 482
Forrest, Stephen ... 332
Forrester, Sydney ... 991
Fortmann, Charles M. ... 22
Foster, Michael ... 587
Foti, Marina ... 1005
Fox, Curtis ... 897
France, Ryan M. ... 71, 341, 394, 475, 994, 1081, 1150
Frasson, Nicola ... 352
Fregosi, Daniel ... 528, 555, 897
French, Roger H. ... 234, 671, 758, 861, 1088, 1485
Friedl, Jared D. ... 220, 275, 279, 1210, 1323
Friedl, Jared ... 683
Friedlmeier, Theresa M. ... 566
Friedlmeier, Theresa Magorian ... 199, 1201
Friedman, Daniel J. ... 394
Friedman, Daniel ... 485
Fritzsche, Helmut ... 476
Frye, Bailey ... 1557
Fthenakis, Vasilis ... 146, 1503
Fu, Oakland ... 377
Fu, Sheng ... 42, 1417
Fukaya, Shohei ... 775
Fürer, Sebastian O. ... 298
Furis, Madalina ... 161
Gabas, Mercedes ... 1433
Gabor, Andrew M. ... 671, 850, 1485
Gaddy, Edward ... 311
Galiazzo, Marco ... 352
Gamboa, Daniel H. ... 761
Gamel, M. ... 942
Ganguly, Subhankar ... 1074
Gao, David Wenzhong ... 840

Gao, Jiaqing	1372
Gao, Munan	1509
Gao, Ningchao	840
Gao, Tina	718
Gao, Zhiyu	822
García, I.	947
García, Iván	1433, 1603
Garcia, Maria Angelica M.	1379
Garcia, R.	1492
Garcia, Rodrigo M.	1068, 1159, 1654
Garcia, Rodrigo	1134
Garín, M.	942
Garrevoet, Jan	651
Gedi, Sreedevi	1549
Gehan, Tim	740
Gehrke, Aaron	578
Geissbühler, Jonas	1
Geistert, Kristina	144
Geisz, John F.	71, 291, 394, 709, 1158
Geisz, John	178, 485
Georghiou, George E.	194, 541, 1270, 1348
Gerardi, Cosimo	1005
Gerber, Andreas	166
Gerger, Andrew	311
Gerton, Jordan	134
Gevorgian, Vahan	360
Geyer-Klingeberg, Jerome	385
Gfroerer, T. H.	1261
Gharabeiki, Sevan	1617
Ghosh, Probir	1366
Ghosh, Sayantani	415
Giacchino, Evan S.	918
Gibbons, Daniel	861, 1088
Gil-Escrig, Lidon	938
Giliberti, Gemma	1413
Ginger, David S.	1020
Giraldo, Sergio	1059
Giridharagopal, Rajiv	1020
Giteau, Maxime	610
Giuri, Antonella	1109
Glatthaar, Markus	1625
Gloeckler, Markus	914
Goga, Adam	51
Gok, A.	737
Golive, Yogeswara Rao	1287
Golubev, Timofey	874
Gomez, Daniel	1433
Gonçalves, Felipe	256
Gong, Yuancai	1059
González, Alanis M. Colón	367
González, Emmanuel J.	416
Good, Brian	245
Goosay, Olivia	727
Gopal, Deepika	126
Gorman, John	922
Gorman, Will	793
Gostein, Michael	254, 295, 1558
Gotoh, Kazuhiro	292, 724, 775
Gottschalg, Ralph	486, 1475
Goubard, Fabrice	1484
Govaerts, Jonathan	48
Graeber, Dietmar	1184
Granello, Pierpaolo	1404
Grassman, Tyler J.	917
Greco, Erminio	1404
Green, Martin	302, 1256
Greenaway, Ann L.	75
Greenhalgh, R. C.	245
Greenhalgh, Rachael C.	981, 1193
Greenhalgh, Rachael	656
Gregory, Christopher	916
Grimm, Benjamin	6
Grisanti, Marco	1567
Grossberg-Kuusk, Maarja	1298, 1432
Grossklaus, Kevin A.	1152
Grover, Sachit	489, 1582
Grovogui, Jann A.	11, 198
Gruenhagen, Philip	662
Grundmann, Marius	12
Gu, Hangyu	366
Gu, Xiaohong	42, 282, 952, 1141
Guc, Maxim	1067
Gudi, Dhanvini	272, 1356
Guennou, M	995
Gueymard, Christian A.	54, 778
Guha, Mousumi	492
Guibin, Shen	1052
Guillemoles, Jean-François	233
Guillemot, Thomas	268
Gulati, Himanshu	259, 1227
Guo, Da	914
Guo, Yonggang	1372
Gupta, Apoorva	644
Gupta, Mool C.	1155, 1522
Gupta, Priya	830
Gurule, Nicholas S.	984
Gütay, Levent	1059
Guthrey, H.	315
Guthrey, Harvey L.	1216
Guthrey, Harvey	939
Gutzler, Rico	432, 1201
Hohn, Oliver	941
Haas, Benedikt	879
Haase, Felix	6, 733

Habisreutinger, Severin 1490
Habte, Aron 474, 587, 600, 778, 1106
Hacene, Benjamin 144
Hacke, Peter 648, 952
Hadi, Sabina Abdul 1471
Hadjipanayi, Maria 194
Hagemann, Johannes 651
Hagendorf, Christian 770
Hages, Charles J. 915, 1241
Hages, Charles 975
Hähnel, Angelika 770
Hajj, Adonis E. 81
Haley, Thomas 928, 1147
Halm, Andreas 1372
Halme, Janne 345
Halsall, Matthew P. 203
Hamadani, Behrang H. 695, 1292
Hameiri, Ziv 5, 227, 472, 828, 1224, 1295, 1468
Hamer, Mike 1359
Hamon, G. 1538
Hamon, Gwenaëlle 133, 781, 1051
Hanif, Muhammad 271, 1564
Hansen, Clifford W. 922, 1110, 1250
Hanuš, Tadeáš 434, 1051
Hao, Xiaojing 1256
Haque, Sirazul 1380
Hara, Tomohiko 724, 1453
Harada, Yukihiro 895
Harder, Nils-Peter 426, 1024
Harder, Ross 1183
Hare, Casey P. 977
Hariskos, Dimitrios 432, 1201
Harper, Jim 14
Harrison, Jason 1002
Harrison, Samuel 352
Härtel, Marlene 945
Hartenstein, Matthew B 475
Hartenstein, Matthew 30
Hartweg, Barry 925
Harvey, Steve P. 946
Harvey, Steven P. 298
Hasan, Arif Yetkin 975
Haselsteiner, Philipp 1428
Hashad, Khaled 505
Hasoon, Falah S. 561
Hatt, Thibaud 1625
Hauch, Jens A. 645
Hauch, Jens 540, 1092
Hauschild, Dirk 91
Hawkins, Nicholas 298
Hayden, Steven 1044
He, Bo 1221

Heath, Garvin 594
Heben, Michael J. 35, 46, 220, 275, 279, 401, 402,
.... 683, 823, 844, 1041, 1193, 1210, 1323, 1376, 1410, 1417
Heben, Michael 804, 910, 1197, 1600
Hegedus, Steven 464
Heidrich, Robert 486
Heilscher, Gerd 1184
Heimsath, Anna 1313
Heinrich, Martin 1060
Heinzle, Nino 1348
Heitmann, Johannes 1266
Helder, Tim 199
Helfer, Eric 931
Helienek, Lubos 1594
Helmers, Henning 941
Helms, Clay 81
Henry, Isaiah 1551
Herasimenka, Stan 110
Heres, Geert C. 622
Hernández, Johann 761
Hernandez, Samuel I. 416
Hernandez-Alvidrez, Javier 984
Herrero, Rebeca 746
Heske, Clemens 91
Hettiaratchy, Elline C. 489
Hidalgo, Juanita 969
Hildreth, Owen J. 400
Hildreth, Owen 333
Hill, Blake 363
Hill, C. 995
Hill, Taylor D. 489
Hill, Taylor 1582
Hillhouse, Hugh W. 340
Hillhouse, Hugh 473
Hinken, David 288
Hinzer, Karin 47, 331, 476, 941
Hirst, Louise C. 1249
Ho, Kevin 1020
Hobbs, William B. 492, 921, 948, 1558
Hodges, Joseph 240
Hoerantner, Maximilian T. 919
Hoex, Bram 1256
Hoffman, Adam 861, 1088
Höger, Ingmar 191
Hoheisel, Raymond 894
Hoke, Andy 840, 1164
Hole, Jarand 749
Holman, Zachary C. 767, 1279, 1379, 1533
Holman, Zachary 925
Holmes-Smith, A Sheila 591
Holmgren, William F. 948
Holzhey, Philippe 289, 298

Hong, Chengjian	195, 1526
Hönig, René	191
Hool, Ryan D.	670, 997
Hörnlein, Stefan	191
Hoss, Jan	795
Hossain, Mohammad I.	27, 1516
Howard, Ian A.	172
Hsieh, Chun-Hao	1443
Hu, Hongjie	377
Hua, Amandee	91
Huaman-Rivera, Anny	477
Huang, Ben	1284
Huang, Jing	285
Huang, Jing-Shun	783
Huang, Jinsong	366, 1044
Huang, Jun-Yu	1443
Hubbard, Seth M.	63, 89, 721, 894
Hübner, Simon	1129
Huddy, Julia E.	1276
Hudry, Damien	996
Hudson, Andrew	1462
Hultqvist, Adam	1151
Huneycutt, Sandra	38
Hunter, Robert	941
Hunwick, Nicholas	981
Huque, Aminul	138, 835, 1096
Hussain, Zulkifl	683
Hwang, Jeong-Mo	179
Hyndman, David W.	1259
Iannascoli, Lorenzo	1404
Ibanez, Eduardo	1393
Ifuji, Yuto	1453
Ilahi, Bouraoui	133, 434, 1051
Im, Kyu-Hyeon	1657
Imaizumi, Mitsuru	76, 463, 721, 1429
Imenes, Anne G.	815
Imperatori, Davide	1404
Imran, Hassan	772
Infante-Ortega, L. C.	245
Infante-Ortega, Luis C.	981, 1497
Inoue, Kazuma	292
Ireton, Scott J	977
Irizarry-Rivera, Agustin	477
Irvine, Stuart J. C.	308, 1497
Irvine, Stuart	245
Irving, Richard	823
Isabella, Olindo	80, 655, 1005, 1398
Ishizuka, Shogo	939
Ital, Donald	38
Ito, Yuta	724
Jacob, David	1024
Jacobs, Janet	203

Jaeger-Waldau, Arnulf	924
Jäger-Waldau, Arnulf	550, 749
Jahandardoost, Mohsen	566
Jahangir, Jabir Bin	626, 1027
Jahelka, Phillip R.	202
Jahelka, Phillip	1157
Jain, Anubhav	1519
Jakob, Leonie	1625
Jamarkattel, Manoj K.	46, 220, 279, 401, 683, 823, 1041, 1197, 1376
Jamarkattel, Manoj	910, 1600
Jannuzzi, Gilberto	256
Jansson, Peter Mark	380
Janz, Stefan	434, 1341
Jaouad, A.	1538
Jaouad, Abdelatif	133
Jarzembowski, Enrico	191
Jaubert, Jean-Nicolas	234, 794
Javier, Gaia Maria N.	472, 828
Jawinski, Tanja	12
Jay, Frédéric	716
Jayaraman, Sreenivas	855
Jeangros, Quentin	1
Jensen, Karissa	42
Jeon, Seokmin	134
Jhang, Song-Syun	282, 952
Ji, Liang	1141
Jia, Huiying	1466
Jiang, C.-S.	315
Jiang, Chun-Shen	1570
Jiang, Chun-Sheng	46, 946
Jiang, Fangyuan	1020
Jiang, Jessica Yajie	302, 345
Jiang, Qi	709, 812, 1158, 1380
Jiang, Yi	1284
Jimenez-Arguijo, Alex	1059
Jin, Hao	1434
Jin, Shuangshuang	240
Jin, Tan	240
Jiyun, Zhang	1092
Jo, Sangmin	436
Joe, Junki	1218
John, Jim Joseph	119
John, Oliver	1588
Johnson, Jay	334
Johnson, Samuel	938
Johnston, Michael B.	298
Johnston, S.	315
Johnston, Steve W.	10, 848
Johnston, Steve	219, 387, 702, 865, 1490, 1570, 1629
Jones, Abigail R.	1250
Jones, C. Birk	455

Jones, James...920
Jones, Luke O. 174, 245, 395, 498
Jones, Luke 656, 1177
Jones, Steve..308
Jordan, Dirk C...................... 219, 1110, 1593
Jordan, Dirk 794, 896, 1145
Jorgensen, Peter Stanley 651
Josepson, Raavo.......................................1432
Jouanneau, Corentin 781
Joyce, Hannah J. 1249
Julien, Arthur..233
Jung, Hyeonjung Tari...................... 1384, 1393
Jung, Kyung Taek 97
Junghänel, Matthias 191
Jurca, Titel..629
Kaaya, Ismail..86
Kabra, Dinesh...1366
Kachman, Dana........................... 272, 1356
Kaewnukultorn, Thunchanok......................464
Kaizuka, Izumi...749
Kakoulaki, Georgia.....................................550
Kalizewski, Lauren M.................................917
Kalpoe, Prashand 305
Kaltenbaugh, Jarod 758
Kaluarachchi, Prabodika N. 402, 683, 844
Kamal, Serene........................... 272, 718
Kambley, Ankur..281
Kamikawa-Shimizu, Yukiko........................939
Kanakkithodi, Arun K. M. 782
Kanaujia, Pawan K. 1522
Kaneko, Ryuji 442
Kanevce, Ana................. 107, 199, 1201
Kang, Min Gu..97
Kanneboina, Venkanna 1410
Karade, Vijay..1556
Karakaya, Sakir...634
Kari, Thøger.................. 448, 1452, 1662
Kartopu, Giray...308
Kasher, Tal...917
Kashkimbayev, Ulan......................................39
Kasik, Camden..1310
Katakumbura, Nadeesha 402, 804
Kauert, Maximilian......................................191
Kauk-Kuusik, Marit..................... 1298, 1432
Kaupmees, Reelika....................................1298
Kaur, Navdeep 606
Kawakami, Mizuto 895
Kaydanik, Katty...783
Kazim, S. 737
Kazmerski, Lawrence L. 94, 1366, 1541
Ke, Cangming 191
Kee, Jared 555

Keller, Jan 887
Kelzenberg, Michael D. 1157
Kempa, Heiko.......................................12, 432
Kempe, Michael D. 794, 1141, 1260
Kendall, Anthony D. 1259
Kenyon, Jacques..................................395, 534
Kern, Dana B.................. 219, 702, 865, 1044, 1570, 1629
Kern, Dana 15, 387, 896
Kerr, Lei 1466
Kersten, Friederike 191
Kessler-Lewis, Emily 894
Kettle, J. 737
Khadka, Dhruba B.............................324, 1401
Khan, M. Ryyan 626
Khenkin, Mark 1066
Khetri, Mahantesh 1522
Khoury, Alexandre 1612
Khulmann, Forrest 363
Khurgin, Daniel..1356
Kikelj, Miha 1
Kile, Kara B. 683
Kim, Bora 670
Kim, Boyoung 436
Kim, Dohyung 97
Kim, Han-Jung 1373
Kim, Hyomin 1549
Kim, Hyungoo 436
Kim, Jeong-Hyeon 1484
Kim, Jin Hyeok 1556
Kim, Jin Young.. 165
Kim, Junhee 1373
Kim, Kiwhan 1556
Kim, Mijung.......................................670, 997
Kim, Munse.. 97
Kim, Sanggyun..1183
Kim, Seul-Gi 1484
Kim, Woo Kyoung 1549
Kim, Yoonkap 1373
Kimura, Keita 1453
Kinfack, J. 1538
King, Bruce H. 188, 254
King, Bruce 353, 409
King, Richard 916
Kinzer, Austin 920
Kipp, Tobias 651
Kirchartz, Thomas..............................199, 1307
Kirmani, Ahmad R 1156
Kirmani, Ahmad 1490
Kita, Takashi 659, 895
Klenk, Reiner 1066
Klenke, Christian 191
Klie, Robert F. 771, 1600

Kline, Michael 492
Klöter, Bernhard 569
Knebel, Kevin J. 769
Knodle, Philip J. 671, 850
Knodle, Philip 1485
Ko, Yohan 1373
Koch, Christoph 879
Kodalle, Tim 969
Koelblin, Pascal 166, 1662
Köhler, Matthias 191
Kojima, Haruki 724
Kojima, Nobuaki 924, 1535
Komoll, Felix 199
Kondzialka, Christoph 1184
Köntges, Marc 733
Kopecek, Radovan 1372
Kopidakis, Nikos 178, 485
Korgel, Brian 1537
Korir, Lilian 1261
Kornienko, Vlad 683
Kornienko, Vladislav 656, 1193
Körtgen, J. 546
Koschier, Linda 749
Koseki, Shuuichi 105
Koskey, Steven 555
Kottantharayil, Anil 1287, 1366
Kottokkaran, Ranjith 641
Kouame, K. 1538
Koumis, Anastasios 541
Krause, Timothy 415
Krebs, Hannes 1630
Kretly, L. C. 1492
Krich, Jacob J 941
Krishna, Anurag 194
Krishnani, Pramod N. 81
Kristensen, Sissel Tind 191
Kroeger, George F. 219
Krückemeier, Lisa 1307
Krustok, Jüri 1298, 1432
Kuba, Austin G. 1545
Kubiniec, Alex 928, 1620
Kuciauskas, Darius 107, 577, 633, 849
Kujovic, Luksa 245, 789, 981, 1497
Kumar, Akash 237, 755
Kumar, Akshay 622
Kumar, Dayanand 327
Kumar, Neetesh 276, 1206, 1437
Kumar, Niranjana Mohan 782, 1634
Kumar, Satyendra 1366
Kumar, Vibhor 1509
Kunkar, Alejandro 708
Kupets, Elaine 594

Kurokawa, Yasuyoshi 292, 775
Kurtz, Sarah R. 783
Kurtz, Sarah 209, 617, 1529
Kusch, G 995
Kwon, Ohjin 191
Lachenal, Damien 1
Lachowicz, Agata 352
Lackner, David 941
Ladd, Anthony J. C. 1241
Lahood, Catherine 22
Lahti, Gabriella D. 1158
Lai, Barry 782, 1183, 1634
Lai, Cheng-Yu 276, 606, 1206, 1437
Lakshmikanth, Balaji Bangolae 126
Lambertz, Andreas 177, 280, 419
Lampa, Nicole 191
Lan, Yucheng 272
Landis, Geoffrey A. 63
Lang, Tom 1002
Lange, Stefan 12, 770
Lao, Yao Y. 11
Lara-Fanego, Vicente 54
Larionova, Yevgeniya 701
Larson, Harry 1551
Lasalvia, V. 1459
Lasalvia, Vincenzo 359
Laufer, Felix 144
Lave, Matthew S. 674, 1170
Law, Adam M. 174, 498, 789, 981
Law, John M. 174
Laws, Nicholas D. 228
Le, Anh Huy Tuan 5
Leccisi, Enrica 146
Lee, Chungho 849, 998, 1008
Lee, Hyunjong 919
Lee, Hyunju 442, 724
Lee, Jehyun 436
Lee, Kyumin 391
Lee, Minjoo L. 670, 997
Lee, Ross 1510
Lee, Sang Hee 97
Lee, Sanghyun 1102
Lee, Songhee 1549
Leever, Benjamin 1466
Leijtens, Tomas 919
Leloux, J. 737
Lemay, AC 1233
Lemire, Amanda 1152
Lemos, Francisco V. E. 1159
Lenert, Andrej 482
Leonardi, Marco 1005
Lerat, Jean-Francois 133

Lestrade, Michel 1645
Leuty, Zachary B. 1533
Lewis, Mandy R. 47
Li, Baojie 1519
Li, Bo 794
Li, Bor 945
Li, Brian D. 670
Li, Can 822
Li, Chongwen 909, 1019
Li, Deng-Bing 275, 279, 823, 1041, 1197, 1353, 1376
Li, Dengbing 46, 804, 910
Li, Dinica 978
Li, Fang 648, 1120
Li, Gan 374, 935
Li, Lulin 1575
Li, Luxi 1183
Li, Mengjie 671, 758, 1485
Li, Minghui 822
Li, Muzhi 938
Li, Ning 1074
Li, Wayne 555, 897
Li, Xinjun 234
Li, Yongxi 332
Li, You 844, 1197, 1417
Li, Zelin 1141
Li, Zhanming S. 1645
Li, Zhen 822
Li, Zhenguo 195, 1526
Li, Zhiqiang 1645
Liao, Weilin 248
Libal, Joris 1372
Libby, Cara 921
Liggett, Max 671
Lightfoote, Stephen 607
Li-Kao, Zacharie Jehl 1059
Lim, Deokoh 436
Lim, Jihun 482
Limodio, Gianluca 305
Lin, Boris 903
Lin, Fen 1052
Lin, Yida 1356, 1575
Lindahl, Johan 749
Lindig, S. 737
Lindig, Sascha 931
Linke, Jonathan 795
Linss, Volker 795
Lipovšek, Benjamin 1
Liu, Baiqiang 248
Liu, Chengfa 1544
Liu, Jiang 1221
Liu, Jie 1221
Liu, Jiqi 234

Liu, Xiaolei 245, 789, 981, 1497
Liu, Xitao 1503
Liu, Yang 1221
Liu, Yangang 386
Livera, Andreas 1270
Lobato, K. 737
Loeding, Adam W. 918
Lombardero, Ivan 1433
Lombardo, Salvatore A. 1500, 1609
Lomuscio, Alberto 1067
Loo, Roger 434, 1390
López, G. 942
Lopez, Hector 1280
Lopez-Becerra, Alan 1283
Lopez-Cardalda, Guillermo 370
López-González, J. M. 942
Lopez-Lorente, Javier 1359
Louks, A. 315
Louks, Amy E 590
Louks, Amy 1570
Lu, Chengchangfeng 1356
Lu, Dingyuan 1197, 1497
Lu, Jianfeng 298
Lu, Junxiong 195, 1526
Lu, Meijun 248
Luderer, Christoph 770
Lüer, Larry 645
Lumb, Matthew P 1217
Luo, Bin 48
Luo, Yanqi 1183
Luque, A. 947
Luque-Heredia, I. 947
Lustig, Zachary 123
Luther, Joseph M. 132, 1156
Luther, Joseph 415, 1490
Lv, Ruirui 1284
Ma, Depu 935
Ma, Jaliu 914
Ma, Jessica 1612
Ma, Yiwei 138
Macalpine, Sara M. 1409
Macdonald, Daniel 1434
Macías, Javier 746
Mack, Charles 178
Mack, Sebastian 795
Mack, Shawn 886
Madonna, Richard G. 1157
Magginetti, David 134
Mahabaduge, Hasitha 1633
Mahaffey, Mason P 1279
Mahamu, Hambalee 659
Mahesh, Suhas 298

Mahmood, Farrukh Ibne 237, 648, 755, 1120
Mahmoudi, Eslam 1134
Mahmud, Rasel 1074
Mahmud, Zabir 617
Mahoney, John 1156
Maiberg, Matthias 432
Mainali, Madan K. 401, 1550
Mainali, Madan 1410
Makita, Kikuo 105
Makrides, Andreas 1348
Makrides, George 541, 1348
Mallajosyula, Arun Tej 1606
Mallick, Rajni 1497
Mamun, Ashraful 134
Manceau, Matthieu 934
Mandic, Vilko 351
Manganiello, Patrizio 80, 655
Mangum, John S. 71, 994, 1150
Mannino, Gaetano 1304, 1567
Mannodi-Kanakkithodi, Arun 1634
Manoukian, Gregory A. 915, 1241
Mansfield, Lorelle M. 35, 1380
Mantel, Claire 16
Mao, Dan 782, 1634
Mapara, Varun N. 161
Maple, Larry 387
Marasini, Ganesh 1096
Marcos, Jesús 888
Mariam, Tamanna 279, 683, 844, 1376, 1417
Mariotti, Davide 281
Mariotti, Silvia 945
Markevich, Vladimir P. 203
Marques, F. C. 1492
Marques, Francisco C. 290, 1068, 1134, 1159, 1654
Marquis, Audrey 295
Marstell, Roderick J. 1024
Martel, Benoit 352
Martín, Francisco 746
Martín, I. 942
Martin, Ina T. 597
Martín, Nazario 804
Martin, P. 947
Martin, Pablo 1433
Martinez, Daniel 15
Martinez-Szewczyk, Michael W. 333, 400
Martinez-Szewczyk, Michael 896
Martins, Giuliano L. 1475
Martír, Pablo 1603
Masi, Sofia 441
Masson, Gaëtan 749
Masuda, Atsushi 442
Masuda, Taizo 924

Matam, Manjunath 671
Mate, Mayank 123, 363
Matera, Fabio 1500, 1609
Mathiak, Gerhard 119
Matsui, Takuya 281, 775
Matthews, Bryan 978
Matthews, David 767
Mayordomo, Alejandra A. 448
Mayyas, Ahmad 299
McAlister, Tom 1620
McCandless, Brian 1299
McCarthy, Robert F. 145
McCarthy, Robert 1446
McCulloch, Manuela 1184
McDanold, Byron 47
McDonald, Calum 281
McGarvey, Elspeth 1393
McGlynn, Ruairi 281
McGott, Deborah L. 308, 633, 915
McKenna, Killian 1074
McKuin, Brandi 209
McMahon, William E. 291, 709, 848, 994, 1150, 1380
McMeekin, David P. 298
McMillon-Brown, Lyndsey 415, 1437, 1490
McNatt, Jeremiah 415
McRae, Mary E. 1510
Medjoubi, Karim 233, 268
Meeker, Rick 1002
Meidanshahi, Reza Vatan 1021
Meier, Rico 689, 970
Meila, Marina 340
Meinhart, Lisa 1630
Melchiorre, Michele 1151, 1617
Meléndez, Cristian R. 420
Mendez, Andres Felipe Castro 969
Méndez-Curbelo, Pablo 717
Mendis, B G 995
Menegassi, Matheus Melati 1588
Meng, Yuhuan 340, 473
Mercimek, Yavuzhan 655
Merkle, Arno P. 291
Mette, Ansgar 191
Metzger, Wyatt K. 1497
Meyer, Abigail 359
Meyers, Bennet 710, 961
Michael, Sheri F 885
Michel, Jesus Ibarra 1434
Micheli, Leonardo 54, 122, 437, 1105
Mihailetchi, Valentin D. 1372
Mikeska, Kurt R. 248
Mikli, Valdek 1298
Mikofski, Mark 81

Miller, Chandler...793, 991
Miller, Clark 916
Miller, David C. 353, 1593
Miller, David W. 1497
Miller, Emily 1557
Miller, Jason 978
Miller, Michael F. 107
Mil'Shtein, Sam 1009
Min, Kwan Hong 179, 1030
Minuto, Alessandro 871
Mirletz, Brian T..228
Mirletz, Heather M. 554
Mirletz, Heather 1416
Mitra, Suchismita...30
Mitterhofer, Stefan..282
Miyano, Kenjiro..324, 1401
Moffitt, Stephanie L.................42, 282, 952, 1141, 1292
Mogannam, Laura..467
Moghadamzadeh, Somayeh 701
Mohammadi, Mahsa 1266
Mohite, Aditya..938
Mohr, William...145
Mohsin, Muhammad Saeed 844
Molinero, R..947
Molto, Cécile 597, 648, 1120
Monahan, Daniele..1462
Monnin, Ryan 914
Montes-Bojorquez, Jose Raul 1283
Montes-Romero, Jesús..1270, 1348
Mood, Thomas C 886, 1217
Mora-Seró, Iván 441
Mordvinkin, Anton 486
Morel, Don..690
Morlier, Arnaud 86
Morris, Kerrie M..1177
Morris, Kerrie 981
Moscoso-Cabrera, Javier A...420
Moser, D..737
Moser-Mancewicz, Nicholas 896
Moses, Paul..665
Motes, Brandon T 590
Mouri, Tasnim K. 91, 1667
Mousumi, Jannatul Ferdous...629
Moutinho, Helio R. 219
Moutinho, Helio...1629
Mu, Teliang..1503
Mugnier, Daniel 749
Mule, Chirag..359
Müller, Jörg W. 191
Muller, Matthew 122, 356, 437, 710
Müller, Matthias..1266
Müller, Thore..1491

Mulloy, Eva M 577
Muñoz, Daniel...1206
Muñoz-Pinzon, Daniel 1437
Munshi, Amit H.................................123, 363, 1234
Munshi, Amit Harenkumar..1537
Munshi, Amit 1337
Murakami, Takurou N. 253
Murphy, Alan 783
Muska, Katri 1298, 1432
Mussakhanuly, Nursultan 227
Muzzillo, Chris..1008
Myneni, Sushmakanth 1234
Nagarajan, Shreyas...81
Nagel, Henning 795
Nagle, Timothy...1497
Nahar, Aayush 1299
Nain, Preeti 743, 1095
Nakado, Takashi 924
Nakamura, Kyotaro..724, 924
Nakamura, Tetsuya...76, 463
Nakano, Yoshiaki...............................342, 374, 935
Nakarmi, Upama 662
Nakka, Laxmi..1052
Nambo, Apolo 38
Nardone, Marco 566, 577, 849
Nascetti, Augusto 1404
Nasser, Michael 498
Navon, David 1292
Nayfeh, Ammar..................... 321, 327, 1320, 1344, 1381, 1471
Nayfeh, Laith 1381
Nayfeh, Leia 1381
Naylor, M 995
Nazeeruddin, Mohammad K. 804
Nazer, Afshin..80
Nazif, Koosha Nassiri 1347
Ndione, Paul F..1158
Ndione, Paul...1629
Neal, Craig J..671
Nekarda, Jan 1588
Nemeth, B. 1459
Nemeth, William 30, 359, 475, 891, 1216
Neto, Pedro O. C. M. 1159
Neubert, Anja 1359
Neuhaus, Dirk H. 1313
Neuhaus, Dirk Holger 1060
Neumann, Anica N. 218, 291, 848, 994, 1130, 1150
Neupane, Ganga R. 695
Neupane, Sabin 46, 220, 279, 1041, 1197, 1210,
....1353, 1376
Neves, M. R. M. 1492
Neves, Mendelsson R. M. 1068, 1159
Newberry, Milton G. 380

Ng, Annie...39, 502
Ni, Chaoying...248
Nicholson, Anthony P. 509
Niebergall, Larissa...191
Nielsen, Michael P. 271, 976, 1564
Nieves, Michael Y. Vazquez................................367
Nigmetova, Gaukhar..39
Nihar, Arafath...758
Nikam, Maitheli..1253
Ninad, Nayeem...155, 1648
Nishihara, Tappei.....................................442, 724
Nishinaga, Jiro..939
Nishioka, Kensuke..924
Nitta, Frederick U. 1347
Noack, Philipp...701
Nocerino, John C...198
Nolde, Jill A..886
Norman, A. 315
Norton, Matthew..194
Nowak, David...1059
Núñez, R. 746
Nuns, Thierry..934
Nuys, Maurice..177
Nyholm, Andrew W. 202
Obrecht, John M. 1578
O'Brien, Colleen......................................794, 1141
Ochoa, Jorge...768, 896, 1533
Ogura, Atsushi..442, 724
Ogut, Mehmet G. 961
Oh, Jaewon...794
Oh, Myeongchan..436
Ohlmann, Jens...434, 1341
Ohshima, Takeshi.......................................76, 1429
Ohshita, Yoshio..............442, 724, 924, 1453, 1535
Ok, Young-Woo 91, 179, 1030, 1657
Okada, Yoshitaka...463
O'Kearney, Felix...1224
Oklobia, Ochai........................245, 308, 1497
Okoli, Fitzgerald C. 918
Okullo, James..1393
Oliver, R A 995
Oltjen, William C...................671, 758, 1485
Oltjen, William..234
Olzhabay, Yerassyl...502
O'Neill, Mark..1446
O'Neill-Carrillo, Efraín....................57, 1170
Onno, Arthur...1279
Ooi, Tzy Wei...855
Opatovsky, Martin..1594
Orejuela, V. 947
Orejuela, Víctor...1603
Oreski, G..737

Oreski, Gernot...931, 1630
O'Rourke, Michelena..330
Ortis, Alessandro..1567
Ortiz, Eduardo I...420
Ortiz, Eduarto I. 416
Ortiz-Rivera, Eduardo I. 525
Ortiz-Rivera, Eduardo.................................370, 1338
Oshima, Ryuji..105
Oshima, Takeshi..463
Osowski, Mark..145
Ossig, Christina......................................435, 651
Ota, Yasuyuki..924
Ottoson, Larry...178
Ovaitt, Silvana..................47, 554, 794, 920, 1416
Owen-Bellini, Michael............952, 1044, 1218, 1260, 1629
Oyewo, Ayobami S...708
Ozaki, Ryo...924
Ozaktas, Ekin Gunes..1334
Ozbeytemur, Josh...240
Ozbolt, Alex...1575
Pacheco, Willian...717
Packard, Corinne E....................................10, 71, 291
Padhamnath, Pradeep...................................1030, 1657
Padmakumar, Govind...305
Padmanaban, Dilli Babu.....................................281
Paesa, Marta Casasola......................................48
Paetel, Stefan.......................................107, 432, 1201
Paetzold, Ulrich Wilhelm..............................144, 701
Page, M. R. 1459
Page, Matthew...30, 359, 475
Pal, Shweta..622
Palacios II, Felipe..1626
Palacios, Felipe...1034
Palekis, Vasilios.....................................690, 1237
Palmer, Jack R. 644
Palmiotti, Elizabeth.......................................353
Palmstrom, A. 315
Palmstrom, Axel F.....................................590, 1158
Palmstrom, Axel.......................................938, 1570
Pamperin, Megan..1393
Pan, Noren...145
Pan, Yida..1434
Panchalogaranjan, Vinushika................................665
Pandey, Ramesh...123
Panzic, Ivana..351
Papaeconomou, Vassilis.....................................1270
Parada, Gabor..981
Paraskeva, Vasiliki..194
Pareek, Devendrá...1059
Paris, Claudio...1404
Park, Chinho...749
Park, Nam-Gyu..1484

Park, Sungeun ... 97
Parke, Tyler ... 1667
Parra, Johan ... 268
Paschen, Jan ... 1588
Passarella, Bianca ... 1005
Patel, Aesha P. ... 402, 823
Patel, Jayeshkumar ... 476
Patel, M. Tahir ... 626
Paul, G. ... 315
Paul, Goutam ... 1570
Paul, Sritoma ... 482
Paupy, Nicolas ... 133, 434
Paviet-Salomon, Bertrand ... 1, 1129
Peaker, Anthony R. ... 203
Pearce, Phoebe M. ... 345, 976
Pearsall, N. ... 737
Pearson, Patrick ... 887
Pechmann, Sabrina ... 194
Peibst, Robby ... 6, 701, 733
Peña, Carlos ... 717
Peng, Fuguo ... 195, 1526
Peng, Hugh ... 727
Penukula, Saivineeth ... 1534
Penukula, Vineeth ... 1109
Perez, Marc ... 285, 662
Perez, Richard ... 662, 928, 1620
Perez-Rodriguez, Paula ... 305
Perini, Carlo A. R. ... 1183
Perini, Carlo Andrea Riccardo ... 969
Perkins, Craig L ... 577, 633
Perna, Allison N. ... 291
Perna, Allison. ... 71
Pernès, Nicolas ... 1129
Perret, Lionel ... 749
Perry, Kirsten ... 710, 1145
Perry, Lakesha N. ... 282
Perullo, Christopher ... 555
Peshek, Timothy J. ... 698
Peshek, Timothy ... 1490
Peter, Christoph ... 1372
Peters, Benjamin ... 634
Peters, Ian Marius ... 540, 645, 1092
Peters, Stefan ... 191
Peterson, Josh ... 600, 1123
Petesic, James ... 1295
Petter, Kai ... 191
Phang, Sieu Pheng ... 1434
Phillips, Adam B. ... 46, 220, 275, 279, 401, 402,
... 683, 823, 1041, 1193, 1210, 1323, 1410, 1417
Phillips, Adam ... 804, 910
Pierce, Benjamin G. ... 512
Pierce, Benjamin ... 101

Pieters, B. E. ... 546
Pieters, Bart E. ... 166
Pikolos, Loucas ... 1348
Pilot, Nicholas ... 14
Pilvet, Maris ... 1298, 1432
Pina, Marissa ... 1667
Pineda, F. Brigham ... 783
Platzer-Björkman, Charlotte ... 887
Ploigt, Hans-Christoph ... 191
Podraza, Nicholas J. ... 1410
Podraza, Nikolas J. ... 401, 1550, 1557
Pogorelov, Kostiantyn ... 1491
Pohl-Hampel, Britta ... 191
Pokhrel, Dipendra ... 683
Polly, Stephen J. ... 63, 721, 894
Polly, Steve J ... 89
Polzin, Jana-Isabelle ... 795
Poortmans, Jef ... 48, 162
Pop, Eric ... 1347
Poplawsky, Jonathan D. ... 46
Porret, Clément ... 434, 1390
Posada, Jorge ... 233, 268
Pothoof, Justin ... 1020
Poulsen, Peter B. ... 448, 1452
Powell, Kaden ... 998
Prabakar, Kumaraguru ... 1164
Pradeep, Nisitaa Karen Clement ... 467
Prasanna, Rohit ... 919
Prathap, Nemalipuri Surya ... 182
Pravettoni, Mauro ... 1188
Prell, Henrik ... 939
Prettl, Michael ... 289
Price, Kent ... 1102
Prot, Aubin JC. M. ... 1067
Provost, Marion ... 268
Prym, Guilherme C. S. ... 1159, 1654
Ptak, Aaron J. ... 10, 71, 291
Puel, Jean-Baptiste ... 233, 268
Pugstaller, Robert ... 1428, 1467, 1481
Pulwin, Ziggy ... 1217
Purkayastha, Atanu ... 1606
Pusch, Andreas ... 976
Qazi, Suleman Sami ... 772
Qian, Chen ... 1256
Qian, Yang ... 134
Qin, Yuan ... 1221
Qiu, Botong ... 1575
Qiu, Feng ... 634
Qu, Minghao ... 195, 1526
Qu, Xiaoyong ... 1372
Quader, Abdul ... 1410
Queck, Martina ... 191

Quinones, Dhamelyz R. S. 1667
Quispe, David 767
Radhakrishnan, Hariharsudan 48
Radu, Daniela R. 276
Radu, Daniela 606, 1206, 1437
Raghoebarsing, A. 737
Ragonesi, Antonino 1005
Rahimi, Amirhossein 844
Rahman, Areefa 482
Rahman, Naveed 1183
Raikar, Subbarao 333, 400
Rajakaruna, Manoj 402, 844, 1417
Raju, Sukhwant 855
Raker, David 402
Ramasubramanian, Deepak 1096
Rametta, Francesco 1005
Ramirez-Iniguez, Roberto 591
Ramos, Wendy Reyes 682
Rampalli, Chaitanya Santosh 1513
Rana, Prem J. S. 1044
Ranalli, Joseph 263, 1242
Rand, BP 1233
Ransome, Steve 1348
Rapp, Jeremy 1259
Rashed, Faisal 356, 1454
Rashkin, Lee 1034
Rasmussen, Mirra 597
Rathgeber, Andreas 385
Rau, U. 546
Rau, Uwe 177, 199, 419, 1307
Raugewitz, Annika 6, 733
Ravello, Magdalena 969
Ravishankar, Sandheep 1307
Reagan, Jeremiah 209
Reddy, K. S. 182
Reddy, Vasudeva Reddy Minnam 1549
Reddy, Yellasiri Bharath Kumar 259, 1227
Reece, Peter J. 976
Reese, Matt 245
Reese, Matthew O. 35, 633, 695, 915, 946, 1008
Reese, Samantha 1416
Reeves, Adam 240
Reich, Carey 849, 1234, 1279
Reich, Gerly 1024
Reichel, Christian 1313
Reinders, A. H. M. E. 737
Reinders, Angèle H. M. E. 809
Reinders, Angele 443
Rendler, Li C. 1313
Rengifo, Fabio Andrade 519
Reno, Matthew J. 13, 334
Repins, Ingrid L. 702, 1260, 1593

Reyes-Colón, Ramón 57
Rey-Stolle, I. 947
Rey-Stolle, Ignacio 1433, 1603
Rezk, Ayman 321, 1320, 1344, 1471
Rhee, Kurt 581
Ribeiro, Andrei C. 1159, 1654
Ribo, Macarena Mendez 554, 1416
Richards, Bryce S. 172, 996
Riedel, Maximillian 1066
Riedl, Gabriel 1428, 1467
Rienäcker, Michael 701
Rigby, O M 995
Rijal, Suman 1197, 1376
Rikhof, Anne 622
Riley, Daniel 101, 409, 1131
Ringel, Steven A. 917
Rippingale, Jan 922
Ritzer, David Benedikt 144
Rivera, Agustín Irizarry 519
Rivera, Eduardo I. Ortiz 630
Rivera-Matos, Yiamar 916
Rizk, Ayman 327
Rizzo, Aurora 1109
Ro, Jason 925
Roberts, Dennice M. 1629
Robles-Rivera, Emmanuel G. 1034
Rock, Nathan 1337
Rockett, Angus A. 489
Rodgers, Marianne 1612
Rodriguez-Cabanas, Lissette 508
Rohatgi, Ajeet 91, 179, 1030, 1155, 1657
Rojas-Gatjens, Esteban 1020
Rojsatien, Srisuda 782, 1634
Rolston, Nicholas 938, 1109, 1534, 1565
Rome, Grace A. 75
Romer, Pascal 1060
Römer, Udo 917
Rong, Eric 272, 1356
Rosales, Bryan 740
Rosenthal, Samuel 1575
Roufberg, Lew 311
Rounsaville, Brian 91, 179
Rousset, Jean 268, 1440
Rout, Bibhudutta 132
Routhier, Alex 916
Rowell, David 145
Roy, Etee Kawna 998
Roy-Layinde, Bosun 482
Ru, Xiaoning 195, 1526
Rudolph, Dominik 1372
Ruhle, Ryan 387
Runkana, Venkataramana 1053

Ruske, Florian 945
Rusnak, Jozef 1594
Russell, Annie C. J. 331
Saavedra-Peña, Nelson E. 1082, 1326
Sacchitella, Elijah 894
Saeed, Muhammad Mohsin 1210
Saenz, Theresa E. 994, 1130, 1150
Sahani, Rishabh 276, 1206, 1437
Sai, Hitoshi 775
Saitta, Federica 305
Saive, Rebecca 348, 622, 996
Sajja, Sunil 855
Saliba, Michael 289
Salles, Caroline Lima 1216
Sampath, Walajabad S. 509, 1537, 1600
Sampath, Walajabad 123, 387, 1279
San José, Luis J. 746
Sanchez-Perez, C. 947
Sanchez-Perez, Clara 1433
Sanci, Sal 978
Sankin, Igor 914
Santala, Annikki L. 919
Santamaría, Rodrigo Del Prado 448, 1452, 1662
Santana, Vinícius Camatta 1541
Santbergen, Rudi 305, 1005, 1398
Santhanam, Lakshmi 126
Santiago, Brian L. Reyes 1338
Santistevan, Kevin 409
Santiwipharat, Chaiwarut 1545
Santos, José 804
Saraswat, Govind 360
Saraswat, Krishna C. 1347
Sargent, Edward H. 1019
Sargent, Ted 518
Sarkisov, Sergey 110
Sartor, Benjamin E. 1008
Sato, Shin-Ichiro 76, 463
Satymov, Rasul 708
Saucedo, Edgardo 1059
Saucedo, Joel 1633
Saw, Min Hsian 1188
Sazzad, Muhammad H. 976
Scarpulla, Michael A. 731
Scarpulla, Michael 1337
Schall, J. W. 315
Schall, Jackson W. 1570
Schall, Jackson 1044
Schaper, Martin 191
Scharf, Jessica 191
Scheer, Roland 12
Scheibner, Michael 415
Scheideler, William J. 1276

Scheidt, Rebecca A. 161, 1156
Scheiman, David 1045
Schelhas, Laura T. 298, 353, 952, 1044, 1260, 1629
Schirone, Luigi 1404
Schlatmann, Rutger 1066
Schlenoff, Tali 1292
Schley, Michael 191
Schmieder, Kenneth J. 886, 1217
Schmitz, J. 737
Schneble, Olivia D. 1150
Schneiderloechner, Eric 795
Schnierer, Branislav 1594
Schönmann, Antje 191
Schramm, Barbara 435
Schreiber, Waldemar 434, 1341
Schropp, Andreas 651
Schüler, Marc Andre 1060
Schulte, Kevin L. 10, 291, 848
Schulte-Huxel, Henning 652
Schulz, Susanne 191
Schurman, Matthew J. 311
Schütze, Matthias 191
Schwab, Andrew J. 977
Schwabedissen, Axel 191
Schwartz, Dakota 1551
Schwung, Julian 64
Schygulla, Patrick 1341
Sciuto, Marcello 1005
Scuto, Andrea 1609
Seal, Sudipta 671
Sehirlioglu, Alp 758
Seibert, Samuel 1353
Seiboth, Frank 651
Seigneur, Hubert P. 671
Seigneur, Hubert 850, 1120, 1485
Sekulic, William 794
Sellers, Ian R. 132, 161, 1156
Selvidge, Jennifer 341
Senaud, Laurie-Lou 1
Sengupta, Manajit 152, 386, 474, 587, 600, 778, 1106
Seo, Seongrok 298
Sepúlveda-Mora, Sergio B. 464
Sepúlveda-Vélez, Fredy A. 1105
Serafini, Patricio 441
Sermarini, Anna Carolina De Paula 256, 858
Serrano, Guillermo 525
Setiawan, Ignatius Andre 1549
Sevillano-Bendezú, Miguel Á. 1066
Seymour, Kyle 928, 1147
Seyrich, Martin 651
Shafarman, William N. 91, 1545
Shafarman, William 1299

Shah, Akash 123
Shan, Ambalanath 1410
Shapiro, Finley R 915
Sharikadze, Saba 641
Shaton, Avishai 861, 1088
Shaw, Daniel 1234
Shen, Heping 1434
Sheppard, Scott 555
Sheyfer, Dina 1183
Shi, Jiahui 1503
Shi, Yangwei 1020
Shimabukuro, Laura 1575
Shimasaki, Takashi 342
Shimpi, Tushar 981
Shiradkar, Narendra 1287, 1366
Shirai, Yasuhiro 324, 1401
Shirazi, Eli 443, 816
Shojaei, D. 942
Shoji, Yasushi 105
Shore, Andrew M. 1292
Shrestha, Bishal 401, 1410
Shrestha, Santosh 881, 1564
Sidhik, Siraj 938
Siebentritt, Susanne 1067, 1151, 1617
Siegneur, Hubert 648
Siepchen, Bastian 656
Silhavy, Jake T. 918
Silva-Acuña, Carlos 1020
Silveira, A. M. C. 1492
Silveira, Allan 1068
Silverman, Timothy J. 106, 921, 1044, 1218, 1260, 1593
Simon, John 10
Sims, Jeremiah D 829
Singh, Luna 718
Singh, Manish K. 1384
Singh, Pritpal 1510
Singh, Rhythm 830, 1230
Sinha, Arpan 1155
Sinha, Parikhit 146, 855
Sinton, Ron 113
Sinton, Ronald A. 798
Sirkisoon, Sarah 22
Sites, James R. 509
Sites, James 1310, 1582, 1600
Sitiriche, Marcel Castro 519
Skoczek, Artur 1594
Slauch, Ian M. 970
Slauch, Ian 689
Slonopas, Andre 508
Smaine, Issam 426
Smets, Arno H. M. 305
Smith, David D. 1024

Smith, Emily 1045
Smith, Ryan 1120
Smith, Soshana 282
Snaith, Henry J. 298
Snell, Jeffrey 1009
Snuggs, Robert 823
Snyder, William 409
Sodabanlu, Hassanet 342, 374, 935
Soeriyadi, Anastasia H. 917
Song, Chang-Yun 432
Song, Hee-Eun 97
Song, Tao 178, 394, 485
Song, Zhaoning ..... 42, 275, 683, 804, 812, 844, 909, ... 1197, 1210, 1323, 1353, 1376, 1417, 1550, 1557
Sood, Mohit 1151
Sourabh, Shashi 132, 161
Sovetkin, E. 546
Sovetkin, Evgenii 166
Spaeth, Bettina 656
Spataru, Sergiu V. 448, 1452, 1662
Springer, Martin 794
Spurgeon, Ben 861, 1088
Sridhar, Seetharaman 554
Stall, Richard 311
Stanley, Bradley 508
St-Arnaud, Louis-Philippe 941
Stein, Joshua S. 188
Stein, Joshua 101
Steinebrunner, Udo 1060
Steiner, Myles A. ..... 10, 75, 218, 291, 341, 394, 848, ... 994, 1081, 1130
Stenzel, Florian 191
Stern, Jillian 467
Stevens, Margaret A 886, 1217
Stevens, Tristan 655
Steyn, Dirk 891
Stid, Jacob T. 743, 1259
Stoffel, Tom 1106
Stoicescu, Liviu 166, 1662
Stradins, P. 1459
Stradins, Paul 30, 475, 891
Stradins, Pauls 1216
Strandberg, Rune 815
Strandins, Pauls 359
Straub-Mueck, Michael 385
Strelow, Christian 651
Stroyuk, Oleksandr 540
Stuckelberger, Michael E. 435, 651
Sturm, Chris 12
Subedi, Kamala Khanal 46, 1019, 1353
Subramanian, Sivakumar 1053
Suemasu, Takashi 1398

Sugaya, Takeyoshi	105, 939
Sugimoto, Hiroki	76
Sugiyama, Masakazu	342, 374, 463, 935
Sulas-Kern, D. B.	315
Sun, Kaiwen	1256
Sun, Yijia	1485
Sun, Yukun	997
Sung, Li-Piin	282
Sunkari, Preetham P.	340
Sunkari, Preetham	473
Sunter, Deborah A.	467
Suppiah, Sam	476
Sutterlueti, Juergen	1348
Svrcek, Vladimir	281
Sweat, Rebekah	330
Syed, Faizan	606
Sytnyk, Mykhailo	1092
Szablewski, M	995
Szábo, Sandor	550
Szyszka, Bernd	945
Taconelli, Mauricio	1654
Taddei, Margherita	1020
Tafur, Lucila D.	927, 1146
Takahashi, Tadatoshi	611
Takamoto, Tatsuya	924
Talavera, Diego L.	1105
Tamizhmani, Govindasamy	237, 648, 755, 794, 1120
Tamuno-Ibuomi, Lewis Osikibo	591
Tan, Jin	840
Tan, Kelvin	85
Tanaka, Taichi	1453
Tanimoto, Tsutomu	924
Tank, Mehul	330
Tao, Meng	85, 1566
Tatavarti, Rao	1429
Taubmann, Rouven	1184
Tawsif, Sheikh Elahi	690
Tayagaki, Takeshi	253
Taylor, Andre D	1008
Taylor, P. Craig	359
Teasley, Corson	555
Teodor, Alexandra H.	11
Teplyakov, Andrew V.	1667
Ternes, Simon	144
Terry, Mason	783
Tervo, Eric J.	341
Terwilliger, Kent	702, 1629
Theelen, M.	737
Theingi, S.	1459
Theingi, San	475
Theocharides, Spyros	541
Theristis, Marios	122, 188, 1250

Thiagarajan, Ramanathan	1164
Thibodeau, Matthew R.	918
Thiel, Christian	924
Thiengi, San	30
Thind, Arashdeep S.	771
Thomas, Adam	682
Thomas, Sinju	432, 939
Thomsen, Vitus B.	16
Thon, Susanna M.	272, 718, 1334, 1356, 1575
Thon, Susanna	695
Tiefenthaler, Martin	1481
Tilli, Francesca	749
Timmo, Kristi	1298, 1432
Timò, Gianluca	871
Timofte, Tudor	1372
Tina, Giuseppe Marco	1304, 1567
Titus, Jochen	919
Tobail, Osama	191
Tobon, Carlos Mario Ruiz	1398
Todaro, Lorenzo	1304
Töfflinger, Jan A.	1066
Togay, Mustafa	245, 789, 981, 1177, 1193, 1497
Toh, Wei Wen	1188
Tokumasu, Takashi	292
Tomita, Yosuke	924
Tonita, Erin M.	331
Topic, Marko	1
Törndahl, Tobias	1151
Tovar, Michael	939
Tracy, Jared	952
Transue, Taos	922, 1250
Tremont-Brito, Rolando J.	1170
Trempa, Matthias	1266
Tresan, Jenner	492
Trimby, Pat	939
Troupe, Anthony T	590
Trupke, Thorsten	227, 472, 828, 1224, 1468
Tsakalids, Anastasios	924
Tse, Yau Yau	656
Tumusange, Marie Solange	1550
Tuomiranta, Arttu	162, 426
Turala, A.	1538
Turcotte, Dave	155
Tutsch, Leonard	1625
Ubukata, Akinori	105
Uddin, Md Aslam	1044
Uene, Naoya	292
Ukaegbu, Ikechi	502
Ulbrich, Carolin	1066
Ulicná, Sona	353, 952, 1044, 1629
Upadhyay, Prashant Kumar	259, 1227
Upadhyaya, Ajay D	1030

Upadhyaya, Ajay 91
Upadhyaya, Vijaykumar D 179, 1030
Upadhyaya, Vijaykumar 91, 1657
Ures, Sandra 888
Urs, Rahul R 299
Usami, Noritaka 292, 724, 775
Vaas, T. S. 546
Valdivia, Christopher E. 331, 476, 941
Valerino, Michael 437
Vallerotto, G. 746
Van De Voorde, Mathis 348, 996
Van Dyck, Rik 48
Van Nijen, David A. 655
Van Sark, W. J. G. H. M. 737
Van Swaaij, René A. C. M. M. 655
Van Velson, Nathan 1114
Van Vuure, Aart Willem 48
Vanderhaegen, Aline 1151
Vandervelde, Thomas E. 1152
Vansant, Kaitlyn 415, 1490
Vargas, Fernando J. 416
Vasconcelos, Cláudia K. B. 94
Vasi, Juzer 1366
Vazquez, Michael Y. 416
Venkat, Sameera Nalin 234
Venkatramanan, D. 1384
Verezhak, Mariana 651
Verkou, Maarten 1005
Verlinden, Pierre 1544
Vignola, Frank 600
Villa-Bracamonte, Maria Fernanda 1283
Villa-Ignacio, Armando 215
Villalva, M. G. 1492
Villalva, Marcelo G 1068, 1159, 1654
Villalva, Marcelo Gradella 1134
Villalva, Marcelo 290
Voarino, Philippe 1423
Vogt, Malte 1005
Volatier, M. 1538
Von Gastrow, Guillaume 1024
Von Wenckstern, Holger 12
Voss, Stephen 607
Vuk, Dragana 351
Wadsworth, Matthew 330
Wagner, Kristen 1620
Wagner, Philipp 945
Wagner-Mohnsen, Hannes 569
Wahl, Tina 1201
Wakamiya, Atsushi 442
Walajabad, Sampath S. 981
Walajabad, Sampath 1234
Walker, Alexandre W 941

Walker, Don 198
Walkons, Curtis 566
Wallner, Gernot M 1467, 1481
Wallner, Gernot 1428
Walls, J. Michael 981, 1193
Walls, John M 245, 498, 789, 1177, 1497
Walls, Michael 656, 683
Wang, Jian 1020
Wang, Jianjian 1114
Wang, Jianming 248
Wang, Jianqiang 1526
Wang, Jing 1074
Wang, Le 1544
Wang, Liwei 240
Wang, Quanzhi 1523
Wang, Tonghui 332
Wang, Wei 690, 1237
Wang, Wenzong 835, 1096
Wang, Yichun 195, 1526
Wang, Yonglei 1221
Wang, Zhaoyu 634
Wanlass, M. W. 1261
Wargulski, Dan R. 879, 939
Warren, Emily L 75, 218, 291, 709, 848, 994, 1130,
.... 1150, 1158, 1380
Wasmer, Sven 569
Watanabe, Kentaroh 342, 374, 935
Watson, Stephanie S. 282
Wattenberg, Bianca 701
Weber, August 1266
Weber, Julian 1060
Weed, Emily 505
Weigand, William J. 1533
Weigand, William 1379
Weihrauch, Anika 191
Weinhardt, Lothar 91
Welch, Liam M 395
Welch, Liam 534
Welser, Roger E. 145, 721
Welser, Roger 1429
Wenner, Scott L. 220
Wenner, Scott 910
Westerhof, Jelle 622
Westraadt, Johan 1625
Whalen, Devin C. 380
Wheeler, Aaron 783
Wheeler, Lance M. 1044
Whiteside, Vincent R 132, 161, 1156
Wibowo, Andree 145, 1429
Wickett, Shelbie 1635
Widrick, Devin 528
Wieghold, Sarah 1183

Wieliczka, Brian 1490
Wieser, Raymond J. 1141
Wieser, Raymond 861, 1088
Wietler, Tobias 6, 652, 733
Wikoff, Hope 1416
Wilcox, Stephen 1106
Williams, Henry J. 505, 727
Williams, Jennifer 415
Williams, Rafell 178
Wilson, Paige 476, 941
Wilson, Samantha S. 607
Wilson, Thomas 90
Wilt, David 1429
Wilterdink, Harrison 113
Winkelmann, Aimo 939
Winkler, Louisa 1060
Winter, Björn Oliver 64
Wirtz, L. 995
Witte, Wolfram 432
Witteck, Robert 709
Wittmann, Ernst 645
Wong, Johnson 978
Woodall, Mark 415
Woodhouse, Michael 1593
Wright, Brendan 1224, 1295
Wu, Xiang 1372
Wu, Yinghui 861, 1088
Wu, Yuh-Renn 1443
Wu, Zhenni 1092
Wyss, Patrick 1129
Xiang, Xiaofeng 578
Xiao, Chuanxiao 46, 822, 1019
Xiao, Yegao 1645
Xie, Tian 195, 1526
Xie, Yu 152, 386, 474, 587
Xiong, Gang 998, 1197, 1497
Xu, Binbin 419
Xu, Jianmei 1544
Xu, Tao 1284, 1523
Xu, Xixiang 195, 1221, 1526
Xu, Yawen 248
Xue, Chaowei 1526
Yagi, Shuhei 926
Yaguchi, Hiroyuki 926
Yaiche, Armelle 1440
Yamaguchi, Masafumi 924, 1535
Yamamoto, Kohei 253
Yan, Di 1434
Yan, Feng 1197
Yan, Yanfa 42, 46, 220, 275, 279, 683, 804, 812,
823, 844, 909, 910, 1019, 1041, 1197, 1210, 1323, 1353, 1376,
1417, 1550, 1557

Yanagida, Masatoshi 324, 1401
Yang, Guangtao 655
Yang, Jaemo 152, 386
Yang, Jie 1434
Yang, Miao 195, 1526
Yang, Ruiquan 975
Yang, Zhaoqing 1020
Yao, Dominique Akissi 758
Yao, Keyi Kang 1575
Ye, Jichun 822
Yelzhanova, Zhuldyz 39
Yermekov, Nurzhan 39
Yi, Chuqi 917
Yildirim, Murat 634
Yilmaz, P. 737
Yin, Shi 195, 1526
Yoon, Heayoung 134, 998
Yoon, Woojun 1045
Yoon, Yohan 134
Yoshita, Masahiro 253
Young, D. L. 1459
Young, David L 106
Young, David 30, 475, 891, 1216
Young, Ethan 921
Young, Matthew R. 1130
Young, Michelle 848
Youtsey, Chris 145, 1446
Yu, Li 1164
Yu, Xuanji 234, 597, 671, 758, 861, 1088, 1141,
.... 1284, 1485
Yu, Yuanjie 794, 1284, 1523
Yu, Zhengshan J. 767
Yu, Zhibin 330
Yuan, Luyao 730
Yun, Changyeol 436
Yun, Jae Ho 1556
Yusuf, Jubair 13
Zabalza, Ruben 1141
Zaka, Awais 1471
Zawisza, Zachary W. 823
Zawisza, Zachary 910
Zech, Matthias 1242
Zehender, M. 947
Zelenina, Anastasia 1067
Zeman, Miro 655, 1005
Zeng, Yiyu 302
Zhang, Changgen 248
Zhang, Fan 1261
Zhang, Guangchun 1284
Zhang, Hongxu 1221
Zhang, K. Max 505, 727
Zhang, Kangping 248

Zhang, Shu 1544
Zhang, Wei 1497
Zhang, Xinyu 1434
Zhang, Yijun 718
Zhang, Yong 1261
Zhang, Zheyu 240
Zhao, Dewei 822
Zhao, Shijia 634
Zhao, Yong 248
Zhao, Zitong 1021
Zhaoning, Song 1019
Zheng, Jian-Yao 622
Zheng, Peiting 1434
Zhong, Ruohan 1030, 1657
Zhu, Kai 812, 1158, 1265, 1484
Zhu, Xitong 809
Zhu, Yan 227
Zilouchian, Ali 1280
Zimányi, Gergely T. 51, 1021
Zimmerman, Jeramy D. 994, 1130, 1150
Zimmermann, Gregor 191
Zimmermann, Iwan 804
Zin, Ngwe 1509
Zinßer, Mario 199
Zoppi, G 995
Zubieta, Diego 597

2023 IEEE 50th Photovoltaic Specialists Conference (PVSC 2023)

San Juan, Puerto Rico, USA
11-16 June 2023

Pages 555-1113

IEEE Catalog Number: CFP23PSC-POD
ISBN: 978-1-6654-6060-6

**Copyright © 2023 by the Institute of Electrical and Electronics Engineers, Inc.
All Rights Reserved**

Copyright and Reprint Permissions: Abstracting is permitted with credit to the source. Libraries are permitted to photocopy beyond the limit of U.S. copyright law for private use of patrons those articles in this volume that carry a code at the bottom of the first page, provided the per-copy fee indicated in the code is paid through Copyright Clearance Center, 222 Rosewood Drive, Danvers, MA 01923.

For other copying, reprint or republication permission, write to IEEE Copyrights Manager, IEEE Service Center, 445 Hoes Lane, Piscataway, NJ 08854. All rights reserved.

*** This is a print representation of what appears in the IEEE Digital Library. Some format issues inherent in the e-media version may also appear in this print version.

IEEE Catalog Number:	CFP23PSC-POD
ISBN (Print-On-Demand):	978-1-6654-6060-6
ISBN (Online):	978-1-6654-6059-0

Additional Copies of This Publication Are Available From:

Curran Associates, Inc
57 Morehouse Lane
Red Hook, NY 12571 USA
Phone: (845) 758-0400
Fax: (845) 758-2633
E-mail: curran@proceedings.com
Web: www.proceedings.com

TABLE OF CONTENTS

3-Terminal Perovskite/Silicon Tandem Modules: A Dead End or a Bright Future of Tandem Based Photovoltaics 1

Miha Kikelj, Laurie-Lou Senaud, Jonas Geissbühler, Damien Lachenal, Derk Baetzner, Benjamin Lipovšek, Marko Topic, Christophe Ballif, Quentin Jeangros, Bertrand Paviet-Salomon

Temperature-Dependent Performance of Ultra-Thin Silicon Heterojunction Solar Cells for Space Applications 5

Anh Huy Tuan Le, Pradeep Balaji, André Augusto, Ziv Hameiri

Utilizing a Soft IZO Sputtering Process to Contact Buffer-Free Semitransparent Perovskite Pin Solar Cells 6

Roland Clausing, Annika Raugewitz, Benjamin Grimm, Marvin Diederich, Tobias Wietler, Felix Haase, Rolf Brendel, Robby Peibst

Effect of Surface Morphology on GaAs Solar Cells Grown on Planarized Spalled (100) GaAs Substrates 10

Anna K. Braun, Jacob T. Boyer, Kevin L. Schulte, John Simon, Steve W. Johnston, Myles A. Steiner, Corinne E. Packard, Aaron J. Ptak

Ultra-Light Environmental Protection for Solar Arrays in Space 11

Pilar Espinet-Gonzalez, Alexandra H. Teodor, Jann A. Grovogui, Yao Y. Lao

Highly Crystalline In2S3:V Thin Films Epitaxially Grown on Sapphire Substrates: A Potential Canditate for Intermediate Band Solar Cells 12

Tanja Jawinski, Chris Sturm, Roland Clausing, Heiko Kempa, Stefan Lange, Roland Scheer, Marius Grundmann, Holger Von Wenckstern

A Model-Free Approach for Estimating Service Transformer Capacity Using Residential Smart Meter Data 13

Joseph A. Azzolini, Matthew J. Reno, Jubair Yusuf

Bifacial PV Fed Electrolysis for Green Hydrogen Generation and Cofiring Hydrogen in an Aeroderivative Gas Turbine 14

Nicholas Pilot, Jim Harper

Reverse-Bias Testing of Perovskite Cells to Inform Bypass-Diode Design 15

Daniel Martinez, Dana Kern, Giles Eperon

Improving Deep Learning-Based Defect Classification in Solar Cells Using Conformal Prediction 16

Vitus B. Thomsen, Claire Mantel, Gisele A. Dos Reis Benatto, Allan P. Engsig-Karup, Søren Forchhammer

Off-The-Shelf Small Scale Photovoltaic Systems for Puerto Rico Sustainable Farms: Assisting Those Who Help Others 22

Catherine Lahood, Sarah Sirkisoon, Charles M. Fortmann

Flexible Photonic Cooler Based on Multi-Stacked Thin Films IR Filters with Anti-Dust Capability for PV-Desert Environment Applications 27

Brahim Aïssa, Mohammad I. Hossain

Loss Analysis and Performance Optimization Pathways of 729-MV V_{OC} Si Solar Cells with Poly-Si on Locally-Etched Dielectric Passivating Contacts 30

Suchismita Mitra, Caroline Lima Anderson, Matthew Hartenstein, William Nemeth, Matthew Page, San Thiengi, David Young, Sumit Agarwal, Paul Stradins

Cadmium Telluride Accelerator Consortium (CTAC) 35

Lorelle M. Mansfield, Matthew O. Reese, Michael J. Heben

Screen Printable Copper Pastes for Silicon Solar Cells 38

Thad Druffel, Ruvini Dharmadasa, Apolo Nambo, Abafriseke Ebong, Sandra Huneycutt, Donald Ital

Optimization of Zinc Oxide Electron Transport Layers for Cs-Based Perovskite Solar Cells 39

Zhuldyz Yelzhanova, Gaukhar Nigmetova, Gulzhan Bizhanova, Nurzhan Yermekov, Ulan Kashkimbayev, Annie Ng

UV Degradation of Formamidinium-Cesium Lead Halide Perovskite Solar Cells 42

Kshitiz Dolia, Abasi Abudulimu, Sheng Fu, Tyler Brau, Karissa Jensen, Stephanie L. Moffitt, Randy J. Ellingson, Xiaohong Gu, Zhaoning Song, Yanfa Yan

Efficient Cd(Se, Te) Solar Cells with Cd(O, S, Se, Te) at the Front Interface 46

Dengbing Li, Sabin Neupane, Sandip Singh Bista, Abasi Abudulimu, Kamala Khanal Subedi, Manoj K. Jamarkattel, Chuanxiao Xiao, Chun-Sheng Jiang, Jonathan D. Poplawsky, David A. Cullen, Adam B. Phillips, Michael J. Heben, Randy J. Ellingson, Yanfa Yan

Energy Yield and Economics of Single-Axis-Tracked Bifacial Photovoltaics with Artificial Ground Reflectors 47

Mandy R. Lewis, Silvana Ovaitt, Byron McDanold, Chris Deline, Karin Hinzer

Advanced Encapsulants for Reduced Thermal Mechanical Stress in Photovoltaic Modules: A Quantitative Analysis Using FBGS 48

Rik Van Dyck, Marta Casasola Paesa, Tine Engelen, Bin Luo, Tom Borgers, Jonathan Govaerts, Hariharsudan Radhakrishnan, Michael Daenen, Jef Poortmans, Aart Willem Van Vuure

TOPCon Solar Cell Degradation Via Pinhole Nucleation 51

Andrew Diggs, Adam Goga, Zachary Crawford, Gergely T. Zimanyi

Soiling Model for PV Applications: Improved Parameterizations 54

Vicente Lara-Fanego, Christian A. Gueymard, Leonardo Micheli

A Sustainable Energy Market Through Community-Based PV Systems 57

Ramón Reyes-Colón, Efraín O'Neill-Carrillo

Radioisotope Thermoradiative Cell Power Generator 63

Stephen J. Polly, Geoffrey A. Landis, Seth M. Hubbard

Robust Detection Method of Low-Voltage Islanding for Grid-Forming Inverters Operated in Conjunction with Existing PV Inverters 64

Björn Oliver Winter, Julian Schwung, Bernd Engel

Improving Performance of III-V Solar Cells Grown on Spalled Germanium with Ex Situ Substrate Planarization 71

John S. Mangum, Anna K. Braun, Allison Perna, John F. Geisz, Aaron J. Ptak, Corinne E. Packard, Ryan M. France

Fabrication and Characterization of III-V Photovoltaic Devices for Use as CO2 Reduction Photoelectrodes 75
  Myles A. Steiner, Grace A. Rome, Ann L. Greenaway, Joel W. Ager, Emily L. Warren

Proton Degradation-Free Flexible Chalcopyrite Solar Cells Without Cover Glass and Adhesive 76
  Hiroki Sugimoto, Tetsuya Nakamura, Mitsuru Imaizumi, Shin-Ichiro Sato, Takeshi Ohshima

Analysis of Hierarchical PV2PV Series Differential Power Processing Configuration for Photovoltaic Applications.......................... 80
  Afshin Nazer, Patrizio Manganiello, Olindo Isabella

Plane of Array Irradiance Cleaning and Generation of Validated POA Readings for Plant Evaluation 81
  Pramod N. Krishnani, Adonis E. Hajj, Clay Helms, Shreyas Nagarajan, Mark Mikofski

A Maximum Current Point Tracking Algorithm for Solar-To-Hydrogen Production......................... 85
  Kelvin Tan, Meng Tao

A Physics Based Approach for PV Lifetime and Degradation Signatures Prediction 86
  Ismail Kaaya, Gofran Chowdhury, Arnaud Morlier

Epitaxial Growth and Testing of 1.1 eV Metamorphic InGaAs/GaAs Laser Power Converters 89
  Katelynn E Fleming, Steve J Polly, Seth M Hubbard

Suitability of GaAsBi as a Candidate Junction in a III-V Multi-Junction Solar Cell 90
  Thomas Wilson, Nicholas Ekins-Daukes

~20% Efficient Si PERC Solar Cell with Emitter Surface Passivated by H2S Reaction 91
  Tasnim K. Mouri, Ajay Upadhyaya, Ajeet Rohatgi, Youngwoo Ok, Amandee Hua, Dirk Hauschild, Lothar Weinhardt, Clemens Heske, Vijaykumar Upadhyaya, Brian Rounsaville, William N. Shafarman, Ujjwal K. Das

Investigations of Snail-Trail and Associated Microcrack Properties and Behavior in Brazil's Tropical Climate.................................... 94
  Antonia Sonia A. C. Diniz, Neolmar De M. Filho, Cláudia K. B. Vasconcelos, Lawrence L. Kazmerski

A Study of POCl3 Deposition Reaction Rate with Residual Gas Analysis Method........................ 97
  Min Gu Kang, Sang Hee Lee, Kyung Taek Jung, Yunae Cho, Dohyung Kim, Sungeun Park, Munse Kim, Hee-Eun Song

Analysis of Photovoltaic Systems Penetration on Demand Curve and Locational Marginal Prices (LMPs) in PJM 98
  Mesude Bayrakci Boz

Single Axis Tracker Performance Modeling on Sloped Terrain 101
  Benjamin Pierce, Joshua Stein, Daniel Riley

Substrate Reuse of Hydride Vapor Phase Epitaxy Grown-GaAs Solar Cells for Low-Cost Photovoltaics 105
  Yasushi Shoji, Ryuji Oshima, Kikuo Makita, Akinori Ubukata, Shuuichi Koseki, Takeyoshi Sugaya

Towards Polymer-Free, Fs Laser Welded Glass/Glass Modules 106
  David L Young, Nick Bosco, Timothy J Silverman

Defects in RbF - Treated $Cu(In_xGa_{1-x})Se_2$ Solar Cells and Their Impact on V_{OC} 107
Michael F. Miller, Ana Kanevce, Alexandra M. Bothwell, Stefan Paetel, Darius Kuciauskas, Aaron R. Arehart

Improving the Space Silicon Solar Cell Efficiency by Adding the Layer Down-Converting UV Light to Visible 110
Alex Fedoseyev, Stan Herasimenka, Sergey Sarkisov

Near-Contactless Production I-V Testing of Silicon Solar Cells 113
Harrison Wilterdink, Ron Sinton, Adrienne Blum, Karoline Dapprich, Nick Degenhart, Wes Dobson

Self-Thermometry of PV Panels...119
Kaushal Chapaneri, Shahzada Pamir Aly, Jim Joseph John, Gerhard Mathiak, Vivian Alberts, Muhammad A. Alam

Inverter Clipping and Its Masking Effect on PV Soiling: Truth Or Myth?..................................... 122
Leonardo Micheli, Matthew Muller, Marios Theristis, Florencia Almonacid, Eduardo F. Fernandez

Effect of Arsenic Doping in Polycrystalline Thin Film CdTe Solar Cells 123
Mayank Mate, Akash Shah, Ramesh Pandey, Zachary Lustig, Walajabad Sampath, Amit H. Munshi

Thermal Modelling of a Renkube Panel 126
Deepika Gopal, Balaji Bangolae Lakshmikanth, Lakshmi Santhanam

Stable High Temperature Operation in Metal Halide Perovskite Solar Cells 132
Hadi Afshari, Shashi Sourabh, Sergio A. Chacon, Vincent R. Whiteside, Bibhudutta Rout, Giles E. Eperon, Joseph M. Luther, Ian R. Sellers

Characteristics of Detachable III-V Solar Cells Grown on Porous Germanium............................. 133
Valentin Daniel, Thomas Bidaud, Jeremie Chretien, Abdelatif Jaouad, Jean-Francois Lerat, Nicolas Paupy, Bouraoui Ilahi, Jinyoun Cho, Kristof Dessein, Christian Dubuc, Gwenaelle Hamon, Abderaouf Boucherif, Maxime Darnon

Impact of Surface Roughness in Measuring Optoelectronic Characteristics of Thin-Film Solar Cells............ 134
David Magginetti, Seokmin Jeon, Yohan Yoon, Ashif Choudhury, Ashraful Mamun, Yang Qian, Jordan Gerton, Heayoung Yoon

Validation of Open-Source Distributed Energy Resources (OpenDER) Model with IEEE 1547-2018 Smart Inverter.. 138
Yiwei Ma, Charles Brewster, Aminul Huque

Upscaling of Perovskite Solar Cell Fabrication Via Slot-Die Coating: In Situ Tracking of the Drying and Crystallization Front During Gas Quenching 144
Kristina Geistert, Simon Ternes, David Benedikt Ritzer, Benjamin Hacene, Felix Laufer, Ulrich Wilhelm Paetzold

Inverted Metamorphic Photovoltaics Utilizing a Distributed Bragg Reflector Compatible with Epitaxial Lift-Off... 145
Robert F McCarthy, David Rowell, Andree Wibowo, William Mohr, Chris Youtsey, Mark Osowski, Martin Drees, Roger E Welser, Noren Pan

Life-Cycle Analysis of Potentially Longer Life Expectancy CdTe PV Modules............................. 146
Vasilis Fthenakis, Enrica Leccisi, Parikhit Sinha

Forecasting Day-Ahead Solar Irradiance for Puerto Rico Using the WRF Model and NSRDB 152
Manajit Sengupta, Jaemo Yang, Yu Xie

Coordinating the Frequency-Droop Controls of Inverter-Based Resources and Diesel Generators in an Isolated Microgrid 155
Mohan Du, Nayeem Ninad, Dave Turcotte

Probing Non-Equilibrium Hot Carrier Dynamics in Metal Halide Perovskite Solar Cells 161
Shashi Sourabh, Hadi Afshari, Vincent R. Whiteside, Giles E Eperon, Rebecca A. Scheidt, Varun N. Mapara, Madalina Furis, Matthew C Beard, Ian R. Sellers

On the Influence of Forced Convection in PV Energy Yield Models 162
Raed I. Bourisli, Bader S. Aldalali, Arttu Tuomiranta, Jef Poortmans

2-Terminal and 3-Terminal Subcell Characterization Platforms for Emerging Tandems 165
Jin Young Kim

Fast Cell Detection and Distortion Correction for Outdoor Electroluminescence Images 166
Evgenii Sovetkin, Bart E. Pieters, Andreas Gerber, Liviu Stoicescu, Pascal Koelblin

Limiting Factors on the Performenace of Luminescent Solar Concentrators for Building Integrated Photovoltaics 172
Bryce S. Richards, Ian A. Howard

3D Printed Transparent Sheet for Solar Panel Encapsulation and Thermal Management 173
Fahad Alam, Nazek El-Atab

Testing the Durability of Fluorine-Free Hydrophobic Coatings vs Porous Silica 174
Luke O. Jones, Adam M. Law, Gary Critchlow, John M. Law

Investigation of the Microstructure of Underdense Hydrogenated Amorphous Silicon Layers for Silicon Heterojunction Solar Cells by Raman Spectroscopy and Hydrogen Effusion 177
Benedikt Fischer, Maurice Nuys, Andreas Lambertz, Weiyuan Duan, Kaining Ding, Uwe Rau

Impact of Irradiation-Induced Filter Temperature Increase on Calibration of Reference Solar Cells with NIR-Longpass Filters 178
Tao Song, Larry Ottoson, Rafell Williams, John Geisz, Charles Mack, Jeremy Brewer, Nikos Kopidakis

Investigation of High Nitrogen Composition SiN_x for Textured Front Surface Passivation of n-Type Silicon Solar Cells in Terms of Light Stability of Injected Negative Charge and Cell Performance 179
Kwan Hong Min, Jeong-Mo Hwang, Christopher Chen, Wook-Jin Choi, Vijaykumar D Upadhyaya, Brian Rounsaville, Ajeet Rohatgi, Young-Woo Ok

Optimizing the Heat Sink for Concentrated Photovoltaic Systems for Different Heat Flux Conditions 182
Nemalipuri Surya Prathap, Harsh Chaurasia, K. S. Reddy

Comparison of Open-Source Photovoltaic Performance Models Against Multi-Year Field Data 188
Lelia Deville, Marios Theristis, Bruce H. King, Terrence L. Chambers, Joshua S. Stein

Q Cells Q.Antum Neo Technology with > 25% Conversion Efficiency Applying Mass-Production Processes 191

Matthias Junghänel, Ingmar Höger, Martin Schaper, Kai Petter, Enrico Jarzembowski, Christian Klenke, Anika Weihrauch, Michael Schley, Hans-Christoph Ploigt, Ohjin Kwon, Antje Schönmann, Osama Tobail, Axel Schwabedissen, Maximilian Kauert, Klaus Duncker, René Hönig, Janko Cieslak, Stefan Hörnlein, Florian Stenzel, Björn Faulwetter-Quandt, Jessica Scharf, Friederike Kersten, Cangming Ke, Sissel Tind Kristensen, Carsten Baer, Martina Queck, Gregor Zimmermann, Matthias Köhler, Nicole Lampa, Britta Pohl-Hampel, Lorenzo Burtone, Larissa Niebergall, Matthias Schütze, Susanne Schulz, Stefan Peters, Ansgar Mette, Fabian Fertig, Markus Fischer, Jörg W. Müller

Outdoor Study of Photovoltaic Mini-Modules with Different Perovskite Compositions 194

Vasiliki Paraskeva, Maria Hadjipanayi, Matthew Norton, Aranzazu Aguirre, Anurag Krishna, Rita Ebner, Tommasso Fontanot, Sabrina Pechmann, Silke Christiansen, George E. Georghiou

Achieving a New World Record Silicon Solar Cell Efficiency of 26.81% Using SHJ Device Structure 195

Xixiang Xu, Minghao Qu, Miao Yang, Xiaoning Ru, Shi Yin, Chengjian Hong, Fuguo Peng, Junxiong Lu, Liang Fang, Zhenguo Li, Yichun Wang, Tian Xie

Utilizing Particle Swarm Optimization for Autocalibration of LED Solar Simulators 198

Jann A Grovogui, John C Nocerino, Don Walker

Analysis of Cu(In, Ga)Se$_2$ Heterojunction Solar Cells in Terms of Their Balance of Thermodynamic Potentials 199

Uwe Rau, Felix Komoll, Tim Helder, Mario Zinßer, Ana Kanevce, Thomas Kirchartz, Theresa Magorian Friedlmeier

Epitaxy-Free, Thin-Film GaAs Solar Cells Fabricated with Diffusion Doping and Mechanical Spalling 202

Phillip R. Jahelka, Andrew W. Nyholm, Harry A. Atwater

Analysis of Impurity-Related Radiative Transitions in Silicon Materials Using Temperature-Dependent Photoluminescence 203

Tarek O. Abdul Fattah, Janet Jacobs, Vladimir P. Markevich, Nikolay V. Abrosimov, Matthew P. Halsall, Iain F. Crowe, Anthony R. Peaker

Reflector Candidates for a Vertical Bifacial Solar Canal 209

Jeremiah Reagan, Brandi McKuin, Sarah Kurtz

Evaluating Leafy Green Production in a Colorado Rooftop Agrivoltaic System 215

Armando Villa-Ignacio, Jennifer Bousselot

Isotropic Wet Etching of Acoustically-Spalled GaAs 218

Anica N Neumann, Myles A Steiner, Pablo G Coll, Mariana Bertoni, Emily L Warren

Degradation-Related Defect Level in Weathered Silicon Heterojunction Modules Characterized by Deep Level Transient Spectroscopy 219

Steve Johnston, Dirk C. Jordan, Dana B. Kern, Kristopher O. Davis, Helio R. Moutinho, George F. Kroeger

Interrogating Dominant Recombination Pathways in CdTe Solar Cells Using Wavelength-Dependent External Radiative Efficiency Measurements 220

Jared D. Friedl, Adam B. Phillips, Manoj K. Jamarkattel, Tyler Brau, Sabin Neupane, Scott L. Wenner, Abasi Abudulimu, Ebin Bastola, Yanfa Yan, Randy J. Ellingson, Michael J. Heben

Dynamic Calibration of Injection Dependent Carrier Lifetime from Time-Resolved Photoluminescence 227
Yan Zhu, Robert Lee Chin, Nursultan Mussakhanuly, Thorsten Trupke, Ziv Hameiri

Impacts of Dispatch Strategies and Forecast Errors on the Economics of Behind-The-Meter PV-Battery Systems 228
Brian T. Mirletz, Nicholas D. Laws

Advanced Characterization and Degradation Analysis of Perovskite Solar Cells Using Machine Learning and Bayesian Optimization 233
Joseph Chakar, Arthur Julien, Karim Medjoubi, Jorge Posada, Jean-François Guillemoles, Jean-Baptiste Puel, Yvan Bonnassieux

Statistical Analysis and Degradation Pathway Modeling of PERC PV Minimodules with Different Packaging Strategies in Indoor Accelerated Exposures 234
Sameera Nalin Venkat, Jiqi Liu, Xuanji Yu, William Oltjen, Xinjun Li, Jean-Nicolas Jaubert, Jennifer L. Braid, Roger H. French, Laura S. Bruckman

Hotspot Endurance of Pristine and Thermal Cycled Glass-Backsheet Photovoltaic Modules 237
Muhammad Afridi, Akash Kumar, Farrukh Ibne Mahmood, Govindasamy Tamizhmani

Integrated Large-Scale Data Management Platform for Photovoltaic Power Conversion Equipment (PCE) Reliability Data 240
Liwei Wang, Buck Brown, Shuan Dong, Tan Jin, Daniel Clemens, Joseph Hodges, Adam Reeves, Josh Ozbeytemur, Shuangshuang Jin, Zheyu Zhang

SnO2 Buffer Layers for High Efficiency CdSeTe/CdTe Devices 245
L. C. Infante-Ortega, Xiaolei Liu, Luksa Kujovic, Mustafa Togay, Luke O. Jones, Ali Abbas, Kieran Curson, R. C. Greenhalgh, Kurt L. Barth, Jake W. Bowers, John M. Walls, Ochai Oklobia, Stuart Irvine, Eric Colegrove, Brian Good, Matt Reese

Contact Interface Morphology of Screen-Printable Front-Side Contacts for Industrial N- TOPCon Crystalline Silicon Solar Cells 248
Meijun Lu, Kurt R. Mikeska, Weilin Liao, Chaoying Ni, Yong Zhao, Jianming Wang, Kangping Zhang, Changgen Zhang, Yawen Xu, Baiqiang Liu

Light-Dark Cycling in Perovskite Solar Cells Studied by MPPT and Ion Migration Current Measurements 253
Takeshi Tayagaki, Kohei Yamamoto, Takurou N. Murakami, Masahiro Yoshita

Field Trial in Progress for Measuring Global, Direct, Diffuse, and Ground-Reflected Irradiance Using a Static Sensor Array 254
Michael Gostein, Bruce H. King

The Proposition of a Public Policy to Stimulate Low-Income Communities' Assess to Distributed Energy Resources 256
Anna Carolina De Paula Sermarini, João Henrique Paulino Azevedo, Vanessa Cardoso De Albuquerque, Rodrigo Flora Calili, Felipe Gonçalves, Gilberto Jannuzzi

PV Plant Performance Review Methodology: Key Performance Indicators (KPI) Estimation 259
Himanshu Gulati, Prashant Kumar Upadhyay, Yellasiri Bharath Kumar Reddy

Validation of Inverter Labeling with Plant Transfer Functions 263
Joseph Ranalli

Long Terms Stability and Metastable Behavior of Perovskite Solar Devices on Outdoor Conditions 268
Karim Medjoubi, Anne Migan Dubois, Jean Castillon, Thomas Guillemot, Johan Parra, Marion Provost, Jean-Baptiste Puel, Jean Rousset, Juan Pablo Medina Flechas, Camille Bainier, Dounya Barrit, Jordi Badosa, Jorge Posada

Ultrafast Dynamics of Photoexcited Carriers and Phonons in Tailored 1D Acoustic Phonon Potentials 271
Muhammad Hanif, Stephen Bremner, Michael P. Nielsen, Milos Dubajic, Gavin J. Conibeer

Improving PbS Colloidal Quantum Dot Solar Cell Performance Via Solution-Phase Engineering 272
Dhanvini Gudi, Arlene Chiu, Dana Kachman, Eric Rong, Serene Kamal, Yucheng Lan, Susanna M. Thon

Understanding and Advancing Bifacial Thin Film Solar Cells Under Dual Illumination 275
Adam B. Phillips, Jared D. Friedl, Zhaoning Song, Abasi Abudulimu, Ebin Bastola, Deng-Bing Li, Yanfa Yan, Randy J. Ellingson, Michael J. Heben

Spray-Assisted Passivation Strategy for Highly Efficient and Stable Perovskite Solar Cells 276
Rishabh Sahani, Neetesh Kumar, Cheng-Yu Lai, Daniela R. Radu

Approaching 19% Efficiency in (InxGa(1-X))2O3/CdSe/CdTe Solar Cells with Improved Front & Back Interfaces 279
Manoj K. Jamarkattel, Adam B. Phillips, Ebin Bastola, Sabin Neupane, Deng-Bing Li, Abasi Abudulimu, Jared D. Friedl, Tamanna Mariam, Yanfa Yan, Randy J. Ellingson, Michael J. Heben

Short-Circuit Current Density Chasing and Breakthroughs in High Efficiency Silicon Heterojunction Solar Cells 280
Weiyuan Duan, Karsten Bittkau, Andreas Lambertz, Kaining Ding

Benefits of Surface Engineered Silicon Quantum Dots in Formamidinium Lead Iodide Perovskite Solar Cells 281
Vladimir Svrcek, Calum McDonald, Dilli Babu Padmanaban, Ruairi McGlynn, Ankur Kambley, Bruno Alessi, Davide Mariotti, Takuya Matsui

Depth Profiling of Glass/POE/Transparent Backsheet Degradation for Bifacial Photovoltaics 282
Xiaohong Gu, Ashlee R. Aiello, Stefan Mitterhofer, Soshana Smith, Stephanie L. Moffitt, Lakesha N. Perry, Song-Syun Jhang, Stephanie S. Watson, Li-Piin Sung

Temporal Downscaling of GHI Clear-Sky Indices Using T-Copula 285
Jing Huang, Marc Perez

Extended FF-VOC Parameterization for Silicon Solar Cells 288
Karsten Bothe, David Hinken, Rolf Brendel

Towards Commercialisation with Lightweight, Flexible Perovskite Solar Cells for Residential Photovoltaics 289
Philippe Holzhey, Michael Prettl, Silvia Collavini, Nathan Chang, Michael Saliba

Thermal Stability of BiI3 Thin Films 290
Natália F. Coutinho, Thais Crestani, Otávio J. De Oliveira, M. M. M. Modesto Ana Paula De, Marcelo Villalva, Francisco C. Marques

In-Situ Smoothing of Facets on Spalled GaAs(100) Substrates During OMPVE Growth of III-V Solar Cells 291
> William E. McMahon, Anna K. Braun, Allison N. Perna, Pablo G. Coll, Kevin L. Schulte, Jacob T. Boyer, Anica N. Neumann, John F. Geisz, Emily L. Warren, Aaron J. Ptak, Arno P. Merkle, Mariana I. Bertoni, I. Bertoni, Corinne E. Packard, Myles A. Steiner

Numerical Simulation Study for Analysis of Hydrogenated Amorphous Silicon/Crystalline Silicon Heterostructure by Reactive Molecular Dynamics Method 292
> Kazuma Inoue, Naoya Uene, Kazuhiro Gotoh, Yasuyoshi Kurokawa, Takashi Tokumasu, Noritaka Usami

Planned Field Test of Soiling and Irradiance Measurement Uncertainties in Bifacial PV Systems Using In-Situ Reference Modules 295
> Michael Gostein, Audrey Marquis, Marine Bila, Robert Campbell

Intermediate-Phase Engineering Via Dimethylammonium Cation Additive for Stable Perovskite Solar Cells 298
> David P. McMeekin, Philippe Holzhey, Sebastian O. Fürer, Steven P. Harvey, Laura T. Schelhas, James M. Ball, Suhas Mahesh, Seongrok Seo, Nicholas Hawkins, Jianfeng Lu, Michael B. Johnston, Joseph J. Berry, Udo Bach, Henry J. Snaith

A Techno-Economic Analysis of Various Grid-Connected Photovoltaic System Configurations for Green Hydrogen Production 299
> Rahul R Urs, Assia Chadly, Ahmad Mayyas

Multi-Layer Dense Antireflection Coatings 302
> Yiyu Zeng, Martin Green, Jessica Yajie Jiang

Transparent Conductive Oxide Bi-Layer as Front Contact for Multijunction Thin-Film Silicon Solar Cells 305
> Federica Saitta, Prashand Kalpoe, Govind Padmakumar, Paula Perez-Rodriguez, Gianluca Limodio, Rudi Santbergen, Arno H. M. Smets

12.3% Efficient Lifted-Off and Reconstructed As-Doped CdTe Thin Film Solar Cell 308
> Ochai Oklobia, Deborah L. McGott, Giray Kartopu, Steve Jones, Stuart J. C. Irvine

Design, Fabrication, Test, and Flight Performance of the Parker Solar Probe Solar Array 311
> Edward Gaddy, Andrew Gerger, Lew Roufberg, Richard Stall, Matthew J. Schurman

Investigating Electric Field and Light Induced Degradation in Perovskite Solar Cells Through Nanometer-Scale Potential Imaging 315
> G. Paul, J. W. Schall, C.-S. Jiang, A. Louks, A. Palmstrom, N. S. Dutta, S. Johnston, H. Guthrey, A. Norman, M. M. Al-Jassim, D. B. Sulas-Kern

2D-MoS$_2$ Nano Structures to Enhance Silicon Solar Cells 321
> Muntaser Abdelrahman Almansoori, Ayman Rezk, Ammar Nayfeh

Effect of Bidentate Ligand Additive in Tin Perovskite Solar Cells 324
> Dhruba B. Khadka, Yasuhiro Shirai, Masatoshi Yanagida, Kenjiro Miyano

Optoelectronic Performance of Solution Processable MoS$_2$ for Application in Photovoltaic Devices 327
> Dayanand Kumar, Ayman Rizk, Ammar Nayfeh, Nazek El-Atab

Mechanical Degradation of Perovskite Thin Films for Photovoltaics: In-Situ Microscopy & Digital Twin Modeling 330
> Melissa A Davis, Mehul Tank, Michelena O'Rourke, Matthew Wadsworth, Zhibin Yu, Rebekah Sweat

Optimal Row Spacing for Monofacial and Bifacial Fixed-Tilt and Tracked Photovoltaic Systems Up to 75°N 331

    Erin M. Tonita, Annie C. J. Russell, Christopher E. Valdivia, Karin Hinzer

Instability of Non-Fullerene Acceptors Used in Organic Solar Cells 332

    Yongxi Li, Tonghui Wang, Aram Amassian, Stephen Forrest

Damp Heat Performance of Silicon Heterojunction Solar Cells with Reactive Silver Ink Metallization 333

    Michael W Martinez-Szewczyk, Steven Digregorio, Subbarao Raikar, Owen Hildreth, Mariana Bertoni

Parametric Analysis of Photovoltaic Inverters Under Balanced and Unbalanced Voltage Phase Angle Jump Conditions 334

    Rachid Darbali-Zamora, Jay Johnson, Matthew J. Reno

Chemical Reaction Kinetics of the Decomposition of Low Bandgap Tin-Lead Halide Perovskite Films and the Effect on the Ambipolar Diffusion Length 340

    Yuhuan Meng, Preetham P. Sunkari, Marina Meila, Hugh W. Hillhouse

Demonstration of Thermoradiative Energy Conversion with InAs Cells 341

    Eric J Tervo, Andrew J Ferguson, Jennifer Selvidge, Myles A Steiner, Ryan M France

InGaP/GaAs/In$_{0.35}$Ga$_{0.65}$As//In$_{0.53}$Ga$_{0.47}$As Four-Junction Solar Cells Integrated by Surface Activated Wafer Bonding 342

    Kentaroh Watanabe, Takashi Shimasaki, Hassanet Sodabanlu, Yoshiaki Nakano, Masakazu Sugiyama

Efficiency Limits for Multi-Junction Coloured Photovoltaics 345

    Phoebe M. Pearce, Janne Halme, Jessica Yajie Jiang, Farid Elsehrawy, Nicholas J. Ekins-Daukes

Microscale, High Aspect Ratio, Effectively Transparent Contacts (ETCs) Fabricated with String Printing 348

    Mathis Van De Voorde, Janis A. Andersons, Rebecca Saive

Optimization of Bulk Heterojunction Layer Constituents in Organic Photovoltaic Device 351

    Vilko Mandic, Dragana Vuk, Radovanovic-Peric Floren, Ivana Panzic

Towards Smart Integration of Cu-Plating for Silver-Free and Edge Passivated SHJ Shingle Modules 352

    Samuel Harrison, Vincent Barth, Benoit Martel, Agata Lachowicz, Nicola Frasson, Marco Galiazzo

A Comparison of Emerging and Industry Benchmark Photovoltaic Backsheets Between Different Outdoor Locations 353

    Elizabeth Palmiotti, Bruce King, Rachael Arnold, Sona Ulicná, Laura T. Schelhas, David C. Miller

Considering the Variability of Soiling in Long-Term PV Performance Forecasting 356

    Matthew Muller, Faisal Rashed

Electron Paramagnetic Resonance Investigation of the Defect Responsible for Light- And Elevated-Temperature-Induced Degradation in Ga-Doped Czochralski Si 359

    Chirag Mule, P. Craig Taylor, Abigail Meyer, William Nemeth, Vincenzo Lasalvia, Matthew Page, Sumit Agarwal, Pauls Strandins

Real-Time Regional PV Spinning Reserve Estimator with AGC Look-Ahead Windows 360
Mengmeng Cai, Govind Saraswat, Vahan Gevorgian

Investigation of Sputtered P-Type Electrical Contacts for Thin Film Cadmium Telluride-Based
Solar Cells 363
Blake Hill, Forrest Khulmann, Mayank Mate, Amit H. Munshi

Perovskite Bafacial Modules-Efficiency, Stability and Upscaling 366
Hangyu Gu, Jinsong Huang

Modular, Array-Mounted Photovoltaic Inspection Robot 367
Michael Y. Vazquez Nieves, Alanis M. Colón González, Jennifer L. Braid

ESSPI as a Fast Tool for Load Prioritization on Microgrids Design 370
Luis Colomba-Colon, Natanael Batista-Alvarez, Guillermo Lopez-Cardalda, Eduardo Ortiz-
Rivera

Investigations on Absorber Type and Junction Position of GaAs Solar Cells 374
Gan Li, Hassanet Sodabanlu, Meita Asami, Kentaroh Watanabe, Masakazu Sugiyama,
Yoshiaki Nakano

Transparent Tedlar® Frontsheet for Lightweight PV Module Designs 377
Hongjie Hu, Stela Chen, Oakland Fu, Michael Demko, Kaushik Roy Choudhury

Photovoltaic Design Projects Increase ECE Student Engagement 380
Devin C. Whalen, Peter Mark Jansson, Milton G. Newberry

Long-Term Degradation Rate of Photovoltaic Modules: A Meta-Analysis.. 385
Michael Straub-Mueck, Jerome Geyer-Klingeberg, Andreas Rathgeber

The Use of a Physics-Based DNI Model to Enhance the National Solar Radiation Database
(NSRDB)... 386
Yu Xie, Jaemo Yang, Manajit Sengupta, Yangang Liu

Novel Module Architecture for Lower CapEx and Improved Recyclability for c-Si PV Modules 387
Ryan Ruhle, Larry Maple, Timothy Delazzer, Steve Johnston, Dana Kern, Walajabad Sampath

Snow Sensing for Photovoltaic Single Axis Tracker Systems... 391
Ayush Chutani, Ana Dyreson, Laurie Burnham, Kyumin Lee

Operando Temperature Measurements of Photovoltaic Laser Power Converter Devices Under
Continuous High-Intensity Illumination... 394
John F. Geisz, Daniel J. Friedman, Myles A. Steiner, Ryan M. France, Tao Song

Optimization of Back-Contact Diffusion Barrier for Solution-Processed CIGS Solar Cells: Case of
MoO_3 and MoN 395
Nada Benhaddou, Jacques Kenyon, Luke O. Jones, Liam M. Welch, Jake W. Bowers

Silicon Heterojunction Cell Metallization with Reactive Silver Inks: Printing Process, Ink Formula,
and Interconnection 400
Steven J. Digregorio, Michael W. Martinez-Szewczyk, Subbarao Raikar, Mariana I. Bertoni,
Owen J. Hildreth

Optical Properties of (InxGal-X)203 Alloys and Evaluation as Emitter Layer in CST PV 401
Bishal Shrestha, Madan K. Mainali, Manoj K. Jamarkattel, Ebin Bastola, Adam B. Phillips,
Michael J. Heben, Nikolas J. Podraza

Temperature Dependent Fill Factor in CdSe/CdTe PV Devices from -20°C to 60°C Under AM1.5G and AM0 Spectra 402

Nadeesha Katakumbura, Prabodika N. Kaluarachchi, Manoj Rajakaruna, Tyler Brau, Aesha P. Patel, Abudulimu Abasi, Ebin Bastola, David Raker, Adam B. Phillips, Michael J. Heben, Sorin Cioc, Randy J. Ellingson

Module Reliability in Winter: Field Analysis of Deflection and Cell Cracking Across Multiple Module Architectures 409

Laurie Burnham, Daniel Riley, Bruce King, William Snyder, Kevin Santistevan, Paul W. Dice

How Useful is a Field-Operable I-V Curve Tracer? 412

Alexander Cimaroli

Results of First Long Duration Space Flight of Hybrid Perovskite Thin Film 415

Lyndsey McMillon-Brown, William Delmas, Samuel Erickson, Jorge Arteaga, Mark Woodall, Michael Scheibner, Timothy Krause, Kyle Crowley, Kaitlyn Vansant, Joseph Luther, Jennifer Williams, Jeremiah McNatt, Sayantani Ghosh

Undergraduate Research Experience in the Design and Construction of a Photovoltaic Inspection Robot 416

Alanis M. Colón, Emmanuel J. González, Fernando J. Vargas, Samuel I. Hernandez, Michael Y. Vazquez, Eduarto I. Ortiz

Development of Gradient Layers to Improve the Efficiency of Transparent Passivating Contact Solar Cells 419

Alexander Eberst, Binbin Xu, Weiyuan Duan, Andreas Lambertz, Uwe Rau, Kaining Ding

A GIS-Based Approach for Prioritization of Photovoltaic Systems with Energy Storage Implementation for Vulnerable Community Resilience 420

Javier A. Moscoso-Cabrera, Edgar E. Cruz, Cristian R. Meléndez, Eduardo I. Ortiz

Position Dependence of the Performance Gain by Selective Ground Albedo Enhancement for Bifacial Installations 426

Nils-Peter Harder, Issam Smaine, Fadi Bourarach, Damien Cosme, Ines Arfaoui, Julien Chapon, Arttu Tuomiranta, Antonios Florakis

Microscopic Origins of Performance Losses in (Ag,Cu)(In,Ga)Se$_2$ Thin-Film Solar Cells 432

Sinju Thomas, Wolfram Witte, Dimitrios Hariskos, Rico Gutzler, Stefan Paetel, Chang-Yun Song, Heiko Kempa, Matthias Maiberg, Daniel Abou-Ras

Overview of Engineered Germanium Substrate Development for Affordable Large-Volume Multi-Junction Solar Cells 434

Jinyoun Cho, Valérie Depauw, Alexandre Chapotot, Waldemar Schreiber, Tadeáš Hanuš, Nicolas Paupy, Valentin Daniel, Guillaume Courtois, Bouraoui Ilahi, Abderraouf Boucherif, Clement Porret, Roger Loo, Jens Ohlmann, Stefan Janz, Kristof Dessein

X-RAYS Meet Neutrons Meet Ions Meet Electrons Meet Lasers Meet Magnets: Combined Access to Multiple Facilities Through EU Project "Remade@ARI" 435

Michael E. Stuckelberger, Christina Ossig, Barbara Schramm, Stefan Facsko

BIPV Market Potential Analysis with Building Shadow Simulation 436

Changyeol Yun, Myeongchan Oh, Boyoung Kim, Jehyun Lee, Hyungoo Kim, Deokoh Lim, Sangmin Jo

An Investigation on the Pollen-Induced Soiling Losses in Utility-Scale PV Plants 437

João Gabriel Bessal, Michael Valerino, Matthew Muller, Mike Bergin, Leonardo Micheli, Florencia Almonacid, Eduardo F. Fernández

The Role of PbS QDs on Strain and Optical Properties in Different Perovskite Matrix 441
  Sofia Masi, Patricio Serafini, Iván Mora-Seró

Zr-Doped In2O3 Film for the Interlayer of Perovskite/Crystalline Silicon Tandem Solar Cells 442
  *Tappei Nishihara, Hyunju Lee, Ryuji Kaneko, Yoshio Ohshita, Atsushi Wakamiya, Atsushi
  Masuda, Atsushi Ogura*

Design with Luminescent Solar Concentrator Photovoltaics in the Built Environment 443
  Eli Shirazi, Wouter Eggink, Angele Reinders

Analysis of Solar Cell Electroluminescence Spectra for Daylight Inspection of c-Si PV Modules 448
  *Gisele A. Dos Reis Benatto, Alejandra A. Mayordomo, Rodrigo Del Prado Santamaria,
  Thøger Kari, Peter B. Poulsen, Sergiu V. Spataru*

Artificial Neural Network and Peer-To-Peer Communications at the Grid-Edge to Mitigate Cyber
Attacks on Distributed Photovoltaic Inverters 455
  C. Birk Jones, Rachid Darbali-Zamora

Improvement of Radiation Tolerance in Solar Cells by Hetero P/N Junction Structure 463
  *Tetsuya Nakamura, Mitsuru Imaizumi, Meita Asami, Masakazu Sugiyama, Hidefumi Akiyama,
  Shin-Ichiro Sato, Takeshi Oshima, Yoshitaka Okada*

Abnormal Responses of Residential Smart Photovoltaic Inverters to Cyberattacks 464
  Thunchanok Kaewnukultorn, Sergio B. Sepúlveda-Mora, Steven Hegedus

Community Influence of Houses of Worship on Rooftop Solar Growth Rates 467
  *Ashley Degen, Laura Mogannam, Nisitaa Karen Clement Pradeep, Jillian Stern, Deborah A.
  Sunter*

Decoupling Open-Circuit Voltage and Series Resistance in Electroluminescence Images Through
Deep Learning 472
  Gaia Maria N. Javier, Priya Dwivedi, Thorsten Trupke, Ziv Hameiri

Decomposition Mechanisms and Kinetics of Perovskite Semiconductors 473
  Hugh Hillhouse, Yuhuan Meng, Spencer Cira, Preetham Sunkari

Development and Evaluation of Typical Plane of Array Year (TPY) for Solar Energy Systems Over
the Americas 474
  Aron Habte, Manajit Sengupta, Grant Buster, Yu Xie

10-Junction Edge-Illuminated Passivated-Contact Silicon Minimodules for Laser Power
Conversion 475
  *Ryan M France, Matthew B Hartenstein, William Nemeth, San Theingi, Matthew Page, Sumit
  Agarwal, David Young, Paul Stradins*

GaAs Betavoltaic Cell Modeling for Light to Medium Element Radiation Conversion into
Electrical Power 476
  *Mathieu De Lafontaine, Gavin Forcade, Paige Wilson, Jayeshkumar Patel, Brian Ellis,
  Helmut Fritzsche, Sam Suppiah, John P. D. Cook, Christopher E. Valdivia, Karin Hinzer*

Residential Electric Energy Storage System to Reduce Voltage and Thermal Violations in
Distribution Lines and Increase PV Integration 477
  Anny Huaman-Rivera, Agustin Irizarry-Rivera

Air-Bridge Cells for Higher Emission Temperatures 482
  *Bosun Roy-Layinde, Areefa Rahman, Jihun Lim, Sritoma Paul, Stephen R. Forrest, Andrej
  Lenert*

On the Accuracy of Spectral Adjustment for Performance Measurements of Multijunction Solar Cells...... 485
Nikos Kopidakis, Tao Song, John Geisz, Daniel Friedman

Spatially Resolved Degradation Analysis of Solar Modules After Combined Accelerated Aging 486
Robert Heidrich, Anton Mordvinkin, Ralph Gottschalg

NiO as a P-Type TCO for Inorganic Thin-Film Photovoltaics 489
Elline C. Hettiaratchy, Angus A. Rockett, Taylor D. Hill, Sachit Grover

Modeled Impacts of Solar Forecast Error on Utility Production Cost 492
William B. Hobbs, Jenner Tresan, Michael Kline, Mousumi Guha, Brent Duncan

Durability Testing of Porous SiO_2 Anti-Reflection Coatings for Solar Cover Glass 498
Adam M Law, Luke O Jones, Michael Nasser, Ali Abbas, John M Walls

Performance Evaluation of Perovskite Solar Cell Modules with Tilt Angle Optimization in BIPV Application: A Case Study for Kazakhstan 502
Yerassyl Olzhabay, Ikechi Ukaegbu, Annie Ng

CFD-Based Machine Learning Model for Agrivoltaic System Design 505
Henry J. Williams, Emily Weed, Khaled Hashad, K. Max Zhang

Towards Highly Stable Metal-Halide Perovskite Materials for a Broad Range of Applications: Film Growth, Degradation Control, and Interfacial Engineering 508
Lissette Rodriguez-Cabanas, Calvin Duong, Bradley Stanley, Andre Slonopas

First-Principles Study of Energy Band Alignment in Pristine $CdTe/TeO_2/Te$ Interfaces 509
Anthony P. Nicholson, James R. Sites, Walajabad S. Sampath

Horizon Profiling Methods for Photovoltaic Arrays 512
Jennifer L. Braid, Benjamin G. Pierce

Field Effect Passivation Enables 2.2 V Open-Circuit Voltage All-Perovskite Tandems 518
Bin Chen, Ted Sargent

Distributed Generation Component Placement and Point of Common Coupling Allocation for Solar Rooftop Microgrid Sizing Costs Minimization 519
Robert A. García Cooper, Marcel Castro Sitiriche, Agustín Irizarry Rivera, Fabio Andrade Rengifo

The Solar Boat: An Academic Research Experience 525
Guillermo Serrano, Erick Aponte, Eduardo I. Ortiz-Rivera

Evaluating the Use of Satellite Data and Machine Learning Models for PV Performance Monitoring 528
Daniel Fregosi, Rabin Dhakal, Devin Widrick

Drying Effects Upon Spin Coating of Solution-Processed Amine-Thiol Thin Film $Cu(In,Ga)(S,Se)_2$ Absorber Fabrication 534
Jacques Kenyon, Nada Benhaddou, Liam Welch, Jake Bowers

Impact of Backsheet Versatility on Inverter Availability 540
Claudia Buerhop-Lutz, Oleksandr Stroyuk, Jens Hauch, Ian Marius Peters

Evaluating the Weather Forecasting Models and the Impact to PV Generation Forecasting 541
Spyros Theocharides, Anastasios Koumis, George Makrides, George E. Georghiou

Plausibility Filtering of PV Outdoor Data 546
  T. S. Vaas, J. Körtgen, E. Sovetkin, U. Rau, B. E. Pieters

The European Solar Communication - Will it Strengthen the Photovoltaic Industry in the European
Union 550
  Arnulf Jäger-Waldau, Anatoli Chatzipanagi, Georgia Kakoulaki, Sandor Szábo

Measuring Sustainability of PV in Energy Transition: Mass, Energy, and Circularity 554
  *Heather M. Mirletz, Silvana Ovaitt, Macarena Mendez Ribo, Seetharaman Sridhar, Teresa M.
  Barnes*

Photovoltaic Site Architecture Estimation Using Performance Data.. 555
  *Steven Koskey, Scott Sheppard, Corson Teasley, Christopher Perullo, Jared Kee, Daniel
  Fregosi, Wayne Li*

Synthesis and Characterization of Bismuth Selenide and Copper Doped Bismuth Selenide Thin
Films by Chemical Bath Deposition.. 561
  Hamda A. Al-Thani, Shifaa M. Al-Baity, Falah S. Hasoon

CIGS Device Stability: A Comparison of Two Different Process Batches... 566
  *Mohsen Jahandardoost, Curtis Walkons, Marco Nardone, Theresa M. Friedlmeier, Shubhra
  Bansal*

Hierarchical Variance Analysis of Solar Cell Production Using Machine Learning and Numerical
Simulations... 569
  Bernhard Klöter, Hannes Wagner Mohnsen, Sven Wasmer

Identifying the Electrical Signature of Snow in Photovoltaic Inverter Data 572
  Emma C. Cooper, Jennifer L. Braid, Laurie M. Burnham

Band Tail Effects on Cd(Se,Te) Device Performance: A Numerical Simulation Approach...................... 577
  Eva M Mulloy, Darius Kuciauskas, Craig L Perkins, Marco Nardone

Understanding the Dopability of as in Selenium-Alloyed Cadmium Telluride Solar Cells...................... 578
  Xiaofeng Xiang, Aaron Gehrke, Scott Dunham

Modeling Transposition for Single-Axis Trackers Using Terrain-Aware Backtracking Strategies 581
  Kurt Rhee

What's New in the NSRDB 587
  Manajit Sengupta, Aron Habte, Grant Buster, Yu Xie, Brandon Benton, Michael Foster

Rapid, Contactless Measurements and Performance Predictions of Photovoltaic Materials.................... 590
  *Brandon T Motes, Anthony T Troupe, Amy E Louks, Axel F Palmstrom, Joseph J Berry, Dane
  W Dequilettes*

Optimizing the Packing Density of Building Integrated Concentrating Photovoltaic Systems for
Improved Performance and Reduced Embodied Carbon Through a Novel Polygonal Concentrator.............. 591
  *Lewis Osikibo Tamuno-Ibuomi, Roberto Ramirez-Iniguez, A Sheila Holmes-Smith, Geraint
  Bevan*

Human Health Risk Assessment for Improper Landfill Disposal of End-Of-Life CdTe PV 594
  Elaine Kupets, Garvin Heath

Accelerating Cycles of Learning for Silicon Heterojunction Architectures: Experimental Design and Data-Driven Degradation Pathway Prediction 597
  Xuanji Yu, Diego Zubieta, Mirra Rasmussen, Chien-Hsuan Chen, Cécile Molto, Mariana Bertoni, Kristopher O. Davis, Laura S. Bruckman, Ina T. Martin

Modeling Reference Cell Performance Using Measured and Modeled Spectral Data 600
  Josh Peterson, Frank Vignola, Afshin Andreas, Aron Habte, Manajit Sengupta

Silver Reflector-Driven Light Harvesting Enhancement in Large Area Dye Sensitized: Solar Cells 606
  Navdeep Kaur, Faizan Syed, Cheng-Yu Lai, Daniela Radu

Survey of Snow Impacts on Bifacial Gain in Commercial Photovoltaic Arrays 607
  Samantha S. Wilson, Stephen Lightfoote, Stephen Voss

New Theoretical Limits for Light Trapping in Solar Cells 610
  Stéphane Collin, Maxime Giteau

Supply Side Management with Agrivoltaics: Feasibility Study of Modeling Methodologies of Solar PV and Crop Response 611
  Tadatoshi Takahashi

Effect of Solar Mounting Configurations on California Zero-Carbon Grid 617
  Zabir Mahmud, Sarah Kurtz

Experimental Demonstration of Diffused Light Collimation in Free Space 622
  Lisanne M. Einhaus, Geert C. Heres, Jelle Westerhof, Shweta Pal, Akshay Kumar, Anne Rikhof, Jian-Yao Zheng, Rebecca Saive

The Planet-Scale Performance Potential of Si-Perovskite Tandem Solar Farms 626
  Jabir Bin Jahangir, M. Tahir Patel, Reza Asadpour, M. Ryyan Khan, M. A. Alam

Investigating the Potential of Hydrogen Plasma Treated ALD-TiOx Films as Hole-Selective Passivating Contacts in Crystalline Silicon Solar Cells 629
  Chien-Hsuan Chen, S. Novia Berriel, Taylor M. Currie, Jannatul Ferdous Mousumi, Titel Jurca, Parag Banerjee, Kristopher O. Davis

The Photovoltaic Exponential Model 630
  Eduardo I. Ortiz Rivera

On the Unappreciated Impact of Se in as-Doped CdSexTe1-X 633
  Deborah L McGott, Darius Kuciauskas, Craig L Perkins, Eric Colegrove, Matthew O. Reese

Leveraging High-Fidelity Sensor Data for Inverter Diagnostics: A Data-Driven Model Using High-Temperature Accelerated Life Testing Data 634
  Sakir Karakaya, Murat Yildirim, Shijia Zhao, Feng Qiu, Jack David Flicker, Benjamin Peters, Zhaoyu Wang

Influence of Interfaces on Stability of Perovskite Solar Cells 641
  Arkadi Akopian, Saba Sharikadze, Ranjith Kottokkaran, Vikram Dalal

A New Route to Facilitate Scaling of Lead-Tin Halide Perovskites: Thin Films Via Solvent Self-Volatilization 644
  Jack R. Palmer, Apoorva Gupta, Sean P. Dunfield, David P. Fenning

Aerial Photoluminescence Imaging of PV Modules 645
  Bernd Doll, Ernst Wittmann, Larry Lüer, Claudia Buerhop-Lutz, Jens A. Hauch, Christoph J. Brabec, Ian Marius Peters

Polarization Type Potential Induced Degradation Under Positive Bias in a Commercial PERC Module 648
Farrukh Ibne Mahmood, Fang Li, Peter Hacke, Cécile Molto, Hubert Siegneur, Govindasamy Tamizhmani

Three-Dimensional and Multimodal X-Ray Microscopy Reveals the Impact of Voids in CIGS Solar Cells 651
Giovanni Fevola, Christina Ossig, Mariana Verezhak, Jan Garrevoet, Martin Seyrich, Dennis Brueckner, Johannes Hagemann, Frank Seiboth, Andreas Schropp, Gerald Falkenberg, Peter Stanley Jorgensen, Christian Strelow, Tobias Kipp, Romain Carron, Jens Wenzel Andreasen, Michael E. Stuckelberger

Optical Characterization and Loss Simulation of Encapsulation Materials and Back Sheets for PERC+ Solar Modules 652
Tim Lukas Brockmann, Henning Schulte-Huxel, Susanne Blankemeyer, Tobias Wietler

Integration of Lateral Power MOSFETs into IBC c-Si Solar Cells with Poly-Si Passivating Contacts 655
David A. Van Nijen, Patrizio Manganiello, Yavuzhan Mercimek, Tristan Stevens, Guangtao Yang, René A. C. M. M. Van Swaaij, Miro Zeman, Olindo Isabella

The Formation of Dendrites in Overtreated CdSeTe/CdTe Solar Cells 656
Vladislav Kornienko, Luke Jones, Kieran Curson, Ali Abbas, Rachael Greenhalgh, Yau Yau Tse, Michael Walls, Christian Drost, Bettina Spaeth, Bastian Siepchen

Carrier Dynamics in $Al_xGa_{1-x}As$/InAs-Based Photon Up-Conversion Solar Cells with a Doubled-Heterointerface 659
Hambalee Mahamu, Shigeo Asahi, Takashi Kita

The Importance of Terrain-Shading Losses in PV Yield Assessment: The Case of Oahu 662
Marc Perez, Upama Nakarmi, Philip Gruenhagen, Richard Perez

Utilizing PSO Technique for Locational-Dependent Feeder PV Hosting Capacity Evaluation 665
Vinushika Panchalogaranjan, Paul Moses

Cryogenic Operation of GaAs Laser Power Converters 670
Bora Kim, Mijung Kim, Brian D. Li, Ryan D. Hool, Minjoo L. Lee

Characterization of Field Exposed Photovoltaic Modules Featuring Signs of Contact Degradation 671
Max Liggett, Dylan J. Colvin, Balaashwin Babu, William C. Oltjen, Xuanji Yu, Manjunath Matam, Hubert P. Seigneur, Mengjie Li, Andrew M. Gabor, Philip J. Knodle, Craig J. Neal, Sudipta Seal, Laura S. Bruckman, Roger H. French, Kristopher O. Davis

Energy Management in a Dynamic Microgrid Using Genetic Algorithms 674
Ricardo Calloquispe-Huallpa, Rachid Darbali-Zamora, Erick E. Aponte-Bezares, Matthew S. Lave

Sn-Based Perovskite Thin Film Solar Cells with Enhanced Stability 682
Wendy Reyes Ramos, Zeying Chen, Adam Thomas, Tara Dhakal

Cadmium Selenide (CdSe) as an Active Absorber Layer for Solar Cells with V_{OC} Approaching 750 mV 683
Ebin Bastola, Adam B. Phillips, Abasi Abudulimu, Vlad Kornienko, Zulkifl Hussain, Manoj K. Jamarkattel, Tamanna Mariam, Prabodika N. Kaluarachchi, Jared Friedl, Dipendra Pokhrel, Kara B. Kile, Zhaoning Song, Yanfa Yan, Michael Walls, Randy J. Ellingson, Michael J. Heben

Mapping Stress in PV Modules: The Influence of Soldering, Tabbing, and Module Architecture 689
Ian Slauch, Rico Meier, Kaushik Roy Choudhury, Mariana Bertoni

Elemental Vapor Transport Deposition of $CdSe_xTe_{1-x}$ Thin Films for n-Type CdTe Solar Cells 690
Wei Wang, Vasilios Palekis, Md Zahangir Alom, Sheikh Elahi Tawsif, Don Morel, Chris
Ferekides

Intra-Grain Local Luminescence Properties of $CdSe_{0.1}Te_{0.9}$ Thin Films 695
Ganga R. Neupane, David S. Albin, Joel N. Duenow, Matthew O. Reese, Susanna Thon,
Behrang H. Hamadani

Electroluminescence Imaging: A Study in the Impact of Microscopic Surface Defects 698
Meghan E. Bush, Timothy J. Peshek

Towards a Three-Terminal Perovskite/Silicon Tandem Solar Cell with Highest Efficiency 701
Michael Rienäcker, Somayeh Moghadamzadeh, Paul Fassl, Yevgeniya Larionova, Philipp
Noack, Bianca Wattenberg, Ulrich Wilhelm Paetzold, Robby Peibst

Luminescence and Thermal Imaging Applied to Half-Cut-Cell and Emitter-Wrap-Through-Cell
Modules 702
Steve Johnston, Dana B. Kern, Kent Terwilliger, Ingrid L. Repins

Role of Solar Photovoltaics for a Sustainable Energy System in Puerto Rico in the Context of the
Entire Caribbean Featuring the Value of Offshore Floating Systems 708
Christian Breyer, Ayobami S Oyewo, Alejandro Kunkar, Rasul Satymov

Energy-Harvesting Efficiency Analysis for Solar Modules Using 2T and 4T Tandem Solar Cells 709
Robert Witteck, William E. McMahon, John F. Geisz, Qi Jiang, Emily L. Warren

A Reproducible Validation of Algorithms for Estimating Array Tilt and Azimuth from Photovoltaic
Power Time Series 710
Kirsten Perry, Bennet Meyers, Kevin Anderson, Matthew Muller

Investigation of P-Type Silicon Heterojunction Radiation Hardness 716
Romain Cariou, Adrien Danel, Nicolas Enjalbert, Frédéric Jay, Sébastien Dubois

Energy Injustice Metrics for Puerto Rico 717
Pablo Méndez-Curbelo, Carlos Peña, Willian Pacheco, Marcel Castro-Sitiriche

Inverse Design of Spectrally-Selective Films for PbS-CQD Tandem Solar Cells 718
Sreyas Chintapalli, Tina Gao, Luna Singh, Serene Kamal, Arlene Chiu, Yijun Zhang, Susanna
M. Thon

AM0 Optimized Dual Junction Quantum Well Solar Cells-Investigation of Radiation Tolerance
Designs and V_{OC} Retention at EOL 721
Brandon M. Bogner, Stephen J. Polly, Seth M. Hubbard, Mitsuru Imaizumi, Roger E. Welser

Evaluation of Process Damage to Crystalline Silicon by Transparent Conductive Oxide Film
Deposition 724
Haruki Kojima, Tappei Nishihara, Yuta Ito, Hyunju Lee, Kazuhiro Gotoh, Noritaka Usami,
Tomohiko Hara, Kyotaro Nakamura, Yoshio Ohshita, Atsushi Ogura

Brownfields to Brightfields: The Potential for Landfill Solar Redevelopment in New York State 727
Henry J. Williams, Hugh Peng, Olivia Goosay, K. Max Zhang

Carbon Footprint of Silicon Photovoltaics Manufacturing in North America 730
Annick Anctil, Angela Farina, Luyao Yuan

The PV Efficiency Vs R&D Effort Learning Curve for Research-Stage Material Technologies 731
Phillip J. Dale, Michael A. Scarpulla

Perovskite/Silicon Tandem Solar Cells with Front Side Metallization Applied Prior to Top Cell
Fabrication Enabling High Curing Temperatures 733
Sara Baumann, Annika Raugewitz, Felix Haase, Tobias Wietler, Robby Peibst, Marc Köntges

Performance and Degradation in Silicon PV Systems Under Outdoor Conditions in Relation to
Reliability Aspects of Silicon PV Modules – Summary of Results of COST Action PEARL PV.................... 737
S. Lindig, J. Ascencio-Vásquez, J. Leloux, D. Moser, M. Aghaei, A. Fairbrother, A. Gok, S.
Ahmad, S. Kazim, K. Lobato, W. J. G. H. M. Van Sark, N. Pearsall, B. G. Burduhos, A.
Raghoebarsing, G. Oreski, J. Schmitz, M. Theelen, P. Yilmaz, J. Kettle, A. H. M. E. Reinders

Progress Towards Scaling Perovskite/Silicon Tandem Modules 740
Chris Eberspacher, Colin Bailie, Tim Gehan, Bryan Rosales, Tom Brenner, Mike Chen, Terry
Banks

Identification of Module Replacements in US Utility-Scale Photovoltaic Installations...................... 743
Chenyang Deng, Jacob T. Stid, Preeti Nain, Annick Anctil

Indoor and Outdoor Evaluation of Curved Modules for VIPV 746
Rebeca Herrero, Ignacio Antón, Francisco Martín, Steve Askins, Javier Macías, Luis J. San
José, G. Vallerotto, R. Núñez, C. Domínguez

Towards an Annual Terrawatt Photovoltaics Market - Comparison of the Social Acceptance in
Various IEA PVPS Countries 749
Arnulf Jäger-Waldau, Georg Altenhöfer-Pflaum, Otto Bernsen, Christian Breyer, Jose
Donoso, Hubert Fechner, Kenn Henrik Bournonville, Jarand Hole, Izumi Kaizuka, Linda
Koschier, Johan Lindahl, Gaëtan Masson, Daniel Mugnier, Chinho Park, Lionel Perret,
Francesca Tilli

Potential Induced Degradation Evaluation of Damp Heat Stressed PV Modules...................... 755
Farrukh Ibne Mahmood, Akash Kumar, Muhammad Afridi, Govindasamy Tamizhmani

Data-Driven Photovoltaic Module Performance Analysis with FAIR Data 758
Mengjie Li, Jarod Kaltenbaugh, Dylan J. Colvin, William C. Oltjen, Arafath Nihar,
Dominique Akissi Yao, Xuanji Yu, Alp Sehirlioglu, Roger H. French, Kristopher O. Davis

Proposed Update of the Colombian Technical Standard NTC 4405 for Evaluating the Efficiency of
Photovoltaic Solar Systems and Their Components...................... 761
Johann Hernández, Daniel H. Gamboa, Juan F. Beltrán

Aerosol-Deposited SnOx as an Electron Contact in Perovskite Solar Cells...................... 767
David Quispe, David Matthews, Zhengshan J. Yu, Zachary C. Holman

The Temperature Dependence of Auger Recombination in Silicon...................... 768
Jorge Ochoa, Simone Bernardini, Mariana Bertoni

Combining Perovskites and Quantum Dots: Application in Solar Cells 769
Jose Raul Montes Bojorquez, Maria Fernanda Villa Bracamonte, Kevin J. Knebel, Arturo A.
Ayon

Nanostructure Analysis of Parasitic Oxides and Contact Resisitivity Degradation During Annealing
of Silicon Heterojunction Solar Cells...................... 770
Stefan Lange, Angelika Hähnel, Christoph Luderer, David Adner, Martin Bivour, Christian
Hagendorf

Impact of Selenium Doping in CdSeTe-Based Solar Cells at the Atomic-Scale 771
    Arashdeep S. Thind, Jack Farrell, Robert F. Klie

Performance of Vertical Bifacial 2T and 3T Perovskite/Silicon Tandem Solar Farms 772
    Syed Usama Bin Afzal, Hassan Imran, Suleman Sami Qazi, Muhammad Ashraful Alam,
    Nauman Zafar Butt

Influence of Insertion Position of a LiF Buffer Layer on Passivation Performance of Crystalline
$Si/SiO_y/TiO_x/Al$ Heterostructures 775
    Shohei Fukaya, Kazuhiro Gotoh, Takuya Matsui, Hitoshi Sai, Yasuyoshi Kurokawa, Noritaka
    Usami

Radiometric Standards and Best Practices: Recent Progress 778
    Aron Habte, Manajit Sengupta, Christian A. Gueymard

Compact and High Efficiency Micro-CPV Module with High Wafer Utilization Rate 781
    Corentin Jouanneau, Thomas Bidaud, Maxime Darnon, Gwenaelle Hamon

Selenium Diffusion During CdCl2 Treatment of CdSeTe Solar Cells 782
    Niranjana Mohan Kumar, Srisuda Rojsatien, Angel De La Rosa, Barry Lai, Arun K. M.
    Kanakkithodi, Maria Chan, Dan Mao, Mariana Bertoni

Analysis of Measured Operating Temperature of Perovskite Modules. 783
    D. Martinez Escobar, Aaron Wheeler, F. Brigham Pineda, Katty Kaydanik, Alan Murphy,
    Jing-Shun Huang, Mason Terry, Sarah R. Kurtz

A Broadband Anti-Reflection Coating for Thin Film CdSeTe/CdTe Solar Cells 789
    Adam M Law, Luksa Kujovic, Mustafa Togay, Xiaolei Liu, Kurt Barth, John M Walls

Can Solar+Storage Keep the Lights On? Assessing Solar+Storage for Backup Power During Long-
Duration Power Interruptions in the US 793
    Will Gorman, Galen Barbose, Juan Pablo Carvallo, Sunhee Baik, Chandler Miller

Close Roof Mounted System Temperature Estimation for Compliance to IEC TS 63126 794
    Michael D. Kempe, Silvana Ovaitt, Martin Springer, Matthew Brown, Dirk Jordan, William
    Sekulic, Colleen O'Brien, Jean-Nicolas Jaubert, Yuanjie Yu, Jaewon Oh, Govindasamy
    Tamizhmani, Bo Li

I-TOPCon Solar Cells Prepared by High Throughput Magnetron Sputtering of In-Situ Doped n-
Type Amorphous Silicon Layers 795
    Eric Schneiderloechner, Tina Dietsch, Jan Hoss, Jonathan Linke, Jana-Isabelle Polzin,
    Sebastian Mack, Henning Nagel, Volker Linss

Optimizing Demand Management to Enable Renewables: Why the Use of a Marginal Emissions
Signal is a Poor Choice 798
    Ronald A. Sinton

Charge Extraction and Recombination Dynamics of CdSe/CdTe Solar Cells Studied with Transient
Photovoltage/Photocurrent Techniques 804
    Abasi Abudulimu, Dengbing Li, Lei Chen, José Santos, Iwan Zimmermann, Nadeesha
    Katakumbura, Tyler Brau, Ebin Bastola, Adam Phillips, Zhaoning Song, Juan Cabanillas,
    Michael Heben, Mohammad K. Nazeeruddin, Nazario Martín, Yanfa Yan, Randy Ellingson

The Feasibility of Luminescent Solar Concentrators Overlays for Conventional Lens 809
    Xitong Zhu, Michael G. Debije, Angèle H. M. E. Reinders

Highly Efficient Bifacial Single Junction Perovskite Solar Cells.. 812
Qi Jiang, Zhaoning Song, Yanfa Yan, Kai Zhu

Efficiency Maps for Tandem Solar Cells Using High Resolution Spectral Data.............................. 815
Rune Strandberg, Anne G. Imenes

Optimal Allocation of Voltage Regulations to Maximize the Hosting Capacity of Distribution
Systems.. 816
Bahman Ahmadi, Eli Shirazi

In-Situ Microscopy Characterization of Light-Induced Phase Segregation in Wide-Bandgap
Perovskite Materials 822
*Fangfang Cao, Liming Du, Zhiyu Gao, Minghui Li, Cong Chen, Dewei Zhao, Can Li, Zhen
Li, Jichun Ye, Chuanxiao Xiao*

Radiation Tolerance Studies of CdSe/CdTe Bilayer Solar Cells on Space-Qualified Cover Glass 823
*Aesha P. Patel, Adam B. Phillips, Ebin Bastola, Abasi Abudulimu, Zachary W. Zawisza,
Robert Snuggs, Manoj K. Jamarkattel, Deng-Bing Li, Richard Irving, Yanfa Yan, Michael J.
Heben, Randy J. Ellingson*

Advanced Production Line Monitoring with Time-Lag Sequential Analysis..................................... 828
Gaia Maria N. Javier, Rhett Evans, Priya Dwivedi, Thorsten Trupke, Ziv Hameiri

NASA GRC Solar Cell Characterization: Facilities 829
Jeremiah D Sims

Application of Noise-Assisted Multivariate Data Analysis for Hour-Ahead GHI Forecasting 830
Priya Gupta, Rhythm Singh

Can Grid-Following DERs Operate in Parallel with Grid-Forming Resources Without
Compromising Microgrid Stability?.. 835
Wenzong Wang, Aminul Huque

Developing Frequency Stability Constraint for Unit Commitment Problem Considering High
Penetration of Renewables 840
Ningchao Gao, Shuan Dong, Xin Fang, Andy Hoke, David Wenzhong Gao, Jin Tan

Precursor Ink Engineering for Scalable Slot-Die Coating of Perovskite Films for Photovoltaic
Mini-Module Production 844
*Manoj Rajakaruna, Jaehoon Chung, You Li, Tamanna Mariam, Muhammad Saeed Mohsin,
Prabodika N. Kaluarachchi, Amirhossein Rahimi, Zhaoning Song, Michael J. Heben, Yanfa
Yan, Randy J. Ellingson*

GaAs Solar Cells Grown on Acoustically-Spalled GaAs Substrates with 27% Efficiency 848
*Kevin L Schulte, Steve W Johnston, Anna K Braun, Jacob T Boyer, Anica N Neumann, William
E McMahon, Michelle Young, Pablo Guimerá Coll, Mariana I Bertoni, Emily L Warren,
Myles A Steiner*

Why Increased CdSeTe Charge Carrier Lifetimes and Radiative Efficiencies Did Not Result in
Voltage Boost for CdTe Solar Cells? 849
*Darius Kuciauskas, Alexandra Bothwell, Carey Reich, Chungho Lee, Eric Colegrove, Marco
Nardone*

Detection and Impact of Cracks Hidden Near Interconnect Wires in Silicon Solar Cells 850
Andrew M. Gabor, Hubert Seigneur, Philip J. Knodle, Dylan J. Colvin, Kristopher O. Davis

Net Zero Water Strategies and Impacts for PV Manufacturing 855
Parikhit Sinha, Sunil Sajja, Tzy Wei Ooi, Sreenivas Jayaraman, Sukhwant Raju

Characterization of Different Groups of Electricity Consumers and Measures Taken to Reduce
Energy Poverty 858
*Anna Carolina De Paula Sermarini, Lucas Aló Rodrigues Araujo Da Silva, Vanessa Cardoso
De Albuquerque, Rodrigo Flora Calili*

Using Neural Network Decomposition to Estimate Field Photovoltaic Performance Loss Rate 861
*Yangxin Fan, Raymond Wieser, Xuanji Yu, Jennifer Braid, Avishai Shaton, Adam Hoffman,
Thevenard Didier, Ben Spurgeon, Daniel Gibbons, Laura S. Bruckman, Yinghui Wu, Roger H.
French*

Capacitance Transients, Photoconductive Decay, and Impedance Spectroscopy on 19% to 22%
Efficient Silicon Solar Cells 865
Steve Johnston, Dana B. Kern

Development of Next Generation Solar Trackers Based on Shape Memory Alloy to Be Integrated in
CPV/PV Hybrid Modules 871
Alessandro Minuto, Edoardo Celi, Gianluca Timò

Quantifying Uncertainty Due to Climate Variability in Vehicle-Integrated Photovoltaic Yield
Predictions 874
Timofey Golubev

Phase Distributions and Local Bandgap Energies in Mixed-Halide Perovskite Nanoparticles 879
*Dan R. Wargulski, Tal Binyamin, Katrina Coogan, Benedikt Haas, Christoph Koch, Lioz
Etgar, Daniel Abou-Ras*

Third Generation Approaches for Low Cost, Radiation Tolerant, Efficienct Space Solar Cells 881
Gavin Conibeer, Santosh Shrestha

Effect of Novel Optimization Algorithm on the Performance of Photovoltaic Devices.................... 885
Sheri F Michael

Towards Transfer Printing GaSb Membranes Using Selective Etchants 886
Margaret A Stevens, Jill A Nolde, Shawn Mack, Thomas C Mood, Kenneth J. Schmieder

Investigating the Role of Ag and Ga Content in the Stability of Wide-Gap (Ag,Cu)(In,Ga)Se2 Thin
Film Solar Cells.. 887
Patrick Pearson, Jan Keller, Charlotte Platzer-Björkman

A Horizontal Single-Axis Tracker Mock-Up to Quickly Assess the Influence of Geometrical
Factors on Bifacial Energy Gain... 888
César Domínguez, Jesús Marcos, Sandra Ures, Steve Askins, Ignacio Antón

A New Method for the Evaluation of Majority and Minority Carrier Contact Resistivity of
Polysilicon on Oxide Contacts 891
Dirk Steyn, William Nemeth, David Young, Paul Stradins, Sumit Agarwal

Simultaneous Solar Power Generation and Bidirectional Data Transmission 894
*Emily Kessler-Lewis, Stephen J. Polly, Elijah Sacchitella, Seth M. Hubbard, Raymond
Hoheisel*

Localized Surface Plasmon Resonance of Quantum Dots in Two-Step Photon Up-Conversion Solar
Cell Structures 895
Yukihiro Harada, Mizuto Kawakami, Shigeo Asahi, Takashi Kita

Understanding the Degradation of Silicon Heterojunction Modules 896
Jorge Ochoa, Michael Martinez-Szewczyk, Nicholas Moser-Mancewicz, Dana Kern, Dirk Jordan, Mariana Bertoni

Enhancing Inverter Reliability: Current Status and Paths to Predictive Maintenance 897
Wayne Li, Rabin Dhakal, Daniel Fregosi, Curtis Fox, Michael Bolen

Improvements on Spectral Correction Predictive Modeling for CdTe Modules 903
Alan J. Curran, Boris Lin, Yuepeng Deng

Modification of PEDOT:PSS Hole Transporting Layer to Improve the Efficiency and Light Stability of Tin-Lead Perovskite Solar Cells 909
Lei Chen, Chongwen Li, Tyler R. Brau, Abasi Abudulimu, Randy J. Ellingson, Zhaoning Song, Yanfa Yan

Improving V_{OC} of CdSe/CdTe Solar Cells Via Incorporating Oxygenated CdS Between Front Buffer and Absorber 910
Abasi Abudulimu, Dengbing Li, Manoj Jamarkattel, Zachary Zawisza, Scott Wenner, Tyler Brau, Ebin Bastola, Adam Phillips, Michael Heben, Yanfa Yan, Randy Ellingson

From Accelerated Life Test to Accurate Degradation Prediction of CdTe PV Devices: A Modeling Approach 914
Da Guo, Jaliu Ma, Samuel Demtsu, Ryan Monnin, Igor Sankin, Jose A. Calderon, Markus Gloeckler

Quantifying Bulk and Surface Recombination in CdSeTe Absorbers by Modeling Terahertz and Photoluminescence Decays 915
Gregory A Manoukian, Calvin Fai, Deborah L McGott, Finley R Shapiro, Matthew O Reese, Charles J Hages, Jason B Baxter

Data Driven Energy Resilience for Low-To Middle-Income Communities in Puerto Rico 916
Christopher Gregory, Angel Echevarria, Yiamar Rivera-Matos, Daniel D. Campo-Ossa, Alex Routhier, Clark Miller, Richard King

Advances in GaAsP Top Cells for Use in GaAsP/Si Tandems 917
Tal Kasher, Lauren M. Kalizewski, Marzieh Baan, Chuqi Yi, Anastasia H. Soeriyadi, Atom Chang, Udo Römer, Gianluca Coletti, Stephen P. Bremner, Tyler J. Grassman, Steven A. Ringel

Key Areas of Due Diligence for Solar PV Project Financing 918
Eric R. Decristofaro, Matthew R. Thibodeau, Fitzgerald C. Okoli, Jake T. Silhavy, Evan S. Giacchino, Adam W. Loeding, Alexander E. Coologeorgen

Convergence of Efficiency, Stability, and Manufacturability in Perovskite Tandem Solar Cells 919
Rohit Prasanna, Tomas Leijtens, Jochen Titus, Laura E. Crowe, Hyunjong Lee, Annikki L. Santala, Maximilian T. Hoerantner, Giles E. Eperon

Validating View-Factor Approach and Spatial Albedo Models for Bifacial and AgriPV Modeling 920
Silvana Ovaitt, Matthew Boyd, Austin Kinzer, James Jones, Chris Deline

Field and Accelerated Aging of Cracked Solar Cells 921
Michael G. Deceglie, Timothy J Silverman, Ethan Young, William B. Hobbs, Cara Libby

Orange Button: Accelerating the Digital Transformation of Distributed Energy 922
Clifford W. Hansen, Jan Rippingale, Taos Transue, Philip Court, John Gorman

Analysis for Effects of Temperature Rise of Solar Cell Modules Upon the Driving Distance of Photovoltaics-Powered Vehicles 924

Masafumi Yamaguchi, Taizo Masuda, Tsutomu Tanimoto, Yosuke Tomita, Yasuyuki Ota, Christian Thiel, Anastasios Tsakalidis, Arnulf Jaeger-Waldau, Takashi Nakado, Kenji Araki, Kensuke Nishioka, Tatsuya Takamoto, Kyotaro Nakamura, Ryo Ozaki, Nobuaki Kojima, Yoshio Ohshita

Laser-Weld Qualification for a Reliable Aluminum Foil Interconnection of Copper-Metallized Back-Contact Silicon Solar Cells 925

Barry Hartweg, Kathryn Fisher, Jason Ro, Zachary Holman

Conversion Efficiency Analysis of Tandem Solar Cells with Intermediate Band Tunnel Connection 926

Shuhei Yagi, Hiroyuki Yaguchi

Effects of Period of Record Extension, Model Diversification, and DHI Measurements on Measure-Correlate-Predict Analyses for On-Site Solar Resource Assessments 927

Lucila D. Tafur, Renn Darawali, Halley Darling

Solar Forecasting: The Value of Using Satellite Derived Irradiance Data in Machine Learning Based Forecasts 928

Alex Kubiniec, Thomas Haley, Kyle Seymour, Richard Perez

Ageing Detection of Encapsulants and Backsheets in the Field Via NIR Spectroscopy 931

Chiara Barretta, Sascha Lindig, Marton Bredács, Alexander Astigarraga, Eric Helfer, Gernot Oreski

In-Situ & Ex-Situ Study of Protons and Electrons Irradiations of Perovskite Solar Cells 934

Carla Costa, Matthieu Manceau, Thierry Nuns, Sophie Duzellier, Romain Cariou

Numerical Investigation on Non-Radiative Recombination in InGaAs Front and Rear Hetero-Junction Solar Cell 935

Depu Ma, Hassanet Sodabanlu, Gan Li, Meita Asami, Kentaroh Watanabe, Masakazu Sugiyama, Yoshiaki Nakano

Strategies to Improve the Mechanical Robustness of Metal Halide Perovskite Solar Cells 938

Muzhi Li, Siraj Sidhik, Lidon Gil-Escrig, Samuel Johnson, Aditya Mohite, Axel Palmstrom, Henk J. Bolink, Nicholas Rolston

Microstructure-Property Relationships in Epitaxial Cu(In, Ga)Se$_2$ Solar-Cell Absorbers 939

Daniel Abou-Ras, Jiro Nishinaga, Takeyoshi Sugaya, Yukiko Kamikawa-Shimizu, Ulrike Bloeck, Henrik Prell, Sinju Thomas, Michael Tovar, Dan R. Wargulski, Harvey Guthrey, Pat Trimby, Aimo Winkelmann, Shogo Ishizuka

High-Performance Multi-Junction C-Band Photonic Power Converters: Calibrated Optoelectronic Model for Next Generation Designs 941

Gavin P Forcade, Meghan N Beattie, Christopher E Valdivia, Henning Helmers, Oliver H?hn, Paige Wilson, Louis-Philippe St-Arnaud, Robert Hunter, David Lackner, Jacob J Krich, Alexandre W Walker, Karin Hinzer

Strategies for High Fill Factor and Open-Circuit Voltage in Low-Doped c-Ge TPV Cells with Partially Contacted Surfaces Using 3D Simulations 942

M. Gamel, D. Shojaei, J. M. López-González, G. López, M. Garín, I. Martín

Minimizing Sputter Damage-Induced Electrical Losses in Monolithic Perovskite/Silicon Tandem Solar Cells During Deposition of the Transparent Front-Electrode 945

Marlene Härtel, Bor Li, Silvia Mariotti, Philipp Wagner, Florian Ruske, Steve Albrecht, Bernd Szyszka

Nanometer-Scale Imaging on Electrical Potential in Absorber of As-Doped CdSeTe Solar Cells 946
Chun-Sheng Jiang, Eric Colegrove, Steve P. Harvey, Joel N. Duenow, Matthew O. Reese

Advanced Germanium TPV Cells for Latent Heat Thermal Batteries 947
A. Luque, P. Martin, R. Molinero, V. Orejuela, C. Sanchez-Perez, M. Zehender, I. García, I.
Luque-Heredia, I. Rey-Stolle

Evaluation of Irradiance Variability Adjustments for Subhourly Clipping Correction 948
William B. Hobbs, Chloe L. Black, William F. Holmgren, Kevin S. Anderson

Evaluation of Beta-Phase Formation in the Failure of PVDF-Based Solar Module Backsheets 952
Stephanie L. Moffitt, Sona Ulicna, Song-Syun Jhang, Michael Owen-Bellini, Peter Hacke,
Jared Tracy, Kaushik R. Choudhury, Laura T. Schelhas, Xiaohong Gu

Development of a Dynamic Photovoltaic Inverter Model with Grid-Support Capabilities for Power
System Integration Analysis 953
Rachid Darbali-Zamora

PV Fleet Modeling Via Smooth Periodic Gaussian Copula 961
Mehmet G. Ogut, Bennet Meyers, Stephen P. Boyd

Unveiling the Structural Formation of Low Dimensional Layers Deposited on Lead Halide
Perovskites by Thermal Evaporation 969
Carlo Andrea Riccardo Perini, Andres Felipe Castro Mendez, Tim Kodalle, Magdalena
Ravello, Juanita Hidalgo, Juan-Pablo Correa-Baena

Ultrasonic Characterization of Ethylene Vinyl Acetate (EVA) Crosslinking for Quality Assurance
and Lamination Process Control 970
Rico Meier, Ian M. Slauch, Mariana I. Bertoni

Detailed Raman Investigation on the Search for the Secondary Phases in the Chalcogenide
Perovskite BaZrS3 975
Arif Yetkin Hasan, Ruiquan Yang, Charles Hages, Phillip J. Dale

A Direct Comparison of Thermoradiative and Thermophotovoltaic Operation of HgCdTe
Photodiodes 976
Michael P. Nielsen, Muhammad H. Sazzad, Andreas Pusch, Phoebe M. Pearce, Peter J.
Reece, Nicholas J. Ekins-Daukes

Solar Simulator Performance Metrics: Balloon Flown Calibration Standards Offer Real Time AM0
Solar Simulation Error Measurements 977
Scott J Ireton, Casey P Hare, Andrew J Schwab

Operation Efficiency Gains from Analyzing Minimal Solar Cells Production Data 978
Johnson Wong, Kissenger Chen, Dinica Li, Bryan Matthews, Jason Miller, Sal Sanci

Extracting Electrical Properties of CdTe, CdSeTe and CdSe Thin Films Using a Parallel Dipole
Line Hall Effect System 981
Mustafa Togay, Rachael C. Greenhalgh, Kerrie Morris, Xiaolei Liu, Luksa Kujovic, Luis C.
Infante-Ortega, Nicholas Hunwick, Adam M. Law, Tushar Shimpi, Sampath S. Walajabad,
Eric Don, Gabor Parada, Kurt L. Barth, J. Michael Walls, Jake W. Bowers

Power Hardware-In-The-Loop Interface Method for Grid Forming Inverters Using a Voltage-
Controlled Power Amplifier 984
Javier Hernandez-Alvidrez, Rachid Darbali-Zamora, Jack D. Flicker, Nicholas S. Gurule

Implications of Battery Storage for Solar Net-Metering Reforms.. 991
Galen Barbose, Sydney Forrester, Chandler Miller

III-V Solar Cells Grown Directly on V-Groove Si Substrates 994
Theresa E. Saenz, Jacob T. Boyer, John S. Mangum, Anica N. Neumann, Myles A. Steiner,
Ryan M. France, William E. McMahon, Jeramy D. Zimmerman, Emily L. Warren

Correlated Mapping of Raman Spectroscopy and Cathodoluminescence of Emerging Absorber
Bournonite (CuPbSbS3) 995
O M Rigby, C Hill, G Kusch, M Naylor, M Guennou, G Zoppi, M Szablewski, R A Oliver, L
Wirtz, P Dale, B G Mendis

Yb3+- Doped CsPbX3 Nanocrystals for Improving Free-Space Luminescent Solar Concentrators............... 996
Mathis Van De Voorde, Damien Hudry, Dmitry Busko, Bryce S. Richards, Rebecca Saive

Rear Heterojunction GaInP Solar Cells for Improved Performance at Elevated Temperatures..................... 997
Mijung Kim, Yukun Sun, Ryan D. Hool, Minjoo L. Lee

Design and Fabrication of PERC-Like CdTe Solar Cells Using Micropatterned Al_2O_3 Layer........................ 998
Etee Kawna Roy, Kaden Powell, Chungho Lee, Gang Xiong, Heayoung Yoon

Issues, Challenges, and Primary Factors in the Estimation of Floating Solar PV Performance 1002
Rick Meeker, Anna Brinck, Tom Lang, Jason Harrison

Partial Shading of Photovoltaic Modules: A Comparison Between Simulated and Measured IV
Characteristics 1005
Bianca Passarella, Maarten Verkou, Marco Leonardi, Fabrizio Coco, Youri Blom, Malte Vogt,
Rudi Santbergen, Agnese Di Stefano, Andrea Canino, Marina Foti, Antonino Ragonesi,
Marcello Sciuto, Francesco Rametta, Miro Zeman, Olindo Isabella, Cosimo Gerardi

Hierarchal Ti3C2Tx MXene and Aluminum Microgrid Back Contacts for Bifacial CdTe PV 1008
Benjamin E Sartor, Matthew O Reese, Chris Muzzillo, Chungho Lee, Andre D Taylor

Design of Electronic Control of PV Tracking Independent of Weather Forecast 1009
Sam Mil'Shtein, Dhawal Athana, Jeffrey Snell, William Brooks

Efficient and Stable All-Lead Perovskite Tandem Solar Cells Enabled by All-Inorganic CsPbI2Br
Top Cells... 1019
Chongwen Li, Chuanxiao Xiao, Kamala Khanal Subedi, Bin Chen, Randy J. Ellingson, Song
Zhaoning, Yanfa Yan, Edward H. Sargent

(3-Aminopropyl)trimethoxysilane Surface Passivation Improves Perovskite Solar Cell Performance
by Reducing Surface Recombination Velocity 1020
Yangwei Shi, Esteban Rojas-Gatjens, Jian Wang, Justin Pothoof, Rajiv Giridharagopal, Kevin
Ho, Fangyuan Jiang, Margherita Taddei, Zhaoqing Yang, Carlos Silva-Acuña, David S.
Ginger

Machine Learning-Based Defect Identification Method at the c-Si/A-Si:H Interface.................................. 1021
Zitong Zhao, Gonglin Chen, Reza Vatan Meidanshahi, Gergely T. Zimányi

Recombination Analysis of Maxeon IBC Production Cells by Time-Resolved Photoluminescence 1024
David Jacob, Guillaume Von Gastrow, Nils-Peter Harder, Luis Buño, Gerly Reich, Maristel
Baldrias, Roderick J. Marstell, Arnold Castillo, David D. Smith, Michael J. Cudzinovic

Can Hierarchical Physics-Based Machine Learning De-Anonymize Solar Farm Locations?........................ 1027
Jabir Bin Jahangir, Amandeep Singh Bhatia, Muhammad Alam

Investigation and Quantitative Understanding of Front Field Passivation in Rear Junction Selective Double-Side TOPCon Solar Cells 1030
Wook-Jin Choi, Young-Woo Ok, Pradeep Padhamnath, Gabby De Luna, Kwan Hong Min, Ruohan Zhong, Sagnik Dasgupta, Vijaykumar D Upadhyaya, Ajay D Upadhyaya, Ajeet Rohatgi

Autonomous Control Strategies for Interconnected DC Microgrids with Geographical Separation 1034
Emmanuel G. Robles-Rivera, Rachid Darbali-Zamora, Erick E. Aponte-Bezares, Jack D. Flicker, Andrew R. R. Dow, Felipe Palacios, Lee Rashkin

Enabling High Efficiency, Flexible, and Lightweight CdTe Solar Cells with a Cadmium Stannate Transparent Conducting Oxide 1041
Manoj K. Jamarkattel, Adam B. Phillips, Ebin Bastola, Sabin Neupane, Deng-Bing Li, Yanfa Yan, Randy J. Ellingson, Michael J. Heben

Post-Mortem Failure Analysis of Metal Halide Perovskite Modules 1044
Sona Ulicna, Nutifafa Y. Doumon, Michael Owen-Bellini, Jackson Schall, Dana B. Kern, Timothy J. Silverman, Lance M. Wheeler, Steven Hayden, Chengbin Fei, Md Aslam Uddin, Prem J. S. Rana, Jinsong Huang, Joseph J. Berry, Laura T. Schelhas

Tuning Device Interfaces for Improved Open Circuit Voltage in Wide-Bandgap Hybrid Perovskite Photovoltaics 1045
Emily Smith, Sarah Brittman, David Scheiman, Woojun Yoon

Multiple-Reuse of Ge Substrates: Towards Cost-Effective and Sustainable III-V Solar Cells Fabrication 1051
Alexandre Chapotot, Bouraoui Ilahi, Tadeáš Hanuš, Gwenaëlle Hamon, Jinyoun Cho, Kristof Dessein, Christian Dubuc, Maxime Darnon, Abderraouf Boucherif

Mitigation of Potential Induced Degradation in Perovskite Solar Cells 1052
Laxmi Nakka, Shen Guibin, Armin G. Aberle, Fen Lin

A Robust Approach for Daily Solar Irradiance Clustering 1053
Roshni Agrawal, Sivakumar Subramanian, Shashank Agarwal, Venkataramana Runkana

On the Role of Sn-Halide Post-Deposition Reactive Annealing for the Passivation of Defective Surfaces in Cu2ZnSnSe4 1059
Alex Jimenez-Arguijo, Yuancai Gong, David Nowak, Devendrá Pareek, Levent Gütay, Lorenzo Calvo-Barrio, Zacharie Jehl Li-Kao, Sergio Giraldo, Edgardo Saucedo

Analysis of Thermal Behavior and Reliability of Bare Die Diodes Embedded Within PV Modules as Bypass Devices 1060
Luis Eduardo Alanis, Julian Weber, Pascal Romer, Marc Andre Schüler, Louisa Winkler, Udo Steinebrunner, Martin Heinrich, Dirk Holger Neuhaus

Measurement and Analysis of Annual Solar Spectra at Different Installation Angles in Central Europe 1066
Guillermo A. Farias-Basulto, Miguel Á. Sevillano-Bendezú, Maximillian Riedel, Mark Khenkin, Jan A. Töfflinger, Rutger Schlatmann, Reiner Klenk, Carolin Ulbrich

Photoluminescence Analysis of the Back Side of Cu(In,Ga)(S,Se)2 Absorbers 1067
Aubin JC. M. Prot, Susanne Siebentritt, Anastasia Zelenina, Hossam Elanzeery, Alberto Lomuscio, Thomas Dalibor, Maxim Guc, Robert Fonoll-Rubio

Effects of Salt Spray on c-Si Photovoltaic Modules in the Brazilian Region 1068
Mendelsson R. M. Neves, Allan Silveira, Hugo S. Alvarez, Rodrigo M. Garcia, Francisco C. Marques, Marcelo G. Villalva

Experimental Analysis of Distribution Network Voltage Regulation Using Smart Inverters 1074
Rasel Mahmud, Subhankar Ganguly, Jing Wang, Killian McKenna, Ning Li

Doped GaInAs/GaP Quantum Well Superlattice Solar Cells with 27.5% Efficiency 1081
Ryan M France, Myles A Steiner

Aggregated Three-Phase Photovoltaic Inverter Model with Sandia Frequency Shift Islanding
Detection 1082
Nelson E. Saavedra-Peña, Rachid Darbali-Zamora, Edgardo Desarden-Carrero, Erick
Aponte-Bezares

Time-Series Imputation Using Graph Neural Networks and Denoising Autoencoders 1088
Raymond Wieser, Yangxin Fan, Xuanji Yu, Jennifer Braid, Avishai Shaton, Adam Hoffman,
Ben Spurgeon, Daniel Gibbons, Laura S. Bruckman, Yinghui Wu, Roger H. French

Cradle to Cradle Recycling of Perovskite Solar Cells 1092
Zhenni Wu, Gülüsüm Babayeva, Zhang Jiyun, Mykhailo Sytnyk, Jens Hauch, Christoph J.
Brabec, Ian Marius Peters

Per- And Polyfluoroalkyl Substances (PFAS) Usage in Solar Photovoltaics 1095
Preeti Nain, Annick Anctil

Preliminary Gap Analysis of Existing IEEE 1547 and IEEE 2800 Standards Towards Grid-Forming
Technology 1096
Ganesh Marasini, Wenzong Wang, Wes Baker, Deepak Ramasubramanian, Aminul Huque,
Jens C. Boemer

Enhanced Bifaciality Factor with Sb_2Se_3 Devices Modeling Cu_2O Back Buffer 1102
Sanghyun Lee, Kent Price

Analysis of the Key Factors Influencing the Economic Competitiveness and Profitability of
Floating Photovoltaics 1105
Leonardo Micheli, Diego L. Talavera, Fredy A. Sepúlveda-Vélez

Towards Integrating Data Quality Assessments and Radiometer Uncertainty for Determining the
Expanded Uncertainty of Three-Component Solar Radiation Measurements 1106
Stephen Wilcox, Tom Stoffel, Aron Habte, Manajit Sengupta

Stable, High-Throughput Production of Robust Perovskites in Open-Air with Polymer Additives 1109
Nicholas Rolston, Carsen Cartledge, Vineeth Penukula, Muneeza Ahmad, Antonella Giuri,
Aurora Rizzo

Clear-Sky Detection Using Time-Averaged, Tilted-Plane Data 1110
Clifford W. Hansen, Dirk C. Jordan

Theoretical Performance Analysis for Thermo-Radiative Assisted Photovoltaic (TRAP™) Cell
Operating in Outer Space 1114
Jianjian Wang, Nathan Van Velson

A Study of Cell Cracks Formation During Freight Shipping : Monitoring Shock and Temperature in
Real-Time & Assessing Damages with Pre and Post-Transit Characterizations of PV Modules 1120
Cécile Molto, Dylan J. Colvin, Farrukh Ibne Mahmood, Fang Li, Ryan Smith, Govindasamy
Tamizhmani, Hubert Seigneur

Quantifying Real-World Sources of Error in Redundant GHI Measurements 1123
Josh Peterson, Julie Chard, Alex Bryan

Sputtering for the Formation of SI-Based Passivating Contacts..1129
Christophe Allebé, Antoine Descoeudres, Patrick Wyss, Nicolas Pernès, Bertrand Paviet-Salomon, Christophe Ballif, Simon Hübner, Torsten Dippell

Copper Metallization for III-V Solar Cells..1130
Theresa E. Saenz, Anica N. Neumann, Matthew R. Young, Jeramy D. Zimmerman, Emily L. Warren, Myles A. Steiner

An Enhanced Snow-Shedding Model: The Module Frame as a Key Variable 1131
Daniel Riley, Laurie Burnham, Paul Dice, Jennifer Braid

The Effect of Tilt and Azimuth Angle Variations on Monthly and Annual Incident Solar Radiations for Locations in Brazil..1134
Eslam Mahmoudi, Tarcio Andre Dos Santos Barros, Hugo Alvarez, Rodrigo Garcia, Francisco C. Marques, Marcelo Gradella Villalva

Rear-Side Irradiance Simulation of Field PV Modules 1141
Zelin Li, Raymond J. Wieser, Xuanji Yu, Stephanie L. Moffitt, Ruben Zabalza, Silvana Ayala, Matthew Brown, Xiaohong Gu, Liang Ji, Colleen O'Brien, Micheal D. Kempe, Laura S. Bruckman, Kenneth P. Boyce

Extreme Weather and PV Performance 1145
Dirk Jordan, Kirsten Perry, Chris Deline

Viability of a Novel Methodology of Measure-Correlate-Predict for Albedo Estimation 1146
Halley C. Darling, Renn Darawali, Lucila D. Tafur

Long Term Soiling Model Tuning for Enhanced PV Cleaning Schedule Optimization 1147
Kyle Seymour, Akanksha Bhat, Thomas Haley

Coalescence of GaP on V-Groove Si for III-V/Si Solar Cells 1150
Theresa E. Saenz, John S. Mangum, Olivia D. Schneble, Anica N. Neumann, Ryan M. France, William E. McMahon, Jeramy D. Zimmerman, Emily L. Warren

Mitigation of Phase Separation in High Ga Cu(In,Ga)S2 Absorbers to Achieve ~ 1 Volt 15.6% Power Conversion Efficiency 1151
Damilola Adeleye, Mohit Sood, Tobias Törndahl, Adam Hultqvist, Aline Vanderhaegen, Michele Melchiorre, Susanne Siebentritt

Germanium-Tin Diode for Thermophotovoltaic Energy Collection..1152
Amanda Lemire, Kevin A. Grossklaus, Thomas E. Vandervelde

Rapid Thermal Annealing of Symmetric p-TOPCon Silicon Test Structures..1155
Arpan Sinha, Sagnik Dasgupta, Ajeet Rohatgi, Mool C. Gupta

Temperature Dependent Carrier Extraction and the Effects of Excitons on Emission and Photovoltaic Performance in Cs0.05FA0.79MA0.16Pb(I0.83Br0.17)3 Solar Cells..1156
Hadi Afshari, Brandon K Durant, Ahmad R Kirmani, Sergio A Chacon, John Mahoney, Vincent R Whiteside, Rebecca A Scheidt, Matthew C Beard, Joseph M Luther, Ian R Sellers

Alba: Testing Emerging Photovoltaic Technologies in Low-Earth Orbit 1157
Michael D. Kelzenberg, Phillip Jahelka, Richard G. Madonna, Harry A. Atwater

Optimizing the Design of 4-Terminal Perovskite/C-Si Tandem Photovoltaics 1158
Paul F. Ndione, John F. Geisz, Qi Jiang, Gabriella D. Lahti, Rosemary C. Bramante, Kai Zhu, Axel F. Palmstrom, Emily L. Warren

Investigation of EMC Tests in Photovoltaic Inverter According to INMETRO Ordinance No. 140.............1159
Andrei C. Ribeiro, Geyciane P. De Lima, Guilherme C. S. Prym, Paulo R. D. R. Da Silva,
Pedro O. C. M. Neto, Romullo R. M. Carvalho, Francisco V. E. Lemos, Mendelsson R. M.
Neves, Tárcio A. S. Barros, Hugo Da S. Alvarez, Rodrigo M. Garcia, Francisco C. Marques,
Marcelo G. Villalva

PV Inverter Testing for Momentary Cessation and Rate-Of-Change-Of-Frequency Events.............1164
Ramanathan Thiagarajan, Kumaraguru Prabakar, Li Yu, Ken Aramaki, Andy Hoke

Microgrid Design Toolkit Cost Optimization for a Rural Community in Puerto Rico.............1170
Rolando J. Tremont-Brito, Rachid Darbali-Zamora, Robert Broderick, Erick E. Aponte-
Bezares, Efrain O'Neill-Carrillo, Matthew S. Lave

Co-Sputtered Sn-Doped ZnO Thin Film n-Type Layers for Incorporation into CdTe Based
Photovoltaics.............1177
Kerrie M. Morris, Mustafa Togay, Luke Jones, John M. Walls, Jake W. Bowers

Investigating the Impact of MACl Doping in FA-Based Perovskites by Multimodal Synchrotron X-
Ray Techniques.............1183
Yanqi Luo, Sanggyun Kim, Carlo A. R. Perini, Naveed Rahman, Luxi Li, Dina Sheyfer, Ross
Harder, Barry Lai, Juan-Pablo Correa-Baena, Sarah Wieghold

Removing Barriers for Participation of Small PV Systems in Balancing Energy Markets by
Utilizing the Established Smart Meter Eco-System.............1184
Christoph Kondzialka, Manuela McCulloch, Rouven Taubmann, Jonas Dierenbach, Dietmar
Graeber, Gerd Heilscher

A Comparative Study of the Reflectance of Commercial Photovoltaic Modules.............1188
Wei Wen Toh, Min Hsian Saw, Erik Birgersson, Mauro Pravettoni

The Microstructure of Thin Film CdSe Following Cadmium Chloride Activation Treatment.............1193
Rachael C. Greenhalgh, Vladislav Kornienko, Mustafa Togay, Ali Abbas, Ebin Bastola, Adam
B. Phillips, Michael J. Heben, Jake Bowers, J. Michael Walls

High Open Circuit Voltage with Organic Hole Transport Layers in Group V Doped CdSeTe Solar
Cells.............1197
Sabin Neupane, Deng Bing Li, Sandip S Bista, Suman Rijal, Zhaoning Song, Alisha Adhikari,
Lei Chen, You Li, Manoj K. Jamarkattel, Abasi Abudulimu, Dingyuan Lu, Xiaomeng Duan,
Feng Yan, Michael Heben, Randy J. Ellingson, Gang Xiong, Yanfa Yan

Thin-Film Tandem Partners Based on Inline-Processed (Ag, Cu)(In,Ga)Se$_2$.............1201
Theresa Magorian Friedlmeier, Rico Gutzler, Tina Wahl, Dimitrios Hariskos, Stefan Paetel,
Erik Ahlswede, Ana Kanevce, Jan-Philipp Becker

Development of 3D/2D Perovskite Solar Cells Using a Spray-Based Sequential Deposition.............1206
Neetesh Kumar, Rishabh Sahani, Daniel Muñoz, Cheng-Yu Lai, Daniela Radu

Numerical Modeling of Bifacial Thin Film Solar Cells.............1210
Briana Dokken, Sabin Neupane, Muhammad Mohsin Saeed, Jared D. Friedl, Adam B.
Phillips, Michael J. Heben, Yanfa Yan, Zhaoning Song

Passivating Contacts with Engineered Pinhole Enabled Transport.............1216
Harvey L Guthrey, Caroline Lima Salles, William Nemeth, Sumit Agarwal, David Young,
Pauls Stradins

InP-Based Tunnel Junctions for Micro-Concentrator Photovoltaics 1217
  Kenneth J Schmieder, Thomas C Mood, Eric A Armour, Mitchell F Bennett, Margaret A
  Stevens, Martin Diaz, Ziggy Pulwin, Matthew P Lumb

Residual Stress Limits Gridline Bridging in Cracked Solar Cells 1218
  Junki Joe, Timothy J Silverman, Michael Owen-Bellini, Nick Bosco

Optimizing the Laser Scribing Process to Achieve a Certified Efficiency of 25.9% for Over 240
cm² Four-Terminal Perovskite/Si Tandem Solar Cells 1221
  Yonglei Wang, Hongxu Zhang, Yang Liu, Yuan Qin, Jie Liu, Jiang Liu, Bo He, Xixiang Xu

Predicting Damp Heat Degradation in Heterojunction PV Modules Using Machine Learning 1224
  Zubair Abdullah-Vetter, Felix O'Kearney, Priya Dwivedi, Robert Lee Chin, Brendan Wright,
  Thorsten Trupke, Ziv Hameiri

Demand Following RE – a Demand Driven Approach for Rapid RE Capacity Addition in India 1227
  Prashant Kumar Upadhyay, Himanshu Gulati, Yellasiri Bharath Kumar Reddy

Perspectives on PV Adoption and Engaging Gen Z and Millennials in the Indian Scenario 1230
  Robins Anto, Rhythm Singh

An Analysis of the Current Status and Future Potential of Rooftop Solar Adoption in the United
States 1233
  AC Lemay, BP Rand

Investigation of Varying Se Vapor Pressure During Deposition of CdSeTe Thin Film PV Devices 1234
  Sushmakanth Myneni, Carey Reich, Daniel Shaw, Sampath Walajabad, Amit H. Munshi

Performance Optimization of the CdSe$_x$Te$_{1-x}$/CdTe Solar Cell 1237
  Md Zahangir Alom, Sheikh Tawsif Elahi, Vasilios Palekis, Wei Wang, Chris Ferekides

Analysis of Optoelectronic Characterization Data Via Bayesian Inference: A Desktop-Scale MCMC
Method 1241
  Calvin Fai, Gregory A. Manoukian, Jason B. Baxter, Anthony J. C. Ladd, Charles J. Hages

Generalizability of Neural Network-Based Identification of PV in Aerial Images 1242
  Joseph Ranalli, Matthias Zech

Modal Analysis of GaAs Nanowire Solar Cells for Optimal Device Design 1249
  Venkata S. A. Chaluvadi, Eduardo Camarillo Abad, Hannah J. Joyce, Louise C. Hirst

Benchmark Tests for IV Fitting Algorithms 1250
  Clifford W. Hansen, Abigail R. Jones, Taos Transue, Marios Theristis

Validation of Photovoltaic Plant Loss Estimation from Monitoring Data: String Faults, Shading and
Degradation 1253
  Karel De Brabandere, Maitheli Nikam, Julien Deckx, Gofran Chowdhury

Single-Junction Bifacial and Semitransparent Sb$_2$(S,Se)$_3$ Solar Cells 1256
  Chen Qian, Kaiwen Sun, Martin Green, Bram Hoex, Xiaojing Hao

The United States Renewable Energy Landscape: Siting, Management, and Potential Impacts 1259
  Jacob T. Stid, Anthony D. Kendall, Annick Anctil, Jeremy Rapp, David W. Hyndman

Interpreting Accelerated Tests on Perovskite Modules: Using Photooxidation of MAPbI3 as an Example 1260
> Ingrid L. Repins, Michael Owen-Bellini, Michael D. Kempe, Michael G. Deceglie, Joseph J. Berry, Nutifafa Y. Doumon, Timothy J. Silverman, Laura T. Schelhas

Charge Carrier Diffusion and Recombination Near Misfit Dislocations in GaAsP/GaInP Heterostructures 1261
> T. H. Gfroerer, A. Edmondson, Lilian Korir, Fan Zhang, Yong Zhang, M. W. Wanlass

Improving Operational Stability of High-Efficiency Inverted Perovskite Solar Cells 1265
> Kai Zhu

Influence of Aluminum Co-Doping on Current-Induced Degradation and Regeneration Kinetics in Boron-Doped Cz PERC Solar Cells 1266
> August Weber, Mahsa Mohammadi, Matthias Trempa, Thomas Buck, Johannes Heitmann, Matthias Müller

Reducing the Photovoltaic Operation and Maintenance Costs Through an Autonomous Control Operation Center 1270
> Andreas Livera, Álvaro Fernández-Solas, Joao G. Bessa, Jesús Montes-Romero, Eduardo F. Fernández, Vassilis Papaeconomou, George E. Georghiou

Large-Area Uniformity Mapping of High-Speed Flexography-Printed Perovskite Solar Cells Via Scanning Photoluminescence 1276
> Julia E. Huddy, William J. Scheideler

Measuring the Doping Concentration of Si and CdTe Absorbers Using Lock-In Amplified Quantitative QSSPL 1279
> Mason P Mahaffey, Arthur Onno, Carey Reich, Adam Danielson, Walajabad Sampath, Zachary C Holman

Peer-To-Peer Energy Trading for PV Prosumers Using Fuzzy Logic Inference Systems 1280
> Hector Lopez, Ali Zilouchian

Carbon Quantum Dots and Their Possible Application in Perovskites Passivation 1283
> Maria Fernanda Villa-Bracamonte, Jose Raul Montes-Bojorquez, Alan Lopez-Becerra, Arturo Ayon

A New Combined Accelerated Stress Test Sequence for Rapid Reliability Screening of Photovoltaic Materials 1284
> Yi Jiang, Xuanji Yu, Ben Huang, Ruirui Lv, Yuanjie Yu, Tao Xu, Guangchun Zhang

Improved Soiling Rate Estimation by Calculating PV Module Temperature Using a Distributed Thermal Model 1287
> Shoubhik De, Yogeswara Rao Golive, Narendra Shiradkar, Anil Kottantharayil

Electrical and Electroluminescence Evaluation of 17 Year Old Monocrystalline Silicon Building Integrated Photovoltaic Modules 1292
> Andrew M. Shore, Tali Schlenoff, Bakary Coulibaly, David Navon, Stephanie L. Moffitt, Brian Dougherty, Behrang H. Hamadani

Automated Photovoltaic Module Quality Assessment: Defect Identification and Classification from Luminescence Images Using Machine Learning 1295
> Brendan Wright, James Petesic, Timothy Dawson, Ziv Hameiri

Influence of Alkali Iodide Fluxes on Cu2ZnSnS4 Monograin Powder Properties and Performance of Solar Cells 1298
  Kristi Timmo, Katri Muska, Maris Pilvet, Mare Altosaar, Valdek Mikli, Mati Danilson, Reelika Kaupmees, Jüri Krustok, Maarja Grossberg-Kuusk, Marit Kauk-Kuusik

Pyrolyzer Assisted Vapor Transport Deposition of Antimony-Doped Cadmium Telluride 1299
  Bin Du, Kevin Dobson, Brian McCandless, Aayush Nahar, Ujjwal Das, Shannon Fields, Aaron Arehart, William Shafarman

Early Degradation Trend Estimation of Bifacial PV, Investigating the Seasonality Effect 1304
  Gaetano Mannino, Giuseppe Marco Tina, Mario Cacciato, Lorenzo Todaro, Agnese Di Stefano, Fabrizio Bizzarri, Andrea Canino

The Rise and the Decay of the Photovoltage in Perovskite Solar Cells 1307
  Uwe Rau, Lisa Krückemeier, Sandheep Ravishankar, Thomas Kirchartz

Cadmium Zinc Telluride as an Electron Reflecting Back-Contact Layer for CdTe Solar Cells 1310
  Camden Kasik, Jennifer Drayton, James Sites

Innovative Layouts for Utility-Scale PV Modules: Module Characteristics, Shading Tolerance, and Electricity Costs 1313
  Li C. Rendler, Christian Reichel, Matthew F. Berwind, Anna Heimsath, Dirk H. Neuhaus

UV Absorption Utilizing a MoS2/Ge Nano-Junction for Solar Applications 1320
  Ayman Rezk, Ammar Nayfeh

Defect Signatures in Admittance Spectroscopy of Perovskite Solar Cells 1323
  Rasha A. Awni, Zhaoning Song, Jared D. Friedl, Abasi Abudulimu, Adam B. Phillips, Randall J. Ellingson, Michael J. Heben, Yanfa Yan

Modeling the Hardware Components of a Power Hardware-In-The-Loop Platform for Photovoltaic Applications 1326
  Edgardo Desarden-Carrero, Rachid Darbali-Zamora, Nelson E. Saavedra-Peña, Erick E. Aponte-Bezares

Effective and Equivalent Refractive Index Models for Patterned Solar Cell Films Via a Robust Homogenization Method 1334
  Ekin Gunes Ozaktas, Sreyas Chintapalli, Susanna M. Thon

Evaluation of Rear Contact Passivation Strategies Via Surface Photovoltage Spectroscopy 1337
  Nathan Rock, Chien-Hsuan Chen, Kristopher Davis, Amit Munshi, Michael Scarpulla

Machine Learning, Unmanned Vehicles, and Energy: A Review 1338
  Brian L. Reyes Santiago, Eduardo Ortiz-Rivera

III-V Epitaxy on Detachable Porous Germanium 4" Substrates 1341
  Waldemar Schreiber, Jens Ohlmann, Patrick Schygulla, Stefan Janz, Jinyoun Cho, Kristof Dessein

Effect of Soiling from Dust Particles on Solar Cell Efficiency in the United Arab Emirates (UAE) 1344
  Muntaser Abdelrahman Almansoori, Rawdah Almannaee, Mariam Aldhefairi, Meerh Alsuwaidi, Ayman Rezk, Ammar Nayfeh

Efficiency Limit of Transition Metal Dichalcogenide Solar Cells 1347
  Koosha Nassiri Nazif, Frederick U. Nitta, Alwin Daus, Krishna C. Saraswat, Eric Pop

Advanced Health-State Data Analytic Workflow for Utility-Scale Photovoltaic Power Plants 1348
Jesus Montes-Romero, Loucas Pikolos, Andreas Makrides, Nino Heinzle, George Makrides, Juergen Sutterlueti, Steve Ransome, George E. Georghiou

Monolithic Bifacial Perovskite-CdSeTe Tandem Solar Cells 1353
Zhaoning Song, Deng-Bing Li, Sabin Neupane, Kamala Khanal Subedi, Samuel Seibert, Randy J. Ellingson, Yanfa Yan

Improving the Performance and Yield of Colloidal Quantum Dot Solar Cells Through Electron
Transport Layer Optimization 1356
Dana Kachman, Arlene Chiu, Dhanvini Gudi, Chengchangfeng Lu, Eric Rong, Sreyas Chintapalli, Yida Lin, Daniel Khurgin, Susanna M. Thon

Uncertainty Considerations in Bifacial Photovoltaic Systems with High Albedo Seasonality 1359
Javier Lopez-Lorente, Anja Neubert, Mike Hamer

India as an Emerging Solar Manufacturing Country 1366
Juzer Vasi, Mrunal Berad, Narendra Shiradkar, Anil Kottantharayil, Dinesh Kabra, Kedar Deshmukh, Aditi Chaubal, Rajeewa Arya, Probir Ghosh, Satyendra Kumar, Lawrence L. Kazmerski

IBC Technology Targeting Fast and Effective Silver Reduction Applying Advanced Screen:
Printing 1372
Radovan Kopecek, Florian Buchholz, Valentin D. Mihailetchi, Joris Libal, Ning Chen, Haifeng Chu, Christoph Peter, Dominik Rudolph, Thomas Buck, Tudor Timofte, Andreas Halm, Yonggang Guo, Xiaoyong Qu, Xiang Wu, Jiaqing Gao, Peng Dong

Development of Machine Vision System for Detection of Wrap-Around in n-TOPCon Solar Cells 1373
Junhee Kim, Han-Jung Kim, Yohan Ko, Yoonkap Kim

Effect of CdS Annealing on the Performance of Antimony Selenosulfide Solar Cells 1376
Alisha Adhikari, Suman Rijal, Sabin Neupane, Manoj K. Jamarkattel, Deng-Bing Li, Tamanna Mariam, Michael J. Heben, Zhaoning Song, Yanfa Yan

20%-Efficient TOPCon Solar Cell with a Silicon Oxide Layer Deposited by Aerosol Impaction-
Driven Assembly 1379
Maria Angelica M Garcia, William Weigand, Zachary C Holman

Understanding Practical Efficiency Limits for Tandem Solar Cells 1380
Emily L. Warren, Sirazul Haque, Qi Jiang, William E. McMahon, Lorelle M. Mansfield

Effect of Angle and Direction of Solar Panels in the Desert Climate of Abu Dhabi, United Arab
Emirates 1381
Laith Nayfeh, Leia Nayfeh, Ammar Nayfeh

Per-Unit Dynamic Models for Grid-Following Photovoltaic Inverters 1384
Hyeonjung Tari Jung, D. Venkatramanan, Manish K. Singh, Sairaj Dhople

'There and Back Again': Reusable Germanium Wafers with Ge-On-Nothing Structures for Triple-
Junction Solar Cells 1390
Valérie Depauw, Guillaume Courtois, Jinyoun Cho, Kristof Dessein, Clément Porret, Roger Loo

Characterizing Capacity Contribution of Renewable Resources Over Time in Renewable-Heavy
Transmission System: MISO Case Study 1393
Hyeonjung Tari Jung, Megan Pamperin, Elspeth McGarvey, Eduardo Ibanez, James Okullo, Armando Figueroa Acevedo, Jordan Bakke

Device Modeling of HTL/BaSi$_2$ Heterojunction Solar Cells... 1398
Sho Aonuki, Carlos Mario Ruiz Tobon, Rudi Santbergen, Olindo Isabella, Takashi Suemasu

Modulating Efficiency and Stability of Methylammonium/Br-Free Perovskite Solar Cells Using Fluoroarene Hydrazine 1401
Dhruba B. Khadka, Yasuhiro Shirai, Masatoshi Yanagida, Kenjiro Miyano

In-Flight Validation of End-Of-Life Optimized Triple Junction Solar Cells Onboard ASTROBIO Cubesat 1404
Luigi Schirone, Pierpaolo Granello, Matteo Ferrara, Matteo Avoli, Davide Imperatori, Nithin Maipan Davis, Lorenzo Iannascoli, Augusto Nascetti, Stefano Carletta, Claudio Paris, Erminio Greco, Roberta Campesato

Evaluation of Module Mismatch Losses and Generation Impact in Utility Scale PV Systems..................... 1409
Sara M. Macalpine, David A. Bowersox

SCAPS-1D Simulations of CdTe Based Solar Cells with an Amorphous Silicon-Based Back Buffer........... 1410
Abdul Quader, Venkanna Kanneboina, Prabin Dulal, Madan Mainali, Bishal Shrestha, Ebin Bastola, Adam B. Phillips, Ambalanath Shan, Nicholas J. Podraza, Michael J. Heben

Impact of Current Collecting Grids on the Scalability of 3-Terminal Perovskite/Silicon Tandems with Bipolar Transistor Architecture 1413
Gemma Giliberti, Federica Cappelluti

Siting Optimization of PV Recycling Plants for Supply Chain Security and Critical Material Recovery... 1416
Macarena Mendez Ribo, Silvana Ovaitt, Hope Wikoff, Heather Mirletz, Samantha Reese

Environmentally Controlled Electroluminescence/Photoluminescence Imaging System with Current Density-Voltage Capabilities for Quantitative Degradation Analysis of Perovskite Thin Film Solar Cells... 1417
Tamanna Mariam, Zahrah S. Almutawah, Adam B. Phillips, Sheng Fu, Jaehoon Chung, You Li, Manoj Rajakaruna, Kshitiz Dolia, Zhaoning Song, Randy J. Ellingson, Yanfa Yan, Michael J. Heben

Method for Evaluating the Silicon Solar Cells Performances Under AM0 Thanks to AM1.5G Spectrum... 1423
Philippe Voarino, Adem Dahi, Romain Cariou

Fatigue Debonding of EVA from Solar Glass at Elevated PV Service Temperatures........................... 1428
Gernot Wallner, Gabriel Riedl, Philipp Haselsteiner, Robert Pugstaller

Demonstration of Dual-Junction ELO Solar Cells with Strain-Balanced and Lattice-Matched Quantum Well Absorbers.. 1429
Rao Tatavarti, Andree Wibowo, Mitsuru Imaizumi, Takeshi Ohshima, David Wilt, Roger Welser

Cu2ZnSnS4 Monograin Layer Solar Cells for Flexible Photovoltaic Applications 1432
Marit Kauk-Kuusik, Kristi Timmo, Maris Pilvet, Katri Muska, Mati Danilson, Jüri Krustok, Raavo Josepson, Maarja Grossberg-Kuusk

Advances in Flexible and Lightweight III-V Multijunction Solar Cells for High Power Density Applications... 1433
Carlos Algora, Ivan Garcia, Clara Sanchez-Perez, Pablo Martin, Pablo Fernandez, Luis Cifuentes, Ivan Lombardero, Daniel Gomez, Mercedes Gabas, Ignacio Rey-Stolle

Improving the Stability of Polycrystalline Silicon Passivated Contacts Using Titanium Dioxide 1434
Di Yan, Jesus Ibarra Michel, Yida Pan, Sieu Pheng Phang, Daniel Macdonald, Heping Shen, Leiping Duan, Kylie Catchpole, Jie Yang, Peiting Zheng, Xinyu Zhang, Hao Jin, James Bullock

Copper Oxide: A Potential Candidate for Hole Transport Material in Perovskite Solar Cells for Space 1437
Daniel Muñoz-Pinzon, Rishabh Sahani, Mateo Ferreira, Neetesh Kumar, Cheng-Yu Lai, Daniela Radu, Lyndsey McMillon-Brown

Innovative Methodology for an Advanced Characterization of Perovskite Systems to Reach Buried Interfaces: In-Depth Profile by Coupling GD-OES and XPS 1440
Mirella Al Katrib, Pia Dally, Armelle Yaiche, Jean Rousset, Muriel Bouttemy

Optimization of Optical and Electrical Properties of 2T Textured Perovskite/Silicon Tandem Solar Cell Structure 1443
Chun-Hao Hsieh, Jun-Yu Huang, Yuh-Renn Wu

Development of an Ultra-Light Curvilinear Prismatic Window Which Mitigates Reflections and Glare for PV Modules and Other Surfaces 1446
Mark O'Neill, Chris Youtsey, Robert McCarthy

Evaluating Multi-Bias Modulation for Diagnostics of PV Modules in Daylight Electroluminescence Inspections 1452
Rodrigo Del Prado Santamaría, Gisele A. Dos Reis Benatto, Thøger Kari, Peter B. Poulsen, Sergiu V. Spataru

Recomibiation Center Defects Induced by TCO Reactive Plasma Deposition in Carrire Selective Contact Solar Cells 1453
Yoshio Ohshita, Tomohiko Hara, Taichi Tanaka, Keita Kimura, Yuto Ifuji

Predicting Site-Specific Adjustments to P50 Energy Production Estimates from Sub-Hourly Irradiance Data 1454
Faisal Rashed

Self-Assembled Monolayer Patterning for PolySi/SiO$_2$ Passivated Contacts 1459
B. Nemeth, D. L. Young, M. R. Page, V. Lasalvia, S. Theingi, P. Stradins

Non-Ionizing Radiation Effects on the Room Temperature Surface Recombination Velocity of Unintentionally Doped AlGaAs/GaAs Heterostructures 1462
Andrew Hudson, Daniele Monahan

Flexible Organic Solar Cells on Ti Foil Substrate 1466
Huiying Jia, Lei Kerr, Benjamin Leever

Novel Approach to Control Environmental Fatigue Tests on Glass/PV Encapsulant Laminates 1467
Gabriel Riedl, Gernot M. Wallner, Robert Pugstaller

Automated Analysis of Internal Quantum Efficiency Measurements of GaAs Solar Cells Using Machine Learning 1468
Zubair Abdullah-Vetter, Priya Dwivedi, N. J. Ekins-Daukes, Thorsten Trupke, Ziv Hameiri

RF-Powered Sputtering of Iron Pyrite for Photovoltaic Applications 1471
Awais Zaka, Ayman Rezk, Sabina Abdul Hadi, Saeed Alhassan, Ammar Nayfeh

Analysis and Identification of Measurement Uncertainty Sources of a LED Sun Simulator with Double-Side Illumination for Bifacial PV Module Power Rating 1475
Sebastian Dittmann, Giuliano L. Martins, Ralph Gottschalg

The Effects of Global Damp Heat Ageing on Debonding of Polyolefin Glass Laminates 1481
Martin Tiefenthaler, Gernot M. Wallner, Robert Pugstaller

Nanographene (NG)-Based Hole Transporter with π- Interface Modifier for Thermally Stable Perovskite Solar Cells 1484
Seul-Gi Kim, Thybault De Monfreid, Jeong-Hyeon Kim, Fabrice Goubard, Joseph J. Berry, Kai Zhu, Thanh-Tuân Bui, Nam-Gyu Park

Automated Workflows for Machine Learning on Photovoltaic Timeseries and UV Fluorescence Image Datasets Using FAIR Principles 1485
William C. Oltjen, Xuanji Yu, Mengjie Li, Dylan J. Colvin, Yijia Sun, Hubert Seigneur, Philip Knodle, Andrew M. Gabor, Laura S. Bruckman, Kristopher O. Davis, Roger H. French

Post-Flight Analysis of Perovskite Solar Cells for NASA Materials International Space Station Experiment (MISSE) 1490
Kaitlyn Vansant, Ahmad Kirmani, Severin Habisreutinger, Steve Johnston, Brian Wieliczka, Joseph Luther, Timothy Peshek, Lyndsey McMillon-Brown

Cleaning Optimization for Photovoltaic Powerplants: A Novel Approach Combining Techno-Economic Modelling with Historic Rain and Soiling 1491
Thore Müller, Kostiantyn Pogorelov, Franco Clandestino

Characterization of Solar Cell Busbar Grid for Different Technologies by Time Domain Reflectometry Simulation: Transmission Line Approach 1492
A. M. C. Silveira, M. R. M. Neves, R. Garcia, H. Alvarez, M. G. Villalva, F. C. Marques, L. C. Kretly

19.5% Efficient CdSeTe/CdTe Solar Cells Using ZnO Buffer Layers 1497
Luksa Kujovic, Xiaolei Liu, Mustafa Togay, Luis C. Infante-Ortega, Kurt L. Barth, Jake W. Bowers, John M. Walls, Ochai Oklobia, Stuart J. C. Irvine, Wei Zhang, David W. Miller, Timothy Nagle, Rajni Mallick, Dingyuan Lu, Wyatt K. Metzger, Gang Xiong

Numerical Evaluation of Optimal Tilt Angle for Energy Production and Minimum Shadowing for Bifacial Solar Modules 1500
Roberto Corso, Fabio Matera, Salvatore A. Lombardo

Optimization of 1-Axis Tracking with N-S Rotating-Axis Orientation 1503
Jiahui Shi, Xitao Liu, Teliang Mu, Xiaotong Feng, Vasilis Fthenakis

A Crucial Role of Spin-Dry Cleaning on the Surface Passivation Quality of Crystalline Silicon 1509
Munan Gao, Vibhor Kumar, Ngwe Zin

Holistic Assessment of Monocrystalline Silicon (mono-Si) Solar Panels with Recycled Content Vs. Virgin-Grade Materials 1510
Christopher C. Bondoc, Ross Lee, Mary E. McRae, Pritpal Singh

Effect of Thickness of Electron Reflector Layer on the Efficiency of CdS/CdTe Heterojunction Thin-Film Solar Cell 1513
Chaitanya Santosh Rampalli, Hamid Fardi

Fabrication Au/TiO_x Nanoislands Systems by a Solid State Thermal Dewetting for Plasmonic Solar Cell Applications 1516
Brahim Aïssa, Mohammad I. Hossain, Adnan Ali

Detection and Analyze of Off-Maximum Power Points of PV Systems Based on PV-Pro Modelling........... 1519
Baojie Li, Xin Chen, Anubhav Jain

Laser Recycling of Silver from Waste Silicon Solar Cells 1522
Mahantesh Khetri, Pawan K. Kanaujia, Mool C. Gupta

Study on Air Gap Effects on Photovoltaic Modules Operating Temperature on Typical Metal
Rooftop Appliation 1523
Quanzhi Wang, Yuanjie Yu, Tao Xu

Development of P-Type Silicon Heterojunction Solar Cells with 26.6% Efficiency 1526
*Xiaoning Ru, Miao Yang, Yichun Wang, Jianqiang Wang, Chaowei Xue, Shi Yin, Chengjian
Hong, Fuguo Peng, Minghao Qu, Junxiong Lu, Liang Fang, Tian Xie, Zhenguo Li, Xixiang
Xu*

Oxy-Fuel Combustion: A Threat Or an Opportunity for Solar? 1529
Mariela Colombo, Sarah Kurtz

High-Throughput In-Line Deposition of Silicon Oxide Passivation Layers in Silicon TOPCon Solar
Cells... 1533
*Zachary B. Leuty, William J. Weigand, Jorge Ochoa, Joe V. Carpenter, Mariana I. Bertoni,
Zachary C. Holman*

Measurement and Control of Mobile Ion Concentration in Halide Perovskites... 1534
Saivineeth Penukula, Nicholas Rolston

2D-GaSe/In$_x$Se$_y$ Layer for Rapid ELO GaAs Technique... 1535
Nobuaki Kojima, Yoshio Ohshita, Masafumi Yamaguchi

NSF Industry-University Cooperative Research Center (IUCRC) for a Solar Powered Future 2050
(SPF2050).. 1537
Amit Harenkumar Munshi, Walajabad S. Sampath, Brian Korgel

Indoor and Outdoor Characterization of III-V/Ge Solar Cells Assembled on Glass Substrate for
Concentrated Photovoltaic Applications.. 1538
*K. Kouame, J. Kinfack, D. Danovitch, P. Albert, T. Bidaud, A. Turala, M. Volatier, V. Aimez, A.
Jaouad, M. Darnon, G. Hamon*

Survey of Module and System Quality in Brazil PV Deployments... 1541
*Lawrence L. Kazmerski, Denio Alves Cassini, Daniel Sena Braga, Suellen C. S. Costa,
Vinícius Camatta Santana, Antonia Sonia A. C. Diniz*

690 WP N-Type i-TOPCon Modules in Mass Production with >25% Efficiency Solar Cells Based
on Large-Area 210 mm Wafers 1544
*Yifeng Chen, Hong Chen, Shu Zhang, Le Wang, Chengfa Liu, Daming Chen, Jianmei Xu,
Pietro Altermatt, Zhiqiang Feng, Pierre Verlinden*

Methylamine Post-Deposition Treatments of Vapor-Deposited Perovskite Thin Films 1545
*Chaiwarut Santiwipharat, Austin G. Kuba, Kevin D. Dobson, Ujjwal K. Das, William N.
Shafarman*

Optimization of Sb2Se3 Thin Films Prepared by Selenization of Sb Metallic Precursors for
Photovoltaic Application 1549
*Woo Kyoung Kim, Vasudeva Reddy Minnam Reddy, Sreedevi Gedi, Salh Alhammadi, Ignatius
Andre Setiawan, Yujeong Ahn, Songhee Lee, Hyomin Kim*

Narrow Bandgap Perovskite Solar Cell Degradation Monitoring by Spectroscopic Ellipsometry 1550
Marie Solange Tumusange, Madan K. Mainali, Lei Chen, Zhaoning Song, Yanfa Yan, Nikolas J. Podraza

Solution Processed N+ CdS/ n-CdTe/ Perovskite Heterojunction Thin-Film Solar Cells 1551
Isaiah Henry, Dakota Schwartz, Harry Larson, Shubhra Bansal

Effect of Double Cation Substitution on Nonradiative Recombination Losses in Cu2ZnSn(S,Se)4 Solar Cells 1556
Vijay Karade, Kiwhan Kim, Jae Ho Yun, Jin Hyeok Kim

Mapping Spatial Variations of Wide Band Gap Perovskite Thin Films.. 1557
Emily Miller, Kshitiz Dolia, Bailey Frye, Yanfa Yan, Zhaoning Song, Nikolas J. Podraza

Exploring Distributed PV Power Measurements for Real-Time Potential Power Estimation in Utility-Scale PV Plants.. 1558
Michael Gostein, William B. Hobbs

Controlling Photoexcited Carrier Relaxation Through Phonon Management in GaAs/AlAs Superlattices 1564
Muhammad Hanif, Milos Dubajic, Stephen P Bremner, Michael P Nielsen, Santosh Shrestha, Gavin J Conibeer

Controlling Residual Stresses for Scalable Open-Air Fabrication of Perovskite Solar Cells........................ 1565
Muneeza Ahmad, Carsen Cartledge, Nicholas Rolston

Silver Recovery Through a Fluoride Chemistry for Solar Module Recycling 1566
Theresa K Chen, Meng Tao

Thermal Models of Monofacial and Bifacial PV Modules: Machine Learning and Physical Estimation Models Comparison 1567
Marco Grisanti, Gaetano Mannino, Giuseppe Marco Tina, Alessandro Ortis, Mario Cacciato, Sebastiano Battiato, Fabrizio Bizzarri, Andrea Canino

In-Situ Photostability Analysis of Perovskite Solar Cells by Time-Evolving Photoluminescence Imaging.. 1570
Jackson W. Schall, Amy Louks, Goutam Paul, Nikita S. Dutta, Steve Johnston, Chun-Shen Jiang, Axel Palmstrom, Mowafak Al-Jassim, Dana B. Kern

Flexible Manufacturing of Colloidal Quantum Dot Solar Cells Via Spray-Casting Techniques.................... 1575
Lulin Li, Botong Qiu, Yida Lin, Laura Shimabukuro, Alex Ozbolt, Keyi Kang Yao, Stephen Farias, Samuel Rosenthal, Susanna M. Thon

Detection of PV Module Temperature Coefficient Using Machine Learning 1578
John M. Obrecht, Julián Ascencio-Vásquez

Widegap CdSe Solar Cells with V_{OC} >750mV 1582
Taylor Hill, Sachit Grover, James Sites

Eliminating the Need for Handling Individual Sub-Cells for Small Appliance PV Modules with Voltage Demands Above 12V.. 1588
Jan Paschen, Andreas Brand, Matheus Melati Menegassi, Oliver John, Jan Nekarda

Setting Priorities for Photovoltaic Reliability Research Using Criticality Analysis..................................... 1593
Ingrid L. Repins, Michael G. Deceglie, Timothy J. Silverman, David C. Miller, Dirk C. Jordan, Michael Woodhouse, Teresa M. Barnes

Uncertainties in PV Power Simulation Chain......................1594

Lubos Helienek, Jozef Rusnak, Branislav Schnierer, Martin Opatovsky, Lukas Dvonc, Vicente Lara Fanego, Artur Skoczek, Tomas Cebecauer

Characterizing TeO_2 Formation in CdTe Devices Using Transmission Electron Microscopy......................1600

John Farrell, Ebin Bastola, Manoj Jamarkattel, Michael Heben, Walajabad S. Sampath, James Sites, Robert F. Klie

Dense Array TPV Modules with Alternating Polarity InGaAs Cells......................1603

Iván García, Aitana Cano, Víctor Orejuela, Pablo Martír, Ignacio Rey-Stolle

Influence of Spectral Albedo on the Performance of Lead-Free Perovskite Bifacial Tandem Solar Cell......................1606

Atanu Purkayastha, Arun Tej Mallajosyula

Outdoor Characterization of a Bifacial Four-Terminal GaAs/Si Mini-Module Under Different Albedo Conditions......................1609

Roberto Corso, Fabio Matera, Andrea Scuto, Salvatore A. Lombardo

Evaluation of PV Snow Loss Models in the East Coast of Canada Using AI Computer Vision......................1612

Jessica Ma, Alexandre Khoury, Marianne Rodgers

Influence of NaF and KF Post-Deposition Treatment on the Sub-Band Gap Absorption of $Cu(In,Ga)Se_2$ Absorber Layers......................1617

Sevan Gharabeiki, Michele Melchiorre, Susanne Siebentritt

The Importance of Data Quality for Reducing the Uncertainty of Site-Adapted Solar Resource Datasets......................1620

Kristen Wagner, Alex Kubiniec, Tom McAlister, Richard Perez

Multifunctional Titanium Oxide Layers in Silicon Heterojunction Solar Cells by Selective Anodization......................1625

Leonie Jakob, Leonard Tutsch, Thibaud Hatt, Johan Westraadt, Markus Glatthaar, Martin Bivour, Jonas Bartsch

Development of an Adaptive Droop Control Method for Interconnected Lunar DC Microgrids Using Power Hardware-In-The-Loop......................1626

Andrew R. R. Dow, Rachid Darbali-Zamora, Felipe Palacios II, Jack D. Flicker, Marc A. Carbone, Jeffrey T. Csank

Assessing Degradation in Bifacial Photovoltaic by Sequential Stress and Outdoor Aging......................1629

Dennice M. Roberts, Sona Ulicna, Michael Owen-Bellini, Paul Ndione, Helio Moutinho, Kent Terwilliger, Steve Johnston, Laura T. Schelhas, Dana B. Kern

Damp Heat Exposure of Glass/Glass Coupons with Different Encapsulants......................1630

Chiara Barretta, Lisa Meinhart, Hannes Krebs, Andreas Brandstätter, Gernot Oreski

Trajectories to Reach 25% Efficiency CdTe Solar Cells with the Implementation of CdTe1-XSex Band Gradient in SCAPs 1-D......................1633

Joel Saucedo, Hasitha Mahabaduge

How Do As-Local Structures in CdSexTe1-X Respond to Bias Conditions Under (X-Ray) Illumination?......................1634

Srisuda Rojsatien, Niranjana Mohan Kumar, Barry Lai, Dan Mao, Arun Mannodi-Kanakkithodi, Maria K. Y. Chan, Mariana Bertoni

Trends in Solar PV Growth in Snowy Climates and Impact on Resource Adequacy 1635
Shelbie Wickett, Ana Dyreson

Modeling of Perovskite/Si Tandem Solar Cell 1645
Yegao Xiao, Michel Lestrade, Zhiqiang Li, Zhanming S. Li

Assessment of a DER Inverter Model for IEEE 1547 Ride-Through Requirements Using a Model
in the Loop Testbed 1648
Nayeem Ninad, Eugene Desjardins Couture

Methodology for the Analysis of Series Arc Fault Algorithms.. 1654
*Paulo R. D. R. Da Silva, Guilherme C. S. Prym, Geyciane P. De Lima, Andrei C. Ribeiro,
Hugo Da S. Alvarez, Rodrigo M. Garcia, Francisco C. Marques, João A. F. G. Da Silva,
Mauricio Taconelli, Tárcio A. Dos S. Barros, Marcelo G. Villalva*

Patterning the Front Polysilicon Contact for Silicon Solar Cells Using Laser Oxidation 1657
*Sagnik Dasgupta, Pradeep Padhamnath, Vijaykumar Upadhyaya, Young-Woo Ok, Ruohan
Zhong, Wookjin Choi, Kyu-Hyeon Im, Ajeet Rohatgi*

Evaluation of Motion-Induced Noise and Pixel-Bleeding in Electroluminescence Field Inspection
of PV Modules.. 1662
*Thøger Kari, Rodrigo Del Prado Santamaria, Gisele A. Dos Reis Benatto, Pascal Koelblin,
Liviu Stoicescu, Sergiu V. Spataru*

Interface Hydrogen and Passivation of Amorphous Silicon / Crystalline Silicon Heterojunction 1667
*Ujjwal K. Das, Tasnim K. Mouri, Marissa Pina, Tyler Parke, Dhamclyz R. S. Quinones,
Andrew V. Teplyakov*

Author Index

Photovoltaic Site Architecture Estimation Using Performance Data

Steven Koskey[1], Scott Sheppard[1], Corson Teasley[1], Christopher Perullo[1], Jared Kee[1], Daniel Fregosi[2], Wayne Li[2]

[1] Turbine Logic, Atlanta, GA, 30308, USA

[2] Electric Power Research Institute, Charlotte, NC, 28262, USA

Abstract — **Most photovoltaic power generation sites schedule maintenance as a result of physical inspections and observations. For example, a site may use aerial infrared imaging to determine the fault status of individual combiner boxes and/or strings. However, the costs to perform aerial scans result in infrequent, typically annual, application. As a result, DC faults can often go unnoticed for months at a time. While this repetitive, expensive task is attractive for automation, the limited granularity of modern sensor suites makes it difficult. The authors' work has enabled continuous, real-time PV anomaly detection using existing, installed sensor suites. This relies on a detailed knowledge of the site layout to correctly predict expected performance. Previously this information was manually codified using available site drawings for each site. However, the manual review and codification of metadata is time-consuming, increasing the investment required for an M&D center to implement the code. This difficulty is exacerbated by the competitiveness of the PV market, which has led to leaner O&M investments. This work presents a new method to estimate the site architecture using performance data and a fraction of the metadata. The setup speed is accelerated by more than a factor of 15, while achieving similar anomaly detection quality to the previous work using manually codified site layouts.**

I. INTRODUCTION

Monitoring and diagnostics (M&D) play an increasingly important role supporting both business and engineering decisions across the power industry. Utility M&D centers leverage the advances associated with modern digitalization – affordable data storage and computational power, advanced analytics, and modern sensor suites – to characterize and improve power plant operations. While the trend towards more data-informed decision-making is universal, the specific goals and implementations vary significantly across power generation methods. In the case of photovoltaic (PV) power plants, M&D centers generally observe electrical properties and weather conditions. In essence, the weather conditions, particularly irradiance measurements, give the operators an idea of how the plant *should be* operating, and the electrical properties give them an idea of how it *is* operating.

Operators at PV plants perform both corrective and preventative maintenance. The former corrects issues which have already occurred, and the latter corrects issues which are expected to occur. Large and/or widespread problems often receive corrective maintenance priority as they impact a plant's power production capabilities more significantly. Preventative maintenance needs to balance the cost of maintenance with the risk of experiencing a future outage [1]. In this case, M&D

centers' data can be used to train risk models that inform preventative maintenance plans.

One significant challenge for M&D centers to address is the scale and remoteness of PV plants. While M&D centers are uniquely situated to handle large amounts of data, due to the costs of new sensors, the need to maintain sensors, and the need to store the data collected by each sensor, it is often cost prohibitive to put sensors on each solar panel or each string of panels. When compared to traditional power generating assets, such as gas turbines, PV plants require one to two orders of magnitude *more* sensors, depending on the size and layout of the plant. Many PV plants collect and record data at the inverter and the combiner box level. Each site analyzed in this paper has between 250 and 2400 combiner boxes, approximately 16 strings per combiner, and approximately 20 modules per string. Instrumenting at the module level would increase the required number of sensors by 320-fold. Each sensor collects its measurements as a function of time, often sending minute-by-minute data back to the M&D center.

Standard sensor suites make it trivial to detect large problems, like inverter outages. Smaller scale sub-inverter issues in the DC collector field, such as string outages, are more difficult to detect even though they can account for approximately 2% of losses from a plant's nameplate capacity [2]. Recent work has attempted to automatically detect and localize these faults using available plant data [3]. Accurately identifying a fault's location within the site is vital because it enables the maintenance team to spend less time finding the faults and more time fixing them. Consider a case in which an anomaly detection routine detects a string outage within an inverter. The maintenance team can find the fault much more efficiently if they know which combiner is faulted, rather than inspecting all combiner boxes within the inverter [5].

If automated, a data-driven anomaly detection method can operate on real-time data to perform continuous fault monitoring. This differs from other fault diagnosis methods, which often involve manual panel inspection or aerial infrared inspections performed by flying an aircraft or drone above the site – both of which are too expensive to perform continuously. Aerial fault status measurements are generally performed as part of annual maintenance routines [1]. As a result, faults can go unnoticed for months on end. Continuous anomaly detection enables operators to prepare maintenance schedules using more up-to-date information.

978-1-6654-6060-6/23 $31.00 © 2023 IEEE

While previous anomaly detection work has shown promising results in terms of accuracy and usability [3], it still requires a significant time investment to configure a model. The competitiveness of the PV market has resulted in lean O&M budgets, making significant time investments to configure new software more difficult. In general, model frameworks are created first, then they are calibrated to specific sites. The variation between sites introduces a tedious problem for the modeling team: configuration. One fundamental variation that all PV modelers have to consider is site layout – that is, *how many combiners are there in each inverter, how many strings are there in each combiner, and how many modules are there in each string*. Traditionally, modelers manually review the site as-built drawings. This is a laborious, time-consuming task that leaves room for user errors.

Third-party modelers, who often interact with data from various utilities, face another hurdle in the site configuration process: they need to translate the tags from the utility's naming convention to one that works across all of the datasets being modeled. Tags are names by which operators and modelers can reference data streams. The tag translation step is needed so third-party code can pull needed data from the source.

The present work introduces two new algorithms to automatically estimate the PV site layout, enabling automated configuration for PV modeling pipelines. The first method automatically translates the tag names used by a utility using a short list of patterns provided by the user. The translation enables modelers to easily access needed data streams regardless of any utility's tag-naming schema – third-party models that interact with data from various utilities benefit most from this functionality. The second method codifies the hierarchical architecture of a site down to the combiner box using the available tags, then estimates the number of strings and the number of modules per string for each combiner box using performance data. The algorithms write code-readable configuration files that can be used in arbitrary PV modeling pipelines.

II. Methodology

The anomaly detection method applied to this work and described in [3] requires an engineer to manually provide (i) tag mappings for each data stream, (ii) a detailed codification of the site architecture, and (iii) individual hardware component specifications. In the authors' experience from doing this for several sites, manual configuration takes at least eight human hours, depending on the size of the plant. The current work simplifies the configuration process by automatically generating the first two items from the list given a reduced set of metadata. The automatic configuration takes approximately 30 minutes and has three high-level steps.

First, the tags are mapped from the utility naming convention to a user-defined convention. Electrical and weather readings are then filtered to only retain clean maximum power point (MPP) data, which is compared with the modules' datasheet values to estimate the combiner-level layout information. Finally, the tag mapping and estimated per-combiner layout are combined into the necessary site configuration files for a fault detection algorithm.

A. Tag Mapping Standardization

Naming conventions vary greatly from utility to utility. For example, imagine a combiner box at hypothetical Site ABC. It may have the following tag: *ABC-1.D.4-A*. From this format, an analyst can infer that it contains amperage data coming from Site ABC, array 1, inverter D, combiner box 4. But an analyst from another utility, or a third-party modeler, can only guess what it means.

Due to the variation between tag naming conventions, it is helpful for the user to translate the tags to their own convention for each of the relevant information types. While relevant information types vary depending on the application, the anomaly detection method discussed in [3] is used to exemplify the process. It relies on the following information types (each as a function of time):

- Inverter current, power, and voltage
- Combiner box current
- Ambient temperature
- Plane of array irradiance
- Wind speed
- Tracker angles (if using solar trackers)

Previously an engineer had to manually map each tag from the site to its standard counterpart. This took an author approximately an hour per site; the specific duration scales significantly with the site's size. The new algorithm relies on user-input encodings to produce a one-to-one mapping between formats.

Table 1 shows a few examples of different tag names for various data typically collected at PV plants. As PV plants are highly modular, it is generally easy to identify regular tag name formats that have been used to identify similar types of information, such as inverter voltages. The new method enables an engineer to list out a much shorter dictionary of tag name formats, such as those in the "Pattern" column of Table 1. So long as the site tags have a regular pattern, that pattern can be used to efficiently translate the tags.

After the user provides a pattern for each information type, the algorithm uses regular expressions to create a mapping dictionary. It first determines which information type a tag contains and then searches each site tag for the components defined in brackets in the pattern dictionary – for example, array, inverter, etc. The algorithm can translate to any user-defined format, using pattern components as specified by the user. For this work, the authors extracted the full inverter ID (i.e., *{Array}.{Inverter}*), the combiner box ID, and the met-station ID for each data stream. After each component is extracted, it is composed into a user-defined standard format. Using this method, site data tags can be translated to the same format regardless of the source format.

978-1-6654-6060-6/23 $31.00 © 2023 IEEE

TABLE 1
EXAMPLE TAG NAMES AND THEIR ARCHETYPICAL PATTERN

Information Type	Example Site Tags	Pattern
Inverter Voltage	ABC-1.D-V	{Site}-{Array}.{Inv.}-V
Inverter Current	ABC-1.D-A	{Site}-{Array}.{Inv.}-A
Combiner Current	ABC-1.D.1-A	{Site}-{Array}.{Inv.}.{Comb.}-A
Combiner Current	ABC-1.D.2-A	{Site}-{Array}.{Inv.}.{Comb.}-A
Ambient Temperature	ABC-1.met1-T	{Site}-{Array}.met{Station ID}-T

B. Data Cleaning

Once the tags have been standardized, the data is cleaned and the meteorological readings are used to determine when a panel can be expected to operate at its MPP. The data filtering described in [3] has been used as a first step to reduce noise in the dataset. Points with low irradiance and low solar elevation angles are removed. Users can provide a list of known bad values (such as historian timeout values) which will also be removed. The cloud detection strategy from [3] was used to remove off-MPP data from each dataset. Additional data quality filters as described in [4] have been used to filter for non-physical values and point-to-point changes in the data streams.

As the final estimation compares measured data to spec sheet data, one additional filtering step has been applied. The PVPRO package has a built-in function to estimate when PV current and voltage data are at the hardware's maximum power point. This functionality is used to filter the dataset to only those points in time.

C. Codification of Site Architecture

PV sites have hierarchical structures similar to Fig. 1. Working inwards, the components are array, inverter, combiner box, string, and module. This hierarchy is reflected in the standard tag mapping – that is, each combiner box tag contains information for the array and inverter under which it is housed, as indicated in Table 1. However, many models and analyses, such as the routine in [3], also rely on the number of strings per combiner box and the number of modules per string. Neither of these numbers is generally encoded in the combiner box tag names. As a result, the automated site setup routine infers these string and panel counts from measured performance data. Given the panel specifications, the current at max power (i_{MP}) and the voltage at max power (V_{MP}) can be estimated for each combiner box using PVPRO [6]. Once i_{MP} and V_{MP} have been calculated, the ratio is taken between the combiner box-level estimate and the single module datasheet values, as suggested in PVPRO's documentation. The ratios are as shown in (1) and (2). The current ratios estimate the number of strings per combiner box and the voltage ratios estimate the number of modules per string.

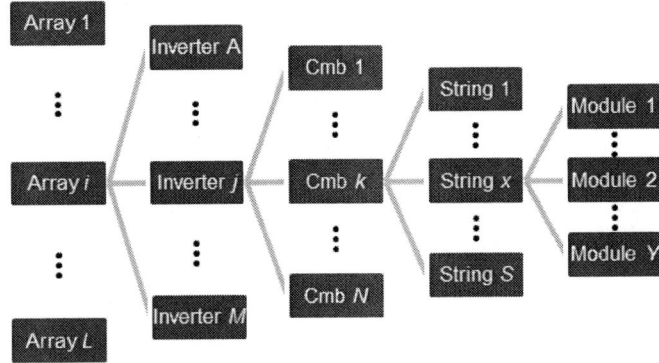

Fig. 1. The hierarchical structure of a PV site can be represented using arrays, inverters, combiner boxes, strings, and modules.

$$n_{strings} = \frac{i_{MP,estimate}}{i_{MP,datasheet}} \quad (1)$$

$$n_{modules} = \frac{V_{MP,estimate}}{V_{MP,datasheet}} \quad (2)$$

The PVPRO method results in a unique estimation of the string count for each combiner box and the number of modules per string for that combiner box. But the precise estimates are not integers, whereas the real sites have integer string and combiner counts. Furthermore, the sites reviewed in this work had a few distinct string counts, and the number of modules per string was constant across the entire site. To convert PVPRO's precise estimates to values more aligned with the real sites, the authors used two separate aggregation methods, one for string counts and the other for module counts. The module estimates are averaged to estimate a single value for each site. The string estimates, on the other hand, are clustered using a kernel density estimate (KDE) method like that described in [7].

Fig. 2(a) shows the essence of the string clustering process used in this work. In this example, the ground truth string counts (as defined by the site drawings) are shown at the top: each combiner box contains either 16, 19, or 22 strings. The combiner-by-combiner estimates are shown next, with the small, filled points. Because the i_{MP} and V_{MP} estimates from PVPRO depend on the actual site performance data, these estimates reflect the current state of the plant and capture any degradation that has grown since the construction of the plant. As a result, the ratio between the estimated and nameplate MPP production does not scale perfectly with the number of strings (or modules) and ratios do not yield whole numbers. The unfilled circles at the bottom visualize how the hypothetical estimates may be clustered by the KDE method. While the clustered estimates are close, they don't always match the blueprint values, as shown by the blue cluster.

Figure 2(b) shows the KDE fit for a real site. In essence, clustering using a kernel density estimate entails: (i) fitting a

978-1-6654-6060-6/23 $31.00 © 2023 IEEE

kernel density estimate to the data and (ii) thresholding using that function's peaks and troughs. The dark blue line indicates the KDE function. The local minima (green) are then selected as bounding thresholds and the local maxima (gray) are selected as cluster centers. Each estimate (yellow) is bounded according to its adjacent thresholds and is assigned the value of the corresponding cluster center.

Interestingly, there is often a small cluster of PVPRO estimates around zero strings per combiner. While relatively unlikely, this can happen if the combiner box is entirely faulted. It is more likely that either the sensor has failed, or the data historian administrator set up combiner box tags that – for whatever reason – do not correspond to a real combiner box collecting data. The authors have observed this at a site with more than 2000 active combiner boxes.

Finally, some light post-processing is performed on both sets of data, such as rounding the estimates to whole numbers (numbers of strings and modules are discrete values). The results are written to configuration files compatible with an anomaly detection pipeline, like that in [3].

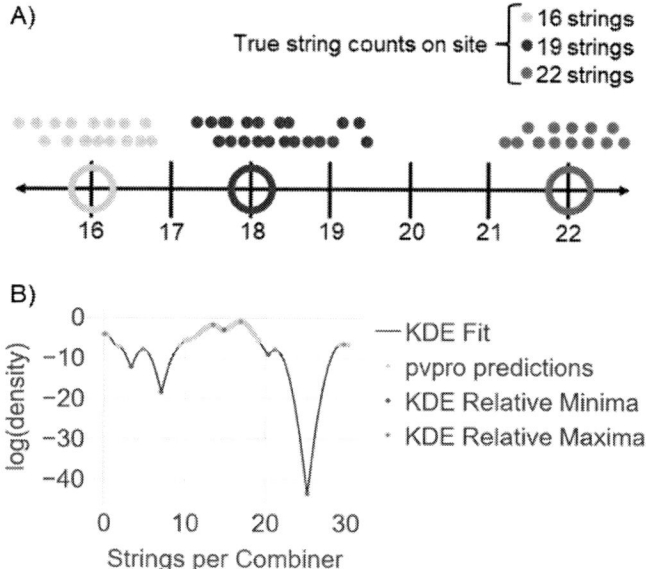

Fig. 2. Overview of the KDE clustering method as it is applied to the strings per combiner estimate clustering. (A) shows hypothetical estimates and true values for a site. (B) shows estimates and thresholds for a real site.

III. RESULTS AND DISCUSSION

A. Configuration Speed

Manually configuring a site takes approximately eight human hours to complete. Using the automated method, configuration takes approximately 20 human minutes and 10 computer minutes. This is achieved by automating the most time-consuming, tedious tasks. Consider the following tasks needed to enable the anomaly detection routine discussed in [3]:

1) Translate 100+ to 1000+ tags (1 hour)
2) Extract site layout from blueprints (4 hours)
3) Create model configuration files (2 hours)
4) Review and fix mistakes (2 hours)

Each of these tasks is automated to some extent in the new method, leaving the following human tasks:
1) Specify ~10 tag name patterns (10 minutes)
2) Specify site specific metadata, like site location and module hardware specification (10 minutes)
3) Run code (10 computer minutes)

Note that the automated method does not require the user to spend nearly two hours finding and debugging mistakes. While repetitive and tedious tasks tend to tire people out, leading to mental lapses and errors, algorithms run precisely the same way each iteration. The automatic method automates each of the first three items in the manual method, making the fourth unnecessary.

B. Accuracy Relative to Blueprints

The automated setup simply translates and codifies existing tags, so the resulting configurations are as accurate as the naming nomenclature and the data collected at the site. In some cases, the automatic approach yields more accurate representations of the site layout (down to the combiner box level) than the as-built drawings. The manually generated configs rely on site drawings to infer the site layout, and while these drawings *should be* perfect representations of the site, supply chain and hardware sourcing issues often lead builders to deviate from the plans. Discrepancies can also result if site operators create tags that correspond to non-existent combiner boxes, which the authors have observed in one instance. In general, however, the two methods generate identical site layouts, down to the combiner box level.

There is more deviation between the two configuration methods for the string and module counts. This is because the automatic method relies on estimates, generated from site operations data, whereas the manual method uses the as-built site drawings. To compare the two values, an error metric is established that is simply the difference between the automatic estimates and the blueprint specifications. A negative error results when the automatic estimate is smaller than the blueprint value. The error – after aggregation – is shown in Figure 3.

Since the automated setup relies on performance data, which is significantly affected by the time-of-year, the configuration was performed using both summer and winter data. It is clear that the performance data seasonality affects the estimates. For three of the four sites investigated, the estimates based on winter data yielded string estimates more similar to what is listed in the as-built bill of materials. Winter data also yielded equal or better estimates for all four sites' module estimates. Several factors impact the accuracy of these results. At high temperatures, PV module performance degrades away from their stated spec sheet performance. The cooler winter temperatures have a slight cooling effect on the modules, lessening the impact of this heat-based performance reduction. The lower overall POA irradiance values in the winter reduce the impact of inverter clipping on the array performance, which artificially shifts the plant away from MPP operation and leads to mischaracterization of the plant by the automatic layout generation algorithm. Finally, curtailment was frequently observed in the summer months. This would have a similar impact as inverter clipping, in which operation is artificially shifted away from the natural MPP of the system.

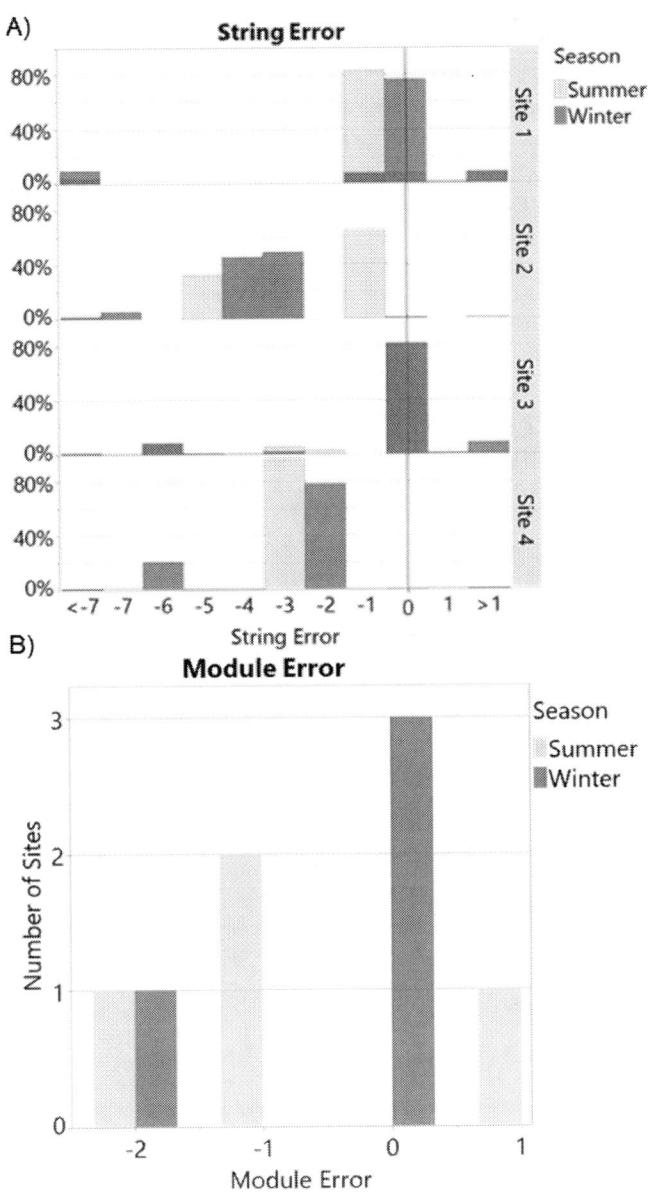

Fig. 3. Overview of the string count and module count estimation error at each site and for different training data.

Figure 3 also shows that the module estimates tend to be far more accurate than the string estimates. This can be explained by the relative stability of voltage signals compared with combiner box signals. Since voltage is measured at the inverter, partial shading will have less of an impact on its signal than it would for a single combiner box. Additionally, the respective aggregation methods likely impact the estimation accuracy. The string estimates rely on KDE clustering to automatically determine the number of unique string counts at a site and which group each combiner box belongs to, which could lead to misassignment for some pieces of hardware. Additionally, since the KDE method is reliant on the point density of the estimates, several outliers could lead to the creation of a highly erroneous cluster value. On the other hand, the module estimates are all clustered into a single group, represented by the average of all estimates. Essentially, the module estimates have more points to determine the single correct cluster-center, so outlier estimates have less of an effect.

Finally, while the bulk of the string estimate errors are within approximately five strings of the site drawing value, they do have a very wide range. This could happen for numerous reasons – for example, imperfections in filtering and modeling noisy data increases the estimates' spread, and sensors flatlining can lead to significant under-predictions. The automatic configuration method is meant to streamline the model configuration process rather than entirely automate it. To mitigate errors of this sort, the authors added a post-processing rule that flags unusually low PVPRO estimates for manual review.

C. Effect on Anomaly Detection

While it is useful to explore the specific differences between the automatic estimates and the as-built site drawings, it does not give a definitive answer about the method's ultimate usefulness. That is, it isn't clear whether methods using the automated site setup work. To this end, the authors assessed the performance of an anomaly detection model derived from that in [3] using both types of site configurations: (i) manual configuration created using the as-built site drawings and (ii) automatic configurations created using the methods in this work. The detection routine first generates features for each combiner box using a physics-based model. It leverages a clustering algorithm to identify the anomalous signals in these features. While the clustering algorithm has a tunable sensitivity, this work uses a single sensitivity to simplify comparisons between sites.

Aerial IR imaging scans from the sites were used to generate ground truth classifications for each site. The ground truth classifications were compared with the automatic workflow's results to calculate the true positive rate (TPR) and false positive rate (FPR) for each of the sites. In discussions with site operators, these are the two most important metrics – the former indicates how many faults the algorithm can catch, and the latter indicates how much time will be wasted inspecting non-faulted hardware. F1 score is a standard classification accuracy metric that incorporates TPR and FPR. Sites 2 and 4 each were each scanned in two separate years, so the routine was evaluated separately for each scan – denoted with (a) and (b).

The impact of the automated site set up on the anomaly detection routine's ability to detect string outages is shown in Fig. 4. This – along with the time to configure a site – is the most important metric of the process's utility. If an automatically configured site could never detect anomalies, the faster setup would be moot.

Generally speaking, automating the model setup slightly decreases anomaly detection performance, but the results vary depending on the site. Automatically configuring the site with winter data mostly performs on par with the manual setup, while the detection success was more noticeably reduced when summer data was used for the architecture estimation. As discussed in the prior section, winter calibration may lead to more accurate results because of the relative infrequency of curtailment and inverter clipping in the winter.

Taking a step back, the main takeaway from Fig. 4 is that the automatic configuration achieves detection rates similar to those achieved with the manual configuration. This demonstrates that any inaccuracies derived from automating the setup process do not jeopardize the overall value proposition of continuous monitoring.

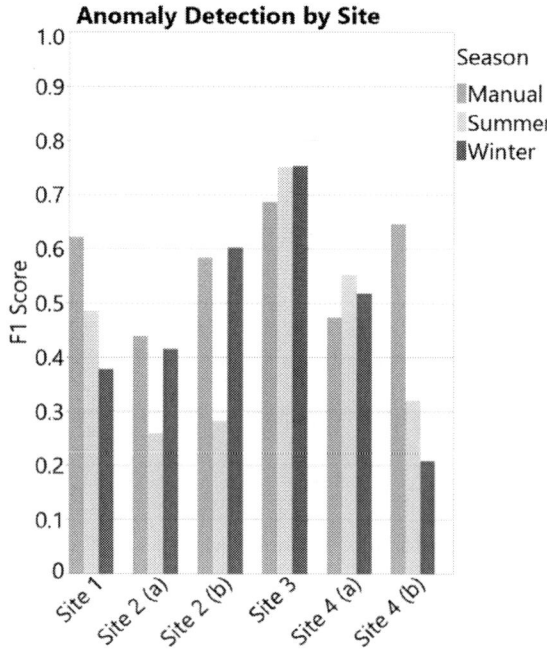

Fig. 4. Automatic configuration achieves detection rates similar to those with manual configuration. Winter mostly outperforms summer.

IV. CONCLUSION

This automated setup method allows simpler implementation of an anomaly detection workflow that detects many DC faults that currently go unnoticed. More broadly, the setup method accelerates model calibration and, as a result, model uptake throughout utility-scale PV applications. Due to the hierarchical nature of PV plants, models that rely on site performance data need to also know the site layout, down to the module level. Otherwise, they cannot scale the results properly. Since data generally is not collected at a finer granularity than the combiner box level, this work introduces a method which estimates the strings per combiner and the modules per string using combiner box level performance data. For higher levels, the tags are used to infer the relationships between inverters and combiner boxes. The tag translation method is particularly well-suited to the large-scale, hierarchical structure of PV sites (and data historians). It can be applied to translate the tags for any dataset, so long as they have a strictly followed schema.

The new workflow is evaluated by the speed to setup a new site, the similarity between automatically generated site layouts and manually generated ones, and the performance of an anomaly detection workflow using each method. Generally speaking, automating the setup leads to slightly different model configurations. Vitally, the anomaly detection pipeline provides usable results with both the manual and the automatic configurations. Continuous anomaly detection provides operators with more up-to-date information, and this algorithm enables widespread adoption by streamlining the setup process.

ACKNOWLEDGEMENTS

This work is funded by the U.S. Department of Energy Solar Technologies Office, under award numbers DE-EE-0008976.

REFERENCES

[1] D. Tansy. "Best practices for operation and maintenance of photovoltaic and energy storage systems; 3rd Edition," Golden, CO, USA. 2018.

[2] N. Vadhavkar, E. Obropta, S. Carey. "Solar Risk Assessment: 2022," kWh analytics, Raptor Maps. 2022.

[3] S. Sheppard, T. Cook, D. Fregosi, C. Perullo, M. Bolen, "Field experience detecting PV underperformance in real time using existing instrumentation," in *2022 IEEE 49th Photovoltaics Specialists Conference (PVSC),* 2022.

[4] "Photovoltaic systems performance – Part 3: Energy Evaluation method," IEC Technical Specification, IEC TS 61724-3, ISBN 978-2-8322-3531-7.

[5] A. Triki-Lahiani, A. Bennani-Ben Abdelghani, I. Slama-Belkhodja, "Fault detection and monitoring systems for photovoltaic installations: A review," *Renewable and Sustainable Energy Reviews,* vol. 82, part 3, pp. 2680-2692, 2018. doi: https://doi.org/10.1016/j.rser.2017.09.101.

[6] DuraMAT, Berkeley, CA. 2022. *PV Production Tools (PV-Pro),* ver 0.0.4.

[7] W.J. Wang, Y.X. Tan, J.H. Jiang, J.Z. Lu, G.L. Shen, R.Q. Yu, "Clustering based on kernel density estimation: nearest local maximum searching algorithm," *Chemometrics and Intelligent Laboratory Systems,* vol 72, iss. 1, pp. 1-8, 2004. Doi: https://doi.org/10.1016/j.chemolab.2004.02.006.

Synthesis and Characterization of Bismuth Selenide and Copper Doped Bismuth Selenide Thin Films by Chemical Bath Deposition

Hamda A. Al-Thani, Shifaa M. Al-Baity, and Falah S. Hasoon

National Energy and Water Research Center (NEWRC), Abu Dhabi, P.O. Box 54111, UAE

Abstract — Bi_ySe_z and $Cu_xBi_ySe_z$ thin films were grown on glass substrates by chemical bath deposition technique at room temperature with up to 24 hours deposition period. The films were deposited from stagnant solutions containing bismuth nitrate (4.16 g/L) complexed with tri-ethanolamine, sodium selenosulfate, and copper nitrate. The bath composition and deposition time were optimized in order to obtain sufficiently thick and crack-free layers for electrical measurements. Four various batches of samples, namely A, B, C, and D which were prepared by varying the chemical bath composition of Selenium (A, and D) or Copper sources (B, and C). The films' thickness is measured by Dektak surface profilometer. X-Ray diffraction (XRD) θ/2θ technique was applied to study the structure of the films, and Electron Probe Micro-Anayser (EPMA) was utilized to investigate the films chemical composition. Atomic Force Microscopy (AFM) was used to examine the films topography and to determine the root-mean-square (RMS) surface roughness. The electrical properties of the films were determined by Hall Effect (HE) measurement technique. X-ray diffraction patterns of the as-deposited films had indicated that the films contain rhombohedral structure of Bi_2Se_3 phases. The RMS film surface roughness was about 30.4 nm and 17.8 nm for $Cu_xBi_ySe_z$ films deposited for 24 hours in baths B and C, respectively. Crack free layers were observed for films deposited in bath A, B, and C. The composition for films deposited for 24 hours in bath B and C was $Cu_{7.55}Bi_{36.04}Se_{56.41}$ and $Cu_{5.28}Bi_{37.79}Se_{56.93}$, respectively. Higher copper content was obtained in films deposited in bath B compared to bath C. Electrical properties for an annealed Bi_ySe_z film deposited in bath A for 24 hours indicated that the film is a n-type semiconductor and the main charge carriers are electrons with an average density of $1.7093 \times 10^{+19}$ /cm² and an average mobility of 1.54 cm²/V-s.

I. INTRODUCTION

Binary and ternary compounds of the V–VI group have attracted wide attention due to their potential applications in thermoelectric devices, optoelectronics and IR spectroscopy [1]. The polyscrystalline Bi_2Se_3 thin film is one of the most promising binary semiconducting compounds for solar cell applications, due to its variable band gap energies in the range of 0.35 to 1.7 eV and its optical properties with a high absorption coefficient in the range of 10^4 cm⁻¹ [1-4]. Several techniques have been employed for the growth of Bi_2Se_3 thin films such as chemical bath deposition (CBD) [3,8-9], electro deposition [2,10], molecular beam epitaxy [11] and SILAR method [12].

The addition of transition-metal cations, such as divalent copper, may convert the n-type Bi_2Se_3 films into p-type semiconductor. Where the Cu sites substitute the Bi sites, due to its smaller ion radius and lower oxidation states.

Consequently, the doped Bi_2Se_3 films become promising candidate as absorber layer in thin film PV devices. For the growth of Cu doped Bi_ySe_z, i.e. $Cu_xBi_ySe_z$ thin films, the CBD technique is commonly applied due to its low cost and its effectiveness to fabricate uniform thin films of large area, with thickness in the range of 0.05-0.3 μm for solar cells applications [9,13].

In this research work, we study the growth of Bi_ySe_z and $Cu_xBi_ySe_z$ films, their structural properties, surface roughness, chemical compositions and electrical properties. The chemical bath deposition experiments for Bi_ySe_z and $Cu_xBi_ySe_z$ films were carried out in stagnant solutions and at room temperature (23 °C) conditions, which are not often applied elsewhere [8,9].

II. EXPERIMENTAL

The CBD technique was used in copper doping of bismuth selenide layers deposited onto glass substrate. Bi_ySe_z and $Cu_xBi_ySe_z$ films were fabricated at room temperature in stagnant reacting solutions. The starting materials used were: (i) aqueous mixture of bismuth (Bi^{3+}) nitrate as Bi Source at 16.67 g/L with Tri-ethanolamine as a complexing agent to dissolve bismuth in water and to prevent the hydrolysis of bismuth (III) nitrate; (ii) Sodium selenosulfate solution as Se Source with initial concentration at 50 g/L and (iii) Copper (Cu^{2+}) nitrate powder as Cu source. Deionized water was added to reach total deposition bath volume of 60 ml. Se source was prepared by dissolving selenium shots in aqueous solution of sodium sulfite and heating the mixture for about 4 hours at 100 °C. Four baths named A, B, C, and D were used for deposition. The final composition of solutions for the films growth is illustrated in Table 1. The initial pH of the reacting solution was in the range of 8.5–9.5. At the end of the deposition, the films were immersed in deionized water and dried by nitrogen.

The films' thickness is measured by Dektak surface profilometer. The overall films' thickness was calculated from three positions in three scans performed at the top edge, the middle and the bottom edge of the deposited films. X-Ray diffraction (XRD) θ/2θ technique was applied to study the structure of the films. Atomic Force Microscopy (AFM) was used to examine the films topography and to determine the root-mean-square (RMS) surface roughness of the films. Also, electron micrograph and the layer elemental composition of the films were obtained using Electron Probe Micro-Analyzer

(EPMA). While, the electrical properties of the films were examined using Hall Effect technique.

TABLE I

COMPOSITION OF THE BATHS A, B, C, AND D USED FOR CBD OF BI_YSE_Z AND $CU_XBI_YSE_Z$ FILMS

Bath/ Film ID	16.67 g/L Bi Source Solution (ml)	50 g/L Se Source Solution (ml)	CuNO₃ Cu Source Powder (mg)
A	15	20	0
B	15	20	10
C	15	20	20
D	15	30	0

III. RESULTS AND DISCUSSION

A. Film Growth

The formation of Bi_2Se_3 in the chemical bath deposition involves (i) the hydrolysis of the selenosulfate compounds which yields selenide anions Se^{2-} and (ii) the dissociation of triethanolamine bismuth (III) complex ions which produces bismuth ions. The film growth mechanisms can occur by ion-by-ion and cluster-by-cluster (or colloidal). In ion-by-ion mechanism, deposition occurs when the ionic product of Bi^{3+} and Se^{2-} is larger than the solubility product of Bi_2Se_3. According to cluster mechanism, absorption and coagulation occurred for the colloidal particles which could be either bismuth selenide or bismuth hydroxide. The hydroxyl group would be replaced by selenide anions, so the bismuth hydroxide will be converted to bismuth selenide [9].

The films thickness measurements and deposition rates per hour are reported in Fig. 1 for films deposited from 1 hour to 24 hours in the four baths. Thickness varied between 55 nm to 609 nm for films deposited in bath A with 20 ml Se source for 1 hour and 24 hours, respectively. The deposition rate decreased from 55 nm/h at 1 hour to 25 nm/h at 24 hours deposition time. Films deposited in bath D with 30 ml Se source show thickness ranging from 84 nm at 1 hour to 874 nm at 24 hours deposition time. The deposition rate was found to start up with 84 nm/h at 1 hour deposition time, followed by a decrease to 36 nm/h at 24 hours deposition time. Also, the thickness for films deposited in bath B with 20 ml Se source and 10 mg Cu source ranged from 24 nm at 1-hour to 106 nm at 24 hours of deposition. The deposition rate was 24 nm/h for 1-hour deposition time and decreased gradually to reach 4 nm/h for 24-hour deposition time. Similarly, thickness of the films deposited in bath C with 20 ml Se source and 20 mg Cu source was found in the range of 37 nm to 136 nm at 1 hour and 24 hours of deposition, respectively. The associated rate of deposition was 37 nm /h and decreased to 6 nm /h for 24 hours deposition time.

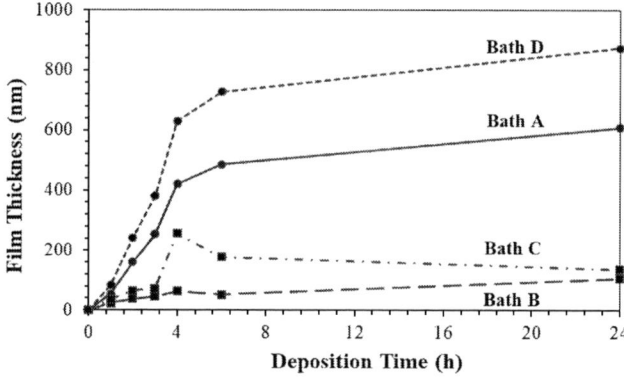

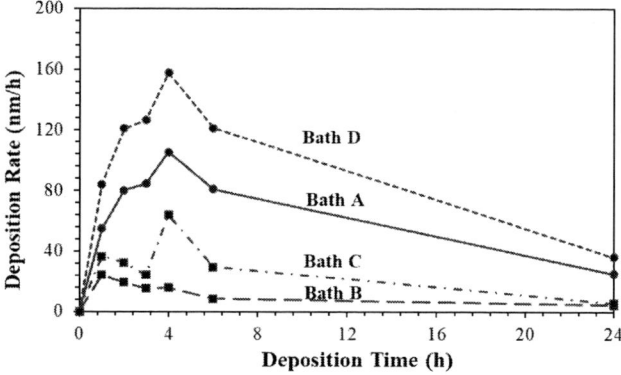

Fig. 1.   (a) Thickness and (b) Deposition rate of Bi_YSe_Z and $Cu_XBi_YSe_Z$ films at room temperature in baths A, B, C and D.

It is evident that thicker films were obtained in baths A and D. In general, addition of Cu source in the bath seems to slow down the rate of deposition (Fig. 1). The deposition occurred in two steps for most of the films. A fast deposition occurred until 4 hours followed by a slow deposition with an optimum rate reached for 4-5 hours. Notice the slow deposition that occurs for films deposited in bath B, which could be due to a difference in environmental factors during the film deposition.

B. Structure of the Films

Using the θ/2θ XRD technique, X-ray patterns were obtained for all films deposited in baths A, B, C, and D for a full period of 24 hours. As illustrated in Fig. 2, the θ/2θ XRD patterns show large peak located at 2θ angle range from 16° to 36.5° that is observed in all XRD patterns which is partially associated to the glass substrate. Also there is a broaden peak located at 2θ angle range from 40° to 46°. PeakFit software [14] was utilized to analyze the diffraction lines that are associated to the deposited films, inspite of their overlapping with the amorphous peak of the glass substrate. Accordingly, the (101), (015), and (110) diffraction lines were determined, at 2θ angle of 24.487°, 29.459°, and 43.626°, respectively. Where these

978-1-6654-6060-6/23 $31.00 © 2023 IEEE

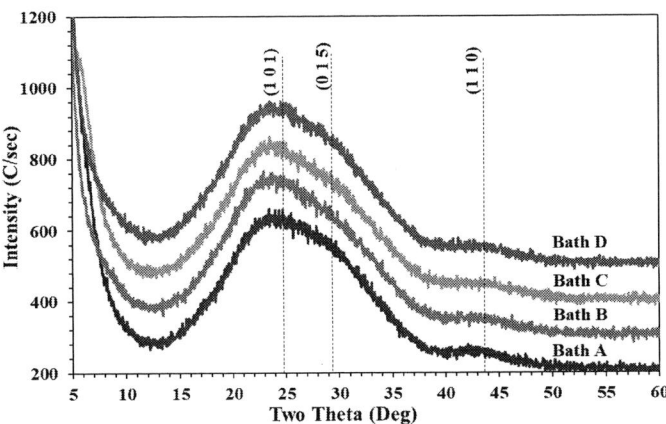

Fig. 2. XRD patterns of the Bi_ySe_z and $Cu_xBi_ySe_z$ films deposited for 24 hours in baths A, B, C and D. The arrow indicates the (101), (015), (110) diffraction lines of Bi_2Se_3 phase (JCPDS 33-0214).

diffraction lines are associated to the rhombohedral structure of Bi_2Se_3 phase (JCPDS 33-0214).

C. RMS Surface Roughness of the Deposited Films

The measured RMS surface roughness for all films is presented in Table 2. It is observed that for the Bi_ySe_z films (A, and D), upon increasing the Se source in the deposition baths from 20 ml (bath A) to 30 ml (bath D), the film surface roughness decreases from 82.3 nm (film A) to 25.4 nm (film D). In comparison, for the $Cu_xBi_ySe_z$ films (B, and D) upon incorporating Cu source while maintaining Se source fixed at 20 ml, the deposited film exhibits a reduction in surface roughness from 30.4 nm (film B) to 17.8 nm (film C).

TABLE II
RMS SURFACE ROUGHNESS OF Bi_ySe_z AND $Cu_xBi_ySe_z$ FILMS

Film	Cu Source (mg) In CBD Bath	RMS Roughness (nm)
A	0	82.3
B	10	30.4
C	20	17.8
D	0	25.4

Fig. 3(a-d) illustrates the AFM topography images for all films A, B, C, and D. For the Bi_ySe_z films (A, and D), it is noticed that grain size decreases in film D, compared to film A upon increasing the Se source in the deposition bath. Which might explains the relative smooth surface of film D compared to rough surface of film A. Furthermore, from the AFM images, for the $Cu_xBi_ySe_z$ films (B, and C), large agglomeration of loose grains on the surface of film B is observed, which is not observed on the surface of film C. Which explains the RMS surface roughness of film B is higher than that of film C.

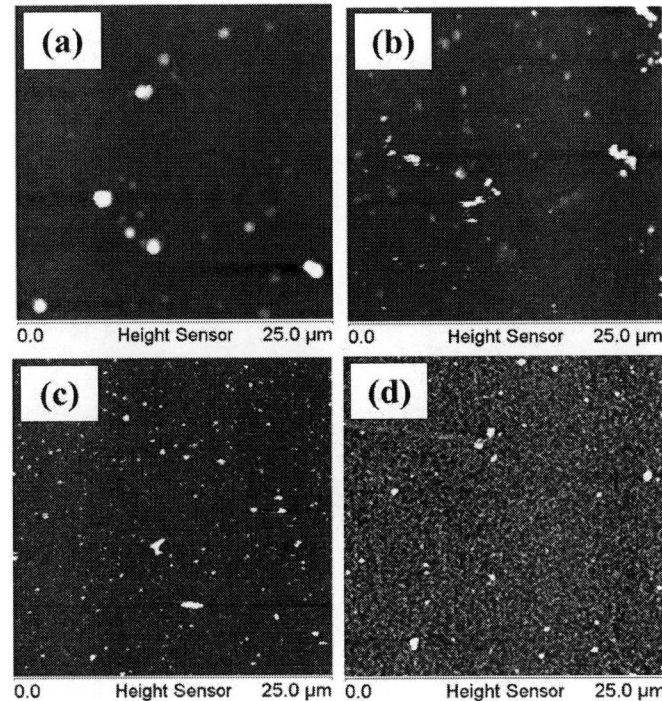

Fig. 3. AFM 2D topography images for (a) Bi_ySe_z film A deposited in bath A for 24 hours, (b) $Cu_xBi_ySe_z$ film B deposited in bath B for 24 hours, (c) $Cu_xBi_ySe_z$ film C deposited in bath C for 24 hours, and (d) Bi_ySe_z film D deposited in bath D for 24 hours.

D. Chemical Composition of the Films

The electron micrographs of the as-deposited films A, B, C, and D is presented in Fig. 4 (a-d) in COMPO imaging mode. The images are obtained at 5 kx magnification. It is clear that the deposited Bi_ySe_z layer D has cracks with the formation of voids (Fig. 4d) that appeared as black dots in the image. On the contrary, Bi_ySe_z film A shows crack-free and dense layer (Fig. 4a). Also, the images reveal two types of $Cu_xBi_ySe_z$ layers in films B and C, i.e. a matrix layer and a powdery layer on the top of the matrix layer, as shown in Fig. 5 (a and b). The matrix layer is dense and crack-free whereas the powdery layer is mainly composed of loose spherical particles. The matrix layer has two main phases with different atomic weight. The dark domain indicates the presence of light atomic weight phase and the brighter dots indicate the presence of heavier atomic weight phase. In contrast, the powdery layer observed in the two films consists of spherical particles with heavy atomic weight phase. Particularly, film B has large agglomerations of spherical particles.

Table 3 illustrates the elemental composition of the as-deposited films and the annealed film A, that was determined by Wavelength Dispersive Spectrometer (WDS) in atomic ratios (at.%) for Bi, Se, and Cu elements. WDS analysis was

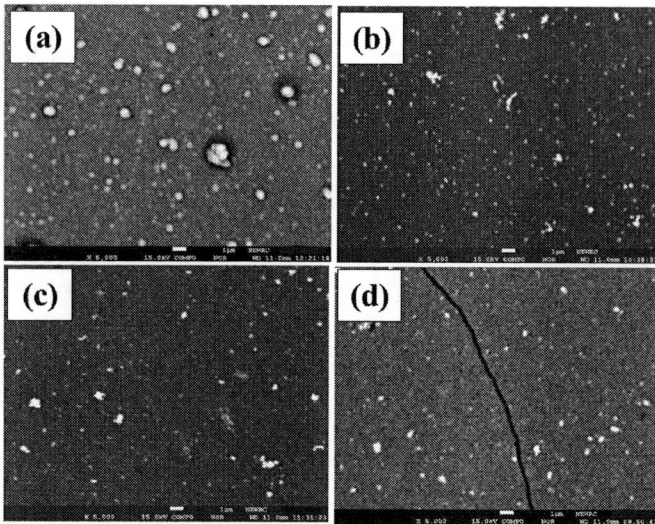

Fig. 4. Electron micrographs (composition images) of the films deposited in (a) bath A, (b) bath B, (c) bath C, and (d) bath D.

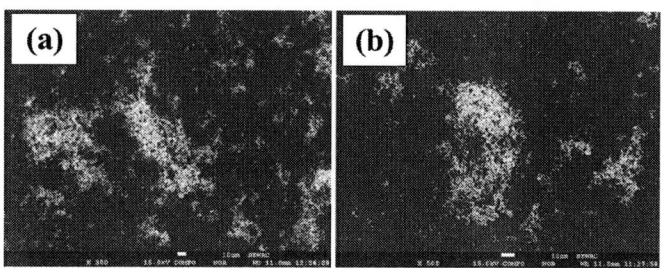

Fig. 5. The EPMA composition images of the powderly layer as observed on top of the matrix $Cu_xBi_ySe_z$ films deposited in (a) bath B, and (b) bath C at 300x and 500x magnification, respectively.

performed in a location on the matrix layer. The obtained compositions of Bi_ySe_z films A and D are $Bi_{45.38}Se_{54.62}$ and $Bi_{42.28}Se_{57.72}$, respectively. It is worth mentioning that the target elemental composition for stoichiometric Bi_2Se_3 phase is $Bi_{40}Se_{60}$. However, even though film D has the higher concentration of Se, the film has cracks that indicate the discontinuity of the film. Therefore, we found that the quality of film A is better, i.e. in terms of morphology and composition, than film D. On the other hand, the obtained composition of deposited $Cu_xBi_ySe_z$ films B and C are $Cu_{7.55}Bi_{36.04}Se_{56.41}$ and $Cu_{5.28}Bi_{37.79}Se_{56.93}$, respectively. In our experimental conditions, i.e. stagnant reacting solutions and no addition of ammonia, increasing copper in the bath C did not increase the concentration of copper in the deposited film C. The lower Cu content in film C may be due to several factors including an equilibrium constant shift toward ionic dissolution because of the higher solubility product Ksp of CuSe at $2x10^{-40}$ [15], when compared to the solubility product K_{sp} of Bi_2Se_3 at 10^{-130} [15]. To prevent ionic dissolution of Cu, complexation of the ion may be required.

TABLE III
WDS ATOMIC RATIOS FOR Bi_ySe_z AND $Cu_xBi_ySe_z$ FILMS DEPOSITED IN BATHS A, B, C, AND D FOR 24 HOURS

Film	Bi (at.%)	Se (at.%)	Cu (at.%)
A	45.38	54.62	-
A- Annealed	53.4	46.6	-
B	36.04	56.41	7.55
C	37.79	56.93	5.28
D	42.28	57.72	-

After annealing film A in air at 100 °C, Se content decreased to 46.6 at.%, as indicated in Table 3. This resulted in a 14.7% loss in selenium for a 2-hour annealing at low temperature.

E. Electrical Properties of the Films

The electrical measurement was performed on as-deposited and annealed films. However, significant results were obtained for the annealed Bi_ySe_z film A due to a better electrical contact with the measurement probes in particular. The sheet resistance of the film was $1.977x10^{+4}$ $\Omega/\square$ at room temperature. Additionally, the Hall measurements were carried out in liquid nitrogen at temperature range of 88 K – 500 K with two-cycle measurement per point and a 10-point measurement per run. The sheet resistance was in the range of 1.07 to 0.607 $\Omega/\square$ and the average Hall coefficient was about -1.38 m^2/C. The results confirm the semiconducting nature of $Bi_{53.4}Se_{46.6}$ film A as n-type semiconductor with an average electron concentration of -$1.7093x10^{+19}$ /cm^2 and an average electron mobility of 1.54 cm^2/V-s.

IV. SUMMARY

Bi_ySe_z and $Cu_xBi_ySe_z$ films were grown on glass substrates by chemical bath deposition method. Most of the films showed low intense peaks which are indexed to the (101), (015), and (110) diffraction lines in the rhombohedral structure of Bi_2Se_3 phase. $Cu_xBi_ySe_z$ film C deposited in bath containing 20 ml Se source and 20 mg Cu source for 24-hour deposition time had the smoother surface with an average roughness of 17.8 nm. The two mechanisms ion-by-ion and cluster-by-cluster were accounted for the films deposition. $Cu_xBi_ySe_z$ film B exhibited preferentially cluster formation and relatively high average surface roughness. WDS analysis indicated that Cu concentration is higher in film B (7.55 at.%) deposited with 10 mg Cu source than in film C (5.28 at.%) deposited with 20 mg Cu source for a long time deposition of 24 hours. The sheet resistance for the annealed film A with composition $Bi_{53.4}Se_{46.6}$ deposited in a bath containing 20 ml Se source was $1.977x10^{+4}$ $\Omega/\square$ at room temperature. The negative hall coefficient for Bi rich film A indicated that the film is n-type semiconductor.

REFERENCES

[1] T. E. Manjulavelli, T. Balasubramanian, and D. Nataraj, Chalcogenide Letters, Vol. 5, No. 11, p.297–302, November 2008.

[2] C. Xiao, J. Yang, W. Zhu, J. Peng, and J. Zhang, Electrochimica Acta 54, pp. 6821–6826, June 2009.

[3] V. M. Garcia, M. T. S. Nair, P. k. Nair, and P. A. Zingaroo, Semicond. Sci. Technol.12, pp. 645–653, 1997.

[4] S. Phok, P. Parilla, R.N. Kini, R. Bhattacharya, B. To, and J. Pankow, Photovoltaic Specialists Conference (PVSC), 2010 35th IEEE, pp. 002946-002951, 20-25 June 2010.

[5] X. L. Qi and S. Zhang, Phys. Today 63, 1, p. 33–38, January 2010.

[6] X. L. Qi, and S. Zhang, Rev. Mod. Phys. 83, 1057–1110, 2011.

[7] Y. S. Hor et al., Physical Review Letters, PRL 104, 057001, p. 057001-1–057001-4 February 2010.

[8] R. H. Bari, and L. A. Patil, Indian Journal of Pure & Applied Physics, vol. 48, Feb. 2010, pp. 127-132, February 2010.

[9] B. Pejova, and I. Grozdanov, Thin Solid Films 408, p. 6-10, 2002.

[10] X. Li, K. Cati, H. Li, L. Wang, and C. Zhou, International Journal of Minerals, Metallurgy and materials, vol. 17, Num. 1, p104, February 2010.

[11] L. He et al, Journal of Applied Physics 109, vol. 103702, 2011.

[12] B. R. Sankapal, R. S. Mane, and C. D. Lokhande, Materials Chemistry and Physics 63, pp. 230-234, 2000.

[13] P. K. Nair et al, Solar Energy Materials & Solar Cells 52, pp. 313-344, 1998.

[14] Peak Separation and Analysis Software, by SYSTAT Software Inc. For more information visit PeakFit – Inpixon – systatsoftware.com; accessed on June 1st, 2023.

[15] G. Hodes, Chemical solution deposition of semiconductor films, Marcel Dekker, Inc, ISBN: 0-8247-0851-2, New York, USA, 2002.

CIGS Device Stability: A Comparison of Two Different Process Batches

Mohsen Jahandardoost[1], Curtis Walkons[1], Marco Nardone[2], Theresa M. Friedlmeier[3], Shubhra Bansal[1,4]

[1] University of Nevada Las Vegas, Dept. of Mechanical Engineering, Las Vegas, NV 89154

[2] Department of Physics and Astronomy, Bowling Green State University, Bowling Green, OH 43403

[3] Zentrum für Sonnenenergie- und Wasserstoff-Forschung Baden-Württemberg, Stuttgart, Germany

[4] School of Mechanical Engineering, School of Materials Engineering, Purdue University, West Lafayette IN 47907

*Email: bansal91@purdue.edu

Abstract — In this work, we compare the light soaking behavior and potential induced degradation in devices from two different process batches (A and C). The two batches of CIGS devices have been prepared by ZSW with different co-evaporation tools and Ga/(Ga+In) depth profiles. Two different Na barriers (SiO_2 or AlO_x–AlN) have been used for low Na devices, and two buffers namely CdS and Zn(O,S) have been used in both batches. Initial device performance before stress indicates an improvement in efficiency and open-circuit voltage (V_{OC}) in the incorporation of Na in the absorber. Replacement of CdS buffer with Zn(O,S) results in a decrease in efficiency due to drop in V_{OC} albeit a slight increase in short-circuit current density (J_{sc}). CdS buffer devices from batch A show more degradation under heat-light soaking (HLS) with short-circuit junction bias compared to batch C. Zn(O,S) devices show similar HLS behavior for both sample batches. Potential induced degradation (PID) tests with 1000 V bias show higher degradation in low Na devices with SiO_2 barrier (batch C) compared to AlO_x-AlN barrier (batch A).

I. INTRODUCTION

CIGS solar cells are a promising solar cell technology with different substrate adaptability, device architecture, and potential for thin-film tandem devices. The current state-of-the-art cell has a graded Ga concentration and holds a record efficiency of 23.35% for the CdS buffer layer[1] and 24.2% for the Zn(O,S) buffer layer.[2] Device structure and growth conditions impact the stability of the CIGS devices. Na has been shown to improve the initial device performance by increasing the V_{OC}.[3], [4] Studies of CIGS/CdS devices indicated an increase in V_{OC} under light soaking due to the alkali ion's effect to reduce donor-like defects.[5]–[7] Increase in V_{OC} was also reported in CIGS devices with Zn(O,S) buffer layer after the HLS stress test.[8]–[10] However, this increase was not observed in our previous work.[11] The current status of the research on PID indicates that preventing the Na migration into the CIGS absorber prolongs the degradation.[12]–[14] Here we investigate the effect of processing on HLS and PID with changes in Na and buffer layers.

II. EXPERIMENTATION

CIGS devices with CdS and Zn(O,S) buffer were fabricated by ZSW in two batches using different in-line CIGS deposition tools. A SiO_2 barrier layer was deposited on a glass substrate, prior to Mo back contact coating for batch C cell (Type 2C), and AlO_x–AlN barrier was used for batch A cell (Type 2A). All samples were sputter-coated with Mo as back contact. The CIGS absorber layers have been grown with a 3-step co-evaporation vacuum method inside the in-line machines to achieve graded bandgap. This process leads to different Ga distributions throughout the CIGS thickness and consequently changes the bandgap energy of the absorber layer where a higher bandgap value achieves by increasing the Ga atomic ratio. A buffer variation has been achieved by chemical bath deposition (CBD) with either the CdS/i-ZnO (Type 1 and 2 A/C) or the ZnOS/ZMO buffer (Type 3 A/C). The tthickness of the buffer layers was kept at 50 nm constant for all devices in batch C but it was changed for the different devices in batch A to reach the maximum efficiency. All samples were finished with a DC-sputtered Al-doped ZnO window layer and an Al/Ni/Al top grid.

Accelerated stress tests (AST) for heat-light soaking have been conducted using an ATLAS XXL+ chamber with AM1.5G spectrum and 1000 W/m^2 illumination at 85 °C, and voltage biases at open-circuit (OC) or short-circuit (SC) conditions. Current density-voltage (J-V) metrics are determined by in-situ current-voltage sweeps every 30 minutes for a duration of 50-100 hours. Potential-induced degradation tests have been performed on unencapsulated devices in dry air by applying a 1000 V voltage bias across the SLG glass substrate with a Keithley 2410 at dark with a temperature of 25 or 85 °C.

978-1-6654-6060-6/23 $31.00 © 2023 IEEE

TABLE I.   AVERAGE AND STANDARD DEVIATION OF INITIAL J-V PARAMETERS FOR CIGS DEVICE TYPES

V_{OC} = OPEN-CIRCUIT VOLTAGE, J_{SC} = SHORT-CIRCUIT CURRENT DENSITY, FF = FILL-FACTOR, Eff = EFFICIENCY, J_0 = DARK SATURATION CURRENT, R_S = SERIES RESISTANCE, G_{SC} = SHUNT CONDUCTANCE.

CIGS Type		V_{OC} (mV)	J_{SC} (mA/cm²)	FF (%)	Eff (%)	J_0 (mA/cm²)	Ideality Factor-A	R_S (ohm-cm²)	G_{SC} (mS/cm²)
Type 1A	Base Na CdS	705±7	31.4±0.8	76.0±2.0	16.9±0.9	8.4×10^{-8}	1.4±0.1	0.8±0.2	0.2±0.1
Type 1C	Base Na CdS	727±11	31.8±0.9	75.0±1.3	17.3±0.5	1.8×10^{-7}	1.5±0.1	0.7±0.2	0.5±0.4
Type 2A	Low Na CdS	664±11	30.4±1.0	72.4±2.6	14.7±1.1	4×10^{-6}	1.7±0.1	0.8±0.2	0.2±0.1
Type 2C	Low Na CdS	717±14	31.8±0.5	73.9±3.3	16.9±1.0	2×10^{-6}	1.4±0.1	1.0 ± 0.2	1.0±1.0
Type 3A	Base Na Zn(O,S)	676±10	32.1±1.0	73.8±1.4	16.0±0.6	2×10^{-4}	1.9±0.5	1.2±0.6	0.3±0.2
Type 3C	Base Na Zn(O,S)	703±8	32.4±0.9	73.9±1.2	16.8±0.4	7×10^{-5}	2.0±0.5	0.7±0.7	1.0±0.4

III. RESULTS

A. Initial Device Performance

The initial mean characteristic parameters for all devices are listed in **Table I**. These values are measured for 10 samples of each device type and the mean value and standard deviation are calculated. We observed from initial measurements before any stress test that devices from batch C have higher efficiency mainly due to the higher open-circuit voltage (V_{OC}). Na improved the device efficiency by increasing the V_{OC} and the replacement of the CdS buffer layer with Zn(O,S) reduced efficiency and V_{OC}, even though there was some increase in J_{SC}. The increase in J_{SC} is related to the higher photon collection of the short-wavelength light by Zn(O,S) buffer layer as observed by QE measurement (not shown here).

B. Accelerated Stress Tests

Devices with baseline sodium (Type 1) and low-Na (Type 2) from both batches (A & C) were light-soaked at 85 °C at open-circuit junction bias which leads to a smaller internal field and higher charge injection, or short-circuit junction bias which results in the larger internal field and lower charge injection. **Figure 1** demonstrates the change in V_{OC} of the devices under OC and SC bias conditions. For CdS buffer devices, V_{OC} shows improvement with OC and a decrease in SC heat-light soaking conditions. SC-HLS stress condition shows a linear decrease in V_{OC}, with ΔV_{OC} being the highest for devices with low-Na in batch A (Type 2A) and minimum for the device with base Na in batch C (Type 1C). Overall, Na proved to be beneficial not only to the initial device performance but also it enhanced device stability.

Figure 2 shows the normalized power production of all cells under PID at different temperatures. The lower temperature required a longer time for the cells to degrade, and the cells had the fastest degradation at 85 °C. All device types completely deteriorated after 60 hours of PID test at 85

°C, however, they were still functional after 1000 hours of PID at 25 °C. Low-Na device from batch A (Type 2A) showed the highest resistance to PID compared to other devices at both temperatures.

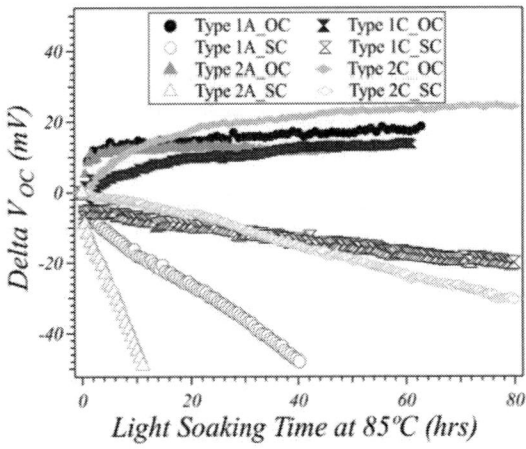

Figure 1. Delta Voc as a function of AST time (hours) for Type 1 and 2 devices that have been processed on different in-line CIGS deposition tools and exposed to 1 sun at 85 °C under open-circuit (OC) or short-circuit (SC) voltage bias.

Figure 3 illustrates the evolution of V_{OC} with time for the HLS stress test and junction voltage biases. Under OC-HLS, an increase in V_{OC} for CdS buffer devices is seen with as much as 20 mV, irrespective of the batch type. Under SC-HLS stress conditions, V_{OC} for baseline CdS devices tends to decrease more rapidly for batch A. For Zn(O, S) buffer devices, on the other hand, V_{OC} decreases under both OC and SC-HLS stress, irrespective of the absorber deposition tools. Both buffer types show a decrease in V_{OC} under SC-HLS stress conditions, but the magnitude of V_{OC} decrease is much higher in Zn(O,S) based devices (Type 3A & C). Zn(O,S) buffer devices show a simultaneous degradation in J_{SC} and FF too (not shown here).

978-1-6654-6060-6/23 $31.00 © 2023 IEEE

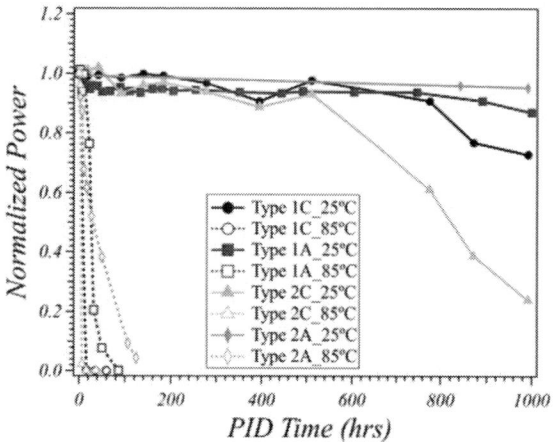

Figure 2. Normalized power versus PID time for Type 1 and 2 devices with different in-line CIGS deposition process and exposed to 1000 V volage bias across the glass substraye at 25 and 85 °C.

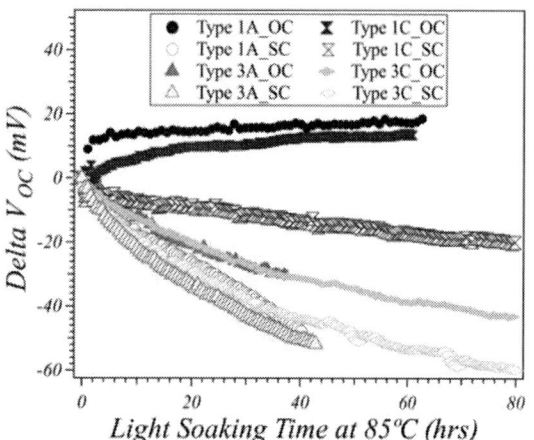

Figure 3. Delta Voc as a function of AST time (hours) for CIGS devices with differnet buffer layer which have been processed on different in-line CIGS deposition tools and exposed to 1 sun at 85 °C under open-circuit (OC) or short-circuit (SC) voltage bias.

IV. CONCLUSION

Two batches of CIGS devices have been processed on different in-line CIGS deposition tools. Initial measurements before any stress test showed that devices from batch C have higher efficiency mainly due to the higher V_{OC}. Na improved the device efficiency by increasing the V_{OC} and the replacement of the CdS buffer layer with Zn(O,S) reduced the efficiency. With CdS buffer, batch C devices with baseline Na exhibit the most stable behavior with heat and light soaking. Preliminary PID tests indicate low Na devices with AlOx-AlN barrier are most resistant to degradation likely due to prevention of Na migration from glass.

ACKNOWLEDGEMENT

This work has been funded by award DE-EE-0007750.

REFERENCES

[1] M. Nakamura, K. Yamaguchi, Y. Kimoto, Y. Yasaki, T. Kato, and H. Sugimoto, "Cd-Free Cu(In,Ga)(Se,S)2 thin-film solar cell with record efficiency of 23.35%," *IEEE J Photovolt*, vol. 9, no. 6, pp. 1863–1867, Nov. 2019.

[2] M. A. Green *et al.*, "Solar cell efficiency tables (Version 60)," *Progress in Photovoltaics: Research and Applications*, vol. 30, no. 7, pp. 687–701, Jul. 2022.

[3] P. T. Erslev, J. W. Lee, W. N. Shafarman, and J. D. Cohen, "The influence of Na on metastable defect kinetics in CIGS materials," *Thin Solid Films*, vol. 517, no. 7, pp. 2277–2281, Feb. 2009.

[4] J. M. Raguse, C. P. Muzzillo, J. R. Sites, and L. Mansfield, "Effects of Sodium and Potassium on the Photovoltaic Performance of CIGS Solar Cells," *IEEE J Photovolt*, vol. 7, no. 1, pp. 303–306, Jan. 2017,.

[5] S. Ishizuka, N. Taguchi, J. Nishinaga, Y. Kamikawa, S. Tanaka, and H. Shibata, "Group III elemental composition dependence of RbF postdeposition treatment effects on Cu(In,Ga)Se2 thin films and solar cells," *Journal of Physical Chemistry C*, vol. 122, no. 7, pp. 3809–3817, Feb. 2018.

[6] J. Matsuura, I. Khatri, T. Y. Lin, M. Sugiyama, and T. Nakada, "Impact of heat-light soaking and heat-bias soaking on NaF-treated CIGS thin film solar cells," *Progress in Photovoltaics: Research and Applications*, vol. 27, no. 7, pp. 623–629, Jul. 2019.

[7] C. Walkons *et al.*, "Behavior of Na and RbF-Treated CdS/Cu(In,Ga)Se2 Solar Cells with Stress Testing under Heat, Light, and Junction Bias," *physica status solidi (RRL) – Rapid Research Letters*, vol. 15, no. 2, p. 2000530, Feb. 2021.

[8] J. Serhan *et al.*, "Investigation of the metastability behavior of CIGS based solar cells with ZnMgO–Zn(S,O,OH) window-buffer layers," *Thin Solid Films*, vol. 519, no. 21, pp. 7606–7610, Aug. 2011.

[9] T. Kobayashi, H. Yamaguchi, and T. Nakada, "Effects of combined heat and light soaking on device performance of Cu(In,Ga)Se2 solar cells with ZnS(O,OH) buffer layer," *Progress in Photovoltaics: Research and Applications*, vol. 22, no. 1, pp. 115–121, Jan. 2014.

[10] W. J. Lee *et al.*, "Behavior of Photocarriers in the Light-Induced Metastable State in the p-n Heterojunction of a Cu(In,Ga)Se2 Solar Cell with CBD-ZnS Buffer Layer," *ACS Appl Mater Interfaces*, vol. 8, no. 34, pp. 22151–22158, Aug. 2016.

[11] M. Jahandardoost, M. Nardone, T. M. Friedlmeier, C. Walkons, and S. Bansal, "Heat- and light-soaking behavior of RbF-treated Cu(In,Ga)Se2 solar cells with two different buffer layers," *J Mater Res*, vol. 37, no. 2, pp. 436–444, Jan. 2022.

[12] V. Fjallstrom *et al.*, "Potential-induced degradation of Cu(In[1-x]Ga[x])Se2 thin film solar cells," *IEEE J Photovolt*, vol. 3, no. 3, pp. 1090–1094, 2013.

[13] C. P. Muzzillo *et al.*, "Potential-Induced Degradation of Cu(In,Ga)Se2 Solar Cells: Alkali Metal Drift and Diffusion Effects," *IEEE J Photovolt*, vol. 8, no. 5, pp. 1337–1342, Sep. 2018.

[14] M. Jahandardoost, T. M. Friedlmeier, M. Nardone, and S. Bansal, "Potential Induced Degradation and Recovery Effects in CdS/CIGS Solar Cells with Na and RbF Treatments," in *IEEE 48th Photovoltaic Specialists Conference (PVSC)*, Jun. 2021, pp. 1693–1696.

Hierarchical variance analysis of solar cell production using machine learning and numerical simulations

Bernhard Klöter, Hannes Wagner-Mohnsen, Sven Wasmer

WAVELABS Solar Metrology Systems GmbH, 04179 Leipzig, Germany

Abstract — **Solar cells are an important source of renewable energy and have become a major industry with annual production capacities above 200 GW. The large amount of data produced during the solar cell production process calls for big data solutions and machine learning based approaches to improve quality and increase efficiency. In this work, we analyzed solar simulator data from 60,000 PERC solar cells using a hierarchical model based on machine learning, and compared the results to a theoretical model to extract missing information and determine the most likely input parameters for each produced solar cell. This approach enabled us to gain insight into the mechanisms that lead to the variance of conversion efficiency.**

I. Introduction

Solar cells are an increasingly important source of renewable energy, and their production has become a major industry reaching annual production capacities of above 200 GW [1]. This corresponds to around 40 billion solar cells with huge amounts of data describing them. Analysis of such large data sets call for big data solutions and machine learning based approaches.

Improving solar cells quality is a main driver for reducing the relative production costs by avoiding production loss. By analyzing data from various stages of the production process, it is possible to identify patterns and trends that can indicate potential problems or areas for improvement. For example, data analysis can be used to identify factors that influence the efficiency of solar cells, such as the purity of the materials used or the conditions under which they are manufactured.

Another motivation for using data analysis and machine learning in solar cell production is to increase the efficiency of the production process itself. By using machine learning algorithms to analyze data from the production process, it is possible to identify bottlenecks or inefficiencies that can be addressed to improve the overall efficiency of the solar cells [2]. This can lead to significant cost savings, as well as reduced energy consumption and a smaller environmental footprint.

Overall, data analysis and machine learning offer a powerful toolkit for optimizing the production processes of solar cells, and their use is likely to become increasingly important in the coming years as the industry continues to grow and evolve.

II. Analysis Setup

Machine learning is particularly useful in situations where complete knowledge about a system or process is lacking. In this case, machine learning algorithms can identify patterns and trends that are not apparent to humans. In the case of solar cells machine learning can be used to analyze data from end-of-line quality control without the knowledge about the complete production process. Missing information can be furthermore extracted by comparing the production results with a theoretical model of the solar cells [3].

In our work we took data of 60,000 PERC solar cells from a state-of-the-art solar simulator (WAVELABS SINUS 300) for further analysis. The data was taken from a typical production day and resembles a data set which is usually analyzed by production experts on a daily base. Besides the commonly used parameters like conversion efficiency (η), short-circuit current (J_{SC}), the fill factor (FF) and open-circuit voltage (V_{oc}) we analyzed additional current-voltage (I-V) characteristics to infer physical properties of these solar cells like the base doping concentration. Furthermore, parameters which might have indirect influence on η and might contain additional information about the solar cells are used in this data set. The data is then analyzed concerning the source of variance of η. Additionally, a theoretical model was set up using the solar cell simulation software Quokka 2.2.5 [4]. By varying the models input parameters within the boundaries of a typical solar cell production, a large set of possible outcomes were calculated. Comparing these outcomes with the measured production outcomes the most likely input parameters for each produced solar cell could be determined. This serves as an additional source of information which we use in the analysis.

III. Results

In Figure 1, the conversion efficiency of the solar cell production is shown. The data resembles a typical production day, and several features are visible, such as outliers to lower efficiency, often appearing at specific times. Also, an overall changing mean value with a maximum between 2 PM and 10PM can be seen. The objective for the following analysis is to gain insight into the mechanisms which lead to the variance of η.

978-1-6654-6060-6/23 $31.00 © 2023 IEEE

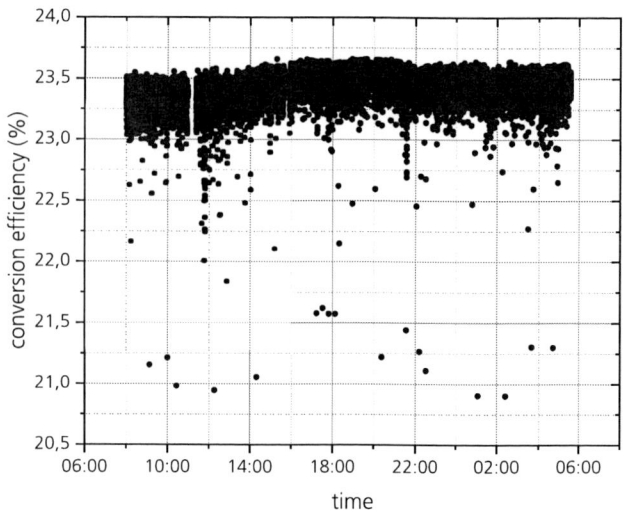

Figure 1: Time series plot of data set for 60k PERC solar cells

The total data set consists of overall 160 parameters per classified cell which can be sorted into three categories. The first category are data points which are directly measured and can be determined from the *I-V*-curve such as η, J_{SC}, FF and V_{oc}.

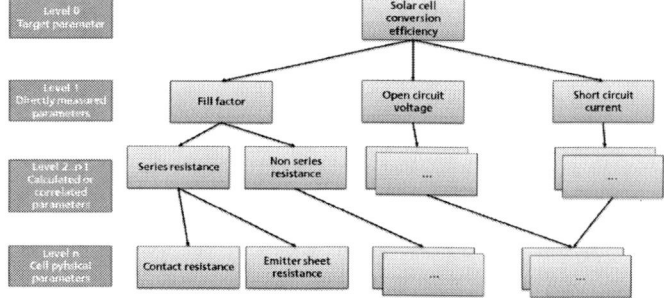

Figure 2: Hierarchical model of analysis

The next category consists of parameters which are calculated from the *I-V*-curve with the help of secondary measurements like the series resistance, the base doping concentration, the cell temperature, or information about the spectral sensitivity. It consists also of parameters describing internal machine states such as internal temperatures, user comments etc. The last category is made up from parameters which can only be inferred by our solar cell simulation model, and which are not easily or not non-destructively measurable. These are e.g., the emitter sheet resistance, the surface recombination velocities, and the contact resistances of the metallized grid. The analysis uses this hierarchy by going through each level separately which allows to reduce simple correlations which e.g., stem from parameters which can be calculated from each other. This also enables a guided root cause analysis which allows us to identify specific sources without losing the big picture.

For the first level, a variance-based sensitivity analysis considering correlations between input variables [5-6] is carried out and which was introduced to PV research in [7]. This shows us the main impacts on η, in this case V_{oc} with a high share of correlated variance. A similar relative high correlation can be seen from the *FF* loss due to R_s. It shows that the root cause for this part of variance influences both V_{oc} and *FF*.

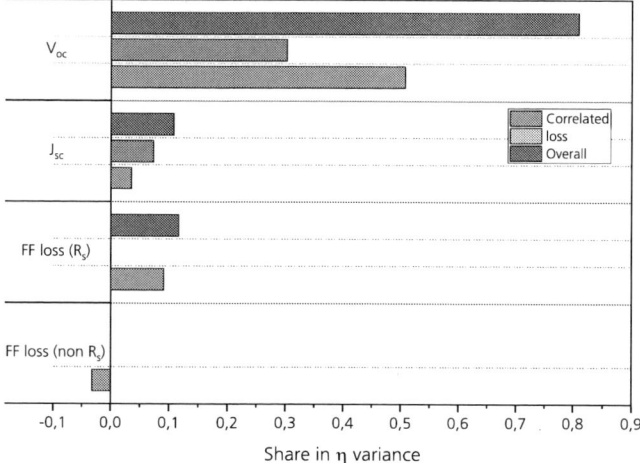

Figure 3: Explained variance for the cell conversion efficiency (η).

As the next level contains most of the data sets parameters a different approach is used to reduce the number of meaningful parameters. For this, a machine learning based model is constructed with all parameters from category level 2 to explain the impacts on η. The individual feature impacts are then extracted based on Shapley values [2] and the main impacts are again analyzed with the sensitivity analysis method. The result is shown in Figure 4. The main impact here can be attributed to the series resistance, the doping density of the bulk material and the total signal of the additional electroluminescence measurement. On first sight, this contradicts the results from the first part of the analysis as both series resistance and doping density are typically attributed with changes in the fill factor. However, further investigation shows that there is indeed a high correlation between the V_{oc} and the *FF*. In this case, R_s is rather a measure for the cell's V_{oc} which could be caused by e.g., an interaction between the contact formation of the electrical terminals and the surface passivation. Furthermore, the doping concentration of the raw silicon wafer is also connected to wafers bulk minority charge carrier lifetime by the ingoting process and is therefore also an indicator of the cell voltage.

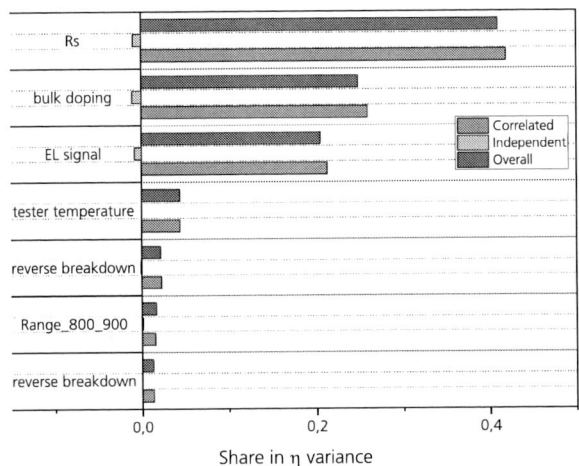

Figure 4: Sensitivity analysis of the main feature impacts extracted by the machine learning model.

For the final layer a simulation based on Quokka 2.2.5 [4] is used. The simulation calculates *I-V*-parameters based on physical inputs like interface and bulk recombination and internal resistances. It then matches the measured cell *I-V*-parameters to the simulated values and returns a most likely set of input parameters for each cell [3]. As before, a sensitivity analysis on these results is done and is shown in Figure 5. From this analysis, the thesis formed in the last part can be confirmed. The simulation shows that the likely root cause for the variance of the cell production is coming from V_{oc} related parameters like the rear and front surface recombination and the bulk lifetime. Together with the information about the correlation between Rs and V_{oc} it can be deduced that the problem is connected to the metallization contacts.

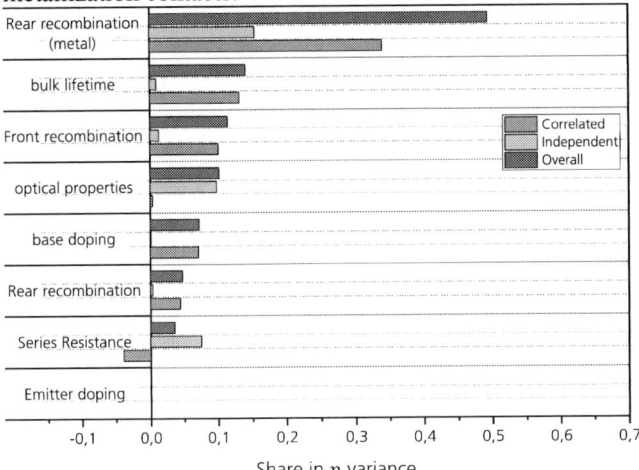

Figure 5 Sensitivity analysis of the numerical cell simulation input parameters

IV. SUMMARY

In summary the hierarchical approach of mass production data of solar cells shows a big potential for understanding root causes of variance. Due to different views on the same data set by using different analysis methods the interpretations can be compared and benchmarked with each other to get a holistic view on the data set. Due to the generic approach this method is not limited to solar cell production data but is universally applicable to all large data set which can be described with a theoretical model

REFERENCES

[1] VDMA, "ITRPV 2022," presented at the PV Celltech Conference, Berlin, Apr. 2022.

[2] S. Wasmer, K. Hübener, and B. Klöter, "Explaining the Efficiencies of Mass-Produced p-Type Cz-Si Solar Cells by Interpretable Machine Learning," Solar RRL, p. 2100477, Oct. 2021

[3] H. Wagner-Mohnsen and P. P. Altermatt, "A Combined Numerical Modeling and Machine Learning Approach for Optimization of Mass-Produced Industrial Solar Cells," in IEEE Journal of Photovoltaics, vol. 10, no. 5, pp. 1441-1447, Sept. 2020, doi: 10.1109/JPHOTOV.2020.3004930.

[4] A. Fell, "A free and fast 3D/2D solar cell simulator featuring conductive boundary and quasi-neutrality approximations," IEEE Transactions on Electron Devices, Vol 60 (2), pp. 733–738, 2012.

[5] T. Most, "Variance-based sensitivity analysis in the presence of correlated input variables", Proc. 5th Int. Conf. Reliable Engineering Computing (REC), Brno, Czech Republic, pp. 335–352, 2012.

[6] G. Li, H. Rabitz, P. E. Yelvington, O. O. Oluwole, F. Bacon, C. E. Kolb, and J. Schoendorf, "Global sensitivity analysis for systems with independent and/or correlated inputs", The journal of physical chemistry. A, vol. 114, no. 19, pp. 6022–6032, 2010.

[7] S. Wasmer, J. M. Greulich, H. Höffler, N. Wöhrle, M. Demant, F. Fertig, S. Rein. "Impact of Material and Process Variations on the Distribution of Multicrystalline Silicon PERC Cell Efficiencies", IEEE J. Photovoltaics, vol. 7, no. 1, pp. 118–128, 2017.

Identifying the electrical signature of snow in photovoltaic inverter data

Emma C. Cooper, Jennifer L. Braid, and Laurie M. Burnham
Sandia National Laboratories, Albuquerque, NM, 87106, USA

Abstract—Snow is a significant challenge for PV plants at northern latitudes, and snow-related power losses can exceed 30% of annual production. Accurate loss estimates are needed for resource planning and to validate mitigation strategies, but this requires accurate snow detection at the inverter level. In this study, we propose and validate a framework for detecting snow in time-series inverter data. We identify four distinct snow-related power loss modes based on the inverter's operating points and electrical properties of the inverter and PV arrays. We validate these modes and identify their associated physical snow conditions using site images. Finally we examine relative frequencies of the snow power loss modes and their contributions to total power loss.

I. INTRODUCTION

As penetration of renewable energy generation increases, seasonal mismatches between renewable production and demand are becoming increasingly problematic [1]. Many studies have demonstrated that snow significantly compromises PV output during winter [2]–[4], often a period of high demand in snowy regions [5], with losses as high as 90% - 100% during winter months for some systems [2], [6], [7]. Quantitative comparisons of designs that promote system resiliency against snow and fast snow shedding are needed, which require accurate estimates of snow losses [8].

Established methods for estimating snow loss can be grouped into data-driven models and physics-based models. The accuracy of stochastic or curve-fitting models has not been rigorously tested, but snow-cover models have been more thoroughly reviewed. Snow cover models can reduce error in generation predictions to as little as 7% during winter months, but reviews have demonstrated that these models tend to significantly underpredict snow losses if they do not reference current snow cover conditions [2]. Timeseries images can be used to provide snow cover conditions for these models, but implementation require advance planning and additional monitoring capabilities.

Over 30% of operable utility-scale fixed-tilt capacity is located in the U.S. at or above 40 degrees of latitude (as of 2021) [9], and represent vast stores of high quality performance and irradiance data collected in cold climates. An accurate method for identifying snow in these utility scale datasets

would make snow loss quantification vastly more accessible to asset owners. Characterizing the electrical signature of snow has been the subject of simulation-based studies, but there has been little research with regard to analysis and identification in field data.

Snow losses were simulated for a single module with varying spatial coverage and snow transmissivity by [10], but models were not validated through field data. IV curves of shaded and unshaded modules connected in series were modeled as a function of shading coverage and validated through field data by [11], but the effect of transmission was not investigated. A recent study [12] developed an algorithm for identifying snow using inverter and onsite ambient temperature data, but comparison with field data demonstrated that the algorithm failed to detect a third of snow events.

In this study, we propose and validate a model for identifying snow in utility-scale PV inverter data with the goal of establishing the foundation for a utility-scale snow loss estimate tool. We use performance models to determine effective transmission and coverage from field data, and corroborate these values with site images. We use these parameters to introduce a framework for behaviors that are distinguishable within utility-scale field data and identify four distinct power-loss modes. We validate the presence of these modes across different system scales and data resolutions. We identify the physical snow cover situations that each mode corresponds to, and note differences across systems. We determine the relative frequencies of each mode and associated fraction of total power loss for each system, and correlate the aggregated results with panel orientation.

II. METHODS

A. Data

A fifteen-minute resolution dataset for a monofacial fixed low-tilt utility-scale site was provided by an electric utility in Northeastern US, including 196 days from December 2020 to February 2022. The site was chosen because of the heavy snow losses observed by its asset owners; over the period of observation, the site experienced over 100 inches of snowfall and persistent snow cover on panels was observed to last for weeks on end.

Available data included inverter-level DC and AC voltage and combiner-level DC current as measured by a Solectria Yakasawa inverter, back-of-module (BOM) temperature, and plane-of-array (POA) irradiance measured by a heated pyranometer. Data was filtered using procedures outlined in

---

Sandia National Laboratories is a multimission laboratory managed and operated by National Technology & Engineering Solutions of Sandia, LLC, a wholly owned subsidiary of Honeywell International Inc., for the U.S. Department of Energy's National Nuclear Security Administration under contract DE-NA0003525. This work was supported by the U.S. Department of Energy's Office of Energy Efficiency and Renewable Energy (EERE) under the Solar Energy Technologies Office under Award 38527.

978-1-6654-6060-6/23 $31.00 © 2023 IEEE

[13]. Additionally, periods of time where AC power output approached the inverter's nameplate limits were excluded to avoid clipped data. Site images showing a string of modules connected to an inverter were collected at 1-hour and 15-minute resolution (Fig. 1). Spatial coverage of snow on the string of modules was determined through an unsupervised pixel clustering algorithm [14] and used as a point of comparison to effective coverage as calculated using electrical data.

Fig. 1. One of the series of site images used to calculate snow cover, where a shedding event occurs between 10 am and 11 am.

B. Snow detection framework

For this work, we assume that a reduction in current or voltage performance index (PI) is due to light-blocking matter (here snow) on the surface of the modules. The electrical configuration of the site was typical in that modules were connected in series strings, which were connected in parallel at combiners. The current is measured at the combiner level, and is equivalent in magnitude to the module current multiplied by the number of module strings in parallel. Combiners are connected in parallel at the inverter, where voltage is measured and is equivalent in magnitude to the voltage of a string of modules. As a result, measured combiner current I_{mp} is representative of the average lowest nonzero current generated by a cell among parallel-connected module strings. Therefore, we estimate the effective transmission of light-blocking matter on the array, T_{eff}, as the fraction of light received by the panels, using the Sandia Array Performance Model (SAPM) [15] with combiner current, and onsite irradiance and BOM temperature data:

$$I_{mp} = N_{strings} \times I_{mp0}(C_0 E_e T_{eff} + C_1 (E_e T_{eff})^2) \\ \times (1 + \alpha_{Imp} (T_{cell} - T_0)) \tag{1}$$

where C is a vector of coeffients specific to the module type, E_e is effective irradiance, T_{cell} is cell temperature, T_0 is 25 °C, I_{mp0} is the nameplate I_{mp} of the module, and $N_{strings}$ is the number of strings connected in parallel.

Inverter voltage was also modeled using the SAPM,

$$V_{mp} = N_{modules} \times [V_{mp0} + C_2 N_s \, \delta \ln(E_e T_{eff}) \\ + C_3 N_s \, (\delta \ln(E_e T_{eff}))^2 \tag{2} \\ + \beta_{Vmp} (T_{cell} - 25)]$$

using T_{eff} and sensor data, where N_s is the number of cells connected in series per module, V_{mp0} is the nameplate V_{mp}

of the module, and $N_{modules}$ is the number of modules in a string.

Given that string operating voltage increases approximately linearly with the number of active (unbypassed) module substrings connected in series, the voltage PI, V_{ratio}, is representative of the average effective snow-free fraction of all PV strings connected to the inverter. The complement of V_{ratio} can be used a measure of effective coverage C_{eff}. In this framework, $V_{ratio} < 1$ corresponds to conditions where there are inactive substrings (activated bypass diodes) in the string, whereas $V_{ratio} \approx 1$ suggests that all substrings are active. A phase diagram of V_{ratio} is plotted against spatial coverage in Figure 2. The 540 V turn-on voltage (V_{turnon}) of site's inverter ensures that the system remains off unless $\approx 73\%$ of modules are operating at maximum power point (MPP), so it should be noted that $V_{ratio} < V_{turnon}/V_{mp0}$ will result in $C_{eff} = 1$. A decrease in transmission appears as a gradient from red to blue and is correlated with spatial coverage; as spatial coverage increases past $(1 - V_{turnon}/V_{mp0}) \times 100\%$, a non-zero V_{ratio} can only be measured for a system where light transmits through the snow and $T_{eff} < 1$.

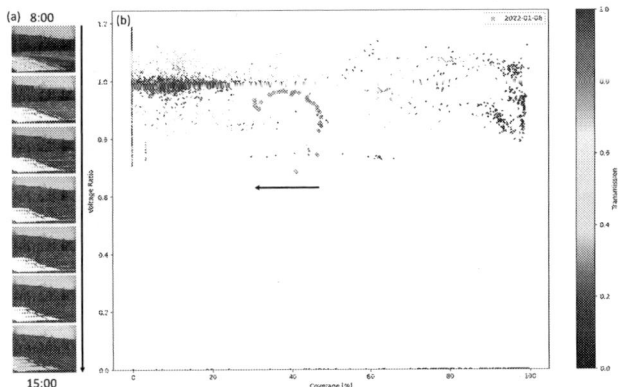

Fig. 2. (a) Images of the utility-scale site from 2022-01-08 where snow recedes from the panels over the course of the day. (b) V_{ratio} plotted against spatial coverage as determined through images. Points are colored by a calculated effective transmission where red indicates $T_{eff} \approx 1$ and blue indicates a lower T_{eff}; data from 2022-01-08 is highlighted to show a decrease in coverage and a corresponding increase in voltage PI.

Four power-loss modes corresponding to the electrically distinguishable behaviors discussed above are identified in Fig. 3. Mode 0 corresponds to a complete outage where the system does not meet the minimum turnon voltage for the inverter. Modes 1 and 2 describe data collected when one or more substrings are bypassed; Mode 1 is the case where the active substrings are limited in current due to less than 100% light transmission of snow, while Mode 2 is for active substrings operating as if uncovered by snow. Mode 3 corresponds to instances where all substrings are active but there is low effective transmission (current is limited due to snow cover). The cutoff values to distinguish between these modes were determined statistically from snow-free data to ensure normal operation was not classified as snow cover.

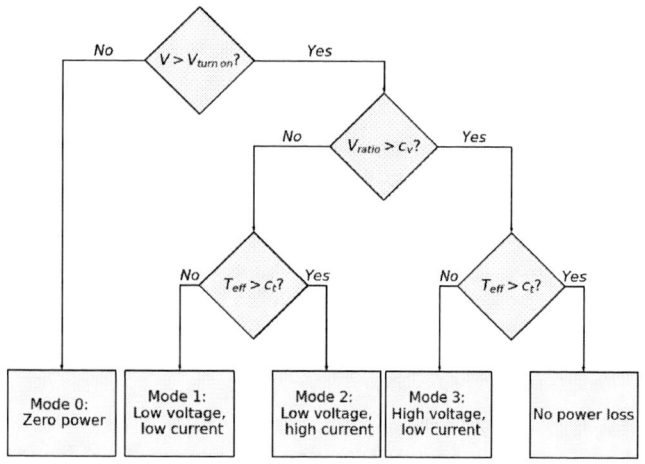

Fig. 3. Flowchart for determining power loss modes. Values for c_v and c_t were determined by identifying the 95th percentile of effective coverage and transmission in snow-free data.

TABLE I
RESEARCH-SCALE VALIDATION SITES

System	String-inverter	Portrait-orientation module	Landscape-orientation module
Voltage resolution	Inverter	Module	Module
Current resolution	Inverter	Module	Module
Orientation	Portrait	Portrait	Landscape
Date range	Nov. 2021 - Apr. 2023	Feb. - May 2023	Feb. - May 2023
Data frequency	1 minute	1 minute	1 minute

C. Validation

Three monofacial research-scale systems at the Michigan Regional Test Center (MI-RTC) were selected for analysis based on their similarity to or differences from the utility-scale system (Table I). Single module systems were selected to verify the presence of power loss modes at a module-level, while a string-inverter system was selected for its structural similarity to the utility-scale system. Both portrait and landscape-orientation systems were included to isolate the effect of orientation on modal frequency. All systems were outfitted with at least one BOM thermocouple and POA irradiance was collected by a heated Kipp & Zonen pyranometer.

Phase diagrams for all three validation sites displayed a correlation between T_{eff} and spatial coverage similar to that seen for the utility-scale site. The two module-scale systems displayed markedly different patterns in voltage ratios; while the plot for the landscape-orientation module shows the number of individual substrings that are online (Figure 4a), the same is not apparent in the plot for the portrait-orientation

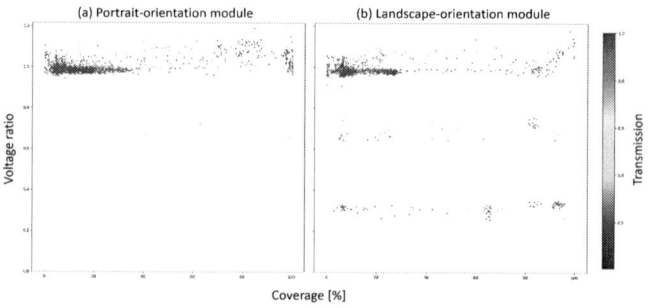

Fig. 4. Relative frequencies of modes for data collected between October and May that includes significant power losses. Data from mode 0 is not collected for the two module-scale systems, but an analysis of missing timestamps found that the portrait-orientation system was offline for ≈ 13 hours longer than the landscape-orientation system.

module (Figure 4b). This is consistent with snow typically shedding down the length of a module, so that individual substrings on the upper half of a landscape orientation module can remain completely uncovered and online even while there is partial snow cover on the bottom portion of the module. The same is not true for a landscape-orientation module, as all substrings remain partially covered even if the upper portion of the panel is uncovered.

III. RESULTS

The electrical/optical characteristics and physical interpretations of power loss modes are summarized in Table II.

The relative frequencies and % power loss of modes across the utility-scale and validation sites are shown for periods of time when a power loss mode is present in Figure 6. For all data collected between the months of October and May during daylight hours, the utility-scale site experienced a snow-related power loss mode for 59% of recorded timestamps and the string-inverter system experienced the same for 52% of timestamps. These statistics are not intended for cross-site comparison, but more so to demonstrate the sheer magnitude of time that these systems experienced snow-related power losses over the period that these datasets was collected.

For the utility-scale and string inverter systems, Mode 0 is achieved when the inverter is unable to reach the turn-on voltage; i.e., too many substrings are covered with effectively opaque snow. We estimate this coverage threshold C_V for each system based on the fraction of the number of substrings operating at $T_{eff} = 1$ necessary to surpass V_{turnon}. Comparisons of modal frequencies and rates of missing timestamps between the two inverter-connected systems and the two single module systems suggest that portrait-orientation systems experience Mode 0 at a higher frequency relative to landscape-orientation modules 4.

Mode 1 describes data with reduced values for both V_{ratio} and T_{eff}. The majority of this data was recorded after snow fell on modules that were already partially covered. We hypothesize that Mode 1 primarily corresponds to full or partial coverage by snow with non-uniform transmission,

TABLE II

SNOW POWER LOSS MODES. EFFECTIVE COVERAGE IS C_{eff}, EFFECTIVE TRANSMISSION IS T_{eff}

	Mode 0	*Mode 1*	*Mode 2*	*Mode 3*	*Mode 4*
C_{eff}	1	$1\text{-}V_{turnon}/V_{mp0} \leq C_{eff} < 1$	$1\text{-}V_{turnon}/V_{mp0} \leq C_{eff} < 1$	≈ 1	0
T_{eff}	0	< 1	≈ 1	< 1	≈ 1
Physical interpretation	Partial or full coverage by opaque snow, minimum inverter voltage not met	Partial coverage by opaque snow, partial coverage by transmissive snow	Partial coverage by opaque snow, partial uncovered	Partial or full coverage by transmissive snow	Zero coverage or partial or full coverage by highly transmissive snow

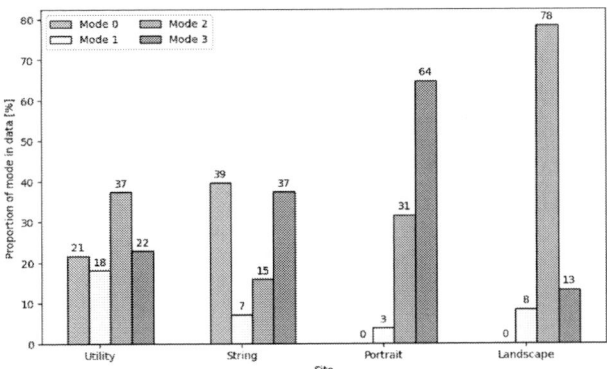

Fig. 5. Relative frequencies of modes for data collected between October and May that includes significant power losses. Data from mode 0 is not collected for the two module-scale systems, but an analysis of missing timestamps found that the portrait-orientation system was offline for ≈ 13 hours longer than the landscape-orientation system.

where $1 - V_{ratio}$ is equivalent to the fraction of array surface area covered by opaque snow and T_{eff} is equivalent of the transmission of non-opaque snow partially or fully covering the remaining area. In this line of thinking, Mode 1 is the product of incomplete shedding and the relative frequency of Mode 1 to other modes may be an indicator of performance by site design in facilitating shedding.

Mode 2 is representative of partial coverage by opaque snow with the remaining portion of the array uncovered. If the system is connected to an inverter, partial coverage by opaque snow must be below the coverage threshold for the inverter to turn on. Based on observations of systems experiencing some kind of snow-related power loss, landscape-orientation systems experience Modes 1 and 2 more frequently than portrait-orientation systems 6. Relative to other modes, portrait-orientation modules are more frequently found to be in Mode 3, which describes data where a full or partial cover of transmissive snow decreases effective transmission but all substrings remain active. Mode 4 was observed to correspond to snow-free production or partial or full coverage by a highly transmissive layer of frost or snow, where in either case, little to no power is lost.

Figure 6 compares modal frequencies with attributable power losses at the utility-scale site. Power losses associated with Mode 2 are minimal, which is consistent with high

transmission values and low partial coverage. Modes 1 and 3, where $T_{eff} < 1$, lead to significant power losses. Some power losses were observed in Mode 4 - these may be attributable to fog or dew - but they were small relative to the frequency of Mode 4.

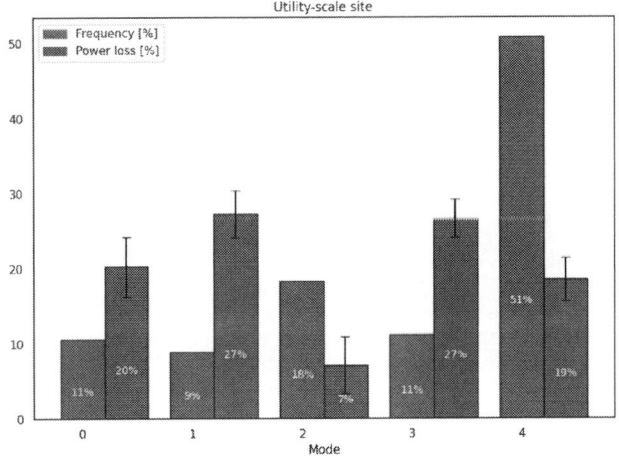

Fig. 6. Frequencies and attributable power losses of modes for all data collected at the utility-scale site between October and May.

CONCLUSION

The introduction and validation of a framework by which snow-related power loss modes can be identified in utility-scale data has the potential to vastly simplify the estimation of power losses and minimize associated requirements for monitoring equipment. Four distinct snow-related power loss modes corresponding to different effective coverage and transmission ranges for snow cover were identified from inverter data at a utility-scale PV site. The presence of snow as detected from electrical data, as well as the physical interpretations of snow cover corresponding to each power loss mode have been verified in time-series images from the site. The modes have been validated using data from inverter-connected and single module research-scale systems, and we observe similar patterns in the relationship between spatial snow cover and snow transmission across systems. We observe significant differences in modal frequencies between landscape and portrait-orientation systems; portrait-orientation systems tend to be completely offline more frequently, while landscape-orientation systems tend remain online despite partial snow coverage. These differences are consistent with the idea that the majority of partial snow covers obscure a nonzero fraction of each substring on a landscape-orientation module, whereas it is typical for at least one substring to be exposed on a landscape orientation module with partial snow cover. For inverter-connected systems, modes with low or zero transmission values have the largest impact on power, while partial coverage has a minimal impact. Future studies should be performed to explore the impact of other system characteristics on modal frequencies such as tilt angle and location.

REFERENCES

[1] "The challenges of achieving a 100% renewable electricity system in the United States - ScienceDirect." [Online]. Available: https://www.sciencedirect.com/science/article/pii/S2542435121001513

[2] R. E. Pawluk, Y. Chen, and Y. She, "Photovoltaic electricity generation loss due to snow – A literature review on influence factors, estimation, and mitigation," *Renewable and Sustainable Energy Reviews*, vol. 107, pp. 171–182, Jun. 2019. [Online]. Available: https://www.sciencedirect.com/science/article/pii/S1364032118308268

[3] E. Andenæs, B. P. Jelle, K. Ramlo, T. Kolås, J. Selj, and S. E. Foss, "The influence of snow and ice coverage on the energy generation from photovoltaic solar cells," *Solar Energy*, vol. 159, pp. 318–328, Jan. 2018. [Online]. Available: https://www.sciencedirect.com/science/article/pii/S0038092X17309581

[4] R. W. Andrews, A. Pollard, and J. M. Pearce, "The effects of snowfall on solar photovoltaic performance," *Solar Energy*, vol. 92, pp. 84–97, Jun. 2013. [Online]. Available: https://www.sciencedirect.com/science/article/pii/S0038092X13000790

[5] "New England Energy Report." [Online]. Available: https://www.eia.gov/dashboard/newengland/electricity

[6] L. Powers, J. Newmiller, and T. Townsend, "Measuring and modeling the effect of snow on photovoltaic system performance," in *2010 35th IEEE Photovoltaic Specialists Conference*, Jun. 2010, pp. 000973–000978, iSSN: 0160-8371.

[7] T. Townsend and L. Powers, "Photovoltaics and snow: An update from two winters of measurements in the SIERRA," in *2011 37th IEEE Photovoltaic Specialists Conference*, Jun. 2011, pp. 003231–003236, iSSN: 0160-8371.

[8] J. Braid, D. Riley, and L. Burnham, "Design Considerations for Photovoltaic Systems Deployed in Snowy Climates," *37th European Photovoltaic Solar Energy Conference and Exhibition*, pp. 1626–1631, Oct. 2020, iSBN: 9783936338737 Publisher: WIP. [Online]. Available: http://www.eupvsec-proceedings.com/proceedings?paper=49817

[9] "Most utility-scale fixed-tilt solar photovoltaic systems are tilted 20 degrees-30 degrees." [Online]. Available: https://www.eia.gov/todayinenergy/detail.php?id=37372

[10] M. B. Øgaard, B. L. Aarseth, F. Skomedal, H. N. Riise, S. Sartori, and J. H. Selj, "Identifying snow in photovoltaic monitoring data for improved snow loss modeling and snow detection," *Solar Energy*, vol. 223, pp. 238–247, Jul. 2021. [Online]. Available: https://www.sciencedirect.com/science/article/pii/S0038092X21003868

[11] A. Dolara, G. C. Lazaroiu, S. Leva, and G. Manzolini, "Experimental investigation of partial shading scenarios on PV (photovoltaic) modules," *Energy*, vol. 55, pp. 466–475, Jun. 2013. [Online]. Available: https://www.sciencedirect.com/science/article/pii/S0360544213003095

[12] Z. DeFreitas, G. Binnard, and T. Delsart, "Using On-site Ambient Temperature and Performance Ratio to Identify Days when Snow Cover Affects PV Plant Production," in *2021 IEEE 48th Photovoltaic Specialists Conference (PVSC)*, Jun. 2021, pp. 1860–1864, iSSN: 0160-8371.

[13] A. Livera, M. Theristis, E. Koumpli, S. Theocharides, G. Makrides, J. Sutterlueti, J. S. Stein, and G. E. Georghiou, "Data processing and quality verification for improved photovoltaic performance and reliability analytics," *Progress in Photovoltaics: Research and Applications*, vol. 29, no. 2, pp. 143–158, 2021, _eprint: https://onlinelibrary.wiley.com/doi/pdf/10.1002/pip.3349.

[Online]. Available: https://onlinelibrary.wiley.com/doi/abs/10.1002/pip.3349

[14] J. L. Braid, D. Riley, J. M. Pearce, and L. Burnham, "Image Analysis Method for Quantifying Snow Losses on PV Systems," in *2020 47th IEEE Photovoltaic Specialists Conference (PVSC)*, Jun. 2020, pp. 1510–1516, iSSN: 0160-8371.

[15] D. L. King, W. E. Boyson, and J. A. Kratochvill, "Photovoltaic Array Performance Model."

Band Tail Effects on Cd(Se,Te) Device Performance: A Numerical Simulation Approach

Eva, M Mulloy, Darius Kuciauskas, Craig, L Perkins, Marco Nardone

Bowling Green State University, Bowling Green, OH, United States

National Renewable Energy Laboratory, Golden, CO, United States

When various types of disorder are present in a semiconductor, the extended states at the edge of the conduction and valence bands can spread into the band gap. Such sub-band gap states, or band tails, can be localized or delocalized depending on their range. Band tails can be detrimental to device performance by enhancing recombination rates and lowering the open-circuit voltage (Voc). They are often observed in photoluminescence (PL) spectra as peak broadening or as a sub-band gap energy peak that is red-shifted away from the band gap measured by absorption. In this work, we describe how to incorporate band tails in device simulations using mathematical models and interpretation of PL data. A rather general model is used allowing for consideration of different microscopic origins of band tails. Occupancy of the band tails is also included, which results in bias-dependent absorption. Inputs to the simulations include the band-tail-corrected absorption coefficient and the reduced band gap that governs the radiative recombination peak (PL peak). Significant Voc losses are observed in device simulations when absorption coefficients including tail effects and band gap shifts are included. It is demonstrated that band tails are increasingly detrimental as the tail extent goes beyond the thermal energy, kT. This approach allows for quantification of band-tail-related Voc and power conversion efficiency losses through a combination of PL, JV, and QE measurements and simulation. Cd(Se,Te) is used as a case study, but the approach can be applied to other technologies.

Understanding the dopability of As in selenium-alloyed cadmium telluride solar cells

Xiaofeng Xiang, Aaron Gehrke, and Scott Dunham

University of Washington, Seattle, WA, 98195, USA

Abstract — Cadmium telluride (CdTe) and its alloy CdTeSe are widely used in optoelectronic devices, such as radiation detectors and solar cells, due to their superior electrical properties. However, the formation of defects and defect complexes in these materials can significantly affect their performance. As a result, understanding the defect formation and recombination processes in CdTe and CdTeSe alloy is of great importance. In recent years, density functional theory (DFT) calculations have emerged as a powerful tool for investigating the properties of defects in semiconductors. In this paper, we use DFT calculations to study the properties of arsenic extrinsic defects in CdTeSe alloy, providing insights into the effects of these defects on the electrical properties of the material. We found that AX centers formation is more favorable in the alloy, which can potentially reduce the dopability of As in CdTeSe alloy. Our findings not only provide new insights into the mechanisms of As extrinsic defects in CdTeSe but also have implications for the development of high-performance solar cells.

I. INTRODUCTION

Cadmium telluride (CdTe) and its alloy with selenium (CdTeSe) have been extensively studied due to their high potential for use in optoelectronic devices, including solar cells and radiation detectors. Compared to CdTe, the addition of Se to the alloy has been shown to improve the material's electronic properties, such as its bandgap and carrier mobility, making it a promising candidate for high-efficiency solar cell applications [1].

The performance of these devices, however, can be significantly impacted by the presence of point defects, which can affect the material's electrical and optical properties. To better understand the effects of defects on the properties of CdTeSe, it is necessary to investigate the formation and recombination of defects and defect complexes in this material.

Arsenic (As) has been identified as an effective dopant in CdTe-based materials, with potential applications in improving *p*-type doping in solar cells [2]. However, the behavior of As in CdTeSe is not well understood, particularly with regards to its effects on defect formation and recombination. In this work, we use density functional theory (DFT) calculations to investigate the properties of As dopants in CdTeSe, with a focus on the formation of As-related point defects and their impact on the material's electronic properties. Our study sheds light on the behavior of As in CdTeSe and provides insights into strategies for optimizing the performance of CdTeSe-based devices.

II. FIRST-PRINCIPLES CALCULATION METHODS

Most DFT calculations apply the generalized gradient approximation (GGA) or local density approximation (LDA) as the exchange-correlation functional. However, it is widely known that DFT with LDA or GGA is likely to underestimate the bandgap of semiconductors. For CdTe, GGA gives us a bandgap of 0.68 eV, which is much smaller than experimental values ($\approx$1.5 eV). This can be explained by the overestimation of the delocalization of Cd-4d electrons, which lifts the valence electron energies (the Te-5p valence band) [3]. There are many approaches to correct the exchange-correlation functional, such as self-interaction correction (SIC) calculations, hybrid functional of Heyd, Scuseria and Ernzerhof (HSE) and spin-orbit coupling (SOC). However, these methods are computationally expensive, which makes them not feasible for study of defects in alloys due to the massive number of possible configurations. DFT with coulomb self-interaction potentials (GGA+U) is another method that has been frequently used to correct the calculated bandgaps [3]. This method combines Hubbard-like model for a portion of states in the system with Coulomb self-interaction potentials (U) to select bands for correction. Non-integer or double occupations of states are described by introducing of two parameters: (1) U, which reflects the intensity of the on-site Coulomb interaction, and (2) J, which adjusts the intensity of the exchange interaction. Typically, for simplicity, an effective parameter $U_{eff} = U - J$ is used. This effective parameter U_{eff} is typically referred to as U.

A. Computational Details

We find that U=12.2 eV for Cd-4d in CdTe gives excellent match to experimental lattice constant and bandgap. We also compare the result with other correction methods, HSE and HSE+SOC. Both approaches can match experimental bandgap. However, the lattice constants from hybrid methods seem to be overestimated with larger error than GGA+U method. We extend this optimized U parameter to zinc blende and wurtzite CdSe. Surprisingly, we found this setting still works well. The bandgap and lattice parameter of zinc blende CdSe shows great match with experiments in

978-1-6654-6060-6/23 $31.00 © 2023 IEEE

TABLE I
Lattice constant, bandgap for CdTe and CdSe obtained from GGA+U HSE06 and experiments.

Method	GGA+U		HSE06		Experiment [4]	
	Bandgap (eV)	a_0 (Å)	Bandgap (eV)	a_0 (Å)	Bandgap (eV)	a_0 (Å)
CdTe (Zinc Blende)	1.50	6.46	1.50	6.58	1.50	6.48
CdSe (Zinc Blende)	1.72	5.95	-	-	1.71	5.98
CdSe (Wurtzite)	1.79	a=b=4.21 c=6.86	1.68	a=b=4.30 c=7.01	1.80	a=b=4.30 c=7.02

Table I. For the wurtzite structure, the lattice parameter is underestimated by GGA+U, but bandgap is in line. The close agreement with experiment using the same U_d for CdTe and CdSe is possibly due to the d–s coupling between Cd–Te and Cd–Se being similar as both Te and Se belong to group VI. GGA+U down shifts the Cd 4d bands and activates strong d–s coupling. At U=12.2 eV, this approach provides appropriate correction for both CdTe and CdSe. As a result, we apply this U value to full composition range of CdTeSe alloy.

The structural optimizations and energy calculations are implemented in the VASP code. The GGA exchange-correlation functional of Perdew, Burke, and Ernzerhof (PBE) has been used for GGA and GGA+U calculations. We also use HSE06 with default parameters for comparison. A plane wave cutoff of 450 eV has been used for the wave functions. The Brillouin zones of all the structures considered were sampled with Γ-centered k-point grids. With this setting, a 2×2×2 k-point grid is generated, which is used in this work.

III. RESULTS AND DISCUSSION

As expected, As can substitute for Te or Se in CdTeSe and act as an acceptor due to its group V nature and having one electron less than Te. In this study, our focus is on the substitutional defect As_{Te}, as other related defects such as As interstitials have relatively higher formation energies [2]. Our calculations reveal that neutral and -1 charged states of AsTe exhibit T_d symmetry, in agreement with previous report [2]. The calculated transition level of As_{Te} is 0.22 eV, indicating a shallow level and the potential application for p-type CdTe. However, As can also form a +1 charged AX center defect, converting it into a donor [2] [5]. As shown in Fig. 1, this asymmetric defect is formed as As moves toward one of the neighbor Te atoms. Then the triply degenerate T_d state will split into two fully occupied states and one empty state as shown in Fig. 2. And each of the two atoms will break one bond with one neighbor Cd. If the band splitting energy ΔE_{split} is higher than bonding energy E_{bond} with Cd, AX defect will be more stable than tetrahedral As substitutional defect. In our calculation, we found AX center has a formation energy of 1.92 eV with "+/-" ionization level 0.08 eV, which is in line with First Solar's report [5] and explains why high

p-type doping in CdTe is possible in experiment [6]. The T_d +1 charge state of AsTe has a formation energy of 1.96 eV in CdTe and 1.96 ~ 1.99 eV in $CdTe_{0.75}Se_{0.25}$, which shows the

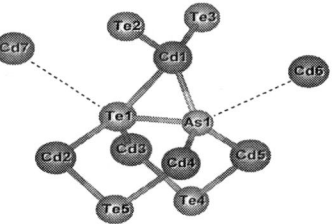

Fig. 1.   Schematic of AX center defect configuration.

T_d +1 defect formation does not vary much with Se alloy. However, the formation energy of the AX defect dramatically decreases with rising Se ratio in CdTeSe alloy. For example. in $CdTe_{0.75}Se_{0.25}$, the formation energy of AX defect ranges from 1.57 eV to 1.79 eV for different Se/Te arrangements, compared to 1.92 eV in CdTe. Therefore, an interaction between Se and the AX defect is likely to exist, which could lower the formation energy of the AX defect and compensate for p-type doping. This helps explain why large compensation effect would happen in front surface of CdTeSe:As solar cell device [7]. As can be seen from Fig. 2, the favorable formation energy of AX defect is not caused by the increase in ΔE_{split}. It is possibly due to nearby Se atom could weaken the bond between Cd and As/Te. In fact, our study of a 64-atom supercell of CdTeSe alloy with only one

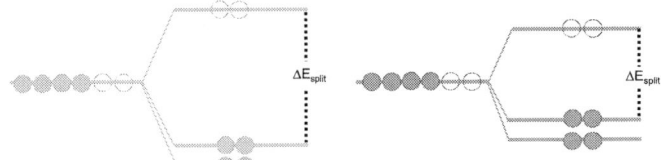

Fig. 2.   Schematic of AX center defect band splitting. Left one indicates AX in CdTe and right one indicates AX in $CdTe_{0.75}Se_{0.25}$. ΔE_{split} is the band splitting energy

Se atom in the group VI lattice revealed that the lowest formation energy of the AX defect was achieved when either the Te4 or Te5 atom was replaced with a Se atom. From Fig.1, we can deduce that when a smaller Se atom is being placed into Te4 and Te5, the nearby Cd atoms will be compressed in the direction of As1-Te1, which will strengthen the bonding

978-1-6654-6060-6/23 $31.00 © 2023 IEEE

TABLE II
Arsenic AX Center Defect vector model

Vector	(0.5, +/-0.5, -1)	(0.5, +/-0.5, 1)	(-0.5, +/-0.5, -1)
ΔE (eV)	-0.100	-0.032	0.045
Vector	(0, +/-1, 2)	(0, 0, 2)	(1.5, +/-0.5, 1)
ΔE (eV)	-0.029	-0.016	-0.017
Vector	(1, 0, 2)	(-1, 0, 2)	(2, 0, 2)
ΔE (eV)	-0.015	0.044	0.013

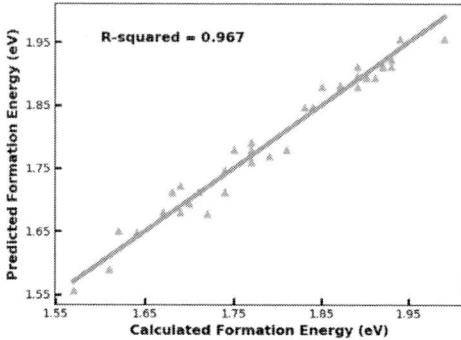

Fig. 3. Defect Vector Model Predication vs Calculation in CdTeSe alloy (Se ratio ranges from 0% to 50%).

As1-Te1 but weaken the bonding As1-Cd6 and Te1-Cd7. Moreover, we found that As atom will only move toward Te atom to form dimer structure in alloy, but moving toward Se atom is not energy favorable.

To investigate the optimal group VI position for AX defect formation, we constructed a new coordinate system for a 64-atom supercell of CdTeSe alloy. An As atom was placed at the origin and a distorted Te atom was positioned in (1, 0, 0), as illustrated in Fig. 4, with all other atoms labeled accordingly. This new coordinate system, comprising 128 atoms, allowed us to systematically vary the number of Se atoms in different vector positions originating from As and quantify the effect on AX defect formation. As the b-direction vectors are symmetrically equivalent, their sign is arbitrary. We calculated approximately 40 formation energies of AX defects in CdTeSe with Se alloy ratios ranging from 0% to 50% and developed a regression model (Table II, Fig.3) to predict the AX defect formation energy based on Se/Te ordering.

IV. Conclusions

In this paper, we have investigated the properties of arsenic extrinsic defects in CdTeSe alloy DFT calculations. We found that AX formation is more favorable in the alloy due to the interaction of selenium and arsenic, which could potentially deteriorate the dopability of As in CdTeSe alloy. Our results indicate that the formation energy of the AX

defect decreases as the Se alloy increases, suggesting an interaction between Se and AX defect that could lower the formation energy and compensate for *p*-type doping.

Furthermore, we built a regression model to quantify the effect of the position of Se atoms on the formation energy of the AX defect by calculating around 40 AX defect formation energies in CdTeSe with Se alloy ratio from 0% to 50%. Lattice Monte Carlo simulations based on the regression model are underway to study the detailed arsenic defect formation.

In conclusion, our study provides insights into the effects of arsenic defects on the electrical properties of CdTeSe alloy. Our findings could help guide the development of new doping strategies for the fabrication of more efficient solar cells based on CdTeSe alloy.

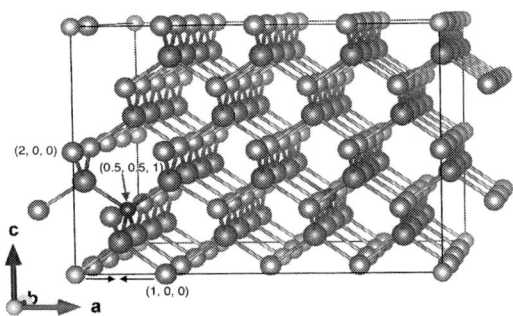

Fig. 4. New coordinate system of CdTeSe 64-atom supercell. Black arrows show As and Te deformation when AX center forms. (Red = Cd, Green = As, Brown = Te and Blue = Se)

References

[1] Fiducia, Thomas AM, et al. "Understanding the role of selenium in defect passivation for highly efficient selenium-alloyed cadmium telluride solar cells." *Nature Energy* 4.6 (2019): 504-511.

[2] Yang, Ji-Hui, et al. "Enhanced p-type dopability of P and As in CdTe using non-equilibrium thermal processing." *Journal of Applied Physics* 118.2 (2015): 025102.

[3] Yelong Wu, Guangde Chen, Youzhang Zhu, Wan-Jian Yin, Yanfa Yan, Mowafak Al-Jassim, and Stephen J Pennycook. LDA+U/GGA+U calculations of structural and electronic properties of CdTe: Dependence on the effective U parameter. *Computational Materials Science*, 98:18–23, 2015.

[4] Landolt, Hans, and Richard Börnstein. Numerical data and functional relationships in science and technology: *Crystal and solid state physics*. Group 3. Vol. 7. Springer-Verlag, 1966.

[5] Krasikov, D., and I. Sankin. "Beyond thermodynamic defect models: A kinetic simulation of arsenic activation in CdTe." *Physical Review Materials* 2.10 (2018): 103803.

[6] Nagaoka, Akira, Darius Kuciauskas, and Michael A. Scarpulla. "Doping properties of cadmium-rich arsenic-doped CdTe single crystals: Evidence of metastable AX behavior." *Applied Physics Letters* 111.23 (2017): 23210

[7] McGott, Deborah L., et al. "Revealing the Importance of Front Interface Quality in Highly Doped CdSe x Te1−x Solar Cells." *ACS Energy Letters* 6.12 (2021): 4203-4208

Modeling Transposition for Single-Axis Trackers Using Terrain-Aware Backtracking Strategies

Kurt Rhee

Nevados Engineering, San Francisco, California, 94107, United States

Abstract—**Nevados Engineering has developed a software program and accompanying modeling methodology which can be used in conjunction with industry standard photovoltaic performance modeling software in order to more accurately estimate the performance of trackers which employ terrain-aware backtracking strategies. Nevados has utilized this software to benchmark its proprietary terrain-aware backtracking algorithm against two other backtracking algorithms at one site in the United States. Results at the site indicate that a standard ground coverage ratio based backtracking algorithm would suffer terrain-related inter-row shadowing losses on the order of 6.7%. A common alternative to standard backtracking entitled artificial ground coverage ratio backtracking was able to recover 33% of this lost energy. The proprietary terrain-aware backtracking algorithm was able to recover around 63% of terrain-related energy losses,**

Index Terms—**horizontal single-axis tracker, tracker, transposition, backtracking, ray casting, terrain**

I. INTRODUCTION

Horizontal single-axis trackers in utility-scale photovoltaic projects have historically been constructed and modeled as flat arrays of repeating tracker rows, with each row indistinguishable from the row next to it. As the solar industry has matured, there has been an effort to construct sites on ground that is not flat. Sites that are constructed on variable terrain differ from flat sites in two main important ways.

In terms of hardware, some tracker manufacturers have added the ability to add angular deflection to the tracker torque-tube. This modification allows trackers to conform to the underlying terrain in the north-south direction, and reduces the amount of grading a site may require. Figure 1 and 2 show real world installation as well as a simplified example of how Nevados trackers change torque tube angles within a given tracker respectively.

In terms of software, some tracker manufacturers have developed terrain-aware backtracking strategies which can be used to reduce terrain related shading losses that would otherwise occur if the system were to be controlled by a standard ground coverage ratio (GCR) based backtracking algorithm. A standard GCR based backtracking algorithm takes into account the spacing between trackers, but does not utilize any information regarding the underlying terrain.

These new features introduced by tracker manufacturers can cause some difficulty for performance modeling groups. First, modeling a solar photovoltaic system sited on terrain as it were flat will not capture row-to-row shading losses if no terrain-aware backtracking algorithm is employed [1] [2] [3]. Second, the number of different terrain-aware backtracking strategies available and their different effects on a performance model are not currently well understood. Third, most performance modeling software programs do not allow for angular deflection within a tracker object. In order to accurately estimate a tracker system on terrain, a performance model not only needs to take into account the unique rotation angles that are generated from the terrain-aware backtracking strategy, it must also take into account the torque tube axis angle deflections that can occur within each tracker object.

At the time of writing, many industry photovoltaic performance modeling software have not yet implemented all forms of terrain-aware backtracking in their internally calculated tracker rotation angles, though this has not stopped users from modeling its effects in an indirect fashion [1]. Some software programs such as Terabase's PlantPredict and DNV's Solar-Farmer give the user the ability to input custom tracker angles which can make the evaluation of terrain-aware backtracking schedules easier.

The main objective of this paper is to illustrate a method for modeling the performance of terrain-aware backtracking algorithms that integrates easily with existing industry performance modeling software. The second objective is to demonstrate the magnitude of performance gain that can be expected from different terrain-aware backtracking algorithms compared to operating a plant with a standard ground coverage ratio based strategy on variable terrain.

II. METHODS

A. Backtracking Model Selection and Preprocessing

Backtracking on flat ground and on mono-slopes can be solved in closed form [7]. Backtracking on variable terrain has not been solved in closed form, and a number of different algorithms exist to solve the problem.

1) Types of Terrain-Aware Backtracking Algorithms: In order to model the performance benefit of a terrain-aware backtracking algorithm over a standard GCR based algorithm, one must first define what type of terrain-aware backtracking algorithm the tracker control system will employ. The simplest type of algorithm is called the "Artificial GCR" method. In this method, the GCR used to control the trackers is set artificially higher than the as-built GCR so that the controls

978-1-6654-6060-6/23 $31.00 © 2023 IEEE

Fig. 1. Nevados trackers being installed on a hill. Angle changes occur along the torque-tube axis at each pile location which allow the Tracker to follow the terrain underneath without the need to level the site.

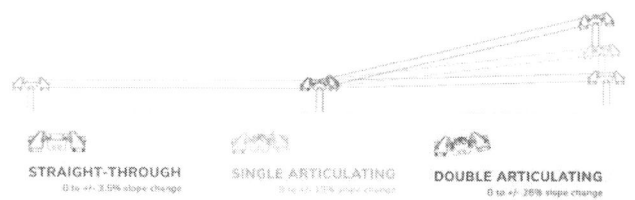

STRAIGHT-THROUGH   SINGLE ARTICULATING   DOUBLE ARTICULATING

Fig. 2. Simplified bearing design for a Nevados tracker. Using different bearings at each pile location allows Nevados trackers to adjust to changes in elevation from pile to pile, while also using the minimum cost bearing at each pile.

system backtracks the trackers earlier and more aggressively than if it were sited on flat land. For example, if a tracker site is constructed with a GCR of 33% on terrain, a tracker manufacturer may control their trackers with an artificial GCR of 36%. In this way, some terrain shade losses are avoided, but because the algorithm does not take the actual terrain into account, it may leave some rows shaded and other rows more backtracked than they need to be.

Another method which can be used to optimize the tracker rotation angles is to utilize a computational geometry engine. In this method, a digital version of the site geometry is created in three dimensions. Then the site geometry is processed via methods borrowed from the graphics processing and computational geometry fields, namely ray-casting and shadow-mapping [5]. Utilising the 3D site geometry allows the algorithm to minimize the angle between the module surface normal vector and the direct insolation coming from the sun, while also avoiding inter-row shadowing for every individual tracker.

It must be mentioned that there are even more possible backtracking algorithms besides the two listed above which can be employed to operate a horizontal single axis tracker site. A tracker company could employ a slope-aware back-

tracking algorithm [7] or use a backtracking algorithm which utilizes machine learning. Comparing these other types of backtracking algorithms is beyond the scope of this paper. Either way, the methods outlined below should provide a framework for comparing diffferent terrain-aware backtracking algorithms should any interested parties choose to extend this work in the future.

2) Preprocessing: Artificial GCR Method: Historically when the artificial GCR method has been deployed, it has been the job of a human operator to manually change the controls GCR set-point until no shading was observed during the commissioning process. Unfortunately, because financing steps come before commissioning, not all developers have been able to take advantage of the performance benefit that this controls strategy may impart on their project.

In order to benchmark the performance of backtracking angles created via a geometry engine to those created by an artificial GCR, Nevados has developed a method which can be used to determine which GCR set-point should be used during modeling. First, an array of rotation schedules is generated using the standard GCR based method. Then, a new array of rotation schedules is created by incrementing the input GCR by 1 percent. For each array of rotation schedules, a metric termed "effective plane of array insolation" is calculated which represents the amount of plane of array insolation we would expect to be converted into electricity.

In the simplified model for effective plane of array insolation used in this paper, the contribution of a shaded tracker "bay" to total plant transposition is reduced to only the diffuse portion of the incident insolation. In a Nevados tracker, a bay represents a subsection of a tracker which has a set of modules which all have the same torque tube axis tilt angle, in other words, a set of modules that exist together before an angular deflection of the torque tube occurs. This model is intended to very roughly mimic the effects of electrical mismatch along the string and though the model is an approximation, the magnitude of the resultant calculations is consistent with more rigorous models [1]. A more robust method would be to complete an energy model considering detailed 3D shading and IV curve mismatch as described and executed in [10].

3) Preprocessing: Geometry Engine Method: Some initial processing is needed to convert solar layouts, usually created in a computer aided design software such as AutoCAD, into a format that is usable by the geometry engine program. Once this conversion is complete, no further preprocessing is needed, the computation engine can create individual tracker rotation schedules bespoke to the terrain at each tracker.

B. Method for Modeling Transposition Utilizing a Terrain-Aware Backtracking Algorithm

1) Average Transposition: Because not all performance modeling software packages have the ability to calculate transposition for the large number of different torque tube axis angles and rotation angles that a site constructed on terrain may contain, it is necessary to first calculate transposed insolation externally. Transposition in this study was created using pvlib [8]. Once transposed insolation is calculated for every bay object, then a weighted average can be created with weights corresponding to the number of modules in a given bay. For example if there was a 3 bay tracker with 1 module, 2 modules and 5 modules in each bay respectively, then the transposition could be averaged as followed:

$$POA = 1/8 \times POA_1 + 2/8 \times POA_2 + 5/8 \times POA_3 \quad (1)$$

Where POA_1 stands for bay 1 transposition, POA_2 stands for bay 2 transposition and POA indicates plant average plane of array insolation.

This method of calculating and averaging transposed insolation can theoretically be broken down into multiple sub-models for enhanced accuracy. For example, instead of averaging all bays at a given site, one could model all bays at a given inverter. Averaging transposition will underestimate mismatch losses for the plant and will therefore cause modeled energy to be higher than actual energy.

2) Retro-Transposition: Some software programs such as PVSyst [13] will retro-transpose plane of array insolation into global horizontal and diffuse horizontal irradiation if plane of array insolation is used as an import. These software programs will then re-transpose the retro-transposed components into the plane of array insolation that is actually used in the simulation model. The results of this retro-transpose, re-transpose process are highly dependent on the underlying transposition models as well as the rotation angles assumed. In PVSyst, the Hay model [14] is used for retro-transposition and the rotation angles are calculated using one of PVSyst's suite of tracking modes.

Because one cannot control the rotation angles of single axis trackers in PVSyst, the only option for approximating the operations of a terrain-aware backtracking strategy is to tune the input meteo data. The input meteo data should be tuned so that once it is transposed, the resulting plane of array insolation matches that of a tracker employing a terrain-aware backtracking strategy. In this paper, TRACE uses the pvlib gti_dirint function with Perez transposition and a standard GCR-based schedule of backtracking angles. This allows the program to back-solve what global horizontal insolation and diffuse horizontal insolation would be necessary to receive the same amount of plane of array insolation as a terrain-aware backtracking strategy on a GCR-based backtracking simulation.

III. RESULTS

A. Terrain

In order to generate models, one real world system located in the north-eastern United States was chosen, built with Nevados trackers. The project is sited on a hill and is sloped in all directions, but most of the site is dominated by a north-eastern aspect. In table I, statistics for torque-tube axis-tilt, cross-axis-tilt and axis-tilt mismatch are reported. Torque-tube axis-tilt is reported in the south to north direction with positive numbers indicating that the modules are inclined towards the southerly horizon. A histogram of torque-tube axis-tilt can be

found in Figure 5. Cross-axis-tilt is reported in the west to east direction with positive numbers indicating that a given bay is lower in elevation than the bay directly east of it. A histogram of cross-axis-tilt can be found in Figure 6. Please note that the process for calculating cross-axis-tilt is not entirely clear for trackers with non-continuous torque-tubes, since a given bay may have a different torque-tube axis angle than the bay directly east or west of it. Additionally, a bay may be offset to the north or south of a bay directly to its east or west. In either case, the center point of each bay was used for comparison, and no difference in torque-tube axis-tilt is assumed. Differences in torque-tube axis angle from a given bay, compared to another bay on another tracker in the cross-axis direction are classified as axis-tilt mismatch. For example, if tracker 1 bay 1 has a torque-tube axis-tilt in the north/south direction of 1 degree and tracker 2 bay 1 directly east has a torque-tube axis-tilt in the north/south direction of 2 degrees, then the resulting axis-tilt mismatch would be −1 degrees. Only bays that are compared for cross-axis-tilt are compared for axis-tilt mismatch. A histogram of axis-tilt mismatch can be found in Figure 7.

TABLE I
SITE TERRAIN CHARACTERISTICS IN DEGREES

Type	Min.	Mean	Max	St. Dev.
N/S Torque-Tube Axis-Tilt	−3.36	−1.07	3.51	1.75
Cross-Axis Slope	−3.98	−1.31	1.41	0.90
Axis-Tilt Mismatch	−0.96	0.00	1.13	0.20

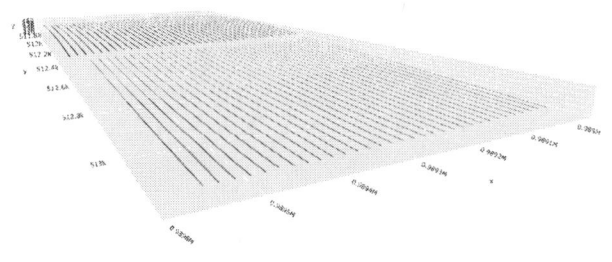

Fig. 3. 3D Model of the site used for schedule generation. Trackers are aligned north to south and are arranged in 4 distinct rows.

B. Backtracking

An hourly TMY weather dataset from the NSRDB's PSMv3 model [12] indicates that the annual average diffuse fraction at the site is 38%, which is relevant in that as the diffuse fraction increases towards a limit of 100%, terrain related inter-row shadowing is expected to be reduced to only the diffuse shading portion of the inter-row shading loss. At the site, backtracking time-steps represent approximately 20% of all daylight tracking time-steps. This 20% represents a smaller impact on overall transposed insolation than true tracking time-steps relative to its proportion, since backtracking is confined to the beginning and end of day when the sun is lower in

978-1-6654-6060-6/23 $31.00 © 2023 IEEE

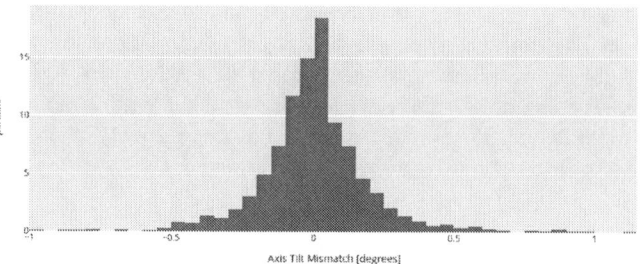

Fig. 7. Histogram of Axis-Tilt Mismatch. The peak centered around 0 degrees indicates that most trackers do not vary in north/south torque-tube axis-tilt from their cross-axis neighbors

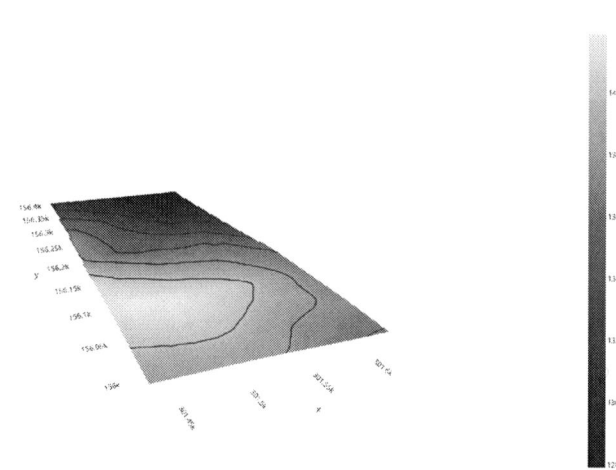

Fig. 4. Site Surface with 2m Contour Lines. Most of the site is oriented towards the north east, though some of the site is oriented towards the south east.

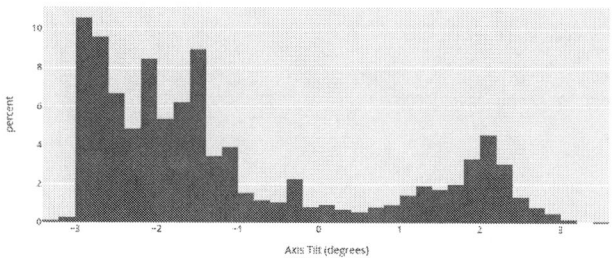

Fig. 5. Histogram of north/south torque-tube axis-tilt for all bays in the system. Evidence of both the dominating northern axis-tilt and smaller southern axis-tilt can be seen as distinct peaks.

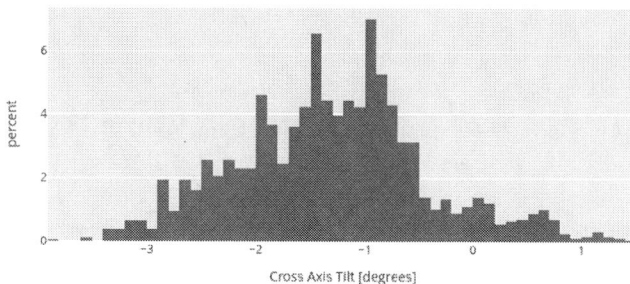

Fig. 6. Histogram of cross-axis-tilt for all bays in the system. The negative peak of the distribution indicates that most of the system's trackers are lower in elevation than the tracker directly to its west.

the sky. Assuming a flat model and Perez transposition [11], backtracking time-steps contribute approximately 16% of total plane of array insolation.

Table II shows the results of 3 different backtracking algorithms which could be used to control the same physical system. Each algorithm's performance in terms of effective transposition gain relative to the annual global horizontal insolation is shown in the "Gain" column. Each algorithm's performance in terms of effective transposed insolation is shown in the "Effective Insolation" column. Effective insolation is the amount of insolation that is usable to the system for the creation of energy that will not be lost to the effects of inter-row shadowing. The standard GCR based backtracking algorithm uses a set-point of 36% which matches the as-built GCR on site. The artificial GCR based backtracking algorithm uses a set-point of 42% which was calculated to be the most optimal GCR via the methodology described earlier in this paper. Results of the parameter sweep to determine the best GCR set-point for the artificial GCR method can be found in Figure 8.

Some important bookends which can be used to understand each algorithm's performance are as follows. The annual global horizontal insolation at the site was 1600.32 kWh/m². If the entire site was graded to flat, the system would have a global plane of array insolation value of 2108.79 kWh/m².

TABLE II
SITE 1: BACKTRACKING POA

Method	Gain (%)	Effective Insolation (kWh/m²)
Standard GCR	25	~2000
Artificial GCR	27.5	~2040
Geometry Engine	29.3	2068.23

POA stands for Plane of Array Irradiance. A ~ symbol is used to remind the reader that these insolation values were created via an approximation described in the Methodologies section of this paper.

By comparing the table above to the results that could have occurred if the site were graded entirely flat, it can be seen that about 6.7% of usable transposed insolation was lost if the tracker system was operated with a standard GCR based backtracking algorithm due to terrain-related inter-row shading. Approximately 2.5% absolute additional effective insolation can be captured if the system were to be controlled with the

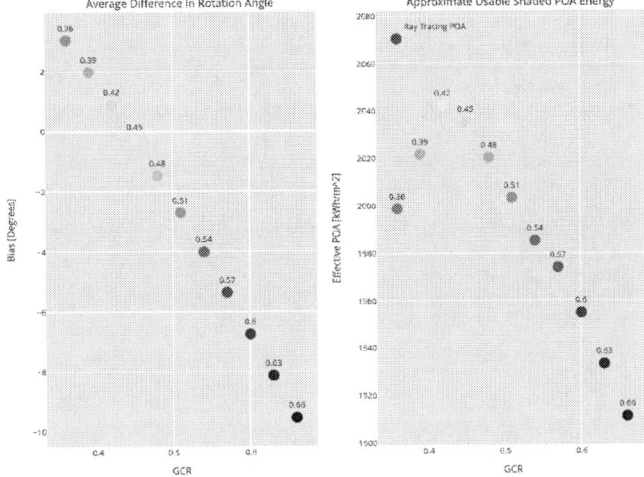

Fig. 8. *Left*: Mean bias in tracker rotation angle when comparing a standard GCR based backtracking algorithm to the most backtracked tracker in the field when calculating rotation angles via the computational geometry engine. Positive values indicate that the standard GCR based backtracking algorithm is less backtracked at that GCR set-point than the most backtracked result from the computation engine. Negative values indicate that the standard GCR based backtracking algorithm is more backtracked at that GCR set-point than the most backtracked result from the computation engine. *Right*: Comparison of effective POA using different GCRs compared to the computational geometry engine (green dot). The results from the computational geometry engine backtracking algorithm outperforms a parameter sweep of all ground coverage ratio based backtracking algorithm set-points at the site.

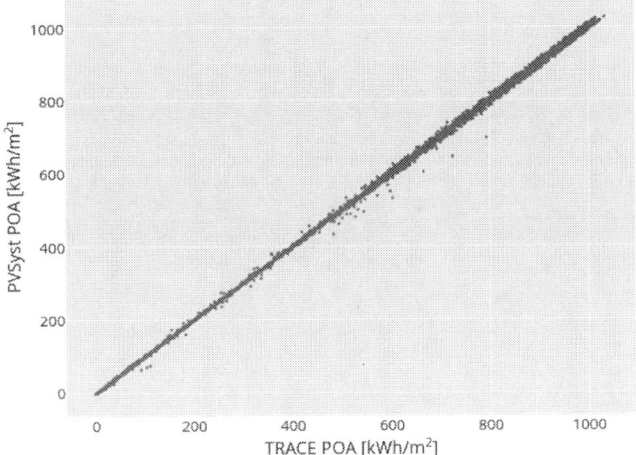

Fig. 9. Plane of array insolation as generated by TRACE on the x-axis compared to plane of array insolation as generated by retro-transposition and re-transposition as generated by PVSyst. The highly correlated nature of the data indicates that the retro-transposition method may be an accurate way to approximate the performance of terrain-aware backtracking strategies in performance modeling software which does not accept custom tracker angles as an input.

artificial GCR algorithm with a GCR set-point of 42%. 4.3% absolute additional effective insolation can be captured by using a computational geometry engine. These modeled results show the large effects that backtracking algorithm choice can have on system performance and also closely match results from field-testing mentioned in prior work [1].

C. Backtracking Meta-Analysis

A few interesting points constituting a meta-analysis of the aforementioned results follow:

1) Contribution of Torque-Tube Axis-Tilt to Transposition Loss: In order to determine, approximately, how much lost transposed insolation can be attributed to the northern aspect of the site, a modeled system was created with all trackers at a 1.07 degree northerly aspect. This aspect represents the average torque-tube axis-tilt for the whole site. Modeling the system with this configuration resulted in a loss of approximately 0.9% global plane of array insolation relative to a flat site. This loss represents 13.4% of the total lost transposed insolation and demonstrates that there is a component of transposed insolation related only to the torque tube axis tilt and not the backtracking rotation angles.

2) Averaging of Backtracking Angles: Another finding of note is that averaging the time series of tracker rotation angles as well as averaging the tracker-axis-tilt before calculating transposition instead of averaging the transposed insolation after the calculation is complete gives nearly the same results. Averaging the rotation angle and torque tube axis angle of every bay at the site for every timestep and then calculating transposed insolation based off of this average gives an annual insolation of 2071.87 kWh/m^2 compared to 2068.23 kWh/m^2 when tranposed insolation is averaged after the fact. This represents a 0.18% difference and falls within the uncertainty expected of transposition modeling in general.

3) Retro-Transposition: The method of retro-transposing the plane of array insolation generated with a terrain-aware backtracking algorithm in order to back-calculate the necessary horizontal insolation components needed to achieve the same plane of array insolation on a modeled single-axis tracker employing a standard ground coverage ratio based backtracking strategy is new to the industry. Attempting this method at this site showed an R-squared correlation of 0.9998 when the target transposed insolation was retro-transposed in pvlib and then re-transposed in PVsyst. Figure 9 shows a scatter-plot comparing plane of array insolation before the retro-transpose, re-transpose process to plane of array insolation re-transposed in PVSyst.

CONCLUSION

Accurately modeling the performance of horizontal single-axis trackers on variable terrain can be a daunting task. A method which can take into account the various intra-tracker torque-tube axis deflections as well as individual tracker rotation angles has been demonstrated. Furthermore, this method was used to benchmark three different backtracking algorithms on one site built on terrain. Using an artificial ground coverage ratio backtracking algorithm can recover around 33% of the losses incurred by terrain-related inter-row shading at the modeled site. Backtracking angle calculated via a computational geometry engine on the other hand can recover around 63% of the losses that may be incurred by terrain-related inter-row shading. Additional work must be done to complete the full performance modeling chain at this site and also to repeat these methodologies at different sites due to the fact that terrain varies widely from site to site.

ACKNOWLEDGMENT

I would like to thank Thang Le, David Spieldenner, and Jesse Milam from Terabase for their great suggestions which have been included in this paper as well as the assistance they have provided in using PlantPredict. At NREL, I would like to thank Kevin Anderson for his help in modeling multi-angle transposition, for all of his help editing this paper, and general mentorship. At DNV, I would like to thank Mark Mikofski for his continued guidance in all aspects of performance modeling. At Primoris I would like to thank Jay DeVilbiss for suggesting the use of alternative backtracking strategies during commissioning. At Nevados I would like to thank Brittanie Jackson for proofreading all of my papers as well as Yezin Taha and Sam Prest for dedicating time and resources towards research and development.

REFERENCES

[1] M. Leung et al., "Tracker Terrain Loss Part Two," in IEEE Journal of Photovoltaics, vol. 12, no. 1, pp. 127-132, Jan. 2022, doi: 10.1109/JPHOTOV.2021.3114599.

[2] A. Kankiewicz, "PV plant performance challenges from near shading and complex terrain," Solar Builder Magazine, May 20, 2021. https://solarbuildermag.com/news/pv-plant-performance-challenges-from-near-shading-and-complex-terrain/.

[3] "Solar Risk Assessment — kWh Analytics", kWh Analytics, 2022. [Online]. Available: https://www.kwhanalytics.com/solar-risk-assessment. [Accessed: 27- Jul- 2022].

[4] K. Passow, L. Ngan, G. Rich, M. Lee and S. Kaplan, "PlantPredict: Solar Performance Modeling Made Simple," 2017 IEEE 44th Photovoltaic Specialist Conference (PVSC), 2017, pp. 600-603, doi: 10.1109/PVSC.2017.8366450.

[5] K. Rhee, "Terrain Aware Backtracking via Forward Ray Tracing," 2022 49th IEEE Photovoltaic Specialists Conference (PVSC), 2022

[6] W.F. Marion and A. P. Dobos, "Rotation angle for the optimum tracking of one-axis trackers," Nat. Renewable Energy Lab., Golden, CO, USA, Tech Rep. NREL/TP-6A20-58891, 2013. [Online]. Available: htps://www.nrel.gov/docs/fy13osti/58891.pdf

[7] K. Anderson and M. A. Mikofski, "Slope-aware back-tracking for single-axis trackers," Nat. Renewable Energy Lab., Golden, CO, USA, Tech. Rep. NREL/TP-5K00-76626, 2020. [Online]. Available: https://www.nrel.gov/docs/fy20osti/76626.pdf

[8] William F. Holmgren, Clifford W. Hansen, and Mark A. Mikofski. "pvlib python: a python package for modeling solar energy systems." Journal of Open Source Software, 3(29), 884, (2018). doi: https://doi.org/10.21105/joss.00884.

[9] K. Passow, K. Lee, S. Shah, D. Fusaro, J. Sharp and L. Creasy, "Strategies to Optimize and Validate Backtracking Performance of Single-Axis Trackers on Sloped Sites," 2021 IEEE 48th Photovoltaic Specialists Conference (PVSC), 2021, pp. 1960-1964, doi: 10.1109/PVSC43889.2021.9518776.

[10] K. Anderson, "Maximizing Yield with Improved Single-Axis Backtracking on Cross-Axis Slopes," 2020 47th IEEE Photovoltaic Specialists Conference (PVSC), 2020, pp. 1466-1471, doi: 10.1109/PVSC45281.2020.9300438.

[11] Perez, Richard, et al. "Modeling daylight availability and insolation components from direct and global insolation." Solar energy 44.5 (1990): 271-289, doi: 10.1016/0038-092X(90)90055-H

[12] Sengupta, M., Y. Xie, A. Lopez, A. Habte, G. Maclaurin, and J. Shelby. 2018. "The National Solar Radiation Data Base (NSRDB)." Renewable and Sustainable Energy Reviews 89 (June): 51-60, doi: 10.25984/1810289

[13] Mermoud, A. Use and validation of PVSYST, a user-friendly software for PV-system design. In W. Freiesleben (Ed.), Thirteenth european photovoltaic solar energy conference (1995). Bedford: H.S. Stephens. Retrieved from https://archive-ouverte.unige.ch/unige:119365

[14] Hay, J.E., Davies, J.A., 1980. Calculations of the solar radiation incident on an inclined surface. In: Hay, J.E., Won, T.K. (Eds.), Proc. of First Canadian Solar Radiation Data Workshop, 59. Ministry of Supply and Services, Canada.

What's New in The NSRDB

Manajit Sengupta[1], Aron Habte[1], Grant Buster[1], Yu Xie[1], Brandon Benton[1] and Michael Foster[2]

1. National Renewable Energy Laboratory, Golden, CO, 80129, USA,

2. University of Wisconsin, Madison, WI, 53715, USA

Abstract — **The National Solar Radiation Database (NSRDB) provides solar resource data across the globe at a high temporal and spatial resolution. This data is primarily used in solar energy modeling. The NSRDB is updated annually for the United States and North, Central and South America and the data is currently available from 1998-2021. In 2022 the NSRDB was updated using the latest version of the underlying Physical Solar Model (PSM). This update includes improved surface albedo and gap-filling of cloud properties. The inclusion of these updates reduced the uncertainty in the data compared to previous versions of the NSRDB. The Himawari and Meteosat Indian Ocean Data Coverage(IODC) satellites were added to the Geostationary Operational Environmental Satellite (GOES) and made our coverage global. While standard data from the GOES continues to be served at an hourly 4km x 4km resolution, full resolution data has also been made available to the user. The NSRDB now contains over 200Tb of data with nearly 40Tb being added annually. We provide significant flexibility for data download depending on the amount of data required by the users. In this paper we provide an update on the current status on the NSRDB.**

I. INTRODUCTION

The NSRDB[1] is a widely used public source of high-quality solar resource and ancillary data used for modeling Photovoltaic (PV) and Concentrated Solar Thermal (CST) generation as well as for grid-integration studies. The NSRDB has over 160,000 annual users and provides a baseline dataset for government, academia and industry for large scale studies and pre-feasibility assessment.

II. METHODOLOGY AND INPUTS

The NSRDB uses a physical approach to satellite-based solar modeling. The underlying Physical Solar Model (PSM) involves modeling cloud-properties using satellite remote sensing and subsequently computing solar radiation using radiative transfer models. The retrieved cloud properties include cloud-mask, cloud-type (i.e. water and ice clouds) cloud optical depth and cloud droplet size[2]. The radiative transfer models require additional input parameters such as aerosol optical properties (AOD), preciptable water vapor, surface albedo, temperature and pressure to accurately model solar radiation. While cloud properties are obtained directly from the geostationary satellites other inputs are obtained from additional source such as the National Aeronautical and Space Administration (NASA) Modern Era Retrospective Analysis for Research and Applications version 2 (MERRA2), the Interactive Multisensor Snow and Ice Mapping System (IMS)

model data from the U.S. National Ice Center and NASA's polar orbiting satellites such as the Moderate Resolution Imaging Spectroradiometer (MODIS) instruments on the Aqua and Terra Plaform. The NSRDB data is provided on a fixed grid that matches both the spatial and temporal resolution of the geostationary satellites. The data from the various geostationary satellites, MERRA2, IMS and MODIS are generally available at unmatched spatial and temporal resolution. Therefore various parameters have to be downscaled to match the resolution of the NSRDB while maintain physical consistency. This requires the use of surface elevation obtained from the United States Geological Survey (USGS). For the NSRDB we use the USGS 30-m Digital Elevation Model (DEM) and downscale temperature, pressure, relative humidity and AOD.

The radiative transfer models for satellite-based solar modeling requires significant computational speed to process millions of pixel for each time. This led to the development of the Fast All-Sky Radiation Model for Satellite Applications (FARMS) model[3]. We also developed the FARMS Narrowband Irradiance on Tilted Surfaces (FARMS-NIT)[4][5] to provide spectral radiation for PV applications. Additionally, the FARMS Direct Normal Irradiance (FARMS-DNI) model[6] was developed to improve Direct Normal Irradiance (DNI) estimates under cloudy conditions. Finally, as PV reliability studies require information about ultraviolet (UV) radiation[7] we developed a model to estimate UV using broadband solar radiation.

As part of the update to the PSM (currently PSM version 3.2.2) two new changes were implemented to the model and input datasets. Satellite-based cloud properties inherently contain missing periods due to various reasons including a failure of the satellite to collect appropriate data and limitations of the physical cloud property algorithms to estimate clouds at high solar zenith angles. A new gap-filling technique using machine learning[8] was implemented to reduce gap-filling uncertainties. With bifacial-PV becoming an important part of utility scale PV deployment there was a need to improve the surface albedo especially during periods with snow on the ground. A new algorithm was implemented to determine snow-albedo based on surface temperature (e.g. Fig. 1).

978-1-6654-6060-6/23 $31.00 © 2023 IEEE

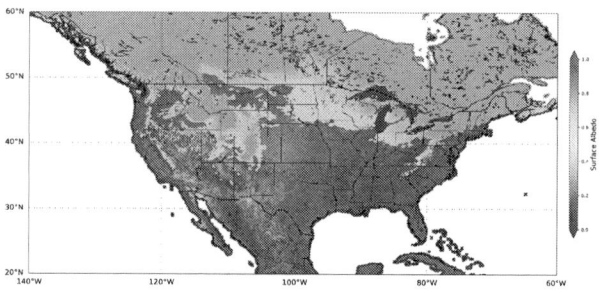

Fig 1: An example of variable surface albedo in snow covered regions.

II. NSRDB DATASETS

The NSRDB provides high-resolution data covering the entire globe (Fig. 2). Data from the GOES, Meteosat IODC and Himawari satellites were used to produce this global coverage.

As the NSRDB seeks to provide data for PV and CSP modeling and grid integration, multiple parameters are provided to meet user needs. This includes solar radiation including Global Horizontal Irradiance (GHI), DNI, Diffuse Horizontal Irradiance (DHI) and UV. Spectral radiation at 2001 wavlengths for user chosen tilt and orientation are made available through an on-demand service that computes the datasets on request. Clear-sky radiation is also available for GHI, DNI and DHI. In addition, ancillary data for surface temperature, surface albedo, surface pressure, relative humidity and 10-m wind speed and wind-direction are provided.

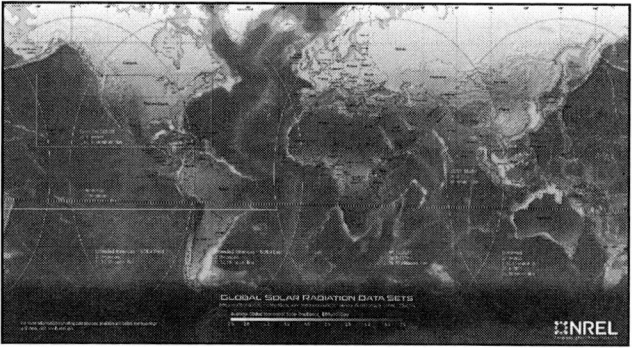

Fig. 2: The NSRDB has global coverage using data from 3 satellites.

The NSRDB also provides information about input parameters that were used to compute the final radiation products. This includes information about AOD and cloud properties and are useful for researchers seeking to study the impact of these various parameters in greater detail. As an example, the NSRDB data is capable of capturing events such as forest fires that severely impact solar generation and variables such as AOD is provided for enabling detailed research in such areas.

Currently data for the U.S. is available for 1998-2021 at a 30-minute temporal 4kmx4km spatial resolution. High-resolution data is also made available from 2018 at 5-minute intervals on a 2kmx2km grid. For other locations the years covered and the temporal and spatial resolutions vary depending on the resolution at which satellite data is collected.

IV. EVALUATION OF THE NSRDB

The NSRDB is continuously evaluated against high-quality ground measurement of solar radiation to assess its accuracy. Ground measurements from NREL Solar Radiation Research Lab (SRRL), US Department of Energy Atmospheric Radiation Measurement (ARM), National Oceanic and Atmospheric Administration (NOAA) Surface Radiation (SURFRAD) and Solar Radiation (SolRad) networks and University of Oregon are used for validation (Fig. 3).

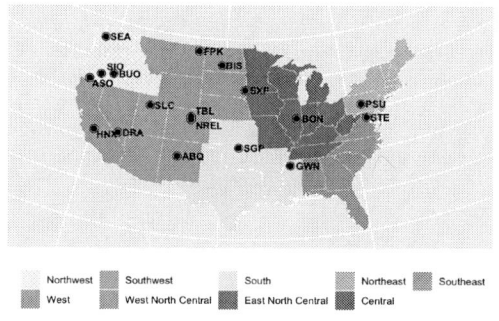

Fig. 3: Solar radiation measurement stations used for validation of the NSRDB.

Evaluation of the NSRDB was conducted for 18 stations and the results are presented in the Fig. 4. The Mean Bias Error (MBE), Mean Absolute Error (MAE) and Root Mean Square Error (RMSE) were computed for both GHI and DNI. The evaluation was conducted for the 1998-2021 period. As the NSRDB provides a cloud mask the evaluation can be separated into clear and cloudy periods. Generally, the MBE lies within plus or minus ±5% for GHI and ±20% for DNI. The RMSE is less than 30% for GHI and 35% for DNI.

V. DATA DISTRIBUTION

An important component of the usefulness of any data is the ease of access to the data. Particular emphasis has been placed on the distribution of the datasets and there are three ways to access the data. Depending on user requirement, primarily depending on the volume required, users can access the data through –

i. Point location or small area through the NSRDB Data Viewer (https://maps.nrel.gov/nsrdb-viewer/).

978-1-6654-6060-6/23 $31.00 © 2023 IEEE

ii.   Application Programming Interface (API) for automated access to larger quantities of data (https://nsrdb.nrel.gov/data-sets/api-instructions.html)

iii. Highly Scalable Data Service (HSDS) hosted on Amazon Web Services for unlimited data (https://nsrdb.nrel.gov/data-sets/nsrdb-data-hsds-demo.html)

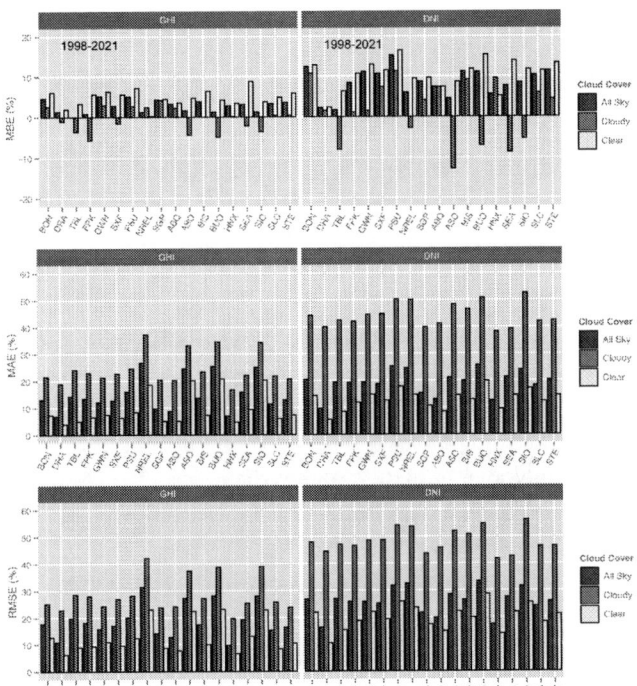

Fig. 4: Evaluation of the NSRDB using 18 ground measurements stations for locations shown in Fig. 2.

V. SUMMARY AND CONCLUSIONS

The NSRDB is a widely used public database of solar radiation providing global data primarily for solar energy applications. The data for the U.S. is available from 1998-2021 at a 30-minute, 4kmx4km resolution with higher resolution data available from 2018. The data from the NSRDB contains both solar radiation information as well as additional atmospheric parameters that enable PV and CSP modeling. The NSRDB is well validated using data from 18 high-quality ground stations. Datasets are easily available to all users through various channels with the data for the US being updated on an annual basis. The PSM is the underlying model that generates the NSRDB. This model is regularly updated based on advancement in the science of remote sensing and atmospheric physics and leads to continuous improvement in the accuracy of the datasets.

REFERENCES

[1] Sengupta, M., Y. Xie, A. Lopez, A. Habte, G. Maclaurin, and J. Shelby. 2018. "The National Solar Radiation Data Base (NSRDB)." Renew. Sustain. Energy Rev., 89: 51–60. https://doi.org/10.1016/j.rser.2018.03.003.

[2] A. Heidinger, M. Foster, A. Walther, X. Zhao, 2014. The pathfinder atmospheres-extended AVHRR climate dataset, Bull Am Meteorol Soc, 95, pp. 909-922

[3] Yu Xie, Manajit Sengupta, Jimy Dudhia. 2016. "Fast All-Sky Radiation Model for Solar applications (FARMS): Algorithm and Performance Evaluation." *Solar Energy*, Vol. 135, 435-445.

[4] Xie, Y., M. Sengupta, 2019, A Fast All-sky Radiation Model for Solar applications with Narrowband Irradiances on Tilted surfaces (FARMS-NIT): Part I. The clear-sky model, Solar Energy, 174, 691-702. https://doi.org/10.1016/j.solener.2018.09.056.

[5] Xie, Y., M. Sengupta, Wang, C., 2019, A Fast All-sky Radiation Model for Solar applications with Narrowband Irradiances on Tilted surfaces (FARMS-NIT): Part II. The cloudy-sky model, Solar Energy, 188, 799-812. https://doi.org/10.1016/j.solener.2019.06.058.

[6] Xie, Y., Sengupta, M., Liu, Y., Long, H., Min, Q., Liu, W., Habte, A., 2020. A physics-based DNI model assessing all-sky circumsolar radiation. iScience 22, https://doi.org/10.1016/j.isci.2020.100893.

[7] Habte, A., M. Sengupta, C.A. Gueymard, R. Narasappa, O. Rosseler, D.M. Burns, (2019), Estimating Ultraviolet Radiation from Global Horizontal Irradiance, IEEE J. Photovolt., 9, 1, 139-146. https://doi.org/10.1109/JPHOTOV.2018.2871780.

[8] Buster, G. Bannister, M., Habte, A., Hettinger, D., Maclaurin, G., Rossol, M., Sengupta, M. and Xie, Y., 2022. Physics-guided machine learning for improved accuracy of the National Solar Radiation Database, Solar Energy, Volume 232, 483-492, https://doi.org/10.1016/j.solener.2022.01.004.

Rapid, Contactless Measurements and Performance Predictions of Photovoltaic Materials

Brandon T Motes, Anthony T Troupe, Amy E Louks, Axel F Palmstrom, Joseph J Berry, Dane W deQuilettes

Optigon Inc, Somerville, MA, United States

Department of Materials Science, Colorado School of Mines, Golden, CO, United States

National Renewable Energy Laboratory, Golden, CO, United States

Department of Physics and Renewable and Sustainable Energy Institute, University of Colorado Boulder, Boulder, CO, United States

Emerging photovoltaic (PV) materials have the potential to add significant value to the PV supply chain, enabling cheap tandem modules, facile manufacturing, and novel form factors. Metrology tools have been critical to the development of PV materials, but current tools are limited in measurement acquisition speeds, form factors, and data interpretation. This has hindered the rate of feedback in the research setting and the application of measurement tools on the production line. To fill this gap, we develop a metrology tool and methodology for perovskite solar cells (PSCs) that measures critical photophysical properties at high-throughput manufacturing speeds and provides live interpretation of these measurements. We demonstrate a compact integrated tool with measurement speeds on the millisecond scale for transmission, time-resolved photoluminescence (TRPL), and spectrally-resolved photoluminescence with an accuracy of 95% compared to commercial-off-the-shelf measurement tools. Then using the measurements of a partially completed device, we derive several key parameters related to charge carrier lifetime, diffusion length, and absorption to predict the current-voltage (JV) curve of the completed device. This capability enables reduced research iteration time with immediate feedback on partially completed devices, accelerates the commercialization and scaling timeline for new materials via faster optimization of pilot lines, and reduces production variation through smarter process control which directly links process parameters to final device performance.

978-1-6654-6060-6/23 $31.00 © 2023 IEEE

Optimizing the Packing Density of Building Integrated Concentrating Photovoltaic Systems for Improved Performance and Reduced Embodied Carbon through a Novel Polygonal Concentrator

Lewis Osikibo Tamuno-Ibuomi[1*], Roberto Ramirez-Iniguez[1], A Sheila Holmes-Smith[1], Geraint Bevan[1]

[1]School of Computing, Engineering and Built Environment, Glasgow Caledonian University, Cowcaddens Road, Glasgow, G4 0BA, United Kingdom

Abstract— Low packing density can influence the performance and embodied energy of building integrated concentrating photovoltaic systems. This paper discusses the software development and optical characterisation of a new family of static low concentrating photovoltaic technology, called hexagonal concentrator (3D Hex). This novel device can improve the packing density of static low concentrating photovoltaic systems by 90.5% when compared to two 3D conventional concentrators. This promises improved performance, reduced module area and lower embodied carbon. The optical analysis shows that the proposed novel design has an optical gain of 3.0, optical efficiency of 91.6% and a maximum half-acceptance angle of +/-20⁰.

Keywords—low packaging density, embodied energy, low concentrating photovoltaic, efficiency, 3D Hexagonal concentrator

I. INTRODUCTION

Static low concentrating photovoltaic (LCPV) systems are gaining attention especially, as they can be integrated within double glazed windows, double skin façades, and skylights of the building envelope. When used in this way, these systems are called building integrated concentrating photovoltaic (BICPV) technology which can replace structural elements of a building, offsetting some of the cost and helping to improve its aesthetics [1]. This could transform the building into an energy efficient infrastructure, able to generate its electricity, reduce carbon and greenhouse gas (GHG) emissions; thus, helping to mitigate global warming and climate change effects [2].

The LCPV system is a type of concentrating photovoltaic (CPV) technology which is basically an assembly of a low-cost optical concentrator and a solar photovoltaic (PV) cell. The CPV unit concentrates sunlight from its wider entrance aperture and directs it to a smaller exit aperture where the PV cell is attached. LCPV devices are preferred in BICPV applications (over the medium concentrating photovoltaic (MCPV) and high concentrating photovoltaic (HCPV) types) because they are smaller in size, less expensive, nonimaging, easy to maintain, and have a wider field of view. In addition, they are static (not requiring any electromechanical sun tracking) and three-dimensional (3D) [3].

Despite their desirable properties and potential of helping to improve energy efficiency and achieve net zero carbon goals in buildings, the LCPV devices have a low packing density (LPD), which can influence the overall performance of the BICPV modules. The presence of large spaces between the concentrators or LCPV devices in a BICPV module, as shown in Fig. 1, is an indication of low packing density which could potentially cause the loss of convertible solar energy [4].

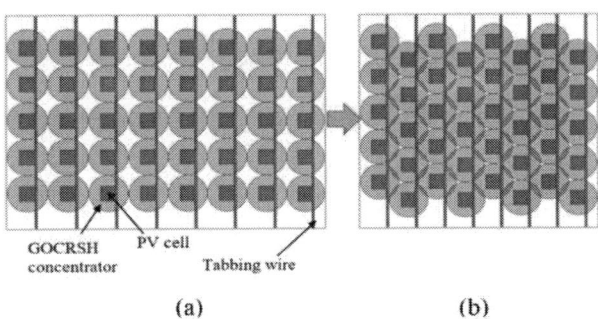

Fig. 1. 3D Genetically optimised circular rotational square hyperboloid concentrator (GOCRSH) module top view (a) with low packing density (b) with enhanced packing density, but overlapped concentrators [4].

In [4], the authors stated that, increasing the packing density could reduce the overall embodied energy and embodied carbon which can be achieved by rearranging the LCPV devices in the BICPV module, but this has not been addressed in other CPV studies.

Only recently, Tamuno-Ibuomi et al. [5], have proposed a novel 3D static low concentrating hexagonal photovoltaic (LCHPV) device, a polygonal BICPV system called 3D Hex concentrator that offers a high packing density (HPD) for the BICPV system. The authors in [5], conducted a theoretical comparative analysis on the packing densities of three 3D BICPV systems with LPD profiles: (i) proposed 3D Hex (ii) 3D genetically optimised circular rotational square hyperboloid (GOCRSH) [4] and (iii) 3D Square Elliptical Hyperboloid (SEH) [2]. It was found that the proposed 3D Hex concentrator

978-1-6654-6060-6/23 $31.00 © 2023 IEEE

can provide a packing efficiency of 89.4% while the packing efficiencies of the 3D GOCRSH and 3D SEH devices were 76.0% and 71.8% respectively at LPD conditions.

Following on with the LPD calculations in [5], this paper now presents the HPD profiles of the proposed 3D Hex, 3D GOCRSH and 3D SEH devices and their packing efficiencies calculations. The software development, simulation and optical performance of the proposed novel 3D Hex are also presented.

II. HPD PROFILES OF THREE 3D CONCENTRATORS

The expected high packing density (HPD) profiles for the three 3D concentrators (Hex, GOCRSH and SEH) are shown in Figs. 2-4 respectively [5].

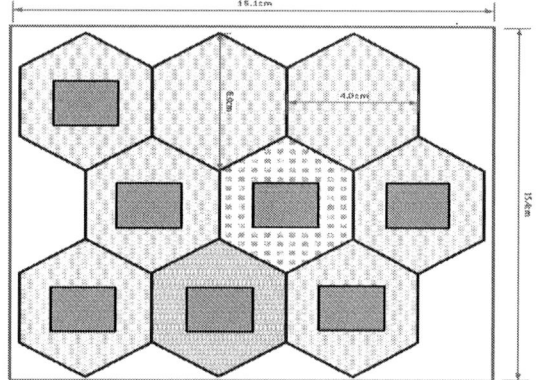

Fig. 2. Proposed HPD profile for the 3-D Hex module [5].

The packing efficiency (eff_{Pd}) for the proposed 3D Hex HPD profile is calculated by using equation (1) below, as explained in [5], and this gives 90.5% for a module area of 232.54cm².

$$eff_{Pd} = (N_{sc} \cdot A_{sc}) / A_{ms} \cdot 100\%. \qquad (1)$$

where N_{sc} is number of solar concentrators, A_{sc} is the entrance aperture area of a concentrator in an LCPV module and A_{ms} is the total module surface area.

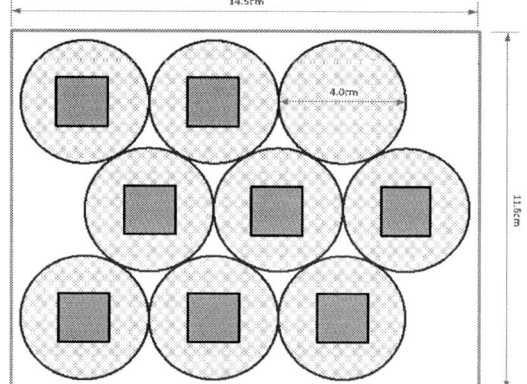

Fig. 3. HPD profile for the 3D GOCRSH module [5].

Similarly, the packing efficiency for the conventional 3D GOCRSH HPD profile is calculated using the approach in [5], and this resulted to 67.3% for a module area of 168.2cm².

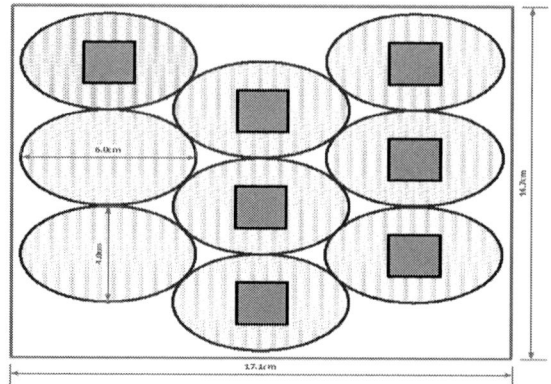

Fig. 4. HPD profile for the 3D SEH module [5].

The same approach in [5] was adopted for determining the packing efficiency for the high packing density (HPD) profile of the conventional 3D SEH concentrator illustrated in Fig. 4, and this gives 67.5% for a module area of 251.37cm².

III. SOFTWARE DEVELOPMENT OF THE PROPOSED 3D HEX CONCENTRATOR

The stages of software development for the proposed 3D Hex concentrator are given in Fig. 5.

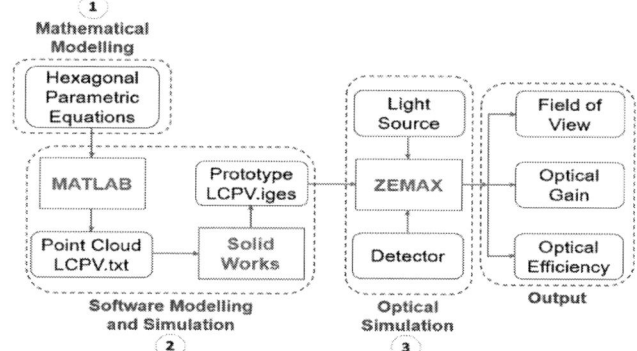

Fig. 5. Modular design structure for the novel 3D Hex LCPV device.

The mathematical modelling and MATLAB design processes have already been discussed in [5]. However, there are three distinct software stages: MATLAB design, computer-aided design (CAD) and optical simulation. The MATLAB point cloud text files defining the 3D Hex concentrator are exported to any suitable CAD software, in this case SolidWorks. AUTOCAD, Rhino-7, and GeoMagic are other CAD software with similar capabilities. The CAD software is used to convert the text files into solid (volume) surfaces, which is then saved in a file format suitable for the optical simulation software, in this case IGES file. STEP files can also be used. The optical simulation and analysis are carried out using an optical software called ZEMAX. Prior to this, the IGES file must be exported and saved within the Objects directory of ZEMAX where it can be accessed and fetched for optical simulation. Also, the light source, number of rays and detector must be properly defined for a successful simulation. The optical performance is aimed at determining the acceptance angle or field of view, optical gain, and optical efficiency of the novel 3D Hex concentrator.

IV. RESULTS AND DISCUSSIONS

A. Packing density evaluation and analysis

Table 1. Summary of packing efficiency calculation

LCPV System	LPD [5]		HPD		Difference in eff_{pd} (%)	Embodied Carbon (kg/cm²)	Inference
	eff_{pd} (%)	Area (cm²)	eff_{pd} (%)	Area (cm²)			
EEA (3D SEH)	71.8	236.22	67.5	251.37	6.2	+15.15 (increase)	Packing density not improved
CEA (3D GOCRSH)	76.0	148.84	67.3	168.2	12.1	+19.36 (increase)	Packing density not improved
HEA (3D Hex)	89.4	235.30	90.5	232.54	1.2	-2.76 (decrease)	Packing density improved

EEA-*elliptical entrance aperture*, **CEA**-*circular entrance aperture* and **HEA**-*hexagonal entrance aperture*.

B. Software Development of novel 3D Hex concentrator

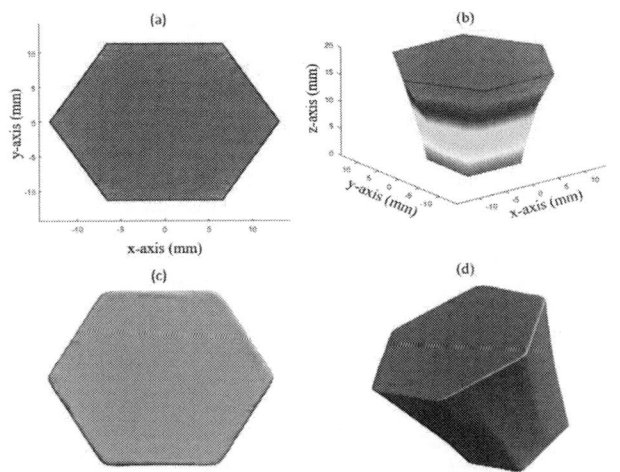

Fig. 6. 3D Hex device: (a) entrance aperture from MATLAB (b) 3D design with MATLAB (c) entrance aperture CAD surface (d) 3D surface design with CAD.

C. Optical characterisation of novel 3D Hex concentrator

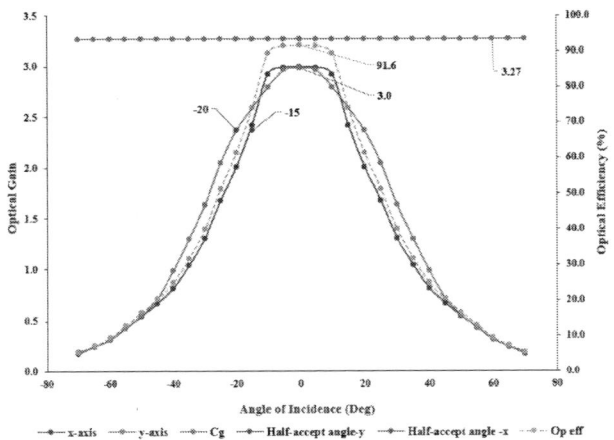

Fig. 7. Optical simulation result of novel 3D Hex concentrator in ZEMAX.

The results in Table 1 show that rearranging the concentrators under the HPD profile for the 3D SEH and 3D GOCRSH modules did not improve the packing density instead the module areas are increased by 15.15cm² and 19.36cm² respectively when compared to the actual module areas under their LPD profiles [5]. This represents an increase in the scale of embodied carbon for the 3D SEH and 3D GOCRSH devices from the LPD to HPD profiles at 15.15kg/cm² and 19.36kg/cm²

respectively. However, for the novel 3D Hex concentrator, there is a significant improvement in the packing density across the LPD to HPD profiles, from 89.4% to 90.5% with a 2.76kg/cm² reduction in both module area and embodied carbon scale. This shows an improved performance, with reduced embodied energy, GHG emission, and material cost.

Fig. 6 gives the results for the software simulation of the novel 3D Hex concentrator using MATLAB and SolidWorks, showing the hexagonal entrance aperture and the 3D designs. Fig. 7 gives the optical performance of the novel polygonal concentrator using ZEMAX. The result shows that the proposed 3D Hex device has an acceptance angle of 30⁰ along the x-axis and 40⁰ along the y-axis, optical efficiency (op eff) of 91.6%, optical gain of 3.0 and a geometrical concentration ratio (Cg) of 3.27 for a compact design with a height of 16mm.

V. CONCLUSIONS

A new family of polygonal LCPV device for BICPV application has been proposed which offers a high packing efficiency of 90.5% which is 23% and 17.4% higher than the packing efficiencies, 71.8% and 76%, of the 3D SEH and 3D GOCRSH concentrators respectively. The 3D Hex concentrator can reduce the embodied carbon, GHG emission and hence, the embodied energy which can contribute to achieve the net zero carbon goals for buildings. Further investigation on the 3D Hex concentrator includes fabrication, experiments, and thermal characterisation.

REFERENCES

[1] A. Alamoudi *et al.*, "Using static concentrator technology to achieve global energy goal," *Sustain.*, vol. 11, no. 11, pp. 1–22, Jun. 2019, doi: 10.3390/su11113056.

[2] N. Sellami and T. K. Mallick, "Optical characterisation and optimisation of a static Window Integrated Concentrating Photovoltaic system," *Sol. Energy*, vol. 91, pp. 273–282, 2013, doi: 10.1016/j.solener.2013.02.012.

[3] D. Freier, R. Ramirez-Iniguez, T. Jafry, F. Muhammad-Sukki, and C. Gamio, "A review of optical concentrators for portable solar photovoltaic systems for developing countries," *Renew. Sustain. Energy Rev.*, vol. 90, no. March, pp. 957–968, 2018, doi: 10.1016/j.rser.2018.03.039.

[4] D. F. Raine, F. Muhammad-Sukki, R. Ramirez-Iniguez, J. A. Ardila-Rey, T. Jafry, and C. Gamio, "Embodied energy and cost assessments of a concentrating photovoltaic module," *Sustain.*, vol. 13, no. 24, pp. 1–15, 2021, doi: 10.3390/su132413916.

[5] L. O. Tamuno-Ibuomi, R. Ramirez-iniguez, A. S. Holmes-Smith, and G. Bevan, "Improving the Packing Efficiency of Building Integrated Concentrating Photovoltaic Systems through a Novel Hexagonal Concentrator," *2022 57th Int. Univ. Power Eng. Conf.*, pp. 1–6, 2022, doi: 10.1109/UPEC55022.2022.9917592.

Human health risk assessment for improper landfill disposal of end-of-life CdTe PV

Elaine Kupets and Garvin Heath

National Renewable Energy Laboratory, Golden, CO, 80401, United States of America

Abstract — The present work is a continuation of the 2020 IEA PVPS Task 12 *Human Health Risk Assessment Methods for PV Part 3: Module Disposal Risks.* The 2020 report performed a human health risk assessment (HHRA) for disposal of a cadmium telluride (CdTe) PV module in an unlined landfill, focusing solely on risks from cadmium. This study extends the 2020 HHRA on CdTe PV, analyzing eleven constituent elements: Cd, Se, Te, Cu, Si, Cr(III), Mo, Sn, Zn, Ni, and Al. The present HHRA was performed through two methods: utilization of the U.S. Environmental Protection Agency's (USEPA) Delisting Risk Assessment Software (DRAS V.4.0) on eight exposure pathways for cancer risk and non-cancer hazards; and comparison of exposure point concentrations to federal standards for groundwater, surface water, air, and soil exposure pathways. Cancer risks and non-cancer hazards posed by elemental leaching through all evaluated exposure pathways, using both methods, were found to be several orders of magnitude below USEPA health-protective thresholds. Cadmium exhibited both the highest risks and lowest uncertainty considering data availability on chemical content, leachate, and federal screening levels.

I. Introduction

Despite recycling being available for CdTe modules through its main manufacturer, First Solar Inc., there is concern that some CdTe modules are being landfilled [1], which can lead to leaching and release of constituent chemical elements to the environment and people. Sinha et al. [2] quantified the risks and hazards of improper PV disposal in an unsanitary landfill scenario for cadmium (Cd) alone. Although the authors found that cadmium was not likely to pose a cancer risk or non-cancer hazard, the study was not a complete HHRA, and thus, inferences regarding the complete risk profile of landfilling CdTe modules could not be drawn. We have now completed a more comprehensive HHRA by analyzing all elements found in CdTe modules (Cd, Se, Te, Cu, Si, Cr(III), Mo, Sn, Zn, Ni, and Al) following the methods outlined in [2], where further details can be found; we also updated as much as possible with respect to First Solar's latest manufactured module, Series 6.

The risk scenario modeled is improper disposal of 10 MWac worth of CdTe modules, crushed and buried in an unsanitary landfill, and then capped after 1 year. An unsanitary landfill is defined as being unlined at its base, allowing leachate to migrate into groundwater and subsurface soils. Analysis of such a disposal scenario serves as a maximum risk scenario, potentially applicable in other jurisdictions.

II. Methods

A. Human health cancer risks and non-cancer hazards

The first of two methods used in this study employed USEPA's DRAS V.4.0 software, which quantifies carcinogenic risks and non-carcinogenic hazards, utilizing USEPA thresholds that trigger requirements for greater regulatory scrutiny set at 1 x 10^{-6} (1 in a million) and 1, respectively [3].

DRAS models the movement of constituents through four surface (water, soil and air) pathways: a) surface water to ingestion, b) air particulate inhalation, c) fish ingestion, and d) soil ingestion—and four subsurface (groundwater) pathways: a) ingestion, b) inhalation (from showers), c) dermal absorption for adults, and d) dermal absorption for children. For the groundwater pathways, acidic landfill conditions cause PV elements to be leached, while wind and water erosion leads to release into the surface pathways [2]. The DRAS model requires three major user-input parameters: (1) leachate values (mg/L) from the USEPA's Toxicity Characterization Leaching Procedure (TCLP); (2) total chemical concentration of the eluted element (mg/kg); and (3) total volume of module waste (m³). The groundwater and surface pathways are affected by the first and second parameter, respectively, while the third parameter affects all pathways.

B. Exposure point concentrations

The benefit to using the regulatory-approved DRAS model is that it directly outputs non-cancer hazards and cancer risks for each leached PV element. Yet, not all CdTe PV elements (Table 1) have the dose-response toxicological values required for the DRAS software. An alternative is calculating exposure point concentrations, which we did for all elements. Exposure point concentrations of elements in groundwater, surface water, ambient air, and soil exposure pathways were back-calculated from DRAS model equations, and then compared to risk-based screening levels unique to each element [3].

To understand whether estimated exposure point concentrations could present health risk, they were compared to two federal guidelines, both of which regulate drinking water: a) USEPA risk-based screening levels (RSLs) and b) (primary and secondary) maximum contaminant levels (MCLs) [4]. It was found, however, that the EPA data on RSLs and MCLs for the evaluated exposure pathways are incomplete for several elements. Upper Tolerance Limits (UTL) established by the U.S. Army Corps Engineers (USACE) were used to supplement missing RSL and MCLs [5], [6]. Nevertheless, thresholds for all the elements were found only for the soil exposure pathway.

C. Data

Data extracted from prior studies of CdTe modules (Table 1) were used as input to both methods. We were interested in leachate values that came from TCLP tests performed on field-representative conditions of landfilled modules—module samples between 3 to 9.5 mm—which eliminated studies that milled or ground modules to unrealistically smaller sizes [7]. Not all elements were tested (N.A. = not available), and of those tested, only cadmium was detected in leachate (N.D. = not detected) [7], [8], which means it is the only element considered for groundwater exposure.

Chemical content of each element was used to estimate chemical concentration (chemical content (mg) / weight of module (kg)). We obtained First Solar's Series 6 chemical content information to align with the source of leachate values (Series 6) [9], [10]. Note that Series 6 modules have eliminated the use of Pb [11], and previous First Solar modules or proxy databases were used for missing Series 6 information [12]–[14]. It is recognized that in doing so, we may overestimate human health risk as newer First Solar modules have decreased content in some elements, e.g., copper [15], and because the area and weight of Series 6 modules (36 kg, 2.47 m^2) differs from the other proxy CdTe modules (12 kg, 0.72 m^2). Our use of maximum chemical concentrations and available TCLP values are considered health-protective assumptions.

Table 1. Major input parameters per CdTe PV element

Metal	Proxy data or Series 6 data?	# of datasets	TCLP values (mg/L)	Element chemical concentration per module (mg/kg)		Waste volume for modules (m³)	Used for DRAS input?
				median	maximum		
Cd	Series 6	7	0.2 – 0.6 (median = 0.42)	649	649	400	yes
Se	Series 6	7	N.D.	61.0	61.0	400	yes
Te	Series 6	7	N.A.	638	638	400	no
Cu	Series 6	7	N.A.	250	250	400	yes
Si	Series 6	7	N.A.	8300	8300	400	no
Cr	Series 6, proxy	1	N.D.	180*	180*	500*	yes
Mo	proxy	1	N.A.	500*	500*	500*	yes
Sn	proxy	3	N.A.	22.6*	45.0*	500*	yes
Zn	proxy	1	N.A.	0.002*	0.002*	500*	yes
Al	proxy	3	N.A.	51.4*	100*	500*	no
Ni	proxy	1	N.A.	1.20*	1.20*	500*	yes

*If chemical content data could not be found for Series 6 modules, data obtained from studies examining other CdTe modules were used, to support chemical concentration and waste volume calculations.

D. Human health cancer risks and non-cancer hazards

DRAS was used to analyze the cancer risks and non-cancer hazards for eight out of the eleven elements of interest and eight exposure pathways. Federal cancer risk determinations have not been made for any element besides Cd and Ni, and our analysis revealed estimated risk for just one pathway: inhalation of particles. Yet the risk estimated—3.27 x 10^{-9} for Cd and 9.14 x

10^{-13} for Ni—falls below USEPA's cancer risk-screening threshold, consistent with [2].

Potential non-cancer hazards could be quantified for every element analyzed through DRAS. Cadmium had the largest potential, aggregate non-cancer hazard among all elements, followed by copper, and is similar to [2] in magnitude. The cumulative non-cancer hazard quotients for all elements across all pathways are found to be below the EPA hazard threshold of 1: 5.28 x 10^{-3} using median chemical concentration and 7.48 x 10^{-3} using maximum chemical concentration (Fig. 1). The aggregate non-cancer hazard quotients of Cu, Mo, Sn, Zn, and Ni were calculated as if these elements had a TCLP value of 0, since DRAS will not differentiate a value of zero from a value that is not detected or not available.

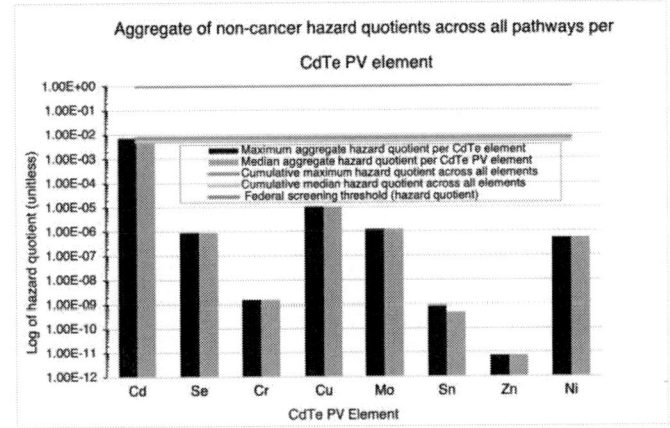

Fig. 1. Non-cancer hazards posed by CdTe elements (median and maximum chemical content) based on DRAS modeling. Note the logarithmic scale. Cumulative hazard quotients are below EPA threshold of 1.

We find that the ranking of exposure point concentration is directly correlated to chemical concentration: Si had the highest concentration values for the surface water, soil, and air pathways at 1.46 x 10^{-7} mg/L, 6.66 x 10^{-3} mg/kg, and 4.72 x 10^{-8} mg/m^3 respectively, followed by Cd, using both maximum and median chemical concentrations. Cadmium alone had a non-zero TCLP value, so it held the greatest groundwater exposure point concentrations for both maximum (1.22 x 10^{-4} mg/L) and median (8.09 x 10^{-5} mg/L) values. Additionally, both Cd and Ni are the only elements with a cancer risk quotient, which the air exposure pathway requires; air exposure point concentrations for all the elements were calculated using the available risk quotient and equations found in [2].

In every case where a federal screening level was available, the exposure point concentrations we estimated were lower (Table 2). Tellurium (Te) has only one screening threshold (for the soil pathway) available, while all other elements had threshold information for at least two pathways.

E. Uncertainties

While results for all pathways and elements with required input data and federal standards demonstrate risks lower than federal risk thresholds, it is important to note that there were many elements that could not be evaluated for specific exposure pathways owing to information gaps.

There are uncertainties associated with the input parameters due to assumptions made, resulting in intentional overestimation of actual risks and hazards; this is done to be health protective. For instance, PV module efficiency and PV project size in this scenario lead to an overestimation in waste volume. As PV module efficiencies continuously increase, the quantity of modules needed for a project will decrease, leading to lower waste volumes [2]. This analysis also assumes that PV disposal occurs at project end, without consideration for cases of PV modules or parts that are reused, refurbished, remanufactured, or recycled [2]. Chemical concentration is perhaps the most prominent source of uncertainty for the surface pathways, followed by waste volume, since data from Series 6 modules and previous generations are used in conjunction; the evolution of First Solar CdTe modules encompasses changed specifications such as dimension, weight, and chemical content of metals [2].

Table 2. Comparison of exposure point concentrations to federal screening thresholds for CdTe PV elements

Metal	Is there a screening threshold indicated for these pathways per element?						Below threshold
	Groundwater		Surface water		Air	Soil	
	RSL	MCL	RSL	MCL	RSL	RSL	
Cd	yes	yes	yes	yes	yes	yes	all pathways
Se	yes	yes	yes	yes	yes	yes	all pathways
Te						yes**	soil
Cu	yes	yes	yes	yes		yes	GW, SW, soil
Si					yes	yes	air, soil
Cr	yes	yes	yes	yes		yes	GW, SW, soil
Mo	yes		yes			yes	all pathways
Sn	yes		yes			yes	GW, SW, soil
Zn	yes	yes*	yes	yes*		yes	GW, SW, soil
Al	yes	yes*	yes	yes*	yes	yes	all pathways
Ni	yes		yes		yes	yes	all pathways

*Secondary MCL was used when Primary MCL was not available.
**The result reported here is based on UTL.

III. CONCLUSION

Eleven elements were investigated in a HHRA, considering eight pathways of exposure from improper disposal of First Solar's Series 6 CdTe modules in an unlined landfill. Using two methods and best obtainable information with health-protective assumptions, we found that no element exhibited cancer risks or non-cancer hazards above available U.S. federal thresholds. This is the most comprehensive HHRA for CdTe modules of which we are aware. Risk assessment of Cd is based on chemical content, leachate data, and federal screening levels. Aside from Cd, uncertainties in information regarding chemical content of Series 6 modules, federal screening levels for certain pathways and elements, available leachate values, and other limitations prevents us from definitively concluding lack of health risk potential. Therefore, additional fundamental research is needed

to establish risk thresholds for more pathways and elements, and regarding PV module data, to increase confidence.

REFERENCES

[1] First Solar, "Solutions: Recycling," *First Solar*, 2022. https://www.firstsolar.com/en/Solutions/Recycling (Accessed: September 9, 2022).

[2] P. Sinha, G. Heath, *et al.*, "Human health risk assessment methods for PV, Part 3: Module disposal risks," International Energy Agency (IEA) PVPS Task 12, T12-16:2020, 2020.

[3] USEPA, "RCRA Delisting Technical Support Document." 2008. https://www.epa.gov/sites/default/files/2016-01/documents/dtsd-20081031-chaps1_6.pdf (Accessed: September 10, 2022).

[4] USEPA, "Regional Screening Levels (RSLs) - User's Guide," 2015. https://www.epa.gov/risk/regional-screening-levels-rsls-users-guide (Accessed: September 15, 2022).

[5] Parsons, "Final 4825 Glenbrook Road Human Health Risk Assessment," Washington, D.C., 2011. https://www.nab.usace.army.mil/Portals/63/docs/SpringValley/4825%20Glenbrook%20Rd.%20Human%20Health%20Risk%20Assesment.pdf (Accessed: September 13, 2022).

[6] ERT, Inc., "The Final SVFUDS Site-Wide Risk Assessment Work Plan for the Spring Valley FUDS Integrated Site-Wide Remedial Investigation/Feasibility Study," Washington, D.C., 2014. https://www.nab.usace.army.mil/Portals/63/docs/SpringValley/Human%20Health%20Risk%20Assessment%20Work%20Plan.pdf (Accessed: September 13, 2022).

[7] G. TamizhMani *et al.*, "Sampling Methods for Toxicity Testing of PV Modules for End-of-Life Decisions," in *2021 IEEE 48th Photovoltaic Specialists Conference (PVSC)*, 2021, pp. 1871–1875. doi: 10.1109/PVSC43889.2021.9518620.

[8] S. Borst, "Analytical Report: Metals," Test America, Irvina, CA, USA, 2019.

[9] V. Fthenakis, H. C. Kim, R. Frischknecht, M. Raugei, and P. Sinha, "Life Cycle Inventories and Life Cycle Assessments of Photovoltaic Systems," International Energy Agency (IEA) PVPS Task 12, T12-04:2015., 2015.

[10] "First Solar Series 6 Module Datasheet." *First Solar*, 2020. https://www.firstsolar.com/en/Resources/Downloads (Accessed: September 10, 2022).

[11] P. Sinha and L. de Rosa, "Alternatives to SnPb solder for First Solar's transition from Series 4 to Series 6 manufacturing," *First Solar*, 2019. https://www.firstsolar.com/-/media/First-Solar/Sustainability-Documents/Alternatives-Assessment-to-SnPb-Solder.ashx (Accessed: September 10, 2022).

[12] A. Domínguez and R. Geyer, "Photovoltaic waste assessment of major photovoltaic installations in the United States of America," *Renewable Energy*, vol. 133, pp. 1188–1200, Apr. 2019, doi: 10.1016/j.renene.2018.08.063.

[13] A. Anctil and V. Fthenakis, "Critical metals in strategic photovoltaic technologies: abundance versus recyclability," *Progress in Photovoltaics: Research and Applications*, vol. 21, no. 6, pp. 1253–1259, 2013, doi: 10.1002/pip.2308.

[14] P. Nain and A. Kumar, "Initial metal contents and leaching rate constants of metals leached from end-of-life solar photovoltaic waste: An integrative literature review and analysis," *Renewable and Sustainable Energy Reviews*, vol. 119, p. 109592, Mar. 2020, doi: 10.1016/j.rser.2019.109592.

[15] K. Pickerel, "First Solar Series 6 CuRe modules use less copper for low degradation rate," *Solar Power World*, 2021. https://www.solarpowerworldonline.com/2021/04/first-solar-series-6-cure-modules-use-less-copper-for-low-degradation-rate/ (Accessed: September 10, 2022).

Accelerating Cycles of Learning for Silicon Heterojunction Architectures: Experimental Design and Data-Driven Degradation Pathway Prediction

Xuanji Yu*, Diego Zubieta*, Mirra Rasmussen*, Chien-Hsuan Chen[†], Cécile Molto[†]
Mariana Bertoni[‡], Kristopher O. Davis[†], Laura S. Bruckman*, Ina T. Martin*
*Case Western Reserve University, Cleveland, OH 44106, USA
[†]University of Central Florida, Orlando, FL 32816, USA
[‡]Arizona State University, Tempe, AZ 85281, USA

Abstract—Advanced crystalline silicon photovoltaic (PV) cell architectures, e.g., heterojunctions, mitigate energy conversion losses present in traditional architectures. However, the use of new materials and processes introduces the potential for new failure modes. Our experimental design includes (1) applying different accelerated aging exposures with combined stressors including light, temperature, and acid to unencapsulated textured silicon heterojunction (SHJ) cells and measuring their electrical performance before and after exposure; (2) applying the same aging conditions to planar non-textured test structures (c-Si/a-Si:H/ITO stacks), in order to conduct parallel materials analyses, including spectroscopic ellipsometry, X-ray photoelectron spectroscopy (XPS), and ToF-SIMS. Using data-driven network structural equation modeling, we will study the contributions of each stressor, deconvolute the potential multiplicative effects from combined stressors that cause changes in electrical performance of cells, and predict future degradation pathways. By further materials analysis and connecting specific performance degradation mechanisms of SHJ cells to the underlying root causes at the materials level, we aim to identify key design constraints to inform a more robust technology. The goal of this research is to establish the basis for a generalized approach to rapid screening of degradation mechanisms in unencapsulated cells.

Index Terms—Silicon heterojunction, accelerated aging, data-driven, degradation pathway network, thin film

I. INTRODUCTION

Advanced crystalline silicon photovoltaic (PV) cell architectures mitigate energy conversion losses present in traditional architectures. Currently, one of the two main technologies expected to replace Passivated Emitter and Rear Contact (PERC) is silicon heterojunction (SHJ) cells. This SHJ cell architecture decouples the metal contact from the absorber to form a passivating, carrier-selective contact, thus limiting contact recombination and achieving higher efficiency. Typically, the heterojunction is formed by depositing a multilayer stack of intrinsic, hydrogenated amorphous silicon (a-Si:H) to passivate dangling bonds at the n-type c-Si absorber surface.

This material is based upon work supported by the U.S. Department of Energy's Office of Energy Efficiency and Renewable Energy (EERE) under Solar Energy Technologies Office (SETO) Agreement Number DE-EE0010250. The views expressed herein do not necessarily represent the views of the U.S. Department of Energy or the United States Government.

This is followed by doped a-Si:H to provide carrier selectivity and an appropriate transparent conductive oxide (TCO), typically indium tin oxide (ITO), to provide lateral transport and ensure good contact with the metal. Moreover, SHJ cells demonstrate a significantly lower temperature coefficient and higher bifaciality than conventional solar cells.

Introducing new materials and processes, however, brings the potential for new failure modes. In the SHJ architecture, the a-Si:H layers provide passivation to the c-Si absorber that enables the record high performance seen in these cells. Despite the criticality of this material for SHJ functionality and performance, there are several degradation processes that affect a-Si:H including the Staebler-Wronski effect and light-induced degradation. [1]–[4] Bertoni's work with surface recombination velocity measurements of c-Si/a-Si:H stacks support that these processes contribute to the formation of electronically active defects, such as dangling Si bonds, that reduce the stability of the layer. [2] The electronically active defects promote carrier recombination at the a-Si:H/c-Si interface, which can lead to device-level decreases in cell performance. However, Sinha et al. compared effect of UV exposure on the performance of different unencapsulated Si architectures and proposed that the degradation of the SHJ was due to an increase in ITO resistance or deterioration of the passivating interfaces. [5] Thus the exact cause of the SHJ cell performance is not well understood; these failure modes differ across materials and technologies and need to be assessed before production levels are ramped up.

Here, we are exposing unencapsulated textured SHJ cells to accelerated aging conditions that encompass heat, moisture, light, and acidic conditions. Device electrical performance will be characterized before and after accelerated aging. Meanwhile, we will apply the same aging conditions to planar non-textured test structures (c-Si/a-Si:H/ITO stacks), in order to conduct parallel materials analyses, including spectroscopic ellipsometry, XPS, and ToF-SIMS. Data-driven degradation pathway network modelings will be used to study the contributions of each stressor, deconvolute the potential multiplicative effects from combined stressors that cause changes in electrical performance of cells, and predict future degradation pathways.

By further combining with the materials analysis to connect specific performance degradation mechanisms of SHJ cells (e.g., optical, recombination, selectivity, contact resistivity) to the underlying root causes at the materials level, we aim to identify key design constraints to inform a more robust technology. Our ultimate goal is to use accelerated aging of unencapsulated stacks and devices to understand, predict, and mitigate avenues of device failure.

II. METHODS

A. Silicon Heterojunction Architectures

Fig. 1 shows samples used in this study:

- textured SHJ cells: monocrystalline n type silicon SHJ cell featuring intrinsic and doped a-Si:H layers, ITO, and Ag contacts on the front and rear, for electrical performance degradation analysis.
- nontextured samples: planar c-Si/a-Si:H/ITO stacks, for parallel materials analyses.

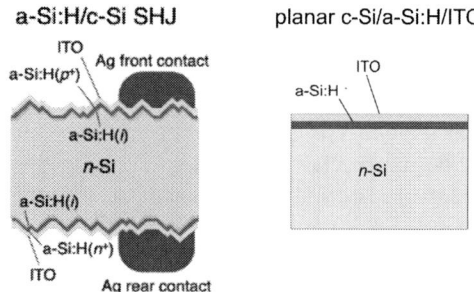

Fig. 1. Silicon heterojunction architecture (left) and planar test structure (right).

B. Accelerated aging exposures

Our previous work [6] demonstrated that unencapsulated SHJ devices are quite stable in damp heat (DH); over 2000 h of exposure, V_{OC}, I_{SC}, and FF values did not change significantly. Signatures of surface degradation of the ITO and the Ag gridlines were present in the XPS data of SHJs exposed to DH, but the ITO/Ag interface was likely not affected, i.e. the oxidation of the gridlines was minor enough to not affect the device measurements. In contrast to DH exposure, acetic acid exposure resulted in both changes to V_{OC}, and degradation of the front contact. From this work, we determined that additional samples, degradation methods, and characterization were necessary to identify trends in the performance of the SHJs. Thus, the following accelerated aging exposures will be used to induce damage in the samples, applying various stressors and stress levels and combinations thereof.

- (UV) light: to determine the effect of light alone; light/heat/temperature (cyclic) to trigger degradation modes that require multiple stressors
- Acetic acid, temperature and concentration: to emulate EVA degradation
- Damp Heat (85°C/85% humidity): sensitivity to heat, moisture

- Na experiment: brush with $NaHCO_3$ and heat, to emulate sodium migration from soda lime glass
- Humidity freeze: cycle temperature and humidity to stress interfaces

C. Electrical Performance Characterizations

Current-voltage (I-V) characteristics of the devices will be collected using an All Real Apollo Solar Simulator to illuminate the samples. Suns-V_{OC} and photoluminescence (PL) imaging at 1 Sun condition will be performed at each step of exposure to track the change in V_{OC}, saturation current density (J_0) and PL counts.

D. Network structural equation modeling (netSEM)

To study the contributions of each stressor and deconvolute the potential multiplicative effects from combined stressors that cause electrical performance reductions, we will use a data-driven tool called Network Structural Equation Modeling (netSEM). netSEM is a statistical approach to perform pathway network analysis in a system composed of multiple variables [7], selecting the best models and retrieving the statistical significance of relationships using p-values and R^2_{adj}. We will use netSEM to illustrate pairwise non-linear relationships between variables, and to predict performance loss under multiple stressors using multiple regression.

E. Advanced Materials Characterizations

We plan to perform non-destructive characterizations including optical profilometry and spectroscopic ellipsometry on the nontextured samples, in order to assess the degradation profiles of the cells. Destructive characterization such as ToF-SIMS and XPS will be performed on the retained untextured samples after the completion of non-destructive characterization.

III. PRELIMINARY RESULTS AND DISCUSSION

Decades of research exist on the structure-property relationships of a-Si:H and ITO for thin film PV applications. [8]–[11] Although the function of a-Si:H differs in SHJ cells, the lessons learned on a-Si:H stability and its relationship to hydrogen content and bonding configuration are highly pertinent to SHJs. As described in the introduction, evidence exists of degradation modes at both the c-Si/a-Si:H and ITO/a-Si:H interfaces in SHJs. Thus, characterization methods that probe the surface, the bulk of all three materials, and their interfaces are desirable. The inclusion of device relevant film stacks in this study will allow characterization of the surface of the ITO (via XPS), and the ITO and a-Si:H films and interfaces (via SE and ToF-SIMS), which will provide complementary information to the degradation measured in parallel on full devices.

Fig. 2 shows the survey XPS spectra of the SHJ front surface, with the measurement encompassing both the ITO and a silver gridline. For the control (as-received) sample (a), In, Sn, O, C, and Ag are detected. The same elements are present in samples exposed to DH (b, c) and acetic acid, but in different proportions. High resolution XPS spectra of the

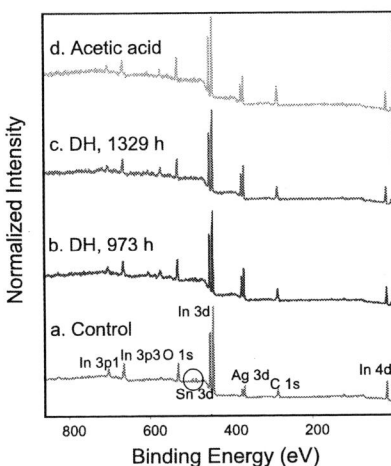

Fig. 2. Survey XPS spectra of the SHJ front surface (ITO and Ag gridline) for a (a) control (as-received) sample (b) after 973 h of DH exposure, (c) 1329 h of DH exposure, and (d) 180 min of acetic acid exposure.

individual elements were also measured, to probe their relative quantities, and chemical binding environments. Fig. 3 shows the high-resolution (a) O 1s and (b) Ag 3d5/2 XPS spectra of the ITO/Ag surface of bare, DH exposed, and acetic acid exposed SHJ cells. For the O 1s spectra, the shift to a higher binding energy is consistent with degradation of the ITO [12]. Both the DH and the acetic acid treatments result in a similar change to the O 1s envelope. Similarly, the Ag spectra also change with both DH and acetic acid exposure. With both exposures, the Ag 3d5/2 spectrum has a new peak at a lower binding energy, consistent with silver oxide formation [13]. Table 1 shows the contributions of the different components

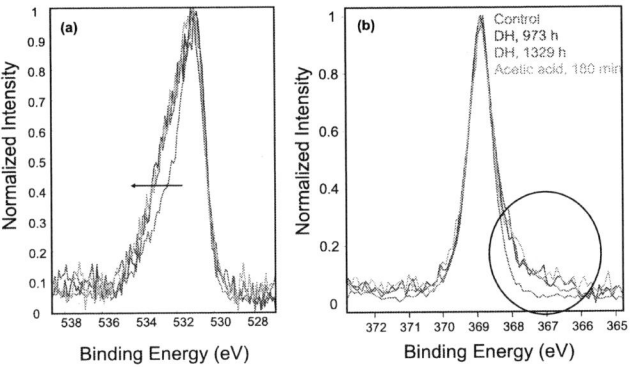

Fig. 3. Survey XPS spectra of the SHJ front surface (ITO and Ag gridline) for a (red) control (as-received) sample (blue) after 973 h of DH exposure, (green) 1329 h of DH exposure, and (orange) 180 min of acetic acid exposure.

of the Ag 3d5/2 spectra. Reported values are the average and standard deviation of three measurements per sample. These data are an example of tracking materials degradation across exposure types.

TABLE I
DIFFERENT COMPONENTS OF THE SILVER XPS SPECTRA

Exposure Type/Time	Ag 3d5/2	
	%Ag^0	%AgO$_x$
None	100	0
DH/970 h	86.8 ± 4.8	13.2 ± 4.8
DH/1329 h	82.0 ± 4.9	18.0 ± 4.9
Acetic acid/180 min	88.8 ± 0.5	11.2 ± 0.5

IV. CONCLUSIONS

We are still preparing samples and gathering data for data-driven analysis.

REFERENCES

[1] R. Vasudevan, I. Poli, D. Deligiannis, M. Zeman, and A. H. M. Smets, "Light-Induced Effects on the a-Si:H/c-Si Heterointerface," *IEEE Journal of Photovoltaics*, vol. 7, no. 2, pp. 656–664, Mar. 2017.

[2] S. Bernardini and M. I. Bertoni, "Insights into the Degradation of Amorphous Silicon Passivation Layer for Heterojunction Solar Cells," *physica status solidi (a)*, vol. 216, no. 4, p. 1800705, 2019.

[3] S. De Wolf, B. Demaurex, A. Descoeudres, and C. Ballif, "Very fast light-induced degradation of a-Si:H/c-Si(100) interfaces," *Physical Review B*, vol. 83, no. 23, p. 233301, Jun. 2011.

[4] P. Mahtani, R. Varache, B. Jovet, C. Longeaud, J.-P. Kleider, and N. P. Kherani, "Light induced changes in the amorphous—crystalline silicon heterointerface," *Journal of Applied Physics*, vol. 114, no. 12, p. 124503, Sep. 2013.

[5] A. Sinha, J. Qian, K. Hurst, S. L. Moffitt, L. T. Schelhas, D. C. Miller, and P. Hacke, "UV-Induced Degradation of High-Efficiency Solar Cells with Different Architectures," in *2020 47th IEEE Photovoltaic Specialists Conference (PVSC)*, Jun. 2020, pp. 1990–1991.

[6] N. Iqbal, N. K. Chockalingam, K. A. Coleman, J. Fina, K. O. Davis, L. S. Bruckman, and I. T. Martin, "Accelerate Cycles of Learning: Unencapsulated Silicon Photovoltaic Cells to Environmental Stressors," in *2022 IEEE 49th Photovoltaics Specialists Conference (PVSC)*, Jun. 2022, pp. 0668–0674.

[7] L. S. Bruckman, N. R. Wheeler, J. Ma, E. Wang, C. K. Wang, I. Chou, J. Sun, and R. H. French, "Statistical and Domain Analytics Applied to PV Module Lifetime and Degradation Science," *IEEE Access*, vol. 1, pp. 384–403, 2013.

[8] A. S. Ferlauto, R. J. Koval, C. R. Wronski, and R. W. Collins, "Phase Diagrams for the Optimization of rf Plasma Enhanced Chemical Vapor Deposition of a-Si:H: Variations in Plasma Power and Substrate Temperature," *MRS Online Proceedings Library*, vol. 664, no. 1, p. 54, Dec. 2000.

[9] I. T. Martin, M. A. Wank, M. A. Blauw, R. A. C. M. M. van Swaaij, W. M. M. Kessels, and M. C. M. van de Sanden, "The effect of low frequency pulse-shaped substrate bias on the remote plasma deposition of a-Si : H thin films," *Plasma Sources Science and Technology*, vol. 19, no. 1, p. 015012, Nov. 2009.

[10] M. Stuckelberger, R. Biron, N. Wyrsch, F.-J. Haug, and C. Ballif, "Review: Progress in solar cells from hydrogenated amorphous silicon," *Renewable and Sustainable Energy Reviews*, vol. 76, pp. 1497–1523, Sep. 2017.

[11] C. R. Wronski and R. W. Collins, "Phase engineering of a-Si:H solar cells for optimized performance," *Solar Energy*, vol. 77, no. 6, pp. 877–885, Dec. 2004.

[12] T. Tohsophon, A. Dabirian, S. De Wolf, M. Morales-Masis, and C. Ballif, "Environmental stability of high-mobility indium-oxide based transparent electrodes," *APL Materials*, vol. 3, no. 11, p. 116105, Nov. 2015.

[13] N. Iqbal, M. Li, T. S. Sakthivel, K. Mikeska, M. Lu, M. Nandakumar, S. Duttagupta, M. Dhamrin, K. Tsuji, S. Bowden, A. Augusto, Y. Guan, S. Seal, and K. O. Davis, "Impact of acetic acid exposure on metal contact degradation of different crystalline silicon solar cell technologies," *Solar Energy Materials and Solar Cells*, vol. 250, p. 112089, Jan. 2023.

978-1-6654-6060-6/23 $31.00 © 2023 IEEE

Modeling Reference Cell Performance Using Measured and Modeled Spectral Data

Josh Peterson[1], Frank Vignola[1], Afshin Andreas[2], Aron Habte[2], Manajit Sengupta[2]

[1]Material Science Institute/University of Oregon, Eugene, Oregon, 97403 (USA)
[2]National Renewable Energy Laboratory, Golden, Colorado, 80401 (USA)

Abstract — The performance of several silicon-based reference cells is examined under clear skies on a horizontal and two-axis tracking surface during the winter of 2022. The ratio of the calculated reference cell output to the measured reference cell output is examined. For each reference cell, when using the measured spectral data, the ratio of the estimated to measured output varies by less than ±0.6% at the P95 level. The analysis was also done using modeled spectral values obtained from the Bird spectrl2 model. The ratio between the estimated reference cell output using the modeled spectral values to the measured reference cell output varies by ±1.1% at the P95 level.

I. INTRODUCTION

Increasingly, reference cells are being used to monitor and evaluate the performance of photovoltaic (PV) systems in the field. Reference cells have several appealing traits that are responsible for their widespread use in industry [1]. Reference cells have a similar spectral response to that of PV modules. The glazing of reference cells yields transmission characteristics similar to those of PV modules. In addition, reference cells are typically less expensive than high-quality pyranometers.

The overarching goal of this project is to understand, characterize, and evaluate the measurements from reference cells. This goal is accomplished by developing and testing a reference cell performance model under a diverse set of experimental conditions. The diversity of conditions includes: different reference cell orientations (horizontal, fixed tilt, one-axis rotation, and two-axis rotation surfaces), multiple experimental locations (Golden, CO, Eugene, OR), and various reference cell makes and models.

This paper expands on previous studies in that the reference cell output is estimated using both measured and modeled spectral data. This approach allows one to evaluate how well estimating the output of reference cells works when only modeled spectral data are available. In most locations where reference cells are used, only modeled spectral irradiance values are available. Results using both measured and modeled spectral data are compared.

To be clear, this paper uses two different models. One model estimates the spectral irradiance using the Bird spectral model. The other model estimates the reference cell output. Efforts have been made to clearly identify which model is being discussed throughout this paper.

This paper is organized as follows. First, descriptions of the differences between the measured and modeled spectrum are presented. A method to quantify the differences between these two spectral sources is discussed. Then a brief discussion on the reference cell model, previously published, is given. Finally, the comparison between the modeled reference cell output and the measured reference cell output is discussed. Comparisons are made when either the measured or modeled spectral data are used as inputs.

II. MEASURED AND MODELED SPECTRAL DATA

A limited number of measured spectral data sets exist because of the expense and maintenance of spectroradiometers. Modeled spectral values are easier to obtain than ground-based measurements or satellite-derived spectral data. As with every model, however, there are some systematic biases. In this section, modeled spectral data from ground-based irradiance values is compared to measured spectral data.

The measured experimental data used in this study were gathered by the National Renewable Energy Laboratory (NREL) at the Solar Radiation Research Laboratory (SRRL) in Golden, CO. For this experiment, both horizontal and two-axis data are analyzed. The time frame of the data collection was from 2022-11-21 through 2022-12-31. The spectral irradiance, I_λ, was measured at 1-nm wavelengths from 290 nm to 1,650 nm using two EKO Weiser spectroradiometers for the horizontal and two-axis tracking surfaces.

The modeled spectral data were generated using the pvlib python spectrl2 library [2] based on the Bird spectrl2 model [3], [4]. This model is frequently used to obtain modeled spectral values. The spectrl2 model is a clear-sky model. The inputs to the spectrl2 model include the solar zenith angle, angle of incidence, tilt of the surface, albedo, air pressure, precipitable water vapor, aerosol optical depth, and ozone. In this study, the aerosol, optical depth, and ozone were set to reasonable constant values. The other variables were time-series values.

A sample comparison of the spectrum from the measured and modeled spectral data sources is shown in Fig. 1. The data shown in Fig. 1. correspond to 2022-11-30 08:00 and 12:00 taken at the SRRL on a two-axis tracking surface as well as on a horizontal surface. During these times the sky was clear.

The upper panels of Fig. 1. correspond to the spectral irradiance values. The vertical scales of the left and right plots are the same. The middle two plots of Fig. 1. show the absolute difference between the modeled and measured data of the upper plots. The lower two plots show the percentage difference of the absolute values between the modeled and measured data of the upper plots.

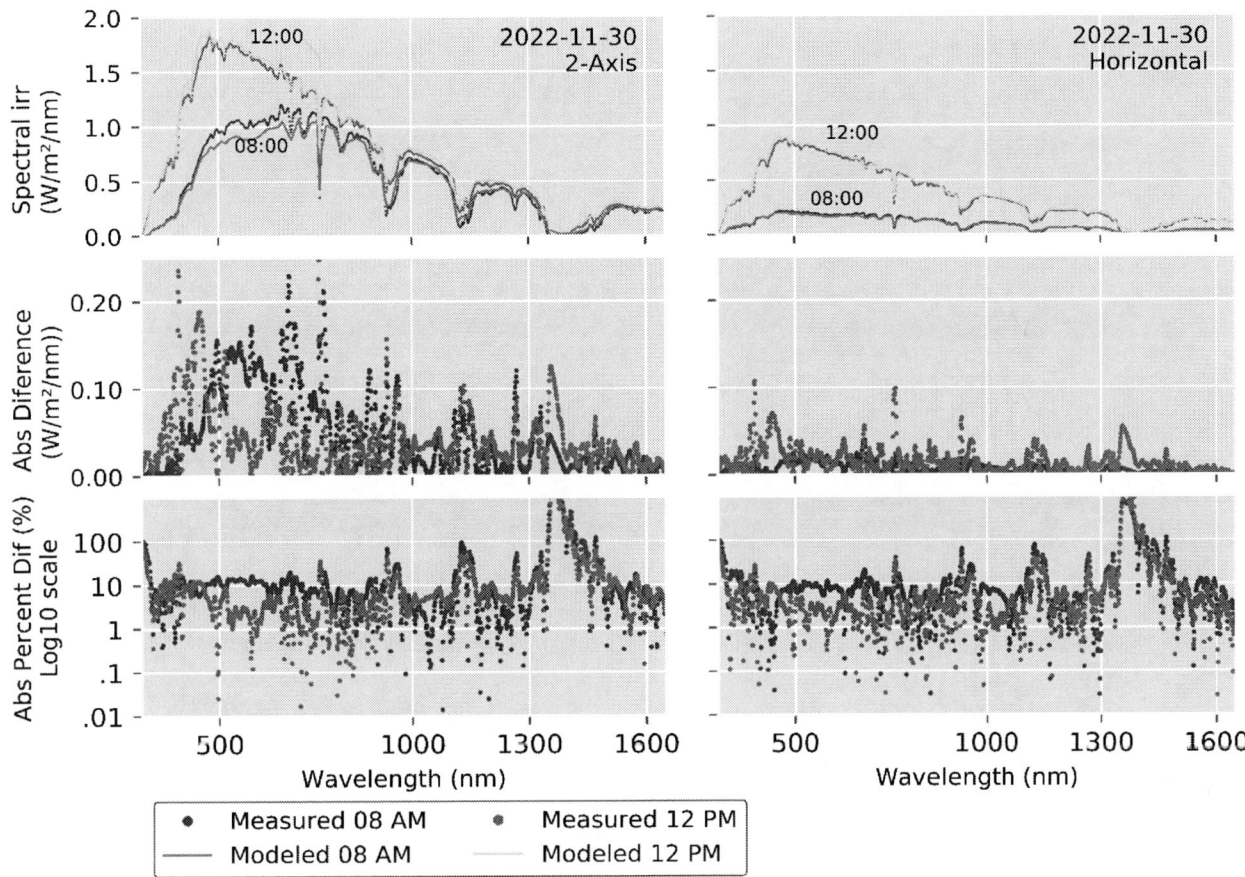

Fig. 1. Sample spectral irradiance curves comparing measured to modeled (top), difference between modeled and measured (center), and percentage difference between modeled and measured (bottom).

The vertical axis of the lower plots is on a Log10 scale. Wavelengths with strong atmospheric absorption effects also have increased percentage difference values. This results from the fact that the percentage differences are exaggerated when the spectral irradiance is significantly decreased, as is the case at these absorption wavelengths.

As shown in Fig. 1, spectral data drastically vary with both time of day as well as across the wavelength range. Because the spectral irradiance exhibits such large variations, standard difference and percentage difference methods to analyze the data are insufficient; therefore, an alternative way to visualize the difference between the measured and modeled spectral values is to plot the percentage differences at selected wavelengths over the day. The selected wavelengths were at 500, 1000, 1300, and 1600 nm because they are not associated with atmospheric absorption effects. These wavelengths are representative samples of the spectrum. The 500-nm wavelength was selected because it is near the peak spectral irradiance value. The 1,000-nm wavelength is representative of the wavelengths near the peak of the spectral responsivity curve of a reference cell. The 1,300-nm wavelength represents the tail

end of the spectral responsivity curve. The 1,600-nm wavelength was selected for completeness.

In Fig. 2, the percentage difference between the measured and modeled spectral data are plotted at the four discrete wavelengths. The data plotted in Fig. 2. correspond to all minutes under clear skies from 2022-11-21 to 2022-12-31, with the individual minute data plotted in blue. The left panels contain data from a two-axis tracking surface. The panels on the right correspond to the horizontal surface. The vertical axis of Fig. 2. is the percentage difference. The horizontal axis of the figure is the solar zenith angle. Because the time frame of the data was only in the winter months, the data do not extend to the smaller solar zenith angle ranges.

A linear fit of the data from the zenith angle range from 60°–80° was computed and is shown as the black line. The equation of the linear fit is given in the text box associated with each plot. Within the range from 60°–80°, the modeled data are within +4 to -10% of the measured data.

Overall, similar trends occur for each wavelength plot between the two orientations. At 500-nm, the percentage difference decreases with solar zenith in both orientations. At

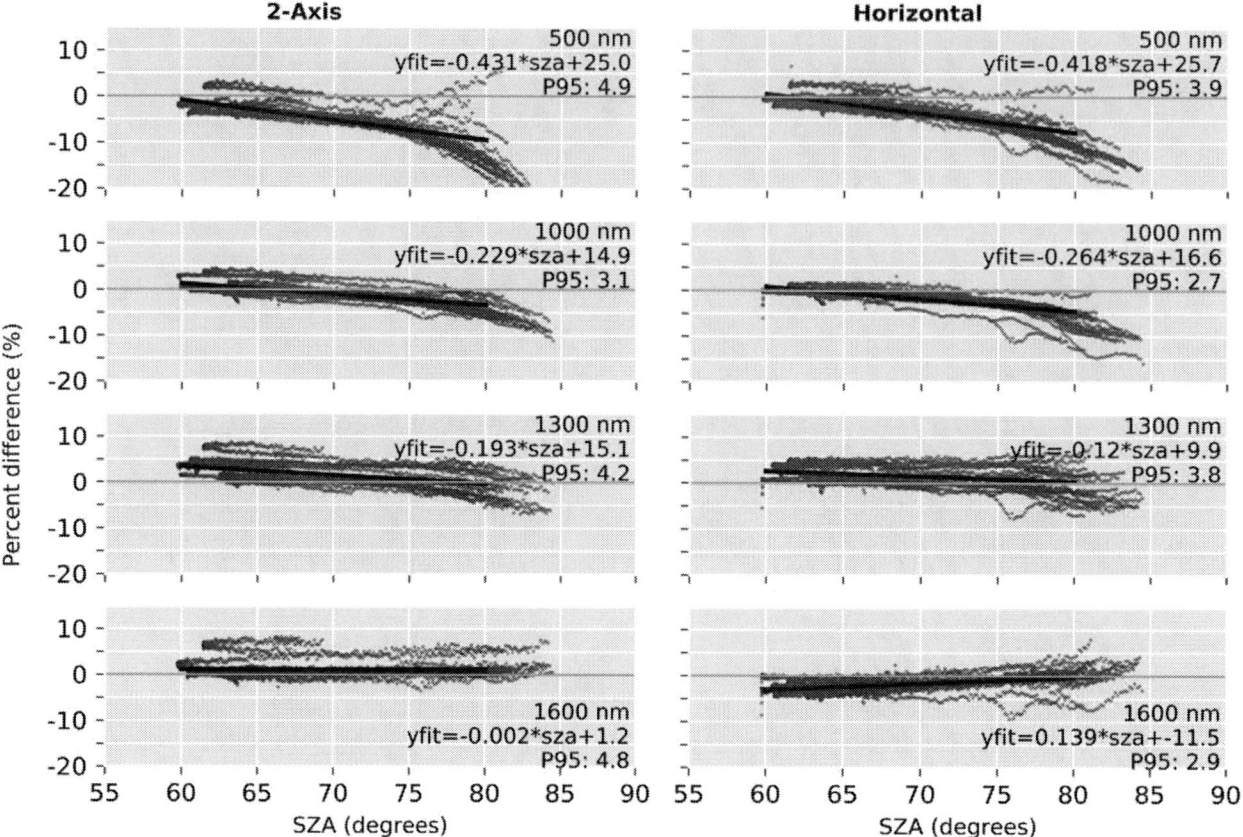

Fig. 2 Percentage differences of modeled to measured spectral data vs. solar zenith angle for the selected wavelengths: 500, 1,000, 1,300, and 1,600 nm.

higher wavelengths, the percentage difference remains constant or slightly increases with zenith angle for both orientations.

The two orientations have very different view factors, but the plots exhibit similar trends with zenith angle. The differences between the modeled and measured data are caused by biases that exist in the two data sets. The difference decreases as the wavelength increases at the 1,000-nm, 1,300-nm, and 1,600-nm wavelengths.

At low sun angles, the modeled spectrum tends to deviate more from the measured spectrum. This likely results from the increased importance in atmospheric effects as the irradiance's path through the atmosphere increases, and any biases in the atmospheric characterization are then amplified.

To compute the scatter in each plot, the following method was used. First, the linear fit of each plot was determined. Then, the absolute difference from the linear fit was computed for each minute of data. In subtracting the data from the fit, the downward sloping trend at 500 nm was taken into account. From the absolute difference, the 95th percentile was calculated. This P95 value is reported in the text associated with each plot. The P95 value can be thought of as the vertical spread in the data. The P95 values range from 2.7% − 4.9% for these

wavelengths and orientations. For comparison, the uncertainties associated with the measured spectral data are between 4% − 8% over these wavelength ranges and incident angles.

III. OVERVIEW OF PREVIOUS MODELED REFERENCE CELL WORK

In previous studies, [5–7], the outputs of the reference cells on various surface orientations were modeled using measured spectral irradiance and the temperature of the reference cells. The results of these studies concluded that four main factors affect the reference cell model:

- The changing spectral distribution of incoming light
- The angle of incidence of the incoming light
- The effect of transmission of light through the glazing
- The spectral responsivity of the reference cell and the temperature effects on the spectral responsivity.

The modeled reference cell output used in this study is given by (1):

$$RC_{model} = F(AOI) \cdot \sum_{\lambda=300nm}^{1300nm} R_\lambda(T) \cdot I_\lambda \qquad (1)$$

where $F(AOI)$ is the average transmission of light through the reference cell glazing and is a function of the angle of incidence of the incoming light. To obtain the angle-of-incidence function, the incident irradiance is separated into the beam irradiance and the diffuse irradiance components in the plane-of-array surface. The diffuse irradiance is further separated into circumsolar, dome, horizon, and ground-reflected components. The various diffuse components are obtained using the Perez model [8]. Finally, the Marion model [9] is applied to the various components to generate an overall $F(AOI)$ term. This model assumes that $F(AOI)$ is spectrally insensitive.

In (1), $R_\lambda(T)$ is the spectral responsivity of the reference cell, which is a function of the reference cell temperature. An adjustment to the spectral responsivity, $R_\lambda(T)$, was determined using the Hishikawa model [10]. I_λ is the spectral irradiance of the incoming light.

Essentially, $R_\lambda \cdot I_\lambda$ is a measure of how much short-circuit current the reference cell will generate at a particular wavelength, λ. The sum in (1) adds all these individual wavelength components to generate a total reference cell short-circuit current. The reference cells used in this study only generate current inside the wavelength range from 300 nm–1,300 nm; however, due to the use of the Hishikawa model, the wavelength range that can generate current extends into the 1,300 nm–1,400 nm range at high temperatures.

Defining a scale factor, K, as the ratio of the modeled output over the measured reference cell irradiance output. The units of RC_{model} are amps. The units of $RC_{measure}$ are in W/m². The units of K are amps/(W/m²). Note: In several previous studies, the inverse of (2) was reported.

$$K = \frac{RC_{model}}{RC_{measure}} \qquad (2)$$

If the model assumptions are valid, then the value of K should be constant under all orientations and weather conditions. The value of K is similar to a responsivity of a pyranometer and is expected to be different for different reference cells. Using (2), the value of K can be determined experimentally. The value of K can be determined under laboratory conditions, but the laboratory spectral irradiance is slightly different than the field spectral irradiance.

In previous studies, the scale factor, K, was analyzed for either one-axis or two-axis orientations; however, the spectroradiometer used to generate the spectral irradiance curves had $5\% - 20\%$ bias in the measurement in the range from 900 nm $-$ 1,650 nm. This biased the spectral irradiance at these higher wavelengths. The spectroradiometers were repaired in the summer to fall of 2022. This study uses spectral data from the period when the upgraded spectroradiometers were installed.

The five reference cells used in this study are: 1. Atonometrics (ATO), 2. EETS (EET), 3. IKS Photovoltaik (IKS), 4. IMT (IMT), and 5. NES (NES). These monocrystalline reference cells are operated at short-circuit current. The quantum efficiency of the cells were measured at the NREL Cell Lab under a standard lamp perpendicular to the reference cell [11]. The spectral responsivities were obtained from the quantum efficiencies by normalizing the responsivity to 1 at the peak response wavelength. Because the responsivity obtained is a relative value, a calibration factor, K, was needed to relate the current from the reference cell to W/m².

IV. Comparing Estimates of Reference Cell Output

Both measured and modeled spectral data were used as input values in (1). These are the spectral values discussed in Section 2. Both two-axis and horizontal orientations were analyzed.

The K values from the IMT reference cell are shown in Fig. 3. The two-axis data are plotted on the left. The horizontal data are plotted on the right. For all plots, the x-axis corresponds to the solar azimuthal angle. Only clear-sky minutes are plotted. The time frame of both data sets was 2022-11-21 to 2022-12-31. The upper plots show the K value of the IMT. The lower plots show the spread in the K values (discussed shortly).

The measured spectral irradiance data were used to generate the black points. The modeled spectral irradiance data were used to generate the blue points. The data in a solar noon window (SNW) is defined as $150° < AZM < 210°$. SNW data are used to analyze the results. The SNW data are highlighted in bold colors.

In the upper plot of the horizontal data set, the K value exhibits a pronounced S-shaped curve. The K value is low in the mornings, and it increases in the afternoon. This might be explained by examining the relative tilt of the two sensors involved in (2). The K value is a function of the spectrum and also the measured reference cell data. If either sensor is tilted relative to the other, the amount of light incident on that sensor will vary over the course of the day. The spectroradiometer is unlikely to be tilted because the modeled and measured spectral data produce similar results. This reasoning can be used to explain the variation throughout the day of the K values. Given the shape of the curve, it can be assumed that the IMT reference cell is slightly tilted to the east. This will generate higher measured reference cell values in the morning, which, in turn, will generate smaller K values in the morning. This is consistent with the data. The K values for the horizontal ATO reference cells do not exhibit such an S-shaped curve (not shown).

To uniformly assess the variance of K between the reference cell output and the modeled reference cell output, the K value data were fit to a third-degree polynomial equation. This was done to both the two-axis as well as the horizontal data sets. The fit was computed only from the data in the solar noon window. The difference from the fit to the measured data was computed. Then the absolute value of this difference was taken. This was done for each minute of data. The third degree polynomial fit is shown in Fig. 3 as the thick solid line.

978-1-6654-6060-6/23 $31.00 © 2023 IEEE

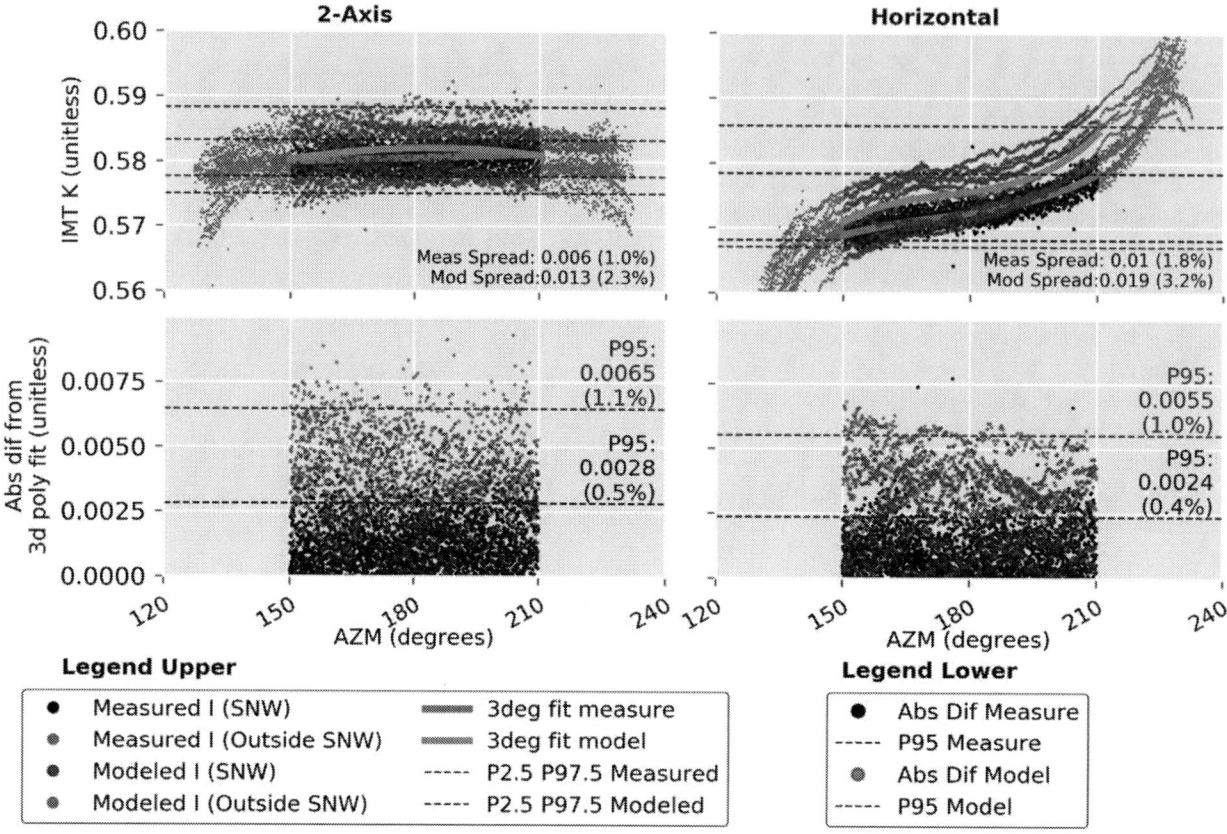

Fig. 3.   IMT reference cell results

The absolute differences for both data sets are plotted in the lower plots of Fig. 3. This adjustment was performed in an effort to minimize the suspected tilt effects present in the horizontal data set.

In the lower plots of Fig. 3, the absolute difference between the fit and the data are plotted along the y-axis. The solar azimuthal angle is plotted along the x-axis. The P95 of each data set is computed. The dashed lines in the lower plots indicate the P95 value. The corresponding P95 percentage difference is listed in the text box as well. The percentage difference is well within the uncertainty associated with the measured spectral data set (4%–8%)

Looking at the differences, in the data derived from the measured spectral data, the uncertainties in the two-axis and horizontal comparisons are very similar (i.e., the two black dashed lines). The same is true for the data derived from the modeled spectral data (i.e., the two blue dashed lines). When comparing the P values of the measured versus the modeled spectral data (i.e., the blue versus black lines), however, there is a factor of two increase for the modeled data.

This increase in uncertainty in the estimated reference cell performance using the modeled spectral data is related to the uncertainty in the spectral values shown in Fig. 2. Because the

modeled spectrum differs from the measured spectrum, larger variations using the modeled spectrum to generate the estimated reference cell output can be expected.

This same method was applied to the other reference cells mounted on the two-axis and horizontal surfaces. Figures similar to Fig. 3. were generated for the other reference cells. The "S" shape characteristic of a tilted sensor was not present in the other two horizontal sensors. This further implies that the IMT reference cell is in fact tilted.

TABLE I

VARIATIONS IN K FOR VARIOUS REFERENCE CELLS

Ref Cell	Surface Orientation	P95 (Measure)	P95 (Model)	P95% (Measure)	P95% (Model)
ATO	2-axis	0.0035	0.0062	0.6%	1.1%
EET	2-axis	0.0026	0.0064	0.5%	1.1%
IKS	2-axis	0.0032	0.0064	0.6%	1.1%
IMT	2-axis	0.0028	0.0065	0.5%	1.1%
NES	2-axis	0.0025	0.0067	0.4%	1.1%
ATO1	Hor	0.0036	0.0066	0.6%	1.1%
ATO2	Hor	0.0028	0.0060	0.5%	1.0%
IMT	Hor	0.0024	0.0055	0.4%	1.0%

The P95 and P95% values for each reference cell are summarized in Table 1. For every reference cell used in this study, the modeled spectral data resulted in twice as much spread in the K values. The spread in the K values was between 0.4% and 0.6% when measured spectral values were used. The spread in the K values was between 1.0% and 1.1% when the modeled spectral values were used. It has been shown here that when modeled spectral is used to generate reference cell output the uncertainty increases by a factor of two.

V. CONCLUSION

In conclusion, a step has been taken in the larger goal of this project to gain a better understanding of reference cells. The key finding of this study is that modeled spectral data can be used to estimate reference cell output to within a P95 value of 1.1% for a two-axis tracking and a horizontal surface.

In the sub-study that was conducted, the modeled spectral values were compared against the measured spectral data in both the two-axis and horizontal orientations. The modeled spectral data were within 10% of the measured spectral data at the discrete wavelengths of 500, 1,000, 1,300, and 1,600 nm. The measurement uncertainty of the spectroradiometer is between 4% and 8% over these wavelength ranges. As expected, the uncertainty of the modeled spectral values are larger than the uncertainty of the measured spectral data. Given the limited inputs of the spectrl2 model, however, the results are encouraging. The comparison between the modeled spectral data and the measured spectral data demonstrated that without more precise inputs into the spectral data model, some systematic biases present in the modeled data increase as the solar zenith angle increases.

The reference cell model incorporates the spectral response of the reference cell and the incoming spectrum of light incident on the reference cell. The reference cell model is confirmed against reference cell measurements through the use of a constant scale factor, K. The more constant the value of K is over all conditions, the better the model performs at predicting the actual reference cell output. Previous studies examined the reference cell model through the use of measured spectral data. In this study, the reference cell model was examined by using modeled spectral values as input in (1) instead of measured spectral data. That is, modeled spectral data were used to generate estimates of reference cell output.

The modeled spectral data generated a variance twice as large in the reference cell output values using measured spectral data. This was tested against five different makes and models of reference cells in the two-axis orientation and three in the horizontal orientation. These are initial results. Considering that the inputs to the spectral model are rather limited, the use of the modeled spectral values in the calculation of reference cell performance, the model spectral irradiance did a remarkably good job at mimicking the reference cell results.

ACKNOWLEDGEMENTS

This work was authored in part by the National Renewable Energy Laboratory, operated by Alliance for Sustainable Energy, LLC, for the U.S. Department of Energy (DOE) under Contract No. DEAC36-08GO28308. Funding provided by U.S. Department of Energy Office of Energy Efficiency and Renewable Energy Solar Energy Technologies Office. The views expressed in the article do not necessarily represent the views of the DOE or the U.S. Government. The U.S. Government retains and the publisher, by accepting the article for publication, acknowledges that the U.S. Government retains a nonexclusive, paid-up, irrevocable, worldwide license to publish or reproduce the published form of this work, or allow others to do so, for U.S. Government purposes.

This work was also sponsored by the Murdoch Family Trust. Additional support for the SRML comes from the Bonneville Power Administration and the Energy Trust of Oregon.

REFERENCES

[1] A. Azouzoute, A.A. Merrouni, E.G. Bennouna, A. Ghennioui, *Accuracy Measurement of Pyranometer vs Reference cell for PV resource assessment* Energy Procedia. 157. 1202-1209. 10.1016/j.egypro.2018.11.286. , 2019

[2] W. Holmgren, C. Hansen, M. Mikofski. *pvlib python: a python package for modeling solar energy systems*. Journal of Open Source Software, 3(29), 884, 2018 [https://doi.org/10.21105/joss.00884]

[3] Bird *Simple Spectral Model: spectrl2_2.c.* [https://www.nrel.gov/grid/solar-resource/spectral.html]

[4] M. Utrillas et al. *A comparative study of SPCTRAL2 and SMARTS2 parameterised models based on spectral irradiance measurements at Valencia, Spain,* Solar Energy, Volume 63, Issue 3, 1998, ISSN 0038-092X F. Vignola, et al., *Reference Cell Performance and Modeling on a One-Axis Tracking Surface* IEEE PVSC 2022

[5] F. Vignola, et al,. *Influence of Diffuse and Ground-Reflected Irradiance on the Spectral Modeling of Solar Reference Cells,* Proceedings of the American Solar Energy Society, Boulder, CO., 2021

[6] F. Vignola, et al., *Improved Field Evaluation of Reference Cell Using Spectral Measurements,* Solar Energy 215, (2021) pp. 482-491

[7] F. Vignola, et. al., *Evaluation of Reference Solar Cells on a Two-Axis Tracking Using Spectral Measurements,* SolarPACES, September 27–October 1, 2020

[8] R. Perez, et al., *Modeling daylight availability and irradiance components from direct and global irradiance,* Solar Energy 44, 271–289, 1990

[9] W. Marion, *Numerical method for angle-of-incidence correction factors for diffuse radiation incident photovoltaic modules,* Solar Energy 147 344–348, 2017

[10] Y. Hishikawa, et al., *Temperature dependence of the short circuit current and spectral responsivity of various kinds of crystalline silicon photovoltaic devices,* Japanese Journal of Applied Physics 57, 08RG17, 2018

[11] M. Sengupta, et al., Solar Radiation Research Laboratory (SRRL) Final Report: Fiscal Years 2019–2021," 2022

Silver Reflector-Driven Light Harvesting Enhancement in Large Area Dye Sensitized Solar Cells

Navdeep Kaur, Faizan Syed, Cheng-Yu Lai, Daniela Radu

Florida International University , Miami, FL, United States

Abstract - Efficient light harvesting by absorber materials in large active area photovoltaic (PV) devices has always been a major challenge as some of the materials have smaller light absorption coefficients and their properties are impacted by scalability attempts. In addition, defects are more difficult to control when active area increases. In the present work, we demonstrated the positive impact of a silver reflecting surface consisting of a micrometer-thick, silver nanoparticles (Ag Nps)-based film on the glass side of platinum (Pt) counter electrode (CE). The films aimed to improve the light-harvesting ability of mesoporous TiO2 photoanodes on large-area dye sensitized solar cells (DSSCs). Although the power conversion efficiency (PCE) of large-area DSSC is lower in comparison to small-area DSSC, the PCE of the large-area Ag reflector based DSSC showed significant PCE improvement relative to reference DSSC. The enhancement is attributable to the improved light harvesting ability of photoanodes demonstrated from the increased short circuit current density (JSC) values of large-area Ag reflector based DSSC, as most of the unused incident light is reflected again due to the improved reflectivity of Pt CE via Ag reflecting film towards the photoanodes and is being reabsorbed by the dye sensitizers. This effect has been further confirmed by the electrochemical impedance spectroscopy (EIS) measurement, where resistances at different interfacial layers in DSSCs have been estimated. Hence, we report on a simple and efficient method of using Ag reflector films to enhance light harvesting in large-area DSSCs and paving the road towards improving performance of other large-area PV devices as well. Keywords: Large area DSSCs, DSSCs, Dye-sensitized solar cells, enhanced light harvesting, silver nanoparticles film

978-1-6654-6060-6/23 $31.00 © 2023 IEEE

Survey of Snow Impacts on Bifacial Gain in Commercial Photovoltaic Arrays

Samantha S. Wilson, Stephen Lightfoote, and Stephen Voss

Power Factors Inc., Brossard, QC, J4Z 1A7, Canada

Abstract — **Bifacial modules are being increasingly deployed in commercial utility arrays worldwide due to their predicted ability to lower the levelized cost of energy versus monofacial modules. The energy gain versus monofacial modules, also known as bifacial gain, is driven primarily by the amount of rear irradiance seen by a module. Data from approximately 100 bifacial sites around the world were obtained and from the Power Factors database and analyzed for bifacial gain based on measured irradiance. The analysis shows that sites have a bifacial gain of 5-10% in the summer. In the winter the value is more variable due to snow fall. New snow can have a large impact on albedo, increasing bifacial gain 2.6-3.2 times. However, this gain quickly degrades and has a half-life of only 3-5 days. For a site to see large gains in albedo in the winter it must experience frequent snow falls of 1 cm snow depth or more.**

I. INTRODUCTION

Bifacial modules are being increasingly deployed in commercial utility scale photovoltaic projects world-wide due to their predicted ability to lower the levelized cost of energy (LCOE) versus monofacial modules. [1] Bifacial modules absorb light through both the front side and back and thus any performance gains associated with bifacial modules are primarily due to additional rear plane-of-array (POA) irradiance. Thus, to truly understand bifacial performance, the most important parameter to understand is the irradiance that is available to the rear of the module. Significant modeling studies have been conducted and individual research arrays have been studied, but there have been limited reviews of rear POA irradiance in the field. [1]-[3]

There are many parameters that can impact the rear irradiance of an array including ground coverage ratio (GCR), post height, table width, diffuse fraction, and ground albedo [2]. Albedo has been found to be the most important parameter for determining rear irradiance [3]. It represents the ratio of reflected light to incident light on a surface. The typical assumption for ground albedo is 0.2, however this can be impacted by local weather conditions. For example, albedo often decreases with rain but increases with snow. The albedo for fresh snow is predicted to be as high as 0.8 [4]. Bifacial gain, or the additional energy generated from the backside of a module, is approximately linearly dependent on ground albedo [3].

Albedo can be directly measured by an inverted pyranometer, often called an albedometer, however albedometers are not as widely deployed in field currently as rear POA pyranometers that directly measure irradiance on the rear of an array. The Power Factors database contains 245 rear pyranometers across

~100 bifacial sites located primarily in the continental US. Approximate site locations are shown in Fig 1. These pyranometers experienced ~300 significant snow events over the past year. Bifacial gain was calculated directly from the pyranometer measurements and indicates that though snow does increase local albedo significantly, the increase in bifacial gain had a half-life of only 3-5 days.

II. METHODS

AC power, plane-of-array irradiance (POA), rear POA irradiance, and solid precipitation data were collected for each qualifying pyranometer in the Power Factors database at a data resolution of 5 minutes using measurements taken from the period of January 2022 – January 2023. Filters were applied to the data for quality issues including flat line, constant slope, minimum intervals, location, and range. All sensors were in the Northern hemisphere. After filtering there were 129 sensors with sufficient data quality for analysis.

Bifacial gain was calculated from the filtered data. For rear sensors it was calculated as:

$$Bifacial\ Gain = \frac{(Irradiance\ Rear)(Bifaciality)}{Irradiance\ POA} \quad (1)$$

A default bifaciality value of 0.7 was used for all sites. Results are shown in Fig 2.

Data was split into two categories based on mounting type, horizontal single axis trackers (HSAT) and fixed tilt arrays. Sites with and without snow were also assessed based on sites that received < 1 cm of solid precipitation in a year.

For sites with snow, we calculated average bifacial gain based on days since a snow event. A snow event was snow fall greater than 1 cm in a day. The rate of decay of bifacial gain

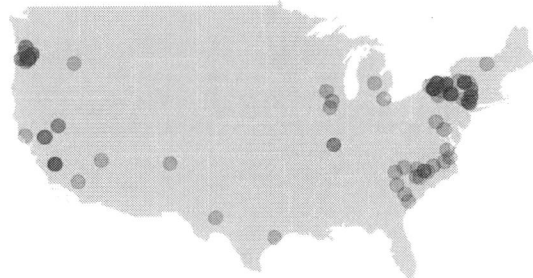

Fig 1. Map of approximate locations of bifacial arrays used in this study.

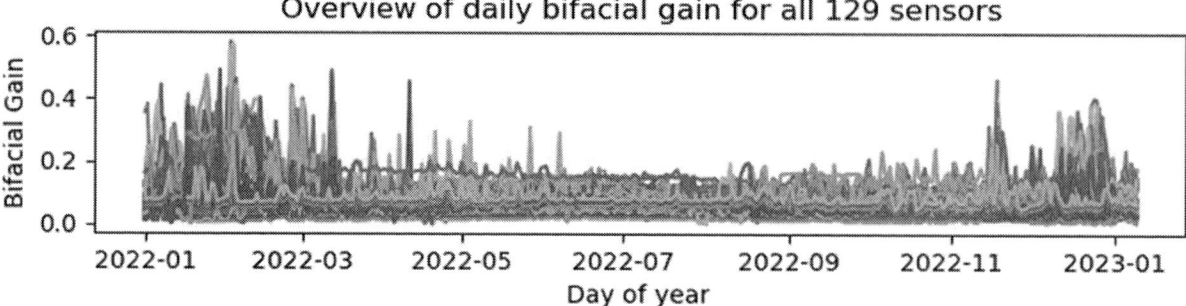

Fig 2. Plot of Bifacial gain for 129 sensors

after snow was calculated by fitting data to a standard exponential decay with a non-zero asymptote:

$$f(x) = a * e^{-b*x} + c \qquad (2)$$

III. RESULTS AND DISCUSSION

Fig 2 shows a plot of bifacial gain recorded by all 129 rear sensors with sufficient data quality for analysis. Bifacial gain in the summer was relatively flat across all sensors which is consistent with low variation in ground albedo and less variation in the diffuse fraction of light. In the winter there are large, sudden spikes in albedo consistent with intermittent snow fall. Fresh snow can have an albedo as high as 0.8 so we would expect spikes to be as high as 0.3 to 0.4 depending on array configuration. Higher spikes could be due to changes in diffuse fraction and albedo happening simultaneously.

Monthly averages for all sensors in Fig 2 are shown in Fig 3. Average bifacial gain across the sensors varied from 5-10% with an average of 7% annually. Variance was higher in the winter due to snow fall. We then split all the data into 4 different categories based on mounting type and if the sensor had experienced any snow events. Snow events were defined as total snow fall greater than 1 cm over the course of 2022. The

sites that experienced snow in the winter had particularly high bifacial gain for December, January and February. March data was a bit lower, and April-November was mostly flat with lower variance. The tracker plants had extremely high bifacial gain in the winter with values as high as 20%. Fixed tilt sites also showed some increase in bifacial gain in the winter, but the difference was not as large. Some possible causes for this disparity are explored later. The fixed tilt sites that didn't experience snow had larger bifacial gains in the summer versus the winter. This could be due to rain in the winter decreasing the ground albedo. It could also be due to low sun angles decreasing the view factor of the ground.

Fig 4 shows a plot of the depth of daily total new snow versus the absolute change in bifacial gain from the day before snow fall to the day of snow fall. There is a large spread in the data which is probably due to variation in the diffuse fraction of light day-to-day. However, there is a slight linear trend to the data. One conclusion of the linear trend is that nearly any amount of new snow can positively impact ground albedo. There is also a slight correlation between the amount of snow and the magnitude of the change in ground albedo. From this data we also concluded that approximately 1 cm of snow can be considered a snow event since the linear fit estimates that 1 cm of snow will lead to +1% absolute change in ground albedo.

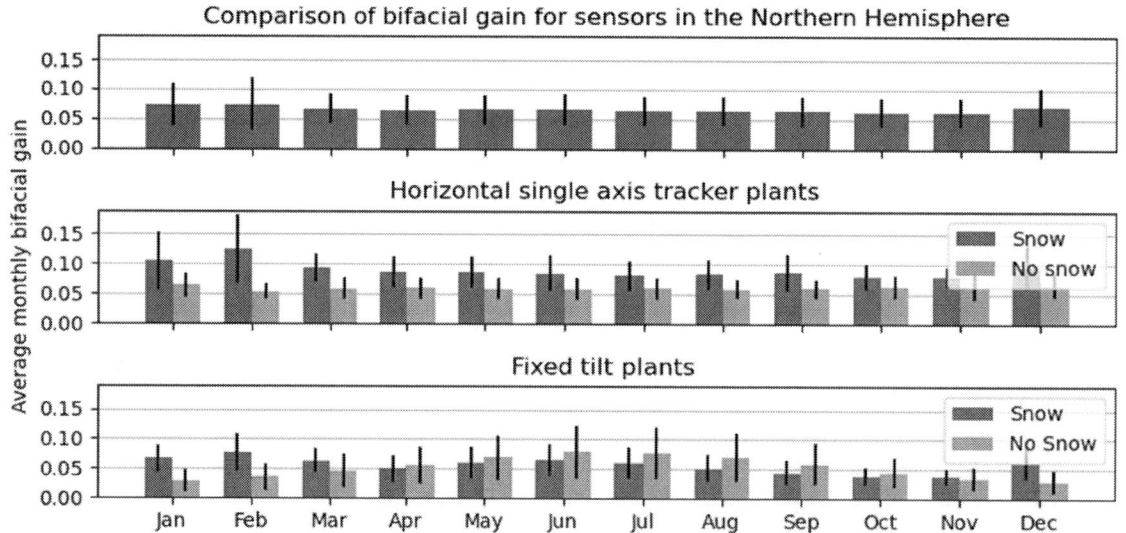

Fig 3. Comparison of average monthly bifacial gain for (a) all sites, (b) sensors mounted on horizontal single axis trackers, and (c) sensors mounted on fixed tilt arrays. Data was also split by sensors that experienced > 1 cm snow and those that did not

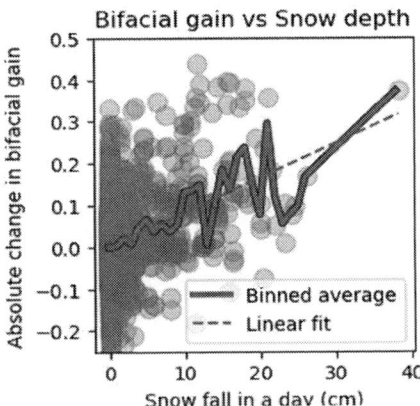

Fig 4. Analysis of depth of new snow required to impact albedo. The linear nature of the data indicates nearly any amount of snow can have an impact.

After establishing what snow depth constitutes a snow event, we then studied how long the impact of new snow on albedo lasted. Data in Fig 5 plots the decay of bifacial gain versus days since last snow event with day 1 being the first snow day. The data was well fit by an exponential decay curve. For horizontal single axis trackers, the half-life of the curve was 4.9 days with an asymptote at 8.4% bifacial gain. The day 1 bifacial gain was 22% which is approximately 2.6 times the asymptote meaning the maximum winter albedo is about 2.6 times the summer values. We also examined the shape of the frontside POA values to determine if the effect was being driven primarily by changes in diffuse fraction. Snow generally falls on very cloudy

TABLE I
COEFFICIENTS FOR EXPONENTIAL DECAY

Fit coefficients	Mounting type	
	HSAT	Fixed tilt
a	0.135	0.109
b	0.142	0.242
c	0.084	0.051

days, so those days tend to be particularly high bifacial gain days both because of the clouds and the new snow.[1] However, except for a large dip on the initial snow day, the frontside POA was relatively flat. This indicates the biggest driver of bifacial gain post snow fall is changes in ground albedo.

The fixed tilt data was also fit to an exponential decay curve. The half-life of the fixed tilt array was 2.9 days, and the asymptote was 5.1% bifacial gain. The day 1 bifacial gain value was 16.0%, which is 3.2 times the summer asymptote. Overall values were lower for fixed tilt plants which could have generally been experiencing fewer diffuse days as the POA for the fixed tilt plants is measurably higher than the HSAT plants for most days. We are continuing to investigate the disparity.

The immediate decay on bifacial gain after snow fall indicates that for a site to experience high bifacial gain for the entire winter it must continuously get fresh snow. Additionally high albedo estimates for the winter in the range of 3-4 times the summer albedo are high risk. Average snow falls measured in this study show gains of 2.6-3.2 times summer albedo and only for short times.

III. CONCLUSIONS

Data from ~100 bifacial sites around the world were obtained and analyzed for bifacial gain based on on-site irradiance measurements. Data shows that sites have a bifacial gain of 5-10% in the summer. In the winter the value is more variable due to snow fall. New snow can have a significant impact on albedo, increasing bifacial gain 2.6-3.2 times. However, this gain quickly degraded and had a half-life of only 3-5 days. For sites to see large gains in albedo in the winter they must experience frequent snow falls of 1 cm or more.

REFERENCES

[1] Deline, C et. al. "Bifacial PV System Performance: separating fact from fiction" PVSC-46, Chicago, IL 2019

[2] Anoma, M et al, 2017. "View Factor Model and Validation for Bifacial PV and Diffuse Shade on Single-Axis Trackers PVSC-44, 2017

[3] Advanced Array Technologies "Field Testing Meets Modeling: Validated Data on Bifacial Solar Performance" https://arraytechinc.com/field-testing-meets-modeling/

[4] Marion, B " Ground Albedo Measurements and Modeling" 2018 Bifacial PV Workshop

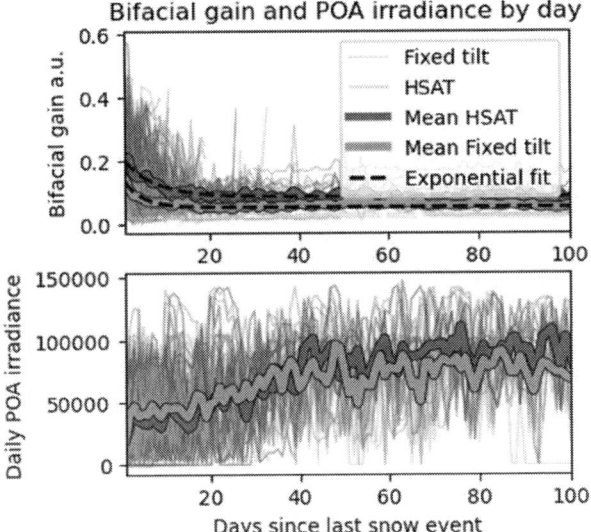

Fig 5. Fresh snow fell on day one, after which bifacial gain immediately began to decay. Half-life for the decay was found to be 3-5 days and the decay reached an asymptote of 5-8%. The daily POA trend shows the trend is driven primarily by changes in ground albedo and not changes in the diffuse fraction.

New theoretical limits for light trapping in solar cells

Stéphane Collin, Maxime Giteau

1Centre de Nanosciences et de Nanotechnologies (C2N), Palaiseau, France

2Institut Photovoltaïque d'Ile-de-France (IPVF), Palaiseau, France

We present a universal model of broadband absorption in a slab of semiconductor. The theoretical framework, based on the description of multiple overlapping resonances, has a very broad domain of validity. We derive simple analytical formulas for reference light-trapping models and absorption upper bounds. Two light-trapping strategies are compared: multi-resonant absorption achieved with a sub-wavelength periodical pattern, and isotropic scattering obtained with random texturing. These new theoretical limits for light-trapping provide guidelines for the design of solar cells, and could be used to revisit the maximum efficiency of single-junction silicon solar cells.

978-1-6654-6060-6/23 $31.00 © 2023 IEEE

Supply Side Management with Agrivoltaics: Feasibility Study of Modeling Methodologies of Solar PV and Crop Response

Tadatoshi Takahashi
Tadatoshi Solar Research
Vancouver, BC, V6B3W7, Canada
tadatoshi@gmail.com

Abstract—Agrivoltaics use a land for both solar photovoltaics (PV) electricity generation and agriculture, increasing overall productivity of the land. However, this can also create competition between solar PV and agriculture for available solar irradiation. Demand of electricity is higher early in the morning than in the middle of the day. Thus, we need to let solar PV modules utilize full solar irradiation early in the morning while higher energy production from solar PV in the middle of the day necessitates curtailment due to inverter clipping in solar balance of system or in order to mitigate Duck Curve effect in Electric Grid by the Utilities. On the other hand, photosynthesis of agricultural crops saturates at certain threshold light intensity, making some crops require full usage of solar irradiation early in the morning while those crops cannot fully utilize solar irradiation in the middle of the day. In order to address this problem, we can look at another aspect. Solar irradiation is naturally intermittent due to weather such as cloud coverage and crops are well adapted to variation of solar irradiation availability. We interpret it as natural Demand Response capability of crops. The objective of this study is to investigate the possibility of shifting solar irradiation usage by crops from early morning to the middle of the day, by changing the light intensity and availability over crops by controlling the solar PV modules coverage over them. The scope of this paper is feasibility study of modeling methodologies for this objective. We incorporated Bayesian analysis to address uncertainty with limited relevant data. The result of this evaluation gives a guideline for further research to realize the objective, taking into account the advantages and disadvantages of proposed modeling approach and agrivoltaics configurations.

Index Terms—Agrivoltaics, Photovoltaics, Agriculture, Supply Side Management, Duck Curve, Photosynthesis, Acclimation in Plants, Soybean, Maize, Bayesian Analysis

I. INTRODUCTION

Agrivoltaics is a combination of solar photovoltaics (PV) and agriculture, sharing a land for both electricity generation and crop cultivation. Agrivoltaics promote efficient way to utilize land. Most of the research on agrivoltaics focus on this aspect, sharing solar irradiation between solar PV and agriculture while minimizing the reduction of electricity generation and agricultural yield compared to the conventional use of land for solar PV only or agriculture only.

There is another factor that is well identified in the field of electricity generation, that is, matching electricity generation to electricity consumption. This is the underlying theme of this study. In particular, the main focus is to supply electricity during the peak electricity consumption period in the morning

when electricity generation by solar PV is low. This factor is not addressed by research on agrivoltaics at the time of this writing, based on our review on research papers on the subject.

Hence, it is important to take into account these three factors, i.e. maximizing electricity generation by solar photovoltaics, maximizing crop yields, and meeting electricity demand.

Conventionally, dynamically adjusting electricity supply in power plant to meet electricity demand is called "supply side management". In the case of solar PV, we cannot perform this supply side management since its electricity generation is dependent on availability of solar irradiation. On the other hand, plants have natural coping mechanism to deal with intermittency of solar irradiation. This study proposes to take advantage of this mechanism, in agrivoltaic setting, in order to produce as much electricity as possible from solar PV when the demand is high in the morning, by shifting solar irradiation over crops to the midday, while optimizing cumulative electricity generation and crop yield. Thus, we borrow the expression, "supply side management", and call our proposed approach "supply side management with agrivoltaics".

One impediment in this approach is in plants' coping mechanism itself, in which plants protect themselves from being damaged by intense solar irradiation in the midday by downregulating photosynthetic capacity. This is called Midday Depression of Photosynthesis [1]. This necessitates full solar irradiation over crops in the morning when full irradiation over PV modules is important in order to produce as much electricity as possible to meet high demand. Agrivoltaics mitigates this problem since PV modules covering over crops protects them from intense irradiation in the midday. The positive impact of agrivoltaics in this aspect is well documented as described in Section I-B below for compilation of literature review.

A. Terminology

Following the convention, we use the terminology, "Irradiation" to mean Energy of sunlight whereas "Irradiance" to mean Power of sunlight. i.e. "Irradiation" to have a unit of Watt hour $[Wh]$ and "Irradiance" to have a unit of Watt $[W]$, unless otherwise noted.

B. Review of Related Research Works

[2] performed systematic literature review on the subject of agrivoltaic systems. It is the most comprehensive review paper on agrivoltaics we found. [3] reviews the installed agrivoltaic systems around the world with variety of coexisting production such as vegetables, beehives, and livestock. It also emphasizes the benefit of such systems based on their positive impact on climate change problem as well as with regard to their Levelized Cost of Electricity (LCOE). [4] reviews several research activities describing the system and structure in detail with figures. [5] reviews agrivoltaic research activities from the perspective of the part of the world with large farmer populations and with arid climate.

In terms of solar irradiance availability in agrivoltaic setting, [6] developed a tool to calculate shading caused by solar PV arrays over the ground as well as Photosynthetically Active Radiation (PAR) on the ground. [7] uses Ray-tracing method of light absorption to evaluate the light availability in agrivoltaic systems. [8] performed a validation of a model comparing with a drone image over experimental farm land with solar PV modules installed over maize. [9] investigated irradiation under agrivoltaic system in seven different climate regions around the world, and found that the optimal design of PV modules installation for crop growth is east-west direction. [10] investigated the effect of the orientation of bifacial modules installation - North/South and East/West orientation by irradiance modeling. In the same setting, [11] investigated the effect of the orientation of bifacial single-axis tracking system. [12] found that optimal PV array configuration for agrivoltaics in terms of decreased shadow duration and insolation homogeneity is East/West tracking system (North/South axis) compared to fixed south-facing system.

In terms of type of solar modules that can be used in agrivoltaic system, [13] analyzed the light spectrum under PV modules and found that the spectrum near edge under bifacial module has more optimal light distribution for crops than the direct solar irradiance.

Various agrivoltaic specific solar PV modules have been designed. [14] used the approach to use a light filter to redirect light to PV module or crop depending on the characteristics of crop's light usage, in order to increase overall productivity. [15] used tinted semi-transparent PV module that uses green and blue spectrum of light while passing red spectrum of light that is used by photosynthesis of plant. It found that under agrivoltaic condition, the ratio of biomass above ground over biomass under ground increased, and amount of protein increased. [16] reports a new solar PV module for agrivoltaics for greenhouses. The focus is on yielding agricultural production in winter and thus energy production by solar PV is not considered in winter.

Related to electric energy production in agrivoltaics, [17] utilizes linear programming to investigate the impact of increased electric power generation by agrivoltaics deployed in agricultural land on electric grid. This research investigated various scenarios including power grid expansion and en-

ergy storage usage. [18] utilizes mathematical optimization in comprehensive aspect of agrivoltaics incorporating crop yield in equations. Their objective function is to maximize Land Equivalent Ratio (LER), minimize power fluctuation (by minimizing standard deviation), and maximize annual electricity production. Their decision variables are Solar Azimuth Angle and Row Distance of installed photovoltaic modules.

In terms of metrics for agrivoltaics performance, [19] gives concise classification and key performance indicator (KPI) definitions. KPI's include Ground Coverage Ratio (GCR), Energy and agricultural yield, Land Equivalent Ratio (LER), and price-performance ratio (ppr) among others. [20] and [21] introduced a new metrics to standardize the effect of PV over various crops's yield - Light Productivity Factor (LPF).

In terms of stress on crops under agrivoltaic setting, [22] investigated the temperature and heat stress on a crop under agrivoltaic condition. [23] investigated if shade-intolerant crops can grow under solar PV modules and found that agrivoltaics can increase crop yields. It points out that light saturation point limits photosynthesis, that sunlight over that point can even damages plant, and that plants have a mechanism to mitigate it which reduces photosynthetic activity. It lists a table of light saturation points for various crops.

In terms of category of crops, C3 and C4 are most notable [24]. [25] proposes a different categorization, based on response to PAR and sunlight duration, categorizing plants and crops in three categories. [26] investigated crop selection methodology for agrivoltaics, based on three climate factors, sunshine, temperature, and precipitation. Crops suitable for different part of agrivoltaic installation area are selected.

Related to agrivoltaics, we reviewed literatures in agriculture investigating the effect of intermittency of sunlight on plant growth. [27] reviews published data on the acclimation of plants under fluctuating irradiance and identifies key points. These include the literatures based on modelings. [28] summarizes the effect of diffuse solar irradiation on photosynthesis, comparing with the case with direct solar irradiation. [29] uses mathematical framework to assess photosynthetic acclimation in fluctuating light and found that longer low light period compared with high light period decreases photosynthetic capacity of plants. [30] uses stochastic mathematical model of fluctuating light on plants in order to reduce computation time. The notable mention this work makes is "longer periods in high light and low light are likely to be less productive than rapid fluctuations." [31] investigated the effect of intercropping on photosynthesis in the case of soybean planted between maizes. They found that while the photosynthetic capacity was limited, efficient interception and absorption of light and carbon gain was observed. [32] investigated another type of intercropping, in which agricultural crop, soybean, is cultivated in between forest species.

Another important factor in the context of agrivoltaics is midday depression of photosynthesis, which is decreased photosynthetic activity around solar noon. [1] investigated this phenomenon and lists related factors, such as light intensity. It identified that the decisive factor is soil water status, i.e. its reduction is the main but indirect cause. One of the causes of midday depression of photosynthesis is high light intensity. [33] investigated various high light stress coping mechanisms, namely, Chloroplast movement, Stomatal response, Photoacclimation, Photoprotection, and Antioxidant defence against reactive oxygen species (ROS).

With regard to conversion factor of parameters used in photovoltaics and agriculture, [34] indicates the conversion factor between solar irradiance and PAR to be 1 [W/m^2] = 4.57 [μmole/(m^2s)] whereas [35] indicates 1 [W/m^2] = 2.02 [μmole/(m^2s)]. The difference seems to be due to the fact that solar irradiance is measured in entire spectrum whereas PAR considers only the wavelength from 400 [nm] to 700 [nm].

[36]- [39] list various values for crops.

Through the reviews, we can emphasize that photosynthetic intensity increase is not linear and that it has saturating point. This is an interesting point in terms of sharing light with solar PV modules whose current intensity increases linearly with light intensity, which leads to relatively linear increase in power production.

C. Research Questions, Hypotheses, and Specific Aims

The research questions in this study are as follows:

1) Is it possible to have sufficient agricultural yield when shifting solar irradiation over crops from the peak energy consumption period in the morning to the midday?
2) What is the optimal solar PV modules configuration to achieve it?
3) What is the good mathematical model to evaluate it, under the constraint that the relevant data are limited?
4) Can we incorporate biofuel crop production under this agrivoltaic configuration in order to produce biofuel to power auxiliary power plant to compensate energy production from solar PV during the peak energy consumption period?

Hypotheses in this research are as follows:

- It is possible to shift solar irradiation over crops from the peak energy consumption period in the morning to the midday, by allowing plants to use its ability to adjust to various environmental conditions.
- Bayesian modeling is an effective mathematical and statistical modeling approach for this agrivoltaic setting, addressing uncertainty with limited relevant data and gradually incorporating data as they become available.

We conceptualized various agrivoltaics configurations for this purpose, based on past research on agrivoltaics. Then, we developed a mathematical and statistical model, and developed software for the model as Open Source. The scope of this report is to report the result of feasibility study on the model, that can help to plan empirical study in the future, i.e. serving as analysis of power of planned research.

II. METHODOLOGY

All the values and calculations are based on per m^2. Hence, solar irradiance is obtained per m^2. Subsequently, electricity generation by solar photovoltaic module is calculated per m^2. For crops, irradiance per m^2 is converted to PAR per m^2, then based on it, overall crop yield is calculated.

A. List of Agrivoltaic Installation Models

Table I lists possible agrivoltaic configurations that can achieve the supply side management approach proposed in this project. Based on it, this project evaluated one simpler model that can be further extended and divided into these three configurations.

B. Diurnal Solar Irradiance Allocation

We used clear sky irradiance obtained from pvlib python library [40]. Peak electricity consumption periods are between 8am and 10am and between 3pm and 6pm. During these periods, the entire solar irradiance is consumed by Solar PV and no irradiance is made available to crops. During the midday (between 10am and 3pm) and the rest of the day, we allocated the maximum irradiance that crops can take to crops and the remaining irradiance to solar PV.

Some crops including soybean we chose has Midday Depression of Photosynthesis. It is also related to "light saturation point" in photosynthesis, beyond which photosynthesis cannot take place any more. Hence, we made sure that irradiance over the crops stay under this light saturation point.

During the time period when irradiance is shared between solar photovoltaics and crops, how much solar irradiance is redirected to solar PV modules and how much to crops is calculated every minute during that period. This calculation is performed by Linear Programming.

C. Crop

This project is expected to be expanded in the future to more specific applications. One application is cultivation of crops for producing biofuel, which is used as a fuel for peak power plant operated during the peak energy consumption periods when solar PV by itself cannot meet the demand. For this reason, this study chose crops that are consumed as biofuel as well as food, i.e. maize (corn – Zea mays) and soybean (Glycine max).

Another interesting fact about selecting these crops is that maize is C4 plant and soybean is C3 plant. This categorization is based on how CO_2 is absorbed through stomata of leaves. C4 plant doesn't lose much water through stomata during the absorption of CO_2, while C3 does, especially under high heat. Thus, C4 plant is more resilient to intense irradiance that happens midday. On the other hand, C4 plant is shade intolerant and C3 plant is shade tolerant. [24] This makes the comparison of maize and soybean under the agrivoltaic configuration of this study an interesting subject for a future study.

Another reason is that soybean has nitrogen-fixing properties and thus regenerates the soil, which can be considered to reduce the usage of chemical fertilizer based on fossil fuel. It will be an interesting future study to assess overall CO_2 emission reduction under the agrivoltaic configuration of this study. Hence, the main focus of this project is on soybean.

978-1-6654-6060-6/23 $31.00 © 2023 IEEE

TABLE I
LIST OF POSSIBLE AGRIVOLTAIC INSTALLATION MODELS FOR SUPPLY SIDE MANAGEMENT

Configuration	Relevant Configuration from other research	PV module type	Modeling or Simulation Tools	Assumptions	Relevant research
Crops between Vertical Bifacial PV modules	Intercropping [31]	Bifacial PV module vertically installed	pvlib python library Bifacial PV module calculation [40], Net Photosynthetic Rate for soybean under fluctuating irradiance from [31]	With intense irradiation around noon, midday depression of photosynthesis is expected. Net Photosynthetic Rate pattern follows that from [31].	[18] [31] [40]
Crops under Agrivoltaic modules equipped with light steering optical film	Agrivoltaic modules from [14]	Monofacial PV module installed with the tilt matching latitude	pvlib python library Monofacial PV module calculation [40], Net Photosynthetic Rate for soybean under shading condition	With the coverage of crops by PV modules, it is expected that air moisture and soil moisture are retained, thus midday depression of photosynthesis is mitigated, and also expected that intensity of light is kept under light saturation point.	[14] [40]
Crops under semi-transparent PV modules	Semi-transparent PV modules from [15]	Monofacial PV module installed with the tilt matching latitude	pvlib python library Monofacial PV module calculation [40]		[15] [40]

D. Photosynthetic Rate

In order to verify the Hypothesis 1), described in Section I-C, the key factor is the change of Net Photosynthetic Rate due to shifting irradiation over crops to the midday.

As theoretical Net Photosynthetic Rate, a non-rectangular hyperbola model (Equation (1) of [29]) is as follows:

$$P(L, P_{max}) = \frac{\phi L + (1+\alpha)P_{max} - \sqrt{(\phi L + (1+\alpha)P_{max})^2 - 4\theta\phi L(1+\alpha)P_{max}}}{2\theta} - \alpha P_{max} \tag{1}$$

where L is Photosynthetic Photon Flux Density (PPFD) $[\mu mol/(m^2 s)]$, P_{max} is maximum photosynthetic capacity $[\mu mol/(m^2 s)]$, ϕ is maximum quantum yield, α is fraction of maximum photosynthetic capacity used for dark respiration, and θ is curvature of light-response curve.

In an empirical study, [31] compared the case of soybeans cultivated between maizes (intercropping) with the case of soybeans only (monocropping). In the intercropping case, soybean is shaded by maize early in the morning and late in the afternoon. The data from this experiment (Fig. 1) serves as a starting point (prior in Bayesian Statistics) since we cover crops with solar PV in a similar way.

We used the modeling approach of fitting (1) to this empirical data (Fig. 1) and then extended the result to obtain the case for agrivoltaics, applying Bayesian analysis, as follows:

Step 1: Apply curve-fitting algorithm to obtain ϕ, α, and θ of (1) by fitting the equation to the empirical data (Fig. 1).

Step 2: Use the result of Step 1 as prior for Bayasian analysis to revise Net Photosynthetic Rate.

Since the Net Photosynthetic Rate based on the agrivoltaic configuration proposed by this project is not available, thus it is being evaluated by model, Bayesian modeling is used to obtain it as the posterior. In other words, it is assumed that Net Photosynthetic Rate of soybean under this project's agrivoltaic configuration is similar to that of Intercropping scenario in [31] where soybean is planted between maize. Based on this incomplete knowledge with uncertainty, we perform Bayesian analysis to revise Net Photosynthetic Rate

with probability distribution. With Bayesian approach, we can iteratively improve the accuracy of expected value.

The proposed Bayesian statistical model is:
- prior: Parameters ϕ, α, and θ assumed to follow Normal Distribution.
- likelihood: Normal Distribution with mean following (1), with Standard Deviation assumed to follow Half Student T distribution with Standard Deviation of net photosynthetic rate, and with data of net photosynthetic rate

We were able to obtain parameters ϕ, α, and θ of (1) taking into account the uncertainty related to expanding the result of intercropping research to agrivoltaics.

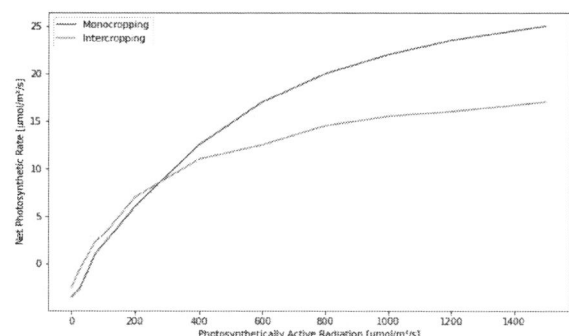

Fig. 1. Net Photosynthetic Rate measured as CO_2 uptake in soybean leaf (Recreated based on Figure 2 of [31])

E. Crop Yield

Crop yield is typically used to assess the agrivoltaic system.

Equation (6) of [18] gives the crop yield as a summation of diurnal biomass production based on PAR in the duration between planting and harvesting:

$$Y_{agri} = HIA \sum_{i=1}^{N}(c_1 \times BE \times PAR_{tot,i} \times (1 - e^{c_2 LAI_i})\gamma_{reg,i} \tag{2}$$

978-1-6654-6060-6/23 $31.00 © 2023 IEEE

where Y_{agri} is agricultural yield $[t/ha]$, HIA is adjusted harvest index at maturity, i is i-th day from planting, N is total number of days from planting to harvesting, c_1 is a coefficient $= 0.001$, BE is biomass energy ratio $[(kg/ha)/(MJ/m^2)]$, $PAR_{tot,i}$ is total daily PAR $[MJ/m^2]$, c_2 is a coefficient $= -0.65$, LAI_i is daily leaf area index $[m^2/m^2]$, and $\gamma_{reg,i}$ is daily crop growth regulating factor.

We used the same equation for calculating produced biomass for PAR per minute. PAR was obtained at the light saturation point (the knee of the curve in Fig. 1) for the Net Photosynthetic Rate per minute obtained in Section II-D above. In this equation, daily crop growth regulating factor needs to be determined. We obtained its relative value based on the result from Section II-D above.

III. RESULTS AND MAIN FINDINGS

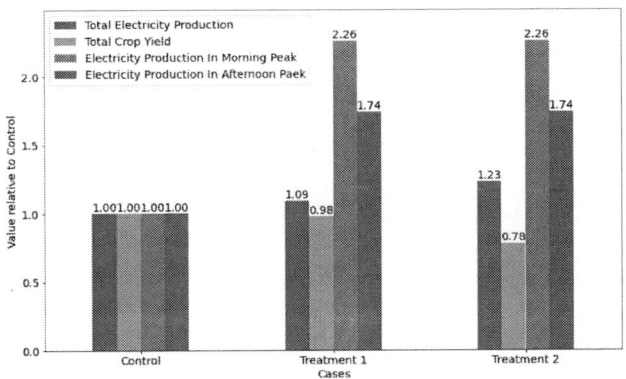

Fig. 2. Preliminary Result: Electricity Production and Crop Yield (Biomass) per day (Relative values to Control)

Fig. 2 shows the preliminary results - electricity production and crop yield per day relative to those values in the control case. The control case (Control) is the agrivoltaic configuration with no shifting of solar irradiation over crops from peak energy consumption period to midday. The first case with shifting of solar irradiation over crops (Treatment 1) is the result of applying only Step 1 in the Section II-D. The second case with shifting of solar irradiation over crops (Treatment 2) is the result of applying both Step 1 and Step 2 in Section II-D.

Comparison between Control and Treatment 1 / Treatment 2 serves for evaluation of Hypothesis 1) described in Section I-C.

Comparison between Treatment 1 and Treatment 2 serves for evaluation of Hypothesis 2) about the validity of Bayesian approach.

Comparison between Control and Treatment 1 indicates that there is no difference in total electricity production and total crop yield, while electricity production during peak energy consumption periods increases significantly, roughly doubling. Comparison between Control and Treatment 2 indicates that total electricity production increases by about 20% and total crop yield decreases by about 20%, while electricity production during peak energy consumption periods increases significantly, roughly doubling. While the result meets overall objective, it also implies the need for more study of the validity of Hypothesis 1).

Comparison between Treatment 1 and Treatment 2 indicates that there is a difference in total electricity production and total crop yield between them. Assumption that prior and likelihood of Bayesian analysis in Treatment 2 follow Normal Distribution needs to be examined. This flexibility of applying various probability distribution to describe uncertainty implies the validity of Hypothesis 2).

IV. DISCUSSION

Since the approach proposed in this study is new and unique encompassing several disciplines, we need to accommodate an ability to improve the model as we find more data and findings. Hence, Bayesian statistics where posterior is updated based on prior is feasible. The modeling performed in this study serves for planning agrivoltaic field experiments to find a way to maximize the electricity production during the peak electricity consumption periods while maintaining sufficient amount of crop yield. The modeling software developed for this study is made available as Open Source [41] [42].

V. SUMMARY

Maximizing crop yield in agrivoltaic setting doesn't have to conflict with solar PV's meeting the electricity consumption pattern. In this study, we entertained the point of view that the factors that necessitate full solar irradiation over crops during peak electricity consumption period, such as Light Saturation Point and Midday Depression of Photosynthesis, can be controlled by covering crops by solar PV modules, in a way of reducing light intensity over crops and keeping air moist around crops. This makes shifting solar irradiation over crops to midday possible, thus, allowing full solar irradiation over solar PV modules at peak electricity consumption period.

Discovering the agrivoltaic configurations to realize this presented a challenge, due to the difficulty in finding relevant literature. We performed extensive literature review and incorporated the findings and the data from the literature to establish models and the way to evaluate the models.

This study presented here serves as a good starting point for more elaborate specific future studies, in supply side management with agrivoltaics.

REFERENCES

[1] D.-Q. Xu and Y.-K. Shen, "External and Internal Factors Responsible for Midday Depression of Photosynthesis," Handbook of photosynthesis, 2nd Ed. CRC Press, Boca Raton, FL, USA, 2005, pp. 287-294.

[2] M. A. A. Mamun, P. Dargusch, D. Wadley, N. A. Zulkarnain, and A. A. Aziz, "A review of research on agrivoltaic systems," Renewable and Sustainable Energy Reviews, vol. 161, p. 112351, 2022, doi: https://doi.org/10.1016/j.rser.2022.112351.

[3] V. Adomavicius,"Review of results of agro-photovoltaic system implementation in agriculture," Proceedings of International Conference on Engineering for Rural Development (ERD), June 2021, pp. 605-610, doi: 10.22616/ERDev.2021.20.TF130.

[4] I. Khele and M. Szabo, "Microclimatic and Energetic Feasibility of Agrivoltaic Systems: State of the Art," Hungarian Agricultural Engineering, No 40/2021 pp. 102-115, Dec. 2021, doi: 10.17676/HAE.2021.40.102

[5] N. C. Giri, R. C. Mohanty, and S. P. Mishra, "Smart Shift from Photovoltaic to Agrivoltaic System for Land-Use Footprint," Ambient science, vol. 8(2), pp. 12–18, Jun. 2021, doi: 10.21276/ambi.2021.08.2.rv01.

[6] H. Wang, H. J. Williams, X. Bu, and K. Max Zhang, "A Combined Shading and Radiation Simulation Tool for Defining Agrivoltaic Systems," in 2022 IEEE 49th Photovoltaics Specialists Conference (PVSC), Jun. 2022, pp. 0384–0386. doi: 10.1109/PVSC48317.2022.9938795.

[7] TI. Zohdi, "A digital-twin and machine-learning framework for the design of multiobjective agrophotovoltaic solar farms," Computational Mechanics, Aug. 2021 68(2), pp. 357-70.

[8] E. K. Grubbs, H. Imran, R. Agrawal, and P. A. Bermel, "Coproduction of solar energy on maize farms — experimental validation of recent experiments," in 2020 47th IEEE Photovoltaic Specialists Conference (PVSC), Aug. 2020, pp. 2071–2075. doi: 10.1109/PVSC45281.2020.9300459.

[9] O. Kwon, J. Kang, M. Trommsdorff, and K. Lee, "Sensitivity Analysis for Optimized Agrivoltaic Designs: An Inquiry on the Trade-off Between Homogenous Light Conditions and Electrical Yield," in 2020 47th IEEE Photovoltaic Specialists Conference (PVSC), Aug. 2020, pp. 1220–1225. doi: 10.1109/PVSC45281.2020.9300677.

[10] M. H. Riaz, H. Imran, and N. Z. Butt, "Optimization of PV Array Density for Fixed Tilt Bifacial Solar Panels for Efficient Agrivoltaic Systems," in 2020 47th IEEE Photovoltaic Specialists Conference (PVSC), Aug. 2020, pp. 1349–1352. doi: 10.1109/PVSC45281.2020.9300670.

[11] H. Imran, M. H. Riaz, and N. Z. Butt, "Optimization of Single-Axis Tracking of Photovoltaic Modules for Agrivoltaic Systems," in 2020 47th IEEE Photovoltaic Specialists Conference (PVSC), Aug. 2020, pp. 1353–1356. doi: 10.1109/PVSC45281.2020.9300682.

[12] A. Perna, E. K. Grubbs, R. Agrawal, and P. Bermel, "Design Considerations for Agrophotovoltaic Systems: Maintaining PV Area with Increased Crop Yield," in 2019 IEEE 46th Photovoltaic Specialists Conference (PVSC), Jun. 2019, pp. 0668–0672. doi: 10.1109/PVSC40753.2019.8981324.

[13] P. M. Jansson, M. G. Newberry, and S. M. Myers, "Agrivoltaics Using Bi-Facial PVs for Permaculture in Utility-Scale Projects," in 2022 IEEE 49th Photovoltaics Specialists Conference (PVSC), Jun. 2022, pp. 0330–0332. doi: 10.1109/PVSC48317.2022.9938476.

[14] C. B. Honsberg, R. Sampson, R. Kostuk, G. Barron-Gafford, S. Bowden and S. Goodnick, "Agrivoltaic Modules Co-Designed for Electrical and Crop Productivity," 2021 IEEE 48th Photovoltaic Specialists Conference (PVSC), 2021, pp. 2163-2166, doi: 10.1109/PVSC43889.2021.9519011.

[15] E. P. Thompson, E. L. Bombelli, S. Shubham, H. Watson, A. Everard, V. D'Ardes, A. Schievano, S. Bocchi, N. Zand, C. J. Howe, P. Bombelli, "Tinted Semi-Transparent Solar Panels Allow Concurrent Production of Crops and Electricity on the Same Cropland," Advanced Energy Materials, vol. 10, no. 35, p. 2001189, Sep. 2020, doi: 10.1002/aenm.202001189.

[16] G. Roccaforte, "Eclipse: A new photovoltaic panel designed for greenhouses and croplands," AIP Conference Proceedings, vol. 2361, no. 1. p. 070002, 2021.

[17] R. A. Gonocruz, "Modeling of large scale integration of agrivoltaic systems: Impact on the Japanese power grid," Journal of Cleaner Production, vol. 363, Jun. 2022, doi: 10.1016/j.jclepro.2022.132545.

[18] P. E. Campana, B. Stridh, S. Amaducci, and M. Colauzzi, "Optimisation of vertically mounted agrivoltaic systems," Journal of Cleaner Production, vol. 325, p. 129091, Nov. 2021, doi: 10.1016/j.jclepro.2021.129091.

[19] B. Uyttkx, B. Uytterhaegen, B. Ronsijn, B. Herteleer, and J. Cappelle, "A standardized classification and performance indicators of agrivoltaic systems.," Oct. 2020. doi: 10.4229/EUPVSEC20202020-6CV.2.47.

[20] M. H. Riaz, H. Imran, H. Alam, M. A. Alam and N. Z. Butt, "Crop-Specific Optimization of Bifacial PV Arrays for Agrivoltaic Food-Energy Production: The Light-Productivity-Factor Approach," IEEE Journal of Photovoltaics, vol. 12, no. 2, pp. 572-580, March 2022, doi: 10.1109/JPHOTOV.2021.3136158.

[21] M. H. Riaz, H. Imran, R. Younas, M. A. Alam and N. Z. Butt, "Module Technology for Agrivoltaics: Vertical Bifacial Versus Tilted Monofacial Farms," in IEEE Journal of Photovoltaics, vol. 11, no. 2, pp. 469-477, March 2021, doi: 10.1109/JPHOTOV.2020.3048225.

[22] N. F. Othman et al., "Modeling of Stochastic Temperature and Heat Stress Directly Underneath Agrivoltaic Conditions with Orthosiphon Stamineus Crop Cultivation," Agronomy, vol. 10, no. 10, p. 1472, Sep. 2020, doi: 10.3390/agronomy10101472.

[23] T. Sekiyama and A. Nagashima, "Solar Sharing for Both Food and Clean Energy Production: Performance of Agrivoltaic Systems for Corn, A Typical Shade-Intolerant Crop," Environments, vol. 6, p. 65, Jun. 2019, doi: 10.3390/environments6060065.

[24] H-W, Heldt, "Plant Biochemistry," Elsevier Academic Press, Burlington, MA, USA, 2005.

[25] D. Wang, Y. Sun, Y. Lin, and Y. Gao, "Analysis of Light Environment Under Solar Panels and Crop Layout," in 2017 IEEE 44th Photovoltaic Specialist Conference (PVSC), 2017, pp. 2048–2053. doi: 10.1109/PVSC.2017.8521475.

[26] D. Wang, Y. Zhang, and Y. Sun, "A Criterion of Crop Selection Based on the Novel Concept of an Agrivoltaic Unit and M-matrix for Agrivoltaic Systems," in 2018 IEEE 7th World Conference on Photovoltaic Energy Conversion (WCPEC) (A Joint Conference of 45th IEEE PVSC, 28th PVSEC & 34th EU PVSEC), Jun. 2018, pp. 1491–1496. doi: 10.1109/PVSC.2018.8547609.

[27] A. Morales, E. Kaiser, "Photosynthetic Acclimation to Fluctuating Irradiance in Plants," Frontiers in Plant Science, mar. 2020, vol. 11, article 268, pp. 357-70, doi 10.3389/fpls.2020.00268.

[28] M. Durand, E. H. Murchie, A. V. Lindfors, O. Urban, P. J. Aphalo, T. M. Robson, "Diffuse solar radiation and canopy photosynthesis in a changing environment," Agricultural and Forest Meteorology, Oct. 2021, vol. 311, doi: 10.1016/j.agrformet.2021.108684.

[29] R. Retkute, S. E. Smith-Unna, R. W. Smith, A. J. Burgess, O. E. Jensen, G. N. Johnson, S. P. Preston, and E. H. Murchie, "Exploiting heterogeneous environments: does photosynthetic acclimation optimize carbon gain in fluctuating light?," Journal of Experimental Botany, May 2015, vol. 66, no. 9, pp. 2437–2447, doi: 10.1093/jxb/erv055.

[30] R. Retkute, A. J. Townsend, E. H. Murchie, O. E. Jensen, S. P. Preston, "Three-dimensional plant architecture and sunlit–shaded patterns: a stochastic model of light dynamics in canopies," Annals of Botany, 2018, vol. 122, pp. 291–302, doi:10.1093/aob/mcy067.

[31] X. Yao, H. Zhou, Q. Zhu, C. Li, H. Zhang, J-J. Wu, and F. Xie, "Photosynthetic Response of Soybean Leaf to Wide Light-Fluctuation in Maize-Soybean Intercropping System," Frontiers in Plant Science, vol. 8, Sep. 2017.

[32] B. O. CARON, J. SGARBOSSA, F. SCHWERZ, E. F. ELLI, E. ELOY, and A. BEHLING, "Dynamics of solar radiation and soybean yield in agroforestry systems," Annals of the Brazilian Academy of Sciences, vol. 90, no. 04, pp. 3799–3812, Oct. 2018, doi: 10.1590/0001-3765201820180282.

[33] R. K. S. V. Kumar, S. Kumar, and B. L. Choudhary, "High Light Stress Response and Tolerance Mechanism in Plant," Interdisciplinary journal of Contemporary Research, vol. 4, no. 1, 2017.

[34] R. W. Langhans and T. W. Tibbits, "Plant Growth Chamber Handbook - Chapter 1 Radiation," in Plant Growth Chamber Handbook, Iowa State University, 1997.

[35] M. Reis and A. Ribeiro, "Conversion factors and general equations applied in agricultural and forest meteorology." Jul. 27, 2019.

[36] Institute of Agriculture and Natural Resources, University of Nebraska–Lincoln, "Cropwatch - Bioenergy Soybeans." [Online]. Available: https://cropwatch.unl.edu/bioenergy/soybeans

[37] Washington State University, "Crop harvest." [Online]. Available: http://modeling.bsyse.wsu.edu/CS_Suite_4/CropSyst/crop_editor/manual/harvest.htm

[38] T. Setiyono, A. Weiss, J. Specht, K. Cassman, and A. Dobermann, "Leaf area index simulation in soybean grown under near-optimal conditions," Field Crops Research, vol. 108, pp. 82–92, Jul. 2008, doi: 10.1016/j.fcr.2008.03.005.

[39] Y. Zhang, Y.-Y. Hu, H.-H. Luo, W. Chow, and W.-F. Zhang, "Two distinct strategies of cotton and soybean differing in leaf movement to perform photosynthesis under drought in the field," Functional Plant Biology, vol. 38, pp. 567–575, Aug. 2011, doi: 10.1071/FP11065.

[40] W. F. Holmgren, C. W. Hansen, and M. A. Mikofski, "pvlib python: A python package for modeling solar energy systems," Journal of Open Source Software, vol. 3, no. 29, p. 884, 2018, doi: 10.21105/joss.00884.

[41] T. Takahashi, "Agrivoltaics Supply Side Management Python source code." GitHub, 2022. [Online]. Available: https://github.com/tadatoshi/agrivoltaics_supply_side_management

[42] T. Takahashi, "Agrivoltaics Supply Side Management Python package." PyPI, 2022. [Online]. Available: https://pypi.org/project/agrivoltaics-supply-side-management/

Effect of Solar Mounting Configurations on California Zero-Carbon Grid

Zabir Mahmud and Sarah Kurtz

University of California Merced, Merced, CA, 95343, United States

Abstract— **Using solar mounting configurations that result in various daily and seasonal solar generation profiles might be impactful in designing a future renewables-driven grid. We study the effects of PV-panel tilt and tracking on California zero-carbon grid modeling and compare results for multiple target years between 2030 and 2045 and for a range of cost assumptions. Using south-facing tilt in the candidate solar projects significantly reduces the annual solar curtailment in 2045. Tilted panels can decrease the total system (grid) cost and average annual cost of delivered solar electricity over time if technological advancement can lower the cost for tilting.**

Keywords—solar, tilting, tracking, curtailment, electricity cost.

I. Introduction

In response to rising global concerns regarding climate change and the escalating frequency of extreme weather events, governments worldwide are accelerating the transition to renewable electricity. California has emerged as one of the leaders in this transition, setting a target to achieve a zero-carbon grid by 2045 [1], albeit without a defined pathway. Given the economic and environmental advantages offered by solar energy, California boasts the largest solar installed capacity in the United States [2] and has long been a champion of solar power. In 2022, solar energy accounted for over 25% of California's electricity generation [2]. However, to meet its energy and climate targets, the state must expand its solar power capacity.

In a grid with high solar penetration, curtailment poses a significant challenge. In a renewables-driven grid like California, seasonal imbalances between load and generation can still occur even with diurnal storage in the grid. To mitigate this, it is crucial to align generation profiles with the demand curve, ensuring a better match between supply and consumption. Here, the mounting configuration of solar panels could play a vital role in minimizing seasonal imbalances and optimizing energy production.

Single-axis tracking is the predominant mounting type for solar panels in the United States. As of 2021, about 90% of newly installed utility-scale photovoltaic (PV) capacity utilized single-axis tracking systems, while the remaining 10% employed fixed-tilt mounting [3]. When conducting capacity expansion planning for a future grid, there is an opportunity to select the desired solar panel mounting configuration. For solar, the commonly chosen configuration today is single-axis tracking with no-tilt mounting setup.

The choice of mounting configurations can shift based on various factors, including location, cost of installation, project specific details, electricity demand, and storage usage in the grid. In the case of California, where renewable generation and load experience seasonal mismatches, the significance of incorporating more winter-dominant generators into the grid becomes apparent [4]. South-facing tilted solar panels can extract maximum energy during the winter months, highlighting the role they play in a renewables-driven grid within the state. Recent trends in California's solar additions have shown the prevailing prominence of rooftop solar installations. From 2017 to 2022, solar installed on rooftops in California have doubled while utility-scale photovoltaic capacity has increased by approximately 70% [5]. When it comes to rooftop solar, the fixed tilt mounting configuration is a preferred option. Understanding the unique implications associated with various solar mounting configurations, we realize the importance of conducting a comparative analysis to assess their respective impacts on a future California zero-carbon grid.

Breyer et al. [6] conducted a study examining the influence of fixed-tilt and single-axis tracking mounting configurations in a 2030 global 100% renewable energy scenario and concluded that the advantages of single-axis tracking PV systems are more pronounced in regions with lower latitudes and a higher proportion of PV generation in the optimal energy mix. However, the analysis may yield different outcomes when applied to regions with mid or high latitudes. Jones et al. [7] proposed that south-facing tilted panels in California's future grid would reduce seasonal imbalances, but they did not create a full grid model to quantify the value.

In this paper, we start by plotting the variations in solar generation profiles resulting from three solar panel mounting configurations. We then implement these profiles in modeling a renewables-driven grid specific to California, with a range of cost assumptions for each orientation. Our analysis focuses on understanding the impact on the total system (grid) cost, annual solar curtailment, and average annual cost of delivered solar electricity across multiple target years.

II. Data and Methodology

A. Solar Profiles

We created solar generation profiles by using the System Advisor Model API [8], which employs the PVWATTS v7 model. These profiles were calculated for seven specific locations within California, with each profile weighted according to the selected AC capacity of the corresponding solar project. We used default input parameters for our analysis, setting the tilt to 0 for no-tilt panels and tilt equal to the latitude facing south for tilted panels.

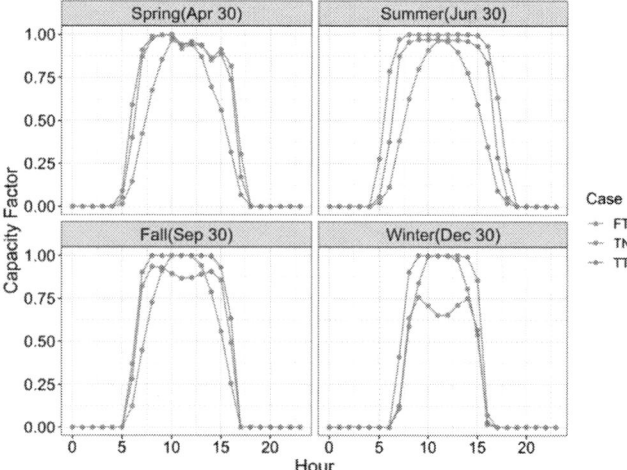

Fig. 1. Hourly solar capacity factor calculated for solar installations in Tehachapi, California for four sunny days in 2007 *using three solar panel mounting configurations.* FT = Fixed Tilt, TNT = Tracked No Tilt, TT = Tracked Tilt.

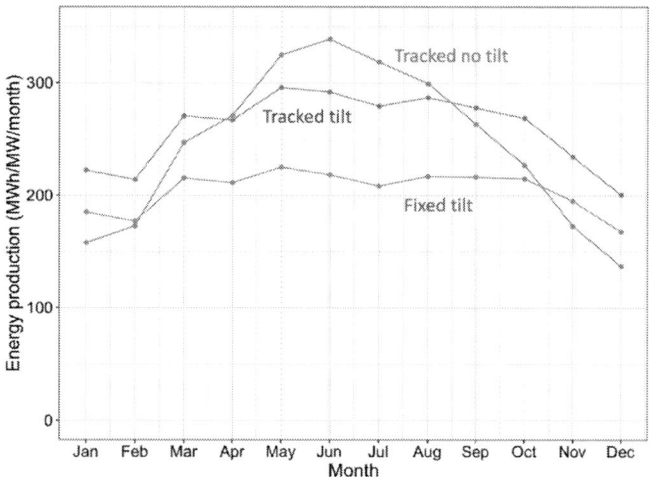

Fig. 2. Monthly electricity-generation profiles for same calculations as Fig. 1.

Our capacity expansion modeling incorporated three different mounting configurations: 1) tilt equal to the latitude without any tracking (referred to as FT - Fixed Tilt), 2) tilt set to 0 with 1-axis tracking (TNT - Tracked No Tilt), and 3) tilt equal to the latitude with 1-axis tracking (TT - Tracked Tilt). These configurations demonstrated both daily and seasonal variations in generation, as illustrated in Figs. 1 and 2, respectively. We selected four days from each season (fall, spring, summer, and winter) for the daily solar profiles depicted in Fig. 1. Our analysis indicated that the utilization of trackers can lead to an increase in daily solar insolation by up to 35%, with an annual increase of around 25%. Notably, during summer, panels without tilt exhibited higher daily generation, while both south-facing tilted configurations performed better during winter. This can be attributed to the positioning of the winter sun in the southern sky, allowing south-facing tilted panels to extract the maximum solar irradiation during midday in fall or winter, which panels without tilt cannot achieve.

The key difference between south-facing tilt and no-tilt solar panels becomes clearly visible when considering the seasonal generation, as depicted in Fig. 2. Without tilting the panels, the winter generation is comparatively lower. On the other hand, by tilting the panels to face south and including trackers (TT), a higher amount of annual solar electricity can be harnessed compared to the tracked no tilt (TNT) mounting configuration. For instance, in the Tehachapi region, the tracked tilted (TT) solar panels exhibit a 36% increase in generation during winter but a 10% decrease in summer resulting in a 6.5% increase in annual generation compared to the tracked no tilt (TNT).

B. Cost Assumptions

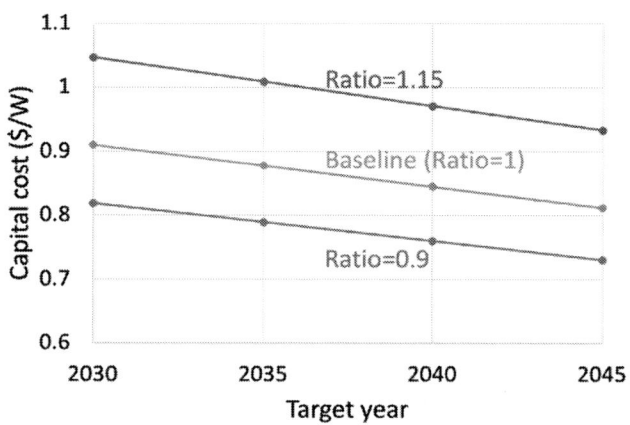

Fig. 3. Capital cost assumption for installed systems as a function of year for three scenarios. Green line shows the baseline scenario. As examples, red and blue lines represent cost assumptions for cost ratios of 0.9 and 1.15, respectively.

The capital cost for the tracked no tilt (TNT) solar mounting configuration was taken based on California's Preferred System Portfolio (PSP) [9], which we assumed as our baseline scenario shown in Fig. 3. We considered the relative cost of fixed tilt (FT) and tracked tilt (TT) systems based on [7]. In their study, Jones et al. [7] provided a detailed breakdown of total PV costs at the utility-scale and found that the capital expenditure (capex) cost of fixed tilt (FT) solar panels can be approximately 7% lower, while the capex cost of tracked tilt (TT) panels can be around 6% higher compared to the tracked no tilt (TNT) configuration.

TABLE I.
RELATIVE ANNUALIZED COST ASSUMPTIONS

Mounting configuration	Relative annualized cost (capex + O&M)
1-axis tracked no-tilt (TNT)	0.9-1.1 (baseline = 1)
1-axis tracked latitude-tilt (TT)	0.95-1.15
Fixed latitude-tilt (FT)	0.85-1.05

Considering that the cost of tilted panels is subject to technological advancements as well as project-specific

978-1-6654-6060-6/23 $31.00 © 2023 IEEE

considerations, we assumed a range of relative annualized new capacity costs as presented in Table 1 and analyzed the sensitivity of the results to these ranges of cost. Additionally, we introduced a 10% variation in the cost of the tracked no tilt (TNT) orientation to assess the impact of different cost assumptions on the outputs, while keeping our baseline scenario as tracked no tilt (TNT) with a cost ratio of 1.

C. Capacity Expansion Modeling

In our analysis, we used RESOLVE [10] as the capacity expansion modeling tool. We optimized total system (grid) cost by identifying the optimal resource build while constraining carbon emissions. To conduct the capacity expansion modeling, we followed the critical timesteps (CTS) technique as explained by Farzan et al. [11]. This approach introduced minimal errors specifically for grids dominated by solar generation. We used the weather data from the year 2007 to generate the input solar profiles and primarily focused on the outputs for the target years of 2030, 2035, 2040, and 2045.

For load profiles, we used the PSP [9], while for electric vehicle (EV) loads, we adopted the 2020 PATHWAYS High Electrification scenario. We set a statewide carbon emission cap of 38MMT by 2030. Our model allowed new lithium batteries but no new pumped hydro to be built. Existing pumped hydro storage was included in the model. We assumed a round-trip efficiency of 85% for batteries, and the annualized new capacity cost of batteries was taken from the PSP [9].

III. RESULTS AND DISCUSSION

A. Annual Solar Curtailment

Solar will be the primary resource in a renewables-driven grid in California as the state has abundant solar. With high levels of solar penetration, there is a risk of increased annual solar curtailment. Therefore, it becomes beneficial to prioritize minimizing curtailment levels, as higher curtailment reduces the overall value of solar generation. The seasonal fluctuations in generation among the three solar mounting configurations have a notable influence on curtailment levels, as illustrated in Fig. 4.

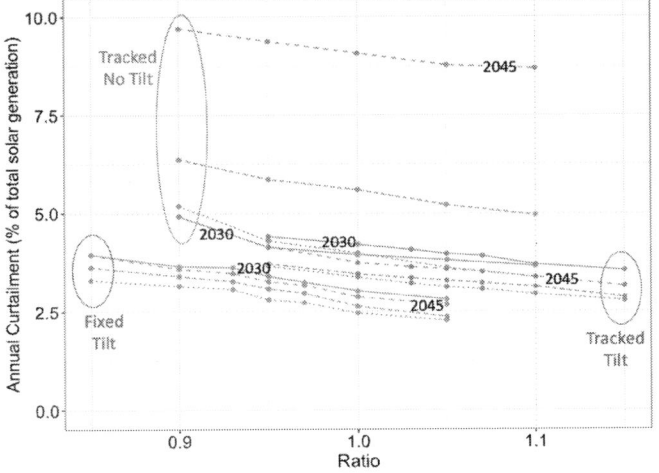

Fig. 4. Annual solar curtailment as a fraction of total solar generation for three solar mounting configurations and four target years

If we consider the no-tilt mounting, the impact is not significant in the initial years. In 2030, the curtailment remains within the range of 3-5% for all mounting types, indicating that south-facing tilt does not provide substantial benefits at that stage. However, when we look at the year 2045, the importance of tilting becomes evident, as solar panels without tilt contribute to a higher level of curtailment, likely due to seasonal variations in generation. In 2045, the curtailment for the tracked no tilt (TNT) mounting configuration reaches 9-10% of the total solar generation, while it stays around 3-4% for the tilted (FT and TT) panels. Given that the California renewables-driven grid may require more generation during winter months, the increased winter-time generation from south-facing tilt panels could prove valuable. This motivates us to consider the adoption of south-facing tilt in solar panels across the state. Minimizing curtailment can stimulate investments in tilting and promote a more reliable and resilient grid infrastructure.

B. Total System Cost

The impact of mounting configurations on the total system (grid) cost can vary and potentially depends on the cost for each specific mounting option. Fig. 5 compares the effect of various cost hypotheses for three solar panel mounting configurations on the grid's total system cost over multiple target years.

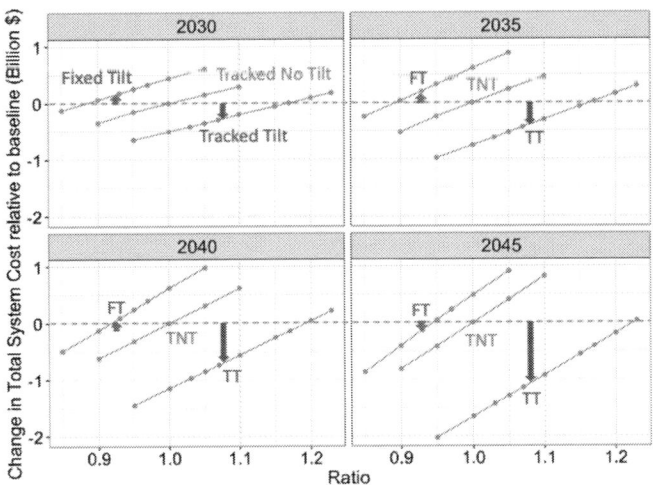

Fig. 5. Change in total optimized cost compared to the baseline scenario as a function of cost assumption for three mounting configurations. Positive y-value represents an increase while negative y-value means decrease in total cost comparing to the tracked no tilt (TNT) scenario with cost ratio=1 (baseline). For specific cost assumptions taken from [7], red and blue arrows show the comparison for fixed tilt (FT) and tracked tilt (TT) respectively.

Due to the grid's heavy reliance on solar energy, the cost of the entire system rises as the price of various mounting configurations increases. The comparison for multiple target years is shown using the red and blue arrows. In comparison with the baseline, if we look at 2030, we can see the red arrow pointing upward suggesting the total system (grid) cost for the fixed tilt (FT) is greater while it is less for the tracked tilt (TT) as indicated by the blue arrow. As we observe for different target years, the reduction in cost of the overall system (grid) achieved by using tracked tilted (TT) panels compared to the baseline scenario increases over time. The decision to use south-facing tilted panels depends on whether the reduction in total system (grid) cost justifies the additional expenses associated with tilting. Although adding tilt to the panels may not lower the overall grid cost today, if the tilting cost can be reduced, it may result in lower system costs in the future.

The added cost of tilting the panels can be brought down through innovation and technological advancement. If it is possible to make the cost of tilted mounting configurations less expensive than we had anticipated, the total system (grid) cost might even be lower. However, the opposite is also possible, and in that circumstance, we might forfeit the benefit of using tilt.

C. Average Annual Cost of Delivered Solar Electricity

The total system costs take into account the perspective of the entire grid. However, it is important to consider the perspective of the solar plant owner who may face the potential challenge of curtailment, as demonstrated in our findings. Fig. 6 shows the average annual cost of delivered solar electricity (the solar electricity that was not curtailed) for the three mounting configurations using their cost assumptions across multiple target years.

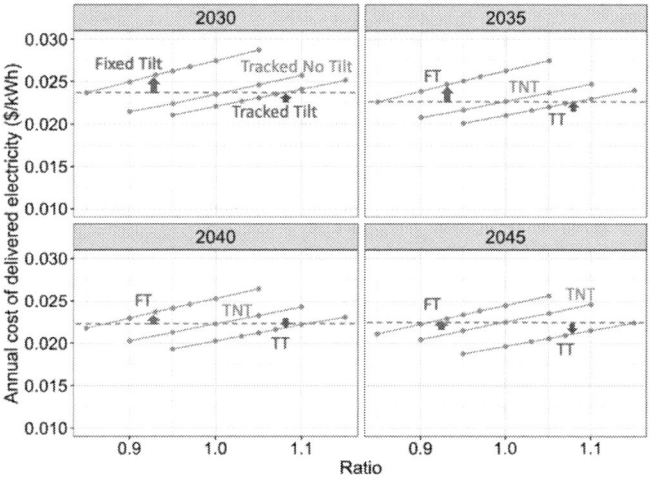

Fig. 6. Effect of three solar mounting configurations as a function of their assumed costs on the average annual cost of delivered solar electricity. For specific cost assumption taken from [7], red and blue arrows show the gradual change in electricity cost over multiple target years for fixed tilt (FT) and tracked tilt (TT), respectively.

Based on our input assumptions of cost reduction for solar installations over time, as shown in Fig. 3, the average annual cost of delivered solar electricity decreases for all mounting configurations from 2030 to 2045. When comparing the 2045 solar electricity cost in Fig. 6, we observe that fixed tilt (FT), with relative annualized new capacity cost 5-10% lower than the baseline assumption, results in a comparable price to the baseline case. On the other hand, assuming a cost increase of 10-15% for tracked tilt (TT) mounting configuration compared to baseline cost could potentially offer the lowest electricity cost by 2045. It's important to note that these effects may vary if the expenses associated with tilting fall outside the range we have assumed.

IV. CONCLUSION

Of the three configurations studied, the tracked, no-tilt mounting configuration experiences greater annual curtailment due to the mismatch between generation and load. Introducing south-facing tilt helps mitigate this mismatch and reduces curtailment to a lower level. While the cost of solar electricity is expected to decrease in the future regardless of the mounting configuration, implementing south-facing tilt may reduce it further. Tilting can also contribute to lowering the overall system cost by reducing the investment needed in solar and storage infrastructure. Today, horizontal single-axis tracking systems are considered optimal. However, in a California zero-carbon grid by 2045, the optimal configuration may shift towards a south-facing tilt. The specifics will depend on advancements and innovations aimed at reducing the costs associated with tilted systems.

ACKNOWLEDGMENTS

We thank R. Go for support with RESOLVE. This work was supported by the California Energy Commission (No. EPC-19-060). This document was prepared as a result of work sponsored by the California Energy Commission. It does not necessarily represent the views of the Energy Commission, its employees, or the State of California. The Energy Commission, the State of California, its employees, contractors, and subcontractors make no warranty, express or implied, and assume no legal liability for the information in this document; nor does any party represent that the use of this information will not infringe upon privately owned rights. This report has not been approved or disapproved by the Energy Commission nor has the Energy Commission passed upon the accuracy of the information in this report.

REFERENCES

[1] "The 100 Percent Clean Energy Act of 2018," Senate Bill 100 (SB 100, De León)

[2] Solar Energy Industries Association (SEIA). Solar State By State. Retrieved Jun 1, 2023, from https://www.seia.org/states-map

[3] M. Bolinger, S. Joachim, C. Warner, and R. Dana. Utility-Scale Solar, 2022 Edition: Empirical Trends in Deployment, Technology, Cost, Performance, PPA Pricing, and Value in the United States. Lawrence Berkeley National Lab.(LBNL), Berkeley, CA (United States), September 2022.

[4] U.S. Energy Information Administration (EIA). Form EIA-923 detailed data with previous form data (EIA-906/920). Retrieved Jun 1, 2023, from https://www.eia.gov/electricity/data/eia923/

[5] U.S. Energy Information Administration (EIA). Electricity Data Browser. Retrieved Jun 1, 2023, from https://www.eia.gov/electricity/data/browser/

[6] S. Afanasyeva, D. Bogdanov, and C. Breyer. Relevance of PV with single-axis tracking for energy scenarios. Solar Energy, 173, pp.173-191. October 2018.

[7] R.K. Jones, and S. Kurtz. Optimizing the Configuration of Photovoltaic Plants to Minimize the Need for Storage. IEEE Journal of Photovoltaics, 12(3), pp.860-870. March 2022.

[8] System Advisor Model Version 2020.11.29 (SAM 2020.11.29). National Renewable Energy Laboratory. Golden, CO. Accessed December 27, 2020. sam.nrel.gov

[9] https://files.cpuc.ca.gov/energy/modeling/2021%20PSP%20RESOLVE%20Package.zip

[10] User Manual RESOLVE Capacity Expansion Model User Manual. (2019). www.ethree.com

[11] F. ZareAfifi, Z. Mahmud, and S. Kurtz. Diurnal, physics-based strategy for computationally efficient capacity-expansion optimizations for solar-dominated grids. Energy, p.128206. September 2023

Experimental demonstration of diffused light collimation in free space

Lisanne M. Einhaus, Geert C. Heres, Jelle Westerhof, Shweta Pal, Akshay Kumar, Anne Rikhof, Jian-Yao Zheng, and Rebecca Saive

University of Twente, Enschede, 7522NB, the Netherlands

Abstract — **We report on a photonic system capable of collimating diffuse light in free space. In this system, photons are absorbed by luminophores (lumogen red) embedded in a PMMA waveguide and subsequently emitted with red-shifted wavelength. These photons are allowed to exit the waveguide if their trajectory falls within the emission cone dictated by a spectro-angular notch filter at the front of the device. We experimentally observed an 18% higher photon emission than an ideal Lambertian reflector at an angle normal to the system surface, thereby confirming our previously established analytical model for the collimation of diffused light in free space.**

I. INTRODUCTION

We envision that the surroundings of solar panels can play a much greater role in collecting sunlight. Recently, researchers and power plant developers have started to pay attention to the albedo, i.e., the ground reflection properties, around (bifacial) solar panels [1-6]. It has been seen that the spectral [7, 8] and angular albedo properties strongly determine the solar energy yield. This raises the question if a material can be designed that performs better than natural materials. Such a material would have to collect light from all angles to operate under diffuse and building integrated (no tracking) conditions and at the same time should exhibit ideal angular and spectral emission properties. We postulate that a collimated beam of photons of energy just above the solar cell absorber band gap directed towards the solar panel would perform best at this task. As depicted in Fig. 1, such a material would enable to redirect light otherwise not usable towards solar panels that are spatially separated. Many areas in the built environment are not efficiently usable due to partial shading or difficult geometries. The collimating material can solve this issue by redirecting the incident photons towards solar panels that are located in more suitable locations, thereby enhancing the yield of those panels.

Recently, we theoretically showed [9, 10] that photon collimation can be performed with a type of luminescent solar concentrator (LSC) that allows photons to exit the waveguide under the designed emission cone, whereas photons outside of this emission cone are recycled and randomized within the waveguide to obtain a further chance for escape within the desired cone. A schematic of the operation principle is shown in the inset of Fig. 1: incoming photons are accepted by the nanophotonic coating within the acceptance cone spanning the full upper hemisphere. Luminophores embedded in a transparent polymer matrix down-shift, i.e. Stokes-shift, the energy of the photons. A Lambertian reflector at the bottom of

Fig. 1.   Schematic of an urban scene in which free space diffused light collimators collect sunlight falling on a house façade and then emit a collimated beam of light towards solar panels positioned across the street and covering canals.

the polymer ensures trapping, i.e. photon recycling, and randomization of the photons. The down-shifted photons falling within the escape cone are allowed to exit the structure whereas photons outside of the escape cone will be reflected back into the structure. The Lambertian reflector recycles and randomizes these photons such that they obtain a chance to exit through the escape cone after a second pass through the structure. Note, that for reciprocity reasons, within the narrow emission wavelength window of the luminophore and outside of the acceptance cone, also incident photons will be reflected off the nanophotonic coating. Contrary to conventional luminescent solar concentrators, these photons should not be regarded as loss, but in our free space configuration might still contribute to the irradiance of the solar panel in the same way a conventional specular mirror would if placed in the vicinity of the panel. With an analytical model we previously predicted light collimation in free space and presented these results during last year's PVSC [10].

This year, we present experimental evidence of free space light collimation using a lumogen red doped PMMA waveguide and an aperiodic spectro-angular notch filter. In measurements of the spectro-angular emission of our best system, optimized to emit vertically, we found 18% more photons than an ideal Lambertian would emit within this solid angle. The results are in good agreement with our theoretical predictions and show that the collimation of diffused light in free space is possible with our system. Further system properties and results were recently published in ACS Photonics [11].

II. SPECTRO-ANGULAR EMISSION MEASUREMENTS

To measure the wavelength and angle dependent emission, we built an optical goniometer in which we can independently set the incident angle of the exciting light source and the angle under which a calibrated fiber coupled spectrometer (Avantes CMOS device, with a bandwidth of 200 to 1100 nm, and accuracy of ±5%) measures the emitted intensity and spectrum. Here, we used a green laser (516 nm) for excitation. To prove the collimating property of the device without convoluting specular reflection, we shone the laser from a 40° angle, which means that the specular reflection peak fell outside or just on the edge of the emission cone and did not contribute to the calculation of the concentration factor. Furthermore, reflected and emitted light could also be deconvoluted through their different wavelengths, the reflected light peaking at 516 nm (laser emission) and the emitted light peaking around 650 nm (dye emission). Note, that all results are reported in photon flux as we would like to determine the systems quantum efficiency. Due to the energy down-shifting, the energy efficiency is lower than the quantum efficiency. In other words, if every green photon was converted into one red photon, our quantum efficiency would be 100%. The energy efficiency would be lower as the

photons lose part of their energy in the process which is thermodynamically necessary to allow for collimation and concentration as explained above.

To enable modular experiments in which parameters can easily be changed while keeping others constant, we designed the system such that the Lambertian back reflector, the Lumogen F Red 305 doped PMMA waveguide and the nanophotonic coating were separate samples mechanically pressed together with a clamp. To avoid total internal reflection within the waveguide, we roughened the back surface of the waveguide and we put an index matching fluid (water) in between the waveguide and the nanophotonic coating.

III. EXPERIMENTAL RESULTS

In Fig. 2, we show photographs of our luminescent waveguides in 4 different scenarios. In Fig. 2a the luminescent waveguide is placed in the setup without the nanophotonic notch filter and the photograph is taken almost under a perpendicular angle. In Fig. 2b, a spectro-angular notch filter was placed in front of the waveguide and index matched with a thin layer of water. In Fig. 2c, the configuration of 2a is rotated to take a picture under more grazing incidence and the same holds for Fig. 2d where the configuration of 2b is rotated to obtain a photograph under a shallow angle. From these photographs, it can already be visually seen, that the notch filter restricts photons from exiting under a shallow incidence angle. From the photograph, it is not possible to see, but in Fig. 2b, more photons are emitted in the center as compared to Fig. 2a. This can be better seen in the quantitative results presented in [11].

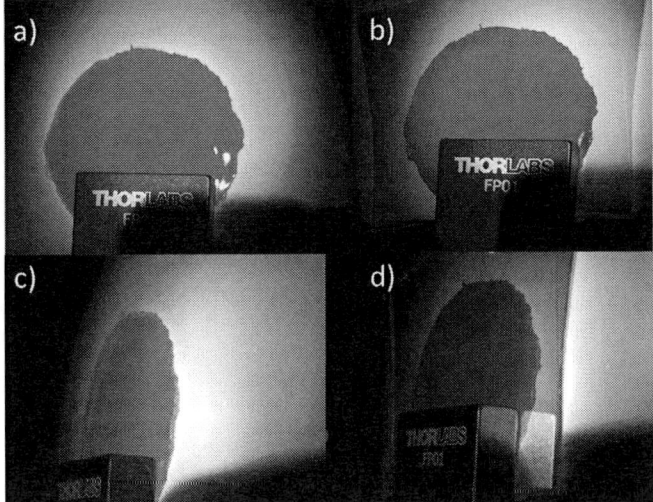

Fig. 2. Photographs of our luminescent waveguides in 4 different scenarios. a) and c) are without, b) and d) with a spectro-angular notch filter. The red disk has a diameter of around 2 inch.

Due to copyright reasons, we are not showing the data in this proceedings papers but we refer to the figures in the respective open access publication [11]. In Fig. 5a several calculated cases of angular photon emission are shown whereas Fig. 5b presents the experimental results and fits. A Lambertian reflector is expected to show cosine behavior and should therefore result in a semicircle when plotted in polar coordinates as can be seen from the black solid line in Fig. 3a. Our measurements (black asterisk in Fig. 5b) confirm the Lambertian nature of the reflector we used and can be fitted with a cosine model (black solid line in Fig. 5b). The luminophore doped waveguide also shows an isotropic emission as seen by the measured data (green asterisk) and fitted curve (green line) in Fig. 5b which would have been intuitively expected also from the photographs in Fig. 2a and 2b. The calculated emission profile for the 40° and 20° emission cone coatings are shown in Fig. 5a as red and blue dashed lines respectively. We considered two distinct cases: an ideal scenario in which all processes have unity efficiency and one scenario in which the parameters were chosen to mimic our case, i.e., 99% luminophore quantum yield, 98% Lambertian back reflector. The resulting concentration factors (C) and system efficiencies (η_sys) as defined above are summarized in Fig. 5c. If no losses were present, we would expect our 40° and 20° emission systems to reach a concentration factor of 1.73 and 1.99 respectively. Including losses, we calculated a concentration factor of 1.28 and 1.30 for the 40° and 20° emission systems respectively. While the loss-free 20° emission system yields a significantly higher concentration, including losses, this benefit almost completely disappears. Figure 3b shows the respective measured data. The 40° emission system surpasses the emission of the Lambertian reflector and yields a concentration factor of 1.18 at 0° emission, thereby experimentally demonstrating the collimation of light in free space. The 20° emission system also shows suppressed emission outside of the emission cone but does not surpass the Lambertian reflector at 0°, the concentration factor is 0.76 and thereby, this system failed to collimate more light within the emission cone than what was sent in. Though, it does have a higher concentration factor than the luminophore doped waveguide (C=0.66).

To shed more light on the reasons behind the difference in performance between simulated and experimental systems, we calculated the system efficiencies for all structures. The system efficiency describes the ratio of outgoing to incoming photons. The number of outgoing photons was obtained by integrating the emission over all wavelengths and over the whole upper hemisphere assuming spherical symmetry. The incoming photon number was calculated by integrating over all wavelengths and angles of the photons reflected by the Lambertian reflector and assuming an efficiency of 98% for the Lambertian. This integration and normalization resulted in 7.1×10^{13} photons/s/cm2 of incoming photons. The efficiencies of all systems are summarized in Fig. 5c.

REFERENCES

[1] Valdivia, C.E., Li, C.T., Russell, A., Haysom, J.E., Li, R., Lekx, D., Sepeher, M.M., Henes, D., Hinzer, K., and P., S.H.: 'Bifacial Photovoltaic Module Energy Yield Calculation and Analysis', IEEE PVSC 2017 conference proceedings, 2017

[2] Russell, A.C., Valdivia, C.E., Bohémier, C., Haysom, J.E., and Hinzer, K.: 'DUET: A Novel Energy Yield Model With 3-D Shading for Bifacial Photovoltaic Systems', IEEE Journal of Photovoltaics, 2022, 12, (6), pp. 1576-1585

[3] Tonita, E.M., Russell, A.C., Valdivia, C.E., and Hinzer, K.: 'Optimal ground coverage ratios for tracked, fixed-tilt, and vertical photovoltaic systems for latitudes up to 75° N', Solar Energy, 2023, 258, pp. 8-15

[4] Tonita, E.M., Valdivia, C.E., Russell, A.C., Martinez-Szewczyk, M., Bertoni, M.I., and Hinzer, K.: 'A general illumination method to predict bifacial photovoltaic system performance', Joule, 2023, 7, (1), pp. 5-12

[5] Ovaitt, S., Brown, M., Deline, C., and Kempe, M.D.: 'Spectral rear irradiance testing and modeling for degradation and performance of solar fields', in Editor (Ed.)^(Eds.): 'Book Spectral rear irradiance testing and modeling for degradation and performance of solar fields' (IEEE, 2022, edn.), pp. 0992-0994

[6] Sun, X., Khan, M.R.H., Amir, Hussain, M.M., and Alam, M.A.: 'The Potential of Bifacial Photovoltaics: A Global Perspective', IEEE PVSC 2017 conference proceedings, 2017

[7] Russell, T., Saive, R., and Atwater, H.A.: 'Thermodynamic Efficiency Limit of Bifacial Solar Cells for Various Spectral Albedos ', in Editor (Ed.)^(Eds.): 'Book Thermodynamic Efficiency Limit of Bifacial Solar Cells for Various Spectral Albedos ' (IEEE, 2017, edn.), pp.

[8] Russell, T.C., Saive, R., Augusto, A., Bowden, S.G., and Atwater, H.A.: 'The Influence of Spectral Albedo on Bifacial Solar Cells: A Theoretical and Experimental Study', IEEE Journal of Photovoltaics, 2017, 7, (6), pp. 1611-1618

[9] Einhaus, L., and Saive, R.: 'Free-space concentration of diffused light for photovoltaics', in Editor (Ed.)^(Eds.): 'Book Free-space concentration of diffused light for photovoltaics' (IEEE, 2020, edn.), pp. 1368-1370

[10] Heres, G.C., Einhaus, L.M., and Saive, R.: 'Analytical Model for the Performance of a Free-Space Luminescent Solar Concentrator', in Editor (Ed.)^(Eds.): 'Book Analytical Model for the Performance of a Free-Space Luminescent Solar Concentrator' (IEEE, 2021, edn.), pp. 1027-1029

[11] Einhaus, L.M., Heres, G.C., Westerhof, J., Pal, S., Kumar, A., Zheng, J.-Y., and Saive, R.: 'Free-Space Diffused Light Collimation and Concentration', ACS photonics, 2023

The Planet-scale Performance Potential of Si-Perovskite Tandem Solar Farms

Jabir Bin Jahangir[1], M. Tahir Patel[1], Reza Asadpour[1], M. Ryyan Khan[2], and M. A. Alam[1]

[1]Electrical and Computer Engineering Department, Purdue University, West Lafayette, IN, USA
[2]Department of Electrical and Electronic Engineering, East West University, Dhaka, Bangladesh

Abstract—Recent developments in monofacial multi-junction Perovskite-Si tandem technology have produced 30% efficient cells under controlled laboratory conditions. A bifacial tandem would further enhance the energy yield potential. However, recent studies have demonstrated that the current matching constraint erases the performance gain of two-terminal (2T) Tandem modules (over single junction HIT cells) due to time-dependent albedo associated with realistic solar farm configurations. Here, in this planet-scale study of the performance potential of tandem solar cells, we show that three or four-terminal modules (3/4T) would achieve the anticipated performance gain (17-23%) despite albedo variation. As such, 3/4T tandem will be the key to realizing/unlocking the next-generation leaps in yield performance and should be the key focus on research/development of next-generation solar modules and solar farms.

Keywords—solar farm, perovskite, Si, tandem, four-terminal, energy yield

I. INTRODUCTION

With commercial single-junction solar cells rapidly reaching their practical limiting performance, $\eta_1 \sim 30\%$ [1], [2] and the industry-wide consensus regarding the transition to bifacial technology, the research efforts have naturally shifted to quantify the remarkable performance potential and economic viability of bifacial tandem solar cells. Indeed, at the thermodynamic limit, an N-junction solar cell with typical albedo, $R \sim 0.3$, reaches an efficiency limit of [3]

$$\frac{\eta_N}{\eta_1(R=0)} = \frac{2(1+R)N}{2+(N-1)(1+R)},$$

so that a 2-junction bifacial tandem promises an astounding efficiency $\eta_2(R=0.3) = \eta_1(R=0) \times 1.57 > 50\%$! In other words, bifacial tandem solar cells could produce 50% more energy compared to single-junction solar cells. This efficiency gain is only possible when cells individually have very high efficiency themselves. That is why there is a broad interest in using perovskite-silicon technology for bifacial tandem [4], [5] modules.

The recent laboratory-scaled developments in 2T tandem solar cell technology have been promising: a *monofacial* Perovskite-Si tandem solar cell with efficiency > 30% was recently reported by EPFL [6]. The efficiency is still significantly lower than the thermodynamic limit for monofacial 2T cells (~42%), due to incomplete absorption and albedo-dependent current matching in a two-terminal (2T) configuration. Further improvement is possible (~33 − 36%) with bifacial modules with moderate albedo, $R_A = 0.3$ [4]. The gain is significant; there is an obvious incentive to develop 2T bifacial solar modules for solar farms worldwide.

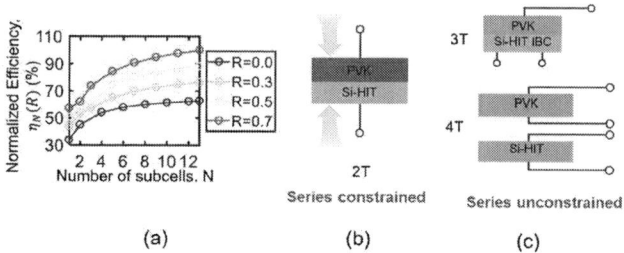

Fig. 1. Multi-junction bifacial cells. (a) Normalized efficiency of multijunction bifacial cells for different albedo (b) Bifacial Perovskite-Si (HIT) two-terminal (2T) tandem cell (c) Bifacial Perovskite-Si (HIT) three-terminal (3T) and four-terminal (4T) tandem cells.

It is well understood that 2T tandems are subject to the current matching constraint. The effect is more pronounced for the bifacial tandems (see. Fig. 1(b)) that also collect irradiance through their rear faces. The bandgaps of the top and bottom cells of a tandem cell are designed for a specific albedo. In practice, however, the rear face collection is determined by the effective albedo (the actual amount of light collected) related to the time-dependent mutual shading between rows of a solar farm. Except for a narrow time window, the effective albedo must necessarily differ from the design albedo. And the corresponding current mismatch would limit the performance benefits of 2T tandem compared to its constituent single-junction HIT in varying albedo conditons. The recent global analysis by Patel et al. [5] found that, compared to constituent bifacial single junction c-Si (HIT) cells, the solar farms employing bifacial 2T tandem modules ($\eta\sim26\%$) show marginally small gains (4-5%) in energy yield at 0.3 R_A, the cell's design albedo. Moreover, some deviation in albedo, e.g., 0.1 R_A, would result in a 2T tandem yielding *less* energy compared to the single junction HIT owing to the current mismatch in the tandem subcells. Moreover, the mismatched energy must be dissipated within the cells; the excess temperature would reduce energy yield and accelerate degradation. Taken together, Patel et al. [6] paint a pessimistic future for the utility-scale deployment of Perovskite-based tandem solar cells. The bifacial 2T tandem would have to achieve high efficiencies with bandgaps tuned for different albedo conditions to realize significant benefits compared to single-junction Si.

Fortunately, the current matching constraint can be obviated by introducing additional terminals (see Fig. 1(c)) in three (3T) or four terminal (4T) configurations. As such, 3/4T tandem, despite their present complexity, will be the key to unlocking the next-generation leaps in yield performance. The goal of this paper is to quantify the planet-scale

978-1-6654-6060-6/23 $31.00 © 2023 IEEE

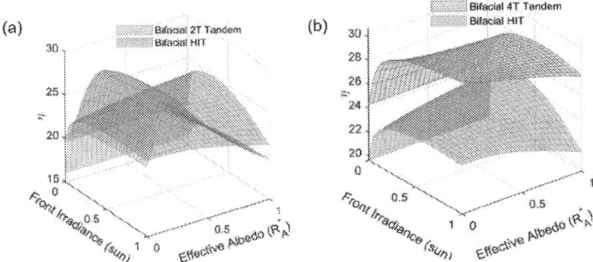

Fig. 2. Comparison of irradiance-efficiency landscapes of bifacial (a) 2T and (b) 3/4T tandem and single junction HIT. Due to current matching in 2T tandem, a greater efficiency than single-junction HIT is obtained only when $R_A^* \sim 0.17$. However, 3/4T is relatively robust to albedo variations.

performance potential of 3/4T perovskite-Si tandem farms. Towards that goal, we developed a detailed physics-based simulation framework for Perovskite-Si tandem cells [5]. The model accounts for the cell-to-farm level characteristics to compute the yearly yield potential of a solar farm. Leveraging the model, the work will address the following questions: (i) *How does variation in albedo condition affect the efficiency landscapes of 2T and 3/4T tandems?* (ii) *How does the gain in energy yield of 3/4T compared to 2T tandem globally?*

II. RESULTS

Using the detailed simulation framework, we have carried out the global scale simulation of solar farms employing the 2T and 3/4T tandem cells. The simulation framework is detailed in Ref. [5]. In these simulations, we consider 2T and 3/4T tandem cells designed to maximize the power output when $R_A = 0.3$. 2T and 3/4T cells have identical material stacks. The only difference is that for the 2T configuration, the system's current is defined by the minimum of the subcell currents, while the current of the 3/4T cells is extracted independently. We have assumed the 3T and 4T output

power to be identical which represents the ideal case. In practice, they do vary owing to practical reasons regarding their fabrication. Nonetheless, the results provide an upper limit of yield performance of these cells. The following sections present the results of the simulations. First, we will use physical simulation of the PVK-Si tandem cells to elucidate how the variation of effective albedo (R_A^*) affects the efficiencies of 2T and 3/4T tandem cells. Then, we will present the results of the worldwide simulation of bifacial solar farms.

A. Irradiance-dependent cell efficiencies

Effective albedo is defined as the ratio of the rear face and front face irradiances. The efficiency (at fixed 25°C) vs. irradiance landscapes of the bifacial 2T and 3/4T cells were determined for varying front and rear face irradiances and shown in Fig. 2(a) and (b). The device opto-electrical simulations were carried out using the method described in Ref. [5]. First, consider the 2T tandem landscape shown in Fig. 2(a). We note that the efficiency of the 2T tandem is only greater in the vicinity of the effective albedo $R_A^* \sim 0.17$, reaching a maximum value of ~30%. Significant variation in albedo, regardless of whether increased or decreased, may result in lower efficiency compared to single junction HIT. The maximum gain in efficiency compared to its constituent high-efficiency HIT cell (24.8%) is ~21%. The figure makes apparent the fundamental challenge of 2T tandem cells in a solar farm setting: bifacial 2T tandems designed for specific albedo conditions, can only be deployed at locations with albedo conditions similar to the design albedo. Modules built for a certain albedo condition may not provide desirable gains at other locations with different natural albedo. Even a natural variation in the albedo condition (e.g., snowfall) could lead to lower performance relative to HIT. By contrast, as shown in Fig. 2(b) the 3/4T

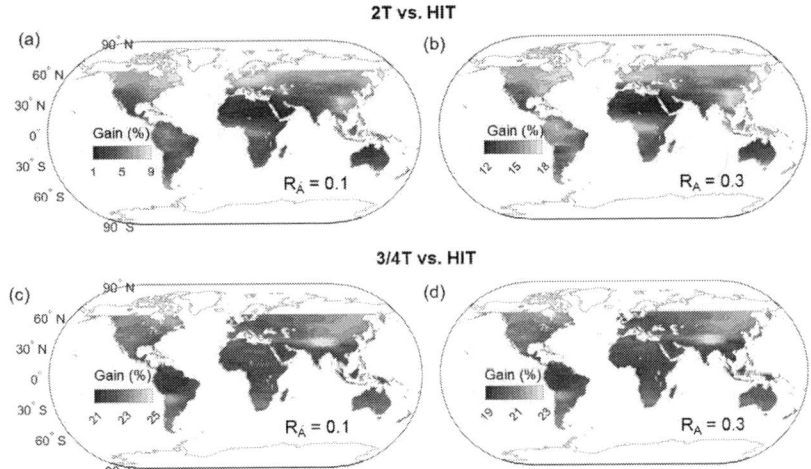

Fig. 3. Worldwide gains in yearly energy yield of single-axis-tracking solar farms employing 2T and 3/4T bifacial Perovskite-Si tandem compared to single-junction HIT. (a) and (b) shows the gains of 2T tandem for albedo 0.1 and 0.3, respectively. (b) and (c) shows the gains of 3/4T tandem for albedo 0.1 and 0.3, respectively. For either albedo condition, 3/4T demonstrates significant gains everywhere.

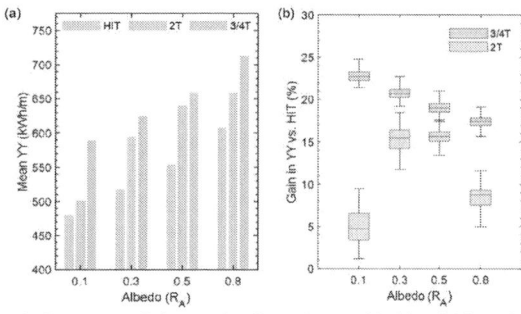

Fig. 4. Summary of the results from the worldwide yield performance simulations. (a) Mean yearly yield (YY) for each technology for different albedo condtions. 3/4T shows the largest yield average irrespective of R_A. (b) The boxplots of the gain distributions across the world for different albedo.

tandem cell is not subject to the current-matching constraint and as a result relatively insensitive to the change in albedo conditions.

B. 3T/4T outperforms its constituent single-junction Si

Worldwide simulation of single-axis-tracking solar farms employing 3/4T tandem farms allows us to paint a detailed picture of their yield potentials. The maps in Fig. 3 compare the gains in yield potential of 2T and 3/4T tandem cells with that of single junction HIT cells at optimal ($R_A = 0.3$) and non-optimal ($R_A = 0.1$) albedo condition. Note that the cells were designed to maximize power output when $R_A = 0.3$. First, consider the 2T tandem gains shown in Fig. 3(b) when $R_A = 0.3$. At this albedo condition, 2T tandem produces a greater gain in energy, ranging between 12-18%, throughout the world as expected from preceding discussion. However, as Fig. 3(a) shows, the gains diminish and range between 1-9% if $R_A = 0.1$. Smaller gains (~1-2%) are observed near the equator at locations receiving predominantly direct insolation. Next, we turn our attention to 3/4T tandem gains shown in Fig. 3(c) and 3(d). We notice that the gains compared HIT are significantly positive for either albedo condition: when $R_A = 0.3$, the gains range between 19-23%, and when $R_A = 0.1$, the gains range between 21-25%. Thus, 3/4T tandem offers a significant advantage in yield potential compared to both single junction HIT and 2T tandem.

Finally, the worldwide results are summarized in Fig. 4. The figure also shows the results when $R_A = 0.5$ and 0.8. Fig. 4(a) shows the mean yearly yield (YY) for the three albedo conditions. Worldwide mean YY increases with increasing albedo for the bifacial system. The boxplots in Fig. 4(b) show that the gains have a relatively narrow distribution for different albedo conditions. The gains of 2T tandem is maximized when the albedo condition match its design albedo. At low (high) albedo the gains decrease as the output power of 2T tandem is limited by current in bottom (top) cell. It is clear that the 3/4T tandem outperforms the 2T counterpart (as well as the single junction bifacial cells) at every location throughout the world.

III. CONCLUSIONS

In this paper, leveraging a detailed and experimentally validated electrical-thermal-optical model of a solar farm, we have analyzed the worldwide yield potentials of tracking solar farms employing 3/4T tandem cells and provided comparisons with 2T tandem and single-junction HIT cells. The key takeaways are:

1. Current-matching constraint in bifacial 2T tandem cells limits performance benefit in practical solar farm configurations if the albedo condition is non-optimal. To achieve any benefit, the bifacial 2T tandem modules must be designed for specific albedo conditions for optimal operation. For a module designed for 0.3 albedo, the gain can be just ~5% gain average compared to HIT cell if the albedo condition is sub-optimal.
2. The current mismatch of 2T solar farms will be dissipated within the module. The corresponding self-heating would decrease yield and reduce lifetime.
3. By avoiding current matching, 3/4T tandems can provide significant performance gains ranging between 19-23%. Additionally, the gains compared to HIT being relatively insensitive to local albedo conditions, 3/4T tandems are more suited for future worldwide deployment in high-performance solar farms.

Deploying 3/4T tandem cells would require us to address the electrical and optical challenges associated with 3/4T tandems. For instance, minimizing optical losses, system integration, etc. Developing 3/4T modules will be expensive, however, the 3/4T farms would provide significant potential economic benefits to offset the cost and reduce LCOE.

REFERENCES

[1] M. A. Alam and M. R. Khan, "Principles of Solar Cells," *Principles of Solar Cells*, Aug. 2022, doi: 10.1142/12139.

[2] M. R. Khan and M. A. Alam, "Thermodynamic limit of bifacial double-junction tandem solar cells," *Appl Phys Lett*, vol. 107, no. 22, 2015, doi: 10.1063/1.4936341.

[3] M. A. Alam and M. Ryyan Khan, "Shockley–Queisser triangle predicts the thermodynamic efficiency limits of arbitrarily complex multijunction bifacial solar cells," *Proc Natl Acad Sci U S A*, vol. 116, no. 48, pp. 23966–23971, 2019, doi: 10.1073/pnas.1910745116.

[4] R. Asadpour, R. V. K. Chavali, M. Ryyan Khan, and M. A. Alam, "Bifacial Si heterojunction-perovskite organic-inorganic tandem to produce highly efficient (ηT* ~ 33%) solar cell," *Appl Phys Lett*, vol. 106, no. 24, p. 243902, Jun. 2015, doi: 10.1063/1.4922375.

[5] M. T. Patel, R. Asadpour, J. bin Jahangir, M. R. Khan, and M. A. Alam, "Current-Matching Erases the Anticipated Performance Gain of Next-Generation Two-Terminal Perovskite-Si Tandem Solar Farms," *Appl Energy*, no. Accepted, 2022.

[6] "Two new world records on perovskite/silicon tandem solar cells - EPFL." https://actu.epfl.ch/news/two-new-world-records-on-perovskitesilicon-tandem-/ (accessed Nov. 07, 2022).

978-1-6654-6060-6/23 $31.00 © 2023 IEEE

Investigating the Potential of Hydrogen Plasma Treated ALD-TiOx Films as Hole-selective Passivating Contacts in Crystalline Silicon Solar Cells

Chien-Hsuan Chen, S. Novia Berriel, Taylor M. Currie, Jannatul Ferdous Mousumi, Titel Jurca, Parag Banerjee, Kristopher O. Davis

1 Department of Materials Science and Engineering, University of Central Florida, Orlando, FL, United States

2 Department of Chemistry, University of Central Florida, Orlando, FL, United States

3 Resilient Intelligent Sustainable Energy Systems Faculty Cluster, University of Central Florida, Orlando, FL, United States

4 Florida Solar Energy Center, University of Central Florida, Orlando, FL, United States

5 CREOL, the College of Optics and Photonics, University of Central Florida, Orlando, FL, United States

The potential of hydrogen plasma treated atomic layer deposited (ALD) titanium oxide (TiOx) films as hole-selective passivating contacts in crystalline silicon (c-Si) solar cells is discussed from three aspects including the passivation quality, carrier selectivity and electrical property. In this work, we also propose an integrated deposition technique through incorporating hydrogen plasma pulses into the thin film deposition. The passivation quality, carrier selectivity and electrical property were studied through photoluminescence (PL) image analysis, Schottky barrier height analysis using dark current-voltage/capacitance-voltage (I-V/C-V) measurements and contact resistivity analysis using transmission line method (TLM). Compared with conventional deposition techniques and post-treatments, high passivation quality, excellent carrier selectivity and low contact resistivity can be achieved through combining our integrated deposition technique with post hydrogen plasma treatment (HPT). Further study will be performed to evaluate the durability and reliability of TiOx contacts in c-Si solar cells. Index Terms-atomic layer deposition, titanium oxide, passivation quality, carrier selectivity, contact resistivity, passivating contact, hole-selective contact, hydrogen plasma

The Photovoltaic Exponential Model

Eduardo I. Ortiz Rivera, *IEEE Senior Member*, eduardo.ortiz7@upr.edu

Abstract—This paper presents the Photovoltaic Exponential Model (PVEM) where the shape, boundary conditions, and performance of the physical PVM are satisfied. The proposed photovoltaic module exponential model is based on the boundary and optimal conditions like open circuit voltage, *Vx*, short circuit current, *Ix*, optimal voltage, *Vop*, optimal current, *Iop*, and maximum power, *Pmax*. Examples to validate the proposed PV Exponential Model are given in the paper using a data sheet for different types of PV Modules.

I. INTRODUCTION

In engineering and sciences, an accurate mathematical modeling for a physical system, object, event, or pattern can determine the behavior and characteristics of the proposed design, saving time, space, money, and materials. Examples of mathematical modeling and simulations are circuit analysis, mechanical systems design, nuclear explosion simulation, power grid simulations, etc. An inaccurate mathematical model can result in serious problems not expected in the system's final design. The performance and behavior of the system can be diminished because of erroneous modeling. One of the most dramatic examples is the Tacoma Narrows Bridge, USA, in 1940, where the natural resonance of the bridge coincided with the frequency of the wind creating a collapse of the bridge, an effect not considered in the original design.

But at the same time, a very complex mathematical model can take time to analyze and is impractical. So a compromise should be taken between the complexity and the number of parameters used to describe a physical system. If the correct assumptions are made, an approximation of the mathematical model that keeps the main properties of the physical system can be obtained. An example is the mathematical model for resistance in circuit analysis, where the temperature effect is neglected on the nominal value for the resistance.

The Underwriters Laboratories has developed a sample of information requirements for photovoltaic modules (www.ul.com/database). In United States, a PV module must comply with the standard UL1703. A PV module datasheet should include the ratings of the short circuit current (*Ix*), open circuit voltage (*Vx*), optimal voltage (*Vop*) and optimal current (*Iop*) of an individual module when operating at maximum power (*Pmax*). This paper presents a photovoltaic module exponential model (PVEM) based on the manufacturer's datasheet. Finally, this paper describes the relationship of the photovoltaic exponential model considering the current, *I*, voltage, *V*, power, *P*, conductance, *G*, or resistance, *R*.

II. PHOTOVOLTAIC MODULE EXPONENTIAL MODEL

The PV Exponential Model (PVEM) takes into consideration the relationship of the current, I, with respect to the voltage, V, effective irradiance level, E_i and temperature, T of operation for the PVM, the characteristic constant for the I-V curves, the short-circuit current and the open-circuit voltage. The relationship of V and I for any photovoltaic module is given in (1) and can be described in terms of the values provided by the manufacturer's data sheet and the standard test conditions. The power is described in (2) and is calculated by multiplying (1) by the voltage, V. Ix is the short circuit current at any given E_i and T, and it can be calculated when the voltage, V is zero as described in (3). Vx is the open circuit voltage at any given E_i and T, and it is the voltage of operation for the PVM when the current, I is zero as described in (4). The range of existence of V will be from 0 to Vx, the range of existence of $I(V)$ will be from 0 to Ix, and for P is from 0 to P_{max} [1].

$$I(V) = \frac{Ix}{1 - \exp\left(-\frac{1}{b}\right)} \cdot \left[1 - \exp\left(\frac{V}{b \cdot Vx} - \frac{1}{b}\right)\right] \qquad (1)$$

$$P(V) = \frac{V \cdot Ix}{1 - \exp\left(-\frac{1}{b}\right)} \cdot \left[1 - \exp\left(\frac{V}{b \cdot Vx} - \frac{1}{b}\right)\right] \qquad (2)$$

$$Ix = I(0) = p \cdot \frac{E_i}{E_{iN}} \cdot \left(Isc + TCi \cdot (T - T_N)\right) \qquad (3)$$

$$Vx = s \cdot \frac{E_{iN}}{E_i} \cdot TCV \cdot (T - T_N) + V_{max} - (V_{max} - V_{min})$$

$$\cdot \exp\left(\frac{E_i}{E_{iN}} \cdot \ln\left(\frac{V_{max} - Voc}{V_{max} - V_{min}}\right)\right) \to I(Vx) = 0 \qquad (4)$$

In order to obtain the maximum power P_{max}, the derivative of P with respect to V as described in (5). Then (5) is set to zero in order to solve for V. The resulting V is called the optimal voltage Vop and is substituted into (1) to obtain the optimal current Iop. Finally the maximum power P_{max} is obtained multiplying Vop by Iop.

$$\frac{\partial P}{\partial V} = \frac{Ix}{1 - \exp\left(-\frac{1}{b}\right)} \cdot \left[1 - \exp\left(\frac{V}{b \cdot Vx} - \frac{1}{b}\right)\right] - \frac{V \cdot Ix}{b \cdot Vx - b \cdot Vx \cdot \exp\left(-\frac{1}{b}\right)}$$

$$\cdot \exp\left(\frac{V}{b \cdot Vx} - \frac{1}{b}\right) = I + \frac{P}{I} \cdot \frac{I - I \cdot \exp\left(-\frac{1}{b}\right) - Ix}{b \cdot Vx - b \cdot Vx \cdot \exp\left(-\frac{1}{b}\right)} \qquad (5)$$

Unfortunately, (5) is not possible to solve it analytically given that it's function without the diffeomorphism property. The Linear Reoriented Coordinates Method (LRCM) can approximate the optimal values for the current and voltage[2]

978-1-6654-6060-6/23 $31.00 © 2023 IEEE

III. PVEM USING THE MANUFACTURER DATA SHEET

This section presents the Photovoltaic Exponential Method using the manufacturer data sheet for a given solar panel. It is essential to mention that the PVEM should keep the exact boundaries, shape, and performance of the I-V and P-V Curves provided by the PV Manufacturer Data Sheet.

Ex. 1. Let's apply the PVEM using the datasheet for the PV Modules SX-5 and SX-10. It's important to mention that the provided PV data sheet is under Standard Test Conditions (i.e. 25oC and 1,000W/m2) where the short circuit current (Ix) is defined as Isc and the open circuit voltage (Vx) as Voc. Fig. 1 shows the datasheet for the given PVM.

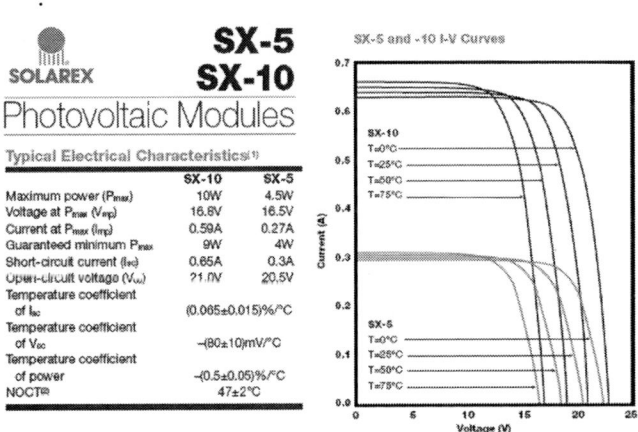

Fig. 1 Typical PV Manufacturer Data Sheet

Using the PVEM and data provided by the manufacturer, it is possible to compare the Manufacturer vs. Simulated I-V Curves under different temperatures for SX-05 and SX-10.

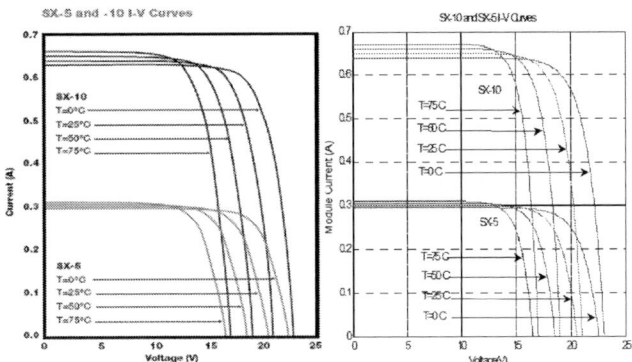

Fig. 2 SX-05 and SX-10 I-V (Manufacturer vs Simulated) Curves under different temperatures.

It's important to remark that the given PVEM could extend the analysis for the PV modules at different I-V, P-V, and R-V Curves at different temperatures and irradiances. An additional added value is that the PVEM considers the variations of the irradiance levels (Ei) and temperatures (T), making it excellent for Real-Time applications. Also, the PVEM can be used for circuit analysis, given the direct relationship between the PV current and the voltage [3].

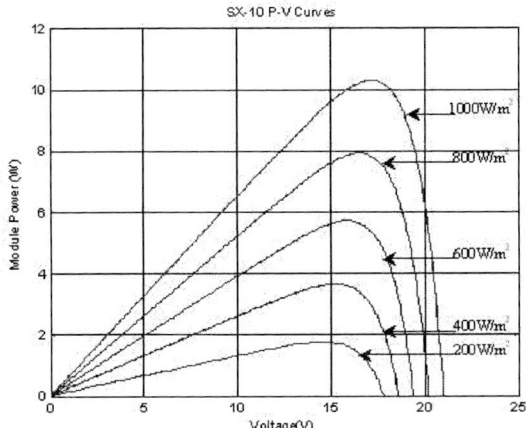

Fig. 3 Simulated SX-05 and SX-10 P-V Curves under irradiance levels.

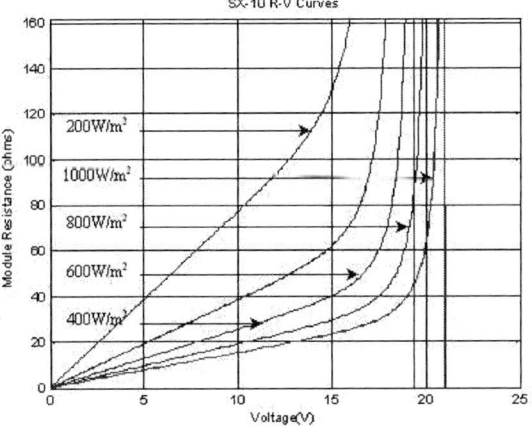

Fig. 4 Simulated SX-05 and SX-10 R-V Curves under irradiance levels.

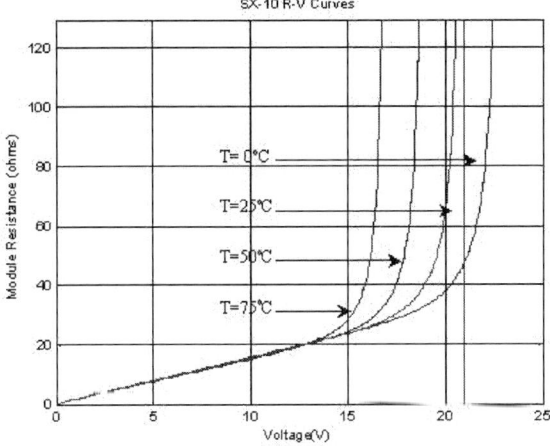

Fig. 5 Simulated SX-05 and SX-10 R-V Curves under different temperatures.

IV. What is the Characteristic Constant?

The characteristic constant, b, is the constant value that the PV will produce the required Vb voltage to change the output current by 63.2% of Isc when the measurements are made under Standard Test Conditions. Given the experimental I-V Curve, b can be estimated. The parameters Voc, Isc, Ib, and Vb can be obtained from the I-V Curve, as shown in Fig. 6. Finally, equation (6) calculates b.

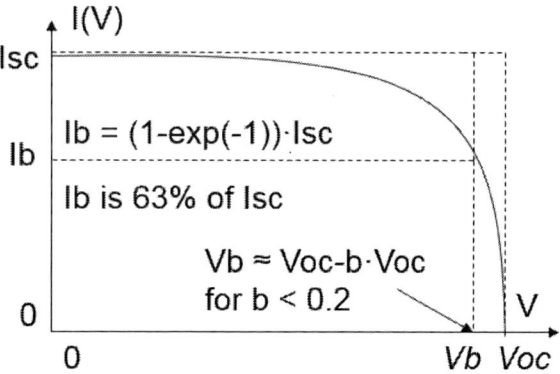

Fig. 6 Graphical method to determine b using the I-V Curve.

$$b \approx 1 - \frac{V_b}{V_{oc}} \tag{6}$$

V. Fill Factor Analysis using the PVEM

The Fill Factor, FF, is a figure of merit for solar panel design. It is defined as the rectangular area covered by $Pmax$ (i.e. Iop multiplied by Vop) divided by the total rectangular area produced by Isc and Voc. Graphically, FF is a measure of the "squareness" of the solar cell and is also the area of the largest rectangle which will fit in the IV curve. In other words, Fill Factor is a parameter that, in conjunction with Voc and Isc, determines the maximum power from a solar cell. The FF is defined as the ratio of the maximum power from the solar cell to the product of Voc and Isc, as shown in (7). As an interesting fact, it is possible to determine the boundaries of the existence of the FF and to prove that the minimum FF is 1/4, as shown in (8).

$$fill\,factor = \frac{P_{max}}{I_{sc} \cdot V_{oc}} = \frac{I_{op} \cdot V_{op}}{I_{sc} \cdot V_{oc}} \tag{7}$$

$$I_{sc} \cdot V_{oc} > \int_0^{V_{oc}} I(V)dV > P_{max} > \frac{1}{4} \cdot I_{sc} \cdot V_{oc} \tag{8}$$

Let's determine the upper and lower boundary conditions using the limits of (2) when b tends to zero and infinite. Calculating the area under the I-V Curve can determine the upper boundary condition.

$$\int_0^{V_{oc}} I(V)dV = \int_0^{V_{oc}} I_{sc} \cdot \frac{1 - exp\left(\frac{V}{b \cdot V_{oc}} - \frac{1}{b}\right)}{1 - exp\left(\frac{-1}{b}\right)} dV = I_{sc} \cdot V_{oc} \cdot \frac{1 - b + b \cdot exp\left(\frac{-1}{b}\right)}{1 - exp\left(\frac{-1}{b}\right)} \tag{9}$$

$$\lim_{b \longrightarrow 0} \left[I_{sc} - I_{sc} \cdot \frac{1 - exp\left(\frac{V}{b \cdot V_{oc}}\right)}{1 - exp\left(\frac{1}{b}\right)} \right] = I_{sc} \tag{10}$$

$$\lim_{b \longrightarrow \infty} \left[I_{sc} - I_{sc} \cdot \frac{1 - exp\left(\frac{V}{b \cdot V_{oc}}\right)}{1 - exp\left(\frac{1}{b}\right)} \right] = I_{sc} - I_{sc} \cdot \frac{V}{V_{oc}} \tag{11}$$

$IL(V)$ is the limit of I(V) when b tends to ∞ at STC where the FF for $IL(V)$ is 1/4, as shown in (11).

$$P(V) = V \cdot I_L(V) = I_{sc} \cdot V - I_{sc} \cdot \frac{V^2}{V_{oc}}$$

$$\Rightarrow \frac{\partial P}{\partial V} = I_{sc} - 2 \cdot I_{sc} \cdot \frac{V}{V_{oc}} = 0$$

$$\Rightarrow V_{op} = \frac{V_{oc}}{2} \Rightarrow P(V_{op}) = \frac{I_{sc} \cdot V_{oc}}{4} \tag{12}$$

Note: b tends to ∞, so I(V) is always bigger than IL(V).

$$I(V) = I_{sc} - I_{sc} \cdot \frac{1 - exp\left(\frac{V}{b \cdot V_{oc}}\right)}{1 - exp\left(\frac{1}{b}\right)} > I_{sc} - I_{sc} \cdot \frac{V}{V_{oc}} = I_L(V) \tag{13}$$

Finally, it is proved that the Fill Factor is more than one quarter and less than the total area inside of the I-V Curve divided by the short circuit current, Isc, and open circuit voltage, Voc, as shown in (13).

$$1 > \int_0^{V_{oc}} \frac{I(V)dV}{I_{sc} \cdot V_{oc}} > fill\,factor > \frac{1}{4} \tag{14}$$

VI. Conclusions

This paper presents the Photovoltaic Module Exponential Model, PVEM. The PVEM considers the information provided by the manufacturer's data sheets. The PVEM considers the irradiance level (Ei) and temperature (T), making it excellent for Real-Time applications. For any PVM, the P-V Curves, I-V Curves, and R-V Curves can be calculated using the proposed model. The PVM model can be used for circuit analysis because it is a continuous and differentiable model with a direct relationship between the current and the voltage. Finally, proof of the existence of the boundary limits for the fill factor for any PVM.

References

[1] Ortiz-Rivera, Eduardo I.; Torres-Feliciano, Yazmin; Sanchez Del Valle, Angelymar*; "Mathematical Models of Renewable Energy Sources developed at UPRM useful for Microgrid Analysis" 48th IEEE Photovoltaic Specialists Conf. (PVSC) from June 20-25, 2021.

[2] Ortiz, E.; Peng, F."Linear Reoriented Coordinates Method", IEEE Int. Conf. Electro/information Technology, May 7-10 2006 pp:459–464

[3] Darbali, Rachid; Ortiz, Eduardo I.; "Optimal Duty Ratio Maximum Power Point Tracking Technique Using the SEPIC Topology for Photovoltaic Systems Applications" 2016 IEEE ANDESCON Andean Council Int. Conf.; Arequipa, Perú, Oct. 19 - 21, 2016

On the unappreciated impact of Se in As-doped CdSexTe1-x

Deborah L McGott, Darius Kuciauskas, Craig L Perkins, Eric Colegrove, Matthew O. Reese

National Renewable Energy Laboratory, Golden, CO, United States

Group-V (e.g., As) doping in CdSexTe1-x solar cells has shown greatly improved absorber hole density, but not increased open-circuit voltages. This is typically attributed to poor interfaces, self-compensation, and sub-bandgap absorption related to group-V dopants. Se, on the other hand, is typically considered to passivate deep defects and improve performance. Here, we demonstrate lifetimes > 1 μs in undoped CdSexTe1-x and identify several processing conditions that influence lifetime. For CdSe0.3Te0.7, the alloy composition commonly used in devices, n-type behavior is shown with substantial sub-bandgap photoluminescence emission that largely disappears with As doping. A detailed analysis reveals potential causes and relates to voltage loss in devices, where Se alloying is found to result in larger losses than As doping.

Leveraging High-Fidelity Sensor Data for Inverter Diagnostics: A Data-Driven Model using High-Temperature Accelerated Life Testing Data

Sakir Karakaya[*][¥], Murat Yildirim[*], Shijia Zhao[†], Feng Qiu[†],
Jack David Flicker[‡], Benjamin Peters[#], Zhaoyu Wang[§]

[*] Wayne State University, Detroit, MI, 48202, USA; [¥]Ministry of Industry and Technology, 06510, Turkey;
[†] Argonne National Laboratories, Lemont, IL, 60439, USA; [‡] Sandia National Laboratories Albuquerque,
NM, 87123, USA; [#] The University of Texas Rio Grande Valley, Edinburg, TX, 78539, USA;
[§] Iowa State University, Ames, IA, 50011, USA

Abstract — Inverters pose substantial reliability risks and significantly impact operations & maintenance costs in photovoltaic (PV) systems. Understanding and predicting inverter failure processes is a key enabler for improving levelized cost of energy and competitiveness of the PV industry. In recent years, there has been a growing interest in harnessing sensor information from inverters to monitor and predict inverter degradation and failure risks. In this paper, we propose a comprehensive diagnostics framework for PV inverters that (i) transforms functional sensor information to time-frequency domain features in an effort to capture both summary statistics and signal dynamics, and (ii) uses the produced signal features to build a diagnostic model that predicts degradation severity in PV inverters. Results using inverter data from an accelerated life testing experiment show that proposed approach offers 91-97% accuracy in predicting degradation severity.

I. INTRODUCTION

Operations and maintenance (O&M) costs account for 17% of the levelized cost of energy (LCOE) for photovoltaic (PV) systems [1], with $9.40/kW/year as an average price for a global utility-scale project [2]. O&M also has a significant impact on operational performance as failure-induced outages cause a 1.6% loss in operational revenues [2]. As the root cause for 43% of PV system failures, PV inverters constitute a significant contributor to the O&M costs and reliability risks within PV systems [2]. Owing to their significance, there has been a growing research studying the reliability risks and failure processes within PV inverters [3]. These failure models typically rely on complex characterizations of degradation and domain-driven failure analysis that provides an in-depth understanding of a specific failure process. While these domain-specific models provide interesting insights, they do not necessarily generalize to other failure processes or inverter types. There is a need for data-driven generalizable models that can effectively harness complex sensor information to detect changes in system behavior as a function of degradation. This research aims to address this need by offering a data-driven diagnostic framework for predicting inverter degradation.

Power systems are witnessing a sensor-driven transformation in recent decades. PV industry is also benefiting from this trend, as a wealth of sensor information is becoming available to PV operators for analysis. In recent years, there has been significant attempts on using sensor data and inspection records for detecting issues in PV panels. Unfortunately, similar developments for PV inverters remained challenging due to the complexity associated with their failure processes. To better understand PV inverter degradation, Sandia National Laboratories conducted extensive accelerated life testing (ALT) experiments on PV inverters [4]. Main motivation for these ALT tests is to continuously observe inverters from brand new stage to failure in an effort to characterize progression of degradation in these systems: i.e., to acquire the degradation curve. These degradation curves are composed of high-fidelity, longitudinal, and multi-stream sensor data from PV inverters.

In this paper, we propose a diagnostic framework for PV inverters that can harness the complex functional sensor data for detecting state-of-health of PV inverters. The proposed framework is composed of two interconnected stages:

- The first stage, called *feature extraction*, transforms high fidelity sensor data from PV inverters (e.g., AC current, AC voltage, power, efficiency etc.), to time and time-frequency domain metrics through the use of signal summary statistics and functional data analysis.
- The second stage refers to *classification*, where multiclass classification model is used to map the time and time-frequency domain features to the degradation conditions, in an effort to train predictive diagnostic models for PV inverters.

We use change point detection algorithms to detect the degradation classes within an inverter lifetime. Results are validated using an inverter accelerated life testing experiment.

978-1-6654-6060-6/23 $31.00 © 2023 IEEE

II. METHOD

In this section, we develop our modeling approach for diagnosing PV inverters. The proposed approach is defined under the data-driven approaches in the related literature, which mainly focuses on exploring the relationship between the current health status of a component (e.g., an inverter in a solar power plant) and sensor data by harnessing machine learning or statistical methods. A typical data-driven approach aims to detect changes or patterns in the data used to monitor some performance parameters that are able to reflect the degradation process. Based on these patterns, the current health of system and time-to-failure are estimated. A typical framework for the data-driven approach includes two stages [5]: (1) data acquisition and processing, and feature extraction, (2) training by a machine learning model and prediction based on the monitored sensor data. Data acquisition and processing involves installing a data collection system based on sensors or physical-inspection, determining which parameters should be monitored (e.g., voltage, current, internal temperature, operational status, weather conditions), and deciding on data resolution, cleaning, and preprocessing of data. In the feature extraction stage, an optimal subset of features that can reflect the health status of a component, system degradation, or failure process is selected based on detected patterns in degradation-fault data. These features can be categorized into three classes: (a) time domain features, (b) frequency domain features, and (c) time-frequency domain features. Time domain features include basic statistical measures such as mean, variance, median, skewness, etc., which are calculated from the signal data (e.g., output/AC current values of an inverter recorded during a certain time interval) and characterized as a time-series. Though these features might be helpful for understanding the behavior of the signal over time, it can be challenging to perform reliable degradation analysis by using time-domain features alone. To capture changes in the signal data, frequency and time-frequency domain features are also used in addition to the time domain features. Frequency domain features include energy, signal-to-noise ratio, entropy, peak frequency etc., which are calculated by transforming time domain signal into frequency-domain signal using techniques such as Fourier transformation or fast Fourier transformation (see [6]-[8]). Time-frequency domain features are also commonly used to transform the time domain signal into time-frequency domain in order to capture anomalies or change-points in a nonstationary signal. The entropy and energy are calculated based on low- and high-frequency decomposition of the signal using techniques such as wavelet transformation (see [9]-[12]) or short-time Fourier transformation (see [13]). After calculating the features of a signal, machine learning algorithms are used to train the model, and failure risk predictions are made based on the degradation level and sensor observations of the component over time in the second stage of the data driven-approach. Inverter specific approaches in this stage are focused on diagnostics, where the emphasis is on fault detection in PV inverters based on sensor data. Artificial neural network [8], k-nearest neighbor [14], random forest [15], Bayesian networks [16] and support vector machine [10] are commonly used approaches in the related literature. For a detailed literature review on the algorithms used, see [17] and [18].

By following the generalized procedure given above, we develop our inverter-focused data-driven approach. In the first stage of our approach, we produce mixed-domain metrics using time domain and time-frequency features obtained by discrete wavelet transformation method. In the second stage, we use these features in a supervised multi-classification algorithm to predict degradation severity in PV inverters. The general framework of our diagnostic modeling approach is displayed in Figure 1.

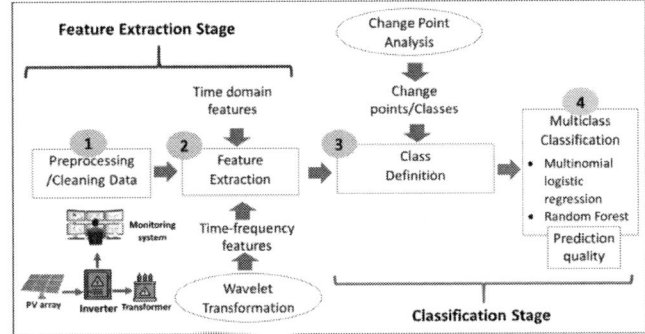

Fig. 1. The General Framework for the Inverter Diagnostics Models

A. Feature Extraction

In this section we produce time, and time-frequency domain features to extract latent degradation features from inverter sensor data. The produced features fall under two categories:

Time Domain Features: We study features including mean, standard deviation, median, minimum, maximum, kurtosis, and skewness values, calculated for signals. We calculate the time domain features and then capture if there are any possible trends or significant changes for each feature as the inverter degrades. In most applications, it may be difficult to perform reliable degradation analysis by using only the time domain features, due to complexity and high-dimensionality of functional data. Therefore, in addition to the time domain features, we also use time-frequency features extracted from discrete wavelet transformation method for our classification-based diagnostic modeling approach.

Time-Frequency Features: Discrete Wavelet Transformation (DWT) is used for signal processing to extract multilevel time-frequency features. DWT decomposes the original signal data into low-frequency (longer time intervals) and high-frequency (shorter time intervals) components, which are considered as input variables to the classification stage of our modeling approach. The DWT mainly allows us to extract both local spectral and temporal information embedded in the original signal data. The theoretical details and applications of DWT can be found in [19]. We elucidate our DWT model as follows:

Based on a transformation process using father scaling function and mother wavelet, we obtain the wavelet representation of the original signa $s(t)$ composed of approximation (low-frequency content, shown as cA) and detail (high-frequency, shown as cD) coefficients as follows:

$$\tilde{s}(t) = \sum_{k \epsilon Z_{j_o}} cA_{j_o,k}\Phi_{j,k}(t) + \sum_{j=j_o}^{J} \sum_{k \epsilon Z_j} cD_{j,k}\psi_{j,k}(t)$$

$$cA_{j_o,k} = \int s(t)\Phi_{j,k}(t)dt$$

$$cD_{j,k} = \int s(t)\psi_{j,k}(t)dt$$

where $\Phi_{j,k}(t) = 2^{-j/2}\Phi(2^{-j}t - k)$ and $\psi_{j,k}(t) = 2^{-j/2}\psi(2^{-j}t - k)$ are father scaling function and mother wavelet, respectively (j: the scaling parameter, k: the translation parameter, z_{j_o}: set of indices for decomposition level of low-frequency component where $j_o = 1$, and z_j: set of indices for decomposition level of high-frequency component). For father scaling function the daubechies-1 (db1) is selected since the data noise level is relatively low. The number of maximum decomposition level is calculated as $J = log_2 T$. After obtaining the coefficients, the energy contained within each decomposition level is quantified as the sum of squared coefficients, and thus the following DWT time-frequency feature array (size of 1 x$(J + 1)$) is obtained.

$$DWT = \left[\sqrt{\sum_{k \epsilon Z_{j_o}} (cA_{j_o,k})^2}, \ \sqrt{\sum_{k \epsilon Z_j} (cD_{J,k})^2}, \ \sqrt{\sum_{k \epsilon Z_{J-1}} (cD_{J-1,k})^2}, \ ..., \ \sqrt{\sum_{k \epsilon Z_{j_o}} (cD_{j_o,k})^2}, \right]$$

In our framework, we divide the whole data into a certain number of slices (i.e., N) with equal sizes (i.e., M), each of which represents a profile observation. n^{th} profile observation at time $t \in T = \{t | t = 0, 1, ..., T\}$ is denoted by $s_n(t)$, $n = 1,2, ..., N$ (the length of $s_n(t)$ is M and the length of whole data, $s(t)$, is NxM). Following the wavelet transformation procedure given above firstly we update the feature array for n^{th} profile observations (slices), DWT^n, as follows:

$$DWT^n = \left[\sqrt{\sum_{k \epsilon Z_{j_o}} (cA_{j_o,k}^n)^2}, \ \sqrt{\sum_{k \epsilon Z_j} (cD_{J,k}^n)^2}, \ \sqrt{\sum_{k \epsilon Z_{J-1}} (cD_{J-1,k}^n)^2}, \ ..., \ \sqrt{\sum_{k \epsilon Z_{j_o}} (cD_{j_o,k}^n)^2} \right]$$

Then, the DWT time-frequency feature matrix for whole signal data is obtained as:

$$DWT = [DWT^1, DWT^2, ..., DWT^N]^T.$$

In addition to DWT features, we also calculate the time domain features, for each profile observation, $n = 1,2, ..., N$. Let TF_i^n be the ith time domain feature, $i = 1,2, ..., P$, calculated for profile observation calculated for n^{th} profile observation. We merge the time domain and time-frequency feature matrixes and find the final matrix, FM, (as shown below) that will be used as the main input for stage two, i.e., classification.

$$FM = \begin{bmatrix} FM^1 \\ \vdots \\ FM^N \end{bmatrix} = \begin{bmatrix} DWT^1 & TF_1^1 & \cdots & TF_P^1 \\ \vdots & \vdots & \ddots & \vdots \\ DWT^N & TF_1^N & \cdots & TF_P^N \end{bmatrix}$$

B. Classification

As shown in Figure 1, the classification stage has two steps. First, we assign class labels to data (i.e., class definition); and then apply multi-classification algorithms to evaluate prediction quality of our approach.

We use change point detection method (a.k.a. anomaly detection, signal segmentation, and regime switching) to define the classes. For our data, $FM^1, FM^2, ..., FM^N$, if a change point exists at a time point $\tau \in T$, then $FM^1, FM^2, ..., FM^\tau$ differs from $FM^{\tau+1}, FM^{\tau+2}, ..., FM^N$ based on selected general distribution measures such as mean, median, rank, etc. An illustrated flowchart of change point detection is displayed in Figure 2.

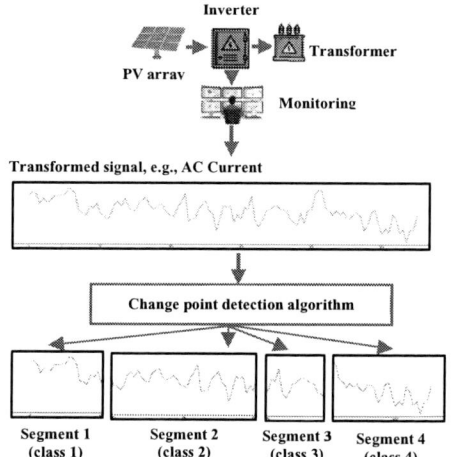

Fig. 2. Flowchart of a Change Point Detection Scheme for a Signal

In this context, the performance of an inverter is monitored with different signals, e.g., AC current (and/or AC voltage, efficiency, power, etc.), which can be used to monitor the degradation levels occurred in the inverter's performance. The resulting signal is divided into segments with similar characteristics via a change point detection algorithm in order to extract insightful features from these segments providing that the boundaries, i.e., change points, of these segments are identified. As shown in Figure 2, to illustrate, the signal is divided into four segments which are considered as classes in our modeling approach.

In our application, we use one of the commonly used change point algorithms called Pruned Exact Linear Time (PELT). It is an optimal detection method yielding exact solutions for segmentation problem. In this method, the number of change points are not given to the algorithm and the objective is to find the optimal change points by minimizing the penalized sum of costs. In this method, a linear penalty function is used as $penalty(T) = \beta|T|$, where $\beta > 0$ is a smoothing parameter. This method handles each point in the data sequentially; and use a pruning rule that provides a point to be or not to be excluded from the set of possible change points. For instance, consider two time-indexes associated to two different data points, t_1 and t_2, where $t_1 < t_2$. The pruning rule is then defined as [20]:

If $\left[\min_{T} C(T, s_{0..t_1}) + \beta|T|\right] + c(s_{t_1..t_2}) \geq \left[\min_{T} C(T, s_{0..t_2}) + \beta|T|\right]$

holds, t_1 cannot be the time of last change point before T.

The cost function is selected so as to detect mean-shifts in the signal. The details of this method can be found in [20].

C. Classification using Multi-Class Algorithms

After defining the class labels, we use these classes in our diagnostic modeling approach. Based on the classes we can predict if a new signal data (profile observation) indicates that the inverter is in its first, second, or m^{th} stage (the last class indicating the stage in which the inverter is highly degraded) of its lifetime. Within this context, we evaluate the performance of our diagnostic model, i.e., *classifier*, through calculating the classification accuracy to validate the prediction quality of the model. To do this, first, we generate two sets from the whole signal data class labels of which are determined: (1) training set used to train the classifier, and (2) testing or validation set used to calculate the prediction quality. Second, we apply multinomial logistic regression as for classification. Third, we calculate the most commonly used measures to evaluate the performance of each method. Finally, based on the importance scores of features, we re-apply the classification methods and report final performance measures. The flowchart of this procedure is summarized in Figure 3.

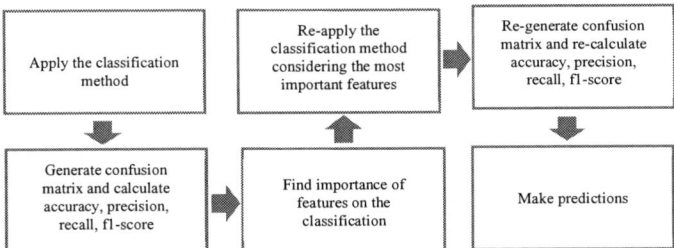

Fig. 3. Flowchart of the Classification Method

Multinomial logistic regression (MNLR) is a classification method, which is an extension of logistic regression (considering only two classes), in which the response variable (the class label of each point of transformed signal) has more than two possible categorical outcomes (i.e., multiple classes). Let the response variable be y_n be the class label of observation profile, n, where $y_n \in M = \{j | j = 1, 2, \ldots, m\}$. Consider there are k number of predictors, $X_1, X_2, \ldots, X_k$ and $X_K = [X_1, X_2, \ldots, X_k]$ is the vector of predictors. Once fitted, logistic regression model predicts the probability of classes. The final prediction becomes the class that has the higher probability.

In contrast to the binary case in which there is a single coefficient vector, MNLR model considers a matrix of coefficients C each of which row vector (C_j) corresponds to

each class j, $C_j = [\beta_{1j}, \ldots, \beta_{kj}]$. And we have also the constant coefficient for each class j as $C_{0,j}$. The aim of the model is now to predict the probability of each class j, $P(y_n = j | X_K)$ as:

$$\hat{\rho}_j(X_K) = \frac{e^{C_{0,j} + X_K C_j}}{\sum_l^{m-1} e^{C_{0,l} + X_K C_l}}$$

Note that $\sum_{j=1}^{m} \hat{\rho}_j(X_K) = 1$ and the last class, m, is considered as the reference or pivot class. As an optimization problem, MNLR with regularization term $r(C)$ minimizes the following cost function:

$$\min_{C} -P \sum_{k=1}^{K} \sum_{j=0}^{m-1} [y_n = j] \log\left(\hat{\rho}_j(XX_K)\right) + r(C)$$

where P is defined as the inverse of regularization strength and $[y_n = j] = 0$ if $y_n = j$ is false, otherwise it takes 1. The regularization term $r(C)$ is selected as:

$$r(C) = \frac{1}{2} \sum_{k=1}^{K} \sum_{j=1}^{m} C_{k,j}^2.$$

Prior to the classification stage, the columns of *FM feature matrix* are standardized to zero mean and unit variance. Then, we use classes determined via change point detection method.

III. CASE STUDY

For our case study, we develop and test a diagnostic model using results from a high static temperature accelerated life testing experiment for inverters, which was conducted at Sandia National Laboratories [4]. Sensor dataset from the test includes AC side signals (AC Current, i.e., ACI) for a micro-inverter tested under a high static temperature level, 125 °C, was continuously monitored (one record per minute). In this test, the inverter was exposed to 125°C for three months until it failed, and during its lifetime ACI values were recorded.

After preprocessing the dataset, we divide the whole data into $N = 130$ slices with equal size, each of which represents a profile observation, $n = 1, 2, \ldots, 130$. Then, the time domain features are calculated for each profile observation and the wavelet transformation is applied based on the procedure given in Section 2, and the *FM* feature matrix for whole signal data is obtained as:

$$FM = \begin{bmatrix} FM^1 \\ \vdots \\ FM^{130} \end{bmatrix} = \begin{bmatrix} DWT^1 & TF_1^1 & \cdots & TF_7^1 \\ \vdots & \vdots & \ddots & \vdots \\ DWT^{130} & TF_1^{130} & \cdots & TF_7^{130} \end{bmatrix}.$$

The number of decomposition level in wavelet transformation is calculated as 9 (so, there are 10 time-frequency features), and thus we have in total 17 features in the *FM* feature matrix. The time domain and time-frequency domain features are displayed in Figure 4 and Figure 5, respectively.

After obtaining the feature matrix, we apply the *PELT* in order to find the change points in the data set. Thus, we identified that the change points occurred at time epochs t = 50, 60, 75, 90, and 115. Based on this result, the signal is divided into 6 segments (as shown in Figure 6) considered as classes based on lifetime percentiles.

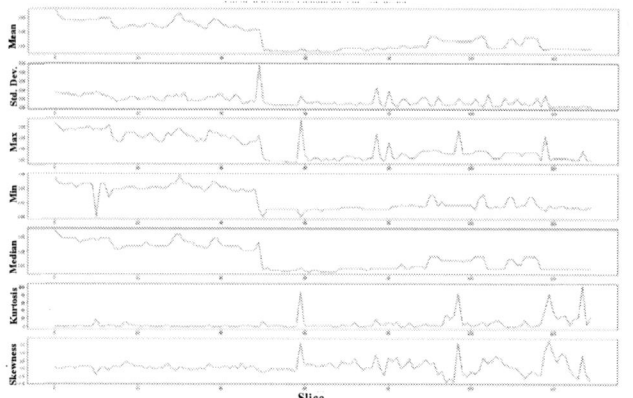

Fig. 4. Time Domain Features for the Inverter Tested, based on ACI Signal Data

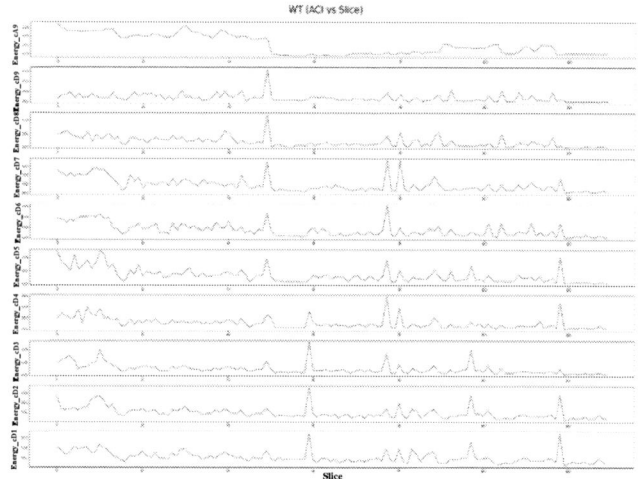

Fig. 5. Time-Frequency Features for the Inverter Tested, based on ACI Signal Data

Class 0	Class 1	Class 2	Class 3	Class 4	Class 5
[1-37]%	[38-47]%	[48-57] %	[58-70]%	[71-89]%	[90-100]%

Fig. 6. Cutoffs for Degradation States based on Lifetime Percentiles

After assigning the associated class labels to each row of the feature matrix, we apply the MNLR method. The results of the predictive diagnostic model is presented in Figure 7 and Figure 8. We evaluate the performance of the MNLR based on the confusion matrix, the classification report that includes the most important measures such as recall, *f1*-score and precision measures, and an accuracy rate (score). The confusion matrix is a *MxM* matrix, where *M* is the number of classes, and each cell of this matrix gives a comparison between the predicted class labels determined by the classifier and the actual class levels defined in the original data. It also allows us to calculate the most important performance measures including accuracy rate, recall, precision, *f1*-score, shown under the classification

report. The accuracy score, which indicates the prediction quality of the classifier, shows the percentage of the correct predictions.

Precision refers to what percentage of predictions labelled as positive class are positive in actual data set; recall refers to what percentage of all positive datapoints are predicted correctly as positive; and *f1*-score is a combined accuracy measure of the harmonic mean between precision and recall.

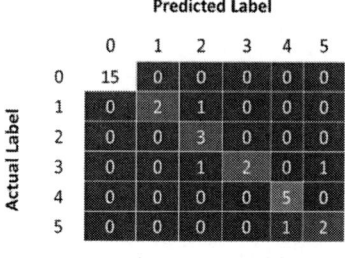

Accuracy score: 0.88

Fig. 7. Predictive Performance of the Proposed Diagnostic Model: Confusion Matrix

	precision	recall	f1-score	support
0	1.00	1.00	1.00	15
1	1.00	0.67	0.80	3
2	0.60	1.00	0.75	3
3	1.00	0.50	0.67	4
4	0.83	1.00	0.91	5
5	0.67	0.67	0.67	3
average	0.85	0.81	0.80	33
weighted avg	0.91	0.88	0.87	33

Fig. 8. Predictive Performance of the Proposed Diagnostic Model: Classification Report

Considering the results given in Figure 8, our proposed model provides a weighted accuracy score of 91% in predicting degradation severity, with an associated recall score of 88%, and *f1*-score of 87%. When the prediction is wrong, the model often confuses between neighboring degradation severities, meaning that even missed predictions do not necessarily imposse significant operational consequences.

In most applications, PV operators will have the opportunity to have multiple observations before making a prediction on inverter degradation. To this end, we will leverage on the predictive diagnostic models based on MNLR and propose a multi-observation prediction model that provides better prediction performance compared to its benchmark given above. Main focus of the model is to get the predictions from two consecutive observations, and use prediction probabilities in a Bayesian framework to predict the degradation class.

More formally, let us assume that we have two consecutive observations at period t_1, t_2. We also assume that these observations are close enough that the degradation state across

978-1-6654-6060-6/23 $31.00 © 2023 IEEE

them does not change. At time t_1, when we run the MNLR, the associated probabilities that the underlying class belongs to i, is denoted as p_{i,t_1}. Given that we have 6 classes, the probabilities at time period t_1 would be given as $\{p_{0,t_1}, p_{1,t_1}, p_{2,t_1}, p_{3,t_1}, p_{4,t_1}, p_{5,t_1}\}$. Likewise, at time period t_2 the corresponding probabilities would be given as $\{p_{0,t_2}, p_{1,t_2}, p_{2,t_2}, p_{3,t_2}, p_{4,t_2}, p_{5,t_2}\}$. Using the observation at time periods t_1, t_2, the following equation provides the probability that the observations at these time periods belong to class i:

$$P_{i,t_1,t_2} = \frac{p_{i,t_1} * p_{i,t_2}}{\sum_{s=0}^{5} p_{s,t_1} * p_{s,t_2}}$$

$$= \frac{\text{Probability that both observations indicate degradation class} = i}{\text{Probability that the degradation class at both times is the same}}$$

In order to find the probability that a certain class should be predicted, the above equation calculates the probability that both observations predict class i, and conditions it on the event that both observations have the same underlying class. The predictor finds this probability for every degradation class, and chooses the maximizer as the predicted degradation class based on the two observations. Using the above equation, we obtain the confusion matrix and the summary results as indicated in Figure 9 and 10, respectively.

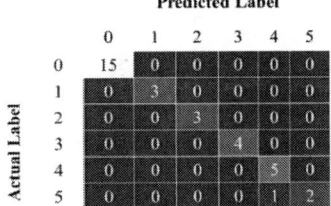

Fig. 9. Predictive Performance of the Proposed Diagnostic Model with Multi-Observations: Confusion Matrix

	precision	recall	f1-score	support
0	1.00	1.00	1.00	15
1	1.00	1.00	1.00	3
2	1.00	1.00	1.00	3
3	1.00	1.00	1.00	4
4	0.83	1.00	0.91	5
5	1.00	0.67	0.80	3
average	0.97	0.94	0.95	33
weighted avg	0.97	0.97	0.97	33

Fig. 10. Predictive Performance of the Proposed Diagnostic Model with Multi-Observations: Classification Report

As shown in Figure 10, the multi-observation model improves precision from 91% to 97%, and improves f1-score from 87% to 97%. Although the multi-observation model provides improvements over its single observation counterpart, it also introduces a lag, i.e., requires multiple observations to take place before it can make a prediction. Thus, the choice of methods needs to be made considering the relative importance of speed vs. accuracy.

IV. CONCLUSION

In this paper, we developed a data-driven diagnostic modeling approach for inverters that uses complex functional sensor data to produce accurate predictions on degradation severity. As a general conclusion, the proposed approach, which uses multinomial logistic regression-based classification method with the classes defined by the change point analysis, can be a valuable tool for determining the state-of-health of inverters. It can be generalized to different degradation processes and inverter types, as it does not require specific domain knowledge on the physics of failure. Since inverters remain a main culprit for PV reliability, the proposed diagnostic approach can be a valuable tool for PV operators that would like to gain more visibility on health condition of their inverter assets, to optimize asset management, and O&M activities.

ACKNOWLEDGMENT

This material is based upon work supported by the U.S. Department of Energy's Office of Energy Efficiency and Renewable Energy (EERE) under the Solar Energy Technologies Office Award Number 38458. The views expressed herein do not necessarily represent the views of the U.S.Department of Energy or the United States Government.

REFERENCES

[1] T. Silverman, M. Deceglie, and K. Horowitz, "NREL Comparative PV LCOE calculator,"

[2] kWh analytics, "Solar Risk Assessment: 2020," 2020.

[3] D. C. Jordan, B. Marion, C. Deline, T. Barnes, and M. Bolinger, "PV field reliability status—Analysis of 100 000 solar systems," Progress in Photovoltaics: Research and Applications, vol. 28, no. 8, pp. 739–754, 2020, doi: https://doi.org/10.1002/pip.3262.

[4] J. Flicker, G. Tamizhmani, M. K. Moorthy, R. Thiagarajan and R. Ayyanar, "Accelerated Testing of Module-Level Power Electronics for Long-Term Reliability," IEEE Journal of Photovoltaics, vol. 7, no. 1, pp. 259-267, 2017. doi:10.1109/JPHOTOV.2016.2621339.

[5] Ren, L., Sun, Y. , Cui, J.,& Zhang, L. (2018). Bearing remaining useful life prediction based on deep autoencoder and deep neural networks. Journal of Manufacturing Systems, 48, pp. 71–77.

[6] Khomfoi, S and Tolbert, L.M. (2007). Fault diagnostic system for a multilevel inverter using a neural network. IEEE Transactions on Power Electronics, 22 (3), pp. 1062–1069.

[7] Kumar, G.K., Parimalasundar, E., Elangovan, D., Sanjeevikumar, P., Lannuzzo, F., & Holm-Nielsen, J. B. (2020). Fault investigation in cascaded H-bridge multilevel inverter through fast fourier transform and artificial neural network approach. Energies, 13 (6).

[8] Sun, Q., Yu, X., &Li, H. (2020). Open-circuit fault diagnosis based on 1D-CNN for three-phase full-bridge inverter. 11th International Conference on Prognostics and System Health Management (PHM-2020 Jinan), Jinan, China, pp. 322-327.

[9] Bhattacharya, M., Saha, S., Khan, D., & Nag, T. (2017). Wavelet based component fault detection in diode clamped multilevel inverter using probabilistic neural network. 2017 2nd International Conference for Convergence in Technology (I2CT), Mumbai, India, pp. 1163-1168.

[10] Bowen, C. and Wei, T. (2019). Switch open-circuit faults diagnosis of inverter based on wavelet and support vector machine. 14th IEEE International

Conference on Electronic Measurement & Instruments (ICEMI), Changsha, China, pp. 1178-1184.

[11] Chowdhury, D., Bhattacharya, M., Khan, D., Saha, S., & Dasgupta, A. (2017). Wavelet decomposition-based fault detection in cascaded H-bridge multilevel inverter using artificial neural network. 2nd IEEE International Conference on Recent Trends in Electronics, Information & Communication Technology (RTEICT), Bangalore, India, pp. 1931-1935.

[12] Kurukuru, V.S.B., Haque, A., Kumar, R., Khan, M.A., & Tripathy, A.K. (2020). Machine learning based fault classification approach for power electronic converters. 2020 IEEE International Conference on Power Electronics, Drives and Energy Systems (PEDES), Jaipur, India, pp. 1-6.

[13] Lu, W.K. and Zhang, Q. (2009). Deconvolutive short-time fourier transform spectrogram. IEEE Signal Processing Letters, 16 (7), pp. 576-579.

[14] Manohar, M., Koley, E., Kumar, Y. , & Ghosh, S. (2018). Discrete wavelet transform and kNN-based fault detector and classifier for PV integrated microgrid. In: Kolhe, M., Trivedi, M., Tiwari, S., Singh, V. (eds) Advances in Data and Information Sciences. Lecture Notes in Networks and Systems, vol 38. Springer, Singapore.

[15] Kou, L., Liu, C., Cai, G.W., Zhou, J.N., De Yuan, Q., & Pang, S.M. (2010). Fault diagnosis for open-circuit faults in NPC inverter based on knowledge-driven and datadriven approaches. IET Power Electronics, 13 (6), pp. 1236–1245.

[16] Cai, B., Zhao, Y., Liu, H., & Xie, M. (2017). A data-driven fault diagnosis methodology in three-phase inverters for PMSM drive systems. IEEE Transactions on Power Electronics, 32 (7), pp.5590–5600.

[17] Malik, A., Haque, A., Kurukuru, V. S. B., Khan, M. A., & Blaabjerg, F. (2022). Overview of fault detection approaches for grid connected photovoltaic inverters. e-Prime - Advances in Electrical Engineering, Electronics and Energy, 2, 100035.

[18] Gedde-Dahl, G. S. (2022). Optimising maintenance operations in photovoltaic solar plants using data analysis for predictive maintenance (Master's thesis, Norwegian University of Life Sciences).

[19] R. Yan, R. X. Gao, and X. Chen. "Wavelets for fault diagnosis of rotary machines: A review with applications." Signal processing 96 (2014): 1-15.

[20] C. Truong, L. Oudre, and N. Vayatis. "Selective review of offline change point detection methods." Signal Processing, vol.167, 2020.

Influence of Interfaces on Stability of Perovskite Solar Cells

Arkadi Akopian, Saba Sharikadze, Ranjith Kottokkaran and Vikram Dalal

Iowa State University, Ames, Iowa 50011, USA

Abstract — This paper examines the influence of organic and inorganic interfaces on the photo-stability of perovskite solar cells. The perovskite cells were both of the hybrid inorganic-organic type (methyl-ammonium iodide) and inorganic type (cesium lead bromide). We measured photostability under continuous illumination for over 100 hours. We also measured cells under different intensities of illumination ranging up to 8X sun. For both organic and inorganic perovskites, it was found that having an organic hole transport layer led to much higher degradation than having an inorganic hole transport layer, even when the organic layer was at the back of the cell and did not receive any UV radiation. An all-inorganic solar cell with no organic layers in either the perovskite or in the electron and hole heterojunction layers showed excellent stability with virtually no degradation, even when exposed to air during illumination The all-inorganic cells were stable under thermal degradation when subjected to 200 ℃, to air exposure for many months, and for continuous AM1.5 illumination under a full spectrum xenon lamp for hundreds of hours. The varying intensity experiment showed that there are two distinct degradation mechanism, one related to native ions in the material that get trapped at the interface, and one related to ions generated in the bulk by light. We will describe a model that explains this behavior.

I. INTRODUCTION

Perovskite solar cells are an important new technology for soalr cells, with excellent conversion efficiencies of >25% achieved in the laboratory [1]. Most perovskite cells are of the hybrid organic-inorganic composition, comprising alloys such as methyl ammonium lead haldies, or methyl-formamidinum lead halide alloys. The electron heterojunction layer is usually an inorganic layer such as tin oxide or titanium dioxide, and the p layer is almost universally an organic layer such as P3HT or SPIRO-Ometad, though occasionally NiO is also used as the p layer. The high efficiency cells almost always use an organic p layer.

While the efficiency achieved in hybid perovskite devics is remarkable, the stability under continuous light generation is not very good, decreasing by at least 10% over a hundred hours or so, even when encapsulated. The thermal stability is also poor, with the cells degrading when subjected to 100°C. The environmental staility when subjected to moist air is rather poor unless encapsulated.

In contrast, we have shown previously that an inorganic perovskite cell made using $CsPbBr_3$ is completely stable under both thermal degradation as well as under moisture degradation[2,3]. In this paper, we examine the stability of both inorganic and hybrid solar cells using either organic or inorganic p-type heterojunction layers. We will show that using a p-type organic heterojunction layer leads to photo-degradation of the solar cell for both inorganic and hybrid perovskites. We will also show that an all inorganic cell which does not contain any organic layer in either the pervoskite or the heterojunction layers is completely stable, under thermal stress, under environmental stress and under illumination. We will also examine the stability of the hybrid solar cell under varying intensuites of illumination, and show that there are two distinct and independent phenomena that lead to photo-instability, one related to the native ions trapped at the interface, and one to photo-gnerated ions.

II. EXPERIMENTAL DETAILS

The inorganic $CsPbBr_3$ solar cells were deposited using thermal evaporation using techniques described previously [2,3]. The hybrid solar cells were deposited using the usual solution growth techniques [4]. The device measurements included the normal I-V measurements conducted under both an inert nitrogen environment and under air exposure. The devices were also measured for quantum efficiency and all current values correspond to the current from QE measurements since for small area cells, significant error arises in current measurements fom I-V curves alone.The p-layers were either P3HT deposited using solution growth, or NiO deposited using e-beam evaporation for hybrid cells, and either P3HT or a Carbon paste for the inorganic perovskite devices. Note that by using a C paste electrode, the inroganic perovskite device was an all-inroganic device with no organic layers anywhere.

For stability testing, we used a chamber which can be evacuated and refilled with either nitrogen or room air. A full-spectrum xenon AM1.5 source from ABET was used to measure the device during test.

III. RESULTS
A. Results on cell performance

Figure 1 shows the I-V curve for a typical inorganic solar cell with a P3HT p-layer initial exposure plus exposure at ambient temp-erature . It had a conversion cell efficiency of 8.8% (the bandgap of the inorganic cell is 2.3 eV). In Fig. 2, we show the changes in the performance when the cell is exposed to room air for 25 days, showing no degradation at all. In Fig. 3, we show the results for a cell with inorganic p layer (C paste). Now the efficiencyis slightly lower, 7.5%.

978-1-6654-6060-6/23 $31.00 © 2023 IEEE

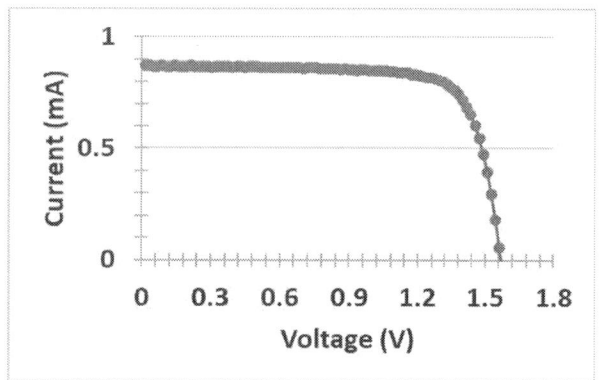

Fig. 1 I-V curve of an inorganic cell with P3HT p-layer. The current densityis 8.8 mA/cm2, the open circuit voltage is 1.64V, and the efficiencyis 8.8%. The bandgap is 2.3 eV.

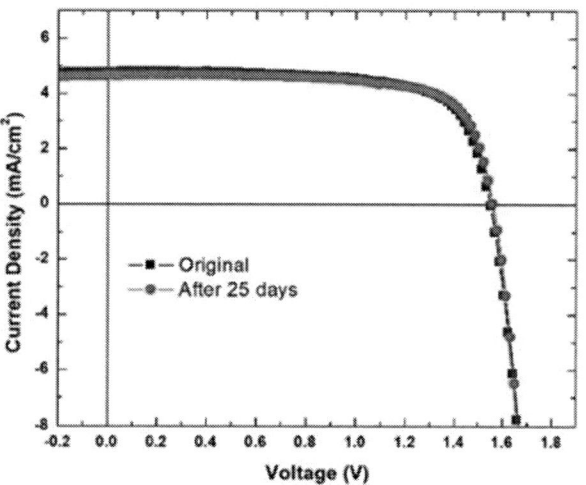

Fig.2 I-V curves of inorganic perovskite cell exposed to air for 25 days. There is no change in the performance.

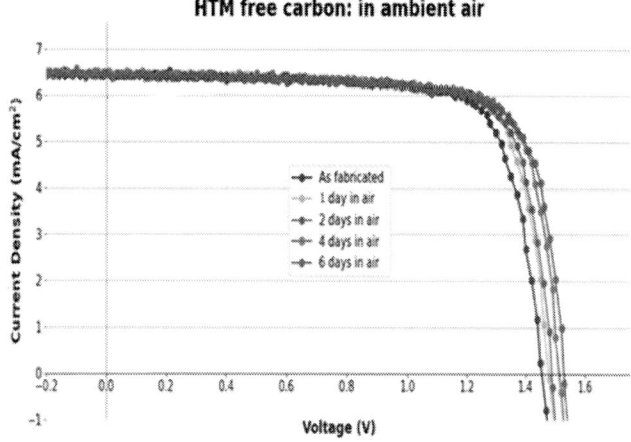

Fig. 3. I-V cuve of a cell with C paste as p layer. The data shows that the performance *improves* upon air exposure. The initial efficiency was 7.1%, increasing to 7.5% after 6 days of exposure.

B.   Results of Photo-stability test

Both types of cells, with organic p alyer, and inorganic p layer, were subjected to 1X full sun illumination from a xenon AM1.5 source, with the samples kept in air, for 60-150 hours.In Fig. 4 we show the results for the performance degradation of the cell with organic(P3HT) p layer, and in Fig. 5 for the cell with inorganic p layer (C paste). Very clearly, the cell with the organic p layer degrades rapidly, but one with the inorganic p layer has not degraded at all. This is a powerful demonstration that the organic heterojunction layers play an improtant role in the photo-induced degradationof perovskite solar cells, even when the cell perovskite material itself is stable under light.

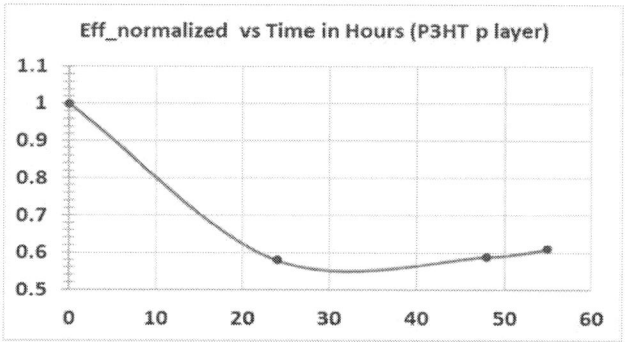

Fig. 4 The degradation of perovskite cells with P3HT p layer exposed to AM 1.5 full spectrum illumination

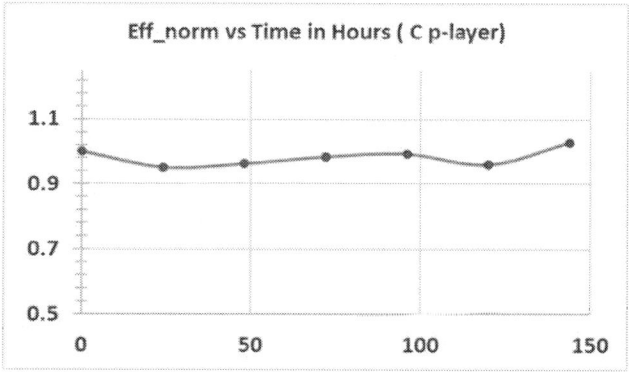

Fig. 5 The degradation of all-inorganic cell with inorganic p layer exposed to full spectrum AM1.5 illumination.

C.Results of stability test under different light intensities on organic hybrid perovskite cells with inorganic p layers.

In this paragraph, we show that there are two distinct mechanisms for degradation of perovskite cells, one related to the interfaces, and one related to generation of ions in the bulk by light. It is well known that there are native ions in the maerial, and that the interfaces trap these ions, leading to a reduction in open circuit votlage which increases upon illumination [5]. It is also known that light decomposes the material and generates more ions which contribute to increased recombination of photo-generated carriers and leads to a reduction in short circuit current[5]. In this work, we separate out these two effects by varying the intensity of light. Intensity only affects the light-generated ion density, but does not change the native defects. We also mathematically model the degradation and show that the two mecahnisms have distinctly different time scales and different saturation densities.

Fig. 6 shows the data for degradation (dots). The data can be modeled using sum of two exponentials, the first one related to interfacial trapped ions, and the second to ions generated by light.

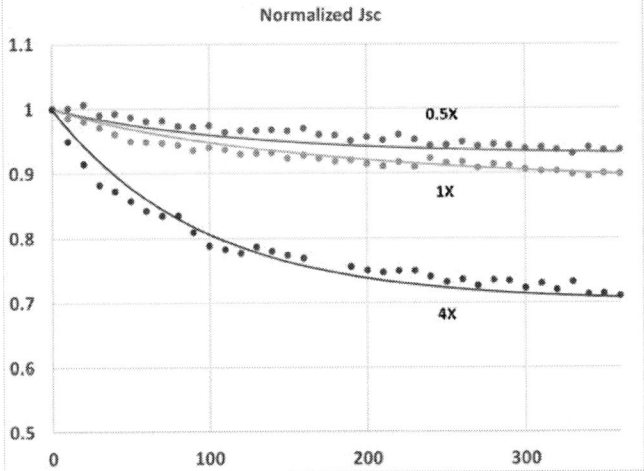

Fig. 6 Results on degradation of a hybrid solar cell vs. time in minutes under three different intensities of light. The dots are the experimental data and the lines are results from a model.

The equations are:

$$J_{sc}(t) = J_{sc.t=0} - \Delta J_{sc}$$

and

$$\Delta Jsc(t) = Jsc,t(0) \ [\ a_i\{ \ 1 - exp(-t/t_i)\} + a_b \ \{1 - exp(-t/t_b)\}]$$

where the subscript i refers to parameters related to traping of ions at the interface and b to bulk ions generated by light. We expect the i parameters to be independent of light intensity, but the b parameters to be functions of light intensity.

In Fig. 6, the lines represent the results of the model equations, and they fit the experimental data (dots) very well. The parameters that fit the data are shown in Table I below.

Table I

Intesity	a_i	a_b	t_i	t_b
0.5	0.07	0.05	120m	3500m
1.0	0.07	0.12	120m	1200m
4.0	0.07	0.23	120m	95m

Note how the interface parameters do not change at different intensities, but the bulk parameters do, with the modulation factor a_b increasing with light intensity, as it must.

We also did a similar experiment to distinguishe between using organic or inorganic p-interfaces for hybrid organic p-i-n cells. One used PTAA as the p layer,and the second one

used NiO as the p layer. In Fig.7, we show the results for the time dependence of degradation of these cells. Very clearly, the time dependence is much faster for the cell on PTAA when compared with the cell on NiO, once again showing that heterojunction layers have a profound influence on photo-stability of the cell.

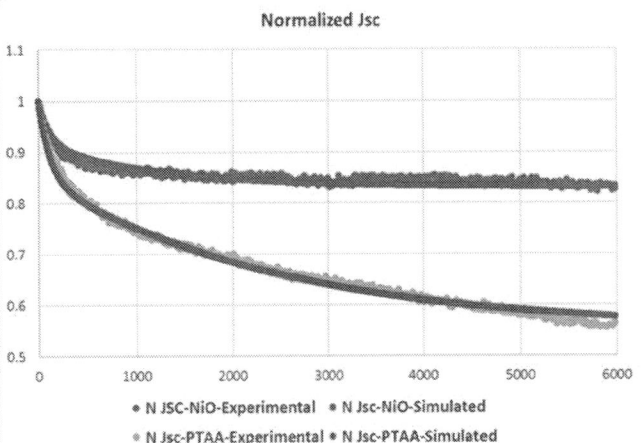

Fig. 7 Different rates of degradation for hybrid organic cells deposited on PTAA(bottom curve) vs. on NiO (top curve). The cell on NiO degrades much less.

IV. DISCUSSION AND CONCLUSIONS

It is clear from the data that heterojunction layers make a significant impact on the photo-stability of perovskite solar cells. Even in totally inorganic perovskite cell materials (CsPbBr3), having an organic interface (p layer) adversely affects the photo-stability. We have also shown that an all-inorganic cell comprising of CsPbBr₃ as the perovskite material and having inorganic n and p layers (TiO₂ and Carbon respectively) leads to a cell which is stable thermally, environmentally in air, and against light induced degradation. We have also developed a new model to explain the transient of light induced degradation which separates out the effects of degradation into an interface ion-trapping related part and a bulk ion generation part.

V.ACKNOLWEDGMENT

This work was supported in part by a grant from NSF. Arkadi Akopian was a Fulbright scholar, supported by U.S. Department of State

VI. REFERENCES

1.D. I. King, J.W.Lee, R. H. Jeong and J-H Boo, Scientific Reports, 12,5980(2022)

2.J.H.Zhu, R. Kottokkaran, S. Sharikadze, M.Noack and V.Dalal, Proc,; of 48th. IEEE PVSC(2021)

3. Harsh Gaonkar, Junhao Zhu, Ranjith Kottokkaran, Behrang Bagheri, Max Noack and Vikram Dalal, ACS Appl. Energy Materials, 3, 3497 (2020)

4. J. H. Im et al, Nature Nanotechnology, 9,927(2014)

5. P. Joshi et al, AIP Advances, 6, 115114 (2016)

A New Route to Facilitate Scaling of Lead-Tin Halide Perovskites: Thin Films via Solvent Self-Volatilization

Jack R. Palmer, Apoorva Gupta, Sean P. Dunfield, David P. Fenning

University of California San Diego, La Jolla, CA, United States

Substantial work is needed to scale metal halide perovskite (MHP) photovoltaics from laboratory research to high-volume manufacturing. In this work, we demonstrate low bandgap, alloyed lead-tin perovskite films and devices prepared using solvent self-volatilization without the use of an antisolvent, gas-quench, or vacuum annealing step. By tuning the ratio of a volatile non-coordinating solvent to a coordinating non-volatile solvent and relying on solvent evaporation to induce crystallization, compact thin-films are produced that achieve optoelectronic and crystalline quality comparable to those produced with traditional antisolvent processing. Device power conversion efficiencies up to 16.8%, short-circuit current densities of 31.2 mA/cm2, fill-factors of 75.1%, and open-circuit voltages (Voc) of 722 mV are achieved. Photoluminescence quantum yield measurements are used to identify the source of Voc losses and offer insight into further device optimization. Overall, results suggest that self-volatilizing inks are tractable for depositing alloyed lead-tin MHP films. Furthermore, self-volatilizing inks may enable a closer R&D feedback loop between large-area deposition and small-scale prototyping by spin coating than traditional processes employing an antisolvent or quenching step.

978-1-6654-6060-6/23 $31.00 © 2023 IEEE

Aerial photoluminescence imaging of PV modules

Bernd Doll[1,2,3], Ernst Wittmann[1], Larry Lüer[2], Claudia Buerhop-Lutz[1], Jens A. Hauch[1,2],
Christoph J. Brabec[1,2], Ian Marius Peters[1]

[1]Forschungszentrum Jülich GmbH, Helmholtz Institute Erlangen-Nürnberg for Renewable Energy,
Erlangen, Germany, 91058
[2]Friedrich-Alexander University Erlangen-Nürnberg, Faculty of Engineering, Materials for Electronics and
Energy Technology, Erlangen, Germany, 91058
[3]Graduate School in Advanced Optical Technologies, Erlangen, Germany, 91058

Abstract — **On-site imaging of photovoltaic systems requires methods that are high-throughput, low-cost and contact free for commercial relevance. Photoluminescence imaging satisfies these requirements, but it has so far not been used for aerial imaging due to a lack of a drone-mounted system. In this study, we present first results captured by our in-house developed PLAI (photoluminescence aerial imaging) setup. The setup consists of a hexacopter aerial drone equipped with an illumination unit and a near infrared camera. The unit is capable of partially illuminating full size modules at night and capturing the photoluminescence response. We show that the setup can be used to detect and identify defects such as cracks and potential-induced degradation with high levels of confidence.**

I. INTRODUCTION

Field characterization of photovoltaic (PV) modules enables the detection of defects causing power loss or safety problems at an early stage. Cost-effective on-site examination is essential for commercial relevance and requires high throughput and low invasiveness. Luminescence imaging is suitable for this purpose, and it is used in the form of electroluminescence imaging (ELI) in the field [1] and in the lab [2]. ELI gives detailed semiconductor information and enables the detection of numerous types of defects [3]. Yet ELI is not ideal as it requires accessing the electrical circuitry of the installation to generate a detectable signal [4], adding effort and cost to the inspection process. A contact-free alternative is photoluminescence (PL) imaging (PLI), for which first outdoor-capable systems are currently developed [5], [6]. A challenge with existing ground-based outdoor PLI systems is the comparably low throughput. To overcome this limitation, we have developed a photoluminescence aerial imaging (PLAI) system, capable of performing PLI acquisition at the same pace as well-established thermal infrared images, yet with greatly improved image resolution compared to IR imaging.

A challenge with the drone-mounted system is a partial and non-homogeneous illumination of the modules, resulting in a superposition of excitation and response, which is potentially detrimental to the identification of module defects. Here, we demonstrate defect detection with the PLAI on five chosen defective PV modules with cracks, shunts and potential induced degradation (PID). The PL images taken with the PLAI setup during our first test flight are compared with indoor EL images

to prove that the mentioned defects and existing anomalies can be detected with high level of confidence.

II. PLAI CONFIGURATION

The aerial setup consists of a custom-made drone frame adjusted by *C5UAV GmbH*, four 100 W broad-white light emitting diode (LED), as visible in Fig. 1, modules and an indium gallium arsenide (InGaAs) detector camera, namely an "OWL 1280" from *Raptor Photonics Limited*. The filter system consists of a custom made long-pass filter in front of the camera with a cut-on wavelength of 970 nm and a short pass filter, namely "KG5" from *Schott*. The camera is controlled by an onboard computer and imaging data can be stored on a solid-state drive (SSD). The video signal is transmitted via *pixhawk* flight control to the remote controller. The take-off weight of the whole setup was about 16 kg. Fig. 2 depicts the PLAI system with the LEDs turned on.

Fig. 1. Aerial PL setup during test flight at day-light without camera unit.

Fig. 2. PLAI setup with four switched-on LED modules and a take-off weight of about 16 kg

The integration time of the InGaAs camera was 5 ms with an aperture of 2.4 and a lens focal length of 35 mm. These parameters enable mobile video recording without motion blur effects at a frame rate of about 10 Hz. The excitation intensity in the PV module plane was estimated at around 10-20 W/m².

III. EXPERIMENTAL SETUP

During the PL flight, different PV modules with different PV technologies were analyzed. Fig. 3 shows the PV modules and their positioning with the corresponding technology and performance parameters, which are summarized in Table I**Fehler! Verweisquelle konnte nicht gefunden werden.**.

Fig. 3.   Positioning of the PV modules A-E at a mounting angle of about 40°.

TABLE I

SUMMARY OF THE PV MODULE A-E HIGHLIGHTING THEIR TECHNOLOGIES AND DEFECTS

PV module	PV technology	Defect
A	mono-crystalline	cracks
B	shingled PERC	backside hits to Si
C	poly-crystalline	PID
D	no gab half cell	crack shunted
E	mono-c PERC	one cracked cell

PV module A, containing monocrystalline solar cells, has long cracks in the upper part of the PV mdule. These cracks can lead to a decrease in open circuit voltage (V_{OC}), which decreases the luminescence intensity, especially at low carrier injection, as is the case for the PLAI excitation. PV module B, with shingled passivated emitter and rear cells (PERC), exhibits strong local scratches in the backsheet of unknown origin to the rear side of the Si material, and three cell parts are affected. PV module C, using polycrystalline solar cells, was artificially degraded with an applied potential and shows signs of PID. PV module D contains half-cut cells connected without gap ("no gap module"). This module was aged with our mechanical stress setup, resulting in severe cell cracking and cell displacement in its center. PV module E, a module with mono-c PERCs, has cracks in one cell, potentially caused by transportation.

IV. COMPARISON BETWEEN PLAI AND EL IMAGES

To determine the detectability of the different defects with the PLAI setup we compared the PL images acquired outdoors to EL images taken in the lab. In Fig. 4 (A-EL to E-EL) these EL images with a low excitation of 100 mA are shown for all of the PV modules. With the EL images taken with a camera resolution of 2048 x 2048 pixels, the defects of the PV modules, as described in Table I, can be identified.

In Fig. 4 all PL images (A-PL to E-PL) of PV modules measured in the outdoor test with the PLAI setup are shown. For each PL image, the contrast was adjusted for optimized image appearance. The contrast for the PL video, accessible through the link in the figure caption or the QR-code in the image, was optimized globally and gives a qualitative comparison of the PL response of the different PV module technologies and their PV module defects. PV module B, D and E have high signal to noise levels and therefore clearly show defective cells. PV module A shows a PID-like pattern, which is generated by strongly cracked cells due to the low excitation intensity. For higher excitation intensities the influence of reduced V_{OC} would not be visible that strongly. The PL image of PV module C, with a low PL intensity level, is noisy, but the typical PID-pattern is clearly visible. During acquisition of the PL video roughly a throughput of 13.6 PV modules per minute and a flight speed of about 1.8 m/s was reached. For PV module B, the effect of the inhomogeneous illumination is apparent, resulting in lower brightness of the lower module part. Yet, the three defective cells are easily detectable. Additionally, the defects and cell intensity distribution is recognizable for the PV modules D and E. PV module C has the lowest PL signal, as expected from its age and type (multi-crystalline Si PV) module. The low PL signal level is highly influenced by the noise of the InGaAs detector. Therefore, to reduce the noise influence, the mean values of each cell were extracted.

All images indicate a high correlation if compared with the EL images. However, the PL images and the PL video indicate significant excitation inhomogeneity during the measurements. These inhomogeneities on one hand present a complication for quantitative evaluation; on the other hand they may be actively used in the future by imaging series- besides parallel resistance alterations, yielding deeper insight in underlying mechanisms. The reason is that the homogeneity adjustments for just four LED modules is difficult to set and reproduce. Especially, a strong difference between the upper and lower half of the PV modules can be determined.

V. SIGNIFICANCE OF THE WORK

Contactless high-throughput imaging techniques bear great potential for reducing the cost of spatially resolved imaging. On-site examination is essential to minimize PV module damage and electrical yield loss caused by additional transport to the laboratory. EL has already proven to be of great value in detecting a variety of defects in solar modules, which are detrimental for reliable and cost-efficient power production.

Since for PL imaging no electrical contacts to the module are necessary, power production losses (compared to EL) are reduced to zero at daytime. The findings of this work show that our PLAI tool provides a path for extended PL analysis in the field, and that it is a very promising tool for detecting performance-relevant defects on-site. A mobile, fully contactless and aerial luminescence method would significantly increase the throughput and is much simpler to use as current EL measurements setups.

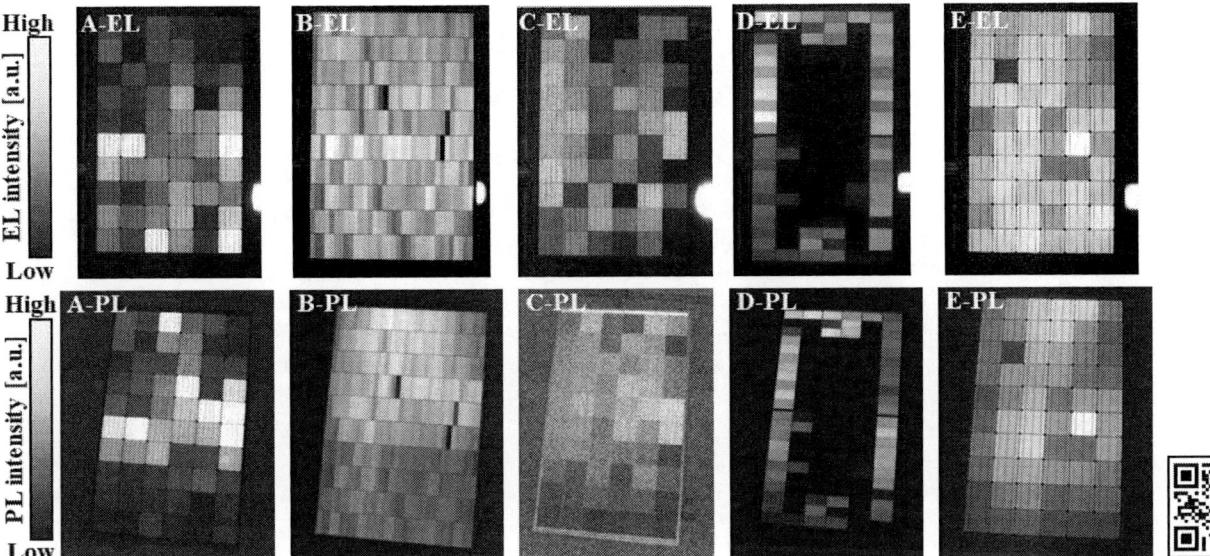

Fig. 4. A-EL to E-EL images for the PV modules A-E with an excitation of 100 mA and an integration time 30 s taken with a Si detector in the lab as ground truth. In the right lower corner of the EL images, a reference cell is visible which is used for quality control. A-PL to E-PL PL image screenshots of all of the PV modules A-E taken with the PLAI system from the PL video and an integration time of 5 ms. The PL video is accessible via QR-code and link https://youtu.be/BayU7MCfsYc.

VI. CONCLUSION

Compared to indoor EL images, similar defects can be detected with our PLAI setup. The examined PV modules showed various defects, such as cracks, PID and scratches in the backsheet. Each defect pattern can be detected with high reliability. Even the low homogeneity of excitation intensity of only 10-20 W/m² does not prevent defect detection. Our PLAI setup is limited due to the take-off weight, using in only four LED lamps for PL excitation, and resulting in low excitation intensity. PV module C shows that higher excitation intensities are required for reliable imaging of low efficiency PV modules.

The PL images presented in this work demonstrate the capability of air-based PL imaging. The advantages of the PLAI setup are that no contacts to the PV modules are required and the setup is not tied to the ground, allowing a potentially high throughput. In our initial experiment, we achieved a rate of 13.6 PV modules measured per minute. Modules in an actual large-scale installation stand closer together, implying a rate of 27 PV modules per minute with our current flight speed of about 1.8 m/s. Increasing flight speed to 5 m/s and image acquisition to 20 fps, which we believe to be feasible without loosing too much image quality, a throughput of up to 300 PV modules per minute (for PV modules with a width of about 1 m) and about 4 images per PV module, which is sufficient for super resolution post-processing, could be achieved. Demonstrating this throughput increase will be our next task.

The proven throughput of 13.6 PV modules per minute is already the highest one among other contactless luminescence imaging methods that we are aware of. The contactless switching modulation methods shown by the research group from the University of New South Wales [5] has a throughput

of about 10 PV modules per minute. Laser induced luminescence, presented by the research group of the Technical University of Denmark [6], shows promising results but was not used at commercial PV installations and therefore a throughput could not be determined.

Right now, we have taken first steps with the PLAI setup, demonstrating that high throughput, similar or better than that for IR imaging should be possible for luminescence imaging. For the five analyzed PV modules, defects like cracks, shunts and PID could be detected at high confidence level when compared to indoor EL images. Hence, PLAI combines the speed of IR aerial imaging with the detailed imaging of luminescence, extending drone based capabilities for outdoor PV-system inspection.

REFERENCES

[1] U. Jahn, et al., "Review on Infrared (IR) and Electroluminescence (EL) imaging for photovoltaic field applications," 2018.

[2] K. G. Bedrich, et al., "1st International Round Robin on EL Imaging: Automated Camera Calibration and Image Normalisation," 35th EUPVSEC., no. 1, pp. 1049–1056, 2018.

[3] T. Fuyuki, H et al., "Photographic surveying of minority carrier diffusion length in polycrystalline silicon solar cells by electroluminescence," Appl. Phys. Lett., vol. 86, no. 26, 2005.

[4] J. A. Tsanakas, et al., "Faults and infrared thermographic diagnosis in operating c-Si photovoltaic modules: A review of research and future challenges," Renew. Sustain. Energy Rev., vol. 62, 2016.

[5] O. Kunz, et al., "High Throughput Outdoor Photoluminescence Imaging via PV String Modulation," in IEEE 48th PVSC, 2021.

[6] G. A. dos Reis Benatto, et al., "Photoluminescence Imaging Induced by Laser Line Scan: Study for Outdoor Field Inspections," in 7th WCPEC, 2018.

Polarization type Potential Induced Degradation under Positive Bias in a Commercial PERC Module

Farrukh ibne Mahmood[1], Fang Li[1], Peter Hacke[2], Cécile Molto[3], Hubert Siegneur[3], GovindaSamy TamizhMani[1]

[1]Photovoltaic Reliability Laboratory, Arizona State University (ASU-PRL), Mesa, AZ, USA
[2]Reliability and System Performance, National Renewable Energy Laboratory (NREL), Golden, CO, USA
[3]Florida Solar Energy Center, University of Central Florida (UCF-FSEC), Cocoa, FL, USA

Abstract — **Potential induced degradation of the polarization type (PID-p) can reduce module performance in a relatively short period of time. PID-p can occur at both voltage polarities, but most studies are focused on degradation under a negative bias. This paper uses commercial bifacial passivated emitter and rear contact (PERC) cells within a monofacial glass-backsheet module construction to evaluate the impact of PID-p under a positive bias on the front side. Using the aluminum-foil (Al-foil) method, the module was stressed for PID in an environmental chamber. After the stress, the maximum power (Pmax) showed a decline of 3.1% at 1000 W/m^2 and 6.2% at 200 W/m^2. Recovery under light was also investigated. Complete recovery was observed at high irradiance, while a partial recovery was seen at lower irradiance. The outcomes of this study can help in understanding PID-p degradation under a positive bias and its recovery under the light.**

Keywords — *potential induced degradation, polarization, positive bias, PERC, recovery*

I. INTRODUCTION

PV modules are connected in series in commercial systems with high absolute system voltages ranging from 1000V to 1500V. The high voltage difference and elevated temperature/humidity cause leakage currents to flow between the grounded module frame and the cell. These leakage currents can be an indicator of PID [1]. Studies have shown that PID can reduce module performance by up to 30% [2]. Three main types of PID phenomenon have been reported in the literature, i.e., PID-shunting (PID-s), PID-polarization (PID-p), and PID-corrosion (PID-c) [1].

When understanding the PID-p effects in PV modules, most studies focus on the degradation under a negative bias on the rear side of the p-PERC bifacial modules. Yamaguchi et al. reported a rapid degradation due to PID-p when PV modules were stressed at -1000V and -1500V [3], [4]. Janssen et al. suggested minimizing PID-p by modifying the dielectric antireflection/passivation stack composition while using stress conditions of -1000V for PID [5]. Similarly, Luo et al. studied the impact of illumination on PID-p in PV modules by stressing them at -1000V [6]. In another study, they stressed the modules at +1000 V but report little to no degradation. Moreover, the samples used in these studies were all one-cell modules instead of commercial ones [7].

Therefore, in this paper, we use a commercial p-PERC PV module to study the effect of PID-p under a positive bias on the module front. Using the Al-foil method in an environmental chamber, the module was stressed for PID. Recovery under sunlight was also carried out after the PID stress. Pre- and post-characterization tests such as Flash IV and electroluminescence (EL) were performed to understand the changes in performance, post-PID and post-recovery.

II. EXPERIMENTAL SETUP

This study used one commercial monofacial glass-backsheet (white backsheet) module employing 144-half-cut bifacial PERC cells. The cell dimensions were 78 x 156 mm, and the module dimensions were 2141 x 994 mm. The encapsulant used in the module was ethylene vinyl acetate (EVA). Kapton tape was used to cover the perimeter of the module frame to prevent contact of the Al-foil with the frame. The Al-foil was placed on the front glass, and 3M Al-tape (0.028 mm) was placed between the foil pieces to ensure electrical connectivity. An additional layer of thick insulating roofing membrane was put on top of the Al-foil to ensure a good contact between the glass and Al foil. The module was placed horizontally on a rack inside the environmental chamber, and extra weight was put on the module front to ensure proper contact between the foil and the front glass.

The module terminals were shorted and connected to the positive end of the power supply, while the front Al-foil was connected to the negative terminal of the power supply. So, the cell was at a positive bias with respect to the front side under stress. The module was stressed at +1500V, 25°C, and 54% relative humidity (RH) (to closely match with typical climatic condition) for 168 hours as per International Electrotechnical Commission (IEC) standard 62804-1. Leakage current was monitored for the duration of the stress. For recovery, after PID, the module was placed under sunlight with an average dosage of 19 kWh/m^2 at open circuit conditions on the front side.

Characterization tests, including Flash IV and EL, were performed pre-PID, post-PID, and post-recovery to measure the change in performance parameters. The Flash IV was done using a Spire 5600 at standard test conditions (STC) (1000W/m^2 and 25°C) and low irradiance of 200W/m^2 (also at 25°C) at SolarPTL. The IV parameters such as Pmax, short circuit current (I_{SC}), open circuit voltage (V_{OC}), fill factor (FF), maximum current (I_{MAX}), and maximum voltage (V_{MAX}) were acquired using the IV sweep. The EL was performed at 100% I_{SC} with an exposure for 30 seconds. The average gray value for the EL images was also obtained using an image analysis tool.

III. RESULTS AND DISCUSSIONS

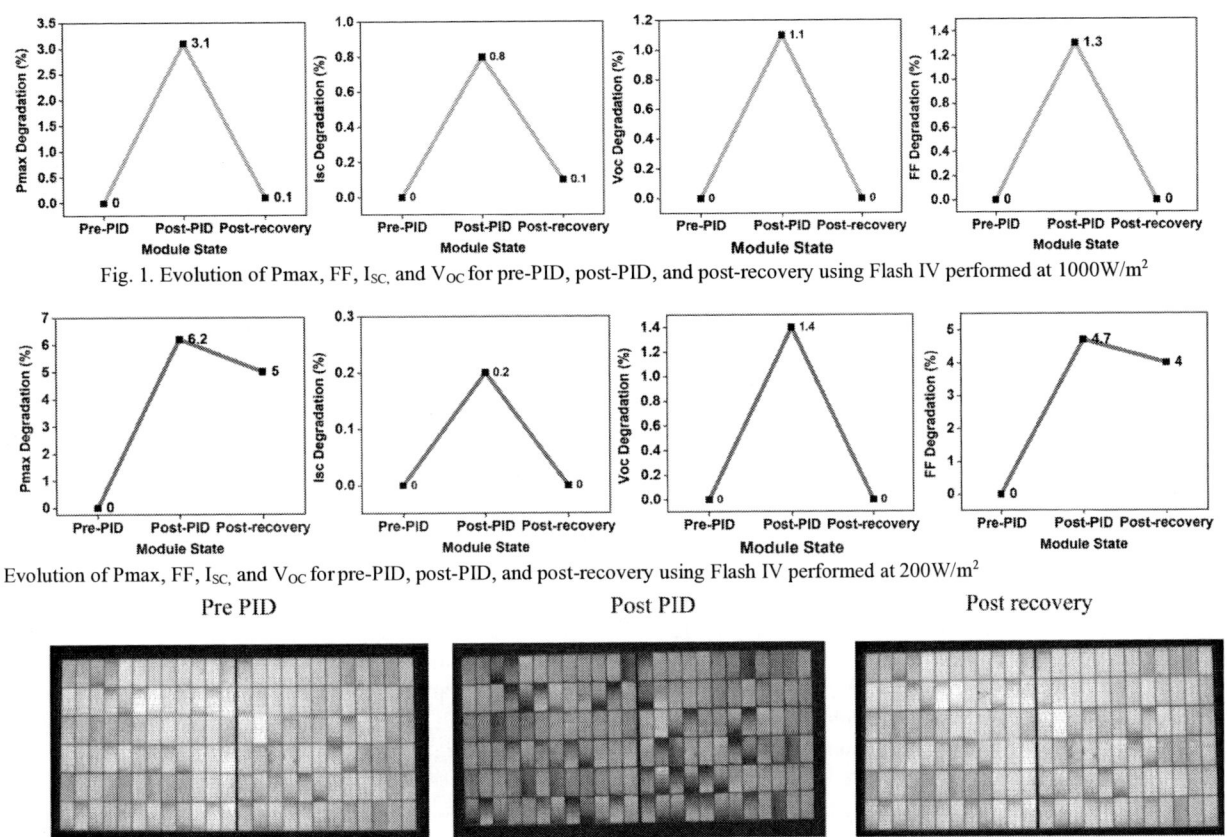

Fig. 1. Evolution of Pmax, FF, $I_{SC,}$ and V_{OC} for pre-PID, post-PID, and post-recovery using Flash IV performed at 1000W/m^2

Fig. 2. Evolution of Pmax, FF, $I_{SC,}$ and V_{OC} for pre-PID, post-PID, and post-recovery using Flash IV performed at 200W/m^2

| Pre PID | Post PID | Post recovery |

EL Gray Value Degradation

| 0% | -26.1% | 0% |

Fig. 3. EL images for pre-PID, post-PID, and post recovery

Fig.1 shows the % degradation in Pmax, I_{SC}, V_{OC}, and FF at high irradiance at different states for the tested module. Similarly, Fig. 2 shows the % change for the same parameters at low irradiance at various states for the module. Higher levels of degradation and partial recovery for Pmax and FF are seen at lower irradiance. Fig. 3 shows the EL images with gray values at 100% I_{SC}. Moreover, post-recovery EL images suggest a complete recovery from PID.

We believe that the degradation in this module under a positive bias is due to PID-p. Since PID-s involve the movement of Na$^+$ ions to the cell junction, the positive ion migration to the positive electrode cannot occur under a positive bias [8]. So, the decline in Pmax should arise either due to PID-p or PID-c. PID-c is not recoverable under light exposure [9]. Since the stressed module almost completely recovers under sunlight, we believe that the observed degradation may be attributed to the PID-p mechanism [9].

The PID-p mechanism occurring in PERC (with AlO$_x$ and SiN$_y$ passivation) cells has been explained by Sporleder et al. [10] using the K center model (for negative bias on the rear side, for positive bias on the front, a mechanism has not been reported in the literature). A Si dangling connection linked to three N atoms forms a K center. K centers (K^0, K$^-$, or K$^+$) can have a neutral, negative, or positive electrical charge. Under a

voltage bias, the K$^+$/K^0 can accept electrons, or K$^-$/K^0 can release electrons. Pre-PID, the passivation layer has a negative charge, and the K-center charge states within SiN$_y$ are arbitrarily dispersed. When the PID stress is applied, K$^-$/K^0 releases electrons causing the K-center charge states within SiN$_y$ to become positive (under positive bias, as in this study, we believe the K-center charge states would become negative). This leads to an approximately zero charge for the passivation layer, increasing the surface recombination velocity (SRV). The increase in SRV reduces the intrinsic field effect passivation of the AlO$_x$ layer (for the module under study, we are not sure about the passivation layer, and therefore the mechanism presented is based on the current literature), and the module power decreases [10]. Positive bias applied to the cell repels positive charge from the front passivating dielectric. Said another way, it attracts net negative charge to the passivating layer. When the net charge in the front passivating layer becomes more negative, it attracts more minority carrier holes from the n$^+$ emitter layer at the front of the n$^+$/p PERC structure. When the industry moves toward more selective emitters, the sensitivity to this effect increases because the selective emitter, with lower phosphorus doping, provides less front surface field screening the minority carriers from cell

978-1-6654-6060-6/23 $31.00 © 2023 IEEE

front recombination at the dielectric interface with the Si emitter.

When exposed to sunlight, almost full recovery is observed because photogenerated electrons are captured by K^+ centers and nullify them. This changes the K^+ charge to a mobile hole that sinks via the SiN_y layer leading to fewer K^+ centers that can draw the minority carrier electrons to the dielectric interface. Hence, the SRV decreases, and recovery is seen [11].

Furthermore, higher degradation level at lower irradiance is linked to the shift in surface recombination injection-level dependency caused by the PID-p progression, which alters the surface charge density of the AlO_x/SiN_y layer [12]. Incomplete recovery of PID at lower irradiance is reported by other works [13], [14]. However, the mechanism is still not clearly understood and will be the topic of future studies.

IV. Conclusions and Future Work

PID-p mechanism can quickly degrade under voltage bias and recover under light in PV modules compared to other types of PID mechanisms. Moreover, most PID studies target degradation under a negative bias, whereas PID-p can also happen under a positive bias in PERC modules, as observed in this study. Using the Al-foil method for PID testing, we see a 3.1% and 6.2% degradation in Pmax at high and low irradiance levels, respectively. Under sunlight, complete recovery is seen when the module is tested at STC. However, recovery is only partial when IV is measured at lower irradiance. The results of this study can help the PV community understand PID-p in PERC cells under a positive bias and its recovery under sunlight.

Future studies may need to be focused on understanding the mechanisms behind incomplete recovery at low irradiance and how outdoor field conditions can propagate or recover PID-p.

Acknowledgement

This material is based upon work supported by the Department of Energy, Office of Energy Efficiency and Renewable Energy (EERE), under Award Number DE-EE-0009345.

References

[1] F. I. Mahmood and G. TamizhMani, "Impact of Anti-soiling Coating on Potential Induced Degradation of Silicon PV modules," in *2022 IEEE 49th Photovoltaics Specialists Conference (PVSC)*, 2022, pp. 1198–1200.

[2] M. Dhimish and G. Badran, "Recovery of photovoltaic potential-induced degradation utilizing automatic indirect voltage source," *IEEE Trans. Instrum. Meas.*, vol. 71, pp. 1–9, 2021.

[3] S. Yamaguchi, K. Nakamura, A. Masuda, and K. Ohdaira, "Rapid progression and subsequent saturation of polarization-type potential-induced degradation of n-type front-emitter crystalline-silicon photovoltaic modules," *Jpn. J. Appl. Phys.*, vol. 57, no. 12, p.

122301, 2018.

[4] S. Yamaguchi, B. B. Van Aken, M. K. Stodolny, J. Löffler, A. Masuda, and K. Ohdaira, "Effects of passivation configuration and emitter surface doping concentration on polarization-type potential-induced degradation in n-type crystalline-silicon photovoltaic modules," *Sol. Energy Mater. Sol. Cells*, vol. 226, no. February, p. 111074, 2021.

[5] G. J. Janssen *et al.*, "Minimizing the polarization-type potential-induced degradation in PV modules by modification of the dielectric antireflection and passivation stack," *IEEE J. Photovoltaics*, vol. 9, no. 3, pp. 608–614, 2019.

[6] W. Luo *et al.*, "Investigation of the Impact of Illumination on the Polarization-Type Potential-Induced Degradation of Crystalline Silicon Photovoltaic Modules," *IEEE J. Photovoltaics*, vol. 8, no. 5, pp. 1168–1173, 2018.

[7] W. Luo *et al.*, "Elucidating potential-induced degradation in bifacial PERC silicon photovoltaic modules," *Prog. Photovoltaics Res. Appl.*, vol. 26, no. 10, pp. 859–867, 2018.

[8] V. Naumann *et al.*, "Explanation of potential-induced degradation of the shunting type by Na decoration of stacking faults in Si solar cells," *Sol. Energy Mater. Sol. Cells*, vol. 120, pp. 383–389, 2014.

[9] K. Sporleder *et al.*, "Local corrosion of silicon as root cause for potential-induced degradation at the rear side of bifacial PERC solar cells," *Phys. status solidi (RRL)--Rapid Res. Lett.*, vol. 13, no. 9, p. 1900163, 2019.

[10] K. Sporleder, V. Naumann, J. Bauer, D. Hevisov, M. Turek, and C. Hagendorf, "Time-resolved investigation of transient field effect passivation states during potential-induced degradation and recovery of bifacial silicon solar cells," *Sol. RRL*, vol. 5, no. 7, p. 2100140, 2021.

[11] B. M. Habersberger and P. Hacke, "Impact of illumination and encapsulant resistivity on polarization-type potential-induced degradation on n-PERT cells," *Prog. Photovoltaics Res. Appl.*, vol. 30, no. 5, pp. 455–463, 2022.

[12] W. Luo *et al.*, "Elucidating potential-induced degradation in bifacial PERC silicon photovoltaic modules," *Prog. Photovoltaics Res. Appl.*, vol. 26, no. 10, pp. 859–867, 2018.

[13] J. Oh, S. Bowden, and G. S. TamizhMani, "Potential-Induced Degradation (PID): Incomplete Recovery of Shunt Resistance and Quantum Efficiency Losses," *IEEE J. Photovoltaics*, vol. 5, no. 6, pp. 1540–1548, 2015.

[14] J. Oh, S. Bowden, and G. TamizhMani, "Application of reverse bias recovery technique to address PID issue: Incompleteness of shunt resistance and quantum efficiency recovery," in *2014 IEEE 40th Photovoltaic Specialist Conference (PVSC)*, 2014, pp. 0925–0929.

Three-dimensional and multimodal X-ray microscopy reveals the impact of voids in CIGS solar cells

Giovanni Fevola, Christina Ossig, Mariana Verezhak, Jan Garrevoet, Martin Seyrich, Dennis Brueckner, Johannes Hagemann, Frank Seiboth, Andreas Schropp, Gerald Falkenberg, Peter Stanley Jorgensen, Christian Strelow, Tobias Kipp, Romain Carron, Jens Wenzel Andreasen, Michael E. Stuckelberger

Deutsches Elektronen-Synchrotron DESY, Hamburg, Germany

Universität Hamburg UHH, Hamburg, Germany

Paul Scherrer Institute PSI, Villigen, Switzerland

Technical University of Denmark DTU, Kgs. Lyngby, Denmark

Empa, Dübendorf, Switzerland

The formation of small voids in polycrystalline absorbers has been repeatedly reported, whether after device fabrication or degradation. Although it certainly increases risk of delamination and complicates deposition of additional top layers and cells, it is not clear whether or how detrimental it is for photovoltaic performance. In our study, using synchrotron imaging, we non-destructively probe local performance deficits attributable to voids and highlight the complex 3D nature of structural defects in high-efficiency thin-film CIGS solar cells. We find that, although possibly detrimental at a local level, voids only have a minor effect at the device level, given the abundance of voids and the high efficiency of the cell. Our quantification of the absolute electron densities at the nanoscale may enable the development of adequate comprehensive models simulating structural and electronic defects. In the full presentation, we will show high-resolution images and 3D renderings of nanostructures in CIGS and put them in the context of spatially resolved performance measurements. Furthermore, we will showcase recent developments in X-ray imaging and give an outlook to technical developments at new X-ray sources that will become available in the coming years and enable a further increase of sensitivity and spatial resolution.

978-1-6654-6060-6/23 $31.00 © 2023 IEEE

Optical characterization and loss simulation of encapsulation materials and back sheets for PERC+ solar modules

Tim Lukas Brockmann[1], Henning Schulte-Huxel[1], Susanne Blankemeyer[1], Tobias Wietler[1,2]

1) Institute for Solar Energy Research Hamelin (ISFH), 38160 Emmerthal, Germany
2) Institute of Electronic Materials and Devices, Leibniz University Hannover, 30167 Hannover, Germany

Abstract — In a PV module the J_{sc} is impacted by multiple factors such as geometrical effects, reflection on interfaces and absorption properties of the module components. A major contributer to parasitic absorption are encapsulation materials which are based on different polymers (ethylene-vinyl acetates and polyolefins) with different UV-properties (UV-absorbing and UV-transparent). This raises the question which material to use and how to find the optimal material for a given solar cell. This can be done with simulations of cell-to-module (CTM) losses. However the optical properties of those encapsulating polymers need to be known.

In this work multiple encapsulants are characterized in a wavelength range from 300 to 1200 nm to determine their optical constants. Those constants are used for ray-tracing simulations of PERC+ silicon half cell modules with glass or backsheet rear sides to identify losses in J_{sc} due to absorption in those materials. To gain further information about optical CTM loss mechanisms of these encapsulation materials different rear sides are simulated. As a main conclusion it is shown that encapsulants and rear sides can be optimized independently to maximize the module J_{sc}. This can reduce the number of required experiments for a number of L encapsulants and M rear sides from L × M to L + M − 1 when optimizing the optical properties of a solar module.

I. INTRODUCTION

A crucial step in the implementation of solar cells into the environment is the module integration. It is used to scale the power output to higher system voltages and also to protect the cells from environmental impacts. [1]

Different encapsulation polymers like ethylene-vinyl acetates (EVA) or polyolefin (PO) are used to build these modules. However EVA shows a tendency to degrade like yellowing or decomposition into acedic acid. [2,3] That is part of the reason why the ITRPV predicts a shift from EVA towards PO over the next years in module module fabrication. [4]

Another major factor in the decision which material to process are the optical properties of these encapsulation materials. Both polymers offer UV-transparent (UVT) and UV-absorbing (UVA) materials with different advantages: UVT materials offer a higher short current density J_{sc} and less yellowing while UVA materials can be used to protect delicate samples like perovskites from UV-radiation. [5, 6]

This work focuses on the differences in optical properties between polymer groups and polymers within the same group. The first step is the optical characterization to determine their opticals constants. Those are used as input parameters for ray-

tracing simulations using DAIDALOS. [7] As an example system PERC+ half cells modules are used. The impact on different types of modules will be assessed by simulating different rear sides: a glass module, a white backsheet module and rear sides with 0 % and 100 % reflection.

II. OPTICAL CHARACTERIZATION

The optical characterization is necessary to aquire the optical constants of encapsulation materials that are used in the optical simulation.

A. Sample fabrication

The unlaminated encapsulation material is cut into 10 x 10 cm² squares with a thickness of 0.45 to 0.55 mm depending on the material. The material is stacked between two non-sticking surfaces and laminated using a laboratory laminator. The adaptation of the lamination process in length and peak temperature is important to ensure that cross-linking encapsulation materials are above the required temperature throughout the sample. A typical process ranges between 120 to 160 °C and 10 to 30 minutes. A small 2 x 2 cm² sample is cut out of the sample to gain a homogenous thickness over the measurement area.

We processed 13 different materials from 8 different manufacturors. Four of these materials are UVT POs, the nine remaining are each three UVA POs, three UVT EVAs and three UVA EVAs.

B. Measurement and unceartainty assessment

We measure transmission, reflection and ellipsometry from 300 to 1700 nm using a photospetrometer (Cary 5000) and the ellipsometer (Wollaam M2000-UI). An optical model is built to fit the transmission and ellipsometry data using the Software WVASE. For the evaluation of measurement a monte-carlo approach is utilized. [5] We generate new data within the unceartinty of the measurement. In 1000 iterations the generated data is fitted using the model and saving the generated data and optical constants for each iteration. To determine the results the mean value and the standard deviation of the iterations are calculated. This way we get the refractive index n and the extinction coefficient k with their corresponding unceartainty.

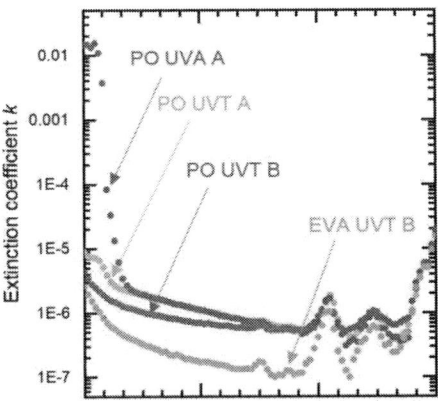

Fig. 1: Extinction coefficients k in dependence of the wavelength for different types of encapsulation materials from different manufacturers (A and B) with UVA and UVT properties

C. Results of encapsulant evaluation

Fig. 1 shows the the extinction coefficient k over the wavelength for four representative samples from two different manufacturers A and B as an example. The polyolefins of manufacturer A show a change in the extinction coefficient for wavelengths below 400 nm, changing over four orders of magnitude. The PO and EVA from manufacturer B show slighty varying extinction coefficients. When comparing the refractive indices n (not shown) of the different manufacturers they show values differing by about 0.02 which corresponds to a change in reflection less than 0.05 %. When estimating the reflection between a glass front and the encapsulation material the additional reflection is about 0.01 %.

III. OPTICAL SIMULATIONS

For the optical simulations the ray-tracer DAIDALOS is used. It uses a multi-domain approach to simulate photons in 3 dimensions over multiple orders of magnitude in size (nm up to m). Table 1 (last page) shows an overview of the input parameters for the simulation. We simulated half cells - the length of the cell and module components is equal to half the width except for the cell-to-cell gap, which is the same for length and width. The reflection, absorption and transmission probability are calculated for each wavelength in every module component. For 10000 photons per wavelength between 300 and 1200 nm with a step width of 10 nm the photon interacts with the module components with their respective probability.

The results are absorptions in each module component and reflection for each interface resoluted by wavelength. The absorption in the cell absorber is intergraded over the 1.5 AMG spectrum, to calculate the J_{sc}.

Loss current densities for parasitic absorption or reflection on interfaces in module components other than the cell are calculated in the same way and are categorized as optical CTM losses. We simulate four different rear sides for each characterized material: one with a glass, one with a white backsheet and one each with 0 % reflection and 100 % reflection respectively.

A. Results of the ray-tracing simulations

In Fig. 2 the J_{sc} and the loss current density in the encapsulation materials calculated from the ray-tracing simulations for different rear sides are shown. The losses are the sum of reflection and absorption caused by the encapsulation above and below the cell.

In Fig. 2a) the J_{sc} of a glass-glass modules over the cut-off wavelength of UVA materials from different manufacturers are shown. For every material except for the PO from manufacturer C the J_{sc} decreases with increasing cut-off wavelength. The PO from manufacturer C shows a lower J_{sc} than the EVAs with the same cut-off wavelength from manufacturer F and G.

The white backsheet rear side in Fig. 2b) shows J_{sc} about 5 % higher than those of the module with the glass rear side in Fig. 2a). The simulation result for the UVT EVA from manufacturer

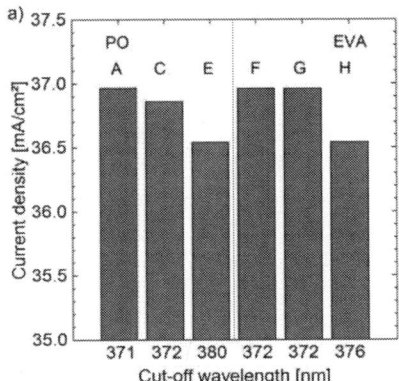
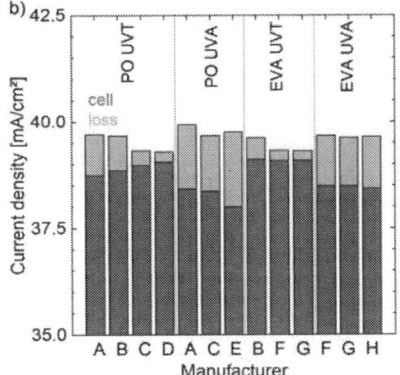
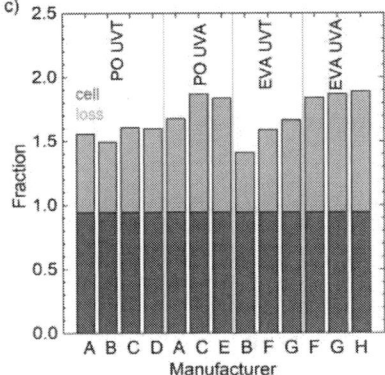

Fig. 1: Results of DAIDALOS simulations, a) J_{sc} for UVA materials with glass rear side over cut-off wavelength, b) J_{sc} and loss current density in encapsulation material for a module with white back sheet, c) ratio of cell current density for a rear side with 100 % reflection and 0 % reflection and lost current density in encapsulation material for a rear side with 100 % reflection and 0 % reflection

B is unexpected, as the loss current density is higher than those of manufacturer F and G, however the module has a higher J_{sc}. The highest J_{sc} for the PO and the highest J_{sc} for the EVA are within each others unceartainty margins.

Fig. 2c) shows the ratio of the two simulations with a reflection of 100 % and one with a reflection of 0 % at the rear side for each encapsulant. The ratio of J_{sc} is constant at 94.4 % for the different materials, varying at most 0.12 %. The ratio of loss current density due to absorption in the encapsulant material varies by up to 50 %.

B. Interpretation of the ray-tracing simulations

Multiple reasons can be found for the higher change in loss current density than the change in J_{sc}. Photons that are not absorbed in the encapsulation material may be absorbed in other layers, reducing their impact on the J_{sc}. Fig. shows that absorption takes place in the wavelength range between 1100 and 1200 nm, where the absorption of silicon decreases, especially a local maximum at 1160 nm. In the wavelength-resolved simulation results the absorption in the encapsulation material for the wavelength between 1100 and 1200 nm increases by a factor of 2 to 8 with a higher reflecting rear side. Thus the 100 % reflection on the rear side mainly increases the propability for photons to be absorbed with multiple passes.

IV. Discussion

The changes in J_{sc} are higher between different manufacturers than between EVAs and POs of the same UV category. However due to the light-scattering properties of some POs their absorption may be overestimated in this work, which could be the case for manufacturer A and B for UVT material and manufacturer C for UVA material. [10]

Changing from UVT to UVA materials decreases J_{sc} which is to be expected. [5] For the module integration of UV-degrading cells different UVA materials offer different cut-off wavelengths and optimization potential. For the optimization of

J_{sc} the adjustment of the encapsulation material and the rear side can be done independently, as shown in this work. The ratio of the J_{sc} for different encapsulation materials is constant for the variation of the rear side of the module. This can reduce the number of required experiments for a number of L encapsulants and M rear sides from L × M to L + M — 1 when optimizing the optical properties of a solar module.

We will publish the optical constants characterized within this work with the proceedings.

References

[1] A. E. Amrani et al., "Solar Module Fabrication", *International Journal of Photoenergy, vol. 2007, Article ID 027610*, 2007

[2] A. Morlier et al. "Ultraviolet fluorescence of ethylene-vinyl acetate in photovoltaic modules as estimation tool for yellowing and power loss", *IEEE 7th WCPEC, 2018, pp. 1597-1602*, 2018

[3] *A. Omazic et al.,* "Relation between degradation of polymeric components in crystalline silicon PV module and climatic conditions : A literature review," *Sol. Energy Mater. Sol. Cells, vol. 192, no. September 2018, pp. 123–133, 2019.*

[4] ITRPV "International Technology Roadmap for Photovoltaic (ITRPV): 2021 Results." *13th edition, 2022*

[5] M. R. Vogt, "Development of physical models for the simulation of optical properties of solar cell modules" *Diss. Technische Informationsbibliothek (TIB)*, 2016

[6] Lang, F., et al., *N. H., Adv. Mater. 2018, 30, 1702905*, 2017

[7] Holst, H., P. P. Altermatt, and R. Brendel. "Daidalos-A plugin based framework for extendable ray tracing." *Proceedings of the 25th EUPVSEC 2150*, 2010

[8] E. D. Palik, "Handbook of Optical Constants of Solids", *p. 350 - 357, Academic Press, Inc.*, 1985

[9] E. Shiles et al. "Self-consistency and sum-rule tests in the Kramers-Kronig analysis of optical data: Appications to aluminum", *Phys. Rev. B 22, 1612 – 1628*, 1980

[10] J. Eymard et al., "Characterization of UV–Vis–NIR optical constants of encapsulant for accurate determination of absorption and backscattering losses in photovoltaics modules", *Solar Energy Materials and Solar Cells, Volume 240*, 2022

TABLE I
Overview of the Parameters for the Optical Simulations

Module component	Thickness	Width	Quantity	Remark	Source
Cell-to-cell gap	-	2 mm		For each module component	-
Front / rear Glass	2 mm	168 mm		With ARC coating	[5]
Front Encapsulation	400 µm	168 mm		Variation with characterized materials	This work
Interconnector	380 µm	380 µm	9	Wire interconnectors	ISFH
Front Finger	12 µm	20 µm	60		[8]
Front Passivation	75 nm	166 mm		Silicon nitride	ISFH
Cell Absorber	150 µm	166 mm		Silicon half cell, length = width/2	[5]
LCO	10 µm	30 µm	75		[5]
Rear Passivation	80 nm + 5 nm			Silicon nitride and aluminum oxide	ISFH
Rear Finger	20 µm	175 µm	75		[9]
Rear Encapsulation	400 µm	168 mm		Variation with characterized materials	This work
Backsheet				Variation with characterized materials	[5]

Integration of lateral power MOSFETs into IBC c-Si solar cells with poly-Si passivating contacts

David A. van Nijen, Patrizio Manganiello, Yavuzhan Mercimek, Tristan Stevens, Guangtao Yang,
René A.C.M.M. van Swaaij, Miro Zeman, Olindo Isabella

TU Delft, Delft, Netherlands

Power electronics (PE) is essential for the optimal operation of photovoltaics (PV). An increasing share of PV systems makes use of module-level PE, and various shade-tolerant PV module designs with sub-module PE have been proposed. Furthermore, PE is used in stand-alone devices powered by PV-storage solutions. One way to facilitate further implementation of PE in PV applications is to integrate PE components into crystalline silicon (c-Si) PV cells. In this contribution, we focus on the integration of transistors, which are crucial for power converters as well as reconfigurable modules, into PV cells. It is important for the cost-effectiveness of the final product that as many fabrication steps as possible apply to both the PV cell and the transistor. Thus, we manufacture lateral power MOSFETs with a process flow highly similar to that of IBC c-Si solar cells with ion-implanted poly-Si passivating contacts. The devices are characterized in both dark and illuminated conditions. Based on these results, we reflect on the remaining challenges for real-world integration of transistors into PV cells. Furthermore, we describe a process flow for the combined fabrication of IBC PV cells and MOSFETs on a single c-Si wafer.

978-1-6654-6060-6/23 $31.00 © 2023 IEEE

The Formation of Dendrites in overtreated CdSeTe/CdTe Solar Cells

Vladislav Kornienko, Luke Jones, Kieran Curson, Ali Abbas, Rachael Greenhalgh,

Yau Yau Tse, and Michael Walls

CREST/Department of Materials, Loughborough University, Loughborough, LE11 3TU, UK

Christian Drost[3], Bettina Spaeth[3], Bastian Siepchen[3]

CTF Solar, Manfred-von-Ardenne-Ring 4, 01099 Dresden, Germany

Abstract — **Aggressive cadmium chloride activation conditions (>450°C) used to activate a FTO/CdSeTe/CdTe device in an oxygen enviroment results in the formation of dendrites at the front and back of the device. Most of the dendrite formation occurs at the front interface (FTO/CdSeTe). Optical Microscopy and X-Ray Photoelectron Spectroscopy (XPS) have been used to study the dendrite morphology and chemical composition. Optical images revealed that the dendrites can grow >500μm in length causing a range of discolourations from oxidation (dark brown/pink). The XPS analysis provides evidence for the presence of cadmium oxychlorides.**

I. INTRODUCTION

The cadmium chloride ($CdCl_2$) activation treatment is an essential step for high efficiency CdTe-based devices. The current record efficiency is 22.1%[1]. The $CdCl_2$ treatment improves the efficiency of CdTe device by removing or passivating defects [2], increasing grain size, grading of the absorber by selenium diffusion[3], passivating grain boundaries at the junction interface [4] and randomizing the absorber texture [5].

The process conditions for the cadmium chloride activation process are critically important. Critical parameters include temperature and annealing time. A narrow optimum process window exists and either side of the optimum, devices can be undertreated or overtreated both resulting in reduced performance. If undertreated, then insufficient chlorine is present for passivation. Grain growth and texture randomisation is incomplete and interdiffusion of Se into CdTe absorber is not optimized. However, overtreatment at high temperatures and long process times (especially in an oxygen environment) can lead to excess chlorine at the interfaces and at the back of the device. Although higher temperatures are desirable for interdiffusion and grain growth, aggressive treatments can result in excess chlorine and void formation at the junction which can then lead to delamination. Cadmium rich regions have been observed at the FTO interface [6].

During over-activation in atmospheric conditions, we have observed a variety of tree-branch shaped defects forming at the front interface of FTO/CdSeTe/CdTe devices. Observation of such defects has not been widely reported and is not well understood. The phenomenon has been reported previously in CdTe nano-particles[7] and has also been observed to occur in CdS/CdTe devices [8], but not in the more recent selenium alloyed CdSeTe/CdTe devices. Using a light microscope, the defects appear as white spots that conglomerate into dendritic patterns. These have been identified elementally by EDS and thought to be cadmium metal or a cadmium rich phase [8]. However, no technique capable of identifying specific chemical bonding, has been used.

A mechanism for dendrite formation has been suggested to be due to an uneven $CdCl_2$ activation or vapor transport of cadmium from the surrounding areas to localised spots[8]. The activation in an oxygen environment may produce a range of oxides. In this study, we investigate the chemical composition of the dendrites using X-ray Photoelectron Spectroscopy (XPS) to determine the chemical composition of the dendrites found in overtreated FTO/CdSeTe/CdTe devices. We show that the dendrites likely contain cadmium oxychlorides.

II. EXPERIMENTAL

The CdSeTe/CdTe devices were fabricated at CTF Solar GmbH. The absorber was deposited by Closed Space Sublimation (CSS). In this study, a 20mg $CdCl_2$ per 100cm^2 of methanol based solution was used for the activation treatment. Overtreatment most commonly occurs, when the sample temperature is reaching >450°C for at least 5 min. The time needed to achieve overactivation is dependent on the absorber thickness. The thinner the absorber, the lower the time for over-activation to occur, as there is less material to treat. The absorber thickness used in the current study was about 3μm. Samples from the deposition were cut down to 10x10 cm^2 area before activation.

Optical images were obtained from both the front side (FTO/CdSeTe) and from the back surface of the films using a Keyence optical microscope. Note that the images taken at the front are through the glass substrate and the FTO. By adjusting the focus, it was possible to examine the different interfaces.

Surface compositional analysis was obtained using a Thermo Scientific K-Alpha X-ray photoelectron spectrometer (XPS) system. The XPS analysis was charge corrected to the C1s adventitious carbon peak at 284.8eV. To access the front interface, the absorber layer was cleaved from the FTO/glass substrate using liquid nitrogen.

III. RESULTS AND DISCUSSION

A. Optical Microscopy

Fig. 1 is an optical image obtained from the FTO/CdSeTe interface through the glass substrate. It clearly shows the damaged film and the dentritic structure at the interface. This is the type of morphology we observed at this interface caused by overactivation. The dendritic patterns shown up as bright yellow with black patterns next to them as seen in Fig.1, with a small less affected area remaining. Further work is needed to disseminate composition between those specific areas.

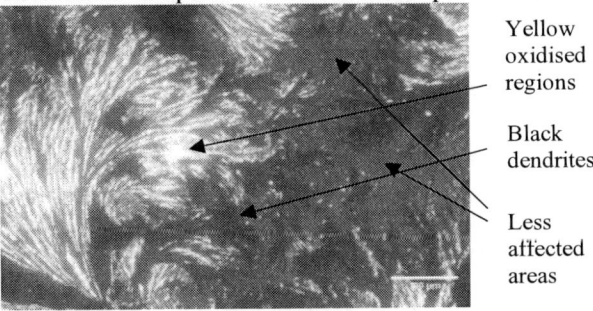

Fig. 1. Optical image taken through the substrate glass with a focus at the FTO/CdSeTe interface showing the formation of dendrites.

To clarify, Fig.2 shows schematic diagram of where the dendrites are located at the interface(not to scale).

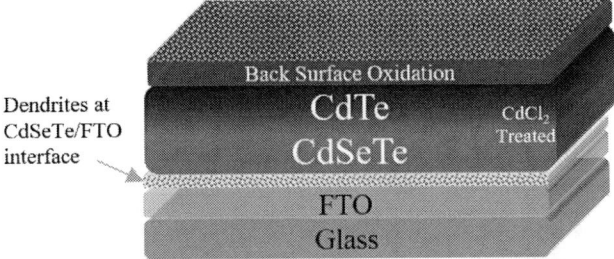

Fig.2 Diagram of the device stack with the location of the dendrite formation.

The size of the dendrites depends on the onset of the nucleation and growth rate of the suspected cadmium oxychlorides or other oxides phases. Some regions of the film are darker and do not show the same morphology as the areas with dendrites, indicating dendrite formation is not uniform in the film. Fig.3 shows typical dendrites of 500µm in length and spreading into branches. Preferred growth orientation of the dendrites could be confirmed with further EBSD work. There are smaller bright patches near the dendrites, which are approximately 1-10µm in size. The chemical composition of such bright spots could be revealed by futher SEM/EDS or TEM/EDS work.

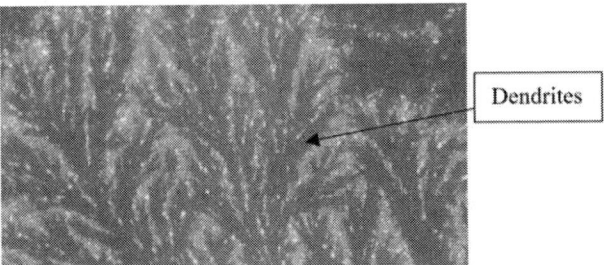

Fig.3. Large dendrites spanning >500µm in length.

Fig.4 (a),(b) and (c) show further examples of other distinct morphologies located at the FTO/CdSeTe interface. Fig. 4 (a) shows a typical dendrite microstructure with pink contrast. Fig 4(b) shows a darker patch forming without the dendrite branches. Fig.4 (c) reveals a small blue coloured circular region about 50µm in diameter. This could be caused by inhomogeneities in the temperatures/conditions inside the activation furnace, as well as to absorber thickness variations and $CdCl_2$ inhomogeneities. However, the aggressive activation in oxygen environment with temperatures exceeding 450°C, even for a short period of time could induce oxide formation. The brown/pink areas could also indicate a form of oxide present. The observed different colour contrast in the optical microscope reflects the variations in the thickness of the oxides.

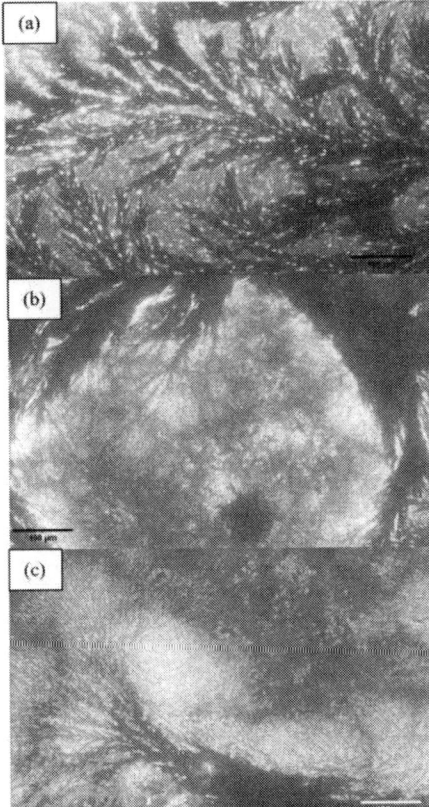

Fig.4. Optical images of dendrites with distinct morphologies and colourations.

Degradation at the back surface is also present following overactivation. Fig.5 shows optical images obtained from the back surface. There is a transition between top of the image which is a degraded region (a) and the area at the bottom (b) which is the unaffected absorber.

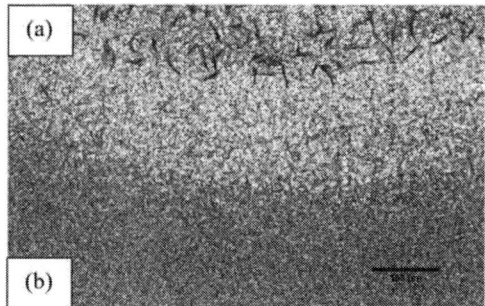

Fig.5. Optical image of an overactivated region at the back surface of the device

B. X-Ray Photoelectron Specroscopy (XPS)

Surface chemical analysis of the front and back interfaces identified the elements Cd, Te, O, and Cl on the sample surface with typical adventitious surface contaminants of C, Na, and Mg.

High resolution scans were performed to identify the chemical state of the dendrites in the substrate to detect the presence of cadmium compounds as opposed to pure cadmium previously suggested to be part of the dendritic growth. Analysing the Cd3d spectra shown in Fig.6 from the back surface, we can identify the peaks at 405.1, 406.4, and 407.6 eV which correspond to CdTe, $CdCl_2$ and an unknown peak which we believe is cadmium oxychloride. Notably, there is no peak at binding energy values less than 405eV which indicates there is no Cd metal present in these films (typical value for Cd metal ~404.6eV).

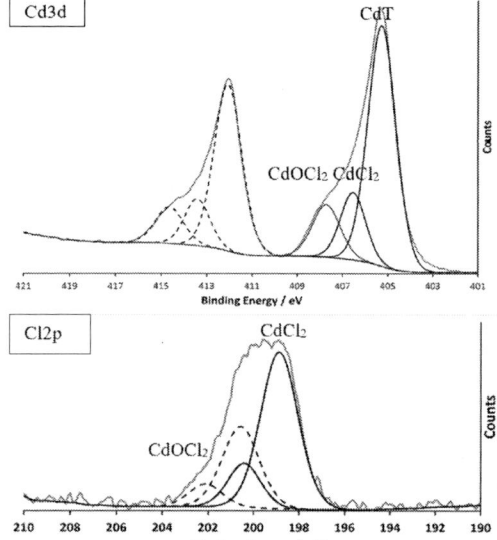

Fig.6: (Top) Cd3d and (Bottom) Cl2p XPS spectra for CST-glass interface. Solid peaks are Cd3d5 or Cl2p3 XPS peaks respectively. Dashed peaks are the corresponding Cd3d3 or Cl2p1 XPS peaks.

To support the assignment to cadmium oxychloride dendrites, the Cl2p spectra reveals evidence of an oxychloride ion bonds at 200.3eV alongside a $CdCl_2$ bond at 198.8eV. It should be noted that these spectra were produced when the spot size of the X-ray was directed towards dendrites, as opposed to the sample bulk, and found primarily on the back of the device. There was only a small quantity of material that was peeled off the substrate for the front interface analysis. The oxychlorides may have remained attached to the substrate. Cadmium oxychlorides salts if dissociated into Cd+ will likely be toxic, but further work is needed to confirm the exact chemistry of the dendrites.

IV. CONCLUSIONS

The aggressive $CdCl_2$ conditions used in this work exceeded a threshold where oxidation takes place to create dendrites. XPS analysis of the dendrites provides evidence for the presence of cadmium oxychloride. The presence of dendrites at the contacts will degrade the device efficiency. Further optimization of the activation process is required to high temperatures (in excess of 450°C) for more interdiffusion and grain growth, but avoidance of dendrite formation. Replacing or reducing the oxygen content during activation may reduce or eliminate the rate of formation of the dendrites. However, this would prevent the activation process to carried out in atmosphere adding cost to an industrial process at scale. Further work to be reported in the full paper will include cross-sectional transmission electron microscopy (TEM) and EDS to provide specific analysis of the interface between the dendrites and the absorber.

REFERENCES

[1] M. A. Green et al., "Solar cell efficiency tables (Version 60)," Progress in Photovoltaics: Research and Applications, vol. 30, no. 7, pp. 687–701, 2022, doi: 10.1002/pip.3595.

[2] P. Hatton, M. J. Watts, A. Abbas, J. M. Walls, R. Smith, and P. Goddard, "Chlorine activated stacking fault removal mechanism in thin film CdTe solar cells: the missing piece," Nat Commun, vol. 12, no. 1, p. 4938, 2021, doi: 10.1038/s41467-021-25063-y.

[3] E. Colegrove et al., "Se diffusion in CdTe thin films for photovoltaics," J Phys D Appl Phys, vol. 54, no. 2, 2021, doi: 10.1088/1361-6463/abbb47.

[4] A. Shah et al., "Understanding the copassivation effect of Cl and Se for CdTe grain boundaries," ACS Appl Mater Interfaces, vol. 13, no. 29, pp. 35086–35096, Jul. 2021, doi: 10.1021/acsami.1c06587.

[5] V. Kornienko et al., "Large Area Survey Grain Size and Texture Optimization For Thin Film CdTe Solar Cells Using Xenon-Plasma Focused Ion Beam (PFIB)," in 2022 IEEE 49th Photovoltaics Specialists Conference (PVSC), 2022, pp. 63–68. doi: 10.1109/PVSC48317.2022.9938662.

[6] A. Abbas et al., "Cadmium chloride assisted re-crystallization of CdTe: The effect of annealing over-treatment," in 2014 IEEE 40th Photovoltaic Specialist Conference, PVSC 2014, Oct. 2014, pp. 701–706. doi: 10.1109/PVSC.2014.6925018.

[7] H. Sun et al., "Self-assembly of CdTe nanoparticles into dendrite structure: A microsensor to Hg2+," Langmuir, vol. 27, no. 3, pp. 1136–1142, Feb. 2011, doi: 10.1021/la104325s.

[8] D. Kim et al., "Thin film CdS/CdTe solar cells fabricated by electrodeposition," AIP Conf Proc, vol. 306, no. 1, pp. 320–328, Jun. 1994, doi: 10.1063/1.45761.

Carrier Dynamics in $Al_xGa_{1-x}As$/InAs-based Photon Up-conversion Solar Cells with a Doubled-heterointerface

Hambalee Mahamu, Shigeo Asahi, and Takashi Kita

Graduate School of Engineering, Kobe University, Kobe, Hyogo, 657-8501, Japan

Abstract — **The concept of two-step photon up-conversion solar cells was proposed aiming to increase the conversion efficiency by the enhancement of the intraband excitation caused by below-bandgap photon absorption. The solar cells mainly consist of two different bandgap semiconductors, namely $Al_{0.3}Ga_{0.7}As$ and GaAs, which create a heterointerface where InAs quantum dots exist. The experimental results indicated that the solar cells enhance photocurrent with additional below-bandgap photoirradiation without degradation of photovoltage. The solar cells, however, showed no photocurrent enhancement when the direct excitation (interband excitation) photon energy is higher than $Al_{0.3}Ga_{0.7}As$ bandgap. In this work, we further developed the solar cells aiming to enhance the intraband transition by adding a larger bandgap semiconductor of $Al_{0.7}Ga_{0.3}As$ forming two heterointerfaces, specifically the heterointerface at $Al_{0.7}Ga_{0.3}As$ / $Al_{0.3}Ga_{0.7}As$ and the heterointerface at $Al_{0.3}Ga_{0.7}As$ /GaAs. The JV characteristics of our solar cells exhibit clear enhancement of short-circuit photocurrent with slight increases in photovoltage. The photocurrent enhancement detected under 484-nm interband excitation proves that the intraband transition occurs at the heterointerface of direct and indirect semiconductors because the 484-nm photons are almost absorbed by $Al_{0.7}Ga_{0.3}As$ and $Al_{0.3}Ga_{0.7}As$. The enhancement detected under 784-nm interband excitation suggests that photogenerated electrons in GaAs absorb two infrared photons in order to overcome the heterointerfaces. This was proved by intraband excitation power-dependent gained photocurrent data. A notable superlinear feature was observed at a small 784-nm intraband excitation power density. The superlinear characteristics suggest that a photogenerated electron provided by GaAs absorbs more than one infrared photon. Therefore, we conclude that the three-step photoexcitation was achieved in our solar cells including one interband excitation and two intraband excitations. This might open the door to further development of photon up-conversion solar cells.**

I. INTRODUCTION

The concept of intermediate band SCs (IBSCs) was proposed in order to overcome Shockley-Queisser limit stating the maximum efficiency of a single junction solar cell cannot exceed 31% under 1 sun illumination [1]. The IBSCs is a single junction SCs with additional energy states, the so-called intermediate band (IB), existing between the conduction band (CB) and the valence band (VB). This creates three photon absorption channels i.e., VB-to-CB, VB-to-IB, and IB-to-CB channels respectively. Despite the weak IB-to-CB transition in actual IBSCs, the maximum theoretical efficiency shows a dramatic achievement of 48.2% under 1 sun illumination [2].

In 2017, the concept of two-step photon up-conversion SCs (TPU-SCs) was proposed theoretically with experimental evidence. The TPU-SCs are based on $Al_xGa_{1-x}As$ with the mole fractions of AlAs of $x = 0.3$ and $x = 0$ respectively. In particular, $Al_{0.3}Ga_{0.7}As$ acts as a wide-bandgap semiconductor (WGS) while GaAs is a narrow-bandgap semiconductor (NGS). Therefore, the TPU-SCs consist of a heterointerface where the different bandgap semiconductors are connected. Here, InAs quantum dots (QDs) was grown in order to achieve the so-called intraband transition which corresponds to below-band gap photon absorption [3]. The quantum states of InAs QDs allow in-plane electronic transition because of the three-dimensional quantum confinement of the QDs. Hence, the TPU-SCs provide three photon absorption channels including two direct transitions in $Al_{0.3}Ga_{0.7}As$ and GaAs, known for interband transition, and one intraband transition at the heterointerface. The calculated efficiency of the TPU-SCs is The experimental results prove that external quantum efficiency can be improved by irradiating 1319-nm below-bandgap photons. This increase in EQE signals indicates the absorption of 1319-nm infrared photons, and hence, the intraband transition. However, the increase in EQE, with 1319-nm infrared photon irradiation, was undetectable when the interband excitation photons are higher energy than $Al_{0.3}Ga_{0.7}As$ bandgap [3]. The reason seems to be understandable because of no heterointerface and InAs QDs on top of the WGS layer.

We, consequently, proposed TPU-SCs with an additional absorber of $Al_{0.7}Ga_{0.3}As$ grown on top of $Al_{0.3}Ga_{0.7}As$ with a layer of InAs QDs. Here, the doubled-heterointerface photon up-conversion SCs (hereafter, DPU-SCs) are based on $Al_xGa_{1-x}As$/InAs with AlAs mole fractions of $x = 0.7$, $x = 0.3$, and $x = 0$ respectively. The heterointerfaces, therefore, are doubled. It is expected that the usage of the additional $Al_{0.7}Ga_{0.3}As$ layer enhances the intraband transition when the interband excitation photon energy is high albeit $Al_{0.7}Ga_{0.3}As$ exhibits an indirect bandgap semiconductor. The results of interband excitation wavelength-dependent JV characteristics indicate that the intraband transition occurs when additional 1319-nm infrared photons are irradiated on the DPU-SCs regardless of the interband excitation wavelength. The interband excitation wavelength-dependent photocurrent difference measured as a function of the infrared photons reveals notably different carrier dynamics compared to the TPU-SCs. The analysis guarantee that the intraband transition

978-1-6654-6060-6/23 $31.00 © 2023 IEEE

can be achieved at the heterointerface of direct and indirect semiconductors as well as the enhancement of the below-bandgap photon absorption which might be useful for the further development of SCs.

II. Experiment

A. Device Fabrication

The DPU-SCs were fabricated by molecular beam epitaxy following the schematic structure indicated in Fig.1. A 400-nm and 150-nm thick p^+-GaAs and p-GaAs, respectively, were alternatively grown on p^+-GaAs (001) substrates under a constant substrate temperature of 550°C. The Be dopant concentrations are shown in Fig.1. A 1400-nm thick i-GaAs, subsequently, was grown and then we deposited 2.1 monolayers of InAs QDs (0.64 nm) under Straski-Krastranov growth condition. Following our recipe, the QD density is about $1 \times 10^{10} \text{cm}^{-2}$. The dimensions are 3 nm height and 20 nm width [3], [5]-[6]. During the growth of the QDs, the substrate temperature was kept at 490°C. Next, a 290-nm i-$Al_{0.3}Ga_{0.7}As$ was grown on top forming a heterointerface with InAs inserted beneath. Here, we denote the heterointerface as HI-2. The process of the QD growth, then, was repeated after the deposition of a 10-nm thick i-GaAs layer to adjust the surface of i-$Al_{0.3}Ga_{0.7}As$ layer. Also, a 10-nm thick i-GaAs capping layer was deposited on top of the QDs. After that, a 200-nm thick i-$Al_{0.7}Ga_{0.3}As$ layer was deposited on top to complete another heterointerface (denoted as HI-1). Then, the substrate temperature was increased to 500°C before the growth of n-layers. A 180-nm thick n-$Al_{0.7}Ga_{0.3}As$, a 30-nm thick n^+-$Al_{0.7}Ga_{0.3}As$, and a 50-nm thick contact layer of n^+-GaAs were grown with Si doping concentration following Fig. 1. A 750-nm thick Au/Au-Zn metal contact was grown on the back side of the DPU-SCs meanwhile 500 nm of Au/Au-Ge was deposited on top. The size of the DPU-SCs is 0.3 mm × 0.37 mm.

Schematic Structure of DPU-SCs

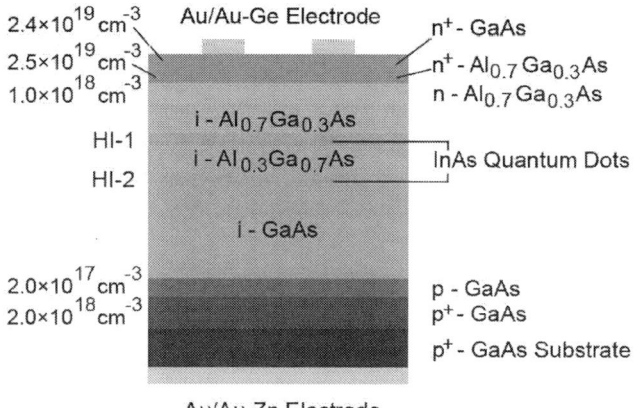

Fig. 1. Schematic illustration of doubled-heterointerface photon up-conversion solar cells (DPU-SCs) structure.

B. Measurement of JV Characteristics and Intraband Excitation Power-dependent Gained Photocurrent

Three different lasers were employed for the interband excitation of DPU-SCs including 784-nm, 660-nm, and 484-nm laser diodes. A 1319-nm infrared laser was used to induce the intraband transition. For interband excitation wavelength-dependent JV characteristics measurement, we used a source measure unit (Keithley 2400) to collect the electrical signals. The interband excitation photon flux was kept at $\sim 10^{16} \text{ cm}^{-3}\text{s}^{-1}$ for each wavelength. For the infrared photon, the intraband excitation power was controlled constantly at 85.5 mW/cm^2. The beam spot diameter was 1.2 mm. The irradiation site on the DPU-SC surface is the same since we employ laser beams using optical fibers.

A similar measurement system was used to measure gained photocurrent at short-circuit conditions (ΔJ_{sc}). The gained photocurrent is defined as the following:

$$\Delta J_{sc} = J_{sc,with\ infrared} - J_{sc,without\ infrared} \qquad (1)$$

where J_{sc} is short-circuit current density.

We measured ΔJ_{sc} as a function of 1319-nm infrared power density under specified excitation power density of monochromatic interband excitation of 784-nm, 660-nm, and 484-nm respectively. The intensity of 1319-nm infrared photons was varied by a variable natural density filter. The two aforementioned measurements were done at room temperature.

The point of using different wavelengths of the interband excitation laser is to selectively study the influence of the heterointerfaces. We pursued simple calculations of optical absorption for each interband excitation wavelength used in our experiments. In the calculation, we used Beer-Lambert law indicated as the following:

$$T \propto \exp(-\alpha x), \qquad (2)$$

where T is the transmittance of photons at a specified wavelength, α is the absorption coefficient depending on the semiconductor at a specified wavelength, and x is the thickness of the material. We incorporated refractive indices provided by Ref. [7] in our calculations. The results predict that 68% of the incident 660-nm photons are absorbed by $Al_{0.3}Ga_{0.7}As$ providing 32% remaining reach GaAs layer generating carriers in both layers. Therefore, the carrier dynamics, in this case, are influenced by both HI-1 and HI-2. Similarly, for the 784-nm interband excitation case, the photons excite merely GaAs providing the photogenerated carriers to be obstructed by HI-2 and HI-1, respectively. On the other hand, a large portion of 484-nm photons are absorbed by $Al_{0.7}Ga_{0.3}As$, in particular 82%. The remaining portions of 17% excite carriers in $Al_{0.3}Ga_{0.7}As$. Therefore, photogenerated carriers provided by GaAs are negligible. The carrier transport, hence, is associated with HI-1. Based on the calculations, we are able to selectively study the influence of the heterointerfaces on the carrier dynamics, specifically the intraband transition.

III. RESULTS AND DISCUSSION

Fig. 2 demonstrates the JV characteristics of DPU-SCs under several interband excitation wavelengths with and without 1319-nm infrared photons. The enhancement of J_{sc} with 1319-nm infrared photoirradiation is clear. In the case of 484-nm interband excitation, the electrons (majority carriers) are generated in $Al_{0.7}Ga_{0.3}As$ and $Al_{0.3}Ga_{0.7}As$. The electrons drift to the n-layers because of the internal electric field. The electrons generated by $Al_{0.3}Ga_{0.7}As$, then, accumulate at HI-1. The enhancement in photocurrent, therefore, is caused by the intraband transition due to the below-bandgap photon absorption. This guarantees that the intraband transition is achieved at the heterointerface formed by direct and indirect semiconductors. On the other hand, 784-nm photons excite carriers in the GaAs layer. The electrons drifting to the n-layer must pass through both HI-2 and HI-1 respectively. Without infrared photons, the photocurrent is detectable because of the thermal activation at the heterointerfaces. With infrared photons, the enhancement is small but significant. Since the electrons are generated in GaAs, the enhancement is supposed to be achieved by the intraband transition at both HI-1 and HI-2. As a result, one electron is expected to absorb two infrared photons in order to overcome the heterointerfaces. Hence, the results suggest a three-step photoexcitation including an interband excitation in GaAs and two intraband excitations at the heterointerfaces.

Wavelength-dependent JV Characteristics

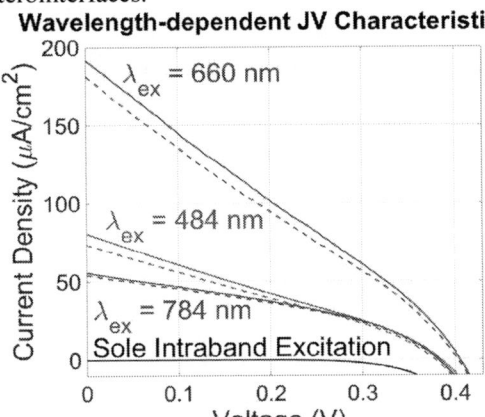

Fig. 2. Interband excitation wavelength-dependent JV characteristics of DPU-SCs with and without additional infrared photons. The dashed lines indicate the characteristics without infrared photons opposite to the solid lines.

Fig. 3 shows the gained photocurrent measured as a function of 1319-nm infrared excitation power density under 784-nm interband excitation. At 4.1 mW/cm^2 of 784-nm interband excitation and small intraband excitation power densities, a notable superlinear feature was observed. We proceed with curve fitting using a simple power function as the following:

$$\Delta J_{sc} \propto P^n, \tag{3}$$

where the power index n corresponds to the number of infrared photons absorbed by one electron generated in GaAs. The superlinear characteristics, $n = 1.137$, suggest that one electron

absorbs two infrared photons at the heterointerfaces. Theoretically, the power index n should be equal to 2. Because of the influence of thermal activation and recombination at the heterointerfaces, the power index n becomes less than we expected. The larger-than-unity power index confirms our speculation that three-step photoexcitation was achieved in the DPU-SCs. With increasing intraband excitation power densities, the power index decreases because of the saturation of quantized states at HI-1. When 784-nm photons excite GaAs layer, photogenerated holes drift to the p-layers providing negligible hole concentration at HI-1. Therefore, electron accumulation at HI-1 seems to be saturated. Although electrons absorb infrared photons, they are hardly extracted as photocurrent. This leads to the reduction of intraband transition, and hence, the power index n.

Excitation Power-dependent Gained Photocurrent

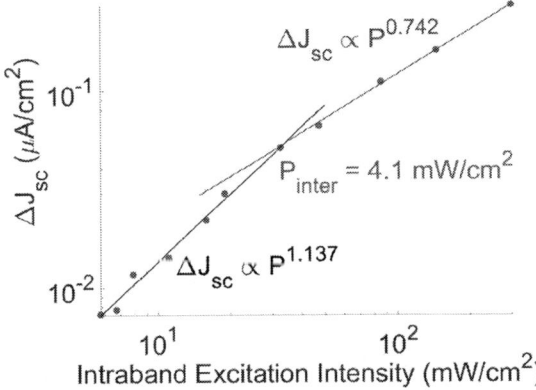

Fig. 3. Intraband excitation power-dependent JV characteristics of DPU-SCs.The blue circles are photocurrent data. The solid lines are fitted curves using power functions indicated in the figure.

IV. CONCLUSION

In conclusion, we successfully observed the intraband transition at the heterointerface of direct/indirect semiconductors. Moreover, we observed a three-step photoexcitation in DPU-SCs. The JV characteristics, measured under 784-nm interband excitation with additional infrared photons, suggest the three-step process. The processes were confirmed by the power index obtained from power-dependent gained photocurrent data despite the saturation behavior at high excitation intensity leading to the reduction of the power index.

REFERENCES

[1] W. Shockley, and H. J. Queisser, "Detailed Balance Limit of Efficiency of *p-n* Junction Solar Cells," *Journal of Applied Physics*, vol. 32, pp. 510, 1961.

[2] T. Kita, Y. Harada, and S. Asahi, *Energy Conversion Efficiency of Solar Cells.* Singapore, Springer Nature Singapore Pte Ltd., 2019.

[3] S. Asahi, H. Teranishi, K. Kusaki, T. Kaizu, and T. Kita, "Two-step Photon Up-conversion Solar Cells," *Nature Communication*, vol. 8, pp. 14962, 2017.

The Importance of Terrain-Shading Losses in PV Yield Assessment: The Case of Oahu

Marc Perez[1], Upama Nakarmi[1], Philip Gruenhagen[1] & Richard Perez[2]

[1]Clean Power Research, Napa, CA, USA
[2]Atmospheric Sciences Research Center, SUNY, Albany, NY, USA

Abstract — Through the case study of the island of Oahu, we investigate the regional impact of terrain-induced shading losses on the regional PV resource production. This impact is characterized as a function of 17 USGS land-use categories. We show that the overall resource impact amounts to ~7% losses over the island. It ranges from less than 1% in estuarine wetlands to over 12% in evergreen forests. When considering only land-use categories where PV is more likely to exist or expand (e.g., developed impervious surfaces), the impact is reduced to about 2-3% losses, i.e., a number that could be considered significant, even before considering the impact of local obstructions.

Keywords—Shading losses, PV yield, land-use, modeling, irradiance

I. BAKGROUND, METHODS

Shading losses, whether from natural terrain (hills, mountains) or local obstructions (neighboring building, trees) follow well-known first principles overall approaches [1-3], entailing:

- the determination of surrounding obstructions' horizon field of view from any point on a surface elevation grid
- the computation of resulting radiative losses at one or more points on that grid, specifically: obstructed direct irradiance (DNI) and reduced diffuse irradiance (DHI) sky view factor.

This first principles methodology underlies numerous PV yield assessment applications and publications such as the Google Project Sunroof [4] or NREL US roof potential evaluation [5] that are concerned with assessing the impact of nearby (building/trees) obstructions.

The resolution of the surface elevation grid and the angular resolution of horizon field of view scans are characteristics that are application-specific, as well as the treatment of diffuse and reflected irradiances.

The shading loss methodology we apply here was initially developed by Clean Power Research (CPR) for the IRENA Solar City Simulator [6] using ultra-high resolution elevation grids from stereoscopic satellite imagery, to precisely capture the impacts of nearby obstructions. For the present regional terrain-induced loss evaluation, we apply a regional elevation grid with a resolution of 0.33 arc seconds (~10 meters) from [7]. For each point, (totaling nearly 16 million) on the island, we scan horizons with a one-degree angular resolution. The treatment of DNI follows the standard first principles methods – i.e., DNI is on, or off if obstructed. The treatment of diffuse irradiance is specific to our methodology as it exploits the Perez diffuse irradiance model framework [8] that includes three diffuse zones: circumsolar, horizon, and isotropic. The circumsolar zone is treated as DNI. The horizon brightening/darkening zone is treated in terms of view factor loss in a region of the sky hemisphere extending up to 20 degrees above the horizon. The isotropic zone, typically accounting for a small fraction of the available diffuse energy, is assumed to be equivalent overall, in terms of integrated radiance, to the mean radiance directly or indirectly reflected by obstructions.

II. PAPER OBJECTIVE

Our objective is to investigate the impact of terrain shading at a regional level, taking the Hawaiian Island of Oahu as a case study. This island is characterized by mountainous terrain and a large existing as well as prospective PV generation resource.

Assuming nominal fixed south-facing latitude-tilt systems and starting from the 0.5 km resolution 15-minute DNI/diffuse irradiance grids from CPR's SolarAnywhere® product [9] spanning multiple years, we superimpose the high-resolution shading methodology (30 meters, 1 degree) specified above to all 16 million grid points on the island. The result is a high-resolution map of terrain-corrected plane-of-array irradiance (POAI) yield that can be compared to the original (unobstructed) yield grid to assess point-specific losses.

In addition to overall regional assessment, we investigate how shading losses are distributed as a function of regional land-use and ground cover from UCSD [10] – i.e., as a function of proxies for where PV is likely to exist or to be deployed in the future. The two-meter resolution of the USCD grid was reduced by averaging to match that of the elevation grid at 10-meter resolution.

III. RESULTS

The maps in Figure 1 compare an unobstructed POAI map (top) to a map accounting for all terrain-induced shading losses (bottom). Figure 2 compares the distribution of gridded irradiance for the 16 million grid points. The overall reduction in total yield integrated over the island is about 7%.

Figure 3 illustrates the repartition and statistical distribution of land cover in Oahu (top) and the distribution of POAI and shading losses we calculated for each of the 17 different land-use categories.

While overall losses relative to uncorrected irradiance amounts to 7%, the losses in land categories more likely to see current and future PV deployment – impervious surfaces, open spaces, impervious surfaces, cultivated land, pasture/hay, grassland, and scrub/shrub – only amounts to 3.9%, and only 2% without counting the latter (scrub/shrub) category that may be too habitat sensitive for PV deployment.

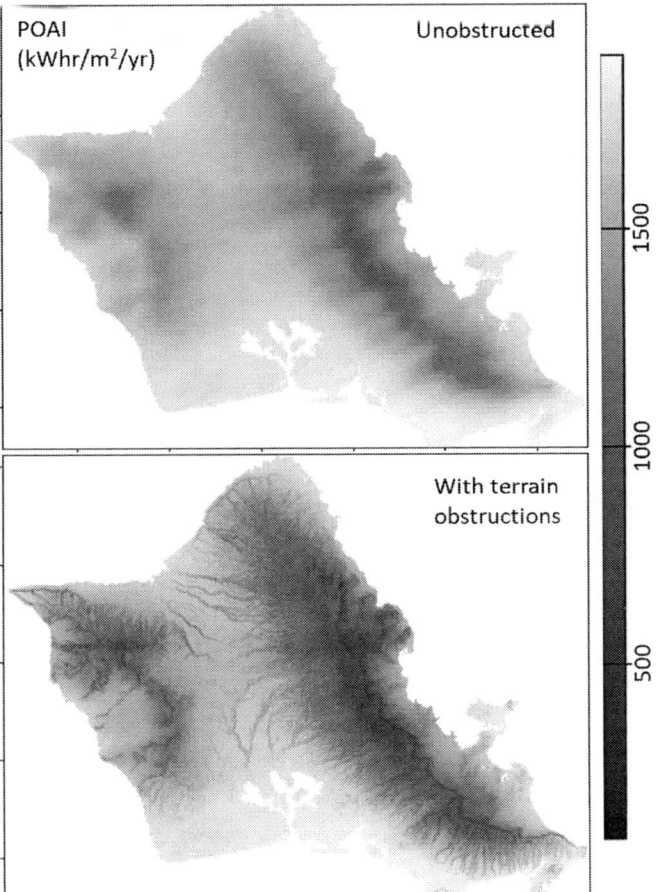

Fig. 1. Comparing annual plane of array irradiance (POAI) without assuming unobstructed horizons (top) and including terrain-induced losses (bottom).

IV. DISCUSSION

The terrain shading analysis presented in this article suggests that, for environments with steep terrain or large elevation changes (including many islands and continental mountainous regions) PV yield losses may not be inconsequential. For regional PV deployment planning purposes, the filter of land-use categories is an effective analytical addition that can put loss numbers in realistic/actionable contexts. This terrain shading detection capability will be incorporated in upcoming versions of CPR's SolarAnywhere® product.

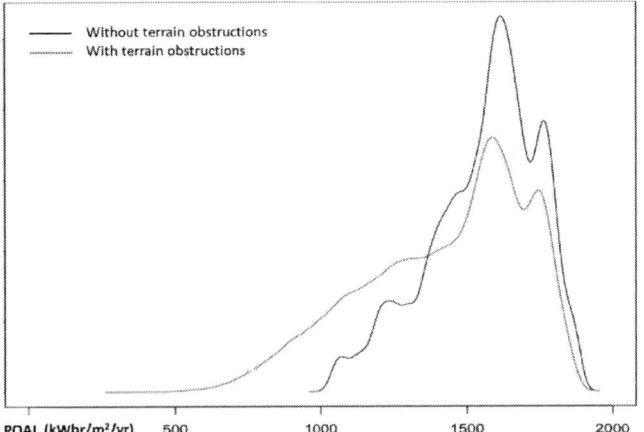

Fig. 2. Comparing unobstructed and actual island wide POAI frequency distributions.

REFERENCES

[1] Ha Nguyen, Joshua Pearce. Incorporating Shading Losses in Solar Photovoltaic Potential Assessment at the Municipal Scale. Solar Energy, 2012, 86 (5), pp.1245-1260. ff10.1016/j.solener.2012.01 .017.hal-00685775

[2] M. Olson (2017): Impacts of Topographic Shading on Surface Energy Balance of High Mountain Asia Glaciers. MS Thesis, Dept. of Geography, University of Utah.

[3] M. Olson and S. Rupper (2018): Impacts of topographic shading on direct solar radiation for valley glaciers in complex topography. The Cryosphere, 13, 29–40, 2019. https://doi.org/10.5194/tc-13-29-2019

[4] Google Project Sunroof (2022): https://sunroof.withgoogle.com/

[5] P. Gagnon, R. Margolis, J. Melius, C. Phillips, and R. Elmore (2016): Rooftop Solar Photovoltaic Technical Potential in the United States: A Detailed Assessment. NREL/TP-6A20-65298

[6] IRENA Solar City Simulator (2022): https://www.irena.org/Energy-Transition/Project-Facilitation/Renewable-potential-assessment/SolarCity-Simulator#:~:text=SolarCity%20is%20a%20web%20application,rooftop%2Dmounted%20solar%20PV%20installations

[7] Pacific Islands Observing System (2022): http://www.pacioos.hawaii.edu/metadata/usgs_dem_10m_oahu.html

[8] R Perez, R Seals, P Ineichen, R Stewart, D Menicucci, (1987): A new simplified version of the Perez diffuse irradiance model for tilted surfaces. Solar energy 39 (3), 221-231

[9] SolarAnywhere (2022): https://www.solaranywhere.com/?gclid=Cj0KCQiAzeSdBhC4ARIsACj36uGJ52EqaY9hFc9JJ0-gRvmhOpgOPAM9VuKqED4kzD7AebMK2MBozPEaAglaEALw_wcB

[10] United States Geological Survey, (2020): National Landcover Database. https://www.usgs.gov/centers/eros/science/national-land-cover-database?qt-science_center_objects=0#qt-science_center_objects

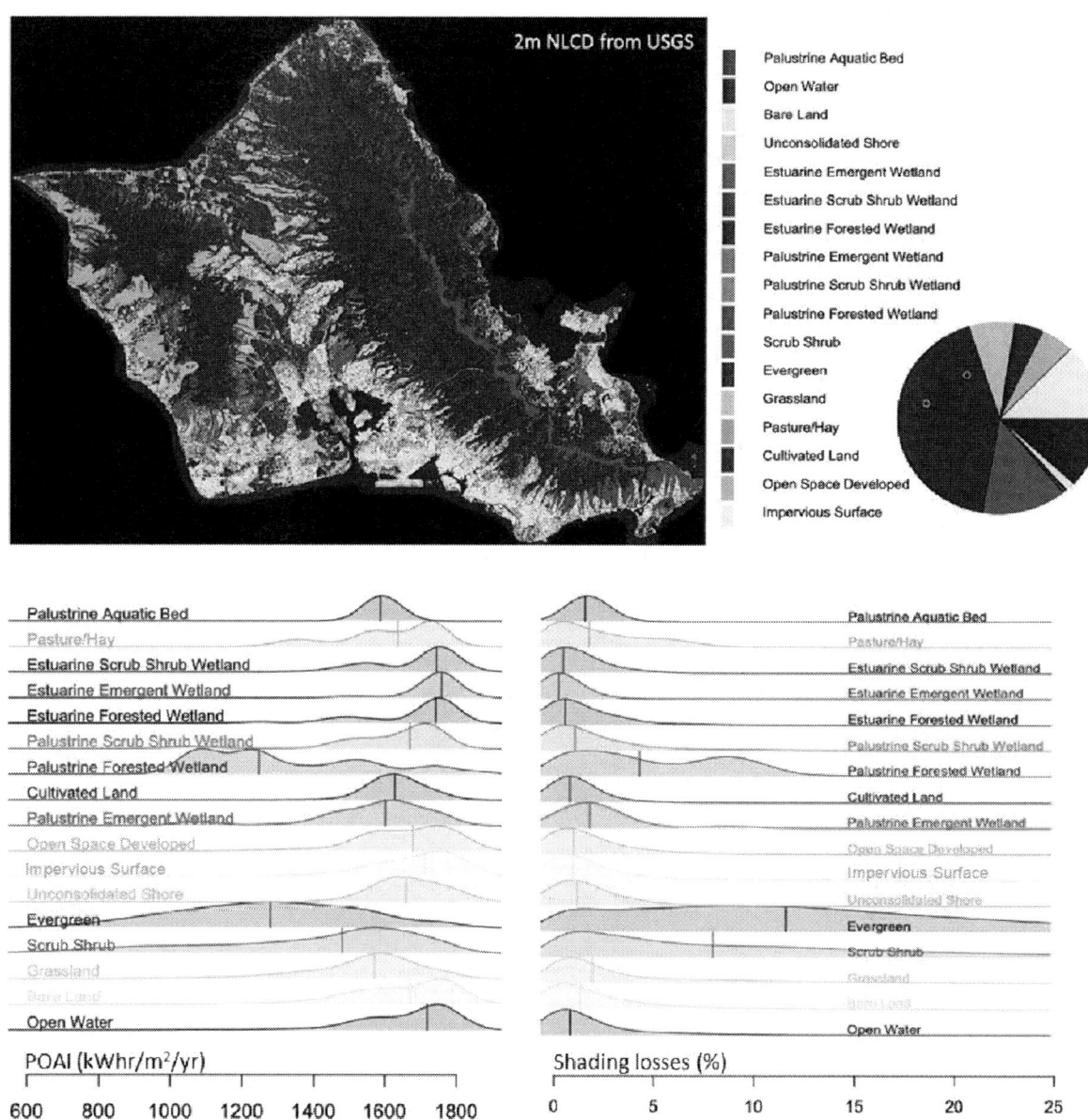

Fig. 3. Geographical and statistical distribution of USGS land-use categories in Oahu (top). Plane of array irradiance and percent shading losses distributions per land-use category (bottom).

Utilizing PSO Technique for Locational-dependent Feeder PV Hosting Capacity Evaluation

Vinushika Panchalogaranjan and Paul Moses

School of Electrical and Computer Engineering, University of Oklahoma, Norman, OK, 73019, USA

Abstract—Integration of renewable energy resources (RER) like solar photovoltaic (PV) power in the distribution system (DS) is one of the key aspects in the evolution of modern power systems. Maximizing the utilization of the RER while adhering to stringent grid standards has become a crucial technical challenge in distributed power systems. The analysis of overall network performance and PV capacity limitations is essential before the installation of a PV source in a distribution feeder. This paper analyses the locational-dependent PV hosting capacity (PVHC) for a central PV plant integrated into the DS using the particle swarm optimization (PSO) method. The proposed methodology is applied in simulations of a real distribution feeder and examines the effectiveness of the framework. The results show that PSO is an effective method to determine the locational-dependent PVHC.

Index Terms—Distribution system, Particle swarm optimization, PV hosting capacity, Renewable energy resources

I. INTRODUCTION

The integration of renewable energy resources (RER) has increased rapidly in the past decade and is expected to keep the trend in the future as well. Distribution side generation with renewable resources such as solar photovoltaic (PV) and wind energy is considered a crucial step in enhancing the power system resiliency and the decarbonization in power generation. The traditional power grid is designed for unidirectional power flow from generation resources to the distribution side through the transmission system. Therefore, integrating a large capacity of RER in the DS causes complications and operational issues. The main issues are voltage violations, network congestion overloads, maloperation of protective devices, and imbalances and unmanaged reverse power flow [1], [2].

The most common distribution side generation resource is solar PV. Therefore, determining the capacity of the PV which can be integrated into the existing distribution system is necessary in DS planning. PVHC is generally defined as the maximum PV capacity that can be accommodated in an existing DS without violating any network operational criteria [3]. The PV systems can be integrated into the distribution system as single or multiple large-scale PV plants or as residential side rooftop PV plants. This paper only considers the integration of single central PV plants into the DS for the PVHC calculation. This PVHC is specifically defined in two ways. The first one is determining the overall PVHC of the DS by providing the minimum and maximum PVHC from [4]. The second one is providing the PVHC for the specific

This material is based on work supported by the National Science Foundation under Grant No. OIA-1946093.

location [5]. This paper focuses on the locational-dependent PVHC of the distribution feeder.

Several methods have been proposed in the literature to determine and analyze the PVHC in the DS for these PV systems. The single PVHC is conventionally determined by increasing the PV capacity in small steps while considering the operational constraints [6]. Usually, utility providers use the detailed distribution network model for planning studies and analysis. Circuit reduction methods are often used to simplify the circuits for the analysis. However, the resultant network would still be substantially large. Therefore, this method will be exhaustive and costly to run the simulation for the whole network. Optimization techniques have been another popular method to determine the PVHC. Authors in [7] proposed a genetic algorithm-based technique to determine the maximum PVHC in the DS. Recent papers use particle swarm optimization (PSO) to determine the PVHC [8]–[10]. Most of these proposed methods are only investigated on the small IEEE test cases thus making it uncertain about the applicability to large distribution networks. Therefore, the effectiveness of the method in a real network needs to be analyzed. In this paper, the PSO method is used to determine the PVHC of the single central PV plant in the DS. This methodology is applied to a real 24.9 kV distribution feeder to examine the applicability of the algorithm.

This paper is organized as follows. Problem formulation is presented in Section II. Section III describes the solution technique with an explanation of PSO and the objective function formulation for the PSO algorithm. Sections IV and V discuss a case study of the proposed method and simulation results. The conclusion is presented in Section VI.

II. PROBLEM FORMULATION

The objective is to determine the maximum PV capacity that a DS can accommodate while keeping the operational constraints within the standard limits. Apart from the operational violations, power losses will occur in the DS. Integrating PV systems can reduce the losses in the DS if connected in a certain capacity range. Therefore maximizing the PV hosting capacity while minimizing the losses is the main objective considered in this paper. The voltage and line loading violations are taken for the operational constraints since those are the essential criteria.

A. Objective Function

The loss function is convex for a single PV plant. Therefore, this multi-objective problem is simply aggregated into a constrained single-objective function optimization problem by summing the weighted total loss and inverse of the PV capacity in one function. Thus, the objective function is defined below.

$$F = \text{Min} \left(w_1 \frac{1}{P_{HC}} + w_2 f_L \right) \quad (1)$$

where P_{HC} is the PV capacity and f_L is the function of loss. w_1 and w_2 are weights that are taken to be 0.5 each by considering similar weights for both objective functions. These weight factors can be varied from 0 to 1 according to the significance of each function.

B. Constraints

The constraints that can be violated and imposed by the user are formulated in this section

1) Voltage: The system voltage should be within the standard regulatory limit in order to avoid voltage instability and voltage quality issues. This constraint is given by:

$$V_{min} \leq V_i \leq V_{max} \quad (2)$$

where V_{min} and V_{max} are minimum and maximum voltage limitations respectively. V_i is the voltage at bus i.

2) Line loading: The distribution system line flow should be kept within the cable rating to prevent thermal overload stresses in the system. This is formulated as:

$$I_{ij} \leq I_R \quad (3)$$

I_{ij} is the current flows from node i to node j while I_R is the rated current for that line.

III. SOLUTION TECHNIQUE

There are several techniques used to solve optimization problems in power systems. The problem defined in Section II can be solved effectively by using metaheuristic techniques since it is a large-scale, highly nonlinear problem. [11].

A. Particle Swarm Optimization

PSO is a swarm intelligence-based method, developed by Eberhart and Kennedy which was motivated by the social behavior of fish schooling and bird flocking [12]. This technique is chosen to solve the optimization since it is not largely affected by the size and nonlinearity of the problem and is easier to implement with fewer parameters to adjust. In this method, a set of particles is assigned to move toward a promising area to reach the global optimum. Each particle keeps track of its best solution, personal best (p_{best}), and the best value of any particle, which is global best (g_{best}). The position of each particle is determined by the vector $x_i \in R^n$ and the velocity of the particles $v_i \in R^n$ which is given below [12].

$$\vec{x_i}(t) = \vec{x_i}(t-1) + \vec{v_i}(t) \quad (4)$$

The velocity is determined by the given equation

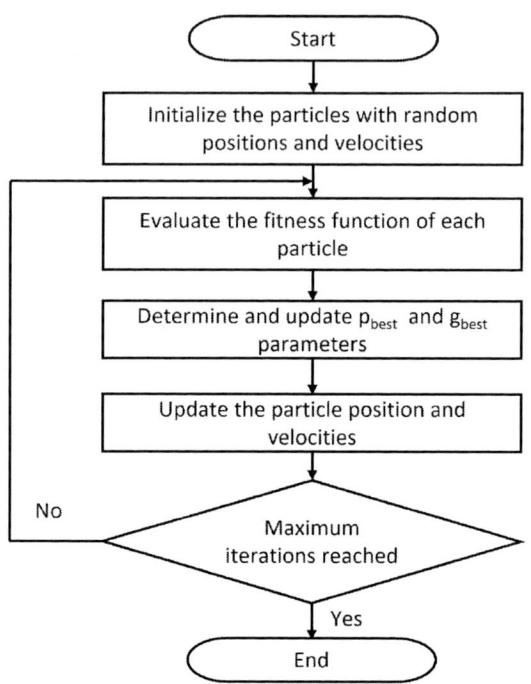

Fig. 1. Flow chart of the PSO algorithm

$$\vec{v_i}(t) = \phi_{ic}\vec{v_i}(t-1) + \phi_1.r_1.(\vec{p_i} - \vec{x_i}(t-1)) \dots$$
$$+\phi_2.r_2.(\vec{p_g} - \vec{x_i}(t-1)) \quad (5)$$

where ϕ_{ic} is referred to as inertia constant, ϕ_1, ϕ_2 are two positive numbers, and r_1, r_2 are two random numbers having a uniform distribution in the range of [0,1]. Vector $\vec{p_i}$ is the best position by the given particle whose corresponding fitness value is the particle's best(p_{best}) while $\vec{p_g}$ best position by any particle whose corresponding fitness value is global best(g_{best}). With the above mathematical formulation, the procedure to implement the PSO algorithm is given in Fig. 1.

B. Objective function for PSO

PSO method uses different techniques to handle the constraints. The popular method is to convert the constrained optimization problem into an unconstrained one by penalizing the infeasible solutions [13]. Therefore, the constraints are incorporated into the objective function in (1) with a penalty and formulated as follows:

$$J = \text{Min} \left(w_1 \frac{1}{P_{HC}} + w_2 f_L + P \times c \right) \quad (6)$$

where the second term in the equation is called the penalty function, where P is the penalty value and c is the number of constraints violated. This equation is referred to as the cost function or fitness function of the PSO method.

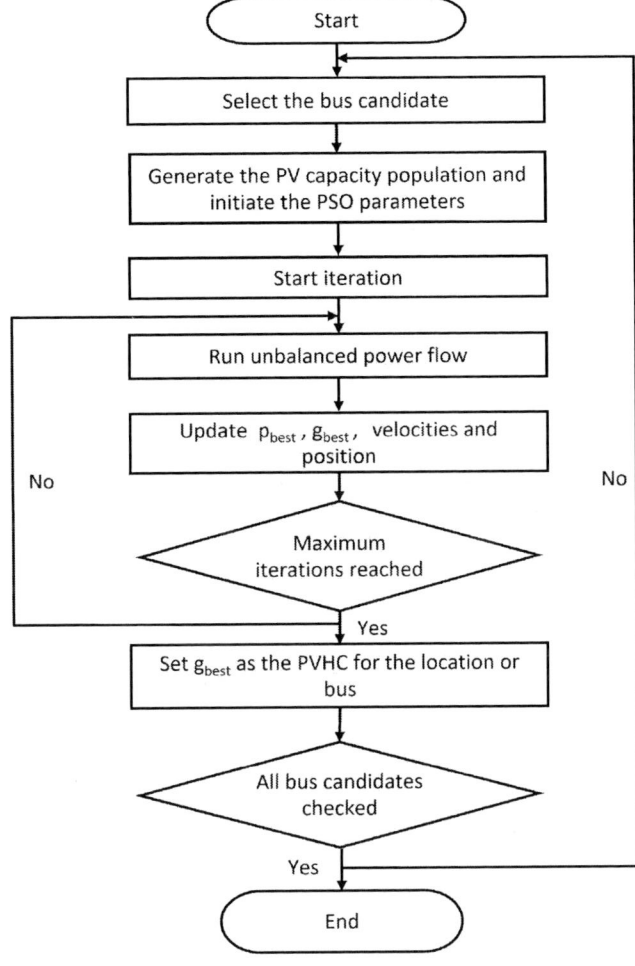

Fig. 2. Flow chart of the methodology to determine PVHC

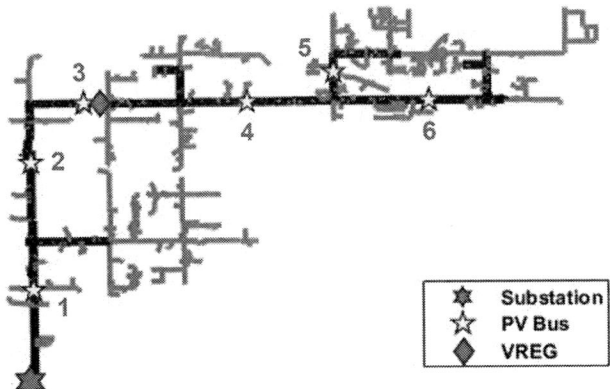

Fig. 3. Feeder network chosen for the analysis

C. Methodology

The large-scale central PV plant is one of the most common ways of integrating generation on the distribution side. As mentioned, the location to integrate the central PV plant has limitations due to the availability of the landscape. Therefore utilities have a limited choice to choose the best location with maximum hosting capacity. The location of the PV plant is to be considered user defined in this paper which makes the PV capacity the only variable. Also, having buses or locations as the variable in this problem will lead to convergence issues in the PSO due to a large set of variables. The simplified methodology to determine the PVHC is shown in the flowchart in Fig. 2.

IV. CASE STUDY

A. Test Network

A real suburban distribution feeder from Oklahoma is chosen to determine the locational-dependent PVHC using the PSO method. This 24.9 kV radial feeder consists of 570 three-phase buses, located in a suburban area with a peak load capacity of 5.2 MW. There is a voltage regulator (VREG)

located in the middle of the network to control the voltage. The network diagram of the feeder is shown in Fig. 3. The three-phase lines are represented in black while the single-phase lines are shown in grey color. The system does not have any existing PV plants. OpenDSS, an open-source three-phase distribution system simulation program, and MATLAB are used to carry out the analysis. GridPV toolbox from MATLAB is used to control OpenDSS through the COM interface [14]. In OpenDSS, the substation is modeled as a voltage source with a load tap changer (LTC) to regulate the voltage. Loads are represented as constant real and reactive power models for each customer separately. The central PV plant injecting power at the unity power factor is considered for the simulation since it is the commonly used configuration.

The worst-case scenario for voltage violation occurs when the minimum loading and high PV generation conditions. The line loading and line losses can get worsened during maximum loading conditions. Therefore, the minimum and maximum loading conditions are considered for the simulation which is chosen to be 50 % and 100 % of the peak load respectively [5]. The maximum allowable PV capacity integrated into the feeder is limited to 5 MW or 100 % to avoid the excessive reverse power flow as well as the limitations with the land and cost. Voltage is considered to be violated if it exceeds the limitation of 0.95 pu and 1.05 pu and the line loading violations are considered if the current flows in the lines exceed the line rating.

B. PSO tuning

In order to produce feasible solutions for the problem, PSO parameters should be tuned accordingly. From the literature and through experiments, the PSO parameters chosen are discussed here. The maximum inertia weight ϕ_{ic}^{max} is taken as 0.9 while the minimum inertia weight ϕ_{ic}^{min} is 0.4 through trial and error. The acceleration constant ϕ_1 and ϕ_2 are chosen to be 2 as suggested in [12]. The penalty P is chosen as a very large value of 10000.

V. SIMULATION RESULTS

For the locational-dependent PVHC calculation, all the three-phase buses except the substation side can be chosen as possible candidates. Six buses from the feeder were chosen to determine the PVHC for the demonstration. As indicated in Fig. 3, buses were chosen from near to far from the substation to estimate the PVHC individually. The hourly simulation is performed during the daytime for 12 hours starting from 7.00 am to 7.00 pm since the PV capacity will be effective in day time only. To obtain the worst cases for the constraints and losses, the 50 % and 100 % of the summer peak load factors of each hour are considered for the simulations. For the PV plant simulation, the clear sky irradiance is taken into account to get the maximum possible capacity occurrence throughout the day.

Conventional and PSO methods are used to obtain the PVHC for comparison and discussion. Conventionally, it is determined by increasing the PV injected from 0 to 5 MW with 0.15 MW steps at each possible location. Hourly static power flow is performed for each scenario with maximum and minimum loading conditions to check for line loading and voltage violations. For each PV capacity step size, the hourly simulation is performed and the maximum of the voltage and line currents are recorded. If any of these values are violated the constraint limits while increasing the PV capacity, then the loop will end. The maximum PV capacity without violations will be the PVHC of that location. Here, the effect of line loss is not considered while determining the PVHC, unlike the PSO method. The hourly simulations were performed for maximum and minimum loading conditions for each particle. The line loss of maximum loading conditions was considered for the fitness functions since the loss is high for this condition.

The plots of total three-phase hourly line losses and maximum voltage recorded with PV capacity are shown in Fig. 4 and Fig. 5 respectively to discuss the results obtained from conventional and PSO methods. Here, only the three-phase line losses were calculated to reduce the simulation. The upper voltage constraint is also shown in the redline in the Fig. 5. It can be observed that the loss is a convex function and the minimum loss is shifted towards the left side when the location moves far from the substation. The total three-phase line loss when there is no PV plant is 379.85 kWh. This shows that the integration of a PV plant reduces the loss since a certain portion of the power flow distance to the loads is reduced when the power is supplied from the PV plant. The reason for the loss increase again with PV capacity is because, power flow begins to reverse to the substation from that point, increasing the losses. The maximum voltage increases with the PV capacity as well as the distance from the substation. Therefore, the PV plant located further from the substation experienced voltage violations early. There were no line loading violations recorded for the range of PV capacity chosen for this DS.

Table I provides the comparison of the results obtained from PSO and exhaustive search. Both methods do not exactly provide the optimal solution from the techniques. The conventional method depends on the step size in providing the accuracy of the solution. Since PSO is a heuristic method, it will only provide an approximately optimal solution. However,

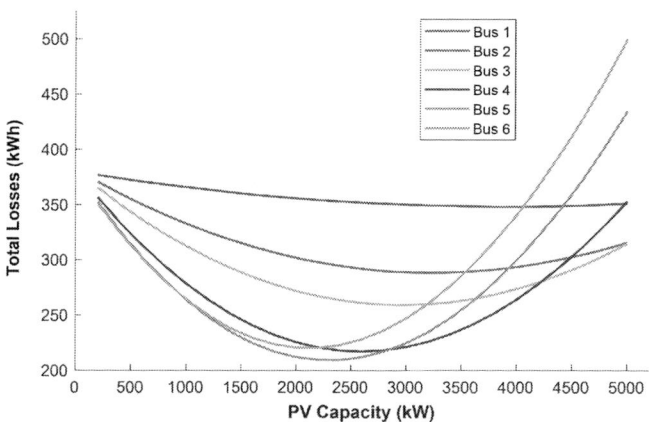

Fig. 4. Variation of three-phase line loss with PV capacity for maximum loading conditions

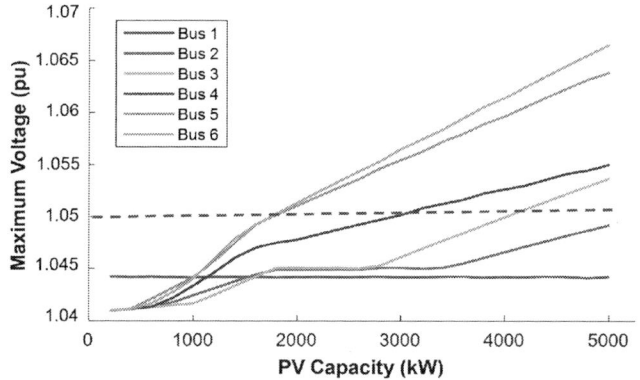

Fig. 5. Variation of maximum voltage with PV capacity for minimum loading conditions

this comparison helps find the variations and the reliability of the PSO. There will be small differences in the PVHC results for some cases with conventional and PSO methods since the effect of loss is considered in the PSO method while the other one did not. The PVHC of Bus 2 with the conventional method is 5 MW while the PSO method is 3.98 MW with a reduced loss of 31.1 kWh. The weights w_1 and w_2 can be adjusted in order to increase the PVHC further in these cases.

Also, the PSO method does not provide the PV capacity with the minimum loss. Instead, the minimum loss is compromised to be a slightly higher value in order to maximize the PVHC. But this small difference in loss can be acceptable since it considerably increases the PVHC. Since Bus 1 is located near the substation, the loss and voltage do not have much variation. Therefore the maximum limited capacity is provided as the PVHC. When the PV location moves away from the substation, the PVHC reduces due to voltage violations. Also, the PSO method considerably reduces the simulation time compared to the conventional method while providing reliable results. The time taken for simulation to determine the PVHC of these six locations for the conventional method is 1 hour and 46 minutes while the PSO method

TABLE I
COMPARISON OF MINIMUM LOSS AND PV HC BETWEEN PSO AND EXHAUSTIVE SEARCH

Location	PSO		Conventional	
	Total losses (kWh)	PVHC (kW)	Total losses (kWh)	PVHC (kW)
Bus 1	352.2	5000	352.2	5000
Bus 2	292.5	3982	323.6	5000
Bus 3	268.1	3762	278.4	4050
Bus 4	219.8	2913	220.3	3000
Bus 5	223.8	1699	224.7	1650
Bus 6	229.4	1699	230.8	1650

took only 38 minutes to simulate. This shows that the PSO method reduces simulation time to approximately one-third of the conventional method. This time will be much less when it is simulated in high-performance computers. However, the time difference and the accuracy show how effective the PSO method can be for this problem.

VI. CONCLUSION

PSO-based technique is proposed to determine the locational-dependent PVHC in an existing radial DS in this paper. The objective function is formulated as maximizing the PVHC while minimizing the DS losses. A real distribution network is used to apply the method for determining the PVHC of a central PV plant. The constraints imposed are voltage and line loading which are the crucial constraints to be considered. The PSO-based method is compared with the conventional method to observe the efficiency of the method. From the case studies, it can be seen that the algorithm is more accurate and faster compared to the conventional methods. Future work will concentrate on assessing the PVHC for the DS in order to integrate the residential rooftop solar PV plants consisting of stochastic variables.

ACKNOWLEDGMENT

The authors are grateful to Oklahoma Electric Cooperative for providing utility data for this analysis.

REFERENCES

[1] M. M. Haque and P. Wolfs, "A review of high PV penetrations in LV distribution networks: Present status, impacts and mitigation measures," *Renewable and Sustainable Energy Reviews*, vol. 62, pp. 1195–1208, 2016.

[2] B. Uzum, A. Onen, H. M. Hasanien, and S. M. Muyeen, "Rooftop solar PV penetration impacts on distribution network and further growth factors—a comprehensive review," *Electronics*, vol. 10, no. 1, p. 55, Dec 2020.

[3] Y.-J. Liu, Y.-H. Tai, Y.-D. Lee, J.-L. Jiang, and C.-W. Lin, "Assessment of PV hosting capacity in a small distribution system by an improved stochastic analysis method," *Energies*, vol. 13, no. 22, p. 5942, Nov 2020.

[4] A. Dubey and S. Santoso, "On estimation and sensitivity analysis of distribution circuit's photovoltaic hosting capacity," *IEEE Transactions on Power Systems*, vol. 32, no. 4, pp. 2779–2789, 2017.

[5] K. Coogan, M. J. Reno, S. Grijalva, and R. J. Broderick, "Locational dependence of PV hosting capacity correlated with feeder load," in *2014 IEEE PES TD Conference and Exposition*, 2014, pp. 1–5.

[6] M. J. Reno, K. Coogan, S. Grijalva, R. J. Broderick, and J. E. Quiroz, "PV interconnection risk analysis through distribution system impact signatures and feeder zones," in *2014 IEEE PES General Meeting — Conference Exposition*, 2014, pp. 1–5.

[7] M. Vatani, D. S. A. . M. J. Sanjari, and G. B. Gharehpetian, "Multiple distributed generation units allocation in distribution network for loss reduction based on a combination of analytical and genetic algorithm methods," vol. 10, 2016.

[8] A. Y. Saber, T. Khandelwal, and A. K. Srivastava, "Fast feeder PV hosting capacity using swarm based intelligent distribution node selection," in *2019 IEEE Power Energy Society General Meeting (PESGM)*, 2019, pp. 1–5.

[9] Y. Alghamdi, A. A. Al-Mehizia, and F. Al-Ismail, "PV hosting capacity calculation using particle swarm optimization," in *2021 North American Power Symposium (NAPS)*, 2021, pp. 1–6.

[10] M. Z. U. Abideen, O. Ellabban, F. Ahmad, and L. Al-Fagih, "An enhanced approach for solar PV hosting capacity analysis in distribution networks," *IEEE Access*, vol. 10, pp. 120 563–120 577, 2022.

[11] G. Chicco and A. Mazza, "Metaheuristic optimization of power and energy systems: Underlying principles and main issues of the 'rush to heuristics'," *Energies*, vol. 13, no. 19, p. 5097, Sep 2020. [Online]. Available: http://dx.doi.org/10.3390/en13195097

[12] Y. del Valle, G. K. Venayagamoorthy, S. Mohagheghi, J.-C. Hernandez, and R. G. Harley, "Particle swarm optimization: Basic concepts, variants and applications in power systems," *IEEE Transactions on Evolutionary Computation*, vol. 12, no. 2, pp. 171–195, 2008.

[13] A. R. Jordehi, "A review on constraint handling strategies in particle swarm optimization," *Neural Computing and Applications*, vol. 26, no. 6, pp. 1265–1275, 2015.

[14] K. Coogan and M. J. Reno, "Grid integrated distributed PV (GridPV)," Sandia National Laboratories SAND2014-20141, Tech. Rep., 2014.

Cryogenic Operation of GaAs Laser Power Converters

Bora Kim, Mijung Kim, Brian D. Li, Ryan D. Hool, Minjoo L. Lee

University of Illinois Urbana-Champaign, Urbana, IL, United States

We present strategies to achieve efficient operation of GaAs laser power converters (LPCs) at cryogenic temperatures under 808 nm high-power laser light. Due to reduced dark current, low temperature (LT)-operation has the potential to enhance performance drastically. However, other changes including reduced absorption coefficient and dopant deactivation can negate the expected efficiency improvements. In this work, we show that heterointerface majority carrier blocking at LT leads to dramatically increased series resistance (RS) and degraded fill factor (FF). We go on to demonstrate a redesigned cell with heterointerfaces that reduce majority carrier barriers while ensuring adequate minority carrier blocking at 83K. The redesigned cell shows a low RS of 0.01 Ωcm-2, a peak efficiency of 62.1%, and a peak FF of 93.3% at 83K.

978-1-6654-6060-6/23 $31.00 © 2023 IEEE

Characterization of Field Exposed Photovoltaic Modules Featuring Signs of Contact Degradation

Max Liggett[1,2], Dylan J. Colvin[2,3], Balaashwin Babu[1], William C. Oltjen[4,5], Xuanji Yu[4,5]
Manjunath Matam[3], Hubert P. Seigneur[3], Mengjie Li[2,3], Andrew M. Gabor[6], Philip J. Knodle[6]
Craig J. Neal[1], Sudipta Seal[1], Laura S. Bruckman[4,5], Roger H. French[4,5], Kristopher O. Davis[1,2,3]

[1] Department of Materials Science and Engineering, UCF, Orlando, FL 32816, USA
[2] Resilient, Intelligent and Sustainable Energy Systems (RISES) Cluster, UCF, Orlando, FL 32816, USA
[3] Florida Solar Energy Center (FSEC), University of Central Florida (UCF), Cocoa, FL 32922, USA
[4] SDLE Research Center, Case Western Reserve University (CWRU), Cleveland, OH 44106, USA
[5] Department of Materials Science and Engineering, CWRU, Cleveland, OH 44106, USA
[6] BrightSpot Automation Automation LLC, Westford, MA 01886, USA

Abstract—As the solar energy industry expands, the reliability and lifespan of photovoltaic (PV) modules have become increasingly important to ensure commercial viability for large-scale applications. To improve reliability and performance, it is necessary to better understand modes of failure through accelerated aging tests, which can identify degradation mechanisms that take a long time to manifest. This work investigates contact corrosion of fielded PV modules using a multi-scale analytical approach. Current-voltage (IV) and Suns-VOC measurements, electroluminescence (EL) imaging, Infrared (IR) imaging, and Ultraviolet Fluorescence (UVF) photography imaging were performed on multicrystalline silicon (multi-Si) and monocrystalline silicon (mono-Si) modules installed in a hot and humid climate. Subsequently, locations of interest were cored from the modules and analyzed using cross-sectional scanning electron microscopy (SEM) and energy-dispersive X-ray spectroscopy (EDS).

Index Terms—

I. INTRODUCTION

With the expansion of the solar energy industry, the lifespan and reliability of PV modules have become increasingly important in order to insure commercial viability for such large-scale applications. This mass adoption of PV technology brings with it a need to better understand modes of failure in order to make more informed decisions regarding design and manufacturing processes that work to directly improve reliability and performance.

Some degradation mechanisms take a long time to manifest and can only be identified rapidly using accelerated aging tests. Accelerated aging tests are therefore a crucial part of establishing the quality of PV modules. The value of accelerated aging tests is heavily reliant on the ability to correlate this indoor testing with the relevant degradation factors for fielded PV modules [1]. The confidence of the correlation of indoor and outdoor data is predicated on the indoor test's ability to accurately represent the environmental stressors the PV device would be subjected to in the field.

Damp heat testing, typically performed by exposing PV modules to 85°C and 85% relative humidity, is one of the most widely adopted accelerated aging tests in the PV industry because the combination of heat and moisture tends to accelerate many known degradation mechanisms. In particular, it is often used to evaluate corrosion of the front silver contacts caused by the formation of acetic acid in PV modules featuring ethylene-vinyl acetate (EVA) encapsulation [2]. The EL features used to qualitatively determine whether your PV module exhibits contact corrosion generally present after 2,000-3,200 hours of damp heat testing [2], [3].

In the case of more complex modes of failure, like contact corrosion, establishing the correlation between indoor accelerated aging tests and data collected from exposed PV modules is more difficult Contact corrosion can occur as a result of a hydrolysis reaction triggered in the encapsulant material, which leads to the formation of acetic acid and subsequent failure of the system. Acetic acid directly corrodes the silver gridlines and reacts with the metal oxides within the glass frit.

The aforementioned complexity of contact corrosion and variations in the parameters used during damp heat testing in research contributes to a reduced level of confidence in how representative indoor data is with respect to field performance.

In this work, we investigate contact corrosion of fielded modules that exhibit front contact degradation as a result of the acetic acid formation using a multi-scale analytical approach. Current-voltage $(I - V)$ and Suns-V_{OC} measurements, EL imaging, IR imaging, and UVF imaging were all performed on the multi-Si aluminum back surface field (Al-BSF) and mono-Si passivated emitter and rear cell (PERC) modules installed in the field in Cocoa, Florida, a hot and humid climate. Subsequently, locations of interest were cored from the modules and analyzed using cross-sectional SEM and EDS techniques [4].

II. EXPERIMENTAL METHODS

For this study, we examine two types of PV modules exposed in the field in Cocoa, Florida for different amounts of time. One group features multi-Si Al-BSF modules installed in 2012 and under field operation for 11 years, and the other group features mono-Si PERC modules in operation for four years and an unexposed control.

A wide range of non-destructive characterization techniques was carried out on all of the modules, including $I-V$ and Suns-V_{OC} measurements, EL imaging, IR imaging, and UVF imaging. These techniques allow power loss to be identified and the potential loss mechanisms identified. EL and IR imaging has long been used by the PV community to detect and identify defects, and UVF imaging has emerged recently as an effective means of identifying damage such as cracks in the cells, delamination, hotspots, or areas of reduced efficiency [5].

To determine whether corrosion of the glass frit in the front silver contacts was present in the modules, portions of the encapsulated cells were cored from the central and near bus bar regions of cells in the module. These regions corresponded to regions of the module featuring high and low EL intensity, respectively (i.e., bright and dark regions of the EL images). We performed cross-sectional SEM imaging and EDS analysis on the cores using a Zeiss Ultra 55 SEM. For the multi-Si Al-BSF module, the front-side encapsulant represented a significant hurdle to the SEM imaging and EDS analysis due to its strong adhesion to the surface PV module. In establishing a suitable removal procedure, it was important to avoid aggressive heat, mechanical, or chemical treatments that may compromise future data collection by changing the cores' material chemistry or introducing fractures The post-coring procedure involved a toluene vapor exposure treatment that served to soften the EVA enough for removal. The cores are raised above the bottom of the container on $1/4$ in thick glass slides to prevent direct contact with the 100-150 mL of liquid organic solvent below. The vapor exposure step lasts for three days, after which, the cores are removed and tweezers are used to peel the EVA from the front side of the cores. Photoluminescence (PL) images are taken before and after the EVA removal in order to observe any significant damage introduced to semiconductor sections of the cores. After the front-side EVA is stripped, the cores are submerged in an acetone bath for 12 hours to remove the posts. These posts were left in place to provide extra structural integrity during the mechanical peeling step of the encapsulant removal.

III. RESULTS AND DISCUSSION

The $I-V$ results from the non-destructive module-level characterization shown in Figure 1 shows that both the multi-Si Al-BSF and mono-Si PERC modules exhibit signs of increased series resistance (R_S) when compared to nameplate. The extent to this R_S is very minor for the mono-Si PERC modules, but absolutely devastating for the multi-Si Al-BSF modules. In addition to the high R_S, EL images of the multi-Si Al-BSF shown in Figure 2 contains features reminiscent of those produced in indoor testing for contact corrosion encouraging further investigation of the state of the contacts with SEM cross-sectional analysis [6].

The EL images for the mono-crystalline PERC module did not share the same features. The UVF images showed the fielded PERC module has an unusually bright fluorescent signal along cell bus bars. The mono-Si PERC module only

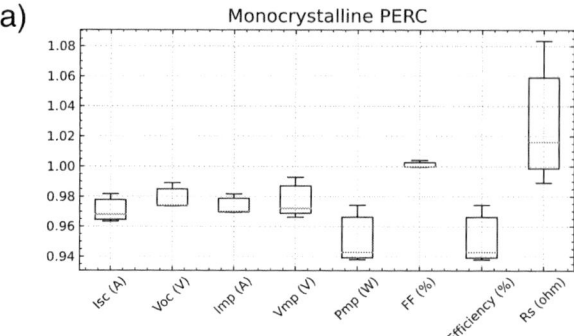

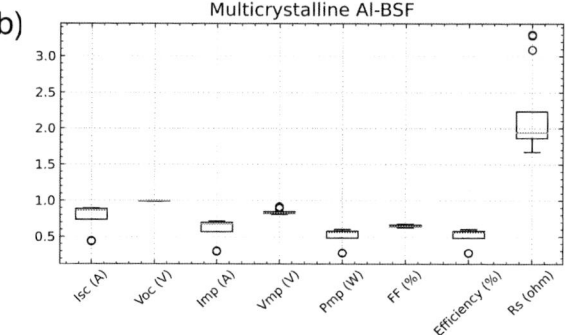

Fig. 1. Box plots of the $I-V$ parameters of the multi-Si Al-BSF and mono-Si PERC modules normalized to their nameplate values.

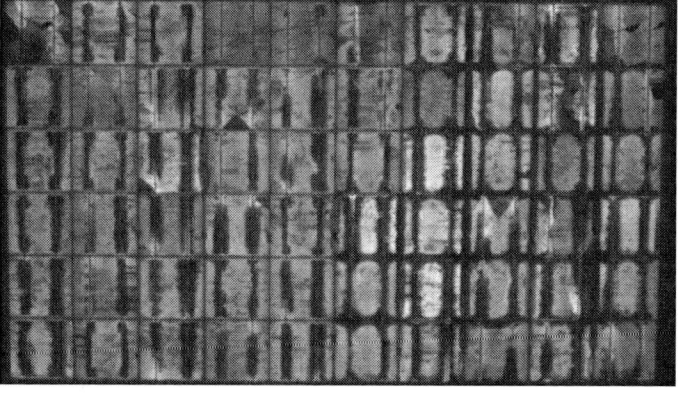

Fig. 2. Electroluminescence image of SLTE module after 10 years of outdoor exposure.

shows minor series resistance losses after 4 years of exposure providing a useful contrast to the severely degraded multi-Si Al-BSF module.

The SEM cross-section for the mono-Si PERC modules in Figure 4 does not show any breakage between the silver and oxide layer or darkening of the glass frit which is often used as indicators for contact corrosion [7], [8].

IV. FUTURE WORK

Future work will be to obtain more chemical information to help to validate the indoor testing used to explore acetic acid contact corrosion. To achieve this, cores from the fielded

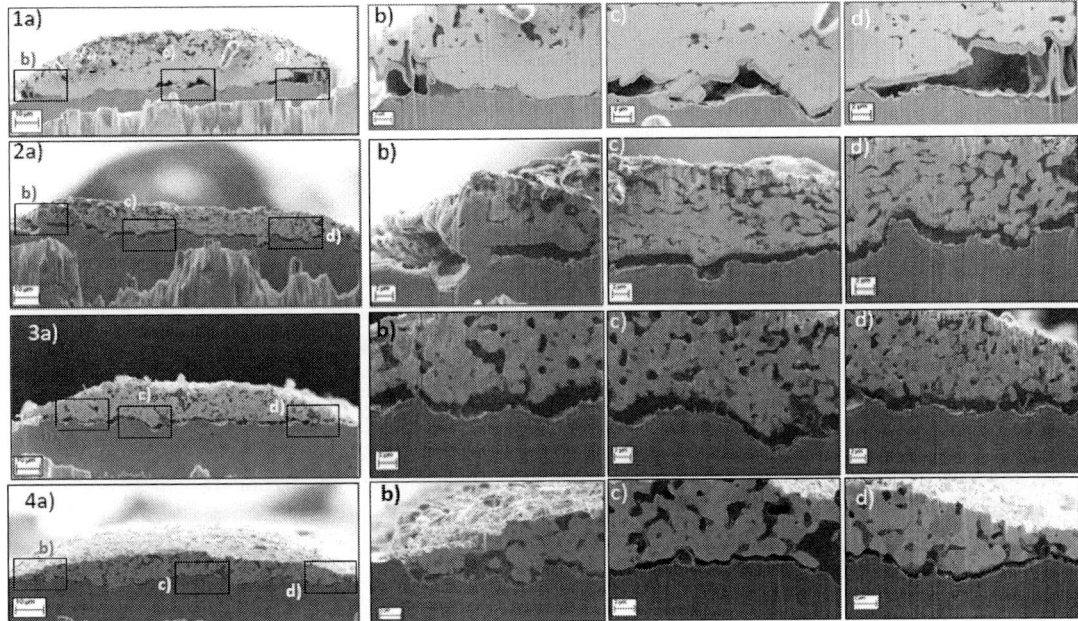

Fig. 3. Central (row 1) and near busbar region (row 2 - row 4) cross-sectional SEM images of the front contact on multi-crystalline Al-BSF PV cores extracted from different cells on a module exposed in the field for 10 years.

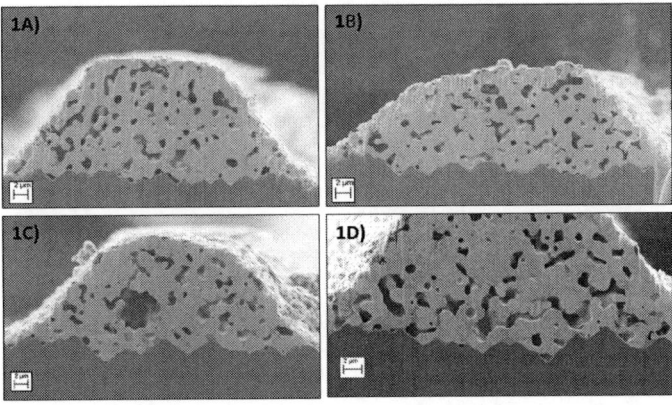

Fig. 4. SEM cross-sectional images of gridlines from central region control core (1A), near busbar region control (1B), central region of field exposed (1C), and the near busbar region of the field exposed module (1D).

modules could be taken and their contacts removed with nitric acid. X-ray photoelectron spectroscopy (XPS) data could then be collected from the glass layer beneath the front contact and compared to that of control cores. Work will be aimed to establish a better understanding of all the factors that influence

ACKNOWLEDGMENT

The authors would like to thank Daniel Riley, Bruce King, Laurie Burnham, Jennifer L. Braid, Joshua S. Stein of Sandia National Laboratories for helpful technical discussions and access to modules. This work is supported by the U.S. Department of Energy's Office of Energy Efficiency and Renewable Energy (EERE) under the Solar Energy Technologies Office Agreement Number DE-EE0009347.

REFERENCES

[1] K.-A. Weiß, E. Klimm, and I. Kaaya, "Accelerated aging tests vs field performance of PV modules," *Progress in Energy*, vol. 4, no. 4, p. 042009, Aug. 2022, publisher: IOP Publishing. [Online]. Available: https://dx.doi.org/10.1088/2516-1083/ac890a

[2] A. Fairbrother, L. Gnocchi, C. Ballif, and A. Virtuani, "Corrosion testing of solar cells: Wear-out degradation behavior," *Solar Energy Materials and Solar Cells*, vol. 248, p. 111974, Dec. 2022. [Online]. Available: https://www.sciencedirect.com/science/article/pii/S0927024822003920

[3] J. Karas, A. Sinha, V. S. P. Buddha, F. Li, F. Moghadam, G. TamizhMani, S. Bowden, and A. Augusto, "Damp Heat Induced Degradation of Silicon Heterojunction Solar Cells With Cu-Plated Contacts," *IEEE Journal of Photovoltaics*, vol. 10, no. 1, pp. 153–158, Jan. 2020, conference Name: IEEE Journal of Photovoltaics.

[4] H. Moutinho, B. To, D. Sulas-Kern, C.-S. Jiang, M. Al-Jassim, and S. Johnston, "Advances in Coring Procedures of Silicon Photovoltaic Modules," in *2020 47th IEEE Photovoltaic Specialists Conference (PVSC)*, Jun. 2020, pp. 1449–1453, iSSN: 0160-8371.

[5] A. Gabor, "Solar panel design factors to reduce the impact of cracked cells and the tendency for crack propagation," 2015.

[6] N. Iqbal, D. J. Colvin, E. J. Schneller, T. S. Sakthivel, R. Ristau, B. D. Huey, B. X. J. Yu, J.-N. Jaubert, A. J. Curran, M. Wang, S. Seal, R. H. French, and K. O. Davis, "Characterization of front contact degradation in monocrystalline and multicrystalline silicon photovoltaic modules following damp heat exposure," *Solar Energy Materials and Solar Cells*, vol. 235, p. 111468, Jan. 2022. [Online]. Available: https://www.sciencedirect.com/science/article/pii/S0927024821005080

[7] T. Tanahashi, N. Sakamoto, H. Shibata, and A. Masuda, "Electrical detection of gap formation underneath finger electrodes on c-Si PV cells exposed to acetic acid vapor under hygrothermal conditions," in *2016 IEEE 43rd Photovoltaic Specialists Conference (PVSC)*, Jun. 2016, pp. 1075–1079.

[8] A. Masuda, N. Uchiyama, and Y. Hara, "Degradation by acetic acid for crystalline Si photovoltaic modules," *Japanese Journal of Applied Physics*, vol. 54, p. 04DR04, Apr. 2015.

Energy Management in a Dynamic Microgrid Using Genetic Algorithms

Ricardo Calloquispe-Huallpa[1], Rachid Darbali-Zamora[2], Erick E. Aponte-Bezares[1], and Matthew S. Lave[2]

[1]University of Puerto Rico-Mayagüez, Mayagüez, Puerto Rico 00682, USA

[2]Sandia National Laboratories, Albuquerque, New Mexico, 87185, USA

Abstract—This paper presents an energy management strategy by opening and closing switches in a dynamic microgrid. This microgrid consists of several groups containing grid following (GFL) photovoltaic (PV) inverters, grid forming (GFM) energy storage and load which can be connected and disconnected from the main microgrid. Optimization for energy management seeks to minimize the energy consumed by the grid with the constraint of continuously supplying energy to a critical load. The algorithm is tasked with energy management is the genetic algorithm. The algorithm evaluates all the parameters of the microgrid and as a result, delivers the states of the switches of all the groups and the power reference for the energy storage. Finally, to observe the dynamic behavior of the microgrid states caused by the opening and closing of switches, the microgrid was modeled in the direct-quadrature frame. For this, the average models of the inverters with voltage control for the GFM group and with current control for the GFL groups were used.

Index Terms—dynamic microgrid, DQ frame, genetic algorithms.

NOMENCLATURE

I_d	Current in the D-axis
I_q	Current in the D-axis
V_d	Voltage in the D-axis
V_q	Voltage in the Q-axis
D_d	Inverter duty cycle in the D-axis
D_q	Inverter duty cycle in the Q-axis
U_{id}	State corresponding to the integrator of the PI current controller in the D-axis
U_{iq}	State corresponding to the integrator of the PI current controller in the Q-axis
U_{vd}	State corresponding to the integrator of the PI voltage controller in the D-axis
U_{vq}	State corresponding to the integrator of the PI voltage controller in the Q-axis
V_{dc}	Inverter dc voltage source
S	Status of the switch (1 or 0)
C_0	Bus 0 capacitance
C	Inverter capacitance
L	Inverter inductance
z	Inverter load
R	Line resistance
K_{ii}	Integral gain of the current controller
K_{pi}	Proportional gain of the current controller
K_{iv}	Integral gain of the voltage controller
K_{pv}	Proportional gain of the voltage controller

S_a	Apparent power
P	Active power
Q	Reactive power
w	System angular frequency
P_{grid}	Grid power
P_{pv}	Photovoltaic power
P_l	Load power
P_{Batt}	Energy storage power
I_{batt}	Energy storage current
I_{max}	Maximum energy storage current
Cap	Energy storage capacity
SoC_i	Initial energy storage state of charge
SoC_f	Final energy storage state of charge
SoC_{ub}	State of Charge upper boundary
SoC_{lb}	State of Charge lower boundary
Δt	Time period
K_g	Grid weight factor
K_b	Energy storage weight factor
K_gr	Groups weight factor

I. INTRODUCTION

THE increase in load demand and the deficit of generation has become a global issue. One alternative to mitigate this is implementing distributed energy resources (DERs). DERs include micro-generators, such as wind turbine generators (WTGs), photovoltaic panels (PVs), fuel cells and diesel generators, as well as energy storage devices, such as batteries, supercapacitors and flywheels [1]. DERs play an important role in microgrids.

Micrgorids are smaller independent grids with the ability to provide on-site generation for local loads. In addition to conventional generation sources, microgrids also incorporate the use of renewable energy such as WTGs and PV. They can also use energy storage for periods when renewable energy is not readily available [2]. A microgrid can connect and disconnect from the grid to enable it to operate in grid-connected or island-mode. Microgrids have the ability to mitigate energy deficiency, increase reliability and resilience of the power system, and decreasing losses and costs.

Microgrids can leverage DERs in order to provide on-site generation, independent from the larger power grid. Due to their dependency to solar and wind resources, DERs power generation is highly variable. This in combination with load variability makes energy management in microgrids complex. Therefore, it is necessary to consider specialized energy management systems for microgrids [3].

978-1-6654-6060-6/23 $31.00 © 2023 IEEE

A way to control the amount of power generated and consumed in a microgrid during the day is through the operation of switches in different sections of the microgrid. These dynamic microgrids allow PV systems to generate enough power to supply certain critical loads [4]. This avoids the use of expensive continuous generators [5].

Hence, this work focuses on the dynamic modeling of a single-phase dynamic microgrid in the direct-quadrature (DQ) frame to observe the variation of voltage, current, and power states in each inverter with varying microgrid size. The size is defined by the optimization goal to always supply energy to a specific critical load. Genetic algorithm (GA) was used to optimize this process and define the states of the switches.

This paper is organized as follows: Section II introduces the model framework of the networked microgrid in the DQ frame, and Section III describes the methodology for the dynamic microgrid approach and optimization stage using genetic algorithms, while Section IV presents the case study. Finally, Section V summarizes the conclusions of the study.

II. FRAMEWORK FOR NETWORKED MICROGRID MODELING IN DQ FRAME

A scalable networked microgrid simulation was developed in MATLAB to study different control and optimization algorithms including their scalability effects. Fig. 1 shows the radial topology of the modeled microgrid. This allows modeling n groups in a microgrid with radial topology, each group with its own inverter.

An inverter with voltage control was used to mimic the behavior of the power grid, in this way, the group with this inverter will fulfill the functions of a grid forming (GFM). This group is responsible for maintaining the voltage of the entire system [6]. Inverters with active and reactive power control were used to model a PV system and fulfill the functions of a grid following (GFL). GFLs receive the power delivered by the PV throughout the day as a reference, thus mimicking the behavior of the PV [7].

These groups are connected to their front bus through a resistor that models the power line and a switch. A first bus labeled 0 was considered so that several groups can be interconnected in parallel at the head-end. Additionally, a capacitor was added to bus 0 to observe the dynamic behavior of the voltage at this bus. It is assumed that the PV and load values are known.

A. Microgrid in the DQ frame

The DQ frame was initially used for the modeling of electrical machines. Currently, it has a wide application in the control of electrical machines, modeling of multi-machines, multi-inverters, simulation of microgrids, the elaboration of phase-locked loops (PLLs), inverter control and others [8]. Inverter voltage and current controller design is generally done on the DQ frame, since the system becomes time-invariant and the control theory for linear time-invariant systems can be straightforwardly applied.

The DQ parameters are obtained by multiplying the parameters of the original system with a transformation matrix. It should be noted that the original system does not need to be a three-phase system. The DQ parameters can also be obtained from single-phase systems in which a second imaginary phase is created to the states that are shifted by 90 degrees [9].

Within the modeling of a microgrid, elements like the electrical lines and loads are straightforward to model. For these, simple elements, such as resistors can be used. According to [10], the most important and complex element to model is the inverter, so it will be discussed in further detail.

Fig. 2 shows the averaged model of the DQ frame inverter used in this work. The advantage of the averaged model is that it does not consider the small ripple that occurs at high frequencies in the currents and voltages due to the energy storage elements. Hence, the high frequency components are neglected, shifting focus to the low frequency components, which are essential for inverter and controlling power flow. Neglecting high-frequency components substantially reduces computational cost. However, averaged models can only be used as long as the system frequency is lower than the switching frequency [11].

The equations of the average interconnected inverter model are given in (1) through (4).

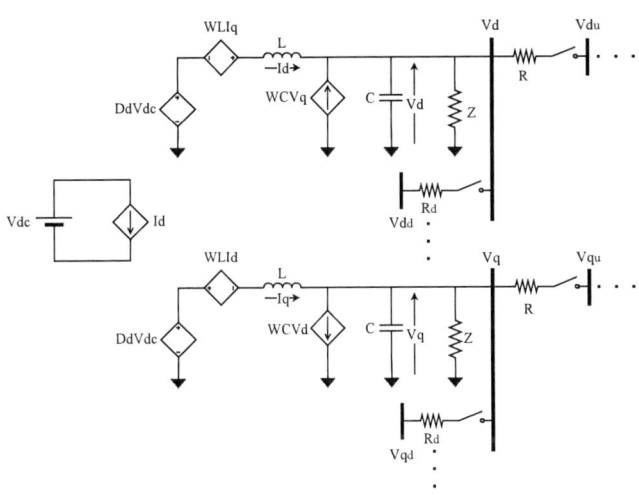

Fig. 2. Average inverter model in the DQ frame.

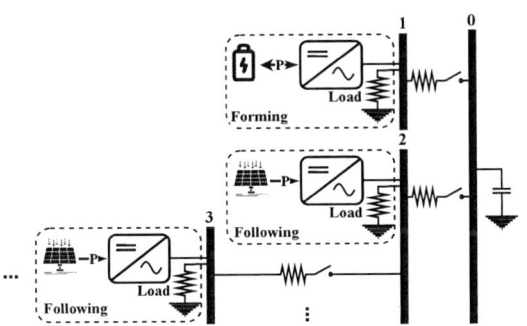

Fig. 1. Microgrid radial topology.

978-1-6654-6060-6/23 $31.00 © 2023 IEEE

$$\frac{d}{dt}Id = \frac{DdVdc}{L} - \frac{Vd}{L} + Iq\omega \tag{1}$$

$$\frac{d}{dt}Iq = \frac{DqVdc}{L} - \frac{Vq}{L} - Id\omega \tag{2}$$

$$\frac{d}{dt}Vd = \frac{Id}{C} - \frac{Vd}{CZ} + Vq\omega - \frac{Vd - Vd_u}{CR} + \sum_{j=1}^{m} S_j \left(\frac{Vd_{d_j} - Vd}{CR_{d_j}} \right) \tag{3}$$

$$\frac{d}{dt}Vq = \frac{Iq}{C} - \frac{Vq}{CZ} - Vd\omega - \frac{Vq - Vq_u}{CR} + \sum_{j=1}^{m} S_j \left(\frac{Vq_{d_j} - Vq}{CR_{d_j}} \right) \tag{4}$$

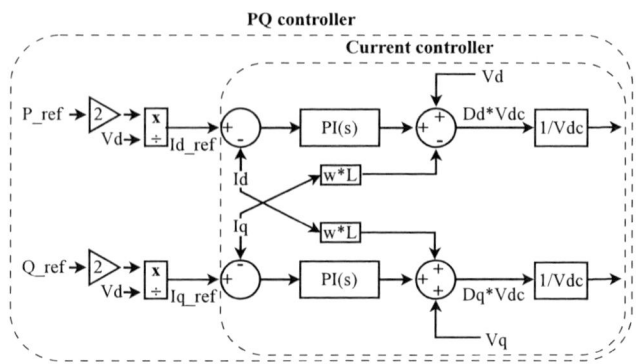

Fig. 4. PQ and current controllers.

where m represents the number of inverters downstream; the sub-index u and d represent the upstream and downstream elements, respectively.

Fig. 3, shows bus 0 with its respective capacitor. The equations corresponding to bus 0 are shown, in (5) and (6).

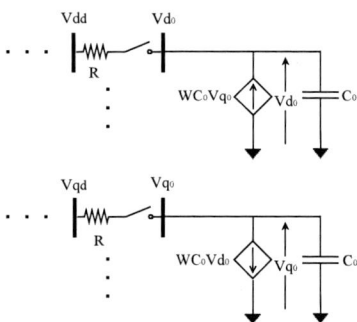

Fig. 3. Bus 0 of the microgrid modeled in the DQ framee.

$$\frac{d}{dt}V_{d0} = V_{q0}\omega + \sum_{i=1}^{m} \left(\frac{S_i(Vd_{d_i} - V_{d0})}{C_0 R_i} \right) \tag{5}$$

$$\frac{d}{dt}V_{q0} = -V_{d0}\omega + \sum_{i=1}^{m} \left(\frac{S_i(Vq_{d_i} - V_{q0})}{C_0 R_i} \right) \tag{6}$$

B. Inverter Control

PV inverters are typically controlled either as PQ sources, in which case the active and reactive powers of the inverter are directly controlled, [12], [13], [14], or as a voltage source, where the voltage amplitude is directly controlled, [9], [15]. These two control schemes are associated with the two main operating modes of an inverter, which are the GFM, and GFL modes. Both control schemes can be implemented using low-level controllers based on the DQ frame [8], [16].

Fig. 4 shows the current control, which outputs the duty cycle for each axis. These two signals, Dd and Dq, are converted back to the stationary frame to obtain the modulating signal to control an inverter in the stationary frame. These two signals will also regulate the controlled voltage sources shown in Fig. 2. A PI controller was implemented for each transfer function. A previous stage of decoupling was added to eliminate the coupling between the two axes [17].

The equations corresponding to the two PI controllers are shown in (7) through (10).

$$\frac{d}{dt}U_{id} = I_{d(ref)}K_{ii} - I_d K_{ii} \tag{7}$$

$$\frac{d}{dt}U_{iq} = I_{q(ref)}K_{ii} - I_q K_{ii} \tag{8}$$

$$D_d V_{dc} = I_{d(ref)}K_{pi} - I_d K_{pi} + U_{id} - I_q \omega L + V_d \tag{9}$$

$$D_q V_{dc} = I_{q(ref)}K_{pi} - I_q K_{pi} + U_{iq} + I_d \omega L + V_q \tag{10}$$

By replacing (9) and (10) in (1) and (2), the equations of the currents in function to the references are:

$$\frac{d}{dt}I_d = \frac{I_{d(ref)}K_{pi}}{L} - \frac{I_d K_{pi}}{L} + \frac{U_{id}}{L} \tag{11}$$

$$\frac{d}{dt}I_q = \frac{I_{q(ref)}K_{pi}}{L} - \frac{I_q K_{pi}}{L} + \frac{U_{iq}}{L} \tag{12}$$

Equations (11), (12), (3), (4), (7) and (8) model a current-controlled inverter. To the four initial states, two additional states were added because of the two current PI controllers.

The PQ controller acts as a current controller with a previous stage. Working with the power equation in DQ frame, for a $V_q = 0$, equation (13) is obtained [18]:

$$S_a = P + jQ = 0.5V_d I_d - 0.5jV_d I_q \tag{13}$$

By varying I_d and I_q, it is possible to control P and Q, respectively:

$$I_{d(ref)} = \frac{2P_{(ref)}}{V_d} \tag{14}$$

$$I_{q(ref)} = -\frac{2Q_{(ref)}}{V_d} \tag{15}$$

Fig. 5 shows the voltage controller. This controller is similar to the current controller and has the reference currents for the D-axis and Q-axis as outputs. These references are then fed to the current controller. This controller is cascaded, where it has an inner current loop and an outer voltage loop [11].

978-1-6654-6060-6/23 $31.00 © 2023 IEEE

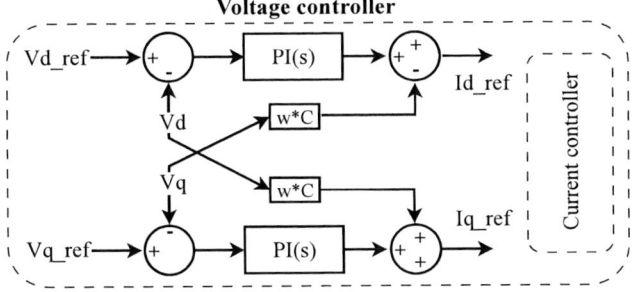

Voltage controller

Fig. 5. Voltage controller.

The equations corresponding to the two PI controllers in the voltage controller are shown in equations (16) through (19).

$$\frac{d}{dt}U_{vd} = V_{d(ref)}K_{iv} - V_dK_{iv} \tag{16}$$

$$\frac{d}{dt}U_{vq} = V_{q(ref)}K_{iv} - V_qK_{iv} \tag{17}$$

$$I_{d(ref)} = V_{d(ref)}K_{pv} - V_dK_{pv} + U_{vd} - V_q\omega C + \frac{V_d}{z} \tag{18}$$

$$I_{q(ref)} = V_{q(ref)}K_{pv} - V_qK_{pv} + U_{vq} + V_d\omega C + \frac{V_q}{z} \tag{19}$$

If (18) is replaced in (7), and (11), and (19) is replaced in (8), and (12), the equations will be obtained as a function of the reference voltages. Therefore, the equations modeling a voltage-controlled inverter will be all the equations of the current controller with the previously specified replacements and equations (16) and (17).

III. DYNAMIC MICROGRID APPROACH

The size of the microgrid will be defined by the load and power available in each group. The algorithm in charge of providing the states of the switches will be a GA.

A. Analysis & Reconfiguration Overview

The control process for microgrid energy management starts with the acquisition of data from the microgrid and ends with the allocation of switch states and power references for the energy storage, as can be seen in Fig. 6.

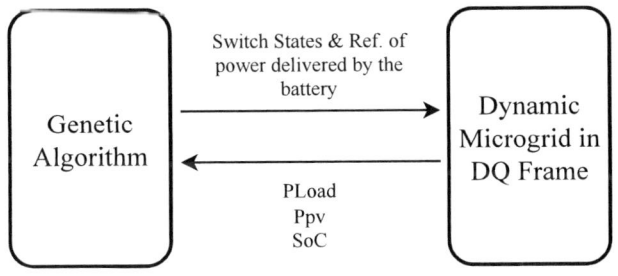

Fig. 6. Information exchange between GA and the microgrid.

It is important to note that the microgrid has a group with a critical load, so one of the constraints of the optimization will always be to provide power to that group. In addition to this, the optimization will have other constraints due to the energy storage. It is not possible to use any derivative-based optimization algorithms for dynamic microgrids, because the operation of the switches causes discontinuous changes in the states of the microgrid, i.e., the derivatives do not behave adequately with discontinuous functions. Therefore, optimization alternatives that do not require the use of derivatives are sought. Among these, is GA. GA make use of a probabilistic search based on natural and genetic selection.

B. Genetic Algorithm

GA is one of the metaheuristic methods available in artificial intelligence. In addition to GA, there are other methods, such as ant colony optimization (ACO) or particle swarm optimization (PSO). All these are inspired by biological evolution. Thus, GA is a global random search optimizer. Based on Darwin's theory of evolution, GA includes the process of selection, crossover and mutation [19], [20].

The reconfiguration of the microgrid will be determined by the optimization performed with GA. The size of the microgrid will depend on the ability to balance the power generated and consumed by the groups. Therefore, the main objective of the optimization will be to minimize the power delivered and received by the grid. A second objective is to minimize the power delivered by the energy storage, thus giving preference to the energy storage to charge and deliver power only in cases where PV generation is not sufficient. Finally, a third objective is to maximize the groups connected to the microgrid. This ensures that the groups that have their generation equal to their load are always connected to the microgrid. These three objectives in the optimization are accompanied by constants that serve as weights, as shown in equations (20) through (27). The higher the weight, the higher the priority given to it by the optimization.

$$min \left(K_g|P_{grid}| + K_bP_{batt} - K_{gr}\sum_{i=1}^{n}(S_i) \right) \tag{20}$$

where:

$$P_{grid} = -P_{Batt} - \sum_{i=1}^{n}[(Ppv_i - Pl_i) * S_i] \tag{21}$$

$$P_{Batt} = I_{batt} * V_{batt} \tag{22}$$

$$I_{batt} = \frac{Cap \times (SoC_i - SoC_f)}{\Delta t} \tag{23}$$

$$st : \quad S_{crit} = 1 \tag{24}$$

$$S_d = S_uS_d \tag{25}$$

$$I_{batt} \leq I_{max} \tag{26}$$

$$SoC_{ub} \geq SoC_f \geq SoC_{lb} \tag{27}$$

978-1-6654-6060-6/23 $31.00 © 2023 IEEE

The proposed optimization with GA is subjected to a variety of constraints. The first constraint of the optimization is that the switch corresponding to the group with the critical load is always closed. The second constraint of the optimization is that the states of each one of the switches of the downstream groups depend on the state of the switch of the upstream group. The third constraint of the optimization is the current limits of the energy storage. Finally, the fourth constraint of the optimization is the operating limits of the energy storage state of charge (SoC).

IV. CASE STUDY

In order to verify the performance of the proposed optimization with GA, a simulation model of a dynamic microgrid in the DQ frame is used. Fig. 7 illustrates a diagram of the dynamic microgrid used to test the optimization with GA. This microgrid consists of 11 groups. Group 1 corresponds to the electrical network and therefore does not have a switch, while the other groups do have their own switch. Notice from this dynamic microgrid model group 2 is the only group with a critical load.

The power generated and consumed by each one of the dynamic microgrid groups is illustrated in Fig. 8. This data corresponds to a period between 5:00 am and 9:00 pm, which is further divided into 32 periods of 30 minutes. The only groups with more generation than consumption is group 2 and group 10. However, for group 10 and to operate, it is necessary for group 8 to operate, which has a higher consumption than the generation of group 10.

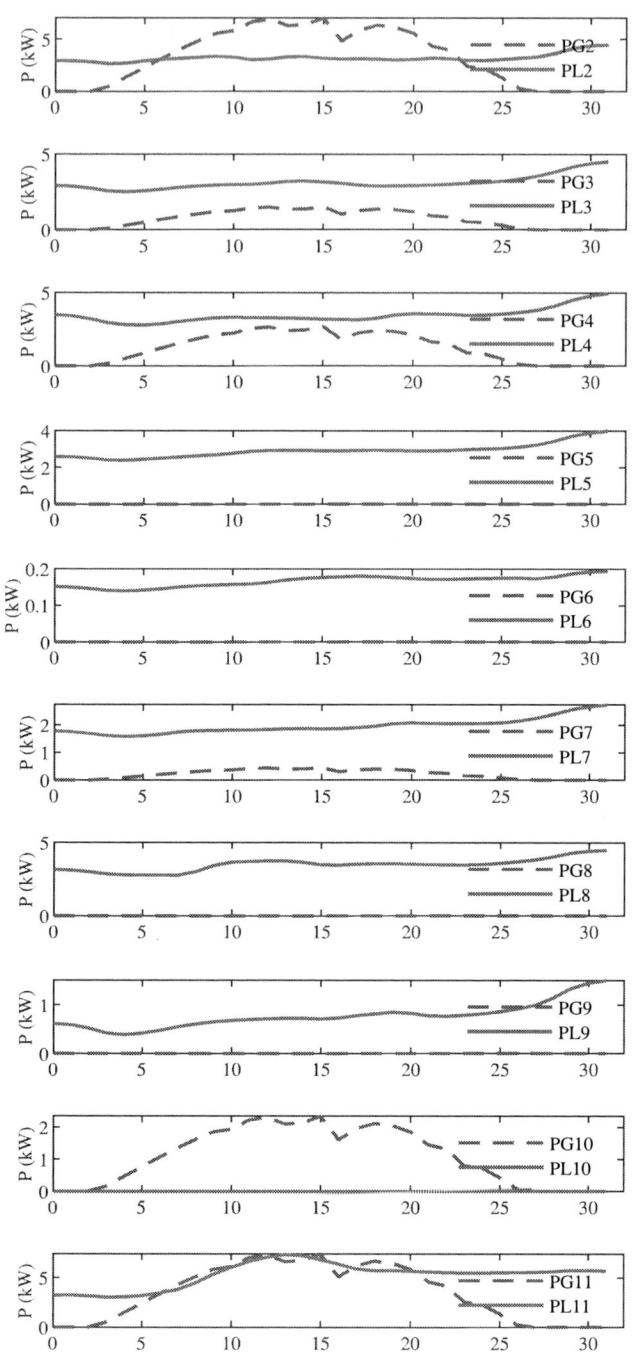

Fig. 8. Power generated and consumed by each group.

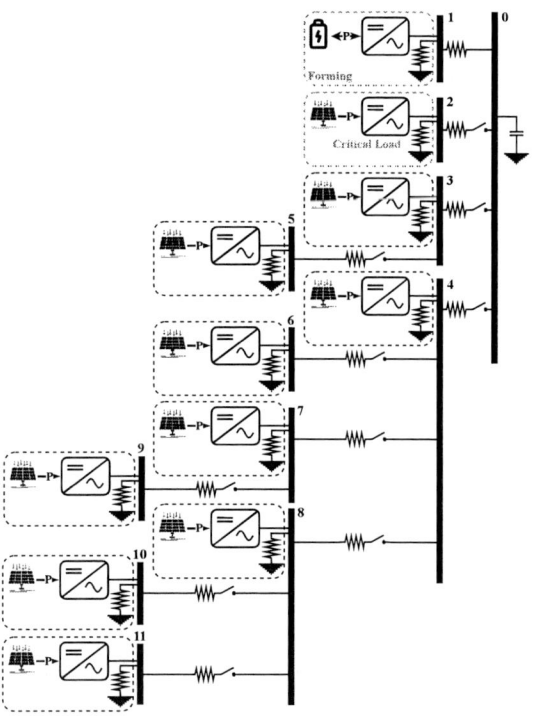

Fig. 7. Diagram of the radial microgrid with 11 groups used to test the GA optimization controls.

Fig. 9 illustrates the simulation results obtained for the total load of the microgrid (including critical and non-critical loads), the total PV generation, and the power provided by the grid. Simulation results demonstrate that the load is higher than the PV generation. Notice that the grid provides approximately 20 kW and 30 kW of power at the beginning and end of the day, respectively. Therefore, the curve of the power delivered by the grid is duck-shaped, because of the high penetration of PV generation.

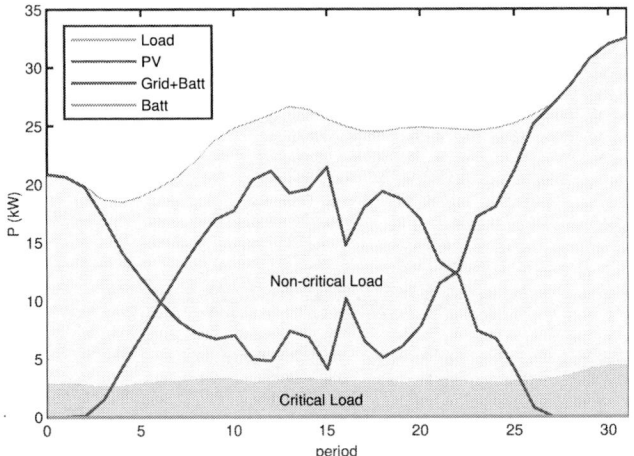

Fig. 9. Total load, PV, grid and energy storage before optimization.

A. Case Scenario 1

For the optimization of this first case, the parameters shown in Table I were considered.

TABLE I
OPTIMIZATION PARAMETERS

Description	Parameter	Value	Unit
Energy storage capacity	Cap	5000	Ah
Maximum energy storage current	I_{max}	20	A
Initial SoC	SoC_i	50%	%
SoC upper boundary	SoC_{ub}	90%	%
SoC lower boundary	SoC_{lb}	20%	%
Grid weight factor	Kg	0.7	
Energy storage weight factor	Kb	0.3	
Groups weight factor	Kgr	300	

Fig. 10 shows the power after the optimization. The load and PV generation have close values. Results show that the critical load remains constant. The power supplied by the grid is less than 3 kW. The energy storage delivers power when the PV generation is lower and charges in periods when there is excess PV. At the beginning and end of the day, the energy storage is limited by the maximum charge and discharge current.

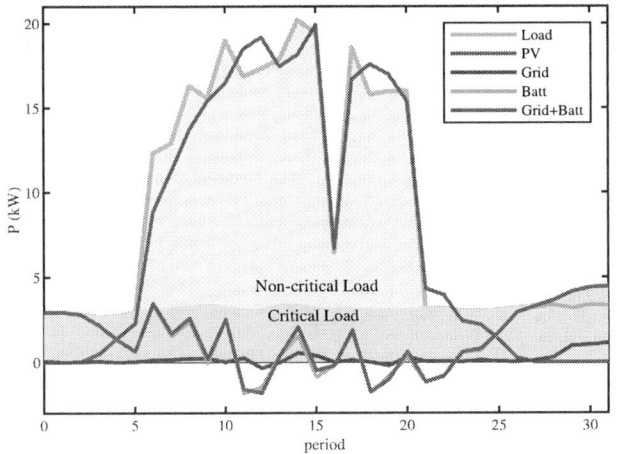

Fig. 10. Total load, PV, grid and energy storage after optimization.

Fig. 11 shows the SoC of the energy storage. The SoC decreases at the beginning and end of the day, while it increases slightly in the middle of the day.

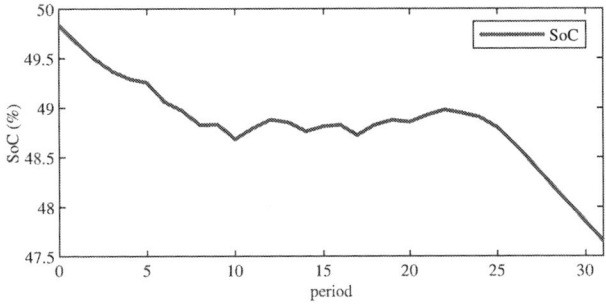

Fig. 11. Energy storage SoC throughout the day.

The values of the microgrid parameters in the DQ frame that were used in the 11 groups are presented in Table II.

TABLE II
MICROGRID PARAMETERS

Group	C(uF)	L(mH)	R(Ω)	Kii	Kpi	Kiv	Kpv
0	1 875	-	-	-	-	-	-
1	43.4	0.72	0.1	2.443E4	6.05	219.71	8.78
2	43.4	0.72	0.1	2.443E4	6.05	-	-
3	26	1.2	0.1	2.172E4	9.89	-	-
4	43.4	0.72	0.1	2.443E4	6.05	-	-
5	26	1.2	0.1	2.172E4	9.89	-	-
6	1.7	18	0.1	5.147E5	179.35	-	-
7	8.7	3.6	0.1	9.998E4	34.83	-	-
8	43.4	0.72	0.1	2.443E4	6.05	-	-
9	8.7	3.6	0.1	9.998E4	34.83	-	-
10	8.7	3.6	0.1	9.998E4	34.83	-	-
11	43.4	0.72	0.1	2.443E4	6.05	-	-

Since group 1 models the power grid and energy storage, it will have a GFM inverter, while all other groups have GFL inverters. To better observe the dynamics of the switching states, the duration of the periods was changed to 0.2 seconds. Fig. 12 shows the microgrid powers dynamically, showing small variations in magnitude. This is mainly because there are now losses in the inverters and the lines.

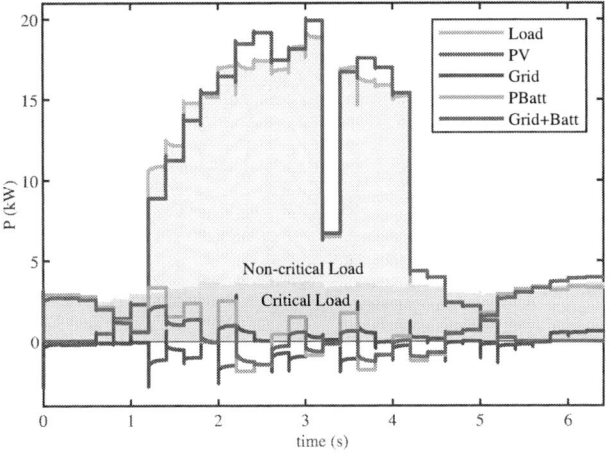

Fig. 12. Total load, PV , grid and energy storage in the DQ frame.

978-1-6654-6060-6/23 $31.00 © 2023 IEEE

The voltages of each group are shown in Fig. 13, while Fig. 14 shows the Vd voltage variations for the 11 groups. The periods in which the groups are connected and disconnected from the main microgrid can be seen clearly. The only group that is always interconnected to group 1 (grid) is group 2 because the critical load is in this group. It is also observed that the voltage control of group 1 fulfills its function of maintaining the voltage at 120 V at its terminals. Due to the voltage drop in the resistors of the lines, there are variations in the voltages of the other groups. This can be seen in more detail in group 2 which is always connected. At the beginning and end of the day, its voltage drops due to the high current drawn from the grid. As the generation of this group increases, the voltage increases because it no longer needs current from the grid to supply its load.

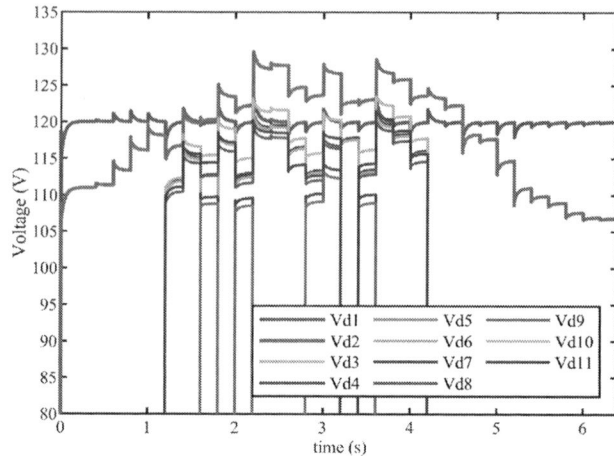

Fig. 14. Vd voltages of the 11 groups as a whole.

Notice that when there is excess PV generation, the current also flows from group 2 to the grid, and again, there is a high current flowing through the line resistor, which causes the group voltage to exceed the nominal system voltage.

B. Case Scenario 2

To observe the effects of varying the weight factors of the optimization results, a second case was proposed. For this case, the Kg value was changed to 0.52 and the Kb value was changed to 0.48; all other parameters remained the same. Fig. 15 shows the results of the power after optimization. The number of groups connected to the dynamic microgrid decreased. In a similar manner, the power delivered by the power grid increased slightly in some periods.

Fig. 16 shows the SoC of the energy storage throughout the entire day. The effects of varying the weight factors of the optimization results are more noticeable in the results for the energy storage behavior. The SoC of the energy storage has a higher charge in periods where there is an excess of PV generation.

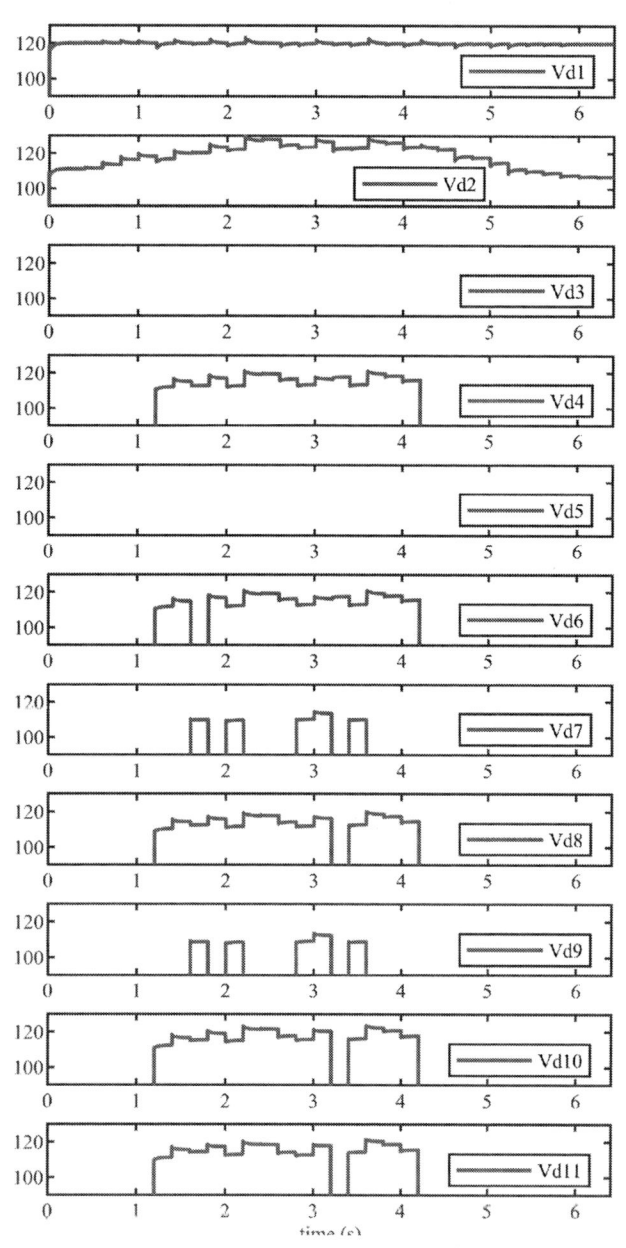

Fig. 13. Vd voltages of the 11 groups.

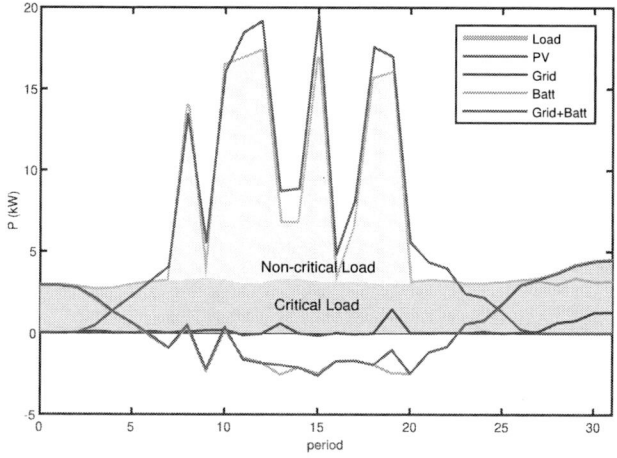

Fig. 15. Total load, total PV generated, grid and energy storage after optimization.

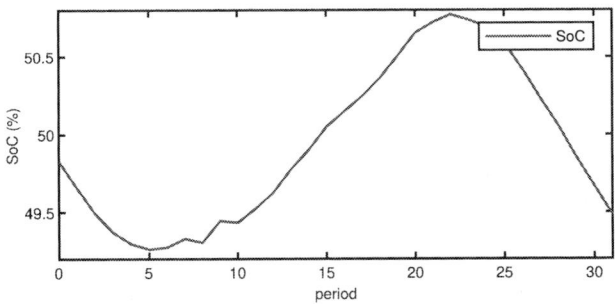

Fig. 16. Energy storage SoC throughout the day.

V. CONCLUSION

In this work, the energy management in a dynamic microgrid was performed through the opening and closing of switches, achieving a great balance between PV generation and load. Several aspects were considered in the optimization, such as the power delivered by the grid and the energy storage, the number of groups that are connected, critical load, microgrid topology, maximum charging and discharging current of the energy storage, and SoC limits. Two cases were proposed in which the weight factors were modified in the objective function, thus modifying the energy storage behavior throughout the day. The results show that the generation provided by the electric grid can be reduced to a great extent, obtaining an almost flat behavior in the power consumption curve of the electric grid. The microgrid model was simulated in DQ frame to observe the dynamic behavior of the states after the opening and closing of the switches. The voltage variation in each group and possible voltage violations due to high penetration of PV systems were observed.

The results obtained with GA were as expected, and an adequate optimization was achieved despite the discontinuity in the objective function of the system, concluding that the algorithms in the metaheuristic methods are effective for optimization of energy management in microgrids with switches.

ACKNOWLEDGMENT

This article has been authored by an employee of National Technology & Engineering Solutions of Sandia, LLC under Contract No. DE-NA0003525 with the U.S. DOE. The employee owns all right, title and interest in and to the article and is solely responsible for its contents. The U.S. Government retains and the publisher, by accepting the article for publication, acknowledges that the U.S. Government retains a non-exclusive, paid up, irrevocable, world-wide license to publish or reproduce the published form of this article or allow others to do so, for U.S. Government purposes. The DOE will provide public access to these results of federally sponsored research in accordance with the DOE Public Access Plan.

This work was sponsored in part by the Consortium for Hybrid Resilient Energy Systems (CHRES) under grant number DE-NA0003982 from the National Nuclear Security Administration part of the U.S. Department of Energy.

REFERENCES

[1] Z. Li, M. Shahidehpour, F. Aminifar, A. Alabdulwahab, and Y. Al-Turki, "Networked microgrids for enhancing the power system resilience," *Proceedings of the IEEE*, vol. 105, pp. 1289–1310, 7 2017.

[2] R. Darbali-Zamora, C. J. Gómez-Mendez, E. I. Ortiz-Rivera, H. Li, and J. Wang, "Solar irradiance prediction model based on a statistical approach for microgrid applications," in *2015 IEEE 42nd Photovoltaic Specialist Conference (PVSC)*, 2015, pp. 1–6.

[3] W. J. Ma, J. Wang, V. Gupta, and C. Chen, "Distributed energy management for networked microgrids using online admm with regret," *IEEE Transactions on Smart Grid*, vol. 9, pp. 847–856, 2018.

[4] R. Darbali-Zamora, C. B. Jones, M. S. Lave, and E. E. Aponte-Bezares, "The capability of a grid-forming inverter to support dynamic microgrids with high penetrations of photovoltaics systems," in *2022 IEEE 49th Photovoltaics Specialists Conference (PVSC)*, 2022, pp. 1091–1098.

[5] C. B. Jones, M. E. Ropp, J. H. Alvidrez, and R. Darbali-Zamora, "Optimized control of distribution switches to balance a low cost photovoltaic microgrid," *2021 IEEE 48th Photovoltaic Specialists Conference (PVSC)*, pp. 0053–0059, 6 2021.

[6] R. Darbali-Zamora, N. S. Gurule, J. Hernandez-Alvidrez, S. Gonzalez, and M. J. Reno, "Performance of a grid-forming inverter under balanced and unbalanced voltage phase angle jump conditions," in *2021 IEEE 48th Photovoltaic Specialists Conference (PVSC)*, 2021, pp. 1409–1416.

[7] R. Darbali-Zamora, J. Johnson, N. S. Gurule, M. J. Reno, N. Ninad, and E. Apablaza-Arancibia, "Evaluation of photovoltaic inverters under balanced and unbalanced voltage phase angle jump conditions," in *2020 47th IEEE Photovoltaic Specialists Conference (PVSC)*, 2020, pp. 1562–1569.

[8] D. Baimel, J. Belikov, J. M. Guerrero, and Y. Levron, "Dynamic modeling of networks, microgrids, and renewable sources in the dq0 reference frame: A survey," *IEEE Access*, vol. 5, pp. 21 323–21 335, 2017.

[9] R. Zhang, M. Cardinal, P. Szczesny, and M. Dame, "A grid simulator with control of single-phase power converters in d-q rotating frame," *2002 IEEE 33rd Annual IEEE Power Electronics Specialists Conference. Proceedings (Cat. No.02CH37289)*, pp. 1431–1436, 2002.

[10] B. Crowhurst, E. F. El-Saadany, L. E. Chaar, and L. A. Lamont, "Single-phase grid-tie inverter control using dq transform for active and reactive load power compensation," *PECon2010 - 2010 IEEE International Conference on Power and Energy*, pp. 489–494, 2010.

[11] A. Roshan, R. Burgos, A. C. Baisden, F. Wang, and D. Boroyevich, "A d-q frame controller for a full-bridge single phase inverter used in small distributed power generation systems," *APEC 07 - Twenty-Second Annual IEEE Applied Power Electronics Conference and Exposition*, pp. 641–647, 2 2007.

[12] B. Li, S. Huang, X. Chen, and Y. Xiang, "A simplified dq-frame current controller for single-phase grid-connected inverters with lcl filters," *2017 20th International Conference on Electrical Machines and Systems (ICEMS)*, pp. 1–5, 8 2017.

[13] J.-C. Liao and S.-N. Yeh, "A novel instantaneous power control strategy and analytic model for integrated rectifier/inverter systems," *IEEE Transactions on Power Electronics*, vol. 15, pp. 996–1006, 11 2000.

[14] D. Reid, "Dq rotating frame pi control algorithm for power inverter voltage regulation modelling and simulation using the openmodelica platform," *SoutheastCon 2015*, pp. 1–4, 4 2015.

[15] J. F. Sultani, "Modelling, design and implementation of d-q control in single-phase grid-connected inverters for photovoltaic systems used in domestic dwellings," *Thesis Ph.D. Montfort University*, pp. 106–148, 2013.

[16] W. Guo and L. Mu, "Control principles of micro-source inverters used in microgrid," *Protection and Control of Modern Power Systems*, vol. 1, p. 5, 12 2016.

[17] A. Roshan, "A dq rotating frame controller for single phase full-bridge inverters used in small distributed generation systems," *Thesis M.S. Virginia Polytechnic Institute and State University*, pp. 25–70, 2006.

[18] M. Restrepo, J. Morris, M. Kazerani, and C. A. Canizares, "Modeling and testing of a bidirectional smart charger for distribution system ev integration," *IEEE Transactions on Smart Grid*, vol. 9, pp. 152–162, 1 2018.

[19] S. Jinlei, L. Wei, T. Chuanyu, W. Tianru, J. Tao, and T. Yong, "A novel active equalization method for series-connected battery packs based on clustering analysis with genetic algorithm," *IEEE Transactions on Power Electronics*, vol. 36, pp. 7853–7865, 7 2021.

[20] S. Zhao, F. Blaabjerg, and H. Wang, "An overview of artificial intelligence applications for power electronics," *IEEE Transactions on Power Electronics*, vol. 36, pp. 4633–4658, 4 2021.

978-1-6654-6060-6/23 $31.00 © 2023 IEEE

Sn-based Perovskite Thin Film Solar Cells with Enhanced Stability

Wendy Reyes Ramos, Zeying Chen, Adam Thomas, Tara Dhakal

Binghamton University, Binghamton, NY, United States

We report Sn-based Perovskite thin film solar cells with enhanced stability. For the perovskite FASnI3 thin film absorber is used with various additives such as SnF2 which exhibited long term stability against oxidization from Sn2+ to Sn4+ ﬂbecause of the thermodynamic stability of this compound. The PL measurement of the absorber film showed the PL maximum around 830 nm, which corresponded to an energy of 1.49 eV. A typical device structure was Glass/ITO/NiOx or PEDOT:PSS/ FASnI3/PCBM/BCP/Ag. The charge transport and stability of the cells were improved using additives and ultrathin buffer layers deposited using atomic layer deposition (ALD) techniques. We will present devices performance together with stability of the cells.

Cadmium Selenide (CdSe) as an Active Absorber Layer for Solar Cells with V_{OC} Approaching 750 mV

Ebin Bastola[1], Adam B. Phillips[1], Abasi Abudulium[1], Vlad Kornienko[2], Zulkifl Hussain[1], Manoj K. Jamarkattel[1], Tamanna Mariam[1], Prabodika N. Kalurachchi[1], Jared Friedl[1], Dipendra Pokhrel[1], Kara B. Kile[1], Zhaoning Song[1], Yanfa Yan[1], Michael Walls[2], Randy J. Ellingson[1] and Michael J. Heben[1]

[1]Wright Center for Photovoltaics Innovation and Commercialization (PVIC), Department of Physics and Astronomy, University of Toledo, Toledo, Ohio, 43606, USA

[2]Centre for Renewable Energy Systems Technology (CREST), Loughborough University, Loughborough, Leicestershire, LE11 3TU, UK

Abstract — **Cadmium Selenide (CdSe) is a semiconductor material with a band gap (1.74 eV) suitable for top cell for the fabrication of tandem devices. Here we explore the optoelectronic properties of evaporated CdSe and the subsequent device performance. The as-deposited CdSe film (thickness ~ 800 nm) has small grains (~200-500 nm) that grow to the order of several microns after cadmium chloride ($CdCl_2$) treatment. In addition, the $CdCl_2$ treatment yielded enhanced photoluminescence (PL) response and long carrier lifetime. However, in addition to a significant band edge PL, we observe a wide peak at energies below the bandgap, suggesting defect states in the absorbance affecting the recombination in the device. The CdSe material was used as an active layer in photovoltaic devices (device structure SnO_2/CdSe/HTLs/Au) and achieved a device efficiency of 2.6% with V_{OC} exceeding 750 mV, FF of 56%, and J_{SC} of 6.1 mAcm^{-2} when illuminated through the thin Au (front) side. The device efficiency can be improved by replacing gold (Au, 10 nm) which has relatively poor transmittance and sheet resistance. We will discuss the comprehensive evaluation of CdSe films and devices for the photovoltaic application.**

Index Terms- **Evaporation, CdSe, $CdCl_2$, Solar Cells.**

I. INTRODUCTION

Cadmium selenide (CdSe), band gap of 1.74 eV, is a suitable material for photovoltaic application for the fabrication of top cell in multijunction solar cells. CdSe related devices were investigated back in 1982 with a device efficiency of about 6% [1] but very few reports exist[2, 3] after that to understand material qualities, defects associated with the absorber layer, and factors limiting the device efficiency; however, there are some recent efforts to fabricate CdSe thin film solar cells.[4, 5]

Cadmium chloride ($CdCl_2$) activation process is very common in CdTe solar cells to enhance the grain size, passivate the grain boundaries, and enhance the opto-electronic properties of the absorber layer.[6, 7] This $CdCl_2$ treatment is found to be effective to improve the grains and electronic properties of CdSe films.[2] However, finding p-type material with good transparency, proper band alignment with CdSe is challenging.[8] Previously, materials such as ZnTe, PEDOT: PSS, and P3HT have been investigated as the hole transport layers (HTLs) for CdSe solar cells[9-11], however, V_{OC} is much lower (< 600 mV) than the band gap of CdSe (~1.74 eV). In a recent study, Tang *et al.* has reported CdSe solar cells with 6% photoconversion efficiency with a V_{OC} of 586 mV.[5] Thus, the improving the V_{OC} of the CdSe solar cells is critical to enhance the device efficiency and use it as the top cell in the tandem devices.

CdSe thin films can be fabricated using various methods including close space sublimation (CSS), thermal evaporation, and sputtering. Here, we report the thermally evaporated CdSe photovoltaic devices with a very long minority carrier lifetime in the order of sub-microsecond (~100 ns). We are using our multisource deposition system to deposit CdSe thin films which is commonly used for fabrication of CdSe/CdTe solar cells.[12] The cadmium chloride ($CdCl_2$) treatment of these CdSe films significantly improved the grain sizes, steady state photoluminescence and charge carrier lifetime. Additionally, the photovoltaic device was fabricated by depositing CdSe film on FTO glass (NSG, TEC™12D) and various HTLs such as ZnTe, ZnTe:Cu, single walled carbon nanotubes (SWCNTs) and Spiro-OMeTAD (Spiro). ZnTe is commonly used in CdTe photovoltaics for commercial production as a back buffer layer, while Spiro is common for perovskite based solar cells. Out of these HTLs, photoconversion efficiency of about 2.6% has been observed with V_{OC} of 755 mV, J_{SC} of 6.1 mAcm^{-2} and FF of 56% on using Spiro as the HTL and thin Au (10 nm) as an electrode. Au contact has very low transmittance, so J_{SC} can be further improved by replacing the Au with another transparent conducting material. Further, we will discuss our preliminary results using indium tin oxide (ITO) as a transparent conducting material to enhance the J_{SC} of the device.

978-1-6654-6060-6/23 $31.00 © 2023 IEEE

II. EXPERIMENTAL

Prior to the deposition of thin film materials, glass substrates were cleaned using a Micro-90 detergent under ultrasonic cleaning and a few rinses with deionized water. CdSe thin films of about 800 nm were deposited on soda lime and TEC12D glasses (NSG Pilkington) for characterization and solar cell devices respectively via thermal evaporation at a base pressure of 4.0×10^{-6} Torr at a substrate temperature of 400 °C. As the thin films cooled down, they were taken out and treated with saturated $CdCl_2$ solution in methanol. About 0.6 mL of $CdCl_2$ solution was dropped on CdSe surface (3"x3") to wet the surface, and then it was allowed to evaporate at room temperature from the surface for about 30 s. Then, the sample was heat treated at 450 °C for 40 mins under dry air environment and rinsed with methanol twice to remove the excess $CdCl_2$. Devices were fabricated by applying various HTLs via evaporation, sputtering, or solution-processing on the top of CdSe film, then gold (Au, 10 nm) was thermally evaporated to complete the devices at a base pressure of about 10^{-6} Torr. Scanning electron microscopy (SEM) images of CdSe films were taken using a Hitachi S-4800 microscope. Transmission spectra were collected using a Perkin-Elmer UV-Vis-NIR 1050 spectrophotometer. Current density-voltage (J-V) characteristics were measured using Keithley 2401 source meter under a simulated AM1.5G spectrum and external quantum efficiencies (EQEs) were measured using PV Instruments system (model IVQE8-C).

III. RESULTS AND DISCUSSION

Fig. 1 shows the surface morphology of CdSe thin films deposited on soda lime glass substrates before and after $CdCl_2$ treatment. As-deposited thin film has grains in the order of 200-300 nm, and after $CdCl_2$ treatments the grains grew to the order of microns (1-5 μm) indicating enormous grains growth due to the $CdCl_2$ treatment.

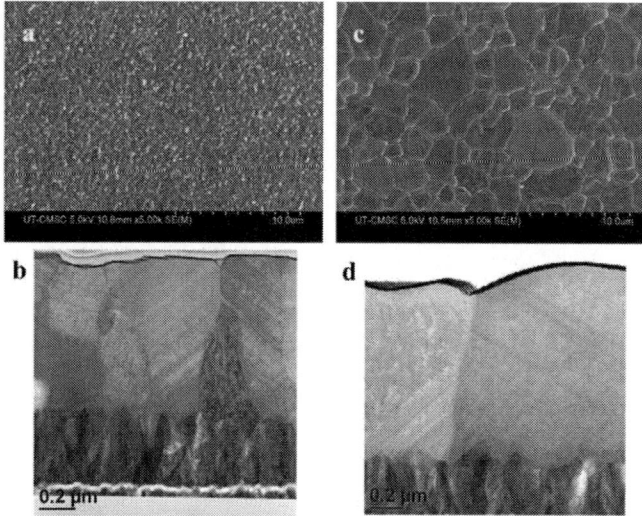

Fig 1. Electron microscopy images of CdSe films w/o (a and b) and w/ (c and d) $CdCl_2$ treatments. The thickness of the CdSe film is ~800 nm, and grains size grow enormously large (microns) after $CdCl_2$ treatment.

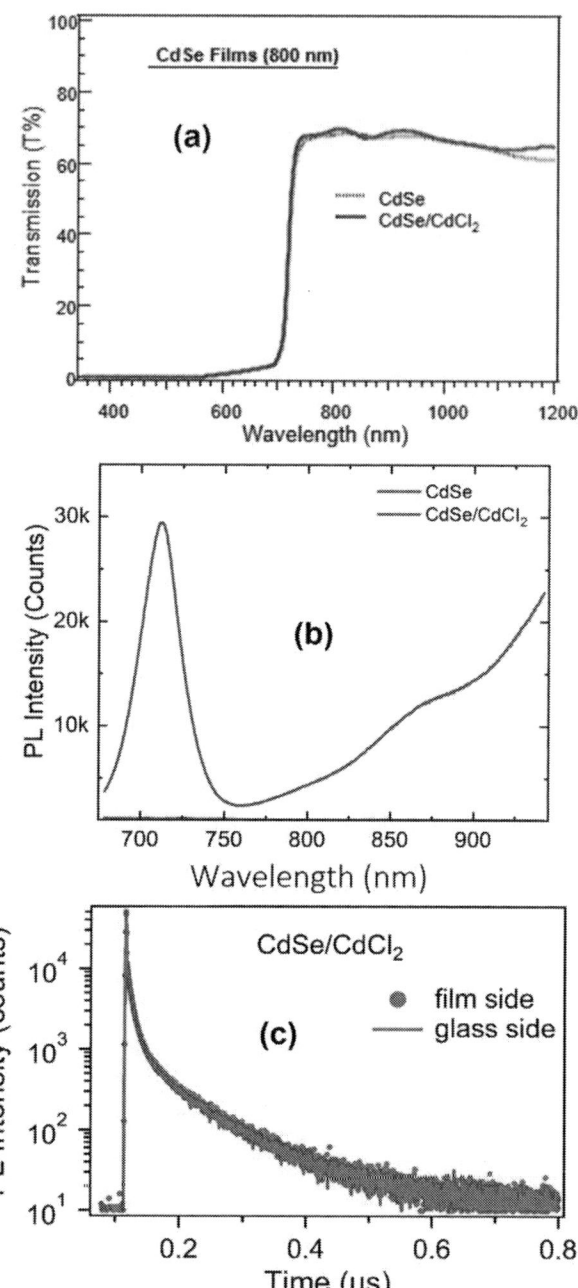

Fig 2. Opto-electronic properties of CdSe films (a) transmission spectra (b) PL and (c) TRPL decay curves.

The untreated sample has many small grains with high density of stacking faults, and these stacking faults were removed with large single grain growing from the substrate. Thus, a $CdCl_2$ treatment improved the material quality, reduced the defect densities which is similar to $CdCl_2$ activation process in CdTe devices.

Similarly, the X-ray diffraction patterns show the highly crystalline nature of these films. Based on the analysis of the XRD patterns of these as deposited and $CdCl_2$ treated films, CdSe films have preferred orientation along (111) plane in mixed phases, wurtzite, and zincblende. The preferred orientation plane remained the same even after CdCl$_2$ treatment of these CdSe films.

The transmission spectra for as deposited and CdCl$_2$ treated films are shown in Fig. 2(a). A sharp transition in the transmission spectra relates to the direct band gap of the CdSe film. The transmission spectrum for as deposited and CdCl$_2$ treated samples show a transition at 710 nm, indicating a band gap of 1.74 eV for these evaporated CdSe films. The steady-state photoluminescence spectra of the CdSe film presented in Fig. 2 (b) shows that there are multiple emission peaks in the wavelength range from 700 nm to 950 nm (and beyond) for the treated film, but as deposited material did not have any PL emission. CdCl$_2$ treatment reduced the defect density, and hence increased the PL emission. Previously, Tang et al. observed the increase in PL emission by 1.5 times with heating at 450 °C for 10 mins.[5] We assign the emission peak at 715 nm (with full width half maximum of 30 nm) to the emission from the band gap, as it matches well with the absorption edge of the CdSe film (Fig 2a). We assume that the difference in photoluminescence and absorption spectra might be related to the sample preparation conditions (e.g. temperature) and structural and morphological properties of the final film. Photoexciting the sample from film side or glass side did not result in many differences in the intensity of the band gap emission but it does on the below band gap emission. The below band gap emission can be attributed to defects (energetic or structural) associated with CdSe layer.

Fig. 2 (c) displays the time-resolved photoluminescence decay traces measured at band gap emission (715 nm) exciting the sample at 633 nm. The analysis of the TRPL decay shows a long carrier lifetime (τ_2) in the order of 100 ns and higher. We did not observe any noticeable differences in the decay kinetics when the samples were excited and measured from glass side or film side. However, we witnessed significant differences between the photoluminescence decay dynamics of the films deposited on the soda-lime glass and on the TEC12D, with the PL lifetime being shorter on the TEC12D

substrate. It can be an indication for quenching of the charges by the TEC12D at the interface.

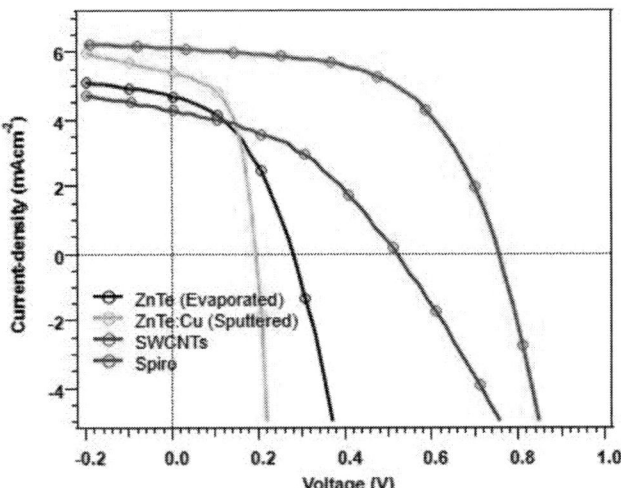

Fig 3. J-V characteristics of CdSe photovoltaic device with various hole transport layers (HTLs).

In our device stack, CdSe is the absorber layer and understanding its excited-state dynamics is one of the key milestones for engineering the devices towards the theoretical performance limit. Fig. 3 shows JVs for our devices TEC12D/CdSe/CdCl$_2$/HTLs/Au (10 nm) which have good diode characteristics under light and dark illuminations. The HTLs tested here include evaporated ZnTe, sputtered ZnTe:Cu, SWCNTs + MoO$_x$ (solution-processed) and spiro-OMeTAD. Table 1 summarizes the deposition conditions and device parameters for all the HTLs investigated to fabricate CdSe solar cells. In the case of ZnTe, V_{OC} is lower compared to SWCNTs and Spiro. For Spiro, V_{OC} is relatively high (759 mV) but still CdSe solar cells have a potential of having high $V_{OC} > 1$ V due to its high band gap (1.74 eV). The high V_{OC} of the device would depend on the finding right p-type partner to fabricate the devices. On all these devices, the p-n junction is

TABLE I

Device Parameters of CdSe Solar Cells with Various Hole Transport Layers (HTLs) Measured from the Front Side of the Devices (Au side)

HTLs		V_{OC} (mV)	J_{SC} (mAcm^{-2})	FF (%)	Eff. (%)
Material	Deposition/Conditions				
ZnTe	Evaporation, 400 C, ~100 nm	279	4.7	42.3	0.55
ZnTe:Cu	Sputtering at 270 C, ~20 nm	195	5.4	55.2	0.58
SWCNTs/MoOx	Spin-coating, 140 C, ~80 nm	518	4.3	40.7	0.90
Spiro-OMeTAD	Spin-coating, RT ~250 nm	759	6.1	51.0	2.4

in between the HTL and CdSe layer, closer towards the Au side of the device. Thin Au (10 nm) reflects larger amount of incident photons, and hence J_{SC} is lower for all the HTLs. The lower V_{OC} in the case of ZnTe could be related to the doping level of these layers.

Similarly, FF is lower (< 60%) in all the cases indicating losses of the charge carries which could be at the interfaces on the devices. Thus, improving the recombination losses is essential to enhance the device performance. The interface between SnO_2/CdSe can also be examined, and possibly improved with other high band gap transparent materials such as magnesium zinc oxide or indium gallium oxide.[13] Additionally, FF (and V_{OC}) can be further improved if the sub band gap emissions below the band gap can be eliminated or reduced, and it would further increase the PL emission at the band edge. It could be possible by incorporating or treating the absorber layer in Se atmosphere to lower Se vacancy in the film or doping with group V elements. Further, the conductivity of 10 nm Au is poor. When a sheet resistance was measured with 10 nm Au deposited on a soda lime glass, it had a sheet resistance of ~255 Ohm/Sq.

Fig. 4 shows JVs and EQEs of slightly better CdSe solar cells for Spiro as an HTL when illuminated from Au and glass sides. The J-V curves show a good diode characteristic of the device under both light and dark illuminations. When the device is illuminated from the Au side, Au reflects a significant amount of incident photons, and short circuit current density is 6.1 $mAcm^{-2}$ only. Similarly, when the device was illuminated from the glass side, it can transmit ~90% of the incident light. However, the device has poor Jsc (3.1 $mAcm^{-2}$) that could be associated with the defects on the absorber layer as suggested by the PL emission spectrum (Fig. 2b). Fig. 4(b) displays the EQEs for the same device when illuminated from Au and glass side of the device, and the dotted line represents the transmission of light from a 10 nm Au film. The poor J_{SC} from Au side illumination is due to (1) low transmission of incident lights from the back electrode (Au) and (2) absorption of blue photons by the hole transport layers (Spiro, band gap ~3.0 eV). The triangular shape EQE when illuminated from the glass side indicates high defect density on the absorber layer, and the p-n junction being far away from the photons absorbed. Usually, triangular shape EQE is common for CdSeTe device when the absorber layer is underdoped with Cu[12], and for bifacial device when it is illuminated from the back contact side and the absorber is thick (> 3 μm).[14] Here, the absorber is only about 800 nm thick, and the possible reason could be the sub-band gap

defects, and weak built-in potential at the p-n junction. The charge carriers do not separate to increase the photocurrent, and being far from the junction, they recombine producing a very small current from glass side. Table 2 summarizes the device parameters for the cells shown in Fig 4 with Spiro HTL. Further, we tested transparent material such as indium tin oxide (ITO) replacing Au to see how the J_{SC} increases with transparent electrode.

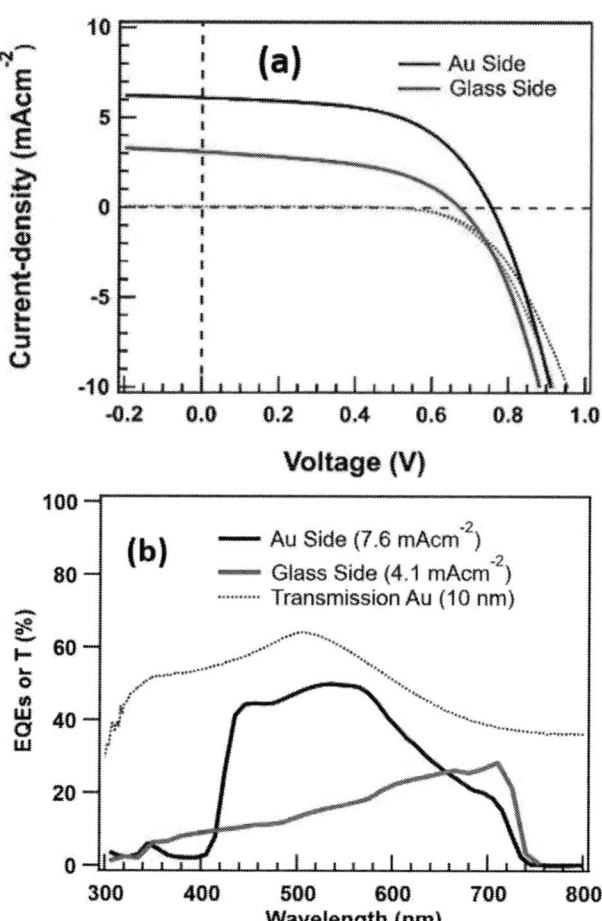

Fig 4. (a) J-V characteristics and (b) EQEs of photovoltaic devices using CdSe as an active absorber layer.

When thin Au is used as an electrode on the sunny side of the device, it transmits only ~60% of light as shown in transmission spectra in Fig 4 (b) and J_{SC} of the device is low.

TABLE II
PV DEVICE PARAMETERS OF CdSe SOLAR CELLS WITH CDSE AS ACTIVE ABSORBER LAYESR

Illumination	V_{OC} (mV)	J_{SC} ($mAcm^{-2}$)	FF (%)	PCE (%)	Rs (Ω cm^2)	Rsh (Ω cm^2)
Au Side	755	6.1	56.3	2.6	30.6	1752
Glass Side	673	3.1	48.5	1.1	57.0	1195

EQE follows the T% spectra of the thin Au layer. Thus, other transparent conducting materials could be an alternative to thin Au (10 nm). Here, we examined CdSe devices with indium tin oxide (ITO) (TEC12D/CdSe/CdCl$_2$/ZnTe/ITO), and the EQEs are shown for thin Au and ITO devices. EQE reaches up to 66% with ITO which is much higher than the Au device (~45%). The integrated J_{SC} is 13.12 mAcm^{-2} for ITO device whereas its value is 8.21 mAcm^{-2} for Au device clearly indicating significant gain in the current collection. However, these devices suffer from very low V_{OC} (~200 mV) and thus the device efficiency is poor (~1%). Probably, this effect can be minimized by using an integrated HTLs using two materials such as Spiro/ZnTe or SWCNTs/ZnTe with an ITO which may help to maintain V_{OC} and high J_{SC} of the device.

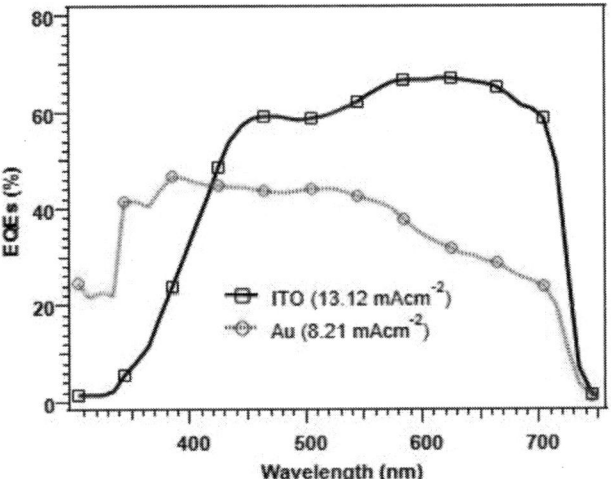

Fig 5. EQEs of CdSe solar cells with the EQE difference for Au (10 nm) and ITO electrodes.

V. CONCLUSIONS

Here, we demonstrated that CdSe solar cells with good diode behavior can be fabricated using a thermally evaporated CdSe absorber layer though there needs much understanding about the defects associated with the absorber layer, and interfaces. V_{OC} as high as 759 mV, and photoconversion efficiency of 2.6% was observed with Spiro as an HTL. The device suffers from low J_{SC} due to low transmittance of thin Au layer. Further, a transparent back electrode needs to be developed to absorb more photons that could reach the absorber layer and increase the photocurrent (J_{SC}). Our preliminary study shows a J_{SC} of > 13 mAcm^{-2} when ITO was used as the transparent electrode. Further, an integrated HTLs would be beneficial to have high V_{OC} and J_{SC} of the CdSe solar cells.

ACKNOWLEDGEMENTS

This material is based on research sponsored by the U. S. DOE's office of Energy Efficiency and Renewable Energy (EERE) under Solar Energy Technologies Office (SETO), Cadmium Telluride Accelerator Consortium (CTAC) (NREL Sub-contract SUB-2021-10715), and Air Force Research Laboratory under agreement number FA9453-19-C-1002. The U.S. Government is authorized to reproduce and distribute reprints for Governmental purposes not withstanding any copyright notation thereon. Disclaimer: The views and conclusions contained herein are those of the authors and should not be interpreted as necessarily representing the official policies or endorsements, either expressed or implied, of the Department of Energy or Air Force Research Laboratory or the U.S. Government.

REFERENCES

[1] E. Rickus, "Photovoltaic behaviour of CdSe thin film solar cells," in *Fourth EC Photovoltaic Solar Energy Conference*, 1982: Springer, pp. 831-835.

[2] B. Bagheri *et al.*, "Influence of post-deposition selenization and cadmium chloride assisted grain enhancement on electronic properties of cadmium selenide thin films," *AIP Advances,* vol. 9, no. 12, p. 125012, 2019, doi: 10.1063/1.5124881.

[3] S. Vakkalanka, C. S. Ferekides, and D. L. Morel, "V_{OC} enhancement in CdSe solar cells using ZnSe$_x$Te$_{1-x}$: N window layers," in *2008 33rd IEEE Photovoltaic Specialists Conference*, 11-16 May 2008 2008, pp. 1-4, doi: 10.1109/PVSC.2008.4922534.

[4] K. Li *et al.*, "Rapid thermal evaporation for cadmium selenide thin-film solar cells," *Frontiers of Optoelectronics,* vol. 14, no. 4, pp. 482-490, 2021.

[5] K. Li *et al.*, "Fabrication and Optimization of CdSe Solar Cells for Possible Top Cell of Silicon-Based Tandem Devices," *Advanced Energy Materials,* vol. 12, no. 26, p. 2200725, 2022, doi: https://doi.org/10.1002/aenm.202200725.

[6] I. Dharmadasa, "Review of the CdCl$_2$ treatment used in CdS/CdTe thin film solar cell development and new evidence towards improved understanding," *Coatings,* vol. 4, no. 2, pp. 282-307, 2014.

[7] P. N. Kaluarachchi *et al.*, "Optimizing CdCl$_2$ Treatment on CdTe Solar Cells Using Spray Deposition Method," in *2022 IEEE 49th Photovoltaics Specialists Conference (PVSC)*, 5-10 June 2022 2022, pp. 0828-0832, doi: 10.1109/PVSC48317.2022.9938796.

[8] J. D. Friedl, R. H. Ahangharnejhad, A. B. Phillips, and M. J. Heben, "Material Requirements for CdSe Wide Bandgap Solar Cells," in *2021 IEEE 48th Photovoltaic Specialists Conference (PVSC)*, 2021: IEEE, pp. 1548-1552.

[9] L.-P. Poly, B. Mace, R. Kottokkaran, B. Bagheri, M. Noack, and V. Dalal, "Novel CdSe Solar Cell," in *2021 IEEE 48th Photovoltaic Specialists Conference (PVSC)*, 2021: IEEE, pp. 0443-0446.

[10] B. Bagheri *et al.*, "Efficient heterojunction thin film CdSe solar cells deposited using thermal evaporation," in *2019 IEEE 46th Photovoltaic Specialists Conference (PVSC)*, 2019: IEEE, pp. 1822-1825.

[11] S. Chanda, R. Anders, C. Ferekides, and D. Morel, "Control of V oc in CdSe solar cells," in *2009 34th IEEE Photovoltaic Specialists Conference (PVSC)*, 2009: IEEE, pp. 001507-001512.

[12] E. Bastola *et al.*, "Understanding the Interplay Between CdSe Thickness and Cu Doping Temperature in CdSe/CdTe Devices," *IEEE Journal of Photovoltaics,* vol. 12, no. 1, pp. 11-15, 2022, doi: 10.1109/jphotov.2021.3110338.

[13] M. K. Jamarkattel *et al.*, "Indium Gallium Oxide Emitters for High-Efficiency CdTe-Based Solar Cells," *ACS Applied Energy Materials,* vol. 5, no. 5, pp. 5484-5489, 2022/05/23 2022, doi: 10.1021/acsaem.2c00153.

[14] D. Pokhrel *et al.*, "Copper iodide nanoparticles as a hole transport layer to CdTe photovoltaics: 5.5% efficient back-illuminated bifacial CdTe solar cells," *Solar energy materials and solar cells,* vol. 235, p. 111451, 2022.

Mapping Stress in PV Modules: The Influence of Soldering, Tabbing, and Module Architecture

Ian Slauch, Rico Meier, Kaushik Roy Choudhury, Mariana Bertoni

Arizona State University, Tempe, AZ, United States

DuPont Photovoltaic and Advanced Materials, Newark, DE, United States

Alternatives to conventional photovoltaic cell tabbing schemes are explored for their promising electrical properties and low cost. When implementing new architectures and bill of materials very little attention is usually paid to the thermomechanical stresses acting on the cell. To this point, even less is understood about the effect these stresses have on failure modes and module lifetime. Here, we use X-ray Topography to image in-situ the stresses in cells with conventional and multi-busbar tabbing with different bill of materials and configurations (glass/glass vs. glass backsheet). We determine that, while residual stresses are reduced with multi bus bar tabbing in glass-backsheet laminates, glass-glass laminates require significantly thicker encapsulant to achieve similar stress levels. At the time of presentation we will complete the set with interdigitated back contact cells and POE vs. EVA.

Elemental Vapor Transport Deposition of CdSe$_x$Te$_{1-x}$ Thin Films for n-type CdTe Solar Cells

Wei Wang, Vasilios Palekis, Md Zahangir Alom, Sheikh Elahi Tawsif, Don Morel, and Chris Ferekides

Electrical Engineering, University of South Florida, Tampa, FL 33620, USA

Abstract — This paper investigated the structural and electrical properties of CdSe$_x$Te$_{1-x}$ thin films grown by the elemental vapor transport (EVT) deposition process. The CdSe$_x$Te$_{1-x}$ films with various selenium concentrations were characterized in terms of optical absorption and atomic concentration, as well as morphology and crystallinity. Incorporating Se into the CdTe films results in a reduction of the bandgap which favors the generation of photocurrents at longer wavelengths. With the increase of Se composition, a phase transition from zinc blende structure to wurtzite structure occurs. SIMS measurements show a uniform distribution of Se in the CdSe$_x$Te$_{1-x}$ film. TRPL lifetime measurements show that the minority carrier lifetimes increase with the increase of Se composition.

Keywords—CdTe solar cell, elemental vapor transport, CdSeTe, minority carrier lifetime

I. INTRODUCTION

Thin-film polycrystalline CdTe-based solar cells have reached a cell efficiency of 22.1% (recently, First Solar announced that the cell efficiency has improved to 22.3%) [1]. The near-ideal band gap (~1.5eV) and high optical absorption coefficient (>10^4/cm) in the visible wavelength range make CdTe one of the most promising materials for the fabrication of low-cost, high-efficiency thin-film solar cells.

The theoretical efficiency limit of CdTe-based solar cells is ~ 33%, which indicates that there is still room to improve cell performance. To further improve cell efficiency, higher carrier concentration and longer minority carrier lifetime are required [2]. One of the most challenging problems that limits the CdTe cell performance is the difficulty to achieve high majority carrier concentration and long minority carrier lifetime at the same time. Selenium was incorporated in CdTe absorbers to improve the short circuit current by lowering the energy bandgap without compensating the open-circuit voltage, which is related to the improvement of the minority carrier lifetime in the absorber [3].

Both p- and n-type CdTe thin films can be achieved by intrinsic or extrinsic doping. N-type CdTe films can be achieved by replacing Cd and Te sites with groups-III (In, Al) and -VII (Cl, I) elements, respectively [4]. In- and Cl-doped n-type CdTe-based solar cells have been studied previously; cell efficiency of 9%, carrier concentration ~ 7E+16 cm^{-3}, and lifetime of 7 ns have been achieved [5] [6] [7] [8]. The incorporation of Se improves the lifetime of the minority carriers of the p-CdTe solar cells. To investigate whether the incorporation of Se has a similar effect on the performance of n-CdTe cells, the material properties of CdSe$_x$Te$_{1-x}$ thin films grown by elemental vapor transport deposition were studied.

II. EXPERIMENTAL

As shown in Fig. 1, the device structure includes Glass/ITO/CdS/CdSe$_x$Te$_{1-x}$. Indium tin oxide (ITO) thin films with a thickness of 4000 Å were deposited by RF sputtering on corning EagleXG glass. Cadmium sulfide (CdS) with a thickness of 250 Å was deposited by the close-spaced sublimation (CSS). Polycrystalline CdSe$_x$Te$_{1-x}$ films were grown by elemental vapor transport deposition (EVT). The EVT process is used to achieve control of the Cd/Te vapor ratios and various Se compositions. Cd, Te, and Se metals with a minimum purity of 5N were used to deposit CdSe$_x$Te$_{1-x}$. alloys. The Se composition can be controlled by varying Se vapor temperature and Se vapor flow rate.

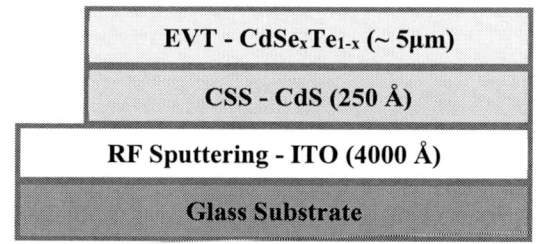

Fig. 1 Device configuration

Spectral response (SR), scanning electron microscope (SEM), energy dispersive spectroscopy (EDS), x-ray diffraction (XRD), secondary-ion mass spectrometry (SIMS), and time-resolved photoluminescence (TRPL) measurements were used to characterize the CdSe$_x$Te$_{1-x}$ thin films and devices. SR measurements were performed using an Oriel monochromator (model 74100) with a light source whose intensity was calibrated using a standard calibrated silicon reference solar cell. SEM and EDS measurements were performed on Hitachi SU-70. XRD measurements were performed on PANalytical X'pert PRO with a Copper (Cu) kα to study the crystallographic properties. SIMS measurements were performed at the National Renewable Energy Laboratory (NREL). TRPL measurements were performed at the University of Florida (UF).

We are thankful to Steve Harvey of the National Renewable Energy Laboratory (NREL) for SIMS measurements, and professor Charles Hages and Calvin Fai of the University of Florida for the TRPL measurements.

III. RESULT AND DISCUSSION

A. Se compositon

Thin films of $CdSe_xTe_{1-x}$ were fabricated by varying the Se gas flowrate (i.e., 0, 5, 10, 20, 40, 60, 100 ccm) while keeping the Se temperature constant at 320°C. Using vapor pressure data from the literature the Se/Te vapor ratios corresponding to the above flow rates were calculated to be 0, 0.15, 0.31, 0.62, 1.24, 1.85, and 3.10, respectively [9].

Figure. 2 shows the normalized relative transmission measurements for the CdTe and $CdSe_xTe_{1-x}$ films mentioned above. The Se composition of these films were measured by EDS, which are shown in Table 1. The transmission edge shifts to the right (longer wavelengths) as the Se gas flow increases. This indicates a reduction in the bandgap as the amount of Se in the films gradually increased from x = 0 to x = 0.21. With the further increase of Se content, the bandgap value then experienced an increase. These results are consistent with the dependence of the $CdSe_xTe_{1-x}$ bandgap on Se composition that exhibits the bowing effect caused by variations in the local structural arrangement, arising from differences in ionic size, electronegativity, and lattice constants among alloys with varying Se content [10]. The film growth at high Se gas flowrates (i.e., 100 ccm) appears to be an outlier.

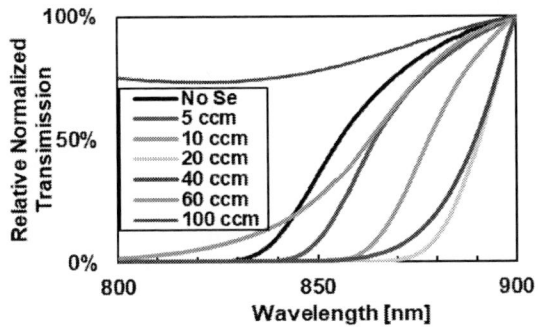

Fig. 2 Normalized relative transmission with various Se vapor flowrates.

According to the Vegard's Law, Se composition can be calculated based on the peak shift from the CdTe cubic peak towards the CdSe cubic peak [11]. For the CdTe and $CdSe_xTe_{1-x}$ films mentioned in Fig. 2, the Se composition calculated from the XRD peak shift and measured by EDS is summarized in TABLE 1. The results are in a good agreement for Se compositions up to ~ x = 0.20. Vegard's law is not applicable at higher Se compositions due to the transition of the film's structural arrangement from cubic to hexagonal. As the Se composition in the CST film increases, the crystal structure changes from zinc blende (CdTe crystal) to wurtzite (CdSe crystal), which can be seen in XRD spectral in Fig. 3. The Se composition increases with the increase of Se vapor flow rate. X

= 0.70 was achieved in $CdSe_xTe_{1-x}$ films. The films become more group-II rich as Se composition increases.

TABLE 1. *SE COMPOSITIONS MEASURED BY EDS AND XRD*

Se flow rate [ccm]	EDS ratio II/VI	Se composition (x)	
		EDS	XRD (estimation)
5	0.92	0.057	0.063
10	0.95	0.115	0.121
20	0.99	0.182	0.184
40	1.35	0.213	0.254
60	1.56	0.440	-
100	1.19	0.698	-

Figure. 3 shows the XRD spectra for the CST films with various Se compositions. Pure CdTe and CdSe with cubic structures have their main peak (111) located at 2θ of 23.759° (Ref. file: 00-015-0770) and 25.354° (Ref. file: 00-019-0191) respectively. CdSe with a hexagonal structure has peaks at (002) located at 2θ = 25.354° and (103) at 2θ = 45.759° (Ref. file: 00-008-0459). The top graph shows the complete range of the XRD spectra, and the bottom left, and right graphs focus on 2θ between 23° - 26° and 23° - 46°, respectively.

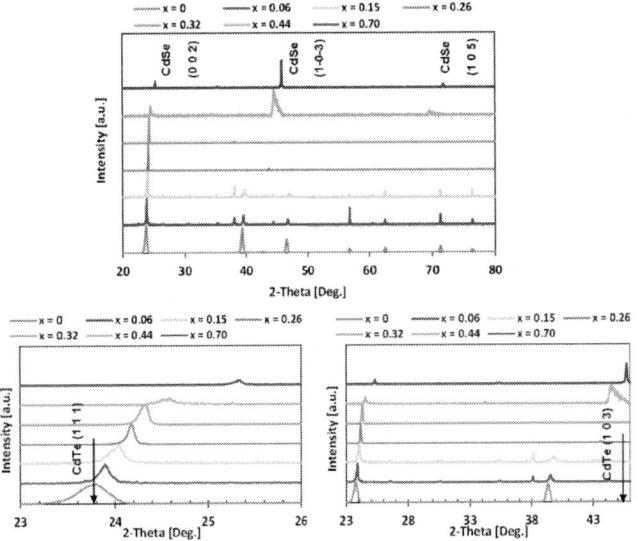

Fig. 3. XRD peak shift with increasing selenium concentrations (from a cubic to hexagonal structure)

The pure CdTe film has the (111) peak located at 2θ = 23.759°. As Se gets incorporated into the films the CdTe (111) peak starts shifting to the right towards the CdSe (111) peak. At x = 0.44 the peak at 2θ = 24.58° decreases and a new peak appears at 2θ = 45.05°. Both peaks are approaching the (002) and (103) CdSe hexagonal structure. It can also be seen from the SEM images (Fig. 4), how the grains have a well-defined columnar shape at x = 0.44 and 0.7. At x = 0.70 the peaks are located at 2θ = 25.335°, 45.735°, and 71.8° which are all orientations of the CdSe hexagonal structure.

The morphology of $CdSe_xTe_{1-x}$ films with various Se compositions is shown in Fig. 4. The Se concentration x shown is from EDS measurements. As the Se concentration increases, significant morphological changes were observed. There is a reduction in grain size as the Se composition increases. The $CdSe_xTe_{1-x}$ films have a cubic structure with x < 0.20. When x = 0.27, the crystal structure appears to be going through a transition. For the higher Se composition films x > 0.3, there is a clear change in the film morphology and the grains appear to have a columnar and low density structure.

Fig. 4. SEM images at x=0, 0.14, 0.27, 0.31, 0.44 and 0.7.

SIMS measurements were performed for a series of $CdSe_xTe_{1-x}$ films with x = 0.07, 0.15, 0.25, and 0.33. The film stack was Al_2O_3/$CdSe_xTe_{1-x}$/Al_2O_3/ITO/glass. SIMS depth profiles for all four Se compositions are similar. In Fig. 5, the SIMS measurements depict the CdSexTe1-x films, demonstrating a relatively constant distribution of Se across the entire film. This uniform distribution of Se was observed across all Se compositions. Fig. 6 presents the extracted Se/Te counts ratios from the SIMS profile in the $CdSe_xTe_{1-x}$ layers shown in Fig. 4. As expected, the Se/Te counts ratios increased with an increase in Se composition. Se/Te counts ratio increased from 1.47 to 3.78 with x increased from 0.07 to 0.33.

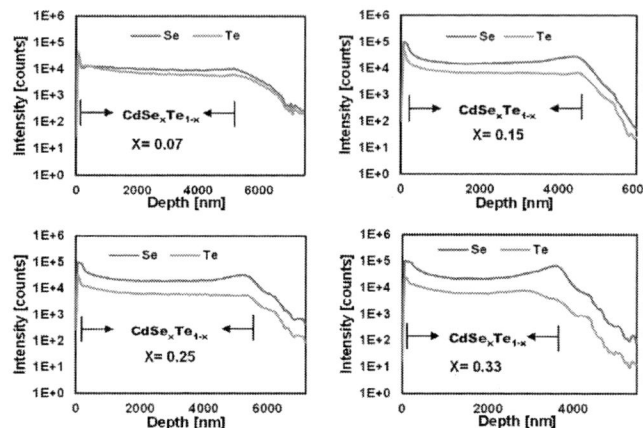

Fig. 5 SIMS depth profile for Se and Te ions in CdSexTe1-x films with x = 0.07, 0.15, 0.25, and 0.33.

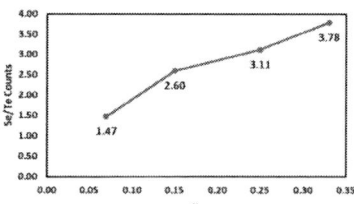

Fig 6. Se/Te counts ratio of the films in Fig 5.

B. TRPL Lifetime Measurements

A series of $CdSe_xTe_{1-x}$ films were fabricated for lifetime measurements to investigate the effects of Se composition. TRPL measurements were used to calculate the minority carrier lifetime of these $CdSe_xTe_{1-x}$ films.

Figure 7 shows the lifetime measurements for the $CdSe_xTe_{1-x}$ films with various Se compositions. The lifetimes are listed on Table 2. As shown in the table, as the amount of Se increased from x = 0 to x = 0.32, the lifetime gradually increased from 8.1 to 85.3 ns. Therefore, it appears that incorporating Se is advantageous as it leads to longer lifetime and further enhances device performance.

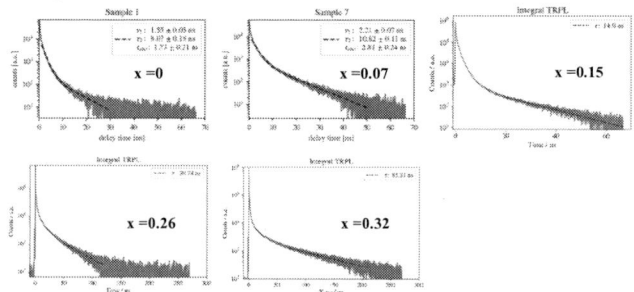

Fig. 7. TRPL lifetime measurements for $CdSe_xTe_{1-x}$ films with various Se compositions.

TABLE 2. LIFETIME VS. SE COMPOSITIONS

Se composition (x)	0	0.07	0.15	0.26	0.32
Lifetime (ns)	8.1	10.6	14.9	39.7	85.3

C. Preliminary Results on Cl-doped CdSe$_x$Te$_{1-x}$ Devices

Chlorine was used as the n-type dopant for CdSe$_x$Te$_{1-x}$ films. In order to investigate the performance of the Cl-doped n-type CdSe$_x$Te$_{1-x}$ solar cell, a full device structure was fabricated, comprising of the following layers: glass/TCO/CdS/ Cl-doped CdSe$_x$Te$_{1-x}$ /ZnTe/Cu/ITO. Films were deposited at different Se and Cl vapor concentrations. Low and high Se compositions, x<0.2 and x>0.4 were chosen. Two Cl/Te gas vapor phase concentrations of 1 and 3 Kppm were used. The film Se composition was measured by EDS measurements.

Figure 8 shows the SR of the cells with various compositions and Cl vapor concentrations. Carrier collection was low across the whole spectrum as seen in Fig. 8. Devices with low Se composition (i.e., x = 0.17) show higher carrier collection at long wavelengths when compared to the high Se composition (i.e., x = 0.55, and 0.52) ones. This is due to the narrowing bandgap for low Se composition. Higher Se concentration cells have a higher bandgap due to the bowing effect in CdSe$_x$Te$_{1-x}$ films. Table 3 shows the net-doping calculated from C-V measurements. Doping is higher >10^{16} cm^{-3} for the higher Se (x > 0.50) devices.

TABLE 3. CL DOPED CST CdSe$_x$Te$_{1-x}$ DEVICES

Device Conditions	Net Doping [cm^{-3}]
x = 0.17 Se – CdCl$_2$ 1K ppm	1.41E+14
x = 0.17 Se – CdCl$_2$ 3K ppm	9.74E+13
x = 0.55 Se – CdCl$_2$ 1K ppm	3.47E+16
x = 0.52 Se – CdCl$_2$ 3K ppm	2.17E+17

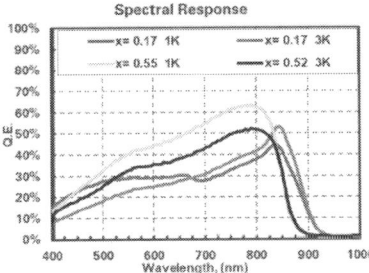

Figure 8. Spectral response for Cl-doped CdSe$_x$Te$_{1-x}$ device with various Se compositions

IV. CONCLUSION

The elemental vapor transport process offers the means to control the vapor pressure of Cd, Te, and Se elements independently. Therefore, CdSe$_x$Te$_{1-x}$ thin films with various Se compositions were fabricated by EVT deposition process. The incorporation of Se results in a reduction in the bandgap of CdSe$_x$Te$_{1-x}$ films. XRD, EDS, and SEM measurements indicate that the structure of CdSe$_x$Te$_{1-x}$ films shifted from cubic to hexagonal structure with increase in the Se compositions. The transition appears to happen around x = 0.44. SIMS measurements demonstrating a uniform distribution of Se for all CdSe$_x$Te$_{1-x}$ films with various Se compositions. TRPL lifetime measurements indicate that minority carrier lifetimes increase with the increase of Se composition. CdSe$_x$Te$_{1-x}$ film with minority lifetime of 85.3 ns was obtained with Se composition of x = 0.32.

ACKNOWLEDGMENT

This work was supported in full by the Department of Energy under award DE-EE0008745. SIMS measurements were performed at the National Renewable Energy Laboratory (NREL). TRPL measurements of CdSe$_x$Te$_{1-x}$ films were performed by professor's Charles Hages group at the University of Florida.

REFERENCES

[1] M. A. Green, E. D. Dunlop, J. Hohl-Ebinger, M. Yoshita, N. Kopidakis and X. Hao, "Solar cell efficiency tables (version 59)," *Progress in Photovoltaics: Research and Applications,* pp. 3-12, 2022.

[2] J. Sites and P. Jun, "Strategies to increase CdTe solar-cell voltage," *Thin Solid Films,* 2007.

[3] A. H. Munshi, J. Kephart, A. Abbas, J. Raguse, J. N. Beaudry, K. Barth, J. Sites, J. Walls and W. Sampath, "Polycrystalline CdSeTe/CdTe absorber cells with 28 mA/cm2 short-circuit current," *IEEE Journal of Photovoltaics,* vol. 8, no. 1, pp. 310-314, 2017.

[4] Y. Marfaing, "Impurity doping and compensation mechanisms in CdTe," *Thin Solid Films,* vol. 387, no. 1-2, pp. 123-128, 2001.

[5] W. Wang, V. Palekis, M. Z. Alom, S. E. Tawsif and C. Ferekides, "Chlorine Doped n-Type CdTe Solar Cells,," in *2022 IEEE 49th Photovoltaics Specialists Conference (PVSC),* Philadelphia, 2022.

[6] V. Palekis, W. Wang, S. E. Tawsif, Z. A. Md and C. Ferekides, "Thin Film Solar Cells with n-type CdTe Absorber and p-type ZnTe Window Layers," in *2021 IEEE 48th Photovoltaic Specialists Conference (PVSC),* 2021.

[7] V. Palekis, I. Khan, S. Collins, C. A. Hsu, S. Misra, M. A. Scarpulla, Y.-H. Zhang, D. Morel and C. Ferekides, "Thin Film Solar Cells Based on n-type Polycrystalline CdTe Absorber," in *2018 IEEE 7th World Conference on Photovoltaic Energy Conversion (WCPEC),* Waikoloa, 2018.

[8] W. Wang, V. Palekis, M. Z. Alom, S. E. Tawsif and C. Ferekides, "Numerical Modeling of n-CdTe/p-ZnTe Thin Film Solar Cells," in *2021 IEEE 48th Photovoltaic Specialists Conference (PVSC),* Fort Lauderdale, 2021.

[9] L. Brooks, "The vapor pressures of tellurium and selenium," *Journal of the American Chemical Society,* vol. 74, no. 1, pp. 227-229, 1952.

[10] S.-H. Wei, S. B. Zhang and A. Zunger, "First-principles calculation of band offsets, optical bowings, and defects in CdS, CdSe, CdTe, and their alloys.," *Journal of applied Physics,* vol. 87, no. 3, pp. 1304-1311, 2000.

[11] A. R. Denton and N. W. Ashcroft, "Vegard's law," *Physical Review A,* vol. 43, no. 6, pp. 3161--3164, 1991.

Intra-grain local luminescence properties of CdSe$_{0.1}$Te$_{0.9}$ Thin Films

Ganga R. Neupane[1,2], David S. Albin[3], Joel N. Duenow[3], Matthew O. Reese[3], Susanna Thon[2], and Behrang H. Hamadani[1]

[1]National Institute of Standards & Technology, Gaithersburg, MD 20899, USA
[2]Johns Hopkins University, Baltimore, MD 21218, USA
[3]National Renewable Energy Laboratory, Golden, CO 80401, USA

We report on the local photoluminescence properties of grains and grain boundaries of CdSe$_{0.1}$Te$_{0.9}$ thin film deposited by the colossal grain growth method using a wide-field hyperspectral imaging technique. We observed significant variations in the photoluminescence intensity of both individual grains and also that of grain boundaries. Multiple sub-bandgap defect peaks were captured in the luminescence spectra in the energy range 1.2 eV to 1.6 eV. The intensity and peak positions of these sub-gap emissions were slightly different among various grains and at the grain boundaries, revealing intra- and inter-grain variations in these polycrystalline thin films. A recently-developed density-of-states based photoluminescence model was extended to include multiple peaks and was fitted to the data. We observed that at a fixed temperature, the quasi-Fermi level splitting energy and a disordered energy parameter can be extracted locally by use of this model.

I. INTRODUCTION

Wide-field, luminescence-based hyperspectral (HS) imaging is a unique and powerful technique that can acquire high resolution images over many continuous spectral bands [1]. Every pixel in the hyperspectral image contains the full spectral information, therefore, making it a useful tool for assessing local charge transport properties of thin-film photovoltaic materials.

The performance of thin-film cadmium telluride (CdTe) solar cells have continued to improve in recent years, owing to refinements in passivation techniques such as alloying of CdTe with Se, select doping and CdCl$_2$ treatment, to reduce defect-mediated non-radiative recombination [2], [3] However, high resolution luminesence imaging that can visualize local recombination phenomena and the role of defects at the grain interiors (GIs) and grain boundaries (GBs) has not been performed extensively. Wide field HS imaging is an attractive technique to quickly and reliably image large areas of the film and investigate local potential fluctuations or variations in defect-related emissions between the grains.

Small-grain polycrystalline CdTe films have poor electro-optical properties. The local charge transport in CdTe is significantly impacted by the size of the grains and the defects associated with them. In addition, it has been found that the carrier lifetime, which correlates well with the open circuit voltage, V$_{oc}$, increases with grain size [2]. So our efforts need to focus on increasing the size of CdTe thin film grains and passivating the local defects and interfaces. In this work, we have used the colossal grain growth (CGG) [4] method to produce large-grain CdSeTe thin films so that they can be used as a template for the epitaxial growth of large-grained CdTe thin films. Studying the absolute photoluminescence (PL) images obtained from these films, we find that there are both intra-grain and inter-grain variations in the luminsence spectra at the micron scale. To understand how these variations lead to local differences in the chemical potential or the disordered energy, we have used and extended a recent single-transition analytical PL model [5] to a multi-transition regime such as seen in data presented here. We show that it is possible to model a complicated multi-peak PL emission spectrum using a single quasi-Fermi level splitting (QFLS) energy and extract the sub-gap absorptivity and the disordered energy of the various sub-gap peaks by use of this model. Explaining local variability through useful quantities such as the QFLS energy can be helpful for materials growth improvements.

II. METHODS

A. Sample preparation

The polycrystalline thin film of CdSeTe was formed on alumina-coated glass substrates. The film was made using a different growth technique called the CGG method. A 3.0 µm CdSeTe precursor film was evaporated onto a 0.7 mm thick alumino-borosilicate glass substrate coated with 100 nm of alumina heated to about 400 °C. Evaporation was from a single-source alumina crucible containing an alloy of CdSe$_{0.1}$Te$_{0.9}$ heated to 660 °C to produce a deposition flux close to 10 A/s. In the subsequent CGG step, the CdSeTe precursor film is then heated to 550 °C while suspended 1 mm, film-side down, above a Se-containing powder (typically, CdSe$_{0.4}$Te$_{0.6}$) in an atmosphere of 100-Torr helium. After the CGG step, the CdSeTe film was then annealed and suspended over a powder of CdCl$_2$ at 550 °C for 10 m in an ambient of 400-Torr helium to reduce recombination and increase luminescence efficiency. Details about the CGG process can be found in [4].

B. Hyperspectral characterization

HS imaging in PL mode was performed in the spectral region from 780 nm to 980 nm using a silicon-based CCD camera system with an appropriate HS grating. A 20X microscope objective was used for capturing the images shown here. For

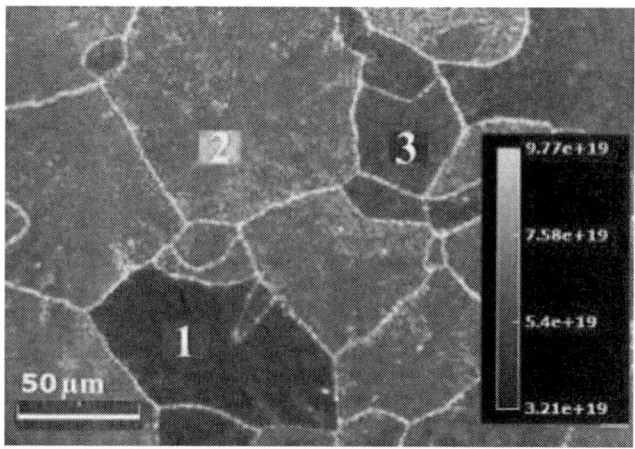

Fig. 1. A PL image of grains and grain boundaries of CdSeTe thin film at 144 K at a fixed intensity of 1776 W/m². The colorbar shows the absolute photon flux in units of photons/m² s eV.

the PL excitation, a 532 nm laser was used to illuminate the entire field of view. Temperature-dependent measurements were performed using a liquid nitrogen flow optical cryostat under vacuum.

III. RESULTS

Fig. 1 shows the PL photon flux emission map of a CdCl₂ treated CdSeTe thin film at 144 K (temperature estimated from the energetic position of the band-to-band peak) taken at 1.36 eV. These measurements have been performed at lower temperatures because the radiative efficiency is too low near room temperature to provide the level of detail and clarity observed here. Surprisingly, the image reveals a higher luminescence intensity at the location of the GBs. This suggests less non-radiative recombination at GBs compared to the GIs. The treatment of as-grown films by CdCl₂ results in overall passivation of both GIs and GBs, but the passivation is stronger at the GBs. Other studies have also reported GB defect passivation using Se and CdCl₂ treatments [6], [7]. In addition, the image also shows a variety of grains with different sizes. The grains are tens of μm in lateral dimension and from the image, it is clear that these grains do not have a uniform defect profile, revealing a remarkable carrier recombination non-uniformity.

Next, we select three grain interiors and one grain boundary to study their spectral emission profile as shown in Fig. 1 labelled as 1, 2 and 3 for the respective grains. These grains are chosen in a such a way that they look different in terms of their size and PL brightness. The PL spectra corresponding to several GIs and one GB are shown in Fig 2. The spectra consist of several peaks in the range of ~ 1.2 to 1.6 eV. The intensity and peak position energies corresponding to each grain are a little different. This is either due to variations in the individual grain structures or variability in the defect distribution across the various grains. Since the band-to-band

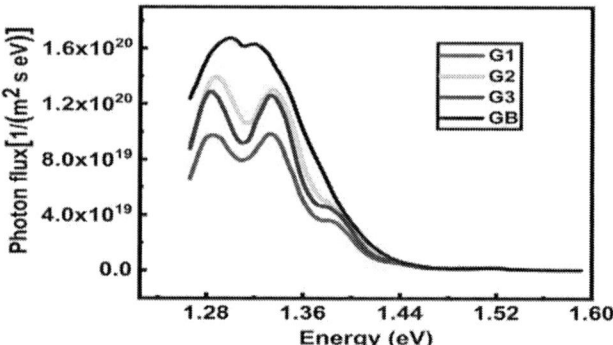

Fig. 2. PL spectra for select grains and a grain boundary at 144 K.

(BB) transition is around 1.52 eV, all these sub-bandgap transitions are related to radiative defects with energies smaller than E_g. These transitions likely stem from donor-acceptor pair defects in this material but further investigation is needed to confirm their characteristics. In addition, the observed multiple PL peaks blend together at the GB sites and form a large broadened peak. This blending of the peaks at the GB sites is due to inherent disorder/ other factors at the boundary. Because the crystalline nature of the material is disrupted at the GB, this energetic broadening is expected. Interestingly, the intensity of the PL emission at the GB is higher across all emitting energies than the grain interiors. Also, we observe (not shown here) that the BB energy is slightly lower at the GBs than GIs, suggesting a slightly higher aggregation of Se at the GBs, which may help with better passivation of GB defects.

To understand the nature of the PL emission spectra better, we extend a single-transition model developed by Katahara and Hillhouse [5] to include multiple optical transitions. Since absolute photon flux spectra are measured in our HS system, this model can be used to estimate important physical parameters such as the quasi-Fermi level splitting energy, $\Delta\mu$, the sub-gap absorption coefficient α and the energy broadening parameter γ (i.e., Urbach energy). The spontaneous radiative emission in terms of the spectral absorptivity under non-equilibrium condition is given by [5], [8]:

$$I_{PL}(E) = \frac{2\pi}{h^3 c^2} \frac{E^2 a(E)}{\exp\left(\frac{E-\Delta\mu}{kT}\right)-1} \qquad (1)$$

where I_{PL} is the photoluminescence intensity, h is the Planck constant, a is the absorptivity, k is the Boltzmann constant, and T is the temperature. The absorptivity is defined in terms of a joint density of states $G_i(E)$ for each emission peak that includes a disordered sub-bandgap tail parameter γ_i, $a(E) \sim \sum \alpha_{0i} G_i(E)$, where α_{0i} is absorption coefficient in the high energy side of the peak. Figures 3 and 4 demonstrate how this model, summed over multiple transitions, can be used to fit our data. For the

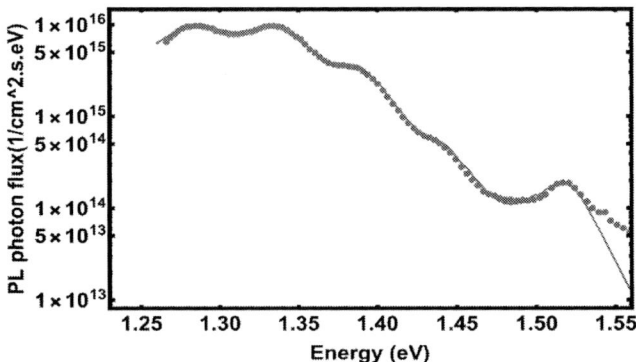

Fig. 3. Log-linear plot of PL spectra of G1 with the multi-peak modeling

simplicity, we choose to show the fitting for G1 and the GB. In the figure, the red symbols are the experimental PL spectra and the solid blue lines are the model-calculated spectra. For a fixed T, the intensity of the PL emission is controlled by both the $\Delta\mu$ and the absorption coefficients α_{0i}. We find that we can fix α_{01} of the BB peak (peak on the furthest right) to a reasonable value of $\sim 10^4$ cm^{-3} and then focus on determing a single $\Delta\mu$ parameter along with α_{0i} for all the below-gap peaks. The disordered energy parameters γ_i can be changed to increase or decrease the low-energy tail broadening of each peak such that the composite curve fits the data. $\Delta\mu$ values for G1, G2, and G3 are 1.265 eV, 1.267 eV, and 1.266 eV respectively and 1.269 eV for the GB. This supports the notion that a higher non-radiative recombination leads to a lower quasi-Fermi level splitting energy or a lower internal voltage. Similarly, we found that the grain boundary is more disordered ($\gamma = 12.5$ meV) than the grain interior ($\gamma = 10.3$ meV) near the band edge energy. This explains our observation of broadening of low energy PL tail of the grain boundary.

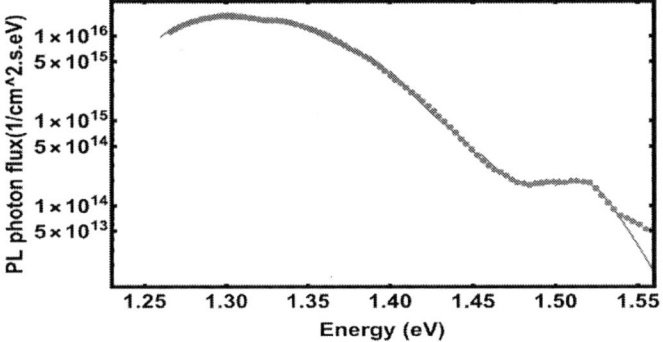

Fig. 4. Log-linear plot of PL spectra of GB with the model fitting.

IV. Summary

We have presented high resolution PL images of CdSeTe thin films using a wide-field hyperspectral imaging

system. Higher defect passivation at the GBs results in a higher luminescence signal there compared to grain interiors. Different sizes of grains and brightness differences reveal remarkable grain-to-grain and grain-to-grain boundary variations which strongly depend on the defect distribution. In addition, a density of states-based model was successfully extended to fit multiple peaks in the PL spectra and different parameters like the quasi-Fermi level splitting energy and the sub-bandgap disorder energy were extracted from each individual peak. The disorder energy at the grain boundaries is higher than grain interiors, causing several defect transitions to blend together.

References

[1] B. H. Hamadani, M. A. Stevens, B. Conrad, M. P. Lumb, and K. J. Schmieder, "Visualizing localized, radiative defects in GaAs solar cells," *Sci Rep*, vol. 12, no. 1, p. 14838, Sep. 2022, doi: 10.1038/s41598-022-19187-4.

[2] H. Nazem, H. P. Dizaj, and N. E. Gorji, "Modeling of J sc and V oc versus the grain size in CdTe, CZTS and Perovskite thin film solar cells," *Superlattices Microstruct*, vol. 128, no. February, pp. 421–427, 2019, doi: 10.1016/j.spmi.2019.02.002.

[3] R. M. Geisthardt, M. Topič, and J. R. Sites, "Status and Potential of CdTe Solar-Cell Efficiency," *IEEE J Photovolt*, vol. 5, no. 4, pp. 1217–1221, Jul. 2015, doi: 10.1109/JPHOTOV.2015.2434594.

[4] D. S. Albin, M. Amarasinghe, M. O. Reese, J. Moseley, H. Moutinho, and W. K. Metzger, "Colossal grain growth in Cd(Se,Te) thin films and their subsequent use in CdTe epitaxy by close-spaced sublimation," *Journal of Physics: Energy*, vol. 3, no. 2, p. 024003, Apr. 2021, doi: 10.1088/2515-7655/abd297.

[5] J. K. Katahara and H. W. Hillhouse, "Quasi-Fermi level splitting and sub-bandgap absorptivity from semiconductor photoluminescence," *J Appl Phys*, vol. 116, no. 17, p. 173504, Nov. 2014, doi: 10.1063/1.4898346.

[6] N. A. Shah, A. Ali, S. Hussain, and A. Maqsood, "CdCl2-treated CdTe thin films deposited by the close spaced sublimation technique," *J Coat Technol Res*, vol. 7, no. 1, pp. 105–110, 2010, doi: 10.1007/s11998-008-9146-0.

[7] R. Wang, M. Lan, and S.-H. Wei, "Enhanced performance of Se-alloyed CdTe solar cells: The role of Se-segregation on the grain boundaries," *J Appl Phys*, vol. 129, no. 2, p. 024501, Jan. 2021, doi: 10.1063/5.0036701.

[8] M. Tebyetekerwa *et al.*, "Quantifying Quasi-Fermi Level Splitting and Mapping its Heterogeneity in Atomically Thin Transition Metal Dichalcogenides," *Advanced Materials*, vol. 31, no. 25, pp. 1–8, 2019, doi: 10.1002/adma.201900522.

Electroluminescence Imaging: A Study in the Impact of Microscopic Surface Defects

Meghan E. Bush and Timothy J. Peshek

Photovoltaic and Electrochemical Systems Branch, NASA Glenn Research Center, Cleveland, Ohio, 44135, United States

Abstract—Electroluminescence imaging can be used as a quantitative characterization method for solar cell performance when combined with image processing. This work aims to leverage this technique to investigate the impact of lunar dust occlusion on solar cells, specifically to identify if dust grain size has any effect on electroluminescence. Currently, this method has proven sensitive enough to differentiate between clean and dusted samples using both Martian and lunar simulants. Additionally, a model has been built to provide a simulated result to compare experimental data and predict what dust grain sizes will have a greater impact on performance.

Index Terms—photovoltaics, electroluminescence, electroluminescence imaging, image processing, lunar dust

I. Introduction

Electroluminescence (EL) imaging is a common solar cell characterization technique that allows for a photograph of the band edge luminescence of a solar cells in forward bias due to radiative electron-hole pair recombination. This technique is often used as a qualitative analysis to quickly identify defects such as cracks or corrosion. However, careful baselining and reproducibility of the imaging device can enable quantitative EL, which, when coupled with computer-based image processing, would allow for expanded insight to be drawn from what is a relatively quick and easy characterization process. Quantitative EL imaging could enable quick analysis of large solar cell data sets and provide a means to derive a relation between surface defects and the resulting power loss. Here, we focus on the utilization of quantitative EL to characterize small surface defects, *i.e.* small grains of dust occluding the surface and create estimates of power loss.

Photovoltaics have been used for in-space power generation for decades due to their hardiness in the space environment, favorable cost/mass ratio, and long lifetimes [1]. Arrays used in surface applications have found a complication in dust, the smallest subset of regolith [2]. On the Moon, Apollo 17 astronauts noted lunar regolith clinging to any exposed surfaces and found it very difficult to remove [3]. Dust accumulation on solar arrays has been a concern in every Mars mission using solar power [7] [8] [9]. While research is currently being done to develop dust mitigation techniques [6], the relationship between dust coverage and solar cell performance still needs to be further investigated.

II. Experimental

We will use EL imaging to quantify dust coverage on the surface of solar cells. Dust occlusion has a direct correlation with power loss [7], but investigation needs to be done to produce a reliable trend. We seek to establish an understanding of how different grain sizes impact occlusion of the solar cell.

Solar cells under study are biased such that a current of approximately equal to the short circuit current was injected, resulting in a high brightness image. A Raspberry Pi camera module will be used to capture all EL images, as the cells under study luminesce in the visible range. The cells will be the same distance away from the camera and the focus will not be adjusted; additionally, the camera will be run in fully manual mode, allowing key parameters such as exposure, white balancing, shutter speed, and aperture to be consistent across all images.

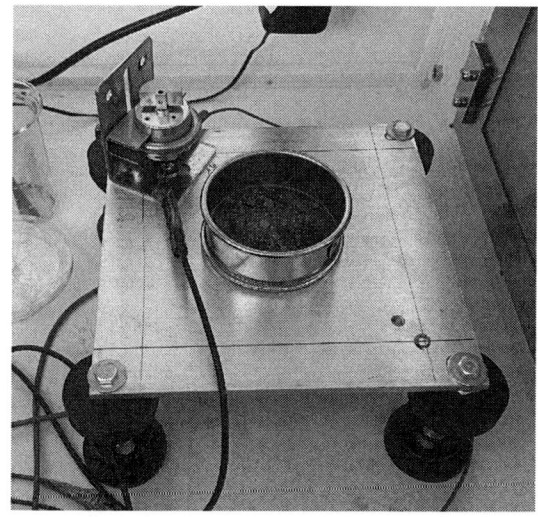

Fig. 1. Major components of the dust deposition system with lunar simulant loaded into the sieve.

Dust was deposited onto a solar cell (Spectrolab XTJ, acquired previously) using a dust deposition system that consists of a vibration motor and a mechanical sieve set into a vibration plate that sits over the sample. Lunar simulant JSC-1A is being used. The sieves being tested have mesh sizes of 20 μm, 75 μm, and 250 μm. Each sample set consists of one pristine and three dusted cells using the same sieve size, so the complete set will use 12 total solar cells. Cells will undergo EL imaging before and after dust deposition to observe the difference. Figure 1 shows the dust deposition system apparatus.

III. Data Analysis and Discussion

Since the dust grains are sufficiently small compared to the pixels of even a high definition camera, we naively expect that

978-1-6654-6060-6/23 $31.00 © 2023 IEEE

the overall brightness of the image is to be reduced at constant current injection. However, the goals of our analytics are to determine approximate coverage of dust to study the effects of dust bonding and removal and to develop a relationship between integrated EL signal and expected power loss.

We developed an algorithmic process for analyzing the images due to the difficulty in obtaining quantifiable metrics from EL imaging. First, the image is cropped to focus on the cell and converted into grayscale. Next, the brightness level of each pixel is calculated and plotted in a histogram. A pristine, defect-free cell will have peaks in the fully bright regions (emitting light) and fully dark regions (electrodes, area around the cell) and appear somewhat bimodal. In the event of a dust-occluded cell, the EL histogram will have a modified shape and appear multimodal, *i.e.* showing more than two peaks. Hypothetically, because lunar dust particles are smaller than the size of a pixel, the presence of grains will not fully block out light but instead dim it locally. This dimming will be evident in the histogram of the electroluminescence image as the bins between the emitting and dark regions. Modeling of the multimodal peaks that are associated with dust coverage can yield insights into the size and coverage density of the particles and therefore allow a first order calculation of short circuit current loss.

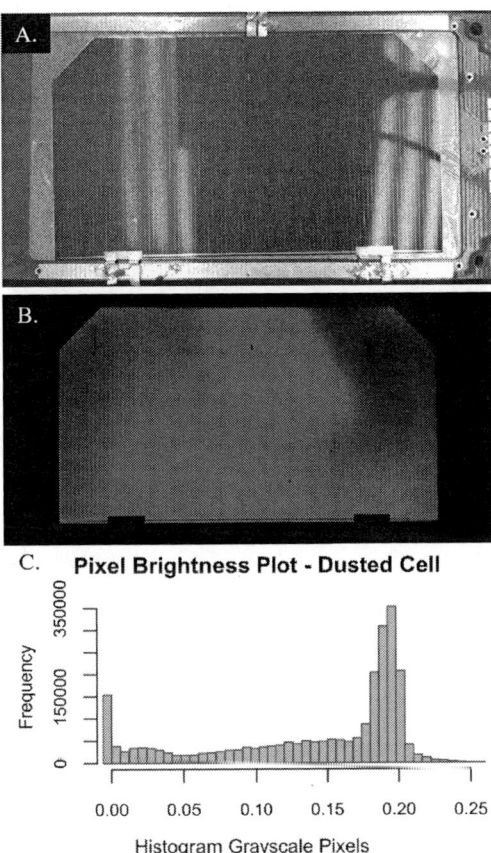

Fig. 3. Dusted solar cell (A) with its EL image (B) and resulting pixel brightness histogram (C).

A pristine cell was imaged by the Raspberry Pi under ambient light and again under bias. Lunar simulant JSC-1A was then deposited on the cell using a $20\,\mu$m sieve and the cell was biased and re-imaged. Visually, the samples looked very similar while illuminated; however, their resulting histograms seen in Figures 2 and 3 showed clear differences once dust was deposited on the surface.

These solar cells luminesce with a dark red light; the brightness peak in the histogram occurs around 0.20. This cell has a dark region in the upper right corner that populates some of the bins in the middle. 0.25 indicates the best-performing, brightest region of the pristine cell under EL. When dust is added, this peak fully disappears and contributes to an increase in the 0.20 grayscale brightness peak and overall population of the rest of the non-zero bins. The fingers in the solar cell become clearer visually as the dust dims the luminescent cell.

A different Spectrolab XTJ solar cell from the same batch was run through the electroluminescence imaging process to compare its histogram with the model-generated result. The experimental and modelled pixel brightness histogram are shown in Figure 4. The experimental histogram has essentially zero population in the bins between 0.00 and 0.10, similar to the histogram in Figure 2; the brightness peaks occur in the same places, though the experimental cell has some

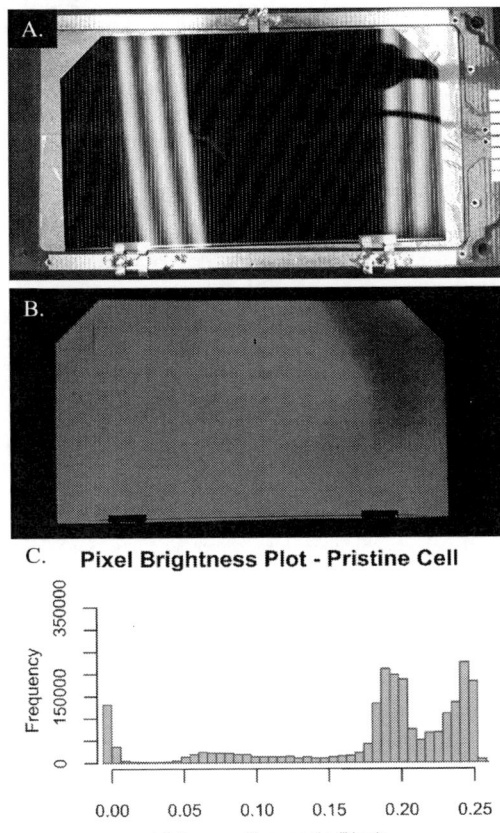

Fig. 2. Pristine solar cell (A) with its EL image (B) and resulting pixel brightness histogram (C).

bin population in the 0.10 - 0.15 region due to defects and shadowing in the EL image. This second cell illustrates the overall difficulty in determining an algorithmic analytical technique for these images, but still shows interesting trends in the histogram that can be feature-modeled.

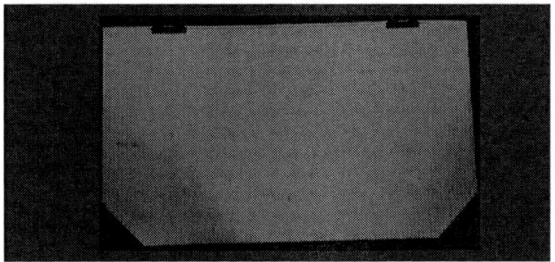

Experimental Brightness Histogram

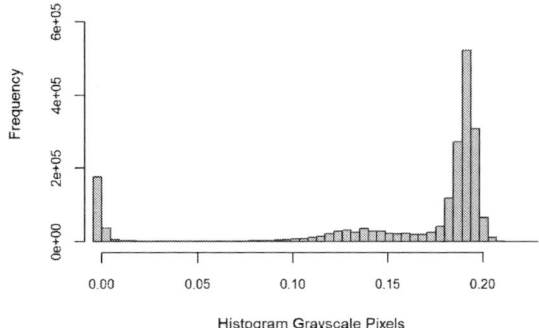

Modelled Brightness Histogram

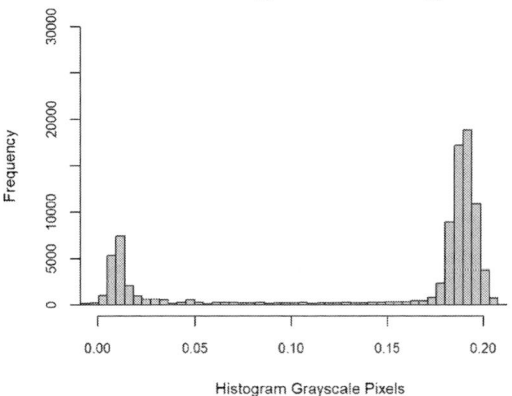

Fig. 4. Luminescent Spectrolab cell and its pixel brightness histogram vs a modelled pristine solar cell.

In order to model the datasets to extract grain size and coverage, a Monte Carlo technique was employed to simulate the randomness of dust grain placement. In this method, a matrix of brightness measures is created that is closely representative of a high quality cell. A random Gaussian noise filter is added to alter the brightness, and a random fluctuation with dark regions is added to simulate regions of poor radiative recombination. The overall histogram for this simulated EL image is closely representative of the histograms we gathered

for pristine cells, see Figure 4. In order to simulate the dust coverage, we applied a reduction in brightness to a random collection of pixels. The collection of pixels with reduced brightness is representative of the amount of dust sticking to the surface. The amount of grains sticking to the surface can be empirically determined by a simple experiment that measured the mass of sieved dust grains deposited by the apparatus per unit time. Using a simple hard sphere approximation, the dust grains are assumed to be fully reflective of approximately 50% of the light normally incident upon them; we ignore edge effects here and utilize a symmetry assumption to suggest that a negligible amount of EL light output is non-normally incident. However, this reduction in brightness metric is relative to the size of the grain, not the pixel. This technique quickly resulted in histograms that well represented the shape profiles observed with dusted samples. Therefore with refined physical models of the dust grain sizes and light scattering behavior, we expect to relate the shape profiles of the dusted histogram regions to estimate surface coverage.

IV. Conclusion

Quantitative electroluminescence imaging provides a means to characterize the impact of surface defects on solar cell performance. We are investigating lunar dust occlusion on solar cells and expect to use this technique broadly to envelope the potential for power loss from dust sticking to solar cells and arrays.

V. Acknowledgements

This work is funded by the Game Changing Development Program (DMFlex ACO - 20- 20 ACO Final - 0020) within NASA's Space Technology Mission Directorate.

References

[1] J. Banik, S. Kiefer, M. LaPointe and P. LaCorte, "On-orbit validation of the roll-out solar array," 2018 IEEE Aerospace Conference, 2018, pp. 1-9, doi: 10.1109/AERO.2018.8396390.

[2] Heiken, Vanniman, French (1991). Lunar Sourcebook. Cambridge University Press. pp. 756. ISBN 978-0-521-33444-0.

[3] T. Stubbs, R. Vondrak, W. Farrell (2005, September 26-30) "Impact of Dust on Lunar Exploration," Dust in Planetary Systems, Kauai, Hawaii.

[4] K. Bothe, et al. (2006, September 4-8), "Electroluminescence imaging as an in-line characterization tool for solar cell production," 21st European Photovoltaic Solar Energy Conference, Dresden, Germany.

[5] A. M. Karimi, J. Fada, et al. "Automated pipeline for photovoltaic module electroluminescence image processing and degradation feature classification," IEEE Journal of Photovoltaics, Vol. 9, No. 5, September 2019, pp 1324-1335. doi: 10.1109/JPHOTOV.2019.2920732

[6] H. Kawamoto et al. (2009, November 16-19) "Mitigation of Lunar Dust on Solar Panels and Optical Elements for Lunar Exploration Utilizing Electrostatic Traveling-Wave", Lunar and Planetary Institute's Lunar Exploration and Analysis Group, Houston, TX, United States.

[7] G. Landis, P. Jenkins. "Dust on Mars: Materials adherence experiment results from Mars Pathfinder," 1997 IEEE Photovoltaics Specialist Conference, 1997, doi: 10.1109/PVSC.1997.654224.

[8] R. D. Lorenz et al. "Scientific Observations with the InSight Solar Arrays: Dust, Clouds, and Eclipses on Mars", Earth and Space Science, Vol. 7, Issue 5, April 2020, doi: 10.1029/2019EA000992.

[9] G. Landis. "Dust Obscuration of Mars Solar Arrays", Acta Astronautica, Vol. 18, No. 11, pp. 885-891, 1996, doi: 10.1016/S0094-5765(96)00088-4

Towards a three-terminal perovskite/silicon tandem solar cell with highest efficiency

Michael Rienäcker, Somayeh Moghadamzadeh, Paul Fassl, Yevgeniya Larionova, Philipp Noack, Bianca Wattenberg, Ulrich Wilhelm Paetzold, Robby Peibst

Institute for Solar Energy Research Hamelin (ISFH), Emmerthal, Germany

Karlsruhe Institute of Technology, Institute of Microstructure Technology, Eggenstein-Leopoldshafen, Germany

Singulus Technologies AG, Kahl am Main, Germany

In this contribution, we aim to progress towards a 3T perovskite/POLO²-IBC tandem cell with highest efficiency. For this purpose, we describe the design considerations for a 3T perovskite/POLO²-IBC tandem cell with respect to the choice of the top and bottom cell configurations and their interconnection. We identify pin perovskite solar cells (PSCs) with a band gap of around 1.6 eV spin-coated on a nano-textured p+/n+ poly-Si tunnel junction front side of a n-type unijunction POLO² IBC bottom cells as a promising cell architecture. We report on the experimental progress with the pin PSCs on a nano-textured surface, POLO junctions optimized for textured and planar surfaces simultaneuously and the realization of the p+/n+ poly-Si tunnel junctions and ITO/n+ poly-Si recombination junctions. We fabricate high performance 3T POLO²-IBC bottom cells with p+/n+ poly-Si and with ITO/n+ poly-Si junctions on a nano-textured front side and find that the pseudo-J-V characteristics of both bottom cells is on par with our filtered single-junction 26.1%-efficient POLO²-IBC cells. Currently, we intergrate the pin PSC on the nano-textured bottom cell' font side and we will present 3T Pk/POLO²-IBC tandem cells at the conference.

978-1-6654-6060-6/23 $31.00 © 2023 IEEE

Luminescence and Thermal Imaging Applied to Half-cut-cell and Emitter-wrap-through-cell Modules

Steve Johnston, Dana B. Kern, Kent Terwilliger, and Ingrid L. Repins

National Renewable Energy Laboratory, Golden, CO, 80401, U.S.A.

Abstract—Imaging techniques provide spatial details and visualization of module defects and degradation mechanisms that affect energy conversion efficiency and performance. We apply photoluminescence, electroluminescence, and dark lock-in thermography imaging techniques to evaluate new modules in their initial state and after applying stresses of damp heat, light-induced-degradation regeneration parameters, thermal cycling, and humidity-freeze cycles. One module uses emitter-wrap-through cells with back contacts connected to a metal-foil backplane, and the other is composed of half-cut cells. Imaging shows examples on non-uniform degradation and damage such as cells that degrade and recover under the applied conditions, cells with cracks and handling damage, and cells with increasing series resistance.

Keywords— Photovoltaic cells, solar panels, degradation, imaging, infrared imaging, photoluminescence, silicon.

I. INTRODUCTION

Silicon-cell-based photovoltaic (PV) modules continue to evolve with more efficient and reliable architectures and materials. Some cell- and module-design innovations target decreases in series resistance and increases in power density per module area. As cells sizes become larger and more efficient, the generated current increases. Cutting the cells to reduce area can reduce series-resistance-related losses. Cells are spaced tightly together to reduce inactive area within the module, so interconnecting metal ribbons must bend sharply, or cells must tilt in a shingle effect to accommodate the tight spacing.

Larger metal grid lines and busbars help decrease series resistance when currents become large. An emitter-wrap-through (EWT) cell architecture creates short paths for current to travel from the cell's front to back while limiting front shadowing. Large metal structures can be used on the back to reduce series resistance, such as a metal foil that doesn't require soldered ribbons to connect cells. Instead, cells can be attached to a metal backplane by using laser welding or electrically conductive adhesives.

We investigate two of these innovative designs, including PV modules with back-contact cells and half-cut cells. We use imaging techniques to visualize spatial features related to PV module performance and any damage or degradation when accelerated stresses are applied. Luminescence techniques such as photoluminescence (PL) imaging and electroluminescence (EL) imaging use direct band-to-band radiative recombination to study cell performance [1]-[3]. PL and EL image intensities often correlate to carrier lifetime and cell voltage such that brighter intensity indicates higher cell voltage and efficiency. Relatively dark areas or dark cells have poorer voltage and

performance that may be attributable to defects or degradation mechanisms. EL imaging shows sensitivity to series resistance, particularly at high current densities since current is used to inject excess carriers through the conductive paths of the cells within the module.

Thermal imaging using dark lock-in thermography (DLIT) shows heating where current flows through cells and their conductive layers with some electrical resistance [4]-[6]. When defects or degradation are present, then current may not be uniformly distributed, such as in the case of cell cracks when increased carrier recombination at unpassivated crack surfaces leads to greater heating. Cell cracks or breaks in the metal grid fingers and increased series resistance can also lead to regions where current cannot flow or current is reduced, and these areas have lower heating and appear relatively cool in the thermal images.

In this study, we show correlated images of PL, EL and DLIT for these advanced module designs. We highlight features such as spatial variation in performance, local defects from processing or handling, patterns from the paths of current flow and localized resistance, and degradation/recovery of carrier lifetime that may be caused by defect instabilities.

II. EXPERIMENT

We apply imaging techniques to both an EWT-cell module, described first, and a half-cut-cell module, described second. We collect PL and EL images using a Princeton Instruments PIXIS 1024BR Si-charge-coupled-device camera. For high resolution and to limit the amount of laser excitation power needed for PL, we image one cell at a time and step the module underneath the camera using automated motion control stages. Two 808-nm fiber-coupled laser diodes provide 0.25-sun intensity (25 mW/cm^2) over the area of a full-size cell or two half-cut cells. A montage of 60 PL images using an exposure time of 0.5 s for each cell image is constructed to display the actual cell positions within the module. When driving 1 A and 10 A of current through the module, low-current and high-current EL-image montages are collected using exposure times of 3 s and 0.2 s, respectively. To prevent temperature from rising more than a couple degrees during imaging, current is turned off during motion movements. PL and EL images for the EWT-cell module are shown in Fig. 1.

DLIT images are collected using a Cedip Silver 660M (FLIR SC5600-M) InSb camera with built-in lock-in data acquisition. For thermal imaging, the modules are imaged all at once using a wide-angle lens. A forward-bias current of 10 A is pulsed on and off at 0.2 Hz, and the images are averaged for 5 minutes to

978-1-6654-6060-6/23 $31.00 © 2023 IEEE

Emitter-wrap-through-cell Module

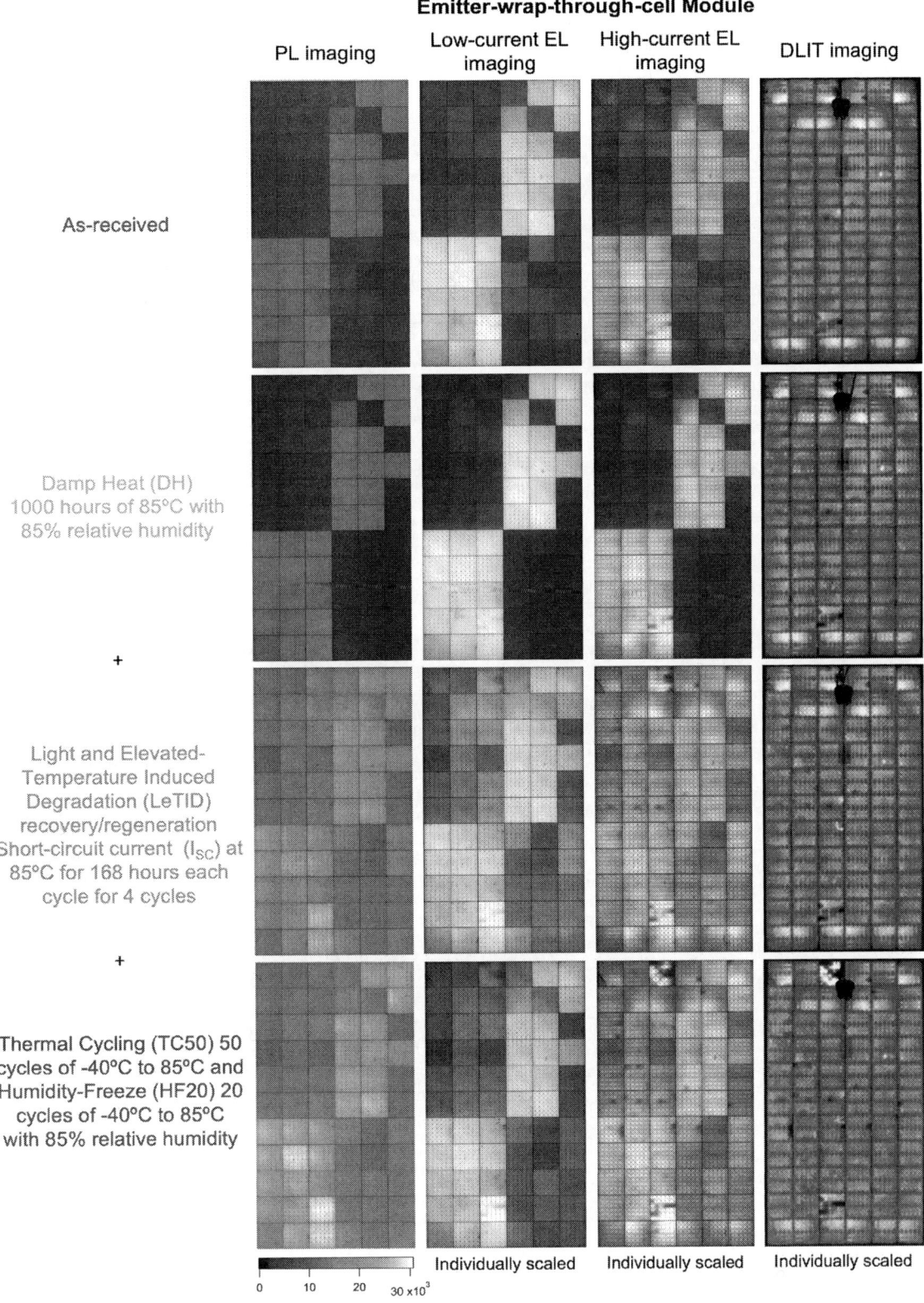

Fig. 1. EWT-cell module: PL (left column), low-current EL (second column), high-current EL (third column) and DLIT (right column) images of the EWT module in the as-received state (top row), after damp heat (second row), after LeTID regeneration (third row), and after thermal cycling and humidity freeze (bottom row).

improve the signal-to-noise ratio. The DLIT images are also acquired through the backsheet to improve signal strength and image resolution, and the images are flipped to show the same orientation as the PL and EL images. DLIT images for the EWT module are also shown in Fig. 1.

The EWT-cell module is flash tested and imaged in its as-received state. The light current-voltage (I-V) curve is plotted in Fig. 2 and gives 326.0 W at the maximum power point. Cells within this module show a distribution of relatively dark cells (mostly top-left and bottom-right quadrants) and bright cells. A cell crack is seen in the top left cell, and a few high resistance contact points are seen where dark spots are present in both the high-current EL and DLIT images, such as in the cells in the third column from the left and the first, sixth, and tenth rows from the top. The DLIT image shows how current flows around the corners and tends to take the inside track of lower resistance as seen by the slightly warmer areas on the corner cells and the end cells where the cell string serpentines up and down in the center columns of the module.

The EWT-cell module is stressed with 1000 hours of damp heat (85°C and 85% relative humidity). The images after damp heat are shown below the as-received images in Fig. 1. After damp heat, the cells that were relatively dark become even darker suggesting that this stress possibly destabilized the cells and induced a degradation mode or promoted degradation in cells that were already unstable. The power reduced by 3.5% to 314.7 W. The heat distribution detected by DLIT shows a subtle increase in the upper right and lower left quadrants of the module after the damp-heat stress. With a constant current flowing through all cells, the ones with a slightly higher voltage as indicated by the PL intensity dissipate a slightly higher power as heat since power is the product of current and voltage.

To address the stability of the changing cells, this module is then subjected to the light- and elevated-temperature induced degradation (LeTID) regeneration conditions of driving I_{SC} (10 A) at 85°C [7]. The cells increase in PL and EL intensity over four cycles of 168 hours each, with the darkest cells showing

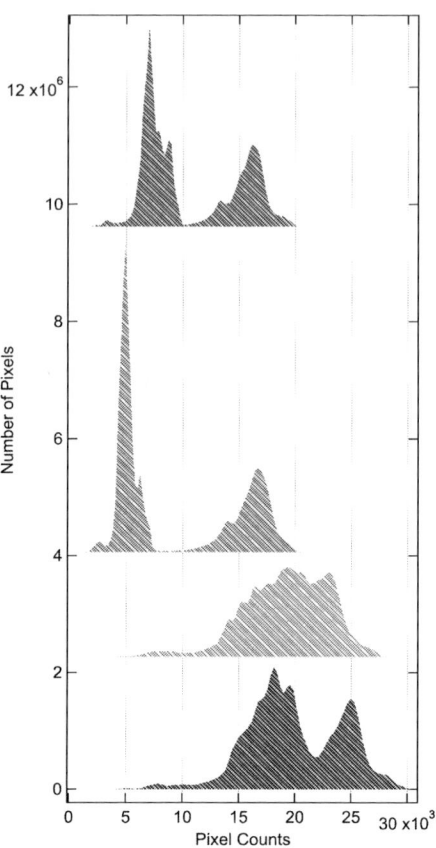

Fig. 3. PL imaging shows an initial bimodal distribution of cells within the module (top, shown in red). After damp heat, PL imaging shows initial dark cells becoming slightly darker (second down, shown in orange). After LeTID regeneration/recovery conditions are applied, PL shows all cells improve in intensity, and the bimodal distribution merges (third down, shown in green). After thermal cycling and humidity freeze stresses, PL and EL imaging shows the bimodal distribution of cells begins to develop again (bottom, shown in blue). Histograms are vertically offset for clarity.

relatively greater improvement (third row of images in Fig. 1). As shown in the PL image histograms for the entire module (Fig. 3), the bimodal distributions of bright and dark cells become more combined into one peak with overall increased intensity after the LeTID regeneration treatment. The corresponding I-V curve shows increased voltage and current, and the power has improved to 327.6 W, which is better than the initial measurement.

The images show a new quarter-circle-shaped, handling-related cell crack near the module's center that was generated between the damp heat and regeneration imaging. After the luminescence improvement, other high-series-resistance contact areas are more easily seen with higher contrast.

The EWT-cell module is additionally stressed with thermal cycling (50 cycles, 40°C to 85°C) and humidity-freeze (20 cycles, 40°C to 85°C with 85% relative humidity). The images after these stresses are shown in the bottom row of Fig. 1. The luminescence intensity begins to split back into a bimodal

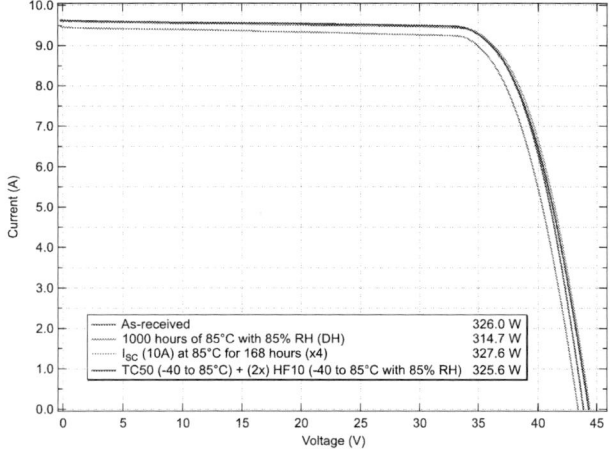

Fig. 2. Flash table I-V curves for the EWT-cell module. Power at the maximum power point initially drops 3.5% after DH, but power recovers after LeTID regeneration conditions are applied. Further stresses of thermal cycling and humidity freeze have not introduced any significant degradation.

distribution as shown in the PL histogram of Fig. 3. No significant changes are seen in the high-current EL and DLIT images which suggests the contacts are robust to thermally-induced mechanical stresses. The flash test power is still near the module's starting value.

The half-cut-cell module is also flash tested and imaged as shown in Figs. 4, 5 and 6. The initial flash-test I-V curve is plotted in Fig. 4 and shows 342.6 W for the maximum power point in the as-received state. The as-received images show several features as seen in Fig. 6. First, there is a checkerboard pattern of relatively dark and bright cells in the top left corner of the module. The cells that are dark in PL and EL have lower voltage and carrier lifetime and would contribute to some mismatch of the cell string. Second, there are well-defined dark features in the bottom-left and top-left corners of the module's PL and EL images. These are seen as bright patterns in the DLIT image. These are cell cracks that likely occurred during module handling when the corner of another module's frame contacted the back of this module during unstacking. There is carrier recombination at the cracks to reduce luminescence and increase heat. Lastly, there are dark features in EL and DLIT in the bottom-right corner that are not apparent in the PL image. These are high series resistance regions due to isolation from metal cracks, poor firing of the contacts, or some form of poor electrical connection. Excess carriers are generated in these regions during PL imaging which leads to uniform emission, but there is low carrier injection during EL and DLIT imaging due to series resistance.

The half-cut-cell module is stressed with 1000 hours of damp heat (85°C and 85% relative humidity). This module shows negligible change for this applied stress, as shown in the second row of images of Fig. 6. The unchanged images correlate to the flash test I-V curve (Fig. 4) that only shows a power reduction of 1.1%. LeTID regeneration cycles also result in very little change in power and imaging (images not shown).

The half-cut-cell module is also additionally stressed with thermal cycling (50 cycles, 40°C to 85°C) and humidity-freeze

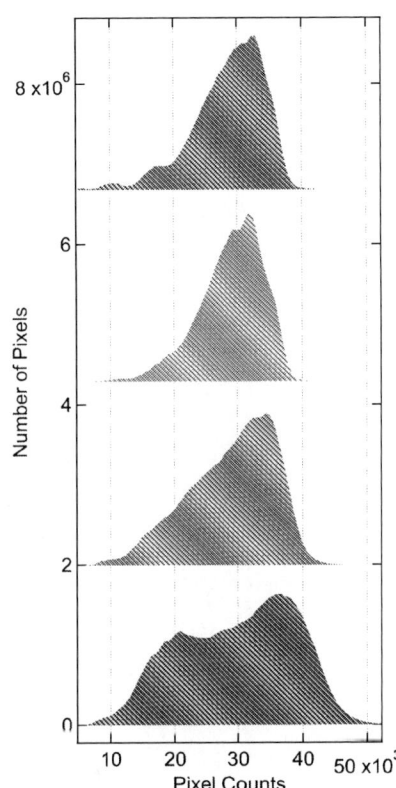

Fig. 5. High-current EL imaging for the half-cut-cell module shows an initial distribution (top, shown in red) that does not change significantly after damp heat stress (second down, shown in orange). After thermal cycling (TC50) and humidity freeze (HF10) stresses, the EL distribution broadens toward lower counts (third down, shown in light blue) that correlate with the formation of dark areas. After an additional stress-cycle of humidity freeze (HF10) EL imaging shows a bimodal distribution beginning to develop (bottom, shown in blue) as high-series-resistance areas continue to increase.

(10 cycles, 40°C to 85°C with 85% relative humidity). The flash test I-V curves begin to show a drop at the knee of the curve corresponding to a loss of fill factor. The power drops 0.9% (to 335.5 W) with thermal cycling and an additional 1.6% (to 330.0 W) with ten cycles of humidity freeze. The images after these stresses are shown in the third row of Fig. 6. The PL images do not show significant changes, thus verifying the degradation is that of series resistance. Some dark areas near the cell edges begin to appear with the low-current EL image but are seen with more contrast in the high-current EL image. There is an especially distinct region of increasing resistance at the edges of the first and second columns of cells on the lower half of the module. The DLIT image also shows cooler edges at the same regions due to the reduced current flowing in these areas.

A second run of ten humidity-freeze cycles further accentuates the degradation. The further loss of fill factor has led to an additional 3.3% loss of power (down to 319.1 W). The high-current EL shows this resistive degradation, and histograms of the high-current EL intensity distribution are plotted in Fig. 5. The initial distribution is shown on the top plot, and the shape and intensity show little change with the damp-

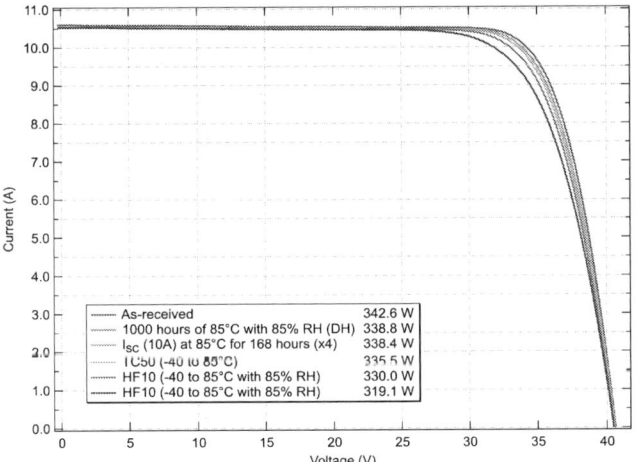

Fig. 4. Flash table I-V curves for half-cut-cell module. Power at the maximum power point initially degrades very slightly (1.1%) after DH and shows no sensitivity to LeTID when regeneration conditions are applied. Stresses of thermal cycling and humidity freeze have introduced more significant additional degradation of 5.7%.

978-1-6654-6060-6/23 $31.00 © 2023 IEEE

Half-cut-cell Module

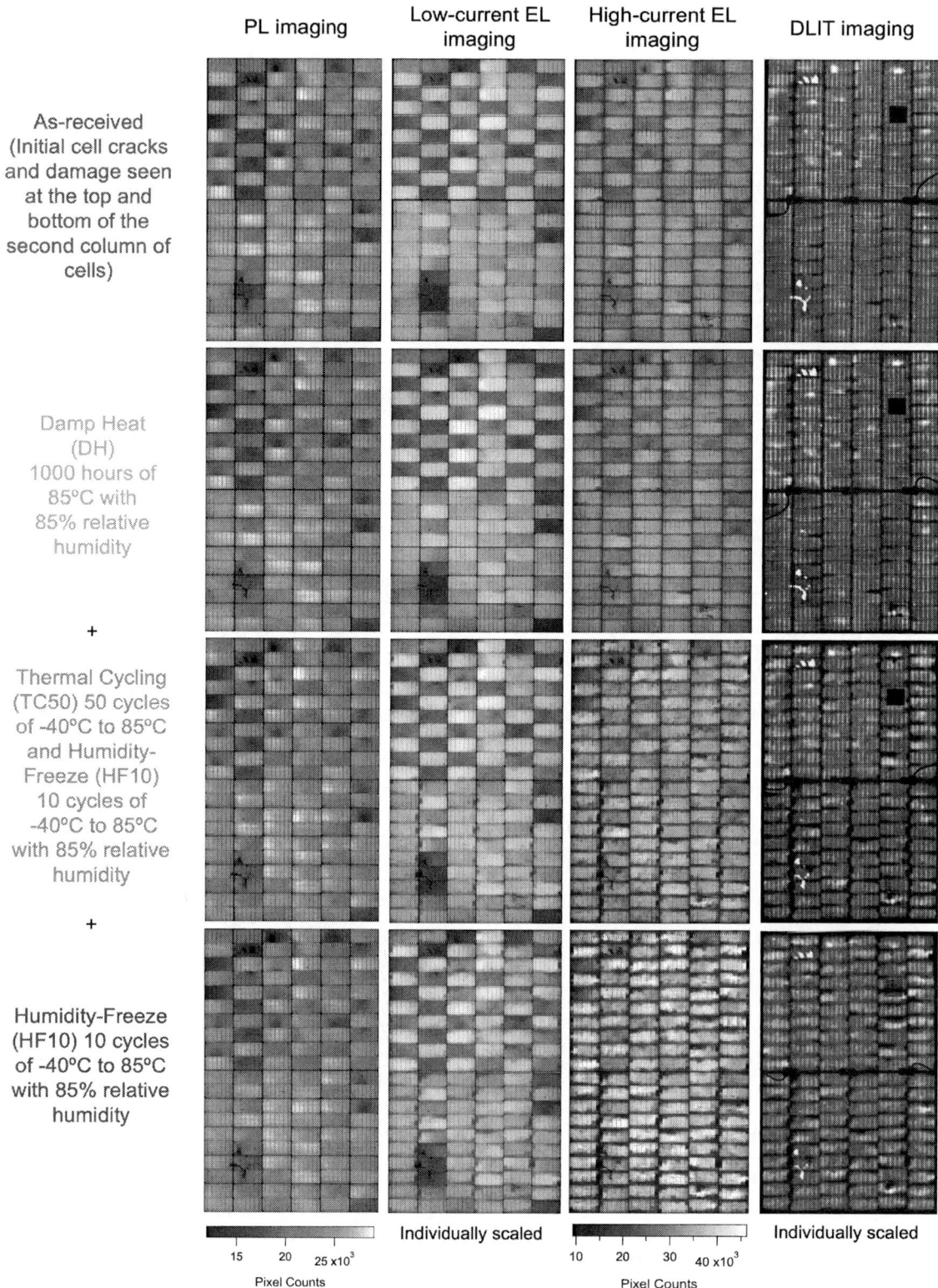

Fig. 6. Half-cut-cell module: PL (left column), low-current EL (second column), high-current EL (third column) and DLIT (right column) images of the half-cut-cell module in the as-received state (top row), after damp heat (second row), after thermal cycling (TC50) and humidity freeze (HF10) (third row), and after additional humidity freeze (HF10) (bottom row).

heat stress shown in the second plot. The distribution broadens toward the lower intensity side as thermal cycling and humidity freeze are applied as shown in the third plot. Lastly, a bimodal distribution is beginning to form as additional humidity freeze stress is applied and the dark regions become more pronounced. A zoomed-in example of one cell is shown in Fig. 7. Here, dark areas are observed in bands between busbars at the top edge of the cell, and these dark regions have increased as temperature-cycle stresses have been applied.

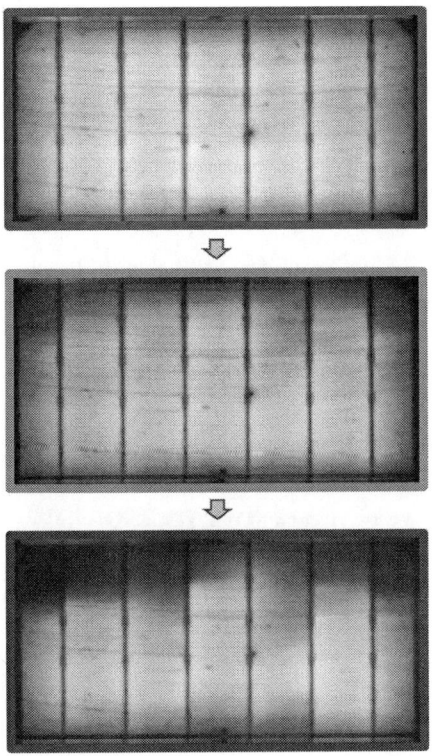

Fig. 7. High-current EL images of one of the cells from the half-cut-cell module are shown. The top image (outlined in red) shows the EL image in the as-received state. Development of dark bands are shown at the top of the cell as thermal cycling (50 cycles) and humidity freeze (10 cycles) are applied, as shown in the middle image (outlined in light blue). The bottom image shows the EL image after an additional 10 cycles of humidity freeze (outlined in dark blue).

III. SUMMARY

We have used imaging techniques to provide spatial characterization and show specific defects and degradation modes on advanced module architectures. PL, EL, and DLIT imaging provide data to distinguish module losses and determine which loss mechanisms are most dominant, such as reduced carrier lifetime and voltage, cell cracks and/or shunting, or series resistance. For the EWT-cell module, the PL and EL images correlate to current, voltage, and maximum power point reductions after the damp-heat stress. The images show that select cells within the module have less stability and greater changes with stress and regeneration. The current, voltage, and maximum power point recover to their initial values or better after the applied regeneration conditions. For the half-cut-cell module, high-current EL and DLIT images show that areas of series resistance develop due to temperature-induced mechanical stresses of thermal cycling and humidity freeze.

As modules change performance over time due to accelerated stress, weathering, or storm or other outdoor events, imaging techniques can provide spatial information to determine the type and extent of degradation and damage. Imaging techniques can be used to characterize varying module architectures, such as back-contact EWT cells and half-cut cells, as shown in these examples. Imaging techniques can be contactless so that they can be applied to modules in the field without disconnecting cables or strings. Thermal imaging can be collected during daylight operation, and PL and light-induced EL [8],[9] are non-contact techniques that can be applied to fielded modules.

ACKNOWLEDGMENT

This material is based upon work supported by the U.S. Department of Energy's Office of Energy Efficiency and Renewable Energy (EERE) under the Solar Energy Technologies Office Award Number 38263. The views expressed herein do not necessarily represent the views of the U.S. Department of Energy or the United States Government.

REFERENCES

[1] T. Fuyuki, H. Kondo, T. Yamazaki, Y. Takahashi, and Y. Uraoka, "Photographic Surveying of Minority Carrier Diffusion Length in Polycrystalline Silicon Solar Cells by Electroluminescence," Appl. Phys. Lett., vol. 86, pp. 262108-1–262108-3, 2005.

[2] T. Trupke, R. A. Bardos, M. C. Schubert, and W. Warta, "Photoluminescence imaging of silicon wafers," Appl. Phys. Lett., vol. 89, pp. 044107-1–044107-3, 2006.

[3] T. Trupke, R. A. Bardos, M. D. Abbott, F. W. Chen, J. E. Cotter, and A. Lorenz, "Fast photoluminescence imaging of silicon wafers," in Proc. 32nd IEEE Photovoltaic Spec. Conf., 4th World Conf. Photovoltaic Energy Convers., pp. 928–931, 2006.

[4] O. Breitenstein, M. Langenkamp, O. Lang, and A. Schirrmacher, "Shunts due to laser scribing of solar cells evaluated by highly sensitive lock-in thermography," Sol. Ener. Mat. Sol. Cells, vol. 65, pp. 55–62, 2001.

[5] O. Breitenstein, J. Bauer, J.-M. Wagner, and A. Lotnyk, "Imaging physical parameters of pre-breakdown sites by lock-in thermography techniques," Prog. Photovolt, Res. Appl., vol. 16, pp. 679–685, 2008.

[6] O. Breitenstein, W. Warta, and M. Langenkamp, Lock-in Thermography: Basics and Use for Evaluating Electronic Devices and Materials, 2nd ed. Berlin, Germany, Springer-Verlag, 2010.

[7] I. L. Repins, F. Kersten, B. Hallam, K. VanSant and M. B. Koentopp, "Stabilization of light-induced effects in Si modules for IEC 61215 design qualification", Solar Energy, vol. 208, pp. 894-904, 2020.

[8] S. Johnston, "Contactless Electroluminescence Imaging for Cell and Module Characterization," Proceedings of the 2015 IEEE 42nd Photovoltaic Specialist Conference (PVSC), 14-19 June 2015, New Orleans, Louisiana 6 pp.

[9] M. Kontges, J. Wagner, M. Siebert, S. Bordihn, C. Schinke, "Applicability of Light Induced Luminescence for Characterization of Internal Series-Parallel Connected Photovoltaic Modules," IEEE Journal of Photovoltaics, 2022, Vol.12 (3), p.805-814.

Role of Solar Photovoltaics for a Sustainable Energy System in Puerto Rico in the Context of the entire Caribbean Featuring the Value of Offshore Floating Systems

Christian Breyer, Ayobami S Oyewo, Alejandro Kunkar, Rasul Satymov

LUT University, Lappeenranta, Finland

The energy transition towards highly sustainable systems is accelerating with solar photovoltaics (PV) being the largest power source by capacity added. The Caribbean and Puerto Rico are lagging in ramping renewable energy (RE) capacities despite high energy supply costs. Energy system transition pathways for Puerto Rico and the Caribbean are analyzed for reaching 100% RE by 2050 for all energy supply in the sectors of power, heat and transport. Islands are often limited in available onshore area for large-scale solar PV applications. For addressing this challenge scenario variations are considered including offshore floating PV. The results for Puerto Rico clearly indicate the enormous benefits of reaching 100% RE, as the levelized cost of electricity (LCOE) can be reduced from more than 100 euro/MWh as of today to 43.5 euro/MWh in 2050, while reaching 100% RE, and the levelized cost of energy including all energy sectors declines from 78 euro/MWh to 51 euro/MWh, respectively. PV reaches 80% of all electricity supply, leading to 32.6 GW installed capacity, thereof 16.7 GW offshore floating PV in case of area limitation. Without area limitation, the total system cost would be about 2.7% lower in cost. Demand for e-fuels for marine and aviation is best to be imported, while this accounts for 10.7% of all energy demand of Puerto Rico and all other needs can be covered by local supply. Total electricity generation is projected to increase from present 21.3 TWh to 72.3 TWh in 2050. Puerto Rico represents about 5.3% of the projected Caribbean population by 2050 but about 11.4% of energy demand due to its high standards of living. The key metrics for the Caribbean development from 2020 to 2050 are as follows: electricity generation from 112 TWh to 632 TWh, PV supply share from 2% to 91%, PV capacity from 1 GW to 307 GW, thereof 20% prosumer, 80% utility-scale with up to 40% offshore floating PV, and LCOE from above 100 euro/MWh to 35.2 euro/MWh. The prosperity of Puerto Rico and the Caribbean is closely related to solar PV, the dominating source of energy in their Solar-to-X Economy.

978-1-6654-6060-6/23 $31.00 © 2023 IEEE

Energy-harvesting efficiency analysis for solar modules using 2T and 4T tandem solar cells

Robert Witteck, William E. McMahon, John F. Geisz, Qi Jiang, Emily L. Warren

National Renewable Energy Laboratory, Golden, CO, United States

We present a framework to compare the energy-harvesting efficiency (EHE) of 2-terminal (2T) and 4-terminal (4T) tandem solar cells integrated into different module architectures. Our study considers perovskite solar cells (PSC) of varying band gaps combined to form all-perovskite tandems as well as PSC/silicon tandems. Examining top cells with varying band gaps shows that spectral variations only marginally impair the EHE of 2T devices if they employ band gaps that make them reasonably current matched. Although situational, we show how the EHE of 2T devices can perform as good or better than 4T devices when the unavoidable optical and series resistance losses for the module integration of 4T cells are included in the analysis. At the conference we will present a ray-tracing-assisted series-resistance model and detailed spectral EHE analysis to find the most suitable module architecture for 2T and 4T devices.

978-1-6654-6060-6/23 $31.00 © 2023 IEEE

A Reproducible Validation of Algorithms for Estimating Array Tilt and Azimuth from Photovoltaic Power Time Series

Kirsten Perry[1], Bennet Meyers[2], Kevin Anderson[1], and Matthew Muller[1]

[1]NREL, Golden, CO, 80401, USA

[2]SLAC National Accelerator Laboratory, Menlo Park, CA, 94025 USA

Abstract—In this research, we assess the viability of four different, publicly available algorithms for estimating the azimuth and tilt parameters of solar photovoltaic systems using only the associated AC power time series data and site latitude-longitude coordinates. In this work, we curated a benchmarking data set of 44 fixed-tilt systems, comprising 275 measured AC power inverter data streams, with known azimuth and tilt parameters. Additionally, we isolated test cases in the data set with real-world issues, including shading and clipping, to determine how algorithm performance varies based on the presence of these phenomena. Using this data set for benchmarking, we evaluated the estimated vs. actual system characteristics for each algorithm, as well as the associated algorithm execution time using a standardized benchmarking process. The two highest performing algorithms were the Solar Data Tools and the PVWatts 5-based methods, which both achieved a median absolute error of approximately 5 and 1 degrees for azimuth and tilt, respectively. During run time analysis, the SDT method was approximately 5 times faster than the PVWatts 5-based method, with the median execution time for a stream varying between 6 and 8 seconds vs. a median run time of 31 seconds for the PVWatts 5-based method.

Index Terms—photovoltaic, azimuth, tilt, system parameters, time series

I. Introduction

The solar industry has experienced rapid growth over the past decade, with 130.9 GWdc of installed total capacity in the United States alone as of Q3 2022 and an average annual growth rate of 33% [1]. The industry is expecting even further growth in the U.S. with the recent passing of the Inflation Reduction Act, with a projected 40% boost, or 62 GWdc, in US solar deployment over the next five years [1]. With this rapid rise in solar investments comes a rise in solar acquisitions. Accurate solar site metadata, including azimuth and tilt, can be lost during transference between fleet owners. Furthermore, site metadata can be recorded incorrectly during manual entry by fleet owners/operators. Accurate site metadata is important for many solar analyses, including degradation analysis and power production forecasting.

In this paper, we present a validation of four methods for estimating solar photovoltaic (PV) array tilt and azimuth, *i.e.*, the angles defining the orientation of plane of the array. Novel to this work, the authors have curated a validation data set of 275 AC power generation data streams from 44 PV systems with known tilt and azimuth parameters. All four

algorithms were validated in a standardized test environment, providing an excellent comparison of the methods. While this data set cannot be made public, this work is intended to be a preview of the *Validation Hub*, which will host the data and test environment described here. This Hub will be available to researchers and practitioners to validate and compare algorithms for accomplishing PV data science tasks, providing a clearinghouse from academia to industry.

II. Related work

The task of estimating PV array tilt and azimuth from a measured powered signal has been previously addressed in the literature, albeit not extensively. Generally, the authors found previously published results to be difficult to recreate due to a lack of publicly available code implementations and non-reproducible test environments. Additionally, many methods require additional data besides a measured power time-series, an important detail that previous studies tend not to emphasize.

A number of authors have proposed methods for directly optimizing the parameters of a 'classic' PV performance model, *e.g.*, PVWatts [2] or the Sandia Array Performance Model [3]. Saint-Drenan et al. [4] make use of brute force simulation and lookup tables to attempts to match system parameters to measured power time series. This approach simultaniously estimates tilt, azimuth, and a factor controlling angular losses, and it requires latitude, longitude, and plane-of-array irradiance in addition to measured power. Haghdadi et. al [5] simultaneously estimate tilt, azimuth and latitude of the PV array. These authors do not assume any exogenous data besides a measured power signal; longitude is required as an input to the method but is estimated from the data independently. Londono-Hurtado et. al [6] followed a similar approach of first estimating longitude independently and then estimating latitude, tilt, and azimuth simultaneously. This approach somewhat uniquely avoids the need to invoke a PV performance model, instead a employing combination of geometric equations from Duffie and Beckman [7, §1.5–1.6] and an adaptive signal decomposition framework [8]. (We note that this method is the basis for two of the implementations tested in this paper.) Recently, authors have attempted to apply deep neural networks to this problem [9]. The authors note that this last approach is ill-suited to the task at hand, and that the

978-1-6654-6060-6/23 $31.00 © 2023 IEEE

results presented in that last paper are of limited operational utility.

III. ALGORITHM BACKGROUND

A. Algorithm Details

This section provides a brief description of the algorithms used in this validation study. Two of the algorithms, both from the open-source Python PVAnalytics package, have not been previously published in the literature. The last two implementations tested are variations of the previously-published Solar Data Tools-based algorithm [10], [11], originally described in Londono-Hurtado et. al [6]. These implementations are discussed briefly:

a) PVWatts 5: The PVWatts 5-based method is publicly available in the PVAnalytics package. Python PVAnalytics is a publicly available package dedicated to quality control and preprocessing of measured solar time series data [12]. In this method, measured PV power is compared against PV power simulated via the PVWatts-5 model using different input parameters, including tilt, azimuth, DC capacity, and DC input limit. The PVWatts-5 simulated power was generated using the associated Python PVLib functions [13], with the methodology described in [14] used. Non-linear least squares optimization is used to find the closest fit between modeled PVWatts v5 power and measured power. Specifically, only azimuth and tilt are returned. A measured PV power signal and the associated system's latitude-longitude coordinates are required as inputs.

b) PV peak: The PV-Peak algorithm is currently being adapted into the PVAnalytics package. In this method, each clear-sky day in a measured PV power time series is fit to a quadratic, and coordinates of the daily peak value, taken as the vertex of the quadratic, are derived. The solar azimuth at this peak and the normalized magnitude of the peak vary systematically of the course of a year as a function of both system tilt and azimuth. This data is assimilated based on day of the year for the entire data set and an initial estimate is derived for azimuth orientation. Then a focused grid search is performed around this azimuth estimate and a range of tilts in 5 degree increments. Clearsky POA irradiance is used to simulate the seasonal variation in azimuth and normalized magnitude at peak performance at each candidate azimuth-tilt pair. The optimal azimuth-tilt pair is chosen by minimizing the sum of squared differences between measured PV performance data and clear-sky irradiance for the two previously described annual fits (the trend for azimuth at peak and the trend for normalized magnitude at peak). This methodology has the same data requirements as the PVWatts 5-based algorithm: a measured PV power signal and the associated system's latitude-longitude coordinates.

c) Solar Data Tools: An implementation of [6] in the Solar Data Tools (SDT) package [10], [11]. This method can adapt to different available data inputs, and two versions of the algorithm were tested—with and without using array latitude and longitude as an input.

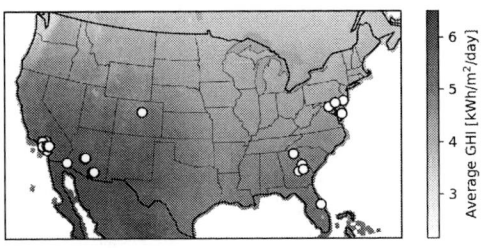

Fig. 1. Map of systems included in the validation set. Color indicates long-term average daily global horizontal insolation taken from the NSRDB [18].

B. Data Preprocessing Steps

It is important to note that for all four methods tested, there was a robust set of preprocessing steps to ensure high PV data quality when passed to the associated function.

For the two methods tested in the PVAnalytics package, several functions from the Python PVAnalytics library were leveraged during pre-processing, including a data/capacity shift detection algorithm [15] and clipping filter algorithm [16] to remove capacity-shifted and clipped power data, respectively. Stale, negative, and outlier data were also removed using existing PVAnalytics functions. Additionally, days with less than 25% data availability were removed. Data streams were further filtered to include clearsky periods only (PVWatts-based algorithm) or clearsky/sunny days only (PV Peak algorithm). Specifically, Physical solar model v3 (PSM3) data was used to derive a clearsky period mask.

During pre-processing, the SDT approach makes use of the SDT DataHandler object for data onboarding and filtering, including built-in modules for clear day detection and inverter clipping detection [17, §4.2.3 and §4.4.4]. The SDT estimation method first identifies and removes days with large operational issues. Then, cloudy *days* and inverter clipping *times* are identified and removed. Finally, values less than 15% of the seasonally adjusted maximum system power output are removed, including night time data.

IV. METHODS

A. Data Sets

AC power data associated with 44 distinct solar systems was collected from the NREL PV Fleets Initiative for the purpose of validating the proposed algorithms [19]. The PV Fleets Initiative is a US Department of Energy-funded project, where operational PV plant data is aggregated into a centralized cloud repository for the purpose of large-scale degradation analysis across the US. This database contains measured time series data for over 3700 sites across the United States, and has over 56 billion rows of solar time series data. In selecting systems, researchers selected sites from a variety of locations across the United States, to ensure algorithm performance was robust to regional variation. Overall, this data set included 275 AC power inverter data streams, where

978-1-6654-6060-6/23 $31.00 © 2023 IEEE

multiple AC power streams could be associated with the same system. Additionally, several data streams within the data set contained common issues associated with measured solar data, including clipping and shading. These phenomena were manually identified, and egregious cases were identified as case studies in the results of this paper. A map of all sites tested is shown in Figure 1, illustrating the diverse range of locations across the United States.

Each system data set consists of 15-minute mean-aggregated AC power time series data, localized to non-DST offset time zone. Associated metadata for each system includes latitude-longitude coordinates, azimuth, and tilt. All systems selected had a fixed-tilt mounting configuration.

B. Testing Environment

In order to accurately and fairly assess all algorithms side-by-side, a standardized Python test script was developed, which allowed each individual algorithm to be submitted and run as a module with standardized inputs and outputs. This standardized method allowed for algorithms to use a specific set of inputs as-needed based on the methodology's data requirements; for example, an algorithm could opt to use a site's latitude and longitude coordinates, but didn't necessarily have to. Standardized module outputs were set as the estimated azimuth and tilt values.

To provide proof of the reproducible test environment and a standardized run time analysis, the validation runs for all software packages were carried out on the same computer, a 2023 Apple M2 MacBook Pro, with 32GB of memory and 12 cores, running macOS 13.3.1. For the Validation Hub, the testing environment will be hosted on a virtual machine and accessible to users through a front-end web interface.

V. RESULTS

A. Algorithm Comparison

For each algorithm, the median absolute error for all 275 data streams was calculated between the estimated and ground-truth azimuth and tilt values for each data stream. Additionally, the median run time in seconds for each algorithm was calculated. Table I displays these results.

The two best performing algorithms were the SDT algorithm with unknown latitude-longitude coordinates and the PVWatts 5-based algorithm, with the lowest median absolute error for azimuth and tilt respectively. Specifically, the SDT algorithm had a median absolute error of 4.94 degrees for azimuth, followed by the PVWatts 5-based model with a median absolute error of 5.19 degrees. For tilt, the PVWatts 5-based model had a median absolute error of 1.21 degrees, followed by the SDT algorithm (unknown latitude-longitude) with a median absolute error of 1.66 degrees.

Unexpectedly, the SDT algorithm with unknown latitude-longitude coordinates outperformed its known latitude-longitude counterpart. The authors believe these results may be due the reduced degrees of freedom of the second model. When latitude and longitude are provided, those values are no longer free parameters, increasing model bias. In many cases,

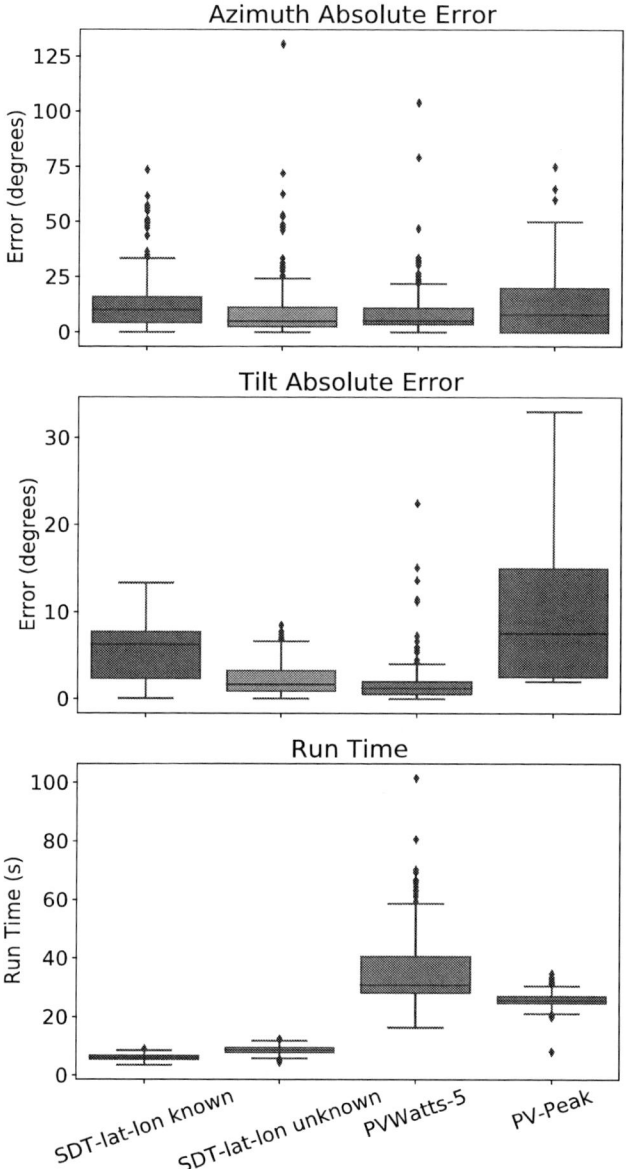

Fig. 2. Box plot depicting the absolute error in azimuth and tilt distributions respectively, as well as the run time distribution for each algorithm tested.

having the extra degrees of freedom evidently provides a better fit to the data.

Overall, the SDT implementations had a significantly faster run time than PV-Peak and the PVWatts 5-based models, with median SDT run time ranging between 6 and 8 seconds. This is approximately 5 times faster than the median PVWatts 5-based model run time, which was 30.87 seconds.

Figure 2 shows the error and run time distributions for individual algorithms. Similar to overall results, the SDT algorithm with unknown latitude-longitude coordinates and the PVWatts 5-based algorithm had the tightest absolute error distributions for tilt and azimuth, respectively. The distributions for these

TABLE I
OVERALL ALGORITHM PERFORMANCE

Algorithm	Data Requirements	Azimuth Median Absolute Error	Tilt Median Absolute Error	Median Run Time (s)
PVWatts	PV power, latitude, longitude	5.19	1.21	30.87
PV Peak	PV power, latitude, longitude	8.0	7.51	25.66
SDT-Known Lat-Long	PV power, latitude, longitude	10.12	6.24	6.02
SDT-Unknown Lat-Long	PV power	4.94	1.66	8.64

two algorithms is promising as it shows that outputs are largely consistent, with the exception of a few outlier cases.

The run time distribution in Figure 2 further illustrates how much faster the SDT implementations are compared to the PVWatts 5-based and PV-Peak algorithms. Both SDT variations had a tight distribution, with most analyses executing within 5 to 15 seconds.

The PVWatts 5-based method had the largest variation in run time, with analyses executing anywhere between 20 to 100 seconds. This is likely because the PVWatts 5-based algorithm requires NSRDB PSM3 data to be pulled for the PVWatts 5 simulation. Consequently, run time is directly impacted by any potential API issues when calling the data. Similarly, the PV-Peak implementation calls the NREL solar position algorithm (SPA) [20] via the Python PVLib package, which may be contributing to slower run times. Furthermore, the PV-Peak algorithm uses a brute force approach for identifying azimuth and tilt, where a total of 468 azimuth-tilt pairs are tested in every fit.

VI. SYSTEM ISSUES

A. Performance on Data Streams with Clipping

A system with heavily clipped data was isolated as a case study for evaluating algorithm performance. Inverter clipping, which occurs when a system's normal operation exceeds an inverter's power conversion limits, prevents inverter overload by operating in a reduced efficiency state [16]. Clipping largely occurs as a result of high DC-to-AC ratio for a system, where the system is overbuilt to increase system output outside of peak irradiance periods. Clipping manifests a flat-line at the peak of an AC inverter's production capacity. Figure 3 illustrates a year of normalized PV power data from the heavily clipped system used in this case study. In this figure, the top of the production profile is largely flat-lined as a result of inverter clipping.

Table II shows how each algorithm performed on PV power data from this heavily clipped system. Results indicate that algorithm performance on heavily clipped system is similar if not slightly higher than median performance, as shown in I. The only exception is the PV-Peak algorithm, with a high absolute tilt error of 23 degrees. All algorithms in this study included a pre-processing step where clipped data was removed before the power series was passed to the associated azimuth-tilt estimation function. This pre-processing step likely helped in producing accurate estimates. For the PV-Peak algorithm in particular, tilt error is higher because daily peak PV data periods are removed in a large proportion of the data set,

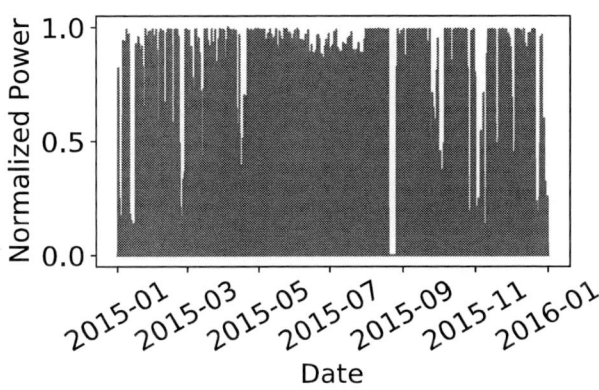

Fig. 3. One year of normalized PV power measurements for a heavily clipped system.

TABLE II
PERFORMANCE ON A SINGLE HEAVILY CLIPPED SYSTEM

Algorithm	Azimuth Absolute Error	Tilt Absolute Error
PVWatts	7.33	2.08
PV Peak	0	23
SDT-Known Lat-Long	5.32	9.49
SDT-Unknown Lat-Long	4.57	2.52

resulting in higher uncertainty when determining peak daily times.

B. Performance on Data Streams with Shading

A heavily shaded system from the data set was selected as a case study for evaluating algorithm performance. This particular system is located at NREL and is mounted to the side of a parking garage with a high tilt angle of 60 degrees. These factors are likely contributing heavily to system shading. A heat map of normalized PV power over time for this particular system is shown in Figure 4. The y-axis of this plot represents time of day in hours. A clear shading trend is evident, with summer months having a significantly lower PV power output compared to winter months.

Table III displays the absolute error for each algorithm on this particular shaded system. Azimuth error for all 4 of the algorithms is low, with algorithm error ranging between 0 and 5 degrees. However, the PVWatts 5-based model has a high tilt error (22.44 degrees). The SDT and PV-Peak algorithm implementations have fairly low tilt error comparatively, with error ranging between 5 and 8 degrees. These results indicate

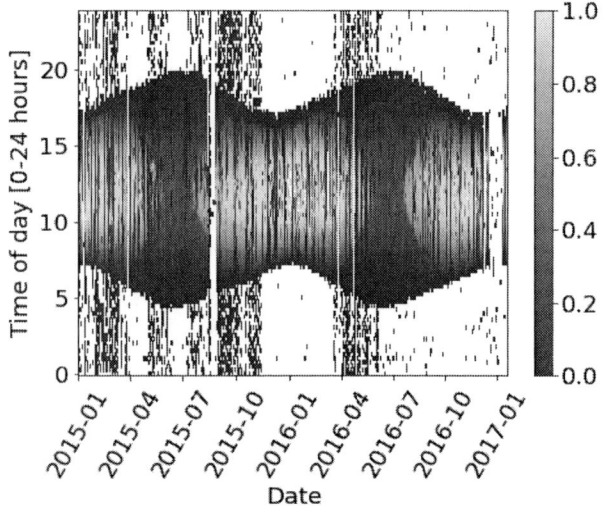

Fig. 4. Heat map depicting normalized PV power for a heavily shaded system on the NREL campus. This particular system is mounted on the side of a parking garage, and has a 60 degree tilt.

TABLE III
PERFORMANCE ON SINGLE SHADED SYSTEM

Algorithm	Azimuth Absolute Error	Tilt Absolute Error
PVWatts	0.41	22.44
PV Peak	0	5
SDT-Known Lat-Long	0.88	7.99
SDT-Unknown Lat-Long	0.45	5.72

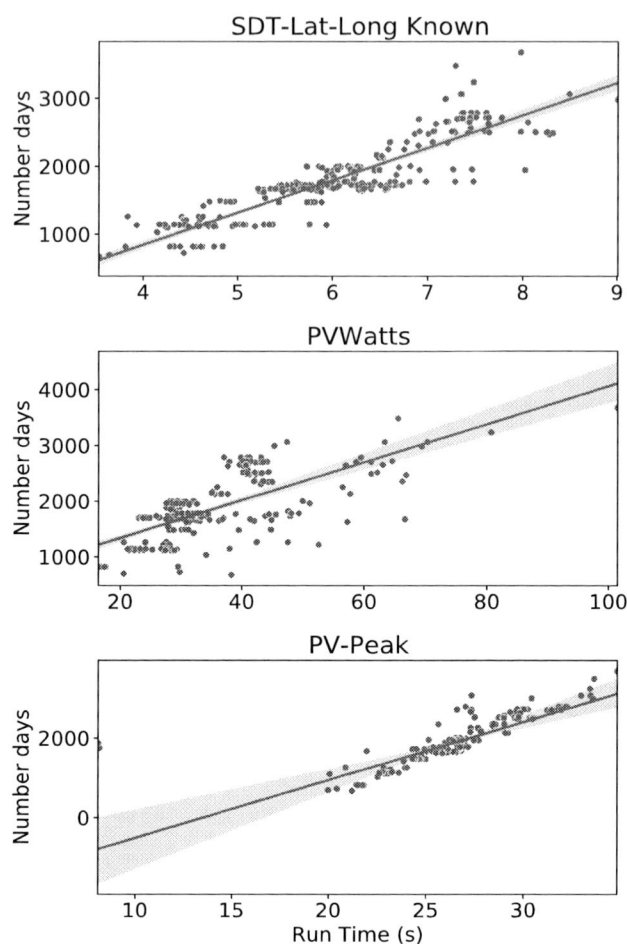

Fig. 5. Execution time vs. data set length for the SDT algorithm (latitude-longitude coordinates known), PVWatts, and the PV-Peak algorithms, respectively.

that all algorithms are less accurate at estimating tilt on heavily shaded system when compared to overall median results, with the PVWatts 5-based model performing especially poorly.

VII. DATA SET LENGTH

The impact of data set length on algorithm run time and azimuth and tilt error was explored. Figure 5 shows the relationship between data set length in days and execution time in seconds for the SDT, PVWatts 5-based, and PV-Peak algorithms, respectively. A clear linear trend is present between data length and run time for the SDT and PV-Peak implementations, with a weaker linear trend for the PVWatts 5-based implementation. This weaker linear relationship is likely a result of the PVWatts 5-based algorithm pulling NSRDB PSM3 data via the API, which may result in some run time variation. These results generally indicate that for all three algorithms, algorithm execution time is linearly correlated to data set length.

VIII. CONCLUSIONS

In this paper, we describe a reproducible methodology for comparing algorithms for estimating array tilt and azimuth from PV power time series. We applied this methodology to four proposed algorithms developed by DOE-funded national

laboratories. To the authors knowledge, this is the first such systematic comparison of algorithms for this purpose in the literature. Such systematic comparisons within a reproducible test environment is critical for the maturation of these PV data science tools and the transfer of this technology from academia to industry.

IX. FUTURE WORK

We intend to incorporate the associated data sets and benchmarking process from this research into the final *Validation Hub* website. When the *Validation Hub* site is launched, other researchers can submit additional algorithms to benchmark against the algorithms presented in this paper. A public leader board, similar to Table I, will display overall algorithm performance, including azimuth and tilt error as well as median run time for each submitted algorithm. Additionally, a private report with in-depth results, similar to the case studies and run time analyses presented in this paper, will be provided to each algorithm submitter. The Hub will include benchmarking routines for additional types of problems facing the PV community, including degradation estimation, soiling detection, and time shift/DST detection and correction, among others. The *Validation Hub* is intended to act as a clearinghouse for easy, side-by-side comparison of different methods for solving specific PV-related time series problems.

ACKNOWLEDGMENT

This work was authored in part by Alliance for Sustainable Energy, LLC, the manager and operator of the National Renewable Energy Laboratory for the U.S. Department of Energy (DOE) under Contract No. DE-AC36-08GO28308. Funding provided by the U.S. Department of Energy's Office of Energy Efficiency and Renewable Energy (EERE) under Solar Energy Technologies Office (SETO) Agreement Numbers 38258 and 38529.

REFERENCES

[1] "US solar market insight," Solar Energy Industries Association, Tech. Rep., 12 2022.

[2] B. Marion, M. Anderberg, R. George, P. Gray-Hann, and D. Heimiller, "PVWATTS version 2—Enhanced spatial resolution for calculating grid-connected PV performance," *NREL Report Number CP-650-30941*, 10 2001. [Online]. Available: https://www.osti.gov/biblio/15000006

[3] J. Kratochvil, W. Boyson, and D. King, "Photovoltaic array performance model." *Sandia Report Number SAND2004-3535*, pp. 1–19, 8 2004. [Online]. Available: https://www.osti.gov/biblio/919131/

[4] Y. Saint-Drenan, S. Bofinger, R. Fritz, S. Vogt, G. Good, and J. Dobschinski, "An empirical approach to parameterizing photovoltaic plants for power forecasting and simulation," *Solar Energy*, vol. 120, pp. 479–493, 2015. [Online]. Available: https://www.sciencedirect.com/science/article/pii/S0038092X15003941

[5] N. Haghdadi, J. Copper, A. Bruce, and I. MacGill, "A method to estimate the location and orientation of distributed photovoltaic systems from their generation output data," *Renewable Energy*, vol. 108, pp. 390–400, 2017. [Online]. Available: https://www.sciencedirect.com/science/article/pii/S0960148117301660

[6] A. Londono-Hurtado, B. Meyers, E. Apostolaki, and R. Flottemesch, "Estimation of photovoltaic system location and orientation from power signals," in *2021 IEEE 48th Photovoltaic Specialists Conference (PVSC)*, 2021, pp. 1807–1812.

[7] J. A. Duffie and W. A. Beckman, *Solar engineering of thermal processes*. John Wiley & Sons, 2013.

[8] B. E. Meyers and S. P. Boyd, "Signal decomposition using masked proximal operators," *Foundations and Trends in Signal Processing*, vol. 17, no. 1, pp. 1–78, 2023. [Online]. Available: http://dx.doi.org/10.1561/2000000122

[9] K. Mason, M. J. Reno, L. Blakely, S. Vejdan, and S. Grijalva, "A deep neural network approach for behind-the-meter residential PV size, tilt and azimuth estimation," *Solar Energy*, vol. 196, pp. 260–269, 1 2020.

[10] B. Meyers, D. Ragsdale, D. Serbetcioglu, J. Goncalves, A. Londono-Hurtado, T. Takahashi, E. Apostolaki, and D. J. F. Rodriguez, "solar-data-tools," May 2022. [Online]. Available: http://dx.doi.org/10.5281/zenodo.6450368

[11] B. Meyers, E. Apostolaki-Iosifidou, and L. Schelhas, "Solar data tools: Automatic solar data processing pipeline," in *2020 47th IEEE Photovoltaic Specialists Conference (PVSC)*, 2020, pp. 0655–0656.

[12] PVLib, "PVAnalytics," https://github.com/pvlib/pvanalytics, 2020.

[13] W. F. Holmgren, C. W. Hansen, and M. A. Mikofski, "pvlib python: a python package for modeling solar energy systems," *Journal of Open Source Software*, vol. 3, no. 29, p. 884, Sep. 2018. [Online]. Available: https://doi.org/10.21105/joss.00884

[14] A. P. Dobos, "PVWatts version 5 manual," National Renewable Energy Laboratory, Golden, CO, Tech. Rep. NREL/TP-6A20-62641, 9 2014.

[15] K. Perry and M. Muller, "Automated shift detection in sensor-based PV power and irradiance time series," in *2022 IEEE 49th Photovoltaics Specialists Conference (PVSC)*, 2022, pp. 0709–0713.

[16] K. Perry, M. Muller, and K. Anderson, "Performance comparison of clipping detection techniques in AC power time series," in *2021 IEEE 48th Photovoltaic Specialists Conference (PVSC)*, 2021, pp. 1638–1643.

[17] B. Meyers, "PVInsight final technical report," Sep 2021. [Online]. Available: https://www.osti.gov/biblio/1897181

[18] M. Sengupta, Y. Xie, A. Lopez, A. Habte, G. Maclaurin, and J. Shelby, "The national solar radiation data base (NSRDB)," *Renewable and Sustainable Energy Reviews*, vol. 89, pp. 51–60, 2018. [Online]. Available: https://doi.org/10.1016/j.rser.2018.03.003

[19] D. C. Jordan, K. Anderson, K. Perry, M. Muller, M. Deceglie, R. White, and C. Deline, "Photovoltaic fleet degradation insights," *Progress in Photovoltaics: Research and Applications*, vol. n/a, no. n/a. [Online]. Available: https://onlinelibrary.wiley.com/doi/abs/10.1002/pip.3566

[20] I. Reda and A. Andreas, "Solar position algorithm for solar radiation applications (revised)." [Online]. Available: https://www.osti.gov/biblio/15003974

Investigation of p-type silicon heterojunction radiation hardness

Romain Cariou, Adrien Danel, Nicolas Enjalbert, Frédéric Jay, Sébastien Dubois

Univ. Grenoble Alpes, CEA, Liten, Campus INES, Grenoble, France

The space sector is facing significant upheavals, in particular in terms of cost reduction challenges, driven by the emergence of Low Earth Orbit constellations. Concerning solar power generation, it opens up perspectives for alternative solar photovoltaics technologies, instead of the highly performant & expensive III-V multi-junction devices. Crystalline silicon solar cells, which have fueled space developments, spark a renewed interest, thanks to their industrial maturity, high efficiencies on p-type substrates & costs of two to three orders of magnitude lower than those of III-V. In this context, we present here the results of electrons radiation hardness studies on p-type (Ga-doped) silicon heterojunction solar cells. Devices with thicknesses down to 60µm are manufactured and then characterized before and after 1MeV electrons irradiations. The best ultra-thin heterojunction cell shows an end-of-life (1.5×10^{14} e/cm2) externally certified efficiency of 15.1% under AM1.5G at room temperature; this translates into ~ 13.4% with AM0 spectrum. The benefits of thickness reduction with respect to radiation hardness are presented, and the cells improvement pathways discussed.

978-1-6654-6060-6/23 $31.00 © 2023 IEEE

Energy Injustice Metrics for Puerto Rico

Pablo Méndez-Curbelo, Carlos Peña, Willian Pacheco, Marcel Castro-Sitiriche

University of Puerto Rico - Mayagüez Campus, Mayagüez, PR, Puerto Rico

According to the statistics census of Puerto Rico, the overall poverty level is approximately 40% and Puerto Rico has been identified as one of the most neglected places worldwide. This means that there is an inherent inequality in how the Puerto Rican population acquires different resources. In terms of electrical energy, this results in a disparity of the electric power service quality indicators. There is no doubt that inequity exists for blue sky scenarios, but it is exacerbated after the passing of a catastrophic phenomenon. After several years of hurricane María the grid is still fragile, revealing the abundant disparity that the Authority of Electrical Energy of Puerto Rico provides. From the electric power service quality indicators, Utuado is the most affected region, with SAIDI peaks of over 5-10 hours from June to October of 2022 and a SAIFI abundance of 1.35 to 1.74 hours from September to October of 2022. After the passage of Hurricane María in 2017 and Hurricane Fiona in 2022, both regions of Ponce and Mayaguez perpetuated the worst levels of CHoLES. The poor management of the network operator on the island is evident when compared to other regions. Although remote areas have a lower density of customers, the allocation of funds for resilience mitigation is not proportional to the CHoLES accumulated in those communities, evading the improvement of a reliable power system. The fair allocation of funds to enable installations of renewable sources in remote areas of Puerto Rico provides a unique opportunity to improve overall resiliency and reliability. While at the same time reducing energy injustice with solar rooftop PV systems with batteries at a lower cost than expected by the power grid.

Inverse Design of Spectrally-Selective Films for PbS-CQD Tandem Solar Cells

Sreyas Chintapalli, Tina Gao, Luna Singh, Serene Kamal, Arlene Chiu, Yijun Zhang, Susanna M. Thon

Johns Hopkins University, Baltimore, MD 21218 USA

Abstract — **We present a method for designing spectrally-selective optoelectronic films with a finite absorption bandwidth. We demonstrate the process by designing a film composed of lead sulfide colloidal quantum dots (PbS-CQDs). Designs incorporate the patterning of absorbing PbS-CQD films into photonic crystal-like slabs which couple incident light into leaky modes within the plane of the absorbing films, modulating the absorption spectrum. Computational times required to calculate optical spectra are drastically decreased by implementing the Fourier Modal Method. Furthermore, a supervised machine-learning-based inverse design methodology is presented which allows tailoring of the PbS-CQD film optical properties for use in a variety of photovoltaic applications, such as tandem cells in which spectral tailoring can enable current-matching flexilibility.**

I. INTRODUCTION

Tandem solar cells (TSC) are an area of large interest to their demonstrated potential to increase solar cell efficiencies beyond current efficiency limits [1]-[2]. In the typical TSC architecture, the ultraviolet-visible cell (VIS Cell) is placed above the infrared absorbing cell (IR cell), closer to the incident illumination. This maximizes photoconversion efficiency by preventing thermalization loss from occurring in the infrared cell, as high energy photons are absorbed prior to exciting charge carriers in the bottom IR cell [3]. Lead sulfide colloidal quantum dots (PbS-CQDs) have a size-tunable bandgap due to quantum confinement, which makes them a promising candidate for use in tandem solar cells as an IR absorber. However, PbS-CQDs, as well as other candidate IR absorbing materials, are unstable at temperatures even as high as 120°C [4]. This places significant processing constraints on device architectures incorporating these films, because after the CQD film is deposited, any further high-temperature processing will degrade device performance. This effectively restricts all layers on top of the IR cell be processed at relatively low temperatures.

Tandem device architectures where the IR cell could be placed on top of existing VIS architectures would provide great flexibility in both material choice for other layers, as well as the ability to quickly leverage existing high efficiency designs for VIS single junction solar cells. To mitigate the thermalization loss, the PbS-CQD film would need to have suppressed absorption in the absorption range of the VIS cell. While all semiconductor-like (SC-like) materials have absorption that continues above some threshold energy, in this work we propose to pattern PbS-CQD films using a photonic crystal-like slab (PhC) structure. In our earlier work, we have shown that patterning PbS-CQD films can modulate their absorption spectra [5]. In this work, we propose a method for designing and patterning films with a finite absorption bandwidth (FAB) which give the desired absorption properties in the IR, as well as desired transmission in the visible spectral range, through incident light coupling into leaky guided modes in the patterned film. This allows for using existing materials and patterning techniques to create custom optical absorption behavior, which can be optimized for many tandem device architectures.

II. METHODOLOGY

A. PbS-CQD Film Characterization

PbS-CQD films were synthesized using the hot-injection method [8], with various target exciton peak wavelengths. The films were measured using a J.A. Woollam UV-NIR Variable Angle Spectroscopic Ellipsometer (VASE) to extract refractive index models. The real and imaginary components are shown in Fig. 1(a) and (b) respectively, with the physical quantum dot sizes calculated from [6] shown in the legend. This model was chosen because it corresponds to $E_g \approx 1.05$ eV, close to the ideal bandgap for a tandem IR cell [9].

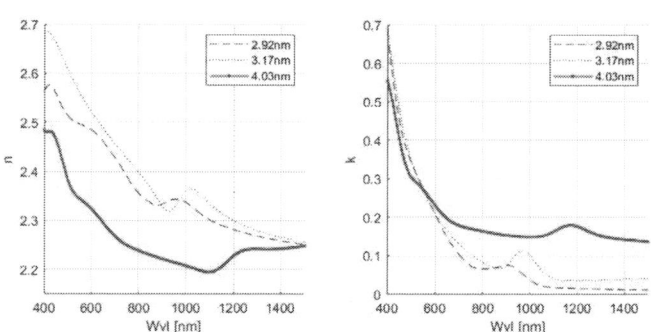

Fig. 1 (a) Real component of refractive index of PbS-CQD Films. (b) Extinction coefficient of PbS-CQD Films. The values in the legend correspond to estimated CQD diameters. The red index model is the model used for the remainder of this work.

B. Traditional Optimization using FMM

In our previous work, we showed that the out-of-plane transmission and reflection spectra of the patterned films couple strongly to in-plane guided modes, even in the presence of absorption [6]. Here, we generate spectra directly from traditional designs We chose a spectral range of 400-1500 nm for the assumed incident light. The design parameters were searched within the following bounds: lattice constant, a, from 100-1500 nm; normalized feature radius, r/a, from 0.05 to 0.5; and normalized slab thickness, t/a, from 0.1 to 2.1. The figure-of-merit (FOM) is given in Equation 1, where w is the weight given to the visible portion of the spectrum during optimization, T_{vis} is the transmission spectrum, A_{IR} is the absorption spectrum, and λ is the wavelength of incidence.

$$\text{FOM} = \max \left(w \int T_{VIS}(\lambda)\, d\lambda + (1-w) \int A_{IR}(\lambda)\, d\lambda \right) \quad (1)$$

This FOM allows us to balance the two design goals between keeping absorbance high in the IR, while allowing the PbS-CQD absorbing film to transmit light to the VIS cell below. Both terms are needed, as the globally optimal solutoin for either the first or second term individually would lead to the thinnest or thickest possible film, respectively. Figure 2(a) shows the absorption and transmission spectra of a hexagonal-hole film compared to an equivalent film with the same volume of PbS-CQD. The shaded region highlights Eg>1.65eV, the

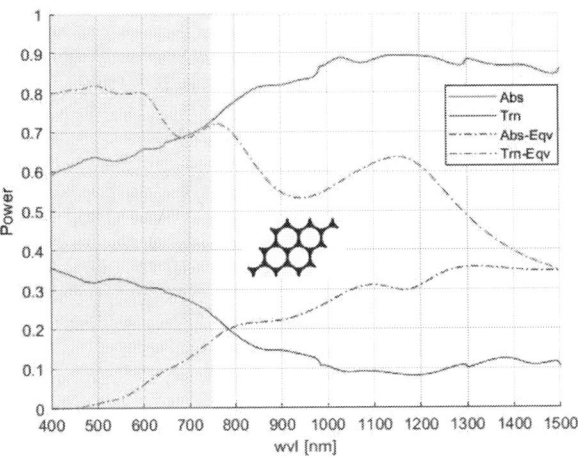

Fig. 2. Optical Spectra of a patterned and unpatterned PbS-CQD film. Red lines are absorption spectra, and blue are transmission spectra. Dashed lines represent the equivalent unpatterned film. The film has a lattice constant of 1500nm, normalized radius of 0.45(675nm), and normalized thickness of 1.3 (1950nm). A few unit cells of the pattern are shown in the center inset, with green representing PbS-CQDs and white representing air.

optimal bandgap for the corresponding VIS cell [8]. With even this simple design, there is a 23.7% increase in the transmission across the visible range, and 10.9% suppression of the absorption.

C. Inverse Design of PbS-CQD Films

Machine learning techniques for the inverse design of metasurfaces and other optical structures have gained prominence in recent year, as these models have the ability to generate high performance, but unintuitive designs [9]-[10]. We use a network architecture inspired by the model used in [11]-[12]. Our model is a multi-input multi-output model that fully parametrizes the design of the patterned film needed for a desired spectrum. Three classes of shapes were used for masks: Bezier polygons, reflectively symmetric, and radially, along with a random scalar lattice constant and thickness, using the same Nannos implementation as the previous section [7]. Spectra were generated for all 30,000 mask-spectra pairs for at 5nm wavelength spacing across 400-1500nm. These spectra were split into proportions of 70%, 20%, and 10% for the training, validation, and testing sets. The entire network, implemented in TensorFlow is available at online at [13], and the random splits are seeded for reproducibility, along with the detailed diagram of the model architecture. The encoder takes each mask and encodes this into a linear 128 dimension feature space similar to [11]. This feature vector concatenates directly with the 2 lattice parameters. The model then goes through a series of fully-connected layers and recovers 665 parameters. The first 663 correspond to each optical spectra (of length 221 each), and the last two elements are the estimators of the recovered lattice parameters. Figure 3 shows examples of the encoding/prediction process. We used the ADAM optimizer [14] with the standard learning rate of 0.001 to minimize the mean-squared error. The process is then reversed and the mask is regenerated from the feature space.

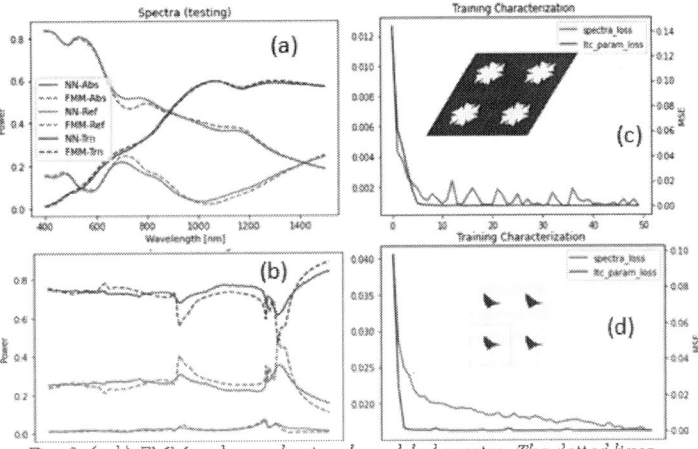

Fig. 3. (a-b) FMM and neural network modeled spectra. The dotted lines represent the FMM model, and the solid lines are the predictions of the encoder. Lattice parameters of (a) were (a, t_a)=(1250, 0.7) with a prediction of (1283, 0.66). Lattice parameters of (b) were (a, t_a)=(1250, 0.7) with a prediction of (1283, 0.66).

(c-d) Training characterization of the model trained on the two lattice types pictures. (a) corresponds to (c) and (b) corresponds to (d). The unit cells passed to the encoder were in the test set, and never seen by the model.

III. Discussion

The tuning of the weights for the loss was critical in the neural network. We use the loss as 0.25 for the spectral encoding, and 0.75 for the lattice parameters Fig 3(a), and as we ed that the mask shape will have a more nuanced effect, as the lattice parameters will strongly define coupling of the incident light leaky modes (in-plane of the crystal as well as the Fabry-Perot modes from the slab thickness). The model is very lightweight, training in less than 6 minutes using the free Google Colab platform, using a batch size of 50 samples over 50 epochs. While the encoding process has relatively high accuracy, and can be further improved with hyperparameter tuning, or drastically increasing the number of variables (and therefore training time), we note that the error from our 'ideal spectrum' for a top IR PbS-CQD absorbing layer is quite high. Unfortunately, it is difficult, if not impossible, to calculate the theoretical minimum of how close one can get to an arbitrary spectrum do to the nearly infinite search space of possible unit cells, but given the constraints of the real index models. As one can see, the 'rod-square' lattice pattern in Fig. 3(d) allows nearly 80% transparency everywhere below the exciton peak of 1200nm of the PbS-CQDs. Here we have used the exact same model on each of the lattice types, which is convenient, but may not be the highest performance. Additionally, we use the exact same features during the encoding and decoding process, which also may not yield the maximum possible FOM.

IV. Summary and Outlook

We have demonstrated a method to improve manufacturing flexibility of tandem cells using PbS-CQDs, as well as the possibility to design custom films to leverage existing VIS cells. In order to further improve these models, as well as realize them in tandem cells, the models can be tuned through various hyperparameter optimizations as well as have enforced manufacturability constraints. The size scale of the unit cells of the inversely designed films are ~600nm, which is within the feature scale of nanoimprint lithography techniques [14] for quantum dots. These methods are also flexible enough that the optical properties of underlying films/substrates can be readily added, due to the flexibility of

V. ACKNOWLEDGEMENTS

We would like to acknowledge the National Science Foundation (ECCS-1846239) for funding this work. This work was carried out at the Advanced Research Computing at Hopkins (ARCH) core facility (rockfish.jhu.edu), which is supported by National Science Foundation (NSF) grant number OAC1920103.

References

[1] T. Ameri, G. Dennler, C. Lungenschmied, and C. J. Brabec, "Organic tandem solar cells: A review," *Energy Environ. Sci.*, vol. 2, no. 4, p. 347, 2009, doi: 10.1039/b817952b.

[2] A. D. Vos, "Detailed balance limit of the efficiency of tandem solar cells," *J. Phys. D: Appl. Phys.*, vol. 13, no. 5, pp. 839–846, May 1980, doi: 10.1088/0022-3727/13/5/018.

[3] T. Todorov, O. Gunawan, and S. Guha, "A road towards 25% efficiency and beyond: perovskite tandem solar cells," *Mol. Syst. Des. Eng.*, vol. 1, no. 4, pp. 370–376, 2016, doi: 10.1039/C6ME00041J.

[4] M. Albaladejo‐Siguan, E. C. Baird, D. Becker‐Koch, Y. Li, A. L. Rogach, and Y. Vaynzof, "Stability of Quantum Dot Solar Cells: A Matter of (Life)Time," *Adv. Energy Mater.*, vol. 11, no. 12, p. 2003457, Mar. 2021, doi: 10.1002/aenm.202003457

[5] B. Qiu, Y. Lin, E. S. Arinze, A. Chiu, L. Li, and S. M. Thon, "Photonic band engineering in absorbing media for spectrally selective optoelectronic films," *Opt. Express*, vol. 26, no. 21, p. 26933, Oct. 2018, doi: 10.1364/OE.26.026933.

[6] I. Moreels et al., "Size-Dependent Optical Properties of Colloidal PbS Quantum Dots," *ACS Nano*, vol. 3, no. 10, pp. 3023–3030, Oct. 2009, doi: 10.1021/nn900863a.

[7] B. Vial and Y. Hao, "Open-Source Computational Photonics with Auto Differentiable Topology Optimization," *Mathematics*, vol. 10, no. 20, p. 3912, Oct. 2022, doi: 10.3390/math10203912.

[8] T. Leijtens, K. A. Bush, R. Prasanna, and M. D. McGehee, "Opportunities and challenges for tandem solar cells using metal halide perovskite semiconductors," *Nat Energy*, vol. 3, no. 10, pp. 828–838, Jul. 2018, doi: 10.1038/s41560-018-0190-4.

[9] S. Molesky, Z. Lin, A. Y. Piggott, W. Jin, J. Vuckovic, and A. W. Rodriguez, "Inverse design in nanophotonics," *Nature Photon*, vol. 12, no. 11, pp. 659–670, Nov. 2018, doi: 10.1038/s41566-018-0246-9.

[10] P. R. Wiecha, A. Arbouet, C. Girard, and O. L. Muskens, "Deep learning in nano-photonics: inverse design and beyond," *Photon. Res.*, vol. 9, no. 5, p. B182, May 2021, doi: 10.1364/PRJ.415960.

[11] T. Christensen et al., "Predictive and generative machine learning models for photonic crystals," *Nanophotonics*, vol. 9, no. 13, pp. 4183–4192, Jun. 2020, doi: 10.1515/nanoph-2020-0197.

[12] A. Nikulin, I. Zisman, M. Eich, A. Yu. Petrov, and A. Itin, "Machine learning models for photonic crystals band diagram prediction and gap optimisation," *Photonics and Nanostructures - Fundamentals and Applications*, vol. 52, p. 101076, Dec. 2022, doi: 10.1016/j.photonics.2022.101076.

[13]. Chintapalli, I (2023). Shaped-PbS-CNN. (Version 1. 1). https://github.com/jhu-nanoenergy/spectral_selectivity

[14] M. J. Hampton, J. L. Templeton, and J. M. DeSimone, "Direct Patterning of CdSe Quantum Dots into Sub-100 nm Structures," *Langmuir*, vol. 26, no. 5, pp. 3012–3015, Mar. 2010, doi: 10.1021/la904787k.

AM0 Optimized Dual Junction Quantum Well Solar Cells – Investigation of Radiation Tolerance Designs and V_{oc} Retention at EOL

Brandon M. Bogner, Stephen J. Polly, Seth M. Hubbard
Rochester Institute of Technology, Rochester, NY, 14623, USA

Mitsuru Imaizumi
Japan Aerospace Exploration Agency, Chōfu, Tokyo, Japan

Roger E. Welser
Independent Consultant, Providence, RI, 02906, USA
*Formerly at Magnolia Optical Technoligies, Inc.

Abstract — **Dual junction upright solar cells composed of a homojunction GaInP top cell and a heterojunction GaInP/GaAs bottom cell incorporating 50 layers of $Ga_{0.9}In_{0.1}As$ quantum wells (QWs) with $GaAs_{0.9}P_{0.1}$ barriers and a 12-pair $Al_{0.9}Ga_{0.1}As/Al_{0.1}Ga_{0.9}As$ distributed Bragg reflector were designed for operation under AM0 and grown by MOVPE. Performance degradation was investigated after irradiance of 1 MeV electrons at a fluence of 1 x 10^{15} cm^{-2}, revealing a Jsc remaining factor of 0.933 in the device with 50 pairs of QWs compared to 0.854 in the 0 QW control. In addition to a 1.7% absolute AM0 efficiency increase in the QW device relative to the control, Voc and fill factor remaining factors were held above 0.86 in both devices, producing equivalent EOL efficiencies of 18%. Modifications to layer and interface designs surrounding the quantum well region will be investigated in an effort to improve EOL V_{oc} and FF in the quantum well device. IV performance under AM0 illumination, layer-specific current collection degradation in the GaAs sub-cell, and sub-cell saturation current densities will be investigated. The device design presented in this work is compatible with the current state-of-the-art GaInP/GaAs/Ge triple-junction architecture. With continued optimization of both subcells, a path to 2J efficiency under AM0 surpassing 27% BOL and 21% EOL may be realized.**

Keywords— Dual junction, Quantum Wells, Distributed Bragg Reflector, Radiation-Induced Degradation, Radiation Tolerance

I. INTRODUCTION

Quantum wells (QW) have been under investigation to increase performance of photovoltaic devices for more than 30 years. These two-dimensional structures with a lower bandgap than the host material allow absorption and collection of photons normally lost to transmission, thus increasing current density. Incorporation into multi-junction devices allows the for greater collection in the QW enhanced subcell, thus increasing the overall current production of the series-connected device, leading to overall increase in BOL efficiency. Space missions, however, expose the PV modules to a harsh environment of high-energy particles. These particles negatively impact the performance of PV devices over time, decreasing power generation at end-of-life (EOL) by over to 20% after a $1x10^{15}$ e/cm^2 fluence of 1 MeV electrons, the equivalent to a 15 year mission in geostationary orbit [1]. In particular, GaAs, the middle sub-cell in triple junction lattice-matched (3J LM) InGaP/GaAs/Ge solar cell, is susceptible to extensive degradation in current collection.

This work builds on previous reports of the integration of InGaAs quantum wells in an upright single junction solar cell incorporating an InGaP/GaAs heterojunction with minimal loss in Voc [2] as well as a dual junction device introducing an InGaP homojunction top subcell [3]. Here 50 layers of strain balanced QWs are added to the InGaP/GaAs heterojunction bottom sub-cell. A distributed Bragg reflector (DBR) was added below the structure centered at the collection energy of the QWs to increase the optical path length of the QW region and further increase subgap current collection. Dual junction device performance under AM0 will be investigated as a function of increasing QW number. The 50 QW and a device without QWs (control) were subjected to 1 MeV electron irradiance at a fluence of 1 x 10^{15} cm^{-2}, where EQE and JV performance was compared before and after to evaluate the effect QWs have on radiation-induced degradation.

II. EXPERIMENTAL

Devices were grown by low pressure metalorganic vapor phase epitaxy (MOVPE) on a 3x2" AIXTRON close-coupled showerhead system at RIT using standard precursors including arsine, phosphine, trimethylgallium, trimethylindium, and trimethylaluminium. Device layer thicknesses were designed using a Drift-diffusion modelling [4], with intentionally lower current generation in the InGaP top subcell in an effort to retain its higher fill factor, as compared to the GaAs bottom cell, in the 2J combination. Specific layer thicknesses, compositions and dopings of the investigated devices were previously reported in [3].

Wafers were fabricated using typical III/V fabrication techniques including contact photolithography and wet chemical mesa isolation etch to define 1x1 cm^2 area solar cells. Front grid contacts designed for 1-Sun current densities were electroplated with Au, and rear contact was thermally evaporated Au/Zn/Au. Antireflective coatings of TiO$_2$, ZnS and MgF$_2$ were deposited by atomic layer deposition and thermal evaporation, respectively.

978-1-6654-6060-6/23 $31.00 © 2023 IEEE

TABLE I
AM0 LIV & EQE RESULTS FOLLOWING 1 MeV 1 x 10^15 cm^-2 ELECTRON IRRADIANCE

Device	Condition	AM0 IV				EQE	
		J_{sc} (mA/cm^2)	V_{oc} (V)	FF (%)	η (%)	InGaP Integrated J_{sc} (mA/cm^2)	GaAs Integrated J_{sc} (mA/cm^2)
0xQW (Control)	BOL	15.23	2.380	88.4	23.5	15.84	17.91
	EOL	13.01	2.191	86.8	18.1	15.6	14.47
	Remaining Factor	0.854	0.920	0.982	0.772	0.985	0.808
50xQW w/ DBR	BOL	17.00	2.349	86.3	25.2	18.25	18.58
	EOL	15.86	2.020	78.1	18.3	18.1	16.17
	Remaining Factor	0.933	0.860	0.905	0.726	0.992	0.870

External and internal quantum efficiency (EQE, IQE) were measured using a Newport IQE-200 system, and 1-Sun light IV was taken with a TS-Space Systems dual source close-match solar simulator calibrated to AM0 using InGaP and GaAs 3J BTJ sub-cells following an NREL AM0 calibration. Electroluminesence-extracted sub-cell JV was taken using a QEPRO spectrometer connected to an integrating sphere, calibrated with an Ocean Insight Tungsten Halogen light source. Samples were irradiated through collaboration with the Japan Aerospace Exploration Agency using a calibrated 1MeV electron beam and subjected to EOL AM0 IV and EQE.

III. RESULTS AND DISCUSSION

Dual junction InGaP / GaAs devices previously reported in [3] were subjected to 1 MeV electron irradiance at a fluence of 1 x 10^15 cm^-2, the equivalent to a 15 year mission in geostationary orbit, and re-evaluated under AM0 illumination. The samples under investigation included a control (0 QW) and a 50 QW device with a DBR coupled to the QW collection range.

Beginning-of-life (BOL) IV performance, shown as solid curves in Figure 1 with figures of merit outlined in Table 1, reveals an AM0 efficiency increase from 23.5% in the control to 25.2% in the 50 QW device, or 1.7% absolute. This efficiency increase is a result of the extended collection range of the QWs in the GaAs subcell from 890 nm to 950 nm (Fig. 2), providing an additional 1.77 mA/cm^2 of sub-cell current while holding open-circuit voltage (V_{oc}) to within 32 mV of the control device.

After exposure to 1x10^15 cm^-2, 1 MeV electron irradiance, end-of-life (EOL) EQE and IV under AM0 were measured to investigate the impact QWs have on radiation-degradation in the multi-junction device. Collection in the InGaP top junction remained unchanged from BOL to EOL in both the control and 50 QW devices, shown in the overlapping quantum efficiency curves and > 0.98 J_{sc} remaining factor. This is typically the case for multi-junction solar cells due to the low displacement damage near the surface of the device at irradiance energies above 150 keV [5].

The GaAs subcells in both samples showed visible degradation in current production and V_{oc}. Again, this is typical since the displacement damage begins to dramatically increase as the electrons scatter and slow down in the GaAs sub-cell [5]. The current degradation in the control GaAs sub-cell spans the entire GaAs collection range, while the damage in the 50 QW sample is held to a smaller collection window of 800 – 900 nm. The resilience in collection below 800 nm and beyond the GaAs band-edge is a consequence of increased i-region thickness, and therefore absorptive region with a high electric field, from the incorporation of QWs. Specifically, the i-region increased from 200 nm in the 0 QW device to 1480 nm in the 50 QW device, to account for the well-barrier absorption region – as well as the inherent radiation resistance of the QWs themselves.

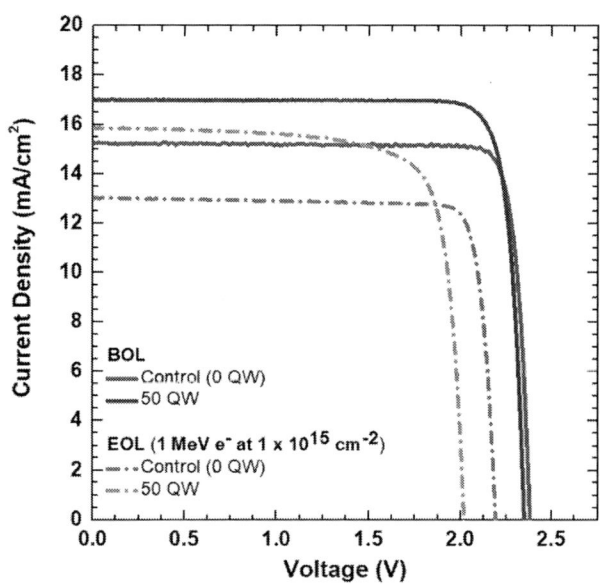

Fig. 1. Illuminated IV results under AM0 at BOL (solid) and EOL (dashed) of the 0 QW and 50 QW dual junction InGaP / InGaP-GaAs devices.

Radiation-induced degradation is primarily known to reduce minority carrier lifetimes through the following relation:

$$\frac{1}{\tau} = \frac{1}{\tau_0} + k_R \phi \tag{1}$$

where τ is the minority carrier lifetime, τ_0 is the BOL lifetime, k_R is the material-dependent damage coefficient and ϕ is the particle fluence [5]. In the 50 QW device, the GaAs subcell maintains effective transport of carriers generated in the i-region. The QW collection remains unchanged between BOL and EOL, shown by the overlapping EQE peaks near 950 nm in Fig. 2 and as integrated J_{sc} of 1.4 and 1.3 mA/cm² from 890-1000 nm at BOL and EOL, respectively. Due to this resilient i-region collection in the 50 QW GaAs subcell, the J_{sc} remaining factor under AM0 is held at 0.933, while the control reported a remaining factor of 0.854.

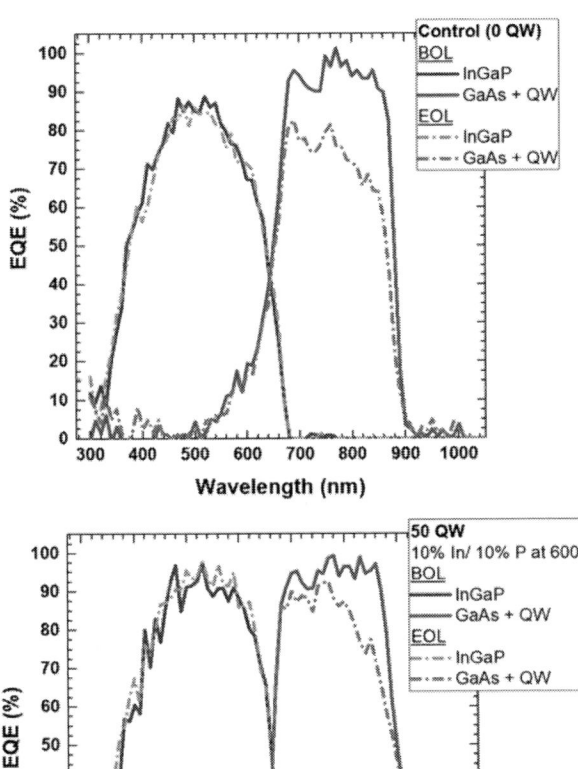

Fig. 2. External Quantum Efficiency at BOL (solid) and EOL (dashed) of the 0 QW (top) and 50 QW (bottom) subcells.

The trade-off of a thicker i-region is seen in V_{oc} and FF degradation. The driving mechanism behind this phenomenon is carrier removal (CR), a process in which radiation-induced crystal defects are formed throughout the device, manifesting as trap states that act as a localized dopant. This process can reduce the abruptness in doping of the i-region/emitter or i-region/base interfaces, producing a dramatic loss in electric field through the junction [6]. With the combined impacts of a lower built-in electric field and an increase in i-region defect states, non-radiative recombination will begin to dominate. As V_{oc} and FF are sensitive to non-radiative recombination, both figures of merit are expected to suffer with increased exposure to high energy electron radiation. This degradation in the 50 QW device is significant, however coupled with the resilience in J_{sc}, EOL efficiency remained 0.2% absolute higher than in the 0 QW device.

This result shows an overall absolute efficiency increase of 1.7% at BOL and 0.2% at EOL through utilization of QWs in a dual junction InGaP / GaAs device. The samples investigated were grown upright on a GaAs substrate, allowing this design to be directly ported to the current state-of-the-art (SOA) GaInP/GaAs/Ge triple-junction architecture.

Dual junction 50 QW devices including potential radiation tolerant design features will also be evaluated in this work at BOL and EOL under AM0. These results will then be used to build a stronger understanding of V_{oc} degradation in QW solar cells, and propose a means of mitigating the previously accepted voltage loss at EOL accompanying QW sub-cells. This work will of course follow device results and growth techniques outlined in [2] and [3], providing design recommendations for space-grade SOA multi-junction QW photovoltaics.

REFERENCES

[1] J. Li, A. Aierken, Y. Liu, Y. ZHaung, X. Yang, J.H. Mo, R.K. Fan, Q.Y. Chen, S.Y Zhang, Y.M. Huang, Q. Zhang, "A Brief Review of High Efficiency III-V Solar Cells for Space Application", Frontiers in Physics, vol. 8, Feb. 2021.

[2] R. E. Welser, S. J. Polly, B. M. Bogner and S. M. Hubbard, "Impact of Well Number on High-Efficiency Strain-Balanced Quantum-Well Solar Cells," IEEE Journal of Photovoltaics, vol. 13, no. 1, pp. 61-69, Jan. 2023, doi: 10.1109/JPHOTOV.2022.3216235.

[3] S. J. Polly et al., "Growth optimization of InGaP/ GaAS dual junction solar cells with quantum wells and a distributed Bragg reflector", Sep. 2022, [online] Available: https://ssrn.com/abstract=4219119.

[4] Lumb, Matthew P. et al. "Extending the 1-D Hovel Model for Coherent and Incoherent Back Reflections in Homojunction Solar Cells." IEEE Journal of Quantum Electronics 49 (2013): 462-470.

[5] S. Sato et al., "Degradation modeling of InGaP/GaAs/Ge triple-junction solar cells irradiated with various-energy protons", Solar Energy Materials & Solar Cells. 93 (2009); doi: 10.1016/j.solmat.2008.09.044

[6] R. Hoheisel et al., " Quantum-Well Solar Cells for Space: The Impact of Carrier Removal on End-of-Life Device Performance", IEEE Journal of Photovoltaics, vol. 4, no. 1, pp. 253-259, Jan. 2014, doi: 10.1109/JPHOTOV.2013.2289935.

Evaluation of process damage to crystalline silicon by transparent conductive oxide film deposition

Haruki Kojima[1], Tappei Nishihara[1,2], Yuta Ito[1], Hyunju Lee[1,5], Kazuhiro Gotoh[3],
Noritaka Usami[3], Tomohiko Hara[4], Kyotaro Nakamura[4], Yoshio Ohsita[4,5], and Atsushi Ogura[1,5]

[1]Meiji University, Scholl of Science and Technology, Kawasaki, Kanagawa, 214-8571, Japan
[2]JSPS Research Fellow DC2, Chiyoda, Tokyo, 102-0083, Japan
[3]Nagoya University, Nagoya, Aichi, 464-8601, Japan
[4]Toyota Technological Institute, Nagoya, Aichi, 468-8511, Japan
[5] Meiji University, Meiji Renewable Energy Laboratory, Kawasaki, Kanagawa, 214-8571, Japan

Abstract — We evaluated the damage to crystalline silicon (c-Si) induced by transparent conductive oxide film (TCO) deposition processes in indium tin oxide (ITO)/ hydrogenated amorphous silicon (a-Si:H)/Si structure. ITO was deposited by reactive plasma deposition (RPD) and sputtering techniques, respectively. After ITO deposition, the post-annealing at 200°C for 30 min in the air atmosphere was also carried out. Carrier lifetime of both samples decreases drastically after ITO deposition, and is significantly recovered by post-annealing both for RPD and sputtering. The plasma processes should produce recombination active defects on the sample surface or in the c-Si substrate resulting in carrier lifetime deterioration. These defects might be relatively small scale. Photoluminescence (PL) spectroscopy revealed the formation of so-called "irradiation-induced defects" that are formed typically after electron beam and ion irradiation with the peak positions at deep levels of 0.767 eV and 0.614 eV in the samples deposited by the RPD technique. Complex defects of impurity carbon and oxygen atoms were considered to be formed in c-Si by the RPD technique. These defects were eliminated after post-annealing. These irradiation-induced defects act as recombination active centers and may cause the deterioration of the conversion efficiency in c-Si solar cells.

I. INTRODUCTION

Silicon heterojunction (SHJ) solar cells are highly promising due to their high conversion efficiency[1, 2]. In SHJ solar cells, an intrinsic hydrogenated amorphous silicon layer (a-Si:H(i)) is used as a passivation layer for crystalline silicon substrate, and an n or p-type hydrogenated amorphous silicon layer (a-Si:H(n, p)) is used as a carrier-selective layer. In addition, a transparent conductive oxide film (TCO) such as tin-doped indium oxide(ITO), tungsten-doped indium oxide (IWO) and zinc-doped indium oxide (IZO) is used to compensate for the high sheet resistance loss of the a-Si:H layer. There have been many studies on improving the manufacturing process to obtain highly efficient SHJ solar cells. In addition, degradation of solar cells due to damage that occurs during the manufacturing process has also been reported [3, 4]. Carrier lifetime in the SHJ solar cell structure decreases after the ITO deposition [5, 6]. This may be caused by induced damage to c-Si as well as degradation of a-Si:H passivation performance. Research on the damage to a-Si:H has been reported extensively [7, 8]. However, the presence or absence of the damage to c-Si has

not been sufficiently clarified yet, which should lead to further improvements in solar cell conversion efficiency.

The purpose of this study is to investigate the damage induced in c-Si during the ITO deposition process of SHJ solar cells. We evaluated the damage induced in c-Si by reactive plasma deposition (RPD) and sputtering deposition ITO, which is one of the most conventional TCO materials.

II. EXPERIMENTAL

We prepared double-sided mirror-polished Czochralski (Cz)-grown n-type c-Si (100) wafers. The resistivity and wafer thickness were 2.6 Ωcm and 250 μm, respectively. These wafers were cleaned following the standard RCA procedure and dipped into 2.5% diluted HF solution for 1 min to remove the native silicon oxide. The substrates were immersed in a 1.5 % H_2O_2 solution for 30 s to form ultra-thin silicon oxide, which prevents crystallization of the a-Si:H layer during annealing for the metallization at the last process in the cell fabrication. Two samples with ITO/a-Si:H(n)/a-Si:H(i)/c-Si(n)/a-Si:H(i)/a-Si:H(n)/ITO structure were prepared by depositing 10 nm thick a-Si:H(i) followed by 10 nm thick a-Si:H(n) on both sides by PECVD and then depositing ITO by RPD or sputtering, respectively.

The ITO (5 wt % Sn) target was used in the RPD technique. The flow rate of argon (Ar) and O_2 gases were 85 and 20 sccm, respectively. The gases were supplied into the plasma chamber and the plasma was guided to the ITO target by the magnetic field. The thickness of deposited ITO was about 91 nm which was evaluated by a stylus profilometer. On the other hand, the 10 wt % Sn doped ITO target is used for sputtering. The flow rate of Ar and O_2 gases were 32 and 1 sccm, respectively. The thickness of deposited ITO was about 100 nm which was evaluated by spectroscopic ellipsometry.

For the carrier lifetime evaluation, we used Sinton WCT-120 instrument and represented the carrier lifetime at the injection level of 1.0×10^{15} cm^{-3}. To evaluate the effect of the post-annealing on the damage, the samples were annealed in the air atmosphere using a hot plate at 200°C for 30 minutes which simulate the thermal budget during the metallization process.

978-1-6654-6060-6/23 $31.00 © 2023 IEEE

In addition, photoluminescence (PL) spectroscopy was performed to evaluate the irradiation-induced defects caused by ITO deposition. The measurement conditions were an excitation wavelength of 532 nm with a beam diameter of approximately 3 mm and the cryostat temperature set at around 4 K. The sample temperature was approximately 12 K estimated from the half width at half maximum of the $I_{TO}(FE)$ peak for the silicon band-edge emission. The detector used in PL spectroscopy was an N_2-cooled indium gallium arsenide (InGaAs) photodiode array with high sensitivity above the 1000–2200 nm spectral range. Furthermore, a spectrum with a high S/N ratio was obtained by integrating the spectra using Lock-in Mode.

III. RESULTS

First, we evaluated the carrier lifetime of samples before and after ITO deposition. Figure 1 shows the change in carrier lifetime for each process, normalized by the value before ITO deposition (about 2 ms). In all samples, the carrier lifetime decreased significantly after ITO deposition and recovered by the post-annealing process. The plasma process should produce recombination-active defects on the sample surface or in the c-Si substrate. In addition, it suggests that these defects are relatively small scale.

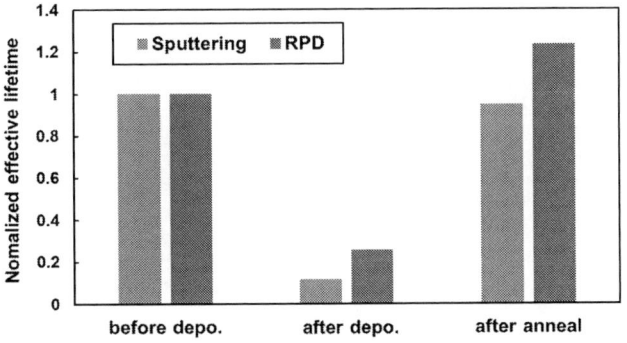

Fig. 1.   Carrier lifetime before and after the plasma process.

Next, we evaluated these defects induced by ITO deposition using PL spectroscopy. Figure 2 shows the PL spectra of samples before and after ITO deposition by the RPD technique. In the sample before ITO deposition, in other words, a-Si:H only, deep level peak, which decreased the carrier lifetime, was not detected. Irradiation-induced defects are not generated in a-Si:H films deposited by PECVD. In the sample after ITO deposition, peaks at 0.767 eV and 0.641 eV were observed. The deep level at 0.767 eV is reported as a defect called P-line, which is related to thermal donor (TD) and generally appears after annealing at temperatures above 450°C [9, 10]. In this study, ITO deposition was performed at room temperature. Therefore, 0.767 eV peak cannot be considered as P-line.

The peak at 0.767 eV was also observed in a sample of ITO deposited by sputtering directly on c-Si and was reported to be the irradiation-induced defects containing impurities such as

carbon and oxygen [4]. And these defects are not due to the plasma emission but due to the ion collisions. In other words, ions during deposition by the RPD technique pass through a-Si:H films and generate defects in c-Si. There are many reports that irradiation-induced defects generated during ITO deposition are composite defects of impurity carbon and oxygen atoms [3, 4]. On the other hand, the origin of the 0.641 eV peak has not yet been reported. We speculated that the peak at 0.641 eV may be one of the composited defects of impurity carbon and oxygen atoms. The observed decrease in carrier lifetime after ITO deposition by the RPD technique and the deep-level emission observed by PL suggested that these peaks are recombination centers in the c-Si band gap and decrease the carrier lifetime.

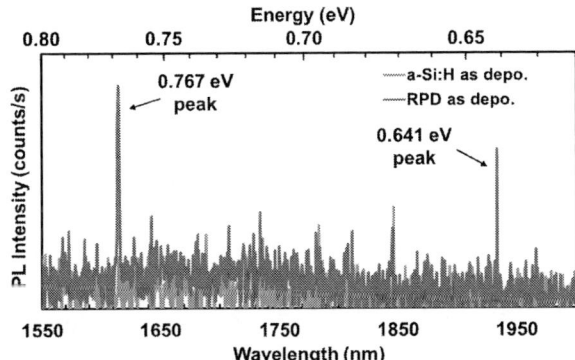

Fig. 2.   PL spectra before and after the RPD process.

Next, we evaluated the effect of post-annealing on irradiation-induced defects. Figure 3 shows the PL spectra of samples before and after annealing. The peaks at 0.767 eV and 0.641 eV disappeared after post-annealing. Combined with the carrier lifetime recovery after the post-annealing process in Fig. 1, the disappearance of these peaks was suggested to be one of the causes of the carrier lifetime recovery. The peak at 0.767 eV observed in the ITO sample deposited by direct sputtering on c-Si did not disappear after annealing at 200°C [4]. This may be due to the ion energy reduction by a-Si:H film, resulting in the formation of defects with a more unstable structure compared to those formed by direct deposition on c-Si.

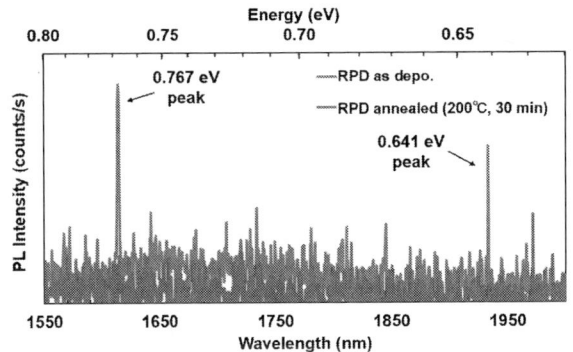

Fig. 3.   PL spectra before and after the post-annealing process.

Finally, we evaluated the defects induced by sputtering ITO deposition. Figure 4 shows the PL spectra of samples before and after ITO deposition by the sputtering technique. In the sample after ITO deposition, a deep-level peak was not observed. This may be due to the ion energy reduction by a-Si:H film or because the laboratory scale deposition equipment used in this study has lower ion energy than industrial deposition equipment used in the previous study [4], and therefore, no luminescence defects were generated. The observed decrease in carrier lifetime after ITO deposition by the sputtering technique and the no deep level emission suggested that the decrease in carrier lifetime by the sputtering technique can be attributed to non-luminescent defects such as point defects caused by inter lattice impurities, vacancies, or their complexes.

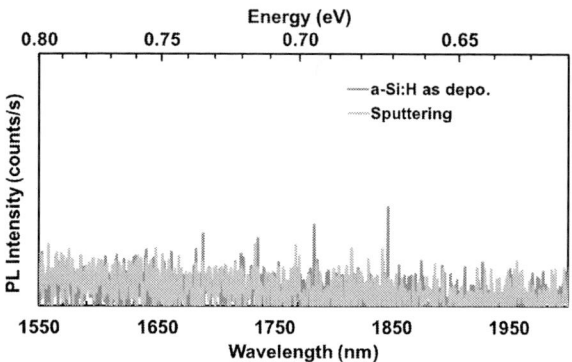

Fig. 4.   PL spectra before and after the sputtering process.

IV. CONCLUSION

We evaluated the damage to c-Si induced by the TCO deposition process in ITO/a-Si:H/c-Si structure. ITO was deposited by RPD and sputtering, respectively. Carrier lifetime of both samples decreases drastically after RPD and sputtering, and significantly recovered by post-annealing. We considered that the plasma process produced recombination active defects on the sample surface or in the c-Si substrate and the carrier lifetime was deteriorated. In addition, it suggested that these defects are relatively small scale.

Furthermore, PL spectroscopy was performed, and the so-called "irradiation-induced defects" were observed at deep levels of 0.767 eV and 0.614 eV in the samples deposited by the RPD technique. We considered that complex defects of impurity carbon and oxygen atoms were formed in c-Si by the RPD technique. These irradiation-induced defects act as recombination active centers and can cause the deterioration of the conversion efficiency in crystalline silicon solar cells. These defects disappeared after post-annealing. The disappearance of these defects by post-annealing is assumed to be one of the causes of lifetime recovery. In the sample after ITO deposition by sputtering technique, a deep level peak was not observed. We considered that the decrease in carrier

lifetime by the sputtering technique can be attributed to non-luminescent defects such as point defects caused by inter lattice impurities, vacancies, or their complexes.

ACKNOWLEDGMENTS

This study is based on the results obtained from a project, JPNP20015, subsidized by the New Energy and Industrial Technology Development Organization (NEDO).

REFERENCES

[1] M. Tanaka, M. Taguchi, T. Matsuyama, T. Sawada, S. Tsuda, S. Nakano, H. Hanafusa, and Y. Kumano, "Development of new a-Si/c-Si heterojunction solar cells: ACJ-HIT (artificially constructed junction-heterojunction with intrinsic thin-layer)," *Jpn. J. Appl. Phys.*, vol. 31, pp. 3518, 1992.

[2] S. De Wolf, A. Descoeudres, Z. C. Holman, and C. Ballif, "High-efficiency silicon heterojunction solar cells: A review," *Green*, vol. 2, pp. 7-24, 2012.

[3] K. Onishi, K. Kinoshita, T. Kojima, Y. Ohshita, and A. Ogura, "Lifetime degradation by oxygen precipitation combined with metal contamination in czochralski silicon for solar cells," *ECS J. Solid State Sci. Technol.*, vol. 8, pp. 4, 2019.

[4] H. Kanai, T. Nishihara, and A. Ogura, "Evaluation of process damage induced by sputtering of transparent conductive oxide films for crystalline silicon solar cells, " *ECS J. Solid State Sci. Technol.*, vol. 10, pp. 035002, 2021.

[5] B. Demaurex, S. D. Wolf, A. Descoeudres, Z. C. Holman, and C. Ballif, "Damage at hydrogenated amorphous/crystalline silicon interfaces by indium tin oxide overlayer sputtering," *Appl.Phys. Lett.*, vol. 101, pp. 171604, 2012.

[6] M. Huang, Z. Hameiri, A. G. Aberle, and T. Mueller, "Influence of discharge power and annealing temperature on the properties of indium tin oxide thin films prepared by pulsed-DC magnetron sputtering," *Vacuum*, vol. 119, pp. 68, 2015.

[7] K. Takeo, O. Keisuke, "Indium tin oxide sputtering damage to catalytic chemical vapor deposited amorphous silicon passivation films and its recovery," *Thin Solid Films*, vol. 635, pp. 73-77, 2017.

[8] A. Le, V. Dao, P. Pham, S. Kim, S. Dutta, C. Nguyen, Y. Lee, Y. Kim and J. Yi, "Damage to passivation contact in silicon heterojunction solar cells by ITO sputtering under various plasma excitation modes," *Solar Energy Materials and Solar Cells*, vol. 192, pp. 36-43, 2019.

[9] M. Tajima, Y. Ishikawa, H. Kiuchi, and A. Ogura, "Origin of room-temperature photoluminescence around C-line in electron-irradiated Si and its applicability for quantification of carbon," *J. Appl. Phys.*, vol. 11, pp. 041301, 2018.

[10] M. Tajima, P. Stallhofer, and D. Huber, "Deep level luminescence ralated to thermal donors in silicon," *Appl. Phys. Lett.*, vol. 22, pp. L586, 1983.

Brownfields to Brightfields: The potential for landfill solar redevelopment in New York State

Henry J. Williams, Hugh Peng, Olivia Goosay, K. Max Zhang

Sibley School of Mechanical and Aerospace Engineering, Cornell University, Ithaca, NY, 14853, USA

Abstract — **Large-scale solar energy development is best suited for flat, open areas, making agricultural land a prime target, particularly in regions dominated by agricultural activity such as New York State (NYS). In order to preserve prime farmland, it is imperative to develop effective policies to maximize the potential of deploying solar farms on marginal, less valuable lands, especially as a growing number of brownfields, and more specifically landfills, remain underutilized in NYS. The concept of repurposing contaminated land for solar energy, known as Brownfields to Brightfields (B2B), has shown promising results in case studies across the United States, but major barriers prevent widespread industry adoption and implementation, including scale limitations and infrastructure integration challenges. This study uses Geographic Information System (GIS)-based tools to demonstrate potential for inactive landfill solar redevelopment in NYS, incorporating important considerations for solar siting and evaluating four possible future infrastructure expansion scenarios. Results indicate that 55 – 67 % of inactive landfill area in NYS demonstrate medium to good suitability for solar development in the four scenarios analyzed, and landfill redevelopment can contribute up to 8.7 % of anticipated solar capacity in 2050 needed to reach statewide climate goals. These findings show the potential for inactive landfills to play a major role in decarbonizing the grid while preserving prime farmland in NYS.**

I. INTRODUCTION

As we accelerate the renewable energy transition, we must emphasize sustainable siting practices for utility-scale solar farms. New York State (NYS) presents a challenging landscape for solar energy, as agricultural land dominates area suitable for solar, and growing public opposition in rural areas creates conflict with developers [1]. Brownfields to Brightfields (B2B) provides a broadly supported strategy to accelerate community solar without jeopardizing local food production, all the while repurposing contaminated land across the state [2]. However, the upfront costs of brownfield redevelopment combined with relatively small land size and infrastructure regulations present significant challenges for developers compared to rural agricultural land. Furthermore, industry lacks sufficient assessment tools to identify appropriate brownfields and estimate risk involved in development.

Brownfields are properties containing hazardous contaminants which complicate redevelopment, including former gas stations, factories, dry cleaners, and waste management sites such as landfills [3]. With such a broad definition, redevelopment strategies must be considered site-by-site, and unknown costs might be incurred in decontamination and redevelopment. Rather than consider all types of brownfields, this study focuses on inactive landfills.

Guidelines exist to provide recommendations for best practices for landfill solar redevelopment, particularly for racking technologies, as traditional pile-driven racking should be replaced with ballasted systems or alternative flexible PV laminate material to avoid cap penetration [4], [5]. With these existing guidelines, developers have the information necessary to redevelop landfill sites. However, there lacks information on the extent to which statewide landfills are suitable for solar, particularly in the context of the NYS Climate Leadership and Community Protection Act (CLCPA), which aims to reduce greenhouse gas emissions by 40% in 2030 and 85% in 2050 from 1990 levels [6]. To achieve these goals, solar PV development is anticipated to reach 60-65 GW capacity by 2050 [6]. With 84% of land available for solar being agricultural, it is important to consider alternative siting locations which can contribute to solar development [7].

A number of studies utilize Geographic Information System (GIS)-based analyses to evaluate solar siting potential based on appropriate criteria including slope, distance to electrical infrastructure, distance to roads, and aspect [7]–[9]. This study aims to use GIS-based methods to evaluate statewide landfill solar redevelopment potential based on sites which demonstrate suitability for solar PV, considering four possible future scenarios of road and substation infrastructure expansion.

II. METHODS

A. Landfill identification

Converting landfills to solar PV sites requires proper capping of the landfill after its decommissioning, and developers should avoid sites capped for less than 2-3 years due to high settlement rate potential [5]. Thus, as a baseline, only inactive landfills should be considered for solar redevelopment. This study

TABLE I
SUITABILITY CRITERIA SCORING

Suitability	Slope (%)	Substation distance (m)	Road distance (m)	Aspect (degrees)
Good (3)	0-3	0-2575	0-15	135-225
Medium (2)	3-7	2575-4025	15-240	90-135, 225-270
Poor (1)	7-13.5	4025-6440	240-770	30-90, 270-330
Unsuitable (0)	>13.5	>6440	>770	0-30, 330-360

utilizes the Inactive Landfill Initiative database obtained from the NYS Department of Environmental Conservation, Division of Materials Management [10]. The dataset includes size and location, among other features, of inactive landfills in NYS as of June 2022. A geospatial dataset containing landfill location and size was created with known coordinates and an average landfill size of 18.4 acres from the original dataset of 2,114 inactive landfills. Thus, a total of 38,803 acres of landfills were used for the GIS analysis, described in the following sections.

B. Criteria selection

Feasibility criteria to determine redevelopment potential includes slope, distance to nearest substation, distance to nearest road, and aspect. These have been presented in literature as important considerations for solar PV siting, in addition to features such as solar radiation availability, wildlife designations, protected areas, and river proximity [11], [12]. This study uses a more focused set of criteria because landfills area assumed to be sited away from protected areas and water bodies, and solar radiation availability does not vary enough throughout NYS to make a significant difference in solar energy generation. Using existing literature specific to NYS, criteria classifications from least suitable (0) to most suitable (3) were determined from ranges shown in Table I [7].

C. Map generation and analysis

Using QGIS, the criteria were applied to NYS overall, then to a masked map including only landfill areas. Each pixel was assigned an equal-weighted sum of the four criteria, for a maximum total score of 12. This score was further reclassified based on the mapping techniques proposed by Katkar et al.: an

original pixel score of 10-12 is reclassified to 3 (good suitability), 8-9 is reclassified to 2 (medium suitability), 1-7 is reclassified to 1 (poor suitability), and 0 remains 0 (unsuitable).

Then, solar PV potential was characterized based on four possible scenarios of future infrastructure expansion. The first scenario assigns a score of 0 only if a pixel scores 0 in all criteria. This is the least restrictive case. The second scenario considers no future expansion of roadway infrastructure, assigning a score of 0 if a pixel scores 0 in the criteria for distance to nearest road. The third scenario considers no future expansion of electrical infrastructure, assigning a score of 0 if a pixel scores 0 in the criteria for distance to nearest substation. The fourth scenario is a combination of scenarios two and three, assigning a score of 0 if a pixel scores 0 in scenario two or scenario three. This is the most restrictive case. Scenarios 2-4 also retain a score of 0 for pixels which score 0 in all criteria.

III. RESULTS AND DISCUSSION

First, overall maps were generated for NYS. These can be used to identify any notable differences in overall solar PV suitability compared to landfill suitability. After generating overall maps for NYS, the data-layer was clipped to inactive landfill areas, and statistics were generated for landfill solar redevelopment suitability. Figure 1 shows the four suitability scenarios with inactive landfill locations plotted over overall maps of NYS. Visually, a large portion of landfills are located within the good (3) range of distance from substation, demonstrating a benefit of landfill redevelopment compared to prime agricultural land, which might be sited in lower-scoring ranges of distance to substation. Results for landfill suitability

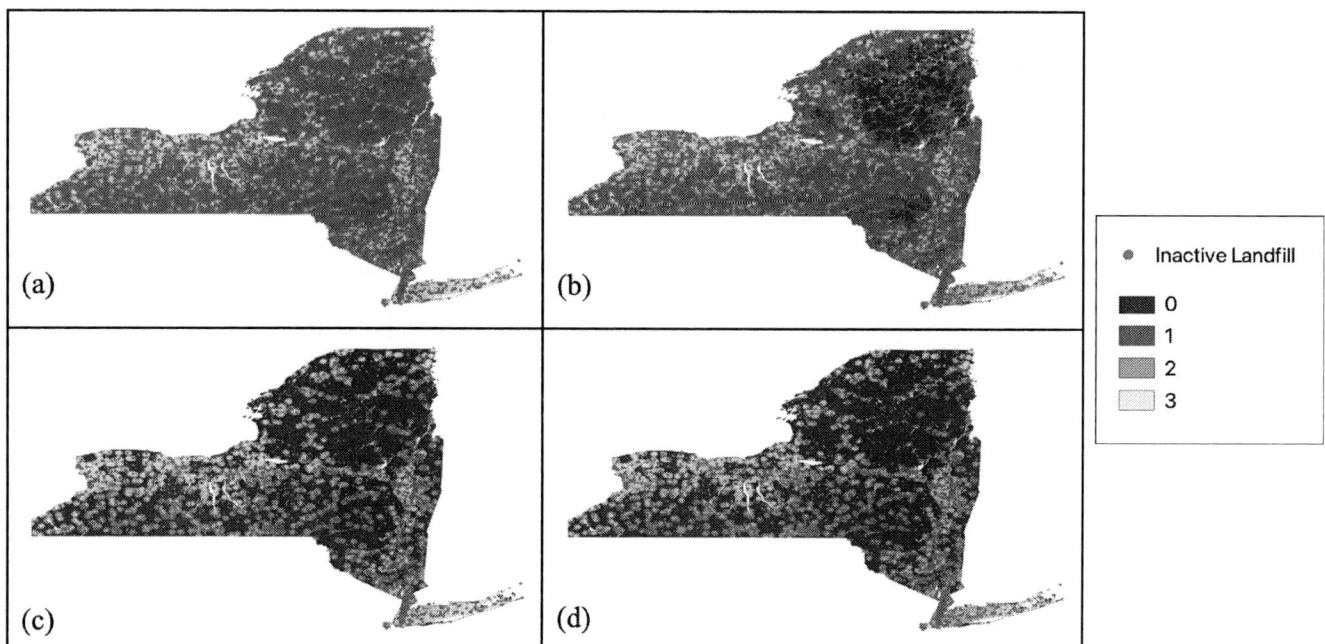

Fig. 1. Solar PV suitability from unsuitable (0) to good (3) for NYS considering inactive landfills in four scenarios: (a) base case, (b) no expansion of roadway infrastructure, (c) no expansion of electrical infrastructure, (d) combination of (b) and (c)

978-1-6654-6060-6/23 $31.00 © 2023 IEEE

in each scenario are presented in Table II. Considering good (3) and medium (2) suitability, 55 – 67 % of total inactive landfill area shows promise for solar redevelopment, from most to least restrictive scenario.

TABLE II
NYS INACTIVE LANDFILL SUITABILITY

Scenario	Good (3) (%)	Medium (2) (%)	Poor (1) (%)	Unsuitable (0) (%)
1	35.3	31.8	32.9	0.01
2	35.3	27.4	28.2	9.10
3	35.3	26.4	15.3	23.1
4	35.3	22.0	12.4	30.4

This study considers a total of 38,803 acres of inactive landfills in NYS. Based on 5 – 7 acres per MW solar capacity [13], the maximum solar PV generation on statewide inactive landfills is 5.5 – 7.8 GW, which would contribute 8.5 – 13 % of the 60 – 65 GW estimated NYS solar PV capacity by 2050 [6]. Table III shows solar PV redevelopment potential for only good (3) and medium (2) suitability landfills, a more realistic representation than all 38,803 acres being redeveloped for solar.

TABLE III
POTENTIAL CAPACITY AND CONTRIBUTION TO CLCPA GOALS

Scenario	Medium (2) and good (3) suitability area (acres)	Contribution to CLCPA 60 – 65 GW 2050 capacity estimate (%)
(a)	26,018	5.7 – 8.7
(b)	24,315	5.3 – 8.1
(c)	23,918	5.2 – 7.9
(d)	22,216	4.9 – 7.4

IV. CONCLUSION AND FUTURE WORK

This study demonstrates the potential for inactive landfill solar redevelopment in NYS based on suitability criteria for four different scenarios corresponding to future infrastructure expansion possibilities. Findings suggest that 55 – 67 % of total inactive landfills in NYS demonstrate medium to good suitability for solar redevelopment, depending on future infrastructure expansion. Based on these suitability criteria, redeveloping statewide landfills can contribute as much as 8.7 % to the anticipated solar capacity needed in 2050 to achieve CLCPA goals. With the rising land-use conflict between solar PV and agriculture, inactive landfills provide an opportunity to maximize the potential of contaminated land while preserving prime farmland and producing clean energy to decarbonize the grid.

Future work can incorporate more types of brownfields, and can guide developers to specific areas where multiple brownfields might be developed together for a single grid interconnection. As it becomes increasingly more time-sensitive to transition to renewable energy resources, these studies are urgently needed to support policy measures and siting decisions for solar PV.

REFERENCES

[1] J. A. Sward *et al.*, "Integrating social considerations in multicriteria decision analysis for utility-scale solar photovoltaic siting," *Appl Energy*, vol. 288, Apr. 2021, doi: 10.1016/j.apenergy.2021.116543.

[2] R. S. Nilson and R. C. Stedman, "Are big and small solar separate things?: The importance of scale in public support for solar energy development in upstate New York," *Energy Res Soc Sci*, vol. 86, Apr. 2022, doi: 10.1016/j.erss.2021.102449.

[3] T. L. Green, "Evaluating predictors for brownfield redevelopment," *Land use policy*, vol. 73, pp. 299–319, Apr. 2018, doi: 10.1016/j.landusepol.2018.01.008.

[4] G. Sampson, "Solar Power Installations on Closed Landfills: Technical and Regulatory Considerations," 2009. [Online]. Available: www.clu-in.org

[5] U. Epa, R.-P. America, and L. Initiative, "Best Practices For Siting Solar Photovoltaics On Municipal Solid Waste Landfills Best Practices For Siting Solar Photovoltaics On Msw Landfills," 2022.

[6] New York State Climate Action Council, "SCOPING PLAN Full Report Preferred Citation," 2022.

[7] V. V. Katkar, J. A. Sward, A. Worsley, and K. M. Zhang, "Strategic land use analysis for solar energy development in New York State," *Renew Energy*, vol. 173, pp. 861–875, Aug. 2021, doi: 10.1016/j.renene.2021.03.128.

[8] J. A. Prieto-Amparán, A. Pinedo-Alvarez, C. R. Morales-Nieto, M. C. Valles-Aragón, A. Álvarez-Holguín, and F. Villarreal-Guerrero, "A regional gis-assisted multi-criteria evaluation of site-suitability for the development of solar farms," *Land (Basel)*, vol. 10, no. 2, pp. 1–19, Feb. 2021, doi: 10.3390/land10020217.

[9] J. R. S. Doorga, S. D. D. V. Rughooputh, and R. Boojhawon, "Multi-criteria GIS-based modelling technique for identifying potential solar farm sites: A case study in Mauritius," *Renew Energy*, pp. 1201–1219, Apr. 2019, doi: 10.1016/j.renene.2018.08.105.

[10] A. Cuomo and B. Seggos, "New York State Inactive Landfill Initiative Comprehensive Plan To Address Priority Solid Waste Sites For Potential Impacts On Drinking Water Quality," 2021.

[11] S. K. Saraswat, A. K. Digalwar, S. S. Yadav, and G. Kumar, "MCDM and GIS based modelling technique for assessment of solar and wind farm locations in India," *Renew Energy*, vol. 169, pp. 865–884, May 2021, doi: 10.1016/j.renene.2021.01.056.

[12] J. Brewer, D. P. Ames, D. Solan, R. Lee, and J. Carlisle, "Using GIS analytics and social preference data to evaluate utility-scale solar power site suitability," *Renew Energy*, vol. 81, pp. 825–836, Sep. 2015, doi: 10.1016/j.renene.2015.04.017.

[13] P. Denholm and R. Margolis, "The Regional Per-Capita Solar Electric Footprint for the United States," 2007. [Online]. Available: http://www.osti.gov/bridge

Carbon Footprint of Silicon Photovoltaics Manufacturing in North America

Annick Anctil, Angela Farina, Luyao Yuan

Michigan State University, East Lansing, MI, United States

There is an unprecedented number of announcements of new PV manufacturing facilities in the US and North America due to tariffs on PV module imports and tax credits to support US manufacturing. This work compares the life-cycle carbon and cumulative energy demand associated with manufacturing modules in North America compared to China and other ASEAN countries (Viet Nam, Thailand, and Malaysia). Specific data on material quality and location for silica sand production is used. The LCA is conducted using country average and regional data for manufacturing in the US, Canada, and Mexico. Scenarios are built considering capacity in various regions and existing partnerships between suppliers and module manufacturers. The lowest possible carbon footprint of PV modules is calculated for PV module production in 2024 for the US and compared with China and ASEAN countries' average production.

The PV Efficiency vs R&D Effort Learning Curve for Research-Stage Material Technologies

Phillip J. Dale[a], Michael A. Scarpulla[b]

[a] *Department of Physics and Materials Science, University of Luxembourg, Belvaux, Luxembourg*
[b] *MSE and ECE Departments, University of Utah, Salt Lake City, UT, USA*

Abstract — The trajectory of PV technologies are frequently presented and discussed in terms of record efficiency versus date, or for commercialized technologies, also in terms of learning curves showing cost or price versus total number of modules. For early-stage PV material technologies, efficiencies are typically low and of course no modules have been produced. We introduce a different learning curve appropriate for assessing the trajectories of PV technologies still in the R & D stage; efficiency as function of total publications which aims to capure total R&D effort for each technology. The premise was that material technologies that were intrinsically better for PV should give more efficiency for lower total effort. Incredibly, the trajectories of many major single-junction cell technologies – Si, CIGSe, CdTe, and halide perovskites – all follow the same curve. The curve is logarithmic, capturing the common-sense idea that each limiting problem for a PV technology is harder than the last (diminishing returns). We present the trajectories of many common single-junction technologies and discuss insights gained from analyzing their trajectories through this different lens. We present this learning curve and give our own discussions and speculations on factors affecting trajectories and implications from the analysis. We hope to spark interesting discussions and further analyses by presenting the work.

Recently, we completed our analysis of the R&D trajectories of different PV material technologies on a plot of record efficiency versus logarithm of number of research publications [1]. This effort was premised on the desire to come up with a method for identifying and assessing promising new PV materials while they were still in their infancy, that is before many papers had been published and before real-world-useful efficiencies had been reached. We noticed that plots of efficiency vs time, while excellent for showing the grand sweep of PV technology development, do not capture the effort put into each technology or the size of the research community dedicated to each. Ideally, we would have data on total person-hours and total budget spent on each technology and the causal relationships of effort leading to both efficiency breakthroguhs and incremental increases. However, since this data is not available, we chose to use publications as a proxy for those variables. The result was Figure 1 reproduced herein. The concept was that an ideal PV material should yield higher effiency for less effort.

Intriguingly, we found that Si, CdTe, CIGSe, and the halide perovskites all seem to follow a common trajectory of 5% absolute efficiency per 10x more papers and reaching 20-24% within 10,000 or fewer papers. This trend holds over 3-4 orders

of magnitude of number of papers. We were surprised to find that the near-commercial halide perovskite technology, despite its metoric rise on the NREL chart of effiency vs time, followed the same learning curve – that is nearly the same amount of total research effort is correlated with efficiency as for other high-efficinecy technologies.

One of our unproven speculations is that the logarithmic dependence of efficiency vs papers ultimately stems from the physics of efficiency; that ultimately the technological trajectory is limited by increases in V_{oc}, which scales as logarithm of external radiative efficiency. That is, how closely the optical and electrical designs asymptotically approach perfection.

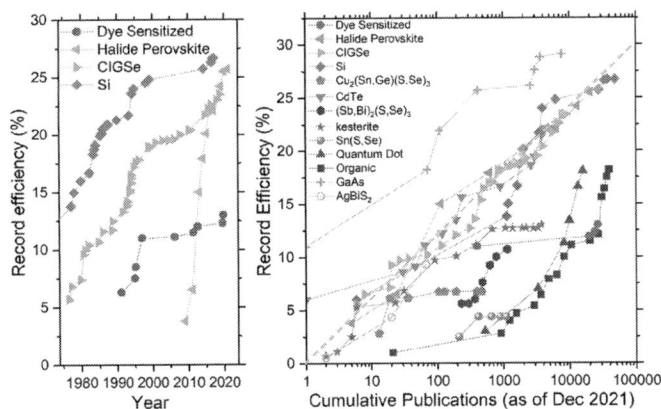

Figure 1 – (right) Efficiency vs. year for representative PV material technologies. (left) Efficiency vs number of publicaitons on log scale.

We point out some predictable factors that will not be captured by this analysis; hidden commercial effort, cross-pollination, and spillover. Record efficiencies may be set by academically-publishing research entities, or by non-publishing commercial entities (especially, but not always, as the technology becomes more mature and heads towards commercialization). Cross-pollination encompasses the transfer of concepts, techniques, and tools from one PV technology to another. For examples, CZTS benefitted heavily from the experience and infrastructure of CIGSe researchers and halide perovskites from those of organic and dye-sensitized cells. Spillover refers to tech transfer from non-PV applications

978-1-6654-6060-6/23 $31.00 © 2023 IEEE

into PV; Si and GaAs benefit from the enormous electronics and optoelectronics industries, while some material technologies like CIGSe are solely used in PV. We hope that ofurther discussions will illuminate and investigate other factors and add understanding of factors affecting these trajectories.

We point out that these technological trajectories are correlations and do not capture causation. From this coarse analysis, we can not resolve questions such as whether technological progress should be viewed as being driven by the entire publishing community, or by only a small group of leaders and seminal papers. We speculate that both the probabilities of breakthroughs and incremental progress scale with the number of people working on a material and the resources brought to bear.

Since we could show that the learning rate for commercial technologies at all stages of development increases logarithmically by 5% absolute per order of magnitude this allows a benchmarking of new materials to see how they compare. Our original analysis investigated the behaviour of 13 material systems and excluded several systems including antimony seleno-sulfoiodide, zinc phosphide, and non-halide containing perovskites. Additionally, we restricted ourselves to single junction technologies, however the rate of progess of tandem photovoltaic devices is of interest. Therefore we will analyse previous historic efficiency data to elucidate the learning curves of the previously unstudied materials as well as tandem devices.

REFERENCES

[1] P.J. Dale and M.A. Scarpulla, Efficiency versus effort: A better way to compare best photovoltaic research cell efficiencies?, Solar Energy Materials and Solar Cells, 251 112097 (2023)

Perovskite/silicon tandem solar cells with front side metallization applied prior to top cell fabrication enabling high curing temperatures

Sara Baumann[1,2], Annika Raugewitz[1], Felix Haase[1], Tobias Wietler[1,3], Robby Peibst[1,2], Marc Köntges[1]

[1]Institute for Solar Energy Research, Hamelin (ISFH), 31860 Emmerthal, Germany, [2]Institute for Electronic Materials and Devices, Leibniz University Hannover, 30167 Hannover, Germany, [3]Institute for Solid State Physics, Leibniz University Hannover, 30167 Hannover, Germany

Abstract — For the industrialization of large area perovskite silicon tandem solar cells a low-resistance and low-cost contact metallization needs to be established. On the one hand, curing or even firing temperatures as applied for screen-printing metallization of Si single junction cells are too high for perovskites. On the other hand, screen-printing pastes with very low curing temperatures have a reduced specific conductivity. Thus, they require wide Ag consuming contacts for a sufficiently low resistance. We present a concept which comprises screen-printing the front contacts onto an isolating layer located on the uppermost layer of the bottom cell before the perovskite top cell is applied. This allows higher curing temperatures and thus lower finger resistivities than in a low curing temperature screen-printing process on a perovskite solar cell. Furthermore, this concept avoids mechanical pressure on the soft perovskite material during cell interconnection. We expect that the latter aspect is beneficial to prevent shunting issues. The concept is shown for perovskite single junction solar cells in a proof-of-concept state. Furthermore, the advantage of higher curing temperatures regarding the Ag finger line resistance and the Ag/ITO contact resistivity is verified by an improvement due to additional sintering to $T \geq 330$ °C and $T \geq 260$ °C after curing, respectively.

I. INTRODUCTION

Perovskite silicon tandem solar cells are a highly promising candidate for the next generation of solar cells as their efficiency surpasses the limit of single junction (SJ) silicon solar cells. Perovskite silicon tandem solar cells have already reached 33.7% efficiency whereas the efficiency record for SJ silicon solar cells is 26.8% [1].

However, the published record efficiencies of perovskite/silicon tandem solar cells are shown on small cells with areas below 1.5 cm² (see [1]), mostly comprising a front side metallization evaporated through a shadow mask. For large cell areas, scalable and cost-efficient front side metallization methods such as screen-printing are needed. When screen-printed on the perovskite solar cell, the drying temperature of the paste is limited to about 140 °C by the thermal instability of perovskite materials [2]. Therefore, screen-printing pastes with a very low curing temperature and thus typically higher resistivity are required. This partially frustrates the benefit of the low current densities in tandem solar cells, i.e. a reduced Ag consumption.

In this work, the screen-printed front metallization can be cured at temperatures way above 200 °C as it is applied onto an isolation layer on the substrate before the perovskite top cell is processed. Thus, screen-printing pastes developed for silicon heterojunction (SHJ) solar cells can be used. Metallization fingers with higher specific conductivities than with low temperature screen-printing might be achieved due to the higher curing temperature [2]. In perspective, a dielectric and thus temperature stable isolation layer in combination with a firing-stable p+ poly-Si / n+ poly-Si tunneling junction [3] instead of a TCO-based recombination layer would even enable "PERC-like" high firing temperatures and thus Al-based front side fingers as in our POLO back junction cell [4].

The isolation between the front metallization and the uppermost layer of the bottom cell possibly has an additional advantage when connecting several cells to a module string. In previous experiments, it was challenging to avoid shunting by the cell interconnecting ribbons due to the mechanical pressure applied to the perovskite solar cell during this process. In our concept, no mechanical pressure is applied on the perovskite layer.

II. SOLAR CELL FABRICATION

The SJ perovskite solar cells are processed on a glass substrate with a 180 nm thin indium tin oxide (ITO) layer on top which has a sheet resistance of about 8 Ω/sq. On top of these 2.5 cm × 2.5 cm substrates solder resist is screen-printed as an isolation layer below the front metallization. The two-component solder resist is dried for 45 min at 130 °C on a hot plate in air. Afterwards the front contact is screen-printed on top of the isolation and the back contact directly onto the ITO within the same printing process. Fig. 1 shows the two front fingers which have a distance of 6.3 mm. The Ag fingers are 98.5 µm ± 8.5 µm wide whereas the solder resist fingers below are 524.3 µm ± 11.9 µm wide. The Ag paste is dried on a hot plate in air for 10 min at 260 °C, whereby the measured temperature on the glass surface is lower and the actual temperature at the Ag paste is assumed to be approximately 190 °C – 210 °C. The drying temperature is not limited by the perovskite but by the thermal stability of the solder resist and the ITO layer. The electrical isolation of the solder resist between the front and back contact is verified for each substrate before perovskite solar cell processing.

After the screen-printing the substrates are cleaned with isopropanol (IPA) and an UV-Ozon process for 15 min at 50 °C. Afterwards a self-assembling monolayer (SAM, MeO-

978-1-6654-6060-6/23 $31.00 © 2023 IEEE

2PACs) is spin-coated and annealed for 10 min at 100 °C to form the hole transport layer (HTL). Then, a perovskite layer (Cs$_{0.05}$(FA$_{0.77}$MA$_{0.23}$)$_{0.95}$Pb(I$_{0.77}$Br$_{0.23}$)$_3$, CsMAFAPbIBr) is spin-coated with 3500 rpm using acetyl acetate as a solvent and annealed for 22 min at 100 °C. This results in a perovskite layer thickness of about 800 nm.

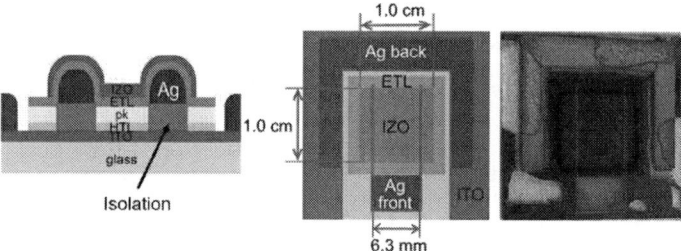

Fig. 1.   SJ perovskite solar cell structure with 0.9 cm² active area given by the IZO on top of ETL (C$_{60}$), perovskite (CsMAFAPbIBr), HTL (MeO-2PACs) and ITO. The HTL is contacted by ITO and the Ag back contact whereas the ETL is contacted by IZO and the Ag front fingers.

As Fig. 2 shows the perovskite material covers only the edges of the solder resist, but does not reach the Ag fingers.

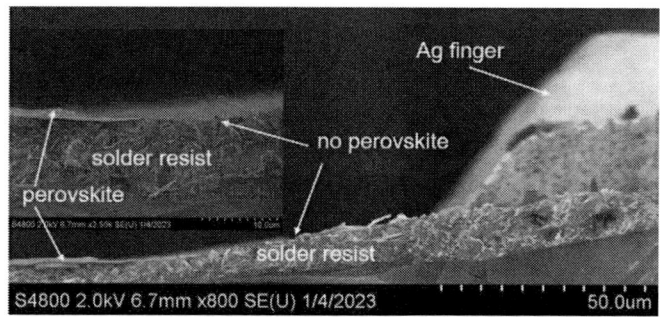

Fig. 2.   Cross sectional view of a scanning electron microscopy image of the solder resist partly covered with spin-coated perovskite material whereas the Ag finger is not.

The electron transport layer (ETL, 23 nm C$_{60}$) is evaporated in an area of 1.3 cm x 1.3 cm (see Fig. 1). Then the perovskite is scraped off along the Ag back contact.

Afterwards 100 nm indium zinc oxide (IZO) with a sheet resistance of 40.1 Ω/sq. are sputtered with a mask on top of the ETL. The IZO area of 1.0 cm $\times$ 1.0 cm defines the perovskite solar cell area. As the solder resist fingers in this first proof-of-concept configuration shade 0.1 cm² of this area, the active solar cell area is 0.9 cm².

The transfer of the samples between the C$_{60}$ and the IZO process in air takes 15 min - 30 min. The IZO layer contacts the Ag front metallization which is isolated from the ITO and the back metallization by the isolation (solder resist) as shown in Fig. 1.

III. *IV* CHARACTERIZATION

An example light *IV* curve, SunsV$_{oc}$ measurement and dark *IV* measurement of a SJ perovskite solar cell measured after 5 min *MPP* tracking is shown in Fig. 3. Thereby the solar cell is illuminated through the IZO and the voltage is varied from - 0.1 V to + 1.2 V with a minimum speed of 150 mV/s. The solar cell has an efficiency of 8.4%, an open circuit voltage V_{oc} of 981 mV and a short circuit current density J_{sc} of 17.2 mA/cm². The fill factor FF is low with 50% which can be explained by a high series resistance R_s of 15 $\Omega \cdot$ cm² that is extracted from linear fits to the dark *IV* curve. The shunt resistance R_{SH} amounts to 12 k$\Omega \cdot$ cm² and the pseudo fill factor pFF is 78%.

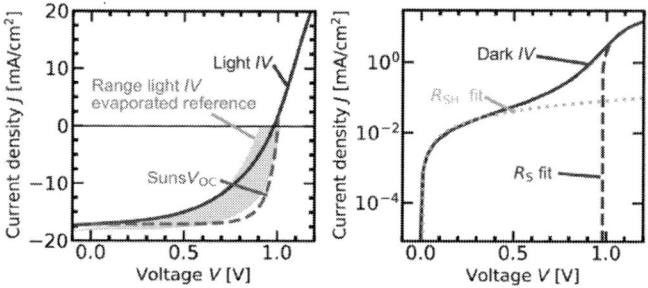

Fig. 3.   Light *IV*, SunsV$_{oc}$, dark *IV* and linear fits at small voltages for the evaluation of R_{SH} and at large voltages for the evaluation of R_S of a SJ perovskite solar cell.

Similar processed solar cells with evaporated contacts, in particular a full-area contact on top of the IZO, with a cell area of only 0.1 cm² show comparable J_{sc} (14 mA/cm² - 18 mA/cm²) and V_{oc} (900 mA $-$ 1010 mV) values and also high series resistances (see grey marked range in Fig. 3 with fill factor FF varying from 46% to 65% for the best solar cell of the batch).

IV. *REDUCTION OF RESISTIVE LOSSES BY MORE EFFICIENT CURING*

The overall resistive loss has several contributions such as the contact resistance between the rear as well as the front transparent conducting oxide (TCO) and the Ag fingers and the finger line resistance. The rear contact (here Ag/ITO) and the Ag finger itself can be cured at higher temperatures in the concept shown here than with usual screen-printing on top of a perovskite solar cell. Therefore, these two resistive contributions are analyzed in the following.

A. *Ag finger line resistance*

For the measurement of the finger line resistance fingers with a distance of 1.6 mm are printed onto glass. For this, the same Ag screen-printing paste as for the perovskite solar cell shown above but a different screen is used to be able to measure the resistance over a finger length of 3 cm with the transfer length method (TLM). The average finger characteristic is determined at 6 fingers. The fingers have an average width of

$(125.7 \pm 2.9)\ \mu m$ and an average height of $(18.3 \pm 1.2)\ \mu m$ which results in an average finger cross section A of $(2300.5 \pm 128.0)\ \mu m^2$.

The whole glass with the Ag fingers on it is cured for 10 min at 220 °C on a hot plate in air. The temperature at the Ag fingers is thereby estimated to be approximately 180 °C.

Afterwards, the glass is divided into strips that contain 6-7 fingers each. Fig. 4 shows the median line resistance R_L and the standard deviation as error bars for 6-7 fingers on these strips before (i.e., after the above-mentioned curing for 10 min at 180 °C) and after additional sintering. Then the strips are individually sintered to various higher temperatures. Therefore, the stripes are put onto the hot plate at about 220 °C and left there until the desired temperature has been kept for 1 min. As the hot plate used for this purpose has a cover, the set and actual temperature T are assumed to be approximately equal.

Fig. 4 compares the median line resistance R_L measured before and after the additional sintering.

The average Ag bulk resistance ρ obtained as the line resistance R_L times the average finger cross section A before sintering is $(6.8 \pm 0.5) \cdot 10^{-6}\ \Omega \cdot cm^2$. Kamino et al. determined $\rho \approx 6.5 \cdot 10^{-6}\ \Omega \cdot cm^2$ for their screen-printing paste after curing at 210 °C [2].

Sintering at $T \geq 330$ °C significantly reduces the line resistance down to $(4.0 \pm 0.2) \cdot 10^{-6}\ \Omega \cdot cm^2$ for sintering at 400 °C. The higher the sintering temperature in the range of 330 °C – 400 °C, the lower the line resistance.

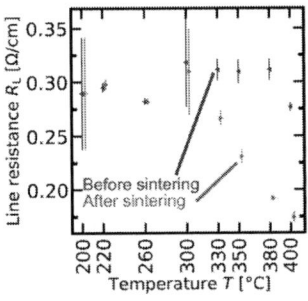

Fig. 4.   Median line resistance after curing for 10 min at 180 °C (blue symbols "before additional sintering") and after additional sintering for 1 min at different temperatures (green symbols). A reduction of Ag finger line resistance upon additional sintering is observed for $T \geq 330$ °C.

B. Ag/ITO contact resistivity

The Ag/ITO contact resistivity is relevant for the back contact of the above shown perovskite SJ solar cells. For its determination the same Ag screen-printing paste as used for the solar cells and the finger line resistance measurement is printed onto wafers with two different ITO layers on top that vary in thickness and charge carrier density with the screen used for the finger line resistance measurement. None of these ITO layers equals the one in the above shown perovskite solar cell

structure. The wafers are cured for 10 min on a hot plate at 220 °C (approximately 200 °C at the Ag fingers) and then cut into 1 cm wide strips perpendicular to the fingers.

By measuring the resistance between fingers in different distances and the finger width the contact resistivity is determined for about 80 fingers per stripe following the transfer length method (TLM).

The strips are additionally sintered for 1 min at temperatures between 200 °C and 400 °C, whereby the actual temperature at the Ag finger is assumed to equal the set temperature of the hot plate as it has a cover. Fig. 5 compares the median Ag/ITO contact resistivity before and after sintering on two different ITO layers. The error bars are given by the standard deviation of all fingers on the corresponding stripe. Although the ITO layer has a major influence on contact resistivity, in both cases the contact resistivity is significantly reduced by sintering at $T \geq 260$ °C to $(41.1 \pm 8.2)\ m\Omega \cdot cm^2$ for ITO 1 and to $(7.3 \pm 1.1)\ m\Omega \cdot cm^2$ for ITO 2, respectively. We therefore assume that sintering improves the Ag/ITO contact resistance in general. In contrast to the line resistance, higher sintering temperatures do not imply a further reduction of the contact resistivity.

Kamino et al. determined an Ag/ITO contact resistivity of $10\ m\Omega \cdot cm^2$ after curing at 210 °C [2] which is comparable to our measurements on ITO 2.

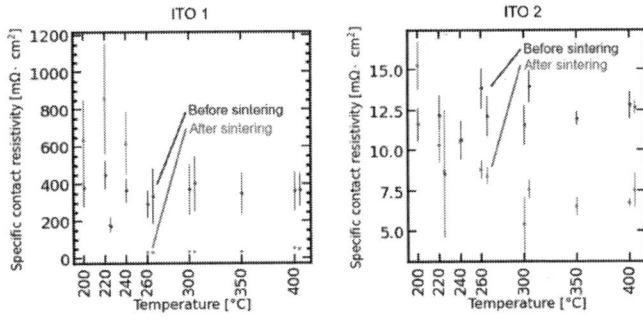

Fig. 5.   Median Ag/ITO contact resistivity on two different ITO layers after curing at 200 °C (blue symbols "before additional sintering") and after additional sintering for 1 min at different temperatures (green symbols). A reduction of Ag/ITO contact resistivity upon sintering is observed for $T \geq 260$ °C on both ITO layers.

V. Discussion

We observe high series resistances for our perovskite solar cells with both contacting methods. One possible reason for this might be a high contact resistance between the Ag contacts and ITO as well as IZO.

In analogy to our measurements, Kamino et al. describe a high series resistance due to high contact resistances between ITO and screen-printed Ag structures, which are higher the lower the curing temperature is [2]. As the curing temperature for the above shown SJ perovskite solar cell (approximately

200 °C) is below the regime in which the Ag/ITO contact resistivity is reduced, further sintering might yield an improvement. However, the Ag/ITO contact resistivity shown in Fig. 5 was determined for different ITO layers than the one used in the perovskite solar cell structure. For the quantification of the exact resistive losses, contact resistivity measurement will be conducted with the same ITO as used for the solar cells.

The IZO/Ag contact formation is performed after the fabrication of the perovskite solar cell. Therefore, it cannot be annealed to as high temperatures as the Ag/ITO contact and thus might be high. This would be relevant for the contact mechanism shown here, and could also apply to evaporated Ag contacts on top of IZO. However, careful annealing below 140°C or local annealing at the contact with a laser could improve the IZO/Ag contact resistivity without harming the perovskite solar cell.

The dark discoloration of the Ag contacts that occurs during the UV-Ozon cleaning process (possibly AgO_2 formation; see Fig. 1) might influence the contact resistance. Therefore, another possible method to optimize the contact resistance between IZO and Ag could be to remove this layer before the IZO process by laser ablation.

Furthermore, the Ag line resistance is directly proportional to the finger series resistance. Therefore, additional sintering of the metallization to 400 °C or even higher could significantly influence the series resistance of the perovskite solar cells.

VI. CONCLUSION

We demonstrated this new contact scheme for perovskite SJ solar cells at a proof-of-concept level. We are convinced that we will achieve higher efficiencies with further improvement of the in-house perovskite solar cell processing. Our concept is fully compatible with perovskite SJ solar cell as well as perovskite silicon tandem cell processing.

The new contacting scheme with a screen-printed front metallization applied before perovskite cell processing results in solar cell parameters comparable to conventional processing with front contacts evaporated on top of the perovskite cell. This shows, that it is suitable for SJ perovskite solar cells and will also be applicable for perovskite silicon tandem solar cells.

Using different isolation layers would allow even higher curing temperatures and might enable Al-based front side fingers as in our POLO back junction cell [4], which could significantly reduce metallization costs. To this end, Al or Cu screen-printing pastes and their contact resistance to ITO and IZO layers should also be analyzed. For contacting the cells with busbars in a solar module the top TCO may be ablated by laser scribing without much effort.

ACKNOWLEDGEMENT

The authors thank the German Federal Environmental Foundation (DBU) and the state of Lower Saxony for their funding. The solar cell processing was carried out within the projects P3T (grant number 03EE1017B), TOP (03EE1080C) and APERO (03EE1113B) supported by the Federal Ministry for Economic Affairs and Climate Action (BMWK). The authors thank L. Brockmann, J. Strey, M. C. Turcu, R. Clausing and R. Winter (all from ISFH) for perovskite solar cell processing.

REFERENCES

[1] NREL: Interactive best research-cell efficiency chart, https://www.nrel.gov/pv/interactive-cell-efficiency.html, last seen 06/25/23

[2] B. A. Kamino, B. Pviet-Salomon, S.-J. Moon, N. Badel, J. Levrat, G. Christmann, A. Walter, A. Faes, L. Ding, J. J. Diaz Leon, A. Paracchino, M. Despeisse, C. Ballif and S. Nicolay: "Low-temperature screen-printed metallization for the scale-up of two-terminal perovskite-silicon tandems", *ACS Appl. Energy Mater.* **2**, pp. 3815-3821 (2019)

[3] R. Peibst, M. Rienäcker, B. Min, C. Klamt, R. Niepelt, T. F. Wietler, T. Dullweber, E. Sauter, J. Hübner, M. Oestreich, R. Brendel: "From PERC to Tandem: POLO- and p+/n+ Poly-Si Tunneling Junction as Interface Between Bottom and Top Cell", *IEEE Journal of Photovoltaics* **9**(1), pp. 49-54, (2019)

[4] B. Min, L. Nasebandt, C. Hollemann, D. Bredemeier, L. Thiemann, T. Brendemühl, K. Bothe, R. Peibst and R. Brendel: "23.1%-efficient POLO back junction solar cells", *Proceedings of the 8th World Conference on Photovoltaic Energie Conversion*, pp. 107-109, Milano, Italy (2022)

978-1-6654-6060-6/23 $31.00 © 2023 IEEE

Performance and degradation in silicon PV systems under outdoor conditions in relation to reliability aspects of silicon PV modules – Summary of results of COST Action PEARL PV

S. Lindig[1], J. Ascencio-Vásquez[2], J. Leloux[3], D. Moser[1], M. Aghaei[4], A. Fairbrother[5], A. Gok[6], S. Ahmad[7], S. Kazim[7,8], K. Lobato[9], W.J.G.H.M. van Sark[10], N. Pearsall[11], B.G. Burduhos[12], A. Raghoebarsing[13], G. Oreski[14], J. Schmitz[13], M. Theelen[15], P. Yilmaz[13,15], J. Kettle[16] and A.H.M.E. Reinders [13,17]

1) EURAC, Bolzano, Bozen – Südtirol, 39100, Italy
2) ASVA Consulting, Santa Cruz de Tenerife, Spain
3) Lucisun, Villers-la-Ville, Walloon Brabant, 1495, Belgium
4) Norwegian University of Science and Technology, Alesund, 6009,Norway
5) École Polytechnique Fédérale de Lausanne (EPFL), Lausanne, Vaud, 1015, Switzerland
6) Gebze Technical University, Gebze, Kocaeli, 41400, Turkey
7) BCMaterials, UPV/EHU Science Park, Bizqaia, Leioa, 48940, Spain
8) IKERBASQUE, Basque Foundation for Science, Bizqaia, Bilbao, 48005, Spain
9) Instituto Dom Luiz, University of Lisbon, Lisbon, Estremadura, 1749-016, Portugal
10) Utrecht University, Utrecht, Utrecht, 3584 CS, The Netherlands
11) Northumbria University, Newcastle upon Tyne, Tyne and Wear, NE1 8ST, United Kingdom
12) Transilvania University of Brasov, Brasov, Transylvania, 500036, Romania
13) University of Twente, Enschede, Overijssel, 7500 AE, The Netherlands
14) Polymer Competence Center Leoben GmbH, Leoben, Styria, 8700, Austria15) TNO Partner in Solliance, Eindhoven, Brabant, 5656 AE, The Netherlands
16) James Watt School of Engineering, University of Glasgow, Glasgow, Scotland, G12 8QQ, United Kingdom
17) Eindhoven University of Technology, Eindhoven, Brabant, 5600 MB, The Netherlands

Abstract — **This paper presents the main results of COST Action PEARL PV, aiming at finding connections between the observed performance of monitored PV systems and degradation causes and failure modes according to literature with a focus on the most dominant technology among installed PV modules, namely silicon PV. It is found that there exists a great potential for performance improvements, though in practice it is difficult to identify exact causes for failure and underperformance.**

I. INTRODUCTION

Understanding of degradation and failure of silicon photovoltaic (PV) modules that have been installed in the field, is one of the key factors to enhance the lifetime of PV systems and, hence, reduce the cost of PV electricity produced. Hereby the basic assumption is that better insights about energy performance, degradation and failure modes can lead to mitigation of causes of less than expected operational lifetime of PV systems. Hence, improving the energy performance and reliability of PV systems has been the focal point of research in COST Action PEARL PV [1] (2017-2022). This project, funded by the COST programme of the European Commission, has been executed by hundreds of PV system experts from 40 countries, which has resulted in a database with monitoring data

of thousands of PV systems, see Figure 1, as well as numerous publications among which 40 in journals [2].

COST Action PEARL PV is one of the few academic networks in the rapidly growing field of PV system research. This research is essential because in the past decade the number of installed PV systems has been steadily growing at 20 to 40% per year, resulting in 1 terawattpeak (TWp) of cumulative installed power worldwide in 2022, contributing about 5% to the global electricity demand.

The PEARL PV network is strongly focused on assessing the actual performance and durability of already installed PV

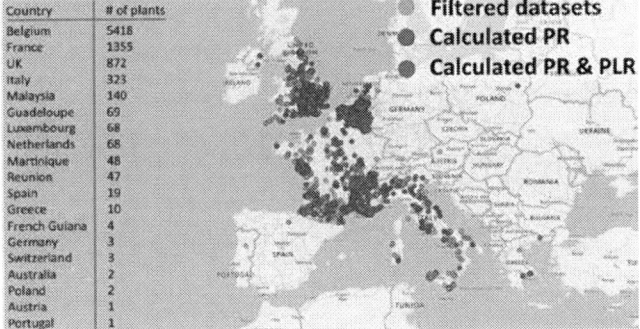

Fig. 1.   Distribution of PV systems of the PEARL PV database.

systems in the short and long run. This task is usually not taken care of by installers, utilities and/or DSOs. This paper will summarize some of the most important results of this COST Action, aiming at finding connections between the observed performance of monitored PV systems and degradation mechanisms and failure modes according to literature. This study will focus on the most prevailing technology among installed PV modules, namely silicon PV. Next, in Section II the research set-up will be described, followed by a summary of results in Section III, and a discussion, conclusions and recommendations for future research in Section IV.

II. RESEARCH SET-UP

The research has a two-fold approach; namely, on the one hand, quantification of the energy performance of a large fleet of PV systems by analytical monitoring and analysis [3], and on the other hand, identification of possible degradation mechanisms and failure modes on the basis of extensive literature studies [4, 5]. These two approaches are connected and discussed, leading to a set of recommendations.

A. Performance analysis of monitored PV systems

To capture degradation in the field, an performance analysis has been executed of monitoring data of over 8400 PV systems [3], see Figure 1. Data have been analyzed regarding the actual monitored long-term performance to quantitatively determine the absolute influences of geographic location, irradiation and ambient temperatures, key system design features such as system components' rated performance and installation types, failure modes, operation and maintenance practices, and performance degradation over service time of these PV systems. The monitoring data have been stored in the PEARL PV database as time series with 10-minutes resolution ranging from one to four years from 2010 to 2016. These time series are covering the following metadata: installed capacity, latitude, longitude, azimuth, and tilt angle of the PV modules. ERA5 reanalysis data are used as irradiance data.

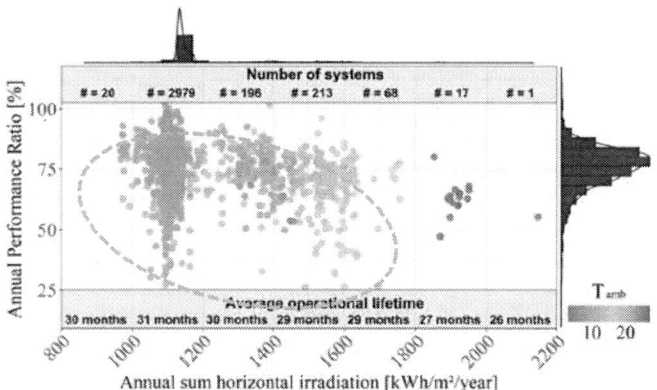

Fig. 2. Annual sum of horizontal irradiation [kWh/m 2/year] versus annual PR [%] categorized by average ambient temperature [∘C] for PV systems in the PEARL PV database as shown in Figure 1.

B. Degradation and failure phenomena

To reduce degradation rates, it is imperative to have insights into degradation and failure phenomena. Hence, an overview is given of the state-of-the-art knowledge on the reliability of silicon PV modules [4, 5] on the basis of a review of about 250 publications. This overview consists of two parts: first, a brief contextual summary about reliability metrics and how reliability is measured, and . second, a summary of the main stress factors and how they influence module degradation.

III. RESULTS

A. Performance analysis of monitored PV systems

The PV plants analyzed are small residential systems with silicon PV modules with a median installed capacity of 6 kWp, primarily installed in Europe, with an average field age of 30.5 months. From the analysis, it is found that the annual mean performance ratio (PR) across all systems was 76.7 %; with an average yield of 954.9 kWh/kWp per year [1] as seen in Fig 2. It can be concluded that the PR is far below the expected range of 80 to 90 % due to outage, failure, shading, and climate conditions, in particular high irradiation (>1400 kWh/m^2/year) and ambient temperatures (> 14 $^\circ$C). Performance loss rates (PLR) were also determined on the basis of an analysis of monitoring data. Average performance losses between 0.74 %/year and 0.86 %/year were calculated depending on the method used

B. Degradation and failure phenomena

An extensive literature review [4,5] has resulted in a summary of the main stress factors defined as the causes of PV modules degradation. Given the findings shown in Section III.A, it is very important to understand how stress factors influence PV module degradation. The review of degradation and failure modes has been focused on the level of individual PV modules' components.

In general, stress factors can be categorized into external and internal stress factors, where external stress factors are related to environmental conditions, such as irradiance, temperature, moisture, mechanical load, soiling and chemicals, while internal stress factors are caused by the PV module architecture and the bill of materials (BOM) of PV modules (including BOM incompatibility) and processing related

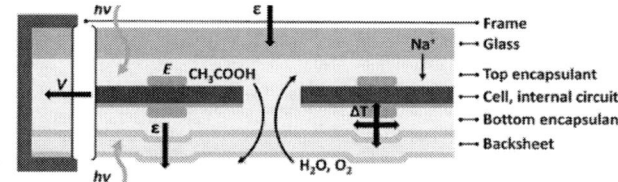

Fig. 3. Some common PV module stressors for a silicon wafer-based PV module, including light (hv), strain (ε), voltage bias (V), chemical diffusion, ingress and egress (CH3COOH, H2O, O2, Na+), electric field (E), and thermomechanical strain (ΔT).

effects. Usually, these stress factors occur simultaneously as seen in Fig. 3. Next, from the existing literature, degradation and failure modes could be identified that generally occur in PV technologies, as shown in Table 1, enabling to draw relationships between stressor, PV module components, failures and effects as shown in Figure 4.

TABLE I
COMMON DEGRADATION AND FAILURE MODES OF PV MODULE COMPONENTS AND THEIR POTENTIAL EFFECTS.

Component	Degradation	Failure Modes	Effects
Frame	Corrosion	Warpage	Increased risk of module damage
Glass	Glass corrosion	Breakage, soiling, abrasion	Reduced current, hotspot formation
Encapsu-lant	Photo-oxidation	Discoloration delamination	Reduced current, increased corrosion
Internal circuit & TCO	Corrosion	Fatigue, cracks	Reduced current, cell isolation, hotspot formation
Solar cells	PID, LID, LETID	Cracks, cell isolation (cracks)	Reduced power, hotspot formation
Back-sheet	Photo-oxidation, hydrolysis	Discoloration, dela-mination, cracks	Increased corrosion, isolation failure
Junction box	---	Arcs delamination	Electrical fault, detachment

IV, DISCUSSION, CONCLUSIONS AND RECOMMENDATIONS

Finally, on the basis of Fig. 4, we dare to draw the conclusion that a mixture of external and internal stress factors makes it difficult to uniquely identify the causes for failure and the effects at the level of PV module output as well as to represent each failure in relation to effects and vice versa. Evenly important, also inverters, electronics and other BOS can show failures, and hence affect the system performance. In practice, for small residential PV systems, average performance losses between 0,74%/year and 0,86%/year have been measured; however, most PV systems don't achieve the expected performance ratio (PR) of 80 to 90%, instead an average PR of 77% is achieved here. Hence, there exists a great potential for performance improvements. Though it is difficult to identify exact causes for failure and underperformance, the present terawatt PV market demands for better insights into long term performance and degradation mechanisms of silicon PV systems. This can be realized by better interaction and data sharing between upstream and downstream. Therefore, it is also advised that more research efforts should be directed towards statistically quantifying and mitigating degradation mechanisms and failure modes of PV systems.

ACKNOWLEDGEMENTS: All volunteers are respectfully acknowledged and thanked for their support to COST Action PEARL PV (CA16235). COST (European Cooperation in Science and Technology) is a funding agency for research and innovation networks. COST Actions help connect research initiatives across Europe and enable scientists to grow their ideas by sharing them with their peers. This boosts their research, career and innovation: www.cost.eu

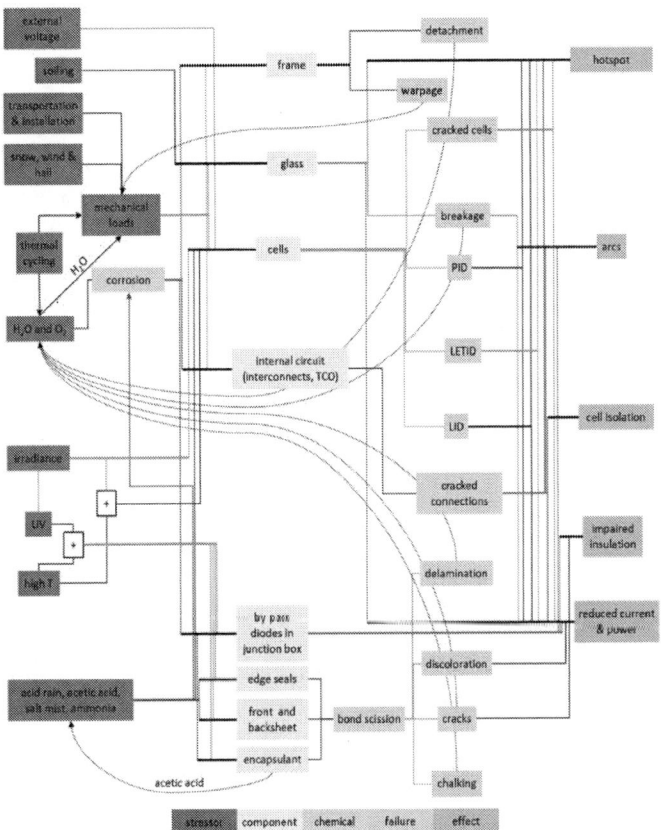

Fig. 4. Flow diagram representing the relationships between stressor, component, failure and effect.

REFERENCES

1. *COST Action PEARL PV*. 2022 09-01-2023]; COST Action PEARL PV "Performance and Reliability of Photovoltaic Systems: Evaluations of Large-Scale Monitoring Data"]. Available from: https://www.pearlpv-cost.eu/.

2. *All publications by COST Action PEARL PV*. 2022 09-01-2023]; Available from: https://www.pearlpv-cost.eu/dissemination/publications/.

3. Lindig, S., et al., *Performance Analysis and Degradation of a Large Fleet of PV Systems*. IEEE Journal of Photovoltaics, 2021. **11**(5): p. 1312-1318.

4. Aghaei, M., et al., *Review of degradation and failure phenomena in photovoltaic modules*. Renewable and Sustainable Energy Reviews, 2022. **159**: p. 112160.

5. Kettle, J., et al., *Review of technology specific degradation in crystalline silicon, cadmium telluride, copper indium gallium selenide, dye sensitised, organic and perovskite solar cells in photovoltaic modules: Understanding how reliability improvements in mature technologies can enhance emerging technologies*. Progress in Photovoltaics: Research and Applications, 2022. **30**(12): p. 1365-1392.

978-1-6654-6060-6/23 $31.00 © 2023 IEEE

Progress towards Scaling Perovskite/Silicon Tandem Modules

Chris Eberspacher, Colin Bailie, Tim Gehan, Bryan Rosales, Tom Brenner, Mike Chen and Terry Banks

Tandem PV, San Jose, CA 95131, USA

Abstract — **Significant progress is being made in scaling metal halide perovskite solar module technology for fabricating durable efficient perovskite submodules suitable for high-efficiency mechanically-stacked perovskite / silicon tandem modules. Semi-transparent 100 cm² submodule efficiencies of 16% have been achieved using scalable non-vacuum processing of perovskite films; refinements in raw materials selection, in superstrate preparation, and in coating and patterning process control have tightened device efficiency distribution. Low-loss monolithic integration has been demonstrated using industrial patterning and contacting tools and techniques; geometric fill factors approaching 95% have been achieved. Device designs and materials selection have increased cell stability and module durability; accelerated tests are underway.**

I. INTRODUCTION

Metal halide perovskite materials are a promising alternative and/or addition to photovoltaic (PV) absorber materials that at present dominate PV power module sales, e.g. crystalline silicon, thin-film CdTe and copper indium gallium selenide (CIGS), etc. As an alternative absorber material, perovskites can be used in single-junction devices, replacing for example CdTe in a monolithic superstrate submodules or CIGS in flexible metal foil substrate cells. As a high-value addition, perovskites can be used in multi-junction devices, as for example a wider-bandgap top cell in a perovskite/silicon or perovskite/CIGS tandem structure. Perovskite-based multi-junction PV can be fabricated in a variety of configurations including perovskite-direct-on-silicon monolithic tandem cells and mechanically-stacked tandem modules. This work focuses on the latter configuration, namely mechanically-stacked perovskite/silicon tandem modules in which a semitransparent superstrate perovskite submodule takes the place of the glass cover sheet of a standard Si-wafer power module to make a high-efficiency tandem module with the form, fit and function of existing high-volume single-junction Si modules.

Whether as an alternative or an addition, perovskites offer up advantages and challenges relative to traditional PV materials. Advantages include small-area champion cell efficiencies above 25% for single-junction cells and above 30% for perovskite/silicon monolithic tandem cells, relatively abundant and cheap raw materials, and functionally simple solution deposition and processing of perovskite thin films. Challenges include susceptibility of some metal halide perovskite materials to thermal and/or water decomposition, and the complexities of translating small-area film processing – e.g. timing-critical spin-coating and anti-solvent rinses – to uniform repeatable processing on commercial power module dimensions. This

work focuses on scaling monolithically-intergrated semi-transparent perovskite superstrate submodules using materials and techniques workable on multi-meter product dimensions, and on selecting submodule design, fabrication sequence and materials of construction to maximize cell stability and in turn module durability.

II. EXPERIMENTAL RESULTS

The design and fabrication of p-i-n superstrate perovskite submodules are summarized as follows. Transparent conductor coated glass is procured from various sources, e.g. 10-15 ohm/sq ITO on 13x13 cm, 1-2 mm thick soda lime glass, and cleaned using industry-standard detergent and water techniques. Thin layers of hole-selective material are deposited by scalable solution processes, e.g. slot die printing. Perovskite films are similarly deposited by preparing proprietary reactant inks and using scalable techniques to deposit, dry and anneal reactant ink layers into high-quality 500-750 nm thick multi-cation, mixed-halide perovskite films. Layers of electron-selective materials and a backside transparent conductor are deposited by physical vapor deposition. Submodule monolithic integration into twenty 0.5x10 cm cells is done by standard three-pattern techniques, e.g. laser patterning of the frontside conductor, of the carrier selective and optical absorber films, and of the backside transparent conductor; geometric fill factors of 90-95% have been achieved. Superstrate submodules are assembled into glass/glass perovskite modules using perimeter polyisobutylene edge seal with nitrogen in the void between the submodule backside surface and the backside cover glass. Figures 1 and 2 show a submodule prior to edge deletion, electrode ribbon attachment and module assembly and a finished perovskite module, respectively.

Fig. 1. Photo of 13x13cm, 20-cell, 100 cm² aperture-area submodule.

Fig. 2. Photo of 13x13cm, 20-cell, 100 cm^2 aperture-area module.

The one-sun aperture-area efficiency of as-made submodules varied widely in early exploratory batches; efficiency losses were dominated by superstrate effects – e.g. inadequate cleaning and surface preparation, perovskite reactant ink composition, and shunt-prone patterning (Figure 3).

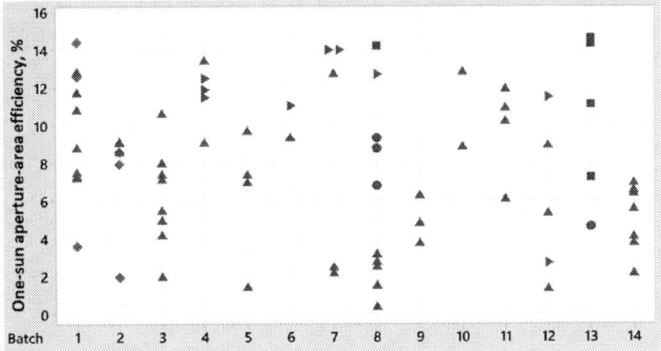

Fig. 3. One-sun, aperture-area efficiency of 100 cm^2 submodules made during exploratory batches.

Key to understanding efficiency losses were Dark Lock-in Thermography (DLIT), Photoluminescence (PL), and Electroluminescence (EL) (Figure 4). A major finding from exploratory batches was that point defects were primarily responsible for the variability observed in and between batches.

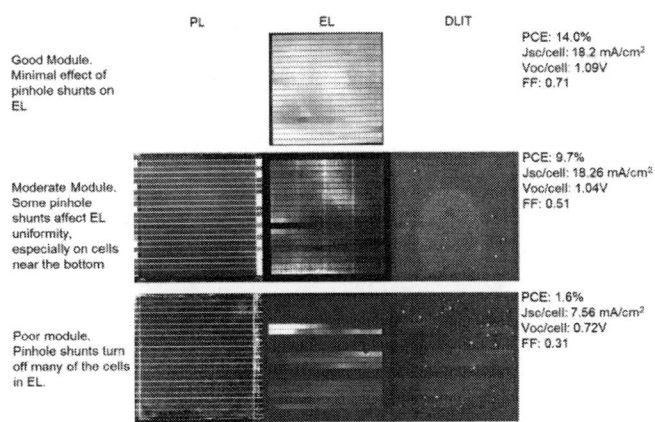

Fig. 4. PL, EL, and DLIT of various submodules and representatve defects failures observed.

Focused effort on reducing the density, size and types of point defects yielded significant improvements in submodule efficiency. An example of a 16% one-sun aperture-area efficient 100 cm^2 perovskite superstrate submodule is shown in Figure 5.

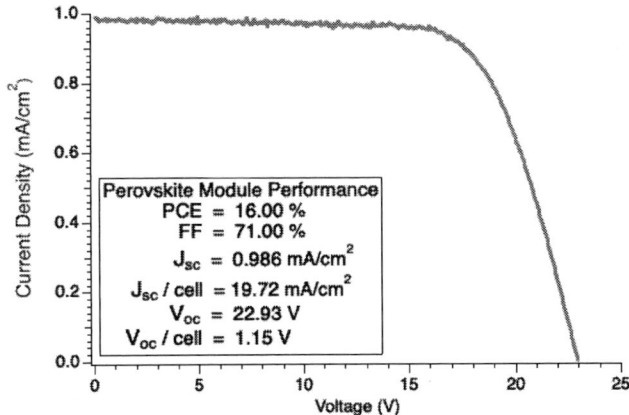

Fig. 5. 16% aperture-area semi-transparent 100 cm^2 submodule

Net of the reduction in losses related to point defects, submodule efficiency significantly tightened as shown in Figure 6. Mitigation of catastrophic point defect losses allowed controlled variations of processing parameters to be better exhibited between submodule batches. The best batches correlated with particular carrier selective layer properties.

As this is being written 100 cm^2 modules are undergoing Highly Accelerated Stress Tests (HAST) including damp heat (DH), humidity freeze (HF) and thermal cycling (TC) as defined in IEC 61215. Early results are promising, but show losses in fill factor related to charge extraction and/or transport. Data will be presented at the PVSC.

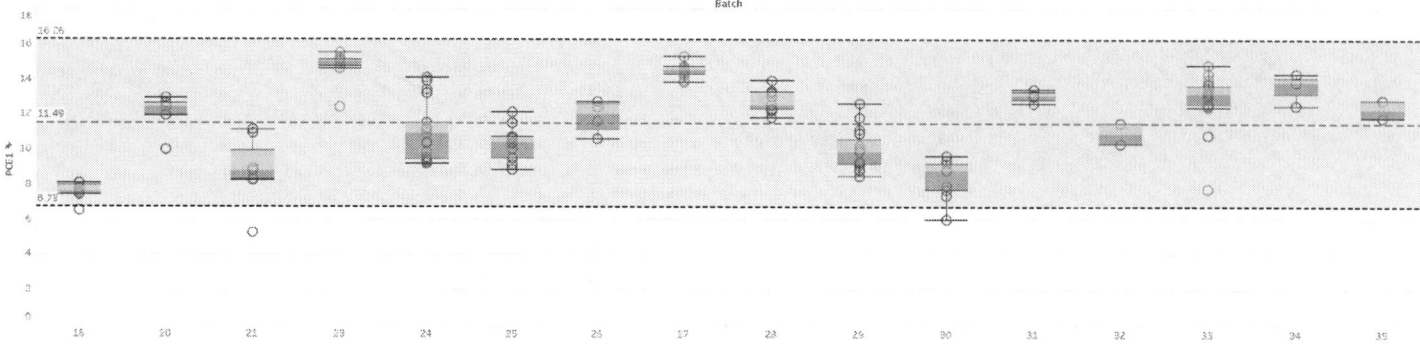

Fig. 6. Second tranche of 100 cm² submodules

III. SIGNIFICANCE AND SUMMARY

Perovskite PV technology holds great promise as a pathway to cheaper, more efficient single-junction power modules capable of being manufactured in volumes sufficient to be of consequence in climate change mitigation. Perovskite PV is particularly well positioned to augment existing PV technologies by facilitating high-efficiency high-value tandem modules, e.g. perovskite / silicon tandems. Key to both single-junction and tandem junction module production is translating small-area cell results to large-area module sizes using manufacturing-ready processes, raw materials and tools.

Significant progress is being made in demonstrating market-relevant efficiencies, and in meeting industry-standard product durability. Notable progress in efficiency, efficiency distribution and durability in 100 cm² peovskite modules will be presented, and a stepwise path to production will be discussed.

IV. PRESENTATION AND PUBLICATION

The authors aim to present a full updated set of results, including durability data, and to publish the results in an expanded version of this three-page abstract.

We believe our contribution is best slotted in sub-area 6.2, but we are open to alternatives as the organizing committee may deem appropriate.

Identification of module replacements in US utility-scale photovoltaic installations

Chenyang Deng[1], Jacob T. Stid[2], Preeti Nain[1], Annick Anctil[1]

[1]Department of Civil and Environmental Engineering, Michigan State University, East Lansing, 48824, MI, USA

[2] Department of Earth and Environmental Science, Michigan State University, East Lansing, 48824, MI, USA

Abstract — **Including replaced modules is crucial to estimate PV waste generation. A new method is proposed to identify past module replacements, which could assist in estimating PV waste. The authors analyzed the variation in the capacity factor (CF) of the US solar plants from 2011 to 2020 to identify possible repowering. A sudden increase in CF is attributed to the possible replacement of old, less efficient modules with higher efficiency modules. The generation and construction data of major PV projects (≥1MW) is collected from US Energy Information Administration and converted into a statistical model to evaluate the capacity factor performance. An algorithmic program is generated that analyses and identify the plants with repowering. Multiple methods, including satellite image, machine learning, and data comparison, are applied to validate and optimize the program. Results show that the method can overall evaluate and monitor the trend of module replacement, although the identification accuracy of a single plant needs further validation. The model will use more parameters, including temperature, location, and irradiance, to improve the success rate.**

Keywords—PV Replacement, PV Repowering, Solar capacity factor, End-of-life, photovoltaic modules

I. INTRODUCTION

Solar photovoltaics (PV) has rapidly developed in the 21st century as an environment-friendly technology. It is considered one of the cleanest renewable energy sources meeting the challenges of increased energy demand with low environmental emissions. In 2021, a large capacity of solar plants was added to the US grids, and utility-scale (29%) and distributed (15%) solar accounted for a combined 45% of the all capacity added to US grids[1]. While the PV modules have low environmental emissions during their operational phase, they still have some environmental impacts during the entire life cycle. Production and end-of-life (EoL) stages contribute maximum to the environmental impacts and human health [2].

To better specify the material flow of the LCA, studies focused on the estimation of PV waste. One general methodology is to track the materials components in PV technologies [3]. The quantification of materials is used to forecast PV waste. The method relies mainly on references from related studies and accounts for crystalline technology and thus, doesn't account for new technologies. Related research typically uses the installation year as the baseline to forecast the

waste after 25-30 years [4]. However, the lifetime of the modules is not as long as manufacturers claim because operators prefer to replace the modules with new ones in the early years to achieve higher electricity generation for economic benefit. Therefore, the 'number/capacity of modules replaced per year' is a critical parameter in the PV waste estimation.

It is economically favorable to replace modules with more efficient ones. The critical condition is that the efficiency gain over installed modules must be large enough to justify the added replacement cost [5]. For crystalline silicon, an increase of 1% in cell efficiency would require the increase of cell production cost to be less than 25% for the process to be accepted [6]. PV manufacturers compete for higher efficiency, continuously developing better-performing modules at a reduced cost. For example, the manufacturer Longi has reached a 26.09% efficiency rate, claiming the materials applied are cheaper and more environmentally friendly [7]. JinkoSolar has released TOPCon solar modules with maximum efficiency of 23.23% [8]. Considering higher electricity generation and financial benefits, utility-scaled solar plant operators are motivated to replace less efficient modules with these new, more efficient ones. The suddenly increased efficiency has provided an excellent opportunity to identify the plants with module replacement.

II. METHODS

The capacity factor (CF) refers to the ratio of the electricity output from a power plant over a particular time to the maximum possible output at its maximum capacity for the entire duration. Assuming that the annual difference of average sunlight time is ignorable for solar energy, the CF is representable for the performance of the solar module. Due to a series of factors, mainly module's degradation, the performance will decrease yearly, as represented by the CF. Since the CF represents the module's efficiency to some extent, an increase in CF from one year to the next is assumed as a replacement (old modules with lower efficiency are replaced with new modules with higher efficiency). The present study applies this approach to utility-scale solar plants (>1 MW) to identify repowered plants.

978-1-6654-6060-6/23 $31.00 © 2023 IEEE

A. Data collection and calculation

The construction data is extracted from EIA 860 forms [9], and the generated data is extracted from EIA 923 forms [10]. The location and other geological information are obtained from EIA Atlas [11]. The CF is calculated as per the following equation:

$$CF = \frac{W}{C \times 24hr \times 365d} \qquad (1)$$

Where W refers to the annual generation, and C refers to the plant's total capacity. Though a solar plant cannot operate for 24 consecutive hours, it is still used to represent the maximum power output. For a plant with multiple arrays which was installed in a different year, the total capacity in a specified year could be represented as:

$$C(year) = \sum C_{1j} + \sum \left(C_{2j} \times \frac{m_j}{12} \right) \qquad (2)$$

C_{1i} refers to the module's capacity installed before the year; C_{2i} refers to the module's capacity installed this year; m_j refers to the entire operating months. Notably, the newly installed modules may not be entirely operated in the initial year, but the overestimation will not impact the results.

B. Identification and correction

The total number of utility-scale solar plants reported by EIA is 4581, and the number of newly operated plants since 2013 is listed in TABLE I.

TABLE I. SUMMARY OF NEWLY INSTALLED PLANTS SINCE 2013

Year	2013	2014	2015	2016	2017	2018	2019	2020
Plants	243	310	370	469	526	473	487	534

The solar plant data includes many abnormal values, including 0 values, negative generation, and non-zero values before the operation year. It is addressed by initial data cleaning and organizing. Also, it is assumed that the fluctuation in annual irradiation has a negligible impact on the CF; therefore not considered. The three considered factors contributing to the yearly variation in capacity factor are:
i)   Degradation of the PV modules
ii)   Module alteration
iii)   Grid fluctuations

A code is developed to detect the variation in capacity factor and reduce the impact of the degradation and grid. Further, a three-class criterion is applied to rank the replacement: i) no replacement, ii) average probability of replacement, and iii) high replacement probability. To reduce the error, the yearly evaluation will depend on the performance for 3 to 5 consecutive years.

C. Validation

To validate the results for the identified replacement in Arizona, satellite images from National Agriculture Imagery Program (NAIP) are used to identify module replacements.

III. RESULTS

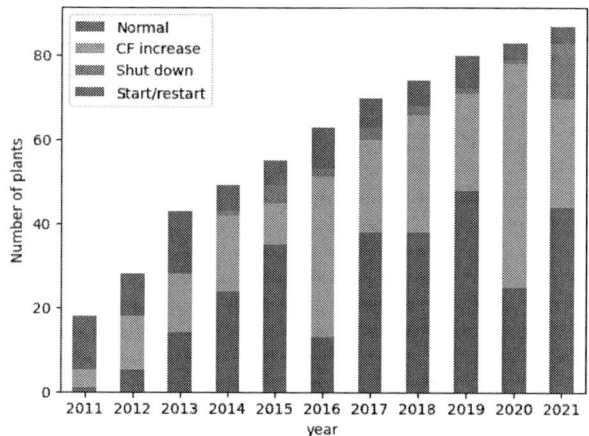

Figure 1: Operation Condition of solar plants in Arizona

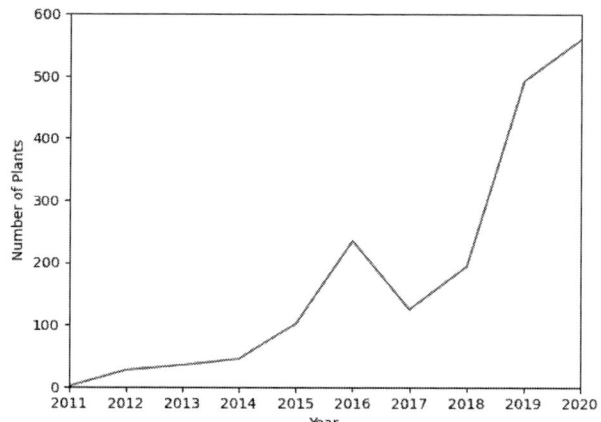

Figure 2: Number of identified repowered plants by year based on the CF change.

Figure 1 shows the capacity factor variation from 2011 to 2020 for plants over 1 MW capacity in Arizona. The total number of solar plants with over 1 MW capacity in Arizona gradually increases. The period from 2011 to 2013 is the fastest growing time, where most plants have more than five years of operation. For increased CF in 2016, the chance of replacement could be anticipated (the overall tendency for all states also happened in 2016).

Figure 2 shows the number of identified repowered plants from the initial algorithm, where the plants are only divided into non-replacement and replacement categories. To reduce the impact of the annual sunlight time, the algorithm detects the replacement based on the performance in a 5-year overall performance to, and the abnormal capacity factor change is excluded. The peak in 2016 is already observed and recorded. This proves the validity of the present method and approach.

Figure 3 shows the solar plant distribution with no replacement, the average probability of replacement, and the high probability of replacement in the Mainland of the US, Hawaii, and Alaska not included. The figure shows that in 2016 most replacements happened in the coastal, and in 2020 northern areas have more replacement than 2016. This could be explained by the start year and location of the solar plants. In coastal regions like California and North Carolina, they have a very early solar plant start, while the north area has a late start following the trends in 2014-2017. Considering the economic benefits, most operators will replace the modules with higher efficient modules for more electricity generation for modules operated over 5 years. Another factor, the location/region, could impact the motivation for replacement. In the northern area, the modules are more frequently impacted by bad weather conditions, mainly the snow. Under such conditions, the modules are easier to be damaged or degraded, causing a shorter lifetime of the module with more frequent replacement.

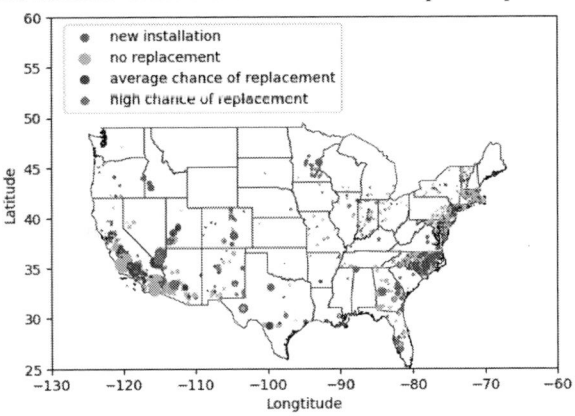

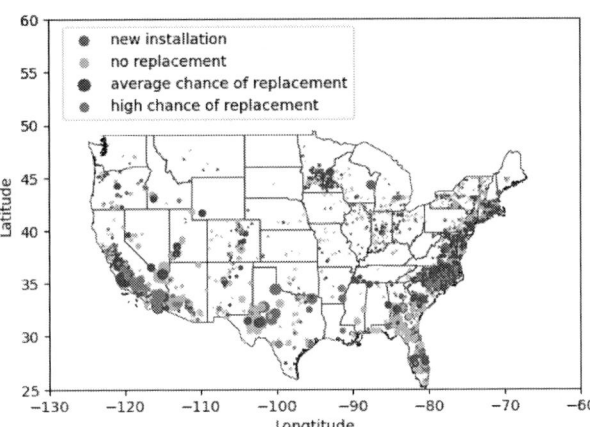

Figure 2: Replacement distribution map of US solar plants in a) 2016; b)2020.

From the figure, there are some patterns about the module replacement related to the location and operation year. If the correlation of the replacement with those parameters could be thoroughly understood, it is possible to build a model covering all the parameters, to explore the relationship between the replacement trends with time, location, temperature, and generation, furtherly predict the capacity of replaced modules. In addition, the machine learning methods could be applied to optimize the algorithm, using the plant information and CF as input and the results from relative studies for training, which is expected to increase the accuracy for the identification.

REFERENCES

[1] Bolinger, M., Seel, J., Warner, C., & Robson, D. (2022). *Utility-Scale Solar, 2022 Edition.* https://emp.lbl.gov/sites/default/files/utility_scale_solar_2022_edition_slides.pdf

[2] Klugmann-Radziemska, E., & Kuczyńska-Łażewska, A. (2020). The use of recycled semiconductor material in crystalline silicon photovoltaic modules production - A life cycle assessment of environmental impacts. *Solar Energy Materials and Solar Cells, 205,* 110259. https://doi.org/10.1016/J.SOLMAT.2019.110259

[3] Domínguez, A., & Geyer, R. (2019). Photovoltaic waste assessment of major photovoltaic installations in the United States of America. *Renewable Energy, 133,* 1188–1200. https://doi.org/10.1016/J.RENENE.2018.08.063

[4] Zhang, L., Chang, S., Wang, Q., & Zhou, D. (2022). Projection of Waste Photovoltaic Modules in China Considering Multiple Scenarios. *Sustainable Production and Consumption, 33,* 412–424. https://doi.org/10.1016/J.SPC.2022.07.012

[5] Jean, J., Woodhouse, M., & Bulović, V. (2019). Accelerating Photovoltaic Market Entry with Module Replacement. *Joule, 3*(11), 2824–2841. https://doi.org/10.1016/J.JOULE.2019.08.012

[6] Benda, V., & Černá, L. (2020). PV cells and modules – State of the art, limits and trends. *Heliyon, 6*(12), e05666. https://doi.org/10.1016/J.HELIYON.2020.E05666

[7] *Longi claims world's highest efficiency for p-type, indium-free HJT solar cells – pv magazine Australia.* (n.d.). Retrieved April 13, 2023, from https://www.pv-magazine-australia.com/2022/12/23/longi-claims-worlds-highest-efficiency-for-p-type-indium-free-hjt-solar-cells/

[8] *JinkoSolar unveils new TOPCon solar products with record efficiency ratings – pv magazine International.* (n.d.). Retrieved February 19, 2023, from https://www.pv-magazine.com/2023/01/10/jinkosolar-unveils-new-topcon-solar-products-with-record-efficiency-ratings/

[9] Form EIA-860 detailed data with previous form data (EIA-860A/860B). (n.d.). Retrieved April 4, 2023, from https://www.eia.gov/electricity/data/eia860/

[10] Form EIA-923 detailed data with previous form data (EIA-906/920). (n.d.). Retrieved April 4, 2023, from https://www.eia.gov/electricity/data/eia923/

[11] Electricity | US Energy Atlas. (n.d.). Retrieved April 9, 2023, from https://atlas.eia.gov/apps/895faaf79d744f2ab3b72f8bd5778e68/explore

[12] Jordan, D. C., Smith, R. M., Osterwald, C. R., Gelak, E., & Kurtz, S. R. (2010). Outdoor PV degradation comparison. Conference Record of the IEEE Photovoltaic Specialists Conference, 2694–2697. https://doi.org/10.1109/PVSC.2010.5616925

[13] Jordan, D. C., Silverman, T. J., Sekulic, B., & Kurtz, S. R. (2017). PV degradation curves: non-linearities and failure modes. Progress in Photovoltaics: Research and Applications, 25(7), 583–591. https://doi.org/10.1002/PIP.2835

Indoor and outdoor evaluation of curved modules for VIPV

Rebeca Herrero[1], Ignacio Antón[1], Francisco Martín[1,2], Steve Askins[1], Javier Macías[1], Luis J. San José[1], G. Vallerotto[1], R. Núñez[1], C. Domínguez[1]

[1]Instituto de Energía Solar-Universidad Politécnica de Madrid (IES-UPM), Madrid, Spain

[2]Solar Added Value (SAV), Madrid, Spain

Abstract — The vehicle integrated photovoltaic (VIPV) technology demands 3D curved PV surfaces to meet the specific design constraints in the automotive industry. The proper characterization of three dimensional PV surfaces requires specific methods and equipment that must be developed. This paper describes the procedures carried out for both outdoor and indoor characterization of modules with curvatures between 1 and 3 meter. IV curves at STC and the angular response of the modules are compared, showing remarkable resemblance. A unique collimated solar simulators has been adapted for indoor characterization while a two axis tracker with programmable deviation of the pointing to the sun has been used outdoors.

I. INTRODUCTION

The deployment of the EV arose the interest of vehicle integrated photovoltaics (VIPV), which consists of integrating PV solar panels in car surfaces such as roof, hood, doors, and even windows [1]. In this market, 3-dimensional (3D) shapes are demanded for the PV surfaces depending on the body car. Thus, VIPV surfaces performance would be dependent on the intrinsic lack of irradiance uniformity proper to the curved shape, and also on the non-stationary working conditions producing, for example, irradiance patterns affected by shadows or fast changing of angle of incident (AOI).

Suitable standards must be developed considering these VIPV particularities not addressed in current ones such as flat IEC 60904 series [2]. In this regard, the standardization body for photovoltaic technology (IEC TC82) has organized an international team (PT600) formed by scientists, engineers, and experts of different disciplines and organizations aimed to developed fair and scientific testing and rating [3]. Among all the activities carried out in the working group, it is worth mentioning the development of an international Round Robin (RR), supported by METI and JET/JEMA (Japan), in which photovoltaic curved samples have been measured by different laboratories and institutions.

Conventional flat PV modules are reliably characterized indoors using solar simulators, which are devices capable of providing irradiance conditions similar to the sunlight in a laboratory environment or production line. Most widely used solar simulators for PV panels are based on a point light source placed far enough from the panel, producing a light pulse long enough to assure a steady-state condition during the IV sweep. This type of solar simulator has demonstrated to achieve the highest quality in terms of spatial non-uniformity of irradiance, spectral mismatch and temporal stability accordingly to [4].

VIPV imposes new constraints for a solar simulator, being the most important related to the curved shape of the PV surfaces. This characteristic causes an inherent non-uniform solar irradiance throughout the surface due to the shape but also dependent on the light source angular distribution. Fig. 1 shows the scheme of collimated solar simulator and conventional one for flat modules. For the case of conventional simulators, the non-uniform irradiance over the curved surface is boosted by the increasing angle of incidence ($\delta > \theta$) throughout the aperture area caused by the divergence of the beam (β). Nevertheless, the collimated solar simulator, with the divergence of the beam ($\pm\alpha$) close to $\pm 0.27°$, reproduces the curved module performance as under sunlight in terms of irradiance over the surface.

This study presents the work carried in the context of the international RR by the Institute of Solar Energy (IES-UPM) focusing on revealing specific considerations in the outdoor and indoor testing of curved surfaces, in particular, analyzing the convenience of a collimated solar simulator for this purpose.

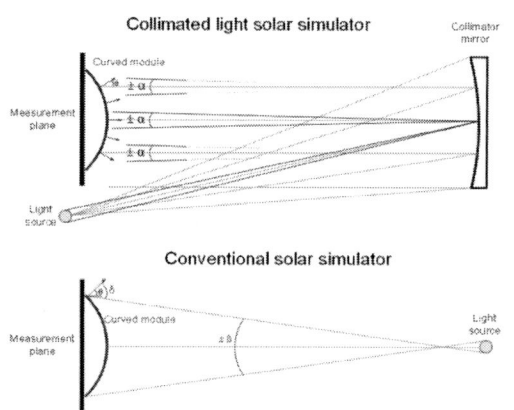

Fig. 1. Scheme of collimated solar simulator compared to a conventional one.

II. EXPERIMENTAL

The measured samples are cilindrical curved modules (1380 mm x 700 mm) manufactured by AGG Energy with different radious of curvature (1 m, 2 m, 3 m, and flat), each module is formed by 8 strings of 4 monocrystalline silicon cells (MWT) series connected, with the inclusion of a bypass diode per string.

The international RR was organized in several routes, one was specifically devoted to research activities for the development of meaningful test procedures for curved PV

978-1-6654-6060-6/23 $31.00 © 2023 IEEE

modules. IES-UPM has participated in this branch, the main goal was the comparison of outdoors and indoors power rating at Standard Test Conditions (GNI=1000 W/m², module temperature of 25ºC, and air mass (AM) of 1.5).

A. Outdoor measurements

The modules under test were installed on a two axis tracker (azimuth-elevation), all in landscape configuration and arranged in the same column avoinding any reflective surfaces at the sides of the modules (see Fig. 2). The IV curve of the 4 modules were measured simultaneously together with temperature probes attached to the rear side of the modules, the GNI irradiance given by a reference monocrystalline silicon cell (calibratred by Fraunhofer ISE CalLab PV Cells) and meteo station data (DNI, GNI, diffuse, ambient temperature, wind, humidity, spectrally resolved DNI 300-1100 nm, etc).

A measurement campaign has been carried out at winter conditions in Madrid at which cold and clear sky days are quite common in this time of the year. The modules performance have been evaluated by means of IV curves measured at steadystate conditions, IV curved measured under thermal transient (to obtain STC) and IV curves measured during 2-axis tracking angular sweeps by following the procedure of [5] (to evaluate the angular performance).

Fig. 2. Curved modules installed in two-axis tracker of different radius of curvature (from top to bottom: 1 m, 2 m, 3m and flat).

These are the steps of the measurement procedure carried out to optain the power rating at Standard Test Condition:
- Calibration of the Isc of each module against the reference cell (GNI) at clear sky conditions.
- Evaluation of modules during thermal transient. First, modules are covered with cardboard, assuring they are at room temperature. After this, the modules are uncovered and IV curves are measured together with the module temperature during warming up, aiming at STC conditions.

B. Indoor Measurements

IES-UPM developed the Helios 3198, a collimated solar simulator devoted for indoor characterization of CPV modules [6-10]. The illumination system consists of a round collimator with a diameter of 2 meters and a focal distance of 6 meters, and a small toroid flash lamp of 6.5 cm in diameter. A black

tunnel for stray light rejection has been designed and built, including the baffles and light trap box and chambers as detailed in Fig. 3.

The modules have been measured at STC (1000W/m² and module temperature of 25ºC), at normal irradiance and with different angle of incidence to determine the rated power of the modules and the angular response.

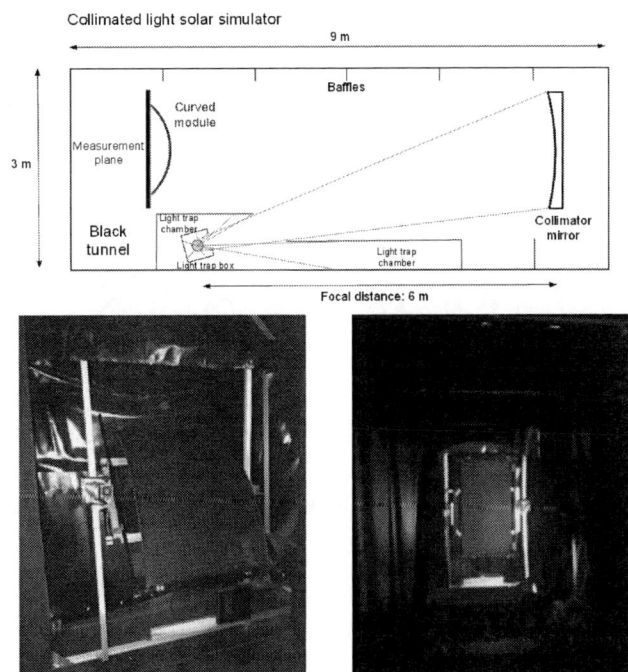

Fig. 3. (up) Scheme of Helios 3198 solar simulator adapted to VIP; (left) module and rotating structure to evaluate angular response: (right) collimator mirror reflecting VIPV modules under test.

III. RESULTS AND CONCLUSIONS

A. Power rating at STC conditions

The IV curves of the 4 modules at STC conditions were obtained both indoors and outdoors. Fig. 4 presents the maximum power (Pmp) measured for the 4 modules under inspection. These Pmp values are related to IV curves measured at steady conditions and IV curve measured during thermal transient (dots inside red circle). Outdoor STC Pmp values (dark dots), and IV curves, are obtained by filtering to achieve measurement conditions as close as possible to STC.

In this study, direct comparison of raw data (IV curves measured at STC both indoors and outdoors) has been carried out. Translation equations and procedures not suitable for curved surfaces due to the steps-shaped IV curve are avoided as much as possible, only minimum irradiance correction has been applied.

978-1-6654-6060-6/23 $31.00 © 2023 IEEE

As example of the achieved results, Fig. 5 shows the IV curves of the module with the larger curvature (R1000). The plot shows the similarity between both indoors and outdoors. The outdoor IV curve was measured at 994 W/m² and module temperature of 24°C. The outdoor IV curve has been corrected to 1000W/m² (only current values were corrected, voltage correction has considered negligible). The rated power, measured in the solar simulator was 127.4 W_p, while outdoors was 126.5 Wp at 994W/m2 (127.3 W_p if corrected to 1000W/m²).

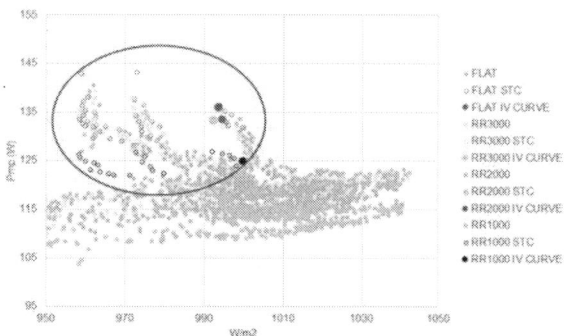

Fig. 4. Pmp vs. GNI for the modules with different radious of curvature (1 m – RR1000, 2 m – RR2000, 3m – RR3000 and flat).

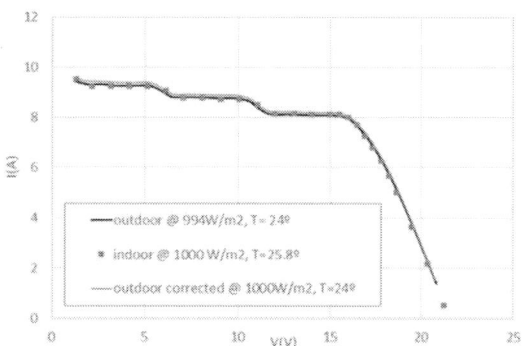

Fig. 5. Measured IV curves (indoor and outdoor) for RR1000 module (curvature radius of 1m).

B. Angular Performance

The angular response (or IAM – Incident Angle Modifier) of VIPV modules is particularly relevant due to both curvature and non-stationary working onditions due to vehicle movement. It must be pointed out that this response depends not only on the curvature, but also on the interconnection topology of the cells comprising the module (cell size, series/parallel connection, bypass diodes…). Fig. 6 shows the results achieved for the module with 3 m of radius of curvature (R3000) both indoors and outdoors and at two bias condition (Isc and Pmp). The cosine response is also plotted for comparison, showing that curved modules do not follow the ideal cosine response.

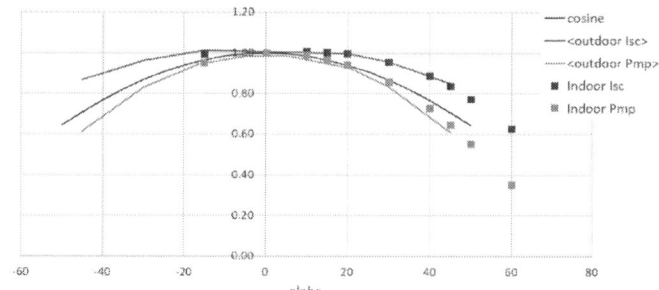

Fig. 6. Measured Pmp vs. GNI for the modules with different radious of curvature (1 m, 2 m, 3m and flat).

ACKNOLEDGMENTS

This work has been supported by the Grant Grant PID2021-128853OB-I00 funded by MCIN/AEI/ 10.13039/ 501100011033 and by "ERDF A way of making Europe". The authors express sincere thanks to the AGG Energy for the manufacturing of the module, and JEMA and JET program for sponsoring the activities with those modules, entrusted by METI (Japan) and carried out under the umbrella of the IEC TC82/PT600 group devoted to Vehicle Integrated PV Systems.

REFERENCES

[1] "State-of-the-Art and Expected Benefits of PV-Powered Vehicles" IEA-PVPS, ISBN 978-3-907281-15-4

[2] K. Araki, et. Al."To Do List for Research and Development and International Standardization to Achieve the Goal of Running a Majority of Electric Vehicles", *Solar Energy Coatings*, vol.8, no. 7, pp, doi: 10.3390/coatings8070251

[3] K. Araki, et. Al. "International collaboration for standardization (testing and rating) for VIPV", in 33rd International Photovoltaic Science and Engineering Conference, 2022.

[4] C. Droz, et Al. "Evaluation of Commercial Large Area Solar Simulator: Features Exceeding the IEC Standard Class AAA" in 25th European Photovoltaic Solar Energy Conference and Exhibition, 2010, pp. 3884 – 3888.

[5] D. Riley et Al. ASME. J. Sol. Energy Eng. June 2015; 137(3): 031008, doi:10.1115/1.4029379

[6] C. Domínguez et Al. Optics Express, vol. 16, no. 19, pp. 14 894–14 901, doi: 10.1364/OE.16.014894

[7] C. Domínguez et Al., "Characterization of five CPV module technologies with the Helios 3198 solar simulator," in *34th IEEE Photovoltaic Specialists Conference* 2009, pp. 001004-001008, doi: 10.1109/PVSC.2009.5411192

[8] C. Domínguez et Al., "Solar simulator for indoor characterization of large area high-concentration PV modules", in *33rd IEEE Photovoltaic Specialists Conference*, 2008, pp. 1-5, doi: 10.1109/PVSC.2008.4922739.

[9] I. Antón et Al., "Characterization Capabilities of Solar Simulators for Concentrator Photovoltaic Modules," *Japanese Journal of Applied Physics*, vol. 51, p. 10ND12, doi: 10.1143/JJAP.51.10ND12

[10] C. Domínguez et Al., "Indoor Characterization of CPV Modules Using the Helios 3198 Solar Simulator," in *24th European Photovoltaic Solar Energy Conference*, 2009, pp. 165-169, doi: 10.4229/24thEUPVSEC2009-1BO.7.6

Towards an Annual Terrawatt Photovoltaics Market
-
Comparison of the social acceptance in various IEA PVPS countries

All authors are members of IEA PVPS Task 1

Arnulf Jäger-Waldau[1]
European Commission
Joint Research Centre (JRC),
Ispra, Italy
arnulf.jaeger-
waldau@ec.europa.eu

Georg Altenhöfer-Pflaum
Projektträger Jülich,
Forschungszentrum Jülich,
Jülich, Germany
g.altenhoefer-pflaum@fz-
juelich.de

Otto Bernsen
Netherlands Enterprise Agency
(RVO), Utrecht, The Netherlands
otto.bernsen@rvo.nl

Christian Breyer
LUT University, Yliopistonkatu
34, 53850 Lappeenranta, Finland
christian.breyer@lut.fi

Jose Donoso
UNEF,
C. de Velázquez, 18,
28001 Madrid, Spain
j.donoso@unef.es

Hubert Fechner
Austrian PV Technology
platform, Vienna, Austria
fechner@technikum-wien.at

Kenn Henrik Bournonville
Frederiksen
Kenergy ApS, 8700 Horsens,
Denmark
kf@kenergy.dk

Jarand Hole
Norwegian Water Resources and
Energy Directorate (NVE), Oslo,
Norway
jho@nve.no

Izumi Kaizuka
RTS Corporation, Tokyo, Japan
kaizuka@rts-pv.com

Linda Koschier
Australian PV Institute, Sydney,
Australia
lindakoschier@gmail.com

Johan Lindahl
Becquerel Institute Sweden AB,
Knivsta, Sweden
johan@becquerelsweden.se

Gaëtan Masson
Becquerel Institute, 146 Rue
Royale, 1000 Brussels, Belgium
g.masson@becquerelinstitute.org

Daniel Mugnier
PLANAIR France, Villeurbanne,
France
daniel.mugnier@planair.fr

Chinho Park
Korea Institute of Energy
Technology (KENTECH), 21
Kentech-gil, Naju 58330, Rep. of
Korea
chpark@kentech.ac.kr

Lionel Perret
PLANAIR SA, Galilée 6,
Yverdon-les-Bains, Suisse
lionel.perret@planair.ch

Francesca Tilli
Gestore dei Servizi Energetici,
GSE S.p.A., Rome, Italy
francesca.tilli@gse.it

Abstract — Over the last two decades, grid-connected solar photovoltaic systems have increased from a niche market to one of the leading power generation capacity additions annually. In spring 2022 the total installed photovoltaic electricity generation capacity broke the 1 TW barrier and it is expected that the annual market will reach 1 TW towards the end of the decade. This development has significant implications for the economy, society and the environment. Social acceptance of the necessary economic activities with regards to raw material extraction, component manufacturing and deployment is becoming a key topic to enable the industry and market to grow to the required multi TW level in the next decade. This paper gives an overview about the current situation in a number of IEA PVPS countries.

I. INTRODUCTION

It took over 40 years until the cumulative installed capacity of PV systems passed the threshold of 100 GWp in 2012. In the next 10 years this capacity increased more than 10-fold and is has reached almost 1.2 TWp by the end of 2022 (Fig. 1) [1, 2].

During this time, annual installations have increased roughly 8-fold from about 31 GWp in 2012 to something between 230 and 260 GWp in 2022 [2, 3, 4,5]. The main growth came from large utility scale PV systems, but some countries also saw an exceptional growth of grid connected residential PV systems.

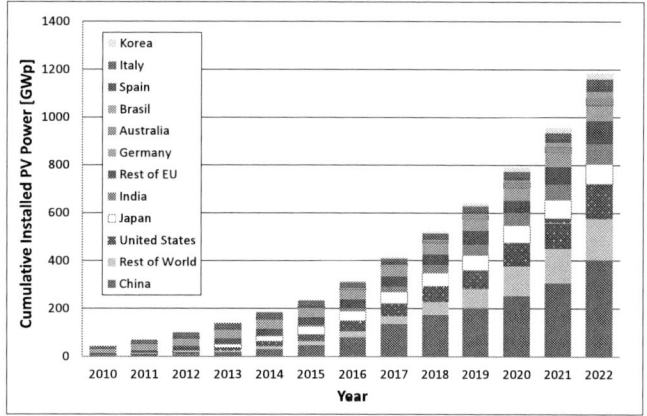

Fig. 1.   Grid-connected PV capacity [1, 2]

The rapid growth of PV systems has not come without a change in the general perception towards the installation of PV systems and the PV industry as a whole.

Various developments over the last two decades have impacted this change.

The social acceptance of the energy transition is a major issue and is becoming a key subject for the development of PV. It is multifaceted: economic, social, societal, and environmental, but also aesthetic.

II. VALUE OF THE PV INDUSTRY

The turnover of the PV sector in 2021 amounted to around 190 Billion USD. This number has been calculated based on the size of the PV market (annual installations and cumulative capacities) and the average price value for installation and Operation & Maintenance (O&M) specific to the different market segments and countries.

IEA PVPS estimates that the PV sector employed an estimated 4.3 million people globally at the end of 2021. An estimated 1.2 million were employed in the upstream part, including materials and equipment, while 3.1 million were active in the downstream part [1]. These numbers have been established based on the IEA PVPS National Survey Reports and additional sources such as the IRENA jobs database. It should be noted that these numbers are strongly dependent on the assumptions and field of activities considered in the upstream and downstream sectors and represent an estimate in the best case.

Besides its direct value for the economy and the jobs that it creates, both making contribution to the prosperity of the countries in which it is being installed and produced, PV entails additional positive implications on the social level if leveraged with appropriate policies.

III. SITUATION IN IEA PVPS COUNTRIES

The IEA Photovoltaic Power Systems Programme (PVPS) is one of the collaborative R&D Agreements established within the IEA and, since its establishment in 1993, the PVPS participants have been conducting a variety of joint projects in the application of photovoltaic conversion of solar energy into electricity. Task 1 aims at promoting and facilitating the exchange and dissemination of information on the technical, economic, environmental and social aspects of PV power systems.

Currently the member countries are: Australia, Austria, Canada, China, Chile, Denmark, Finland, France, Germany, Israel, Italy, Japan, Korea, Malaysia, Mexico, Morocco, the Netherlands, Norway, Portugal, Spain, South Africa, Sweden, Switzerland, Thailand, Turkey and the United States. In addition, the European Commission, Indonesia, Solar Power Europe (SPE), the Solar Industries Association (SEIA), the Solar Electric Power Association (SEPA), the Solar Energy Research Institute of Singapore (SERIS) and the Copper Alliance participate in the IEA PVPS programme.

Australia has now over 30 GWp of installed PV capacity. Historically dominated by rooftop PV we now see the proportion of rooftop approximately 65% with utility scale solar emerging at 35% in 2022.

The utility scale market started around 2013 and has faced numerous challenges along the way including political opposition and insufficiencies in grid infrastructure. With over

978-1-6654-6060-6/23 $31.00 © 2023 IEEE

40,000 km of high voltage transmission lines spanning over 5,000 km from top to bottom, the infrastructure in Australia requires large investment to accommodate the transition to renewables.

Australian projects tend to be large in scale in the hundreds of MW and multiple parcels of land are required for these projects to come to fruition. Social acceptance to date has been evidenced by the large uptake of rooftop PV with over 3 million rooftop installations and an average of 33% of households with PV on the roof. Utility scale projects require careful stakeholder management and numerous guides have been published by governments across Australia on how best to share benefits and manage community engagement and therefore social acceptance.

In Austria, the political target is fixed with 100% RES electricity until 2030 with +11 TWh PV electricity (based on 2020 values). The annual national PV market, exceeded the 1 GW mark for the first time in 2022, leading to a total of about 3.8 GWp. Various and sometimes high bureaucratic hurdles in all nine federal states of the country as well as a lack of electricity grid capacities represent current hurdles. After many years of almost exclusively rooftop PV systems, the first larger ground-mounted systems were only implemented in 2021.

Large investors are trying to secure agricultural land, which leads frequently to conflicts with agricultural use and landowners. Strong restrictions regarding aesthetics and nature compatibility lead to the designation of certain PV suitable zones by the local authorities. Architecturally designed BIPV solutions, the combination with agricultural use but also the increased use of PV in the transport sector are met with far greater approval.

Denmark has reached around 3.9 GWp installed at the end of 2022 now covering 6.2% of the yearly consumption. PV on rooftops in Denmark are in general highly accepted by the public and are only commented if the design and detachment of the systems does not match an architecturally good practice.

The first MW ground mounted PV systems where build in 2013 and mostly in areas without people living nearby a plant so they did not receive much attention. In the last 2-3 years several larger +50 MW systems have been build and several plants over 100 MW are being planned. Resistance against these larger PV plants now comes from local communities (NIMBY effect) and different organisations argument against the use of farmland for PV. However, using undammed land and land covering underground water resources is more accepted. In addition, some argue that there are roof areas enough for the PV needed to achieve the RE goals of Denmark.

Beside a national legislation, which stated that the developer of a ground mounted PV plant must compensate the municipality with a onetime fee of 40.000 DKK/MW installed and the nearby private neighbours with a smaller fee some developers now offer local ownership models of ground mounted PV plants.

The development of PV in Finland has been quite dynamic in recent years, starting from a very low level. Finland did not introduce relevant support schemes for PV and the relatively low yield of about 800-950 kWh/kWp led to a late market ramping [6]. The installations in 2022 are estimated to about 150-200 MWp to reach about 600 MWp in total. Annual installations in 2025 may reach 1 GWp.

Barriers for PV can be structured into technological, economic, behavioural and institutional and political ones [7]. The economic barriers have been largely overcome in recent years, with very high profits in 2022, as PV electricity feed-in is remunerated with the wholesale market price, which was on historic high levels in 2022 and led to a very short amortisation period. PV is the second cheapest cost source for electricity in Finland, next to wind power [8]. Political and behavioural barriers are still existent due to vested public investment in traditional sources of energy and practical experience across stakeholders due to the very late market ramping, which practically only started in 2016.

In France, solar PV development had always nearly equal shares of rooftop and ground mounted systems. Various policies favouring a centralised framed playground have strongly limited the market. Since 2021, the national market has increased from nearly 1 GWp/y to nearly 3 GWp realising that more efforts are needed to reach the national targets.

However, the ground mounted segment is facing new barriers due to various competing laws of land use management. In addition, especially in French rural areas, large PV projects are more facing reluctance of acceptance from citizens. Counter measures are more public information to show the economic advantages of such projects for local authorities and financial participation of local citizens.

In Germany, PV installations reached the 1 GWp/year threshold in 2007 and grew to a maximum of 8.2 GWp (around 100 Wp/inhabitant) in 2012 mainly triggered by a feed in tariff (FiT) model. Due to a significant reduction of the FiT, installations went back to 1.2 GWp in 2014 and since then are slowly recovering to a value of above 7 GWp in 2022. PV now covers more than 10% of the German electricity production and is approaching a total of 1 kWp installed per inhabitant with more than 65 GWp installed at the end of 2022.

During all this time, PV was accepted by a vast majority of the population as an important part of electricity production. A strong lobby against PV has never (been) formed. Nevertheless discussions arise around different regulatory and technical aspects. In the earlier years the public discourse concerned e.g. the social justice (only prosperous people own a roof and can profit from the FiT) or the energetic and financial pay-back-time of PV. By now, topics like maximum PV penetration, grid stability, storage (and storage cost), land consumption or recycling issues are taking over. Looking at the ambitious goals of the government (215 GWp in 2030, 400 GWp in 2040) with a share of each 50% for ground mounted and building attached systems, it can be expected that future acceptance of the development depends on convincing solutions for the present challenges.

Italy at the end of 2022 has reached around 25 GWp with 2.5 GWp of PV capacity installed in 2022. The Integrated National Energy and Climate Plan foresees an increase of RES electricity share in consumption, rising from 34% in 2017 to about 55% by 2030. The main contribution is expected from PV production, with a target of cumulative PV capacity of about 52 GWp by 2030 and a corresponding electricity production of 73 TWh/year. With the approval of the European 'Fit for 55' package by the European Parliament and the Member States [9], it is expected that these targets will be exceeded, leading to a cumulative PV capacity installed exceeding 70 GWp in 2030.

Gestore dei Servizi Energetici, within its participation in international projects like IEA PVPS TCPs, over the years 2009 – 2022 analyzed the main barriers to PV deployment in Italy, thus approaching the topic of social acceptance. The analysis is based on answers to questionnaires/interviews to stakeholders of the PV market such as industry, associations, consultancy firms, financial institutes, universities, research institutes, utilities, specialized press, and private citizens (including some PV/BIPV owners).

From the 2000s on, policymakers have been strongly committed to the dissemination of this technology. It is important to highlight the different paths related to ground PV plants and to PV on building (both BAPV and BIPV).

Initially, ground mounted PV systems were welcomed by citizens and local politicians due to the higher local wealth creation compared to common agricultural activities. However, after the massive growth of ground PV installation of 2009/2010 due to generous incentive scheme (especially in the south of Italy), local administrators and citizens complained about the missing respect for the cultural heritage of the territory and for the traditional cultivation of arable land.

In 2022 the NIMBY effect for this kind of plants is quite strong, resulting in long authorization processes, even if in the last two years several administrative simplifications have been adopted for RES plants installation aimed to reduce the time for authorization.

On the contrary, a PIMBY (Please In My Back Yard) effect for roof-top installations can be observed from citizens thanks to a "PV on the roof" attitude. Some opposition from local authorities still exist, especially in (historical) centers of Italian cities. In this regard, it is important to highlight that most of the capacity installed in 2022 is not on the ground, due also to profitable tax deduction schemes for building energy efficiency. Thus, it is quite clear that, probably, there is a growing awareness about the need of the city skyline modification with the new PV/BIPV material.

In Japan, almost 6.5 GWp of PV systems were installed in 2022 and cumulative installed capacity in the end of the year is expected to reach about 85 GWp. The Sixth Strategic Energy Plan increased the share of renewable energy in the energy mix from 22-24% to 36-38% by 2030. PV is expected to account for 14-16%, and the target PV installed capacity by 2030 was set at 117.6 GW_{AC}. To achieve the target, Japan needs to maximize installation on rooftops and find areas for utility scale projects. After the FIT program started in 2012, the significant growth of utility scale market resulted in opposition from local residents because of environmental damage and concern about safety issues and other reasons. Especially, several accidents of PV installations by Typhoons created negative impacts. To address the safety issues and harmonization of PV projects with local communities, the Ministries of Economy, Trade and Industry, Agriculture, Forestry and Fisheries, Land, Infrastructure, Transport and Tourism and Environment jointly published Guidelines of PV projects development, which covers issues at pre-development stage as well as treatment of end-of-life PV modules. Social acceptance in the Netherlands is tied to different aspects such as societal goals, participation, aesthetics, social justice, carbon footprint, multifunctional land use of the limited space available and lately human rights. As the costs and benefits in the energy transition do not fall to the same parties, and most of the (grid) costs are socialised, this was always going to be a pressing issue down the road.

The aesthetics of rooftop mounted solar systems were from the start a concern and not only on the historic building but in general. The roll out on roof-tops of this patch work of solar panels was addressed early on in the already closed (2001) task 7 of the IEA PVPS. It mentions an additional number of problems that hinder social acceptance; the lack of independent and reliable information, mistrust of planners and project developers, the lack of regulation, guidelines, certification, unclear liabilities and lack of after sales support. Some of these issues have been solved by the solar sector and policy makers but not all and some can be found again in the current task 15 on BIPV.

After the first serious signs of grid congestion occurred in the Dutch electricity grid in 2018 a framework of priorities was adopted in parliament for the deployment of solar panels in 2019. Today a steady amount of over 3 GWp is installed annually, divided almost equally between roof and ground mounted systems. However, there is an increasing back log of not realized solar parks due to congestion in the distribution grids. At the moment solar runs the risk of being identified as a "problem" and not the lack of planning and anticipation by the parties involved. While almost every municipality has an energy cooperative, their total share in the installed solar capacity is less than 2%. Cooperatives of house owners in apartment blocks often lack the necessary information and expertise to reach consensus to implement their sustainable plans.

978-1-6654-6060-6/23 $31.00 © 2023 IEEE

In Norway, PV deployment is still modest with around 300 MWp installed by the end of 2022. However, half of this was installed in 2022, with yearly installations increasing by 350 percent from 2021 to 2022. Preliminary data from 2023 suggest installation rates are still increasing. The discussion of social acceptance of PV in Norway is very much coloured by the intense movement against wind power which got a foothold in Norway in 2019. The PV market has until now been dominated by rooftop solar, but now ground mounted projects are being developed. As the share of agricultural land in Norway is relatively small, many of the projects are planned in forest areas. One of the main discussions in Norway going forward will be weighing the value of ground mounted PV against the value of the forest, with both alternative area uses, biodiversity and carbon emissions etc. In addition, many of the topics mentioned from other countries in this paper are also relevant for Norway, such as aesthetics.

The developers of PV projects are openly talking about social acceptance as key for their success, referring to what happened to wind power development, which came to a full stop in 2019. Some of the largest projects under planning in Norway are agri-PV, where PV is to be combined with sheep grazing. One additional argument that might increase acceptance of ground mounted PV in rural areas, is that these kinds of agri-PV projects could possibly reduce to the loss of sheep to predators such as wolverine, lynx and wolves, as the PV plant needs fencing.

South Korea's cumulative PV installation has grown from 650 MWp at the end of 2010 to more than 24 GWp at the end of 2022. This high speed market growth has caused serious problems in public acceptance due to several incidents related with PV installations in a relatively small land size and a high population density, although most of the installations were made properly. Examples include installations in mountainous areas where trees were cut (landslides), installations on farm land near the ocean, degrading it through sea water penetration as well as ugly installations on roof-tops and balconies. A few PV installers took advantage of naïve farmers and fishermen like the wild-west bonanza seekers did. Grid connection of the installed PV systems also became a serious issue in southern part of the peninsula due to the imbalance between available grids and nearby PV power plants, causing more than 1 GW of curtailments.

Recently empirical research and demonstration projects are under way with governmental support to provide criteria and regulations for PV installations in different locations and environment, and this will help provide new certification standards (KS-mark) in PV installations such as roof-top PV, BIPV, Agri-PV and Floating PV. Replacements of old PV inverters and introduction of medium voltage (70 kV) lines are being undertaken to resolve the grid connection issues.

In Spain, the installation of ground-mounted photovoltaic plants has undergone significant development in recent years. Although the majority perception is still positive, in certain areas where there is a higher concentration of plants, there are significant opposition movements. This opposition, which we could call neo-negationist, is a broad coalition of competing economic interests (real estate, rural tourism, hunters), environmental groups without a global vision of environmental issues, second home owners, anti-capitalists and political opportunists.

The best way to tackle the problem is to increase communication at the local level and to improve the behavioral habits of enterprises. The Spanish Industry Association has created a Certificate of Excellence for companies working on the concept of fully reversible plants that are an integral reserve for nature while leaving a significant positive economic impact at the local level. UNEF has also worked on the concept of bio-agrovoltaics.

From the start, the Swedish market for distributed PV has been driven by self-consumption, as there has never existed a feed-in-tariff in Sweden. Between 2009 and 2021, a capital subsidy programme, where the support level was decreased as the market grew and prices fell, existed. From 2015 on, the capital subsidy was complemented by feed-in premium scheme available for PV systems with a fuse below 100 amperes.

As of 2022 no subsidies exist except for the private domestic PV market segment, which still have access to the feed-in premium and the possibility to deduce 19% of the PV system cost by a tax reduction scheme for green technology.

So far only 8% of the grid-connected capacity are ground-mounted centralized PV parks, but this market segment is expanding fast as the prices since the beginning of the decade has made PV parks economical attractive even without subsidies [5]. With the rapid roll out of PV parks the problem with local social acceptance has risen in Sweden, which originates from the two common opposition motives; (1) The "not in my back yard" (NIMBY) resistance from local residents who do not want to see their picturesque countryside changed, and (2) the national interest of keeping agricultural land productive for food production autonomy.

Local Swedish authorities have not been prepared for the now exploding interest from farmers to lease their land to PV park projects, with long permit processes for PV park projects on agriculture land as a result. However, many PV parks have, and are being built, on other type of less valuable land types, such as industrial sites next to major highways, former landfills, grass areas next to airports, etc. [10]. These projects have in general not faced any social opposition, but rather a positive reception from the local community that believes that it is good that this type of land can be used for something positive such as production of renewable energy.

In Switzerland, the PV market was fairly stable from 2013 to 2019 with around 300 MWp, almost exclusively on roofs. During this phase where subsidies were reduced in the same proportion as the fall in system costs, solar installations grew in reputation. The good image of solar energy thus obtained has made it possible to place solar energy as a future pillar of the Swiss energy supply, with a national target 2050 raised to 34 TWh or about 38 GWp. Therefore, the objective is to move to a market of 1.1 GWp per year, or even more with the recent European energy industry.

This strong acceleration of the market requires new weightings of interest on protected buildings, with judgments in the direction of limiting restrictions. With the growth of the market, unserious offers appear, encouraging the strengthening of the national label for audited professional. "Les Pros du Solaire/ Solar Profi" to ensure the quality of the sector. In order to reduce the risk of supply in winter, an emergency law has also been validated to allow the installation of large ground installations at altitude in the Alps. These installations of at least 7 MWp then require additional land use and now create associations for the protection of the landscape and opposition from certain ecological circles. For these installations, the international teachings of ground-based PV plants, or wind, will be valuable in improving their acceptance.

III. CONCLUSIONS

PV is a major contributor on the road to sustainability: the nature of the energy transformation, and the acceptance of change are essential elements in the success of this revolution: dealing with the number of jobs concerned, the impact on the environment and the social aspects linked to the development of PV has become unavoidable. Ensuring a local development of the PV industry and improving the use of resources is part of the response to the need for PV to be more virtuous than the energy sources that it replaces. Now, with the challenge to develop a multi TWp market, the focus of public scrutiny has more shifted to the area requirements for installations as well as the environmental impacts of securing the raw materials needed as well as the actual manufacturing of the PV components needed.

In 2022, photovoltaic technology has become increasingly a source of affordable, local, and low-carbon energy. In the context of geopolitical tensions and resource scarcity, PV could become a stabilization element, promoting peace through reduced tensions in energy markets while accelerating the development of the world.

Disclaimer: Arnulf Jäger-Waldau works for the Joint Research Centre of the European Commission. The scientific output expressed is based on the current information available to the author, and does not imply a policy position of the European Commission.

REFERENCES

[1] IEA PVPS, Trends in PV applications 2022, ISBN 978-3-907281-35-2
[2] A. Jäger-Waldau, Snapshot of Photovoltaics – May 2023, EPJ Photovoltaics
[3] TrendForce, New PV Installations Worldwide Will Grow by More Than 50% YoY to 351GW for 2023 Thanks to Rising Demand, Press release 16 February 2023, https://www.trendforce.com/presscenter/news/20230216-11568.html
[4] IEA PVPS, Snapshot of Global PV Market 2023, ISBN 978-3-907281-43-7:
[5] Bloomberg New Energy Finance, 1Q 2023 Global PV Market Outlook, 28 February 2023
[6] IEA-PVPS, 2020. National survey report of PV power applications in Finland 2019
[7] Child M. et al, The Role of Solar Photovoltaics and Energy Storage Solutions in a 100% Renewable Energy System for Finland in 2050, *Sustainability* **2017**, *9*(8), 1358, doi 2071-1050/9/8/1358
[8] Vartiainen E. et al Impact of weighted average cost of capital, capital expenditure, and other parameters on future utility-scale PV levelised cost of electricity, Progress in photovoltaics Volume 28, Issue 6, Pages 439-453, doi 10.1002/pip.3189
[9] European Council, Fit for 55%, https://www.consilium.europa.eu/en/policies/green-deal/fit-for-55-the-eu-plan-for-a-green-transition/
[10] Lindahl J. et al, Economic analysis of the early market of centralized photovoltaic parks in Sweden, Renewable Energy, Volume 185, February 2022, Pages 1192-1208, doi 60148121018012

Potential Induced Degradation Evaluation of Damp Heat Stressed PV Modules

Farrukh ibne Mahmood, Akash Kumar, Muhammad Afridi, GovindaSamy TamizhMani

Photovoltaic Reliability Laboratory, Arizona State University (ASU-PRL), Mesa, Arizona, USA

Abstract — Potential induced degradation (PID) mode is one of the most critical degradation modes in the photovoltaic (PV) modules. The PID issue is highly dependent on the conductivity of the materials (especially encapsulant and glass surface) and the adhesion strengths of interfaces (glass/encapsulant, encapsulant/cell and encapsulant/backsheet). In the field-aged modules, the material conductivity typically increases due to aging of materials, weakened interface and cemented soiling of glass surface. Currently, the PID tests are performed on the fresh modules as per IEC standards' requirements. In the fresh modules, the encapsulant conductivity is very low (compared to the field-aged modules) and the adhesion strengths of interfaces are high. Therefore, the PID loss in the current tests represent only the loss which would happen in the fresh modules and young modules, not in the long-term field-aged modules. In the current work, we have used two pre-stressed modules to represent the long-term field-aged PV modules. The pre-stressing was done on two glass/polymer modules for 2000 hours at 85°C/85%RH according to IEC 61215 standard. After the damp-heat pre-stressing, these two modules were subjected to PID (+ve PID on one module; -ve PID on the other module) in an environmental chamber at 1000 V, 60°C, and 85% RH, according to IEC 62804-1. These modules experienced 6.3-7.5% power degradation which is beyond the allowed limit of 5% by the standard. Based on these results, we believe that the PID tests may need to be performed on the unstressed/fresh and pre-stressed modules to appropriately represent the PID issue corresponding to the short-term and long-term field-aged modules, respectively.

I. INTRODUCTION

PV modules in the field are under multiple stresses simultaneously, which can lead to unforeseen degradations and failure modes. Most of these issues arise as conventional testing methods are designed to apply single stress at a time or the stress duration is short to observe actual field failure [1]. Additionally, stressors applied in traditional testing methods may not be administered in the appropriate sequence to mimic environmental impact accurately [1].

Typical PID testing methods are designed to test fresh modules which may not be field representative [2], [3]. Fresh modules usually have high interfacial adhesion strengths, due to which moisture ingress is lowered. However, as the module ages, these bonds weaken over time, leading to higher levels of moisture penetration. High moisture ingress in PV modules is frequently associated with a greater PID susceptibility [3], [4]. In outdoor operation, PV modules experience a variety of issues, including electrical strains, mechanical pressures, moisture intrusion, UV radiation, and drastic temperature changes. Therefore, it takes more than one stress test to

determine if modules are prone to degradation. To evaluate the level of deterioration modules may encounter in the field, mixing various stressors is optimal [5]. Additionally, it is recommended to perform electrical stress (PID) on DH-stressed modules rather than fresh ones to imitate actual operating conditions [6].

In this study, two commercial GB modules were tested to understand the impact of PID due to reduced interfacial adhesion strengths (25+ years of field operation). Both modules undergo DH stress for 2000 hours (to imitate 25+ years of operation), and then the same modules are subjected to PID (one under negative and one under positive bias). Various pre- and post-characterization tests such as light IV, electroluminescence (EL), and dark IV are carried out before and after the stress testing to determine the change in performance parameters.

II. EXPERIMENTAL SETUP

For this investigation, two identical commercial GB modules were used. The GB module is composed of a glass-encapsulant-cell-encapsulant-backsheet-frame and weighs 23 kg, with dimensions 2004 mm x 996 mm by 35 mm. It is a 144-cell half-cut mono-facial module with a rated power of 380 W. The cells are PERC and are 78 mm x 156 mm in size. The front glass is 3.2mm coated tempered glass, the frame is anodized aluminum (Al) alloy, and the encapsulant used is EVA.

Both modules were stressed under DH for 2000 hours at 85°C and 85% relative humidity (RH) under short circuit conditions in an indoor environmental chamber as per International Electrotechnical Commission (IEC) 61215-2. After the DH stress, the modules were subjected to PID in an indoor environmental chamber at 1000 V, 60 °C, and 85% RH, according to IEC 62804-1. One module was stressed under a negative polarity for PID stress by supplying a negative voltage to the cell through shorted module connections, and a positive voltage was applied to the frame. The second module was stressed under a positive bias by supplying a positive voltage to the cell through shorted module leads, and a negative voltage was applied to the frame. To apply voltage to the frame, the anodized Al frame was removed to expose the conductive Al layer beneath.

PID stress was applied twice, with the first cycle lasting 96 hours. As little degradation was observed during the first round, a second round of 192 hours of PID stress was also conducted, totaling 288 hours of PID stress. Leakage current was also monitored for the duration of PID. To ascertain the

978-1-6654-6060-6/23 $31.00 © 2023 IEEE

change in performance characteristics for each module, pre- and post-characterization tests were carried out. These included outdoor light IV, which was used to measure the IV data (results were converted to standard test conditions (STC) 1000 W/m^2 and 25°C). EL (at I_{SC} and 60-sec exposure) was utilized for EL image/gray value calculation, and dark IV was performed to calculate shunt resistance (R_{SH}) and series resistance (R_S).

III. RESULTS AND DISCUSSION

A. Impact of DH Stress

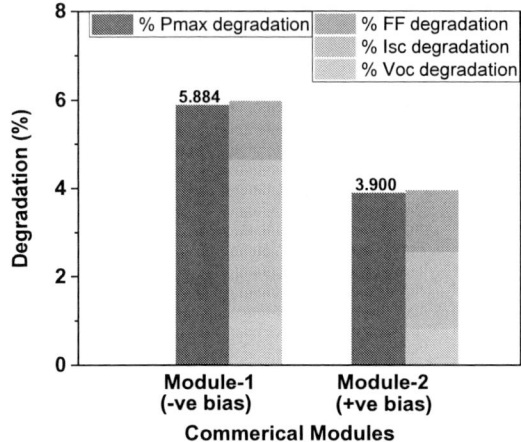

Fig. 1. Change in IV parameters due to DH stress (2000h)

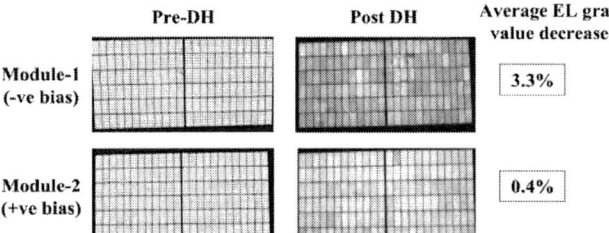

Fig. 2. Pre and post (DH) EL captured at I_{SC} and 60s exposure with gray value change impact of PID on DH stressed module

Fig. 1 shows the change in IV parameters after the DH stress, and Fig.2 shows EL images with EL gray value degradation after the DH stress. The results show that modules experience 4 to 6% degradation in Pmax after the DH stress.

Moisture can seep through the backsheet during the DH stress, causing water damage to the module. The acetic acid that develops from EVA is the main factor for cell darkening and power loss in these modules after DH stress. Acetic acid can erode the glass in the area between the silicon cell and the silver paste, which can lead to a significant increase in the grid-contact resistance [7].

B. Impact of PID on DH stressed modules

For the PID stress, module-1 is stressed under a negative bias, and module-2 is stressed under a positive bias. The PID results in this section are combined for both rounds (96+192h) for the total 288h PID duration since minimal degradation was seen during the first 96h. Fig. 4 shows the change in IV parameters after PID on DH-stressed modules. Fig. 4 shows that EL images with gray value change after PID in both modules, and Fig. 5 combines the decline in R_{SH} and R_S with the gray value change. Table.1 displays the absolute values for R_{SH} and R_S. The leakage current plot for the PID stress is also shown in Fig. 5.

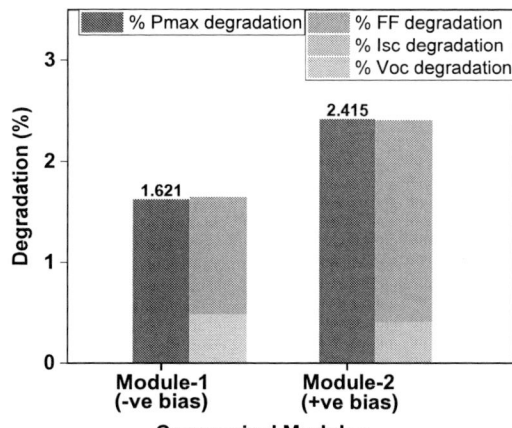

Fig. 3. Change in IV parameters due to PID (288h) after DH (2000h) stress

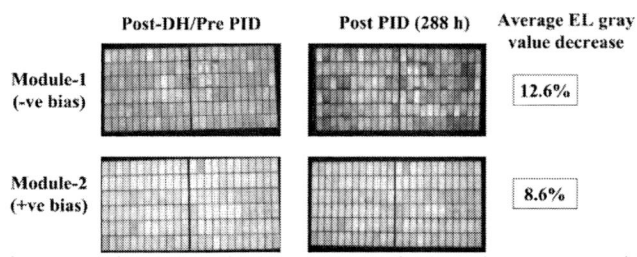

Fig. 4. Pre and post (PID after DH) EL captured at I_{SC} and 60s exposure with gray value change

Table 1. Pre and post-PID (after DH) data for R_{SH} and R_S for both modules obtained from dark IV

Module	R_{SH} pre-PID / Ω	R_{SH} post-PID / Ω	R_S pre-PID / Ω	R_S post-PID / Ω
1	12029	3316	0.533	0.541
2	5000	5000	0.511	0.528

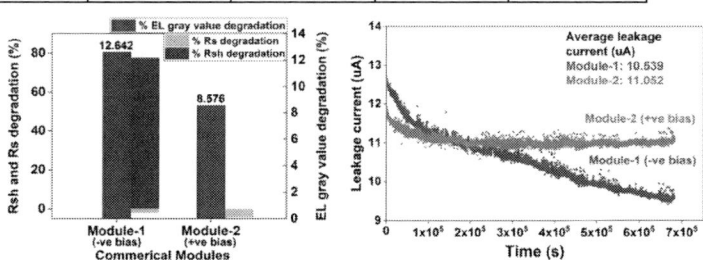

Fig. 5. % EL gray value degradation with R_{SH} and R_S change (left), Time series plot of leakage current for 192h of PID stress (right)

The power loss in module-1 mainly arises from a loss in FF and R_{SH} (more EL darkening), as observed in Fig. 3,4, and 5. It could suggest that PID-shunting (PID-s) is the degradation mechanism. Under a negative bias, sodium ions in the glass can migrate towards the cell and diffuse into the PN, junction leading to a decline in the FF and R_{SH} [8].

The power loss in module-2 comes from a drop in FF linked to a rise in R_S (less EL darkening), as seen in Fig. 3,4, and 5. The leakage current for module-2 is also slightly higher than module-1. These results conform to the findings of Kern et al. [9]. They see a corresponding increase in R_S that results in a decrease in FF in GB modules under a positive bias. They attribute this loss to PID-corrosion (PID-c). According to their explanation, the metallization of the module was corroded electrochemically. Due to a positive bias, the Silicon Nitride layer can be damaged due to moisture ingress and the presence of acetic acid evolving from EVA. Hence, electrochemical oxidation is triggered by water entering the module and coming into contact with the metallization leading to PID-c of grid fingers [9].

C. Combined impact of sequential DH and PID stress

Due to sequential DH and PID, module-1 declined in Pmax by 7.5% and module-2 by 6.3%. A decline in EL gray value of 15.9% was seen for module-1 and 9.0% for module-2.

Fig. 6 shows the infra-red (IR) images taken after both DH and PID stress. The IR images are taken on a bright sunny day with an irradiance of 1062 W/m^2 and 23.5°C ambient temperature under short circuit conditions. ΔT is computed using (1) and compared with the control module, which was unstressed.

$$\Delta T = Max\ module\ temp - Avg\ module\ temp$$

The ΔT for the control module is 11.8°C, whereas a ΔT of 10.6°C and 16°C for module-1 and module-2, respectively. There is no change in ΔT for module-1. The minor rise in ΔT for module-2 could be due to the rise in R_S. The difference is negligible mainly because the backsheet allows for heat removal. Moreover, the half-cut cell structure leads to lower I^2R losses.

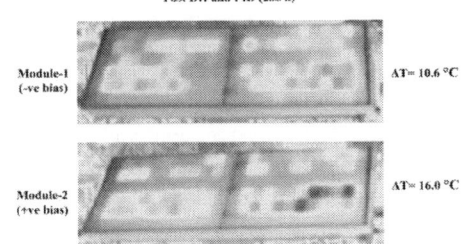

Fig. 6. Outdoor IR for both modules after DH and PID stress

IV. CONCLUSIONS

The GB modules under sequential DH and PID stress exhibit 4 to 6% DH-related deterioration and an additional 1.5 to 2.5% PID-related degradation. Due to sequential testing of PID after DH, GB modules exhibit 6 to 7% deterioration.

Moisture ingress through the backsheet is the leading cause of DH stress deterioration. Due to reduced interfacial adhesion after the DH stress, the modules exhibit some deterioration during the PID stress. PID-s reduces the power in module-1 under a negative bias because FF accounts for most of the power loss. Due to a decrease in the FF and a rise in the R_S, module-2 experiences PID-c under a positive bias.

The results presented in this work are statistically insignificant and cannot be associated with all GB modules.

ACKNOWLEDGEMENT

This material is based upon work supported by the Department of Energy, Office of Energy Efficiency and Renewable Energy (EERE), under Award Number DE-EE-0008565.

REFERENCES

[1] M. Owen-Bellini et al., "Advancing reliability assessments of photovoltaic modules and materials using combined-accelerated stress testing," Prog. Photovoltaics Res. Appl., vol. 29, no. 1, pp. 64–82, 2021.

[2] "IEC TS 62804-1: 2015 Photovoltaic (PV) modules-Test methods for the detection of potential induced degradation-Part 1: Crystalline silicon," IEC--International Electrotech. Comm. Ed, vol. 1, 2015.

[3] S. R. V. Tatapudi, "Potential induced degradation (PID) of pre-stressed photovoltaic modules: effect of glass surface conductivity disruption," Arizona State University, 20152.

[4] P. Hacke et al., "System voltage potential-induced degradation mechanisms in PV modules and methods for test," 2011 37th IEEE Photovolt. Spec. Conf., pp. 000814–000820, 2011.

[5] Y. Kobayashi, H. Morita, K. Mori, and A. Masuda, "Effect of barrier property of backsheet on degradation of crystalline silicon photovoltaic modules under combined acceleration test composed of UV irradiation and subsequent damp-heat stress," Jpn. J. Appl. Phys., vol. 57, no. 12, 2018.

[6] G. TamizhMani and J. Kuitche, "Accelerated Lifetime Testing of Photovoltaic Modules Solar America Board for Codes and Standards," 2013.

[7] C. Peike et al., "Origin of damp-heat induced cell degradation," Sol. Energy Mater. Sol. Cells, vol. 116, pp. 49–54, 2013.

[8] J. Bauer, V. Naumann, S. Großer, C. Hagendorf, M. Schütze, and O. Breitenstein, "On the mechanism of potential-induced degradation in crystalline silicon solar cells," Phys. status solidi (RRL)–Rapid Res. Lett., vol. 6, no. 8, pp. 331–333, 2012.

[9] K. Brecl, M. Bokalic, and M. Topic, "Examination of Photovoltaic Silicon Module Degradation under High-Voltage Bias and Damp Heat by Electroluminescence," J. Sol. Energy Eng. Trans. ASME, vol. 139, no. 3, pp. 1–6, 2017.

Data-Driven Photovoltaic Module Performance Analysis with FAIR Data

Mengjie Li[1,2], Jarod Kaltenbaugh[3], Dylan J. Colvin[1,2], William C. Oltjen[4,6], Arafath Nihar[4,6], Dominique Akissi Yao[4,6], Xuanji Yu[4,6], Alp Sehirlioglu[5,6], Roger H. French[4,6], Kristophor O. Davis[1,2,3]

[1] Florida Solar Energy Center (FSEC), University of Central Florida (UCF), Cocoa, FL 32922, USA
[2] Resilient, Intelligent and Sustainable Energy Systems (RISES) Cluster, UCF, Orlando, FL 32816, USA
[3] Department of Materials Science and Engineering, UCF, Orlando, FL 32816, USA
[4] SDLE Research Center, Case Western Reserve University (CWRU), Cleveland, OH, 44106, USA
[5] Electro-ceramics Group, CWRU, Cleveland, OH, 44106, USA
[6] Department of Materials Science and Engineering, CWRU, Cleveland, OH, 44106, USA

Abstract—**Due to rapid growth of the photovoltaic (PV) market, a huge amount of data is generated everyday. However, networking and data exchange remains challenging due to the scarcity of interoperability and due to the fact that most datasets are stored locally. The "Findable, Accessible, Interoperable, Reusable" (FAIR) Data Principles have been developed to provide guidelines to improve the management of digital assets, so that metadata and data are both human-readable and machine-readable. This work showcases an example where data collected from current and voltage ($I-V$) measurements of over 1,500 photovoltaic (PV) modules of varying types (e.g., monocrystalline, multicrystalline; aluminium back surface field (Al-BSF), passivated emitter rear cell (PERC), heterojunction technology (SHJ) and interdigitated back contact module (IBC) and histories of exposure (e.g., as manufactured, installed in the field and exposed to accelerated aging) can be FAIRified and stored as linked data. This is an important step towards unified documentation of data with multiple registries. Subsequently, the FAIR data is used to build a data-driven PV module performance analysis model. The $I-V$ characteristics of a PV module carry a huge amount of information. Traditional practice of $I-V$ analysis involves curve fitting with predefined mathematical equations. The simplified equation however could create obstacles in understanding the data and the actual fault and degradation happening within the PV module could be overlooked. Data-driven model on the other hand will be able to extract information that is carried by the real data. This work adapts the previous developed data-driven $I-V$ analysis model and implements additional features to collectively analyse dark $I-V$ curves, light $I-V$ curves and pseudo $I-V$ curves. The integration of data FAIRification and use of data-driven models play an important role in long-term scalability and maintainability in PV research.**

Index Terms—**FAIRification, data-driven, light $I-V$ curves, dark $I-V$ curves, PV reliability and durability, degradation analysis**

I. Introduction

By the end of third quarter of 2022, the U.S. reached a total of 135.7 GW_{dc} of installed solar capacity [1]. Solar has accounted for 45% of all new electricity-generating capacity added in the U.S. through the first three quarters of 2022. In 2021, solar photovoltaic (PV) employed around 3.4 million people globally [2]. With the rapid growth of the PV market

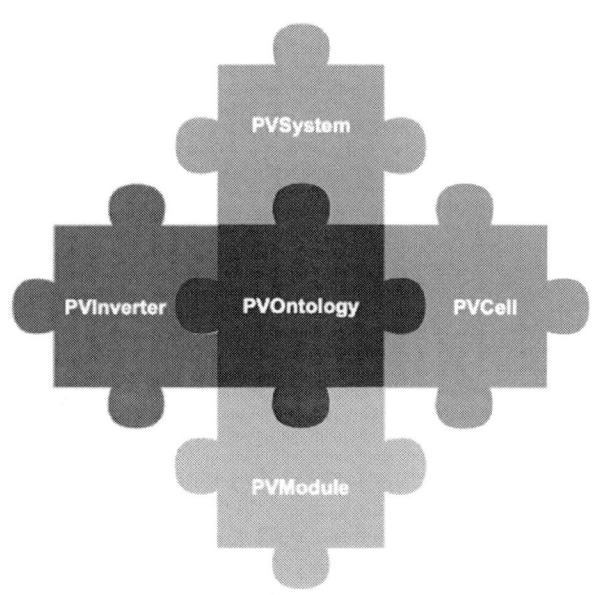

Fig. 1. The structure of PV ontology, where PVSystem, PVInverter, PVModule and PVCell are defined as their own sub-ontologies. The sub-ontologies are designed to be modular and interconnected.

and the vast number of people working within the field, a huge amount of data is generated everyday within the PV community. A good data management approach is the key conduit leading to innovation and scientific discovery, and to subsequent knowledge integration and data reuse by the community. Unfortunately, the existing data management practices, or lack thereof, often prevents us from maximising the benefits of existing research. Data sharing can often be challenging due to the fact that various naming conventions exist and most data are stored locally. The "Findable, Accessible, Interoperable, Reusable" (FAIR) Data Principles [3] are proposed and intended to serve as guidelines to improve the data management and data reusability. The goal is to create structured data that can be easily accessed by both humans and machines. By

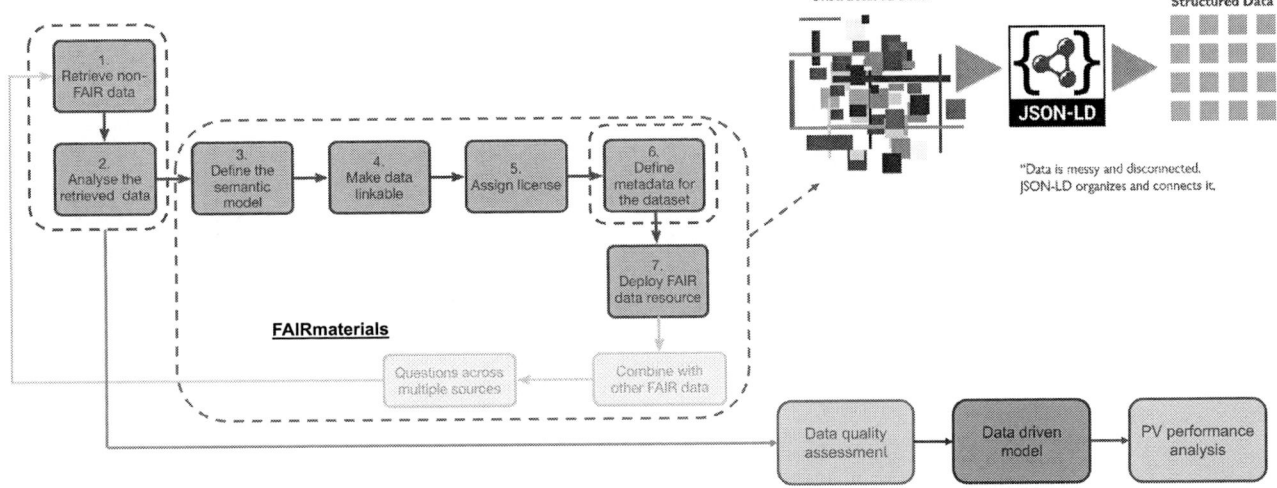

Fig. 2. Workflow: Raw data (non FAIR data) is unstructured data. Raw data is first retrieved and passed through FAIRmaterials [4], [5] package. Within FAIRmaterials, it defines the semantic model, makes the data linkable, assign license and defines metadata for the dataset. The package automatically generates a FAIRified JSON-LD file based off the PV Ontology. The FAIR data is assessed for quality before passed into data-driven model for PV performance analysis. The FAIR data is also write back to the database for future reuse.

adopting the FAIR principles, FAIRmaterials [4], [5] packages have been created to help the data FAIRification process. In this work, we propose to demonstrate an example of using the FAIRmaterials to FAIRify a huge dataset of current-voltage ($I - V$) curves measured on various types of PV modules. Additionally, building upon previously developed data-driven $I - V$ analysis tool [6], [7], we present the implementation of dark $I - V$, light $I - V$ and pseudo $I - V$ analysis to extract various key performance parameters of PV modules.

II. FAIRIFICATION

The FAIR principles were first published in 2016 [8]. The authors intended to provide guidelines to improve the Findability, Accessibility, Interoperability and Reuse of digital assets. Due to the increase in volume, complexity and creation speed of data, humans are relying more and more on machines for computational support. The FAIR principles aim to create structured data accompanied with clearly defined metadata, so that the (meta)data is easily human-readable as well as machine-readable. The most important step in the FAIRification process is coming up with a set of globally unique and persistent identifiers that will be used to describe the (meta)data. To create the identifiers that are unique yet globally accepted and used in PV research, we have created the PV Ontology (Figure 1). An ontology is a formal dictionary of concepts and categories in a subject area or domain that describes their properties and the relations between them. The PV Ontology carefully defined the entities that are commonly involved in PV research and the (meta)data associated with them. Specific terms used in PV Ontology have referenced Orange Button [9]. Within PV Ontology, there are sub-ontologies currently including PVSystem, PVModule, PVInverter and PVCell. The sub-ontologies are designed in a way that they are modular, yet interconnected. When an ontology is filled with real data, it

TABLE I
SUMMARY OF THE $I - V$ DATASET

No. of curves	Dark IV	Light IV	Pseudo IV
Mono-AlBSF	598	598	598
Multi-AlBSF	779	779	779
Mono-PERC	298	298	298
HJT	77	77	77
IBC	13	13	13

becomes a knowledge graph. Previous works have showcased how PV system time series data can be FAIRified and analysed [10], [11]. This work will be focused on FAIRification of PV module $I - V$ curves. The FAIRification is realized with the FAIRmaterials packages, currently available online in both Python [4] and R [5]. After FAIRification, the (meta)data is stored as a standardized Javascript Object Notation for Linked Data (JSON-LD) file. The data used in this study is not made public yet at the time of submission, but it will be made accessible on OSF.io and figshare.com.

III. DATA-DRIVEN $I - V$ ANALYSIS

Figure 2 shows the workflow, where raw data was first FAIRified with FAIRmaterials [4], [5]. Then, the FAIR data was passed through data quality assessment and then into data driven model. The dataset of $I - V$ curves are collected at Florida Solar Energy Center, located in Cocoa, Florida, U.S. FSEC is a research institute of the University of Central Florida (UCF), located on a 20-acre (approx. 81,000 m^2) research complex on Florida's Space Coast at UCF's Cocoa satellite campus. FSEC is the largest and most active state-supported renewable energy and energy efficiency research,

978-1-6654-6060-6/23 $31.00 © 2023 IEEE

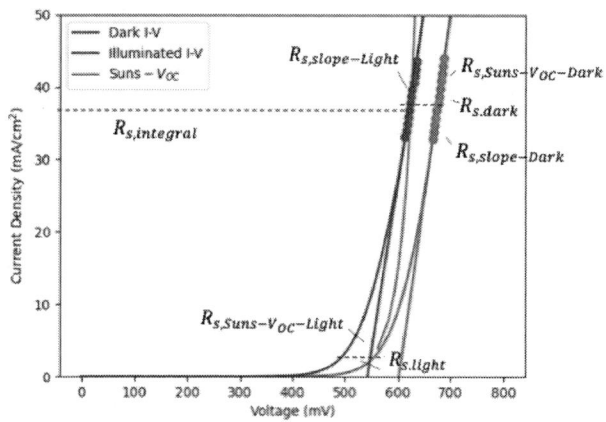

Fig. 3. Different methods summarized in CurveRs [12] to calculate series resistance R_s with input of dark $I - V$, light $I - V$ and $SunsV_{oc}$ curves.

training, testing and certification institute in Florida. FSEC manages and maintains PV performance data of over 100 systems via SunSmart emergency shelters program and the field test site monitors over 850 modules of 17 different types installed between 2002 and 2018. The dataset of $I - V$ curves studied in this work includes, dark $I - V$ curves, light $I - V$ curves and $SunsV_{oc}$ (pseudo $I - V$) curves measured on various type of modules. Table I summarized the number of each type of $I - V$ curves available within this dataset. Within the time frame of 2002 - 2022, monocrystalline and multicrystalline modules with aluminium back surface field (Al-BSF) cells, passivated emitter rear cell (PERC), heterojunction technology (SHJ) cells and interdigitated back contact module (IBC) cells have been periodically measured at FSEC. These modules have a combination of histories of exposure (e.g., as manufactured, installed in the field and exposed to accelerated aging). In total, about 5,295 entries of $I - V$ curves measured on about 1,765 modules will be used in the data-driven model.

The $I - V$ characteristics of PV modules carry a huge amount of information. Especially when collectively analysing the dark $I - V$ curves, light $I - V$ curves and pseudo $I - V$ curves, key performance parameters and indicators of the PV module degradation behavior can be obtained. For example, open circuit voltage (V_{oc}) and short circuit current (J_{sc}), can be directly extracted from the $I - V$ curves. On top of that, fill factor (FF), maximum power output (P_{mpp}), shunt resistance (R_{sh}) and series resistance (R_s) can be calculated from the $I - V$ curves. In previous studies [12] the authors have reported different approaches on R_s calculation (Figure 3). In this work, the results obtained with the different approaches will be compared with the data-driven model.

IV. CONCLUSION

There is an urgent need to improve the management of research data generated within the PV community. The data FAIRification process can benefit a wide range of stakeholders, including: academia, industry, funding agency and publishers. This work attempts to present an example of the process of FAIRifying a huge dataset of $I - V$ curves measured on various types of PV modules and subsequently using the FAIR data to build a data-driven model for PV performance analysis. By creating the PV Ontology, we would like to propose standards for the naming and ways of organising the data generated within the PV community. With a data-driven model built with a huge dataset, we would like to demonstrate a new way of PV performance analysis.

ACKNOWLEDGMENT

This work is supported by the U.S. Department of Energy (DOE) - Office of Energy Efficiency and Renewable Energy (EERE) under Solar Energy Technologies Office (SETO): [DE-EE0009347] and Department of Energy (DOE) - National Nuclear Security Administration (NNSA): [DOE-NNSA-B6477887].

REFERENCES

[1] U.S. Solar Market Insight, https://www.seia.org/us-solar-market-insight, Dec 13 2022.
[2] Solar PV Employed Around 3.4 Million People In 2021, https://www.pv-magazine.com/2022/09/19/solar-pv-employed-around-3-4-million-people-in-2021/, Sep 19 2022.
[3] Go FAIR, https://www.go-fair.org.
[4] FAIRmaterials, https://pypi.org/project/fairmaterials/, Jan 10 2023.
[5] FAIRmaterials, https://cran.r-project.org/web/packages/FAIRmaterials/index.html, Jan 10 2023.
[6] X. Ma et al., "Data-Driven $I - V$ Feature Extraction for Photovoltaic Modules," in IEEE Journal of Photovoltaics, vol. 9, no. 5, pp. 1405-1412, Sept. 2019, doi: 10.1109/JPHOTOV.2019.2928477.
[7] ddiv, https://cran.r-project.org/web/packages/ddiv/index.html, Apr, 14, 2021.
[8] Wilkinson, M., Dumontier, M., Aalbersberg, I. et al. The FAIR Guiding Principles for scientific data management and stewardship. Scientific Data, Nature 3, 160018 (2016). https://doi.org/10.1038/sdata.2016.18
[9] Orange Button, https://orangebutton.io.
[10] A. Nihar et al., "Toward Findable, Accessible, Interoperable and Reusable (FAIR) Photovoltaic System Time Series Data," 2021 IEEE 48th Photovoltaic Specialists Conference (PVSC), Fort Lauderdale, FL, USA, 2021, pp. 1701-1706, doi: 10.1109/PVSC43889.2021.9518782.
[11] W. C. Oltjen et al., "FAIRification, Quality Assessment, and Missingness Pattern Discovery for Spatiotemporal Photovoltaic Data," 2022 IEEE 49th Photovoltaics Specialists Conference (PVSC), Philadelphia, PA, USA, 2022, pp. 0796-0801, doi: 10.1109/PVSC48317.2022.9938523.
[12] CurveRs, https://github.com/ucf-photovoltaics/PVRs.

Proposed Update of the Colombian Technical Standard NTC 4405 for Evaluating the Efficiency of Photovoltaic Solar Systems and Their Components

Johann Hernández[1], Daniel H. Gamboa[1], and Juan F. Beltrán[2]

[1]Grupo LIFAE. Universidad Distrital Francisco José de Caldas. Bogotá, Colombia.
[2]Northern Alberta Institute of Technology. Alberta, Canada.

Abstract — **This study introduces the updated version of the NTC 4405 standard, which was last revised in 1998 and no longer aligns with the modern equipment being used. Furthermore, it fails to address crucial stages, such as energy conversion and inversion. By thoroughly examining national and international regulations, a comprehensive methodology has been developed to accurately measure the total efficiency of off-grid solar photovoltaic systems. This methodology encompasses every stage of the system, including solar panels, charge controllers, battery banks, and power converters/inverters.**

I. INTRODUCTION

Solar photovoltaic systems offer an environmentally friendly alternative and a viable solution to meet energy demands. These systems utilize clean and renewable solar energy to generate electricity, significantly reducing greenhouse gas emissions and other negative environmental impacts associated with conventional power generation [1].

One type of these systems is Off-Grid, which provides great benefits in areas without access to the conventional electrical grid. They can drive social and economic development while promoting environmental sustainability. Due to the efficiency criteria that Off-Grid systems must meet, their measurement becomes an essential aspect in the development of these projects [2].

Therefore, it is crucial to have a technical regulation that defines and specifies the components and procedures, including pre-conditioning and current connection diagrams, to effectively measure the efficiency of photovoltaic systems. As there is no international technical standard that addresses the measurement methodology, and the existing Colombian technical standard lacks updated information, it is essential to compile the relevant data on efficiency measurement in photovoltaic systems into a document. This compilation should consider the new technologies implemented in component manufacturing and system design.

II. JUSTIFICATION

A detailed study was conducted on the Colombian technical standard NTC 4405 (*Evaluation of the efficiency of off-grid solar photovoltaic systems and their components*) [3], which presents an incomplete model of off-grid photovoltaic systems. It only describes solar panel or module components, charge controllers, and battery banks, without including all the equipment necessary for modern systems that incorporate different technologies. Although the standard references the measurement and calculation of component efficiency through various tests, some information is different and not directly related to the standard's efficiency measurement objectives. Instead, the standard indicates several manufacturer measurements and procedures, and in some cases, mentions that the manufacturer should attach data sheets.

III. NTC 4405 STANDARD ANALYSIS

This standard consists of 8 chapters. The first four chapters focus on the formal structure of the technical standard, which is defined in the introduction, objectives, reference standards, and nomenclature. The following four chapters address the measurement of efficiency in the equipment that comprises the photovoltaic system. This system is primarily composed of three equipment components: solar panels, charge controllers, and batteries. Although these components are presented in the diagram shown in Figure 1, the standard does not refer to any specific diagram.

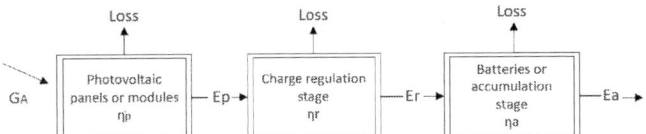

Fig. 1. OFF-GRID Solar System blocks' diagram presented in the NTC 4405 standard [3]

Furthermore, with the mentioned diagram, the formula for the system efficiency is the one presented below:

$$\eta = \eta_p * \eta_r * \eta_a * \eta_c \quad (1)$$

From the above, it can be observed that the description of the photovoltaic system presents an inconsistency in the number of

978-1-6654-6060-6/23 $31.00 © 2023 IEEE

components. According to Eq. 1, the system is composed of four components, while the diagram only shows three.

The standard analysis showed spelling, referential, procedural, and information presentation errors. The most significant errors found in the standard were those in the chapter referring to calculating and measuring efficiency in the load regulation stage.

In Chapter 6, a model of the solar cell is presented, including its characteristic diode curve, the operating principle of the solar cell, equations for measuring efficiency, instrumentation requirements, and a connection diagram for testing.

In Chapter 7, tests for measuring and calculating the efficiency of charge controllers are described. These tests include the maximum load current test, overcurrent test, voltage drop test in the fully loaded controller, power consumption test in the absence of radiation, maximum load current measurement, reconnection voltage measurement, and power consumption test during the day. However, these tests are of the Factory Acceptance Test (FAT) type and are not directly focused on measuring the efficiency of charge controllers.

On the other hand, Chapter 8, titled "Test of Lead-Acid Accumulators for Photovoltaic Installations," specifies that the batteries must comply with the provisions of the NTC 2959 standard [4]. However, it does not describe the procedures, connection diagrams, instruments, or any relevant information that would lead to measuring the efficiency of the batteries.

III. RESULTS

Among the identified errors, the proposed update focused on defining a precise measurement methodology and all that it entails, such as pre-conditioning, connection diagrams, instrumentation, and variable descriptions. Additionally, presenting the obtained data, separating each stage of the system into chapters, creates a methodology that understands the components of an off-grid photovoltaic system. Through the corresponding input and output variables, it becomes possible to determine the total system efficiency by measuring the output variable against the input variable.

Chapter 5 presents the generalities of the efficiency assessment. In summary, the changes are described in Figure 2. This chapter also includes the energy conversion/inversion stage. With this addition, the system becomes adaptable to different types of loads that can be connected, eliminating the limitation, as shown in Figure 1, to only DC loads and limited by the output voltage level of the battery bank. This new equipment provides versatility in delivering power to both DC and AC loads with different voltage levels.

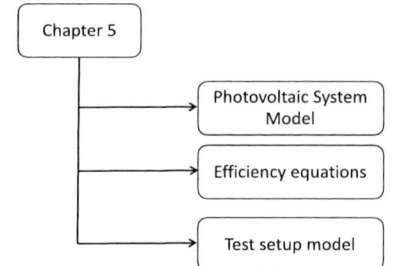

Fig. 2. Changes made on chapter 5

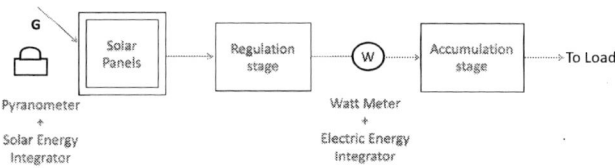

Fig. 3. Tests' Blocks' diagram model presented in standard NTC 4405 [3]

As a final point, the information presented in the connection setup for the tests is updated:

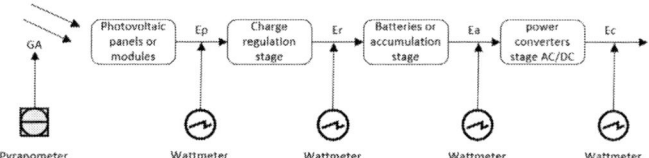

Fig. 4. Updated block diagram of off-grid photovoltaic solar system (own model)

From the above, the representation of each stage in Equation Eq. 1 is described as follows:

$$\eta_p = \frac{E_p}{G_A} = \frac{Useful\ energy\ delivered\ by\ the\ field\ of\ panels\ or\ modules}{Incident\ solar\ energy\ in\ the\ field\ of\ panels\ or\ modules} \quad (2)$$

$$\eta_r = \frac{E_r}{E_p} = \frac{Useful\ energy\ delivered\ by\ the\ regulation\ stage}{Useful\ energy\ delivered\ by\ the\ field\ of\ panels\ or\ modules} \quad (3)$$

$$\eta_a = \frac{E_a}{E_r} = \frac{Useful\ energy\ delivered\ by\ the\ accumulator}{Useful\ energy\ delivered\ by\ the\ regulation\ stage} \quad (4)$$

$$\eta_c = \frac{E_c}{E_a} = \frac{Useful\ energy\ delivered\ to\ the\ load}{Useful\ energy\ delivered\ by\ the\ accumulator} \quad (5)$$

Regarding the photovoltaic panel or module model, the chapter 6 of the standard proposes a thorough analysis of the electricity generation process through semiconductors. The objective is to provide greater clarity on the panel's functioning and its capacity to generate energy by utilizing the polarizing region of the semiconductor, as shown in Figure 6.

While it is true that this information can be found in general literature, we believe it is not essential to include it in this standard as it could distract the reader from the measurement process. Our main goal is to offer a document that is easily understandable for those with limited experience in the subject, considering that these systems may be installed in areas where the study of such systems is limited. The changes made in Chapter 6 are described in Figure 5.

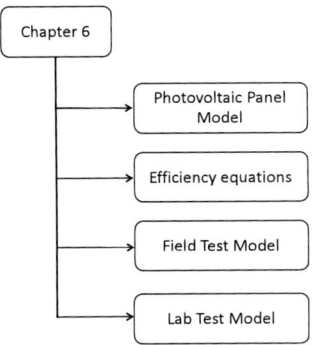

Fig. 3 changes made in Chapter 6

This chapter provides a detailed description of the parameters necessary to determine the output power of solar panels through current and voltage measurements, as well as how to identify the maximum power point to determine the solar panel efficiency. It also includes the measurement of incident solar energy. Furthermore, the presentation of information is updated, clearly specifying the measured variables and ensuring the verification of the maximum power point by measuring both voltage and current.

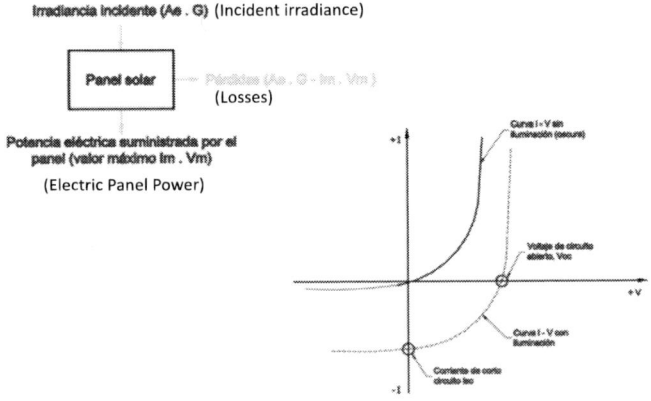

Fig. 4. Solar Cell Model and Dark I-V Curve Graph of the Semiconductor

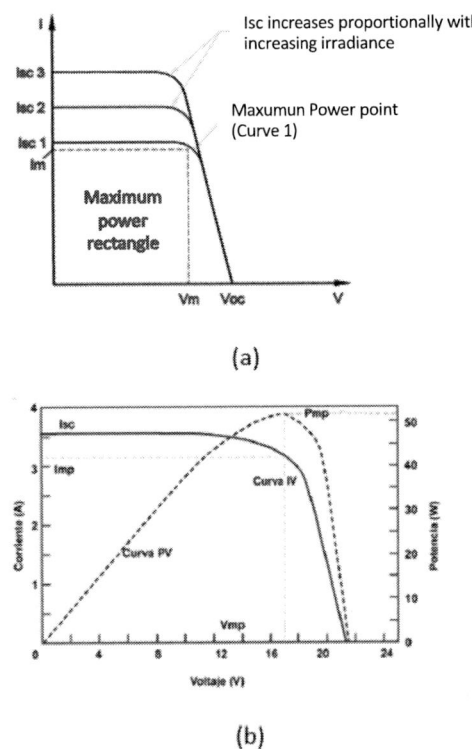

Fig. 7. Determination of the maximum power point (a) in the NTC 4405 standard [3] (b) Own model

Additionally, the equations that were previously established with measurement unit errors in the standard have been updated.

TABLE I.   EFFICIENCY EQUATION MODEL

Efficiency equation model un the NTC 4405 standard	
Equations NTC 4405	**Units**
$\eta = \dfrac{I_{sc} * V_{oc}}{G_A}$	$\eta = \dfrac{A * V}{\dfrac{W}{W+h}} = \dfrac{W}{\dfrac{W+h}{m^2}} = \dfrac{m^2}{h}$
Model of the efficiency equation in the proposal	
Updates Equations	**Units**
$\eta = \dfrac{I_{sc} * V_{oc}}{G * A}$	$\eta = \dfrac{A * V}{\dfrac{W}{m^2} * m^2} = \dfrac{W}{\dfrac{W}{m^2} * m^2}$ $= p.u$

Finally, the connection diagrams in both field and laboratory settings have been updated, illustrating the connections of the measurement equipment.

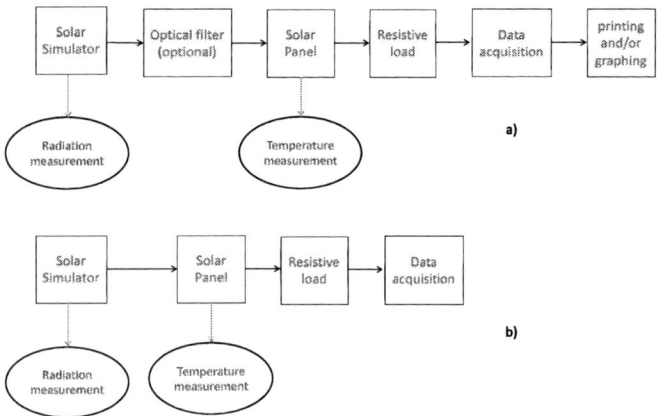

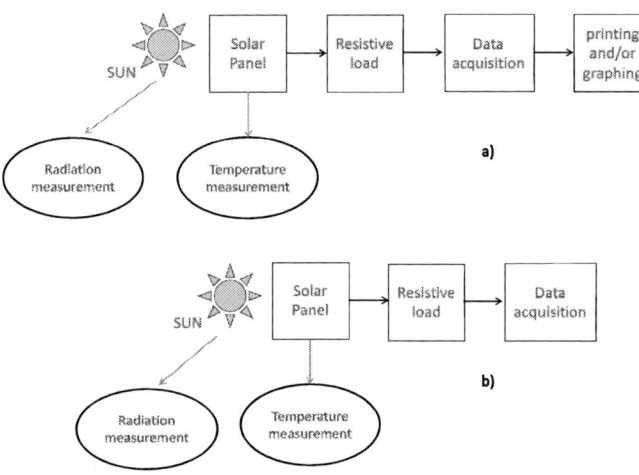

Fig. 8. a) Laboratory test setup in the NTC 4405 standard b) Custom-designed laboratory test setup

Fig. 9. (a) Field test setup in the NTC 4405 standard (b) Custom-designed field test setup

Regarding Chapter 7 on charge controllers, the presented information focused on manufacturer-specific characteristics for Factory Acceptance Tests (FAT) or Site Acceptance Tests (SAT) to verify system parameters. However, it did not have a specific focus on measuring efficiency. The content covered in this chapter is as follows:

- Maximum load current test
- Overcurrent test
- Voltage drop test in the regulator under full load
- Power consumption test in the absence of radiation
- Measurement of maximum load current
- Reconnection voltage measurement
- Daytime power consumption test

The changes made to this chapter are presented as follows:

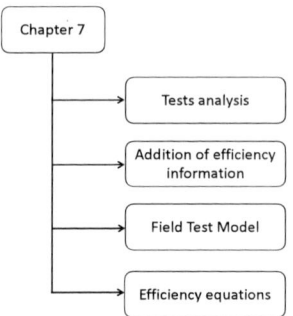

Fig. 10. Changes made to chapter 7

Based on the IEC 62509 standard, which states that this test should be conducted for a load range between 10% and 100%, several factors should be considered:

1. Self-consumption: When the battery voltage is equivalent to 2.1 V/Cell ±2% and the ambient temperature is 25°C ±2 °C.

TABLE II.   MAXIMUM SYSTEM AUTO-CONSUMPTION

Nominal load current	Maximum self-consumption
< 5 A	5 mA
5 A < I < 50 A	0.1% of rated load current
> 50 A	50 mA

2. Configuration and pre-conditioning:

- Solar simulator:
- $V_{OC} \geq 2 * V_{BAT-NOM}$
- $I_{SC} \geq 1,25 * I_{BCC-IN}$
- Controlled voltage source:
- $V \geq 2 * V_{BAT-NOM}$
- $I \geq 1,25 * I_{BCC-IN}$
- Battery bank:
- $V \geq 1,4 * V_{BAT-NOM}$
- $I \geq 1,25 * I_{BCC-OUT}$

3. Charge cycle configuration:
- Solar simulator:

$$V_{PV-PSU} = 1,25 * V_{BAT-MAX}$$
$$I_{PV-PSU} = 10\% \ of \ the \ nominal \ PV \ input \ current$$

The voltage drops across R_s should be between 10% and 15%

$$\frac{0,1 * V_{PV-PSU}}{I_{PV-PSU}} \leq R_S \leq \frac{0,15 * V_{PV-PSU}}{I_{PV-PSU}}$$

Therefore, the minimum required power dissipation of R_s is given by:

$$P_{R_S} = I_{PV-PSU}^2 * R_S$$

- Battery bank:

$$V_{BAT-PSU} = V_{BAT-TEST}$$
$$I_{BAT-PSU} = 1,3 * I_{CHG-MAX}$$

The value of the battery capacitance should be 0.1 F ± 20% (C_B).

R_B It is a fixed-value resistor that dissipates the load current plus the current from the power supply used as a battery:

$$R_B = \frac{V_{BAT-TEST}}{1,15 * I_{CHG-MAX}} \pm 10\%$$
$$P_{R_B} \geq 1,3 * V_{BAT-TEST} * I_{CHG-MAX}$$

4. Efficiency measurement

The voltage drop and efficiency of the regulator must be measured; the efficiency in the charging state is given by the following equation:

$$\eta_R = \left(\frac{P_O}{P_I}\right) * 100$$

Finally, a connection diagram is presented for the test, which is as follows:

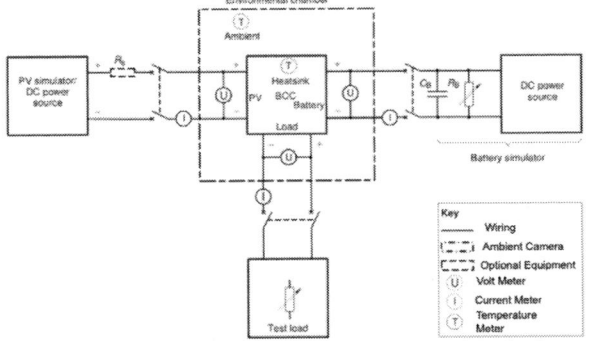

Fig. 11. Proposed setups for efficiency measurement in the regulation stage

The following equipment is the battery bank, and in the standard, it only references the NTC 2959 standard. The first change is in the title, where it is presented as "Testing of Lead-Acid Accumulators for Photovoltaic Installations", and it is changed to the following: "Evaluation of Batteries or Battery Banks", which, in a general sense, refers to the different current technologies. The consulted standards that provided relevant information about this equipment were the NTC 2959 and CAN/CSA F382-M89. With these standards, a methodology is structured to measure the efficiency of this equipment, as follow:

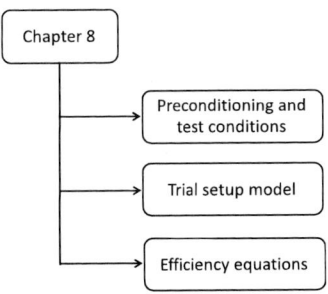

Fig. 12 Changes made to chapter 8

According to NTC 2959 [4], the preconditioning process should comply with the following:

"Before conducting the efficiency test, the battery should be discharged and charged ten times at 25 °C ± 3 °C with a C/10 rate, using the final voltages established in Tables 3 and 4, or until the capacities measured in two successive discharges differ by less than 2.5%."

The test conditions should be Standard Test Conditions (STC); however, it is at the discretion of the designer if different temperatures are required.

The voltage levels are as follows:

TABLE III. CHARGE CONDITIONS

DISCHARGE REGIME AT 25°C	ACID - LEAD (V)	NICKEL/LITHIUM (V)
C/10	1.75	1.1
C/100	1.9	1.2
C/500	1.9	1.2

TABLE IV. DISCHARGE CONDITIONS

CHARGE REGIME C/20. TEMP, °C	ACID - LEAD (V)	NICKEL/LITHIUM (V)
40	**2.12**	**1.41**
30	2.25	1.43
20	2.2	1.47
10	2.27	1.5
0	2.34	1.54
-10	2.45	1.57
-20	2.5	1.6
-30	2.55	1.63
-40	2.57	1.65

The proposed connection diagram without charge controller is as follows:

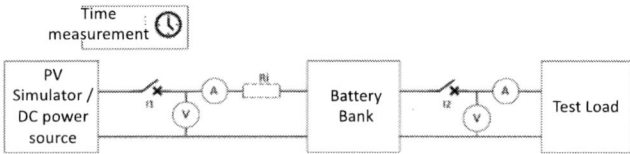

Fig. 13. Diagram of battery test setup without charge controller.

The efficiency measurement is established as follows:

$$Eficiencia\ (\eta c) = \frac{Capacidad\ (C)}{Ic\ (A)\ por\ la\ duracion\ hasta\ la\ tension\ final\ de\ carga\ (h)} * 100$$

$$\eta c = \frac{I_D * t_p}{I_C * t_c} * 100$$

The proposal includes a new equipment, the power inverter/converter, which allows the connection of loads with different levels of DC and AC voltage. The presented information is based on the IEC 61683 standards [9]. The proposed tests are as follows:

- The minimum nominal input voltage of the equipment as defined by the manufacturer.
- The nominal voltage of the inverter or the average of its nominal input range.
- 90% of the maximum input voltage of the inverter.

The efficiency will be determined by:

$$\eta_R = \left(\frac{P_O}{P_I}\right) * 100 \qquad \text{Nominal output efficiency}$$

$$\eta_{par} = \left(\frac{P_{OP}}{P_{IP}}\right) * 100 \qquad \text{Partial output efficiency}$$

Now, the connection diagram and the efficiency tolerance for the test are described in the following figure:

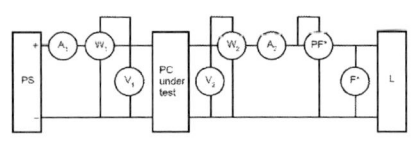

L	Load
PS	Voltage variable Source (DC)
F	Frequency meter
A1	DC Current Meter
V1	DC Volt Meter
A2	AC Current Meter
V2	AC Volt Meter
W1	DC Watt Meter
PF	Power Factor Meter
W2	AC Watt Meter

Fig 14. Proposed setups for testing inverters/converters

IV. Summary

Taking national and international standards as a reference, the changes generated are gathered in a document as a proposal for updating the NTC 4405 standard and which are presented for each stage as follows:

- Solar Panels: some descriptions were reassigned in the measurement processes.
- Regulation stage: The methodology for measuring the efficiency of this equipment and everything related to the measurement process was updated.
- Accumulation stage or battery bank: the necessary tests for measurement and prior conditioning are specified, as well as the updating of the new technologies.
- Inversion/conversion stage: information regarding the connection diagram, measurement methodology, equipment conditioning, and variables to consider.

V. References

[1] Colombian Regulatory Framework and Renewable Energy. "Colombian Regulatory Framework for ZNI and Renewable Energy Sources." Retrieved from https://ipse.gov.co/documento_prensa/documento/documentos_de_investigacion/Articulo_MundoElectrico.pdf.

[2] "New Regulations for the Provision of Electric Service in Non-Interconnected Areas, in Law 2099 of 2021 - Blog Del Sector Minero-Energetico." Retrieved from https://boletinmineroenergetico.uexternado.edu.co/nuevas-regulaciones-para-la-prestacion-del-servicio-electrico-en-las-zonas-no-interconectadas-en-la-ley-2099-de-2021/ (January 16, 2023).

[3] ICONTEC. 1998. "NTC 4405." Icontec.

[4] ICONTEC. 2009. "NTC 2959." Icontec (571).

[5] IEC. 2010. "INTERNATIONAL STANDARD IEC Standard International Standard Battery Charge Controllers for Photovoltaic Systems - Battery Charge Controllers Performance and Operation for Photovoltaic Systems - Performance and Operation." Retrieved from http://solargostaran.com.

[6] "CSA F382 - Characterization of storage batteries for photovoltaic systems | ↑ Engineering360." Retrieved from https://standards.globalspec.com/std/14207805/csa-f382 (January 16, 2023).

[7] "IEC 62620:2014 | IEC Webstore | Rural electrification, energy storage, battery, energy efficiency, smart city, LVDC." Retrieved from https://webstore.iec.ch/publication/7268 (January 16, 2023).

[8] IEC 61427-1:2013 | IEC Webstore | Rural electrification, energy storage. Retrieved from https://webstore.iec.ch/publication/5449 (January 16, 2023).

[9] IEC 61683:1999. "Photovoltaic Systems - Power Conditioners - Procedure for Measuring Efficiency." Retrieved from www.iec.ch.

Aerosol-deposited SnOx as an electron contact in perovskite solar cells

David Quispe, David Matthews, Zhengshan J. Yu, Zachary C. Holman

Arizona State University, Tempe, AZ, United States

Metal oxides are excellent candidates as carrier-selective contacts for perovskite solar cells due to their tunable surface work functions. In particular, the use of solution-based SnOx has realized perovskite solar cells with >20% efficiency but its application has been limited to structures with SnOx deposited before the perovskite. On the other hand, atomic layer deposition (ALD) is used to enable >20% efficient cells with SnOx deposited after the perovskite on top of buffer layers. Thus, perovskite/silicon tandem solar cells utilizing SnOx have been limited to the use of ALD. To provide an alternative to ALD, we investigate the potential for a novel aerosol-based tool to deposit SnOx, at room temperature, on top of the perovskite (with and without buffer layers). We find that the SnOx is not selective enough to electrons, regardless of the two precursors used, resulting in fill factors < 59%. For one SnOx precursor, a champion efficiency of 10.9% was achieved usng a C60/BCP buffer and a 100 °C post-annealing treatment. For the other SnOx precursor, a champion efficiency of 8.0% was achieved with no buffer layers or post-treatments. Additionally, we find that the work function of this SnOx can be reduced to 4.05 eV by incorporating oxygen in the deposition. This provides a path towards potentially improving the electron-selectivity of the SnOx and the use of this aerosol-deposited SnOx in perovskite/silicon tandem solar cells.

The Temperature Dependence of Auger Recombination in Silicon

Jorge Ochoa, Simone Bernardini, Mariana Bertoni

Arizona State University, Tempe, AZ, United States

For Silicon based solar cells with high carrier lifetimes, the carrier density at maximum power point sits at injection levels above 5x1015 cm-3 where the intrinsic Auger lifetime represents the next boundary to the overall device performance. Recent work has proposed revised and improved parametrizations for the Auger recombination, however these studies are restricted to room temperature. In this work we experimentally evaluate the Auger recombination and calculate its ambipolar coefficient across a range of temperatures from 303 to 453 K from photoconductance-based lifetime measurements. Results in n-type FZ silicon suggest the Auger coefficient is temperature-independent when temperature-dependent mobility models are used for the lifetime calculation. We also show that the enhancement factors geeh and gehh are the temperature-dependent parameters governing the temperature behavior of the Auger lifetime.

Combining perovskites and quantum dots: application in solar cells.

Jose Raul Montes Bojorquez, Maria Fernanda Villa Bracamonte, Kevin J. Knebel, Arturo A. Ayon

University of Texas at San Antonio, San Antonio, TX, United States

Perovskite solar cells (PSC) have experience remarkable progress over the last years and exhibit the steepest growth in power conversion efficiency (PCE) among photovoltaic (PV) technologies. Despite reaching competitive efficiencies, with potential for higher performance, durability is currently the largest technological risk for perovskite PV commercialization. Perovskites are inherently vulnerable to moisture, high temperature and UV light, thus improving their stability is one of the main unsolved issues on the field. On the other hand, quantum dots (QDs) are highly attractive materials for a variety of applications. Their tunable optoelectronic properties, easy maneuverability and high functionality offers several advantages as light harvesters that can be complementary to other technologies when employed in hybrid architectures. This work explores the combination of perovskites and luminescent down-shifting quantum dots as a strategy to simultaneously increase the UV response of PSC while improving their stability by suppressing UV-light-induced degradation.

Nanostructure Analysis of Parasitic Oxides and Contact Resisitivity Degradation during Annealing of Silicon Heterojunction Solar Cells

Stefan Lange, Angelika Hähnel, Christoph Luderer , David Adner, Martin Bivour, Christian Hagendorf

Fraunhofer Center for Silicon Photovoltaics CSP, Halle (Saale), Germany

Robert Bosch GmbH, Reutlingen, Germany

Fraunhofer Institute for Solar Energy Systems ISE, Freiburg (Breisgau), Germany

Within the last years, the dominating charge carrier transport barrier in silicon heterojunction (SHJ) solar cell contacts could be identified to be located at the interface of the indium-tin oxide (ITO) to the doped amorphous silicon (a-Si). The formation of a parasitic oxide during high-temperature annealing was hypothesized as source of contact degradation due to increased contact resistivity. However, no experimental proof could be obtained so far. In this contribution, we simultaneously investigate the contact resistivity and the interfacial structure of the electron contact of a SHJ solar cell after annealing between 140 °C and 240 °C with the help of micro transfer length measurements (μ-TLM) and highly-resolved analytical (scanning) transmission electron microscopy ((S)TEM). Contact degradation for temperatures higher than 170 °C is observed. This correlates with a densification of a parasitic silicon oxide layer at the ITO/a-Si junction and an increase of Si oxidation state within this layer. Furthermore, Ag from the metallization diffused into the Si, possibly inducing deep acceptor trap states.

978-1-6654-6060-6/23 $31.00 © 2023 IEEE

Impact of selenium doping in CdSeTe-based solar cells at the atomic-scale

Arashdeep S. Thind, Jack Farrell, Robert F. Klie

Department of Physics, University of Illinois Chicago, Chicago, IL, United States

The high cell efficiency of ~ 22% and cost-effective commercially available modules make thin-film polycrystalline CdTe-based solar cells promising alternatives to conventional energy sources. The photoconversion efficiency of CdTe-based photovoltaics can be further improved by Se doping in the absorber layer. The superior performance of $CdSe_{1-x}Te_x$ (CdSeTe)-based solar cells is attributed to a lower band gap of the CdSeTe layer in the absorber, which leads to increased carrier lifetime, and simultaneous passivation of grain interiors and grain boundaries. Se doping in the absorber layer also suppresses non-radiative recombination due to the reduction of mid-gap defect states. Despite conclusive evidence of enhanced device performance upon Se doping in the absorber layer, direct atomic-scale observations of changes in the bulk, defect, and electronic structures for CdSeTe are still lacking. Moreover, the diffusion of Se through the absorber layer, especially at low temperatures, is not well understood. Here, we have used atomic-resolution scanning transmission electron microscopy and first-principles density functional theory calculations to understand the role of Se doping on the atomic and electronic structure of CdSeTe-based photovoltaic devices. We observe that the smaller grains (< 1μm) in the CdSeTe layer are accompanied by an increased density of dislocations and dislocation cores. We find that the Se concentration profile changes across the absorber layer upon aging for 100' of hours at operational temperatures. We show that the presence of planar defects in the absorber layer leads to heterogeneities in the chemical distribution of Se. Overall, our findings provide atomic-scale insights into the role of Se doping, which can be further leveraged to improve the photovoltaic efficiency of CdSeTe-based solar cells.

978-1-6654-6060-6/23 $31.00 © 2023 IEEE

Performance of Vertical Bifacial 2T and 3T Perovskite/Silicon Tandem Solar Farms

[1]Syed Usama Bin Afzal, [2]Hassan Imran, [3]Suleman Sami Qazi, [4]Muhammad Ashraful Alam, [1]Nauman Zafar Butt[*]

[1]Department of Electrical Engineering, School of Science and Engineering, LUMS, Lahore, Pakistan
[2]Department of Electrical Engineering, School of Engineering and Applied Science, GIFT, Gujranwala, Pakistan
[3]Department of Electrical Engineering, University of Engineering and Technology, Lahore, Pakistan
[4]School of Electrical and Computer Engineering, Purdue University, West Lafayette, IN, USA
[*]Corresponding Author: nauman.butt@lums.edu.pk

Abstract— **Tandem solar cells based on Silicon Heterojunction (SHJ) and Perovskite (PVK) are being widely explored for high efficiency solar modules beyond the thermodynamic/practical limits of single junction solar cell. While a two terminal tandem (2TT) cell configuration is generally the most convenient in terms of fabrication, the performance is hampered by the current-matching constraint for the series connected sub-cells. A four terminal tandem (4TT), on the other hand, is affected by contact shading and fabrication challenges. The three terminal tandem (3TT) cell with the bottom Interdigitated Back Contact (IBC) is an attractive choice as it avoids isolation between sub-cells and the lateral current collection. Recently, there has been a number of studies on PVK/IBC tandem cells and farm level configurations, but their performance has not been explored for vertical solar farms which are attractive due to their morning/afternoon peaks and for agrivoltaic applications. Here, we present a detailed computational study on vertical bifacial solar farms made of 2TT and 3TT PVK/IBC tandem cells and their relative performance as compared to the vertical farms made of single junction HIT solar cells. We show that the annual energy production for East/West faced 3TT PVK/IBC vertical farm is 37% more as compared to 2TT series connected PVK/HIT vertical farm and 16% more as compared to single junction bifacial HIT vertical farm for the latitude of $31.5°$ N at Lahore.**

Keywords—*3TT Configuration, Albedo, IBC, Perovskite, SHJ, Solar Module, Vertical Bifacial Farm, VM Strings*

I. INTRODUCTION

Single junction silicon solar cells and modules currently have a dominant market share with the technology being highly mature and reliable with efficiency approaching the physical limits of ~29%. A remarkable progress on Perovskite (PVK) solar cells over last decade has however prompted broad interest on PVK/Si tandem cells for efficiencies beyond 29%. A variety of options for the terminal configuration of PVK/Si tandem cells have been explored including two terminal tandem (2TT) and four terminal tandem (2TT), but the monolithic solution based on three terminal tandem (3TT) using an interdigitated back contact (IBC) silicon cell has shown the best potential [1]. While 2TT can be the most convenient and cost-effective choice, the output current is limited by the spectral mismatch and albedo variation. The 4TT solar cells allow independent maximum power point tracking for each sub-cell but at the cost of increased fabrication complexity and it remains to be seen if they can reach the anticipated high performance due to contact shading, lateral transport losses and added parasitic absorption

owing to the requirement of at least three transparent electrical contacts [2],[3]. The 3TT IBC tandem configuration can have many advantages as it provides an extra terminal at the back to extract carriers which would otherwise be potentially lost to recombination in a series connected current-matching constraint.

The promising cost-performance tradeoff for 3TT IBC solar cells has recently inspired a lot of research. Tockhorn et al. [4] reported tandem solar cells with Perovskite top sub-cell and IBC-SHJ bottom sub-cell with a cited potential efficiency of 27% approximately. Since working modes of 3TT solar cells can have a broad range, many authors have used TCAD [4]-[8] for the cell level characterization and efficiencies. Innovative tandem architectures such as Bandgap Engineered Smart Three-Terminal (BESTT) solar cell [9] has also been proposed. A comprehensive understanding of various 3TT tandem configurations is provided in [10] by developing a taxonomy for all possible combinations depending on the relative doping sense for the sub-cells.

The string connections for 3TT cells in a module for an optimum performance requires an interlacing of individual sub-cells to create voltage matched (VM) strings [11]. The sub-cells are connected in series-parallel combinations with optimal VM ratio at the string level so that both top and bottom sub-strings in a module could operate near their maximum power points. The I-V curve of the overall module for a given number cells can be compared for 2TT vs. 3TT configurations for a relative comparison as detailed in Section III. The VM strings however have intrinsic end losses due to some unused sub-cells at the ends of the string within a module [12]. The interlacing of sub-cells being unique to 3TT can have some fabrication challenges but the recent interconnection technology based on bipolar junction can enable lean PERC-like fabrication process [13].

Although the cell/string level characteristics for 3TT PVK/IBC tandem has been well-characterized, the farm level performance as compared to conventional configurations has not been quantified. The vertical bifacial configuration, in particular, is of interest as it allows morning/afternoon peaks which can often be matched to the local demands. More recently, innovative applications such as agrivoltaics have sown a great interest in the vertical farms [14]. Here, we present a detailed computational study to quantify the performance of PVK/Si tandem solar farms with 3TT IBC bottom cell in comparison with the single junction HIT and the current matched 2TT PVK/HIT for the vertical module configuration.

978-1-6654-6060-6/23 $31.00 © 2023 IEEE

II. MODELING FRAMEWORK

Individual sub-cells for the tandem PVK/Si solar cells are simulated using Silvaco ATLAS tool [12]. The sub-cell characteristics are then concatenated at the module level for a given VM ratio to simulate the complete string. The cell results are then post processed for a given number of cells, V_M ratio and cell area which are taken to be 72, 3/2 and 15.6cm×15.6cm, respectively for illustration of a typical panel [15]. Cell level results are then fed into the farm level modeling framework to simulate the performance of vertical bifacial solar PV arrays.

A. Tandem Cell Configurations

The nomenclature described in [5] is used which simplifies the classification of various 2TT/3TT cell configurations. The 2TT cell used here is a 2T PVK/s/HIT(p/n) as shown in Figure 1(a) whereas the 3TT cell in figure 1(b) is a 3T PVK/r/nuIBC, similar to the one used by Tockhorn et al. [4] but albedo optimized back contacts to effectively capture the albedo light for the bottom sub-cell [2].

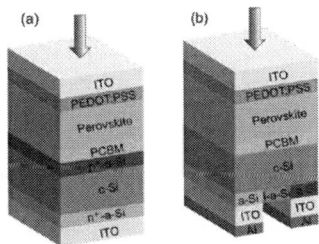

Fig. 1. Structure of the (a) Two Terminal PVK/HIT Tandem (b) Three Terminal PVK/HIT IBC Tandem

The 2TT is a forward series connected configuration where the Front is the hole collecting contact and the Back is the electron collecting contact. The 3TT is a reverse connected configuration with both, one of the Back as well as the Front contacts being hole collecting contacts whereas the extra contact at the back is the electron collecting contact responsible for solely collecting for both, the top as well as the bottom sub-cell ensuring maximized carrier extraction by avoiding current mismatch, implying higher efficiency for 3TT.

The solar illumination on the front of the sub-cell is hindered by ITO only which serves as the Front contact whereas the albedo is hindered by ITO as well as blocked by metallic fingers of the Back and Extra contacts. It is worthwhile to notice that a traditional tandem bottom SHJ sub-cell has only one contact at the bottom to hinder albedo but for an IBC sub-cell as in 3TT there are two contacts at the bottom for each doping, leading to an inherent loss in albedo photocurrent compared to 2TT. Therefore, although the third terminal is beneficial in eliminating the current matching constraint, it results in loss of albedo photocurrent too due to extra shading.

The 3TT cells are connected using VM strings which are the most cost-efficient way to implement 3TT for voltage-matching to avoid redundant use of external power converters [12]. The VM string in 3TT module in this study consists of reverse connected 72 tandem cells out of which 2 cells are lost due to end losses. The 2TT module, on the other hand, has no end losses for 72 series connected sub-cells which do not have VM interlacing. At optimal albedo, 2TT string shows a higher efficiency as compared to 3TT as shown in Fig. 2.

The individual V_{mp} for top and bottom sub-cells is 0.9V and 0.57V, respectively, implying an ideal VM ratio of 1.579. This may accurately be realized by 158/100 top/bottom sub-cells in a string would put a constraint of a large number of tandem cells in the string. Taking VM to be 1.5 or 3/2 in this study provides a close match to the required VM of 1.579 and allow using only 72 tandem cells in the string. Although a closer match could be achieved by using VM ratio of 5/3, the end losses for 5/3 are twice as much as that of 3/2 [12], therefore practically the latter VM ratio gives higher efficiency.

B. Vertical Bifacial Solar Farm Model

The PV farm consists of bifacial PV panels of height h which are installed vertically in East/West orientation and are separated by row-to-row pitch p, as discussed in Section III. To compute the energy produced by the PV farm, we use a 2D view factor approach as discussed in our earlier work [14]. We use Sandia's PV library [16] to calculate the Sun's trajectory described by zenith and azimuth angles [17]. The Global Horizontal Irradiance (GHI) is first calculated using Haurwitz clear sky model [18], which is then renormalized based on the monthly average data for 22 years obtained from NASA Surface Meteorology and Solar Energy Database [19]. The GHI is then split into direct normal (DNI) and diffuse horizontal (DHI) irradiance components by using Orgill and Hollands model [20]. To compute energy produced (IPV_{dir}) by direct irradiance, angle of incident of the sun with respect to PV panels is computed. The energy produced due to diffuse (IPV_{diff}) or albedo (IPV_{Alb}) irradiance components are then calculated using the view factor approach [14].

III. RESULT AND DISCUSSION

Figure 2 shows the trend of 2TT and 3TT cells with respect to the albedo where a generally increasing trend in efficiency is observed due to increased photogeneration as a direct effect and a secondary effect of increased V_{mp} with increased I_{mp}.

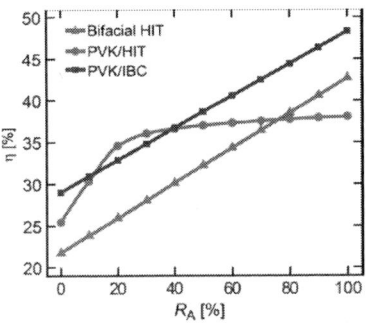

Fig. 2. The efficiency of Two Terminal PVK/HIT (2TT), Three Terminal PVK/HIT IBC (3TT), and bifacial HIT as a function of albedo.

Initially, for 2TT the overall current is limited by the photocurrent of bottom SHJ sub-cell, the overall 2TT efficiency therefore initially increases with albedo and is higher than bifacial HIT as observed previously [21]. This increasing trend however saturates once the current is limited by the photocurrent of the perovskite sub-cell. This occurs at 21.5% albedo, which is the point of current matching for top and bottom sub-cells, as displayed in figure 2. At the optimal albedo, the efficiency for 2TT is 1.83% higher than 3TT i.e., 35% as opposed to 33.17%. Thereafter, further increase in albedo predominantly increases the photocurrent of bottom HIT sub-cell which is no longer the limiting current for 2TT. A major part of the potential gain due to albedo is hence lost beyond the optimal albedo and a mere 2% increase in efficiency is observed as albedo is increased from 30% to 100% for 2TT.

978-1-6654-6060-6/23 $31.00 © 2023 IEEE

For the case of 3TT, the efficiency increases in a quasi-linear manner as the albedo is increased from 0 to 100%. Since there is no current matching constraint in 3TT, the excess carriers are collected by their respective contacts leading to a constant increase in efficiency. The efficiency reaches almost 50% at 100% albedo which indicates a higher potential for 3TT for high reflecting ground surfaces. 3TT provides an exceptional performance for a broad range of albedo in comparison to 2TT and is hence a spectrally robust bifacial tandem configuration.

2TT design is optimized for the albedo of ~21.5%, around which, it outperforms the 3TT. The relative losses for the latter are quantified as 0.2% due to shading of the extra back contact, 1.8% due to voltage mismatch of the parallel strings, and 0.94% due to end loss for the VM string. A further optimization to reduce these losses can enable efficiency beyond 36% for 3TT.

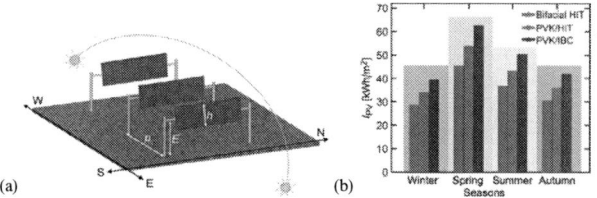

Fig. 3. (a) Illustration of vertical bifacial farm, (b) The seasonal energy output per square meter for the 2TT (PVK/HIT), 3TT (PVK/IBC), and bifacial HIT technologies

The cell/string level differences for 2TT vs. 3TT eventually determine the output at the module level. The farm with vertical bifacial solar panels having one-meter vertical dimension and row to row separation of two-meter between adjacent arrays is illustrated in Figure 3(a). The seasonal output is highest in spring, followed by summer and autumn and is lowest in winter, as illustrated in Figure 3(b). For all seasons, power output of 3TT outperforms single junction HIT while 2TT underperforms relative to both 3TT and HIT which has been pointed previously too as a concern [51]. The yearly energy production for 3TT, 2TT, and HIT are 194,000 kWh, 167,000 kWh, and 142,000 kWh, respectively.

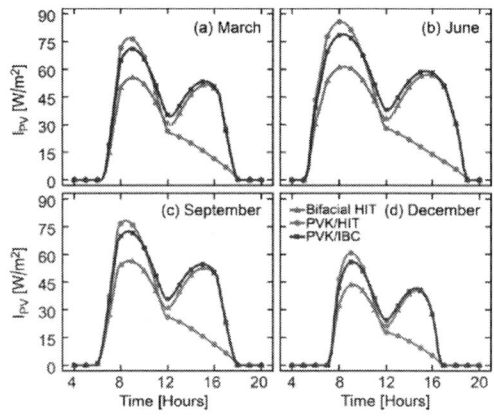

Fig. 4. The power output in a 24-hour period for four different months of the quarter sampled on the 15th of each given month

The daily power output trends for 3TT, 2TT, and HIT are shown in Figure 4, for various months. The top sub-cell for 3TT and 2TT is facing East. The daily profile for 3TT shows a peak in the morning and a smaller peak in the afternoon. The peak for 3TT is higher than that for HIT before noon when the front of the tandem is receiving direct sunlight. The 2TT, on the other hand, although has the highest peak in the morning since the

albedo is kept constant to be optimal for 2TT, leading to current matching and optimum power output. However, during the afternoon, 2TT performs poorly due to current mismatch as the lack of direct light at the top sub-cell bottlenecks the output leading to a reduced overall daily for 2TT.

IV. CONCLUSION

The performance of vertical bifacial solar farms made of three terminal tandem (3TT) perovskite (PVK)/IBC cells is explored using TCAD and detailed farm level physical models. The 3TT configuration is shown to be significantly superior as compared to the series connected two terminal PVK/HIT tandem (2TT) solar farms as well as to the single junction bifacial HIT solar farms. The annual energy production for East/West facing 3TT PVK/IBC vertical farm is 37% more as compared to 2TT series connected PVK/HIT vertical farm and 16% more as compared to single junction bifacial HIT vertical farm for the latitude of 31.5 N at Lahore. The morning and afternoon peaks for 3TT PVK/IBC have a slight skew which can be utilized effectively when aligned with the load demand. The 3TT PVK/IBC has significant advantages over 2TT PVK/HIT and the single junction HIT and should therefore be the desired technology for vertical bifacial tandem solar farms.

REFERENCES

[1] T. Nagashima et al., IEEE Xplore, Sep. 01, 2000
[2] U. B. Qasim et al., Solar Energy, vol. 203, pp. 1–9, Jun. 2020
[3] G. W. P. Adhyaksa et al., Nano Letters, vol. 17, no. 9, pp. 5206–5212, Aug. 2017
[4] P. Tockhorn et al., ACS Applied Energy Materials, vol. 3, no. 2, pp. 1381–1392, Jan. 2020
[5] E. L. Warren et al., ACS Energy Letters, vol. 5, no. 4, pp. 1233–1242, Mar. 2020
[6] E. L. Warren et al., Sustainable Energy & Fuels, vol. 2, no. 6, pp. 1141–1147, 2018
[7] Y. Zou et al., IEEE Xplore, Jun. 01, 2018
[8] R. Santbergen et al., IEEE Journal of Photovoltaics, vol. 9, no. 2, pp. 446–451, Mar. 2019
[9] Z. Djebbour et al., Progress in Photovoltaics: Research and Applications, vol. 27, no. 4, pp. 306–315, Jan. 2019
[10] M. Rienäcker et al., Progress in Photovoltaics: Research and Applications, vol. 27, no. 5, pp. 410–423, Feb. 2019
[11] H. Schulte-Huxel et al., IEEE Journal of Photovoltaics, vol. 8, no. 5, pp. 1370–1375, Sep. 2018
[12] W. E. McMahon et al., IEEE Journal of Photovoltaics, vol. 11, no. 4, pp. 1078–1086, Jul. 2021
[13] Rienäcker, M., et al. "Three-terminal bipolar junction bottom cell as simple as PERC: Towards lean tandem cell processing." 2019 IEEE 46th Photovoltaic Specialists Conference (PVSC). IEEE, 2019
[14] M. H. Riaz et al., IEEE Journal of Photovoltaics, vol. 11, pp. 469-467, 2021
[15] M. A. Alam et al., PRINCIPLES OF SOLAR CELLS, 2022
[16] PV performance modeling collaborative | an industry and national laboratory collaborative to improve photovoltaic performance modeling. https://pvpmc.sandia.gov/, 2016. [Online Accessed: 19-December-2022]
[17] I. Reda et al., Solar energy, vol. 76, pp. 577-589 2004
[18] B. Haurwitz, Journal of Meteorology, vol. 2, pp, 154-166, 1945
[19] P.W. Stackhouse, "Surface Meterorology and Solar Energy-A renewable energy resource website," NASA, POWER :(release 60) https://power.larc.nasa.gov/data-access-viewer/
[20] J. Orgill et al., Solar energy, vol. 19, pp. 357-359, 1977
[21] R. Asadpour et al., Applied Physics Letters, vol. 106, no. 24, p. 243902, Jun. 2015
[22] M. T. Patel et al., Applied Energy, vol. 329, p. 120175, Jan. 2023

Influence of insertion position of a LiF buffer layer on passivation performance of crystalline Si/SiO$_y$ /TiO$_x$/Al heterostructures

Shohei Fukaya[1], Kazuhiro Gotoh[1], Takuya Matsui[2], Hitoshi Sai[2], Yasuyoshi Kurokawa[1] and Noritaka Usami[1,3]

[1]Graduate School of Engineering, Nagoya University, Nagoya, 464-8603, Japan

[2]National Institute of Advanced Industrial Science and Technology, Ibaraki, 305-8568, Japan

[3]Institutes of Innovation for Future Society, Nagoya University, Nagoya, 464-8601, Japan

Abstract —**We report on the effect of the lithium fluoride (LiF) buffer layer on passivation performance for crystalline Si (c-Si)/silicon oxide (SiO$_y$)/titanium oxide (TiO$_x$) heterostructure studied by experiment and band lineup simulation. For the samples with LiF, the passivation performance and electrical contact characteristics were significantly improved. Furthermore, by changing the LiF insertion position, improvement of passivation performance was confirmed while maintaining good electrical contact characteristics for a c-Si/LiF/TiO$_x$/Al structure. However, the band lineup simulation did not show the marked dependence of the band alignment on LiF layer position but rather a trend toward lower performance for the c-Si/LiF/TiO$_x$/Al structure. This suggests that the difference in passivation performance depending on LiF layer position is largely due to the formation of an Al oxide interface layer that is thought to act as an Al diffusion barrier, which was not considered in the simulations. These results are expected to contribute to the elucidation of the mechanism behind the passivation degradation of c-Si/TiO$_x$ interface after Al deposition.**

I. INTRODUCTION

To further improve the power conversion efficiency of crystalline silicon (c-Si) heterojunction solar cells, new carrier-selective contacts (CSCs) are being explored. Hydrogenated amorphous silicon (a-Si:H)-based CSCs provide high-efficiency solar cells, but the band gap energy (E_g) of a-Si:H is about 1.7 eV, which causes photocurrent losses due to parasitic absorption [1]. Therefore, searching for practical materials with wide E_g is attractive for further improvement of the conversion efficiency.

Titanium oxide (TiO$_x$) is a candidate for novel materials for CSC with a wide E_g of 3.3 eV. At the TiO$_x$/n-type c-Si (n-c-Si) heterointerface, a small conduction band offset below ~0.05 eV and a large valence band offset above ~2 eV are created[2], which is a preferable band lineup for selective transport of electrons. High-performance c-Si heterojunction solar cells with TiO$_x$ have already been reported [3,4].

Although TiO$_x$ electron-selective contact has the advantages mentioned above, it has not yet been used in practical devices due to several issues. For example, the passivation performance is degraded after aluminium (Al) electrode deposition [5]. This degradation is thought to be due to chemical reactions such as the formation of Al oxides at the TiO$_x$/Al interface and the reduction of TiO$_x$. The insertion of lithium fluoride (LiF), which has a low work function and high stability, between TiO$_x$ and Al is known to improve passivation performance[6]. However, the underlying mechanism for this improvement has not yet been clarified.

In this research, we investigated the effect of the presence of LiF on the passivation performance of TiO$_x$/Al contacts. Marked differences in passivation performance were observed depending on the LiF insertion position. The underlying mechanism is discussed with reference to the band bending predicted from simulations.

II. EXPERIMENTS

All experiments were performed using double-side polished, floating zone (FZ) grown n-type c-Si (n-c-Si) as a substrate. The resistivity and thickness were 1.0-5.0 Ω·cm and 280 μm, respectively. Prior to deposition of TiO$_x$, the substrates were dipped into 5 % HF solution to remove the native oxide layer and then immersed in a SC-2 (H$_2$O$_2$: HCl: deionized water = 1:1:6) solution at 60 °C for 10 min for the formation of 1-nm-thick silicon oxide (SiO$_y$). After that, 1-nm-thick TiO$_x$ was deposited on the front side by thermal ALD (GEMStar-6, Arradiance Inc.). Tetrakis (dimethylamido) titanium (TDMAT) and H$_2$O were used as a titanium precursor and an oxidant, respectively. Subsequently, for the rear side, 1-nm-thick TiO$_x$ and 1-nm-thick LiF were deposited in different orders by ALD and thermal evaporation, respectively. After that, Al was deposited on each sample. The schematic structure is depicted in Fig. 1.

To measure the passivation performance, we used the implied open-circuit voltage calculated from the photoluminescence (PL) intensity (i-$V_{OC, PL}$) [7] based on the experimentally calibration curve between PL intensity and implied open-circuit voltage (i-V_{OC}). For measuring the *I-V* properties, TiO$_x$, SiO$_y$ and LiF stacks were prepared on the front side of the n-c-Si wafers, and Al contact dots with diameter ϕ = 2.5 mm, were deposited by thermal evaporation using a metal mask. After that, antimony-doped gold (Au:Sb) was evaporated at the rear side,

978-1-6654-6060-6/23 $31.00 © 2023 IEEE

and post-deposition annealing was carried out in vacuum to form ohmic contact by the diffusion of Sb atoms into c-Si between the rear Au:Sb layer and c-Si.

We simulated the band lineup by using AFORS-HET ver 2.5 to estimate the band alignment assuming the ideal interfaces in each sample.[8] In the simulation, the material properties were referred to the literature [9].

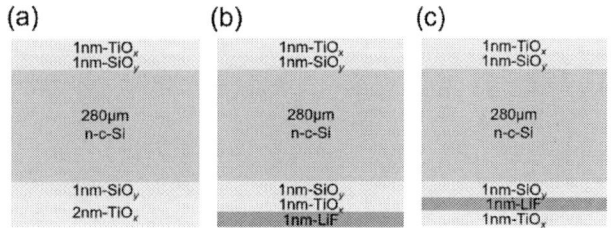

Fig. 1.   Schematic structure with the samples. All samples have 1nm-SiO_y and 1nm-TiO_x layer on the front side, and (a) n-c-Si/SiO_y/2nm-TiO_x, (b) n-c-Si/SiO_y/1nm-TiO_x/LiF and (c) n-c-Si/SiO_y/LiF/1nm-TiO_x for rear side.

III. RESULTS AND DISCUSSION

Fig. 2 shows (a) the i-$V_{OC, PL}$ mapping and (b) box and swarm plots of variation of i-$V_{OC, PL}$ of the samples with n-Si/SiO_y/TiO_x, n-Si/SiO_y/TiO_x/LiF and n-Si/SiO_y/LiF/TiO_x for the rear side before and after Al deposition. It is seen that all the samples showed i-$V_{OC, PL}$ degradation after Al deposition, especially for the samples without the LiF layer. On the other hand, for the

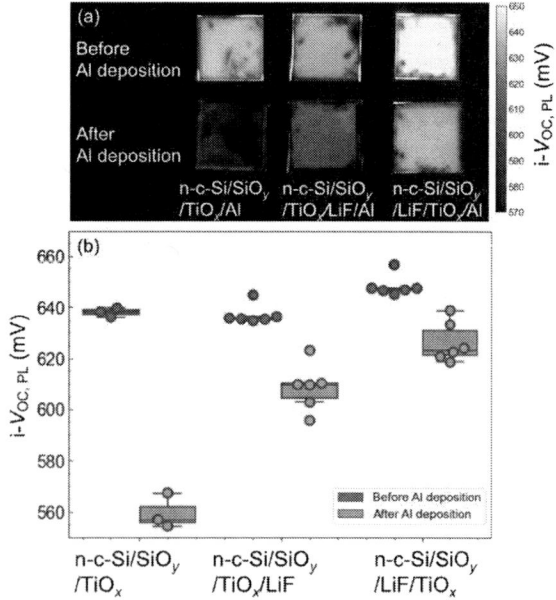

Fig. 2.   (a) i-$V_{OC, PL}$ mapping of the samples with n-Si/SiO_y/TiO_x, n-Si/SiO_y/TiO_x/LiF and n-Si/SiO_y/LiF/TiO_x for rear side before and after Al deposition and (b) box and swarm plots of variation of i-$V_{OC, PL}$ before and after Al deposition.

samples with the LiF layer, the degradation of i-$V_{OC, PL}$ was suppressed. In addition, among the samples with the LiF layer, the samples with n-c-Si/SiO_y/LiF/TiO_x structure showed low passivation degradation after Al deposition compared to the samples with n-c-Si/SiO_y/TiO_x/LiF structure.

Fig. 3 shows the I–V curves of the samples with n-c-Si/SiO_y/TiO_x/Al, n-c-Si/SiO_y/TiO_x/LiF/Al, n-c-Si/SiO_y/LiF/TiO_x/Al heterojunctions with TiO_x layer thickness of 1.0 nm under dark condition. Both samples with n-Si/SiO_y/TiO_x/LiF and n-Si/SiO_y/LiF/TiO_x showed Ohmic properties, whereas rectification property was observed for the sample without the LiF layer. This could be explained by the downward band bending due to the low work function of LiF. I-V curves of the n-Si/SiO_y/TiO_x/LiF and n-Si/SiO_y/LiF/TiO_x heterostructures are almost identical, suggesting that the position of the LiF layer has no significant effect on the field-effect passivation on n-c-Si. These results suggest that the chemical passivation is dependent on the position of the LiF layer while the field-effect passivation is not dependent on the position of the LiF layer.

Fig. 4 shows the simulated band lineups of the samples with (a) n-c-Si/TiO_x/Al, (b) n-c-Si/TiO_x/LiF/Al, and (c) n-c-Si/TiO_x/LiF/Al heterojunctions by AFORS-HET. The SiO_y layer is modeld to act as a perfect passivation layer with no thickness. In addition, we assumed no defects in bulk n-c-Si. For the structure with the LiF layer, the band of n-c-Si bends downward as shown in Fig. 4(b), leading to good field-effect passivation. On the other hand, a slight upward band bending is observed in Fig.4(c). These results suggest that the position of the LiF layer affects the band bending in n-c-Si, and n-c-Si/TiO_x/LiF/Al is preferable to improve the field-effect passivation.

However, this does not agree with the experiments to show the better passivation performance for the sample with n-c-Si/LiF/TiO_x/Al junction. These could be due to the presence of structural defects, which are not considered in the simulation. In fact, it is reported that the formation of Al oxides at Al/TiO_x

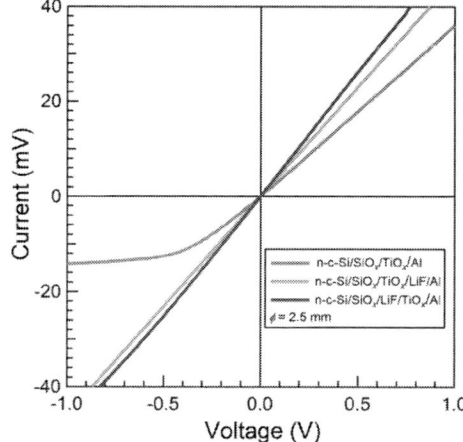

Fig. 3.   I–V curves of the samples with n-c-Si/SiO_y/TiO_x/Al, n-c-Si/SiO_y/TiO_x/LiF/Al, n-c-Si/SiO_y/LiF/TiO_x/Al heterojunctions under dark condition. The electrode diameter (ϕ) is 2.5 mm.

interface[3], and also suggested that Al-O-Ti layer acts the barrier layer to prevent Al diffusion[6]. Therefore, to suppress the degradation of passivation performance after Al electrode deposition, having TiO$_x$/Al contact and LiF layer is considered necessary.

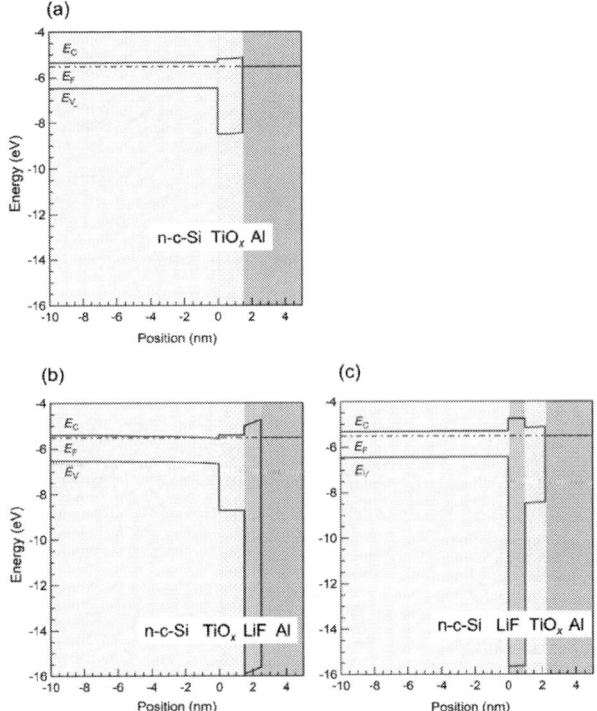

Fig. 4. Band lineups for (a) n-c-Si/TiO$_x$/Al, (b) n-c-Si /TiO$_x$/LiF/Al and (c) n-c-Si/LiF/TiO$_x$/Al simulated by AFORS-HET.

IV. SUMMARY

We investigated the effect of inserting the LiF layer for passivation performance after Al deposition. The structure with the LiF layer showed the suppression of passivation degradation after Al deposition compared to without samples.

In addition, we confirmed the passivation difference for the samples after Al deposition by varying the LiF position. From the results of both of experiments and band lineup simulations, the presence of the LiF layer and Al oxide layer formed at TiO$_x$/Al interface would be the key to obtain the superior passivation performance by suppressing Al diffusion into n-c-Si.

ACKNOWLEDGMENT

This work was supported by New Energy and Industrial Technology Development Organization (NEDO) and MEXT, Grants-in-Aid for Scientific Research on Innovative Areas "Hydrogenomics," JP18H05514.

REFERENCES

[1] Z. C. Holman, A. Descoeudres, L. Barraud, F. Z. Fernandez, J. P. Seif, S. D. Wolf, and C. Ballif, Christophe, IEEE J. Photovolt. 2, 7 (2012)

[2] S. Avasthi, W. E, McClain, G. Man, A. Kahn, J. Schwartz, and J. C. Sturm, Appl. Phys. Lett., 102, 203901 (2013).

[3] X. Yang, Q. Bi, H. Ali, K. Davis, W. V. Schoenfeld and K. Weber, Adv. Mater. 28, 5891 (2016).

[4] T. Matsui, M. Bivour, M. Hermle, and H. Sai, ACS Applied Materials and Interfaces, 12(44), 49777 (2020).

[5] V. Titova and J. Schmidt, Phys. Status Solidi Rapid Res. Lett., 15 (9), 18 (2021).

[6] W. Liang et al., Phys. Status Solidi RRL, 2200304 (2022).

[7] S. Fukaya, K. Gotoh, T. Matsui, H., Sai, Y., Kurokawa and N., Usami, "Quantitative Evaluation of Passivation Performance after Electrode Deposition on TiO$_x$/Si Heterostructures by Photoluminescence Imaging," 12th International Conference on Crystalline Silicon Photovoltaics, Konstanz, Germany, March, 2022.

[8] R. Varache, C. Leendertz, M. E. Gueunier-Farret, J. Haschke, D. Muñoz, and L. Korte., Solar Energy Materials and Solar Cells 141, 14 (2015).

[9] M.Q. Khokhar, S.Q. Hussain, D.P. Pham, M. Alzaid, A. Razaq, I. Sultana, Y. Kim, Y.H. Cho, E.C. Cho, and J. Yi, Mater. Sci. Semicond. Process. 134, 105982 (2021).

Radiometric Standards and Best Practices: Recent Progress

Aron Habte[1], Manajit Sengupta[1], and Christian A. Gueymard[2]

[1]National Renewable Energy Laboratory, Golden, Colorado, 80401, U.S.A
[2]Solar Consulting Services, Colebrook, New Hampshire, U.S.A

Abstract — International standards and best practices for solar resource assessments provide assurance for traceable measurements associated with a low uncertainty. This benefits the solar energy industry by reducing the investment risks through heightened confidence in the solar resource information. NREL, in collaboration with international organizations, such as subcommittees G03 of ASTM (Radiometry) and TC180/SC1 of ISO (Climate measurement and data), recently revised some standards that are widely used by the solar industry. These include radiometric standards that assist in (i) maintaining calibration traceability; (ii) uncertainty analysis; (iii) measurement quality assurance, and (iv) establishing reference spectral irradiance distributions. The latter are widely used by the solar community to evaluate the actual absorptance, reflectance, and transmittance of solar energy materials, or the performance of solar energy devices and systems, relative to standard conditions. This paper provides a summary of the recent changes brought to these standards as a way to better support the solar energy industry.

I. INTRODUCTION

International standards play a notable role in all phases of the development of solar energy projects, including laboratory measurements under standard test conditions, methods and instrumentation, accelerated testing, and service lifetime of materials and systems. Moreover, the design and performance monitoring of solar energy systems depend on the quality of radiometric solar resource data. Radiometric standards contribute to better characterize the actual resource available to solar energy systems. Moreover, they help develop and improve computer models of the geographical distribution of the solar resource over the world, particularly for areas where solar radiation measurements are scarce (e.g., ASTM G222 [1]). In summary, international standards support the present and future deployment of solar conversion systems.

NREL, in collaboration with national and international collaborators, develops and updates such radiometric standards through international standard organizations. These include the International Standard Organization (ISO) through the ISO/TC180/SC1 Climate-Measurement and Data subcommittee, ASTM through both the G03.09 subcommittee on Radiometry and the E44 committee on Solar, Geothermal and Other Alternative Energy Sources. As stated, these standards provide procedures for solar resource calibrations, measurements, and modeling approaches of known data quality.

This paper provides some details on five new, revised or forthcoming standards, two from ASTM and three from ISO.

II. RECENT REVISED RADIOMETRIC STANDARDS

The ISO/TC180/SC1 subcommittee already updated, and/or in the process of updating, five standards. Similarly, the ASTM G03.09 subcommittee has one impactful standard for the solar energy community, which is now in the balloting process. The subsections below summarize which standards have been recently updated or are still in the updating process.

Most importantly, radiometric standards are essential for the proper determination of photovoltaic (PV) cell efficiency and module performance. This paper provides an overview of current international radiometric standards and, as illustrated in Fig. 1, their contribution to maintaining traceability, uncertainty analysis, measurement data quality assurance, and ultimately support the solar energy industry.

A. ASTM G173 - Standard Tables for Reference Solar Spectral Irradiances: Direct Normal and Hemispherical on 37° Tilted Surface

The revised G173 standard includes recent advancements in science and knowledge, resulting in new tables and figures based on the Simple Model of the Atmospheric Radiative Transfer of Sunshine (SMARTS) software, which was recently updated [2]. The most recent version (v2.9.9) includes significant improvements over v2.9.2, which was used to define the reference spectra in the current G173 standard, initially promulgated in 2003. The revised standard based on this SMARTS revision also includes recommended editorial and technical changes, as well as updated references for the SMARTS model and its inputs. Overall, the standard's revision benefits from these improvements:

- New solar constant: Based on a 42-year time series of recalibrated irradiance data, a revised solar constant of 1361.1 W•m^{-2} was recently accepted as part of standard ASTM E490-22 (see [3] for details). This value contrasts with the older World Meteorological Organization (WMO) value of 1367 W•m^{-2}, which was embedded in the original version of G173.
- New revised Extraterrestrial Spectrum (ETS): It is now based on a recalibration and merging of various sources of spaceborne ETS observations [4] and is included in SMARTS v2.9.9. This new ETS distribution forms the basis of standard E490-22 [5] and differs by up to a few percent (depending on spectral band) compared to the earlier ETS used for the original

978-1-6654-6060-6/23 $31.00 © 2023 IEEE

version of G173. This affects the UV and visible spectral bands of the resulting modeled terrestrial spectra, in particular.

- Various model improvements in the spectral calculation of atmospheric effects, particularly affecting the UV.

The various modifications described above are documented in [2–4] and are expected to improve the experimental realization of reference modeled spectra, as suggested by Fig. 2. The standard is expected to be published by ASTM in 2023

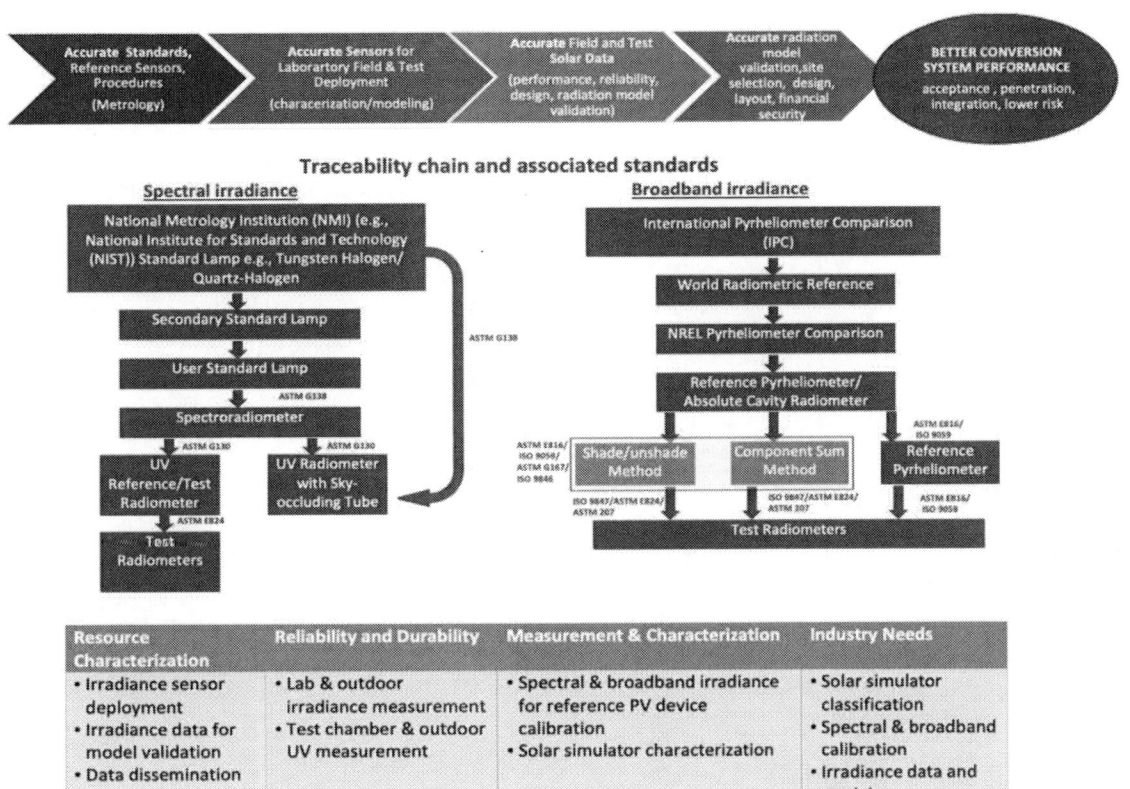

Fig. 1. Example of radiometric standards calibration traceability chain for both spectral and broadband irradiance, and for use cases in solar energy applications.

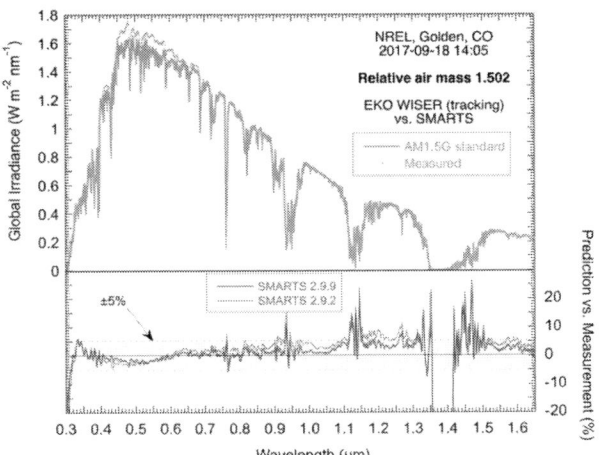

Fig. 2. Experimental realization of the AM1.5G standard reference spectrum at Golden on 2017-09-18. Top panel: Comparison between AM1.5G and the spectroradiometer measurement of global normal irradiance. Bottom panel: percent difference between the observations and predictions from SMARTS v2.9.9 or v2.9.2.

B. ASTM G222-21- Standard Practice for Estimation of UV Irradiance Received by Field-Exposed Products as a Function of Location

The ASTM G03 subcommittee recently published a new standard on methods to estimate the total UV irradiance, based on an empirically-derived model that assumes that measured or modeled data of global horizontal irradiance (GHI) are available. This standard is expected to help understand the degradation of materials exposed outdoors. In particular, the UV irradiance is a key input to obtain reliable assessments of the service life of many products. Unfortunately, the vast majority of locations that monitor and/or model solar irradiance only have instrumentation/models for GHI, and thus completely lack information on UV irradiance. As a remedy, this standard promulgates a mathematical model that provides a simple way to estimate the contribution of UV to GHI through the UV/GHI ratio.

The model estimates the global UV irradiance in two widely-used ranges (280–400 nm and 295–385 nm) from GHI (≈280–

4000 nm). SMARTS is used to evaluate the UV/GHI ratio under a large variety of clear-sky conditions, and examines the impacts of various atmospheric constituents, such as aerosols, precipitable water vapor, or ozone, and of the local surface characteristics (albedo), on the predicted UV. Using this large database of SMARTS predictions, the clear-sky UV/GHI model is created by fitting the predictions of this ratio to a 4th-order polynomial as a function of air mass only:

$$\frac{UV_{clear}}{GHI_{clear}} = \sum_{i=0}^{4} m_i \, AM^i$$

where:

UV = extended global UV irradiance, $W \cdot m^{-2}$
GHI = global horizontal irradiance, $W \cdot m^{-2}$
m_i = location-specific numerical coefficient obtained from the fitting of SMARTS results
AM = air mass coefficient.

The ratio above is then applied to calculate the global all-sky horizontal UV from measured or modeled all-sky GHI:

$$UV_{mod\_all\_sky} = GHI_{all\ sky} * \frac{UV_{clear}}{GHI_{clear}}$$

The model was validated on multiple locations; Fig. 3 shows one example result of the modeled versus ground measured data for NREL's station in Golden, Colorado. Considering multiple sites, the overall mean absolute error of the model is within 8% and the details about the accuracy are stated in [1].

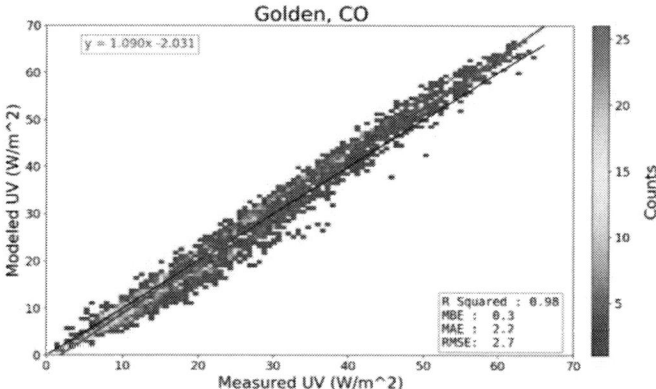

Fig. 3. Scatterplot of modeled and measured global UV (280–400 nm) at the NREL location.

Parallel to the work conducted at ASTM, the ISO TC180/SC1 committee has revised or is still revising the following standards:

• *ISO/TR 9901:2021 - Solar Energy — Field Pyranometers — Recommended practice for use.* After a significant revision, this standard was published in 2021. The summary of the updates includes consideration of recommended practices for use of modern digital pyranometers equipped with internal diagnostic techniques. Further, the standard provides recommended practices for the use of pyranometers to measure two quantities of importance for the for the PV community: plane-of-array irradiance and ground-reflected radiation.

• *ISO 9847: 2023 - Solar energy — Calibration of pyranometers by comparison to a reference pyranometer.* This standard underwent significant updating compared to the previous version. The main changes consist of a new section on indoor calibration, which is now used by manufacturers and calibration service providers. The terms and definitions are updated and harmonized with recently published standards, such as ISO 9060:2018 [6] and ISO Guide 99 [7]. The publication of this revised standard is expected for mid-2023.

• *ISO 9845: 2022 - Solar energy — Reference solar spectral irradiance at the ground at different receiving conditions — Part 1: Direct normal and hemispherical solar irradiance for air mass 1.5.* This standard was thoroughly revised and published in 2022. The revision includes the implementation of the SMARTS model to generate additional reference spectral irradiance distributions. Most notably, 171 subordinate spectra for hemispherical tilted irradiance are provided for a variety of atmospheric conditions and receiver geometries.

REFERENCES

[1] ASTM G222-21, "Standard Practice for Estimation of UV Irradiance Received by Field-Exposed Products as a Function of Location," ASTM International, West Conshohocken, PA.

[2] Gueymard, C.A., "The SMARTS spectral irradiance model after 25 years: New developments and validation of reference spectra," Solar Energy, 187, 2019, pp. 233–253, Corrigendum, *Solar Energy*, 236, 222, pp. 906–907.

[3] Gueymard, C.A., "A reevaluation of the solar constant based on a 42-year total solar irradiance time series and a reconciliation of spaceborne observations," *Solar Energy*, 168, 2018, pp. 2 9.

[4] Gueymard, C.A., "Revised composite extraterrestrial spectrum based on recent solar irradiance observations," *Solar Energy*, 169, 2018, pp. 434–440.

[5] ASTM E490-22, "Solar Constant and Zero Air Mass Solar Spectral Irradiance Tables," ASTM International, West Conshohocken, PA.

[6] ISO 9060:2018. 2018, "Solar energy — Specification and classification of instruments for measuring hemispherical solar and direct solar radiation," International Standard.

[7] BIPM IEC, ILAC IFCC, IUPAC ISO, and OIML IUPAP. 2012. JGCM 200, ISO Guide 99, The international vocabulary of metrology—basic and general concepts and associated terms (VIM), JCGM 200:2008.

Compact and high efficiency micro-CPV module with high wafer utilization rate.

Corentin Jouanneau, Thomas Bidaud, Maxime Darnon, Gwenaelle Hamon

Institut interdisciplinaire d'innovation technologique, sherbrooke, QC, Canada

Laboratoire nanotechnologies nanosystemes, sherbrooke, QC, Canada

Compact and high efficiency micro-CPV module with high wafer utilization rate Corentin Jouanneau[1,2], Thomas Bidaud[1,2], Maxime Darnon[1,2], Gwenaelle Hamon[1,2] [1]Institut Interdisciplinaire d'innovation Technologique (3iT), Sherbrooke, Québec, J1K 0A5, Canada [2]Laboratoire Nanotechnologies Nanosystèmes (LN2) - CNRS IRL-3463, Institut Interdisciplinaire d'Innovation Technologique(3iT), Sherbrooke, Québec, J1K 0A5, Canada   Micro-CPV (μ-CPV) modules promise to overcome the limitations of CPV such as compactness. Miniaturization involves new challenges for micro-fabrication, packaging, and optics. In this paper, we fabricate sub-millimetric solar cells with high performances and low kerf losses (< 10%). A new interconnection method for these cells is also presented allowing to reduce the shading of the cells by 10 compared to the wire bondable cell. We propose the design of a 350× one-stage optics to fabricate modules with manufacturing tolerances on cell misalignment reaching ±50 μm.

Selenium diffusion during CdCl2 treatment of CdSeTe solar cells

Niranjana Mohan Kumar, Srisuda Rojsatien, Angel De La Rosa, Barry Lai, Arun K. M. Kanakkithodi, Maria Chan, Dan Mao, Mariana Bertoni

Arizona State University, Tempe, AZ, United States

Argonne National Lab, Lemont, IL, United States

Purdue University, West Lafayette, IN, United States

First Solar, Perrysburg, OH, United States

Selenium alloying of the absorber is a key component in achieving high efficiency polycrystalline CdSeTe solar cells. Much work has been carried out in characterizing the macroscopic electrical, optical, and structural properties of this material system. However, the movement of Se in the absorber layer and the role of CdCl2 in this process are still not fully understood. In this study, nano-scale correlative X-ray microscopy has been utilized to study ex-situ and in-situ Se-alloyed CdTe samples before, during and after CdCl2 treatment. Changes in the Se distribution were mapped by nano x-ray fluorescence and detailed information about the Se local environment in the absorber layer was collected through x-ray absorption spectroscopy ex-situ and in-situ during CdCl2 treatment. Complementary nano-Xray diffraction grain maps were collected to properly assess the evolution of stress/strain through the recrystallization process.

Analysis Of Measured Operating Temperature Of Perovskite Modules.

D. Martinez Escobar,[1] Aaron Wheeler,[1] F. Brigham Pineda,[2] Katty Kaydanik,[2] Alan Murphy,[2] Jing-Shun Huang[2], Mason Terry,[2] and Sarah R. Kurtz[1]

[1]University of California Merced, Merced, CA, 95343, USA. [2]Caelux Corporation, Pasadena, CA, 91107, USA.

Abstract — **This work analyzes the operating temperature of perovskite mini-modules compared with silicon solar modules. Experimental data from single-junction perovskite modules, tandem (perovskite + Si) structures, and different-sized silicon modules were analyzed. The results showed that single-junction perovskite modules operated at temperatures 3 to 8 degrees Celsius cooler than silicon modules. Interestingly, we observed that one of the tandem modules operated at a similar temperature as the single-junction perovskite modules, while the other exhibited a temperature increase of approximately 5 degrees Celsius, close to the operating temperature of the silicon modules. This discrepancy in temperature can be attributed to several factors, variances in their assembly, optical properties, and conversion efficiency. The reduced operating temperature has important implications for the overall performance of perovskite modules.**

I. INTRODUCTION

In the past couple of years, there has been a remarkable improvement in the energy conversion efficiency of perovskite-based solar cells. Quantifying the performance of perovskite solar cells outdoors is the next step. Therefore, researchers must conduct exhaustive testing to understand this type of solar cell's performance. Various testing methods have been used, primarily indoor testing [1]–[3] at different conditions, and some groups have examined the performance of the devices under outdoor conditions [4]–[8].

Outdoor testing of solar modules holds significant importance in assessing their thermal performance and degradation characteristics. As perovskite technology gains prominence as a viable alternative to conventional silicon, it becomes essential to thoroughly evaluate how they perform under real-world conditions and comprehend the impact of outdoor factors. The thermal performance of solar modules, including temperature fluctuations, heat dissipation, and potential degradation mechanisms, directly affects their energy conversion efficiency and long-term reliability. Outdoor testing of perovskite modules allows researchers and industry experts to monitor and analyze the performance in response to varying temperatures, environmental stressors, and seasonal changes. By studying the thermal performance, we can refine the design, optimize material selection, and develop effective strategies to mitigate potential degradation risks. Understanding the behavior of perovskite modules in outdoor environments is crucial for enhancing their performance, extending their lifespan, and facilitating the widespread adoption of emerging technology for different applications.

In this work, we aim to add to the literature the results of the operating temperature of single-junction and tandem perovskite-silicon modules in outdoor conditions. We analyze three modules and operate them in maximum power point bias for this experiment. In addition, we included three different silicon modules for reference to compare the operating temperatures with existing silicon technology. To quantify the difference in operating temperature, we use two different metrics; first, we use the distribution of the difference between modules and ambient temperature using filtered data (irradiance between 950 and 1020 W/m^2) and report the mean of the distribution as an approximation of the temperature above the ambient. The second way is calculating the average daily delta temperature for the 96 days of data using the same filtered data and reporting the result as another approximation of the operational temperature.

II. EXPERIMENTAL METHODS

A. Outdoor Set up

We conducted our experiment using an outdoor testing facility near Atwater, California, USA, which falls within a Mediterranean climate zone. This region experiences hot and dry summers. The average annual temperature in Merced ranges from around 54°F (12°C) to 85°F (29°C). Summers are typically characterized by abundant sunshine and high temperatures often exceeding 90°F (32°C). In terms of humidity, Merced generally has a relatively low levels, with average relative humidity ranging from 40% to 60%. The site experiences moderate to high solar irradiance levels, making it an ideal location for studying solar module performance. In addition, Merced also exhibits moderate wind speeds, which can impact the cooling and heat dissipation of solar modules.

The outdoor setup is configured as an open rack at a fixed tilt angle of 35 south facing (see Fig. 1). In total, we exposed seven modules from which, two are single-crystal Si modules from commercially available technologies labeled as "SC1" and "SC2", a medium-sized single-crystal Si module labeled as "Si_ref" (green rectangle), two single-junction perovskite mini-module (red rectangle), and two tandem perovskite mini-modules (represented by the yellow rectangles in Fig. 1). For this work, Caelux Corporation fabricated Methylammonium-containing perovskite mini-modules with a 4-terminal architecture where the perovskite is located the top and silicon

978-1-6654-6060-6/23 $31.00 © 2023 IEEE

at the bottom. We present the single-junction perovskite and tandem perovskite mini-module results labeled as "PK_single2", "PK_single3", "PK_tandem2" (PERC silicon) and "PK_tandem3" (TOPCON silicon). For the tandem cell, we only present the results of the perovskite measurement.

Fig. 1.   Outdoor rack used for this experiment. with red rectangles highlighting the single-junction perovskite mini-modules, yellow rectangles representing the tandem perovskite used in this study, and a green rectangle representing the reference silicon.

B.   Data Analysis

In this work, our analysis is based on data collected over a period of 96 days during outdoor exposure. We gathered comprehensive data that includes power output measurements of each module, temperature performance, and the corresponding weather conditions. For the first part of the results, we present the non-filtered distribution of the data to provide a complete overview of the dataset including both low and high irradiance records. Subsequently, we applied a series of filters to clean the distribution of data, focusing on capturing near one-sun conditions, which will be explained further in the following subsection.

To accurately monitor the thermal performance of each module, we attached a T-type thermocouple to the back of every module, located at the center of the cell for the min-modules and in a middle cell for the commercial-grade modules. This measurement point is referred to as the module temperature. Additionally, we recorded the ambient conditions, including plane-of-array irradiance using a thermopile pyranometer and silicon reference cell, ambient temperature, a cup anemometer for measuring both wind speed and direction, and Omega HX71-V2 sensor for measuring relative humidity. Throughout the experiment, all modules were biased at Maximum Power Point to simulate real operating condition. A Daystar MT5 multi-tracer controlled the bias and recorded all variables with a 5-minute interval.

C.   Data cleaning procedure

In the data cleaning procedure, we specifically addressed outliers in the measurements due to disconnection from the sensor and/or weather phenomena such as from cloud reflections and high wind speeds. We developed a set of Python scripts to automate the data cleaning process before doing the main analysis. This code allowed us to efficiently apply a series of filters to clean the dataset.

To handle sensor disconnection, we filter the data using the delta T calculation and removed all the records where the delta T was less than 5 degrees while the irradiance levels were greater than 500 Wm^2. We identified the 5-degree filter by analyzing the distribution of the delta T for various irradiance levels. Additionally, for some of the results, we present a filtered dataset to include only irradiance between 950-1020 W/m^2, wind speed measurements between $0.5 - 1.5$ m/s and ambient temperature between 25-35 °C.

D.   Temperature modeling

In addition to the experimental work, we modeled the temperature above ambient temperature for the perovskite solar modules using the thermal balance equation shown in equation (1) from [4], [9]

$$
\begin{aligned}
(\alpha - \eta) \times G = \ & h_{cf} \times (T_{mod} - T_{amb}) \\
& + \varepsilon_g \times \sigma(T_{mod}^4 - R_{sky}) \\
& + \varepsilon_b \times \sigma(T_{mod}^4 - T^4{}_{gnd})
\end{aligned} \tag{1}
$$

where α is the module absorption coefficient, G is the solar irradiance in W/m^2, h_{cf} is the module convection coefficient for the rack configuration, T_{mod} is the module temperature, T_{amb}, is the ambient temperature, $\varepsilon_g = 0.84$ is the emissivity of glass, $\varepsilon_b = 0.89$ is the emissivity of the back sheet and finally R_{sky} is the sky's thermal downwelling radiation that can be estimated by using the humidity measurement and ambient temperature [9].

Using the thermal balance equation, we plugged in the experiment measurements and solved for the convective transfer coefficients to obtain a model to predict the mini-module temperature. We used the measured data to obtain the thermal energy entering the solar cell plus the calculated conversion efficiency of the perovskite solar cell. Also, we assumed some optical properties of the absorption coefficient of the perovskite. We assumed that the broad-band absorption coefficient remained constant. We did not explore the change of this coefficient due to changing spectrum, degradation of the material, or delamination.

With all the measured parameters, we solved for the convection coefficient h_{cf} by using an adapted version from equation (1) from [6] and also presented in [9]

$$
h_{cf} = [(h_1 \times v + h_2)^3 + h_3^3]^{\frac{1}{3}} \tag{2}
$$

where v is the wind speed $h1$, $h2$, and $h3$ are the forced-convection and free-convection coefficients. To calculate these coefficients numerically, we designed a small optimization problem that calculated the value of the sum of the squares of equation (1) + equation (2) and constrained the range of all the h coefficients to be in the range of already reported values [9]. With the fitted coefficients, we now use them to obtain the module temperature by solving equation (1).

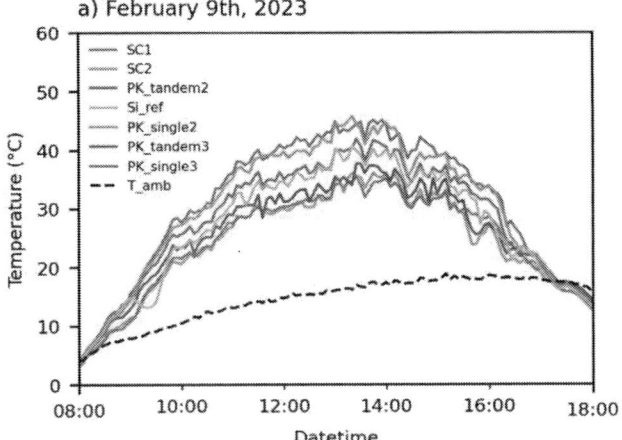

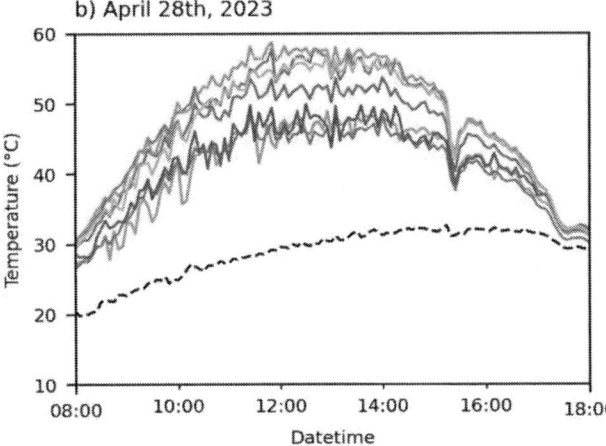

Fig. 2.   Temperature measurements for all the modules in the experiment for two selected days a) cold but sunny day on February 9th 2023, and b) hot and sunny day on April 28th 2023.

III. RESULTS

A. *Thermal performance of perovskite modules*

Fig. 2 illustrates the temperature measurements for each module for two selected days. In Fig. 2a we show the measurements for a cold but sunny day with a maximum ambient temperature of 19 °C and a peak irradiance of 960 W/m^2. During this day, we observed that the single-junction perovskite and the PK_tandem3 had the lowest temperature readings from all the modules. Interestingly, the PK_tandem2 had temperature like the Si_ref but lower than the commercial modules. In contrast, Fig. 2b shows the temperature readings for a hot and sunny day with a maximum ambient temperature of 32 °C and a peak irradiance of 1032 W/m^2. The temperature

difference between the single-junction perovskite and the silicon modules grew considerably. However, we obtained mixed results for the tandem-perovskite and single-junction perovskite where the mini-module PK_tandem3 had a similar temperature reading to its single-junction counterpart, but the PK_tandem2 had higher temperature readings.

To explain these results, we suggest that there are two main factors that contribute to the variation of the temperature readings of the modules. Firstly, as shown in the methodology, there is a difference in size between the commercial modules and the perovskite mini-modules and we obtained –as expected– that modules with larger active areas heat up more throughout the day than their smaller counterpart mini-modules. Second, to explain the difference between the hotter tandem module and single-junction perovskite module is the energy band gap of the material. With a higher band gap, the perovskite module heated up less while the lower band gap silicon was hotter as shown in Fig. 2. Also, we found that when the ambient temperature is higher (around 13 °C), the difference in temperatures starts to increase considerably, up to 10 °C hotter. From our knowledge, we expected that the tandem modules would have an operating temperature similar to the silicon modules, but the area effect in addition to the properties and construction yielded a lower operating temperature. Moreover, we did not expect to obtain that the PK_tandem3 had similar temperature readings to the single-junction modules while PK_tandem2 was hotter consistently throughout the experiment. After some filtering, we could not identify the main reason, but we recognize that further testing is required with different perovskite modules to fully understand the nature of these contradictory results.

From the experiment, we obtained a consistent temperature difference between the single-junction perovskite and PK_tandem2 and the silicon modules. Fig. 3 shows the average daily temperature above ambient over the 96 days of exposure. From the data presented in Fig. 3, we confirm that perovskite modules do run cooler than silicon modules and that single-perovskite run cooler than a tandem configuration mini-module. These results show that the average temperature for the single-junction perovskite is 14-15 °C, whereas the tandem mini-modules were 15-20 °C for PK_tandem3 and PK_tandem2, respectively. From our results, the differences in temperature are less than what was previously reported for other technologies such for a comparison between GaAs and silicon modules [5], [9].

We also noticed that the PK_tandem2 had lower conversion efficiency in comparison with PK_tandem3. This could explain why this mini-module runs hotter. Another possibility might be a change in the optical properties of the material. While we were unable to measure optical properties such as reflection, conducting a further analysis is necessary to gain a better understanding of why similar modules exhibit different temperature characteristics.

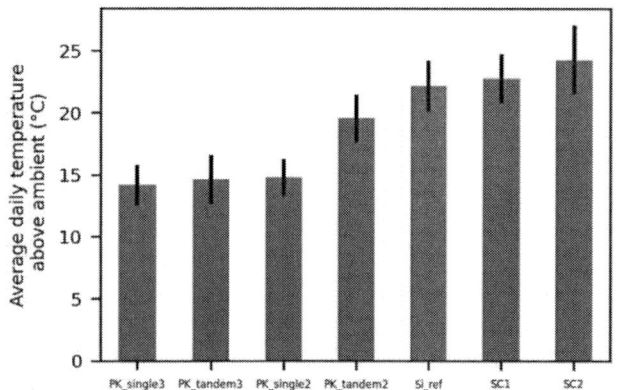

Fig. 3. Average daily temperature above ambient measured at high irradiance. Data were calculated by taking the daily average and averaging it for the 96 days of exposure. Error bars represent the standard deviation of the daily average. Data were filtered for irradiance between 950-1020 W/m².

Additionally, we include a visualization of the correlation between the single-junction perovskite modules and the other modules as shown in Fig. 4. We filtered the data to include only irradiance values that range between 950 and 1020 W/m² to observe conditions near one sun and plotted the temperature above ambient observed for the PK_single2 (the coldest single-junction perovskite module) and compared with the other single-junction perovskite module and the tandem-silicon modules. We observed that most of the temperature data using these filters followed a normal distribution and were consistent with the previous results.

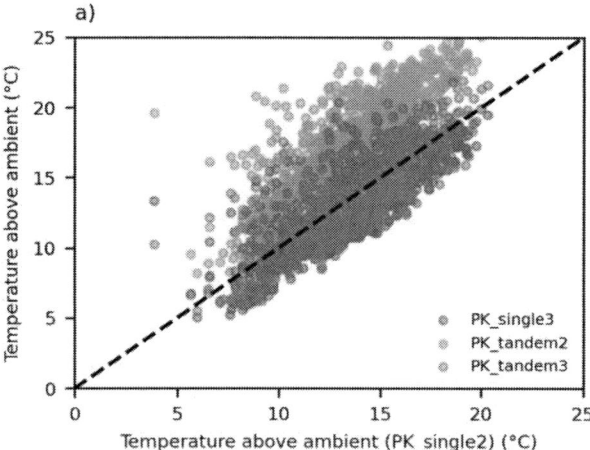

Fig. 4. Temperature correlation between single-junction perovskite mini-module against tandem-silicon perovskite. Data above the dotted line were hotter than the reference single-junction perovskite and data below the line were colder.

B. Thermal modeling and convection coefficient fitting

For the final part of the work, we endeavor to model the temperature of the perovskite mini-modules using equation (1) and adjusting for the measured weather variables and the efficiency of the perovskite. Although, the focus of this paper is the thermal performance and modeling, we calculated the efficiency for all the modules that we used to calculate the electricity delivered (ηG). This parameter plays an important role while calculating the convection coefficients and therefore must be treated with caution. Since we are analyzing emerging perovskite materials that could experience degradation, the method that we implemented to obtain the module temperature might not be valid if the efficiency changes over time since this will yield different values for h. We assumed that there is no loss in efficiency over time and calculated the h coefficient for the single-junction perovskite and the tandem configuration. The convection coefficient fit results are shown in Table 1.

Table 1. Fitted convection coefficients for the different solar mini-modules. Calculated using a sum of squares constrained optimization by using equation (1) and (2) and bounding the range of each h coefficient. Data reported with two significant figures and the units are W/(m²K).

	PK_single2	PK_single3	PK_tandem_2	PK_tandem3
h_1	5.3	5.1	6.5	6.5
h_2	4.3	3.2	8.5	13
h_3	7.4	7.4	11	14

Although the fitted values from Table 1 are straightforward to calculate, we warn the reader about the interpretations of these coefficients and their usefulness outside of using them for equation (1). Various factors like, but not limited to, size, wind speed, absorption coefficient, and humidity may have different convection coefficients. The weather conditions, especially the sky radiation, also affect the calculated coefficients. Nevertheless, it is useful to tabulate them from the measured data. With the fitted coefficients, we solved the equation (1) to calculate the module temperature for the different mini-modules. We show the model validation (Fig. 5) for the single-junction perovskite module (Fig. 5a) and for the tandem module (Fig. 5b).

Fig. 5. shows the modeled module temperature using the coefficients from Table 1 for all 96 days of the experiment. For the single-junction perovskites (Fig. 5a) we obtained a good fit overall, except for some measurements that the model underestimated the temperature for a temperature range from 10-30 degrees. There is no substantial difference in terms of the model accuracy between the two single-perovskite modules. The same analysis was applied to the tandem modules (Fig. 5b). Interestingly, the model for the tandem module is more consistent than for the perovskite-only modules.

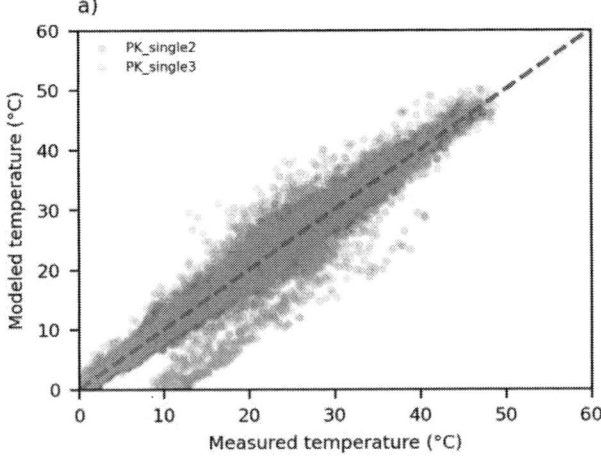

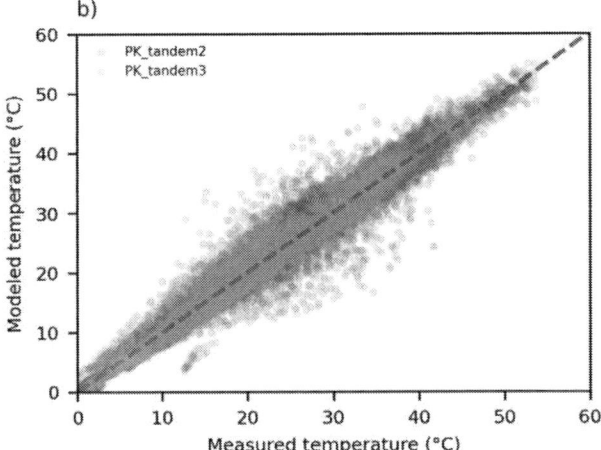

Fig. 5. Temperature model verification using the thermal balance equation from equation (1) with the coefficients from Table 1 versus the measured temperature a) for the single-junction perovskites, and b) for the tandem perovskites mini-modules. Data includes 96 days of data without filtering.

IV. CONCLUSIONS

In conclusion, our analysis shows that perovskite mini-modules exhibited a different temperature than silicon modules. The temperature differentials observed between the two technologies were significant, with perovskite modules consistently operating at lower temperatures. Moreover, within our sample modules, the single-junction perovskite mini-modules displayed lower temperatures than one of the tandem perovskite-silicon mini-modules.

It is important to note that the variability of the data and the differences in module sizes limited our ability to draw definitive quantitative conclusions. The temperature variations observed among the different module configurations may have been influenced by various factors, including variations in module dimensions, band gaps, optical properties, and conversion efficiencies. These factors collectively contribute to differences in heat dissipation and absorption characteristics, thus influencing the operating temperatures.

Additionally, it is important to acknowledge the potential bias introduced by our calculation of the convection coefficient, which could have implications for the accuracy of our results. The convection coefficient plays a crucial role in determining the heat transfer between the module surface and the surrounding environment. Any inaccuracies or assumptions made in estimating this coefficient can impact the modeled temperatures.

ACKNOWLEDGEMENT

We would like to express our gratitude to M. McGehee for the suggestion to document the operating temperature. We also extend our thanks to P. Sanchez for his assistance in developing scripts and discussion for the study. Their contributions were instrumental in the successful completion of this project.

REFERENCES

[1] T. Abzieher et al., "Photovoltaic Devices: Electron-Beam-Evaporated Nickel Oxide Hole Transport Layers for Perovskite-Based Photovoltaics (Adv. Energy Mater. 12/2019)," *Advanced Energy Materials*, vol. 9, no. 12, p. 1970035, 2019, doi: 10.1002/aenm.201970035.

[2] S. Seo, S. Jeong, C. Bae, N.-G. Park, and H. Shin, "Perovskite Solar Cells with Inorganic Electron- and Hole-Transport Layers Exhibiting Long-Term (≈500 h) Stability at 85 °C under Continuous 1 Sun Illumination in Ambient Air," *Advanced Materials*, vol. 30, no. 29, p. 1801010, 2018, doi: 10.1002/adma.201801010.

[3] Z. Guo, A. K. Jena, and T. Miyasaka, "Halide Perovskites for Indoor Photovoltaics: The Next Possibility," *ACS Energy Lett.*, vol. 8, no. 1, pp. 90–95, Jan. 2023, doi: 10.1021/acsenergylett.2c02268.

[4] R. Gehlhaar, T. Merckx, W. Qiu, and T. Aernouts, "Outdoor Measurement and Modeling of Perovskite Module Temperatures," *Global Challenges*, vol. 2, no. 7, p. 1800008, 2018, doi: 10.1002/gch2.201800008.

[5] A. Wheeler, M. Leveille, I. Anton, M. Limpinsel, and S. Kurtz, "Outdoor Performance of PV Technologies in Simulated Automotive Environments," in *2019 IEEE 46th Photovoltaic Specialists Conference (PVSC)*, Jun. 2019, pp. 3103–3110. doi: 10.1109/PVSC40753.2019.8981352.

[6] T. J. Silverman, M. G. Deceglie, B. Marion, S. Cowley, B. Kayes, and S. Kurtz, "Outdoor performance of a thin-film gallium-arsenide photovoltaic module," in *2013 IEEE 39th Photovoltaic Specialists Conference (PVSC)*, Jun. 2013, pp. 0103–0108. doi: 10.1109/PVSC.2013.6744109.

[7] W. Tress et al., "Performance of perovskite solar cells under simulated temperature-illumination real-world operating conditions," *Nat Energy*, vol. 4, no. 7, Art. no. 7, Jul. 2019, doi: 10.1038/s41560-019-0400-8.

[8] N. Irvin et al., "Deleterious Effect of Light Trapping on the Temperatures of Solar Modules," Jun. 2022, pp. 1059–1059. doi: 10.1109/PVSC48317.2022.9938767.

[9] N. P. Irvin, D. M. Escobar, A. Wheeler, R. R. King, C. B. Honsberg, and S. R. Kurtz, "Thermal Impact of Rear Insulation, Light Trapping, and Parasitic Absorption in Solar Modules,"

IEEE Journal of Photovoltaics, vol. 12, no. 4, pp. 1043–1050, Jul. 2022, doi: 10.1109/JPHOTOV.2022.3173785.

[10] M. Muller, B. Marion, and J. Rodriguez, "Evaluating the IEC 61215 Ed.3 NMOT procedure against the existing NOCT procedure with PV modules in a side-by-side configuration," in *2012 38th IEEE Photovoltaic Specialists Conference*, Jun. 2012, pp. 000697–000702. doi: 10.1109/PVSC.2012.6317705.

A Broadband Anti-reflection Coating for Thin Film CdSeTe/CdTe Solar Cells

Adam M Law, Luksa Kujovic, Mustafa Togay, Xiaolei Liu, Kurt Barth and John M Walls
CREST, Wolfson School of Mechanical and Manufacturing Engineering,
Loughborough University, Loughborough, LE11 3TU, United Kingdom

Abstract—Thin film cadmium telluride (CdTe) photovoltaics (PV) is the most successful second generation PV technology, with a current market share of ~5% and deployment predominantly at utility scale. Improvements in the buffer layer of the device, switching from cadmium sulphide (CdS) to tin oxide (SnO_2) or magnesium-doped zinc oxide (MZO), and the addition of selenium to the absorber layer, have expanded the wavelength range over which CdTe devices operate, from 400-850nm to 350-900nm, resulting in increased efficiency. As a result, an optimised anti-reflection (AR) coating design is required to improve the efficiency even further. A 6-layer AR coating of SiO_2 and ZrO_2, building on a previous 4-layer design for CdTe devices, has been designed, modelled, and fabricated on 3.8mm thick FTO-coated TEC15 substrates, reducing reflection by 3.38% absolute. Electrical measurements of a CdSeTe/CdTe device before and after addition of the AR coating show a relative increase in short-circuit current density, J_{sc}, of 3.45%, and a relative increase in efficiency of 3.54%. The use of this AR coating on CdSeTe/CdTe PV devices, which is stable under the high processing temperatures required in device manufacturing, will enable significantly higher conversion efficiencies.

Index Terms—solar, photovoltaics (PV), anti-reflection (AR), coatings, thin film, CdTe

I. INTRODUCTION

Thin film CdTe is the most commercially successful second-generation solar photovoltaics (PV) technology. High efficiency modules are manufactured at scale and at low cost, and their use is predominantly in large scale solar utilities. Laboratory devices have reached efficiencies of 22.1% [1], while module efficiencies can be as high as 19.5% [2]. The high efficiency of CdTe PV is the result of a number of significant advances in device structure, the most recent being the addition of CdSe(1-x)Te(x) to the CdTe absorber layer. The addition of CdSeTe lowers the bandgap at the front of the absorber from 1.5eV (CdTe) to 1.41eV (CdSeTe/CdTe), extending the spectral response to longer wavelengths, increasing photon absorption and leading to higher current. The addition of selenium to the absorber also passivates defects in the absorber and increases carrier lifetime, further improving device performance [3], [4].

PV modules experience a reflection loss of just over 4% from the cover glass surface due to the difference in refractive index between glass and air, although this can be significantly reduced by the use of an anti-reflection (AR) coating placed at the glass/air interface. Typically, a single layer of ~125nm thick porous SiO_2 is used [5], as the porous structure allows a lower refractive index, resulting in lower reflection, although this lowers their resistance to abrasion. Multilayer antireflection (MAR) coatings are used to overcome some of the limitations of single layer AR coatings, giving broadband anti-reflection and higher durability than the porous SiO_2 coatings currently used in the field [6], [7]. Such coatings have previously been designed for silicon, CdTe, and other thin film PV technologies [8].

Traditional CdTe devices operate from 400-850nm, however state-of-the-art CdSeTe/CdTe devices now operate over a range of 350-900nm. The addition of a tin oxide (SnO_2) or magnesium-doped zinc oxide (MZO) buffer layer to replace the previously used cadmium sulphide (CdS) increases photon absorption around 350nm [9], while the incorporation of CdSeTe into the absorber shifts the absorption edge from 850 to 900nm. This improvement in the wavelength range utilised by CdTe devices requires existing AR coating designs to be adapted, to take into account the additional availability of photons at either edge. Previous MAR coatings for CdTe PV have been designed to operate from 400-850nm [10], with reflection values increasing rapidly beyond this range, making them less suitable for improved CdSeTe/CdTe devices.

In this work we report on the design, fabrication, and testing of a multilayer broadband AR coating optimised for CdSeTe/CdTe devices. We provide details of the coating design process using a commercially available optical modelling program, Essential Macleod [11], as well as the coating deposition on fluorine-doped tin oxide (FTO) coated TEC15 substrates via pulsed-DC reactive magnetron sputtering. We also include current-voltage (I-V) and efficiency measurements of CdSeTe/CdTe devices with and without the AR coating to show the improvements in device performance.

II. COATING DESIGN & MODELLING

Previous work on a MAR coating design for CdTe devices was optimised for the 400-850nm range using a 4-layer coating, but state-of-the-art CdSeTe/CdeTe devices operate over 350-900nm, so the design needs to be modified to accommodate the longer wavelength range. MAR coatings can be easily adapted for different wavelength ranges by tuning the layer thicknesses and adding more layers if needed, and this is done using the optimisation and refinement process in Essential Macleod. Further details on the design process

978-1-6654-6060-6/23 $31.00 © 2023 IEEE

and software can be found in previous work [7], [8].

The previous design consisted of 4 layers of alternating SiO_2 and ZrO_2 and is the starting point for the new coating design. SiO_2 is the low index layer, and ZrO_2 the high index layer in the alternating high/low index structure necessary for MAR coatings. Essential Macleod uses an iterative process to optimise layer thicknesses for low reflection across a target wavelength range, making small adjustments in layer thickness and re-calculating reflectance until performance improvements are realised. To begin with, the existing 4-layer design was re-optimised for the longer wavelength range, although it was found that the new 4-layer design offered no improvement in weighted average reflectance (WAR) when compared to the original CdTe design. A comparison including uncoated glass is shown in Table II. Therefore, a 6-layer design is required to improve performance in the extended wavelength range. Although more layers are required, the total thickness for the 6-layer design is actually lower than the initial 4-layer design, resulting in ~5% less material use which would lower production costs. Generally, the addition of more layers with optimised thickness in a MAR coating will reduce reflection further, although this needs to be weighed against the potential increase in materials and/or production costs.

TABLE I
THE MATERIALS USED IN EACH COATING LAYER AND THEIR RESPECTIVE THICKNESSES.

Layer	Material	Thickness (nm)
Medium	Air	
1	SiO_2	91
2	ZrO_2	28
3	SiO_2	11
4	ZrO_2	94
5	SiO_2	22
6	ZrO_2	22
Substrate	Glass	

Table I shows the modelled coating structure and optimised thicknesses for each layer in the multilayer coating design. The 6-layer coating consists of alternating SiO_2 and ZrO_2 layers and has a total thickness of 268nm.

TABLE II
THE NUMBER OF LAYERS, TOTAL THICKNESS, AND WAR VALUES FOR EACH OF THE DESIGNS INVESTIGATED

Design	Layers	Thickness (nm)	WAR (%)
Glass			4.22
CdTe AR	4	282	1.25
CdSeTe AR	4	264	1.24
CdSeTe AR	6	268	1.08

Fig. 1 shows the modelled reflectance of both the previous 4-layer AR coating design for CdTe devices, and the new

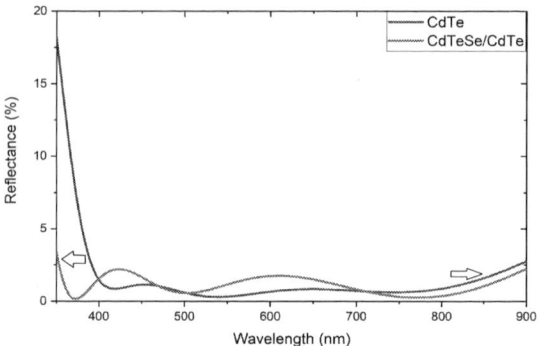

Fig. 1. Modelled reflectance of the new 6-layer AR coating design compared to the previous 4-layer design

6-layer design for CdSeTe/CdTe. In this case only the front surface reflection is shown. The reflection edge at both sides has been shifted to capture the increased availability of photons between 350-400nm and 850-900nm. The shift is much more apparent from 400nm down to 350nm. This shift of both edges does result in slightly increased reflectance at ~600nm, although the WAR is lower, at 1.08% for the 6-layer design compared to 1.25% for the original 4-layer design. The re-optimised 4-layer design has a WAR of 1.24%, offering no improvement compared to the previous design. The front surface of uncoated, 1mm thick soda-lime glass has a WAR of 4.22% across the same wavelength range, so the 6-layer AR coating design offers an absolute improvement of 3.14%.

III. EXPERIMENTAL DETAILS

The 6-layer AR coating comprising alternating SiO_2 and ZrO_2 layers has been deposited on both 1mm thick 5cm x 5cm soda-lime and 3.8mm thick 5cm x 5cm TEC15 substrates (NSG Pilkington) via pulsed-DC reactive magnetron sputtering, using a MyCoat optical coating system (VisionEase, Ramsey MN). Further details on the deposition process, as well as the materials used, can be found in previous reports [7]. Substrates were cleaned using an Ossila UV-Ozone cleaner for 15 minutes prior to deposition.

Reflection measurements have been taken across a wave-length range of 300-1000nm using a Varian Cary5000 UV-VIS-NIR spectrophotometer with integrating sphere attach-ment. WAR values have been calculated using Equation (1), where Φ is the photon flux and R is the percent reflectance at each given wavelength, λ. The calculations were performed using a custom Matlab script.

$$WAR(\lambda_{min}, \lambda_{max}) = \int_{\lambda_{min}}^{\lambda_{max}} \frac{\Phi \cdot R}{R} d\lambda \qquad (1)$$

I-V measurements were obtained for a CdSeTe/CdTe device, before and after addition of the AR coating, using a Wacom solar simulator. The Wacom is a steady-state dual lamp solar simulator with halogen and xenon sources which exceeds the requirements for class AAA as defined in IEC

60904-2-(2020). Electrical measurements were taken with a Keithley source meter. I-V measurements were taken from 4 cells on the device, with 3 measurements taken at each point, and the average value calculated. The cell measurement area was 0.44cm^2, and short-circuit current (I_{sc}) values are divided by this area to give short-circuit current density (J_{sc}) values. The CdSeTe/CdTe device was deposited under similar conditions to those described in [12], and the AR coating was deposited onto the uncoated glass side of the TEC15 superstrate of the device after the initial electrical measurements.

IV. RESULTS & DISCUSSION

A. Reflectance Measurements

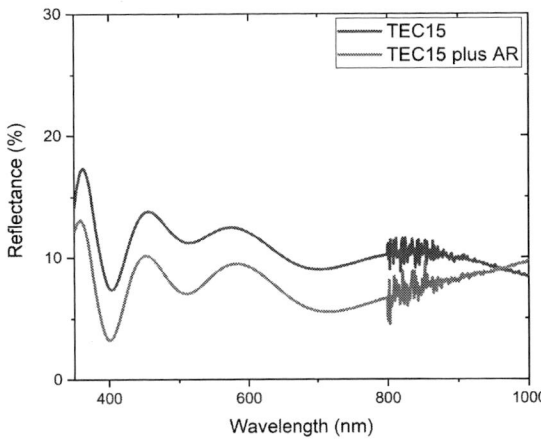

Fig. 3. Measured reflectance of uncoated and AR-coated TEC15 substrates

uncoated TEC15 is 10.85%, which is reduced to 7.47% by the addition of the AR coating, an absolute reduction of 3.38%.

B. Cell Measurements

A summary of the electrical measurements for the CdSeTe/CdTe device before and after the addition of the AR coating is given in Table III. The average short-circuit current density (J_{sc}) across the device increases from 28.25mAcm^{-2} to 29.23mAcm^{-2} after addition of the AR coating, an increase of 3.45%, which is consistent with the reduction in WAR provided by the AR coating. Small variations in measurement may occur as a result of contacting different parts of each cell in the before and after measurements, although almost all of the change can be attributed to the increased photocurrent provided by the AR coating. The fill factor (FF) and open-circuit voltage (V_{oc}) are unchanged by the addition of the AR coating, as they are cell-dependent parameters. The increase in J_{sc} results in an increase in the efficiency of the device from an average of 16.93% to 17.53%, a 3.54% relative increase. This results in a significant gain in power output for the device.

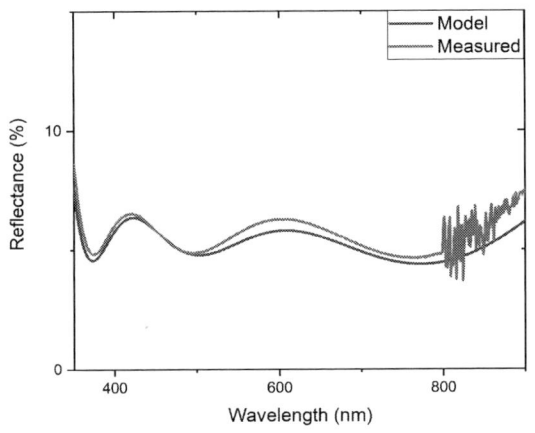

Fig. 2. Measured vs modelled reflectance of the 6-layer AR coating deposited on 1mm thick glass substrates

Fig. 2 shows the measured reflectance of the deposited 6-layer AR coating on 1mm thick soda-lime glass compared to the initial model, with back surface reflectance from the uncoated glass included. The measured reflectance is generally in good agreement with the model, especially at shorter wavelengths. Noise in the measured data at 800nm and above is caused by a changeover in the detector used by the spectrophotometer.

After verifying the design by depositing on 1mm thick soda lime glass, the 6-layer AR coating has then been deposited on 3.8mm thick TEC15 substrates. These substrates are coated with a layer of FTO, a transparent conductor used in CdSeTe/CdTe devices, on one side. CdSeTe/CdTe devices use the superstrate configuration, with the absorber and intermediate layers deposited on glass. A schematic showing a typical CdSeTe/CdTe device structure is shown in [12]. Fig. 3 shows the measured reflectance of FTO-coated TEC15 substrates with and without the 6-layer AR coating, and the back-surface reflectance is included. The reflectance is significantly reduced across all useful wavelengths. The WAR across 350-900nm for

V. SUMMARY & CONCLUSIONS

Improvements in both the buffer layer and absorber structure of CdSeTe/CdTe thin film solar cells have resulted in increased light absorption between 350-400nm, and expanded the absorption edge from 850nm to 900nm, resulting in the need for a newly optimised AR coating to further improve light collection at these wavelengths. A multilayer coating comprising 6 layers of alternating SiO$_2$ and ZrO$_2$ has been designed and optimised across the wavelength range of 350-900nm and has been deposited on both 1mm thick soda-lime glass and 3.8mm thick FTO-coated TEC15 substrates by pulsed-DC reactive magnetron sputtering. Addition of the AR coating to TEC15 substrates lowers weighted average reflectance by 3.38% compared to uncoated glass. I-V measurements of CdSeTe/CdTe devices with and without AR-coated TEC15 substrates show an increase in J_{sc} of almost 1mAcm^{-2}, a relative improvement of 3.45%, and an efficiency

Cell no.	Initial	With AR	Initial	With AR	Initial	With AR	Initial	With AR
	$\mathbf{J}_{sc}$ (mAcm^{-2})		FF (%)		$\mathbf{V}_{oc}$ (mV)		Eff. (%)	
1	28.4	29.4	79.4	79.3	780	783	17.6	18.3
2	28.0	29.1	76.2	76.3	771	770	16.5	17.1
3	28.0	28.9	73.5	73.6	770	772	15.8	16.5
4	28.6	29.5	76.7	76.1	811	808	17.8	18.2
Average	28.25	29.23	76.45	76.33	783	783.25	16.93	17.53
% Change	-	3.45	-	-0.16	-	0.03	-	3.54

TABLE III

ELECTRICAL MEASUREMENTS FOR CDSETE/CDTE CELLS BEFORE AND AFTER AR COATING. DATA COURTESY OF DR M. TOGAY AND A. SMITH

improvement of 3.54%. High temperature stability is a further advantage of the all dielectric stack used in the multilayer design [13]. This allows AR coated superstrates to be used in the cell manufacturing process even if high temperatures are used in the absorber deposition and activation processes. This is not possible with conventional porous SiO_2 AR coatings.

REFERENCES

[1] Martin A Green, Ewan D Dunlop, Joel Hohl-Ebinger, Masahiro Yoshita, Nikos Kopidakis, Karsten Bothe, David Hinken, Michael Rauer, and Xiaojing Hao. Solar cell efficiency tables (version 60). *Prog. Photovoltaic Res. Applic.*, 30(7):687–701, 2022.

[2] NREL. Champion photovoltaic module efficiency chart, 2022.

[3] Thomas AM Fiducia, Budhika G Mendis, Kexue Li, Chris RM Grovenor, Amit H Munshi, Kurt Barth, Walajabad S Sampath, Lewis D Wright, Ali Abbas, Jake W Bowers, and John M Walls. Understanding the role of selenium in defect passivation for highly efficient selenium-alloyed cadmium telluride solar cells. *Nature Energy*, 4(6):504–511, 2019.

[4] Amit H Munshi, Jason Kephart, Ali Abbas, John Raguse, Jean-Nicolas Beaudry, Kurt Barth, James Sites, John Walls, and Walajabad Sampath. Polycrystalline cdsete/cdte absorber cells with 28 ma/cm 2 short-circuit current. *IEEE Journal of Photovoltaics*, 8(1):310–314, 2017.

[5] Klemens Ilse, Charlotte Pfau, Paul-T Miclea, Stephan Krause, and Christian Hagendorf. Quantification of abrasion-induced arc transmission losses from reflection spectroscopy. In *2019 IEEE 46th Photovoltaic Specialists Conference (PVSC)*, pages 2883–2888. IEEE, 2019.

[6] Jimmy M Newkirk, Illya Nayshevsky, Archana Sinha, Adam M Law, QianFeng Xu, Bobby To, Paul F Ndione, Laura T Schelhas, John M Walls, Alan M Lyons, et al. Artificial linear brush abrasion of coatings for photovoltaic module first-surfaces. *Solar Energy Materials and Solar Cells*, 219:110757, 2021.

[7] Adam M Law, Farwah Bukhari, Luke O Jones, Patrick JM Isherwood, and John M Walls. Multilayer antireflection coatings for cover glass on silicon solar modules. *IEEE Journal of Photovoltaics*, 12(5):1205–1210, 2022.

[8] Piotr M Kaminski, G Womack, and John M Walls. Broadband anti-reflection coatings for thin film photovoltaics. In *2014 IEEE 40th Photovoltaic Specialist Conference (PVSC)*, pages 2778–2783. IEEE, 2014.

[9] Taowen Wang, Shengqiang Ren, Chunxiu Li, Wei Li, Cai Liu, Jingquan Zhang, Lili Wu, Bing Li, and Guanggen Zeng. Exploring window buffer layer technology to enhance cdte solar cell performance. *Solar Energy*, 164:180–186, 2018.

[10] Gerald Womack, Piotr M Kaminski, Ali Abbas, Kenan Isbilir, Ralph Gottschalg, and John Michael Walls. Performance and durability of broadband antireflection coatings for thin film cdte solar cells. *Journal of Vacuum Science & Technology A: Vacuum, Surfaces, and Films*, 35(2):021201, 2017.

[11] Angus Macleod and Christopher Clark. Optical coating design with the essential macleod. 2012.

[12] Wyatt K Metzger, S Grover, D Lu, E Colegrove, John Moseley, CL Perkins, X Li, R Mallick, W Zhang, R Malik, et al. Exceeding 20% efficiency with in situ group v doping in polycrystalline cdte solar cells. *Nature Energy*, 4(10):837–845, 2019.

[13] Gerald Womack, PM Kaminski, and JM Walls. High temperature stability of broadband anti-reflection coatings on soda lime glass for solar modules. In *2015 IEEE 42nd Photovoltaic Specialist Conference (PVSC)*, pages 1–6. IEEE, 2015.

978-1-6654-6060-6/23 $31.00 © 2023 IEEE

Can solar+storage keep the lights on? Assessing solar+storage for backup power during long-duration power interruptions in the US

Will Gorman, Galen Barbose, Juan Pablo Carvallo, Sunhee Baik, Chandler Miller

Lawrence Berkeley National Lab, Berkeley, CA, United States

Recent market trends reveal rapid growth in the adoption of paired behind-the-meter (BTM) solar photovoltaic and energy storage systems (PVESS). Those trends have been driven in part by customer concerns over electric system reliability and demand for backup power, which are likely to become even more pronounced as wildfire, hurricane, and other climate-driven risks rise over the coming decades. But what can customers expect, and what can installers promise, in terms of how well these systems might actually perform in providing backup power during long-duration power interruptions? The presentation will highlight key findings from Berkeley Lab' examination of BTM PVESS in backup power applications. The analysis is based on simulating PVESS backup performance during both a set of synthetic power interruption events as well as a set of 10 historical long-duration power interruption events. The analysis evaluates performance across a wide range of outage conditions and across thousands of individual building models, capturing both different building types and variations in the existing building stock. The results show how even a relatively small PVESS can provide backup power to a basic set of critical loads, while also highlighting some of the key considerations and constraints in providing backup to electric heating and cooling loads. The analysis illustrates both the challenges and opportunities associated with electrification, in terms of PVESS backup power capabilities.

978-1-6654-6060-6/23 $31.00 © 2023 IEEE

Close Roof Mounted System Temperature Estimation for Compliance to IEC TS 63126

Michael D. Kempe, Silvana Ovaitt, Martin Springer, Matthew Brown, Dirk Jordan, William Sekulic, Colleen O'Brien, Jean-Nicolas Jaubert, Yuanjie Yu, Jaewon Oh, Govindasamy Tamizhmani, Bo Li

National Renewable Energy Laboratory (NREL), Golden, CO, United States

Underwriters Laboratory (UL), Fremont, CA, United States

Canadian Solar, Suzhou, China

Arizona Stat University (ASU), Mesa, AZ, United States

Intertek, Lake Forest, CA, United States

University of North Carolina at Charlotte, Charoltte, NC, United States

When photovoltaic modules are installed on rooftops, the temperature depends on the geographic location and the mounting configuration. If the mounting structure does not provide sufficient airflow in a hot environment, the 98th percentile temperature will exceed 70°C, which according to IEC TS 63126, requires higher levels of thermal stability testing. However, there is no clear way to determine the temperature level needed for a particular location and system design. In this work, we identify a relationship between the module standoff to the rooftop and the module temperature, and propose methods to describe a minimum standoff for a given location. Because system designs vary in more ways than just the standoff, we show how to determine an effective standoff that can be applied to generic calculations. Lastly, we show measurements and calculations from several systems to demonstrate how this method could work.

i-TOPCon solar cells prepared by high throughput magnetron sputtering of in-situ doped *n*-type amorphous silicon layers

Eric Schneiderloechner[1], Tina Dietsch[1], Jan Hoss[2], Jonathan Linke[2], Jana-Isabelle Polzin[3], Sebastian Mack[3], Henning Nagel[3], Volker Linss[1]

[1] VON ARDENNE GmbH, Am Hahnweg 8, D-01328 Dresden,
[2] International Solar Energy Research Center Konstanz, Rudolf-Diesel-Straße 15, D-78467 Konstanz,
[3] Fraunhofer Institute for Solar Energy Systems ISE, Heidenhofstrasse 2, D-79110 Freiburg

Abstract — We report on the progress of developing an industrially relevant in-situ doped sputtering process for depositing 100-150 nm thick aSi(*n*) layers to passivate the rear contact of i-TOPCon solar cells. All relevant properties of the poly-Si layers have been optimized using symmetrical lifetime samples. The sputtered aSi(*n*) layers do not show blistering effects and can be doped with Phosphorus concentrations larger than 1E21 at/cm³ which has been evaluated via calibrated SIMS measurements. ECV measurements of optimized crystallized layers confirm the necessary electrically active concentration $C_P \sim$ 1-3E20 at/cm³. After tuning the wet chemically or thermally grown interfacial oxide layer and optimizing the PVD process parameters as well as the annealing process implied open circuit voltages $iV_{OC} >$ 735 mV could be achieved at one sun injection density on planar 1-3 Ωcm n-type Cz-lifetime samples. The best sputtered i-TOPCon solar cell produced in this work has a confirmed efficiency of 23.6 %.

I. INTRODUCTION

Over the span of the last few years passivated contacts based on tunnel oxides and highly doped poly-silicon thin films, known as tunnel oxide passivated contacts - TOPCon [1] or poly-Si on oxide - POLO [2], have been identified as the most likely technology [3,4] to overcome the efficiency limitations of industrially manufactured PERC solar cells. The underlying principles of this type of passivated contacts were investigated and published already in the 1980s [5,6] and today the Industry has started to roll out this technology into multi-GW scale.

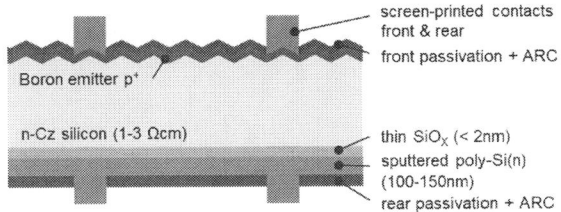

Fig. 1   Sketch of the industrial TOPCon cell structure.

As published in [7] and thereafter elsewhere [8, 9] sputtering can be used to deposit doped aSi layers for passivating contacts. Here, we report on the development of an industrial relevant, in-situ doped DC-sputtering process for depositing 100-150 nm

thick aSi(*n*) layers to be used as passivating rear contacts in TOPCon solar cells as depicted in Fig. 1.

Another important goal was to demonstrate that the sputter process enables a 100% aSi(*n*)-coverage of the rear surface without creating any wrap-around of the aSi(*n*) layer, resulting in low reverse currents. This allows to omit the costly and yield problematic back etching step as it is needed when using LPCVD or PECVD for depositing the thin silicon film.

II. EXPERIMENTAL AND RESULTS

All relevant properties of the poly-Si layers have been optimized using symmetrical FZ- and Cz-silicon lifetime samples with a 1-2 nm thin thermal interface-SiO$_X$ on either side. Other than aSi from PECVD deposition, the sputtered aSi(*n*) layers show no blistering effects as there is no significant amount of hydrogen in the process even for low processing temperatures. Sputtered aSi(*n*) layers can be loaded with Phosphorus up to a concentration of 2E21 at/cm³ which has been evaluated via calibrated SIMS measurements (see Fig. 3). ECV measurements (see Fig. 4) of layers after a high temperature anneal at 850-980 °C for 10-30 min find the necessary electrically active concentration $C_P \sim$ 1-3E20 at/cm³.

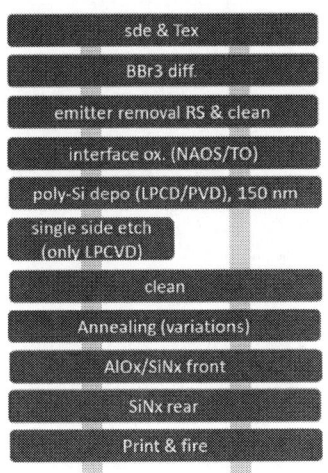

Fig. 2   Process sequence to manufacture i-TOPCon solar cells. Please note that the PVD sequence (right) is without single side etch of silicon layer wrap-around from the front surface.

978-1-6654-6060-6/23 $31.00 © 2023 IEEE

After tuning the thickness of the tunnel oxide layer, the PVD process parameters (T, p, P) as well as the annealing process implied open circuit voltages $iV_{OC} > 735$ mV have been measured (by QSSPC) on symmetrical lifetime structures at about one sun injection density on 1-3 Ωcm n-type Cz-wafers.

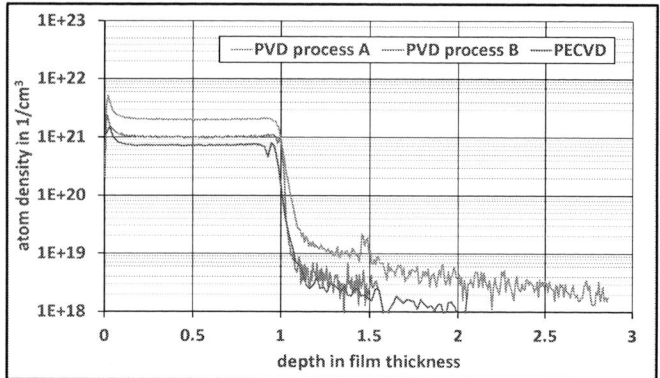

Fig. 3   SIMS depth profiles of Phosphorus of one PECVD- and two PVD-deposited aSi(n) layers.

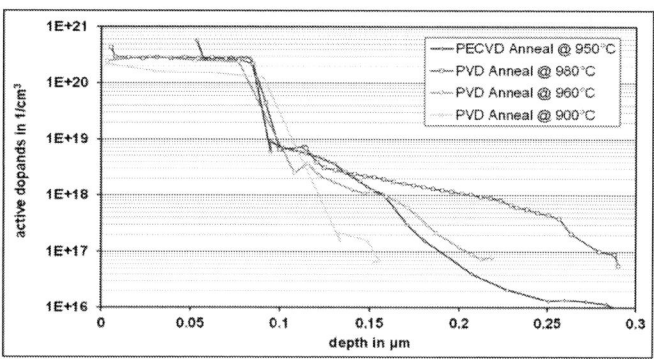

Fig. 4   Active dopant concentration depth profiles (ECV) of one PECVD- & three PVD aSi(n) layers after annealing for 10 min, at plateau temperatures in the range of 900-980 °C.

In cooperation with two different R&D centers our optimized in-situ doped sputtered aSi(n) layers have been integrated in n-type M2-Cz i-TOPCon solar cells with diffused boron emitter and screen-printed contacts. The process sequence is depicted in

Fig. 2 above. At the first R&D center the best solar cells featuring the sputtered layers reach an efficiency of 22.4 % which is 0.4% lower as the already optimized LPCVD references, which is mostly due to the optimized firing of the screen-printed metallization. Applying the laser enhanced contacting technique (LECO, [10]) that allows to post process the screen-printed contacts locally to reduce the specific contact resistivity without changing other solar cell properties, the eta deficit is reduced to 0.3% with the PVD performing at 22.6%.

Most remarkably though is the fact that no etching of any residual aSi-layer on the front side had to be performed for the sputtered i-TOPCon solar cells to reach the reverse current

characteristics $I_{Rev2} < 0.2$ A measured at -12 V as shown in Fig. 5. This truly single sided deposition is realized since sputtering is a directional deposition method. This feature allows for a leaner TOPCon process with respect to CVD based poly-Si deposition methods.

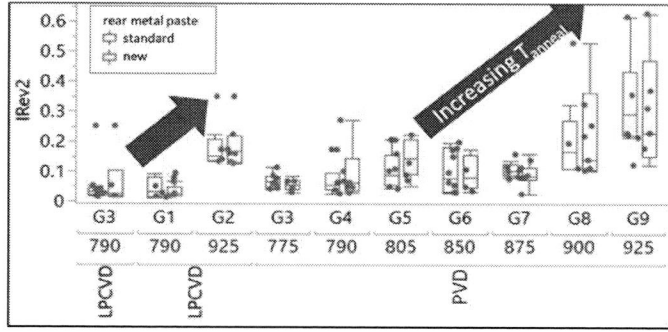

Fig. 5   Reverse current I_{Rev2} @-12V of the M2-format i-TOPCon solar cells measured at R&D center 1. Both, LPCVD- as well as PVD-groups show increasing I_{Rev2} with increasing anneal temperature. Effect is under investigation.

Our latest results achieved in cooperation with R&D center 2 are shown in Fig. 6. Again, there was no back etching step involved in the processing sequence. To optimize the solar cell performance the PVD parameters as well as the annealing temperature have been varied. Also, there were two different tunnel oxides prepared for the PVD groups, thermally grown with a thickness of about 1.5 nm and also wet chemically grown oxides using an ozone-based method. Again, all IV-parameters depicted in Fig. 6 were measured after LECO processing of the screen-printed metal grids on either side.

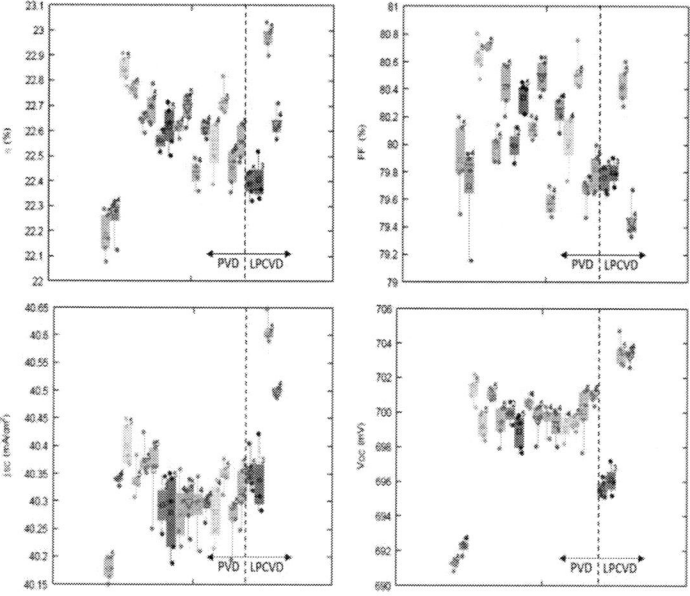

Fig. 6 IV-Data of the differently processed PVD- and LPCVD-groups of i-TOPCon solar cells measured on an industrial cell tester at R&D center 2 after LECO processing.

It can be seen that the V_{OC} and J_{SC} levels of the LPCVD references are slightly higher than that of our PVD-processed samples which can be explained by mail-transporting the PVD samples after tunnel oxide deposition back and forth between the two locations (V_{OC}) and due to the fact that the PVD-layers are 140nm thick compared to 100nm for the references (J_{SC}). Still the best PVD groups are as good as some of the LPCVD groups and the best PVD group reaches on average 22.85% efficiency compared to 22.95% for the best LPCVD group.

The best PVD-processed solar cell was sent to Fraunhofer ISE CalLab and the measured cell efficiency was 23.6%.

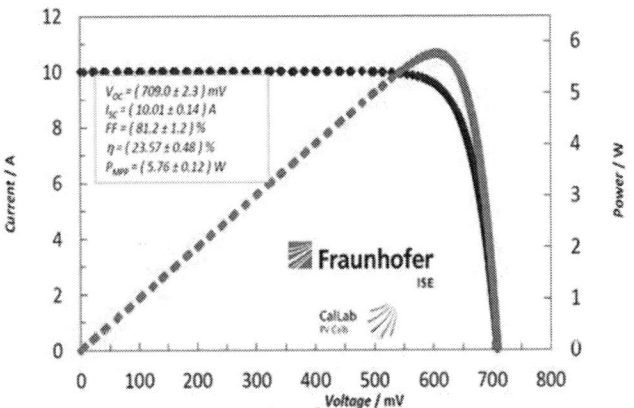

Fig. 7 *IV*-curve of the best PVD co-processed i-TOPCon solar cell measured at *Fraunhofer ISE CalLab PV Cells* on a reflecting chuck.

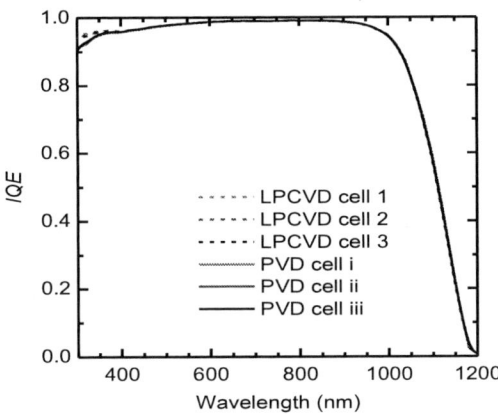

Fig. 8 *IQE* of three PVD processed and three LPCVD reference solar cells of an earlier solar cell batch, (*IV*-data not depicted here). The effective minority carrier diffusion length L_{eff} fitted to the IQE is about 4000 μm.

Finally, Fig. 8 compares the internal quantum efficiency (*IQE*) of solar cells deposited with different PVD processes to LPCVD reference cells. There is no performance difference visible in the near infrared range which confirms that the optical properties of the sputtered poly-silicon layers are perfectly suited for application in TOPCon solar cells, i. e. the free carrier absorption is low.

III. SUMMARY AND OUTLOOK

It has been shown that in-situ *n*-doped, inline sputtered, amorphous silicon films are suitable for high-efficiency TOPCon solar cells. The used inline sputter process can be adapted for high volume production. Based on the processes developed a cost-effective sputter tool with a productivity of more than 8000 M10-wafers per hour can be engineered, similar to the ones already available for the TCO deposition of SHJ-solar cells [11].

Our ongoing work includes the comparison of the crystallinity between the annealed PVD-processed silicon layers and LPCVD and PECVD based references using XRD measurements.

REFERENCES

[1] F. Feldmann, M. Bivour, C. Reichel, M. Hermle, S.W. Glunz, "A passivated rear contact for high-efficiency n-type Si solar cells enabling high Voc's and FF>82 %", *Proceedings of 28th EU-PVSEC*, pp. 988–992, 2013.

[2] U. Römer, R. Peibst, T. Ohrdes, B. Lim, E. Bugiel, T. Wietler, R. Brendel, "Recombination behavior and contact resistance of n+ and p+ poly-crystalline Si/mono-crystalline Si junctions", *Solar Energy Materials and Solar Cells,* vol. 131, pp. 85–91, 2014.

[3] J. Schmidt, R. Peibst, R. Brendel, "Surface passivation of crystalline silicon solar cells: present and future". *Solar Energy Materials and Solar Cells*, vol.187, pp. 39-54, 2014.

[4] A.Richter, R. Müller, J.Benick et.al., "Design rules for high efficiency both-sides-contacted silicon solar cells with balanced charge carrier transport and recombination losses", *Nature Energy* Vol 6, pp. 429-438, 2014.

[5] E. Yablonovitch, R.M. Swanson, Y.H. Kwark, "A 720 mV open circuit voltage SiOx:c-Si:SiOx double heterostructure solar cell", *Appl. Phys. Lett., vol* 47, pp. 1211–1213, 1985.

[6] J.Y. Gan, R.M. Swanson, "Polysilicon emitters for silicon concentrator solar cells", *Proceedings of the 21st IEEE Photovoltaic Specialists Conference*, pp. 245–250, 1990.

[7] D. Yan, A. Cuevas, S.P. Phang, Y. Wan, D. Macdonald, "23% efficient p-type crystalline silicon solar cells with hole-selective passivating contacts based on physical vapor deposition of doped silicon films", *Applied Physics Letters*. vol. 113, 061603, 2018.

[8] C. Allebé, J.J. Diaz Leon, A. Descoeudres, T. Dippell, B. Paviet-Salomon, C. Ballif, "Sputtered poly-Si for the formation of passivating contacts in n-PERT c-Si Solar Cells", *11ᵗʰ Intern. Conf. on Crystalline Silicon Photovoltaics*, 2022.

[9] L. Nasebandt, B. Min, C. Hollemann, S. Hübner, T. Dippell, R. Peibst, R. Brendel, „Sputtered phosphorus-doped poly-Si on Oxide Contacts for Si solar cells" Solar RRL, vol. 6 (9), 2022.

[10] R. Mayberry, K. Myers, V. Chandrasekaran, A. Henning, H. Zhao, E. Hofmüller. Laser enhanced contact optimization a novel technology for improved cell efficiency. *36th European Photovoltaic Solar* Energy Conference and Exhibition, 2019.

[11] M. Dimer, A. Cruz, S. Janke, S. Wendlandt, B. Stannowski, E. Schneiderlöchner, Potential of Sputtered AZO Layers for the Manufacturing of HJ-Solar Cells, 8th WCPEC, pp. 83-87, 2022.

Optimizing Demand Management to Enable Renewables: Why the Use of a Marginal Emissions Signal is a Poor Choice

Ronald A. Sinton

Sinton Instruments, Boulder, CO, 80305

Abstract—**Public Service Company of Colorado has been rapidly transforming from a coal-dominant utility to a wind and solar-dominant utility as they develop and implement plans for 80% renewables by 2030. Above 50% renewables with 15% PV and 39% wind, the utility projects high levels of PV curtailment that effectively begin to limit future PV additions. Although storage is being installed, significant demand shaping and flexibility is not anticipated by the utility before 2030. In analyzing the possibilities to incorporate flexible demand, this paper uses monthly-average hourly projections of load, renewables, and emissions through 2042 from utility regulatory proceedings and hourly data from the Energy Information Agency to conclude that average emissions provide a better signal for demand flexibility to enable renewables than marginal emissions for this specific case of Colorado. Marginal emissions are due to the power plant on the margin providing incremental power when there is an increment in load. Prior to frequent renewable curtailments, this is usually a fossil fuel plant in this balancing area. This provides a weak signal for the cleanest hours with no signal at all indicating renewables availability until renewables are on the margin due to curtailment. In contrast, average emissions per MWh anticipate future renewables in this balancing area at levels currently on the grid. This can be used to further modify demand in anticipation of future renewables. The delayed renewables signal for marginal emissions patterns relative to average emissions is about 5-6 years in this case. The effect is very similar to recent discussion on the advantages of using long-run marginal emissions rates in preference to short-run marginal emission rates to evaluate the effectiveness of various demand profiles on reducing future grid emissions and costs.**

Keywords—solar photovoltaics, wind, utility-resource planning, demand management, demand flexibility

I. INTRODUCTION

In 2019, the U.S. state of Colorado legislature passed a bill to require an 80% reduction in CO_2 by 2030 from the major utilities relative to a 2005 baseline. In 2021, one utility (Public Service Company of Colorado, PSCO) proposed a resource plan to accomplish this goal through a phase-out of coal with additions of wind and PV[1,2,3]. The regulatory proceedings resulted in detailed modeling of the projected monthly load, marginal emissions, and renewables as a function of the hour of the day for 2022 through 2042[4]. Flexible load, such as quickly increasing future EV charging, was assumed to primarily occur at night through 2042, based on the time-of-use rates designed for the 2022 fossil grid when there was excess coal and gas capacity, and therefore lowest marginal costs, at night [2]. However, the renewables fraction is on a steeply rising curve with time, with PV increasing from 9% of load on an annual basis in 2022 to 15% of load in 2023 with 39% wind. Clearly, there could be a role for aligning demand with solar generation in this scenario, where the utility projects significant curtailment of PV during daytime beginning in 2023. In filings based on the assumptions for load through 2030, curtailments of renewables due to incremental PV additions are projected to be 32% of additions in 2025 and 52% in 2030 [5]. In the resource plan and subsequent regulatory proceedings on demand-side-management, the company proposes that no flexible demand will be introduced prior to 2030 that is significant enough to modify resource additions such as combustion turbines [6]. The modeling does include a limited amount of battery and pumped storage. Under these assumptions, the PV additions are extremely limited due to the increasing daytime curtailments reducing the apparent value of PV.

It seems clear that Xcel's TOU rates could be used to shape demand towards the renewables availability to enable more PV with less curtailment. The effectiveness of this demand flexibility depends on how it is designed. Because the utility is a regulated monopoly, there is no price transparency on the system. If the goal is to minimize CO_2 emissions on the system, would it be better to specify a marginal emissions signal to control demand flexibility, or some other metric?

The answer seems intuitively obvious. As a simple example, if electric vehicle (EVs) are added to the system (as already planned) and charged primarily during daytime synchronized with renewable availability, then the next resource addition to serve this new load will be new PV due to the low-cost and low CO_2 from PV. This PV can be added with no additional curtailment. However, the incumbent technology for demand-response signals uses marginal emission rates based primarily on which fossil fuel plant is supplying power on the margin [7]. This is quite a different signal from the intuitive one described above. Marginal emission signals only respond to renewables once curtailments are very common long after resource plans determined the generation mix. Average emissions can have

978-1-6654-6060-6/23 $31.00 © 2023 IEEE

their own issues [8]. If the cleanest average emissions rates of the day are at night due to low demand and available large hydro and nuclear, it wouldn't make sense to encourage a lot of new load at night if this nuclear or large hydro could not meet new demand in the desired time frame. In the case of Colorado, EIA statistics indicate that there is no nuclear and less than 3% hydroelectric generation in Colorado.

A better solution to the issues with marginal and average emissions rates is to use a Long-Run Marginal Emissions Rate, LRMER, instead of the Short-Run Marginal Emissions Rate described above. LRMER is calculated by running capacity expansions models for each load intervention over time to include the new generation resources enabled by the new load profile[8,9]. Once this is done for the entire planning period, the emissions for the new demand profiles can be evaluated. For the simple EV example above, the intuitive result is accomplished. The new EV charged during daytime is clean over the majority of its lifetime and the new PV is subsequently installed with the minimum of new curtailments since it is meeting the demand created by the new EV. Determining LRMER is calculation intensive in the context of a utility resource plan and it isn't really a real-time or day-ahead signal, but more an assessment of which demand profiles integrate more future clean energy efficiently. What is more easily available as a day-ahead or week-ahead signal for optimizing demand response that could enable renewables in both the short and the long term for the particular case of Colorado?

II. METHODOLOGY

This paper will use the hourly marginal emissions rate data calculated monthly by PSCO from utility regulatory proceedings (CO2/MWh), along with load and renewables which was provided for all 24 hours on a monthly basis from January 2022 up until December 2042[4]. The possible use of marginal emissions (from internal PSCO calculations) will be compared to the use of an approximation of average emissions that can be constructed from the data as the fraction of fossil fuel used in each hour, which would be the load with the renewables subtracted out as a fraction of the total load, Fossil-generated MWh/total MWh.

A second method will be compared to this first. The Public Service Company of Colorado runs a balancing authority (BA) that reports hourly data to the Energy Information Agency (EIA). The PSCO data reported in regulatory proceedings comprises 81% of the PSCO balancing area. This publicly available hourly data will be used to calculate a large-signal marginal emissions rates based on the change in generation from coal and from gas between each hour in response to changes in the load. The marginal emissions for each hour will be assigned as

$$0.67 + 0.33 * |\Delta coal| / (|\Delta coal| + |\Delta gas|) \qquad (1)$$

The absolute magnitude of changes in coal and gas in this equation are in MWh. The changes in generation are for 3 hours and assigned to the middle hour. This results in a relative carbon intensity varying from gas (at 0.67) to coal (at 1.0) weighted by their contributions to increases or decreases in generation. The ratio of coal to gas CO_2 emissions rate is taken as 1.5 for simplicity in this calculation. At the relatively-low renewable penetration for 2022 for PSCO, renewables never exceed the load and coal and gas will always be on the margin in this calculation. For the data in 2022, this emissions rate is generally skewed towards gas, with the frequency distribution indicating gas on the margin more than twice as often as coal, with a monotonic trend between the two extremes. In this data, coal appears constrained to have both a higher minimum run in MW than gas and a lower maximum output.

For all of these cases, a map of the cleanest hours for each day or month will be displayed, calculated as the hours cleaner than the average of the cleanest hour and the median hour for each day. This algorithm determines the cleanest 6 or 7 hours of each 24 for potential use as preferred hours to shift load into.

Finally, these results will be compared to calculations from the NREL Cambium modeling for the entire state of Colorado to provide perspective and corresponding Long-Run Marginal Emission Rates, Short-Run Marginal Emission Rates, Average Emission Rates, and cost calculations for two specific choices of potential demand interventions corresponding to the case of charging EVs or other flexible load during daytime vs. night.

III. RESULTS

Fig. 1 plots a color map of the marginal emissions, as calculated by PSCO, for each hour by month from January 2022 through December 2031 reading from the top row of data to the bottom row. On the right, the same is shown for the fraction of fossil fuels. This fraction of fossil fuels was calculated from PSCO data for total load and renewables.

While the fraction of fossil fuel map clearly shows the increase in renewables during daytime starting in 2023, the marginal emission rates do not. No specific patterns can be seen in the marginal emission rates until renewable curtailments become very prevalent in the signal by 2029 placing these renewables on the margin. August of each year is flagged from early afternoon towards 9PM as a dirty black streak that corresponds to high air-conditioning loads during this month.

Fig. 2 shows an interpretation of the data in Fig. 1 in terms of a potential demand response signal. Here, a simple algorithm was used to flag the cleanest 6 or 7 hours of each day as potential times to shift flexible power use into. The issues with marginal emissions rates are evident (left map). There is no clear signal to shift power use to better integrate the PV until about 6 years after this would be clear from the nominal average emissions shown on the right. In contrast, the use of the average emissions signal would systematically ratchet the system towards a cleaner grid due to the fact that even a small amount of PV (9% in 2022) begins to differentiate daytime from night. Data through 2042 looks similar to the 2031 data in both cases. In addition to being cleaner, shifting power use into daytime has been shown to be the preferred behavior for grid reliability and costs in the western USA to minimize grid cost and maximize reliability[10].

978-1-6654-6060-6/23 $31.00 © 2023 IEEE

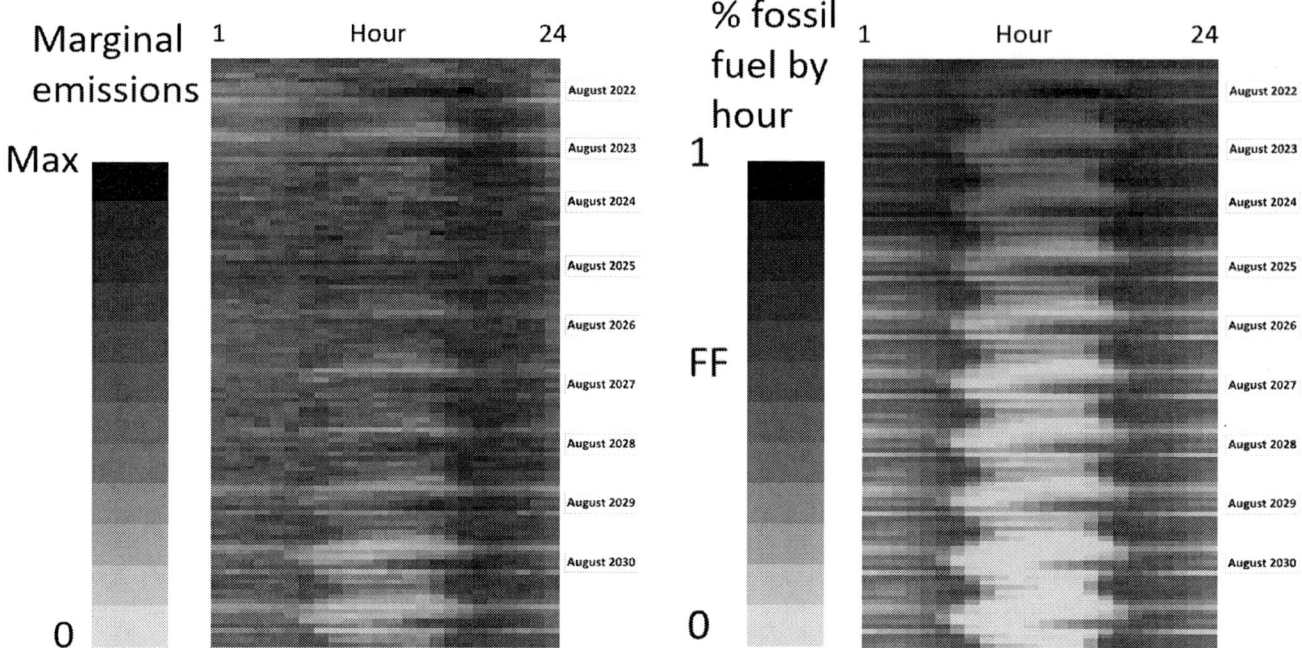

Fig. 1. On left is the marginal emissions for each of 24 hours (Columns left to right) and months (rows, January 2022 (top) to Dec. 2031 (bottom). At right, the same plot for fraction of fossil fuel in each hour, a proxy roughly indicative of relative average emissions per MWh.

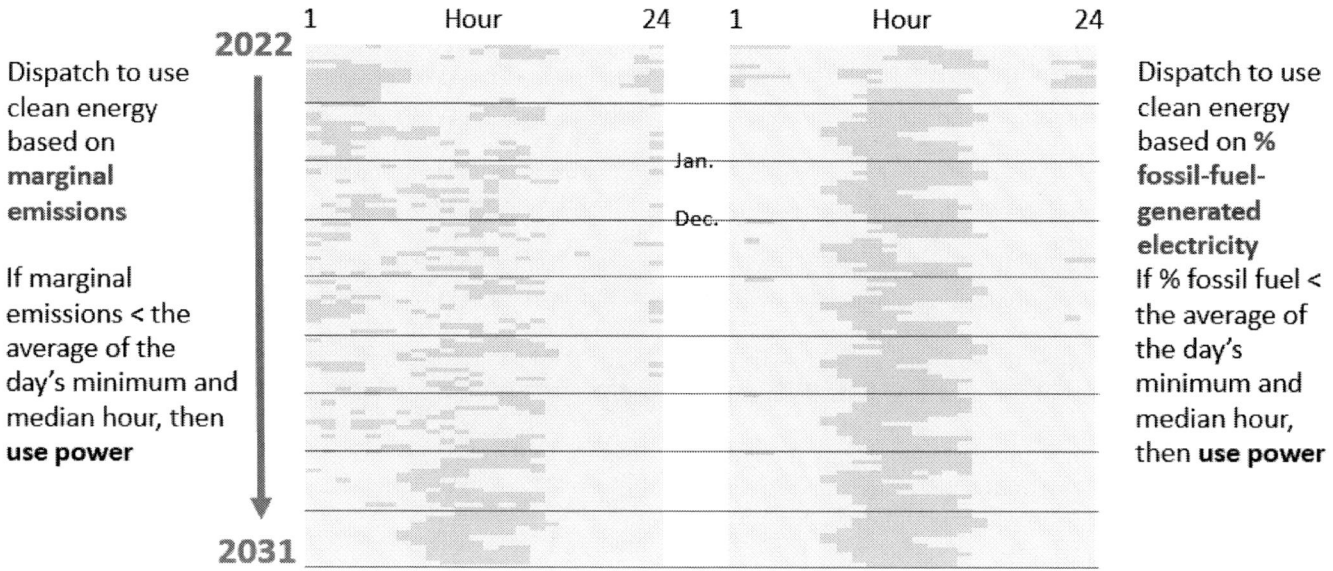

Fig. 2. For each hour in Fig. 1, an algorithm that chooses the 6 or 7 cleanest hours in the day using the marginal emissions (left) or the fraction of fossil fuel (right). The cleanest hours are defined as those between the minimum in an hour and the median for all hours for the 24 hours in the day.

IV. HOURLY MARGINAL EMISSIONS BASED ON EIA DATA

Based on this data, the average emissions rate is better suited to enabling renewables in this case of Colorado, with its anticipated mix of PV and wind as the clean electricity sources. This data was comprised of monthly averages. Normally, marginal emissions would be based on hourly resolution or

better. Hourly data was not available from the utility on a non-confidential level. However, the same geographic area with an overlap of 81% of generation can be obtained from the EIA database as the PSCO balancing area. The large-signal average and marginal emissions are shown in Fig. 3 for 2022. Coal was

2022 Day of Year (1-365)

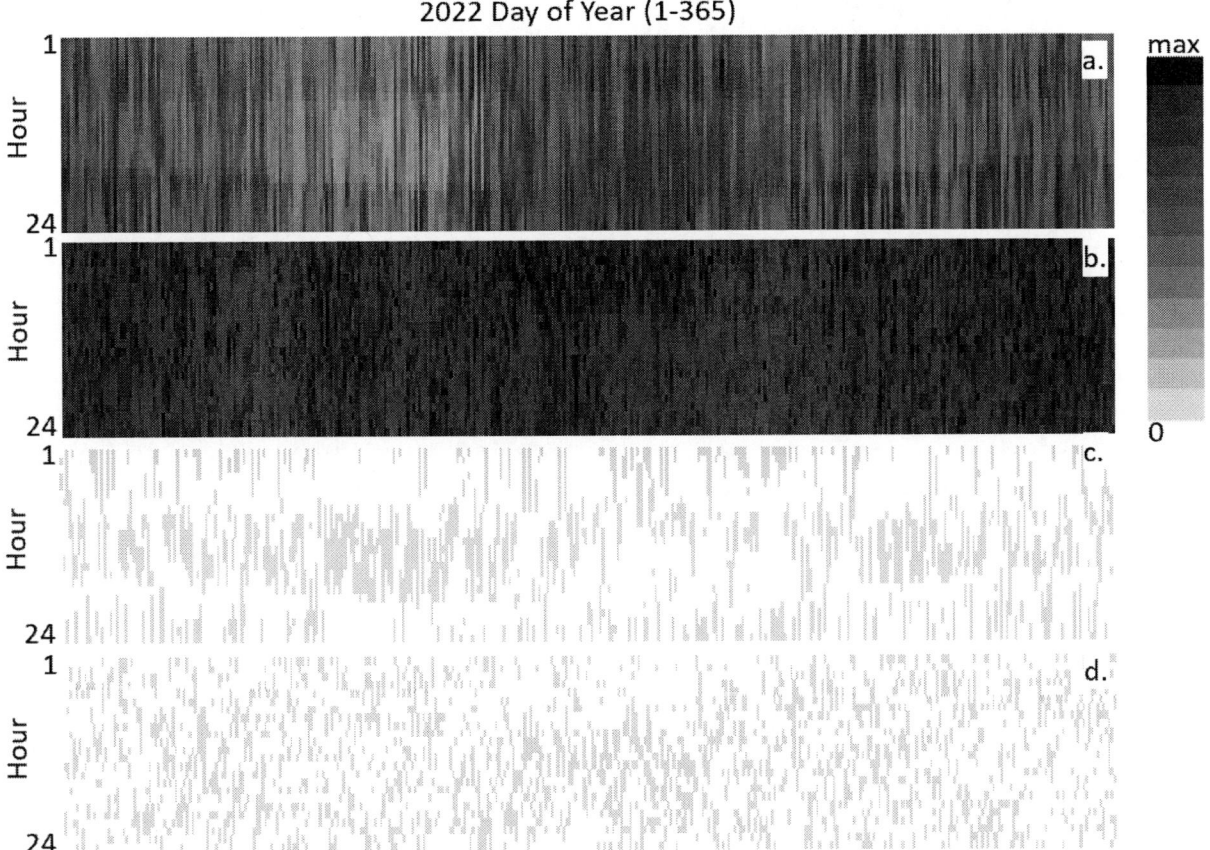

Fig. 3. a.) Average emissions for each of 8760 hours in 2022 for the PSCO balancing area. b.) Large signal marginal emissions calculated from this EIA data using changes in MWh from coal and gas between hours. c.) The cleanest hours determined from average emissions. d.) The cleanest hours from the large-signal marginal emissions calculations.

considered to have 1.5 times the emissions of gas in this calculation.

In Fig. 3, many of the same trends are seen in this hourly data as in the monthly averages in Fig. 1 for 2022. Specifically, the relatively clean hours are seen in the data presented as percentage of fossil fuel (3a and 3c) even in 2022 prior to the much higher levels of PV in 2023 that can be seen in Fig. 1. The hourly data is much better to show the behavior at night. Individual windy nights can be clean but calm nights are dirty. This is not as obvious in the monthly average data in Fig. 1. The clean hours from marginal emissions shown in Fig. 3b and 3d indicate no clear patterns that would shift power use into the daytime hours. Determining demand shifting based on the cleanest hours of marginal emissions (Fig. 3d) essentially dithers between times with coal on the margin or gas on the margin, but does not shift power use towards the most renewable times of the day. In contrast, the cleanest times of day calculated based on average emissions (Fig. 3c) will be indicative of future additions of these resources as also seen in Fig. 1.

There is a notable discrepancy between the marginal emissions plotted in Figure 1 and Fig. 3. The afternoon hours in July and August appear very dirty in the PSCO calculations

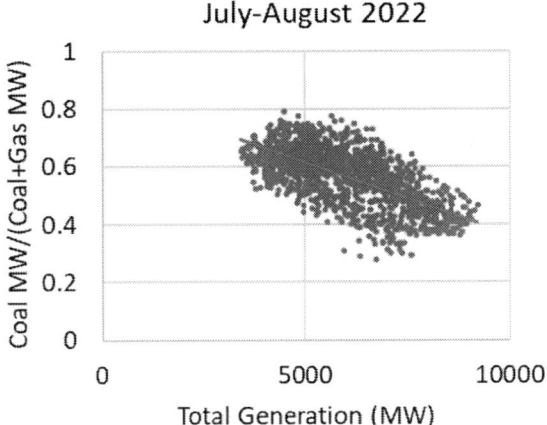

Fig. 4. From the EIA data shown in Fig. 3, the ratio of coal to total fossil fuel (Coal + Gas) is shown here as a function of total generation for the summer months of July and August 2022. During times of high total generation, the mix shifts towards gas since the coal has limited dynamic range to meet summer demand peaks.

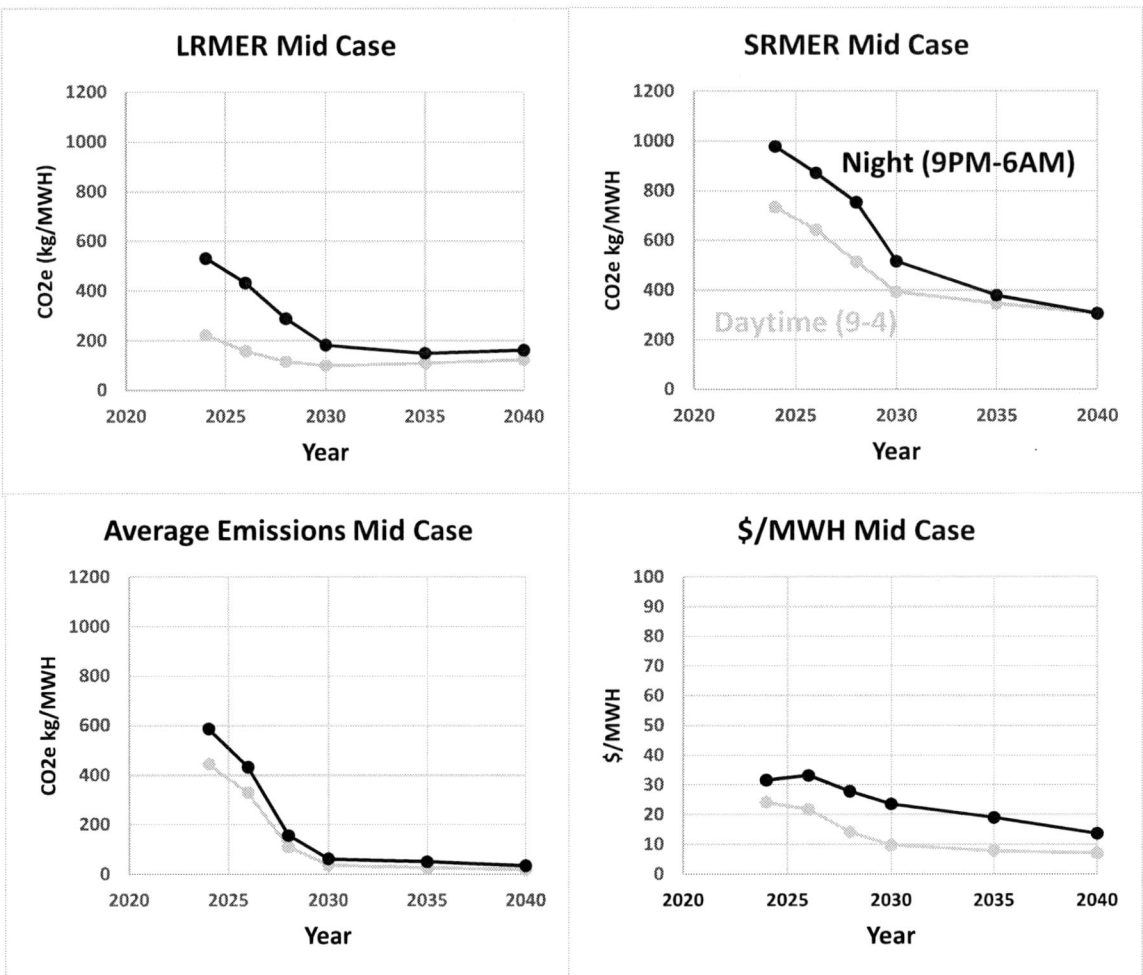

Fig. 5. NREL Cambium calculations for specific demand interventions contrasting potential daytime and night charging of EVs. Long-run marginal emissions are shown upper left, Short-run marginal emissions upper right. Average emissions rate lower left, and total cost at lower right. Green corresponds to a daytime charging schedule and black corresponds to a night schedule. Both the daytime and night schedules exclude peak-demand hours of 4-9PM.

of marginal emissions, but relatively clean in the large-signal marginal emissions calculated from hourly EIA data. For the EIA data, the reason is clear, and illustrated in Fig. 4. Demand peaks in summer are met primarily with gas after the coal has reached its upper bound. Figure 4 shows the ratio of coal as a fraction of the total and coal and gas (MWH) for July and August as a function of the total generation. The mix shifts towards gas at times of high demand resulting in relatively cleaner power when calculated with the method of eq. 1.

V. NREL CAMBIUM ANALYSIS

To have a broader perspective on this topic, the National Renewable Energy Laboratory Cambium analysis can be applied to the geographic area of Colorado to model Long-Run Marginal Emissions Rates as well as average and Short-Run Emission Rates and costs[8]. The main results in the upper panels and the lower left panel in Fig. 5, for example, are the difference between daytime and nighttime emission rates. The Cambium analysis was applied to a simplified case to capture this main effect that is indicated in figs. 1, 2, and 3. What would

be the difference between implementing extra load during daytime (9AM-4PM) vs. night (9PM-6AM)? 9PM to 6AM corresponds to the most popular EV charging program option currently promoted by PSCO, and the 9AM to 3PM would roughly capture the cleanest hours evident in the fraction of fossil fuel generation map in Figs. 1 and 3. NREL maintains "standard scenarios" based on specific sets of assumptions. The results shown here are for the "mid case", described in documentation as using "central inputs for inputs such as technology costs, fuel prices and demand growth"[11]. All of the emission rates indicate that daytime electricity use will be preferable to minimize emissions. In addition, the total cost of addressing the new load will be lower if this load is during the daytime hours rather than night hours.

VI. CONCLUSION

This paper presents results from multiple analysis methods to start to understand how emissions rates might be used to influence demand shifting to enable renewables and renewables planning for Colorado. Utility data from the largest utility in

978-1-6654-6060-6/23 $31.00 © 2023 IEEE

Colorado, Public Service Company of Colorado, indicates that as renewables are increasing on the grid, average emission rates are more indicative of future renewable availability than marginal emission rates. Since marginal emission rates dither between coal and gas for low-penetration of renewables, a strong signal suitable for demand response to integrate and anticipate renewables is delayed by about 6 years until there is large curtailment of renewables. This general conclusion was validated using more fine-resolution hourly data from the EIA for the slightly larger balancing area that includes this utility.

Simulations of the grid using the NREL Cambium package also validate these general conclusions with a more complete analysis including the long-run marginal emissions that can be used for planning the effect of demand profile interventions. The conclusions in this paper are in contrast to the current utility planning that assumes predominantly demand growth during night to serve EV charging long into the future based on historical demand profiles and costs for the fossil grid with excess gas and coal capacity at night.

VII. Acknowledgments

The author would like to thank Elaine Hale at NREL for the modeling results for the day vs. night demand profiles for Colorado using the Cambium models developed and maintained by NREL.

References

[1] Direct testimony of Alice K. Jackson, Proceeding number 21A-0141E of the Colorado Public Utilities Commission, Plan overview and appendix 2. (Unpublished)

[2] R. A. Sinton, "Visualizing the promise and difficulty of demand shifting during a clean-energy transition, in *Proc. 2020 IEEE Photovoltaic Specialist Conference*, pp. 2564-2566.

[3] R. A. Sinton, "Observations on a Colorado electric-utility resource plan for increasing renewables from 55-80% by 2030", *Proc. 2022 IEEE Photovoltaic Specialists Conference*, https://doi.org/10.1109/PVSC48317.2022.9938890.

[4] Xcel 2023 DSM Plan (22A-0315EG), Nick Mark Direct Testimony, NCM-1, DSM Plan, at pages 499-520 (Unpublished)

[5] Rebuttal Testimony and Attachments of Steven W. Wishart, Proceeding No. 21A-0625EG, pg. 13. (Unpublished)

[6] Direct Testimony of Jon T. Landrum, Proceeding number 22A-0309EG of the Colorado Public Utilities Commission, pg. 21, July 2022 (Unpublished).

[7] D. S. Callaway, M. Fowlie, and G. McCormick. "Location, location, location: The variable value of renewable energy and demand-side efficiency resources." Journal of the Association of Environmental and Resource Economists vol. 5.1, pp. 39-75, 2018

[8] Pieter J. Gagnon, and Wesley J. Cole. "Planning for the evolution of the electric grid with a long-run marginal emission rate." Iscience 25.3 (2022): 103915. https://www.sciencedirect.com/science/article/pii/S2589004222001857

[9] Pieter J. Gagnon, John E. T. Bistline, Marcus H. Alexander, and Wesley J. Cole. "Short-run marginal emission rates omit important impacts of electric-sector interventions", PNAS 2022 Vol. 119 No. 49 e2211624119 https://doi.org/10.1073/pnas.2211624119

[10] S. Powell, G. V. Cezar, L. Min, I. M. L. Azevedo, and R Rajagopal, "Charging infrastructure access and operation to reduce the grid impacts of deep electric vehicle adoption" *Nature Energy* , vol. 7, October 2022, pp. 932-945, Available: https://www.nature.com/articles/s41560-022-01105-7.

[11] Pieter Gagnon, Brady Cowiestoll, and Marty Schwarz. 2023. *Cambium 2022 Scenario Descriptions and Documentation*. Golden, CO: National Renewable Energy Laboratory. NREL/TP-6A40-84916. https://www.nrel.gov/docs/fy23osti/84916.pdf

Charge Extraction and Recombination Dynamics of CdSe/CdTe Solar Cells Studied with Transient Photovoltage/Photocurrent Techniques

Abasi Abudulimu[1], Dengbing Li[1], Lei Chen[1], José Santos[2], Iwan Zimmermann[3], Nadeesha Katakumbura[1], Tyler Brau[1], Ebin Bastola[1], Adam Phillips[1], Zhaoning Song[1], Juan Cabanillas[2], Michael Heben[1], Mohammad K. Nazeeruddin[3], Nazario Martín[2,4], Yanfa Yan[1], Randy Ellingson[1]

[1] Wright Center for Photovoltaics Innovation and Commercialization (PVIC), Department of Physics and Astronomy, The University of Toledo, OH 43606 USA. [2] IMDEA Nanociencia, 28049 Madrid, Spain. [3] EPFL VALAIS, 1951 Sion, Switzerland. [4] Facultad de Ciencias Químicas, Universidad Complutense de Madrid, 28040 Madrid, Spain

Abstract — **Significant advancements have been made in the development of high-performance cadmium telluride (CdTe)-based thin film solar cells. However, studies examining the transient excited-state charge dynamics, which determine the final steady-state device performance, are relatively scarce, particularly under device operando conditions. In this work, we investigated charge recombination and extraction dynamics of CdTe solar cells, in comparison with MAPbI3 and narrow bandgap perovskite-based solar cells, using bias-light intensity-dependent transient photovoltage (TPV) and transient photocurrent (TPC) techniques. We found that trap-assisted recombination is the dominant mechanism in the CdTe device at open-circuit, even at 1 sun illumination. Parameters such as charge density, carrier lifetime, and recombination order were extracted from the TPV/TPC data and successfully used to reproduce the open-circuit voltage and short-circuit current densities of the devices measured at steady-state. Consequently, we conclude that these techniques are effective for characterizing charge recombination and extraction dynamics of CdTe solar cells under operando conditions.**

I. INTRODUCTION

The function of solar cells is primarily described by the dynamics of charge carriers within the device. These dynamics include the generation of charge in the active layer (absorber) upon light absorption, the transport of charge to adjacent charge-selecting layers, and the collection of charge at the electrodes. Device performance largely depends on the balance between these beneficial photophysical processes and the undesirable loss processes, such as radiative and non-radiative carrier recombination. These processes occur on a timescale ranging from femtoseconds (charge generation) to microseconds (charge collection) [1-2].

The undesired loss processes are highly dependent on film quality—uniformity, crystallinity, grain size, impurities, and defects [3-4]. They are also influenced by energy level alignment among the layers in the device stack and their conductivity [5-6]. Ultrafast spectroscopies [7] commonly study charge generation and recombination processes in the active layer and transfer to charge-selecting layers, as these processes often elude electrical measurement techniques due to their nanosecond time resolution.

Evaluating solar cells, however, primarily focuses on power conversion efficiency (PCE) obtained from measurements of short-circuit current (*Jsc*), open-circuit voltage (*Voc*), and fill factor (*FF*) under one sun illumination. While fast spectroscopy techniques significantly contribute to understanding material properties [8-9], it is challenging to directly correlate these observed charge dynamics with the electrical parameters of the device measured at steady-state.

In this paper, we employ transient photovoltage (TPV) and transient photocurrent (TPC) techniques to examine charge extraction and recombination dynamics in cadmium telluride-based solar cells (hereafter referred to as CdTe), in comparison with MAPbI3 and narrow bandgap perovskite solar cells (referred to as low_Eg).

While TPV and TPC techniques are familiar to organic and perovskite solar cell communities, they haven't been widely used in the CdTe community, to our knowledge. TPV (TPC) is a technique that observes the decay of Voc (Jsc) after a light pulse, providing information about carrier recombination (extraction) in an operational device [10].

In an open-circuit condition, the device is continuously illuminated by bias-light to generate a steady-state charge carrier density (n) that creates a *Voc*. A weak pulsed light is then introduced on top of the bias-light to produce additional charge carriers (Δn), leading to an increase in open-circuit voltage (ΔVoc). This voltage will decay after the light pulse, revealing the recombination dynamics of Δn,

$$\frac{d\Delta V_{oc}}{dt} \propto \frac{d\Delta n}{dt} = -\frac{\Delta n}{\tau_{\Delta n}} \qquad (1)$$

978-1-6654-6060-6/23 $31.00 © 2023 IEEE

where t is the time and $\tau_{\Delta n}$ is the lifetime of Δn (referred to as TPV lifetime (τ_{TPV}) hereafter).

II. Experimental Methods

CdTe Device: An 80 nm CdSe layer was deposited onto a fluorine-doped tin oxide (Tech 12D, NSG) substrate using radio-frequency (RF) magnetic sputtering. The process took place under a 2% oxygen and 98% argon flow at room temperature. This step was followed by the deposition of a 3.5 μm CdTe layer through close-space sublimation, conducted at source and substrate temperatures of 560°C and 500°C, respectively, at 1 Torr. The films underwent CdCl2 treatment at 400°C for 50 min in dry air. Subsequently, 2 mg/ml CuSCN, dissolved in 30 wt% ammonium hydroxide, was spin-coated at 2000 rpm for 30 s, followed by rapid thermal annealing at 180°C. The process concluded with the evaporation of a 40 nm Au layer onto the film stack, done through a shadow mask with a cell area of 0.08 cm². Further details can be found in [11-12].

MAPbI3 Perovskite Device: The MAPbI3 perovskite solar cells followed an FTO/cTiO2/mTiO2/MAPbI3/capping-layer/Spiro-OMeTAD/Au structure. Complete details of the device fabrication process are available elsewhere [6].

Narrow Bandgap (Low_Eg) Perovskite Device: PEDOT:PSS films were spin-coated onto ITO substrates at 4000 rpm for 30 s and dried at 150°C for 20 min. A 1.8M perovskite precursor, composed of 240.8 mg FAI, 79.5 mg MAI, 26 mg CsI, 373 mg SnI2, 462 mg PbI2, and 15.6 mg SnF2 dissolved in DMF and DMSO (3:1 ratio), was spun onto the ITO/PEDOT:PSS film through a two-step procedure. The process involved spinning at 1000 rpm for 10 s, and then at 5000 rpm for 50 s, with 400 μl of toluene being dropped 20 s before the end of the second spin-coating step. The films were then annealed at 100°C for 8 min, treated with C2H8N2·2HI (in IPA/Toluene) once cooled, and annealed again at 100°C for 5 min. The devices were completed by sequentially evaporating 25 nm C60, 6 nm BCP, and 70 nm Ag onto the perovskite films.

TPV/TPC Measurements: Measurements were made using a home-built system equipped with a 532 nm excitation laser (with a 400 ps pulse width), a high-power white LED (as the bias light), and a picoscope (5444D, Pico Technology).

III. Results And Discussion

Figure 1(a) presents the TPV data of a CdTe solar cell measured at varying bias light intensities. The TPV signal decay rate increases with the bias-light intensity. This result can be attributed to the radiative recombination of photogenerated carriers, which causes τ_{TPV} to decrease as the stationary n, controlled by the bias-light and manifested as Voc, increases. This relationship is described by the equation:

$$\tau_{TPV} = \tau_0 e^{-\beta V_{oc}} \qquad (2)$$

where τ_0 is τ_{TPV} when $Voc = 0$, and β is a constant.

However, the TPV decay can also be influenced by the device's geometrical capacitance, shunts, and traps. Geometrical capacitance becomes significant if the device's active layer is excessively thin or the material has a high dielectric constant. It can dominate TPV decay even under 1 sun bias-light intensity [13-14]. The capacitance characteristic lifetime can be expressed as [13]:

$$\tau_c = \frac{n_{id} C k_B T}{e J_{sc}} \qquad (3)$$

where n_{id} is the ideality factor, C is capacitance, k_B is the boltzman constant, T is temperature, and e is the elementary charge.

The TPV of CdTe, MAPbI3, and Low_Eg perovskite devices measured at 1-sun bias-light (Fig. 1b) suggests that carrier recombination is fastest in the CdTe device. Figure 1c displays the Voc-dependent τ_{TPV}, indicating the presence of multiple dominant mechanisms. At low Voc, TPV decay is predominantly determined by the device's capacitance (dotted lines). Notably, the flat region at low bias-light intensities ($Voc < 0.85$ V) for the MAPbI3 device is due to shunts.

At moderate bias-light intensities (Voc of around 0.77 V for CdTe, 0.75 V for MAPbI3, and 1.0 V for Low_Eg), TPV decays are primarily governed by Shockley-Read-Hall (SRH) recombination in all three devices. Since the trap-assisted carrier lifetime (τ_{trap}) does not rely on the density of photogenerated carriers, it appears as a flat region in Fig. 1c (dashed lines). As TPV probes carrier dynamics via the device's electrodes, the current dataset does not allow us to attribute τ_{trap} to either front or back interfaces or the bulk. Hence, we refer to τ_{trap} as the effective trap-assisted carrier lifetime. The effective τ_{trap} for CdTe, MAPbI3, and Low_Eg perovskite devices are approximately 680 ns, 1500 ns, and 2800 ns, respectively.

Radiative recombination (τ_{rad}) begins to dictate the TPV decay of perovskite devices around 1 sun bias-light intensity

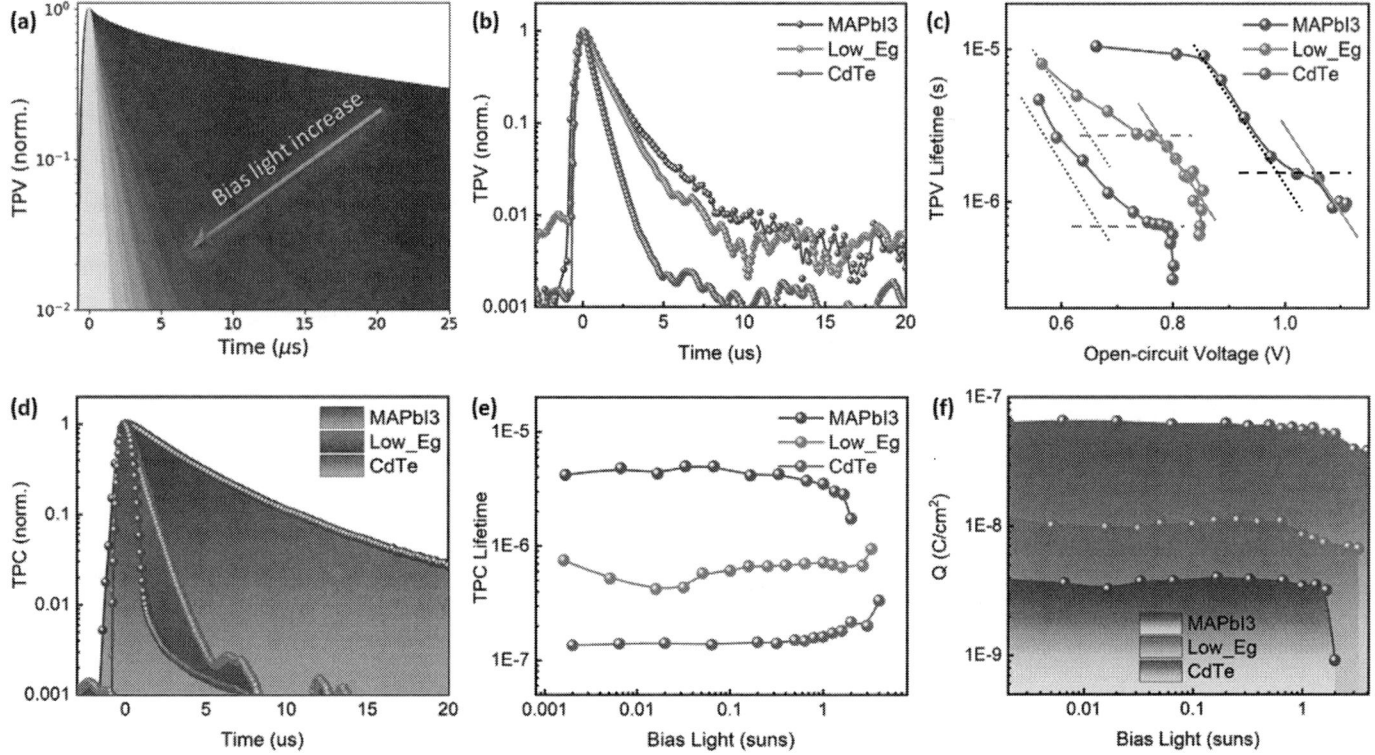

Fig. 1 (a) TPV data of CdTe device measued at a set of bias-light intensities; (b) comparison of TPV data of CdTe, MMAPbI₃, and the Low_Eg perovskite devices measured at 1 sun bias-light intensity; (c) TPV lifetimes as a function of bias-light intensity, where dotted lines = τ_c, dashed lines = τ_{trap}, and solid green lines = τ_{rad}; (d) TPC data of CdTe, MMAPbI₃, and the Low_Eg perovskite devices measured at 1 sun bias-light intensity; (e) TPC lifetime as a function of bias-light intensity; (f) collected charge density (integral of TPC over time) as a function of bias-light intensity.

(solid green lines in Fig. 1c), causing the TPV lifetime to decrease exponentially as per equation (2). Such a transition from trap-assisted to radiative recombination is not observed for this particular CdTe device. The steep decline in τ_{TPV} in the last few data points (for CdTe and Low_Eg devices) is due to the flattening of Voc above 2 suns, attributed to either the decrease of absorber bandgap with increasing cell temperature, high surface recombination, or both. We will investigate this further in future work. Hence, we limit our analysis to measurements taken at 0.02 suns ≤ bias-light ≤ 1 sun.

Contrarily, the TPC data (Fig. 1d-e) reveals that charge extraction in the CdTe device is much faster than in the perovskite devices, and it is not affected by the n generated by the bias-light below 1 sun. For the CdTe and perovskite devices, charge extraction (TPC) slows down at bias-light > 1 sun, while the MAPbI₃ device exhibits the opposite behavior. A slowdown in TPC suggests charge accumulation in the device, while an acceleration indicates the influence of charge recombination, resulting in fewer collected charges (Fig. 1f). Fig. 1f, calculated as ($\int_0^t TPC\, dt = Q$ (the shaded

area in Fig. 1d), shows constant charge collection in the devices below 1 sun bias-light.

The information presented so far concerns the charge recombination/extraction dynamics of a small charge density Δn, generated by the weak pulsed light (see equation (2)). However, to explain the device performance at steady-state, this needs to be correlated with the total charge density n and lifetime τ, as follows:

$$n = \frac{1}{Aed} \int_0^{Voc} \frac{Q}{\Delta V_{oc}} dV = n_0 e^{\gamma V_{oc}} \qquad (4)$$

$$\tau(V_{oc}) = \tau_{TPV}(\beta/\gamma + 1) \qquad (5)$$

where n_0 is the charge carrier density at $Voc = 0$, and γ is a constant. The recombination current J_{rec} and V_{oc} can then be calculated as:

$$J_{rec} = \frac{edn(V_{oc})}{\tau(V_{oc})} \qquad (6)$$

$$V_{oc} = \frac{1}{\gamma(\beta/\gamma+1)} \ln\left(\frac{J_{rec}}{J_{rec0}}\right) \qquad (7)$$

978-1-6654-6060-6/23 $31.00 © 2023 IEEE

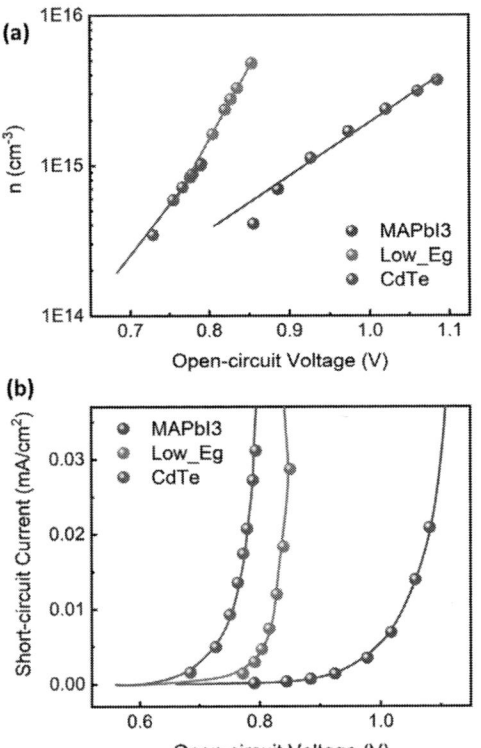

Fig. 2 Total charge density *n* as a function of measured *Voc* using equation (4); (b) measured and reproduced *Jsc* and *Voc* (from TPV/TPC using equation (6) and (7)).

Figure 2a displays the total charge density of the device as a function of *Voc*, derived from TPV/TPC data measured at various bias-light intensities. As expressed in equation (4), *n* grows exponentially with *Voc* for all three types of solar cells. The differences in the slope (*γ*) depend on the intrinsic properties of CdTe, MAPbI₃, and Low_Eg perovskite, such as the definition of the conduction and valence band edges, the density of trap states, and the presence of Fermi level pinning in the device [15], among others.

Figure 2b shows that the measured *Jsc* and *Voc* at different bias-light intensities are well-reproduced via equations (6) and (7), using parameters extracted from TPV/TPC data. From these observations, we conclude that this technique is applicable for studying CdTe devices.

IV. CONCLUSIONS

In this study, we conducted TPV/TPC measurements on a CdSe/CdTe-based solar cell, a MAPbI₃ cell, and a low bandgap perovskite cell. As anticipated, our findings revealed that trap-dominated charge recombination is prevalent in the CdTe device, which is also the case for the

two perovskite solar cells below 1 sun illumination. This recombination in the CdTe device is significantly faster than in the perovskite devices. Interestingly, this pattern is also true for charge extraction, which aligns with the high short-circuit current that CdTe devices frequently produce.

By reproducing the steady-state measured Jsc and open-circuit voltage using parameters obtained from TPV/TPC, such as charge density and carrier lifetime, we have demonstrated the applicability of TPV/TPC techniques for studying charge extraction and recombination dynamics in CdTe-based devices under operando conditions. This work presents valuable insights and offers a reliable approach for further exploration of CdTe solar cells and other photovoltaic technologies.

ACKNOWLEDGMENT

This work is based on research sponsored by Air Force Research Laboratory under agreement number FA9453-19-C-1002, and by the U.S. DOE's Office of Energy Efficiency and Renewable Energy (EERE) under the Solar Energy Technologies Office (SETO), through Agreement DE-EE0008974 and through the Alliance for Sustainable Energy, LLC, Managing and Operating Contractor for the National Renewable Energy Laboratory for the U.S. Department of Energy, under Award Number 37989. The U.S. Government is authorized to reproduce and distribute reprints for Governmental purposes notwithstanding any copyright notation thereon. The views and conclusions contained herein are those of the authors and should not be interpreted as necessarily representing the official policies or endorsements, either expressed or implied, of the Air Force Research Laboratory or the U.S. Government.

REFERENCES

[1] Shi, Jiangjian, et al. "From Ultrafast to Ultraslow: Charge-Carrier Dynamics of Perovskite Solar Cells." *Joule,* 2018

[2] Abdulimu, A, et al. "Single - Walled Carbon Nanotubes as an Additive in Organic Photovoltaics: Effects on Carrier Generation and Recombination Dynamics." *Solar RRL* 6.4, 2101010, 2022.

[3] Seok, Sang Il, et al. "Methodologies toward Highly Efficient Perovskite Solar Cells." *Small* 14.20, 1704177, 2018

[4] Abudulimu, Abasi, et al. "Photophysical Properties of CdSe/CdTe Bilayer Solar Cells: A Confocal Raman and Photoluminescence Microscopy Study." *IEEE 49th Photovoltaics Specialists Conference (PVSC),* 1088-1090, 2022. doi: 10.1109/PVSC48317.2022.9938717.

[5] Abudulimu, Abasi, et al. "Crucial role of charge transporting layers on ion migration in perovskite solar cells." *Journal of Energy Chemistry* 47, 132-137, 2020.

[6] Abudulimu, A, et al. "Hole transporting materials for perovskite solar cells and a simple approach for determining the performance limiting factors." *Journal of Materials Chemistry A* 8.3, 1386-1393, 2020.

[7] Shi, Junqing, et al. "Designing high performance all-small-molecule solar cells with non-fullerene acceptors: comprehensive studies on photoexcitation dynamics and charge separation kinetics." Energy & Environmental Science 11.1, 211-220, 2018.

[8] Abudulimu, Abasi, et al. "Chirality specific triplet exciton dynamics in highly enriched (6, 5) and (7, 5) carbon nanotube networks." The Journal of Physical Chemistry C 120.35 (2016): 19778-19784.

[9] Sandoval-Torrientes, Rafael, et al. "Minimizing geminate recombination losses in small-molecule-based organic solar cells." *Journal of Materials Chemistry C* 7.22, 6641-6648, 2019.

[10] C. G. Shuttle, B. O'Regan, A. M. Ballantyne, J. Nelson, D. D. Bradley, J. De Mello, and J. R. Durrant, *Appl. Phys. Lett.* 92, 80, 2008.

[11] L, Deng-Bing, et al. "20%-efficient polycrystalline Cd (Se, Te) thin-film solar cells with compositional gradient near the front junction." *Nature Communications* 13.1, 1-8, 2022.

[12] Li, Deng-Bing, et al. "Oxygen Management to Avoid Photo-Inactive Cd (S, Se) for Efficient Cd (Se, Te) Solar Cells." *ACS Energy Letters* 8.3, 1529-1534, 2023.

[13] Wang, Zi Shuai, et al. "Transient photovoltage measurements on perovskite solar cells with varied defect concentrations and inhomogeneous recombination rates." *Small Methods* 4.9, 2000290, 2020.

[14] Kiermasch, David, et al. "Unravelling steady-state bulk recombination dynamics in thick efficient vacuum-deposited perovskite solar cells by transient methods." *Journal of Materials Chemistry A* 7.24, 14712-14722, 2019.

[15] A. Maurano, C. G. Shuttle, R. Hamilton, A. M. Ballantyne, J. Nelson, W. Zhang, M. Heeney, and J. R. Durrant, *J. Phys. Chem. C*, 115, 5947-5957, 2011

The Feasibility of Luminescent Solar Concentrators Overlays for Conventional Lens

Xitong Zhu[1], Michael G. Debije[2], Angèle H.M.E. Reinders[1,3]

1) Energy Technology Group, Department of Mechanical Engineering, Eindhoven University of Technology, 5612 AE Eindhoven, The Netherlands

2) Stimuli-responsive Functional Materials & Devices Group, Department of Chemical Engineering and Chemistry, Eindhoven University of Technology, 5612 AE Eindhoven, The Netherlands

3) Department of Design, Production and Management, Faculty of Engineering Technology, University of Twente, 7522 NB Enschede, The Netherlands

Abstract — **The vast majority of LSC devices studied to date have been flat. To enhance their integration potential in buildings and vehicles, this study assesses the performance of non-planar LSC PV devices with integrated lenses. Four different combinations of lenses and LSCs have been designed in Solidworks and simulated by means of ray tracing in LightTools software. These devices have a diameter (D) of 20mm and a focal length (f) of 25.3 mm. All parts are made of PMMA, with the LSC sections doped with Lumogen Red 305 dye at a concentration of 110 parts per million (ppm). A circular silicon solar cell (diameter 10 mm) has been placed 15 mm below the lens and a long silicon solar cell to the circular edge of the top LSC. The simulation results show that under diffuse irradiance conditions, a 1 mm LSC cover layer on the top of the lens results in a power conversion efficiency above 7%, compared with 6% for a bare lens. Additional advantages are a red shift of the incoming spectrum which improves the power conversion efficiency of silicon PV cells. Although LSCs in combination with lenses do not always result in better PCEs than devices without LSC parts, the better performance under diffuse irradiance conditions lets LSC technology has a good potential to be combined with lenses.**

I. INTRODUCTION

Using luminescent solar concentrator photovoltaic (LSC PV) modules to harvest solar energy was originally proposed in the 1970s[1]. The main body of the LSC is a lightguide made of plastic or glass with luminescent materials embedded inside, or applied as a separate layer on top and/or bottom of the lightguide. The luminescent materials can be organic fluorescent dyes, inorganic phosphors, or quantum dots[2]. Sunlight penetrates the top surface of the lightguide and is absorbed by embedded luminescent materials, and re-emitted at longer wavelengths. Due to total internal reflection (TIR), a fraction of the re-emitted light is guided towards small PV cells attached to the edges of the lightguide. There are advantages to using LSCs: the lightguide elements can be colorful and manufactured in various sizes, and the device performs equally well under both direct and indirect illumination[3]. Due to its exciting design features, such as customized coloring and

formability, LSC PV is an attractive technology for integration in buildings and products such as vehicles [4].

For the past decades, almost all LSC PV research and applications have been limited to flat devices. In some literature, flat LSC-PV devices already have achieved a reasonable performance in simulation or practical measurement [5], [6]. A small body of research has suggested that the potential performance of curved LSC PV devices [7] or non-planar LSCs could equal to or better than the planar LSC PV devices [8]. This study therefore evaluates the performance of several non-planar LSC PV in combination with lenses by means of ray-tracing simulations.

A converging lens can have a high concentration ratio (in the order of 100x), which can reduce the cost of high-efficiency PV cells [9]. However, a high-precision lens is easily damaged by the external environment [10], so using an inexpensive LSC layer to protect the lens and at the same time using the LSC layer as a light collector converting incoming irradiance to a more reddish spectrum are two other underlying motivationsof this study. Moreover, LSCs has the advantage of concentrating diffuse light which would improve the poor performance of lens under diffuse light.

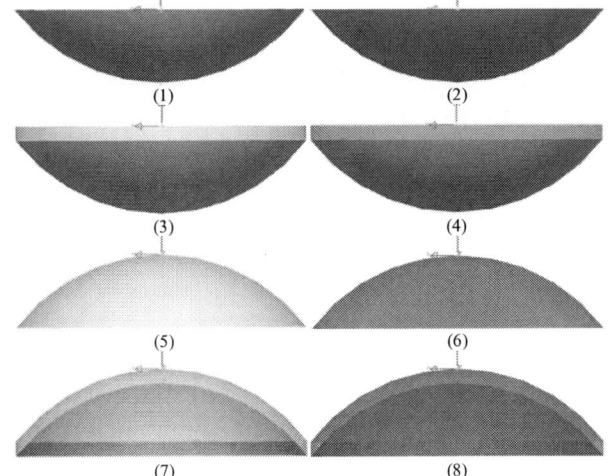

TABLE I
STRUCTURE AND MATERIAL OF THE MODELS

Case	Lens material	Convex surface	Cover material
1	PMMA	Lower	/
2	LSC	Lower	/
3	PMMA	Lower	PMMA
4	PMMA	Lower	LSC
5	PMMA	Upper	/
6	LSC	Upper	/
7	PMMA	Upper	PMMA
8	PMMA	Upper	LSC

Fig. 1. Eight models used in this study, the red parts are LSC, and the gray parts are PMMA.

II. METHOD

Table I lists the structure and materials of all 8 different models used in this study. Four different models (cases 2, 4, 6, 8) have been created which combine LSC PVs with convex lenses. As Fig. 1. shows, two LSC plano-convex lenses models (cases 2 and 6) were used to explore the performance of LSC lenses, and the two models of PMMA plano-convex covered by a 1 mm LSC layer (cases 4 and 8) were used to explore the performance and possibility of the LSC protective layer a the normal lens. The simulations are based on a PMMA plano-convex lens with a diameter of 20 mm ($D = 20\ mm$), 5mm thickness ($T = 5\ mm$), 12.5 mm convex radius ($R_1 = 12.5\ mm$), and a 25 mm focal length ($f = 25.3mm$). Cases 7 and 8 are PMMA lenses covered by a 1 mm spherical cap layer which has the same curvature of 0.08 as the convex surface of the lens. Cases 3 and 4 are PMMA lenses covered by a 1mm flat layer.

All the LSC parts use a PMMA lightguide containing Lumogen Red 305 dye at a concentration of 110 ppm; The lens and LSC modules have been designed in Solidworks and simulated by means of ray tracing in LightTools. As a reference, lens modules without dyes and pure PMMA covering layer modules have been simulated as well.

The light source was set as a direct sun and a diffused sky. In order to explore how the performance varies under different lighting conditions, and whether LSC can improve the efficiency under diffuse light as expected. We simulated only direct light, only diffuse light, and three different ratios of direct light to diffuse light, namely 70% direct light with 30% diffuse light, 50% direct light with 50% diffuse light, and 30% direct light with 70% diffuse light. All light sources used the AM 1.5 spectral distribution.

The LSC PV performance has been quantified by the optical collection efficiency (η_{opt}) and power conversion efficiency (η_{PCE}). The optical collection efficiency is calculated by the following formula:

$$\eta_{opt} = \frac{\sum \dot{Q}_{PV}}{\dot{Q}_{in}} \tag{1}$$

Where:
$\dot{Q}_{in}$ is the radiant flux which falls on the aperture area (W)
$\dot{Q}_{PV}$ is the radiant flux which falls on each cell (W)
The power conversion efficiency which is used to evaluate the electrical performance is calculated by the followed formula:

$$\eta_{PCE} = \frac{\sum P}{\dot{Q}_{in}}$$
$$= \frac{\sum_{i=1}^{n=total\ PV\ cells} \int_{\lambda=0}^{4500\ nm} S_{PV}^i(\lambda)\ SR(\lambda)\ FF\ Voc\ A_{PV}^i\ d\lambda}{\int_0^\infty S_{ap}(\lambda) A_{ap}\ d\lambda} \tag{2}$$

where:
P is the power production of each cell (W)
S_{PV} is the spectral distributed irradiance on each solar cell (W/m²nm)
SR is the spectral response of the spectral distribution (A/W)
FF is the fill factor of the solar cell
Voc is the open-circuit voltage of the solar cell (V)
A_{PV} is the area of the solar cell (m²)
S_{ap} is the spectral distributed irradiance on the aperture area (W/m²nm)
A_{ap} is the area of the aperture area (m²)

TABLE II
SUMMARY OF SIMULATION RESULT

Case	Direct light only		70% direct light +30% diffuse light		50% direct light +50% diffuse light		30% direct light +70% diffuse light		Diffuse light only	
	Optcal collection efficiency	PCE of whole devices	Optcal collection efficiency	PCE of whole devices	Optcal collection efficiency	PCE of whole devices	Optcal collection efficiency	PCE of whole devices	Optcal collection efficiency	PCE of whole devices
1	60.9%	12.1%	43.9%	8.7%	32.6%	6.5%	21.3%	4.2%	4.4%	0.9%
2	47.1%	9.5%	34.1%	6.9%	25.3%	5.1%	16.6%	3.4%	3.5%	0.7%
3	54.9%	10.9%	39.6%	7.9%	29.4%	5.9%	19.2%	3.8%	3.9%	0.8%
4	53.0%	10.9%	39.4%	8.2%	30.4%	6.4%	19.2%	6.8%	7.9%	1.9%
5	90.1%	17.9%	64.4%	12.8%	47.2%	9.4%	30.1%	6.0%	4.4%	0.9%
6	72.8%	14.6%	52.1%	10.5%	38.2%	7.7%	24.4%	4.9%	3.7%	0.7%
7	81.3%	16.2%	58.0%	11.5%	42.5%	8.5%	27.0%	5.4%	3.7%	0.7%
8	75.2%	15.1%	54.8%	11.1%	41.2%	8.4%	25.4%	7.7%	7.2%	1.6%

978-1-6654-6060-6/23 $31.00 © 2023 IEEE

III. SIMULATION RESULT

Simulation results are shown in Table II. and Fig. 2.; Considering only under a mixed light source, case 5 (a PMMA lens without dyes) has the best performance in the cases without LSC part, namely has a high efficiency of 12.8% for 70% direct irradiance, while case 8 (case 5 with LSC cover) results in 7.7% efficiency for 70% diffuse irradiance and 11.1% for 70% direct irradiance.

It can be noticed that under the direct light source, the bare PMMA lens has great optical efficiency reach to 90.1% and PCE reaches 17.9% (case 5 under direct only light). However, under the situation with diffuse light only, the optical efficiency and PCE have decreased to 4.4% and 0.9%. On the other hand, the optical efficiency and PCE of the PMMA lens covered by the LSC layer (case 8) under the direct light source are 75.2% and 15.1%, but under the situation with diffuse light only, the optical efficiency and PCE could reach to 7.2% and 1.6% which are obviously better than bare PMMA lens.

In the case 2 and 5 with other cases, the performance of the LSC lens is always not ideal; this is because, the thickness of the LSC lens is great, and the light is more affected by the fluorescent particles. The irradiance emitted by the fluorescent particles would be semi-random in the LSC lens and due to the change of direction, cannot be focused to the focal point. Additionally, some of the emission wouldn't even escape the LSC lens due to trapping by TIR.

As the results of cases 1, 3, 4 and 5, 7, 8 show, compared with the 1 mm PMMA protective layer (cases 3 and 7), which only reduces the overall performance and efficiency, the LSC protective layer (case 4 and 8) improves the performance under diffuse light with a small sacrifice in the performance under direct light. Under cloudy conditions dominated by diffuse light (30% direct light +70% diffuse light), the LSC protective layer can maintain the efficiency of the overall device at about 7% with a small impact under more direct lighting.

IV. CONCLUSIONS AND RECOMMENDATIONS

This study shows that LSC technology has a good potential to be combined with lenses to improve lens performance under diffuse light and supply a protective layer for high precision and high-cost lenses at the same time, although LSCs in combination with lenses do not always result in better PCEs than devices without LSC parts. With the same 20 mm in diameter top view area, the 1 mm LSC layer would maintain the PCE of the whole device above 7%, compared with 6% of total power conversion efficiency for bare lenses under cloudy skies. Moreover, the LSC layer also acts as a protective layer for the lens. Thus, combining LSCs with lenses could be a feasible design direction for LSC PV devices.

Since this abstract covers ongoing work, the final paper will cover the distribution of irradiance intensity behind the lens in the x-y plane, newer simulations to be executed for different thickness LSC layers, and combined LSC with Fresnel lens.

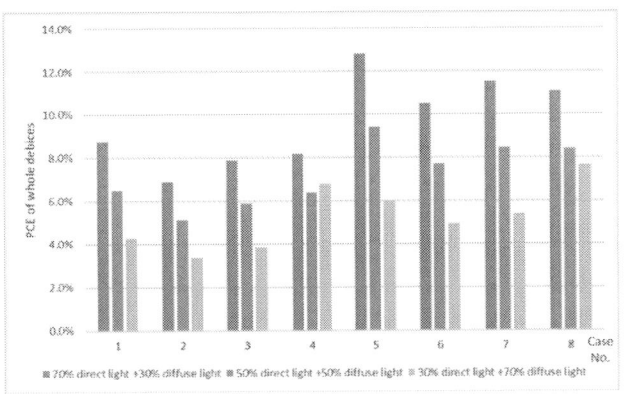

Fig. 2. PCE of whole devices simulation results of all cases under three natural possible different light conditions.

REFERENCES

[1] A. Goetzberger, "Fluorescent Planar Collector-Concentrators: a Review," *Sol. Cells*, vol. 4, pp. 3–23, 1980.

[2] S. J. Gallagher, B. Norton, and P. C. Eames, "Quantum dot solar concentrators: Electrical conversion efficiencies and comparative concentrating factors of fabricated devices," *Sol. Energy*, vol. 81, no. 6, pp. 813–821, 2007, doi: 10.1016/j.solener.2006.09.011.

[3] M. Rafiee, S. Chandra, H. Ahmed, and S. J. McCormack, "An overview of various configurations of Luminescent Solar Concentrators for photovoltaic applications," *Opt. Mater. (Amst).*, vol. 91, no. March 2018, pp. 212–227, 2019, doi: 10.1016/j.optmat.2019.01.007.

[4] A. Reinders, *Designing with Photovoltaics*. 2020.

[5] M. G. Debije and P. P. C. Verbunt, "Thirty years of luminescent solar concentrator research: Solar energy for the built environment," *Adv. Energy Mater.*, vol. 2, no. 1, pp. 12–35, 2012, doi: 10.1002/aenm.201100554.

[6] M. Aghaei, X. Zhu, M. Debije, W. Wong, T. Schmidt, and A. Reinders, "Simulations of Luminescent Solar Concentrator Bifacial Photovoltaic Mosaic Devices Containing Four Different Organic Luminophores," *IEEE J. Photovoltaics*, vol. 12, no. 3, pp. 771–777, 2022, doi: 10.1109/JPHOTOV.2022.3144962.

[7] B. Vishwanathan *et al.*, "A comparison of performance of flat and bent photovoltaic luminescent solar concentrators ScienceDirect A comparison of performance of flat and bent photovoltaic luminescent solar concentrators," *Sol. ENERGY*, vol. 112, no. January 2019, pp. 120–127, 2015, doi: 10.1016/j.solener.2014.12.001.

[8] X. Zhu, M. G. Debije, and A. H. M. E. Reinders, "Ray Tracing of Bent Applications of Luminescent Solar Concentrator PV Modules," pp. 0089–0091, 2022, doi: 10.1109/pvsc48317.2022.9938486.

[9] W. Sweatt, G. Nielson, and M. Okandan, "Concentrating photovoltaic systems using micro-optics," *Opt. InfoBase Conf. Pap.*, 2011, doi: 10.1364/ose.2011.srwc6.

[10] P. Li, X. Liu, Q. Cheng, and Z. Liang, "Long-term photovoltaic performance of thin-film solar cells with diffractive microlens arrays on glass substrates," *Results Phys.*, vol. 21, p. 103841, 2021, doi: 10.1016/j.rinp.2021.103841.

Highly efficient bifacial single junction perovskite solar cells.

Qi Jiang[1], Zhaoning Song[2], Yanfa Yan[2], Kai Zhu[1]

1. National Renewable Energy Laboratory, Golden, CO, 80401, USA

2. University of Toledo, Toledo, OH, 43606, USA

Abstract — We report on efficient, single-junction bifacial perovskite solar cells (PSCs) that simultaneously exhibit high front-side-illumination power conversion efficiency (PCE) (over 22%) and high bifaciality (over 91%), which represents a significant advancement for single junction bifacial PSC development. The bifacial PSCs exhibit substantial performance boost under concurrent front and rear illumination conditions. The stabilized power output goes up to 25.3 mW/cm² under albedos of 0.2, which is comparable to the state-of-the-art monofacial single-junction PSCs. Our work demonstrates the value and potential of bifacial single junction PSCs as an attractive direction for future scientific study and commercialization.

I. INTRODUCTION

The past decade has witnessed the revolution of perovskite photovoltaics (PV). The certified power conversion efficiency (PCE) of laboratory-sized perovskite solar cells (PSCs) has rapidly increased to more than 25%, comparable to or greater than the records of more established PV technologies on the market[1]. This rapid progress was propelled by the countless innovations in the design and processing of PSCs as well as the global research activities devoted to developing a fundamental understanding of perovskite materials[2, 3]. To further increase the power output per unit area, perovskite-based tandem solar cells have received significant attention, with substantial progresses being made in recent years[4-7]. Another effective approach is to use a bifacial configuration to enhance the utilization of direct, reflected, and diffused sunlight from both sides of a PSC[8-11]. A bifacial single-junction PSC does not significantly alter the production processes; thus, it has a manufacturing complexity and cost similar to that of the conventional monofacial architecture.

Among thin-film PV technologies, PSC is uniquely suited to a bifacial structure, owing to its high absorption coefficients, long carrier lifetimes, benign and readily passivated surfaces, and proper band alignment with charge selective contacts[3, 8]. With the albedo light in the range of 0.1–0.5 (albedo is the ratio of rear light intensity to front light intensity), the equivalent bifacial efficiency (or the output power density) of a typical bifacial PSC can, in principle, be about 10.2%–51.5% greater than that of its monofacial counterpart[12]. Over the past several years, bifacial PSCs have seen gradual performance increases, due to efforts to improve the transparent electrodes, charge transport layers, and perovskite absorbers[9, 11, 13, 14]. However, bifacial PSC performance is still inadequate in comparison to its monofacial PSC[3, 11]. As for single-junctionbifacial PSCs, it is very hard to achieve high bifaciality and high front PCE simultaneously. Thus, significant efforts are required to fully demonstrate the potential of bifacial PSCs. An ideal bifacial single-junction PSC should: (1) have a front-side-illumination efficiency close to that of state-of-the-art monofacial cells; (2) have a high (close-to-unity) bifaciality (efficiency ratio between the rear-side and front-side illuminations); and (3) function well under simultaneous illumination (up to an equivalent 2× sunlight) from both sides of the device.

Here, we demonstrate efficient bifacial single-junction PSCs, guided by technic optimization. When measured under one-side illumination, the bifacial PSC reached front-side and rear-side efficiencies of ~22.8% and ~21.1%, respectively; the bifaciality was within the range of ~91%–93%. When measured under concurrent illumination from both sides, the bifacial PSC generated stable power outputs of 25.3 mW/cm² under the albedo of 0.2, highlighting the potential of bifacial PSCs. Our work shows the promising future of bifacial single junction PSCs.

II. BIFACIAL DEVICE CONFIGURATION

The most critical part affects the front and rear-side PCE of bifacial PSC is the transparent conducting rear electrode, which often involves a transparent conducting oxide layer. Here in this study, we adopted indium zinc oxide (IZO), and the perovskite composition is also following our previous study as

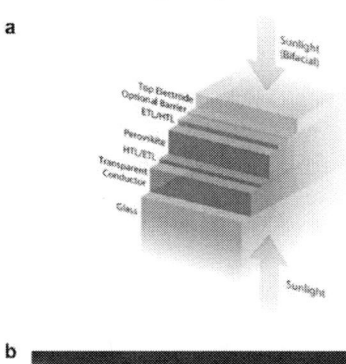

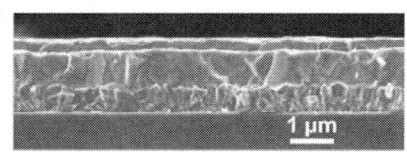

Fig. 1. a) Schematic device structure of a bifacial PSC. b) A representative cross-section scanning electron microscope (SEM) image of a bifacial perovskite solar cells.

$Rb_{0.05}Cs_{0.05}MA_{0.05}FA_{0.85}Pb(I_{0.95}Br_{0.05})_3$, in which MA stands for methylammonium and FA is formamidinium. A schematic device stack is shown in Fig. 1a, and a representative cross-sectional scanning electron microscopy image is shown in Fig. 1b. In this study, the TCO is indium zinc oxide (IZO), and the optimized thickness is ~210-240 nm. The optimized perovskite layer thickness is about 850 nm, which is sufficient to absorb >95% of photons with a wavelength smaller than 830 nm.

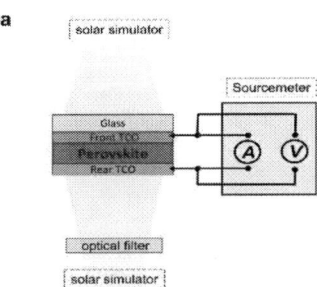

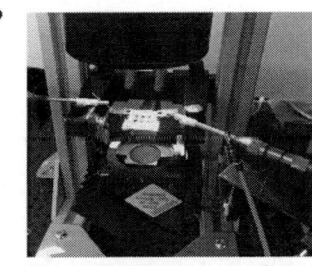

Fig. 2. a) Schematic measurements setup for a bifacial illumination using two simulators. b) A photo of real set-up for bifacial solar cells measurement under concurrent bifacial illumination.

III BIFACIAL DEVICE RESULTS AND DISCUSSION

Accurate assessment of bifacial solar cell performance for indoor characterization plays a key role in promoting the deployment of bifacial PV technologies[15]. To evaluate the efficiency gain under bifacial illumination, we placed bifacial PSCs between two AM1.5G solar simulators, the front illumination was set to an intensity of 100 mW/cm^2 and the rear illumination was set to be 20 mW/cm^2 to simulate realistic albedo light, as a preliminary test (albedo=0.2). Note that albedo is the ratio of rear light intensity to front light intensity. The schematic setup is shown in Fig. 2a and the actual measurement setup in the lab is shown in Fig. 2b. The bifacial device's front and back sides were both masked during the concurrent illumination measurement.

Fig. 3a, b show the J-V curves of a bifacial PSC illuminated separately from the front (glass) and back (IZO) sides. When illuminated from the front side, the bifacial device showed a PCE of 22.8% (current density (J_{sc})=24.34 mA/cm^2, open circuit (V_{oc}) =1.147 V, fill factor (FF)=81.6%) from the reverse scan, and a PCE of 21.2% (J_{sc}=24.34 mA/cm^2, V_{oc}=1.135 V and FF=76.6%) from the forward scan direction.

When illuminated from the back side, the reverse- and forward-scan PCEs are reduced to 21.1% (J_{sc}=22.31 mA/cm^2, V_{oc}=1.149 V and FF=82.1%) and 19.8% (J_{sc}=22.26 mA/cm^2, V_{oc}=1.132 V and FF=78.5%), respectively, corresponding to the bifaciality of ~92%-93%. The difference in PCE between the front-side and back-side illumination is mainly caused by the different in J_{sc} (~2 mA/cm^2).

Fig. 3b shows the corresponding external quantum efficiency (EQE) spectra of the bifacial PSC measured seprately from the front and back sides. The integrated J_{sc} values from the EQE spectra are 24.19 mA/cm^2 for front illumination and 21.8 mA/cm^2 for rear illumination; these values are consistent with the J_{sc} results from the J-V curves. The optical loss at short wavelengths can be attributed mainly to the parasitic absorption of the IZO and C$_{60}$ layers.

Fig. 3 c, d and Table 1 show the J-V curves and stable power outputs (SPOs) (or the equivalent steady-state bifacial efficiencies) of another representative bifacial device under albedo=0 and albedo=0.2. We obtained an efficiency of 22.39% (reverse scan, or rev)/20.79% (forward scan, or fwd)/21.8% (SPO) for front-side illumination without rear illumination (albedo = 0). When we added 20 mW/cm^2 (or albedo = 0.2) of rear illumination, we achieved bifacial equivalent efficiencies of 26.4% (rev)/25.15% (fwd)/25.3% (SPO), corresponding to an increase of ~4% in comparison to the situation without albedo light. Note that the albedo of 0.2—a value commonly used by the International Electrotechnical Commission (IEC) standard to approximate naturally reflected surroundings[11, 15, 16]. With an albedo of 0.2, our bifacial device performance is comparable to our opaque metal electrode PSC efficiency of ~25%[3].

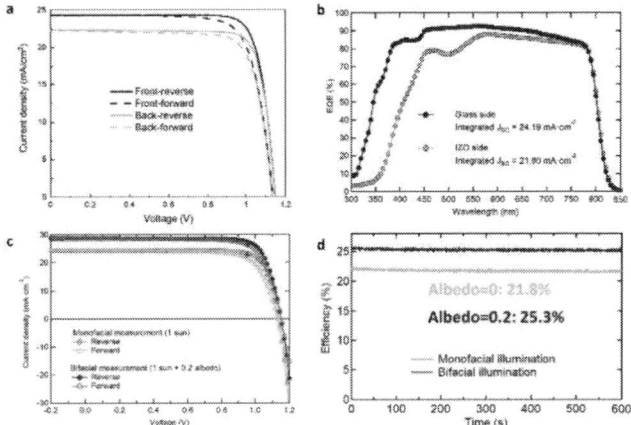

Fig. 3 a) Current density-Voltage (J-V) curves with both reverse and forward scans and b) corresponding EQE spectra with integrated J_{sc} value of a bifacial PSC under front-side and rear-side 1-sun illumination, measured separately. c) J-V curves with reverse and forward scans and d) corresponding stable power output (SPO) of the bifacial device without albedo and under concurrent illumination, with 1-sun intensity from the front side (100 mW/cm^2) and back/rear side illumination of 20 mW/cm^2 (albedo=0.2).

TABLE I
BIFACIAL PSC PERFORMANCE UNDER DIFFERENT ALBEDOS FROM THE REAR SIDE

Albedo	Scan	Voc (V)	Jsc (mA/cm²)	FF (%)	PCE (%)	SPO (%)
0	Reverse	1.136	24.41	80.7	22.4	21.8
	Forward	1.129	24.28	75.8	20.8	
0.2	Reverse	1.153	28.57	80.1	26.4	25.3
	Forward	1.152	28.68	76.1	25.2	

IV. SUMMARY AND OUTLOOK

Bifacial solar cells harvest solar irradiance from both their front and rear surfaces, boosting energy conversion efficiency to maximize their electrical power production. For single-junction perovskite solar cells (PSCs), the performance of bifacial configurations is still far behind that of their state-of-the-art monofacial counterparts. Here, we report on highly efficient, bifacial, single-junction PSCs based on the p-i-n (or inverted) architecture. We optimized our transparent conducting rear electrode for bifacial PSCs to enable optimized efficiency under concurrent illumination conditions. The bifaciality of the PSCs was about 92%–93%. Under concurrent bifacial measurement conditions, we obtained equivalent, stabilized bifacial power output densities of 25.3 mW/cm² under albedos of 0.2. Our work shows that single-junction bifacial PSCs are promising for the commercialization of perovskite PV.

REFERENCES

[1] "National Renewable Energy Laboratory, Best research-cell efficiency chart; www.nrel.gov/pv/cell-efficiency.html.."

[2] Y. Zhou, L. M. Herz, A. K. Y. Jen, and M. Saliba, "Advances and challenges in understanding the microscopic structure–property–performance relationship in perovskite solar cells," *Nature Energy*, vol. 7, pp. 794-807, 2022.

[3] Q. Jiang, J. Tong, Y. Xian, R. A. Kerner, S. P. Dunfield, C. Xiao, R. A. Scheidt, D. Kuciauskas, X. Wang, M. P. Hautzinger, R. Tirawat, M. C. Beard, D. P. Fenning, J. J. Berry, B. W. Larson, Y. Yan, and K. Zhu, "Surface reaction for efficient and stable inverted perovskite solar cells," *Nature*, vol. 611, pp. 278-283, 2022.

[4] M. A. Green, E. D. Dunlop, G. Siefer, M. Yoshita, N. Kopidakis, K. Bothe, and X. Hao, "Solar cell efficiency tables (Version 61)," *Progress in Photovoltaics: Research and Applications*, vol. 31, pp. 3-16, 2022.

[5] R. Lin, J. Xu, M. Wei, Y. Wang, Z. Qin, Z. Liu, J. Wu, K. Xiao, B. Chen, S. M. Park, G. Chen, H. R. Atapattu, K. R. Graham, J. Xu, J. Zhu, L. Li, C. Zhang, E. H. Sargent, and H. Tan, "All-perovskite tandem solar cells with improved grain surface passivation," *Nature*, vol. 603, pp. 73-78, 2022.

[6] Q. Jiang, J. Tong, R. A. Scheidt, X. Wang, A. E. Louks, Y. Xian, R. Tirawat, A. F. Palmstrom, M. P. Hautzinger, S. P. Harvey, S. Johnston, L. T. Schelhas, B. W. Larson, E. L. Warren, M. C. Beard, J. J. Berry, Y. Yan, and K. Zhu, "Compositional texture engineering for highly stable wide-bandgap perovskite solar cells," *Science*, vol. 378, pp. 1295-1300, 2022.

[7] A. Al-Ashouri, E. Köhnen, B. Li, A. Magomedov, H. Hempel, P. Caprioglio, J. A. Márquez, A. B. Morales Vilches, E. Kasparavicius, J. A. Smith, N. Phung, D. Menzel, M. Grischek, L. Kegelmann, D. Skroblin, C. Gollwitzer, T. Malinauskas, M. Jošt, G. Matič, B. Rech, R. Schlatmann, M. Topič, L. Korte, A. Abate, B. Stannowski, D. Neher, M. Stolterfoht, T. Unold, V. Getautis, and S. Albrecht, "Monolithic perovskite/silicon tandem solar cell with >29% efficiency by enhanced hole extraction," *Science*, vol. 370, pp. 1300-1309, 2020.

[8] Z. Song, C. Li, L. Chen, and Y. Yan, "Perovskite solar cells go bifacial—mutual benefits for efficiency and durability," *Advanced materials*, vol. 34, p. 2106805, 2022.

[9] P. Kumar, G. Shankar, and B. Pradhan, "Recent progress in bifacial perovskite solar cells," *Applied Physics A*, vol. 129, p. 63, 2022.

[10] F. Fu, T. Feurer, T. Jäger, E. Avancini, B. Bissig, S. Yoon, S. Buecheler, and A. N. Tiwari, "Low-temperature-processed efficient semi-transparent planar perovskite solar cells for bifacial and tandem applications," *Nature communications*, vol. 6, p. 8932, 2015.

[11] C. Zhang, M. Chen, F. Fu, H. Zhu, T. Feurer, W. Tian, C. Zhu, K. Zhou, S. Jin, and S. M. Zakeeruddin, "CNT-based bifacial perovskite solar cells toward highly efficient 4-terminal tandem photovoltaics," *Energy & Environmental Science*, vol. 15, pp. 1536-1544, 2022.

[12] Z. Song, C. Chen, C. Li, S. Rijal, L. Chen, Y. Li, and Y. Yan, "Assessing the true power of bifacial perovskite solar cells under concurrent bifacial illumination," *Sustainable Energy & Fuels*, vol. 5, pp. 2865-2870, 2021.

[13] H. Li, Y. Wang, H. Gao, M. Zhang, R. Lin, P. Wu, K. Xiao, and H. Tan, "Revealing the output power potential of bifacial monolithic all-perovskite tandem solar cells," *eLight*, vol. 2, p. 21, 2022.

[14] B. Chen, Z. Yu, A. Onno, Z. Yu, S. Chen, J. Wang, Z. C. Holman, and J. Huang, "Bifacial all-perovskite tandem solar cells," *Science Advances*, vol. 8, p. eadd0377, 2022.

[15] T. S. Liang, M. Pravettoni, C. Deline, J. S. Stein, R. Kopecek, J. P. Singh, W. Luo, Y. Wang, A. G. Aberle, and Y. S. Khoo, "A review of crystalline silicon bifacial photovoltaic performance characterisation and simulation," *Energy & Environmental Science*, vol. 12, pp. 116-148, 2019.

[16] R. Guerrero-Lemus, R. Vega, T. Kim, A. Kimm, and L. Shephard, "Bifacial solar photovoltaics–A technology review," *Renewable and sustainable energy reviews*, vol. 60, pp. 1533-1549, 2016.

Efficiency maps for tandem solar cells using high resolution spectral data

Rune Strandberg, Anne G. Imenes

Universitetet i Agder, Grimstad, Norway

Spectral data have been collected every minute for a full year at our test station in Grimstad, Norway. These spectra have been used to model how spectral variation impacts the production from multi-junction solar cells and to produce efficiency maps that show how the efficiency varies with the band gaps. We find that there is good agreement between the efficiency under the AM1.5-spectrum and the annual efficiency found using the collected spectra. The main difference is that a slight blue-shift favors larger band gaps. Although series-connected tandem cells are more sensitive to spectral variation than four-terminal devices, the penalty experienced by the former cells due to spectral variations throughout the year is found to be small.

978-1-6654-6060-6/23 $31.00 © 2023 IEEE

Optimal Allocation of Voltage Regulations to Maximize the Hosting Capacity of Distribution Systems

Bahman Ahmadi[1] and Eli Shirazi[2]

[1]Faculty of EEMCS, University of Twente, Enschede, the Netherlands
Email: b.ahmadi@utwente.nl
[2]Faculty of Engineering Technology, University of Twente, Enschede, Netherlands.
Email: e.shirazi@utwente.nl

Abstract—This paper proposes a novel approach for enhancing the hosting capacity (HC) of distributed generators (DGs) in a grid by optimizing the allocation problem of DGs besides the placement and operation of voltage regulators (VRs). The proposed model aims to maximize the installed capacity of DGs while ensuring the bus voltage/current stays within operating limits and minimizing the investment costs of VRs. The optimization model is implemented in the dynamic hunter leadership optimization algorithm, and the effectiveness of the proposed method is demonstrated using IEEE 33-bus and 69-bus distribution systems.

Index Terms—Power system planning, Photovoltaic systems, Voltage Regulators, Simulation, Distribution networks, Dynamic Hunting Leadership, Optimization

I. INTRODUCTION

Distributed generation (DG) based on renewable energy sources is gaining popularity worldwide as a way to tackle energy and environmental issues. However, high penetration of DG into the distribution network (DN) can lead to problems such as voltage rise, reverse power flow, increased losses, and decreased power quality [1]. The variability of renewable energy sources makes it challenging to predict the power output of DG systems [2]. To overcome these challenges, it is crucial for planners to adopt a methodical approach to determine the most suitable capacity for DG installations within a DN. This approach must consider uncertainties associated with DG power output, such as those from wind turbines (WTs) and photovoltaic (PV) systems, as well as load consumption. Additionally, it is essential to avoid any technical limitations that may arise in the process.

The amount of DG capacity that a DN can handle without violating operational constraints is known as the DG hosting capacity [3]. This capacity is influenced by various factors, such as the type of DG, the characteristics of the DN, and restrictions set by the network operator. Limitations to DG penetration may arise from issues such as voltage violations, current limits, control schemes, and impacts on the network's protection systems. To optimize the integration of DG units into a system, several methods have been proposed for maximizing hosting capacity (MHC) by using battery energy storage systems [4], reactive power compensation [3], and network reconfiguration [5].

Several studies have investigated ways to address voltage quality issues when integrating PV systems into DNs. In [6], low voltage grids can accommodate a higher relative penetration of PV systems than medium voltage grids, but medium grids manage to experience fewer voltage problems such as over-voltages compared. A separate research effort aimed to control the reactive and active power of PV units using inverters in order to increase the number of PV units that can be connected to the grid [7]. The study proposed new strategies for Volt-VAr and Volt-Watt coordination of PVs to minimize the impact of PV output on the voltage level of the system. In other words, by regulating the extra active power of PV systems during over-voltage incidents and adjusting the reactive power control during under-voltage incidents, the voltage profiles and the power quality of DNs can be improved. Overall, it is crucial to develop techniques that maximize PV integration while keeping the grid's operational constraints within limits [8].

The use of voltage regulators (VRs) as a solution to technical problems in DNs is becoming increasingly popular due to decreasing costs and longer lifetimes. VRs are mainly used for voltage control, which can affect the MHC of the distribution system. There have been many studies on the operation and placement of VRs in DNs. The most commonly used methods are heuristic optimization algorithms to determine the best operation of VRs for different goals. In [9], the optimal tap positions of the VRs are determined to increase the hosting capacity of PV units. In [10], the goal is to minimize voltage violations in the DNs using VRs. Usually, the optimal operation of the VRs used for MHC is only applied to scenarios that increase the size of existing DG units [9] or with optimal allocation of the DG units but not the optimal allocation of the VRs [3].

Several research gaps in the field are found based on the literature review. While previous studies have investigated ways to address voltage quality issues when integrating PV systems into DNs, there is still a need to develop techniques that maximize multi-type DG integration while keeping the grid's operational constraints within limits. Additionally, although the use of VRs is becoming increasingly popular as a solution to technical problems in DNs, there is still a lack of research on the optimal placement of VRs to enhance the optimal size and location of DG units to increase the hosting capacity for the system. Furthermore, while heuristic optimization algorithms are commonly used to determine the best parameters of DG and/or VRs allocation for different goals, there is still a need to apply more reliable methods

978-1-6654-6060-6/23 $31.00 © 2023 IEEE

that can deal with mixed integer non-linear program (MINLP) problems.

In this paper, a model for the optimal placement of VRs is proposed to enhance the optimal size and location of DG units to increase the hosting capacity for the system while ensuring bus voltage stays within operating limits. The model is formulated as an optimization problem, where the optimal number and placement for VR units and optimal sizes and locations of DG units are determined for the new objective function formulated for MHC. Dynamic hunter leadership (DHL) algorithm [11] is used to handle the MINLP problem. The DG output and load demand operation scenarios are determined using historical data. The effectiveness of the proposed model is verified through testing on modified IEEE 33-bus and 69-bus DNs.

The paper presents several key contributions, including:

- Proposes an efficient method for improving DG hosting capacity (DGHC), which plays an essential role in determining the capability of a DN to support more DG generation.
- Explores the possible advantages of optimal operation and placement of VRs to improve DGHC by eliminating voltage and current violation problems due to large-scale DG penetration.
- adapts the DHL algorithm for the proposed problem and extends it with an additional stopping criterion to improve the convergence properties.
- Determining the optimal size and placement of DG units beside the optimal placement and operation of the VRs.

The structure of the paper is as follows. The statement of the problem, including the objective functions and constraints, is described in II. Section III is devoted to the DHL method and implementation for the problem. Section IV presents the simulation results for test system applications. Conclusions are summarized in Section V.

II. FORMULATION OF THE PROBLEM

The problem of optimizing the allocation of DG (WT and/or PV units) and VR units in a grid can be represented mathematically as follows:

$$\max_{w.r.t \ \vec{X}} F(\vec{X}), \tag{1}$$

where $\vec{X} = \{\vec{L}, \vec{S}, \vec{O}\}$ is a vector representing the control variables of the problem, $\vec{S}$ and $\vec{L}$ represent the size and location vectors of the units, $\vec{O}$ denotes the operation strategy of the VR units, and $F(\vec{X})$ is the objective function to be maximized.

The optimization is subject to the following constraints:

$$\text{subject to :} \begin{cases} g_i(\vec{X}) \geq 0, i = 1, 2, ..., m \\ h_j(\vec{X}) = 0, j = 1, 2, ..., p \end{cases}$$

where $g(\vec{X})$ and $h(\vec{X})$ represent the inequality and equality constraints, respectively, for the control variable vector $\vec{X}$, and m and p are the numbers of inequality and equality constraints, respectively.

The locations, sizes, and operating strategies of the DG and VR units are optimized to maximize the DGHC of the grid. The objective function for the problem is defined below.

A. DGHC objective function

The proposed objective function to maximize the size of DG units, including WT and PV in the network, is defined as follows:

$$F(\vec{X}) = \frac{\sum_{n=1}^{N_{DG}} S_{DG}(\vec{X})}{1 + \sum_{i=1}^{N_T}(\sum_{j=1}^{N_{Bus}} K_{i,j}(\vec{X}) + \sum_{n=1}^{N_{Line}} Y_{i,n}(\vec{X}))} \tag{2}$$

where N_T and N_{BUS} denote the total number of time steps and the number of buses in the network, respectively. S_{DG} is the size of DG units, and $K_{i,j}$ and $Y_{i,j}$ are penalty terms used to penalize the objective function for the voltage magnitudes and current that violate a reference value. The penalty terms are defined as follows:

$$K_{i,j}(\vec{X}) = \begin{cases} 0, & \text{if } 0.95V_{\text{ref}} < V_{i,j}(\vec{X}) < 1.05V_{\text{ref}} \\ k \cdot |V_{i,j}(\vec{X}) - V_{\text{ref}}|, & \text{if other} \end{cases} \tag{3}$$

$$Y_{i,n}(\vec{X}) = \begin{cases} 0, & \text{if } \underline{I}_{i,n} < I_{i,n}(\vec{X}) < \overline{I}_{i,n} \\ y, & \text{if other} \end{cases} \tag{4}$$

where $V_{i,j}$ represents the voltage magnitude at bus j and time step i. k and y are fixed constants used to increase the penalty value, and $I_{i,n}$ is the current in the n_{th} line of the DN.

B. Constraints

The proposed formulations for the network inequality and equality constraints, including voltage magnitude limitations, nodal power balance equations, DG output limits, and VR unit constraints, are given as follows:

1) Power balance:

$$\begin{aligned} P_{\text{MG}_i} + P_{\text{DG}_i} - P_{\text{load}_i} - P_{\text{losses}_i} &= 0 \\ Q_{\text{MG}_i} - Q_{\text{load}_i} - Q_{\text{losses}_i} &= 0 \\ i = 1, 2, \cdots, N_T \end{aligned} \tag{5}$$

where P_{MG} is the active and Q_{MG} is the reactive power supplied from the main grid at the slack bus. P_{DG} is the total active power generated by DGs. The active and reactive system loads and line losses are represented by P_{load}, P_{losses}, Q_{load}, and Q_{losses}.

2) Generation constraints:

$$P_{\text{MG}_i} \leq P_{\text{MG}_{max}} \tag{6}$$

$$Q_{\text{MG}_i} \leq Q_{\text{MG}_{max}} \tag{7}$$

$$P_{\text{DG}} \leq P_{\text{DG}_{max}} \tag{8}$$

III. SOLUTION METHODOLOGY AND PROBLEM IMPLEMENTATION

The proposed solution methodology for maximizing the hosting capacity of a grid addresses the problem outlined in the previous section and is presented in this section by utilizing the DHL algorithm as the main tool. Additionally, the mathematical formulation of the DHL algorithm in the optimization process was modified to be applicable to our specific optimization problem. Finally, we describe how we have implemented the DHL algorithm in maximizing the DG hosting capacity. These sub-sections are outlined as follows:

A. Dynamic hunter leadership

The DHL [11] algorithm was introduced based on inspiration from the cooperative teamwork observed in the hunting process. The DHL algorithm is a computational method that models a social hierarchy and hunting strategy in order to search for optimal solutions in a given problem space. The algorithm involves coordination between leaders and hunters in order to encircle potential solutions. The number of leaders involved in following a prey(s) is reduced over time, allowing for a decrease in the exploration of the search space to increase exploitation and subsequently focus on the best target. The optimization process in the DHL algorithm is comprised of the following steps.

1) Initializition: Initially, a collection of potential solutions is randomly generated.

$$\mathbf{X} = \vec{X}_m : m = 1, 2, ..., N_h \tag{9}$$

$$\vec{X}_m = \{X_{m,1}, ..., X_{m,D}\} \tag{10}$$

$$LB_n \leq X_{m,n} \leq UB_n \tag{11}$$

where N_h represents the number of solutions in the set $\mathbf{X}$ and D represents the number of control parameters within each solution vector. The lower and upper bounds for the nth component of the solutions are represented by LB_n and UB_n, respectively.

Each solution is ranked using the objective function value, and a subset of N_l best solutions is selected as the leaders for the evaluation of the hunters in the next step.

$$F^* = F_\mu : \mu = 1, 2, .., N_l \tag{12}$$

2) Main loop of the evaluation: The objective of the main loop in the DHL algorithm, as presented in [11], is to improve the value of the objective function (increase the value for the maximization problem) for each hunter by updating the solutions based on the guidance of N_l leaders. The number of leaders in each iteration of the optimization process is dynamically updated according to a discriminate function, with the aim of balancing the exploration and exploitation phases of the search process.

By gradually reducing the number of leaders, DHL prioritizes exploitation around a promising solution and limits the exploration of the solution space. The exploration phase focuses on searching the entire solution space to identify promising candidates and prevent getting trapped in locally optimal solutions. In contrast, the exploitation phase concentrates on refining the candidate solutions identified during the exploration phase by exploring the local areas of the solution space. The dynamic number of leaders helps simulate a balance between the exploration and exploitation phases of the evaluation process.

At each iteration of the main loop, the evaluation of solutions is guided by the positions of N_l leaders. The value of N_l at each iteration is updated using a dynamic approach. A history of the objective function values for the leaders (F^*) is maintained in a repository for reference. It is worth noting that in the first iteration, the value of N_l is equal to N_h.

The decision to decrease the number of leaders is made by comparing the objective value of the worst leader (i.e. the leader with the lowest F value in the set F^* compared to other leaders) with its value in a predefined percentage of previous iterations. If the improvement in F is less than a specified amount, that leader is removed from the set of leaders. This methodology is then applied to the remaining leaders, and they are removed one by one until only one leader remains at the end of the optimization process.

$$N_l = \begin{cases} N_l - 1, & if \ |F_{it} - F_{(it-(\beta \cdot \text{Max}_{it}))}| < \epsilon \\ N_l, & if \ \text{ otherwise} \end{cases} \tag{13}$$

$$\beta = \frac{\varphi \cdot \text{Max}_{it}}{100} \tag{14}$$

where ϵ denotes a small number serving as tolerance, and φ represents a predefined number between 1 and 100. F_{it} represents the objective value of the worst leader in the set.

In each iteration of the hunting phase, based on the N_l leaders found in (13), the positions of the other hunters ($\vec{X}_m$) are then updated based on the position variables of the leaders using the following set of equations:

$$\vec{X}_{m,it+1} = \frac{\sum_{i=1}^{N_l}(\vec{X}_{l,it}^i - \rho \cdot |\vec{r}_2 \cdot \vec{X}_{l,it}^i - \vec{X}_{m,it}|)}{N_l} \tag{15}$$

$$\rho = 2 \cdot (1 - \frac{it}{\text{Max}_{it}}) \cdot (2 \cdot \vec{r}_1 - 1) \tag{16}$$

where $\vec{X}_{l,it}^i$ represents the position of the ith leader in the set of leaders at the current iteration of the main loop, while $\vec{X}_{m,it}$ represents the position of m_{th} hunter in that same iteration. In each iteration of the optimization process, after the hunters' positions are updated, the set of leaders is updated with the top-performing hunters. This process continues until the pre-defined stopping criteria for the main loop have been met, indicating the optimization has reached a satisfactory solution.

The specific stopping criteria for the DHL algorithm include *maximum number of iterations* and *convergence*. The algorithm stops the main loop of the optimization after a specified number of iterations. In addition, the algorithm stops the process when the change in the objective function value for the best leader from one iteration to the next is below a pre-defined tolerance level, indicating that the optimization process has converged to a solution.

B. Implementation of DHL algorithm for hosting capacity problem

It is important to acknowledge that the nonlinear VR and DG allocation and sizing problem is a nonconvex problem that commercial solvers are unable to solve. To address this issue, the DHL algorithm can be utilized to solve the hosting capacity problem of DG units by considering the location and operation of the VRs. By following a series of steps, the DHL algorithm can efficiently find near-optimal solutions to this complex problem. A flowchart of the process for determining near-optimal solution can be found in Fig. 1.

IV. TEST SYSTEMS AND RESULTS

A. Test systems and DG and VR modeling

The IEEE 33 and 69 bus distribution systems are used in the simulations. Within these models, the main modifications are the addition of DG systems and VR at best busses of the systems found by the optimization algorithm. Real data concerning sun irradiation and load curve behavior are used to

978-1-6654-6060-6/23 $31.00 © 2023 IEEE

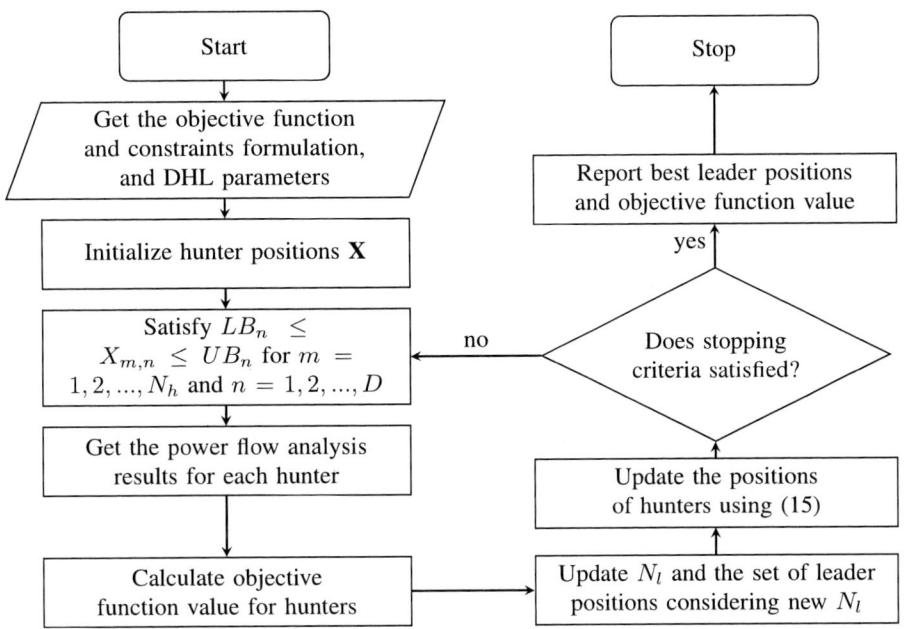

Fig. 1. Flowchart of the hosting capacity optimization based on the DHL algorithm.

model the PV yield and load characteristics. The details of the line data, peak load data, and daily load curves for residential, commercial, and industrial loads are taken from [3]. The load curves for the three illustrative days are shown in Fig. 2. Note that the curve's peaks and valleys are critical for considering under and over-voltage problems, respectively.

Seasonal averages are used for the power outputs of DG units. The scaled values for DG outputs [3] are illustrated in Fig. 3 and 4. According to yearly PV outputs, the average capacity factor for PV units was found as 34% while for WT units, it is 33%. The same output patterns and capacity factors are used for all DG units since the length of the feeder is short enough to consider identical atmospheric conditions.

Assume a VR is installed in between node a and b. Then, the voltage transmitted along the branch at any individual time interval can be described as,

$$\delta_r V_a = V_b, \tag{17}$$

$$\delta_r = 1 + \Delta_{\text{tap}} V_{\text{step}} , \forall i \in N \tag{18}$$

$$\Delta_{\text{tap}} \in [-16, 16], \text{integer} \tag{19}$$

$$V_{\text{sep}} = 0.00625 \text{ p.u.} \tag{20}$$

where V_a and V_b are the voltage magnitude of the nodes a and b, δ represents the turns ratio of the VR, and V_{step} denote the tap-position of the same VR. Note that Δ_{tap} for each VR and for the proposed time interval of the optimization is shown with $\vec{O}$.

B. Results and discussion

To implement the optimization algorithms in the proposed problem, a computer with 16 GB RAM, an Intel Core i7-7700 3.6 GHz processor configuration, and MATLAB version 2021b is chosen to handle the calculations. The DHL parameters used for the simulations are as follows: N_h is set to 25 hunters, MAX$_{it}$ is set to 5000 iterations with enabling second stopping criteria after 200 iterations. The number of control

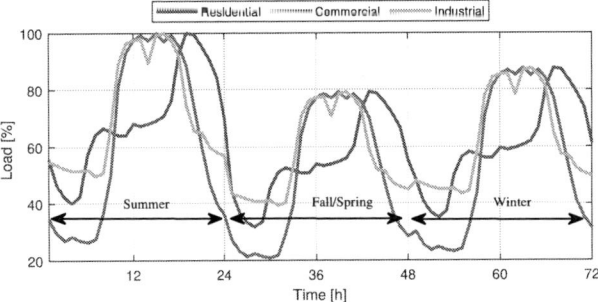

Fig. 2. Scaled seasonal load curve for the residential, commercial, and industrial load.

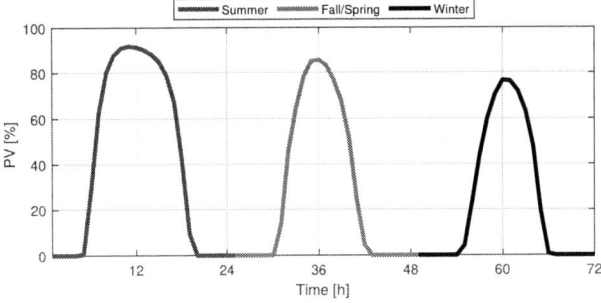

Fig. 3. Scaled seasonal PV output.

variables D depends on the maximum number of DG and VR units in each scenario.

The base case scenario (*BC*) is defined as the system with maximum hosting capacity for the DG units without utilizing VRs. Each scenario for finding the maximum hosting capacity of DG units while improving it (compared to the base case) based on finding optimal parameters of X VR units is named as *B-X*. Table I shows the near-optimal hosting capacity values

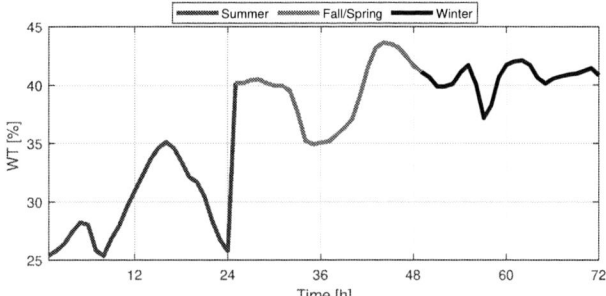

Fig. 4. Scaled seasonal WT output.

TABLE I
MAXIMUM HOSTING CAPACITY DETERMINED BY THE OPTIMIZATION ALGORITHM.

		BC	C-1	C-2	C-3	C-4	C-5
F	33-bus	11.8	14.6	14.9	16.8	16.8	16.8
	69-bus	12.9	14.8	15.3	17.8	18.2	18.2

TABLE II
THE OPTIMAL LOCATION, SIZE, AND TYPE OF DG UNITS FOR *C-3* IN 33-BUS AND *C-4* FOR 69-BUS SYSTEMS.

DG #	IEEE 33-bus			IEEE 69-bus		
	Location	Size [kW]	Type	Location	Size [kW]	Type
1	7	1000	WT	17	1000	WT
2	10	1000	PV	23	250	WT
3	12	1000	WT	30	1000	PV
4	16	900	PV	36	1000	WT
5	17	870	WT	40	1000	PV
6	18	1000	WT	43	1000	WT
7	21	1000	WT	50	1000	PV
8	22	1000	PV	52	1000	WT
9	24	1000	WT	53	1000	WT
10	25	1000	PV	56	1000	WT
11	26	1000	WT	59	1000	WT
12	27	1000	WT	62	1000	WT
13	29	1000	WT	63	1000	PV
14	30	1000	WT	64	1000	WT
15	31	1000	PV	65	1000	PV
16	32	1000	WT	66	1000	WT
17	33	1000	PV	67	1000	WT
18				68	1000	PV
19				69	1000	WT

TABLE III
THE OPTIMAL PLACEMENT OF VR UNITS FOR *C-3* IN 33-BUS AND *C-4* FOR 69-BUS SYSTEMS.

VR #	33-bus		69-bus	
	From	To	From	To
1	1	2	1	2
2	5	6	5	6
3	8	9	10	11
4			55	56

for different scenarios of the test systems. As can be seen from the table, the maximum hosting capacity values for all scenarios are higher than the base case scenario, indicating that the use of VRs can significantly improve the integration of DG units into the network. The results of the table suggest that the installation of more VR units can lead to a significant increase in the hosting capacity of the grid for the integration of DG units. The optimal selection of parameters (location and tap positions) for VRs can further enhance their effectiveness in improving the hosting capacity of DG while also effectively addressing voltage and/or current violation issues caused by DG integration.

The *F* value in Table I comprises the total size of the DG units integrated into the system and the penalty term for showing the violations. Based on the results for the base case and different scenarios, the penalty term for the solutions is found as zero. It means that all the voltage/current violations in the near-optimal solutions are solved by finding the optimal parameters of the optimization problem, and so the F value is the total size of the DG units.

Comparing the results, it is observed that the use of VRs in the *C-1* scenario can increase the DGHC by a significant amount of 23.7% and 14.7% for the 33-bus and 69-bus systems, compared to the *BC* scenario. Furthermore, the increase in the DGHC continues as the number of VRs increases. In the 33-bus system, the DGHC increases from 11.8 MW in the *BC* scenario to 16.7 MW in the *C-3* to *C-5*, which is a 41.5% improvement. The optimal number of VRs in the grid is three units, and more than this number will not improve the DGHC value. In the 69-bus system, the DGHC increases from 12.9 MW in the *BC* scenario to 18.2 MW in the *C-4* and *C-5* scenario, which is a 41.1% improvement. Also, the result indicates that the optimal number of VRs to maximize the DGHC is 4 units, and after that, no improvement is observed. These results indicate that the use of VRs can significantly enhance the hosting capacity of the grid, enabling greater DG integration while ensuring the system operates within safe voltage/current limits.

Table II displays the location, size, and type of DG units for the optimal scenario. In the IEEE 33-bus system, there are

a total of 6 PV units with a cumulative size of 5.90 MW and 11 WT units with a total size of 10.87 MW. The IEEE 69-bus system, on the other hand, has 6 PV units and 13 WT units, with a total size of 6 MW and 12.25 MW, respectively.

Table III shows the optimal placement of VRs in both systems. Note that VRs are positioned between two nodes, and their placement is listed from node to node in the table. To assess the effectiveness of the proposed optimization approach, the voltage profile of the systems based on seasonal data is illustrated in Fig. 5 and Fig. 6. These figures demonstrate that the proposed approach has led to removing all the voltage violations and a significant improvement in the voltage profile of both systems. Specifically, in the IEEE 33-bus system, the voltage deviations expected by WTs and PVs have been reduced, and the maximum and the minimum voltage magnitude are 0.957 and 1.05 p.u. in winter and 0.960 and 1.05 p.u. in summer. Similarly, the voltage deviations in the IEEE 69-bus system have been reduced, and maximum and minimum voltage magnitudes for the system are 0.966 and 1.047 p.u.

These results demonstrate the effectiveness of the proposed optimization approach in determining the optimal location, size, and type of DG units, as well as the placement and operation of VRs, to enhance the voltage profile and increase the hosting capacity of grids.

Table IV presents the outcomes of the DHL algorithm for the near-optimal values of the objective functions in the IEEE 33 and 69 bus systems for the *BC* and the best scenario of VR number. Additionally, the table offers a comparison between the results obtained by the DHL method and those obtained by other optimization methods, such as AOA [12], GWO [13], and PSO [14], using the same Max_{it} number and dynamic stopping criteria. Note that the algorithm parameters are given

978-1-6654-6060-6/23 $31.00 © 2023 IEEE

TABLE IV
COMPARISON OF THE RESULTS OBTAINED BY DHL TO AOA, GWO, AND PSO RESULTS.

			DHL			AOA			GWO			PSO		
			mean	std.	best	mean	std.	best	mean	std.	best	mean	std.	best
IEEE 33-bus	BC	F	**11.759**	0.032	**11.800**	11.733	0.035	11.800	11.732	0.033	11.787	11.753	0.019	11.789
		ET	87	39	30	86	49	37	76	34	32	87	39	30
	C-4	F	**16.660**	0.21	**16.800**	16.657	0.23	16.800	16.629	0.13	16.642	16.658	0.49	16.670
		ET	1362	6.1	1353	1364	4.9	1357	1364	5.5	1355	1374	9.6	1355
IEEE 69-bus	BC	F	12.847	0.31	**12.896**	12.834	0.26	12.880	12.491	2.02	12.896	**12.855**	0.25	12.890
		ET	145	22	106	148	23	105	144	23	106	145	22	106
	C-4	F	**18.144**	7.1	**18.237**	18.095	7.2	18.227	18.116	8.3	18.229	18.128	6.9	18.219
		ET	3632	78.3	3511	3645	69.2	3509	3618	75.6	3510	3654	88.9	3507

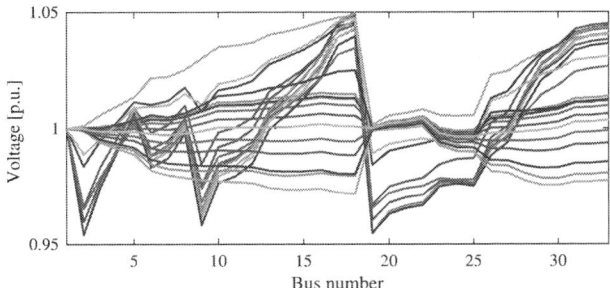

Fig. 5. The voltage profile of IEEE 33-bus system in C-3.

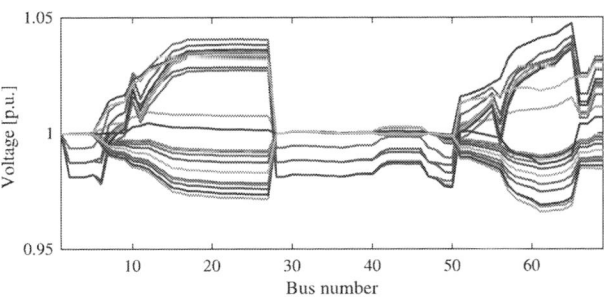

Fig. 6. The voltage profile of IEEE 69-bus system in C-4.

in [15]. The performance of these different methods is evaluated based on the mean, standard deviation (STD), and best objective function values, computed over 50 independent runs. Moreover, the convergence speed of the heuristic methods is assessed in terms of computational time, expressed in seconds (s) and denoted by ET in the table.

Based on the results, it seems that the DHL method performs the best for both systems in terms of both objective function values and execution time.

V. CONCLUSION

We proposed a model to optimize the placement and operation of VRs and DGs in DNs. The objective is to increase hosting capacity while maintaining the voltage and current of the grid within operating limits. We used the DHL algorithm to solve the optimization problem. Our model outperforms existing methods in terms of DGHC, voltage stability, and technical constraints. Our proposed method increased the optimal capacity of DG installations up to 41% by considering VR units. Future work includes testing on more complex networks and optimizing the allocation problem based on multi-objective optimization problems.

ACKNOWLEDGMENT

This research is funded by EU HORIZON 2020 project SERENE, grant agreement No 957682.

REFERENCES

[1] A. Rabiee and S. M. Mohseni-Bonab, "Maximizing hosting capacity of renewable energy sources in distribution networks: A multi-objective and scenario-based approach," *Energy*, vol. 120, pp. 417–430, 2017.

[2] E. Shirazi, J. Lemmens, M. Gofran Chowdhury, A. Tuomiranta, F. Catthoor, E. Voroshazi, and I. Gordon, "Cloud detection for pv power forecast based on colour components of sky images," in *2021 IEEE 48th Photovoltaic Specialists Conference (PVSC)*, 2021, pp. 2389–2391.

[3] B. Ahmadi, O. Ceylan, and A. Ozdemir, "Reinforcement of the distribution grids to improve the hosting capacity of distributed generation: Multi-objective framework," *Electric Power Systems Research*, vol. 217, p. 109120, 2023.

[4] S. Zhang, Y. Fang, H. Zhang, H. Cheng, and X. Wang, "Maximum hosting capacity of photovoltaic generation in sop-based power distribution network integrated with electric vehicles," *IEEE Transactions on Industrial Informatics*, vol. 18, no. 11, pp. 8213–8224, 2022.

[5] E. Kazemi-Robati, M. S. Sepasian, H. Hafezi, and H. Arasteh, "Pv-hosting-capacity enhancement and power-quality improvement through multiobjective reconfiguration of harmonic-polluted distribution systems," *International Journal of Electrical Power & Energy Systems*, vol. 140, p. 107972, 2022.

[6] T. Aziz and N. Ketjoy, "PV penetration limits in low voltage networks and voltage variations," *IEEE Access*, vol. 5, pp. 16 784–16 792, 2017.

[7] E. Kazemi-Robati, M. S. Sepasian, H. Hafezi, and H. Arasteh, "Pv-hosting-capacity enhancement and power-quality improvement through multiobjective reconfiguration of harmonic-polluted distribution systems," *International Journal of Electrical Power & Energy Systems*, vol. 140, p. 107972, 2022.

[8] S. Hashemi and J. Østergaard, "Methods and strategies for overvoltage prevention in low voltage distribution systems with pv," *IET Renewable power generation*, vol. 11, no. 2, pp. 205–214, 2017.

[9] B. Ahmadi, O. Ceylan, and A. Ozdemir, "Enhancing photovoltaic hosting capacity in distribution networks by optimal allocation and operation of static var compensators," in *2022 57th International Universities Power Engineering Conference (UPEC)*. IEEE, 2022, pp. 1–6.

[10] A. Bedawy, N. Yorino, K. Mahmoud, Y. Zoka, and Y. Sasaki, "Optimal voltage control strategy for voltage regulators in active unbalanced distribution systems using multi-agents," *IEEE Transactions on Power Systems*, vol. 35, no. 2, pp. 1023–1035, 2019.

[11] B. Ahmadi, J. S. Giraldo, and G. Hoogsteen, "Dynamic hunting leadership optimization: Algorithm and applications," *Journal of Computational Science*, 2023 (submitted).

[12] L. Abualigah, A. Diabat, S. Mirjalili, M. Abd Elaziz, and A. H. Gandomi, "The arithmetic optimization algorithm," *Computer methods in applied mechanics and engineering*, vol. 376, 2021.

[13] X. Sun, Z. Jin, Y. Cai, Z. Yang, and L. Chen, "Grey wolf optimization algorithm based state feedback control for a bearingless permanent magnet synchronous machine," *IEEE Transactions on Power Electronics*, vol. 35, no. 12, 2020.

[14] J. Kennedy and R. Eberhart, "Particle swarm optimization," in *Proceedings of ICNN'95-international conference on neural networks*, vol. 4. IEEE, 1995.

[15] B. Ahmadi, J. S. Giraldo, G. Hoogsteen, M. E. T. Gerards, and J. L. Hurink, "A decentralized control strategy for voltage regulators and energy storage devices in active unbalanced distribution systems," *2022 International Conference on Smart Energy Systems and Technologies (SEST)*, pp. 1–6, 2022.

In-situ Microscopy Characterization of Light-induced Phase Segregation in Wide-Bandgap Perovskite Materials

Fangfang Cao, Liming Du, Zhiyu Gao, Minghui Li, Cong Chen, Dewei Zhao, Can Li, Zhen Li, Jichun Ye, Chuanxiao Xiao

Ningbo Institute of Materials Technology and Engineering, Chinese Academy of Sciences, Ningbo, China

Ningbo University, Ningbo, China

Northwestern Polytechnical University and Shaanxi Joint Laboratory of Graphene, Xi'an, China

Sichuan University, Chengdu, China

Ningbo New Materials Testing and Evaluation Center CO., Ltd, Ningbo, China

Tandem solar cells using perovskite materials with wide-bandgap show great potential for improving photovoltaic technology efficiency, but stability issues remain a significant challenge for commercialization. One technique to adjust the perovskite bandgap is by mixing bromine and iodine, but this can result in phase segregation under illumination, leading to crystal and interface defects, nonradiative recombination, and voltage loss. The lack of understanding of the dynamic process of phase segregation and defects limits progress. This work uses high-resolution in-situ Kelvin probe force microscopy (KPFM) to spatially resolve the topography and surface potential of phase segregation in wide-bandgap perovskite materials. On topography mappings, we observed 100-200 nm chucks appear and continue to break into smaller particles during light soaking, and it is not fully reversible in dark storage. The surface potential mappings provided rich information about iodine- and bromine-related materials evolution. The initial surface has a chuck with lower potential, but the potential contrast gradually decreases and becomes neglectable after light soaking. And the light drove phase segregation into spikes with higher and lower potential. The potential can return back similar to its initial stage, but the grain boundaries become higher potential than the grain interiors. Our comprehensive KPFM results offer valuable information to further our understanding of phase segregation, but further analysis is needed to fully comprehend the findings.

Radiation Tolerance Studies of CdSe/CdTe Bilayer Solar Cells on Space-Qualified Cover Glass

Aesha P. Patel, Adam B. Phillips, Ebin Bastola, Abasi Abudulimu, Zachary W. Zawisza, Robert Snuggs, Manoj K. Jamarkattel, Deng-Bing Li, Richard Irving, Yanfa Yan, Michael J. Heben, Randy J. Ellingson

Wright Center for Photovoltaics Innovation and Commercialization, Department of Physics and Astronomy, The University of Toledo, 2801 Bancroft Street, Toledo, OH 43606, USA

Abstract — Here, we report on the fabrication of CdSe/CdTe bilayer solar cells, based on multi-source evaporation, directly onto space-qualified cover glass. The aim was to optimize the device structure to achieve maximum device performance under AM0 atmospheric conditions with minimal losses compared to that under the AM1.5G condition. Two sets of optimized solar cells underwent proton irradiation—one from the film-side, the other from the glass-side—at a fluence of 5 x 10^{13} cm^{-2} and varying energies. We utilized SRIM calculations to assess the potential impact and penetration depth of protons at three different energies (50 keV, 150 keV, and 240 keV) in the CdSe/CdTe filmstack. Evaluations of device performance changes pre- and post-radiation exposure were conducted through current density-voltage (*J-V*) characteristics and external quantum efficiency (EQE) measurements. Additionally, scanning electron microscopy (SEM) was employed to study surface morphology and determine layer thicknesses. Glass-side irradiated samples showed almost no degradation compared to those irradiated from the film-side, indicating the importance of using space-qualified cover glass for space PV.

Keywords—CdTe, cover glass, proton radiation, space PV, thin-film

I. INTRODUCTION

To employ solar cells in space, achieving high specific power (W/kg) and radiation hardness is crucial. A high power-to-weight ratio is required for the ease of launching lighter satellites into space. One approach to improve the specific power of solar cells is to substitute the conventional ~3 mm thick glass substrates with ultra-thin and lightweight substrates. Solar cells in space are subject to high-intensity radiation (electron, proton, and UV) which causes the development of color centers (darkening) within conventional glass substrates leading to current loss. Hence, cerium-doped space-qualified cover glasses are recommended for space-based solar cells [1]. CdTe, a II-VI thin film solar cell, with its high light absorption and lower production cost could be a good candidate for space application. The highest recorded efficiency for flexible CdS/CdTe solar cells on ultra-thin glass superstrate has been 16.4% [2]. However, the maximum efficiency achieved by CdTe under terrestrial STC is 22.1% at the laboratory scale [3]. Replacing CdS with CdSe as the window layer as enabled higher current collection at short and long wavelengths and has significantly improved device efficiency [4]. While CdS/CdTe

solar cells have been extensively researched in terms of performance degradation and to a lesser extent, radiation tolerance [5], the higher efficiency architecture based on CdSe/CdTe remains relatively unexplored for space application.

Here, we demonstrate the fabrication of CdSe/CdTe bilayer solar cell directly onto the 150 μm space-qualified cover glass with a novel n-type emitter layer indium gallium oxide (IGO) [6] and report device efficiencies under AM1.5G and AM0 illuminations along with preliminary studies on proton radiation tolerance of CdSe/CdTe solar cells.

II. EXPERIMENTAL DETAILS

The space-qualified cover glass (Corning 0214) - 0.75% cerium oxide doped borosilicate glass in use was provided by Martin Materials Solutions, Inc. ~290 nm of aluminum-doped zinc oxide (AZO) was RF sputtered at 250°C as transparent conducting oxide (TCO). 75 nm thick indium gallium oxide (IGO) (E$_g$ = 4.02 eV) film was deposited as an emitter layer by co-sputtering from In$_2$O$_3$ and Ga$_2$O$_3$ targets at 250°C under Ar/O$_2$ environment [6]. CdSe/CdTe layers were sequentially grown in a multisource evaporation chamber at a substrate temperature of 400°C and a base pressure of 1.5 x 10^{-6} Torr [4]. CdSe/CdTe bilayer film stack was activated by CdCl$_2$ treatment using a saturated CdCl$_2$-methanol solution and annealing at 400°C for 30 min in nitrogen environment. Standard Cu/Au back contact was fabricated using CuCl$_2$ treatment at 250°C for 10 min in a hot oven and completed with 80 nm of evaporated Au with a device area of 0.08 cm^2.

Proton radiation tests were carried out by bombarding singly ionized hydrogen ions (H$^+$) using a 330 kV positive ion accelerator at the Toledo heavy ion accelerator (THIA) at The University of Toledo [7]. THIA can achieve proton energies and fluences between 50 keV and 240 keV, and 1 x 10^{12} cm^{-2} and 1 x 10^{15} cm^{-2}, respectively. Two sets of samples (three samples each) were irradiated from the film-side as well as the glass-side (one set each) at three different proton energies (50 keV, 150 keV, and 240 keV) at a constant fluence of 5 x 10^{13} cm^{-2} to monitor the device performance degradation caused by protons of lower energy range and the importance of incorporating space cover glass in device fabrication. Stopping and Range of Ions in Matter (SRIM) was used to carry out

978-1-6654-6060-6/23 $31.00 © 2023 IEEE

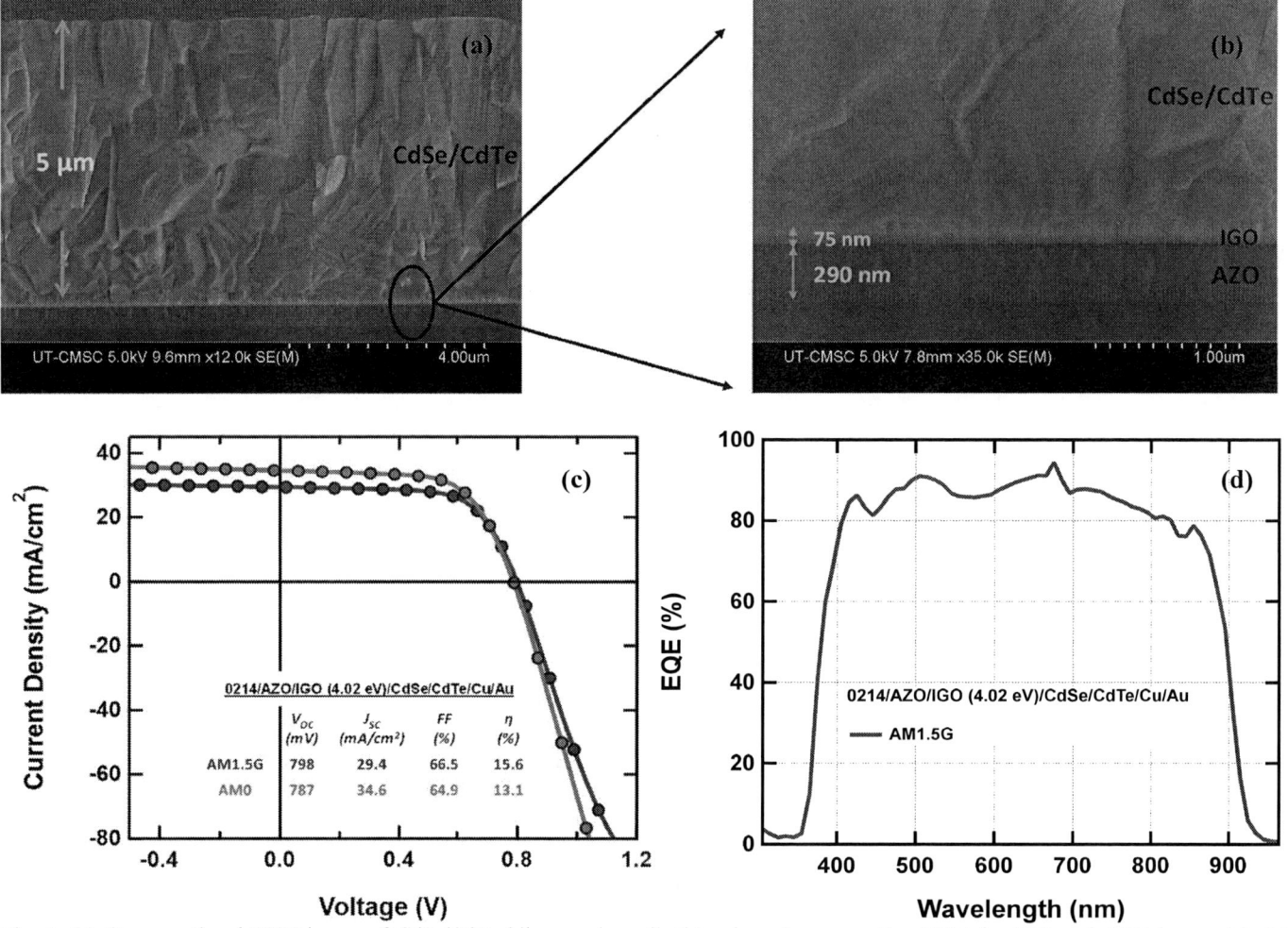

Fig. 1. (a) Cross-sectional SEM image of CdSe/CdTe bilayer solar cell, (b) enlarged cross-section SEM for IGO and AZO layers, (c) *J-V* measurements for 0214/AZO/IGO/CdSe/CdTe/Cu/Au device under AM1.5G and AM0 illuminations, and (d) EQE for CdSe/CdTe bilayer solar cell under AM1.5G illumination

Monte Carlo simulations of particle path and collisions with ion beam at normal incidence. Simulations were carried out for glass-side as well as film-side irradiation to predict the ion penetration depth range through the cover glass and Au, respectively. The layers used for simulations were as follows:

- 80 nm Au (ρ=19.31 g cm^{-3})
- 3.5 μm CdTe (1:1 ratio, ρ=5.85 g cm^{-3})
- 150 μm Ce-doped borosilicate cover glass (ρ=2.53 g cm^{-3})

Current density-voltage (*J-V*) curves under AM1.5G illumination were measured using a Keithley 2440 digital source meter and a solar simulator. AM0 illumination-based *J-V* measurements were conducted on a led-based Sunbrick solar simulator (G2V Optics, Inc.). External quantum efficiency (EQE) was measured using the PV Instruments system (model IVQE8-C). Scanning electron microscopy (SEM) was carried out using a Hitachi S-4800 UHR scanning electron microscope.

III. RESULTS AND DISCUSSION

Cross-sectional SEM images (Fig. 1 (a, b)) for CdSe/CdTe film stack were recorded post-CdSe/CdTe deposition. AZO (~290 nm), IGO (~75 nm), and CdSe/CdTe (~5 μm) layer thicknesses were determined from these images. Though it is difficult to differentiate between CdSe and CdTe, it is evident that the CdSe/CdTe bilayer thickness seems to be slightly higher than the target thickness of ~3.5 μm. One of the reasons for such discrepancy could be the sticking coefficient of the thinner glass. This SEM image was recorded before CdCl$_2$ treatment, thus, the CdTe grains are small.

Fig. 1 (c, d) summarizes the *J-V* characteristics of 0214/AZO/IGO/CdSe/CdTe/CdCl$_2$/CuCl$_2$/Au (standard) device under AM1.5G and AM0 conditions. The champion device of η = 15.6% (J_{SC} = 29.4 mA/cm^2, V_{OC} = 798 mV, and FF = 66.5%) and 13.1% (J_{SC} = 34.6 mA/cm^2, V_{OC} = 787 mV, and FF = 64.9%) was obtained under AM1.5G and AM0

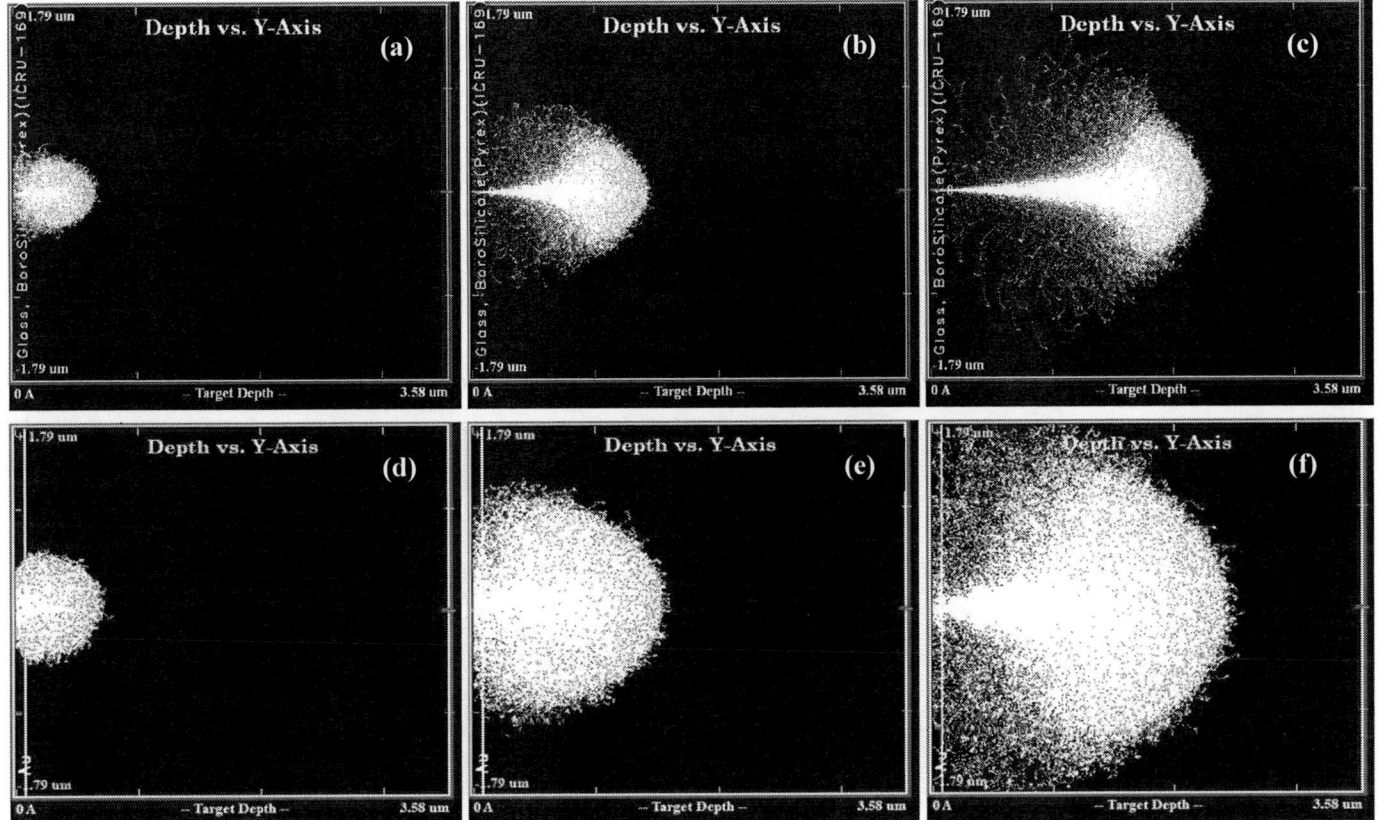

Fig. 2. SRIM calculations of ion trajectories for proton energies of 50 keV (a, d), 150 keV (b. e), and 240 keV (c, f). Simulations were carried out for glass-side irradiation through 150 μm borosilicate cover glass (a-c), and film-side irradiation through 80 nm Au (d-f)

illumination, respectively. A significant increase in J_{SC} is observed with a slight decrease in FF under AM0 condition. However, a V_{OC} decrease of 11 mV indicates recombination losses which would require further investigation.

Fig. 2 shows the SRIM simulations for glass-side (Fig. 2 (a-c)) as well as film-side proton irradiation (Fig. 2 (d-f)) at proton energies of 50 keV (Fig. 2 (a, d)), 150 keV (Fig. 2 (b, e)), and 240 keV (Fig. 2 (c, f)). All calculations were completed using 99,999 ions to obtain relatively robust statistics. It is evident that the ion penetration depth ranges between ~0.3-2 μm for proton energies ranging from 50-240 keV, respectively, through both the glass-side as well as film-side. Hence, theoretically, these low energy protons would barely penetrate through the first few microns of glass during glass-side irradiation and through 80 nm Au and a few microns of CdTe from the back during film-side irradiation.

Fig. 3 shows the relative J-V characteristics (expressed as the ratio of post-irradiated (i) to pre-irradiated (0) device parameters) of the above fabricated CdSe/CdTe solar cell onto the space-qualified cover glass. The cells were irradiated at a constant proton fluence of 5 x 10^{13} cm^{-2} and proton energies of 50 keV, 150 keV, and 240 keV from the glass-side as well as the film-side. Overall, irrespective of the proton energy, degradation in glass-side irradiated cells is almost negligible, indicating the importance of encapsulation using space-

qualified cover glass in space PV applications. However, the device performance of film-side irradiated cells drops significantly post-irradiation. The large drop in efficiency for film-side irradiated cells at all 3 proton energies (Fig. 3 (a)), could be attributed to a significant decrease in J_{SC}, as can be seen in Fig. 3 (b). The V_{OC} (Fig. 3 (d)) does not show such a large decrease for any proton energy and an even smaller decrease for FF than J_{SC} and V_{OC} can be seen in Fig. 3 (c). Since these low energy protons do not reach the cover glass in the front, when irradiated through film-side, efficiency losses for film-side irradiated cells could be due to Au diffusion and possible structural and/or optical property changes due to damages caused by the impact of protons in bulk CdTe (~1-2 μm from the back).

IV. CONCLUSION

The CdSe/CdTe bilayer solar cells have been successfully deposited directly onto the 150 μm thick space-qualified cover glass by using AZO and IGO materials as TCO and n-type emitter layers, respectively, under superstrate configuration and device results under AM1.5G and AM0 illuminations have been recorded. SEM images showed thicker than expected CdSe/CdTe bilayer (~5μm) deposition and IGO non-uniformity. The best device efficiencies of 15.6% (AM1.5G)

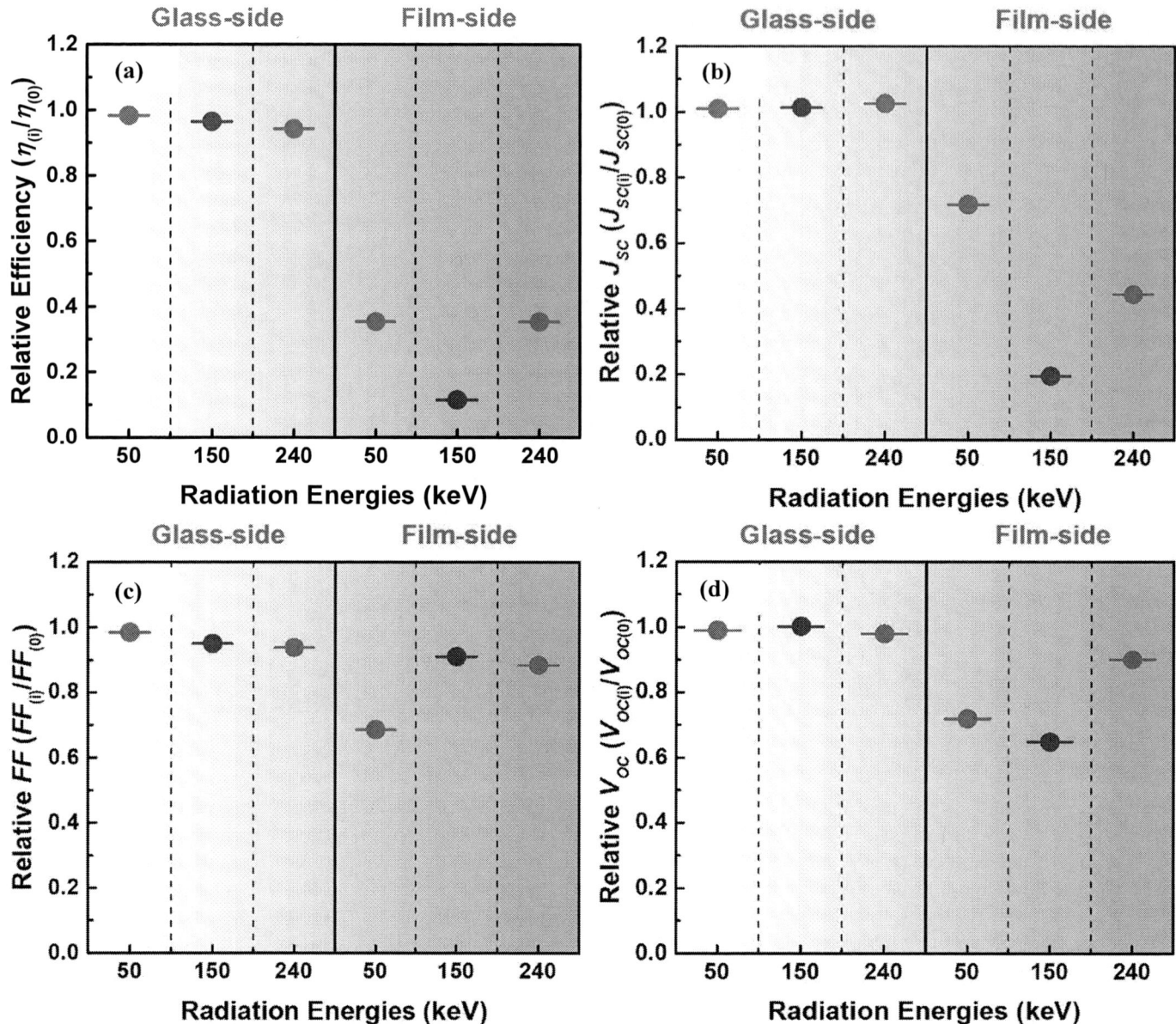

Fig. 3. Relative *J-V* parameters of samples expressed as a ratio of post (i)- to pre (0)-proton irradiation device performance (a) efficiency (η), (b) J_{SC}, (c) *FF*, and (d) V_{OC}

and 13.1% (AM0) show great promise for improvement under further optimization. We are collecting data from further experiments to optimize the IGO bandgap, CdCl$_2$ treatment conditions, and Cu doping to obtain maximum device efficiency. Proton tolerance tests were carried out on CdSe/CdTe bilayer devices at THIA to study the degradation mechanisms under space conditions. SRIM calculations, based on Monte Carlo simulations, were carried out to determine the ion trajectories and ion penetration depths at different proton energies. From SRIM simulations, it is evident that irrespective of the cell irradiation side, ions with proton energies ranging from 50-240 keV barely penetrate through ~0.3-2 μm of the layers. The device measurements of glass-side irradiated cells

showed almost no degradation. Minor efficiency losses could be due to the cell exposure to ambient air during sample transfer to THIA. SRIM calculations suggest that the ions barely reach 1-2 μm bulk CdTe for film-side irradiation. However, a decrease in device efficiency for film-side irradiated cells resulted from a significant J_{SC} loss. Hence, further investigation is required to study the structural and/or optical property changes in CdTe post-proton radiation. It is difficult to determine a trend in the device performance degradation of film-side irradiated cells. Hence, further experimentation is required where cells are irradiated at higher proton energies and difference fluences to develop a better understanding of proton radiation tolerance in CdSe/CdTe bilayer solar cells.

ACKNOWLEDGMENT

This material is based on research sponsored by Air Force Research Laboratory under agreement number FA9453-21-C-0056. The U.S. Government is authorized to reproduce and distribute reprints for Governmental purposes notwithstanding any copyright notation thereon. The views expressed are those of the authors and do not reflect the official guidance or position of the United States Government, the Department of Defense or of the United States Air Force. The appearance of external hyperlinks does not constitute endorsement by the United States Department of Defense (DoD) of the linked websites, or the information, products, or services contained therein. The DoD does not exercise any editorial, security, or other control over the information you may find at these locations. Approved for public release; distribution is unlimited. Public Affairs release approval #AFRL-2023-0188.

REFERENCES

[1] A. Romeo, D. L. Bätzner, W. Hajdas, H. Zogg, and A. N. Tiwari, "Potential of CdTe Thin Film Solar Cells for Space Application," *Proceedings of 17th European Photovoltaic Solar Energy Conference and Exhibition*, vol. 3, no. October, pp. 2183–2186, 2002.

[2] H. P. Mahabaduge, W. L. Rance, J. M. Burst, M. O. Reese, D. M. Meysing, C. A. Wolden, J. Li, J. D. Beach, T. A. Gessert, W. K. Metzger, S. Garner, and T. M. Barnes, "High-efficiency, flexible CdTe solar cells on ultra-thin glass substrates," *Appl Phys Lett*, vol. 106, no. 13, pp. 3–7, 2015, doi: 10.1063/1.4916634.

[3] M. A. Green, E. D. Dunlop, G. Siefer, M. Yoshita, N. Kopidakis, K. Bothe, and X. Hao, "Solar cell efficiency tables (version 61)," *Progress in Photovoltaics: Research and Applications*, vol. 31, no. 1, pp. 3–16, 2022, doi: 10.1002/pip.3646.

[4] E. Bastola, A. B. Phillips, G. Barros-King, M. K. Jamarkattel, DB. Li, A. Quader, D. Pokhrel, J. Friedl, J. M. Gibbs, X. Mathew, Y. Yan, R. J. Ellingson, and M. J. Heben, "Understanding the Interplay between CdSe Thickness and Cu Doping Temperature in CdSe/CdTe Devices," *IEEE Journal of Photovoltaics*, vol. 12, no. 1, pp. 11-15, 2022, doi: 10.1109/jphotov.2021.3110338.

[5] D. A. Lamb, C. I. Underwood, V. Barrioz, R. Gwilliam, J. Hall, M. A. Baker, and S. J.C. Irvine, "Proton irradiation of CdTe thin film photovoltaics deposited on cerium-doped space glass," *Progress in Photovoltaics: Research and Applications*, vol. 25, no. 12, pp. 1059–1067, 2017, doi: 10.1002/pip.2923.

[6] M. K. Jamarkattel, A. B. Phillips, I. Subedi, A. Abasi, E. Bastola, DB. Li, X. Mathew, Y. Yan, R. J. Ellingson, N. J. Podraza, and M. J. Heben, "Indium Gallium Oxide Emitters for High-Efficiency CdTe-Based Solar Cells," *ACS Appl Energy Mater*, 2022, doi: 10.1021/acsaem.2c00153.

[7] R. R. Haar, D. J. Beideck, L. J. Curtis, T. J. Kvale, A. Sen, R. M. Schectman, and H. W. Stevens, "The Toledo heavy ion accelerator," *Nuclear Instruments and Methods in Physics Research Section B: Beam Interactions with Materials and Atoms*, vol. 79, no. 1, pp. 746-748, 1993/06/02/ 1993, doi: https://doi.org/10.1016/0168-583X(93)95458-II

Advanced production line monitoring with time-lag sequential analysis

Gaia Maria N. Javier, Rhett Evans, Priya Dwivedi, Thorsten Trupke, Ziv Hameiri

The University of New South Wales, Sydney, Australia

Conventional methods for monitoring variations in solar cell production are often subjective and suffer from various limitations. In this study, a novel statistical technique is proposed to evaluate these variations using time series measurements. A metric based on lag-sequential analysis is developed to infer probable batching, degree of randomness, and stability in production. The metric, validated through simulations and measurements from an industrial production line, was proven to be effective in detecting random and non-random variations in different parameters using current-voltage measurements and electroluminescence (EL) images. The technique not only unlocks new capabilities for EL images, but can also be easily extended to photoluminescence images, enabling early detection of anomalies in production lines. Furthermore, this technique can be adapted for real-time monitoring. The method can be readily implemented, enabling powerful monitoring of process variations, an important capability for improving production yield and thereby contributing to the further reduction in the cost of photovoltaic energy.

NASA GRC Solar Cell Characterization Facilities

Jeremiah D Sims

NASA Glenn Research Center, Cleveland, OH, United States

Abstract- NASA GRC maintains various equipment to calibrate and characterize photovoltaic devices to Air Mass Zero standards and different space-simulated environments. These characterization capabilities are critical to developing and advancing photovoltaic devices, allowing advancements in various NASA-related projects (Artemis, Lunar Gateway, Human Landing System, etc.). Our facilities include an X-25 Triple source Solar Simulator, G2V Sunbrick, Angstrom Designs pLEDss (Programmable LED Solar Simulator), Thermal Balance facility, ER-2 High Altitude aircraft calibration.

Application of noise-assisted multivariate data analysis for hour-ahead GHI forecasting

Priya Gupta[1], Rhythm Singh[1]

[1]Department of Hydro and Renewable Energy, Indian Institute of Technology Roorkee, Roorkee – 247667, Uttarakhand, India

Abstract — This paper proposed a hybrid model based on a noise-assisted data analysis technique and a less time-complex stacked (ensemble) model for 1-h ahead GHI forecasting. Data analysis techniques reduce the non-stationary issues of time series data. Noise-assisted multivariate empirical mode decomposition (NA-MEMD) proposed in this work is an updated form of the standard MEMD technique. The former can overcome the drawback (mode mixing) of the latter technique. The ensemble model combines the prediction of three simple ML models (decision tree regressor, k-nearest neighbor, and ridge regressor). The results are analyzed for two locations in India under different climate zones. The results show that after replacing MEMD with NA-MEMD, the forecasting error has further reduced. The corresponding reduced values in terms of %RMSE (%MAE) are as follows: 23.87% (26.84%) at Delhi and 20.65% (25.64%) at Nainital. Interestingly, MEMD took around 232 sec, whereas NA-MEMD consumed around 180 sec to decompose the time series. Further, the proposed model's performance is visualized against some popular and complex ML algorithms. The minimum %RMSE reduction of 58.85%, 61.04%, 59.76%, and 63.36% has been reported after replacing stacked model LSTM, RF, ANN, and (kNN-DTR-Ridge), respectively, with the proposed NA-MEMD-stacked model.

Keywords—Global horizontal irradiance, noise-assisted multivariate empirical mode decomposition, ridge regressor, decision tree regressor, k nearest neighbor

I. INTRODUCTION

Developing accurate solar irradiance forecasting models is of great importance for the efficient functioning of grid-integrated solar energy systems. Due to more effective nonlinear mapping capabilities of modern ML algorithms (tree-based ensemble, neural networks, deep learning, etc.), simple techniques (DTR, kNN, ridge regression, etc.) are becoming less popular [1], [2]. However, this is achieved with high time and model complexity. Developing increasingly precise models while ignoring time and model complexity is not particularly practicable. Moreover, learning the complex time series of solar energy is difficult even for modern and popular ML algorithms. Combining multiresolution analysis tools for pre-data analysis with forecasting models is one of the emerging areas of research in solar forecasting [3]. Empirical mode decomposition (EMD), Ensemble EMD (EEMD), complete EEMD adaptive noise (CEEMDAN), etc., are popular univariate data analysis techniques [4]. In contrast, MEMD [5] emerges as a promising approach for multivariate time series due to the dependency of solar energy on various weather variables. While the data demarcation techniques improve forecasting accuracy, they increase the model complexity even more and necessitate significant time for data decomposition that should not be neglected.

Based on the above discussion, the prior study developed a hybrid model that balances accuracy and complexity [6]. Combining MEMD with an ensemble of simple ML techniques accomplished the stated aim. The developed ensemble or stacked model combines three simple ML algorithms (kNN, DTR, and Ridge regressor) as base models, whereas Linear regressor is a meta-model. The idea of this model is that if the data analysis technique has already performed the important task of mapping nonlinearity and non-stationarity from a complex time series, then simpler ML models may perform the accurate prediction. Although a significant improvement was observed with the combination of MEMD, the main limitation of MEMD is that it suffers from mode mixing. Mode mixing alters the physical meaning of subseries (intrinsic mode functions (IMFs)) extracted from complex and intermittent time series. In contrast, a new noise-assisted version of MEMD (NA-MEMD) [7] may overcome this limitation and improve the accuracy of the model even more. This paper presents the potential of combining NA-MEMD with the stacked model (kNN-DTR-Ridge), comparing the performance of two multivariate data demarcation techniques, i.e., MEMD and NA-MEMD.

II. NOISE-ASSISTED MULTIVARIATE EMPIRICAL MODE DECOMPOSITION

NA-MEMD is an updated version of the adaptive multivariate decomposition technique, i.e., MEMD. The former is designed to reduce the drawback of the latter technique. Mode mixing arising due to signal intermittency is the problem with MEMD that distorts the physical meaning of IMFs and erroneously suggests 1) the availability of different physical processes in a single IMF or 2) the presence of similar properties in two or more IMFs. The steps involved in this algorithm are as follows:

1) An m-dimensional noise signal $((S_m(t)_{t=1}^T)$ is added to the original n-variate time series $((X_n(t)_{t=1}^T)$.

2) In place of the original signal, the resultant signal $((R_{m+n}(t)_{t=1}^T)$ is utilized in NA-MEMD in further steps.

3) The same methodology of the MEMD technique, as discussed in [8] is applied to decompose the resultant signal.

978-1-6654-6060-6/23 $31.00 © 2023 IEEE

4) The n-channel IMFs related to the original signal are retrieved after decomposition, whereas the m-channel IMFs corresponding to the noise can be discarded. This is because adding separate noise channels does not overlap with the original signal.

III. THE PROPOSED WORK

A. Data Collection

Five years of data on GHI, clear-sky GHI, and four weather variables with hourly time resolution are collected from the 'National Solar Radiation Database' (NSRDB). The data is collected at two locations in India, viz—Delhi (composite climate zone) and Nainital (cold climate zone) [9]. The weather variables include solar zenith angle (SZA, degree), wind speed (WS, m/sec), temperature (T, ̊C), and relative humidity (RH, %).

B. Methodology

The present work is based on analyzing the potential of the new NA-MEMD technique with the stacked model developed in our previous research [6]. Fig.1. displays the block diagram of the present work. Firstly, the proposed decomposition technique decomposed six variate time series of solar data. For this, two noise channels are added with six-variate solar data, resulting in an eight-variate noise-added time series. Generally, adding two to three noise channels is sufficient for reducing the mode mixing. The steps discussed in Section 2 have been applied to decompose the resultant signal and extract the IMFs of the solar data. Secondly, the decomposed IMFs have been divided into 80% for training and 20% for testing. The base models, i.e., kNN, DTR, and ridge regressor, are trained using training data, and the best parameters of these models are selected using 10-fold cross-validation. Next, the meta-model, i.e., linear regressor, assigns weight to the prediction result of each base model by optimizing the loss function. Finally, the trained stacked model has been validated on the test dataset. Readers may refer to our previously published article [6] for a deep understanding of the designed stacked (ensemble) model.

The following are the main contributions of the present work:

1) This is the first application of NA-MEMD in solar resource forecasting.

2) The forecast precision of an ensemble of simple and old ML algorithms has been improved even more after replacing MEMD with NA-MEMD.

3) It is reported that NA-MEMD consumed less time to decompose the data than the MEMD algorithm.

4) The proposed NA-MEMD-stacked model (N-M-stacked) outperformed modern and complex ML techniques, i.e., LSTM, ANN, and RF, at two locations in India lying under different climate zones.

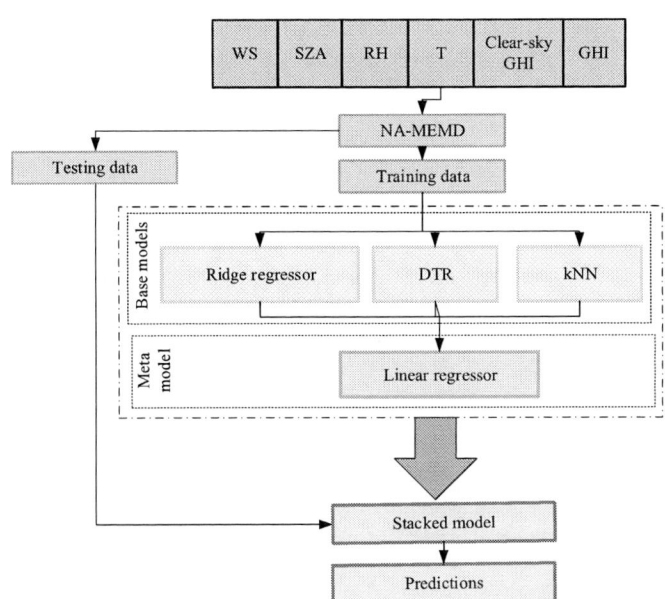

Fig.1. Methodology of the proposed work

IV. RESULTS AND DISCUSSION

A. Models' comparison based on statistical metrics

Table 1 shows that the modern algorithms (RF, ANN, and LSTM) perform better than the designed stacked model. This is because of the better ability of modern algorithms to map nonlinear features of complex time series. As discussed in Section 1, this is achieved with high training time complexity. The training time of these algorithms is as follows: around 2 sec for the stacked model, 23 sec for RF, 40 sec for ANN, and 60 sec for LSTM. The minimum training time of the designed ensemble (stacked) model (kNN-DTR-ridge) motivated us to analyze its potential with the MEMD technique, as discussed in [6]. A noticeable reduction (increment) in the RMSE and MAE (R^2 score) of the stacked model can be visualized in Table 1 after integrating it with the MEMD technique. Although, the main limitation of MEMD is its mode mixing and aligning problem [7]. However, a noise-assisted version of MEMD can overcome this limitation. This inspires us to replace MEMD with NA-MEMD and analyze the potential of the latter technique with the stacked model. After replacing MEMD with NA-MEMD, the RMSE (MAE) has reduced significantly by 23.87% (26.84%) at Delhi and 20.65% (25.64%) at Nainital (Fig.1). Further, the model's is visualized against some modern ML algorithms. A minimum %RMSE (%MAE) reduction across two locations after replacing competing models with NA-MEMD-stacked is as follows: 61.04% (55.89%), 59.76% (56.19%), 58.85% (50.77%), and 63.36% (60.34%) compared to RF, ANN, LSTM, and stacked model, respectively (Fig.1).

TABLE 1: NUMERICAL ANALYSIS OF DEVELOPED FORECASTING MODELS

Forecasting Model	RMSE (W/m²)	MAE (W/m²)	R² score
Delhi			
NA-MEMD-stacked	28.22	18.51	0.989
MEMD-stacked	37.07	25.30	0.981
Stacked model	77.06	46.67	0.918
LSTM	71.08	37.60	0.930
ANN	74.95	42.26	0.921
RF	76.70	41.97	0.918
Nainital			
NA-MEMD-stacked	34.65	22.74	0.982
MEMD-stacked	43.67	30.58	0.972
Stacked model	94.56	62.04	0.873
LSTM	84.21	49.61	0.899
ANN	86.10	53.23	0.895
RF	88.93	55.40	0.887

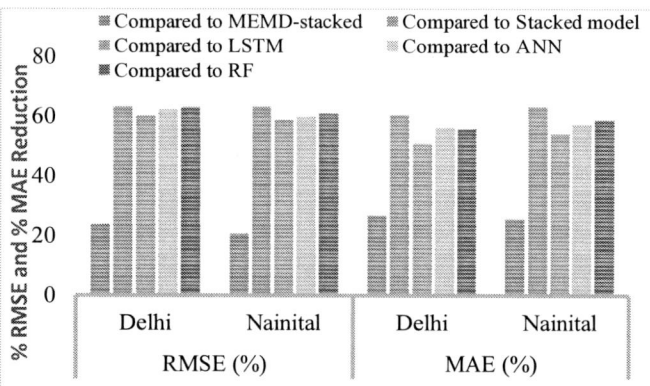

Fig.2. % RMSE and MAE reduction after replacing competing models with the proposed model

B. Models' comparison based on the graphical representation

For graphical visualization, the performance of the models is assessed using Taylor's diagram and box plot in Fig. 2 a) and b), respectively, for Delhi. Similar results are obtained for Nainital and displayed in Figure 3. Taylor diagram displays the closeness of predicted and observed patterns in terms of three indicators: 1) standard deviation (SD, dotted blue contour), 2) Pearson correlation coefficient (r, dotted black contour), and 3) centered RMSE (cyan contour). The blue point on the x-axis represents the ideal scenario, where the error is zero, and the predicted and observed patterns are identical. This is termed a reference point. It is seen from Fig. 2 a) and 3 a) that the model close to this reference point is NA-MEMD-stacked (green point). This reflects the superiority of the proposed model with high r, low RMSE, nearly equal SD to a reference point, and high similarity between the modeled and measured patterns.

Further, the box plot (Fig. 2. b and Fig. 3. b) represents the models' analysis based on absolute forecast error (|FE|). |FE| is the absolute difference between the predicted and real GHI values at time instant t. The box plot displays the lower and upper quartiles and median (orange line). The suggested model's lowest quartile cutoff indicates its suitability for hourly GHI forecasting.

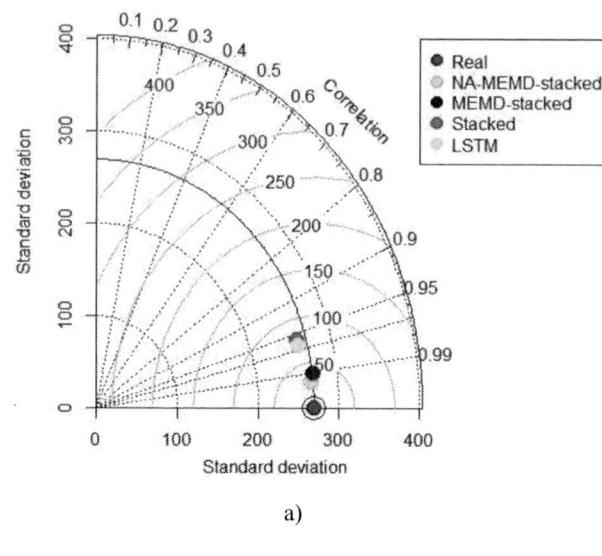

a)

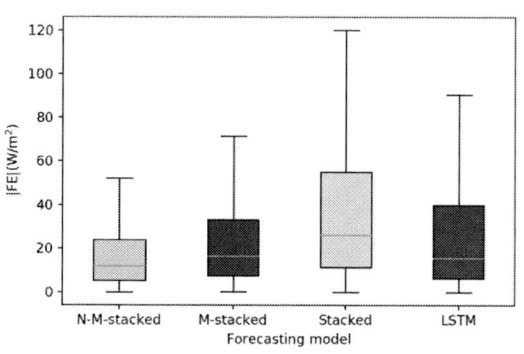

b)

Fig.3. Graphical representation of forecasting models at Delhi in terms of a) Taylor plot and b) box plot

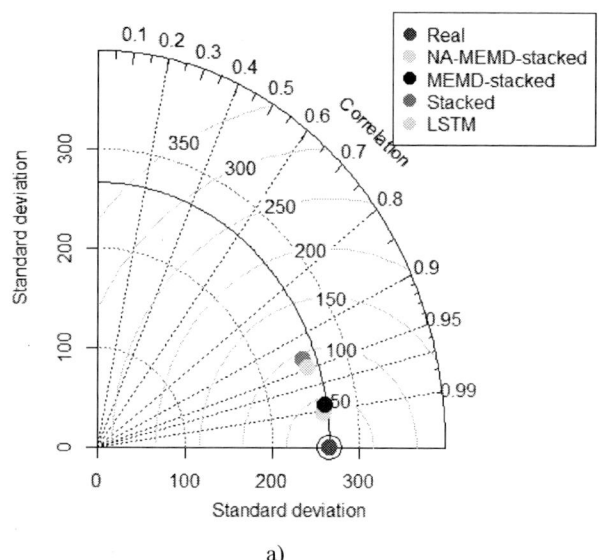

a)

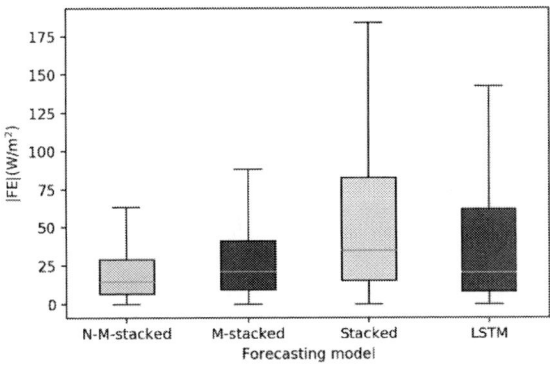

b)

Fig. 4. Performance analysis at Nainital in terms of graphical representation: a) Taylor plot and b) box plot

C. Models' comparison based on decomposition time complexity

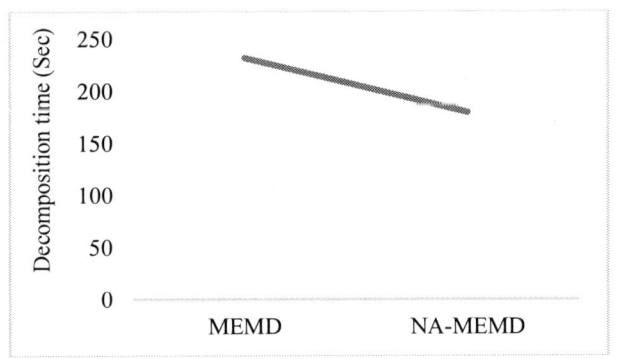

Fig.5. Decomposition time consumed by two considered multivariate decomposition techniques

Both MEMD and NA-MEMD are multivariate data demarcation techniques. NA-MEMD can overcome the limitation of MEMD by incorporating sufficient noise channels. In this work, two noise channels are added with six-variate solar energy data. Intel core i7-11700@2.50GHz processor has been used to run all the developed models. The decomposition time of two data analysis techniques, MEMD and NA-MEMD, was around 232 sec and 180 sec, respectively (Fig. 5). Interestingly, along with high-performance accuracy, NA-MEMD consumes less time for data decomposition compared to MEMD. This may be because noise channels provide a uniformly distributed reference scale, which enhances the learning capability of MEMD and minimizes the time needed for decomposition.

V. CONCLUSION

This work presents a modification to the previously published work [6]. The multivariate data demarcation technique used in the previous work has a limitation of mode mixing, which alters the physical meaning of obtained subseries. This work suggests replacing MEMD with its updated version, i.e., NA-MEMD. NA-MEMD alleviates the drawback of MEMD by adding sufficient noise channels with original multivariate data. Integrating NA-MEMD significantly enhanced the performance of a less complex ensemble of some simple ML techniques. As a result of their inability to appropriately map nonlinearity from complex time series data, these techniques are utilized less frequently today. However, combining NA-MEMD overcomes this limitation. Additionally, this work demonstrates the suitability of the proposed model over modern and complex ML techniques. Following are the main findings of the present work:

1) MEMD significantly reduced the error of the stacked model, and replacing MEMD with NA-MEMD has reduced forecasting error even more. The corresponding reduced %RMSE values after replacing MEMD-stacked with NA-MEMD-stacked are 23.87% at Delhi and 20.65% at Nainital. Overall, incorporating NA-MEMD with a stacked model has reduced the RMSE (MAE) of the stacked model by a minimum of 63.36% (60.34%) across studied locations.

2) The analysis showed the proposed model's accuracy better than modern and complex ML models for 1-h ahead GHI forecasting. The minimum percentage RMSE reduction across two locations after replacing RF, ANN, and LSTM with the proposed model is as follows: 61.04%, 59.76%, and 58.85%, respectively.

3) The Taylor diagram and box plot validate the superiority of the proposed model based on three indicators and quartiles cutoff, respectively.

4) Interestingly, NA-MEMD enhanced (reduced) the performance accuracy (error) and consumed less time for data decomposition. MEMD took around 232 sec, whereas NA-MEMD consumed almost 180 sec in decomposition.

This work suggests the use of the proposed model in several other fields, such as forecasting air pollution, PV power, wind speed, stock prices, fault diagnosis, etc. Besides this, incorporating data analysis techniques with the proposed simple ML techniques reported remarkable improvement in performance. These data analysis techniques consume significant time for decomposing the data, which can not be ignored. Still, after considering time complexity and performance accuracy jointly, it can be said that combining data demarcation techniques may open a path for simple and less complex ML techniques, which are less commonly used nowadays. Moreover, NA-MEMD is more suitable than MEMD as a multivariate data analysis technique.

ACKNOWLEDGMENT

This work was made possible with the fellowship support provided by the Ministry of Education, Government of India. The first author is also grateful to the Science and Engineering Research Board (SERB), Government of India, and Dean of Resources and Alumni Affairs (DORA), IIT Roorkee for providing the travel grant to present this paper.

REFERENCES

[1] P. Gupta and R. Singh, "PV power forecasting based on data-driven models: a review," *International Journal of Sustainable Engineering*, pp. 1–23, Oct. 2021, doi: 10.1080/19397038.2021.1986590.

[2] C. Voyant *et al.*, "Machine learning methods for solar radiation forecasting: A review," *Renew Energy*, vol. 105, pp. 569–582, 2017, doi: 10.1016/j.renene.2016.12.095.

[3] E. Dokur, N. Erdogan, M. E. Salari, C. Karakuzu, and J. Murphy, "Offshore wind speed short-term forecasting based on a hybrid method: Swarm decomposition and meta-extreme learning machine," *Energy*, vol. 248, Jun. 2022, doi: 10.1016/j.energy.2022.123595.

[4] P. Gupta and R. Singh, "Univariate model for hour ahead multi-step solar irradiance forecasting," pp. 494–501, 2021, Accessed: Aug. 25, 2022. [Online]. Available: 10.1109/PVSC43889.2021.9519002

[5] N. Rehman and D. P. Mandic, "Multivariate empirical mode decomposition," *Proceedings of the Royal Society A: Mathematical, Physical and Engineering Sciences*, vol. 466, no. 2117, pp. 1291–1302, 2010, doi: 10.1098/rspa.2009.0502.

[6] P. Gupta and R. Singh, "Combining simple and less time complex ML models with multivariate empirical mode decomposition to obtain accurate GHI forecast," *Energy*, vol. 263, p. 125844, Jan. 2023, doi: 10.1016/j.energy.2022.125844.

[7] N. Ur Rehman and D. P. Mandic, "Filter bank property of multivariate empirical mode decomposition," *IEEE Transactions on Signal Processing*, vol. 59, no. 5, pp. 2421–2426, May 2011, doi: 10.1109/TSP.2011.2106779.

[8] P. Gupta and R. Singh, "Combining a deep learning model with multivariate empirical mode decomposition for hourly global horizontal irradiance forecasting," *Renew Energy*, vol. 206, pp. 908–927, Apr. 2023, doi: 10.1016/j.renene.2023.02.052.

[9] Bureau of Indian Standards, *NATIONAL BUILDING CODE OF INDIA 2005*. New Delhi, 2005.

Can Grid-Following DERs Operate in Parallel with Grid-Forming Resources without Compromising Microgrid Stability?

Wenzong Wang and Aminul Huque

Electric Power Research Institute, Knoxville, TN, USA

Abstract — **In feeder-level multi-customer microgrid design, there can be cases where the capacity of existing distributed energy resources (DERs), e.g., solar photovoltaic (PV) plants can potentially be leveraged and hence reduce the need for new generation capacity. A question to be addressed under this scenario is how much grid-following DER capacity can operate in parallel with grid-forming plant(s) without compromising stability of the microgrid. To shed light on this, this paper investigates how the size, location and control of grid-following inverter-based plants affect the stability of the islanded microgrid, and subsequently evaluates the maximum ratio between grid-following and grid-forming capacity that can be allowed in an example microgrid circuit under different conditions.**

Keywords — **Grid-following, grid-forming, islanding detection, microgrid, stability.**

I. Introduction

Intentional islanding or microgrid operation is an effective way to improve resilience of a section of power distribution grid. Among the different scales of microgrids, feeder-level multi-customer microgrid which involves utility medium voltage (MV) feeder and loads/generations at different locations, is of particular interest to utility operators. A number of pilot feeder-level microgrids are already set up and running at different locations across the US and around the world [1].

With the increasing penetration of inverter-based distributed energy resources (DERs), such as solar photovoltaic (PV) and battery energy storage system (BESS), there is a growing interest to design fully inverter-based microgrids without the need for desiel generators. However, present-day inverter control technologies are mainly designed to operate in the presence of (or parallel with) a stiff power system, which are often termed as "grid-following (GFL) inverter control". Such inverter control approaches face challenges to provide grid services and maintain stability in inverter-dominated systems, such as in a fully inverter-based microgrid [2].

To deal with the challenges, improvement of the present-day control design and various new forms of inverter controls have been proposed [3]. These controls are termed as "grid forming (GFM) inverter control" due to their capability to form grid voltage and operate without relying on synchronous machines [4]. In today's inverter-based microgrid design, the capacity of the GFM DERs (typically BESS) is often selected large enough to supply the entire microgrid loads during islanded operation. Moreover, the GFL DERs (e.g., behind-the-meter or utility-scale PV) inside the microgrid footprint may be required to be offline during the microgrid islanded operation due to the concern that they may jepodize the microgrid stability, especially when the aggregated capacity of the GFL DERs are similar to that of the GFM DERs.

However, there is a clear benefit of operating the GFL PV plants in parallel with the GFM BESS plant(s), as they can help supply the load and hence may reduce the BESS capacity needed. Moreover, the BESS may be charged during times with excess PV generation, such that the microgrid can operate longer, thereby further increasing grid resilience.

Therefore, it is important to understand and identify the maximum capacity of GFL DERs that can operate in parallel with GFM DERs without causing instability of the microgrid. Towards this goal, this paper investigates the key factors affecting the capacity of the GFL DERs that can be allowed online. In particular, leveraging the generic GFL and GFM inverter models developed by EPRI [4], [5], how the location, grid support functions, and active islanding detection (AID) of GFL inverter-based DERs affect the microgrid stability is investigated through electromagnetic transient (EMT) simulation on a real microgrid circuit.

It is important to notice that stability depends heavily on inverter control and load dynamics, among other factors. Since generic GFM and GFL inverter models are used in the study and the loads are modeled as constant impedance loads, the exact numbers (e.g., inverter settings, plant size) at which the microgrid becomes unstable should not be generalized. Instead, the impacting factors and the trend of the impact should be focused on.

II. Microgrid and Inverter Modeling

A. Microgrid Modeling

A real microgrid circuit is considered in this study which is part of a utility feeder. When there is a power outage at the substation (say for example due to extreme weather), part of the MV feeder can be isolated and operates as a microgrid, powered by a single BESS plant as the only GFM resource. The peak load of this microgrid is around 3000 kW with an average power factor of 0.88 (absorbing reactive power). The microgrid circuit is modeled in PSCAD for EMT simulation studies, where the loads are modeled as static constant impedance load. The topology of the microgrid area is illustrated in Fig. 1, where the red boxes denote the reclosers/breakers in the system. Besides the GFM plant with a rating of 4 MVA, one GFL plant

978-1-6654-6060-6/23 $31.00 © 2023 IEEE

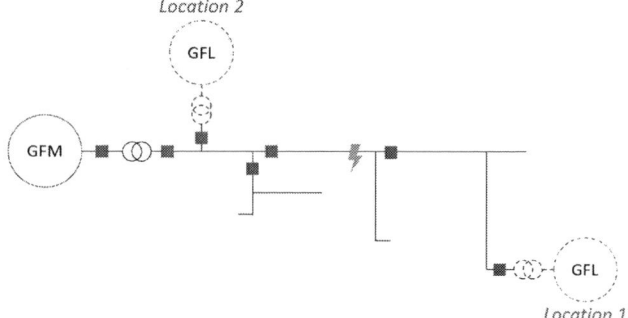

Fig. 1. Illustration of the microgrid topology and plant locations.

is connected to the feeder at one of the two locations. Location 1 is at the end of the feeder where location 2 is close to the GFM plant. Note that the GFL plant is fictitous as the microgrid under consideration does not have significant amount of GFL DERs inside.

B. GFM Inverter Modeling

The generic three-phase GFM inverter EMT model developed by EPRI is utilized which has both positive sequence and negative sequence control [4], [6]. As illustrated in Fig. 2, four types of positive sequence control methods are implemented in the model: phase locked loop (SRF-PLL) based GFM control, droop based GFM control, virtual synchronous machine (VSM) based GFM control, and dispatchable virtual oscillator (dVOC) based GFM control. Similarly, two different control methods are implemented for negative sequence outer loop control (V_{neg} elimination and Kfactor). The V_{neg} elimination control utilizes a proportional integral (PI) controller and eliminates the negative sequence voltage when the current output is within the limit, whereas Kfactor control emulates a constant negative sequence impedance. For the simulation studies in this paper, droop based control and V_{neg} elimination are selected for positive and negative sequence, respectively. The GFM control is generic in the sense that it does not represent a particular product, but the generic behavior expected from GFM inverters.

C. GFL Inverter Modeling

In this study, the EPRI developed generic white-box EMT model for a three-phase solar PV inverter is leveraged [5]. The model is developed to closely represent an inverter that complies with IEEE Std 1547™-2018, including grid support functions, unintentional islanding detection, and protection functions defined in the standard.

III. SIMULATION CASE STUDIES AND RESULTS

A. Description of the Case Studies

To identify the key factors that affect the amount of GFL capacity that can be online without compromising microgrid stability, different case studies are constructed, which are summarized in Table 1. Case a1 is the base case where neither

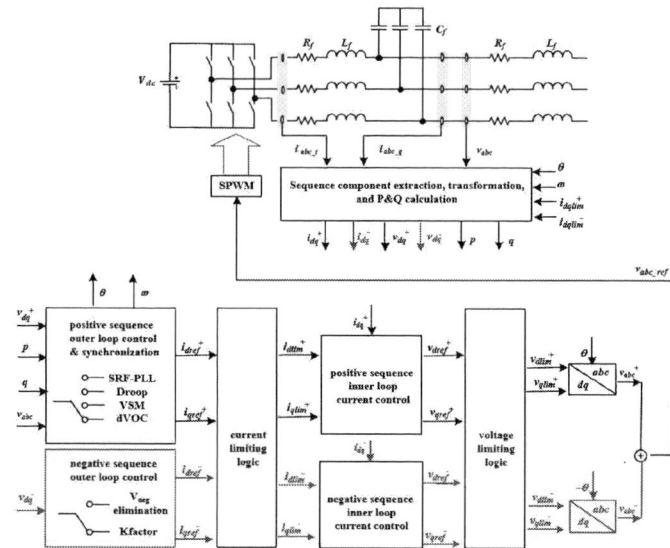

Fig. 2. Block diagram of the GFM inverter model.

AID nor grid support functions are enabled. Cases a2 through a5 then analyze the impact of volt-var and/or freq-watt functions with different settings. Cases b1 through b5 repeat the previous case studies but with AID enabled. The active AID method considered is a group 2c method which injects/absorbs reactive power based on the frequency deviation [7]. Finally, case c1 is constructed to show the impact of GFL plant location.

For each case study, the rating of the GFM plant is fixed at 4 MVA and the GFL plant size is gradually increased (with a step size of 1 MW) until instability is observed. The maximum GFL plant rating before reaching instability over the GFM plant rating is the maximum GFL/GFM ratio shown in Table 1.

B. Base Case with both Grid Support Function and AID Disabled

As a base case study, the GFL plant has its islanding detection disabled. Moreover, the plant is working at unity power factor with an active power limit of 0.8 pu, without any grid support functions enabled. Note that the maximum power point (MPP) for the GFL plant is at 1pu. However, the output power is limited to 0.8 pu such that it has headroom for under frequency response when freq-watt function is enabled in later case studies.

Simulation results when the GFL plant is rated at 6 MW is shown in Fig. 3. The active power output, reactive power output, voltage, and frequency of the GFM plant are shown in (a) whereas those for the GFL plant are shown in (b).

Events in the system are labeled with the circled numbers and they correspond to:

(1) $t = 0s$ GFM inverter black start
(2) $t = 0.5s$ All switches closed, transformer and load energization starts
(3) $t = 1.45s$ Voltage and frequency at the GFL plant meet enter service criteria, GFL plant active power generation starts ramping up

978-1-6654-6060-6/23 $31.00 © 2023 IEEE

TABLE I

SUMMARY OF THE CASE STUDIES

Case study #	AID	Grid Support Functions		GFL plant location	Maximum GFL/GFM ratio
		Volt-Var	Freq-Watt		
a1	No	No	No	1	1.5
a2	No	Yes, 1547-2018 Category B default settings	Yes, 1547-2018 Category III default settings	1	1.5
a3	No	Yes, aggressive settings	No	1	1.75
a4	No	No	Yes, aggressive settings	1	1.75
a5	No	Yes, aggressive settings	Yes, aggressive settings	1	0.75
b1	Yes	No	No	1	0.75
b2	Yes	Yes, 1547-2018 Category B default settings	Yes, 1547-2018 Category III default settings	1	0.75
b3	Yes	Yes, aggressive settings	No	1	1.25
b4	Yes	No	Yes, aggressive settings	1	1.25
b5	Yes	Yes, aggressive settings	Yes, aggressive settings	1	0.5
c1	Yes	No	No	2	1.25

(4) $t = 3s$ Three-phase fault occurs

(5) $t = 3.5s$ Three-phase fault clears

As can be seen, when the GFL plant has a rating of 6 MW (1.5 times the GFM plant rating), it can ride-through the fault and the system is stable at the steady states before and after the fault as well as during the fault.

However, when the rating of the GFL plant is increased to 7 MW, as shown in Fig. 4, even though the microgrid is stable before the fault, the GFL control becomes unstable during the fault, as indicated by the large frequency deviation from the GFM plant frequency. Due to the instability, the GFL plant is tripped by over frequency protection right after the fault clearance. The large transients due to instability during the fault may trigger system protection and cause (partial) black out of the microgrid. Thefore, the maximum GFL/GFM ratio in this case is identified as 1.5.

C. Impact of AID of GFL Plant

When active islanding detection methods are enabled at the GFL plant, it is expected that it would negatively impact the microgrid stability. To evaluate this, case b1 is constructed to compare with case a1. The results of case b1 is shown in Fig. 5 and Fig. 6, where the AID is activated right after the GFL plant starts to ramp up its active power following the enter service delay. As can be seen, when the size of the GFL plant is 3 MW, there are some oscillations right after the AID is activated but the damping is sufficiently high. However, when the GFL plant size is increased to 4 MW, the AID starts causing sustained oscillations after it is activated.

Decreasing the gain of the islanding detection algorithm can improve the damping and may resolve the oscillation if the gain is sufficiently low. Since disabling AID is an extreme case of that, and the impact has been analyzed in case a1, the effect of reducing the islanding detection gain is not further demonstrated here.

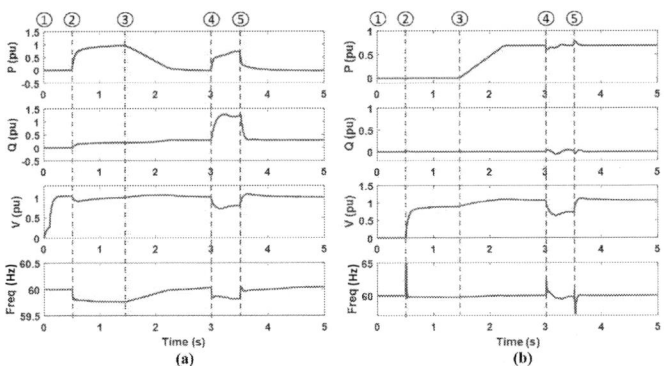

Fig. 3. Response of the GFM plant (a) and 6MW GFL plant (b) in a1.

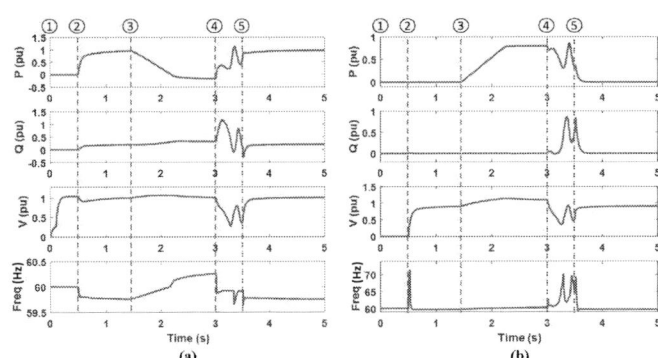

Fig. 4. Response of the GFM plant (a) and 7MW GFL plant (b) in a1.

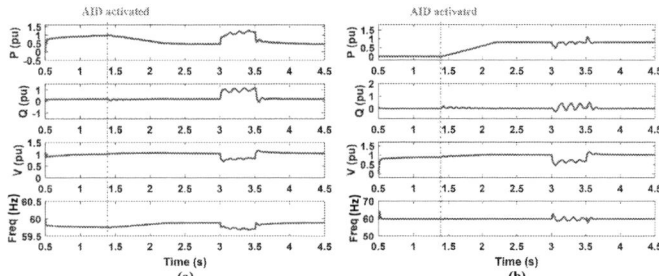

Fig. 5. Response of the GFM plant (a) and 3MW GFL plant (b) in b1.

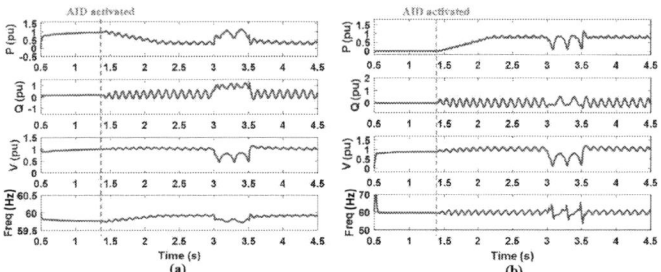

Fig. 6. Response of the GFM plant (a) and 4MW GFL plant (b) in b1.

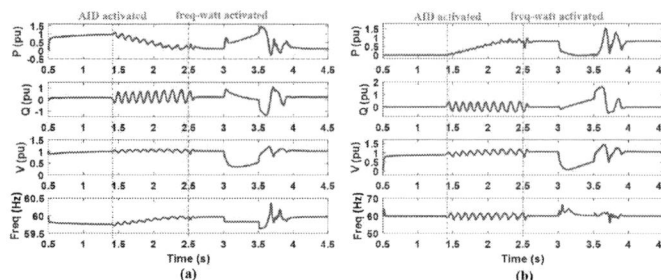

Fig. 7. Response of the GFM plant (a) and 5 MW GFL plant (b) in b4.

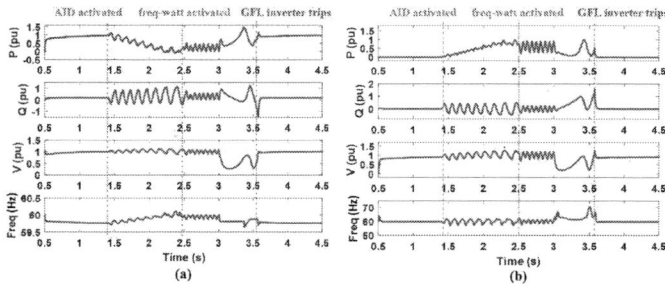

Fig. 8. Response of the GFM plant (a) and 6MW GFL plant (b) in b4.

Because of the oscillations that can be caused by AID, the maximum GFL/GFM ratio that can be hosted in the microgrid without losing stability is significantly lower than in the case without AID, as can be seen comparing results from case b1 and case a1.

D. Impact of Grid Support Functions

Since AID may induce oscillation in the microgrid, the next question to address is whether grid support functions can help resolve the oscillation or make it worse. To shed light on that, in case b2, the scenario where both volt-var and frequency-watt functions are activated with the IEEE Std 1547™-2018 settings is investigated. Results indicate that volt-var and frequency-watt functions with the default settings do not affect the system response in any significant way. The default 5s open loop response time makes the grid support functions too slow to affect the ~10 Hz oscillations caused by AID.

Moreover, even when the open loop response times are set to the lowest values defined in IEEE Std 1547™-2018 (1s for volt-var and 0.2s for frequency-watt), the impact of grid support functions is still negligible. Therefore, the open loop response times for both functions are further reduced to 20 ms to evaluate the impact, while using the steepest droop slope settings allowed in IEEE Std 1547™-2018 (case b3, b4 and b5).

Results from case b4 with aggressive frequency-watt functions are shown in Fig. 7 and Fig. 8. Comparing with case b1, aggressive freq-watt function with 20 ms open loop response time can stabilize the oscillation caused by AID right after it is activated at 2.5s, when the GFL plant rating is 5 MW. With a larger GFL plant (6 MW), higher frequency (~16 Hz) oscillation occurs after the freq-watt control is activated. Moreover, in the case with 6 MW GFL plant, the GFL inverter becomes large signal unstable during the fault and is tripped due to over frequency right after the clearance of the fault.

Similar results were obtained in case b3 with aggressive volt-var function alone. The volt-var function stabilizes the oscillations caused by AID when the GFL plant rating is 5 MW but induced higher frequency oscilllations when the rating is increased to 6 MW.

In case b5, activating volt-var and freq-watt functions together and both with aggressive settings leads to unstable oscillations when the GFL plant rating is 3 MW, as shown in Fig. 9. The GFL/GFM capacity ratio is therefore limited to 0.5, which is even lower than that in case b1 (0.75). This case study suggests that even though aggressive volt-var and freq-watt functions can help improve microgrid stability to some extent and allow more GFL capacity online when activated individually, having both activated at the same time with aggressive settings may cause oscillations and affect system stability adversely and therefore should be carefully studied.

E. Impact of GFL Plant Location

As a GFM control provides stabilizing effects to the microgrid operation, it is expected that the shorter the electric distance between the GFM and the GFL plant, the easier the GFL plant dynamics can be stabilized. Notice that for all the previous case studies, the GFL plant is connected at location 2,

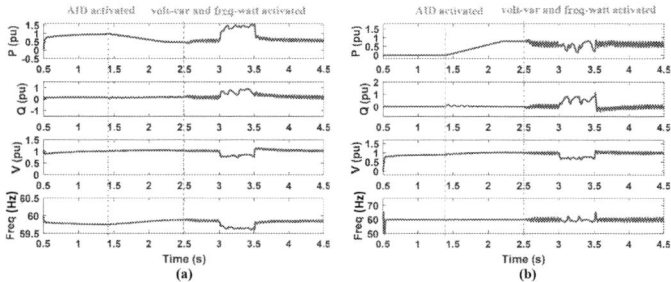

Fig. 9. Response of the GFM plant (a) and 3MW GFL plant (b) in b5.

978-1-6654-6060-6/23 $31.00 © 2023 IEEE

which has the longest electric distance from the GFM plant inside the microgrid.

To investigate the impact of GFL plant location on system stability, in case c1, the GFL plant is connected at location 1 which is closer to the GFM plant. The GFL inverter has its AID enabled while the grid support functions are disabled, which is the same as in case b1.

The results from this case study are shown in Fig. 10 and Fig. 11. As can be seen, when the GFL plant has shorter electric distance from the GFM plant, the GFL plant size can go up to 1.25 times the size of the GFM plant. In comparison, in case b1 when the GFL plant is farther away, the ratio is 0.75. This demonstrates that a microgrid can have more GFL capacity online with shorter electric distance between the GFL and the GFM plant(s). One thing to note is that as the GFL plant is closer to the GFM plant, during black start, the voltage at the GFL plant enters the enter service region faster and this explains why the GFL plant starts ramping up its power before 1s, instead of before 1.5s as in the previous cases.

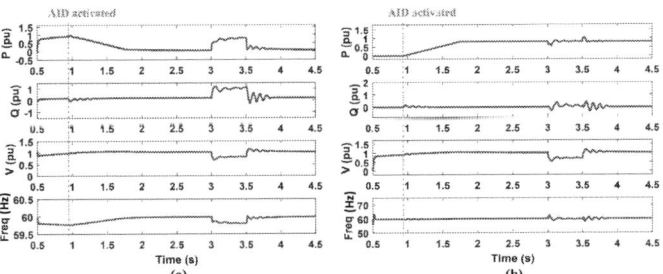

Fig. 10. Response of the GFM plant (a) and 5MW GFL plant (b) in c1.

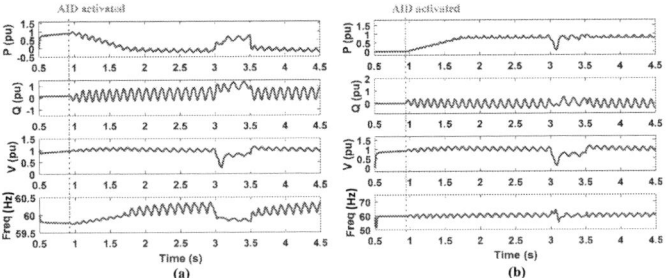

Fig. 11. Response of the GFM plant (a) and 6MW GFL plant (b) in c1.

IV. CONCLUSION

Based on simulation case studies on one real microgrid circuit, this paper reveals that a microgrid is less stable with more GFL capacity online in parallel with the GFM resources. The instability can manifest in voltage and frequency oscillations within the normal operating range, or it may cause large voltage/frequency excursions and cause GFL or GFM plants to trip following a large disturbance (e.g., a short circuit fault). AID of the GFL plants tends to negatively affect the microgrid stability by introducing unstable voltage and frequency oscillations, and causing large voltage/frequency transients that can trip the GFL plants following a fault condition. Moreover, with IEEE Std 1547™-2018 default settings, volt-var and freq-watt functions from the GFL plants has minimum impact on microgrid stability. When the volt-var and freq-watt functions have faster open loop response time than that specified in IEEE Std 1547™-2018, they may help stabilize the oscillations caused by GFL plant AID and improve the stability of the GFL plants during a fault. However, caution should be exercised as the aggressive volt-var and freq-watt functions may also introduce unstable oscillations, especially if the two functions are activated simultaneously. In addition, it is found that the shorter the electric distance between the GFL and the GFM plants, the more GFL capacity can be online without causing instability.

REFERENCES

[1] *Grid Forming Inverters: EPRI Tutorial (2022)*. EPRI, Palo Alto, CA: 2022. 3002025483.

[2] W. Wang, G. M. Huang, D. Ramasubramanian and E. Farantatos, "Transient stability analysis and stability margin evaluation of phase-locked loop synchronised converter-based generators," IET Generation, Transmission & Distribution, vol. 14, no. 22, pp. 5000-5010, Nov. 2020.

[3] P. Unruh, M. Nuschke, P. Strauß, and F. Welck, "Overview on Grid-Forming Inverter Control Methods," Energies, 2020, 13, 2589.

[4] *Performance Requirements for Grid Forming Inverter Based Power Plant in Microgrid Applications: Second Edition*. EPRI, Palo Alto, CA: 2022. 3002024431.

[5] *White-box Inverter EMT Model in PSCAD with IEEE 1547-2018 Functions*. EPRI, Palo Alto, CA: 2022. 3002025896.

[6] *Grid Forming Inverter Models*, PSCAD Knwledge Base, https://www.pscad.com/knowledge-base/article/894.

[7] *Taxonomy for Inverter Island Detection Methods*. EPRI, Palo Alto, CA: 2021. 3002022455.

Developing Frequency Stability Constraint for Unit Commitment Problem Considering High Penetration of Renewables

Ningchao Gao[1,3], Shuan Dong[1], Xin Fang[2], Andy Hoke[1], David Wenzhong Gao[3], and Jin Tan[1]

1. *National Renewable Energy Laboratory*, Golden, CO, 80401, US

2. *Department of Electrical and Computer Engineering, Mississippi State University*, Starkville, MS, 39762, US

3. *Department of Electrical and Computer Engineering, University of Denver*, Denver, CO, 80210, US

Abstract—As zero-carbon electricity systems become the trend of future grid, the system inertia provided by conventional synchronous generators (SGs) keeps decreasing. The resultant lower system inertia will inevitably cause frequency stability problem, especially in the first few seconds following disturbance. To tackle this challenge, this paper proposes a frequency stability constraint for power systems unit commitment problem by considering the fast frequency responses (FFRs) from inverter-based resources (IBRs). Our developed frequency stability constraint is grounded on an analytical frequency nadir estimation framework that considers both SG and IBR dynamics. The accuracy of our frequency nadir estimation framework is validated by most severe N-1 contingency simulation result in a real island system. Then, the adaptive inertia frequency stability constraint is derived by performing sensitivity analysis with our frequency nadir estimation framework. Finally, we demonstrate the effectiveness of our developed frequency stability constraint with one year day-ahead unit commitment results of the island system.

Index Terms—Fast frequency response, frequency nadir, inverter-based resources, stability constraint, unit commitment.

I. INTRODUCTION

THE increasing penetration of Inverter-Based Resources (IBRs) in power grids, particularly in islanded systems, is necessitating new considerations in unit commitment (UC) problems. The traditional role of synchronous generators (SGs) in providing inherent inertia to resist frequency changes is being disrupted by the influx of IBRs, which have different dynamical response characteristics compared with SGs. However, recent advancements in control strategies and technological solutions have opened up possibilities for IBRs to contribute positively to frequency stability.

With these advancements, understanding how to best incorporate frequency stability constraints into UC while considering IBRs' FFR becomes a complex and challenging problem. It requires the development of advanced mathematical models and optimization algorithms that can handle the complex dynamics and uncertainties associated with these resources, while ensuring the cost-effectiveness and reliability of power system operations [1]–[3]. With the trend towards renewables, how to consider IBRs' contributions in power system scheduling problems like UC is indeed becoming a paramount topic in power systems research and practice [4]–[8].

The authors in [9], [10] employed differential algebraic equations (DAEs) to capture the dynamic frequency response of the system, providing a more accurate and realistic representation of system dynamics than traditional steady-state optimal power flow models. Since only a single type of SG turbine governor model is considered, this limits the general applicability of the model. This is because the real-world power systems typically include various types of turbine governor models, each with unique dynamic characteristics.

Thus, this paper proposes an adaptive inertia frequency stability constraint for the UC problems, which considers the largest contingency in power systems analytically. Our key contributions are as follows:

1) Validate the accuracy of the dynamic frequency nadir prediction framework in predicting the island system's frequency nadir following disturbances.
2) Develop an adaptive inertia frequency stability constraint for UC problem.
3) Include our developed stability constraint to the NREL-developed framework Multi-Timescale Integrated Dynamic and Scheduling (MIDAS) and validate its effectiveness on the island system [11].

The rest of this paper is organized as follows: Section II introduces the frequency nadir estimation framework, which predicts the largest time constant of different governor and IBR models, and computes the aggregated parameters of a real island system. Section III proposes adaptive inertia frequency stability constraint. Section IV performs the case study to demonstrate the effectiveness of the added frequency stability constraint in improving frequency dynamics. Section V concludes the paper.

II. FREQUENCY NADIR ESTIMATION

The authors in [12] propose a frequency nadir estimation framework which is summarized in (1)-(3) below:

$$f_{nadir} = f_n + \frac{P_{sus} - P_{genmax}}{D_\Sigma + R_g^{-1}} - \frac{T_g R_g^{-\frac{1}{2}} M e^{\alpha - \phi - \pi} \cot\phi}{(t_2 - t_1)(D_\Sigma + R_g^{-1})^{\frac{3}{2}}},$$
$$(1)$$

$$\zeta = \frac{1}{2}\left(\frac{D_\Sigma}{2H_\Sigma} + \frac{1}{T_g}\right)\sqrt{\frac{2T_g H_\Sigma}{D_\Sigma + R_g^{-1}}} = \cos\phi, \qquad (2)$$

978-1-6654-6060-6/23 $31.00 © 2023 IEEE

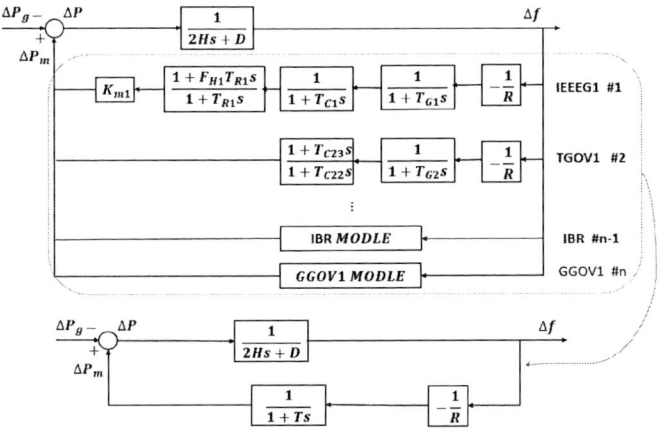

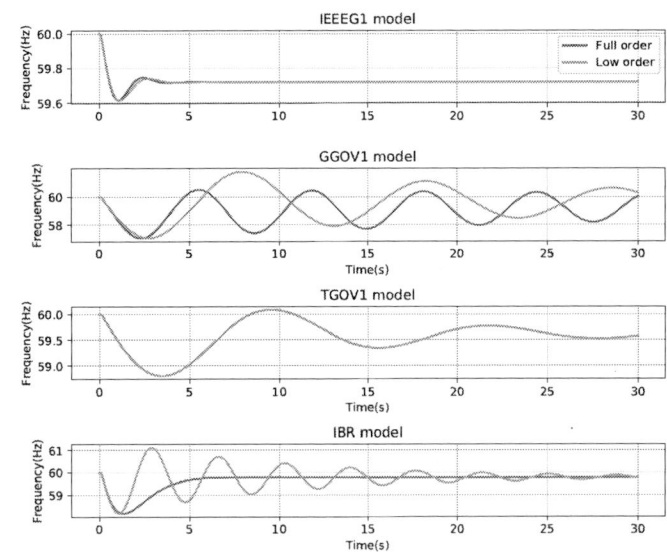

Fig. 1. Low- and Full-order SFR models of Governor and IBR

$$m(t) = P_{sus}e^{\zeta\omega_n t_2}\sin(\omega_d(t-t_2)-\beta-\phi) - P_{sus}e^{\zeta\omega_n t_1}$$
$$\sin(\omega_d(t-t_1)-\beta-\phi) - \Delta P_{load}\omega_n(t_2-t_1)\sin(\omega_d t-\beta),$$
$$(3)$$

where f_{nadir} is the grid frequency nadir in most severe N-1 contigency, f_n is the rated grid frequency, P_{sus} is the IBR step response, P_{genmax} is the ative-power output of the largest generation unit, D_Σ is the aggregated damping constant, R_g is the aggregated droop constant, T_g represent the grid's aggregated time constant, and H_Σ is the aggregated system inertia constant [12].

A. Low-Order Approximated SG Governor and IBR Models

In the island system model, there are three types of SG governors, i.e., IEEEG1, TGOV1, and GGOV1. Among them, TGOV1 and GGOV1 have particularly complex control models, and this precludes us from analyzing them. In [13] and [14], a low-order system frequency response (SFR) model has been proposed, which neglects nonlinearities and all but the most significant time constants. This low-order SFR model provides a simple but accurate method to estimate complex generator models [15]. This paper uses the low-order SFR model to predict the most prominent time constant of SG governors and the IBR model. As shown in Fig. 1, we build both low- and full-order SFR models of governor and IBR in PSCAD and adjust the most prominent time constant in low order SFR model. In so doing, we expect the frequency response of low-order model to approximate that of the full-order model. From Fig. 2, we can get the low-order model of IEEEG1, and TGOV1 can precisely predict the full-order model dynamics. We note that the low-order model of GGOV1 cannot track the full-order model's steady-state frequency perfectly but can predict the frequency nadir with high precision.

B. Frequency Nadir Estimation

Here, we validate the accuracy of the frequency nadir prediction method with the island system model. To achieve this, we use one-day real-time economic dispatch (RTED) results with 288 snapshots, trip the largest generator, and

Fig. 2. Comparison between full- and low-order system frequency response.

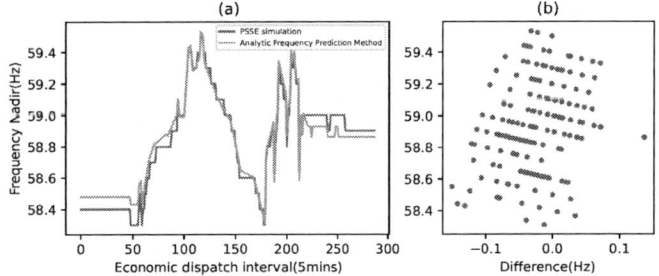

Fig. 3. Comparison between simulated and predicted frequency nadirs following most severe N-1 congitency in island power systems.

compare the frequency nadirs between PSS/E dynamic simulation results and our analytic frequency nadir predictions. Figure 3(a) shows that our analytical prediction of frequency nadir (orange trace) aligns well with PSS/E simulation results (blue trace). Also, based on Fig. 3(b), the frequency nadir difference between PSS/E simulation and analytic prediction method is very small, and 90 percent of these differences are limited within ± 0.1 Hz.

III. FREQUENCY STABILITY CONSTRAINT

This section leverages the sensitivity analysis method to analyze all the aggregated system frequency response (ASFR) model parameters for the island system. We note that two ASFR model parameters, i.e., the largest generation unit loss $P_{genloss}$ and system inertia H_Σ, have a high correlation with frequency nadir.

In Fig. 4, the blue points reflect the relationship between $P_{genloss}$ (P_{genmax}) and system inertia H_Σ when the post-disturbance frequency nadir is 59 Hz. By visually checking those blue points, we can find that while increasing the largest output of the generator, a larger system inertia is required to secure the frequency nadir above 59 Hz. So, we select the upper bound of these blue data points (red line) as the

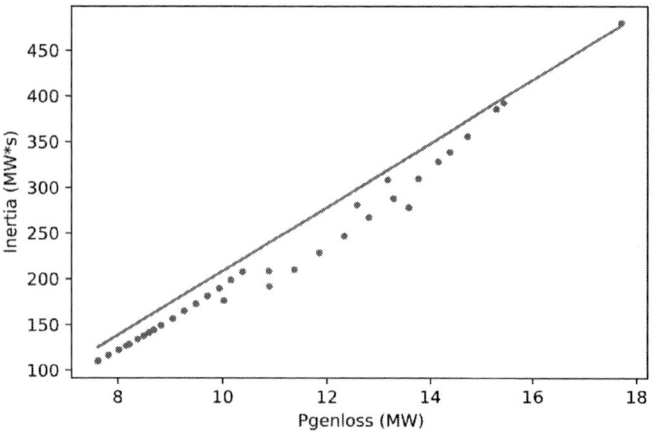

Fig. 4. Approximate linear frequency stability constraint obtained by taking the upper bound of original data points computed from (1)–(3).

frequency stability constraint, which is also expressed in (4) below.

$$H_\Sigma \le k \cdot P_{genmax} + b, \qquad (4)$$

where k and b is obtained from the sensitivity analysis of the frequency nadir estimation framework. For the case in Fig. 4, k and b, respectively, take 34.8 and -140. Recall that our frequency stability constraint is developed based on the points when the frequency nadir is precisely 59 Hz. Thus, by including our developed frequency constraint (red trace), we expect that the post-disturbance frequency nadir will not be lower than 59 Hz.

With our developed linear frequency nadir stability constraint (4) in place, we can include it into the UC model in MIDAS framework developed by NREL [11].

IV. CASE STUDIES

In this section, we conduct two case studies to evaluate the effectiveness of our developed frequency stability constraints: the base case without our constraint and the Freq. Const. case with our constraint. In both cases, the SG capacity is 167 MW, and the governor model includes IEEEG1, TGOV1, and GGOV1. The total renewable capacity is 375 MW, and the penetration level of renewable is 70% . Figure 5 shows the PV input, wind input, and load time series curve in our two designed case studies. We note that for illustrative purpose, Fig. 5 only plot the first 336 hours (two weeks) of time series data to show the detail.

The proposed adaptive inertia frequency stability constraint unit commitment is considered in day-ahead unit commitment (DAUC) problem. After solving one-year DAUC, we get 8760 generation scheduling results, in which the largest output of generator and battery for base case and Freq. constr. case and are shown in Fig 6. Based on Fig. 6, it is evident that the largest generator output in the base case is larger than that in the Freq. constr. case. This is because based on our developed frequency stability constraint in (4), the system inertia restricts the generator largest output P_{genmax}.

We plot the N-1 contingency PSS/E simulation result in Fig. 7. Based on Fig. 7(a), 99% of the frequency nadirs in

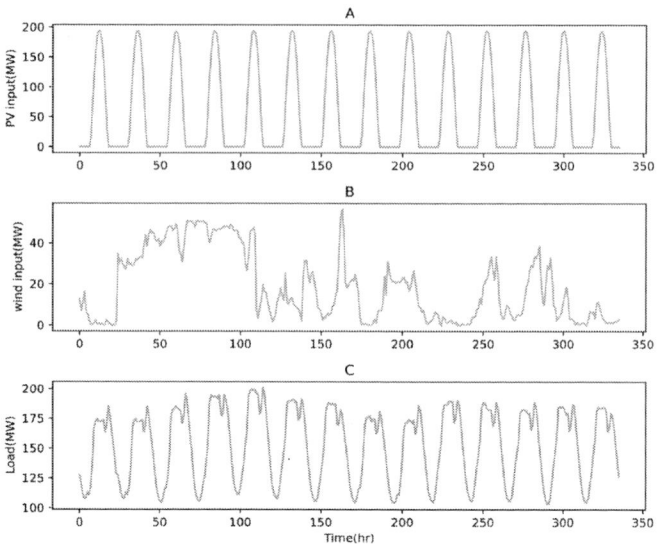

Fig. 5. Time-series data of available PV sources, wind source, and Load demand within two weeks.

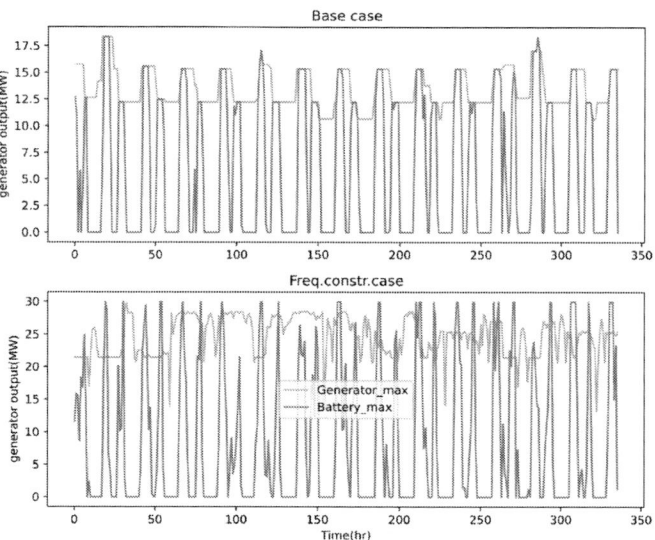

Fig. 6. Largest outputs of generators and batteries in base case and Freq. Const. case.

the base case (blue trace) are below 59 Hz in the period of 8760 hours. While in the freq. constr. case, 98% of frequency nadirs remain above 59 Hz. Specially, figure 7(b) depicts the frequency nadirs within first 100 hours. Similarly, we can find that 97% of the frequency nadirs in the base case are below 59 Hz, but all the frequency nadirs in the Freq. Constr. case are above 59 Hz. In addition, as shown in Table I, the Freq. constr. case's total generation cost increases by 20% compared with the base case. This is because our frequency constraint limits the largest output from the conventional generators and renewables.

V. CONCLUSION

This paper proposes an Adaptive Inertia Frequency Stability Constraint for UC problems. Our developed constraint aims

978-1-6654-6060-6/23 $31.00 © 2023 IEEE

TABLE I
GENERATION COST COMPARISON OF BASE CASE AND FREQUENCY CONSTRAINT CASE FOR ONE YEAR DAY-AHEAD UNIT COMMITMENT (DAUC)

Case	Generation Cost ($)
Base Case	135,839,782
Freq. Constr. Case	168,386,186

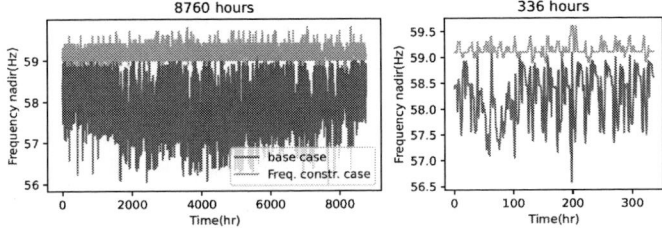

Fig. 7. Comparison of post-disturbance frequency nadir results between base case (blue trace) and Freq. Const. case (orange trace).

to contains the frequency nadir above certain threshold value following the largest N-1 contingency. Then, we validate our frequency stability constraint in DAUC scheduling problems through N-1 PSS/E dynamic simulation results. Specifically, the result shows that our constraint can guarantee 98% of the system frequency nadirs are above the under-frequency load shedding setting point (59 Hz) following most severe N-1 contingency. Compared with the base case, our constraint improves the frequency stability significantly. But we note that the total generation cost will increase by 20% while including our frequency nadir stability constraint. In future work, we will explore how to optimally dispatch more IBRs to provide inertia response instead of pushing more SGs online.

ACKNOWLEDGMENT

This work was authored in part by the National Renewable Energy Laboratory, operated by Alliance for Sustainable Energy, LLC, for the U.S. Department of Energy (DOE) under Contract No. DE-AC36-08GO28308. This material is based upon work supported by the U.S. Department of Energy's Office of Energy Efficiency and Renewable Energy (EERE) under the Solar Energy Technologies Office Award Number 37772. The U.S. Government retains and the publisher, by accepting the article for publication, acknowledges that the U.S. Government retains a nonexclusive, paid-up, irrevocable, worldwide license to publish or reproduce the published form of this work, or allow others to do so, for U.S. Government purposes. The views expressed herein do not necessarily represent the views of the U.S. Department of Energy or the United States Government.

REFERENCES

[1] X. Fang, Q. Hu, F. Li, B. Wang, and Y. Li, "Coupon-based demand response considering wind power uncertainty: A strategic bidding model for load serving entities," *IEEE Transactions on Power Systems*, vol. 31, no. 2, pp. 1025–1037, 2016.

[2] X. Liu, J. Xie, X. Fang, H. Yuan, B. Wang, H. Wu, and J. Tan, "A comparison of machine learning methods for frequency nadir estimation in power systems," in *2022 IEEE Kansas Power and Energy Conference (KPEC)*, 2022, pp. 1–5.

[3] W. Wang, X. Fang, H. Cui, F. Li, Y. Liu, and T. J. Overbye, "Transmission-and-distribution dynamic co-simulation framework for distributed energy resource frequency response," *IEEE Transactions on Smart Grid*, vol. 13, no. 1, pp. 482–495, 2022.

[4] X. Zhao, H. Wei, J. Qi, P. Li, and X. Bai, "Frequency stability constrained optimal power flow incorporating differential algebraic equations of governor dynamics," *IEEE Trans. Power Syst.*, vol. 36, no. 3, pp. 1666–1676, 2021.

[5] H. Ahmadi and H. Ghasemi, "Security-constrained unit commitment with linearized system frequency limit constraints," *IEEE Trans. Power Syst.*, vol. 29, no. 4, pp. 1536–1545, 2014.

[6] J. Restrepo and F. Galiana, "Unit commitment with primary frequency regulation constraints," *IEEE Trans. Power Syst.*, vol. 20, no. 4, pp. 1836–1842, 2005.

[7] N. Gao, D. W. Gao, and X. Fang, "Manage real-time power imbalance with renewable energy: Fast generation dispatch or adaptive frequency regulation?" *IEEE Transactions on Power Systems*, pp. 1–12, 2022.

[8] X. Fang, K. S. Sedzro, H. Yuan, H. Ye, and B.-M. Hodge, "Deliverable flexible ramping products considering spatiotemporal correlation of wind generation and demand uncertainties," *IEEE Transactions on Power Systems*, vol. 35, no. 4, pp. 2561–2574, 2020.

[9] S. G. Vennelaganti and N. R. Chaudhuri, "Stability criterion for inertial and primary frequency droop control in mtdc grids with implications on ratio-based frequency support," *IEEE Trans Power Syst.*, vol. 35, no. 5, pp. 3541–3551, 2020.

[10] H. Ahmadi and H. Ghasemi, "Security-constrained unit commitment with linearized system frequency limit constraints," *IEEE Trans. Power Syst.*, vol. 29, no. 4, pp. 1536–1545, 2014.

[11] J. Tan et al., "Final technical report: Multi-timescale integrated dynamics and scheduling for solar (midas-solar)," National Renewable Energy Lab.(NREL), Golden, CO (United States), Tech. Rep., 2023.

[12] S. Dong, X. Fang, J. Tan, X. Cui, and A. Hoke, "Analytical frequency nadir prediction considering inverter-based fast frequency responses," *https://arxiv.org/abs/2209.09413*, 2022.

[13] D. L. H. Aik, "A general-order system frequency response model incorporating load shedding: analytic modeling and applications," *IEEE Trans. Power Syst.*, vol. 21, no. 2, pp. 709–717, 2006.

[14] P. Anderson and M. Mirheydar, "A low-order system frequency response model," *IEEE Trans. Power Syst.*, vol. 5, no. 3, pp. 720–729, 1990.

[15] Q. Shi, F. Li, and H. Cui, "Analytical method to aggregate multi-machine sfr model with applications in power system dynamic studies," *IEEE Trans. Power Syst.*, vol. 33, no. 6, pp. 6355–6367, 2018.

978-1-6654-6060-6/23 $31.00 © 2023 IEEE

Precursor Ink Engineering for Scalable Slot-Die Coating of Perovskite Films for Photovoltaic Mini-Module Production

Manoj Rajakaruna, Jaehoon Chung, You Li, Tamanna Mariam, Muhammad Saeed Mohsin, Prabodika N. Kaluarachchi, Amirhossein Rahimi, Zhaoning Song, Michael J. Heben, Yanfa Yan, Randy J. Ellingson

Wright Center for Photovoltaics Innovation and Commercialization (PVIC), Department of Physics and Astronomy, University of Toledo, Toledo, Ohio, 43606, USA

Abstract — As the efficiency of small-area perovskite solar cells has reached more than 25% in recent years, it is now essential to accelerate the commercialization of perovskite photovoltaics. Developing low-cost, reliable, and automatic large-area coating techniques is a crucial step toward the industrial production of perovskite solar modules. For large-area solution-based deposition of perovskite films, solvent extraction from the wet precursor film is critical to producing pinhole-free, uniform, and high-quality perovskite films. Here, we report a scalable slot-die coating method for perovskite photovoltaic mini-module production. A vacuum quenching technique is developed to assist the formation of a perovskite intermediate phase by extracting solvents from the precursor films uniformly throughout the coating area. We study the impact of N-Methyl-2-Pyrrolidone (NMP) as a coordinating solvent in the perovskite precursor ink on the quality of slot-die-coated perovskite films and the photovoltaic performance of corresponding devices. We demonstrate small-area cells and 40 cm² perovskite mini-modules with power conversion efficiencies of up to 19% and 17%, respectively. Furthermore, external radiation efficiency (ERE) measurement carried out to characterize the uniformity of mini-modules and study the perovskite degradation.

I. INTRODUCTION

As a result of developing organic-inorganic hybrid perovskites, solar cells have been certified to achieve ~24% power conversion efficiency, indicating a promising future for using large-scale perovskite solar cells. One of the strongest advantages of perovskite solar cells (PSCs) is their low-cost solution method for fabricating the light absorbers. Due to their high efficiency and ability to be produced in high volumes by printing, metal halide perovskite-based solar cells (PSCs) have been identified as an exciting emerging PV technolog [1]. Moreover, perovskite thin-film PV technology can be manufactured on lightweight and flexible substrates to achieve high specific powers (power-to-weight ratio) [2] and exhibit high cosmic radiation resistance [3], making it a promising candidate for space photovoltaics.

There are several fabrication methods available for large-area perovskite solar cell fabrication, including spray coating, slot-die coating, vacuum deposition, blade coating, etc. Among them, slot-die coating has been considered one of the options with the greatest commercial potential due to its high industrial readiness for scaling up and automatization. However, several technical challenges remain in the large-area slot-die coating of perovskite films. The most critical issue is how to produce pinhole-free, uniform, and high quality perovskite film over a large area.

Here, we develop a scalable slot-die coating method for perovskite photovoltaic mini-module production. We use a vacuum quenching approach to replace the typical air quenching method to prepare perovskite precursor films uniformly over a large coating area. Furthermore, we engineer the coordinating solvent, N-Methyl-2-Pyrrolidone (NMP), used in the perovskite precursor ink to enable the coating of compact, uniform, and pinhole-free perovskite films. Using this method, we demonstrate 40 cm² perovskite mini-modules with a power conversion efficiency of up to 17%.

II. METHODOLOGY

For the perovskite ink, 1.2 M $Cs_{0.07}(FA_{0.85}MA_{0.15})_{0.93}PbI_3$ perovskite composition was dissolved in a solvent mixture of DMF (Dimethylformamide) and NMP. Here, DMF is the main solvent, and NMP is the coordinating solvent. Five perovskite precursors were made by changing the NMP: DMF ratio.

Fig. 1.   Five perovskite precursors prepare with different amount of NMP: DMF ratio.

After making precursor solutions, slot-die coating was used to deposit thin layers of perovskite solution on cleaned 3" × 3" FTO glass substrates. 0.2 m/min and 45 µl/min were the coating and pumping speeds, respectively. Upon coating, the wet film on the glass substrate was quickly transferred into a vacuum chamber and vacuum quenched for 40 s until the wet film reached its intermediate phase. The glass substrate was then transferred into a vacuum oven to anneal at 115 °C for 15 min to convert the precursor to the perovskite phase.

Perovskite solar cells and mini-modules were fabricated on cleaned FTO glass. A compact TiO_2 blocking layer was first deposited by spray pyrolysis, followed by slot-die coating of SnO_2 nanoparticles as the electron transport layer and $Cs_{0.07}(FA_{0.85}MA_{0.15})_{0.93}PbI_3$ as the absorber layer. A thin layer of OABr (octylammonium bromide in Chloraform) was deposited as a passivation layer, using the spin coating method. Then, the spiro-OMeTAD hole transport layer was spin-coated afterward. Finally, 60 nm gold was deposited by thermal evaporation to complete the devices.

III. RESULTS AND DISSCUSSION

To obtain high-quality, pinhole-free, and compact perovskite films, it is essential to form a stable perovskite intermediate phase [6]. NMP, GBL (gamma-Butyrolactone), DMSO (Dimethyl sulfoxide), CHP (N-Cyclohexyl-2-pyrrolidone) are widely used co-solvents in perovskite precursors as they have high boiling points and low vapor pressures (Figure 1) [4] [5]. The surface morphology of perovskite film can be significantly affected by adding a high boiling point co-solvent to retard the nucleation of perovskite intermediates. This phenomenon is important for large-area coating to obtain uniform crystal growth across the whole coating area.

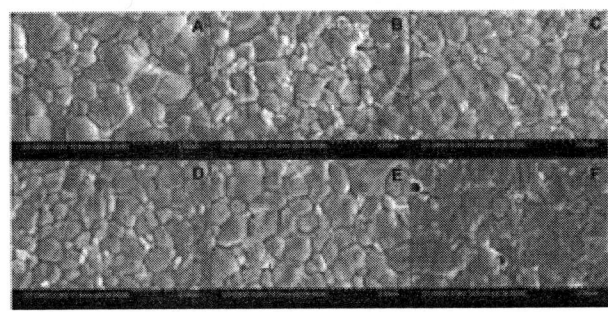

Fig. 2. Structure and boiling point of different co-solvents used in perovskite precursors.

We select NMP to modify our perovskite precursor ink for slot-die coating. Figure 2 shows SEM images of each perovskite film obtained by slot die coating with different NMP: DMF concentrations. Starting from Figure 2 (A) to (E), the amount of NMP was reduced from 1:1 to 1:10. It is observed that after adding 1: 10 NMP: DMF ratio, film morphology can be improved significantly. In contrast, in the perovskite film prepared using the solution without NMP (NMP: DMF = 0:1), one observes some pinholes and grain structures without any clear grain boundaries (Figure 2 (F)).

Since NMP has a higher boiling point than DMF, solvent evaporation during vacuum quenching was retarded. Therefore, intermediate perovskite films obtained with high NMP concentrations had a thick wet film before annealing.

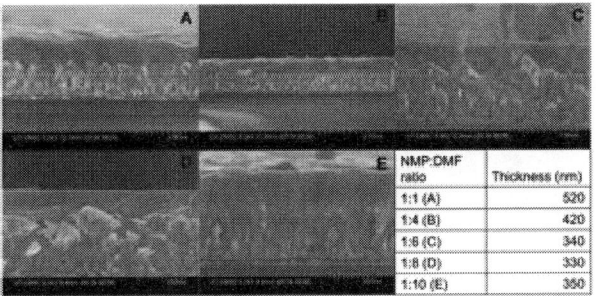

Fig. 3. Surface SEM images of perovskite films with different NMP: DMF ratios (A) 1:1 (B) 1:4 (C) 1:6 (D) 1:8 (E) 1:10 (F) 0:1.

According to Figure 4, the thickness of the final film highly depends on the amount of NMP concentration in the precursor solution. Thickness decreased from 520 nm to 350 nm when decreasing NMP: DMF ratio from 1:1 to 1:10. It is required to have at least 750 nm of perovskite film thickness to have a decent perovskite mini-module. This can be achieved by changing slot die coating parameters accordingly.

NMP:DMF ratio	Thickness (nm)
1:1 (A)	520
1:4 (B)	420
1:6 (C)	340
1:8 (D)	330
1:10 (E)	350

Fig. 4. Cross section SEM images of perovskite films with different NMP:DMF ratios (A) 1:1 (B) 1:4 (C) 1:6 (D) 1:8 (E) 1:10.

X-ray diffraction (XRD) analysis was carried out to evaluate the crystallinity of perovskite films prepared using each NMP concentration. XRD results show a dominant perovskite phase in all films regardless of NMP concentrations (Figure 5). A minority PbI_2 phase was also identified in the film, agreeing with the white particles on the boundaries of grain domains of perovskites, as shown in the SEM images. A small amount of PbI_2 is beneficial to passivate defects on the surface and grain boundaries. Among all the samples, the one with a 1:10 NMP: DMF ratio exhibits the most intense perovskite peak, indicating the best crystallinity.

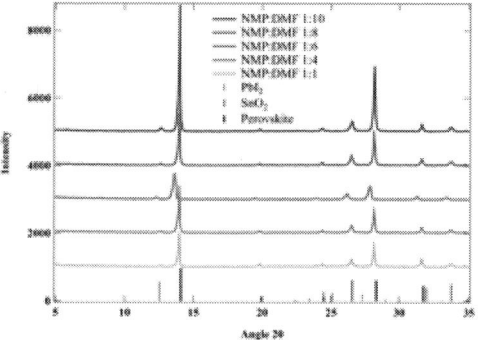

Fig. 5. XRD patterns of perovskite films with different NMP: DMF ratios.

Based on perovskite film characterization results, 1:6, 1:8, and 1:10 NMP: DMF concentration solvents were selected to make unit cells to test performance. Figure 7 shows the PV parameters of these devices. Devices made using the 1:10 ratio precursor solution show the best open circuit voltage and efficiency among the group, delivering ~19% efficiency.

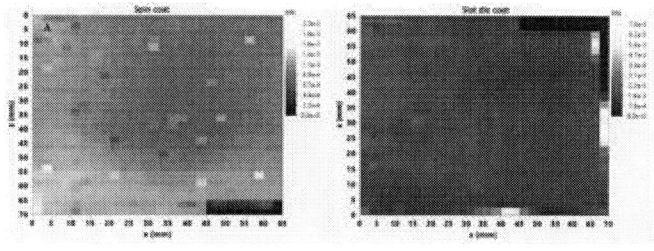

Fig. 6. Percentage change in external radiative efficiency of 3" by 3" substrate with perovskite film deposited by (A) spin (B) slot die coating.

To validate the uniformity of the slot die coated perovskite film, here we use the concept of External Radiative Efficiency (ERE), which is defined as the fraction of radiative emission from the total recombination that takes place in the cell mapped as a function of spatial position as a quantitative characterization method to compare the fabricated films' uniformity using spin and slot-die coating techniques. 2mW laser (~634 nm) was used to shine the film through the glass side at a 2 mm step size.

The sample fabricated using the spin coating technique gave 0.098 % ERE with a standard deviation of 2.4×10^{-4}. 95% of the data ranged between zero and 2×10^{-3}. ERE values were low radially close to the spinning axis and increased when distant from the axis, resulting in a dark circular area in the middle of the ERE map. This signifies the film is thicker and possesses better quality, radially close to the spinning axis, and fades as going away across the 3" × 3" inch sample. The sample fabricated using the slot die coating method gave 0.18% ERE with a standard deviation of 3.2×10^{-4} and 99% of data ranged

between 1 to 2.2×10^{-3} resulting in a uniform film across the 3" × 3" sample.

In conclusion, it can be stated that the sample fabricated using slot die coating method has a higher ERE response and a tighter spread of data when compared to the sample fabricated using spin coating method. Hence Slot die coating technique can be identified as a uniform fabrication method for thin films.

Finally, 1:10 NMP: DMF concentration perovskite precursor was used to make 3" × 3" mini-modules by air and vacuum quenching. It is clear that the vacuum-quenched module shows a higher current density than the air-quenched one (Figure 9 and Table 1). This is mainly due to the thickness difference between the two modules (figure 8). During the air-quenching process, the wet film is forced to evaporate solvent by putting pressure throughout the film. Due to this phenomenon, thickness control is a limitation of air quenching.

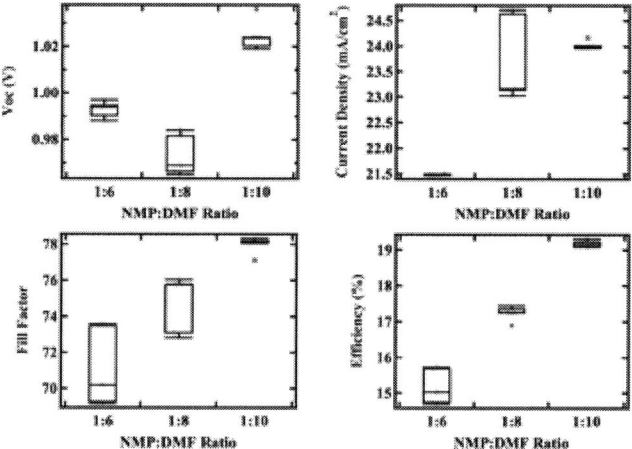

Fig. 7. Statistical distribution of photovoltaics parameters of unit cells.

To achieve a high fill factor and current density, the absorber of perovskite mini-modules needed to be sufficiently thick (>750 nm). For vacuum-quenching, there is no influence on wet film thickness as solvent extraction is done in a low-pressure environment.

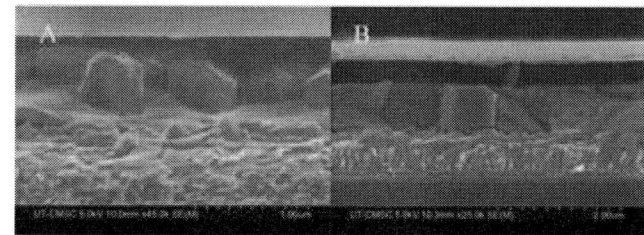

Fig. 8. Cross-sectional SEM of two films with different quenching methods. (A) Air quenching (B) Vacuum quenching.

Therefore, a thick perovskite absorber is feasible. Nonetheless, a 17% efficiency module was demonstrated using the slot-die coating with vacuum quenching.

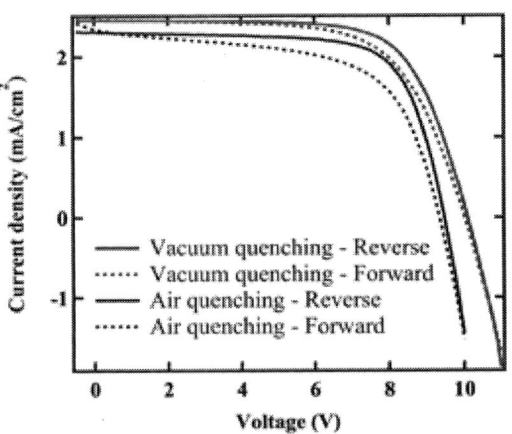

Fig. 9.   J-V curves of two mini-modules prepared by air and vacuum quenching.

Table 1.   Photovoltaic parameters of two modules

Module	Voc (V)	Jsc (mA/cm^2)	Fill Factor (%)	Efficiency (%)	P Max (mW)	HI
Vacuum	10.02	2.47	65.54	16.20	552.35	3%
	10.12	2.46	69.74	17.37	592.19	
Air	9.35	2.35	60.65	13.33	536.00	8%
	9.60	2.31	70.46	15.66	629.45	

IV. CONCLUSION

We demonstrated the importance of tuning the coordinating solvent in $Cs_{0.07}(FA_{0.85}MA_{0.15})_{0.93}PbI_3$ perovskite ink to enable large-area coating of perovskite mini-modules. To find the optimal NMP content, we change the NMP: DMF concentration until we find an intermediate perovskite phase that forms a good perovskite film upon annealing. Vacuum quenching resulted in an optimal intermediate phase when the NMP and DMF ratio was 1:10 using this NMP: DMF ratio, we obtained over 19% efficiency for unit cells and 17% for 3" × 3"

mini-module. Moreover, ERE mapping (PL mapping) shows high uniformity and high PL response of slot die-coated perovskite films.

ACKNOWLEDGMENT

This material is based on research sponsored by Air Force Research Laboratory under agreement number FA9453-21-C-0056. The U.S. Government is authorized to reproduce and distribute reprints for Governmental purposes notwithstanding any copyright notation thereon. The views expressed are those of the authors and do not reflect the official guidance or position of the United States Government, the Department of Defense or of the United States Air Force. The appearance of external hyperlinks does not constitute endorsement by the United States Department of Defense (DoD) of the linked websites, or the information, products, or services contained therein. The DoD does not exercise any editorial, security, or other control over the information you may find at these locations. Approved for public release; distribution is unlimited. Public Affairs release approval #AFRL-2023-0186.

REFERENCES

[1] J. Chung et al., "Engineering Perovskite Precursor Inks for Scalable Production of High-Efficiency Perovskite Photovoltaic Modules," Advanced Energy Materials, vol. n/a, no. n/a, p. 2300595, doi: https://doi.org/10.1002/acnm.202300595.

[2] Y.-B. Cheng, A. Pascoe, F. Huang, and Y. Peng, "Print flexible solar cells," Nature, vol. 539, no. 7630, pp. 488-489, 2016/11/01 2016, doi: 10.1038/539488a.

[3] Y. Tu et al., "Perovskite Solar Cells for Space Applications: Progress and Challenges," Advanced Materials, vol. 33, no. 21, p. 2006545, 2021,

[4] T. Bu et al., "Lead halide–templated crystallization of methylamine-free perovskite for efficient photovoltaic modules," Science, vol. 372, no. 6548, pp. 1327-1332, 2021.

[5] J. W. Yoo et al., "Efficient perovskite solar mini-modules fabricated via bar-coating using 2-methoxyethanol-based formamidinium lead tri-iodide precursor solution," Joule, vol. 5, no. 9, pp. 2420-2436, 2021/09/15/ 2021, doi: https://doi.org/10.1016/j.joule.2021.08.005.

[6] T. Bu et al., "Modulating crystal growth of formamidinium–caesium perovskites for over 200 cm2 photovoltaic sub-modules," Nature Energy, vol. 7, no. 6, pp. 528-536, 2022/06/01 2022, doi: 10.1038/s41560-022-01039-0.

GaAs solar cells grown on acoustically-spalled GaAs substrates with 27% Efficiency

Kevin L Schulte, Steve W Johnston, Anna K Braun, Jacob T Boyer, Anica N Neumann, William E McMahon, Michelle Young, Pablo Guimerá Coll, Mariana I Bertoni, Emily L Warren, Myles A Steiner

National Renewable Energy Laboratory, Golden, CO, United States

Colorado School of Mines, Golden, CO, United States

Crystal Sonic Inc., Phoenix, AZ, United States

Arizona State University, Tempe, AZ, United States

We report the growth of high-efficiency GaAs solar cells grown by organometallic vapor phase epitaxy on non-flat substrate surfaces created by acoustic spalling, or "Sonic Lift-off" (SLO). SLO is a potentially low-cost source of III-V epitaxial growth substrates, but surface facets formed during the SLO process can impact the performance of subsequently-grown devices. We show that non-linear shunts can form in regions where the surface contains facets with 2-3 μm peak-to-valley height. These defects degrade the device performance via a reduction in open-circuit voltage, despite quantum efficiency measurements that suggest that the bulk material quality is only slightly affected by the surface roughness. We present evidence from electrical device measurements and structural analyses that these shunts form at regions of non-conformal coating of the epitaxial layers over surface features, which we hypothesize leads to the formation of Schottky diodes where the front contact grid-lines contact the p-type base of the n-on-p diode structure. We demonstrate that these defects can be mitigated or eliminated by planarizing the surface using wet chemical etching and/or growth. Using a combination of etching and growth planarization, we demonstrate 0.25 cm2 devices with 26.94% photovoltaic conversion efficiency under the one-sun AM1.5G spectrum grown on an acoustically spalled substrate. These results show that the growth of high performance III-V devices is possible on rougher, non-traditional substrates that offer the potential for reduced cost.

Why increased CdSeTe charge carrier lifetimes and radiative efficiencies did not result in voltage boost for CdTe solar cells?

Darius Kuciauskas, Alexandra Bothwell, Carey Reich, Chungho Lee, Eric Colegrove, Marco Nardone

NREL, Golden, CO, United States

CSU, Fort Collins, CO, United States

First Solar, Santa Clara, CA, United States

BGSU, Bowling Green, OH, United States

We show that trapping impacts charge carrier dynamics in undoped CdSeTe, undoped CdSeTe/CdTe bilayers, and in As-doped CdSeTe/CdTe solar cells. Trapping has much smaller influence in CdTe-only (no Se, no As) films. Electrostatic potential fluctuation model (amplitude γ > 30 meV) applies to As-doped CdSeTe/CdTe, and defect model (defect activation energy Ea = 0.2 eV) applies to undoped CdSeTe. Unusually, undoped CdSeTe with trap states can have high radiative efficiencies and TRPL lifetimes (10-15 μs). As a result, thermodynamic CdTe solar cell analysis (radiative voltage, implied voltage, external radiative efficiency metrics) needs to be used with caution. As a metric for trap states, PL emission spectroscopy and charge carrier lifetimes at low temperatures provide distinct trap signatures. Trapping impact on devices can be evaluated in modeling and assessed in measurements outlined in this paper, perhaps providing a verifiable hypothesis when we search how to overcome voltage bottleneck in CdTe solar cells.

Detection and Impact of Cracks Hidden Near Interconnect Wires in Silicon Solar Cells

Andrew M. Gabor*, Hubert Seigneur§, Philip J. Knodle*, Dylan J. Colvin§, and Kristopher O. Davis§

* BrightSpot Automation LLC, Boulder, CO 80303, USA

§ University of Central Florida, Orlando, FL 32816, USA

Abstract — The thermal stresses associated with the soldering of interconnect wires onto the busbars of solar cells is one of the leading causes of cracks in silicon solar cells. Cracks will often branch outward from the busbar region so that they are easily seen in an electroluminescence (EL) image. However, since the wires are often wider than the busbar metallization, cracks can be located underneath or close to the wires and be "hidden" within the EL image. If the cracks remain beneath the busbar metallization, they may cause no reliability problems. However, if they propagate along the side of busbars and remain hidden under or next to the wires, they can prevent continuity of the gridlines to the busbars. The cracks may cause minimal problems in a new solar panel, but over time they can open up with thermal cycling and cyclic loading in the field. We demonstrate how these hidden cracks may be detected with the technique of UV Fluorescence, and we show examples of their signature in EL images. It is our observation that many groups are not familiar with these EL signatures, and do not consider that hidden cracks may be the cause of many gridline interruptions. We also show how gridline corrosion is strongly linked to hidden cracks where moisture can penetrate through the cracks. We hope that these techniques and improved understanding can lead to improved testing and feedback that accelerate product development and improve panel reliability.

I. INTRODUCTION

Electroluminescence (EL) is the main technique used to image defects in silicon solar cells. EL is now commonly performed in solar panel factories at the following stages for every solar panel: 1) After interconnecting the cells with soldered wires, today commonly forming a 10 or 12-cell string of rectangular cells; 2) At the layup table after placing strings of cells on glass/encapsulant and interconnecting the strings with bussing wire; 3) After lamination prior to packaging. Commonly seen defects are 1) Long cracks that commonly start and end at either a cell edge or a wire location, 2) short "V-cracks" most commonly at the tips of wires or where wires cross a cell edge, 3) short "X-cracks" due to sharp impacts, 4) dark bands along a portion of the wire due to cold solder joints between the wires and silver busbars, 4) dark bands along an entire wire due to a cold solder joint between the bussing wire and the interconnect wire, 5) dark gradient bands along a silver gridline due to a break in the gridline from a screen printing defect where the band is darkest near the break and lightest near the wires, and 6) dark bands along a silver gridline where the band is darkest next to a wire. The origin of this last defect is commonly attributed to a break in the gridline metallization near the wire, but the understanding of how these breaks occur

and the risks they pose is not commonly understood. Kang [1] referred to these as *DR* (Dark Rectangular) defects and showed how they became more severe with an increasing number of thermal shock cycles, and hypothesized about how the damage may be linked to changes in the microstructure of the solder, despite showing an image where the break in a gridline clearly was located over a crack in the silicon. Earlier, Lin [2] referred to these as *GFIB* (GridFinger Interruption at Busbar) defects, and found them to be the main contributor to P_{max} degradation in Humidity Freeze testing, but did not speculate on the details of how the interruptions occurred. Earlier still, Chaturvedi [3] studied the dark rectangular defects and accredited them to stresses between the soldered wires and gridlines that caused cracks in the gridlines, but not the underlying silicon. The oldest publication from Q-Cells [4] suggested the term *GICS* (Gridlines Interrupted Caused by Soldering) and correlated some of these interruptions to microchips in the silicon. A microchip is a scallop shaped crack that is largely parallel to the face of the cell and which does not extend through the thickness of the cell but which can intersect the top cell surface along a line. Where this line intersects the gridlines, cracks in the gridlines can occur. We find the Q-Cells analysis the most convincing of the literature we surveyed.

In this paper we aim to reintroduce and expand upon the *GICS* defect concept and explain further how different types of cracks hidden under or near the interconnect wire can leads to defects seen within EL images. We revisit older environmental chamber data and temperature-effect data that can lead to crack opening/closing, and we show new UV Fluorescence data that clearly shows the location of cracks through the thickness of the silicon that are hidden beneath or adjacent to the interconnect wires.

II. MODEL FOR GRIDLINE-BUSBAR DISCONTINUITY

Figure 1 shows an example of monocrystalline cells within a panel with polymer backsheet that we placed under static and cyclic mechanical loading stresses. These EL images and subsequent images from our groups were captured with a BrightSpot Automation EL camera system with 24 Megapixel resolution and with the panel biased at $-I_{sc}$. The cells show a variety of cracks across which the metallization is discontinuous to varying degrees. The red outlined regions show examples of *GICS* defects that may be due hidden cracks adjacent to a) the left busbar, and b) to both the left and right busbars.

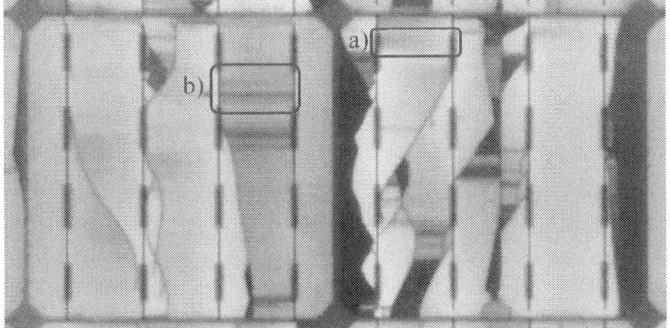

Fig. 1. EL image of monocrystalline cells within a panel that has undergone mechanical load testing and which shows various types of cracks. The red circled regions show examples of GICS defects that may be due hidden cracks adjacent to a) the left busbar, and b) to both the left and right busbars

The drawings in Figure 2 show how a crack adjacent to a busbar could be hidden from EL imaging by the interconnect wire. Such long cracks may be propagated from sub-millimeter microcracks originating from the differential contraction between the copper wire and the silicon during cooling during the soldering process [5]. The crack may wander, sometimes also falling under the busbar or even slightly beyond the wire and still be obscured from EL imaging. Cracks may propagate through the entire thickness of the silicon, or they may be roughly parallel to the surface as in the Q-Cells microchip example [4].

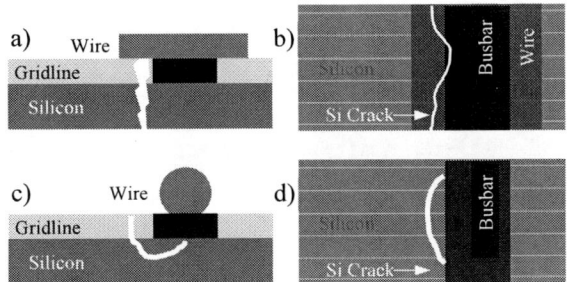

Fig. 2. Drawings of the cross section and top view of solar cells with cracks near the busbar. In a) and b) a flat ribbon wire and a crack running through the thickness of the cell are shown, while in c) and d) a round wire and a microchip crack running roughly parallel to the cell are shown.

If the entire interface between the busbar and the wire is well soldered, then a crack under the busbar location should not cause a gridline discontinuity since the current can enter the wire before being blocked by an open crack. However, the solder might only be bonded well to discrete "solder pad" locations along the busbar, leaving some percentage of the busbar sensitive to cracks under the busbar, leading to GICS defects. It is unclear how well the solder may bond directly to the silver gridlines, and this likely varies significantly among

manufacturers, but well bonded gridlines would provide good resistance to GICS defects for cracks that propagate along the wire/gridline interface region. In modern panels with wire array interconnects using 9 or more round wires, there may be little wire/gridline interface region where cracks could be completely hidden, but cracks just alongside the busbars may still be obscured from EL imagery. In addition, the busbars on the rear side of the cells often have wider bonding pad regions as can be seen in Figure 1. The higher surface recombination in these regions cause the EL image to be darker above the bonding pads, making it more challenging to see cracks in those regions.

Our older publication [5] shows some evidence for the microchip variation of *GICS*. Figure 3 shows how a wire pull test can pull up chunks of silicon with relatively little force on badly soldered cells, presumably due to the existence of crack planes in the silicon that do not penetrate through the entire thickness of the wafer. Figure 3 also shows an SEM cross-section image where a crack roughly parallel to the cell surface can be seen.

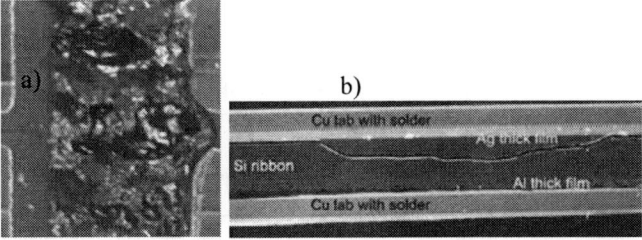

Fig. 3. a) An optical microscope image of the busbar region after a wire has been pulled off, taking chunks of silicon with it, and b) an SEM cross section of a soldered cell showing a scallop crack roughly parallel to the surface of the cell. Both images taken from [5].

III. TEMPERATURE EFFECTS ON GRIDLINE DISCONTINUITY

Figure 4 shows the crack closure and improvement in metallization continuity that occurs when heating a panel by applying a forward bias current [6]. Silverman has also seen changes in the continuity of metallization by heating and cooling effects and has measured the change in the width of gap across the cracks with changes in temperature [7]. It is not clear how the change in the state of *GICS* defects could be explained in the absence of cracks.

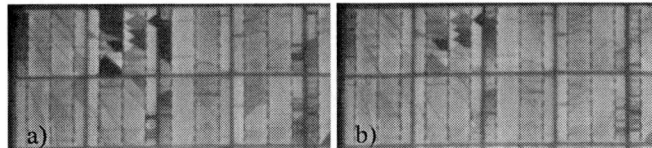

Fig. 4. EL images shown previously in [6] showing crack closure after forward bias current heating from a) 20.6 C to b) 35.3 C. Some *GICS* defects improve significantly.

IV. MECHANICAL LOADING EFFECTS ON GRIDLINE DISCONTINUITY

We have published previously on the concept of crack opening with front side mechanical loading and crack closure with rear side loading [8,9]. Figure 5 shows EL images of a panel with many cracks and *GICS* defects that we previously had placed under rear side load with a *RailPad* brace to place the cells into compressive stress and close previously open cracks. Again, it is not clear how the change in the state of *GICS* defects could be explained in the absence of cracks.

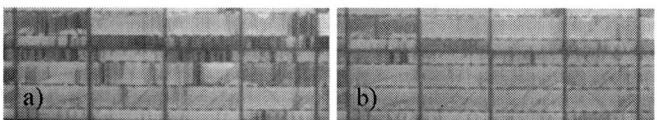

Fig. 5.   EL images shown previously in [8] showing crack closure after rear side loading with a *RailPad* brace. Most *GICS* defects improve significantly.

V. MACHINE LEARNING DETECTION OF GRIDLINE DISCONTINUITIES

We have developed and trained a machine learning model to automatically detect a range of different defects related to cracks and soldering/wire problems. The *GICS* defect can often present itself as a narrow dark line that might be confused by the model as a short "non-branching crack" defect. However, these defects are always perpendicular to the busbars and thus show little confusion with other defect types, and our model handles these inferences well as can be seen below in Figure 6. The challenge for automating the detection of the *GICS* defect is related to the wide variability in darkness contrast with the surrounding region. If one annotates every minor occurrence of the *GICS* defect in the training data, then the operator may be overwhelmed with inference detection events in cases where this defect type is common.

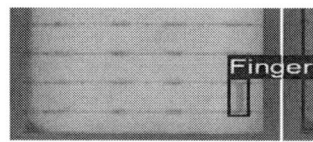

Fig. 6.   Successful machine learning inference of a GICS defect signature.

VI. UV FLUORESCENCE OF PANELS WITH CRACKS NEAR BUSBARS

UV Fluorescence (UVF) is a panel characterization technique that can be used to image a wide variety of defects in field aged or environmental-chamber aged panels [10,11]. In panels with polymer backsheets, it is particularly powerful at imaging cell cracks since oxygen can diffuse through the polymer backsheet and through the cracks and spread laterally a few mm's to quench the fluorescence in the front encapsulant such that a wide dark line is easily seen around the crack locations. Since the dark lines are much wider than the wires, cracks near or under the wire locations that would normally be hidden in an EL image are easily seen in the UVF image. In this work, we captured UVF images of fielded panels with a BrightSpot Automation pole mounted UVF camera system with a UV flash.

Figure 7 shows UVF images of fielded modules that indicate extensive cell cracking near the busbar locations on most cells. In some cases, the cracks appear to be very short, resulting in more of a dot shape than a line, while in many cases, the crack or series of shorts cracks cause the entire busbar region to be dark. While most cracks remain confined to the busbar regions, some cracks have propagated at an angle into the region between busbars. In the SolarWorld panel, narrow dark lines parallel to gridlines emanate from the busbars which we attribute to gridline corrosion. It is significant that the corrosion only spreads from busbars with cracks, indicating that the hidden cracks play an important role in the corrosion. We hypothesize that moisture can penetrate the cracks and react with the front encapsulant to form acetic acid that attacks the silver metallization and gradually works its way down the length of the gridlines. Such corrosion can result in high contact resistance between the silver and the silicon [12] and reduced line resistance along the length of the gridlines.

Fig. 7.   UV Fluorescence images of panels with extensive cell cracking near busbars: a) Schott Solar after 11 years in CO; b) SolarWorld after 10 years in MA where gridline corrosion is seen.

While the classic UVF image exhibits dark lines over the crack regions, some panel designs exhibit a white line over the crack regions. This presumably is related to rear encapsulants with a high concentration of fluorophores that can diffuse through the cracks to the top encapsulant layer. The competing kinetics of oxygen diffusion through the crack to quench the fluorescence is apparently insufficient to overpower the strong diffusion of the fluorophores in these cases. The UVF image shown in Figure 8 shows interesting variations in the brightness of 3 cracks. The left crack along the edge of the isolated area fluoresces white in a narrow band, while a wider dark band from the oxygen quenching behavior is seen where the crack crosses the strongly fluorescing ring. This bright ring is due to fluorophore diffusion from the gap region between cells. The middle crack only shows the dark band. It is unclear why different cracks have such different behavior, but it is perhaps related to the width of the gap in the silicon across the crack. The right crack is very close to the wire and leads to strongly fluorescing wire regions, presumably due to the diffusion of the fluorophores over the wire and to the higher reflectivity of the wires. Smaller regions where the left crack is near the wire also show this behavior. Thus, we can conclude that strongly fluorescing wires also are an indicator of cracks near to or underneath the wires.

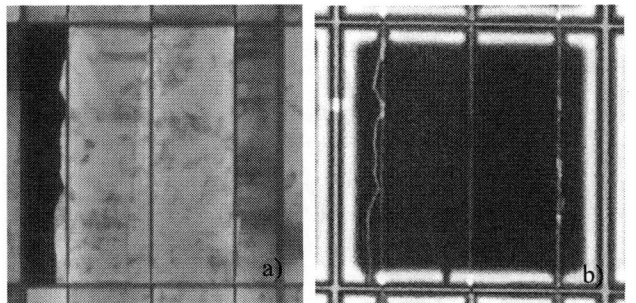

Fig. 8. a) EL image of a cell from a Jinko JKM305P-72 panel showing 3 cracks; b) the corresponding UV Fluorescence image where the left crack fluoresces white, the middle crack is easily seen only where it causes a quenching of the fluorescence in the ring region, and the right crack causes white fluorescence above the wire.

VII. DISCUSSION

The temperature effect and mechanical loading data suggest that many of the *GICS* defects are due to cracks in the silicon. It seems more likely that these observed cracks are of the type that penetrate the full thickness of the wafer rather than the microchip variety. More characterization is needed to show the relative prevalence of cracks in *GICS* defects and to demonstrate whether there are cases where such defects are present without cracks in the silicon. By studying the prevalence of such defects in panels from different manufacturers, we can better understand the design and process factors that make panels resilient to these defects.

Based on the UVF data, we see that there are cases where most busbar regions have cracks in the silicon that have propagated through the thickness of the wafer, even if the cracks rarely propagate into the region between busbars where they can be imaged by EL. Indeed, these may be the most common type of crack in the silicon PV industry. The degradation risks are that 1) these cracks open up over time, leading to visible *GICS* defects in the EL images, 2) that these cracks propagate away from the busbars under mechanical load or thermal cycling, leading to isolated regions, and 3) that they enable gridline corrosion.

Panel manufacturers have internal quality control guidelines based on EL testing at various stages, where a detected crack may prompt rework prior to lamination, or where a certain frequency of cracks in the panel post-lamination results in a downgrade in the class and selling price of a panel. The fact that we may be missing a large percentage of cracks by EL imaging raises concerns for panel reliability. Additionally, a misinterpretation of the dark finger bands seen in EL images can mean that manufacturers are not accurately following their own internal quality metrics. For example, a GICS defect that occurs between a busbar and the edge of the cell or when gridlines are not connected to any busbar due to GICS defect at each neighboring busbar, the dark region in the EL image is close in nature to an "isolated area," and the grading score for a crack defect would be very different than that of a screen printing defect. For panel grading criteria such as the MBJ Solar Module Judgment Criteria [13], commonly used in field EL testing, the grader is tasked to assume that any crack will continue to propagate in the direction it is heading, and the implications of the *GICS* defects being due to cracks could have a severe impact on panel grading scores.

During product development, panels normally undergo extensive environmental chamber testing. Thermal and/or UV exposure can cause the evolution of fluorescence in the front encapsulant when illuminated by UV light. This raises the possibility of using UVF during product development to help detect the hidden cracks, at least for panels with polymer backsheets. Many modern panels use a front encapsulant without UV absorbing additives, and such panels exhibit "ring pattern" fluorescence where it is more challenging to see cracks over the whole cell area. By using front encapsulant with UV absorbing additives for special product development work, the hidden cracks may be more easily detected by UVF.

VIII. CONCLUSIONS

Cracks within the silicon wafer near the busbar regions are common within the PV industry, and these cracks are often obscured from detection by EL imaging. We suggest reimplementing the old *GICS* defect (Gridlines Interrupted Caused by Soldering) terminology proposed by Q-Cells [4] to refer to the dark bands in EL images parallel to the gridlines. The silicon cracks can be of the *microchip* variety and can cause breaks in the gridlines where the cracks intersect the gridlines. The cracks can alternatively propagate through the entire thickness of the silicon, and we present UV Fluorescence (UVF) data that confirms the widespread presence of such cracks. These dark rectangular EL defects are presently not well understood within the PV community, and definitively assigning them to a subcategory of crack defect has potentially significant financial implications in terms of how panels are graded for quality, factory yield, and insurance/warranty claims.

We demonstrate how gridline corrosion appears to be well correlated to the presence of these hidden cracks, and we propose a model whereby moisture can penetrate the cracks in the busbar regions to initiate the reactions with the encapsulant to form acetic acid and attack the silver gridlines and their interface with the silicon. Reducing soldering induced damage could thus reduce these corrosion risks.

We also propose the use of UVF and front encapsulant layers with UV absorbing additives to enable easy imaging of the hidden cracks during product development. Such imaging can be a powerful tool to enable the optimization of materials, designs, equipment, and processes. We are especially concerned that modern high-speed soldering equipment has insufficient pre-heat and post-heat zones [2,4,5] to minimize soldering induced damage, and that the equipment vendors and panel manufacturers are "flying blind" without such imaging to see the hidden cracks.

ACKNOWLEDGMENT

This material is based upon work supported by the U.S. Department of Energy's Office of Energy Efficiency and Renewable Energy (EERE) under the Solar Energy Technologies Office Agreement Number DE-EE0009347.

REFERENCES

[1] M.-S. Kang, Y.-J. Jeon, D.-S. Kim, and Y.-E. Shin, "Comparison of the 60Sn40Pb and 62Sn2Ag36Pb Solders for a PV Ribbon Joint in Photovoltaic Modules Using the Thermal Shock Test," *Energies (Basel)*, vol. 10, no. 4, p. 529–, 2017, doi: 10.3390/en10040529.

[2] K. Lin et al., "Detection of soldering induced damages on crystalline silicon solar modules fabricated by hot-air soldering method," Renewable energy, vol. 83, pp. 749–758, 2015, doi: 10.1016/j.renene.2015.05.017.

[3] P. Chaturvedi, B. Hoex, and T. M. Walsh, "Broken metal fingers in silicon wafer solar cells and PV modules," Solar energy materials and solar cells, vol. 108, pp. 78–81, 2013, doi: 10.1016/j.solmat.2012.09.013.

[4] J. Wendt et. al., "The Link Between Mechanical Stress Induced by Soldering and Micro Damage in Silicon Solar Cells." *24th European Photovoltaic Solar Energy Conference*, Hamburg, Germany, September (2009), p. 3420.

[5] A. M. Gabor, et. al., "Soldering induced damage to thin Si solar cells and detection of cracked cells in modules." *21st European Photovoltaic Solar Energy Conference*, Dresden, Germany, September (2006), p. 4.

[6] H. Seigneur, A. M. Gabor, E. Schneller, J. Lincoln, "Electroluminescence-Testing Induced Crack Closure in PV modules," *Proceedings 46th IEEE PVSC*, 2019.

[7] T. Silverman and S. Huang, "Temperature dependent electroluminescence of a commercial mc-Si module," *NREL PV Module Reliability Workshop*, Lakewood CO, USA, Feb 2018.

[8] A. M. Gabor *et al.*, "Compressive Stress Strategies for Reduction of Cracked Cell Related Degradation Rates in New Solar Panels and Power Recovery in Damaged Solar Panels," *2018 IEEE 7th World Conference on Photovoltaic Energy Conversion (WCPEC) (A Joint Conference of 45th IEEE PVSC, 28th PVSEC & 34th EU PVSEC)*, Waikoloa, HI, USA, 2018, pp. 2820-2825, doi: 10.1109/PVSC.2018.8547207.

[9] A. M. Gabor, et. al., "Mechanical load testing of solar panels — Beyond certification testing," *2016 IEEE 43rd Photovoltaic Specialists Conference (PVSC)*, Portland, OR, USA, 2016, pp. 3574-3579, doi: 10.1109/PVSC.2016.7750338.

[10] M. Köntges, A. Morlier, G. Eder, E. Fleiß, B. Kubicek, and J. Lin, "Review: Ultraviolet Fluorescence as Assessment Tool for Photovoltaic Modules," *IEEE Journal of Photovoltaics*, vol. 10, no. 2, pp. 616 - 633, 2020.

[11] A. M. Gabor and P. Knodle, "UV Fluorescence for Defect Detection in Residential Solar Panel Systems," *2021 IEEE 48th Photovoltaic Specialists Conference (PVSC)*, 2021, pp. 2575-2579, doi: 10.1109/PVSC43889.2021.9518884.

[12] N. Iqbal *et al.*, "Accelerate Cycles of Learning: Unencapsulated Silicon Photovoltaic Cells to Environmental Stressors," *2022 IEEE 49th Photovoltaics Specialists Conference (PVSC)*, Philadelphia, PA, USA, 2022, pp. 0668-0674, doi: 10.1109/PVSC48317.2022.9938492.

[13] MBJ Solutions, *MBJ Solar Module Judgment Criteria, Revision 5.0*. Nov. 10, 2022. [Online]. Available: www.mbj-solutions.com/fileadmin/cont_solutions/downloads/MBJ_PV-Module_Judgement_Criteria_rev5.0.pdf

Net Zero Water Strategies and Impacts for PV Manufacturing

Parikhit Sinha[1], Sunil Sajja[2], Tzy Wei Ooi[3], Sreenivas Jayaraman[2], and Sukhwant Raju[2]

[1]First Solar, Tempe, AZ, 85281, USA, [2]First Solar, Perrysburg, OH, 43551, USA, [3]First Solar, Kulim, Kedah, Malaysia

Abstract — Net zero water is a sustainable development strategy for manufacturing in water-stressed locations. A case study in Tamil Nadu, India shows that sustainable net zero water PV manufacturing can be achieved by a) utilizing on-site wastewater treatment and zero liquid discharge units to maximize the usage of onsite reclaimed water b) using offsite reclaimed water to meet the remaining water demand and c) implementing continuous improvement in water conservation. Net zero water can be combined with net zero electricity to reduce the life cycle water footprint of PV modules by ~half while also reducing the life cycle carbon footprint by ~40%. While crucial for managing local water and energy resources, net zero strategies have a relatively small (~15%) impact on reducing the total multi-criteria product footprint. Adding a third strategy of high value recycling with semiconductor recovery can achieve up to ~65% reduction in the multi-criteria PV module product environmental footprint covering health, ecosystem, and natural resource impact categories.

I. INTRODUCTION

While net zero concepts have focused primarily on energy [1] and greenhouse gas emissions [2], net zero water concepts are also of strategic importance in water-stressed locations. A net zero water facility is defined as one where the amount of alternative water used and water returned to the original water source is greater than or equal to the facility's total water usage (Eq. 1) [3].

$$W_T \leq W_A + W_R \qquad (1)$$

where:
- W_T: total water usage
- W_A: alternative water usage which is not derived from fresh surface or ground water sources (e.g., harvested rainwater, reject water from water purification, reclaimed wastewater)
- W_R: water returned to original water source (e.g., wastewater treated on-site and returned to original water source, stormwater infiltrated to the original water source through green infrastructure)

Therefore, a net zero water facility can be designed by:
- Minimizing W_T
- Maximizing W_A
- Minimizing wastewater discharge from the facility and maximizing W_R

A case study approach is utilized that evaluates a new thin film cadmium telluride (CdTe) photovoltaics (PV) module manufacturing facility under construction near Chennai in Tamil Nadu, India [4]. New facilities provide the opportunity to design with net zero strategies, including access to reclaimed water.

While Chennai has a wet climate with approximately 1400 mm of annual rainfall [5], it has experienced hydrological extremes in recent years of flooding in 2015 and water shortage in 2019. Therefore, industrial water management with regards to water supply and wastewater discharge is a priority for the area.

II. METHODS

An approximate water balance is developed for the Tamil Nadu facility to understand design elements that contribute to Eq. 1. The water balance considers facility water usage (cooling tower), process water usage for manufacturing, wastewater treatment, and flows of reclaimed water. De eminimis uses of sanitary and irrigation water are not included. Life cycle assessment (LCA) is also conducted to evaluate the water footprint of the PV modules produced in the Tamil Nadu facility and the associated improvements from progress toward net zero water practices.

Cradle-to-gate LCA has been conducted with Simapro (V. 9.4.0.2) software and UVEK DQRv2:2018 background unit processes. Water withdrawal estimates (per kWp) are based on all Simapro raw material water categories (groundwater, lake, river, ocean, salt, cooling, unspecified, etc.) excluding water used in running hydroelectric turbines (turbine use) because it is in-stream (not off-stream) use. LCA follows ISO 14040/14044 and IEA Task 12 LCA methodology guidelines [6] and is based on First Solar Series 7 life cycle inventory (LCI) with average module conversion efficiency of 18.7% (187 W_p per m^2) manufactured in India (50%) and USA (50%).

In order to understand the impacts of net zero strategies on a broader set of environmental impact categories, LCA impacts were also assessed in accordance with PV product environmental footprint (PEF) category rules [7] based on the PEF Guide [8]. Impacts were assessed with the ILCD 2011 Midpoint impact method (long term emissions excluded [7]), normalized [9], and equally weighted, and compared to corresponding product stage (cradle-to-gate) impacts for CdTe PV in the PEF screening study [10].

978-1-6654-6060-6/23 $31.00 © 2023 IEEE

III. RESULTS AND DISCUSSION

A. Net zero water strategies

Water usage is roughly evenly divided between process water for manufacturing and facility water for air conditioning (cooling tower) (Fig. 1). Most of the water consumption by the facility is due to evaporation from the cooling tower, where water consumption is defined as the amount of water removed from the immediate water environment.

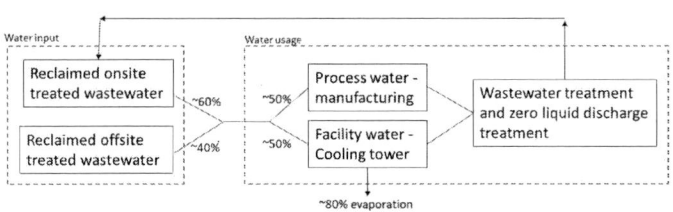

Fig. 1. Simplified net zero water flows for Tamil Nadu facility.

Within the facility, an ultra pure water unit is used to treat raw water for use in manufacturing, whereas water for the cooling tower does not require purification. After use in manufacturing, process water is treated onsite in a wastewater treatment unit. Treated process water and any non-evaporated water from the cooling tower is finally sent to a zero liquid discharge (ZLD) treatment unit.

Water reclaimed from the ZLD unit is useimpplemented at d as a raw input water for the Tamil Nadu facility and accounts for most (~60%) of facility's demand. The remaining ~40% is provided by an alternative water source, an offsite tertiary treatment reverse osmosis (TTRO) plant that provides reclaimed wastewater to industrial facilities.

As a result of minimizing wastewater discharge and using onsite and offsite reclaimed water, the facility is designed to approximately achieve a net zero water balance (Figs.1-2). Not shown in Fig. 1 is de minimis use of tanker fresh water for sanitary water. After use, sanitary water is treated onsite and reused for irrigation water (Fig. 2).

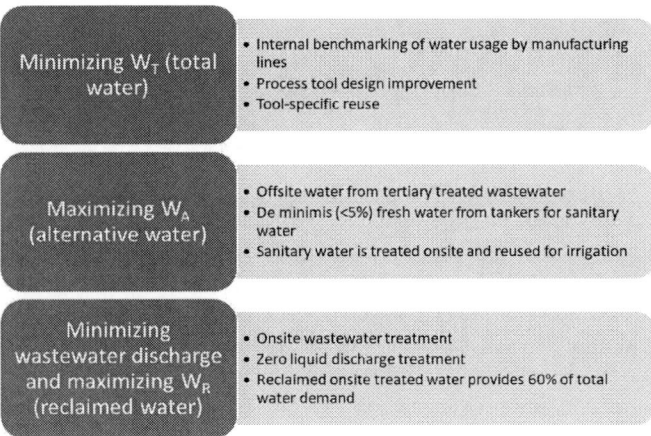

Fig. 2. Net zero water strategies for Tamil Nadu facility corresponding to Eq. 1

Achieving a net zero water balance is also aided by continuous improvement in water conservation (minimizing W_T). Continuous improvement requires establishing submeters for monitoring water usage in manufacturing. Specifically, internal benchmarking of water usage by washing tools in manufacturing process lines is used to identify best practices and opportunities for improvement. Process tool design improvement initiatives are also used to eliminate unused functions and reduce water carry-over and associated losses between sections of the manufacturing line. Tool-specific reuse is also used to treat some of the process wastewater locally (next to the process tool) and reuse at the same tool.

B. Net zero water impacts

Cradle-to-gate life cycle water withdrawal is shown in Fig. 3 for production of Series 7 PV modules in India and USA. The contribution of direct water usage at the manufacturing facility is < 5%. Therefore, while net zero water strategies are important for local water management in water stressed locations, they do not make a significant impact on the product water footprint. Life cycle water withdrawal is primarily due to embodied water in grid electricity as well as water intensive product components (glass, steel frame, encapsulant, and junction box).

In addition to pursuing net zero water strategies, the Series 7 PV manufacturing facilities are targeting net zero electricity by year 2028 in support of a corporate RE100 commitment. Because of the large contribution of grid electricity to the product water footprint, a net zero electricity strategy would result in a significant (~50-60%) reduction in the product water footprint (Fig. 3; last column). It should be noted that net zero water strategies can impact net zero electricity implementation as some water management tools increase electricity demand (e.g., evaporators for zero liquid discharge).

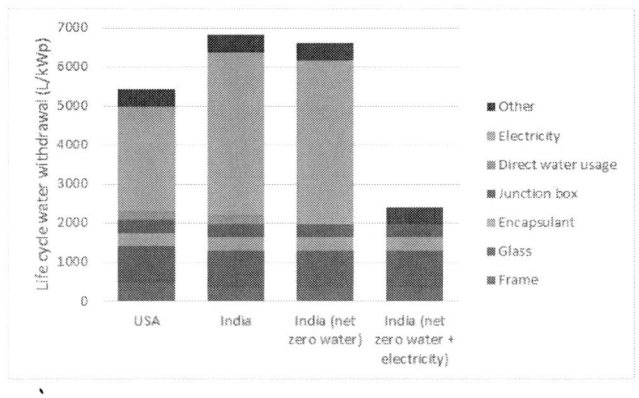

Fig. 3. Life cycle water withdrawal of Series 7 PV module production in USA and India, including net zero water and electricity strategies

C. Multi-criteria impacts

While net zero water and electricity strategies are crucial for managing water and energy resources, they can be viewed as part of a larger set of environmental strategies that consider health, ecosystems, and natural resources. Multi-criteria life cycle assessment of Series 7 PV module production indicates that net zero water and electricity strategies have a relatively small (~15%) impact on reducing the total Series 7 multi-criteria product footprint (Fig. 4; difference between second and third columns).

A third environmental strategy of high-value recycling is needed to significantly reduce the multi-criteria product footprint which is largely attributable to resource depletion impacts (~60%; second column of Fig. 4 in light green). High-value recycling which recovers semiconductor materials as well as bulk materials (e.g., glass, frame, cables) reduces the resource depletion impact which is primarily associated with the use of semiconductor materials (CdTe).

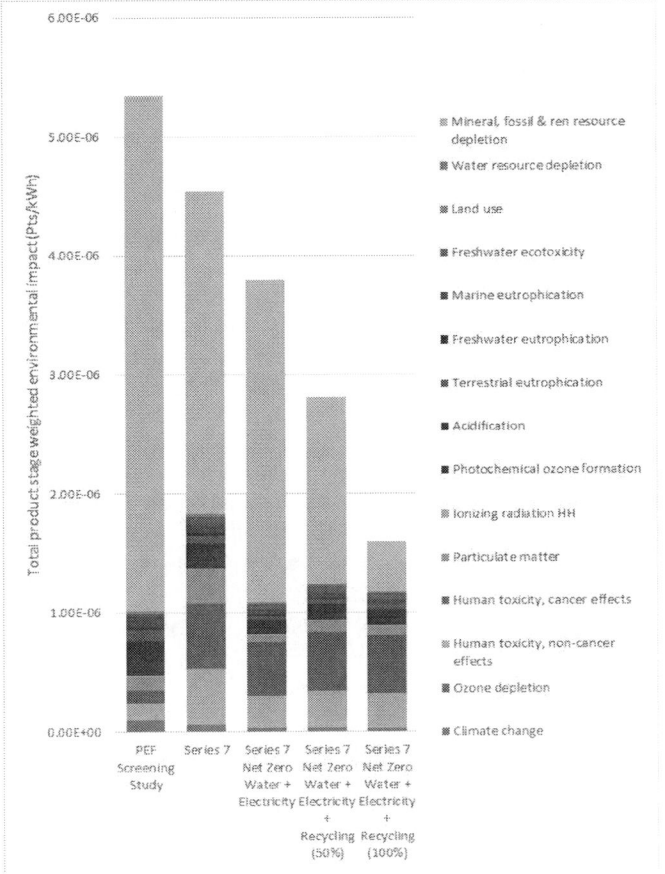

Fig. 4. Series 7 total environmental impact (normalized [9] and equally weighted) per kWh of electricity produced (based on PEF category rules of European deployment with average annual energy yield of 975 kWh/kWp and 30 year lifetime). Impacts are also shown for the PEF screening study CdTe PV benchmark and for Series 7 scenarios with net zero water and electricity and with high value recycling (including semiconductor recovery) with 50% and 100% avoided burden credits.

The Series 7 multi-criteria product footprint is about 15% lower than the CdTe PV benchmark in the PEF screening study. Impacts for human toxicity are higher than the benchmark due to the use of a steel frame in Series 7, whereas the benchmark was for an earlier generation frameless CdTe PV module. The Series 7 impacts for particulate matter are also higher than the benchmark due to the Indian electricity grid which was not part of the screening study.

The particulate matter impacts are mitigated by net zero electricity strategies (Fig. 4), and the carbon footprint is also reduced by ~40% with net zero strategies. Total environmental impacts are reduced by up to 65% relative to the Series 7 baseline when all all three strategies are utilized (net zero water, net zero electricity, high value recycling) (Fig. 4). The recycling benefits are shown with both 50% and 100% of the avoided burden credits from recycling, with 50% as the default assumption under the PEF category rules.

REFERENCES

[1] Climate Group and CDP, "RE100," available at: https://www.there100.org/
[2] SBTi, "Science based targets," available at: https://sciencebasedtargets.org/
[3] U.S. Department of Energy, "Net zero water building strategies," available at: https://www.energy.gov/eere/femp/net-zero-water-building-strategies
[4] First Solar, "First Solar Intends to Expand Global Manufacturing Footprint with New 3.3 GW Facility in India," available at: https://investor.firstsolar.com/news/press-release-details/2021/First-Solar-Intends-to-Expand-Global-Manufacturing-Footprint-with-New-3.3-GW-Facility-in-India/default.aspx
[5] India Meteorological Department, "Climatological normal 1981-2010," available at: https://imdpune.gov.in/library/public/1981-2010%20CLIM%20NORMALS%20%28STATWISE%29.pdf
[6] R. Frischknecht, P. Stolz, G. Heath, M. Raugei, P. Sinha, M. de Wild-Scholten, "Methodology Guidelines on Life Cycle Assessment of Photovoltaic Electricity, 4th edition," Report IEA-PVPS T12-18:2020, 2020.
[7] Technical Secretariat PEF Pilot PV Electricity Generation, "Product Environmental Footprint Category Rules: Photovoltaic Modules Used in Photovoltaic Power Systems for Electricity Generation," Version 1.1, 2019.
[8] European Commission, "Commission Recommendation of 9 April 2013 on the use of common methods to measure and communicate the life cycle environmental performance of products and organisations." Official Journal of the European Union, vol. L 124, pp. 1-210, 2013.
[9] L. Benini, L. Mancini, S. Sala, S. Manfredi, E. M. Schau, and R. Pant, "Normalisation method and data for Environmental Footprints," European Commission Joint Research Centre, Institute for Environment and Sustainability, Ispra, Italy, 2014.
[10] P. Stolz, R. Frischknecht, F. Wyss, and M. de Wild-Scholten, "PEF screening report of electricity from photovoltaic panels in the context of the EU Product Environmental Footprint Category Rules (PEFCR) Pilots, v. 2.0," treeze Ltd. and SmartGreenScans, Uster, Switzerland, 2016.

Characterization of different groups of electricity consumers and measures taken to reduce energy poverty

Anna Carolina de Paula Sermarini[1], Lucas Aló Rodrigues Araujo da Silva[1], Vanessa Cardoso de Albuquerque[1], Rodrigo Flora Calili[1]

[1]Pontifical Catholic University of Rio de Janeiro, Rio de Janeiro/RJ, 22451-900

ABSTRACT — The concept of energy poverty has been discussed in recent years, being multidisciplinary, subjective and different according to the reality of each territory. In any case, to understand how to solve the problem and propose appropriate measures, it is necessary to evaluate the reality of each community. The adoption of solar photovoltaics in communities can help reduce energy poverty and ensure access to basic energy services, being aligned with ODS 7. Thus, to identify and characterize different consumers in low-income communities in Rio de Janeiro, a survey was conducted on 310 households in the communities of Babilônia and Chapéu mangueira. Aiming at the fair promotion of the benefits of clean and renewable energy, a renewable energy cooperative was created in the same locality. This paper aims to present the survey results and the main impacts obtained after a little over a year of the creation of the cooperative.

I. INTRODUCTION

Energy is fundamental to ensuring human well-being and quality of life. The essential energy sources for the residential class come primarily from electricity and natural gas. Although access to safe and reliable energy services is paramount to economic development, especially in slums and low-income communities, the instability of the grid and the low quality of service provided in these areas, together with the high costs of electricity, tend to lead these residents to irregularity, causing the so-called Non-Technical Losses (NTL), one of the consequences of energy poverty [1][2].

Energy poverty (EP) is a subjective and multidisciplinary theme that depends on the territory and context in which it is being analyzed, as exposed by [3]. In one of the most widespread theories, introduced by [4], it is considered a precursor as to how to measure EP. According to the theory, a family is considered energy poor if it spends more than 10 % of its household budget on energy. Other authors believe that not having basic needs fulfilled, such as decent access to energy, configures a situation of EP [5]. It is worth noting that according to [6], in communities and slums in developing countries, the percentage of income commitment can be much higher than 10 %.

The 17 Sustainable Development Goals (SDGs), established in 2015 by 193 United Nations (UN) member states, organized in the so-called Agenda 2030, were created to promote sustainable development inclusively and fairly and improve people's quality of life [7]. To achieve the established goals, especially for SDG 7 - Clean and Affordable Energy - and SDG 11 - Sustainable Cities and Communities, the growth of renewable energy sources, such as solar photovoltaics, has become essential, as shown [8].

In practice, it has been observed that the use of photovoltaic systems for self-generation of energy (PV) is much more common in families with higher purchasing power [9][10]. However, some authors [8][11] explain that the adoption of PV in low-income families and communities contributes to the fight against EP, mainly by ensuring access to basic energy services and fair promotion of the benefits of clean and renewable energy, being in line with another SDG, 10 - Reduction of inequalities.

Thus, the present work aims to identify and characterize the different energy users through a survey applied to 310 households in the communities of Babilônia and Chapéu Mangueira in Rio de Janeiro, as well as to characterize the initiative of a Solar Energy Cooperative implemented in the same locality as a possible solution to reduce EP.

II. BACKGROUND

There are some studies [8][11] that expose that the adoption of PV systems by the low-income population is a necessary means to achieve energy justice [12], ensuring access to essential energy services and the fair distribution of Distributed Energy Resources (DER). And [13] shows that the main factor for low-income consumers to be favourable to Distributed Generation (DG) is the potential reduction of their electricity bills. However, the initial cost of installing these systems is one of the leading causes that keep them from using DER.

A widespread way to improve a community's socio-economic welfare, and promote its development, is the cooperative, a society of people acting towards a common goal. In Brazil, they are managed by the National Cooperative Policy (Law no. 5.764/1971). Besides strengthening community ties, cooperatives observe the needs of the place where they are inserted and guarantee the necessary goods and services [14]. In the context of DG, Law 14.300/2022, which established the legal framework of DG in Brazil and the Electricity Compensation System (ECS), allows such credit compensation to be made in the shared generation modality through cooperatives and other similar organizations.

In this way, considering the concept of EP previously exposed and the advent of DER, especially DG, creating a cooperative is a possible solution in the quest to introduce this technology in low-income locations.

III. METHODOLOGY

Low-income families, usually living in more precarious territories such as favelas, can be considered the primary victims of EP in Brazil. Therefore, to identify and characterize such residents, it is fundamental to analyze the local reality so that it is possible to design any proposal for the problem. Thus, for this purpose, a survey was applied in Babilônia and Chapéu Mangueira, located in the South Zone of the City of Rio de Janeiro, in February 2022.

The survey used as a data collection instrument a questionnaire divided into four parts, the first with information about the socio-economic profile, the second about access to electricity and level of education, the third contained information about consumption habits and participation in the energy transition, and in the fourth, there was information about quality, price, perceived value and fairness about the services provided by the electricity utility.

After selecting the questions for each stage, the questionnaire was reviewed by a local community resident to verify that the language and questions used were appropriate to the community and to avoid as much evasion as possible, mitigating possible conflicts and false answers.
To define the sample size, first, the number of inhabitants of the communities was observed, then determining a sample confidence interval of 95 % with a sampling error of 5 %, reaching the number of 310 respondents. Finally, the sample was divided according to the proportion of residents living in each area of the communities (Fig. 1).

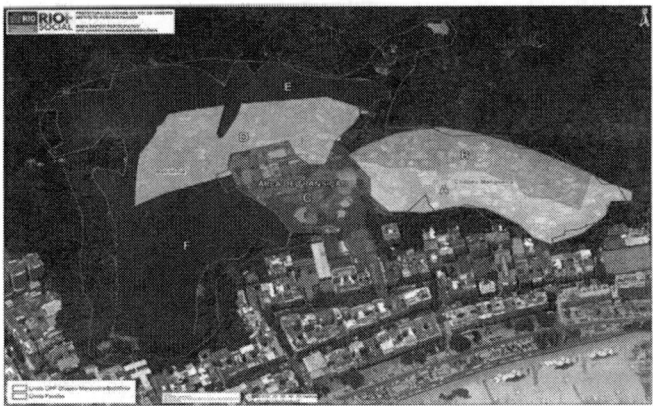

Fig. 1.   Division of the communities of Babilônia and Chapéu Mangueira into areas for distribution of the questionnaires.

Before the survey application, there was a training of interviewers, and local residents, so the respondents felt greater confidence, reducing the number of unanswered questions and false and discrepant answers. A pre-test was applied to validate the questionnaire and train the interviewers, considering 10 % of the sample. After collecting the data in digital format (.xlsx) with the validated answers, they were treated and analyzed according to the objective of this work. The following section brings the main results found.

IV. RESULTS AND DISCUSSIONS

The entire process of designing the survey, validating the questions with a local resident, and applying and analyzing the results, was supported by Revolusolar. This non-profit organization promotes sustainable development in low-income communities through solar energy.

A. Survey Results

Most of the residents consulted consider themselves black or brown (68.3 %), and only 32.7 % have completed high school or have an incomplete college education. Of the families surveyed, 56.1 % are male heads of household, and, on average, almost three people live in each house, with the majority of the population being young (67.6 %) up to 39 years old. The average monthly family income was R$ 1,865.30.

Most residents (71.3 %) do not use the social tariff (Brazilian policy of discounts on energy bills); of these, only 6.4 % know it. Of those who do not have the benefit, 39 % have or had in the last three months all appliances in the house consuming energy irregularly.

About the reasons for having or having already had some irregular electricity connection, 23 % of the residents say it is because they don't have money to pay, followed by 21 % who consider the price unfair.

Another significant result portrayed by the survey is that 38.2 % of the residents would be willing to replace the energy source with another type to save energy, such as solar.

Regarding the commitment of income to the cost of electricity, of the 153 respondents who informed the value of family income and the cost of the last electricity bill, 20.3 % committed greater than or equal to 10 % of their revenue with electricity alone (energy poverty), disregarding the costs of other energy sources. Of these, 97 % find the price charged for energy services high (42 %) and very high (55 %).

As a way of combating the problem and contributing to the reduction of low-income residents' expenses with electricity, Revolusolar has created a solar energy cooperative.

B. Solar energy cooperative in Babilônia and Chapéu Mangueira

In line with the UN's SDGs, especially with SDG 7, which is aligned with the fight against EP through photovoltaics solar energy, the NGO Revolusolar, located in Rio de Janeiro and based in the slums of Babilônia and Chapéu Mangueira, was created in 2015. In 2021, through grants, sponsorships, and collective fundraising, the organization formed the first solar energy cooperative in slums in Brazil, called Percília and Lúcio Renewable Energy Cooperative, named after two historical residents of the communities. The cooperative has 60 photovoltaic panels installed on the roof of the Association of Residents of Babylon, resulting in a 26 kWp power plant, benefiting about 30 local families that have their expenses with electricity reduced through solar generation.

The cooperative families were selected according to some established criteria, such as mandatory regulations and compliance with the local utility. Each month, each member's electricity bill is reduced due to the plant's solar power generation. Half of this discount returns to the cooperative in the form of monthly fees to cover expenses such as energy bills with the utility and taxes. The other half stays with the member, who now freely disposes of this value.

Over one year of operation (from September 2021 to December 2022), the plant generated 38,762 kWh, resulting in R$ 30,481 in total savings (about R$ 1,000 per cooperated family) and avoided 2.39 tCO2, equivalent. Another significant result to be mentioned is the average compliance rate regarding the payment of monthly fees by the cooperative members, which is currently 92 %.

Due to the selection criteria cited above, some families in EP identified by the research could not become cooperative members, so the average degree of impact on income before participating in the energy cooperative was 6 %, reduced to 4 % after using the PV. There are also significant cases related to the reduction of EP, in which the cooperative members had a reduction of up to 8 % in electricity expenses, considering their monthly fees.

It is worth mentioning that a redistribution of the plant's credits, that is, a redefinition of the percentage of generated credits allocated to each cooperative member, is in the process of implementation by the concessionaire, in addition to the performance of new PV plants, aiming to benefit more families in the community and improve the results obtained in this first year of operation.

V. FINAL CONSIDERATIONS

Naturally, each community has its specificity and characteristics. Understanding the profile of electricity consumers is essential to develop actions aimed at these territories, especially regarding the supply of quality and fair electricity.

The satisfactory results that have been observed with the implementation of a solar energy cooperative in a slum in Rio de Janeiro corroborate the position of some authors regarding the benefits of adopting PV in low-income families and communities to combat EP, mainly by guaranteeing access to basic energy service and by the fair promotion of the benefits of clean and renewable energy.

The importance of this study is evidenced, which can be used as a tool for formulating policies aimed at the implementation of PV in low-income communities, as well as an aid for public agencies and companies in the sector to work on the subject. As future work, it is suggested to analyze all the data from the survey and follow up and monitor the results of the Percília e Lúcio Renewable Energy Cooperative and its expansion capacity.

ACKNOWLEDGEMENT

The authors acknowledge NGO Revolusolar for support during field research, and Brazilian institutes FAPERJ and CNPq for financial support.

REFERENCES

[1] CAUSONE, Francesco; TATTI, Anita. Solar Technologies as a Driver to Limit Energy Poverty in the Rocinha Favela. In: **Environmental Performance and Social Inclusion in Informal Settlements.** Springer, Cham, 2020. p. 55-86.

[2] LEE, J.; SHEPLEY, M. M. Benefits of solar photovoltaic systems for low-income families in social housing of Korea: Renewable energy applications as solutions to energy poverty. **Journal of Building Engineering**, v. 28, 2020.

[3] CARVALHO, F. R. Mapeamento de potencial fotovoltaico em contexto de favela. Lisboa, 2021. 116 p. Dissertação (mestrado) - Universidade Nova de Lisboa, Engenharia das Energias Renováveis, 2021.

[4] BOARDMAN, B. **Fuel poverty: from cold homes to affordable warmth.** Belhaven Press, 1991.

[5] PEREIRA, M. G.; FREITAS, M. A. V.; SILVA, N. F. Rural electrification and energy poverty: Empirical evidences from Brazil. **Renewable and Sustainable Energy Reviews**, v. 14, n. 4. p. 1229-1240, 2010.

[6] BUTERA, F. M. et al. Energy access in informal settlements. Results of a wide on site survey in Rio De Janeiro. **Energy Policy**, v. 134, 2019.

[7] UNITED NATIONS. United Nations: Peace, dignity and equality on a healthy planet. Página inicial. Disponível em: <https://www.un.org/en/>. Acesso em: 7 de jan. de 2023.

[8] MILČIUVIENĖ, S.; KIRŠIENĖ, J.; DOHEIJO, E.; URBONAS, R.; MILČIUS, D. The Role of Renewable Energy Prosumers in Implementing Energy Justice Theory. **Sustainability**, 11, 5286, 2019.

[9] O'SHAUGHNESSY, E., BARBOSE, G., WISER, R. ET AL. The impact of policies and business models on income equity in rooftop solar adoption. **Nat Energy**, v. 6, p. 84–91, 2021.

[10] MACIEL, L. S. B.; BONATTO, B. D.; ARANGO, H.; ARANGO, L. G. Evaluating Public Policies for Fair Social Tariffs of Electricity in Brazil by Using an Economic Market Model. **Energies**, v. 13, article number 4811, 2020.

[11] SIGRIN, B.; SEKAR, A.; TOME, E. The solar influencer next door: Predicting low income solar referrals and leads. **Energy Research & Social Science**, v. 86, 102417, ISSN 2214-6296, 2022.

[12] SOVACOOL, B.K.; DWORKIN, M.H. Energy justice: Conceptual insights and practical applications. **Applied Energy**, v. 142, ISSN 0306-2619, p. 435-444, 2015.

[13] ILLIOPOULOS, N.; ESTEBAN, M.; KUDO, S. Assessing the willingness of residential electricity consumers to adopt demand side management and distributed energy resources: A case study on the Japanese market. **Energy Policy**, v.137, article number 111169, 2020.

[14] MAJEE, W.; HOYT, A. Cooperatives and Community Development: A Perspective on the Use of Cooperatives in Development. **Journal of Community Practice**, p. 48-61, 2011.

Using Neural Network Decomposition to Estimate Field Photovoltaic Performance Loss Rate

Yangxin Fan*, Raymond Wieser*, Xuanji Yu*, Jennifer Braid†, Avishai Shaton‡, Adam Hoffman§, Thevenard Didier¶,
Ben Spurgeon‖, Daniel Gibbons**, Laura S. Bruckman*, Yinghui Wu*, Roger H. French*

*SDLE Research Center, Case Western Reserve University, Cleveland, OH, USA
†Sandia National Laboratories, Albuquerque, NM, USA
‡SolarEdge Technologies, Herzliya, Israel
§Maxeon Solar Technologies, Singapore
¶Canadian Solar Inc., Guelph, ON, Canada
‖Brookfield Renewable U.S., New York City, NY, USA
**Bay4 Energy, Tucson, AZ, USA

Abstract—Estimation of Photovoltaic (PV) Performance Loss Rate (PLR) becomes increasingly important in the stage of rapid growth of PV industry. Accurate PLR estimation benefits PV users by providing real-time monitoring of the PV modules' performance. Explainable PLR estimation can help PV manufacturers study and improve performance of their products. However, traditional PLR estimations, based on statistical models, have some major disadvantages. First, they need user knowledge and decisions. Second, they tend to be less robust to non-uniform and low-quality data. To address these issues, we propose the NN-PLR, a Neural Network decomposition method for field PV PLR estimations. NN-PLR decomposes the power timeseries data into seasonality, trend, and remainder components and uses trend to estimate PLR. Decompositions used to derive PLR can reveal how PLR vary and change across different PV systems. Using digital power plant datasets, we have demonstrated that NN-PLR can produce comparable PLR estimation results to traditional PLR estimation methods while needing much less input tweaks.

Index Terms—Neural Network, Performance Loss Rate, Timeseries Decomposition

I. INTRODUCTION

As Photovoltaics (PV) have become an increasingly dominant energy sector over past few decades [1], the estimation of Performance Loss Rate (PLR) of PV systems becomes a recent focus of research [2]–[4]. Such methods could potentially provide a PV manufacturer a competitive edge since the manufacturer can estimate the performance of PV systems without time-consuming experiments and consistent monitoring of sampled systems. The growth of PV market has pushed the demand for degradation analysis for a large number of PV systems which have spatio-temporal correlations that can be used to improve model performance.

There are many challenges towards performing accurate field PV PLR estimation. First, field PV timeseries data used to derive PLR may be low-quality. They typically bear

This material is based upon work supported by the U.S. Department of Energy's Office of Energy Efficiency and Renewable Energy (EERE) under Solar Energy Technologies Office (SETO) Agreement Number DE-EE0009353. The views expressed herein do not necessarily represent the views of the U.S. Department of Energy or the United States Government.

missingness issue (with missing data and incorrect data from faulty sensors) [5]. Addressing the data quality issue is an essential part of data preprocessing, which itself may take more time than the PLR estimation model. We have proposed an automated framework for detecting missingness data and impute them using spatio-temporal graph neural networks [6]. Second, traditional PLR estimations need human-guided input tweaks such as data filtering, thresholding and external data sources like satellite data [7]. Due to the lack of a successful automated PLR estimation framework, the current approaches may take more time and tend to be less robust in estimating the PLR of "out-of-sample" PV systems.

Neural Networks and its deep learning variants have been successfully applied in a wide range of PV research problems, from fault detection and diagnosis to power forecasting [8], [9]. In this paper, we propose **NN-PLR**, an automated end-to-end PLR estimation model. NN-PLR uses feed-forward neural network, which conveys information in one direction through input nodes. Information continues to be processed in this single direction until it reaches the output layer. Traditional PLR methods require user decisions, they tend to generate inconsistent results and bias [7]. Neural Network-based method (NN-PLR) generates more consistent results and needs minimal user knowledge. Our experiments has shown that NN-PLR achieves comparable results to state-of-the-art traditional PLR estimation methods while needing much less input tweaks.

II. METHODS

A. Traditional PLR Estimation

Traditional PLR estimation was conducted using the R package **PVplr** [10]–[12]. Reported PLR values are the result of a 500 iteration bootstrap with a sampling rate of 80% of the original data. Each power predictive model is run with multiple irradiance thresholds and temporal aggregations. Irradiance thresholds were set to: 0, 200, 400, 600, and 800 ($\frac{W}{m^2}$). Temporal aggregations summarized the 5-minute interval data

978-1-6654-6060-6/23 $31.00 © 2023 IEEE

into periods of days, weeks, or months. Thus, each PLR estimation is ensemble value of 6000 PLR realizations.

B. Fourier Transform for Timeseries Decomposition

Statistical methods decompose timeseries into trend, seasonality, and residual elements. Fourier transform (FT) can be used to extract seasonal component [13]. Recall that a sine function $y(t)$ can be defined as:

$$y(t) = A \sin(wx + \phi). \tag{1}$$

where A is the magnitude, w is the frequency, and ϕ refers to the phase (also called time shift). FT decomposes a timeseries into the frequency domain. Simply put, an seismic wave in the time domain is decomposed into a set of frequencies and their corresponding and magnitudes. An important variant of FT is Discrete Fourier Transform (DFT). The input of DFT can be a sequence of numerical values (e.g., an AC power timeseries of a PV inverter). The output is a series of fourier coefficients with one value of amplitude for each frequency. If the amplitude of a particular frequency is large, the seasonality corresponding to such frequency is important.

C. Proposed NN-PLR Framework

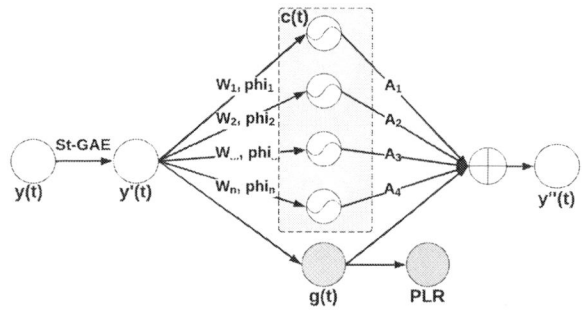

Fig. 1. Proposed NN-PLR Framework for PLR Estimation

Inspired by the prior work on neural decomposition for timeseries data [14], we propose an automated framework NN-PLR to estimate field PV performance loss rate, shown in Fig. 1. Neural decomposition is a neural network technique for extrapolation of time-series. The input of our framework is the inverter alternating current power timeseries, denoted by $y(t)$. However, the real-world timeseries normally include missingness values, i.e., missing and incorrect values. NN-PLR detects and then imputes missingness values in $y(t)$ using our proposed St-GAE (*denoised spaio-temporal graph autoencoder*) model [6]. St-GAE leverages spatio-temporal coherence within PV systems and value dependencies from domain knowledge to impute missingness data. It incorporates a domain-knowledge aware data augmentation module and integrates spatiotemporal graph convolution layers and denoising autoencoder to improve imputation accuracy. The imputed time series data $y'(t)$ serves as the high-quality input to a downstream *decomposition neural network*. NN-PLR decomposes $y'(t)$ using a three-layer feed-forward neural network. The decomposition process is shown in Eq. 2.

$$y''(t) = \sum_{i=1}^{n} (A_i sin(w_i t + \phi_i)) + g(t). \tag{2}$$

In the hidden layer, NN-PLR decomposes $y'(t)$ into two components $c(t)$ and $g(t)$. $c(t)$ is the seasonality component, consisting of n neurons. n is the number of samples in $y'(t)$. $g(t)$ denotes the non-periodical component, or in our case a linear trend, which may include one or more neurons. $y''(t)$ is the reconstructed timeseries. The model is trained by minimizing the loss between reconstructed $y''(t)$ and $y'(t)$. NN-PLR further estimates PLR from $g(t)$ using Eq. 3.

$$PLR = \frac{g(n) - g(1)}{n/N} \times 100\%. \tag{3}$$

Where N is the number of samples per year. In our framework, A_i, w_i, and ϕ_i are all trainable parameters. NN-PLR applies a $L1$ regularization to the amplitude A_i in the output layer. In this way, NN-PLR can force the amplitude weights of trivial sine signals to be zero and only keep relevant frequencies of sines that contribute most to seasonal component $c(t)$. Hence, learned weights of amplitudes can be used to rank what are the most important seasonal information in the timeseries.

III. DATASETS DESCRIPTION

We select four EDF digital power plant timeseries datasets from our previous work [7] to evaluate the performance of NN-PLR. Detailed information about datasets are shown in Table. I. All datasets include same set of attributes, including timestamp, Global Plane of Array Irradiance (GPOA), ambient temperature, wind speed, AC power, and DC power. Since these are simulated data, we know the "ground truth" of PLR for each PV systems. Two of them have no degradation at all and the other two have -4.89%/annual degradation. We can acquire fair comparison of PLR estimation accuracy between NN-PLR and traditional PLR methods using these datasets.

TABLE I
DIGITAL POWER PLANT DATASETS USED FOR PLR ESTIMATION

Dataset	Meteo	Degradation
Rennes_SameMeteo_5ans_Deg0	Same every year	0.00%/a
Rennes_SameMeteo_5ans_Degx	Same every year	-4.89%/a
DPP_Rennes_HC5ansELM_DEG0	HelioClim+1 cold year	0.00%/a
DPP_Rennes_HC5ansELM_DEGx	HelioClim+1 cold year	-4.89%/a

IV. PFCRELIMINARY RESULTS

A. Traditional PLR Estimation Results

The accuracy of the traditional PLR estimation methods varied greatly across different hyper-parameters. Traditional methods performed the best on simulated data with no degradation and cyclic weather patterns, as shown in Table. II. All power predictive models over predict PLR value for the real world weather simulated data. **XbX + UTC** and **PVUSA** power models have the smallest ensemble MAE compared to the ground truth value of 0% per year. For the digital power

978-1-6654-6060-6/23 $31.00 © 2023 IEEE

TABLE II
TRADITIONAL PLR ESTIMATION RESULTS (DATASET: RENNES_SAMEMETEO_5ANS_DEG0)

Est. Method	Power Model	$\overline{PLR}$	Std. $\overline{PLR}$ (% per a)	95% Confidence Interval
Regression	6K	-0.887	0.381	(-1.103,-0.671)
Regression	PVUSA	0.045	0.123	(-0.025,0.114)
Regression	XbX	0.044	0.118	(-0.023,0.111)
Regression	XbX + UTC	-0.107	0.099	(-0.162,-0.051)
YoY	6K	0.155	0.305	(-0.018,0.327)
YoY	PVUSA	-0.004	0.011	(-0.011,0.002)
YoY	XbX	-0.010	0.026	(-0.025,0.005)
YoY	XbX + UTC	-0.005	0.044	(-0.031,0.02)

plants with simulated degradation PLR estimation methods under predicted the truth value of PLR. The presence of real world weather signals increased the MAE of the PLR estimation when compared to 4.89% per year.

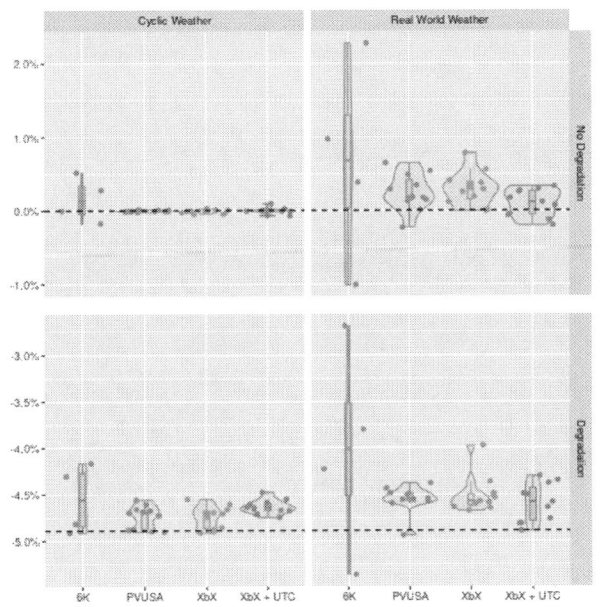

Fig. 2. Violin plots of PLR Estimated by Traditional PLR Estimation methods (Dataset: EDF digital Power Plants).

B. NN-PLR Results

Our preliminary results are acquired using Rennes_SameMeteo_5ans_Deg0 dataset (see Fig. 3). We set the epoch to be 100, $L1$ regularization rate to be 0.01, batch size to be 8, and validation_length to be 24 in our experiments. NN-PLR decomposes original timeseries into $c(t)$ and $g(t)$. We then use the Eq. 3 to estimate the PLR. To generate Fig. 3, we have run NN-PLR model with same parameter setting 500 times. It represents the cumulative distribution function (cdf) of PLR by NN-PLR. Both mean and median of PLR estimations are close to 0.00%/a, i.e.: the "ground truth" of this dataset. Mean of the 500 runs of NN-PLR is 0.0084%/a with standard deviation 0.067%/a. The 95% confidence interval is $(-0.013, -0.004)$. Compared to the state-of-the-art traditional methods, NN-PLR outperformed

all regression based methods and achieves comparable results to best performed YoY methods (please refer to Table. II).

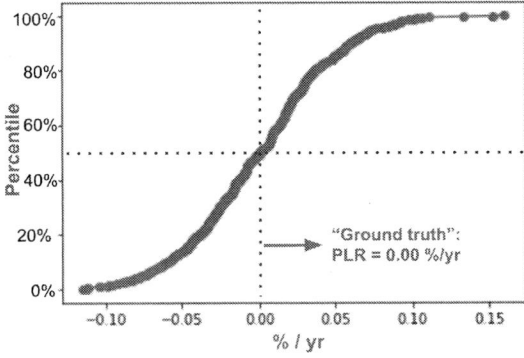

Fig. 3. Cumulative Distribution Plot of PLR Estimated by NN-PLR (Dataset: Rennes_SameMeteo_5ans_Deg0)

V. CONCLUSIONS

In this paper, we propose the NN-PLR, a dataset-agnostic and explainable Neural Network decomposition method for field PLR estimations. NN-PLR decomposes the power time-series data into seasonality (daily and yearly), trend, and remainder components. It uses trend derived from decomposition to estimate PLR. Decompositions used to derive PLR can help to calculate how PLR vary and change across different PV systems. Using digital power plant datasets, we have shown that NN-PLR can generate comparable PLR estimation results of state-of-the-art traditional PLR estimation methods.

REFERENCES

[1] J. Christiansen, "Global Market Outlook for Solar Power," SolarPower Europe, Tech. Rep., 2021.

[2] M. Theristis, A. Livera, C. B. Jones, G. Makrides, G. E. Georghiou, and J. S. Stein, "Nonlinear photovoltaic degradation rates: Modeling and comparison against conventional methods," *IEEE Journal of Photovoltaics*, vol. 10, no. 4, pp. 1112–1118, 2020.

[3] A. J. Curran, C. B. Jones, S. Lindig, J. Stein, D. Moser, and R. H. French, "Performance loss rate consistency and uncertainty across multiple methods and filtering criteria," in *2019 IEEE 46th Photovoltaic Specialists Conference (PVSC)*. IEEE, 2019, pp. 1328–1334.

[4] I. Kaaya, M. Koehl, A. P. Mehilli, S. de Cardona Mariano, and K. A. Weiss, "Modeling outdoor service lifetime prediction of pv modules: effects of combined climatic stressors on pv module power degradation," *IEEE Journal of Photovoltaics*, vol. 9, no. 4, pp. 1105–1112, 2019.

[5] S. Lindig, D. Moser, A. J. Curran, K. Rath, A. Khalilnejad, R. H. French, M. Herz, B. Müller, G. Makrides, G. Georghiou *et al.*, "International collaboration framework for the calculation of performance loss rates: Data quality, benchmarks, and trends (towards a uniform methodology)," *Progress in Photovoltaics: Research and Applications*, vol. 29, no. 6, pp. 573–602, 2021.

[6] Y. Fan, X. Yu, R. Wieser, D. Meakin, A. Shaton, J.-N. Jaubert, R. Flottemesch, M. Howell, J. Braid, L. Bruckman, R. French, and Y. Wu, "Spatio-temporal denoising graph autoencoders with data augmentation for photovoltaic timeseries data imputation," in *2023 ACM SIGMOD Conference on Management of Data (SIGMOD)*. ACM, 2023.

[7] R. H. French, L. S. Bruckman, D. Moser, E. Lindig, M. van Iseghem, J. S. Stein, M. Richter, M. Herz, W. van Sark, F. Baumgartner *et al.*, "Assessment of performance loss rate of pv power systems," 2021.

[8] W. Chine, A. Mellit, V. Lughi, A. Malek, G. Sulligoi, and A. M. Pavan, "A novel fault diagnosis technique for photovoltaic systems based on artificial neural networks," *Renewable Energy*, vol. 90, pp. 501–512, 2016.

[9] A. M. Karimi, Y. Wu, M. Koyutürk, and R. H. French, "Spatiotemporal graph neural network for performance prediction of photovoltaic power systems." in *AAAI*, 2021, pp. 15 323–15 330.

[10] A. Curran, T. Burleyson, S. Lindig, D. Moser, R. French, and SDLE Research Center, "PVplr: Performance Loss Rate Analysis Pipeline," Oct. 2020, tex.ids: a.j.curranPVplrSDLEPerformance2020, curran-PVplrPerformanceLoss2020. [Online]. Available: https://CRAN.R-project.org/package=PVplr

[11] A. J. Curran, C. Birk Jones, S. Lindig, J. Stein, D. Moser, and R. H. French, "Performance Loss Rate Consistency and Uncertainty Across Multiple Methods and Filtering Criteria," in *2019 IEEE 46th Photovoltaic Specialists Conference (PVSC)*. Chicago, IL, USA: IEEE, Jun. 2019, pp. 1328–1334.

[12] D. Moser, D. Bertani, A. J. Curran, R. H. French, M. Herz, and S. Lindig, "International Collaboration Framework for the Calculation of Performance Loss Rates: Data Quality, Benchmarks, and Trends," in *36th European Photovoltaic Solar Energy Conference and Exhibition*, Oct. 2019, pp. 1266–1271, citation Key Alias: moserInternationalCollaborationFramework2019a.

[13] A. C. Jackson and S. Lacey, "Seasonality and anomaly detection in rare data using the discrete fourier transformation," in *2019 First International Conference on Digital Data Processing (DDP)*. IEEE, 2019, pp. 13–17.

[14] L. B. Godfrey and M. S. Gashler, "Neural decomposition of time-series data," in *2017 IEEE International Conference on Systems, Man, and Cybernetics (SMC)*. IEEE, 2017, pp. 2796–2801.

978-1-6654-6060-6/23 $31.00 © 2023 IEEE

Capacitance Transients, Photoconductive Decay, and Impedance Spectroscopy on 19% to 22% Efficient Silicon Solar Cells

Steve Johnston and Dana B. Kern

National Renewable Energy Laboratory, Golden, CO, 80401, U.S.A.

Abstract—High efficiency silicon solar cells are characterized using current-voltage curves, electroluminescence imaging, impedance spectroscopy, capacitance transients, microwave photoconductive decay, and time-resolved photoluminescence imaging. The sample set is composed of cells from different manufacturers and includes an n-type silicon heterojunction (SHJ), an n-type passivated emitter rear totally diffused (PERT), and five different p-type passivated emitter rear contact (PERC) cells. Carrier lifetimes, both photoconductivity and photoluminescence, are measured co-located with the light excitation pulse and within the cell but away from the light spot. Luminescence intensity and excess carrier lifetimes correlate to cell voltage. The capacitance transient time constants correlate to the capacitance values extracted from impedance spectroscopy.

Keywords— Photovoltaic cells, charge carrier lifetime, imaging, silicon, impedance spectroscopy, capacitance, photoconductivity.

I. INTRODUCTION

Silicon solar cells continue to increase in efficiency as designs, materials, and manufacturing are improved. Transitions include a migration from p-type bulk with aluminum back surface field to high quality p- and n-type monocrystalline wafers with enhanced passivation such as n-type silicon heterojunction (SHJ), n-type passivated emitter rear totally diffused (PERT), and p-type passivated emitter rear contact (PERC) which has a localized back-surface field. Characterization can include several techniques that measure efficiency and cell parameters, provide spatial resolution for imaging defects and degradation, and determine time constants such as excess carrier lifetime. We apply electroluminescence (EL) imaging [1] and impedance spectroscopy [2]-[5] and collect transients of capacitance, microwave photoconductive decay (μ-PCD), and photoluminescence (PL) imaging [6],[7] for a set of efficient solar cell types.

II. EXPERIMENT

Selected SHJ, PERT, and PERC cells have been collected from several different manufacturers, and these have been initially screened for defects or abnormalities using EL imaging. The EL images are collected using a Princeton Instruments PIXIS 1024BR silicon charge-coupled-device camera cooled to -60°C. A current of 9 A (~short-circuit current) is driven through the cells, and the camera exposure time is set to 0.5 s, except for the SHJ cell using 0.05 s. The EL images are shown in Fig. 1, and the cells appear mostly uniform with no obvious cell cracks or areas of damage or non-uniformity.

The cells are characterized by measuring their current-voltage (I-V) curves with a light-emitting-diode-based G2V

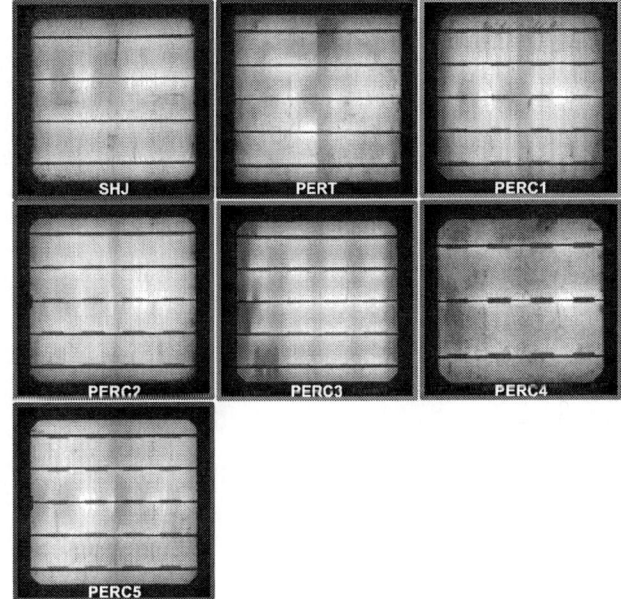

Fig. 1. EL images of the SHJ, PERT, and PERC cells are individually scaled in contrast and brightness to display the cell features.

Sunbrick having a simulated 1-Sun-intensity AM1.5 spectrum. The cells are mounted on a temperature-controlled metal plate

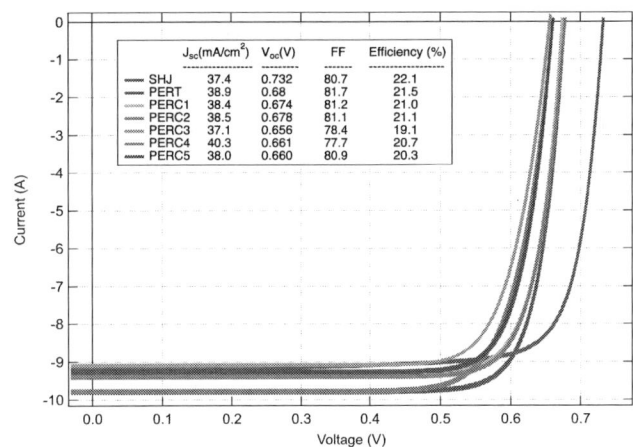

	J_{sc}(mA/cm^2)	V_{oc}(V)	FF	Efficiency (%)
SHJ	37.4	0.732	80.7	22.1
PERT	38.9	0.68	81.7	21.5
PERC1	38.4	0.674	81.2	21.0
PERC2	38.5	0.678	81.1	21.1
PERC3	37.1	0.656	78.4	19.1
PERC4	40.3	0.661	77.7	20.7
PERC5	38.0	0.660	80.9	20.3

Fig. 2. Light I-V curves, collected using a 1-Sun AM1.5 simulated spectrum, are used to extract parameters of J_{sc}, V_{oc}, FF, and efficiency.

(25°C), and narrow-bar, multi-pin probes on each busbar are used to contact the cells. The I-V curves are plotted in Fig. 2, where cells have 19 to 22% efficiency. Cell performance parameters include short-circuit current density (J_{SC}), open-circuit voltage (V_{OC}), and fill factor (FF).

Impedance spectroscopy (IS) is collected using a Zurich Instruments MFIA impedance analyzer. The real and imaginary impedances are collected as frequency is swept from 0.5 Hz to 1 MHz using 200 data points spread out logarithmically. The applied bias is 0 V_{dc}, and the measurement signal has an amplitude of 300 mV_{ac}. The cells are similarly mounted as the I-V measurements using a metal plate and bar probes on the top busbars. The cells are measured in the dark. The results are plotted in a Nyquist-plot format [3],[4] where imaginary impedance (not inverted, in this case) is plotted against real impedance, as shown in Fig. 3. The half-circle shape is characteristic of an impedance that consists of a parallel combination of capacitance and resistance, and those values are extracted from the data. The capacitance values are shown in the legend, and the shunt resistances are the widths of the semicircles ranging from 200 to 2000 Ω. The values are modeled as parallel components and the resulting impedances are shown as black dotted-line fits. Any offset of the half-circle in the x-axis from the origin would be due to series resistance, and those values are negligible. The equivalent circuit time constant is given by $\tau_{RC} = RC$, where R is the shunt resistance and C is the cell capacitance.

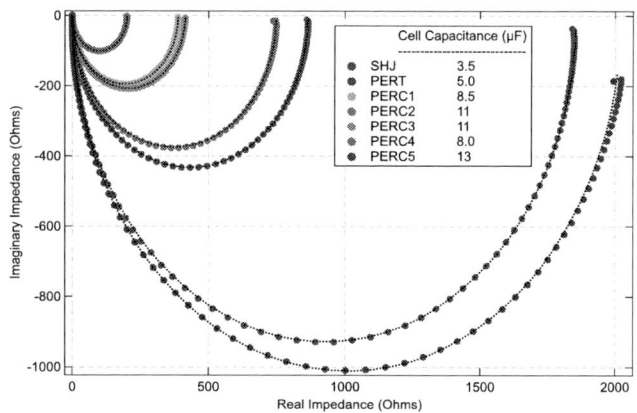

Fig. 3. Nyquist plots generated from impedance spectroscopy data.

Using the same impedance analyzer and cell probing setup, we collect capacitance transients by pulsing the applied voltage between 0 V and -0.2 V_{dc} (reverse bias). The applied reverse bias expands the solar cell built-in depletion width which reduces the capacitance of the cell. The voltage is applied as a square wave with a 20 ms period. The transients are collected using 1000 averages and a frequency of 100 kHz. As the dc voltage is pulsed to 0 V, the capacitance increases over time as shown in Fig. 4. When voltage is pulsed back to -0.2 V, capacitance exponentially decays back to values for that voltage. In Fig. 5, capacitance decays are plotted on a log scale, and exponential fits are shown and give the capacitance transient time constants, τ_{CAP}.

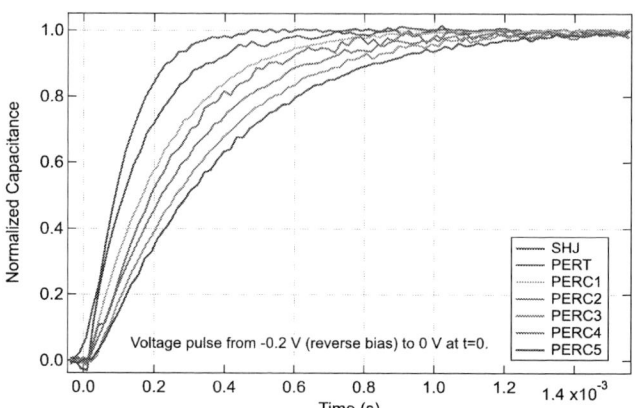

Fig. 4. Capacitance transients as voltage is stepped from reverse bias of -0.2 V to 0 V. Peak capacitance values are normalized to 1 for comparison of all data and respective transition times.

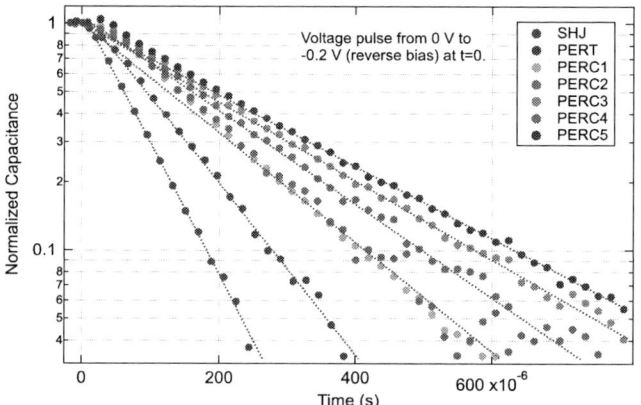

Fig. 5. Capacitance transients plotted on a log scale when voltage is stepped from 0 V to a reverse bias of -0.2 V. Exponential fits give the capacitance time constants.

Excess carrier lifetimes are measured by collecting transients of microwave photoconductive decay (μ-PCD). The excitation source is a ns-pulse Nd:YAG laser with optical parametric oscillator tuned to 1100 nm. Microwave power at 20 GHz is directed to the front side of the cell using an open-ended waveguide, and reflections are collected, amplified, and rectified to display the PCD signal on an oscilloscope. We label the data as 'on-spot' when the microwave signal is measured at the same area where the laser pulses excite carriers. The PCD data for on-spot measurements are shown in Fig. 6. The decay shape initially shows a relatively long lifetime at the highest injection followed by a shorter lifetime which is labeled as mid-injection-level region. This shape of this decay is typically associated with defect-related recombination where the majority-carrier capture cross section is smaller than that for minority carriers. The knee of the decay curve would then be associated with an injection level near the bulk doping of the silicon wafers. However, since these samples are cells with a built-in electric field, the high-injection-region is associated with an increase in photoconductivity where excess carriers are generated and separated by the built-in junction. Carrier recombination still counteracts the build-up of excess carriers and voltage, but the field-effect separation also decreases total

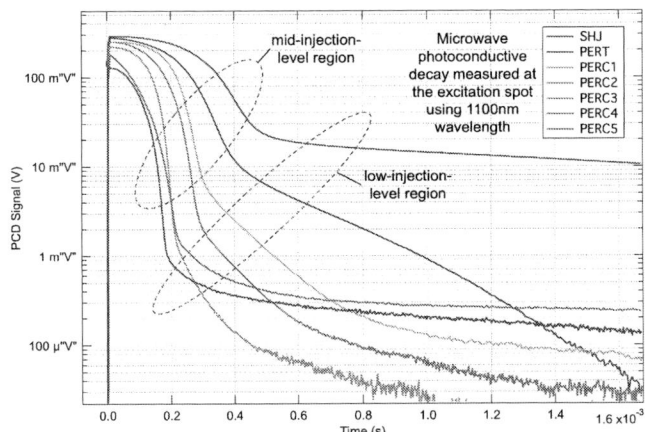

Fig. 6. Microwave photoconductive decay transients measured at the same area as the laser excitation (on-spot).

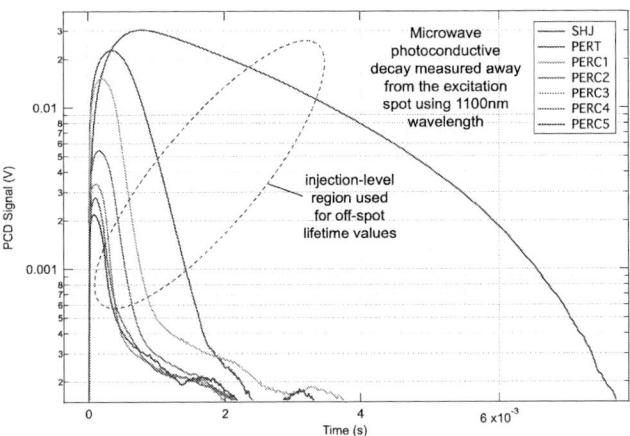

Fig. 7. Microwave photoconductive decay transients measured at an opposite corner of the cell than the corner of laser excitation (off-spot).

recombination. The mid-injection-level region then corresponds to a combination of recombination of carriers and voltage spreading out over the entire cell area which is connected by the highly-doped and metalized structures of the cell.

The PCD signals then transition to a longer lifetime as labeled in Fig. 6 as the low-injection-level region. This region of injection-level-related recombination might typically be associated with carrier trapping and emission of carriers in a bulk silicon sample. However, for the finished cells measured here, the built-in junction sweeps carriers to the open-circuited base and emitter sides of the cell. The low-injection-level lifetimes are associated with the full-cell equilibration of carriers and voltage.

Localized light on the solar cell initially generates a localized voltage, and that voltage spreads out across the cell using the conductive contact structures. The voltage then induces excess carriers similarly to externally applying a forward bias. This is visualized later with luminescence imaging. We define PCD measurements where the laser spot and microwave waveguide are not co-located as 'off-spot'. The off-spot PCD data are shown in Fig. 7. The curves show a rounded rise-time peak

which is the transition for the laser light to generate excess carriers and build up a voltage that spreads throughout the cell structure. The high- and mid-injection-level regions of the "on-spot" μ-PCD curves (Fig. 6) match this rise time in the "off-spot" curves of Fig. 7. After cell voltage equilibrium is established, the excess carriers recombine with very similar rates at both the "on-spot" and "off-spot" locations, as shown by comparing the low-injection-level region of Fig. 6 to the circled injection-level region of Fig. 7.

Excess carrier lifetimes are also measured by imaging luminescence of the cells when using the pulsed laser spot for carrier excitation. The laser is tuned to a 900 nm excitation wavelength. PL images are collected using a Princeton Instruments NIRvana InGaAs camera with a ~950 nm long-pass filter to block reflected laser light. The camera uses a 100 μs exposure time, and 100 images are averaged for each delay time.

The camera is triggered at fixed delay times with respect to the laser pulse. For short times, images are collected every 20 μs up to 640 μs. These steps are shown in Fig. 8 on the logarithmic-scaled time axis where the red tick marks show the image

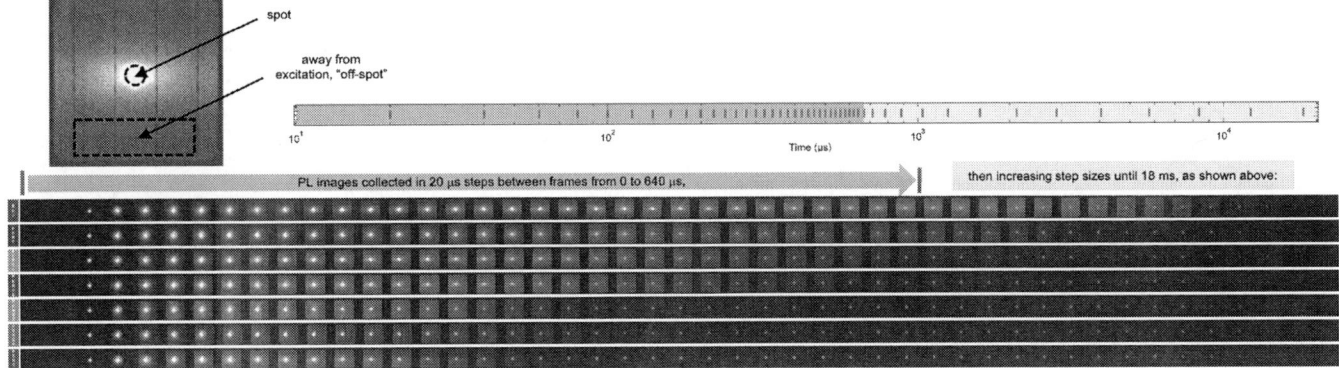

Fig. 8. PL imaging collects an image of the full cell but uses a localized laser excitation spot near the center of the cell. PL images are collected using a 100 μs exposure time, and the image collection time begins at fixed steps in time with respect to the laser pulse. Images are collected in fixed steps of 20 us from just before the laser pulse until about 640 μs after the laser pulse. Then, images are collected with increasing steps in time from 640 μs to 18 ms.

978-1-6654-6060-6/23 $31.00 © 2023 IEEE

collection times. After 640 μs, images are collected with increasing times steps up to a final time of 18 ms.

The example PL image of Fig. 8 shows the laser excitation area is nominally near the center of the wafer but kept between the front busbars. This round spot is the location where excess carriers are initially generated. Soon after the laser pulse, the voltage induced by the excess carriers builds up and spreads throughout the cell area. The strips of images in Fig. 8 show how the voltage spreads over the cells in time and induces a contactless, light-induced EL [8]-[11]. The colored boxes at the left of each image strip in Fig. 8 correlate to the cell types listed in the legends of the other figures. The n-type cells (upper image strips) show extended decay times of full-cell luminescence compared to the cells labeled as PERC3 through PERC5 (lower image strips).

The PL intensity from each image is measured by averaging within a circular area from the direct laser excitation, and this value is plotted against the respective image delay time. This "on-spot" PL data is plotted in Fig. 9. The voltage-induced luminescence intensity is measured away from the excitation spot by averaging within a rectangular area from the bottom portion of the cell as shown in Fig. 8. This "off-spot" luminescence data is plotted in Fig. 10.

For each of the measurements performed on the set of cells, the data are summarized and plotted in the graphs of Fig. 11. The data points are color coded as labeled in the previous figures. The EL intensity is scaled to an equivalent camera exposure time by subtracting dark background counts and multiplying by factors for the exposure time correction. The EL intensity is plotted on a logarithmic scale since EL intensity typically varies exponentially with cell voltage. The full-cell-area average intensity shows a correlation to the cells' V_{OC} values as plotted in Fig. 11(a).

The PL transients are fit with exponential decay time constants to extract an excess carrier lifetime. These lifetime values are fit for the PL transients that are directly illuminated by the laser spot and plotted in Fig. 11(b). These "on-spot" PL lifetimes show a direct correlation to the cell V_{OC}. Luminescence intensity is also fit in the bottom area of the cell where the voltage-induced luminescence quickly builds and then decays as shown in Figs. 8 and 10. These "off-spot" luminescence decay lifetimes are plotted in Fig. 11(c). and show a direct correlation to the PL lifetimes. Since they are directly proportional, the "off-spot" luminescence decay times also directly correlate to V_{OC}.

The μ-PCD excess carrier lifetimes represented by decay fits of the mid-injection-level PCD of Fig. 6 also correlate to V_{OC} as shown in Fig. 11(d). The on-spot PCD curves show two regimes of decay where the mid-level is affected by local carrier dynamics where both recombination and spreading voltage compete to reduce excess carriers within the original generation area. The low-injection-level region is dominated by full-cell-area dynamics that have equilibrated at the longer times. There is correlation of these two injection-level regimes as seen in Fig. 11(e). This indicates that both are affected by carrier recombination even if carriers are contributing to voltage spread in parallel. As previously described, the on-spot low-level decay lifetimes match the off-spot decay lifetimes very well, as shown in Fig. 11(f) since these both occur after the initial localized injection of carriers has equilibrated to full cell voltage and excess carrier distribution.

The capacitance decay rates from Fig. 5 show an inversely proportional relationship to V_{OC}, as plotted in Fig. 11(g). Using the IS data, capacitance and resistance values are extracted from the real and imaginary impedances of the Nyquist plot. As shown in Fig. 11(h), the capacitance decay times show a directly proportional relationship to the extracted cell capacitance values. These relationships then give an inverse correlation of cell capacitance to V_{OC}. An equivalent circuit of a parallel resistance and capacitance fit the Nyquist data well (Fig. 3), and the values are also used to calculate a circuit time constant where $\tau_{RC} = RC$. The transient capacitance decay rates are compared to these RC time constants in Fig. 11(i), and no strong correlation is observed. While the capacitance transients have decay rates that correlate to the cell capacitance, the cell resistance values do not give equivalent-circuit time constants that correlate to the capacitance transients of the cells.

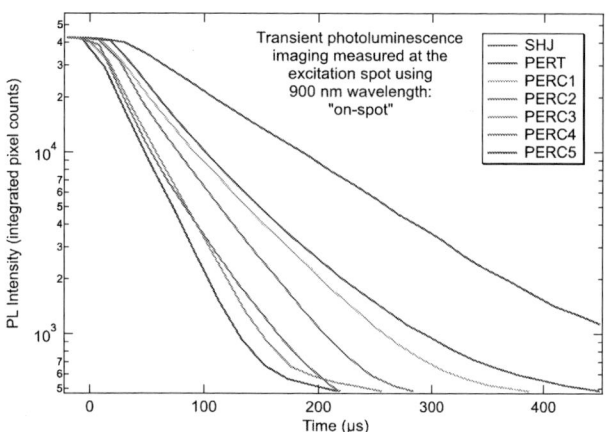

Fig. 9. PL decay transients measured at the same area as the laser excitation (on-spot).

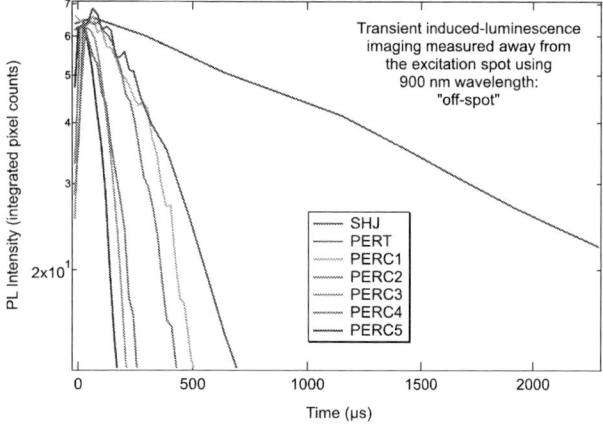

Fig. 10. Voltage-induced luminescence decay transients measured at a bottom area of the cell away from the cells' center laser excitation areas (off-spot).

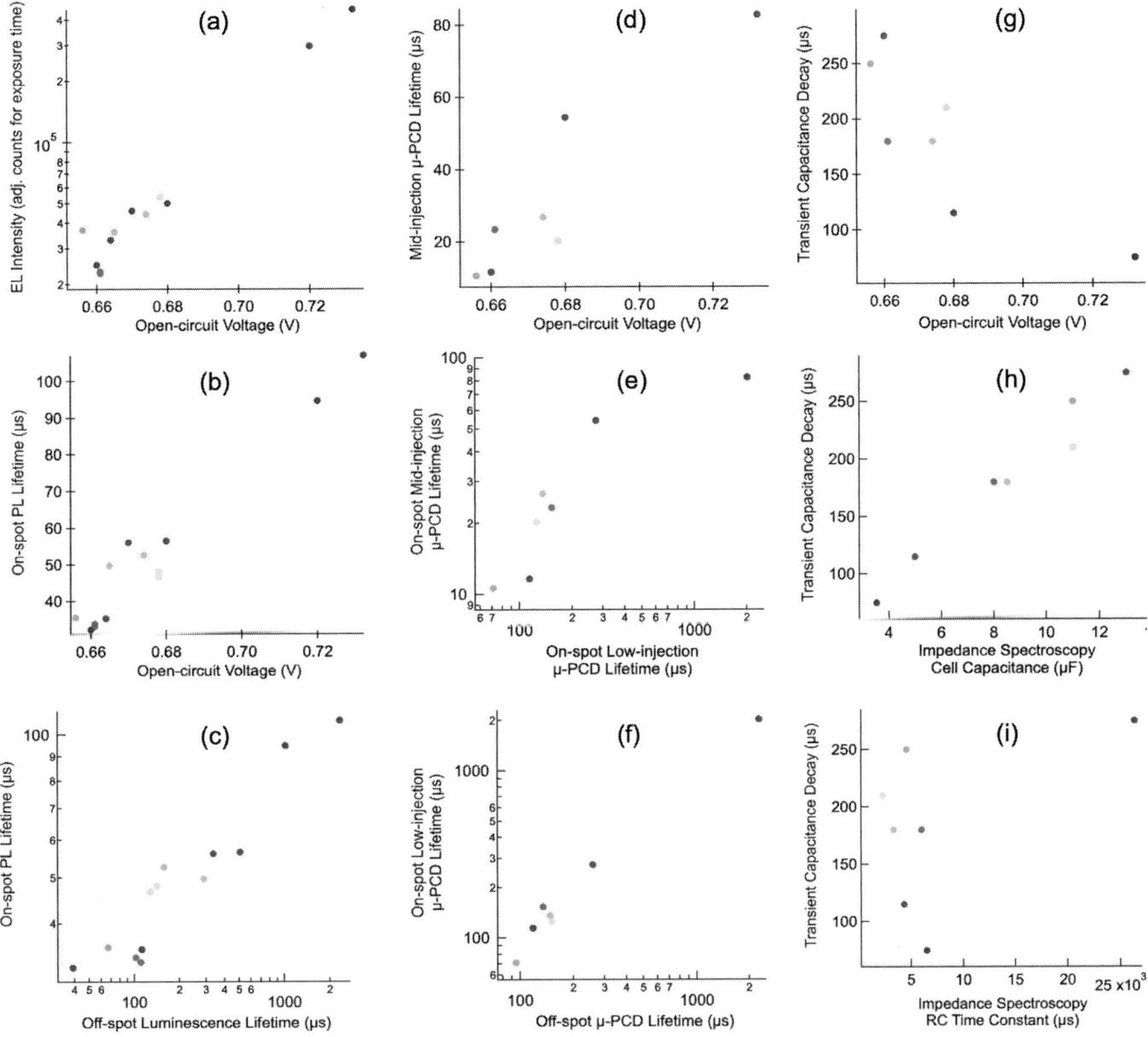

Fig. 11. Measured parameters are plotted to show interactive relationships. EL image intensity shows a direct correlation to V_{OC} (a). On-spot PL lifetime shows a direct correlation to V_{OC} (b) and to off-spot luminescence lifetime (c). Mid-injection-level μ-PCD shows a direct correlation to V_{OC} (d) and to low-injection-level μ-PCD (e). Low-injection μ-PCD matches off-spot μ-PCD (f). Transient capacitance time constants show an inversely proportional relationship to V_{OC} (g). Transient capacitance time constants show a direct correlation to measured cell capacitance extracted from IS (h). Transient capacitance time constants do not show a strong relationship to the equivalent circuit time constants, τ_{RC}, which are calculated using the equivalent circuit R and C values (i).

III. Summary

High efficiency silicon SHJ, PERT, and PERC solar cells are characterized using I-V curves, EL imaging, impedance spectroscopy, microwave photoconductive decay, and photoluminescence decay. Electroluminescence intensity and excess carrier lifetimes correlate to cell voltage. Carrier lifetimes are measured either co-located with the light excitation pulse or at a location of the cell away from the light spot, and each are affected by recombination and correlate to cell voltage. The capacitance values extracted from impedance spectroscopy correlate to the exponential time constants of the capacitance transients. The capacitance transient decay rates show inversely proportional relationships to cell voltages and excess carrier lifetimes.

ACKNOWLEDGMENT

We thank Peter Hacke for providing varieties of cells for this study.

This work was authored by the National Renewable Energy Laboratory, operated by Alliance for Sustainable Energy, LLC, for the U.S. Department of Energy (DOE) under Contract No. DE-AC36-08GO28308. Funding provided by U.S. Department of Energy Office of Energy Efficiency and Renewable Energy Solar Energy Technologies Office. The views expressed in the article do not necessarily represent the views of the DOE or the U.S. Government.

The U.S. Government retains and the publisher, by accepting the article for publication, acknowledges that the U.S. Government retains a nonexclusive, paid-up, irrevocable, worldwide license to publish or reproduce the published form of this work, or allow others to do so, for U.S. Government purposes.

This material is based upon work supported by the U.S. Department of Energy's Office of Energy Efficiency and Renewable Energy (EERE) under Solar Energy Technologies Office (SETO) Agreement Number 38263.

REFERENCES

[1] T. Fuyuki, H. Kondo, T. Yamazaki, Y. Takahashi, and Y. Uraoka, "Photographic Surveying of Minority Carrier Diffusion Length in Polycrystalline Silicon Solar Cells by Electroluminescence," *Appl. Phys. Lett.*, vol. 86, pp. 262108-1–262108-3, 2005.

[2] A. Blank and N. Suhareva, "Admittance Spectra of Silicon Photocells in Dark Mode." *Sensors and actuators. A. Physical.* 331 (2021): 112909–.

[3] D. K. Sharma, K. Pareek, and A. Chowdhury, "Investigation of Solar Cell Degradation Using Electrochemical Impedance Spectroscopy." *International journal of energy research* 44, no. 11 (2020): 8730–8739.

[4] A. Bouzidi, W. Jilani, I. S. Yahia, and H. Y. Zahran, "Impedance Spectroscopy of Monocrystalline Silicon Solar Cells for Photosensor Applications: Highly Sensitive Device." *Physica. B, Condensed matter* 596 (2020): 412375–.

[5] M. M. Shehata, T. N. Truong, R. Basnet, H. T. Nguyen, D. H. Macdonald, and L. E. Black, "Impedance Spectroscopy Characterization of c-Si Solar Cells with SiOx/ Poly-Si Rear Passivating Contacts." *Solar energy materials and solar cells* 251 (2023): 112167–.

[6] T. Trupke, R. A. Bardos, M. D. Abbott, F. W. Chen, J. E. Cotter, and A. Lorenz, "Fast photoluminescence imaging of silicon wafers," in *Proc. 32nd IEEE Photovoltaic Spec. Conf.*, 4th World Conf. Photovoltaic Energy Convers., pp. 928–931, 2006.

[7] T. Trupke, R. A. Bardos, M. C. Schubert, and W. Warta, "Photoluminescence imaging of silicon wafers," *Appl. Phys. Lett.*, vol. 89, pp. 044107-1–044107-3, 2006.

[8] D. B. Sulas, S. Johnston, and D. C. Jordan, "Comparison of Photovoltaic Module Luminescence Imaging Techniques: Assessing the Influence of Lateral Currents in High-Efficiency Device Structures." *Solar energy materials and solar cells* 192 (2019): 81–87.

[9] D. B. Sulas, S. Johnston, and D. C. Jordan, "Imaging Lateral Drift Kinetics to Understand Causes of Outdoor Degradation in Silicon Heterojunction Photovoltaic Modules," Article No. 1900102, 2019, *Solar RRL* 3, 8, (August 2019), 6 pp.

[10] S. Johnston, "Contactless Electroluminescence Imaging for Cell and Module Characterization," *Proceedings of the 2015 IEEE 42nd Photovoltaic Specialist Conference (PVSC)*, 14-19 June 2015, New Orleans, Louisiana 6 pp.

[11] M. Kontges, J. Wagner, M. Siebert, S. Bordihn, C. Schinke, "Applicability of Light Induced Luminescence for Characterization of Internal Series-Parallel Connected Photovoltaic Modules," *IEEE Journal of Photovoltaics*, 2022, Vol.12 (3), p.805-814.

Development Of Next Generation Solar Trackers Based On Shape Memory Alloy To Be Integrated In CPV/PV Hybrid Modules

Alessandro Minuto, Edoardo Celi and Gianluca Timò

RSE, Piacenza, Italy, Alessandro.minuto@rse-web.it

Abstract — **A solar tracker based on shape memory alloy (SMA) actuators has been developed with the purpose of its future integration within a CPV/PV hybrid module. The use of SMAs allows simplifying the technology of solar trackers, drastically reducing their costs, as well as their environmental impact, while increasing their reliability. To demonstrate the integrated solar tracker (IST) concept, a first functional unit (FU) of CPV module has been developed. It consists of four fixed optical lenses that focus the solar light on four CPV multijunction solar cells installed on a mobile plate that is translated in X and Y direction by SMA actuators. A new closed-loop control system that uses a low-cost LEDs matrix as a feedback sensor completes the FU. Remarkably, the movable plate has been manufactured with a weight of 7 kg to speed up the development efforts towards future commercial modules made up of hundreds of CPV cells. Despite the well-known difficulty to smoothy control SMA actuators, the performed preliminary outdoor tests of the FU successfully showed that in 90% of the experimental cases the absolute misalignment error of the light spot concentrated by the lenses has been less than 0.1 mm, by considering a total stroke of 65 mm. These results are very promising for the development of the new generation of low cost-high concentration, environmentally friendly, CPV/PV hybrid modules with IST.**

I. INTRODUCTION

The Concentrator Photovoltaic technology (CPV) is based on high efficiency multijunction solar cells, that allow, for the same surface area, generating significantly higher power peaks than the flat plate PV one. Nevertheless, the requirement of using high precision biaxial solar trackers has been so far a bottleneck for the diffusion of CPV technology, in particular, due to reliability, cost, weight and size issues [1]. For these reasons, a new generation of hybrid CPV/PV modules with integrated micro-trackers (IMT) have been proposed [2]. In such modules the optical lens array is fixed, while there is a movable plane actuated by micro-motors, installed on the back side of the module, in order to keep the alignment of the concentrated light over the CPV cells along the time. However, the IMT actuated by micro-motors are based on conventional control scheme which could be affected by maintenance and cost issues. To overcome these limitations, we have started developing a new integrated solar tracker (IST) concept (preliminarily described in [3]), in which the electric motors are replaced by shape memory alloy (SMA) springs. SMA are composed on NiTi nickel-titanium alloys which, following a considerable plastic deformation with respect to the nominal length of the springs, once heated (for example, by injecting electric current) can contract, resuming their initial shape. The main advantages of the SMA actuators compared to conventional ones (e.g., electric motors) are their low cost, high reliability and durability, high force density (i.e., the ability to generate high force values in relation to its weight and volume), with consequent ease of integration within the module and lower environmental impact. So far, the precision applications of these actuators have been limited to the displacement of very light loads (a few grams) and excursions of a few millimetres (for example the stabilization system of the mobile phone camera). In applications where the weight of the load is much higher (over 1 kg) and where the excursions are wider (a few centimetres), these actuators have been mostly used as switches, in ON/OFF mode, without requiring smoothy control. This is due to the strong non-linear behaviour of the force generated by the SMA actuators as a function of temperature (which includes a hysteresis behaviour) [4][5]. Solar tracking systems based on SMA actuators have already been reported: in [6], (not integrated in the module), where high precision was not required, as they weren't designed for the CPV application and in [7], where a first IST concept based on SMA suitable for the micro-CPV application was presented, however, in this case, the strokes and weights were very limited.

With reference to the previous work [3], in this contribution we describe the progress related to the development of the new IST control system based on SMA actuators, in which the weight and the stroke are respectively in the range of kilograms and several centemeters, therefore suitable to develop next generation CPV/PV hybrid modules with size comparable to the current PV modules.

II. DEVELOPMENT OF THE NEW INTEGRATED SOLAR TRACKER BASED ON SMA ACTUATORS

To reduce costs and increase the reliability of the IST, in addition of replacing the motors with SMA springs, we have designed a new tracking concept based on a reduced number of components which, since, already massive widespread in other industrial sectors, are very cheap. With reference to the schemes reported in Fig.1 and Fig.2, thanks to the wide acceptance angle of the sun pointing device (SPD) (already described in [3]), the new IST concept uses few components, as it allows eliminating: a) the precision clock with its buffer battery; b) the encoders for reading the current position of the tracking plane; c) the communication ports for entering input data (astronomical coordinates and roof orientation) by the

978-1-6654-6060-6/23 $31.00 © 2023 IEEE

installer; d) the reference switches with the encoders. Furthermore, thanks to the use of SMA actuators, gearmotors and limit switches are not required. The difficulty to smoothy control the SMA actuators has been managed both by improving the SMA spring thermo-mechanical parameters (by applying a proper thermal process during their manufacturing) and by adopting a new control scheme, reported in Fig.3. Referring to such scheme, each SMA spring can be represented by a block with three inputs: the ambient temperature T_{air}, the movable plane position $x_m(t)$, the electric current $I_1(t)$, and only one output that is the force value F1.

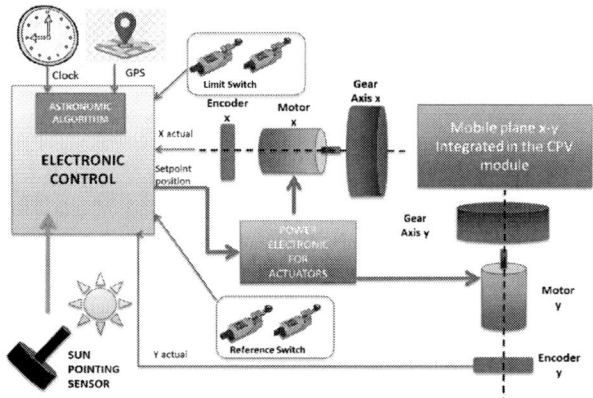

Fig. 1.   Block scheme of a conventional IST system.

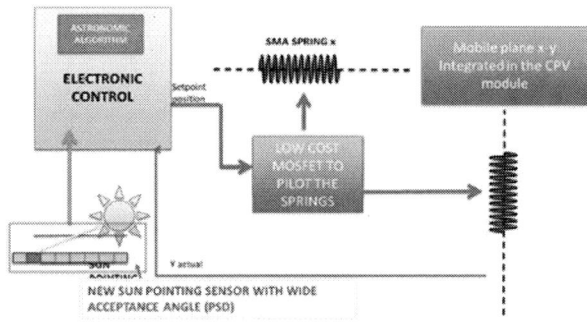

Fig. 2.   Block scheme of the new IST based on SMA springs.

The position $x_m(t)$ of the movable plate is given by the equilibrium point obtained between the forces produced by SMAs (F1 of SMA1 and F2 of SMA2) and by the mechanical friction. Every time the movable plate moves, the new position $x_m(t)$ involves a new springs elongation with consequent variation of the value of their force. The PSD, receiving the sun light from two slits positioned on the front side of the module, can directly measure the alignment error $e_a(t)$, that is, the difference between the position of the movable plane $x_m(t)$, which can be seen as the position of a generic CPV cell placed on the movable plane, and the position of the sunlight spot concentrated by the lenses, $x_s(t)$. According to the control

scheme, since the reference setpoint error $e_r(t)$ is zero, the control error $e_c(t)$, assumes a value opposite to the alignment error $e_a(t)$.

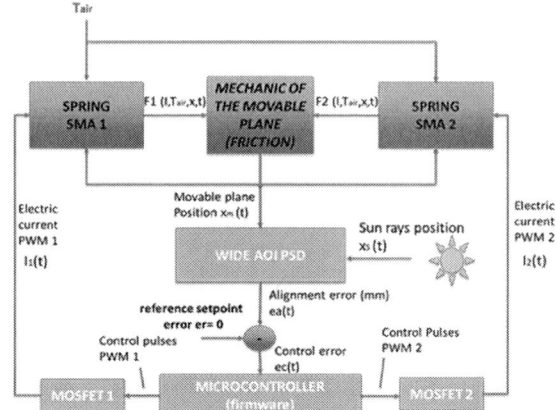

Fig. 3. Control scheme of the new IST based on SMA springs.

The control logic, implemented in the firmware of a microcontroller, adopts a digital proportional-integral-derivative (PID) algorithm which continuously calculates the thermal power to be injected into the SMA springs based on the sum of three signals which are, respectively, proportional, the integral and the derivative of $e_c(t)$. In particular, the integral action allows the system self-adapting as the ambient temperature varies, while the derivative action reduces the power when the SMA actuator spring contracts too quickly to avoid large oscillations of the movable plate (a situation which clearly affects the optimal operation of CPV cells). The PID technique has been optimized in order to manage the non-linearity of the SMA actuators. Besides, the control logic evaluates which of the two SMA springs must be powered considering that when the system moves one SMA spring mainly acts as an actuator (i.e., it is heated) while the second one mainly acts as an antagonist (i.e., it is cold). The electrical power to be injected, calculated by the microcontroller, is then transferred from a power supply to the SMA springs by controlling the switching of proper mosfets, by using the pulse width modulation (PWM) technique. To prove the IST concept a first hybrid module FU, which includes only the CPV cells, and the IST system based on SMA actuators (see Fig. 4) has been developed. On the movable frame, four-9 mm² - CPV cells able to convert the direct sunlight focused by four lenses have been installed. In addition to the CPV cells, the movable frame contains the electronic board with the microcontroller and the PSD (whose details are shown in Fig.5). The module FU includes PV cells on the front side just to show that, in the future version of the module, the IST will be powered by these cells, avoiding any external power demand. The movable plate has been loaded up to 7 kg in order to test the IST concept with a weight that is comparable to that of a future commercial module

978-1-6654-6060-6/23 $31.00 © 2023 IEEE

made with hundreds of CPV cells and including PV cells for the hybrid CPV/PV concept.

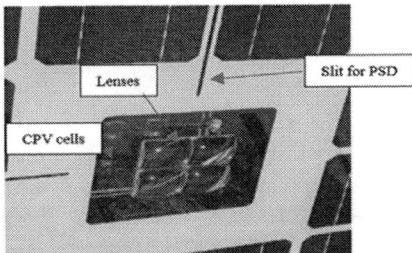

Fig. 4. Topside view of the CPV module FU where the slit for the PSD, the CPV cells and lenses are shown.

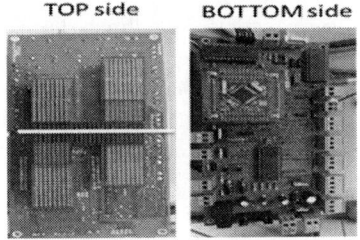

Fig. 5. Electronic board which includes the LED matrices of the PSD sensor (top side) and the electronic components (bottom side).

III. EXPERIMENTAL TEST AND RESULTS

Fig. 6 shows the trends of the misalignment $e_d(t)$ and the power P_{x1} and P_{x2} used to operate the springs SMA1 and SMA2, during a preliminary outdoor test of few hours in which the CPV plane has been moved along the x-axis to track the concentrated light spot.

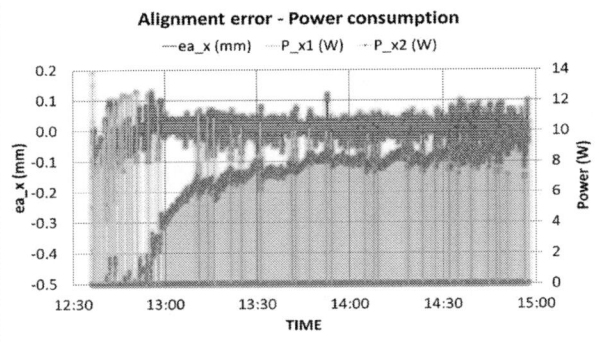

Fig. 6. Graphs of alignment error (in blue) and electrical power for actuating springs SMA1 (in yellow) and SMA2 (in red), obtained during an outdoor x-axis test.

The logic of the control scheme reported in Fig.3 requires that in order to track the sun trajectory along the x axis, the SMA2 spring should work mainly as an actuator while the SMA1 spring as an antagonist. However, in the first part of the test

(transient), from 12:30 to 12:45, it is possible to point out that the IST microcontroller mainly activated the SMA1 spring. The reason can be understood as the following. At the start of the test, the SMA2 was already elongated and had a strong mechanic energy due to its elastic behavior which was able to pull the plate. To smooth such movement and stay within an error lower than 0.2 mm, the control logic automatically activated the SMA1. At about 12:45 the SMA2 extension was decreased such as its elastic force couldn't pull the plate anymore. Then, to properly track the sun, the SMA2 spring started to be heated by the IST microcontroller and acquired a force due to the SMA material transformation. From 12:45 until the end of the test, the SMA2 mainly worked as actuator while the SMA1 spring as an antagonist, except in some cases in which it was activated to compensate some overshoots. Overall, statistical analysis allows evaluating that the absolute value of error has been less than 0.1 mm and 0.2 mm respectively in 90% and 99% of the time of the experimental test.

IV. CONCLUSIONS

A new IST concept based on SMA springs has been preliminary validated by a CPV module FU tested in outdoor operating condition. In 90% of the experimental cases the absolute alignment error has been less than 0.10 mm. This accuracy is sufficient to develop CPV modules with cells of 4 mm^2 area and geometrical concentration factors up to 400 suns. The accuracy result is relevant also by considering that the total stroke of the SMA spring was 65mm and that the movable plate has been specially loaded up to 7 kg in order to simulate the possible weight of a future commercial module made with hundreds of CPV cells. The new solar tracking system based on SMA actuators, the new control logic and PSD is a promising solution to develop a new generation of low cost-high concentration, environmentally friendly, CPV/PV hybrid modules with IST.

REFERENCES

[1] H. Zsiborács, N. H. Baranyai, " The Impacts of Tracking System Inaccuracy on CPV Module Power," *Processes* v 8, 1278, 2020.

[2] S. Askins, N. Jost, and I. Antón, " Performance of Hybrid Micro-Concentrator Module with Integrated Planar Tracking and Diffuse Light Collection," in *in 46th IEEE PVSC, Chicago, 2019.*

[3] A.Minuto, E.Celi, G. Timò « A New CPV/PV Hybrid Module with a Self-Powered, Integrated, 3D Meso Solar Tracker Based on Shape Memory Alloys,» 17th international conference on concentrator photovoltaic systems (CPV-17), 2022.

[4] A. U. Attanasi, «Theoretical and Experimental Investigation on SMA Superelastic Springs,» Journal of Materials Engineering and Performance, 2011.

[5] S. Kannan, «Modeling and control of Shape Memory Alloy Actuator,» HAL archives, 2011.

[6] Patent N. US7692091B2,

[7] S. Grede, Alex J. Brulo at al., "Motorless Microtracking for Rooftop CPV", *IEEE 7th World Conference on Photovoltaic Energy Conversion, WCPEC 2018,* vol2. pp. 1648–1651.

978-1-6654-6060-6/23 $31.00 © 2023 IEEE

Quantifying Uncertainty Due to Climate Variability in Vehicle-Integrated Photovoltaic Yield Predictions

Timofey Golubev

ThermoAnalytics, Inc., Calumet, MI, 49913, USA

Abstract—This work uses a surrogate modeling approach to quantify the uncertainty due to climate variability in vehicle-integrated photovoltaic (VIPV) energy yield predictions. An artificial neural network (ANN) is trained to predict VIPV yield from weather data using results from thermal-electrical simulations of a VIPV system. The ANN is then used with 23 years of historical weather data to evaluate the variability of the VIPV system's yield at 260 locations in North America.

Index Terms—vehicle-integrated photovoltaics, uncertainty quantification, energy yield, surrogate modeling, artificial neural network, electric vehicles, thermal-electrical

I. Introduction

The rapid development of electric vehicles combined with increasing efficiencies and reducing costs of photovoltaic cells has resulted in increased interest in vehicle-integrated photovoltaics (VIPVs) [1], [2]. Climate conditions can have a significant impact on VIPV system energy production [3]. Therefore, evaluating VIPVs through simulation is necessary to predict VIPV yield under a wide range of environmental conditions. However, detailed physics-based VIPV models that calculate the incident irradiance on curved surfaces and consider thermal effects are computationally expensive, making them impractical to use directly for comprehensive studies. To address this limitation, we developed an approach for training a surrogate model on physics-based simulations for rapid prediction of VIPV energy yields [4]. In this work we use our surrogate modeling approach to quantify uncertainty due to climate variability in VIPV yield predictions for 260 locations in North America.

II. Methodology

A. Surrogate Model

We train an artificial neural network (ANN) to predict VIPV yield using results from a coupled thermal-electrical model of a vehicle with on-board PVs and the surrogate modeling approach developed in our previous work [4]. We briefly summarize the surrogate modeling approach in this section. The thermal-electrical model was created in the commercial heat transfer software TAITherm [5], which uses a numerical, finite volume method based on first principles physics to solve for heat transfer due to conduction, convection, and radiation. A PV equivalent circuit electrical model was implemented in Python and coupled with the thermal model. To simulate the climatic conditions, we applied typical meteorological year (TMY) weather data from the National Solar Radiation Database (NSRDB) [6] as boundary conditions. As an example model, we simulated the VIPV energy production from 2.8

m^2 of SunPower Maxeon Gen III solar cells integrated into the roof of a sports utility vehicle. Equivalent circuit parameters for the solar cells were derived from the manufacturer datasheet [7] and are listed in our previous work [4].

To generate the training dataset for the surrogate model creation, we ran year-long thermal-electrical simulations of the VIPV system with hourly time-steps. We use the following model inputs (i.e. features): global horizontal irradiance (GHI), direct normal irradiance (DNI), direct horizontal irradiance (DHI), solar azimuth, solar zenith, and ambient air temperature. The target variable is the power output of the VIPV in each hour of the year. For the surrogate model, we use the best performing model architecture from our previous work, a multi-layer perceptron neural network from the Scikit-Learn Python package (version 0.24.2) [8] with a logistic activation function, 3 hidden layers with 100 neurons in each, an L2 penalty parameter of 10^{-5}, and the `lbfgs` solver with 5000 maximum iterations. In our previous work, we trained the surrogate models using a single year-long TAITherm simulation of the VIPV system in Seattle, Washington. In this work, we retrained the ANN using TAITherm results from two locations instead of one: Seattle, Washington and Phoenix, Arizona. The second location was added to improve the model accuracy for predictions at new locations by providing the ANN with training data that includes a wider range of climate conditions. After training, we tested the ANN's performance by comparing its predictions to TAITherm results for six additional locations in the United States.

B. Sensitivity Analysis

To evaluate the sensitivity of the VIPV yield prediction on the model inputs, we performed a Saltelli sensitivity analysis with the OpenTURNS Python package [9]. Historical weather data from multiple locations in the U.S. was used to estimate non-parametric probability distributions of the inputs. These input distributions were then sampled 80,000 times and Sobol indices [10], which quantify the sensitivity of the VIPV yield to each model input, were calculated.

C. Uncertainty Quantification

We used the ANN and historical weather data from the NSRDB (PSMV3 dataset for 1998-2020) to quantify the variability in annual VIPV yield at 260 locations in or near the United States. The ANN model was run to predict the hourly VIPV energy production for each of the 5980 years (260 locations x 23 years) of weather data. The ANN model

978-1-6654-6060-6/23 $31.00 © 2023 IEEE

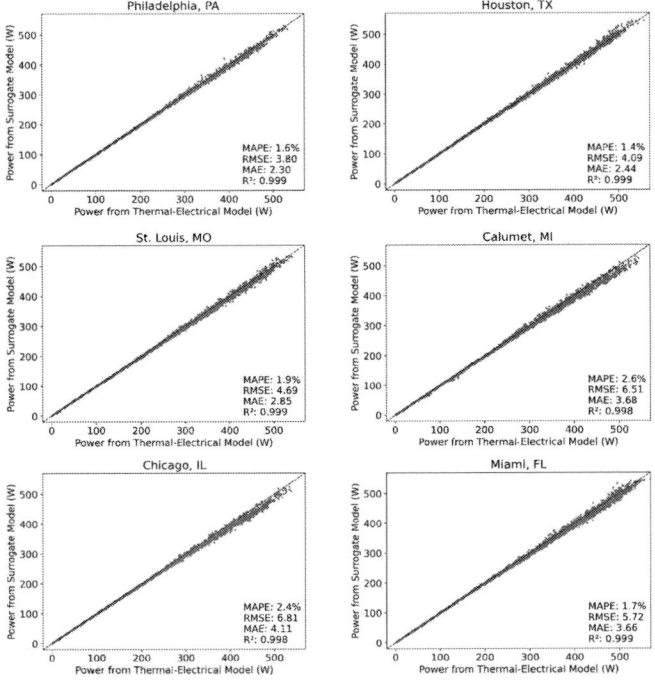

Fig. 1. Comparison of the VIPV power output prediction from the surrogate model versus the thermal-electrical model for six test locations.

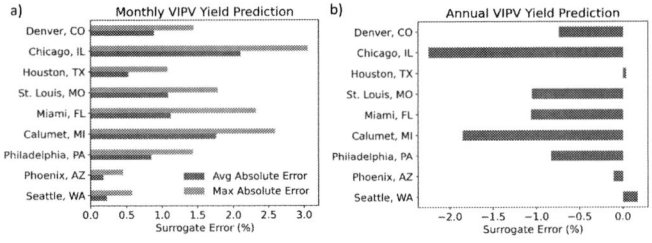

Fig. 2. Surrogate model's deviation from thermal-electrical model results for monthly and annual total VIPV yield predictions.

evaluation took only 0.18 seconds per year (on an Intel i9-10900K). The hourly predictions were then summed for each year at each location. The variability in annual VIPV yield at each location was quantified by the percent difference between the median annual yield and the P90 annual yield (i.e. yield that was predicted to be exceeded in 90% of the historical years) [11].

III. RESULTS AND DISCUSSION

A. Surrogate Model Validation

Fig. 1 shows comparisons of the hourly PV power output prediction from the ANN and the thermal-electrical model on the test locations. The mean absolute error (MAE), root mean square error (RMSE), mean absolute percent error (MAPE), and coefficient of determination (R^2) were used to quantify model performance. The surrogate model's hourly power output predictions are on average within 1.4-2.6% (depending on location) of the thermal-electrical model results, indicating

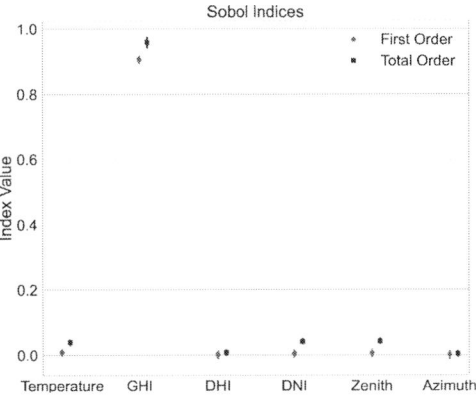

Fig. 3. Sobol sensitivity indices.

good prediction accuracy. Fig. 2 displays the error between the surrogate and thermal-electrical models' total monthly and annual VIPV yield predictions. We find that the ANN has good prediction accuracy for the monthly and annual total yields with predictions within 3% of the thermal-electrical results. These results validate that the surrogate model is able to replicate the full thermal-electrical model's predictions with sufficient accuracy to be used for sensitivity analysis and uncertainty quantification.

B. Sensitivity Analysis

Fig. 3 quantifies the sensitivity of the model to its input parameters in terms of Sobol indices where parameters with larger index values correspond to a greater sensitivity to that parameter. As expected, the VIPV energy production is most sensitive to the GHI. The VIPV yield also has smaller non-zero sensitivity to air temperature, DNI, and zenith angle. The first order indices for temperature, DNI, and zenith angle are near zero while the total order are non-zero indicating that the VIPV yield depends on combinations of these parameters, instead of directly on each parameter individually.

C. Uncertainty Quantification

As an example of the interannual yield distributions at individual locations, Fig. 4 shows the distributions of predicted VIPV yield in 10 locations from 2001-2020 and Fig. 5 shows the monthly yields for four of these locations. We see that there can be significant variability in both annual and monthly energy production from year to year. In Fig. 5, three of the four locations exhibit strong seasonality in VIPV yield with summer energy production being several times greater than winter production. Calumet, MI, the northernmost location of these four, exhibits the strongest seasonality with summer yields approaching those of locations much further south such as St. Louis and Miami, while winter yields are very low. Miami, the southernmost location of the four, has the least seasonality.

Fig. 6a shows the medians of the predicted annual VIPV yields from 1998-2020 for 260 locations. The median VIPV annual yield has similar trends to the annual solar GHI at these

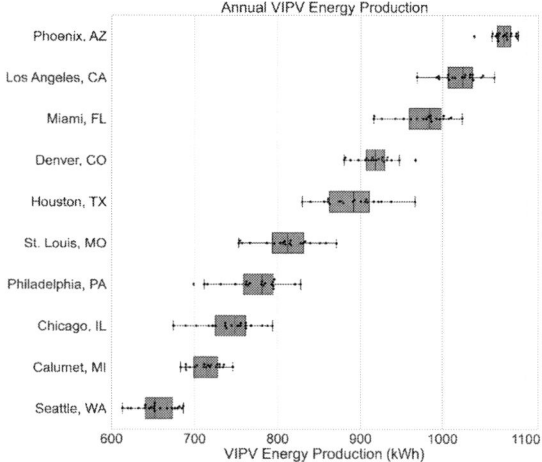

Fig. 4. Uncertainty quantification of annual VIPV yield in ten locations using historical weather data from 2001-2020.

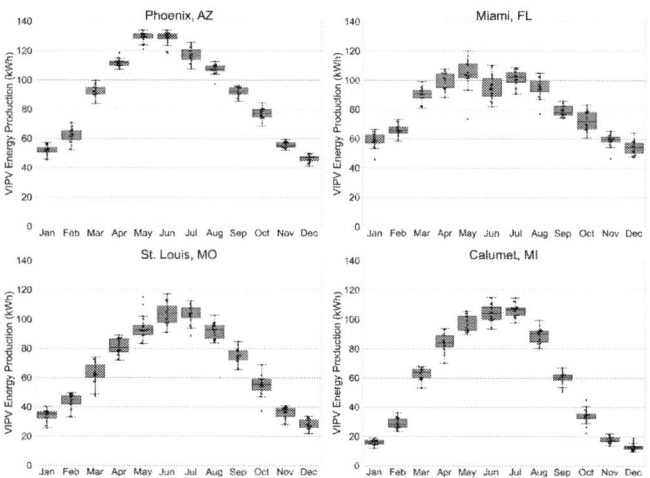

Fig. 5. Uncertainty quantification of monthly VIPV yield in four locations using historical weather data from 2001-2020.

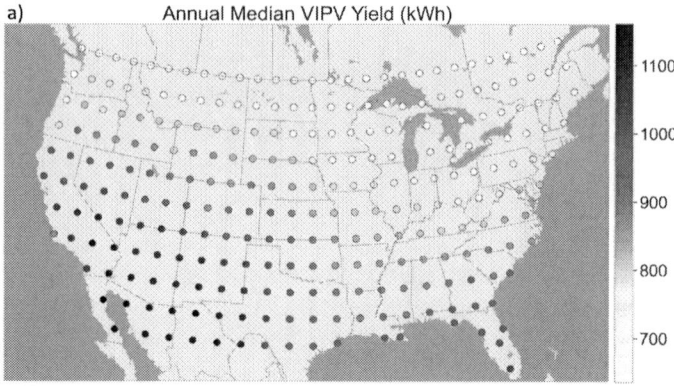

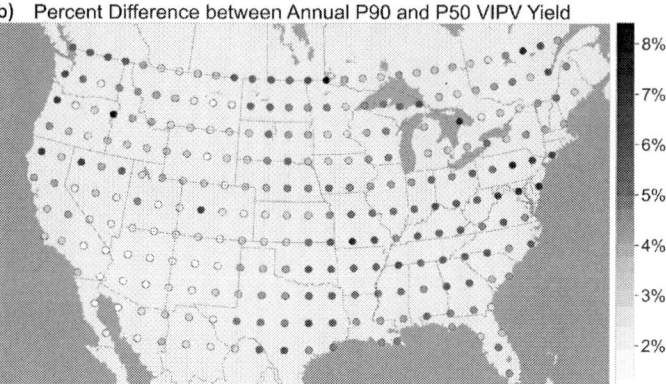

Fig. 6. Summary of VIPV yield predictions for 1998-2020. a) Median annual VIPV yields. b) Interannual variability of VIPV yields quantified as percent difference between P90 and median (P50) values.

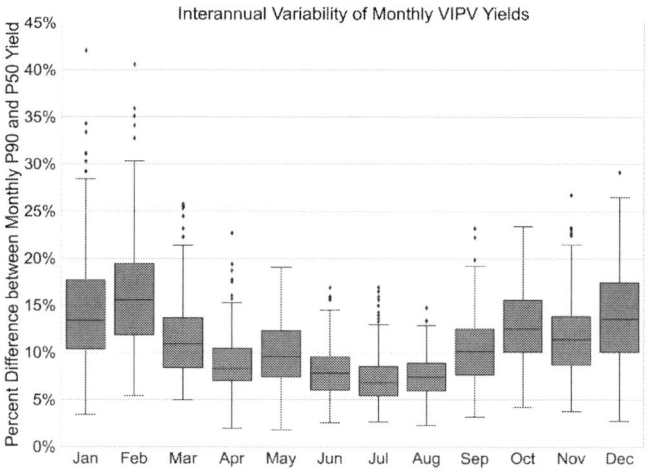

Fig. 7. Statistical summary of interannual variability of monthly total VIPV yields for all 260 locations quantified as percent difference between P90 and median (P50) values for each location.

locations with the highest annual yields being in the southwest of the region studied and the lowest yields being in the north [6]. Fig. 6b shows the variability in predicted annual VIPV yields in terms of the percent difference between the P90 and median (P50) values for each location. The U.S. Southwest and portions of Wyoming and Montana (just east of the Rocky Mountains) are predicted to have the least variability in VIPV energy yield with interannual fluctuations expected to be only 1-3% from the median. Locations in the U.S. Southeast, Mid-Atlantic, and Northwest Coast are predicted to have in-general the greatest variability in yield with fluctuations of 5-8.4%. The median of the annual variabilities for all locations is 4.1%.

We also calculated the uncertainty due to interannual climate variability in monthly VIPV yields for all locations. Fig. 7 presents a statistical summary of the results. In general, the uncertainty is highest in the winter months and lowest in the summer months. The winter months have the largest range of values for uncertainties in different locations, with some

outlier locations having greater than 30% difference between P90 and P50 values in January and February.

Fig. 8 shows a summary of the medians of the monthly VIPV yields for 1998-2020 for the 260 locations considered in this study. The trends are consistent with monthly trends in

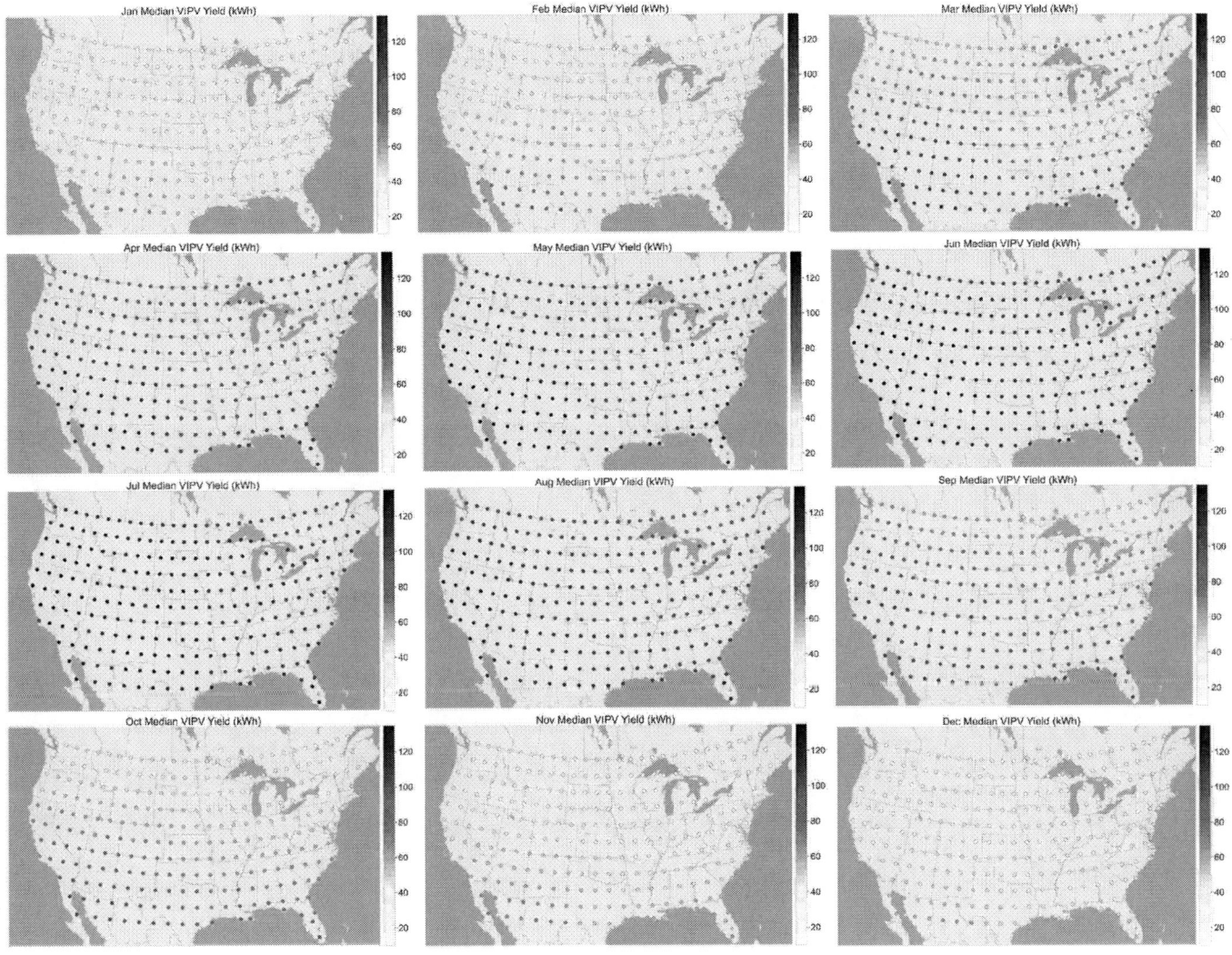

Fig. 8. Summary of median monthly VIPV yields predicted for 1998-2020.

the GHI for each location. Like seen with the annual median yields, the highest yields during any given month tend to be in the southwest of the region studied, while the lowest yields are in the north.

Fig. 9 shows the variability in predicted monthly VIPV yields. Overall, the variability in monthly yields is significantly larger in the winter months than in the summer months. The southwest is predicted to have the lowest variabilities in the winter months. In the summer months, the yield variabilities of the majority of locations are reduced and the variabilities are more similar throughout the entire region.

IV. CONCLUSION

This work uses an ANN surrogate model, which was trained on physics-based thermal-electrical simulations, to quantify uncertainty due to climate variability in VIPV energy yield predictions. The high computational efficiency of the surrogate model improves the practicality of performing uncertainty quantification, which often requires thousands of model runs.

The ANN was found to have less than 3% error in hourly, monthly total, and yearly total yield predictions when compared to the physics-based simulations, thus validating that it can be used reliably for uncertainty quantification. A sensitivity analysis showed that the uncertainty in VIPV yield is most sensitive to GHI with smaller non-zero sensitivities due to the combined effects of temperature, DNI, zenith angle, and GHI. The ANN was applied to predict the expected variability in VIPV energy production at 260 locations in North America. Areas with generally lower or higher interannual variability were identified. Interannual variability of total monthly yields of four locations were evaluated as a demonstration of the impact of geographic location on VIPV yield and its uncertainty. In future work, the surrogate VIPV model could be combined with electric vehicle energy consumption simulations to predict the solar-powered driving range in different locations.

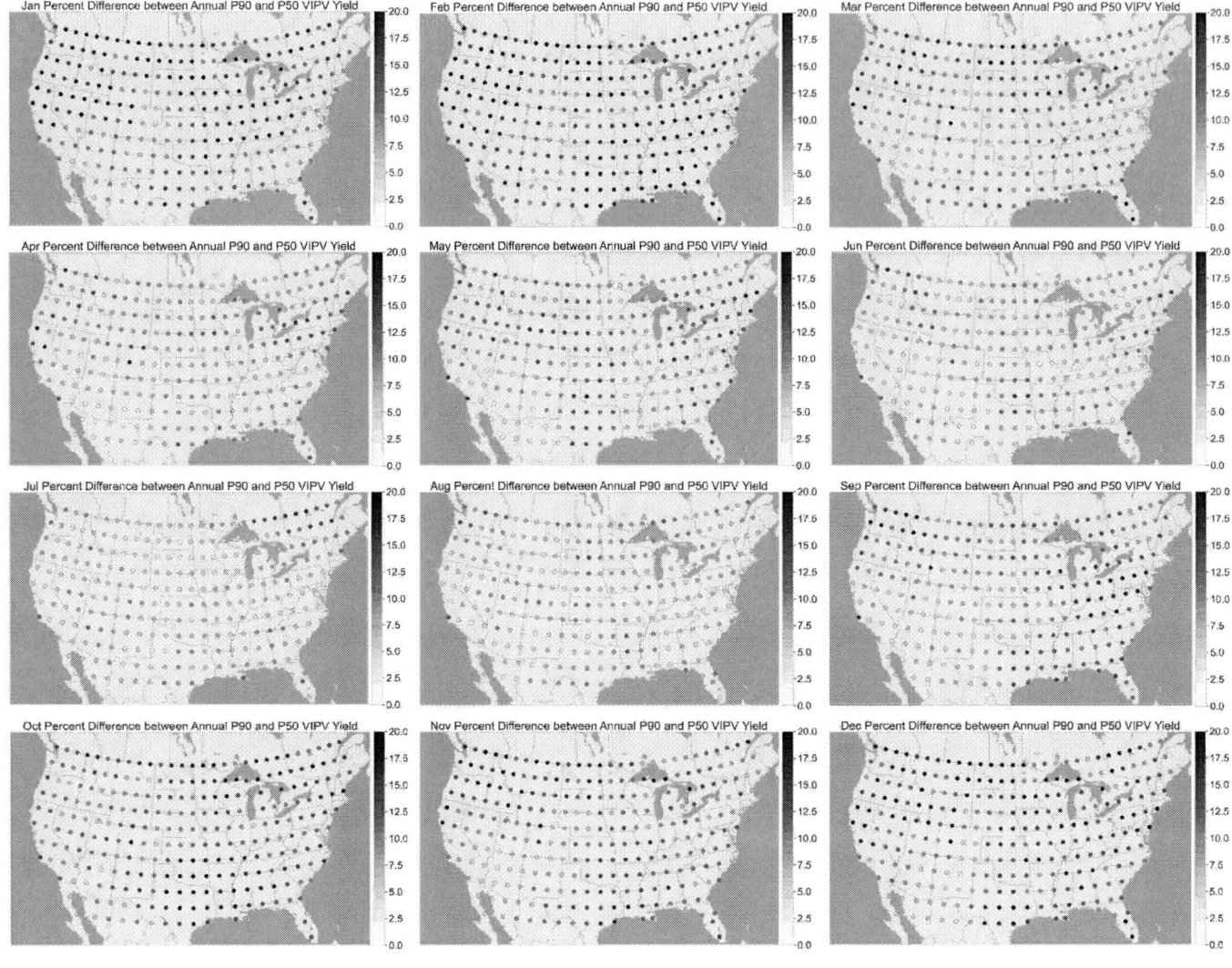

Fig. 9. Summary of interannual variability of monthly VIPV yields predicted for 1998-2020.

REFERENCES

[1] B. Commault, T. Duigou, V. Maneval, J. Gaume, F. Chabuel, E. Voroshazi, "Overview and perspectives for vehicle-integrated photovoltaics," *Appl. Sci.*, vol. 11, 2021, doi:10.3390/app112411598.

[2] M. Heinrich et al., "Potential and challenges of vehicle integrated photovoltaics for passenger cars," in 37th European PV Solar Energy Conference and Exhibition (EU PVSEC), 2020.

[3] C. Thiel et al., "Impact of climatic conditions on prospects for integrated photovoltaics in electric vehicles," *Renew. Sustain. Energy Rev.*, vol. 158, 2022, doi:10.1016/j.rser.2022.112109.

[4] T. Golubev, "Surrogate modeling for rapid prediction of energy yield from vehicle-integrated photovoltaics," in *49th IEEE Photovoltaic Specialists Conference*, 2022, doi:10.1109/PVSC48317.2022.9938713.

[5] *TAITherm* (2022.2.0). ThermoAnalytics, Inc.

[6] M. Sengupta, Y. Xie, A. Lopez, A. Habte, G. Maclaurin, and J. Shelby, "The national solar radiation database (NSRDB)," *Renew. Sustain. Energy Rev.*, vol. 89, pp. 51–60, 2018, doi:10.1016/j.rser.2018.03.003.

[7] Sunpower. "Maxeon 3 BLK". [Online]. Available: https://sunpower.maxeon.com/int/sites/default/files/2020-09/sp_mst_MAX3-375BLK_355BLK_ds_en_a4_mc4_532497.pdf [Accessed: 21-Nov-2021].

[8] F. Pedregosa et al., "Scikit-learn: Machine learning in python," *J Mach Learn Res.*, vol. 12, pp. 2825–2830, 2011.

[9] M. Baudin, A. Dutfoy, B. Iooss, A. Popelin, "OpenTURNS: An industrial software for uncertainty quantification in simulation," *Handbook of Uncertainty Quantification*, 2017, doi: 10.1007/978-3-319-12385-1_64.

[10] I. M. Sobol, "Global sensitivity indices for nonlinear mathematical models and their Monte Carlo estimates," *Mathematics and Computers in Simulation*, vol. 55, pp. 271-280, 2001, doi:10.1016/S0378-4754(00)00270-6.

[11] D. Ryberg, J. Freeman, N. Blair, "Quantifying interannual variability for photovoltaic systems in PV Watts," NREL Technical Report, 2015, doi:10.2172/1226165.

Phase Distributions and Local Bandgap Energies in Mixed-Halide Perovskite Nanoparticles

Dan R. Wargulski[1], Tal Binyamin[2], Katrina Coogan[3], Benedikt Haas[3], Christoph Koch[3], Lioz Etgar[2] & Daniel Abou-Ras[1]

[1]Helmholtz-Zentrum Berlin für Materialien und Energie, Germany
[2]Casali Center for Applied Chemistry, Hebrew University of Jerusalem, Israel
[3]Department of Physics & IRIS Adlershof, Humboldt-Universität zu Berlin, Germany

Abstract—Monochromated, aberation-corrected scanning transmission electron microscopes enable the investigation of chemical distribution and optical properties such as the bandgap energy. We conducted valence and core electron energy-loss spectroscopy on nanoparticles made of the inorganic double-halide perovskites $CsPb(I_xBr_{1-x})_3$ (x=0-1) to investigate the distribution of halogen anions as well as Cs and Pb cations inside the nanoparticles and correlated the results with bandgap energy mappings. Changes of the chemical distribution with continuous electron beam irradiation were observed.

Keywords—*valence-loss EELS, core-loss EELS, halide perovskites, nanoparticles, monochromated STEM, photovoltaic*

I. INTRODUCTION

For several years, halide perovskites have received a lot of attention as absorber layers for solar cells or as active films in light-emitting diodes. One reason is the tunability of the band-gap energy, which can be varied e.g., by mixing halide ions in the compounds. This tunability enables the development of customized, widegap semiconductors as absorber layers in top cells for tandem devices.

However, mixing of halides may invoke the "Hoke effect"[1,2], a light-induced segregation of halides, which leads to a substantial loss in open-circuit voltage and power conversion efficiency of the corresponding solar cells. To improve the understanding of halide mixing in perovskites and their phase stabilities, we investigated $CsPb(I_xBr_{1-x})_3$ (x=0-1) nanoparticles (NPs) at atomic scales in a dedicated scanning transmission electron microscope (STEM) by means of high-resolution imaging combined with electron energy-loss spectroscopy (EELS).

II. EXPERIMENTAL DETAILS

Halide NPs with various halide mixtures of the type $CsPb(I_xBr_{1-x})_3$ were synthesized and dispersed in hexane. Particle sizes of about 40-80 nm were produced to avoid quantum confinement effects at sizes below 20 nm and to provide a sufficient electron transparency, which is not guaranteed at larger sizes. The NPs in solution were dripped onto 8-nm-thin Si_3N_4 membranes. The drop-cast particles were dried in a vacuum oven at 50 °C for 8 h to remove solvent residues. Subsequently, plasma cleaning in Ar/O_2 atmosphere was performed for 1 min, to prevent contamination during measurements. Imaging and EELS analyses were conducted in a monochromated and aberration-corrected Nion Hermes 200 kV STEM equipped with a Dectris ELA direct electron detector.

III. RESULTS AND DISCUSSION

A. Imaging

We obtained high-resolution images of $CsPb(I_xBr_{1-x})_3$ NP without observing electron beam damage during the acquisitions (Figure 1). The images confirmed that indeed, the desired halide-perovskite phase[3] was obtained. Furthermore, the particle sizes and halide ratios were determined and correlated with the EELS results.

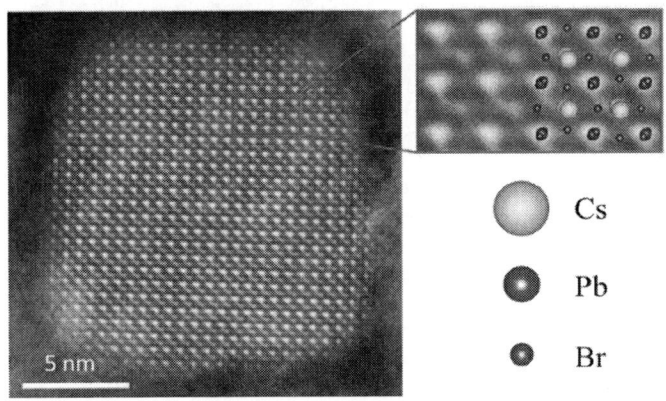

Figure 1: High-resolution image of a $CsPbBr_3$ nanoparticle in the [001] projection.

B. Core-loss EELS

Elemental distributions in the NPs were analyzed, providing the chemical compositions and revealing inhomogeneous cation and halide distributions in the $CsPb(I_xBr_{1-x})_3$ NPs. Moreover, image and core-loss EEL spectrum image series were acquired during continuous electron irradiation of several NPs. In some NPs, the image series featured changes in contrast suggesting interdiffusion from/to the center of the NPs to/from their edges. The evaluation of the EELS series is still ongoing.

978-1-6654-6060-6/23 $31.00 © 2023 IEEE

C. Valence-loss EELS

EEL spectra were acquired in the low-energy range of several eV on various $CsPb(I_xBr_{1-x})_3$ NPs. The resulting spectra exhibited onsets at energies (1.7-2.4 eV) which are the expected band-gap energies for the corresponding NP compositions. We will correlate the band-gap energies with the elemental distributions obtained by core-loss EELS.

IV. CONCLUSIONS

$CsPb(I_xBr_{1-x})_3$ NPs with various sizes and chemical compositions were analyzed by means of high-resolution imaging as well as by EELS. Local band-gap energies in NPs were determined successfully by means of valence EELS. Moreover, electron irradiation for more than 10 min did not result in a full decomposition of the NPs but seemed to induce interdiffusion processes with some of the analyzed NPs.

ACKNOWLEDGMENTS

The present work was supported by the Helmholtz-International Research School "HI-SCORE" (HIRS-008) and by the Graduate School HyPerCells.

REFERENCES

1. Hoke, E. T. *et al.* Reversible photo-induced trap formation in mixed-halide hybrid perovskites for photovoltaics. *Chem. Sci.* **6**, 613–617 (2015).

2. Unger, E. L. *et al.* Roadmap and roadblocks for the band gap tunability of metal halide perovskites. *J. Mater. Chem. A* **5**, 11401–11409 (2017).

3. Yu, Y., Zhang, D. & Yang, P. Ruddlesden–Popper Phase in Two-Dimensional Inorganic Halide Perovskites: A Plausible Model and the Supporting Observations. *Nano Lett* **17**, 22 (2017).

Third Generation approaches for low cost, radiation tolerant, efficienct Space solar cells

Gavin Conibeer and Santosh Shrestha

School of Photovoltaics and Renewable Energy Engineering, University of New South Wales, Sydney, NSW 2052 Australia

Abstract— The large increase in space missions in the next decade represents a large increase in space PV demand. These cells must be relatively cheap but also radiation resistant. Some existing solar cell materials offer good radiation resistance, but third generation concepts involving multiple energy levels offer great promise to de-couple optical absorption from power generation.

Keywords—space solar cells, radiation environment, tandem cells, hot carrier cells, up/down conversion, luminescent concentrators.

I. INTRODUCTION

The large increase in space missions at present and over the next several decades, dwarfs previous space activity. There are 10,352 satellites in space out of 1 ever launched. [1] These satellites are in a large range of Earth orbits, as illustrated in Fig. 1, although the vast majority are in low earth orbit (LEO) and engaged in earth observation, geolocation, telecommunications and other communication and earth information gathering applications, many of which are important to attaining the UN Sustainable Development Goals (SDGs). [1]

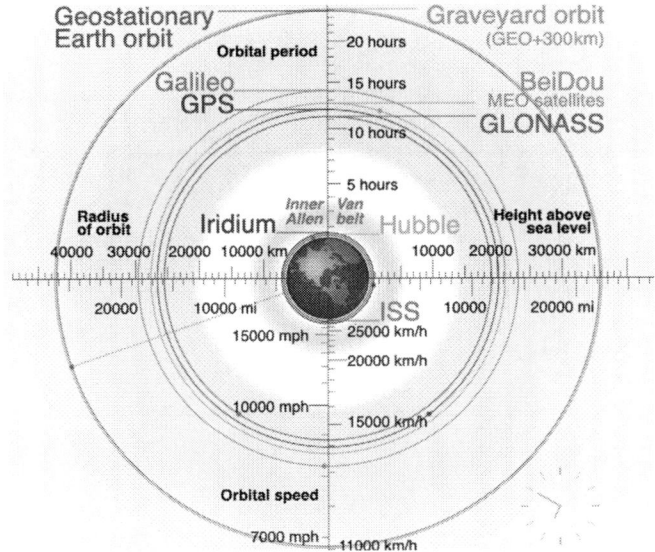

Figure 1: Earth satellite navigation orbits. [2] As of Jan 2023 > 10,000 satellites in Earth orbit – most in LEO - each with 400W to 4kW of PV. [1]

There were 1274 satellites launched in 2020, 1809 in 2021, 2397 in 2022 and so far 129 so far in 2023. They are driven by emerging markets in global information systems and need low cost components and require a mean of about 600W of reliable power. Hence, the space market is already growing and is set to grow further very rapidly, particularly with the deployment of constellation satellites each of which will have between 400Wp and 4kWp of PV. With current orders and planned deployments there will be at least 8,000 constellation satellites by 2024 rising from about 2,000 today. [3] The obvious and almost always used power source is photovoltaics, but the current primarily used photovoltaic technology (triple junction GaInP/InGaAs/Ge cells on active Ge substrates) is very expensive to manufacture and in terms of raw materials (>\$250/W). In addition, future supply is unlikely to be able to keep up with the large numbers of Ge substrate and III-V growth facilities required to maintain this technology as the most widely used. So, there is much scope for the development of lower cost photovoltaics that nonetheless also provide high power outputs and long-lived operation.

There is therefore now a new set of parameters required for solar cells for space applications: reasonably lower cost than existing triple junction space cells; tolerance to the severe radiation in the space environment; a high power output per kg of mass (specific power) rather than area, because of the high costs of launching 1 kg into LEO; and reasonably high efficiency (per area) to optimize the generating capacity even at lower costs.

These somewhat modified parameters compared to terrestrial cells, suggest it is worth looking again at some the alternative approaches to photovoltaics that have been suggested – 'third generation' concepts. The moderate cost and radiation tolerance in particular make some of the third generation approaches certainly worth re-visiting and quite likely superior to their application for terrestrial devices.

II. SPACE SOLAR CELL REQUIREMENTS

Space solar cell properties that do not generally need to be considered for terrestrial cells:

Radiation tolerance – minimize the decrease in performance on exposure to high energy proton and electron fluxes – the degree of tolerance required depends on the intended lifetime of mission and the orbit and altitudes of the mission – the most important parameter being the end of life (EOL) efficiency. LEO orbits have moderate radiation environments, with a combination of high energy electrons around 1 MeV and high

978-1-6654-6060-6/23 \$31.00 © 2023 IEEE

energy protons around 100keV. [4] These are highly detrimental to electronic devices, creating a series of defects as high energy particles cascade through the material depositing energy as they go – principally displacing atoms to create vacancies, interstitials and di-vacancies that act as efficient recombination centres and can even lead to type change of the semiconducting material. Higher orbits tend to have higher proportions of high energy electrons peaking with the extremely aggressive environment of the Upper van Allen belts. Geostationary orbit at 38,000 km is slightly less aggressive but still around 10x the radiation of LEO. [4] Hence the need for protection from this radiation is important for any electronic device including solar cells. Cover glass (doped with high mass elements such as Ce) provides reasonable protection from moderate energy protons (which paradoxically create more damage than high energy protons – because they stop in the material) but do little to attenuate high energy electrons. Cover glass is also expensive – around 10% the cost of triple junction cells - which becomes a problem if the cost of cells is significantly reduced.

Cost – the costs of cells needs to be reduced from the very expensive >\$250/W for III-V multijunction cells to around a few \$s per Watt for the new Space 2.0 missions, which represent a much larger PV market that must be delivered at much lower costs per Watt and per kg.

Specific power – very important from a launch mass consideration, the specific power in W/kg should be as high as possible.

Efficiency – the high efficiencies of triple junction III-V cells on Ge substrate (~34% AM0) should be approached in Space 2.0 cells, but lower efficiencies can be tolerated if they come at much lower cost per Watt. Total system sizing and economics and mission duration and orbit need to be considered in deciding on the EOL efficiencies required.

Of these properties, radiation tolerance and lower cost are the two most important factors.

III. TANDEM CELLS

Tandem cells: series connected tandem or multijunction cells need to be current matched, or at least their maximum current is the one generated by the lowest current of the cells in the stack (with some modifications due to luminescent coupling). Most commercial tandem cells are indeed series connected cells because these are much easier to fabricate in a largely planar way as they minimize the number of photolithographic steps. However, unless radiation damage on each of the cells in a stack gives exactly the same decrease in current through increased non-radiative recombination, the current matching is likely to drift out and dramatically reduce the efficiency of the overall tandem stack.

Two approaches can mitigate this somewhat: The first of these is that standard triple junction cells on Ge active substrate, typically GaInP on GaAs (or InGaAs) on Ge are not as prone to this effect because the Ge bottom cell has too low a bandgap and is capable of producing a lot more current. On degradation by radiation it is the Ge bottom cell which has the largest reduction in current, so there is a certain amount of 'spare' current loss that can be accommodated before the overall current and efficiency of the cell is compromised significantly. [5,6]

However, this Ge bottom cell is problematic for two other reasons, firstly it gives a lower voltage than an otherwise more optimally matched cell, such as Si, would give, thus reducing the maximum efficiency of the initial tandem cell. Secondly it is very expensive and contributes to the very high cost of these triple junction cells. Si would be better or other approx. 1 eV bandgap materials, but Ge is almost perfectly lattice matched to the InGaAs and InGaP layers grown on top, and this has a greater effect on reducing threading dislocations and other defects than the effect of radiation.

Secondly, it is certainly possible to design cells to be still close to lattice matched but less optimized for relative bandgaps for the AM0 solar spectrum, i.e. poorly current matched. If this is done in such a way that it is the cells most tolerant to radiation that generate the lowest currents (and thus determine the overall current of the tandem cell), then as the cell degrades under radiation, the overall cell will become more current matched and EOL performance will only be slightly lower than 'beginning of life' (BOL) performance and as mentioned above, it is EOL that is most important in determining system sizing. However, such a design is not easy and will vary depending on the exact mission duration and altitude (i.e. on the radiation environment and fluence to be encountered).

Mechanically stacked or multi-terminal tandems also avoid this current matching problem by ensuring the individual cells are electrically independent. Hence degradation of one cell to a greater extent does not have such a non-linear effect on the overall stack efficiency. It does however, introduce more complexity into fabrication, and does require more complex external management of the power generated, but it also frees up the choice of materials. Essentially individual cells of relatively abundant materials can be manufactured relatively cheaply and then assembled with decreasing bandgaps on top of each other and connected independently. The extra complexity may well be offset by the less effects of radiation and hence better long term performance (EOL efficiency) and relatively lower costs.

Tandem cells on active silicon cell substrates are probably the most relevant cell type in this context. There is of course much current work on tandems on silicon for terrestrial applications. [7,8] Adaption of this work to the space environment require modification of the silicon cell to be more radiation tolerant (see next section) and deposition or assembly of tandem cell elements optimized for AM0. It is also probable that stacked tandems with independent connections will give the best EOL efficiencies in the space environment. Furthermore, it is possible that same bandgap tandems will become attractive because of their potentially lower costs.

IV. RADIATION TOLERANCE

Tolerance to radiation can be achieved more directly by modifying the material or the device to be less prone to the defects caused by radiation damage. This can be achieved by reducing or recovering defects caused by radiation – passivation, shallow defects in first place. [9]

Passivation of the defects caused by radiation can occur fairly naturally in some materials. This is most notable for perovskite solar cells, which tend to have very shallow defects because of the large metal atoms in the centre of their octahedra, with large atoms bonding with relatively weak bond strength and hence shallow defects for out of place interstitial or substitutional defects. These shallow defects are one of the reasons for the high efficiencies achievable with relatively simple fabrication and hence relatively highly defected material, (although the relatively weak bonds is also one of the contributing factors to the relative instability of hybrid organic perovskite materials). These shallow 'native' defects also mean that many of those defects caused by radiation damage will also tend to be shallow and hence have a lesser influence on reducing efficiency. There is some evidence for this in the radiation tolerance of some perovskite solar cell materials. [10,11]

Cadmium telluride cells [12] show good and CIGS cells [13] even better radiation tolerance largely because of their high absorption coefficients. Similarly, nanowire solar cells [14] also show excellent radiation tolerance because of their low displacement damage density under radiation. Hence these are all attractive for space applications.

Silicon cells were the main type used in space for the first three decades of space missions. They gave way to III-V cells and then to multi-junction III-V cells because of the latters' both greater efficiency and increased tolerance to radiation – primarily because the greater absorption coefficient of direct bandgap of III-Vs results in a much thinner cell and hence less damage caused by radiation for a given area, because this is a volumetric effect. There has been much work in the last two decades on terrestrial silicon solar cells that has resulted in much thinner cells, better handling of defects and much better light management in for instance PERC, TOPCon and heterojunction achitectures. [15] Some of these approaches can be exploited to increase the radiation tolerance of silicon cells significantly, both for use as a bottom cell in tandems and as single bandgap cells for space applications. [16]

V. HIGHER RADIATION TOLERANCE THIRD GENRATION CELLS

The light collecting regions of some third generation type concepts will tend to have a lower susceptibility to radiation damage than a solar cell that is light harvester and voltage generating material in one, and which hence requires excellent optical AND electronic transport properties.

These include up-conversion [17] and down-conversion [18] devices that operate purely optically. Their important properties are optical, luminescent and highly localized. Whilst they will suffer from enhanced localized direct recombination dues to localized radiation defects as this does not require long distance electronic transport (i.e. they can have very low electron/hole mobilities) the action of enhanced defects on transport will be much more limited.

Similarly a material designed to capture the excess energy of hot carriers – a hot carrier absorber – must also work on a very local level (it must do because of the extremely short time available - several ps - to capture excess energy in the form of optical or acoustic phonons). [19] It too will not be so badly affected by detrimental defects that reduce transport. This is particularly true of the "optical hot carrier cell" [20], [21], which decouples the light and hot carrier energy collector, which must be highly luminescent, from the single bandgap solar cell via a monochromatic filter that can double as a radiation shield, see Fig. 2. Potentially, this filter could be configured as a waveguide to change the direction of illumination and allow a robust radiation shield in front of the solar cell.

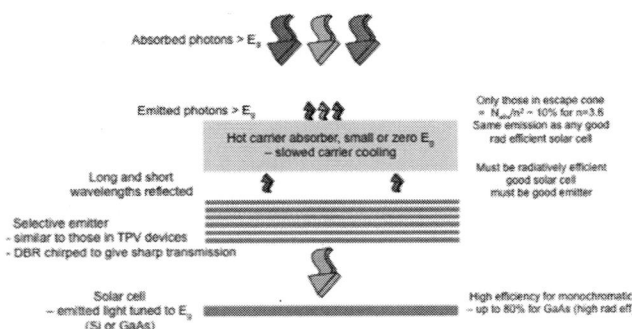

Figure 2: Optical Hot Carrier Cell showing decoupling of optical absorption and power generation via a monochromatic filter that can double as a radiation shield, potentially via a waveguide. [21]

VI. DECOUPLING OF LIGHT GENERATION FROM VOLTAGE GENERATION

Alternatively radiation tolerance of the overall device can be tolerated by moving the photovoltaic cell element away from the source of radiation. This can be done by decoupling the light collecting surface from the electronic power generated part. An example of this is a luminescent solar concentrator – in which a large area collects light and then transports laterally to small cells out of line of sight and potentially behind radiation shielding. [22] The original idea for LSCs was to reduce the area of expensive high efficiency solar cell required, but this required large area luminescent panels, which even at 99% re-absorption/re-emission efficiency lead quite quickly to unacceptable losses before the small area cells are reached. However, for a radiation shielded version the luminescent collectors need not be large area and so not suffer from large amounts of parasitic absorption. Geometric concentration ratios of 10x or less would be quite sufficient to provide the shielding needed of the high efficiency cells, rather than the few 100x of earlier iterations. [23]

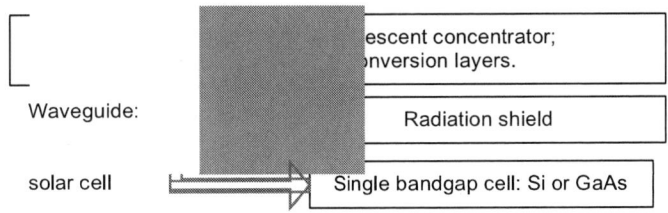

Figure 3: Generic decoupling approach, allowing the solar cell to be behind a radiation shield.

The overall idea of such a decoupled approach is shown in Fig. 3 and can be applied to other solar cells types and different light collectors, such as Hot Carrier absorbers, Luminescent concentrators or up or down converting layers. For instance the optical hot carrier cell and up/down conversion devices discussed above in Fig 2, with the decoupling filter configured as a waveguide to change the direction of illumination and allow a robust opaque radiation shield in front of the sensitive photovoltaic part of the device.

VII. CONCLUSIONS

The very large increase in space missions over the next decade and the consequent large increase in demand for space solar cells, is leading to a demand for much lower cost, radiation tolerant solar cells of reasonable efficiency. Radiation tolerance is one of the major properties needed for such cells which must be designed for their end of life performance.

Improvement in the radiation tolerance of existing materials for solar cells offers some potential, with annealing of defects or inherently low damage materials, but the possibility of re-visiting the third generation concepts of multi-energy level devices is very attractive in this context. Tandem cells are the obvious choice and series connected and mechanically stacked tandems on germanium and on silicon substrates are looking promising. Alternatively separating the absorption of light in a relatively radiation tolerant material from the power generation part in a radiation sensitive but potentially shielded solar cell is also an attractive approach that will be developed more thoroughly at the conference.

REFERENCES

1. UN: United Nations Office of Outer Space Affairs, accessed Jan '23:
   https://www.unoosa.org/oosa/index.html
2. Comparison satellite navigation orbits.svg, https://commons.wikimedia.org/w/index.php?
3. Curzi: G Curzi, D Modenini, P Tortoa, Aerospace 2020, 7(9), 133, Large Constellations of Small Satellites: A Survey of Near Future Challenges and Missions
   https://www.mdpi.com/2226-4310/7/9/133/htm
4. M. Yamaguchi, et al., Progress in Photovoltaics, 29 (2021) 98-108 Analysis for nonradiative recombination loss and radiation degradation of Si space solar cells.
5. M. Yamaguchi, F. Dimroth, J.F. Geiz, N.J. Ekins-Daukes, Journal of Applied Physics 129, 240901 (2021). Multi-junction solar cells paving the way for super high-efficiency.
6. A. Baiju, M. Yarema, Front. Energy Res., Sec. Solar Energy 10 (2022). Status and challenges of multi-junction solar cell technology.
   https://doi.org/10.3389/fenrg.2022.971918
7. R. Cariou et al., Nature Energy, 3 (2018) 326-333. III-V-on-silicon solar cells reaching 33% photoconversion efficiency in two terminal configuration.
8. S. Akhil, et al., Materials & Design, 211 (2021) 110138. Review on perovskite silicon tandem solar cells: Status and prospects 2T, 3T and 4T for real world conditions
9. A.W.Y. Ho-Baillie, et al, Advanced Material Technologies, 2022, 7, 2101059. Deployment Opportunities for Space Photovoltaics and the Prospects for Perovskite Solar Cells.
10. J. Huang, H. A. Atwater, et al., 2017 IEEE 44th Photovoltaic Specialist Conf. (PVSC), IEEE, 2017, pp. 1248–1252.
11. J. Barbé, et al., Solar RRL, 3 (2019) 1900219.
12. G. Yang, et al., Energy Technology. Radiation-Hard and Ultralightweight Polycrystalline Cadmium Telluride Thin-Film Solar Cells for Space Applications
    https://onlinelibrary.wiley.com/doi/10.1002/ente.201600346
13. S. Kawakita, M. Imaizumi, H. Kusawake, Cambridge University Press (2015). Space Environments and Effects on CIGS Solar Cells and Modules.
14. P. Espinet-Gonzalez, H.A. Atwater, et al., ACS Nano 2019, 13, 11, 12860–12869. Radiation Tolerant Nanowire Array Solar Cells.
15. M.A. Green, Solar Photovoltaics Efficiency limits for PERC, TOPCon, HJT and IBC - PV CellTech Conference online 2021 as a Keynote presentation. https://www.youtube.com/watch?v=RGQ1Q_V0-EU
16. G. Li, et al., Asia-Pacific Solar Research Conference, Necastle Austraila (nov 2022). Potential Application of Silicon Solar Cells in Low Earth Orbit Space Scenario
17. B. Richards, et al., Chem. Rev. 2021, 121, 15, 9165–9195. Photon Upconversion for Photovoltaics and Photocatalysis
18. M.B. de la Mora, et al., Sol. Ener. Mats. Sol. Cells, 45 (2017) 59-71. Materials for downconversion in solar cells. [Down conversion refs]
19. G.J. Conibeer, D. König, M.A. Green, J-F. Guillemoles, "Slowing of carrier cooling in Hot Carrier solar cells", Thin Solid Films, 516, 6948-6952 (2008).
20. D.J. Farrell, et al., Appl. Phys. Lett. 99, 111102 (2011). A hot-carrier solar cell with optical energy selective contacts
21. J. Yang, et al., Physical Review Applied, 3 (2015) 044006. Working Principle in an Optically Coupled Hot-Carrier Solar Cell
22. A. Goetzberger, Appl. Phys., 16 (1978), p. 399, Fluorescent solar energy collectors: operating conditions with diffuse light.
23. B.C. Rowan, L.R. Wilson, B.S. Richards, IEEE J. Sel. Top. Quant. Electron., 14 (2008), p. 1312. Advanced material concepts for luminescent solar concentrators.
24. M.A. Hermandez-Rodriguez, S.F.H. Correla, R.A.S. Ferreira, L.D. Carlos, Journal of Applied Physics 131, 140901 (2022). A perspective on sustainable luminescent solar concentrators.

Effect of Novel Optimization Algorithm on the Performance of Photovoltaic Devices

SheriF Michael

Naval Postgraduate School,, Monterey, CA, United States

A new method for developing a realistic and accurate physical model of any type of solid state device has been previously introduced by the author. Application to model advanced multi-junction solar cells; Thermo-photovoltaics; sensors; as well as other novel solid state devices were previously presented.. The primary goal of multijunction solar cell design is to maximize the output power for a given solar spectrum. The construction of multijunction or tandem cells places the individual junction layers in series, thereby limiting the overall output current to that of the junction layer producing the lowest current. The solution to optimizing a multijunction design involves both the design of individual junction layers which produce an optimum output power and the design of a series-stacked configuration of these junction layers which yields the highest possible overall output current and voltage. This presentation demonstrates and compares the use of different optimization techniques to achieve that goal. The first approach is to use Genetic Algorithm in a two-part process to refine a given multijunction solar cell design for near-optimal output power for a desired light spectrum. Consequently, a Novel Space-Filling Experimental Design optimization technique is also utilized and comparison of the results is presented. These approaches can similarly be utilized to optimize the parameters of any Solid state device to yield any desired performance.

Towards Transfer Printing GaSb Membranes Using Selective Etchants

Margaret A Stevens, Jill A Nolde, Shawn Mack, Thomas C Mood, Kenneth J. Schmieder

U.S. Naval Research Laboratory, Washington, DC, United States

Jacobs, Hanover, MD, United States

To support micro-transfer printing of antimonide photovoltaics, we characterized the effectiveness of two selective etchants. Using 2 μm GaSb membranes as a stand-in for a photovoltaic device, we utilized stacks of lattice matched InAsSb and AlGaAsSb to create etch-release layers for a hydrofluoric acid based etch and a citric acid based etch. Nomarski, infrared, and scanning electron microscopy was used to monitor the progress of the undercut etch underneath the GaSb membrane. We found the hydrofluoric acid based etch will undercut the mesas, but needs thicker etch-stop layers or alternative photoresist tethering schemes to be compatible with micro-transfer printing. The citric acid based etch is much slower, requiring 3 hours for mesa undercutting to begin, but AlGaAsSb is found to be a very stable etch-stop layer in this material system.

Investigating the Role of Ag and Ga Content in the Stability of Wide-Gap (Ag,Cu)(In,Ga)Se2 Thin Film Solar Cells

Patrick Pearson, Jan Keller, Charlotte Platzer-Björkman

Uppsala University, Uppsala, Sweden

Thin film solar cells spanning a wide range of compositions within the (Ag,Cu)(In,Ga)Se2 system are fabricated and characterised using current-voltage, external quantum efficiency and capacitance-based measurements. The stability of the devices over time, after dry annealing and after lightsoaking is evaluated, and the role of Ag and Ga content is explored. Ag-free CuInSe2 and Cu(In,Ga)Se2 are observed to be stable, however high-Ag, high-Ga (Ag,Cu)(In,Ga)Se2 is observed to exhibit fluctuations in short-circuit current. High-Ag (Ag,Cu)InSe2 is instead observed to have stable current, but open-circuit voltage and fill-factor are strongly responsive. Thus, it is indicated that high fractions of Ag in the material lead to significant stability concerns, with high Ga fractions affecting the way in which degradation manifests. Discussion of probable causes and mechanisms follows.

978-1-6654-6060-6/23 $31.00 © 2023 IEEE

A Horizontal Single-Axis Tracker Mock-Up to Quickly Assess the Influence of Geometrical Factors on Bifacial Energy Gain

César Domínguez[1,2], Jesús Marcos[1,2], Sandra Ures[1,2], Steve Askins[1] and Ignacio Antón[1]

[1]Instituto de Energía Solar, Universidad Politécnica de Madrid, Madrid, 28040, Spain

[2]ETSI Diseño Industrial, Universidad Politécnica de Madrid, 28012, Spain

Abstract — Bifacial photovoltaic modules on horizontal single-axis trackers have become the mainstream type of utility-scale solar power plant. Their optimal design involves many trade-offs between different geometric properties, which are difficult to capture via modeling tools or underestimated by research-scale test beds. This article presents the design and preliminary demonstration of a scaled test bed for monitoring irradiance non-uniformity on both sides of the panel and realistically assess the influence of geometrical parameters on the bifacial energy gain achieved. The mock-up developed includes an open-loop control system with back-tracking actuated by stepper motors. This approach is intended to dramatically reduce the cost of benchmarking new HSAT designs for bifacial PV systems.

I. INTRODUCTION

The market share of bifacial solar cells has surpassed 50%[1]. Bifacial photovoltaic modules on horizontal single-axis trackers (HSAT) have become the mainstream type of utility-scale solar power plant, as a means of increasing the energy yield (5%-15% typ.) and minimizing the levelized cost of the electricity (LCOE) generated[2]. Bifacial solar cells are able to collect radiation from both front and rear sides, which increases the input solar resource. Besides, HSATs increase the solar energy collected with respect to a fixed-tilt installation by rotating panels around a horizontal axis, thus minimizing the angle of incidence of direct irradiation. However, tracker mounting structures can partially shade the diffuse or ground reflected irradiance incident on the rear side of PV modules. Shading not only reduces available irradiance (optical losses), but also increases non-uniformity of light on the rear side (electrical mismatch losses).

Thus, optimal tracker configuration (geometry, tracking algorithm, module mounting scheme, layout of trackers in the field) requires a careful estimation of the potential energy gain produced by the bifacial modules via some optical and electrical model of the complete solar plant and the available solar resource. Many parameters have to be considered: PV module electrical interconnection, tracker geometry, layout of tracking structures, albedo and scattering pattern from the ground, or the angular properties of direct and diffuse irradiation over the typical meteorological year of the site.

Tracker manufacturers and solar plant developers can rely either on experimental field tests or modeling tools to optimize their designs. Modeling tools can be either simplistic (e.g. the view factors approach used by PVsyst [3], which assumes infinite tracker rows and Lambertian ground scattering) and

unable to consider the actual tracker geometry, or require a great amount of computing power and/or simulation time (RADIANCE-based approach [4]). Even in the latter case, the results are very dependent on the scattering pattern from the ground.

On the other hand, field tests are based on side-by-side comparisons with some reference designs, but they are limited by the costly investments required (in terms of materials or available land) to compare many different configurations. Furthermore, small research-scale systems have been found to be dominated by edge effects: the view factor of modules not surrounded by other tracking rows is unrealistically high. Thus, translating results from small sites to actual performance in large-scale power plants requires further comprehensive models built upon field experience. As in many other fields, a way to circumvent the cost limitation is to build mock-ups of the system to be studied. In [5], a scaled test bed was built for quickly comparing different bifacial models to experimental data.

This work presents the design and preliminary assessment of a fully functional 1:10 scale model of a horizontal single-axis tracker with tunable geometrical and mounting properties that serves as a convenient testbed for rapid evaluation of the energy gain produced by bifacial modules under different geometrical configurations. This mock-up replicates the optical and cinematic properties of horizontal single-axis trackers to provide experimental results that are actually representative of real-size large-scale HSAT power plants (in contrast with mock-ups for wind tunnels, light behaves identically at meter and centimeter scales). The system developed includes irradiance and temperature sensors on both front and rear side planes to provide a direct measurement of the non-uniformity of the irradiance and an open-loop control system with back tracking actuated by a (micro)stepper motor. This approach is intended to dramatically reduce the cost of finding an optimized HSAT design for bifacial PV systems, in terms of maximum energy yield and minimum LCOE.

II. DESIGN OF A SCALED HSAT TEST BENCH

Bifacial energy gain is the main figure of merit for comparing bifacial system configurations, which is defined as the difference in energy produced by a bifacial module (Y_{bifa})

compared to a monofacial module (Y_{mono}) under the same geometrical configuration:

$$BG_E = \frac{Y_{f,bifa} - Y_{f,mono}}{Y_{f,mono}} = \frac{\varphi_{Pmax} G_{rear}}{G_{front}} \left(1 - L_{rearNU}\right) \qquad (1)$$

where G_{front} and G_{rear} are the front- and rear-side irradiance, respectively, φ_{Pmax} is the bifaciality factor of the solar cells in the module (the fraction of the incident irradiance that is actually collected by the back-side of the module, compared to the front side, ranging from 60%-70% of modules with standard p-type PERC cells to >90% in heterojunction-cell modules), and L_{rearNU} accounts for the electrical mismatch losses produced by the non-uniformity of the light on the rear side of the panel. BG_E rises with gap between modules, distance between tracker rows or axis height, but those reduce the energy yield per unit land area or increase costs. Many trade-offs hard to fine tune.

A. Mechanical design

The mock-up has to allow varying key geometrical characteristics of a HSAT array:

- Height of the torque tube (hub) and aspect ratio (defined as axis height divided by collector width).
- Orientation of panels: portrait (short side parallel to the axis) or landscape.
- Number of modules stacked perpendicularly to the axis: e.g. 2P (2V) means two modules in portrait (vertical) position, one at each side of the torque tube.
- Gap over torque tube between stacked modules to reduce the shading loss on the rear side of the module.
- Purling length and width, which has an influence on the unshaded irradiance and rear-side non-uniformity.
- Row-to-row distance (between axes) and the resulting ground coverage ratio (GCR, defined as the collector width divided by row-to-row distance).
- Albedo (ground reflectivity and scattering properties)

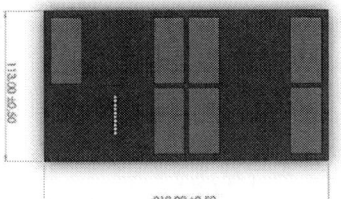 

Figure 1. *Left*: PCB mock-up of a bifacial PV module with surface-mount solar cells that monitor incident irradiance in different locations of the aperture area. Solar cells can be soldered at 8 different positions on both sides. *Right*: Threaded post with adjustable height to modify HSAT aspect ratio easily.

A 1:10 mock-up of a 30-m HSAT row has been designed following these requirements using mostly off-the-shelf mechanical components plus a few custom Aluminum pieces that can be easily machined. GCR and aspect ratio are known to be the largest influence in BG_E, and they are insensitive to

the actual value of height or row-to-row distance as long as the proportions to collector width are preserved. Module mock-ups with 1:10 scale have been created as well using printed circuit boards (PCB).

Bearing posts are made of threaded rod to easily vary axis height and hence aspect ratio (see Fig. 1). A modular wooden platform supports the pillars through small guiding holes, which provide a means of varying GCR. It supports as much as 5 tracker rows with 1P configuration (or 3 rows for 2P configuration) to mimic a realistic view factor and avoid edge effects at the center of the model (otherwise, the small dimensions allow to add other similar mock-ups easily). The support platform can be coated with different types of terrains (e.g. using sand, gravel, grass basing from railroad models) to widely alter albedo properties (both reflectivity and scattering pattern) quickly.

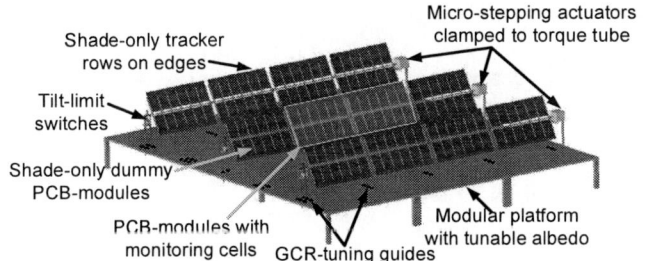

Fig. 2. Final mock-up design in 2P configuration.

This non-uniformity can be characterized by different metrics like the coefficient of variation (CoV) or the relative mean absolute difference (RMAD) between the values measured at different locations of the module. However, its influence on module power loss (L_{rearNU}) depends on the electrical layout of the module. The majority of utility-scale power plants employ PV modules with half cells because its sensitivity to partial shading is lower (the module height is divided in two sub-matrices connected in parallel). Thus, the monitoring of irradiance has been designed to study these halves separately.

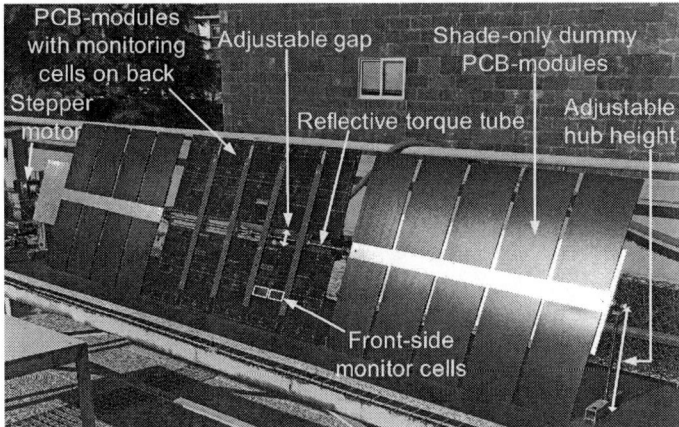

Fig. 3. Preliminary version of the mock-up design.

III. EXPERIMENTAL

A preliminary version of the mock-up design has been set up in order to quickly assess the idea and provide feedback to improve the design (see Fig. 5) It has been installed outdoors

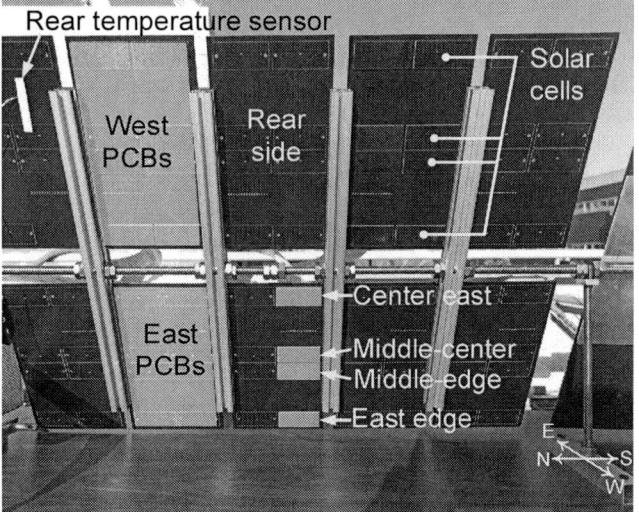

Fig. 5. Distribution of irradiance and temperature sensors in the preliminary mock-up design.

Solar cells have been placed in the center PCB-modules, at both East and West sides of the axis to investigate the non-uniformity of the irradiance. The test bed has been installed at the rooftop of the Solar Energy Institute for a short-term monitoring campaign, along with all the relevant solar resource and weather quantities (global, diffuse and direct irradiance, ambient temperature, wind speed, etc.). A large irradiance variability has been found between both edges of the module (perpendicular to the axis), and a sharp intensity transition at noon between sides: non-intuitive results that are telling about the value of the test bed to provide realistic results hard to predict with a single unit of a real-sized system or via software.

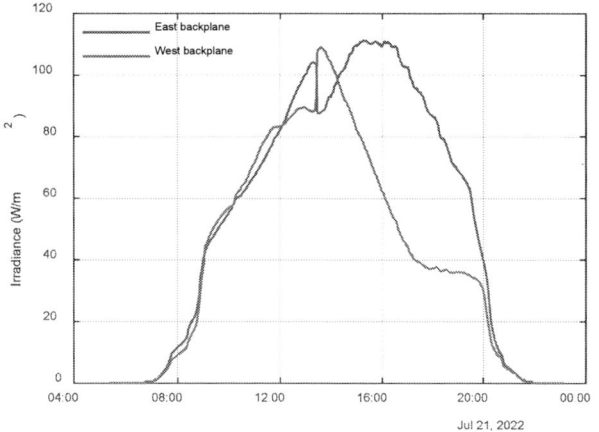

Fig. 6. Average rear-side irradiance in East and West-side modules, showing a sharp transition between them close to noon, probable due to the reflection of direct light at the torque tube.

IV. CONCLUSIONS

A mock-up for monitoring irradiance non-uniformity on both sides of bifacial modules installed on HSAT systems has been presented, which allows varying the main drivers of bifacial energy gain variability. A final version of the system will be set up and monitored at the Solar Energy Institute to evaluate the gain produced by different reference tracker designs and types of albedo.

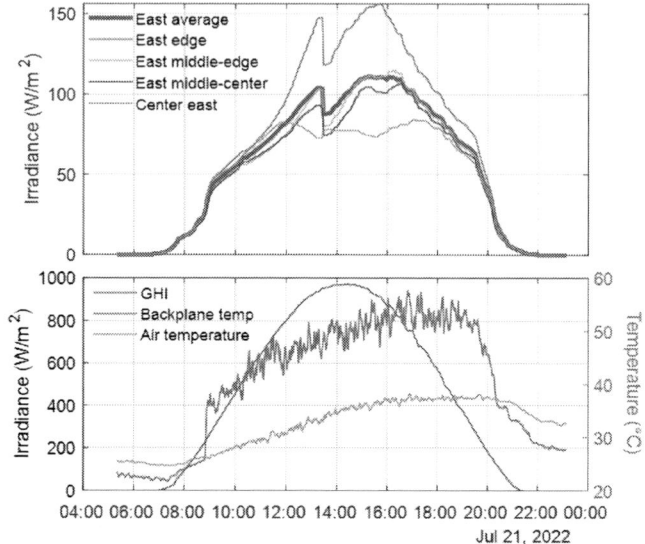

Fig. 7. Monitored rear-side irradiance at different locations of East-side PCB-modules for a clear-sky summer solstice day in Madrid.

ACKNOWLEDGEMENTS

Project supported by grant TED2021-130920B-C21 funded by MCIN/AEI/10.13039/501100011033 and by the "European Union NextGenerationEU/PRTR"

REFERENCES

[1] M. Fischer, M. Woodhouse, S. Herritsch, and J. Trube, "International Technology Roadmap for Photovoltaic, 2021 Results," Frankfurt, Germany: VDMA eV, 2022.

[2] C. D. Rodríguez-Gallegos *et al.*, "Global Techno-Economic Performance of Bifacial and Tracking Photovoltaic Systems," *Joule*, vol. 4, no. 7, pp. 1514–1541, 2020.

[3] B. Wittmer and A. Mermoud, "Yield simulations for horizontal axis trackers with bifacial PV modules in PVsyst," 2018.

[4] S. A. Pelaez, C. Deline, P. Greenberg, J. S. Stein, and R. K. Kostuk, "Model and Validation of Single-Axis Tracking with Bifacial PV," *IEEE J. Photovoltaics*, vol. 9, no. 3, pp. 715–721, May 2019

[5] S. A. Pelaez, C. Deline, S. M. Macalpine, B. Marion, J. S. Stein, and R. K. Kostuk, "Comparison of Bifacial Solar Irradiance Model Predictions with Field Validation," *IEEE J. Photovoltaics*, vol. 9, no. 1, pp. 82–88, Jan. 2019.

A New Method for the Evaluation of Majority and Minority Carrier Contact Resistivity of Polysilicon on Oxide Contacts

Dirk Steyn[1,2], William Nemeth[2], David Young[2], Paul Stradins[2], Sumit Agarwal[1]

[1] Colorado School of Mines, Golden CO, 80401, USA, [2] National Renewable Energy Laboratory, Golden CO, 80401, USA

Abstract — In this contribution we present a novel technique for characterizing the majority and minority carrier contact resistivity in polysilicon on oxide passivating contacts. Traditionally, contact resistivity is measured through the Transfer Length Method (TLM). This method presents difficulties when measuring the minority carrier contact due to the presence of a blocking diode which prevents current flow between the pads. Our novel method is performed by creating a unipolar device which has the same contact on both sides (either symmetric *n*-type or symmetric *p*-type) and metalizing the device with a grid on one side, and a full metal contact on the back (the same pattern that will be used in a final cell). We then measure the resistance across the device at increasing illumination. Since the resistivity of the wafer is a function of carrier density, when increasing illumination, the wafer resistivity decreases. We can then fit modeling equations to the measured resistivity and extrapolate the fit to infinite illumination where only the contact resistivity is measured to accurately measure both the minority and majority carrier contact resistivity.

I. INTRODUCTION

Polysilicon on oxide passivated contacts have accumulated a great deal of interest [1-3] in recent years. These contacts are both selective in that they only allow the transport of one type of charge carrier (electrons in the case of *n*-type contacts and holes in the case of *p*-type contacts), and passivating in that they allow for very low surface recombination. In developing better contacts for solar cells, contacts need to be both very conductive and have a low surface recombination parameter J_0. The contact resistivity is traditionally measured with the Transmission Length Method (TLM). Measuring the minority carrier contact resistivity with TLM is difficult due to the presence of a blocking diode formed by the p-n junction. To measure the minority carrier contact resistivity then, the base wafer is replaced by a wafer of the opposite type. For example, to measure the resistivity of a *p*-type contact on an *n*-type wafer, the same *p*-type contact will be created on a *p*-type wafer. Since it is not obvious that this new contact will have the exact same resistivity, it is desirable to be able to directly measure the *p*-type contact on an *n*-type wafer.

In our novel tehcnique, the measurement is done by first creating symmetric unipolar structures with the same polysilicon on oxide contact on both sides (shown in Fig. 1). These contacts will only allow one type of current (either electron or hole) to pass through since they are selective, allowing for the measurement of electron or hole current separately depending on the contact used. The structures will then be metalized using the same metallization geometry as a solar cell, with a metal grid in the front, and a full metal rear contact.

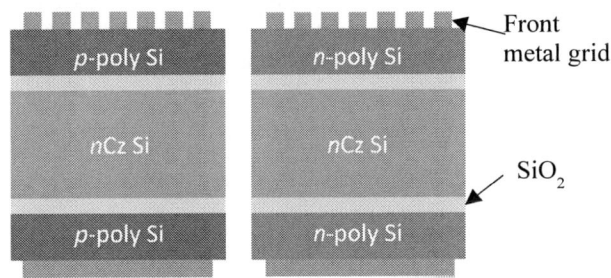

Fig. 1. Cross sectional view of symmetric structures created.

For the minority carrier contact, the effect of the blocking diode disappears under illumination. This is because the band bending induced by illumination decreases the strength of the electric field. We have confirmed this and shown that under illumination these devices exhibit ohmic behavior. Under illumination the total resistance through the structure will be:

$$R_{tot} = R_{wafer} + R_{cont,Total} \tag{1}$$

where R_{tot}, R_{wafer}, and $R_{cont,Total}$ are the total resistance, resistance of the wafer, and the total contact resistance (the sum of the front and rear contact resistance), respectively. This expression is derived by modeling the entire structure as 3 resistors in series (the two contact resistances and the bulk wafer resistance). R_{wafer} can be expressed as:

$$R_{wafer} = \frac{W}{\sigma A_c} \tag{2}$$

where W, A_c, and σ is the thickness of the wafer, cross-sectional area of the device, and conductivity, respectively. For a *n*-type wafer under illumination, the conductivity is different for *n*-type and *p*-type contacts (σ_n and σ_p, respectively). Two different conductivities are expected because the symmetric *n*-type contact will only pass electron current and the symmetric *p*-type contact will only pass hole current. The conductivities can be expressed as

$$\sigma_n = e\mu_n(N_d + g\tau) \tag{3}$$

$$\sigma_p = e\mu_p(g\tau) \tag{4}$$

978-1-6654-6060-6/23 $31.00 © 2023 IEEE

where σ_n and σ_p are the electron and hole conductivities, respectively; and μ_n and μ_p are the electron and hole conductivities, respectively; g is the generation rate and τ is the lifetime of the carriers. Under illumination, the base wafer conductivity will increase as g increases, causing the wafer resistance to drop to zero at "infinite" illumination. This can be seen by combining (1)-(4) into one equation:

$$R_{tot,n} = \frac{W}{e\mu_n(N_d+g\tau)A_c} + R_{cont,Total} \tag{5}$$

$$R_{tot,p} = \frac{W}{e\mu_p(g\tau)A_c} + R_{cont,Total} \tag{6}$$

Where (5) and (6) describe the resistance for a symmetric n and p-type contacts respectively. Using these equations, we can plot the total resistance as a function of the inverse of illumination. We should expect to see the following response from a symmetric n-type contact structure.

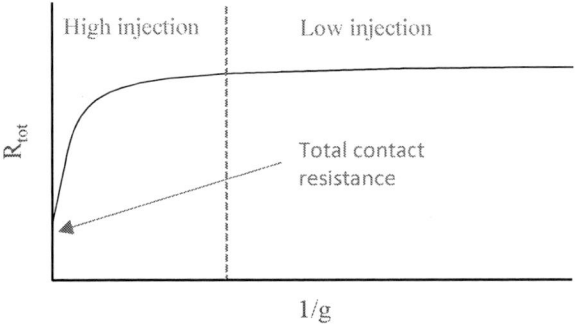

Fig. 2.   Total resistance vs inverse illumination for a symmetric n-type contact.

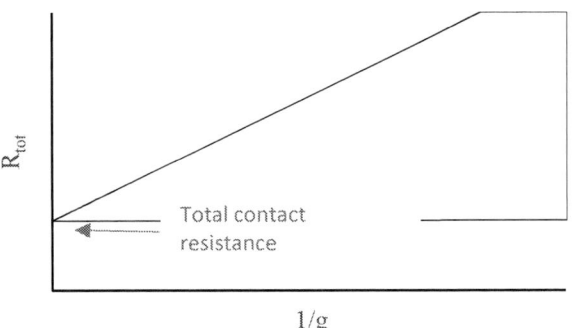

Fig. 3.   Total resistance vs inverse illumination for a symmetric p-type contact.

In these graphs, the y-intercept will correspond to the total contact resistance. On the right side of Fig. 2, we can see the low injection behaviour of the symmetric n-type structure. This occurs because the excess carriers provided by the dopant atoms dominate the bulk resistance. Once enough light-induced carriers are injected into the wafer to overcome the dopant atoms, we see a sharp decrease in the total resistance. This

behaviour should not occur in the symmetric p-type contacts since only hole current is measured and the dopant atoms do not provide excess holes. Therefore a linear response is expected.

II. EXPERIMENTAL PROCEDURE

The structures were created using 170 µm thick n-type Cz Si wafers. The wafers were cut into 30x50 mm samples before undergoing a saw damamge removal step in 22.5% potassium hydroxide solution. The samples then underwent a standard RCA clean. Then a thin (1-1.5 nm) tunneling oxide was thermally grown on the surface. After this, highly doped amorphous silicon was deposited via Plasma Enhanced Chemical Vapor Deposition (PECVD), and crystallized into highly doped polysilicon at 850°C to form the contacts. The samples then had a 15 nm layer of alumina deposited via Atomic Layer Deposition (ALD) and annealed in forming gas to inject hydrogen and passivate the interface. Then a front metal grid with 20 µm thick fingers with a 1 mm spacing and a full area back contact was deposited via the thermal evaporation of aluminum. To measure the resistance of the structures, kelvin probes were placed at the front and rear of the device, and a Kiethly 2440 was used to collect current-voltage curves at varying illumination. The illumination was provided by an in-house solar simulator, calibrated by a reference solar cell. To calculate the resistance of the samples, the slope of the current-voltage curve was taken at zero volts.

III. RESULTS

The results of both the symmetric n-type and symmetric p-type are shown in Fig. 4 and Fig. 5 respectively. As predicted, the n-type sample shows a leveling out of the total resistance as g^{-1} approaches infinity, while the p-type samples show an approximately linear behavior.

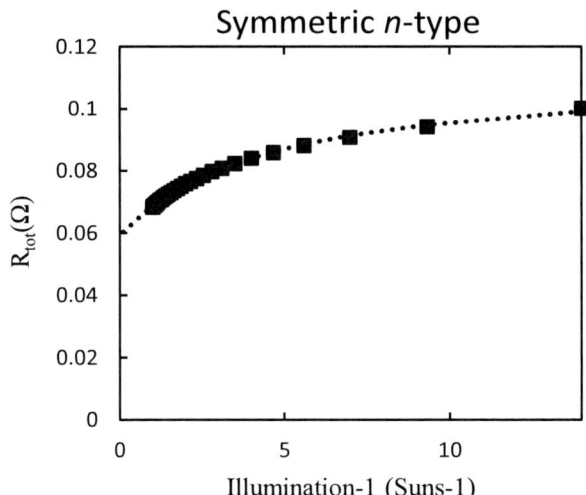

Fig. 4.   Measured total resistance vs. inverse illumination for an n-type contact structure

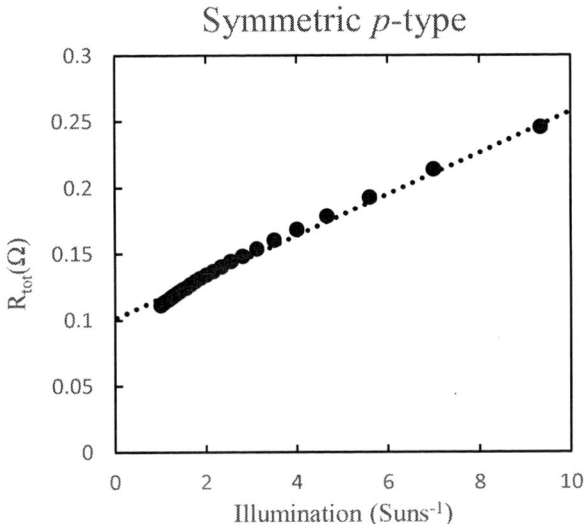

Fig. 5. Measured total resistance vs. inverse illumination for a symmetric p-type contact structure

As can be seen by Fig. 5, the response is not perfectly linear. This is likely due to the effect of Auger-Mitner recombination in the device decreasing the lifetime of the carrier. As injection increases in the device, the lifetime decreases and therefore the number of excess carriers decreases. This leads to a weaker response to incident light than expected, which can be seen in (6). We can then fit the data collected to (5) and (6) and extract the contact resistance. After fitting the equations, a total contact resistance of 101 mΩ for the symmetric p-type 59.3 mΩ for the symmetric n-type was measured. To convert this number to contact resistivity, the following equation is used.

$$\rho_c = \frac{R_{cont,Total}}{\frac{1}{A_{front}} + \frac{1}{A_{rear}}} \tag{5}$$

Here, ρ_c is the contact resistivity and A_{front} and A_{rear} are the front and rear metallization areas respectively. Applying this equation we find that the contact resistivity is 5.49 mΩ-cm² and 9.4 mΩ-cm² for the symmetric n and p-type contact structures, respectively. This value was confirmed with TLM measurements for the symmetric n-type structure. The symmetric p-type structure was not verified due to TLM not being able to measure the contact resistivity, however this contact resistivity value aligns with literature values [2], as well as approximate values from TLM that we have measured.

In the fitting of these equations, it was observed that the wafer bulk resistance was approximately 2.5x higher than the resistance values measured by a 4-point probe. This effect can most clearly be seen in the dark, as a higher resistance is measured through the device than would be expected. We propose that this discrepancy is a result of spreading resistance due to current crowding into the fingers of the solar cell. We have hypothesized this because the higher resistance measured

does not apply uniformly throughout all illuminations, and the correct value for contact resistance is still predicted by the fitting equations. This shows that the spreading resistance is illumination dependent. We will present Quokka 3 simulations that have confirmed this theory. Modeling the effect of spreading resistance is achieved in the modeling equations by adding a prefactor a (which is a function of metal geometery) that is multiplied by the bulk wafer resistance. Including this prefactor the final equation for total resistance in a symmetric n-type contact structure is as follows.

$$R_{tot,n} = a\frac{w}{e\mu_n(N_d + g\tau)A_c} + R_{cont,Total} \tag{5}$$

This method of measuring the contact resistance then also has the added benefit of showing spreading resistance and current transport effects of differing metal geometries.

IV. SUMMARY AND CONCLUSIONS

In this contribution, we have presented a new method for measuring the contact resistivity of both p and n-type contacts on the same wafer that will be used in the final solar cell device. This will allow for the accurate characterization of contact resistance for both contact types, which has not been demonstrated before. This method also has the benefit of providing information on the current transport in the final solar cell, and could potentially be used to optimize metallization geometry, which could be particularly useful in the design of bifacial solar cells.

In addition to the results shown, we will be presenting Quokka 3 simulations that show the effect of geometry on this spreading resistance, and how the current crowding affects carrier transport through the device. We will also show some theoretically derived equations that describe the prefactor a and how it is dependent on geometry.

IV. REFERENCES

[1]   Brendel et al., 2016 IEEE J. Photovoltaics, vol. 6, no. 6, pp. 1413–1420
[2]   Yan et al., 2022 Prog. Photovoltaics Res. Appl., pp. 1- 17
[3]   Feldmann et al., 2014 Sol. Energy Mater. Sol. Cells, vol. 120, pp. 270–274

Simultaneous Solar Power Generation and Bidirectional Data Transmission

Emily Kessler-Lewis, Stephen J. Polly, Elijah Sacchitella, Seth M. Hubbard, Raymond Hoheisel

Rochester Institute of Technology, Rochester, NY, United States

Blacksky Aerospace LLC, Arlington, VA, United States

Photovoltaics (PV) provide an enabling platform for integrating data transmission capabilities. A four-terminal, mechanically stacked hybrid PV device/electroabsorption modulator (EAM) is presented, combining both solar power generation and data transmission via free space optical at 1.55 μm in one package. A 23 % AM0 efficient dual junction, InGaP/GaAs solar cell was bonded to an InGaAs/InAlAs segmented EAM with a designed cutoff frequency approaching 1 MHz. Simultaneous power generation and data transmission will be demonstrated.

Localized Surface Plasmon Resonance of Quantum Dots in Two-Step Photon Up-Conversion Solar Cell Structures

Yukihiro Harada, Mizuto Kawakami, Shigeo Asahi, Takashi Kita

Kobe University, Kobe, Japan

We theoretically studied the electron density dependence of the electric field enhancement effect caused by the localized surface plasmon resonance in heavily-doped InAs/GaAs quantum dots. The resonant wavelength of the field enhancement factor shows a redshift with increasing electron density at mid-to-near infrared wavelengths. Furthermore, the multiple resonant wavelengths appear in semi-ellipsoid InAs/GaAs quantum dots due to the lowered symmetry, which is promising for the electric field enhancement at the intraband transition wavelength in the two-step photon up-conversion solar cells consisting of $Al0.3Ga0.7As/GaAs$ hetero-structure.

978-1-6654-6060-6/23 $31.00 © 2023 IEEE

Understanding the Degradation of Silicon Heterojunction Modules

Jorge Ochoa, Michael Martinez-Szewczyk, Nicholas Moser-Mancewicz, Dana Kern, Dirk Jordan, Mariana Bertoni

Arizona State University, Tempe, AZ, United States

National Renewable Energy Laboratory, Golden, CO, United States

With silicon heterojunction (SHJ) solar cells being close to their practical efficiency limit, long-term reliability, and stability are the lowest hanging fruit to improve their market adoption. However, the passivation degradation over time under field operating conditions at amorphous silicon (a-Si:H) and crystalline silicon (c-Si) interface remains a concern. Herein, we compare our damp heat accelerated tests and fielded module results to the aging behaviour of the passivation quality of a-Si:H/c-Si. We extract surface recombination velocity (SRV) using the temperature- and injection- dependent lifetime spectroscopy technique and model the expected cell performance over time. Our results show a degradation of the a-Si/c-Si interface through an increase in SRV originating from failing chemical passivation exhibited by an increase in defect density at the interface. The field passivation for the a-Si:H/c-Si stack remains the same through the time of these experiments. Moreover, microstructural analysis shows increase in defect density originates due to changes at the interface and not in the bulk of a-Si:H.

Enhancing Inverter Reliability: Current Status and Paths to Predictive Maintenance

Wayne Li, Rabin Dhakal, Daniel Fregosi, Curtis Fox and Michael Bolen

EPRI, Palo Alto, California, 94304 USA

Abstract — In large-scale PV plants, inverters have consistently been the leading cause of corrective maintenance and downtime. Improving inverter reliability is critical to increasing solar photovoltaic (PV) affordability and overall plant reliability. This study combines a literature review with field diagnostics to better understand inverter failure modes, and to identify opportunities for improving inverter reliability and developing predictive maintenance practices for inverters. It is an important step towards the longer-term goal of developing and demonstrating methodologies for predictive inverter maintenance and establishing best practices for inverter-focused inventory management and retrofitting at large-scale PV plants.

I. INTRODUCTION

Inverters are a critical component for generating PV electricity. For nearly a decade they have been identified as the largest source of corrective maintenance and downtime in large-scale PV plants [1]. Seminal research on documenting inverter failures indicates that electrical components are, cumulatively, the biggest culprit [2]-[4]. Extending component lifetime towards improving inverter reliability is important for numerous reasons, including: continuing to reduce the levelized cost of energy (LCOE) from PV plants and increasing reliance on inverter-based resources to provide grid support as penetration levels increase.

Preliminary findings suggest that inverter availability averages approximately 97% after infant mortality issues are handled [5]. This translates to 3% lost energy and a commensurate reduction to LCOE. This is an important difference compared to lost module power, such as from degradation, upon which the solar industry has typically focused and is often not directly proportional with energy loss due to plant design decisions, such as dc:ac ratios well over 1. Eliminating a persistent 3% energy loss over 50-years has a similar benefit to lifetime energy production as reducing module degradation from 0.5%/year to 0.2%/year [6] .

Inverters are trending towards modularity, enabled by tends such as improved power density and larger nameplates for string inverters, the aggregation of many string inverters to form a virtual central inverter, and components that are easier to replace. Fig. 1 shows the historical and forecasted inverter technologies being deployed, focusing primarily on those used in large-scale PV plants in the blue-hued chart elements [7]. These trends can improve plant uptime by reducing diagnostic and repair time, however, the frequency of failures and repairs have deleterious impact on LCOE. Increasing the reliability of the components is needed to reduce LCOE.

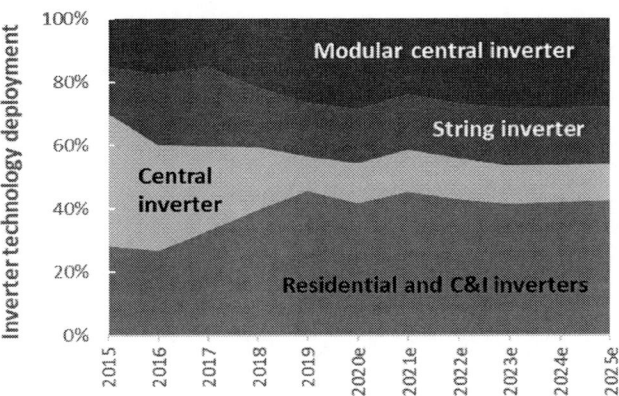

Fig. 1. Inverters are becoming more modular to improve plant uptime [7].

Previous and on-going research efforts to broadly characterize inverter reliability and failure generally use two data sets: time-series performance data and maintenance logs. From first-hand experience, there are limitations to both approaches. Time-series data – like those being collected and analyzed in the PV Fleet Initiative by the National Renewable Energy Laboratory (NREL) – can quantify inverter downtime and lost energy but does not provide insights into what caused the inverter to trip offline. Maintenance logs are highly inconsistent in their information quality, quantity, and accuracy. EPRI has worked with Sandia National Laboratory (SNL) for a decade on collecting and analyzing maintenance and failure-related data from large-scale PV plants. Maintenance tickets rarely contain sufficient information to ascertain understanding about failures, let alone the cause. For example, corrective maintenance tickets may be as rudimentary as "inverter broke, technician fixed". When the original equipment manufacturer (OEM) of an inverter sends their own technicians to fix inverters, which is a common occurrence, the data about what broke becomes proprietary and is not shared. More controlled and rigorous studies are needed to determine the causes of failure in inverters and identify opportunities to improve reliability.

II. METHODOLOGY

This study combines a literature review with field diagnostics to better understand inverter failure modes and to identify opportunities for improving inverter reliability. As shown in Fig. 2, a combination of field-based and lab-based testing will be conducted afterwards to develop service lifetime estimates and predictive maintenance for PV inverters, as well as best practices for inverter-focused retrofitting at large-scale PV plants. Some inverters will be fitted with additional sensors that are tied to a monitoring package with standalone, self-sufficient data acquisition and transmitting capabilities. The instrumented inverters are meant to monitor component health during field and lab testing that is above-and-beyond the intrinsic self-reporting capabilities of most off-the-shelf inverters. Field-based testing would occur at a large-scale, commercially operating PV plant in the U.S. that has been operational for over 7 years and has hundreds of existing 3-phase string inverters. The OEM of the existing inverters (OEM A) has exited the business and the inverters are entering the wear-out phase of the reliability bathtub curve. Additional inverter OEMs (OEM A & OEM B) will be explored for retrofitting to the existing site. Lab-based testing includes performing accelerated lifetime test until failure.

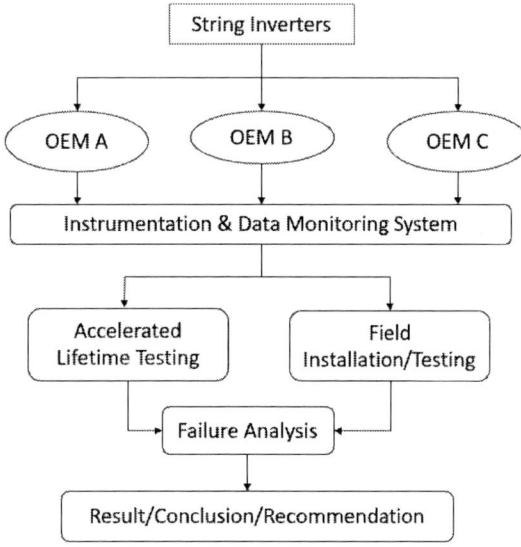

Fig. 2. Research methodology

III. PRELIMINARY RESULTS

Inverters comprise roughly one-third of annual operating expenses, which includes corrective maintenance, preventative maintenance, and inverter replacement reserve accounts [8]. Most plant owners and/or maintenance providers follow a time-based preventative maintenance approach detailed by the inverter OEM to maintain a warranty. More sophisticated approaches, such as predictive maintenance, detect signatures of impending failure and send an alert to provide sufficient time to fix the impending failure before it happens, increasing uptime and reducing maintenance costs. Lessons learned from inverter failure modes and service life estimates can be built upon to predict how rectifying known deficiencies may improve reliability. For example, using the failure stressors and

kinetics of more robust technology versus the commonly failing technology to predict how reliability is improved.

A. Litreature Survey Result

Investigation and generalization of inverter failures have been hampered by the complexity of inverters, number of OEMs, and short product lifecycles. Therefore, it is crucial to identify and study the most vulnerable electronic components or function blocks. Fig. 3 depicts the relative frequency of inverter failures by merging multiples data sources [2]. It shows that inverter downtime is primarily caused by software and component issues.

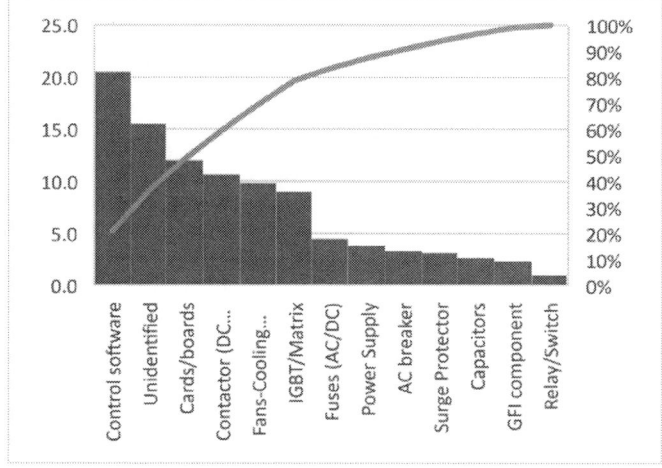

Fig. 3. Relative frequency of inverter component failures, primarily for central inverters. Replotted from Hacke et al.[2].

The components that are most likely to fail are cards and boards, contactors, fans or cooling systems, and insulated-gate bipolar transistor (IGBT) matrix. In addition, capacitors have been identified as a vulnerable component based on various reports [19]-[22] and expert insights. As a result, this study focused on these five components to better understand their failure mechanisms, mitigation measures, and failure prediction means that have been proposed or tested.

TABLE I

FAILURE MECHANISM, MITIGATION APPROACHES AND FAILURE PREDICTION METHODS FOR DIFFERENT COMPONENTS [9]-[24]

Cards and Borads	
Failure Mechanism	• Overheating & moisture between wafer layers causing delamination and cracks • Thermal expansion coefficient mismatch between components and PCB • Vibration damaging solder joints • Environmental contamination or ion migration causing short circuit bridges
Mitigation Approach	• Proper board design using prediction model • Improving sealing of Inverter components and PCB • Consistent reliability and QA program
Prediction Methods	• Regular visual inspection & board tests • Temperature and humidity monitoring • Strain measurement

IGBTs

Failure Mechanism	• Overvoltage, overcurrent, electrostatic discharge led to hot carrier current leaking through damaging gate oxide
	• High operating temperature causing fast degradation
	• Thermal runaway due to excessive reverse recovery charge and die-attach delamination
	• Thermal mismatch between the components and/or moisture causing:
	   − cracks and delamination in solder layers,
	   − wire bonding cracks / wire lift off
	   − voids in encapsulation and insulation failure
	• Surrounding circuitry failures

Mitigation Approach	• High quality designs using prediction model • High voltage rating for worst case junction temperature • Lowering thermal resistance between IGBT and heat sink • Improve chip thickness and bonding technology • Using high wire material purity with larger diameter • Operating IGBT power module at temperature <150°C • Using snubber, low inductance bus etc.
Prediction Methods	• Heatsink sensor monitoring with performance data • Collector/emitter or source/drain voltage measurement • Gate signal monitoring • Output current waveform measurement

Capacitors

Failure Mechanism	• High operating temperature accelerate chemical process • Electrolytic evaporation for electrolytic capacitors • Polymeric film/Al oxide fail under applied voltage for polymeric /ceramic capacitors • Moisture, oxygen, halogen causing degradation • Leakage current cause localize heating, ion transport
Mitigation Approach	• Proper design with extra margin on voltage, current, and hot spot temperature • Design with proper overload protection scheme • Use thin film capacitors
Prediction Methods	• Health monitoring of Equivalent Series Resistance (ESR) • Frequent measurement of tan delta • Temperature measurement of capacitor

Fans/Cooling System

Failure Mechanism	• Mechanical damage of cage, bearing, and lubrication deterioration • Environmental stress from excessive vibration, dust and dirt on fan and heat sink causing degradation • Electrical overstress, loose wire, wiring errors from installation. • Coolant leakage
Mitigation Approach	• A high-quality design • Improving serviceability • Periodic maintenance such as lubrication and filters
Prediction Methods	• Temperature sensors for cooling monitoring and protection • Precursor monitoring (such as acoustic noise, current, and shaft speed) • Heatsink sensor monitoring with performance data

Contactors

Failure Mechanism	• Contactor coil aging, mechanical wear, over voltage and current causing contactor fail to open/close • Rust deposits, oxidation and polluted contacts increasing resistance • Water ingress from manufacturing process • Arcing weld the contacts
Mitigation Approach	• Proper design for protection • Design with extra margin on power rating • Electrical life prediction based on contact erosion loss and electrical load stress

	• Consistent reliability and quality assurance program
Prediction Methods	• Periodic visual inspection • Preventive maintenance to carry out contactor electrical test • In-situ contact resistance measurement • Current to open contactor & contactor temperature

Table I provide a summary of the reported outcomes of five major components likely to fail in the inverter. The main conclusions are:

- Most components failures are caused by environmental stress such as high temperature, moisture, and thermal cycling, as well as electrical stress, such as over voltage and over current.
- Failure mitigation measures may include high-quality upfront design with extra margin on power rating. With fast development and a constant cost reduction effort in the industry, a consistent reliability and quality assurance (QA) program is essential to improve inverter reliability. Potential improvements may run counter to pervasive downward cost pressure trends in the industry.
- Inverter sealing improvement, operation temperature reduction, and preventative maintenance can also mitigate inverter failures.
- Some researchers focused on inverter lifetime prediction using a physical model. This can be used at the design stage to improve component selection.
- Predictive maintenance was frequently discussed, and many approaches were proposed and tested in laboratory. But few were tested and validated in the field.

B. Failure Inspection Result

In this study, failure frequency of different components is also collected through inspection of the failed string inverters. The inspected string inverters are from the commercially operative PV plant where the field based testing is proposed. The inspection is conducted to identify failed electronics compoennts such as capacitor, transistor, and boards. This investigation was conducted to build confidence in placing sensor on the components likely to fail. 40 invertes of OEM A, which failed during 7 years of operation of the PV plant, were investigated through visual examination method. The visual examination consists of a series of tasks from disassembling each components of the inverter, followed by examining for physical damage, to documenting the failures. The details of the inspected inverter are presented in Table II, and the individual compoennts such as fuse blocks, swithes, DC MOVs, DC filters, main control boards, inductors, IGBTs, capacitors, fans, contactors, current and voltage transformers, communication boards, terminal boards, etc., are presented in Fig. 4 .

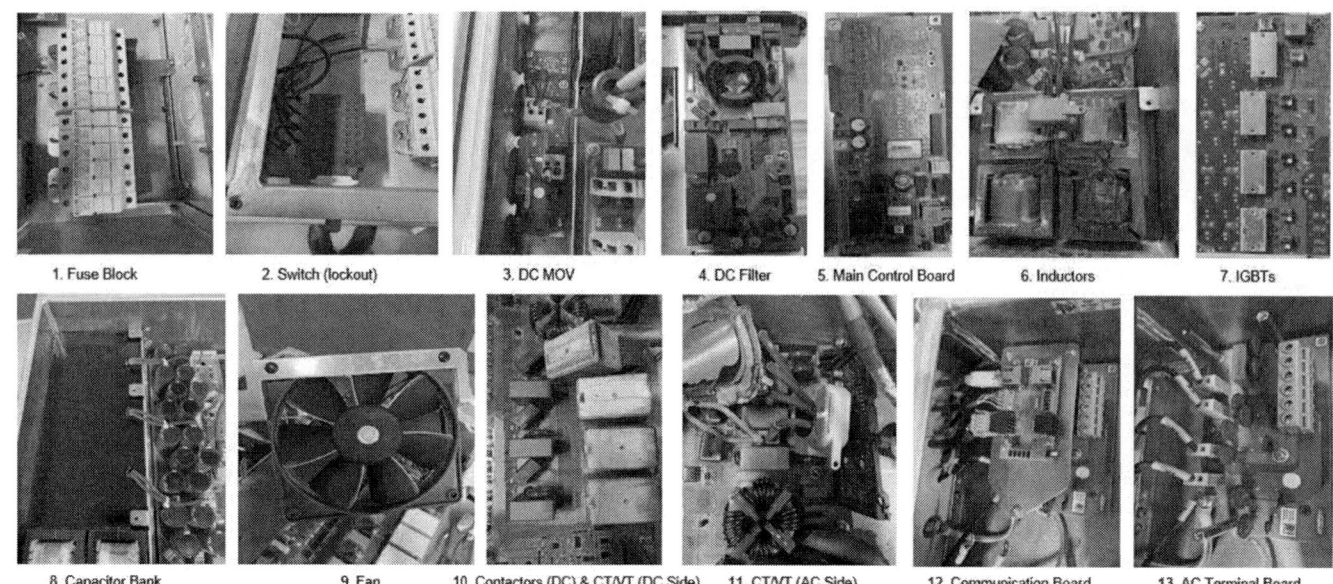

Fig. 4. Component investigated in the visual inspection

TABLE II
FAILED STRING INVERTERS DETAILS

PV inverter information	
Inverter Rating	23 kW
Approximate DC:AC ratio	1.2
DC input design	
Number of modules in series	19
Number of strings per inverter	6 strings (3 strings per input)
Fuse/OCPD rating	15 amp/1000 VDC
AC output design	
Output breaker rating	40A AC panel has 6 inverters 400A feed and 250A switch
Output Voltage	480 V balanced 3 phase
AC collection ground system	Wye-grounded
Disconnets between inverter and panel	No

The result of visual examination study shows that the failed inverter from the commercially operated PV plant had missing components. This could indicate failures or their utilization for repair/maintenance of other working inverters. Table III shows the details of the missing components. The missing components are the high value, high failure elements. They are most likely pulled for spares.

TABLE III
INVERTERS AND MISSING COMPONENTS DETAILS

Number of Inverter Inspected	40
Missing Components Details	
Fuse Block	9
Main Control Board	8
Communication Board	1
AC Terminal Board	1

The overall result of the failure inspection shows that the major components that failed are the current and voltage sensing devices. They are followed by inductors, contactors, IGBT and capacitor bank. The missing components presented in table III is not included in the summary of Fig. 5. As those missing components may be related to inverter failure caused by other components failure, and they were replaced in other working inverter as a part of the repair and maintannce process in the plant.

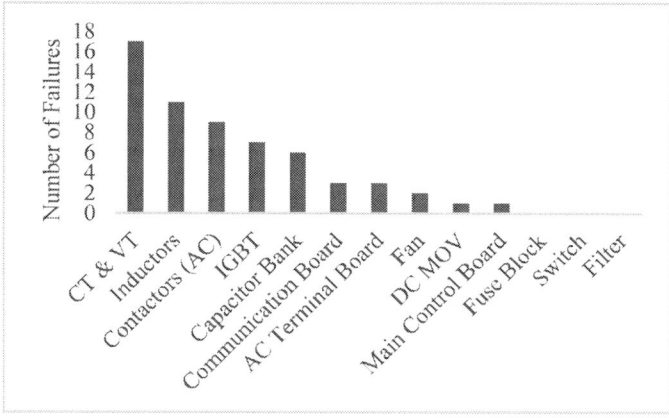

Fig. 5. Summary of the failure inspection on string inverters from operational commercial PV plant.

The method to identify failure is based on observation of the discoloration, burn out components, offset, distortion etc. Based on initial discussion, the causes of failure of those components appear to be overheating, manufacturing defects, arcing, short circuit etc. Further investigation and discussion on the failure analysis are pending with the owner of the commercial PV plant and other stakeholders working on repair and maintenance of the string inverters.

IV. FUTURE WORKS

This study has so far established a baseline for the study of component level predictive maintenance of inverters. The next phase of this project will involve the installation of sensors and monitoring platform, followed by laboratory and field based testing. The current project plan will involve additional type of inverters (OEM B and OEM C) along with the existing inverter (OEM A).

Fig. 6 outlines a single inverter laboratory mock up of parameters that can be sensed while the inverter is operated under various conditions controlled with an environmental chamber. The sensors that are required for sensing the inverter component parameters including ambient environment variables have been identified.

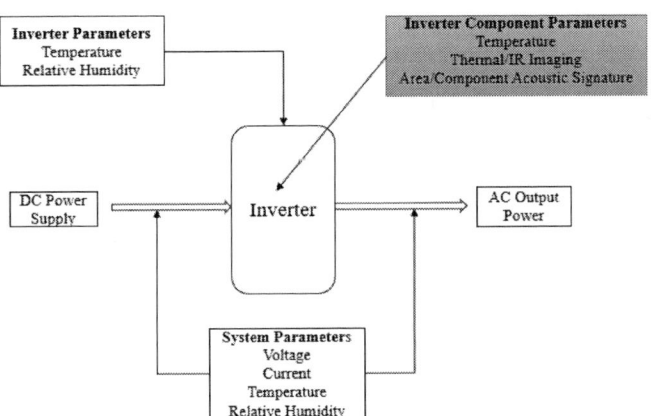

Fig. 6. Laboratory mockup of experimental set up and sensing parameters

The Data Acquisition System (DAS) envisioned for one or more inverter sensing systems is also designed and can be modified to support different options of sensor package that is decided later in the study. Specifications for each sensor have been identified and system compatibility with the envisioned DAS is being evaluated.

In the instrumented inverters, accelerated life time testing and field based testing will be conducted. The data from lab and field is monitored and collected for the further analysis. The failed inverters from lab and field based testing will be analyzed to identify the cause of failure. Information is sought to identify what failed – electronics component(s) like a capacitor or transistor, solder bond, and so on.

The final result of the study is shared among stakeholders working in the field for discussion. Insights will be collected via expert elicitation from the organizations and relevant staff involved in these monitoring and diagnosis of inverter reliability. The goal is to identify common themes and opportunities for improvement from the individual cases. This is of particular importance since extending component lifetime and improving inverter reliability may exacerbate challenges, assuming the historical rapid product lifecycle and lack of reverse compatibility persists of the life of a 50-year plant.

IV. CONCLUSION

Inverters remain the largest source of corrective maintenance and downtime at the PV palnts. The components that are most likely to fail are cards/boards, contactors, fans/cooling systems, IGBT/matrix, and capacitors. High-quality upfront design with extra margin on power rating, adopting consistent realiablity and quality assurance program and sealing improvement, temperature reduction and presventative maintenance can be method to increase reliability of the inverters. In Summary, there are limited data on practices to measure and predict component failures. Most studies at the laboratory scale may be overly complicated for field application. Some promising techniques for failure prediction need further validation. New techniques need to be developed for certain component failures.

ACKNOWLODGEMENT

This work is funded in part by the U.S. Department of Energy Solar Energy Technologies Office, under award number DE-EE-0009826.

REFERENCES

[1] D. Jordan, et al. PV field reliability status—Analysis of 100 000 solar systems. Prog Photovolt Res Appl. 2020; 28: 739– 754.

[2] P. Hacke, et al. A status review of photovoltaic power conversion equipment reliability, safety, and quality assurance protocols. Renewable and Sustainable Energy Reviews 2018, 82: 1097-1112.

[3] A. Golnas, PV System Reliability: An Operator's Perspective. IEEE JPV 2013; 3: 416-421.

[4] T. Gunda, et al. A Machine Learning Evaluation of Maintenance Records for Common Failure Modes in PV Inverters, IEEE Access 2020; 8: 211610-211620.

[5] C. Deline, et al. PV Fleet Performance Data Initiative: Performance Index–Based Analysis. NREL. TP-5K00-78720. Golden, CO: 2021.

[6] Modeled energy with System Advisor Model using 1.3 dc:ac ratio, single-axis tracking, sited at Kansas City.

[7] Global solar PV inverter and MLPE landscape 2020: Prices, forecasts, market shares, trends, and vendor profiles. Wood Mackenzie. Edinburgh, Scotland: 2020.

[8] 2020 Solar Technology Status, Cost, and Performance. EPRI, Palo Alto, CA: 2020. 3002018729.

[9] R. Grinberg et. al, Analysis of failure rate prediction by using part stress method for printed circuit board assemblies in power electronics building block. IET Conference 2012. 1-5.

[10] A. Wileman et al, Physics of Failure (PoF) Based Lifetime Prediction of Power Electronics at the Printed Circuit Board Level. Appl. Sci. 2021, 11, 2679.

[11] M. Khalil, Reliability assessment of PV inverters, 14th IMEKO TC10 Workshop Technical Diagnostics New Perspectives in Measurements, Tools and Techniques for system's reliability, maintainability and safety, Milan, Italy, June 27-28, 2016

[12] Predictive Maintenance of solar PV plants: The time is now, DNV GL, 2021

[13] A. Nagarajan et. al, Photovoltaic Inverter Reliability Assessment. Golden, CO: National Renewable Energy Laboratory. NREL/TP-5D00-74462

[14] S. Atcitty, Utility-Scale Grid-Tied PV Inverter Reliability Workshop Summary Report, SAND2011-4778, 2011

[15] H. Leobardo et al. Early fault detection in SiC-MOSFET with application in boost converter. Rev.fac.ing.univ. Antioquia, Medellín, 87: 8-15, June 2018

[16] M. S. Khanniche, Fault Detection and Diagnosis of 3-Phase Inverter System. Rev. Energ. Ren.: Power Engineering (2001) 69-75

[17] B.H. Kwon, et al. Fault Diagnosis of Open-Switch Failure in a Grid-Connected Three-Level Si/SiC Hybrid ANPC Inverter. Electronics 2020, 9, 399.

[18] F. Gonzalez-Hernando et al. Wear-Out Condition Monitoring of IGBT and mosfet Power Modules in Inverter Operation. IEEE Transactions on Industry Applications, vol. 55, no. 6, pp. 6184-6192, Nov.-Dec. 2019

[19] Quality and Reliability Assessment of Select Photovoltaic Inverters. EPRI, Palo Alto, CA: 2012. 1024366

[20] J. Lenz et al. Benchmarking of capacitor power loss calculation methods for wear-out failure prediction in PV inverters. Microelectronics Reliability, 100-101, 113491

[21] A. Lahyani et al. Failure Prediction of Electrolytic Capacitors During Operation of a Switchmode Power Supply. IEEE TRANSACTIONS ON POWER ELECTRONICS, VOL. 13, NO. 6, NOVEMBER 1998

[22] P. Hacke et. al., Evaluation of the DC bus link capacitors and power transistor modules in the qualification testing of PV inverters. Prog Photovolt Res Appl. 2020;1–9.

[23] H. Oh et al. Precursor monitoring approach for reliability assessment of cooling fans. J Intell Manuf 23, 173–178 (2012)

[24] K. Li et al. Electrical Performance Degradation Model and Residual Electrical Life Prediction for AC Contactor. IEEE Transactions on Components, Packaging and Manufacturing Technology. PP. 1-1.

Improvements on Spectral Correction Predictive Modeling for CdTe Modules

Alan J. Curran, Boris Lin, and Yuepeng Deng

First Solar, Perrysburg OH, 43551, USA

Abstract—**An updated spectral prediction model is proposed for CdTe modules which includes predictive capabilities for cloud cover while maintaining the simplicity required for plant performance simulations. Clear sky index is used to quantify cloud cover given it can be calculated using existing solar plant simulation software like PVsyst and Plant Predict. Initial fitting results show the new model has an irradiance weighted testing RMSE 29.4% lower in the GHI orientation and 56.7% lower in the 1-Axis tracking orientation compared to the prior 2-parameter spectral model for CdTe. Monthly and yearly spectral correction comparisons show in all locations except for Miami that the new model predicts expected spectral correction more accurately than the existing model which underpredicts spectral impact.**

Keywords—*Solar, CdTe, Energy prediction, Spectral correction, Modeling*

I. INTRODUCTION

The power production of photovoltaic (PV) systems are fundamentally dependent on the immediate weather conditions, most notably solar irradiance and temperature, but numerous other effects including the position of the sun, the composition of the atmosphere, location of the modules, angle of incidence, etc. will all influence the energy production of a system. As such, monitoring an active system or predicting the performance of a future system requires accurate data collection and prediction as well as a strong understanding of the impact from each different mechanism. Software used to design and simulate PV systems, such as PVsyst and PlantPredict, have built-in functionality to calculate the energy generation performance and evaluate the quantitative impact of different mechanisms.

Solar irradiance (incident power from the sun normalized by area) is typically measured with pyranometers to use as a reference to the power output of a solar system. The standard for solar spectrum is the ASTM G173 AM 1.5 spectrum [1], which is defined between 280nm and 4000nm of photon wavelengths. Pyranometers measure the irradiance across the entire spectral range, the integral of which is the final irradiance value. The spectral range of solar panels are defined by their quantum efficiency (QE) curves which denote efficiency as a function of wavelength. The spectral range of a QE curve is far lower than that of a pyranometer, which creates a potential for disparity between the irradiance reported by a pyranometer and the irradiance experienced by a solar panel. The QE curves of crystalline silicon (c-Si) modules generally range from 350nm-1100nm and those of cadmium telluride (CdTe) modules have a tighter range of 280nm-950nm. Fig. 1 shows the AM 1.5 spectra overlayed with representative QE curves (normalized to peak)

for both c-Si and CdTe modules. Spectral correction refers to adjustments made to pyranometer measured irradiance to correct for the differences caused by the disparate spectral ranges of the pyranometer and the solar panel. Spectral correction (M) is defined by (1) where E_R is the reference spectra, E_S is the measured spectra, SR_R is the spectral response of the spectroradiometer or pyranometer, and SR_T is the spectral response of the module [2].

$$M = \frac{\int_{280}^{4000} E_R SR_R d\lambda}{\int_{280}^{4000} E_S SR_R d\lambda} \times \frac{\int_{280}^{4000} E_S SR_T d\lambda}{\int_{280}^{4000} E_R SR_T d\lambda} \qquad (1)$$

The resultant M is a proportionality between the irradiance reported by the pyranometer and the irradiance observed within the solar panels QE range. The QE curve difference between c-Si and CdTe leads to disparities between their spectral corrections. In particular, the CdTe QE range falls short of two significant water absorption bands (highlighted in Fig. 1) between 900 and 1200nm which fall under typical c-Si QE ranges. This effect in particular makes CdTe more susceptible to spectral variations as cloud cover and humidity conditions affecting these bands would be captured by a pyranometer, captured in part by a c-Si module, but not captured by a CdTe module. This has made spectral correction predictions for CdTe modules both more extreme and harder to predict than c-Si modules [3].

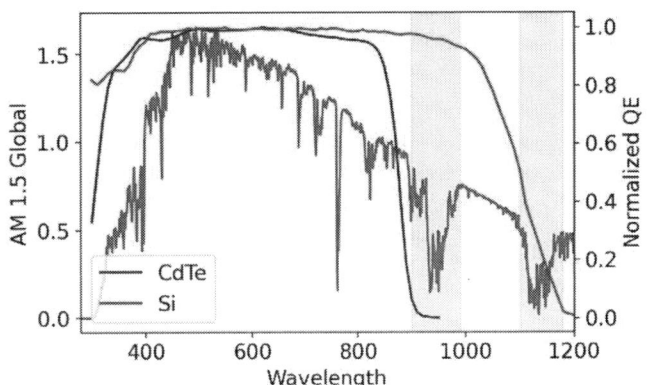

Figure 1. AM 1.5 spectra overlayed with examples of typical c-Si and CdTe module normalized QE curves. The curves were normalized by dividing each one by its maximum values. Two significant water absorption bands have been highlighted that fall outside the CdTe QE range but inside the c-Si QE range.

Spectral variations have been studied at First Solar and shown to significantly affect both short- and long-term energy yield performance of CdTe PV systems [3]. Historically, the Simple Model of the Atmospheric Radiative Transfer of Sunshine (SMARTS) [4] was applied to predict the spectrum and quantify its impact on monthly and yearly energy yield performance of CdTe PV modules. However, SMARTS can be computationally expensive to run and remained practically difficult to incorporate into any commercially available energy simulation tools. In an effort to approximate the SMARTS output results for use in PV prediction software, Nelson et al. [3] firstly published a one-variable parameterization (Spectral 1.0) for the spectral correction on CdTe PV systems as an exponential function form of precipitable water content (P_{wat}), for First Solar module products prior to Series 4-2. Subsequently, the same functional form with updated coefficients was presented to reflect the improved QE curves on First Solar Series 4-2 modules [5], as follows:

$$M \approx 1.266 - 0.091\exp(1.199(P_{wat} + 0.5)^{-0.210}) \quad (2)$$

Lee et al. [6] developed a two-variable parameterization (Spectral 2.0) as a function of both precipitable water content (P_{wat}) and absolute air mass (AM_a), which captured the secondary dependence on AM_a for the spectral sensitivity of CdTe PV modules. Spectral 2.0 demonstrated an improved and better correlation between the measured and predicted spectral correction values through analysis of publicly available outdoor test data from three National Renewable Energy Laboratory (NREL) test sites in Golden, Colorado; Cocoa, Florida; and Eugene, Oregon, compared to Spectral 1.0 using P_{wat} only [6]. The functional form for Spectral 2.0 is given in (3):

$$M = b_0 + b_1 \cdot AM_a + b_2 \cdot p_{wat} + b_3 \cdot \sqrt{AM_a} +$$
$$b_4 \cdot \sqrt{p_{wat}} + b_5 \cdot \frac{AM_a}{\sqrt{p_{wat}}} \quad (3)$$

Correspondingly, the module-specific b_n coefficients were determined through training a multivariate sensitivity analysis in SMARTS for First Solar module products up to Series 6 as well as including Mono-Si and Poly-Si modules [6]. Spectral 2.0 has been implemented in PVsyst and Plant Predict.

Both Spectral 1.0 and Spectral 2.0 were built based on simplifying the SMARTS simulation outputs of M values, while intrinsically speaking, the applicable scope of SMARTS is only under clear sky conditions [4]. Therefore, a fundamental limitation exists that so far there has never been a model with a simple functional form that is capable of quantitatively addressing the impact of cloud cover on the spectral effects of CdTe PV energy yield performance. PVsyst has implemented a CREST model [7] to account for the spectral correction of amorphous silicon (a-Si) modules exclusively, which incorporates its dependence on the relative air mass for the atmospheric path length and an air mass independent clearness index for the cloud cover. The CREST model correlates an indirect spectral characterization parameter, the average photon energy (APE) [8], through a parametric function of the relative air mass (AM) and an AM corrected clearness index (kT*) [9],

while its set of coefficients were fit against a year of outdoor spectral irradiance data from Loughborough, UK [7]. The clearness index (kT), calculated as the fraction of Global Horizontal Irradiance (GHI) to its corresponding extraterrestrial irradiance at the top of the atmosphere, is an older measure of cloud cover than the clear sky index (CSI) used in this study. The spectral correction of a-Si modules was then correlated from APE with linear or polynomial relationships [7]. Therefore, cloud cover should be reasonably convinced as a nonnegligible contributing factor towards the M values of CdTe modules, given that a-Si modules also have a narrower QE range than c-Si modules.

As an additional source of spectral irradiances data and an effort to fit the proposed spectral prediction model (Spectral 3.0) in this paper for 1-Axis tracking systems, this study incorporates a state-of-start radiative transfer model developed at the NREL, named as Fast All-sky Radiation Model for Solar applications with Narrowband Irradiances on Tilted surfaces (FARMS-NIT) [10]. FARMS-NIT is a major progression of spectral irradiances computation model from the SMARTS, profoundly under the cloudy-sky conditions. FARMS-NIT analytically and effectively computes the spectral irradiances in 2002 narrow-wavelength bands from 280 to 4000 nm, over inclined PV panels for both fixed tilt and 1-Axis tracking systems [10], where the integral of spectral irradiances across the wavelengths equals to the plane-of-array (POA) irradiances, for both clear-sky and cloudy-sky conditions. The validation analysis has also been performed by NREL to confirm that FARMS-NIT shows substantial improvement in both computational accuracy and efficiency for spectral irradiances on both ground and inclined surfaces than many other radiative transfer models [11].

II. METHODOLOGY

A. Data Aquisition

Global Horizontal Irradiance (GHI) data for the spectral study in this paper are collected from Kipp and Zonen CMP11 broadband pyranometers at three First Solar operated test sites located in Arizona, Ohio, and Malaysia. These three sites show a large diversity in global locations and climates which allows fitting over significantly different spectral conditions. Broadly speaking, Arizona is hot and dry, Ohio is cold and mildly humid, and Malaysia is hot and humid. Density plots of the relevant metrology data are shown in Fig. 2 for each test site location, as well as Miami which is used in 1-Axis tracking modeling. Trends show less cloud cover (stronger clear sky index with peak around 1) in Miami and Arizona, and higher precipitable water measurements in Malaysia and Miami. At each test site there is a Spectrafy SolarSIM-G, a software-augmented multi-filter radiometer [12], in a GHI orientation measuring 280-4000nm spectra with data collection starting from 2018. The spectra are merged with GHI, solar zenith angle, ambient temperature and relative humidity data from on-site pyranometers and meteorological stations with a resolution of 15 minutes. Precipitable water content and absolute air mass are determined by applying Gueymard's precipitable water model [13] and PVLib sun-earth geometry functions [14] where on-site measured solar zenith angle, ambient temperature and relative humidity serve as inputs. The spectral corrections are then calculated using (1) with the GHI calculated from the integral of the spectra rather than a local pyranometer. Using the measured spectra as a GHI source rather than a pyranometer prevents potential errors due to instrumentation offsets caused by calibration differences, small orientation offsets, different local cloud cover, and collection timing. Spectral correction factors from local instrumentations and from different models can then be calculated and compared over several years, on hourly, daily, monthly and yearly time scales.

1-Axis tracking spectral data was not available from test site spectroradiometers at the time of this study for these locations, however 1-Axis tracking systems represent the most relevant commercial orientation. Therefore, it is important to compare 1-Axis model fitting results alongside GHI model fitting results. To obtain the 1-Axis spectral data which most closely represents the field measurements, the FARMS-NIT model [10] was used as the reference of hourly spectral irradiances data across 280-4000nm spectra in 1-Axis tracking conditions. FARMS-NIT spectral data were accessed and collected from the National Solar Radiation Database (NSRDB) [15], developed by the NREL, where an hourly spectral dataset is able to be generated and downloaded after the user specifies a geographical location, year, time interval, fixed tilt angle and azimuth angle or 1-Axis tracking. As the Malaysia region was not available for FARMS-NIT spectral data at the time of this study, Miami, Florida is chosen as a substitute for 1-Axis tracking model fitting being a high temperature and high humidity region.

B. Variable Selection

To maintain modeling simplicity, clear sky index (CSI) is chosen as a metric to quantify the cloud cover. This allows the proposed Spectral 3.0 to maintain the computationally light and universally applicable usage that Spectral 2.0 has. CSI is defined as a ratio of measured irradiance over modeled clear sky irradiance. The capability to calculate typical metrological year (TMY) CSI is already built into common PV modeling software such as PVsyst and PlantPredict, so it is not a disruptive inclusion in their existing algorithms.

$$M = \beta_0 CSI + \beta_1 P_{wat} + \beta_2 AM_a + \beta_3 \quad (4)$$

Precipitable water content and absolute air mass are also maintained as predictive variables from Spectral 2.0. The initial iteration of the Spectral 3.0 model is a multiple linear model of each significant variable, shown in (4), where M is the spectral correction factor (unitless), CSI is the clear sky index (unitless), P_{wat} is precipitable water content (cm), and AM_a is absolute airmass (unitless). The simplified Solis clear sky irradiance model from PVLib [14] is used in conjunction with ground sensor measurements to calculate clear sky index for each of the locations.

C. Data Selection and Model Evaluation

Model fitting is completed through a training and testing schema, with the coefficients for the model are fit against the training data and then predicted and evaluated over the testing

dataset to prevent overfitting. In each case 1 year of data is selected as the training set and a different year is selected as the testing set to prevent blindness to seasonal spectral trends at each location. For the GHI model fitting from the test sites, the training data set is July to July from 2020-2021 which is a continuous year with a high degree of data quality across all sites. The testing data set is the 2022 year, with the first 4 months excluded from Arizona due to data quality issues. The 1-Axis tracking model is fit against FARMS-NIT data rather than the measured spectra, so data collection quality is not a concern. The training set for the 1-Axis data is the year 2020 and the testing set is the year 2021 of FARMS-NIT modeled spectra for each location with a time interval of 1 hour, which is currently the highest available resolution from the NSRDB [15].

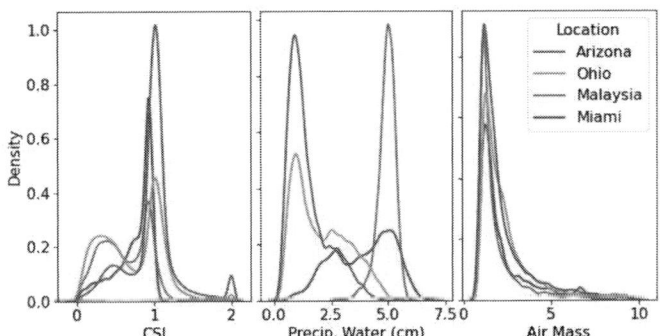

Figure 2. Density plots of the metrology data used for fitting the spectral models grouped by location. Ohio and Malaysia show a more significant bimodal distribution for clear sky index given they are cloudier regions compared to Miami and Arizona which have stronger peaks around 1. Miami and Malaysia show significantly higher precipitable water measurements as well.

The models are not fit by individual locations, rather the site level data are combined and a single set of Spectral 3.0 coefficients are fit and evaluated, for GHI and 1-Axis tracking models respectively. These are referred to as universal coefficients. In a typical use scenario, it is impractical to have a unique set of coefficients for each location as the end goal of the proposed model is to extrapolate spectral predictions to any given location. Performance is monitored by site to ensure that the models do not show significantly different performance at locations with different weather conditions and spectral adjustments.

Modeling is done with linear regression weighted to the inverse of the absolute value of irradiances to bias the fitting towards higher irradiance regions. The most extreme spectral corrections tend to occur at low irradiance regions which reduce their overall impact on net energy production. In a practical setting it is more important to accurately predict spectral corrections during high energy production periods while errors during low energy production are less consequential. Biasing the models ensures that the models will fit best during high irradiance periods, preventing low irradiance regions from acting as influential outliers. Coefficients are not given in this paper as they are considered to be preliminary and need to be further validated with additional field data before being released, however they are available upon request if there is an interest in testing this model against other data sources.

III. RESULTS

A. GHI Spectral Modeling

Equivalence plots for the spectral models and measured spectrometer results are shown right in Fig. 3 for each of the three test sites. In all cases the Spectral 2.0 model shows difficulty in predicting spectral corrections above 1.05. This is not a set filter, rather an intrinsic behavior of the model and given coefficients. In most cases Spectral 2.0 falls at or below the line of equivalence, indicating a general behavior of underperformance. Spectral 3.0 shows a better adherence to the line of equivalence until high spectral correction conditions where it begins to tail off and starts to underpredict. This has a minor impact on overall predictions as these high spectral correction conditions normally occur under high cloud cover and low irradiance which amount to low contributions to the total insolation. This tailing trend is exacerbated by the weighted regression prioritizing high irradiances and therefore near AM 1.5 spectra regions, however this is beneficial for the net accuracy of the model where total insolation is considered. Testing Root Mean Squared Error (RMSE) for each model vs. the measured results are shown in Table I. RMSE is shown as both unweighted and weighted, where the weighted RMSE is the spectral correction error multiplied by the irradiance to bias towards predicting at high irradiance values which are the most impactful for system performance. The unweighted and weighted RMSE values are numerically incomparable and can only be compared between themselves. In all cases the Spectral 3.0 model shows improved unweighted RMSE and weighted RMSE results compared to the Spectral 2.0 model. Spectral 3.0 shows a 51.8% reduction in unweighted RMSE and 29.4% reduction in weighted RMSE compared to Spectral 2.0 in the GHI orientation.

The equivalence plots alone are not a sufficiently good indicator of overall spectral impact as there is an irradiance dependence on spectral correction factor with higher irradiances generally trending towards a correction of 1, meaning there is an additional weight component to consider when looking at the net impact. Monthly spectral correction comparisons in the year of 2021 are shown in Fig. 4 for the GHI results across all sites. The lowest spectral correction is in Arizona where influences like cloud cover and precipitable water are minimal. Consequently, this is also the location with the lowest discrepancy between the 2.0 model and the spectrometer measurements. Deviations between the 2.0 model and the 3.0 model and spectrometer results generally occur in months and regions where cloud cover becomes more significant, most notably in the winter months of Ohio. The yearly Spectral 3.0 model deviates from the spectrometer results by -0.3, -0.16, and 0.18 percentage points in Arizona, Ohio, and Malaysia respectively, compared to Spectral 2.0 model which is -1.85, -2.47, and -3.03 percentage points in Arizona, Ohio, and Malaysia.

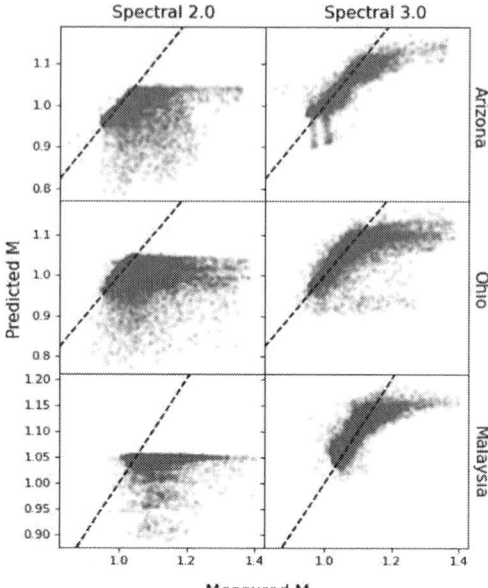

Figure 3. Equivalence plots of measured vs. modeled spectral correction for the Spectral 2.0 and 3.0 models at each of the three data collection locations. The black lines on each plot are lines of equivalence (slope of 1, intercept of 0). The Spectral 3.0 model shows a much more accurate prediction of the measured spectral correction.

B. 1-Axis Tracking Spectral Modeling

The equivalence plots for the 1-Axis tracking spectral modeling (Fig. 5) show similar trends to the GHI results for both spectral models. For Spectral 2.0, there is still an upper threshold of 1.05 that the model cannot predict above, although this effect is less pronounced as the 1-Axis correction results are generally closer to 1 compared to the GHI results. For the 3.0 model the same tail behavior is observed at high correction with much better equivalence between the FARMS-NIT and Spectral 3.0 results prior to this region. An additional outlier cluster in the 3.0 results can be seen in each of the three locations where the 3.0 model predicts a near 1 spectral correction but the FARMS-NIT results are below 1.

Monthly spectral boost/derate plots are given in Fig. 6 for each of the locations. Similar to the GHI results, 2.0 shows a trend of underperformance compared to the FARMS-NIT and 3.0 results, most notably in winter months when cloud impact is at highest. Miami is a notable exception however, where Spectral 2.0 model predicts higher correction values in the summer periods than the 3.0 model and the FARMS-NIT results in all months except June. RMSE values for the models (Table I) show error reduction in 3.0 by 47.9% unweighted and 56.7% weighted compared to 2.0. Yearly spectral impacts show percentage point deviations from FARMS-NIT results of 0.2, -0.58, and -0.54 for Spectral 3.0 and -1.88, -1.71, -0.21 for Spectral 2.0 in Arizona, Ohio, and Miami respectively. Miami represents the best fit of the 2.0 model but also the only scenario where the 2.0 model outperforms the 3.0 model in yearly correction predictions, however the 3.0 model is still within 1 percentage point of the FARMS-NIT results.

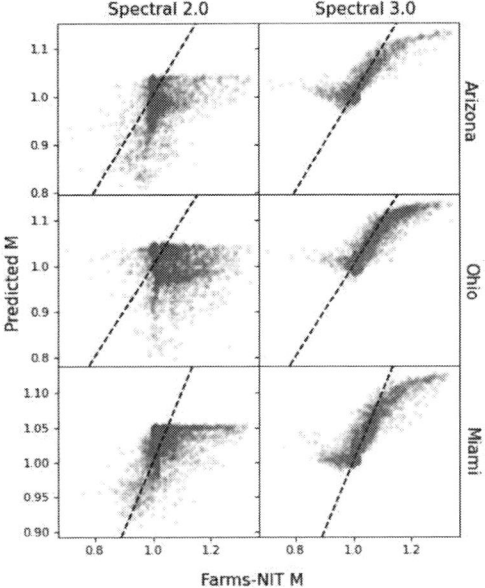

Figure 5. Equivalence plots of FARMS-NIT modeled vs. Spectral 2.0 and 3.0 modeled spectral correction at each of the three data collection locations. The black lines on each plot are lines of equivalence (slope of 1, intercept of 0). The Spectral 3.0 model shows better alignment to the lines of equivalence in predicting the FARMS-NIT results than the 2.0 model.

TABLE I. TESTING RMSE RESULTS FOR THE SPECTRAL 2.0 AND 3.0 MODELS, INCLUDING IRRADIANCE WEIGHTED RMSE.

	Spectral Model	Testing RMSE	Weighted Testing RMSE
GHI	2.0	0.085	14.04
	3.0	0.041	9.91
1-Axis	2.0	0.071	21.23
	3.0	0.037	9.20

IV. DISCUSSION

From a fitting perspective the Spectral 3.0 model offers clear advantages over the Spectral 2.0 model. Unweighted RMSE and weighted RMSE both improve for the 3.0 model and the equivalence trend is far more stable with the 3.0 model for both GHI and 1-Axis tracking orientations. In both of the 3.0 fitting cases, the tailing behaviors at high spectral correction conditions demonstrate the importance of the irradiance weighted regression in maintaining the modeling performance during the highest insolation periods. Despite these low irradiance outliers, monthly and yearly predictions of spectral correction are still accurate. Adjustments to the model can be made to include non-linear terms if it becomes important to fit these regions more accurately, however more validation against the field data is needed to determine if this is necessary. The yearly 1-Axis tracking spectral corrections are lower and closer

to the ideal case of 1 for each location, demonstrating the usefulness of fitting the model to multiple orientations.

In all GHI cases, the 3.0 model outperforms the 2.0 model both from a fitting standpoint and from total spectral impact prediction, being within the yearly spectrometer measurements by at most 0.3 percentage points. The 2.0 model was at best -1.85 percentage points below the spectrometer which aligns with the expectation of general underperformance given that the 2.0 model was only fit against clear sky data and cannot account for cloud cover. While this is a promising result it does not represent the practical use cases of most modern commercial systems, which are in 1-Axis tracking orientation.

The 1-Axis tracking modeling results are less consistent than those of GHI. While both unweighted and weighted RMSEs improve, the yearly spectral impact of the Spectral 3.0 model prediction deviates further from the FARMS-NIT results than the 3.0 model does in GHI from the spectrometer results. Additionally, Miami is unique among all the locations in that it is the only case where the 2.0 model is closer to the FARMS-NIT results than the 3.0 model. In Miami, the 3.0 model significantly underpredicts the FARMS-NIT result, while the 2.0 model is off by only 0.21 percentage points on a yearly scale. There appears to be a deviation between the expected trends in the FARMS-NIT and spectrometer data at high humidity conditions. In Malaysia, both models predict high

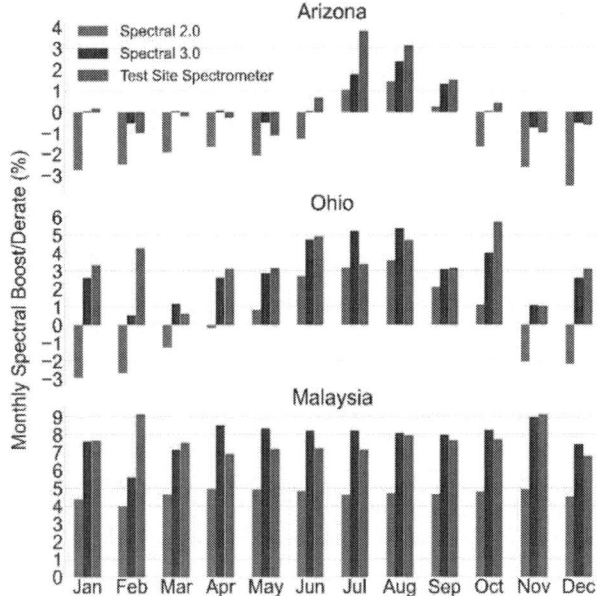

Figure 4. Bar plots of monthly spectral boost or derate percentage in the year of 2021 for the Spectral 2.0 and 3.0 models and spectrometer results at each of the three data collection locations, in the GHI orientation. On the yearly basis, in Arizona, Spectral 2.0 is observed as -1.08%, and Spectral 3.0 is observed as 0.47%, whereas the spectrometer is observed as 0.77%; in Ohio, Spectral 2.0 is observed as 1.02%, and Spectral 3.0 is observed as 3.33%, whereas the spectrometer is observed as 3.49%; in Malaysia, Spectral 2.0 is observed as 4.66%, and Spectral 3.0 is observed as 7.87%, whereas the spectrometer is observed as 7.69%.

correction, over 4%. Miami in the summer shows similar precipitable water readings to Malaysia which is reflected in the

2.0 model predictions being much higher than in other months, however the FARMS-NIT and 3.0 predictions lag behind in this period. It is unknown if this is due to the FARMS-NIT predictions not fully accounting for the impact of precipitable water, or if this is an intrinsic behavior of 1-Axis tracking spectral data compared to GHI considering the lower ratio of direct to diffuse light in the GHI orientation.

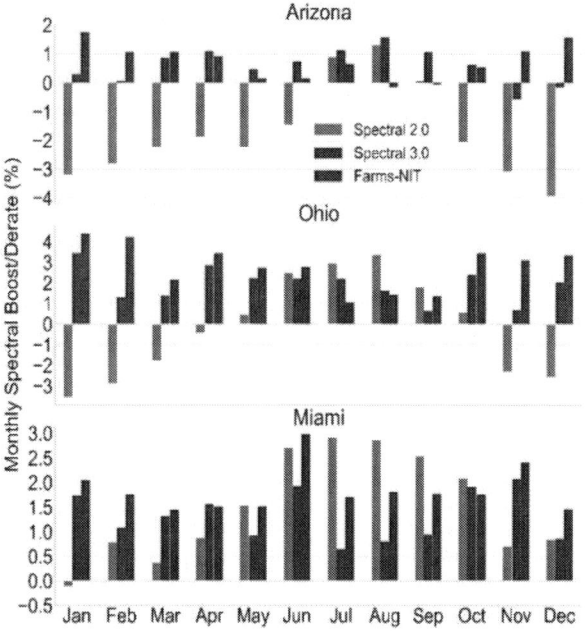

Figure 6. Bar plots of monthly spectral boost or derate percentage in the year of 2021 for the Spectral 2.0 and 3.0 and FARMS-NIT models at each of the three data collection locations, for 1-Axis tracking systems. On the yearly basis, in Arizona, Spectral 2.0 is observed as -1.31%, and Spectral 3.0 is observed as 0.77%, whereas the FARMS-NIT is observed as 0.57%; in Ohio, Spectral 2.0 is observed as 0.73%, and Spectral 3.0 is observed as 1.86%, whereas the FARMS-NIT is observed as 2.44%; in Miami, Spectral 2.0 is observed as 1.60%, and Spectral 3.0 is observed as 1.27%, whereas the FARMS-NIT is observed as 1.81%.

Overall, clear sky index appears to be a promising and easily accessible parameter in predicting CdTe spectral correction. Despite concerns about the 1-Axis tracking FARMS-NIT results not fully capturing the influence of precipitable water, the functional form of a multiple linear model of clear sky index, precipitable water content, and absolute air mass demonstrates improved fitting over the existing 2.0 model in almost all cases. Further work is needed to collect and validate the 3.0 model against additional field measured data, most notably in the 1-Axis tracking orientation.

V. CONCLUSION

This study proposes and demonstrates the initial iteration of an updated CdTe spectral prediction model (Spectral 3.0) as a progression from the currently used Spectral 2.0 model. The goal of the updated model is to be more accurate than the existing 2.0 model while still being computationally lightweight and built on widely available metrology data for use in PV simulation software over a wide geographical range. The new model expands on the 2.0 model by including clear sky index

(CSI) in the model as a numeric indicator of cloud cover which is already accessible in common PV plant simulation software, along with precipitable water content and absolute air mass from Spectral 2.0. Coefficients are fit to field measured spectral data spanning several years from spectrometers in GHI orientation and FARMS-NIT modeled spectra in the 1-Axis tracking orientation. Results show that the 3.0 model has an irradiance weighted testing RMSE 29.4% lower in the GHI orientation and 56.7% lower in the 1-Axis tracking orientation compared to the 2.0 model. With the exception of Miami, the 3.0 model is able to better predict the spectrometer or FARMS-NIT spectral model results at each tested location. The new functional form demonstrates a promising capability of fitting spectral trends, however there is a need to further validate the model with field measured spectra, particularly in the 1-Axis tracking orientation.

REFERENCES

[1] ASTM G173-03 (2020), "Standard Tables for Reference Solar Spectral Irradiances: Direct Normal and Hemispherical on 37° Tilted Surface."

[2] G. Eason, B. Noble, and I. N. Sneddon, "On certain integrals of Lipschitz-Hankel type involving products of Bessel functions," Phil. Trans. Roy. Soc. London, vol. A247, pp. 529–551, April 1955.

[3] L. Nelson, M. Frichtl, and A. Panchula, "Changes in cadmium telluride photovoltaic performance due to spectrum," IEEE Journal of Photovoltaics, vol. 3, No. 1, pp. 488-493, 2013.

[4] C. Gueymard, "SMARTS, a simple model of the atmospheric radiative transfer of sunshine: Algorithms and performance assessment" Florida Sol. Energy Center, Cocoa, FL, Tech. Rep. FSEC-PF-270-95, 1995.

[5] M. Lee, L. Ngan, W. Hayes, and A.F. Panchula, "Understanding Next Generation Cadmium Telluride Photovoltaic Performance due to Spectrum", New Orleans, 2015.

[6] M. Lee and A.F. Panchula, "Spectral Correction for PV Performance Based on Air Mass and Precipitable Water," 43rd IEEE Photovoltaic Specialists Conference (2016).

[7] T. R. Betts, D. G. Infield, and R. Gottschalg, "Spectral irradiance correction for PV system yield calculations," presented at the 19th European Photovoltaic Solar Energy Conference, Paris, France, 2004, pp. 2533–2536.

[8] T. R. Betts, R. Gottschalg, and D. G. Infield, Proceedings of the 3rd World Conference in Photovoltaic Energy Conversion, Vol. B (2003) 1756.

[9] J. Merten and J. Andreu, "Clear separation of seasonal effects on the performance of amorphous silicon solar modules by outdoor *I/V*-measurements", Solar Energy Materials and Solar Cells 52 (1998) 11.

[10] Xie, Y., Sengupta, M., Wang, C., 2019. "A Fast All-sky Radiation Model for Solar applications with Narrowband Irradiances on Tilted surfaces (FARMS-NIT): Part II. The cloudy-sky model". Sol. Energy 188, 799-812.

[11] Xie, Y., Sengupta, M., Dooraghi, M., 2018. "Assessment of uncertainty in the numerical simulation of solar irradiance over inclined PV panels: New algorithms using measurements and modeling tools". Sol. Energy 165, 55-64.

[12] Viktar Tatsiankou, Karin Hinzer, Henry Schriemer, Stelios Kazadzis, Natalia Kouremeti, J. Gröbner, Richard Beal. 2018. "Extensive validation of solar spectral irradiance meters at the World Radiation Center". Solar Energy. 166. 80-89.

[13] C. Gueymard, "Analysis of Monthly Average Atmospheric Precipitable Water and Turbidity in Canada and Northern United States," Solar Energy vol 53(1), pp. 57-71, 1994.

[14] Sandia Corporation, "PV Performance Modeling Collaborative: PV_Lib Toolbox," 2023. https://pvpmc.sandia.gov/applications/pv_lib-toolbox/.

[15] Sengupta, M., Y. Xie, A. Lopez, A. Habte, G. Maclaurin, and J. Shelby. 2018. "The National Solar Radiation Data Base (NSRDB)." Renewable and Sustainable Energy Reviews 89 (June): 51-60.

Modification of PEDOT:PSS hole transporting layer to improve the efficiency and light stability of tin-lead perovskite solar cells

Lei Chen, Chongwen Li, Tyler R. Brau, Abasi Abudulimu, Randy J. Ellingson, Zhaoning Song, Yanfa Yan

Department of Physics and Astronomy, and Wright Center for Photovoltaics Innovation and Commercialization, University of Toledo, Toledo, OH, United States

Mixed tin (Sn)-lead (Pb) halide perovskites are promising photovoltaic materials for efficient single- and multi-junction solar cells. However, the power conversion efficiency (PCE) and light stability of Sn-Pb perovskite solar cells (PSCs) are limited by the PEDOT:PSS hole transport layer (HTL). In this work, we introduce potassium citrate (PC) additives to modify PEDOT:PSS HTL. The performance of mixed Sn-Pb PSCs is boosted significantly after PC modification. Using this strategy, we demonstrated the best PCE of 22.67% with enhanced open circuit voltage (Voc) and fill factor (FF). The characterization for films and devices revealed that PC molecules significantly suppress non-radiative recombination and thus improve device performance. Owing to the reduced non-radiative recombination, the device with PC modification also exhibited improved light stability.

Improving V_{OC} of CdSe/CdTe Solar Cells via Incorporating Oxygenated CdS Between Front Buffer and Absorber

Abasi Abudulimu, Dengbing Li, Manoj Jamarkatel, Zachary Zawisza, Scott Wenner, Tyler Brau, Ebin Bastola, Adam Phillips, Michael Heben, Yanfa Yan, Randy Ellingson

Wright Center for Photovoltaics Innovation and Commercialization (PVIC), Department of Physics and Astronomy, The University of Toledo, Toledo, OH 43606 USA

Abstract — The recent trend in enhancing the performance of cadmium telluride (CdTe) solar cells involves alloying CdTe with selenium (Se) and substituting the window layer, cadmium sulfide (CdS), with magnesium zinc oxide (MZO). However, the stability and reproducibility of MZO raise concerns. Recent studies have proposed the use of commercial SnO_2 in place of MZO and introducing oxygenated CdS between SnO_2 and the absorber, achieving an efficiency of over 20%. This significant V_{OC} gain is ascribed to the photoactive bandgap gradient region, Cd(O,S,Se,Te), formed at the device's front interface, which avoids a detrimental interface. To assess the general applicability of this finding, we analyzed CdSe/CdTe devices' performance using MZO and indium gallium oxide (IGO) compared to SnO_2. We discovered that incorporating the oxygenated CdS layer notably improved the V_{OC} of devices across all three buffers, achieving 876 mV, which closely aligns with the previously simulated value of 882 mV for SnO_2. Steady-state and time-resolved photoluminescence measurements indicate that the Voc improvement results from significantly reduced carrier recombination at the front of the device.

I. INTRODUCTION

Cadmium telluride (CdTe)-based thin film solar cells present a compelling technology in the solar cell market, offering low production costs, scalability, stability, and high efficiency. Recent advancements in CdTe solar cell research, including the bandgap grading of the absorber by alloying CdTe with selenium (Se) via the cadmium chloride (CdCl$_2$) treatment process and the removal of the cadmium sulfide (CdS) window layer, have moved this technology into a new development phase. These advances have allowed for increased light absorption at longer wavelengths and an enhancement of the short-circuit-current density (J_{SC}), inching it closer to the Shockley-Queisser limit [1].

However, the improvement in JSC did not significantly enhance the power conversion efficiency (PCE) compared to conventional CdS window layer-based CdTe devices. The PCE breakthrough came with the introduction of magnesium zinc oxide (MZO) as the front buffer [2]. Despite this, MZO's sensitivity to humidity and instability in open air pose substantial obstacles for commercializing these devices [3].

To address these issues, Li et al. proposed using commercially available SnO_2 as the front buffer and incorporating an oxygenated CdS layer between the front buffer and the absorber (Cu doped) [4-5]. This approach significantly increased the open-circuit voltage (V_{OC}) to 865 mV, leading to a record PCE of 20.03% for CdTe devices fabricated in academic institutions.

In this study, we explore the application of an oxygenated CdS layer for devices using different front buffer materials, namely indium gallium oxide (IGO), MZO, and commercial SnO2. Our findings reveal a considerable increase in VOC for all devices, regardless of the front buffer material used, when the oxygenated CdS layer is incorporated between the absorber and the front buffer layer. We demonstrate through steady-state and time-resolved photoluminescence spectroscopy measurements that this V_{OC} improvement is partly due to reduced carrier recombination at the front of the devices.

II. EXPERIMENTAL METHODS

We fabricated devices using cleaned fluorine-doped tin oxide (FTO) substrates (Tech 12, NSG USA), while Tech 12 coated with a 30 nm tin oxide layer (TEC12D, NSG USA) was employed for the devices with SnO_2 buffer layers. For MZO devices, 75 nm MZO films were deposited via radio-frequency (RF) magnetron sputtering from an MZO target containing 8 wt % magnesium oxide (MgO) at room temperature under 10 mTorr of Ar pressure [6]. IGO devices were prepared with 75 nm IGO films, produced by co-

978-1-6654-6060-6/23 $31.00 © 2023 IEEE

sputtering from In_2O_3 and Ga_2O_3 targets under 4 mTorr pressure in an Ar/O_2 environment, onto substrates maintained at 250 °C [7].

We deposited the oxygenated CdS (60 nm) and CdSe layers (80 nm) using RF sputtering with a 2-inch target under 2% oxygen and 98% argon at room temperature. The CdTe absorber layer (3.5 μm) was then added by close-space sublimation at source and substrate temperatures of 560 and 500 °C, respectively, at 1 Torr. This was followed by a wet CdCl2 treatment at 400 °C for 50 min in dry air. CuSCN (~50 nm) was applied via spin-coating, using a 2 mg/ml solution in 30 wt% ammonium hydroxide, at a spin speed of 2000 rpm for 30 seconds. Then, rapid thermal annealing at 180 °C facilitated Cu diffusion. After this treatment, a 40 nm Au layer was evaporated onto the back surface, resulting in a cell area of 0.08 cm².

We measured J-V characteristics of the completed devices at room temperature in air using a Keithley 2440 source meter and an LED solar simulator (Newport Minisolo model LHS-7320) under simulated 1-sun AM1.5G solar irradiance spectrum conditions. We conducted external quantum efficiency (EQE) measurements on a PV Instruments (model IVQE8-C) system. Steady-state PL measurements were conducted using a custom-built system equipped with a Horiba Symphony-II CCD detector and Horiba iHR320 monochromator, utilizing 633 nm continuous wave lasers as the excitation source (2.5×10^{17} cm^{-2} s^{-1}). Time-resolved PL measurements were done using a Becker & Hickl TCSPC system equipped with a Fianium supercontinuum laser at 633 nm (10^{11} cm^{-2} pulse^{-1}).

III. RESULTS AND DISCUSSION

The J-V characteristics of the highest-performing devices from each configuration are displayed in Fig. 1. Our results demonstrate that integrating an oxygenated CdS layer between the absorber and front buffer increases V_{oc} for all devices, irrespective of the type of front buffer used. The peak V_{oc} values achieved were 860 mV, 873 mV, and 876 mV for devices made with MZO, IGO, and SnO₂ buffer layers, respectively. These values are even higher than that of the previously reported champion device (20.3% PCE) with oxygenated CdS. It should be noted that these devices have not yet been systematically optimized. This is evidenced by the rollover of the J-V curves of the devices at bias voltages above V_{oc} (see Fig. 1b), low J_{sc}, and low device efficiency (table in Fig. 1).

Devices fabricated without a front buffer layer were also tested and, as expected, their performance was subpar. Interestingly, the performance loss was primarily in V_{oc},

while FF and J_{sc} values were comparable to devices with a front buffer layer. This suggests that there are no significant energetic barriers at the front interface of the device for charge extraction but there is an issue with charge selectivity. This could also be attributed to poor interface formation with a high defect density [8].

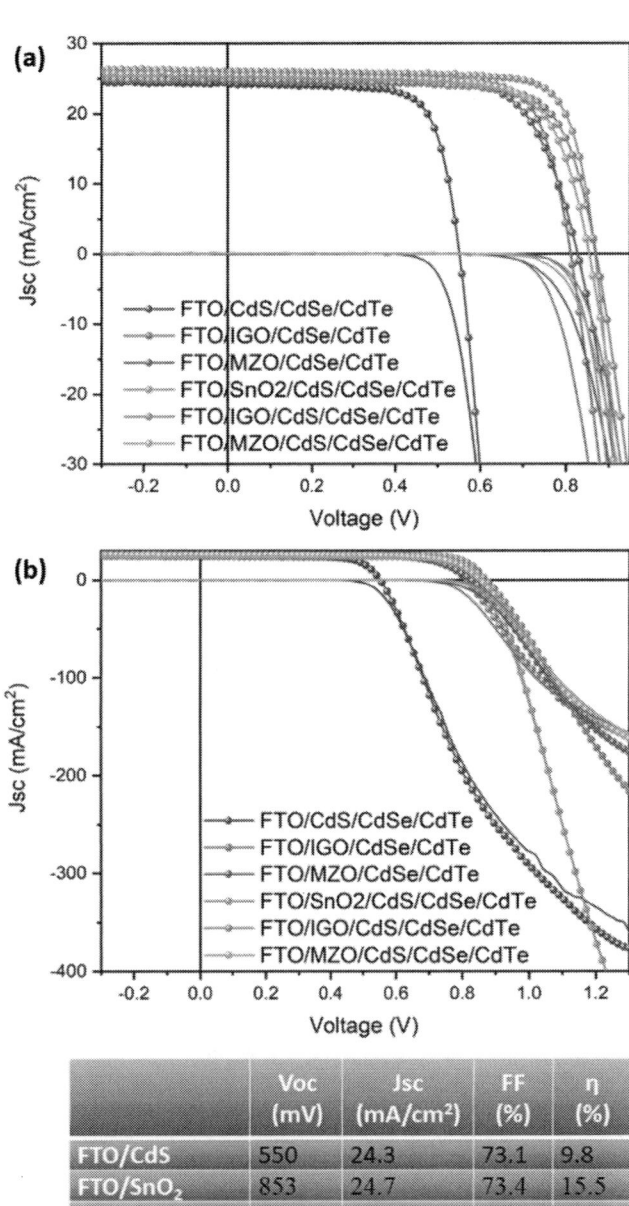

	Voc (mV)	Jsc (mA/cm²)	FF (%)	η (%)
FTO/CdS	550	24.3	73.1	9.8
FTO/SnO₂	853	24.7	73.4	15.5
FTO/IGO	811	25.4	75.9	15.6
FTO/MZO	834	25.6	71.1	15.2
FTO/SnO₂/CdS	868	26.0	78.3	17.7
FTO/IGO/CdS	870	24.8	73.8	15.9
FTO/MZO/CdS	854	25.4	72.2	15.7

Fig. 1 J-V characteristics of the best-performing devices. Subfigure (b) replicates subfigure (a) but features a broader voltage scan range.

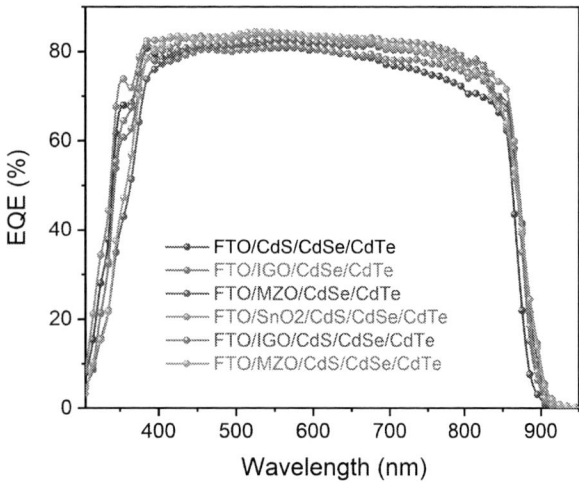

Fig. 2   EQE characteristics of the best-performing devices.

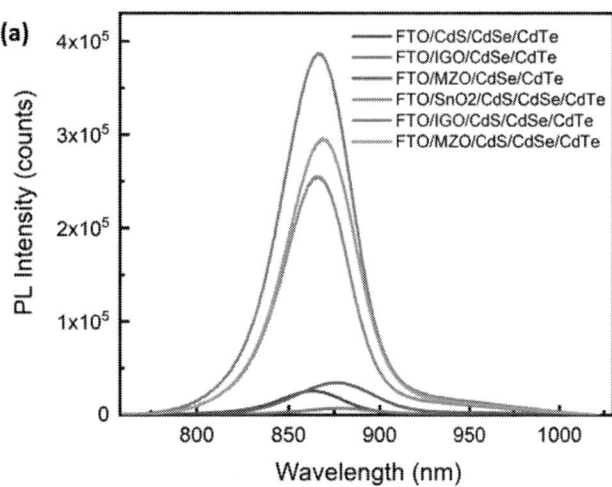

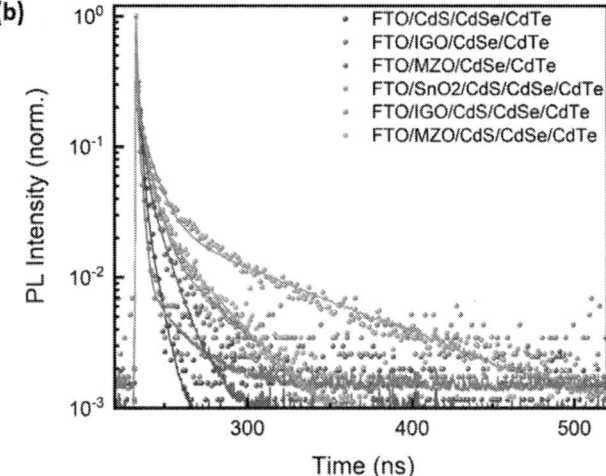

Fig. 3   (a) Steady-state and (b) time-resolved PL spectra of the best-performing devices measured from the glass side.

As depicted in Fig. 1, the J_{SC} of the devices is uniformly low for all buffer layers, resulting in an approximate 2% efficiency cost, as we typically observe 28-29 mA/cm^2 for devices constructed with a front buffer layer [4-5]. The low J_{SC} is further corroborated by external quantum efficiency (EQE) measurements in Fig. 2. The EQE indicates no significant charge collection loss due to optical loss from CdS light absorption, which would generally occur in the wavelength region below 600 nm. Therefore, CdS is not the cause of the low J_{SC}. This low J_{SC} issue likely stems from the absorber and could be associated with the non-optimum deposition of CdSe/CdTe layers, and/or non-optimal post-treatments (CdCl$_2$ and Cu). The devices with MZO collected fewer charges below 400 nm (with or without the CdS layer) indicating an MZO related optical loss.

Steady-state PL spectra, excited and measured from the glass side, reveal that the PL intensity of the devices equipped with the oxygenated CdS layer dramatically increased compared to their counterparts (see Fig. 3a). The surge in PL intensities correlates to an implied V_{OC} increase of 45 mV, 92 mV, and 55 mV for devices with SnO$_2$, IGO, and MZO buffer layers incorporating a CdS layer, respectively. Although the measured V_{OC} did improve with the inclusion of CdS, the V_{OC} gain was noticeably less than that implied by the PL emission intensity (that is also confired by external radiative efficiency measurements). This suggests that processes beyond those detectable by the PL spectroscopy may influence the resulting V_{OC}, such as energetic band alignment between the absorber and the front buffer.

It's worth noting that, unlike others, devices with MZO and IGO with the CdS layer exhibited a more pronounced blue shift in the PL peak position. This might be an effect of CdS on the distribution of Se, as Se concentration regulates the bandgap of the absorber (CdSeTe alloy). This might be an effect of CdS on the distribution of Se, since the Se concentration governs the bandgap of the absorber (CdSeTe alloy). However, this requires further investigation. Additionally, the time-resolved PL data, presented in Fig. 3b, indicated that the carrier lifetime of the devices significantly increased with the integration of the oxygenated CdS layer. This implies a reduction in carrier recombination at the front of the devices when the oxygenated CdS layer is present.

IV. CONCLUSIONS

Building on recent work [4], we introduced oxygenated CdS between the absorber (CdSe/CdTe) and three different front buffer layers (MZO, IGO, and SnO$_2$). We observed a significant increase in V_{OC} for each device, regardless of the buffer used, compared to their counterparts without the CdS layer. The highest V_{OC} values of 876 mV and 873 mV were achieved with devices featuring SnO_2 and IGO buffers, respectively.

Through steady-state and time-resolved PL measurements, we attribute these improvements in *Voc* to a reduction in carrier recombination at the front of the devices. However, JSC is noticeably low for all the devices without a CdS layer, which diminishes the expected efficiency. EQE confirmed there were no significant charge collection losses associated with the CdS layer, and the flatness of the J-V curve around 0 bias voltage suggested no significant charge extraction barrier within the devices. Therefore, we suspect the low Jsc might be related to the deposition of the absorber or the its post treatments such as $CdCl_2$ or Cu.

Given that these devices are not yet systematically optimized, our future work will explore the causes behind the low *Jsc* and investigate whether a trade-off exists between high *Voc* and low *Jsc*.

ACKNOWLEDGMENT

This work is based on research sponsored by Air Force Research Laboratory under agreement number FA9453-19-C-1002, and by the U.S. DOE's Office of Energy Efficiency and Renewable Energy (EERE) under the Solar Energy Technologies Office (SETO), through Agreement DE-EE0008974 and through the Alliance for Sustainable Energy, LLC, Managing and Operating Contractor for the National Renewable Energy Laboratory for the U.S. Department of Energy, under Award Number 37989. The U.S. Government is authorized to reproduce and distribute reprints for Governmental purposes notwithstanding any copyright notation thereon. The views and conclusions contained herein are those of the authors and should not be interpreted as necessarily representing the official policies or endorsements, either expressed or implied, of the Air Force Research Laboratory or the U.S. Government.

REFERENCES

[1] R, Sven. "Tabulated values of the Shockley–Queisser limit for single junction solar cells." *Solar energy* 130, 139-147, 2016.

[2] Munshi, Amit H., et al. "Polycrystalline CdSeTe/CdTe absorber cells with 28 mA/cm 2 short-circuit current." *IEEE Journal of Photovoltaics* 8.1, 310-314, 2017.

[3] B, Francesco, et al. "Degradation of Mg-doped zinc oxide buffer layers in thin film CdTe solar cells." *Thin Solid Films* 691, 137556, 2019.

[4] L, Deng-Bing, et al. "20%-efficient polycrystalline Cd (Se, Te) thin-film solar cells with compositional gradient near the front junction." *Nature Communications* 13.1, 1-8, 2022.

[5] Li, Deng-Bing, et al. "Oxygen Management to Avoid Photo-Inactive Cd (S, Se) for Efficient Cd (Se, Te) Solar Cells." *ACS Energy Letters* 8.3 (2023): 1529-1534.

[6] Jamarkattel, M. K., et al. "High vacuum heat-treated MZO: Increased n-type conductivity and elimination of S-kink in MZO/CdSe/CdTe solar cells." *MRS Advances* 7, 713-717, 2022.

[7] Jamarkattel, M. K., et al. "Indium Gallium Oxide Emitters for High-Efficiency CdTe-Based Solar Cells." *ACS Applied Energy Materials* 5.5, 5484-5489, 2022.

[8] Abudulimu, Abasi, et al. "Photophysical Properties of CdSe/CdTe Bilayer Solar Cells: A Confocal Raman and Photoluminescence Microscopy Study." *2022 IEEE 49th Photovoltaics Specialists Conference (PVSC)*. IEEE, 1088-1090, 2022. doi: 10.1109/PVSC48317.2022.9938717.

978-1-6654-6060-6/23 $31.00 © 2023 IEEE

From Accelerated Life Test to Accurate Degradation Prediction of CdTe PV Devices: A Modeling Approach

Da Guo, Jaliu Ma, Samuel Demtsu, Ryan Monnin, Igor Sankin, Jose A. Calderon, Markus Gloeckler

First Solar Inc, Perrysburg, OH, United States

An accurate prediction of solar module performance during long-term field operation has been one of the key system-level challenges since the beginning of Photovoltaic (PV) technologies commercialization. In this study, we develop an orthogonal model that predicts long-term metastable behaviors of thin-film CdTe-based PV devices under varied climate conditions. The model can be parametrized using data from a multi-temperature ALT. We describe both the model and the parametrization procedure, and confirm the prediction accuracy by comparing forecasted device performance against service-life measured data collected from First Solar operated field tests. This modeling approach could have the potential to be adopted for other types of PV technologies.

Quantifying Bulk and Surface Recombination in CdSeTe Absorbers by Modeling Terahertz and Photoluminescence Decays

Gregory A Manoukian, Calvin Fai, Deborah L McGott, Finley R Shapiro, Matthew O Reese, Charles J Hages, Jason B Baxter

Drexel University, Philadelphia, PA, United States

University of Florida, Gainesville, FL, United States

National Renewable Energy Laboratory, Golden, CO, United States

We demonstrate a technique for estimating critical recombination parameters Shockley-Read-Hall lifetime (τ_{SRH}) and front and back surface recombination velocities (SF and SB) in Cd(Se)Te. We evaluated uniform single heterostructures of glass/Al2O3/CdSexTe1-x/air (x = 0 & 0.2) using time resolved terahertz spectroscopy (TRTS) and time resolved photoluminescence (TRPL) with variable excitation wavelengths and powers as well as front vs back-side illumination. The addition of TRTS to traditional TRPL enhances analysis by providing insight into the first nanoseconds after photoexcitaton when TRPL is dominated by redistribution of carriers. Data were fit using numerical simulations of the semiconductor and Poisson' equations with best-fit parameters determined by Bayesian inference. Our CdSe0.2Te0.8 films had τ_{SRH} ~50 ns, compared with ~26 ns for CdTe. τ_{SRH} values extracted from our model were approximately double those derived from multi-exponential fit of TRPL decays, which do not rigorously account for surface recombination. Se alloying also decreased SB from 2x105 to 6x103 cm/s. Upper limits of 104 and 103 cm/s were determined for SF for CdTe and CdSe0.2Te0.8, respectively. Precise determination of recombination velocity at the well-passivated Al2O3/ CdSexTe1-x interface was limited by the dominant bulk and/or back surface recombination in these films. This study was made on passivated heterostructures of absorber layers to for the primary purpose of evaluating new methods of parameter determination. Further development of the models toward full devices can lead to rapid correlation of changes in bulk lifetime and surface recombination rates with processing and could enable evaluation of effects of electric fields and other phenomena that cannot be captured using conventional approaches.

Data Driven Energy Resilience for Low- to Middle-Income Communities in Puerto Rico

Christopher Gregory, Angel Echevarria, Yiamar Rivera-Matos, Daniel D. Campo-Ossa, Alex Routhier, Clark Miller, Richard King

Arizona State University, Tempe, AZ, United States

University of Puerto Rico, Mayaguez, PR, United States

A solar and battery system model was built to study energy use and energy resilience in communities in Puerto Rico. The model utilizes data from NREL' NSRDB and PV-RDBPR. The findings show that energy use habits, including time of day and amount of energy used, dominate how well the system performs. This means smaller, inexpensive systems may be used very effectively if used during the correct times of day.

Advances in GaAsP Top cells for use in GaAsP/Si Tandems

Tal Kasher, Lauren M. Kalizewski, Marzieh Baan, Chuqi Yi, Anastasia H. Soeriyadi, Atom Chang, Udo Römer, Gianluca Coletti, Stephen P. Bremner, Tyler J. Grassman, Steven A. Ringel

The Ohio State University, Columbus, OH, United States

University of New South Wales, Sydney, Australia

Oxford University, Oxford, United Kingdom

TNO Energy Transitions, Petten, Netherlands

To date, monolithic, epitaxially-integrated III-V/Si tandem solar cells, including the current record-holders, remain limited by excessive threading dislocation densities (TDD) in the metamorphic III-V top cell(s). Our recent development of GaAs$_y$P$_{1-y}$/Si virtual substrates with TDD in the low-10^6 cm-2 range paves a pathway toward significant improvement in total device efficiencies for such architectures. In this work, the combination of improved heteroepitaxial processes and growth structures with more optimal subcell design has resulted in a significant increase in GaAs$_{0.75}$P$_{0.25}$-on-Si top cell performance. This new subcell now outperforms our previous best by ~4.4% absolute AM1.5G efficiency, with increases in fill factor, JSC, and WOC of about 4.9% absolute, 1.9 mA/cm2, and 0.05 V, respectively.

Key Areas of Due Diligence for Solar PV Project Financing

Eric R. DeCristofaro, Matthew R. Thibodeau, Fitzgerald C. Okoli, Jake T. Silhavy, Evan S. Giacchino, Adam W. Loeding, Alexander E. Coologeorgen

Sargent & Lundy , Chicago, IL, United States

Financing plays a key role in the successful implementation of solar PV projects. To obtain financing solar PV projects are subject to a thorough due diligence process typically conducted by an independent engineering firm in support of the financiers. In the past five years Sargent & Lundy has provided independent engineering services for the renewable industry in support of financing for over 300 renewable power projects. This includes support for the financing of dozens of solar PV projects for major tax equity investors and debt lenders. Sargent & Lundy has developed a presentation that touches on key areas of due diligence that solar PV project financiers are most interested in as well as key recommendations and lessons learned from our experience supporting the financing process as an independent engineer. The presentation covers the topics outlined below: 1. Review of Project Parties 2. Site Suitability Assessment 3. Key Technology Review (PV Module, Power Conversion System, Main Power Transformer, Mounting System) a. Manufacturer Overview b. Manufacturing Quality System Review c. Product Review / Installation Statistics d. Site Suitability Considerations e. Warranty Review 4. System Design Review a. PV Array Design b. Electrical Design c. Civil Review i. Geotechnical Conditions ii. Hydrology and Flood Risk d. Structural Design 5. Energy Production a. Solar Resource Assessment i. Analysis of Available Solar Resource Data ii. Data Correlation (Measure-Correlate-Predict) b. Independent Energy Yield Assessment using PVsyst c. Probabilistic Analysis 6. Environmental and Permitting Review 7. Contracts and Agreements Review a. Engineering, Procurement, and Construction Contract b. Supply Agreements c. Power Purchase Agreements d. Interconnection Agreement e. O&M Agreement 8. Financial Model Review a. Revenue, Capacity and General Technical Inputs b. Energy Output and Revenue Projections c. Project Operating Cost d. Useful Life

Convergence of efficiency, stability, and manufacturability in perovskite tandem solar cells

Rohit Prasanna, Tomas Leijtens, Jochen Titus, Laura E. Crowe, Hyunjong Lee, Annikki L. Santala, Maximilian T. Hoerantner, Giles E. Eperon

Swift Solar, San Carlos, CA, United States

Perovskite-silicon tandem solar cells offer a clear step past the best silicon single junction solar cells in the roadmap of photovoltaics. Having recently reached over 32% efficiency at an R&D scale, they have long-term stability and manufacturability as important things to prove. In this talk, I will discuss Swift Solar' development of perovskite deposition processes that achieve excellent visual and compositional uniformity over standard M6 wafer areas. I will discuss unique challenges to ensuring high yield while fabricating the perovskite thin film device stack on silicon bottom cells, which are rougher than glass surfaces even if planarized silicon is used, and of course have significantly larger roughness in the case of textured surfaces that are typical of commercial silicon solar cells today. To ensure long term stability, I will highlight the photolysis of lead iodide as an important degradation mechanism of the perovskite material under illumination at elevated temperature. Lastly, I will present effects of the chosen contact material in determining mechanical and chemical stability of the contact layer interfaces. With a stable perovskite composition with a tandem-relevant band gap and appropriately chosen contact layer, we demonstrate perovskite solar cells that maintain 97% of their initial performance after over 2000 hours of aging at 70 C under illumination.

978-1-6654-6060-6/23 $31.00 © 2023 IEEE

Validating view-factor approach and spatial albedo models for Bifacial and AgriPV modeling

Silvana Ovaitt, Matthew Boyd, Austin Kinzer, James Jones, Chris Deline

National Renewable Energy Laboratory, Golden, CO, United States

University of Maryland, Maryland, MD, United States

View factor models are used in due diligence software to calculate rear irradiance for bifacial modules. An intermediate step in this calculation is the irradiance at the ground level, which can be leveraged for evaluating the Photo Active Radiation available for crops in Agrivoltaic setups. This paper presents the metrics and modifications to the model for ground irradiance study with the view factor approach, compares it to the raytracing method, and validates it with field measurements of ground irradiance. It is found that for the clearances, row-to-row setups, and tilts studied, the view factor method matches with raytracing results within 2% MBD. The comparison is performed for the nine most common agriPV configurations using high-performance computing and the NSRDB database for the whole US, with results and data made available open-source on the InSPIRE AgriPV website.

Field and Accelerated Aging of Cracked Solar Cells

Michael G. Deceglie, Timothy J Silverman, Ethan Young, William B. Hobbs, Cara Libby

National Renewable Energy Laboratory, Golden, CO, United States

Southern Company, Birmingham, AL, United States

Electric Power Research Institute, Palo Alto, CA, United States

Cracks can form in silicon solar cells in an otherwise intact module due to mechanical stresses like rough handling or hail. The immediate impact on power loss due to these cracks can be readily measured, but it is also known that the cracks can worsen over time, as has been previously shown with accelerated tests. However, it is not clear how to predict the extent of future degradation of modules with cracked cells in the field. Predicting future power loss requires a calibrated accelerated test that can be performed on modules with cracked cells. In this presentation we describe progress towards such a test. In particular, we report on the outdoor ageing of modules with cracked cells for nearly two years. We find that modules with cracked cells degraded in the field an average of 0.5% absolute more than uncracked modules over a period of 21 months. We achieved comparable degradation using a novel accelerated test, dynamic mechanical acceleration (DMX) to apply thousands of pressure cycles at pressures < 200 Pa which are relevant to the wind-driven pressure cycles experienced outdoors. We also characterize the modules with multi-temperature electroluminescence and find that the degradation is associated with cell fragments that become increasingly electrically isolated.

Orange Button: Accelerating the Digital Transformation of Distributed Energy

Clifford W. Hansen[1], Jan Rippingale[2], Taos Transue[2], Philip Court[3], John Gorman[3]

[1] Sandia National Laboratories, Albuquerque, NM USA
[2] Blu Banyan, Berkeley, CA USA
[3] Ecogy Energy, Brooklyn, NY USA

Abstract — Data processing adds substantial soft costs to distributed energy systems. These costs are incurred primarily as labor necessary to collect, normalize, store and communicate data. The open-source Orange Button data exchange standard comprises data taxonomies, common data sources, and interoperable software tools which together can dramatically reduce these costs and thereby accelerate the deployment of distributed energy systems. We describe the data taxonomies and datasets, and the software enabled by these capabilities.

I. INTRODUCTION

Costs associated with data processing contribute substantially to the soft costs associated with solar power projects. A recent analysis quantified these soft costs, comprising developer overhead, permitting, inspection and engineering, procurement and construction (EPC) overhead, to amount to roughly 30% of total project cost for commercial scale ground-mount PV systems [1]. Anecdotally, much of these costs relate to data processing.

In 2016 the U.S. Department of Energy (US DOE) launched the Orange Button Initiative[1] to foster development of resources and tools that would reduce the soft costs associated with data processing. One grant awarded to SunSpec Alliance, resulted in the first release of the Orange Button Taxonomy implemented using xBRL and focused on data exchange needs for financial institutions. A follow-on grant to Sandia National Laboratories re-implemented the taxonomy using OpenAPI, a data exchange standard more readily usable for a broader set of users and use cases spanning the lifecycle of solar power systems, from sales through commissioning and operations. At the conclusion of the DOE development grants, the Open Orange Button taxonomy was released as open-source and free-to-use[2].

We describe the taxonomy, summarize the principles underlying its design and on-going development processes, and illustrate its use. We summarize data sources and software tools enabled by the taxonomy's capabilities.

II. OPEN ORANGE BUTTON TAXONOMY

A taxonomy is fundamentally a dictionary of common, readable, well-defined terms with associated keywords that facilitate the use of the terms in software. For example, in the Open Orange Button (OB) taxonomy, one term is "Energy (AC) measured for a period of time" which has the associated keyword EnergyAC. Some terms (e.g., EnergyAC) are defined generally to encourage re-use in different contexts. Terms definitions are carefully written to distinguish between similar, but different, data; for example, because EnergyAC is specific to "measured" energy, the taxonomy also contains EnergyACModeled for output from a model, and EnergyACCurtailed, for the energy that could have been produced except for directed inverter curtailment.

Data elements, i.e., individual terms, are often grouped into objects with related terms to provide convenient capability to report a body of data. For example, EnergyAC is grouped with EnergyACCurtailed, Insolation, and RevenueAccrued, and several other terms to form an OperatingReport object. Objects can be placed within other objects to provide hierarchical context and can be used to compose arrays of objects. For example OperatingReport contains an object PerformanceRatios, which is an array of PerformanceRatio objects, which in turn contains EnergyAC, EnergyACModeled, and PerfRatio, among others.

Each element has a fixed set of six primitives where data are stored as well as metadata about the data:

- Value: the data itself, which can be float, string, integer, Boolean, or arrays of these types.
- Unit: a string identifying one unit from an included unit registry. For example, the element EnergyAC's Unit may be 'kWh', 'MWh' or 'Wh'.
- StartTime and EndTime. These ISO8601-compliant timestamp strings provide a means to associate time intervals with a data value.
- Decimals and Precision: these primitives are available to communicate significant digits in numerical data.

For many elements, e.g., string-valued elements, only the Value primitive would be used. For numeric data, all primitives may be relevant.

We chose to implement the taxonomy consistent with the OpenAPI 3.0 specification (aka Swagger). This specification:

---

[1] https://www.energy.gov/eere/solar/orange-button-solar-bankability-data-advance-transactions-and-access-sb-data

[2] https://github.com/Open-Orange-Button/Orange-Button-Taxonomy

978-1-6654-6060-6/23 $31.00 © 2023 IEEE

"…defines a standard, language-agnostic interface to RESTful APIs which allows both humans and computers to discover and understand the capabilities of the service without access to source code, documentation, or through network traffic inspection. When properly defined, a consumer can understand and interact with the remote service with a minimal amount of implementation logic." [3]

As an OpenAPI 3.0-compliant taxonomy, OB is designed to mediate data exchange between applications, as illustrated in Figure 1. The taxonomy source is coded as a JSON schema.

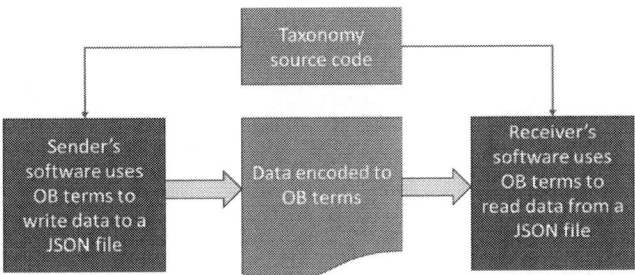

Figure 1. Illustration of OB taxonomy's use by software.

To enable development and use of the taxonomy, a browser-based editor is available[4]. The editor operates on the source JSON code, provides search capabilities, and displays the taxonomy in a human-friendly form. Figure 2 illustrates the editor display. EnergyProductions is an array of the EnergyProduction object (indicated by the brackets). The EnergyProduction object contains two elements: SiteID (an identifier for the location or system) and an array of the object EnergyMeasurement. Each EnergyMeasurement object contains the elements DeviceID (an identifier for the device providing the measurements, which could be a meter or inverter) and EnergyAC. Finally, the primitives are displayed for EnergyAC.

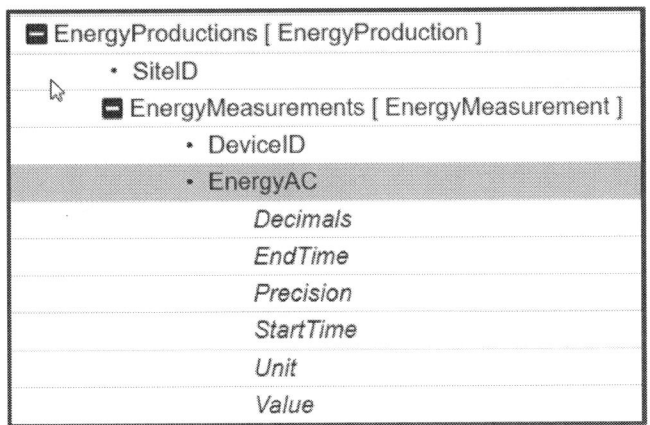

---
[3] https://swagger.io/specification/
[4] https://openobeditor.sunspec.org/

Figure 2. Screenshot of OB taxonomy editor.

III. DATA SOURCES LEVERAGING OPEN ORANGE BUTTON

During taxonomy development, it became clear that a frequent and labor-intensive task for solar developers is identifying the jurisdictions having authority (AHJs) at a given location. These jurisdictions specify building, electrical and other codes with which solar power systems must comply. Armed with the Open Orange Button taxonomy, a database of AHJs, and digital census maps, a look-up service called the AHJ Registry[5] was rapidly developed and deployed. A product registry is currently in development.

IV. SOFTWARE TOOLS USING OPEN ORANGE BUTTON

A number of software tools and services are already leveraging the data normalization capability of the OB taxonomy. At the front end of the PV system lifecycle, Blu Banyan's Solar Success™ provides an integrated set of business tools spanning the sales-to-commissioning process. An OB-compliant API is in development for SolarAPP+, a DOE-sponsored tool to automate permitting[6]. Taxonomy development currently focuses on data exchange needs of operations and maintenance (O&M) service providers. In each case, interfaces between tools and external-facing APIs are built on the OB taxonomy.

V. SUMMARY

Our conference presentation will describe the taxonomy and examples of its use in code. We will demonstrate the AHJ Registry and Product Registry data services. Finally, we will briefly survey software tools which use Orange Button to provide interoperability.

ACKNOWLEDGEMENT

Sandia National Laboratories is a multimission laboratory managed and operated by National Technology and Engineering Solutions of Sandia, LLC., a wholly owned subsidiary of Honeywell International, Inc., for the U.S. Department of Energy's National Nuclear Security Administration under contract DE-NA-0003525.

REFERENCES

[1] V. Ramasamy, J. Zuboy, E. O'Shaughnessy, D. Feldman, J. Desai, M. Woodhouse, P. Basore, and R. Margolis. U.S. Solar Photovoltaic System and Energy Storage Cost Benchmarks, With Minimum Sustainable Price Analysis: Q1 2022. NREL/TP-7A40-83586. National Renewable Energy Laboratory, Golden CO, USA. 2022.

---
[5] https://ahjregistry.sunspec.org/
[6] https://solarapp.nrel.gov/

Analysis for effects of temperature rise of solar cell modules upon the driving distance of photovoltaics-powered vehicles

Masafumi Yamaguchi, Taizo Masuda, Tsutomu Tanimoto, Yosuke Tomita, Yasuyuki Ota, Christian Thiel, Anastasios Tsakalids, Arnulf Jaeger-WAldau, Takashi Nakado, Kenji Araki, Kensuke Nishioka, Tatsuya Takamoto, Kyotaro Nakamura, Ryo Ozaki, Nobuaki Kojima, Yoshio Ohshita

Toyota Tech. Inst., Nagoya, Japan

Toyota Motor Co., Susono, Japan

Nissan Motor Co., Yokosuka, Japan

Miyazaki Univ., Miyazaki, Japan

EC-JRC, Ispra, Italy

Toyota Motor Co., Toyota, Japan

Sharp Co., Nara, Japan

The development of vehicles powered by photovoltaics (PV) is desirable and very important for reducing CO2 emissions from the transport sector to realize a decarbonized society. Although long-distance driving of PV-powered vehicles without electricity charging is expected in a sunny region, driving distance of PV-powered vehicles is affected by climate conditions such as solar irradiation and temperature rise of solar cell modules. In this paper, analytical results for effects of climate conditions such as solar irradiation and temperature rise of solar cell modules are presented by using our test data for Toyota Prius and Nissan Van demonstration cars installed with high-efficiency InGaP/GaAs/InGaAs 3-junction solar cell modules with a module efficiency of more than 30%. The potential of PV-powered vehicles to be deployed in major cities in the world is also analyzed. Mild weather cities such as Sydney in Australia are thought to have long driving ranges even under high sunshine conditions. In addition, the importance of heat dissipation of solar cell modules and the development of high-efficiency solar cell modules with better temperature coefficients are suggested. Further studies are necesary because there are additional losses for PV-powered vehicles.

978-1-6654-6060-6/23 $31.00 © 2023 IEEE

Laser-weld qualification for a reliable aluminum foil interconnection of copper-metallized back-contact silicon solar cells

Barry Hartweg, Kathryn Fisher, Jason Ro, Zachary Holman

Arizona State University, Tempe, AZ, United States

Sunflex Solar, Phoenix, AZ, United States

Laser welding is a low-temperature processing method to interconnect high-efficiency Cu-metallized interdigitated back-contact cells using low-cost Al foil. This interconnection technique is relatively new and, thus, requires detailed vetting of its reliability before being adopted commercially. In this study, we observe that the laser-weld adhesion, module fill factor, and reliability through thermocycling are all highly correlated to each other. We find that 94.4% of the modules fabricated using laser welds that had a mean 90o peel energy above 0.8 mJ lost less than 5% relative of their initial power after 200 thermocycles, and 90.0% of modules fabricated with an initial fill factor above 77% passed thermocycling. These metrics can be used in the further development of new laser settings and as quality control parameters in the manufacturing of these modules.

Conversion Efficiency Analysis of Tandem Solar Cells with Intermediate Band Tunnel Connection

Shuhei Yagi, Hiroyuki Yaguchi

Saitama University, Saitama, Japan

We propose a novel device structure of intermediate band solar cell (IBSC) in which the intermediate band (IB) state is connected to the valence band (VB) of the adjacent bottom cell through a tunnel junction. In this structure, the top cell is an IBSC while the bottom cell is a conventional single gap cell having the identical gap energy to that of top cell conduction band (CB)-IB gap. Detailed balance analysis demonstrated that such IB-tunnel connected tandem cell (IB-TC2) does not require significant CB-IB absorption, and its conversion efficiency less suffers from small shunt resistance connected to the CB-IB diode in the equivalent circuit compared with conventional IBSC with the same bandgap configuration. This new device configuration could be an attractive option for IBSC development in the face of unavoidable origin of shut resistance such as non-optical electron relaxation from the CB and thermal electron escape form the IB.

978-1-6654-6060-6/23 $31.00 © 2023 IEEE

Effects of Period of Record Extension, Model Diversification, and DHI Measurements on Measure-Correlate-Predict Analyses for On-site Solar Resource Assessments

Lucila D. Tafur, Renn Darawali, Halley Darling

UL Solutions, Albany, NY, United States

For solar PV projects, on-site irradiance measurements are used with historical satellite models to develop long-term solar resource estimates. Current industry practices rely on one year of measurements correlated to a single satellite model and are commonly performed only on GHI. This study will explore three areas in which current industry practices may want to add further consideration for Measure-Correlate-Predict (MCP) analyses: (1) extending the period of record of the measurement campaign (2) including a diverse set of modeled long-term datasets, and (3) including DHI measurements in the campaign. Industry-standard collection periods for a typical solar resource campaign are a minimum of one year. More extended periods of record, however, could lower the uncertainty of MCP analyses for prospective projects. Other common industry practices include the use of only one long-term satellite dataset for correlation. Reliance on a single long-term satellite dataset for correlation may result in increased sensitivity and uncertainty, as satellite models across the industry exhibit varying estimations of interannual variability. Lastly, while GHI measurements have historically been the primary focus of solar measurement campaigns, Diffuse Horizontal Irradiance (DHI) measurements are an integral component in the calculation of Plane of Array (POA), the main driver in energy simulations. Using over 10 test cases across various regions of the US, this study evaluates the potential uncertainty benefits of using a 1+ year campaign, a diverse set of satellite models, and DHI measurements thereby mitigating the financial risk of preconstruction projects.

978-1-6654-6060-6/23 $31.00 © 2023 IEEE

Solar Forecasting: The value of using satellite derived irradiance data in machine learning based forecasts.

Alex Kubiniec[1], Thomas Haley[1], Kyle Seymour[1], and Richard Perez[2]

[1] Clean Power Research, Kirkland, WA, 98033, USA
[2] SUNY Albany, NY, 12222, USA

Abstract — **Solar forecasts lower the cost of solar power and reduce the barriers to firm power generation. Solar forecasting conventionally has relied on advecting near real time observations and numerical weather predictions (NWPs). Observation based methods have limitations in cost and operational feasibility. NWPs generally have coarse spatial and temporal resolution, resulting in forecasts that may be overly general for a solar plant's location. Machine learning (ML) based forecasts have the potential to extract and blend observations and NWP data in an optimal blend, adding forecast skill. A primary drawback is ML based forecasts usually require training data. This paper will quantify the ML based forecast skill of using satellite derived irradiance data in lieu of ground, and the relationship between length of input training data and trained forecast skill gained. Climate and regional effects will be investigated by testing sites across the globe. Importantly these ML based forecasts will be compared to persistence forecasts and current NWP forecasts as a baseline.**

I. INTRODUCTION

To decarbonization energy systems, low carbon, renewable energy sources are needed to meet 100% renewable targets in 2035 [1]. Solar power is an excellent choice due to its low cost and global availability. Forecasts further improve solar power as an energy source. Knowledge of future power production lowers costs and helps grid managers ensure grid reliability. Higher quality solar power forecasts can reduce costs [2]. A solar plant operator can use a solar power forecast to sell all generated solar power more consistently. A solar forecast that under predicts, results in solar power that may not be sold, with possible results including, curtailment or energy being wasted. Overprediction adds cost to the plant operator, assuming the operator has promised to deliver more energy than the plant is currently producing, requiring energy to be procured at cost and potentially from non-renewable sources.

Conventional solar forecasts are based on physically defined models and observations of current conditions advected into the future. Conventional solar forecasting methods have useful forecast skill. However, recent advances in ML, processing power, and data availability, make ML approaches, practical to research and apply operationally. Generally, observation-based forecasts that rely on onsite sensors may have skill advantages in short term forecasts but have the drawback of increased cost and maintenance requirements. Forecast approaches that utilize ground observations, include auto regressive based forecasts requiring ground based sensors that measures global horizontal irradiance (GHI). Numerical

weather predictions (NWPs) are physically based models which use large amounts of observations, satellite data, and atmospheric soundings, combined with large computing power. The drawback is typically the spatial and temporal resolution are lower. The NWP skill globally generalizes well. NWPs have difficultly forecasting site-specific solar plants. ML has the potential to excel in this area. ML approaches can be tailored a specific location and offer skill over a conventional NWP forecast [5]. A limitation is the need for ground measurements or training data. A new site might also have little training data available. This paper will attempt to answer the question, how much training data is enough, and can a satellite derived irradiance model be substituted for true ground measurements.

The methods used to make a solar prediction are dependent on the time frame of desired forecast lead time. Conventional solar forecasts have used, in order of shortest lead time to longest, real time measurements, sky imagers, satellite images, and numerical weather predictions as inputs. The relationship between input data and forecast lead time is shown in Figure 1.

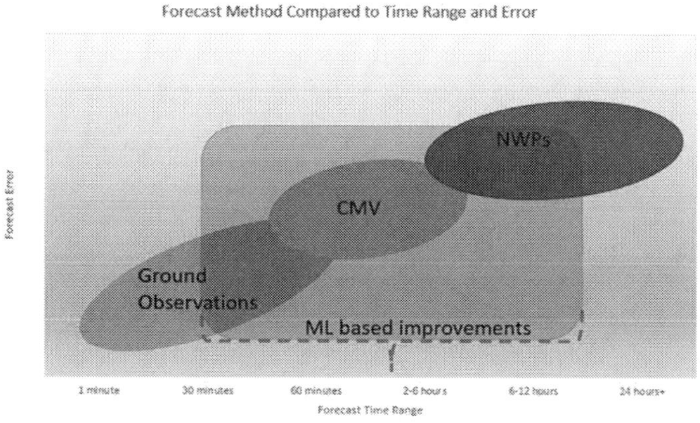

Fig. 1.   Visual illustration of forecast method, time range, and forecast error. The area of possible improvement using a ML approach is outlined in red. Forecast error increases as forecast horizon or time range increases. Forecast error also increases as the forecast method becomes more removed from observations at the ground site. This figure is for illustration purposes only, and axis values are only to be interpreted in a relative or conceptual manner.

978-1-6654-6060-6/23 $31.00 © 2023 IEEE

Machine learning based forecasts have the potential to bridge the gap between observation-based forecasts and NWPs, with higher skill than current methods. Some possible benefits of a ML based approach include a more custom or site-specific forecast, due to the trained nature of a ML forecast. This is timely due to the increasingly operational feasibility of ML based forecasts. A drawback or limitation is the need for training data, which this paper will attempt to address by quantifying the performance using modelled irradiance data in place of ground-based measurements for training.

II. METHODS

The input data for this work includes, satellite derived irradiance data, in the form of SolarAnywhere V3.6 (SA) [3]. Ground measured GHI, sourced from the SURFRAD / BSRN network [4]. NWP data is sourced from the National Digital Forecast Database (NDFD) and the Global Forecast System (GFS). Cloud motion vector (CMV) based forecasts are sourced from SolarAnywhere forecasts, using geostationary satellite imagery.

The sample period for results shown, cover 2021-01-01 to 2021-12-31, and a longer period of assessment is planned as a future task.

Variables or features during the sample periods, are quality controlled, temporally aligned, and preprocessed for use training a ML based forecast model. The two primary ML models assessed are a Random Forest and TensorFlow Keras (Keras) [6]. A random forest was chosen, because it trains quickly, easy to implement, and robust to hyper parameter permutation. The specific implementation is from the Python Scikit-learn library [7]. A Keras model will also be assessed. Benefits of Keras are its more sophisticated than a random forest, and generally should be more appropriate than a random forest-based approach.

Other models will be considered time and computation permitting. However, the goal of this work is to not benchmark

different ML algorithms against each other, instead to focus on input data.

The procedure is as follows:
1. Select and prepare GHI data for the sample period (in this paper the period is 2021-01-01 to 2021-12-31, this will be expanded to 2019-2022 in the coming months.
   a. Data including ground measurements, model GHI, and forecasted GHI will all be quality controlled, and temporally aligned.
   b. Sites in various locations and climates will be trained and assessed.
2. Train a ML model (Random forest in the case of this paper, Keras and additional models are planned as future work.) Varying the number of months and the use of ground data or modelled irradiance data, to test for trained forecast skill sensitivity to the variables of training data input type, and length of data. For purposes of testing SA vs ground with respect to training all other training inputs will be held constant.
   a. Time permitting other training features will also be tested, including additional forecast variables such windspeed, cloud cover, relativity humidity, various forecast lead times etc.
3. Compare results to benchmark/baseline forecasts such as persistence and current NWPs. It is important to put ML skill into a practical context so this is a task for future work as NWP data needs to be retrieved. Current work examines the forecasted first 6 hours, future work will examine longer forecast horizons, with more comprehensive error metrics.

Using this simple framework, many variables or features for training can be tested. The primary focus will be to test the difference in trained forecast skill using models trained using ground observations, and models trained using satellite derived irradiance data, in the form of SA. The goal is to quantify the difference in trained forecast skill if any.

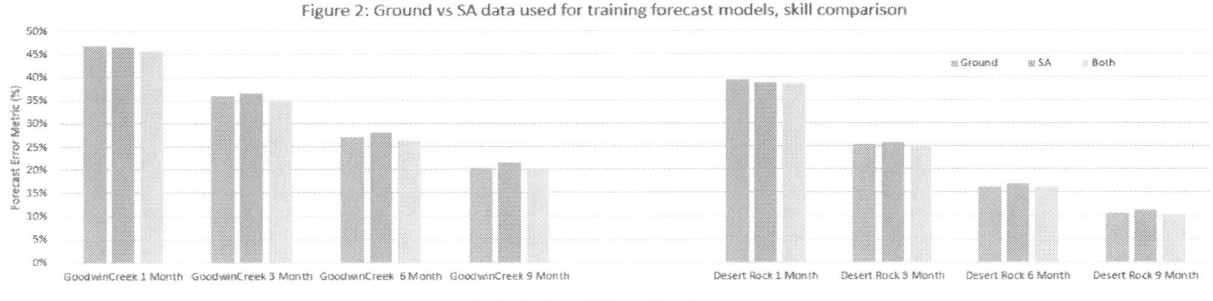

Figure 2: Compares 3 training approaches. One, using ground data as the variable input, two using SA model data as the variable input, and three using both. Note, that performance using ground, model, or both result in similar skill, and the controlling factor is the length of data used. Trained forecast skill is strongly related to the length of input data. Two sites are tested in this figure, however, all BSRN sites will be included in future work.

978-1-6654-6060-6/23 $31.00 © 2023 IEEE

A second goal is to test the sensitivity to data length in trained model forecast skill, as well as insights into various additional variables that may add forecast skill.

III. RESULTS AND DISCUSSION

Preliminary results suggest that using SA in lieu of ground is feasible. Trained forecast model skill using ground for training or using SA results in similar performance. 2 sites were tested, with a plan to expand to all BSRN sites reporting in real time, or in recent years. Additionally, determining the length of input data that results in no further trained skill gain is planned for future work. It is computationally intensive to test many locations, input lengths, and model types. The preliminary results in figure 2 suggest that at the very least 9 months vs 1 month result in dramatic trained forecast skill increases.

Specifically, the average percent difference in skill between a random forest based forecast using ground GHI as a training input vs. using SA as a training input is 3.2%. Which may at first appear to be a large difference, but when compared to the percent difference in using 1 month of training data vs. 9 months of training data, which is approximately 96%. This suggests that the difference between using ground data as the training input, and satellite derived irradiance data in the form of SA is very small relative to impact of a longer period of record. Detailed results are shown in figure 2.

It is worth noting that, when using both ground and SA model data as training inputs, along with NWP forecast data, which is same for all training scenarios, as to not give any additional information or advantage, the forecast skill is generally slightly better. These results suggest that using as much model and

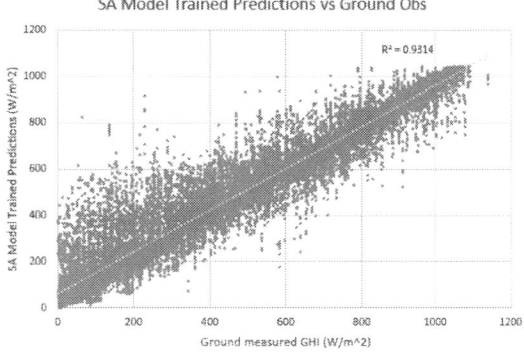

Figure 3: Ground measured GHI compared to forecast predictions based on a random forest trained using SA model data. Note the similarity to figure 4, both in data shape and correlation coefficient. Both are very similar, further suggesting model data and ground data can be used interchangeably for training.

ground data as possible will generally result in the best forecast skill trainable, but this is not novel conclusion, using model data in lieu ground data allows for future work exploring the sensitivity of trained forecast approaches on a global scale.

Actual predictions were also considered to confirm that predictions based on ground or model data had similar characteristics, not only forecast skill metrics.

Figures 3 and 4 confirm that actual predicted forecasted GHI values are very similar. Correlation coefficients are effectively the identical. Further suggesting that ground or modelled SA

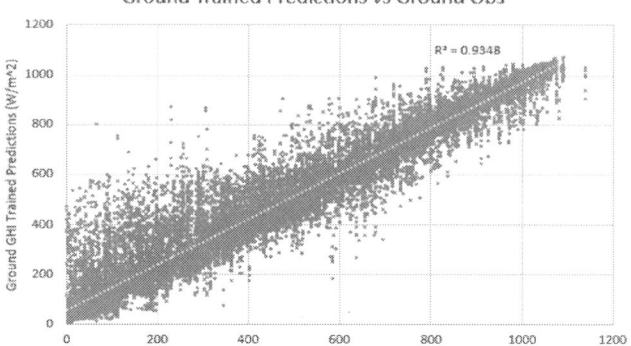

Figure 4: Note the similarity to figure 3. Goal in presenting these figures is to confirm trained predictions similarity independent of training inputs.

data can be used interchangeably for training-based forecasting.

IV. PRELIMINARY CONCLUSIONS / FUTURE WORK

This paper summarizes the basic framework for comparing the use of ground data as an input for ML based forecasts vs. satellited derived irradiance GHI. These two sources of data can be used interchangeably, with appreciable loss of forecast skill. Use of both results in the highest forecast skill. Forecast skill increases as longer periods are used for training, future work will determine the ideal amount of data to use to get maximum trained forecast skill. Climate's relationship to training skill will also be explored in future work. The bulk of outstanding results and work require significant computation and data compilation; however, the core method and comparisons are established. Those being, ground data and SA model data can be used interchangeably for machine learning based forecasts, and using the framework outlined in this paper an ideal length of training data for maximum machine learning based forecast skill can be determined, given computation time.

REFERENCES

[1] Douglas J. Arent, et. al; "Challenges and opportunities in decarbonizing the U.S. energy system", Renewable and Sustainable Energy Reviews, Volume 169, 2022.

[2] Yuhan Wang, et. al;"The cost of day-ahead solar forecasting errors in the United States", Solar Energy, Volume 231, 2022.

[3] Perez, Richard, et al. "Detecting calibration drift at ground truth stations a demonstration of satellite irradiance models' accuracy." 2017 IEEE 44th Photovoltaic Specialist Conference (PVSC). IEEE, 2017.

[4] Augustine, John A., et al. "SURFRAD—A National Surface Radiation Budget Network for Atmospheric Research." Bulletin of the American Meteorological Society, vol. 81, no. 10, 2000,

[5] Jesus Lago, et. al, "Short-term forecasting of solar irradiance without local telemetry: A generalized model using satellite data", Solar Energy, Volume 173, 2018

[6] Chollet, F., & others. (2015). Keras. GitHub. Retrieved from https://github.com/fchollet/keras

[7] Scikit-learn: Machine Learning in Python, Pedregosa et al., JMLR 12, pp. 2825-2830, 2011.

Ageing Detection of Encapsulants and Backsheets in the Field via NIR Spectroscopy

Chiara Barretta[1], Sascha Lindig[2], Márton Bredács[1], Alexander Astigarraga[2], Eric Helfer[1], Gernot Oreski[1]

[1] Polymer Competence Center Leoben GmbH (PCCL), Leoben, 8700, Austria, [2] EURAC Research, Bolzano, 39100, Italy

Abstract — **Photovoltaic (PV) modules and systems degrade during operation. It is necessary to understand how those mechanisms develop over time and to find precursors that allow an early detection of potential fatal faults. In this work, we investigated the evolution of performance loss rates (PLR) of 6 PV systems exposed in northern Italy for about 11 years, and we characterized 19 PV modules belonging to the systems via near infrared (NIR) spectroscopy to identify encapsulant and backsheet types. Additionally, we analyzed the spectral data with multivariate data analysis (MVDA) to detect hidden signs of polymer degradation that are not visible with the naked eye.**

I. INTRODUCTION

Photovoltaic (PV) modules that are exposed in the field undergo degradation processes that depend on environmental and operating conditions. Appearing degradation modes can affect one or more components of the PV modules as well as balance of system (BoS) components with effects on the electrical performance. One of the most challenging tasks is to understand what is happening in a system during operation and to identify the root causes of the faults.

Non-destructive measurements techniques are preferable, if not necessary, to investigate PV systems without damaging and affecting them. The most common procedure that allows non-destructive assessment of the status of a PV module is visual inspection. Degradation modes such as, e.g., polymer discoloration, delamination, snail trails, backsheet cracks and junction box faults can be usually detected visually [1]–[2]. In addition to visual inspection, several imaging techniques are available to detect degradation and malfunctioning in the field. Some examples are electroluminescence, ultraviolet (UV) fluorescence and infrared (IR) imaging, which are useful to detect potential induced degradation (PID), cell cracks, corrosion, failures of interconnections and solder joints, and encapsulant degradation [3]–[4]. Near infrared (NIR) spectroscopy is an analytical method that allows polymer identification and characterization. It is routinely used in quality control for several applications and in mechanical recycling of polymers. In photovoltaics, it can be used in the field to identify the type of encapsulant and backsheet materials [5]-[6]. In this work, we performed NIR spectroscopy measurements on modules from 6 PV systems exposed for about 11 years in northern Italy. The objective was to (I) show that NIR spectroscopy can be a powerful tool to detect ageing of polymers in PV modules and to (II) correlate observed polymer degradation to performance data of modules and systems, and to observed degradation modes.

II. EXPERIMENTAL

The subject of the investigation were 19 PV modules belonging to 6 systems that are exposed in the field since about 11 years. The modules are located at a test site at the airport in Bolzano/Italy (ABD), in a climate classified as temperate with medium irradiation [7]. The PV technologies analyzed are based on crystalline silicon solar cells encapsulated in a glass/polymer backsheet structure.

Performance loss rates (PLRs) were calculated on system level using performance data from the inverters. The PLR of a PV system describes the overall decline in performance due to reversible and irreversible performance losses such as degradation, soiling, snow, BoS degradation, etc. [8]. Techniques such as NIR might be a useful tool to subdivide the overall PLR into its respective components. The encapsulants and backsheet materials were identified and characterized via NIR spectroscopy using a hand-held NIR spectrophotometer from trinamiX GmbH (Germany). At least 5 spectra were acquired by putting the spectrometer in contact with the glass and with the backsheet (as shown in Fig. 1) to get information about the encapsulant and the backsheet, respectively.

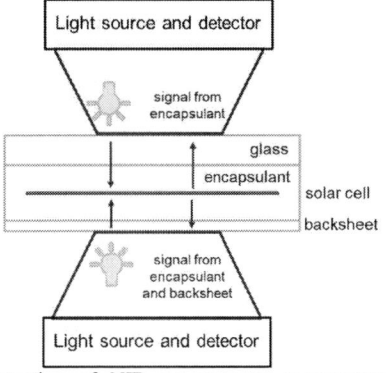

Fig. 1. Illustration of NIR spectroscopy measurements on PV modules in the field (not in scale).

The light beam can go through all the layers of the backsheet and through the encapsulant, the signal is reflected back to the detector by the solar cell, which allows to obtain information about the molecular composition of all the layers encountered. A non-exposed module was available for each PV system and is considered as a reference. The spectra were baseline corrected and normalized using standard normal variate (SNV) transformation to perform Multivariate Data Analysis (MVDA) using the software Unscrambler X 10.5 (CAMO Software AS., Oslo, Norway). NIR spectra show overtone absorption bands and combination bands of fundamental molecular vibrations, whereas IR spectra show fundamental molecular vibrations, meaning that interpretation of variations in the NIR spectra due to ageing is more difficult with respect to IR spectra. MVDA as well Principal Component Analysis (PCA) can be useful tools to extract hidden chemical information in NIR spectra.

III. RESULTS AND DISCUSSION

The systems analyzed showed PLR between −0.89 %/year (pc-Si B) and −1.56 %/year (pc-Si A). The high PLR for the system pc-Si A might be due to the particular glass surface structure that traps soling, thus reducing the power generated. Visual inspection of the modules highlights the presence of several degradation modes such as backsheet cracking, chalking, cell cracks, moisture ingress and corrosion of various metallic components. More detailed information are given in Table 1. The NIR spectrophotometer is coupled with a software that compares the measured spectra with a backsheet and

encapsulant library, thus allowing a fast identification of the polymers. Although the polymer identification showed quite reliable results, further destructive tests will be carried out on the backsheets to analyze their composition in detail. The encapsulant material was identified as ethylene vinyl acetate (EVA) for all the modules, as shown in Fig. 2. The peaks observable at about 1730 nm and 1760 nm represent the C−H stretching vibrations of methyl and methylene units of the polyethylene (PE) moieties [9] in EVA. Regarding the backsheets, it was possible to identify at least 3 different types. The modules pc-Si A and pc-Si B have a polyamide (PA) based backsheet. Additionally, all modules from the pc-Si B system showed the typical squared cracks between the cells that were already reported in literature for PA based backsheets [10]. The modules from the system pc-Si A showed strong chalking, which is a typical precursor of cracking phenomena for this type of backsheet. The backsheets from the modules mc-Si A, pc-Si C, pc-Si D, pc-Si E are characterized by a polyethylene terephthalate (PET) core layer, as it is shown by the presence of the peak at about 1650 nm, corresponding to the first overtone of the aromatic C−H stretch and methylene stretching [11]. The backsheet of the module pc-Si E was identified as a polyvinylidene fluoride (PVDF)/PET backsheet. The spectra of the modules pc-Si C, pc-Si D and pc-Si E are quite similar and additional destructive analysis will be performed to validate the composition prediction.

TABLE 1. SUMMARY OF RESULTS OF THE PV SYSTEMS AND MODULES ANALYZED

PV systems	Performance loss rate (PRL) (2011-2019) %/year	Visual observations	PV modules analyzed	Bill of materials (BOM)		Multivariate data analysis (MVDA)	
				Encapsulant	Backsheet	Encapsulant	Encapsulant + Backsheet
mc-Si A	−1.13	Dirt spots on backsheets	mc-Si A_ref, mc-Si A_1, mc-Si 2, mc-Si 3	EVA	PET core	Weak model	Strong model
pc-Si A	−1.56	Strong chalking, soiling on the glass surface	pc-Si A_ref, pc-Si A_1, pc-Si A_2	EVA	PA	Medium model	Strong model
pc-Si B	−0.89	Backsheet chalking, cell cracks, humidity ingress, corrosion	pc-Si B_ref, pc-Si B_1, pc-Si B_2	EVA	PA	Weak model	Strong model
pc-Si C	−1.10	Polymer coming out of frame	pc-Si C_ref, pc-Si C_1, pc-Si C_2	EVA	PET core	Medium model	Strong model
pc-Si D	−1.23	−	pc-Si D_ref, pc-Si D_1, pc-Si D_2	EVA	PET core	Weak model	Weak model
pc-Si E	−1.05	Cell cracks, humidity ingress and corrosion	pc-Si E_ref, pc-Si E_1, pc-Si E_2	EVA	PVDF/PET core	Medium model	Strong model

The measured spectra were then analyzed by means of MVDA and an example of the results obtained can be seen in Fig. 3. The plot shows the principal components (PCs) of the model that describe the relationship between the data for the system pc-Si C. PC1 allows, in this case, to describe 85% of the variance between the data and the plot shows a clear difference between the aged and the reference modules measured from the backsheet side.

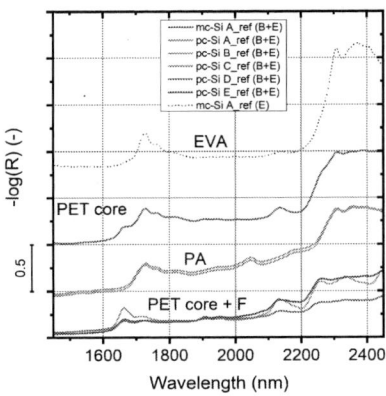

Fig. 2.   NIR spectroscopy measurements on the reference modules from the glass side (mc-Si A_ref) and from the backsheet side (curves marked with B+E).

Even though the model that describes the data is rather robust, it is not necessarily correlated to significant differences in physical and chemical properties of the materials. Looking at the loading plots (not shown here), there are clear peaks that can describe one group or the other. Additional destructive characterization will be necessary to determine whether the difference corresponds to significant physical or chemical degradation.

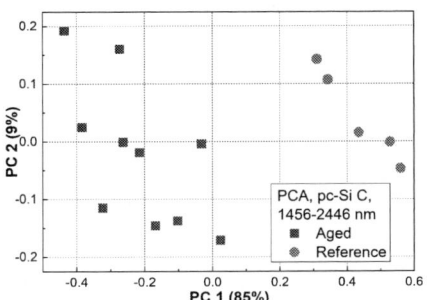

Fig. 3.   Plot of Principal Components (PC) of the model that describes the differences between the measurements on the aged (blue squares) and reference samples (red circles) of the pc-Si C modules (E+B).

IV. CONCLUSIONS

In this work, the performance loss rates of six PV systems, installed in the north of Italy and operational for roughly 11 years, were calculated. NIR spectroscopy was applied to randomly selected modules of these systems and was proven to be a valuable tool to detect polymer ageing in the field. MVDA analysis showed that there is a significant difference between exposed PV modules and non-exposed references, and that the difference is more pronounced for the backsheets than for the encapsulants. Further destructive characterization will be performed on the tested modules to identify physical and chemical root causes for the spectral changes and to correlate them to observed degradation modes and the calculated performance loss rates. NIR spectroscopy could be effectively used in the field to monitor ageing of modules in PV plants for future operation and maintenance.

REFERENCES

[1] A. Virtuani et al., "35 years of photovoltaics: Analysis of the TISO-10-kW solar plant, lessons learnt in safety and performance—Part 1," *Progress in Photovoltaics: Research and Application*, vol. 27, no. 4, pp. 328–339, 2019.

[2] M. Aghaei, A. Fairbrother, A. Gok, S. Ahmad, S. Kazim, K. Lobato, G. Oreski, A. Reinders, J. Schmitz, M. Theelen, P. Yilmaz, J. Kettle, *Review of degradation and failure phenomena in photovoltaic modules*, Renewable and Sustainable Energy Reviews, Volume 159, 2022, 112160, ISSN 1364-0321, https://doi.org/10.1016/j.rser.2022.112160.

[3] U. Jahn, M. Herz, M. Koentges, D. Parlevliet, and M. Paggi, "Review on Infrared and Electroluminescence Imaging for PV Field Applications," *Report IEA-PVPS T13-10:2018, IEA-PVPS Task 13*, 2018. [Online]. Available: https://iea-pvps.org/key-topics/review-on-ir-and-el-imaging-for-pv-field-applications/

[4] M. Köntges, A. Morlier, G. Eder, E. Fleis, B. Kubicek, and J. Lin, "Review: Ultraviolet Fluorescence as Assessment Tool for Photovoltaic Modules," *IEEE J. Photovoltaics*, vol. 10, no. 2, pp. 616–633, 2020, doi: 10.1109/JPHOTOV.2019.2961781.

[5] G. C. Eder, Y. Lin, Y. Voronko, and L. Spoljaric-Lukacic, "On-site identification of the material composition of PV modules with mobile spectroscopic devices," *Energies*, vol. 13, no. 8, 2020, doi: 10.3390/en13081903.

[6] O. Stroyuk, C. Buerhop-Lutz, A. Vetter, J. Hauch, and C. J. Brabec, "Nondestructive characterization of polymeric components of silicon solar modules by near-infrared absorption spectroscopy (NIRA)," *Solar Energy* Materials and Solar Cells, vol. 216, p. 110702, 2020, doi: 10.1016/j.solmat.2020.110702.

[7] I. Kaaya, S. Lindig, K-A. Weiss, A. Virtuani, M. Sidrach de Cardona Ortin, D. Moser. "Photovoltaic lifetime forecast model based on degradation patterns". *Prog Photovolt Res Appl*. 2020; 28: 979– 992. https://doi.org/10.1002/pip.3280.

[8] S. Lindig, M. Theristis, D. Moser, "Best practices for photovoltaic performance loss rate calculations". *Prog. Energy*. Vol. 4, no. 2, 2022. DOI 10.1088/2516-1083/ac655f.

[9] J. J. Workman, Jr., "Near-Infrared Spectroscopy of Polymers and Rubbers," in *Encyclopedia of Analytical Chemistry*, R. Meyers and T. Provder, Eds.

[10] G. C. Eder et al., "Error analysis of aged modules with cracked polyamide backsheets," *Solar Energy Materials and Solar Cells*, vol. 203, p. 110194, 2019, doi: 10.1016/j.solmat.2019.110194.

[11] G. Lachenal, "Characterization of Poly(Ethylene Terephthalate) Using Near and Far FTIR Spectroscopy," *International Journal of Polymer Analysis and Characterization*, vol. 3, no. 2, pp. 145–158, 1997, doi: 10.1080/10236669708032760.

In-situ & ex-situ study of protons and electrons irradiations of perovskite solar cells

Carla Costa, Matthieu Manceau, Thierry Nuns, Sophie Duzellier, Romain Cariou

CEA INES, Le Bourget du Lac, France

ONERA, Toulouse, France

The need for low cost photovoltaic solutions is becoming more and more important with the ongoing NewSpace revolution. In this context, alternative solar cell technologies are under the spotlight, in particular perovskites which can reach high specific power. In this study, we investigate the electron and proton radiation hardness of multi-cation mixed halide perovskite cells $Cs_xFA_{1-x}Pb(I_yBr_{1-y})_3$. The proton irradiations demonstrate an excellent radiation hardness of the perovskite material but also highlight the degradation of the HTL layer. And the in-situ IV measurements under vacuum following the electron irradiations reveals a self-healing phenomena.

Numerical Investigation on Non-radiative Recombination in InGaAs Front and Rear Hetero-Junction Solar Cell

Depu Ma[1], Hassanet Sodabanlu[2], Gan Li[1], Meita Asami[1], Kentaroh Watanabe[2], Masakazu Sugiyama[1, 2], Yoshiaki Nakano[1]

[1]Department of Electrical Engineering, University of Tokyo, Tokyo, 1138656 Japan

[2]RCAST, University of Tokyo, Tokyo, 1538904 Japan

Abstract — Front hetero-junction (FHJ) and rear hetero-junction (RHJ) InGaAs solar cells were grown by metal-organic vapor phase epitaxy (MOVPE). Compared with RHJ solar cell, FHJ InGaAs solar cell shows a better short circuit current density (Jsc), but a worse open circuit voltage (Voc). Based on simulation results, non-radiative recombination in FHJ and RHJ InGaAs solar cells has been compared. FHJ has a better carrier diffusion length, but greater carrier loss at the interface resulted to worse Voc. RHJ has a less carrier loss in the interface but worse carrier diffusion length resulted to worse Jsc.

I. INTRODUCTION

The fabrication of $In_{0.53}Ga_{0.47}As$ with high crystal quality and lattice matched to InP substrate has been investigated in many researches. Besides, several structures including front hetero-junction (FHJ) and rear hetero-junction (RHJ) have been proposed to suppress non-radiative recombination and improvement of open-circuit voltage (Voc) in solar cells [1]. However, there are few reports about advantages and disadvantages about FHJ and RHJ in InGaAs solar cell application. In the present work, experiments and SCAPS-1D simulation software has been used to analyze the non-radiative recombination in FHJ and RHJ InGaAs cells. The results revealed different influences of RHJ and FHJ on the interface and bulk recombination.

II. THEORY

Recombination in solar cell devices includes radiative recombination and non-radiative one, the former is mainly determined by the basic property of the material, and the latter is mainly determined by the crystal quality of the material. Radiative, Auger and Schockley-Read-Hall (SRH) recombination are considered as main bulk recombination in our simulation, which are calculated by (1), (2), (3), respectively.

$$R_{radiative} = K(np - n_i^2) \tag{1}$$

$$R_{Auger} = (c_n^A n + c_p^A n)(np - n_i^2) \tag{2}$$

$$R_{SRH} = \frac{np - n_i^2}{\tau_{n,SRH}(p + p_t) + \tau_{p,SRH}(n + n_t)} \tag{3}$$

where K is the radiative recombination coefficient, c_n^A and c_p^A are the Auger capture coefficients of electron and hole respectively, $\tau_{n,SRH}$ and $\tau_{p,SRH}$ are the SRH recombination-dependent lifetime of electron and hole respectively.

Interface recombination is another source for non-radiative recombination, which heavily affects the performance of solar cell. Recombination in interface states is modeled by the Pauwels-Vanhoutte theory [2], which is an extension of the SRH theory.

The numerical calculation is performed by the simulation software, SCAPS-1D, which is capable of self-consistently solving the Poisson equation, current continuity equation and photon rate equations with the proper boundary conditions. Most basic materials parameters of InP and $In_{0.53}Ga_{0.47}As$ are set after previous experiments results [3]. To be noted, SRH bulk recombination and interface recombination parameters are set to have a good agreement with our experiment results, as shown in Table I. Sunlight wavelength-dependent absorption and reflection of materials are based on previous experimental data [3] and AM1.5G spectrum (same with experiment measurement condition) has been used for simulation.

II. EXPERIMENT

In this experiment, both of FHJ and RHJ InGaAs cells were grown on p-type (Zn-doped) exactly oriented (001) InP substrates. MOVPE reactor (AIXTRON 2000HT) using standard H_2 carrier gas was used for materials growth. Low-toxic tertiarybutylarsine (TBAs) and tertiarybutylphosphine (TBP) are used as group-V precursors, trimethylgallium (TMGa) and trimethylindium (TMIn) as group-III metal-organics. Dimethylzinc (DMZn) and hydrogen sulfide (H_2S) were utilized for p-type and n-type doping, respectively. The FHJ and RHJ structures are shown in Fig. 1 (a) and (b).

978-1-6654-6060-6/23 $31.00 © 2023 IEEE

TABLE I
Basic recombination parameters of InP and $In_{0.53}Ga_{0.47}As$

materials	Radiative Recombination coefficient [2] ($cm^{-3}s^{-1}$)	Auger Electron/Hole Capture Coefficient [2] ($cm^{-6}s^{-1}$)	Electron/Hole Capture Section of bulk SRH Recombination (cm^2)	Electron/Hole Capture Section of Interface Recombination (cm^2)
InP	1.2×10^{-10}	$1.5\times10^{-29}/1.5\times10^{-29}$	$6\times10^{-11}/6\times10^{-11}$	
P-$In_{0.53}Ga_{0.47}As$	9.6×10^{-11}	$5\times10^{-31}/5\times10^{-31}$	$8.5\times10^{-12}/6.5\times10^{-12}$	$4\times10^{-11}/4\times10^{-11}$
N-$In_{0.53}Ga_{0.47}As$	9.6×10^{-11}	$5\times10^{-31}/5\times10^{-31}$	$1.5\times10^{-11}/2.5\times10^{-11}$	

Electro-chemical capacitance voltage (ECV) was used for obtaining the desired doping concentration. More detailed growth optimization (including growth temperature, V-III ration and gas flow rate) for InGaAs solar cells has been reported in a previous work of our laboratory [4].

III. RESULTS AND DISCUSSION

The numerical calculation and experiment results of the J-V curve and external quantum efficiency (EQE) are shown in Fig.1 (c) and (d), respectively. Our experiment and simulation results have achieved good agreement, which shows that our simulation results can reflect the real situation of the experimental samples to a certain extent. Compared with RHJ structure, FHJ InGaAs solar cell shows a better short circuit current density (J_{sc}), but a worse V_{oc}. The J_{sc} depends on the diffusion length of the minority carriers, thus, based on the simulation results, we extracted the base layer minority carrier diffusion lengths in the FHJ and RHJ, respectively (shown in Fig.1 (d)). The diffusion lengths of the minority carrier from base layer in the FHJ and RHJ structures are larger and smaller than the length of the base region, respectively. Therefore, the J_{sc} is larger in the FHJ structure. To explain the difference in V_{oc} between FHJ and RHJ, we first briefly introduce the reciprocity relation. According to reciprocity relation [5], V_{oc} is expressed as (5), where V_{oc}, V_{oc}^{rad} and $\eta_{ext}(V_{oc})$ are actual V_{oc}, ideal V_{oc} and non-radiative dependent external radiative efficiency. Thus, more serious non-radiative recombination in FHJ should be the reason for a smaller V_{oc}.

$$V_{oc} = V_{oc}^{rad} + \frac{KT}{q}\ln(\eta_{ext}(V_{oc})) \qquad (5)$$

In order to show the non-radiative recombination in FHJ and RHJ, numerical calculation of non-radiative recombination rate in FHJ and RHJ when voltage equal to 0 and V_{oc} has been calculated in Fig.2 (b) and (c). It can be clearly seen that in the case of voltage equal to V_{oc}, the non-radiative recombination in the FHJ is greater than that in

the RHJ, thus, larger V_{oc} in RHJ. Dark current (See Fig.2(a)) also suggests the same conclusion as well. Because the voltage-dependent dark current can be expressed by (6) [6], the former part is the current related to radiative recombination, and the latter part is the current related to non-radiative recombination. Part of the reason for the increase in dark current in the FHJ is due to the increase in non-radiative recombination current and thus, by reciprocity relation, the degradation of V_{oc}.

$$J_{dark}(V) = J_{em,0}(\exp(\frac{qV}{KT})-1) + J_{nr}(V) \qquad (6)$$

FHJ has a better carrier diffusion length, but why does it have a poorer Voc? To explain this problem, We calculated the non-radiative recombination current densities under the conditions of V=0 and V=Voc in FHJ and RHJ respectively, and displayed them in Table II. In our simulation results, Auger recombination, SRH recombination as well as interfacial recombination are considered. It can be seen that the Auger recombination current density is much smaller than the other two recombination current densities, so only SRH recombination and interface recombination will be discussed below. It can be clearly seen that in the FHJ structure, the interface recombination current density accounts for a larger proportion of the total non-radiative recombination current, indicating that the carriers have a greater loss at the interface. In the RHJ structure, the SRH recombination current density in the base layer accounts for a larger proportion of the total non-radiative recombination current, indicating that the carriers have a greater loss in the base region, which also is consistent with worse carrier diffusion length conclusion. In the future, the improvement of the crystal quality of the hetero-junction interface and the base region may be the key to improve the performance of FHJ and RHJ InGaAs solar cells, respectively.

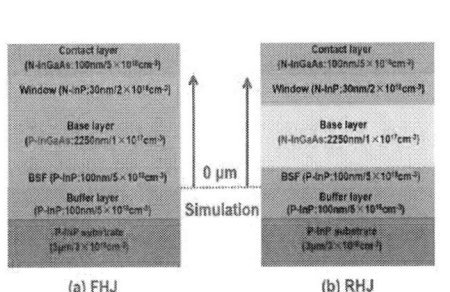
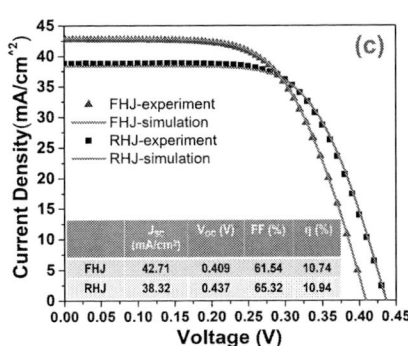
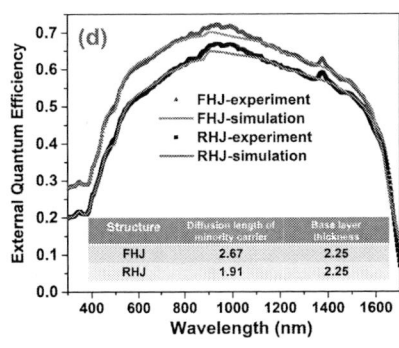

Fig.1 The structure of (a) FHJ and (b) RHJ InGaAs solar cell; the numerical calculation and experiment results of (c) J-V curve and (d) EQE under AM1.5G illumination.

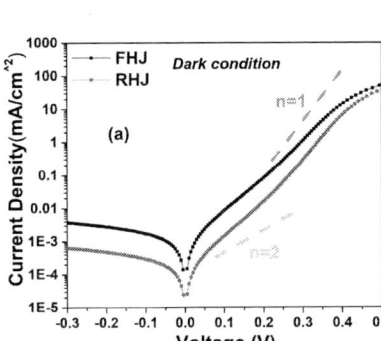
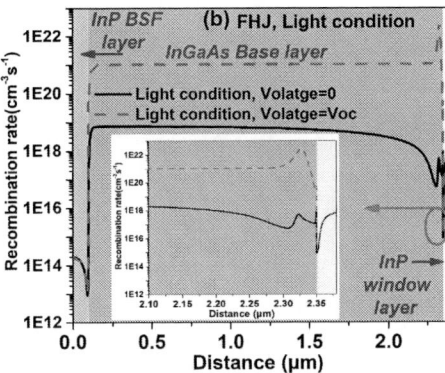
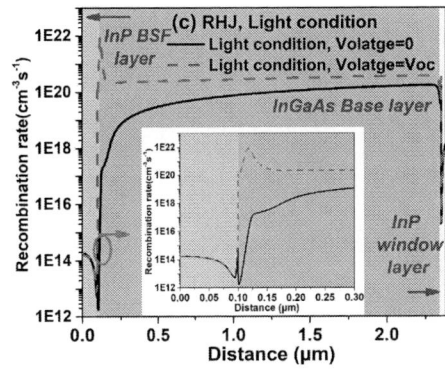

Fig. 2 The dark J-V of (a) FHJ and (b) RHJ InGaAs solar cell; the numerical calculation and experiment results of non-radiative recombination rate in (a) FHJ and (b) RHJ with voltage=0 and V_{oc}.

TABLE II
Non-radiative recombination current density in FHJ and RHJ solar cell

Structure	Condition	Total Non-radiative recombination current density J_{tot} (mA/cm^2)	Interface recombination current density J_{INT} (mA/cm^2), J_{INT}/J_{tot}(%)	SRH recombination current density J_{SRH} (mA/cm^2), J_{SRH}/J_{tot}(%)	Auger recombination current density J_{AUG} (mA/cm^2), J_{AUG}/J_{tot}(%)
FHJ	V=0	2.35	1.74, 74.1	0.57, 24.2	0.04, 1.7
	V=V_{oc}	26.67	14.29, 53.6	11.01, 41.3	1.36, 5.1
RHJ	V=0	4.67	0.87, 18.6	3.69, 79.1	0.11, 2.3
	V=V_{oc}	24.41	4.21, 17.2	18.31, 75.0	1.90, 7.8

IV. SUMMARY

Based on experiment samples and simulation results, non-radiative recombination in FHJ and RHJ InGaAs solar cells have been discussed. Based on simulation results, non-radiative recombination in FHJ and RHJ InGaAs solar cells has been compared. FHJ has a better carrier diffusion length, but greater carrier loss at the interface resulted in worse Voc. RHJ has a less carrier loss at the interface but worse carrier diffusion length resulted in worse Jsc. In the future, the improvement of the crystal quality of the hetero-junction interface and the base region may be the key to improve the performance of FHJ and RHJ InGaAs solar cells, respectively.

REFERENCES

[1] R. Yokota et al., "Photovoltaics Reaching for the Shockley-Queisser Limit", in 68h Spring meeting of Japan Society of Applied Physics, 2021, 16p-Z02-6.

[2] H.J. Pauwels et al., "Influence of interface states and energy barriers on efficiency of heterojunction solar-cells", J. Phys. D-Appl. Phys., vol. 11, pp. 649-667, 1978.

[3] S. Adachi et al., Physical Properties of III-Y Semiconductor compounds, John Wiley and Sons, 1992.

[4] Hassanet Sodabanlu et al., "Growth of InGaAs(P) using tertiarybutylarsine and tertiarybutylphosphine for photovoltaic applications",vol. 57, 08RD09, 2018.

[5] Uwe Rau et al., "Reciprocity relation between photovoltaic quantum efficiency and electroluminescent emission of solar cells", Phys. Rev. B vol. 76, 085303, 2007.

[6] J. F. Geisz et al., "Enhanced external radiative efficiency for 20.8% efficient single-junction GaInP solar cells", Appl. Phys. Lett. vol.103, 041118, 2013.

Strategies to Improve the Mechanical Robustness of Metal Halide Perovskite Solar Cells

Muzhi Li, Siraj Sidhik, Lidon Gil-Escrig, Samuel Johnson, Aditya Mohite, Axel Palmstrom, Henk J. Bolink, Nicholas Rolston

Arizona State University, Tempe, AZ, United States

Rice University, Houston, TX, United States

University of Valencia, Valencia, Spain

University of Colorado Boulder, Boulder, CO, United States

National Renewable Energy Laboratory, Golden, CO, United States

Here, we investigate the mechanical properties of high-efficiency perovskite solar cells with different chemical components by measuring fracture energy (Gc) of films and devices. With the help of both macroscopic and microscopic techniques, we identify the locations where the fractures take place in the devices (either adhesive or cohesive failure) with various structures. We propose strategies that can improve the fracture energy of PSCs based on the measured Gc: replacing 3D with 2D perovskites and either avoidance of small-molecule charge transport layer or the addition of a tin oxide layer for mechanical reinforcement. Our findings offer a pathway to rationally study the mechanical properties of perovskite-based solar cells and enable such cells to be more mechanically robust to reach commercial viability.

Microstructure-property relationships in epitaxial Cu(In,Ga)Se₂ solar-cell absorbers

Daniel Abou-Ras[1], Jiro Nishinaga[2], Takeyoshi Sugaya[2], Yukiko Kamikawa-Shimizu[2], Ulrike Bloeck[1], Henrik Prell[1], Sinju Thomas[1], Michael Tovar[1], Dan R. Wargulski[1], Harvey Guthrey[3], Pat Trimby[4], Aimo Winkelmann[5], and Shogo Ishizuka[2]

1. Helmholtz-Zentrum Berlin für Materialien und Energie GmbH, 14109 Berlin, Germany
2. National Institute of Advanced Industrial Science and Technology, Umezono, Tsukuba, Ibaraki, Japan
3. National Renewable Energy Laboratory (NREL), Golden, CO, USA
4. Oxford Instruments Nanoanalysis, High Wycombe, Buckinghamshire HP12 3SE, UK
5. ST Development GmbH, 33102 Paderborn, Germany

Abstract—Epitaxially grown Cu(In,Ga)Se₂ (CIGS) absorber layers were analyzed by various techniques in scanning electron microscopy in order to reveal microstructure-property relationships in these thin films. Owing to their epitaxial nature, these CIGS absorber layers do not contain any grain boundaries, but only anti-phase domains (APDs) and dislocations. By combining electron channeling contrast imaging, electron backscatter diffraction, and cathodoluminescence (CL), in some cases on identical specimen positions of polished cross-sections of CIGS/Mo/glass stacks, it was possible to correlate the presence and orientations of APDs and dislocations with the lateral distributions of the CL intensity and emission-peak energy. We studied CIGS layers with three different [Ga]/([Ga]+[In]) ratios as well as with and without NaF/KF treatments. Considerable differences between the CIGS layer properties in the microstructure-property relationships were found, depending on the growth parameters. Dislocations in the epitaxial CIGS layers do not tend to exhibit strong CL intensity decreases, which contrasts with the situation in numerous other semiconductor materials.

Keywords—epitaxial, Cu(In,Ga)Se₂, anti-phase domains, dislocations, electron channeling-contrast imaging, electron backscatter diffraction, cathodoluminescence

I. INTRODUCTION

Thin-film solar cells with epitaxially grown Cu(In,Ga)Se₂ (CIGS) absorber layers have reached conversion efficiencies of more than 20% [1]. Since the epitaxial CIGS layers do not contain any grain boundaries, they are appropriate to study the impact of dislocations on the optoelectronic properties. Apart from dislocations, epitaxially grown CIGS thin films also feature anti-phase domains (APDs). It is known that their boundaries are connected with a dislocation core in ordered alloys [2]. Thus, these CIGS layers contain dislocations in the bulk and further dislocation cores linked to the APD boundaries.

For the present study, a series of epitaxial CIGS solar cells with three different [Ga]/([Ga]+[In]) (GGI) ratios, 0.2, 0.6, and 0.8, with and without NaF/KF treatments, were fabricated. Thus, it was possible using various techniques in scanning electron microscopy in addition to X-ray diffraction (XRD) to study the effects of the GGI ratio as well as of the alkali treatment on the microstructure-property relationships of the epitaxial CIGS thin films.

II. EXPERIMENTAL DETAILS

Epitaxial CIGS layers were grown on GaAs (001) p+-type substrates by means of molecular beam epitaxy. For the present study, CIGS layers with three different [Ga]/([Ga]+[In]) ratios with and without NaF/KF treatments were produced. Solar cells were completed by the chemical-bath deposition of a CdS buffer as well as a sputtered ZnO/ZnO:Al window layer. Specimens for electron microscopy were prepared by gluing two pieces of solar cells or CIGS/Mo/glass face-to-face together using epoxy glue and by polishing the cross-sections of these stacks mechanically and by an Ar-ion beam. Electron channeling-contrast imaging (ECCI), electron backscatter diffraction (EBSD), and cathodoluminescence (CL) were performed using Zeiss UltraPlus, Gemini 460, and Merlin scanning electron microscopes equipped with a dedicated backscattered electron detector, an Oxford Instruments Symmetry EBSD detector, and a DELMIC SPARC CL system. Also, all CIGS layers were analyzed by means of XRD using a Bragg-Brentano and a grazing-incidence setup.

III. RESULTS AND DISCUSSION

The presence and relative orientations of APDs are different for the three different GGI ratios. At GGI=0.2, no APDs were found in the epitaxial CIGS films, and only dislocations were present. Their density seems to be smaller in the CIGS layer with NaF/KF treatments. At GGI=0.6, both, vertical and diagonal APDs, were detected, while their orientations with respect to the substrate were only diagonal at GGI=0.8.

In addition, the texture of the CIGS films also changes with changing GGI, which can be attributed to the deviation from the pseudocubic symmetry on one hand side and to lattice match with the GaAs(001) substrate on the other. We provide a model describing the different film textures, orientations of APDs, and crystal facets at APD boundaries for the various GGIs.

By means of ECCI and EBSD, it was possible to detect dislocations and to characterize their densities and Burgers vectors. In all the six studied CIGS layers, no strong CL intensity contrasts related to the presence of dislocations were detected, as reported before in a precedent work [3]. Reduced or enhanced CL signals on APDs can be linked rather to different channeling conditions between matrix and APDs.

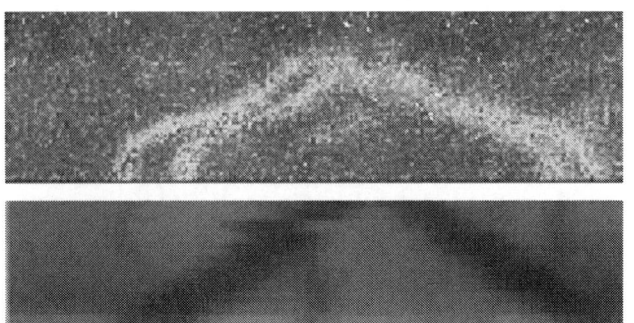

Fig. 1. (Top) EBSD map highlighting geometrically necessary dislocations located at the boundaries of a APD. (Bottom) CL intensity distribution from the identical position showing reduced signals at the positions of APDs, which can be explained by a channeling effect on the CL signals. Height of the maps about 2 μm.

IV. CONCLUSIONS

The microstructures of epitaxial CIGS thin films were found to depend substantially on their GGI ratios. The presence and relative orientations of APDs varies considerably when changing the GGI ratio from 0.2 and 0.8. Alkali treatments do not affect the properties of APDs. Dislocations do not induce strong decreases in CL intensity in any of the analyzed CIGS thin films.

ACKNOWLEDGMENTS

The present work was supported by the Graduate School "MatSEC", the Helmholtz-International Research School "HI-SCORE" (HIRS-0008), the BMWK-funded project "EFFCIS-II" (03EE1059B), and the New Energy and Industrial Technology Development Organization "Research and Development Program for Promoting Innovative Clean Energy Technologies Through International Collaboration" under the Ministry of Economy, Trade and Industry. Moreover, this work was authored, in part, by the National Renewable Energy Laboratory, operated by the Alliance for Sustainable Energy, LLC, for the U.S. Department of Energy (DOE) under Contract No. DE-AC36-08GO28308. The views expressed in the article do not necessarily represent the views of the DOE or the U.S. Government. The U.S. Government retains and the publisher, by accepting the article for publication, acknowledges that the U.S. Government retains a nonexclusive, paid-up, irrevocable, worldwide license to publish or reproduce the published form of this work, or allow others to do so, for U.S. Government purposes.

REFERENCES

[1] J. Nishinaga, T. Nagai, T. Sugaya, H. Shibata, S. Niki, Single-crystal Cu(In,Ga)Se2 solar cells grown on GaAs substrates, Appl. Phys. Express 11 (2018), 082302.

[2] A.H. Cottrell, Interactions of dislocations and solute atoms, in: Relation of properties to microstructure, American Society of Metals (1954).

[3] D. Abou-Ras, A. Nikolaeva, M. Krause, L. Korte, H. Stange, R. Mainz, E. Simsek Sanli, P.A. van Aken, T. Sugaya, J. Nishinaga, Optoelectronic inactivity of dislocations in Cu(In,Ga)Se2 thin films, phys. stat. sol. (RRL) 15 (2021) 2100042, doi: 10.1002/pssr.202100042

High-Performance Multi-Junction C-Band Photonic Power Converters: Calibrated Optoelectronic Model for Next Generation Designs

Gavin P Forcade, Meghan N Beattie, Christopher E Valdivia, Henning Helmers, Oliver Höhn, Paige Wilson, Louis-Philippe St-Arnaud, Robert Hunter, David Lackner, Jacob J Krich, Alexandre W Walker, Karin Hinzer

Department of Physics, University of Ottawa, Ottawa, ON, Canada

SUNLAB, School of Electrical Engineering and Computer Sciences, Ottawa, ON, Canada

Fraunhofer Institute for Solar Energy Systems ISE, Freiburg, Germany

Advanced Electronics and Photonics Research Centre, National Research Council of Canada, Ottawa, ON, Canada

Predicting behavior of optoelectronic devices is critical for device design and optimization. Such predictions can be made with a calibrated drift-diffusion model. Recently, photovoltaics using InGaAs as the absorber material, lattice matched to InP, have shown excellent performance in many applications. Further enhancements may be possible by optimizing the designs with an optoelectronic model, calibrated to results from experimental devices. We characterize fabricated InGaAs photonic power converters (PPCs) that have 60 nm, 180 nm, and 540 nm absorber layer thicknesses, using experimental results to calibrate our drift-diffusion model. The calibrated model predicts external quantum efficiencies to better than 1% accuracy and overestimates the open-circuit voltage under 1540 nm illumination by less than 2%. Using this calibrated model, we predict a realistic 1-junction PPC efficiency of up to 46% for a 1540 nm laser, with efficiency limited by carrier collection and series resistance. Improved junction architecture and cell segmentation could help to mitigate both issues. We also fabricated and characterized a 10-junction PPC made of series-connected InGaAs subcells and measured a maximum efficiency of 44% under 1540 nm illumination at 33 W/cm2 power with an open-circuit voltage above 5 V, providing both high efficiency and the voltage required to power electronic circuitry.

Strategies for high Fill Factor and Open-Circuit Voltage in low-doped c-Ge TPV cells with partially contacted surfaces using 3D simulations

M. Gamel, D. Shojaei, J.M. López-González, G. López, M. Garín and I. Martín

Departament d'Enginyeria Electrònica, Universitat Politècnica de Catalunya, Gran Capità s/n, Mòdul C4, 08034 Barcelona, Spain. Pho: +34 93 405 4193, e-mail: mansur.gamel@upc.edu

Abstract —**Dielectric passivated surfaces provide excellent surface recombination properties while point-like contacts are defined to locally extract the current. In this solution, the distance between contacts, or pitch, is critical because it reveals the trade-off between the series resistance and the surface passivation, i.e., between fill factor (*FF*) and open circuit voltage (*V$_{oc}$*). A priori, this trade-off is of paramount importance for low-doped substrates especially under thermophotovoltaic (TPV) conditions, where current densities in the range of 1 A/cm^2 can be obtained. In this study, we investigate strategies to simultaneously achieve high *FF* and *V$_{oc}$* in low-doped c-Ge TPV devices whose rear surface is partially contacted. We use 3D device simulations (Silvaco Atlas) to accurately study the effect of rear pitch under TPV illumination conditions where the photovoltaic device is placed close to a 1473 K thermal emitter. Firstly, we develop a model to introduce surface recombination in the 3D simulations which consists of defining a narrow (1 nm) region in bulk close to the surface with an equivalent lifetime. Next, the 1.2 Ω·cm c-Ge TPV device model with a rear partially contacted surface is investigated. Preliminary results show that optimum pitch can be extended from 75 to 125 µm by improving the saturation current density of the electron selective contact on the front surface. By doing so, apart from the obvious improvement in *V$_{oc}$*, substrate resistivity is reduced due to the increase in excess carrier density at the maximum power point. As a consequence, the trade-off between *FF* and *V$_{oc}$* is relaxed permitting high values for both.**

I. Introduction

Thermophotovoltaic (TPV) energy conversion, *i.e.,* conversion of heat into electricity via radiated photons, has emerged as a promising technology. Typically, most thermal applications radiate at temperature lower than 1200 °C whose emission spectrum is mainly concentrated in the infrared. To harvest infrared photons, TPV cell must utilize narrow bandgap semiconductors which are typically based on III-V compounds. For example, the highest TPV system efficiencies, in the range of 40% [1], are based on InGaAs and AlGaInAs absorbers. The commercial exploitation of this technology is hindered by its high-cost per device area unit (€/cm^2) mainly related to the use of complex epitaxial techniques on III-V costly substrates.

Recently, c-Ge substrates have arisen as a cost-effective alternative for TPV devices with approximately 6 times lower cost than III-V semiconductors [2-3]. An industrially feasible approach to create hole selective contacts that has been successfully developed in p-type c-Si is laser processing aluminum oxide/silicon carbide (Al$_2$O$_3$/SiC) stacks [4]. The surface passivation of low-doped p-type c-Ge by Al$_2$O$_3$ films has been recently demonstrated [5-6] and the transfer of the laser process technique to p-type c-Ge devices is currently underway in our research group. In particular, we use p-type c-Ge substrates with doping concentration and thickness of 1.2 Ω·cm (N_A =2×10^{15} cm^{-3}) and 170 µm, respectively. Figure 1 shows a cross-section of the double-sided c-Ge TPV cell with point-like contact.

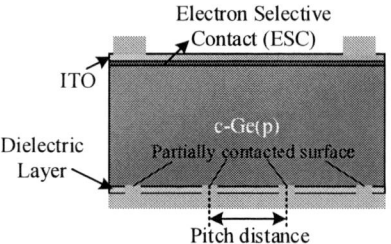

Fig. 1. Double-sided Ge cell, with partially contacted surface.

Generally, point-like contacted surfaces have a trade-off between dielectric passivation (high *V$_{oc}$*) and parasitic series resistance (low *FF*). For good passivation, relatively low-doped substrates, as the one used in this work, are needed. A priori, the optimum rear pitch must be very short to keep R_s under control, especially for TPV devices where typical photogenerated currents can easily reach 1 A/cm^2. However, device resistivity depends not only on the nominal resistivity (doping), but also on the excess carrier density that the device can maintain at the maximum power point. The precise calculation of this effect and its impact on the optimum rear pitch demands 3D device simulations, which will allow us to identify strategies to simultaneously get high *V$_{oc}$* and *FF* values.

II. Discussion and Results

A. Surface recombination in 3D simulations with Silvaco Atlas

Before going into the device simulations, we must introduce surface recombination in the 3D model which is not straightforward because in Silvaco Atlas a surface recombination velocity can be only defined in 2D simulations. For the 3D approach, we define a narrow region with 1-10 nm thickness with an equivalent lifetime so that recombination within this layer equals the surface recombination rate given by the following equation:

$$U_s = \Delta n \cdot S_{eff} \qquad (1)$$

where Δn is the excess carrier density and S_{eff} is the effective surface recombination velocity. To determine the equivalent bulk lifetime in the narrow region, we need to integrate the bulk recombination (U_b) along the region and make it equal to U_s, as shown in equation (2). Figure 2 shows U_b with a relatively low lifetime in a 10 nm region (c-Ge surface at $x = 0$) extracted from the simulator. Notice that the mesh used is 2 nm step (symbols), and consequently, the integral of the bulk recombination should be extended up to 12 nm, including the linear interpolation executed by the simulator between 10 and 12 nm U_b values.

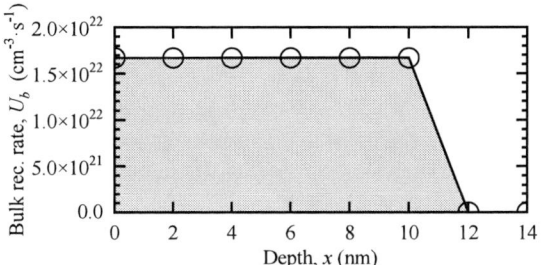

Fig. 2. Bulk recombination distribution calculated by the 3D simulation in a 10 nm region close to the c-Ge surface ($x = 0$).

It is obvious that this integral is the area below the curve indicated in the figure and it is calculated as follows:

$$\Delta n \cdot S_{eff} = \int_{0nm}^{12nm} U_b(x)dx = 11\text{nm} \cdot U_b = 11\text{nm} \cdot \frac{\Delta n}{\tau_b} \quad (2)$$

where τ_b is the bulk lifetime, and we have assumed constant Δn in the 10 nm region. From this equation, we reach an equivalent τ_b that represents the S_{eff} value as $\tau_b = 11\text{nm}/S_{eff}$. In general, the equivalency must include the thickness of the region (d_{reg}) plus half of the next mesh step (m_{step}):

$$\tau_b = \left(d_{reg} + \frac{m_{step}}{2}\right) \cdot \frac{1}{S_{eff}} \quad (3)$$

To validate this equation, we carry out 3D simulations of a 170 μm thick piece of 1.2 Ω·cm p-type c-Ge where narrow regions define the front and rear surface recombination with the corresponding lifetime from equation (3). We simulate the following combination of region widths and mesh steps: 10 nm with 2 nm step, 4 nm with 2 nm step, and 1 nm with 1 nm step. We also change the surface recombination velocity for every simulation using 1, 10, and 100 cm/s, respectively. The excess carrier density is changed by means of the light intensity of a monochromatic beam at 1800 nm, which leads to a constant photogeneration rate along the device. The simulated results are compared to the theoretical expression of the effective surface recombination velocity that considers the surface recombination dependence on excess carrier density. With only interface states at the intrinsic energy level and equal surface recombination velocities for electrons and holes (S) [7]:

$$U_S = \Delta n \cdot S_{eff} = \frac{\Delta n \cdot (N_A + \Delta n) - n_i^2}{S \cdot (\Delta n + n_i) + S \cdot ((N_A + \Delta n) + n_i)} \quad (4)$$

In figure 3, we show the simulated (symbols) and theoretical (line) results with an excellent match, validating the proposed way to introduce surface recombination in the 3D model.

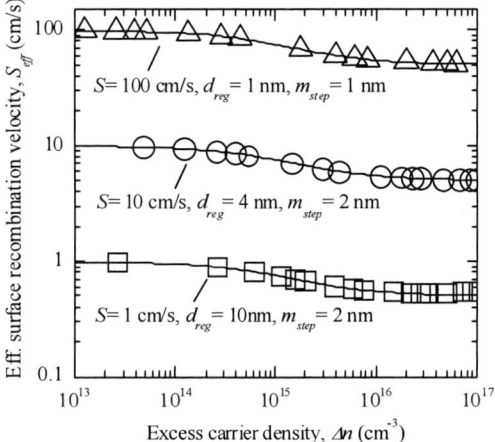

Fig. 3. Simulated (symbols) and theoretical (lines) values for S_{eff} vs. Δn for different S values defined at narrow regions close to the surface.

B. Impact of the pitch distance

The surface recombination model explained in the previous section is integrated into the 3D simulation of the c-Ge device at the dielectric passivated rear surface. In particular, we introduced $S_{eff} = 8$ cm/s while the pointed-like contact is a 10 μm side square with a surface recombination velocity of 1700 cm/s, both experimentally determined. On the other hand, an electron selective contact is defined at the front surface contacted by a 170 nm ITO layer. The ITO's thickness is determined to maximize the photogenerated current from a 1473 K blackbody thermal emitter with a view factor of 0.3. The electron selective contact (ESC) consists of a 50 nm n-type c-Ge material whose quality can be changed by means of the donor density (N_D), leading to the values of hole saturation current density (J_{op}) shown in Table I.

TABLE I
ELECTRON SELECTIVE CONTACT CHARACTERISTICS.

ESC quality	N_D (cm⁻³)	J_{op}(A·cm⁻²)
Low	$2 \cdot 10^{19}$	$1.47 \cdot 10^{-6}$
Medium	$4 \cdot 10^{19}$	$0.47 \cdot 10^{-6}$
High	$6 \cdot 10^{19}$	$0.28 \cdot 10^{-6}$

Figure 4 shows the effect of varying the pitch distance and ESC quality on the Ge TPV parameters (V_{oc}, J_{sc}, FF and P_{out}). Focusing on the J_{sc}, we see that its value is almost constant with pitch except for the longest ones, where the high series resistance reduces it. Regarding V_{oc} and FF, we can see the trade-off between them, but a strong effect of the ESC quality can be seen. The V_{oc} values increase with better ESC quality, as a consequence of the reduction in J_{op}, and the decrease in FF with pitch is smoother. As a result of this relaxed trade-off, the optimum pitch for higher ESC qualities also shifts to 125 μm.

978-1-6654-6060-6/23 $31.00 © 2023 IEEE

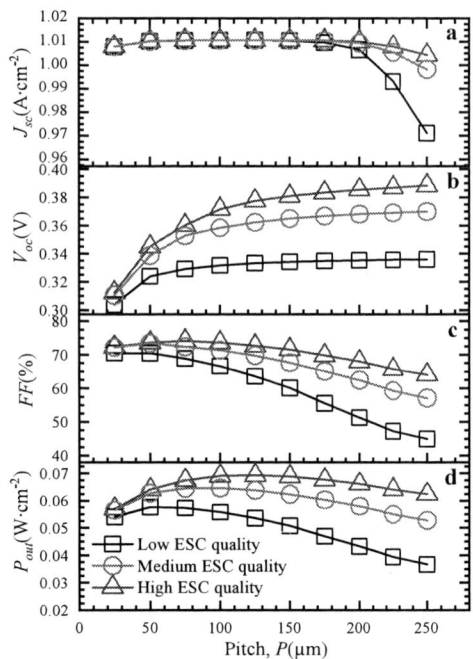

Fig. 4. (a). J_{sc}, (b). V_{oc}, (c). FF and (d). P_{out} as a function of the pitch distance and for the three qualities of ESC explored in this work.

TABLE II
PHOTOVOLTAIC FIGURES WITH THE OPTIMUM PITCH FOR THE THREE ESC QUALITIES EXPLORED

ESC quality	Pitch (μm)	V_{oc} (V)	FF (%)	J_{sc} (A/cm^2)	P_{out} (W/cm^2)
Low	75	0.329	68.92	1.01	0.0573
Medium	100	0.358	71.35	1.01	0.0646
High	125	0.378	72.65	1.01	0.0693

A summary of the obtained photovoltaic figures is shown in Table II. As we can see, the improvement of ESC quality not only leads to better V_{oc} but also better FF, even with longer pitches. This striking result can be linked to the Δn maintained at the maximum power point. The higher ESC performance leads to higher Δn, which reduces substrate resistivity, from its nominal value. This behaviour allows us to obtain high V_{oc} (well-passivated surfaces on low-doped substrates) combined with high FF (low substrate resistivity due to high Δn).

In order to get a deeper look into this effect, we extract the series resistance at the maximum power point by fitting the simulated I-V curve with a simple one-diode model and a lumped series resistance (R_s). Using the 1D model proposed by Fischer [8], we can calculate the theoretical R_s for every pitch depending on the substrate resistivity. In figure 5, we can observe the theoretical curve for the nominal resistivity of the substrate (1.2 $\Omega\cdot$cm). We also plot the R_s values obtained from simulations (symbols) for the three ESC qualities resulting in much lower values. The 1D model can reproduce the observed trends with substrate resistivities of 0.53, 0.31 and 0.20 $\Omega\cdot$cm for low, medium and high ESC qualities, respectively, revealing the reduction in R_s.

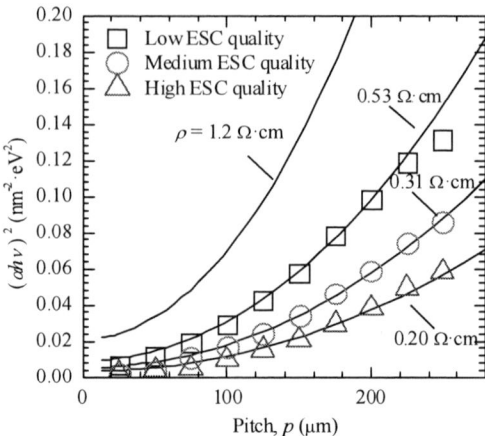

Fig. 5. Evolution of R_s with pitch for the nominal wafer resistivity (ρ= 1.2 $\Omega\cdot$cm) and the equivalent resistivities that fit the simulated values.

IV. CONCLUSION

In this preliminary work, we have demonstrated how the improvement in the electron selective contact not only increases V_{oc} but also positively impacts the FF relaxing the trade-off between them. In the conference, accurate results of the simulations including the effect of substrate resistivity will be reported, with detailed explanations about the identified trends.

ACKNOWLEDGEMENTS

This work has been supported by the Spanish government under project PID2020–115719RB-C21 (GETPV) funded by MCIN/ AEI/10.13039/501100011033.

REFERENCES

[1] A. LaPotin *et al.*, "Thermophotovoltaic efficiency of 40%," *Nature*, vol. 604, no. 7905, pp. 287–291, 2022.

[2] J. Fernández *et al.*, "Back-surface optimization of germanium TPV cells", AIP Conference Proceedings 890, 190–197, 2007.

[3] A. Jiménez, "Development of laser-based diffusion process in Ge for the manufacturing of low-cost thermophotovoltaic cells" (PhD thesis), Universidad Politécnica de Madrid, 2022.

[4] I. Martín *et al.*, " Laser processing of Al$_2$O$_3$/a-SiC$_x$:H stacks: a feasible solution for the rear surface of high-efficiency p-type c-Si solar cells", *Progr. Photovolt: Res. App.* 21, pp. 1171-5, 2013.

[5] W.J.H. Berghuis *et al.*, "Surface passivation of germanium by atomic layer deposited Al$_2$O$_3$ nanolayers", *J. Mater. Res.* 36, 571, 2021.

[6] I. Martín *et al.*, "Effect of the thickness of amorphous silicon carbide interlayer on the passivation of c-Ge surface by aluminium oxide films," *Surf. Interfaces* 31, 102070, 2022.

[7] Martin A. Green, "Silicon solar cells: advanced principles & practice", Centre Photovoltaic Devices and Systems, 179, 1995.

[8] B. Fischer, "Loss analysis of crystalline silicon cells using photoconductance and quantum efficiency measurements" (PhD Thesis), University of Konstanz, 2003.

Minimizing sputter damage-induced electrical losses in monolithic perovskite/silicon tandem solar cells during deposition of the transparent front-electrode

Marlene Härtel, Bor Li, Silvia Mariotti, Philipp Wagner, Florian Ruske, Steve Albrecht, Bernd Szyszka

Helmholtz-Zentrum Berlin, Berlin, Germany

Technische Universität Berlin, Berlin, Germany

The front electrode of monolithically integrated perovskite/silicon tandem solar cells commonly consists of a transparent conductive oxide (TCO). TCOs are usually deposited using the well-established method of magnetron sputtering. High particle energies, however, can cause sputter damage to sensitive substrates during the deposition process. Therefore, a SnO_2 buffer layer is used in all current perovskite top-cell designs with p-i-n polarity and competitive efficiencies. Here, we propose a methodology to identify electrical losses in perovskite solar cells (PSC) caused by sputter damage during the TCO deposition. We also show a simple method for minimizing sputter damage to the PSC, which enables SnO_2 buffer layer-free devices. Evaluation of the ideality factor and pseudo-current density-voltage (J-V) curves, reconstructed from light intensity-dependent J-V measurements on tandem top-cell-equivalent semi-transparent single-junction PSCs, revealed that sputter damage causes transport and non-radiative recombination losses. These losses result in a lower open-circuit voltage (V_{OC}) and fill factor (FF), limiting the PSC performance. By lowering the sputter power density, we reduced the impact of sputter damage. This resulted in improved V_{OC}s (~13 mV) and FFs (~3%) of the semi-transparent PSCs, which is a direct consequence of the reduced electrical losses[1]. Finally, we applied our low-damage sputter approach on SnO_2 buffer layer-free monolithic perovskite/silicon tandem devices. Compared to tandem devices with a SnO_2 buffer layer, the SnO_2 buffer layer-free devices were optically superior, resulting in a device current density improvement of 0.52 mA/cm² due to increased current densities in both sub-cells[1]. This is an important development for further optical performance optimization of tandem devices. [1]   M. Härtel et al., 'Reducing sputter damage-induced recombination losses during deposition of the transparent front-electrode for monolithic perovskite/silicon tandem solar cells', vol. 252, no. October 2022, pp. 1-7, 2023, doi: 10.1016/j.solmat.2023.112180.

Nanometer-Scale Imaging on Electrical Potential in Absorber of As-doped CdSeTe Solar Cells

Chun-Sheng Jiang, Eric Colegrove, Steve P. Harvey, Joel N. Duenow, Matthew O. Reese

National Renewable Energy Laboratory, Golden, CO, United States

CdTe technology has been largely improved by adding Se in region of CdTe absorber near the front interface. Replacing the low Cu-doping by group-V elements has become a major focus for further improving Voc. Indeed, average carrier concentrations have reached ~1016/cm3. However, the best devices remain significantly Voc deficient. Group-V doping can induce defects and affect the front interface in different ways from Cu-doping. In this contribution, we report on a nm-scale electrical potential imaging throughout the As-doped CdSeTe absorber using Kelvin probe force microscopy (KPFM). The potential imaging was conducted both laterally and vertically on beveled films using ion-milling in a small glancing angle. KPFM images electrical potential on the uniformly beveled surface and assesses defect-charging in the subsurface region within a screening length from the beveled surface. We found that grain boundaries are positively charged and there are significant potential fluctuations in both grain boundary versus grain interior and intragrain. We further found that these potential fluctuations decrease significantly toward the front interface. Time of flight secondary ion mass spectrometry imaging shows Se content increases toward the front interface, consistent with Se-passivation of defects. The potential fluctuation was induced by defect-charging, with positive charges by donor defect levels above the Fermi level and with negative charges by acceptor defect level below the Fermi level. Furthermore, the results elucidate different details of defect configurations and grain structures of the films with different CdCl2 treatment temperatures. The defect configurations in the region near the front interface can be a main factor contributing to the device performance difference. Our potential imaging provides insights about the defects throughout the absorber films, and shows that the potential fluctuation has a direct correlation to the Voc deficit.

Advanced Germanium TPV Cells for Latent Heat Thermal Batteries

A. Luque, P. Martin, R. Molinero, V. Orejuela, C. Sanchez-Perez, M. Zehender, I. García, I. Luque-Heredia, I. Rey-Stolle

SILBAT Energy Storage Solutions, Madrid, Spain

Universidad Politécnica de Madrid, Madrid, Spain

Latent heat thermal batteries constitute a novel electricity storage technology that, in combination with thermophotovoltaic cells, can offer drastic cost reductions at elevated energy densities. In these devices, incoming electricity is used to melt a material with a high fusion point, which stores energy in the form of latent heat. When needed, the stored heat is turned back to electricity by a TPV converter. InGaAs TPV cells have produced the highest efficiencies so far, but their cost is so high that hampers the volume development of these batteries. In this work, we present a cost-effective Ge TPV cell design with improved performance because of 1) better PV parameters from improved MOVPE growth and p/n junction passivation; 2) lower series resistance resulting from contact engineering; and 3) better sub-bandgap photon management through improved infrared mirrors and wafer thinning. With these changes, theoretical models forecast that ~30% TPV efficiencies are within reach.

Evaluation of irradiance variability adjustments for subhourly clipping correction

William B. Hobbs[1], Chloe L. Black[1], William F. Holmgren[2], and Kevin S. Anderson[3]

[1]Southern Company, Birmingham, AL, 35203, USA
[2]DNV, San Diego, CA, 92123, USA
[3]Sandia National Laboratories, Albuquerque, NM, 87123, USA

Abstract—Subhourly changes in solar irradiance can lead to energy models being biased high if realistic distributions of irradiance values are not reflected in the resource data and model. This is particularly true in solar facility designs with high inverter loading ratios (ILRs). When resource data with sufficient temporal and spatial resolution is not available for a site, synthetic variability can be added to the data that is available in an attempt to address this issue. In this work, we demonstrate the use of anonymized commercial resource datasets with synthetic variability and compare results with previous estimates of model bias due to inverter clipping and increasing ILR.

Index Terms—photovoltaic, inverter, clipping, modeling, high-frequency, subhourly, irradiance, variability, synthetic, satellite

I. INTRODUCTION

In previous work [1], we showed that typical photovoltaic (PV) system performance modeling workflows based on hourly resource datasets suffer from overprediction bias that increases with increasing inverter loading ratio (ILR). This ILR-dependent model bias was shown to be consistent with the bias expected from subhourly clipping, a normal behavior of real PV systems that hourly simulations are inherently incapable of modeling directly (although adjustment-based workarounds have been explored [2]–[5]). Direct simulation of this effect is possible with the combination of high-frequency resource datasets and performance modeling tools capable of making use of them. Sub-hourly simulation tools have been available for some time in the form of various open-source [6], [7] and proprietary software packages, but high-frequency resource datasets have historically been limited to the few locations with ground-based resource monitoring stations.

In more recent years, enabled by advancements in weather satellite imagery, the "native" time resolution of many satellite-based resource datasets has improved to 5-minute intervals [8]. While this represents a significant improvement, it does not offer a complete solution to direct modeling of subhourly clipping. One problem is that even intervals as short as 5 minutes do not fully capture the fastest components of real-world irradiance fluctuation, which in many climates require minutely or even sub-minutely data to accurately represent. Another limitation is that the current models for producing satellite-based irradiance datasets do not account for some complex secondary phenomena like cloud edge enhancement

and therefore underestimate the upper tail of the distribution of irradiance data [1].

To address these issues, statistical/stochastic adjustments are sometimes applied after the primary irradiance modeling process to generate plausible irradiance signals with higher time resolution and more desirable variability characteristics. Existing methods include the use of Markov chains with transition probability tables [9], implemented in modeling software PV*SOL [10] and resource data from Meteonorm [11]. Markov chains are also used in a subhourly resource data product from Solargis [12]. Random noise based on regional parameters adds variability and clear-sky exceedence to a SolarAnywhere product [13], [14].

However, it remains to be seen whether these variability adjustments successfully address the limitations mentioned previously. In this work, we evaluate the suitability of five such datasets for the purpose of directly simulating the effect of subhourly clipping.

II. METHODS

To evaluate some of these enhanced resource datasets with synthetic variability adjustments, we use the six PV systems from [1], with specifications shown in Table I. We modeled the plants with NREL SAM, version 2022.11.21 [6], matching plant specifications as closely as possible with available models and equipment in SAM, using default options otherwise, and using the Perez transposition model with DNI and GHI as inputs. For each of the sites we used five commercially-available enhanced resolution datasets. In each case, we first modeled a year of plant operation using the enhanced resource dataset at its native time resolution (e.g., 1 or 5 minutes). Next, we averaged the resource dataset to 60 minute intervals and modeled it again. Finally, we compared the two resulting annual energy values. This is analogous to "clip, then average" and "average, then clip" bias described in [15]. We refer to this as simulation bias:

$$\text{bias}_{\text{sim}} = \frac{E_{\text{sim, AtC}} - E_{\text{sim, CtA}}}{E_{\text{sim, CtA}}} \quad (1)$$

where $E_{\text{sim, AtC}}$ is total energy modeled in SAM using resource data that was first averaged to 60 minute intervals and $E_{\text{sim, CtA}}$ is energy from the same model but using the resource data at its native resolution.

978-1-6654-6060-6/23 $31.00 © 2023 IEEE

TABLE I
SUMMARY OF PV SYSTEM CONFIGURATIONS

System	Size [kW$_{dc}$]	ILR	Rack	Location	Year	SFACZ[1]	SVZ[2]
NIST	271.0	1.04	Fixed Tilt	Gaithersburg, Maryland	2018	7	Moderate (low)
NREL	204.1	0.82	Fixed Tilt	Golden, Colorado	2020	4	Moderate (high)
SSRC 30S	2.4	0.80	Fixed Tilt	Birmingham, Alabama	2019	6	Moderate (low)
SSRC 1-Axis	2.4	0.80	Single-Axis Tracking	Birmingham, Alabama	2019	6	Moderate (low)
Commercial Plant 1	4609.2	1.15	Single-Axis Tracking	Southeast US	2018	6	Moderate (low)
Commercial Plant 2	594.4	1.19	Single-Axis Tracking	Southeast US	2020	7	Low

[1] Solar Forecast Arbiter Climate Zone [16]
[2] Solar Variability Zone [17]

We then increased the number of strings behind the inverter in the model, and in some cases modified the inverter model, to evaluate ILRs of 1.2 to 2.0, with intervals of 0.1.

We note that the enhanced resource datasets we used are not from the same calendar year as the previous analyses, however, we expect that interannual variability in subhourly clipping error is small relative to the magnitude of the error based on [18].

As a reference to compare the simulation bias to, we calculated an empirical bias using real power measurements from an inverter. We used the same methods as in [1], with the exception that we did not apply rescaling coefficients in the current work, as the mismatched calendar years between measured power and the enhanced resource datasets make tuning to the modeled output impractical. The empirical approach of using measured power from low ILR systems lets us estimate subhourly clipping bias that varies with ILR without risk of model error, controlling for other effects outside of ILR. Empirical subhourly clipping bias is calculated similarly to simulation bias, using the "average, then clip" vs "clip, then average" method:

$$\text{bias}_{\text{emp}} = \frac{E_{\text{meas,AtC}} - E_{\text{meas,CtA}}}{E_{\text{meas,CtA}}} \qquad (2)$$

where $E_{\text{meas,AtC}}$ is total energy calculated by averaging measured data to 60 minute intervals and then artificially clipping (analogous to what a conventional hourly simulation model would calculate) and $E_{\text{meas,CtA}}$ is energy calculated by first clipping at 1 minute intervals and then averaging to 60 minutes (analogous to what a real system would do). For each system, this bias is evaluated at ILRs of 1.2 through 2.0 in increments of 0.1.

III. RESULTS

Simulated bias results for five datasets (V1-V5) are shown in Fig. 1, along with empirical biases calculated in the same way as in [1]. Tabular data are in Table II. For V1 across all sites except NREL, simulated bias is higher than empirical bias at lower ILR values, and simulated bias is lower than empirical bias at higher ILR values. For V2 across all sites, simulated bias is very low across all ILR values, and V2 bias is lower than V1 bias for all sites and ILR values, with one exception (NIST site at 2.0 ILR, where the two are approximately equal).

For V3-V5 there is a broad range of results, with V3 being significantly higher than other datasets and empirical bias, V4 is similar to V2, and V5 is generally higher than V2 but lower than empirical bias.

IV. CONCLUSION

Our analysis of five enhanced resource datasets using five PV systems indicates that there is a broad range of results for subhourly model bias, with some trends evident. All of the tested subhourly datasets modeled at all ILRs yield lower energy than is obtained from the hourly model, which is the expected general behavior. However, most datasets tend to underestimate subhourly losses and resulting bias at higher ILRs. Exceptions were V3, which notably overestimated bias at all ILRs and most sites, and V1, which tended to slightly overestimate bias at ILRs up to 1.3–1.5, both with the exception of the NREL site. Models using V2 have similar bias to the empirical model at low ILR for most projects, but significantly under predict bias at high ILR. Models using V4 underpredict bias at all ILR greater than 1.2–1.3. Models using V5 also underpredict bias for all but the lowest ILR, but are generally closer to the empirical bias than V4.

Regarding sites analyzed, NREL was an exception in a number of areas. As previously noted in [1], the NREL site had unexplained performance issues, which could account for some differences there. It appears, however, based on empirical and simulated biases being approximately equal to or lower than other sites, that the higher Solar Variability Zone (SVZ) [17] classification does not necessarily indicate higher bias in hourly modeling.

Based on the sites and datasets evaluated here, the authors recommend that users seek validation of synthetic variability datasets. Ideal validation would include multiple relevant sites and ILRs. As an example to support needing validation at multiple sites, V5 matches empirical data well at the NREL site, but appears to overestimate significantly at the other sites. And in support of needing multiple ILRs, for example, V1 and V5 appear to match well for several sites at ILRs of 1.4-1.5, but they underestimate at high ILRs and often overestimate at low ILRs.

Regarding applicability of this work, we recognize that some datasets may be intended for things like ramp rates analysis for battery sizing and grid integration studies, which

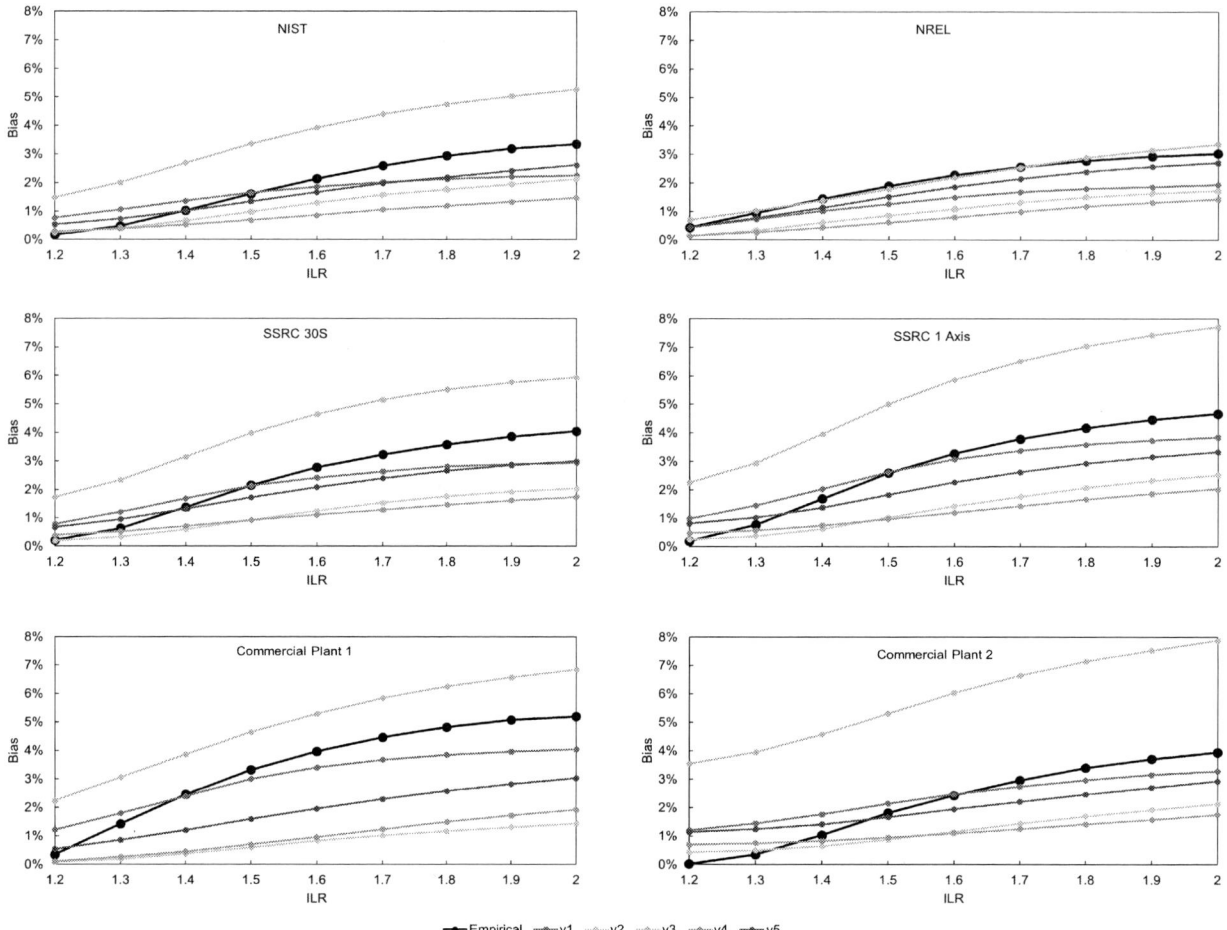

Fig. 1. Comparison of the ILR-dependent simulated and empirical biases for five sites, using between two and five enhanced resource datasets with synthetic variability adjustments, anonymously denoted V1-V5. See Table II for tabular data.

is not something we looked at here. And specific to sub-hourly model bias, we note that inverter-level effects are the primary focus here, and phenomena like plant-level clipping with inverter ac overbuilds are a separate consideration.

Future work could include comparing enhanced resource datasets with comparable standard datasets from the same source (rather than post-processing to get hour averages ourselves) and exploring the reasons for different bias results in the different datasets. We also expect the vendor algorithms to significantly evolve over the next few years and that reevaluation will be needed.

ACKNOWLEDGMENT

This work was supported by the U.S. Department of Energy's Office of Energy Efficiency and Renewable Energy (EERE) under the Solar Energy Technologies Office Award Number 38267. Sandia National Laboratories is a multimission laboratory managed and operated by National Technology & Engineering Solutions of Sandia, LLC, a wholly owned subsidiary of Honeywell International Inc., for the U.S. Department of Energy's National Nuclear Security Administration under contract DE-NA0003525. This paper describes objective technical results and analysis. Any subjective views or opinions that might be expressed in the paper do not necessarily represent the views of the U.S. Department of Energy or the United States Government.

REFERENCES

[1] K. S. Anderson, W. B. Hobbs, W. F. Holmgren, K. R. Perry, M. A. Mikofski, and R. A. Kharait, "The effect of inverter loading ratio on energy estimate bias," in *2022 IEEE 49th Photovoltaics Specialists Conference (PVSC)*, 2022, pp. 0714–0720. [Online]. Available: https://doi.org/10.1109/PVSC48317.2022.9938632

[2] A. Villoz, B. Wittmer, A. Mermoud, M. Oliosi, and A. Bridel-Bertomeu, "A model correcting the effect of sub-hourly irradiance fluctuations on overload clipping losses in hourly simulations," in *8th World Conference on Photovoltaic Energy Conversion*, 2022.

[3] K. Anderson and K. Perry, "Estimating subhourly inverter clipping loss from satellite-derived irradiance data," in *2020 47th IEEE Photovoltaic Specialists Conference (PVSC)*. IEEE, Jun. 2020. [Online]. Available: https://doi.org/10.1109/pvsc45281.2020.9300750

[4] A. Parikh, K. Perry, K. Anderson, W. B. Hobbs, R. Kharait, and M. A. Mikofski, "Validation of subhourly clipping loss error corrections," in *2021 IEEE 48th Photovoltaic Specialists Conference (PVSC)*. IEEE, Jun. 2021. [Online]. Available: https://doi.org/10.1109/pvsc43889.2021.9518564

[5] M. Ghiz, "Hourly modeling corrections for PV energy assessments," *2022 15th PV Performance Modeling Workshop Salt Lake City, Utah*

TABLE II

TABULAR BIAS RESULTS. SEE FIG. 1 FOR A GRAPHICAL VERSION.

Site	Dataset	ILR								
		1.2	1.3	1.4	1.5	1.6	1.7	1.8	1.9	2
NIST	V1	0.77%	1.06%	1.36%	1.65%	1.85%	2.02%	2.12%	2.21%	2.24%
	V2	0.25%	0.40%	0.67%	0.98%	1.29%	1.58%	1.76%	1.94%	2.12%
	V3	1.49%	2.00%	2.69%	3.36%	3.92%	4.40%	4.74%	5.04%	5.27%
	V4	0.29%	0.39%	0.52%	0.69%	0.86%	1.06%	1.18%	1.33%	1.47%
	V5	0.54%	0.73%	1.01%	1.34%	1.66%	1.98%	2.19%	2.41%	2.62%
	Empirical	0.17%	0.48%	1.03%	1.60%	2.14%	2.59%	2.94%	3.19%	3.35%
NREL	V1	0.44%	0.73%	1.02%	1.26%	1.50%	1.68%	1.79%	1.86%	1.94%
	V2	0.14%	0.34%	0.61%	0.86%	1.09%	1.31%	1.50%	1.64%	1.73%
	V3	0.71%	1.03%	1.38%	1.78%	2.18%	2.54%	2.89%	3.14%	3.34%
	V4	0.15%	0.27%	0.43%	0.61%	0.80%	0.99%	1.17%	1.32%	1.43%
	V5	0.44%	0.77%	1.13%	1.51%	1.86%	2.15%	2.39%	2.57%	2.70%
	Empirical	0.44%	0.95%	1.45%	1.89%	2.28%	2.56%	2.77%	2.93%	3.03%
SSRC S30	V1	0.78%	1.21%	1.69%	2.14%	2.41%	2.63%	2.81%	2.88%	2.94%
	V2	0.19%	0.34%	0.58%	0.91%	1.25%	1.54%	1.76%	1.92%	2.04%
	V3	1.74%	2.34%	3.15%	3.98%	4.64%	5.15%	5.50%	5.75%	5.93%
	V4	0.39%	0.51%	0.71%	0.92%	1.11%	1.29%	1.47%	1.61%	1.74%
	V5	0.67%	0.94%	1.32%	1.73%	2.08%	2.40%	2.66%	2.85%	2.99%
	Empirical	0.21%	0.63%	1.38%	2.15%	2.78%	3.22%	3.57%	3.86%	4.04%
SSRC 1-Axis	V1	0.98%	1.46%	2.03%	2.62%	3.07%	3.37%	3.59%	3.74%	3.84%
	V2	0.25%	0.37%	0.61%	1.03%	1.43%	1.76%	2.08%	2.32%	2.51%
	V3	2.25%	2.94%	3.95%	5.00%	5.86%	6.51%	7.04%	7.43%	7.71%
	V4	0.47%	0.57%	0.74%	0.97%	1.20%	1.43%	1.67%	1.87%	2.03%
	V5	0.81%	1.02%	1.38%	1.83%	2.27%	2.62%	2.92%	3.15%	3.33%
	Empirical	0.19%	0.77%	1.69%	2.59%	3.27%	3.78%	4.17%	4.46%	4.67%
Commercial Plant 1	V1	1.21%	1.79%	2.39%	3.00%	3.40%	3.67%	3.84%	3.96%	4.04%
	V2	0.09%	0.18%	0.37%	0.59%	0.83%	1.01%	1.16%	1.30%	1.43%
	V3	2.24%	3.06%	3.86%	4.64%	5.29%	5.83%	6.24%	6.56%	6.84%
	V4	0.10%	0.26%	0.45%	0.70%	0.95%	1.23%	1.49%	1.72%	1.92%
	V5	0.53%	0.85%	1.20%	1.59%	1.95%	2.29%	2.58%	2.82%	3.03%
	Empirical	0.34%	1.42%	2.45%	3.32%	3.97%	4.46%	4.81%	5.07%	5.20%
Commercial Plant 2	V1	1.20%	1.45%	1.76%	2.14%	2.47%	2.74%	2.97%	3.15%	3.28%
	V2	0.43%	0.50%	0.64%	0.87%	1.15%	1.44%	1.69%	1.92%	2.13%
	V3	3.54%	3.95%	4.57%	5.31%	6.03%	6.64%	7.14%	7.54%	7.89%
	V4	0.70%	0.75%	0.83%	0.95%	1.09%	1.25%	1.42%	1.58%	1.75%
	V5	1.13%	1.24%	1.41%	1.66%	1.94%	2.20%	2.47%	2.71%	2.93%
	Empirical	0.02%	0.35%	1.03%	1.81%	2.43%	2.96%	3.40%	3.71%	3.94%

USA, 2022. [Online]. Available: https://pvpmc.sandia.gov/download/8540/

[6] P. Gilman, A. Dobos, N. DiOrio, J. Freeman, S. Janzou, and D. Ryberg, "SAM photovoltaic model technical reference update," National Renewable Energy Laboratory, Golden, CO, Tech. Rep. NREL/TP-6A20-67399, 2018. [Online]. Available: https://doi.org/10.2172/1429291

[7] W. F. Holmgren, C. W. Hansen, and M. A. Mikofski, "pvlib python: a python package for modeling solar energy systems," *Journal of Open Source Software*, vol. 3, no. 29, p. 884, 2018. [Online]. Available: http://doi.org/10.21105/joss.00884

[8] M. Sengupta, A. Habte, Y. Xie, G. Buster, B. Benton, and M. Bannister, "Recent updates to the national solar radiation database (NSRDB)," in *EMS Annual Meeting 2022, Bonn, Germany*. Copernicus GmbH, Sep. 2022. [Online]. Available: https://doi.org/10.5194/ems2022-611

[9] M. Hofmann, S. Riechelmann, C. Crisosto, R. Mubarak, and G. Seckmeyer, "Improved synthesis of global irradiance with one-minute resolution for PV system simulations," *International Journal of Photoenergy*, vol. 2014, 2014. [Online]. Available: https://doi.org/10.1155/2014/808509

[10] S. Lindemann, "PV*SOL overview for PV modeling," Sandia PV Performance Modeling Collaborative, 2017. [Online]. Available: https://www.slideshare.net/sandiaecis/11-presentation-valentinsoftware20170509

[11] J. Remund, "Comparison of radiation models generating minute values," Sandia PV Performance Modeling Collaborative, 2017. [Online]. Available: http://www.slideshare.net/sandiaecis/generation-of-one-minute-data

[12] "Stochastic generation of 1-minute solar radiation data," 2019. [Online].

Available: https://solargis.com/docs/methodology/1-minute-solar-data

[13] J. Huang, R. Perez, J. Schlemmer, M. Perez, A. Bhat, P. Keelin, and A. Kubiniec, "Enhancing temporal variability of 5-minute satellite-derived solar irradiance data," in *2022 IEEE 49th Photovoltaics Specialists Conference (PVSC)*, 2022, pp. 0314–0318. [Online]. Available: https://doi.org/10.1109/PVSC48317.2022.9938759

[14] "SolarAnywhere® High-resolution data with True Dynamics™," Jul 2022. [Online]. Available: https://www.solaranywhere.com/support/historical-data/high-resolution/

[15] J. O. Allen and W. B. Hobbs, "The effect of short-term inverter saturation on modeled hourly PV output using minute DC power measurements," *Journal of Renewable and Sustainable Energy*, vol. 14, no. 6, p. 063503, 2022. [Online]. Available: https://doi.org/10.1063/5.0130265

[16] C. W. Hansen, W. F. Holmgren, A. Tuohy, J. Sharp, A. T. Lorenzo, L. J. Boeman, and A. Golnas, "The solar forecast arbiter: An open source evaluation framework for solar forecasting," in *2019 IEEE 46th Photovoltaic Specialists Conference (PVSC)*, 2019, pp. 2452–2457. [Online]. Available: https://doi.org/10.1109/PVSC40753.2019.8980713

[17] M. Lave, R. J. Broderick, and M. J. Reno, "Solar variability zones: Satellite-derived zones that represent high-frequency ground variability," *Solar Energy*, vol. 151, pp. 119–128, 2017. [Online]. Available: https://doi.org/10.1016/j.solener.2017.05.005

[18] M. Bolen and J. O. Allen, "Improved photovoltaic (PV) plant energy production prediction (phase 1): The effect of short term inverter saturation on PV performance modeling," EPRI, Palo Alto, CA, Tech. Rep. 3002014718, 2018. [Online]. Available: https://www.epri.com/research/products/000000003002014718

978-1-6654-6060-6/23 $31.00 © 2023 IEEE

Evaluation of beta-phase formation in the failure of PVDF-based solar module backsheets

Stephanie L. Moffitt, Sona Ulicna, Song-Syun Jhang, Michael Owen-Bellini, Peter Hacke, Jared Tracy, Kaushik R. Choudhury, Laura T. Schelhas, Xiaohong Gu

National Institute of Standards and Technology, Gaithersburg, MD, United States

National Renewable Energy Laboratory, Golden, CO, United States

E. I. du Pont de Nemours and Company, Wilmington, DE, United States

There are concerns about the potential for polyvinylidene fluoride (PVDF)-based backsheets to crack and fail prematurely. Several studies have suggested that polymer phase change may play a role in the failure mechanism. Here we utilize wide-angle X-ray scattering maps to show that alpha- to beta-phase transformations occurs at cracks tips in aged PVDF-based backsheets. In addition, our work demonstrates that beta-phase formation is not required for crack growth and only occurs at crack tips when plastic deformation of the PVDF polymer has occurred.

Development of a Dynamic Photovoltaic Inverter Model with Grid-Support Capabilities for Power System Integration Analysis

Rachid Darbali-Zamora

Sandia National Laboratories, Albuquerque, New Mexico, 87185, USA

Abstract – As the interest in distributed energy resources (DERs) grows and more photovoltaic (PV) inverters are connected into the power grid, standards are being developed to tackle the high penetration of DERs. Newer DERs are required to provide grid-support functionality (GSF) to aid in regulating both voltage and frequency. With these advances in PV inverter technology, there exists a need for developing reliable dynamic models that emulate the behavior of these commercial devices with GSF capabilities. Herein, this paper presents a PV inverter model with GSF for power system analysis. The proposed model is composed of a dynamic mathematical PV module model, a state-space equations-based DC/DC converter tasked with maximum power point tracking and active power curtailment and a current controlled average PV inverter model. The model is designed to provide Volt-Var Curve and Frequency-Watt Curve control as well as to operate with voltage and frequency ride-through capabilities. The proposed PV inverter model is simulated utilizing *MATLAB/Simulink*. Simulation results are demonstrated for the proposed PV inverter model operating under various GSFs.

Index Terms – photovoltaic inverter, distribution systems, modelling, circuit analysis, grid-support, simulation

I. INTRODUCTION

The use of variable generation from distributed energy resources (DERs) can have adverse effects on the electric grid, causing fluctuations in both frequency and voltage [1]. These adverse effects can cause power system reliability issues. With recent advancements in grid-support functions (GSF) and interoperability standardization, devices such as photovoltaic (PV) inverters can manage active and reactive power to help compensate any frequency and voltage deviations caused by renewable energy sources (RES). Worldwide, efforts are being made to incorporate additional GSF that will allow voltage and frequency regulation. Grid codes and interconnection standards such as IEEE 1547-2018 require DERs to provide GSF [2], [3]. There has been extensive research in PV inverters GSF that provide ancillary services for voltage and frequency regulation utilizing techniques such Frequency-Watt Curve (FWC) and Volt-Var Curve (VVC) control [4], [5], [6]. These techniques allow a PV inverter to provide active or reactive power based on the measured grid frequency or voltage level at the point of common coupling (PCC). To avoid sudden loss of generation and increase system reliability, standards are also requiring DERs to operate even when subjected to temporary disturbances. Voltage ride-through (VRT) and frequency ride-through (FRT) enable DERs to continue operating even when voltage and frequency stray away from normal operating conditions [7], [8].

As PV inverter technology becomes more advanced, there is a need for developing dynamic models that emulate the behavior of commercial devices, including GSF. In some cases, PV inverter models with GSF capabilities do not represent the PV module (PVM) and utilize ideal sources, without considering PVM power limitations [9], [10]. When these models are implemented in power systems integration studies, the DC-side dynamics of the PVM are completely neglected, and it's assumed that changes in irradiance do not significantly alter the behavior of the PV inverters output, e.g. voltage, current, and power. Thus, there is no representation of the PVM. Hence, most PV inverter models don't utilize irradiance profile as an input variable [11], [12], although the effects of resource variability can be represented by varying the active power provided to the PV inverter. This omits the Maximum Power Point Tracking (MPPT) portion altogether, assuming that the voltage reference signal of the PV inverter is equal at all times to the Maximum Power Point (MPP) value at any given irradiance. Previous PV inverter models have been developed to provide active and reactive power based on a reference set points [13], and GSF for voltage regulation [14], but did not consider the PVM. Leveraging these prior works and a PVM model [15], a dynamic PV inverter model with GSFs as specified by IEEE 1547-2018 is proposed.

Herein, this paper presents a PV inverter model with GSF capabilities for power systems analysis. The proposed PV inverter model is composed of a dynamic PVM model, state-space equations-based DC/DC converter tasked with executing MPPT and active power curtailment (APC) algorithms, a three-phase current controlled average PV inverter model with GSF capabilities and a three-phase RLC filter. The proposed PV inverter model is simulated utilizing *MATLAB/Simulink*. Simulation results are presented for the PV inverter model operating under MPPT, VVC with reactive power priority (RPP), FWC with APC, VRT, and FRT conditions to demonstrate the various GSF capabilities.

II. PROPOSED PHOTOVOLTAIC INVERTER MODEL

The proposed PV inverter model is composed of a PVM, a state-space based DC/DC converter, an average three-phase inverter, and a three-phase RLC filter. The dynamic PVM mathematical model uses irradiance as well as temperature as input variables. The state-space equation-based DC/DC converter allows regulating the PVM voltage to control its output power. Two algorithms are employed to either operate the PVM at MPPT, extracting the utmost available power from the PVM or operating in APC, commanding a desired reduced active power value. The PV inverter model is able to provide VVC, FWC, VRT and FRT functionality. Fig. 1 illustrates the diagram of the proposed PV inverter model.

978-1-6654-6060-6/23 $31.00 © 2023 IEEE

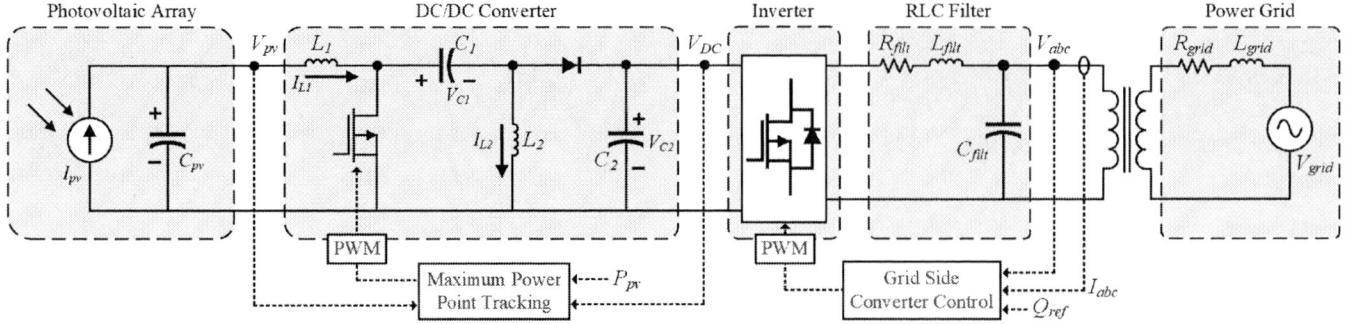

Fig. 1. Block Diagram of the Proposed Photovoltaic Inverter Simulation Model.

III. PHOTOVOLTAIC MODULE MATHEMATICAL MODEL

The power for the PV inverter is provided by a mathematical PVM model. The PVM is emulated using a dynamic exponential equation [16], as shown in equation (1).

$$I_{pv}(V) = \frac{p \cdot I_x}{1 - exp\left(-1/b\right)} \cdot \left[1 - exp\left(\frac{V_{pv}}{s \cdot b \cdot V_x} - \frac{1}{b}\right)\right] \quad (1)$$

In this equation, the variable V is the PVM voltage, and the variable b is the PV characteristic constant. The algorithm that generates the characteristic constant b is given by equation (2).

$$b_{n+1} = \frac{V_{op} - V_{oc}}{V_{oc} \cdot ln\left[1 - \frac{I_{op}}{I_{sc}} \cdot \left(1 - exp\left(-1/b_n\right)\right)\right]} \quad (2)$$

The constants p and s represent the number of PVM connected in series and in parallel, respectively. The variables V_x and I_x are obtained using the manufacturer's datasheet. The expression used to obtain the V_x is shown in equation (3).

$$V_x = \frac{E_{iN}}{E_i} \cdot TCv \cdot (T - T_N) + V_{max} - (V_{max} - V_{min}) \cdot$$
$$exp\left[\frac{E_i}{E_{iN}} \cdot ln\left(\frac{V_{max} - V_{oc}}{V_{max} - V_{min}}\right)\right] \quad (3)$$

In this equation, the variable V_{max} is the open-circuit voltage at 25°C at an irradiance of 1,200 W/m². The variable V_{min} is the open-circuit voltage at 25°C obtained at an irradiance of less than 200 W/m². V_{max} is approximately $1.03 \cdot V_{oc}$ and V_{min} is approximately $0.85 \cdot V_{oc}$. E_i is the effective irradiance in W/m². The variable TCv is the temperature coefficient of V_{oc} measured in V/°C. The variable T is the PV temperature in °C. The constants T_N is 25 °C and the nominal effective irradiance, E_{iN} is 1,000 W/m². The variable V_{oc} is the open-circuit voltage at a temperature level of 25 °C and irradiance of 1,000 W/m². Variable I_x can be calculated using equation (4).

$$I_x = \frac{E_i}{E_{iN}} \cdot [I_{sc} + TCi \cdot (T - T_N)] \quad (4)$$

In the previous equation, TCi is the temperature coefficient of the I_{sc} in A/°C. The multiplication of the PVM voltage with equation (1) yields the PVM power as shown in equation (5).

$$P(V) = \frac{V_{pv} \cdot p \cdot I_x}{1 - exp\left(-1/b\right)} \cdot \left[1 - exp\left(\frac{V_{pv}}{s \cdot b \cdot V_x} - \frac{1}{b}\right)\right] \quad (5)$$

The maximum power, P_{max} can be obtained by a PVM when it is operating at the optimal voltage, V_{op}. The expression for maximum available power that can be extracted from a PVM is shown in equation (6).

$$P_{max}(V_{op}) = \frac{p \cdot V_{op} \cdot I_x}{1 - exp\left(-1/b\right)} \cdot \left[1 - exp\left(\frac{V_{op}}{s \cdot b \cdot V_x} - \frac{1}{b}\right)\right] \quad (6)$$

These simulated results are compared with experimental results from a *mono-Si* PVM. The data sheet provides the values for V_{oc} and the I_{sc} of an individual PVM, 21.70 V and 8.429 A, respectively. The characteristic constant b of the PVM used is 0.04. Fig. 2 (a) shows the P-V curve obtained from the PVM at varying irradiance. Fig. 2 (b) shows the I-V curve obtained from the PVM at varying irradiance.

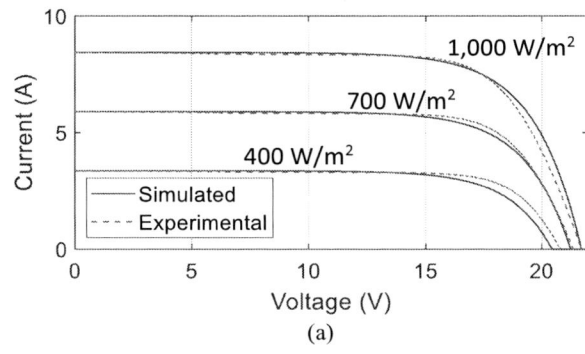

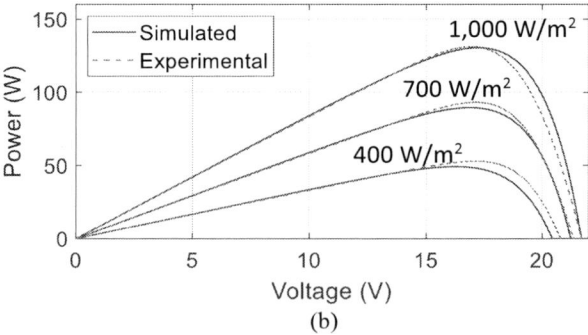

Fig. 2. Photovoltaic Module Mathematical Model at Varying Irradiance conditions. (a) I-V Curve. (b) P-V Curve.

IV. DC/DC CONVERTER TOPOLOGY

For the proposed PV inverter model, the Single Ended Primary Inductor Converter (SEPIC) is used to either perform the MPPT or to curtail active power. The state-space equations that describe the SEPICs behavior are shown in equation (7).

$$
\begin{bmatrix} \frac{dI_{L1}}{dt} \\ \frac{dI_{L2}}{dt} \\ \frac{dV_{C1}}{dt} \\ \frac{dV_{C2}}{dt} \end{bmatrix} = \begin{bmatrix} 0 & 0 & \frac{S-1}{L_1} & \frac{S-1}{L_1} \\ 0 & 0 & \frac{S}{L_2} & \frac{(S-1)}{L_2} \\ \frac{(S-1)}{C_1} & -\frac{S}{C_1} & 0 & 0 \\ \frac{(S-1)}{C_2} & \frac{(S-1)}{C_2} & 0 & \frac{-1}{RC_2} \end{bmatrix} \cdot \begin{bmatrix} I_{L1} \\ I_{L2} \\ V_{C1} \\ V_{C2} \end{bmatrix} + \begin{bmatrix} \frac{1}{L_1} \\ 0 \\ 0 \\ 0 \end{bmatrix} \cdot (V_{in}) \quad (7)
$$

In these equations, S is the status of the switch, I_{L1} is the current through the inductor L_1 while I_{L2} is the current through the inductor L_2. V_{C1} is the voltage across the capacitor C_1 and V_{C2} is the voltage across the capacitor C_2. If the duty cycle, D is an averaged S, the relationship between the input and output voltage can be derived, shown in equation (8).

$$
V_{out} = V_{in} \cdot \left(\frac{D}{1-D}\right) \quad (8)
$$

In this expression, V_{out} is the voltage at the output terminal and V_{in} is the input voltage.

A. Maximum Power Point Tracking (MPPT)

The optimal duty ratio (ODR) MPPT method is used to obtain the optimal duty cycle of a DC/DC converter [17]. To operate at the MPP, the PVM must be at the optimal voltage V_{op}. as shown in equation (9).

$$
D_{op} = \frac{V_{out}}{V_{out} + V_{op}} \quad (9)
$$

To calculate the optimal voltage V_{op}, the linear reoriented coordinate method (LRCM) is employed [18]. The LRCM is used as a MPPT technique to obtain the optimal voltage for a PVM and is defined by equation (10).

$$
V_{op} = V_x + b \cdot V_x \cdot \ln\left(b - b \cdot exp\left(-\frac{1}{b}\right)\right) \quad (10)
$$

Evaluating the optimal voltage in the PV current equation (1), yields the optimal current I_{op}, shown in equation (11).

$$
I_{op} = \frac{I_x}{1 - exp\left(\frac{-1}{b}\right)} \cdot \left[1 - b + b \cdot exp\left(\frac{-1}{b}\right)\right] \quad (11)
$$

The maximum power is obtained by multiplying the optimal voltage and current values, as shown in equation (12).

$$
P_{max}(V_{op}) = \frac{I_x \cdot V_x}{1 - exp\left(\frac{-1}{b}\right)} \cdot \left[1 - b + b \cdot exp\left(\frac{-1}{b}\right)\right]
$$
$$
\cdot \left[V_x + b \cdot V_x \cdot \ln\left(b - b \cdot exp\left(\frac{-1}{b}\right)\right)\right] \quad (12)
$$

B. Active Power Curtailment

When curtailing active power, the MPPT is halted and a voltage corresponding to the curtailed active power is used [19]. APC is required when there is not enough headroom to provide the requested reactive power due to the PV inverter operating at its rated power at maximum irradiance. This may also occur when the PV inverter is not at rated power but the active power and requested reactive power Q_{ref}, exceeds the PV inverter power rating S_{rtd}. The curtailed active power can be calculated from equation (13).

$$
P_{curt} = \sqrt{S_{rtd}^2 - Q_{ref}^2} \quad (13)
$$

To curtail active power, the duty cycle for the DC/DC converter must operate at the curtailed voltage, V_{curt} that yields the PV inverters curtailed active power, as shown in equation (14).

$$
D_{curt} = \frac{V_{out}}{V_{out} + V_{curt}} \quad (14)
$$

To calculate the curtailed voltage V_{curt}, that will yield the curtailed active power can be obtained from equation (15).

$$
V_{curt} = V_{op} + \left[\frac{(V_{op} \cdot I_{op} - P)}{(V_{op} \cdot I_{op})} \cdot (V_x - V_{op})^\beta\right]^{1/\beta} \quad (15)
$$

The variable β can be obtained from equation (16).

$$
\beta = \frac{(V_{op} - V_x)}{(V_{op} \cdot I_{op})} \cdot \left[\frac{\partial P(V_x)}{\partial V}\bigg|_{V = V_x}\right] \quad (16)
$$

The partial derivative of the PV power shown in equation (5), in terms of the PV voltage is defined in equation (17).

$$
\frac{\partial P(V)}{\partial V} = I(V) + V \cdot \frac{\partial I(V)}{\partial V} \quad (17)
$$

The partial derivative of the PV current shown in equation (1), in terms of to the PV voltage is defined in equation (18).

$$
\frac{\partial I(V)}{\partial V} = \frac{I(V)}{b \cdot V_x} + \frac{I_x}{b - b \cdot exp\left(-1/b\right)} \quad (18)
$$

Replacing equation (18) in equation (17) yields equation (19).

$$
\frac{\partial P(V)}{\partial V} = I(V) + \frac{V \cdot I(V)}{b \cdot V_x} + \frac{V \cdot I_x}{b - b \cdot exp\left(-1/b\right)} \quad (19)
$$

Note that when replacing V_x, the PV current $I(V)$ is zero as shown in equation (20).

$$
\frac{\partial P(V_x)}{\partial V}\bigg|_{V = V_x} = -\frac{I_x}{b - b \cdot exp\left(-1/b\right)} \quad (20)
$$

Evaluating the curtailed voltage V_{curt}, in equation (1), yields the corresponding curtailed current, as shown in equation (21).

$$I_{curt}(V_{curt}) = \frac{I_x}{1 - exp\left(-\frac{1}{b}\right)} \cdot \left[1 - exp\left(\frac{V_{curt}}{b \cdot V_x} - \frac{1}{b}\right)\right] \quad (21)$$

The curtailed active power can be obtained by multiplying the curtailed voltage and current values, as shown in equation (22).

$$P_{curt}(V_{curt}) = \frac{V_{curt} \cdot I_x}{1 - exp\left(-\frac{1}{b}\right)} \cdot \left[1 - exp\left(\frac{V_{curt}}{b \cdot V_x} - \frac{1}{b}\right)\right] \quad (22)$$

V. PHOTOVOLTAIC INVERTER MODEL

The PV inverter models uses an equivalent per phase circuit of a PV inverter model with RLC output filter is derived. The control is represented using the direct (d) and quadrature (q) axis reference frame, and a current controller [20], [21]. Equation (23) illustrates the PV inverter models dq control.

$$\begin{bmatrix} \dot{I_d} \\ \dot{I_q} \end{bmatrix} = \begin{bmatrix} \frac{-R}{L} & 0 \\ 0 & \frac{-R}{L} \end{bmatrix} \cdot \begin{bmatrix} I_d \\ I_q \end{bmatrix} + \frac{1}{L} \cdot \begin{bmatrix} M_d \\ M_q \end{bmatrix} \quad (23)$$

In equation (23), the filter effects are included in the modulation voltages M_d and M_q, shown in equation (24) and equation (25), respectively.

$$M_d = V_d + \omega \cdot L \cdot I_q - V_{ds} \quad (24)$$

$$M_q = V_q - \omega \cdot L \cdot I_d - V_{qs} \quad (25)$$

Adjusting the d and q axis currents allows controlling the active and reactive power, respectively. The active and reactive setpoints are determined by the specified GSF settings.

A. Volt-Var Curve

VVC is an autonomous control that provides voltage regulation based on the relationship between voltage and reactive power. The VVC is generated by interpolating between the voltage measured at the PCC V_{pcc}, and available reactive power as shown in equations (26).

$$\left.\begin{array}{ll} Q_{ref} = Q_1 = Q_{max} & v_1 \leq v_{pcc} \\ Q_{ref} = \frac{(Q_2 - Q_1)}{(v_2 - v_1)} \cdot (v_{pcc} - v_1) + Q_1 & v_1 < v_{pcc} < v_2 \\ Q_{ref} = 0 & v_2 \leq v_{pcc} \leq v_3 \\ Q_{ref} = \frac{(Q_4 - Q_3)}{(v_4 - v_3)} \cdot (v_{pcc} - v_3) + Q_3 & v_3 < v_{pcc} < v_4 \\ Q_{ref} = Q_4 = Q_{min} & v_4 \leq v_{pcc} \end{array}\right\} \quad (26)$$

Fig. 3 illustrates an example of a standard VVC. Notice that the slope of the VVC determines the aggressiveness of the reactive power that can be delivered and absorbed by the PV inverter.

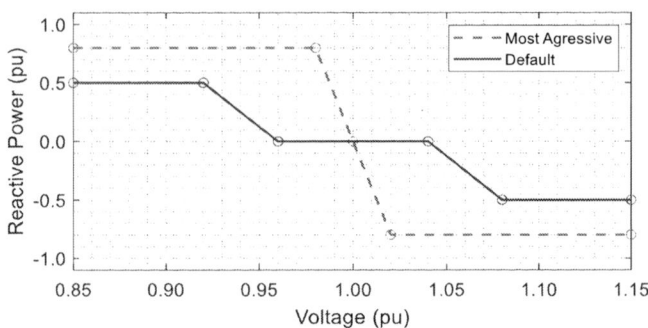

Fig. 3. Diagram of a VVC Illustrating a Default (Blue) and Most Aggressive (Red) Curves.

B. Frequency-Watt Curve

FWC is an autonomous control that provides frequency regulation based on the relationship between frequency and active power. The FWC is generated by interpolating between the frequency measured at the PCC f_{pcc}, and available active power as shown in equations (27).

$$\left.\begin{array}{ll} P_{ref} = P_2 = P_{max} & f_1 \leq f_{pcc} \\ P_{ref} = \frac{(P_3 - P_2)}{(f_3 - f_2)} \cdot (f_{pcc} - f_2) + P_2 & f_1 < f_{pcc} < f_2 \\ P_{ref} = 0 & f_2 \leq f_{pcc} \leq f_3 \\ P_{ref} = \frac{(P_5 - P_4)}{(f_5 - f_4)} \cdot (f_{pcc} - f_4) + P_4 & f_3 < f_{pcc} < f_4 \\ P_{ref} = P_6 = P_{min} & f_4 \leq f_{pcc} \end{array}\right\} \quad (27)$$

Fig. 4 illustrates an example of a standard FWC. Notice that the slope of the FWC determines the aggressiveness of the active power that can be delivered by the PV inverter.

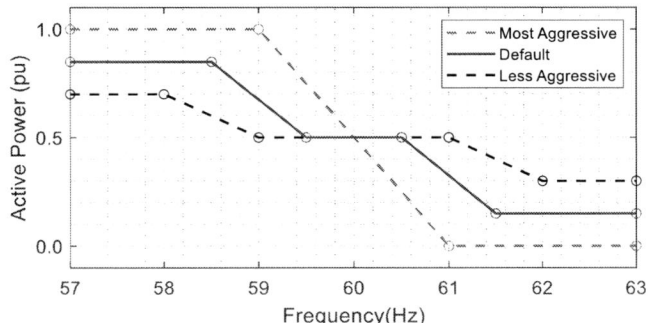

Fig. 4. Diagram of a FWC Illustrating a Default (Blue), Less Aggressive (Black) and Most Aggressive (Red) Curves.

C. Voltage Ride-Through

The objective of the VRT is to allow the PV inverter to continue operating during momentary voltage disturbances that stray outside of the PV inverter's normal operating conditions. This requirement was implemented to avoid a sudden loss of PV inverter generation under tolerable voltage conditions. Fig. 5 illustrates a diagram of the VRT characteristics specified by the IEEE Std. 1547-2018.

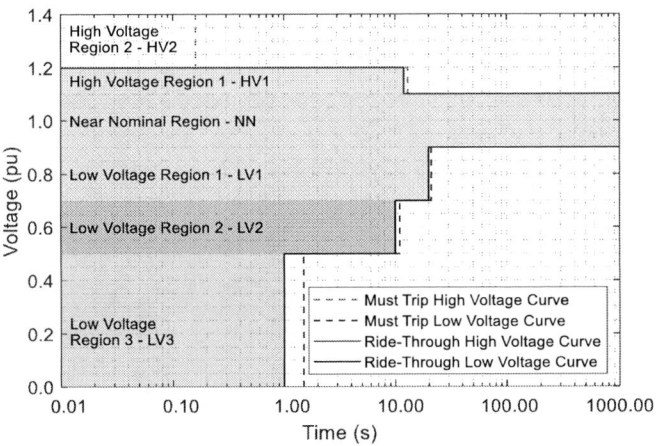

Fig. 5. Diagram of the IEEE Std. 1547-2018 VRT Requirements.

Note that voltage deviations tend to propagate throughout an entire distribution system. Both low voltage ride-through (LVRT) and high voltage ride-through (HVRT) functions are implemented by measuring the voltage at the PCC of the PV inverter and initiating a counter that compares the stored counter time with the voltage ride-through times specified by IEEE Std. 1547.1. When the PV inverter senses a change in voltage at the PCC, depending on the specified IEEE Std. 1547.1 settings, it will either go into momentary cessation, ride-through the event or disconnect from the utility.

D. Frequency Ride-Through

The FRT allows the PV inverter to continue operating during momentary frequency disturbances. Note that in most cases frequency is effectively uniform across an entire distribution system. Fig. 6 illustrates a diagram of the FRT characteristics specified by the IEEE Std. 1547-2018. The FRT characteristic ride-through trip time requirements are summarized in Table II. Both low frequency ride-through (LFRT) and high voltage ride-through (HFRT) functions are implemented by measuring the frequency at the PCC of the PV inverter and initiating a counter that compares the stored counter time with the times specified by IEEE Std. 1547.1. When the PV inverter senses a change in frequency at the PCC, it will either go into momentary cessation, ride-through the event or disconnect from the utility.

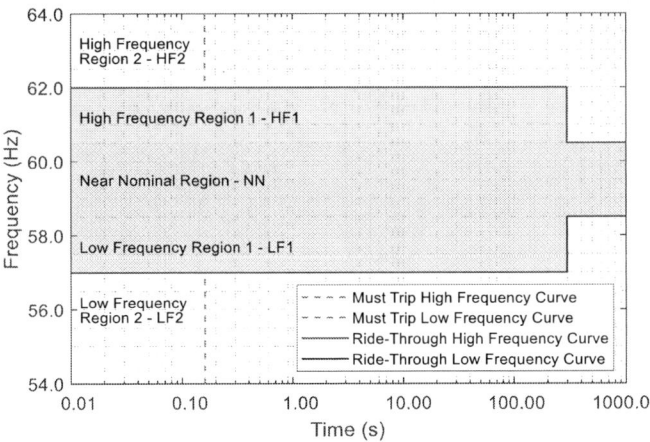

Fig. 6. Diagram of the IEEE Std. 1547-2018 FRT Requirements.

E. Reactive Power Priority

The RPP functionality curtails active power, prioritizing reactive power generation or absorption, achieved by adhering to the power relationship between active, reactive, and apparent power as shown in equation (28).

$$S_{rtd} = \sqrt{\left(P_{ref}\right)^2 + \left(Q_{ref}\right)^2} \qquad (28)$$

In this equation, the variables P_{ref} and Q_{ref}, are the reference active and reactive powers commanded by the enabled GSF, respectively. Solving for the active power yields the RPP relationship, shown in equation (29).

$$P_{ref} = \sqrt{(S_{rtd})^2 - \left(Q_{ref}\right)^2} \qquad (29)$$

F. Active Power Priority

The APP functionality allows the PV inverter to provide a certain degree of reactive power, while prioritizing active power generation. This means that there is no curtailment of active power to provide reactive power. Solving for the reactive power yields the APP relationship, shown in equation (30).

$$Q_{ref} = \pm\sqrt{(S_{rtd})^2 - \left(P_{ref}\right)^2} \qquad (30)$$

VI. SIMULATION RESULTS

This section summarizes the simulation results obtained from testing the PV inverters GSF. The voltage and frequency at the PCC of the PV inverter model are varied to test the dynamic performance of developed model. Tables I through III summarize the parameters for the PVM, the DC/DC converter, and the PV inverter, respectively.

TABLE I: PHOTOVOLTAIC MODULE PARAMETERS

Parameter	Description	Value	Unit
I_{sc}	Short-Circuit Current	5.69	A
V_{oc}	Open-Circuit Voltage	64.2	V
b	Characteristic Curvature Constant	0.4	-
s	Modules in Series	5	-
p	Modules in Parallel	66	-

TABLE II: DC/DC CONVERTER PARAMETERS

Parameter	Description	Value	Unit
C_o	Input Capacitor	10	mF
C_1	Coupling Capacitor	1	mF
C_2	Output Capacitor	25	mF
L_1	Input Inductor	0.05	mH
L_2	Output Inductor	0.05	mH

TABLE III: PHOTOVOLTAIC INVERTER PARAMETERS

Parameter	Description	Value	Unit
P_{nom}	Nominal Power	100	kW
F_{nom}	Nominal Frequency	60	Hz
V_{AC}	Terminal AC Voltage	480	V
V_{DC}	Nominal DC Voltage	500	V
R_{filt}	Filter Resistance	1.9	mΩ
L_{filt}	Filter Inductance	100	uH
C_{filt}	Filter Capacitance	20	μF

A. Maximum Power Point Tracking Under Varying Irradiance

This simulation demonstrates that the ODR is able to calculate the reference MPP values in order for the DC/DC converter to execute the MPPT. Fig. 7 shows PVM power. Fig. 8 shows the PVM voltage. Fig. 9 shows the PV inverter RMS current. Fig. 10 shows the PV inverter RMS voltage. Fig. 11 shows the PV inverter active power.

B. Volt-Var Curve Under Varying Voltage

In this simulation the VVC is enabled with RPP to demonstrate that the PV inverter can regulate voltage. Fig. 12 shows PVM power. Fig. 13 shows the PVM voltage. Fig. 14 shows the PV inverter RMS current. Fig. 15 shows the PV inverter RMS voltage. Fig. 16 and Fig. 17 shows the PV inverter active and reactive power, respectively.

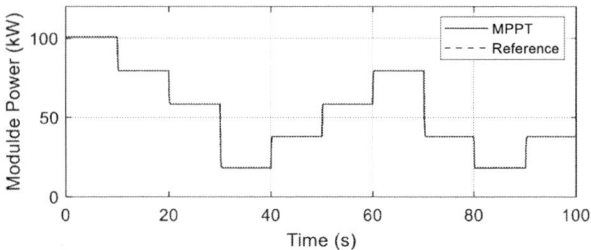

Fig. 7. Results for the Photovoltaic Module Power.

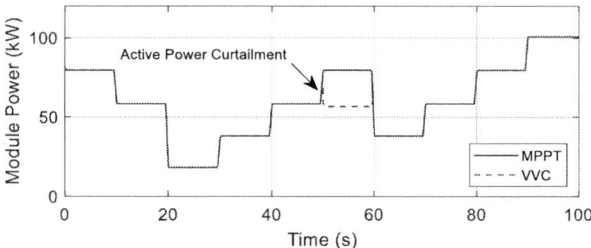

Fig. 12. Results for the Photovoltaic Module Power.

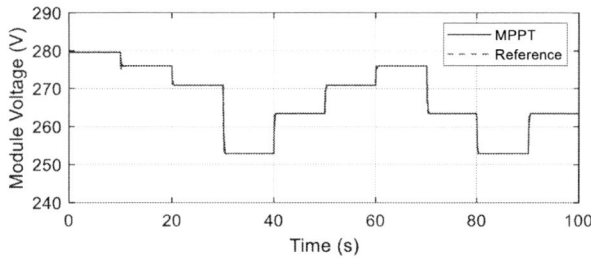

Fig. 8. Results for the Photovoltaic Module Voltage.

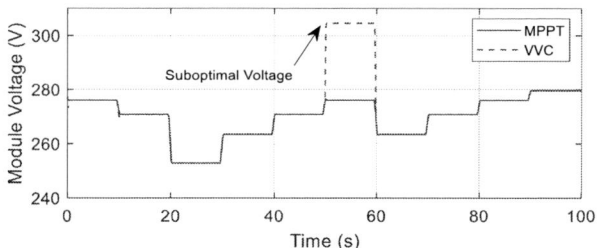

Fig. 13. Results for the Photovoltaic Module Voltage.

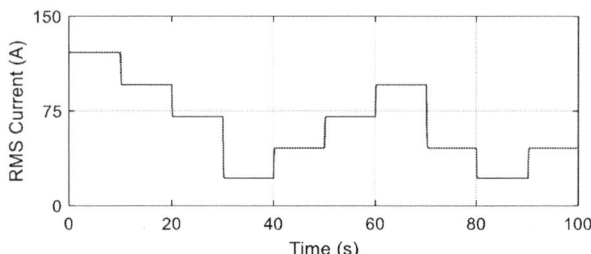

Fig. 9. Results for the Photovoltaic Inverter RMS Current.

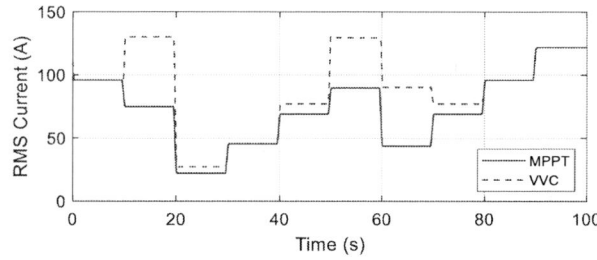

Fig. 14. Results for the Photovoltaic Inverter RMS Current.

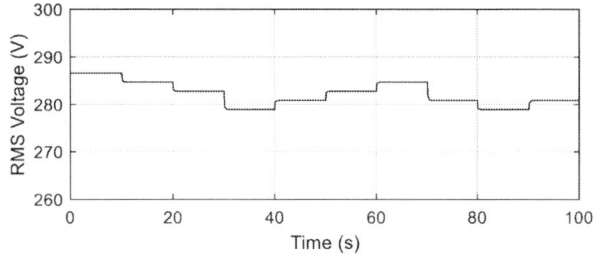

Fig. 10. Results for the Photovoltaic Inverter RMS Voltage.

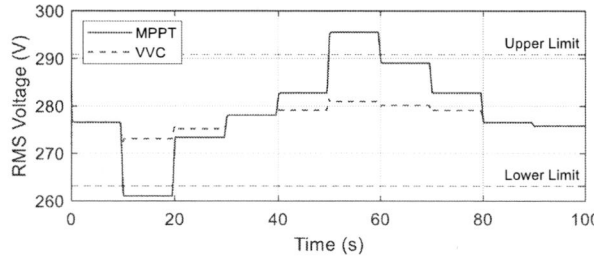

Fig. 15. Results for the Photovoltaic Inverter RMS Voltage.

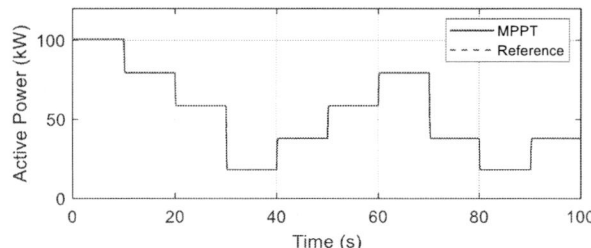

Fig. 11. Results for the Photovoltaic Inverter Active Power.

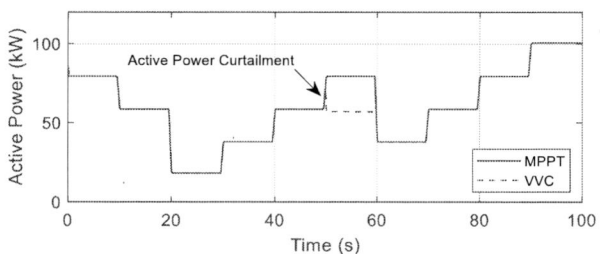

Fig. 16. Results for the Photovoltaic Inverter Active Power.

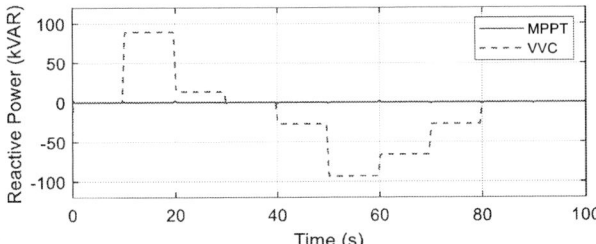

Fig. 17. Results for the Photovoltaic Inverter Reactive Power.

C. Frequency-Watt Curve Under Varying Frequency

In this simulation, the FWC is enabled with APC at a constant 75 kW to demonstrate that the PV inverter can provide frequency regulation. Fig. 18 shows PVM power. Fig. 19 shows the PVM voltage. Fig. 20 shows the PV inverter RMS current. Fig. 21 illustrates the PV inverter frequency. Fig. 22 shows the PV inverter active power.

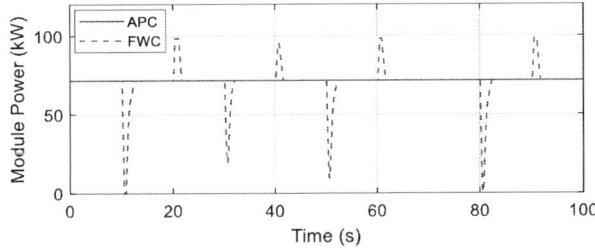

Fig. 18. Results for the Photovoltaic Module Power.

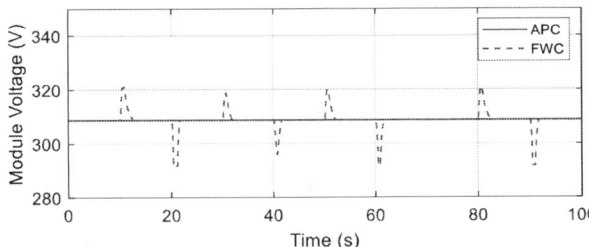

Fig. 19. Results for the Photovoltaic Module Voltage.

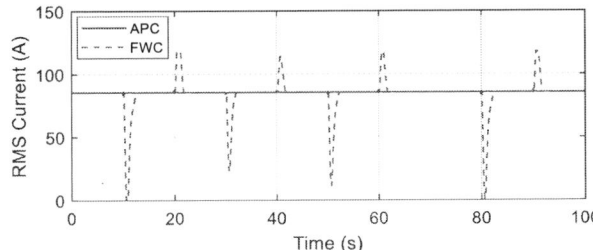

Fig. 20. Results for the Photovoltaic Inverter RMS Current.

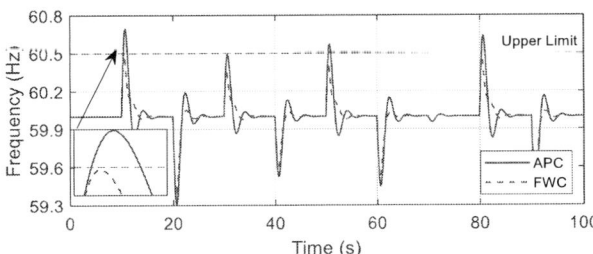

Fig. 21. Results for the Photovoltaic Inverter Frequency.

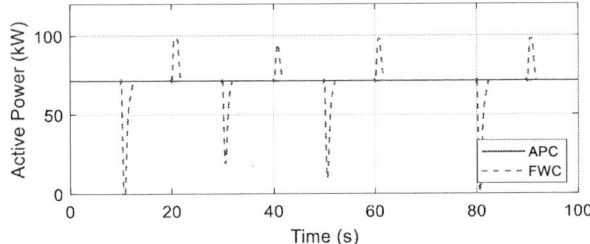

Fig. 22. Results for the Photovoltaic Inverter Active Power.

D. Voltage Ride-Through at Varying Voltage

In this simulation, the PV inverter is subjected to varying voltage conditions to demonstrate VRT capabilities. Fig. 23 illustrates the PV inverter RMS voltage. In these results, the PV inverter initially operates at NN (277.1 V) and is then subjected to LV1 (207.8 V), LV2 (180.1 V) and HV1 (218.7 V) voltages. Fig. 24 illustrates the PV inverters active power. When the PV inverter is subjected to LV1 (trips in 20 s), LV2 (trips in 10 s) and HV1 (trips in 12.9 s), the PV inverter is able to perform the required VRT before tripping. Results demonstrate that when the PV inverter is subjected to momentary voltage deviations, the PV inverter performs the VRT, but when subjected to voltage deviations for a prolonged time, the PV inverter disconnects.

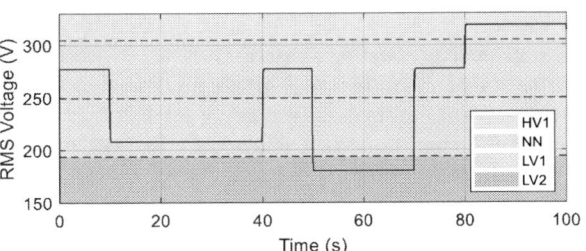

Fig. 23. Results for the Photovoltaic Inverter RMS Voltage.

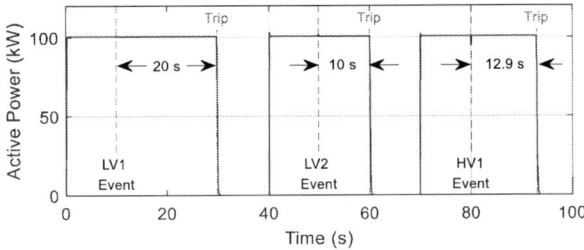

Fig. 24. Results for the Photovoltaic Inverter Active Power.

E. Frequency Ride-Through at Varying Frequency

In this simulation, the PV inverter is subjected to varying frequency conditions in order to demonstrate FRT capabilities. Fig. 25 illustrates the PV inverter frequency. In these results, the PV inverter initially operates at NN (60 Hz) and is then subjected to LF1 (58 Hz) and HF2 (62.1 Hz) frequencies. Fig. 26 illustrates the PV inverters active power. When the PV inverter is subjected to LF1 (trips in 299 s) and HF2 (trips in 015 s), the PV inverter is able to perform the required FRT before tripping. Results demonstrate that when the PV inverter is subjected to momentary frequency deviations, the PV inverter performs the FRT, but when subjected to frequency deviations for a prolonged time, the PV inverter disconnects.

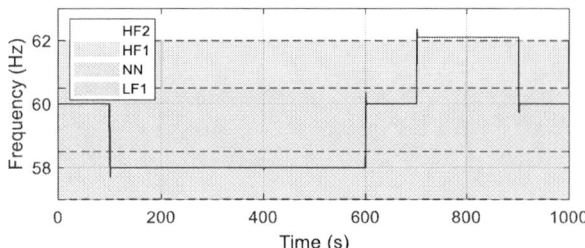

Fig. 25. Results for the Photovoltaic Inverter Frequency.

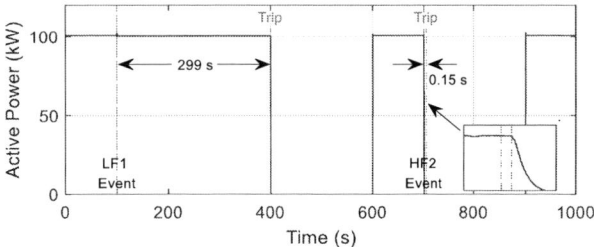

Fig. 26. Results for the Photovoltaic Inverter Active Power.

VII. CONCLUSION

This paper presents a dynamic PV inverter model with the ability of executing various GSF capabilities under varying grid conditions for power system analysis. The proposed PV inverter model is composed of a dynamic mathematical PVM model, state-space equations-based DC/DC converter and current controlled *dq*-frame three-phase inverter model. The mathematical PVM has the ability to apply at varying irradiance and temperature conditions. The state-space average model of the DC/DC SEPIC is able to execute the optimal duty ratio MPPT. A mathematical method for determining the suboptimal voltage that results in the desired curtailment of active power is also presented. This approach calculates the suboptimal duty ratio for the DC/DC SEPIC which yields the curtailed active power. The three-phase average inverter model utilizes a current controlled *dq*-frame scheme. The presented simulation results demonstrate that the PV inverter is able to perform the MPPT, extracting the maximum power from the PVM. Simulation results demonstrate how the PV inverter model is able to regulate voltage and frequency, utilizing VVC with RPP and FWC with APC, respectively. Simulation results demonstrate that the PV inverter model is able to perform VRT and FRT as specified by IEEE 1547 Std.

ACKNOWLEDGEMENT

This article has been authored by an employee of National Technology & Engineering Solutions of Sandia, LLC under Contract No. DE-NA0003525 with the U.S. DOE. The employee owns all right, title and interest in and to the article and is solely responsible for its contents. The U.S. Government retains and the publisher, by accepting the article for publication, acknowledges that the U.S. Government retains a non-exclusive, paid-up, irrevocable, world-wide license to publish or reproduce the published form of this article or allow others to do so, for U.S. Government purposes. The DOE will provide public access to these results of federally sponsored research in accordance with the DOE Public Access Plan.

REFERENCES

[1] R. Darbali-Zamora *et al.*, "Evaluation of Photovoltaic Inverters Under Balanced and Unbalanced Voltage Phase Angle Jump Conditions", *47th IEEE Photovoltaic Specialists Conference (PVSC)*, 2020, pp. 1562-1569.

[2] "IEEE Standard for Interconnection and Interoperability of Distributed Energy Resources with Associated Electric Power Systems Interfaces", IEEE Std 1547-2018 (Revision of IEEE Std 1547-2003), pp.1-138, Apr. 2018.

[3] "IEEE Standard Conformance Test Procedures for Equipment Interconnecting Distributed Energy Resources with Electric Power Systems and Associated Interfaces", IEEE Std 1547.1-2020, pp.1-282, Jan. 2020.

[4] J. Johnson *et al.*, "Photovoltaic Frequency–Watt Curve Design for Frequency Regulation and Fast Contingency Reserves", *IEEE Journal of Photovoltaics*, vol. 6, no. 6, pp. 1611-1618.

[5] J. Johnson, S. Gonzalez, and D. B. Arnold, "Experimental Distribution Circuit Voltage Regulation using DER Power Factor, Volt-Var, and Extremum Seeking Control Methods", *44th IEEE Photovoltaic Specialist Conference (PVSC)*, 2017.

[6] N. Ninad *et al.*, "PV Inverter Grid Support Function Assessment using Open-Source IEEE P1547.1 Test Package", *47th IEEE Photovoltaic Specialists Conference (PVSC)*, 2020, pp. 1138-1144.

[7] M. Ropp, D. Schultz, J. Neely, and S. Gonzalez, "Effect of grid support functions and VRT/FRT capability on autonomous anti-islanding schemes in photovoltaic converters", *43rd IEEE Photovoltaic Specialists Conference (PVSC)*, 2016, pp. 1853-1856.

[8] E. Desarden-Carrero, R. Darbali-Zamora and E. E. Aponte-Bezares, "Analysis of Grid Support Functionality Dynamics under Ride-Through Requirements Using Power-Hardware-in-the-Loop Implementation", *48th IEEE Photovoltaic Specialists Conference (PVSC)*, 2021, pp. 1795-1802.

[9] W. Du *et al.*, "Voltage-Source Control of PV Inverter in a CERTS Microgrid", *IEEE Transactions on Power Delivery*, vol. 29, no. 4, pp. 1726-1734, Aug. 2014.

[10] N. Guruwacharya *et al.*, "Modeling Inverters with Grid Support Functions for Power System Dynamics Studies", *IEEE Power & Energy Society Innovative Smart Grid Technologies Conference (ISGT)*, 2021, pp. 1-5.

[11] D. Ramasubramanian, V. Vittal, and J. M. Undrill, "Transient stability analysis of an all converter interfaced generation WECC system", *2016 Power Systems Computation Conference (PSCC)*, 2016, pp. 1-7.

[12] H. N. V. Pico and B. B. Johnson, "Transient Stability Assessment of Multi-Machine Multi-Converter Power Systems", *IEEE Transactions on Power Systems*, vol. 34, no. 5, pp. 3504-3514, Sept. 2019.

[13] R. Darbali-Zamora *et al.*, "Distribution Feeder Fault Comparison Utilizing a Real-Time Power Hardware-in-the-Loop Approach for Photovoltaic System Applications", *46th IEEE Photovoltaic Specialists Conference (PVSC)*, 2019, pp. 2916-2922

[14] J. Johnson *et al.*, "Distribution Voltage Regulation Using Extremum Seeking Control with Power Hardware-in-the-Loop", *IEEE Journal of Photovoltaics*, vol. 8, no. 6, pp. 1824-1832, Nov. 2018.

[15] R. Darbali-Zamora *et al.*, "Exponential Phase-Locked Loop Photovoltaic Model for PHIL Applications", *IEEE ANDESCON*, 2018, pp. 1-6.

[16] E. I. Ortiz-Rivera and F. Z. Peng, "Analytical Model for a Photovoltaic Module using the Electrical Characteristics provided by the Manufacturer Data Sheet", *36th IEEE Power Electronics Specialists Conference*, 2005, pp. 2087-2091.

[17] R. Darbali-Zamora and E. I. Ortiz-Rivera, "Optimal duty ratio maximum power point tracking technique using the SEPIC topology for photovoltaic systems applications", *IEEE ANDESCON*, 2016.

[18] E. I. Ortiz Rivera and F. Z. Peng, "Linear Reoriented Coordinates Method", *IEEE International Conference on Electro/Information Technology*, 2006, pp. 459-464.

[19] Y. León-Ruiz *et al.*, "Fast dc Link Voltage Controller under Large Irradiance Variations in a PV Single- Phase Grid-Connected Inverter", *13th IEEE International Conference on Power Electronics and Drive Systems (PEDS)*, 2019, pp. 1-6.

[20] A. Yazdani; R. Iravani, "Two-Level, Three-Phase Voltage-Sourced Converter" *Voltage-Sourced Converters in Power Systems: Modeling, Control, and Applications*, IEEE, 2010, pp.115-126.

[21] N. E. Saavedra-Peña *et al.*, "Development of Photovoltaic Inverter Model with Islanding Detection Using the Sandia Frequency Shift Method", *49th IEEE Photovoltaics Specialists Conference (PVSC)*, 2022, pp. 0398-0404.

978-1-6654-6060-6/23 $31.00 © 2023 IEEE

PV Fleet Modeling via
Smooth Periodic Gaussian Copula

Mehmet G. Ogut[1], Bennet Meyers[2], and Stephen P. Boyd[1]

[1] Stanford University, Stanford, CA, 94305, USA

[2] SLAC National Accelerator Laboratory, Menlo Park, CA, 94025, USA

Abstract—We present a method for jointly modeling power generation from a fleet of photovoltaic (PV) systems. We propose a white-box method that finds a function that invertibly maps vector time-series data to independent and identically distributed standard normal variables. The proposed method, based on a novel approach for fitting a smooth, periodic copula transform to data, captures many aspects of the data such as diurnal variation in the distribution of power output, dependencies among different PV systems, and dependencies across time. It consists of interpretable steps and is scalable to many systems. The resulting joint probability model of PV fleet output across systems and time can be used to generate synthetic data, impute missing data, perform anomaly detection, and make forecasts. In this paper, we explain the method and demonstrate these applications.

Index Terms—photovoltaic systems, photovoltaic fleet modeling, distributed power generation, power generation planning, forecasting, convex optimization, copula method, probability distributions, forecast uncertainty

I. Introduction

Modeling the power output of a fleet of photovoltaic (PV) systems is of great importance for digital operations and maintenance in the PV sector, which is now a multi-billion dollar industry [1]. Applications include predicting power production, detecting anomalies, and making informed decisions about when and where to send workers to service a site. In recent years, there has been a significant increase in the deployment of PV systems, making it necessary to develop scalable models that can handle thousands of systems simultaneously, while staying robust to real-world data challenges, such as the ability to handle missing data.

In this paper we propose a method to estimate the joint probability distribution of the power outputs of a fleet of PV systems, modeling all relevant correlations in the data—across individual PV systems and across time. As a copula method, we first develop a novel set of nonlinear marginal transforms that map the power from each system to a scalar Gaussian, and then develop a set of linear transformations that model the marginally transformed data as a large joint Gaussian distribution. We interact with that model to carry out various applications including synthetic data generation, data imputation, anomaly detection, and forecasting.

II. Prior work

a) Fleet models: Modeling PV systems for operations and maintenance (O&M) purposes based on measured data

has a long history [2]–[7]. These techniques focus on predicting either the maximum power point or the full current-voltage relationship of a PV system under a given set of environmental conditions. O&M tasks are then carried out using the model. Fault detection is performed by comparing actual system power generation to the predicted power from the model. Forecasting future PV system power output is done by running the models on predicted weather trends, such as those generated by NOAA [8]. 'Fleet modeling' is the practice of constructing an independent, bespoke model for each system in a PV fleet and is quite labor intensive. More recently, researchers have attempted to reduce the effort of fleet modeling using systematic approaches such as learning algorithms and machine inference [9]. These approaches tend to be more task dependent, *e.g.*, focusing on the task of anomaly detection [10]–[12] or forecasting [13]–[19]. While these methods reduce the human effort to create a PV system model, they do not provide *joint* models of system behavior in a fleet. Several recent papers have explored models to predict aggregate quantities of fleets, such as temporal variability and maximum feed-in power [20]–[23].

b) Copula models and Gaussianization: Our proposed method draws on previous work on both copula models and Gaussianization methods.

Copulas are tools for modeling dependence of several random variables, first proposed by Abe Sklar in 1959 [24] and recently translated into English in [25]. The basic idea is to apply a nonlinear invertible mapping to each component of a random variable so it has some standard distribution such as uniform or Gaussian, and then model the dependence of these transformed variables [26]–[28]. Copulas have been applied in many domains [29]–[33], including PV data analysis [34]. Copula models may be based on theoretical constructions (*e.g.*, the multivariate Gaussian copula) or may be learned directly from data [27]. The *quantile transform* is a typical choice for data-driven models of the marginal distributions, and various options exist [35], [36]. Our method includes an autoregressive component in the copula model, an approach that has been explored by other authors [37].

An alternative approach to modeling multivariate joint distributions is *Gaussianization*, also called 'normalizing flows'. These methods seek an invertible mapping under which the transformed variable has a standard (jointly) Gaussian distribution [38]–[40]. The transformations are typically built in steps, so the transformation is a composition of multiple

978-1-6654-6060-6/23 $31.00 © 2023 IEEE

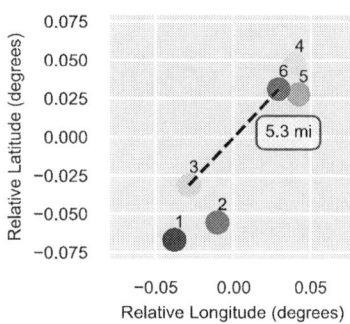

Fig. 1: Relative location of the six PV systems.

transformations, with the distribution of mapped variables getting closer to Gaussian as more layers or steps are added. We can think of a Gaussian copula as a simple first step in such a normalizing flow. While a normalization method can in principle model any probability distribution, Gaussian copula models cannot. On the other hand a Gaussian copula model makes several operations such as conditioning on some known values very easy, involving just basic linear algebra. (These computions can be done for more complex normalizing flows, but they are much more involved, *e.g.*, requiring Monte Carlo or other sampling type methods.)

c) Modeling via convex optimization: Our method relies on convex optimization in every step. This guarantees efficient algorithms for finding global solutions [41], and mature tools exist to easily specify convex optimization problems in code [42], [43]. Many traditional statistical models rely on convex optimization for fitting such as regression [44, Chap. 12], auto-regressive (AR) models [44, Ch. 13], and fitting Gaussian distributions to data [41, §3.5], among many others. Our method is inspired by recent work on the trade-off of fit versus roughness in stratified Gaussian models [45], [46], as well as work on convex optimization based signal decomposition [47].

III. DATA

We will illustrate our method on PV fleet data provided by SunPower Corporation under a nondisclosure agreement. We select six PV systems located in Southern California, with three grouped in Santa Ana, CA and three grouped in the hills to the east in Tustin, CA. The relative locations are shown in figure 1. This choice of system locations was intentional, as we wanted to verify that our model captures similarity of power profiles for nearby systems.

The data consist of 15-minute (average) power values (in kW) for each of the six systems, recorded from 3/1/2017 to 3/31/2017. Figure 2 depicts the power output of the six systems over the three day period between 3/4/2017 and 3/6/2017. At night the power output is zero; during daylight hours, we see different types of power profiles. On 3/6/2017 we see clear-sky behavior, characterized by a smooth increase until noon

followed by a gradual decrease until evening. On 3/5/2017, however, we see the power generation curves with multiple dips and peaks throughout the day, which can be attributed to weather factors such as passing clouds. The maximum power output of the systems varies, with system 4 peaking at around 9 kW, and the other systems peaking at around 2 kW.

We denote the data as $y_t \in \mathbf{R}^d$, with $d = 6$, and the time index running from $t = 1$ to $t = T = 2976$. This particular data set does not have any missing data. However, our method gracefully handles missing values, and indeed, relies on this ability to choose hyper-parameters by cross-validation.

We also use data for the following 2 weeks, from 4/1/2017 to 4/14/2017 as our test set for validating our models and applications. This data was not used to fit our model. The test data set has index running from $t = 1$ to $T^{\text{test}} = 1344$.

IV. METHOD

We propose a method for fitting the given data $y_1, \ldots, y_T$ to a smooth 24-hour-periodic stochastic process. We apply a sequence of three invertible transformations so that the transformed data is approximately a standard Gaussian. These transformations, which respect periodicity and are constructed to vary smoothly across time, are applied in the three steps shown in figure 3. First we use a smooth periodic nonlinear transform to make the data approximately marginally Gaussian. This allows us to model the changing distribution over a day, in addition to the differing maximum values seen across systems. In the second step we use an autoregressive (AR) model to account for dependencies across time. This results in a residual that is approximately uncorrelated across time. Finally, we fit a smooth periodic Gaussian distribution to the residual of the AR model; from this we can whiten the residual so that it is approximately a standard Gaussian. In the language of copula modeling, our first step is our marginal transformation and the final two steps constitute our copula (*i.e.*, 'linking') function.

A. Fitting a smooth periodic model

Here we describe the general technique, used in steps 1 and 3 of our method, for fitting a smooth P-periodic parameter, given by $\theta_1, \ldots, \theta_T \in \Theta \subseteq \mathbf{R}^m$ to some data, where Θ is a convex set of allowed parameter values. We will use a Fourier series with K harmonics to represent θ,

$$\theta_t = \sum_{k=0}^{K} \left(\cos\left(\frac{2\pi kt}{P}\right) \alpha_k + \sin\left(\frac{2\pi kt}{P}\right) \beta_k \right), \quad (1)$$

for $t = 1, \ldots, P$, where $\alpha_k, \beta_k \in \mathbf{R}^m$ are the (vector) coefficients that define θ.

We take as a measure of smoothness the Dirichlet energy,

$$\mathcal{D} = \frac{(2\pi)^2}{P} \sum_{k=1}^{K} k^2 \left(\|\alpha_k\|_2^2 + \|\beta_k\|_2^2 \right).$$

The loss function has the form

$$\mathcal{L} = \sum_{t=1}^{T} \ell_t(\theta_t),$$

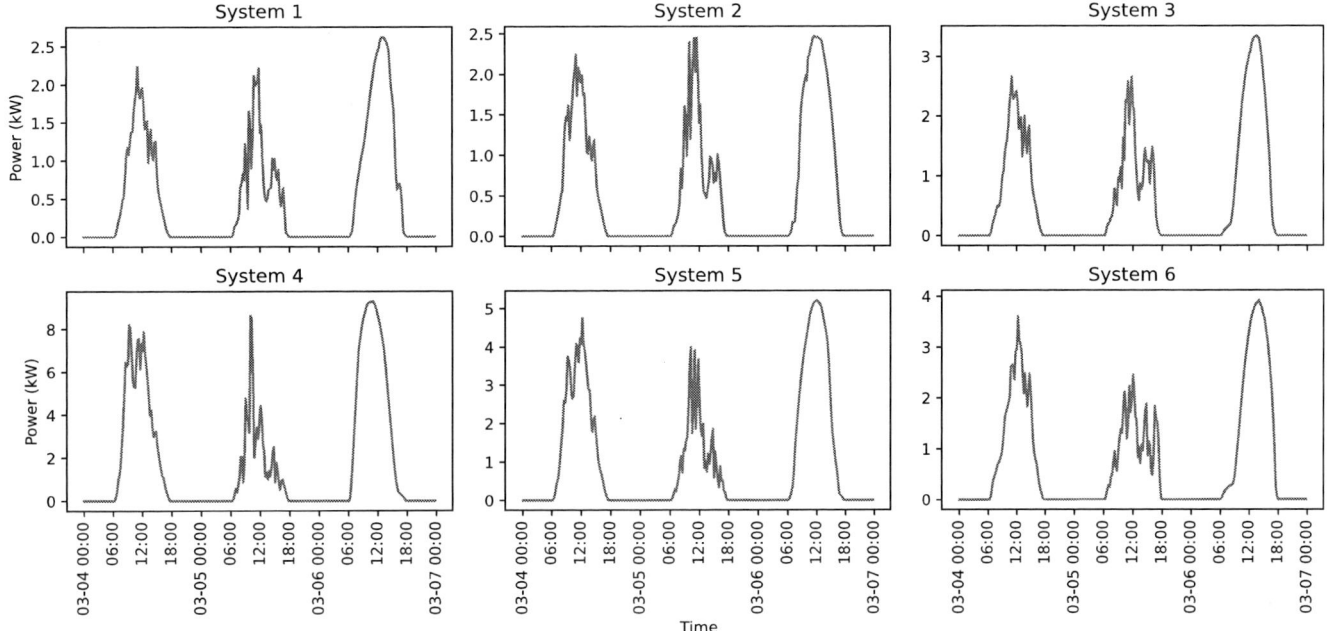

Fig. 2: 15 minute power output data for six PV systems from 3/4/2017 to 3/6/2017.

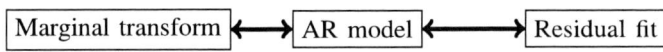

Fig. 3: Three invertible tranformations used in the proposed method.

where $\ell_t : \Theta \to \mathbf{R}$ is a convex loss function that depends on some data at time t. (If some data are missing, then the sum is only over t for which the data are available.)

Our generic fitting method takes θ as a solution of the convex optimization problem

$$
\begin{aligned}
\text{minimize} \quad & \mathcal{L} + \lambda \mathcal{D}, \\
\text{subject to} \quad & \theta_t \in \Theta, \quad t = 1, \ldots, P,
\end{aligned}
\tag{2}
$$

where $\lambda > 0$ is the smoothing regularizer hyper-parameter. The variables are the m-vectors $\alpha_0, \ldots, \alpha_K$ and $\beta_1, \ldots, \beta_K$. This is a convex optimization problem, and readily solved.

This generic fitting method contains the hyper-parameter λ (and possibly others), but good values of these can be found automatically using cross-validation [48, §7.10], [44, §13.2], so the method is essentially hyper-parameter free and automatic.

Two steps of our method solve a different instance of the problem (2) with λ chosen using cross-validation. This method of fitting a smooth periodic parameter is a special case of a Laplacian regularized stratified model [45], [46] with the underlying graph a cycle representing the periodicity. It can also be thought of as a signal decomposition problem [47].

B. Marginal transforms

In this first step, we seek continuous increasing functions $\varphi_{t,i} : \mathbf{R} \to \mathbf{R}$. We allow these functions to change with t but enforce that they are periodic and smooth. Our goal is for the transformed values,

$$
x_{t,i} = \varphi_{t,i}(y_{t,i}), \quad \text{for } t = 1, \ldots, T,
$$

to have an approximately Gaussian (marginal) distribution for each $i = 1, \ldots, d$. Typical quantile transforms are static; but our method defines a transformation that changes smoothly in time and is periodic.

We carry out this step for each component of the original data, so the method described in this section is carried out separately for each $i = 1, \ldots, d$. To lighten the notation in this section, we drop the component index i, and consider the original data y_t to be scalar.

Our first step is to estimate a set of quantiles $0 \leq \eta_1 < \cdots < \eta_r \leq 1$ of the data. By default we take these to be 2nd percentile, the 98th percentile, and the 10 deciles, so $r = 11$ and

$$
\eta = (0.02, 0.10, 0.20, \ldots, 0.80, 0.90, 0.98),
$$

but our method is general and any other choice of quantiles could be used. We denote the estimated η_i quantile of y_t as $q_{t,i}$, $t = 1, \ldots, T$, $i = 1, \ldots, r$. We assume these are P-periodic and smooth.

To estimate these quantiles from the data we use standard quantile regression [49], [50], which relies on the so-called pinball loss, defined as

$$
\ell^{\mathrm{pin}}(u; \eta) = \max\{(1 - \tau)u, \tau u\} = (\tau - 1/2)|u| + (1/2)u,
$$

for quantile $\tau \in [0, 1]$. We take our loss function to be

$$
\ell(q_t) = \ell^{\mathrm{pin}}(q_{t,1} - y_t; \eta_1) + \cdots + \ell^{\mathrm{pin}}(q_{t,r} - y_t; \eta_r), \tag{3}
$$

978-1-6654-6060-6/23 $31.00 © 2023 IEEE

the sum of the pinball losses associated with each of our r quantiles. To estimate the quantiles we solve the generic problem (2), with loss function (3), and constraint set

$$\Theta = \{q \mid q_1 \leq \cdots \leq q_r\},$$

which enforces that the quantiles are consistent. This simultaneously estimates the r quantiles of y_t for each t, with the quantiles being smooth and periodic, and always satisfiying the consistency constraint. The hyper-parameter λ, which controls how smooth the quantile estimates are, can be chosen automatically via cross-validation.

Given these periodic smooth quantile estimates, we construct nonlinear mappings $\varphi_t : \mathbf{R} \to \mathbf{R}$ as continuous piecewise linear functions, with knot-points given by the quantiles, and values at those points given by the associated value of a standard scalar Gaussian for the same quantiles, *i.e.*,

$$\varphi_t(q_{t,j}) = \Phi^{-1}(\eta_j), \quad j = 1, \ldots, r, \quad (4)$$

where Φ is the cumulative distribution function (CDF) of a standard Gaussian. This gives the marginally transformed data $x_{t,i} = \varphi_t(y_{t,i})$.

C. Autoregressive model

The time series $x_1, \ldots, x_T$ has entries with approximately standard Gaussian marginal distribution, but there are dependencies between the components, as well as across time. Our next step is to handle the dependency across time. We fit a vector autoregressive (AR) model to the marginally Gaussianized data x_t. The model is

$$x_t = A_1 x_{t-1} + \cdots + A_M x_{t-M} + v_t, \quad (5)$$

where v_t is a process noise or residual, M is the memory of the AR model, and $A_1, \ldots, A_M \in \mathbf{R}^{d \times d}$ are the coefficients. We could fit these AR coefficients as smooth and periodic, but we have found that a constant AR model does just as well as a more complex time-varying one.

We fit these coefficients by minimizing the mean-squared error, the average of

$$\ell_t = \|x_t - A_1 x_{t-1} + \cdots + A_M x_{t-M}\|_2^2$$

over those entries where all x_t are known. We add ridge regularization to this average loss,

$$\lambda^{\text{ridge}} \left(\|A_1\|_F^2 + \cdots + \|A_M\|_F^2 \right),$$

where $\lambda^{\text{ridge}} > 0$ is a hyper-parameter that scales the regularization, and $\| \cdot \|_F^2$ denotes the square of the Frobenius norm, *i.e.*, the sum of squares of entries. The hyperparameter λ^{ridge} can be chosen by cross-validation.

We denote the AR residual as

$$v_t = x_t - (A_1 x_{t-1} + \cdots + A_M x_{t-M}),$$

defined when $x_t, \ldots, x_{t-M}$ are all known. Note that in the special case $M = 0$, which corresponds to the model that x_t are approximately uncorrelated for different t, the residual reduces to x_t.

D. Smooth periodic residual fit

Our last step is fit a smoothly varying periodic Gaussian distribution to the residual v_t, $v_t \sim \mathcal{N}(\mu_t, \Sigma_t)$, where we assume that v_s and v_t are independent for $s \neq t$. We model $v_t \in \mathbf{R}^d$ as smooth periodic Gaussian,

$$v_t \sim \mathcal{N}(\mu_t, \Sigma_t), \quad t = 1, \ldots, T \quad (6)$$

where Σ_t and μ_t are smooth and periodic. We expect μ_t to be small.

Our loss ℓ_t will be the negative log-likelihood of the Gaussian model (6), which is

$$\ell_t = \frac{d}{2} \log(2\pi) - \sum_{j=1}^{d} \log\left(\mathbf{diag}\left(L_t\right)_j \right) + \frac{1}{2} \|L_t^T v_t - \nu_t\|_2^2,$$

with variables $L_t \in \mathbf{R}^{d \times d}$ and $\nu_t \in \mathbf{R}^d$. Here, L_t is the Cholesky factor of Σ_t^{-1} and $\nu_t = L_t^{-T}\mu_t$. This change of variables makes the loss a convex function. We can recover Σ_t and μ_t as

$$\Sigma_t = (L_t L_t^T)^{-1}, \qquad \mu_t = L_t^{-1}\nu_t.$$

once we solve the optimization problem to find L_t and ν_t. Here too the smoothness hyper-parameter λ is found by cross-validation.

Our final whitened signal is given by

$$z_t = \Sigma_t^{-1/2}(x_t - \mu_t) = L_t^T x_t - \nu_t,$$

defined when x_t is. According to our model, these are independent identically distributed (IID) with $z_t \sim \mathcal{N}(0, I)$.

E. The whole model

From our first two steps, we see that our model of y_t is a stationary periodic Gaussian process x_t, mapped entrywise through a smoothly periodic transformation. Using all three steps, we interpret it as IID Gaussians z_t, passed through an AR filter to obtain y_t, and then mapped entrywise.

Such a model allows us to carry out several operations. We can generate samples from the model. We can evaluate the density at a sequence y_t. We can condition on a set of known values of some of the components, as well as computing conditional marginal quantiles for each unknown entry. These allow us to carry out imputation, *i.e.*, guessing missing values, by evaluating the conditional median of a missing entry given the known ones. (We also can get error bars, *e.g.*, the 10th and 90th conditional marginal quantiles.) We can also do anomaly detection, where we detect known entries that do not fit the model. To do this we compute the conditional quantile of each known entry, given the other known entries but not that particular value; conditional quantile values that are either near zero or one are then flagged as suspicious.

These operations (and others) can be carried out for many types of statistical models, for example using Monte Carlo sampling. But due to the specific structure of our model, we can carry them out using simple linear algebra, which makes the operations fast and reliable. Details of how we implement these operations will be given in a forthcoming paper.

978-1-6654-6060-6/23 $31.00 © 2023 IEEE

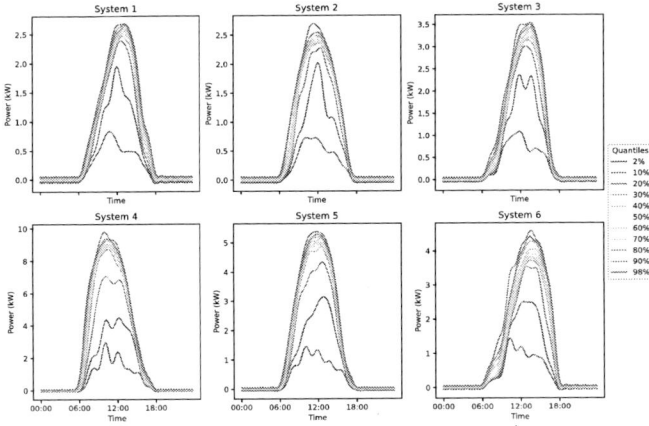

Fig. 4: Smooth periodic quantiles.

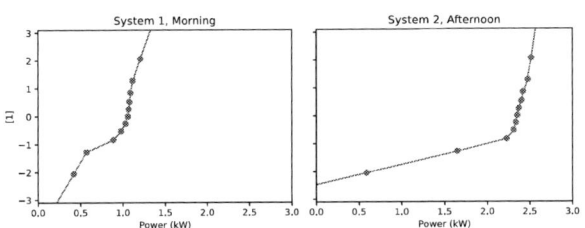

Fig. 5: Time aware copula transform.

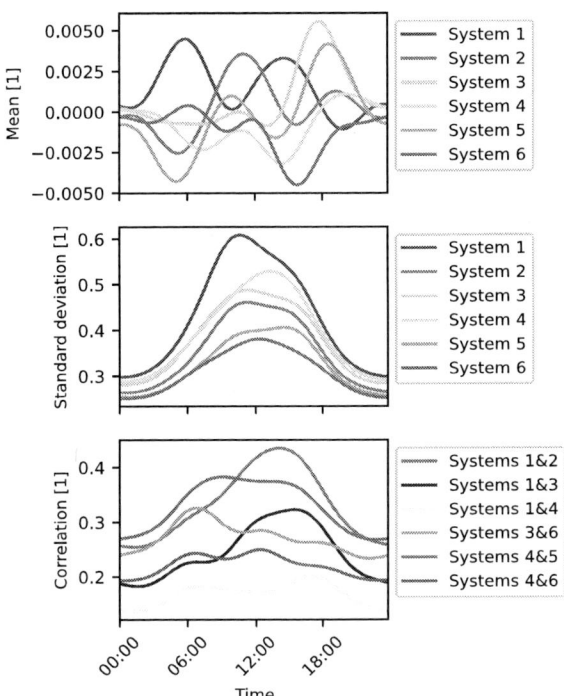

Fig. 6: Means, standard deviations, and selected correlations of AR residuals.

V. RESULTS

Here we show the results of our modeling method on the PV data described above, using default parameters. Estimated quantiles for each system are shown in figure 4. Note that the 98th percentile serves as an effective statistical clear-sky model. The estimated quantiles collapse to zero at night as expected. The spread between the upper quantiles is narrower than that of the lower quantiles, especially around noon. Quantiles for each system exhibit distinct characteristics, with the systems physically near each other showing similar shaped quantiles.

A few samples of the associated piecewise linear copula transforms are shown in figure 5. We see that the same power value is mapped to different points based on the time of day and the system. This shows how our time-aware copula transform adapts to both the time of day and the unique characteristics of different systems as opposed to a standard fixed copula transform, which does not.

We use AR memory $M = 3$, with coefficients. We observe several interesting phenomena in these coefficient matrices. First, the entries of A_1 are generally bigger than those of A_2 and A_3, showing that the previous period plays a larger role in predicting the current values than the previous two values. We also see that the diagonal entries are generally larger than the off-diagonal ones, meaning that the previous values for each system play a larger role in predicting the current value than the previous values of the other systems. However,

the many non-zero off-diagonal elements in the coefficient matrices show that the predictions for each system do depend on the previous values of the other systems.

Finally we fit a smoothly varying periodic Gaussian distribution to the residuals of the AR model, shown in figure 6. The top plot shows the means, which are indeed small, as expected. The middle plot shows the standard deviations of the residuals. These are smaller than one, which is approximately the standard deviation of the entries of x_t, which means we are able to predict the current values using previous values better than simply guessing $x_t = 0$, i.e., treating them as uncorrelated across time. We can see that the residual standard deviations vary considerably across systems and time of day. Roughly speaking, the residual of system 1 has almost twice the standard variation of the residual of system 6. We also see that the residual standard deviation is smaller at noon than in the morning and afternoon. The bottom plot shows the correlations of selected pairs of residuals. These correlations are generally around 30%, but we can see that systems that are physically near each other are more highly correlated. We can also see variation of the correlation over the day.

VI. APPLICATIONS

A. Generating simulated data

We generate simulated data from our model by simulating data from the periodic Gaussian stochastic process, and then applying the inverse nonlinearities $\phi_{t,i}^{-1}$ to the entries of these samples. Figure 7 shows two simulations of fake data for

978-1-6654-6060-6/23 $31.00 © 2023 IEEE

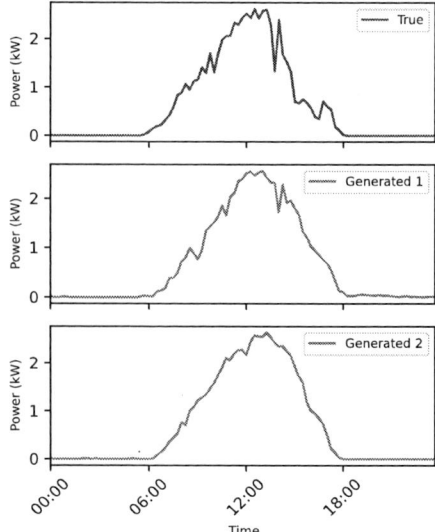

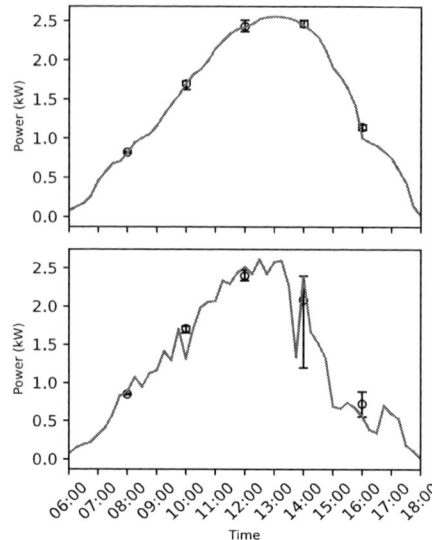

Fig. 7: *Top.* System 1 on 4/7/2017. *Middle and bottom.* Two simulations from our model.

Fig. 8: 10th, 50th, and 90th marginal conditional quantiles at five times, conditioned on values of all systems at other times. *Top.* Clear day. *Bottom.* Cloudy day.

TABLE I: Anomaly detection example.

Time	True value	Perturbed value	Conditional quantile
08:30	0.9919	0.8431	0.0001
10:00	1.7280	1.9872	0.9999
11:30	2.3633	2.7178	0.9999
13:00	2.5890	2.9773	0.9999
14:15	2.4294	2.0650	0.0001
15:30	1.6576	1.4090	0.0013

system 1, with the actual data for the specific day 4/7/2017 shown at top for reference. They appear quite similar.

B. Conditional marginal quantiles

We can compute the marginal quantiles of any entry, conditioned on all other known entries, in time and across systems. When the entry is unknown, this gives us a sophisticated method for imputing or guessing what the missing value might have been. We can use the conditional marginal median (50th percentile) as the imputed value, with the 10th and 90th percentiles defining an uncertainty interval. Figure 8 shows the marginal conditional quantiles for system 1 at 5 times on two days, one clear and one partially cloudy. We observe that the model correctly adapts the uncertainty bounds to the weather, with tighter bounds on clear days. Additionally, we note that the uncertainty bounds are asymmetric, with decreases in output (say, due to clouds) more likely than increases. The predictions themselves, shown as the circle representing the conditional median, are good.

C. Anomaly detection

We can use marginal conditional quantiles to identify anomalous entries in our data. To estimate whether a given known entry is an anomaly we pretend that it is unknown, compute its conditional marginal CDF given all other known values, and evaluate it at the known value. We can flag an entry as anomalous if this quantile value is less than ϵ or more than $1 - \epsilon$, where ϵ is a threshold value such as 10^{-2}. With this threshold, we would expect a false positive rate around 2ϵ.

To illustrate this, we consider system 2 on 4/1/2017. We introduce synthetic anomalies by perturbing the power values at 8:30, 10:00, 11:30, 13:00, 14:15 and 15:30, by randomly increasing or decreasing the true values by 15%. Table I shows

true values, perturbed values and conditional marginal quantiles of perturbed values, clipped to the range [0.0001, 0.9999].

With threshold $\epsilon = 0.01$, we detect all of the artificial anomalies. We also have three false positives, *i.e.*, times when a true value is flagged as an anomaly. The conditional marginal quantiles for this day are shown in figure 9. The vertical axis shows $\min\{q, 1 - q\}$, with the threshold $\epsilon = 0.01$ shown as the darker horizontal line. True negatives are shown as blue circles, and true positives are shown as blue squares. False positives are shown as orange circles. The three false positives are all cases where the true power was low compared to our predicted median. This is not suprising; clouds can easily reduce power output unexpectedly by 15% or more.

D. Forecast

Here we forecast the values of system 2 from 13:15 on, using data from all systems up through 13:00. We show the forecast, which is the median or 50th conditional marginal quantile, along with the conditional marginal 10th and 90th quantiles, which give us confidence bands for the forecast values. This is illustrated in figure 10 on the clear day 4/9/2017 and the cloudy day 4/6/2017. Our forecast on the clear day is very good, with tight uncertainty bands. Our forecast on the

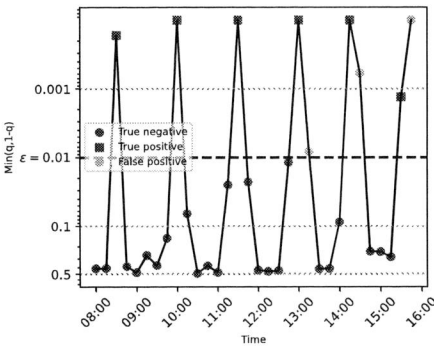

Fig. 9: Marginal conditional quantiles for system 2 on 4/1/2017, with six artificial anomalies added.

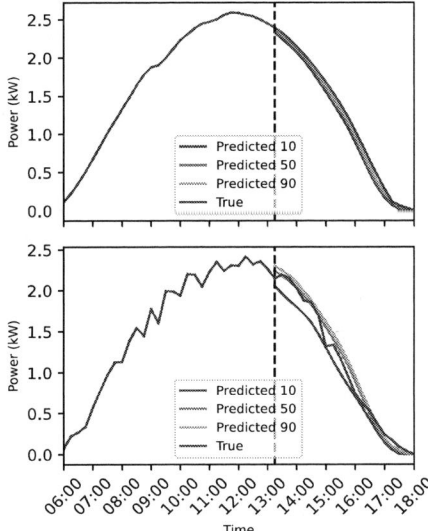

Fig. 10: Forecast for system 2 on the clear day 4/9/2017, and the cloudy day 4/6/2017. We forecast from 13:15 on, given values for all systems up through 13:00.

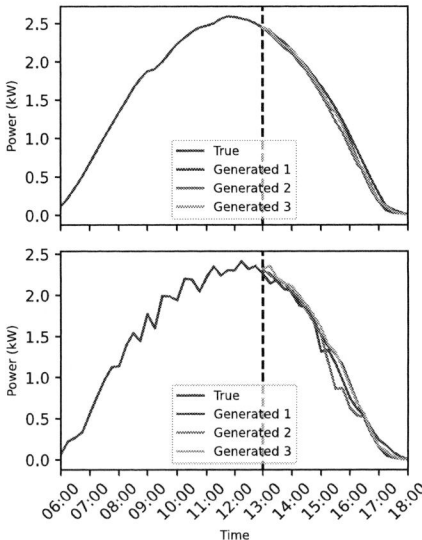

Fig. 11: Forecast for system 2 on the clear day 4/9/2017, and the cloudy day 4/6/2017. We forecast from 13:15 on, given values for all systems up through 13:00.

VII. CONCLUSIONS

We presented a novel approach to modeling and analyzing fleets of PV systems based on a smooth periodic Gaussian copula transform, and illustrated some of its applications. While we have demonstrated the method on a small example, it can scale gracefully to much larger problems; details will be given in a forthcoming paper.

ACKNOWLEDGMENT

This material is based on work supported by the U.S. Department of Energy's Office of Energy Efficiency and Renewable Energy (EERE) under the Solar Energy Technologies Office Award Number 38529. Stephen Boyd's work was funded in part by the AI Chip Center for Emerging Smart Systems (ACCESS).

REFERENCES

[1] L. G. da Fonseca, "The state of digital O&M for the solar market," *Greentech Media*, October 2019, accessed May 24 2023. [Online]. Available: https://www.greentechmedia.com/articles/read/the-state-of-digital-om-for-the-solar-market1

[2] D. L. King, J. A. Kratochvil, and W. E. Boyson, "Measuring solar spectral and angle-of-incidence effects on photovoltaic modules and solar irradiance sensors," *IEEE 26th Photovoltaic Specialist Conference*, pp. 1113–1116, 1997.

[3] ——, "Temperature coefficients for PV modules and arrays: measurement methods, difficulties, and results," *IEEE 26th Photovoltaic Specialist Conference*, pp. 1183–1186, 1997.

[4] D. L. King, W. E. Boyson, and J. A. Kratochvil, "Photovoltaic array performance model." *Sandia Report No. 2004-3535*, vol. 8, pp. 1–19, August 2004. [Online]. Available: https://www.osti.gov/servlets/purl/919131/

[5] B. Marion, J. Adelstein, K. Boyle, H. Hayden, B. Hammond, T. Fletcher, B. Canada, D. Narang, A. Kimber, L. Mitchell, G. Rich, and T. Townsend, "Performance parameters for grid-connected PV systems," *IEEE 31st Photovoltaic Specialist Conference*, pp. 1601–1606, 2005.

[6] W. D. Soto, S. A. Klein, and W. A. Beckman, "Improvement and validation of a model for photovoltaic array performance," *Solar Energy*, vol. 80, pp. 78–88, January 2006.

cloudy day is reasonable, but with much wider uncertainty bands.

In addition to forecasting marginal quantiles of a single system, we can generate joint forecasts using all systems. We illustrate this in figure 11, where we forecast the values of system 2 from 13:15 on, using data from all systems up through 13:00. We show three different forecasts, sampled from the full joint conditional distribution. We see that the forecasts agree with the marginal forecasts, in the sense that generated instances are within the 10–90% confidence bands of the marginal forecasts. We also see that the forecasts are reasonable, since we observe that for cloudy days, the forecasts are more volatile than for clear days with the latter having a more stable, smooth behavior.

978-1-6654-6060-6/23 $31.00 © 2023 IEEE

[7] W. F. Holmgren, R. W. Andrews, A. T. Lorenzo, and J. S. Stein, "PVLIB Python 2015," *IEEE 42nd Photovoltaic Specialist Conference*, pp. 1–5, June 2015.

[8] R. G. Miller, "GEM: A statistical weather forecasting procedure," *NOAA Technical Report NWS-28*, November 1981. [Online]. Available: https://www.weather.gov/media/owp/oh/hdsc/docs/TR28.pdf

[9] S. Quintarelli, "Let's forget the term AI. let's call them systematic approaches to learning algorithms and machine inferences (SALAMI)," *Quinta's weblog*, November 2019, accessed May 10 2023. [Online]. Available: https://blog.quintarelli.it/2019/11/lets-forget-the-term-ai-lets-call-them-systematic-approaches-to-learning-algorithms-and-machine-inferences-salami/

[10] Y. Zhao, Q. Liu, D. Li, D. Kang, Q. Lv, and L. Shang, "Hierarchical anomaly detection and multimodal classification in large-scale photovoltaic systems," *IEEE Transactions on Sustainable Energy*, vol. 10, pp. 1351–1361, July 2019.

[11] F. Aziz, A. U. Haq, S. Ahmad, Y. Mahmoud, M. Jalal, and U. Ali, "A novel convolutional neural network-based approach for fault classification in photovoltaic arrays," *IEEE Access*, vol. 8, pp. 41 889–41 904, 2020.

[12] V. Veerasamy, N. I. A. Wahab, M. L. Othman, S. Padmanaban, K. Sekar, R. Ramachandran, H. Hizam, A. Vinayagam, and M. Z. Islam, "LSTM recurrent neural network classifier for high impedance fault detection in solar PV integrated power system," *IEEE Access*, vol. 9, pp. 32 672–32 687, 2021.

[13] P. Li, K. Zhou, X. Lu, and S. Yang, "A hybrid deep learning model for short-term PV power forecasting," *Applied Energy*, vol. 259, p. 114216, February 2020.

[14] Q. Huang and S. Wei, "Improved quantile convolutional neural network with two-stage training for daily-ahead probabilistic forecasting of photovoltaic power," *Energy Conversion and Management*, vol. 220, p. 113085, September 2020.

[15] S. Ding, R. Li, and Z. Tao, "A novel adaptive discrete grey model with time-varying parameters for long-term photovoltaic power generation forecasting," *Energy Conversion and Management*, vol. 227, p. 113644, January 2021.

[16] A. A. du Plessis, J. M. Strauss, and A. J. Rix, "Short-term solar power forecasting: Investigating the ability of deep learning models to capture low-level utility-scale photovoltaic system behaviour," *Applied Energy*, vol. 285, p. 116395, March 2021.

[17] W. Zhao, H. Zhang, J. Zheng, Y. Dai, L. Huang, W. Shang, and Y. Liang, "A point prediction method based automatic machine learning for day-ahead power output of multi-region photovoltaic plants," *Energy*, vol. 223, p. 120026, May 2021.

[18] Q. Li, X. Zhang, T. Ma, C. Jiao, H. Wang, and W. Hu, "A multi-step ahead photovoltaic power prediction model based on similar day, enhanced colliding bodies optimization, variational mode decomposition, and deep extreme learning machine," *Energy*, vol. 224, p. 120094, June 2021.

[19] F. Najibi, D. Apostolopoulou, and E. Alonso, "Enhanced performance Gaussian process regression for probabilistic short-term solar output forecast," *International Journal of Electrical Power & Energy Systems*, vol. 130, p. 106916, September 2021.

[20] T. E. Hoff and R. Perez, "Quantifying PV power output variability," *Solar Energy*, vol. 84, pp. 1782–1793, October 2010.

[21] ——, "Modeling PV fleet output variability," *Solar Energy*, vol. 86, pp. 2177–2189, August 2012.

[22] G. Wirth, E. Lorenz, A. Spring, G. Becker, R. Pardatscher, and R. Witzmann, "Modeling the maximum power output of a distributed PV fleet," *Progress in Photovoltaics: Research and Applications*, vol. 23, pp. 1164–1181, September 2015.

[23] J. Marcos, I. de la Parra, M. García, and L. Marroyo, "Simulating the variability of dispersed large PV plants," *Progress in Photovoltaics: Research and Applications*, vol. 24, pp. 680–691, May 2016.

[24] A. Sklar, "Fonctions de répartition à n dimensions et leurs marges," *Publications de l'Institut de Statistique de l'Université de Paris*, vol. 8, pp. 229–231, 1959.

[25] B. V. Vliet, "Abe Sklar's 'Fonctions de répartition à n dimensions et leurs marges': The original document and an English translation," *SSRN Electronic Journal*, 2023.

[26] T. Schmidt, "Coping with copulas," in *Copulas: From theory to application in finance*, J. Rank, Ed.   University of California, 2007, pp. 3–34.

[27] A. Charpentier, J.-D. Fermanian, and O. Scaillet, "The estimation of copulas: Theory and practice," in *Copulas: From theory to application in finance*, J. Rank, Ed.   University of California, 2007, pp. 35–61.

[28] A. J. Patton, "Copula-based models for financial time series," in *Handbook of Financial Time Series*.   Springer, 2009, pp. 767–785.

[29] Y. Stander, D. Marais, and I. Botha, "Trading strategies with copulas," *Journal of Economic and Financial Sciences*, vol. 6, no. 1, pp. 83–107, 2013.

[30] P. Xu, D. Wang, V. P. Singh, Y. Wang, J. Wu, H. Lu, L. Wang, J. Liu, and J. Zhang, "Time-varying copula and design life level-based nonstationary risk analysis of extreme rainfall events," *Hydrology and Earth System Sciences Discussions*, pp. 1–59, 2019.

[31] Y. Zhao and M. Udell, "Missing value imputation for mixed data via Gaussian copula," in *Proceedings of the 26th ACM SIGKDD International Conference on Knowledge Discovery & Data Mining*, 2020, pp. 636–646.

[32] J. L. Schafer, "NORM users guide: Multiple imputation of incomplete multivariate data under a normal model," *University Park: The Methodology Center, Penn State*, 1999. [Online]. Available: https://scholarsphere.psu.edu/collections/v41687m23q

[33] Y. Zhao and M. Udell, "Matrix completion with quantified uncertainty through low rank Gaussian copula," *Adv. Neural Inf. Process. Syst.*, vol. 33, pp. 20 977–20 988, 2020.

[34] J. Munkhammar and J. Widén, "An autocorrelation-based copula model for producing realistic clear-sky index and photovoltaic power generation time-series," in *2017 IEEE 44th Photovoltaic Specialist Conference (PVSC)*.   IEEE, 2017, pp. 3067–3072.

[35] W. Härdle and O. Linton, "Applied nonparametric methods," in *Handbook of Econometrics*, 4th ed., R. Engle and D. McFadden, Eds.   Springer, 1994, pp. 767–785.

[36] A. Pagan and A. Ullah, *Nonparametric Econometrics*.   Cambridge University Press, 1999.

[37] E. C. Brechmann and C. Czado, "COPAR-multivariate time series modeling using the copula autoregressive model," *Applied Stochastic Models in Business and Industry*, vol. 31, pp. 495–514, July 2015.

[38] S. Chen and R. Gopinath, "Gaussianization," *Advances in Neural Information Processing Systems*, vol. 13, 2000.

[39] C. W. Huang, R. T. Q. Chen, C. Tsirigotis, and A. Courville, "Convex potential flows: Universal probability distributions with optimal transport and convex optimization," *arXiv preprint*, 2020.

[40] H. Liao and J. He, "Jacobian determinant of normalizing flows," *arXiv preprint arXiv:2102.06539*, 2021.

[41] S. Boyd and L. Vandenberghe, *Convex optimization*.   Cambridge University Press, 2009.

[42] S. Diamond and S. Boyd, "CVXPY: A Python-embedded modeling language for convex optimization," *Journal of Machine Learning Research*, vol. 17, no. 83, pp. 1–5, 2016.

[43] A. Agrawal, R. Verschueren, S. Diamond, and S. Boyd, "A rewriting system for convex optimization problems," *Journal of Control and Decision*, vol. 5, no. 1, pp. 42–60, 2018.

[44] S. Boyd and L. Vandenberghe, *Introduction to Applied Linear Algebra*. Cambridge university press, 2018.

[45] J. Tuck and S. Boyd, "Fitting Laplacian regularized stratified Gaussian models," *SSRN Electronic Journal*, vol. 1, pp. 1–24, 2020.

[46] J. Tuck, S. Barratt, and S. Boyd, "A distributed method for fitting Laplacian regularized stratified models," *Journal of Machine Learning Research*, vol. 22, pp. 1–37, 2021.

[47] B. Meyers and S. Boyd, "Signal decomposition using masked proximal operators," *Foundations and Trends in Signal Processing*, vol. 17, no. 1, pp. 1–78, 2023.

[48] T. Hastie, R. Tibshirani, and J. Friedman, *The Elements of Statistical Learning*, ser. Springer Series in Statistics.   New York, NY: Springer New York, dec 2009. [Online]. Available: https://hastie.su.domains/ElemStatLearn/

[49] R. Koenker, *Quantile Regression*.   Cambridge University Press, July 2005.

[50] ——, "Quantile regression: 40 years on," *Annual Review of Economics*, vol. 9, pp. 155–176, August 2017.

Unveiling the structural formation of low dimensional layers deposited on lead halide perovskites by thermal evaporation

Carlo Andrea Riccardo Perini, Andres Felipe Castro Mendez, Tim Kodalle, Magdalena Ravello, Juanita Hidalgo, Juan-Pablo Correa-Baena

Georgia Institute of Technology, Atlanta, GA, United States

Lawrence Berkeley National Laboratory, Berkeley, CA, United States

In this work, we reveal differences in the formation of capping layers based on large organic cations via thermal evaporation and solution deposition. We show that the use of solvents, such as isopropanol, causes the formation of defects at the treated interface that lower the open circuit voltage of solar cells. The addition of large cations heals those defects and facilitates the conversion of the 3D perovskite surfaces into Ruddlesden-Popper (RP) phases. We observe formation of a dominant n = 2 RP both via solution and via vapor, and solvent-mediated formation of n = 1. Deposition via vapor prevents the formation of aggregates of the organohalide salts upon deposition and enables better phase purity and narrower solar cell performances after passivation with respect to solution. This study shines light on the formation dynamics of capping layers deposited via solution and vapor routes, which enables the design of more efficient passivation.

Ultrasonic Characterization of Ethylene Vinyl Acetate (EVA) Crosslinking for Quality Assurance and Lamination Process Control

Rico Meier [a→b], Ian M. Slauch [a], Mariana I. Bertoni [a]

[a]School of Electrical, Computer and Energy Engineering, Arizona State University, Tempe, AZ, 85287, USA
[b]School of Engineering: Energy and Information, HTW Berlin - University of Applied Sciences, Berlin, 12459, Germany

Abstract — To increase throughput and reduce cost, module manufacturers seek to minimize lamination processing times. However, too short lamination result in poor Ethylene-Vinyl Acetate (EVA) properties due to unfinished crosslinking and a high content of residual aggressive reaction starters in the completed module. These modules are prone to aging, delamination, mechanical fatigue, corrosion, and yellowing, severely reducing long-term reliability. Identification of unfinished crosslinking is crucial for lamination process surveillance, but since material and process conditions vary locally within the module (and over time), destructive characterization techniques (e.g. differential scanning calorimetry (DSC), Soxhlet extraction or peel testing) on some selected samples only provide limited insight.

For that reason, we propose a non-destructive method that uses ultrasound to characterize crosslinking within the solar module. High-frequency longitudinal ultrasound is transmitted through the thickness of the solar module and analyzed. By knowing the thickness of the polymer, the speed of sound and the frequency-dependent ultrasonic attenuation, the degree of crosslinking can be determined. In this paper, we demonstrate an experimental correlation between the DSC degree of crosslinking and the speed of sound, as well as the frequency-dependent attenuation coefficient. After an initial drop in velocity during initial melting (or humidity evaporation), sound velocity and attenuation increased with curing duration. This correlation can be utilized to determine the degree of crosslinking. However, a calibration of the material-specific model with conventional methods is required (e.g. via DSC analysis). The proposed ultrasonic method then allows a non-destructive characterization of the module lamination process, capable of detecting inhomogeneities and deficiencies in crosslinking, to ultimately drive process optimizations, increase production yield and reduce the LCOE.

I. INTRODUCTION

Deficiencies in the lamination process conditions (curing temperature, pressure, duration, and homogeneity of these conditions) can cause poor Ethylene-Vinyl Acetate (EVA) properties, which are often not identified by initial testing because the power losses are small at this stage. However, these issues can become much more critical during operation in the field and significantly decrease the long-term power yield, and thus increase the LCOE. Typical module degradation modes caused by deficient lamination process conditions include:

1. Cell breakage: Sharp changes in pressure or temperature incorporate thermomechanical stress, increasing cell breakage.

2. Corrosion: A reduction of encapsulants' moisture barrier due to weakened adhesive bonds may accelerate corrosion of front-, and backside metallization, leading to yellowing, snail trails, and an increase in series resistance [1].

3. Delamination: Weakened adhesive bonding promotes delamination of the encapsulant from module components under accelerated testing and field loading [2]. Delamination increases the humidity vulnerability (adding to 2) and results in optical losses due to additional reflections and scattering.

4. Local Inhomogeneities: Lamination conditions in the module and the additives in the foil are often not homogenously distributed [1, 3]. This can lead to local variations of properties, stress concentration, and failure.

Identifying unfinished crosslinking is critical for lamination process surveillance, but since crosslinking can be inhomogeneous within the module, local destructive characterization techniques (differential scanning calorimetry (DSC), Soxhlet extraction, peel testing) give limited insight. We propose a non-destructive method using ultrasound to characterize the crosslinking. On a scientific level, ultrasound has been tested as a crosslinking characterization tool in PV before [4-6], but was not transferred to industry due to a lack of sensitivity and/or non-industrial sample requirements. We want to change that.

II. SAMPLE PREPARATION

Three sets of Glass/Ethylene vinyl acetate/Backsheet laminates were prepared. Laminates measuring 300 x 300 x 3.2 mm³ were made using low-iron sodalime, EVA encapsulant, and backsheet in an NPC model LM-110x160-S laminator. Prior to lamination, the glass was cleaned by rinsing both surfaces with isopropanol and drying with an N_2 spray gun. The lamination chamber was preheated to operating temperature before loading the lay-up and beginning the lamination process. Each lamination required a 4.5-minute vacuum step, where the chamber pressure was reduced to $< 0.1\ kPa$, a $90\ s$ initial press, where the bladder pressure was increased from near vacuum to atmospheric pressure, and a final press, during which the EVA was cured. A single layer of EVA ($\sim 400\ \mu m$ thickness) was laminated between the glass and backsheet with a variable final press time of 5, 6, 10, 15, 18, or 50 minutes.

978-1-6654-6060-6/23 $31.00 © 2023 IEEE

II. INITIAL DIFFERENTIAL SCANNING CALORIMETRY

Differential scanning calorimetry (DSC) analysis was performed on EVA extracted from the final laminates to establish a correlation between the crosslinking state of the EVA and the final press duration. DSC is a standard characterization technique to measure the extent of crosslinking reaction [1-4]. The reaction enthalpy ΔH was obtained from the DSC measurement using Equation (1). This involves integrating the heat flow Q over time from the onset T_1 to the ending temperature T_2 of the cross-linking reaction peak, as depicted in Fig. 1. A baseline heat flow $Q_{baseline}$ is determined by linear interpolation of the heat flow values at T_1 and T_2. This baseline heat flow is subtracted from the measured heat flow profile $Q_{measured}$. The obtained heat released during cross-linking is normalized to the sample mass m_{sample}.

$$\Delta H = \frac{1}{m_{sample}} \int_{t(T_1)}^{t(T_2)} (Q_{measured}(t) - Q_{baseline}(t))dt \quad (1)$$

To calculate the extent of crosslinking, the DSC Degree of Crosslinking X_{DSC}, the reaction enthalpy of a partially cured sample ΔH is compared to that of an uncured sample ΔH_0 by

$$X_{DSC} = (\Delta H_0 - \Delta H)/\Delta H_0. \quad (2)$$

During the first run, two endothermic peaks, which are characteristic to crystal melting, are observed within the temperature range from 30 to 80 °C. At 135 °C, an exothermic peak emerges, indicating the heat release from the crosslinking reaction. In a second run, there was a straight line at the former crosslinking peak location, indicating that the crosslinking reaction was fully completed after the first run.

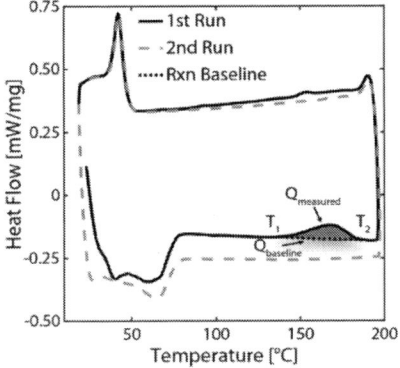

Fig. 1. The DSC plot shows heat flow versus temperature for an EVA sample (10-minute final press). The first run shows a crosslinking peak, while the second run does not. A black dotted line highlights the baseline of the reaction peak (red).

The results of the DSC analysis are shown in Fig. 2. As expected, the degree of crosslinking increases with lamination time. There is a sharp increase in the degree of crosslinking at short lamination times, followed by a convergence to a plateau later. EVA laminated for 50 minutes showed no further reaction, indicating that the sample is fully crosslinked.

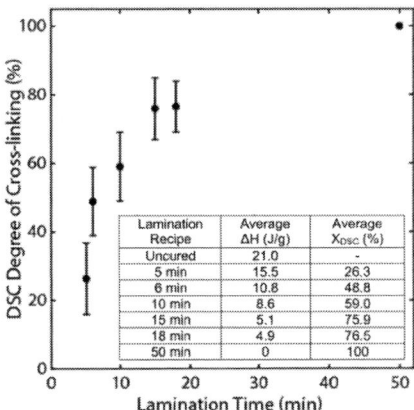

Lamination Recipe	Average ΔH (J/g)	Average X_{DSC} (%)
Uncured	21.0	-
5 min	15.5	26.3
6 min	10.8	48.8
10 min	8.6	59.0
15 min	5.1	75.9
18 min	4.9	76.5
50 min	0	100

Fig. 2. DSC degree of crosslinking vs. lamination time. Error bars indicate the standard deviations of multiple measurements.

II. ULTRASONIC TRANSMISSION EXPERIMENTS AT ROOM TEMPERATURE

Short ultrasonic pulse durations and a high signal-to-noise ratio are critical to distinguish individual pulses in those samples. Low frequencies were not sensitive enough to characterize thin EVA layers, whereas frequencies that were too high were highly absorbed, resulting in a low signal quality for further analysis. 10 MHz Olympus transducers were selected to excite and receive the most sensitive, yet detectable signals. The sending transducer was fixed to the glass and excited by a high-voltage pulse, while the receiving transducer was placed on the other side of the module, i.e. the backsheet (transmission measurement). Ultrasound was coupled to the glass-EVA-backsheet sample, propagated through it, before being received by the other transducer. The electric signal generated by the receiving transducer was amplified and recorded with a transient recorder (TiePie HS5). The first six echoes were identified in the time signal (Fig. 3), and the times-of-flight, as well as the signal amplitudes, were evaluated. The mean longitudinal velocities were calculated based on the time-of-flight of the peaks of each echo and sample thickness. We identified a slight increase in mean longitudinal wave speed with lamination duration in those samples.

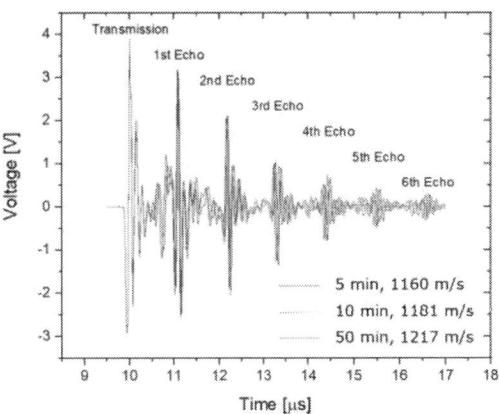

Fig. 3. Transmitted time signal for the different laminated samples.

Furthermore, we analyzed the spectra of the received signals (Fig. 4). The center frequencies of the eigenfrequencies slightly shifted to higher values with increasing lamination time, which coincides with a higher overall stiffness. Amplitudes were normalized to the highest peak ($U_{ref} = 7.5\ MHz$) to take coupling losses into account.

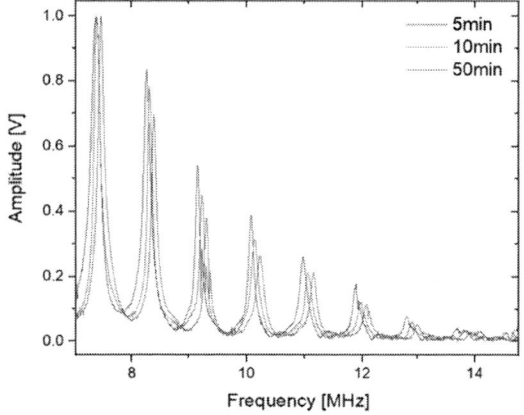

Fig. 4.   Frequency spectra of the received signals from Fig. 2. The amplitudes where normalized to the highest peak (at 7.5 MHz).

Attenuation was modeled as a polynomial in the Beer-Lambert law: $U(f) = U_{ref}\ exp(-\alpha x)$, where U and U_{ref} are the signal amplitudes, x is the sample thickness, and α the attenuation coefficient, which was modelled via $\alpha = -a \cdot f^b$. As expected, Fig. 4 shows higher attenuation at higher frequencies. Furthermore, a linear relationship between the logarithm of the amplitude and frequency was observed (in Fig. 5, $b \approx 1$).

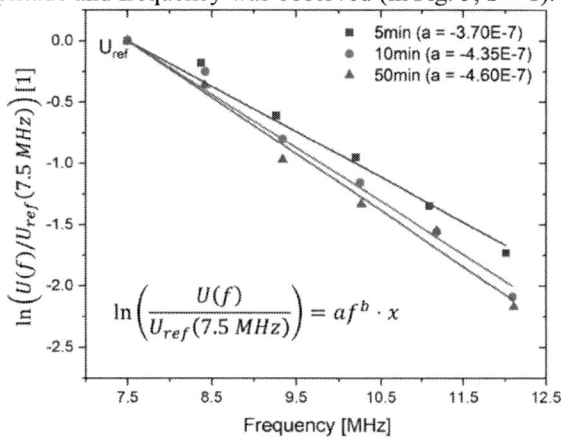

Fig. 5.   The normalized amplitudes of the eigenfrequencies were linearly fitted (since $b \approx 1$).

This is typical for viscoelastic materials in the ultrasonic frequency range from 20 kHz to 20 MHz when scattering mechanisms can be ignored, and the attenuation arises from viscous damping or hysteretic behavior [7]. Additionally, longer-cured samples show a higher attenuation prefactor. This behavior cannot be explained by a decrease in EVA thickness and is a result of the crosslinking reaction. Coupled with a thickness gauge, these measurements allow a local mapping of

the crosslinking. In Fig. 6 the evaluated sound velocities and attenuation prefactors are compared to the obtained DSC degree of crosslinking. Such an empirical model allows a determination of crosslinking only by the ultrasound transport properties (after being calibrated).

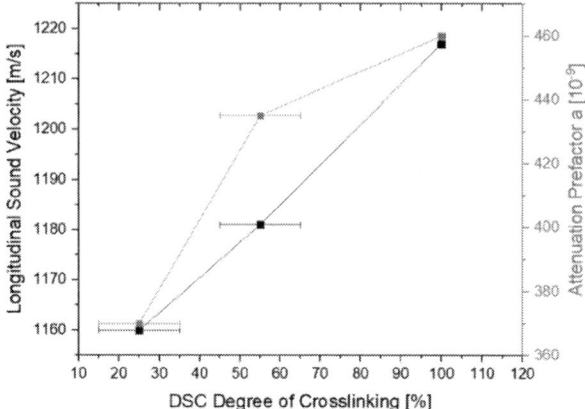

Fig. 6.   Correlating the ultrasound properties to the DSC degree of crosslinking. The error bars indicate the standard deviations of multiple DSC measurements.

The observed decrease in time-of-flight, as well as the shift in eigenfrequencies (but not the increase in attenuation prefactor), at higher lamination times could also be explained by a decrease in EVA thickness during lamination (due to the melting of the EVA combined with the applied lamination pressure). To check this, we analyzed the crosslinking progression of a partially pre-cured sample (laminated for 5 minutes) in an oven at 150°C. Since no vacuum pressure was applied, the thickness was assumed to be constant. Based on thermocouple measurements, 5-minute heating intervals were chosen to allow the EVA to reach crosslinking temperatures ($T > 140°C$). After each interval, the sample was cooled to room temperature, and the ultrasonic transmission properties of the sample were characterized.

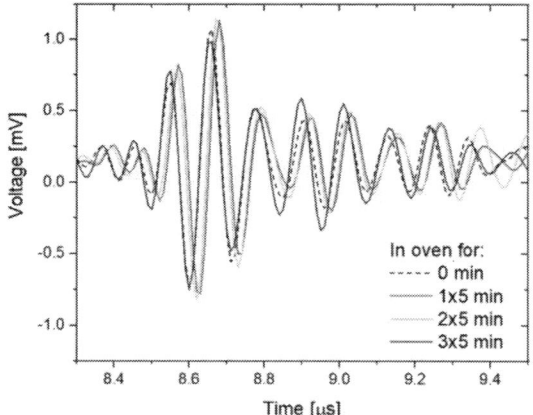

Fig. 7.   Received time signals after different curing durations. The reference (0 min) sample was 5 min pre-laminated. Afterwards additional 5 min heating cycles (without pressure) were added.

In Fig. 7, the time signals of the measured transmitted signals are shown. As described in the literature [4], at the beginning of the crosslinking reaction, the speed of sound decreases (between 0 and 5 min). The reference signal (0 min) arrives before the signal of a sample being 5 min in the oven. This is a result of either initial melting [4] or initial internal moisture, which evaporates during the first heating cycle. Additional heating intervals (each of 5 min) increase the sound velocity in the EVA. The growing polymer network increases the stiffness of the material and, therefore, the measured sound velocity. This is in accordance with the previously observed increase in sound velocity with crosslinking and agrees with the literature [4].

To apply this method to full modules, ultrasound needs to be transmitted through the entire module stack. It is crucial to determine the thickness and temperature of the polymer, which may vary significantly [8], but can be obtained through independent approaches such as optical or capacitive sensors or mechanical thickness gauges. Especially ribbons, texture, busbars, and backside metallization lead to inhomogeneous ultrasonic scattering of high frequencies, complicating the signal evaluation. For that reason, the method is better suited for module locations around the cells or within cell gaps. In those locations, the module represents a glass-EVA-backsheet laminate, and the method described in this paper can be directly applied.

III. ULTRASONIC TRANSMISSION EXPERIMENTS DURING CROSSLINKING REACTION

To test the method at relevant lamination temperatures, high-temperature piezo ceramics measuring 5.00 x 1.02 mm² (with an operation range up to 160°C) were implemented in a 3D printed mold. To prepare for future measurements, we constructed very thin transducers (<2 mm) as we intend to install a thin sheet containing an array of these transducers on both sides of the PV module within the module laminator. At this point, measurements were carried out on 5-minute partially pre-cured samples inside an oven in order to maintain a constant EVA thickness (see chapter II). Unlike the previously mentioned measurements, the samples were analyzed directly during the heating step (crosslinking process) instead of being tested sequentially at room temperature after oven intervals. The time-of-flight of the ultrasonic pulse was determined by the cross-correlation of the complex-conjugated time signal received $u^*(n, t)$ with a reference signal $u_{ref}(n, t_0)$, which was recorded at room temperature before the experiment. In this notation, n represents the n-th sample of the recorded transient, and t indicates the time when the transient was recorded. By utilizing the location τ_{max} of the maximum of the cross-correlation function

$$(u * u_{ref})(\tau, t) = \sum_{n_{initial}}^{n_{final}} u^*(n, t)\, u_{ref}(n + \tau, t_0)$$

and the sampling frequency of the transient recorder f_s, the time-of-flight of the received ultrasonic pulse at time t was determined via $ToF(t) = \tau_{max}(t)/f_s$.

In Fig. 8, the time-of-flight of the received ultrasonic pulse is shown over the measurement time. Prior to the measurement, the sample was heated to 85°C. At 15:32, the oven temperature was increased to 145°C. The rise in temperature led to a drastic reduction in stiffness (elastic modulus) in all the module components (glass, EVA, backsheet), which decreased the speed of sound and thus increased the time-of-flight of the ultrasonic transmission signal. The drastic rise at the beginning converged when a uniform temperature was reached within the sample. At 15:44, an increase in noise was observed in the signal. At the same time, the recorded transient (time signal) visibly changed, leading to disturbances in the cross-correlation signal. These changes could indicate structural changes inside the EVA due to crosslinking, which influence the ultrasonic signal. A noticeable decrease in time-of-flight is evident at 15:51, which corresponds with our previous research at room temperature (chapter II) and literature [4], showing an increase in elastic modulus during crosslinking. The recorded reduction in time-of-flight reached a plateau at 15:57, indicating that the modulus has stopped changing significantly and the crosslinking process is finished. The oven was switched off, and the sample cooled down at 16:02. The time-of-flight was significantly reduced due to an increase in elastic moduli and sound velocities in all module components during the cooling process. It should be emphasized that the impact of the crosslinking reaction on time-of-flight (and hence ultrasonic velocity) is considerably less significant than that of the temperature change. A well-controlled temperature profile is therefore crucial for accurate measurements. In that case, examining the plateau region in the measured time-of-flight (or sound velocity) could be a valuable technique for tracking the crosslinking progress during lamination. This is particularly useful as no prior calibration step is needed for a qualitative analysis. Additional samples and materials will be used to confirm these initial findings. A more comprehensive comparison to DSC results at various positions on the plateau will establish the method's sensitivity.

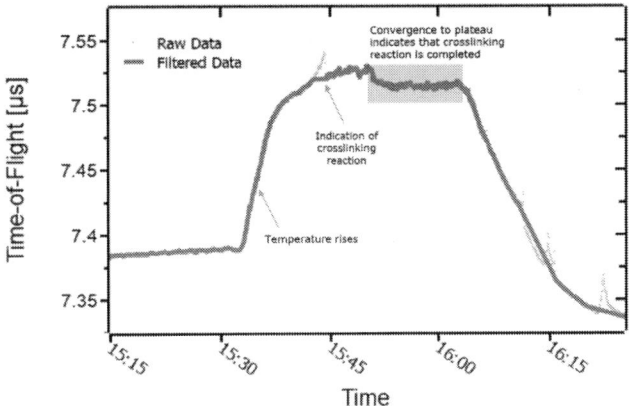

Fig. 8. The time-of-flight (ToF) of the transmitted ultrasonic pulse measured during a heating step profile to 150°C. Changes in temperature highly affected the ToF. The crosslinking reaction led to a smaller but significant decrease in ToF until a plateau was reached, indicating that the crosslinking reaction was completed.

IV. CONCLUSION

To investigate changes in the degree of crosslinking of the encapsulant, high-frequency longitudinal ultrasound was transmitted through the thickness of the solar module and analyzed. By knowing the thickness of the polymer, the speed of sound, and the frequency-dependent ultrasonic attenuation, we determined a correlation between the DSC degree of crosslinking and the speed of sound, as well as the frequency attenuation coefficient. The obtained results are in accordance with literature findings [4], where sound velocity and attenuation were analyzed during the crosslinking reaction of thick EVA samples in a specially designed high-temperature setup. After an initial drop in velocity during initial melting (or humidity evaporation), sound velocity and attenuation increased with curing time. An initial material-specific calibration of a sound velocity or attenuation vs. crosslinking model allows for an ultrasonic determination of the degree of crosslinking.

In a second study, we examined the times-of-flight of high-frequency ultrasonic pulses transmitted through solar module laminates at lamination temperature (which are related to the speed of sound). Temperature had a significant impact on the stiffness of the materials, which in turn affected the times-of-flight. As the EVA crosslinking reaction occurred, there was a noticeable but smaller decrease in time-of-flight until it reached a plateau, indicating that the crosslinking reaction was completed. Analyzing this plateau region either in time-of-flight or sound velocity could be an effective method for non-destructively monitoring the progress of crosslinking during lamination. This approach could be particularly beneficial as no prior calibration step is needed for qualitative analysis.

In conclusion, we were able to identify and quantify changes in the degree of crosslinking of the encapsulant by analyzing high-frequency longitudinal ultrasound that was transmitted through the thickness of the solar module. This shows enormous potential for the development of an ultrasonic instrument that can non-destructively characterize the progress of crosslinking during the production of PV modules.

IV. ACKNOWLEDGMENT

This work was supported by U.S. Department of Energy's Office of Energy Efficiency and Renewable Energy (EERE) under Award Number DE-EE0008987. The transmission measurements at high temperature (chapter III) were part of the IFAF project "GG-XLink" under supervision of the Berlin Institute for Applied Research (IFAF) and funded by the Senate Chancellery for Science and Research of the Mayor of Berlin.

REFERENCES

[1] J. Zhu, T. R. Betts, R. Gottschalg, "Effects of lamination condition on durability of PV module packaging and Performance", *Proceedings of the 10th Photovoltaic Science Applications and Technology Conference C96 (PVSAT-10)*, 2014.

[2] A. K. Öz et al. "The Impact of the Lamination Process on the Adhesion Properties at the Glass-Encapsulant Interface and Damp Heat Stability of PV Modules" *Proc. 38th European Photovoltaic Solar Energy Conference and Exhibition (EUPVSEC)*, 2021

[3] S.-H. Schulze et al. "Encapsulation polymers – a key issue in module reliability", Photovoltaics International Volume 11, Mar 2011.

[4] W. Stark et al., "Investigation of the crosslinking behaviour of ethylene vinyl acetate (EVA) for solar cell encapsulation by rheology and ultrasound", *Polymer Testing 31*, pp. 904–908, 2012.

[5] Ch. Hirschl et al. "Determining the degree of crosslinking of ethylene vinyl acetate photovoltaic module encapsulants—A comparative study", *Solar Energy Materials & Solar Cells 116*, pp. 203–218, 2013.

[6] M. Sander, M. Ebert. "Vibration analysis of PV-modules by laser-doppler vibrometry", *Proc. 24th European Photovoltaic Solar Energy Conference and Exhibition (EUPVSEC)*, 2009

[7] K. Ono. "A Comprehensive Report on Ultrasonic Attenuation of Engineering Materials, Including Metals, Ceramics, Polymers, Fiber-Reinforced Composites, Wood, and Rocks", Appl. Sci. 2020, 10, 2230

[8] A. Pfreundt et al. "Post-processing thickness variation of pv module materials and its impact on temperature, mechanical stress and power", *Proc. 36th European Photovoltaic Solar Energy Conference and Exhibition (EUPVSEC)*, 2019

Detailed Raman Investigation on the Search for the Secondary Phases in the Chalcogenide Perovskite BaZrS3

Hasan Arif Yetkin, Ruiquan Yang, Charles Hages, Phillip J. Dale

University of Luxembourg, Physics and Materials Science Research Unit, Belvaux, Luxembourg

University of Florida, Department of Chemical Engineering, Gainesville, FL, United States

Remarkable progress in photovoltaic has been made by the halide perovskites, paving the way for overcoming the Shockley-Queisser-limit of 33.7 % of a single junction solar cell by in tandem photovoltaic applications. However, the biggest challenge for the commercialization is the stability of the halide perovskites which are rather prone to degradation through air, light, heat, and electric field mechanisms. Alternatively, highly stable chalcogenide perovskite materials, containing S, Se instead of halides might maintain their remarkable optoelectronic properties. Among these materials, BaZrS3 with a bandgap of around 1.9 eV and strong absorption coefficient attracts most attention due to the possible integration as a top cell in tandem applications. However, the biggest hurdle seems to be the synthesis of these materials at moderate temperatures on conductive substrates in order to get phase-pure material. Here, we investigate the formation of secondary phases in the low-temperature synthesis of BaZrS3 at 550 °C, compatible with device integration, with various annealing durations. Raman analysis reveals secondary phases such as BaS3 and ZrS3 which cannot be readily identified from X-ray diffraction. Longer annealing times improve the phase purity of BaZrS3.

A direct comparison of thermoradiative and thermophotovoltaic operation of HgCdTe photodiodes

Michael P. Nielsen, Muhammad H. Sazzad, Andreas Pusch, Phoebe M. Pearce, Peter J. Reece, Nicholas J. Ekins-Daukes

School of Photovoltaic and Renewable Energy Engineering, University of New South Wales, Sydney, Australia

School of Physics, University of New South Wales, Sydney, Australia

The thermoradiative diode (TRD) is the symmetric counterpart to the photovoltaic solar cell that generates power via the net emission rather than absorption of light. While the TRD has enticing applications in night-sky power generation, there are also opportunities for power generation via waste heat recovery. However, while the theoretical limits for power generation are promising, the current technological limits have not been explored. Here we compare the electro-optical characteristics of HgCdTe photodiodes in operating in both thermoradiative and thermophotovoltaic (TPV) modes, supported by optical modelling. By contrasting thermoradiative and TPV operation using the same devices, we set realistic expectations for power generation using mid-infrared semiconductors.

Solar Simulator Performance Metrics: Balloon Flown Calibration Standards Offer Real Time AM0 Solar Simulation Error Measurements

Scott J Ireton, Casey P Hare, Andrew J Schwab

Angstrom Designs, Inc., Santa Barbara, CA, United States

Solar simulators are critical for photovoltaic power systems. Solar simulators are used in research, development, evaluation, and manufacturing. Solar Simulator metrics are defined by the ASTM standard (E927-10), which registers the highest capability as class AAA. However, class AAA was defined with single junction silicon in mind and has never been an adequate metric for multi-junction solar cells allowing for stacking of error in ways that can lead to large uncertainties. However, modern LED solar simulators and low-cost balloon flown calibration standards allow for direct measurement of the error in every junction in every cell of an array. This measurement is called a Zone Error measurement. Next generation AM0 solar simulator calibration standards are made by flying solar cell isotypes on Angstrom Designs small balloons. Balloon flown calibration provides a more accurate measurement of AM0 solar cell performance. These cells are then returned to earth and brought back to the lab to tune out solar simulator spectral mismatch between AM0 irradiance and solar simulator irradiance. The cost-effective nature of flying small balloons allows for solar simulator standards to be generated for each type of cell technology a solar simulator encounters. This saves time and improves measurement accuracy by preventing the need to measure solar simulator spectrum and solar cell quantum efficiency to perform a spectral mismatch calculation. This talk will provide an overview of modern LED solar simulator performance as well as small balloon flight capabilities before describing in detail how the direct measurement of zone errors using flown isotypes can decrease both measurement error and cost.

Operation Efficiency Gains from Analyzing Minimal Solar Cells Production Data

Johnson Wong, Kissenger Chen, Dinica Li, Bryan Matthews, Jason Miller, Sal Sanci

Aurora Solar Technologies Inc, Vancouver, British Columbia, V7P3N4, Canada

Abstract — On the typical solar cell factory floor, end-of-line I-V measurements and batch-level tracking data alone contain sufficient information about the underlying process tool performances for a variety of meaningful analysis that will lead to improved plant operational efficiency. In one PERC cell manufacturing site, an evaluation over two months data concluded that there is potentially 0.045-0.058% average efficiency gain by taking advantage of big-data analysis to perform preferential batch routing. When implemented to detect tool performance excursions, simulations for a larger 2-3 GW PERC plant show that an abrupt tool performance drop that negatively impacts cell efficiency by 0.05%, can be detected and traced to the correct problematic tool (out of 88 possible process tools) within 24 hours. This anomaly detection and root cause tracing can be further improved if other end-of-line and mid-stream measurements are also taken into consideration. Finally, the accuracy of the analysis technique in the face of tracking errors (imperfect batch integrity) is examined.

I. INTRODUCTION

Solar cell manufacturing is typically a batch based process, consisting of a handful of steps, where at each step each batch is routed to one of many identically configured process tools. There is some inherent randomness that results in a generally non-deterministic process path taken by each batch. Each day, there's also a large variety of process paths and it is not straightforward to glean information from them. This creates a situation where even if the process path and end-of-line efficiency of each batch were perfectly known, it would be difficult to trace issues detected at the end-of-line backwards to the problem source using simple heuristics. To get around this, some manufacturers divide the factory floor into "lines" where each line is one zone that the routing of batches is confined to on a best-effort basis. The problems with this system are 1) issues can be quickly narrowed down to the tools within a line, but not more specific than that, 2) such restriction hampers throughput because it is harder to have perfect throughput matching of all steps within each zone, 3) when throughput is the highest priority, the line confinement cannot be enforced strictly and there is significant crossing of batches from one line ot another.

To tackle the problem of delineating tool performance in this production environment, reference [1] introduced a factorial analysis method that can decouple the influence of process tools of different steps to the batch efficiency. This is a big-data analysis technique that is not only useful for problem tracing, but also for providing general quantification of tool performance so that resources can be focused on the lowest performing tools, and also that batches can be preferentially routed to the better performing tools when there is spare production throughput at any step. This paper presents results based on both actual production data as well as simulated data, to quantify the operational efficiency gains enabled by the factorial analysis technique.

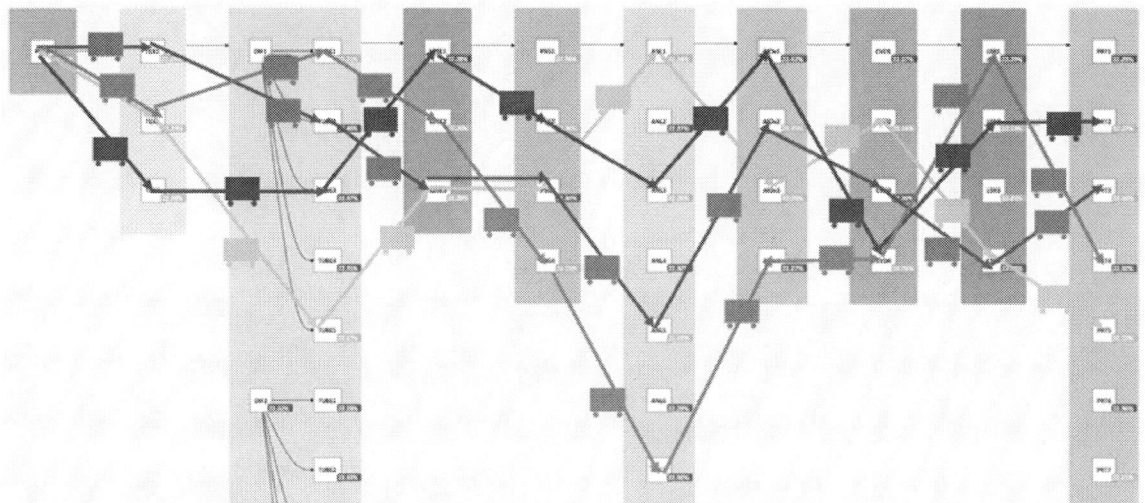

Fig. 1. Factorial analysis to survey entire factory production data to delineate performance differences in all tools. Different colour lines represent different cell batches that took on different process routes.

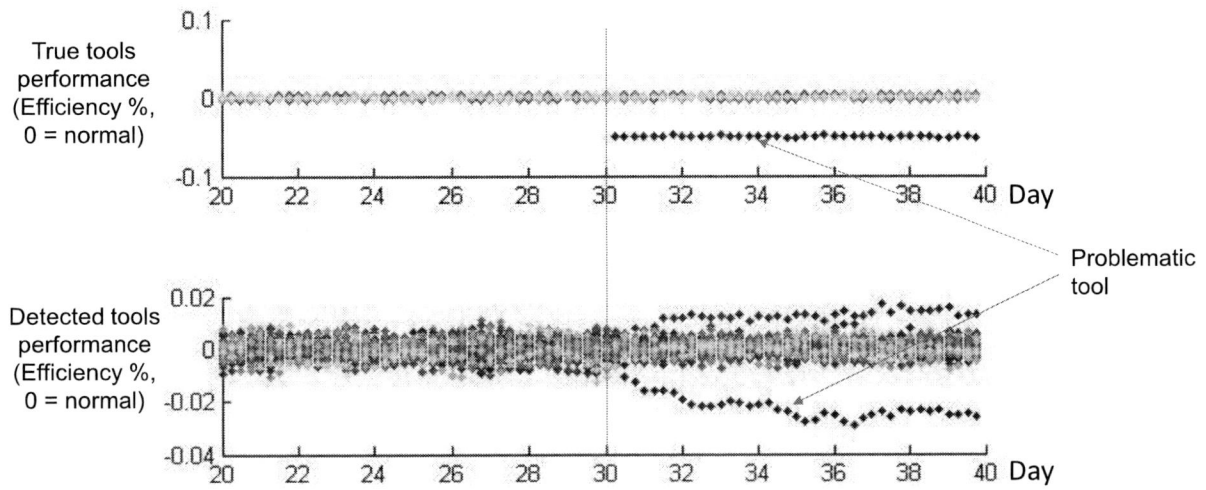

Fig. 2. Simulation results of true tool performance trends in time, versus the inferred tool performance in time from factorial analysis. There is built-in inherent batch-to-batch variation in efficiency of 0.04% when all tools of each process steps have identical performance. The problematic tool performance experiences a 0.05% drop on the 30th day.

II. RESULTS AND DISCUSSIONS

Firstly, in simulation we create a virtual 2-3 GW production line that consists of 9 process steps and a total of 88 process tools. There is some built-in inherent fluctuations in the tool performances such that, even if all the tools of each step were at stable and identical average performance, the batch to batch variation in mean efficiency would be about 0.04%. This is in line with what is typically observed at a PERC cell factory. Then, the simulation proceeds by generating 576 batches each day over 40 days. Each batch mean efficiency depends on the performance of the tools which produced it. The tool performance is quantified in terms of its impact on the efficiency, with zero being no net impact. For the first 30 days, all of the tools have zero impact, and then from the 30th day onwards one tool performance drops by 0.05%. This time trend is depicted on the upper half of Figure 2. The lower half of the figure illustrates the factorial analysis determination of the tool performance, based on the resultant batch efficiencies and their process paths. It can be seen that the underperforming tool is correctly determined to have lower than average performance that becomes clear after roughly 24 hours. While this may not be the optimal amount of time to trace an issue, it should be noted that human analysis of the raw data that arises from up to 88 machines and 576 batches each day, for instance by simple examination of batches filtered by tools, will have significantly worse resolution and will be far more prone to erroneous conclusions. Moreover, manual verification of guesses using diagnostic batches (experiments that compare two batches routed identically except at the process step of interest, where one batch splits off and is processed by the suspected low performing tool) is also prone to erroneous conclusion if the

tool performance degradation is about 0.05% but the batch to batch natural variation is 0.04%. Therefore, it is far cheaper, expedient and accurate to perform a one-step process of detection and root cause tracing using big-data methods, than to do a two-step process of first examining aggregate data to make guesses, and then doing verification using diagnostic batches.

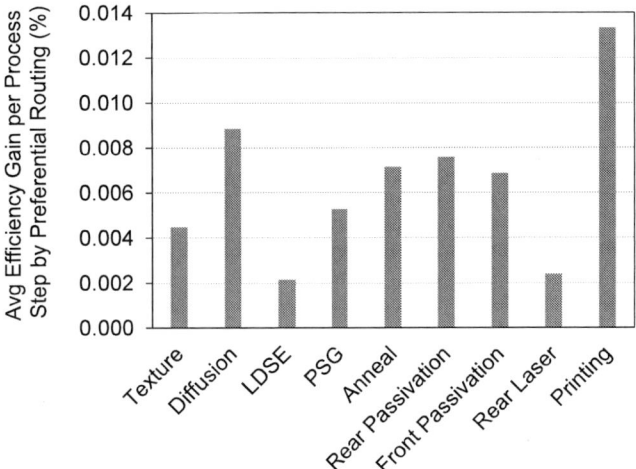

Fig. 3. Potential average efficiency gain per process step by preferentially routing batches to better performing tools when a process step has spare throughput (i.e. is not the bottle neck). The results are based on analyzing real production data (actual process paths, batch efficiencies, throughput, and capacity of machines).

Next, we make use of actual production data from a PERC cell manufacturing site, to determine the potential gain in average cell efficiency if batches were routed to better performing tools (as determined by the factorial analysis) over

978-1-6654-6060-6/23 $31.00 © 2023 IEEE

the course of two months. Figure 3 shows the breakdown of the average efficiency gain potential, by process step. The total average efficiency gain is 0.058% which is very substantial. Evidently, routing batches to better performing print lines (the last step) stands to gain the most, but the caveat is that because each print line is tied to a unique I-V tester, part of the efficiency gain may be attributed to the testers having different bias in efficiency measurement. Nevertheless, even if we discount the disparity in print line performance differences, preferential routing at the other steps still result in 0.045% average efficiency gain.

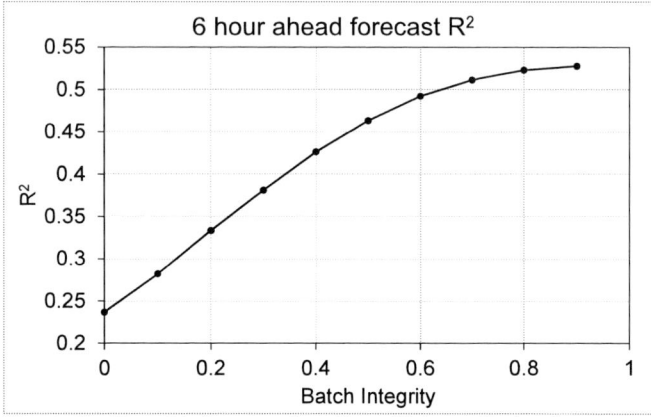

Fig. 4. Quality of inference of median efficiency for batches up to 6 hours after tool performance determination from factorial analysis, calculated based on the batches process paths. The x-axis denotes tracking error in the form of imperfect batch integrity.

In most factories today, there is bound to be some level of tracking errors, manifesting in some fraction of some batches not being traced to the correct process tools. The most common reason for tracking errors is the imperfection of batch integrity, or the assumption that the content (cells) of each batch remains the same from start to finish. To assess the impact of imperfect batch integrity, we have performed retrospective analysis on the PERC plant data by examining the consequence if some cells are shuffled between batches that are in close proximity when they are loaded into the printers. This leads to errors in the batch efficiency medians because now each batch contains a certain fraction of cells that don't actually belong in it. Figure 4 shows the quality of "6 hour ahead forecast" of batch efficiency at different levels of batch integrity. The forecasting is done by applying factorial analysis on batches paths and efficiency data up to time t, assuming that the solved tool performance persist unchanged into the future, and applying them on batches whose process paths are provided between time $t - t + 6$ hours to infer their median efficiency. The R^2 of agreement between the inferred and actual efficiency is then an indicator of the quality of the inference. We do not expect R^2

close to 1 to be achievable because the batch path information does not explain all the efficiency variance in the production environment. Evidently, the R^2 drops steeply only when the batch integrity drops below 0.7-0.8. This threshold of batch integrity can in fact be attained at most plants.

Figure 5 shows the potential average efficiency gain by preferential routing at the PERC plant, again at different levels of batch integrity. Once again, there is only steep drop in the potential gain when the batch integrity drops below 0.7 or so. This assures us that the factorial analysis method is fairly tolerant to imperfect tracking. In contrast, alternative methods that do not use stable statistics like batch median, such as anomaly detection based on electroluminescence (EL) defect features that occur in < 1% of cells, will be far more prone to imperfect tracking.

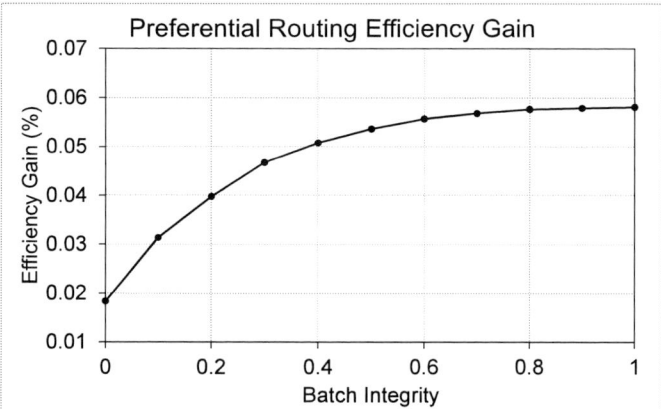

Fig. 5. Potential average efficiency gain by preferentially routing batches to better performing tools at the PERC plant. The x-axis denotes tracking error in the form of imperfect batch integrity.

III. CONCLUSIONS

Factorial analysis is a robust method that relies on only the most easily obtained production data in the solar cell plant, namely the cell efficiency and batch process paths. It can be used for anomaly detection and root cause analysis, as well as for preferential routing with significant efficiency potential gains. It is quite tolerant to tracking errors. Therefore it can be a ubitquitous and reliable method in any big data analysis system for solar cell production, including ones that may incorporate other kinds of data like electroluminescence (EL) images or midstream measurements.

REFERENCES

[1] J. Wong, D. Li, G. Deans, "Data Mining of Solar Cells Production Data Using Factorial Analysis", in *49th IEEE Photovoltaic Specialist Conference*, 2022.

Extracting electrical properties of CdTe, CdSeTe and CdSe thin films using a Parallel Dipole Line Hall effect system

Mustafa Togay[1], Rachael C. Greenhalgh[1], Kerrie Morris[1], Xiaolei Liu[1], Luksa Kujovic[1], Luis C. Infante-Ortega[1], Nicholas Hunwick[1], Adam M. Law[1], Tushar Shimpi[2], Sampath S. Walajabad[2], Eric Don[3], Gabor Parada[4], Kurt L. Barth[1], J. Michael Walls[1] and Jake W. Bowers[1]

[1]CREST, Wolfson School of Mechanical, Electrical and Manufacturing Engineering, Loughborough University, Loughborough, Leicestershire, LE11 3TU, United Kingdom

[2]NSF I/UCRC for Next Generation Photovoltaics, Colorado State University, Fort Collins, CO 80526 United States

[3]SemiMetrics Ltd., PO Box 36, Kings Langley, WD4 9WB, United Kingdom

[4]Semilab Co. Ltd., Prielle Kornélia u. 4/A. H-1117 Budapest, Hungary

Abstract — **Electrical properties of as-deposited CdSe, as-deposited and CdCl₂ treated CdTe, and CdCl₂ treated CdSeTe/CdTe films were investigated using a high sensitivity parallel dipole line (PDL) Hall effect system. The PDL Hall offers AC, and traditional DC magnetic field modes with an increased signal to noise ratio (S/N) for more reliable and accurate measurements. The steps used in the measurement process for the PDL Hall effect measurements are provided in detail. Carrier concentration, mobility, resistivity, sheet resistance and majority charge carrier type (p or n) of the films can be extracted from the Hall measurements. Schottky type behaviour has been observed from the resistivity measurements during the I-V scans of CdTe thin films. More suitable material for sample contacts, such as gold will be used to eliminate this non-ohmic behaviour. This will provide an improved signal to extract reliable results.**

I. INTRODUCTION

Thin film solar cells (TFSCs) based on CdTe and CIGS absorbers are devices that can provide low-cost and efficient large-scale solar energy conversion, offering a wide variety of choices in device structure, fabrication, and applications. The most fundamental measurement technique which provides information on the overall performance of the device under standard test conditions (STC) is the current density-voltage (J-V) measurement. For more advanced electrical properties of TFSC materials such as carrier concentration, mobility, resistivity, and majority charge carrier type (p or n) which are not highlighted in standard electrical characterization, the Hall effect is used as a measurement approach to cover these properties. Such properties provide valuable insights to help understand the overall device performance and assists the exploration of the various options of materials in development

of new emerging solar cell materials (window layer, absorber, buffer, back-contact) for device application.

A standard conventional Hall effect system operates with a constant magnetic field (DC) using the Van der Pauw configuration [1]. These measurements are used to determine the carrier concentration and mobility by measuring the Hall voltage at constant magnetic field. Many polycrystalline materials have very low mobility, very high resistance (such as thin films) or very low resistance (such as metals). It is difficult to measure a reliable Hall voltage due to the low signal to noise ratio (S/N). The S/N can be improved by increasing the magnetic field or drive current or by reducing the overall measurement noise. In parallel dipole line (PDL) Hall measurements, the magnetic field is varied using two simultaneously rotated magnets [2]. A Lock-in signal detection is used to filter the desired Hall signal (X) from the electrical noise (Y). A variable (AC) magnetic field along with modulated drive currents, and a Lock-in detection of the Hall signal can enhance the S/N to provide more reliable measurements.

In this work, we utilize the PDL Hall effect system to measure the electrical properties of different TFSC materials that are deposited on an insulating substrate. Materials such as CdTe, CdSe and CdSeTe (CST) have been studied to extract the carrier concentration, mobility, resistivity, and majority charge carrier type of the films. Such materials are problematic and/or impossible to measure with a DC Hall measurement due to the low Hall signal. IIence, the AC Hall measurement is used for these materials. In addition, the effect of the CdCl₂ activation treatment on electrical properties has been studied for the CdTe and CdSeTe absorbers.

A. Film deposition on glass substrates

Hall measurements require the thin film to be deposited on an insulating surface. The CdSe films used for this study were deposited on soda lime glass (SLG) substrates, and the CdTe films were deposited on ZrO_2 coated Abrisa Eagle XG® glass substrates at Loughborough University, UK. A thin layer of ZrO_2 coating promotes better adhesion of CdTe onto the substrate. The CdTe absorbers were deposited in a closed-spaced sublimation (CSS) system [3], and the CdSe films were deposited using pulsed DC magnetron sputtering [4]. The thicknesses of the as-deposited films were: ~ 4 μm of CdTe and ~ 1 μm of CdSe. The as-deposited CdTe absorbers were also $CdCl_2$ treated to measure its effect on the individual film properties. The CdSeTe/CdTe films were fabricated on NSG-Pilkington TEC™ SB glass substrates using an in-line vacuum deposition system at Colorado State University (CSU) [5]. The film thickness for as-deposited CST/CdTe is about 3.5 μm. These films were measured before and after the $CdCl_2$ treatment.

B. Semilab PDL Hall effect system

The Hall effect measurements were performed using a high sensitivity PDL Hall effect system from SemiLab, Hungary [6]. The PDL system is capable of working in both AC and conventional DC measurement modes. The AC field can be used for materials with mobility below 0.1 cm²/Vs, and it can measure highly resistive materials by providing an enhanced S/N for reliable measurements. The system can measure very high carrier concentrations over 1×10^{21} cm^{-3} and very low carrier concentration down to 1×10^{12} cm^{-3}.

Fig. 1. The experimental setup used for Semilab PDL Hall effect system.

Fig. 1 shows the experimental setup used for the PDL Hall effect system. It uses two magnets, a driver (at the bottom) and a follower magnet (at the top) with a large magnetic field, approximately 2.5 T peak to peak. The driver magnet is rotated by a motor and the follower magnet mirrors its rotation. The sample is placed between the two magnets using a sample holder. The system is capable of measuring resistivity, sheet resistance, mobility, carrier concentration and carrier type of the material. The system also has the capability to measure the concentration of both majority and minority carriers, as well as the recombination lifetime, diffusion length and recombination

coefficient using carrier-resolved photo-Hall effect measurements when implemented with additional hardware [7].

C. Sample preperation and Hall effect measurements

To prepare samples for Hall measurements, 0.5 cm x 0.5 cm samples were cut from the substrates, and four metal alloy contacts were put at the corners using an ultrasonic soldering iron. Then, samples were mounted on a sample holder with thin aluminum wires for each contact using a soldering iron. The sample holder has 12 tin coated pads and 4 of them are used for the Van der Pauw configuration when performing Hall effect measurements. The measurement follows 3 important steps to extract reliable and accurate results, the contact check, the sheet resistance measurement, and the Hall resistance measurement.

In step 1, contacts between the sample and the sample holder are checked to ensure there is a good connection established. 2-point resistance of each contact is measured. This is used to check the linearity of I-V response from each measurement for each contact, and if there is no linear I-V response, then the connections should be improved before moving to the next step.

In step 2, the sheet resistance and sample resistivity measurements are performed using the Van der Pauw configuration. The I-V response of the signal is measured. The linearity of the I-V response is assessed using the correlation coefficient (R) of the measurements. The measurement with a correlation coefficient equal or higher than 0.98 (maximum R=1) indicates a linear response. Any non-linear I-V response will affect the final result from step 3 when extracting the carrier concentration and the mobility, and so the importance of contacts with no potential barriers between the metal and semiconductor (such as a Schottky contact) is clear.

In the final step, the Hall voltage is recorded simultaneously from the Van der Pauw configuration with changing magnetic field as the two magnets rotate at a constant angular speed. Moreover, numerical Lock-in detection is performed, and Fourier spectra of the magnetic field and the raw Hall signal is extracted to determine the carrier concentration, its carrier type and mobility.

D. Check list for a successful Hall effect measurement

Preparation of samples is critical as the samples are required to be very small and square to obtain a uniformly distributed magnetic field. However, there are other variables to take in consideration. For a successful and reliable Hall effect measurements the following conditions are required:

- Linear I-V responses from the contact check (step 1) and conductivity measurements (step 2)

- From the Hall resistance measurements (step 3)

  - at least 10 cycles of magnetic field

  - a visible Hall signal variation which follows the signal of the magnetic field (and this is difficult to detect for highly resistive samples)

  - clear Fourier Transform (FT) peaks from both the magnetic field and Hall signal, and both FT peaks are aligning with each other

- o a stabilized Lock-in detection (where the magnitude of Lock-in X > magnitude of Lock-in Y)
- o results from the Lock-in and the Fourier analysis are close to each other

In order to meet the requirements, there are many parameters that can be controlled to achieve successful measurements. The measurement duration time, drive current, and the speed of the rotating magnets are the most important parameters. The speed of the rotating magnets can be varied from 0.1-2 rpm. The duration time for the measurement depends on the Hall signal intensity, and it can be in the range from 30 minutes to 10 hours. Measuring the Hall effect on films which are highly resistive can be quite challenging as it is difficult to detect visible Hall signal variation with the magnetic field. Carrier concentration can be extracted from both FT analysis and the Lock-in detection. The sign of Lock-in X indicates the carrier type of the material.

II. Preliminary Results and Discussion

TABLE I. Extracted parameters from the AC PDL Hall effect measurements

Samples	Sheet Resistance [$\Omega/\square$]	Sample Resistivity [$\Omega.cm$]	Carrier Concentration & Type [cm^{-3}]	Mobility [cm^2/Vs]
as-deposited CdSe	5.43×10^3	5.75×10^{-1}	1.98×10^{18}, N	6.2
as-deposited CdTe	4.95×10^9	4.95×10^6	1.10×10^{12}, P	17.6
CdCl$_2$ treated CdTe	3.96×10^8	3.96×10^5	2.80×10^{14}, P	0.1
CdCl$_2$ treated CST/CdTe	8.72×10^9	1.65×10^6	1.63×10^{14}, P	4.5

Table 1 shows the extracted parameters from PDL Hall measurements for as-deposited CdSe, as-deposited and CdCl$_2$ treated CdTe and CdCl$_2$ treated CdSeTe/CdTe films. All films provided linear I-V responses from step 1, indicating a good connection between the sample and the sample holder. However, for step 2 only the as-deposited CdSe sample provided a linear I-V response, and it then passed all the checks in step 3. The other films displayed Schottky type behaviour in the I-V response from step 2, and no clear FT peaks were detected. When no linear I-V response from step 2 is found, the mobility value is the one of the parameters most affected as it is calculated from this step. Thus, in this case the final extracted value for carrier concentration and mobility are not reliable. Although the measurement is not successful, the extracted carrier concentration for CdTe absorbers is around 10^{14} cm^{-3} which is in the range of what has been previously reported for undoped CdTe absorbers [8]. An increase of carrier concentration by 2 orders of magnitude after CdCl$_2$ treatment was observed. This was expected as the CdCl$_2$ treatment results in the removal of defects, increase in grain size and passivation of grain

boundaries. These improvements in microstructure enhance the electrical properties of the film [9].

III. Future Work

For highly resistive p-type absorbers especially CdTe and CdSeTe films, gold contacts will be used for the preparation of samples. The gold contacts will be deposited using a metal evaporator through a specially designed mask to deposit 4 even contacts at the corners of each sample. A wire bonder will be used along with a new sample holder, consisting of gold coated pads replacing tin to enable secure bonding with a thin gold wire. Measurements providing a non-linear I-V response will be repeated using the new configuration. In addition to CdTe related absorbers, results will be reported from various metal and alloy buffer layers and CIGS based absorbers using the PDL Hall effect measurement system. In addition, the effects of a series of light soaking experiments will be reported to investigate metastable behaviour that occur on electrical properties on thin film PV absorbers.

Acknowledgements

The authors are grateful to EPSRC grant for funding through EP/S017690/1, EP/V013858/1, EP/P02484X/1 and EP/N510014/1 to support this work.

References

[1] D. Schroeder, "Semiconductor Material and Device Characterization Third Edition," John Wiley & Sons Inc., 2006.

[2] O. Gunawan, Y. Virgus, and K. F. Tai, "A parallel dipole line system," Appl. Phys. Lett., vol. 106, no. 6, 2015.

[3] C. Potamialis, F. Lisco, J. W. Bowers, and J. M. Walls, "Fabrication of CdTe Thin Films by Close Space Sublimation," in PVSAT-12 UoL, 2016.

[4] R. C. Greenhalgh, V. Kornienko, K. Morris, A. Abbas, J. W. Bowers and J. M. Walls, "High Rate Deposition of CdSe Thin Films by Pulsed DC Magnetron Sputtering," 2020 47th IEEE Photovoltaic Specialists Conference (PVSC), Calgary, AB, Canada, pp. 2132-2135, 2020.

[5] D. E. Swanson, J. Kephart, P. S. Kobyakov, K. E. Walters, K. C. Cameron, K. L. Barth, W. S. Sampath, J. A. Drayton and J. R. Sites, "Single vacuum chamber with multiple close space sublimation sources to fabricate CdTe solar cells," J. Vac. Sci. Technol. A Vacuum, Surfaces, Film., vol. 34, no. 2, p. 21202, 2016.

[6] G. Paráda, F. Korsós, A. Bojtor, J. W. G. Bos, E. Don, J. W. Bowers, and M. Togay, "Exploiting Bi-Modulated Magnetic Field and Drive Current Modulation to Achieve High-Sensitivity Hall Measurements on Thermoelectric Samples." MRS Advances, vol. 7, no. 0123456789, pp. 608-613, March, 2022.

[7] O. Gunawan, S. R. Pae, D. M. Bishop, Y. Virgus, J. H. Noh, N. J. Jeon, Y. S. Lee, X. Shao, T. Todorov, D. B. Mitzi, and B. Shin, "Carrier-resolved photo-Hall effect," Nature, vol. 575, no. 7781, pp. 151–155, 2019.

[8] C. Li, Y. Wu, J. Poplawsky, T. J. Pennycook, N. Paudel, W. Yin, S. J. Haigh, M. P. Oxley, A. R. Lupini, M. A. Jassim, S. J. Pennycook and Y. Yan, "Grain-boundary-enhanced carrier collection in CdTe solar cells," Phys. Rev. Lett., vol. 112, no. 15, pp. 1–5, 2014.

[9] A. H. Munshi, J. M.Kephart, A. Abbas, A. Danielson, G. Gelinas, J. N. Beaudry, K. L. Barth, J. M. Walls, and W. S. Sampath, "Effect of CdCl$_2$ passivation treatment on microstructure and performance of CdSeTe/CdTe thin-film photovoltaic devices," Sol. Energy Mater. Sol. Cells, vol. 186, pp. 259–265, 2018.

Power Hardware-in-the-Loop Interface Method for Grid Forming Inverters Using a Voltage-Controlled Power Amplifier

Javier Hernandez-Alvidrez, Rachid Darbali-Zamora, Jack D. Flicker, and Nicholas S. Gurule

Sandia National Laboratories, Albuquerque, New Mexico, 87185, USA

Abstract – **Incorporating grid-forming inverters (GFMI) into a power hardware-in-the-loop (PHIL) setup presents critical stability challenges if the GFMI operates in isochronous mode. Attempting to use the ideal transformer method (ITM) to interface a GFMI may pose the risk of catastrophic damage to either: the GFMI, or the power amplifier. This is caused by the lack of a synchronization mechanism in some commercially available GFMI designed to operate only in isochronous mode. One possible solution is to use a power amplifier that operates in current-control mode. However, most of the power amplifiers available in the market operate only under voltage-control mode. This paper presents a novel yet simple method to convert a voltage-controlled power amplifier into a current-controlled device by using the computational capabilities of a real-time simulator to embed the current control. This way, a stable synchronization between the GFMI and the power amplifier can be achieved, and the PHIL setup stability can be maintained.**

Keywords – Grid-forming, photovoltaic inverters, real-time simulation, power hardware-in-the-loop, blackstart, stability analysis.

I. INTRODUCTION

As the proliferation of the interfacing of grid-forming inverters (GFMI) into distribution and transmission systems continues, it is essential to test the dynamics of such GFMI by the use of rigorous and automated tests that emulate abnormal scenarios such as faults, loss of generation, islanding, and blackstarts. Such tests must facilitate the understanding of stability benefits and reliability impacts when the GFMI interacts with either large-scale grids or microgrids. This can test how a GFMI system interacts with the larger grid or how a GFMI inverter may work in a distribution system with significant Photovoltaics (PV) penetration. Furthermore, the testing of GFMI should also focus on the compliance of such devices with the industry standards currently under development [1]. One common way to test power devices is by incorporating them into a power hardware-in-the-loop (PHIL) setup.

Initially, the first PHIL interfaces were related to grid-following inverters (GFLI), which require a voltage reference for their internal synchronization mechanism, usually a phase-locked-loop (PLL), to operate [2]. Once synchronized, the GFLI interfaces with the reference by regulating the amount of real and reactive power injected into it. Under this functionality, the PHIL interface with a GFLI allows a direct interconnection between the power amplifier and the GFLI, as depicted in Fig. 1. The PHIL interfacing method represented in Fig. 1 is called the ideal transformer method (ITM), and it is widely used to test GFLI [3]-[4].

II. BACKGROUND

The interfacing of GFMI into PHIL setups presents its challenges when the GFMI operates in isochronous mode and without a synchronization mechanism. Under this operation mode, the GFMI regulates the voltage and frequency at its terminals, but it does not utilize a PLL to synchronize with the grid before interconnection. This can pose a hazard if a direct connection between the GFMI and a voltage source (such as a power amplifier) is attempted. Since both sources will most likely be out of synchronization, the amplitude and phase differences between the two sources will cause an unregulated flow of real and reactive power that could potentially damage one of the devices involved. Therefore, the ITM must not be used if the GFMI does not have a synchronization mechanism, which is the case for some commercially available GFMI [5],[6],[7]. Because of this rationale, a novel method to interface a GFMI into a PHIL setup is presented.

This method aims to incorporate an external synchronization mechanism, such as a PLL, to regulate the frequency and phase angle of the power amplifier used in a PHIL setup. Furthermore, the output current of the power amplifier is controlled using the classic *dq*-frame current control [2]. Including the PLL intrinsically forces the power amplifier to be driven by current signals, a feature that very few commercially available power amplifiers include nowadays. Therefore, once the power amplifier is synchronized with the GFMI, both devices can be interconnected, and the power flow between them can be regulated using conventional control schemes like proportional-integral (PI) controllers.

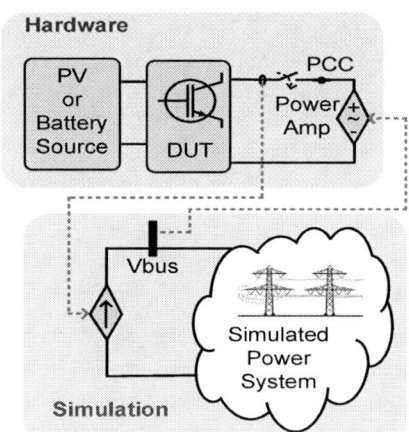

Fig. 1. Ideal transformer method used to interface a GFLI.

III. Proposed PHIL Interface Method

The critical aspect of the proposed method is to control the power amplifier using current signals instead of voltage signals. To accomplish this functionality, a PLL needs to be incorporated in the real-time simulator, as depicted in Fig. 2 The physical elements section of the setup is composed of: the GFMI, the power amplifier, and an external interface inductor (L) that avoids direct interconnection between the GFMI and the power amplifier. The control section of the setup is also implemented in the real-time simulator, and it has the following elements: a dq-frame based PLL, an abc to dq transformation block, a couple of PI controllers to regulate the current of the system, and a dq to abc transformation block that brings back the time-varying signals that will drive the power amplifier.

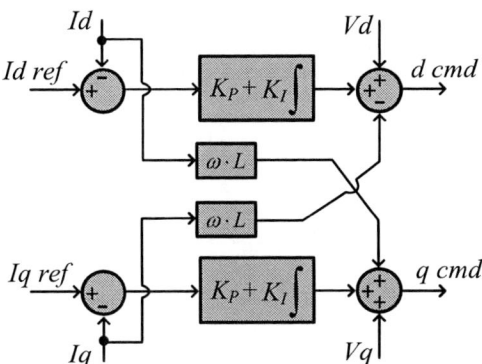

Fig. 3.   Detailed diagram of the dq current control scheme

Notice that the sensed three-phase voltage signals (V_{abc}) not only drive the PLL but also are used to drive the voltage-controlled voltage source (V_{sim}) in the simulated power system, which for this particular case is only a load bank (P, Q). The simulation currents (I_{sim}) are transformed into the dq domain, and these transformed values become the PI controllers' reference values ($I_{dq\ ref}$). If such controllers are appropriately tuned, the physical currents I_{abc} will track the simulation currents I_{sim}, and the real and reactive power delivered by the GFMI to the power amplifier (P_{inv} and Q_{inv}) will match the real and reactive power that V_{sim} provides to the load bank (P_{sim} and Q_{sim}).

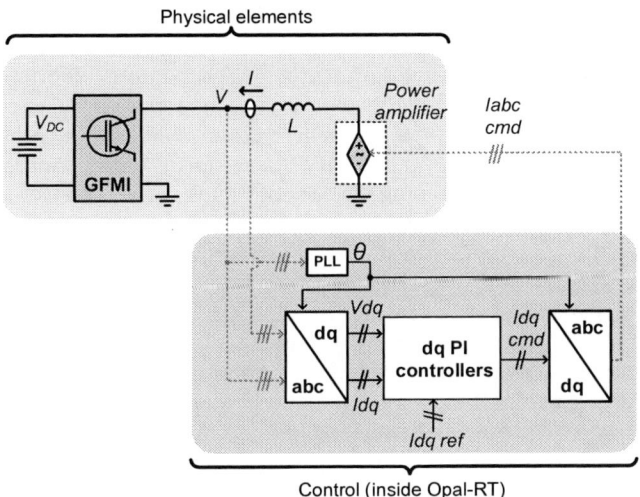

Fig. 2. Converting the power amplifier into a current-controlled device.

The PLL provides the angle reference (θ) for the dq transformation blocks. The system's control variable is the sensed current signal (I) at the terminals of the power amplifier. The current is transformed into its d and q components. Each current component is then compared to a current reference value provided by the simulated system that will be interfaced with the GFMI. A more detailed diagram of the dq current control scheme is shown in Fig. 3. The use of cross-coupling terms (ωL) helps to decouple each control loop from each other, facilitating the design and tuning of the PI controllers [2]. Also, notice the use of the feedforward signals Vd and Vq, which are the d and q components of the reference voltage provided by the GFMI. The outputs of the current control scheme are the commanded dq-signals that are transformed back into a set of three-phase positive-sequence currents, which drive the analog inputs of the power amplifier.

As of yet, the description of the proposed method has focused merely on controlling the power amplifier with current-regulation schemes. Another important use of the real-time simulator is to host the simulated power system interacting with the GFMI. The reference current signals provided to the PI controllers ($I_{d\ ref}$ and $I_{q\ ref}$) come from the simulated power system, as depicted in Fig. 4, which shows the proposed PHIL interfacing method applicable for a three-phase GFMI.

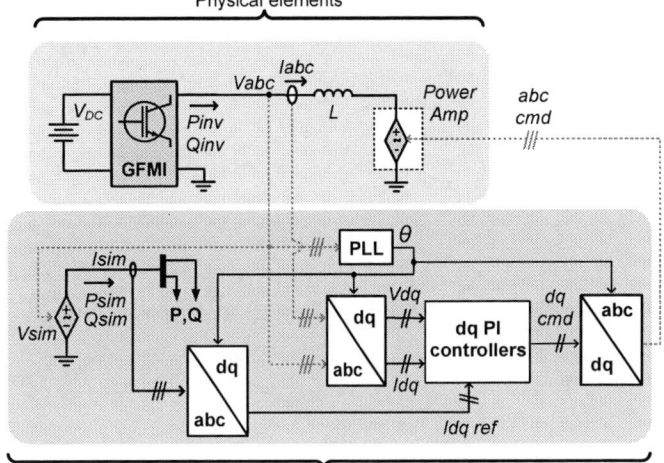

Fig. 4.   Detailed diagram of the proposed PHIL setup.

IV. Simulation Results

To demonstrate the concept of the proposed method, the aforementioned physical section was also emulated using the internal FPGA of the real-time simulator, which provides the capability of smaller time steps [8]. Such capability becomes very useful when simulating switching devices, like the IGBTs used in the H-Bridge of the GFMI. The fully emulated method is depicted in Fig. 5, where the FPGA is assigned to: the switching devices of the GFMI, the inductor L, and the power amplifier. The control scheme of the GFMI is the well-known CERTS control [9], and it is implemented in one of the CPU cores of the real-time simulator because the FPGA section does

978-1-6654-6060-6/23 $31.00 © 2023 IEEE

not allow either control blocks or transfer functions in it. Using the system depicted in Fig. 5, a blackstart scenario was tested using the simulation parameters described in Table I.

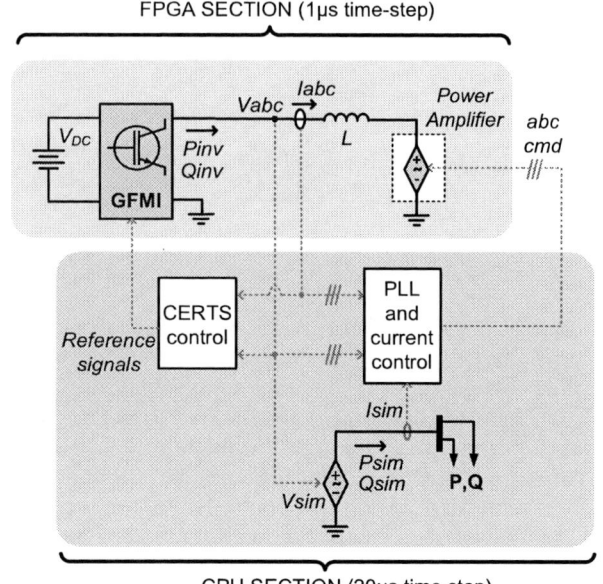

Fig. 5. Proposed PHIL interface method with emulated physical elements.

TABLE I. GFMI SIMULATION PARAMETERS.

Parameter	Value	Units
GFMI rating	100	kVA
GFMI rated voltage	480	V
Interface Inductor	1	mH
CERTS voltage droop	5	%
CERTS frequency droop	1	%
Simulated load (P)	70	kW
Simulated load (Q)	30	kVAR

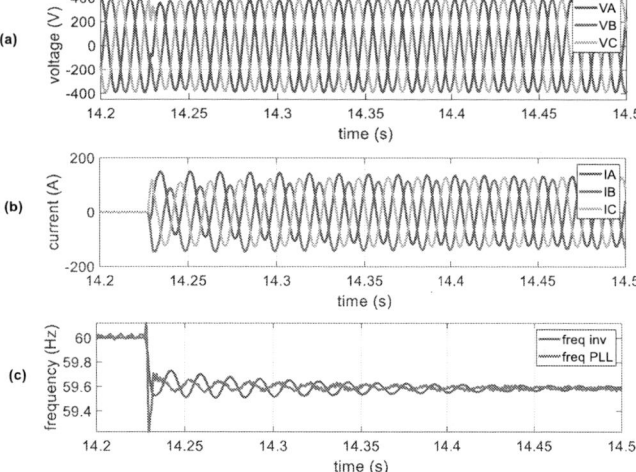

Fig. 6. Simulation results of a blackstart scenario. (a) Voltage. (b) Current. (c) Frequency.

The results from the simulation with emulated physical components are shown in Fig. 6. The voltage and current traces are depicted in Fig. 6-(a) and Fig. 6-(b), respectively. Notice

that the voltage traces show a small transient when the load is enabled. The traces of the currents also show a DC transient component that resembles the sub-transient dynamics of a synchronous generator. Fig. 6-(c) compares the GFMI and PLL synthesized frequencies. Notice that under steady state, both traces converge to the same value, which is dictated by the droop characteristic.

V. EXPERIMENTAL RESULTS WITH PLL

Before implementing the complete control scheme depicted in Fig. 2, designing the PLL while considering an appropriate bandwidth and phase margin is essential. The PLL scheme used for this method is the classic Synchronous Reference Frame PLL (SRF-PLL). The diagram of a single-phase SRF-PLL is shown in Fig. 7, where the parameters K_{PPLL} and K_{IPLL} determine the bandwidth and phase margin of the SNR-PLL.

The dynamics of the SRF-PLL are inherently nonlinear. Still, it is a common practice to linearize it around an equilibrium point on which the q component is regulated to a set value of zero. This forces the d component to correspond with the peak value of the voltage at the GFMI's terminals. Furthermore, linearization allows tuning the correct values of K_{PPLL} and K_{IPLL} to a point where the desired open loop frequency response of the PLL can be obtained. For this particular SRF-PLL, the linearization and frequency response plots for amplitude gain and phase angle were obtained using the procedure in [10]. This response is depicted in Fig. 8, showing a bandwidth of 20 Hz and a phase margin of 65 degrees. The chosen bandwidth is low to prevent voltage transients from interfering with the SRF-PLL dynamics and potentially making the system prone to instability [10]. The phase margin was designed according to the recommended linear control techniques explained in [11] to guarantee a low overshoot response to step changes.

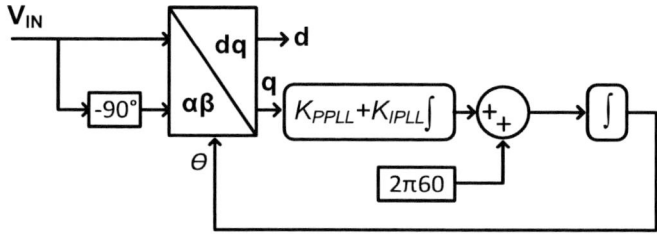

Fig. 7. SRF-PLL diagram.

To assess the stability of the SRF-PLL, the experimental setup shown in Fig. 9 was used. The GFMI is a commercially available device that operates in split-phase mode. For this reason, two interfacing inductors were used (one for each phase). TABLE II summarizes the values of the components, simulation parameters, and the inverter ratings. The PLL uses per-unit voltage signals at its input. It is a reduced setup since it does not have the current controllers, but the SRF-PLL still helps to synchronize the user-given voltage components V_{dcmd} and V_{qcmd} with the angle reference provided by the GFMI. Since time delays are associated with the sensed voltage, the parameter δ helps compensate for those delays by adding a constant angular value to the synthesized phase angle θ.

TABLE II. PARAMETERS OF THE EXPERIMENTAL SETUP WITH PLL ONLY.

Parameter	Value	Units
GFMI rating	5	kVA
GFMI rated voltage	240	V
Interface Inductors	1	mH
K_{PPLL}	182	Hz/V
K_{IPLL}	9155	Hz/V

Before closing the contactor between the GFMI and the power amplifier in Fig. 9, the PLL was engaged in the simulation until the synthesized frequency was 60 Hz. Then, V_{dcmd} was set to the peak value of the GFMI's rated voltage (339 V), whereas V_{qcmd} was set to zero. This way, the voltage traces of the GFMI and the power amplifier are the same right before the contactor is commanded to close. Under this equal voltage condition, the power the GFMI provides is zero. Later, the variable δ was varied in small increments that delayed the phase angle of the voltage of the power amplifier with respect to the phase angle of the GFMI. This way, the real power provided by the GFMI can be controlled. Since the main control objective is to regulate the system's current, it is essential to understand the current dynamics under different values of real power delivered by the GFMI.

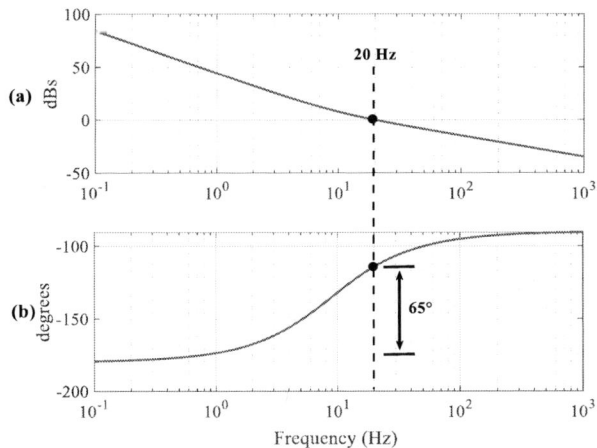

Fig. 8. Open loop SRF-PLL's frequency responses. (a) Magnitude Gain. (b) Phase angle.

Fig. 10 shows the current trace of the GFMI for the PHIL case using only the SRF-PLL. During the experiment, the angle δ was varied, so the GFMI delivered 500 W, 1500 W, and 5000 W. Notice how the amplitude levels of the current trace correlate with such increasing power levels in the form of a staircase. A magnified portion of the time-varying current trace is shown during steady state at rated power (5 kW). The harmonic content in the magnified current trace resulted from a mismatch between the nominal values (1mH) of the physical inductors used in the experiment. This also translates into significantly noisy dq current components. Ideally, the dq-current components must be a DC quantity during steady state. Still, the harmonics caused by the inductors' mismatches introduce significant noise into the DC value of the d and q components of the current, as depicted in Fig. 11 and Fig. 12. From Fig. 11, notice in the magnified portion of the current how the noise oscillates around the corresponding steady-state value of $I_d=25$ A. In order to address the issue with the noise, stringent

low-pass filters must be included, but the current controller in the dq-frame must account for the dynamics and phase margin reduction introduced by the filters.

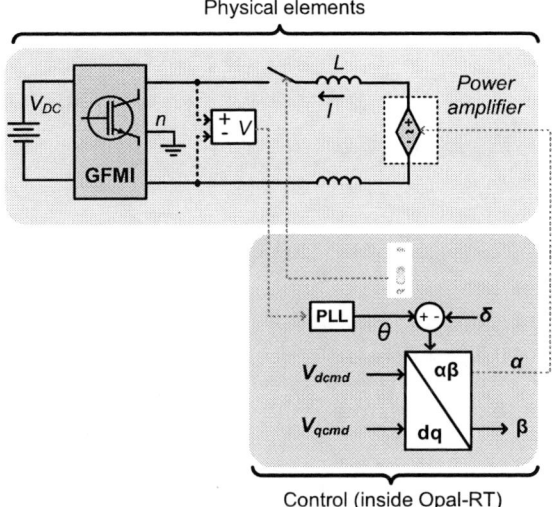

Fig. 9. Experimental setup to test the SRF-PLL.

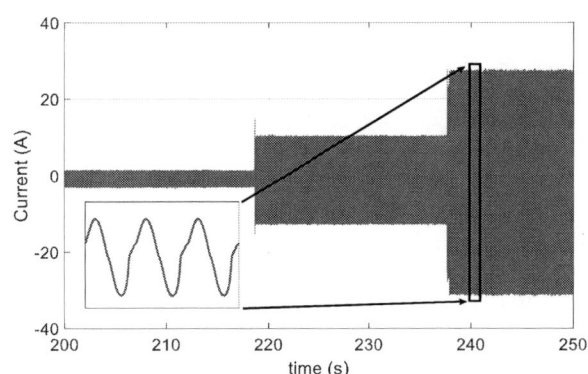

Fig. 10. Current traces for PHIL interfacing using only the SRF-PLL.

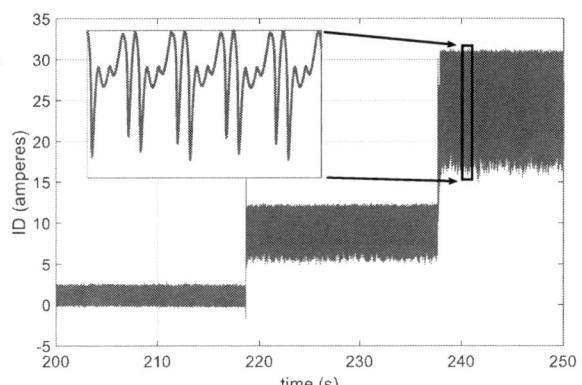

Fig. 11. d-component of the current trace.

978-1-6654-6060-6/23 $31.00 © 2023 IEEE

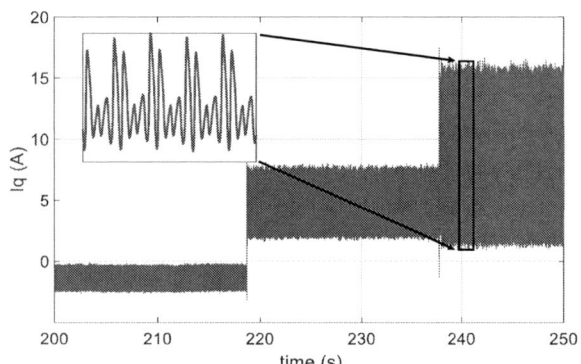

Fig. 12. *q*-component of the current trace

The preceding experiment qualifies as an interfacing method of GFMIs into PHIL testbeds that could be considered an extension of the method proposed in [5], but with the synchronization reinforced by the SRF-PLL. Furthermore, this experiment could play an essential role in characterizing the current's open-loop frequency response while incorporating the dynamics of the SRF-PLL. Such characterization could be performed by injecting small signal sinusoids around an equilibrium point (linearization) of V_{dcmd} and V_{qcmd} by injecting small signal sinusoids (v_{ds} and v_{qs}) and measuring the current outputs of the same frequency (i_{ds} and i_{qs}) to obtain the amplitude gain and phase difference. The diagram of the experimental setup for such characterization with the injection and measuring points is shown in Fig. 13. It is important to bear in mind that this experiment operates on the *dq* components of the current, but given the amount of noise on such components, the characterization for this specific inverter could not be performed. Nevertheless, if the harmonic content of the current permits, it is beneficial to attempt this plant characterization since it could provide insightful information about the current components' frequency response, which can provide the bandwidth and phase margin for each control loop. Knowing these two parameters can make the initial tuning of the current PI regulators less prone to instability [11],[10].

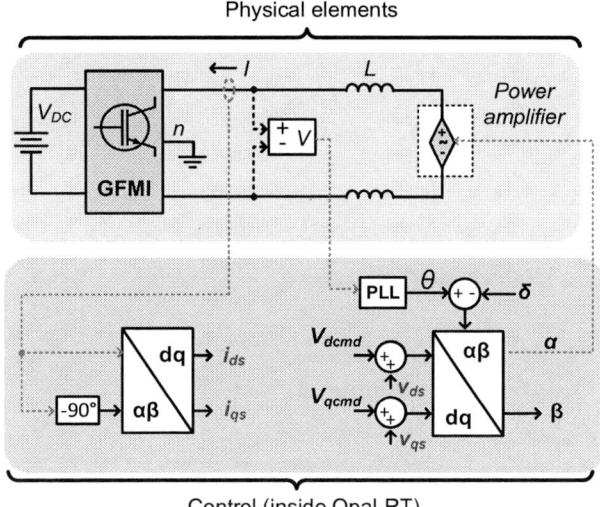

Fig. 13. Experimental setup for plant characterization.

VI. Experimental Results with Current Control

The proposed interfacing method that includes the current PI regulators in the *dq*-frame (see Fig. 2.) was also implemented using the same commercially available GFMI. Due to the high noise content in the *dq* current components shown in the previous experiment, the foregoing plant characterization method could not be achieved. Instead, for the first experiment, we relied on the PI tuning parameters of the emulated circuit as the initial parameters for the first test setup, which was used to assess stability. Such initial parameters are summarized in Table III. For this experiment, both PI regulators were engaged while their correspondent references were set to zero. Subsequently, $I_{d\,ref}$ was stepped up to a value of 5 A while $I_{q\,ref}$ was kept at zero. Fifteen seconds later, $I_{d\,ref}$ was brought back to zero. The test's results are summarized in Fig. 14, where the four most relevant traces are shown, with each one pointed to its corresponding location in the control diagram for ease of reference. The two plots on the left show the traces of the current's *d* and *q* components (green colored) and their corresponding set values (black dashed traces). Notice how the I_d component tracked the steps (up and down) applied at $I_{d\,ref}$ but with significant overshoot and slowly damped low-frequency oscillations of about 0.75 Hz. The same dynamics occurred on the I_q component while tracking the corresponding reference value $I_{q\,ref}$, which was set to zero during the entire test. Moreover, the other two plots on the right of Fig. 14 depict the traces of the outputs of each PI regulator, where it can be noticed that the same low-frequency oscillations appeared again.

From a reference tracking perspective, the dynamics in Fig. 14 show a stable system with very poorly damped responses. Efforts to withstand the oscillations are discussed next.

TABLE III. Initial Tuning Parameters of the Current-Control Experiment.

Parameter	Value	Units
K_P of current regulators	1	V/I (Ω)
K_I of current regulators	3	V/I (Ω)
K_{PPLL}	182	Hz/V
K_{IPLL}	9155	Hz/V

VII. Discussion

Correcting the transient dynamics shown in Fig. 14 represents the second most challenging part after achieving the system's stability. However, it is important to highlight the importance of the simulation with the emulated physical devices since it provided stable tuning parameters for both the PLL and the current PI regulators. The best further approach to improve the transient response is to tune the parameters K_P and K_I of the current regulators [12]. As an initial approach, both values were reduced (K_P=0.4 and K_I=2.4), but the system reached instability immediately, causing damage to the overvoltage circuit board of the inverter.

The system also showed slower oscillations (0.8 Hz). Since this test was unsuccessful to the point of damaging a protection circuit, it is always recommended to use low-rating converters during the initial tests, this way, the economic impact of the failure is minimized.

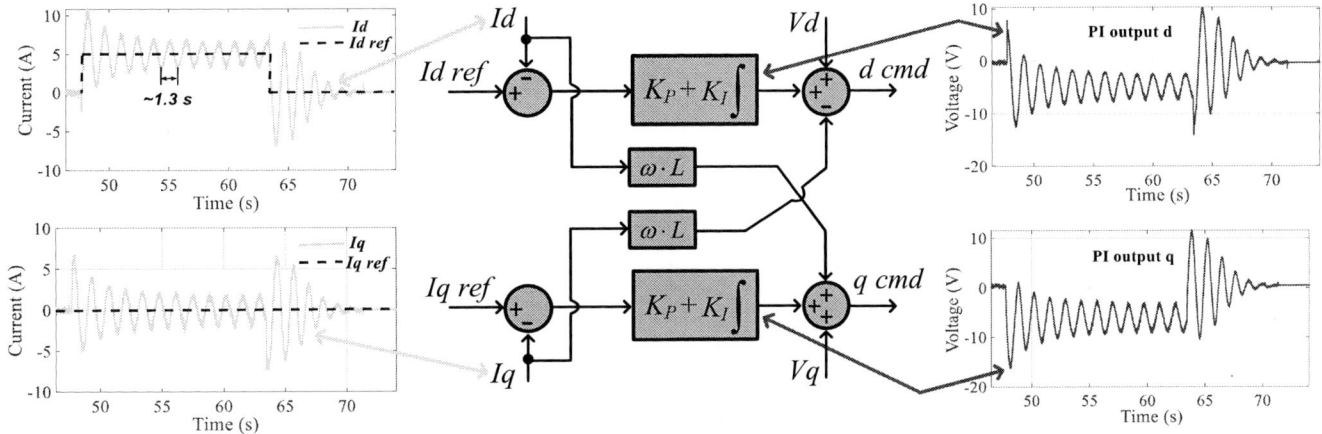

Fig. 14. Experimental results using *dq* current control scheme.

Further tuning steps will involve increasing the values of *KP* and KI, but this time putting more stringent operational limits on the signal that controls the power amplifier and the current of the system. Another approach to withstand the large overshoot will be to use an alternative PI regulator structure, where the proportional control is directly applied to the *dq* current components instead of the feedback error, whereas the integral control is still applied to the feedback error [12]. Such structure is shown in Fig. 15.

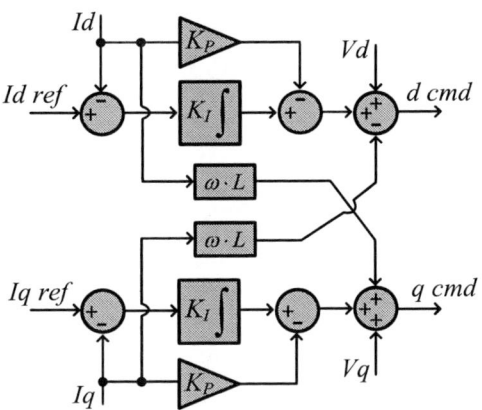

Fig. 15. Modified PI regulator structure to compensate for overshoot.

VIII. CONCLUSIONS

A novel method to interface a GFMI into a PHIL is proposed. The main feature of the method relies upon using the computational resources of the real-time simulator to incorporate a current control scheme whose output drives the power amplifier. The method aims to help research institutions that do not have a power amplifier unit with current control capabilities but do have a voltage control unit. The proposed method's stability was first validated by simulations that emulated the physical instruments where the transient responses of the current were satisfactory. Experimental results also validated the method's stability, with transient responses showing significant overshoot and oscillations.

ACKNOWLEDGMENT

This article has been authored by an employee of National Technology & Engineering Solutions of Sandia, LLC under Contract No. DE-NA0003525 with the U.S. Department of Energy (DOE). The employee owns all right, title and interest in and to the article and is solely responsible for its contents. The United States Government retains and the publisher, by accepting the article for publication, acknowledges that the United States Government retains a non-exclusive, paid-up, irrevocable, world-wide license to publish or reproduce the published form of this article or allow others to do so, for United States Government purposes. The DOE will provide public access to these results of federally sponsored research in accordance with the DOE Public Access Plan https://www.energy.gov/downloads/doe-public-access-plan.

REFERENCES

[1] H. H. Alhelou, N. Mohammed, and B. Bahrani, *Grid-Forming Power Inverters*. Boca Raton: CRC Press, 2023. doi: 10.1201/9781003302520.

[2] R. Teodorescu, M. Liserre, and P. Rodríguez, *Grid Converters for Photovoltaic and Wind Power Systems*, First. Wiley, 2011.

[3] S. Chakraborty, A. Nelson, and A. Hoke, "Power hardware-in-the-loop testing of multiple photovoltaic inverters' volt-var control with real-time grid model," in *2016 IEEE Power & Energy Society Innovative Smart Grid Technologies Conference (ISGT)*, IEEE, Sep. 2016, pp. 1–5. doi: 10.1109/ISGT.2016.7781160.

[4] RTDS Technologies, "Power Hardware in the Loop Simulations (PHIL report)," 2018. [Online]. Available: https://www.rtds.com/applications/power-hardware-in-the-loop/

[5] J. Hernandez-Alvidrez, N. S. Gurule, M. J. Reno, J. D. Flicker, A. Summers, and A. Ellis, "Method to Interface Grid-Forming Inverters into Power Hardware in the Loop Setups," in *47th IEEE Photovoltaic Specialists Conference.*, 2020.

[6] H. Kikusato *et al.*, "Verification of power hardware-in-the-loop environment for testing grid-forming inverter," *Energy Reports*, vol. 9, pp. 303–311, May 2023, doi: 10.1016/J.EGYR.2022.12.126.

[7] Z. Feng, A. Alassi, M. Syed, R. Pena-Alzola, K. Ahmed, and G. Burt, "Current-Type Power Hardware-in-the-Loop Interface for Black-Start Testing of Grid-Forming Converter," in *IECON 2022 – 48th Annual Conference of the IEEE Industrial Electronics Society*, IEEE, Oct. 2022, pp. 1–7. doi: 10.1109/IECON49645.2022.9968517.

[8] Opal RT, "OPAL-RT News, Innovation & Breakthroughs," 2019.

[9] W. Du *et al.*, "Modeling of Grid-Forming and Grid-Following Inverters for Dynamic Simulation of Large-Scale Distribution Systems," *IEEE Transactions on Power Delivery*, vol. 36, no. 4, pp. 2035–2045, Aug. 2021, doi: 10.1109/TPWRD.2020.3018647.

[10] T. Suntio, T. Messo, and J. Puukko, *Power Electronic Converters*, First edition. Wiley, 2018.

[11] J. C. Doyle, B. A. Francis, and A. R. Tannenbaum, *Feedback Control Theory*, 1st ed. Mineola, NY: Dover Publications, INC, 2009.

[12] L. Wang, *PID Control System Design and Automatic Tuning using MATLAB/Simulink*, First Edition. Wiley-IEEE Press, 2020.

Implications of Battery Storage for Solar Net-Metering Reforms

Galen Barbose, Sydney Forrester, and Chandler Miller

Lawrence Berkeley National Laboratory, Berkeley, CA, 94720, USA

Abstract — **Compensation structures for residential solar PV are evolving toward a model that incentivizes the use of battery storage to maximize solar self-consumption. Using metered data from 1,800 residential customers across six U.S. utilities, we show that batteries operated solely in this manner often provide no grid value, due to misalignment with market prices. Incentivizing customers to discharge storage in response to market prices, particularly on infrequent peak load days would greatly enhance storage dispatch value. However, doing so requires consideration of local distribution network impacts. We illustrate a net billing design that yields a storage dispatch value equal to 50-70% of its maximum potential market value, without materially degrading solar self-consumption levels or increasing local grid stress.**

I. INTRODUCTION

Historically, the predominant compensation structure for distributed solar PV in the United States has been net metering, which credits all exported generation at the full retail electricity price. This has become a significant source of contention in many states, as utilities and others have raised concerns that net metering fails to recover fixed utility infrastructure costs borne to serve solar customers, and as a result tends to shift those costs onto non-solar customers [1].

In response, many states are considering transitions away from net metering, often to a "net billing" structure that provides lower levels of compensation for grid exports, while continuing to allow solar PV customers to directly offset their own usage contemporaneously [2]. The defining feature of this approach is its *asymmetric pricing structure*, with higher (retail) prices paid for solar PV that directly serves onsite load and lower (grid export) prices for solar PV exported to the grid.

Simultaneous with these compensation reforms, increasing numbers of residential customers are installing battery storage alongside solar PV [3]. Those trends have been spurred in large part by declining battery costs and rising demand for backup power. However, battery storage is also ideally suited to managing solar PV grid exports under net billing structures. In effect, customers can use storage to arbitrage between retail and grid export prices, thereby retaining retail rate compensation for much of their PV production, similar to net metering.

But does this arbitrage behavior provide commensurate value to the electric system? If not, then PV customers' use of storage will ultimately undermine the intended objective of net metering reforms, raising the spectre of prolonging the contentious debates around distributed PV compensation.

This paper evaluates the aforementioned question through an empirical approach relying on historical hourly load from a diverse sample of residential customers across six U.S. utility service territories, paired with wholesale electricity market pricing data for the corresponding locations and time periods. We model the energy+capacity value of storage when dispatched in response to net billing pricing structures and compare to the maximum potential value if it were instead dispatched directly against energy and capacity market prices.

II. DATA AND METHODS

The analysis relies on three sets of time-series data: (1) *Load Data*, consisting of metered hourly interval load data collected from roughly 1,800 residential customers without PV or storage, located in six utility service territories, over the period from 2012-2013; (2) *Solar production profiles*, generated using the National Renewable Energy Laboratory's System Advisor Model, for the same locations and time periods as the load data; and (3) *Wholesale market price data*, consisting of day-ahead energy market prices and balancing authority system loads for the same locations and time period as the load data.

TABLE I
BATTERY DISPATCH SCENARIOS

	Scenario	Consumption Price	Grid Export Price
1	Net billing with time-invariant pricing	Annual average retail price	Annual average solar market value
2	Net billing with hourly pricing, but no grid charging or discharging allowed	Hourly energy + capacity + T&D adder	Hourly energy + capacity
3	+ Grid charging allowed	Same as above	Same as above
4	+ Limited grid discharging allowed	Same as above	Same as above
5	+ Unlimited grid discharging allowed	Same as above	Same as above
6	- T&D adder	Hourly energy + capacity	Same as above

Using these data, we simulate storage dispatch and compute the market value of the resulting dispatch profile for each customer, across the series of scenarios outlined in Table 1. The scenarios progress in a step-wise fashion from net billing with time-invariant pricing in Scenario 1 to full market-based dispatch in Scenario 6. As the first intermediate step between those bookends, flat pricing is replaced with hourly varying pricing in Scenario 2, but consumption and export prices remain asymmetrical due to the the volumetric T&D adder included in

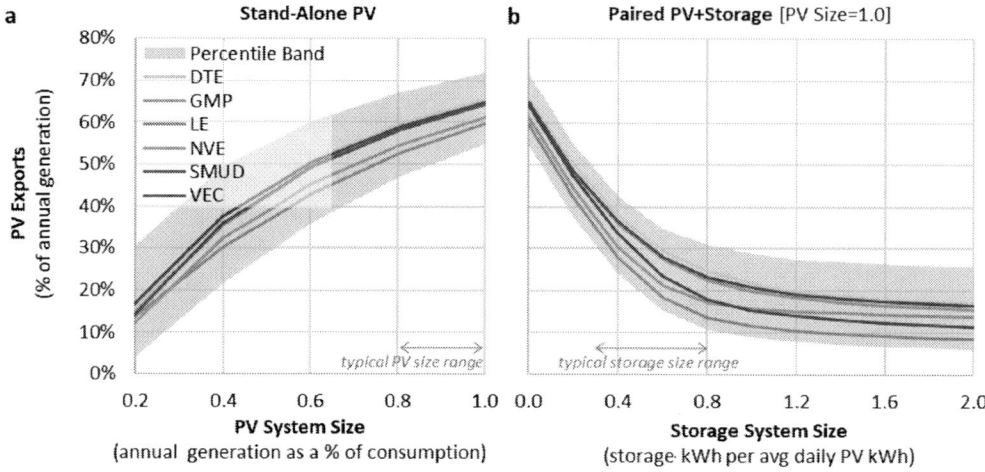

Fig. 1. Solar PV Grid Exports with and without Battery Storage. *Utility abbreviations: Detroit Edison (DTE), Green Mountain Power (GMP), Lakeland Electric (LE), Nevada Energy (NVE), Sacramento Municipal Utilities District (SMUD), and Vermont Electric Cooperative (VEC).*

the consumption price. Next, the limits on grid charging and discharging are sequentially relaxed in Scenarios 3-5. "Limited grid charging" in Scenario 4 refers to a case where the combined hourly exports from PV+storage are limited to the PV nameplate capacity, as is sometimes stipulated in interconnection agreements. Finally in Scenario 6, the T&D adder is removed, resulting in symmetric hourly prices for both consumption and exports. This is equivalent to the price signal faced by storage connected in front of the customer meter, responding directly to wholesale market prices.

In each scenario, storage is dispatched to maximize customer bill savings under the specified set of tariff prices and constraints. PV and storage system sizes are stipulated for each customer based on typical sizes currently observed in the market, with the base-case PV sized to generate 100% of the customer's annual consumption and storage sized at 50% of average daily PV production.

III. RESULTS

Before delving into the market value analysis, Fig. 1 first shows how PV export levels vary with system sizing, and how increasing amounts of battery storage can reduce those export levels. As shown in Fig. 1a, PV systems within a typical size range and without storage generally export roughly 50-75% of annual PV generation over the course of the year. Storage operated to maximize self-consumption of solar PV can reduce export to roughly 10-30% of annual PV generation, for storage systems at the upper end of sizes typically observed in the market today. Larger batteries reduce exports further, but with diminishing returns, as a result of limits on the amount of nighttime load to serve with stored solar energy. Notably, these results are highly consistent across all six utilities.

Fig. 2 compares the market value of the storage dispatch profile under each scenario for each utility. The values plotted

in the bar charts are the averages across all customers in the sample for each utility, and are disaggregated into the energy and capacity components.

The market value of the storage dispatch profile under net billing with flat rates (Scenario 1) is approximately zero across all utilities. This reflects the poor alignment between the times of storage charging and discharging and the times when energy and capacity prices are lowest or highest, respectively. This compares to a maximum potential market value of roughly $15-30 per kWh of storage capacity, annually, if storage were dispatched directly against market prices in Scenario 6.

Replacing flat prices with hourly varying prices in Scenario 2, but prohibiting grid charging or discharging with storage, increases storage dispatch value only negligibly. A much more significant jump in market value occurs in Scenario 4 where storage is allowed to discharge to the grid, albeit with limits based on nameplate PV capacity. Much of that jump is associated specifically with capacity value, which is concentrated in a relatively small number of high-priced hours each year. Depending on the utility, storage dispatch value under Scenario 4 equalas roughly 50-70% of its maximum potential market value. Notably, fully relieving grid discharge constraints in Scenario 5 has negligible incremental impacts on dispatch value, indicating that the discharge constraints in Scenario 4 are rarely binding during high-priced hours.

Finally, removing the fixed T&D adder on consumption prices in Scenario 6 eliminates the asymmetry between consumption and export prices, thereby fully aligning storage dispatch with market value. The value gap between Scenarios 5 and 6 is, effectively, a measure of the inefficiency attributable directly to assymetric pricing, and depending on the utility, equates to roughly 20-50% of the potential market value. The underyling source of that gap is that the fixed T&D adder tends to suppress routine daily arbitrage that would otherwise occur in response to diurnal variation in energy market prices.

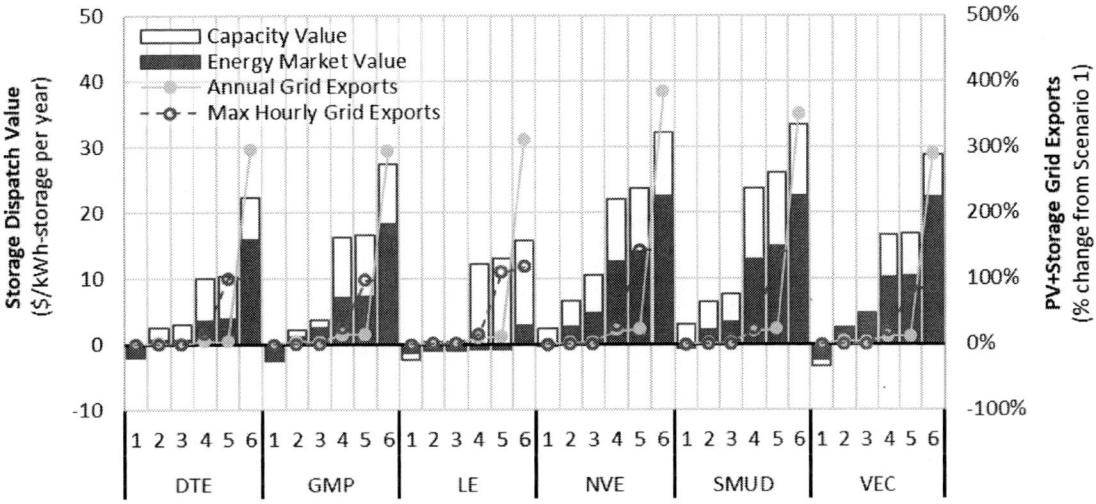

Fig. 2. Storage Dispatch Value and Grid Exports under Sequential Tariff Scenarios.

In addition to the dispatch value, Fig. 2 also shows both the annual and maximum hourly grid exports under each tariff scenario, presented as a percent change from Scenario 1. Two notable findings emerge. First, the maximum hourly grid exports increase dramatically from Scenarios 4 to 5, when grid discharge limits are fully lifted, yet market value is largely unchanged between those two scenarios. This is significant, because large amounts of instantaneous exports could create stress on the local distribution network. Scenario 4 thus represents a potential "sweet spot", insofar as it induces storage dispatch that realizes a significant portion of market value, without severely stressing the local network.

The second key finding related to grid exports is that annual grid exports remain virtually flat until Scenario 6. Removing the asymmetry between consumption and export prices in scenario 6 incentivizes a large volume of relatively low value (though still economical) grid exports.

IV. CONCLUSIONS

The most basic take-away from this analysis is that net billing is suboptimal from the utility system perspective, in terms of encouraging customers to operate storage in a manner that benefits the grid; however, the degree of suboptimality can be managed through rate design and interconnection rules.

Under many circumstances—particularly with time-invariant pricing—battery storage dispatched in response to net billing structures yields virtually no market value. The implications of this are three-fold. First, it suggests the potential for significant deadweight loss, if PV customers make large capital outlays for storage equipment solely or primarily for the purpose of arbitrage between retail and grid export prices. Second, battery storage could undermine the intent of net-metering reforms: moving solar grid exports back behind the meter maintains the same sales and revenue erosion issues as with net-metering reform, but potentially without providing any commensurate cost savings for utility ratepayers. Third, net billing structures could perpetuate some of the same inequities that have been leveled at net metering, insofar as wealthier customers will be better positioned to purchase the storage equipment necessary to capture the net billing arbitrage opportunity.

The above notwithstanding, the results show how these issues can be mitigated through well-designed net billing tariffs. Most important is to ensure that customers with PV and storage are incentivized, allowed, and capable of discharging storage to the grid during the highest value hours over the year, which typically coincides to times of peak demand. Doing so, however, may create stress on local distribution systems, if large numbers of customers on a single circuit are discharging to the grid in unison. It is therefore essential to consider those potential impacts when developing tariff reforms for net metering, given the increasing prevalence of paired PV and storage.

REFERENCES

[1] Sergici, S., Yang, Y., Castaner, M., Faruqui, A. "Quantifying net energy metering subsidies." Electricity Journal, 2019.

[2] NC Clean Energy Technology Center. *The 50 States of Solar: 2020 Policy Review and Q4 2020 Quarterly Report*. Raleigh, NC: North Carolina State University, 2021

[3] G. Barbose, N. Darghouth, E. O'Shaughnessy, and S. Forrester. *Tracking the Sun: 2022 Edition*. Berkeley, CA: Lawrence Berkeley National Laboratory, 2022.

III-V Solar Cells Grown Directly on V-groove Si Substrates

Theresa E. Saenz, Jacob T. Boyer, John S. Mangum, Anica N. Neumann, Myles A. Steiner, Ryan M. France, William E. McMahon, Jeramy D. Zimmerman, Emily L. Warren

National Renewable Energy Lab, Golden, CO, United States

Colorado School of Mines, Golden, CO, United States

We report on the development of GaAs solar cells directly grown on nanopatterned V-groove Si substrates by metalorganic vapor-phase epitaxy (MOVPE). A low threading dislocation density (TDD) of 3×10^6 cm-2 was achieved in the GaAs through a combination of thermal cycle annealing and InGaAs dislocation filter layers. Front junction GaAs solar cells were then grown on these low-TDD substrates, but preliminary devices produced a conversion efficiency of only 6.6% without an anti-reflection coating. Electron channeling constast imaging measurements on this cell showed a high density of misfit dislocations at the interface between the AlInP/GaInP window layer and GaAs absorber, likely causing poor surface passivation and thus poor performance. The source of these misfit dislocations will be discussed, as well as mitigation strategies to improve solar cell performance.

978-1-6654-6060-6/23 $31.00 © 2023 IEEE

Correlated Mapping of Raman Spectroscopy and Cathodoluminescence of Emerging Absorber Bournonite (CuPbSbS3)

O M Rigby, C Hill, G Kusch, M Naylor, M Guennou, G Zoppi, M Szablewski, R A Oliver, L Wirtz, P Dale, B G Mendis

Durham University, Durham, United Kingdom

Luxembourg Insitute of Science and Technology, Esch-sur-Alzette, Luxembourg

University of Cambridge, Cambridge, United Kingdom

Northumbria University, Newcastle, United Kingdom

University of Luxembourg, Esch-sur-Alzette, Luxembourg

Bournonite ($CuPbSbS_3$) is an emerging absorber material for solar cells. There is little literature on some of its fundamental structural and optical properties. We report the first complete Raman spectra of bournonite as well as the first luminescence signal. A chosen area has been mapped with both Raman spectra and cathodoluminescence (CL) to investigate the structural and optical uniformity. Raman mapping reveals a new secondary phase, tetrahedrite, but otherwise a uniform compositional thin-film. CL mapping reveals slight redshift at grain boundaries (GBs). Preliminary devices made with a bournonite absorber layer show rectification under standard test conditions.

Yb3+- doped CsPbX3 Nanocrystals for Improving Free-space Luminescent Solar Concentrators

Mathis Van de Voorde, Damien Hudry, Dmitry Busko, Bryce S. Richards, Rebecca Saive

University of Twente, Enschede, Netherlands

Institute of Microstructure Technology, Karlsruhe Institute of Technology, Karlsruhe, Germany

Light Technology Institute, Karlsruhe Institute of Technology, Karlsruhe, Germany

Free-space luminescent solar concentrators (FSLSCs) are optical structures that can enhance photovoltaic yield by efficiently concentrating diffuse light into a small cone in free space. To do so, a luminophore-doped waveguide captures the incoming light, and down-shifts the photons to an energy level which can escape the waveguide through a specific emission cone. The emission cone is tuned via a nanophotonic coating, consisting of multiple aperiodic bilayers of high and low refractive index materials, deposited on top of the waveguide. Non-ideality of the luminophores severely cripple the output of the FSLSCs. Therefore, we are investigating state-of-the-art CsPbX3 (X=Cl,Br) perovskite nanocrystals. The combination of these nanoparticles with rare-earth dopants (i.e., Yb3+) make excellent candidates for this technology due to their attractive optical properties such as large Stokes shifts (difference between absorption and emission peaks), narrow emission profiles, and high brightness. Here, we focus on Yb3+-doped CsPbCl3. We highlight initial results of our experimental investigation, including thorough characterization of the structural (STEM, EDX and XRD) and optical properties (PLQY, absorption/emission spectra and PL lifetimes) of the synthesised nanocrystals. Our approach contributes to the understanding of rare-earth incorporation in the perovskite matrix and will shed light on the role of state-of-the-art luminophores on the concentration of diffuse light using free space luminescent solar concentrators.

Rear heterojunction GaInP solar cells for improved performance at elevated temperatures

Mijung Kim, Yukun Sun, Ryan D. Hool, Minjoo L. Lee

University of Illinois Urbana-Champaign, Urbana, IL, United States

We present the characteristics of n-on-p front-junction (FJ) and rear-heterojunction (RHJ) GaInP solar cells at operating temperatures (T) up to 400°C. Photovoltaic cells with efficient operation at high T may be important for satellite missions near the sun or as laser power converters for sensors operating in harsh environments, such as gas turbines or hypersonic vehicles. In this work, we show that the time-resolved photoluminescence lifetime (tTRPL) in both lattice-matched (LM) n-$Ga_{0.51}In_{0.49}P$ and metamorphic (MM) n-$Ga_{0.37}In_{0.63}P$ double heterostructures (DHs) increases significantly with T. In contrast, the tTRPL values in their p-type counterparts are lower and decrease with T. We go on to demonstrate both LM and MM solar cells in FJ and RHJ configurations. For both LM and MM cells, the internal quantum efficiency (IQE) of the RHJ cells increases significantly up to T = 300 °C due to the increase in both tTRPL and the linear increase in diffusivity with T. In contrast, the IQE of MM FJ cells drops slightly with increasing T, while the IQE of LM cells drops sharply. RHJ cells maintain higher VOC and fill factor than their FJ counterparts, leading to a significant efficiency advantage at T = 200-400°C. Taken together, our work shows that MM cells perform well at elevated temperatures and that RHJ cells are promising for high-T operation.

Design and Fabrication of PERC-Like CdTe Solar Cells Using Micropatterned Al₂O₃ Layer

Etee Kawna Roy[1], Kaden Powell[1], Chungho Lee[2], Gang Xiong[2], and Heayoung Yoon[1]

[1]Department of Electrical and Computer Engineering, University of Utah, Salt Lake City, UT 84112, USA

[2]California Technology Center, First Solar Inc., Santa Clara, CA, USA.

Abstract —Recent studies have investigated novel strategies to further improve the limited V_{oc} of CdTe solar cells via increased carrier lifetime and doping density of CdTe thin films. Among various metal oxides, aluminum oxide (Al₂O₃) is a promising passivation candidate, where the negatively charged Al₂O₃ layer repels the minority carrier in CdTe and Al₂O₃ provides a chemically passivating interface, increasing the carrier lifetime. Despite the continuing efforts, an optimized back-contact architecture to improve the V_{oc} while maintaining high J_{sc} and FF is still under development. In this work, we report the design, fabrication, and characterization of PERC-like CdTe solar cells, where an Al₂O₃ passivation layer is patterned using laser-beam lithography. Our process enables reproducible patterning on a rough surface CdTe while maintaining the size of the array in the design. Analysis of CdTe PERC devices (As-doped) shows a notably different V_{oc} trend compared to FF and J_{sc}, independent of the patterned array structures used in this study. The subsurface electronic structure of CdTe and the interplay between carrier selectivity and collection of the patterned Al₂O₃ could be responsible for the observed PV characteristics.

I. INTRODUCTION

Thin-film CdTe solar cells are a competitive photovoltaic (PV) technology that can meet the rapidly growing societal demand for energy owing to cost-effective manufacturing and relatively short energy payback time [1, 2]. At a record efficiency of 22.1 %, significant attention has been devoted to further improving the power conversion efficiencies of CdTe-based solar cells [3-5]. An optimized front contact that leads the J_{sc} over 31 mA/cm² and a fill factor (FF) above 79 % were reported [6, 7]. An open-circuit voltage (V_{oc}) over 1 V was achieved with As-doped CdTe single crystals. Despite the continuing efforts, high V_{oc} (> 1 V) has not yet been observed in thin-film polycrystalline CdTe PVs. The underlying physical mechanisms responsible for the V_{oc} loss are not presently well understood.

Recent studies have demonstrated novel device architectures and passivation strategies to enhance the V_{oc} via increasing carrier lifetime and doping density of CdTe thin films. Among various metal oxides, Al₂O₃ is a promising passivation candidate, where the negatively charged Al₂O₃ layer (fixed charge density of $10^{12} \sim 10^{13}$ cm⁻²) could repel the minority carrier (electrons) in CdTe, increasing the carrier lifetime [7, 8]. Previous studies also suggested that Al₂O₃ could passivate the CdTe surface via chemical reactions during the fabrication processes. This configuration is similar to a passivated emitter and rear contact (PERC) design, frequently used in Si photovoltaic (PV) technology [9, 10]. An improved V_{oc} was reported with a conformal Al₂O₃ (< 5 nm) on CdTe, yet the J_{sc} and FF suffer from the tunneling barrier [11-18]. On the other hand, Kephart and co-workers used a 20 nm-thick Al₂O₃ with micropatterning. A remarkably high lifetime ($\tau_2 > 400$ ns) was measured for double heterostructures, but no consistent improvement of V_{oc} was observed for the Cu-doped CdTe solar cells [11].

This work reports the design and fabrication of PERC-like CdTe solar cells (CdTe PERC), where the Al₂O₃ passivation layer is patterned via laser-beam lithography. We optimize the beam dose to polymerize the photoresist on a rough surface CdTe while maintaining the size of the array in the design. We measure the dark and light I-Vs of As-doped CdSeTe PERC having different hole diameters and pitches (i.e., the distance between adjacent holes). Quantitative analysis shows the impact of Al₂O₃ patterns on V_{oc} is notably different from other device parameters of J_0, J_{sc}, and FF. We discuss a similar V_{oc} trend observed after annealing (70°C for 2.5 hours) of the complete CdTe PERC. Our results indicate the electrically-active Al₂O₃ patterns, in conjunction with the defect chemistry in CdTe bulk, could play an essential role in determining the V_{oc} of advanced CdTe-based solar cells.

II. EXPERIMENT

Figure 1(a) displays a schematic of a CdTe PERC device consisting of a thin-film CdTe solar cell and a patterned Al₂O₃. The As-doped CdSeTe/CdTe absorber layer was synthesized on a stack of buffer/TCO [transparent conductive oxide]/glass substrate. A scanning electron microscopy (SEM) image shows a representative Al₂O₃ hole array on CdTe fabricated in this work (Figure 1b). The same-size individual holes are uniformly

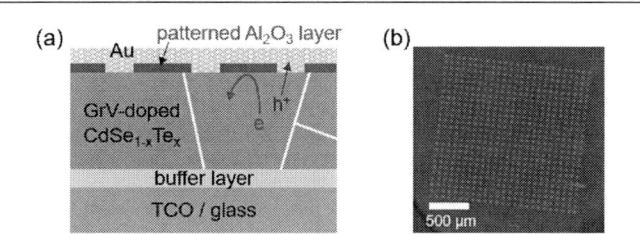

Figure 1. (a) Schematic of a PERC-like CdTe solar cell, (b) Representative scanning electron microscopy (SEM) image of patterned Al₂O₃ layer on CdTe.

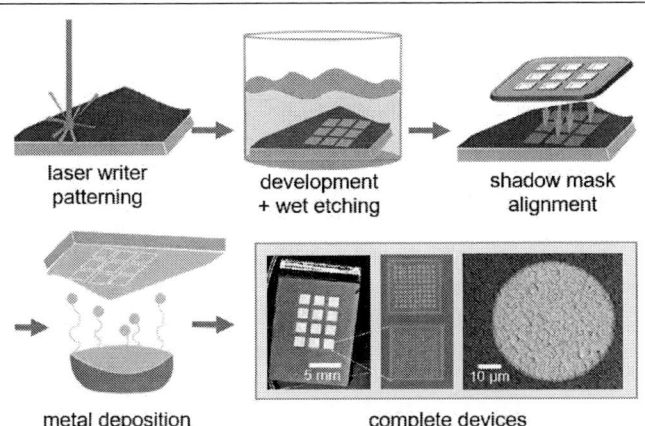

Figure 2. Schematics illustrating the fabrication process. The lower right panel displays the complete CdTe PERC devices. The back-scattered electron (BSE) image shows exposed CdTe (bright hole area) after the selective wet etching processes (AZ 1:1 developer, AZ 300 MIF).

distributed in a square pattern across the entire device. This study used three different hole sizes (10 μm, 20 μm, and 40 μm). The distance between the adjacent holes ranges from 10 μm to 320 μm, introducing different sizes of the exposed CdTe area to metal contact.

Figure 2 illustrates the fabrication procedures to produce patterned Al_2O_3 back contacts. As-doped CdSeTe/CdTe solar cells were obtained from First Solar that was coated with a 20 nm Al_2O_3 layer. The samples were cleaned with acetone and isopropyl alcohol (IPA), and blown dry with nitrogen (N_2). The samples were baked on a hot plate at 100°C for 60 seconds and cooled down before spin coating of photoresist. A thin layer of positive photoresist (Shipley 1813) was coated on the sample at a spin speed of 3,000 rpm (revolution per minute) for 60 seconds, followed by soft baking for another 60 seconds on the hot plate at 100°C. A laser writer (Heidelberg μPG 101) was used to pattern the hole array design (L-Edit) on the photoresist-coated Al_2O_3 on CdTe.

We performed a series of control experiments to optimize the laser dose that could polymerize the photoresist on a rough surface CdTe while maintaining the size of the array of the design [19]. We found a relatively high beam energy (an effective dose of 13.5 mW) is needed for CdTe compared to traditional Si flat samples (10 mW). The photoresist was developed in a solution (AZ Developer 1:1) for 60 seconds and soaked in DI water. We used another solution (AZ 300 MIF) to selectively etch the exposed Al_2O_3 on CdTe after the photoresist development. This chemical contains a small amount of tetramethyl ammonium hydroxide (< 3 % TMAH), which has a faster etch rate for Al_2O_3 than photoresist. The Al_2O_3 etching was conducted for 20 minutes, followed by the photoresist removal and cleaning using acetone, IPA, deionized water, and N_2-blown dried.

The metal contact for each device was designed to be 2 mm × 2 mm in size. We fabricated a shadow mask using a stainless-steel foil (40 μm) containing a square array (3 × 4) with a spacing of 1 mm. This shadow mask was aligned to the hole array patterns on Al_2O_3/CdTe (top right in Figure 2). The sample was placed on a holder in an electron beam evaporator. Thin films of Cu/Au (3 nm/80 nm) were deposited on the patterned Al_2O_3/CdTe through the shadow mask, which served as a back contact. A front indium contact was formed on TCO after removing the CdTe segment using a razor blade. Figure 2 displays the complete CdTe PERC devices.

The dark and light current-voltage characteristics were measured using a probe station (dark enclosure) connected to a semiconductor analyzer (Agilent 4145C). Each probe tip (25 μm diameter W-tip) was placed on top and bottom metal contact while recording the I-Vs using the LabVIEW program. Before the measurements, the solar simulator (G2V) was calibrated to 1-sun using a reference cell (Newport; 91150V).

III. RESULTS AND DISCUSSION

We analyze the dark and light I-Vs measured from 36 CdTe PERC devices having various hole array patterns. All devices consisted of a patterned area in the center (1 mm × 1 mm) of the metal contact (2 mm × 2 mm). For a statistical comparison, we estimate the exposed CdTe area of each device from the design, where the Cu/Au metal is directly deposited on CdTe without the Al_2O_3 layer. Figure 3 summarizes the parameters (I_0, I_{sc}, FF, and V_{oc}) of CdTe PERC.

The distribution of leakage current of our devices (Figure 3a) is relatively constant, supporting the fidelity of the fabrication processes developed in this work. The magnitudes of I_{sc} and FF

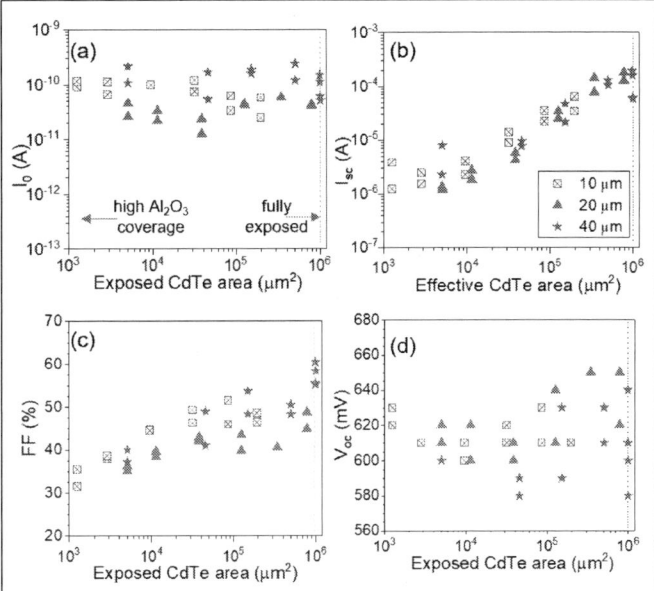

Figure 3. Comparison of I_0, I_{sc}, FF, and V_{oc} extracted from the I-Vs of 36 CdTe PERC devices. The diameter of hole arrays is 10 μm, 20 μm, or 40 μm, and the distance between adjacent holes ranges from 10 μm to 320 μm.

proportionally increase with the exposed CdTe area. As seen in Figure 3(b), the I_{sc} of the CdTe PERC having an exposed CdTe area of 1 % is 1 µA. This current increases to 10 mA when the Al_2O_3 layer is fully removed in the patterned area (1 mm × 1 mm). The FF increases to a factor of two (approximately 30 % to 60 %) with the Al_2O_3 removal. These trends are expected as the electrically resistive Al_2O_3 layer (20 nm thick) can interfere with the photocarrier collection (hole carriers) from CdTe to metal contact.

Interestingly, the V_{oc} trend (Figure 3d) for the As-doped CdTe PERC is notably different from that of I_{sc} and FF. The V_{oc} ranges from 580 mV to 660 mV, which seems independent to the exposed CdTe areas. In a close look, the V_{oc} slightly decreases from $\approx$ 640 mV to $\approx$ 590 mV when the CdTe area increases from 10^3 µm^2 (1 % of the total device area) to 10^4 µm^2 (10 %). The V_{oc} increases as high as 660 mV with the increase of the exposed CdTe area (40 % ~ 80 %). The V_{oc} of the control devices, which have the fully open CdTe area (100 %), ranges from 580 mV to 640 mV, still below the highest V_{oc} observed with the patterned Al_2O_3 PERC devices. We note that the V_{oc} trend of As-doped PERC is also significantly different from our Cu-doped CdTe PERC, where the V_{oc} decreases with the increased CdTe exposed area (not shown in the data). Our results indicate the electrically-active Al_2O_3 patterns could play

an important role in determining the V_{oc} of advanced CdTe-based solar cells.

We have examined the performance of the As-doped CdTe PERC solar cells under heating. We annealed the complete PERC devices at 70 °C for 2.5 hours in this preliminary experiment. The diameter of the patterned Al_2O_3 is 10 µm, and the adjacent hole distance ranges from 10 µm to 320 µm. Figures 4 (a, b) shows the light I-Vs collected "before" and "after" the heating. Overall, the PERC devices preserve good diode behaviors after the heating.

Figures 4 (c ~ f) compare the device parameters of the saturation current (I_s), I_{sc}, FF, and V_{oc} extracted from the I-Vs. The magnitudes of the I_s and V_{oc} are relatively constant, while the I_{sc} and FF increase with the exposed CdTe area, with similar trends observed in Figure 3. The magnitudes of I_s, I_{sc}, and FF show only slight variation after the heating at 70 °C. In contrast, the V_{oc} of the CdTe PERC increases after annealing with a magnitude as high as $\approx$ 10 % of its initial value (e.g., 600 mV to 650 mV). Presumably, the group-V dopant (As) in the CdTe bulk could be more activated, or the defects near Al_2O_3 may be further passivated during the annealing [20]. Further studies are in progress to gain a better understanding of such heating effects and back-contact passivation.

IV. CONCLUSIONS

In summary, we have demonstrated the fabrication of CdTe PERC solar cells with a patterned Al_2O_3 passivation layer. Our process based on laser-beam lithography enables reproducible patterning on a rough surface CdTe while maintaining the size of the array in the design. Quantitative analysis of the As-doped CdTe PERC shows a proportional increase of FF and J_{sc} with an exposed CdTe area. We have observed that the complete CdTe PERC annealing increases the V_{oc} by $\approx$ 10 % of their initial magnitudes, whereas the I_s, I_{sc}, and FF show negligible changes. Our results indicate that the subsurface electronic properties of CdTe and the interplay between carrier selectivity and collection of the patterned Al_2O_3 could be responsible for the observed PV characteristics.

ACKNOWLEDGEMENT

The authors thank B. Baker, A. Hurlbut, P. Perez, D. Albin, A. Chowdhury, and D. Magginetti for experimental assistance and valuable discussions in this work. This research was supported by the U.S. Department of Energy's Office of Energy Efficiency and Renewable Energy (EERE) under the DE-FOA-0002064 program award number DE-EE0008983. We acknowledge support in part by the National Science Foundation (NSF) CAREER Award No. 2048152.

REFERENCES

[1] G. M. Wilson *et al.*, "The 2020 photovoltaic technologies roadmap," (in English), *Journal of Physics D-Applied Physics*, vol. 53, no. 49, Dec 2 2020.

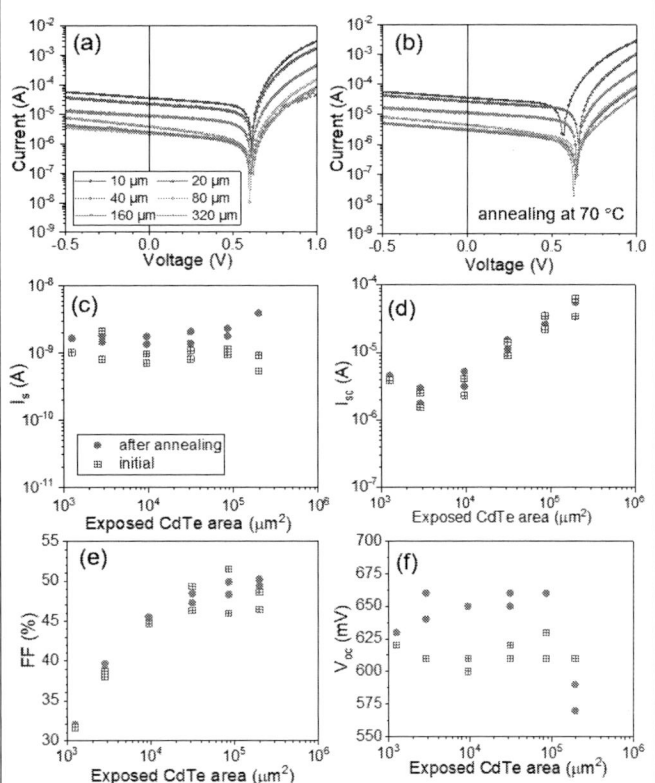

Figure 4. Light I-Vs of 24 CdTe PERC "before" and "after" heating at 70 °C for 2.5 hours. The hole diameter of the arrays is 10 µm (pitch: 10 µm to 320 µm). Comparison of the parameters of (c) saturation current (d) short-circuit current (e) fill-factor and (f) open-circuit voltage.

[2] "Cadmium Telluride," Energy.gov. https://www.energy.gov/eere/solar/cadmium-telluride (accessed Oct. 10, 2022).

[3] K. J. Hsiao and J. R. Sites, "Electron reflector to enhance photovoltaic efficiency: Application to thin-film CdTe solar cells," Progress Photovolt., Res. Appl., vol. 20, no. 4, pp. 486–489, 2012

[4] G. K. Liyanage, A. B. Phillips, F. K. Alfadhili, R. J. Ellingson, and M. J. Heben, "The Role of Back Buffer Layers and Absorber Properties for >25% Efficient CdTe Solar Cells," ACS Appl. Energy Mater., vol. 2, no. 8, pp. 5419–5426, Aug. 2019, doi: 10.1021/acsaem.9b00367.

[5] A. Kanevce and T. Gessert, "Optimizing CdTe solar cell performance: Impact of variations in minority-carrier lifetime and carrier density profile," IEEE J. Photovolt., vol. 1, no. 1, pp. 99–103, Jul. 2011.

[6] Kuciauskas, D.; Moseley, J.; Šč ajev, P.; Albin, D. Radiative Efficiency and Charge-Carrier Lifetimes and Diffusion Length in Polycrystalline CdSeTe Heterostructures. Phys. Status Solidi RRL 2020, 14, No. 1900606.

[7] L. Wu et al., "CdTe surface passivation by electric field induced at the metal-oxide/CdTe interface," Solar Energy, vol. 225, pp. 83–90, Sep. 2021, doi: 10.1016/j.solener.2021.07.015.

[8] Li, X., Shen, K., Li, Q., Deng, Y., Zhu, P., Wang, D., 2018b. "Roll-over behavior in currentvoltage curve introduced by an energy barrier at the front contact in thin film CdTe solar cell," Sol. Energy 165, 27–34.

[9] M. A. Green, A. W. Blakers, and J. Shi, "19.1% efficient silicon solar cell," Appl. Phys. Lett. 44 (1984) 1163-1164.

[10] A.W. Blakers, A. Wang, A.M. Milne, J. Zhao, M.A. Green, "22.8% Efficient Silicon Solar Cell," Appl. Phys. Lett. 55 (1989) 1363–1365.

[11] J. M. Kephart et al., "Sputter-Deposited Oxides for Interface Passivation of CdTe Photovoltaics," IEEE J. Photovoltaics, vol. 8, no. 2, pp. 587–593, Mar. 2018, doi: 10.1109/JPHOTOV.2017.2787021.

[12] J. Liang et al., "Rectification and tunneling effects enabled by Al 2 O 3 atomic layer deposited on back contact of CdTe solar cells," Appl. Phys. Lett., vol. 107, no. 1, p. 013907, Jul. 2015, doi: 10.1063/1.4926601.

[13] F. K. Alfadhili et al., "Back-Surface Passivation of CdTe Solar Cells Using Solution-Processed Oxidized Aluminum,"

ACS Appl. Mater. Interfaces, vol. 12, no. 46, pp. 51337–51343, Nov. 2020, doi: 10.1021/acsami.0c12800.

[14] Y. Su et al., "Band Alignment for Rectification and Tunneling Effects in Al 2 O 3 Atomic-Layer-Deposited on Back Contact for CdTe Solar Cell," ACS Appl. Mater. Interfaces, vol. 8, no. 41, pp. 28143–28148, Oct. 2016, doi: 10.1021/acsami.6b07421.

[15] A. Kanevce, M. O. Reese, T. M. Barnes, S. A. Jensen, and W. K. Metzger, "The roles of carrier concentration and interface, bulk, and grain-boundary recombination for 25% efficient CdTe solar cells," Journal of Applied Physics, vol. 121, no. 21, p. 214506, Jun. 2017, doi: 10.1063/1.4984320.

[16] D. Kuciauskas, J. M. Kephart, J. Moseley, W. K. Metzger, W. S. Sampath, and P. Dippo, "Recombination velocity less than 100 cm/s at polycrystalline Al 2 O 3 /CdSeTe interfaces," Appl. Phys. Lett., vol. 112, no. 26, p. 263901, Jun. 2018, doi: 10.1063/1.5030870.

[17] Q. Lin et al., "A novel p-type and metallic dual-functional Cu–Al 2 O 3 ultra-thin layer as the back electrode enabling high performance of thin film solar cells," Chem. Commun., vol. 52, no. 71, pp. 10708–10711, 2016, doi: 10.1039/C6CC04299F.

[18] A. H. Munshi, A. H. Danielson, A. Kindvall, K. Barth, and W. Sampath, "Investigation of Sputtered Oxides and p+ Back-contact for Polycrystalline CdTe and CdScTe Photovoltaics," in 2018 IEEE 7th World Conference on Photovoltaic Energy Conversion (WCPEC) (A Joint Conference of 45th IEEE PVSC, 28th PVSEC & 34th EU PVSEC), Jun. 2018, pp. 3009–3012. doi: 10.1109/PVSC.2018.8548203.

[19] K. M. Powell, Y.-L. Hsu, E. K. Roy, D. J. Magginetti, and H. P. Yoon, "Fabrication of Microscale Back-Contact Arrays for Local Charge Transport Measurements," in 2022 IEEE 49th Photovoltaics Specialists Conference (PVSC), Philadelphia, PA, USA, Jun. 2022, pp. 1–4. doi: 10.1109/PVSC48317.2022.9938877.

[20] D. J. Elliott, "Annealing and Planarizing," in Ultraviolet Laser Technology and Applications, Elsevier, 1995, pp. 209–250. doi: 10.1016/B978-0-12-237070-0.50011-X.

Issues, Challenges, and Primary Factors in the Estimation of Floating Solar PV Performance

Rick Meeker[1], Anna Brinck[1], Tom Lang[2], and Jason Harrison[3]

[1]Nhu Energy, 1736 W. Paul Dirac Dr., Tallahassee, FL 32310, [2]Floating Solar Energy, Lake Mary, FL, [3]AccuSolar, 1790 SW 13th Ct, Pompano Beach, FL 33069

Abstract — Floating solar photovoltaic (FPV) systems are beginning to play an increasing role worldwide in solar generation's overall contribution to decarbonization and clean energy. There are unique aspects of FPV design, installation, operation, and maintenance when compared to ground-mount solar PV systems that introduce uncertainty into estimating and predicting the performance of these systems using current tools and methods. This work, part of a major project underway to de-risk, validate and field a new FPV system design, describes these uncertainties and challenges, identifies and examines important factors in estimating FPV performance, and illustrates with existing solar PV performance prediction tools the sensitivity of key performance metrics to selected factors.

I. INTRODUCTION

Global floating solar photovoltaic (FPV) generation capacity is growing steadily, projected to reach 4.8 GW by 2026 [1]. With several unique advantages, FPV can play an important role supporting the overall expansion of solar to meet decarbonization goals. These advantages include:

- Providing the option to site solar generation where land is unavailable, too costly, or better utilized for other purposes.

- No land acquisition costs.

- Characteristics beneficial to certain water bodies, including reduction of evaporation losses, algae growth, wave formation, and coupled erosion effects [2].

- Improved PV module efficiency, on the order of 10-15% [3], due to cooling effects over water.

- Synergies with marine and hydrokinetic energy sources including hydroelectric dams and reservoirs [4].

There is estimated to be potential for 400 GW of FPV capacity worldwide (double the overall worldwide total existing solar PV capacity) [5] from 400,000 sq. km of man-made reservoirs. And, it is estimated there is capacity to provide 10% of US generation needs from FPV deployed on just 27% of suitable man-made water bodies in the US [6].

While there is clearly the potential for FPV to contribute significantly to a clean energy transformation, it is still on the early part of the technology learning curve and is not yet experiencing the same exponential growth rates as land-based solar has. One of the reasons for this is a greater perceived technical and financial risk. Compared to land-based solar, FPV presently has lower "bankability", the ease with which FPV projects can secure financing. Because greater uncertainty in estimating life-cycle performance represents increased risk for project financiers, there has been recent research by U.S. Dept. of Energy (DOE) national laboratories to better quantify these uncertainties [7]. Several first and second-order factors that are sources of uncertainty in current state of the art modeling and estimation tools and methods [7][8] are even less well-understood and modeled for FPV performance prediction.

There are various components that go into determining a solar PV generation plant's life-cycle cost and return on investment. Categorically, challenges considered here fall into two areas:

1. Optimizing FPV system design for performance and cost.

2. Predicting FPV system performance, including energy output, capacity factor, and energy yield.

II. OPTIMIZING FPV DESIGN

There are design considerations and constraints unique to FPV systems that impact life-cycle performance and cost. Some of the most impactful are related to stability and anchoring requirements.

Providing sufficiently stable and reliable anchoring for FPV plants is more complex than for ground-mount or roof-mount solar. Like land-based solar, FPV systems are designed for particular wind-loads. In addition, FPV systems must also be designed for wave height, current (water flow), and water level variation. Wind and wave loading requirements have considerable effect on the physical dimensions, geometry, and weight of the system and, consequently, the structural and anchoring requirements and cost. In addition to anchoring and mooring systems, FPV systems derive stability from having all of the PV arrays and their flotation systems connected together to form the entire FPV plant. This is another important difference compared to land-based systems that ends up constraining key parameters affecting system performance.

Two such parameters that are particularly important in determining FPV plant energy output, capacity factor and energy yield are 1.) PV array tilt, and 2.) Ground Coverage Ratio (GCR). Tilt is the angle up from horizontal of the array of PV panels, and, GCR is the ratio of the length of a side of the PV array, 'a', to the row spacing, 'b', (Fig 1).

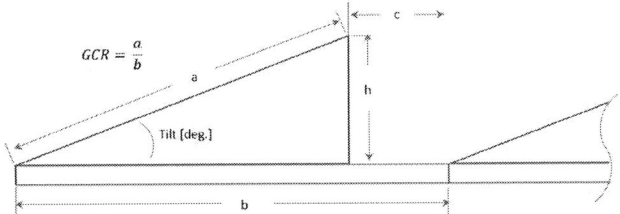

Figure 1. Key FPV installation dimensions and geometry (side elevation view)

PV panel specifications and the arrangement of the panels on the floatation system will constrain the geometry and dimensions, which are then further constrained by design maximum wind speed. Increasing design maximum wind speed increases the stress the floatation and PV assembly itself must withstand and increases the stress the anchoring or mooring system must withstand in keeping the system stationary and stable. The bathymetry of the water body and the shoreline and bottom characteristics further determine what the anchoring and mooring options are. These additional interdependent design factors associated with FPV systems, add to the challenge and complexity of optimizing tilt and GCR to maximize energy output, capacity factor, and energy yield. The impacts and sensitivity will be examined with a realistic example case.

III. METHODOLOGY AND CASE STUDY

Impacts of PV array tilt and GCR on key system performance metrics were analyzed utilizing the System Advisor Model (SAM) techno-economic software [9] developed by the National Renewable Energy Laboratory (NREL). Dimensional parameters, geometry and physical arrangement of the FPV system are based on a realistic design prototype and a conceptual 40 MW$_{DC}$ FPV plant design for installation on a large reservoir located in the Southeastern U.S. Parameter sweeps are informed by known design constraints associated with the FPV prototype. The system design is summarized in Table 1.

Table 1. System Design

Plant Power	40 MW$_{DC}$, 33 MW$_{AC}$
PV Modules	61,440 Canadian Solar BiHiKu7 650W, bifacial
Inverters	22 Dynapower CPS 1500 600V$_{DC}$
Strings	1,920 strings, 32 modules / string
DC:AC Ratio	1.21
FPV Units	7,680 FPV units in 128 rows of 60 each
FPV Rating	5.2 kW / unit
Array	Landscape orientation, 2 high x 4 across / unit
Array Tilt	Fixed, 15 degrees
Life	25 years (analysis period)

IV. DESIGN PERFORMANCE CASE: TILT

Array tilt angle becomes techno-economically constrained due to increasing wind force as tilt angle increases and given the objective to minimize the height of the FPV unit. It is desirable to design FPV units so PV panel arrays can be stowed in a zero deg. tilt position in high winds. This reduces the maximum forces throughout the structure and anchoring and mooring systems, and, therefore the system cost. Seasonal average optimal tilt for a south-facing system in the northern hemisphere increases with latitude. To reduce wind load and improve stability, FPV systems will tend to have lower tilt than land-based PV. At a tilt of only 15 deg., which is sub-optimal for most of the continental U.S. and Canada, transverse wind forces are roughly 10 times greater than when the array is stowed.

For the Case Study described, SAM was used to examine impact on performance of varying tilt angle for the FPV prototype design. Parametric sweeps over a range of 0 to 45 degrees were performed to find the optimum tilt based on energy yield in kWh/kW at different GCR's. Optimal tilt and energy yield are shown in Table 2 for GCR's of 0.6, 0.76 and 0.84.

Table 2. Optimal FPV array tilt for different GCR's

GCR	Optimal Tilt [deg.]	Energy Yield [kWh/kW]
0.6	20	1470
0.76	10	1407
0.84	5	1374

For the Case Study installation, which is a site located at a latitude of 35 deg. N., the optimal tilt is 20 deg. at 0.6 GCR. Reducing the tilt by 5 deg., to 15 deg., at the same GCR reduces energy yield by a relatively small amount, 7 kWh/kW or about a 0.5% reduction. Reducing tilt has benefits due to reduction in wind-loading. Conversely, increasing tilt above the optimum would reduce yield and increase wind-loading and cost. This provides a bases for a system cost trade study between tilt angle, yield, and revenue and wind-loading and FPV unit installed cost, which can aid in the techno-economic optimization of the FPV design.

V. DESIGN PERFORMANCE CASE: GCR

As can be seen in Fig. 1, increasing GCR for a given tilt comes from decreasing row spacing and visa versa. Decreasing row spacing (increasing GCR) increases self-shading, reducing energy yield, though possibly increasing power and energy density on a per area (land or water) basis. Typical GCR values for FPV are in the range of 0.6 to 0.85, compared to 0.4 to 0.5 for fixed tilt ground-mount PV and 0.25 to 0.4 for single-axis tracking ground-mount PV [10]. The higher GCR for FPV is due to the rows of solar arrays on associated floating racking systems being close coupled, connected together to form a contiguous system that comprises the entire solar PV generating plant. Ground mount systems, conversely, have no physical connection between rows and, in fact, often have a physical separation between rows of approximately 2 to 3 meters for ground and system maintenance. Access walkways between rows in FPV systems can be merely 0.5 meters wide.

For the Case Study described, SAM was used to examine impact on energy yield by parametric search of the solution space varying GCR over a range of 0.1 to 0.9 and tilt over a range of 0 to 45 degrees. Figure 2 shows Energy Yield for various combination of GCR and tilt over the specified ranges for the study case. It can be seen that for this FPV and plant design, energy yield suffers considerably as GCR increases (smaller row spacing) and optimal tilt to maximize energy yield declines as GCR increases.

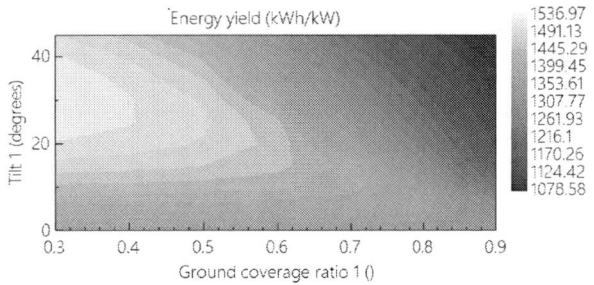

Figure 2. *Energy yield versus tilt and GCR for the study case.*

VI. CHALLENGES PREDICTING FPV PERFORMANCE

Existing solar PV performance prediction tools that are available are sophisticated and comprehensive when it comes to land-based solar PV. Some aspects of these tools extend adequately to FPV systems. Some factors identified in [7] that are particular sources of uncertainty for FPV in model accuracy for predicting performance are *irradiance transposition* (the need to account for irradiance over water for mono- and bi-facial panels and impact of float structure), *standard test conditions* (greater variation from these over water), and *cell temperature* (cooling effect from being over water not modeled well in current state of the art).

A further consideration is that, in some cases, there are additional beneficial or potentially detrimental impacts from FPV on the water body that are useful to have quantifiable metrics for and reliable means to estimate. Examples of these are *evaporation rate* (the expected benefit being reduction in evaporation loss), and *water temperature* (e.g. the impact of FPV on cooling water canals).

VII. CONCLUSIONS AND FUTURE WORK

Optimizing FPV system design for performance and estimating system performance for a given FPV system design, plant design, and location has complexities and special considerations not present in the optimization and analysis of land based solar PV systems. The unique challenges of keeping a system stable, stationary, and intact in the presence of wind, wave, current, and water level changes, while floating on the surface of the water drives dimensional and geometric constraints and trade-offs that differ considerably compared to land-based systems. Some key design optimization factors involved in these constraints and trade-offs are PV array tilt and ground coverage ratio (GCR), with relevant examples examined herein. A key performance estimation factor that is not yet well-modeled and accounted for in existing tools is temperature, in particular translating ambient temperature to panel temperature over water and the impact on PV efficiency.

There is also a need for additional performance metrics. Since advantages may include effect of PV on the water body itself, predicting effects, such as reduction in evaporation rate, is an area for further future research.

As part of a multiyear project underway to validate, demonstrate, and test a new FPV design, the authors plan additional research on improving methods for modeling and accounting for PV panel temperature when analyzing and estimate FPV performance prediction. The project team will also be instrumenting multiple FPV test sites to acquire field data to use in validating FPV design optimization and system performance estimation methods.

ACKNOWLEDGEMENT

This material is based upon work supported by the U.S. Department of Energy's Office of Energy Efficiency and Renewable Energy (EERE) under the Solar Energy Technologies Office (SETO) Award Number DE-EE0009641.

REFERENCES

[1] Kennedy, R., "Floating PV could reach 4.8 GW globally by 2026", *PV Magazine*, January 19, 2022.

[2] Perez, M.J.R., "Deploying Effectively Dispatchable Floating PV on Reservoirs: Comparing Floating PV to Other Renewable Technologies", *Solar Energy*, Oct. 2018.

[3] Haggerty, J., "Floating solar nearing price parity with land-based US solar", *PV Magazine,* Oct. 7, 2020.

[4] Zahid, Herman, Abdullah Altamimi, Syed Ali Abbas Kazmi, Zafar A. Khan, and Abdulaziz Almutairi. "Floating solar photovoltaic as virtual battery for reservoir based hydroelectric dams: A solar-hydro nexus for technological transition." *Energy Reports* 8, 2022, 610-621.

[5] "Where Sun Meets Water: Floating Solar Market Report", *World Bank*, 2018.

[6] Spencer, R.S., et al, "Floating Photovoltaic Systems: Assessing the Technical Potential of Photovoltaic Systems on Man-Made Water Bodies in the Continental United States", Environmental Science & Technology, Dec. 2018.

[7] Prilliman, M. J., Hansen, C.W., Keith, J.M.F., Janzou, S., Theristis, M., Scheiner, A., and Ozakyol, E., "Quantifying Uncertainty in PV Energy Estimates Final Report. No. NREL/TP-7A40-84993", National Renewable Energy Lab (NREL), Golden, CO (United States), 2023.

[8] Reise, C., Muller, B., Moser, D., Belluardo, G., Ingenhoven, P., "Uncertainties in PV System Yield Predictions and Assessments", *IEA PVPS Task 13, Subtasks 2.3 & 3.1, Report IEA-PVPS T13-12:2018*, Intl. Energy Agency (IEA), April 2018.

[9] NREL System Advisor Model (SAM), https://sam.nrel.gov/

[10] Bolinger, M., Bolinger, G., "Land Requirements for Utility-Scale PV: An Empirical Update on Power and Energy Density*", IEEE Journal of Photovoltaics, Vol. 12, No.2*, IEEE, March 2022.

Partial shading of photovoltaic modules: a comparison between simulated and measured IV characteristics

Bianca Passarella(1), Maarten Verkou(2), Marco Leonardi(1), Fabrizio Coco(3), Youri Blom(4), Malte Vogt(4), Rudi Santbergen(4), Agnese Di Stefano(3), Andrea Canino(3), Marina Foti(1), Antonino Ragonesi(1), Marcello Sciuto(1), Francesco Rametta(1), Miro Zeman(2,4), Olindo Isabella(2,4), Cosimo Gerardi(1)

1. Enel Green Power S.p.A, 3SUN, Contrada Blocco Torrazze Zona Industriale 95121, Catania, Italy
2. PVWorks, Mekelweg 4, 2628CD, Delft, the Netherlands
3. Enel Green Power S.p.A, Innovation Lab Contrada Passo Martino Zona Industriale 95121, Catania, Italy
4. Delft University of Technology, PVMD, Mekelweg 4, 2628CD, Delft, the Netherlands

Abstract — **Photovoltaic (PV) technology is raising attention as a low-cost green energy source. It mainly finds applications in solar fields, on building facades and on rooftop. One of the main issues that can occur is the shading of solar cells inside the photovoltaic module which could affect the maximum power output of the PV panel and the lifetime of the cell itself. In order to predict the behaviour of PV panels in partial shading conditions, simulations and then measurements on two different photovoltaic modules have been carried out and compared. Data have shown that the maximum power output of the panels under 1sun illumination can be predicted by simulation with a 3% discrepancy from measured values, independently from the type of technology and interconnections of the PV module.**

Keywords: partial shading, simulation, modeling, photovoltaic

I. INTRODUCTION

Photovoltaic (PV) technology is used to convert solar light into electricity. It is a renewable energy source widely chosen for green energy production [1] thanks to its reduced levelized cost of generated electricity [2]. Photovoltaic panels find their application both in power plants or on rooftop where partial shading (PS) represents an issue. Solar cells in the photovoltaic panel can be shaded due to light-blocking objects like trees, buildings or chimneys and can experience issues due to reverse bias and high temperature [3]. In addition, partial shading causes a deterioration of the maximum power point (MPP) and sometimes generates multiple MPPs of the solar panel output making the maximum power point tracking (MPPT) system very complicated. In this scenario, it becomes necessary to predict the behaviour of solar panel in shading conditions by proper modeling. There exist several models in literature to simulate shading conditions on cells, panels and also on power plants [4].

In this work, the model proposed by Atia et al. [5] has been implemented in the Photovoltaic Materials and Devices Toolbox (PVMD Toolbox) [6] to predict the behaviour of two photovoltaic panels which differ for the type of interconnections. In order to validate the model and the

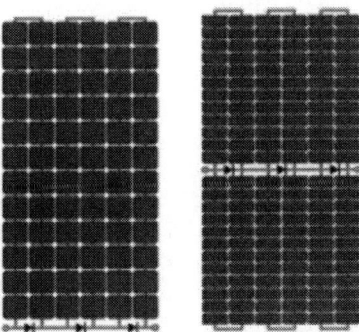

Figure 1: a) Standard module having 72 cells and 3 strings; b) butterfly module having 144 half cut cells divided in 6 substrings.

simulation tool used, the two photovoltaic panels have been then tested under the typical AM1.5G 1sun condition, with different shading patterns.

II. EXPERIMENTAL SECTION

Two different PV modules have been considered for both simulations and measurements. The first is a standard module composed by 72 M2 cells (individual cell area: 244.43 cm^2) connected in series. It can be divided into three strings each one connected to one bypass diode (Figure 1.a). The second one is a butterfly module and is made by 144 half M10 cells (individual cell area: of 165.35 cm^2). In this case, the bypass diode is placed in the middle of the string, forming two substrings connected in parallel (Figure 1.b). The IV curve of an individual cell of the module was calculated using a 1-diode model [7]. The model parameters saturation current (J_0), series resistance (R_s), shunt resistance (R_{sh}), and ideality factor (n_{id}) change with temperature and irradiance. For the model to work properly the shading factor was set to 99.5% instead of setting the incident irradiance to zero.

978-1-6654-6060-6/23 $31.00 © 2023 IEEE

Standard module used for electrical measurements have been processed in EGP 3SUN factory in Catania and is made of heterojunction (HJT) solar cells. Butterfly module has been fabricated by Jinko and is made by Passivated Emitter and Rear Cells (PERC). Measurements at 1sun AM1.5G have been carried out in Enel Green Power under Pasan Solar simulator.

A. Shading patterns

For both the photovoltaic modules, the same shading patterns have been considered, meaning that two half-cut cells are shaded in the butterfly module for each whole cell in the standard module.

In Figure 2, four different patterns are represented: pattern 1 and 2 have been chosen to compare the effect of shading on a single string and on a single diode, while pattern 3 and 4 have been chosen to compare the effect of shading on two different strings connected on two diodes.

B. Electrical characterization

Simulation and measurements under 1 sun illumination have been carried out on 3Sun standard module and on Jinko butterfly module in order to study the effects of shading. Every IV curve of the shaded module has been compared with the electrical output of the unshaded module. In Figure 3 the simulated and measured IV characteristics under 1 sun illumination have been reported.

For the standard module, the output IV curves given by Patterns 1 and 2 are overlapping since cells are connected in series. This means that covering one cell of the string or the whole string gives the same output IV curve. In Patterns 1 and 2 the open circuit voltage (V_{oc}) of the shaded module is equal to 2/3 of the $V_{oc,unshaded}$ since only 2/3 of the PV module is actually working and the maximum power (P_{max}) is 64% of $P_{max,unshaded}$ (Table 1). Also for Patterns 3 and 4 the IV output curves are overlapping independently from the covered area. In this case, the Voc of the shaded module is 1/3 the $V_{oc,unshaded}$ and the P_{max} is 27% of $P_{max,unshaded}$ (Table 1). Looking at the IV output curves of the butterfly module, a difference with respect to the standard configuration can be appreciated: Patterns 1 and 2 do not give rise to the same output. Indeed, in the case of parallel connection of the string, current keeps flowing into the unshaded substring [8]. In Patterns 1, 3 and 4 where the string is not totally covered, the IV curve has a hump which generates a double P_{max} peak (a global and a local P_{max}). Even if there is this P_{max} double peak, for Patterns 1 and 2 the global P_{max} is the same and corresponds to 64% of $P_{max,unshaded}$ (Table 1) as for the standard module.

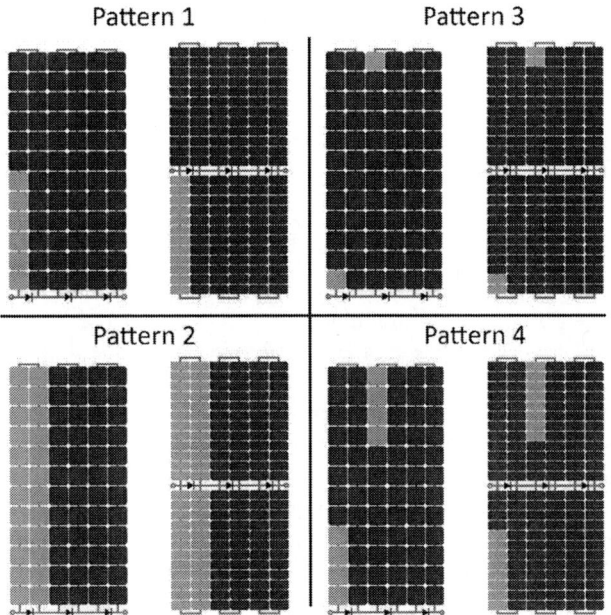

Figure 2: Shading patterns reported for both standard and butterfly module. Pattern 1 and 2 affects one only bypass diode while pattern 3 and 4 affect two bypass diodes.

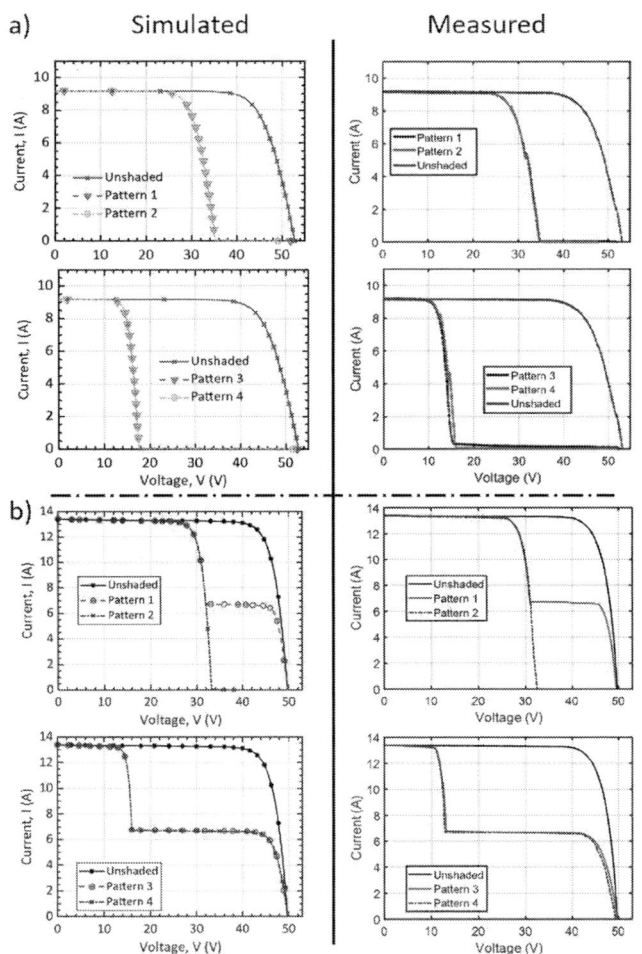

Figure 3: simulated and measured IV characteristic curves under 1 sun illumination for a) standard and b) butterfly modules.

TABLE I
MEASURED MAXIMUM POWER POINT

	$P_{max,STD}$ [W]	$P_{max,BUTTERFLY}$ [W]
Unshaded	364.9	543
Pattern 1	233.7	351
Pattern 2	233.7	351
Pattern 3	99.4	279
Pattern 4	102.1	277

TABLE II
SIMULATED MAXIMUM POWER POINT

	$P_{max,STD}$ [W]	$P_{max,BUTTERFLY}$ [W]
Unshaded	364.2	543.4
Pattern 1	242.8	362.3
Pattern 2	242.8	362.3
Pattern 3	121.4	284.7
Pattern 4	121.4	282.4

In Patterns 3 and 4, the presence of the hump in the IV curves gives an output P_{max} equal to the 51% of the $P_{max,unshaded}$ (Table 1) which means that, if MPPT capable in finding the global P_{max} is present, the butterfly configuration is more convenient in case of PS of the string.

Comparing the measured and simulated IV curves for both standard and butterfly modules it can be noticed that with the chosen model, the behaviour of the PV panels can be predicted in a reliable way. Simulated and measured IV curves are in accordance and for the four presented patterns they differ of about 3% in terms of maximum power since there is a slight overestimation of the voltage at the maximum power point (V_{mpp}).

III.CONCLUSIONS

In this work, an experimental investigation of partial shading influence on PV panels has been performed and then compared to simulated results. Measurements under 1sun illumination have been carried out on two different PV modules: a standard module with HJT cells and a butterfly module with PERC cells. Measured and simulated IV characteristics were in accordance for both the PV panels independently from the kind of connections and the cells technology and a difference of around 3% in the P_{max} has been calculated.

REFERENCES

[1] A. G. Olabi and M. A. Abdelkareem, "Renewable energy and climate change," *Renewable and Sustainable Energy Reviews*, vol. 158, Apr. 2022, doi: 10.1016/j.rser.2022.112111.

[2] M. A. Green, "Photovoltaic technology and visions for the future," *Progress in Energy*, vol. 1, no. 1. Institute of Physics, Jul. 01, 2019. doi: 10.1088/2516-1083/ab0fa8.

[3] M. Dhimish, V. Holmes, P. Mather, and M. Sibley, "Novel hot spot mitigation technique to enhance photovoltaic solar panels output power performance," *Solar Energy Materials and Solar Cells*, vol. 179, pp. 72–79, Jun. 2018, doi: 10.1016/j.solmat.2018.02.019.

[4] F. Bayrak and H. F. Oztop, "Effects of static and dynamic shading on thermodynamic and electrical performance for photovoltaic panels," *Appl Therm Eng*, vol. 169, Mar. 2020, doi: 10.1016/j.applthermaleng.2020.114900.

[5] A. Atia, F. Anayi, and M. Gao, "Influence of Shading on Solar Cell Parameters and Modelling Accuracy Improvement of PV Modules with Reverse Biased Solar Cells," *Energies (Basel)*, vol. 15, no. 23, Dec. 2022, doi: 10.3390/en15239067.

[6] M. R. Vogt *et al.*, "Introducing a comprehensive physics-based modelling framework for tandem and other PV systems," *Solar Energy Materials and Solar Cells*, vol. 247, Oct. 2022, doi: 10.1016/j.solmat.2022.111944.

[7] M. A. Koondhar, M. I. Jamali, A. S. Channa, and I. A. Laghari, "PARTIAL SHADING EFFECT ON THE PERFORMANCE OF PV PANEL AND ITS DIFFERENT CIRCUIT TOPOLOGIES BASED MITIGATION TECHNIQUES: A REVIEW," *Article ID: IJARET_12_04_003 International Journal of Advanced Research in Engineering and Technology (IJARET)*, vol. 12, no. 4, pp. 15–23, 2021, doi: 10.34218/IJARET.12.4.2021.003.

[8] R. G. Vieira, F. M. U. de Araújo, M. Dhimish, and M. I. S. Guerra, "A comprehensive review on bypass diode application on photovoltaic modules," *Energies*, vol. 13, no. 10. MDPI AG, May 01, 2020. doi: 10.3390/en13102472.

Hierarchal Ti3C2Tx MXene and aluminum microgrid back contacts for bifacial CdTe PV

Benjamin E Sartor, Matthew O Reese, Chris Muzzillo, Chungho Lee, Andre D Taylor

New York University, New York, NY, United States

National Renewable Energy Lab, Golden, CO, United States

First Solar, Santa Clara, CA, United States

A hierarchal transparent back contact leveraging an AlGaOx passivating layer, high work function Ti3C2Tx MXenes, and a transparent cracked film lithography (CFL) templated aluminum microgrid is demonstrated on copper-free CdTe devices. AlGaOx improves device VOC but leads to reduced fill factor when CFL is used. The includion of Ti3C2Tx interlayer improves fill factor, removes detrimental Schottky barriers, and enables metallization with low cost aluminum. The bifacial performance of an AlGaOx / Ti3C2Tx / CFL Al contact is evaluated, reaching 1.9% backside efficiency under 1-sun illumination.

Design of Electronic Control of PV Tracking Independent of Weather Forecast

Sam Mil'shtein[1], Dhawal Athana[1], Jeffrey Snell[1], William Brooks[1,2]

[1]Advanced Electronic Technology Center

ECE Dept., UMass Lowell, MA, USA

[2]Aerodyne Research, Inc., Billerica, MA, USA

Abstract. With over 28 Terawatt-hours of annual electricity produced globally, only 2% is generated by photovoltaics (PV). Solar tracking is one of the most efficient methods to increase the amount of clean energy produced by PV panels. It was shown in experimental studies that the solar day could become 30% longer, i.e., 30% more solar energy collected daily, if proper single-axis (or dual-axis) tracking systems are installed. However, many commercial systems are controlled by input from local weather forecasts, which rarely take the specific location of a solar farm into account. Thus, usage of pyranometers positioned close to solar farms produce accurate solar irradiance information, despite their delayed response and drawbacks with predicting solar production. In the current study we describe the design and operation of a miniature, low cost, electronic system which tracks the sun and instantly prescribes the optimal orientation of PV panels [1] for any local weather. The system consists of two stepper motors, where each carry three small photodetectors connected to its shaft. The total sensitivity of sensors covers the entire spectrum of silicon solar cells. Stepper motors control the movement of optical sensors scanning the skies in the vertical (elevation) and horizontal (azimuth) directions. The programmable microcontroller stores information of complete elevation and azimuthal scans. The algorithm then predicts the amount of power that could be generated by large PV panels, if their orientation replicates positions of motor shafts. Experimental testing of the novel system supports our method of efficient sun tracking.

1. Introduction

Among natural sources of energy usually converted into electricity such as coal, gas, oil, and solar radiation, PV has the best ratio of Energy Returned (ER) to Energy Invested (EI). In simple words, the ratio of energy obtained from the energy production per unit consumed in the process. This ER/EI ratio was offered by Prof. M. Fermeglia of Italian Trieste University [2] and discussed further with respect to different solar cell technologies [3]. Even with the added maintenance cost to PV manufacturing, the conversion of solar energy to electricity is characterized by ER/EI = 25:1, compared to fossil fuel electricity generation varying in the range of ER/EI = 7:1 - 13:1. This ratio will only improve as many existing static solar farms invest in tracking arrays to improve production [4-5].

The primary controllable factor that affects PV output is the incident angle of the sun's rays on the surface of the panel. Maximum PV efficiency is achieved when the solar rays are perpendicular to the panel surface, which is the motivation behind solar tracking. Single-axis tracking systems (SATS) are the most common type of utility-scale PV system in the United States, compared to dual-axis tracking systems (DATS) and static systems. The improved efficiency of DATS over static systems varies from 24.91% on an overcast day to 82.12% on a clear day [6]. SATS are able to generate up to 90% of the energy produced by DATS, and due to lower costs, are preferred for most PV farm usage.

Regarding the cost-effectiveness and surplus power generation, single axis trackers are becoming increasingly popular and adoptable in the industry [5, 7]. With the consequent capital cost reductions associated with them, it will soon be advisable for fixed tilt solar PV installations located northwards of certain latitudes to be revamped with single axis trackers [7]. As a result, real time assessment of the optimal angular orientation of the PV panels has come across as a major challenge [8-10]. Constantly varying sky conditions, changing proportions of the different components of irradiance, and location-specific factors further convolute this challenge.

Existing solutions in the industry rely on computational algorithms that receive inputs from multiple sources of data like real-time satellite data, weather forecast, analysis of solar equations, and pyranometers installed at dedicated measuring stations. While these solutions have been successful in tracking the sun and optimally orienting solar

978-1-6654-6060-6/23 $31.00 © 2023 IEEE

arrays in different conditions, they have not been able to sufficiently address the local topographic factors that determine the availability of reflected and diffused components of available irradiance at the site [11-13].

Among these irradiance detecting tools, the pyranometer instrument is currently the most popular and standardized solution [14-15]. Pyranometers operate based on the signals produced from a thermocouple by means of the Seeback phenomenon. The two major figures of merit associated with these devices include relatively fast response time, uniform spectral response, and wide range spectral response. The best response time achieved by pyranometers is at least 8 seconds, which is the minimum time required for a thermocouple to heat up. With the omnidirectional response in space, these devices also require complex installation and orientation structures to assess the spatial distribution of available solar resource, as seen in Fig. 1. As a result, there is an urgent need for measurement solutions that are scalable enough to be adopted at distributed generation (DG) scale PV installation sites.

Figure 1: A pyranometers setup strategically oriented for measuring diffuse and direct solar irradiance [after 15].

One of the issues associated with pyranometer's uniform spectral response being used as a figure of merit in PV production is the fact that solar PV cells operate at different efficiencies in response to different wavelengths in the solar spectrum [16-19]. Pyranometers are typically designed to cover the entire solar spectrum, generally ranging from 0.3 μm to 3.0 μm. PV production depends on the presence of a specific range of wavelengths in the sun spectra which is largely defined by the type of semiconductor material used in the solar cell. Therefore, the wide spectral sensitivity of pyranometers means they are inherently inaccurate at predicting PV production at a location.

Operation of tracking systems are strongly defined by weather conditions and controlled by direct forecast signals in the area, or localized signals of pyranometer platforms, described above. This study presents the development of a miniature electronic system which provides localized information about weather, with an algorithm to help SATS and/or DATS track sun position. This novel system provides instant information of the incoming sun power and predicts the amount of energy that the sun tracking PV array would produce in any tested orientation. The low cost of our system is incomparable to the expensive cost of physically sizable pyranometers.

2. Conventional Tracking System Control

Conventional tracking systems rely on algorithms for control over the actuators. There are two main types of algorithms that are used by tracking solar farms today. Open-loop algorithms, also known as schedule trackers, use only a microprocessor for mathematical calculations and do not require any sensors for feedback. Closed-loop algorithms on the other hand, use electro-optic sensors to determine the direction of highest intensity light. These sensors perform well in clear sky conditions, but often fail to accurately measure the solar position in overcast or intermittent cloudy conditions [20]. Numerous studies found that schedule-based trackers can outperform optical tracking system's production by 2.1%-4.2% over the span of a few days [20-21].

3. Weather dependance of existing tracking systems

The performance of various tracking strategies is most dependent on the weather and atmospheric conditions at a given location. Irradiance hitting a PV panel in most cases is broken into three different components: direct beam irradiance, diffuse irradiance, and reflected irradiance. Direct normal radiation is the primary type which arrives directly from the sun when there are no obstructions between the sun and ground surface. Diffuse radiation is the component that is scattered throughout the atmosphere from cloud cover and other aerosols and hits ground surfaces from all directions. Reflected irradiance is radiation that gets reflected off objects and surfaces with high albedo. This is generally a small component of the total radiation hitting a tilted surface, unless the ground is highly reflective, like the case with water or snow in front of the panels. In any given location, the prominent type of radiation changes based on these countless factors that are constantly changing in the sky and surroundings. Therefore, to maximize PV production, optimal tracking algorithms need to be developed that continuously consider location, weather, and seasonal changes [22].

Kelly et al. [23] proved that during days of heavy overcast cloud conditions, horizontally positioned panels are the optimal orientation for solar energy capture at up to 50% more than a sun tracking panel. Kelly et al. later proposed and modeled a hybrid algorithm that compared the maximum electrical output of a horizontal fixed system versus a DATS. Over an entire cloudy day, the model confirmed that an additional 50% of energy can be collected by utilizing a horizontal panel position during periods of heavy overcast weather. When modeled in Detroit, where up to half of the days are cloudy, this new algorithm produced an average of 1% additional energy over one year compared to a normal DATS [24]. Overall, the gained energy from the horizontal position on cloudy days has a diminished impact on energy collection over the span of months or a year. This is due to the significantly less irradiance that's available for collection on cloudy days in the first place. However, improving energy output on cloudy days becomes important for large systems, which need to produce sufficient electrical power on the worst solar days. In some cases, improving the worst-case energy production scenario can lead to a cheaper and smaller system.

In a similar study, Saymebetov et al. [25] compared a hybrid algorithm to a DATS using a schedule-based tracking algorithm. The adaptive hybrid algorithm compared the output power from a small horizontal panel and a small tracking panel to determine the optimal position of the larger system. The hybrid tracker generated 66-69% more energy on cloudy days compared to the traditional DATS and overall, 1.5% more energy over the span of three months [25]. These studies prove that adaptive algorithms that can recognize cloudy conditions provide valuable energy gains during these conditions. However, developing an algorithm that can quickly and accurately determine when to position horizontally proves challenging.

Antonanzas et al. [26] developed a prediction algorithm model that works in real time and another that predicts the position ahead of time based on global horizontal irradiance forecasts issued twice per day. The first model compared the irradiance on a horizontal surface to a tracking panel surface to determine the best position for the larger panel. This algorithm increased daily energy gains between 0.16% and 3.01% compared to SATS in the southernmost and northernmost sites respectively. The predictive algorithm using forecasted data was less effective for every site that was tested, mainly due to the presence of false positive decisions to switch positions [26]. Forecasted tracking algorithms for PV systems remain difficult to implement due to the inaccuracies of forecasts and often insufficient data for rural locations.

With the intent of solving these challenges, this study presents a novel measuring device which carries two platforms, each with three different photodetectors attached. Each photodetector is sensitive to a different portion of the solar spectrum. Each platform is installed on a shaft of a miniature stepper motor, which rotate through the elevation and azimuth planes respectively. Stepper motors are used to rotate these shafts across discrete angular steps to record the output signals from the photodetectors in each position. In the experiments that follow, the signals from these photodetectors are compared to the predicted output power from PV panels oriented at prescribed angular positions. Additionally, the system was programed and tested alongside regression models and the methodology for assessment of available solar resource and evaluation of PV panel performance is discussed. Consequently, the eligibility of the novel device for applications such as PV production forecasting, smart sun tracking and assessment of local solar resource is inferred.

978-1-6654-6060-6/23 $31.00 © 2023 IEEE

4. Design of novel system

This novel experimental system investigates the performance of a closed-loop algorithm that evaluates the intensity of solar radiation throughout the elevation and azimuth plane. The system can identify the orientation for optimal PV power and determine whether the sky is sufficiently overcast to warrant switching from sun tracking to horizontal positioning of PV arrays. The block diagram below in Fig. 2 illustrates how the different components of the system interact with each other.

The microcontroller used to implement the algorithm was an Arduino Nano ATmega328P. The Arduino program controls the two motor drivers which operate the stepper motors to the desired number of steps. The motors rotate the elevation platform and the azimuthal platform, both of which have three solar sensors on them pointing outwards from the motor shaft. The elevation platform completes a scan by rotating through 180° in the elevation plane. It begins with the platform and sensors pointed due west and travels up to the zenith then back down to due east to capture the sun's general path, similar to SATS. The second platform rotates horizontally 360° around the azimuth plane on a separate axis. The stepper motor moves each platform in discrete 14.4° steps, with a brief pause at each step while the controller records the sensor's measurements. This totals 13 steps in the elevation plane and 25 steps in the azimuth plane.

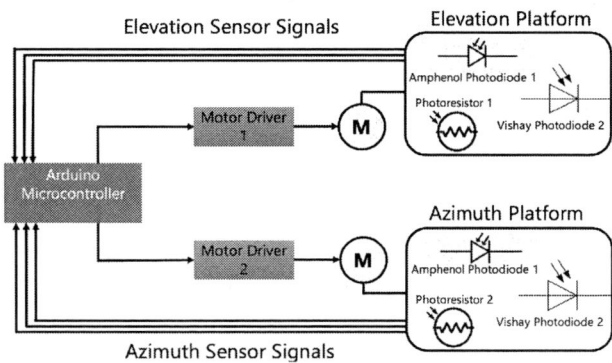

Figure 2: Block diagram of prototype system. Not shown are the photodiode reverse biases and the stepper motor's power supply.

The novelty of this system is the use of three different solar sensors that simultaneously respond to the intensity of solar irradiance. Each platform has two photodiodes and a cadmium sulfide (CdS) photoresistor, each of which produce a signal for the microcontroller to read. These three sensors have spectral sensitivities that combined cover the entire solar spectrum in which silicon solar cells are responsive to. This combined spectral response, shown in Fig. 3, allows for better prediction of silicon PV production, especially in overcast conditions where sensor-based tracking tends to not perform well. Each photodiode is also reverse biased with either 5V or 6V to increase the responsivity to light intensity. The overall footprint of the system is smaller than a laptop.

At each step of the scan, the controller reads the voltage from the voltage divider for each sensor circuit and converts the values to the current generated by the sensors. Both the elevation and azimuth scan occur simultaneously and when finished, the algorithm identifies the position exposed to the highest intensity of solar energy based on the combined sensor outputs. At the end, both platforms rotate back to point in the orientation with maximal power for visualization purposes. In a real-world setting, the program can be adjusted to perform scans at any interval of time desired for the application.

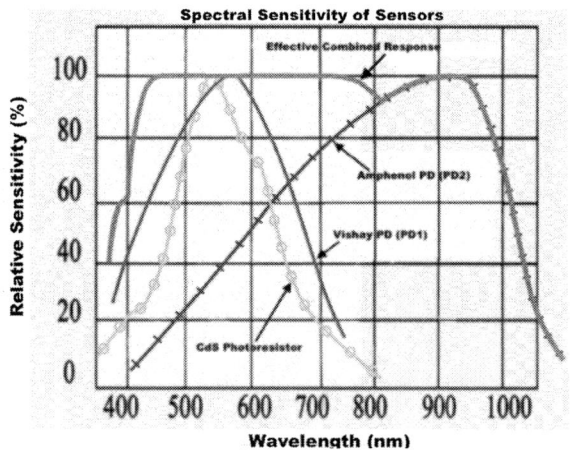

Figure 3: Sensitivity of the spectral responses from each sensor. Combined they cover most of the spectrum from 400 to 1100 nanometers, corresponding to the sensitivity of most solar cells.

5. Experimental measurements and testing

The results of the prototype testing alongside the model are shown below with an example elevation scan in sunny conditions shown in Fig. 4.

The main standard for model accuracy was the Mean Absolute Percent Error (MAPE), which averages the absolute error between the predicted performance and actual PV power for every step of the scan. R^2-correlation coefficients were also computed for the two clear sky cases but disregarded for the overcast scans due to the inherently low variance in values. For the elevation scan in overcast conditions, the MAPE was 17.1%. The azimuth scan had a MAPE of 20.9%. For clear sky conditions, the elevation scan had a MAPE of 28.8% and R^2 correlation coefficient of 0.85. The clear sky scan in the azimuth plane had a MAPE of 43.3% and R^2 correlation coefficient of 0.82. The model was also tested against randomly selected data points in all types of sky conditions. The test used 80% of the total available data for training the model and the remaining 20% for testing. This overall test case resulted in an R^2 of 0.89 and a MAPE of 36.4%.

It is important to note that this system needs to be independently calibrated at any new location with the chosen solar array that it will be used with. As discussed throughout the paper, location-based factors, type of panel design, and parameters of the semiconductor materials being used, all need to be considered in the specific characteristics of the site. As a result, the following scans should only be used as a reference for what the system is capable of. Results will certainly vary, and parameters should be fine-tuned from site to site.

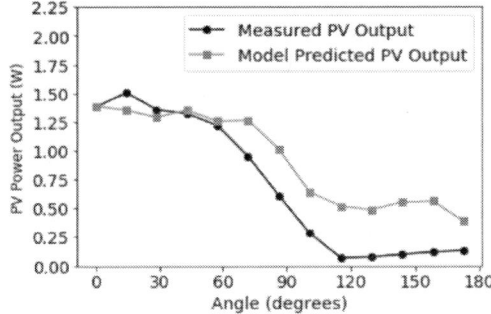

Figure 4: The actual and predicted values vs elevation angle is shown from a sunny day scan. The model predicts PV power based on sensor outputs from the scan. Test taken at 5:00PM on September 2.

6. Discussion

Some experimental curves (figures) collected from testing under different weather conditions were excluded, but verbally described in this discussion are the most important results.

It is well established that horizontal positioning of PV panels during heavy overcast conditions can result in short term production gains by up to 80%, and monthly gains of up to 1.5% [23-25]. This makes it imperative for solar farms to have systems that can not only track the sun to maximize direct irradiance, but also identify sufficiently cloudy conditions to optimize the collection of diffuse irradiances. From the results above, the two overcast scans have less error than the scans in clear conditions, a trend seen throughout testing. This is mostly due to the inherently lower variance of values in overcast scans. The overcast azimuth curve is entirely flat, and the overcast elevation prediction curve only varies by a tenth of 1 Watt, compared to the clear sky scans which have a high variance in PV production throughout each scan. As a result, the error for the overcast scans revealed every predicted point with less than 55% error and over half with less than 20% error. It is also important to notice that the overcast azimuth scan has such low power output that just a fluctuation of 0.05W in the predicted output leads to an error of 25%, showing that the error values are very sensitive to small mispredictions.

Based on the flat predictions in the two overcast scans, the sensors are not sensitive to minute changes in PV power in low light conditions. This was expected to an extent, as numerous articles highlight that schedule-based trackers are more efficient than sensor-based trackers in overcast and partly cloudy conditions [20-21]. Despite the sensor's failure to predict maximal PV power point in these low light conditions, this ability is trivial when the implementation of horizontal positioning is considered. An analysis of meteorological data for Detroit, MI and Phoenix, AZ showed a significant percentage of cloudy or partly cloudy days throughout the year with 79% in Detroit and even 42% in Phoenix [23]. This proves that horizontal positioning during overcast conditions is an extremely useful technique that any tracking array in any location can take advantage of, even in highly sunny areas such as Phoenix. As a result, it is unnecessary for sensor-based tracking systems to be as accurate in cloudy conditions, as long as the algorithm being used can still identify when arrays should stop tracking and switch to horizontal orientation.

The tested elevation and azimuth scans taken in sunny conditions are shown above in Fig. 5. As expected, the power output in both the elevation and azimuth scans had a wider range than the overcast scans. Both sunny scans had a higher MAPE than the overcast condition scans with 28.8% in the elevation and 43.3% in the azimuth. The overall error in the clear sky elevation scans is comparable to the overcast scans with 69% of the points below 20% error. However, the measured PV power at $90°$ elevation drops significantly and appears to be an outlier measurement, leading to a prediction error of over 100% at that point. The clear sky azimuth scan has a wide error range with only 44% of the points having less than 20% error and 44% of the points with greater than 50% error.

In the elevation scans, the sensor's predicted optimal position was one step off from the actual maximum power position. The sensor's predicted the maximum elevation angle to be $115°$ from the initial position, but the PV panel produced the highest output closer to $130°$. In the azimuth scan the sensor's predicted optimal position were also off from the correct orientation by ~$30°$. Throughout testing, the elevation scan was successful at predicting the optimal orientation about 67% of the time in relatively clear conditions. 100% of the predictions were within $14.4°$ (one step) of the optimal orientation with an average error of $4.8°$. The azimuth scan's ability to predict the optimal power orientation was overall poor. For more precision, step size could be decreased to produce a higher resolution scan at the cost of scan time. Consequently, it is proposed that alongside the tested algorithm, a schedule-based tracking algorithm could be used congruently, due to their extremely high precision tracking the sun. The proposed prototype system could be primarily used to determine when to switch from scheduled tracking to horizontal positioning and predict how much power will be generated with that orientation, while the schedule-based tracker keeps track of the exact sun position.

One major difference found between the elevation and azimuth plane was the higher presence of noise in the azimuth scans, leading to curves less smooth than in the elevation. The majority of this noise appeared as random jumps in value and are primarily attributed to reflections along the horizon. Tests were conducted in a relatively open field, but there was still a nearby parking garage to the west along with some surrounding fences and metal bleacher seats. In the elevation, the sensors are pointed toward the sky for most of the scan, seeing primarily direct and diffuse irradiance. The sensors in the azimuth plane, however, rarely experience significant direct irradiance and are mostly influenced by the diffuse and reflected components. Reflections appeared particularly prevalent in the azimuth on clear sky tests, resulting in lots of noise in sensor values when pointed away from the sun and towards surrounding objects. For example, the predicted values between $120°$ and $250°$ in the azimuth scan in Fig. 4 show lots of sensitivity to minute changes that the larger PV panel does not respond to. It is also likely that the elevation scan sensors could be affected by reflections when within $30°$ of the horizon. Upon further analysis of the sunny azimuth scans, it was typically large jumps in the photodiode 1 current which caused these fluctuations in value.

Compared to the predicted sunny azimuth power curves, the actual PV panel had an extremely flat response curve when pointed away from the sun in the azimuth, proving that the solar cells are not as affected by the random reflections on the horizon. This is logical considering the surface area of the photodetectors is smaller than the PV panel by a factor of ~10,000. A pinpoint reflection will have a much larger effect on the response of a 1 mm^2 solar cell compared to an entire solar panel containing numerous solar cells. As a whole, solar panels are typically not heavily impacted by reflected irradiance considering the aim is to point towards the sun's rays, whereas a large portion of reflected radiation emanates from the opposite side of the panel's face. This reflected irradiance component is only significant if bifacial panels are being used or if the ground in front of the PV array has a high albedo.

The final model test was completed with randomly selected data points from the collection of scans. Data from all different types of weather conditions were used to train the model and then it was tested on 20% of the total data, shown above in Fig. 6. The R^2 correlation coefficient of 0.89 reveals that the model predictions correlate very well with the actual power, regardless of the weather conditions. The 36.4% MAPE is on par with the mean error in the sunny azimuth scan, which was the highest error of the four conditions. However, half of the predicted points had less than 10% error. The power range that had the largest uncertainty was the lower power values from 0.25W to 0.5W. This further confirms that the sensors are incapable of picking up small changes in irradiance in overcast conditions.

To minimize error in PV measurements in future correlation tests, it is recommended to use a PV tracking array scanning the same positions as the prototype system with a microcontroller collecting power data. In this study, PV power was recorded by hand by measuring the short circuit current and open circuit voltage at every orientation. It was initially underestimated how quickly sky conditions could change while calibrating the system, especially with moving clouds. As a result, it proved difficult at times to collect PV measurements while clouds were moving. For best calibration, PV measurements should be recorded simultaneously with the system scans or immediately after in order to preserve the exact sky conditions at the given moment.

One of the major implications of including the idea of smart tracking, is real-time estimation of output power from PV panels at a specific location using photocurrent data or estimated irradiance from the miniature system. This system provides the PV farm operator with an added degree of freedom as the interval between two consecutive scans of the miniature system can also be customized to vary from the recommended value of 30 minutes. As an embodiment of this idea, the energy required to move the solar panels to a new angular position can be estimated prior to finalizing the decision to change positions. An example evaluation of this can be seen in Table 1.

With a variety of recent developments taking place within the PV power generation industry, absence of thoroughly defined design standards is one the challenges for estimating the energy required to rotate a PV array [27]. As a result, the assessment

Table 1: Mechanical specifications of dual axis solar tracking systems

S/No.	Quantity	Value
1	Weight of a single solar panel	42 lbs=19 kg or 190 N
2	Wind loading on a single panel	2067 N/m²
3	Max force due to wind on a single panel	4000 N
4	Area of a single commercial panel	2 m²
5	Maximum torque single panel exerts about the axis of rotation	320 Nm (8-inch separation for the axis of rotation)
6	Max extension length of linear actuators	0.45 m
7	Stroke value for 15°	150 mm
8	Rated Force of linear actuators	150000 N
9	Energy spent in moving by 15°	2250 J
10	Number of solar panels serviced by a single actuator	12-16
11	Power capacity handled by a single actuator	4.8 kW to 6.4 kW
12	Energy needed by linear actuators to move by 15°	351.56 J/kW to 468.75 J/kW
13	**Max. energy to move linear actuators by 15° for SATS**	**0.097 Wh/kWh to 0.130 Wh/kWh of installed capacity**
14	**Max. energy to move linear actuators by 15° for DATS**	**0.194 Wh/kWh to 0.260 Wh/kWh of installed capacity**

presented is an approximation based on average values [28-32] and specifications for the following:

- The dimensions of the mounting frames used are from commercial scale dual axis PV systems
- Wind loading on the surface of the solar panels can be a significant factor in tracking. This parameter is set to vary significantly depending on geographic conditions
- Energy spent in moving the loaded solar tracker by 15°. This value must be assumed as constant because the total span of rotation along the elevation and azimuth axes varies according to geography and other factors like budget, scale of output power generation, etc.

Despite the low torque exerted by the weight of the solar panel about the mounting axis, there can be erratic levels of torque about the rotating axis due to wind loading. The result for maximum value of energy required, presented in Table 1, was evaluated by using the value of rated force of the linear actuators used for the tracking systems. This value was multiplied by the stroke length to estimate moving the panels by 15°. A similar calculation to the one conducted can be used to determine if a change in PV position is warranted, by comparing the energy required for movement with the estimated gain in energy from the system scan.

6. Conclusions

a) Simultaneous scanning for available sunlight energy in vertical (elevation) direction and diffused light in the horizontal (azimuthal) direction provides instant information of the total available sunlight. Moreover, the miniature scanner, with improvements in the calibration method, will precisely define the optimal position for the solar panel to collect the maximum amount of solar energy without the need to consider the weather forecast. The DATS farm can use both scanning results and SATS farm can use results of elevation scanning.

b) The scanning methods described above should be calibrated with the specific type of solar cell design and semiconductor material installed at the given farm taken into consideration as well the quality of the solar panel's surface. Silicon solar cells absorb sunlight and generate different amounts of electricity in comparison, for example, with CdS or other semiconductor materials. The presence of corrugated surfaces and/or antireflection coating enhance efficiency of the solar cell. These surface features and some other specific factors of solar cell design would justify calibration with data generated by our instant sun tracking.

c) Available solar energy predicted from instant scanning should be compared to the estimated energy needed for the movement of PV panels. Then, if the expected energy accumulation is higher than the energy to be spent in the mechanical movement, than the PV array can make a move to the new position.

References

[1] Sam Mil'shtein, Dhawal Asthana, Jeffrey Snell "Design and Method of Operation of Miniature System Controlling Sun Tracking by PV", Patent appl., December 2022.

[2] Maurizio Fermeglia." How to Avoid the Perfect storm: the Role of Energy and Photovoltaics" Invited talk, MRS Webinar on Solar Energy and the Circular Economy, February, 2021.

[3] S. Mil'shtein, D. N. Asthana, "Brief Comparison of Cascaded and Tandem Solar Cells". Proceed. of IEEE PVSC 48, pp. 2260-2263, 2021.

[4] Zinaddinov, M., Mil'shtein, S. "Solar Tracking with Anti-Tracking Support for Ancillary Service". IEEE PVSC Conf. Proceed., (46), 2019.

[5] Samson Mil'shtein, Dhawal Asthana, "Harvesting Solar Energy: Efficient Methods and Materials", Springer Briefs in Materials, ISBN 978-3-030-93379-1, ISSN 2192-1091, 2022.

[6] Lee, J. F., Rahim, N. A., & Al-Turki, Y. A." Performance of Dual-Axis Solar Tracker versus Static Solar System by Segmented Clearness Index in Malaysia". Intern. Journ. of Photoenergy, 1-13, 2013. https://doi.org/10.1155/2013/820714

[7] M. T. Patel, H. Imran, M. S. Ahmed, N. Z. Butt, M. A. Alam and M. R. Khan, "When and Where to Track: A Worldwide Comparison of Single-axis Tracking vs. Fixed Tilt Bifacial Farms", 2020 47th IEEE Photovoltaic Specialists Conference (PVSC), (2020), pp. 1735-1737.

[8] Kurnianto, Rudi & Hiendro, Ayong & Yusuf, Muhammad Ismail. Optimum Orientation Angles for Photovoltaic Arrays to Maximize Energy Gain at Near Equator Location in Indonesia. International Review of Automatic Control (IREACO) 2017.

[9] Abid, Eman. The optimum tilt angle and orientation for Solar panels. Journal Port Science Research. 2. 259-263 (2019).

978-1-6654-6060-6/23 $31.00 © 2023 IEEE

[10] Thuillier, G., Zhu, P., Snow, M. et al. Characteristics of solar-irradiance spectra from measurements, modeling, and theoretical approach. Light Sci Appl 11, 79 (2022). https://doi.org/10.1038/s41377-022-00750-7

[11] Al-Aboosi, F.Y. Models and hierarchical methodologies for evaluating solar energy availability under different sky conditions toward enhancing concentrating solar collectors use: Texas as a case study. Int Journ of Energy Environ Eng 11, 177–205 (2020). https://doi.org/10.1007/s40095-019-00326-z

[12] J. F. Weaver, How we fail at solar power generation projections, Green Dealflow news, July 2020: Available at: https://greendealflow.com/https-greendealflow-com-how-we-fail-at-solar-power-generation-projections, Last Accessed: September 2022.

[13] Y. Wanga, D. Millstein, A. D. Mills, S. Jeong, A. Ancella, "The cost of day-ahead solar forecasting errors in the United States" Solar Energy Vol. 231 pp. 846-856, (2022).

[14] K Tohsing et al. A development of a low-cost pyranometer for measuring broadband solar radiation. J. Phys.: Conf. Ser. Vol. 1380 012045, (2019).

[15] Datasheet: Total Solar Pyranometer TSP-700, Yankee Environmental Systems Inc. Last Accessed: July 2022.

[16] Andersen, Elsa & Nielsen, Kristian & Dragsted, Janne & Furbo, Simon. Measurements of the Angular Distribution of Diffuse Irradiance. Energy Procedia. 70. 729-736. 10.1016/j.egypro.2015.02.182. (2015).

[17] B. Ramkiran, C.K. Sundarabalan, K. Sudhakar, Performance evaluation of solar PV module with filters in an outdoor environment, Case Studies in Thermal Engineering, Volume 21, (2020), 100700, https://doi.org/10.1016/j.csite.2020.100700.

[18] Ogherohwo E.P. , Barnabas .B.. Investigating the Wavelength of Light and Its Effects on the Performance of a Solar Photovoltaic Module International Journal of Innovative Research in

Computer Science & Technology (IJIRCST), 3, no. 4 (July, 2015): 61-65.

[19] Evaldo C. Gouvêa, Pedro M. Sobrinho and Teófilo M. Souza. Response of Polycrystalline Silicon Photovoltaic Cells under Real-Use Conditions. Energies, Vol. 10 (2017), pp. 1178.

[20] Chowdhury, M. E. H., Khandakar, A., Hossain, B., & Abouhasera, R. "A Low-Cost Closed-Loop Solar Tracking System Based on the Sun Position Algorithm". Journ. of Sensors, 2019, 1–11. https://doi.org/10.1155/2019/3681031

[21] Kuttybay, N., Saymbetov, A., Mekhilef, S., Nurgaliyev, M., Tukymbekov, D., Dosymbetova, G., Meiirkhanov, A., & Svanbayev, Y. "Optimized Single-Axis Schedule Solar Tracker in Different Weather Conditions. Energies", 13(19), 5226, 2020. https://doi.org/10.3390/en13195226

[22] Kafka, J. L., & Miller, M. A., "A Climatology of Solar Irradiance and its Controls Across the United States: Implications for Solar Panel Orientation. Renewable Energy", 135, 897–907, 2019. https://doi.org/10.1016/j.renene.2018.12.057

[23] Kelly, N. A., & Gibson, T. L. "Improved Photovoltaic Energy Output for Cloudy Conditions with a Solar Tracking System". Solar Energy, 83(11), 2092–2102, 2009. https://doi.org/10.1016/j.solener.2009.08.009

[24] Kelly, N. A., & Gibson, T. L., "Increasing the Solar Photovoltaic Energy Capture on Sunny and Cloudy Days". Solar Energy, 85(1), 111–125, 2011. https://doi.org/10.1016/j.solener.2010.10.015

[25] Saymbetov, A., Mekhilef, S., Kuttybay, N., Nurgaliyev, M., Tukymbekov, D., Meiirkhanov, A., Dosymbetova, G., & Svanbayev, Y, "Dual-axis Schedule Tracker with an Adaptive Algorithm for a Strong Scattering of Sunbeam", Solar Energy, 224, 285–297, 2011. https://doi.org/10.1016/j.solener.2021.06.024

[26] Antonanzas, J., Urraca, R., Martinez-de-Pison, F., & Antonanzas, F, "Optimal Solar Tracking Strategy to Increase Irradiance in the Plane of Array Under Cloudy Conditions: A study across Europe", Solar Energy, 163, 122–130, 2018. https://doi.org/10.1016/j.solener.2018.01.080

[27] J. H. Wohlgemuth "Standards for PV Modules and Components – Recent Developments and Challenges" 27th European Photovoltaic Solar Energy Conference and Exhibition, NREL/CP-5200-56531, 2012.

[28] Contributed by Alibaba.com "Hot Sale Solar Tracking System Dual Solar Tracker System" Available at: www.Alibaba.com, Last Accessed: July 2022.

[29] Joseph H. Cain, D. Banks, Principal, Cermak Peterka Petersen "Wind Loads on Utility Scale Solar PV Power Plants" SEAOC CONVENTION PROCEEDINGS, 2015.

[30] Anna Sapinga "Calculating Power Consumption of Electric Linear Actuators" Available at: https://www.progressiveautomations.com/blogs/how-to/calculating-power-consumption-of-linear-actuators, 2021.

[31] Contributed by MECA Enterprises "How to Find Wind Pressure on Solar Panels" Available at: https://www.mecaenterprises.com/how-to-solar-panel windpressure/#:~:text=p%20%3D%2043.191%20 psf,isn't%20usually%20the%20case. Last Accessed: 2022.

[32] Contributed by Progressive Automations "Solar Tracker Linear Actuator" Available at: https://www.progressiveautomations.com/products /solar-hall-effect-actuator?variant=19371633606723. Last Accessed: 2022.

Efficient and Stable All-Lead Perovskite Tandem Solar Cells Enabled by All-Inorganic CsPbI2Br Top Cells

Chongwen Li, Chuanxiao Xiao, Kamala Khanal Subedi, Bin Chen, Randy J. Ellingson, Song Zhaoning, Yanfa Yan, Edward H. Sargent

The Edward S. Rogers Department of Electrical and Computer Engineering, University of Toronto, Toronto, ON, Canada

Department of Physics and Astronomy, and Wright Center for Photovoltaics Innovation and Commercialization, University of Toledo, Toledo, OH, United States

Materials Science Center, National Renewable Energy Laboratory, Golden, CO, United States

Fabricating all-perovskite tandem solar cells (APTSCs) is promising to overcome the thermodynamic limit of single-junction perovskite solar cells. Here, we demonstrate efficient and stable APTSCs based on Pb-based perovskites - a 1.91 eV cesium lead iodide bromide (CsPbI2Br) and a 1.51 eV formamidinium cesium lead triiodide (FA0.95Cs0.05PbI3). We introduce poly[3-(4-methylamine carboxybutyl)thiophene] (P3CT-N) to modify NiOx hole transport layer (HTL) for the wide-bandgap CsPbI2Br cells. The HTL surface modification enhances charge extraction and suppresses the formation of the detrimental interfacial layer, leading to high performance and stability of wide-bandgap CsPbI2Br solar cells. These advances allow us to demonstrate efficient and stable all-Pb APTSCs consisting of CsPbI2Br/FA0.95Cs0.05PbI3 subcells with power conversion efficiencies (PCEs) exceeding 22% and an operational lifetime of more than 1,000 hours under one sun illumination at ~65 Â°C.

(3-Aminopropyl)trimethoxysilane Surface Passivation Improves Perovskite Solar Cell Performance by Reducing Surface Recombination Velocity

Yangwei Shi, Esteban Rojas-Gatjens, Jian Wang, Justin Pothoof, Rajiv Giridharagopal, Kevin Ho, Fangyuan Jiang, Margherita Taddei, Zhaoqing Yang, Carlos Silva-Acuña, David S. Ginger

Department of Chemistry, University of Washington, Seattle, WA, United States

School of Chemistry and Biochemistry, Georgia Institute of Technology, Atlanta, GA, United States

School of Physics, Georgia Institute of Technology, Atlanta, GA, United States

School of Materials Science and Engineering, Georgia Institute of Technology, Atlanta, GA, United States

Halide perovskites have attracted tremendous research interest due to their potential applications in photovoltaics. In this work, we demonstrate reduced surface recombination velocity (SRV) and enhanced power-conversion efficiency (PCE) in mixed-cation mixed-halide perovskite solar cells by using a surface passivator of (3-aminopropyl)trimethoxysilane (APTMS). We show that APTMS passivates defects at the perovskite surface, while also decoupling the perovskite from detrimental interactions at the C60 interface. We measure an increase of ~100 meV in quasi-Fermi level splitting (QFLS) in APTMS-passivated devices compared to the control devices. We use time-resolved photoluminescence (TRPL) and excitation-correlation photoluminescence (ECPL) spectroscopy to show that APTMS passivation effectively suppresses non-radiative recombination. Furthermore, we show that APTMS improves both the fill factor (FF) and open-circuit voltage (VOC), increasing VOC from 1.03 V for control devices to 1.09 V for APTMS-passivated devices, and leads to a PCE increase from 15.90% to 18.03% on average. We attribute the enhanced performance to reduced defect density resulting in suppressed nonradiative recombination and lower SRV at the perovskite/transport layer interface. Finally, we use scanning probe microscopy (SPM) techniques, revealing that the APTMS polymerizes heterogeneously at the perovskite surface but still ensures efficient extraction of charge carriers.

Machine Learning-based defect identification method at the c-Si/a-Si:H interface

Zitong Zhao[1], Gonglin Chen[2], Reza Vatan Meidanshahi[3], and Gergely T. Zimányi[1]

[1]Physics Department, University of California Davis, Davis, CA, 95616, USA

[2]Department of Computer Science, University of Southern California, Los Angeles, CA, 90089, USA

[3]School of Electrical, Computer and Energy Engineering, Arizona State University, Tempe, AZ, 85287, USA

Abstract — **Generation of defects leads to performance degradation of silicon solar cells. However, *ab initio* simulations of defect structures are constrained by system size, while the representative simulations of amouphous silicon (aSi) require a large amount of computation because of the wide distribution of relevant local structures. We propose a machine learning-based approach to replace the quantum mechanical *ab initio* calculations to identify defect formation in aSi-related structures. We demonstrate that with a correlated set of descriptors and a representative set of training data, the machine learning method can identify localized defects in structures containing Si-H structures with high fidelity.**

I. INTRODUCTION

Performance degradation is ubiquitous to all commercial solar cells, with an averaged rate of at least 0.2%/year, reaching 0.7%/year in the recently-developed silicon heterojunction (Si HJ) cells. Such decrease in the efficiencies significantly impacts the total energy produced over the lifetime of the solar cell, and thus the levelized cost of solar energy. Experimental studies have characterized the performance degradation by the decrease of minority carrier lifetime over time, suggesting an increase in the recombination rate as the solar cell structure evolves. To gain insight into the microscopic origin of the change in electronic properties, one must resort to atomistic modeling.

In our recent study, we identified a potential driver for the degradation in Si HJ: the breaking of the passivating Si-H bonds at the crystalline/amorphous interface, followed by the drift of the H atoms away from the interface. Our study relied on Molecular Dynamics (MD) and Density Functional Theory (DFT) to obtain realistic structural configurations and the electronic defects that lie within. The MD simulations utilized the Gaussian Approximation Potential (GAP) [1], a machine learning (ML)-trained interatomic potential, that reproduced DFT-energies with a remarkable 4 meV/atom accuracy while reducing computational complexity from $O(N^3)$ to $O(N)$. We determined the statistics of the bond-breaking and drift barriers from more than 2000 MD processes. However, identifying which Si-H bond-breaking process generated electronic defects required the use of DFT calculations. The high computational cost of DFT calculations severely limited the size and number of the simulated systems.

Much effort has been devoted to accelerating DFT with ML, such as the GAP and the deep potential molecular dynamics (DPMD) [2]. Many ML models seek to predict the correct total energy for any given structural configuration, but few focus on extracting the other key information generated in a single DFT calculation, such as electron and spin density, partial charges, and energies and spatial structure of the Kohn-Sham orbitals. In this work we present a proof-of-concept ML method that predicts whether electronic orbits get localized on a given atom, based only on the local atomic environment. We show that a trained ML model is capable of mapping the input to the output of a DFT calculation in a data-driven approach with high fidelity. Our ML model can replace the iterative process of solving the Kohn-Sham Hamiltonian each time for similar inputs. Such approach can be easily generalized to other atomic systems and to predict other results of a DFT calculation.

II. METHODS

The structures we study contain Si and H atoms in both the crystalline and amorphous phase of Si. All the DFT calculations were performed by the Quantum Espresso 6.2.1 software package using the Perdew-Burke-Ernzerhof exchange-correlation functional [3][4]. We define the presence of electronic defected states on a given atom using the Inverse Participation Ratio (IPR) of the Kohn-Sham wavefunction. The details of the IPR defect-identification technique were already described in our previous study [5]. An IPR_k value close to 1 indicates that the wavefunction is localized onto an atomic orbital of atom k. If the IPR_k is $\ll 1$, then the orbital is extended. We constructed our ML model by transforming the inputs of the DFT, the structural configurations, and the IPR_k results into the descriptors and labels of the ML model.

A. Labels

First, we identify the maximum IPR_k values of each atom from the Kohn-Sham orbitals whose energies lie within the band gap of amorphous Si. Then, we set a threshold of 0.4 to divide all atoms into two categories: atoms with $IPR_k > 0.4$ are identified as having a defect orbital; whereas atoms with IPR_k < 0.4 do not have defect orbitals. We train the model to assign either of the two labels to each "target" atom.

B. Descriptors

Since the amplitude of a localized orbital decays exponentially with distance, its presence should strongly correlate with the local atomic configuration. Therefore, the presence or absence of a localized defect orbital should strongly correlate with the local atomic environment. Motivated by this, we provide the model only information regarding the 10 nearest neighbors of the target atom instead of the entire structure. We characterize the local atomic configuration using the positions and bond angles, while conserving the rotational symmetry. We sort the neighbor 10 atoms by distance to the target atom, and capture their relative positional information with the following descriptors per target atom (111 in total per atom):

- Bond length: distance to target atom (10 descriptors)
- 3-body angle: bond angle between the current and the other 9 atoms with respect to the target atom (55 descriptors)
- 4-body angle: a more complex structural angle involving four atoms (45 descriptors)
- Atom species (11 descriptors, distinguishing Si or H)

C. Dataset

Machine learning models in general are excellent at interpolating rather than extrapolating, so it is essential for the training set to cover a large enough range of inputs. We include structures from the following categories to obtain a diverse set of local configurations:

1) Snapshots of crystalline Si (cSi) in 2x2x2 unit cell with 1,2,3 vacancies, during local annealing at 1,200K and 3,000K (2,500 structures/131,000 atoms).

2) cSi/aSi interface structures in 3x3x5 unit cell with local annealing (1,500 structures/555,000 atoms).

3) cSi/aSi:H interface structures with 12% and 15% H in 3x3x5 unit cell at equilibrium and during the Si-H bond breaking process (500 structures/160,000 atoms).

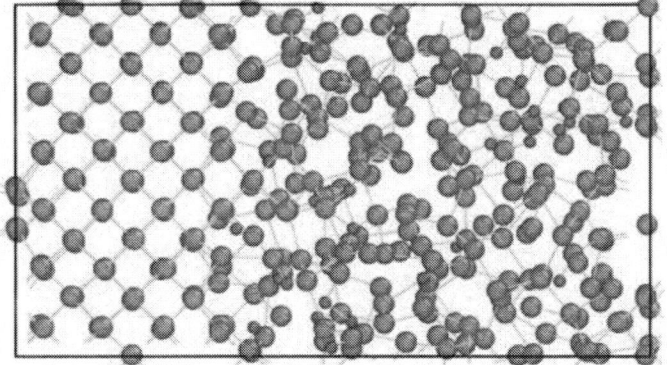

Fig. 1. Example of a cSi/aSi:H interface structure.

Fig. 1 shows an example of a typical cSi/aSi:H interface structure. **Fig. 2** shows the distribution of the IPR_k values of all

atoms of all simulated structures. The defect atoms comprise only around 1% of the total atoms. Therefore, to prevent the model from becoming biased towards the majority category of npon-localized orbitals, we select randomly a small portion of non-defected atoms such that the number of atoms in each category remains the same.

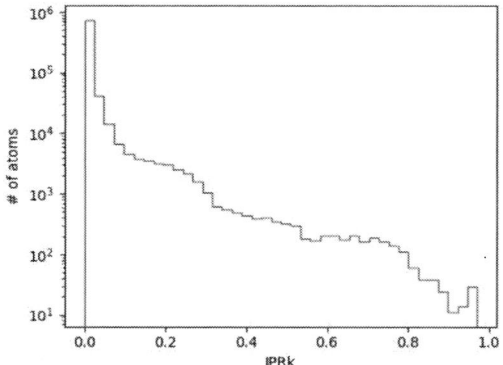

Fig. 2. Histogram of IPR_k values on all atoms in the dataset.

D. Neural network model

The model processed the inputs by passing them through multiple layers of a Tensorflow neural network. The architecture of the model consisted of four hidden layers with tanh activation, and an output layer with SoftMax activation to predict whether an electronic orbital on the target atoms is defected or not. The parameters are optimized using an Adam optimizer with respect to a categorical cross-entropy loss function, using a learning rate of 0.001.

The dataset was divided into training and validation sets. We performed validation by comparing the predicted labels to target labels. We chose as an accuracy metrics the percentage of correct predictions, and used it as a measure of the ML model's performance. The training and prediction of the model only required a few minutes on a regular desktop computer.

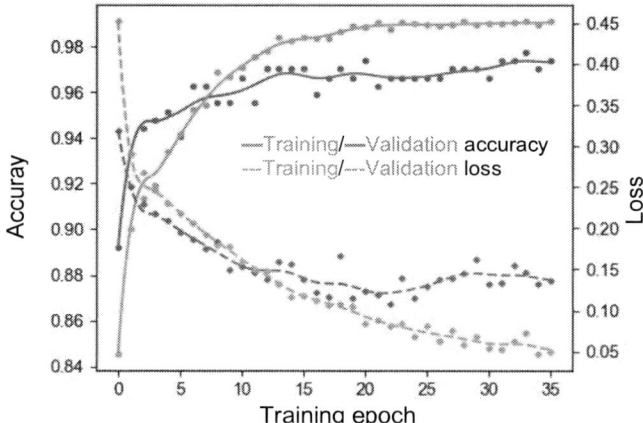

Fig. 3. Loss and accuracy of the training and validation set during the training process of the model, eventually reaching 97% accuracy.

III. RESULTS AND DISCUSSION

We performed training on a series of combinations of IPR_k threshold, descriptor set and datasets. **Table I** shows that we experimented with different subsets of descriptors on different subsets of training structures. As visisble, our ML model was able to achieve a remarkable validation accuracy of 97% on the first category of cSi-based training structures, where we used the bond lengths and 3-body angles as descriptors, and an IPR_k threshold of 0.4. **Fig. 3** illustrates the training process of the optimal model. **Table I** details the predictive accuracies of other descriptor and training set pairings.

TABLE I
SUMMARY OF MODEL CONFIGURATIONS

Descriptors \ Dataset	Training/validation accuracy		
	cSi	cSi/aSi	cSi/aSi:H
Bond lengths only	68%/68%	61%/61%	60%/60%
3-body angles only	97%/94%	85%/78%	77%/77%
Bond lengths + subset of 3-body angles	98%/95%	75%/75%	76%/76%
Bond lengths + 3-body angles	98%/97%	90%/81%	78%/78%
Bond lengths + 3-body angles + 4-body angles	99%/97%	93%/83%	80%/78%

The performance of the model improved when we gave a larger and more complete set of descriptors of the local environment, indicating that it is indeed uncovering correlations between the classical description of the atomic structures and the quantum mechanical information regarding the electrons. Our ML model was able to correctly identify a distinct set of local configurations of neighbor atoms that caused the formation of defect orbitals. The 97% validation accuracy persisted through cross validation of the model, giving us confidence of the model's high fidelity.

We further speculate that the ML model can become able to learn how to characterize the degree of localization of the electornic orbitals. As a preliminary study, we changed the IPR_k threshold that defined the defected atoms in 0.1 steps from 0.1 to 0.7. We found that the accuracy of the predictive power of the model dropped only by 5% at most. This suggested that our model has the potential to use the IPR_k thresholds to describe the degree of localization of the defect orbitals.

We note that the model in its current formulation delivered an 83% accuracy for amorphous and hydrogenated amorphous silicon, where the local configurations of the atoms became more complex. Motivated by this modest step-down, our project is now focused on systematically enlarging the training set, as well as identifying more suitable descriptors to the model.

IV. SUMMARY

In this paper we reported the development of a machine learning model that takes classical mechanical descriptions of atomic environments as input, and produces quantum mechanical information of the electronic structures as output. The model was trained on data already generated with reasonable computational expense during previous projects. We demonstrated that once trained with such dataset, the model was indeed capable of inferring the quantum mechanical information whether an elecotrnic state was localized or not based only on classical structural information, without needing expensive DFT calculations. The model required only negligible computational effort. We believe that our approach is easily generalizable to other types of materials and other quantities derived from *ab initio* calculations.

This work was supported by the U.S. Department of Energy's Office of Energy Efficiency and Renewable Energy (EERE) under the Solar Energy Technologies Office Award Numbers DE-EE0008979 and DE-EE0009835. The views expressed herein do not necessarily represent the views of the U.S. Department of Energy or the United States Government.

REFERENCES

[1] D. Unruh, R. V. Meidanshahi, S. M. Goodnick, G. Csányi, and G. T. Zimányi, "Gaussian approximation potential for amorphous Si : H," *Phys. Rev. Mater.*, vol. 6, iss. 6, p. 065603, 2022.

[2] L. Zhang, J. Han, H. Wang, R. Car and W. E, "Deep potential molecular dynamics: a scalable model with the accuracy of quantum mechanics," *Phys. Rev. Lett.*, vol. 120, iss. 14, p. 143001, 2018.

[3] P. Giannozzi, O. Andreussi, T. Brumme, O. Bunau, M. Buongiorno Nardelli, M. Calandra, R. Car, C. Cavazzoni, D. Ceresoli, M. Cococcioni, et al. "Advanced capabilities for materials modelling with QUANTUM ESPRESSO," *J. Phys.: Condens. Matter*, vol. 29, p. 465901, 2017

[4] J. P. Perdew, K. Burke, M. Ernzerhof, "Generalized gradient approximation made simple," *Phys. Rev. Lett.*, vol. 77, pp. 3865-3868, 1996

[5] D. Unruh, R. V. Meidanshahi, C. Hansen, S. Manzoor, M. I. Bertoni, S. M. Goodnick, and G. T. Zimányi, "From femtoseconds to gigaseconds: the SolDeg platform for the performance degradation analysis of silicon heterojunction solar cells," *ACS Applied Mater. & Interfaces*, vol. 13, pp. 32424-32434, 2021

Recombination analysis of Maxeon IBC production cells by time-resolved photoluminescence

David Jacob, Guillaume von Gastrow, Nils-Peter Harder, Luis Buño III, Gerly Reich, Maristel Baldrias, Roderick J. Marstell, Arnold Castillo, David D. Smith, Michael J. Cudzinovic

Maxeon Solar Technologies, San Jose, CA, 94601, USA

Abstract — **Photoluminescence is a powerful technique to analyze solar cell recombination across the manufacturing steps. However, this analysis requires accurate modelling of the phenomena caused by injection-dependent lifetimes. We present a transient photoluminescence technique that we apply to our current new-generation production cells. We extract the saturation current J_o and the bulk lifetime, in a refined version of the method developed by Kane and Swanson. We include band-gap narrowing effects occurring at the large operating voltages typical of our back-contact cells. Our approach results in excellent fittings of the carrier density decay (R^2=99.9%). We find an area averaged J_o of 6.2 fA/cm² and an inverse bulk Shockley-Read-Hall lifetime of 14 s⁻¹ in a production cell measured after metal etching.**

I. INTRODUCTION

Photoluminescence (PL) techniques are used extensively in the solar industry to measure spatially resolved physical properties of the test wafers and finished cells [1]-[3]. They can be contactless and suited for non-metallized cells. Photoluminescence measurements can be performed either in quasi-steady-state (QSSPL) or in transient mode (time-resolved PL or TRPL). The TRPL approach has the advantage of being independent of the photogeneration and of the intensity of the excitation.

The simplest way to perform a TRPL lifetime analysis is to model the PL transient as an exponential decay. However, this approach is an approximation that assumes no dependence of the lifetime on the injection level [4], which is often not verified. For instance, high-efficiency cells have nowadays such high voltages that part of their current-voltage characteristic is in high injection, particularly for moderately- or lowly-doped material. In such case of high-injection, the 'Jo-type' emitter recombination leads to an injection-dependent effective lifetime. Additionally, bandgap-narrowing effects occurring at these voltages influence the intrinsic lifetime, which cannot be ignored for an accurate determination of the low J_o values of high-efficiency cells.

In this paper, we propose an improved TRPL analysis compatible with high-efficiency solar cells. We use a general Kane&Swanson approach including band-gap narrowing to extract images of crucial recombination parameters, namely the saturation current J_o and the inverse bulk lifetime.

II. EXPERIMENTAL DETAILS

The experimental apparatus we use, PixEL, developed and manufactured by Tau Science Corporation, consists of 4 visible lasers illuminating a cell and an InGaAs camera in a dark chamber. The cell sits on a temperature-controlled chuck that is stabilized at 25±0.5 °C. The chuck has a reflectance below 10% in the near infrared, and is able to measure cell current and voltage in parallel of the imaging. We tune the laser intensity to a level of 0.65 suns and set a time-modulation to produce square pulses with a duration of 5 ms, to reach a high carrier-density. We chose an exposure of 100 μs for all our devices, much shorter than the typical carrier lifetime. By changing the delay between the laser pulse cutoff and the camera acquisition, between 0 ms to 15 ms, we obtain a series of 17 images from the initial high-injection state down to low-injection. In order to increase the signal-to-noise ratio in the images, we repeat every measurement 100 times. We produce image sets on Interdigitated-Back-Contact (IBC) production cells in about 1min/cell, which is suited for R&D and offline measurements in a factory, but does not lend itself for inline production quality monitoring in high unit-per-hour environments.

III. CALIBRATION PROCEDURE

We first perform a PL-Voc calibration on metallized devices [5]. For finished solar cells with good passivation uniformity, we observe a quasi-uniform photoluminescence image, with less than 5% relative variation within the cell area. This indicates that the combination of open-circuit conditions and the high lateral conductivity provided by the metal produces a very homogeneous carrier concentration, even in the presence of minor passivation or lifetime inhomogeneities. We record the PL signal in steady-state and calculate the average intensity I_{PL}^{avg} over the surface of the cell. We simultaneously measure the cell open-circuit voltage V_{oc} and convert it into a minority carrier density Δp, taking into account band-gap narrowing [6]. This procedure is repeated several times at different light intensities. Using the expression $I_{PL} = A\Delta p(\Delta p+N_{dop})$ [7], we

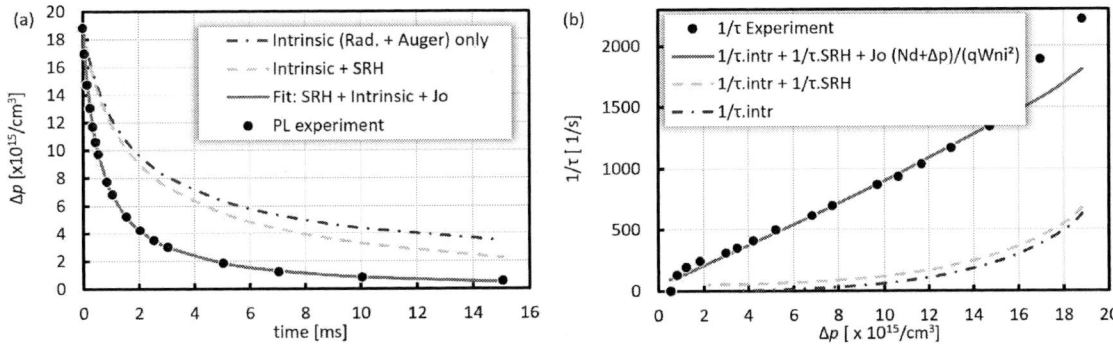

Fig. 1. (a) experimental time-resolved decay of carriers (black dots), fitted model (red line) with J_o=9.3±0.2 fA/cm2 and $1/\tau_{SRH,bulk}$=48±10 s^{-1}, model with intrinsic recombination alone (blue dash-dotted line) and model with intrinsic and SRH,bulk recombination (yellow dashed line). The fit parameters are displayed in the upper left. (b) experimental inverse lifetime, fitted $1/\tau$ using the parameters from (a) (red line), model with intrinsic recombination alone (blue dash-dotted line) and model with intrinsic and SRH,bulk recombination (yellow dashed line).

extract the calibration factor A. The metal on the cell that is used for calibration acts as a rear-side reflector, so we need to adapt the calibration constant by a simple optical analysis for samples without metal. We measure the signals $I_{PL}^{avg,low-refl}$ and $I_{PL}^{avg,high-refl}$ of a test wafer using two back reflectors, respectively lowly- (bare chuck) and highly-reflective, with similar optical properties as the metallization layer. We then correct the factor A as follows: $A_{corr}=A \times I_{PL}^{avg,low-refl}/I_{PL}^{avg,high-refl}$. This accounts for the difference in optical properties [8] between the calibration with a (metallized) cell and non-metallized test samples. For lifetime samples with notably different optical properties, such as non-textured wafers, a calibration with an optically similar cell has to be provided.

IV. FITTING THE CARRIER DECAY EQUATION

In transient mode, the evolution of excess carriers follows the continuity equation [9]:

$$\frac{\partial \Delta p(r,t)}{\partial t} = -\frac{\Delta p(r,t)}{\tau(r,\Delta p)} + D\nabla^2[\Delta p(r,t)] + G(t) \qquad (1)$$

where D is the diffusion constant and G(t) the carrier generation rate. When the generation is off, G(t)=0. We have not yet incorporated in our evaluation method the technique of [10] for lateral diffusion. While this is of notable numerical importance in case of rapid lateral variation of material or passivation quality, our cells of interest for detailed analysis are typically rather homogeneous, which justifies neglecting lateral diffusion. The first term in (1) is the loss of carriers with time due to the effective lifetime $\tau(r)$. The inverse of the injection-dependent (effective) lifetime can be written as the sum of the Shockley-Read-Hall recombination rate in the bulk (SRH,bulk), the surface recombination rate and the intrinsic Auger and radiative recombination rates, as described in [11]. For the two latter terms, we use the advanced parametrization for the intrinsic lifetime τ_{intr} developed in [12]. The expression for inverse effective lifetime is:

$$\frac{1}{\tau(r,\Delta p)} = \frac{1}{\tau_{SRH,bulk}} + \frac{qWJ_o(\Delta p + N_{dop})}{n_i^2(\Delta p)} + \frac{1}{\tau_{intr}(\Delta p)} \qquad (2)$$

where q is the elementary charge, W the wafer thickness, n_i the intrinsic carrier concentration and N_{dop} the doping concentration. The two parameters $\tau_{SRH,bulk}$ and J_o depend on the defect density in the bulk and on the passivation at the position r, respectively. We analyze the lifetime in the high-injection regime, which implies that the SRH,bulk component of lifetime is independent of injection level [11]. Because of the non-trivial dependencies of n_i and τ_{intr} with carrier density, there is no analytical solution for (1), we thus solve this equation numerically. For each pixel j in the image, we use an error minimization procedure with the fitting parameters $1/\tau_{SRH,bulk,j}$ and $J_{o,j}$. We obtain the resulting images in about two minutes with a definition of 40 pixels × 40 pixels.

V. RESULTS AND DISCUSSION

We first investigate a n-type monocrystalline silicon wafer, with texture on the front, passivation and laser-contacts but no metallization, with an average lifetime of 5 ms at a minority carrier density of 2×10^{15} cm^{-3}. We study the decay of the center area of this cell, averaged into an effective single region. In Fig. 1. we show the experimental decay of carriers (a) and the corresponding total inverse lifetime and its components (b). The fit parameters on this decay are J_o=9.3±0.2 fA/cm2 and $1/\tau_{SRH,bulk}$=48±10 s^{-1}, with R^2=99.9%. In this example, the surface recombination dominates the early decay, shown by a steep drop of carriers. Once the carriers are depleted to about one-tenth of the density at saturation, the bulk and intrinsic losses become more significant.

Fig. 2. shows a full-cell mapping example. In our factory, we sample devices at various stage of their fabrication process, and monitor their $1/\tau_{SRH,bulk}$ and J_o recombination maps. In production, we track the change of either bulk or surface

978-1-6654-6060-6/23 $31.00 © 2023 IEEE

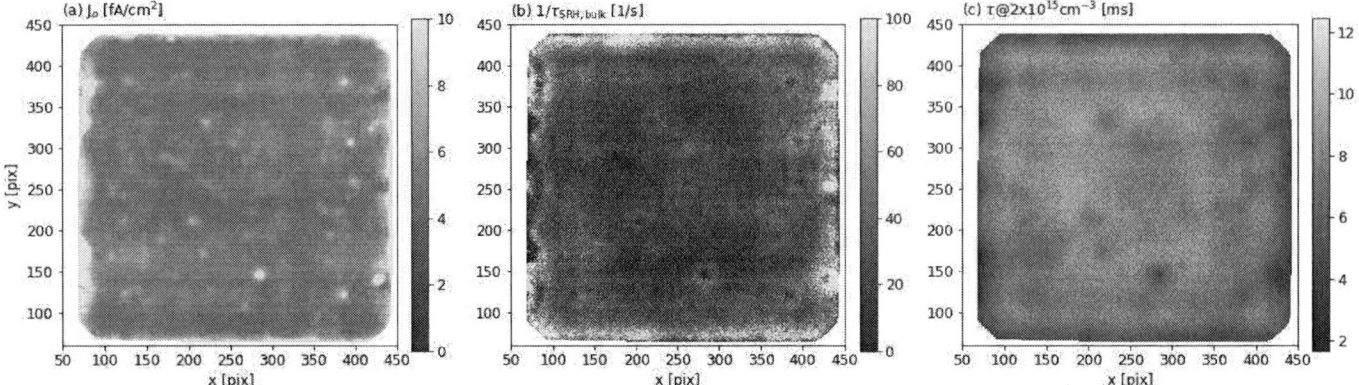

Fig. 2. Recombination analysis maps of a production cell with etched-off metal. (a) J_o map, average over the cell area is 6.2 fA/cm^2. (b) $1/\tau_{SRH,bulk}$ map, average is 14 s^{-1}. (c) τ map at of 2×10^{15} cm^{-3}, average is 7.8ms. The 3 maps have a resolution of 300µm/pixel.

recombination after a single process step, which allows the fine optimization of our process parameters. The maps we show in Fig. 2. come from a measurement of a production cell, after metal etching, that reached an efficiency of 25.5%.

Except at the edges and in a few spots in the center, the lifetime is quite uniform, and we see that is dominated by the surface recombination losses, with an average of 6.2 fA/cm^2. Note that at the edges, a more rigorous approach including diffusion and an edge surface recombination velocity (SRV) would lead to more accurate evaluation of the local $1/\tau_{SRH,bulk}$ and J_o. We can nevertheless use this measurement to quantify the decrease of efficiency due to lifetime non-uniformity and calculate the efficiency gain on the finished devices that could be achieved by improving the lower lifetime regions. In this example, the average lifetime would increase by about 20% if we were able to get the entire cell to reach the same level as the highest 10% lifetimes. This corresponds to an increase of about 0.3% absolute efficiency, which would bring the efficiency up to 25.8%. With these maps, we can analyze and quantify precisely the impact of non-uniform lifetime, locate the defects and trace back to the processes they originate from.

V. CONCLUSION

In this paper, we have reported on a method to extract the different components of recombination in a high-efficiency solar cell: surface, bulk and intrinsic recombination. We have implemented a numerical method using an injection-dependent lifetime model fitted on experimental TRPL data. This method can be used at any step of the manufacturing process. We are pursuing refinement of this approach to include lateral diffusion for the analysis of large defects, edge-SRV, efficiency and series-resistance mapping.

REFERENCES

[1] T. Trupke, B. Mitchell, J. Weber, W. McMillan, R. Bardos, and R. Kroeze, "Photoluminescence imaging for photovoltaic applications," Energy Procedia, vol. 15, pp. 135–146, 2012, international Conference on Materials for Advanced Technologies 2011, Symposium O.

[2] P. Wuerfel, T. Trupke, T. Puzzer, E. Schaeffer, W. Warta, and S. W. Glunz, "Diffusion lengths of silicon solar cells from luminescence images," Journal of Applied Physics, vol. 101, no. 12, p. 123110, 2007.

[3] Z. Hameiri, P. Chaturvedi, M. K. Juhl, and T. Trupke, "Spatially resolved emitter saturation current by photoluminescence imaging," in 2013 IEEE 39th Photovoltaic Specialists Conference (PVSC), 2013, pp. 0664–0668.

[4] S. Parola, M. Daanoune, A. Kaminski-Cachopo, M. Lemiti, and D. Blanc-P´elissier, "Time-resolved photoluminescence for self-calibrated injection-dependent minority carrier lifetime measurements in silicon," Journal of Physics D: Applied Physics, vol. 48, no. 3, p. 035102, dec 2014.

[5] T. Trupke, R. A. Bardos, M. D. Abbott, and J. E. Cotter, "Suns-photoluminescence: Contactless determination of current-voltage characteristics of silicon wafers," Applied Physics Letters, vol. 87, no. 9, p. 093503, 2005.

[6] A. Schenk, "Finite-temperature full random-phase approximation model of band gap narrowing for silicon device simulation," Journal of Applied Physics, vol. 84, no. 7, pp. 3684–3695, 1998.

[7] T. Trupke and R. Bardos, "Photoluminescence: a surprisingly sensitive lifetime technique," in Conference Record of the Thirty-first IEEE Photovoltaic Specialists Conference, 2005., 2005, pp. 903–906.

[8] H. T. Nguyen, F. E. Rougieux, S. C. Baker-Finch, and D. Macdonald, "Impact of carrier profile and rear-side reflection on photoluminescence spectra in planar crystalline silicon wafers at different temperatures", IEEE Journal of Photovoltaics, vol. 5, no. 1, pp. 77–81, 2015.

[9] S. Sze and K. Ng, Physics of Semiconductor Devices. Wiley, 2006.

[10] S. P. Phang, H. C. Sio, and D. Macdonald, "Carrier de-smearing of photoluminescence images on silicon wafers using the continuity equation" Applied Physics Letters, vol. 103, no. 19, p. 192112, 2013.

[11] R.M. Kane, D.E. Swanson, "Measurement of the emitter saturation current by a contactless photoconductivity decay method" 1985.

[12] A. Richter, S. W. Glunz, F. Werner, J. Schmidt, and A. Cuevas, "Improved quantitative description of auger recombination in crystalline silicon," Phys. Rev. B, vol. 86, p. 165202, Oct 2012

Can hierarchical physics-based machine learning de-anonymize solar farm locations?

Jabir Bin Jahangir, Amandeep Singh Bhatia, and Muhammad Alam

Elmore Family School of Electrical and Computer Engineering, Purdue University, West Lafayette, IN, 47907, USA

Abstract—In the last decade, machine learning techniques powered by big data have revolutionized data-driven methods to explore and exploit underlying patterns in data. For private entities sharing data, some guarantee of anonymity is essential. In this paper, we discuss a scheme to de-anonymize solar farms only based on their monthly yield profiles. By hierarchically training physics-guided convolution neural networks in a federated setting, the models attain above 90% accuracy when mapping yield profiles to geographical regions. However, the maximum accuracy attainable is limited. We argue that although the monthly yield profile can indicate a farm's latitude, it does not provide sufficient information to uniquely identify its location.

Index Terms—solar farm, machine learning, de-anonymization, convolutional neural network, privacy

I. INTRODUCTION

In the last decade, machine learning techniques powered by big data have revolutionized data-driven methods to explore and exploit underlying patterns in data. The PV industry also stands to benefit tremendously from these advances; from material design to energy yield forecasting, machine learning techniques are being applied extensively. However, the success of machine learning is predicated on the availability of copious amounts of data. To produce and curate this data, cross-regional collaborations like DuraMAT, and PEARL PV will play an important role. However, ensuring privacy is critical when sharing data at any scale—and de-anonymization is a potential threat. The goal of a de-anonymization attack is to re-identify the original source or owner of the data. The de-anonymization threats regarding shared consumer data have been extensively discussed in the literature [1]–[3]. In this work, we address the issue of de-anonymization of PV power stations sharing yield data.

Previously, we have shown that climatic similarities between different regions on earth can be exploited to data-efficiently solve the *forward* problem of determining the worldwide energy yield potential of a PV technology with monthly data from only five carefully selected sites [4]. Now, we consider the *inverse* problem: *given only the yield profile of the solar farm can the location of the farm be determined?* The answer to this question has implications from a privacy perspective: the entity sharing the data may wish to remain anonymous (and their hide their plant identity) for various legal and/or competitive reasons.

In this paper, we demonstrate that the power generation profile of a solar farm naturally bears a location's signature. Obviously, the information contained depends on the temporal resolution of the data. Time integration leads to levels of obfuscation. Yearly data (high degree of uncertainty, less information) is less re-identifiable than hourly data (lower degree of uncertainty, more information). From a physical perspective, yearly yield data is challenging to de-anonymize because many regions of the earth receive similar integrated irradiance—and thus obfuscates its original location. However, monthly data over a year bears the signature of a location's sun-earth relation—and therefore provides information about the location's latitude. But, the monthly yield profile itself cannot provide information about the location's longitude. Therefore naturally, there is some uncertainty associated with geo-locating PV systems only based on monthly yield. A finer resolution yield data (e.g., hourly) will provide more information about the system and may be correlated with additional covariates to reveal its location more accurately. In this work, we restrict ourselves to the scenario where we only possess a farm's monthly yield profile.

To identify the location's latitude signature and attempt to solve the inverse problem, we employ machine learning techniques. In recent years, federated learning (FL) techniques have been introduced as a solution that enhances data privacy without any raw data leaving the devices. Since the introduction of Google's federated learning-based android app called Gboard, it has received a significant response to solve the data island problem [5]. It has been applied in several domains for collaborative learning while protecting data privacy such as smart cities, finance, and healthcare. FL supports the researchers on a common objective: to propose a model that learns from the shared data in the cloud and generalizes to private data at local nodes without transferring private data explicitly. Each client sends its parameters to the global model at each round of communication. Finally, the global model performs aggregation and sends the updated model weights to clients for further training. The process is repeated until the global model achieves the intended accuracy.

The paper is structured as follows: Section II discusses the physical intuition behind how a farm's yield profile may reveal its location. Section III discusses the convolution neural network-based hierarchical learning scheme. The experimental results and discussions are presented in Section IV. Finally, conclusions are drawn in Section V.

II. PV-SPECIFIC CLIMATE ZONES AND YIELD PROFILES

A. *PV climate map to partition the earth*

Several climate classification systems, such as the Koppen-Geiger system, the Thornthwaite system, etc., exist to classify

978-1-6654-6060-6/23 $31.00 © 2023 IEEE

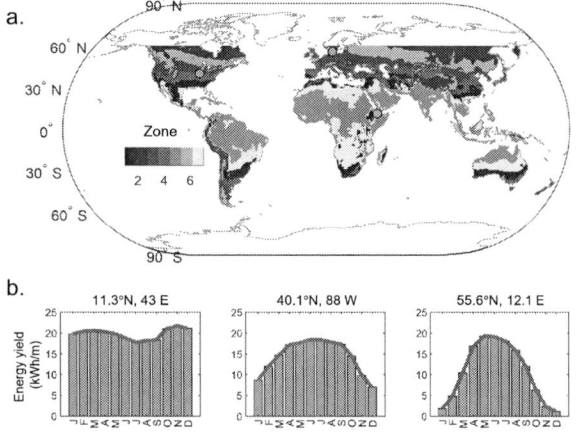

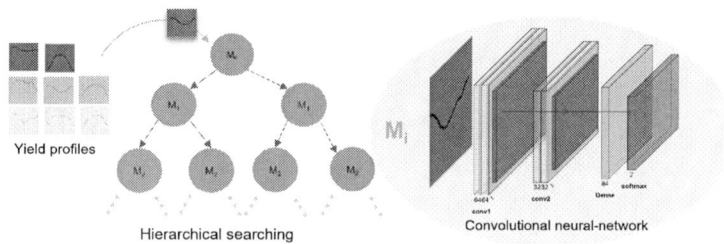

Fig. 2. Hierarchically search algorithm to determine the original geographic region.

Fig. 1. PV climate map and yield profiles. (a) The PV-climate map was constructed through cluster analysis of the global gridded climatology dataset. (b) Monthly energy yield profiles of simulated solar farms.

the earth's climate based on some shared climatic features. However, not all climatic variables are equally important in determining the energy yield potential of a PV system at a given location. The yield potential at any location is largely determined by the Global Horizontal Irradiance (GHI), temperature (T_{amb}), and clearness index (k_t). Therefore to understand trends in energy yield potential, it is more instructive to construct a climate map based on shared similarities in these variables specifically relevant to PV. Fig. 1 shows such a map. To construct this map, we partitioned the world into seven ($k = 7$) zones using the k-means algorithm. The global gridded climatological data was obtained from NASA's POWER database [6]. We observe that the zones are largely latitudinally distributed.

B. PV Yield Profiles

The yield profile of a PV system is a record of the energy generated throughout the year at specific time intervals, e.g., hourly, monthly, etc. An optimally designed PV system's energy yield potential largely depends on the GHI experienced at the location; it will generate the most energy during the summer months GHI is the highest. However, the variance in GHI experienced within a year strongly depends on the location's latitude. A location near the equator receives similar monthly GHI throughout the year, this is reflected in the relatively flat yield profile shown in Fig. 1(b). But, at high latitudes, the monthly average GHI has a larger variance resulting in bell-shaped yield profiles at these locations as shown in Fig. 1(b). Although the absolute yield of PV systems will vary with system capacity across the world, the *shape* of the yield profile will bear the signature of the latitude. Based on this observation, we devise a scheme to determine the region on earth the data originated from. It is again worth noting that a profile with finer temporal resolution will contain more information about the location.

III. Geo-locating solar farms

Our goal is to map a given yield profile to climate zones shown in Fig. 1(a). Casting the problem as an image recognition problem, we are able to leverage existing sophisticated machine-learning techniques, namely the convolutional neural networks. To train the neural networks, the training data, i.e., labeled yield profiles were generated by detailed physical simulations of solar farms [7]. The systems were assumed to be optimally designed and consist of monofacial modules with identical efficiencies across the world. But, training AI systems need access to a large amount of data, and security presents AI with a new set of challenges.

A. Training Convolutional neural network in FL settings

In our proposed system, we trained several convolutional neural networks collaboratively without sharing the data using Tensorflow federated framework (TFF), TensorFlow (TF) 2.0, and Keras libraries for the classification task. The grayscale images (shown in Fig. 2) of energy yield with a size 64×64 are used to train the centralized model. To demonstrate the effectiveness of federated CNN, the dataset is distributed non-independently and non-identically (non-iid) among the clients. Each dataset is split into a 70:30 ratio of training and testing. We considered the number of clients to be four and the number of communication rounds to be 500.

The architecture of the federated CNN used for the classification task has two 3×3 convolution layers (conv1 and conv2), consisting of 64, 32 filters and 2 × 2 max pooling layers. Then, a fully connected dense layer with 64 units, followed by the output layer with a softmax activation function, is shown in Fig. 2.

B. Algorithm to hierarchically geo-locate solar farms

To determine the potential originating region of the yield profile, our algorithm (shown in Fig. 2) carries out a search in multiple stages in a manner akin to a binary search. Our current implementation consists of three stages of inference, M_i where $i = 0, 1, 2$. In the first stage M_0, the model determines the hemisphere the profile belongs to. In each subsequent stage, the hemispheres are divided into 2^i regions and classified with numerical labels $L_i \in \{0, ..., i - 1\}$. Once the hemisphere is determined, the next stages of the algorithm determine the subregion of the hemisphere the profile may belong to. After 3 stages of evaluation, the sequence of labels

978-1-6654-6060-6/23 $31.00 © 2023 IEEE

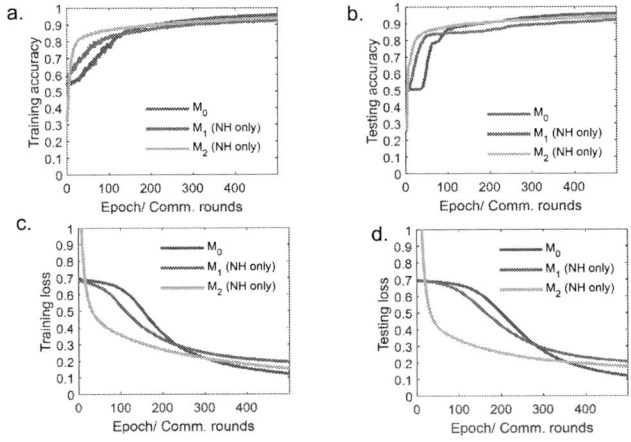

Fig. 3. (a) and (b) shows the training and testing accuracy of the model, respectively. During the training and testing, the model reaches accuracies greater than 90% after 500 epochs. (c) and (d) shows the training and testing losses of the model, respectively.

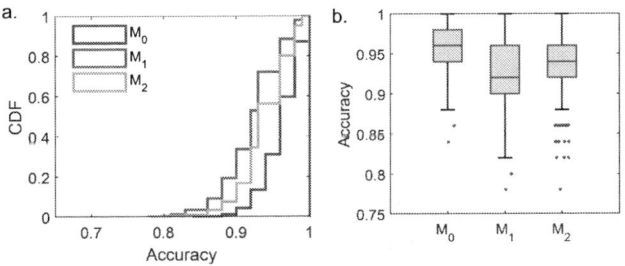

Fig. 4. Distribution of accuracy after 1000 inference trials. (a) Cumulative density functions of accuracy. (b) Boxplot of the model accuracies. In each trial, 50 images were classified. The distributions show that in random trials the models demonstrate high mean accuracy of 95%.

$L_0 L_1 ... L_i$ identifies the region. Although additional stages may be introduced to the algorithm, the accuracy of the model diminishes with increasing stages as the profiles from the subregions become indistinguishable. This is encouraging from a privacy perspective.

IV. RESULTS

We perform a set of experiments using convolutional neural networks and show how well our model performs to determine the location of solar farms. The training and testing accuracies as a function of communication rounds are shown in Fig. 3 (a-b). After 500 epochs, each model reaches both training and testing accuracy above 90%. For optimization, the stochastic gradient descent (SGD) optimizer is used with a learning rate of 0.01. To gauge the models' accuracy in practice, we carried out repetitive inference trials on the models. After training of CNN model, 50 yield profiles were randomly selected from the test dataset for the testing model during each trial. The results in Fig. 4 show that each model maintains high accuracy in trials consisting of random profiles from the test data set. The mean accuracy after 1000 trials is approximately 95%. The boxplots show that the distributions are quite narrow about the

mean. The CNN models' loss function is shown in Fig. 3 (b-c). Each model has converged well and has a reasonable loss on all three stages of inference. The application of the CNN algorithm to identify the locations of solar farms appears to be very promising.

V. CONCLUSIONS

In this work, we have shown how one can leverage the physical understanding of the PV yield profiles coupled with machine-learning techniques to potentially de-anonymize solar farm data. While the method does not give the exact coordinate of the system, it allows one to narrow down to the geographic region the solar farm may belong to. The conclusions are the following:

1) PV systems cannot be exactly located only based on their monthly yield profile. This is because the profiles cannot be distinguished from profiles in a large geographical region.
2) It follows that the companies owning solar farms may share yearly, and monthly yield data anonymously. Federated learning offers another layer of security and reduces the cost of transmitting data over the network.

REFERENCES

[1] L. Sweeney, "k-anonymity," *International Journal of Uncertainty, Fuzziness and Knowledge-Based Systems*, vol. 10, pp. 557–570, 10 2002.

[2] M. Jawurek, M. Johns, and K. Rieck, "Smart metering de-pseudonymization," *ACM International Conference Proceeding Series*, pp. 227–236, 2011.

[3] A. Narayanan and V. Shmatikov, "Robust de-anonymization of large sparse datasets," *Proceedings - IEEE Symposium on Security and Privacy*, pp. 111–125, 2008.

[4] J. B. Jahangir, M. T. Patel, and M. A. Alam, "Physics-Guided Machine Learning Identifies 5 Optimum Test Locations to Predict Global PV Energy Yield for Arbitrary Farm Topologies," pp. 0843–0846, 11 2022.

[5] H. Brendan McMahan, E. Moore, D. Ramage, S. Hampson, and B. Agüera y Arcas, "Communication-Efficient Learning of Deep Networks from Decentralized Data," *Proceedings of the 20th International Conference on Artificial Intelligence and Statistics, AISTATS 2017*, 2 2016.

[6] "POWER. Surface meteorology and solar energy: a renewable energy resource web site (release 6.0); 2017.¡https://eosweb.larc.nasa.gov/cgi-bin/sse/sse.cgi? ¿."

[7] M. R. Khan, M. T. Patel, R. Asadpour, H. Imran, N. Z. Butt, and M. A. Alam, "A review of next generation bifacial solar farms: Predictive modeling of energy yield, economics, and reliability," *Journal of Physics D: Applied Physics*, vol. 54, no. 32, 2021.

Investigation and Quantitative Understanding of Front Field Passivation in Rear Junction Selective Double-side TOPCon Solar Cells

Wook-Jin Choi[1], Young-Woo Ok[1], Pradeep Padhamnath[2], Gabby De Luna[2], Kwan Hong Min[1], Ruohan Zhong[1], Sagnik Dasgupta[1], Vijaykumar D Upadhyaya[1], Ajay D Upadhyaya[1] and Ajeet Rohatgi[1]

[1]Georgia Institute of Technology, Atlanta, Georgia, 30332, USA
[2]Solar Energy Research Institute of Singapore, 7 Engineering Drive 1, 117574, Singapore

Abstract — This paper investigated the opportunity and challenge in fabrication of rear junction (RJ) selective double side tunnel oxide passivated contact (DS-TOPCon) cell, featuring full-area p-TOPCon on the back and a selective-area n-TOPCon on the front of n-type c-Si wafer. 2D device simulation showed that this cell structure, which exploits the potential of TOPCon on both sides, possesses the potential to reach cell efficiency higher than 25.0 %, provided that the passivation on front field region between the metal grid can attain a J_0 value of < 5 fA/cm². It is shown in this work that Al_2O_3/SiN_X:H passivation stack can achieve J_0 of ~4 fA/cm² on the front field region. However, the J-V parameters of fabricated selective DS-TOPCon cell showed a significant cell efficiency degradation, presumably due to the formation of inversion layer, which is induced by a negative fixed charge in Al_2O_3 layer. This causes minority carrier leakage via tunneling or shunting at the p^+ inversion layer on the field region and n^+ region under the metal-Si contact. A second passivation scheme, SiN_X:H layer, gave much higher J_0 of ~20 fA/cm² due to poor chemical passivation at the SiN_X:H/Si interface. Our recent experimental result showed that growth of 15 nm of thermal SiO_2 capped with SiN_X:H can achieve the J_0 of ~10 fA/cm² with potential of further improvement.

I. INTRODUCTION

The photovoltaic industry has rapidly adopted the advanced solar technology since the high efficiency solar cells are the key for cost reduction and widespread use of PV. Among the advanced technologies, the concept of tunnel oxide passivated contact (TOPCon) composed of ultra-thin interfacial oxide (iOx) and heavily doped poly-Si has been widely received industrial acceptance, since it can effectively mitigate the recombination losses at the diffused and metallized regions by physically displacing these regions from the Si absorber [1].

In this work, we report on the quantitative understanding and requirements of making high efficiency screen-printed RJ selective DS-TOPCon solar cells (Fig. 1). This cell structure is composed of full-area iOx/p^+ poly-Si (p-TOPCon) on the back, but selective-area iOx/n^+ poly-Si (n-TOPCon) on the front side to minimize the parasitic light absorption losses in poly-Si layer [2-4]. This paper focusses on the requirement and challenges in passivating the front field region of selective DS-TOPCon cell to achieve high cell efficiency. Two different passivation schemes, Al_2O_3/SiN_X:H and SiN_X:H, were investigated to minimize the recombination current density (J_0) in the field region. Various symmetric test structures were fabricated to

assess the J_0 values of the passivated field regions. In addition, we fabricated the selective DS-TOPCon cell using two different front surface passivation schemes.

II. EXPERIMENTAL DETAILS

A. Preparation of symmetric test structures

Various symmetric test structures (a)-(d), shown in Fig. 2, were fabricated on n-type substate to evaluate their passivation quality.

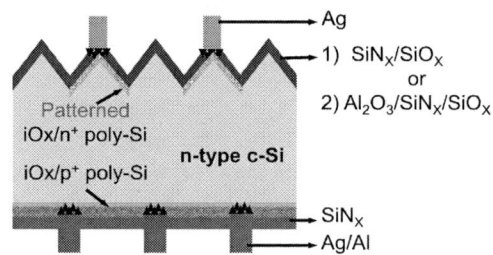

Fig. 1. Schematic diagram of selective DS-TOPCon solar cell featuring patterned front iOx/n^+ poly-Si and rear iOx/p^+ poly-Si.

After saw damage etching (SDE) in 20 %$_{wt}$ potassium hydroxide (KOH) solution at 80 °C for 9 minutes, both sides of the wafers were textured using the standard texturing process. A stack of ultra-thin interfacial oxide (iOx) and intrinsic poly-Si of desired thickness were deposited on both planar and textured samples in a single step two-stage process in low pressure chemical vapor deposition (LPCVD) system. The ex-situ doping of the poly-Si layers was performed in an atmospheric tube diffusion furnace), using liquid phosphorus oxychloride (POCl₃) and boron-tribromide (BBr₃) as dopant sources, resulting in symmetric n-type and p-type poly-Si structures, respectively. Up to this process stage, a batch of symmetric textured n-TOPCon (90 nm) and planar p-TOPCon (250 nm) samples were fabricated and ready for the next process.

A proportion of these symmetric textured n-TOPCon samples were dipped in 9 %$_{wt}$ KOH solution at 40 °C to etch off the n^+ poly-Si. This etching process removed the entire n^+

poly-Si, but was stopped its reaction exactly at iOx interface, which was demonstrated in our previous work [3]. Due to the etch protection by iOx, a weakly phosphorus (P) in-diffused layer under the iOx, formed during the ex-situ doping process, was preserved, which imitates the field region of selective DS-TOPCon cell (Fig. 1). Then, for the fabrication of structure (c) in Fig. 2, a 30 Å of Al_2O_3 film was deposited using plasma-enhanced atomic layer deposition (ALD) system. Finally, both sides of all the samples were passivated with SiN_X:H using plasma-enhanced chemical vapor deposition (PECVD) system. High temperature firing process was performed in an industrial belt furnace.

B. Characterization details

The recombination current density (J_0) of all test structures was measured at an injection level of 5×10^{15} cm^{-3} using a contactless photoconductance decay measurement tool (WCT-120) [5]. The current-voltage (J-V) characteristics of the fabricated selective DS-TOPCon solar cells were measured using a flash tester (FCT-450).

III. RESULTS AND DISCUSSIONS

A. Preliminary simulation: Impact of front $J_{0,field}$ on efficiency of selective DS-TOPCon cell

To start with, a 2D device simulation was performed for selective DS-TOPCon cell structure using Quokka 2 [6] with practically achievable device parameters to find out the target J_0 value for the front field region, which can attain ~25.0 % cell efficiency. As shown in Fig. 3, the front $J_{0,field}$ should not exceed 5 fA/cm^2 to cross the 25.0 % cell efficiency with selective DS-TOPCon cell structure. Also, it shows that the cell efficiency of rear junction device is extremely sensitive to the front field

passivation, projecting nearly 1.3 % absolute efficiency degradation as the $J_{0,field}$ rises from 5 to 30 fA/cm^2.

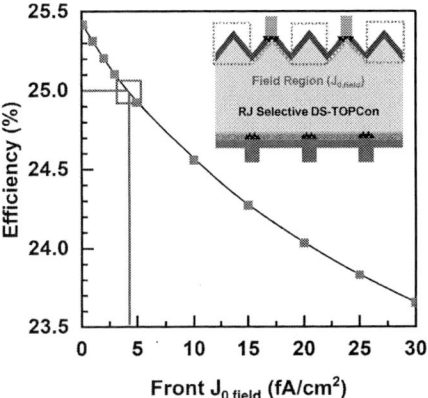

Fig. 3. Simulated cell efficiency of RJ selective DS-TOPCon cell structure as a function of front $J_{0,field}$.

B. Investigation of passivation property of each region of selective DS-TOPCon solar cells

To evaluate the feasibility of the proposed selective DS-TOPCon cell structure (Fig. 1) and identify the potential performance-limiting factors, the passivation property of each region of the cell was individually monitored. As summarized in Fig. 2, the J_0 of each test samples was measured before and after high temperature screen-printed contact firing step (~740 °C). The symmetric textured n-TOPCon (structure (a)) and planar p-TOPCon (structure (d)) samples showed excellent passivation quality after simulated firing, resulting in full-area passivated J_0 values of ~9 fA/cm^2 and ~4 fA/cm^2 respectively.

Since, in a real device, textured n-TOPCon is replaced by dielectric passivated textured n-type c-Si, therefore, we also

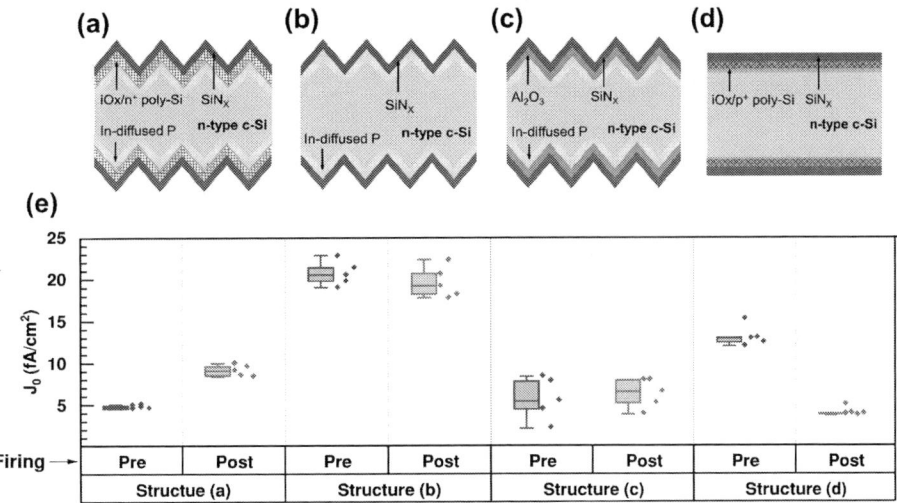

Fig. 2. (a)-(d) Schematic diagrams of symmetric test structures and (e) the summary of measured J_0 before and after high temperature firing

studied dielectric passivated fields region on the front. We investigated and compared the passivation quality of SiN_X:H layer and Al_2O_3/SiN_X:H stack. Fig. 2 shows that Al_2O_3/SiN_X:H stack passivated the surface extremely well, resulting in textured field J_0 value as low as 4 fA/cm^2 ,which is comparable to n-TOPCon and is consistent with the requirement for 25.0 % cell efficiency (Fig. 3). This excellent passivation is attributed to the large negative fixed charge density ($\sim$1x10^{13}/cm^2) in Al_2O_3 layer which provides excellent passivation via formation of inversion layer at interface of AlO_3/n-Si. On the other hand, SiN_X:H passivation resulted in a relatively high J_0 of 18-22 fA/cm^2. This may be due to the poor chemical passivation at the SiN_X:H/n-Si interface and the positive fixed charge density in SiN_X:H is not enough to create strong accumulation layer to lower the J_0 to the desired value of <5 fA/cm^2.

B. J-V characteristic of fabricated selective DS-TOPCon cell

After monitoring the recombination behaviors of each layer (J_0 assessment), we made an attempt to fabricate the selective DS-TOPCon cell. Table I summarizes J-V characteristics of fabricated selective DS-TOPCon solar cell with above two different front field surface passivation schemes. Contrary to our expectation, we found that the performance of the cell passivated with SiN_X layer was superior to the cell passivated with Al_2O_3/SiN_X stack. Further C-V analysis revealed that Al_2O_3 layer provided superior surface passivation quality because it has larger negative fixed charge density ($\sim$1x10^{13}/cm^2) than the positive charge density ($\sim$1x10^{12}/cm^2) in the SiN_X:H. However, the cell performance of Al_2O_3/SiN_X:H stack passivated cell was relatively inferior because the depletion and inversion layers underneath the dielectric layer attracts minority carriers to the surface, which aggravated the carrier collection. In addition, the positively charged p$^+$ inversion layer in the field and the n$^+$ region underneath the metal contacts form a tunnel junction to trigger tunneling or leakage of minority carriers. This causes a substantial drop in the cell V_{OC} and FF, as seen in the Table I. Thus, neither SiN_X:H nor Al_2O_3/SiN_X:H passivation schemes are sufficient for very high efficiency solar cells. Recently we have conducted the experiment with 15 nm thermal SiO_2 capped with SiN_X:H on the field region. A preliminary optimization on this passivation scheme has given J_0 values of $\sim$ 10 fA/cm^2, which getting close to the target $J_{0,\text{field}}$ value. Further dielectric optimization and cell fabrication is in progress.

IV. CONCLUSIONS

In this work, we have investigated the opportunity and the challenge in fabrication of RJ selective DS-TOPCon solar cells. We found that the performance of selective DS-TOPCon cell structure is largely dependent on the front surface passivation since vast majority of carriers are photo-generated near the top surface and the collecting junction is located at the back. A 2D device simulation showed that in order to cross the 25.0 % cell, the passivated $J_{0,\text{field}}$ should be below 5 fA/cm^2. We have seen from the symmetric test samples that Al_2O_3/SiN_X:H passivation stack can achieve the J_0 value as low as 4 fA/cm^2, which satisfies the high efficiency requirement from the basis of J_0 value. To demonstrate and verify the feasibility of Al_2O_3/SiN_X:H passivation stack, a selective DS-TOPCon cells were fabricated and passivated its front surface with Al_2O_3/SiN_X:H stack. However, due to the existence of P in-diffusion on the front field region, this passivation stack formed strong inversion layer in the field region, which detrimentally affected the J-V parameters of the finished cells. Both Al_2O_3/SiN_X:H and SiN_X:H passivation schemes investigated in this work were concluded unsatisfactory for different reasons. Finding out the passivation scheme which can achieve $J_{0,\text{field}} <$ 5 fA/cm^2 and at the same time induce the strong accumulation layer on the field region will be a key to cross the 25.0 % cell efficiency with selective DS-TOPCon cell structure.

ACKNOWLEDGEMENT

This material is based upon work supported by the U.S. Department of Energy's Office of Energy Efficiency and Renewable Energy (EERE) under Solar Energy Technologies Office (SETO) Agreement Number DE-EE0009350.

REFERENCES

[1] S. W. Glunz et al., "Passivating and Carrier-selective Contacts - Basic Requirements and Implementation," in 2017 IEEE 44th Photovoltaic Specialist Conference (PVSC), 25-30 June 2017 2017, pp. 2064-2069, doi: 10.1109/PVSC.2017.8366202.

[2] W. J. Choi, Y. W. Ok, K. Madani, S. Duttagupta, and A. Rohatgi, "Development of a co-anneal process for double-side TOPCon precursor fabricated by ex-situ POCl3 and APCVD boron diffusion," in 2022 IEEE 49th Photovoltaics Specialists Conference (PVSC), 5-10 June 2022 2022, pp. 1068-1068, doi: 10.1109/PVSC48317.2022.9938560.

[3] S. Dasgupta et al., "Novel Process for Screen-Printed Selective Area Front Polysilicon Contacts for TOPCon Cells Using Laser Oxidation," IEEE Journal of Photovoltaics, pp. 1-7, 2022, doi: 10.1109/JPHOTOV.2022.3196822.

[4] Y. Y. Huang, A. Jain, W. J. Choi, K. Madani, Y. W. Ok, and A. Rohatgi, "Modeling and Understanding of Rear Junction Double-Side Passivated Contact Solar Cells with Selective Area TOPCon on Front," in 2021 IEEE 48th Photovoltaic Specialists Conference (PVSC), 20-25 June 2021 2021, pp. 1971-1976, doi: 10.1109/PVSC43889.2021.9518628.

[5] R. A. Sinton, A. Cuevas, and M. Stuckings, "Quasi-steady-state photoconductance, a new method for solar cell material and device characterization," ed, 1996, pp. 457-460.

TABLE I
SUMMARY OF J-V PARAMETERS OF FABRICATED SELECTIVE DS-TOPCON SOLAR CELLS

Passivation stack	V_{OC} [mV]	J_{SC} [mA/cm^2]	FF [%]	ɳ [%]	pFF [%]	J_{02} [nA/cm^2]
SiN_X passivation	687	38.7	79.3	21.1	82.2	8
Al_2O_3/SiN_X passivation	675	37.5	78.5	19.8	80.6	18

[6] A. Fell, "A Free and Fast Three-Dimensional/Two-Dimensional Solar Cell Simulator Featuring Conductive Boundary and Quasi-Neutrality Approximations," IEEE Transactions on Electron Devices, vol. 60, no. 2, pp. 733-738, 2013, doi: 10.1109/TED.2012.2231415.

Autonomous Control Strategies for Interconnected DC Microgrids with Geographical Separation

Emmanuel G. Robles-Rivera[1], Rachid Darbali-Zamora[2], Erick E. Aponte-Bezares[1], Jack D. Flicker[2], Andrew R. R. Dow[2], Felipe Palacios II[2], Lee Rashkin[2]

[1]University of Puerto Rico-Mayagüez, Mayagüez, Puerto Rico 00682, USA
[2]Sandia National Laboratories, Albuquerque, New Mexico, 87185, USA

Abstract – **Distributed Energy Resources (DER) with renewable energy sources are propagating quickly and as such, research for integrating these new technologies more efficiently and reliably is paramount to a well working grid. A significant amount of energy is produced and stored in DERs as Direct Current (DC). DC microgrids are being studied as an alternative to Alternating Current (AC) microgrids due to their increased efficiency and simplicity of control. Microgrids are usually controlled by broadband communications, but some forms of autonomous control have been developed as contingency for loss of communications scenarios. Previously developed autonomous controls have been shown in practice to work well for close proximity interconnected grids. However, geographical separation of assets is generally not considered. This paper presents simulation results for an autonomous control strategy to interconnect geographically extended microgrids through aerial and underground power lines, with the use of averaged bidirectional boost converters.**

Keywords – renewable energy, DC microgrids, autonomous control, bidirectional converter

I. INTRODUCTION

Microgrids, which are small scale, controllable versions of interconnected electrical grids [1], are increasingly in demand due to their ability to provide increased resilience and reliability when used in conjunction with renewable energy sources. Most modern grids and microgrids work with alternating current (AC), which historically has had major benefits, including that most traditional generators are AC and transformers have long been available for voltage magnitude changes allowing for decreased power line losses, especially as line distance increases. However, AC microgrids also have drawbacks in terms of controlling and maintaining power quality and system stability, such as when synchronizing several generators together, causing issues with reactive power and frequency regulation. Alternatively, direct current (DC) does not pose such issues. A significant number of loads and distributed energy resources (DERs), such as photovoltaics (PV) and energy storage systems (ESS) are DC in nature, making DC microgrids an attractive alternative [2], [3]. Because of the existence of power electronics converters, the integration between different renewable energy sources is possible. Bidirectional boost converters are a good example of such devices used in DC microgrids, since they can be used to change voltage magnitudes (i.e., for transmission and distribution, much like transformers in an AC system).

Most microgrid converters have multiple control layers, each with a specific type of regulation or goal. The use of a control strategy to enable energy management or economic dispatch is common. However, this type of tertiary controls usually needs broadband communications, which can sometimes be interrupted, especially in resiliency scenarios. A primary control strategy called Droop is also commonly used in microgrids. It is a type of control which applies a change in the asset's output proportional to a certain measured deviation from nominal values. It provides voltage regulation and also provides proportional load sharing between microgrids. Several other microgrid controls have been developed to either decrease the reliance on communications for control [4] or utilize autonomous methods when communications are lost and grid functionality is required [5], [6], [7]. This article focuses on autonomous controls for emergency scenarios without communications applied to interconnected DC microgrids with geographical separation.

When interconnecting several microgrids together, the complexity of control variables significantly increases, since every asset must be coordinated to maintain system stability as well as bus voltage. However, existing autonomous controls frequently do not consider geographical separation between microgrids, which means that the addition of a power line interconnections between microgrid buses may affect the stability of the system, depending on line characteristics and length. Under emergency scenarios, a possible solution to potential instability introduced by aerial or underground power lines between microgrids under autonomous control is the use of bus-tie (transactional) converters with fixed duty cycle. This approach for the use of DC bus-ties will function in a similar manner to the operation of a transformer in a traditional distribution system.

While using this method, system stability relies on enhanced droop compensation controls utilized by the dispatchable units connected to the local microgrid but will enable the bus-tie converters to transfer power between grids autonomously through power lines.

In this paper, several dynamic scenarios of interconnected microgrids with dispatchable and non-dispatchable assets are presented. The interconnected microgrids are simulated utilizing state-space average models of their power electronic converters. The interconnection between microgrids consists of aerial or underground power lines with DC bus-tie converters (bidirectional boost converters), that increase voltage magnitude (max 500 V to avoid high voltage considerations) for the power lines. Each simulation uses different line lengths or configurations. Results for aerial lines are presented for a preset conductor separation of 1 ft to make use of existing manufacturer's parameters tables [8]. The line values used for underground conductors will presume they are in close proximity and inside of a conduit, so line capacitance, resistance and inductance will only be affected by line length. Additionally, a set of parametric line distance vs. power transfer curves were developed to provide a sense of maximum distances and power transfer boundaries for the selected voltage.

II. System Description

A. Microgrid Architecture and Configuration

For this analysis, two microgrid topologies are simulated, as shown in Fig. 1 and Fig. 2. Fig. 1 represents a two-bus microgrid system and Fig. 2 represents a three-bus microgrid system, each with their respective assets and loads. This first diagram shows the interconnection between DC Microgrid A and DC Microgrid C. Both microgrids contain a single bus, load and a DC bus tie converter. Microgrid A has two dispatchable units, and Microgrid C has two units, one dispatchable and one non-dispatchable. Each dispatchable unit has an enhanced droop control containing a voltage correction term to improve regulation of the common bus voltage. In contrast to a conventional droop method, the enhanced droop compensation used adds a shift voltage term into the droop expression of the proportional controller [9]. This term enhances voltage regulation usually set by traditional droop methods by compensating for line resistances. A non-dispatchable unit is configured with power control, to simulate the maximum power point tracking function for the PV.

Fig. 2 shows a three-bus system with the interconnection between DC Microgrids A, C and C1. Both microgrids A and C remain unchanged from the two-bus system, and microgrid C1 is an exact copy of microgrid C. It is important to notice that this interconnection and autonomous method is not limited to a two microgrid interconnection, and loads can be shared proportionally between microgrids with the desired droop constants. In this case, droop compensations are equal between microgrids.

The objective for each simulation is to demonstrate how autonomous (without communications), fixed duty cycle DC bus-tie converters can achieve stable power transfer between interconnected microgrids when traditional controls would not work. As previously mentioned, fixed duty converters emulate the behavior of traditional distribution transformers. In the event of communications loss, fundamental power system operations with local control signals and measures take over and provide a well-known framework to maintain system operations with the remaining energy in the system. This is a very desirable situation to support continued microgrid operation and careful evaluation is required to explore possible limitations.

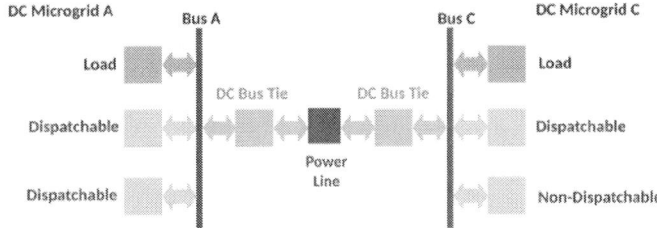

Fig. 1. System description for two-bus interconnected microgrids via single power line with DC bus-tie converters.

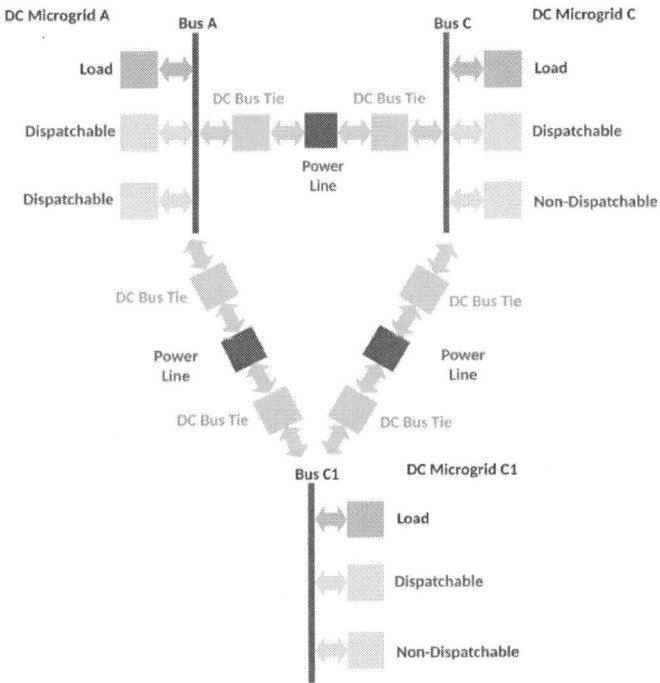

Fig. 2. System description for three-bus interconnected microgrids via power lines and DC bus-tie converters.

For every test shown, each microgrid will start with a constant load that is locally supplied by its dispatchable and non-dispatchable units. At 0.5 seconds, Microgrid A will have an overload event, which will occur due to a local abrupt load change. For the two-bus system, this will force Microgrid C to supply approximately 2 kW through the power line via the bus-tie converters in order to constrain the overloaded bus to a 5% voltage drop from nominal, which is 250 V. For the three-bus system simulation, Microgrid A will also experience this overload event, but compensation will be provided equally between both microgrids C and C1.

B. Power Line Models

As the system is DC, in steady state, only a resistive value for the power line is needed, and capacitive and inductive line values will not affect the final result. However, to support transient response scenarios, capacitive and inductive line values are included in the power line model in order to obtain accurate results.

i. Aerial Copper Conductor

To obtain line values for this model, a mathematical model that represents a single-phase aerial copper conductor in a T line configuration was used. The equations for the aerial power line inductance and capacitance are shown in equation (1) and (2), respectively.

$$L = 2 \cdot 10^{-7} \cdot ln\left(\frac{D}{r}\right) \cdot 1609.39 \quad (H/miles) \quad (1)$$

$$C = \frac{2 \cdot \pi \cdot \varepsilon}{ln\left(\frac{D}{r}\right)} \cdot 1609.39 \quad (F/miles) \quad (2)$$

In these equations, the variable D represents the distance between conductors (in meters), r represents the Geometric Mean Radius (GMR) of the conductor in meters and ε is the permittivity of free space (8.85×10-12 F/m). Inductive and Capacitive values are obtained per mile of total conductor (For DC systems the conductor distance D is twice the physical distance between microgrids). Three different power line lengths were tested, including 0.5, 1, and 2 miles. The line characteristics are shown in Table I.

TABLE I.
2/0 AWG AERIAL COPPER CONDUCTOR LINE PARAMETERS

Parameter	Variable	Value	Unit
GMR	r	0.003816	meters
Resistance	R_{DC}	0.44	Ω/miles
Inductance	L	1.41	mH/mile
Capacitance	C	22.01	nF/mile

*when 1 ft apart (0.3048 m)

ii. Underground Copper Conductor

An underground belted paper insulated copper conductor model in a nominal T-model was chosen to compare and contrast versus the aerial alternative [5]. Similar to the aerial conductor, line characteristics for the underground conductor are affected by line length, however in this case, the distance between conductors is not part of the model equations, as underground conductors are generally in very close proximity to each other. The equations for the underground power line inductance and capacitance are shown in equation (3) and equation (4), respectively.

$$L = \frac{X_L}{2 \cdot \pi \cdot f} \quad (H/miles) \quad (3)$$

$$C = \frac{1}{2 \cdot \pi \cdot f \cdot X_C} \quad (F/miles) \quad (4)$$

The variable f represents the parameter's nominal frequency in Hertz (Hz), X_L is the inductive reactance in Ohms/mile, and X_C represents the capacitive reactance in Ohms/mile. Because this model will only be used with a DC system, it could be noted that the nominal frequency to estimate the parameters do not apply to the DC simulations. It's included to extract inductance

and capacitance values from standard AC (60Hz) tables. Obtaining DC specific line parameters for underground copper conductors is considered beyond the scope of this paper. The line characteristics are shown in Table II.

TABLE II.
2/0 AWG UNDERGROUND BELTED PAPER
INSULATED COPPER CONDUCTOR LINE PARAMETERS

Parameter	Variable	Value	Unit
Resistance	R_{DC}	0.495	Ω/miles
Inductance	L	0.366	mH/mile
Capacitance	C	947.4	nF/mile

iii). Line Lengths vs. Ampacity

Given the relatively novel approach that using DC microgrids encompasses, the following section was developed to present general guidelines for choosing the appropriate underground line caliber and length for a certain ampacity, a reference graph was developed to be used for DC interconnections at 500V nominal.

Fig. 3 shows the maximum line lengths to be used per total line ampacity percentage, while not exceeding the 10% Voltage drop (steady state) from nominal during doing power transfers. Maximum line ampacity percentages were established according to the National Electrical Code for insulated copper conductors at 90°C.

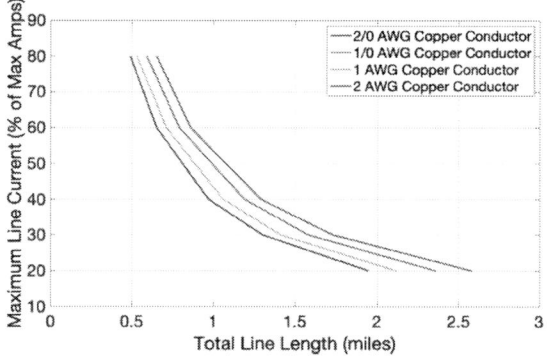

Fig. 3. Line Ampacities vs. Lengths for Underground Copper Conductors at 500 V nominal.

The copper conductors do not exceed 80% of their total ampacity rating. The maximum total line resistance was calculated using equation (5), while the maximum total underground length was calculated using equation (6).

$$R_{MaxTotal} = \frac{V_{loss}}{i_{line}} \quad (5)$$

$$MaxLinelength(miles) = \frac{R_{MaxTotal}}{Line \frac{\Omega}{mile}} \quad (6)$$

In this equation, the variable $R_{MaxTotal}$ represents the total conductor resistive value in Ohms (including positive and negative conductors) for it not to exceed the 10% Voltage drop threshold. V_{loss} represents the 50V maximum voltage drop when the line is at 500V nominal. The line current, i_{line}, will be a variable in the system that will change while doing power transfer, and the Line Ω/mile will represent the maximum allowable total line length for it not to exceed the chosen voltage drop. The values for Fig. 3 are shown in Table III through Table VI.

TABLE III.
2/0AWG AMPACITY VALUES VS. MAX LENGTH

Line Ampacity Percentage*	Current (A)	Maximum Resistance (Ω)	Max Total Line Length (miles)
20%	39	1.28	2.58
40%	78	0.64	1.29
80%	156	0.32	0.65

*Max 2/0AWG Cable Ampacity 195 A (NEC 310-16)

TABLE IV.
1/0AWG AMPACITY VALUES VS. MAX LENGTH

Line Ampacity Percentage*	Current (A)	Maximum Resistance (Ω)	Max Total Line Length (miles)
20%	34	1.47	2.36
40%	68	0.74	1.19
80%	136	0.367	0.59

*Max 1/0AWG Cable Ampacity 170 A (NEC 310-16)

TABLE V.
1AWG AMPACITY VALUES VS. MAX LENGTH

Line Ampacity Percentage*	Current (A)	Maximum Resistance (Ω)	Max Total Line Length (miles)
20%	30	1.67	2.12
40%	60	0.833	1.06
80%	120	0.416	0.53

*Max 1AWG Cable Ampacity 150 A (NEC 310-16)

TABLE VI.
2AWG AMPACITY VALUES VS. MAX LENGTH

Line Ampacity Percentage*	Current (A)	Maximum Resistance (Ω)	Max Total Line Length (miles)
20%	26	1.92	1.95
40%	52	0.96	0.97
80%	104	0.48	0.49

*Max 2AWG Cable Ampacity 130 A (NEC 310-16)

As it was chosen to use 500V as a nominal line voltage for all tests, reasonable line lengths had to be chosen to represent practical uses for power transfer between DC microgrids. As seen in Fig. 3, reasonable line distances range at around 0.5 to 0.75 miles for 80% of maximum cable ampacity, and extend up to 2.5 miles of total conductor length while using 20% of total cable ampacity. It would be necessary to increase line voltages if longer line lengths are needed while not exceeding the 10% of voltage loss across the line.

It is also important to note that the values on Fig. 3 do not consider the transient behavior of the line and are valid for steady state operation. Depending on the case and use, transient behavior should be considered for determining the appropriate line length, caliber, and configuration, as this can decrease maximum line lengths further.

C. Boost Converter Model

In order to avoid switching frequency and ripple considerations in the converters, state-space averaged models for the bus-tie equations where chosen, as shown in equation (7) and equation (8) [10], [11], [12].

$$\frac{di_L}{dt} = \frac{1}{L} \cdot (v_i - D \cdot v_c - i_L \cdot R_s) \tag{7}$$

$$\frac{dv_c}{dt} = \frac{1}{C} \cdot (D \cdot i_L - i_o) \tag{8}$$

In these equations, the variables i_L and v_i represent the boost converter inductor current and input voltage, respectively. v_c is the capacitor voltage at the converter's output, R_s is a small source resistance at the input, and i_o is the output current for the converter. The variable D is the average of the duty cycle value for the converter. These equations can also be written in matrix form, as shown in equation (9) [13].

$$\begin{bmatrix} \dfrac{di_L}{dt} \\ \dfrac{dv_c}{dt} \end{bmatrix} = \begin{bmatrix} \dfrac{-R_S}{L} & \dfrac{-D}{L} \\ \dfrac{D}{C} & \dfrac{-1}{R_S \cdot C} \end{bmatrix} \cdot \begin{bmatrix} i_L \\ V_c \end{bmatrix} + \begin{bmatrix} \dfrac{1}{L} \\ 0 \end{bmatrix} \cdot v_i \tag{9}$$

The boost converter is operated to have a gain of two so that the output voltage connecting to the power line is twice that of the input, around 500 V. A fixed duty cycle of $D = 0.5$ achieves this result. Table VII summaries the boost converter parameters.

TABLE VII.
BOOST CONVERTER PARAMETER

Parameter	Variable	Value	Unit
Inductance	L	0.013	mH
Capacitance	C	1	mF
Series Resistance	R	0.06	Ω

D. Droop Parameters for Dispatchable Units

Each microgrid contains dispatchable and non-dispatchable units that will provide power to the existing loads. The enhanced droop controls for the dispatchable units provide improved voltage regulation when compared to traditional droop control. Equation (10) presents the current reference (i_{ref}) equation using the droop with an enhanced voltage compensation term. V_{nom} corresponds to the nominal microgrid bus voltage, which was chosen to be 250V, while V_{bus} is the feedback from the local bus voltage. The voltage compensation term is represented as ΔV_{corr} (12). The variable k_{droop} is the droop coefficient for the dispatchable unit and is calculated as

978-1-6654-6060-6/23 $31.00 © 2023 IEEE

shown in equation (11). Also in this equation, the variables V_{max} and V_{min} are the acceptable 5% deviation from the nominal voltage. The complete droop scheme us implemented by a variable P-Controller, where the variables *kpi* and *ki* are the proportional and integral gains for the control voltage compensation correction, respectively (12).

$$i_{ref} = \frac{V_{nom} - V_{bus} + AV_{corr}}{k_{droop}} \qquad (10)$$

$$k_{droop} = \frac{V_{max} - V_{min}}{P_{rated}} \qquad (11)$$

$$AV_{corr} = \left(kpi + \frac{ki}{s}\right)(V_{ref} - V_{bus}) \qquad (12)$$

III. SIMULATION RESULTS

Simulation results are presented for the two and three-bus systems. These simulations results consider microgrids with dispatchable and non-dispatchable assets, and a single variable load per microgrid (as shown in Fig. 1). The interconnection between the microgrids was implemented via aerial or underground power lines and DC bus ties (bidirectional boost converters) that increase voltages to twice the bus voltage. Each simulation utilizes different line lengths.

A. Two-Bus System Results

Fig. 4 shows that when Bus A is overloaded at 0.5 seconds, the compensation response from Bus C supplies enough power through the power line and bus tie converters to maintain the receiving bus voltage within 5% of nominal. Fig. 5 shows the power line voltage drop. The simulation results of Fig. 5 show that when increasing the power line length, the voltage drop in steady state across the line also increases, since only the resistive value for the line is in effect, while current flow is unchanged at around 4 A.

It is important to note that this autonomous control for the bus-tie converters supports scenarios of current flow between microgrids and, scenarios where both microgrids are being supplied by their local assets (no transfers). Dispatchable units from both microgrids are utilizing the same enhanced droop control with the same parameters and gains, so load sharing only occurs when a voltage difference between microgrids occur. As seen in Fig. 5, when there is no current across the power line before 0.5 seconds, the line is maintained at 500 V with no voltage drop. Fig 6 shows the simulation results obtained for the current flow from microgrid C to microgrid A.

Since the non-dispatchable unit of Bus C was set to produce a fixed power value, its dispatchable unit with enhanced droop control was able to produce the necessary power to maintain both bus voltages. The current flow remains at approximately 4 A since the local microgrids control does not have a feedback mechanism to compensate for voltage losses at the far end of the power line.

Simulation results demonstrated that both aerial and underground conductors can present several technical

difficulties for their use in the field. Parameters such as line voltages, line lengths and distance between conductors plays a significant role in system stability and can greatly affect system performance. Notice from these results that underground conductors, like the one used for the following results, tend to have significantly higher capacitive values per unit length in comparison to its aerial counterpart, since the distance between underground conductors is minimal and depends entirely on the conductor insulation. This increase in capacitance could affect system stability and should be considered prior to its implementation.

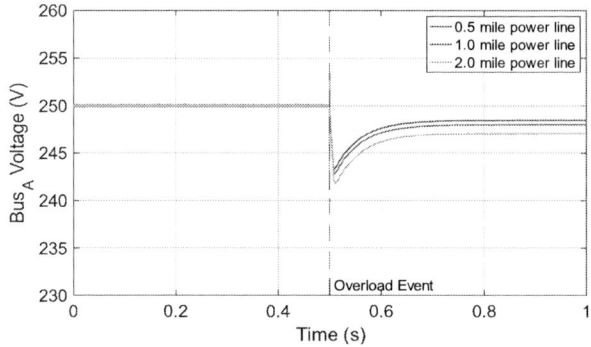

Fig. 4. Bus A (power receiving bus) Voltage response through a 2/0 AWG aerial copper conductor (1 ft apart at 500 V).

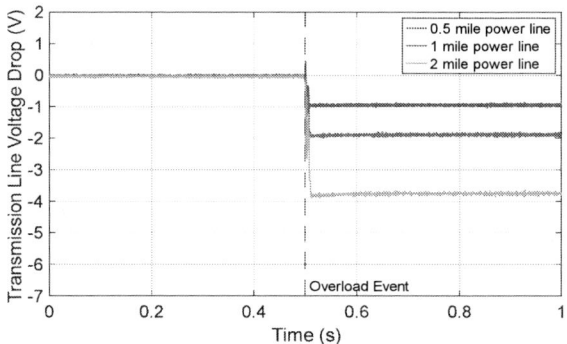

Fig. 5. Power line Voltage drop through a 2/0 AWG aerial copper conductor (1 ft apart at 500 V).

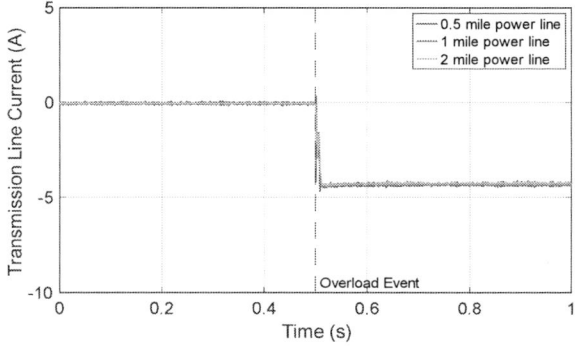

Fig. 6. Current flow from microgrid C to microgrid A through a 2/0 AWG aerial copper conductor (1 ft apart at 500 V).

As shown in Fig. 7, the voltage response for Bus A at 0.5 seconds is similar to its aerial counterpart, as all three-line lengths show less than 5% of voltage drop from nominal on its transitory period, and the voltage drop at steady state only responds to the line's resistive value.

When observing the power line voltage drop for the underground conductor, it can be noted that the transitory response right after the overload event is slightly more abrupt than the aerial conductor. This may be because the underground conductor has around 43 times more capacitance value than the aerial line.

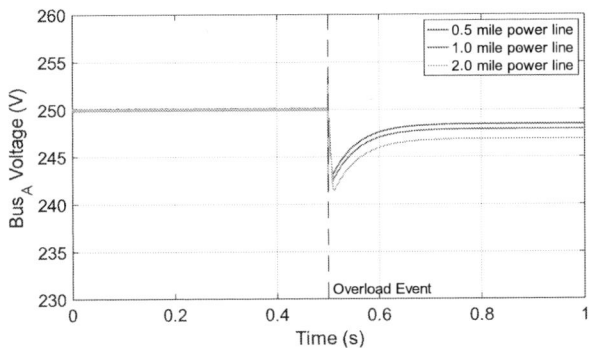

Fig. 7. Bus A (power receiving bus) Voltage response through a 2/0 AWG underground copper conductor (at 500 V).

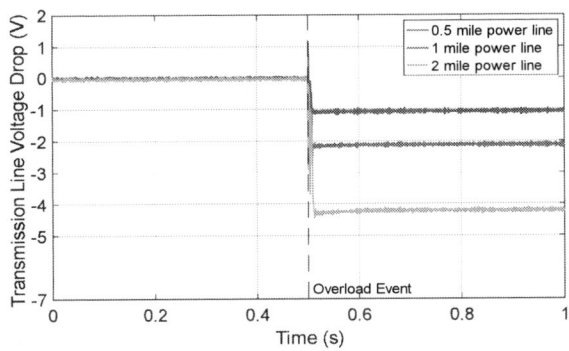

Fig. 8. Power line Voltage drop through a 2/0 AWG underground copper conductor (at 500 V).

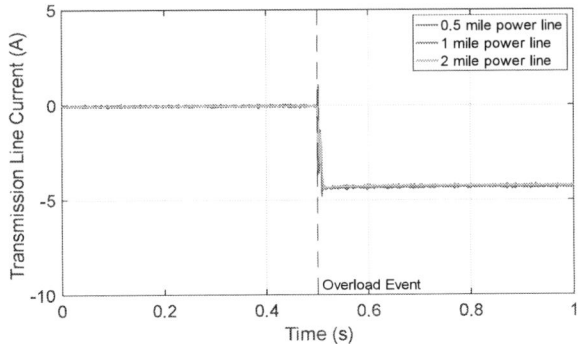

Fig. 9. Current flow from microgrid C to microgrid A through a 2/0 AWG underground copper conductor (at 500 V).

These results demonstrate that the increase in capacitance values for the underground conductor are significantly higher than its aerial counterpart and obtaining the step response to the underground line alone will produce a second order transient not observed on the aerial conductors. Regardless, this increase in capacitance did not significantly affect the system stability, since the results generated from both lines are similar.

B. Three-Bus System Results

The main objective for the three-bus system was to demonstrate how loads would be shared equally and proportionally between multiple interconnected microgrids, while using a fixed duty cycle in a transactional converter. A similar scenario was tested where only Bus A is being affected by the same overload event as presented in the previous section, but in this case, it will be supported by both microgrids C and C1, which will not experience an overload event locally, and will remain with the same constant loads as they were used with the two-bus cases. As can be observed in Fig. 10, the Bus A voltage response is identical to what is observed in Fig. 7. Notice that the maximum voltage drop does not exceed the 5% of nominal as well for the 3 tested line lengths. In contrast to Fig. 7, the overloaded Bus A is now being supplied by both Bus C and C1.

For Fig. 11 and Fig. 12, it was chosen to show only the 2-mile simulations, as they represent the worst-case scenario for these tests. Because microgrids C and C1 are identical to each other, it can be noted that there is no current flow on the power line between these buses, and this identical behavior can be confirmed when seeing that there is no power line voltage drop (as illustrated in Fig. 11) between C and C1 before or after the overload event occurs. Notice that the voltage response and total current supplied externally for the overloaded bus does not significantly change when it is being supplied proportionally by one or more microgrids, as the current gets divided equally, as can be observed in Fig. 12 when the current response and magnitudes from C to A and from C1 to A are identical. While the overload event occurring in microgrid A could be externally supplied with only one microgrid, the interconnection between multiple DC microgrids could increase resiliency and local bus voltage stability if greater overload events would occur.

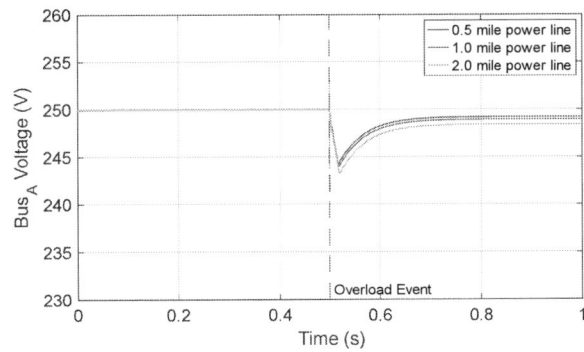

Fig. 10. Bus A (power receiving bus) Voltage response, being supplied by Buses C and C1 through 2/0 AWG underground copper conductors with different line lengths (at 500 V).

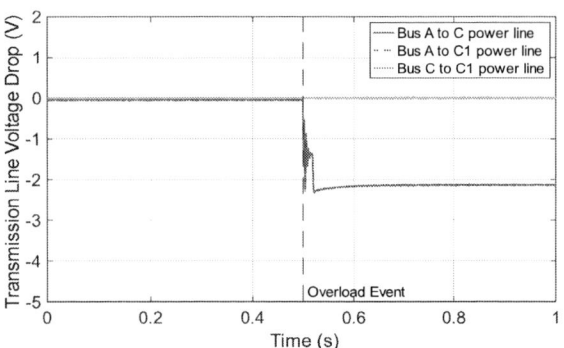

Fig. 11. Power line Voltage drops through 2/0 AWG underground copper conductors with 2 miles (at 500 V).

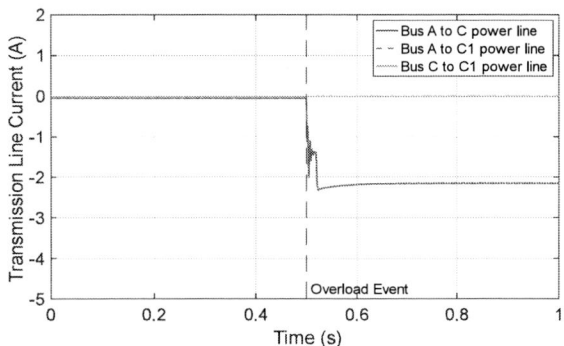

Fig. 12. Current flow from microgrid C to microgrid A, from microgrid C1 to A, and from C to C1 through a 2/0 AWG underground copper conductor with 2 miles (at 500 V).

IV. CONCLUSIONS

The simulations and system presented shows autonomous control solutions for interconnecting geographically distant DC microgrids through aerial and underground power lines with bus-ties (transactional) converters. When local microgrid assets are implemented with enhanced voltage droop control strategies, fixed duty cycle bus-tie converter can be used as transactional converters between microgrids. This enables load sharing between microgrids when local power generation is not sufficient. Simulation results showed that, to choose a DC aerial power line for interconnection, it is important to note that distances between lines greatly affect the transient stability of the system, since capacitive and inductive values will change. Similarly, when choosing an underground conductor, it is important to note that the increase in capacitance for this line can generate second order transients while power sharing, and its system stability will greatly depend on how robust the compensation for the local microgrid assets is. Under a fixed duty cycle, overloaded buses were able to maintain voltage regulation within 5% of nominal when sharing 2 kW through either power line. The two-bus and three-bus cases showed that averaged model DC microgrid simulation with fixed duty on transactional converters point to a positive outlook under contingency scenarios, when communications between microgrids fail and dispatch commands are unavailable.

ACKNOWLEDGEMENT

This article has been authored by an employee of National Technology & Engineering Solutions of Sandia, LLC under Contract No. DE-NA0003525 with the US Department of Energy (DOE). The employee owns all right, title and interest in and to the article and is solely responsible for its contents. The US government retains and the publisher, by accepting the article for publication, acknowledges that the US government retains a non-exclusive, paid-up, irrevocable, world-wide license to publish or reproduce the published form of this article or allow others to do so, for US government purposes. The DOE will provide public access to these results in accordance with the DOE Public Access Plan.

REFERENCES

[1] R. Darbali-Zamora, C. J. Gómez-Mendez, E. I. Ortiz-Rivera, H. Li, and J. Wang, "Solar irradiance prediction model based on a statistical approach for microgrid applications", *2015 IEEE 42nd Photovoltaic Specialist Conference (PVSC)*, 2015, pp. 1-6.

[2] R. Darbali-Zamora, J. E. Quiroz, J. Hernández-Alvidrez, J. Johnson and E. I. Ortiz-Rivera, "Viability Assessment of a Real-Time Simulation Model for a Residential DC Microgrid Network to Compensate Electricity Disturbances in Puerto Rico",*2018 IEEE ANDESCON*, 2018.

[3] R. Darbali-Zamora, J. E. Quiroz, J. Hernández-Alvidrez, J. Johnson and E. I. Ortiz-Rivera, "Implementation of a Dynamic Real Time Grid-Connected DC Microgrid Simulation Model for Power Management in Small Communities", *2018 IEEE 7th World Conference on Photovoltaic Energy Conversion (WCPEC) (A Joint Conference of 45th IEEE PVSC, 28th PVSEC & 34th EU PVSEC)*, 2018, pp. 1179-1184.

[4] M. S. Sadabadi, N. Mijatovic and T. Dragičević, "A Robust Cooperative Distributed Secondary Control Strategy for DC Microgrids With Fewer Communication Requirements", *IEEE Transactions on Power Electronics*, vol. 38, no. 1, pp. 271-282, Jan. 2023.

[5] M. S. Sadabadi, N. Mijatovic and T. Dragičević, "A Robust Cooperative Distributed Secondary Control Strategy for DC Microgrids With Fewer Communication Requirements", *IEEE Transactions on Power Electronics*, vol. 38, no. 1, pp. 271-282, Jan. 2023.

[6] X. Li, W. Jiang, J. Wang, P. Wang, and X. Wu, "An Autonomous Control Scheme of Global Smooth Transitions for Bidirectional DC-DC Converter in DC Microgrid", *IEEE Transactions on Energy Conversion*, vol. 36, no. 2, pp. 950-960, Jun. 2021.

[7] C. O. Gonzalez-Candelario, R. Darbali-Zamora, J. D. Flicker, L. Rashkin, J. Neely, and E. E. Aponte-Bezares, "Autonomous Control Strategies for Interconnected DC Microgrid Applications with Multiple DER Resource Penetration", *2021 IEEE 48th Photovoltaic Specialists Conference (PVSC)*, 2021, pp. 2194-2199.

[8] J. Zaborszky and J. W. Rittenhouse, *Electric power transmission;* New York: Ronald Press Co, 1954.

[9] G. Spiazzi, "Modeling approaches for switching converters", University of Padova-Department of Information Engineering.

[10] R. Darbali-Zamora, A. R. R. Dow, F. Palacios, J. D. Flicker and D. Bauer, "Development of a State-Space Average Nanogrid Model for DC Microgrid Power Management Applications", *2023 IEEE Power & Energy Society Innovative Smart Grid Technologies Conference (ISGT)*, 2023, pp. 1-5

[11] R. Darbali-Zamora and E. I. Ortiz-Rivera, "A State Space Average Model for Dynamic Microgrid Based Space Station Simulations", *2017 IEEE 44th Photovoltaic Specialist Conference (PVSC)*, 2017, pp. 2957-2962.

[12] A. R. R. Dow, R. Darbali-Zamora, J. D. Flicker, F. Palacios, and J. T. Csank, "Development of Hierarchical Control for a Lunar Habitat DC Microgrid Model Using Power Hardware-in-the-Loop", *2022 IEEE 49th Photovoltaics Specialists Conference (PVSC)*, 2022, pp. 0754-0760.

[13] R. Darbali-Zamora and E. I. Ortiz-Rivera, "An Overview into the Effects of Nonlinear Phenomena in Power Electronic Converters for Photovoltaic Applications", *2019 IEEE 46th Photovoltaic Specialists Conference (PVSC)*, 2019, pp. 2908-291.

Enabling High Efficiency, Flexible, and Lightweight CdTe Solar Cells with a Cadmium Stannate Transparent Conducting Oxide

Manoj K. Jamarkattel, Adam B. Phillips, Ebin Bastola, Sabin Neupane, Deng-Bing Li, Yanfa Yan, Randy J. Ellingson, and Michael J. Heben.

Wright Center for Photovoltaic Innovation and Commercialization, Department of Physics and Astronomy, University of Toledo, Toledo, OH, 43606, USA

Abstract—**Fabrication of high-efficiency solar cells using a light weight, flexible substrate has been a challenge due to the high-temperature device fabrication processes. One substrate that can withstand the process temperatures is Corning® Willow® glass; however, this material does not include a transparent conduction oxide (TCO), which is critical to high performance CdTe-based solar cells typically fabricated in the superstrate configuration. Here, we investigate the properties of cadmium stannate (CTO) as a TCO layer for CdTe solar cells. To improve the transmission, conductivity, and crystallinity of the as-deposited CTO layer, the Willow®/CTO films were put in contact with a CdS/substrate film and annealed at 600 °C under high vacuum condition. After this proximity anneal, the device was fabricated by depositing an indium gallium oxide (IGO) emitter, CdSe, and CdTe, followed by device processing. The efficiency of the devices improved to 16.1% with a specific power density of ~ 700 W/kg compared to devices in which the CTO did not undergo the proximity anneal have efficiency of 12.5%. These results CTO is a promising TCO for fabricating high efficiency, lightweight, flexible CdTe solar cells which could be applicable to space applications.**

I. INTRODUCTION

Cadmium telluride (CdTe) has an optical band gap of 1.5 eV with a high absorption coefficient which made it an excellent material for solar cells. Low weight and flexibility are key aspects for solar cells to achieve high specific power (total power/ total weight) and flexible installation capability which are basic requirements for space applications. One of the simplest ways to fabricate lightweight and flexible solar cells is by using flexible and lightweight substrates. Efforts have been done to fabricate lightweight and flexible CdTe solar cells using different substrates such as polymer, ceramic substrate, or thin glass substrates [1-5]. Polymers substrates are limited to low-temperature processing which limits the efficiency of the devices. Ceramic substrate can resist high-temperature processing but are currently fabricated in such a way to make them highly reflective, thereby limiting current generation [2]. Thin glass substrates like Corning® Willow® glass can withstand high temperature processing and are optically clear. As a result, Willow® has been used to fabricate lightweight flexible devices [3, 4]. However, CdTe solar cell technology has greatly improved [6, 7] since Willow® was last used as a device substrate, and improvements to those device structures could yield efficiencies greater than the highest reported value of 16% [3].

In the previous studies, the device structure was TCO/buffer/CdS/CdTe, where the n-type CdS used to create the junction was photo-inactive, resulting in a loss in photocurrent collection at shorter wavelengths [4] . Since those studies, CdS has been replaced with CdSe, which intermixes with CdTe to form a photo-active p-type layer and moves the junction to the buffer layer [8]. In our previous work, we fabricated CdTe solar cells using CTO (CTO/i-ZnO/CdS:O/CdTe) on a 100-microns thick Willow® glass substrate and achieved efficiencies of ~14% [4]. In these devices, the transparency of the CTO films was ~0 % to 65% from 350 nm to 500 nm wavelength region, but the parasitic absorption in the CdS dominated the current losses in these devices. However, moving to a CdSe layer, it is now critical to improve the underlying CTO. This has been done by annealing the CTO at high temperatures ($560 - 700$ °C) under a number of conditions such as in an inert atmosphere [9], under a CdS layer, or a so-called "proximity anneal" in which the CTO is in contact with CdS layer on another substrate during heating [10].

Here, we investigate how the properties of CTO affect the performance of Willow®/CTO/IGO/CdSe/CdTe devices. The optical and electrical properties of the CTO films are measured before and after various processing conditions, and the proximity anneal yields the most promising TCO. Efficiency greater than 16% was achieved for devices fabricated with Willow®/CTO substrates after undergoing a proximity anneal. With this efficiency, these solar cells have a specific power density of ~ 700 W/kg (without packaging), which is encouraging results for space applications.

II. EXPERIMENTAL DETAILS

Devices were fabricated on 100 μm Willow® glass that was ultrasonically cleaned using Micro-90 detergent and deionized water. A 300 nm CTO film was sputtered as a TCO layer in a 20% O_2 and 80% Ar environment at room temperature. As-deposited CTO films underwent a vacuum anneal or proximity anneal at 600 °C for 30 min under a vacuum of 10^{-6} Torr. After CTO processing, 75 nm of Indium Gallium Oxide (IGO) with a bandgap of 4.02 eV or 4.1 eV was RF co-sputtered as described in our previous work [11] followed by sputtering of 90 nm CdSe. 3.5 micron CdTe was deposited using a high vacuum close-space sublimation system (CSS) [12]. $CdCl_2$ and $CuCl_2$ were performed as described in our previous work [13]. Anti-reflecting (AR) layer was coated on the front surface of the Willow® substrate.

A Perkin Elmer Lambda 1050 UV/Vis/NIR spectrophotometer was used to measure optical properties, and X-ray diffraction patterns were recorded with a Rigaku Ultima III X-ray diffractometer. Current density-voltage (JV) characteristics were measured using a Keithley 2440 digital source meter and a solar simulator with AM1.5G illumination.

III. RESULTS AND DISCUSSIONS

To study the electrical and optical properties of CTO films after processing, we deposited CTO layers on SLG glass, and samples were 1) annealed and 2) proximity annealed separately at 600 °C for 30 min under 1×10^{-6} Torr pressure. Four-point probe, spectrophotometry, and energy dispersive x-ray (EDX) were performed to measure sheet resistance, transmission, and the Cd-to-Sn compositional ratio in CTO films.

TABLE.1. Sheet resistance and Cd:Sn ratios of CTO layer

CTO Films	Sheet resistance ($\Omega/\square$)	Cd:Sn ratio
As deposited CTO	4953 ± 739	2:1
HT CTO	101 ± 22	1:1
Proximity HT, CTO	11.3 ± 2.8	2:1

As seen in TABLE.1, as-deposited CTO film is highly resistive and has a Cd:Sn ratio of 2:1. XRD shows that as-deposited CTO film is amorphous and has low transmission in 350-500 nm wavelength region (Figure.1) After annealing between graphite susceptors inside the CSS, the sheet resistance of CTO films decreases by an order of magnitude but still too resistive for a high-quality TCO layer. EDX analysis shows that the compositional ratio between Cd and Sn is 1:1, indicating Cd loss during annealing and that the CTO films are not Cd_2SnO_4. To overcome this Cd loss during annealing, proximity annealing with a CdS source was performed [10] for separate sample. For this, ~300 nm CdS was deposited on Willow® glass as a source material for Cd and brought in contact with SLG/CTO films, then annealed at 600 °C for 30 min under high vacuum. After the proximity anneal, CTO films have a composition ratio of 2:1 (Cd:Sn) with sheet resistance values of 11 $\Omega/\square$. The transmission improved significantly in the 350-500 nm wavelength region indicating a "blue shift" in the absorption peak on the transmission curve (Figure.1.)[9]. Crystallization of CTO films occurs after proximity annealing with a spinal phase of Cd_2SnO_4 [9, 10].

To investigate the properties of underlying CTO films during the fabrication of actual CdTe solar cells, we deposited IGO, CdSe, and CdTe on SLG/proximity annealed CTO substrate and performed $CdCl_2$ and $CuCl_2$ treatments as described in our previous work [13]. Lift-off of layers at different fabrication steps has been done using epoxy and liquid nitrogen [14]. We did not notice any significant changes

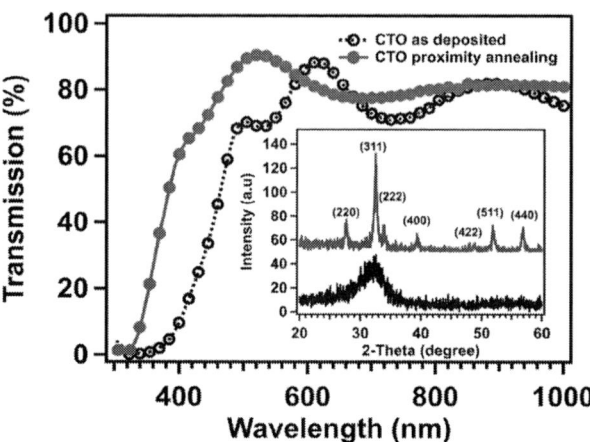

Figure.1. Transmission and XRD data (inset) of as-deposited and proximity annealed CTO films.

in the position of an absorption peak in the transmission data or sheet resistance of the underlying CTO films, indicating that CTO films are thermally stable and suitable for CdTe-based solar cell fabrication.

In an attempt to fabricate high-efficiency lightweight and flexible CdTe devices using high-quality CTO as a TCO layer, we fabricated devices with a device stack of AR/Willow®/CTO/IGO/CdSe/CdTe/CdCl$_2$/CuCl$_2$/Au. Here, a 4.02 eV bandgap IGO film is used as the emitter layer to minimize the carrier recombination at the front interface [11]. As seen in the JV curves (Figure.2.), devices using CTO without a proximity anneal (red graph) have a short circuit current density (Jsc) of ~ 26 mAcm^{-2}, fill factor (FF) ~ 65 %, and device efficiency of 12.5 %. Both the low Jsc (~ 26 mAcm^{-2}) and FF (< 65%) values can be due to poor quality CTO with low transmission through the film and high resistance. However, for devices with a CTO that has undergone proximity annealing, the current collection

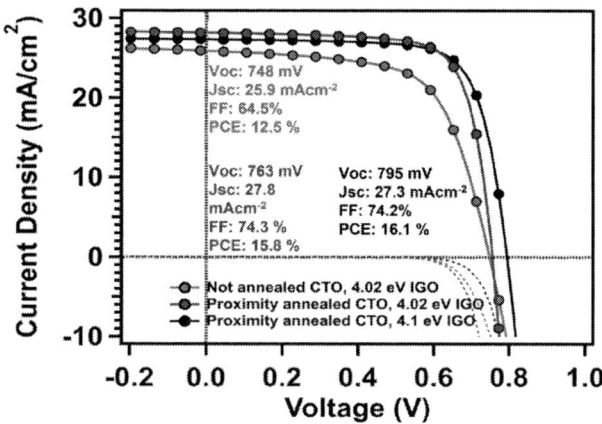

Figure.2. JV graphs of champion cells

improves to 27.8 mAcm^{-2}, FF increases to ~ 75% and open circuit voltage (Voc) increases to 760 mV with a champion cell of 15.8% efficiency. Here, the V_{OC} is lower than expected, and may be due to misalignment of band energies between IGO and absorber layer [11]. To test this possibility, we fabricated another device with a wider bandgap IGO (4.1 eV) while keeping all the other layers and parameters identical to that of a 4.02 eV bandgap IGO device. As the JV curve shows (black graph, Figure.2), increasing the IGO bandgap increases the Voc to ~790 mV, and the FF of ~75% is maintained to yield a champion cell with an efficiency of 16.1% and a specific power density of ~700 W/kg without packaging. By changing the IGO bandgap by 0.08 eV, a Voc gain of ~ 30 mV is achieved. Further investigation of the role of IGO in CTO films is currently underway. A champion cell of 16.1% efficiency and a specific power density of ~700 W/kg are promising for space applications, but other requirements for space deployment, such as radiation hardness testing of each device layer, is required to determine the potential of CdTe-based solar cells for space applications.

IV. CONCLUSION

In this work, we fabricated high-efficiency, lightweight, and flexible CdTe solar cells using a proximity annealed CTO as a TCO layer. With champion device efficiency of 16.1% and a specific power density of ~700 W/kg, these results could lead to a new pathway for CdTe-based solar cells for space applications.

ACKNOWLEDGEMENT

This report is based on research sponsored by the U.S. DOE's Office of Energy Efficiency and Renewable Energy (EERE) under Solar Energy Technologies Office (SETO) Agreement DE-EE0008974, through the Alliance for Sustainable Energy, LLC, Managing and Operating Contractor for the National Renewable Energy Laboratory for the U.S. Department of Energy, under Award Number 37989, and Air Force Research Laboratory under agreement numbers FA9453-21-C-0056 and FA9453-19-C-1002. Approved for public release, distribution is unlimited. Public Affairs release approval #AFRL-2023-0291. The U.S. Government is authorized to reproduce and distribute reprints for Governmental purposes not withstanding any copyright notation thereon. The views and conclusions contained herein are those of the authors and should not be interpreted as necessarily representing the official policies or endorsements, either expressed or implied, of Air Force Research Laboratory or the U.S. Government. The Authors like to thank Dr. Sean Garner and Corning Inc. for providing the Corning® Willow® glass.

REFERENCES

[1] A. N. Tiwari, A. Romeo, D. Baetzner, and H. Zogg, "Flexible CdTe solar cells on polymer films," *Progress in Photovoltaics: Research and Applications,* vol. 9, no. 3, pp. 211-215, 2001, doi: 10.1002/pip.374.

[2] M. K. Jamarkattel *et al.*, "Ultra-Thin and Lightweight CdS/CdTe Solar Cell Fabricated on Ceramic Substrate for Space Applications," in *2022 IEEE 49th Photovoltaics Specialists Conference (PVSC),* 5-10 June 2022 2022, pp. 0348-0350, doi: 10.1109/PVSC48317.2022.9938514.

[3] H. P. Mahabaduge *et al.*, "High-efficiency, flexible CdTe solar cells on ultra-thin glass substrates,"

Applied Physics Letters, vol. 106, no. 13, 2015, doi: 10.1063/1.4916634.

[4] G. K. Liyanage *et al.*, "RF-sputtered Cd2SnO4 for flexible glass CdTe solar cells," in *2016 IEEE 43rd Photovoltaic Specialists Conference (PVSC)*, 5-10 June 2016 2016, pp. 0450-0453.

[5] S. S. Bista *et al.*, "Water-Assisted Lift-Off Process for Flexible CdTe Solar Cells," *ACS Applied Energy Materials,* vol. 6, no. 2, pp. 885-891, 2023/01/23 2023, doi: 10.1021/acsaem.2c03287.

[6] D.-B. Li *et al.*, "20%-efficient polycrystalline Cd(Se,Te) thin-film solar cells with compositional gradient near the front junction," *Nature Communications,* vol. 13, no. 1, p. 7849, 2022/12/21 2022, doi: 10.1038/s41467-022-35442-8.

[7] D.-B. Li *et al.*, "Oxygen Management to Avoid Photo-Inactive Cd(S,Se) for Efficient Cd(Se,Te) Solar Cells," *ACS Energy Letters,* pp. 1529-1534, 2023/02/20 2023, doi: 10.1021/acsenergylett.3c00141.

[8] N. R. Paudel and Y. Yan, "Enhancing the photo-currents of CdTe thin-film solar cells in both short and long wavelength regions," *Applied Physics Letters,* vol. 105, no. 18, p. 183510, 2014.

[9] Z. Du, X. Liu, Y. Zhang, and Z. Zhu, "High-quality cadmium stannate annealed in N2 atmosphere for low-cost thin film solar cell," *RSC Advances,* 10.1039/C7RA00394C vol. 7, no. 30, pp. 18545-18552, 2017, doi: 10.1039/C7RA00394C.

[10] D. M. Meysing *et al.*, "The influence of cadmium sulfide and contact annealing configuration on the properties of high-performance cadmium stannate," *Sol. Energy Mater. Sol. Cells,* vol. 117, pp. 300-305, 2013/10/01/ 2013, doi: https://doi.org/10.1016/j.solmat.2013.06.009.

[11] M. K. Jamarkattel *et al.*, "Indium Gallium Oxide Emitters for High-Efficiency CdTe-Based Solar Cells," *ACS Applied Energy Materials,* vol. 5(5), pp. 5484-5489, 2022/04/26 2022.

[12] M. K. Jamarkattel *et al.*, "Improving CdSeTe Devices With a Back Buffer Layer of Cu$_x$AlO$_y$," *IEEE Journal of Photovoltaics,* vol. 12, no. 1, pp. 16-21, 2022.

[13] M. K. Jamarkattel *et al.*, "Reduced Recombination and Improved Performance of CdSe/CdTe Solar Cells due to Cu Migration Induced by Light Soaking," *ACS Applied Materials & Interfaces,* 2022/04/22 2022, doi: 10.1021/acsami.1c23937.

[14] D. L. McGott, E. Colegrove, J. N. Duenow, C. A. Wolden, and M. O. Reese, "Revealing the Importance of Front Interface Quality in Highly Doped CdSexTe1−x Solar Cells," *ACS Energy Letters,* vol. 6, no. 12, pp. 4203-4208, 2021/12/10 2021, doi: 10.1021/acsenergylett.1c01846.

Post-Mortem Failure Analysis of Metal Halide Perovskite Modules

Sona Ulicna, Nutifafa Y. Doumon, Michael Owen-Bellini, Jackson Schall, Dana B. Kern, Timothy J. Silverman, Lance M. Wheeler, Steven Hayden, Chengbin Fei, Md Aslam Uddin, Prem J. S. Rana, Jinsong Huang, Joseph J. Berry, Laura T. Schelhas

National Renewable Energy Laboratory, Golden, CO, United States

Department of Applied Physical Sciences, University of North Carolina, Chapel Hill, NC, United States

Metal halide perovskite (MHP) solar cells have reached remarkable device performances in a very short period of time. However, the majority of research focus to date has been at the cell level, and questions regarding their reliability in real-world operating conditions remain. At this stage of research, module-level reliability testing of MHP modules is needed to advance these PV technologies to large scale deployment. Learning from the long history of silicon module reliability testing, accelerated stress testing protocols and post-mortem analysis methodologies are being developed to match the complex nature of MHP-based devices and reveal degradation and failure modes these modules suffer from in the field. MHP modules encapsulated in a polymer-free glass-glass package are evaluated after indoor accelerated stress testing and outdoor exposure. Indoor accelerated stress testing is designed to decouple various environmental stressors (UV, temperature, visible light), probing weaknesses of the perovskite devices as well as module packaging. To catch up to the fast device development, field performance characterization and degradation prediction tools are being developed in parallel. Here, we present an initial strategy for post-mortem failure analysis of aged MHP modules. The analysis is focused on morphological, chemical, and structural degradation. Fielded and stress tested modules are used for non-destructive and destructive characterizations to understand the degradation mechanisms caused by module aging. The results serve to validate the accelerated testing protocols against field failure and improve perovskite module performance stability and reliability. The preliminary results of this work suggest that module failure was related to either package failure, or module specific fabrication defects.

978-1-6654-6060-6/23 $31.00 © 2023 IEEE

Tuning Device Interfaces for Improved Open Circuit Voltage in Wide-bandgap Hybrid Perovskite Photovoltaics

Emily Smith[1], Sarah Brittman[2], David Scheiman[2], Woojun Yoon[2]

[1]NRC Postdoctoral Research Associate residing at the Naval Research Laboratory, Washington, D.C. 20375, United States

[2]Naval Research Laboratory, Washington, D.C. 20375, United States

Wide-bandgap hybrid organic-inorganic perovskite solar cells (PSCs) are of interest to the research community because of their potential for high performance, low-cost and lightweight integration as persistent photovoltaic (PV) sources for tandem PV, space, expeditionary, and underwater power generation. However, despite recent advancements in hybrid PSCs with bandgaps close to 1.75 eV, there are fewer research efforts exploring single-junction PSCs with band gaps 1.8 – 2.1 eV. Here, we show the solution-processed fabrication of a wide-bandgap hybrid PSCs utilizing a formamidinium-based mixed iodide/bromide perovskite composition (band gap ~1.84 eV) as the active layer. Additionally, we employ solution-processable electron and hole transport materials, SnO_2, and poly[bis(4-phenyl)(2,4,6-trimethylphenyl)amine] (PTAA). Our work illustrates the pivotal role that interfaces between the transport materials and the perovskite layer play on the open circuit voltage (V_{oc}) of these devices. For example, we observe an increase in the V_{oc} of devices when we add a C_{60}-SAM (4-(1$'$,5$'$-dihydro-1$'$-methyl-2$'$ H-[5,6]fullereno-C_{60}-I$_h$-[1,9-c]pyrrol-2$'$-yl)benzoic acid) interlayer between the SnO_2 and perovskite layer, as well as utilizing effective p-type doping of the hole transport layer. We hypothesize that these results are due to an improved energetic band alignment and/or improved charge extraction efficiencies at those interfaces. Through careful manipulation of perovskite/transport material interfaces, we can achieve wide-bandgap hybrid PSCs with a V_{oc} greater than 1.1 V. This study is one of only a few detailed works into single-junction PSCs with a bandgap > 1.75 eV utilizing a hybrid formamidinium-based perovskite composition.

I. INTRODUCTION

Hybrid perovskites are an emerging class of photovoltaic (PV) materials that have the potential to be implemented in high efficiency, cost-effective, and flexible solar cells. In particular, the composition of these materials can be tuned to achieve a wide array of bandgaps spanning from 1.5 – 2.3 eV [1], [2], [3] making them ideal candidates for wide-bandgap PV (bandgap >1.7 eV) applications. There has been substantial work to optimize PSCs with a bandgap ~1.5 – 1.75 eV due to their optimal bandgap for perovskite/Si, perovskite/CIGS, and perovskite/perovskite tandem cells [4], [5], [6]. However, there has been limited research conducted on hybrid perovskite compositions with bandgaps >1.8 eV, which could have applications for PV tailored to the underwater solar spectrum, all-perovskite tandem, and extraterrestrial PV applications [7], [8]. Thus, there is a significant need for continued exploration into these wide-bandgap perovskite compositions.

Traditional hybrid perovskite materials take on the stoichiometry ABX_3 where 'A' is a organic cation, typically methylammonium (MA), or formamidinium (FA), 'B' is a divalent metal, typically Pb, or Sn, and X is a halide, typically I, or Br. The prototypical hybrid perovskite PSCs often utilize some mixed composition that takes the form $MA_xFA_{1-x}PbI_yBr_{1-y}$. However, due to poor thermal stability of narrow-bandgap MA-containing materials [4], [7], we opted to focus our study on hybrid materials that do not incorporate MA.

The bandgap can be tuned to higher energies directly through the shortening of the Pb-X bond via either the replacement of I with Br or Cl [2], or the replacement of the 'A' site ion for a smaller sized cation [9]. Wide-bandgap compositions in the 1.8 – 2.1 eV range will typically consist of a composition containing >30% Br [2], [9], [10]. Importantly, mixed-halide perovskite materials in this region of interest undergo light induced halide phase segregation, leaving unanswered questions regarding their long-term stability [4], [7], [11]. However, the incorporation of some Cs into the 'A' site of ammonium-based perovskites has been identified as a promising strategy for bandgap tuning and enhanced stability in wide-bandgap materials [9].

'All-inorganic' perovskites, where the organic ion at the 'A' site is entirely replaced with a metal cation, such as Cs, have shown some promise in recent studies, e.g. $CsPbI_2Br$ (bandgap ~1.91) [10]. However, Cs salts suffer from poor solubility in common solvents and require higher processing temperatures, which can limit their compatibility and tunability with solution-based processing methods [7]. Additionally, in other bandgap regions (1.5 – 1.75 eV), the optoelectronic properties of Cs-based all-inorganic PSCs tend to lag behind their organic counterparts [12]. Thus, we reason that there should be further exploration into hybrid wide-bandgap materials for single-junction PSCs in the 1.8 – 2.1 eV bandgap range.

Herein, we illustrate the development of a formamidinium-based PSC (bandgap ~1.84 eV) with the nominal composition $FA_{0.8}Cs_{0.2}PbI_{1.5}Br_{1.5}$. We first characterize the perovskite material using powder X-ray diffraction (PXRD), Ultraviolet-visible light absorption (UV-Vis), scanning electron microscopy (SEM), and photoluminescence (PL). We then utilized this material in a single-junction PSC with a base architecture of ITO/SnO_2/Perovskite/PTAA/Au, where SnO_2 is the electron transport material, and PTAA (poly[bis(4-

978-1-6654-6060-6/23 $31.00 © 2023 IEEE

phenyl)(2,4,6-trimethylphenyl)amine]) is the hole transport material. We found that these devices were sensitive to the manipulation of interfaces between the perovskite and charge transport materials.

The doping level of the PTAA is instrumental in improving device open-circuit voltage (V_{OC}). The device V_{OC} can also be enhanced by modifying the interface between the SnO$_2$ and perovskite with a interlayer of C$_{60}$-SAM (4-(1′,5′-dihydro-1′-methyl-2′H-[5,6]fullereno-C$_{60}$-I$_h$-[1,9-c]pyrrol-2′-yl)benzoic acid). These enhancements are likely attributed to better energetic band alignment, improved charge extraction, or both.

II. RESULTS & DISCUSSION

We first developed a fabrication method to solution process the perovskite material FA$_{0.8}$Cs$_{0.2}$PbI$_{1.5}$Br$_{1.5}$. Then we tested this composition in a PV device to assess its properties.

A. Material characterization

We tested various conditions to determine optimal fabrication parameters for these perovskite films. We settled on a one-step spin-coating method performed inside an nitrogen filled glovebox. We deposited the thin-film from a 1.5 M precursor solution containing the precursor salts CsBr, FAI, PbBr$_2$, and PbI$_2$ in a stoichiometric ratio, dissolved in a solvent mixture of dimethylformamide (DMF), and dimethyl sulfoxide (DMSO) in a 9:1 (v:v) ratio [13]. The precursor salts were stored in a nitrogen filled glovebox and massed in ambient conditions. The solutions were prepared from anhydrous solvents in a nitrogen glovebox and were stirred at rt for ~12 hours before spin-coating the films.

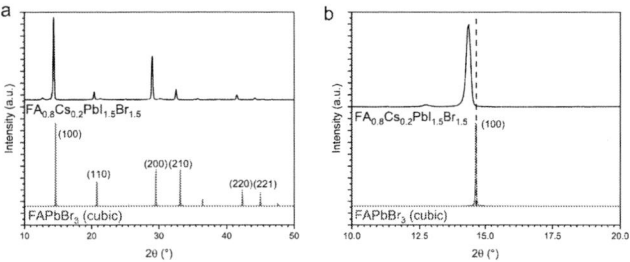

Fig. 1.   (a) Powder x-ray diffraction pattern of the perovskite film in this study (black) compared to a calculated pattern for the cubic phase of FAPbBr$_3$ (red) obtained from [15] (b) patterns from (a) focused in on the region 2θ between 10 - 20° to illustrate the shift to larger lattice spacing as expected for incorporation of iodide.

After spin-coating this solution onto ITO- and colloidal SnO$_2$- coated glass substrates, we included a five second immersion of the films in an antisolvent bath of chlorobenzene, then we annealed the films at 130 °C to complete the perovskite fabrication [13], [14].

We characterized the structure of our optimized material using PXRD and found evidence of a 3D perovskite phase with characteristic diffraction peaks at 14.4° and 29.0° 2θ (Fig. 1a).

We compared our sample to a calculated diffraction pattern for FAPbBr$_3$ [15] in a cubic phase. Diffraction peaks in our sample shifted to smaller degree 2θ, consistent with a lattice expansion due to the addition of larger iodide ions in this composition (Fig. 1b). The exact crystal structure and phase of this perovskite material is still under investigation.

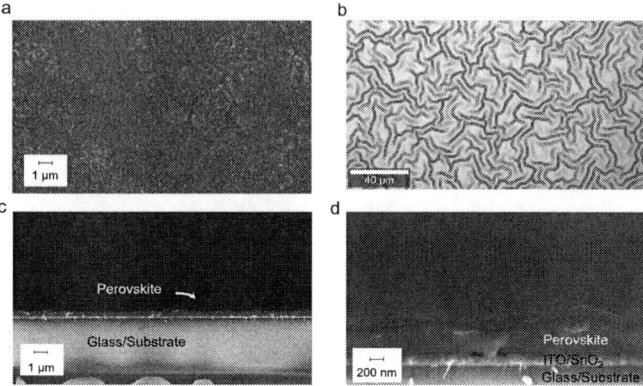

Fig. 2.   (a) SEM image of the surface of a perovskite film deposited on ITO/SnO$_2$. The scale bar represents 1 micron. (b) optical microscopy image of the surface of a perovskite film deposited on ITO/SnO$_2$. The scale bar represents 40 microns. (c) Cross-sectional SEM of the perovskite film deposited on ITO/ SnO$_2$. The scale bar represents 1 micron. (d) Cross-sectional SEM of the perovskite film deposited on ITO/SnO$_2$. The scale bar represents 200 nm.

We estimated the apparent grain size of the resulting films to be ~50 – 100 nm by SEM (Fig. 2a). Interestingly, we observed that films prepared by this method showed evidence of folding and wrinkling at the surface when viewed in an optical microscope. (Fig. 2b). In cross-sectional SEM images of pristine films coated on a substrate of ITO/SnO$_2$ we could observe the wrinkled morphology as a "hilly" landscape characterized by regions of variable thickness (Fig. 2c,d). Folding and wrinkling in perovskite films with similar mixed-halide compositions has been previously reported and is usually attributed to film stress during crystallization [16]. Film wrinkling has implications on the optoelectronic properties of perovskite films; however, reports indicate that films that contain wrinkling can still be effectively deployed in high efficiency PV cells [17].

In addition to the folding and wrinkling, we observed some pinholes in perovskite films at the interface between the perovskite and the ITO/SnO$_2$ (Fig. 2d). This could be due to the inadequate removal of precursor solvents during the anti-solvent submersion step, leading to the trapping of solvents and subsequent appearance of voids at the buried interface of the film after annealing [18]. It's likely that this effect could be mitigated by tuning the solvent precursor ratio and/or the anti-solvent processing procedure.

We characterized the optical properties of the thin-films by PL and UV-Vis absorption. Tauc analysis taken from UV-Vis experiments allowed us to estimate a bandgap of ~1.84 eV, assuming a direct, allowed transition (Fig. 3a). We estimated a film thickness of 450 nm for Tauc analysis by taking an average

978-1-6654-6060-6/23 $31.00 © 2023 IEEE

thickness over both thin and thicker portions of the wrinkled film. To account for scattering effects, we determined the bandgap using the intersection of two linear fits; one at the absorption edge and the other in the lower energy region (~1.6 – 1.7 eV) with no absorption.

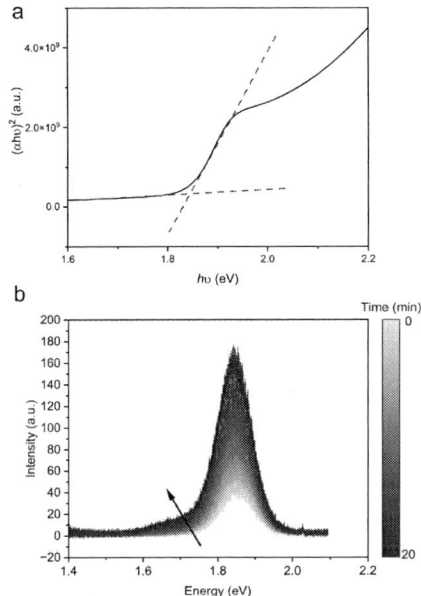

Fig. 3. (a) Tauc analysis of the perovskite film taken from the UV-Vis absorption assuming a direct, allowed transition, and a film thickness of 450 nm. (b) Time-dependent photoluminescence (PL) of the film conducted in ambient conditions with an excitation wavelength of 405 nm and an irradiation power of 12 mW cm^{-2}. The PL scans were taken sequentially every 20 s for a total of 20 min.

We additionally assessed the PL emission and illumination stability of films (Fig. 3b). For PL experiments, samples were measured in ambient atmosphere under constant 405 nm laser irradiation at a power density of 12 mW cm^{-2}. To assess phase stability, multiple PL spectra were taken sequentially every 20 seconds for a total of 20 minutes. For this experiment, we prepared perovskite films on a glass substrate coated with ITO and a thin layer of colloidal SnO$_2$, and illuminated the film from its top, air-exposed surface. Thus, some charge extraction by the conducting substrate and non-radiative interfacial recombination is expected to weaken the measured PL of the samples.

We observed PL from the films centered around 1.85 eV, consistent with the estimated bandgap of 1.84 eV (Fig. 3b). However, we observed that under these conditions, the PL showed characteristic red shifting of the emission to ~1.67 eV under constant illumination, consistent with halide phase segregation (Fig. 3b). Halide-segregation is implicated as a factor in overall device stability and V_{OC} losses [19]. These experiments were conducted in ambient conditions. It has been reported that water exposure in similar perovskite compositions accelerates film degradation and segregation kinetics [19]. Thus, we observed the rates of phase segregation in our films

were highly variable based on the overall environment of measurement.

B. Device Characterization

Devices were fabricated on a glass substrate with the architecture of glass/ITO/SnO$_2$/Perovskite/PTAA/Au (Fig. 4). All layers in the film were solution processed using procedures adapted from the literature [8], [13], [14]. Devices were then transferred to a custom-built air free sample holder and transported to a separate glovebox where the Au contact was evaporated using a thermal evaporator. We measured all devices under simulated one sun (AM1.5G solar irritation with 100 mW cm^{-2} at 25 °C). We determined that the V_{OC} (< 0.5 V) of our initial devices was much lower than the V_{OC} predicted by the detailed balance limit for PVs based on this bandgap (~1.549 V). We hypothesized this discrepancy was due to defect-mediated recombination occurring at device interfaces due to inadequate material optimization. To test our hypothesis, we implemented two interventions in our device configuration at the electron transport material and the hole transport material.

To address the interface between SnO$_2$ and the perovskite layer, we deployed an interlayer material that has shown success in similar wide-bandgap solar cells, C$_{60}$-SAM (4-(1′,5′-dihydro-1′-methyl-2′H-[5,6]fullereno-C$_{60}$-I$_h$-[1,9-c]pyrrol-2′-yl)benzoic acid) [8]. We found that the inclusion of the C$_{60}$-SAM layer enhanced the V_{OC} of devices to ~810 mV (Table 1). The average power conversion efficiency (PCE) for this device stack was 1.35% (standard deviation = 1.03%, n=9) in the forward scan direction, and 1.53% (standard deviation = 1.27%, n=9) in the reverse direction. We hypothesize that this observation can be attributed to an energetically improved electron extraction at this interface, passivation of non-radiative recombination at this interface, or both [20], [21].

TABLE I
DEVICE METRICS WITH C$_{60}$-SAM INTERLAYER AND UN-DOPED PTAA

Forward Scan		Reverse Scan	
J_{sc} (±SD) (mA/cm^2)	V_{oc} (±SD) (mV)	J_{sc} (±SD) (mA/cm^2)	V_{oc} (±SD) (mV)
4.98 ± 3.62	817 ± 209	4.90 ± 3.59	862 ± 227

At the hole transport material, we sought to dope PTAA with 2,3,5,6-tetrafluoro-tetracyanoquinodimethane (F4TCNQ) to improve the poor intrinsic conductivity of PTAA [22] and enhance charge extraction at that interface. Table II depicts the device metrics for PVs that contained PTAA with four F4TCNQ doping concentrations, 52% (n=15), 39% (n=5), 26% (n=13), and 13% (n=11). These devices additionally employ the C$_{60}$-SAM layer. Doping concentrations reflect the amount of F4TCNQ added to the polymer solution before spin coating. We observed that devices that deployed both the doped PTAA layers and the C$_{60}$-SAM interlayer showed enhancement in the V_{OC}, with some devices displaying voltages >1.1 V. The effect of PTAA doping concentration did not uniformly influence

TABLE II
DEVICE METRICS AT DIFFERENT PTAA:F4TCNQ % DOPING

F4TCNQ doping (%)	Forward Scan			Reverse Scan		
	J_{sc} ($\pm$SD) (mA/cm^2)	V_{oc} ($\pm$SD) (mV)	PCE ($\pm$SD) (%)	J_{sc} ($\pm$SD) (mA/cm^2)	V_{oc} ($\pm$SD) (mV)	PCE ($\pm$SD) (%)
52	11.48 $\pm$ 5.66	650 $\pm$ 37	2.95 $\pm$ 1.44	11.83 $\pm$ 5.70	617 $\pm$ 60	2.39 $\pm$ 0.97
39	11.12 $\pm$ 3.64	612 $\pm$ 39	1.93 $\pm$ 0.48	11.66 $\pm$ 4.16	632 $\pm$ 87	1.84 $\pm$ 0.47
26	11.76 $\pm$ 5.05	872 $\pm$ 132	3.98 $\pm$ 2.48	11.9 $\pm$ 4.61	913 $\pm$ 101	4.05 $\pm$ 2.09
13	10.29 $\pm$ 3.57	1072 $\pm$ 19	4.32 $\pm$ 1.44	10.35 $\pm$ 3.53	1039 $\pm$ 28	5.01 $\pm$ 1.63

device properties. We found that devices displayed optimal V_{OC} in the case of F4TCNQ solution of 13%; an average V_{OC} of 1.072 V and an average J_{sc} of 10.29 mA cm^{-2} were obtained (Table II).

We observed substantial variability in devices with this perovskite composition across multiple experiments conducted over the course of many months (see more discussion in section C). Importantly, at this juncture, we are careful not to directly compare the relative enhancement of the device V_{OC} independently with C$_{60}$-SAM and doped PTAA. While our data support the conclusion that both the addition of C$_{60}$-SAM and doping PTAA resulted in enhancements to device V_{OC}, we cannot determine which intervention was more impactful.

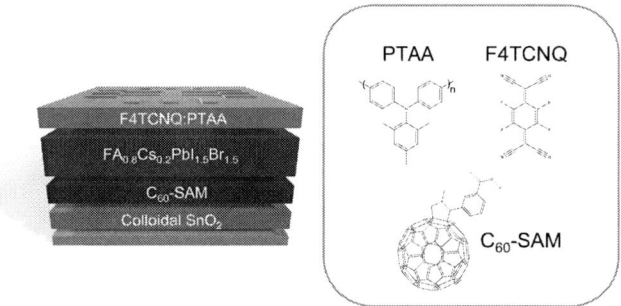

Fig. 4. A schematic of the fully optimized device in this work, and chemical structures of the transport materials

We note, it is also possible that there is interplay between both interfaces such that modifying one is only influential (or measurable) when extraction at both interfaces is sufficient. For example, in one experiment, devices that contained doped PTAA and *no* C$_{60}$-SAM displayed a very low V_{OC} of 298 mV (standard deviation = 106 mV). When the same devices included both doped PTAA and C$_{60}$-SAM, the V_{OC} increased to 970 mV (standard deviation = 122 mV). In a later experiment, we showed that, in the presence of C$_{60}$-SAM, tuning the doping density of PTAA can further refine the V_{OC} (Table II).

One possible explanation is that devices that do not contain C$_{60}$-SAM are entirely limited by poor extraction and recombination at the perovskite/SnO$_2$ interface. In this case, any effects of the relative doping of PTAA may not be captured by the observed device performance. Therefore, doping PTAA

might only be effective as a strategy for V_{OC} enhancement when the perovskite/SnO$_2$ interface is sufficiently addressed first. It's possible that the properties of the SnO$_2$ contact could be more influential for overall device characteristics because the SnO$_2$ constitutes the transparent electrode, therefore, electronic carrier generation is higher closer to that contact.

From this, we conclude that comparisons across devices that contain any number of interventions should be interpreted cautiously. Especially given the highly variable nature of perovskite properties in thin-films of this composition, we are careful to conclude only that both interfacial modifications appear to influence device output.

C. Challenges and Future Work

While we observed enhancements to device performance utilizing these interfacial modifications, challenges remain regarding developing these PV cells further. The observed halide phase segregation under illumination will be a challenge for the overall stability and V_{OC} of our PVs. Nonetheless, we anticipate that halide phase segregation effects can be mitigated through the careful optimization of device interfaces, defect engineering, and appropriate encapsulation [19].

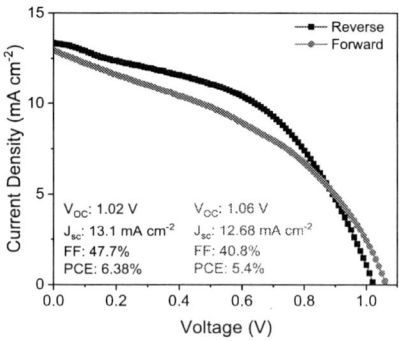

Fig. 5. Representative JV scans of a device with the architecture ITO/SnO$_2$/C$_{60}$-SAM/Perovskite/PTAA:F4TCNQ (13%)/Au

Additionally, these devices are characterized by low fill factors (< 50%) and J_{SC} (Fig. 5 & Table II). The maximum J_{SC} predicted by the detailed balance limit for this bandgap is 18.2 mA cm^{-2}. We are currently working to optimize film fabrication

978-1-6654-6060-6/23 $31.00 © 2023 IEEE

conditions, thicknesses, and interfacial energy alignments to enhance these metrics in our PV devices. We are additionally exploring the use of precursor and interfacial additives to further enhance device metrics.

Finally, we have observed that these films display a high degree of variability in overall properties and PV performance when we attempt to replicate process conditions. While reproducibility is a well-documented challenge in perovskite literature, this nominal composition appears to be especially sensitive to overall conditions of fabrication. We are currently working to better understand the role of fabrication environment, precursor material purity, and precursor stoichiometry on the crystalline and optoelectronic properties of these films.

III. CONCLUSIONS

We showed the fabrication and characterization of a hybrid formamidinium-based perovskite material with the band gap 1.84 eV, which is substantially wider than the band gap of materials used in the most widely investigated perovskite PVs (up to 1.75 eV). We successfully deployed this material for use in single-junction PVs. We discovered that the operation of resulting devices was significantly influenced by the interfacial contacts of devices. By adjusting the doping levels of PTAA, we could adjust the V_{OC} and J_{sc} of devices to achieve improved performance. We also observed significant enhancement of the V_{OC} on the addition of a C_{60}-SAM interlayer between the SnO_2 and the perovskite layer. Further analysis into these devices will yield insights that will allow us to continue to optimize these materials and resulting PV devices.

ACKNOWLEDGMENT

This research was performed while E. Smith held a National Research Council (NRC) Research Associateship Award (RAP) at the U.S. Naval Research Laboratory. The Office of Naval Research (ONR) is acknowledged for financial support of this work.

REFERENCES

[1] H. Kim, C. Lee, J. Im, K. Lee, T. Moehl, A. Marchioro, S. Moon, R. Humphry-Baker, J. Yum, J. Moser, M. Gratzel, and N. Park, "Lead Iodide Perovskite Sensitized All-Solid-State Submicron Thin Film Mesoscopic Solar Cell with Efficiency Exceeding 9%," *Scientific Reports*, 2, pp. 591, 2012.

[2] J. Hong Noh, S. Hyuk Im, J. Hyuck Heo, T. Mandal†, and S. Seok, "Chemical Management for Colorful, Efficient, and Stable Inorganic–Organic Hybrid Nanostructured Solar Cells," *Nano Letters*, 13, pp. 1764-1769, 2013.

[3] W. Yoon, J. Boercker, M. Lumb, J. Tischler, P. Jenkins and R. Walters, "Vapor deposition of organic-inorganic hybrid perovskite thin-films for photovoltaic applications," *2014 IEEE 40th Photovoltaic Specialist Conference (PVSC)*, pp. 1577-1580, 2014.

[4] J. Tong, Q. Jiang, F. Zhang, S. Kang, D. Kim, and K. Zhu, "Wide-Bandgap Metal Halide Perovskites for Tandem Solar Cells," *ACS Energy Letters*, 6, pp. 232-248, 2020.

[5] W. Yoon, Z. Song, C. Chen, D. Scheiman and Y. Yan, "21.1% Efficient Space Perovskite/Si Four-Terminal Tandem Solar Cells," *2020 47th IEEE Photovoltaic Specialists Conference (PVSC)*, pp. 1552-1556, 2020.

[6] W. Yoon, D. Scheiman, Y. Ok, Z. Song, C. Chen, G. Jernigan, A. Rohatgi, Y. Yan, and P, Jenkins, "Sputtered indium tin oxide as a recombination layer formed on the tunnel oxide/poly-Si passivating contact enabling the potential of efficient monolithic perovskite/Si tandem solar cells," *Solar Energy Materials and Solar Cells,* 210, pp. 110482, 2020.

[7] Y. Tong, A. Najar, L. Wang, L. Liu, M. Du, J. Yang, J. Li, K. Wang, S. Liu "Wide-Bandgap Organic-Inorganic Lead Halide Perovskite Solar Cells," *Advanced Science*, 14, pp. 2105085, 2022.

[8] Z. Song, C. Li, C. Chen, J. McNatt, W. Yoon, D. Scheiman, P. Jenkins, R. Ellingson, M. Heben, and Y. Yan, "High Remaining Factors in the Photovoltaic Performance of Perovskite Solar Cells after High-Fluence Electron Beam Irradiations," *Journal of Physical Chemistry C*, 2, pp. 1330-1336, 2019.

[9] K. Bush, K. Frohna, R. Prasanna, R. Beal, T. Leijtens, S. Swifter, and M. McGehee, "Compositional Engineering for Efficient Wide Band Gap Perovskites with Improved Stability to Photoinduced Phase Segregation," *ACS Energy Letters*, 3, pp. 428-435, 2018.

[10] R. He, S. Ren, C. Chen, Z. Yi, Y. Luo, H. Lai, W, Wang, G Zeng, X. Hao, Y. Wang, J. Zhang, C. Wang, L. Wu, F. Fu, and D. Zhao, "Wide-bandgap organic–inorganic hybrid and all-inorganic perovskite solar cells and their application in all-perovskite tandem solar cells," *Energy & Environmental Science*, 11, pp. 5723-5759, 2021.

[11] N. Kotulak, Z.Almutawah, S. Watthage, A. Phillips, D. Zhao, K. Knipling , P.Jenkins, R. Walters, Y. Yan, M. Heben, and W. Yoon, "3D imaging compositional map in one-step growth of $CH_3NH_3PbI_3$," *2018 IEEE 7th World Conference on Photovoltaic Energy Conversion (WCPEC) (A Joint Conference of 45th IEEE PVSC, 28th PVSEC & 34th EU PVSEC)*, pp. 0485-0489, 2018.

[12] Y. Yuan, G. Yan, R. Hong, Z. Liang, and T. Kirchartz, "Quantifying Efficiency Limitations in All-Inorganic Halide Perovskite Solar Cells," *Advanced Materials*, 34, pp. 2108132, 2022.

[13] Q. Jiang, L. Zhang, H. Wang, X. Yang, J. Meng, H. Liu, Z. Yin, J. Wu, X. Zhang, and J. You, "Enhanced electron extraction using SnO_2 for high-efficiency planar-structure $HC(NH_2)_2PbI_3$-based perovskite solar cells," *Nature Energy*, 2, pp. 16177. 2017.

[14] Y. Zhou, M. Yang, W. Wu, A. Vasiliev, K. Zhu, and N. Padture, "Room-temperature crystallization of hybrid-perovskite thin films via solvent–solvent extraction for high-performance solar cells," *Journal of Materials Chemistry A*, 3, pp. 8178-8184, 2015.

[15] E. Mozur, J. Trowbridge, A. Maughan, M. Gorman, C. Brown, T. Prisk, J. Neilson, "Dynamical Phase Transitions and Cation Orientation-Dependent Photoconductivity in $CH(NH_2)_2PbBr_3$," *ACS Materials Letters*, pp. 260-264, 2019.

[16] S. Kim, J. Kim, P. Ramming, Y. Zhong, K. Schötz, S. Kwon, S. Huettner, F. Panzer, N. Park, "How antisolvent miscibility affects perovskite film wrinkling and photovoltaic properties," *Nature Communications*, pp. 1554, 2021.

[17] S. Braunger, L.. Mundt, C. Wolff, M. Mews, C. Rehermann, M. Jošt, A. Tejada, D. Eisenhauer, C. Becker, J. Guerra, E. Unger, L. Korte, D. Neher, M. Schubert, B. Rech, S. Albrecht, "$Cs_xFA_{1-x}Pb(I_{1-y}Br_y)_3$ Perovskite Compositions: the Appearance of Wrinkled Morphology and its Impact on Solar Cell

Performance," *Journal of Physical Chemistry C*, pp. 17123-17135, 2018.

[18] S. Chen, X. Dai, S. Xu, H. Jiao, L. Zhao, J. Huang, "Stabilizing perovskite-substrate interfaces for high-performance perovskite modules," *Science*, pp. 902-907, 2021.

[19] A. Knight, L. Herz, "Preventing phase segregation in mixed-halide perovskites: a perspective," *Energy & Environmental Science*, pp. 2024-2046, 2020.

[20] A. Abrusci, S. Stranks, P. Docampo, H. Yip, A. K.-Y. Jen, H. Snaith, "High-Performance Perovskite-Polymer Hybrid Solar Cells via Electronic Coupling with Fullerene Monolayers," *ACS Nano*, pp. 3124-3128, 2013.

[21] K. Wojciechowski, S. Stranks, A. Abate, G. Sadoughi, A. Sadhanala, N. Kopidakis, G. Rumbles, C. L, R. Friend, A. K.-Y. Jen , H. Snaith, "Heterojunction Modification for Highly Efficient Organic–Inorganic Perovskite Solar Cells," *ACS Nano*, pp. 12701-12709, 2014.

[22] Q. Wang, C. Bi, J. Huang, "Doped hole transport layer for efficiency enhancement in planar heterojunction organolead trihalide perovskite solar cells," *Nano Energy*, pp. 275-280, 2015.

Multiple-reuse of Ge Substrates: Towards Cost-effective and Sustainable III-V Solar Cells Fabrication.

Alexandre chapotot, Bouraoui Ilahi, Tadeáš Hanuš, Gwenaëlle Hamon, Jinyoun Cho, Kristof Dessein, Christian Dubuc, Maxime Darnon, Abderraouf Boucherif

Institut Interdisciplinaire d′Innovation Technologique (3IT), Université de Sherbrooke, 3000 Boulevard Université, J1K 0A5, Sherbrooke, QC, Canada

Laboratoire Nanotechnologies Nanosystèmes (LN2) - CNRS IRL-3463, Institut Interdisciplinaire d′Innovation Technologique (3IT), Université de Sherbrooke, 3000 Boulevard Université, J1K 0A5, Sherbrooke, QC, Canada

Umicore Electro-Optic Materials, Watertorenstraat 33, 2250, Olen, Belgium

Saint-Augustin Canada Electric Inc. 75 rue d'Anvers, G3A 1S5, Saint-Augustin, QC, Canada

High efficiency solar cells based on III-V layers on Ge substrates are commonly used for space devices, but their high cost, partly due to Ge substrate, make them less suitable for terrestrial photovoltaic applications. The porous lift-off process offers a solution for detaching a thin, flexible, and lightweight cell and reusing the substrate, potentially making these solar cells more cost-effective. To date, only one reuse of the substrate has been proven. However, further examination of multiple reuse cycles is crucial in understanding the trends in the surface of the wafer over cycles. The HF-based wet chemistry employed in this study, allows for the substrate to be reused three times by removing the broken pillars that remain after each porous lift-off cycle. The reusability of the substrate is guaranteed due to the smooth and blemish-free surfaces with low RMS roughness achieved through reconditioning. Even after three porous lift-off cycles, the epitaxial layer remains monocrystalline with a roughness of only 3 nm, making it suitable for solar cell growth. These findings open up the possibility for a more affordable and sustainable method of producing high efficiency solar cells, reducing costs by a factor of 4 and Ge consumption by a factor of 19 compared to traditional manufacturing methods.

Mitigation of Potential Induced Degradation in Perovskite Solar Cells

Laxmi Nakka, Shen Guibin, Armin G. Aberle, Fen Lin

Department of Electrical and Computer Engineering, National University of Singapore, Singapore, Singapore

Solar Energy Research Institute of Singapore, National University of Singapore, Singapore, Singapore

Potential induced degradation (PID) is one of the failure mechanisms of photovoltaic (PV) modules which arises when the devices are exposed to high relative potential causing leakage current and device failure. This problem is of serious concern in perovskite solar cells (PSCs), due to the prevalent ion migration and halide segregation under high-voltage stress. However, there is so far very limited research on this topic. In this work, for the first time, we demonstrate the mitigation of PID in PSCs by introducing NiOx as a blocking layer between the ITO and the self-assembled monolayer. Our results show that the modified devices are less susceptible to PID and retain over 65% of the initial PV efficiency, whereas the control devices with PID only retain about 27%. We find that NiOx behaves as a blocking layer suppressing the Na+ ion migration from the glass into the active solar cell layers. The results of this work are useful for the development of reliable and PID-free perovskite solar cells, possibly providing a pathway for future mass applications.

A Robust Approach for Daily Solar Irradiance Clustering

Roshni Agrawal
TCS Research, TRDDC
Tata Consultancy Services
Pune, India
◎ 0000-0002-7669-1309

Sivakumar Subramanian
TCS Research, TRDDC
Tata Consultancy Services
Pune, India
◎ 0000-0003-0284-4744

Shashank Agarwal
TCS Research, TRDDC
Tata Consultancy Services
Pune, India
◎ 0000-0003-4835-9592

Venkataramana Runkana
TCS Research, TRDDC
Tata Consultancy Services
Pune, India
◎ 0000-0002-3609-5529

Abstract—Variability in solar irradiance is a major concern for power utilities and island grids. Clustering of historical irradiance data can help develop a better understanding of the patterns in a given location. This work uses a simple, yet effective scalar metric named Daily Irradiance Index (DII) obtained by comparing the measured irradiance level with the clear sky irradiance level. We explore a two-layered clustering approach for a data set from NREL's Solar Radiation Research Laboratory for Golden, Colorado. The first layer is designed to identify the seasons and in the second layer seasonal days are clustered based on the irradiance levels as measured by DII. For the first layer of clustering, we find the use of attributes from clear sky irradiance is sufficient to clearly identify the seasons. In the second layer of clustering, based on the cloud activities, days are clustered into high, medium and low irradiance levels corresponding to clear, intermittent, and cloudy days. The quality of the clusters is monitored through the well-known Silhouette score, Calinski-Harabasz score and Davies-Bouldin score. The effectiveness of employing clear sky irradiance, specifically sunshine duration, as a means of identifying solar seasons was demonstrated.

Index Terms—Daily Global Horizontal Irradiance; Daily clear sky irradiance; K-Means Clustering; Daily Irradiance Index

I. INTRODUCTION

The variability in solar irradiance has a direct impact on the generated power from a solar plant. This variability in solar irradiance at ground level is caused by various factors, such as water vapor in the air, aerosol/dust levels and cloud activity. Among these, the cloud plays a dominant role in interacting with radiation.

The variability in the incident irradiance not only causes fluctuations in photovoltaic (PV) power generation, but also poses challenges in terms of power quality concerns in voltage instability, unscheduled frequent outages that are difficult to plan and compensate. Hence, accurate prediction of PV generation is desirable for better operation and management of the grid. To deal with the uncertainty in solar power generation, support from ancillary services such as storage capacities are suggested. For island grids with significant reliance on solar power, robust estimates of the power availability become more critical to plan for energy storage or procurement from alternative sources.

Forecasting of solar irradiance is particularly difficult due to its tight coupling to local and global weather-related parameters such as cloud cover, water vapor, wind speed,

temperature and aerosols. Clustering of the past measured irradiance profiles helps in grouping similar days and analyze them for building robust forecasting models [1]–[3]. In this work, we recommend an elegant and a simple metric, referred here as Daily Irradiance Index (DII or β) to characterize the observed daily irradiance level. The β compares the profile of the measured irradiance curve to the corresponding clear sky irradiance (CSI) profile taking the area under profiles for quantifying them. As we shall see, this normalized index works well for both seasonal and annual clustering analysis.

We adopt a two-layered clustering framework. In the first layer of clustering, seasonal boundaries are established, and in the second layer, DII-based clustering is used to group the daily irradiance levels in a season. The automation of the season identification through an unsupervised clustering algorithm removes any need for rule-based season definitions. For this purpose, clear sky irradiance profiles are employed instead of measured irradiance which has been used in many of the earlier works [4], [5]. The main reason being variations observed in the measured irradiance normally makes this process more difficult. The season identification while in itself might be of some interest, also has the benefit of reducing the computational complexity of similarity metrices (distances) such as Euclidean distance (ED), and Dynamic Time Warping (DTW) in the second layer of clustering as only the days in a season need to be compared. Typically, three seasons were identified, namely 'Winter', 'Transition' and 'Summer'. Transition is used to mark both the periods between Winter to Summer and Summer to Winter.

The second layer of clustering that primarily depends on DII groups days into 'High' (H), 'Medium' (M) and 'Low' (L) levels of irradiance in each season. The performance of clustering is evaluated using three performance metrics, namely Silhoutte score, Calinski-Harabasz score and Davies-Bouldin score [6]. The irradiance patterns of these clusters can be useful in building better irradiance forecasting models factoring periods of significant variability. Thus, the clustering analysis is useful in developing an understanding of the power generation capabilities of a selected site, planning for their upgrade and sizing storage capacities to meet a given or projected power demand.

The paper is organized as follows. Section II discusses

978-1-6654-6060-6/23 $31.00 © 2023 IEEE

selected prior art from the clustering literature, categorizes them based on their approach and presents their merits and demerits. Section III provides details of the irradiance database and representation used. Section IV and section V presents the two-layer clustering framework and its results. Finally, we conclude the paper in section VI.

II. LITERATURE REVIEW

The irradiance intensity has been analyzed in numerous studies previously. The deterministic solar geometry and non-deterministic atmospheric extinction are considered separately in the two most commonly used transmissivity measures such as clearness index and clear-sky index. These are instantaneous, normalized, and dimensionless indicators of solar irradiation at the ground-level. The clearness index is defined as the ratio of measured irradiance at ground level to the approximated extra-terrestrial irradiance and has been used in [7], [8]. The effect of diurnal and annual cycles on solar irradiance data is reduced by using the clearness index. Considering the impact of atmospheric extinction, the clear sky index is defined as a ratio of measured irradiance to the modeled irradiance that is beam irradiance of a clear sky day. However, the changes in the mean clear sky index are not always linearly related to cloud activity. Clear sky index has been used mostly to process measured irradiance to a normalized time-series in [1], [3], [9].

The daily clearness index (CI), is the ratio of the area under the measured irradiance and the area under the clear sky irradiance, and the variability index (VI), is the ratio of the length of the measured irradiance profile to the length of clear sky irradiance profile. These are used in [10]. The classification rules are formed based on the combinations of CI and VI to find five weather groups. Similarly, the same approach is applied and validated in [11]. To identify weather types using clustering analysis, Irradiance index is used in [12], which is same as the daily clearness index discussed earlier. They attempted to build forecasting models for each seasonal cluster. However, this work does not provide sufficient details on the clustering results, its strengths and performance. It can be observed that to signify the same index, different authors have chosen to use different names and it leads to confusion. Later in Section IV, we refer to this index as Daily Irradiance Index (DII or β) taking into consideration the daily irradiance level indicated by the index.

Researchers have clustered the irradiance using various approaches, and they can be categorized into four types as follows. The category (a) and (b) differs based on their approach to clustering – either clustering the daily profiles based on time-series distance or clustering the features derived from daily irradiance time-series, category (c) and (d) differs based on their approach to categorize the irradiance database in seasonal groups, – either based on season definitions or automated from clustering. Here, the daily profiles refer to daily measured irradiance time-series or daily indexed time-series such as daily clearness index and, daily clear sky index.

A. Clustering the daily profiles using time-series distance

This is a direct and commonly used approach for clustering the daily time series irradiance profiles. When there are N days in the irradiance dataset, the similarity measure is computed for each pair of days resulting in $N \times N$ matrix using distance metrics such as ED, MD or DTW. Later, unsupervised clustering is performed on the similarity matrix to cluster the days based on the distances. K-Means clustering is used to cluster the daily temporal cloud cover in [13] and daily direct fraction profiles in [14]. Daily Clearness Index profiles are clustered using K-Medoids clustering in [2] and using self-organizing maps and hierarchical clustering in [7].

The key drawback of this approach is the huge computational complexity associated with computing similarity matrix for the entire dataset. Moreover, depending upon the sunshine duration, the day-time daily profiles have different day-time samples. In such cases, the DTW distance metric is usually the preferred choice [1]. It has to be borne in mind that not only the number of DTW distance computations increases to the square of number of days to be clustered but also its dimension of confusion matrix increases with the increase in the temporal resolution. Hence, with DTW, the search for optimal time-warping path becomes more compute-intensive. Comparatively, ED as a similarity measure is more straightforward to compute.

B. Clustering based on features derived from daily profiles

To overcome some of the computational complexities in using the daily profiles, one may alternately use features extracted from daily profiles for clustering. Here, the days are clustered using an unsupervised clustering from a $(N \times M)$ feature matrix, where N is the number of days and M is the number of features. Typically, $M \ll N$. The success of this approach of clustering is sensitive to the features used. This approach was used to cluster similar geographies of solar potential in [8] using cloud cover dataset. A list of 13 features was clustered using K-Means, K-Medoids and Hierarchical clustering in [5]. The efforts should be directed to design features to represent the variability of irradiance time series well.

C. Two-layer clustering with fixed seasonal clusters

It is well known that solar irradiance shows systematic seasonal changes over the year. Thus, it may be used to partition the datasets into seasonal groups to reduce any ambiguity in the clustering of the irradiance profiles. Therefore, the first step should be to implement clustering to identify irradiance profiles representing seasons [12]. Later, these profiles can aid in building forecasting models and PV performance monitoring models. Here, the dataset is grouped into initial clusters of seasons based on the calendar rule for geography of interest. [4] split the irradiance database into four seasonal subsets based on the calendar rule and divided them into four clusters in each season using five irradiance features such as mean, standard deviation, skewness, kurtosis, and moving fluctuation intensity. Similarly, [9] partitioned the data into four seasons

978-1-6654-6060-6/23 $31.00 © 2023 IEEE

and each season's daily clear sky index profiles were clustered using Fuzzy C-Means clustering. K-Means clustering was used further for categorizing daily irradiance index of four seasons into three clusters in [12]. It is important to note that the seasonal calendar rules are location-specific and need careful inputs when dealing with varied locations.

D. Two-layer clustering with automated seasonal clusters

In this approach, clustering methods are applied in two layers. In the first layer, dataset is clustered to find seasonal or weather groups. Later, another clustering method is implemented in the second layer on the groups resulting from the earlier stage. This two-layer clustering methodology has the ability to automate seasonal distribution of groups in a dataset compared to manual or pre-defined categorization based on the calendar rule of a given geography. Moreover, these seasonal boundaries vary every year with astronomical conditions. These drawbacks have been overcome in this approach. Three seasons were identified using K-Means++ clustering on a feature, that is, total daily irradiance in [15]. Further, the variability time series of each season were clustered into three variability types using K-Means clustering on DTW distance-based similarity matrix. A 25-d vector, i.e. probability of occurrence, was used to categorize three seasons and then seasons were further categorized into four clusters each using hierarchical clustering in [7].

III. IRRADIANCE DATABASE

The day-time measured irradiance data for this study was obtained from Baseline Measurement System (BMS) located at National Renewable Energy Laboratory (NREL) Solar Radiation Research Laboratory (SRRL) in Golden, Colorado (39.742 North, 105.18 West, 1828.8 AMSL Elevation). In the present study, 1-minute sampled Global Horizontal Irradiance data from 1 Jan 2015 to 31 Dec 2018 is considered. These were obtained using Global CMP22 with a Kipp & Zonen pyranometer sensor [16]. The clear sky irradiance was computed using PVLIB Toolbox using *Location.getclearsky(times)* method with geographical location and date range as inputs. The PVLIB python package is open-source software that was initially developed at Sandia National Laboratories [17].

Fig. 1(a) shows the clear sky irradiance at Golden, Colorado for the period January 2015 to December 2018. The seasonal cycles are clearly visible in this figure. It can be observed that the irradiance intensity in a day and sunshine duration vary throughout the year. The measured irradiance shown in Fig. 1(b) shows significant variation due to factors such as cloud, for the same period for the location. Both the clear sky and measured irradiance data were collected with a sampling time of 1-minute.

Interestingly, on some of the days, the measured irradiance exceeds the clear-sky irradiance. The observed drop in irradiance, and the time of occurrence varies. In the selected data set, the maximum (of peaks in) clear sky irradiance and measured irradiance are approximately 1058 W/m² and 1488 W/m², respectively. This difference could be explained

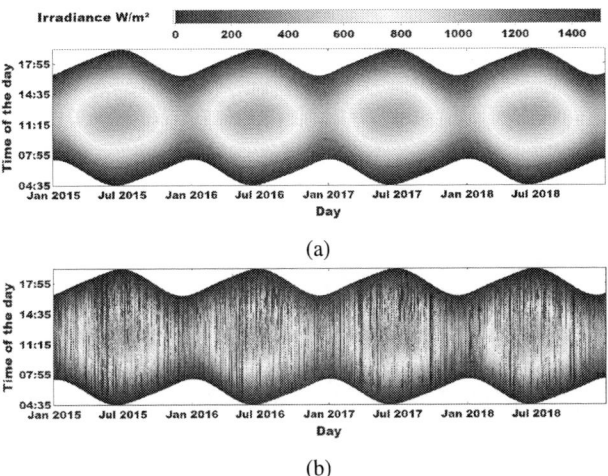

(a)

(b)

Fig. 1: Instantaneous irradiance for Golden Colorodo for the period of Jan 2015 to Dec 2018 with a sampling frequency one minute: (a) Clear sky irradiance, and (b) Measured irradiance at SRRL BMS

by cloud enhancement or irradiance enhancement. Due to cloud enhancement, instantaneous irradiance levels can be occasionally higher than they are on clear sky irradiance that is computed without considering the effect of passing clouds. This has been attributed to the reflection of radiation on the edges of the clouds [15].

IV. METHODOLOGY

As discussed earlier, in this work, a two-layer clustering approach is adopted. In the first layer of clustering, seasonal boundaries are identified using features extracted from clear sky irradiance. Features of clear sky irradiance, specifically sunshine duration and area under the clear sky irradiance profile that represent the total irradiance are used. In the second layer of clustering, we employ Daily Irradiance Index (DII or β) as a measure to capture the irradiance level on a day. It is defined as:

$$\beta = \frac{\text{Area under the daily measured irradiance}}{\text{Area under the daily clear sky irradiance}}$$

The β values are expected to range from 0 to 1, the limits representing a fully cloudy day and a clear day, respectively. A larger difference between the area under clear sky irradiance and measured irradiance typically denotes a drop in solar potential due to cloud-related factors. As noted earlier, the upper limit of 1 represents a soft limit. On some days, the location may receive more irradiance than what is normally predicted using the clear sky model. It is hypothesized that β captures the irradiance level of any given day better than other aggregate measures, such as maximum, mean, and standard deviation. of measured irradiance profiles. Moreover, β is computed for each day separately and independently of other days in a dataset unlike distance-based similarity measures such as ED or DTW. In this aspect, it is similar to any other metric extracted from daily measured irradiance profile such as mean, and maximum. Note that its complement $(1 - \beta)$ may be used to indicate the cloud level on a given day. It is worth noting that [10], [11] had used a measure same as β along with the variability index to classify days into sunny, cloudy, and so on. Applying simple empirical rules, they have given ranges for classifying the days into five categories. In this work, β is used in conjunction with clustering algorithms to arrive at recommendations that shall help in drawing better boundaries for 'High', 'Medium', and 'Low' irradiance levels representing 'Clear', 'Intermittent', and 'Cloudy' days for a location.

In this work, K-Means clustering was used in both layers. This clustering method was employed by others for clustering solar irradiance, for example, [1], [3]. Here, the distance of an input daily object to all the centroids are computed and the daily object is assigned to a centroid with the minimum distance. As the iteration proceeds, the centroids are updated as a mean of all daily objects in its cluster.

In this work, we evaluate the performance of clustering using three metrics, namely Silhoutte (Silh) score, Calinski-Harabasz (CH) score and Davies-Bouldin (DB) score. Silhouette width quantifies both clustering cohesion and separation. It is defined as the average of the Silhouette coefficients of all the days. Silhouette coefficients for a daily object was calculated based on the distance between a daily object and other objects within the same cluster, and the distance between the same daily object and the objects in the nearest neighboring cluster. Silh score varies in the interval [-1,1]. When the clusters are apart from each other and well-defined, Silh score is closer to 1 [6]. Calinski-Harabasz (CH) index is also known as variance ratio criterion. It is the ratio of the sum of inter-clusters dispersion and of intra-cluster dispersion for all clusters, where dispersion is defined as the sum of distances squared. Larger values of CH scores denote well defined clusters. This measure is normally better suited for problems when the clusters are dense and well separated [6]. The DB score is defined as the average similarity measure of each cluster with its most similar cluster, where similarity is the ratio of within-cluster distances to between-cluster distances. Thus, clusters with better separation will have lesser similarity and it will result in a score close to zero, with a lower score indicating better clusters [6]. Better clusters will yield a Silh score closer to one, a larger CH score and a DB score close to zero.

V. Results and Discussion

In this section, results from the two levels of clustering for solar irradiance data of Golden, Colorado using the methodology described was presented and discussed.

A. Seasonality Detection

The daily irradiance data of 1460 days was grouped into three solar seasons using K-Means clustering. Sunshine duration and area under the daily clear sky irradiance were used as the features for clustering. Their spread is shown in Table I. It can be observed that the sunshine duration and area under

TABLE I: Monthly averages of clustering features from clear sky irradiance used in automated seasonality detection

Features	Jan	Feb	Mar	Apr	May	Jun	Jul	Aug	Sep	Oct	Nov	Dec
Sunshine Duration [min]	580	639	715	793	859	892	875	817	742	664	595	560
Area (Daily CSI) [kWh/m²]	3.12	4.38	6.06	7.63	8.59	9.04	8.8	7.85	6.39	4.76	3.36	2.73

the daily CSI range from 560 to 896 minutes and 2.73 to 9.04 kWh/m², respectively.

The solar seasons that can be categorized as 'Winter' (W), 'Summer' (S) and 'Transition' (T) are distinctly identified by the algorithm as shown in Fig. 2. This calendar view clearly identifies the contiguous sequence of seasons, namely W, T, S, T, and W with this pattern repeating year-after-year. This approach is useful in automating the detection of solar seasons in a year for a location. Our attempts to identify the seasons using the features obtained from measured irradiance resulted in ambiguous or multiple boundaries as shown in Fig. 2(b).

The clusters are clearly delineated when presented in the features space of the area under the clear sky irradiance profile and the sunshine duration as shown in Fig. 3(a). When the corresponding measured irradiance features were used the spread in the area under the profile showed significant variability in the area attribute (figure not shown due to space limitation). Interestingly, sunshine duration was able to guide the clusters clearly. When the area under the curve is replaced by the daily mean of measured irradiance, the clusters become ambiguous as can be seen in Fig. 2(b) and Fig. 3(b). These observations emphasize the need for adequate care in selecting the set of features when performing feature-based clustering.

B. Clustering with daily irradiance index (β)

First of all, the β values for each day in the data set was computed. For a selected few days covering the observed range of β, the daily measured irradiance profile, the clear sky profile along with the computed β are shown in Fig. 4(a). It is clear from these graphs that the measure captures the extent of irradiance level on any given day well with higher levels of β indicating clearer days. The β distribution is presented in Fig. 4(b) for the entire data set. The β varies from 0.1 to 1.2 with more frequently observed values ranging from 0.7 to 1.1. When monthly means of β were analyzed (figure not shown due to space constraints), they were found to vary in a narrower band of 0.7 to 0.9. Interestingly, the month of May showed the lowest mean β value perhaps due to higher level of cloud activity.

Using K-Means clustering, the β values were grouped for each season into three clusters indicating irradiance level as 'High' (H), 'Medium' (M), and 'Low' (L) representing clear, intermittent and cloudy days, respectively. The mean of irradiance profile of each of these clusters for the three seasons are shown in Fig. 5. The labels used in this plot follow the convention with the first letter indicating the seasonal cluster and the second letter indicating the irradiance level cluster. The mean profiles were constructed by taking the mean values of the irradiance levels of all the days in a cluster for every sampling instant (minute here).

978-1-6654-6060-6/23 $31.00 © 2023 IEEE

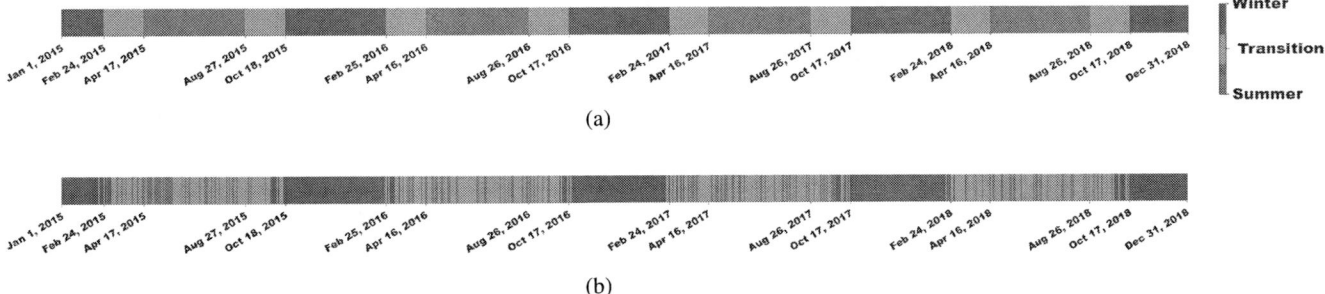

(a)

(b)

Fig. 2: Seasonal boundaries identified in Layer-1 with features: (a) Sunshine duration and area under the curve of CSI or the corresponding values from Measured Irradiance, and (b) Sunshine duration and daily mean irradiance

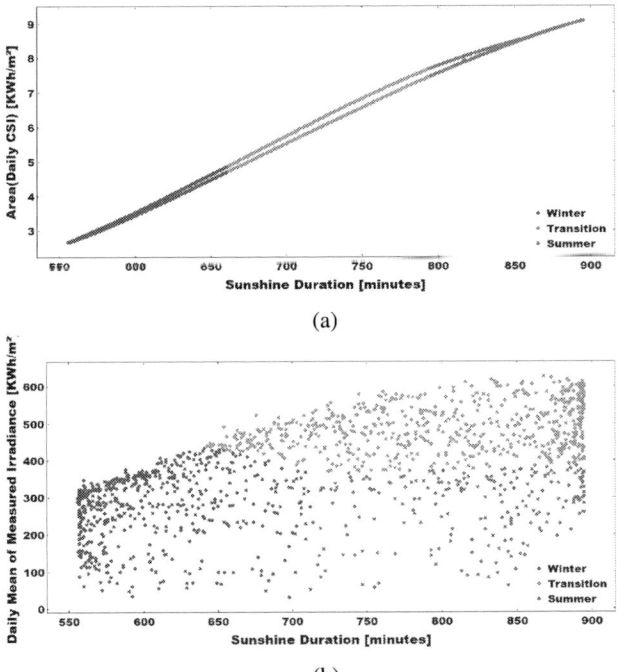

Fig. 3: The first layer of clustering identified seasonal clusters as shown in respective colors

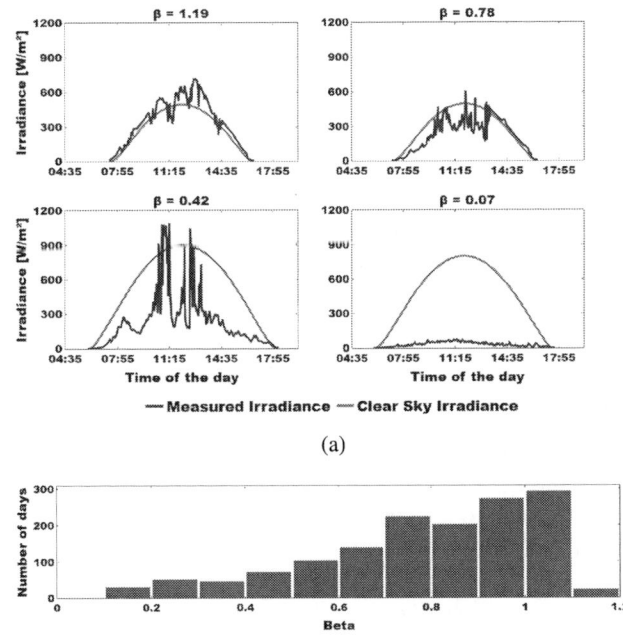

Fig. 4: (a) Comparison of CSI and measured irradiance for a few selected days in the data set along with computed β, (b) Distribution of beta values in the data set

As expected, days in 'Winter' have a shorter sunshine duration and days in 'Summer' have a longer sunshine duration. It is interesting to note that, peak irradiance 'WH' is lesser than 'SH', 'SM', 'TH', and 'TM'. This denotes the limited power generation capabilities of a clear day in Winter and such days might require additional grid support and storage capacities to meet the demand. As expected, the clear days in each season with a 'High' irradiance level had a smoother curve with limited variation. The mean profiles of cloudy days represented by their respective 'Low' irradiance level profiles are useful in forecasting typical solar power generation when the incident irradiance is limited for a given season.

The range of β for the second layer of clusters for the three

TABLE II: Range of β index in irradiance level-based clusters

Irradiance level	Winter	Transition	Summer	All seasons
High	$0.88 \leq \beta \leq 1.19$	$0.83 \leq \beta \leq 1.13$	$0.8 \leq \beta \leq 1.04$	$0.84 \leq \beta \leq 1.19$
Medium	$0.55 < \beta < 0.88$	$0.49 < \beta < 0.83$	$0.5 < \beta < 0.8$	$0.51 < \beta < 0.84$
Low	$0.1 \leq \beta \leq 0.55$	$0.07 \leq \beta \leq 0.49$	$0.1 \leq \beta \leq 0.5$	$0.07 \leq \beta \leq 0.51$

seasons is shown in Table II. Being a single variable-based clustering, the ranges specify non-overlapping definitions of these clusters. Considering that β is a normalized index, we attempted to use it to cluster the days into 'High', 'Medium', and 'Low' irradiance levels for all the days ignoring the seasons. Interestingly, the β ranges for the new clusters are not very different as shown in the table (column titled 'All

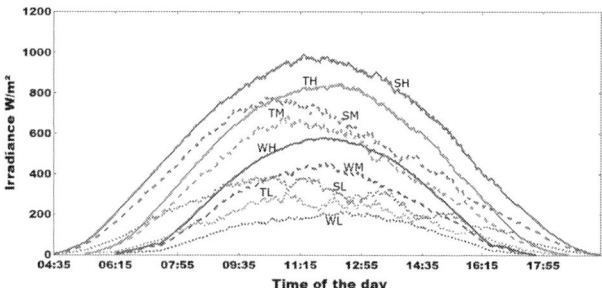

Fig. 5: Irradiance profiles representing time-series instantaneous mean of 9 clusters

TABLE III: Clustering performance metrics

Clustering Layer	Features	Name of group	Silhouette Score	CH Score	DB Score
Layer 1	Sunshine duration, Area CSI	W, T, S	0.63	7964.95	0.46
Layer 1	Sunshine duration, Area Measured Irr	W, T, S	0.63	7949.71	0.46
Layer 1	Sunshine duration, Daily Mean Irr	W, T, S	0.45	1557.82	0.83
Layer 2: Winter	β	WH, WM, WL	0.63	2102.35	0.48
Layer 2: Transition	β	TH, TM, TL	0.63	1401.87	0.48
Layer 2: Summer	β	SH, SM, SL	0.58	1619.80	0.51
All seasons	β	H, M, L	0.60	4893.63	0.49

seasons'). This demonstrates that β can be reliably used either to cluster irradiance levels for a season or for the complete year.

The calculated clustering scores are presented in Table III for both layers of clustering. The first layer clustering with sunshine duration and daily mean of irradiance as features have resulted in a poorer score. On the other hand, using sunshine duration with the area under measured irradiance as features in clustering shows better performance. It should be noted that sunshine duration combined with the area under the clear sky irradiance as features in clustering have shown improved performance as well as clearly delineated clusters. In layer-2 clustering for each season, the clusters were found to be well separated as evidenced by the scores presented.

After two layers of clustering, the irradiance level of the day is categorized into 'High', 'Medium' and 'Low' in each season. The information on the irradiance level of consecutive days is useful in deriving the transition probability of irradiance level in each season and each year. The 'Winter' season has lesser variability and has a higher probability of a 'High' irradiance level on the next day. The transition probability could be useful in building a forecasting model with some weightage to previous days' irradiance levels.

VI. CONCLUSION

The daily irradiance index, a direct measure that compares clear sky irradiance to the measured irradiance, was used to successfully cluster solar irradiance obtained from NREL SRRL in Golden, Colorado. This approach offers a more robust set of clusters when compared to the time-series based measures such as ED, DTW and MD that have been employed widely in the past. The efficacy of using clear sky irradiance, and in particular sunshine duration for solar seasons

identification was also brought out. The presented approach is generic and is expected to work for other locations.

REFERENCES

[1] C. S. Lai, Y. Jia, M. D. McCulloch, and Z. Xu, "Daily clearness index profiles cluster analysis for photovoltaic system," *IEEE Transactions on Industrial Informatics*, vol. 13, no. 5, pp. 2322–2332, 2017.

[2] B. Zhou, W. Zhao, X. Su, S. Lu, T. Wang, W. Yao, P. Xie, T. Mao, L. Guan, and Y. Lv, "Pv power characteristic modeling based on multi-scale clustering and its application in generation prediction," in *2018 IEEE power & energy society general meeting (PESGM)*. IEEE, 2018, pp. 1–5.

[3] P. Govender, M. J. Brooks, and A. P. Matthews, "Cluster analysis for classification and forecasting of solar irradiance in durban, south africa," *Journal of Energy in Southern Africa*, vol. 29, no. 2, pp. 51–62, 2018.

[4] X. Fu, F. Gao, J. Wu, X. Guan, X. Li, P. Liu, and P. Li, "A simulation method of solar irradiance data based on feature clustering and markov transition probability matrix," in *2018 13th World Congress on Intelligent Control and Automation (WCICA)*. IEEE, 2018, pp. 1741–1746.

[5] C. Feng, M. Cui, B.-M. Hodge, S. Lu, H. F. Hamann, and J. Zhang, "An unsupervised clustering-based short-term solar forecasting methodology using multi-model machine learning blending," *arXiv preprint arXiv:1805.04193*, 2018.

[6] Scikit-learn. Clustering metrics. [Online]. Available: https://scikit-learn.org/stable/modules/classes.html#module-sklearn.metrics

[7] T. Watanabe, K. Oka, and Y. Hijioka, "Assessment of characteristics of surface solar irradiance on consecutive days using a self-organizing map and clustering methods," *Meteorological Applications*, vol. 28, no. 2, p. e1984. [Online]. Available: https://rmets.onlinelibrary.wiley.com/doi/abs/10.1002/met.1984

[8] T. Watanabe, T. Takamatsu, and T. Y. Nakajima, "Evaluation of variation in surface solar irradiance and clustering of observation stations in japan," *Journal of Applied Meteorology and Climatology*, vol. 55, no. 10, pp. 2165–2180, 2016.

[9] C. S. Lai, Y. Jia, M. D. McCulloch, and Z. Xu, "Daily clearness index profiles cluster analysis for photovoltaic system," *IEEE Transactions on Industrial Informatics*, vol. 13, no. 5, pp. 2322–2332, 2017.

[10] C. Trueblood, S. Coley, T. Key, L. Rogers, A. Ellis, C. Hansen, and E. Philpot, "Pv measures up for fleet duty : Data from a tennessee plant are used to illustrate metrics that characterize plant performance," *IEEE Power and Energy Magazine*, vol. 11, no. 2, pp. 33–44, 2013.

[11] B. Hartmann, "Comparing various solar irradiance categorization methods–a critique on robustness," *Renewable Energy*, vol. 154, pp. 661–671, 2020.

[12] H. He, R. Hu, Y. Zhang, Y. Zhang, and R. Jiao, "A power forecasting approach for pv plant based on irradiance index and lstm," in *2018 37th Chinese control conference (CCC)*. IEEE, 2018, pp. 9404–9409.

[13] P. Govender and V. Sivakumar, "Investigating diffuse irradiance variation under different cloud conditions in durban, using k-means clustering," *Journal of Energy in Southern Africa*, vol. 30, no. 3, p. 22–32, Sep. 2019. [Online]. Available: https://energyjournal.africa/article/view/6314

[14] P. Jeanty, M. Delsaut, L. Trovalet, H. Ralambondrainy, J. Lan-Sun-Luk, M. Bessafi, P. Charton, and J.-P. Chabriat, "Clustering daily solar radiation from reunion island using data analysis methods," in *Proceedings of the International Conference on Renewable Energies and Power Quality (ICREPQ'13), Bilbao, Spain. https://doi. org/10.24084/repqj11*, vol. 340, 2013.

[15] I. Santiago, J. L. Esquivel-Martin, D. Trillo-Montero, R. J. Real-Calvo, and V. Pallarés-López, "Classification of daily irradiance profiles and the behaviour of photovoltaic plant elements: The effects of cloud enhancement," *Applied Sciences*, vol. 11, no. 11, p. 5230, 2021.

[16] N. R. E. Laboratory. Srrl bms daily plots and raw data files. [Online]. Available: https://midcdmz.nrel.gov/apps/day.pl?BMS

[17] W. F. Holmgren, C. W. Hansen, and M. Mikofski, "Pvlib python: a python package for modeling solar energy systems," *J. Open Source Softw.*, vol. 3, p. 884, 2018.

On the role of Sn-halide post-deposition reactive annealing for the passivation of defective surfaces in Cu2ZnSnSe4

Alex Jimenez-Arguijo, Yuancai Gong, David Nowak, Devendra Pareek, Levent Gütay, Lorenzo Calvo-Barrio, Zacharie Jehl Li-Kao, Sergio Giraldo, Edgardo Saucedo

Fundació Institut de Recerca Energética de Catalunya(IREC), Barcelona, Spain

Universitat Politècnica de Catalunya (UPC), Barcelona, Spain

University of Oldenburg, Oldenburg, Germany

University of Barcelona, Barcelon, Spain

The volatility of Sn-Se compounds can lead to detrimental surface termination of Cu2ZnSnSe4. The possible causes can be a high density detrimental defects or a surface decomposition during the cooling down stage. Furthermore, it has been recently proposed that the complexing agents for Cd2+ used in the CdS chemical bath deposition process can also degrade the surface of CZTSe, explaining the need of a specific CdS bath for a specific kesterite surface. The exploration of processes that can tune and passivate the surface and junction of kesterite based devices is of uttermost importance for their mass production. The use of Sn-Halides during a low temperature post deposition thermal treatment is proposed to passivate the surface of CZTSe as well as tuning the grain boundary composition and the hole density.

Analysis of Thermal Behavior and Reliability of Bare Die Diodes Embedded Within PV Modules as Bypass Devices

Luis Eduardo Alanis[1], Julian Weber[1], Pascal Romer[1], Marc Andre Schüler[1], Louisa Winkler[1], Udo Steinebrunner[2], Martin Heinrich[1], and Dirk Holger Neuhaus[1]

[1]Fraunhofer Institute for Solar Energy Systems, Heidenhofstr. 2, 79110 Freiburg, Germany
[2]Diotec Semiconductor AG, Kreuzmattenstr. 4, 79423 Heitersheim, Germany

Abstract— The installation of bypass devices embedded within a photovoltaic (PV) module is an increasingly attractive option for simplifying module design and manufacturing, offering an alternative to the current standard of connecting and gluing to the laminated module one or more external boxes containing the bypass diodes. However, certain technical challenges require further exploration for a reliable integration. We analyze the role of the geometry of the diode, and of the module's bill of materials (BOM) in dissipating heat when partial shading occurs on a PV array, putting the diode in forward bias. Critical parameters for module-integrated bare die diodes are defined such that its maximum operating temperature remains below 95 °C. Different approaches are explored experimentally and a Finite Element Method (FEM) model is created to predict said maximum temperatures. It is found that while the chip size is one main factor in dissipating heat efficiently, other methods for modifying the geometry of the diode-connectors subassembly can aid considerably in reducing the temperature of the device during operation, all without substantially increasing the size of the semiconductor. Connecting in parallel two or more small diodes, increasing the width of the connectors, and adding flat heat sinks are some of the proposed approaches. Furthermore, performance and reliability tests are executed to understand how degradation of the integrated diodes might affect the lifetime of the PV module, showing that when operation temperatures above 130 °C are maintained even for a relatively short time, some degradation of the electrical contacting of the diode will occur.

Keywords— *bypass diode, bare die diode, heat dissipation, partial shading, diode temperature, hot spot, shading*

I. INTRODUCTION

To avoid land use conflicts for PV installations and to utilize more available surfaces for renewable energy generation, there is a growing interest in integrating PV modules into infrastructure such as buildings, roads, or vehicles. This results in specific requirements for the modules in terms of design, reliability and safety in operation [1–9].

Partial shading of single solar cells within a PV module can cause hot-spots, and bypass diodes are a well-known approach to prevent damage in the event of shading [10, 11]. Comprehensive literature has been collected including investigations regarding the working principle, different diode configurations and their power characteristics [11].

However, the standard approach for incorporating bypass diodes into photovoltaic modules through an external junction box glued to its rear side might not be a practical solution for all integrated PV concepts due to space [11, 12] and module layout. This growing need for customization in module design, together with the emergence of new cell technologies, and the reduction in the size of electronic components, have made the embedding of diodes and bypass devices into a PV module an increasingly attractive option to simplify module design and manufacturing [12, 13].

While case studies claim failure rates of bypass diodes are neglectable when integrated in the junction box [14], others have seen them as one of the dominant failures in the first years of operation in PV plants [13, 15]. In this way, it is essential to understand the technical constraints and to establish a set of design guidelines that ensure the safety, resilience, and long-term functionality of integrated bypass devices and, therefore, the longevity of the module [11].

Knowing that the operating temperature of diodes is one of the main drivers of their functional lifespan, the thermal behavior of bare die diodes is assessed to obtain insights into their functionality and reliability for PV integration.

Due to the absence of casing, these unpackaged semiconductor chips can be manufactured for a relative low cost and with thicknesses that are close to that of a solar cell. These characteristics give bare die diodes great potential for being utilized as integrated bypass devices in PV modules, as they might offer a solution that is more affordable than alternative devices, and can also be implemented without altering the module's BOM and lamination parameters.

Different approaches are explored experimentally to limit the diode's heating when operating under critical conditions so that its maximum temperature does not reach the diode's breaking point or cause degradation of the surrounding materials.

A FEM model is created to predict the diode temperature under forward bias. This is done by calculating the power dissipated by each type of diode based on its measured current-voltage (IV) characteristic curve. A temperature coefficient of -1.6 mV/°C for the diode's forward voltage (V_f), given by the manufactuer, is used in the calculations to account for the temperature dependency of its electrical behavior.

The simulation model, which can be adapted to different diode geometries and configurations, BOMs, and boundary conditions, is adjusted to the different testing setups for validation of the predicted temperatures.

Lastly, functional parameters of the devices are assessed and compared before and after prolonged operation at high temperatures.

II. METHODS

The testing and experimental activities were conducted on bare die p-n junction Si diodes of various sizes with a rated thickness of 165 μm and a typical V_f of 0.85 V @ 10 A. These were supplied by Diotec Semiconductor AG and manufactured specially for this investigation.

According to the manufacturer, the junction temperature of the diode under long-term operation should not exceed 95 °C. Nevertheless, the device should be able to withstand up to 150 °C for a limited time. It must be noticed that 150 °C is within range of the processing temperature of PV module encapsulation materials, therefore, besides the possibility of damage to the diode assembly, degradation of the laminate could also be expected if one component within the stack reaches this temperature level.

Due to the small thickness of the metallization layer on the surface of the p-n diodes, which covers most of their area, a full contact with the connectors must be ensured. For this, two approaches are considered:

- Manual soldering with Pb-Sn solder paste
- Gluing with lead-free Electrically Conductive Adhesive (ECA)

The diode subassemblies are laminated in a glass-foil configuration and placed on top of a heating plate to simulate a specific module operating temperature. All measurements are recorded with the heating plate set to 25 °C and 75 °C, which are testing temperatures referenced in the section MQT18.1 of the test standard IEC 61215-2 that addresses bypass diodes [16].

The laminated devices are connected in forward mode to a power source that is set to drive a constant current, which is applied for approximately 20 minutes while the temperature is recorded.

The measuring procedure involves an IR camera for thermal imaging that senses the temperature on top of the laminate and its surroundings. Based on the observation that the diode's temperature reaches its peak and becomes stable between 5 and 13 minutes after starting the power source, the reported maximum temperatures are acquired after 15 minutes of testing.

The reported temperatures obtained via thermal imaging take into account the emissivity of the ETFE foil and thermal resistance of the front layers, hence offering a close approximation of the diode's junction temperature.

A. Single diode

For the first iteration in this series of tests, four square diodes with a chip size of 4.6 mm per side (21.16 mm²) are contacted

using the abovementioned approaches to tin-plated copper ribbons with a cross section of 5 x 0.3 mm².

Furthermore, half of the diodes are attached to a tinned copper plate with a nominal thickness of 100 μm and an area of 1,350 mm² that is intended to serve as a heat sink.

The contacted diodes are subsequently laminated in the configuration shown in Fig. 1. This is done in a plate-membrane laminator using a standard process for EVA, with a target temperature of 160 °C and 850 mbar of membrane pressure.

In the case of the diodes contacted with ECA, the glue is placed manually using a dispenser and cured only during the lamination process.

The testing is performed with currents of 3 A, 6 A, and 9 A.

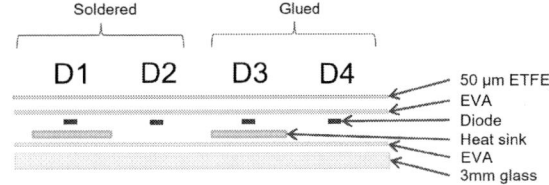

Fig. 1. Glass-foil lamination configuration for temperature measurements of single diodes.

A sample image of the temperature data recorded during the testing of the laminated diodes is shown in Fig. 2.

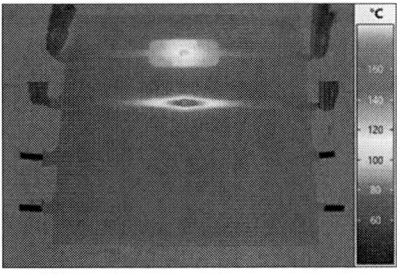

Fig. 2. Thermal image of diodes in forward bias with 9 A of current applied.

The geometry of the assembly, electrical characteristics of the diode, and material parameters are set on the FEM model to resemble the setup shown in Fig. 1, and the maximum temperature of the chip is computed. Additionally, a glass-glass configuration that replaces the front layer with a 3 mm glass is also simulated.

B. Parallelized diodes

It is anticipated that by connecting two or more diodes in parallel, the maximum operating temperature will decrease as the electric current will be split between the devices. Furthermore, the total dissipated power will be lower due to the resulting IV characteristic of the parallelized chips. This can be seen in Fig. 3, which exemplarily shows the IV curve of several individual diodes and the one resulting from parallelizing them.

978-1-6654-6060-6/23 $31.00 © 2023 IEEE

In this approach, however, it is required that all diodes to be connected in parallel have a similar V_f characteristic, otherwise, it is possible that one of the chips begins to dissipate more power than the rest, potentially leading to overload in forward direction. For this reason, a set of 6.2 x 6.2 mm² (38.44 mm²) uncased p-n junction diodes, pre-characterized and sorted by Diotec Semiconductor AG, is used for this experiment.

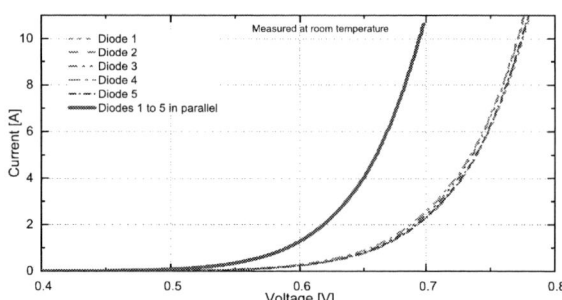

Fig. 3. IV characteristic curve of 5 individual 38.44 mm² diodes and their resulting combined curve after parallelization.

Groups of three diodes are contacted in parallel via soldering. The distance (d) between the diodes is taken as an additional parameter to consider, as well as the width (w) of the ribbon connector below the diode. Additionally, a piece of Kapton tape is placed in the space between diodes to avoid short circuit between the ribbons as shown in Fig. 4.

TABLE 1.
GEOMETRICAL PARAMETERS OF PARALLEL DIODE SUB-ASSEMBLIES AS SHOWN IN FIG. 4

Sample	d [mm]	w [mm]
01	3	5
02	5	5
03	5	10

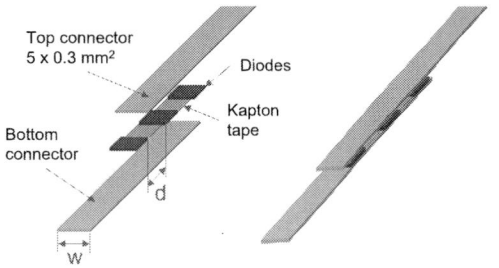

Fig. 4. Parallel diode sub-assembly in exploded (left) and condensed view (right).

Three sample assemblies are built with the geometrical parameters shown in Table 1, referenced to the assembly shown in Fig. 4, and subsequently laminated in the same glass-foil

configuration. Their temperature is then measured with the same method and current levels as on the single diode test.

The individual temperature of each diode is acquired from the thermal data (exemplified in Fig. 5).

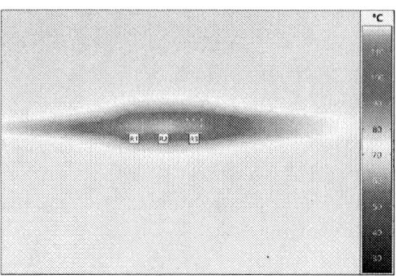

Fig. 5. Thermal image of parallel diodes (Sample 03) in forward bias with 9 A at 75°C.

These three variations are also simulated via the FEM model. Additionally, the single diode case is also simulated for this chip size (6.2 x 6.2 mm²) for a direct comparison.

C. Variation of chip size

The potential of reducing the heating of the diode by increasing its dimensions instead of the number of chips is considered in this test.

Samples of square bare die diodes a side length of 10 mm and 25 mm (100 mm² and 625 mm², respectively), also provided by Diotec Semiconductor AG, are contacted with manually dispensed ECA, which in this case is precured before lamination with applied heat and pressure to improve the quality of the connection.

Due to the size of the chips, standard PV ribbon connectors would not allow for a full contact with the metallization layer of the chips, therefore special connectors are designed and laser-cut from tin-plated copper foil with a width of 7 mm and a nominal thickness of 100 μm.

Furthermore, a heat sink with a flat area of 625 mm² is designed and cut from the same material and attached to one 10 mm diode with the same method. This provides a direct comparison with the area of the 25 mm diode.

All pre-contacted diodes are characterized, laminated with the same BOM from the previous experiments, and put through the same testing and temperature logging procedures. However, in this case, the experiments are performed with forward currents of 6 A, 9 A and 12 A to increase the rigor of the testing.

D. Full-size module integration

To investigate the behavior of integrated bare die diodes in a setup that resembles real-life operation, three chips with a length of 20 mm per side are laminated behind the cell area of a full-size module featuring shingle matrix interconnection technology [9, 17]. An EPE layer is used as insulation between the cells and bypass circuits and diodes, and a PET-based back sheet is used as a rear cover for the module. The connectors are

978-1-6654-6060-6/23 $31.00 © 2023 IEEE

rigged so that the diodes can be externally connected and disconnected for individual testing and characterization.

All diodes are characterized, and the module is brought up to 75 °C. The temperature behind the module is recorded via the attachment of themocouples while the diodes are forwardly biased at the short-circuit current (I_{SC}) of the module for one hour. Afterwards, the current is raised to the equivalent of 1.25 times the I_{SC} and the temperature is logged for one more hour.

III. RESULTS AND DISCUSSION

A. Single diode

For the laminated diodes at 25 °C and 75 °C, the maximum measured temperatures are entered in Table 2. The measurements for Diode 4 (glued with no heat sink) are interrupted due to heat-related failure while testing at 9 A.

TABLE 2.
MAXIMUM DIODE TEMPERATURE IN GLASS-FOIL
CONFIGURATION AT 25 °C AND 75 °C

Current [A]	Max. Temperature [°C]							
	D1		D2		D3		D4	
	At 25 °C	At 75 °C	At 25 °C	At 75 °C	At 25 °C	At 75 °C	At 25 °C	At 75 °C
3	49	88	63	104	51	87	64	102
6	72	108	101	138	74	122	111	ND
9	99	127	130	162	114	149	ND	ND

It is observed that the chips contacted using ECA tend to heat up more than their soldered counterparts, which is seen as a possible indicator of lower quality of the connection and differences in the thermal properties of the materials.

The measurements show that for this diode size, the maximum junction temperature is consistently surpassing the manufacturer's recommendation of 95 °C when operating in forward mode at 9 A, regardless of module temperature and addition of heat sinks, and in most cases when the initial module temperature is set to 75 °C.

It is seen, however, that the heat sink has a significant effect in removing heat from the diode, in some cases helping reduce the temperature of the stack by over 30 °C.

The simulation for the glass-glass configuration shows that the temperature of the diode without heat sink is lower (with a difference between 3 °C and 9 °C) than in glass-foil. This is due to the low thermal conductivity of the ETFE foil as compared to that of glass, which in this case prevents more heat dissipation towards the surrounding air, while the glass conducts heat away from the diode more efficiently and, due to its mass and heat capacity, aids in heat dissipation.

Fig. 6 shows a direct comparison between experimental and simulated maximum temperatures of a 4.6 x 4.6 mm² diode without heat sink.

It is seen that the simulated maximum temperatures are close to the values obtained experimentally, with a variation no larger than 5 °C between corresponding datapoints, which shows a high level of accuracy from the model.

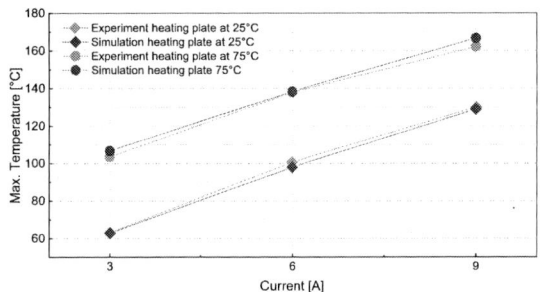

Fig. 6. Comparison between experimental and simulated maximum temperature reached by single 4.6 mm diode in glass-foil configuration without heat sink.

B. Parallelized diodes

The temperature of each individual diode is obtained from the thermal imaging and, as expected, it is observed that the temperature of the chips in the central position (see Fig. 5) is marginally higher (between 1 °C and 4 °C difference) throughout all the iterations of the experiment, as these are simultaneously influenced by the heat generated by two neighboring diodes.

An average temperature of all three diodes is calculated and used for further comparisons.

The maximum temperatures measured for the bundles of three diodes connected in parallel, while conducting 3 A, 6 A and 9 A at 25 °C and 75 °C, are summarized on Table 3 for the sample configurations detailed in Table 1.

TABLE 3.
MAXIMUM DIODE TEMPERATURE FOR THREE PARALLEL DIODES
IN GLASS-FOIL CONFIGURATION AT 25 °C AND 75 °C

Current [A]	Max. Temperature [°C]					
	Sample 01		Sample 02		Sample 03	
	At 25 °C	At 75 °C	At 25 °C	At 75 °C	At 25 °C	At 75 °C
3	47	86	46	83	45	81
6	66	104	62	97	62	97
9	92	116	79	113	79	108

The results of the simulation are consistent with what is seen on the experiment, with both the larger distance between diodes (comparing Sample 01 with Sample 02) and the additional width of the connector (comparing Sample 02 with Sample 03) having the desired effect of lowering the diode temperature. The increase in width of the connector does not seem to have a large influence at low module temperature, but at higher set temperature it may lead to more efficient heat dissipation.

Comparing the 3 parallel diodes to the single diode case, an improvement is seen after the parallelization, but the overall change in temperature is not much more significant than that achieved when using a heat sink.

The obtained temperatures for the three configurations shown in Table 1 are plotted in Fig. 7 together with the experimental results of Sample 01 for reference.

The maximum temperatures found via measurements and simulation differ from each other to various degrees depending on the case, but in all cases the deviation is below 12 °C, which is within the expected variability due to noncontrolled variables such as ambient temperature and humidity. In most cases the simulated values are higher than those found experimentally.

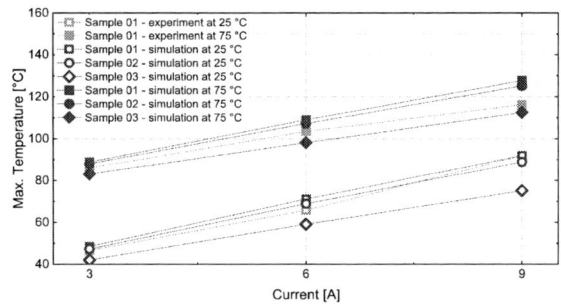

Fig. 7. Maximum simulated temperatures of different geometrical configurations for bundles of three 6.2 mm diodes in parallel.

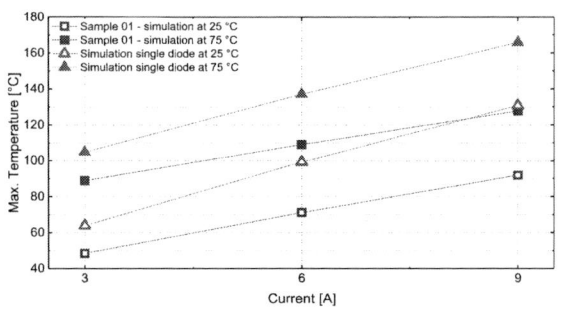

Fig. 8. Maximum simulated temperatures of three 6.2 mm diodes in parallel and of a single diode of the same size.

The thermal behavior of a single 6.2 mm diode is simulated at the same three levels of current, with the same BOM, and no additional aid in heat dissipation. The resulting maximum temperatures are plotted in Fig. 8 together with the simulated results of Sample 01 for reference.

C. Variation of chip size

The maximum temperatures obtained experimentally for the 100 mm² and 625 mm² diodes are shown in Table 4.

The 10 mm diode without heat sink stopped functioning during the experiment, therefore no reliable experimental data is given for the 9 A and 12 A cases at 75 °C.

TABLE 4.
MAXIMUM DIODE TEMPERATURE FOR LAMINATED DIODES WITH 10 AND 25 MM SIDE LENGTH

Current [A]	Max. Temperature [°C]					
	10 mm no heat sink		10 mm with heat sink		25 mm no heat sink	
	At 25 °C	At 75 °C	At 25 °C	At 75 °C	At 25 °C	At 75 °C
6	68	97	61	104	60	89
9	83	ND	70	133	64	100
12	103	ND	93	158	79	113

It is found that the 25 mm diode with the heating plate set to 25 °C manages to remain below 95 °C even at 12 A. However, at 75 °C only this one device succeeded in doing so at 6 A, and all exceeded this recommendation with higher currents. This is supported by the results from the simulation, shown in Fig. 9 for a heating plate temperature of 75 °C. Additionally, the maximum temperature for a 7.5 mm diode with and without a 625 mm² heat sink is simulated and also shown in Fig. 9.

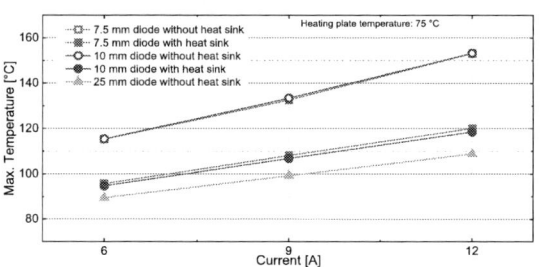

Fig. 9. Maximum simulated temperatures of different diode sizes with and without heat sink.

In most cases, the FEM model successfully predicts the maximum temperature with a deviation of under 10 °C from the experimental data, but some outliers are observed for the 10 mm diode without heat sink, with margins as wide as 18 °C.

D. Full-size module integration

The temperatures of the full size module sensed on the back sheet on the position of the diodes go as high as 131 °C with I_{SC} applied. In the second part of the measurements, at 1.25 I_{SC}, the temperature goes up to 157 °C.

After the testing procedure, the diodes are characterized and the IV characteristic curves before and after testing are compared in Fig. 10, where a change is noticeable.

Although the device is still operational, this shift indicates an increase in the series resistance of the diode assembly, which is an indicator of degradation of the diode interconnection. Due to this, it is expected that the power dissipated by the device will increase in future shading events, leading to larger power losses and diode temperatures.

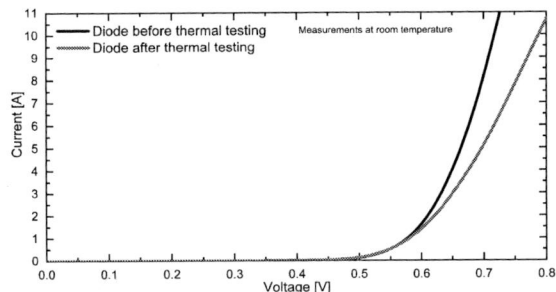

Fig. 10. Change of the characteristic curve of one of the diodes integrated into a full-size PV module after rigurous thermal testing.

IV. CONCLUSIONS AND OUTLOOK

The integration of diodes within PV modules is a promising way to simplify module design and manufacturing, and the use of bare die diodes for this application offers great potential and the possibility of reducing production costs. However, their IV characteristic leads to a higher power being dissipated in the device as compared of that of Schottky diodes, hence the effective removal of heat is necessary so that the lifetime and functionality of the device is not compromised.

Different methods such as the addition of heat sinks, the parallel connection of multiple chips, or the increase of the diode's area itself, are proven to have the desired effect of significantly lowering its temperature. However, it is observed that, under critical conditions, reaching the objective of a maximum junction temperature of 95 °C is quite challenging, and therefore a case-by-case analysis and the combination of different methods might be necessary.

Future investigations require further optimization of the geometry of the diode-connectors subassembly and contacting methodes, as well as a cost-benefit study of the abovementioned approaches. With the built and verified FEM model for the thermal behavior of bare die bypass diodes, we are now able to to design and prototype bare die bypass diodes which fullfill the requirements and which can be tested in further experments.

ACKOWLEDGEMENT

We would like to thank the German Ministry of Economic Affairs and Climate Action (FKZ 03EE1115A "Baldachin") for their funding.

REFERENCES

[1] Patrina Eiffert, Ph.D. and Gregory J. Kiss: NREL, "Building-Integrated Photovoltaic Desings for Commerical and Institutional Structures: A Sourcebook for Architects,"
[2] T. Bauer, "BIPV - a future with thin-film modules," in *Proceedings of the Intersolar Europe*, Munich, Germany, 2011.
[3] T. E. Kuhn, C. Erban, M. Heinrich, J. Eisenlohr, F. Ensslen, and D. H. Neuhaus, "Review of Technological Design Options for

Building Integrated Photovoltaics (BIPV)," *accepted for publication in the Energy and Buildings special issue review articles from the editors*, 2020.
[4] M. Heinrich, C. Kutter, F.Basler, M. Mittag, E. Alanis, D. Eberlein, A. Schmid, C. Reise, T. Kroyer, H. Neuhaus, H. Wirth, "Potential and Challenges of Vehicle Integrated Photovoltaics for Passenger Cars," in *37th EU PVSEC 2020*, 2020.
[5] M. A. Schüler, C. Kutter, L. E. Alanis, L. Saroch, C. Barz, J. Bornwasser, S. Reichert, H. Neuhaus, M. Heinrich, "High-Voltage Vehicle Integrated Photovoltaic System Concept and Demonstrator Truck: Presented at WCPEC-8, 8th World Conference on Photovoltaic Energy Conversion, 26-30 September 2022, Milan, Italy,"
[6] J. Forster, "Design Aspects and Constraints Regarding Hotspot Resilient Interconnection Topologies for Integrated Photovoltaic Modules," Master Thesis, University of Freiburg, 2021.
[7] H. Wirth, M. Vehse, Rau, Björn, Peibst, Robby, A. Colsmann, A. Stephan, and P. Lechner, "Integrierte Photovoltaik: Aktive Flächen für die Energiewende," Berlin, Oct. 23 2019.
[8] C. Kutter, M. Mittag, C. Ferrara, B. Bläsi, T. Kuhn, T. Kroyer, and O. Höhn, "Decorated building integrated photovoltaic modules: power loss, color appearance and cost analysis," in *Proceedings of the 35th EU PVSEC; Brussels, Belgium*, 2018.
[9] P. Baliozian, N. Klasen, N. Wöhrle, C. Kutter, H. Stolzenburg, A. Münzer, P. Saint-Cast, M. Mittag, E. Lohmüller, T. Fellmeth, M. Al-Akash, A. Kraft, M. Heinrich, A. Richter, A. Fell, A. Spribille, D. Neuhaus, and R. Preu, "PERC-based shingled solar cells and modules at Fraunhofer ISE,"*Photovoltaics International*, no. 43, pp. 129–145, 2019.
[10] C. E. Clement, J. P. Singh, Y. S. Khoo, A. Halm, D. Tune, and E. Birgersson, "Design of shading and hotspot resistant shingled modules,"*Prog. Photovolt: Res. Appl.*, 2021.
[11] R. G. Vieira, F. M. U. de Araújo, M. Dhimish, and M. I. S. Guerra, "A Comprehensive Review on Bypass Diode Application on Photovoltaic Modules," *Energies*, vol. 13, no. 10, p. 2472, 2020.
[12] K. M. Coetzer, A. J. Rix, and P. G. Wiid, "The Measurement and SPICE Modelling of Schottky Barrier Diodes Appropriate for Use as Bypass Diodes within Photovoltaic Modules," *Energies*, vol. 15, no. 13, p. 4783, 2022.
[13] C. Xiao, P. Hacke, S. Johnston, D. B. Sulas-Kern, C.-S. Jiang, and M. Al-Jassim, "Failure analysis of field-failed bypass diodes," *Prog. Photovolt: Res. Appl.*, vol. 28, no. 9, pp. 909-918, https://onlinelibrary.wiley.com/doi/10.1002/pip.3297, 2020.
[14] L. Lillo-Sánchez, G. López-Lara, J. Vera-Medina, E. Pérez-Aparicio, and I. Lillo-Bravo, "Degradation analysis of photovoltaic modules after operating for 22 years. A case study with comparisons," *Progress in Solar Energy 1*, vol. 222, pp. 84–94, 10.1016/j.solener.2021.04.026, 2021.
[15] M. Herz, G. Friesen, U. Jahn, M. Koentges, S. Lindig, and D. Moser, "Identify, analyse and mitigate—Quantification of technical risks in PV power systems," (en), *Prog. Photovolt: Res. Appl.*, 2022.
[16] Deutsche Kommission Elektrotechnik, Elektronik, Informationstechnik, *DIN EN IEC 61215-2 (VDE 0126-31-2), Terrestrische Photovoltaik(PV)-Module - Bauarteignung und Bauartzulassung. Teil 2, Prüfverfahren (IEC 61215-2:2021): = Terrestrial photovoltaic (PV) modules - design qualification and type approval. Part 2, Test procedures (IEC 61215-2:2021)*, 61215th ed. Berlin: VDE Verlag GmbH, 2022.
[17] A. Mondon, N. Klasen, M. Mittag, M. Heinrich, and H. Wirth, "Comparison of Layouts for Shingled Bifacial PV Modules in Terms of Power Output, Cell-to-Module Ratio and Bifaciality," in *Proceedings of the 35th European Photovoltaic Solar Energy Conference and Exhibition (EU PVSEC)*, Brussels, Belgium, 2018, pp. 1006–1010.

Measurement and analysis of annual solar spectra at different installation angles in central Europe

Guillermo A. Farias-Basulto, Miguel Á. Sevillano-Bendezú, Maximillian Riedel, Mark Khenkin, Jan A. Töfflinger, Rutger Schlatmann, Reiner Klenk, Carolin Ulbrich

PVcomB/ Helmholtz-Zentrum Berlin für Materialien und Energie GmbH , Berlin, Germany

Pontificia Universidad Católica del Perú, Department of Sciences, Physics Section, Lima, Peru

HTW Berlin - University of Applied Sciences, Berlin, Germany

Real world solar spectra are rarely equal to those under standard test conditions, which points to the importance of measuring, analyzing and understanding the implications of solar spectra variations in photovoltaic performance. This work presents and analyzes one year of solar spectra measured with installed spectrometers at optimum and vertical (building-integrated-photovoltaics relevant) angles in central Europe, reporting for the first time experimental datasets measured simultaneously over such timespan. In addition, we quantify the differences between these datasets with known key performance indicators and discuss the implications of spectral changes on the maximum current density of ideal single-junction and tandem devices for both installation angles.

Photoluminescence analysis of the back side of Cu(In,Ga)(S,Se)2 absorbers

Aubin JC. M. Prot, Susanne Siebentritt, Anastasia Zelenina, Hossam Elanzeery, Alberto Lomuscio, Thomas Dalibor, Maxim Guc, Robert Fonoll-Rubio

University of Luxembourg, Luxembourg, Luxembourg

Avancis, Munich, Germany

Catalonia Institute for Energy Research, Barcelona, Spain

Record efficiency in chalcopyrite based solar cells Cu(In,Ga)(S,Se)2 (CIGSSe) is achieved using the standard gallium gradient structure to increase the band gap of the absorber towards the back side. Although this structure has reduced the recombination at the back side, we demonstrate that in sequentially processed absorbers with intentional gallium gradient, the back side is a source of non-radiative recombinations. Photoluminescence measurements performed on both the front and back sides reveal two main radiative recombination paths, the dominant one corresponding to band-to-band recombination at the band gap minimum and the second one attributed to a secondary phase, mainly located towards the back. A linear relation between the non-radiative voltage losses and the contribution of the secondary PL peak is observed and suggests that reducing the amount of secondary phase could decrease these losses by up to 180mV. Additionally, we show that as the back side of the absorber is etched away until complete removal of the gallium gradient, the photoluminescence intensity increases by 1 order of magnitude, which translates into an increase of 60meV in quasi Fermi level splitting or voltage.

Effects of Salt Spray on c-Si Photovoltaic Modules in the Brazilian Region

Mendelsson R. M. Neves[1], Allan Silveira[2], Hugo S. Alvarez[4], Rodrigo M. Garcia[4], Francisco C. Marques[3] e
Marcelo G. Villalva[1]

[1]School of Electrical and Computer Engineering, Dept. of Systems and Energy - DSE, Campinas, Brazil
[2]School of Electrical and Computer Engineering, Communications Department - DECOM, Campinas, Brazil
[3]Institute of Physics Gleb Wataghin, Dept. of Applied Physics - DFA, Campinas, Brazil
[4]Build Your Dreams Company - BYD, Dept. Research and Development - R&D, Campinas, Brazil

Abstract — **Brazil has great potential for this type of energy generation due to its geographic location, allowing the development of viable photovoltaic (PV) projects in several regions. its use in places close to the sea has increased, with its use on boats and even resorts and hotels. This proximity to the sea requires attention to the local salinity, more precisely to the saline mist. This article will describe the methodology used to carry out the salinity resistance test of PV modules, choosing a specific classification of corrosive atmosphere according to the brazilian environment on the coast where the module will be placed in real conditions.**

I. INTRODUCTION

Photovoltaic solar energy has grown a lot in recent years, either because of its vast potential or because of the non-emission of greenhouse gases in the operation of plants. Brazil has great potential for this type of energy generation due to its geographic location, allowing the development of viable photovoltaic projects in several regions.

As it is renewable energy available anywhere and at any time, photovoltaic solar energy can be applied in different ways. Its use in places close to the sea has increased, with its use on boats, resorts, and hotels. This proximity to the sea requires attention to the local salinity, more precisely to the saline mist (Salt Spray).

The salt spray has the effect of enhancing the corrosion of metal components. Metallic corrosion involves the loss of electrons, which, in the form of ions, react with other species existing in the corrosive medium, forming oxides, salts, and hydroxides, known as corrosion products [1]–[3].

The range of salt spray can vary greatly due to the movement of droplets containing chlorine ions (Cl-) and local characteristics such as topography and meteorological parameters, especially the direction of wind speed.

Photovoltaic (PV) modules are electrical devices in the PV system that undergo continuous exposure throughout their lifetime. In this way, places close to the ocean or marine environments have a lot of salt and water vapor mixed with acidic and basic substances, acting directly on the surface of the module. These chemical agents can eventually degrade some components of the photovoltaic module (corrosion of metallic parts and deterioration of non-metallic components due to the assimilation of salts), causing degradation of the module's power and even the electrical safety of the module, impairing its operation. Fig. 1 shows how salt deposition on the module surface can occur.

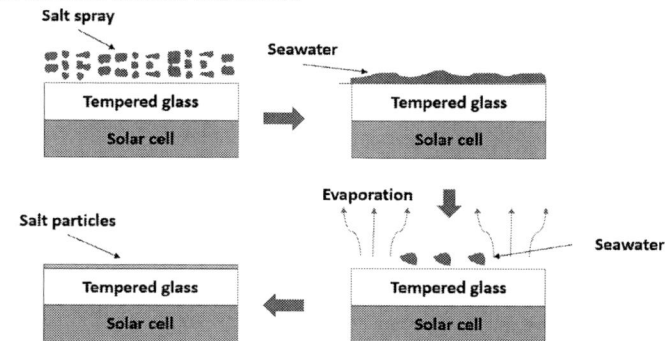

Fig. 1.   Process of changing salt spray and sea water. Adapted from [1].

Corrosion tests are mostly carried out in closed laboratory climatic chambers, as they allow simulation of an aggressive environment under which the module will be exposed in a short period, compared to when it will be in the field. With this method, it is possible to obtain results similar to real conditions.

To accelerate the corrosion tests and make it more aggressive, sodium chloride (NaCl) or copper chloride is used as a solution, including factors such as drying and humidification cycles and temperature increase, among others [1], [3]. For this, there is no need to faithfully reproduce all the corrosion agents involved as long as it is possible to reproduce the test quickly, obtain data related to field exposure data and be executable in chambers and other environments test.

Variations of salt spray test solutions depend on the materials to be tested. We can divide the Salt Spray test into Neutral Salt Spray (NSS), Cuproacetic Salt Spray (CASS), and Acetic Acid Salt Spray (AASS).

Neutral Salt Spray (NSS) is a test of spraying a sodium chloride solution at neutral pH. It is used for various parts and coatings, allowing faster results than actual exposure, although some modified tests may be more recommended for specific materials and surface treatments [4].

978-1-6654-6060-6/23 $31.00 © 2023 IEEE

Cuproacetic Salt Spray (CASS) is widely used as a rapid test on copper/nickel/chromium or nickel/chromium, chrome coatings on iron and zinc for relatively severe use, and aluminum, particularly when anodized. When carried out after the acetic salt spray test, it proved to be more efficient due to the addition of CuCl2 in the solution and higher test temperature [4].

The Acetic Acid Salt Spray (AASS) is the salt spray test with acidic pH due to the addition of glacial acetic acid. It is recommended for aluminum parts, copper/nickel, copper/nickel/chrome coatings, and anodized aluminum [4].

In this article, the methodology used to carry out the salinity resistance test of PV modules will be described, choosing a specific classification of corrosive atmosphere according to the environment in which the module will be placed in real conditions. Small faults were drilled into the modules to test the effect of salinity when the module's external components were compromised. The back sheet was compromised in one of the modules, while the other passed the static and dynamic mechanical load test, suffering cracks in its cells.

II. METHODOLOGY

IEC 61701 (Photovoltaic modules – Salt mist corrosion testing) is the international standard that sets a test standard for salt spray resistance. Tests described in the IEC 61215-2 (Terrestrial photovoltaic modules – Design qualification and type approval – Part 2: Test procedures) and IEC 60068-2-52 (Environmental testing – Part 2-52: Tests – Test Kb: Salt mist, cyclic (sodium chloride solution) standards were included, which combined provide means to evaluate possible failures caused in photovoltaic modules when operating under humid atmospheres with a high concentration of dissolved salt (NaCl).

Depending on the specific nature of the PV module in the surrounding atmosphere when put into actual operation, various test methods can be applied as per IEC 60068-2-52. In addition, IEC 61701 provides a test rating based on the severity of the corrosive atmosphere in which the PV module will be installed.

Corrosivity ratings C1 to CX based on ISO 9223(Corrosion of metals and alloys – Corrosivity of atmospheres – Classification, determination and estimation) provide an informative estimate of the corrosivity category based on knowledge of the local environmental situation. This classification can be seen in Table I.

The percentage of humidity time, from the English ToW, is defined as the number of hours during the year in which the relative humidity (RH) is equal to or greater than 80% and the temperature is greater than 0 °C, divided by the total and hours in a year.

Table I shows C1 and C2 represent environments with low corrosivity, generally in dry climates. C1 environments are rare, but C2 environments are common in dry areas far from

saltwater bodies. C3 environments are common in industrialized areas with distances of 2 to 10 km, or more than 10 km from saltwater sites and areas with high humidity during much of the year.

TABLE I

DETERMINATION OF CORROSIVE RATING ACCORDING TO ISO 9223 AND TEST METHODS CORRELATED TO A YEAR OF CORROSIVITY. ADAPTED FROM [5]

Module location corrosivity rating	Characteristics of the locality		Mass long range (g/m²) in one year of bare steel	60068-2-52 test method equivalent to one year of corrosivity
	Distance from salt water (km)	Percent Time of Wetness (ToW)		
C1	-	-	< 10	-
C2	≥ 10	< 25%	10 – 200	2,3
C3	≥ 10	≥ 25%	200 – 400	4 (14 days)
	2 a 10	< 25%		
C4	2 a 10	≥ 25%	400 – 650	1 (28 days)
	< 2	< 25%		5 (28 days)
C5	< 2	≥ 25%	650 – 1500	6 (56 days)
CX	Maritime	-	1500 – 5500	7 (90 days)
				8 (70 days)

The C4 and C5 environments are highly corrosive and are associated with proximity to salt water. The extreme corrosiveness category CX has been added to the other classifications to represent marine (offshore) environments, such as those experienced by floating module installations or platforms used for offshore oil and gas extraction.

As for the test methods used, we must follow what is recommended by IEC 60068-2-52, divided into eight methods, and in some of them, the processes involved in each cycle are the same, changing only the number of cycles for each method. The processes involved in each cycle are shown in detail in Fig. 2.

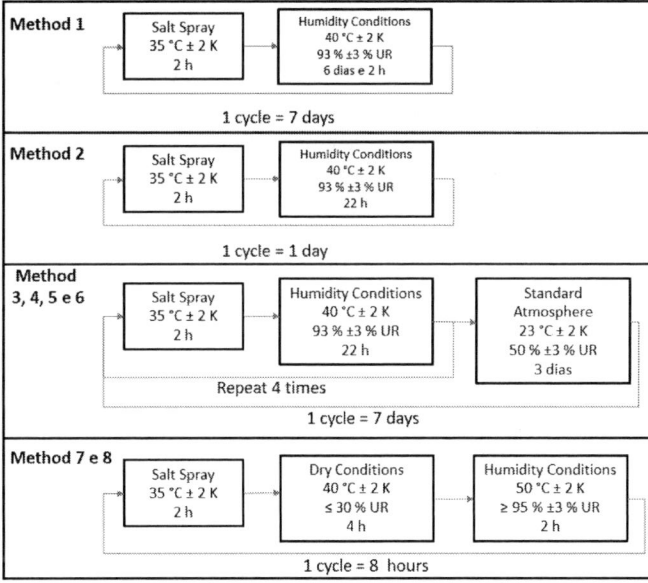

Fig. 2.   Test cycles for test methods 1 to 8. Adapted from [6].

Fig. 3 shows the atmospheric corrosion map applicable in several countries and environmental conditions, bringing data related to corrosion based on the categories of environments described in ISO 9223. We can see in the legend of the map the categories with extremely severe corresponding to C4 and C5, severe to C3, moderate to C2, and mild to C1, which can be related to Table I. With the help of this map, we can design an environment category to test the photovoltaic modules for the Brazilian environment. The National Association of Corrosion Engineers (NACE) is responsible for producing these maps in Fig. 3, which are based on grip data from different research projects related to aerodynamic traction.

To prepare the solution, an amount of NaCl was dissolved in deionized water with a concentration of 50 g/L, the total prepared solution was 20 liters, totaling a total mass of 1000 g of NaCl used. Using a pH meter, the pH of the solution was measured to ensure that the solution had a neutral pH between 6.5 and 7.2 for a temperature of 25 °C, according to ISO 9227 (Corrosion tests in artificial atmospheres – Salt spray tests).

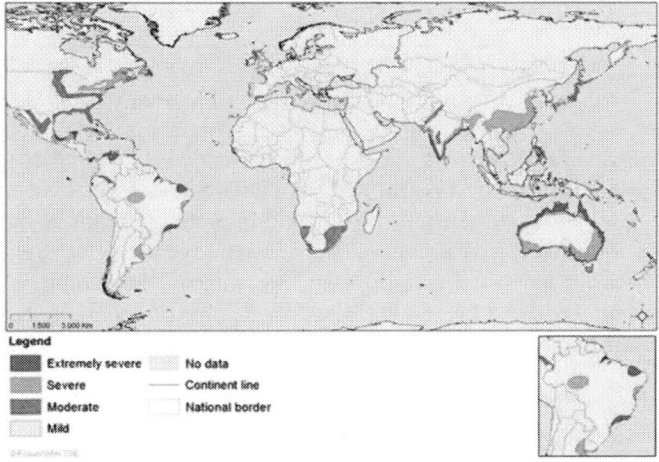

Fig. 3. Map of the atmosphere of saline corrosion in Brazil and the world. Adapted from [3].

The solution is prepared directly in a solution tank of the saline mist machine itself with the addition of NaCl, choosing the Salt Spray test with Neutral pH. The corrosive environment category chosen was C3 because it is an environment with a medium level of corrosion. Following Table I, it is recommended that Test Method 4 be carried out, which presents three operating modes of the climate chamber, namely: Salt Spray, Humidity Conditions and Standard Atmosphere [4]. Fig. 4 shows the climatic chamber used.

In Fig. 4, the test aims to evaluate the corrosion resistance of a photovoltaic module due to a saline mist. Water must be deionized because minerals present in untreated water can interfere with the final result. The solution is stored in the chamber tank. To start the test, the solution must be stirred through a command in the control center. The modules must be carefully positioned in the portrait position to be mechanically stable throughout the test. The solution

dispersion tower, responsible for emitting the saline spray in the chamber, must be superimposed on the module so that its flow occurs in full. The solution collection funnel must be installed in such a way that its performance is also not impaired.

In the Salt Spray condition, the sample is eroded by a complex electrochemical or chemical reaction with neutral saline solution. The saline solution forms a thin electrolyte film on the test surface of the sample. This can initiate corrosion and allow it to proceed.

When switching from Salt Spray to Wet Condition, the existing moisture on the sample surface at the end of the salt spray period is maintained without excessive dilution of the solution that may result in moisture condensation.

Standard Atmosphere is when the test sample is equilibrated with standard laboratory conditions, which results in gradual drying and stabilization of most corrosion reactions. The period of dry atmosphere can occur, in practice, during breaks in operation [4], [5], [7]. However, this may apply to other types of equipment. Since the module is almost always in operation. Thus, the inclusion of such a dry period can lead to corrosion mechanisms that can be quite different from those under constant wet conditions.

Fig. 4. Salt spray climate chamber model Serie SS – FINAME 2044830 from Equilam®.

In Fig. 5 we have the flowchart of tests that are followed for the corrosion test as stated at the beginning of this work. The IEC 61215-2 standard was used for the characterization of the samples. MQT 01 – Visual Inspection, MQT 19.1 – Initial Stabilization, and MQT 02 – Maximum Power Determination correspond to module quality tests according to IEC 61215. MQT 01 is intended to detect any visual defects in the module. The MQT 19.1 is intended to electrically stabilize the modules, with this procedure being repeated and power measured until the module has reached an electrically stable power output level. MQT 02 is intended to determine the module's maximum power after stabilization, as well as before and after various environmental stress tests.

Still, in Fig. 5, the electroluminescence (EL) image test was added to the original flowchart of the test adopted by IEC 61701. The inclusion of the EL is because possible degradations that are not seen with the naked eye can be visualized. This gives us a more detailed visual inspection of the sample for a qualitative and quantitative comparison of damage to the modules.

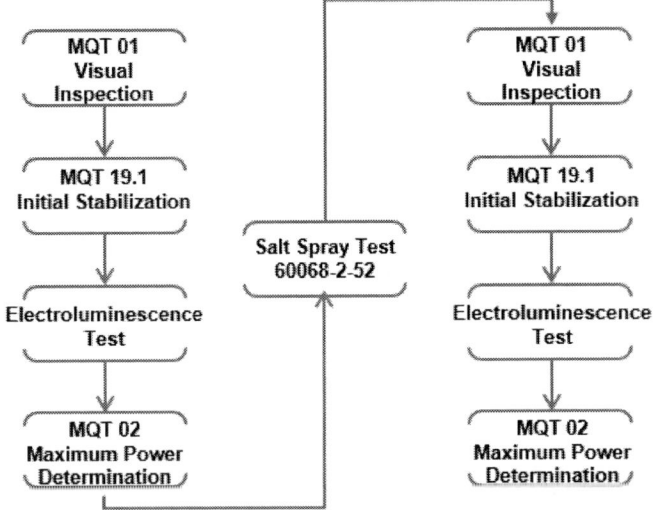

Fig. 5. Corrosion test sequence for salt spray tests for PV modules.

step is repeated three times, in the initial state of the module and always after the module suffers some degradation. This allows you to track the parameters that were affected by the degradation. In the process of provoked failures, we have the mechanical load test tested following the IEC 61215-2 standard. In the salt spray corrosion process, the test applied is Method 4. To carry out the MQT 02 test, we used a table-type solar simulator model XJCM-11A+ from Gsola®. It features class A+A+A+, a long pulse duration ranging from 10ms to 150ms, and can test crystalline silicon modules and thin film. The simulator presents irradiance instability <1%, irradiation non-uniformity <1% and repeatability 0.3%

In salt spray corrosion tests previously performed in the laboratory, completely healthy modules without any failure were tested. However, the module did not show any degradation in its performance. In this way, it was thought about what could happen if the module suffered a failure before the corrosion test and how much this failure could influence the performance of the module. Thus, a failure was caused in each sample, which can occur when the module is in the field.

The failure caused by the mechanical load test is related to the mechanical efforts that the modules can suffer from the wind load applied to them, represented by the static load test. And the vibration caused to the module also by the wind effect

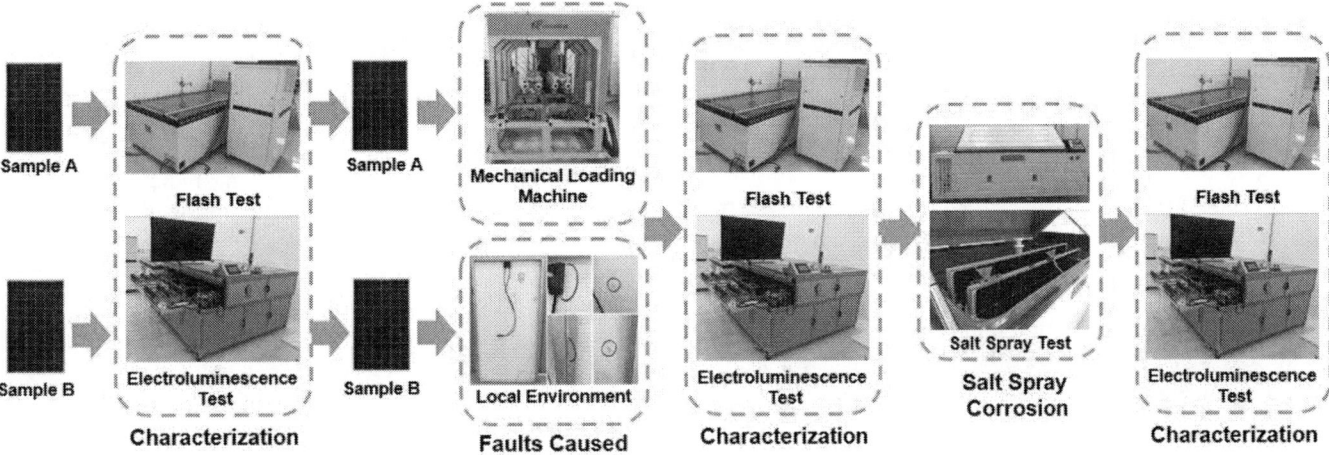

Fig. 6. Details of the steps of the experimental procedure.

In Fig. 6, we detail the tests performed at each stage of the testing process in a more explanatory way. At first, we have two samples of commercial PV modules. Manufacturers will not be mentioned in this work. Sample A and B are modules with 5BB (busbar) monocrystalline cell Passivated Emitter and Rear Cell (PERC) technology, with 72 full-cell cells. According to their commercial labels, samples A and B have 400 Wp and 380 Wp, respectively.

Also, in Fig. 6, we have the representation of the module characterization process, in which the measurements of the electrical parameters of the PV module are performed. This

is represented by the dynamic load test. To carry out the mechanical load test, we used a Hototech® mechanical simulator model HTPV-08 for a positive test load of 5400 Pa and 2400 Pa for a negative test load.

Failure related to backsheet wear represents a situation in which there may be poor handling during module installation, a problem caused in the module transport stage, the use of low-quality backsheets, or any other action that may cause wear of the backsheet.

After going through the first module characterization test, Sample A suffered from a failure caused by the mechanical load test (static and dynamic), with its cells damaged by microcracks. After this provoked failure process, the module is characterized again. The module is placed in the salt spray climate chamber for two weeks and finally characterized again, as illustrated in Fig. 6.

Sample B likewise goes through the characterization processes of sample A, and the difference is in the failure caused. While sample A suffered degradation due to mechanical stress, sample B suffered backsheet wear. After this differentiated process underwent the same testing processes as sample A, as illustrated in Fig. 6.

III. RESULTS AND DISCUSSIONS

Table II presents the results of the electrical parameters obtained in each stage of module characterization by the Flash test. Fig. 7 shows the data from the Table II for comparative visualization of degradation.

TABLE II
ELECTRICAL PARAMETERS OF SAMPLES A AND B IN STC (AM 1.5; 1000 W/M²; 25 °C)

Parameter	Sample A			Sample B		
	Initial	Post failure	Post Salt Spray	Initial	Post failure	Post Salt Spray
P_{max} (W)	390.93	359.90	355.09	378.92	378.92	364.22
I_{sc} (A)	9.87	9.81	9.78	9.52	9.52	9.66
V_{oc} (V)	49.35	48.56	48.77	49.96	49.96	48.769
FF (%)	80.27	75.58	74.46	79.65	79.65	77.26
R_s (Ω)	0.45	0.59	0.64	0.53	0.53	0.58
R_{sh} (Ω)	235.96	364.76	420.11	263.55	263.55	149.09

A. Sample A

By analyzing the electrical data obtained from sample A from its initial state to the post-failure state, we can see that there were significant changes in module power (P_{max}), series resistance (R_s), and shunt resistance (R_{sh}). The maximum power loss of the module after the faults were triggered was a 7.93% drop in power. The series resistance increased by 29.24%, a considerable increase that may indicate that when placed in the field, the module may present hot spots mainly due to the increase in the loss due to the Joule effect in the busbars and fingers of the cell.

In Fig. 8, the image of EL A.ii, it is possible that the module cells were damaged with cracks in all their extensions due to the mechanical effect. This analysis is only possible thanks to the EL image since the appearance of the module to the naked eye was not changed, being possible to explain the increase in the series resistance of the module.

Still, in sample A, we have the shunt resistance that increased by 54.59%, which does not necessarily harm the module since it is the low shunt resistance that causes energy losses in the photovoltaic cells, providing an alternative path for the generated current by light. This bypass reduces the

current flow through the solar cell junction and the solar cell voltage.

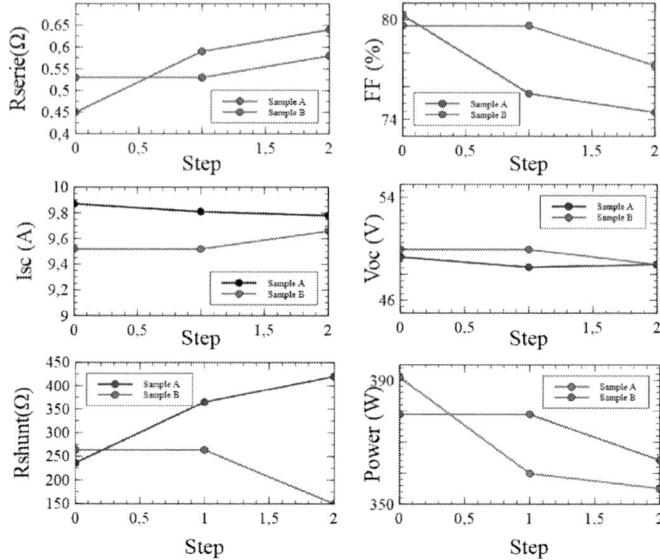

Fig. 7. Graphical representation of module degradation after each step of the experiment.

Following the flow of tests shown in Fig. 6, a third characterization of sample A was carried out. Comparing the data after the salt spray test, we have changes in the electrical data shown in Table II. The module's maximum power after the salt spray test was reduced by 1.33%. Series resistance increased by 8.55%, and shunt resistance increased by 42.58%. A high shunt resistance indicates that the vast majority of free carriers produced by solar radiation generate energy [8], [9]. However, the losses due to the effects caused by photovoltaic cells with cracks outweigh the gains that may accompany the increase in shunt resistance to energy generation by free-charge carriers.

B. Sample B

After carrying out the measurements in its initial state before suffering the provoked failures, the module of sample B did not present any alteration in the measurements or the EL. This can be seen in Fig. 8, in the images of EL B.i and B.ii. This characterization was necessary to verify that, when there was forced wear on the backsheet of the module, it did not cause it to show false corrosions in the EL image after the salt spray test.

Again, let's analyze the parameters of P_{max}, R_s, and R_{sh} in the same way as sample A. The maximum power loss of the module after the faults were triggered was a 3.88% drop in power. Series resistance increased by 10.53%, and shunt resistance decreased by 43.43%. In this case, we can consider the energy losses generated by the low shunt resistance of the module.

When comparing the images of EL B.ii and B.iii, in Fig. 8, we have the darkening of the cells in the regions that caused the flaws, with cells A4 and D6 darkened at specific points and between cells C4 and D4, these degradations did not reach the expected degradation, which could erode the cells.

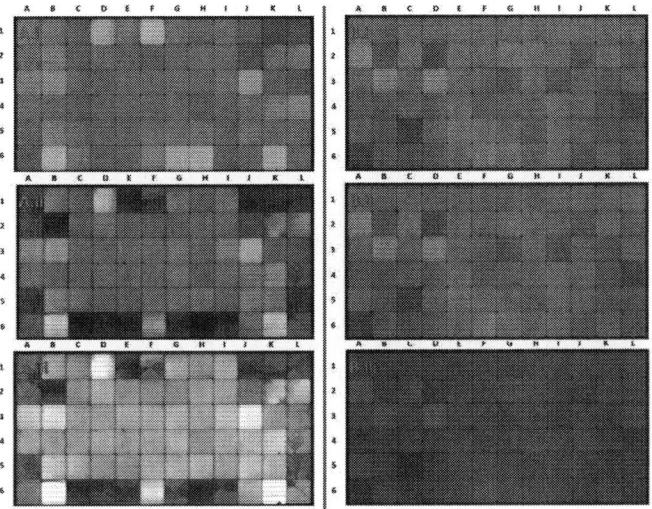

Fig. 8.   EL image of the modules in their initial state and after each fault process executed. The sequence on the left corresponds to sample A. The sequence on the right corresponds to sample B.

IV. CONCLUSION AND FINAL CONSIDERATIONS

The influence of salt spray on the PV module is a dynamic process that can affect the solar irradiance and the temperature received by the solar cell by changing the conditions of the surface and the photovoltaic module, thus changing the electrical characteristics.

In real marine environments, salt spray and seawater often coexist. The water film of salt spray formation, the evaporation of water, and the precipitation of salt particles occur simultaneously. The resulting changes in PV module output are often very complex and can lead to increases or decreases in PV module output. Meanwhile, considering the rain, seawater salinity, and other factors, the salt deposition will be slower, affecting the output of the PV module. More research is needed.

From the experiments, we can conclude that the neutral salt spray test (NSS) does not harm the module so severely, considering the 14-day cycle performed. Longer cycle times and even test switching to a more acidic salt spray may show more severe corrosion on both the anodized aluminum frame and module backsheet. For future tests, it is intended to use Acetic Acid Saline Mist (AASS) or even Cuproacetic Saline

Mist (CASS) to be able to increase the corrosion of the module components and to be able to estimate its durability when placed in more acidic, such as the large centers close to the coast.

ACKNOWLEDGMENT

The authors gratefully acknowledge support from BYD Energy Brazil through the PADIS (Brazilian Program for the Technological Development of the Semiconductor Industry)/MCTI program, Project No. 5779/FUNCAMP. Also thanks to CNPq (National Council for Scientific and Technological Development) for granting of scholarship to the author.

REFERENCES

[1]   Y. Zhang e C. Yuan, "Effects of marine environment on electrical output characteristics of PV module", *Journal of Renewable and Sustainable Energy*, vol. 13, nº 5, p. 053701, set. 2021, doi: 10.1063/5.0060201.

[2]   A. Hasan e I. Dincer, "A new performance assessment methodology of bifacial photovoltaic solar panels for offshore applications", *Energy Conversion and Management*, vol. 220, p. 112972, set. 2020, doi: 10.1016/j.enconman.2020.112972.

[3]   K. Slamova, C. Schill, S. Wiesmeier, M. Köhl, e R. Glaser, "Mapping atmospheric corrosion in coastal regions: methods and results", *JPE*, vol. 2, nº 1, p. 022003, jun. 2012, doi: 10.1117/1.JPE.2.022003.

[4]   International Organization for Standardization (ISO), "ISO 9227:2017 Corrosion tests in artificial atmospheres — Salt spray tests". 2017.

[5]   *ISO 9223 - Corrosion of metals and alloys: corrosivity of atmospheres - classification, determination and estimation*. London: British Standards Institution, 2012.

[6]   International Electrotechnical Commission(IEC), "IEC 60068-2-52:2018 Environmental testing - Part 2: Tests - Test Kb: Salt mist, cyclic (sodium chloride solution)". 2018.

[7]   International Organization for Standardization (ISO), "IEC 61701:2020 Photovoltaic (PV) modules — Salt mist corrosion testing". 2020.

[8]   A. D. Dhass, E. Natarajan, e L. Ponnusamy, "Influence of shunt resistance on the performance of solar photovoltaic cell", em *2012 International Conference on Emerging Trends in Electrical Engineering and Energy Management (ICETEEEM)*, dez. 2012, p. 382–386. doi: 10.1109/ICETEEEM.2012.6494522.

[9]   T. J. McMahon, T. S. Basso, e S. R. Rummel, "Cell shunt resistance and photovoltaic module performance", em *Conference Record of the Twenty Fifth IEEE Photovoltaic Specialists Conference - 1996*, maio 1996, p. 1291–1294. doi: 10.1109/PVSC.1996.564369.

Experimental Analysis of Distribution Network Voltage Regulation Using Smart Inverters

Rasel Mahmud[1], Subhankar Ganguly[1], Jing Wang[1], and Killian McKenna[1] Ning Li[2]

Abstract—Smart inverters (SIs) have demonstrated their potential to provide grid services for both transmission and distribution systems. One of these grid services, distribution network voltage regulation by SIs, has the potential to improve network voltage regulation through controlling the reactive and active power output of the SIs. Voltage regulation by SIs will be distributed and might be better suited to controlling local conditions to complement traditional voltage-regulating assets, e.g., tap-changing transformers, capacitor banks, and line voltage regulators. There is a gap in the literature on comparing the SI response characteristics when the SIs are controlled by a local controller or external control signals. This paper presents an experimental study to characterize SI reactive power regulation responses to two different control methods: autonomous control and remote dispatch. We found that SI reactive power regulation responses exhibit important differences between these methods in terms of delays and ramp rate. Finally, power-hardware-in-the-loop (PHIL) tests were conducted to evaluate the performance of these two methods. The PHIL test results show that the SI response characteristics for autonomous control and remote dispatch need to be considered when planning for distribution network voltage regulation using SIs.

I. INTRODUCTION

Utilities have traditionally employed legacy equipment (such as on-load tap-changing transformers, capacitor banks, and line voltage regulators) for distribution network voltage regulation. This voltage-regulating equipment is usually installed at either the substation or along the feeder primary [1] to help reduce voltage drop along the network and ensure that end-of-line voltages remain within service limits. The integration of distributed energy resources (DERs)—in particular, solar photovoltaic (PV) systems with smart inverters (SIs)—brings challenges and opportunities to voltage control. DERs are typically interconnected along the length of the distribution circuit and and can provide grid-edge voltage control by modulating reactive or active power in response to local voltage conditions [2], [3].

In general, voltage regulation by SIs can be realized in two ways: *i)* by remote dispatch method and *ii)* by autonomous SI

This work was authored in part by the National Renewable Energy Laboratory, operated by Alliance for Sustainable Energy, LLC, for the U.S. Department of Energy (DOE) under Contract No. DE-AC36-08GO28308. Support for the work was also provided by Utilidata, Inc., under CRD-20-16909. The views expressed in the article do not necessarily represent the views of the DOE or the U.S. Government. The U.S. Government retains and the publisher, by accepting the article for publication, acknowledges that the U.S. Government retains a nonexclusive, paid-up, irrevocable, worldwide license to publish or reproduce the published form of this work, or allow others to do so, for U.S. Government purposes.

[1]Authors are with the National Renewable Energy Laboratory, Golden, Colorado
[2]Authors are with the Utilidata, Inc., Providence, RI

control. In remote dispatch method, the SI receives external control signals, potentially from a centralized DER management solution, using appropriate communication channels. The external controller could be a centralized [4] or distributed controller [5] that coordinates multiple assets in the network to achieve some control objective, e.g., distribution network voltage regulation, and it might or might not use the SI local voltage measurements in the decision-making process. To leverage the communication-based DER management solution, DERs need to have interoperability with the external controller and access to the communication networks [6]. In distributed control, DERs can act autonomously based on local information (e.g., voltage, frequency, active and reactive power) and predefined functions (e.g., volt-var, frequency-watt) without requiring any communication with any external controller [7].

Voltage regulation using autonomous SI control has been extensively investigated in both simulations and field tests [8]. The configuration, operation, and demonstration of autonomous SI control has been extensively tested and analyzed in the laboratory [9], [10]; however, there is lack of literature on the response and characterization of DERs controlled by remote dispatch method and contrasted with autonomous SI control. Currently, few DERs installed in the field are connected with utility communication networks, and the lack of understanding of centralized control has been a barrier to DER utilization in voltage regulation. Many inverters installed in the field can be connected with the appropriate communication systems using necessary gateways to leverage the benefits of remote dispatch method. It is necessary to continue the discussion on establishing interoperable communications between existing, field-deployed SIs and external controllers to understand the potential of using DERs in centralized voltage regulation schemes.

To address these issues, we developed an experimental setup to investigate and characterize the SI inverter responses for both control modes. The experimental tests were designed to focus on establishing, debugging, and characterizing remote dispatch method and autonomous SI control. The objectives were as follows:

- Demonstrate the successful operation of communication-enabled remote dispatch method for SI control.
- Propose characterization parameters and evaluate the performance of autonomous and remote dispatch control methods of SIs.
- Compare the SI responses in remote dispatch method versus autonomous SI control operation.

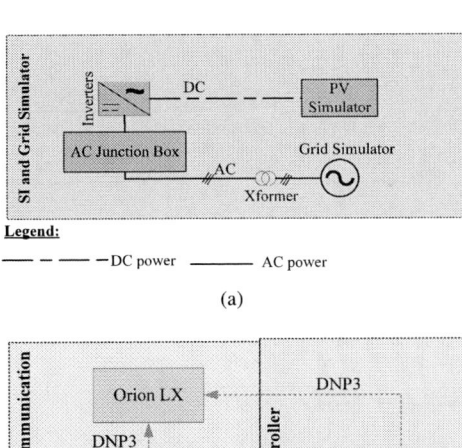

(a)

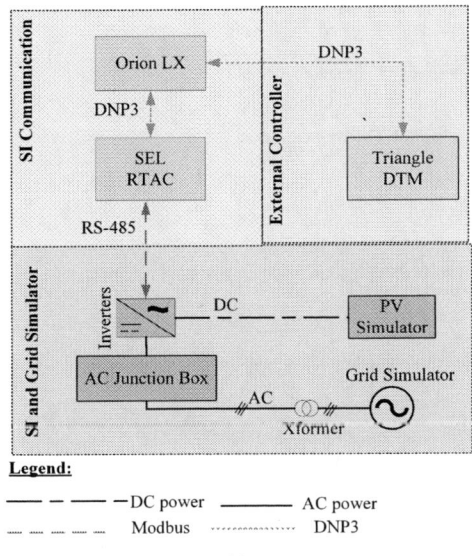

(b)

Fig. 1. Experimental setup to evaluate the performance of grid services from smart inverters: a) autonomous SI control, b) remote dispatch method.

- Perform a scenario analysis using remote dispatch method and autonomous SI control.

II. TESTING PROTOCOL AND METHODOLOGY

A. Overview of Experimental Setup

An experimental setup was developed using a commercially available, off-the-shelf 125-kW PV inverter to test the autonomous SI control mode of operation, as shown in Fig. 1a. In this setup, the inverter was connected with a 270-kW rated, three-phase grid simulator on the AC side through a 480-V/600-V transformer and a 250-kW rated PV emulator on the DC side. AC-side measurements were taken on the inverter side of the transformer. A setup similar to that shown in Fig. 1b was developed to evaluate the remote dispatch method. For the test setup shown in Fig. 1b, the Triangle Microworks, Inc., Distributed Test Manager (DTM) [11] was used as the remote server to emulate the remote dispatch controller. The communications between the SI and the remote server are described in II-B.

B. Communications Between SI and Remote Server

The communications between the SI and the DTM are shown in Fig. 1b to represent a potential field installation scenario. The SEL Real-Time Automation Controller (RTAC)

was represented as a data concentrator for the SIs communicating over serial using the RS-485/EIA-485 interface. The data points exchanged over serial were mapped to Distributed Network Protocol 3 (DNP3) [12] points in the RTAC. Also, the RTAC was configured as a DNP3 outstation device to interchange data points with OrionLX using the DNP3 protocol. OrionLX was included in the setup to represent a utility substation data gateway. Finally, the remote server functionality was simulated in the DTM. The DNP3 protocol was also used for the communications between the DTM and OrionLX. A basic heartbeat logic was developed in the RTAC using the IEC 61131 [13] programming language to check the integrity of the communications between the SIs and the DTM. If any communication failure with the remote server was detected, the logic was configured to restore the SI set points to the values stored before the initialization of the communication session. The algorithm for the remote server functionality was developed in the DTM using Javautonomous SI controlript [14]. The remote server was programmed to initiate and terminate the communication session with the SIs and to send active and reactive power set points to the SIs. The set points were calculated based on the measured SI terminal voltages and the volt-var curve programmed in the DTM and automatically dispatched to the SIs.

C. SI Characterization

A quantitative comparison of the two SI control methods can be obtained based on the inverter responses to different situations. The Triangle DTM was programmed with volt-var curves to remotely sense the inverter terminal voltages and issue the corresponding reactive power dispatch. This allowed for validating the proposed communication paths and characterizing the response times. Both open-loop and closed-loop tests were conducted, defined as follows:

- Open-loop: The DTM directly issues reactive power set points to the inverter.
- Closed-loop: The DTM receives the inverter terminal voltage, calculates the reactive power based on a pre-defined volt-var curve, and issues the reactive power set points to the inverter.

The volt-var functions described in IEEE 1547-2018 specify categories with parameter values for defining the volt-var curve and the allowable ranges for those parameters; however, it is common in the industry to use non-default volt-var curves [15] to meet the requirements of the grid considering local conditions. Considering all these factors, the test setup was developed to demonstrate the following capabilities:

- Program the inverter for custom volt-var curves.
- Send inverter terminal voltage measurements to a remote server
- Issue active/reactive power dispatch command from a remote server to the inverter

Same volt-var curve was programmed in the inverter when in autonomous SI control mode of operation and in the remote server (DTM) when in remote dispatch method mode of

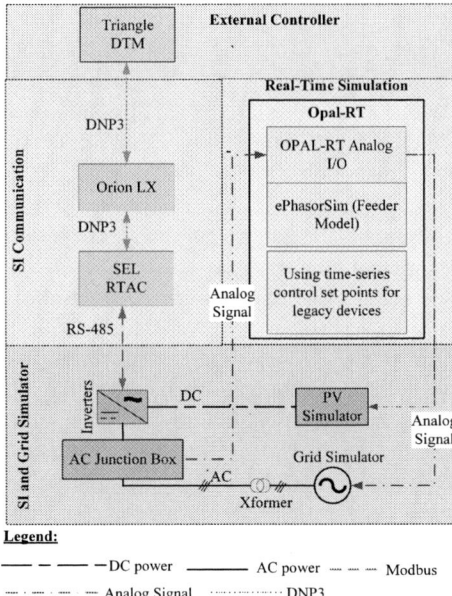

Fig. 2. Experimental setup for scenario analysis using real-time network model.

operation so that the responses can be compared with each other.

D. PHIL set-up for scenario analysis

The test setup in Fig. 1b was expanded by including a network model running in real time to regulate the inverter point of common coupling (PCC) voltage, as shown Fig. 2. This power-hardware-in-the-loop (PHIL) setup was used to run different loading and PV generation scenarios. Eventually, the performance of the two different inverter control modes were evaluated in a PHIL testing framework.

E. Real-time Network Model With Legacy Voltage Regulation Equipment

A real-time distribution network model was used in the PHIL test to experimentally evaluate the performance of remote dispatch method and autonomous SI control in distribution network voltage regulation. Separate profiles for all the load and PV generation for the simulated PV units were used in the real-time simulation. Time-series control set points (e.g., on/off status of capacitor banks, line regulator tap positions) for the legacy devices were used to emulate the operation of those devices. The PV simulator was programmed with an irradiance profile for the duration of the tests. The inverter PCC voltage obtained from the real-time network model was used to drive the voltage controlled by the grid simulator. Three scenarios were tested for the inverter control, which are described in Section IV-F.

III. PARAMETERS TO CONSIDER WHEN CHARACTERIZING SI RESPONSE

The parameters to consider when characterizing the SI responses are illustrated in Fig.3b for a generic connection

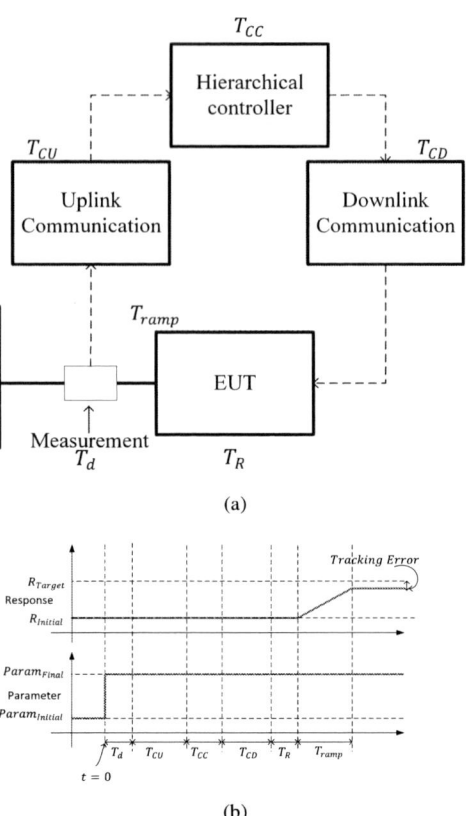

(a)

(b)

Fig. 3. Grid services from inverter: a) generic connection and signal flow diagram for the SIs, b) typical response from SIs.

diagram of the equipment under test (EUT), as shown Fig.3a. The local measurements from the EUT were transmitted to an external controller using an uplink communication channel. The external controller processed that information and generated new control set points for the EUT. The control set points were then transmitted to the EUT using a downlink communication channel. The EUT needs some time to receive the control signal, perform the calculation in the local controller, and execute the control set points. The following list summarizes the process and defines the associated parameters:

- Detection time (T_d): Time needed to measure the parameter (e.g., PCC voltage)
- Uplink communication latency (T_{CU}): Time needed to transmit the measured parameter from the EUT terminal to the external controller
- External controller computation time (T_{CC}): Time needed for the external controller to compute the control signal for the EUT based on the measured parameter
- Downlink communication latency (T_{CD}): Time needed to transmit the external controller generated control signal to the EUT terminal
- EUT response delay (T_R): Time needed by the EUT to respond to the control signal
- Ramping time (T_{ramp}): Time needed by the EUT to

change the response from one step to another step
- Communication and computational time (T_L): Total time needed to measure the EUT parameter, communicate the parameter to the external controller, compute the external control signal, and transmit the control signal to the EUT
- Tracking error: Difference between the control signal and the actual EUT response.

IV. EXPERIMENTAL RESULTS

The SIs under the autonomous SI control and remote dispatch method modes of operation were characterized by running several experiments using the testing protocols and methodologies described in Section II. Table I provides a summary of these experiments.

A. Remote Dispatch Method (Open-Loop): Reactive Power Ramp Rate

Open-loop tests using remote dispatch method by sending reactive power set points from the DTM to the inverter were carried out to determine the reactive power ramp rate in response to an external control signal. Fig. 4a shows the SI responses when the external control signal requests a change in reactive power from a low value to a high value, and Fig. 4b shows a similar change but in the opposite direction. In this test, the positive reactive power ramp rate was found to be 4.16%/second, whereas the negative reactive power ramp rate was found to be -14.24%/second. The ramping was observed to be linear.

B. Remote Dispatch Method (Closed-Loop): Communication and Computation Delay

Closed-loop tests involving the inverter, the DTM, and the corresponding communication setup were performed to determine the communication and computational loop times (T_L). In these tests, the DTM was programmed with a volt-var function. In this test as shown in Fig. 5, for the step change in the voltage from high to low, the T_L was found to be 11.243 seconds and for the step change in the voltage from low to high, the T_L was found to be 16.1 seconds. Note that the measured T_L reported here were based on two tests. A statistical analysis of T_L from additional measurements is presented in Section IV-E.

C. Remote Dispatch Method: volt-var Dispatched Through DTM

Test case similar to those described in Section IV-B but including step changes in the PCC voltage were used to characterize the closed-loop communication delay; the test results are shown in Fig. 6. We found that the closed-loop communication delay was not constant. One interesting observation here is that there were voltage fluctuations in the PCC voltage near 1.08 per unit (p.u.). The grid simulator voltage was set to stiff in this test case. So, this voltage fluctuation was caused by the transformer impedance and the fluctuation reactive power (closer to maximum reactive power, Q_{max}, output from the inverter) from the inverter.

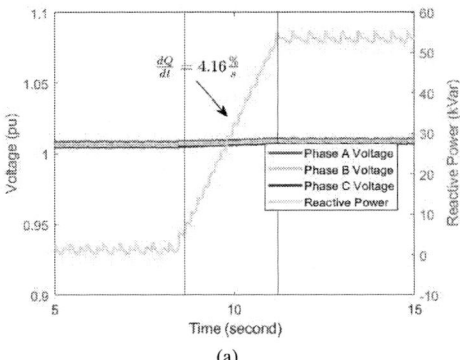

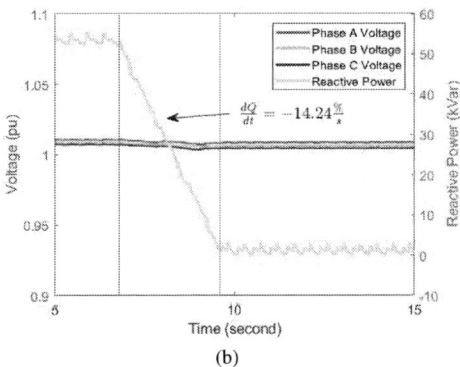

Fig. 4. Test result to characterize the reactive power ramp rate when the inverter was controlled by remote dispatch method: a) positive ramp rate, b) negative ramp rate.

D. Autonomous SI Control: volt-var Programmed at Inverter Local Controller

To compare the performance of the remote dispatch method, the volt-var function with the same parameters from the previous test cases was programmed in the inverter in autonomous SI control. Experimental results are shown in Fig. 7. The mean ramping time (Tramp) was found to be 24.6 seconds with a standard deviation of 2.244 seconds. The reactive power ramping was following a first-order pattern instead of the linear ramp of the reactive power in response to the external control signal.

E. Summary of the Characterization Experiments

The observation from all the characterization test results are summarized in Table II. Significant differences are noticeable between the remote dispatch method and autonomous SI control for the same grid voltage conditions. Based on the data shown in Table II, the observed ramp rates and delays for the two control methods need to be considered when planning and designing voltage regulation using SIs. Though fast response from the inverters might be helpful to address disturbances in the grid quickly, there could be potential risks of unintended consequences associated with fast response [7]. Similarly, the inverter response types, e.g. linear, step changes, or PI type responses, might have considerable impact on the overall performance of the voltage regulations when large

TABLE I
CHARACTERIZATION TEST MATRIX

Case #	Description	Test Step	Objective
1	Remote dispatch method: Open-loop DTM to inverter Q command*	Send Q1=0kVar and Q2=55kVar** from DTM to inverter Send Q1=55kVar and Q2=0kVar** from DTM to inverter	1. Determine the ramping time (Tramp).
2	Remote dispatch method: Closed-loop DTM to inverter Q command* (volt-var in DTM)	Change AC PCC voltage from high to low and low to high.	1. Test the interoperability performance between the DTM and the inverters using the applied communication method. 2. Characterize the inverter response in remote dispatch method. 3. Identify any stability and/or unknown issue for inverter control from an external controller.
3	Autonomous SI control: volt-var in inverter	Change AC PCC voltage from high to low and low to high.	1. Characterize the volt-var function in autonomous SI control.

*The Q command refers to reactive power command from the DTM to the inverter.
**Q1 and Q2 refers to different reactive power set points from the DTM to the inverter.

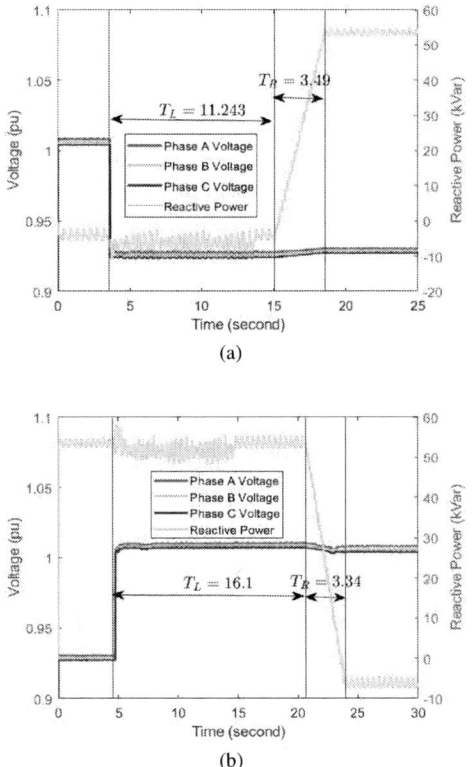

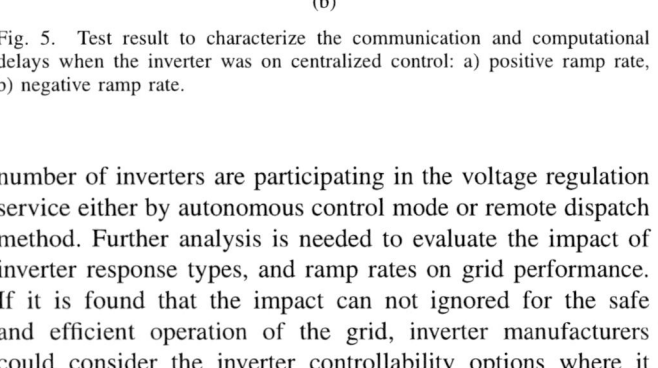

Fig. 5. Test result to characterize the communication and computational delays when the inverter was on centralized control: a) positive ramp rate, b) negative ramp rate.

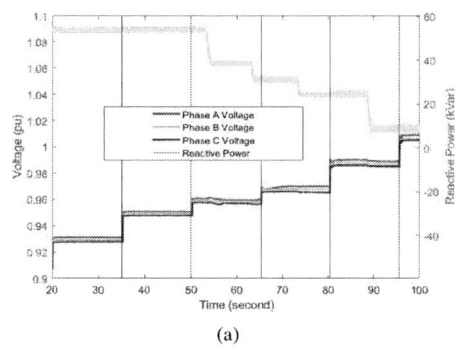

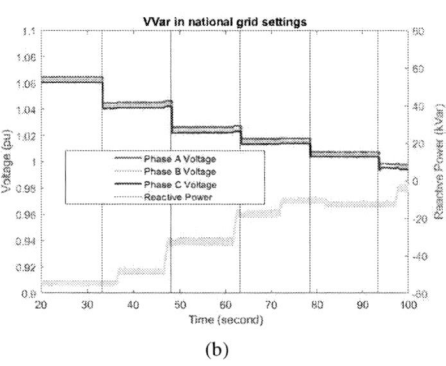

Fig. 6. Performance of the inverter in remote dispatch method with volt-var hosted in the DTM: a) voltage changes from low to high in steps; b) voltage changes from high to low in steps.

number of inverters are participating in the voltage regulation service either by autonomous control mode or remote dispatch method. Further analysis is needed to evaluate the impact of inverter response types, and ramp rates on grid performance. If it is found that the impact can not ignored for the safe and efficient operation of the grid, inverter manufacturers could consider the inverter controllability options where it is possible to set the way in which inverters respond and the rate/timing/character of response to a change in set-point. Such inverter response controllability options can also be incorporated in the future version of IEEE Std 1547 which currently has limited requirements on inverter response types, and ramp rates.

978-1-6654-6060-6/23 $31.00 © 2023 IEEE

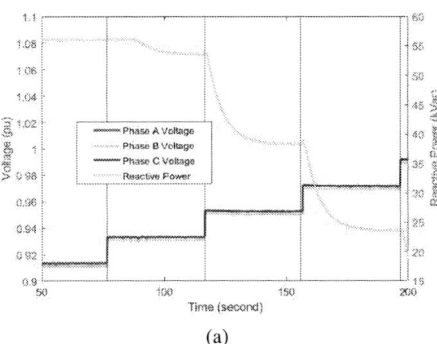

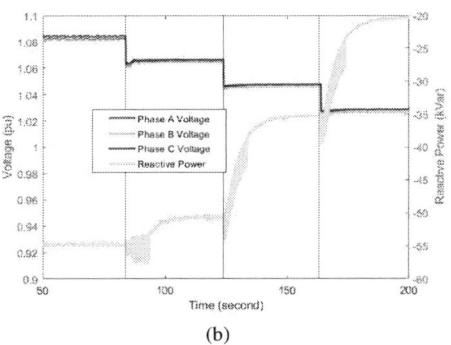

Fig. 7. Test result to characterize the inverter volt-var response in autonomous SI control: a) voltage changes from low to high; b) voltage changes from high to low.

TABLE II
QUANTITATIVE COMPARISON OF REMOTE DISPATCH METHOD TO AUTONOMOUS SI CONTROL RESPONSE

Characterization Parameter	Statistical Parameter	Remote Dispatch	Autonomous Control
T_L [1] Unit: second	Mean	11.74	0.77
	Std. deviation	4.78	0.31
	Maximum	20.16	1.20
	Minimum	2.80	0.20
Positive ramp rate [2] Unit: %/second	Mean	10.87	0.45
	Std. deviation	3.68	0.13
	Maximum	16.26	0.58
	Minimum	6.37	0.15
Negative ramp rate [3] Unit: %/second	Mean	-10.02	-0.49
	Std. deviation	2.96	0.01
	Maximum	-4.50	-0.47
	Minimum	-14.05	-0.51

[1] For autonomous SI control: $T_{CU} = 0$, $T_{CC} = 0$, $T_{CD} = 0$
[2] Rate of change of reactive power from low value to high value
[2] Rate of change of reactive power from high value to low value

F. Scenario Analysis Using the PHIL Test Platform

Three different scenarios as explained in Table III were executed using the PHIL setup described in Section II-D. The experimental results from these scenario analyses are illustrated in Fig. 8. These tests demonstrate that voltage regulation can be accomplished by both the autonomous SI control and remote dispatch method modes of operation. The response delay to a changing PCC voltage (e.g., T_L) as well as the reactive power ramping rate impact the network voltage regulation; however, these tests mainly focused on i) the interoperability of the external controller and the inverter and ii) the demonstration of an experimental setup (via the scenario analysis) to examine the performance of autonomous SI control and remote dispatch method for voltage regulation. Detailed performance analysis of any autonomous SI control and remote dispatch method using the setup will need additional experiments.

TABLE III
SCENARIOS TESTING USING PHIL SETUP

Scenario	Reactive Power Mode	Control Model
1	Unity pf	autonomous SI control
2	VVar	autonomous SI control
3	External signal	remote dispatch method

V. CONCLUSION

This paper reports an experimental study demonstrating and validating the communication setup from a remote controller using intermediate communication gateways to an SI. The inverter was able to receive the external control signal for the remote dispatch of the reactive power set points. The study compared the response times of the inverter for remote dispatch method and autonomous SI control. For remote dispatch method, the computational and communication time was found to be within the range from 2–20.16 seconds, and the inverter ramp rates were observed to be within the ranges from 6.3%/second–16.26%/second and from -4.5%/second–14.05%/second. The ramp rates of the inverter in autonomous SI control included the inverter embedded control response and were slower than remote dispatch method. The reactive power ramp rates for autonomous SI control mode averaged 0.45%/second and -0.49%/second. The testing demonstrated the communication paths for a remote dispatch method-controlled inverter to dispatch inverters using remote dispatch controller via the intermediate communication gateways and communication channel. The delays in the communication and response times might be of concern for voltage control, and there might be advantages to using a hybrid communication architecture (i.e., remote control and dispatch but with the ability to receive inverter measurements to update the remote settings from a centralized controller).

978-1-6654-6060-6/23 $31.00 © 2023 IEEE

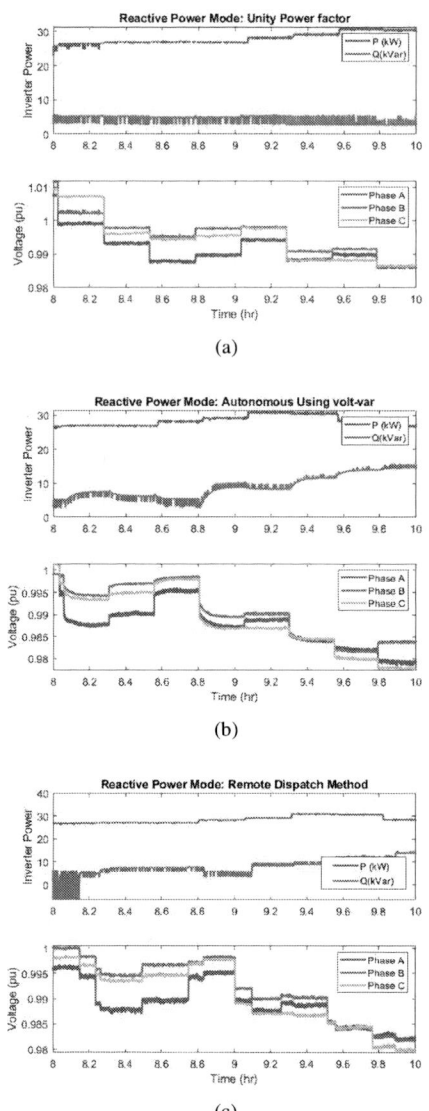

Fig. 8. PHIL test result to evaluate the applicability of remote dispatch method and autonomous SI control: a) Scenario 1, b) Scenario 2, 3) Scenario 3.

REFERENCES

[1] D. Narang, R. Mahmud, M. Ingram, and A. Hoke, "An overview of issues related to ieee std 1547-2018 requirements regarding voltage and reactive power control," 9 2021. [Online]. Available: https://www.osti.gov/biblio/1821113

[2] M. Shahin, E. Topriska, M. Gormley, and M. Nour, "Design and field implementation of smart grid-integrated control of PV inverters for autonomous voltage regulation and VAR ancillary services," *Electric Power Systems Research*, vol. 208, p. 107862, jul 2022.

[3] R. Mahmud, A. Hoke, and D. Narang, "Validating the test procedures described in UL 1741 SA and IEEE P1547.1," in *2018 IEEE 7th World Conference on Photovoltaic Energy Conversion (WCPEC) (A Joint Conference of 45th IEEE PVSC, 28th PVSEC & 34th EU PVSEC)*. IEEE, jun 2018, pp. 1445–1450. [Online]. Available: https://ieeexplore.ieee.org/document/8547346/

[4] J. Wang, H. Padullaparti, F. Ding, M. Baggu, and M. Symko-Davies, "Voltage Regulation Performance Evaluation of Distributed Energy Resource Management via Advanced Hardware-in-the-Loop Simulation †," 2021. [Online]. Available: https://doi.org/10.3390/en14206734

[5] R. Mahmud and G.-S. Seo, "Blockchain-Enabled Cyber-Secure Microgrid Control Using Consensus Algorithm; Blockchain-Enabled Cyber-Secure Microgrid Control Using Consensus Algorithm," 2021.

[6] M. Ingram, R. Mahmud, and D. Narang, "Informative background on the interoperability requirements in ieee std 1547-2018," 7 2021. [Online]. Available: https://www.osti.gov/biblio/1807667

[7] A. Hoke, R. Mahmud, A. Nelson, D. Pattabiraman, M. Asano, D. Arakawa, B. Pierre, M. Elkhatib, J. Tan, V. Gevorgian, C. Antonio, and E. Ifuku, "Fast grid frequency support from distributed energy resources," 3 2021. [Online]. Available: https://www.osti.gov/biblio/1772978

[8] J. I. Giraldez Miner, A. F. Hoke, P. Gotseff, N. D. Wunder, M. Emmanuel, A. Latif, E. Ifuku, M. Asano, T. Aukai, R. Sasaki, and M. Blonsky, "Advanced Inverter Voltage Controls: Simulation and Field Pilot Findings," Tech. Rep., 2018.

[9] R. Mahmud, A. Hoke, and D. Narang, "Fault Response of Distributed Energy Resources Considering the Requirements of IEEE 1547-2018," in *2020 IEEE Power & Energy Society General Meeting (PESGM)*. IEEE, aug 2020, pp. 1–5. [Online]. Available: https://ieeexplore.ieee.org/document/9281878/

[10] R. Mahmud, A. Hoke, and L. Yu, "Characterization of DER Momentary Cessation and Rate-of-Change-of-Frequency Response," in *The IEEE Photovoltaic Specialists Conference*, Philadelphia, PA, 2022.

[11] "Distributed Test Manager - Overview." [Online]. Available: https://www.trianglemicroworks.com/products/testing-and-configuration-tools/dtm-pages/overview

[12] "Distributed Network Protocol3- Overview." [Online]. Available: https://www.dnp.org/About/Overview-of-DNP3-Protocol

[13] "IEC-61131-3:2013." [Online]. Available: https://webstore.iec.ch/publication/4552

[14] "JavaScript." [Online]. Available: https://www.ecma-international.org/publications-and-standards/standards/ecma-262

[15] X. Zhu, R. Mahmud, B. Mather, P. Mishra, and A. Meintz, "Grid Voltage Control Analysis for Heavy-Duty Electric Vehicle Charging Stations," in *2021 IEEE Power & Energy Society Innovative Smart Grid Technologies Conference (ISGT)*. IEEE, 2021, pp. 6–10.

Doped GaInAs/GaP quantum well superlattice solar cells with 27.5% efficiency

Ryan M France, Myles A Steiner

National Renewable Energy Laboratory, Golden, CO, United States

Record efficiency III-V multijunction devices were recently demonstrated by incorporating GaInAs quantum wells (QWs) to increase absorption below the GaAs band edge, resulting in a more optimal multijunction bandgap combination. The QW cell had excellent performance by overcoming limitations to material quality including interface quality, stress-balancing, and strain-driven surface fluctuations, enabling an optically-thick quantum well solar cell by incorporating many quantum wells into a thick intrinsic region. However, the thick intrinsic region still results in large depletion region recombination, observed in JV curves as a low FF and large J02 depletion region recombination. With very thin GaAsP barriers in a strain-balanced superlattice (SLS) solar cell, tunneling begins to dominate transport, and carrier collection via drift in an intrinsic region may not be needed. Here, we experimentally investigate doping in GaInAs SLS solar cells that utilize very thin, ~2 nm, GaP barriers for stress-balancing. Doping reduces J02 depletion region recombination and improves the solar cell FF to over 86%, but high doping reduces the carrier collection, leading to a tradeoff in efficiency. We show that the barrier thickness also plays an important role in carrier collection, and we demonstrate doped SLS devices with 27.5% under the AM1.5 global spectrum and 23.9% under the AM0 space spectrum are demonstrated, with pathways toward further improvement.

Aggregated Three-Phase Photovoltaic Inverter Model with Sandia Frequency Shift Islanding Detection

Nelson E. Saavedra-Peña[1], Rachid Darbali-Zamora[2] Edgardo Desarden-Carrero[1], and Erick Aponte-Bezares[1]

[1]University of Puerto Rico-Mayagüez, Mayagüez, Puerto Rico 00682, USA
[2]Sandia National Laboratories, Albuquerque, New Mexico, 87185, USA

Abstract – **Aggregated models of inverters have previously been proposed to represent the reality of bulk power systems where high penetration of photovoltaic (PV) inverters is present. High penetration can affect dynamics of electric power system (EPS) in general and power quality especially during peak solar hours where PV inverters will tend to dictate the performance of EPS. In this work, an approach for evaluating a multi-inverter aggregated model for islanding detection application is taken. A system of parallel-connected inverters is modeled as a single aggregated inverter model. The objective is to study the effects of scaling (aggregation) of this PV inverter model to the islanding detection method, in this case the SFS.**

Keywords – *aggregated inverter model, unintentional islanding, islanding detection, sandia frequency shift, photovoltaic inverter.*

I. INTRODUCTION

Electric power systems (EPS) are expected to be reliable and resilient to meet the energy demands of the modern world [1]. Irregular modes of operation as well as improper interconnection of distributed energy resources (DERs) can affect the reliability and resilience of modern EPS. With increasing penetration of DERs, especially photovoltaic (PV) inverters, behavior of EPS must be analyzed [2]. DERs must comply with current standards and regulations in order to provide satisfactory performance and be safe in overall. PV power systems are a type of DER that, by definition, converts solar energy into electric energy in a form that is usable and appropriate for the application intended [3]. PV power systems have become an accessible and cost-effective way of meeting energetic demands for applications involving households, grid utilities, and industrial environments. A common type of power electronic converter is a PV inverter. A PV inverter transforms the DC electric energy produced from PV panels into AC electric energy. Various models for inverters have been proposed to describe and analyze the behavior of the power converter dynamics. A common topology for a voltage source inverter (VSI) is the H-bridge. Usually, cascaded H-bridges (CHB) are used to provide multiple inverter levels and reduce the introduction of unwanted harmonics created by switching phenomena [4]. Multi-level inverters can operate without need of a harmonic output filter [5]. Two main classifications for grid-connected inverters exist: grid-forming (GFM) and grid-following (GFL). GFM inverters are designed to operate in a similar manner to a synchronous machine. Control schemes specifically designed for GFM inverters have been developed. to provide stability [6], [7], [8], as well as to regulate system frequency, and voltage [9], [10], [11]. This type of inverter is not part of the scope of this work, as we will only be dealing with GFL inverters. GFL inverters rely on grid presence and phase lock loop (PLL) mechanisms to operate correctly. GFL inverters are commonly programmed with current controllers and power controllers based of the measurements taken from the grid. When the grid is not present, GFL inverters will most likely terminate operation or in worst case scenario operate incorrectly. Aggregated models of inverters have been proposed to represent the reality high penetration of PV inverters [12]. High penetration can affect dynamics of an EPS in general and power quality especially during peak solar hours where PV inverters will tend to dictate the performance of the EPS. Newer PV inverters have the ability to provide grid-support functionalities (GSF) to regulate both frequency and voltage [13], [14]. Proposed models include GSF, frequency droop control and power factor correction but are limited when representing islanding scenarios, tripping and ride through procedures as described in the IEEE 1547 standard [15], [16].

Islanding is a condition in which a part of an EPS is energized by at least one DER but is disconnected from the rest of the EPS [17]. Unintentional islanding (UI) is when islanding occurs without coordinated action or planning. In case of an UI, the DER is required to detect and terminate operation within 2 seconds of the formation of such island [18]. By analyzing the performance of different UI detection methods, the relative effectiveness of these methods can be determined by defining a non-detection zone parameter (NDZ). The NDZ is defined as an operational region of the detection method in which the method fails to detect the island. Small NDZs implicate that the method is very likely to detect an islanding scenario. However, large NDZs suggest that the DER is less probable to detect the formation of an island [19].

In this work, an approach for evaluating a multi-inverter aggregated model for islanding detection application is taken. A system of parallel-connected inverters is modeled as single aggregated inverter model. The aggregated inverter model is programmed with the Sandia frequency shift (SFS) islanding detection method. The objective is to study the effects of scaling (aggregation) of this PV inverter model to the islanding detection method, in this case the SFS.

II. SANDIA FREQUENCY SHIFT (SFS) DETECTION METHOD

The SFS is an active islanding detection method that was developed by Sandia National Laboratories. The SFS injects a disturbance signal into inverter current phase angle, defined in equation (1), that is proportional to the error between PV inverter frequency and nominal operating frequency [20], [21]. When grid-connected, the SFS will have no noticeable effect on PV inverter operation due to small error between inverter frequency, f_{PCC}, and nominal system frequency, f_{nom}. Also,

978-1-6654-6060-6/23 $31.00 © 2023 IEEE

the grids stiffness contributes enormously to the stability of the system. On the other hand, when islanded, inverter frequency is determined by loading conditions and will differ from nominal system frequency. The SFS then amplifies this error by a predetermined gain, K, and inject this disturbance into the PV inverter current phase angle. The SFS phase angle disturbance is injected via a phase angle transformation operation. This phase angle transformation creates a dependance in the references of the inverter that is proportional to the difference between PCC frequency and nominal frequency (60 Hz).

$$\theta_{sfs} = \frac{\pi}{2} \cdot K \cdot (f_{PCC} - f_{nom})] \tag{1}$$

The SFS disturbance signal is injected via a phase angle transformation operation. The SFS phase angle transformation, shown in equation (2), makes the commanded current references of the PV inverter a function of the difference between PCC frequency and nominal frequency (60 Hz).

$$\begin{bmatrix} I_{d_{ref}}^* \\ I_{q_{ref}}^* \end{bmatrix} = \begin{bmatrix} \cos(\theta_{sfs}) & -\sin(\theta_{sfs}) \\ \sin(\theta_{sfs}) & \cos(\theta_{sfs}) \end{bmatrix} \cdot \begin{bmatrix} I_{d_{ref}} \\ I_{q_{ref}} \end{bmatrix} \tag{2}$$

III. PHOTOVOLTAIC INVERTER MODEL

For the three-phase GFL PV inverter model, an LCL output filter was selected as the harmonic filter to reduce noise created by switching components (e.g., FETs, IGBTs) when modulating DC voltage into AC voltage. Fig. 1 shows a diagram of LCL PV inverter model. Using the two-axis reference frame (dq0) theory from [22], the three-phase GFL PV inverter was represented in the equivalent dq circuits shown in Fig. 2, and Fig. 3, respectively.

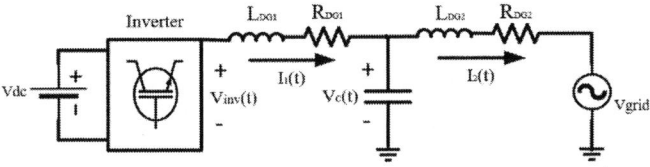

Fig. 1. One-line diagram of LCL PV inverter model used.

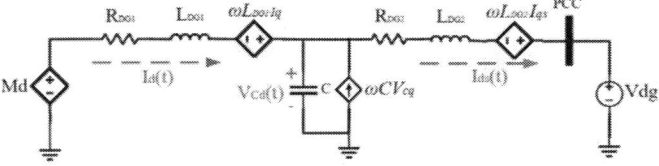

Fig. 2. Direct ('d') axis PV inverter cqt. with LCL filter.

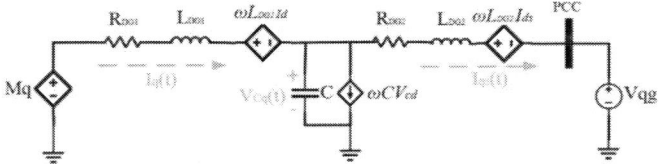

Fig. 3. Quadrature ('q') axis PV inverter cqt. with LCL filter.

The equations for total inverter active and reactive power injected to the grid using the Park's transformation are shown in equation (3) and equation (4), respectively.

$$P_{inv} = 1.5 * \left(V_d * I_{ds} + V_q * I_{qs} \right) \tag{3}$$

$$Q_{inv} = 1.5 * \left(V_q * I_{ds} - V_d * I_{qs} \right) \tag{4}$$

IV. PHOTOVOLTAIC INVERTER SCALING AND AGGREGATION

The inverter scaling was done using the method described in [23]. A power scaling factor, shown in equation (5), is introduced to aggregate the system of inverters into a reduced order inverter model. The aggregated PV inverter model represents the dynamics of multiple inverters by scaling associated controller and filter parameters by this power scaling factor, $\widehat{k_p}$. The variable k_s, represents the individual contribution of each inverter to the aggregated inverter model thru the power scaling factor. This variable is calculated by dividing the power rating of the aggregated model by the rating of the individual inverter used as base (reference) for scaling, as shown in equation (6).

$$\widehat{k_p} = \sum_{s=1}^{N} k_s \tag{5}$$

$$k_s = \frac{P_{rated}}{P_{base}} \tag{6}$$

The specific parameters of the inverter model that were scaled are: filter inductance (L_{DG}), filter resistance (R_{DG}), filter capacitor (C_{DG}), and controller gains (K_p, K_i). The inverter parameters mentioned before were scaled as follows: $L_{DG}/\widehat{k_p}$, $R_{DG}/\widehat{k_p}$, $C_{DG} * \widehat{k_p}$, $K_p/\widehat{k_p}$, $K_i/\widehat{k_p}$. By substituting the base system parameters with the scaled parameters, a reduced-order aggregated model of N parallel connected inverter systems can be represented using this scaling theory. A 24 kVA, and 33 kVA PV inverters were connected in parallel and compared with a reduced-order 57 kVA PV inverter model using the scaling theory described. The developed aggregated 57 kVA PV inverter model was then compared with the two PV inverter models using the same PCC and load parameters. This was done in order to compare the performance of the aggregated PV inverter model to the two PV inverter system dynamics. The parameters for the 24 kVA, and 33 kVA PV inverter models are shown in Table I and Table II, respectively

TABLE I:
SFS 24 KVA PV INVERTER PARAMETER DESCRIPTION

Parameter	Description	Value	Units
R_{DG1}	Parasitic resistance of L_{DG1}	0.8	Ω
L_{DG1}	Inverter Side Inductor	2.5	mH
R_{DG2}	Parasitic resistance of L_{DG2}	0.8	Ω
L_{DG2}	Grid Side Inductor	2.5	mH
C	Filter Capacitor	4.7	μF
f_{sw}	Switching Frequency	20	kHz
f_{nom}	Nominal Frequency	60	Hz
V_{LL}	Line to line voltage	480	V_{rms}

978-1-6654-6060-6/23 $31.00 © 2023 IEEE

TABLE II:
SFS 33 kVA PV INVERTER PARAMETER DESCRIPTION

Parameter	Description	Value	Units
R_{DG1}	Parasitic resistance of L_{DG1}	0.8	Ω
L_{DG1}	Inverter Side Inductor	1.4	mH
R_{DG2}	Parasitic resistance of L_{DG2}	0.8	Ω
L_{DG2}	Grid Side Inductor	1.4	mH
C	Filter Capacitor	9.19	μF
f_{sw}	Switching Frequency	20	kHz
f_{nom}	Nominal Frequency	60	Hz
V_{LL}	Line to line voltage	480	V_{rms}

The scaled parameters for the aggregated 57 kVA PV inverter are shown in Table III. Table IV summarizes the SFS gains selected for PV inverter models. The gains were empirically selected such that the SFS method would detect islanding in less than the 2 second regulation specified by IEEE 1547 Std. As the system used as base (reference) for the aggregated model was the 33 kVA, the SFS gains are defined equally. The power scaling factor in this case is 1.72.

TABLE III:
SFS SCALED PV INVERTER PARAMETER DESCRIPTION

Parameter	Description	Value	Scaled Value	Units
K_i	Integral Gain	400	231.58	V/A·s
K_p	Proportional Gain	0.85	0.49	V/A
L_{DG1}	PV Side Inductor	1.4	0.81	mH
R_{DG1}	L_{DG1} Resistance	0.8	0.46	Ω
L_{DG2}	Grid Side Inductor	1.4	0.81	mH
R_{DG2}	L_{DG2} Resistance	0.8	0.46	Ω
C	Filter Capacitor	9.19	15.87	μF

TABLE IV:
SFS GAINS FOR PV INVERTER MODELS

Parameter	Description	Value
K_{SFS1}	33 kVA Model SFS Gain	0.07
K_{SFS2}	24 kVA Model SFS Gain	0.12
K_{SFS3}	57 kVA Model SFS Gain	0.07

V. RESULTS

A resonant island was created for the simulation to represent a case in which no islanding detection method was programmed into the PV inverter controller, the PV inverter will continue to operate. The active and reactive power setpoint were 57 kW and 0 kVARS, respectively. The SFS PV inverter models were subjected to islanding conditions using the distributed system under test (DSUT), shown in Fig. 4. Two PV inverter models (24 kVA and 33 kVA) were connected to the same PCC. The aggregated model of 57 kVA, using scaled controller and filter parameters, was connected to another PCC with same grid and loading conditions as the two PV inverter system. The goal was to compare the islanding detection times between the two PV inverter system and aggregated model when using the SFS as the islanding detection method. The under frequency (UF) and over frequency (OF) load parameters for the resonant island are described in Table V and Table VI, respectively.

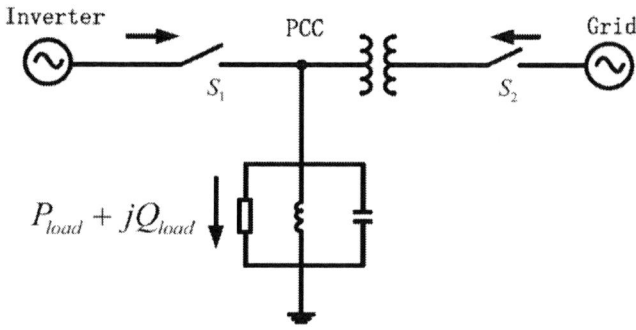

Fig. 4. DSUT for aggregated and two PV inverter system simulations.

TABLE V:
UF LOAD PARAMETER DESCRIPTION

Parameter	Description	Value	Units
R_{Load}	Resistive Load	4.14	Ω
C_{Load}	Capacitive Load	808	μF
L_{Load}	Inductive Load	9.1	mH
f_0	Resonant Frequency	58.7	Hz
P_{Load}	Active Load Power	55.6	kW
Q_{Load}	Reactive Load Power	-3.03	kVARS
S_{Load}	Apparent Load Power	55.7	kVA

TABLE VI:
OF LOAD PARAMETER DESCRIPTION

Parameter	Description	Value	Units
R_{Load}	Resistive Load	4.14	Ω
C_{Load}	Capacitive Load	808	μF
L_{Load}	Inductive Load	8.68	mH
f_0	Resonant Frequency	60.09	Hz
P_{Load}	Active Load Power	55.6	kW
Q_{Load}	Reactive Load Power	0.224	kVARS
S_{Load}	Apparent Load Power	55.6	kVA

A. Under Frequency Scenario Results

Fig. 5 illustrates the PCC frequency for both, two PV inverter system (24 kVA and 33 kVA) and the aggregated 57 kVA PV inverter model. Notice from these results that once the islanding condition is created, the SFS islanding detection method drives the frequency of the PV inverter into an UF scenario by amplifying the difference between nominal frequency and actual island frequency once the resonant load gets disconnected from the grid (utility). In this case the limit for the UF is set to 52 Hz. As soon as the PCC frequency reaches this limit, the passive OFP/UFP of the PV inverter models controller detects the islanding scenario and ceases to operate. The SFS was able to detect the island by tripping UF protection after 0.68 seconds from the start of the islanding condition.

The same islanding detection times for both systems are observed. Fig. 6 and Fig. 7 illustrates the active (P), and reactive (Q) power, respectively. The active and reactive power for the model deviate from setpoint power during the non-detection period of the islanding scenario. This is directly related to the way that the SFS phase angle transformation injects the disturbance into the PI controller. Although the SFS has the same gain, dynamics of the systems are slightly different in response. The main goal is to develop an aggregated model that detects the island at the same time as the multiple PV system. In order to detect the island at the same time (same Hz/s) for systems with different output filters, and power ratings, the controller for the aggregated PV inverter has to supply slightly less reactive power to yield the same detection time.

Fig. 8 illustrates the instantaneous voltages and currents of both PV inverter systems. Only phase A voltage and currents are shown as we assume a three-phase balanced system. As we can observe from the plots, the SFS method creates the deviation in frequency by introducing a phase shift between the PCC voltage and PV inverters output current being injected into the load. In a PV inverter system, usually, the PV inverter operates at unity power factor (Q=0). This condition changes throughout the islanding detection period (until the UF limit is reached) as the SFS phase angle transformation introduces some direct axis component into the quadrature axis reference and vice versa. Fig. 9 illustrates the instantaneous voltages and current of the aggregated PV inverter model. For the UF scenario, instantaneous current will tend to lag the phase voltage until the UF limit is reached. This is in essence the way that the SFS drives the frequency into UFP/OFP. The voltage at the PCC will also begin to sag from nominal value as the grid (utility) is the one that is in charge of providing voltage regulation. As we can observe from both the aggregated and two PV inverter systems, the behavior of the current/voltage dynamics are very similar in both the aggregated and two PV inverter system when islanding occurs. This is due to the fact that dynamics for both of the systems are governed by the SFS detection method, and the gains selected.

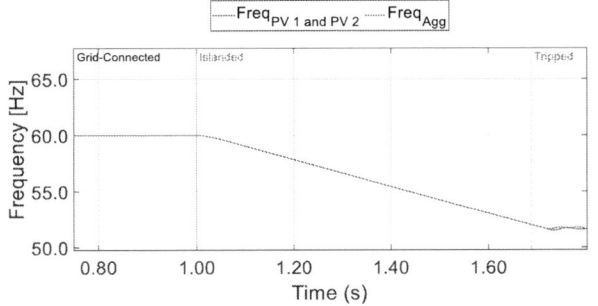

Fig. 5. Frequency at the PCC for the two PV inverters (33 kVA and 24 kVA) and the aggregated 57 kVA PV inverter model for an UF scenario.

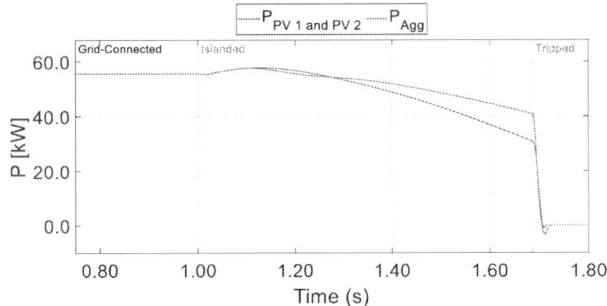

Fig. 6. Active power at the PCC for the two PV inverters (33 kVA and 24 kVA) and the aggregated PV inverter model for an UF scenario.

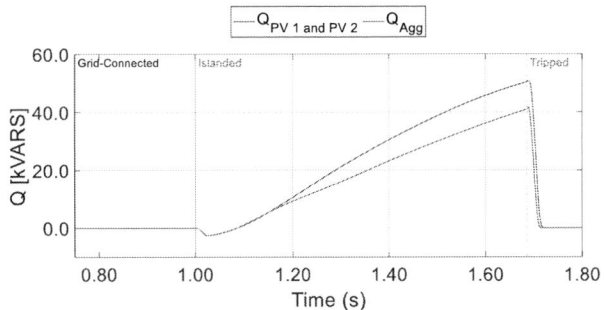

Fig. 7. Reactive power at the PCC for the two PV inverters (33 kVA and 24 kVA) and the aggregated PV inverter model for an UF scenario.

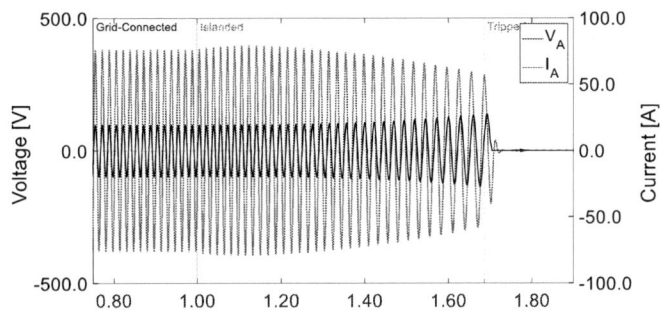

Fig. 8. Instantaneous voltage and current of the multiple PV inverters connected in parallel (33 kVA and 24 kVA) for an UF scenario.

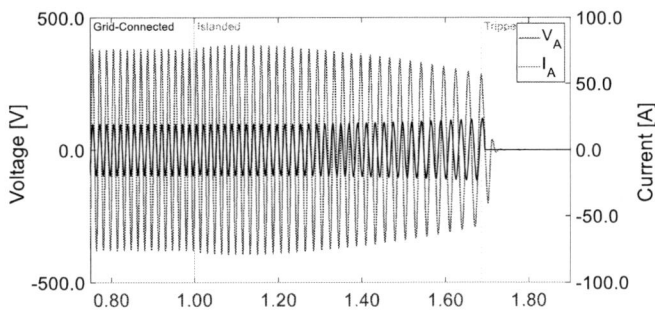

Fig. 9. Instantaneous voltage and current of the aggregated PV inverter model (57 kVA) for an UF scenario.

B. Over Frequency Scenario Results

For the OF scenario the same DSUT was used for comparison between the aggregated and two PV inverters. The only change was in the inductance value of the load. This was done in order to create a resonant frequency above nominal frequency (60 Hz). Fig. 10 illustrates the frequencies of the two PV inverter system (33 kVA and 24 kVA) and the aggregated 57 kVA inverter model. Notice from these results that once the islanding condition is created in both PV inverters, the SFS deviates the frequency towards the OF limits (68 Hz). The frequency for both PV inverters is deviated with the same Hz/s slope due to the controlling action of SFS, once again. The OF protection is able to detect the island and cease operation after around 0.68 second of the creation of the island, complying with applicable standards and regulations for grid-connected PV systems.

Fig. 11 shows the active power (P) of both PV inverters. As in the UF scenario, the active power gets deviated by the SFS by the same co-dependence between quadrature axis current reference and direct axis current reference. The SFS begins to change these references as a function of the deviation of PCC frequency. Comparing the UF and the OF cases shows that the PV inverter absorbs reactive power instead of supplying it (as was seen in the UF scenario), in order to drive PCC frequency into the OF limit, as shown in Fig. 12.

Fig. 13 and Fig. 14 show the instantaneous voltage and current of the two PV inverter system and the aggregated PV inverter system, respectively, for the OF scenario. In this case the current leads the phase voltage as the SFS takes controlling action over the dynamics of the PV inverter. This statement is true in both systems. As in the UF scenario, the PCC voltage starts to sag because of absence of regulation from the grid.

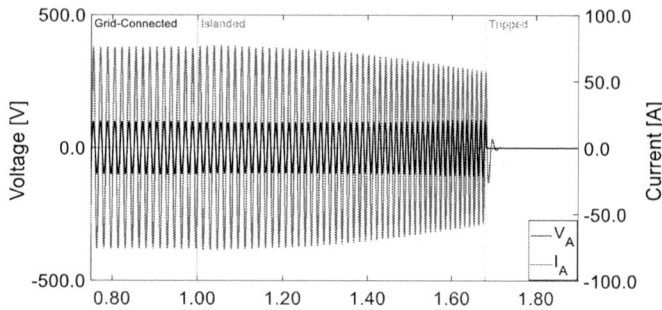

Fig. 13. Instantaneous voltage and current of the multiple PV inverters connected in parallel (33 kVA and 24 kVA) for an OF scenario.

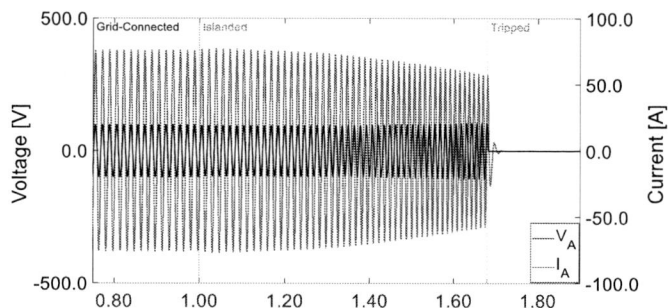

Fig. 14. Instantaneous voltage and current of the aggregated PV inverter model (57 kVA) for an OF scenario.

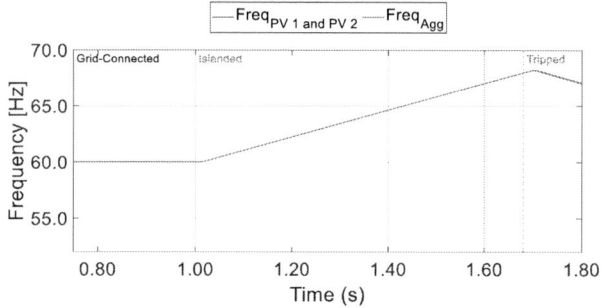

Fig. 10. Frequency at the PCC for the two PV inverters (33 kVA and 24 kVA) and the aggregated 57 kVA PV inverter model for an OF scenario.

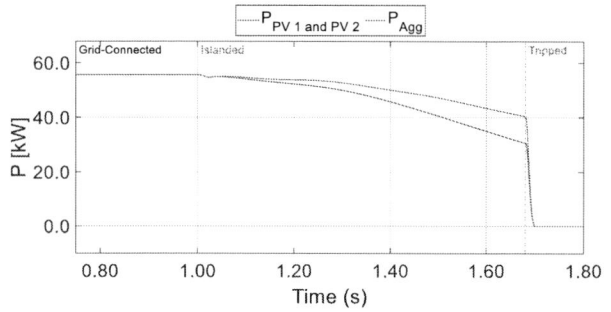

Fig. 11. Active power at the PCC for the two PV inverters (33 kVA and 24 kVA) and the aggregated PV inverter model for an OF scenario.

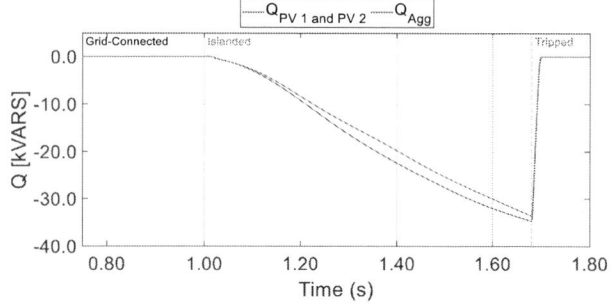

Fig. 12. Reactive power at the PCC for the two PV inverters (33 kVA and 24 kVA) and the aggregated PV inverter model for an OF scenario.

VI. CONCLUSIONS

In this work, an approach for modeling a multi-inverter aggregated model for islanding detection applications is presented. A simulation model of a system consisting of PV inverters connected in parallel is modeled as single aggregated inverter model using MATLAB/Simulink. The developed aggregated PV inverter model is programmed with the SFS method to provide islanding detection functionality. The objective is to study the effects that scaling has on the performance of both the PV inverter model and the SFS islanding detection method. Two PV inverters simulation models, both rated at 24 kVA, and 33 kVA were connected in parallel and compared with the aggregated 57 kVA PV inverter simulation model. The developed aggregated 57 kVA PV inverter model was then compared with the two PV inverter models using the same PCC and load parameters. This was done in order to compare the performance of the aggregated PV inverter model to the two PV inverters dynamics.

Results demonstrate that the proposed aggregated PV inverter model is able to detect both UF and OF islanding scenario at the same time as the two PV inverter system. This is significant since it demonstrates that the aggregated PV inverter model can represent the islanding detection times and frequency behavior of the two individual PV inverter system but with a reduced order system when using scaled control and filter parameters. Results show no dependency on load parameters such as load resonant frequency, and quality factor. Additional analysis is required to characterize scaling factor limitations in order to yield the same islanding detection times for the larger scaling factors.

ACKNOWLEDGEMENT

This article has been authored by an employee of National Technology & Engineering Solutions of Sandia, LLC under Contract No. DE-NA0003525 with the US Department of Energy (DOE). The employee owns all right, title and interest in and to the article and is solely responsible for its contents. The US government retains and the publisher, by accepting the article for publication, acknowledges that the US government retains a non-exclusive, paid-up, irrevocable, world-wide license to publish or reproduce the published form of this article swill provide public access to these results in accordance with the DOE Public Access Plan.

This work was sponsored in part by the Consortium for Hybrid Resilient Energy Systems (CHRES) under grant number DE-NA0003982 from the National Nuclear Security Administration part of the US DOE.

REFERENCES

[1] R. Darbali-Zamora, C. J. Gómez-Mendez, E. I. Ortiz-Rivera, H. Li, and J. Wang, "Solar irradiance prediction model based on a statistical approach for microgrid applications", *2015 IEEE 42nd Photovoltaic Specialist Conference (PVSC)*, 2015, pp. 1-6.

[2] R. Darbali-Zamora, J. Hernandez-Alvidrez, A. Summers, N. S. Gurule, M. J. Reno and J. Johnson, "Distribution Feeder Fault Comparison Utilizing a Real-Time Power Hardware-in-the-Loop Approach for Photovoltaic System Applications", *2019 IEEE 46th Photovoltaic Specialists Conference (PVSC)*, 2019, pp. 2916-2922.

[3] "IEEE Recommended Practice for Utility Interface of Photovoltaic (PV) Systems", *IEEE Std 929-2000*, pp. 1-32, Jan. 2000.

[4] P. Kongsuk, S. Boontua, V. Kinnares and J. Boonseng, "Cascaded H-bridge Multilevel Inverter for Induction Motor Drive with Improved Grid Current Quality", *2021 18th International Conference on Electrical Engineering/Electronics, Computer, Telecommunications, and Information Technology (ECTI-CON)*, 2021, pp. 802-805.

[5] E. Desarden-Carrero, R. D. Zamora, E. Aponte-Bezares, and E. I. Ortiz-Rivera, "Seven-Level Cascaded H-Bridge Multilevel Single-Phase Inverter Implemented with an ATMEGA Microprocessor", *2022 IEEE 49th Photovoltaics Specialists Conference (PVSC)*, 2022, pp. 0916-0922.

[6] P. Marinakis and N. Schofield, "Grid Forming Control for Power Systems with up to 100% Inverter Based Generation", *The 9th Renewable Power Generation Conference (RPG Dublin Online 2021)*, Online Conference, 2021, pp. 143-148.

[7] R. Darbali-Zamora, C. B. Jones, M. S. Lave and E. E. Aponte-Bezares, "The Capability of a Grid-Forming Inverter to Support Dynamic Microgrids with High Penetrations of Photovoltaics Systems", *2022 IEEE 49th Photovoltaics Specialists Conference (PVSC)*, 2022, pp. 1091-1098.

[8] R. Darbali-Zamora, N. S. Gurule, J. Hernandez-Alvidrez, S. Gonzalez and M. J. Reno, "Performance of a Grid-Forming Inverter Under Balanced and Unbalanced Voltage Phase Angle Jump Conditions", *2021 IEEE 48th Photovoltaic Specialists Conference (PVSC)*, 2021, pp. 1409-1416.

[9] M. E. Elkhatib, W. Du and R. H. Lasseter, "Evaluation of Inverter-based Grid Frequency Support using Frequency-Watt and Grid-Forming PV Inverters", *2018 IEEE Power & Energy Society General Meeting (PESGM)*, 2018, pp. 1-5.

[10] R. A. Mastromauro, "Voltage control of a grid-forming converter for an AC microgrid: A real case study", *3rd Renewable Power Generation Conference (RPG 2014)*, 2014, pp. 1-6.

[11] R. Darbali-Zamora, J. Johnson, N. S. Gurule, M. J. Reno, N. Ninad, and E. Apablaza-Arancibia, "Evaluation of Photovoltaic Inverters Under Balanced and Unbalanced Voltage Phase Angle Jump Conditions", *2020 47th IEEE Photovoltaic Specialists Conference (PVSC)*, 2020, pp. 1562-1569.

[12] Electric Power Research Institute (EPRI), "The New Aggregated Distributed Energy Resources (der_a) Model for Transmission Planning Studies: 2019 Update", pp. 1-35, Mar. 2019.

[13] E. Desarden-Carrero, R. Darbali-Zamora and E. E. Aponte-Bezares, "Analysis of Grid Support Functionality Dynamics under Ride-Through Requirements Using Power-Hardware-in-the-Loop Implementation", *2021 IEEE 48th Photovoltaic Specialists Conference (PVSC)*, 2021, pp. 1795-1802.

[14] N. Ninad *et al.*, "PV Inverter Grid Support Function Assessment using Open-Source IEEE P1547.1 Test Package", *2020 47th IEEE Photovoltaic Specialists Conference (PVSC)*, 2020, pp. 1138-1144.

[15] "IEEE Standard for Interconnection and Interoperability of Distributed Energy Resources with Associated Electric Power Systems Interfaces", *IEEE Std 1547-2018 (Revision of IEEE Std 1547-2003)*, pp.1-138, Apr. 2018.

[16] "IEEE Standard Conformance Test Procedures for Equipment Interconnecting Distributed Energy Resources with Electric Power Systems and Associated Interfaces", *IEEE Std 1547.1-2020*, pp.1-282, Jan. 2020.

[17] N. E. Saavedra-Peña, R. Darbali-Zamora, E. Desarden-Carrero and E. Aponte-Bezares, "Towards an Islanding Detection Method Using a Digital Twin Concept", *2021 IEEE 48th Photovoltaic Specialists Conference (PVSC)*, 2021, pp. 0410-0415.

[18] E. Desarden-Carrero, R. Darbali-Zamora, N. S. Gurule, E. Aponte-Bezares and S. Gonzalez, "Evaluation of the IEEE Std 1547.1-2020 Unintentional Islanding Test Using Power Hardware-in-the-Loop", *2020 47th IEEE Photovoltaic Specialists Conference (PVSC)*, 2020, pp. 2262-2269.

[19] E. Desardén-Carrero, R. Darbali-Zamora and E. E. Aponte-Bezares, "Analysis of Commonly Used Local Anti-Islanding Protection Methods in Photovoltaic Systems in Light of the New IEEE 1547-2018 Standard Requirements", *2019 IEEE 46th Photovoltaic Specialists Conference (PVSC)*, 2019, pp. 2962-2969.

[20] M. J. Mukarram and S. V. Murkute, "Sandia Frequency Shift Method for Anti-Islanding Protection of a Gridtied Photovoltaic System", *2020 IEEE International Students' Conference on Electrical,Electronics and Computer Science (SCEECS)*, 2020, pp. 1-5.

[21] P. C. Krause; O. Wasynczuk; S. D. Sudhoff, and S. Pekarek, *"Analysis of Electric Machinery and Drive Systems"*, Wiley-IEEE Press, pp.1- 632, 2013.

[22] N. E. Saavedra-Peña, R. Darbali-Zamora, E. Desarden-Carrero and E. Aponte-Bezares, "Development of Photovoltaic Inverter Model with Islanding Detection Using the Sandia Frequency Shift Method", *2022 IEEE 49th Photovoltaics Specialists Conference (PVSC)*, 2022, pp. 0398-0404.

[23] V. Purba, S. V. Dhople, S. Jafarpour, F. Bullo and B. B. Johnson, "Reduced-order structure-preserving model for parallel-connected three-phase grid-tied inverters", *2017 IEEE 18th Workshop on Control and Modeling for Power Electronics (COMPEL)*, 2017, pp. 1-7.

978-1-6654-6060-6/23 $31.00 © 2023 IEEE

Time-series Imputation using Graph Neural Networks and Denoising Autoencoders

Raymond Wieser*, Yangxin Fan*, Xuanji Yu*, Jennifer Braid[†], Avishai Shaton[‡], Adam Hoffman[§],
Ben Spurgeon[‖], Daniel Gibbons[**], Laura S. Bruckman*, Yinghui Wu*, Roger H. French*

*SDLE Research Center, Case Western Reserve University, Cleveland, OH, USA
[†]Sandia National Laboratories, Albuquerque, NM, USA
[‡]SolarEdge, Herzliya, Israel
[§]Maxeon Solar Technologies, Singapore
[‖]Brookfield Renewable, Toronto, Ontario, Canada
[**]Bay4 Energy, Tucson, AZ, USA

Abstract—Robust system monitoring and prediction relies on having accurate and complete records of system performance. However, real world sensors and monitoring equipment will fail eventually leading to loss of data. This "missingness" effects the stability and accuracy of performance loss rate estimation. To this end, different missingness scenarios were developed to simulate the real world data-stream interruptions. Multiple state of the art and classical imputation methods are bench-marked with different missingness % and missingness type. Accuracy and the precision of the imputation methods was measured against the measured values of AC power from the uncorrupted dataset. It was found that the commonly used interpolation methods such as Linear Interpolation (LI) and Mean interpolation (MI) were significantly outperformed by cutting edge imputation methods such as K-Nearest-Neighbor Imputation (KNN) and a novel graph based neural network imputation (ST-GNN).

Index Terms—Graph Neural Network, Imputation, Timeseries Decomposition, Linear Interpolation, Timeseries, Statistical Analysis, Deep Learning

I. INTRODUCTION

Photovoltaic (PV) power generation is a critical component of a sustainable energy generation ecosystem. The feasibility of using PV as a renewable energy source is dependent on distributing the environmental cost of production across a long service lifetime. In addition, the cost competitiveness of PV technologies is also dependent on system longevity. Due to rapid production of novel PV technologies and the in-feasibility of certifying modules under lifetime exposures, research into the durability of PV components must rely on pre-existing implementations of system monitoring. The most basic forms of system monitoring include measurements of inverter level power production and locally measured metrology data. By leveraging these data-streams, data-driven models of system performance can be generated to evaluate the performance loss rate (PLR) of fleets of PV systems. The best models of PLR depend on the quality of the time-series dataset and the length of time interval. Data sets with large amounts of "missingness" (i.e missing values or anomalous measurements) or with small time intervals can

lead to calculation of misleading PLR values [1]–[3]. Thus, data imputation modules are critical for augmenting time-series datasets with large amounts of missingness or short time windows [4], [5]. Various methods for time-series imputation have been adopted into PV modeling, while novel techniques are still being developed. Imputation methods rely on physical models of PV systems [6]–[8] or can purely data driven [9].

PV imputation presents a specific set of challenges that include, different severity of missingness, different missing types, and lack of high quality training data. Additionally, the large daily and yearly seasonality makes it difficult to have a single model perform well for all time scales. Finally, the spatial location of the PV system inevitably changes its performance and prevents the creation of a single model for all exposure environments. Despite these challenges, the spatio-temporal coherence of PV system data presents a valuable opportunity to introduce novel methods that leverage these inherent attributes to increase predictive accuracy. Spatio-temporal imputation frameworks use measured data that is similar to the attributes of missing points to impute missing values. The intuitive understanding of these methods is that inverters that have a high degree of "similarity" (i.e distance, climate type, cell type, module brand, etc...) should be more strongly influencing than sites that have low "similarity". These methods not only can be used to recover missingness but check to see if measured values are outliers or otherwise faulty.

Graph Neural Networks (GNN) are a machine learning framework built to analysis spatial-temporal problems. The have shown promise in real-time imputation problems [10]–[14]. The GNN can be trained to minimize reconstruction loss from the original signal or to include reconstruction loss terms that are based on physical models of PV systems, allowing for a hybrid data-driven physically informed imputation module. GNNs have also been used to forecast PV system data [15]. This work presents a novel Spatio-Temporal Graph Auto Encoder (ST-GAE) for the purposes of identifying and recovering

978-1-6654-6060-6/23 $31.00 © 2023 IEEE

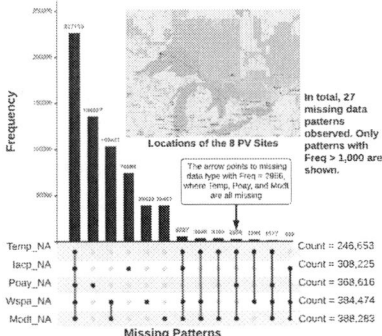

Fig. 1. Location of the PV inverters and their most common missingness patterns. A full measurement blackout is the most common missingness pattern. Adapted from "Spatio-Temporal Denoising Graph Autoencoders with Data Augmentation for Photovoltaic Timeseries Data Imputation" by Fan et al., 2023 [22]. Copyright SIGMOD publishing 2023.

missingness from real world PV datastreams.

II. METHODS

A. Missingness and Imputation

Data quality can a substantial impact on the results of performance calculations. Poor data quality is not limited to having instruments that fail to read measurements, but also includes instrument drift, faulty sensors, or lack of proper maintenance. These errors are substantially harder to detect because they may appear to be within the range of normal readings but bias the analysis of the data. Different types of missingness were injected into the raw data. Block Missing represents the condition where continuous periods of time-series data is concurrently missing. Missing Completely At Random (MCAR), has random missing datum points located uniformly throughout the timeseries.

Several state of the art imputation methods were tested on the PV dataset. Imputation methods include classical statistical methods like Linear Interpolation (LI) [16], Multiple Imputation by Chained Equations (MICE) [17], and Mean Imputation [18]. K-nearest neigbhors (KNN) [19], Low-Rank Tensor Completion Truncated Nuclear Norm (LRTC-TNN) [20], and Multiple Imputation by Denoising Autoencoders (MIDA) [21] represent cutting edge machine learning models.

B. Dataset

Historical timeseries data from 98 inverters installed in a **Dfa** climate zone was acquired for the entire year of 2015 1. The data included onsite measurements of inverter AC power (**IACP**) and site metrology data. The total percentage of missing data was 2.074% for **IACP** .

C. ST-GAE Framework

1) Graph Autoencoders: Graph Autoencoders (GAEs) are unsupervised learning frameworks, consisting of a graph encoder and a graph decoder. The graph encoder learns "important" features by mapping inverters into a latent representation, while the graph decoder learns to reconstruct the data from the encoded counterparts. It aggregates signals from neighboring nodes to learn embeddings for each node:

$$\bar{X} = \sigma(\tilde{D}^{-\frac{1}{2}}\tilde{A}\tilde{D}^{-\frac{1}{2}}XW) \quad (1)$$

where X is the inverter attribute matrix, $\tilde{A}$ is the adjacency matrix with self loop, W is a trainable weight matrix, $\tilde{D}$ is the degree matrix with $D_{i,i} = \sum_j W_{i,j}$, and σ denotes the activation function.

Previous studies in GAEs have shown their outstanding performance in learning from corrupted data, which is the natural extension of the problem of missing data imputation [23]. ST-GAE integrates GAEs to learn the distribution of node and topological features to handle complex missing scenarios.

2) Denoising Autoencoders: Denoising Autoencoder (DAE) is a variant of AutoEncoder (AE), with a goal to recover data from corrupted input. AE tends to over-fit for data imputation, since it minimizes the reconstruction loss between the input and the reconstructed counterpart. Instead, DAE introduces noise (corruption) to the input. Input data can be corrupted by noises added to the input vector in a stochastic manner [21]. The model is then trained to minimize the reconstruction loss between the recovered data and the uncorrupted counterpart.

D. PLR estimation

PLR estimation was conducted using PVplr [24]. Power predictive models used include: **XbX, XbX + UTC, PVUSA,** and **6K**. Irradiance threshold values where: 200, 400, 800 $\frac{W}{m^2}$, and PLR estimations were calculated for daily, weekly, and monthly aggregated data. Three different missingness scenarios were simulated at 25%, 50% and 75% missing values. PLR estimation was conducted using both **YoY** and **Weighted Regression**. Power models were bootstrapped re-sampled at a 60% data threshold for 100 iterations for each model. A total of 21,168 realizations of PLR values were estimated.

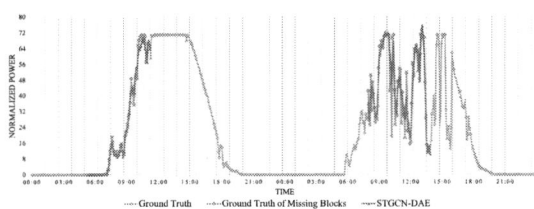

Fig. 2. Imputation of 6 hrs of block missing data. The STD-GAE not only imputed the correct shape for a severe case of missingness, but also captures the natural variance of real world measurements. Adapted from "Spatio-Temporal Denoising Graph Autoencoders with Data Augmentation for Photovoltaic Timeseries Data Imputation" by Fan et al., 2023 [22]. Copyright SIGMOD 2023.

III. RESULTS

A. Missingness Imputation

Missingness imputation was conducted on a variety of missingness scenarios and severity. Traditional imputation methods performed differently depending on the type of missingness scenario. Block missing imputation had significantly larger

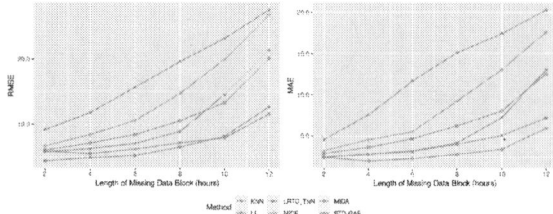

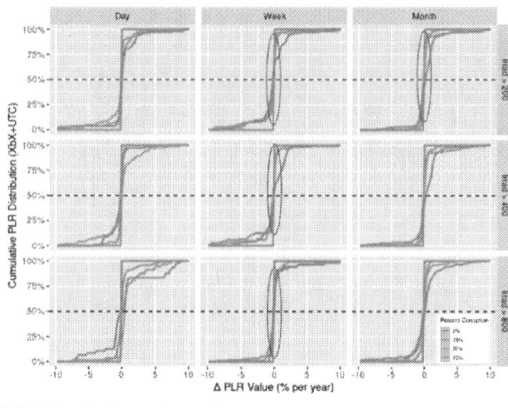

Fig. 3. Imputation accuracy of a variety of imputation methods. STD-GAE outperforms all models for all types of missingness and severity. Adapted from "Spatio-Temporal Denoising Graph Autoencoders with Data Augmentation for Photovoltaic Timeseries Data Imputation" by Fan et al., 2023 [22]. Copyright SIGMOD publishing 2023.

Fig. 4. Cumulative Distribution Curves of ΔPLR from raw 2% missingness dataset with **XbX + UTC** with YoY regression. A vertical curve at $\Delta PLR = 0$ indicates good agreement between recovered PLR and raw PLR. Red ovals highlight good agreement between imputation and raw PLR values. At large missingness percentages, with high irradiance filters and low temporal aggregation, their exists a significant portion large outliers. However, recovery of any information at missingness percentages of 75% is remarkable.

MAE and RMSE compared to Random Missing patterns. This is due to the severity of the lost of information. However, the ST-GAE performed consistently across all severity and missingness scenarios. The lower RMSE and MAE of the ST-GAE shows that the GNN imputation module is not only fitting the observed values accurately but also is robust enough to reproduce the random high frequency fluctuation present in PV time-series data. The ST-GAE is also able to capture the different effects of cloud coverage (Fig. 2). Two separate days where injected with 6hrs BM, one day representing a clear sky and the other a cloudy day. The ST-GAE was not only able to correctly impute the clear sky data, but near perfectly regenerated the turbulent signal of a overcast day.

B. PLR Estimation

After the recovery of missingness from the time-series data, PLR for the systems was estimated. In order to assess the affects of data imputation on PLR estimation, baseline PLR estimation was conducted for the uncorrupted dataset of 98 inverters and again on datasets with different missingness severity. PLR was estimated again after imputation. A total of 21,168 individual PLR values were estimated, across different power predictive models, filtering constraints and aggregation levels. Although the true value of PLR for each inverter is not known, we can compare the estimated PLR after imputation to the original PLR estimation.

PLR estimation requires large lengths of continuous time-series data. As the severity of missingness increases, the loss of information prevents the accurate modeling of seasonal effects leading to the failure of PLR estimation. This trend is especially significant for PLR estimations using high irradiance filters and low data aggregation. After imputation, PLR estimation was successfully run on missingness severity of up to 75% and even increased the number of successful models for the baseline 2% missingness data

Imputation increases the number of successful calculations of PLR, but does not necessarily correlated to the estimation of the true PLR value. In order to assess the performance of the imputed data, PLR was calculated for the raw 2% missing dataset of 98 inverters. The imputation performance is then base-lined as a deviation from the raw PLR value. Large variation from a ΔPLR of zero indicates the imputed

data does not represent the original dataset well. Cumulative distribution curves of ΔPLR have been generated to evaluate the fleet level performance of imputation. A perfect imputation algorithm would result in a vertical line CDF centered at zero. Another important aspect to consider is the variance of the PLR estimation. A high-quality imputation method not only matches the mean value well, but should be able to replicate the variance of the measurements as well.

Imputation with ST-GAE allowed for the estimation of PLR for datasets with up to 75% missingness. The bulk performance of imputation shows good agreement with the baseline values. Larger aggregation periods and lower irradiance filters performed the closest to their uncorrupted PLR value. Missingness percentages exceeding 50% and with short term aggregation, and large irradiance thresholds displayed large outliers, but the mean performance was still in agreement with the raw PLR.

IV. CONCLUSIONS

PV system data has inherent spatio-temporal correlations. This information is critical for expanding PV system modeling from a myopic single system paradigm to a more natural understanding of PV systems as a system of interconnected pieces. GNNs and spatio-temporal modeling can leverage existing information to increase model accuracy. These novel techniques have been incorporated into a imputation module for PV system data. ST-GAE have been shown to outperform all state of the art imputation techniques for highly-variable, large fluctuation time-series data streams. PLR estimation on the reconstructed data showed good agreement with PLR estimation on the uncorrupted dataset. Information loss is inherent for large scale systems but, by leveraging fleet level datastreams the consequences to performance monitoring can be mitigated.

REFERENCES

[1] A. Livera, M. Theristis, E. Koumpli, S. Theocharides, G. Makrides, J. Sutterlueti, J. S. Stein, and G. E. Georghiou, "Data processing and quality verification for improved photovoltaic performance and reliability analytics," *Progress in Photovoltaics: Research and Applications*, p. pip.3349, Oct. 2020.

[2] G. Makrides, M. Theristis, J. Bratcher, J. Pratt, and G. E. Georghiou, "Five-year performance and reliability analysis of monocrystalline photovoltaic modules with different backsheet materials," *Solar Energy*, vol. 171, pp. 491–499, Sep. 2018.

[3] C. R. Osterwald, J. Adelstein, J. A. del Cueto, B. Kroposki, D. Trudell, and T. Moriarty, "Comparison of Degradation Rates of Individual Modules Held at Maximum Power," in *2006 IEEE 4th World Conference on Photovoltaic Energy Conference*, vol. 2, May 2006, pp. 2085–2088.

[4] S. Lindig, M. Theristis, and D. Moser, "Best practices for photovoltaic performance loss rate calculations," p. 17, 2022.

[5] S. Lindig, A. Louwen, D. Moser, and M. Topic, "Outdoor PV System Monitoring—Input Data Quality, Data Imputation and Filtering Approaches," *Energies*, vol. 13, no. 19, p. 5099, Sep. 2020.

[6] D. L. King, W. E. Boyson, and J. A. Kratochvil, "Photovoltaic array performance model." Sandia National Laboratories (SNL), Albuquerque, NM, and Livermore, CA (United States), Tech. Rep. SAND2004-3535, Aug. 2004, tex.ids= kratochvilPhotovoltaicArrayPerformance2004, kratochvilPhotovoltaicArrayPerformance2004a. [Online]. Available: https://www.osti.gov/biblio/919131

[7] J. Kratochvil, W. Boyson, and D. King, "Photovoltaic array performance model." Tech. Rep. SAND2004-3535, 919131, Aug. 2004. [Online]. Available: https://www.osti.gov/servlets/purl/919131/

[8] D. Faiman, "Assessing the outdoor operating temperature of photovoltaic modules," *Progress in Photovoltaics: Research and Applications*, vol. 16, no. 4, pp. 307–315, 2008. [Online]. Available: https://onlinelibrary.wiley.com/doi/abs/10.1002/pip.813

[9] R. Perez, T. Cebecauer, and M. Šúri, "Semi-Empirical Satellite Models," in *Solar Energy Forecasting and Resource Assessment*. Boston: Academic Press, 2013, pp. 21–48. [Online]. Available: http://www.sciencedirect.com/science/article/pii/B9780123971777000024

[10] Z. Diao, X. Wang, D. Zhang, Y. Liu, K. Xie, and S. He, "Dynamic spatial-temporal graph convolutional neural networks for traffic forecasting," in *AAAI*, 2019.

[11] B. Yu, H. Yin, and Z. Zhu, "Spatio-temporal graph convolutional networks: A deep learning framework for traffic forecasting," in *Proceedings of the 27th International Joint Conference on Artificial Intelligence*, ser. IJCAI'18. AAAI Press, 2018, p. 3634–3640.

[12] X. Zhang, C. Huang, Y. Xu, and L. Xia, "Spatial-temporal convolutional graph attention networks for citywide traffic flow forecasting," in *Proceedings of the 29th ACM International Conference on Information and Knowledge Management*, ser. CIKM '20. New York, NY, USA: Association for Computing Machinery, 2020, p. 1853–1862. [Online]. Available: https://doi.org/10.1145/3340531.3411941

[13] A. Cini, I. Marisca, and C. Alippi, "Multivariate time series imputation by graph neural networks," *arXiv preprint arXiv:2108.00298*, 2021.

[14] Y. Ye, S. Zhang, and J. J. Q. Yu, "Spatial-temporal traffic data imputation via graph attention convolutional network," in *Artificial Neural Networks and Machine Learning – ICANN 2021*, I. Farkaš, P. Masulli, S. Otte, and S. Wermter, Eds. Cham: Springer International Publishing, 2021, pp. 241–252.

[15] A. M. Karimi, Y. Wu, M. Koyuturk, and R. H. French, "Spatiotemporal graph neural network for performance prediction of photovoltaic power systems," in *Proceedings of the AAAI Conference on Artificial Intelligence*, vol. 35, no. 17, 2021, pp. 15323–15330.

[16] T. Blu, P. Thévenaz, and M. Unser, "Linear interpolation revitalized," *IEEE Transactions on Image Processing*, vol. 13, no. 5, pp. 710–719, 2004.

[17] P. Royston and I. R. White, "Multiple imputation by chained equations (mice): implementation in stata," *Journal of statistical software*, vol. 45, pp. 1–20, 2011.

[18] A. R. T. Donders, G. J. Van Der Heijden, T. Stijnen, and K. G. Moons, "A gentle introduction to imputation of missing values," *Journal of clinical epidemiology*, vol. 59, no. 10, pp. 1087–1091, 2006.

[19] R. Malarvizhi and A. S. Thanamani, "K-nearest neighbor in missing data imputation," *International Journal of Engineering Research and Development*, vol. 5, no. 1, pp. 5–7, 2012.

[20] X. Chen, J. Yang, and L. Sun, "A nonconvex low-rank tensor completion model for spatiotemporal traffic data imputation," *Transportation Research Part C: Emerging Technologies*, vol. 117, p. 102673, 2020.

[21] L. Gondara and K. Wang, "Mida: Multiple imputation using denoising autoencoders," in *Pacific-Asia conference on knowledge discovery and data mining*. Springer, 2018, pp. 260–272.

[22] Yangxin Fan, Xuanji Yu, Raymond Wieser, David Meakin, Avishai Shaton, Jean-Nicolas Jaubert, Robert Flottemesch, Michael Howell, Jennifer Braid, Laura S.Bruckman, Roger H.French, and Yinghui Wu, "Spatio-Temporal Denoising Graph Autoencoders with Data Augmentation for Photovoltaic Timeseries Data Imputation," in *Proceedings of the 2023 ACM SIGMOD International Conference on Management of Data*. Seattle, WA, USA: Association for Computing Machinery, Jun. 2023.

[23] R. C. Pereira, M. S. Santos, P. P. Rodrigues, and P. H. Abreu, "Reviewing autoencoders for missing data imputation: Technical trends, applications and outcomes," *Journal of Artificial Intelligence Research*, vol. 69, pp. 1255–1285, 2020.

[24] A. Curran, T. Burleyson, S. Lindig, D. Moser, R. French, and SDLE Research Center, "PVplr: Performance Loss Rate Analysis Pipeline," Oct. 2020, tex.ids: a.j.curranPVplrSDLEPerformance2020, curranPVplrPerformanceLoss2020. [Online]. Available: https://CRAN.R-project.org/package=PVplr

Cradle to cradle Recycling of Perovskite Solar Cells

Zhenni Wu[1,2], Gülüsüm Babayeva[2], Zhang Jiyun[1,2], Mykhailo Sytnyk[1], Jens Hauch[1,2],
Christoph J. Brabec[1,2], Ian Marius Peters[1]

[1] Forschungszentrum Jülich, IEK 11, Helmholtz Institute Erlangen-Nürnberg for Renewable-Energies
HI ERN, 91058 Erlangen, Germany

[2] Friedrich-Alexander-University Erlangen-Nuremberg (FAU), Faculty of Engineering, Department of
Material Science, Materials for Electronics and Energy Technology (i-MEET), Erlangen 91058 Germany

Abstract — **Cradle-to-cradle recycling plays an important role in resolving one of the major upcoming challenges for photovoltaics in the energy transition – resource management. As photovoltaic production continues to grow, it absorbs a vast amount of resources. Cradle-to-cradle recycling is essential to keep these resources available perpetually and minimize waste generation. To achieve this, a new design paradigm for solar panels that puts recycling in its center is needed. In this study, we show first results of a perovskite solar cell that was created using such a design paradigm. We created a device with 18.1% efficiency and an architecture that facilitates disassembly of the functional layers. We show that the absorber material, MaPbI₃ was used for the test device, can be recycled without causing a loss of solar cell efficiency.**

I. Introduction

Cradle-to-cradle (C2C) recycling becomes inevitable for photovoltaic technology at the terawatt (TW) scale. On the one hand, one terawatt of today's silicon solar panels has a mass of roughly 50 million tons, and will create waste accordingly. On the other hand, solar panel production will become a major consumer of every material involved in its production. The photovoltaic industry is already a major consumer of silicon, flat glass, silver, aluminum and certain polymers, and scaling todays manufacturing capacity by, following Verlinden [1], a factor of about ten, will cement the industries dominant consumer role. Hence, resource and supply chain management will be among the greatest upcoming challenges for PV. C2C recycling has the potential to alleviate waste- as well as resource management greatly, especially once PV installations have reached the end of their growth phase.

Assuming no recycling at all, one TW of today's silicon panels (350W, 1 x 1.67 m², 40mm thick, 18kg mass) equals roughly 50 million tons of waste with a volume, assuming panels are stacked, of approximately 200 million m³. Verlinden suggests a global annual production of 3TW in 2040, meaning that maybe twice this amount - there will be improvements in the power per mass ratio of modules- would become waste, probably in the early 2070s. This compares to an annual global landfill generation of 500 million tons in 2022, estimated by the World Bank. Hence, if all solar panels were dumped they would constitute a notable share of global landfill waste, and a significant fraction of electronic waste. While managing such an amount of waste would not be impossible, it would mean a significant and unnecessary loss of resources.

Today, the PV industry consumes about 10% of globally mined silver [2], and a similar amount of flat glass [3]. Scaling capacities by a factor of ten will entail massive transitions for the way the PV industry works. When the industry became the dominant market for silicon and started to overtake information technology, new polysilicon factories were tailored entirely to solar cell production. Today, photovoltaics is the dominant consumer for silicon, and the silicon for PV supply chain is entirely linked to solar cell production. A similar trend is already observable for glass. New photovoltaic production lines, like the ones of First Solar [4], are constructed alongside glass manufacturing. As PV glass is already a specialized product, we expect to see stronger links between glass and module manufacturing. While for silicon and glass, resource availability is not a principle issue; the situation is different for silver. Competition with other consumers and rising prices are bound to become a challenge for the PV industry, motivating strong efforts to reduce or replace silver [2].

As we continue to scale PV capacity, we have to invest the resources needed to achieve the targeted 50 to 80 TW [1]. During this time, recycling can aid to reduce material demand and help reduce waste. Yet, by the time the target capacity is installed, C2C recycling has to be in place to minimize the need for additional material to maintain this capacity. The sheer amount will likely make mining end-of-life panels the most attractive option for new materials. Yet, to be able to do this, a paradigm shift is needed that includes recyclability into PV module design. The new design should have the goal to make resources perpetually usable (see Figure 1). In this work, we introduce first results of our attempt to create a completely cradle-to-cradle recyclable module, using perovskite technology.

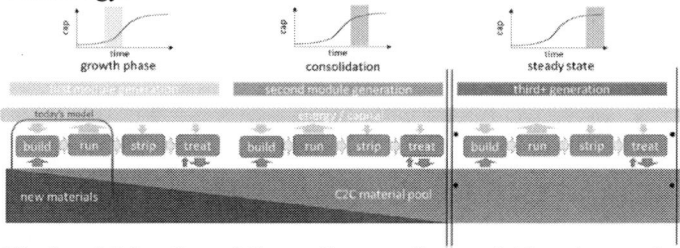

Fig. 1. Vision for a fully cradle-to-cradle recyclable solar panel. Rather than designing a device with a focus on a single operation period only, we want to develop a design that enables a perpetual use of all involved components.

978-1-6654-6060-6/23 $31.00 © 2023 IEEE

II. SOLAR CELL DESIGN AND CONSTRUCTION

Our choice to use perovskites was motivated by the fact that all functional layers of this solar cell are deposited using solution processing, which enables the use of solvents with complementary solubility for subsequent layers. Our initial goal was to construct a solar cell using a reversible sequence of processing steps, materials, and solvents so that this cell could be disassembled in the same way as it was produced. Our first test structure consisted of an ITO glass substrate with SnO_2 electron transport layer, a $MaPbI_3$ absorber layer deposited using DMF and NMP, and a hole transport layer made of PDCBT deposited using 1,2-dichlorobenzene coupled with a layer of PTAA-BCF. The test stricter employs contacts of evaporated gold. Image, structure, and performance of this test structure are displayed in Figure 2. Shown is the result for a champion cell out of a batch of 18 with an efficiency of 18.1%.

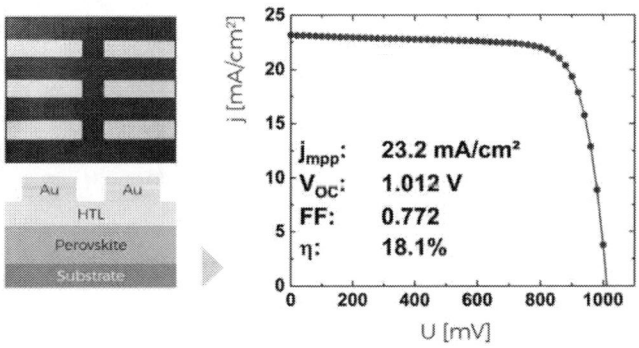

Fig. 2. Structure and current voltage characteristics of the initially produced solar cell. .

III. DISASSEMBLY AND REASSEMBLY

To recover a sufficient amount of material for characterization and treatment, we produced batches of 50 solar cells with matching architecture. On all 50 cells, we peeled of the gold contacts, and used the same solvents as for the deposition to remove the functional layers. When characterizing the recovered $MaPbI_3$, we observed that the recovered material exhibited material- and phase impurities, requiring an additional treatment step to reproduce the initial material quality. This treatment step consisted of i) recycled perovskite solution concentration with rotovap; ii) perovskite crystallization and precipitation with a non-solvent; iii) centrifugation; iv) vacuum drying. The processing and the improvement of material quality during treatment are shown in Figure 3.

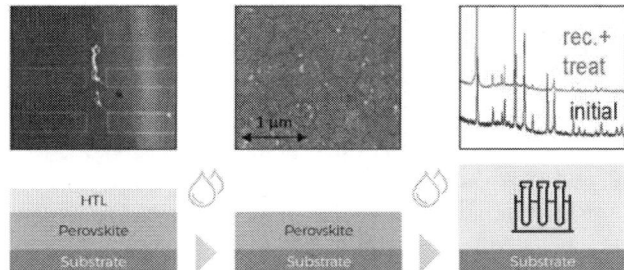

Fig. 3. Disassembly process and recovery of materials.

In a next step, we used the recovered $MaPbI_3$ to fabricate a new solar cell. While we recovered all components of the previous cell, we didn't produce new cells from only recycled materials to investigate the impact of every recycled material on cell efficiency. Hence, the cell with recovered $MaPbI_3$ included a new substrate with ETL, a newly made HTL, and new contacts. For this cell with recycled $MaPbI_3$, we were able to reproduce the same efficiency as with the initial cell. The result is shown in Figure 4, and, again, displays the champion cell of a batch of 18. While there is a variation in the characteristics between the initial cell and the one with recycled $MaPbI_3$, this is within the typical variations we obtain between cell batches. We take the reproduction of the initial efficiency as confirmation that the recycled $MaPbI_3$ does not limit solar cell efficiency differently from the initial cell.

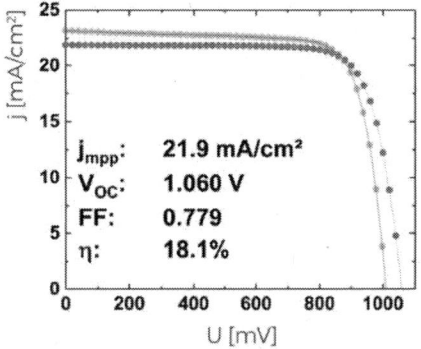

Fig. 4. Current voltage characteristics of the solar cell manufactured with recycled $MaPbI_3$.

In follow-up experiments, we will sequentially replace more and more components with recycled ones until we are able to make a solar cell entirely of recycled materials. The next step will be the creation of a solar cell with recycled substrate and recycled $MaPbI_3$. The challenge in this is not so much achieving a good device performance, but doing so while keeping the amount of energy, capital and new materials to restore the quality of recycled components at a minimum. For this purpose, we track energy and material flows during this process.

IV. SUMMARY & CONCLUSIONS

To avoid waste and reduce the challenges with providing sufficient resources for PV in the energy transition, we need photovoltaic modules that are C2C recyclable. Current silicon PV modules are produced within a "design for immortality" paradigm, making it challenging to recycle them efficiently. Our aim is to introduce a new "design for circularity" paradigm. In this paradigm, every step during assembly has to be accompanied by a corresponding disassembly step at the end of life. In this way, PV module manufacturing becomes reversible. As a first vehicle to demonstrate this concept, we have chosen perovskite solar cell, because solution processing provides a straightforward way to recover all functional materials that requires designing the use of material stack and solvents. In a first experiment, we designed a solar cell with a reversible assembly sequence, measured its efficiency, disassembled the cell, and used the recovered absorber, MaPbI$_3$, to create a new solar cell. We achieved matching efficiencies for the initial cell and the recycled cell, demonstrating that perovskite based absorbers are capable of circular recycling. The potential to recover perovskites has also been shown, for example, by [5] and [6]. Yet, a full circular recycling of a perovskite PV module is, to the best of our knowledge, still an open task.

Finally, it should be noted that silicon and not perovskite technology will most likely carry the majority of the capacity that will be installed in the next two decades. Neverthelss, developing recycling concepts for perovskites has the potential to impact photovoltaic technology on a larger scale. On the one hand, and directly, perovskite recycling will be relevant for the recycling of perovskite on silicon tandem solar cells, and indirectly, recycling concepts for the module package may be transferable to silicon and CdTe technology.

REFERENCES

[1] N. Haegel et a., "Terawatt-scale photovoltaics: Transform global energy", *Science*, vol. 364 pp. 836-838, 2019.

[2] B. Hallam, M. Kim, Y. Zhang, L. Wang, A. Lennon, P. Verlinden, P. P. Altermatt, P. R. Dias, „The silver learning curve for photovoltaics and projected silver demand for net-zero emissions by 2050", *Progress in Photovoltaics*, 2022.

[3] J. C: Goldschmidt, L. Wagner, R. Pietzacker and L. Freidrich, "Technological learning for resource efficient terawatt scale photovoltaics", *Energy & Environmental Science*, vol. 14, pp 5147 − 5160, 2021.

[4] https://www.firstsolar.com/en-Emea/Technology/Manufacturing

[5] J. M. Kadro, N. Pellet, F. Giordano, A. Ulianov, O. Müntener, J. Maier, M. Grätzel, and A. Hagfeldt, "Proof-of-concept for facile perovskite solar cell recycling", *Energy & Environmental Science*, vol. 9, pp 3172 − 3179, 2016.

[6] B. Chen, C. Fei, S. Chen, et al. „Recycling lead and transparent conductors from perovskite solar modules". *Nat Commun* vol. 12, pp. 5859ff, 2021.

Per- and polyfluoroalkyl substances (PFAS) usage in solar photovoltaics

Preeti Nain, Annick Anctil

Michigan State University, East Lansing, MI, United States

Per- and polyfluoroalkyl substances (PFAS) are increasingly used in the renewable energy sector, from photovoltaic (PV) modules to batteries. There are increasing concerns from the community about PFAS land contamination after PV solar farm construction, but no comprehensive studies on the subject. In terms of potential PFAS use in solar modules, there are no known benefits for using PFAS in the active layer. However, PFAS could be used in outer layers to increase the power conversion efficiency and module lifetime. This study provides an overview of PFASs usage in solar modules and reviews the literature to understand PFAS type and amount used in solar modules. Within the published literature, the information on PFAS for solar panels is limited and mainly includes PFAS usage as anti-reflective and self-cleaning coatings. While there are limited studies, but it can't be argued PFAS certain usage in soalr modules. Most PFAS identified in solar modules belong to the fluoropolymers category, which is a distinct group of PFAS and is considered a polymer of low concern. Academic research on PFAS quantification methods specific to solid matrices could help end-of-life decisions and identify treatment options.

Preliminary Gap Analysis of Existing IEEE 1547 and IEEE 2800 Standards towards Grid-Forming Technology

Ganesh Marasini[1], Wenzong Wang[2], Wes Baker[2], Deepak Ramasubramanian[2], Aminul Huque[2], Jens C. Boemer[2]

[1]University of Central Florida, Orlando, FL, USA

[2]Electric Power Research Institute, Knoxville, TN, USA

Abstract — **Grid-forming inverter control has been proposed as a key enabler of future inverter-dominated power system, due to its capability to provide grid services without relying on synchronous machines. However, before the wide deployment of grid-forming inverters in power transmission and distribution systems, the respective generation interconnection rules should be examined such that they do not impose inadvertent requirements that may prevent the interconnection of grid-forming inverter based power plants. In this paper, preliminary gap analysis is conducted to examine some of the clauses in both IEEE Std 1547™-2018 and IEEE Std 2800™-2022. The steady-state and dynamic behavior of a generic grid-forming inverter based power plant model is checked against the requirements in the standards. The gaps identified may provide inputs to future revision of the standards to better accommodate grid-forming technology.**

Keywords — **Grid-forming, interconnection requirements, IEEE Std 1547™-2018, IEEE Std 2800™-2022.**

I. INTRODUCTION

Present-day inverter control technologies are mainly designed to operate in the presence of (or parallel with) a stiff power system, which are often termed as "grid-following (GFL) inverter control". Such inverter control approaches face challenges to provide grid services and maintain stability in inverter-dominated systems [1], [2].

To deal with the challenges, improvement of the present-day control design and various new forms of inverter controls have been proposed [3]. These controls are termed as "grid forming (GFM) inverter control" due to their capability to form grid voltage and operate without relying on synchronous machines [1]. The benefits of utilizing GFM inverter control have been shown in power transmission [4] and distribution systems [5] especially when the inverter-based resource (IBR) penetration level is high and (part of) the power system becomes weak.

For a new IBR to be interconnected with power system, it has to go through interconnection review and commissioning test to make sure that its performance complies with the interconnection requirements. Since today's interconnection requirements may not fully consider the chariceritics and needs of the future inverter-dominated power system, they might impose inadvertent barrier which may then prevent the interconnection of a GFM IBR. This aspect requires careful examination and the inadvertent barriers (if any) need to be resolved before GFM IBRs can be widely deployed.

Towards this goal and as an initial step, this paper examines certain clauses in IEEE Std 1547™-2018 [6] and IEEE Std 2800™-2022 [7], which are widely cited in interconnection rules in power distribution (IEEE 1547) and transmission (IEEE 2800) systems across the North America.

Since an industry acceptable consistent and uniform method of defining the performance requirements for GFM IBR is presently lacking, instead of comparing IEEE Std 1547™-2018 and IEEE Std 2800™-2022 against a GFM standard, they are checked against the behavior observed from a generic GFM IBR model with a particular set of control parameters. The GFM IBR model is generic in the sense that it does not represent a particular product. Instead, it represents generic behavior expected from GFM inverters. The model has been tuned in various past studies and it demonstrates GFM properties such as maintaining system stability in weak grids and after tripping of the last synchronous machine, enabling fast fault recovery, etc.

A set of tests (e.g., frequency and voltage step change) is designed to check the GFM IBR behavior against the standards, and a gap is identified when the GFM IBR behavior is outside the range required by a certain clause of the standards. It should be noted that the set of tests described in this paper are not complete and more tests are being developed to check whether the GFM behavior violates the requirements of the corresponding standard.

Moreover, as the tests are carried out with one GFM model, its behavior is expected to be a subset of the GFM behavior. Whether the full spectrum of GFM behavior violates the standards will be evaluated as future work.

II. TEST SYSTEMS AND INVERTER MODELING

A. Description of the Test Systems

Fig. 1 shows the single line diagram of the test system constructed for IEEE Std 1547™-2018 related tests. The 1MVA GFM plant (including the inverter and interconnection transformer inside the generic GFM model box) is connected to a controllable AC source with negligible impedance. The reference point of applicability (RPA) of the plant is at the system side of the interconnection transformer. Since the performance requirements to be evaluated in IEEE Std 1547™-

978-1-6654-6060-6/23 $31.00 © 2023 IEEE

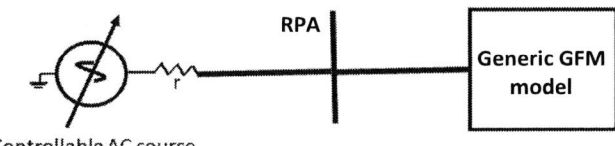

Fig. 1. Illustration of the test system for IEEE Std 1547™-2018 tests

2018 are defined by open loop response of the distributed energy resource (DER) at its RPA, in the test system, the DER is connected to an infinite bus without impedance between the RPA and the ideal AC source.

Fig. 2 shows the single line diagram of the test system constructed for IEEE Std 2800™-2022 related tests. The GFM inverter and IBR unit transformer are modeled inside the generic GFM model box. The 10 MVA IBR plant also includes the collector system (modeled by PI section) and the plant interconnection transformer. The point of measurement (POM) where the IEEE Std 2800™-2022 requirements apply is at the high voltage side of the interconnection transformer. For the initial investigation presented in this paper, the impedance inside the controllable AC source is configured to represent a short circuit ratio (SCR) of 20 at the POM.

B. GFM Inverter Modeling

The generic three-phase GFM inverter EMT model developed by EPRI is utilized which has both positive sequence and negative sequence control [8], [9]. As illustrated in Fig. 3, four types of positive sequence control methods are modeled: phase locked loop (SRF-PLL) based GFM control, droop based GFM control, virtual synchronous machine (VSM) based GFM control, and dispatchable virtual oscillator (dVOC) based GFM control. All four types of controls are simulated in the gap analysis. The objective is not to determine pros and cons of each particular type of GFM. Rather, it is to observe whether the four types of GFM control have similarities in performance. Similarly, two different control methods are implemented for negative sequence outer loop control (V_{neg} elimination and Kfactor). The V_{neg} elimination control utilizes a proportional integral (PI) controller and eliminates the negative sequence voltage when the current output is within the limit, whereas Kfactor control emulates a constant negative sequence impedance. The Kfactor approach is analyzed since it aligns closer with the IEEE Std 2800™-2022 requirements.

C. GFL Inverter Modeling

In this study, the EPRI developed generic white-box EMT models for transmission and distribution connected solar PV

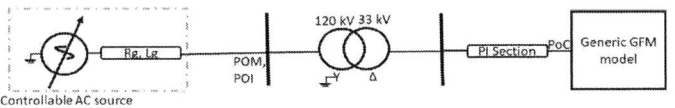

Fig. 2. Illustration of the test system for IEEE Std 2800™-2022 tests

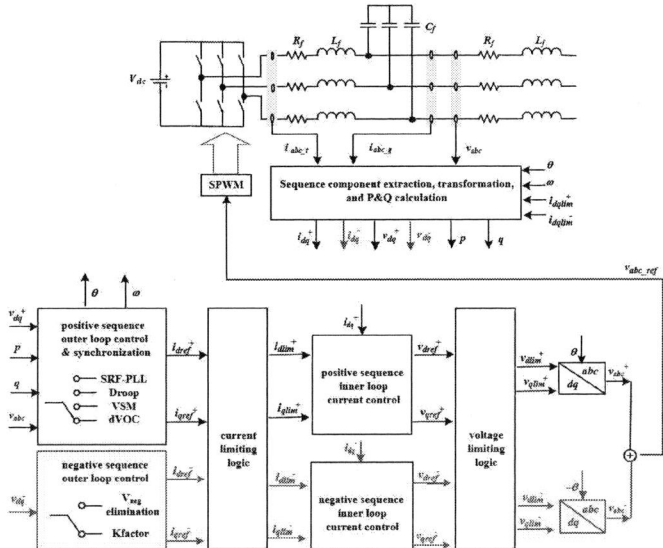

Fig. 3. Illustration of the GFM inverter model

IBRs are leveraged. The transmission connected IBR model is developed to closely represent an IBR that complies with IEEE Std 2800™-2022, while the DER model represents an IEEE Std 1547™-2018 compliant DER. The GFL models are leveraged to better visualize the response defined in the standards. It is not to compare the capability of GFL inverter with GFM inverter.

III. SIMULATION BASED GAP ANALASYS AND RESULTS

A. Reactive Power-Voltage Control Test Against IEEE Std 1547™-2018

In this test, following a step voltage change at the plant RPA, the reactive power response of the GFM plant is checked against the voltage-reactive power (volt-var) control mode requirements in section 5.3.3 of IEEE Std 1547™-2018. The evaluation metrices that are of interest are the open loop response time (OLRT) and the steady state operating characteristics. The OLRT is the time between the step change in the RPA voltage and when the reactive power output (also measured at the RPA) of the DER reaches 90 % of the final steady state value, before any overshoot. Note the response is characterized as "open loop" if the RPA voltage is not affected by power injection of the DER, which is the case in the developed test system.

IEEE Std 1547™-2018 requires that the OLRT of volt-var control function to be between 1s and 90s, and it specifies a range of steady state volt-var curve settings, both of which are checked against in this test.

The reactive power response of the GFM plant to a RPA voltage change from 1pu to 0.9pu at 4s is shown in Fig. 4. Note the test is repeated with four different types of GFM control as described before, as well as the IEEE Std 1547™-2018 compliant GFL control (with volt-var control and OLRT set to

978-1-6654-6060-6/23 $31.00 © 2023 IEEE

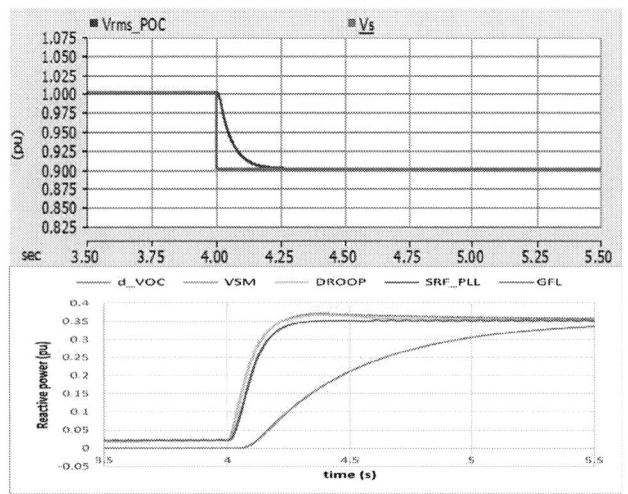

Fig. 4. Response of generic GFM and GFL model for a step change in AC source voltage

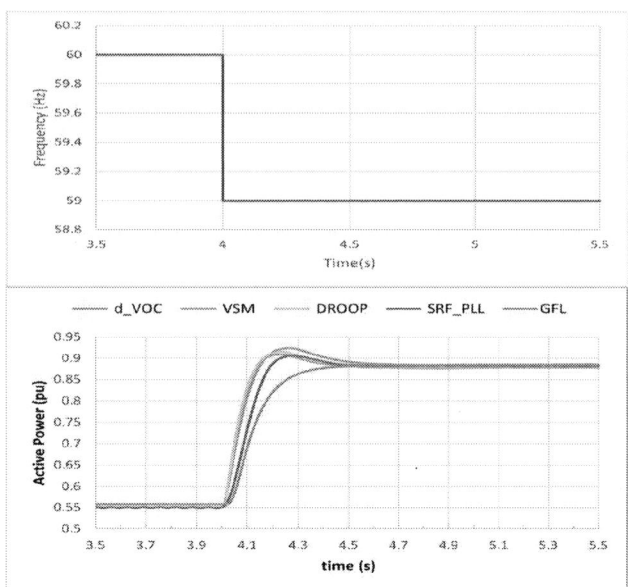

Fig. 5. Response of generic GFM and GFL model for a step change in the grid frequency

1s). As can be seen, the responses of the GFM controls are faster compared to the response of the GFL inverter. The OLRT of the GFM controls (around 0.3s) is less than the fastest response allowed in IEEE Std 1547™-2018 (1s). Therefore, a gap is identified as requiring OLRT according to the standard can prevent the interconnection of this particular GFM plant. Note this does not imply GFM behavior in general violates the volt-var OLRT requirement in IEEE Std 1547™-2018. For example, a GFM plant may be able to utilize additional slower reactive power control to satisfy the requirements. This aspect is under investigation.

Since the reactive power-voltage droop of the GFM plant is configurable and can be set within the range allowed by IEEE Std 1547™-2018, no violation is identified regarding the steady state volt-var characteristic at the plant RPA.

B. Active Power-Frequency Control Test Against IEEE Std 1547™-2018

This test is used to compare the performance of the generic GFM model with the requirements in section 6.5.2.7 of IEEE Std 1547™-2018, which lays out the active power-frequency (freq-watt) control requirements for a DER for a change in grid frequency.

The test criteria for this test are the steady state active power and the open loop response time for a step change in grid frequency. The open loop response time in the context of this test is the time between the step change in the grid frequency and when the active power output of the DER reaches 90 % of the final steady state value, before any overshoot. Note that the standard has different range of allowable settings for different DER categories. For Category III DER, the range of allowable settings for open loop response time is the widest (0.2s to 10s).

The reactive power response of the GFM plant to a frequency change at the AC voltage source from 60 Hz to 59 Hz at 4s is shown in Fig. 5. Similar to the volt-var test, responses from the four types of GFM control as well as the IEEE Std 1547™-2018 compliant GFL control (with freq-watt control and OLRT set to 0.2s) are captured.

Results indicate that the response of the generic GFM model is faster than that of the GFL model. In fact, the open loop response time is less than the minimum value of the allowable ranges for Category I, II and III DERs. However, foot note b under Table 24 in IEEE Std 1547™-2018 indicates that response times may be set to lower values than the minimum values shown in the table (1s for Category I and II, and 0.2s for Category III). Therefore, the response time of the GFM model to a frequency disturbance is allowed by IEEE Std 1547™-2018.

Moreover, the frequency-watt droop characteristic of the GFM model is a straight line (without deadband) with a slope slightly less than 5%. The slope is inside the range of allowable settings for all the three DER performance categories. Given that foot note c under Table 24 in IEEE Std 1547™-2018 states that a deadband of less than 0.017 Hz shall be permitted for all three DER performance categories, the steady state frequency-watt droop characteristic of the GFM model is allowed by IEEE Std 1547™-2018 and hence no gap is identified.

C. Reactive Power-Voltage Control Test Against IEEE Std 2800™-2022

This test is related to the voltage control mode requirements set forth in section 5.2.2 of IEEE Std 2200™-2022. This section lays out the requirements for the dynamic reactive power response of the IBR plant to a step change in the applicable voltage within the continuous operation region. As per the requirement, the IBR plant shall operate in closed-loop automatic voltage control mode to regulate the steady-state

978-1-6654-6060-6/23 $31.00 © 2023 IEEE

voltage at the RPA to the reference value as adjusted by the droop function.

The evaluation metrices that are of interest are the reaction time, maximum step response time and damping ratio of the reactive power output response of the IBR for a step change in applicable voltage. Requirements set forth in the standard are summarized in Table I.

The test is carried out by applying a step voltage decrease as well as a step voltage increase, and simulate the GFM model response. The time domain response to the voltage increase event are shown in Fig. 6 and the metrices are summarized in Table II. As can be seen, the response of the GFM model is within the allowable range of IEEE Std 2200™-2022 and no gap is identified.

TABLE I
PERFORMANCE SPECIFICATION FOR VOLTAGE CONTROL MODE IN IEEE Std 2800™-2022

Parameter	Performance target	Notes
Reaction time	Less than 200 ms	
Maximum step response time	As required by transmission system operator	Typical range: 1s to 30s
Damping ratio	No less than 0.3	Depends on grid strength

TABLE II
REACTION TIME, RESPONSE TIME AND DAMPING RATIO OF THE GFM RESPONSE

GFM control modes	Reaction time (ms)	Response time (s)	Damping ratio
SRF-PLL	11	0.08	1
Droop	8	0.065	0.61
VSM	7	0.065	0.62
dVOC	6	0.06	0.59

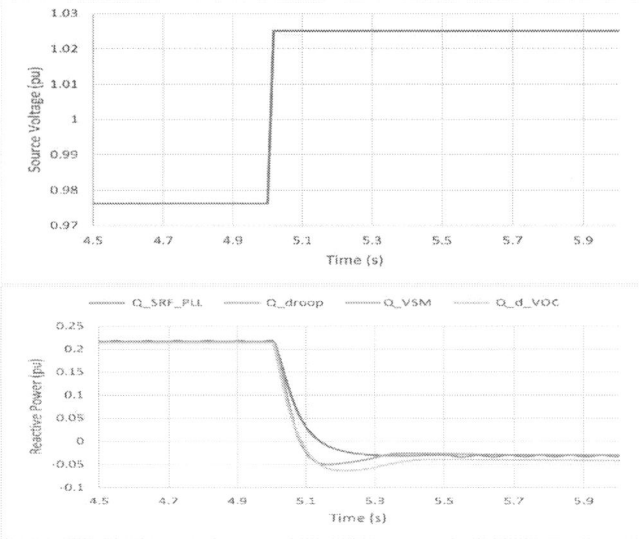

Fig. 6. Generic GFM model response for a step increase in source voltage at t = 5 s

D. Active Power-Frequency Control Test Against IEEE Std 2800™-2022

This test is designed to compare the performance of generic GFM model with the response requirements of an IBR to a change in system frequency as detailed in section 6 of IEEE Std 2800™-2022. The requirements related to primary frequency response (PFR) and fast frequency response (FFR1) are considered.

The test criteria for PFR test are the reaction time, rise time, settling time and damping ratio. The reaction time in the context of PFR test is the time between the step change in the applicable frequency and when the active power output of the system starts changing in the direction of the control. The rise time is the duration between the time when active power response reaches 10% of the steady state value and when it reaches 90% of the steady state value. The settling time is the duration taken by the response to settle within the allowable settling band. The damping ratio is the ratio of the actual damping of the response to the damping level at the critical damping. The baseline metrics (defined in IEEE Std 2800™-2022) for evaluation are given in Table III.

A test case which simulates a decrease in applicable frequency with rate of change of frequency (ROCOF) of 3 Hz/s is created. The response of the GFM model is shown in Fig. 7 and the values of the evaluation metrics are given in Table IV. The values of evaluation metrics are within the allowable range. Furthermore, since the response time is less than 1 s, the requirement of FFR1 is also fulfilled. Therefore, the behavior of the GFM model satisfies the PFR requirements in IEEE Std 2800™-2022 and no gap is identified.

E. Low Voltage Ride-Through Test Against IEEE Std 2800™-2022

This test compares the response of the generic GFM model

TABLE III
PERFORMANCE SPECIFICATION FOR PFR IN IEEE Std 2800™-2022

Parameter	Units	Default value	Minimum	Maximum
Reaction time	Seconds	0.50	0.20	1
Rise time	Seconds	4.0	2.0	20
Settling time	Seconds	10.0	10	30
Damping ratio	Unitless	0.3	0.2	1

TABLE IV
VALUES OF EVALUATION METRICS UNDER FREQUENCY DECREASE

Control mode	Reaction time (ms)	Rise time (s)	Settling time (s)	Damping ratio
SRF-PLL	11	0.139	0.48	0.6137
Droop	9	0.0756	0.49	0.514
VSM	9	0.09	0.48	0.48
d_VOC	9	0.0756	0.49	0.514

978-1-6654-6060-6/23 $31.00 © 2023 IEEE

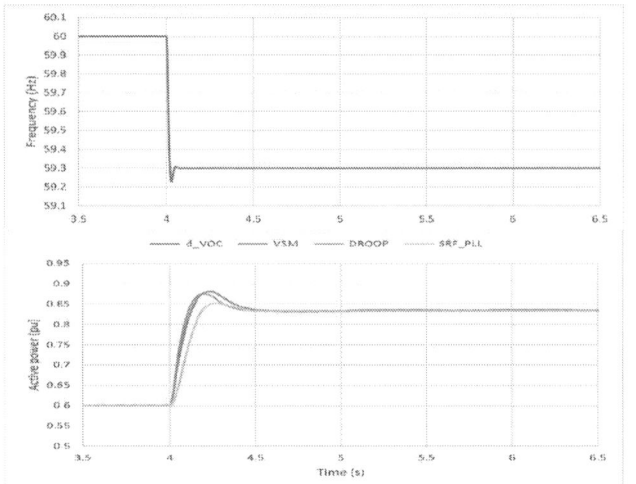

Fig. 7. Response of GFM model for decrease in applicable frequency with ROCOF of 3 Hz/s

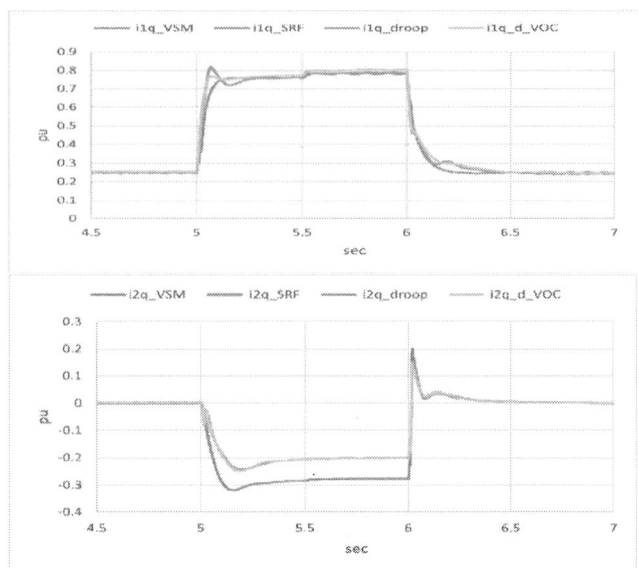

Fig. 8. Positive and negative sequence reactive currents (i1q, i2q) during single line to ground fault

with the low and high voltage ride-through response requirements set forth in section 7.2.2.3 of IEEE Std 2800™-2022. This section describes the required response capability of the IBR when in the mandatory operating region defined in IEEE Std 2800™-2022 section 7.2.2. This work has specifically compared the performance of the GFM model with the current injection requirements specified in section 7.2.2.3.4. The default RPA for current injection during ride through mode is the POC.

The standard requires the IBR unit to inject reactive current dependent on the IBR unit terminal voltage for balanced faults, and inject negative sequence current dependent on the IBR unit terminal negative sequence voltage that leads the negative sequence voltage by 90-100 degrees (in addition to the positive sequence reactive current). Note the standard intentionally does not specify the magnitude of incremental positive sequence and negative sequence reactive current and leaves those to be specified by system operators.

The low voltage ride through behavior of the GFM model has been tested under balanced and different unbalanced fault conditions. As an example, Fig. 8 shows the results under a single line to ground fault, where the positive and negative sequence reactive currents are denoted as i1q and i2q, respectively. In terms of current injection, the GFM IBR injects reactive current in both positive and negative sequence during the fault, and the negative sequence current is leading the

negative sequence voltage by close to 90 degrees (not shown in the figure), which satisfies the standard. However, it was found that the step response time and settling time of the GFM response is slower than what is required and hence a potential gap exists. Whether this violation is specific to this one GFM model and its parameterization is under investigation.

IV. CONCLUSION

This paper examines several cluases in IEEE Std 1547™-2018 and IEEE Std 2800™-2022 to identify inadvertent barriars that may potentially prevent the interconnection of GFM plants into power distribution and transmission systems. Preliminary results reveal that for a voltage disturbance inside the continuous operation region, the response of the generic GFM model with the specific parameterization considered is faster than what is allowed in IEEE Std 1547™-2018, which is a potential gap. For transmission connected IBRs, the tests conducted show that the response of the generic GFM model mostly aligns with the IEEE Std ™ 2800-2022 requirements regarding reactive power-voltage control, active power-frequency control, and voltage disturbance ride through, except that the step response time and settling time under voltage ride through are slower than the requirements. For the potential gaps identified, whether they are specific to the investigated GFM model and its parameterization or applies to all GFM IBRs is under study.

ACKNOWLEDGEMENT

This paper is based upon work supported by the U.S. Department of Energy's Office of Energy Efficiency and Renewable Energy (EERE) under the Solar Energy Technologies Office Award Number 38637.

TABLE V
PERFORMANCE SPECIFICATION FOR CURRENT INJECTION DURING
VOLTAGE RIDE THROUGH IN IEEE Std 2800™-2022

Parameter	IBR units except Type III WTGs
Step response time	No more than 2.5 cycles
Settling time	No more than 4 cycles
Settling band	-1.5%/+10% of IBR unit maximum current

978-1-6654-6060-6/23 $31.00 © 2023 IEEE

REFERENCES

[1] *Grid Forming Inverters: EPRI Tutorial (2022)*. EPRI, Palo Alto, CA: 2022. 3002025483.

[2] W. Wang, G. M. Huang, D. Ramasubramanian and E. Farantatos, "Transient stability analysis and stability margin evaluation of phase-locked loop synchronised converter-based generators," IET Generation, Transmission & Distribution, vol. 14, no. 22, pp. 5000-5010, Nov. 2020.

[3] P. Unruh, M. Nuschke, P. Strauß, and F. Welck, "Overview on Grid-Forming Inverter Control Methods," Energies, 2020, 13, 2589.

[4] D. Ramasubramanian, "Differentiating between plant level and inverter level voltage control to bring about operation of 100% inverter based resource grids," Electric Power Systems Research, vol. 205, no. 107739, Apr 2022.

[5] W. Wang, et al., "Benefit of Fast Reactive Power Response from Inverters in Weak Distribution Systems," 2022 IEEE Rural Electric Power Conference, Savannah, GA, USA, 2022.

[6] IEEE Std 1547™-2018, "IEEE Standard for Interconnection and Interoperability of Distributed Energy Resources with Associated Electric Power Systems Interfaces," IEEE standard, 2018.

[7] IEEE Std 2800™-2022, "IEEE Standard for Interconnection and Interoperability of Inverter-Based Resources (IBRs) Interconnecting with Associated Transmission Electric Power Systems," IEEE standard, 2022.

[8] *Performance Requirements for Grid Forming Inverter Based Power Plant in Microgrid Applications: Second Edition.* EPRI, Palo Alto, CA: 2022. 3002024431.

[9] *Grid Forming Inverter Models*, PSCAD Knwledge Base, https://www.pscad.com/knowledge-base/article/894.

Enhanced Bifaciality Factor with Sb$_2$Se$_3$ devices modeling Cu$_2$O back buffer

Sanghyun Lee[a], Kent Price[b]

[a]Department of Engineering Technology, University of Kentucky, Lexington, KY 40506, USA

[b]Department of Physics, Morehead State University, Morehead, KY 40351, USA

Abstract — We have investigated bifacial Ge-incorporated Sb$_2$Se$_3$ thin-film solar cells by modeling and simulating a back buffer layer including Cu$_2$O to optimize the performance of bifacial devices. We proposed Ge-incorporated Sb$_2$Se$_3$ absorber layers for bifacial devices, which were used as modeling input parameters after characterizing the absorber film. With extracted parameters (bandgap, 1.23 eV and the absorption coefficient, 3 x 10^5 cm^{-1} at 600 nm), we investigated the impact of various back buffer layers on bifacial device performance. The simulated device structure is ZnO:Al/i-ZnO/CdS/Ge-Sb$_2$Se$_3$/back buffer layer/TCO. First, the impact of conduction and valence band offsets was investigated, showing the significant dependence of device performance on valence band offset. To minimize the Schottky hole barrier at the back contact region, high back buffer doping concentration (1 x 10^{15} cm^{-3} or higher) is required to improve device performance, whereas low absorber doping concentration (<1 x 10^{14} cm^{-3}) enables higher efficiency due to the enlarged energy band bending. With a Cu$_2$O back buffer layer, the best efficiency of front-side illumination is 19.7 %, Voc (744.4 mV), Jsc (40.14 mA/cm2), and FF (66.1 %) and that of the rear-side illumination is 13.0 %, Voc (724.5 mV), Jsc (31.6 mA/cm2), and FF (56.7 %). Consequently, the bifaciality factor is 66 %.

Keywords — *bifacial solar cells, Sb$_2$Se$_3$, Ge, modeling*

I. INTRODUCTION

Thin-film solar cells based on an Antimony Selenide (Sb$_2$Se$_3$) absorber layer are a promising candidate for next-generation photovoltaic devices. A Sb$_2$Se$_3$ absorber has 1.1 eV bandgap, a high absorption coefficient at visible light (>10^5 cm^{-1}), good carrier mobility (<15 cm^2/Vs), long carrier lifetime (<67 ns), and simple binary compound with high vapor pressure and low melting point (550 °C) [1-3].

Due to the versatility of Sb$_2$Se$_3$ thin-films, Sb$_2$Se$_3$ has been studied by varying compositions and concentrations of incorporated elements for various applications. In particular, Ge-incorporated Sb$_2$Se$_3$ thin-films (hereafter, Ge-Sb$_2$Se$_3$) have been reported as a good polycrystalline absorber candidate with Ge concentration <15 % [2-3].

To further improve the efficiency of polycrystalline Sb$_2$Se$_3$ devices, both front and rear-side illumination could be captured by utilizing the bifacial device configuration of Sb$_2$Se$_3$ devices. However, developing bifacial devices has stagnated in thin-film photovoltaic technologies due to the short carrier lifetime (<100 nS) as compared to polysilicon counterparts [4]. For instance, the record efficiency of rear-side illumination is reported as 9.2 % for CIGS, 8.0 % for CdTe, and 9.0 % for Kesterite solar cells,

respectively [5-6]. Designing a bifacial photovoltaic structure has been studied by several research groups [4-6] and recently, a few studies of thin-film devices with the thin absorber (<2 um) such as CIGS, CdTe, and Kesterite have been reported [4-6]. By carefully selecting a transparent back buffer layer coupled with a transparent conducting back contact, a bifacial device configuration needs to be optimized in thickness and doping concentration to maximize a bifaciality factor, which is the ratio of rear-to-front efficiency to the same irradiation.

In this work, we have theoretically studied bifacial Ge-Sb$_2$Se$_3$ devices after fabricating Ge- Sb$_2$Se$_3$ absorber thin-films, followed by characterizing the optical properties. The optical properties of a Ge-Sb$_2$Se$_3$ absorber layer were used as input parameters for modeling and numerical simulation of the bifacial device configuration. We have proposed new bifacial devices to improve the overall device performance and optimized developed device models with various parameters.

II. EXPERIMENTAL RESULTS

1. Ge-Sb$_2$Se$_3$ absorber film

Sb$_2$Se$_3$ films were deposited on NSG Fluorine-doped Tin Oxide (FTO) soda-lime glass produced by Pilkington Group. FTO glass substrate show approximately 6 ohm/sq. with the thickness of ~250 nm. The substrates were cleaned in an ultrasonic cleaning bath, followed by acetone, isopropanol, and N$_2$ drying in sequence. Ge- Sb$_2$Se$_3$ films were deposited using Vapor Transfer Deposition processes. The temperature for Sb$_2$Se$_3$ substrates and sources was kept at 500 °C and the temperature for Ge sources was separately maintained at 900 °C for 5 min with the ramp rate of 37 °C/min, resulting in ~1 μm overall thickness. Upon film deposition, the furnace was

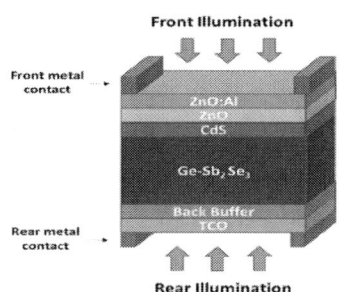

Fig. 1. Simulated Ge-Sb$_2$Se$_3$ device structure.

naturally cooled down before optical characterization.

2. Modeling Methodology

Cu₂O (Back Buffer)	Value/Unit
Thickness	0.05 um
CB effective density of state	8.0×10^{18} cm⁻³ (various)
VB effective density of state	1.8×10^{19} cm⁻³ (various)
Electron mobility	40 cm²/V/s (various)
Hole mobility	20 cm²/V/s (various)

Table I: Key parameters used in this work

To investigate the bifacial Ge-Sb₂Se₃ thin-film devices, we emphasize our study around a transparent back buffer layer, which is central to capturing the incoming light from the back surface. In Fig. 1, the proposed device structure considered in this study presents ZnO:Al/i-ZnO/CdS/Ge Sb₂Se₃/back buffer layer/Transparent conductive oxide (TCO) between front and rear metal grid contacts. TCO for the front-side is ZnO:Al(AZO)/intrinsic-ZnO (i-ZnO) and Indium Tin Oxide (ITO) is used for the back-side.

For simulation of Ge-Sb₂Se₃ devices, we began with our key baseline input parameters and models in Table 1 [2,3]. With our extensive previous studies around the back contact region, we updated models of interfaces, defects, and interlayers [7-10].

III. RESULTS AND DISCUSSION

With UV-Vis characterization, the bandgap energy of the Ge-Sb₂Se₃ absorber is 1.23 eV and an absorber coefficient is 3×10^5 cm⁻¹ at 600 nm (2.1 eV), which were used in the simulation.

1. Band alignment and back buffer doping

The band alignment with the absorber is crucial to minimize the effective hole Schottky barrier at the interface between the absorber and the transparent back buffer interlayer [7-10]. At the same time, reflecting electrons toward the front conduct junction is as important as the valence band alignment. For high p-type doping concentration with Fermi level near valence band, zero or smaller positive valence band offset ($E_{V,absorber} - E_{V,back\ buffer} = \Delta E_V$) are essential to minimize hole barriers while the negative conduction band offset ($E_{C,absorber} - E_{C,back\ buffer} = \Delta E_C$) are preferred [8-10]. To systematically investigate factors affecting bifacial device performance, we conducted our study through two phases: impact of different band alignment through conduction and valence band offsets and optimization of the selected back buffer by varying input parameters.

As the band alignment to the absorber plays a crucial role, we investigated conduction and valence band offsets with two cases: no valence band offset ($\Delta E_V = 0.0$ eV, ideal), and positive valence band offset ($\Delta E_V = 0.51$ eV). In Fig. 2, the valence band of the back buffer is aligned with the absorber ($\Delta E_V = 0.0$ eV) without a hole Schottky barrier, whereas negative conduction band ($\Delta E_C = -0.5$ eV) enables electron reflection (an ideal case). Without a Schottky barrier to holes, high back buffer doping concentration ($1 \times 10^{15} - 1 \times 10^{18}$ cm⁻³) improves the device performance in Fig 2. However, lower buffer doping concentration (1×10^{14} cm⁻³) reduces the absorber energy band

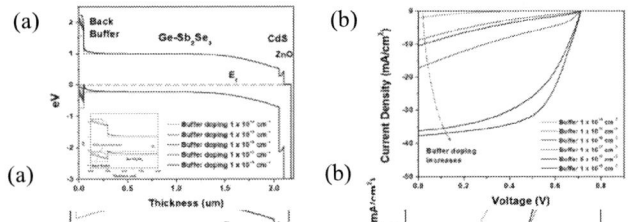

Fig. 3. (a) Band diagram of back buffer layers with conduction (-0.56 eV) and valence band (0.51 eV) offsets: an inset for back contact area zoom-up, (b) current-voltage characteristics

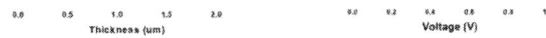

Fig. 2. (a) Band diagram of back buffer layers with conduction (-0.5 eV) and no valence band offsets: an inset for back contact area zoom-up, (b) current-voltage characteristics

near the back buffer region, hampering hole movement and acting like a hole barrier, leading to clamping hole current near the back buffer region in current-voltage (JV) curves.

However, the second case study is practical with the positive valence band offset of the back buffer layer ($\Delta E_V = 0.51$ eV), and negative conduction band offset ($\Delta E_C = -0.51$ eV) by varying back buffer doping concentration (Fig. 3). The shape of the hole barrier changes from near-rectangle to triangle, increasing the probability for holes to tunnel through the barrier, improving device performance. Hence, high enough back buffer doping concentration ($>1 \times 10^{17}$ cm⁻³) is preferred with minimal positive valence and negative conduction band offsets for the efficiency improvement of Ge-Sb₂Se₃ devices.

Given the high doping concentration of a back buffer (3×10^{18} cm⁻³), Fig. 4 shows the dependency of device performance on conduction and valence band offsets. Efficiency is dependent on valence band offset with the absence of conduction band offset dependency.

2. Absorber doping concentration

Now we investigate the impact of the absorber doping concentration, coupled with various defect concentration of the back buffer layer and valence band offset. As reported [7-10], an effective Schottky hole barrier is increased by higher defect concentration ($>1 \times 10^{15}$ cm⁻³) of the back buffer between the absorber and TCO. Fig. 5 shows various absorber doping concentrations at two buffer doping concentrations (1×10^{14} and 1×10^{15} cm⁻³). Firstly, efficiency strongly depends on the valence band offset throughout the simulation. Secondly,

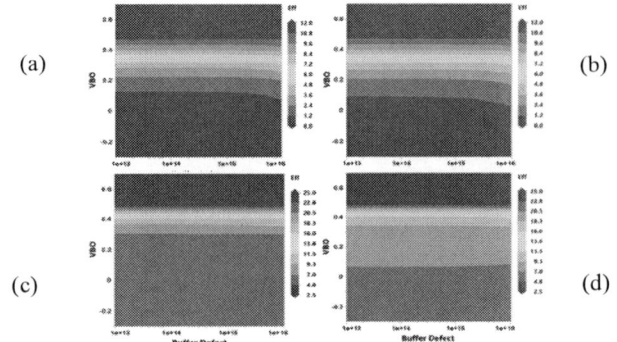

Fig. 5. Efficiency for absorber and buffer doping concentrations and valence band offset (VBO): absorber/buffer (cm^{-3}), (a) $1 \times 10^{13}/1 \times 10^{14}$, (b) $1 \times 10^{14}/1 \times 10^{14}$, (c) $1 \times 10^{13}/1 \times 10^{15}$, (d) $1 \times 10^{14}/1 \times 10^{15}$

efficiency generally improves for the same buffer defect and doping concentration, and the valence band offset as the absorber doping concentration decreases due to enlarged absorber band bending (Fig. 6). Hence, lower absorber and higher back buffer doping concentrations greatly enhance efficiency for the given valence band offset of the buffer layer.

3. Bifacial efficiency with Cu₂O back buffer

We consider the dependency of thickness and doping concentration of the Ge-Sb₂Se₃ absorber on front and rear-side illuminated bifacial efficiency (Fig. 7). For front-side illumination, efficiency increases as the absorber thickness increases near lower side doping concentration ($<5 \times 10^{14}$ cm^{-3}), whereas rear-side illumination slightly decreases near $<1 \times 10^{15}$ cm^{-3} for increasing thickness. For absorber thickness dependency, rear-side illuminated efficiency increases as the thickness is reduced. Consequently, the optimum thickness of the absorber is >0.7 um with the absorber doping concentration, $<1 \times 10^{14}$ cm^{-3}. With optimized input parameters, front-side illumination yields conversion efficiency, 19.7 %, Voc (744.4 mV), Jsc (40.14 mA/cm²), and FF (66.1 %), and rear side illumination shows efficiency, 13.0 %, Voc (724.5 mV), Jsc (31.6 mA/cm²), and FF (56.7 %). The bifacility factor is 66 %.

III. CONCLUSIONS

We have investigated bifacial Ge-incorporated Ge-Sb₂Se₃ thin-film solar cells. Valence band offset to the Ge-Sb₂Se₃ absorber is crucial in selecting an appropriate back buffer layer, whereas the impact of conduction band offset is insignificant. As the Schottky hole barrier is minimized with zero or small positive valence band offset, hole current clamping could be

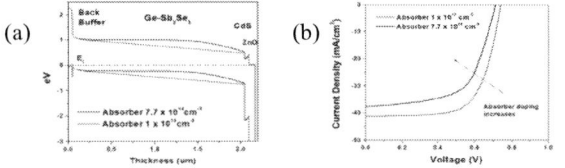

Fig. 6. (a) Band diagram for two absorber doping concentration. (b) Current-voltage characteristics

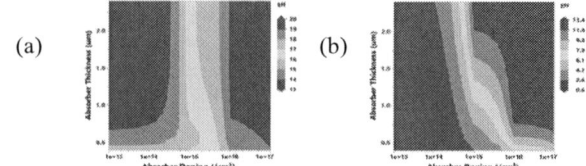

Fig. 7. Front (a) and rear-side (b) illuminated bifacial efficiency on thickness and absorber doping concentration with a Cu₂O back buffer.

reduced at the interface between the absorber and the back buffer layer. With satisfactory band alignment with the absorber, Cu₂O back buffer is used to investigate device performance with various critical input parameters for the bifacial devices such as absorber thickness, doping concentration, and defect concentration near the back buffer interface. With optimized parameters, front-side illumination yields conversion efficiency, 19.7 %, Voc (744.4 mV), Jsc (40.14 mA/cm²), and FF (66.1 %) and rear side illumination shows efficiency, 13.0 %, Voc (724.5 mV), Jsc (31.6 mA/cm²), and FF (56.7 %). All in one, the bifaciality factor is 66 %.

REFERENCES

[1] X. Hu, J. Tao, Y. Wang, J. Xue, 5.91 %-efficient Sb₂Se₃ solar cells with a radio-frequency magnetron-sputtered CdS buffer layer, Appl. Mater. Today 16 (2009) 367–374

[2] S. Lee, M. McInerney, Modeling of Ag incorporated Sb(S$_x$Se$_{1-x}$) bi-layer devices: Enhanced bi-layer absorber configuration, Solar Energy 238 (2022) 363–370.

[3] S. Lee, M. McInerney, Theoretical Study of the Improved bi-layer absorber of Silver incorporated Sb₂Se₃ devices, Proc. of 8th WCPEC, 1 (2022) 276

[4] R. Lemus, R. Vega, T. Kim, A. Kimm, L. Shephard, Bifacial solar photovoltaics - A technology review, Renewable Sustainable Energy Rev. 60 (2016) 1533–1549.

[5] K. Subedi, A. Phillips, N. Shrestha, F. Alfadhili, Enabling Bifacial Thin Film Devices by Developing a Back Surface Field using CuAlO$_y$, Nano Energy 83 (2021) 105827

[6] H. Deng, Q. Sun, Z. Yang, W. Li, Novel symmetrical bifacial flexible CZTSSe thin film solar cells for indoor photovoltaic applications, Nat. Comm. 12 (2021) 3107

[7] S. Lee, K. Price, E. Saucedo, Interface models of CZTSe: Ge bilayers solar cells with Impedance Spectroscopy, *Proc. of IEEE Photovoltaic Spec. Conf., 47th*, 1 (2020) 1-3

[8] S. Lee, K. Price, E. Saucedo, S. Giraldo, Engineering of effective back-contact barrier of CZTSe: Nanoscale Ge solar cells – MoSe₂ defects implication, Solar Energy, 194 (2019) 114-120

[9] S. Lee, K. Price, DH Kim, Two local built-in potentials of H₂S processed CZTSSe by complex impedance spectroscopy, Solar Energy, 225 (2021) 11-18

[10] S. Lee, K. Price, E. Saucedo, Estimation of front and back junctions of CZTSe: Ge solar cells by combined modulus and impedance spectroscopy, J Phys D Appl Phys, 54 (2021)

Analysis of the key factors influencing the economic competitiveness and profitability of floating photovoltaics

Leonardo Micheli, Diego L. Talavera, Fredy A. Sepúlveda-Vélez

DIAEE, Sapienza University of Rome, Rome, Italy

IDEA Research Group, University of Jaén, Jaén, Spain

Electronics and Automation Engineering Department, University of Jaén, Jaén, Spain

If modules are installed on water, PV can alleviate the land competition arising with its growing deployment. However, this solution, known as floating photovoltaics (FPV), still represents a small percentage of the global PV capacity. Because of the relatively young age and the lack of a developed supply chain, the costs associated with FPV are still higher compared to those of land-based PV (LPV). Despite that, specific economic conditions can favor and/or hinder the deployment of this technology, independently of the cost of LPV. This work assesses the competitiveness and profitability of FPV from an economic standpoint in various countries. This way, conditions more or less favorable to the development of FPV are identified and discussed.

Towards Integrating Data Quality Assessments and Radiometer Uncertainty for Determining the Expanded Uncertainty of Three-Component Solar Radiation Measurements

Stephen Wilcox[1], Tom Stoffel[1], Aron Habte[2], and Manajit Sengupta[2]

[1]Solar Resource Solutions, LLC, Louisville, CO 80401 USA

[2]National Renewable Energy Laboratory, Golden CO 80204 USA

Abstract — Accurate solar irradiance data are fundamental for determining the design and performance characteristics of photovoltaic systems. The uncertainty of solar irradiance measurements depends on many factors including radiometer design, calibration, installation, maintenance, and operational environment. The key contributors to this uncertainty can be classified as the measurement uncertainty of a particular radiometer and the operational uncertainty determined for the time of measurement. Radiometer measurement uncertainty estimates (U_R) can be based on well-established methods used as part of the radiometer calibration process. Estimates of operational uncertainties (U_O) require consideration of additional site-specific factors that affect data quality. A method is needed for establishing the accuracy of solar irradiance data by integrating an existing data quality process and measurement uncertainty estimates for specific radiometers. An algorithm has been developed to integrate data quality analyses and measurement uncertainty estimates for three-component solar irradiance data: global horizontal (total hemispheric) irradiance, direct normal (beam) irradiance, and diffuse horizontal (sky) irradiance collected at one- to 60-minute intervals. The algorithm has been tested using one-minute irradiance measurements. The goal of the project is to distribute a user-friendly software package based on the new algorithm.

I. Introduction

Acquiring solar resource data with known uncertainty directly supports the goal of making solar energy systems more cost competitive with other forms of energy. In the past, the uncertainty of a solar irradiance data set has frequently been solely represented by either the manufacturer's stated instrument uncertainty or the uncertainty from a calibration process. The historic approach fails to acknowledge many additional sources of error in a field measurement [1]. This approach incorporates methods that conform with the *Guide to the Expression of Uncertainty in Measurements* [2] (GUM).

The uncertainty of a data set is determined by several factors identified in established best practices [3], including:

- Design and manufacturing characteristics of a measuring instrument
- Configuration and installation of a measurement station
- Quality of the instrument calibration and the uncertainty of the reference instruments
- Uncertainty of the data logging equipment and associated electronic infrastructure

- Errors introduced during ongoing measurement operations.

The last item, which we are calling *Operational Uncertainty* (U_O), is difficult to ascertain because many uncontrolled factors affect a field measurement, such as frequency of instrument maintenance, degradation, failure of supporting equipment, and multiple environmental and weather impacts. An estimate of U_O is derived by examining interrelated data from measurement station instruments to detect errors in field measurements.

The operational uncertainty method described here requires the use of simultaneous measurements of global horizontal irradiance (GHI), direct normal irradiance (DNI), and diffuse horizontal irradiance (DHI) and assumes that data were carefully collected using best practices protocols. It is not intended to assign uncertainty to data collected under deficient or unknown conditions.

II. Determining Operational Uncertainty

U_O is derived from the three-component solar irradiance measurements of GHI, DNI, and DHI through the functionality of SERI QC [4], a data quality assessment software package developed by the National Renewable Energy Laboratory (NREL). These values are converted by SERI QC to Clearness Indices in K-space, a normalized representation of a measurement independent of the attenuating effects of the atmosphere and station location. For K-space, each measurement is normalized (divided) by the like component as if observed at the top of the atmosphere without any atmospheric attenuation, a value here referred to as extraterrestrial irradiance.

The direct normal extraterrestrial irradiance (ETRn) is computed from the date and time information:

$$ETRn = TSI * (R/Ro)^2 \quad (1)$$

where:

TSI = total solar irradiance (1360.8 ±0.5 W/m2)
R = sun-Earth distance at the time of interest
Ro = annual mean sun-Earth distance

The global horizontal (total hemispheric) extraterrestrial irradiance (ETR) is calculated as:

$$ETR = ETRn * \cos(SZA) \quad (2)$$

where:

SZA = solar zenith angle at the location and time of interest.

Thus, in K-space the total, normal, and diffuse clearness indices are computed:

$$Kt = GHI\, /\, ETR \qquad (3)$$
$$Kn = DNI\, /\, ETRn \qquad (4)$$
$$Kd = DHI\, /\, ETR \qquad (5)$$

These values are related in K-space by the coupling equation:

$$Kt = Kn + Kd \qquad (6)$$

In the absence of a recognized measurement reference at the monitoring station, (6) and algebraic rearrangements are used to determine U_O by establishing a field reference for each component through the other two components in the coupling equation. An operational uncertainty *ratio* can then be calculated, which with "perfect data" will equal one. Any measurement error would result in a ratio less than or greater than one. A percent error is calculated to determine the operational uncertainties for the three components:

$$U_O\,Kt = \left(\frac{Kt}{Kn+Kd} - 1\right) \cdot 100 \qquad (7)$$

$$U_O\,Kn = \left(\frac{Kn}{Kt-Kd} - 1\right) \cdot 100 \qquad (8)$$

$$U_O\,Kd = \left(\frac{Kd}{Kt-Kn} - 1\right) \cdot 100 \qquad (9)$$

(8) and (9) are undefined if Kt=Kd or Kt=Kn. Certain other measurement conditions will result in exaggerated and unrealistic results from the same equations. Additionally, measurements that occur at high zenith angles (near sunrise and sunset) can result in similar unrealistic values from ratios between small measurement values. Thus, it was determined that (8) and (9) can be applied only in specifically restricted data scenarios.

Further analysis has shown that $U_O Kt$ (7) will reveal errors from the other two K-space components, and it can be used nearly universally to represent a reasonable operational uncertainty in all three components [5]. This approach does not provide specific uncertainties for each component, but in all cases described, regardless of method of calculation, the nature of the uncertainty equations and the use of field data are such that each is subject to crossover errors contributed by the other two components. This is what we call the *ambiguity of fault*, or the general inability to determine which of the three components causes an imbalance in an uncertainty equation. Nonetheless, $U_O Kt$ as an overarching uncertainty can be applied to the entire data record, a value that embodies a valuable contribution to estimating the uncertainty of a data set.

III. DERIVING THE EXPANDED UNCERTAINTIES

The first step relies on the fixed instrument uncertainty from the NREL radiometer uncertainty tool [1] (external to this process). This value represents the bounds of measurement error based on the design performance specifications of a radiometer operated under *well-controlled conditions*, such as a calibration scenario. These radiometer uncertainties (U_R) for the three-component data set are defined as:

Uncertainty of GHI: $U_R GHI$
Uncertainty of DNI: $U_R DNI$
Uncertainty of DHI: $U_R DHI$.

Beyond these uncertainties, we expect adverse environmental or operational field conditions to add to the uncertainty of a measurement. Through the uncertainty equations (7)-(9), we have an estimate of all uncertainties in a field measurement, and as noted previously, we will rely on $U_O Kt$ to represent errors in all three components. For this evaluation, we will re-label $U_O Kt$ as a *system uncertainty*, $U_O SYS$, which has relevance to all three components:

$$U_O SYS = U_O Kt \qquad (10)$$

In the context of field measurements, $U_O SYS$ is comprised of two discrete, non-overlapping components: 1) the radiometer expanded uncertainties (U_R), and 2) uncertainties attributable to field operations. To separate the field uncertainty from the radiometer expanded uncertainties, we first establish the collective expanded uncertainty for the radiometers (U_{RADS}) by combining the three radiometer uncertainties in quadrature.

$$U_{RADS} = 2 \cdot \sqrt{\left(\frac{UrGHI}{2}\right)^2 + \left(\frac{UrDNI}{2}\right)^2 + \left(\frac{UrDHI}{2}\right)^2} \qquad (11)$$

(11) is consistent with GUM principles [2] in that the three U_R values are assumed to have normal distributions, i.e., each expanded uncertainty is divided by 2 to represent the standard uncertainties and combined in the quadrature to represent combined uncertainty, and ultimately a coverage factor (k) of 2 is applied to the radical to yield an expanded uncertainty.

We next define $U_O Field$, the operational uncertainty attributable to environmental and operational effects, as the difference between $U_O SYS$ and U_{RADS}.

$$U_O Field = MAX[U_O SYS - U_{RADS}, 0] \qquad (12)$$

The MAX function in (12) sets negative values of $U_O Field$ to zero. In this arrangement, $U_O Field$ represents the uncertainty in $U_O SYS$ beyond that of the radiometer uncertainties. Combining $U_O Field$ with each radiometer uncertainty in turn, we arrive at estimates of the expanded uncertainty for each of the three measured irradiance components:

$$U_{95}GHI = 2 \cdot \sqrt{\left(\frac{UrGHI}{2}\right)^2 + \left(\frac{UoField}{2}\right)^2} \qquad (13)$$

$$U_{95}DNI = 2 \cdot \sqrt{\left(\frac{UrDNI}{2}\right)^2 + \left(\frac{UoField}{2}\right)^2} \qquad (14)$$

$$U_{95}DHI = 2 \cdot \sqrt{\left(\frac{UrDHI}{2}\right)^2 + \left(\frac{UoField}{2}\right)^2} \qquad (15)$$

IV. APPLICATION OF THE METHOD

A preliminary evaluation of the method was performed using three-component one-minute solar data from the NREL Solar Radiation Research Laboratory (SRRL) for all of 2021.

Values for $U_R GHI$, $U_O SYS$, U_{RADS}, and the final value for $U_{95}GHI$ are plotted in Figure 1. The average of values plotted in Figure 1 are shown in Table I, along with the percent of times $U_{95}GHI$ values exceed $U_R GHI$. Results for $U_{95}DNI$ and $U_{95}DHI$ are similar. The values for $U_R GHI$, $U_R DNI$, and $U_R DHI$ derived from the NREL Radiometer Uncertainty Tool using specifics of the radiometer manufacturer and model are shown in Table II.

TABLE I
AVERAGE UNCERTAINTY VALUES FOR ONE YEAR OF DATA

Parameter	Value (%)
$U_{95}GHI$	3.58
$U_O SYS$	1.04
$U_O Field$	0.12
U_{RADS}	4.39
$U_{95}GHI$ exceeds $U_R GHI$	2.3

TABLE II
SRRL RADIOMETER UNCERTAINTIES (U_R)

Parameter	Instrument	Value (%)
$U_R GHI$	Kipp & Zonen CMP22	3.5
$U_R DNI$	Kipp & Zonen CHP1	2.3
$U_R DHI$	Kipp & Zonen CM22	3.5

V. CONCLUSIONS

A new algorithm consistent with GUM has been developed to assess the uncertainty of three-component solar irradiance measurements using SERI QC, an existing data quality assessment tool, and radiometer measurement uncertainties to determine the contribution of operational uncertainty to the overall uncertainty of data used for PV. The method has been evaluated using one-minute solar irradiance measurements collected during 2021 according to accepted best practices at the SRRL in Golden, Colorado. The resulting distributed software package will be based on the algorithm.

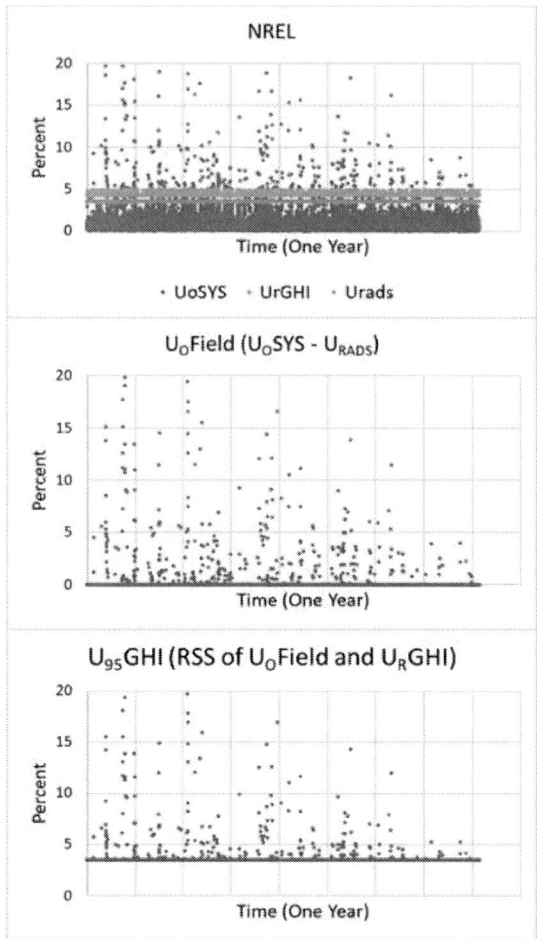

Fig. 1. Constituent parameters of the $U_{95}GHI$ calculations.

REFERENCES

[1] A. Habte, "Spreadsheet for estimating radiometer uncertainties", NREL Measurement & Instrumentation Data Center. Golden, CO: National Renewable Energy Laboratory. 2014 https://midcdmz.nrel.gov/radiometer_uncert.xlsx

[2] International Organization for Standardization (ISO). 2008. ISO/IEC Guide 98-3:2008(E): *Uncertainty of measurement – Part 3: Guide to the expression of uncertainty in measurement* (GUM: 1995). Geneva, Switzerland.

[3] M. Sengupta, A. Habte, S. Wilbert, C. Gueymard, and J. Remund, 2021. *Best Practices Handbook for the Collection and Use of Solar Resource Data for Solar Energy Applications: Third Edition* (NREL TP-5D00-77635). Golden, CO: National Renewable Energy Laboratory.

[4] E. Maxwell, S. Wilcox, and M. Rymes. *Users Manual for SERI QC Software—Assessing the Quality of Solar Radiation Data* (NREL/TP-463-5608). Golden, CO: National Renewable Energy Laboratory, 1993.
https://www.nrel.gov/docs/legosti/old/5608.pdf.

[5] S. Wilcox, and T. Stoffel, *Developing the Proof of Concept for SERI QC Flag Translation*. Golden, CO: National Renewable Energy Laboratory, 2023 [in press].

Stable, High-Throughput Production of Robust Perovskites in Open-Air with Polymer Additives

Nicholas Rolston, Carsen Cartledge, Vineeth Penukula, Muneeza Ahmad, Antonella Giuri, Aurora Rizzo

Arizona State University, Tempe, AZ, United States

CNR Nanotec, Lecce, Italy

Metal halide perovskites hold the potential for making the next generation of high-efficiency, low-cost solar cells through solution processing of Earth abundant materials. The U.S Department of Energy has identified four major challenges facing PVSKs: stability, scalability, manufacturability, and accelerated test protocols for reliability validation. Typical processing utilizes an antisolvent or gas quenching step. Unmodified perovskite inks designed for spin coating produce pinholed, uneven, and shunted films when processed in air or with scalable processing methods. Novel polymeric modifiers have recently been used to improve film uniformity and density. These additives not only act as rheological modifiers by increasing ink viscosity, thus allowing longer periods for crystallization and smooth films, but the long chains of the polymer create an organizational scaffold that increases resistance to humidity. Furthermore, additives from the food industry, such as cornstarch, have been found to enhance mechanical integrity and the operational lifetime of devices in addition to inducing spherulitic domains that can be tuned in size by temperature and precursor concentration to increase performance. In this work, the effects of gellan gum, a nontoxic polymeric modifier from the food industry, are studied on improving PVSK stability using scalable, one-step processing in open air. The use of a nontoxic solvent system further extends the wetting period, thus improving crystallization control. We measure the film stress to be tunable and compressive based on the additive concentration. We find that the additive enables reproducible and scalable perovskite devices to be produced in ambient conditions that match state-of-the-art performance on areas >25 cm2 with efficiencies exceeding 15%. The devices are stable under both thermal cycling and continuous operation (1000 hours). We attribute the compressive stress in the films along with significantly reduced ion migration based on impedance spectroscopy and transient photocurrent measurements as key reasons for the improved device stability. We validate the effect through aging perovskites with different film stresses and additive concentrations for mechanistic understanding of how film stress influences ion migration and device stability.

978-1-6654-6060-6/23 $31.00 © 2023 IEEE

Clear-sky detection using time-averaged, tilted-plane data

Clifford W. Hansen[1], Dirk C. Jordan[2]

1. Sandia National Laboratories, Albuquerque, NM 87185-1033

2. National Renewable Energy Laboratory, Golden, CO 80401

Abstract — **A method is presented to detect clear-sky periods for plane-of-array, time-averaged irradiance data that is based on the algorithm originally described by Reno and Hansen. We show this new method improves the state-of-the-art by providing accurate detection at longer data intervals, and by detecting clear periods in plane-of-array data, which is novel. We illustrate how accurate determination of clear-sky conditions helps to eliminate data noise and bias in the assessment of long-term performance of PV plants.**

I. INTRODUCTION

Methods are available which can identify clear-sky conditions using time series of global horizontal irradiance (GHI). These methods are designed and validated using short-interval (e.g., one minute) data. To date, no method is published for identifying clear-sky conditions using tilted-plane or time-averaged data.

PV systems are often monitored by recording concurrently the system's output power and irradiance in the PV system's plane-of-array (POA) and are time-averaged to, e.g., 15 minutes, to reduce the data storage requirements. Therefore, a method to identify clear-sky conditions using these data—time-averaged plane-of-array irradiance (Gpoa)—is of great interest for determination of PV system health.

We present a method to identify periods of time with clear-sky conditions, using a time series of time-averaged POA irradiance. A precise definition of clear-sky condition relies to some extent on judgment. For our purposes we use the term "clear-sky" to refer to conditions when the irradiance received by the measurement device does not differ from irradiance received in cloudless sky conditions to any meaningful degree.

II. BACKGROUND

Gueymard et al. [1] surveys and evaluates available methods with the intent to identify those best suited for detection of cloudless sky conditions. From that analysis, Bright et al. [2] propose a globally applicable method termed Bright-Sun for detection of cloudless sky conditions using 1-minute time series of GHI and either diffuse horizontal irradiance (DHI) or direct normal irradiance (DNI). The Bright-Sun algorithm builds on earlier work of Reno and Hansen [3] that proposed a method to identify clear-sky conditions using 1-minute time series of GHI. The Reno method essentially quantifies congruence of the GHI profile with a clear-sky model, by computing statistics on moving windows that quantify magnitude and variability. A point in the GHI time series is labeled as clear-sky if the statistics for every window containing the point agree within tolerance with the statistics computed for the same windows applied to the time series from the clear-sky model. Ellis [4] showed that the parameters of the Reno method could be adjusted to reliably identify clear-sky conditions in time series subsampled to 15-, 30- or 60-minute intervals from a base time series of 1-minute values. The Ellis technique does not, however, identify clear-sky conditions in data averaged to these longer intervals. Engerer and Mills [5] showed that the same statistics used for GHI in the Reno method can be applied to direct normal irradiance (DNI) to strengthen the evidence for the identified clear-sky conditions. In an evaluation of models for clear-sky DNI, Inman et al. [6] applied the Reno method (with one minor change) to synchronous one-minute GHI and DNI and give parameter values for each statistic.

A different approach to labeling clear-sky conditions is described by Meyers et al. [7]. In that work, a quantile regression technique is applied to time series of PV system power to extract daily profiles of maximum system output. This technique, termed "Statistical Clear Sky Fitting", circumvents the need for irradiance measurements, but requires a relatively long time series of system power e.g., several years. Moreover, the profile produced is not synonymous with clear-sky conditions, as the profile reflects any recurring external or internal factors that act to limit PV system output, such as shading from nearby structures or inverter clipping, although this is still an active field of development.

III METHODOLOGY

We extend the Reno method by finding optimized parameters for different data-averaging intervals. The parameters for the Reno method comprise a data window length and a threshold value for each of six statistics. For each data-averaging interval, we found optimized parameters (Table I.) by a combination of Design of Experiment (DOE) and Monte Carlo (MC) simulation [8]. The default parameters for the Reno method are validated only for one-minute data frequency.

As a test object we used a 1 kW crystalline silicon (c-Si) PV system at the National Renewable Energy Laboratory (NREL) in Golden, CO that has been described in more detail before [9]. The system was equipped with a broad-spectrum pyranometer

978-1-6654-6060-6/23 $31.00 © 2023 IEEE

TABLE I
PARAMETERS FOR THE RENO METHOD AND OPTIMIZED PARAMETERS (THIS WORK) FOR EACH DATA FREQUENCY.

Method	Data frequency (min)	Window length (min)	Mean difference	Maximum difference	Lower line length	Upper line length	Variance difference	Slope deviation
Reno	1	10	75	75	-5	10	0.005	8
This work	1	50	75	60	-45	80	0.005	50
This work	5	60	75	65	-45	80	0.010	60
This work	15	90	75	75	-45	80	0.032	75
This work	30	120	75	90	-45	80	0.070	96

to measure Gpoa and an ambient temperature (Tamb) sensor. In addition, an adjacent weather station provided GHI, DNI and DHI. Times with Gpoa below 100 W/m2 were removed to exclude possible shading of either the array or irradiance instruments at high incident angles. All data were averaged to a frequency of 1 minute (min) from 5 second (s) data logger increments. Clear-sky Gpoa was modeled using the Ineichen clear-sky model and Hay-Davies transposition model in pvlib-python [10]. Having measured Gpoa at the same natural data frequency allowed us to compare directly modeled and measured Gpoa.

A two-year measured data set and corresponding modeled Gpoa were inspected day-by-day to manually identify and label clear-sky periods. We chose two years characterized by the lack of interruptions due to repairs, maintenance, and the lack of outliers during the test time period. This inspection resulted in a "ground-truth" data set for quantifying accuracy of clear-sky detection methods. We note that even visual inspection can result in a small percentage of misclassified periods because of thin cloud coverage. To investigate the impact of various clear-sky detection algorithms at different data intervals, the 1 min data were averaged subsequently to 5 min, 15 min and 30 min data, carefully aligning the time stamps to avoid any unintended offsets that may have resulted from the averaging procedure.

IV. RESULTS

Having measured Gpoa in its native 1-min form permits evaluation of different clear-sky detection methods at different data frequencies. Several methods to detect clear-sky conditions were applied to the two-year data at each frequency. The first method was a simple filter implemented in RdTools, an open-source Python software co-developed by the National Renewable Energy Laboratory [11] that defines any point as clearsky if measured Gpoa is within 15% of modeled Gpoa. Second, we applied the Reno method [3] as implemented in the pvlib-python function pvlib.clearsky.detect_clearsky [10] with its default parameters. Third, we applied the Reno method using alternate parameter values as described by Ellis [4]. Finally, we apply the Reno method using parameters developed in this work.

Figures 1 and 2 show an example of detecting clear periods in a mostly clear day, and a day with mixed conditions,

respectively, using the 15 min averaged Gpoa data. The native 1 min data are represented by a continuous light blue line. The 15 min averaged data are shown by the black diamonds, correctly identified clear-sky points by green triangles, incorrectly identified clear-sky points (false positives) by red circles and missed clear-sky points (false negatives) by dark blue squares. The 15 min data set is chosen because the most common data frequency in large scale PV data collection such as the PV Fleet initiative is 15 min (averaged) and at this frequency differences between the different clear-sky detection algorithms are more readily visible [12].

It is not surprising that the Reno method with default (one-minute) parameters performs poorly. The present work surpasses the Ellis method whose parameters were determined for subsampled, rather than time-averaged, data. In mixed conditions, the present work mostly avoids the false positives that are evident in the Rdtools method's results.

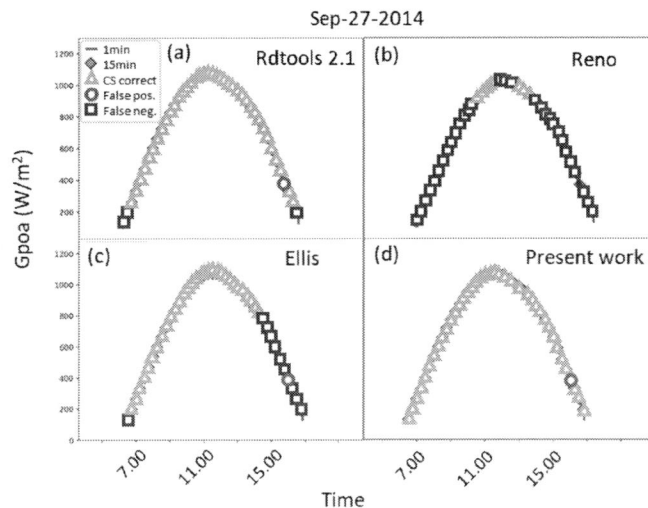

Fig. 1. Example mostly clear day with clear-sky periods in 15-minute averaged data identified by different algorithms: RdTools(a); Reno with default parameters (b); Ellis (c) and present work (d).

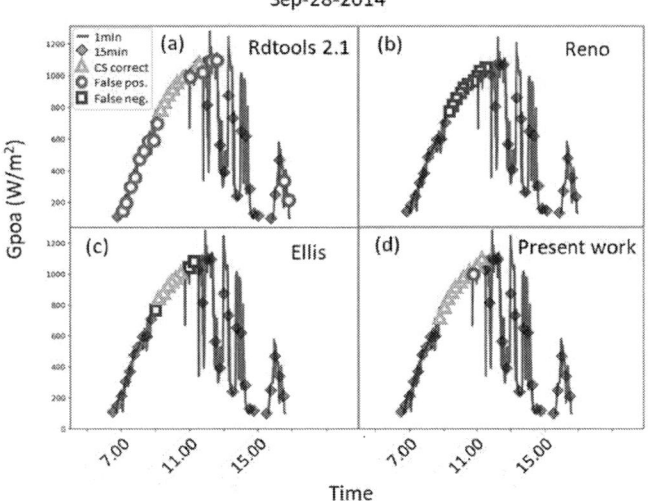

Fig. 2.   Example day of early clear and later cloudy conditions, with clear-sky periods in 15-minute averaged data identified by different algorithms: RdTools(a); Reno with default parameters (b); Ellis (c) and present work (d).

Figure 3 shows detection accuracy across the entire two-year data set. Data for plane-of-array irradiance is indicated by open symbols while GHI is indicated with filled symbols. The symbol shape represents the detection method. The percentage of correctly identified data points is represented by green data points, false positive by red and false negative points by blue symbols.

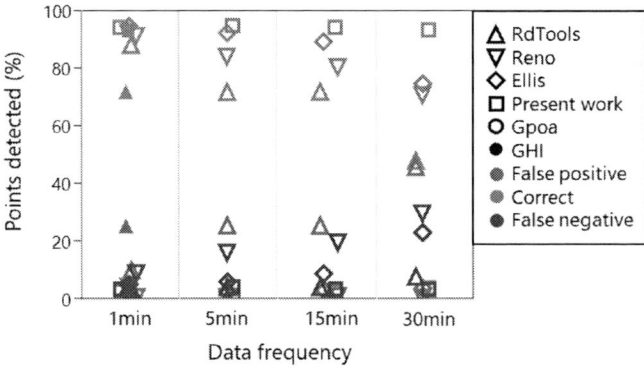

Fig. 3.   Detection accuracy using different clear-sky detection methods for different data frequencies.

As already seen in the example days, the RdTools clear-sky filter (triangles) performs worst for both GHI and Gpoa at all data intervals, having the lowest true positive and highest false positive rates. For 1-minute GHI or Gpoa data the methods of Reno (inverted triangles), Ellis (diamonds) and this work (squares) all perform well, correctly identifying more than 90% of labeled clear-sky points. More differentiation among the

methods is seen as the data interval increases. At 5-min intervals the Ellis method and the present work perform essentially equally well. The Reno method is next but also shows a large percentage of false negatives. As the data interval is increased to 15 min and lastly to 30 min parameters from the present work perform the best. The Reno and Ellis method miss many clear-sky periods at the 30-min data frequency.

DATA AVAILABILITY

Supplementary data can be found online through the Duramat datahub, DOI: 10.21948/1874985: https://datahub.duramat.org/dataset/labeled-clearsky-period-data-set

ACKNOWLEDGEMENT

Sandia National Laboratories is a multimission laboratory managed and operated by National Technology and Engineering Solutions of Sandia, LLC., a wholly owned subsidiary of Honeywell International, Inc., for the U.S. Department of Energy's National Nuclear Security Administration under contract DE-NA-0003525.

REFERENCES

[1]   C.A. Gueymard, J.M. Bright, D. Lingfors, A. Habte, M. Sengupta, A posteriori clear-sky identification methods in solar irradiance time series: Review and preliminary validation using sky imagers, Renewable and Sustainable Energy Reviews, 109, 2019, pp. 412-427, https://doi.org/10.1016/j.rser.2019.04.027.

[2]   J. M. Bright, X. Sun, C.A. Gueymard, B. Acord, P. Wang, N.A. Engerer, Bright-Sun: A globally applicable 1-min irradiance clear-sky detection model, Renewable and Sustainable Energy Reviews, 121, 2020, 109706, https://doi.org/10.1016/j.rser.2020.109706.

[3]   M.J. Reno, C. W. Hansen, Identification of Periods of Clear Sky Irradiance in Time Series of GHI Measurements, Renewable Energy, 90, 2016, pp. 520-531, https://doi.org/10.1016/j.renene.2015.12.031.

[4]   B.H. Ellis, M. Deceglie, A. Jain. Automatic detection of clear-sky periods using ground and satellite based solar resource data, 45th IEEE Photovoltaic Specialist Conference, Waikoloa, HI, 2018. 2018. p. 2293–8.

[5]   N.A. Engerer, F.P. Mills, Validating nine clear sky radiation models in Australia, Solar Energy, 120, 2015, pp. 9-24, https://doi.org/10.1016/j.solener.2015.06.044.

[6]   R. H. Inman, J.G. Edson, C.F.M. Coimbra, Impact of local broadband turbidity estimation on forecasting of clear sky direct normal irradiance, Solar Energy, 117, 2015, pp. 125-138, https://doi.org/10.1016/j.solener.2015.04.032.

[7]   B. Meyers, M. Tabone, E. C. Kara, Statistical Clear Sky Fitting Algorithm, in Proc. of the 45th IEEE Photovoltaic Specialist Conference, Waikoloa, HI, 2018.

[8]   D.C. Jordan and C.W. Hansen. Clear-sky detection for PV degradation analysis using multiple regression, Renewable Energy, 209, 2023, pp. 393-400, https://doi.org/10.1016/j.renene.2023.04.035.

[9]   D.C Jordan, C. Deline, S. Johnston, S.R. Rummel, B. Sekulic, P. Hacke, S.R. Kurtz, K.O. Davis, E.J. Schneller, X. Sun, M.A.

Alam, R.A. Sinton., Silicon Heterojunction System Field Performance, Journal of Photovoltaics, 8(1), 2018, pp. 177–182.

[10] W.F. Holmgren, C.W. Hansen, M.A. Mikofski. pvlib python: a python package for modeling solar energy systems, Journal of Open Source Software, 3(29), 884, (2018). https://doi.org/10.21105/joss.00884

[11] M.G. Deceglie, A. Nag, A. Shinn, G. Kimball, D. Ruth, D. Jordan, J. Yan, K. Anderson, K. Perry, M. Mikofski, M. Muller, W. Vining, C. Deline RdTools, version {2.1}, Computer Software, https://github.com/NREL/rdtools.

[12] D.C. Jordan, K. Anderson, K. Perry, M. Muller, M. Deceglie, R. White, C. Deline, Photovoltaic fleet degradation insights, Progress in PV, April 2022.

AUTHOR INDEX

Abad, Eduardo Camarillo 1249
Abasi, Abudulimu 402
Abbas, Ali 245, 498, 656, 1193
Abdullah-Vetter, Zubair 1224, 1468
Aberle, Armin G. 1052
Abou-Ras, Daniel 432, 879, 939
Abrosimov, Nikolay V. 203
Abudulimu, Abasi 42, 46, 220, 275, 279, 683, 804,
.... 823, 909, 910, 1197, 1323
Acevedo, Armando Figueroa 1393
Adeleye, Damilola 1151
Adhikari, Alisha 1197, 1376
Adner, David 770
Afridi, Muhammad 237, 755
Afshari, Hadi 132, 161, 1156
Afzal, Syed Usama Bin 772
Agarwal, Shashank 1053
Agarwal, Sumit 30, 359, 475, 891, 1216
Ager, Joel W. 75
Aghaei, M. 737
Agrawal, Roshni 1053
Aguirre, Aranzazu 194
Ahlswede, Erik 1201
Ahmad, Muneeza 1109, 1565
Ahmad, S. 737
Ahmadi, Bahman 816
Ahn, Yujeong 1549
Aiello, Ashlee R. 282
Aimez, V. 1538
Aïssa, Brahim 27, 1516
Akiyama, Hidefumi 463
Akopian, Arkadi 641
Al Katrib, Mirella 1440
Alam, Fahad 173
Alam, M. A. 626
Alam, Muhammad A. 119
Alam, Muhammad Ashraful 772
Alam, Muhammad 1027
Alanis, Luis Eduardo 1060
Al-Baity, Shifaa M. 561
Albert, P. 1538
Alberts, Vivian 119
Albin, David S. 695
Albrecht, Steve 945
Aldalali, Bader S. 162
Aldhefairi, Mariam 1344
Alessi, Bruno 281
Algora, Carlos 1433

Alhammadi, Salh 1549
Alhassan, Saeed 1471
Ali, Adnan 1516
Al-Jassim, M. M. 315
Al-Jassim, Mowafak 1570
Allebé, Christophe 1129
Almannaee, Rawdah 1344
Almansoori, Muntaser Abdelrahman 321, 1344
Almonacid, Florencia 122, 437
Almutawah, Zahrah S. 1417
Alom, Md Zahangir 690, 1237
Alsuwaidi, Meerh 1344
Altenhöfer-Pflaum, Georg 749
Altermatt, Pietro 1544
Al-Thani, Hamda A. 561
Altosaar, Mare 1298
Alvarez, H. 1492
Alvarez, Hugo Da S. 1159, 1654
Alvarez, Hugo S. 1068
Alvarez, Hugo 1134
Aly, Shahzada Pamir 119
Amassian, Aram 332
Anctil, Annick 730, 743, 1095, 1259
Anderson, Caroline Lima 30
Anderson, Kevin S. 948
Anderson, Kevin 710
Andersons, Janis A. 348
Andreas, Afshin 600
Andreasen, Jens Wenzel 651
Anto, Robins 1230
Antón, Ignacio 746, 888
Aonuki, Sho 1398
Aponte, Erick 525
Aponte-Bezares, Erick E. 674, 1034, 1170, 1326
Aponte-Bezares, Erick 1082
Araki, Kenji 924
Aramaki, Ken 1164
Arehart, Aaron R. 107
Arehart, Aaron 1299
Arfaoui, Ines 426
Armour, Eric A. 1217
Arnold, Rachael 353
Arteaga, Jorge 415
Arya, Rajeewa 1366
Asadpour, Reza 626
Asahi, Shigeo 659, 895
Asami, Meita 374, 463, 935
Ascencio-Vásquez, J. 737

Ascencio-Vásquez, Julián 1578
Askins, Steve 746, 888
Astigarraga, Alexander 931
Athana, Dhawal 1009
Atwater, Harry A. 202, 1157
Augusto, André 5
Avoli, Matteo 1404
Awni, Rasha A. 1323
Ayala, Silvana 1141
Ayon, Arturo A. 769
Ayon, Arturo 1283
Azevedo, João Henrique Paulino 256
Azzolini, Joseph A. 13
Baan, Marzieh 917
Babayeva, Gülüsüm 1092
Babu, Balaashwin 671
Bach, Udo 298
Badosa, Jordi 268
Baer, Carsten 191
Baetzner, Derk 1
Baik, Sunhee 793
Bailie, Colin 740
Bainier, Camille 268
Baker, Wes 1096
Bakke, Jordan 1393
Balaji, Pradeep 5
Baldrias, Maristel 1024
Ball, James M. 298
Ballif, Christophe 1, 1129
Banerjee, Parag 629
Banks, Terry 740
Bansal, Shubhra 566, 1551
Barbose, Galen 793, 991
Barnes, Teresa M. 554, 1593
Barretta, Chiara 931, 1630
Barrit, Dounya 268
Barros, Tárcio A. Dos S. 1654
Barros, Tárcio A. S. 1159
Barros, Tarcio Andre Dos Santos 1134
Barth, Kurt L. 245, 981, 1497
Barth, Kurt 789
Barth, Vincent 352
Bartsch, Jonas 1625
Bastola, Ebin 220, 275, 279, 401, 402, 683, 804,
.... 823, 910, 1041, 1193, 1410, 1600
Batista-Alvarez, Natanael 370
Battiato, Sebastiano 1567
Baumann, Sara 733
Baxter, Jason B. 915, 1241
Beard, Matthew C 161, 1156
Beattie, Meghan N 941
Becker, Jan-Philipp 1201

Beltrán, Juan F. 761
Benatto, Gisele A. Dos Reis 16, 448, 1452, 1662
Benhaddou, Nada 395, 534
Bennett, Mitchell F. 1217
Benton, Brandon 587
Berad, Mrunal 1366
Bergin, Mike 437
Bernardini, Simone 768
Bernsen, Otto 749
Berriel, S. Novia 629
Berry, Joseph J. 298, 590, 1044, 1260, 1484
Bertoni, I. 291
Bertoni, Mariana I. 291, 400, 848, 970, 1533
Bertoni, Mariana 218, 333, 597, 689, 768, 782, 896, 1634
Berwind, Matthew F. 1313
Bessa, Joao G. 1270
Bessal, João Gabriel 437
Bevan, Geraint 591
Bhat, Akanksha 1147
Bhatia, Amandeep Singh 1027
Bidaud, T. 1538
Bidaud, Thomas 133, 781
Bila, Marine 295
Binyamin, Tal 879
Birgersson, Erik 1188
Bista, Sandip S 1197
Bista, Sandip Singh 46
Bittkau, Karsten 280
Bivour, Martin 770, 1625
Bizhanova, Gulzhan 39
Bizzarri, Fabrizio 1304, 1567
Black, Chloe L. 948
Blankemeyer, Susanne 652
Bloeck, Ulrike 939
Blom, Youri 1005
Blum, Adrienne 113
Boemer, Jens C. 1096
Bogner, Brandon M. 721
Bojorquez, Jose Raul Montes 769
Bolen, Michael 897
Bolink, Henk J. 938
Bondoc, Christopher C. 1510
Bonnassieux, Yvan 233
Borgers, Tom 48
Bosco, Nick 106, 1218
Bothe, Karsten 288
Bothwell, Alexandra M. 107
Bothwell, Alexandra 849
Boucherif, Abderaouf 133
Boucherif, Abderraouf 434, 1051
Bourarach, Fadi 426
Bourisli, Raed I. 162

Bournonville, Kenn Henrik...749
Bousselot, Jennifer..215
Bouttemy, Muriel..1440
Bowers, Jake W.............................. 245, 395, 981, 1177, 1497
Bowers, Jake.. 534, 1193
Bowersox, David A..1409
Boyce, Kenneth P..1141
Boyd, Matthew 920
Boyd, Stephen P..961
Boyer, Jacob T. 10, 291, 848, 994
Boz, Mesude Bayrakci..98
Brabec, Christoph J..645, 1092
Bracamonte, Maria Fernanda Villa 769
Braga, Daniel Sena 1541
Braid, Jennifer L. 234, 367, 512, 572
Braid, Jennifer 861, 1088, 1131
Bramante, Rosemary C. 1158
Brand, Andreas 1588
Brandstätter, Andreas...1630
Brau, Tyler R. 909
Brau, Tyler 42, 220, 402, 804, 910
Braun, Anna K. 10, 71, 291, 848
Bredács, Marton...931
Bremner, Stephen P..917, 1564
Bremner, Stephen...271
Brendel, Rolf 6, 288
Brenner, Tom 740
Brewer, Jeremy 178
Brewster, Charles 138
Breyer, Christian 708, 749
Brinck, Anna 1002
Brittman, Sarah 1045
Brockmann, Tim Lukas..652
Broderick, Robert 1170
Brooks, William 1009
Brown, Buck 240
Brown, Matthew 794, 1141
Bruckman, Laura S. 234, 597, 671, 861, 1088, 1141, 1485
Brueckner, Dennis 651
Bryan, Alex 1123
Buchholz, Florian 1372
Buck, Thomas..1266, 1372
Buerhop-Lutz, Claudia..540, 645
Bui, Thanh-Tuân 1484
Bullock, James...1434
Buño, Luis 1024
Burduhos, B. G. 737
Burnham, Laurie M. 572
Burnham, Laurie 391, 409, 1131
Burtone, Lorenzo 191
Bush, Meghan E...698
Busko, Dmitry 996

Buster, Grant 474, 587
Butt, Nauman Zafar..772
Cabanillas, Juan 804
Cacciato, Mario...1304, 1567
Cai, Mengmeng..360
Calderon, Jose A. 914
Calili, Rodrigo Flora..256, 858
Calloquispe-Huallpa, Ricardo 674
Calvo-Barrio, Lorenzo 1059
Campbell, Robert 295
Campesato, Roberta 1404
Campo-Ossa, Daniel D..916
Canino, Andrea..1005, 1304, 1567
Cano, Aitana 1603
Cao, Fangfang 822
Cappelluti, Federica 1413
Carbone, Marc A. 1626
Cariou, Romain...716, 934, 1423
Carletta, Stefano..1404
Carpenter, Joe V. 1533
Carron, Romain 651
Cartledge, Carsen 1109, 1565
Carvalho, Romullo R. M. 1159
Carvallo, Juan Pablo 793
Cassini, Denio Alves 1541
Castillo, Arnold 1024
Castillon, Jean 268
Castro-Sitiriche, Marcel 717
Catchpole, Kylie..1434
Cebecauer, Tomas 1594
Celi, Edoardo 871
Chacon, Sergio A. 132, 1156
Chadly, Assia...299
Chakar, Joseph 233
Chaluvadi, Venkata S. A...1249
Chambers, Terrence L. 188
Chan, Maria K. Y. 1634
Chan, Maria...782
Chang, Atom 917
Chang, Nathan..289
Chapaneri, Kaushal..119
Chapon, Julien 426
Chapotot, Alexandre..434, 1051
Chard, Julie 1123
Chatzipanagi, Anatoli 550
Chaubal, Aditi 1366
Chaurasia, Harsh 182
Chen, Bin...518, 1019
Chen, Chien-Hsuan..597, 629, 1337
Chen, Christopher 179
Chen, Cong 822
Chen, Daming 1544

Chen, Gonglin 1021
Chen, Hong 1544
Chen, Kissenger 978
Chen, Lei 804, 909, 1197, 1550
Chen, Mike 740
Chen, Ning 1372
Chen, Stela 377
Chen, Theresa K 1566
Chen, Xin 1519
Chen, Yifeng 1544
Chen, Zeying 682
Chin, Robert Lee 227, 1224
Chintapalli, Sreyas 718, 1334, 1356
Chiu, Arlene 272, 718, 1356
Cho, Jinyoun 133, 434, 1051, 1341, 1390
Cho, Yunae 97
Choi, Wookjin 1657
Choi, Wook-Jin 179, 1030
Choudhury, Ashif 134
Choudhury, Kaushik R. 952
Choudhury, Kaushik Roy 377, 689
Chowdhury, Gofran 86, 1253
Chretien, Jeremie 133
Christiansen, Silke 194
Chu, Haifeng 1372
Chung, Jaehoon 844, 1417
Chutani, Ayush 391
Cieslak, Janko 191
Cifuentes, Luis 1433
Cimaroli, Alexander 412
Cioc, Sorin 402
Cira, Spencer 473
Clandestino, Franco 1491
Clausing, Roland 6, 12
Clemens, Daniel 240
Coco, Fabrizio 1005
Colegrove, Eric 245, 633, 849, 946
Coletti, Gianluca 917
Coll, Pablo G. 218, 291
Coll, Pablo Guimerá 848
Collavini, Silvia 289
Collin, Stéphane 610
Colomba-Colon, Luis 370
Colombo, Mariela 1529
Colón, Alanis M. 416
Colvin, Dylan J. 671, 758, 850, 1120, 1485
Conibeer, Gavin J. 271, 1564
Conibeer, Gavin 881
Coogan, Katrina 879
Cook, John P. D. 476
Coologeorgen, Alexander E. 918
Cooper, Emma C. 572

Cooper, Robert A. García 519
Correa-Baena, Juan-Pablo 969, 1183
Corso, Roberto 1500, 1609
Cosme, Damien 426
Costa, Carla 934
Costa, Suellen C. S. 1541
Coulibaly, Bakary 1292
Court, Philip 922
Courtois, Guillaume 434, 1390
Coutinho, Natália F. 290
Couture, Eugene Desjardins 1648
Crawford, Zachary 51
Crestani, Thais 290
Critchlow, Gary 174
Crowe, Iain F. 203
Crowe, Laura E. 919
Crowley, Kyle 415
Cruz, Edgar E. 420
Csank, Jeffrey T. 1626
Cudzinovic, Michael J. 1024
Cullen, David A 46
Curran, Alan J. 903
Currie, Taylor M. 629
Curson, Kieran 245, 656
Da Silva, João A. F. G. 1654
Da Silva, Lucas Aló Rodrigues Araujo 858
Da Silva, Paulo R. D. R. 1159, 1654
Daenen, Michael 48
Dahi, Adem 1423
Dalal, Vikram 641
Dale, P. 995
Dale, Phillip J. 731, 975
Dalibor, Thomas 1067
Dally, Pia 1440
Danel, Adrien 716
Daniel, Valentin 133, 434
Danielson, Adam 1279
Danilson, Mati 1298, 1432
Danovitch, D. 1538
Dapprich, Karoline 113
Darawali, Renn 927, 1146
Darbali-Zamora, Rachid 334, 455, 674, 953, 984,
.... 1034, 1082, 1170, 1326, 1626
Darling, Halley C. 1146
Darling, Halley 927
Darnon, M. 1538
Darnon, Maxime 133, 781, 1051
Das, Ujjwal K. 91, 1545, 1667
Das, Ujjwal 1299
Dasgupta, Sagnik 1030, 1155, 1657
Daus, Alwin 1347
Davis, Kristopher O. 219, 597, 629, 671, 758, 850, 1485

Davis, Kristopher 1337
Davis, Melissa A 330
Davis, Nithin Maipan 1404
Dawson, Timothy 1295
De Albuquerque, Vanessa Cardoso 256, 858
De Brabandere, Karel 1253
De La Rosa, Angel 782
De Lafontaine, Mathieu 476
De Lima, Geyciane P. 1159, 1654
De Luna, Gabby 1030
De Monfreid, Thybault 1484
De Oliveira, Otávio J. 290
De, M. M. M. Modesto Ana Paula 290
De, Shoubhik 1287
Debije, Michael G. 809
Deceglie, Michael G. 921, 1260, 1593
Deckx, Julien 1253
Decristofaro, Eric R. 918
Degen, Ashley 467
Degenhart, Nick 113
Delazzer, Timothy 387
Deline, Chris 47, 920, 1145
Delmas, William 415
Demko, Michael 377
Demtsu, Samuel 914
Deng, Chenyang 743
Deng, Yuepeng 903
Depauw, Valérie 434, 1390
Dequilettes, Dane W 590
Desarden-Carrero, Edgardo 1082, 1326
Descoeudres, Antoine 1129
Deshmukh, Kedar 1366
Dessein, Kristof 133, 434, 1051, 1341, 1390
Deville, Lelia 188
Dhakal, Rabin 528, 897
Dhakal, Tara 682
Dharmadasa, Ruvini 38
Dhople, Sairaj 1384
Di Stefano, Agnese 1005, 1304
Diaz, Martin 1217
Dice, Paul W. 409
Dice, Paul 1131
Didier, Thevenard 861
Diederich, Marvin 6
Dierenbach, Jonas 1184
Dietsch, Tina 795
Diggs, Andrew 51
Digregorio, Steven J. 400
Digregorio, Steven 333
Ding, Kaining 177, 280, 419
Diniz, Antonia Sonia A. C. 94, 1541
Dippell, Torsten 1129

Dittmann, Sebastian 1475
Dobson, Kevin D. 1545
Dobson, Kevin 1299
Dobson, Wes 113
Dokken, Briana 1210
Dolia, Kshitiz 42, 1417, 1557
Doll, Bernd 645
Domínguez, C. 746
Domínguez, César 888
Don, Eric 981
Dong, Peng 1372
Dong, Shuan 240, 840
Donoso, Jose 749
Dougherty, Brian 1292
Doumon, Nutifafa Y. 1044, 1260
Dow, Andrew R. R. 1034, 1626
Drayton, Jennifer 1310
Drees, Martin 145
Drost, Christian 656
Druffel, Thad 38
Du, Bin 1299
Du, Liming 822
Du, Mohan 155
Duan, Leiping 1434
Duan, Weiyuan 177, 280, 419
Duan, Xiaomeng 1197
Dubajic, Milos 271, 1564
Dubois, Anne Migan 268
Dubois, Sébastien 716
Dubuc, Christian 133, 1051
Duenow, Joel N. 695, 946
Dulal, Prabin 1410
Duncan, Brent 492
Duncker, Klaus 191
Dunfield, Sean P. 644
Dunham, Scott 578
Duong, Calvin 508
Durant, Brandon K 1156
Dutta, N. S. 315
Dutta, Nikita S. 1570
Duzellier, Sophie 934
Dvonc, Lukas 1594
Dwivedi, Priya 472, 828, 1224, 1468
Dyreson, Ana 391, 1635
Eberspacher, Chris 740
Eberst, Alexander 419
Ebner, Rita 194
Ebong, Abafriseke 38
Echevarria, Angel 916
Edmondson, A. 1261
Eggink, Wouter 443
Einhaus, Lisanne M. 622

Ekins-Daukes, N. J. 1468
Ekins-Daukes, Nicholas J. 345, 976
Ekins-Daukes, Nicholas 90
Elahi, Sheikh Tawsif 1237
Elanzeery, Hossam 1067
El-Atab, Nazek 173, 327
Ellingson, Randall J. 1323
Ellingson, Randy J. 42, 46, 220, 275, 279, 402,
...... 683, 823, 844, 909, 1019, 1041, 1197, 1353, 1417
Ellingson, Randy 804, 910
Ellis, Brian 476
Elsehrawy, Farid 345
Engel, Bernd 64
Engelen, Tine 48
Engsig-Karup, Allan P. 16
Enjalbert, Nicolas 716
Eperon, Giles E. 132, 161, 919
Eperon, Giles 15
Erickson, Samuel 415
Escobar, D. Martinez 783
Espinet-Gonzalez, Pilar 11
Etgar, Lioz 879
Evans, Rhett 828
Facsko, Stefan 435
Fai, Calvin 915, 1241
Fairbrother, A. 737
Falkenberg, Gerald 651
Fan, Yangxin 861, 1088
Fanego, Vicente Lara 1594
Fang, Liang 195, 1526
Fang, Xin 840
Fardi, Hamid 1513
Farias, Stephen 1575
Farias-Basulto, Guillermo A. 1066
Farina, Angela 730
Farrell, Jack 771
Farrell, John 1600
Fassl, Paul 701
Fattah, Tarek O. Abdul 203
Faulwetter-Quandt, Björn 191
Fechner, Hubert 749
Fedoseyev, Alex 110
Fei, Chengbin 1044
Feng, Xiaotong 1503
Feng, Zhiqiang 1544
Fenning, David P. 644
Ferekides, Chris 690, 1237
Ferguson, Andrew J 341
Fernández, Eduardo F. 122, 437, 1270
Fernandez, Pablo 1433
Fernández-Solas, Álvaro 1270
Ferrara, Matteo 1404

Ferreira, Mateo 1437
Fertig, Fabian 191
Fevola, Giovanni 651
Fields, Shannon 1299
Filho, Neolmar De M. 94
Fischer, Benedikt 177
Fischer, Markus 191
Fisher, Kathryn 925
Flechas, Juan Pablo Medina 268
Fleming, Katelynn E 89
Flicker, Jack D. 984, 1034, 1626
Flicker, Jack David 634
Florakis, Antonios 426
Floren, Radovanovic-Peric 351
Fonoll-Rubio, Robert 1067
Fontanot, Tommasso 194
Forcade, Gavin P 941
Forcade, Gavin 476
Forchhammer, Søren 16
Forrest, Stephen R. 482
Forrest, Stephen 332
Forrester, Sydney 991
Fortmann, Charles M. 22
Foster, Michael 587
Foti, Marina 1005
Fox, Curtis 897
France, Ryan M. 71, 341, 394, 475, 994, 1081, 1150
Frasson, Nicola 352
Fregosi, Daniel 528, 555, 897
French, Roger H. 234, 671, 758, 861, 1088, 1485
Friedl, Jared D. 220, 275, 279, 1210, 1323
Friedl, Jared 683
Friedlmeier, Theresa M. 566
Friedlmeier, Theresa Magorian 199, 1201
Friedman, Daniel J. 394
Friedman, Daniel 485
Fritzsche, Helmut 476
Frye, Bailey 1557
Fthenakis, Vasilis 146, 1503
Fu, Oakland 377
Fu, Sheng 42, 1417
Fukaya, Shohei 775
Fürer, Sebastian O. 298
Furis, Madalina 161
Gabas, Mercedes 1433
Gabor, Andrew M. 671, 850, 1485
Gaddy, Edward 311
Galiazzo, Marco 352
Gamboa, Daniel H. 761
Gamel, M. 942
Ganguly, Subhankar 1074
Gao, David Wenzhong 840

Gao, Jiaqing 1372
Gao, Munan 1509
Gao, Ningchao 840
Gao, Tina 718
Gao, Zhiyu 822
García, I. 947
García, Iván 1433, 1603
Garcia, Maria Angelica M. 1379
Garcia, R. 1492
Garcia, Rodrigo M. 1068, 1159, 1654
Garcia, Rodrigo 1134
Garín, M. 942
Garrevoet, Jan 651
Gedi, Sreedevi 1549
Gehan, Tim 740
Gehrke, Aaron 578
Geissbühler, Jonas 1
Geistert, Kristina 144
Geisz, John F. 71, 291, 394, 709, 1158
Geisz, John 178, 485
Georghiou, George E. 194, 541, 1270, 1348
Gerardi, Cosimo 1005
Gerber, Andreas 166
Gerger, Andrew 311
Gerton, Jordan 134
Gevorgian, Vahan 360
Geyer-Klingeberg, Jerome 385
Gfroerer, T. H. 1261
Gharabeiki, Sevan 1617
Ghosh, Probir 1366
Ghosh, Sayantani 415
Giacchino, Evan S. 918
Gibbons, Daniel 861, 1088
Gil-Escrig, Lidon 938
Giliberti, Gemma 1413
Ginger, David S. 1020
Giraldo, Sergio 1059
Giridharagopal, Rajiv 1020
Giteau, Maxime 610
Giuri, Antonella 1109
Glatthaar, Markus 1625
Gloeckler, Markus 914
Goga, Adam 51
Gok, A. 737
Golive, Yogeswara Rao 1287
Golubev, Timofey 874
Gomez, Daniel 1433
Gonçalves, Felipe 256
Gong, Yuancai 1059
González, Alanis M. Colón 367
González, Emmanuel J. 416
Good, Brian 245

Goosay, Olivia 727
Gopal, Deepika 126
Gorman, John 922
Gorman, Will 793
Gostein, Michael 254, 295, 1558
Gotoh, Kazuhiro 292, 724, 775
Gottschalg, Ralph 486, 1475
Goubard, Fabrice 1484
Govaerts, Jonathan 48
Graeber, Dietmar 1184
Granello, Pierpaolo 1404
Grassman, Tyler J. 917
Greco, Erminio 1404
Green, Martin 302, 1256
Greenaway, Ann L. 75
Greenhalgh, R. C. 245
Greenhalgh, Rachael C. 981, 1193
Greenhalgh, Rachael 656
Gregory, Christopher 916
Grimm, Benjamin 6
Grisanti, Marco 1567
Grossberg-Kuusk, Maarja 1298, 1432
Grossklaus, Kevin A. 1152
Grover, Sachit 489, 1582
Grovogui, Jann A. 11, 198
Gruenhagen, Philip 662
Grundmann, Marius 12
Gu, Hangyu 366
Gu, Xiaohong 42, 282, 952, 1141
Guc, Maxim 1067
Gudi, Dhanvini 272, 1356
Guennou, M 995
Gueymard, Christian A. 54, 778
Guha, Mousumi 492
Guibin, Shen 1052
Guillemoles, Jean-François 233
Guillemot, Thomas 268
Gulati, Himanshu 259, 1227
Guo, Da 914
Guo, Yonggang 1372
Gupta, Apoorva 644
Gupta, Mool C. 1155, 1522
Gupta, Priya 830
Gurule, Nicholas S. 984
Gütay, Levent 1059
Guthrey, H. 315
Guthrey, Harvey L. 1216
Guthrey, Harvey 939
Gutzler, Rico 432, 1201
Hohn, Oliver 941
Haas, Benedikt 879
Haase, Felix 6, 733

Habisreutinger, Severin 1490
Habte, Aron 474, 587, 600, 778, 1106
Hacene, Benjamin 144
Hacke, Peter 648, 952
Hadi, Sabina Abdul 1471
Hadjipanayi, Maria 194
Hagemann, Johannes 651
Hagendorf, Christian 770
Hages, Charles J. 915, 1241
Hages, Charles 975
Hähnel, Angelika 770
Hajj, Adonis E. 81
Haley, Thomas 928, 1147
Halm, Andreas 1372
Halme, Janne 345
Halsall, Matthew P. 203
Hamadani, Behrang H. 695, 1292
Hameiri, Ziv 5, 227, 472, 828, 1224, 1295, 1468
Hamer, Mike 1359
Hamon, G. 1538
Hamon, Gwenaëlle 133, 781, 1051
Hanif, Muhammad 271, 1564
Hansen, Clifford W. 922, 1110, 1250
Hanuš, Tadeáš 434, 1051
Hao, Xiaojing 1256
Haque, Sirazul 1380
Hara, Tomohiko 724, 1453
Harada, Yukihiro 895
Harder, Nils-Peter 426, 1024
Harder, Ross 1183
Hare, Casey P 977
Hariskos, Dimitrios 432, 1201
Harper, Jim 14
Harrison, Jason 1002
Harrison, Samuel 352
Härtel, Marlene 945
Hartenstein, Matthew B 475
Hartenstein, Matthew 30
Hartweg, Barry 925
Harvey, Steve P. 946
Harvey, Steven P. 298
Hasan, Arif Yetkin 975
Haselsteiner, Philipp 1428
Hashad, Khaled 505
Hasoon, Falah S. 561
Hatt, Thibaud 1625
Hauch, Jens A. 645
Hauch, Jens 540, 1092
Hauschild, Dirk 91
Hawkins, Nicholas 298
Hayden, Steven 1044
He, Bo 1221

Heath, Garvin 594
Heben, Michael J. 35, 46, 220, 275, 279, 401, 402,
.... 683, 823, 844, 1041, 1193, 1210, 1323, 1376, 1410, 1417
Heben, Michael 804, 910, 1197, 1600
Hegedus, Steven 464
Heidrich, Robert 486
Heilscher, Gerd 1184
Heimsath, Anna 1313
Heinrich, Martin 1060
Heinzle, Nino 1348
Heitmann, Johannes 1266
Helder, Tim 199
Helfer, Eric 931
Helienek, Lubos 1594
Helmers, Henning 941
Helms, Clay 81
Henry, Isaiah 1551
Herasimenka, Stan 110
Heres, Geert C. 622
Hernández, Johann 761
Hernandez, Samuel I. 416
Hernandez-Alvidrez, Javier 984
Herrero, Rebeca 746
Heske, Clemens 91
Hettiaratchy, Elline C. 489
Hidalgo, Juanita 969
Hildreth, Owen J. 400
Hildreth, Owen 333
Hill, Blake 363
Hill, C. 995
Hill, Taylor D. 489
Hill, Taylor 1582
Hillhouse, Hugh W. 340
Hillhouse, Hugh 473
Hinken, David 288
Hinzer, Karin 47, 331, 476, 941
Hirst, Louise C. 1249
Ho, Kevin 1020
Hobbs, William B. 492, 921, 948, 1558
Hodges, Joseph 240
Hoerantner, Maximilian T. 919
Hoex, Bram 1256
Hoffman, Adam 861, 1088
Höger, Ingmar 191
Hoheisel, Raymond 894
Hoke, Andy 840, 1164
Hole, Jarand 749
Holman, Zachary C. 767, 1279, 1379, 1533
Holman, Zachary 925
Holmes-Smith, A Sheila 591
Holmgren, William F. 948
Holzhey, Philippe 289, 298

Hong, Chengjian 195, 1526
Hönig, René 191
Hool, Ryan D. 670, 997
Hörnlein, Stefan 191
Hoss, Jan 795
Hossain, Mohammad I. 27, 1516
Howard, Ian A. 172
Hsieh, Chun-Hao 1443
Hu, Hongjie 377
Hua, Amandee 91
Huaman-Rivera, Anny 477
Huang, Ben 1284
Huang, Jing 285
Huang, Jing-Shun 783
Huang, Jinsong 366, 1044
Huang, Jun-Yu 1443
Hubbard, Seth M. 63, 89, 721, 894
Hübner, Simon 1129
Huddy, Julia E. 1276
Hudry, Damien 996
Hudson, Andrew 1462
Hultqvist, Adam 1151
Huneycutt, Sandra... 38
Hunter, Robert 941
Hunwick, Nicholas 981
Huque, Aminul 138, 835, 1096
Hussain, Zulkifl 683
Hwang, Jeong-Mo.. 179
Hyndman, David W. 1259
Iannascoli, Lorenzo.. 1404
Ibanez, Eduardo 1393
Ifuji, Yuto 1453
Ilahi, Bouraoui 133, 434, 1051
Im, Kyu-Hyeon 1657
Imaizumi, Mitsuru 76, 463, 721, 1429
Imenes, Anne G. 815
Imperatori, Davide... 1404
Imran, Hassan 772
Infante-Ortega, L. C. 245
Infante-Ortega, Luis C. 981, 1497
Inoue, Kazuma 292
Ireton, Scott J 977
Irizarry-Rivera, Agustin 477
Irvine, Stuart J. C. 308, 1497
Irvine, Stuart 245
Irving, Richard 823
Isabella, Olindo 80, 655, 1005, 1398
Ishizuka, Shogo 939
Ital, Donald 38
Ito, Yuta 724
Jacob, David 1024
Jacobs, Janet 203
Jaeger-Waldau, Arnulf...................................... 924
Jäger-Waldau, Arnulf................................. 550, 749
Jahandardoost, Mohsen 566
Jahangir, Jabir Bin.................................. 626, 1027
Jahelka, Phillip R. 202
Jahelka, Phillip 1157
Jain, Anubhav 1519
Jakob, Leonie 1625
Jamarkattel, Manoj K. 46, 220, 279, 401, 683, 823,
.... 1041, 1197, 1376
Jamarkattel, Manoj 910, 1600
Jannuzzi, Gilberto 256
Jansson, Peter Mark 380
Janz, Stefan 434, 1341
Jaouad, A. 1538
Jaouad, Abdelatif 133
Jarzembowski, Enrico 191
Jaubert, Jean-Nicolas................................. 234, 794
Javier, Gaia Maria N. 472, 828
Jawinski, Tanja.. 12
Jay, Frédéric 716
Jayaraman, Sreenivas 855
Jeangros, Quentin 1
Jensen, Karissa 42
Jeon, Seokmin 134
Jhang, Song-Syun.................................... 282, 952
Ji, Liang... 1141
Jia, Huiying 1466
Jiang, C.-S. 315
Jiang, Chun-Shen 1570
Jiang, Chun-Sheng.................................... 46, 946
Jiang, Fangyuan 1020
Jiang, Jessica Yajie.................................... 302, 345
Jiang, Qi 709, 812, 1158, 1380
Jiang, Yi 1284
Jimenez-Arguijo, Alex 1059
Jin, Hao 1434
Jin, Shuangshuang... 240
Jin, Tan 240
Jiyun, Zhang.. 1092
Jo, Sangmin 436
Joe, Junki 1218
John, Jim Joseph 119
John, Oliver 1588
Johnson, Jay 334
Johnson, Samuel... 938
Johnston, Michael B. 298
Johnston, S. 315
Johnston, Steve W. 10, 848
Johnston, Steve.............. 219, 387, 702, 865, 1490, 1570, 1629
Jones, Abigail R. 1250
Jones, C. Birk 455

Jones, James...920
Jones, Luke O. 174, 245, 395, 498
Jones, Luke 656, 1177
Jones, Steve 308
Jordan, Dirk C. 219, 1110, 1593
Jordan, Dirk 794, 896, 1145
Jorgensen, Peter Stanley 651
Josepson, Raavo.................................... 1432
Jouanneau, Corentin 781
Joyce, Hannah J. 1249
Julien, Arthur 233
Jung, Hyeonjung Tari...................... 1384, 1393
Jung, Kyung Taek 97
Junghänel, Matthias 191
Jurca, Titel 629
Kaaya, Ismail 86
Kabra, Dinesh 1366
Kachman, Dana.................................272, 1356
Kaewnukultorn, Thunchanok....................464
Kaizuka, Izumi..749
Kakoulaki, Georgia.....................................550
Kalizewski, Lauren M...................................917
Kalpoe, Prashand 305
Kaltenbaugh, Jarod 758
Kaluarachchi, Prabodika N. 402, 683, 844
Kamal, Serene.....................................272, 718
Kambley, Ankur 281
Kamikawa-Shimizu, Yukiko 939
Kanakkithodi, Arun K. M. 782
Kanaujia, Pawan K. 1522
Kaneko, Ryuji 442
Kanevce, Ana...................... 107, 199, 1201
Kang, Min Gu...97
Kanneboina, Venkanna 1410
Karade, Vijay 1556
Karakaya, Sakir...634
Kari, Thøger 448, 1452, 1662
Kartopu, Giray 308
Kasher, Tal 917
Kashkimbayev, Ulan 39
Kasik, Camden 1310
Katakumbura, Nadeesha 402, 804
Kauert, Maximilian.......................................191
Kauk-Kuusik, Marit 1298, 1432
Kaupmees, Reelika 1298
Kaur, Navdeep 606
Kawakami, Mizuto 895
Kaydanik, Katty 783
Kazim, S. 737
Kazmerski, Lawrence L.............. 94, 1366, 1541
Ke, Cangming 191
Kee, Jared 555

Keller, Jan 887
Kelzenberg, Michael D. 1157
Kempa, Heiko 12, 432
Kempe, Michael D. 794, 1141, 1260
Kendall, Anthony D. 1259
Kenyon, Jacques.................................395, 534
Kern, Dana B...............219, 702, 865, 1044, 1570, 1629
Kern, Dana 15, 387, 896
Kerr, Lei 1466
Kersten, Friederike 191
Kessler-Lewis, Emily 894
Kettle, J. 737
Khadka, Dhruba B..............................324, 1401
Khan, M. Ryyan 626
Khenkin, Mark 1066
Khetri, Mahantesh 1522
Khoury, Alexandre 1612
Khulmann, Forrest 363
Khurgin, Daniel 1356
Kikelj, Miha 1
Kile, Kara B. 683
Kim, Bora 670
Kim, Boyoung 436
Kim, Dohyung...97
Kim, Han-Jung 1373
Kim, Hyomin 1549
Kim, Hyungoo 436
Kim, Jeong-Hyeon 1484
Kim, Jin Hyeok 1556
Kim, Jin Young..165
Kim, Junhee 1373
Kim, Kiwhan 1556
Kim, Mijung...................................... 670, 997
Kim, Munse 97
Kim, Sanggyun..1183
Kim, Seul-Gi 1484
Kim, Woo Kyoung 1549
Kim, Yoonkap 1373
Kimura, Keita 1453
Kinfack, J. 1538
King, Bruce H. 188, 254
King, Bruce 353, 409
King, Richard 916
Kinzer, Austin 920
Kipp, Tobias 651
Kirchartz, Thomas.............................199, 1307
Kirmani, Ahmad R 1156
Kirmani, Ahmad 1490
Kita, Takashi 659, 895
Klenk, Reiner 1066
Klenke, Christian 191
Klie, Robert F. 771, 1600

Kline, Michael 492
Klöter, Bernhard 569
Knebel, Kevin J. 769
Knodle, Philip J. 671, 850
Knodle, Philip 1485
Ko, Yohan 1373
Koch, Christoph 879
Kodalle, Tim 969
Koelblin, Pascal 166, 1662
Köhler, Matthias 191
Kojima, Haruki 724
Kojima, Nobuaki 924, 1535
Komoll, Felix 199
Kondzialka, Christoph 1184
Köntges, Marc 733
Kopecek, Radovan 1372
Kopidakis, Nikos 178, 485
Korgel, Brian 1537
Korir, Lilian 1261
Kornienko, Vlad 683
Kornienko, Vladislav 656, 1193
Körtgen, J. 546
Koschier, Linda 749
Koseki, Shuuichi 105
Koskey, Steven 555
Kottantharayil, Anil 1287, 1366
Kottokkaran, Ranjith 641
Kouame, K. 1538
Koumis, Anastasios 541
Krause, Timothy 415
Krebs, Hannes 1630
Kretly, L. C. 1492
Krich, Jacob J 941
Krishna, Anurag 194
Krishnani, Pramod N. 81
Kristensen, Sissel Tind 191
Kroeger, George F. 219
Krückemeier, Lisa 1307
Krustok, Jüri 1298, 1432
Kuba, Austin G. 1545
Kubiniec, Alex 928, 1620
Kuciauskas, Darius 107, 577, 633, 849
Kujovic, Luksa 245, 789, 981, 1497
Kumar, Akash 237, 755
Kumar, Akshay 622
Kumar, Dayanand 327
Kumar, Neetesh 276, 1206, 1437
Kumar, Niranjana Mohan 782, 1634
Kumar, Satyendra 1366
Kumar, Vibhor 1509
Kunkar, Alejandro 708
Kupets, Elaine 594

Kurokawa, Yasuyoshi 292, 775
Kurtz, Sarah R. 783
Kurtz, Sarah 209, 617, 1529
Kusch, G 995
Kwon, Ohjin 191
Lachenal, Damien 1
Lachowicz, Agata 352
Lackner, David 941
Ladd, Anthony J. C. 1241
Lahood, Catherine 22
Lahti, Gabriella D. 1158
Lai, Barry 782, 1183, 1634
Lai, Cheng-Yu 276, 606, 1206, 1437
Lakshmikanth, Balaji Bangolae 126
Lambertz, Andreas 177, 280, 419
Lampa, Nicole 191
Lan, Yucheng 272
Landis, Geoffrey A. 63
Lang, Tom 1002
Lange, Stefan 12, 770
Lao, Yao Y. 11
Lara-Fanego, Vicente 54
Larionova, Yevgeniya 701
Larson, Harry 1551
Lasalvia, V. 1459
Lasalvia, Vincenzo 359
Laufer, Felix 144
Lave, Matthew S. 674, 1170
Law, Adam M. 174, 498, 789, 981
Law, John M. 174
Laws, Nicholas D. 228
Le, Anh Huy Tuan 5
Leccisi, Enrica 146
Lee, Chungho 849, 998, 1008
Lee, Hyunjong 919
Lee, Hyunju 442, 724
Lee, Jehyun 436
Lee, Kyumin 391
Lee, Minjoo L. 670, 997
Lee, Ross 1510
Lee, Sang Hee 97
Lee, Sanghyun 1102
Lee, Songhee 1549
Leever, Benjamin 1466
Leijtens, Tomas 919
Leloux, J. 737
Lemay, AC 1233
Lemire, Amanda 1152
Lemos, Francisco V. E. 1159
Lenert, Andrej 482
Leonardi, Marco 1005
Lerat, Jean-Francois 133

Lestrade, Michel	1645
Leuty, Zachary B.	1533
Lewis, Mandy R.	47
Li, Baojie	1519
Li, Bo	794
Li, Bor	945
Li, Brian D.	670
Li, Can	822
Li, Chongwen	909, 1019
Li, Deng-Bing	275, 279, 823, 1041, 1197, 1353, 1376
Li, Dengbing	46, 804, 910
Li, Dinica	978
Li, Fang	648, 1120
Li, Gan	374, 935
Li, Lulin	1575
Li, Luxi	1183
Li, Mengjie	671, 758, 1485
Li, Minghui	822
Li, Muzhi	938
Li, Ning	1074
Li, Wayne	555, 897
Li, Xinjun	234
Li, Yongxi	332
Li, You	844, 1197, 1417
Li, Zelin	1141
Li, Zhanming S.	1645
Li, Zhen	822
Li, Zhenguo	195, 1526
Li, Zhiqiang	1645
Liao, Weilin	248
Libal, Joris	1372
Libby, Cara	921
Liggett, Max	671
Lightfoote, Stephen	607
Li-Kao, Zacharie Jehl	1059
Lim, Deokoh	436
Lim, Jihun	482
Limodio, Gianluca	305
Lin, Boris	903
Lin, Fen	1052
Lin, Yida	1356, 1575
Lindahl, Johan	749
Lindig, S.	737
Lindig, Sascha	931
Linke, Jonathan	795
Linss, Volker	795
Lipovšek, Benjamin	1
Liu, Baiqiang	248
Liu, Chengfa	1544
Liu, Jiang	1221
Liu, Jie	1221
Liu, Jiqi	234

Liu, Xiaolei	245, 789, 981, 1497
Liu, Xitao	1503
Liu, Yang	1221
Liu, Yangang	386
Livera, Andreas	1270
Lobato, K.	737
Loeding, Adam W.	918
Lombardero, Ivan	1433
Lombardo, Salvatore A.	1500, 1609
Lomuscio, Alberto	1067
Loo, Roger	434, 1390
López, G.	942
Lopez, Hector	1280
Lopez-Becerra, Alan	1283
Lopez-Cardalda, Guillermo	370
López-González, J. M.	942
Lopez-Lorente, Javier	1359
Louks, A.	315
Louks, Amy E	590
Louks, Amy	1570
Lu, Chengchangfeng	1356
Lu, Dingyuan	1197, 1497
Lu, Jianfeng	298
Lu, Junxiong	195, 1526
Lu, Meijun	248
Luderer, Christoph	770
Lüer, Larry	645
Lumb, Matthew P	1217
Luo, Bin	48
Luo, Yanqi	1183
Luque, A.	947
Luque-Heredia, I.	947
Lustig, Zachary	123
Luther, Joseph M.	132, 1156
Luther, Joseph	415, 1490
Lv, Ruirui	1284
Ma, Depu	935
Ma, Jaliu	914
Ma, Jessica	1612
Ma, Yiwei	138
Macalpine, Sara M.	1409
Macdonald, Daniel	1434
Macías, Javier	746
Mack, Charles	178
Mack, Sebastian	795
Mack, Shawn	886
Madonna, Richard G.	1157
Magginetti, David	134
Mahabaduge, Hasitha	1633
Mahaffey, Mason P	1279
Mahamu, Hambalee	659
Mahesh, Suhas	298

Mahmood, Farrukh Ibne 237, 648, 755, 1120
Mahmoudi, Eslam 1134
Mahmud, Rasel 1074
Mahmud, Zabir 617
Mahoney, John 1156
Maiberg, Matthias 432
Mainali, Madan K. 401, 1550
Mainali, Madan 1410
Makita, Kikuo 105
Makrides, Andreas 1348
Makrides, George 541, 1348
Mallajosyula, Arun Tej 1606
Mallick, Rajni 1497
Mamun, Ashraful 134
Manceau, Matthieu 934
Mandic, Vilko 351
Manganiello, Patrizio 80, 655
Mangum, John S. 71, 994, 1150
Mannino, Gaetano 1304, 1567
Mannodi-Kanakkithodi, Arun 1634
Manoukian, Gregory A. 915, 1241
Mansfield, Lorelle M. 35, 1380
Mantel, Claire 16
Mao, Dan 782, 1634
Mapara, Varun N. 161
Maple, Larry 387
Marasini, Ganesh 1096
Marcos, Jesús 888
Mariam, Tamanna 279, 683, 844, 1376, 1417
Mariotti, Davide 281
Mariotti, Silvia 945
Markevich, Vladimir P. 203
Marques, F. C. 1492
Marques, Francisco C. 290, 1068, 1134, 1159, 1654
Marquis, Audrey 295
Marstell, Roderick J. 1024
Martel, Benoit 352
Martín, Francisco 746
Martín, I. 942
Martin, Ina T. 597
Martín, Nazario 804
Martin, P. 947
Martin, Pablo 1433
Martinez, Daniel 15
Martinez-Szewczyk, Michael W. 333, 400
Martinez-Szewczyk, Michael 896
Martins, Giuliano L. 1475
Martír, Pablo 1603
Masi, Sofia 441
Masson, Gaëtan 749
Masuda, Atsushi 442
Masuda, Taizo 924

Matam, Manjunath 671
Mate, Mayank 123, 363
Matera, Fabio 1500, 1609
Mathiak, Gerhard 119
Matsui, Takuya 281, 775
Matthews, Bryan 978
Matthews, David 767
Mayordomo, Alejandra A. 448
Mayyas, Ahmad 299
McAlister, Tom 1620
McCandless, Brian 1299
McCarthy, Robert F. 145
McCarthy, Robert 1446
McCulloch, Manuela 1184
McDanold, Byron 47
McDonald, Calum 281
McGarvey, Elspeth 1393
McGlynn, Ruairi 281
McGott, Deborah L. 308, 633, 915
McKenna, Killian 1074
McKuin, Brandi 209
McMahon, William E. 291, 709, 848, 994, 1150, 1380
McMeekin, David P. 298
McMillon-Brown, Lyndsey 415, 1437, 1490
McNatt, Jeremiah 415
McRae, Mary E. 1510
Medjoubi, Karim 233, 268
Meeker, Rick 1002
Meidanshahi, Reza Vatan 1021
Meier, Rico 689, 970
Meila, Marina 340
Meinhart, Lisa 1630
Melchiorre, Michele 1151, 1617
Meléndez, Cristian R. 420
Mendez, Andres Felipe Castro 969
Méndez-Curbelo, Pablo 717
Mendis, B G 995
Menegassi, Matheus Melati 1588
Meng, Yuhuan 340, 473
Mercimek, Yavuzhan 655
Merkle, Arno P. 291
Mette, Ansgar 191
Metzger, Wyatt K. 1497
Meyer, Abigail 359
Meyers, Bennet 710, 961
Michael, Sheri F. 885
Michel, Jesus Ibarra 1434
Micheli, Leonardo 54, 122, 437, 1105
Mihailetchi, Valentin D. 1372
Mikeska, Kurt R. 248
Mikli, Valdek 1298
Mikofski, Mark 81

Miller, Chandler...............................793, 991
Miller, Clark...916
Miller, David C...............................353, 1593
Miller, David W....................................1497
Miller, Emily..1557
Miller, Jason..978
Miller, Michael F....................................107
Mil'Shtein, Sam....................................1009
Min, Kwan Hong.........................179, 1030
Minuto, Alessandro................................871
Mirletz, Brian T....................................228
Mirletz, Heather M.................................554
Mirletz, Heather...................................1416
Mitra, Suchismita.....................................30
Mitterhofer, Stefan.................................282
Miyano, Kenjiro............................324, 1401
Moffitt, Stephanie L............42, 282, 952, 1141, 1292
Mogannam, Laura..................................467
Moghadamzadeh, Somayeh.....................701
Mohammadi, Mahsa..............................1266
Mohite, Aditya......................................938
Mohr, William.......................................145
Mohsin, Muhammad Saeed......................844
Molinero, R...947
Molto, Cécile.......................597, 648, 1120
Monahan, Daniele.................................1462
Monnin, Ryan.......................................914
Montes-Bojorquez, Jose Raul.................1283
Montes-Romero, Jesús....................1270, 1348
Mood, Thomas C............................886, 1217
Mora-Seró, Iván....................................441
Mordvinkin, Anton................................486
Morel, Don...690
Morlier, Arnaud......................................86
Morris, Kerrie M..................................1177
Morris, Kerrie......................................981
Moscoso-Cabrera, Javier A.....................420
Moser, D...737
Moser-Mancewicz, Nicholas....................896
Moses, Paul..665
Motes, Brandon T...................................590
Mouri, Tasnim K............................91, 1667
Mousumi, Jannatul Ferdous.....................629
Moutinho, Helio R.................................219
Moutinho, Helio...................................1629
Mu, Teliang...1503
Mugnier, Daniel....................................749
Mule, Chirag..359
Müller, Jörg W.....................................191
Muller, Matthew................122, 356, 437, 710
Müller, Matthias..................................1266
Müller, Thore.......................................1491

Mulloy, Eva M......................................577
Muñoz, Daniel.....................................1206
Muñoz-Pinzon, Daniel...........................1437
Munshi, Amit H....................123, 363, 1234
Munshi, Amit Harenkumar......................1537
Munshi, Amit.......................................1337
Murakami, Takurou N.............................253
Murphy, Alan.......................................783
Muska, Katri................................1298, 1432
Mussakhanuly, Nursultan.........................227
Muzzillo, Chris....................................1008
Myneni, Sushmakanth............................1234
Nagarajan, Shreyas..................................81
Nagel, Henning.....................................795
Nagle, Timothy....................................1497
Nahar, Aayush......................................1299
Nain, Preeti..................................743, 1095
Nakado, Takashi....................................924
Nakamura, Kyotaro.........................724, 924
Nakamura, Tetsuya............................76, 463
Nakano, Yoshiaki....................342, 374, 935
Nakarmi, Upama....................................662
Nakka, Laxmi......................................1052
Nambo, Apolo..38
Nardone, Marco.....................566, 577, 849
Nascetti, Augusto.................................1404
Nasser, Michael....................................498
Navon, David......................................1292
Nayfeh, Ammar...........321, 327, 1320, 1344, 1381, 1471
Nayfeh, Laith......................................1381
Nayfeh, Leia.......................................1381
Naylor, M...995
Nazeeruddin, Mohammad K......................804
Nazer, Afshin..80
Nazif, Koosha Nassiri...........................1347
Ndione, Paul F.....................................1158
Ndione, Paul.......................................1629
Neal, Craig J..671
Nekarda, Jan.......................................1588
Nemeth, B..1459
Nemeth, William..............30, 359, 475, 891, 1216
Neto, Pedro O. C. M.............................1159
Neubert, Anja......................................1359
Neuhaus, Dirk H...................................1313
Neuhaus, Dirk Holger............................1060
Neumann, Anica N.........218, 291, 848, 994, 1130, 1150
Neupane, Ganga R..................................695
Neupane, Sabin.............46, 220, 279, 1041, 1197, 1210,
....1353, 1376
Neves, M. R. M....................................1492
Neves, Mendelsson R. M...................1068, 1159
Newberry, Milton G................................380

Ng, Annie..39, 502
Ni, Chaoying...248
Nicholson, Anthony P. 509
Niebergall, Larissa..191
Nielsen, Michael P. 271, 976, 1564
Nieves, Michael Y. Vazquez.............................367
Nigmetova, Gaukhar..39
Nihar, Arafath..758
Nikam, Maitheli..1253
Ninad, Nayeem.....................................155, 1648
Nishihara, Tappei...................................442, 724
Nishinaga, Jiro...939
Nishioka, Kensuke...924
Nitta, Frederick U. 1347
Noack, Philipp...701
Nocerino, John C 198
Nolde, Jill A...886
Norman, A. 315
Norton, Matthew...194
Nowak, David...1059
Núñez, R. 746
Nuns, Thierry..934
Nuys, Maurice...177
Nyholm, Andrew W. 202
Obrecht, John M. 1578
O'Brien, Colleen...................................794, 1141
Ochoa, Jorge......................768, 896, 1533
Ogura, Atsushi.....................................442, 724
Ogut, Mehmet G. 961
Oh, Jaewon...794
Oh, Myeongchan...436
Ohlmann, Jens.....................................434, 1341
Ohshima, Takeshi....................................76, 1429
Ohshita, Yoshio.............442, 724, 924, 1453, 1535
Ok, Young-Woo 91, 179, 1030, 1657
Okada, Yoshitaka..463
O'Kearney, Felix...1224
Oklobia, Ochai.....................245, 308, 1497
Okoli, Fitzgerald C. 918
Okullo, James..1393
Oliver, R A 995
Oltjen, William C.......................671, 758, 1485
Oltjen, William..234
Olzhabay, Yerassyl..502
O'Neill, Mark..1446
O'Neill-Carrillo, Efraín 57, 1170
Onno, Arthur...1279
Ooi, Tzy Wei...855
Opatovsky, Martin.......................................1594
Orejuela, V. 947
Orejuela, Víctor..1603
Oreski, G. 737
Oreski, Gernot.....................................931, 1630
O'Rourke, Michelena......................................330
Ortis, Alessandro..1567
Ortiz, Eduardo I. 420
Ortiz, Eduarto I. 416
Ortiz-Rivera, Eduardo I. 525
Ortiz-Rivera, Eduardo.................370, 1338
Oshima, Ryuji..105
Oshima, Takeshi...463
Osowski, Mark...145
Ossig, Christina....................................435, 651
Ota, Yasuyuki..924
Ottoson, Larry...178
Ovaitt, Silvana................47, 554, 794, 920, 1416
Owen-Bellini, Michael.............952, 1044, 1218, 1260, 1629
Oyewo, Ayobami S...708
Ozaki, Ryo...924
Ozaktas, Ekin Gunes.....................................1334
Ozbeytemur, Josh...240
Ozbolt, Alex..1575
Pacheco, Willian...717
Packard, Corinne E......................10, 71, 291
Padhamnath, Pradeep.....................1030, 1657
Padmakumar, Govind......................................305
Padmanaban, Dilli Babu..................................281
Paesa, Marta Casasola......................................48
Paetel, Stefan...................107, 432, 1201
Paetzold, Ulrich Wilhelm..................144, 701
Page, M. R. 1459
Page, Matthew.........................30, 359, 475
Pal, Shweta..622
Palacios II, Felipe.......................................1626
Palacios, Felipe..1034
Palekis, Vasilios....................................690, 1237
Palmer, Jack R. 644
Palmiotti, Elizabeth.......................................353
Palmstrom, A. 315
Palmstrom, Axel F. 590, 1158
Palmstrom, Axel.....................................938, 1570
Pamperin, Megan...1393
Pan, Noren...145
Pan, Yida...1434
Panchalogaranjan, Vinushika.............................665
Pandey, Ramesh...123
Panzic, Ivana..351
Papaeconomou, Vassilis.................................1270
Parada, Gabor..981
Paraskeva, Vasiliki..194
Pareek, Devendrá...1059
Paris, Claudio..1404
Park, Chinho...749
Park, Nam-Gyu...1484

Park, Sungeun 97
Parke, Tyler 1667
Parra, Johan 268
Paschen, Jan 1588
Passarella, Bianca 1005
Patel, Aesha P. 402, 823
Patel, Jayeshkumar 476
Patel, M. Tahir 626
Paul, G. 315
Paul, Goutam 1570
Paul, Sritoma 482
Paupy, Nicolas 133, 434
Paviet-Salomon, Bertrand 1, 1129
Peaker, Anthony R. 203
Pearce, Phoebe M. 345, 976
Pearsall, N. 737
Pearson, Patrick 887
Pechmann, Sabrina 194
Peibst, Robby 6, 701, 733
Peña, Carlos 717
Peng, Fuguo 195, 1526
Peng, Hugh 727
Penukula, Saivineeth 1534
Penukula, Vineeth 1109
Perez, Marc 285, 662
Perez, Richard 662, 928, 1620
Perez-Rodriguez, Paula 305
Perini, Carlo A. R. 1183
Perini, Carlo Andrea Riccardo 969
Perkins, Craig L 577, 633
Perna, Allison N. 291
Perna, Allison. 71
Pernès, Nicolas 1129
Perret, Lionel 749
Perry, Kirsten 710, 1145
Perry, Lakesha N. 282
Perullo, Christopher 555
Peshek, Timothy J. 698
Peshek, Timothy 1490
Peter, Christoph 1372
Peters, Benjamin 634
Peters, Ian Marius 540, 645, 1092
Peters, Stefan 191
Peterson, Josh 600, 1123
Petesic, James 1295
Petter, Kai 191
Phang, Sieu Pheng 1434
Phillips, Adam B. 46, 220, 275, 279, 401, 402,
.... 683, 823, 1041, 1193, 1210, 1323, 1410, 1417
Phillips, Adam 804, 910
Pierce, Benjamin G. 512
Pierce, Benjamin 101

Pieters, B. E. 546
Pieters, Bart E. 166
Pikolos, Loucas 1348
Pilot, Nicholas 14
Pilvet, Maris 1298, 1432
Pina, Marissa 1667
Pineda, F. Brigham 783
Platzer-Björkman, Charlotte 887
Ploigt, Hans-Christoph 191
Podraza, Nicholas J. 1410
Podraza, Nikolas J. 401, 1550, 1557
Pogorelov, Kostiantyn 1491
Pohl-Hampel, Britta 191
Pokhrel, Dipendra 683
Polly, Stephen J. 63, 721, 894
Polly, Steve J. 89
Polzin, Jana-Isabelle 795
Poortmans, Jef 48, 162
Pop, Eric 1347
Poplawsky, Jonathan D. 46
Porret, Clément 434, 1390
Posada, Jorge 233, 268
Pothoof, Justin 1020
Poulsen, Peter B. 448, 1452
Powell, Kaden 998
Prabakar, Kumaraguru 1164
Pradeep, Nisitaa Karen Clement 467
Prasanna, Rohit 919
Prathap, Nemalipuri Surya 182
Pravettoni, Mauro 1188
Prell, Henrik 939
Prettl, Michael 289
Price, Kent 1102
Prot, Aubin JC. M. 1067
Provost, Marion 268
Prym, Guilherme C. S. 1159, 1654
Ptak, Aaron J. 10, 71, 291
Puel, Jean-Baptiste 233, 268
Pugstaller, Robert 1428, 1467, 1481
Pulwin, Ziggy 1217
Purkayastha, Atanu 1606
Pusch, Andreas 976
Qazi, Suleman Sami 772
Qian, Chen 1256
Qian, Yang 134
Qin, Yuan 1221
Qiu, Botong 1575
Qiu, Feng 634
Qu, Minghao 195, 1526
Qu, Xiaoyong 1372
Quader, Abdul 1410
Queck, Martina 191

Quinones, Dhamelyz R. S. 1667
Quispe, David 767
Radhakrishnan, Hariharsudan 48
Radu, Daniela R. 276
Radu, Daniela 606, 1206, 1437
Raghoebarsing, A. 737
Ragonesi, Antonino 1005
Rahimi, Amirhossein 844
Rahman, Areefa 482
Rahman, Naveed 1183
Raikar, Subbarao 333, 400
Rajakaruna, Manoj 402, 844, 1417
Raju, Sukhwant 855
Raker, David 402
Ramasubramanian, Deepak 1096
Rametta, Francesco 1005
Ramirez-Iniguez, Roberto 591
Ramos, Wendy Reyes 682
Rampalli, Chaitanya Santosh 1513
Rana, Prem J. S. 1044
Ranalli, Joseph 263, 1242
Rand, BP 1233
Ransome, Steve 1348
Rapp, Jeremy 1259
Rashed, Faisal 356, 1454
Rashkin, Lee 1034
Rasmussen, Mirra 597
Rathgeber, Andreas 385
Rau, U. 546
Rau, Uwe 177, 199, 419, 1307
Raugewitz, Annika 6, 733
Ravello, Magdalena 969
Ravishankar, Sandheep 1307
Reagan, Jeremiah 209
Reddy, K. S. 182
Reddy, Vasudeva Reddy Minnam 1549
Reddy, Yellasiri Bharath Kumar 259, 1227
Reece, Peter J. 976
Reese, Matt 245
Reese, Matthew O. 35, 633, 695, 915, 946, 1008
Reese, Samantha 1416
Reeves, Adam 240
Reich, Carey 849, 1234, 1279
Reich, Gerly 1024
Reichel, Christian 1313
Reinders, A. H. M. E. 737
Reinders, Angèle H. M. E. 809
Reinders, Angele 443
Rendler, Li C. 1313
Rengifo, Fabio Andrade 519
Reno, Matthew J. 13, 334
Repins, Ingrid L. 702, 1260, 1593

Reyes-Colón, Ramón 57
Rey-Stolle, I. 947
Rey-Stolle, Ignacio 1433, 1603
Rezk, Ayman 321, 1320, 1344, 1471
Rhee, Kurt 581
Ribeiro, Andrei C. 1159, 1654
Ribo, Macarena Mendez 554, 1416
Richards, Bryce S. 172, 996
Riedel, Maximillian 1066
Riedl, Gabriel 1428, 1467
Rienäcker, Michael 701
Rigby, O M. 995
Rijal, Suman 1197, 1376
Rikhof, Anne 622
Riley, Daniel 101, 409, 1131
Ringel, Steven A. 917
Rippingale, Jan 922
Ritzer, David Benedikt 144
Rivera, Agustín Irizarry 519
Rivera, Eduardo I. Ortiz 630
Rivera-Matos, Yiamar 916
Rizk, Ayman 327
Rizzo, Aurora 1109
Ro, Jason 925
Roberts, Dennice M. 1629
Robles-Rivera, Emmanuel G. 1034
Rock, Nathan 1337
Rockett, Angus A. 489
Rodgers, Marianne 1612
Rodriguez-Cabanas, Lissette 508
Rohatgi, Ajeet 91, 179, 1030, 1155, 1657
Rojas-Gatjens, Esteban 1020
Rojsatien, Srisuda 782, 1634
Rolston, Nicholas 938, 1109, 1534, 1565
Rome, Grace A. 75
Romer, Pascal 1060
Römer, Udo 917
Rong, Eric 272, 1356
Rosales, Bryan 740
Rosenthal, Samuel 1575
Roufberg, Lew 311
Rounsaville, Brian 91, 179
Rousset, Jean 268, 1440
Rout, Bibhudutta 132
Routhier, Alex 916
Rowell, David 145
Roy, Etee Kawna 998
Roy-Layinde, Bosun 482
Ru, Xiaoning 195, 1526
Rudolph, Dominik 1372
Ruhle, Ryan 387
Runkana, Venkataramana 1053

Ruske, Florian..............945
Rusnak, Jozef..............1594
Russell, Annie C. J...........331
Saavedra-Peña, Nelson E..........1082, 1326
Sacchitella, Elijah..........894
Saeed, Muhammad Mohsin..........1210
Saenz, Theresa E..........994, 1130, 1150
Sahani, Rishabh..........276, 1206, 1437
Sai, Hitoshi..........775
Saitta, Federica..........305
Saive, Rebecca..........348, 622, 996
Sajja, Sunil..........855
Saliba, Michael..........289
Salles, Caroline Lima..........1216
Sampath, Walajabad S...........509, 1537, 1600
Sampath, Walajabad..........123, 387, 1279
San José, Luis J..........746
Sanchez-Perez, C...........947
Sanchez-Perez, Clara..........1433
Sanci, Sal..........978
Sankin, Igor..........914
Santala, Annikki L...........919
Santamaría, Rodrigo Del Prado..........448, 1452, 1662
Santana, Vinícius Camatta..........1541
Santbergen, Rudi..........305, 1005, 1398
Santhanam, Lakshmi..........126
Santiago, Brian L. Reyes..........1338
Santistevan, Kevin..........409
Santiwipharat, Chaiwarut..........1545
Santos, José..........804
Saraswat, Govind..........360
Saraswat, Krishna C...........1347
Sargent, Edward H...........1019
Sargent, Ted..........518
Sarkisov, Sergey..........110
Sartor, Benjamin E..........1008
Sato, Shin-Ichiro..........76, 463
Satymov, Rasul..........708
Saucedo, Edgardo..........1059
Saucedo, Joel..........1633
Saw, Min Hsian..........1188
Sazzad, Muhammad H...........976
Scarpulla, Michael A...........731
Scarpulla, Michael..........1337
Schall, J. W...........315
Schall, Jackson W...........1570
Schall, Jackson..........1044
Schaper, Martin..........191
Scharf, Jessica..........191
Scheer, Roland..........12
Scheibner, Michael..........415
Scheideler, William J...........1276

Scheidt, Rebecca A..........161, 1156
Scheiman, David..........1045
Schelhas, Laura T...........298, 353, 952, 1044, 1260, 1629
Schirone, Luigi..........1404
Schlatmann, Rutger..........1066
Schlenoff, Tali..........1292
Schley, Michael..........191
Schmieder, Kenneth J...........886, 1217
Schmitz, J...........737
Schneble, Olivia D...........1150
Schneiderloechner, Eric..........795
Schnierer, Branislav..........1594
Schönmann, Antje..........191
Schramm, Barbara..........435
Schreiber, Waldemar..........434, 1341
Schropp, Andreas..........651
Schüler, Marc Andre..........1060
Schulte, Kevin L...........10, 291, 848
Schulte-Huxel, Henning..........652
Schulz, Susanne..........191
Schurman, Matthew J...........311
Schütze, Matthias..........191
Schwab, Andrew J..........977
Schwabedissen, Axel..........191
Schwartz, Dakota..........1551
Schwung, Julian..........64
Schygulla, Patrick..........1341
Sciuto, Marcello..........1005
Scuto, Andrea..........1609
Seal, Sudipta..........671
Sehirlioglu, Alp..........758
Seibert, Samuel..........1353
Seiboth, Frank..........651
Seigneur, Hubert P...........671
Seigneur, Hubert..........850, 1120, 1485
Sekulic, William..........794
Sellers, Ian R...........132, 161, 1156
Selvidge, Jennifer..........341
Senaud, Laurie-Lou..........1
Sengupta, Manajit..........152, 386, 474, 587, 600, 778, 1106
Seo, Seongrok..........298
Sepúlveda-Mora, Sergio B...........464
Sepúlveda-Vélez, Fredy A...........1105
Serafini, Patricio..........441
Sermarini, Anna Carolina De Paula..........256, 858
Serrano, Guillermo..........525
Setiawan, Ignatius Andre..........1549
Sevillano-Bendezú, Miguel Á...........1066
Seymour, Kyle..........928, 1147
Seyrich, Martin..........651
Shafarman, William N...........91, 1545
Shafarman, William..........1299

Shah, Akash 123
Shan, Ambalanath 1410
Shapiro, Finley R 915
Sharikadze, Saba 641
Shaton, Avishai 861, 1088
Shaw, Daniel 1234
Shen, Heping 1434
Sheppard, Scott 555
Sheyfer, Dina 1183
Shi, Jiahui 1503
Shi, Yangwei 1020
Shimabukuro, Laura 1575
Shimasaki, Takashi 342
Shimpi, Tushar 981
Shiradkar, Narendra 1287, 1366
Shirai, Yasuhiro 324, 1401
Shirazi, Eli 443, 816
Shojaei, D. 942
Shoji, Yasushi 105
Shore, Andrew M. 1292
Shrestha, Bishal 401, 1410
Shrestha, Santosh 881, 1564
Sidhik, Siraj 938
Siebentritt, Susanne 1067, 1151, 1617
Siegneur, Hubert 648
Siepchen, Bastian 656
Silhavy, Jake T. 918
Silva-Acuña, Carlos 1020
Silveira, A. M. C. 1492
Silveira, Allan 1068
Silverman, Timothy J. 106, 921, 1044, 1218, 1260, 1593
Simon, John 10
Sims, Jeremiah D 829
Singh, Luna 718
Singh, Manish K. 1384
Singh, Pritpal 1510
Singh, Rhythm 830, 1230
Sinha, Arpan 1155
Sinha, Parikhit 146, 855
Sinton, Ron 113
Sinton, Ronald A. 798
Sirkisoon, Sarah 22
Sites, James R. 509
Sites, James 1310, 1582, 1600
Sitiriche, Marcel Castro 519
Skoczek, Artur 1594
Slauch, Ian M. 970
Slauch, Ian 689
Slonopas, Andre 508
Smaine, Issam 426
Smets, Arno H. M. 305
Smith, David D. 1024

Smith, Emily 1045
Smith, Ryan 1120
Smith, Soshana 282
Snaith, Henry J. 298
Snell, Jeffrey 1009
Snuggs, Robert 823
Snyder, William 409
Sodabanlu, Hassanet 342, 374, 935
Soeriyadi, Anastasia H. 917
Song, Chang-Yun 432
Song, Hee-Eun 97
Song, Tao 178, 394, 485
Song, Zhaoning 42, 275, 683, 804, 812, 844, 909,
.... 1197, 1210, 1323, 1353, 1376, 1417, 1550, 1557
Sood, Mohit 1151
Sourabh, Shashi 132, 161
Sovetkin, E. 546
Sovetkin, Evgenii 166
Spaeth, Bettina 656
Spataru, Sergiu V. 448, 1452, 1662
Springer, Martin 794
Spurgeon, Ben 861, 1088
Sridhar, Seetharaman 554
Stall, Richard 311
Stanley, Bradley 508
St-Arnaud, Louis-Philippe 941
Stein, Joshua S. 188
Stein, Joshua 101
Steinebrunner, Udo 1060
Steiner, Myles A. 10, 75, 218, 291, 341, 394, 848,
.... 994, 1081, 1130
Stenzel, Florian 191
Stern, Jillian 467
Stevens, Margaret A 886, 1217
Stevens, Tristan 655
Steyn, Dirk 891
Stid, Jacob T. 743, 1259
Stoffel, Tom 1106
Stoicescu, Liviu 166, 1662
Stradins, P. 1459
Stradins, Paul 30, 475, 891
Stradins, Pauls 1216
Strandberg, Rune 815
Strandins, Pauls 359
Straub-Mueck, Michael 385
Strelow, Christian 651
Stroyuk, Oleksandr 540
Stuckelberger, Michael E. 435, 651
Sturm, Chris 12
Subedi, Kamala Khanal 46, 1019, 1353
Subramanian, Sivakumar 1053
Suemasu, Takashi 1398

Sugaya, Takeyoshi	105, 939
Sugimoto, Hiroki	76
Sugiyama, Masakazu	342, 374, 463, 935
Sulas-Kern, D. B.	315
Sun, Kaiwen	1256
Sun, Yijia	1485
Sun, Yukun	997
Sung, Li-Piin	282
Sunkari, Preetham P.	340
Sunkari, Preetham	473
Sunter, Deborah A.	467
Suppiah, Sam	476
Sutterlueti, Juergen	1348
Svrcek, Vladimir	281
Sweat, Rebekah	330
Syed, Faizan	606
Sytnyk, Mykhailo	1092
Szablewski, M	995
Szábo, Sandor	550
Szyszka, Bernd	945
Taconelli, Mauricio	1654
Taddei, Margherita	1020
Tafur, Lucila D.	927, 1146
Takahashi, Tadatoshi	611
Takamoto, Tatsuya	924
Talavera, Diego L.	1105
Tamizhmani, Govindasamy	237, 648, 755, 794, 1120
Tamuno-Ibuomi, Lewis Osikibo	591
Tan, Jin	840
Tan, Kelvin	85
Tanaka, Taichi	1453
Tanimoto, Tsutomu	924
Tank, Mehul	330
Tao, Meng	85, 1566
Tatavarti, Rao	1429
Taubmann, Rouven	1184
Tawsif, Sheikh Elahi	690
Tayagaki, Takeshi	253
Taylor, Andre D	1008
Taylor, P. Craig	359
Teasley, Corson	555
Teodor, Alexandra H.	11
Teplyakov, Andrew V.	1667
Ternes, Simon	144
Terry, Mason	783
Tervo, Eric J.	341
Terwilliger, Kent	702, 1629
Theelen, M.	737
Theingi, S.	1459
Theingi, San	475
Theocharides, Spyros	541
Theristis, Marios	122, 188, 1250

Thiagarajan, Ramanathan	1164
Thibodeau, Matthew R.	918
Thiel, Christian	924
Thiengi, San	30
Thind, Arashdeep S.	771
Thomas, Adam	682
Thomas, Sinju	432, 939
Thomsen, Vitus B.	16
Thon, Susanna M.	272, 718, 1334, 1356, 1575
Thon, Susanna	695
Tiefenthaler, Martin	1481
Tilli, Francesca	749
Timmo, Kristi	1298, 1432
Timò, Gianluca	871
Timofte, Tudor	1372
Tina, Giuseppe Marco	1304, 1567
Titus, Jochen	919
Tobail, Osama	191
Tobon, Carlos Mario Ruiz	1398
Todaro, Lorenzo	1304
Töfflinger, Jan A.	1066
Togay, Mustafa	245, 789, 981, 1177, 1193, 1497
Toh, Wei Wen	1188
Tokumasu, Takashi	292
Tomita, Yosuke	924
Tonita, Erin M.	331
Topic, Marko	1
Törndahl, Tobias	1151
Tovar, Michael	939
Tracy, Jared	952
Transue, Taos	922, 1250
Tremont-Brito, Rolando J.	1170
Trempa, Matthias	1266
Tresan, Jenner	492
Trimby, Pat	939
Troupe, Anthony T	590
Trupke, Thorsten	227, 472, 828, 1224, 1468
Tsakalids, Anastasios	924
Tse, Yau Yau	656
Tumusange, Marie Solange	1550
Tuomiranta, Arttu	162, 426
Turala, A.	1538
Turcotte, Dave	155
Tutsch, Leonard	1625
Ubukata, Akinori	105
Uddin, Md Aslam	1044
Uene, Naoya	292
Ukaegbu, Ikechi	502
Ulbrich, Carolin	1066
Ulicná, Sona	353, 952, 1044, 1629
Upadhyay, Prashant Kumar	259, 1227
Upadhyaya, Ajay D	1030

Upadhyaya, Ajay 91
Upadhyaya, Vijaykumar D 179, 1030
Upadhyaya, Vijaykumar 91, 1657
Ures, Sandra 888
Urs, Rahul R 299
Usami, Noritaka 292, 724, 775
Vaas, T. S. 546
Valdivia, Christopher E 331, 476, 941
Valerino, Michael 437
Vallerotto, G. 746
Van De Voorde, Mathis 348, 996
Van Dyck, Rik 48
Van Nijen, David A. 655
Van Sark, W. J. G. H. M. 737
Van Swaaij, René A. C. M. M. 655
Van Velson, Nathan 1114
Van Vuure, Aart Willem 48
Vanderhaegen, Aline 1151
Vandervelde, Thomas E. 1152
Vansant, Kaitlyn 415, 1490
Vargas, Fernando J. 416
Vasconcelos, Cláudia K. B. 94
Vasi, Juzer 1366
Vazquez, Michael Y. 416
Venkat, Sameera Nalin 234
Venkatramanan, D. 1384
Verezhak, Mariana 651
Verkou, Maarten 1005
Verlinden, Pierre 1544
Vignola, Frank 600
Villa-Bracamonte, Maria Fernanda 1283
Villa-Ignacio, Armando 215
Villalva, M. G. 1492
Villalva, Marcelo G 1068, 1159, 1654
Villalva, Marcelo Gradella 1134
Villalva, Marcelo 290
Voarino, Philippe 1423
Vogt, Malte 1005
Volatier, M. 1538
Von Gastrow, Guillaume 1024
Von Wenckstern, Holger 12
Voss, Stephen 607
Vuk, Dragana 351
Wadsworth, Matthew 330
Wagner, Kristen 1620
Wagner, Philipp 945
Wagner-Mohnsen, Hannes 569
Wahl, Tina 1201
Wakamiya, Atsushi 442
Walajabad, Sampath S. 981
Walajabad, Sampath 1234
Walker, Alexandre W 941

Walker, Don 198
Walkons, Curtis 566
Wallner, Gernot M. 1467, 1481
Wallner, Gernot 1428
Walls, J. Michael 981, 1193
Walls, John M. 245, 498, 789, 1177, 1497
Walls, Michael 656, 683
Wang, Jian 1020
Wang, Jianjian 1114
Wang, Jianming 248
Wang, Jianqiang 1526
Wang, Jing 1074
Wang, Le 1544
Wang, Liwei 240
Wang, Quanzhi 1523
Wang, Tonghui 332
Wang, Wei 690, 1237
Wang, Wenzong 835, 1096
Wang, Yichun 195, 1526
Wang, Yonglei 1221
Wang, Zhaoyu 634
Wanlass, M. W. 1261
Wargulski, Dan R. 879, 939
Warren, Emily L. 75, 218, 291, 709, 848, 994, 1130,
.... 1150, 1158, 1380
Wasmer, Sven 569
Watanabe, Kentaroh 342, 374, 935
Watson, Stephanie S. 282
Wattenberg, Bianca 701
Weber, August 1266
Weber, Julian 1060
Weed, Emily 505
Weigand, William J. 1533
Weigand, William 1379
Weihrauch, Anika 191
Weinhardt, Lothar 91
Welch, Liam M. 395
Welch, Liam 534
Welser, Roger E. 145, 721
Welser, Roger 1429
Wenner, Scott L. 220
Wenner, Scott 910
Westerhof, Jelle 622
Westraadt, Johan 1625
Whalen, Devin C. 380
Wheeler, Aaron 783
Wheeler, Lance M. 1044
Whiteside, Vincent R 132, 161, 1156
Wibowo, Andree 145, 1429
Wickett, Shelbie 1635
Widrick, Devin 528
Wieghold, Sarah 1183

Wieliczka, Brian 1490
Wieser, Raymond J. 1141
Wieser, Raymond 861, 1088
Wietler, Tobias 6, 652, 733
Wikoff, Hope 1416
Wilcox, Stephen 1106
Williams, Henry J. 505, 727
Williams, Jennifer 415
Williams, Rafell 178
Wilson, Paige 476, 941
Wilson, Samantha S. 607
Wilson, Thomas 90
Wilt, David 1429
Wilterdink, Harrison 113
Winkelmann, Aimo 939
Winkler, Louisa 1060
Winter, Björn Oliver 64
Wirtz, L. 995
Witte, Wolfram 432
Witteck, Robert 709
Wittmann, Ernst 645
Wong, Johnson 978
Woodall, Mark 415
Woodhouse, Michael 1593
Wright, Brendan 1224, 1295
Wu, Xiang 1372
Wu, Yinghui 861, 1088
Wu, Yuh-Renn 1443
Wu, Zhenni 1092
Wyss, Patrick 1129
Xiang, Xiaofeng 578
Xiao, Chuanxiao 46, 822, 1019
Xiao, Yegao 1645
Xie, Tian 195, 1526
Xie, Yu 152, 386, 474, 587
Xiong, Gang 998, 1197, 1497
Xu, Binbin 419
Xu, Jianmei 1544
Xu, Tao 1284, 1523
Xu, Xixiang 195, 1221, 1526
Xu, Yawen 248
Xue, Chaowei 1526
Yagi, Shuhei 926
Yaguchi, Hiroyuki 926
Yaiche, Armelle 1440
Yamaguchi, Masafumi 924, 1535
Yamamoto, Kohei 253
Yan, Di 1434
Yan, Feng 1197
Yan, Yanfa 42, 46, 220, 275, 279, 683, 804, 812, 823, 844, 909, 910, 1019, 1041, 1197, 1210, 1323, 1353, 1376, 1417, 1550, 1557

Yanagida, Masatoshi 324, 1401
Yang, Guangtao 655
Yang, Jaemo 152, 386
Yang, Jie 1434
Yang, Miao 195, 1526
Yang, Ruiquan 975
Yang, Zhaoqing 1020
Yao, Dominique Akissi 758
Yao, Keyi Kang 1575
Ye, Jichun 822
Yelzhanova, Zhuldyz 39
Yermekov, Nurzhan 39
Yi, Chuqi 917
Yildirim, Murat 634
Yilmaz, P. 737
Yin, Shi 195, 1526
Yoon, Heayoung 134, 998
Yoon, Woojun 1045
Yoon, Yohan 134
Yoshita, Masahiro 253
Young, D. L. 1459
Young, David L 106
Young, David 30, 475, 891, 1216
Young, Ethan 921
Young, Matthew R. 1130
Young, Michelle 848
Youtsey, Chris 145, 1446
Yu, Li 1164
Yu, Xuanji .... 234, 597, 671, 758, 861, 1088, 1141, ... 1284, 1485
Yu, Yuanjie 794, 1284, 1523
Yu, Zhengshan J. 767
Yu, Zhibin 330
Yuan, Luyao 730
Yun, Changyeol 436
Yun, Jae Ho 1556
Yusuf, Jubair 13
Zabalza, Ruben 1141
Zaka, Awais 1471
Zawisza, Zachary W. 823
Zawisza, Zachary 910
Zech, Matthias 1242
Zehender, M. 947
Zelenina, Anastasia 1067
Zeman, Miro 655, 1005
Zeng, Yiyu 302
Zhang, Changgen 248
Zhang, Fan 1261
Zhang, Guangchun 1284
Zhang, Hongxu 1221
Zhang, K. Max 505, 727
Zhang, Kangping 248

Zhang, Shu..1544
Zhang, Wei..1497
Zhang, Xinyu..1434
Zhang, Yijun...718
Zhang, Yong...1261
Zhang, Zheyu...240
Zhao, Dewei..822
Zhao, Shijia...634
Zhao, Yong..248
Zhao, Zitong...1021
Zhaoning, Song...1019
Zheng, Jian-Yao...622
Zheng, Peiting...1434
Zhong, Ruohan 1030, 1657
Zhu, Kai...........................812, 1158, 1265, 1484
Zhu, Xitong..809
Zhu, Yan...227
Zilouchian, Ali..1280
Zimányi, Gergely T. 51, 1021
Zimmerman, Jeramy D. 994, 1130, 1150
Zimmermann, Gregor 191
Zimmermann, Iwan.......................................804
Zin, Ngwe...1509
Zinßer, Mario..199
Zoppi, G..995
Zubieta, Diego 597

2023 IEEE 50th Photovoltaic Specialists Conference (PVSC 2023)

San Juan, Puerto Rico, USA
11-16 June 2023

Pages 1114-1669

IEEE Catalog Number: CFP23PSC-POD
ISBN: 978-1-6654-6060-6

**Copyright © 2023 by the Institute of Electrical and Electronics Engineers, Inc.
All Rights Reserved**

Copyright and Reprint Permissions: Abstracting is permitted with credit to the source. Libraries are permitted to photocopy beyond the limit of U.S. copyright law for private use of patrons those articles in this volume that carry a code at the bottom of the first page, provided the per-copy fee indicated in the code is paid through Copyright Clearance Center, 222 Rosewood Drive, Danvers, MA 01923.

For other copying, reprint or republication permission, write to IEEE Copyrights Manager, IEEE Service Center, 445 Hoes Lane, Piscataway, NJ 08854. All rights reserved.

*** This is a print representation of what appears in the IEEE Digital Library. Some format issues inherent in the e-media version may also appear in this print version.

IEEE Catalog Number:	CFP23PSC-POD
ISBN (Print-On-Demand):	978-1-6654-6060-6
ISBN (Online):	978-1-6654-6059-0

Additional Copies of This Publication Are Available From:

Curran Associates, Inc
57 Morehouse Lane
Red Hook, NY 12571 USA
Phone: (845) 758-0400
Fax: (845) 758-2633
E-mail: curran@proceedings.com
Web: www.proceedings.com

TABLE OF CONTENTS

3-Terminal Perovskite/Silicon Tandem Modules: A Dead End or a Bright Future of Tandem Based Photovoltaics 1
Miha Kikelj, Laurie-Lou Senaud, Jonas Geissbühler, Damien Lachenal, Derk Baetzner, Benjamin Lipovšek, Marko Topic, Christophe Ballif, Quentin Jeangros, Bertrand Paviet-Salomon

Temperature-Dependent Performance of Ultra-Thin Silicon Heterojunction Solar Cells for Space Applications.. 5
Anh Huy Tuan Le, Pradeep Balaji, André Augusto, Ziv Hameiri

Utilizing a Soft IZO Sputtering Process to Contact Buffer-Free Semitransparent Perovskite Pin Solar Cells 6
Roland Clausing, Annika Raugewitz, Benjamin Grimm, Marvin Diederich, Tobias Wietler, Felix Haase, Rolf Brendel, Robby Peibst

Effect of Surface Morphology on GaAs Solar Cells Grown on Planarized Spalled (100) GaAs Substrates 10
Anna K. Braun, Jacob T. Boyer, Kevin L. Schulte, John Simon, Steve W. Johnston, Myles A. Steiner, Corinne E. Packard, Aaron J. Ptak

Ultra-Light Environmental Protection for Solar Arrays in Space..11
Pilar Espinet-Gonzalez, Alexandra H. Teodor, Jann A. Grovogui, Yao Y. Lao

Highly Crystalline In2S3:V Thin Films Epitaxially Grown on Sapphire Substrates: A Potential Canditate for Intermediate Band Solar Cells 12
Tanja Jawinski, Chris Sturm, Roland Clausing, Heiko Kempa, Stefan Lange, Roland Scheer, Marius Grundmann, Holger Von Wenckstern

A Model-Free Approach for Estimating Service Transformer Capacity Using Residential Smart Meter Data.. 13
Joseph A. Azzolini, Matthew J. Reno, Jubair Yusuf

Bifacial PV Fed Electrolysis for Green Hydrogen Generation and Cofiring Hydrogen in an Aeroderivative Gas Turbine.. 14
Nicholas Pilot, Jim Harper

Reverse-Bias Testing of Perovskite Cells to Inform Bypass-Diode Design 15
Daniel Martinez, Dana Kern, Giles Eperon

Improving Deep Learning-Based Defect Classification in Solar Cells Using Conformal Prediction 16
Vitus B. Thomsen, Claire Mantel, Gisele A. Dos Reis Benatto, Allan P. Engsig-Karup, Søren Forchhammer

Off-The-Shelf Small Scale Photovoltaic Systems for Puerto Rico Sustainable Farms: Assisting Those Who Help Others 22
Catherine Lahood, Sarah Sirkisoon, Charles M. Fortmann

Flexible Photonic Cooler Based on Multi-Stacked Thin Films IR Filters with Anti-Dust Capability for PV-Desert Environment Applications 27
Brahim Aïssa, Mohammad I. Hossain

Loss Analysis and Performance Optimization Pathways of 729-MV V_{OC} Si Solar Cells with Poly-Si on Locally-Etched Dielectric Passivating Contacts 30
    Suchismita Mitra, Caroline Lima Anderson, Matthew Hartenstein, William Nemeth, Matthew Page, San Thiengi, David Young, Sumit Agarwal, Paul Stradins

Cadmium Telluride Accelerator Consortium (CTAC) 35
    Lorelle M. Mansfield, Matthew O. Reese, Michael J. Heben

Screen Printable Copper Pastes for Silicon Solar Cells 38
    Thad Druffel, Ruvini Dharmadasa, Apolo Nambo, Abafriseke Ebong, Sandra Huneycutt, Donald Ital

Optimization of Zinc Oxide Electron Transport Layers for Cs-Based Perovskite Solar Cells 39
    Zhuldyz Yelzhanova, Gaukhar Nigmetova, Gulzhan Bizhanova, Nurzhan Yermekov, Ulan Kashkimbayev, Annie Ng

UV Degradation of Formamidinium-Cesium Lead Halide Perovskite Solar Cells 42
    Kshitiz Dolia, Abasi Abudulimu, Sheng Fu, Tyler Brau, Karissa Jensen, Stephanie L. Moffitt, Randy J. Ellingson, Xiaohong Gu, Zhaoning Song, Yanfa Yan

Efficient Cd(Se, Te) Solar Cells with Cd(O, S, Se, Te) at the Front Interface 46
    Dengbing Li, Sabin Neupane, Sandip Singh Bista, Abasi Abudulimu, Kamala Khanal Subedi, Manoj K. Jamarkattel, Chuanxiao Xiao, Chun-Sheng Jiang, Jonathan D. Poplawsky, David A. Cullen, Adam B. Phillips, Michael J. Heben, Randy J. Ellingson, Yanfa Yan

Energy Yield and Economics of Single-Axis-Tracked Bifacial Photovoltaics with Artificial Ground Reflectors 47
    Mandy R. Lewis, Silvana Ovaitt, Byron McDanold, Chris Deline, Karin Hinzer

Advanced Encapsulants for Reduced Thermal Mechanical Stress in Photovoltaic Modules: A Quantitative Analysis Using FBGS 48
    Rik Van Dyck, Marta Casasola Paesa, Tine Engelen, Bin Luo, Tom Borgers, Jonathan Govaerts, Hariharsudan Radhakrishnan, Michael Daenen, Jef Poortmans, Aart Willem Van Vuure

TOPCon Solar Cell Degradation Via Pinhole Nucleation 51
    Andrew Diggs, Adam Goga, Zachary Crawford, Gergely T. Zimanyi

Soiling Model for PV Applications: Improved Parameterizations 54
    Vicente Lara-Fanego, Christian A. Gueymard, Leonardo Micheli

A Sustainable Energy Market Through Community-Based PV Systems 57
    Ramón Reyes-Colón, Efraín O'Neill-Carrillo

Radioisotope Thermoradiative Cell Power Generator 63
    Stephen J. Polly, Geoffrey A. Landis, Seth M. Hubbard

Robust Detection Method of Low-Voltage Islanding for Grid-Forming Inverters Operated in Conjunction with Existing PV Inverters 64
    Björn Oliver Winter, Julian Schwung, Bernd Engel

Improving Performance of III-V Solar Cells Grown on Spalled Germanium with Ex Situ Substrate Planarization 71
    John S. Mangum, Anna K. Braun, Allison Perna, John F. Geisz, Aaron J. Ptak, Corinne E. Packard, Ryan M. France

Fabrication and Characterization of III-V Photovoltaic Devices for Use as CO2 Reduction
Photoelectrodes 75
    Myles A. Steiner, Grace A. Rome, Ann L. Greenaway, Joel W. Ager, Emily L. Warren

Proton Degradation-Free Flexible Chalcopyrite Solar Cells Without Cover Glass and Adhesive 76
    Hiroki Sugimoto, Tetsuya Nakamura, Mitsuru Imaizumi, Shin-Ichiro Sato, Takeshi Ohshima

Analysis of Hierarchical PV2PV Series Differential Power Processing Configuration for
Photovoltaic Applications 80
    Afshin Nazer, Patrizio Manganiello, Olindo Isabella

Plane of Array Irradiance Cleaning and Generation of Validated POA Readings for Plant Evaluation 81
    Pramod N. Krishnani, Adonis E. Hajj, Clay Helms, Shreyas Nagarajan, Mark Mikofski

A Maximum Current Point Tracking Algorithm for Solar-To-Hydrogen Production 85
    Kelvin Tan, Meng Tao

A Physics Based Approach for PV Lifetime and Degradation Signatures Prediction 86
    Ismail Kaaya, Gofran Chowdhury, Arnaud Morlier

Epitaxial Growth and Testing of 1.1 eV Metamorphic InGaAs/GaAs Laser Power Converters 89
    Katelynn E Fleming, Steve J Polly, Seth M Hubbard

Suitability of GaAsBi as a Candidate Junction in a III-V Multi-Junction Solar Cell 90
    Thomas Wilson, Nicholas Ekins-Daukes

~20% Efficient Si PERC Solar Cell with Emitter Surface Passivated by H_2S Reaction 91
    *Tasnim K. Mouri, Ajay Upadhyaya, Ajeet Rohatgi, Youngwoo Ok, Amandee Hua, Dirk
    Hauschild, Lothar Weinhardt, Clemens Heske, Vijaykumar Upadhyaya, Brian Rounsaville,
    William N. Shafarman, Ujjwal K. Das*

Investigations of Snail-Trail and Associated Microcrack Properities and Behavior in Brazil's
Tropical Climate 94
    *Antonia Sonia A. C. Diniz, Neolmar De M. Filho, Cláudia K. B. Vasconcelos, Lawrence L.
    Kazmerski*

A Study of POCl3 Deposition Reaction Rate with Residual Gas Analysis Method 97
    *Min Gu Kang, Sang Hee Lee, Kyung Taek Jung, Yunae Cho, Dohyung Kim, Sungeun Park,
    Munse Kim, Hee-Eun Song*

Analysis of Photovoltaic Systems Penetration on Demand Curve and Locational Marginal Prices
(LMPs) in PJM 98
    Mesude Bayrakci Boz

Single Axis Tracker Performance Modeling on Sloped Terrain 101
    Benjamin Pierce, Joshua Stein, Daniel Riley

Substrate Reuse of Hydride Vapor Phase Epitaxy Grown-GaAs Solar Cells for Low-Cost
Photovoltaics 105
    *Yasushi Shoji, Ryuji Oshima, Kikuo Makita, Akinori Ubukata, Shuuichi Koseki, Takeyoshi
    Sugaya*

Towards Polymer-Free, Fs Laser Welded Glass/Glass Modules 106
    David L Young, Nick Bosco, Timothy J Silverman

Defects in RbF - Treated $Cu(In_xGa_{1-x})Se_2$ Solar Cells and Their Impact on V_{OC} 107
Michael F. Miller, Ana Kanevce, Alexandra M. Bothwell, Stefan Paetel, Darius Kuciauskas, Aaron R. Arehart

Improving the Space Silicon Solar Cell Efficiency by Adding the Layer Down-Converting UV Light to Visible 110
Alex Fedoseyev, Stan Herasimenka, Sergey Sarkisov

Near-Contactless Production I-V Testing of Silicon Solar Cells 113
Harrison Wilterdink, Ron Sinton, Adrienne Blum, Karoline Dapprich, Nick Degenhart, Wes Dobson

Self-Thermometry of PV Panels..119
Kaushal Chapaneri, Shahzada Pamir Aly, Jim Joseph John, Gerhard Mathiak, Vivian Alberts, Muhammad A. Alam

Inverter Clipping and Its Masking Effect on PV Soiling: Truth Or Myth?.. 122
Leonardo Micheli, Matthew Muller, Marios Theristis, Florencia Almonacid, Eduardo F. Fernandez

Effect of Arsenic Doping in Polycrystalline Thin Film CdTe Solar Cells 123
Mayank Mate, Akash Shah, Ramesh Pandey, Zachary Lustig, Walajabad Sampath, Amit H. Munshi

Thermal Modelling of a Renkube Panel 126
Deepika Gopal, Balaji Bangolae Lakshmikanth, Lakshmi Santhanam

Stable High Temperature Operation in Metal Halide Perovskite Solar Cells.................................... 132
Hadi Afshari, Shashi Sourabh, Sergio A. Chacon, Vincent R. Whiteside, Bibhudutta Rout, Giles E. Eperon, Joseph M. Luther, Ian R. Sellers

Characteristics of Detachable III-V Solar Cells Grown on Porous Germanium................................ 133
Valentin Daniel, Thomas Bidaud, Jeremie Chretien, Abdelatif Jaouad, Jean-Francois Lerat, Nicolas Paupy, Bouraoui Ilahi, Jinyoun Cho, Kristof Dessein, Christian Dubuc, Gwenaelle Hamon, Abderaouf Boucherif, Maxime Darnon

Impact of Surface Roughness in Measuring Optoelectronic Characteristics of Thin-Film Solar Cells............ 134
David Magginetti, Seokmin Jeon, Yohan Yoon, Ashif Choudhury, Ashraful Mamun, Yang Qian, Jordan Gerton, Heayoung Yoon

Validation of Open-Source Distributed Energy Resources (OpenDER) Model with IEEE 1547-2018 Smart Inverter.. 138
Yiwei Ma, Charles Brewster, Aminul Huque

Upscaling of Perovskite Solar Cell Fabrication Via Slot-Die Coating: In Situ Tracking of the Drying and Crystallization Front During Gas Quenching 144
Kristina Geistert, Simon Ternes, David Benedikt Ritzer, Benjamin Hacene, Felix Laufer, Ulrich Wilhelm Paetzold

Inverted Metamorphic Photovoltaics Utilizing a Distributed Bragg Reflector Compatible with Epitaxial Lift-Off.. 145
Robert F McCarthy, David Rowell, Andree Wibowo, William Mohr, Chris Youtsey, Mark Osowski, Martin Drees, Roger E Welser, Noren Pan

Life-Cycle Analysis of Potentially Longer Life Expectancy CdTe PV Modules............................... 146
Vasilis Fthenakis, Enrica Leccisi, Parikhit Sinha

Forecasting Day-Ahead Solar Irradiance for Puerto Rico Using the WRF Model and NSRDB 152
    Manajit Sengupta, Jaemo Yang, Yu Xie

Coordinating the Frequency-Droop Controls of Inverter-Based Resources and Diesel Generators in
an Isolated Microgrid 155
    Mohan Du, Nayeem Ninad, Dave Turcotte

Probing Non-Equilibrium Hot Carrier Dynamics in Metal Halide Perovskite Solar Cells 161
    Shashi Sourabh, Hadi Afshari, Vincent R. Whiteside, Giles E Eperon, Rebecca A. Scheidt,
    Varun N. Mapara, Madalina Furis, Matthew C Beard, Ian R. Sellers

On the Influence of Forced Convection in PV Energy Yield Models 162
    Raed I. Bourisli, Bader S. Aldalali, Arttu Tuomiranta, Jef Poortmans

2-Terminal and 3-Terminal Subcell Characterization Platforms for Emerging Tandems 165
    Jin Young Kim

Fast Cell Detection and Distortion Correction for Outdoor Electroluminescence Images 166
    Evgenii Sovetkin, Bart E. Pieters, Andreas Gerber, Liviu Stoicescu, Pascal Koelblin

Limiting Factors on the Performenace of Luminescent Solar Concentrators for Building Integrated
Photovoltaics 172
    Bryce S. Richards, Ian A. Howard

3D Printed Transparent Sheet for Solar Panel Encapsulation and Thermal Management 173
    Fahad Alam, Nazek El-Atab

Testing the Durability of Fluorine-Free Hydrophobic Coatings vs Porous Silica 174
    Luke O. Jones, Adam M. Law, Gary Critchlow, John M. Law

Investigation of the Microstructure of Underdense Hydrogenated Amorphous Silicon Layers for
Silicon Heterojunction Solar Cells by Raman Spectroscopy and Hydrogen Effusion 177
    Benedikt Fischer, Maurice Nuys, Andreas Lambertz, Weiyuan Duan, Kaining Ding, Uwe Rau

Impact of Irradiation-Induced Filter Temperature Increase on Calibration of Reference Solar Cells
with NIR-Longpass Filters 178
    Tao Song, Larry Ottoson, Rafell Williams, John Geisz, Charles Mack, Jeremy Brewer, Nikos
    Kopidakis

Investigation of High Nitrogen Composition SiN_x for Textured Front Surface Passivation of n-Type
Silicon Solar Cells in Terms of Light Stability of Injected Negative Charge and Cell Performance 179
    Kwan Hong Min, Jeong-Mo Hwang, Christopher Chen, Wook-Jin Choi, Vijaykumar D
    Upadhyaya, Brian Rounsaville, Ajeet Rohatgi, Young-Woo Ok

Optimizing the Heat Sink for Concentrated Photovoltaic Systems for Different Heat Flux
Conditions 182
    Nemalipuri Surya Prathap, Harsh Chaurasia, K. S. Reddy

Comparison of Open-Source Photovoltaic Performance Models Against Multi-Year Field Data 188
    Lelia Deville, Marios Theristis, Bruce H. King, Terrence L. Chambers, Joshua S. Stein

Q Cells Q.Antum Neo Technology with > 25% Conversion Efficiency Applying Mass-Production Processes 191

Matthias Junghänel, Ingmar Höger, Martin Schaper, Kai Petter, Enrico Jarzembowski, Christian Klenke, Anika Weihrauch, Michael Schley, Hans-Christoph Ploigt, Ohjin Kwon, Antje Schönmann, Osama Tobail, Axel Schwabedissen, Maximilian Kauert, Klaus Duncker, René Hönig, Janko Cieslak, Stefan Hörnlein, Florian Stenzel, Björn Faulwetter-Quandt, Jessica Scharf, Friederike Kersten, Cangming Ke, Sissel Tind Kristensen, Carsten Baer, Martina Queck, Gregor Zimmermann, Matthias Köhler, Nicole Lampa, Britta Pohl-Hampel, Lorenzo Burtone, Larissa Niebergall, Matthias Schütze, Susanne Schulz, Stefan Peters, Ansgar Mette, Fabian Fertig, Markus Fischer, Jörg W. Müller

Outdoor Study of Photovoltaic Mini-Modules with Different Perovskite Compositions 194

Vasiliki Paraskeva, Maria Hadjipanayi, Matthew Norton, Aranzazu Aguirre, Anurag Krishna, Rita Ebner, Tommasso Fontanot, Sabrina Pechmann, Silke Christiansen, George E. Georghiou

Achieving a New World Record Silicon Solar Cell Efficiency of 26.81% Using SHJ Device Structure 195

Xixiang Xu, Minghao Qu, Miao Yang, Xiaoning Ru, Shi Yin, Chengjian Hong, Fuguo Peng, Junxiong Lu, Liang Fang, Zhenguo Li, Yichun Wang, Tian Xie

Utilizing Particle Swarm Optimization for Autocalibration of LED Solar Simulators 198

Jann A Grovogui, John C Nocerino, Don Walker

Analysis of Cu(In, Ga)Se$_2$ Heterojunction Solar Cells in Terms of Their Balance of Thermodynamic Potentials 199

Uwe Rau, Felix Komoll, Tim Helder, Mario Zinßer, Ana Kanevce, Thomas Kirchartz, Theresa Magorian Friedlmeier

Epitaxy-Free, Thin-Film GaAs Solar Cells Fabricated with Diffusion Doping and Mechanical Spalling 202

Phillip R. Jahelka, Andrew W. Nyholm, Harry A. Atwater

Analysis of Impurity-Related Radiative Transitions in Silicon Materials Using Temperature-Dependent Photoluminescence 203

Tarek O. Abdul Fattah, Janet Jacobs, Vladimir P. Markevich, Nikolay V. Abrosimov, Matthew P. Halsall, Iain F. Crowe, Anthony R. Peaker

Reflector Candidates for a Vertical Bifacial Solar Canal 209

Jeremiah Reagan, Brandi McKuin, Sarah Kurtz

Evaluating Leafy Green Production in a Colorado Rooftop Agrivoltaic System 215

Armando Villa-Ignacio, Jennifer Bousselot

Isotropic Wet Etching of Acoustically-Spalled GaAs 218

Anica N Neumann, Myles A Steiner, Pablo G Coll, Mariana Bertoni, Emily L Warren

Degradation-Related Defect Level in Weathered Silicon Heterojunction Modules Characterized by Deep Level Transient Spectroscopy 219

Steve Johnston, Dirk C. Jordan, Dana B. Kern, Kristopher O. Davis, Helio R. Moutinho, George F. Kroeger

Interrogating Dominant Recombination Pathways in CdTe Solar Cells Using Wavelength-Dependent External Radiative Efficiency Measurements 220

Jared D. Friedl, Adam B. Phillips, Manoj K. Jamarkattel, Tyler Brau, Sabin Neupane, Scott L. Wenner, Abasi Abudulimu, Ebin Bastola, Yanfa Yan, Randy J. Ellingson, Michael J. Heben

Dynamic Calibration of Injection Dependent Carrier Lifetime from Time-Resolved Photoluminescence 227
    Yan Zhu, Robert Lee Chin, Nursultan Mussakhanuly, Thorsten Trupke, Ziv Hameiri

Impacts of Dispatch Strategies and Forecast Errors on the Economics of Behind-The-Meter PV-Battery Systems 228
    Brian T. Mirletz, Nicholas D. Laws

Advanced Characterization and Degradation Analysis of Perovskite Solar Cells Using Machine Learning and Bayesian Optimization 233
    Joseph Chakar, Arthur Julien, Karim Medjoubi, Jorge Posada, Jean-François Guillemoles, Jean-Baptiste Puel, Yvan Bonnassieux

Statistical Analysis and Degradation Pathway Modeling of PERC PV Minimodules with Different Packaging Strategies in Indoor Accelerated Exposures 234
    Sameera Nalin Venkat, Jiqi Liu, Xuanji Yu, William Oltjen, Xinjun Li, Jean-Nicolas Jaubert, Jennifer L. Braid, Roger H. French, Laura S. Bruckman

Hotspot Endurance of Pristine and Thermal Cycled Glass-Backsheet Photovoltaic Modules 237
    Muhammad Afridi, Akash Kumar, Farrukh Ibne Mahmood, Govindasamy Tamizhmani

Integrated Large-Scale Data Management Platform for Photovoltaic Power Conversion Equipment (PCE) Reliability Data 240
    Liwei Wang, Buck Brown, Shuan Dong, Tan Jin, Daniel Clemens, Joseph Hodges, Adam Reeves, Josh Ozbeytemur, Shuangshuang Jin, Zheyu Zhang

SnO2 Buffer Layers for High Efficiency CdSeTe/CdTe Devices 245
    L. C. Infante-Ortega, Xiaolei Liu, Luksa Kujovic, Mustafa Togay, Luke O. Jones, Ali Abbas, Kieran Curson, R. C. Greenhalgh, Kurt L. Barth, Jake W. Bowers, John M. Walls, Ochai Oklobia, Stuart Irvine, Eric Colegrove, Brian Good, Matt Reese

Contact Interface Morphology of Screen-Printable Front-Side Contacts for Industrial N- TOPCon Crystalline Silicon Solar Cells 248
    Meijun Lu, Kurt R. Mikeska, Weilin Liao, Chaoying Ni, Yong Zhao, Jianming Wang, Kangping Zhang, Changgen Zhang, Yawen Xu, Baiqiang Liu

Light-Dark Cycling in Perovskite Solar Cells Studied by MPPT and Ion Migration Current Measurements 253
    Takeshi Tayagaki, Kohei Yamamoto, Takurou N. Murakami, Masahiro Yoshita

Field Trial in Progress for Measuring Global, Direct, Diffuse, and Ground-Reflected Irradiance Using a Static Sensor Array 254
    Michael Gostein, Bruce H. King

The Proposition of a Public Policy to Stimulate Low-Income Communities' Assess to Distributed Energy Resources 256
    Anna Carolina De Paula Sermarini, João Henrique Paulino Azevedo, Vanessa Cardoso De Albuquerque, Rodrigo Flora Calili, Felipe Gonçalves, Gilberto Jannuzzi

PV Plant Performance Review Methodology: Key Performance Indicators (KPI) Estimation 259
    Himanshu Gulati, Prashant Kumar Upadhyay, Yellasiri Bharath Kumar Reddy

Validation of Inverter Labeling with Plant Transfer Functions 263
    Joseph Ranalli

Long Terms Stability and Metastable Behavior of Perovskite Solar Devices on Outdoor Conditions............. 268
Karim Medjoubi, Anne Migan Dubois, Jean Castillon, Thomas Guillemot, Johan Parra, Marion Provost, Jean-Baptiste Puel, Jean Rousset, Juan Pablo Medina Flechas, Camille Bainier, Dounya Barrit, Jordi Badosa, Jorge Posada

Ultrafast Dynamics of Photoexcited Carriers and Phonons in Tailored 1D Acoustic Phonon
Potentials 271
Muhammad Hanif, Stephen Bremner, Michael P. Nielsen, Milos Dubajic, Gavin J. Conibeer

Improving PbS Colloidal Quantum Dot Solar Cell Performance Via Solution-Phase Engineering 272
Dhanvini Gudi, Arlene Chiu, Dana Kachman, Eric Rong, Serene Kamal, Yucheng Lan, Susanna M. Thon

Understanding and Advancing Bifacial Thin Film Solar Cells Under Dual Illumination 275
Adam B. Phillips, Jared D. Friedl, Zhaoning Song, Abasi Abudulimu, Ebin Bastola, Deng-Bing Li, Yanfa Yan, Randy J. Ellingson, Michael J. Heben

Spray-Assisted Passivation Strategy for Highly Efficient and Stable Perovskite Solar Cells 276
Rishabh Sahani, Neetesh Kumar, Cheng-Yu Lai, Daniela R. Radu

Approaching 19% Efficiency in (InxGa(1-X))2O3/CdSe/CdTe Solar Cells with Improved Front &
Back Interfaces 279
Manoj K. Jamarkattel, Adam B. Phillips, Ebin Bastola, Sabin Neupane, Deng-Bing Li, Abasi Abudulimu, Jared D. Friedl, Tamanna Mariam, Yanfa Yan, Randy J. Ellingson, Michael J. Heben

Short-Circuit Current Density Chasing and Breakthroughs in High Efficiency Silicon
Heterojunction Solar Cells 280
Weiyuan Duan, Karsten Bittkau, Andreas Lambertz, Kaining Ding

Benefits of Surface Engineered Silicon Quantum Dots in Formamidinium Lead Iodide Perovskite
Solar Cells 281
Vladimir Svrcek, Calum McDonald, Dilli Babu Padmanaban, Ruairi McGlynn, Ankur Kambley, Bruno Alessi, Davide Mariotti, Takuya Matsui

Depth Profiling of Glass/POE/Transparent Backsheet Degradation for Bifacial Photovoltaics 282
Xiaohong Gu, Ashlee R. Aiello, Stefan Mitterhofer, Soshana Smith, Stephanie L. Moffitt, Lakesha N. Perry, Song-Syun Jhang, Stephanie S. Watson, Li-Piin Sung

Temporal Downscaling of GHI Clear-Sky Indices Using T-Copula.. 285
Jing Huang, Marc Perez

Extended FF-VOC Parameterization for Silicon Solar Cells 288
Karsten Bothe, David Hinken, Rolf Brendel

Towards Commercialisation with Lightweight, Flexible Perovskite Solar Cells for Residential
Photovoltaics 289
Philippe Holzhey, Michael Prettl, Silvia Collavini, Nathan Chang, Michael Saliba

Thermal Stability of BiI3 Thin Films 290
Natália F. Coutinho, Thais Crestani, Otávio J. De Oliveira, M. M. M. Modesto Ana Paula De, Marcelo Villalva, Francisco C. Marques

In-Situ Smoothing of Facets on Spalled GaAs(100) Substrates During OMPVE Growth of III-V Solar Cells 291
  William E. McMahon, Anna K. Braun, Allison N. Perna, Pablo G. Coll, Kevin L. Schulte, Jacob T. Boyer, Anica N. Neumann, John F. Geisz, Emily L. Warren, Aaron J. Ptak, Arno P. Merkle, Mariana I. Bertoni, I. Bertoni, Corinne E. Packard, Myles A. Steiner

Numerical Simulation Study for Analysis of Hydrogenated Amorphous Silicon/Crystalline Silicon Heterostructure by Reactive Molecular Dynamics Method 292
  Kazuma Inoue, Naoya Uene, Kazuhiro Gotoh, Yasuyoshi Kurokawa, Takashi Tokumasu, Noritaka Usami

Planned Field Test of Soiling and Irradiance Measurement Uncertainties in Bifacial PV Systems Using In-Situ Reference Modules 295
  Michael Gostein, Audrey Marquis, Marine Bila, Robert Campbell

Intermediate-Phase Engineering Via Dimethylammonium Cation Additive for Stable Perovskite Solar Cells 298
  David P. McMeekin, Philippe Holzhey, Sebastian O. Fürer, Steven P. Harvey, Laura T. Schelhas, James M. Ball, Suhas Mahesh, Seongrok Seo, Nicholas Hawkins, Jianfeng Lu, Michael B. Johnston, Joseph J. Berry, Udo Bach, Henry J. Snaith

A Techno-Economic Analysis of Various Grid-Connected Photovoltaic System Configurations for Green Hydrogen Production 299
  Rahul R Urs, Assia Chadly, Ahmad Mayyas

Multi-Layer Dense Antireflection Coatings 302
  Yiyu Zeng, Martin Green, Jessica Yajie Jiang

Transparent Conductive Oxide Bi-Layer as Front Contact for Multijunction Thin-Film Silicon Solar Cells 305
  Federica Saitta, Prashand Kalpoe, Govind Padmakumar, Paula Perez-Rodriguez, Gianluca Limodio, Rudi Santbergen, Arno H. M. Smets

12.3% Efficient Lifted-Off and Reconstructed As-Doped CdTe Thin Film Solar Cell 308
  Ochai Oklobia, Deborah L. McGott, Giray Kartopu, Steve Jones, Stuart J. C. Irvine

Design, Fabrication, Test, and Flight Performance of the Parker Solar Probe Solar Array 311
  Edward Gaddy, Andrew Gerger, Lew Roufberg, Richard Stall, Matthew J. Schurman

Investigating Electric Field and Light Induced Degradation in Perovskite Solar Cells Through Nanometer-Scale Potential Imaging 315
  G. Paul, J. W. Schall, C.-S. Jiang, A. Louks, A. Palmstrom, N. S. Dutta, S. Johnston, H. Guthrey, A. Norman, M. M. Al-Jassim, D. B. Sulas-Kern

2D-MoS$_2$ Nano Structures to Enhance Silicon Solar Cells 321
  Muntaser Abdelrahman Almansoori, Ayman Rezk, Ammar Nayfeh

Effect of Bidentate Ligand Additive in Tin Perovskite Solar Cells 324
  Dhruba B. Khadka, Yasuhiro Shirai, Masatoshi Yanagida, Kenjiro Miyano

Optoelectronic Performance of Solution Processable MoS$_2$ for Application in Photovoltaic Devices 327
  Dayanand Kumar, Ayman Rizk, Ammar Nayfeh, Nazek El-Atab

Mechanical Degradation of Perovskite Thin Films for Photovoltaics: In-Situ Microscopy & Digital Twin Modeling 330
  Melissa A Davis, Mehul Tank, Michelena O'Rourke, Matthew Wadsworth, Zhibin Yu, Rebekah Sweat

Optimal Row Spacing for Monofacial and Bifacial Fixed-Tilt and Tracked Photovoltaic Systems Up to 75°N 331

Erin M. Tonita, Annie C. J. Russell, Christopher E. Valdivia, Karin Hinzer

Instability of Non-Fullerene Acceptors Used in Organic Solar Cells 332

Yongxi Li, Tonghui Wang, Aram Amassian, Stephen Forrest

Damp Heat Performance of Silicon Heterojunction Solar Cells with Reactive Silver Ink Metallization 333

Michael W Martinez-Szewczyk, Steven Digregorio, Subbarao Raikar, Owen Hildreth, Mariana Bertoni

Parametric Analysis of Photovoltaic Inverters Under Balanced and Unbalanced Voltage Phase Angle Jump Conditions 334

Rachid Darbali-Zamora, Jay Johnson, Matthew J. Reno

Chemical Reaction Kinetics of the Decomposition of Low Bandgap Tin-Lead Halide Perovskite Films and the Effect on the Ambipolar Diffusion Length 340

Yuhuan Meng, Preetham P. Sunkari, Marina Meila, Hugh W. Hillhouse

Demonstration of Thermoradiative Energy Conversion with InAs Cells 341

Eric J Tervo, Andrew J Ferguson, Jennifer Selvidge, Myles A Steiner, Ryan M France

InGaP/GaAs/In$_{0.35}$Ga$_{0.65}$As//In$_{0.53}$Ga$_{0.47}$As Four-Junction Solar Cells Integrated by Surface Activated Wafer Bonding 342

Kentaroh Watanabe, Takashi Shimasaki, Hassanet Sodabanlu, Yoshiaki Nakano, Masakazu Sugiyama

Efficiency Limits for Multi-Junction Coloured Photovoltaics 345

Phoebe M. Pearce, Janne Halme, Jessica Yajie Jiang, Farid Elsehrawy, Nicholas J. Ekins-Daukes

Microscale, High Aspect Ratio, Effectively Transparent Contacts (ETCs) Fabricated with String Printing 348

Mathis Van De Voorde, Janis A. Andersons, Rebecca Saive

Optimization of Bulk Heterojunction Layer Constituents in Organic Photovoltaic Device 351

Vilko Mandic, Dragana Vuk, Radovanovic-Peric Floren, Ivana Panzic

Towards Smart Integration of Cu-Plating for Silver-Free and Edge Passivated SHJ Shingle Modules 352

Samuel Harrison, Vincent Barth, Benoit Martel, Agata Lachowicz, Nicola Frasson, Marco Galiazzo

A Comparison of Emerging and Industry Benchmark Photovoltaic Backsheets Between Different Outdoor Locations 353

Elizabeth Palmiotti, Bruce King, Rachael Arnold, Sona Ulicná, Laura T. Schelhas, David C. Miller

Considering the Variability of Soiling in Long-Term PV Performance Forecasting 356

Matthew Muller, Faisal Rashed

Electron Paramagnetic Resonance Investigation of the Defect Responsible for Light- And Elevated-Temperature-Induced Degradation in Ga-Doped Czochralski Si 359

Chirag Mule, P. Craig Taylor, Abigail Meyer, William Nemeth, Vincenzo Lasalvia, Matthew Page, Sumit Agarwal, Pauls Strandins

Real-Time Regional PV Spinning Reserve Estimator with AGC Look-Ahead Windows 360
Mengmeng Cai, Govind Saraswat, Vahan Gevorgian

Investigation of Sputtered P-Type Electrical Contacts for Thin Film Cadmium Telluride-Based
Solar Cells 363
Blake Hill, Forrest Khulmann, Mayank Mate, Amit H. Munshi

Perovskite Bafacial Modules-Efficiency, Stability and Upscaling 366
Hangyu Gu, Jinsong Huang

Modular, Array-Mounted Photovoltaic Inspection Robot 367
Michael Y. Vazquez Nieves, Alanis M. Colón González, Jennifer L. Braid

ESSPI as a Fast Tool for Load Prioritization on Microgrids Design 370
Luis Colomba-Colon, Natanael Batista-Alvarez, Guillermo Lopez-Cardalda, Eduardo Ortiz-Rivera

Investigations on Absorber Type and Junction Position of GaAs Solar Cells 374
Gan Li, Hassanet Sodabanlu, Meita Asami, Kentaroh Watanabe, Masakazu Sugiyama, Yoshiaki Nakano

Transparent Tedlar® Frontsheet for Lightweight PV Module Designs 377
Hongjie Hu, Stela Chen, Oakland Fu, Michael Demko, Kaushik Roy Choudhury

Photovoltaic Design Projects Increase ECE Student Engagement 380
Devin C. Whalen, Peter Mark Jansson, Milton G. Newberry

Long-Term Degradation Rate of Photovoltaic Modules: A Meta-Analysis 385
Michael Straub-Mueck, Jerome Geyer-Klingeberg, Andreas Rathgeber

The Use of a Physics-Based DNI Model to Enhance the National Solar Radiation Database
(NSRDB) 386
Yu Xie, Jaemo Yang, Manajit Sengupta, Yangang Liu

Novel Module Architecture for Lower CapEx and Improved Recyclability for c-Si PV Modules 387
Ryan Ruhle, Larry Maple, Timothy Delazzer, Steve Johnston, Dana Kern, Walajabad Sampath

Snow Sensing for Photovoltaic Single Axis Tracker Systems 391
Ayush Chutani, Ana Dyreson, Laurie Burnham, Kyumin Lee

Operando Temperature Measurements of Photovoltaic Laser Power Converter Devices Under
Continuous High-Intensity Illumination 394
John F. Geisz, Daniel J. Friedman, Myles A. Steiner, Ryan M. France, Tao Song

Optimization of Back-Contact Diffusion Barrier for Solution-Processed CIGS Solar Cells: Case of
MoO_3 and MoN 395
Nada Benhaddou, Jacques Kenyon, Luke O. Jones, Liam M. Welch, Jake W. Bowers

Silicon Heterojunction Cell Metallization with Reactive Silver Inks: Printing Process, Ink Formula,
and Interconnection 400
Steven J. Digregorio, Michael W. Martinez-Szewczyk, Subbarao Raikar, Mariana I. Bertoni, Owen J. Hildreth

Optical Properties of (InxGal-X)2O3 Alloys and Evaluation as Emitter Layer in CST PV 401
Bishal Shrestha, Madan K. Mainali, Manoj K. Jamarkattel, Ebin Bastola, Adam B. Phillips, Michael J. Heben, Nikolas J. Podraza

Temperature Dependent Fill Factor in CdSe/CdTe PV Devices from -20°C to 60°C Under AM1.5G and AM0 Spectra 402

Nadeesha Katakumbura, Prabodika N. Kaluarachchi, Manoj Rajakaruna, Tyler Brau, Aesha P. Patel, Abudulimu Abasi, Ebin Bastola, David Raker, Adam B. Phillips, Michael J. Heben, Sorin Cioc, Randy J. Ellingson

Module Reliability in Winter: Field Analysis of Deflection and Cell Cracking Across Multiple Module Architectures 409

Laurie Burnham, Daniel Riley, Bruce King, William Snyder, Kevin Santistevan, Paul W. Dice

How Useful is a Field-Operable I-V Curve Tracer? 412

Alexander Cimaroli

Results of First Long Duration Space Flight of Hybrid Perovskite Thin Film 415

Lyndsey McMillon-Brown, William Delmas, Samuel Erickson, Jorge Arteaga, Mark Woodall, Michael Scheibner, Timothy Krause, Kyle Crowley, Kaitlyn Vansant, Joseph Luther, Jennifer Williams, Jeremiah McNatt, Sayantani Ghosh

Undergraduate Research Experience in the Design and Construction of a Photovoltaic Inspection Robot 416

Alanis M. Colón, Emmanuel J. González, Fernando J. Vargas, Samuel I. Hernandez, Michael Y. Vazquez, Eduarto I. Ortiz

Development of Gradient Layers to Improve the Efficiency of Transparent Passivating Contact Solar Cells 419

Alexander Eberst, Binbin Xu, Weiyuan Duan, Andreas Lambertz, Uwe Rau, Kaining Ding

A GIS-Based Approach for Prioritization of Photovoltaic Systems with Energy Storage Implementation for Vulnerable Community Resilience 420

Javier A. Moscoso-Cabrera, Edgar E. Cruz, Cristian R. Meléndez, Eduardo I. Ortiz

Position Dependence of the Performance Gain by Selective Ground Albedo Enhancement for Bifacial Installations 426

Nils-Peter Harder, Issam Smaine, Fadi Bourarach, Damien Cosme, Ines Arfaoui, Julien Chapon, Arttu Tuomiranta, Antonios Florakis

Microscopic Origins of Performance Losses in (Ag,Cu)(In,Ga)Se$_2$ Thin-Film Solar Cells 432

Sinju Thomas, Wolfram Witte, Dimitrios Hariskos, Rico Gutzler, Stefan Paetel, Chang-Yun Song, Heiko Kempa, Matthias Maiberg, Daniel Abou-Ras

Overview of Engineered Germanium Substrate Development for Affordable Large-Volume Multi-Junction Solar Cells 434

Jinyoun Cho, Valérie Depauw, Alexandre Chapotot, Waldemar Schreiber, Tadeáš Hanuš, Nicolas Paupy, Valentin Daniel, Guillaume Courtois, Bouraoui Ilahi, Abderraouf Boucherif, Clement Porret, Roger Loo, Jens Ohlmann, Stefan Janz, Kristof Dessein

X-RAYS Meet Neutrons Meet Ions Meet Electrons Meet Lasers Meet Magnets: Combined Access to Multiple Facilities Through EU Project "Remade@ARI" 435

Michael E. Stuckelberger, Christina Ossig, Barbara Schramm, Stefan Facsko

BIPV Market Potential Analysis with Building Shadow Simulation 436

Changyeol Yun, Myeongchan Oh, Boyoung Kim, Jehyun Lee, Hyungoo Kim, Deokoh Lim, Sangmin Jo

An Investigation on the Pollen-Induced Soiling Losses in Utility-Scale PV Plants 437

João Gabriel Bessal, Michael Valerino, Matthew Muller, Mike Bergin, Leonardo Micheli, Florencia Almonacid, Eduardo F. Fernández

The Role of PbS QDs on Strain and Optical Properties in Different Perovskite Matrix 441
  Sofia Masi, Patricio Serafini, Iván Mora-Seró

Zr-Doped In2O3 Film for the Interlayer of Perovskite/Crystalline Silicon Tandem Solar Cells 442
  Tappei Nishihara, Hyunju Lee, Ryuji Kaneko, Yoshio Ohshita, Atsushi Wakamiya, Atsushi
  Masuda, Atsushi Ogura

Design with Luminescent Solar Concentrator Photovoltaics in the Built Environment 443
  Eli Shirazi, Wouter Eggink, Angele Reinders

Analysis of Solar Cell Electroluminescence Spectra for Daylight Inspection of c-Si PV Modules 448
  Gisele A. Dos Reis Benatto, Alejandra A. Mayordomo, Rodrigo Del Prado Santamaria,
  Thøger Kari, Peter B. Poulsen, Sergiu V. Spataru

Artificial Neural Network and Peer-To-Peer Communications at the Grid-Edge to Mitigate Cyber
Attacks on Distributed Photovoltaic Inverters 455
  C. Birk Jones, Rachid Darbali-Zamora

Improvement of Radiation Tolerance in Solar Cells by Hetero P/N Junction Structure 463
  Tetsuya Nakamura, Mitsuru Imaizumi, Meita Asami, Masakazu Sugiyama, Hidefumi Akiyama,
  Shin-Ichiro Sato, Takeshi Oshima, Yoshitaka Okada

Abnormal Responses of Residential Smart Photovoltaic Inverters to Cyberattacks 464
  Thunchanok Kaewnukultorn, Sergio B. Sepúlveda-Mora, Steven Hegedus

Community Influence of Houses of Worship on Rooftop Solar Growth Rates 467
  Ashley Degen, Laura Mogannam, Nisitaa Karen Clement Pradeep, Jillian Stern, Deborah A.
  Sunter

Decoupling Open-Circuit Voltage and Series Resistance in Electroluminescence Images Through
Deep Learning 472
  Gaia Maria N. Javier, Priya Dwivedi, Thorsten Trupke, Ziv Hameiri

Decomposition Mechanisms and Kinetics of Perovskite Semiconductors 473
  Hugh Hillhouse, Yuhuan Meng, Spencer Cira, Preetham Sunkari

Development and Evaluation of Typical Plane of Array Year (TPY) for Solar Energy Systems Over
the Americas 474
  Aron Habte, Manajit Sengupta, Grant Buster, Yu Xie

10-Junction Edge-Illuminated Passivated-Contact Silicon Minimodules for Laser Power
Conversion 475
  Ryan M France, Matthew B Hartenstein, William Nemeth, San Theingi, Matthew Page, Sumit
  Agarwal, David Young, Paul Stradins

GaAs Betavoltaic Cell Modeling for Light to Medium Element Radiation Conversion into
Electrical Power 476
  Mathieu De Lafontaine, Gavin Forcade, Paige Wilson, Jayeshkumar Patel, Brian Ellis,
  Helmut Fritzsche, Sam Suppiah, John P. D. Cook, Christopher E. Valdivia, Karin Hinzer

Residential Electric Energy Storage System to Reduce Voltage and Thermal Violations in
Distribution Lines and Increase PV Integration 477
  Anny Huaman-Rivera, Agustin Irizarry-Rivera

Air-Bridge Cells for Higher Emission Temperatures 482
  Bosun Roy-Layinde, Areefa Rahman, Jihun Lim, Sritoma Paul, Stephen R. Forrest, Andrej
  Lenert

On the Accuracy of Spectral Adjustment for Performance Measurements of Multijunction Solar Cells...... 485
Nikos Kopidakis, Tao Song, John Geisz, Daniel Friedman

Spatially Resolved Degradation Analysis of Solar Modules After Combined Accelerated Aging 486
Robert Heidrich, Anton Mordvinkin, Ralph Gottschalg

NiO as a P-Type TCO for Inorganic Thin-Film Photovoltaics 489
Elline C. Hettiaratchy, Angus A. Rockett, Taylor D. Hill, Sachit Grover

Modeled Impacts of Solar Forecast Error on Utility Production Cost........ 492
William B. Hobbs, Jenner Tresan, Michael Kline, Mousumi Guha, Brent Duncan

Durability Testing of Porous SiO_2 Anti-Reflection Coatings for Solar Cover Glass........ 498
Adam M Law, Luke O Jones, Michael Nasser, Ali Abbas, John M Walls

Performance Evaluation of Perovskite Solar Cell Modules with Tilt Angle Optimization in BIPV Application: A Case Study for Kazakhstan 502
Yerassyl Olzhabay, Ikechi Ukaegbu, Annie Ng

CFD-Based Machine Learning Model for Agrivoltaic System Design 505
Henry J. Williams, Emily Weed, Khaled Hashad, K. Max Zhang

Towards Highly Stable Metal-Halide Perovskite Materials for a Broad Range of Applications: Film Growth, Degradation Control, and Interfacial Engineering........ 508
Lissette Rodriguez-Cabanas, Calvin Duong, Bradley Stanley, Andre Slonopas

First-Principles Study of Energy Band Alignment in Pristine $CdTe/TeO_2/Te$ Interfaces........ 509
Anthony P. Nicholson, James R. Sites, Walajabad S. Sampath

Horizon Profiling Methods for Photovoltaic Arrays........ 512
Jennifer L. Braid, Benjamin G. Pierce

Field Effect Passivation Enables 2.2 V Open-Circuit Voltage All-Perovskite Tandems........ 518
Bin Chen, Ted Sargent

Distributed Generation Component Placement and Point of Common Coupling Allocation for Solar Rooftop Microgrid Sizing Costs Minimization 519
Robert A. García Cooper, Marcel Castro Sitiriche, Agustín Irizarry Rivera, Fabio Andrade Rengifo

The Solar Boat: An Academic Research Experience........ 525
Guillermo Serrano, Erick Aponte, Eduardo I. Ortiz-Rivera

Evaluating the Use of Satellite Data and Machine Learning Models for PV Performance Monitoring........ 528
Daniel Fregosi, Rabin Dhakal, Devin Widrick

Drying Effects Upon Spin Coating of Solution-Processed Amine-Thiol Thin Film $Cu(In,Ga)(S,Se)_2$ Absorber Fabrication........ 534
Jacques Kenyon, Nada Benhaddou, Liam Welch, Jake Bowers

Impact of Backsheet Versatility on Inverter Availability 540
Claudia Buerhop-Lutz, Oleksandr Stroyuk, Jens Hauch, Ian Marius Peters

Evaluating the Weather Forecasting Models and the Impact to PV Generation Forecasting........ 541
Spyros Theocharides, Anastasios Koumis, George Makrides, George E. Georghiou

Plausibility Filtering of PV Outdoor Data 546
T. S. Vaas, J. Körtgen, E. Sovetkin, U. Rau, B. E. Pieters

The European Solar Communication - Will it Strengthen the Photovoltaic Industry in the European
Union 550
Arnulf Jäger-Waldau, Anatoli Chatzipanagi, Georgia Kakoulaki, Sandor Szábo

Measuring Sustainability of PV in Energy Transition: Mass, Energy, and Circularity 554
Heather M. Mirletz, Silvana Ovaitt, Macarena Mendez Ribo, Seetharaman Sridhar, Teresa M.
Barnes

Photovoltaic Site Architecture Estimation Using Performance Data 555
Steven Koskey, Scott Sheppard, Corson Teasley, Christopher Perullo, Jared Kee, Daniel
Fregosi, Wayne Li

Synthesis and Characterization of Bismuth Selenide and Copper Doped Bismuth Selenide Thin
Films by Chemical Bath Deposition 561
Hamda A. Al-Thani, Shifaa M. Al-Baity, Falah S. Hasoon

CIGS Device Stability: A Comparison of Two Different Process Batches 566
Mohsen Jahandardoost, Curtis Walkons, Marco Nardone, Theresa M. Friedlmeier, Shubhra
Bansal

Hierarchical Variance Analysis of Solar Cell Production Using Machine Learning and Numerical
Simulations 569
Bernhard Klöter, Hannes Wagner-Mohnsen, Sven Wasmer

Identifying the Electrical Signature of Snow in Photovoltaic Inverter Data 572
Emma C. Cooper, Jennifer L. Braid, Laurie M. Burnham

Band Tail Effects on Cd(Se,Te) Device Performance: A Numerical Simulation Approach 577
Eva M Mulloy, Darius Kuciauskas, Craig L Perkins, Marco Nardone

Understanding the Dopability of as in Selenium-Alloyed Cadmium Telluride Solar Cells 578
Xiaofeng Xiang, Aaron Gehrke, Scott Dunham

Modeling Transposition for Single-Axis Trackers Using Terrain-Aware Backtracking Strategies 581
Kurt Rhee

What's New in the NSRDB 587
Manajit Sengupta, Aron Habte, Grant Buster, Yu Xie, Brandon Benton, Michael Foster

Rapid, Contactless Measurements and Performance Predictions of Photovoltaic Materials 590
Brandon T Motes, Anthony T Troupe, Amy E Louks, Axel F Palmstrom, Joseph J Berry, Dane
W Dequilettes

Optimizing the Packing Density of Building Integrated Concentrating Photovoltaic Systems for
Improved Performance and Reduced Embodied Carbon Through a Novel Polygonal Concentrator 591
Lewis Osikibo Tamuno-Ibuomi, Roberto Ramirez-Iniguez, A Sheila Holmes-Smith, Geraint
Bevan

Human Health Risk Assessment for Improper Landfill Disposal of End-Of-Life CdTe PV 594
Elaine Kupets, Garvin Heath

Accelerating Cycles of Learning for Silicon Heterojunction Architectures: Experimental Design and Data-Driven Degradation Pathway Prediction 597
    Xuanji Yu, Diego Zubieta, Mirra Rasmussen, Chien-Hsuan Chen, Cécile Molto, Mariana Bertoni, Kristopher O. Davis, Laura S. Bruckman, Ina T. Martin

Modeling Reference Cell Performance Using Measured and Modeled Spectral Data 600
    Josh Peterson, Frank Vignola, Afshin Andreas, Aron Habte, Manajit Sengupta

Silver Reflector-Driven Light Harvesting Enhancement in Large Area Dye Sensitized: Solar Cells 606
    Navdeep Kaur, Faizan Syed, Cheng-Yu Lai, Daniela Radu

Survey of Snow Impacts on Bifacial Gain in Commercial Photovoltaic Arrays 607
    Samantha S. Wilson, Stephen Lightfoote, Stephen Voss

New Theoretical Limits for Light Trapping in Solar Cells 610
    Stéphane Collin, Maxime Giteau

Supply Side Management with Agrivoltaics: Feasibility Study of Modeling Methodologies of Solar PV and Crop Response 611
    Tadatoshi Takahashi

Effect of Solar Mounting Configurations on California Zero-Carbon Grid 617
    Zabir Mahmud, Sarah Kurtz

Experimental Demonstration of Diffused Light Collimation in Free Space 622
    Lisanne M. Einhaus, Geert C. Heres, Jelle Westerhof, Shweta Pal, Akshay Kumar, Anne Rikhof, Jian-Yao Zheng, Rebecca Saive

The Planet-Scale Performance Potential of Si-Perovskite Tandem Solar Farms 626
    Jabir Bin Jahangir, M. Tahir Patel, Reza Asadpour, M. Ryyan Khan, M. A. Alam

Investigating the Potential of Hydrogen Plasma Treated ALD-TiOx Films as Hole-Selective Passivating Contacts in Crystalline Silicon Solar Cells 629
    Chien-Hsuan Chen, S. Novia Berriel, Taylor M. Currie, Jannatul Ferdous Mousumi, Titel Jurca, Parag Banerjee, Kristopher O. Davis

The Photovoltaic Exponential Model 630
    Eduardo I. Ortiz Rivera

On the Unappreciated Impact of Se in as-Doped CdSexTe1-X 633
    Deborah L McGott, Darius Kuciauskas, Craig L Perkins, Eric Colegrove, Matthew O. Reese

Leveraging High-Fidelity Sensor Data for Inverter Diagnostics: A Data-Driven Model Using High-Temperature Accelerated Life Testing Data 634
    Sakir Karakaya, Murat Yildirim, Shijia Zhao, Feng Qiu, Jack David Flicker, Benjamin Peters, Zhaoyu Wang

Influence of Interfaces on Stability of Perovskite Solar Cells 641
    Arkadi Akopian, Saba Sharikadze, Ranjith Kottokkaran, Vikram Dalal

A New Route to Facilitate Scaling of Lead-Tin Halide Perovskites: Thin Films Via Solvent Self-Volatilization 644
    Jack R. Palmer, Apoorva Gupta, Sean P. Dunfield, David P. Fenning

Aerial Photoluminescence Imaging of PV Modules 645
    Bernd Doll, Ernst Wittmann, Larry Lüer, Claudia Buerhop-Lutz, Jens A. Hauch, Christoph J. Brabec, Ian Marius Peters

Polarization Type Potential Induced Degradation Under Positive Bias in a Commercial PERC Module 648

    Farrukh Ibne Mahmood, Fang Li, Peter Hacke, Cécile Molto, Hubert Siegneur, Govindasamy Tamizhmani

Three-Dimensional and Multimodal X-Ray Microscopy Reveals the Impact of Voids in CIGS Solar Cells 651

    Giovanni Fevola, Christina Ossig, Mariana Verezhak, Jan Garrevoet, Martin Seyrich, Dennis Brueckner, Johannes Hagemann, Frank Seiboth, Andreas Schropp, Gerald Falkenberg, Peter Stanley Jorgensen, Christian Strelow, Tobias Kipp, Romain Carron, Jens Wenzel Andreasen, Michael E. Stuckelberger

Optical Characterization and Loss Simulation of Encapsulation Materials and Back Sheets for PERC+ Solar Modules 652

    Tim Lukas Brockmann, Henning Schulte-Huxel, Susanne Blankemeyer, Tobias Wietler

Integration of Lateral Power MOSFETs into IBC c-Si Solar Cells with Poly-Si Passivating Contacts 655

    David A. Van Nijen, Patrizio Manganiello, Yavuzhan Mercimek, Tristan Stevens, Guangtao Yang, René A. C. M. M. Van Swaaij, Miro Zeman, Olindo Isabella

The Formation of Dendrites in Overtreated CdSeTe/CdTe Solar Cells 656

    Vladislav Kornienko, Luke Jones, Kieran Curson, Ali Abbas, Rachael Greenhalgh, Yau Yau Tse, Michael Walls, Christian Drost, Bettina Spaeth, Bastian Siepchen

Carrier Dynamics in $Al_xGa_{1-x}As$/InAs-Based Photon Up-Conversion Solar Cells with a Doubled-Heterointerface 659

    Hambalee Mahamu, Shigeo Asahi, Takashi Kita

The Importance of Terrain-Shading Losses in PV Yield Assessment: The Case of Oahu 662

    Marc Perez, Upama Nakarmi, Philip Gruenhagen, Richard Perez

Utilizing PSO Technique for Locational-Dependent Feeder PV Hosting Capacity Evaluation 665

    Vinushika Panchalogaranjan, Paul Moses

Cryogenic Operation of GaAs Laser Power Converters 670

    Bora Kim, Mijung Kim, Brian D. Li, Ryan D. Hool, Minjoo L. Lee

Characterization of Field Exposed Photovoltaic Modules Featuring Signs of Contact Degradation 671

    Max Liggett, Dylan J. Colvin, Balaashwin Babu, William C. Oltjen, Xuanji Yu, Manjunath Matam, Hubert P. Seigneur, Mengjie Li, Andrew M. Gabor, Philip J. Knodle, Craig J. Neal, Sudipta Seal, Laura S. Bruckman, Roger H. French, Kristopher O. Davis

Energy Management in a Dynamic Microgrid Using Genetic Algorithms 674

    Ricardo Calloquispe-Huallpa, Rachid Darbali-Zamora, Erick E. Aponte-Bezares, Matthew S. Lave

Sn-Based Perovskite Thin Film Solar Cells with Enhanced Stability 682

    Wendy Reyes Ramos, Zeying Chen, Adam Thomas, Tara Dhakal

Cadmium Selenide (CdSe) as an Active Absorber Layer for Solar Cells with V_{OC} Approaching 750 mV 683

    Ebin Bastola, Adam B. Phillips, Abasi Abudulimu, Vlad Kornienko, Zulkifl Hussain, Manoj K. Jamarkattel, Tamanna Mariam, Prabodika N. Kaluarachchi, Jared Friedl, Dipendra Pokhrel, Kara B. Kile, Zhaoning Song, Yanfa Yan, Michael Walls, Randy J. Ellingson, Michael J. Heben

Mapping Stress in PV Modules: The Influence of Soldering, Tabbing, and Module Architecture 689
    Ian Slauch, Rico Meier, Kaushik Roy Choudhury, Mariana Bertoni

Elemental Vapor Transport Deposition of $CdSe_xTe_{1-x}$ Thin Films for n-Type CdTe Solar Cells 690
    Wei Wang, Vasilios Palekis, Md Zahangir Alom, Sheikh Elahi Tawsif, Don Morel, Chris Ferekides

Intra-Grain Local Luminescence Properties of $CdSe_{0.1}Te_{0.9}$ Thin Films 695
    Ganga R. Neupane, David S. Albin, Joel N. Duenow, Matthew O. Reese, Susanna Thon, Behrang H. Hamadani

Electroluminescence Imaging: A Study in the Impact of Microscopic Surface Defects 698
    Meghan E. Bush, Timothy J. Peshek

Towards a Three-Terminal Perovskite/Silicon Tandem Solar Cell with Highest Efficiency 701
    Michael Rienäcker, Somayeh Moghadamzadeh, Paul Fassl, Yevgeniya Larionova, Philipp Noack, Bianca Wattenberg, Ulrich Wilhelm Paetzold, Robby Peibst

Luminescence and Thermal Imaging Applied to Half-Cut-Cell and Emitter-Wrap-Through-Cell Modules 702
    Steve Johnston, Dana B. Kern, Kent Terwilliger, Ingrid L. Repins

Role of Solar Photovoltaics for a Sustainable Energy System in Puerto Rico in the Context of the Entire Caribbean Featuring the Value of Offshore Floating Systems 708
    Christian Breyer, Ayobami S Oyewo, Alejandro Kunkar, Rasul Satymov

Energy-Harvesting Efficiency Analysis for Solar Modules Using 2T and 4T Tandem Solar Cells.................. 709
    Robert Witteck, William E. McMahon, John F. Geisz, Qi Jiang, Emily L. Warren

A Reproducible Validation of Algorithms for Estimating Array Tilt and Azimuth from Photovoltaic Power Time Series........................ 710
    Kirsten Perry, Bennet Meyers, Kevin Anderson, Matthew Muller

Investigation of P-Type Silicon Heterojunction Radiation Hardness 716
    Romain Cariou, Adrien Danel, Nicolas Enjalbert, Frédéric Jay, Sébastien Dubois

Energy Injustice Metrics for Puerto Rico 717
    Pablo Méndez-Curbelo, Carlos Peña, Willian Pacheco, Marcel Castro-Sitiriche

Inverse Design of Spectrally-Selective Films for PbS-CQD Tandem Solar Cells........................ 718
    Sreyas Chintapalli, Tina Gao, Luna Singh, Serene Kamal, Arlene Chiu, Yijun Zhang, Susanna M. Thon

AM0 Optimized Dual Junction Quantum Well Solar Cells-Investigation of Radiation Tolerance Designs and V_{OC} Retention at EOL 721
    Brandon M. Bogner, Stephen J. Polly, Seth M. Hubbard, Mitsuru Imaizumi, Roger E. Welser

Evaluation of Process Damage to Crystalline Silicon by Transparent Conductive Oxide Film Deposition 724
    Haruki Kojima, Tappei Nishihara, Yuta Ito, Hyunju Lee, Kazuhiro Gotoh, Noritaka Usami, Tomohiko Hara, Kyotaro Nakamura, Yoshio Ohshita, Atsushi Ogura

Brownfields to Brightfields: The Potential for Landfill Solar Redevelopment in New York State 727
    Henry J. Williams, Hugh Peng, Olivia Goosay, K. Max Zhang

Carbon Footprint of Silicon Photovoltaics Manufacturing in North America................................ 730
    Annick Anctil, Angela Farina, Luyao Yuan

The PV Efficiency Vs R&D Effort Learning Curve for Research-Stage Material Technologies 731
Phillip J. Dale, Michael A. Scarpulla

Perovskite/Silicon Tandem Solar Cells with Front Side Metallization Applied Prior to Top Cell
Fabrication Enabling High Curing Temperatures 733
Sara Baumann, Annika Raugewitz, Felix Haase, Tobias Wietler, Robby Peibst, Marc Köntges

Performance and Degradation in Silicon PV Systems Under Outdoor Conditions in Relation to
Reliability Aspects of Silicon PV Modules – Summary of Results of COST Action PEARL PV 737
*S. Lindig, J. Ascencio-Vásquez, J. Leloux, D. Moser, M. Aghaei, A. Fairbrother, A. Gok, S.
Ahmad, S. Kazim, K. Lobato, W. J. G. H. M. Van Sark, N. Pearsall, B. G. Burduhos, A.
Raghoebarsing, G. Oreski, J. Schmitz, M. Theelen, P. Yilmaz, J. Kettle, A. H. M. E. Reinders*

Progress Towards Scaling Perovskite/Silicon Tandem Modules 740
*Chris Eberspacher, Colin Bailie, Tim Gehan, Bryan Rosales, Tom Brenner, Mike Chen, Terry
Banks*

Identification of Module Replacements in US Utility-Scale Photovoltaic Installations 743
Chenyang Deng, Jacob T. Stid, Preeti Nain, Annick Anctil

Indoor and Outdoor Evaluation of Curved Modules for VIPV 746
*Rebeca Herrero, Ignacio Antón, Francisco Martín, Steve Askins, Javier Macías, Luis J. San
José, G. Vallerotto, R. Núñez, C. Domínguez*

Towards an Annual Terrawatt Photovoltaics Market - Comparison of the Social Acceptance in
Various IEA PVPS Countries 749
*Arnulf Jäger-Waldau, Georg Altenhöfer-Pflaum, Otto Bernsen, Christian Breyer, Jose
Donoso, Hubert Fechner, Kenn Henrik Bournonville, Jarand Hole, Izumi Kaizuka, Linda
Koschier, Johan Lindahl, Gaëtan Masson, Daniel Mugnier, Chinho Park, Lionel Perret,
Francesca Tilli*

Potential Induced Degradation Evaluation of Damp Heat Stressed PV Modules 755
Farrukh Ibne Mahmood, Akash Kumar, Muhammad Afridi, Govindasamy Tamizhmani

Data-Driven Photovoltaic Module Performance Analysis with FAIR Data 758
*Mengjie Li, Jarod Kaltenbaugh, Dylan J. Colvin, William C. Oltjen, Arafath Nihar,
Dominique Akissi Yao, Xuanji Yu, Alp Sehirlioglu, Roger H. French, Kristopher O. Davis*

Proposed Update of the Colombian Technical Standard NTC 4405 for Evaluating the Efficiency of
Photovoltaic Solar Systems and Their Components 761
Johann Hernández, Daniel H. Gamboa, Juan F. Beltrán

Aerosol-Deposited SnOx as an Electron Contact in Perovskite Solar Cells 767
David Quispe, David Matthews, Zhengshan J. Yu, Zachary C. Holman

The Temperature Dependence of Auger Recombination in Silicon 768
Jorge Ochoa, Simone Bernardini, Mariana Bertoni

Combining Perovskites and Quantum Dots: Application in Solar Cells 769
*Jose Raul Montes Bojorquez, Maria Fernanda Villa Bracamonte, Kevin J. Knebel, Arturo A.
Ayon*

Nanostructure Analysis of Parasitic Oxides and Contact Resisitivity Degradation During Annealing
of Silicon Heterojunction Solar Cells 770
*Stefan Lange, Angelika Hähnel, Christoph Luderer, David Adner, Martin Bivour, Christian
Hagendorf*

Impact of Selenium Doping in CdSeTe-Based Solar Cells at the Atomic-Scale 771
    Arashdeep S. Thind, Jack Farrell, Robert F. Klie

Performance of Vertical Bifacial 2T and 3T Perovskite/Silicon Tandem Solar Farms 772
    Syed Usama Bin Afzal, Hassan Imran, Suleman Sami Qazi, Muhammad Ashraful Alam,
    Nauman Zafar Butt

Influence of Insertion Position of a LiF Buffer Layer on Passivation Performance of Crystalline
Si/SiO_y /TiO_x/ Al Heterostructures 775
    Shohei Fukaya, Kazuhiro Gotoh, Takuya Matsui, Hitoshi Sai, Yasuyoshi Kurokawa, Noritaka
    Usami

Radiometric Standards and Best Practices: Recent Progress 778
    Aron Habte, Manajit Sengupta, Christian A. Gueymard

Compact and High Efficiency Micro-CPV Module with High Wafer Utilization Rate 781
    Corentin Jouanneau, Thomas Bidaud, Maxime Darnon, Gwenaelle Hamon

Selenium Diffusion During CdCl2 Treatment of CdSeTe Solar Cells 782
    Niranjana Mohan Kumar, Srisuda Rojsatien, Angel De La Rosa, Barry Lai, Arun K. M.
    Kanakkithodi, Maria Chan, Dan Mao, Mariana Bertoni

Analysis of Measured Operating Temperature of Perovskite Modules. 783
    D. Martinez Escobar, Aaron Wheeler, F. Brigham Pineda, Katty Kaydanik, Alan Murphy,
    Jing-Shun Huang, Mason Terry, Sarah R. Kurtz

A Broadband Anti-Reflection Coating for Thin Film CdSeTe/CdTe Solar Cells 789
    Adam M Law, Luksa Kujovic, Mustafa Togay, Xiaolei Liu, Kurt Barth, John M Walls

Can Solar+Storage Keep the Lights On? Assessing Solar+Storage for Backup Power During Long-
Duration Power Interruptions in the US 793
    Will Gorman, Galen Barbose, Juan Pablo Carvallo, Sunhee Baik, Chandler Miller

Close Roof Mounted System Temperature Estimation for Compliance to IEC TS 63126 794
    Michael D. Kempe, Silvana Ovaitt, Martin Springer, Matthew Brown, Dirk Jordan, William
    Sekulic, Colleen O'Brien, Jean-Nicolas Jaubert, Yuanjie Yu, Jaewon Oh, Govindasamy
    Tamizhmani, Bo Li

I-TOPCon Solar Cells Prepared by High Throughput Magnetron Sputtering of In-Situ Doped n-
Type Amorphous Silicon Layers 795
    Eric Schneiderloechner, Tina Dietsch, Jan Hoss, Jonathan Linke, Jana-Isabelle Polzin,
    Sebastian Mack, Henning Nagel, Volker Linss

Optimizing Demand Management to Enable Renewables: Why the Use of a Marginal Emissions
Signal is a Poor Choice 798
    Ronald A. Sinton

Charge Extraction and Recombination Dynamics of CdSe/CdTe Solar Cells Studied with Transient
Photovoltage/Photocurrent Techniques 804
    Abasi Abudulimu, Dengbing Li, Lei Chen, José Santos, Iwan Zimmermann, Nadeesha
    Katakumbura, Tyler Brau, Ebin Bastola, Adam Phillips, Zhaoning Song, Juan Cabanillas,
    Michael Heben, Mohammad K. Nazeeruddin, Nazario Martín, Yanfa Yan, Randy Ellingson

The Feasibility of Luminescent Solar Concentrators Overlays for Conventional Lens 809
    Xitong Zhu, Michael G. Debije, Angèle H. M. E. Reinders

Highly Efficient Bifacial Single Junction Perovskite Solar Cells 812
Qi Jiang, Zhaoning Song, Yanfa Yan, Kai Zhu

Efficiency Maps for Tandem Solar Cells Using High Resolution Spectral Data 815
Rune Strandberg, Anne G. Imenes

Optimal Allocation of Voltage Regulations to Maximize the Hosting Capacity of Distribution
Systems 816
Bahman Ahmadi, Eli Shirazi

In-Situ Microscopy Characterization of Light-Induced Phase Segregation in Wide-Bandgap
Perovskite Materials 822
*Fangfang Cao, Liming Du, Zhiyu Gao, Minghui Li, Cong Chen, Dewei Zhao, Can Li, Zhen
Li, Jichun Ye, Chuanxiao Xiao*

Radiation Tolerance Studies of CdSe/CdTe Bilayer Solar Cells on Space-Qualified Cover Glass 823
*Aesha P. Patel, Adam B. Phillips, Ebin Bastola, Abasi Abudulimu, Zachary W. Zawisza,
Robert Snuggs, Manoj K. Jamarkattel, Deng-Bing Li, Richard Irving, Yanfa Yan, Michael J.
Heben, Randy J. Ellingson*

Advanced Production Line Monitoring with Time-Lag Sequential Analysis 828
Gaia Maria N. Javier, Rhett Evans, Priya Dwivedi, Thorsten Trupke, Ziv Hameiri

NASA GRC Solar Cell Characterization: Facilities 829
Jeremiah D Sims

Application of Noise-Assisted Multivariate Data Analysis for Hour-Ahead GHI Forecasting 830
Priya Gupta, Rhythm Singh

Can Grid-Following DERs Operate in Parallel with Grid-Forming Resources Without
Compromising Microgrid Stability? 835
Wenzong Wang, Aminul Huque

Developing Frequency Stability Constraint for Unit Commitment Problem Considering High
Penetration of Renewables 840
Ningchao Gao, Shuan Dong, Xin Fang, Andy Hoke, David Wenzhong Gao, Jin Tan

Precursor Ink Engineering for Scalable Slot-Die Coating of Perovskite Films for Photovoltaic
Mini-Module Production 844
*Manoj Rajakaruna, Jaehoon Chung, You Li, Tamanna Mariam, Muhammad Saeed Mohsin,
Prabodika N. Kaluarachchi, Amirhossein Rahimi, Zhaoning Song, Michael J. Heben, Yanfa
Yan, Randy J. Ellingson*

GaAs Solar Cells Grown on Acoustically-Spalled GaAs Substrates with 27% Efficiency 848
*Kevin L Schulte, Steve W Johnston, Anna K Braun, Jacob T Boyer, Anica N Neumann, William
E McMahon, Michelle Young, Pablo Guimerá Coll, Mariana I Bertoni, Emily L Warren,
Myles A Steiner*

Why Increased CdSeTe Charge Carrier Lifetimes and Radiative Efficiencies Did Not Result in
Voltage Boost for CdTe Solar Cells? 849
*Darius Kuciauskas, Alexandra Bothwell, Carey Reich, Chungho Lee, Eric Colegrove, Marco
Nardone*

Detection and Impact of Cracks Hidden Near Interconnect Wires in Silicon Solar Cells 850
Andrew M. Gabor, Hubert Seigneur, Philip J. Knodle, Dylan J. Colvin, Kristopher O. Davis

Net Zero Water Strategies and Impacts for PV Manufacturing 855
Parikhit Sinha, Sunil Sajja, Tzy Wei Ooi, Sreenivas Jayaraman, Sukhwant Raju

Characterization of Different Groups of Electricity Consumers and Measures Taken to Reduce
Energy Poverty 858
*Anna Carolina De Paula Sermarini, Lucas Aló Rodrigues Araujo Da Silva, Vanessa Cardoso
De Albuquerque, Rodrigo Flora Calili*

Using Neural Network Decomposition to Estimate Field Photovoltaic Performance Loss Rate 861
*Yangxin Fan, Raymond Wieser, Xuanji Yu, Jennifer Braid, Avishai Shaton, Adam Hoffman,
Thevenard Didier, Ben Spurgeon, Daniel Gibbons, Laura S. Bruckman, Yinghui Wu, Roger H.
French*

Capacitance Transients, Photoconductive Decay, and Impedance Spectroscopy on 19% to 22%
Efficient Silicon Solar Cells 865
Steve Johnston, Dana B. Kern

Development of Next Generation Solar Trackers Based on Shape Memory Alloy to Be Integrated in
CPV/PV Hybrid Modules 871
Alessandro Minuto, Edoardo Celi, Gianluca Timò

Quantifying Uncertainty Due to Climate Variability in Vehicle-Integrated Photovoltaic Yield
Predictions 874
Timofey Golubev

Phase Distributions and Local Bandgap Energies in Mixed-Halide Perovskite Nanoparticles 879
*Dan R. Wargulski, Tal Binyamin, Katrina Coogan, Benedikt Haas, Christoph Koch, Lioz
Etgar, Daniel Abou-Ras*

Third Generation Approaches for Low Cost, Radiation Tolerant, Efficienct Space Solar Cells 881
Gavin Conibeer, Santosh Shrestha

Effect of Novel Optimization Algorithm on the Performance of Photovoltaic Devices 885
Sheri F Michael

Towards Transfer Printing GaSb Membranes Using Selective Etchants 886
Margaret A Stevens, Jill A Nolde, Shawn Mack, Thomas C Mood, Kenneth J. Schmieder

Investigating the Role of Ag and Ga Content in the Stability of Wide-Gap (Ag,Cu)(In,Ga)Se2 Thin
Film Solar Cells 887
Patrick Pearson, Jan Keller, Charlotte Platzer-Björkman

A Horizontal Single-Axis Tracker Mock-Up to Quickly Assess the Influence of Geometrical
Factors on Bifacial Energy Gain 888
César Domínguez, Jesús Marcos, Sandra Ures, Steve Askins, Ignacio Antón

A New Method for the Evaluation of Majority and Minority Carrier Contact Resistivity of
Polysilicon on Oxide Contacts 891
Dirk Steyn, William Nemeth, David Young, Paul Stradins, Sumit Agarwal

Simultaneous Solar Power Generation and Bidirectional Data Transmission 894
*Emily Kessler-Lewis, Stephen J. Polly, Elijah Sacchitella, Seth M. Hubbard, Raymond
Hoheisel*

Localized Surface Plasmon Resonance of Quantum Dots in Two-Step Photon Up-Conversion Solar
Cell Structures 895
Yukihiro Harada, Mizuto Kawakami, Shigeo Asahi, Takashi Kita

Understanding the Degradation of Silicon Heterojunction Modules 896
  Jorge Ochoa, Michael Martinez-Szewczyk, Nicholas Moser-Mancewicz, Dana Kern, Dirk Jordan, Mariana Bertoni

Enhancing Inverter Reliability: Current Status and Paths to Predictive Maintenance 897
  Wayne Li, Rabin Dhakal, Daniel Fregosi, Curtis Fox, Michael Bolen

Improvements on Spectral Correction Predictive Modeling for CdTe Modules 903
  Alan J. Curran, Boris Lin, Yuepeng Deng

Modification of PEDOT:PSS Hole Transporting Layer to Improve the Efficiency and Light
Stability of Tin-Lead Perovskite Solar Cells 909
  Lei Chen, Chongwen Li, Tyler R. Brau, Abasi Abudulimu, Randy J. Ellingson, Zhaoning Song, Yanfa Yan

Improving V_{OC} of CdSe/CdTe Solar Cells Via Incorporating Oxygenated CdS Between Front
Buffer and Absorber 910
  Abasi Abudulimu, Dengbing Li, Manoj Jamarkattel, Zachary Zawisza, Scott Wenner, Tyler Brau, Ebin Bastola, Adam Phillips, Michael Heben, Yanfa Yan, Randy Ellingson

From Accelerated Life Test to Accurate Degradation Prediction of CdTe PV Devices: A Modeling
Approach 914
  Da Guo, Jaliu Ma, Samuel Demtsu, Ryan Monnin, Igor Sankin, Jose A. Calderon, Markus Gloeckler

Quantifying Bulk and Surface Recombination in CdSeTe Absorbers by Modeling Terahertz and
Photoluminescence Decays 915
  Gregory A Manoukian, Calvin Fai, Deborah L McGott, Finley R Shapiro, Matthew O Reese, Charles J Hages, Jason B Baxter

Data Driven Energy Resilience for Low-To Middle-Income Communities in Puerto Rico 916
  Christopher Gregory, Angel Echevarria, Yiamar Rivera-Matos, Daniel D. Campo-Ossa, Alex Routhier, Clark Miller, Richard King

Advances in GaAsP Top Cells for Use in GaAsP/Si Tandems 917
  Tal Kasher, Lauren M. Kalizewski, Marzieh Baan, Chuqi Yi, Anastasia H. Soeriyadi, Atom Chang, Udo Römer, Gianluca Coletti, Stephen P. Bremner, Tyler J. Grassman, Steven A. Ringel

Key Areas of Due Diligence for Solar PV Project Financing 918
  Eric R. Decristofaro, Matthew R. Thibodeau, Fitzgerald C. Okoli, Jake T. Silhavy, Evan S. Giacchino, Adam W. Loeding, Alexander E. Coologeorgen

Convergence of Efficiency, Stability, and Manufacturability in Perovskite Tandem Solar Cells 919
  Rohit Prasanna, Tomas Leijtens, Jochen Titus, Laura E. Crowe, Hyunjong Lee, Annikki L. Santala, Maximilian T. Hoerantner, Giles E. Eperon

Validating View-Factor Approach and Spatial Albedo Models for Bifacial and AgriPV Modeling 920
  Silvana Ovaitt, Matthew Boyd, Austin Kinzer, James Jones, Chris Deline

Field and Accelerated Aging of Cracked Solar Cells 921
  Michael G. Deceglie, Timothy J Silverman, Ethan Young, William B. Hobbs, Cara Libby

Orange Button: Accelerating the Digital Transformation of Distributed Energy 922
  Clifford W. Hansen, Jan Rippingale, Taos Transue, Philip Court, John Gorman

Analysis for Effects of Temperature Rise of Solar Cell Modules Upon the Driving Distance of Photovoltaics-Powered Vehicles 924

Masafumi Yamaguchi, Taizo Masuda, Tsutomu Tanimoto, Yosuke Tomita, Yasuyuki Ota, Christian Thiel, Anastasios Tsakalids, Arnulf Jaeger-Waldau, Takashi Nakado, Kenji Araki, Kensuke Nishioka, Tatsuya Takamoto, Kyotaro Nakamura, Ryo Ozaki, Nobuaki Kojima, Yoshio Ohshita

Laser-Weld Qualification for a Reliable Aluminum Foil Interconnection of Copper-Metallized Back-Contact Silicon Solar Cells 925

Barry Hartweg, Kathryn Fisher, Jason Ro, Zachary Holman

Conversion Efficiency Analysis of Tandem Solar Cells with Intermediate Band Tunnel Connection 926

Shuhei Yagi, Hiroyuki Yaguchi

Effects of Period of Record Extension, Model Diversification, and DHI Measurements on Measure-Correlate-Predict Analyses for On-Site Solar Resource Assessments 927

Lucila D. Tafur, Renn Darawali, Halley Darling

Solar Forecasting: The Value of Using Satellite Derived Irradiance Data in Machine Learning Based Forecasts 928

Alex Kubiniec, Thomas Haley, Kyle Seymour, Richard Perez

Ageing Detection of Encapsulants and Backsheets in the Field Via NIR Spectroscopy 931

Chiara Barretta, Sascha Lindig, Marton Bredács, Alexander Astigarraga, Eric Helfer, Gernot Oreski

In-Situ & Ex-Situ Study of Protons and Electrons Irradiations of Perovskite Solar Cells 934

Carla Costa, Matthieu Manceau, Thierry Nuns, Sophie Duzellier, Romain Cariou

Numerical Investigation on Non-Radiative Recombination in InGaAs Front and Rear Hetero-Junction Solar Cell 935

Depu Ma, Hassanet Sodabanlu, Gan Li, Meita Asami, Kentaroh Watanabe, Masakazu Sugiyama, Yoshiaki Nakano

Strategies to Improve the Mechanical Robustness of Metal Halide Perovskite Solar Cells 938

Muzhi Li, Siraj Sidhik, Lidon Gil-Escrig, Samuel Johnson, Aditya Mohite, Axel Palmstrom, Henk J. Bolink, Nicholas Rolston

Microstructure-Property Relationships in Epitaxial Cu(In, Ga)Se$_2$ Solar-Cell Absorbers 939

Daniel Abou-Ras, Jiro Nishinaga, Takeyoshi Sugaya, Yukiko Kamikawa-Shimizu, Ulrike Bloeck, Henrik Prell, Sinju Thomas, Michael Tovar, Dan R. Wargulski, Harvey Guthrey, Pat Trimby, Aimo Winkelmann, Shogo Ishizuka

High-Performance Multi-Junction C-Band Photonic Power Converters: Calibrated Optoelectronic Model for Next Generation Designs 941

Gavin P Forcade, Meghan N Beattie, Christopher E Valdivia, Henning Helmers, Oliver H?hn, Paige Wilson, Louis-Philippe St-Arnaud, Robert Hunter, David Lackner, Jacob J Krich, Alexandre W Walker, Karin Hinzer

Strategies for High Fill Factor and Open-Circuit Voltage in Low-Doped c-Ge TPV Cells with Partially Contacted Surfaces Using 3D Simulations 942

M. Gamel, D. Shojaei, J. M. López-González, G. López, M. Garín, I. Martín

Minimizing Sputter Damage-Induced Electrical Losses in Monolithic Perovskite/Silicon Tandem Solar Cells During Deposition of the Transparent Front-Electrode 945

Marlene Härtel, Bor Li, Silvia Mariotti, Philipp Wagner, Florian Ruske, Steve Albrecht, Bernd Szyszka

Nanometer-Scale Imaging on Electrical Potential in Absorber of As-Doped CdSeTe Solar Cells.................. 946
Chun-Sheng Jiang, Eric Colegrove, Steve P. Harvey, Joel N. Duenow, Matthew O. Reese

Advanced Germanium TPV Cells for Latent Heat Thermal Batteries.. 947
A. Luque, P. Martin, R. Molinero, V. Orejuela, C. Sanchez-Perez, M. Zehender, I. García, I. Luque-Heredia, I. Rey-Stolle

Evaluation of Irradiance Variability Adjustments for Subhourly Clipping Correction................................ 948
William B. Hobbs, Chloe L. Black, William F. Holmgren, Kevin S. Anderson

Evaluation of Beta-Phase Formation in the Failure of PVDF-Based Solar Module Backsheets...................... 952
Stephanie L. Moffitt, Sona Ulicna, Song-Syun Jhang, Michael Owen-Bellini, Peter Hacke, Jared Tracy, Kaushik R. Choudhury, Laura T. Schelhas, Xiaohong Gu

Development of a Dynamic Photovoltaic Inverter Model with Grid-Support Capabilities for Power
System Integration Analysis 953
Rachid Darbali-Zamora

PV Fleet Modeling Via Smooth Periodic Gaussian Copula.. 961
Mehmet G. Ogut, Bennet Meyers, Stephen P. Boyd

Unveiling the Structural Formation of Low Dimensional Layers Deposited on Lead Halide
Perovskites by Thermal Evaporation.. 969
Carlo Andrea Riccardo Perini, Andres Felipe Castro Mendez, Tim Kodalle, Magdalena Ravello, Juanita Hidalgo, Juan-Pablo Correa-Baena

Ultrasonic Characterization of Ethylene Vinyl Acetate (EVA) Crosslinking for Quality Assurance
and Lamination Process Control... 970
Rico Meier, Ian M. Slauch, Mariana I. Bertoni

Detailed Raman Investigation on the Search for the Secondary Phases in the Chalcogenide
Perovskite BaZrS3 975
Arif Yetkin Hasan, Ruiquan Yang, Charles Hages, Phillip J. Dale

A Direct Comparison of Thermoradiative and Thermophotovoltaic Operation of HgCdTe
Photodiodes 976
Michael P. Nielsen, Muhammad H. Sazzad, Andreas Pusch, Phoebe M. Pearce, Peter J. Reece, Nicholas J. Ekins-Daukes

Solar Simulator Performance Metrics: Balloon Flown Calibration Standards Offer Real Time AM0
Solar Simulation Error Measurements.. 977
Scott J Ireton, Casey P Hare, Andrew J Schwab

Operation Efficiency Gains from Analyzing Minimal Solar Cells Production Data 978
Johnson Wong, Kissenger Chen, Dinica Li, Bryan Matthews, Jason Miller, Sal Sanci

Extracting Electrical Properties of CdTe, CdSeTe and CdSe Thin Films Using a Parallel Dipole
Line Hall Effect System 981
Mustafa Togay, Rachael C. Greenhalgh, Kerrie Morris, Xiaolei Liu, Luksa Kujovic, Luis C. Infante-Ortega, Nicholas Hunwick, Adam M. Law, Tushar Shimpi, Sampath S. Walajabad, Eric Don, Gabor Parada, Kurt L. Barth, J. Michael Walls, Jake W. Bowers

Power Hardware-In-The-Loop Interface Method for Grid Forming Inverters Using a Voltage-
Controlled Power Amplifier 984
Javier Hernandez-Alvidrez, Rachid Darbali-Zamora, Jack D. Flicker, Nicholas S. Gurule

Implications of Battery Storage for Solar Net-Metering Reforms 991
    Galen Barbose, Sydney Forrester, Chandler Miller

III-V Solar Cells Grown Directly on V-Groove Si Substrates 994
    Theresa E. Saenz, Jacob T. Boyer, John S. Mangum, Anica N. Neumann, Myles A. Steiner, Ryan M. France, William E. McMahon, Jeramy D. Zimmerman, Emily L. Warren

Correlated Mapping of Raman Spectroscopy and Cathodoluminescence of Emerging Absorber Bournonite (CuPbSbS3) 995
    O M Rigby, C Hill, G Kusch, M Naylor, M Guennou, G Zoppi, M Szablewski, R A Oliver, L Wirtz, P Dale, B G Mendis

Yb3+- Doped CsPbX3 Nanocrystals for Improving Free-Space Luminescent Solar Concentrators 996
    Mathis Van De Voorde, Damien Hudry, Dmitry Busko, Bryce S. Richards, Rebecca Saive

Rear Heterojunction GaInP Solar Cells for Improved Performance at Elevated Temperatures 997
    Mijung Kim, Yukun Sun, Ryan D. Hool, Minjoo L. Lee

Design and Fabrication of PERC-Like CdTe Solar Cells Using Micropatterned Al₂O₃ Layer 998
    Etee Kawna Roy, Kaden Powell, Chungho Lee, Gang Xiong, Heayoung Yoon

Issues, Challenges, and Primary Factors in the Estimation of Floating Solar PV Performance 1002
    Rick Meeker, Anna Brinck, Tom Lang, Jason Harrison

Partial Shading of Photovoltaic Modules: A Comparison Between Simulated and Measured IV Characteristics 1005
    Bianca Passarella, Maarten Verkou, Marco Leonardi, Fabrizio Coco, Youri Blom, Malte Vogt, Rudi Santbergen, Agnese Di Stefano, Andrea Canino, Marina Foti, Antonino Ragonesi, Marcello Sciuto, Francesco Rametta, Miro Zeman, Olindo Isabella, Cosimo Gerardi

Hierarchal Ti3C2Tx MXene and Aluminum Microgrid Back Contacts for Bifacial CdTe PV 1008
    Benjamin E Sartor, Matthew O Reese, Chris Muzzillo, Chungho Lee, Andre D Taylor

Design of Electronic Control of PV Tracking Independent of Weather Forecast 1009
    Sam Mil'Shtein, Dhawal Athana, Jeffrey Snell, William Brooks

Efficient and Stable All-Lead Perovskite Tandem Solar Cells Enabled by All-Inorganic CsPbI2Br Top Cells 1019
    Chongwen Li, Chuanxiao Xiao, Kamala Khanal Subedi, Bin Chen, Randy J. Ellingson, Song Zhaoning, Yanfa Yan, Edward H. Sargent

(3-Aminopropyl)trimethoxysilane Surface Passivation Improves Perovskite Solar Cell Performance by Reducing Surface Recombination Velocity 1020
    Yangwei Shi, Esteban Rojas-Gatjens, Jian Wang, Justin Pothoof, Rajiv Giridharagopal, Kevin Ho, Fangyuan Jiang, Margherita Taddei, Zhaoqing Yang, Carlos Silva-Acuña, David S. Ginger

Machine Learning-Based Defect Identification Method at the c-Si/A-Si:H Interface 1021
    Zitong Zhao, Gonglin Chen, Reza Vatan Meidanshahi, Gergely T. Zimányi

Recombination Analysis of Maxeon IBC Production Cells by Time-Resolved Photoluminescence 1024
    David Jacob, Guillaume Von Gastrow, Nils-Peter Harder, Luis Buño, Gerly Reich, Maristel Baldrias, Roderick J. Marstell, Arnold Castillo, David D. Smith, Michael J. Cudzinovic

Can Hierarchical Physics-Based Machine Learning De-Anonymize Solar Farm Locations? 1027
    Jabir Bin Jahangir, Amandeep Singh Bhatia, Muhammad Alam

Investigation and Quantitative Understanding of Front Field Passivation in Rear Junction Selective Double-Side TOPCon Solar Cells 1030
Wook-Jin Choi, Young-Woo Ok, Pradeep Padhamnath, Gabby De Luna, Kwan Hong Min, Ruohan Zhong, Sagnik Dasgupta, Vijaykumar D Upadhyaya, Ajay D Upadhyaya, Ajeet Rohatgi

Autonomous Control Strategies for Interconnected DC Microgrids with Geographical Separation 1034
Emmanuel G. Robles-Rivera, Rachid Darbali-Zamora, Erick E. Aponte-Bezares, Jack D. Flicker, Andrew R. R. Dow, Felipe Palacios, Lee Rashkin

Enabling High Efficiency, Flexible, and Lightweight CdTe Solar Cells with a Cadmium Stannate Transparent Conducting Oxide 1041
Manoj K. Jamarkattel, Adam B. Phillips, Ebin Bastola, Sabin Neupane, Deng-Bing Li, Yanfa Yan, Randy J. Ellingson, Michael J. Heben

Post-Mortem Failure Analysis of Metal Halide Perovskite Modules 1044
Sona Ulicna, Nutifafa Y. Doumon, Michael Owen-Bellini, Jackson Schall, Dana B. Kern, Timothy J. Silverman, Lance M. Wheeler, Steven Hayden, Chengbin Fei, Md Aslam Uddin, Prem J. S. Rana, Jinsong Huang, Joseph J. Berry, Laura T. Schelhas

Tuning Device Interfaces for Improved Open Circuit Voltage in Wide-Bandgap Hybrid Perovskite Photovoltaics 1045
Emily Smith, Sarah Brittman, David Scheiman, Woojun Yoon

Multiple-Reuse of Ge Substrates: Towards Cost-Effective and Sustainable III-V Solar Cells Fabrication 1051
Alexandre Chapotot, Bouraoui Ilahi, Tadeáš Hanuš, Gwenaëlle Hamon, Jinyoun Cho, Kristof Dessein, Christian Dubuc, Maxime Darnon, Abderraouf Boucherif

Mitigation of Potential Induced Degradation in Perovskite Solar Cells 1052
Laxmi Nakka, Shen Guibin, Armin G. Aberle, Fen Lin

A Robust Approach for Daily Solar Irradiance Clustering 1053
Roshni Agrawal, Sivakumar Subramanian, Shashank Agarwal, Venkataramana Runkana

On the Role of Sn-Halide Post-Deposition Reactive Annealing for the Passivation of Defective Surfaces in Cu2ZnSnSe4 1059
Alex Jimenez-Arguijo, Yuancai Gong, David Nowak, Devendrá Pareek, Levent Gütay, Lorenzo Calvo-Barrio, Zacharie Jehl Li-Kao, Sergio Giraldo, Edgardo Saucedo

Analysis of Thermal Behavior and Reliability of Bare Die Diodes Embedded Within PV Modules as Bypass Devices 1060
Luis Eduardo Alanis, Julian Weber, Pascal Romer, Marc Andre Schüler, Louisa Winkler, Udo Steinebrunner, Martin Heinrich, Dirk Holger Neuhaus

Measurement and Analysis of Annual Solar Spectra at Different Installation Angles in Central Europe 1066
Guillermo A. Farias-Basulto, Miguel Á. Sevillano-Bendezú, Maximillian Riedel, Mark Khenkin, Jan A. Töfflinger, Rutger Schlatmann, Reiner Klenk, Carolin Ulbrich

Photoluminescence Analysis of the Back Side of Cu(In,Ga)(S,Se)2 Absorbers 1067
Aubin JC. M. Prot, Susanne Siebentritt, Anastasia Zelenina, Hossam Elanzeery, Alberto Lomuscio, Thomas Dalibor, Maxim Guc, Robert Fonoll-Rubio

Effects of Salt Spray on c-Si Photovoltaic Modules in the Brazilian Region 1068
Mendelsson R. M. Neves, Allan Silveira, Hugo S. Alvarez, Rodrigo M. Garcia, Francisco C. Marques, Marcelo G. Villalva

Experimental Analysis of Distribution Network Voltage Regulation Using Smart Inverters 1074
    Rasel Mahmud, Subhankar Ganguly, Jing Wang, Killian McKenna, Ning Li

Doped GaInAs/GaP Quantum Well Superlattice Solar Cells with 27.5% Efficiency 1081
    Ryan M France, Myles A Steiner

Aggregated Three-Phase Photovoltaic Inverter Model with Sandia Frequency Shift Islanding
Detection 1082
    Nelson E. Saavedra-Peña, Rachid Darbali-Zamora, Edgardo Desarden-Carrero, Erick
    Aponte-Bezares

Time-Series Imputation Using Graph Neural Networks and Denoising Autoencoders................................ 1088
    Raymond Wieser, Yangxin Fan, Xuanji Yu, Jennifer Braid, Avishai Shaton, Adam Hoffman,
    Ben Spurgeon, Daniel Gibbons, Laura S. Bruckman, Yinghui Wu, Roger H. French

Cradle to Cradle Recycling of Perovskite Solar Cells 1092
    Zhenni Wu, Gülüsüm Babayeva, Zhang Jiyun, Mykhailo Sytnyk, Jens Hauch, Christoph J.
    Brabec, Ian Marius Peters

Per- And Polyfluoroalkyl Substances (PFAS) Usage in Solar Photovoltaics 1095
    Preeti Nain, Annick Anctil

Preliminary Gap Analysis of Existing IEEE 1547 and IEEE 2800 Standards Towards Grid-Forming
Technology 1096
    Ganesh Marasini, Wenzong Wang, Wes Baker, Deepak Ramasubramanian, Aminul Huque,
    Jens C. Boemer

Enhanced Bifaciality Factor with Sb_2Se_3 Devices Modeling Cu_2O Back Buffer 1102
    Sanghyun Lee, Kent Price

Analysis of the Key Factors Influencing the Economic Competitiveness and Profitability of
Floating Photovoltaics 1105
    Leonardo Micheli, Diego L. Talavera, Fredy A. Sepúlveda-Vélez

Towards Integrating Data Quality Assessments and Radiometer Uncertainty for Determining the
Expanded Uncertainty of Three-Component Solar Radiation Measurements 1106
    Stephen Wilcox, Tom Stoffel, Aron Habte, Manajit Sengupta

Stable, High-Throughput Production of Robust Perovskites in Open-Air with Polymer Additives................ 1109
    Nicholas Rolston, Carsen Cartledge, Vineeth Penukula, Muneeza Ahmad, Antonella Giuri,
    Aurora Rizzo

Clear-Sky Detection Using Time-Averaged, Tilted-Plane Data 1110
    Clifford W. Hansen, Dirk C. Jordan

Theoretical Performance Analysis for Thermo-Radiative Assisted Photovoltaic (TRAP™) Cell
Operating in Outer Space 1114
    Jianjian Wang, Nathan Van Velson

A Study of Cell Cracks Formation During Freight Shipping : Monitoring Shock and Temperature in
Real-Time & Assessing Damages with Pre and Post-Transit Characterizations of PV Modules 1120
    Cécile Molto, Dylan J. Colvin, Farrukh Ibne Mahmood, Fang Li, Ryan Smith, Govindasamy
    Tamizhmani, Hubert Seigneur

Quantifying Real-World Sources of Error in Redundant GHI Measurements... 1123
    Josh Peterson, Julie Chard, Alex Bryan

Sputtering for the Formation of SI-Based Passivating Contacts..1129
  Christophe Allebé, Antoine Descoeudres, Patrick Wyss, Nicolas Pernès, Bertrand Paviet-
  Salomon, Christophe Ballif, Simon Hübner, Torsten Dippell

Copper Metallization for III-V Solar Cells..1130
  Theresa E. Saenz, Anica N. Neumann, Matthew R. Young, Jeramy D. Zimmerman, Emily L.
  Warren, Myles A. Steiner

An Enhanced Snow-Shedding Model: The Module Frame as a Key Variable 1131
  Daniel Riley, Laurie Burnham, Paul Dice, Jennifer Braid

The Effect of Tilt and Azimuth Angle Variations on Monthly and Annual Incident Solar Radiations
for Locations in Brazil..1134
  Eslam Mahmoudi, Tarcio Andre Dos Santos Barros, Hugo Alvarez, Rodrigo Garcia,
  Francisco C. Marques, Marcelo Gradella Villalva

Rear-Side Irradiance Simulation of Field PV Modules 1141
  Zelin Li, Raymond J. Wieser, Xuanji Yu, Stephanie L. Moffitt, Ruben Zabalza, Silvana Ayala,
  Matthew Brown, Xiaohong Gu, Liang Ji, Colleen O'Brien, Micheal D. Kempe, Laura S.
  Bruckman, Kenneth P. Boyce

Extreme Weather and PV Performance 1145
  Dirk Jordan, Kirsten Perry, Chris Deline

Viability of a Novel Methodology of Measure-Correlate-Predict for Albedo Estimation 1146
  Halley C. Darling, Renn Darawali, Lucila D. Tafur

Long Term Soiling Model Tuning for Enhanced PV Cleaning Schedule Optimization 1147
  Kyle Seymour, Akanksha Bhat, Thomas Haley

Coalescence of GaP on V-Groove Si for III-V/Si Solar Cells 1150
  Theresa E. Saenz, John S. Mangum, Olivia D. Schneble, Anica N. Neumann, Ryan M. France,
  William E. McMahon, Jeramy D. Zimmerman, Emily L. Warren

Mitigation of Phase Separation in High Ga Cu(In,Ga)S2 Absorbers to Achieve ~ 1 Volt 15.6%
Power Conversion Efficiency 1151
  Damilola Adeleye, Mohit Sood, Tobias Törndahl, Adam Hultqvist, Aline Vanderhaegen,
  Michele Melchiorre, Susanne Siebentritt

Germanium-Tin Diode for Thermophotovoltaic Energy Collection..1152
  Amanda Lemire, Kevin A. Grossklaus, Thomas E. Vandervelde

Rapid Thermal Annealing of Symmetric p-TOPCon Silicon Test Structures..............................1155
  Arpan Sinha, Sagnik Dasgupta, Ajeet Rohatgi, Mool C. Gupta

Temperature Dependent Carrier Extraction and the Effects of Excitons on Emission and
Photovoltaic Performance in Cs0.05FA0.79MA0.16Pb(I0.83Br0.17)3 Solar Cells......................1156
  Hadi Afshari, Brandon K Durant, Ahmad R Kirmani, Sergio A Chacon, John Mahoney,
  Vincent R Whiteside, Rebecca A Scheidt, Matthew C Beard, Joseph M Luther, Ian R Sellers

Alba: Testing Emerging Photovoltaic Technologies in Low-Earth Orbit 1157
  Michael D. Kelzenberg, Phillip Jahelka, Richard G. Madonna, Harry A. Atwater

Optimizing the Design of 4-Terminal Perovskite/C-Si Tandem Photovoltaics 1158
  Paul F. Ndione, John F. Geisz, Qi Jiang, Gabriella D. Lahti, Rosemary C. Bramante, Kai Zhu,
  Axel F. Palmstrom, Emily L. Warren

Investigation of EMC Tests in Photovoltaic Inverter According to INMETRO Ordinance No. 140..............1159
Andrei C. Ribeiro, Geyciane P. De Lima, Guilherme C. S. Prym, Paulo R. D. R. Da Silva,
Pedro O. C. M. Neto, Romullo R. M. Carvalho, Francisco V. E. Lemos, Mendelsson R. M.
Neves, Tárcio A. S. Barros, Hugo Da S. Alvarez, Rodrigo M. Garcia, Francisco C. Marques,
Marcelo G. Villalva

PV Inverter Testing for Momentary Cessation and Rate-Of-Change-Of-Frequency Events..........................1164
Ramanathan Thiagarajan, Kumaraguru Prabakar, Li Yu, Ken Aramaki, Andy Hoke

Microgrid Design Toolkit Cost Optimization for a Rural Community in Puerto Rico....................................1170
Rolando J. Tremont-Brito, Rachid Darbali-Zamora, Robert Broderick, Erick E. Aponte-
Bezares, Efrain O'Neill-Carrillo, Matthew S. Lave

Co-Sputtered Sn-Doped ZnO Thin Film n-Type Layers for Incorporation into CdTe Based
Photovoltaics 1177
Kerrie M. Morris, Mustafa Togay, Luke Jones, John M. Walls, Jake W. Bowers

Investigating the Impact of MACl Doping in FA-Based Perovskites by Multimodal Synchrotron X-
Ray Techniques 1183
Yanqi Luo, Sanggyun Kim, Carlo A. R. Perini, Naveed Rahman, Luxi Li, Dina Sheyfer, Ross
Harder, Barry Lai, Juan-Pablo Correa-Baena, Sarah Wieghold

Removing Barriers for Participation of Small PV Systems in Balancing Energy Markets by
Utilizing the Established Smart Meter Eco-System 1184
Christoph Kondzialka, Manuela McCulloch, Rouven Taubmann, Jonas Dierenbach, Dietmar
Graeber, Gerd Heilscher

A Comparative Study of the Reflectance of Commercial Photovoltaic Modules..........................1188
Wei Wen Toh, Min Hsian Saw, Erik Birgersson, Mauro Pravettoni

The Microstructure of Thin Film CdSe Following Cadmium Chloride Activation Treatment.....................1193
Rachael C. Greenhalgh, Vladislav Kornienko, Mustafa Togay, Ali Abbas, Ebin Bastola, Adam
B. Phillips, Michael J. Heben, Jake Bowers, J. Michael Walls

High Open Circuit Voltage with Organic Hole Transport Layers in Group V Doped CdSeTe Solar
Cells..1197
Sabin Neupane, Deng Bing Li, Sandip S Bista, Suman Rijal, Zhaoning Song, Alisha Adhikari,
Lei Chen, You Li, Manoj K. Jamarkattel, Abasi Abudulimu, Dingyuan Lu, Xiaomeng Duan,
Feng Yan, Michael Heben, Randy J. Ellingson, Gang Xiong, Yanfa Yan

Thin-Film Tandem Partners Based on Inline-Processed (Ag, Cu)(In,Ga)Se$_2$ 1201
Theresa Magorian Friedlmeier, Rico Gutzler, Tina Wahl, Dimitrios Hariskos, Stefan Paetel,
Erik Ahlswede, Ana Kanevce, Jan-Philipp Becker

Development of 3D/2D Perovskite Solar Cells Using a Spray-Based Sequential Deposition 1206
Neetesh Kumar, Rishabh Sahani, Daniel Muñoz, Cheng-Yu Lai, Daniela Radu

Numerical Modeling of Bifacial Thin Film Solar Cells 1210
Briana Dokken, Sabin Neupane, Muhammad Mohsin Saeed, Jared D. Friedl, Adam B.
Phillips, Michael J. Heben, Yanfa Yan, Zhaoning Song

Passivating Contacts with Engineered Pinhole Enabled Transport 1216
Harvey L Guthrey, Caroline Lima Salles, William Nemeth, Sumit Agarwal, David Young,
Pauls Stradins

InP-Based Tunnel Junctions for Micro-Concentrator Photovoltaics 1217
Kenneth J Schmieder, Thomas C Mood, Eric A Armour, Mitchell F Bennett, Margaret A Stevens, Martin Diaz, Ziggy Pulwin, Matthew P Lumb

Residual Stress Limits Gridline Bridging in Cracked Solar Cells 1218
Junki Joe, Timothy J Silverman, Michael Owen-Bellini, Nick Bosco

Optimizing the Laser Scribing Process to Achieve a Certified Efficiency of 25.9% for Over 240 cm² Four-Terminal Perovskite/Si Tandem Solar Cells 1221
Yonglei Wang, Hongxu Zhang, Yang Liu, Yuan Qin, Jie Liu, Jiang Liu, Bo He, Xixiang Xu

Predicting Damp Heat Degradation in Heterojunction PV Modules Using Machine Learning 1224
Zubair Abdullah-Vetter, Felix O'Kearney, Priya Dwivedi, Robert Lee Chin, Brendan Wright, Thorsten Trupke, Ziv Hameiri

Demand Following RE – a Demand Driven Approach for Rapid RE Capacity Addition in India 1227
Prashant Kumar Upadhyay, Himanshu Gulati, Yellasiri Bharath Kumar Reddy

Perspectives on PV Adoption and Engaging Gen Z and Millennials in the Indian Scenario 1230
Robins Anto, Rhythm Singh

An Analysis of the Current Status and Future Potential of Rooftop Solar Adoption in the United States 1233
AC Lemay, BP Rand

Investigation of Varying Se Vapor Pressure During Deposition of CdSeTe Thin Film PV Devices 1234
Sushmakanth Myneni, Carey Reich, Daniel Shaw, Sampath Walajabad, Amit H. Munshi

Performance Optimization of the CdSe$_x$Te$_{1-x}$/CdTe Solar Cell 1237
Md Zahangir Alom, Sheikh Tawsif Elahi, Vasilios Palekis, Wei Wang, Chris Ferekides

Analysis of Optoelectronic Characterization Data Via Bayesian Inference: A Desktop-Scale MCMC Method 1241
Calvin Fai, Gregory A. Manoukian, Jason B. Baxter, Anthony J. C. Ladd, Charles J. Hages

Generalizability of Neural Network-Based Identification of PV in Aerial Images 1242
Joseph Ranalli, Matthias Zech

Modal Analysis of GaAs Nanowire Solar Cells for Optimal Device Design 1249
Venkata S. A. Chaluvadi, Eduardo Camarillo Abad, Hannah J. Joyce, Louise C. Hirst

Benchmark Tests for IV Fitting Algorithms 1250
Clifford W. Hansen, Abigail R. Jones, Taos Transue, Marios Theristis

Validation of Photovoltaic Plant Loss Estimation from Monitoring Data: String Faults, Shading and Degradation 1253
Karel De Brabandere, Maitheli Nikam, Julien Deckx, Gofran Chowdhury

Single-Junction Bifacial and Semitransparent Sb$_2$(S,Se)$_3$ Solar Cells 1256
Chen Qian, Kaiwen Sun, Martin Green, Bram Hoex, Xiaojing Hao

The United States Renewable Energy Landscape: Siting, Management, and Potential Impacts 1259
Jacob T. Stid, Anthony D. Kendall, Annick Anctil, Jeremy Rapp, David W. Hyndman

Interpreting Accelerated Tests on Perovskite Modules: Using Photooxidation of MAPbI3 as an Example.. 1260
    Ingrid L. Repins, Michael Owen-Bellini, Michael D. Kempe, Michael G. Deceglie, Joseph J. Berry, Nutifafa Y. Doumon, Timothy J. Silverman, Laura T. Schelhas

Charge Carrier Diffusion and Recombination Near Misfit Dislocations in GaAsP/GaInP Heterostructures... 1261
    T. H. Gfroerer, A. Edmondson, Lilian Korir, Fan Zhang, Yong Zhang, M. W. Wanlass

Improving Operational Stability of High-Efficiency Inverted Perovskite Solar Cells 1265
    Kai Zhu

Influence of Aluminum Co-Doping on Current-Induced Degradation and Regeneration Kinetics in Boron-Doped Cz PERC Solar Cells 1266
    August Weber, Mahsa Mohammadi, Matthias Trempa, Thomas Buck, Johannes Heitmann, Matthias Müller

Reducing the Photovoltaic Operation and Maintenance Costs Through an Autonomous Control Operation Center 1270
    Andreas Livera, Álvaro Fernández-Solas, Joao G. Bessa, Jesús Montes-Romero, Eduardo F. Fernández, Vassilis Papaeconomou, George E. Georghiou

Large-Area Uniformity Mapping of High-Speed Flexography-Printed Perovskite Solar Cells Via Scanning Photoluminescence 1276
    Julia E. Huddy, William J. Scheideler

Measuring the Doping Concentration of Si and CdTe Absorbers Using Lock-In Amplified Quantitative QSSPL 1279
    Mason P Mahaffey, Arthur Onno, Carey Reich, Adam Danielson, Walajabad Sampath, Zachary C Holman

Peer-To-Peer Energy Trading for PV Prosumers Using Fuzzy Logic Inference Systems 1280
    Hector Lopez, Ali Zilouchian

Carbon Quantum Dots and Their Possible Application in Perovskites Passivation 1283
    Maria Fernanda Villa-Bracamonte, Jose Raul Montes-Bojorquez, Alan Lopez-Becerra, Arturo Ayon

A New Combined Accelerated Stress Test Sequence for Rapid Reliability Screening of Photovoltaic Materials.. 1284
    Yi Jiang, Xuanji Yu, Ben Huang, Ruirui Lv, Yuanjie Yu, Tao Xu, Guangchun Zhang

Improved Soiling Rate Estimation by Calculating PV Module Temperature Using a Distributed Thermal Model 1287
    Shoubhik De, Yogeswara Rao Golive, Narendra Shiradkar, Anil Kottantharayil

Electrical and Electroluminescence Evaluation of 17 Year Old Monocrystalline Silicon Building Integrated Photovoltaic Modules.. 1292
    Andrew M. Shore, Tali Schlenoff, Bakary Coulibaly, David Navon, Stephanie L. Moffitt, Brian Dougherty, Behrang H. Hamadani

Automated Photovoltaic Module Quality Assessment: Defect Identification and Classification from Luminescence Images Using Machine Learning.. 1295
    Brendan Wright, James Petesic, Timothy Dawson, Ziv Hameiri

Influence of Alkali Iodide Fluxes on Cu2ZnSnS4 Monograin Powder Properties and Performance of Solar Cells 1298
Kristi Timmo, Katri Muska, Maris Pilvet, Mare Altosaar, Valdek Mikli, Mati Danilson, Reelika Kaupmees, Jüri Krustok, Maarja Grossberg-Kuusk, Marit Kauk-Kuusik

Pyrolyzer Assisted Vapor Transport Deposition of Antimony-Doped Cadmium Telluride 1299
Bin Du, Kevin Dobson, Brian McCandless, Aayush Nahar, Ujjwal Das, Shannon Fields, Aaron Arehart, William Shafarman

Early Degradation Trend Estimation of Bifacial PV, Investigating the Seasonality Effect 1304
Gaetano Mannino, Giuseppe Marco Tina, Mario Cacciato, Lorenzo Todaro, Agnese Di Stefano, Fabrizio Bizzarri, Andrea Canino

The Rise and the Decay of the Photovoltage in Perovskite Solar Cells 1307
Uwe Rau, Lisa Krückemeier, Sandheep Ravishankar, Thomas Kirchartz

Cadmium Zinc Telluride as an Electron Reflecting Back-Contact Layer for CdTe Solar Cells 1310
Camden Kasik, Jennifer Drayton, James Sites

Innovative Layouts for Utility-Scale PV Modules: Module Characteristics, Shading Tolerance, and Electricity Costs 1313
Li C. Rendler, Christian Reichel, Matthew F. Berwind, Anna Heimsath, Dirk H. Neuhaus

UV Absorption Utilizing a MoS2/Ge Nano-Junction for Solar Applications 1320
Ayman Rezk, Ammar Nayfeh

Defect Signatures in Admittance Spectroscopy of Perovskite Solar Cells 1323
Rushu A. Awni, Zhaoning Song, Jared D. Friedl, Abasi Abudulimu, Adam B. Phillips, Randall J. Ellingson, Michael J. Heben, Yanfa Yan

Modeling the Hardware Components of a Power Hardware-In-The-Loop Platform for Photovoltaic Applications 1326
Edgardo Desarden-Carrero, Rachid Darbali-Zamora, Nelson E. Saavedra-Peña, Erick E. Aponte-Bezares

Effective and Equivalent Refractive Index Models for Patterned Solar Cell Films Via a Robust Homogenization Method 1334
Ekin Gunes Ozaktas, Sreyas Chintapalli, Susanna M. Thon

Evaluation of Rear Contact Passivation Strategies Via Surface Photovoltage Spectroscopy 1337
Nathan Rock, Chien-Hsuan Chen, Kristopher Davis, Amit Munshi, Michael Scarpulla

Machine Learning, Unmanned Vehicles, and Energy: A Review 1338
Brian L. Reyes Santiago, Eduardo Ortiz-Rivera

III-V Epitaxy on Detachable Porous Germanium 4" Substrates 1341
Waldemar Schreiber, Jens Ohlmann, Patrick Schygulla, Stefan Janz, Jinyoun Cho, Kristof Dessein

Effect of Soiling from Dust Particles on Solar Cell Efficiency in the United Arab Emirates (UAE) 1344
Muntaser Abdelrahman Almansoori, Rawdah Almannaee, Mariam Aldhefairi, Meerh Alsuwaidi, Ayman Rezk, Ammar Nayfeh

Efficiency Limit of Transition Metal Dichalcogenide Solar Cells 1347
Koosha Nassiri Nazif, Frederick U. Nitta, Alwin Daus, Krishna C. Saraswat, Eric Pop

Advanced Health-State Data Analytic Workflow for Utility-Scale Photovoltaic Power Plants 1348
  Jesus Montes-Romero, Loucas Pikolos, Andreas Makrides, Nino Heinzle, George Makrides,
  Juergen Sutterlueti, Steve Ransome, George E. Georghiou

Monolithic Bifacial Perovskite-CdSeTe Tandem Solar Cells 1353
  Zhaoning Song, Deng-Bing Li, Sabin Neupane, Kamala Khanal Subedi, Samuel Seibert,
  Randy J. Ellingson, Yanfa Yan

Improving the Performance and Yield of Colloidal Quantum Dot Solar Cells Through Electron
Transport Layer Optimization 1356
  Dana Kachman, Arlene Chiu, Dhanvini Gudi, Chengchangfeng Lu, Eric Rong, Sreyas
  Chintapalli, Yida Lin, Daniel Khurgin, Susanna M. Thon

Uncertainty Considerations in Bifacial Photovoltaic Systems with High Albedo Seasonality 1359
  Javier Lopez-Lorente, Anja Neubert, Mike Hamer

India as an Emerging Solar Manufacturing Country 1366
  Juzer Vasi, Mrunal Berad, Narendra Shiradkar, Anil Kottantharayil, Dinesh Kabra, Kedar
  Deshmukh, Aditi Chaubal, Rajeewa Arya, Probir Ghosh, Satyendra Kumar, Lawrence L.
  Kazmerski

IBC Technology Targeting Fast and Effective Silver Reduction Applying Advanced Screen:
Printing 1372
  Radovan Kopecek, Florian Buchholz, Valentin D. Mihailetchi, Joris Libal, Ning Chen,
  Haifeng Chu, Christoph Peter, Dominik Rudolph, Thomas Buck, Tudor Timofte, Andreas
  Halm, Yonggang Guo, Xiaoyong Qu, Xiang Wu, Jiaqing Gao, Peng Dong

Development of Machine Vision System for Detection of Wrap-Around in n-TOPCon Solar Cells 1373
  Junhee Kim, Han-Jung Kim, Yohan Ko, Yoonkap Kim

Effect of CdS Annealing on the Performance of Antimony Selenosulfide Solar Cells 1376
  Alisha Adhikari, Suman Rijal, Sabin Neupane, Manoj K. Jamarkattel, Deng-Bing Li, Tamanna
  Mariam, Michael J. Heben, Zhaoning Song, Yanfa Yan

20%-Efficient TOPCon Solar Cell with a Silicon Oxide Layer Deposited by Aerosol Impaction-
Driven Assembly 1379
  Maria Angelica M Garcia, William Weigand, Zachary C Holman

Understanding Practical Efficiency Limits for Tandem Solar Cells 1380
  Emily L. Warren, Sirazul Haque, Qi Jiang, William E. McMahon, Lorelle M. Mansfield

Effect of Angle and Direction of Solar Panels in the Desert Climate of Abu Dhabi, United Arab
Emirates 1381
  Laith Nayfeh, Leia Nayfeh, Ammar Nayfeh

Per-Unit Dynamic Models for Grid-Following Photovoltaic Inverters 1384
  Hyeonjung Tari Jung, D. Venkatramanan, Manish K. Singh, Sairaj Dhople

'There and Back Again': Reusable Germanium Wafers with Ge-On-Nothing Structures for Triple-
Junction Solar Cells 1390
  Valérie Depauw, Guillaume Courtois, Jinyoun Cho, Kristof Dessein, Clément Porret, Roger
  Loo

Characterizing Capacity Contribution of Renewable Resources Over Time in Renewable-Heavy
Transmission System: MISO Case Study 1393
  Hyeonjung Tari Jung, Megan Pamperin, Elspeth McGarvey, Eduardo Ibanez, James Okullo,
  Armando Figueroa Acevedo, Jordan Bakke

Device Modeling of HTL/BaSi$_2$ Heterojunction Solar Cells 1398
  Sho Aonuki, Carlos Mario Ruiz Tobon, Rudi Santbergen, Olindo Isabella, Takashi Suemasu

Modulating Efficiency and Stability of Methylammonium/Br-Free Perovskite Solar Cells Using
Fluoroarene Hydrazine 1401
  Dhruba B. Khadka, Yasuhiro Shirai, Masatoshi Yanagida, Kenjiro Miyano

In-Flight Validation of End-Of-Life Optimized Triple Junction Solar Cells Onboard ASTROBIO
Cubesat 1404
  *Luigi Schirone, Pierpaolo Granello, Matteo Ferrara, Matteo Avoli, Davide Imperatori, Nithin
  Maipan Davis, Lorenzo Iannascoli, Augusto Nascetti, Stefano Carletta, Claudio Paris,
  Erminio Greco, Roberta Campesato*

Evaluation of Module Mismatch Losses and Generation Impact in Utility Scale PV Systems 1409
  Sara M. Macalpine, David A. Bowersox

SCAPS-1D Simulations of CdTe Based Solar Cells with an Amorphous Silicon-Based Back Buffer 1410
  *Abdul Quader, Venkanna Kanneboina, Prabin Dulal, Madan Mainali, Bishal Shrestha, Ebin
  Bastola, Adam B. Phillips, Ambalanath Shan, Nicholas J. Podraza, Michael J. Heben*

Impact of Current Collecting Grids on the Scalability of 3-Terminal Perovskite/Silicon Tandems
with Bipolar Transistor Architecture 1413
  Gemma Giliberti, Federica Cappelluti

Siting Optimization of PV Recycling Plants for Supply Chain Security and Critical Material
Recovery... 1416
  Macarena Mendez Ribo, Silvana Ovaitt, Hope Wikoff, Heather Mirletz, Samantha Reese

Environmentally Controlled Electroluminescence/Photoluminescence Imaging System with
Current Density-Voltage Capabilities for Quantitative Degradation Analysis of Perovskite Thin
Film Solar Cells... 1417
  *Tamanna Mariam, Zahrah S. Almutawah, Adam B. Phillips, Sheng Fu, Jaehoon Chung, You
  Li, Manoj Rajakaruna, Kshitiz Dolia, Zhaoning Song, Randy J. Ellingson, Yanfa Yan, Michael
  J. Heben*

Method for Evaluating the Silicon Solar Cells Performances Under AM0 Thanks to AM1.5G
Spectrum... 1423
  Philippe Voarino, Adem Dahi, Romain Cariou

Fatigue Debonding of EVA from Solar Glass at Elevated PV Service Temperatures 1428
  Gernot Wallner, Gabriel Riedl, Philipp Haselsteiner, Robert Pugstaller

Demonstration of Dual-Junction ELO Solar Cells with Strain-Balanced and Lattice-Matched
Quantum Well Absorbers 1429
  *Rao Tatavarti, Andree Wibowo, Mitsuru Imaizumi, Takeshi Ohshima, David Wilt, Roger
  Welser*

Cu2ZnSnS4 Monograin Layer Solar Cells for Flexible Photovoltaic Applications 1432
  *Marit Kauk-Kuusik, Kristi Timmo, Maris Pilvet, Katri Muska, Mati Danilson, Jüri Krustok,
  Raavo Josepson, Maarja Grossberg-Kuusk*

Advances in Flexible and Lightweight III-V Multijunction Solar Cells for High Power Density
Applications.. 1433
  *Carlos Algora, Ivan Garcia, Clara Sanchez-Perez, Pablo Martin, Pablo Fernandez, Luis
  Cifuentes, Ivan Lombardero, Daniel Gomez, Mercedes Gabas, Ignacio Rey-Stolle*

Improving the Stability of Polycrystalline Silicon Passivated Contacts Using Titanium Dioxide 1434
  Di Yan, Jesus Ibarra Michel, Yida Pan, Sieu Pheng Phang, Daniel Macdonald, Heping Shen, Leiping Duan, Kylie Catchpole, Jie Yang, Peiting Zheng, Xinyu Zhang, Hao Jin, James Bullock

Copper Oxide: A Potential Candidate for Hole Transport Material in Perovskite Solar Cells for Space 1437
  Daniel Muñoz-Pinzon, Rishabh Sahani, Mateo Ferreira, Neetesh Kumar, Cheng-Yu Lai, Daniela Radu, Lyndsey McMillon-Brown

Innovative Methodology for an Advanced Characterization of Perovskite Systems to Reach Buried Interfaces: In-Depth Profile by Coupling GD-OES and XPS 1440
  Mirella Al Katrib, Pia Dally, Armelle Yaiche, Jean Rousset, Muriel Bouttemy

Optimization of Optical and Electrical Properties of 2T Textured Perovskite/Silicon Tandem Solar Cell Structure 1443
  Chun-Hao Hsieh, Jun-Yu Huang, Yuh-Renn Wu

Development of an Ultra-Light Curvilinear Prismatic Window Which Mitigates Reflections and Glare for PV Modules and Other Surfaces 1446
  Mark O'Neill, Chris Youtsey, Robert McCarthy

Evaluating Multi-Bias Modulation for Diagnostics of PV Modules in Daylight Electroluminescence Inspections 1452
  Rodrigo Del Prado Santamaría, Gisele A. Dos Reis Benatto, Thøger Kari, Peter B. Poulsen, Sergiu V. Spataru

Recomibiation Center Defects Induced by TCO Reactive Plasma Deposition in Carrire Selective Contact Solar Cells 1453
  Yoshio Ohshita, Tomohiko Hara, Taichi Tanaka, Keita Kimura, Yuto Ifuji

Predicting Site-Specific Adjustments to P50 Energy Production Estimates from Sub-Hourly Irradiance Data 1454
  Faisal Rashed

Self-Assembled Monolayer Patterning for PolySi/SiO$_2$ Passivated Contacts 1459
  B. Nemeth, D. L. Young, M. R. Page, V. Lasalvia, S. Theingi, P. Stradins

Non-Ionizing Radiation Effects on the Room Temperature Surface Recombination Velocity of Unintentionally Doped AlGaAs/GaAs Heterostructures 1462
  Andrew Hudson, Daniele Monahan

Flexible Organic Solar Cells on Ti Foil Substrate 1466
  Huiying Jia, Lei Kerr, Benjamin Leever

Novel Approach to Control Environmental Fatigue Tests on Glass/PV Encapsulant Laminates 1467
  Gabriel Riedl, Gernot M. Wallner, Robert Pugstaller

Automated Analysis of Internal Quantum Efficiency Measurements of GaAs Solar Cells Using Machine Learning 1468
  Zubair Abdullah-Vetter, Priya Dwivedi, N. J. Ekins-Daukes, Thorsten Trupke, Ziv Hameiri

RF-Powered Sputtering of Iron Pyrite for Photovoltaic Applications 1471
  Awais Zaka, Ayman Rezk, Sabina Abdul Hadi, Saeed Alhassan, Ammar Nayfeh

Analysis and Identification of Measurement Uncertainty Sources of a LED Sun Simulator with Double-Side Illumination for Bifacial PV Module Power Rating 1475
Sebastian Dittmann, Giuliano L. Martins, Ralph Gottschalg

The Effects of Global Damp Heat Ageing on Debonding of Polyolefin Glass Laminates 1481
Martin Tiefenthaler, Gernot M. Wallner, Robert Pugstaller

Nanographene (NG)-Based Hole Transporter with π- Interface Modifier for Thermally Stable Perovskite Solar Cells 1484
Seul-Gi Kim, Thybault De Monfreid, Jeong-Hyeon Kim, Fabrice Goubard, Joseph J. Berry, Kai Zhu, Thanh-Tuân Bui, Nam-Gyu Park

Automated Workflows for Machine Learning on Photovoltaic Timeseries and UV Fluorescence Image Datasets Using FAIR Principles 1485
William C. Oltjen, Xuanji Yu, Mengjie Li, Dylan J. Colvin, Yijia Sun, Hubert Seigneur, Philip Knodle, Andrew M. Gabor, Laura S. Bruckman, Kristopher O. Davis, Roger H. French

Post-Flight Analysis of Perovskite Solar Cells for NASA Materials International Space Station Experiment (MISSE) 1490
Kaitlyn Vansant, Ahmad Kirmani, Severin Habisreutinger, Steve Johnston, Brian Wieliczka, Joseph Luther, Timothy Peshek, Lyndsey McMillon-Brown

Cleaning Optimization for Photovoltaic Powerplants: A Novel Approach Combining Techno-Economic Modelling with Historic Rain and Soiling 1491
Thore Müller, Kostiantyn Pogorelov, Franco Clandestino

Characterization of Solar Cell Busbar Grid for Different Technologies by Time Domain Reflectometry Simulation: Transmission Line Approach 1492
A. M. C. Silveira, M. R. M. Neves, R. Garcia, H. Alvarez, M. G. Villalva, F. C. Marques, L. C. Kretly

19.5% Efficient CdSeTe/CdTe Solar Cells Using ZnO Buffer Layers 1497
Luksa Kujovic, Xiaolei Liu, Mustafa Togay, Luis C. Infante-Ortega, Kurt L. Barth, Jake W. Bowers, John M. Walls, Ochai Oklobia, Stuart J. C. Irvine, Wei Zhang, David W. Miller, Timothy Nagle, Rajni Mallick, Dingyuan Lu, Wyatt K. Metzger, Gang Xiong

Numerical Evaluation of Optimal Tilt Angle for Energy Production and Minimum Shadowing for Bifacial Solar Modules 1500
Roberto Corso, Fabio Matera, Salvatore A. Lombardo

Optimization of 1-Axis Tracking with N-S Rotating-Axis Orientation 1503
Jiahui Shi, Xitao Liu, Teliang Mu, Xiaotong Feng, Vasilis Fthenakis

A Crucial Role of Spin-Dry Cleaning on the Surface Passivation Quality of Crystalline Silicon 1509
Munan Gao, Vibhor Kumar, Ngwe Zin

Holistic Assessment of Monocrystalline Silicon (mono-Si) Solar Panels with Recycled Content Vs. Virgin-Grade Materials 1510
Christopher C. Bondoc, Ross Lee, Mary E. McRae, Pritpal Singh

Effect of Thickness of Electron Reflector Layer on the Efficiency of CdS/CdTe Heterojunction Thin-Film Solar Cell 1513
Chaitanya Santosh Rampalli, Hamid Fardi

Fabrication Au/TiO$_x$ Nanoislands Systems by a Solid State Thermal Dewetting for Plasmonic Solar Cell Applications 1516
Brahim Aïssa, Mohammad I. Hossain, Adnan Ali

Detection and Analyze of Off-Maximum Power Points of PV Systems Based on PV-Pro Modelling........... 1519
  Baojie Li, Xin Chen, Anubhav Jain

Laser Recycling of Silver from Waste Silicon Solar Cells 1522
  Mahantesh Khetri, Pawan K. Kanaujia, Mool C. Gupta

Study on Air Gap Effects on Photovoltaic Modules Operating Temperature on Typical Metal
Rooftop Appliation 1523
  Quanzhi Wang, Yuanjie Yu, Tao Xu

Development of P-Type Silicon Heterojunction Solar Cells with 26.6% Efficiency 1526
  Xiaoning Ru, Miao Yang, Yichun Wang, Jianqiang Wang, Chaowei Xue, Shi Yin, Chengjian
  Hong, Fuguo Peng, Minghao Qu, Junxiong Lu, Liang Fang, Tian Xie, Zhenguo Li, Xixiang
  Xu

Oxy-Fuel Combustion: A Threat Or an Opportunity for Solar? 1529
  Mariela Colombo, Sarah Kurtz

High-Throughput In-Line Deposition of Silicon Oxide Passivation Layers in Silicon TOPCon Solar
Cells.. 1533
  Zachary B. Leuty, William J. Weigand, Jorge Ochoa, Joe V. Carpenter, Mariana I. Bertoni,
  Zachary C. Holman

Measurement and Control of Mobile Ion Concentration in Halide Perovskites............................. 1534
  Saivineeth Penukula, Nicholas Rolston

2D-GaSe/In$_x$Se$_y$ Layer for Rapid ELO GaAs Technique... 1535
  Nobuaki Kojima, Yoshio Ohshita, Masafumi Yamaguchi

NSF Industry-University Cooperative Research Center (IUCRC) for a Solar Powered Future 2050
(SPF2050)... 1537
  Amit Harenkumar Munshi, Walajabad S. Sampath, Brian Korgel

Indoor and Outdoor Characterization of III-V/Ge Solar Cells Assembled on Glass Substrate for
Concentrated Photovoltaic Applications.. 1538
  K. Kouame, J. Kinfack, D. Danovitch, P. Albert, T. Bidaud, A. Turala, M. Volatier, V. Aimez, A.
  Jaouad, M. Darnon, G. Hamon

Survey of Module and System Quality in Brazil PV Deployments... 1541
  Lawrence L. Kazmerski, Denio Alves Cassini, Daniel Sena Braga, Suellen C. S. Costa,
  Vinícius Camatta Santana, Antonia Sonia A. C. Diniz

690 WP N-Type i-TOPCon Modules in Mass Production with >25% Efficiency Solar Cells Based
on Large-Area 210 mm Wafers 1544
  Yifeng Chen, Hong Chen, Shu Zhang, Le Wang, Chengfa Liu, Daming Chen, Jianmei Xu,
  Pietro Altermatt, Zhiqiang Feng, Pierre Verlinden

Methylamine Post-Deposition Treatments of Vapor-Deposited Perovskite Thin Films 1545
  Chaiwarut Santiwipharat, Austin G. Kuba, Kevin D. Dobson, Ujjwal K. Das, William N.
  Shafarman

Optimization of Sb2Se3 Thin Films Prepared by Selenization of Sb Metallic Precursors for
Photovoltaic Application 1549
  Woo Kyoung Kim, Vasudeva Reddy Minnam Reddy, Sreedevi Gedi, Salh Alhammadi, Ignatius
  Andre Setiawan, Yujeong Ahn, Songhee Lee, Hyomin Kim

Narrow Bandgap Perovskite Solar Cell Degradation Monitoring by Spectroscopic Ellipsometry 1550
Marie Solange Tumusange, Madan K. Mainali, Lei Chen, Zhaoning Song, Yanfa Yan, Nikolas J. Podraza

Solution Processed N+ CdS/ n-CdTe/ Perovskite Heterojunction Thin-Film Solar Cells 1551
Isaiah Henry, Dakota Schwartz, Harry Larson, Shubhra Bansal

Effect of Double Cation Substitution on Nonradiative Recombination Losses in Cu2ZnSn(S,Se)4
Solar Cells 1556
Vijay Karade, Kiwhan Kim, Jae Ho Yun, Jin Hyeok Kim

Mapping Spatial Variations of Wide Band Gap Perovskite Thin Films 1557
Emily Miller, Kshitiz Dolia, Bailey Frye, Yanfa Yan, Zhaoning Song, Nikolas J. Podraza

Exploring Distributed PV Power Measurements for Real-Time Potential Power Estimation in
Utility-Scale PV Plants 1558
Michael Gostein, William B. Hobbs

Controlling Photoexcited Carrier Relaxation Through Phonon Management in GaAs/AlAs
Superlattices 1564
Muhammad Hanif, Milos Dubajic, Stephen P Bremner, Michael P Nielsen, Santosh Shrestha, Gavin J Conibeer

Controlling Residual Stresses for Scalable Open-Air Fabrication of Perovskite Solar Cells 1565
Muneeza Ahmad, Carsen Cartledge, Nicholas Rolston

Silver Recovery Through a Fluoride Chemistry for Solar Module Recycling 1566
Theresa K Chen, Meng Tao

Thermal Models of Monofacial and Bifacial PV Modules: Machine Learning and Physical
Estimation Models Comparison 1567
Marco Grisanti, Gaetano Mannino, Giuseppe Marco Tina, Alessandro Ortis, Mario Cacciato, Sebastiano Battiato, Fabrizio Bizzarri, Andrea Canino

In-Situ Photostability Analysis of Perovskite Solar Cells by Time-Evolving Photoluminescence
Imaging 1570
Jackson W. Schall, Amy Louks, Goutam Paul, Nikita S. Dutta, Steve Johnston, Chun-Shen Jiang, Axel Palmstrom, Mowafak Al-Jassim, Dana B. Kern

Flexible Manufacturing of Colloidal Quantum Dot Solar Cells Via Spray-Casting Techniques 1575
Lulin Li, Botong Qiu, Yida Lin, Laura Shimabukuro, Alex Ozbolt, Keyi Kang Yao, Stephen Farias, Samuel Rosenthal, Susanna M. Thon

Detection of PV Module Temperature Coefficient Using Machine Learning 1578
John M. Obrecht, Julián Ascencio-Vásquez

Widegap CdSe Solar Cells with V_{OC} >750mV 1582
Taylor Hill, Sachit Grover, James Sites

Eliminating the Need for Handling Individual Sub-Cells for Small Appliance PV Modules with
Voltage Demands Above 12V 1588
Jan Paschen, Andreas Brand, Matheus Melati Menegassi, Oliver John, Jan Nekarda

Setting Priorities for Photovoltaic Reliability Research Using Criticality Analysis 1593
Ingrid L. Repins, Michael G. Deceglie, Timothy J. Silverman, David C. Miller, Dirk C. Jordan, Michael Woodhouse, Teresa M. Barnes

Uncertainties in PV Power Simulation Chain.. 1594
  Lubos Helienek, Jozef Rusnak, Branislav Schnierer, Martin Opatovsky, Lukas Dvonc, Vicente Lara Fanego, Artur Skoczek, Tomas Cebecauer

Characterizing TeO_2 Formation in CdTe Devices Using Transmission Electron Microscopy....................... 1600
  John Farrell, Ebin Bastola, Manoj Jamarkattel, Michael Heben, Walajabad S. Sampath, James Sites, Robert F. Klie

Dense Array TPV Modules with Alternating Polarity InGaAs Cells... 1603
  Iván García, Aitana Cano, Víctor Orejuela, Pablo Martír, Ignacio Rey-Stolle

Influence of Spectral Albedo on the Performance of Lead-Free Perovskite Bifacial Tandem Solar Cell... 1606
  Atanu Purkayastha, Arun Tej Mallajosyula

Outdoor Characterization of a Bifacial Four-Terminal GaAs/Si Mini-Module Under Different Albedo Conditions... 1609
  Roberto Corso, Fabio Matera, Andrea Scuto, Salvatore A. Lombardo

Evaluation of PV Snow Loss Models in the East Coast of Canada Using AI Computer Vision 1612
  Jessica Ma, Alexandre Khoury, Marianne Rodgers

Influence of NaF and KF Post-Deposition Treatment on the Sub-Band Gap Absorption of $Cu(In,Ga)Se_2$ Absorber Layers 1617
  Sevan Gharabeiki, Michele Melchiorre, Susanne Siebentritt

The Importance of Data Quality for Reducing the Uncertainty of Site-Adapted Solar Resource Datasets 1620
  Kristen Wagner, Alex Kubiniec, Tom McAlister, Richard Perez

Multifunctional Titanium Oxide Layers in Silicon Heterojunction Solar Cells by Selective Anodization 1625
  Leonie Jakob, Leonard Tutsch, Thibaud Hatt, Johan Westraadt, Markus Glatthaar, Martin Bivour, Jonas Bartsch

Development of an Adaptive Droop Control Method for Interconnected Lunar DC Microgrids Using Power Hardware-In-The-Loop.. 1626
  Andrew R. R. Dow, Rachid Darbali-Zamora, Felipe Palacios II, Jack D. Flicker, Marc A. Carbone, Jeffrey T. Csank

Assessing Degradation in Bifacial Photovoltaic by Sequential Stress and Outdoor Aging........................... 1629
  Dennice M. Roberts, Sona Ulicna, Michael Owen-Bellini, Paul Ndione, Helio Moutinho, Kent Terwilliger, Steve Johnston, Laura T. Schelhas, Dana B. Kern

Damp Heat Exposure of Glass/Glass Coupons with Different Encapsulants 1630
  Chiara Barretta, Lisa Meinhart, Hannes Krebs, Andreas Brandstätter, Gernot Oreski

Trajectories to Reach 25% Efficiency CdTe Solar Cells with the Implementation of CdTe1-XSex Band Gradient in SCAPs 1-D... 1633
  Joel Saucedo, Hasitha Mahabaduge

How Do As-Local Structures in CdSexTe1-X Respond to Bias Conditions Under (X-Ray) Illumination? 1634
  Srisuda Rojsatien, Niranjana Mohan Kumar, Barry Lai, Dan Mao, Arun Mannodi-Kanakkithodi, Maria K. Y. Chan, Mariana Bertoni

Trends in Solar PV Growth in Snowy Climates and Impact on Resource Adequacy 1635
Shelbie Wickett, Ana Dyreson

Modeling of Perovskite/Si Tandem Solar Cell 1645
Yegao Xiao, Michel Lestrade, Zhiqiang Li, Zhanming S. Li

Assessment of a DER Inverter Model for IEEE 1547 Ride-Through Requirements Using a Model
in the Loop Testbed 1648
Nayeem Ninad, Eugene Desjardins Couture

Methodology for the Analysis of Series Arc Fault Algorithms 1654
*Paulo R. D. R. Da Silva, Guilherme C. S. Prym, Geyciane P. De Lima, Andrei C. Ribeiro,
Hugo Da S. Alvarez, Rodrigo M. Garcia, Francisco C. Marques, João A. F. G. Da Silva,
Mauricio Taconelli, Tárcio A. Dos S. Barros, Marcelo G. Villalva*

Patterning the Front Polysilicon Contact for Silicon Solar Cells Using Laser Oxidation 1657
*Sagnik Dasgupta, Pradeep Padhamnath, Vijaykumar Upadhyaya, Young-Woo Ok, Ruohan
Zhong, Wookjin Choi, Kyu-Hyeon Im, Ajeet Rohatgi*

Evaluation of Motion-Induced Noise and Pixel-Bleeding in Electroluminescence Field Inspection
of PV Modules 1662
*Thøger Kari, Rodrigo Del Prado Santamaria, Gisele A. Dos Reis Benatto, Pascal Koelblin,
Liviu Stoicescu, Sergiu V. Spataru*

Interface Hydrogen and Passivation of Amorphous Silicon / Crystalline Silicon Heterojunction 1667
*Ujjwal K. Das, Tasnim K. Mouri, Marissa Pina, Tyler Parke, Dhamelyz R. S. Quinones,
Andrew V. Teplyakov*

Author Index

Theoretical Performance Analysis for Thermo-Radiative Assisted Photovoltaic (TRAP™) Cell Operating in Outer Space

Jianjian Wang, Nathan Van Velson

Advanced Cooling Technologies, Inc., Lancaster, PA, 17601, U.S.A.

Abstract — **In the paper, we propose a thermo-radiative assisted photovoltaic (TRAP™) cell to simultaneously harvest incoming sunlight and outgoing thermal radiation when it is operating in outer space. The TRAP™ cell device consists of three layers: a conventional space photovoltaic (PV) cell as the top layer, a mid-infrared transparent solar absorber as the middle layer, and a thermo-radiative (TR) cell as the bottom layer. When producing electrical power, the conventional photovoltaic cell needs to face the sun, while the thermo-radiative cell needs to face the ultra-cold deep space. To address the challenges when combining them together, we introduce a mid-infrared transparent solar absorber layer between the solar cell and the thermo-radiative cell. Due to the mid-infrared transparent nature of the solar cell (top layer) and solar absorber (middle layer), the bottom thermo-radiative cell layer is only radiatively coupled to the ultra-cold deep space, while the top and middle layers will utilize the entire solar spectrum by either generating electricity directly or converting unused solar radiation into heat to provide thermal energy for the bottom layer. We analyzed the theoretical performance of the TRAP™ cell using detailed balance principle when the cell is operating in the outer space from near the Mercury orbit to near Jupiter orbit, calculated the maximum achievable temperature for the TRAP™ cell under only solar illumination without any other active heating, and investigated the effect of an imperfect solar absorber on the performance of the TRAP™ cell.**

I. INTRODUCTION

In the space environment the thermal balance of a spacecraft is dominated by radiation. This radiation includes solar radiation as well as the thermal radiation. The incoming solar irradiation and outgoing thermal radiation usually occupy different spectral ranges. The sunlight is in the wavelength range spanning from ultraviolet (UV) to approximately 3 μm, whereas the thermal radiation is usually in the mid-infrared range. Therefore, it is possible to harvest both solar radiation and thermal radiation [1] using the same device in different operation modes.

It is well-known that solar radiation can be harvested using the photovoltaic cell. Thermo-radiative cell, as a type of emissive energy harvester (EEH) [2], was proposed by Strandberg [3] to convert heat into electricity and reject the unused heat to the heat sink via thermal radiation. In deep space, the extremely cold universe (3 K) provides a robust heat sink. Even for a heat source with a temperature below 373 K, the corresponding Carnot efficiency can be more than 99%. Therefore, for outer space applications, the thermo-radiative cell is one of the most promising devices to harvest the outgoing thermal radiation by utilizing the ultracold deep space temperature [4-5].

There are two primary challenges that need to be overcome in order to integrate photovoltaic cells and thermo-radiative cells into a single device to simultaneously harvest solar energy and outgoing thermal radiation. First, the thermo-radiative cell should not have radiative communication with the sun (e.g., receive substantial solar energy from the sun), otherwise it will just behave as a low-efficiency solar cell. Second, to harvest the outgoing thermal radiation, the thermo-radiative cell must be able to radiatively communicate with ultracold deep space. Therefore, if there are layers on top of the thermo-radiative cell, these layers must be transparent in the mid-infrared region of typical thermal radiation, similar to the "atmospheric window" in the terrestrial EEH.

To address these two challenges, the layers on top of the thermo radiative cell should be able to utilize the entire solar spectrum by either converting it into electricity or absorbing it as heat, but should also be transparent in the mid-infrared to provide a "path" for the thermal radiation between the thermo-radiative cell and the ultracold deep space. GaAs is commonly used to make space photovoltaic cells to convert a portion of the solar spectrum into electricity (as well as some waste heat during thermalization process). It is also transparent in the range of 1 to 16 μm. An additional layer is required to absorb the rest of solar spectrum as heat, while remaining transparent in the mid-infrared spectral region, i.e., a mid-infrared transparent solar absorber (such as Ge, GaSb, $In_xGa_{1-x}As_ySb_{1-y}$). Consequently, even though the entire device faces the sun during the operation, the strong solar radiation only interacts with the first two layers, while the bottom thermo-radiative cell layer utilizes the heat from the top two layers to generate electricity and the net thermal radiation of thermo-radiative cell goes outwards since it only "sees" the ultracold deep space. Therefore, such thermo-radiative assisted photovoltaic (TRAP™) cell technology could achieve the simultaneous harvesting of incoming sunlight and outgoing thermal radiation in the space environment. Figure 1 shows the structure of TRAP™ cell and its operation mechanism.

978-1-6654-6060-6/23 $31.00 © 2023 IEEE

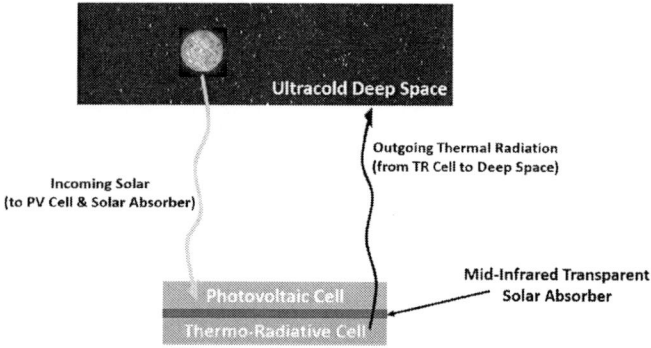

Figure 1: TRAP™ cell for simultaneous energy harvesting. The PV cell converts a portion of the solar spectrum into electricity and waste heat, while the solar absorber converts the rest of the solar spectrum into heat. These two layers utilize the entire solar spectrum and both are transparent in the mid-infrared region. Hence the bottom TR cell will not be affected by the sun during the outgoing thermal radiation harvesting process.

II. THEORETICAL PERFORMANCE ANALYSIS FOR TRAP™ CELL

For theoretical efficiency and power density limits analysis, the temperatures of the sun T_s and the universe T_u are assumed to be 5777 K and 3 K, respectively. The top layer of the TRAP™ cell is a photovoltaic cell with a bandgap E_g^{PV}. The theoretical efficiency limit of a single-junction photovoltaic cell is well-established as the Shockley-Queisser limit [6], in which the cell has unity absorptance for photons with energy higher than the bandgap and zero absorptance for photons with energy lower than the bandgap. Each photon (with $h\nu > E_g^{PV}$) can only excite one electron-hole pair, and the portion of photon energy higher than the bandgap ($h\nu - E_g^{PV}$) will be quickly dissipated as heat. The middle layer of the TRAP™ cell is a mid-infrared transparent solar absorber. Under ideal conditions, it will convert the rest of the solar spectrum into heat. Both the PV cell and the solar absorber layers have unity transmissivity below their bandgaps. The bandgaps of the solar absorber and the TR cell are E_g^{abs} and E_g^{TR}, respectively.

The maximum possible efficiency and power output of the TRAP™ cell can be achieved when non-radiative recombination is eliminated. Using the principle of detailed balance, the total generated current in the PV or TR cell layer is [7-8]:

$$I(\mu) = q[N_{abs}(\theta_{max}, T, \mu) - N_{emit}(\theta_{max}, T, \mu)]$$

where $\mu = qV_{PV}$ or qV_{TR}, q is the elementary charge, N is the photon flux absorbed or emitted by the cell, θ_{max} is the maximum angle for absorption or emission, T is the temperature of the ambient (sun or universe) for absorption or the cell for emission. The total power output of the TRAP™ cell P_e can thus be evaluated as:

$$P_e = P_e^{PV} + P_e^{TR} = I_{PV}(\mu_{PV})V_{PV} + I_{TR}(\mu_{PV})V_{TR}$$

The maximum power P_e^{max} can be achieved when optimized loads are connected to the PV cell and TR cell. The efficiencies

of the TRAP™ cell η_{TRAP} and the PV cell η_{PV} are evaluated at the maximum power output conditions, i.e.,

$$\eta_{TRAP} = \frac{P_e^{max}}{P_{in}^{solar}} = \frac{I_{PV}^{opt}V_{PV}^{opt} + I_{TR}^{opt}V_{TR}^{opt}}{\frac{\Omega_s}{\pi}\sigma T_s^4}$$

and

$$\eta_{PV} = \frac{P_{e\ max}^{PV}}{P_{in}^{solar}}$$

where P_{in}^{solar} is the solar energy incident on the TRAP™ cell, I_{PV}^{opt}, I_{TR}^{opt}, V_{PV}^{opt}, V_{TR}^{opt} are the currents and voltages in the PV and TR cells respectively when optimized loads are connected to the PV cell and TR cell, Ω_s is the solid angle subtended by the sun which depends on the radius of the sun and distance between the sun and the TRAP™ cell, and σ is the Stefan-Boltzmann constant.

III. RESULTS

When evaluating the theoretical performance of the TRAP™ cell, we assume: 1) the bandgap of the PV cell varies from 0.8 eV to 2.5 eV; 2) the bandgap of the solar absorber is around 0.41 eV ($\lambda_g^{abs} = 3$ μm which covers the solar spectrum); 3) the bandgap of the TR cell is 0.1 eV. The maximum output power density and efficiency of the TRAP™ cell as a function of operating temperature and PV cell bandgap are shown in Figure 2.

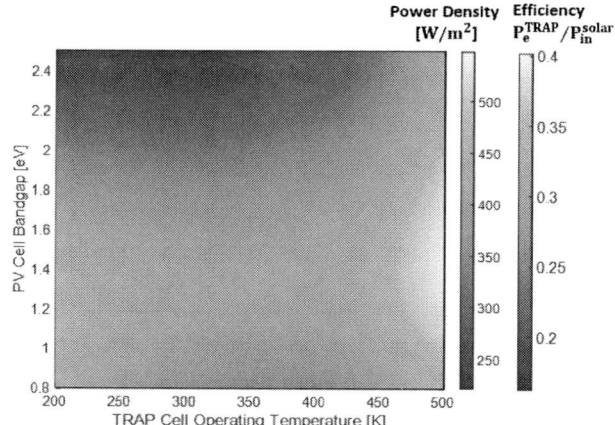

Figure 2: TRAP™ cell maximum output power density and efficiency as a function of operating temperature and PV cell bandgap. The solar absorber and TR cell are assumed to have bandgaps 0.41 eV and 0.1 eV, respectively.

Figure 2 shows that at either lower temperature range (in which additional cooling is required) or higher temperature range, the TRAP™ cell shows better performance, i.e., higher output power density and efficiency. The bandgap of the PV cell is then fixed to be 1.42 eV, which corresponds to the bandgap of GaAs. The maximum output power density from the PV cell and TR cell are plotted in Figure 3 for comparison. The calculations show that the power density of the PV cell decreases linearly with the temperature increase, while the power density of the TR cell increases rapidly with the temperature increase.

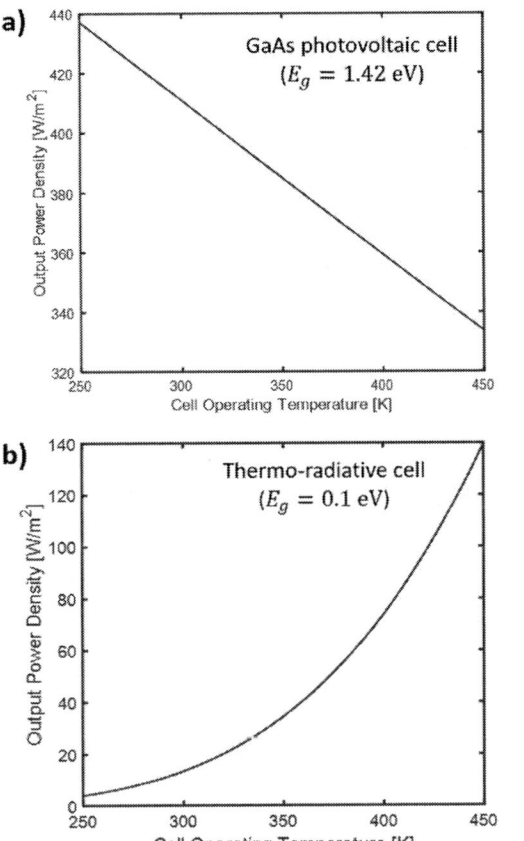

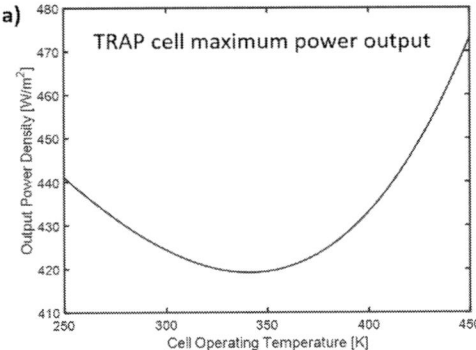

Figure 3: TRAP™ cell power output analysis as a function of temperature. The solar absorber is assumed to have a bandgap of 0.41 eV. a) Maximum power output from the PV cell component of the TRAP™ cell as a function of cell temperature; b) Maximum power output from the TR cell component of the TRAP™ cell as a function of cell temperature.

Using the results from Figure 3, the overall efficiency and total output power density of the TRAP™ cell can be evaluated as a function of temperature, as shown in Figure 4.

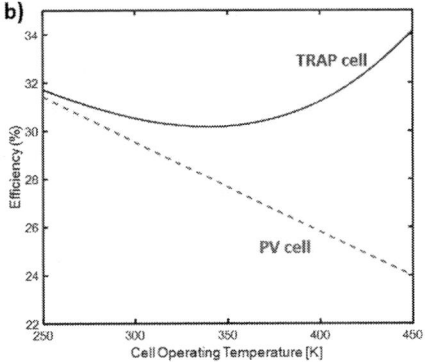

Figure 4: a) Maximum total power output of the TRAP™ cell as a function of cell temperature; b) Efficiency of the TRAP™ cell *vs.* PV cell as a function of cell temperature.

Since the performance of PV cell and TR cell show opposite trends with the temperature, the total power output from the TRAP™ cell decreases with the temperature first, then increases with the temperature. As indicated by Figure 4b, using the TRAP™ cell structure can significantly increase the total efficiency of the traditional PV cell at higher temperatures.

As we know, PV cell panels tend to increase the temperature under solar illumination. Therefore, it would be very challenging to keep the PV or TRAP™ cell at very low temperature (e.g., ~200 K) without the addition of active cooling system. In the range of higher temperature, the TRAP™ cell performance increases rapidly with the temperature. An interesting question arises when considering the applications of TRAP™ cell in deep space: what is the maximum achievable temperature for the TRAP™ cell under solar illumination and without active heating? We conducted the energy balance analysis for the TRAP™ cell, as shown in Figure 5.

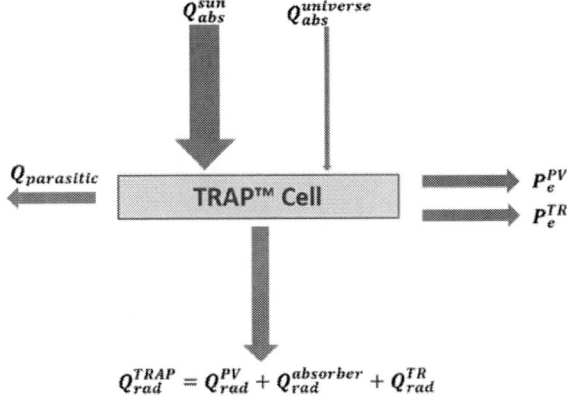

Figure 5: Energy balance of the TRAP™ cell. Incoming energy flows include the absorbed radiation from the sun Q_{abs}^{sun} at 5777 K and from the universe $Q_{abs}^{universe}$ at 3 K. Outgoing energy flows include the electrical power generation P_e ($= P_e^{PV} + P_e^{TR}$), outgoing thermal radiation from the TRAP™ cell Q_{rad}^{TRAP}, and the heat flow from the TRAP™ cell to the other parts of the spacecraft $Q_{parasitic}$.

As long as $Q_{parasitic} > 0$, the maximum power output conditions for the TRAP™ cell is achievable without any active heating to the TRAP™ cell. As the temperature continuously increases, both the total electrical power generation P_e and the outgoing thermal radiation from the TRAP™ cell increase. When $Q_{parasitic} = 0$, the temperature of the TRAP™ cell reaches maximum when there is no active heating. The temperature of the TRAP™ cell can continue rising if active heating is allowed, i.e., $Q_{parasitic} < 0$ which means the direction of heat transfer is now from the other part of the spacecraft to the TRAP™ cell. When the TRAP™ cell is used for applications near the orbits of different planets, the maximum achievable temperature varies as the solar intensity changes (due to the change of Ω_s). The maximum achievable temperature for the TRAP™ cell when its location varies from near Mercury to near Jupiter is calculated and shown in Figure 6.

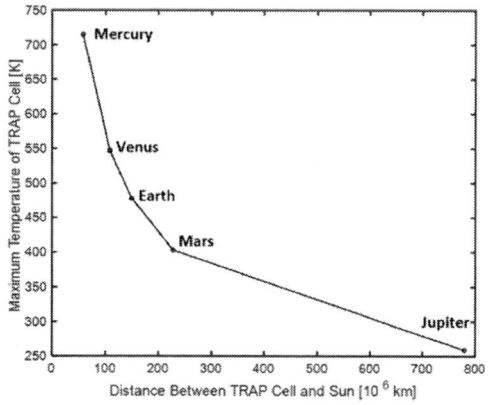

Figure 6: TRAP™ cell maximum achievable temperature without active heating as a function of distance between the cell and the sun.

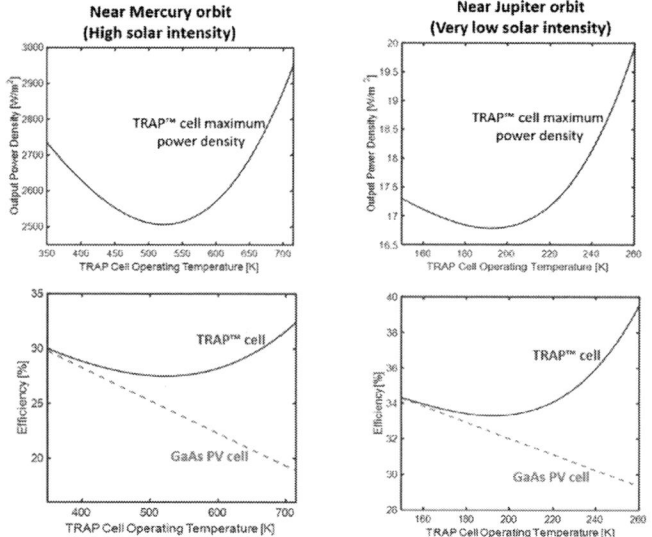

Figure 7: TRAP cell maximum output power density and efficiency as a function of operation temperature when the cell is located near Mercury orbit (upper left & lower left figures) and near Jupiter orbit (upper right & lower right figures).

We then calculated the performance (power density and efficiency) of the TRAP™ cell near Mercury orbit (high solar intensity) and near Jupiter orbit (very low solar intensity), shown in Figure 7. The trend of the TRAP™ cell performance vs. operating temperature is similar, despite the fact the cell is operating in different temperature ranges with quite different output power densities.

We have assumed the bandgap of the mid-infrared solar absorber layer is about 0.41 eV. An absorber with such band gap could absorb solar spectrum up to 3 μm which can be thought as an ideal solar absorber. Ideally, the absorption band of the solar absorber layer (determined by E_g^{abs} or λ_g^{abs}) should cover the solar spectrum. If λ_g^{abs} is too small (e.g., < 2 μm), a small portion of the solar radiation will penetrate the TR cell and reduce the performance of TR cell. If λ_g^{abs} is too large, the width of the spectral transparent window (between λ_g^{abs} and λ_g^{TR}) for the TR cell will be reduced. Potential solar absorber candidates can be germanium, GaSb, or $In_xGa_{1-x}As_ySb_{1-y}$. It is worthwhile to understand the effect of an imperfect solar absorber on the TRAP™ cell performance. We compared the behavior of two mid-infrared solar absorbers. One absorbs infrared radiation up to 1.6 μm (imperfect solar absorber), and the other one absorbs infrared radiation up to 3 μm (perfect solar absorber). Figure 8 shows the net photon flux from the thermo-radiative cell surface between λ_g^{abs} and λ_g^{TR}. Positive values mean the net photon flux is leaving out of the TR cell surface. For an imperfect solar absorber ($\lambda_g^{abs} = 1.6$ μm), the net photon flux becomes positive when the temperature is above about 290 K. Since thermal radiation is proportional to the 4th power of temperature, while the incoming solar radiation in the spectral window between λ_g^{abs} and λ_g^{TR} is fixed, the bottom TR cell layer should be able to function normally when the device temperature is higher than 300 K.

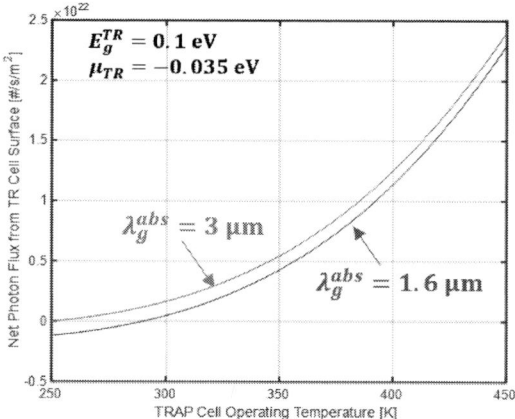

Figure 8: Net photon flux from the thermo-radiative cell surface between λ_g^{abs} and λ_g^{TR}. Positive values mean the net photon flux is leaving out of the TR cell surface.

Figure 9 shows the TRAP™ cell maximum output power density as a function of cell temperature with an ideal mid-infrared solar absorber $\lambda_g^{abs} = 3$ μm and an imperfect mid-infrared solar absorber $\lambda_g^{abs} = 1.6$ μm. As indicated by Figure 9, the performance degradation of using an imperfect mid-infrared transparent solar absorber seems small. A solar absorber that can absorb the solar spectrum up to 1.8-2.5 μm should be good enough.

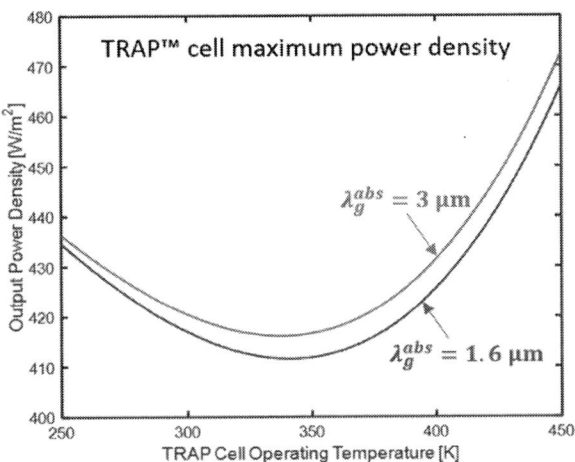

Figure 9: TRAP™ cell maximum output power density as a function of cell temperature with an ideal mid-infrared solar absorber $\lambda_g^{abs} = 3$ μm (red curve) and an imperfect mid-infrared solar absorber $\lambda_g^{abs} = 1.6$ μm (blue curve).

For the operation of the PV cell, the incoming sunlight can only come from one side of the cell for simple geometry configuration. However, outgoing thermal radiation can come from both sides of the cell. If the thermo-radiative cell layer can be designed to radiate into both directions when operating in outer space, it is possible to further increase the power generation by the TR cell part, which also increases the total output power by the TRAP™ cell. Figure 10 shows that when the TR cell can radiate from both sides, the total power generation and efficiency can be significantly improved.

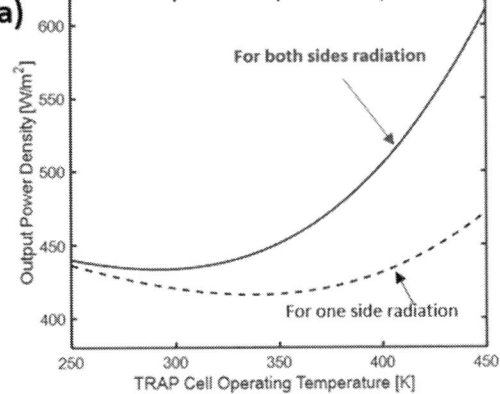

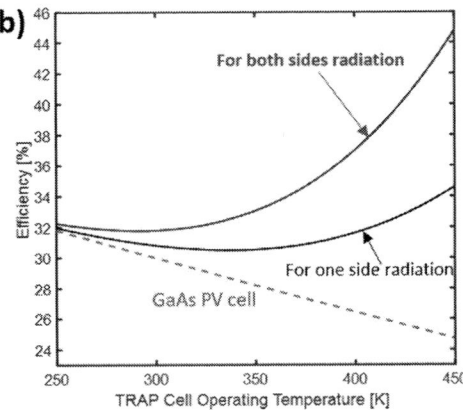

Figure 10: a) Maximum output power density of the TRAP™ cell when the TR cell radiates into deep space from only one side and from both sides; b) Efficiency of the TRAP™ cell when the TR cell radiates into deep space from only one side and from both sides, compared with only GaAs PV cell.

IV. CONCLUSIONS

In summary, we investigated the theoretical efficiency and maximum output power density of the TRAP™ cell when operating in outer space using detailed balance principle. The performance of a TRAP™ cell is significantly better than a PV cell alone, especially at higher operating temperatures. The maximum achievable operating temperature for the TRAP™ cell under only normal solar illumination varies from 715 K (near Mercury) to ~260 K (near Jupiter). The temperature-dependent behavior of the TRAP™ cell is consistent at different distances from the sun (i.e., from near Mercury to near Jupiter), while operating temperature range and output power density range vary with distance. The use of an imperfect solar absorber only slightly reduces the TRAP™ cell performance. When the TR cell component is able to radiate in both directions, the performance of the TRAP™ can be significantly enhanced. However, the theoretical analysis in the paper considers only radiative recombination process, neglecting all non-radiative processes. One of future work could be investigating the practical performance of TRAP™ cell by considering the non-radiative recombination processes in the device and using experimental optical, electrical, and thermal properties of materials. Fabrication of such TRAP™ cell device can be quite challenging. Developing a high-quality high-performance thermo-radiative cell (e.g., using InSb for TR cell fabrication, see ref. [9]) should be the focus of next step research.

V. ACKNOWLEDGMENT

This work was supported by NASA SBIR funding under contract No. 80NSSC22PA921. We would also like to thank Dr. Geoffrey Landis from NASA Glenn Research Center and Prof. Jamie Phillips from University of Delaware for fruitful discussions.

REFERENCES

[1] Chen, Z., Zhu, L., Li, W., and Fan, S. "Simultaneously and Synergistically Harvest Energy from the Sun and Outer Space," Joule, 3, 101-110, 2019.

[2] Byrnes, S.J., Blanchard, R., and Capasso, R. "Harvesting renewable energy from earth's mid-infrared emissions," Proc. Natl. Acad. Sci. USA, 111, 3927-3932, 2014.

[3] Strandberg, R. "Theoretical efficiency limits for thermoradiative energy conversion," J. Appl. Phys., 117, 055105, 2015.

[4] Wang, J.J., Chen, C.-H., Bonner, R. and Anderson, W.G. "Thermo-radiative Cell - A New Waste Heat Recovery Technology for Space Power Applications," AIAA Propulsion & Energy Forum 2019

[5] Santhanam, P. and Fan, S. "Thermal-to-electrical energy conversion by diodes under negative illumination," Phys. Rev. B, 93, 161410, 2016

[6] Shockley, W., and Queisser, H.J. "Detailed balance limit of efficiency of p-n junction solar cells," J. Appl. Phys. 32, 3, 510–519, 1961

[7] Hsu, W.-C., Tong, J.K., Liao, B., Huang, Y., Boriskina, S.V., and Chen, G. "Entropic and near-field improvements of thermo-radiative cells," Sci. Rep., 6, 34837, 2016

[8] Xu, Y., Gong, T. & Munday, J.N. "The generalized Shockley-Queisser limit for nanostructured solar cells," Sci. Rep., 5, 13536, 2015

[9] Wang, J.J., Van Velson, N., Moon, E., Lentz, R. and Phillips, J. "Development of InSb Thermo-Radiative Cell for Waste Heat Recovery of Radioisotope Power Systems," 50th Intl. Conf. on Environ. Sys., July 12-15, 2021

A study of Cell Cracks Formation During Freight Shipping : Monitoring Shock and Temperature in Real-Time & Assessing Damages With Pre and Post-Transit Characterizations of PV Modules

Cécile Molto[1], Dylan J. Colvin[1], Farrukh ibne Mahmood[2], Fang Li[2], Ryan Smith[3], Govindasamy TamizhMani[2], Hubert Seigneur[1]

[1]Florida Solar Energy Center University of Central Florida (UCF-FSEC), Cocoa, FL, 32922, USA
[2]Photovoltaic Reliability Laboratory, Arizona State University (ASU-PRL), Mesa, AZ, 85212, USA
[3]Pordis LLC, Austin, TX, 78729, USA

Abstract — The solar photovoltaic (PV) industry often experiences module damage during transportation. PV modules stacked horizontally and strapped on wooden pallets may develop failures impacting their efficiency (glass shattering, cracks, frame indentation, etc.), especially when modules are not packed by professionals. The appearance of failures during the shipment from one laboratory to another can be detrimental to research studies. In this study, we tested a reusable plastic pallet designed to ship modules vertically. Five technologies of commercial PV modules were shipped together from Arizona to Florida including: glass/glass bifacial modules (framed and frameless), glass/transparent backsheet bifacial modules and glass/white backsheet modules (framed). Each module technology has a different physical size. Some modules had pre-existing cell-cracks allowing for the study of cell-crack formation and propagation during shipment. Dark current-voltage (I-V), light I-V and electroluminescence characterization were performed before and after shipment. A datalogger was used to monitor shocks and ambient temperature throughout the shipment and identify when the modules were more susceptible to damage. The modules were tilted upon receipt and the pallet were damaged but with no shattered modules. Time series shock data provided by the datalogger were used to determine potential responsible events. Our results reveal a few new post-transportation cracks for some modules. Modules parameters before and after shipment are also compared regarding the module constructions and their position in the pallet.

Keywords — Crystalline silicon, crack, bifacial commercial modules, electroluminescence, shipment.

I. INTRODUCTION

Photovoltaic (PV) modules can be shipped several times during their lifetime. From the manufacturer to the field where they will be installed, from the manufacturer to the laboratory where they will be studied or from a laboratory to another for research purposes. It remains challenging to minimize/eliminate the risk of failure (cracks, glass shattering, frame indentation etc.) during shipment, in particular when modules are not carefully packaged inside a wooden crate with an appropriate amount of cushioning material. To significantly reduce the cost and efforts related to wooden or cardboard crates, wooden pallets can be used along with a variety of packaging materials including cardboard, plastic wrap, straps, spacers or a combination of these. This approach would be cheaper but less effective to protect the PV modules from being damaged. The problem would become worse for frameless modules which are expected to have a market share of about 20% by 2028 [1] as they are less robust to mechanical stresses [2], [3]. When PV modules are damaged during shipment, it might impact their performance and reliability, in particular when cell cracks are formed [4]–[10]. It can also compromise their mechanical/electrical integrity and could become a safety hazard. If modules were shipped as part of ongoing research experiments, damage might result in the loss of invaluable sample sets.

In this study, we tested a new type of pallet to ship modules from Arizona to Florida. The pallet is reusable compared to a wooden pallet that can be reused only for a limited amount of time. It is also made of recyclable plastic while wooden pallets are not environmentally friendly. Also, it allows to ship vertically a greater number of modules per pallet and pallets can be stacked on top of each others which saves a significant amount of space. Five different commercial module technologies were shipped together (framed/frameless, glass-glass/glass-backsheet) to determine if any specific module construction is more susceptible to damage during shipment. Modules were characterized (dark current-voltage (I-V), light I-V, electroluminescence) before and after the shipment to monitor for the cell cracks formation or the propagation of existing cracks and evaluate the impact on the modules' performance. A datalogger was placed inside the pallet to monitor the shocks and temperature throughout the shipment and identify when the modules were most likely damaged.

II. EXPERIMENT

A. Module technologies

A total of 19 modules comprising full size and half-cut bifacial Passivated Emitter and Rear Cell (PERC) cells were shipped. They have different module constructions: glass/transparent backsheet (G/TBS) framed (manufacturer 1), glass/glass (G/G) framed (manufacturer 2), glass/white backsheet (G/WBS) framed (manufacturer 3), G/G frameless (manufacturer 4), G/TBS framed (manufacturer 4). Details are

978-1-6654-6060-6/23 $31.00 © 2023 IEEE

given in Table 1. Some modules have pre-existing cracks while others do not to study crack formation and propagation.

TABLE I
MODULES CONSTRUCTION DETAILS

Technology (number)	Manu-facturer	Cell technology	Module construction
A (5)	M1	PERC-144 half cut	G/TBS Framed
B (2)	M2	PERC-144 half cut	G/G Framed
C (4)	M3	PERC-144 half cut	G/WBS Framed
D (4)	M4	PERC-72 full cells	G/G Frameless
E (4)	M4	PERC-144 half cut	G/TBS Framed

B. Packing process

The sidewalls of the pallet can be adjusted to fit most commercial PV modules. Modules were loaded vertically in the following order: Technology C (x4)/Technology E (x4)/Technology B (x2)/Technology D (x4)/Technology A (x5). Frameless modules were placed with corner protections and foam on both sides. Foam was placed on both pallet sides with straps as shown in Fig. 1. A MSR175 shock and temperature datalogger placed inside the pallet and protected from humidity.

Fig. 1.   Picture of the pallet ready for shipment.

C. Characterization techniques

Modules were characterized before and after the shipment with I-V curves, Dark I-V curves and electroluminescence (EL). We used a Sinton FMT-350 flash tester to obtain multi-irradiance I-V and suns-Voc. Dark I-V curves were obtained from 0A to I_{sc} current. EL images were measured using a modified consumer DSLR with a 950 nm longpass filter.

III. PRELIMINARY RESULTS AND DISCUSSION

We found it easier to load the modules in this reusable plastic pallet compared to a wooden pallet. Pictures of the pallet upon receipt are given in Fig. 2. The original green straps were broken, modules had slid and a piece of the pallet was broken. These events could have occurred in a different sequence; still, the positioning of the panels during delivery strongly suggests

that these events are related. The freight company added the yellow strap to keep the modules from completely falling off. Despite these events, we did not experience any catastrophic module breakage (e.g. shattered glass, broken frame, etc.).

Fig. 2.   Pallet upon delivery showing that the modules slid. Straps and a piece of the pallet were also broken.

An analysis of the shocks and temperature data is given in Fig. 3. The different shock events groups have been framed in black and red. In black (10 groups) we were able to find a connection with the transit operations: 1) pallet loaded with the modules; 2) departure from Arizona (AZ); 3) 1st transit at Albuquerque (NM); 4) 2nd transit at Amarillo (TX); 5) 3rd transit at Dallas (TX); 6) 4th transit at Memphis (TN); 7) 5th transit at Conley (GA); 8) 6th transit at Orlando (FL); 9) 7th transit at Cocoa; 10) delivery at the laboratory. The events framed in red do not match with any documented event.

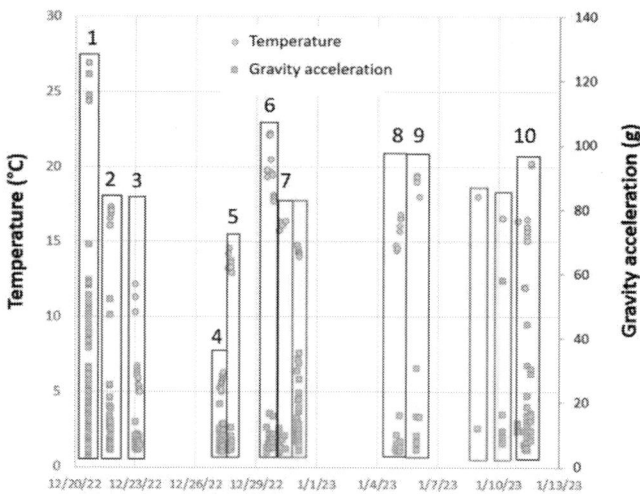

Fig. 3.   Shocks recorded during shipment and corresponding temperatures.

Modules did not slid at events 1, 2 and 10 as a researcher was witnessing the operations. The pallet might have been broken and modules could have slid during any other event. However, we can hypothesized this happened during the shock event group number 7. This latter is followed by another shock event group (framed in red) that could correspond to the repositioning

978-1-6654-6060-6/23 $31.00 © 2023 IEEE

and restrapping of the modules by the carrier as it does not match with any transit step. This is support by the high shock magnitude which was also observed during modules loading (shock event group 1). The temperature ranged from 5°C to 26 °C and some studies have demonstrated that cold temperatures can favorise cracks formation [11], [12].

EL measurements revealed that some modules have additional cracks after shipment. An example is given in Fig. 4 for a G/WBS module.

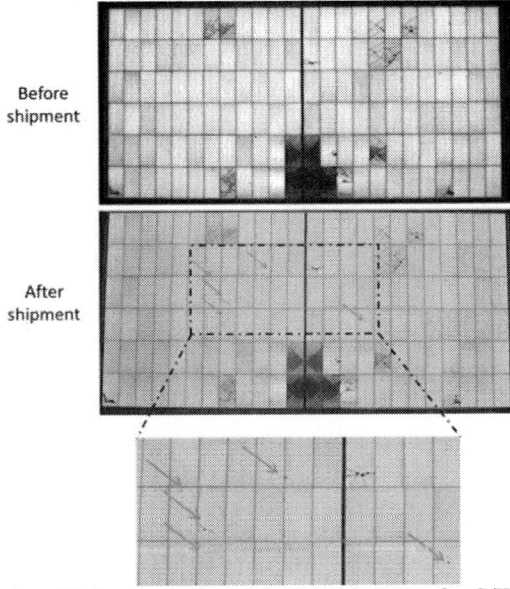

Fig. 4.   EL images before and after shipment of a G/WBS module with pre-existing cracks. New cracks are indicated by orange arrows.

EL images analysis is in progress to identify potential propagation of pre-existing cracks. We are currently comparing the modules parameters before and after shipment with I-V and dark I-V measurements. Results will be analyzed based on the module constructions and their position in the pallet.

IV. CONCLUSION AND FUTURE WORK

The preliminary results of this study indicate that modules were not broken; only a few new cell-cracks have developed despite the fact that the pallet underwent a significant shock event (pallet broken and modules slid). The datalogger allowed us to match the shock events with the different transit steps and we hypothesized during which one the pallet might have been broken. This study is still in progress to evaluate the potential relationship between crack formation and module construction. The impact of the module position in the pallet will be also studied. Monitoring shock events during transportation could be a solution for the PV module owners to figure out which transportation company is responsible when damages happen. For a future work, a Global Positioning System (GPS), displacement sensors, or even a camera could be also used for more accuracy.

REFERENCES

[1]   "ITRPV Roadmap 2021 - Twelfth edition - 2020 results." Accessed: Feb. 07, 2019. [Online]. Available: https://itrpv.vdma.org/

[2]   A. J. Beinert, M. Ebert, U. Eitner, and J. Aktaa, "Influence of Photovoltaic Module Mounting Systems on the Thermo-Mechanical Stresses in Solar Cells by FEM Modelling," in *32nd European Photovoltaic Solar Energy Conference and Exhibition; 1833-1836*, Munich, Germany, 2016, p. 4 pages, 7163 kb. doi: 10.4229/EUPVSEC20162016-5BV.1.14.

[3]   G. Oreski *et al.*, "Designing New Materials for Photovoltaics: Opportunities for Lowering Cost and Increasing Performance through Advanced Material Innovations.," SAND2021-4837R, 1779380, 695676, Apr. 2021. doi: 10.2172/1779380.

[4]   M. Bdour, Z. Dalala, M. Al-Addous, A. Radaideh, and A. Al-Sadi, "A Comprehensive Evaluation on Types of Microcracks and Possible Effects on Power Degradation in Photovoltaic Solar Panels," *Sustainability*, vol. 12, no. 16, p. 6416, Aug. 2020, doi: 10.3390/su12166416.

[5]   H. Seigneur, K. Ogutman, E. Schneller, K. O. Davis, and W. V. Schoenfeld, "Effect of cracks on spatially resolved c-Si solar cell parameters," in *2016 IEEE 43rd Photovoltaic Specialists Conference (PVSC)*, Portland, OR, USA, Jun. 2016, pp. 0704–0707. doi: 10.1109/PVSC.2016.7749692.

[6]   E. J. Schneller, R. Frota, A. M. Gabor, J. Lincoln, H. Seigneur, and K. O. Davis, "Electroluminescence Based Metrics to Assess the Impact of Cracks on Photovoltaic Module Performance," in *2018 IEEE 7th World Conference on Photovoltaic Energy Conversion (WCPEC) (A Joint Conference of 45th IEEE PVSC, 28th PVSEC & 34th EU PVSEC)*, Waikoloa Village, HI, Jun. 2018, pp. 0455–0458. doi: 10.1109/PVSC.2018.8547636.

[7]   C. Buerhop *et al.*, "Evolution of cell cracks in PV-modules under field and laboratory conditions," *Prog. Photovolt. Res. Appl.*, vol. 26, no. 4, pp. 261–272, Apr. 2018, doi: 10.1002/pip.2975.

[8]   C. M. Whitaker, B. G. Pierce, A. M. Karimi, R. H. French, and J. L. Braid, "PV Cell Cracks and Impacts on Electrical Performance," in *2020 47th IEEE Photovoltaic Specialists Conference (PVSC)*, Calgary, AB, Canada, Jun. 2020, pp. 1417–1422. doi: 10.1109/PVSC45281.2020.9300374.

[9]   M. Dhimish, V. Holmes, B. Mehrdadi, and M. Dales, "The impact of cracks on photovoltaic power performance," *J. Sci. Adv. Mater. Devices*, vol. 2, no. 2, pp. 199–209, Jun. 2017, doi: 10.1016/j.jsamd.2017.05.005.

[10] M. Köntges, I. Kunze, S. Kajari-Schröder, X. Breitenmoser, and B. Bjørneklett, "The risk of power loss in crystalline silicon based photovoltaic modules due to micro-cracks," *Sol. Energy Mater. Sol. Cells*, vol. 95, no. 4, pp. 1131–1137, Apr. 2011, doi: 10.1016/j.solmat.2010.10.034.

[11] H. Seigneur *et al.*, "Microcrack Formation in Silicon Solar Cells during Cold Temperatures," in *2019 IEEE 46th Photovoltaic Specialists Conference (PVSC)*, Chicago, IL, USA, Jun. 2019, pp. 1–6. doi: 10.1109/PVSC40753.2019.9198968.

[12] E. J. Schneller, H. Seigneur, J. Lincoln, and A. M. Gabor, "The Impact of Cold Temperature Exposure in Mechanical Durability Testing of PV Modules," in *2019 IEEE 46th Photovoltaic Specialists Conference (PVSC)*, Chicago, IL, USA, Jun. 2019, pp. 1521–1524. doi: 10.1109/PVSC40753.2019.8980533

Quantifying Real-World Sources of Error in Redundant GHI Measurements

Josh Peterson, Julie Chard, Alex Bryan

GroundWork Renewables, Inc., Sand City, CA 93955, USA

Abstract — Many sources of pyranometer measurement uncertainty can be explicitly studied in the lab. These pointed, indoor studies work well to characterize some aspects of the instrument. However, contributions to measurement uncertainty of real-world conditions, such as instrument cleanliness and alignment, maintenance practices, and the inherent noise in the measurement are difficult to quantify. This study quantifies these sources of error by comparing the outputs of paired, collocated pyranometers in the field over the two-year period from 2021-2022 at each of 290 stations spread over a wide range of geographical locations in the United States.

I. Introduction

Pyranometer-based irradiance measurements are the backbone of a solar resource assessment. They are also a critical input to models that predict and evaluate the performance of a PV power plant. Alongside the irradiance value is a corresponding estimate of the irradiance measurement uncertainty.

The uncertainty of pyranometer-based measurements is addressed in the literature [1]-[4]; however, there is no international consensus on pyranometer measurement uncertainty. The uncertainty is typically calculated from a combination of individual sources such as directional response, calibration, temperature response, tilt response, non-linearity, spectral response, zero offsets, aging, and datalogger accuracy. Some sources are correlated, or melded, with other sources (e.g., non-linearity and spectral response are included in the directional and calibration error, respectively [4]). Others are either corrected for (e.g., temperature response) or negligible (e.g., tilt response ~ 0.2%). According to the ASTM G0213-17 standard [1], the directional response and calibration uncertainties account for more than 50% of the overall measurement uncertainty.

All of the above aspects of pyranometer uncertainty can be explicitly studied in the lab. An additional critical component of the uncertainty budget is the maintenance-related, or real-world uncertainty which is difficult to quantify in the lab. Contributions to measurement uncertainty due to operation in the field include instrument cleanliness and alignment, maintenance practices, and inherent noise in the measurement. Users are directed by [1] to estimate the maintenance uncertainty to the best of their ability.

Here we quantify these real-world sources of error by comparing the measurements made by paired, collocated pyranometers in the field for a large number of field campaigns. The data for this study originates from the large network of GroundWork Renewables, Inc. (GroundWork) measurement stations with redundant global horizontal irradiance (GHI) measurements.

GroundWork is a privately owned solar monitoring company operating primarily across North America. At any given time, GroundWork has over 150 client-owned stations in its maintenance and monitoring program. Each station is deployed for at least one year. During operation GroundWork collects data, performs quality control tests, and delivers data to clients on a routine basis. A local technician is contracted to perform weekly maintenance visits at each station. Most stations include redundant GHI measurements along with rainfall, air temperature, wind speed and direction, relative humidity, and air pressure.

The stations included in this study encompass most of the continental United States with a higher concentration in the southern portion of the country. This study also includes ~10 stations south of the Continental United States (Hawaii, Mexico, Puerto Rico, Columbia) and ~10 stations that were operated in southern Canada.

This large number of stations spanning a broad geographical area with a diversity of maintenance technicians allows statistical analysis of the inherent error in measurements made due to operation under real-world conditions.

This paper is organized as follows. First the GHI data from two sensors for a single station is investigated as a function of both time and solar elevation. Throughout this paper, the relationship between the two sensors' measurements is presented in both "difference" and "ratio" forms as defined in the following section. Next, the directional response bias of the sensors is discussed. A directional response adjustment is applied to the data to minimize this bias. A second adjustment is applied to normalize minor calibration-related differences. Following these two adjustments, the difference and ratio of all stations are investigated by grouping all days and stations together as an overall measure of the typical deviation between two sensors resulting from operating in the field under real-world conditions.

II. Time Series Irradiance Data

All 290 stations included in this study were instrumented with redundant Hukseflux SR30 pyranometers [4]. The two sensors on each station were mounted east and west of one another on the same crossarm and oriented horizontally for measurement of GHI. The GHI measurements were temperature-corrected

978-1-6654-6060-6/23 $31.00 © 2023 IEEE

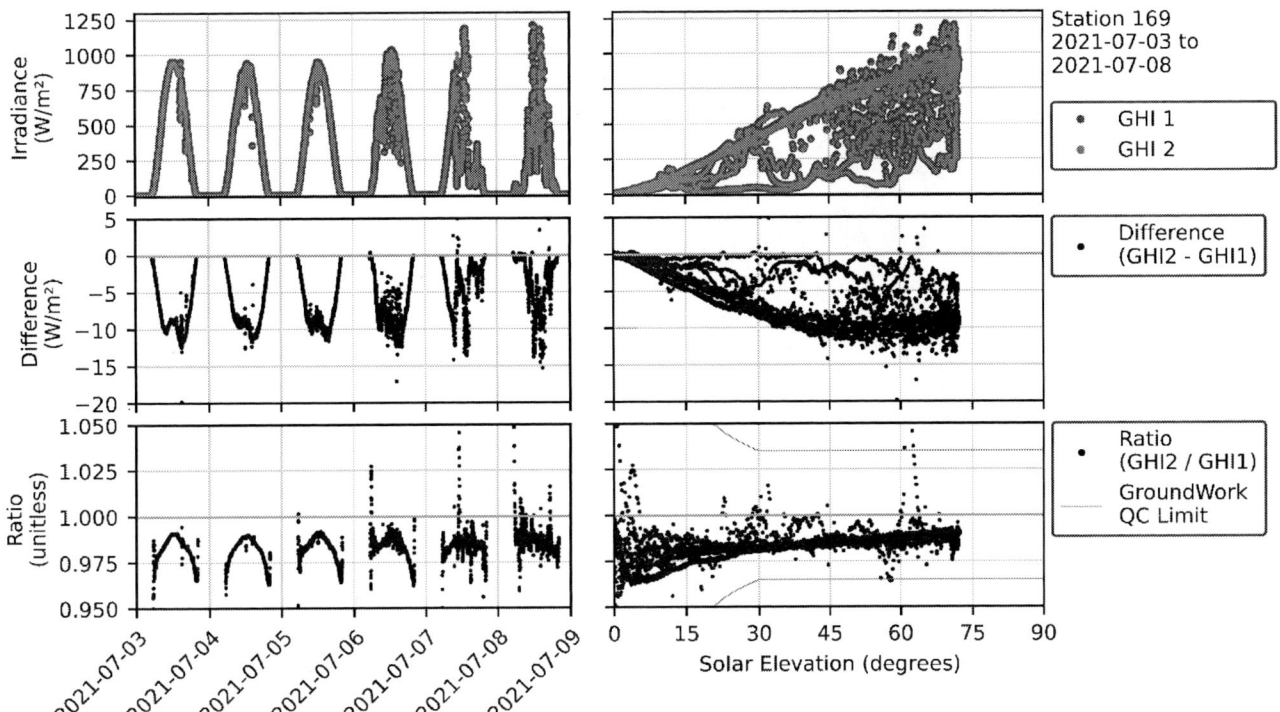

Fig. 1.   Sample time series data set (left) and sample data set plotted as a function of solar elevation (right). GHI (upper), Difference (middle), Ratio (lower). In the upper plots GHI1 is shown in blue. GHI2 is shown in red.

using sensor-specific coefficients supplied by the manufacturer. Temperature-corrected measurements were sampled at 3-second intervals and averaged to one-minute data points. The data in this study spans the period from January 1, 2021 to December 31, 2022. There are more than 100 million one-minute data points over this two-year period.

From the two GHI measurements, the difference between the two sensors' measurements (1) and the ratio of the two sensors' measurements (2) are computed. Since both sensors are identical, except for the directional response differences addressed in section 3, the assignment of GHI1 and GHI2 is arbitrary.

$$\Delta GHI = GHI_2 - GHI_1 \qquad (1)$$

$$ratio = \frac{GHI_2}{GHI_1} \qquad (2)$$

In Fig. 1, a sample time series of redundant GHI measurements is plotted. The data in Fig. 1 was taken in the summer in Indiana, USA. The left panels are time series plots, the right panels plot the data vs solar elevation angle.

The upper plots show the GHI data directly. In the upper plot, GHI1 is shown in blue and GHI2 is shown in red. The middle plots show the difference between the two measurements. The lower plots show the ratio of the two measurements.

Inspection of the upper plot appears to show excellent agreement between the two sensors. Only upon examination of the difference and ratio plots can any disparities be seen.

Differences are greatest at high irradiance values, up to ±10 W/m² between the two sensors. Ratios diverge further from 1 as light levels decrease, up to ±4% around dawn and dusk.

In the time series plots of Fig. 1 (left), there are repeating patterns in the difference and ratio plots from one day to the next. This is especially true in the first three days with clear skies. This pattern could be caused by several sources related to one or both of the sensors: sensor cleanliness, minor tilt-related issues, or differences in the directional responses of the two sensors.

The same irradiance data is plotted on the right in Fig. 1 as a function of solar elevation angle. The data for the clear sky days can be seen to overlap with each other in all three plots. The ratio plot is roughly constant around 0.985, indicating a slight bias of up to ~1.5% between the two pyranometers. Any slope in the trend and/or the scatter around the trend may be attributable to the sources of error mentioned above. These sources of discrepancy will be discussed in the upcoming sections.

III. Directional Response Adjustments

The directional response (DR) of a pyranometer is a measure of how the sensor operates when light is incident from different locations [4]. The directional response of each sensor is unique to that sensor and is characterized by the manufacturer upon construction. The directional response is a known bias that

978-1-6654-6060-6/23 $31.00 © 2023 IEEE   1124

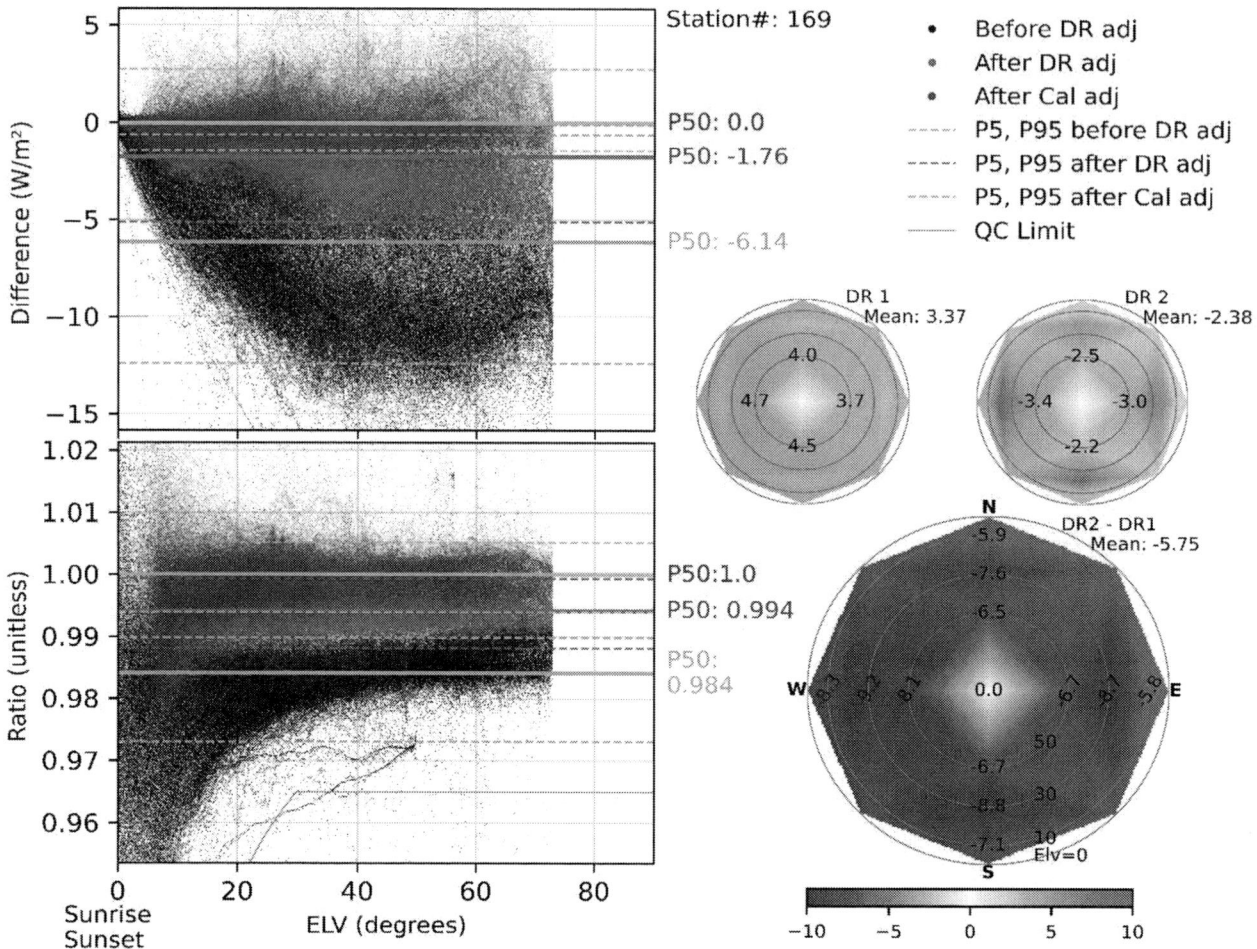

Fig. 2. Difference and ratio before any adjustments (black), after the directional response adjustment (green), and after the calibration adjustment (purple). The directional response of the two sensors is plotted in the lower right.

exists in each sensor. This section discusses this bias and presents a method to correct for it.

The DR of a pyranometer is determined using (3), by comparing the measured irradiance with its expected irradiance at a known zenith (Z) and azimuthal (A) angle.

$$DR(Z, A) = I_{Measured}(Z, A) - I_{Expected}(Z, A) \quad (3)$$

The manufacturer reports 16 DR values; at zenith angles of 40, 60, 70, 80°, and azimuthal angles of 0, 90, 180, 270°. The DR at Z=0 is zero by definition of the calibration process. To generate a continuous DR at all Z and A values a Python interpolating grid function is used. With this function, the DR at any Z and A value is known.

The challenge in applying (3) to GHI measurements is that the light is incident on the sensor from many different locations. Moreover, the ratio of diffuse (DHI) light to direct (DNI) light varies with sky conditions. To overcome this challenge, the GHI measurements are separated into direct and diffuse components using the Erbs model [5, 6].

With proxy values of DNI and DHI for each minute as well as DR grid values for each sensor, the measured GHI is adjusted using (4). The adjustment has a DNI component which is directly correlated with the zenith and azimuth of the sun (Z, A). The adjustment also has a DHI term.

$$GHI_{Adj} = GHI_{meas} - \frac{DR(Z,A) * DNI}{1000} - \frac{DR_{Mean} * DHI}{1000 * cos(45)} \quad (4)$$

Since the DHI is incident on the sensor from all locations, a mean directional response was used for the DHI adjustment term. To generate a mean DR value for each sensor, the grid of DR values computed in the previous step are separated into 1-degree concentric rings of increasing Z. For each ring, the grid points inside this ring are selected and the mean of these points is computed. This process is performed on all rings from 0 to 90°. To compute the mean DR for the sensor, the mean of all the rings is then computed. This is done in an effort to weight the relatively small central rings just as heavily as the large outside rings.

The cosine(45) in the denominator of the DHI term of (4) is needed to translate the normal irradiance directional response calculation to a horizontal measurement. The value of 45 was selected as an average between 0 and 90°. More work is needed to optimize the DR_{Mean} calculation as well as the value of 45°. However, since the DHI value is typically much less than 1000 W/m², the DHI adjustment is relatively small. Therefore, the approximations of DR_{Mean} and 45° are considered acceptable for the purposes of this study.

The adjustment of (4) was applied to all data in the measurement campaign for a single station. This was done to the GHI1 and GHI2 sensors independently since the DR of a sensor is specific to that sensor. A sample of the DR adjustment results is shown in Fig. 2. The three heat map plots in the lower right show the DR of GHI1, GHI2, and the difference between the two DR values (DR2 − DR1). Shades of red indicate a positive DR. Shades of blue represent a negative DR. The DR difference between GHI1 and GHI2 is largely negative, so it can be expected that the difference between the sensors' measurements should also be negative. The DR adjustment aims to minimize this negative bias.

For each minute of the measurement campaign shown in Fig. 1, the difference and ratio are plotted in the left panels of Fig. 2. The difference before the DR adjustment is plotted in black. Notice that the difference trends visible in the right panels of Fig. 1 are also present in Fig. 2. This indicates that the effects seen in Fig. 1 occur throughout the measurement campaign.

The green points in Fig. 2 represent the differences and ratios after the DR adjustments have been applied. The median (P50) difference over all elevation angles is identified as the solid gray (before DR adjustment) and solid green (after DR adjustment) horizontal lines. Notice that before the DR adjustment was applied the median difference was -6.14 W/m². After the DR adjustment was applied the median difference shifted to up to -1.76 W/m². This implies that the DR adjustment helped to minimize the difference between the two sensors. The dashed lines represent the P5 and P95 values. Likewise, the ratio of the two sensors before the DR adjustment had a median value of 0.984 (1.6%). After the DR adjustment the median ratio rose to 0.994 (0.6%).

Ideally the directional response adjustment would shift the data such that after the adjustment, the median difference would be very close to zero and the median ratio would be very close to 1. Some amount of residual noise due to calibration and to operation under real-world conditions is to be expected for an individual station. This reasoning can be used to explain why the adjustments in Fig. 2 did not perfectly correct the issue. However, when analyzing many stations, if the DR is performing as expected, statistically one expects that overall the median difference after the adjustment should be near zero for the bulk of stations.

The directional response adjustment was applied to all 290 stations. Fig. 3 summarizes the median differences for all stations before the adjustment (x axis) and after the adjustment

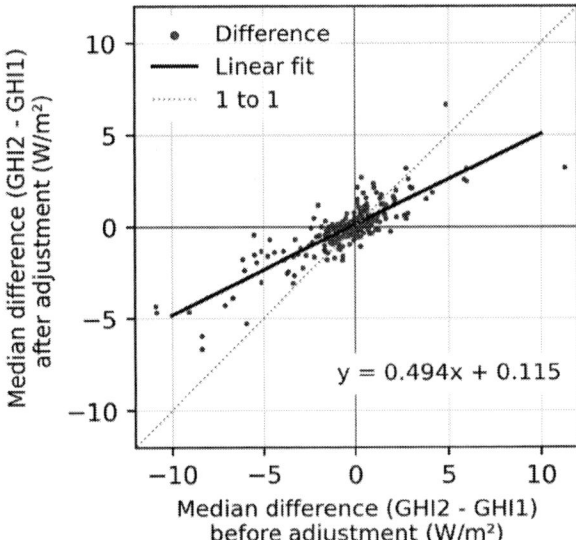

Fig. 3. Median difference between the two sensors before and after the DR adjustment for all 290 stations used in this study.

(y axis). Positive differences before the adjustment generally remain positive after the adjustment. Likewise for the negative differences. A linear fit of the data is plotted in Fig. 3. Without the DR correction the slope of the regression line would be 1. With a perfect DR correction, the slope would be 0. The slope of the fit line is 0.494, an improvement of ~50%, illustrating that the directional response adjustment helps but does not go far enough. Note that most stations have median differences less than 5 W/m² both before and after adjustment.

IV. CALIBRATION ADJUSTMENT

Pyranometer calibrations have some level of uncertainty associated with them. This means that the numerical value of the sensitivity reported on the calibration certificate may not be the actual sensitivity of the sensor. With relation to this study, this can have the effect of one sensor measuring higher than the other by a consistent multiplier. This will reveal itself in the difference plot as an increasing difference as irradiance increases, and the ratio plot being shifted away from 1 by a consistent amount regardless of irradiance. As with directional response biases, calibration-related biases can be taken into account and corrected for.

A second adjustment, applied to the GHI2 sensor, is performed according to (5). This adjustment forces the median ratio of $GHI2_{Adj2}$ and $GHI1_{Adj}$ to be 1 and the median difference to be zero, where $GHI1_{Adj}$ and $GHI2_{Adj}$ are the adjusted GHI values obtained from the directional response adjustment step.

$$GHI2_{Adj2} = GHI2_{Adj} \Big/ Median\left(\frac{GHI2_{Adj}}{GHI1_{Adj}}\right) \tag{5}$$

The data after the calibration adjustment is plotted in Fig. 2 as the purple points. As required by (5), the median ratio is 1.0

after the second adjustment. Likewise, the median difference is 0.0 after the second adjustment.

In Fig. 2, the spread between P95 and P5 decreased significantly after the DR adjustment. The spread did not decrease significantly when the calibration adjustment was applied. However, when the calibration adjustment was applied the P5 and P95 values became more symmetric about the difference =0, ratio =1 lines.

V. DIFFERENCE AND RATIO OF ALL STATIONS

The adjustments discussed in the previous sections were applied to all 290 stations used in this study. After the adjustments, the difference and ratio values were computed for all pairs of sensors. In this section, (1) and (2) were modified to look at the absolute differences and ratios by prescribing GHI2 as the larger sensor. This results in the difference always being positive, and the ratio being greater than or equal to 1.

Fig. 4 plots the results. In the left panels of Fig. 4, the data is plotted vs solar elevation angle. In the middle panels, the data is plotted vs GHI. In the right panels, the data is plotted vs date. A random sample of individual minute level data points are plotted as small blue points. Only after the large collection of data is plotted can the trends in the data be seen.

For all six plots of Fig. 4 the P99, P95, P90, and P50 values were computed. The colored regions of the plot correspond to different bounds when binning by the corresponding x-axis parameter. P values were computed only from elevation angles greater than 10° to exclude erroneous horizon edge effects.

In the left panels of Fig. 4, showing the difference and ratio vs solar elevation, P values were computed for each 1-degree elevation angle. P values were not computed at elevation angles less than 10°. As expected, differences increase with solar elevation. This is because more light is available at higher sun elevations. The opposite is true for the ratio. At low sun elevations, less light exaggerates the ratio. At very high elevation angles the apparent noise in the P values results from a reduced number of available data points. Ratios decrease below 2% at elevation angles greater than ~20° 95% of the time (orange-red boundary).

For the central plots of Fig. 4 (differences and ratios vs GHI), the x-axis GHI value was obtained by averaging the GHI1 and GHI2 data sets after both adjustments. P values were computed on 10-W/m² bins. (Likewise, solar elevations less than 10° were excluded). Similar to the solar elevation plots, P values tend to increase for the differences and decrease for the ratios, as GHI and solar elevation are typically correlated. 95% of the ratios are below 2% for GHIs greater than 250 W/m², representing most of the daylight hours.

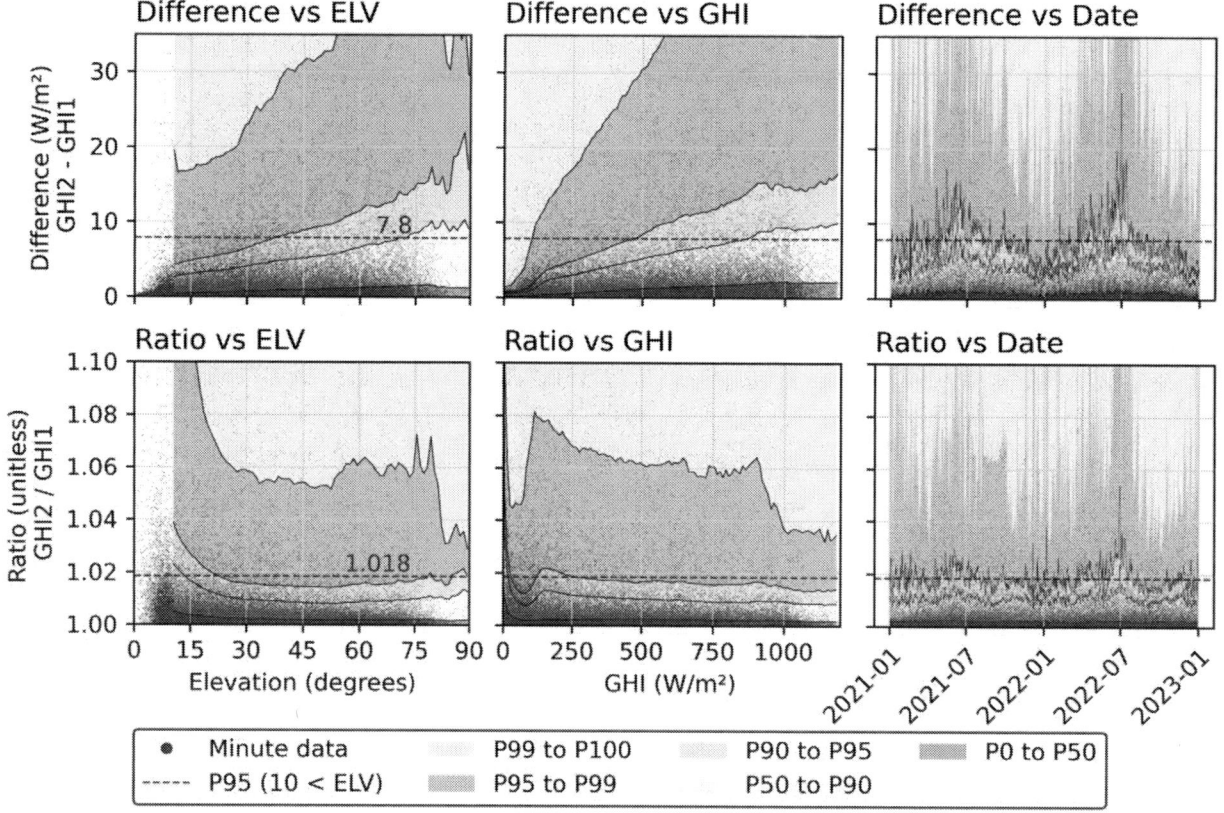

Fig. 4. Difference and ratio results for all stations. Data is plotted vs solar elevation (left), GHI (center), date (right).

Finally, in the right panels of Fig. 4 (differences and ratios by date), P values are computed on a daily basis. There is a visible seasonality in the differences with greater differences during the summer and smaller during the winter. This is consistent with the previous two plots, correlating summer months with higher irradiance values. Ratios, on the other hand, while not as strong, also show mild seasonality toward peaking in the summer, in contrast with trends in the previous two columns. This may be due to soiling and other real-world factors having a larger impact at higher irradiance due to greater attenuation, and/or more soiling during the summer months.

The peak in both difference and ratio that exists around 2022-07-01 is caused by a temporary increase in the number of stations with at least one soiled, unlevel, or non-functioning pyranometer. Issues with a small number of stations can significantly influence the higher P values (e.g., P95), however the P50 is quite well behaved, all things considered.

For all data with elevation angles greater than $10°$, the P50 (median) difference between the two sensors was 0.8 W/m^2 (not shown). The median ratio between the two sensors was 1.002 (0.2%). This 'typical divergence' is the boundary between the yellow and blue regions in Fig. 4.

Likewise, the P95 value was computed for both the difference and the ratio. The P95 of the absolute value of the difference is a measure of how much the two sensors diverged throughout the entire campaign after most of the major known sources of uncertainty are accounted for. This is represented in Fig. 4 as the dashed purple lines. The numerical value of the P95 difference is 7.8 W/m^2. The numerical value of the P95 ratio is 1.018 (1.8%). The P95 values quantify the divergence, at most, attributable to operation under real-world conditions.

These results can be summarized as follows: *For a GroundWork-operated station with weekly maintenance, at solar elevations greater than $10°$, at any time of year, one-minute differences between two collocated, redundant GHI pyranometers due to operation under real-world conditions are less than 7.8 W/m^2, or less than 1.8%, 95% of the time.*

Without applying the DR and calibration adjustments, the P95 difference and ratio values would have been 11.1 W/m^2 and 1.025, respectively. If only the DR correction were applied, the P95 difference and ratio values would have been 9.1 W/m^2 and 1.021, respectively.

The difference and ratio values would be further improved by omitting data flagged during the GroundWork quality control process. For the purpose of this study, quantifying errors due to operation in real-world conditions, no data was excluded.

These results may include other biases which are discrete inputs to the uncertainty budget of a pyranometer. These biases were either accounted for (directional response and calibration) or assumed to be negligible or equal for paired sensors [1, 4]. Some sources of error may be double-counted in the uncertainty calculation, but the effect of double-counting on the overall uncertainty is expected to be small.

VI. CONCLUSION

The differences between redundant GHI measurements for 290 solar monitoring stations operated and maintained by GroundWork Renewables were examined.

The irradiance values from the pyranometer were adjusted using a novel directional response adjustment algorithm in combination with manufacturer-specified DR values and errors were significantly reduced. More work is needed to fine-tune the adjustment. Future work includes refining the DR correction and providing DR-adjusted GHI data to GroundWork clients to further reduce measurement uncertainty.

A second adjustment was applied to account for minor discrepancies between the calibrations of the two sensors. The difference and ratio of the two sensors were computed for all data after the two adjustments were applied.

From the difference and ratio calculations, the P95 difference for weekly-maintained stations was computed. The P95 difference between sensors for all one-minute data at solar elevations greater than $10°$ for all 290 stations was 7.8 W/m^2. The P95 ratio was 1.018 (1.8%).

In summary, having a quantitative understanding of the noise in a GHI measurement due to operation under real-world conditions is a step toward quantifying the 'maintenance' contribution to the uncertainty budget of a pyranometer. Other groups operating solar monitoring networks are encouraged to follow the procedures outlined here to derive a "typical" maintenance uncertainty for the stations in their networks. Through this group exercise, the contribution of various maintenance regimes to pyranometer measurement uncertainty can be obtained and improvements to maintenance best practices may be implemented.

This work was sponsored by GroundWork. Data presented in this study is the property of GroundWork's clients and is used by GroundWork on an aggregated and anonymized basis and is not publicly available. We thank Jörgen Konings and Chiel Donkers at Hukseflux for their guidance in this work.

REFERENCES

[1] American Society for Testing and Materials (ASTM) G213-17, Standard guide for evaluating uncertainty in calibration and field measurements of broadband irradiance with pyranometers and pyrheliometers.

[2] World Meteorological Organization, Guide to meteorological instruments and methods of observation, eighth edition, 2021.

[3] Reda (2011) Method to Calculate Uncertainties in Measuring Shortwave Solar Irradiance Using Thermopile and Semiconductor Solar Radiometers.

[4] https://www.hukseflux.com/uploads/product-documents/SR30-M2-D1_manual_v2203.pdf.

[5] D. G. Erbs, S. A. Klein and J. A. Duffie, Estimation of the diffuse radiation fraction for hourly, daily and monthly-average global radiation, Solar Energy 28(4), pp 293-302, 1982. Eq. 1

[6] William F. Holmgren, Clifford W. Hansen, and Mark A. Mikofski. "pvlib python: a python package for modeling solar energy systems." Journal of Open Source Software, 3(29), 884, (2018).

SPUTTERING FOR THE FORMATION OF SI-BASED PASSIVATING CONTACTS

Christophe Allebé, Antoine Descoeudres, Patrick Wyss, Nicolas Pernès, Bertrand Paviet-Salomon, Christophe Ballif, Simon Hübner, Torsten Dippell

CSEM, Neuchatel, Switzerland

Singulus Technologies AG, Kahl am Main, Germany

Passivating contacts based on Silicon on Passivating Oxide activated during a high temperature treatment have demonstrated excellent passivation and carrier extraction properties, enabling to reach performances above 26% on laboratory scale devices. At industrial scale, several industrial players have built pilot lines delivering solar cells integrating a n-type rear side passivating contact in n-PERT cell architecture, with conversion efficiencies above 24.5%. A key process step of this technology is the poly-Si deposition. Low pressure CVD has been widely investigated at industrial level, despite the non-directional nature of its deposition, leading to wrap-around deposition and resulting p-n junction shunting; consequently, requiring an additional etch-back of the poly-Si from the unwanted side of the solar cell. Plasma-enhanced CVD and sputtering - a physical vapor deposition technique - do not suffer from this limitation thanks to the directional nature of their deposition. In addition, deposition of poly-Si by sputtering has several other advantages including the fact that it can be performed at room temperature and does not involve hazardous gases. We investigate the development of passivating contacts based on poly-Si layer deposited by sputtering and their implementation at the rear side of n-PERT solar cells. In this contribution, the impact of the poly-Si layer deposition parameters, the doping approaches and doping levels on the passivation quality and carrier extraction properties is being studied. Optimized layers have enabled to reach on symmetrical structures implied open circuit voltage values up to 730 mV and specific contact resistivity down to 4m$\Omega \cdot cm2$. The optimized conditions for sputtered Si passivating contacts have been integrated at the rear side of full-area 6-inch n-PERT solar cells, leading to cell' Voc of 697 mV, FF of 80.5 %

978-1-6654-6060-6/23 $31.00 © 2023 IEEE

Copper Metallization for III-V Solar Cells

Theresa E. Saenz, Anica N. Neumann, Matthew R. Young, Jeramy D. Zimmerman, Emily L. Warren, Myles A. Steiner

National Renewable Energy Lab, Golden, CO, United States

Colorado School of Mines, Golden, CO, United States

We study the effect of Cu front contacts on the performance of a GaAs solar cell. No initial degradation was observed compared to a baseline cell with Au contacts, but annealing at 200Â°C for 30 minutes caused a 20 mV drop in open circuit voltage. SIMS measurement showed that the Cu diffuses readily across the GaAs surface on the scale of millimeters, and that Cu concentrations are higher in GaInP layers than surrounding GaAs layers. We will also discuss potential Cu diffusions barriers and electroplated Cu contacts.

An Enhanced Snow-Shedding Model: the Module Frame as a Key Variable

Daniel Riley[1], Laurie Burnham[1], Paul Dice[2], Jennifer Braid[1]

Sandia National Laboratories[1], Albuquerque, NM, USA
Michigan Technological University[2], Houghton, MI, USA

Abstract — Snow sliding is one of the primary methods of natural snow removal from PV modules. In this paper, we evaluate and add to an existing model for predicting snow sliding from PV modules and the rate at which that sliding occurs. Specifically, we look at the impact of the module frame on snow shedding, presenting shedding data from two side-by-side arrays that are identical except one set of modules are frameless while the other set has aluminum frames. This team has previously shown that frameless PV modules shed snow faster than framed PV modules, but the difference in the rate of snow shedding was based on a single winter of data (2018-2019). This study augments that earlier work with data from the 2021-2022 winter season and will also include data from the 2022-2023 winter season.

I. INTRODUCTION

As photovoltaic (PV) installations in northern regions accelerate, the effect of snow accumulation on PV systems and the resulting loss of energy production have become increasingly popular areas of research [1-2]. Marion, et. al developed one of the first models to estimate the rate of snow shedding from PV modules via sliding as a function of temperature, irradiance, tilt angle, and mounting configuration [3]. More recently, Riley et. al. proposed that sliding rates are also affected by the presence or absence of a metallic module frame and had data showing frameless modules shed snow at a rate approximately 50% faster than framed modules [4]. This paper not only validates the accuracy of Marion's original model but increases the model's accuracy by adding the module frame as a key parameter.

In 2019, this team compared snow shedding rates from two PV systems of identical modules, differing only in the presence or absence of a frame. These systems were installed at Sandia's Vermont Regional Test Center (RTC) in Willison, VT. In the spring of 2020, the Vermont RTC closed and, in 2021, the two PV systems were slightly reconfigured and reinstalled at the Michigan RTC in Calumet, MI [5]. The systems were similarly instrumented and monitored with a camera to capture images of snow coverage on the PV modules every 15 minutes. The two PV systems are shown in Fig. 1 with the frameless PV modules on the right side of the image, and the framed PV modules on the left side of the image.

Fig. 1. The PV systems under test in Calumet, Michigan. The systems are identical except for the lack of a frame on the right modules and the presence of a black-anodized frame on the left modules.

The data shown in this evaluation abstract was collected from the 2021-2022 winter but will be expanded with data from the 2022-2023 winter by the time of presentation.

II. METHODOLOGY

A camera installed south of the PV systems, collected module images at 15-minute time intervals. These images were used to determine the snow coverage (in percent) of each module within each system over time. Each system image was cropped and planar-indexed into 36 module images (18 modules of each type). The red, green, blue (RGB) pixel values from the module images were automatically classified into two categories via k-means clustering, which were then labeled them as 'snow' and 'non-snow'. Then, the percent coverage for each module was quantified as the percentage of 'snow' pixels. The individual module coverage was averaged to produce the average snow coverage of each PV system.

Additionally, ambient temperature and plane-of-array irradiance values were continuously measured and used as inputs to the snow-shedding model to predict the percent of each module that would be covered with snow. The model was run in both the ordinary form, which was developed for framed PV modules, and was also run in the modified form to estimate shedding rates for frameless PV modules.

III. RESULTS AND DISCUSSION

A. Snow shedding model performance

Throughout the 2021-2022 winter period, snowfall occurred almost every day [6]. Several of these snowfall events are presented here. The Marion model for estimating snow shedding accounts for temperature and irradiance (among others) as primary determinants of snow sliding on a PV module. Because cold and low-irradiance conditions are expected to cause less sliding than warm and high-irradiance conditions, the Marion model tends to correctly model lack-of-sliding, as shown in Fig 2 where no snow sliding was observed in images and the model predicted that snow would not slide.

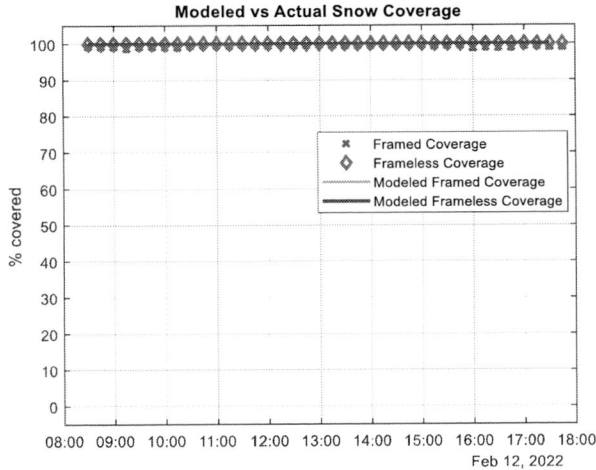

Fig. 2. Under low irradiance, low temperature conditions, snow did not slide from either PV system, and the Marion model accurately predicts this lack of sliding.

However, there are also events where the model did not predict snow sliding, but snow *did* slide from the modules, as captured in images and shown in Fig. 3. Events such as these are expected to occur given that there are both false negatives (snow slides without model predicting sliding) and false positives (snow doesn't slide, but is predicted to slide) in Fig. 5 of [3]. Note that in Fig. 3, the snow slides most quickly from the frameless PV as was expected in [4].

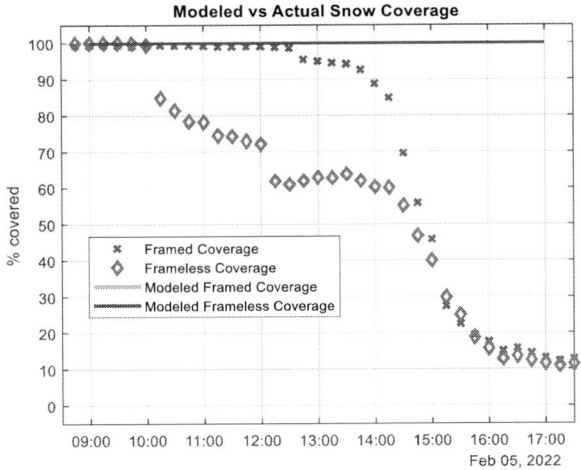

Fig. 3. Snow did slide from both PV systems, but was not expected to slide by the model.

B. Enhanced snow shedding model performance

The increased sliding rate for frameless PV modules was captured in numerous snow shedding events. In shedding events where the shedding rate between framed and frameless PV modules differed, shedding was three times more likely to be higher on frameless PV modules thand framed PV modules. An example is shown in Fig. 4, where Riley's earlier demonstration of a 50% increase in shedding rate for frameless modules is accurately reflected in measured data from February, 2022. Improvements to the model are still needed to account for the *earlier* shedding of frameless PV, rather than solely predicting the increased shed rate, but that data is forthcoming.

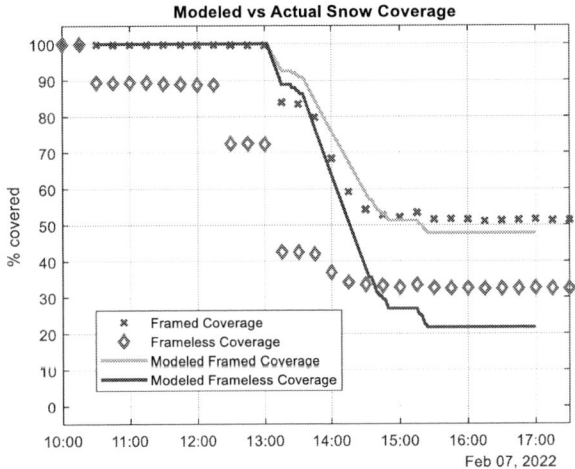

Fig. 4. Snow sheds more quickly from frameless PV modules than framed PV modules. This occurred in 3 out of 4 shedding events where the shedding rates differed. The modeled 50% shedding rate increase for frameless PV seems to predict this well.

It must be noted, however, that occasionally the framed PV modules shed snow more quickly than the frameless PV modules, as shown in Fig. 5. We do not know the cause of these discrepancies, although we do know that snow shedding is a complex chemical-physical phenomenon with difficult-to-model interactions between snow particles, PV module temperatures, and surface interactions.

978-1-6654-6060-6/23 $31.00 © 2023 IEEE

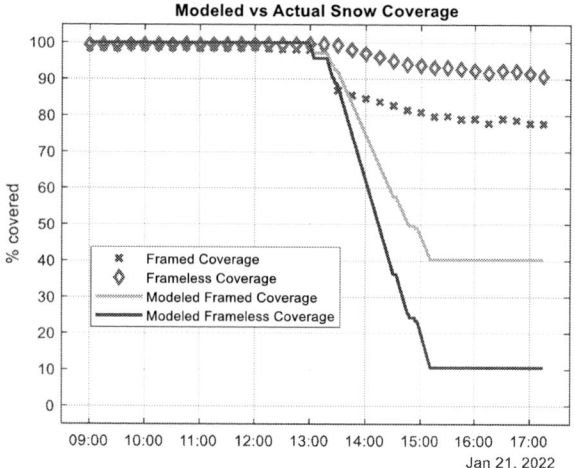

Fig. 5. Here, snow sheds more quickly from the framed PV modules than the frameless PV modules. This occurred in 1 of 4 shedding events where the shedding rates differed.

IV. CONCLUSIONS AND FUTURE WORK

Snow shedding (via sliding) from PV modules is influenced by many factors and therefore challenging to model. Even so, the relatively simplistic Marion model, which takes into account only a limited number of variables, seems to perform admirably in estimating the coverage of PV by snow.

As our study shows, however, the assumption that a module's architecture has no impact on shedding rates, needs to be corrected. As we have now validated, under most circumstances, frameless PV modules shed snow more quickly than framed PV modules. We speculate the difference is due to the additional inertia required for snow to slide over the physical barrier posed by a frame. The actual rate of shedding for frameless modules, however, is difficult to determine as there is wide variability in the actual versus modeled shedding of snow.

Our proposal of a 50% increase in shedding rate for frameless modules produces modeled shedding rates that approximate the measured shedding rates. However, we do note that simply increasing the rate at which the model predicts snow shedding may not fully account for the differences between framed and frameless PV. Frameless PV modules not only have faster shedding rates but they tend to shed snow slightly earlier than framed modules, indicating that the snow-shedding model could be further improved to better account for the snow shedding of frameless PV modules in winter.

Future research on this topic will incorporate data from the 2022-2023 winter and years to come. We are also monitoring other PV systems with design differences that do not include the presence or absence of a frame. Some factors being studied are bifaciality, module orientation (landscape or portrait), and backsheet color (black or white). While data evaluating these other factors is being collected, this paper will focus only on the performance of the standard snow model in predicting shedding from framed modules, expanded to included data from the latest winter, and the performance of the enhanced model in predicting shedding for frameless PV.

ACKNOWLEDGMENTS

Sandia National Laboratories is a multimission laboratory managed and operated by National Technology & Engineering Solutions of Sandia, LLC, a wholly owned subsidiary of Honeywell International Inc., for the U.S. Department of Energy's National Nuclear Security Administration under contract DE-NA0003525. This material is based upon work supported by the U.S. Department of Energy's Office of Energy Efficiency and Renewable Energy (EERE) under the Solar Energy Technologies Office Award Number 34367.

REFERENCES

[1] R.E. Pawluk, Y. Chen, Y.She, "Photovoltaic electricity generation loss due to snow—a literature review on influence factors, estimation and mitigation,",2019, Renewable and Sustaiunable Energy Reviews 107:171-182.

[2] J. Bogenrieder, C. Camu, M.Huttner, P. Offermann, J. Hauch, C. Brabec, 2018. "Technology-dependent analysis of the snow melting and sliding behavior on photovoltaic modules," Journal of Renewable and Sustainable Energy 10, 021005.

[3] B. Marion, R. Schaefer, H. Caine, G. Sanchez, "Measured and modeled photovoltaic system energy losses from snow for Colorado and Wisconsin locations," in *Solar Energy*, vol. 97, pp. 112-121, 2013.

[4] D. Riley, L. Burnham, B. Walker, J. Pearce, "Differences in Snow Shedding in Photovoltaic Systems with Framed and Frameless Modules," in *46th IEEE Photovoltaic Specialist Conference*, 2019.

[5] L. Burnham, D. Riley, B. King, J. Braid, P. Dice, A. Dyreson, W. Snyder, C. Pike. Dedicated cold-climate field laboratory for photovoltaic system and component studies: the Michigan Regional Test Center as a case study, *49th IEEE Photovoltaic Specialist Conference*, 2022.

[6] https://www.keweenawcountyonline.org/snowfall2.php

The Effect of Tilt and Azimuth Angle Variations on Monthly and Annual Incident Solar Radiations for Locations in Brazil

Eslam Mahmoudi [a*], Tarcio Andre dos Santos Barros [a], Hugo Alvarez [b], Rodrigo Garcia [b], Francisco C. Marques [c], Marcelo Gradella Villalva [a]

[a]Department of Systems and Energy, Faculty of Electrical and Computer Engineering, University of Campinas, Brazil

[b] BYD energy, Campinas, Brazil

[c] Gleb Wataghin Institute of Physics, University of Campinas, Brazil

Abstract—This paper analyzes the effect of the surface's azimuth and tilt angle changes on the incident solar irradiation for every location in Brazil. The monthly and annual incident solar radiations on tilted surfaces under different azimuth and inclination angles are estimated to generate graphs as color contour maps, which show the fraction (%) of the maximum irradiation availability at each location. A simulation tool is developed in Python to generate the graphs by computing the incident solar radiations on inclined surfaces based on the Brazilian global horizontal solar irradiation data and the locations' geographic coordinates. Simulation results show that the changes in azimuth angle did not generally result in significant incident solar radiation losses for the slope angle of up to around 20 degrees. Hence, mounting the PV modules under this tilt angle limitation and any azimuth angle results in maintaining a high level of annual incident radiation. This simulation tool can be used by designers to quickly evaluate the effect of specific tilt and azimuth angles on the monthly and yearly incident solar irradiation for a particular location.

I. INTRODUCTION

Energy is a crucial component of every country's economic and social progress. Its presence significantly enhances the overall quality of life, whether derived from oil, gasoline, nuclear power, or any form of renewable energy [1]. Among all the renewable energy resources, solar energy is rapidly emerging as one of the most promising green energy technologies worldwide [2]. Solar energy is environment-friendly and does not produce harmful emissions that contribute to the greenhouse effect or disrupt the planet's delicate ecological balance [3]. Photovoltaic (PV) solar energy conversion has grown significantly in recent years, particularly in its applications integrated with public electricity grids [4]. However, architects and building planners do not tend to consider PV panels in their design plans because they often cannot find an optimal orientation for PV installation on the buildings [5]. Nevertheless, it is possible to achieve satisfactory performance even on non-optimal PV orientations to still benefit from using PV technology [6]. A rule-of-thumb for optimal orientation of PV modules states that the tilt angle should be equal to the location's latitude, with the azimuth angle towards the south for the northern hemisphere and toward the north for the southern hemisphere [7]. However, although the rule-of-thumb method may be appropriate for specific locations, it may lead to increased costs due to oversizing systems if considered without detailed analysis [8].

Furthermore, owing to the architectural consideration in PV module integration in buildings, orientations other than optimal are also may of interest for installing PV modules at the same angle as a sloping roof, even if it leads to a decrease in the received solar radiation [9]. Considering that a solar tracking system may not be preferred because of its capital and maintenance costs, many studies [10]–[13] express that optimal tilt and azimuth angles can be found for every location on Earth with different radiation characteristics for the best solar energy reception. Since non-optimal tilt and azimuth angles can also produce acceptable energy generation amounts, the location-specific losses need to be assessed to identify the limitations of the tilt and azimuth angles [14]. Several studies have been conducted on the architectural aspects of PV in buildings [15]–[18] to evaluate the annual and seasonal generated energy PV energy under different architectural compositions. The most important factor influencing the performance of integrated PV systems is the maximization of the incoming solar irradiation on a monthly or annual basis. Since the seasonal and annual solar irradiation levels

978-1-6654-6060-6/23 $31.00 © 2023 IEEE

highly depend on the PV's location, the optimal PV module orientation and tilt angles can also be very much location specific [4]. Hence, the architects and PV system designers should consider site specific data and irradiance calculation methods to assess the performance of integrated PV systems. As such, regarding the available irradiance data at the desired locations, the decomposition and transposition methods should be considered to estimate the plane-of-array (POA) irradiance at the locations in accordance with the azimuth and tilt angle of the PV systems. Accordingly, the losses in the PV generation in different orientations can be estimated for specific locations for evaluating the PV installation at non-ideal orientations and tilt angles. Therefore, the PV system designers and installers require location-specific monthly and annual irradiation calculations for possible variations in the module's orientation and slope angles. Simulation tools can be used to evaluate the effect of a specific surface orientation in a particular geographic location to know how much monthly and annual incident solar radiations are reduced for non-optimal surface orientations. In this context, this paper proposes a simulation tool to understand how the modules' different positioning can influence the received radiation. The proposed tool is developed in Python to estimate the monthly and annual irradiance incident on surfaces with different tilt and azimuth angles in every location of Brazil. The simulation tool generates graphs demonstrating the potential losses that occur in the variations of inclination and azimuth angles of the positioning of the modules. These graphics are helpful for PV system designers, installers, and architects to determine the modules' installation position.

II. Modeling Incident Solar Irradiation on Inclined Surfaces

A. Solar radiation database

The Global Horizontal Irradiance (GHI) data for Brazilian locations are extracted from the Brazilian Atlas of Solar Energy - 2nd Edition [19]. For this 2nd edition, more than 17 years of satellite data were used, and several advances were implemented in the parameterizations of the BRASIL-SR radiative transfer model, aiming to improve further the reliability and accuracy of the database produced and made available for public access. This dataset provides Brazil's monthly average Global Horizontal Irradiance (GHI), which was validated by data collected in solar stations. The dataset comprises irradiation data with a spatial resolution of $0.1° \times 0.1°$ (approximately 10 km × 10 km). As such, Brazil is divided into 72272 cells, in which each cell's center is specified by certain latitude and longitude, as shown in Fig.1.

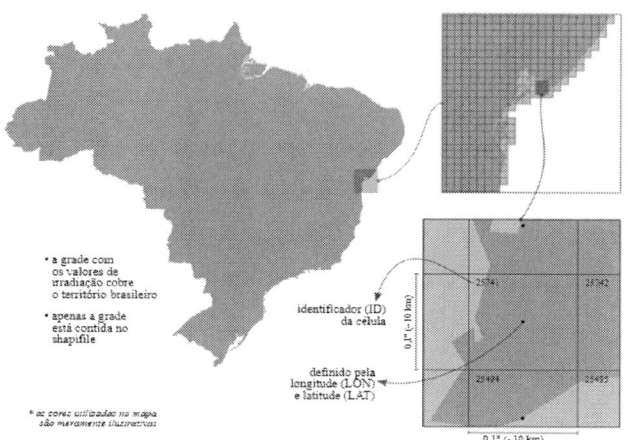

Fig. 1. Division of Brazil into cells of $0.1° \times 0.1°$

B. POA irradiance calculation

This paper aims to generate graphs to analyze the effect of azimuth and tilt angle variations on the POA incident on inclined surfaces. Thus, according to the available data, which is the monthly average GHI, the monthly average POA is calculated for each pair of azimuth and tilt angles. To this end, the Klein and Theilacker (KT) model [20] is employed to estimate the monthly average POA on sloped surfaces. In contrast to the models that restricted to the surfaces facing the equator, the KT model developed a more general form that is valid for any surface azimuth angle γ. The KT model estimates the monthly average daily solar radiation ($\overline{H}_T$) incident on a tilted surface with slop angle β based on the following equation:

$$\overline{H}_T = \overline{H}\overline{R} \tag{1}$$

Where $\overline{H}$ is the measured monthly average GHI, and the term $\overline{R}$ that is given in equation (1) is formulated as the sum of direct, diffuse, and ground-reflected radiations incident on the tilted surface.

$$\overline{R} = D + \left\{\frac{\overline{H}_D}{\overline{H}} \times \left(\frac{1+\cos\beta}{2}\right)\right\} + \left\{\rho_g\left(\frac{1-\cos\beta}{2}\right)\right\} \tag{2}$$

Where D is defined as the direct solar radiation on an inclined surface that is as:

$$D = \begin{cases} max(0, G(\omega_{ss}, \omega_{sr})) & \text{if } \omega_{ss} \geq \omega_{sr} \\ max(0, [G(\omega_{ss}, -\omega_s) + G(\omega_s, \omega_{sr})]) & \text{if } \omega_{ss} < \omega_{sr} \end{cases} \tag{3}$$

Where,

$$G(\omega_1, \omega_2) = \frac{1}{2d}\left[\left(\frac{bA}{2} - a'B\right)(\omega_1 - \omega_2)\frac{\pi}{180} + (a'A - bB)(\sin\omega_1 - \sin\omega_2) - a'C(\cos\omega_1 - \cos\omega_2) + \right.$$

$$\left(\frac{bA}{2}\right)(\sin \omega_1 \cos \omega_1 - \sin \omega_2 \cos \omega_2) + \left(\frac{bC}{2}\right)(\sin^2 \omega_1 -$$
$$\sin^2 \omega_2)] \tag{4}$$

$$a' = [0.4090 + 0.5016 \sin(\omega_s - 60)] - \left(\frac{\overline{H}_D}{\overline{H}}\right) \tag{5}$$

$$a = 0.6609 - 0.4767 \sin(\omega_s - 60) \tag{6}$$

$$d = \sin \omega_s - ((\pi/180)\, \omega_s \cos \omega_s) \tag{7}$$

The ω_1 and ω_2 are correspond to ω_{ss} and ω_{sr} and ω_s, in which the ω_s is the sunset hour and is calculated as

$$\omega_s = \cos^{-1}(-\tan \varphi \, \tan \delta) \tag{8}$$

The δ is the declination angle that is determined from the equation (9).

$$\delta = 23.45 \sin(360 \tfrac{284+num}{365}) \tag{9}$$

Where *num* is the number of days in the year in Julian's calendar, the number of the representative day for each month of the year from January to December are 17, 47, 75, 105, 135, 162, 198, 228, 258, 288, 318, 344, respectively.

The ω_{ss} and ω_{sr} as the sunset and sunrise hour angles for direct radiation on a tilted surface are calculated by the following equations.

$$|\omega_{sr}| = min\left[\omega_s, \cos^{-1}\left(\frac{AB+C\sqrt{A^2-B^2+C^2}}{A^2+C^2}\right)\right] \tag{10}$$

$$\omega_{sr} = \begin{cases} -|\omega_{sr}| & \text{if}(A > 0 \ and \ B > 0) \ or \ (A \geq B) \\ +|\omega_{sr}| & \text{otherwise} \end{cases} \tag{11}$$

$$|\omega_{ss}| = min\left[\omega_{ss}, \cos^{-1}\left(\frac{AB-C\sqrt{A^2-B^2+C^2}}{A^2+C^2}\right)\right] \tag{12}$$

$$\omega_{ss} = \begin{cases} +|\omega_{ss}| & \text{if}(A > 0 \ and \ B > 0) \ or \ (A \geq B) \\ -|\omega_{ss}| & \text{otherwise} \end{cases} \tag{13}$$

$$A = \cos \beta + \tan \varphi \cos \gamma \sin \beta \tag{14}$$

$$B = \cos \omega_s \cos \beta + \tan \delta \sin \beta \cos \gamma \tag{15}$$

$$C = (\sin \beta \sin \gamma)/\cos \varphi \tag{16}$$

In equations (10) and (12), the value within the square root can be negative under specific orientations where the solar incidence angle is less than 90 at sunrise or an incidence angle more than 90 at sunset at all times. In these cases, a boundary has been imposed by setting the ω_{ss} and ω_{sr} to $-\omega_s$ and $+\omega_s$.

$\overline{H}_d$ is the monthly average daily diffuse irradiation on the horizontal surface. Since the measurement of $\overline{H}_d$ is rarely available, it must be estimated based on $\overline{H}$. In the KT model, the equation of $\overline{H}_d/\overline{H}$ proposed by Liu and Jordan [21] is used, which is:

$$\frac{\overline{H}_D}{\overline{H}} = 1.390 - 4.027\overline{K}_T + 5.531\overline{K}_T{}^2 - 3.108\overline{K}_T{}^3 \tag{17}$$

Where,

$$\overline{K}_T = \frac{\overline{H}}{\overline{H}_0} \tag{18}$$

$\overline{H}_0$ is the monthly average extraterrestrial radiation, which is calculated as:

$$\overline{H}_0 = \frac{1}{(m_2-m_1)} \sum_{n=m_1}^{m_2} (H_0)_n \tag{19}$$

Where m_1 and m_2 are the days of the year at the start and end of the month, respectively. $(H_0)_n$ is the extraterrestrial radiation on a horizontal surface for the *n-th* day of the year that is given by:

$$(H_0)_n = \frac{24}{\pi} I_{sc} \left[1 + 0.033 \cos\left(\frac{360n}{365}\right)\right] \times$$
$$[\cos \varphi \cos \delta \sin \omega_s + (\omega_s 2\pi/360) \sin \varphi \sin \delta] \tag{20}$$

Where I_{sc} is the solar constant, and *n* is the day of the year given for each month (i.e., *n* each month of the year from January to December are 17, 47, 75, 105, 135, 162, 198, 228, 258, 288, 318, 344, respectively).

ρg is the ground reflectance factor, assumed to be 0.2 for all months and locations due to the Brazilian weather having less climatic changes and no snow cover throughout the year.

Therefore, the KT model is used to calculate the monthly average daily POA at various azimuth and tilt angles from July to December as well as the annual average daily POA for each cell of Brazil based on the cell's center latitude/longitude and the monthly average GHI.

III. SIMULATION RESULTS AND DISCUSSION

The proposed model is used to generate the graphs of the monthly and annual average daily POA for each location in Brazil. To this end, based on the latitude/longitude of the desired location and the GHI data of the cell where location is located, the KT model estimates the average monthly and annual POAs for each pair of tilt and azimuth angles. Thus, by changing the tilt angle from 0° to 90° and the azimuth angle in the range from -180° to 180° with an interval of 1°, twelve matrices of monthly POAs and one matrix of annual POA are obtained for each location. According to the obtained monthly and annual POA matrices, the graphs can be generated as a color contour map, in which 100% indicates the maximum received average daily POA in the locations. for instance, Fig. 2 to Fig. 13 show the graphs of annual POA for twelve Brazilian capital cities.

978-1-6654-6060-6/23 $31.00 © 2023 IEEE

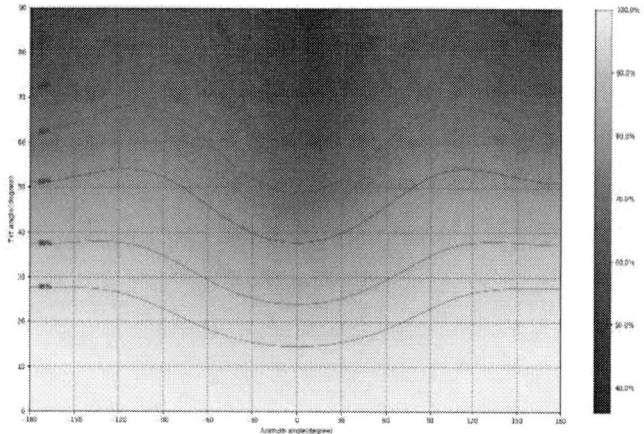

Fig. 2. Annual average daily POA, Aracaju-SE
$(100\%=5.49\text{kWh}/m^2)$

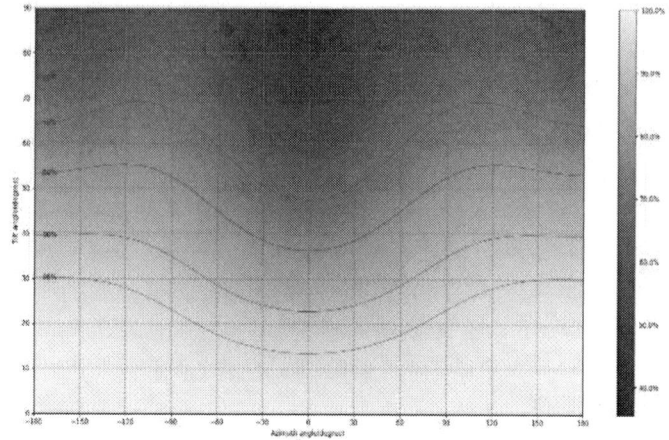

Fig. 5. Annual average daily POA, Salvador - BA
$(100\%=5.36\text{kWh}/m^2)$

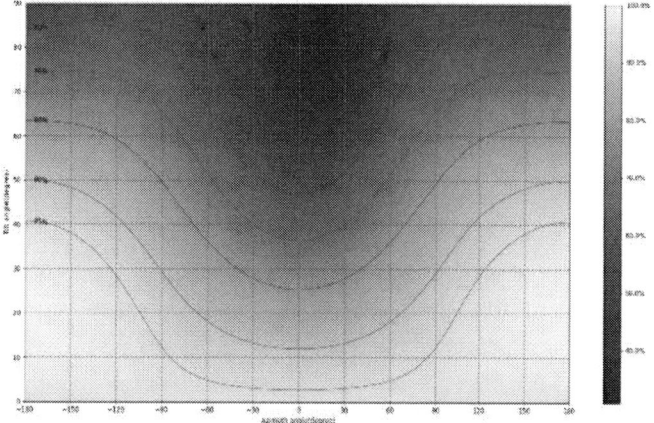

Fig. 3. Annual average daily POA, Belo Horizonte - MG
$(100\%=5.29\text{kWh}/m^2)$

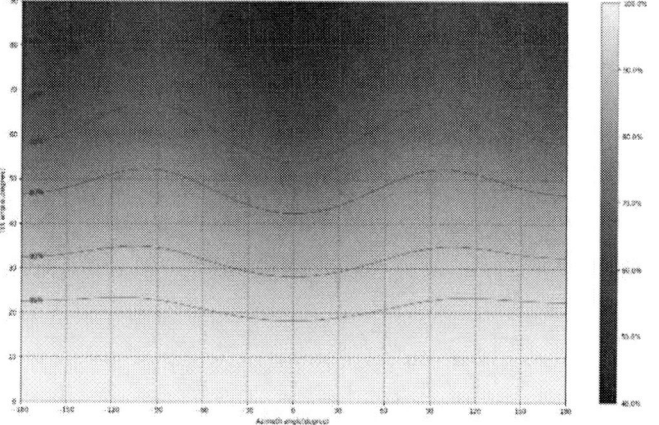

Fig. 6. Annual average daily POA, São Luís - MA
$(100\%=5.2\text{kWh}/m^2)$

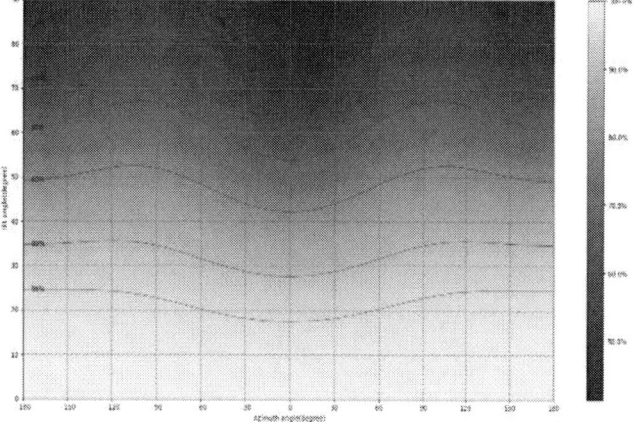

Fig4. Annual average daily POA, Boa Vista - RR
$(100\%=4.48\text{kWh}/m^2)$

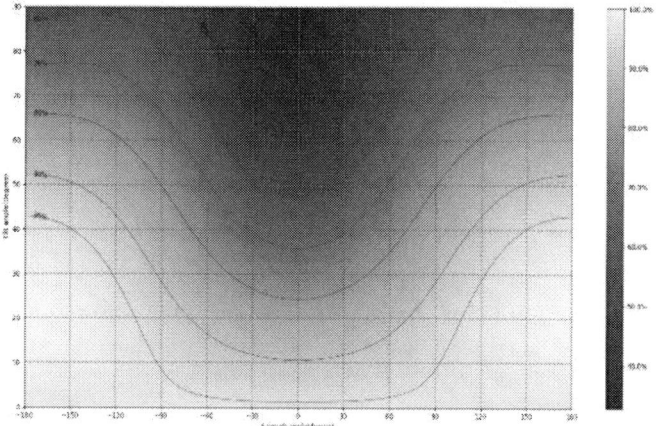

Fig. 7. Annual average daily POA, Porto Alegre - RS
$(100\%=4.6\text{kWh}/m^2)$

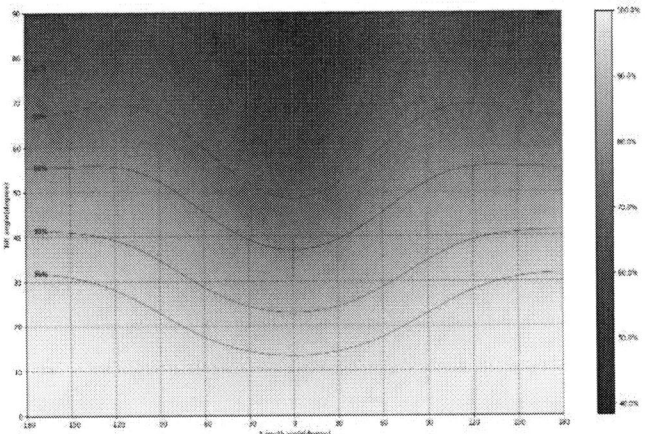

Fig. 8. Annual average daily POA, Porto Velho - RO
($100\%=4.48$kWh/m^2)

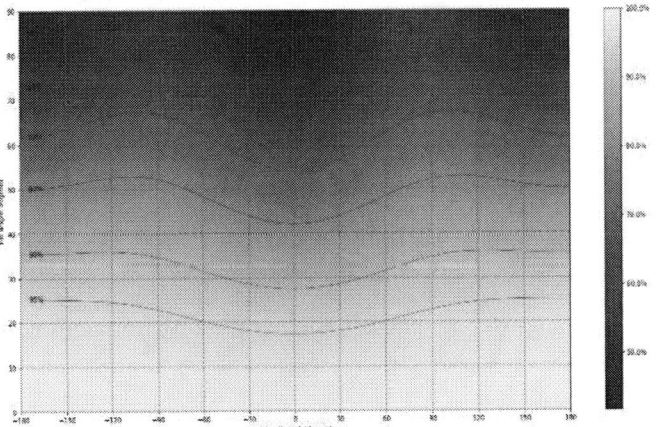

Fig. 9. Annual average daily POA, Curitiba - PR
($100\%=4.33$kWh/m^2)

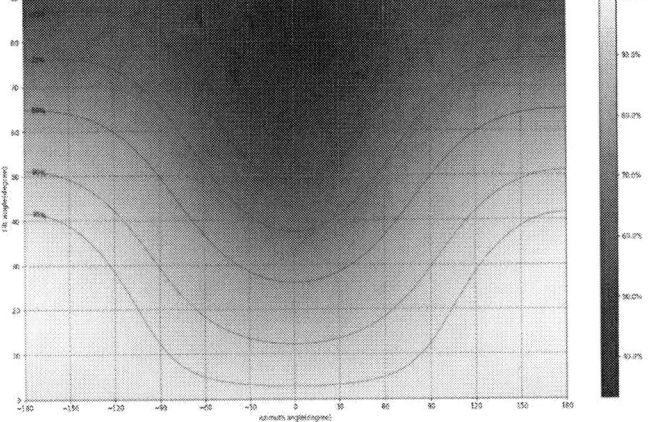

Fig. 10. Annual average daily POA, São Paulo- SP
($100\%=4.6$kWh/m^2)

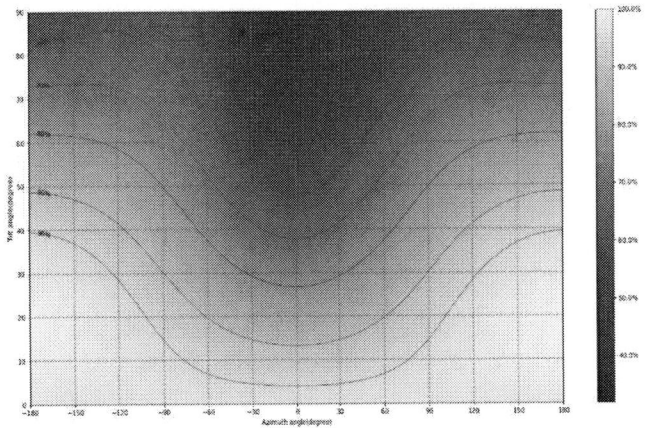

Fig. 11. Annual average daily POA, Brasília- DF
($100\%=5.39$kWh/m^2)

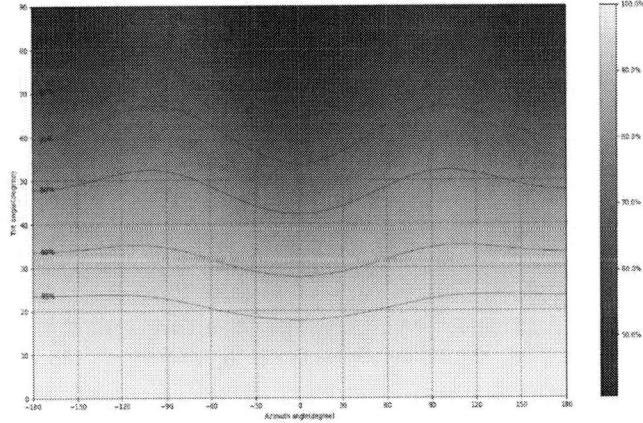

Fig. 12. Annual average daily POA, Belém- PA
($100\%=4.87$kWh/m^2)

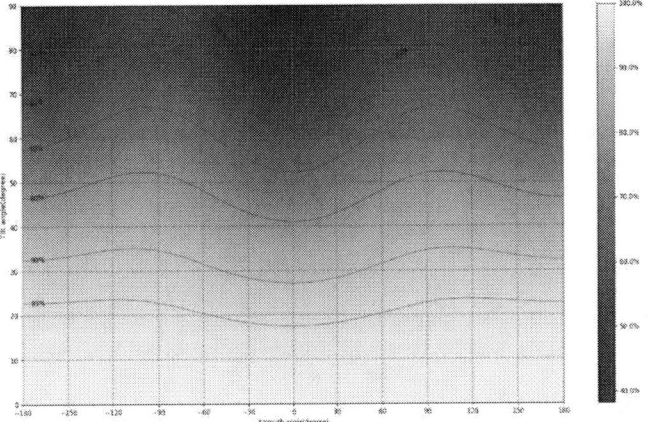

Fig. 13. Annual average daily POA, Goiânia- GO
($100\%=5.77$kWh/m^2)

978-1-6654-6060-6/23 $31.00 © 2023 IEEE

The generated graphs for locations across Brazil demonstrated that in the southern states of Brazil, the highest radiation levels only occur in positions very close to the ideal, with an inclination similar to latitude and north orientation. As observed in the graphs, there cannot be variations in azimuth and inclination, suggesting to the designer to choose the position of the modules prioritizing either the orientation or the inclination of the modules. In the northern states, the possibilities of positioning with good radiation reception are greater so that in these places, it is possible to use any position of azimuth, provided that an inclination of up to 20° is maintained. Moreover, there is also the prevalence of east and west orientation (in the north) when the installation is vertical.

V. CONCLUSION

This study developed a simulation tool to generate graphs that allow a quick and simplified evaluation of the performance of PV generators located in Brazilian locations. The results demonstrate that variations in azimuth or slope do not lead to significant annual irradiation losses, even up to tilt angles of approximately 20 degrees. It can be concluded that installing PV modules at slopes ranging from horizontal to the latitude tilt, and even up to 10° above the latitude tilt, has a minimal impact on the POA for azimuthal deviations from 0° to 90°, provided that the latitude of the site is below 15°. However, if thelocation's latitude is above 15°, small losses occur only when there are variations of plus or minus 10° in PV module tilt angles, as long as surface azimuthal deviations are limited to below 60 °. These findings can inform PV system planners that, under these reasonably flexible conditions, it is feasible to install PV panels on any orientation while maintaining high levels of annual irradiation. With the proposed simulation tool, the designer can evaluate the best PV integration for their projects or even the best geographic location for their investments. Furthermore, it facilitates a rapid analysis of PV retrofitting for building-integrated PV systems, aiding in the search for the most appropriate building surfaces for seamless PV integration.

IV. AKNOWLAGMENT

This work was supported by BYD Energy Brazil (through the PADIS/MCTI program, Funcamp project N.5779) and Total Energies (through ANP - Brazilian Oil & Gas Agency, Funcamp project N.6002).

REFERENCES

[1] Chenic AȘ, Cretu AI, Burlacu A, Moroianu N, Vîrjan D, Huru D, Stanef-Puica MR, Enachescu V. Logical Analysis on the Strategy for a Sustainable Transition of the World to Green Energy—2050. Smart Cities and Villages Coupled to Renewable Energy Sources with Low Carbon Footprint. *Sustainability*. 2022; 14(14):8622.

[2] Ali O M Maka , Jamal M Alabid, Solar energy technology and its roles in sustainable development, *Clean Energy*, Volume 6, Issue 3, June 2022, Pages 476–483.

[3] Hayat, M. B., Ali, D., Monyake, K. C., Alagha, L., & Ahmed, N. (2019). Solar energy—A look into power generation, challenges, and a solar-powered future. International Journal of Energy Research, 43(3), 1049-1067.

[4] Lujean Ahmad, Navid Khordehgah, Jurgita Malinauskaite, Hussam Jouhara, Recent advances and applications of solar photovoltaics and thermal technologies, Energy, Volume 207, 2020, 118254.

[5] Jia Liu, Xi Chen, Sunliang Cao, Hongxing Yang, Overview on hybrid solar photovoltaic-electrical energy storage technologies for power supply to buildings, Energy Conversion and Management, Volume 187,2019, Pages 103-121.

[6] Nikolaos Skandalos, Dimitris Karamanis, An optimization approach to photovoltaic building integration towards low energy buildings in different climate zones, Applied Energy, Volume 295, 2021,117017.

[7] H.Z. Al Garni, A. Awasthi, M.A.M. Ramli, Optimal design and analysis of grid connected photovoltaic under different tracking systems using HOMER, Energy Convers. Manag. 155C, 42-57.2018.

[8] Guo, M.; Zang, H.; Gao, S.; Chen, T.; Xiao, J.; Cheng, L.; Wei, Z.; Sun, G. Optimal Tilt Angle and Orientation of Photovoltaic Modules Using HS Algorithm in Different Climates of China. Appl. Sci., 7, 1028. 2017

[9] Hummon M, Denholm P, Margolis R. Impact of photovoltaic orientation on its relative economic value in wholesale energy markets. Prog Photovolt Res Appl 2012.

[10] Christensen, Craig & Barker, G. Effects of tilt and azimuth on annual incident solar radiation for United States locations. International Solar Energy Conference.2001.

[11] Hafez, AZ; Soliman, A.; El-Metwally, K.A.; Ismail, I.M. Tilt and azimuth angles in solar energy applications—A review. Renew. Sustain. Energy Rev. 77, 147–168. 2017.

[12] Nfaoui, M.; El-Hami, K. Optimal tilt angle and orientation for solar photovoltaic arrays: Case of Settat city in Morocco. Int. J. Ambient. 1–18, 2018.

[13] Hassan Z. Al Garni, Anjali Awasthi, David Wright, Optimal orientation angles for maximizing energy yield for solar PV in Saudi Arabia, Renewable Energy, Volume 133, 538-550.2019.

[14] Perpiñan O, Lorenzo E, Castro MA, Eyras R. On the complexity of radiation models for PV energy production calculation. Sol Energy. 82, 125-31, 2008.

[15] Hagemann I. PV in buildings e the influence of PV on the design and planningprocess of a building. Renew Energy 1996;8:467e70.

[16] Hussein HMS, Ahmad GE, El-Ghetany HH. Performance evaluation of photovoltaic modules at different tilt angles and orientations. Energ Convers Manag 2004;45:2441-52.

[17] Prasad DK, Snow M. Examples of successful architectural integration of PV: Australia. Prog Photovolt Res Appl 2004;12:477-83.

[18] Hestnes AG. Building integration of solar energy systems. Sol Energy 1999;67: 181-7.

[19] Pereira, E. B. Martins, F. R. Goncalves, A. R. Costa, R. S. Lima, F. L. Ruther, R. Abteu, S. L.Tiepolo, G. M.; Pereira,V.,SouzG. Atlas brasileiro de energia solar.2.ed. São José dos Campos: INPE.80.2017

[20] Klein, S.A., Theilacker, J.C.: An algorithm for calculating monthly-average radiation on inclined surfaces. J. Sol. Ener. Eng. 103(1), 29–33. 1981.

[21] S.A. Klein, Calculation of monthly average insolation on tilted surfaces, Solar Energy, Volume 19, Issue 4, 1977, Pages 325-329.

Rear-side Irradiance Simulation of Field PV Modules

Zelin Li*, Raymond J. Wieser*, Xuanji Yu*, Stephanie L. Moffitt[†], Ruben Zabalza[‡], Silvana Ayala[§],
Matthew Brown[§], Xiaohong Gu[†], Liang Ji[‡], Colleen O'Brien[‡],
Micheal D. Kempe[§], Laura S. Bruckman* , Kenneth P. Boyce[‡]

* Case Western Reserve University, Cleveland, Ohio, 44106, USA
[†] National Insitute of Standards and Technology, Gaithersburg, Maryland, 20878, USA
[‡] UL Solutions, Northbrook, Illinois, 60062, USA
[§] National Renewable Energy Laboratory, Golden, Colorado, 80401, USA

Abstract—**Assessing the durability of photovoltaic (PV) module backsheet is critical to increasing module lifetime. Historically, laboratory-based accelerated testing has insufficient in predicting large-scale failures of commercial polymeric materials. Additionally, there is growing concern that standard condition tests do not reflect non-uniformities in field exposure, and that certain modules experience more severe degradation due to their location. Anisotropy in field exposures is installation-dependent and reflects different levels of exposure to irradiance due to mounting geometry, ground surface albedo, and climatic zone. "Bificial_Radiance" [1], simulates the amount of reflected irradiance on the Polyethylene Naphthalate (PEN) backsheet of a PV array located in Maryland. In our work, site specific weather data are gathered for the entire length of exposure for the modules. The total full-spectrum dose for the incident irradiance on the backsheet was then determined. The simulation results are integrated with historical field survey data to better understand realworld outdoor degradation.**

Index Terms—**Backsheet, Degradation, Simulation, Field Survey**

I. INTRODUCTION

The adoption of Photovoltaic (PV) energy has continues to increase. Meanwhile, the cost of solar photovoltaic energy is rapidly declining [2]. PV represents one of the most critical alternative energy technologies to replace no-renewable sources of energy. PV power has the benefit of being more stable than other renewables, like wind energy and hydropower, under a uncertain climatic future [3], [4]. To futher the development of solar energy, the reliability of PV modules in the field is a key point to control the cost. [5]. Preventative maintance in the field serves to lower the risk of module failure and help research institutions to improve the protocols of lab-accelerated aging tests.

In recent research, most of studies put emphasize on the degradation of PV cells and encapsulants [6]. However, the polymeric material in the Backsheets of PV modules also play a role in degradation. Backsheets always experience various synergistic stressors, including but not limited to humidity, air temperature, irradiance, abrasion, wind speed [7]–[15]. For backsheets materials, Ultraviolet (UV) light could cause chemical and structural changes induced by the photooxidation and accumulated applied stress, which will produce visible crack, see Figure 1 , on the outer layer and failure finally [8], [9], [15]. These stressors, especially irradiance, will also change spatially with different array geometry, ground albedo, and shading of external objects. Thus, the distribution of irradiance is not uniform, which has been observed to cause non-uniform degradation of backsheet materials in a single mounting structure. Furthermore, with the aging or exposure increased, the variance among modules will be more significant. In fact, it was found that the edge modules have a higher degradation rate [16], [17].

Fig. 1. Visible crack on backsheet in Maryland Field

In order to accurately model the total exposure of backsheet materials, a specially resolved rear-side irradiance simulation was needed. A raytracing tool, "Bifacial_radiance" [1], canb be used to simulate the irradiance of the real world site from rear side of PV modules [18]. The original purpose of this software is to simulate the exposure of bifacial modules from light reflected by creating a digital twin of the installation.

This material is based upon work supported by the U.S. Department of Energy's Office of Energy Efficiency and Renewable Energy (EERE) under Solar Energy Technologies Office (SETO) Agreement Number DE-EE-0008748. The views expressed herein do not necessarily represent the views of the U.S. Department of Energy or the United States Government. This work made use of the Rider High Performance Computing Resource in the Core Facility for Advanced Research Computing at Case Western Reserve University.

For the rear side simulation, Direct Normal Irradiation (DNI) and Diffuse Horizontal Irradiance (DHI), Global Horizontal Irradiance (GHI), and spectral ground reflectance for various ground types are the main components of the modeling. In order to improve the accuracy of the simulation, the parameter of field, like size of the modules, rows, tubes, are required as well.

In this paper, Simulation results are compared to field degradation in a PV array in Maryland, MD (SS-16-8). In this field, PEN is determined by FTIR as the backsheet material. Compared to raw weather data, the simulation produces data that are more representative of the relationship between irradiance and degradation. The reason is that simulation can generate the distribution of irradiance on the modules, rather than the data collected from the real sensors in the field or weather station nearby.

II. METHODOLOGY

A. Field Survey Protocol

A field survey protocol was adopted to standardize the measurement locations and the measurement types recorded to produce a uniform data set [19]. Additionally, to observe the 'edge effect', additional modules are surveyed on the row ends. Each module is measured in 6 locations across the surface of its backsheet. Data is collected on the YI, Gloss , and FTIR spectra of every module surveyed without cleaning the modules. The exposure conditions of the field are also noted. In this study, simulation requires not only the irradiance data but also the information from sites, including but not least the location, azimuth, tilt angle, size of modules and cells.

YI measurments follow the standard ATSM E313. MiniScan EZ 4000 with Diffuse/8°Geometry , produced by HunterLab, is used to measure the YI. It is suited for reading textured and non-uniform samples, which fit the situation of backsheet in the fields.

Gloss measurements follow the standard ASTM D523. Micro-TRI-gloss, produced by BYK, is used to measure the gloss. It combines 20°, 60°and 85°three angles in one glossmeter, but values of 60°are usually more stable and less variance than the other two angles. Thus, in this paper, values at 60°are consider as the extend of degradation.

FTIR is measured for each module by 4300 handheld FTIR Spectrometer with Diamond Attenuated Total Reflectance(D-ATR), produced by Agilent. In this paper, it is used to figure out the type of materials, Polyethylene Naphthalate (PEN) in SS-16-8, by characteristic peaks in the FTIR spectrum. Furthermore, the spectrum could reflect the proportion of different functional groups in the material, before and after exposure.

B. Data

Weather data used for the simulation is Typical Meteorological Year (TMY) sourced from NSRB [20]. The data consisted of DNI, DHI, GHI, temperature, relative humidity, wind speed, et.al. SolarGIS is an alternative data could be used in simulation as well. Both types of climate data was obtained from the closet available weather station to the fields, and then ingested into a High-Performance Computing Cluster before we used.

For the data collected from the field survey, they will be saved in the same High-Performance Computing Cluster after they are cleaned and merged with meta-data by scripts. All data processing is performed in RStudio by R Program.

C. Rear-Side Irradiance Simulation

In contrast to the data processing, the simulation for the rear-side of modules is performed in a python package, "Bifacial_radiance", by Jupyter Lab [1]. As mentioned before, climate data and site information are required.

The total length of exposure for the surveyed site was ten years, it was assumed that the TMY exposure did not significantly change during the course of the material exposure. The site information, used in the simulation, is shown as Table I.

Site name		SS-16-8
Location	Latitude	39.131895
	longitude	-77.214147
Subrows/Numpanels		4/5
Albedo		0.14
Azimuth (°)		180
Module	x (m)	1.64
	y (m)	0.994
	x gap (m)	0.0286
	y gap (m)	0.0254
	z gap (m)	/
Sites	Pitch (m)	9.779
	Clearance(m)	0.6604
	nMods	48
	nRows	5
	Tilt angle (°)	20
I-Beams	Count	6
	Length (m)	80

TABLE I
SITE INFORMATION OF SS-16-8

III. RESULTS / DISCUSSION

A. Irradiance Distribution

The map of Irradiance (See Figure2) could be simulated for each panel since it built. Sixty sensors are set to simulating for each module, which corresponds to a sensor placed in the center of each PV cell. By assembling the individual modules into rows, the non-uniform distribution of irradiance appears. Comparing from the horizontal direction, two edges absorb more irradiance than the middle. Comparing from the vertical direction, the position close to the I-Beams endure more exposure than others, which might come from the reflection of I-Beams.

B. Irradiance and Degradation

Combined the result from YI and Simulation, degradation and irradiance show a positive correlation, though the YI start to decrease after saturated exposure, which might be coursed by the material failure inducing microcrack.

978-1-6654-6060-6/23 $31.00 © 2023 IEEE   1142

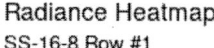

Radiance Heatmap
SS-16-8 Row #1

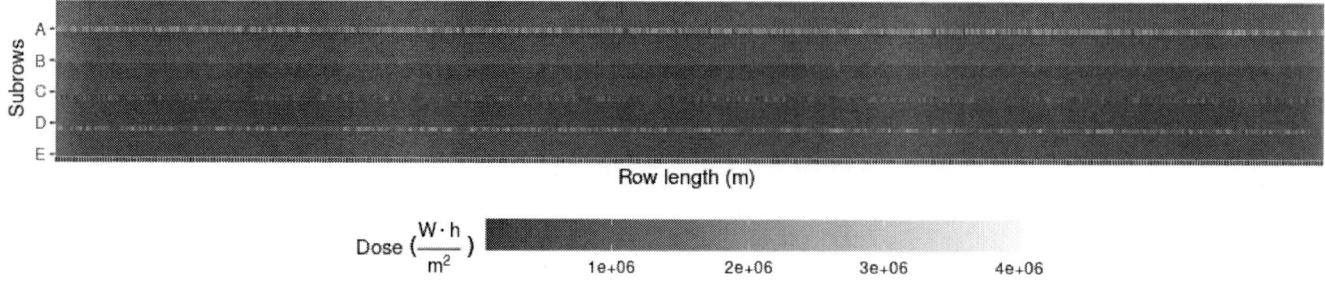

Fig. 2. The irradiance heatmap of Row #1 in the field in Maryland

IV. CONCLUSIONS

A 1 MW field exposed to **Cfa** climate for ten years has been surveyed and the total exposure simulated. The material of backsheet is PEN, determined by the FTIR. With information from the Site and Weather, Simulations of Irradiance for each module in the sites are generated by the "Bifacial_Radiance" package. The simulations display the distribution of dose of exposure since it built, which is not uniformed. Combined with the YI, the relationship between degradation and exposure is regard as positive. In future work, the weather data could be replaced by SolarGis, including real weather data for each year. Since the difference of various material, the relevance between degradation and irradiance could be different. The results could correlate to the Indoor accelerating aging test as well, which will help us understand what the difference between indoor and outdoor exposure is based on materials.

REFERENCES

[1] S. A. Pelaez and C. Deline, "bifacial_radiance: a python package for modeling bifacial solar photovoltaic systems," *Journal of Open Source Software*, vol. 5, no. 50, p. 1865, Jun. 2020. [Online]. Available: https://joss.theoj.org/papers/10.21105/joss.01865

[2] IRENA, *Renewable Power Generation Costs in 2021*, International Renewable Energy Agency, Abu Dhabi, 2022.

[3] M. A. Russo, D. Carvalho, N. Martins, and A. Monteiro, "Forecasting the inevitable: A review on the impacts of climate change on renewable energy resources," *Sustainable Energy Technologies and Assessments*, vol. 52, p. 102283, Aug. 2022. [Online]. Available: https://www.sciencedirect.com/science/article/pii/S2213138822003356

[4] A. I. Osman, L. Chen, M. Yang, G. Msigwa, M. Farghali, S. Fawzy, D. W. Rooney, and P.-S. Yap, "Cost, environmental impact, and resilience of renewable energy under a changing climate: a review," *Environmental Chemistry Letters*, Oct. 2022. [Online]. Available: https://doi.org/10.1007/s10311-022-01532-8

[5] R. A. Kharait, P. Stiles, J. Carriere, and L. McClung, "Impact of Degradation Rates on Solar PV Financing for Projects Located in the United States," in *2017 IEEE 44th Photovoltaic Specialist Conference (PVSC)*, Jun. 2017, pp. 2833–2835.

[6] R. Dubey, S. Chattopadhyay, S. Zachariah, S. Rambabu, H. K. Singh, A. Kottantharayil, B. M. Arora, K. Narasimhan, N. Shiradkar, and J. Vasi, "On-Site Electroluminescence Study of Field-Aged PV Modules," in *2018 IEEE 7th World Conference on Photovoltaic Energy Conversion (WCPEC) (A Joint Conference of 45th IEEE PVSC, 28th PVSEC & 34th EU PVSEC)*, Jun. 2018, pp. 0098–0102, iSSN: 0160-8371.

[7] W. Gambogi, Y. Heta, K. Hashimoto, J. Kopchick, T. Felder, S. MacMaster, A. Bradley, B. Hamzavytehrany, L. Garreau-Iles, T. Aoki, K. Stika, T. J. Trout, and T. Sample, "A Comparison of Key PV Backsheet and Module Performance from Fielded Module Exposures and Accelerated Tests," *IEEE Journal of Photovoltaics*, vol. 4, no. 3, pp. 935–941, May 2014, conference Name: IEEE Journal of Photovoltaics.

[8] Y. Lyu, J. H. Kim, A. Fairbrother, and X. Gu, "Degradation and Cracking Behavior of Polyamide-Based Backsheet Subjected to Sequential Fragmentation Test," *IEEE Journal of Photovoltaics*, vol. 8, no. 6, pp. 1748–1753, Nov. 2018, conference Name: IEEE Journal of Photovoltaics.

[9] C.-C. Lin, Y. Lyu, L.-C. Yu, and X. Gu, "Correlation between mechanical and chemical degradation after outdoor and accelerated laboratory aging for multilayer photovoltaic backsheets," in *Reliability of Photovoltaic Cells, Modules, Components, and Systems IX*, vol. 9938. SPIE, Sep. 2016, pp. 57–67. [Online]. Available: https://www.spiedigitallibrary.org/conference-proceedings-of-spie/9938/99380H/Correlation-between-mechanical-and-chemical-degradation-after-outdoor-and-accelerated/10.1117/12.2238216.full

[10] A. Omazic, G. Oreski, M. Halwachs, G. C. Eder, C. Hirschl, L. Neumaier, G. Pinter, and M. Erceg, "Relation between degradation of polymeric components in crystalline silicon PV module and climatic conditions: A literature review," *Solar Energy Materials and Solar Cells*, vol. 192, pp. 123–133, Apr. 2019. [Online]. Available: https://www.sciencedirect.com/science/article/pii/S0927024818305956

[11] M. D. Kempe and J. H. Wohlgemuth, "Evaluation of temperature and humidity on PV module component degradation," in *2013 IEEE 39th Photovoltaic Specialists Conference (PVSC)*, Jun. 2013, pp. 0120–0125, iSSN: 0160-8371.

[12] E. Wang, H. E. Yang, J. Yen, S. Chi, and C. Wang, "Failure Modes Evaluation of PV Module via Materials Degradation Approach," *Energy Procedia*, vol. 33, pp. 256–264, Jan. 2013. [Online]. Available: https://www.sciencedirect.com/science/article/pii/S1876610213000763

[13] N. Kim, H. Kang, K.-J. Hwang, C. Han, W. S. Hong, D. Kim, E. Lyu, and H. Kim, "Study on the degradation of different types of backsheets used in PV module under accelerated conditions," *Solar Energy Materials and Solar Cells*, vol. 120, pp. 543–548, Jan. 2014. [Online]. Available: https://www.sciencedirect.com/science/article/pii/S092702481300514X

[14] W. Gambogi, Y. Heta, K. Hashimoto, J. Kopchick, T. Felder, S. MacMaster, A. Bradley, B. Hamzavytehraney, V. Felix, T. Aoki, K. Stika, L. Garreau-Illes, and T. J. Trout, "Weathering and durability of PV backsheets and impact on PV module performance," in *Reliability of Photovoltaic Cells, Modules, Components, and Systems VI*, vol. 8825. SPIE, Sep. 2013, pp. 80–90. [Online]. Available: https://www.spiedigitallibrary.org/conference-proceedings-of-spie/8825/88250B/Weathering-and-durability-of-PV-backsheets-and-impact-on-PV/10.1117/12.2024491.full

[15] F. Liu, L. Jiang, and S. Yang, "Ultra-violet degradation behavior of polymeric backsheets for photovoltaic modules," *Solar Energy*, vol. 108, pp. 88–100, Oct. 2014. [Online]. Available: https://www.sciencedirect.com/science/article/pii/S0038092X14003260

[16] A. Fairbrother, M. Boyd, Y. Lyu, J. Avenet, P. Illich, Y. Wang,

978-1-6654-6060-6/23 $31.00 © 2023 IEEE

M. Kempe, B. Dougherty, L. Bruckman, and X. Gu, "Differential degradation patterns of photovoltaic backsheets at the array level," *Solar Energy*, vol. 163, pp. 62–69, Mar. 2018. [Online]. Available: https://www.sciencedirect.com/science/article/pii/S0038092X18300938

[17] Y. Wang, W.-H. Huang, A. Fairbrother, L. S. Fridman, A. J. Curran, N. R. Wheeler, S. Napoli, A. W. Hauser, S. Julien, X. Gu, G. S. O'Brien, K.-T. Wan, L. Ji, M. D. Kempe, K. P. Boyce, R. H. French, and L. S. Bruckman, "Generalized Spatio-Temporal Model of Backsheet Degradation From Field Surveys of Photovoltaic Modules," *IEEE Journal of Photovoltaics*, vol. 9, no. 5, pp. 1374–1381, Sep. 2019, conference Name: IEEE Journal of Photovoltaics.

[18] S. Ovaitt, M. Brown, C. Deline, and M. D. Kempe, "Spectral rear irradiance testing and modeling for degradation and performance of solar fields," in *2022 IEEE 49th Photovoltaics Specialists Conference (PVSC)*, Jun. 2022, pp. 0992–0994.

[19] R. J. Wieser, Z. Z. Li, S. L. Moffitt, R. Zabalza, E. Boucher, S. Ayala, M. Brown, X. Gu, L. Ji, C. O'Brien, A. W. Hauser, G. S. O'Brien, X. Yu, R. H. French, M. D. Kempe, J. Tracy, K. R. Choudhury, W. J. Gambogi, L. S. Bruckman, and K. P. Boyce, "Spatiotemporal Modeling of Real World Backsheets Field Survey Data: Hierarchical (Multilevel) Generalized Additive Models," in *2022 IEEE 49th Photovoltaics Specialists Conference (PVSC)*, Jun. 2022, pp. 0255–0260.

[20] "National Solar Radiation Data Base: Typical Meteorological Year 3 (TMY3) between 1991 and 2005," https://rredc.nrel.gov/solar/old_data/nsrdb/1991-2005/tmy3/.

Extreme Weather and PV Performance

Dirk Jordan, Kirsten Perry, Chris Deline

NREL, Golden, CO, United States

NREL, Golden, CO, United States

NREL, Golden, CO, United States

The impact of extreme weather events on PV performance was studied by comparing the National Oceanic and Atmospheric Administration database on severe weather with the PV Fleet database on continuous PV performance. We identified 170 systems that were immediately impacted by weather events. These severe weather events lead to a median loss of only % of annual production. However, flooding and high wind events were found to have an extremely long tail extending to 60 % loss showing that these discrete events can pose a substantial risk to PV systems. Besides the short-term impact of lost production due to outages, we also found a statistically significant increased performance loss rate (PLR) for high wind events comparing PLR before and after these weather events. In addition, hail events caused a higher PLR for 2 out of 3 systems. More data are required to better quantify the impact, but these first results illustrate the substantial risk these events pose short-and long-term.

Viability of a Novel Methodology of Measure-Correlate-Predict for Albedo Estimation

Halley C. Darling, Renn Darawali, Lucila D. Tafur

UL Solutions, Albany, NY, United States

As utility scale solar facilities employ bifacial panels in larger market proportions and across various biomes, greater scrutiny is required in determining the impacts of albedo estimation on energy gains. Current industry practices evaluate projects using a singular satellite model for albedo. Selecting any singular source can lead to high uncertainty as multiple available satellite modeled albedo datasets often differ significantly at a given location. An alternative to satellite-modeled albedo is direct use of ground measurements from on-site albedometers. Use of these on-site measurements increases confidence in albedo magnitude for a specific time frame, however, may lack coverage for year-to-year albedo variability from snow or vegetation. In this study, UL investigates (1) differences between several satellite-modeled albedo sources and on-site measurements by comparing site-specific estimates and corresponding net energy differences for a hypothetical bifacial system, (2) impacts on energy estimation due to interannual albedo variability by simulating long-term estimates, (3) the viability of measure-correlate-predict (MCP) analyses on albedo to improve accuracy by testing a sample analysis method performed on one year of a longer-term ground measured dataset against the remaining period of that dataset.

978-1-6654-6060-6/23 $31.00 © 2023 IEEE

Long Term Soiling Model Tuning for Enhanced PV Cleaning Schedule Optimization

Kyle Seymour, Akanksha Bhat, and Thomas Haley

Clean Power Research, Kirkland, WA, 98003, United States

Abstract—We use soiling ratio time series data collected by Fracsun stations to tune the HSU soiling model using satellite-derived particulate matter and precipitation data as inputs at four sites. Soiling history from 2003 through 2022 is inferred using the tuned models for each site. A revenue optimized threshold based soiling mitigation strategy is presented to exemplify cost savings potential for a 50 MWdc PV system at each site.

Keywords—soiling, optimization, modeling, photovoltaic, energy

I. BACKGROUND AND MOTIVATION

Given the increasing proliferation of PV power generation, an important issue that can significantly decrease revenue is soiling, which refers to the accumulation of dust, dirt, and other particles on the surface of the panels. This can lead to a reduction in the amount of solar energy that is transmitted to the PV cells, causing a decrease in power output and an increase in production uncertainty [1]. The effects of soiling on PV performance have been studied extensively in recent years with research showing that local environmental conditions can lead to soiling losses as high as 50% [2] in arid regions and annual losses in the United States typically in the range of 0-6% [3].

A common method of measuring the effects of soiling directly is with soiling stations comprised of two identical PV cells, where one is regularly cleaned and the other is allowed to naturally soil. The power loss due to soiling can be represented by the ratio of outputs from the cells (e.g. short circuit currents). Where measurement is not possible, such as during the preconstruction phase, it is necessary to utilize a model to estimate soiling losses [4]. Model development has been aided by studies focused on characterizing environmental parameters on dust deposition [5] or annual soiling losses [6], with a key result being that atmospheric particulate matter and rainfall were found to be some of the most determinant predictor of annual soiling losses [7]. A method developed by Coello et al. leverages these parameters to estimate soiling accumulation at finer timescales [8].

Direct soiling measurement leads to the most accurate evaluations of soiling losses and can capture the effect of localized factors on particulate deposition. It also has the added benefit of enabling real-time monitoring to help guide cleaning decisions. On the other hand, soiling models can assist in economic assessment of projects or optimization of cleaning schedules at little to no cost. Without the need for a temporally constrained measurement campaign, a model can produce soiling estimates on large time scales given the right set of inputs.

In this work, the benefits of both are combined by leveraging soiling measurements to tune a long-term soiling model. Local soiling kit measurements are used to tune an empirical model to location-specific conditions, leading to improved characterization of historic soiling patterns. The key contribution, long-term soiling characterization tuned by local short-term measurements, is then used for a cleaning schedule optimization whereby the cost of lost revenues due to soiling is optimally balanced with the cost of cleanings.

II. DATA SOURCES

SolarAnywhere provides satellite-derived particulate matter (PM10, PM2.5) and precipitation data to support soiling loss modeling. The self-cleaning array-mounted soiling stations produced by Fracsun provide realistic measurements of daily losses due to soiling

A. MERRA-2 particulate matter

SolarAnywhere particulate matter (PM) data is based on NASA Modern-Era Retrospective analysis for Research and Applications (MERRA-2) reanalysis and has a nominal spatial resolution of 50km x 62.5km (0.5 degrees x 0.625 degrees). The data is available in all geographic locations with SolarAnywhere v3.5 and newer data versions and is updated monthly with SolarAnywhere historical data (approximately 1-2 months lag from real-time). SolarAnywhere particulate matter data is created in two steps. First, the surface aerosol concentrations are extracted from the MERRA-2 dataset. These include key constituents of particulate matter such as dust, sulfate, organic carbon, black carbon and sea salt. Thereafter, a weighted combination of surface aerosols is used to generate the total PM10 and PM2.5 concentrations at a location. This process, referred to as particulate matter reconstruction, leverages previous research comparing MERRA-2 data with ground-based measurements. Previous studies by NASA and others have found that MERRA-2 PM concentrations generally agree well with observations [9], [10].

B. SNODAS precipitation

In addition to particulate matter data, precipitation data is used as a key input in PV soiling loss modeling. Rainfall events can remove accumulated dirt, ash, etc. from the PV modules. SolarAnywhere provides time-series data of solid and liquid precipitation in kilograms per meters squared. In the contiguous United States, precipitation measurements are received as daily totals from the National Oceanic and Atmospheric Administration's (NOAA) Snow Data Assimilation System (SNODAS) dataset at a high spatial resolution of 1km. They are

978-1-6654-6060-6/23 $31.00 © 2023 IEEE

then divided by the number of observations included in each day (varies based on requested temporal resolution). Outside the United States, precipitation values are retrieved at an hourly temporal resolution and a spatial resolution of 25km. Validation efforts indicated that daily SNODAS precipitation totals have good correlation with other data sources such as the European Centre for Medium-Range Weather Forecasts (ECMWF) ERA5 reanalysis product.

C. Fracsun soiling measurements

The soiling stations produced by Fracsun consist of two identical coplanar PV cells that can be mounted on a fixed tilt or tracking array. One cell is cleaned daily by an automated spray system while the other is left to soil naturally. Instantaneous soiling ratio is calculated from the irradiance readings of the two cells.

III. MODEL FITTING

The satellite-derived PM and precipitation data can be conveniently coupled with the method developed by Coello et al. [8] to estimate soiling accumulation (henceforth referred to as the "HSU model"). In the HSU model, the soiling ratio is calculated as a function of accumulated particulate matter mass on the PV panels. Particulate accumulation during each time interval is a linear function of the coincident near-surface atmospheric concentration of PM2.5 and PM10. Accumulated mass is assumed to be completely cleared from the panels when some specified 24-hour precipitation threshold (cleaning threshold) is surpassed. Using measurements of the soiling ratio from a soiling kit, a PM multiplier (which scales the PM concentrations) and the cleaning threshold are tuned to improve the specificity of the model to a particular site.

A. Soiling measurements

Fracsun soiling kits sample irradiance of the dirty and clean panels at five-minute intervals. A daily soiling ratio is calculated from the Fracsun system measurements in accordance with methods utilized in previous works [7], in which measurements taken between 11:00 AM and 1:00 PM local under clear sky conditions were utilized to calculate the daily mean soiling ratio. Cleaning events were identified using an adaptation of the quartile method described by [11]. As rainfall is the only cleaning mechanism represented by the HSU model, cleaning events identified on days with no corresponding rainfall in the precipitation data were determined to be unexplainable and were flagged as inferred washings.

B. Model tuning

The HSU model was fit to the measurements by simultaneously optimizing the PM multiplier and cleaning threshold to minimize the Root Mean Squared Error (RMSE) of the daily modeled soiling ratio relative to the measurement-derived daily soiling ratio. On inferred washing days, the model output was replaced by the measured soiling ratio to avoid erroneous overfitting. With this method, the PM multiplier and cleaning threshold are effectively tuned to align the soiling model with measurements between consecutive inferred washing events.

The results of the model tuning process are shown for four sites in Fig. 1. Sites A, B, and C are in California's Central Valley in regions classified as temperate or arid (Köppen-Geiger classifications of Csa, Bsk, and BSk, respectively). Site D is in a hot arid desert region of Nevada (BWh). By tuning the PM multiplier, the slope of the soiling loss line, representing daily soiling rate, was approximated well by the model. Discrete drops in soiling loss are captured by the model as a consequence of precipitation exceeding the tuned cleaning threshold.

IV. CLEANING ANALYSIS CASE STUDY

The ability to estimate soiling behavior over a period of many years enables a robust analysis of the effects of different cleaning strategies. In this section, we present the results of performing such an analysis and demonstrate the cost-savings opportunities made available by doing so. For this case study, power simulations were performed for Site C using SolarAnywhere rainfall and PM data from 2003 through 2022 with the objective of identifying a cost-optimal soiling threshold at which the panels should be cleaned. The soiling threshold is optimized by minimizing the cost of soiling, which is defined as net revenue loss due to reduced energy production plus the cost of cleaning the panels.

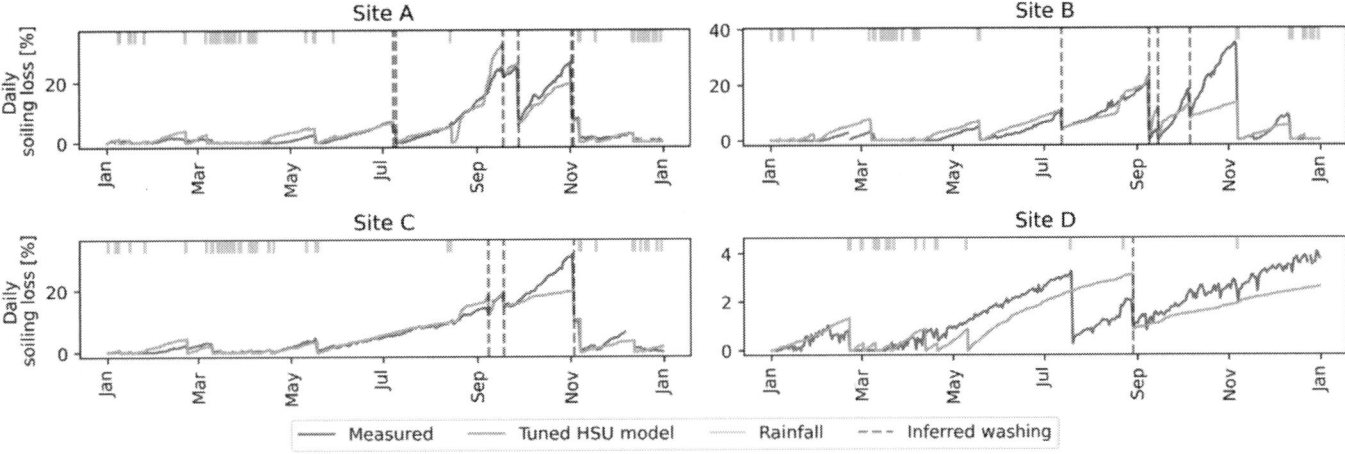

Fig. 1. Results of tuning the HSU model to measured soiling ratio at four sites in 2020. On inferred washing days, the measured soiling ratio replaces the modeled value so only natural soiling and cleaning behavior is learned.

978-1-6654-6060-6/23 $31.00 © 2023 IEEE

The results of the analysis are shown for Site C in Fig. 2. A 50 MWdc single-axis tracking PV system is modeled with a cost per cleaning of $50,000 and an energy price of 0.13$/kWh [12]. Cleaning the panels every time the soiling ratio reaches 0.93 leads to the best balance of cleaning cost and lost energy production revenue due to soiling. Given the 20-year meteorological history of this site, this would result in approximately 1.8 cleanings per year. At this soiling ratio threshold, the cost of soiling is estimated to be approximately 4.3% of annual revenue. Allowing the panels to soil more between cleanings would lead to an additional 1% of revenue loss due to diminished energy production. Cleaning the panels more frequently would result in cleaning costs that sub-optimally outweigh the avoided revenue loss benefit.

Based on the 20-year modeled soiling history, cleaning the panels every time the soiling ratio reaches 0.93 could avoid $191,000 annual revenue loss due to soiling with a net annual benefit of $101,000 after cleaning costs are subtracted. This result and those associated with optimization of 50 MWdc systems at the other three sites are presented in Table I. While the net benefit of cleaning optimization of Site D in Nevada is relatively small, the three sites in California could achieve benefits in the tens of thousands of dollars annually by long term model-derived optimization cleaning schedules.

V. CONCLUSIONS

The technique developed leverages the benefits of both soiling modeling, which can make use of long-term remote sensing data, and direct measurement, which provides the most accurate assessment of localized soiling. A study of this analysis on four sites in the western United States demonstrated that potential cost savings due to long-term cleaning optimization could surpass $100,000 annually for a 50 MWdc PV plant.

Further development of soiling models, such as regression or artificial neural networks as in [13], is needed to make the cleaning analysis results more robust and reduce uncertainty. While some studies identified PM and rainfall as the dominant predictors of soiling, others suggest wind or other meteorological parameters can also have a strong impact [5]. Thus, future work will also focus on incorporating more model parameters from a wider set of sources.

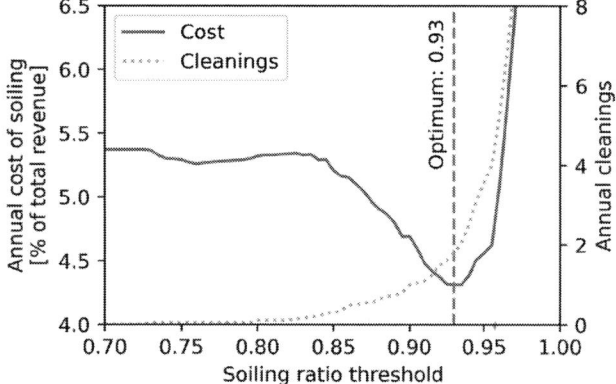

Fig. 2. Cost optimization of cleaning PV panel cleaning for Site C in California's Central Valley. Cleaning the panels when the soiling ratio reaches 0.93 leads to a minimization of the annual cost of soiling.

TABLE I
COST SAVINGS RESULTING FROM CLEANING OPTIMIZATION OF FOUR HYPOTHETICAL 50 MWDC PV PLANTS

	Site A	Site B	Site C	Site D
Optimum soiling ratio threshold	0.89	0.90	0.93	0.97
Annual revenue loss avoided by cleaning	$85k	$157k	$191k	$7k
Annual cleaning expenditure	$42k	$70k	$90k	$5k
Net annual optimization benefit	$43k	$87k	$101k	$2k

ACKNOWLEDGMENT

The authors gratefully acknowledge C.Mattheis, B.Fisher and the Fracsun team for providing ground measured soiling data for this study.

REFERENCES

[1] N. Reich, J. Zenke, B. Muller, K. Kiefer, and B. Farnung, "On-site performance verification to reduce yield prediction uncertainties," *2015 IEEE 42nd Photovoltaic Specialist Conference, PVSC 2015*, pp. 1–6, 2015, doi: 10.1109/PVSC.2015.7355614.

[2] X. Li, D. L. Mauzerall, and M. H. Bergin, "Global reduction of solar power generation efficiency due to aerosols and panel soiling," *Nat Sustain*, vol. 3, no. 9, pp. 720–727, 2020, doi: 10.1038/s41893-020-0553-2.

[3] A. Kimber, L. Mitchell, S. Nogradi, and H. Wenger, "The effect of soiling on large grid-connected photovoltaic systems in California and the Southwest Region of the United States," *Conference Record of the 2006 IEEE 4th World Conference on Photovoltaic Energy Conversion, WCPEC-4*, vol. 2, pp. 2391–2395, 2006, doi: 10.1109/WCPEC.2006.279690.

[4] A. Kimber, L. Mitchell, S. Nogradi, and H. Wenger, "The effect of soiling on large grid-connected photovoltaic systems in California and the Southwest Region of the United States," *Conference Record of the 2006 IEEE 4th World Conference on Photovoltaic Energy Conversion, WCPEC-4*, vol. 2, pp. 2391–2395, 2006, doi: 10.1109/WCPEC.2006.279690.

[5] B. Figgis, B. Guo, W. Javed, S. Ahzi, and Y. Rémond, "Dominant environmental parameters for dust deposition and resuspension in desert climates," *Aerosol Science and Technology*, vol. 52, no. 7, pp. 788–798, Jul. 2018, doi: 10.1080/02786826.2018.1462473.

[6] L. Micheli, M. Muller, and S. Kurtz, "Determining the effects of environment and atmospheric parameters on PV field performance," *2017 IEEE 44th Photovoltaic Specialist Conference, PVSC 2017*, no. 3, pp. 619–625, 2017, doi: 10.1109/PVSC.2017.8366376.

[7] L. Micheli and M. Muller, "An investigation of the key parameters for predicting PV soiling losses," *Progress in Photovoltaics: Research and Applications*, vol. 25, no. 4, pp. 291–307, 2017, doi: 10.1002/pip.2860.

[8] M. Coello and L. Boyle, "Simple Model for Predicting Time Series Soiling of Photovoltaic Panels," *IEEE J Photovolt*, vol. 9, no. 5, pp. 1382–1387, Sep. 2019, doi: 10.1109/JPHOTOV.2019.2919628.

[9] C. A. Randles, "The MERRA-2 aerosol assimilation," *NASA Technical Report*, vol. 45, no. December, pp. 1–153, 2016, [Online]. Available: https://gmao.gsfc.nasa.gov/pubs/docs/Randles887.pdf

[10] R. D. Koster et al., "Technical Report Series on Global Modeling and Data Assimilation, Volume 45 The MERRA-2 Aerosol Assimilation," 2016.

[11] M. G. Deceglie, L. Micheli, and M. Muller, "Quantifying Soiling Loss Directly From PV Yield," *IEEE J Photovolt*, vol. 8, no. 2, pp. 547–551, Mar. 2018, doi: 10.1109/JPHOTOV.2017.2784682.

[12] D. L. Alvarez, A. S. Al-Sumaiti, and S. R. Rivera, "Estimation of an Optimal PV Panel Cleaning Strategy Based on Both Annual Radiation Profile and Module Degradation," *IEEE Access*, vol. 8, pp. 63832–63839, 2020, doi: 10.1109/ACCESS.2020.2983322.

[13] W. Javed, B. Guo, and B. Figgis, "Modeling of photovoltaic soiling loss as a function of environmental variables," *Solar Energy*, vol. 157, no. July, pp. 397–407, 2017, doi: 10.1016/j.solener.2017.08.046.

Coalescence of GaP on V-groove Si for III-V/Si Solar Cells

Theresa E. Saenz, John S. Mangum, Olivia D. Schneble, Anica N. Neumann, Ryan M. France, William E. McMahon, Jeramy D. Zimmerman, Emily L. Warren

National Renewable Energy Lab, Golden, CO, United States

Colorado School of Mines, Golden, CO, United States

Here, we study the morphology and strain relaxation of GaP as it coalesces over V-groove Si. V-groove nanopatterning is a promising way to create substrates that enable the grow of high quality III-V on Si, and it offers a route to III-V growth on low-cost Si substrates rougher than those typically required for epitaxial growth. We find that even at optimized growth conditions, careful control of the geometry of the nanopatterns is needed to promote a smooth morphology of the III-V material. We also find that strain relaxation, which resulted in an elevated threading dislocation density, occurs during the coalescence process. With further dislocation reduction, the templates described here are promising for further development of III-V solar cells on Si. Additional work on growing these templates on low-cost, wet-polished Si will also be presented.

Mitigation of phase separation in high Ga Cu(In,Ga)S2 absorbers to achieve ~ 1 volt 15.6% power conversion efficiency

Damilola Adeleye, Mohit Sood, Tobias Törndahl, Adam Hultqvist, Aline Vanderhaegen, Michele Melchiorre, Susanne Siebentritt

University of Luxembourg, Esch-sur-Alzette, Luxembourg

Uppsala University, Uppsala, Sweden

The use of Cu(In,Ga)S2 as a top cell in tandem solar cell, despite having suitable properties for such an application, is hampered by a high open-circuit voltage (VOC) deficit. The deficit arises from a poor optoelectronic quality of the absorbers - engendered by phase separation - and the inadequate translation of the optoelectronic quality of the absorber into device VOC. In this work, we report the role of first stage substrate temperature in the mitigation of phase separation and optimized Cu-excess during growth in Cu(In,Ga)S2, which leads to reduced VOC deficit, resulting in a device with 15.6 % PCE with a VOC of ~ 981 mV when completed with atomic layer deposited (Zn,Sn)O and Al:ZnMgO transparent conductive oxide.

Germanium-Tin Diode for Thermophotovoltaic Energy Collection

Amanda Lemire, Kevin A. Grossklaus, and Thomas E. Vandervelde

Tufts University, Medford, MA, 02155, USA

Abstract — The bandgap of germanium-tin alloys extends from 0.8 eV into the metallic as the Sn composition increases, which overlaps the strongly absorbing range of the dominant III-V systems used for thermophotovoltaic (TPV) power collection and for narrow bandgap diodes in multijunction PV cells. Si and Ge wafers are more physically resilient than III-V wafers and material abundance is higher than indium-containing cells. Si wafers are also available in larger formats, with a broader array of processing technologies. In fully relaxed GeSn films with greater than 6% Sn, the bandgap transitions to direct; in compressively strained films, the transition is higher. For high-stress and cost-limited applications, Group-IV devices have the potential to be a useful substitute for existing options. We designed a $Ge_{1-x}Sn_x$ device and simulated its performance in Silvaco Atlas using material properties obtained from preliminary growths. We deposited the $Ge_{1-x}Sn_x$ TPV device by molecular beam epitaxy, targeting 3% Sn to achieve a similar bandgap to GaSb. Boron and antimony were used as p and n dopants. Material quality was characterized by XRD, spectroscopic ellipsometry, photoluminescence, and microscopy techniques. Component material electrical characterization was performed via Hall effect.

I. INTRODUCTION

Commercial and experimental thermophotovoltaic (TPV) technology is dominated by III-V compound semiconductors. There are, however, applications for which Group-IV alloys may be equally or more appropriate. The natively direct, narrow bandgap of GaSb enables the manufacture of efficient cells by simple manufacturing methods, such as Zn diffusion and ion implantation. Consequently, GaSb is the predominant material used for TPVs presently deployed for waste heat recovery [1]. Harvesting waste heat can be applied to many manufacturing processes, such as steel smelting, that generate waste heat in the 1300-1600K range. The bandgap of GaSb at 300K is 0.72 eV, which limits its efficiency for converting low-temperature blackbody emissions to electricity. Without incorporating additional alloying elements, or manipulating the strain by growing GaSb heteroepitaxially, the bandgap cannot be reduced to take advantage of lower energy/temperature sources.

$In_xGa_{1-x}As$ is a common choice for a tunable bandgap material with an appropriate absorption range for TPV, being adjustable from 1.42 to 0.36 eV by varying the composition. Using a composition that is lattice-matched to InP, which is a readily available substrate, $In_{0.53}Ga_{0.47}As$ has a direct bandgap of 0.74 eV, similar to GaSb. Epitaxial growth methods allow for the manufacture of mismatched compositions, so InGaAs can be used to extract power from lower temperature materials than GaSb. The improved absorption at lower energies exceeds the losses from lattice defects as strain increases, but material quality is a limiting factor in InGaAs efficiency [1].

Some studies have considered Ge as a potential TPV material, with an indirect gap of 0.66 eV [2]. Ge is particularly advantageous for mixed applications like multijunction cells that use Si, owing to their shared diamond lattice structure, type-II band alignment, and high miscibility. As with III-V compound semiconductors, Ge's bandgap may be tuned by alloying and by strain manipulation. Si and Sn are leading alloying candidates. In fully relaxed $Ge_{1-x}Sn_x$ films with greater than 6 atomic% Sn, the bandgap transitions to direct; in films that are compressively strained to Ge, the transition point may be as high as 25% [3]. Compressively strained GeSn is limited by the same sort of growth defects as InGaAs. However, Ge is 6 times more abundant than In, and produced at 100 times the rate worldwide [4], [5]. Ge and Sn are also less potentially toxic under breakdown conditions than As-containing cells. The Si wafers on which GeSn is often grown are more physically resilient than III-V wafers, making Group-IV cells sturdier for applications that put cells under large movement stresses. They are available in 300 mm wafer formats, twice the size of largest III-V wafer, due both to the economy of scale in Si processing and to the greater strength of the material. The long history of Si device processing has also created a suite of better-characterized processing technologies which have been applied successfully to mixed SiGe and SiGeSn devices.

GeSn has been investigated for infrared light emission and photodetection for decades. Effective detectors have been made by chemical vapor deposition with as little as 2% Sn [6] and as much as 7% [7]. The highest reported Sn incorporation is 27%, in a film grown by molecular beam epitaxy (MBE) and limited to a thickness of 30nm [8]. A TPV cell using fully relaxed $Ge_{0.83}Sn_{0.17}$ has recently been proposed, with a bandgap of 0.29 eV [9]. That ambitious work suggests that power conversion efficiencies of 9% could be achieved for 2 μm thick layers of high quality material under a black blody radiator at 1500K. Their results indicate that thick material layers substantially reduce the impact of Shockley-Read-Hall recombination on the device performance.

We propose in this work to design and grow a GeSn TPV diode close in bandgap to GaSb and $In_{0.53}Ga_{0.47}As$, so that its efficiency can be compared to the dominant technologies. $Ge_{0.97}Sn_{0.03}$ has an indirect bandgap of 0.61 eV when strained to a Ge underlayer. The theoretical critical thickness of a 3% Sn layer is 2 μm, which allows us to grow a thick device without a high density of defects [10]. A comparison of the bandgap to several black body radiation curves is shown in Fig. 1.

978-1-6654-6060-6/23 $31.00 © 2023 IEEE

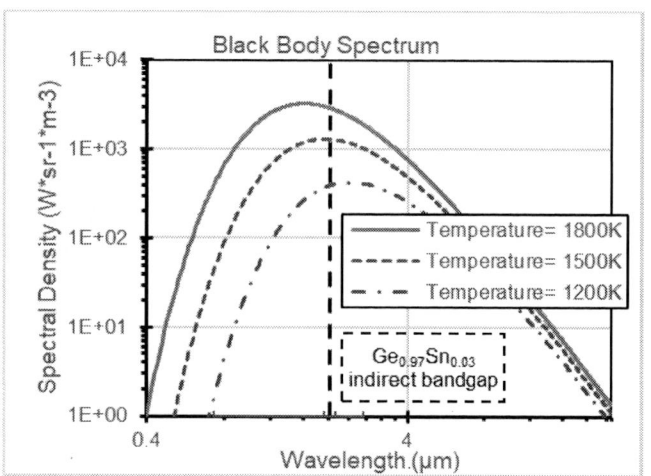

Fig. 1. Smallest bandgap of proposed TPV diode contrasted against the black body spectra for three temperatures in the range of interest.

II. EXPERIMENTAL DETAILS

We design our diode in the simulation software Silvaco Atlas, optimizing for efficiency using a 1500K black body heat source. We vary design parameters including layer thicknesses, doping concentrations, and n-p order to enhance quantum efficiency and power output. Optical properties for strained GeSn are taken experimentally from samples previously grown in the lab. We assume a composition of 3% Sn, fully strained to Ge, which from our previous growths and by theory should have an indirect band gap of 0.61 eV and a direct bandgap of 0.72 eV [3]. Electrical properties are taken from experimental results, both from our work and the literature [2], [9].

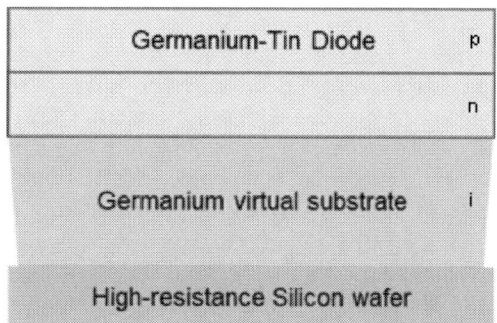

Fig. 2. Simplistic design for the GeSn TPV material stack. Dopings are indicated on the right side of layers to be grown, as p, n, and i. The intrinsic doping of Ge in our system is 4×10^{16} cm^{-1}.

Our material samples are grown on a custom designed VG-90 MBE system, with Ge evaporated by electron beam and Sn, Sb, and B evaporated by solid source effusion cells. Sb is used as the n-dopant, and B as the p-dopant. Growth temperature is monitored by a thermocouple located behind the wafer, held at 175°C to delay the onset of relaxation. The critical thickness for relaxation was determined by previous experiment on the growth effects of Sb, and considered in the design of the diode. The diode is grown on top of a fully relaxed Ge virtual substrate, as shown in Fig. 2.

The conformation of strain and composition to design is confirmed by high-resolution x-ray diffraction (HRXRD) and variable-angle spectroscopic ellipsometry (VASE). The bandgap is confirmed by VASE and photoluminescence spectroscopy (PL). TEM is employed to examine material defectivity and its impact on the device structure. Hall effect measurements of separately grown material layers are compared to infrared VASE data to determine the doping and mobilities of the diode layers.

III. SIGNIFICANCE OF WORK

A Group-IV TPV cell is designed for direct comparison to existing GaSb and InGaAs TPV technologies. Previous experimental works have focused on unalloyed Ge for TPV; this work represents an early foray into narrower bandgap Group-IV materials, which are preferable for incorporation into Si multijunction cells and for high-stress applications.

IV. SUMMARY OF THE WORK

A thermophotovoltaic cell comprising a Ge$_{0.97}$Sn$_{0.03}$ diode, a Ge virtual substrate, and a high-resistance Si wafer is designed in Silvaco. Its dimensions are optimized for a 1500K black body radiator. The material stack is grown by molecular beam epitaxy, and its real electrical characteristics are assumed from equivalent layer data collected by spectroscopic ellipsometry and Hall effect. These are fed back into the model to improve the accuracy of comparison between its performance, and GaSb and InGaAs TPV cells with similar bandgaps.

REFERENCES

[1] M. M. A. Gamel, H. J. Lee, W. E. S. W. A. Rashid, P. J. Ker, L. K. Yau, M. A. Hannan, and M. Z. Jamaludin, "A Review on Thermophotovoltaic Cell and Its Applications in Energy Conversion: Issues and Recommendations," *Materials*, vol. 14, no. 17, Art. no. 17, Jan. 2021, doi: 10.3390/ma14174944.

[2] M. M. A. Gamel, K. P. Jern, W. E. Rashid, L. K. Yau, and M. Z. Jamaludin, "The Effect of Illumination Intensity on the Performance of Germanium Based-Thermophotovoltaic Cell," in *2019 IEEE Regional Symposium on Micro and Nanoelectronics (RSM)*, Aug. 2019, pp. 129–132. doi: 10.1109/RSM46715.2019.8943521.

[3] N. S. Fernando, R. A. Carrasco, R. Hickey, J. Hart, R. Hazbun, S. Schoeche, J. N. Hilfiker, J. Kolodzey, and S. Zollner, "Band gap and strain engineering of pseudomorphic Ge$_{1-x-y}$Si$_x$Sn$_y$ alloys on Ge and GaAs for photonic applications," *Journal of Vacuum Science & Technology B, Nanotechnology and Microelectronics: Materials, Processing, Measurement, and Phenomena*, vol. 36, no. 2, p. 021202, Mar. 2018, doi: 10.1116/1.5001948.

[4] C. S. Anderson, "Indium," US Geological Survey, Jan. 2022. Accessed: Apr. 10, 2023. [Online]. Available: https://pubs.usgs.gov/periodicals/mcs2022/mcs2022-indium.pdf

[5] A. C. Tolcin, "Germanium," US Geological Survey, Jan. 2022. Accessed: Apr. 10, 2023. [Online]. Available: https://pubs.usgs.gov/periodicals/mcs2022/mcs2022-germanium.pdf

[6] H. H. Tseng, H. Li, V. Mashanov, Y. J. Yang, H. H. Cheng, G. E. Chang, R. A. Soref, and G. Sun, "GeSn-based p-i-n photodiodes with strained active layer on a Si wafer," *Appl. Phys. Lett.*, vol. 103, no. 23, p. 231907, Dec. 2013, doi: 10.1063/1.4840135.

[7] T. Pham, W. Du, H. Tran, J. Margetis, J. Tolle, G. Sun, R. A. Soref, H. A. Naseem, B. Li, and S.-Q. Yu, "Systematic study of Si-based GeSn photodiodes with 2.6 μm detector cutoff for short-wave infrared detection," *Opt. Express, OE*, vol. 24, no. 5, pp. 4519–4531, Mar. 2016, doi: 10.1364/OE.24.004519.

[8] D. Imbrenda, R. Hickey, R. A. Carrasco, N. S. Fernando, J. VanDerslice, S. Zollner, and J. Kolodzey, "Infrared dielectric response, index of refraction, and absorption of germanium-tin alloys with tin contents up to 27% deposited by molecular beam epitaxy," *Appl. Phys. Lett.*, vol. 113, no. 12, p. 122104, Sep. 2018, doi: 10.1063/1.5040853.

[9] G. Daligou, R. Soref, A. Attiaoui, J. Hossain, M. R. M. Atalla, P. Del Vecchio, and O. Moutanabbir, "Group IV Mid-Infrared Thermophotovoltaic Cells on Silicon." arXiv, Feb. 21, 2023. Accessed: Mar. 11, 2023. [Online]. Available: http://arxiv.org/abs/2302.10742

[10] W. Wang, Q. Zhou, Y. Dong, E. S. Tok, and Y.-C. Yeo, "Critical thickness for strain relaxation of Ge$_{1-x}$Sn$_x$ ($x \leq$ 0.17) grown by molecular beam epitaxy on Ge(001)," *Appl. Phys. Lett.*, vol. 106, no. 23, p. 232106, Jun. 2015, doi: 10.1063/1.4922529.

Rapid Thermal Annealing of Symmetric p-TOPCon Silicon Test Structures

Arpan Sinha, Sagnik Dasgupta, Ajeet Rohatgi, Mool C. Gupta

University of Virginia, Charlottesville, VA, United States

Georgia Institute of Technology, Atlanta, GA, United States

Thermal annealing of tunnel oxide passivated contacts (TOPCon) solar cell devices is an important step in obtaining high efficiency. The furnace annealing is carried out at 850 0C for 30 minutes. Shorter thermal annealing times are desired, and rapid thermal annealing (RTA) process could achieve this goal. The RTA also provides an opportunity to improve our understanding of the thermal annealing of TOPCon device structures as it allows the variation of heating rate, high-temperature hold time, and cooling rate. Systematic investigation of time-dependent kinetics of thermal annealing as well as the optimization of the annealing time for boron-doped poly-Si based p-TOPCon Si test structures was carried out. The photoluminescence and the sheet resistance were monitored to evaluate the passivation quality. The fast-heating rate of ~80 °C/s or similar cooling rates led to irreversible degradation in PL peak intensity and surface passivation quality, and forming gas annealing (FGA) did not improve the performance. Significant improvement in the PL peak intensity was observed for a slower heating rate of ~13 °C/s (high-temperature hold time of 100 s) and cooling rates of ~2.6 °C/s. A further increase was observed after FGA treatment, and the results were comparable to the furnace annealing. A low sheet resistance was obtained under fast heating and cooling rates, indicating that poly-Si films rapidly crystallize, and dopant activation is relatively fast compared to the improvement in open circuit voltage. Currently, the best results were obtained with a 60 s heating time, a 60 s hold time at 825 °C, and a 300 s cooling time. The PL peak intensity was similar to the value obtained with furnace annealing at 875 °C for 30 min. So, the RTA high-temperature annealing times are much shorter than the currently used furnace annealing times with similar performance. Further work is in progress.

Temperature Dependent Carrier Extraction and the Effects of Excitons on Emission and Photovoltaic Performance in Cs0.05FA0.79MA0.16Pb(I0.83Br0.17)3 Solar Cells

Hadi Afshari, Brandon K Durant, Ahmad R Kirmani, Sergio A Chacon, John Mahoney, Vincent R Whiteside, Rebecca A Scheidt, Matthew C Beard, Joseph M Luther, Ian R Sellers

University of Oklahoma, Norman, OK, United States

National Renewable Energy Laboratory (NREL), Golden, CO, United States

The photovoltaic parameters of triple cation perovskite (Cs0.05FA0.79MA0.16Pb(I0.83Br0.17)3) solar cells are investigated focusing on the electro-optical properties and differences in performance at low and high temperatures. The signature of a parasitic barrier to carrier extraction is observed at low temperatures, which results in a loss of performance at T < 200 K. Intensity dependent measurements indicate extraction across this parasitic interface is limited by a combination of the exciton binding energy and thermionic emission. Loss of solar cell performance is observed to be strongly correlated to an increase in photoluminescence intensity, indicating inhibited carrier extraction results in strong radiative recombination. However, the photovoltaic performance of the device is recovered at low intensity - where the photo-carrier generation rate threshold is lower than the thermionic extraction rate. These systems do not appear to be limited by significant thermally activated non-radiative processes. Evidence of limited carrier extraction due to excitonic effects is also observed with a strong anti-correlation in photoluminescence and carrier extraction observed at lower temperatures.

978-1-6654-6060-6/23 $31.00 © 2023 IEEE

Alba: Testing Emerging Photovoltaic Technologies in Low-Earth Orbit

Michael D. Kelzenberg, Phillip Jahelka, Richard G. Madonna, Harry A. Atwater

California Institute of Technology, Pasadena, CA, United States

Abstract-Caltech' SSPD-1 mission seeks to investigate three key technologies relevant for the development of space-based solar power: ultralight radiation-tolerant photovoltaics, wireless power transfer, and compact lightweight deployable structures. The mission comprises three experimental modules with shared central avionics, integrated as a hosted payload on the Momentus Vigoride 5 vehicle. The photovoltaics module is Alba. It features (32) small-area research cells with accompanying instrumentation capable of recording I-V sweep data as well as temperature and sun angle. The mission launched successfully to low-earth orbit aboard a SpaceX Falcon 9 Transporter flight on January 3, 2023, and is currently undergoing initialization activities, which will be followed by approximately six months of data collection prior to deorbit. We plan to present data collected to-date from the flight.

978-1-6654-6060-6/23 $31.00 © 2023 IEEE

Optimizing the Design of 4-Terminal Perovskite/c-Si Tandem Photovoltaics

Paul F. Ndione, John F. Geisz, Qi Jiang, Gabriella D. Lahti, Rosemary C. Bramante, Kai Zhu, Axel F. Palmstrom, Emily L. Warren

National Renewable Energy Laboratory, Golden, CO, United States

Parasitic optical losses from interlayers in 4-Terminal Perovskite/c-Si tandem photovoltaics device reduce charge generation and impact the device performance. Here, optical simulations of the experimental device are performed, and detailed optical analysis help identify the source of losses and mitigate them by optimizing the thickness of the TCO layers from the tandem top cell.

Investigation of EMC Tests in Photovoltaic Inverter according to INMETRO Ordinance No. 140

Andrei C. Ribeiro[1, †], Geyciane P. de Lima[1], Guilherme C. S. Prym[1], Paulo R. D. R. da Silva[1], Pedro O. C. M. Neto[1], Romullo R. M. Carvalho[1], Francisco V. E. Lemos[1], Mendelsson R. M. Neves[1], Tárcio A. S. Barros[1], Hugo da S. Alvarez[2], Rodrigo M. Garcia[2], Francisco C. Marques[1], and Marcelo G. Villalva[1]

[1]State University of Campinas, Campinas, São Paulo, Brazil, †a223987@dac.unicamp.br
[2]Build Your Dreams (BYD), Campinas, São Paulo, Brazil

Abstract—This paper presents the results of electromagnetic interference observed during radiated and conducted emission tests in a 5 kW photovoltaic inverter commercialized in the Brazilian market. The main objective is to assess whether this equipment meets the EMC specifications now required with the entry into force of INMETRO (National Institute of Metrology, Quality and Technology) Ordinance No. 140 - 2022. Before the results, some criteria and basic definitions of electromagnetic compatibility are briefly introduced. The article lists the principal regulations and standards and describes how the tests can be performed.

Index Terms—PV inverter tests, radiated emission, conducted emission, EMC compliance.

I. INTRODUCTION

According to [1], a system is electromagnetically compatible (EMC) when it satisfies three criteria: it does not cause interference to other systems, is not susceptible to emissions from other equipment and does not interfere with itself.

The first criterion requires that the functioning of the systems does not affect the functioning of other systems. The second criterion demands that the system's functioning in question is not compromised by the noise generated by other equipment. Finally, the third criterion specifies that the noise generated during equipment operation does not alter its performance. In Brazil, ABNT NBR / IEC CISPR 11 (2020) quantifies these standards, delimiting the permitted limits of electromagnetic interference (EMI) in both radiated and conducted forms.

INMETRO Ordinance No. 140 (2022) is an update of Ordinance No. 4 (2011). This update protects users and the national market from unsafe and low-quality products. This objective will be achieved through the necessary technical improvement of equipment for generating, conditioning, and storing electricity in photovoltaic systems [2]. The main standards regarding electromagnetic compatibility (EMC) are listed in Table I.

Consider a photovoltaic system like the one shown in Fig. 1. There are two ways to propagate electromagnetic noise. In radiation form (high frequency), the electromagnetic noise can create antennas even in internal tracks. These antennas may, at specific frequencies, impair the operation of other equipment. DC/AC converter switching or clock signals are mainly responsible for this noise.

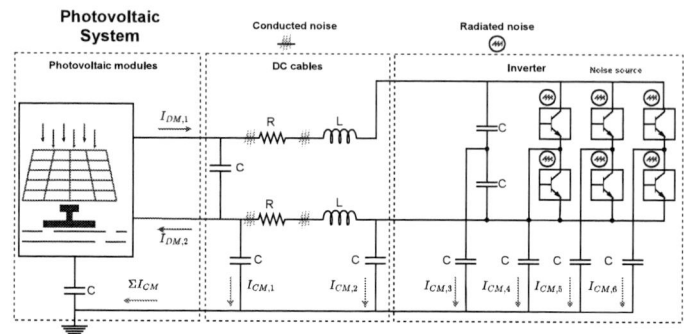

Fig. 1: Equivalent circuit of a photovoltaic system emphasizing electromagnetic noise radiated from switching and conducted propagated through the cables.

In conducted form (low frequency), electromagnetic noise uses cables which can be DC or AC, to propagate throughout the system. The terms high frequency and low frequency are justifiable, as conducted electromagnetic interference is more worrisome in the frequency range 150 kHz to 30 MHz. On the other hand, radiated electromagnetic interference demands more attention in frequency bands above 30 MHz.

Although conducted electromagnetic noise can be carried by DC or AC cables, generally, DC cables have a greater length. In power plants, these cables can be in the order of tens or even hundreds of meters in length and thus form inductances. In addition, capacitance is emerging in the case of power plants, with generation demanding large territorial areas of photovoltaic modules. A resonant circuit is formed for various frequencies with unwanted inductances and capacitances.

The larger the module array area and the DC cables' length, the lower the resonant frequency. Thus, the cables can act as an antenna. In this scenario, disturbances previously only conducted may also be radiated.

II. EMC TESTS AND THEIR RESULTS

There are several devices used to measure EMI, e.g., electromagnetic field meter, spectrum analyzer, oscilloscope, carrier wave meter, and line impedance meter.

TABLE I: Technical Quality Regulation (TQR) related to electromagnetic compatibility in photovoltaic systems according to annexes of Ordinance 140.

RTQ	Title
IEC 61000-6-3:2006 [3]	Electromagnetic compatibility (EMC) - Part 6-3: Generic standards - Emission standard for residential, commercial and light-industrial environments.
IEC 61000-6-3:2006/AMD1:2010 [4]	Amendment 1 - Electromagnetic compatibility (EMC) - Part 6-3: Generic standards - Emission standard for residential, commercial and light-industrial environments.
IEC 61000-6-3:2020 [5]	Electromagnetic compatibility (EMC) - Part 6-3: Generic standards - Emission standard for equipment in residential environments.
IEC 61000-6-4:2006 [6]	Electromagnetic compatibility (EMC) - Part 6-4: Generic standards - Emission standard for industrial environments.
IEC 61000-6-4:2006/AMD1:2010 [7]	Amendment 1 - Electromagnetic compatibility (EMC) - Part 6-4: Generic standards - Emission standard for industrial environments.
IEC 61000-6-4:2018 [8]	Electromagnetic compatibility (EMC) - Part 6-4: Generic standards - Emission standard for industrial environments.
IEC 61000-3-11:2000 [9]	Electromagnetic compatibility (EMC) - Part 3-11: Limits - Limitation of voltage changes, voltage fluctuations and flicker in public low-voltage supply systems - Equipment with rated current 75 A and subject to conditional connection.
IEC 61000-3-3:2013 [10]	Electromagnetic compatibility (EMC) - Part 3-3: Limits - Limitation of voltage changes, voltage fluctuations and flicker in public low-voltage supply systems, for equipment with rated current 16 A per phase and not subject to conditional connection.
IEC 62920:2017 [11]	Photovoltaic power generating systems - EMC requirements and test methods for power conversion equipment.
IEC 62920:2017/AMD1:2021 [12]	Amendment 1 - Photovoltaic power generating systems - EMC requirements and test methods for power conversion equipment.
ABNT NBR IEC/CISPR 11:2020 [13]	Industrial, scientific and medical equipment - Radio-frequency disturbance characteristics - Limits and methods of measurement.

The first two devices are used to measure radiated electromagnetic interference, while the others are used for conducted electromagnetic interference. In the case of photovoltaic inverters, the tests must be based on [11].

A. Radiated Emission Test

The radiated emission test setup was assembled following the recommendations of the IEC 62920:2017 standard, with the configuration as shown in Fig. 2. The setup was implemented in the semi-anechoic chamber (SAC) with internal dimensions of 6.4m x 10m x 19m at the Eldorado Research Institute.

Inside the SAC, the antennas detect and measure horizontal and vertical emissions, varying the position between 100 cm and 400 cm at each test frequency. The equipment under test was placed on a table that rotates 360 degrees.

B. Conducted Emission Test

The conducted emission test setup was assembled in accordance with the IEC 62920 (2017) standard with the suggested configuration, as shown in Fig. 3.

The main differences about the previous test were that the conducted emission test was carried out in a shielded room and had artificial DC and AC power supply networks in the test apparatus. The equipment under test received a voltage of 220 V on its AC side.

On the DC side, two power supplies were used together to deliver to the inverter its nominal power (5 kW) with a voltage equal to 300 V. These values were chosen to simulate normal conditions of use.

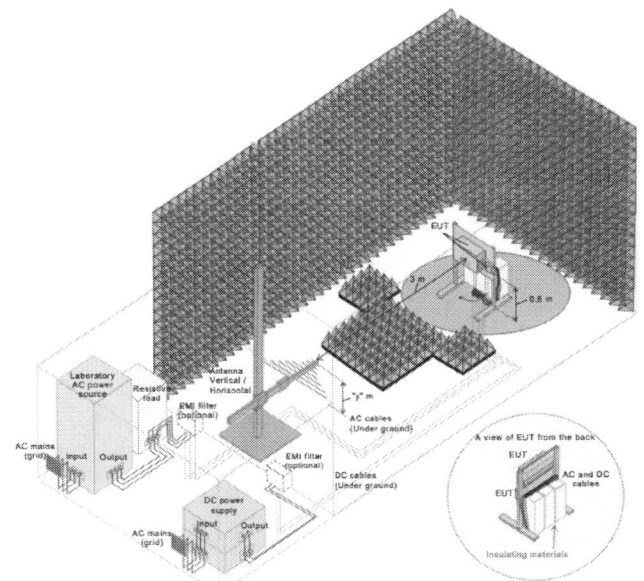

Fig. 2: Setup recommended by IEC 62920 (2017) for radiated emission tests [11].

III. RESULTS AND DISCUSSION

A. Radiated Emission Results

The highest Quasi-Peak (QPK) measurement values of the scans are recorded and shown in Fig. 4. ABNT CISPR 11 requires that at least the six largest disturbances above (L - 10 dB), where L is the level of the specified limit, are sampled. Each sampled disturbance measurement should also show the antenna polarization, height, and turntable rotation position in degrees.

978-1-6654-6060-6/23 $31.00 © 2023 IEEE

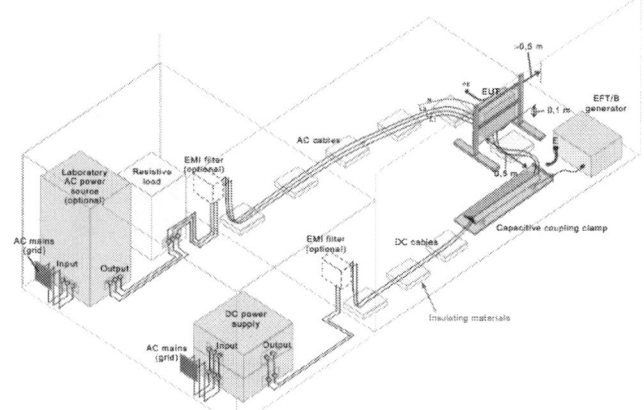

Fig. 3: Setup recommended by IEC 62920 (2017) for conducted emission tests [11].

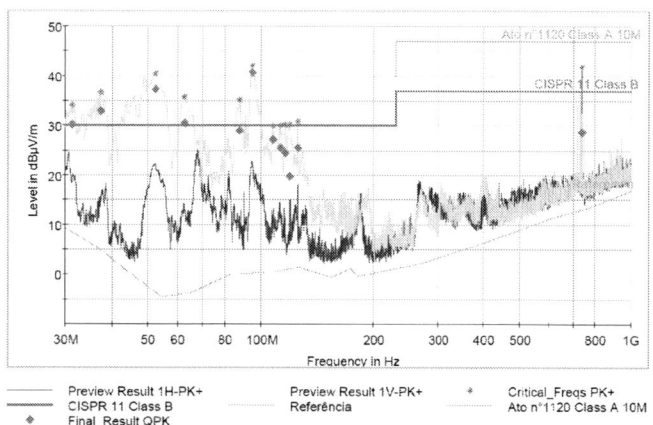

Fig. 4: Radiated emission test results.

The equipment under test must meet Quasi-Peak limits when a Quasi-Peak detector is used to pass the radiated emission test. According to Fig. 4 and [13], the photovoltaic inverter sold in the Brazilian market would be disapproved by the requirements of INMETRO Ordinance No. 140.

Another interesting observation was the presence of a pulse of unexpected electromagnetic disturbance picked up by the antenna. Such an occurrence happened near the frequency of 750 MHz. A mobile device near the test site likely influenced this measurement.

B. Conducted Emission Results

The highest Quasi-Peak (blue marker) and Average (green marker) measurement values of the scans were recorded and are highlighted in Fig. 5a - 6b. NBR CISPR 11 requires that at least the six largest disturbances above (L - 20 dB), where L is the specified threshold level, are recorded.

To pass the conducted emission test, the equipment under test must meet the specified mean limit for measurements with an averaging detector and the specified quasi-peak limit

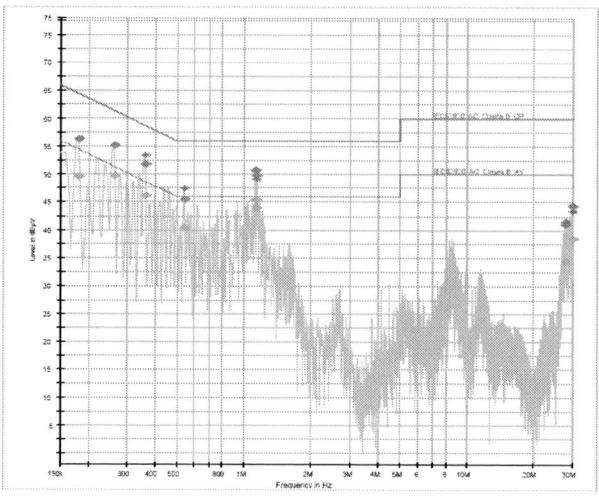

(a) Input 1 - AC.

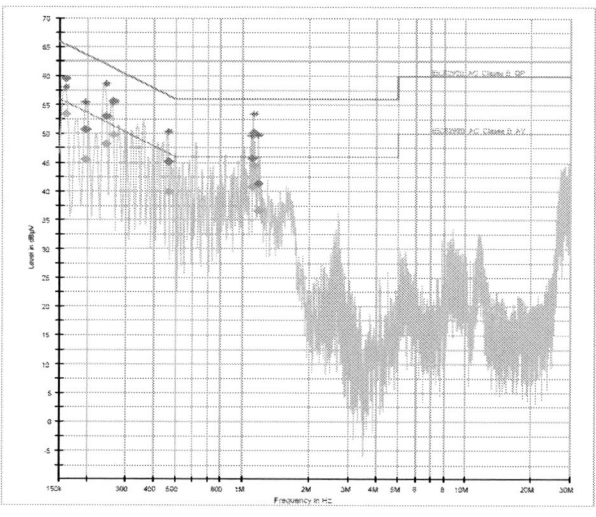

(b) Input 2 - AC.

Fig. 5: Conducted Emission Results in AC inputs

for measurements with a quasi-peak detector or the average threshold using a quasi-peak detector.

From Fig. 6 to 6, and considering the acceptable limits of the conducted emission test [13], the photovoltaic inverter would be approved, as the equipment respects the limits on the DC and AC inputs, despite the small margin in some frequencies.

IV. APPLIED SOLUTIONS

A. Generic recommendations

It is primarily recommended to minimize the range and spectrum of the EMI (Electromagnetic Interference) source as much as possible. This is achieved by using structures that naturally generate less EMI, by increasing switch times (rise time and fall time), using slower components, and sizing the system to work at lower frequencies or with a smaller duty cycle.

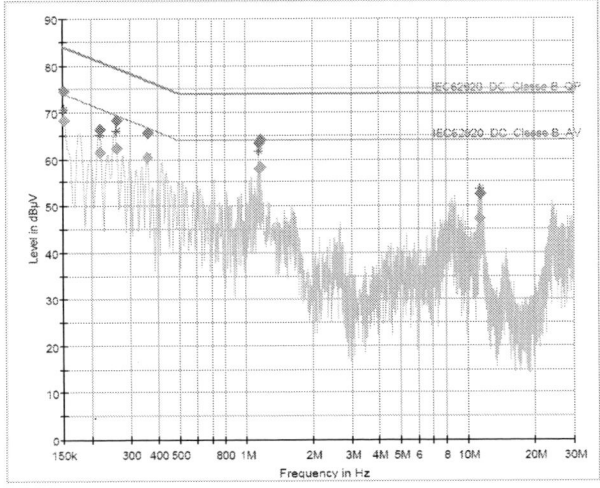

(a) DC Power Supply A.

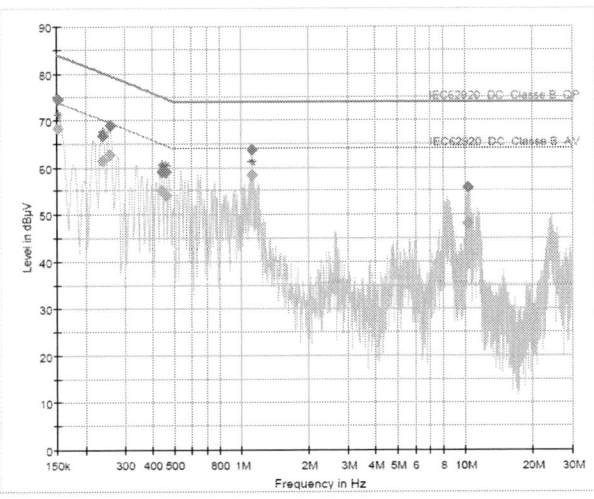

(b) DC Power Supply B.

Fig. 6: Conducted Emission Results in DC inputs.

Although combating interference problems at the source of the EMI is the most suitable approach, it is not always possible to implement this due to project specifications. In these cases, the generation of EMI must be accepted, and efforts should be made to combat the propagation of EMI. This is achieved by reducing the influence of parasitic elements, providing paths of lower impedance for EMI containment, and using line filters and shielding.

B. Practices to mitigate EMI applied to construction and installation of PV inverters

In addition to the previously mentioned recommendations, Figure 7 illustrates good practices for the construction and installation of photovoltaic inverters. The use of grounding for EMI control is not essential and seldom used in most applications, such as in electronic equipment used in medical sciences. However, in photovoltaic systems, as shown in Figure 7a, solid grounding plays a more significant role and can be

used to significantly reduce conducted emission. Solid grounding provides a return path with sufficiently low impedance for frequencies of interference of interest.

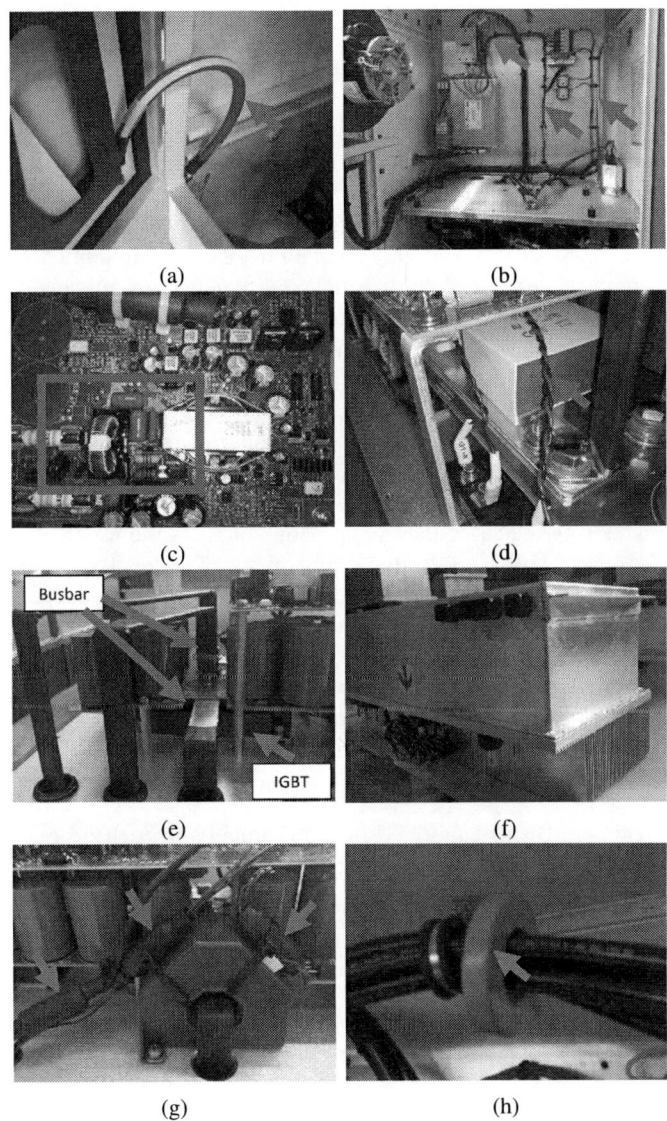

Fig. 7: Practical recommendations for reducing EMI (Electromagnetic Interference) emission (both radiated and conducted) and susceptibility (both radiated and conducted) in photovoltaic inverters: (a) solid grounding, (b) proper routing of cables, (c) use of filters and PCB layout focused on EMI reduction, (d) twisting cables, (e) stacking to reduce area and hence parasitic elements, (f) shielding, (g) use of choke for noise and EMI suppression, (h) high power wiring on the DC side for EMI shielding [14].

It is possible that electronic devices subjected to EMI may experience communication failures among themselves. In addition, the reduction in the lifespan of components and unforeseen operation are common in electromagnetically incompatible environments. The proper distribution of cabling, as shown in Figure 7b, reduces the area of the magnetic field present in the installation and combats effects of conducted magnetic interference. It's also worth noting the use of protection techniques such as cable twisting, indicated in Figure 7d, and the use of ferrites (chokes), as indicated in Figure 7g. These act as high-frequency noise suppressors and should be considered to minimize the effects of EMI.

V. CONCLUSION

A photovoltaic inverter must be designed to meet technical and economic requirements. In the case of grid tie inverters, they must comply with the regulations for connection to the electrical grid and the regulation against unintentional islanding. There are other essential requirements as a guarantee of reliability, high efficiency, reduced volume and weight, and electromagnetic compatibility, whose limits are established by NBR IEC CISPR 11. With INMETRO Ordinance No. 140, an evolution is expected in terms of the Brazilian's quality of photovoltaic projects. This evolution secures the market for unsafe and low-quality products by increasing the number of tests required in equipment for generating, conditioning, and storing electric energy in photovoltaic systems. A 5-kW power inverter already commercialized in the Brazilian market was submitted to radiated emission and conducted emission tests. The equipment failed in radiated emission test. This diagnosis can enable further research in Brazil on electromagnetic compatibility and changes in projects since solving the problem of electromagnetic interference at the source is always more desirable.

ACKNOWLEDGMENT

This work was supported by BYD Energy Brazil (through the PADIS/MCTI program, Funcamp project N.5779) and Total Energies (through ANP - Brazilian Oil & Gas Agency, Funcamp project N.6002).

REFERENCES

[1] C. R. Paul, *Introduction to electromagnetic compatibility.* John Wiley & Sons, 2006.

[2] Brazil, "INMETRO Ordinance No. 140 from march 21th," *Official Gazette of the Federative Republic of Brazil*, 2022. [Online]. Available: https://www.in.gov.br/en/web/dou/-/portaria-n-140-de-21-de-marco-de-2022-389587680

[3] International Electrotechnical Commission (IEC), "Electromagnetic compatibility (EMC) - Part 6-3: Generic standards - Emission standard for residential, commercial and light-industrial environments," International Electrotechnical Commission, Standard 61000-6-3, 2006.

[4] ——, "Amendment 1 - Electromagnetic compatibility (EMC) - Part 6-3: Generic standards - Emission standard for residential, commercial and light-industrial environments," International Electrotechnical Commission, Standard 61000-6-3:2006/AMD1, 2010.

[5] ——, "Electromagnetic compatibility (EMC) - Part 6-3: Generic standards - Emission standard for equipment in residential environments," International Electrotechnical Commission, Standard 61000-6-3, 2020.

[6] ——, "Electromagnetic compatibility (EMC) - Part 6-4: Generic standards - Emission standard for industrial environments," IEC, IEC Standard 61000-6-4:2006, 2006.

[7] ——, "Amendment 1 - Electromagnetic compatibility (EMC) - Part 6-4: Generic standards - Emission standard for industrial environments," IEC, IEC Standard 61000-6-4:2006/AMD1:2010, 2010.

[8] ——, "Electromagnetic compatibility (EMC) - Part 6-4: Generic standards - Emission standard for industrial environments," IEC, IEC Standard 61000-6-4, 2018.

[9] ——, "Electromagnetic compatibility (EMC) - Part 3-11: Limits - Limitation of voltage changes, voltage fluctuations and flicker in public low-voltage supply systems - Equipment with rated current 75 A and subject to conditional connection," IEC, IEC Standard 61000-3-11:2000, 2000.

[10] ——, "Electromagnetic compatibility (EMC) - Part 3-3: Limits - Limitation of voltage changes, voltage fluctuations and flicker in public low-voltage supply systems, for equipment with rated current 16 A per phase and not subject to conditional connection," IEC, IEC Standard 61000-3-3:2013, 2013.

[11] ——, "Photovoltaic power generating systems - EMC requirements and test methods for power conversion equipment," International Electrotechnical Commission, Standard IEC 62920, 2017.

[12] ——, "Amendment 1 - Photovoltaic power generating systems - EMC requirements and test methods for power conversion equipment," IEC, IEC Standard 62920:2017/AMD1, 2021.

[13] ABNT, "Industrial, scientific and medical equipment - Radio-frequency disturbance characteristics - Limits and methods of measurement," ABNT, Standard NBR IEC CISPR 11, 2020.

[14] S. Hong and M. Zuercher-Martinson, "Harmonics and noise in photovoltaic (PV) inverter and the mitigation strategies," *Solectria Renewables, Lawrence, MA, USA, Tech. Rep*, 2010.

PV Inverter testing for Momentary Cessation and Rate-of-Change-of-Frequency events

Ramanathan Thiagarajan[1], Kumaraguru Prabakar[1], Li Yu[2], Ken Aramaki[2], and Andy Hoke[1]

[1]Power System Engineering Center, National Renewable Energy Laboratory, Golden, CO-80401, USA
[2]Hawaiian Electric Company, Honolulu, HI, 96813, USA

Abstract — To understand the power system stability and develop better electromagnetic transient (EMT) models of field deployed photovoltaic (PV) inverters, it is important to characterize inverters' response to abnormal voltage and frequency scenarios. Because EMT models are not typically available for small distribution-connected PV inverters, and because inerterconnection standards historically did not specify desired ride-through behaviors, we tested two such inverters in the lab to characterize their responses to severe undervoltage events and high rate-of-change-of-frequency (ROCOF) conditions. The inverters tested were pre-IEEE 1547-2018 residential PV inverters widely used in the Hawaiian Electric territory and many other areas. The testing results for undervoltage scenarios showed that the inverter from one vendor exhibited momentary cessation while the inverter from the other vendor did not exhibit momentary cessation behavior or tripping for most of the events below the 120 ms undervoltage trip threshold duration set by IEEE 1547-2003. The testing results for ROCOF scenarios showed that the inverter from one vendor temporarily lost synchronization during ROCOF conditions while the inverter from the other vendor did not lose synchronization or cease generation for any ROCOF conditions. Both the inverters were also tested for EMT-simulated grid events with severe changes in frequency and voltage. The observed responses of the inverters were different from the simulated response of PV inverters represented the best available assumptions from pre-existing information. The results from these experiments can be used to update the inverter models used in bulk power system studies.

I. INTRODUCTION

With the increasing presence of inverter interfaced distributed energy resources (DERs) in the power grid, it is important to understand and model the responses of DER inverters to abnormal voltage and frequency conditions. Various studies have been performed for systems with very high presence of DER such as the ones in Hawaii [1][2]. Under the IEEE 1547-2018 standard [3] for interconnection and interoperability of DERs, the newly installed category 3 DERs (those with the greatest ride-through capabilities) must respond to abnormal voltages below 0.5 per unit (p.u.) by momentary cessation [4] of active current output and restore active current to normal after such conditions. However, vast majority of the inverters installed in Hawaiian Electric territory are pre-1547-2018 inverters, so system modelers have to make conservative guesses on how the legacy inverters in the field respond to abnormal voltage and frequency conditions [5]. A previous study by some of the coauthors [6] mapped the dominant population of installed inverters in Hawaiian Electric territory

and provided some information on their responses to abnormal conditions based on a meta-analysis of past study results. Other studies have also attempted to characterize responses to abnormal conditions of residential-scale DER inverters [7], [8]. In this paper, the authors verify the response of legacy PV inverters from two different vendors to momentary reductions in voltage and high ROCOF. The inverters under test from the two vendors will be referred to as inverter 1 and inverter 2.

The characterization of these dominant legacy PV inverters used in the Hawaiian Electric territory is performed by exposing them to test profiles designed to investigate momentary cessation and ROCOF responses. Additionally, some grid events with abnormal voltages and frequency were captured from utility stability studies and were reproduced at the inverter terminals to understand the inverters' responses to these events. These tests were performed at the Energy Systems Integrations Facility (ESIF) using a grid simulator to vary the voltage and frequency and PV simulators to vary the input power to the PV inverter. To replicate field behavior, the inverters tested at ESIF were programmed under the same settings as the legacy inverters installed in the Hawaiian Electric territory. Section II presents the inverters' responses to momentary cessation tests. Section III presents the inverters' responses to ROCOF tests. Section IV shows the inverters' responses to emulated grid events.

II. MOMENTARY CESSATION TESTS

The momentary cessation tests were performed with various levels of available DC input power while varying the voltage applied to the grid side (AC) of the inverter. The change in input DC power was applied using a TerraSAS PV emulator. On the grid side, the momentary cessation test profiles were provided using a grid simulator controlled by an analog output from a

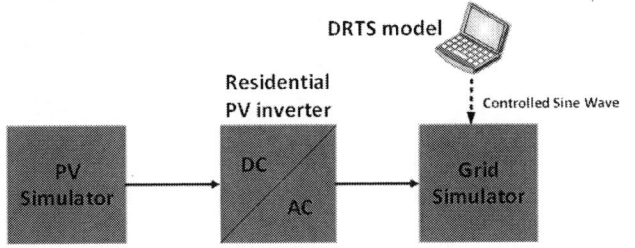

Figure 1. Hardware setup for DER testing for momentary cessation and ROCOF

978-1-6654-6060-6/23 $31.00 © 2023 IEEE

digital deal time simulator (DRTS). The voltage, current, and power measurements were captured at a high sampling rate via an oscilloscope. This setup is shown in Figure 1.

To test the inverters' responses to momentary voltage reduction, a test profile was developed in the DRTS environment. At the start of the experiment, the voltage and frequency are at the nominal values of 1 p.u.. The inverter is synchronized to the grid and is exporting power to the grid. Then the momentary voltage reduction is applied, where the rise time, t_{rise}; fall time, t_{fall}; and the idle time, t_{idle}; can be configured, as shown in Figure 2.

To understand the inverters' responses to the momentary voltage reduction, the low voltage value for testing, V_{low}, was varied between 0.8 p.u. to 0.05 p.u. The idle time and rise time parameters were chosen such that the sum of these two times is equal to 120 ms, the undervoltage threshold trip time set by IEEE 1547-2003 standard for interconnection of DERs [9]. The parameters used in the experiment are shown in Table 1. The total number of tests per inverter was 108.

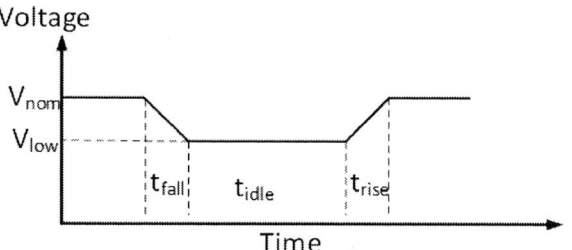

Figure 2. Test profile to emulate momentary voltage reduction

The inverters were tested to capture any momentary cessation behavior under the provided voltage sag scenario. As stated earlier, the two dominant legacy residential inverters from [4] were the devices under test. The inverters under test had the same protection settings as the inverters installed in the field.

Table 1. Parameters for testing momentary voltage cessation in inverters

Parameter	Range of values
V_{nom}	1 p.u.
V_{low}	{0.8, 0.7, 0.6,0.5,0.4,0.3,0.2,0.1,0.05} p.u.
t_{fall}	1 ms
t_{rise}	20 ms
t_{idle}	100 ms
Frequency	60 Hz
Input power	{1, 0.66, 0.33} p.u.

From the tests, the following observations were made for inverter 1:

1. In response to the various low voltages, inverter 1 did not exhibit any momentary cessation.
2. In response to the momentary cessation profile of voltage at all values of V_{low}, the inverter 1 current increased as an initial response to the reduced voltage. The current increased as a result of the voltage reduction to attempt to keep the power output constant. This behavior was

recorded in all instances by varying the input power between 33%, 66% and 100%. This behavior by inverter 1 is shown in Figure 3.

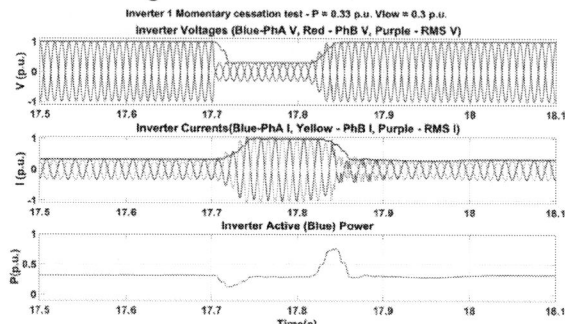

Figure 3. Increase in current in response to momentary reduction in voltage to 0.3 p.u.

3. After the restoration of voltage to the nominal value, inverter 1 exhibited distorted output current temporarily for a few seconds before restoring normal output in 36% of the test cases. This behavior is shown in Figure 4.

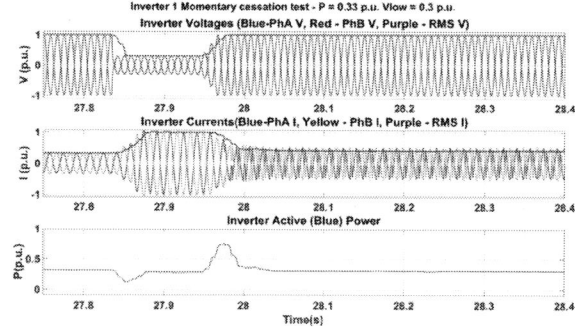

Figure 4. Abnormal recovery dynamic with increased distortion of current waveform after the momentary reduction in voltage to 0.3 p.u.

4. Inverter 1 tripped for voltage sags at or below 0.3 p.u. in two of the 108 cases. No tripping was observed at higher voltages. In these cases, the inverter tripped as shown in Figure 5, and the connection was only restored after the five-minute reset countdown for nominal grid conditions. It should be noted that inverter 1 did continue functioning normally for the remaining instances at or below 0.3 p.u. voltage.

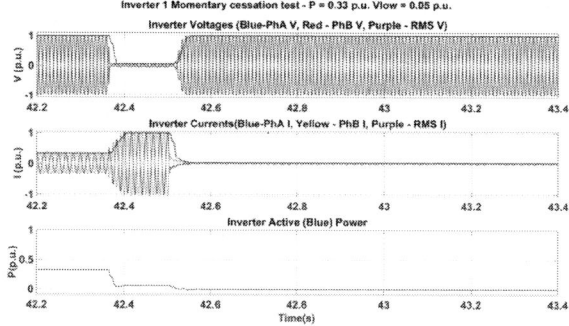

Figure 5. Inverter tripping after the momentary voltage reduction to 0.05 p.u., observed by current waveform going to 0A

5. To understand the variations in behavior of the inverter under identical momentary voltage conditions, the same test profile was applied starting at four different poinst on the sine wave of the voltage waveform. This was performed to check if the inverter exhibited different recovery dynamics or tripping behavior to the same test profile. The variation between tests noted above for Inverter 1 may be related to the point-on-wave timing of the event. An example is shown in Figure 6.

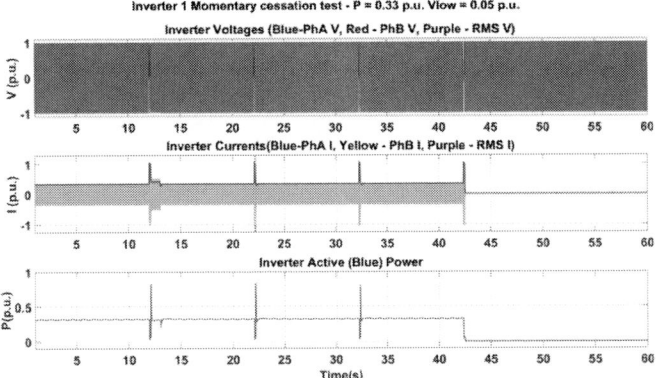

Figure 6. Difference in recovery dynamics and tripping for the same momentary voltage reduction profile at various points on wave for V_{low} of 0.05 p.u. The inverter rides through the first three 0.05 pu voltage sags but trips on the fourth such sag.

From the tests, the following observations were made for inverter 2:

1. In response to low voltages at or above 0.6 p.u., no momentary cessation or tripping was observed. The current increased during the voltage sag to attempt to maintain constant power output, as shown in Figure 7.

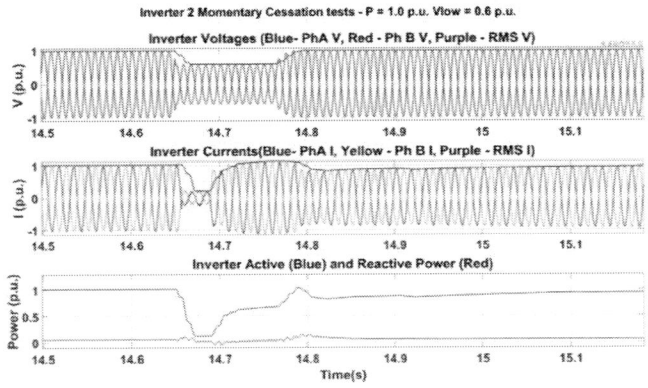

Figure 7. Increase in current in response to momentary reduction in voltage to 0.6 p.u.

2. Inverter 2 exhibited near-zero current for a period of only 2-3 AC cycles (~30-50 ms) for voltages at or above 0.4 p.u. and less than 0.6 p.u. for all the 24 test cases, as shown in Figure 8. For voltage at 0.4 p.u., the inverter exhibited a momentary cessation lasting 6 cycles for all the 12 cases at various power setpoints, as shown in Figure 9.

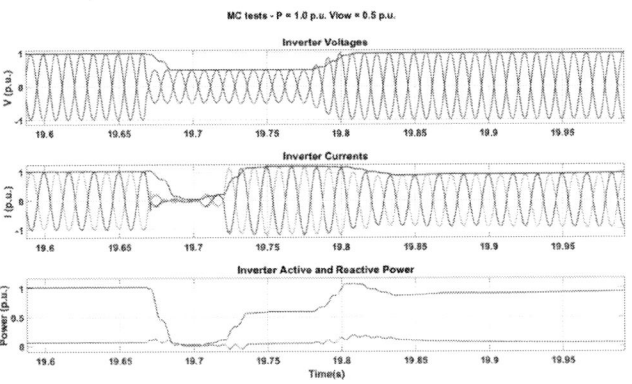

Figure 8. Short duration zero current, lasting 2-3 AC cycles in response to momentary reduction in voltage to 0.5 p.u.

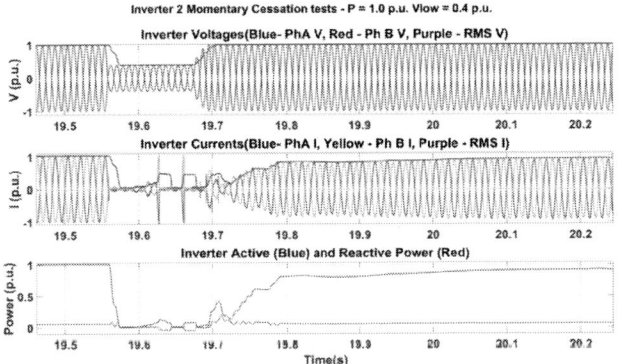

Figure 9. Short duration momentary cessation, lasting 6 AC cycles in response to momentary reduction in voltage to 0.4 p.u.

3. For voltages at or below 0.3 p.u., a momentary cessation lasting approximately 1.5 seconds was consistently observed in inverter 2. This behavior was recorded in all instances for voltages between 0.3 p.u. and 0.05 p.u. with tests repeated at four different point on wave while varying the input power between 33%, 66%, and 100%, shown in Figure 10 and Figure 11.

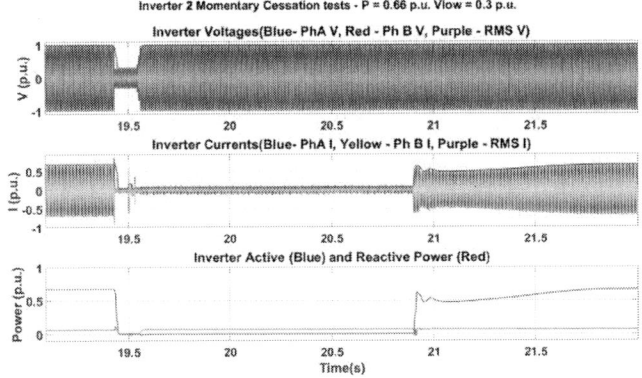

Figure 10. Momentary cessation lasting 1.5 seconds in response to momentary reduction in voltage to 0.3 p.u. at 66% rated inverter power

978-1-6654-6060-6/23 $31.00 © 2023 IEEE

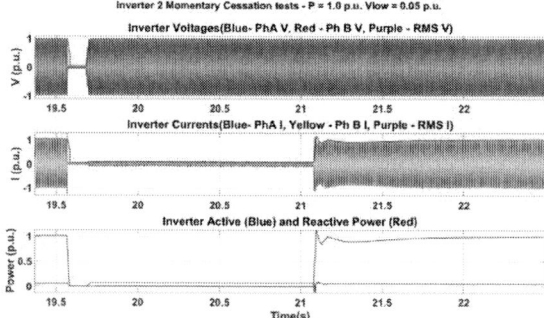

Figure 11. Momentary cessation lasting 1.5 seconds in response to momentary reduction in voltage to 0.05 p.u. at 100% rated inverter power

4. Inverter 2 generally exhibited a nearly instantaneous return to near the pre-event current level once it resumed current export. It also exhibited somewhat variable recovery dynamics behavior depending on the event point of the wave, similar to the behavior of inverter 1.

III. ROCOF TESTS

To characterize the response of PV inverters to different ROCOFs, a test profile was developed to vary the frequency and the rate of change in frequency, as shown in Figure 11. Similar to momentary cessation tests, the inverter under test had the same settings as the field inverter. The inverters were programmed to have an underfrequency setpoint of 56 Hz to emulate most field inverter conditions in Hawaii.

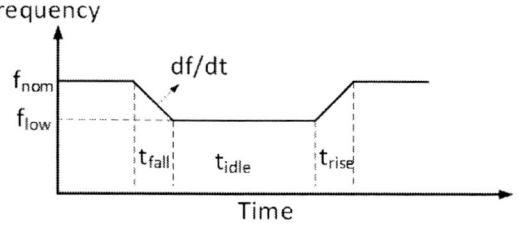

Figure 12. Test profile to characterize ROCOF response.

From the tests, the following observations were made for inverter 1:

1. Inverter 1 did not trip or cease output for ROCOFs up to 60 Hz/s. No loss of generation was observed. The ROCOF events tested lasted between 50 ms and 1000 ms.

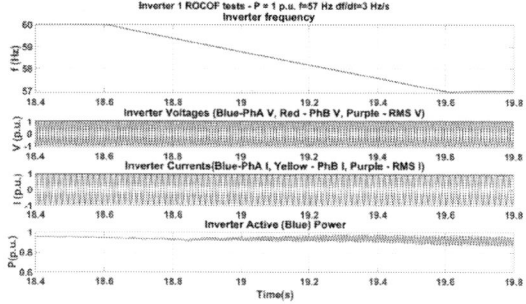

Figure 13. Inverter 1 test results for a ROCOF of 3 Hz/s for a period of 1 second for a change in frequency from 60 Hz to 57 Hz

2. Some results from the tests are shown in Figure 13 and Figure 14. It can be seen that inverter 1 did not trip or reduce generation for a ROCOF of 3 Hz/s lasting 1 second, for a change in frequency from 60 Hz to 57 Hz, as shown in Figure 13. Similarly, inverter 1 did not trip or reduce output for a ROCOF of 60 Hz/s lasting 50 ms, for a change in frequency from 60 Hz to 57 Hz, as shown in Figure 14.

Similarly, observations were made from the ROCOF tests for inverter 2:

1. Inverter 2 did not trip for ROCOFs up to 60 Hz/s.

2. Inverter 2 did reduce power temporarily during a ROCOF of 3 Hz/s for a change in frequency from 60 Hz to 57 Hz, as shown in Figure 15. No tripping of the inverter was observed.

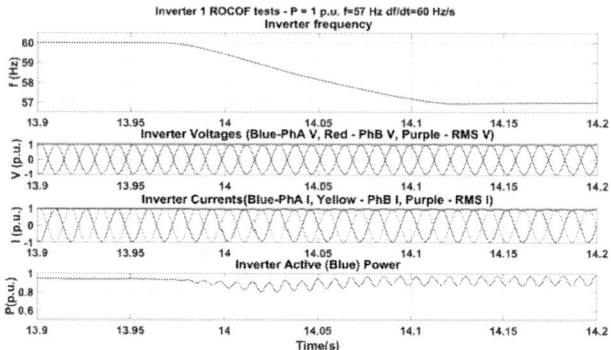

Figure 14. Inverter 1 test results for a ROCOF of 60 Hz/s for a period of 50 ms for a change in frequency from 60 Hz to 57 Hz

3. Inverter 2 lost power momentarily for about 2.5 seconds for a ROCOF of 4 Hz/s for a change in frequency from 60 Hz to 57 Hz, as shown in Figure 16. Similar to the test above, the inverter did not trip itself.

4. With the observed change in behavior between a ROCOF of 3 Hz/s to 4 Hz/s, inverter 2 was tested for ROCOF values incrementally in steps of 0.1 Hz/s from 3 Hz/s, to determine the threshold value at which inverter 2 reduces power output to near zero. This ROCOF value was determined to be 3.6 Hz/s, at which the output power reduces to near zero, as shown in Figure 17. This may be due to the inverter controls temporarily losing synchronism with the grid.

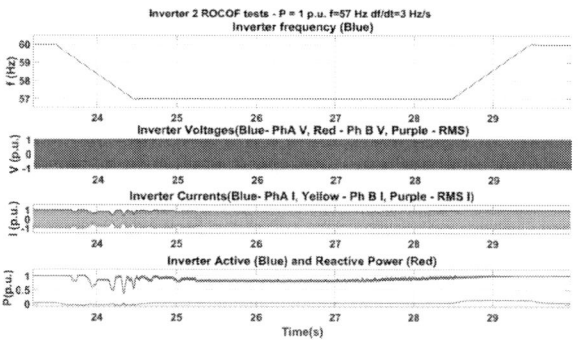

Figure 15. Inverter 2 test results for a ROCOF of 3 Hz/s for a period of 1 second for a change in frequency from 60 Hz to 57 Hz

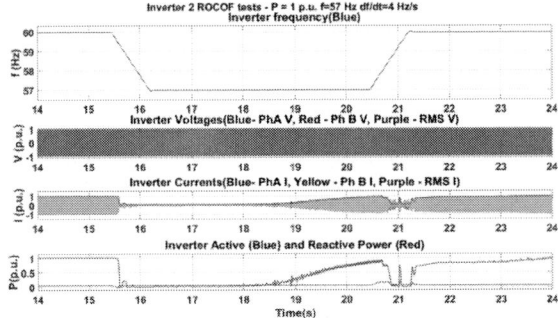

Figure 16. Inverter 2 test results for a ROCOF of 4 Hz/s for a period of 750 ms for a change in frequency from 60 Hz to 57 Hz

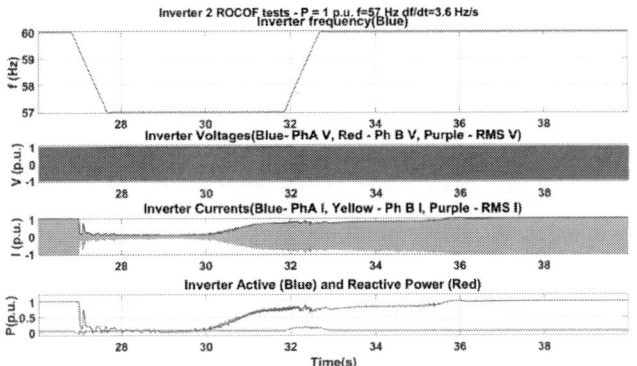

Figure 17. Inverter 2 test results for a ROCOF of 3.6 Hz/s for a change in frequency from 60 Hz to 57 Hz

IV. GRID EVENT RESPONSE TESTS

Utilities perform security and stability studies to understand different contingencies to the grid, such as loss of a transmission line, loss of a generator, or a a fault. During these studies, while looking at the contingencies, certain voltage and frequency changes can be observed. With a large amount of PV inverters in the grid, it is important to understand how the inverters respond to such events. Characterizing the response of the inverters to such events is the motivation for the tests in this section.

Transmission stability studies typically do not simulate the distribution voltages where DER inverters are connected. Thus, the voltage waveforms have to be normalized from (sub)transmission voltages (e.g. 69 kV or 138 kV) to nominal voltages at which residential PV inverters are operated.

Additionally, these studies are performed using EMT software, with the total event duration lasting less than 20 seconds. PV inverters typically require 5-7 minutes to get connected to the grid and start generating power. This necessitates appending nominal voltage waveform data for approximately 10 minutes sampled at the same frequency as the grid event, prior to the 20 second grid event. The data manipulation to ensure practical implementation of the test cases was performed using MATLAB.

A total of 4 simulated cases were tested for inverter response for inverters 1 and 2. The following behaviors were observed for cases 1-4:

1. For case 1, following a fault, the voltage and frequency values reduced well beyond the trip settings of the inverters, causing inverter 1 (shown in Figure 18) and inverter 2 (shown in Figure 19) to trip as a result of exceeding the undervoltage limits and frequency limits.

2. For case 2, following a contingency event, the voltage and frequency were restored to the nominal values, allowing the inverters to resume generating power following the event. For inverter 1, no momentary cessation was observed, as shown in Figure 20. The current increased in response to the momentary voltage reduction. For inverter 2, momentary cessation was observed with the inverter resuming power generation after the momentary cessation, as shown in Figure 21. No tripping was observed.

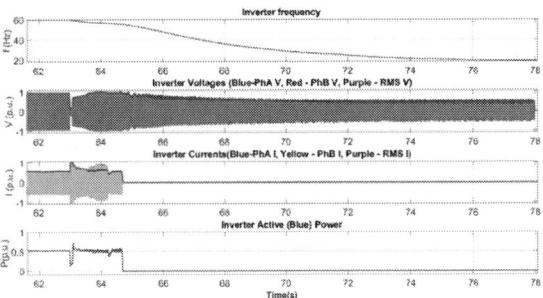

Figure 18. Inverter 1 response to PSCAD trace 1 at input power of 0.5 p.u.The inverter trips.

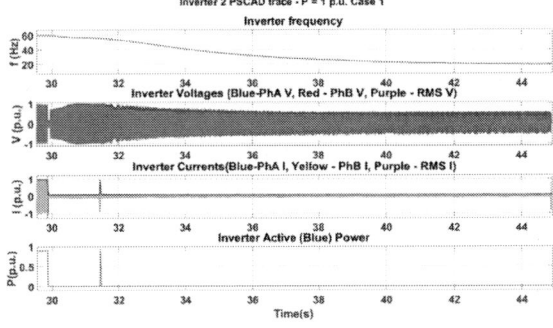

Figure 19. Inverter 2 response to PSCAD trace 1 at input power of 1 p.u. The inverter trips.

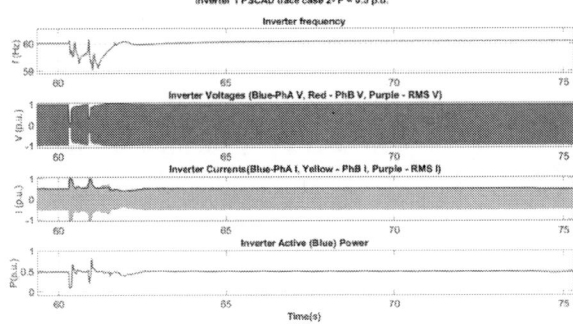

Figure 20. Inverter 1 response to PSCAD trace 1 at input power of 0.5 p.u. The inverter increases current during the faults and does not enter momentary cessation.

3. Two additional cases, cases 3 and 4, were tested for inverter response. Case 3 caused inverters 1 and 2 to trip due to exceeding of the undervoltage durations limits, similar to the inverter response in case 1. Case 4 had the inverters 1 and 2 responding similar to that of case 2, with inverter 1 exhibiting no momentary cessation versus the inverter 2 exhibiting momentary cessation.

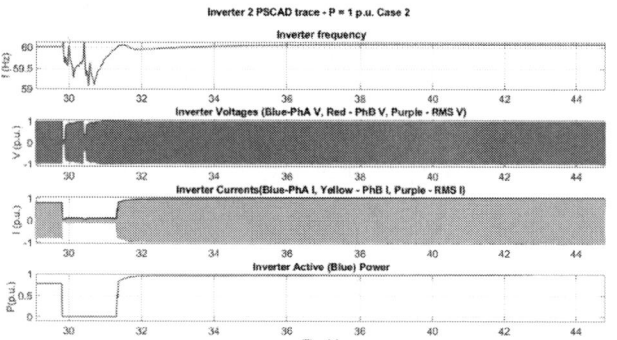

Figure 21. Inverter 2 response to PSCAD trace 1 at input power of 1 p.u. Momentary cessation was observed during voltage drip with full restoration to pre-disturbance power after the event.

IV. CONCLUSION

Following up from the work in [4], the authors tested dominant legacy inverters from two vendors used in Hawaiian Electric territory. For testing the inverters, the authors developed testing profiles for undervoltage response and ROCOF response. The test results show that one legacy inverter exhibits momentary cessation without tripping for severe undervoltages, and reduces output power to near zero for ROCOFs exceeding 3.6 Hz/s but does not trip for even very severe and long ROCOF events. The other legacy inverter from a different vendor did not exhibit any momentary cessation or loss of generation during fast and long ROCOF events or for undervoltage events with residual voltage above 0.3 p.u.; below 0.3 p.u., the inverter tripped. Results from these tests can be utilized to develop better representative power system aggregate models of DERs' responses to momentray voltage reduction and high ROCOF grid events for transmission studies.

V. ACKNOWLEDGEMENTS

This work was authored in part by the National Renewable Energy Laboratory, operated by Alliance for Sustainable Energy, LLC, for the U.S. Department of Energy (DOE) under Contract No. DE-AC36-08GO28308. Funding provided by Hawaiian Electric Company. The views expressed in the article do not necessarily represent the views of the DOE or the U.S. Government. The U.S. Government retains and the publisher, by accepting the article for publication, acknowledges that the U.S. Government retains a nonexclusive, paid-up, irrevocable, worldwide license to publish or reproduce the published form of this work, or allow others to do so, for U.S. Government purposes.

The authors would also like to express their thanks to ESIF research engineers, Kurtis Buck and Peter Gotseff, for their assistance in developing and debugging the test setup through the testing.

REFERENCES

[1] Hoke, Andy, Julieta Giraldez, Bryan Palmintier, Earle Ifuku, Marc Asano, Reid Ueda, and Martha Symko-Davies. "Setting the smart solar standard: Collaborations between hawaiian electric and the national renewable energy laboratory." *IEEE Power and Energy Magazine* 16, no. 6 (2018): 18-29.

[2] Thiagarajan, Ramanathan, Peter Gotseff, Andy Hoke, and Earle Ifuku. "Inverter testing for verification of Hawaiian Electric Rule 14H." In *2019 IEEE 46th Photovoltaic Specialists Conference (PVSC)*, pp. 0906-0910. IEEE, 2019.

[3] "IEEE Standard for Interconnection and Interoperability of Distributed Energy Resources with Associated Electric Power Systems Interfaces," *IEEE Std 1547-2018* (Revision of IEEE Std 1547-2003) , pp.1-138, 6 April 2018, doi: 10.1109/IEEESTD.2018.8332112.

[4] *1,200 MW Fault Induced Solar Photovoltaic Resource Interruption Disturbance Report*. Technical Report. North American Reliability Corporation. https://www. nerc. com/pa/rrm/ea/1200_MW_Fault_Induced_Solar_Photovoltaic_ Resource_/1200_MW_Fault_Induced_Solar_Photovoltaic_Reso urce_Interruption_Final. pdf, 2017.

[5] Modeling Notification: Recommended Practices for Modeling Momentary Cessation Initial Distribution, NERC, 2018.

[6] Mahmud, Rasel, Li Yu, and Andy Hoke. "Characterization of DER Momentary Cessation and Rate-of-Change-of-Frequency Response." In *2022 IEEE 49th Photovoltaics Specialists Conference (PVSC)*, pp. 1000-1002. IEEE, 2022.

[7] L. Callegaro, G. Konstantinou, C. A. Rojas, N. F. Avila and J. E. Fletcher, "Testing Evidence and Analysis of Rooftop PV Inverters Response to Grid Disturbances," in *IEEE Journal of Photovoltaics*, Nov. 2020.

[8] A. Ahmad, H. D. Tafti, G. Konstantinou, B. Hredzak and J. E. Fletcher, "Analysis on the Behavior of Grid-Connected Single-Phase Photovoltaic Inverters Under Voltage Phase-Angle Jumps," *2021 IEEE 12th Energy Conversion Congress & Exposition - Asia (ECCE-Asia)*, Singapore, 2021.

[9] "IEEE Standard for Interconnecting Distributed Resources with Electric Power Systems," IEEE Std 1547-2003 , pp.1-28, 28 July 2003, doi: 10.1109/IEEESTD.2003.94285.

Microgrid Design Toolkit Cost Optimization for a Rural Community in Puerto Rico

Rolando J. Tremont-Brito[1], Rachid Darbali-Zamora[2], Robert Broderick[2], Erick E. Aponte-Bezares[1], Efrain O'Neill-Carrillo[1] and Matthew S. Lave[2]

[1]University of Puerto Rico-Mayagüez, Mayagüez, Puerto Rico 00681, USA
[2]Sandia National Laboratories, Albuquerque, New Mexico, 87185, USA

Abstract – **Hurricanes Irma and Maria had a devastating effect on communities all around Puerto Rico and highlighted the lack of resilience of the island's power system. After months without electric service, and being left with unreliable service years later, community leaders are looking for alternatives to fulfill their energy needs. Three microgrid design alternatives for a rural community that consider centralized and decentralized diesel generation, energy storage, and photovoltaic (PV) generation are evaluated in this paper. Design basis threats and performance metrics are defined with community feedback and requirements. The Microgrid Design Toolkit (MDT) was used to run multi-objective optimization, including cost, performance, and sizing optimization. The trade-off space of solutions for the three alternatives is presented in the form of a Pareto frontier. Optimization results for the centralized alternative slightly favor diesel and big storage. However, for less aggregated alternatives, the optimization algorithm favors solar generation leading to higher renewable penetration.**

Keywords – rural microgrid, trade-offs analysis, cost optimization, design basis threats, microgrid sizing.

I. INTRODUCTION

Puerto Rico, an archipelago located in the Caribbean, hosts a population of approximately 3.194 million (as of 2020). Among its various communities, Corcovada is a small rural community in the mountainous region of the municipality of Añasco, located to the West of Puerto Rico. This community has been operating its aqueduct system for 42 years. In 2017, Puerto Rico was impacted by two hurricanes in the span of two weeks. On September 7th, Hurricane Irma passed close to the northeast of Puerto Rico as a category 5 hurricane. Two weeks later, on September 20th, Hurricane Maria made landfall, as a strong category 4. According to leaders of Corcovada, after Hurricane María, they lacked electric service for up to 5 months. To this day, they suffer from constant power outages and voltage fluctuations that affect their appliances and well-being. The power system collapsed [1], and in some areas, it took over a year to fully recover [2]. The system struggles with unreliable electrical service years after, especially in harder-to-reach places like rural communities.

Currently, there is interest in integrating distributed energy resources (DER) and the application of electrical microgrids emerge as possible solutions to these rural communities in Puerto Rico [3]. Microgrids are independent grids of smaller sizes that can provide power to rural areas locally [4]. These systems incorporate multiple forms of localized generation, including conventional generation such as diesel generators,

and renewable sources such as wind turbine generators (WTG) and solar photovoltaics (PV). However, energy storage is essential to ensure a consistent power supply. A power grid system that incorporates microgrids would still have conventional centralized energy sources, but many more users would have local sources of energy [4]. This provides on-site generation for local loads in both grid-tied and islanded modes of operation.

The Microgrid Design Toolkit (MDT) is a sophisticated decision-support software tool designed to aid microgrid designers during the early stages of the design process [5]. The software employs powerful search algorithms to identify and characterize a trade-off space of microgrid design decisions based on user-defined metrics, including cost, performance, and reliability. In this paper, an MDT model is developed for a distribution system, serving as a representative case scenario for a rural community in a mountainous area with predominantly residential energy consumption. For this analysis, three different scenarios are considered, featuring combinations of centralized and decentralized solar, diesel, and storage resources. The MDT evaluates these cases, yielding results for the Pareto frontier and optimization metrics.

This paper is organized in the following manner: Section II provides an assessment of Corcovada's power distribution system. Section III explains the methodology to develop the microgrid model in MDT. Results are presented in Section IV. The conclusion and future works are presented in Section V.

II. MICROGRID DESCRIPTION

The community of Corcovada is located in the western region of Puerto Rico in the municipality of Añasco. The community of Corcovada has a population of approximately 627 (as of 2010). They have been operating its own aqueduct system for 42 years. Fig. 1 illustrates a map of the main island of Puerto Rico.

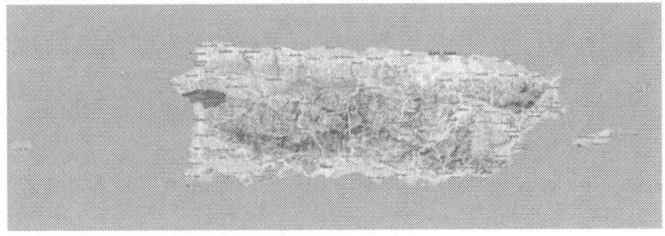

Fig. 1. Map of the archipelago, Puerto Rico (Añasco shown in purple). Map data ©2015 Google, QGIS project.

978-1-6654-6060-6/23 $31.00 © 2023 IEEE

Corcovada's energy needs are supplied mainly by a 2.40/4.16kV distribution feeder from a substation over 13 km away at the municipality of San Sebastian, east of Añasco. Fig. 2 illustrates the feeder that supplies the community, in red, and the area of interest (AoI), highlighted in orange. The AoI serves as the boundary for the microgrid [6]. There's a secondary connection from the municipality of Añasco, in blue feeder, shown at the top left corner of Fig. 2. The community has a total transformer capacity of 1 MW and an existing 21 kW PV system with 13.6/200 kW/kWh of energy storage, yellow star in Fig. 2 [7]. This system helps supply the community-built aqueduct's water pump. Electrical loads in Corcovada are mostly residential with around 162 of the 195 residential buildings in the community being occupied at the time of the study. Nonetheless, all 195 residential buildings will be modeled.

Electrical loads were assigned from billing information provided by community residents. Buildings with unknown data were randomly assigned to one of three different consumption groups: low (5 − 7 kWh average daily energy consumption, 52.475% of buildings), medium (11 − 13 kWh average daily energy consumption, 38.119% of buildings), or high (17 − 19 kWh average daily energy consumption, 9.406% of buildings). The three consumption groups generated follow the same distributions from the known consumption data of the community. In total, this system is modeled with 195 residential loads, 1 commercial load, the community center, and 7 other potential shelters (such as churches). Also modeled as an electrical load is the water pump of the community-built aqueduct system. The water pump is a 5-hp pump that represents approximately 3.72 kW of power consumption.

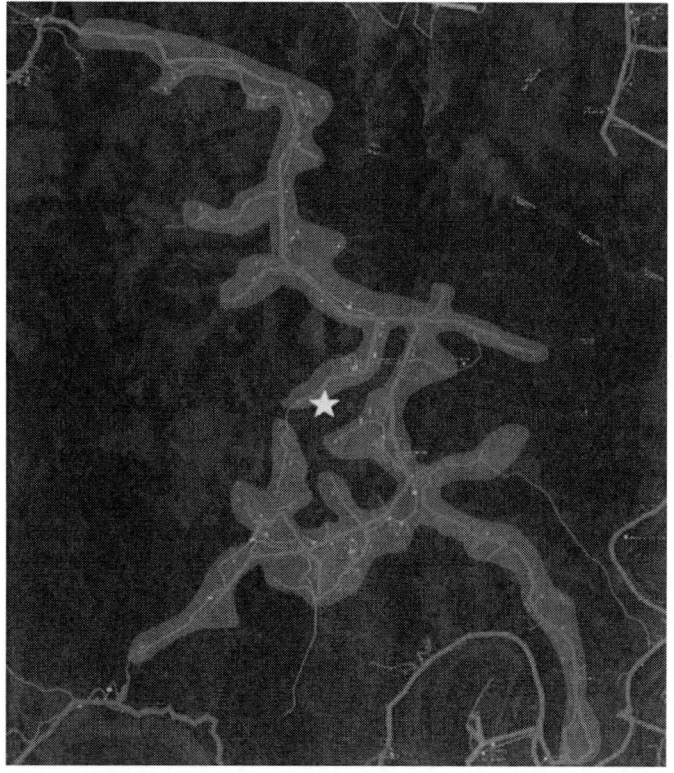

Fig. 2. Community of Corcovada shown in Orange. Map data ©2015 Google, QGIS project.

III. Microgrid Design Toolkit Modelling

An MDT model is developed for a power distribution system representing the community of Corcovada in Añasco, Puerto Rico. The system serves as a case scenario for a microgrid representative of a rural community in a mountainous area with mostly residential energy consumption and various critical load demands that is susceptible to constant weather-related events such as hurricanes and storms. The first step in the model development process is performing an aerial assessment of the power system using geographic information system (GIS) data. This data included power system parameters necessary for developing the MDT model of the community of Corcovada. Latest public version 1.3.1902.0 of MDT was used.

The community center and 7 other potential shelters are defined as critical loads based on discussions with the community members. The aqueduct's water pump is another critical load not included in the loads summarized in Table I. Table I summarizes the minimum, maximum, average, and daily energy consumption for the residential, commercial, and shelter (critical) loads in the Corcovada MDT model. The objective of the model is to accurately represent energy usage. Power demand extremes are affected by the choice of hourly period load models. Higher and lower extreme values are to be expected. The existing water pump load is increased to 100 kW for modeling purposes. The pump runs for periods of 4 hours. For this analysis, commercial loads are defined as priority due to the resilience service they provide (e.g., gas, food), while residential are classified as non-priority [6].

TABLE I:
Summary of the Corcovada Loads

Load	Min (kW)	Max (kW)	Avg (kW)	Daily Energy (kWh)	Priority
Residential	34.3	310.3	75.6	2,894.1	Non-Priority
Commercial	5.2	25.0	14.0	337.0	Priority
Shelters	31.2	84.0	149.6	3,095.2	Critical
Total	70.7	419.3	239.2	6326.2	N/A

The estimated hourly power consumption profiles for an average day for the community of Corcovada are shown in Fig. 3. These power consumption profiles consider residential, commercial loads as well as critical infrastructure consisting of community center & shelters and the community water pump (modeled as a 100 kW load).

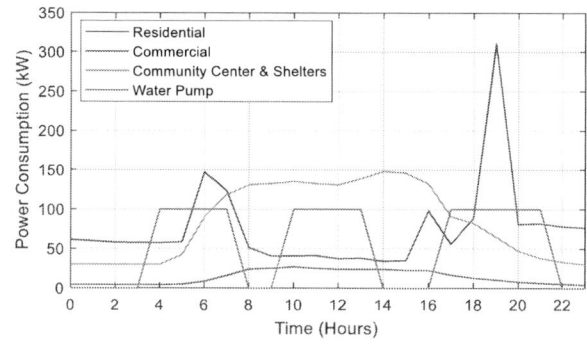

Fig. 3. Estimated Power Consumption for the Community of Corcovada.

A set of design basis threats (DBTs) and performance metrics are defined for the system. DBTs must be defined for the system as part of the optimization. DBTs in MDT refers to events, such as hurricanes, which can cause utility failures forcing the microgrid to operate in islanded mode. The optimizer runs during the islanded operation. If DBTs are not defined, MDT could not perform any optimization. For this system, three DBTs were defined. Due to its geographical location, Puerto Rico is highly susceptible to hurricanes and tropical storms. Puerto Rico electric grid is outdated and lacks maintenance resulting in constant power outages for rural communities like Corcovada. Table II shows the DBTs used for the model with their respective time between occurrences and duration of the outage caused by the event.

TABLE II:
CORCOVADA DESIGN BASIS THREATS

Event	Time Between Occurrence (Hours)	Duration (Hours)
Power Outages	72	3
Tropical Storm	8,760	72
Category 3+ Hurricane	87,600	2,880

MDT's optimization objective also relies on user-defined metrics, which include total purchase cost, energy availability for the different load tiers, renewable energy penetration, and others as needed. These metrics are numerical measures of microgrid performance [6] defined to minimize or maximize with desired limits and objective values. The performance metrics defined for Corcovada are shown in Table III. The solver is set to optimize for cost (metric #1) versus performance (metrics #2 − #6). Metrics #7 and #8 are detached from the optimization and used for measurement purposes only.

TABLE III:
OPTIMIZATION METRICS

#	Name	Action	Limit	Obj.
1	Cost ($ in Millions)	Min.	5	5
2	Energy Availability – Priority (%)	Max.	80	95
3	Energy Availability – Non-Priority (%)	Max.	75	90
4	Energy Availability – Critical (%)	Max.	95	99
5	Average Renewable Penetration (%)	Max.	15	40
6	Diesel Fuel Used (Gal/Hour of Outage)	Min.	5	0
7	Average Renewables Energy Supplied (%)	Max.	0	100
8	Average Renewable Energy Spilled (kWh)	Min.	1000	0

The MDT model is given inputs for the technologies to be considered according to the case scenario being evaluated. The first case scenario considers centralized assets options for diesel generation, solar generation (ground-mounted solar farms), and BESS (large, centralized storage; 800 kW/3 MWh – Lithium-ion units). The second case scenario considers asset options for centralized diesel generation with options for distributed solar

(residential roof-mounted) and BESS (Tesla Powerwall 2; Lithium-Ion − 2.8 hrs.). The third case scenario considers centralized diesel generation and storage, and distributed options for solar. Parameters associated with each technology are disclosed in Table IV.

TABLE IV:
TECHNOLOGY DESCRIPTION

Asset Type	Range (MW)	Steps (kW)	MTBF	MTTR	Capital Cost ($/kW)
Diesel Generator	0.1-2	100	15,000	34	1,150
Centralized PV	0.1-2	100	8,468	55	3,200
Centralized Batteries	0.8-4	800	5,500	168	1,488-1,624
Distributed PV	0.1-2	100	8,468	55	2,650
Distributed Batteries	0.05-0.5	50	5,500	168	2,100

To better represent the dynamics of the electrical loads throughout the day, a profile of power consumption per hour is applied to the loads according to load type. Fig. 4 illustrates the four residential load profiles [8]. The first load profile represents a typical residential load profile for residences in Puerto Rico, Residential 1. The other three load profiles are generated from various consumption surveys from community residents. Fig. 5 illustrates the two load profiles used for commercial buildings generated from [9] and [10]. Profile labeled Commercial 1 is used for the commercial building in the community. Profiles Commercial 1 and 2 are assigned randomly to the community center and the shelters.

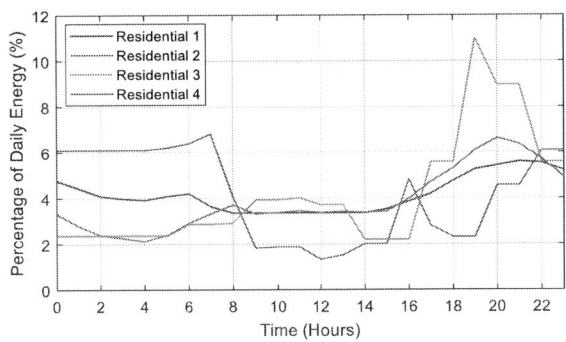

Fig. 4. Residential Load Profiles for the Community of Corcovada.

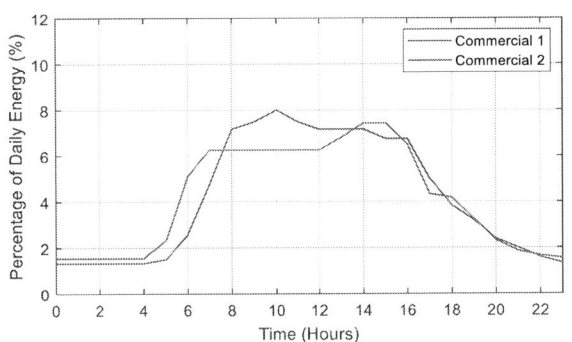

Fig. 5. Commercial Load Profiles for the Community of Corcovada.

IV. SIMULATION RESULTS

For this analysis, three different case scenarios are considered. Each consisting of a combination of centralized and distributed solar, diesel and storage. The different cases are:

- Case 1: Centralized solar, diesel, and storage
- Case 2: Centralized diesel, with distributed solar and storage
- Case 3: Centralized diesel and storage with distributed solar

These cases are evaluated using MDT and results are obtained for the Pareto frontiers, summary of the technologies and optimization metrics for the three scenarios. The pareto frontiers displays various dot colors. Red dots represent the "high cost, high performance" group of feasible solutions. Blue dots represent the "low cost, low performance" group of feasible solutions. The yellow dots represent the "knee" or optimal solutions space. The "knee" is defined to be a standard deviation divided by a factor b (σ/b) away from the mean of each axis. Green dots in the pareto represent other feasible solutions within 1 standard deviation of both means. The factor b is used for visualization purposes, and it varies depending on the solutions set to accommodate 6 − 7 solutions at the "knee". Both red and blue groups are one standard deviation away from the mean of both axes, highlighting both extremes.

A. Case 1: Centralized Solar, Diesel, and Storage

For the first case scenario considering centralized solar, diesel, and storage, a wide range of solutions appear feasible. Fig. 6 illustrates a smooth Pareto frontier, rescaled to show only feasible solutions. The optimization yielded 29 feasible solutions, each one being a different combination of the technologies considered. Factor b had a value of 2.12 for this result set to show six solutions at the "knee".

Fig. 7 illustrates the costs for the average installed capacity low group, "knee" and high group solutions for the diesel, PV, and BESS for the centralized microgrid alternative. Fig. 8 illustrates the cost for the solutions for the lowest (1082), six knee solutions and highest (874) solutions for the centralized solar, diesel, and storage.

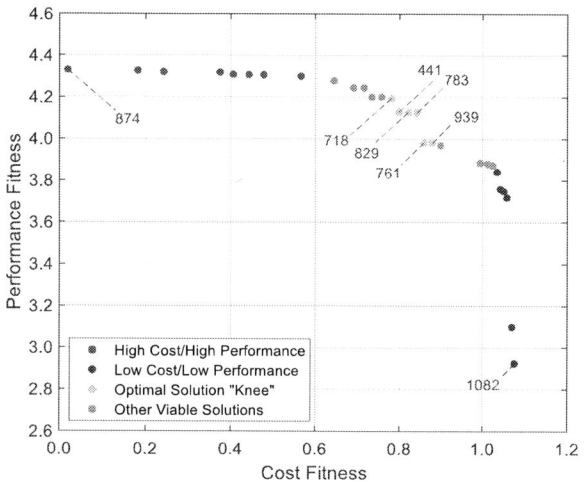

Fig. 6. Results Obtained for the Pareto Frontier.

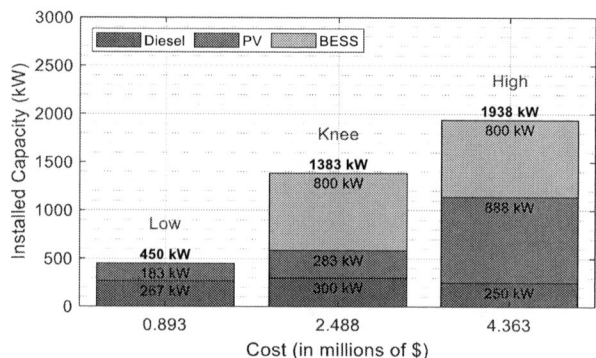

Fig. 7. Results for the average installed capacity solutions.

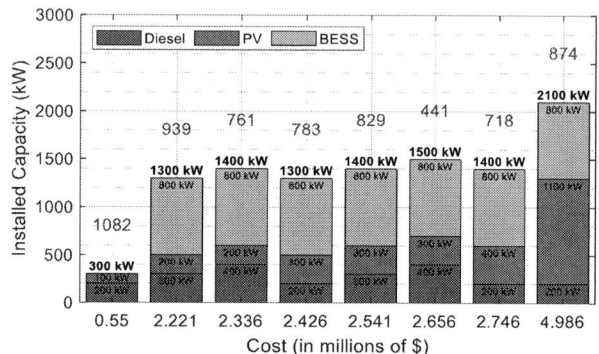

Fig. 8. Results for the installed capacity solutions at "knee".

All but ten of the feasible solutions opted for 1 unit of centralized storage. This is seen in all the "knee" solutions in Fig. 8. Solutions shown around the knee of the pareto range from 1.2 − 1.4 MW of capacity with either 200 kW or 300 kW in central solar and 200 kW, 300 kW and 400 kW in central diesel generation. Solutions at the knee cover over 98 % of the energy for all tiers throughout the simulation, yet results show high dependency on diesel usage. Table V summarizes the optimization results for the lowest (1082), the knee (829) and highest (874) solutions for the centralized solar, diesel, and storage. Energy availability has a noticeable increase for Priority and Non-Priority loads from the low solution (1082) to the "knee" solution (892). However, from the "knee" to the high solution (874), this increase is not as pronounced.

TABLE V:
PIVOT TABLE – OPTIMIZATION METRICS RESULTS

Metric	1082	829	874
Average Energy Supplied (kWh)	0	0	0
Average Energy from Renewables (%)	100	100	100
Average Renewable Penetration (%)	19.94	33.92	43.89
Diesel Fuel Used (Gal/Hour of Outage)	9.27	12.83	5.8
Diesel Generator Utilization (%)	46.56	82.50	75.45
Energy Availability – Critical (%)	99.52	99.98	99.99
Energy Availability – Priority (%)	87.2	99.48	99.77
Energy Availability – Non-Priority (%)	85.82	98.49	99.51
New Central BESS	0	800	800
New Central Diesel	200	300	200
New Central PV	100	300	1100
Purchase Cost (Millions of $)	0.550	2.541	4.985

B. Centralized Diesel, Solar and Distributed Storage

The second case scenario for the community presents an alternative that considers centralized diesel generation with distributed PV generation and small storage. This specific configuration is notable for its finer granularity in the energy storage options (small, distributed storage, 5 kW/13.5 kWh – Lithium-ion units), which is apparent in the increased number of feasible solutions identified through the optimization process. Specifically, a total of 40 feasible solutions were identified, as depicted in Fig. 9. With a factor b of 2, 7 solutions emerge as part of the "knee" group, representing the optimal trade-off space between cost and performance.

A key aspect observed in this configuration is the significant decrease in the required BESS installed capacity for these smaller, distributed systems. Illustrated in Fig. 10, the average installed capacity for the low, "knee", and high groups. The "knee" shows an average of 193 kW (540 kWh) BESS capacity needed, compared to the 800 kW (3 MWh) required in the first case scenario. This decrease remains prevalent even in the high group, where a higher investment is expected, signifying the cost-effectiveness of the distributed BESS configuration. In contrast, a notable increase of 142% in installed solar capacity is observed in the "knee" of this distributed PV scenario compared to the first case scenario, primarily due to its relatively lower cost compared to its centralized counterpart.

Fig. 11 illustrates the cost for the solutions for the lowest (1791), seven "knee" solutions and highest (957) solutions for the centralized diesel with distributed solar and storage. Focusing on the "knee", solutions range from 950 kW to 1250 kW total installed capacity with ranges of 150 kW (30 Tesla Powerwall 2s) to 250 kW (50 Tesla Powerwall 2s) for storage, 600 kW to 800 kW in solar, and 200 kW of diesel across all solutions. Solutions in the "knee" group cover over 98% of the energy needs for all load tiers.

These results underscore the value of the decentralized approach in microgrid design and the MDT's potential in identifying optimized solutions. The increased number of feasible solutions and the optimized allocation of resources highlight the flexibility and resilience offered by distributed microgrid designs.

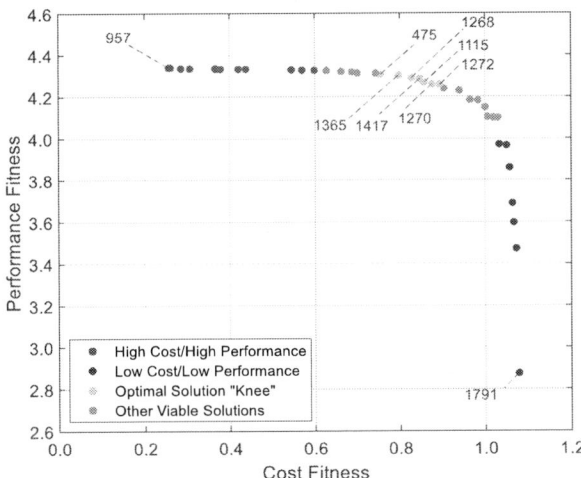

Fig. 9. Results Obtained for the Pareto Frontier.

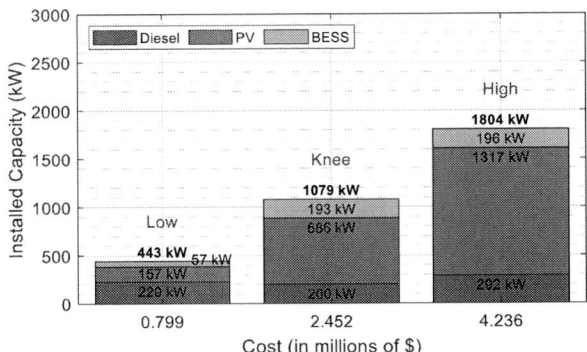

Fig. 10. Results for the average installed capacity solutions.

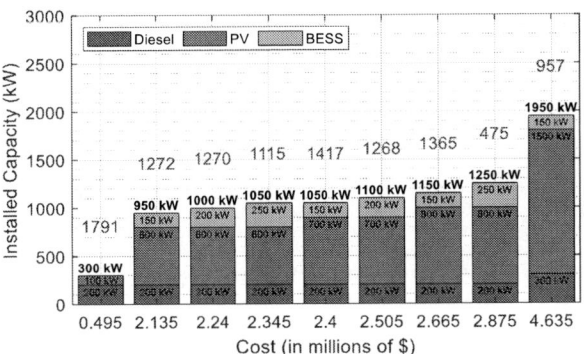

Fig. 11. Results for the installed capacity solutions at "knee".

Table VI shows the optimization metrics results, summarizing the low (1791) and high (957) solutions. Additionally, a representative solution from the "knee" group (1417) is presented. Similar to the first case scenario, the changes in terms of energy availability are consistent. However, there are differences in capacity allocation for storage and solar. Notice the decrease in storage capacity, highlighting the benefits of smaller distributed storage assets. In the solar capacity there is a substantial increase due to the decrease in cost per kW it brings. This translates in an increase in the average renewable penetration metric, bringing it over 40% for solutions around the "knee" in this case scenario compared to the first one. Solutions around the "knee" are consistent with the high levels of solar energy integration.

TABLE VI:
PIVOT TABLE – OPTIMIZATION METRICS RESULTS

Metric	1791	1417	957
Average Energy Supplied (kWh)	0	0	0
Average Energy from Renewables (%)	100	100	100
Average Renewable Penetration (%)	20.12	41.03	45.15
Diesel Fuel Used (Gal/Hour of Outage)	9.24	6.93	6.72
Diesel Generator Utilization (%)	46.61	45.85	35.48
Energy Availability – Critical (%)	99.24	99.93	99.97
Energy Availability – Priority (%)	86.76	99.42	99.78
Energy Availability – Non-Priority (%)	85.47	98.75	99.49
New Distributed BESS	0	150	150
New Central Diesel	200	200	300
New Distributed PV	100	700	1500
Purchase Cost (Millions of $)	0.495	2.400	4.635

C. Centralized Diesel and Storage with Distributed Solar

The third case scenario evaluated for the community considers centralized diesel generation and large, centralized storage units while keeping solar in a distributed form factor. Like in the first case scenarios, the results show a decrease in installed PV. However, costs remain in a similar range to that of the first and second case scenarios. For this simulation run, 30 solutions were found to be feasible within the trade-off space. To display six solutions at the "knee", a factor b of 1.92 was used, as shown by the Pareto frontier in Fig. 12.

Fig. 13 breaks down the average installed capacity of the different technologies considered in the model for the low, high, and "knee" solution groups. Averages in total installed capacity for the "knee" solutions are close to those of the first case scenario but with a small increase in PV capacity due to its lower cost. The higher average installed capacity, compared to the second case scenario, highlights the impact that the large, centralized storage has on the optimizer's ability to invest in other assets. Although the average installed solar capacity increased by almost 30% (367 kW vs. 283 kW on the first case scenario), the increase is small, compared to the second case scenario. This is due to the high investment needed for centralized storage in this centralized storage scenario. The average installed capacity for diesel decreased by 5.67% to 283 kW compared to the first case scenario, however, both averages are higher than the second case scenario at 200 kW for solutions at the "knee".

Fig. 14 shows the installed capacity for centralized diesel and storage, and distributed solar for the lowest (822), highest (1403), and six "knee" solutions. The total installed capacity for the "knee" solutions is within the 1,300 kW to 1,600 kW range. Similar to the centralized case scenario, optimal solutions opted for 1 unit of centralized storage with solar and diesel installed capacities range from 300 kW to 500 kW and 200 kW to 400 kW respectively. This represents a wider range in total installed capacity compared to the first case scenario, however, higher than the total installed capacities for the "knee" solutions in the second case scenario. It is noticeable the impact of having big, centralized assets, in this case energy storage, has on the ability to invest in other aspects such as PV generation to increase renewable penetration.

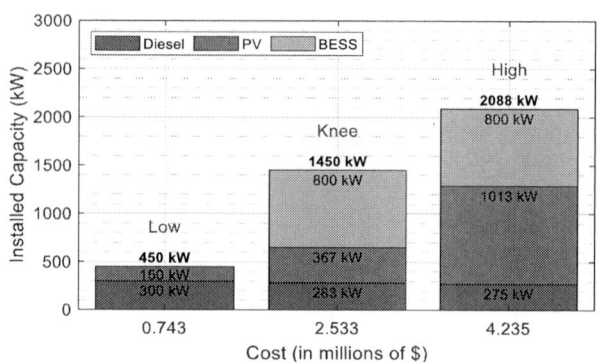

Fig. 13. Results for the average installed capacity solutions.

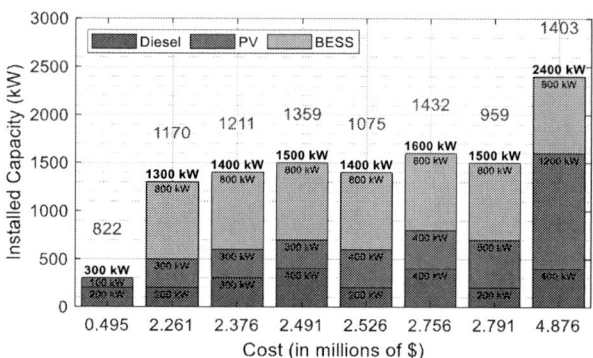

Fig. 14. Results for the installed capacity solutions at "knee".

Table VII illustrates the optimization metrics for the centralized diesel and large storage with distributed solar generation. Metric solutions for the lowest (822), highest (1403) and representative "knee" (1072) solutions are presented. Solutions in the "knee" group can provide over 99% of the energy needs of the community. There is a balance of diesel, solar and energy storage capacities at costs comparable to those of the two other case scenarios. With the 30% increase in installed solar capacity, the renewable penetration reaches 35% for solutions around the "knee". It is also noticeable the high diesel generator utilization, even though diesel is still used, it's being operated at what can be assumed to be a high-efficiency level. A decrease in distributed PV generation capacity is noticeable compared to the second case scenario.

TABLE VII:
PIVOT TABLE – OPTIMIZATION METRICS RESULTS

Metric	822	1075	1403
Average Energy Supplied (kWh)	0	0	0
Average Energy from Renewables (%)	100	100	100
Average Renewable Penetration (%)	19.90	36.79	44.31
Diesel Fuel Used (Gal/Hour of Outage)	9.29	8.24	10.18
Diesel Generator Utilization (%)	46.55	89.96	56.63
Energy Availability – Critical (%)	99.65	99.98	100
Energy Availability – Priority (%)	87.29	99.57	99.81
Energy Availability – Non-Priority (%)	85.98	99.22	99.58
New Central BESS	0	800	800
New Central Diesel	200	200	400
New Distributed PV	100	400	1200
Purchase Cost (Millions of $)	0.495	2.526	4.876

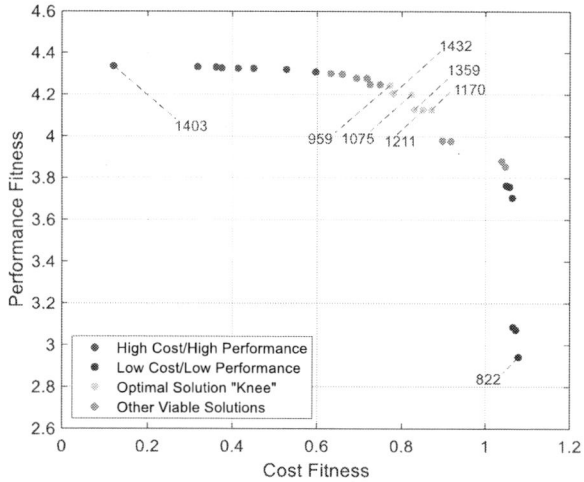

Fig. 12. Results Obtained for the Pareto Frontier.

V. CONCLUSIONS

This paper shows the trade-off space of the optimization of three microgrid design alternatives for a rural community in a mountainous region highly susceptible to hurricanes and storms. The three case scenarios explored offer a range of feasible solutions that balance the initial investment cost with microgrid performance metrics, such as energy availability to various load types. At this stage, stakeholder feedback will be crucial in refining these solutions and ensuring they meet the community's needs effectively. In the first case scenario, it is shown that centralized solar, diesel, and storage systems can provide a range of feasible solutions. However, this scenario exhibited a high dependency on diesel usage. In the second case scenario, the results demonstrate that a decentralized approach, with centralized diesel generation, and distributed solar and energy storage can provide a significant number of feasible solutions, emphasizing the cost-effectiveness of distributed solar and storage configurations. The results also highlighted the increased renewable energy penetration, exceeding 40% for solutions around the "knee" of the Pareto frontier.

In the third case scenario, results suggested that the combination of centralized diesel and storage with distributed solar generation offers a set of feasible solutions with a slightly increased PV capacity due to its lower cost, compared to the first case scenario. This analysis highlights the potential benefits of distributed energy resources in rural settings and the importance of considering multiple alternatives in designing microgrids. A key next step is to engage with the stakeholders and use this feedback to refine and further optimize the microgrid model. It's worth emphasizing that this approach allows for a detailed evaluation of the energy needs and priorities of the specific community in question. The use of MDT modeling offers an in-depth assessment that goes beyond just the cost-effectiveness of each solution, it also considers local factors such as critical load demands, regional threats like hurricanes and tropical storms, and the different load profiles within the community. Optimization results for the centralized alternative slightly favor diesel and high-capacity storage. However, for less aggregated alternatives, the optimization algorithm results favor a balance between storage and solar generation leading to higher renewable penetration.

As mentioned in the paper, the objective of the model is to accurately represent energy usage. Using smaller periods, such as 15 minutes or even less, for the power consumption load profiles could help better represent the fast power demand extremes that residential loads experience (e.g., turning on a water heater) in future modeling. Also, considering the hazards presented by the DBTs (e.g., wind speeds, flooding) and their impact on the components evaluated in the MDT model is an aspect that will be explored in the future. MDT allows modeling the hazards with probability parameters similar to those of the DBT. Component fragility is modeled as a fragility curve that represents the probability of failure of the component due to the hazard intensity and its mean time to repair after such failure. This can help not only on the optimization of the installed capacity of the microgrid but could also help in the decision-making of mitigation alternatives (e.g., undergrounding lines vs new distribution poles) providing a more detailed analysis for the community.

ACKNOWLEDGEMENT

This article has been authored by an employee of National Technology & Engineering Solutions of Sandia, LLC under Contract No. DE-NA0003525 with the U.S. DOE. The employee owns all right, title and interest in and to the article and is solely responsible for its contents. The U.S. Government retains and the publisher, by accepting the article for publication, acknowledges that the U.S. Government retains a non-exclusive, paid-up, irrevocable, world-wide license to publish or reproduce the published form of this article or allow others to do so, for U.S. Government purposes. The DOE will provide public access to these results of federally sponsored research in accordance with the DOE Public Access Plan.

This work was sponsored in part by the Consortium for Hybrid Resilient Energy Systems (CHRES) under grant number DE-NA0003982 from the National Nuclear Security Administration part of the U.S. DOE. This material is based upon work supported by the U.S. DOE's Office of Electricity under agreement with the FEMA. The authors would also like to acknowledge the leadership and support of the Corcovada Communal Committee Inc. and its community members.

REFERENCES

[1] C. B. Jones, C. J. Bresloff, R. Darbali-Zamora, M. S. Lave and E. E. A. Bezares, "Geospatial Assessment Methodology to Estimate Power Line Restoration Access Vulnerabilities After a Hurricane in Puerto Rico", *IEEE Open Access Journal of Power and Energy*, vol. 9, pp. 298-307, 2022.

[2] E. Parés-Atiles, E. O'Neill-Carrillo and F. Andrade, "Best Practices for Microgrids Applied to a Case Study in a Community", *2020 IEEE Power & Energy Society Innovative Smart Grid Technologies Conference (ISGT)*, 2020, pp. 1-5.

[3] "Fundamentals of Advanced Microgrid Design: Coursebook for Advancing Caribbean Energy Resilience Workshop", *Energy Transitions Initiative*, Sandia National Laboratories, 2019.

[4] R. Darbali-Zamora, C. J. Gómez-Mendez, E. I. Ortiz-Rivera, H. Li, and J. Wang, "Solar irradiance prediction model based on a statistical approach for microgrid applications", *2015 IEEE 42nd Photovoltaic Specialist Conference (PVSC)*, 2015, pp. 1-6.

[5] J. P. Eddy, and S. Gilletly, "Microgrid Design Toolkit (MDT) User Guide", Software Version 1.3., 2020.

[6] R. Broderick, B. M. Garcia, S. E. Horn, and M. S. Lave, "Microgrid Conceptual Design Guidebook", *Energy Transition Initiative*, Sandia National Laboratories, 2022.

[7] R. Darbali-Zamora, C. Birk Jones, and E. E. Aponte-Bezares, "Grid-Forming Inverter's Capability to Support Dynamic Microgrids with High Penetrations of Photovoltaics Systems", *49th IEEE Photovoltaic Specialist Conference (PVSC)*, June 5-10, 2022.

[8] A. A. Shcherbakova *et al.*, "Power Consumption of Typical Apartments of Multi-Storey Residential Buildings", *2020 International Youth Conference on Radio Electronics, Electrical and Power Engineering (REEPE)*, 2020, pp. 1-4.

[9] DSO Electric Cooperative "Managing Energy Costs in Convenience Stores", *pamphlet licensed for distribution to the customers of E SOURCE members*, 2009.

[10] P. Mincu and T. Boboc, "Load profiles in Smart Cities", *2017 International Conference on Energy and Environment (CIEM)*, 2017, pp. 480-484.

978-1-6654-6060-6/23 $31.00 © 2023 IEEE

Co-sputtered Sn-doped ZnO thin film n-type layers for incorporation into CdTe based photovoltaics.

Kerrie M. Morris, Mustafa Togay, Luke Jones, John M. Walls and Jake W. Bowers

Centre for Renewable Energy Systems Technology (CREST), Loughborough University, Loughborough, UK

Abstract—**Tin dioxide (SnO_2) and zinc oxide (ZnO) could be used to replace cadmium sulfide (CdS) in cadmium telluride (CdTe) photovoltaic cells. CdS undergoes parasitic absorbance due to a low bandgap and does not have a favourable band alignment. SnO_2 and ZnO were co-sputtered to form a Sn-doped ZnO film. The bandgap of the film is tunable depending on the amount of SnO_2 present within the structure. This was analysed by depositing SnO_2 at varying power densities. The co-sputtered film at 2% oxygen in argon process gas, substrate temperature 500°C and a deposition power density of 2.97 W/cm^2 for SnO_2 and ZnO at 3.96 W/cm^2 had the most promising window layer characteristics with high transmission, a bandgap of 3.82 eV, a high carrier concentration of 2.56×10^{17} cm^{-3} and low resistivity. However when full devices were made with CdTe it was seen the sample with the highest resistivity (10^6) and lowest carrier concentration (10^{12}) made the most efficient device with an efficiency of 6.2%**

Index Terms—**zinc oxide, tin dioxide, thin film, RF magnetron sputtering, cadmium telluride**

I. INTRODUCTION

Cadmium Telluride (CdTe) thin-film photovoltaics saw a large increase in efficiency several years ago, up to 22.1% [1]. This success has been partly attributed to including cadmium selenide (CdSe) within the device structure [2]. Progress has now slowed down and other parts of the device structure are under investigation, such as the interfaces, the n-type layer and the back contact to see if this can improve the efficiency further.

The theoretical limit for the maximum efficiency of a CdTe solar cell is around 30% [3], so there is much scope for improvement throughout the entire device structure. This work has aimed to replace the n-type cadmium sulfide (CdS) layer that is classically used in conjunction with p-type CdTe. Cadmium is a toxic material so working towards removing this from the device structure as much as possible would be advantageous.

CdS has conventionally been grown by several methods including chemical bath deposition as a popular choice [4] [5]. The film is usually around 100 nm thick and must be uniform enough to avoid any short circuits [6]. The crystal structure of CdTe is cubic and this results in a lattice mismatch of 9.7% with CdS [7] due to its most stable structure being hexagonal. This mismatch or strain between the crystalline phases introduces defects that increase recombination velocity [8].

The crystallinity of the CdTe however is affected by the application of a CdS film and an intermixed layer occurs

$CdTe_{1-x}S_x$ where the sulfur substitutes onto Te sites and reduces lattice mismatch [9], the interface is homogenized [10] during this intermixing. A small amount of band bending also occurs which allows light at longer wavelengths to be absorbed [11]. So careful consideration must be made when selecting replacement materials as several mechanisms occur that work in favour of improving the interface between CdS and CdTe layers.

The band alignment of a pn junction can have several configurations, either a spike, cliff, or flat. A spike has a positive ΔE_c and a cliff alignment has a negative ΔE_c. If the positive spike is between 1 ev and 3 eV a favourable effect occurs [12], a hole barrier is created at the interface region due to an absorber inversion occurring. This results in a reduction of the recombination velocity as the holes are blocked which increases the V_{oc}. A spike of 4 eV or more however blocks electron transport and causes a drop in efficiency which is due to a reduction in photocurrent and fill factor (FF). A cliff configuration allows holes to pool at the junction so a large supply of these increase the recombination velocity and will reduce the V_{oc}. CdS exhibits a cliff-type alignment when paired with CdTe so finding an alternative partner that would form a spike alignment of the correct ΔeV would improve the overall efficiency.

The relatively small band gap of CdS with a value of 2.4 eV [13] is also problematic. Any light with energy above the bandgap of 2.4 eV is absorbed as parasitic absorbance and only light with energy below can pass through and be transmitted to the absorber layer. It is anticipated that parasitic losses will be reduced by using a material with a higher band gap than that of CdS for the window layer.

Zinc oxide (ZnO) has previously been investigated as an n-type layer used to form a heterojunction with CdTe [14] and has high thermal and chemical stability. It has a higher band gap than CdS of 3.2 eV and relatively low resistivity of 1×10^8 - 10^{13} Ω.cm.

Tin dioxide (SnO_2) also has a much larger band gap of around 3.3 - 3.7 eV and has also been incorporated within CdTe cells previously [15]. These materials will be deposited by co-sputtering to form an intermixed layer with (SnO_2) acting as a dopant within the ZnO bulk material. By varying the sputtering powers and deposition times the bandgap can be tailored for a favourable band alignment with CdTe and increase the transmission of light and reduce parasitic absorbance.

978-1-6654-6060-6/23 $31.00 © 2023 IEEE

II. EXPERIMENTAL

A. SnO2 and ZnO Thin Film Preparation

Soda lime glass (SLG), NSG Tec™10C (Tec10) and sodium-free eagle XG glass (XG) substrates were first cleaned in a 1:10 ratio of detergent to deionised (DI) water for 30 minutes followed by a 1:5 acetone to DI water for 30 minutes and finally 1:5 IPA to DI for 30 minutes in an ultrasonic bath. They were then stored in DI water until required and then dried and cleaned in a UV ozone cleaner for 5 minutes.

Firstly both SnO_2 and ZnO films were deposited by radio frequency (RF) magnetron sputtering with an Orion 8 HV magnetron sputtering system (AJA International, USA). Oxygen content within the oxygen in argon (O_2/Ar) process gas was varied from 0 - 4% and substrate temperature was varied from room temperature (RT) to 500°C during deposition. Each deposition was completed for 15 minutes at a sputtering power of 3.96 W/cm² which is the recommended maximum power.

Deposition rates for SnO_2 were calculated with powers varying from 75, 50 and 25% of the maximum allowed sputtering power (2.96, 1.97 and 0.99 W/cm²). The SnO_2 and ZnO were co-sputtered using the deposition rates calculated to control the amount of SnO_2 within the ZnO films.

B. Preparation of ZnO/SnO2/CdTe Devices

The devices were completed on Tec10 with a 2-4 μm CdTe layer deposited using close space sublimation (CSS) with a substrate temperature of 515°C and a source temperature of 630°C for 3 minutes with a pressure of 1 Torr and 6% oxygen in argon gas mix. This was followed by evaporating $CdCl_2$ onto the surface and annealing in air at 420°C for 1 minute. 85 nm of gold was then deposited using thermal evaporation to form the back contacts.

C. Film and Device Characterisation Techniques

The optical properties were analysed using a Varian Cary 5000 UV–VIS–NIR spectrophotometer and the film thickness was measured using an Ambios XP-2 profilometer. Hall measurements were taken using a Semilab PDL Hall system. JV Measurements were taken using a bespoke built solar system using a 1000 W Newport xenon lamp. EQE measurements were taken using a Bentham PVE300 system.

III. RESULTS AND DISCUSSION

A. SnO2 Thin Film Characterisation

Transmission data in Fig. 1 was obtained for the SnO_2 film with varying percentages of O_2/Ar. The highest transmission occurred at 2% oxygen in argon. This also has the highest bandgap according to the Tauc plot [16] in Fig. 2 at 3.8 eV. The film sputtered at 2% O_2/Ar was then sputtered at increasing substrate temperatures. The transmission spectra were similar with a slight advantage at wavelengths between 550 - 800 nm and an average transmission between 400 - 550 nm for the film sputtered at 500°C. This also had the highest bandgap at 3.9 eV so saw a slight increase with increasing temperature when compared to the baseline film, see Fig. 4.

Transmission data was also obtained for SnO_2 sputtered at different powers see Fig. 5. As anticipated a reduction in power improved the transmittance of the films as it resulted in thinner films. The bandgap was maintained at 3.9 eV for the highest power of 2.96 W/cm² and slightly reduced to 3.8 ev for 1.97 and 0.99 W/cm² Fig. 6.

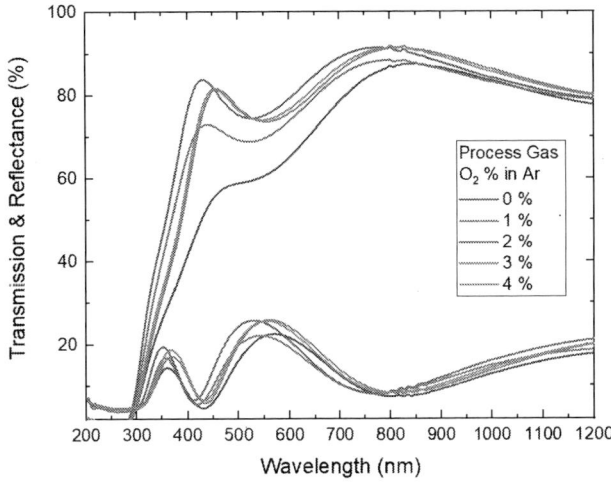

Fig. 1. Transmission and reflection data of SnO_2 sputtered at increasing oxygen content.

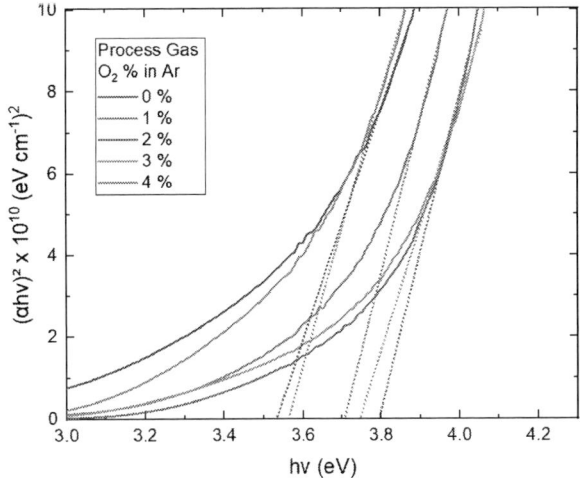

Fig. 2. Tauc plot of SnO_2 sputtered at increasing oxygen content.

Hall measurements seen in Table (I) for the varying percentage of O_2/Ar depositions all had low resistivity and good carrier concentrations in the 10^{19} and 10^{20}. The film sputtered at 2% O_2/Ar also had the highest mobility at 15 cm²/Vs. Due to this also showing the highest bandgap and highest transmission it was selected for repeating at increasing temperatures.

Good carrier concentration was still seen with the increase in temperature of the substrates during deposition up to 500°C. Resistivity was still low for all of the samples with a slight increase to 0.99 Ω.cm for the substrate sputtered at 400°C.

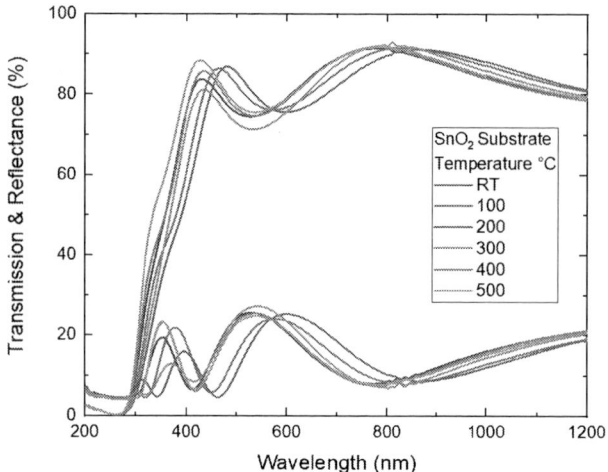

Fig. 3. Transmission and reflection data of SnO$_2$ sputtered at increasing temperatures.

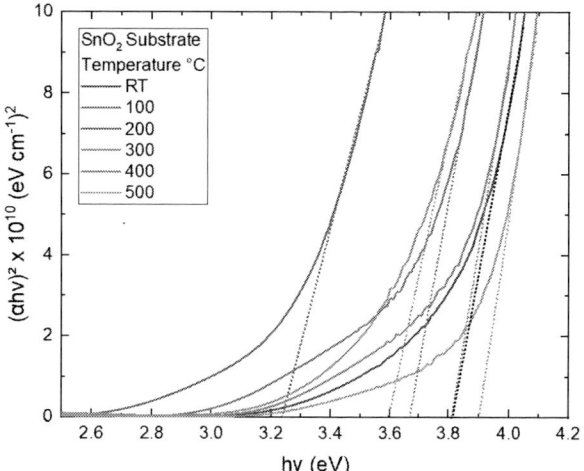

Fig. 4. Tauc plot of SnO$_2$ sputtered at increasing temperatures.

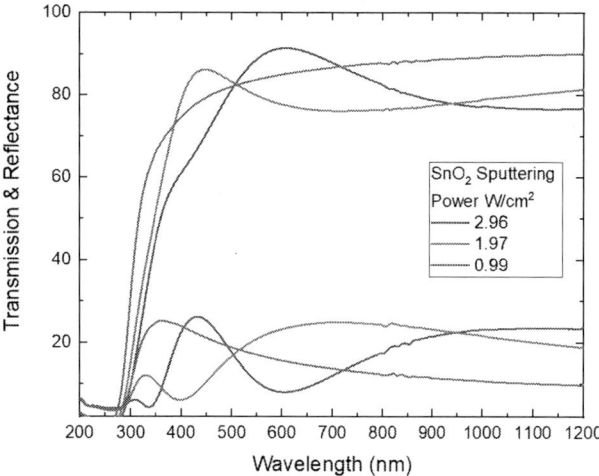

Fig. 5. Transmission and reflection data of SnO$_2$ sputtered at decreasing powers.

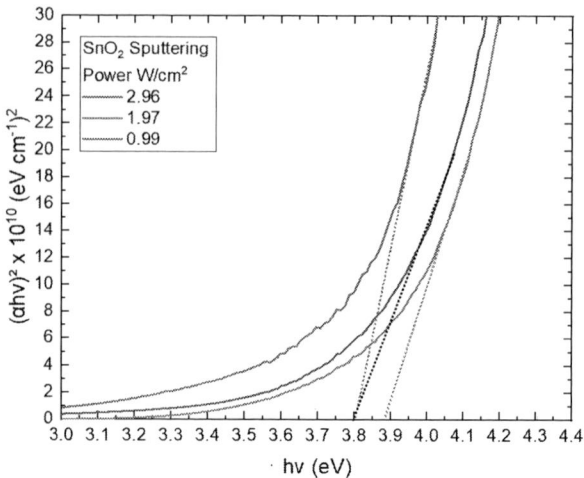

Fig. 6. Tauc plot of SnO$_2$ sputtered at decreasing powers.

Mobility was maintained for the substrates at 200 and 500°C, reduced at 100°C and increased at 300 and 400°C at a maximum of 17.5 cm^2/Vs. The results for varying temperatures were less definitive in choosing an ideal temperature but the higher bandgap and possibility of better crystallisation at higher substrate temperatures led to a temperature of 500°C being the favoured deposition temperature.

TABLE I
HALL DATA AND THICKNESS MEASUREMENTS FOR SnO$_2$ THIN FILMS

	Value	Thickness (nm)	Resistivity (Ω.cm)	Carrier Conc (cm^{-3})	Mobility (cm^2/Vs)
O$_2$%	0	225	0.023	9x10^{19}	2.9
	1	209	0.06	1.3x10^{20}	7.4
	2	223	0.012	3.2x10^{19}	15
	3	207	0.012	4.8x10^{19}	10.5
	4	210	0.039	1.6x10^{19}	9.3
Temp °C	100	190	0.01	6x10^{19}	10
	200	197	0.011	3.5x10^{19}	15.1
	300	202	0.01	3.6x10^{19}	16.8
	400	190	0.99	3.6x10^{19}	17.5
	500	204	0.079	5.3x10^{19}	14.9

B. ZnO Thin Film Characterisation

Varying the percentage of O$_2$/Ar during deposition had very little effect on the transmission and reflection data for ZnO films. Increasing the substrate temperature had the highest transmission for RT, 400 and 500°C see Fig. 7. A Tauc plot see Fig. 8 showed a very narrow bandgap range between 3.23 and 3.26 eV. Hall data could not be measured as the films were too resistive.

When examined by eye, the films did not show any discernible tint when compared to the glass substrate. As the films showed no wide variations with increasing percentage of O$_2$ and had the highest transmission at 500°C it was suitable to use the optimal conditions chosen for SnO$_2$ at 500°C and 2% O$_2$Ar for co-sputtered depositions.

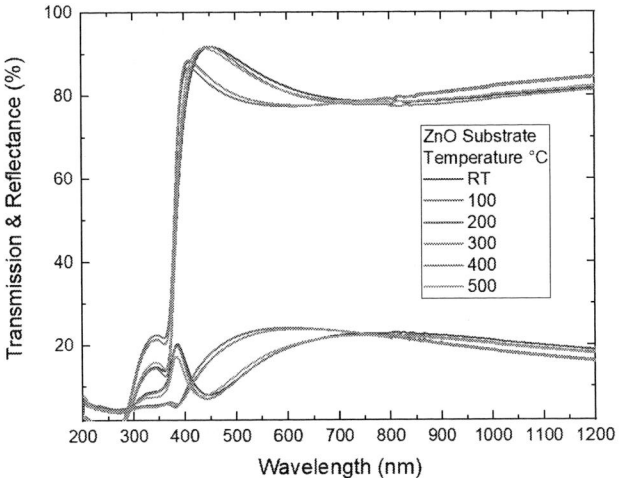

Fig. 7. Transmission and reflection data of ZnO sputtered at varying substrate temperatures.

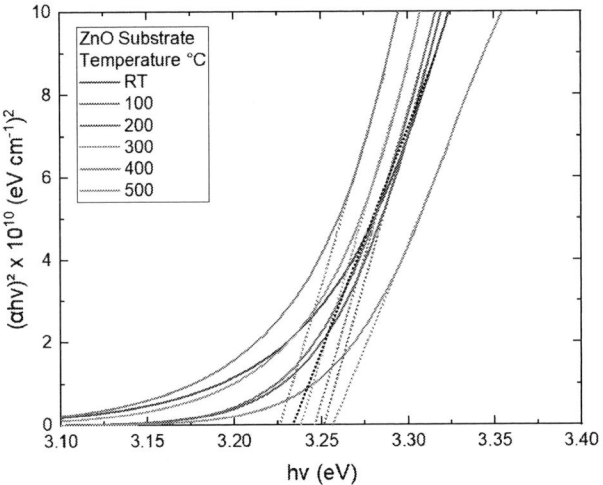

Fig. 8. Tauc plot of ZnO sputtered at varying temperatures.

C. ZnO/SnO₂ Co-sputtered Thin Film Characterisation

The deposition times were calculated using the deposition rates for the SnO₂ and ZnO sputtered at differing powers.

R1 = SnO₂ deposition rate.
R2 = ZnO deposition rate.

$$100 \div (R1 + R2) = time(s) \qquad (1)$$

TABLE II
ZnO/SnO₂ THIN FILM CO-SPUTTERING DEPOSITION POWERS AND TIMES

ZnO Sputtering Power (W/cm²)	SnO₂ Sputtering Power (W/cm²)	Time (s)
3.95	2.96	282
3.95	1.97	312
3.95	0.99	467

A film of 100 nm was determined to be an appropriate layer thickness and the times were calculated to reflect this (1), see Table II.

Transmission data was obtained for the co-sputtered films which indicated the films sputtered with SnO₂ at the highest power had the greatest transmission from 400 - 550 nm Fig. (9). From 550 - 1200 nm The film with SnO₂ sputtered at the lowest power had the highest transmission. A Tauc plot taken from the transmission data showed a decrease in bandgap as the sputtering power of SnO₂ was decreased. This would be indicative of a reduced SnO₂ content which was anticipated. The bandgaps ranged from 3.66 - 3.82 eV Fig (10).

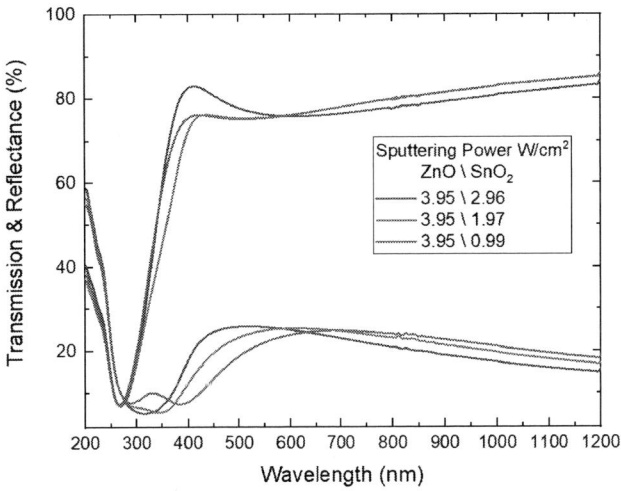

Fig. 9. Transmission and reflection data of ZnO/SnO₂ co-sputtered films at reducing powers.

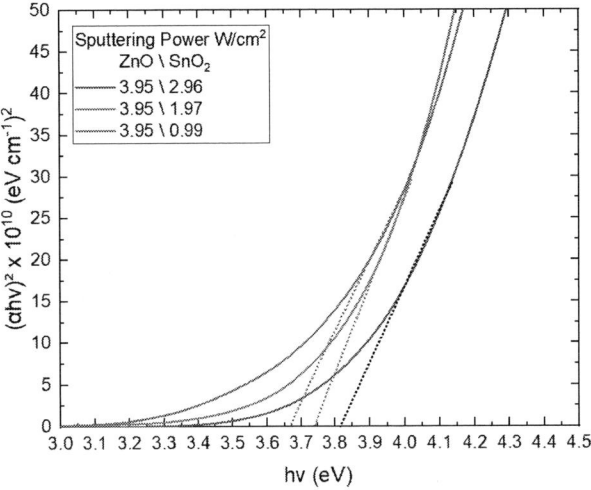

Fig. 10. Tauc plots of ZnO/SnO₂ co-sputtered films at reducing powers.

XPS data also confirmed that the SnO₂ content decreased as the power decreased. 2.96 W/cm² had a SnO₂ content of

11.7%. 1.97 W/cm^2 had a SnO$_2$ content of 8.7% and 0.99 W/cm^2 had a SnO$_2$ content of 5.1%.

Hall Data indicated that all films had a reduction in carrier concentration and mobility and an increased resistivity The SnO$_2$ sputtered at the lowest power having the most substantial changes. This film had the lowest carrier concentration at 4.91x10^{12} cm^{-3}, the lowest mobility at 2.7 cm^2/Vs and a higher resistivity at 2.36x10^{12} Ω.cm Table (III)

TABLE III
HALL DATA AND THICKNESS MEASUREMENTS FOR ZnO/SnO$_2$
CO-SPUTTERED THIN FILMS

Power (W/cm^2) ZnO/SnO$_2$	Thickness (nm)	Resistivity (Ω.cm)	Carrier Conc (cm^{-3})	Mobility (cm^2/Vs)
3.95 / 2.96	119	8.61	2.56x10^{17}	2.8
3.95 / 1.97	86	17.3	1.14x10^{17}	3.2
3.95 / 0.99	98	2.36x10^6	4.91x10^{12}	2.7

D. ZnO/SnO$_2$ Co-sputtered Device Characterisation

JV measurements were taken of the devices see Fig (11), and summerised in Table (IV) as sputtering power decreased for the SnO$_2$ the efficiency increased with the device with a SnO$_2$ sputtering power of 0.99 W/cm^2 having the highest efficiency of 6.2 %. At the lowest power rollover was also improved. Rollover is a common effect seen in CdTe photovoltaic cells due to the high work function of CdTe and the formation of a Schottky barrier forming at the back contact. All devices were made uniformly with regards to CdTe deposition, post-deposition treatments and contacts therefore an improvement can therefore be partly attributed to an improved band alignment between the window layer and CdTe [17].

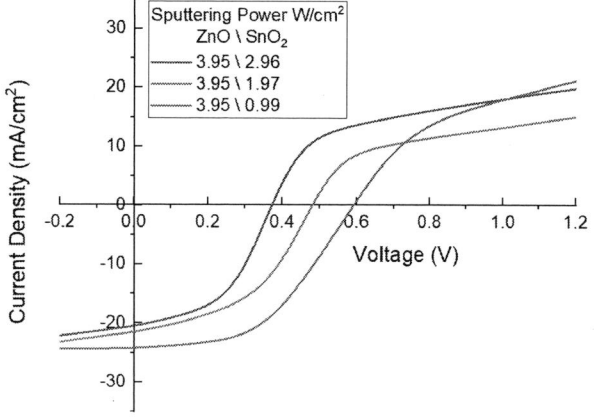

Fig. 11. JV plots of ZnO/SnO$_2$/CdTe devices.

The EQE of the devices, see Fig. (12) show a similar response for the films produced from the higher SnO$_2$ sputtering powers with a slight advantage at the lower wavelengths from 300 - 400 nm due to the higher bandgap allowing higher energy photons to pass through. They also show degradation of EQE response at higher wavelengths due to recombination losses. From 400 nm to 850 nm the SnO$_2$ sputtered at 0.99

TABLE IV
EFFICIENCY (η), SHORT CIRCUIT CURRENT DENSITY (J$_{SC}$), OPEN CIRCUIT
VOLTAGE (V$_{OC}$), AND FILL FACTOR (FF) OF THE ZnO/SnO$_2$/CDTE
DEVICES

Power (W/cm^2) ZnO/SnO$_2$	η (%)	J$_{sc}$ (mA/cm^2)	V$_{oc}$ (mV)	FF (%)
3.95 / 2.96	3.2	20.1	357	44
3.95 / 1.97	4.5	21.3	473	44
3.95 / 0.88	6.2	24.2	581	44

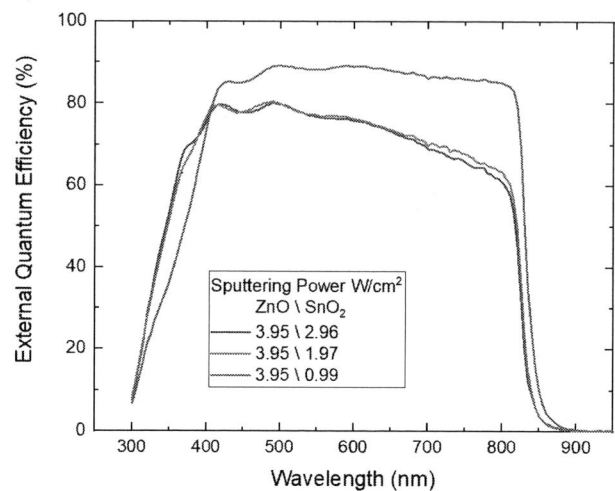

Fig. 12. EQE spectra of ZnO/SnO$_2$/CdTe devices.

W/cm^2 is much improved, this could be due to lower reflection at wavelengths over 600 nm and less surface recombination when compared to the higher power sputtered devices.

E. ZnO/SnO$_2$ Thin Films with SnO$_2$ Sputtered at Further Reduced Powers

In other work it has been shown a film thickness of 50 nm is ideal. The co-sputtered films were repeated at the same powers at the new thickness. Two more thin films were produced and analysed at SnO$_2$ sputtering powers of 60 and 40 W/cm^2, whilst maintaining the sputtering power of ZnO, to reduce the SnO$_2$ doping percentage further.

Transmission and reflection spectra of the new 50 nm films see Fig. 13 were more narrowly clustered than the previous thicker films. A Tauc plot see Fig. 14 followed the same trend as the thicker films and as the sputtering power of the SnO$_2$ decreased the bandgap also decreased. This indicated the SnO$_2$ content had reduced as seen in the previous section.

IV. CONCLUSIONS AND FURTHER WORK

Co-sputtering SnO$_2$ and ZnO resulted in an intermixed layer that had a range of bandgaps that increased as the tin content of the thin film increased. The thin film to have the most promising results was obtained when the oxygen content was 2%, substrate temperature of 500°C, SnO$_2$ deposition power density of 2.96 W/cm^2 and ZnO deposition power density of 3.95 W/cm^2. This resulted in a high transmission in the

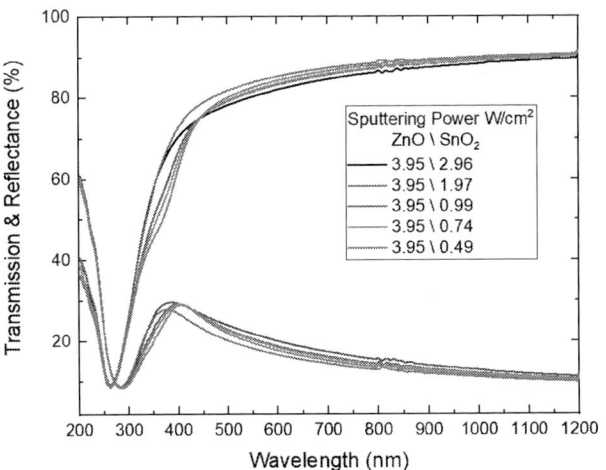

Fig. 13. Transmission and reflection data of ZnO/SnO₂ co-sputtered films at reducing powers, 50 nm film thickness, including lower SnO₂ sputtering powers of SnO₂, 60 and 40 W/cm².

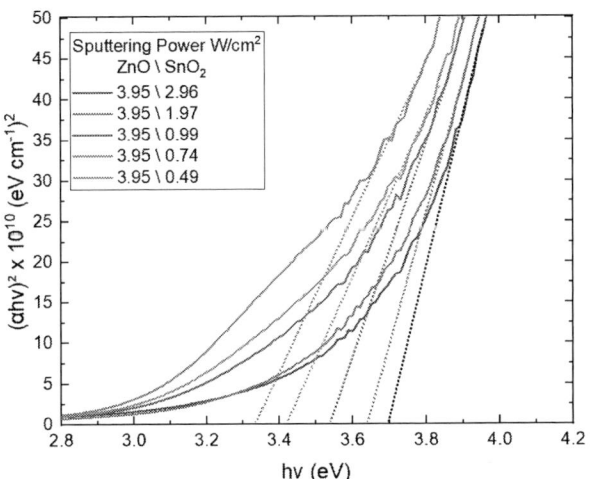

Fig. 14. Tauc plots of ZnO/SnO₂ co-sputtered films at reducing powers, 50 nm film thickness, including lower SnO₂ sputtering powers of SnO₂, 60 and 40 W/cm².

desired wavelength spectrum along with a bandgap of 3.82 eV, a carrier concentration of 2.56×10^{17} cm⁻³ and a resistivity of 8.61 Ω.cm.

However when made into a full stack with CdTe the device with the SnO₂ deposited at power density of 0.99 W/cm² did in fact perform better with an efficiency of 6.2%. This could be due to better band alignment with CdTe forming a spike of between 0.1 and 0.3 eV which is known to improve efficiency.

Surface recombination could also be responsible for the reduced EQE resulting in a low current in the devices made from the higher-power sputtered films.

New co-sputtered films were prepared of 50 nm film thickness and lower SnO₂ sputtering power of 60 and 40 W/cm² and these were assessed to determine if they followed the same pattern of decreasing bandgap with decreasing power which

was seen to be the case.

The ability to tailor the bandgap whilst utilising the electrical and optical properties of the individual materials would be advantageous when trying to improve the overall efficiency by adjusting and optimising the band alignment of the window layer with CdTe. Further work will encompass making the new films into full CdTe devices to determine the efficiency. The device parameters will be used in SCAPS modeling software to estimate band alignment.

REFERENCES

[1] M. A. Green, E. D. Dunlop, G. Siefer, M. Yoshita, N. Kopidakis, K. Bothe, and X. Hao, "Solar cell efficiency tables (Version 61)," *Progress in Photovoltaics: Research and Applications*, vol. 31, pp. 3–16, Jan 2023.

[2] R. M. Geisthardt, M. Topic, and J. R. Sites, "Status and potential of CdTe solar-cell efficiency," *IEEE Journal of Photovoltaics*, vol. 5, pp. 1217–1221, Jul 2015.

[3] W. Shockley, H. J. Queisser, and R. Ell, "Detailed balance limit of efficiency of pn junction solar cells," *Journal of Applied Physics*, vol. 32, p. 510, 1961.

[4] N. B. Chaure, S. Bordas, A. P. Samantilleke, S. N. Chaure, J. Haigh, and I. M. Dharmadasa, "Investigation of electronic quality of chemical bath deposited cadmium sulphide layers used in thin film photovoltaic solar cells," *Thin Solid Films*, vol. 437, pp. 10–17, Aug 2003.

[5] I. M. Dharmadasa and A. A. Ojo, "Unravelling complex nature of cds/cdte based thin film solar cells," *Journal of Materials Science: Materials in Electronics*, vol. 28, pp. 16598–16617, Nov 2017.

[6] S. G. Kumar and K. S. Rao, "Physics and chemistry of cdte/cds thin film heterojunction photovoltaic devices: fundamental and critical aspects," *Energy Environmental Science*, vol. 7, pp. 45–102, Dec 2013.

[7] A. J. Strauss, "The physical properties of cadmium telluride," *Revue de Physique Appliquee*, vol. 12, pp. 167–184, Feb 1977.

[8] J. M. Burst, J. N. Duenow, D. S. Albin, E. Colegrove, M. O. Reese, J. A. Aguiar, C.-S. Jiang, M. K. Patel, M. M. Al-Jassim, D. Kuciauskas, S. Swain, T. Ablekim, K. G. Lynn, and W. K. Metzger, "CdTe solar cells with open-circuit voltage breaking the 1 V barrier," *Nature Energy*, vol. 1, pp. 1–7, Mar 2016.

[9] K. Nakamura, M. Gotoh, T. Fujihara, T. Toyama, and H. Okamoto, "Influence of cds window layer on 2-m thick cds/cdte thin film solar cells," *Solar Energy Materials and Solar Cells*, vol. 75, pp. 185–192, Jan 2003.

[10] J. Major and K. Durose, "Study of buried junction and uniformity effects in cdte/cds solar cells using a combined obic and eqe apparatus," *Thin Solid Films*, vol. 517, no. 7, pp. 2419–2422, 2009. Thin Film Chalcogenide Photovoltaic Materials (EMRS, Symposium L).

[11] X. Mathew, J. S. Cruz, D. R. Coronado, A. R. Millán, G. C. Segura, E. R. Morales, O. S. Martínez, C. C. Garcia, and E. P. Landa, "Cds thin film post-annealing and te–s interdiffusion in a cdte/cds solar cell," *Solar Energy*, vol. 86, pp. 1023–1028, Jul 2012. ISRES 2010.

[12] T. Song, A. Kanevce, and J. R. Sites, "Emitter/absorber interface of CdTe solar cells," *Journal of Applied Physics*, vol. 119, p. 233104, Jun 2016.

[13] M. Ichimura, F. Goto, and E. Arai, "Structural and optical characterization of cds films grown by photochemical deposition," *Journal of Applied Physics*, vol. 85, p. 7411, May 1999.

[14] M. S. Tomar, "Photovoltaic properties of zno/p-cdte thin film heterojunctions," *Thin Solid Films*, vol. 164, pp. 295–299, Oct 1988.

[15] P. Zhan, J. Chen, and L. Chen, "Influence of sno2, zno and tio2 layer on the performance of cigs and cdte solar cells," *IOP Conference Series: Earth and Environmental Science*, vol. 781, p. 042069, May 2021.

[16] J. Tauc, "Optical properties and electronic structure of amorphous Ge and Si," *Materials Research Bulletin*, vol. 3, pp. 37–46, Jan 1968.

[17] X. Li, K. Shen, Q. Li, Y. Deng, P. Zhu, and D. Wang, "Roll-over behavior in current-voltage curve introduced by an energy barrier at the front contact in thin film cdte solar cell," *Solar Energy*, vol. 165, pp. 27–34, May 2018.

Investigating the Impact of MACl Doping in FA-Based Perovskites by Multimodal Synchrotron X-ray Techniques

Yanqi Luo, Sanggyun Kim, Carlo A. R. Perini, Naveed Rahman, Luxi Li, Dina Sheyfer, Ross Harder, Barry Lai, Juan-Pablo Correa-Baena, Sarah Wieghold

Advanced Photon Source, Argonne National Laboratory, Lemont, IL, United States

School of Materials Science and Engineering, Georgia Institute of Technology, Atlanta, GA, United States

Formamidinium (FA)-based lead iodide perovskites are intrinsically unstable and the unfavorable non-perovskite yellow phase readily forms because of the large FA cations. Here, we investigate the impact of methylammonium (MA) chloride additives in cesium FA-based lead iodide perovskite devices on the overall performance. We use a multimodal synchrotron-X-ray microscopy approach to reveal the nanoscale composition and optoelectronic performance in functional devices. We find a critical doping percentage of MACl upon which a dendritic growth of a Cs-rich phase is observed. Complemented by X-ray beam-induced current and X-ray excited optical luminescence measurements, we illustrate that the cesium-rich clusters are photoactive but with a reduction in extractable current. These multimodal correlative results provide an in-depth understanding of the undesired Cs-rich dendrites and pave the way for realizing the stoichiometric balance in preparing phase pure and high-performing hybrid perovskite devices.

Removing barriers for participation of small PV systems in balancing energy markets by utilizing the established smart meter eco-system

Christoph KONDZIALKA[1], Manuela MCCULLOCH[1], Rouven TAUBMANN[1], Jonas DIERENBACH[1], Dietmar GRAEBER[1], Gerd HEILSCHER[1]

[1]Ulm University of Applied Sciences, Albert-Einstein-Allee 53, 89081 Ulm, Germany
christoph.kondzialka@thu.de

Abstract — With the number of PV systems installed in Germany rising again since 2018, the provision of system services is increasingly coming into focus. On sunny off-peak days, the use of PV systems to provide balancing power will be indispensable in the future power grid [1]. This paper is based on a joint project, which is funded by the Ministry of Science, Research and the Arts of Baden-Württemberg, with the goal to develop a solution that will enable a large proportion of all photovoltaic (PV) systems to participate in the balancing power market [2]. In the context of the European harmonization of energy markets, the balancing energy market is therefore also undergoing change.

In principle, PV plants have a very high potential for providing negative balancing power. Small PV systems with an active power limitation (peak power capping) of 70% can additionally provide considerable amounts of positive control power. Technically, cost-effective control of small-scale PV systems via the German smart meter infrastructure is a very attractive option based on the findings presented in this paper. Regulatory hurdles, such as prequalification or market conditions in the balancing energy market can also be overcome in the foreseeable future, in the view of the project partners. However, stronger support at the political level is necessary for their timely elimination. An important step for a cost-efficient technical solution for the control of small PV plants results from the implementation obligation from the German federal law for the digitalization of the energy transition [3]. This conversion to smart metering represents an essential metrological basis for participation in the balancing energy market. As an additional benefit, the introduced intelligent metering system (iMSys) can also be used to control PV systems. The communication channel available for iMSys can essentially be shared here for controlling local plants without much additional effort. This paper proposes a concept that connects all systems involved to realize such an optimised plant controllability. For this purpose, a local control device (so-called CLS-gateway) connects the PV inverter with the remote control system of the balancing service provider (BSP). For communication, the secured channel provided by the iMSys is used, which is encrypted by the highly secured Smart Meter PKI. For the communication between the field device and the BSP control system, the MQTT protocol based on the publish-subscribe scheme is used [4]. The data models used are based on IEC 61850-7-4/420, but in contrast to the standard, they are mapped in JSON and not with MMS.

I. INTRODUCTION

This publication is based on a published article in the German-language energy economics journal "Energiewirtschaftliche Tagesfragen [5] and was prepared within the scope of the research project "Use of small PV systems to provide balancing power", which was carried out by the Smart Grids Research Group at Ulm University of Applied Sciences, the transmission system operator TransnetBW GmbH and the distribution system operator Stadtwerke Ulm/Neu-Ulm Netze GmbH. The aim of the project, which is funded by the Baden-Württemberg Ministry of Science, Research and Arts, is to develop technical and regulatory solutions that will also enable small PV systems to participate successfully in the balancing energy market. In this context, small systems were defined as those with an output of less than 100 kWp.

In the predecessor project "PV control", which was completed at the beginning of 2019 with the participation of TransnetBW as part of the funding initiative "Future-proof power grids" of the German government, proof has already been provided that small PV systems can basically be used to provide the different control power types (FCR, aFRR, mFRR) in accordance with the requirements of the transmission system operators [2]. Along the power curve, a majority of the inverters installed in Germany can adjust the power generation output within one or a few seconds. The "PV control" project focused in particular on the provision of negative manual Frequency Restoration Reserve (mFRR). For this purpose, it could be determined that over large parts of the year during the day, the demand can be covered to a large extent by PV systems [1]. In principle, this also applies to the negative control energy of the control power types Frequency Containment Reserve (FCR) and automatic Frequency Restoration Reserve (aFRR).

II. TECHNICAL AND ECONOMIC POTENTIAL

The economic potential of PV plants participating in the balancing power market was estimated. It was assumed that small PV plants are combined with other technical units to form a reserve group in order to theoretically fulfill the technical prequalification conditions, such as the temporal provision period or a symmetrical provision of power at FCR. Under these general conditions, the research project determined potential revenues for FCR and aFRR - exemplified by a 9.99 kWp system with commissioning in 2019, 70% active power limitation and 1,250 full load hours. It was assumed that the PV system in a reserve group is always used as a priority as soon as a power generation output is available or the active power limitation takes effect.

A. Technical, regulatory and economic barriers

Despite the high technical potential for providing balancing power from PV plants, none participate in the German balancing power market so far. The main reason for this is that PV systems have not yet been approved for prequalification [12] and are thus de facto excluded from the balancing power market. In addition, the EEG also excludes participation in the balancing energy market for PV plants that claim a feed-in tariff, according to § 21 (2) [6]. This prevents participation in the balancing energy market, especially for plants below 100 kWp, since direct marketing is still the absolute exception here.

The long supply time slices continue to be an obstacle. The switch to calendar-day tenders for balancing energy in July 2018 at least allows more accurate forecasts of fluctuating generators for balancing power offers. The current six four-hour blocks are a step toward integrating smaller PV systems. However, these improvements are not sufficient for integrating large numbers of PV plants into the balancing power markets. This would require a further shortening of the time slices [1].

Another obstacle to the participation of small PV plants in the balancing energy market is the cost of technical connection and control. Due to the limited economic potential, these may only be a few euros per year. Previous solutions, such as those used for the technical connection of wind power plants in a balancing power supply, usually cost several hundred euros or more per year. One focus of the research project is therefore to design cost-effective technical solutions

B. Cost-efficient control via the smart meter infrastructure

An important step for a cost-efficient technical solution for the control of small PV plants results from the conversion obligation from the smart meter rollout. In this process, existing meters of PV systems will also be successively exchanged for a smart meter. In addition, owners of PV systems of 7 kWp or more will have a smart meter gateway installed as a communication unit in the future, which, together with the smart meter, will form the so-called intelligent metering system (iMSys) [3]. This conversion to smart metering represents an essential basis in terms of measurement technology for participation in the balancing energy market.

In addition, iMSys can also be used to control PV systems, which is necessary for the provision of balancing power. The communication channel set up for the iMSys can essentially be used here for plant control. The fact that it is possible to implement such a system for controlling small PV plants via an iMSys has already been successfully demonstrated in previous research projects [7]. Furthermore, the requirements of the interface description for the provision of control power of the transmission system operator [8] were taken as a basis. Figure 1 shows the proposed concept, which connects all involved components to implement the controllability of the plants using the iMSys.

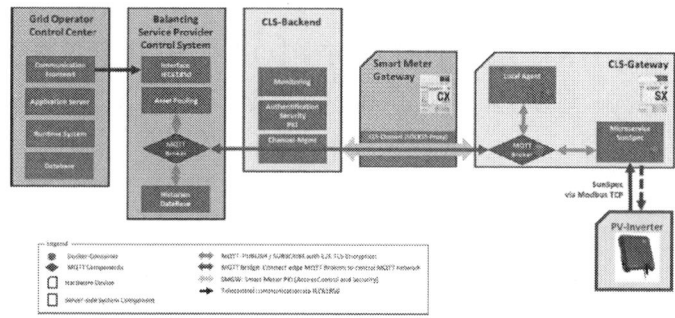

Fig. 1 Proposed system architecture for the communication via the german intelligent metering system (iMSys) of a Balancing Service Provider Control System with decentralized energy systems utilizing IEC61850 in combination with the MQTT protocol

An additional local control unit is used, the so-called CLS gateway, which connects the PV inverter to the control system of the balancing service provider (BSP control system). The BSP control system is responsible for the aggregation of the individual PV systems and combines them into an aggregated and thus marketable unit. A communication channel provided by the iMSys (so-called CLS channel) serves as the communication channel, the security of which is secured as an additional benefit by encryption using the official Smart Meter PKI. Other market participants such as direct marketers or the BSP itself can then retrieve the control power offered via the interfaces provided by the BSP control system. The open standard protocol SunSpec is used to transmit the status and the measured value query between the inverter and the CLS gateway. This is already implemented on most inverters and is based on the Modbus fieldbus protocol.

III. Technical implementation of the telecontrol link using IEC61850 and MQTT

The proposed technical concept uses data models based on IEC 61850-7-4/420, but in contrast to the IEC61850 standard, they are not transmitted using MMS but mapped in JSON. The protocol used between the CLS gateway and the BSP control system is MQTT based on the publish-subscribe scheme [4]. This allows the system design to respond to the special requirements of communication with decentralized power generation plants [9]. The use of proven communication protocols from the IoT environment is crucial here, as this can increase the reliability, availability and security of the controllability of decentralized renewable plants [10]. Furthermore, the aspects of cost efficiency are important, for example, in the minimization of data transmission as well as the use of established hardware as well as software modules from the IoT environment instead of especially newly developed processes. Last but not least, this ensures higher maintainability of the rolled-out systems [11], which benefits cyber security as well [12]. The proposed system architecture (see Figure 1) combines the advantages of the data model defined in the IEC61850 standard with the use of MQTT as a modern network protocol optimized for use in limited networks. As a first step, the desired parameters are queried by the CLS gateway from the PV inverter, which is done using the established and open industry standard SunSpec [13]. The values obtained are transformed into an IEC61850 data model and mapped to JSON format [14]. This JSON object is then further processed as the payload of an MQTT topic and provided using the local MQTT broker. The open-source software library Mosquito 2.0.14 is used, which in particular has implemented the functionality of an MQTT bridge. Here, the entire content of a local system is transmitted to the balancing service provider control system at once. Thus, a complete and consistent image of the local unit is always available at this central unit. For this purpose, the central MQTT broker subscribes to the entire content of the local MQTT broker using the bridge functionality. In addition, this has the advantage that the transmission of the data can be considered separately from the content, which improves the maintainability of the software.

This data transmission is implemented in terms of communication technology by means of the use of the CLS channel of the smart meter infrastructure. To this end, the local CLS gateway initiates the establishment of an encrypted TLS transport channel by the smart meter gateway. This process is based on the use of certificates of the official Smart Meter PKI, thus a mutual authentication of the involved components is always given. In addition, the data transport is also content-encrypted in accordance with the TLS protocol. The counterpart of this communication link is a CLS backend, which assumes the role of an active external market participant. This CLS backend is part of the Smart Meter PKI and makes the established communication channels available to the participating peripheral systems. On the side of the balancing service provider control system, the received data is transferred to the internal data model and included in the plant pooling. The control power provided by the pooled plants can then be called up by grid operators, for example. Technically, this is implemented in the project by means of an IEC61850 interface. Parallel to this process, the exchanged messages can be logged in a database.

IV. CONCLUSION AND OUTLOOK

In principle, small PV systems can additionally provide significant amounts of control power. This makes them an attractive option for maintaining system stability. However, a number of technical, economic, and regulatory barriers still need to be removed in order to leverage their potential to provide control power. It should be discussed to what extent requirements for the provision of control reserve can be simplified across all technologies in order to further reduce unnecessary market barriers, especially for fluctuating and small plants.

It is advisable to enter the test phase already in the upcoming pilot phase with the conditions actually required. In this way, it can be determined whether the PV plants can achieve the final conditions. In addition, it would make sense to extend the pilot phase to all three control energy markets FCR, aFRR and mFRR [15]. The proven procedure for determining the potential feed-in for wind energy should be adopted to the peculiarities of PV plants and adapted accordingly if necessary.

The administrative effort for prequalification with an increasing number of smaller technical units should be limited by standardization and type prequalification. It would be conceivable that the prototypical solution could be approved for type prequalification after successful tests. In this way, an attempt can be made to reduce the technical, economic and regulatory burden on market players.

REFERENCES

[1] D. Premm et al., 'Providing Control Reserve with PV Systems - Goals and Results of the Research Project PV-Regel', in International ETG Congress 2015; Die Energiewende - Blueprints for the new energy age, Nov. 2015, pp. 1–7.

[2] T. Bülo, M. Bünemann, B. Osterkamp, S. Poehling, and M. Stark, 'PV-Regel: Entwicklung von Konzepten und Lösungen zur Regelleistungserbringung mit Photovoltaikanlagen', SMA Solar Technology AG; Technische Universität Braunschweig; GEWI AG, Niestetal, Final Report, 2019. [Online]. Available: https://doi.org/10.2314/KXP:1679249207

[3] 'MsbG - Gesetz über den Messstellenbetrieb und die Datenkommunikation in intelligenten Energienetzen', Jul. 16, 2021. https://www.gesetze-im-internet.de/messbg/BJNR203410016.html (accessed Jan. 08, 2022).

[4] A. Banks, E. Briggs, K. Borgendale, and R. Gupta, 'MQTT Version 5.0 Committee Specification 02'. OASIS Message Queuing Telemetry Transport (MQTT) TC, May 15, 2018. [Online]. Available: https://docs.oasis-open.org/mqtt/mqtt/v5.0/cs02/mqtt-v5.0-cs02.html

[5] M. McCulloch, K. Wiedemann, F. Ebe, J. Dierenbach, D. Graeber, and C. Kondzialka, 'Einsatz von kleinen PV-Anlagen zur Erbringung von Regelleistung', Energiewirtschaftlichen Tagesfragen, pp. 61–64, Nov. 2021.

[6] 'Drittes Gesetz zur Neuregelung energiewirtschaftsrechtlicher Vorschriften (Änderung EEG 2012) | Clearingstelle EEG|KWKG'. https://www.clearingstelle-eeg-kwkg.de/gesetz/2219 (accessed Jan. 07, 2022).

[7] A. Reuter, O. Langniß, B. Haller, and Nicolas Spengler, 'Schlussbericht C/sells - das Energiesystem der Zukunft im Solarbogen Süddeutschlands', smartgrids-bw, Stuttgart, 2021.

[8] TransnetBW, 'Schnittstellenbeschreibung - Anbindung PRL/SRL/MRL/AbLa', Jun. 2017.

[9] H.-J. Jun and H.-S. Yang, 'Performance of the XMPP and the MQTT Protocols on IEC 61850-Based Micro Grid Communication Architecture', Energies, vol. 14, no. 16, p. 5024, Aug. 2021, doi: 10.3390/en14165024.

[10] K. Diwold et al., 'Grid watch dog: a stream reasoning approach for lightweight SCADA functionality in low-voltage grids', in Proceedings of the 8th International Conference on the Internet of Things, Santa Barbara California USA, Oct. 2018, pp. 1–8. doi: 10.1145/3277593.3277601.

[11] A. Krishna, M. Le Pallec, R. Mateescu, L. Noirie, and G. Salaun, 'IoT Composer: Composition and Deployment of IoT Applications', in 2019 IEEE/ACM 41st International Conference on Software Engineering: Companion Proceedings (ICSE-

Companion), Montreal, QC, Canada, May 2019, pp. 19–22. doi: 10.1109/ICSE-Companion.2019.00028.

[12] H. Ghadeer, 'Cybersecurity Issues in Internet of Things and Countermeasures', in 2018 IEEE International Conference on Industrial Internet (ICII), Seattle, WA, Oct. 2018, pp. 195–201. doi: 10.1109/ICII.2018.00037.

[13] SunSpec Alliance, 'SunSpec Device Information Model Specification'. Apr. 20, 2021. [Online]. Available: http://sunspec.org/specifications/#

[14] ECMA International, 'ECMA-404 JSON'. ECMA International. Accessed: Jan. 12, 2022. [Online]. Available: https://www.ecma-international.org/publications-and-standards/standards/ecma-404/

[15] T. Schlüter, Zukünftige Bereitstellung von Regelleistung unter Berücksichtigung technischer und marktwirtschaftlicher Potenziale. Düren: Shaker Verlag, 2019.

A Comparative Study of the Reflectance of Commercial Photovoltaic Modules

Wei Wen Toh, Min Hsian Saw, Erik Birgersson, and Mauro Pravettoni

Solar Energy Research Institute of Singapore, National University of Singapore, 117574 Singapore

Abstract—**Reflective glare of urban photovoltaic systems is a challenge for engineers and urban planners due to the potential operational hazard it poses. Glare considerations have been raised for building-integration photovoltaics, concerning the effect of reflections on nearby buildings, pedestrians, and traffic. Similar warnings are also often posed for installations in other areas, e.g. airports, or in combination with agriculture. The quantification of reflectance is therefore important to aid researchers in the selection of solar modules. In this work we characterise the reflectance of various commercial solar modules to analyse the impact of antiglare coatings or films, soiling, and of silicon-based versus thin-film technologies**

Keywords—*reflectance, glare, building integrated PV (BIPV), angular transmittance, antireflection coating*

I. INTRODUCTION

Due to growing concerns about global warming, countries worldwide are switching from conventional energy sources to renewable ones [1-2]. Solar energy displays a huge potential as it is renewable and clean, with virtually zero greenhouse gas emissions during its operation [3]. It is also scalable, making it versatile for industrial and household use. Moreover, the efficiency of photovoltaic (PV) modules continues to improve in recent years, from 15-18% in the 2010s to more than 20% nowadays, with highly efficient modules up to 23%.

In Singapore, solar energy is proven to be the most feasible renewable energy to be deployed in the city-state [4]. As Singapore faces land scarcity with a growing population size, other renewable energy sources such as wind and hydro energy are not viable. Hence, researchers have been studying the potential of building-integrated photovoltaics (BIPV) in urban districts, floating PV at reservoirs, as well as rooftop solar installations at airports. However, the installation of PV systems in most of these areas may present an operational hazard if the PV modules used are too "glaring" [5].

II. MOTIVATION AND OBJECTIVES

Glare is the visual discomfort experienced in the presence of bright light from direct or reflected light sources. In scientific terms, reflected glare refers to veiling reflection, which translates to regular reflection, superimposed on diffuse reflection from an object, which partially or entirely obscures the details to be seen by lowering the contrast.

Reflection is the process by which radiation is returned by a medium without a change in the frequency of its monochromatic component. Reflection can be categorised into two groups:

specular (or regular) reflection and diffuse reflection. In specular reflection, the angle of incidence is equivalent to the angle of reflection. On the other hand, in diffuse reflection, the incident radiation is scattered at multiple random angles. Mixed reflection is a combination of partial regular and partial diffuse reflection. Although PV modules are manufactured to have high absorbance and minimal reflectance, they are often highly reflective even at low angles due to their optical characteristics. PV glare affects operational safety in urban areas and in the proximity of airports. The presence of PV glare may cause distraction and/or temporary flash blindness to pedestrians and vehicles on the streets, airports, or in the sea. Thus, it is important to study the reflectance of PV modules in the visible range and minimise their disturbance to pedestrians.

This study aims at characterising the reflectance of various types of commercially available PV modules, by analysing the fraction of irradiance reflected in the visible and compared to the reflectance of a standard reference mirror.

III. METHODOLOGY

A. Test Samples

Table I lists the six PV modules that have been sampled for this study.

TABLE I. PV SAMPLES UNDER TEST.

ID	Cell type	Module characteristics
1	poly-Si (benchmark)	60 cells (5 busbars), glass-backsheet, white backsheet
2	poly-Si (soiled)	50 cells (2 busbars), glass-backsheet, white backsheet, aged module (~20% I_{sc} loss due to soiling)
3	CdTe	216 cells (thin-film), glass-glass, with anti-reflective coating.
4	poly-Si PERC (antiglare)	60 cells (5 busbars), polymer-backsheet, with antiglare texturing, white backsheet
5	*p*-type mono-Si PERC (black)	120 cells (half-cut, 9 busbars), glass-backsheet, black backsheet
6	Interdigitated Back-Contact (IBC)	96 cells, glass-backsheet, white backsheet

All samples were commercially available PV modules, although only sample 5 (a *p*-type mono-Si PERC module, with 120 half-cut cells, 9 busbars, glass-backsheet, with black backsheet) is still commercially available at the time of writing. Sample 1 was chosen as a benchmark for the reflectance of sample 2, a soiled poly-Si module of the early 2010s (50 cells, 2 busbars, glass-backsheet, with white backsheet): this module belongs to a group of 70 modules decommissioned from a 10-year building installation in central Singapore for reliability studies, resulting as the most soiled modules among others (for a more comprehensive degradation and soling analysis, the interested reader may refer to previous works [6-7]). Sample 4 is a lightweight glass-free module (poly-Si PERC, 6.0 kg, 60 cells, 5 busbars), with a textured polymeric front-sheet that aims at providing excellent antiglare properties, particularly focusing on building integration. Figure 1 shows pictures of the six samples taken with the camera flashlight on, to qualitatively compare the potential glare effect of the various cases.

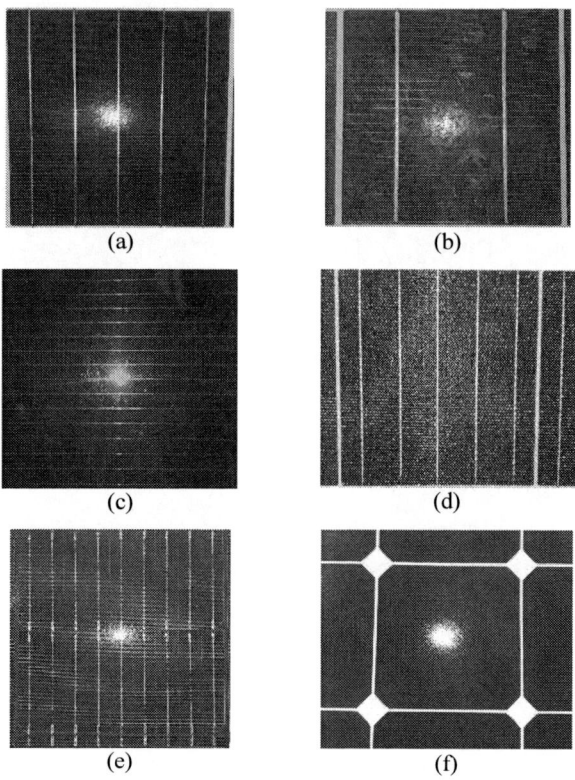

Fig. 1. Visual imaging of the effect of glare for the six modules tested: (a) poly-Si (benchmark); (b) poly-Si (soiled); (c) CdTe; (d) poly-Si PERC (antiglare); (e) mono-Si PERC (black); (f) IBC. All photos were taken with the same flash configuration of the camera, at approximately the same distance to the target.

B. Regular Spectral Reflectance as a Function of the Angle of Incidence

Fig. 2 shows the experimental setup for measurements of regular spectral reflectance as a function of the angle of incidence θ. A customized angled fiber probe holder (Fig. 2a), capable of coupling incident and reflected beams at angles varying from ±10° to ±80° (step 10°) is placed on the testing sample; the incident beam is provided by a xenon light source (Fig. 2b, with variable total irradiance in the range 100-1000 W/m^2); a 3-channel Avantes spectroradiometer (Fig. 2c) detects the reflected spectral irradiance in the 200-1750 nm range (resolution: 0.4 nm in the 200-1095 nm band; 6 nm in the 900-1750 nm band; detectors: 2 CMOS linear arrays with 2048 pixels, bands 200-715 nm and 600-1095 nm; an InGaAs linear

photodiode array with 256 pixels in the 900-1750 band). A standard BK7 Al+MgF$_2$ mirror tile is used as reference, with nominal 85% spectral reflectance in the visible band.

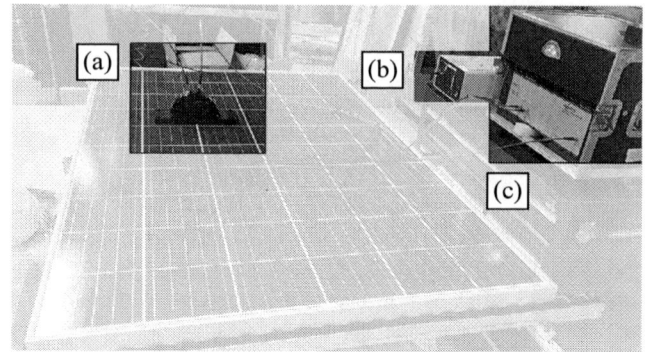

Fig. 2. Setup for measurements of spectral regular reflectance as a function of the angle of incidence θ on the benchmark poly-Si module: (a) customized optical fiber probe holder (from ±10° to ±80°, step 10°), to optically couple the incident and reflected beam driven by optical fibers; (b) a 1000 W/m^2 Xe light source; and (c) a 3-channel spectroradiometer by Avantes (2 CMOS detectors from 250 to 1095 nm, plus an InGaAs photodiode array detector from 900 to 1750 nm). Relative regular spectral reflectance is measured with respect to a standard reflector with nominal 85% reflectance in the visible band.

From the measured regular spectral reflectance at various angles of incidence θ, the relative regular reflectance at $\lambda = 620$ nm, $R(620, \theta)$, is selected as representative of the central part of spectral responsivity of the tested modules. The following reflectance difference is then calculated

$$y(\theta) = 1 - \frac{R(620,\theta) - R(620,10°)}{1 - R(620,10°)} \quad (1)$$

where $R(620,10°)$ is the relative spectral reflectance at the angle of incidence $\theta = 10°$ where the angular transmittance of the testing modules is assumed to be $\tau(10°) = \tau(0°) = 1$ (i.e. cosine response). Under the assumptions that the angular dependence of light absorption by the encapsulant is negligible, that the spectral differences in light transmittance are also negligible, and that the reflectance is purely regular (i.e. negligible diffusive reflectance), $y(\theta)$ defined in (1) is expected to be equivalent to the angular transmittance $\tau(\theta)$.

C. Reflectance and Glare

The glare analysis is performed by measuring the regular reflectance of the six samples of Table 1 at $\theta = 10°$ relative to the calibrated reflectance of a reference mirror (0.869).

Since glare is an expression of visual discomfort, its effect is evaluated in photometric units as follows. The measured relative regular spectral reflectance $R(\lambda, 10°)$, in normalized units, is used to calculate the (regularly) reflected spectral irradiance (in W/m^2/nm) at standard AM1.5g spectral conditions

$$E_R(\lambda, 10°) = E_{AM1.5g}(\lambda) \cdot R(\lambda, 10°). \quad (2)$$

where $E_{AM1.5g}(\lambda)$ is the standard AM1.5g spectral irradiance.

The reflected spectral illuminance (in lm/m^2/nm) can then be obtained by

$$\Phi_R(\lambda, 10°) = K_m V(\lambda) E_R(\lambda, 10°) \qquad (3)$$

where $K_m = 683$ lm/W is the maximum luminous efficiency for photopic vision and $V(\lambda)$ is the dimensionless photopic spectral luminous efficiency function [8-9]. The effect of glare is finally quantified via the ratio of the total reflected illuminance over the incidence total illuminance

$$\frac{\int_{380}^{770} \Phi_R(\lambda, 10°) d\lambda}{\int_{380}^{770} \Phi_{AM1,5}(\lambda, 10°) d\lambda} \qquad (4)$$

where $\Phi_{AM1,5}(\lambda, 10°) = K_m V(\lambda) E_{AM1.5g}(\lambda, 10°)$ is the incident spectral illuminance at AM1.5g. According to the study by Moereke et al. [9], glare is prevented when the total reflected illuminance ratio is less than 0.01%.

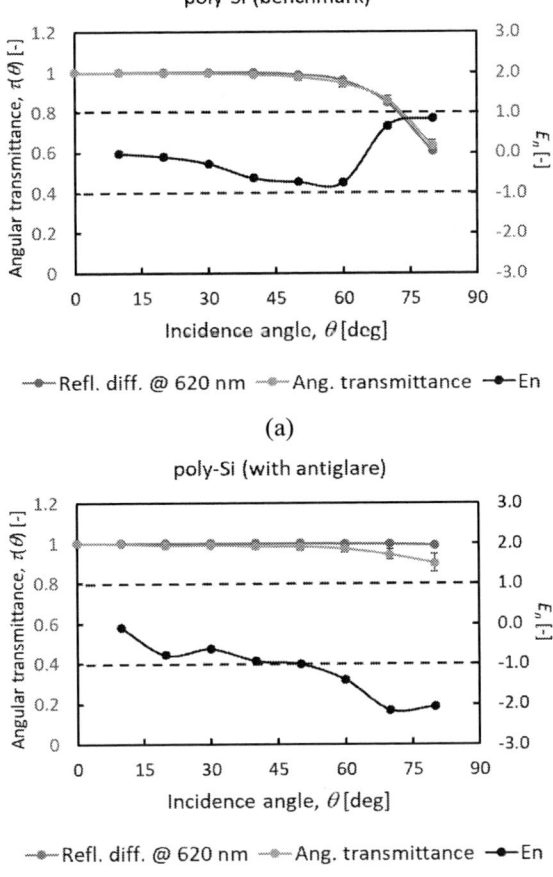

Fig. 3. Reflectance and angular transmittance: Comparison between the relative difference in spectral reflectance at 620 nm (see Equation (1)) and the measured angular transmittance $\tau(\theta)$, measured according to IEC 61853-2 [10]: (a) poly-Si (benchmark); and (b) poly-Si PERC (antiglare). The charts show also the calculated En value for method validation.

IV. RESULTS

A. Reflectance and Angular Transmittance

The charts in Fig. 3 show the comparison between the angular transmittance $\tau(\theta)$ derived from (1) under the assumptions of section III.B, and the measured $\tau(\theta)$, as described in IEC 61853-2 [10]. The charts also show the result of method validation analysis in terms of E_n numbers, i.e. the difference between the two methods normalized over the expanded uncertainty ($k = 2$, level of confidence of approximately 95%): standard practice indicates that method validation is successful when $|E_n| < 1$ (area within the horizontal dashed lines).

The assumptions of section III.B, thus the use of (1) as a practical method for the measurement of $\tau(\theta)$, is satisfactorily validated for the poly-Si benchmark module (Fig. 3a). However, for the poly-Si module with antiglare front sheet, (1) does not provide an accurate estimate of $\tau(\theta)$, thus the assumptions of section III.B seem not to be satisfied. This is not surprising, since the reflectance of the module with the antiglare frontsheet is dominated by the diffuse component, which is not negligible.

B. Glare

Fig. 4 and 5 show the results of glare analysis, comparing the incident spectral illuminance at AM1.5g $\Phi_{AM1,5}(\lambda, 10°)$ (black lines, in lm/m²/nm, vertical axis on the right) with the reflected spectral illuminance (coloured lines, also in lm/m²/nm, vertical axis on the left), for the six selected samples: the soiled poly-Si module (Fig. 4a), the (unsoiled) poly-Si benchmark (Fig. 4b), the CdTe thin-film module (Fig. 4c), the poly-Si PERC module with antiglare front-sheet (Fig. 5a), the p-type mono-Si PERC black backsheet module (Fig. 5b), and the IBC module (Fig. 5c). The charts also show the corresponding picture of Fig. 1, for easy reference, and the ratio of the total reflected illuminance over the incidence total illuminance, calculated using (4).

The sample showing the highest glare is the poly-Si benchmark, which reflects 4.9% of the incident illuminance; then come, in decreasing order of glare, the black backsheet p-type mono-Si PERC (2.1%), and the high-efficiency IBC module (1.4%); the soiled poly-Si sample (1.0%) and the thin-film CdTe sample (0.9%) show nearly the same reflectance. Not surprisingly the antiglare poly-Si module has negligible reflectance (0.01%), being the only sample that satisfies the requirement of Moereke et al. [9] to qualify as "antiglare".

The ranking from highest to lowest glare is consistent with the order from lowest to highest efficiency, with the interesting exception of CdTe (which, although has lower efficiency than IBC, shows a lower glare), and the soiled module. This can be qualitatively accepted by the fact that the modules with the highest efficiency generally have a higher spectral responsivity, hence reflecting less light than the lower efficiency modules (in fact appearing "darker", compared to the "bluer" poly-Si samples).

The CdTe sample uses an anti-reflective coated glass [11], that provides excellent performance in minimizing glare.

Soiling, although generally considered a severe degradation mechanism, is confirmed to reduce the effect of glare, due to the scattering of light of the particles on the module surface.

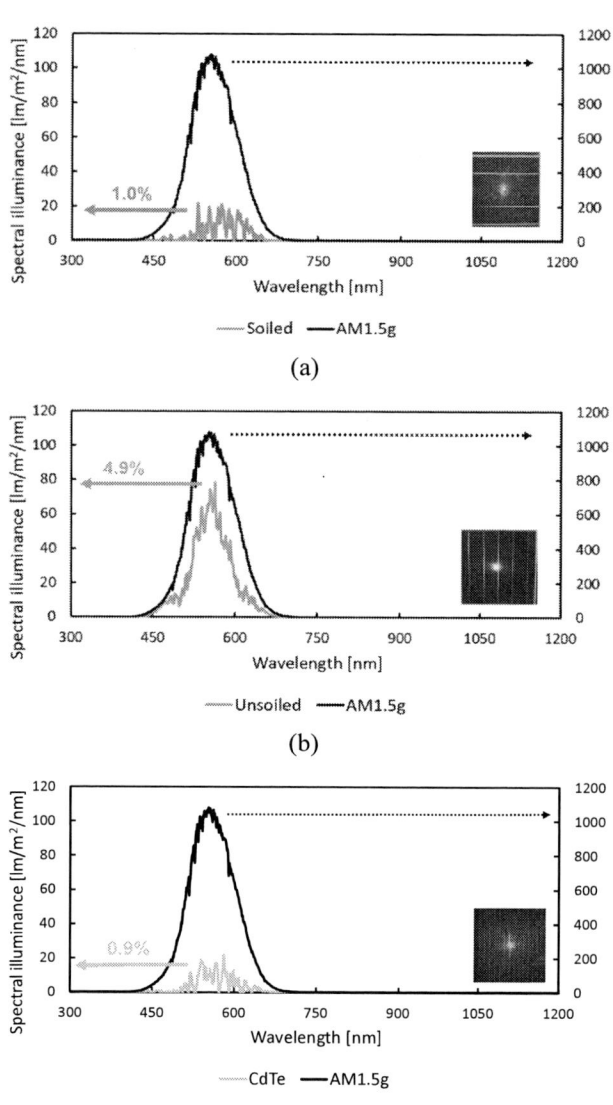

Fig. 4. Effect of glare: Measured spectral reflected illuminance of the tested samples (left axis), compared to the incident illuminance at AM1.5g spectrum (right axis): (a) poly-Si (soiled); (b) poly-Si (benchmark); (c) CdTe. The charts also indicate the fraction of reflected illuminance (as a percentage of the total incident illuminance).

V. Conclusions

A study of reflectance and its effect on glare is often requested for urban PV applications, however, a robust standard method for glare analysis is still missing. In this work, we studied regular reflectance measurements at different incidence angles. The results proved to be potentially a practical method for the angular responsivity (angular transmittance) of commercial modules that have negligible diffuse reflectance. We also quantified the percentage of irradiance reflected in the visible (illuminance), comparing 6 different samples of different technologies, including a soiled module. The results highlighted the impact of anti-glare treatments and showed that only the sample designed with antiglare coating fulfils the reported 0.01% reflectance indication to prevent glare.

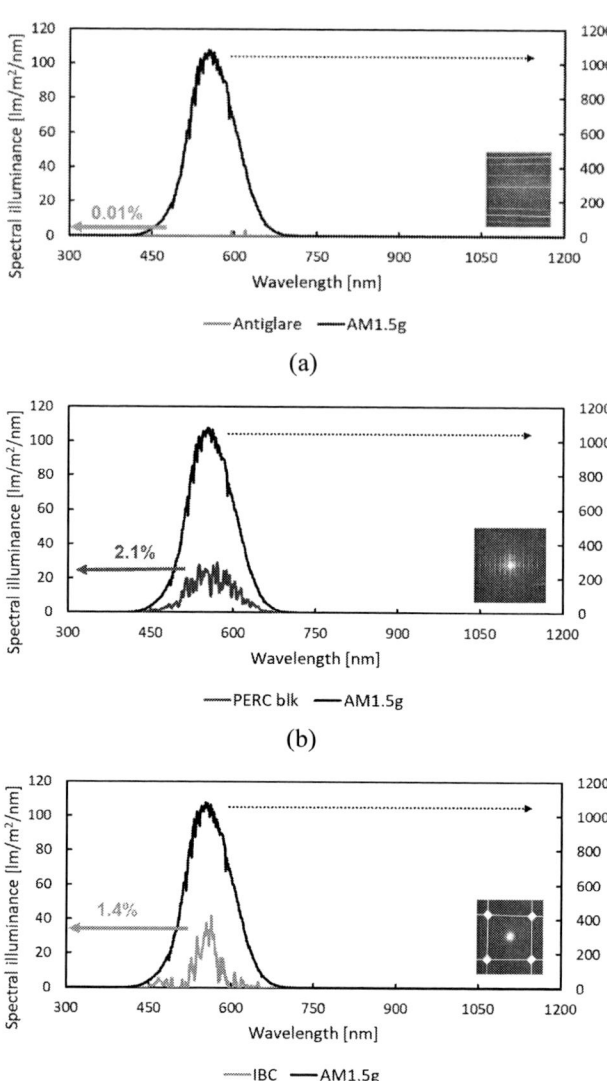

Fig. 5. Effect of glare *(continued from Fig. 4)*: (a) poly-Si PERC (antiglare); (b) mono-Si PERC (black); (c) IBC.

Acknowledgment

SERIS is a research institute at the National University of Singapore (NUS). SERIS is supported by NUS, the National Research Foundation Singapore (NRF), the Energy Market Authority of Singapore (EMA), and the Singapore Economic Development Board (EDB).

References

[1] IEA, "Global Energy Review: CO2 Emissions in 2021," International Energy Agency, Paris, 2022.

[2] United Nations , "The Paris Agreement," United Nations, [Online]. Available: https://unfccc.int/process-and-meetings/the-paris-agreement/the-paris-agreement/.

[3] IRENA, "Renewable Energy Statistics 2022," International Renewable Energy Agency, Abu Dhabi, 2022.

[4] Singapore Green Plan, https://www.greenplan.gov.sg/.

978-1-6654-6060-6/23 $31.00 © 2023 IEEE

[5] "Daylight Reflectance Design Guide", edited by Building Construction Authority and National University of Singapore, Version 1, September 2022.

[6] M. Pravettoni, M. Mahesh S/O S. Das, A. S. Rajput, W. Li, and S. Valliappan, "A Study of Module Degradation from a 10-year Installation in the Urban Tropical Environment," 8th World Conference on Photovoltaic Energy Conversion, pp. 594–598, October 2022.

[7] W. Li, A. S. Rajput, S. Valliappan, and M. Pravettoni, SiliconPV 2022, 28 March – 1 April 2022, in press.

[8] Bureau International des Poids et Mesures (BIPM), "Principles governing Photometry", 05/2019 2nd edition

[9] J. Moereke, P. Borowski, S. Grünsteidl, J. Palm, S. Babu Sapkota, and T. Dalibor, "Light Reflection Analysis of PV Modules: Comparison to Building Facades and Assessing the Possibility of Glare", 8th World Conference of Photovoltaic Energy Conversion, Milan, p.1050-1056 (2022).

[10] IEC 61853-2:2016, "Photovoltaic (PV) module performance testing and energy rating - Part 2: Spectral responsivity, incidence angle and module operating temperature measurements", Ed. 1.0 (2016).

[11] https://www.firstsolar.com/-/media/First-Solar/Technical-Documents/User-Guides/Series-4-Module-User-Guide-Global.ashx?la=en

The Microstructure of thin film CdSe following Cadmium Chloride Activation Treatment

[1]Rachael C. Greenhalgh[1], Vladislav Kornienko[1], Mustafa Togay[1], Ali Abbas[1], Ebin Bastola[2], Adam B. Phillips[2], Michael J. Heben[2], Jake Bowers[1], J. Michael Walls[1]

[1]Centre for Renewable Energy Systems Technology (CREST), Loughborough University, Loughborough, LE11 3TU, UK

[2]Wright Center for Photovoltaics Innovation and Commercialization (PVIC), Department of Physics and Astronomy, University of Toledo, Toledo, Ohio, 43606, USA

Abstract — **Thin film CdSe is an important precursor layer for CdSeTe/CdTe photovoltaic devices and as a potential future top cell in multijunction devices. It is also used in photodetectors. In this paper we report on the improvements in the microstructure of CdSe thin films caused by the cadmium chloride (CdCl2) treatment. Using high resolution cross-sectional HRTEM and EBSD, we show that the CdCl2 treatment leads to recrystallisation, grain growth, texture randomization and defect removal. These improvements in microstructure result in a dramatic increase in luminescence and carrier lifetime as measured using PL and TRPL.**

Keywords—CdSe, Solar Photovoltaics, Top Cell, Tandem, CdTe, Thin Film, Photoactive, Microstructure, CdSeTe, HRTEM

I. INTRODUCTION

Cadmium Selenide (CdSe) thin films have become important in the fabrication of CdSeTe/CdTe solar cells. Thin film CdSe is used as a precursor film in the fabrication of CdSeTe solar cells [1]. As deposited cells consist of a CdSe and CdTe bilayer. CdSe/CdTe devices undergo a heat treatment with $CdCl_2$ during processing into a CdSeTe/CdTe cell. Se diffuses from the CdSe layer into the CdTe during the cadmium chloride activation treatment, grading the bandgap from 1.4 eV in CdSeTe to 1.5 eV at the back [2]. This increases the photocurrent, but the open circuit Voltage is unaffected. Depositing the CdSe/CdTe stack is superior to depositing the CdSeTe alloy because it produces a more consistent and repeatable compositional gradient. CdSe thin films have also been used to produce solar devices with a 6% efficiency. The bandgap of CdSe is ~1.7 eV, which is the optimal bandgap for a potential top cell on Si devices [3]–[5]. CdSe has also been examined for use as a photodetector [6]. Selenium is an important addition to CdTe absorbers in photovoltaic devices. It improves the lifetime [7], passivates grain interiors [8] and passivates grain boundaries [9]. CdSe films have been deposited by various deposition methods including chemical bath [10], RF sputtering [11] DC sputtering [12], vapor transport deposition [13] and evaporation [14]. The effect of the $CdCl_2$ activation treatment has been extensively analyzed for thin film CdTe devices. $CdCl_2$ activation causes drastic recrystallization, Cl decoration of the grain boundaries [15], removal of stacking faults [15] and increase in the photoactivity of the film [16]. However, few studies have been performed on the effects of the $CdCl_2$ treatment on thin film CdSe. It is known that the $CdCl_2$ treatment improves the photoactivity of the device by incorporating chlorine in the grain boundaries [17]. The treatment improves electrical conductivity and reduces strain in the film [18]. In this paper we will present new results obtained using high resolution HRTEM to reveal the effects of the treatment at the atomic scale.

II. EXPERIMENTAL

800 nm thick CdSe films were deposited on TEC12D™ (NSG Pilkington) by evaporation of CdSe powder (99.999%, Alfa Aesar) at a substrate temperature of 400 °C at a base pressure of ~ 4 x 10^{-6} Torr. $CdCl_2$ treatment of the CdSe was performed using a saturated CdCl2 solution in methanol, and then it,was annealed at 450 C for 40 mins in dry air environment The $CdCl_2$ treated CdSe film was then rinsed with methanol and dried with N_2. Transmission Electron Microscopy (TEM) was performed on a FEI Tecnai F20 at 200 kV for the high-resolution TEM images. Specimens for TEM were prepared by Focused Ion Milling (FIB). Electron Backscatter Diffraction (EBSD) was performed on a clean edge of the film, prepared using the G4 Xenon Plasma focused ion beam (PFIB) combined with an Oxford Symmetry CMOS detector. X-Ray Diffraction (XRD) structural analysis of the films was obtained using a Bruker-D2 benchtop XRD using Cu Kα X-rays (X = 1.542Å). The Hall Effect measurement was taken on a Semilab PDL Hall System. These samples required an insulating surface, so the Hall Effect measurement was performed on a soda lime glass substrate. Photoluminescence (PL) and Time Resolved Photoluminescence (TRPL) were performed on an in-house

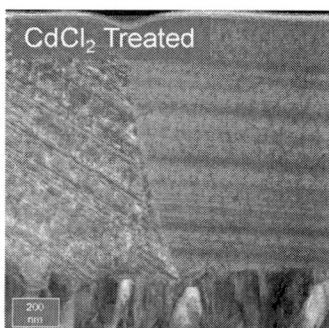

Figure 1 TEM cross sections of CdSe before and after $CdCl_2$ treatment showing large, full thickness grains after the $CdCl_2$ treatment.

978-1-6654-6060-6/23 $31.00 © 2023 IEEE

built system [19] between the wavelengths 600-900 nm with a 1 nm step size and a 640 nm excitation laser.

III. RESULTS AND DISCUSSION

The TEM image in Figure 1 shows untreated evaporated CdSe with full thickness columnar grains After CdCl$_2$ treatment the CdSe has broad, full thickness grains. This shows the CdCl$_2$ treatment of CdSe has led to the removal of cubic defects and the reduction of grain boundaries to leave a purely hexagonal CdSe film. This removal of cubic defects is analogous to the of the removal of hexagonal defects observed in CdTe treated with CdCl$_2$ [15] resulting in a purely cubic structure. The High-Resolution TEM (HRTEM) cross sections in Figure 2 show stacking faults present in the untreated CdSe. After CdCl$_2$ treatment the stacking faults are removed, although twin boundaries are observed. EDX elemental maps of the TEM cross sections (Figure 3) show chlorine decorating the grain boundaries. This is also observed following the CdCl$_2$ treatment of CdTe, and sputtered CdSe treated with CdCl$_2$ [15], [17].CdCl$_2$ treatment of CdSe leads to improvements of the film quality and changes the texture to a more random texture. As deposited CdSe is hexagonal with a high density of stacking faults of cubic phase. After treatment, the CdSe is purely hexagonal. This is observed in in the EBSD data shown in Figure 4 and the XRD data shown in Figure 5. EBSD cross sections are shown in Figure 4, the grain size increase can be clearly observed before and after CdCl$_2$ treatment. A grain size analysis was performed on the EBSD cross section images. As deposited the CdSe grains were 0.3 μm ±0.3 μm, and after

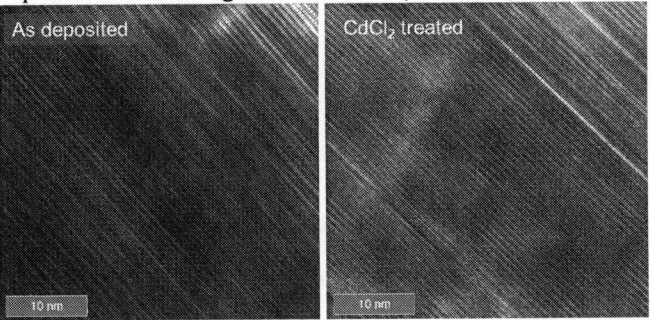

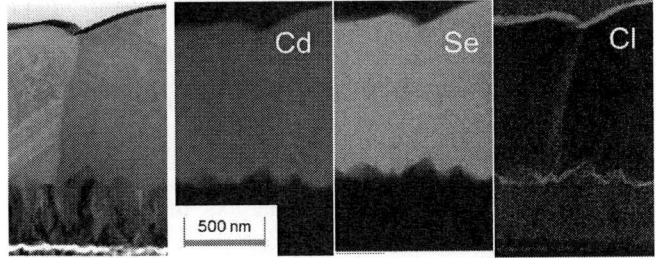

Figure 3 High-Resolution TEM images of CdSe as deposited and after CdCl$_2$ treatment. High density twins are present after treatment

CdCl$_2$ treatment the CdSe thin film has an average grain size of

Figure 5 EDX data for CdCl$_2$ treated CdSe film showing Cl at the grain boundaries and front CdSe/FTO interface.

1.8 μm ±0.5 μm. The XRD pattern shows a small peak at 25° and a broad peak at 46° for the untreated sample. The peak at 46° relates to the (103)$_h$ hexagonal Miller plane. After CdCl$_2$ treatment the (103)$_h$ peak has a much greater intensity and narrower FWHM. The peak at 25° has also increased in intensity, this relates to the (002)$_h$ hexagonal Miller plane. Photoluminescence spectra of evaporated CdSe films before and after CdCl$_2$ treatment are presented in Figure 6. As deposited CdSe has a small, broad peak at 720 nm. After CdCl$_2$ treatment the CdSe peak is much more intense at the 719 nm bandgap. A small shoulder at 650 nm is an artefact from the measurement laser. This increase in the PL intensity is due to

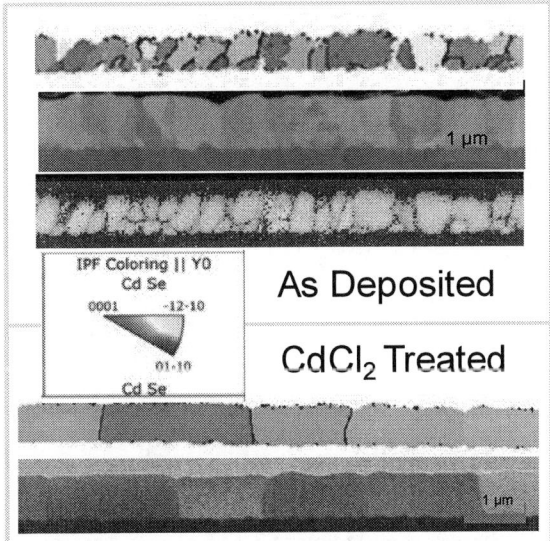

Figure 2 EBSD data for evaporated CdSe before and after CdCl$_2$ treatment. Texture changes observed and dramatic grain growth after CdCl$_2$ treatment.

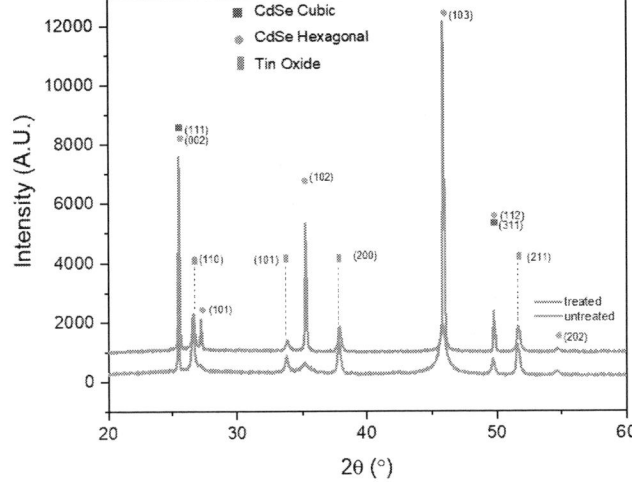

Figure 4. XRD patterns of CdSe deposited by evaporation and treated with a CdCl$_2$ treatment. The CdSe is hexagonal as deposited and after CdCl$_2$ treatment, with more resolved peaks after treatment.

978-1-6654-6060-6/23 $31.00 © 2023 IEEE

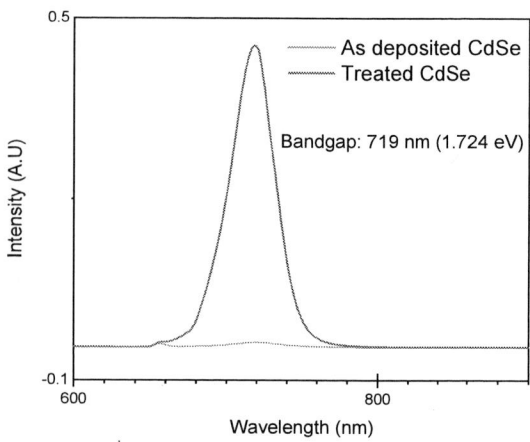

Figure 6. PL spectra for CdSe untreated and treated. After treatment the luminescence intensity at the bandgap increased due to defect removal.

the removal of non-radiative defects and the reduction in the number of grain boundaries due to the larger grain size. The lifetime of the as deposited CdSe in the TRPL has one very fast decay mode – $\tau 1$ at 0.26 ns and cannot be separated from the instrument response. After $CdCl_2$ treatment the CdSe lifetime has two decay modes $\tau 1$ 0.812 ns at and $\tau 2$ at 3.243 ns. The $CdCl_2$ treatment has caused a longer lifetime in the CdSe film, indicating the removal of recombination centres within the grains.

IV. CONCLUSIONS

The $CdCl_2$ treatment of thin film CdSe causes recrystallisation and reorientation. The grain size is increased dramatically. The treatment causes chlorine to decorate and passivate the grain boundaries. HRTEM analysis shows that high densities of stacking faults are removed and EBSD shows the treated films are purely hexagonal. As a result of the improvements in microstructure, the films become much more photoactive and minority carrier lifetimes increase significantly. The $CdCl_2$ treatment of CdSe/CdTe bilayers is an important process in achieving high efficiency in resulting CdSeTe/CdTe devices.

V. ACKNOWLEDGEMENTS

The authors from Loughborough University are grateful to UKRI and EPSRC for funding the project through EPW00092X/1. The authors from University of Toledo acknowledge the research sponsored by the U. S. DOE's office of Energy Efficiency and Renewable Energy (EERE) under Solar Energy Technologies Office (SETO), Cadmium Telluride Accelerator Consortium (CTAC) (NREL Sub-contract SUB-2021-10715), and Air Force Research Laboratory under agreement number FA9453-19-C-1002. The U.S. Government is authorized to reproduce and distribute reprints for Governmental purposes not withstanding any copyright notation thereon. Disclaimer: The views and conclusions contained herein are those of the authors and should not be interpreted as necessarily representing the official policies or endorsements, either expressed or implied, of the Department of Energy or Air Force Research Laboratory or the U.S. Government.

REFERENCES

[1] N. R. Paudel and Y. Yan, "Enhancing the photo-currents of CdTe thin-film solar cells in both short and long wavelength regions," *Appl. Phys. Lett.*, vol. 105, no. 18, p. 183510, Nov. 2014,

[2] X. Zheng, E. Colegrove, J. N. Duenow, J. Moseley, and W. K. Metzger, "Roles of bandgrading, lifetime, band alignment, and carrier concentration in high-efficiency CdSeTe solar cells," *J. Appl. Phys.*, vol. 128, no. 5, p. 053102, Aug. 2020,

[3] T. Leijtens, K. A. Bush, R. Prasanna, and M. D. McGehee, "Opportunities and challenges for tandem solar cells using metal halide perovskite semiconductors," *Nat. Energy 2018 310*, vol. 3, no. 10, pp. 828–838, Jul. 2018,

[4] K. Li *et al.*, "Fabrication and Optimization of CdSe Solar Cells for Possible Top Cell of Silicon-Based Tandem Devices," *Adv. Energy Mater.*, vol. 12, no. 26, pp. 1–11, 2022,

[5] J. D. Friedl, R. H. Ahangharnejhad, A. B. Phillips, and M. J. Heben, "Material Requirements for CdSe Wide Bandgap Solar Cells," *Conf. Rec. IEEE Photovolt. Spec. Conf.*, pp. 1548–1552, Jun. 2021,

[6] N. T. Shelke, S. C. Karle, and B. R. Karche, "Photoresponse properties of CdSe thin film photodetector," *J. Mater. Sci. Mater. Electron.*, vol. 31, no. 18, pp. 15061–15069, 2020,

[7] D. Kuciauskas, J. Moseley, P. Ščajev, and D. Albin, "Radiative Efficiency and Charge-Carrier Lifetimes and Diffusion Length in Polycrystalline CdSeTe Heterostructures," *Phys. status solidi – Rapid Res. Lett.*, vol. 14, no. 3, p. 1900606, Mar. 2020,

[8] T. A. M. Fiducia *et al.*, "Understanding the role of selenium in defect passivation for highly efficient selenium-alloyed cadmium telluride solar cells," *Nat. Energy*, vol. 4, no. 6, pp. 504–511, 2019,

[9] T. Fiducia *et al.*, "Selenium passivates grain boundaries in alloyed CdTe solar cells," *Sol. Energy Mater. Sol. Cells*, vol. 238, p. 111595, May 2022,

[10] K. M. Morris, C. Potamialis, F. Bittau, J. W. Bowers, and J. M. Walls, "Chemical bath deposition of thin film CdSe layers for use in Se alloyed CdTe solar cells," in *2019 IEEE 46th Photovoltaic Specialists Conference (PVSC)*, Jun. 2019, pp. 1857–1862.

[11] T. Baines *et al.*, "Incorporation of CdSe layers into CdTe thin film solar cells," *Sol. Energy Mater. Sol. Cells*, vol. 180, no. February, pp. 196–204, Jun. 2018,

[12] R. Greenhalgh *et al.*, "The Origins of Void formation in Sputtered CdSe," in *2021 IEEE 48th Photovoltaic Specialists Conference (PVSC)*, Jun. 2021, pp. 886–889.

[13] T. Hussain, M. F. Al-Kuhaili, S. M. A. A. Durrani, and H. A. Qayyum, "Effect of collision during vapor transport between Cd and X (X = Te2, Se2, or S2) molecules on the properties of thermally evaporated CdTe, CdSe, and CdS thin films," *Results Phys.*, vol. 8, pp. 988–1000, 2018,

[14] S. Mathuri, K. Ramamurthi, and R. Ramesh Babu, "Effect of Sb incorporation on the structural, optical, morphological and electrical properties of CdSe thin films deposited by

electron beam evaporation technique," *Thin Solid Films*, vol. 660, pp. 23–30, Aug. 2018,

[15] A. Abbas *et al.*, "The effect of a post-activation annealing treatment on thin film CdTe device performance," in *2015 IEEE 42nd Photovoltaic Specialist Conference (PVSC)*, Jun. 2015, pp. 1–6.

[16] J. D. Poplawsky *et al.*, "Direct imaging of Cl and Cu induced short-circuit efficiency changes in CdTe solar cells," *Adv. Energy Mater.*, vol. 4, no. 15, pp. 1–8, 2014,

[17] R. Greenhalgh *et al.*, "Effect of Microstructure on the Photoactivity of Thin Film CdSe," in *2022 IEEE 49th Photovoltaics Specialists Conference (PVSC)*, 2022, pp. 0900–0903.

[18] S. Kumari, G. Chasta, R. Sharma, N. Kumari, and M. S. Dhaka, "Phase transition correlated grain growth in CdSe thin films: Annealing evolution to cadmium chloride activation," *Phys. B Condens. Matter*, vol. 649, p. 414422, Jan. 2023,

[19] V. Tsai, F. Bittau, C. Potamialis, M. Bliss, T. R. Betts, and R. Gottschalg, "Combined Electrical and Optical Characterisation of Recombination Mechanisms and Minority Carrier Lifetime in Solar Cells," in *Conference Proceedings of the 14th Photovoltaic Science, Applications and Technology Conference (PVSAT-14), London, UK, 18-19 April 2018.*, 2018.

High Open Circuit Voltage with Organic Hole Transport Layers in Group V Doped CdSeTe Solar Cells

Sabin Neupane[1], Deng Bing Li[1], Sandip S Bista[1], Suman Rijal[1], Zhaoning Song[1], Alisha Adhikari[1], Lei Chen[1], You Li[1], Manoj K. Jamarkattel[1], Abasi Abudulimu[1], Dingyuan Lu[2], Xiaomeng Duan[3], Feng Yan[3], Michael Heben[1], Randy J. Ellingson[1], Gang Xiong[2], and Yanfa Yan[1].

[1] Wright Center for Photovoltaic Innovation and Commercialization, Department of Physics and Astronomy, University of Toledo, Toledo, OH, 43606, USA

[2] California Technology Center, First Solar Inc. 1035 Walsh Ave, Santa Clara, CA, USA

[3] School for Engineering of Matter, Transport, and Energy, Arizona State University, Tempe, AZ, USA

Abstract— **Effective hole transport layer (HTL) is one of the key factors for the highly efficient cadmium selenium telluride (CdSeTe) solar cells. Here, we present the effect of different organic hole transport layers as HTL in CdSeTe solar cells, including Spiro-OMeTAD, PTAA, PEDOT: PSS, and P3CT. With all these HTLs, the device shows significantly improved open circuit voltages ($V_{OC}s$). Especially, with P3CT, Spiro-OMeTAD, and PTAA as HTLs, the $V_{OC}s$ can reach a maximum of 866, 857, and 869 mV, respectively, which are greater than the control sample without any HTL.**

Keywords— CdSeTe, hole transport layer, Spiro: OmeTAD, PEDOT: PSS, PTAA, P3CT

I. INTRODUCTION

Solar energy is one of the most reliable green energies. In the solar market, there are different types of solar cells like crystalline silicon, copper indium gallium selenide (CIGS), cadmium telluride (CdTe), etc. Among the commercialized solar technologies, CdTe-based solar cell is one of the most cost-effective technologies. The popularity of CdTe solar cells are near the zenith as these cells can be made in both rigid and flexible substrates with a reasonably high efficiency [1, 2]. CdTe-based solar cells with copper (Cu) as the doping element has reached 20% PCE with 863 mV V_{OC} using copper thiocyanate (CuSCN) as HTL in academic institute [3-6]. The leading CdTe solar cell commercializing company First Solar has reported more than 22% PCE with 887 mV V_{OC} for copper-doped CdTe solar cells with ZnTe as back contact [7]. However, these solar cells are still well below their theoretical limit above 30%. This is mainly because the V_{OC} is much lower than the maximum V_{OC} of 1.156 V as per theoretical calculations, resulting in a large V_{OC} deficit [8].

To reduce the V_{OC} deficit, one approach is to replace the Cu dopant with group V elements, the other is to use a new HTL with better band alignment with CdTe to reduce the back barrier height. Recently, first solar reported 22% efficient in-situ As-doped CdTe device [9]. However, the activation of in-situ doped As requires careful engineering [10] of the absorber and could potentially induce front-interface recombination. There is a trade-off between a high carrier (hole) concentration and a long carrier lifetime. A low-temperature ex-situ group V chloride doping recently was reported to produce efficient, copper-free devices [11]. Although the ex-situ group V doping preserves the front interface, due to the limited doping concentration, the V_{OC} reported through ex-situ group V doping is only 863 mV, on the bar with the state-of-the-art CdTe-based solar cells and much lower than the maximum V_{OC} of the theoretical calculations [8]. This large deficit implies a big room for improving V_{OC}. To avoid this V_{OC} mismatch, hole transport layers with better band alignment with CdTe are promising to further improve the V_{OC}. In CdTe solar cells, different inorganic HTLs have been investigated for this purpose but the V_{OC} is still limited below 862 mV [12-15]. Comparing with inorganic HTLs, organic HTLs have rarely been investigated in CdTe-based solar cells. Here, we investigated different organic HTLs in group V doped CdSeTe solar cells and demonstrated a maximum V_{OC} of 869 mV.

II. EXPERIMENTAL DETAILS

We did the ex-situ group V doping on the CdCl₂ treated sample prepared by First Solar Inc. Then the sample was treated with different organic HTLs. For the 1″×1″ sample, 30 μL of Spiro-OMeTAD was spread on the surface. Then, it is spin-coated at a speed of 2,000 RPM for 30 Seconds. For the devices with PEDOT as HTL, 100 μL PEDOT: PSS solution was spin-coated dynamically at a speed of 2,000

RPM for 30 seconds. For the sample with P3CT as HTL, 60 μL P3CT was spin-coated at a spin speed of 3,000 RPM for 30 sec. For the sample with PTAA as HTL, we used dynamic spin-coating by taking 30μL of PTAA at a speed of 3,00RPM for 30 sec. Then, all the samples were dried for 10-20 minutes at 100°C. Gold (Au) was evaporated as the electrode to finish the device. The sample was soaked under 1 sun illumination from the back side for 30 minutes before performing the characterization.

III. RESULTS AND DISCUSSION

To confirm the improvement of the carrier collection efficiency, steady-state photoluminescence (PL) spectra measurements were performed from the film side of the sample without and with different HTLs (Fig. 1). From PL spectra, we can see two peaks for all five samples with the weak emission at 825 nm and the dominant one at 875 nm, which are band to band emissions of pure CdTe and CdSeTe, respectively. For both peaks, the intensities are higher in the device without HTL than that with HTLs. Among all the samples with HTLs, the one with PTAA has the lowest emission peak at 875 nm, indicating the beneficial quenching effect and highest carrier collection efficiency.

Fig 1: PL intensity of the different HTLs and control

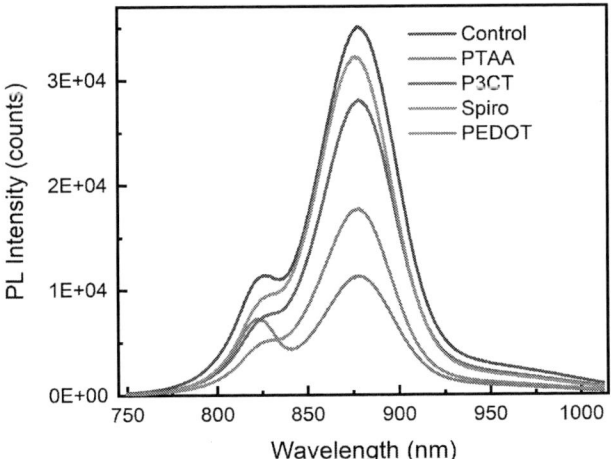

In figure 2, we have shown the analysis of JVT measurement. The barrier height is minimum with 398mV for the device with spiro-OMeTAD as HTL which explains a good Voc and a not bad fill factor for that device. Also, it is the maximum for PEDOT at 491mV. The control device without any HTL has a barrier height of 425mV. The device with PTAA as HTL has a barrier height of 439 mV.

To demonstrate the improved carrier collection efficiencies in the devices with HTLs, five devices were fabricated without and with different HTLs. The performance of all the devices were summarized in Table 1. We can see that the V_{OC} increases in the devices with the use of HTLs. Unfortunately, the FFs decreased for all the HTL-caped devices. This is probably due to the low conductivity of the

HTL materials, which might be amended by reducing the HTL thickness or improving the conductivity through doping optimization. Among all the devices, the highest V_{oc} of 869 mV was obtained in the device with PTAA as HTL, but the fill (FF) of the device decreased. Thereby, the overall device performance is lower than the control device without HTL. Next step, we will try to reduce the PTAA layer thickness or optimize the doping by lithium to improve the FF. Comparing the device with different HTLs, the devices with PTAA show decent V_{OC} and FF with overall improved device performance.

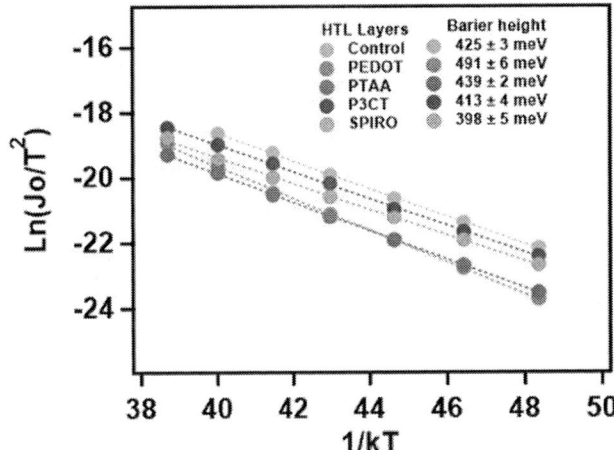

Fig. 2: JVT measurement and analysis

HTL	Voc		Jsc		PCE		FF	
	Max mV	Avg mV	Max mA/cm²	Avg mA/cm²	Max %	Avg %	Max %	Avg %
No HTL (Control)	856	849±4	28.64	27.48±0.88	16.24	15.46±0.49	69.74	66.26±2.08
SPIRO:OMeTAD	857	854±4	27.44	26.53±0.85	16.30	15.18±1.00	69.29	66.92±2.28
P3CT	866	852±1	28.96	28.54±0.34	15.70	14.75±0.46	63.34	60.98±1.26
PTAA	869	863±6	28.35	27.54±0.44	16.07	15.57±0.35	66.61	65.49±0.71
PEDOT:PSS	860	849±8	29.45	26.44±4.70	16.35	14.65±2.47	67.78	65.44±1.22

Table 1: Statistical performance of the devices without (control) and with different HTLs: (a) V_{OC}, (b) J_{SC}, (c) Efficiency, and (d) Fill Factor.

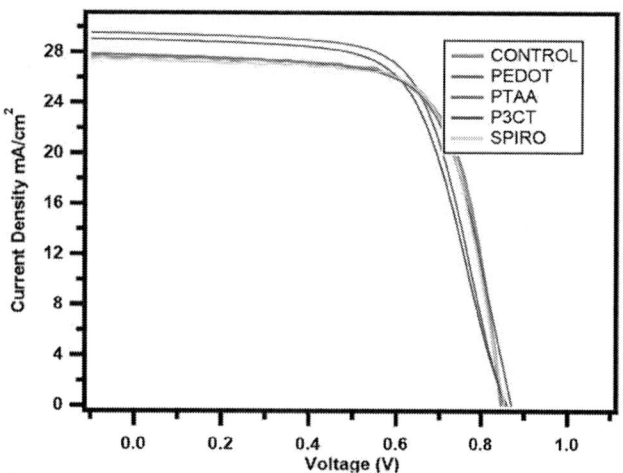

Fig.3: JV curves of the devices without HTL (control) and with different HTLs.

The performance of the devices without and with different HTLs can further be demonstrated from the JV curves in Fig. 3. The device without the HTLs produces the lowest V_{OC} of 856 mV but the highest FF of ~70%. While the devices with PEDOT: PSS, P3CT, and spiro-OMeTAD produce improved V_{OC}s of 860, 866, and 857 mV and decreased FFs of 68%, 69%, and 63%, resulting in overall reduced average PCEs of 14.65%, 15.18%, and 14.75%, respectively, comparing with the device without HTL. Only the device with PTAA show improved device performance with only slightly reduced FF. To further improve the performance of the devices with organic HTLs, the conductivity and/or thickness of the HTL layer should be further optimized.

IV. CONCLUSION

Through the comparison of the devices without and with different organic HTLs, we demonstrated that organic HTLs can further improve the V_{OC} of group V-doped CdTe solar cells, demonstrating an impressive maximum V_{OC} of 869 mV. Although the overall performance of the devices with organic HTLs doesn't show significant improvement compared with the device without HTL due to the reduced FFs, the exploration of organic HTLs is promising with improved carrier collection at the back surface. After systematical optimization of the HTL material conductivity and/or the deposition thickness, the application of organic HTLs has the potential to make a breakthrough contribution to the CdTe community.

ACKNOWLEDGMENTS

This work is sponsored by the U.S. DOE's Office of Energy Efficiency and Renewable Energy (EERE) under the Solar Energy Technologies Office (SETO), through Agreement DE-EE0008974 and DE-EE0009832, and the Alliance for Sustainable Energy, LLC, Managing and Operating Contractor for the National Renewable Energy Laboratory for the U.S. Department of Energy, under Award Number 37989. The U.S. Government is authorized to reproduce and distribute reprints for Governmental purposes notwithstanding any copyright notation thereon. The views expressed are those of the authors and do not reflect the official guidance or position of the United States Government, the Department of Defense, or the United States Air Force. The appearance of external hyperlinks does not constitute an endorsement by the United States Department of Defense (DoD) of the linked websites, or the information, products, or services contained therein. The DoD does not exercise any editorial, security, or other control over the information you may find at these locations.

REFERENCES

[1] S. Bista, "Water-Assisted Lift-Off Process for Flexible CdTe Solar Cells," *ACS Applied Energy Materials,* 2023.

[2] S. S. Bista *et al.*, "Flexible and Lightweight CdS/CdTe Solar Cells via a Water-Assisted Lift-Off Process," in *2022 IEEE 49th Photovoltaics Specialists Conference (PVSC)*, 2022: IEEE, pp. 0464-0466.

[3] D.-B. Li *et al.*, "20%-efficient polycrystalline Cd(Se,Te) thin-film solar cells with compositional gradient near the front junction," *Nature Communications,* vol. 13, no. 1, p. 7849, 2022/12/21 2022, doi: 10.1038/s41467-022-35442-8.

[4] S. S. Bista *et al.*, "Effects of Cu Precursor on the Performance of Efficient CdTe Solar Cells," *ACS Applied Materials & Interfaces,* vol. 13, no. 32, pp. 38432-38440, 2021/08/18 2021, doi: 10.1021/acsami.1c11784.

[5] D. B. L. e. all, "D. -B. Li et al., "Fabricating Efficient CdTe Solar Cells: The Effect of Cu Precursor,," *IEEE Specialists conference,* vol. 48th, 2021.

[6] D.-B. Li *et al.*, "Oxygen Management to Avoid Photo-Inactive Cd (S, Se) for Efficient Cd (Se, Te) Solar Cells," *ACS Energy Letters,* vol. 8, no. 3, pp. 1529-1534, 2023.

[7] M. A. Green *et al.*, "Solar cell efficiency tables (Version 61)," *Progress in Photovoltaics: Research and Applications,* 2003.

[8] R. M. Geisthardt, M. Topič, and J. R. Sites, "Status and potential of CdTe solar-cell efficiency," *IEEE Journal of photovoltaics,* vol. 5, no. 4, pp. 1217-1221, 2015.

[9] W. K. Metzger *et al.*, "As-Doped CdSeTe Solar Cells Achieving 22% Efficiency With− 0.23%/° C Temperature Coefficient," *IEEE Journal of Photovoltaics,* vol. 12, no. 6, pp. 1435-1438, 2022.

[10] M. K. Jamarkattel *et al.*, "Incorporation of Arsenic in CdSe/CdTe Solar Cells During Close Spaced Sublimation of CdTe:As," in *2020 47th IEEE Photovoltaic Specialists Conference (PVSC)*, 15 June-21 Aug. 2020 2020, pp. 2605-2608, doi: 10.1109/PVSC45281.2020.9300772.

[11] D.-B. Li *et al.*, "Low-temperature and effective ex situ group V doping for efficient polycrystalline CdSeTe solar cells," *Nature Energy,* vol. 6, no. 7, pp. 715-722, 2021.

[12] S. S. Bista *et al.*, "Solution-Processed Copper Selenium Oxide (CuSeO_{3}) as Hole Transport Layer for CdS/CdTe Solar Cells," in *2022 IEEE 49th Photovoltaics Specialists Conference (PVSC)*, 2022: IEEE, pp. 1170-1172.

[13] K. P. Bhandari *et al.*, "Thin film iron pyrite deposited by hybrid sputtering/co-evaporation as a hole transport layer for sputtered CdS/CdTe solar cells," *Solar Energy Materials and Solar Cells,* vol. 163, pp. 277-284, 2017.

[14] D. Pokhrel *et al.*, "Copper iodide nanoparticles as a hole transport layer to CdTe photovoltaics: 5.5% efficient back-illuminated bifacial CdTe solar cells," *Solar Energy Materials and Solar Cells,* vol. 235, p. 111451, 2022.

[15] S. Rijal *et al.*, "Post - annealing Treatment on Hydrothermally Grown Antimony Sulfoselenide Thin Films for Efficient Solar Cells," *Solar RRL,* 2022.

Thin-Film Tandem Partners based on Inline-Processed (Ag,Cu)(In,Ga)Se$_2$

Theresa Magorian Friedlmeier, Rico Gutzler, Tina Wahl, Dimitrios Hariskos, Stefan Paetel, Erik Ahlswede, Ana Kanevce, Jan-Philipp Becker

Zentrum für Sonnenergie- und Wasserstoff-Forschung Baden-Württemberg (ZSW), Stuttgart, Germany

Abstract — (Ag,Cu)(In,Ga)Se$_2$-based solar cells can be optimized for different applications by adjusting the composition. A wide range of compositions has been explored in a production-relevant deposition line with minor adjustments to the process. The best efficiencies are achieved for the (Ag,Cu)(In,Ga)Se$_2$ material optimized for single-junction applications, where most optimization efforts have been directed. For tandem devices the achievable bandgap range is suitable for both top and bottom cells. In both cases, material quality must be improved and further modifications to the growth process are necessary. For the bottom cell, standard and reduced-Ga compositions could already be successfully implemented in high-efficiency two-terminal tandems with solution-processed perovskite top cells. In this contribution we review our R&D line with mostly in-line deposition steps, discuss the performance of chalcopyrite solar cells at different compositions, and present modifications which are beneficial for the development of tandem devices. Best results for completely in-house processing are over 18 % for two-terminal perovskite/(Ag,Cu)(In,Ga)Se$_2$ tandems and over 26.6 % for the four-terminal configuration.

Index Terms—ACIGS, (Ag,Cu)(Ga,In)Se$_2$, perovskites, thin-film solar cells, tandem solar cells

I. INTRODUCTION

Thin-film solar cells and modules based on the chalcopyrite compound family (Ag,Cu)(In,Ga)Se$_2$ (ACIGS) have been developed to achieve very high efficiencies as laboratory devices and a high degree of manufacturability as full-size modules [1]. The current records are 23.6 % for cells [2], and 20.3 % for small modules [3] and 17.6 % for large modules [4].

One advantage of the ACIGS materials family is the wide range of bandgaps (E_g) which are accessible by adjusting the composition: from ca 1.0 eV for CuInSe$_2$ to ca 1.7 eV for CuGaSe$_2$, with a further slight increase accompanying substitution of Cu by Ag up to 1.78 eV for AgGaSe$_2$ [5]. Composition gradients with increased E_g at the front and back interfaces are employed to reduce recombination while providing a lower E_g near the space-charge region to improve absorption of lower-energy photons. Most research efforts to date have been directed at the development of ACIGS absorber layers for single-junction cells and modules. The highest efficiencies for laboratory ACIGS solar cells are for integral compositions with GGI = [Ga]/([Ga]+[In]) near 0.3 and ACGI = ([Ag]+[Cu])/([Ga]+[In]) in the range of 0.8 to 0.9.

In tandem solar cells, two or more subcells with different bandgaps are stacked on top of each other to boost overall cell efficiency beyond the maximum achievable by a single-junction cell by a more efficient use of the solar spectrum. For the most optimal performance, the bottom cell should have a bandgap around 1.0 eV to 1.2 eV and top cell a bandgap of 1.6 eV to 1.8 eV [6].

Development of ACIGS absorber layers in tandem structures requires optimization of the deposition process for suitable compositions – either in the high-E_g region for the top cell, or in the low-E_g region for the bottom cell. In both cases the optimization is not just for the cell itself, but together with the intended tandem partner device. Further considerations include the choice of either two-terminal (2T) or four-terminal (4T) tandem configuration. In this contribution we will briefly describe the device processing, addressing the performance for different compositions. Then we will review the current status and challenges involved with application of ACIGS absorbers in the top cell of a tandem, followed by a closer look at the technological developments involved in optimizing ACIGS as a bottom cell with solution-deposited perovskite as a top cell. In all cases the ACIGS deposition and process development is performed in a high-throughput inline coating tool, ensuring good transferability to industry.

II. EXPERIMENTAL

In this work, the standard layer stack for ACIGS solar cells is: soda-lime glass (3 mm)/Mo (550 nm)/ACIGS (2000 to 2500 nm)/CdS or Zn(O,S) (30 to 50 nm)/high-resistive layer of undoped ZnO or (Zn,Mg)O (80 nm)/TCO (200 to 300 nm)/Ni-Al-Ni grid (2.5 µm). The cell size is defined by mechanical scribing to 0.5 cm^2. The standard TCO (transparent conductive oxide) is Al-doped ZnO (AZO), with optional indium zinc oxide (IZO) or H-doped In$_2$O$_3$ (IO:H). Most baseline depositions are in inline equipment dimensioned for sample sizes up to 30×30 cm^2. Specifically, the sputtered contact layers Mo and AZO, the sputtered second buffer ZMO, as well as the absorber layer which is coated by coevaporation of the elements in a multistage process with RbF post-deposition treatment as an option are all deposited in industrially relevant inline systems. The batch processes are the buffer layer by chemical bath deposition and the grid by e-gun evaporation through a

978-1-6654-6060-6/23 $31.00 © 2023 IEEE

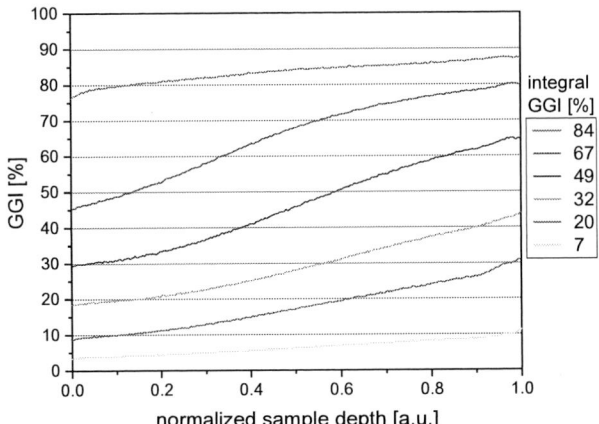

Fig. 1. Examples for GGI composition profiles in ACIGS layers (here without Ag) generated by the inline coevaporation process and resulting in the labeled integral GGI values. The data is acquired by GDOES and calibrated with XRF data. Sample depth 0 refers to the front interface to the buffer layer, and 1 to the Mo back contact.

mask. Consistent labelling of sample IDs and reference windows for all processes support a stable baseline despite the wide variety of experimental research. Ref. [7] provides a detailed description of the laboratory production line and processing steps.

The ACIGS coater is a special proprietary design with linear evaporation sources in a top-down configuration for uniform coating over 30 cm width. It is designed to be as close as possible to a production coater while still allowing research flexibility. In this system, the sources for the ACIGS deposition are distributed between two chamber sections to enable a multi-stage process similar to the commonly used static three-stage deposition process. Typical nominal substrate temperatures are 450 °C for the first stage, and 580 to 650 °C for ACIGS in the second and third stages, with the lower temperatures employed for processes including Ag. Separated by units for temperature adjustment, an additional chamber is set up for optional finishing steps like post-deposition treatment (PDT) by evaporation of RbF and Se or a capping layer.

Fig. 1 illustrates examples for the final GGI composition profile of the ACIGS layer for different integral GGI values (this data is without Ag, which tends to flatten the profile), as measured by GDOES (glow discharge optical emissions spectroscopy) and calibrated by XRF (X-ray fluorescence spectroscopy). The resulting profile is largely but not solely determined by the relative flux rates of the elements during the deposition process as well as the substrate temperatures. Modification of the front gradient is accomplished by adjusting the relative flux rates for In and Ga in the third stage.

The final ACIGS layer is typically 2.2 μm thick which can be modified by adjusting the flux rates and/or the speed of substrate transportation through the coater, with ACIGS growth typically finishing in 26 min. Up to 8000 nominally identical cells can be produced in one ACIGS coater run, with up to 80 test substrates with 10 cells on each of 10 carriers. Two reference devices at the front and back of each carrier document any intended or unintended variations. For example, short breaks between carrier passage can accommodate process adjustments within one run, e.g. for a monotone variation of composition or a comparison of samples with and without RbF post-deposition treatment while limiting other process-related variability.

Reference samples and devices are generated at various stages of the ACIGS processing line and analyzed for quality-control purposes. Typically, every experiment is tagged with a characterization of the individual processes, with the assumption that other samples in the same deposition run have the same properties as the references. This includes conductivity and, where applicable, transmittance of the sputter contact and high-resistive layers, and integral composition of the ACIGS layer by XRF, as well as current-voltage (IV) performance at standard testing conditions (STC).

For 2T tandems, wide-gap perovskite cells are deposited directly on the ACIGS device (no grid). AZO exhibits strong free-carrier absorption in the infrared so IZO or IO:H are used as the TCO for ACIGS bottom cells. For the perovskite cell, we start with a NiO_x/(2-(3,6-Dimethoxy-9H-carbazol-9-yl)ethyl)phosphonic acid (MeO-2PACz) hole conductor and transparent recombination contact, followed by a monolayer of SiO_x nanoparticles for improved wetting of the solutions applied to form the $FA_{0.83}Cs_{0.17}Pb(I_{0.87}Br_{0.13})_3$ perovskite absorber layer with a bandgap of 1.63 eV. A phenethylammonium iodide (PEAI) defect passivation layer

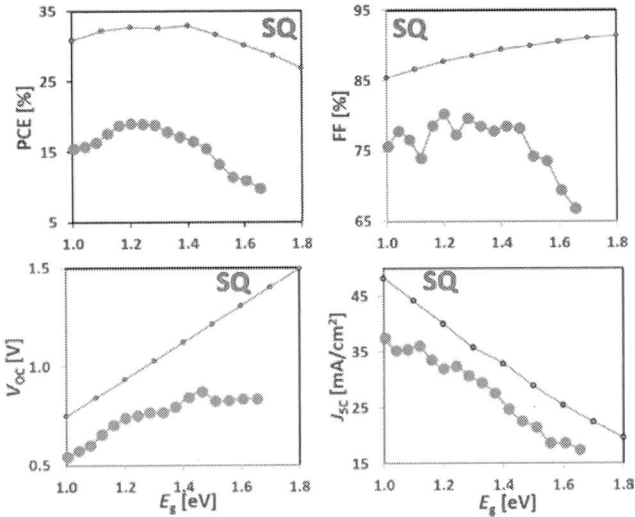

Fig. 2. The solar parameters PCE, fill factor, J_{SC} and V_{OC} for the best-performing 0.5-cm² devices for each bandgap E_g as calculated from the integral GGI as measured by XRF (orange circles). Special features like Ag content or RbF-PDT are included in the data when they lead to the highest efficiencies. The theoretical maximum values as calculated from the Shockley-Queisser (SQ) model are included for comparison (green circles, from Ref. [8]).

follows and the device is finished with [6,6]-Phenyl-C61-butyric acid methyl ester (PCBM) and bathocuproine (BCP) electron transport layers as well as a sputtered IO:H transparent front contact. The layers are deposited by spin-coating, doctor-blading, or slot-die coating, except for NiO$_x$ (ALD) and the TCO (sputtered). The contact grid improves current transport and an anti-reflection coating of MgF$_2$ is included for best devices (both evaporated).

II. RESULTS

A. Single Junction Devices

Reference devices are generated for every carrier of the ACIGS coating process, keeping all other processes within standard windows. The orange circles in Fig. 2 demonstrate the solar parameters open-circuit voltage (V_{OC}), short-circuit current density (J_{SC}), fill factor (FF), and power conversion efficiency (PCE) of the best devices for each GGI, representing the best of nearly 50,000 devices measured at standard testing conditions (STC). Most of these highest-efficiency devices do not contain Ag. For comparison, the maximal values for single-junction solar cells according to the Shockley-Queisser model are included with small green symbols and using the right-side axis [8]. Well-known trends linked to the change in bandgap like the decrease of J_{SC} and the corresponding increase of V_{OC} to higher GGI, and also that V_{OC} drops again for GGIs exceeding ca. 0.7, are apparent in this dataset. This drop is generally attributed to reduced material quality and non-optimal band alignment with the buffer layer. Loss analysis of our experiments with ACIGS absorber layers across all GGIs indicates increased non-radiative losses [7]. The best STC efficiencies of 19.0 % without anti-reflective coating are for cells with GGI around 0.3 and ACGI around 0.88 and employing RbF post-deposition treatment.

B. Top Cell for Tandems

This dataset spans the bandgap range of ca 1.0 eV to 1.7 eV, thus including materials suitable for the top and/or bottom partner of a tandem photovoltaic device. A suitable widegap ACIGS top cell must be highly efficient and at the same time transmit sufficient photons for conversion by the bottom cell. The ACIGS technology will need significant developments to perform well as a top cell device as the efficiencies tend to be lower for high GGI (see Fig. 2) while also employing a transparent back contact that is subjected to the ACIGS deposition process. Recently, promising results for bifacial ACIGS cells employing indium tin oxide as a transparent back contact have been published [9]. The authors reported that including Ag in the chalcopyrite layer allowed substrate temperature reduction while maintaining high film quality and thus reduced impact on the transparent back contact, especially reducing or preventing the formation of GaO$_x$ at the ACIGS/ITO interface. They achieved 19.77 % with a V_{OC} of 708 mV with front illumination and 10.89 % with rear illumination. Results of 12 % on an IO:H transparent back contact with an E_g of 1.45 eV and a V_{OC} of 835 mV have also been reported [10].

Besides transferring the ACIGS deposition process to transparent back contacts, a successful top-cell development requires high efficiencies at high bandgaps. V_{OC} losses to higher GGI is related to interface recombination which can be mitigated by appropriately aligned band offsets and the junction interface. Guidelines for materials selection are presented in Ref. [11].

These promising results indicate the potential for ACIGS top cells which could partner e.g. with a standard silicon wafer-based cell or another ACIGS cell but with low bandgap. Such tandem configurations with mature anorganic technologies have advantages in terms of long-term stability.

C. Bottom Cell for Tandems

In contrast to the top cell, the bottom cell of a tandem device can have an opaque back contact, so the well-studied processes with a Mo back contact and Na supply through the glass can be applied. As shown in Fig. 2, the ACIGS efficiencies also maintain a relatively high level at lower GGI. The combination of low-bandgap ACIGS and high-bandgap perovskite solar cells is already now suitable for application in tandem devices, taking advantage of the high efficiencies achievable with high-bandgap perovskite materials.

The bottom cell of a tandem device can only convert the remaining sunlight which has passed through the top cell

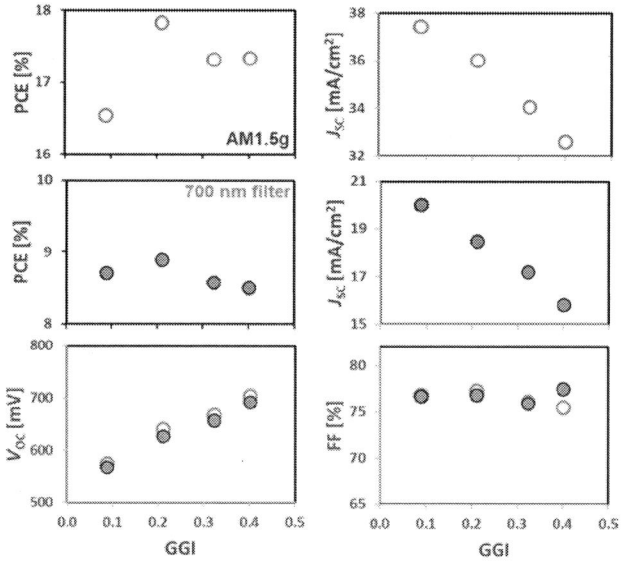

Fig. 3. Solar cell parameters for efficiency PCE, FF, J_{SC} and V_{OC} for a series of devices with different Ga contents GGI. The open blue symbols indicate values at STC, the filled orange ones are measured under a 700-nm longpass filter to mimic a perovskite top cell (but PCE still referenced to AM1.5g).

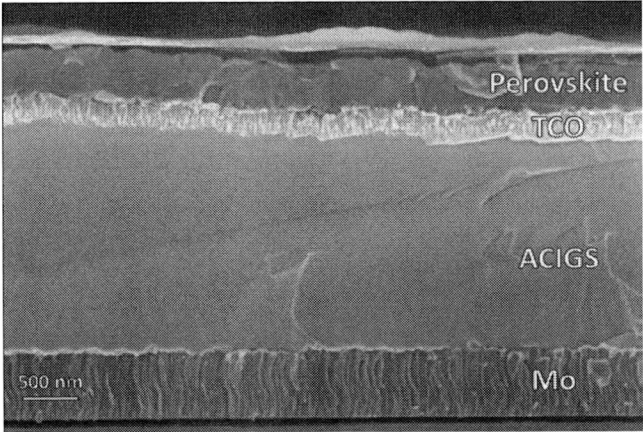

Fig. 4. Cross-section SEM image of a perovskite/ACIGS tandem solar cell with a particularly large-grained and flat ACIGS layer.

(unless in bifacial configuration). The infrared performance is therefore crucial. In order to extend the absorption range towards the infrared, the bandgap of the ACIGS layer is decreased by reducing the gallium content. Fig. 3 shows results from a sample series processed in one ACIGS coating run, without Ag and with a GGI variation. The solar parameters are measured at STC and with a 700-nm longpass cut-off filter to simulate the optical absorption in the top cell and thus obtain bottom cell performance. These devices have a Zn(O,S) buffer layer for higher efficiency at low GGI and an AZO front contact, leaving room to improve photocurrent by switching to IZO instead. The expected behavior of decreasing J_{SC} and increasing V_{OC} with increased GGI and therefore increased bandgap is observed. The PCE is highest for the GGI of 0.21 in this dataset. Filtered illumination leads to a large reduction in J_{SC} of ca 17 mA/cm^2 since fewer photons are available to convert, and this in turn causes a slight decrease in V_{OC} due to reduced Fermi-level splitting. The fill factor is unaffected by filtering. Thus, the performance of the devices under filtered light is reduced by the same amount as the photocurrent, regardless of GGI in this range. STC performance is therefore a good indication of bottom-cell performance. Looking at Fig. 2 it is apparent that low-GGI ACIGS solar cells underperform in both V_{OC} and J_{SC}, so both material quality and optical efficiency need to be improved.

For the 2T tandem devices with a perovskite top cell, the morphology of the bottom cell can disturb the top-cell deposition when using non-conformal, solution-based processes. The roughness of the ACIGS grains plays a large role. One improvement is to include Ag in the absorber layer because it leads to larger and flatter grains. However, too much Ag can lead to phase separation during the growth process. This should be avoided as it leads to the formation of crevices and causes short-circuiting between the cells, see Ref [12]. Fig. 4 shows a scanning electron microscope (SEM) image of a two-terminal perovskite/ACIGS tandem solar cell in cross-section (cleaved). In this case the device stack ist Mo/ACIGS/CdS (35 nm)/i-ZnO/IZO (105 nm), followed by NiO$_x$/MeO-

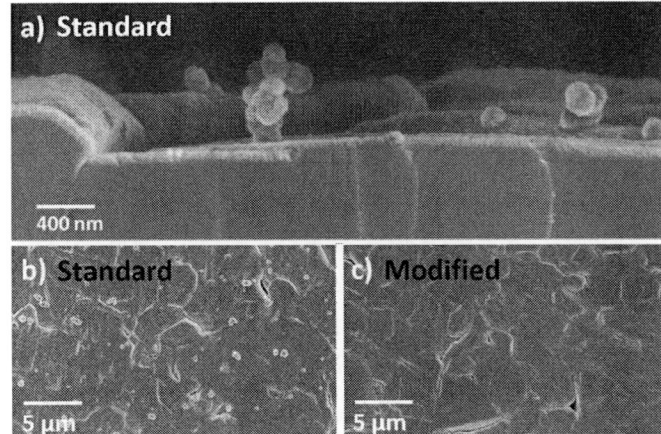

Fig. 5. SEM images of an ACIGS layer after CdS deposition. a) Cross-section of standard process, b) top view of standard process, c) top view of modified process with reduced CdS colloid clusters.

2PACz/SiO$_x$ nanoparticles/FA$_{0.83}$Cs$_{0.17}$Pb(I$_{0.87}$Br$_{0.13}$)$_3$/PCBM /BCP/ IZO/grid.

A further morphology-related optimization of ACIGS bottom cells for 2T devices is needed. As shown in the SEM image of an ACIGS/CdS cross-section in Fig. 5a, and the top view in Fig. 5b, clusters of little ball shapes are distributed on the top of the devices. These are caused by colloid formation during the standard CdS buffer layer deposition process. These small particles stick to the surface and are then amplified by the subsequent growth of window layers around them. They have negligible impact on single-junction ACIGS performance, since the layer is closed and any pinholes caused by colloid removal are insulated by the high-resistive layer. However, they lead to short circuits in the 2T tandem cells because the height of these clusters is about the same as the perovskite film thickness.

A two-fold approach was taken to reduce the problem of colloid clusters in the cell stack. One was to adjust the bath chemistry to slow the reaction by a factor of two, which promotes CdS growth on the substrate surface as opposed to particle formation in the bath solution. The growth rate was slowed to 3 to 4 nm/min by reducing the concentrations of Cd salt and thiourea by 30 %. The other modification was to reduce sticking of the colloid particles to the surface by performing the deposition in an ultrasonic bath, thereby strongly moving the solution and the particles. A final cleaning of the CdS-coated ACIGS layer was also performed with isopropanol in an ultrasonic bath in order to loosen and remove any colloids that stuck to the layer despite the other measures. The top view SEM image in Fig. 5c illustrates the colloid-free surface as compared to the standard situation shown in Fig. 5b.

With these modifications to the ACIGS bottom cell, a 2T tandem cell efficiency of 18.2 % (V_{OC} = 1681 mV, J_{SC} = 15.9 mA/cm^2, FF = 68.1 %, 0.5 cm^2, with anti-reflective coating) was obtained with all processing in house. In the 4T configuration without current-matching limitations a PCE exceeding 27.0 % is achieved.

IV. CONCLUSION

Production-relevant deposition tools and processes designed for single-junction ACIGS solar cells and modules are suitable for manufacturing top or bottom cells for tandem applications. However, efficiencies need to be improved. Inclusion of Ag expands the processing range to lower substrate temperatures, enabling progress with transparent back contacts required for top cell development. Ag alloying is also advantageous for bottom-cell ACIGS by reducing film roughness which is problematic for the subsequent perovskite top cell layers deposited from solution. Modifications to the CdS process already reduced a source of roughness-induced short circuits in 2T tandem solar cells, but the ACIGS layer from inline deposition still needs to be less rough and crevice-free. Material quality as apparent in the V_{OC} deficit will likely improve as more development efforts for inline coating are applied to the high-gap and low-gap ACIGS compositions for tandem applications.

ACKNOWLEDGEMENTS

We thank the teams at ZSW for solar cell preparation and characterization. This work was funded by the German Federal Ministry for Economic Affairs and Climate Action (BMWK) under contract numbers 03EE1078 (ODINCIGS), 03EE1059A (EFFCIS-II), 0324353A (CIGS-TheoMax) and 03EE1038B (CAPITANO).

REFERENCES

[1] M. Powalla, S. Paetel, E. Ahlswede, R. Wuerz, C. D. Wessendorf, and T. Magorian Friedlmeier, "Thin-film solar cells exceeding 22% solar cell efficiency: An overview on CdTe-, Cu(In,Ga)Se 2 -, and perovskite-based materials," *Appl. Phys. Rev.*, vol. 5, no. 4, p. 41602, 2018.

[2] *Interactive Best Research-Cell Efficiency Chart.* [Online] Available: https://www.nrel.gov/pv/interactive-cell-efficiency.html. Accessed on: May 15 2023.

[3] AVANCIS press release, "CIGS efficiency limit of 20 % breached," https://www.avancis.de/en/magazine/avancis-press-release-cigs-efficiency-world-record-20p3.

[4] PV Magazine, "Nice Solar Energy sets new world record for CIGS efficiency," https://www.pv-magazine.com/2019/12/04/nice-solar-energy-sets-new-world-record-for-cigs-efficiency/.

[5] J. Boyle, G. Hanket and W. Shafarman, "Optical and quantum efficiency analysis of (Ag,Cu)(In,Ga)Se2 absorber layers," *2009 34th IEEE Photovoltaic Specialists Conference (PVSC)*, Philadelphia, PA, USA, 2009, pp. 001349-001354.

[6] T. J. Coutts, J. S. Ward, D. L. Young, K. A. Emery, T. A. Gessert, and R. Noufi, "Critical issues in the design of polycrystalline, thin-film tandem solar cells," *Prog Photovolt Res Appl*, vol. 11, no. 6, pp. 359–375, 2003.

[7] R. Gutzler, W. Witte, A. Kanevce, D. Hariskos, and S. Paetel, "V_{oc} -losses across the band gap: Insights from a high-throughput inline process for CIGS solar cells," *Prog. Photovolt. Res. Appl. (Progress in Photovoltaics: Research and Applications)*, 2023.

[8] S. Rühle, "Tabulated values of the Shockley–Queisser limit for single junction solar cells," *Sol. Energy*, vol. 130, pp. 139–147, 2016.

[9] S.-C. Yang, T.-Y. Lin, M. Ochoa, H. Lai, R. Kothandaraman, F. Fu, A. N. Tiwari, and R. Carron, "Efficiency boost of bifacial Cu(In,Ga)Se$_2$ thin-film solar cells for flexible and tandem applications with silver-assisted low-temperature process," *Nat Energy*, vol. 8, no. 1, pp. 40–51, 2023.

[10] J. Keller, L. Stolt, O. Donzel-Gargand, T. Kubart, and M. Edoff, "Wide-Gap Chalcopyrite Solar Cells with Indium Oxide-Based Transparent Back Contacts," *Sol. RRL*, p. 2200401, 2022.

[11] J. Keller, K. V. Sopiha, O. Stolt, L. Stolt, C. Persson, J. J. Scragg, T. Törndahl, and M. Edoff, "Wide-gap (Ag,Cu)(In,Ga)Se$_2$ solar cells with different buffer materials - A path to a better heterojunction," *Prog. Photovolt.: Res. Appl. (Progress in Photovoltaics: Research and Applications)*, vol. 28, no. 4, pp. 237–250, 2020.

[12] S. Essig, S. Paetel, T. M. Friedlmeier, and M. Powalla, "Challenges in the deposition of (Ag,Cu)(In,Ga)Se$_2$ absorber layers for thin-film solar cells," *J. Phys. Mater.*, vol. 4, no. 2, p. 24003, 2021.

Development of 3D/2D Perovskite Solar Cells using a Spray-based Sequential Deposition

Neetesh Kumar, Rishabh Sahani, Daniel Muñoz, Cheng-Yu Lai and Daniela Radu*

Department of Mechanical and Materials Engineering, Florida International University, Florida, 33174
U.S.A.

Abstract — The 3D-perovskite halides have gained a considerable reputation as solar absorbers versus their counterpart semiconductor materials since they achieved a remarkably high-power conversion efficiency of 25.2% within a decade. However, lattice degradation and sensitivity against moisture, oxygen, and strong irradiation still hinder perovskite solar cells mass production and commercialization. Herein, we report a new sequential spray deposition method to deposit bulky organic cations (BOCs) on top of the 3D perovskite layer toward forming a protective 3D/2D layer. With controlled thickness and composition, this layer protects the perovskite grains from moisture penetration. The formation of the 2D layer was confirmed by X-ray diffraction, scanning electron microscopy, atomic force microscopy, photoluminescence, and absorption spectroscopy measurements.

Keywords—2D-perovskite, Spray coating, bulky-cation.

I. INTRODUCTION

The revolutionary raise in the power conversion efficiency (PCE) of solution-processed three-dimensional(3D)-perovskite solar cells (PSCs) has been demonstrated by using several strategies such as passivation of grain boundaries, improving the interfaces between the perovskite absorber and the charge transport layers.[1,2] Most of the 3D perovskite absorbers have degradation problems on exposure to the moisture, oxygen, and strong irradiation. Two-dimensional (2D) halide perovskites have emerged as efficient passivation layers, being the most effective to date in improving the device performance and stability. These 2D perovskites are commonly grown by spin-coating of bulky-organic cations (BOCs) dispersed in isopropyl alcohol or chloroform on top of 3D perovskite layer.[3,4] The spin coating of these materials using isopropyl alcohol (IPA) removes some of the excess lead iodide (PbI_2) from the 3D perovskite layer to then form heterogeneous 2D phases or ultrathin layers of wide-bandgap 2D perovskite.[5,6] Notably, the exposure of the 3D perovskite with these solvents degrades the perovskite layer. The effective way is to fabricated 2D perovskite layer via solvent free methods or use a new fabrication technique with minimal exposure of 3D perovskite with solvents. Jang et al., reported a solid-state solvent-free growth of the 2D perovskite on the 3D film by controlling the pressure, temperature, and time.[7] However, such solid-state in-plane growth is difficult to scale to large areas. Thus, the fabrication of solution-processed heterostructures of 3D/2D

with the desired energy levels, thickness, and orientation has been lacking. We report a spray deposition technique for fabricating 3D/2D perovskite heterophase structures with desired film thickness on a large-area. The process is designed such that it has minimal exposures of solvent to the 3D layer and effectively deposits the bulky-cations and leading to in-situ growth of 2D perovskite phase at room temperature without post-annealing. The thickness of 2D perovskite can be controlled by varying the concentration of bulky-cation in the spray solution and by increasing the spray cycles.

II. EXPERIMETAL DETAIL

The 3D perovskite films were fabricated using two-set sequential deposition technique. For this, first-step 1.5 M PbI_2 solution (DMF-DMSO 9:1 V%) was spin coated at 1500 rpm for 40 s and annealed at 70 ℃ for 1 min under the N_2 atmosphere. In second-step organic salt solution (FAI:MABr:MACl, 90:9:9 mg/mL in anhydrous IPA) was spin coated at 2000 rpm for 30 s. For the growth of perovskite phase the substrate were annealed at 150 ℃ for 20 min in ambient condition at a relative humidity of 40-45%. For the fabrication of 2D layer, 10 mg of phenethylammonium iodide (PEAI) bulky cation was dissolved in 10 mL of IPA solvent. The spray solution was pumped into the nozzle at a flow rate of 3.0 mL/min and was atomized by the pressurized (0.5 mPa) N_2 flow at the flow rate of 14 L/min. The spray cycles were varied from 4, 6 and 8 to achieve different thickness. X-ray diffractometer (XRD, Rigaku MiniFlex) with CuKα radiation (λ = 1.5405 Å) was used for crystallographic measurements. The surface and cross-sectional morphology and surface topography was analyzed by a field emission electron microscope (FESEM, Jeol JSM-F100) and an atomic for microscope (Dantom) respectively. For optical absorbance and transmittance measurements, a Shimadzu UV-3600 Plus, spectrophotometerwas used. The steady state photoluminescence (PL) spectra of perovskite films were measured with an Edinburgh Instruments FS5 spectrofluorometer.

III. RESULTS AND DISCUSSION

Structural and Morphological characterizations

PEAI, a non-fluorinated BOC derivative with short-chain ligands, is used as a surface passivation agent to fine-tune the morphology and dimensionality of 3D perovskite film. For PEAI passivation, PEAI solution was sprayed on 3D perovskite with 4, 6, 8, and 10 cycles. Figure 1 depicts the basic-difference between the spin coating and spray deposition process. In spin coating process the solvent infiltrates onto the 3D perovskite through the grain boundaries and leading the non-uniform growth of 2D perovskite, whereas in the spray process the solvent containing BOC precursor get deposited uniformly on the top of 3D perovskite and produces uniform growth of 2D perovskite.

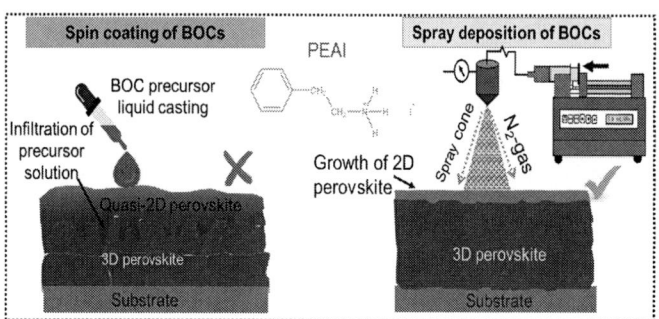

Fig. 1. Spin and spray deposition of BOCs on 3D perovskite for the growth of 2D perovskite layer.

Previous studies have reported that the BOC atop 3D-perovskite film can either exist in pristine form or react with remnant PbI_2 to form 2D perovskite, depending on the fabrication condition.[1] To confirm this, X-ray diffraction (XRD) was used to investigate the state and crystal phase of the PEAI atop the 3D perovskite film. The XRD compilation in Figure 2 (a) show that all of the perovskite films exhibited a dominant diffraction α-phase 3D perovskite peak at 2θ of 13.98° and other subsidiary α-perovskite peaks at 2θ positions of 14.9°, 24.3°, 28.8°, 31.6°, 34.78°, 40.4° and 42.9°. However, PEAI (4-8 cycles) passivated films are displaying an additional diffraction peak at 2θ of 5.3°, and this peak is particularly pronounced as spray cycles increased from 4 to 8. To identify the origin of this peak, we compared XRD of PEAI powder and the self-synthesized PEA_2PbI_4 2D perovskite film. As shown in the XRD compilations in Figure 2 (b), the PEAI powder displayed a representative intense diffraction peak at 2θ ~ 4.56°, whereas the 2D PEA_2PbI_4 perovskite film produced a dominant characteristic peak at 2θ of 5.3°. This confirmed that the additional diffraction in perovskite films passivated with PEAI (4-8 spray cycles) are displaying an additional diffraction peak at 2θ of 5.3°, which is corresponded to the PEA_2PbI_4 2D perovskite, instead of the PEAI salt. The XRD pattern of the 2D PEA_2PbI_4 perovskite agreed closely with results reported by Liu et al. [8] After PEAI passivation, the content of the remnant PbI_2 in the PEAI-treated films quenched (8 cycles), indirectly confirming that the PEAI salt reacted with remnant PbI2 on the perovskite film surface, resulting in in situ formation of 2D PEA_2PbI_4 perovskite. Therefore, PEAI-passivated perovskite film is believed to possess a unique 3D/2D

multidimensionality. For 8-spray cycles PbI_2 peaks intensity increased and additional peak at 11.75° appeared, which is correspond to the delta perovskite. This is due to degradation of perovskite for longer spray cycles.

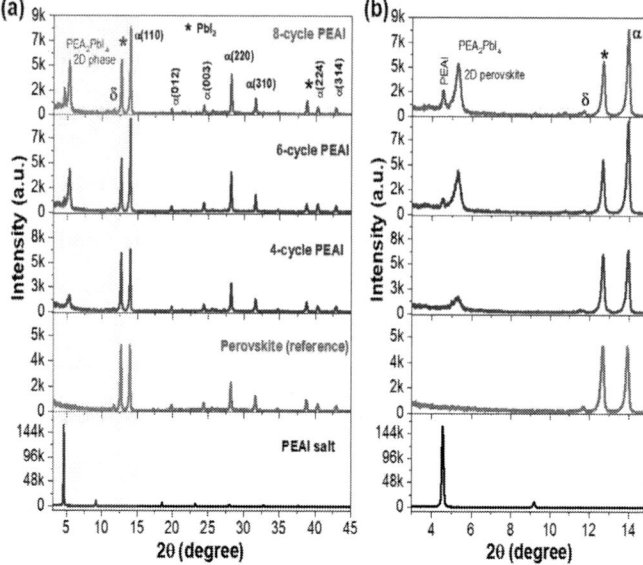

Fig. 2. (a) XRD patterns of perovskite films passivated with PEAI with different spray cycles (4-8) showing the formation 2D perovskite phase. (b) Zoom out image (3-15 degree) of XRD pattern showing the presence of 2D perovskite.

Simultaneously, the optical properties of the pristine and PEAI-passivated perovskite films were investigated. As depicted in Figure 3(a), the optical absorption of perovskite film in the UV-visible region was unaffected by the 2D layer. In addition, there is no deviation in the absorption band edge of the PEAI-passivated perovskite film, indicating that the formation of the PEA_2PbI_4 layer has no effect on the optical band gap of the perovskite film. This finding supported our hypothesis that the PEAI passivation effect is limited to the surface of a perovskite film due to the bulky phenyl ring and chain ligands of the $PEA+$ ions. The increment in absorption around 550 nm may due to formation of 2D perovskite. Figure 3(b) depicts the photoluminescence spectra of perovskite films. Reference films exhibits PL emission in the wavelength range of 780 to 880 nm with peak at 823 nm.

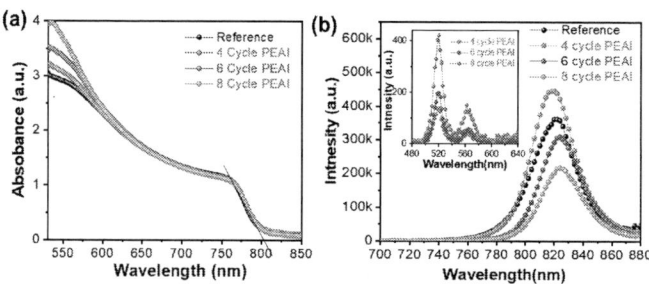

Fig. 3. (a) Absorption and (b) photoluminescence spectra of perovskite films passivated with PEAI with different spray cycles (4-8). Inset image showing PL spectra for 2D perovskite phase. (a)

Absorption and (b) photoluminescence spectra of of perovskite films passivated with PEAI with different spray cycles (4-8). Inset image showing PL spectra for 2D perovskite phase.

Form PL emission we can estimated the band gap of perovskite to be 1.5 eV. For 4-cycles of PEAI spray the PL intensity increased by 25% due surface defect passivation. For higher spray cycles PL intensity quenched due to the formation 2D perovskite on the top of 3D perovskite. The PL emission correspond to 2D perovskite is observed at 530 and 560 nm as shown in inset image of Figure 3 (b). The formation of 2D

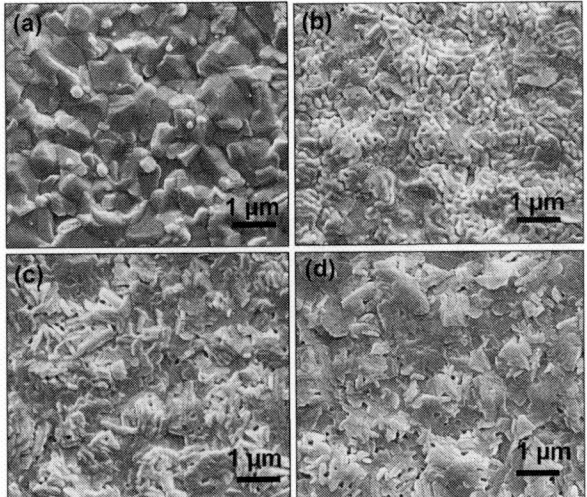

pcrovskite is further confirmed by the FE-SEM measurements. Fig. 4. FE-SEM images of perovskite films passivated with PEAI with different cycles. (a) Reference, (b) 4 cycles, (c) 6 cycles, (d) 8-cycles.

Fig. 4 depicts the surface morphology evolution of the perovskite films following PEAI (4-8 cycles) of PEAI spray. As can be seen in pristine (untreated) perovskite film exhibited pinhole-free, densely-packed grain morphology. Each grain and grain boundary on the surface are clearly visible. A dramatic change in the morphology of the PEAI-passivated perovskite film is evident in Figure 4(b-d). The PEAI passivation appears to have healed the grain boundaries on the perovskite film surface and rendered the perovskite grains indistinguishable. On increasing the spray cycles fine microstructure in more pronounced which is PEA_2PbI_4 2D perovskite in a sheet or rod like morphology. Similarly, atomic force microscopy (AFM) revealed that perovskite film surfaces became smoother after PEAI treatment.

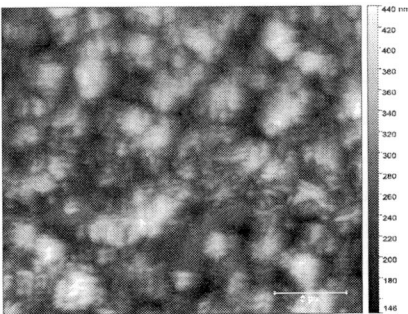

Fig. 5. AFM micrograph of perovskite film passivated with 8 cycles of PEAI.

As displayed in Figure 5, the PEA_2PbI_4 2D perovskite film has a much smaller grain size and more grain boundaries than the 3D perovskite film. The difference in the surface morphology of the PEA_2PbI_4 2D perovskite and the PEAI-modified 3D/2D perovskite film suggests that the former and the latter have entirely different crystal structures, as reflected by the XRD patterns in Figure 2.

IV. SUMMARY AND FUTURE STEPS

These preliminary results suggest that spray deposition technique is very promising for the fabrication 2D perovskite layer on the top of 3D perovskite with damaging or degrading it. As, XRD results reviled that as spray cycles increases the amount of 2D perovskite increased and simultaneously unreacted PEAI also increased. The simultaneously presence of 2D and PEAI can effectively passivate the 3D perovskite and it can work as a protective layer against the moisture related degradation. By optimizing the spray parameters these films be applied for the fabrication of high efficiency perovskite with high stability, for on large-area substrates.

ACKNOWLEDGEMENT

The authors acknowledge the funding from NASA, Award # 80NSSC19M0201, and Award # 80NSSC21M0310, DoD Office of Naval Research Award # N00014-20-1-2539, and NSF Award # DMR-2122078

REFERENCES

[1] Qi Jiang, Yang Zhao, Xingwang Zhang, Xiaolei Yang, *et.al.*, "Surface passivation of perovskite film for efficient solar cells," *Nature Photonics,* vol. 13, pp. 460–466, 2019.

[2] Fei Zhang, So Yeon Park, Canglang Yao, Haipeng Lu, "Metastable Dion-Jacobson 2D structure enables efficient and stable perovskite solar cells," *Science,* vol. 375, pp. 71-76, 2022.

[3] Efat Jokar, Po-Yuan Cheng, Chia-Yi Lin, Sudhakar Narra, Saeed Shahbazi and Eric Wei-Guang Diau, "Enhanced performance and stability of 3D/2D tin perovskite solar cells fabricated with a sequential solution deposition," *ACS Energy Lett.*, vol. 6, pp. 485–492, 2021.

[4] Jason J. Yoo, Sarah Wieghold, Melany C. Sponseller, Matthew R. Chua, *et al.* "An interface stabilized perovskite solar cell with high stabilized efficiency and low voltage loss," *Energy Environ. Sci.*, vol.12, pp. 2192-2199, 2019.

[5] Tiankai Zhang, Mingzhu Long, Minchao Qin, Xinhui Lu, *et al.* "Stable and Efficient 3D-2D Perovskite-perovskite planar heterojunction solar cell without organic hole transport layer., *Joul*, vol. 2, pp. 2706-2721, 2019.

[6] Mengqi Jin, Huilin Li, Qiang Lou, Qing Du, *et. al.* "Toward high-efficiency stable 2D/3D perovskite solar cells by incorporating multifunctional $CNT:TiO_2$ additives into 3D perovskite layer," *EcoMat*, vol 4 1-13, 2022.

[7] Woongchan Lee, Young Jin Yoo, Jinhong Park, *et al.*, "Perovskite microcells fabricated using swelling-induced crack propagation for colored solar windows, *Nat. Commun.*, vol. 13, pp. 1-10, 2022.

[8] Yucheng Liu, Yunxia Zhang, Zhou Yang, Haochen Ye, et al. "Multi-inch single-crystalline perovskite membrane for high-detectivity flexible photosensors," *Nat. Commun.* vol. 5302, pp. 1-11, 2018.

Numerical Modeling of Bifacial Thin Film Solar Cells

Briana Dokken[1,2], Sabin Neupane[1], Muhammad Mohsin Saeed[1], Jared D. Friedl[1], Adam B. Phillips[1], Michael J. Heben[1], Yanfa Yan[1], and Zhaoning Song[1]

[1.] Wright Center for Photovoltaics Innovation and Commercialization, Department of Physics and Astronomy, The University of Toledo, 2801 W. Bancroft St., Toledo, OH 43606 USA

[2.] Department of Physics, University of Minnesota Morris, Morris, MN 56267, USA

Abstract— **Bifacial thin film solar cells are a promising photovoltaic (PV) technology with the potential to generate more electrical power than their monofacial counterparts with the same footprint. However, bifacial thin film solar cells based on polycrystalline inorganic absorber layers typically exhibit inferior efficiencies under back illumination than under front illumination. Understanding the parameters that govern the back-illumination performance of bifacial thin film solar cells is essential to improving their bifaciality and overall performance. In this study, we used SCAPS software to evaluate the impact of some critical factors, including contact work function, back surface recombination velocity, minority carrier mobility, total defect density, shallow acceptor density, and absorber layer thickness on the PV parameters of a pseudo-ideal bifacial cadmium telluride thin film solar cell. We show that back surface recombination current is the most dominant factor that determines the bifacial PV performance of a thin-film solar cell.**

I. INTRODUCTION

As energy demands continue to increase and the need for renewable energy remains pertinent, thin film solar cells (TFSCs) seem to be a promising solution. Thin film photovoltaic (PV) technologies in general take less material and energy consumption than traditional Si cells, making them less expensive to produce. Unlike monofacial TFSCs, bifacial TFSCs can absorb and convert light from both their front and back surfaces into electrical energy [1]. Therefore, bifacial TFSCs can generate more energy than regular monofacial devices while using similar amounts of materials and space, further reducing the cost of solar electricity.

A current challenge in creating efficient bifacial TFSCs is improving the bifaciality factor by increasing back-illumination short-circuit current density (J_{SC}) and efficiency [2]. The conventional inorganic TFSCs, such as cadmium telluride (CdTe) and copper indium gallium selenide (CIGS), rely on a built-in electric field formed by a p-n heterojunction to efficiently separate photoexcited charge carriers [2]. These TFSCs typically exhibit poor back-illumination performance due to the lack of a proper electric field and surface passivation at the back interface, resulting in a bifaciality factor of less than 50% [3-5], much lower than perovskite and Si solar cells [6, 7]. Particularly, the downward band bending and high back surface recombination velocity (BSRV) of CdTe severely limit the performance of back-illuminated CdTe solar cells [8]. The low diffusion length of minority carriers determined by the low absorber minority carrier mobility, high absorber total defect density, and high shallow acceptor density can also hinder the performance of bifacial TFSCs under back illumination.

To better understand the working principle of bifacial TFSCs and gain insights into the parameters that are crucial to the bifacial performance of TFSCs, here we simulate the bifacial performance of CdTe solar cells and systematically evaluate the impact of some critical factors, including contact work function, back surface recombination velocity, minority carrier mobility, total defect density, shallow acceptor density, and absorber layer thickness on the PV parameters and bifaciality of bifacial CdTe TFSCs. Through the analysis, we conclude that back surface recombination current is the most dominant factor that determines the bifacial PV performance of a TFSC.

II. EXPERIMENTAL DETAILS

In this study, we used SCAPS [9] to simulate the PV performance of a pseudo-ideal CdTe (bandgap = 1.5 eV) solar cell consisting of a simple P-N homojunction. Fig. 1(a) illustrates the baseline model that has the structure of back contact/p-type absorber/n-type buffer/front contact. Fig. 1 (b) depicts the corresponding band diagram of a 4 μm p-type ($N_a = 10^{14}$ cm^{-3}) absorber and a 10 nm n-type ($N_d = 10^{17}$ cm^{-3}) buffer layer. For the ideal model, flat bands (4.6 and 5.6 eV) are assumed for both front and back contacts. The defect density of the absorber is set at 10^{10} cm^{-3} to minimize the non-radiative recombination for the pseudo-ideal absorber layer. Electron and hole mobilities of the absorber layer are 320 and 40 cm^2/Vs. The radiative recombination coefficient is 10^{-12} cm^3/s. Other parameters are adopted from the literature [10]. This pseudo-ideal benchmark model allows the efficiency of the thin-film CdTe cell to approach its detailed balance limit. As a result, the real impact of variables can be better assessed.

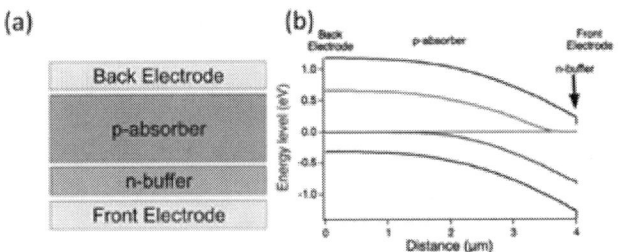

Fig. 1. (a) Schematic structure and (b) band diagram of our baseline cell.

We simulated the PV performance of bifacial TFSCs with different thicknesses, including 1, 2, and 4 μm. First, we varied the back contact WF and BSRV of the cell under the front and back illumination. The back electrode work function varied from 4.6 to 5.7 eV, corresponding to a Fermi level offset of -1.3 to 0.2 eV, while the BSRV of electrons also varied from 1 to 10^9 cm/s. Short-circuit current density (J_{SC}), Open-circuit voltage (V_{OC}), and power conversion efficiency (PCE) values of the cell under both front and back illumination were simulated.

After simulating cells with varying contact WFs and BSRVs, we screened the impact of the electron (minority carrier) mobility (μ_e), absorber layer total defect density (N_t), and absorber layer shallow acceptor density (N_a) of the absorber layer on the performance of the cell. We varied the BSRV of electrons from 1 to 10^9 cm/s while also varying the electron mobility in the absorber from 10^{-3} to 10^4 cm²/Vs. Then, the BSRV of electrons was varied from 1 to 10^9 cm/s, and the total defect density of the absorber was varied from 10^{10} to 10^{19} cm^{-3}. Finally, the BSRV of electrons was varied from 1 to 10^9 cm/s while the shallow acceptor density was varied from 10^{12} to 10^{19} cm^{-3}. All three of these tests were also done under front and back illumination and PV parameters were recorded. J-V and QE curves for each of these trials were also graphed.

III. RESULTS AND DISCUSSION

A. Contact Work Function and BSRV

We first studied the impact of back WF and BSRV. Fig. 2 compares the J_{SC}, V_{OC}, and PCE of the back-illuminated cells as a function of back contact WF and BSRV of electrons. Under back illumination, the minority carrier (electron) BSRV has the biggest impact on the output of the cell. Increasing BSRV significantly limits the J_{SC} and PCE of the cell. For the 4 μm absorber layer, above a BSRV of about 10^5 cm/s, the cell has a J_{SC} and PCE around zero. The thinner absorber layers tolerate higher BSRVs, allowing back-illuminated J_{SC} of more than 20 mA/cm² and back-illuminated PCE of ~10%. As the back contact WF increases, the V_{OC} of the back-illuminated cell increases. Higher back WFs and lower BSRVs lead to a higher J_{SC}, V_{OC}, and PCE for the back-illuminated cells.

Fig. 3 compares the J_{SC}, V_{OC}, and PCE of the same cells under front illumination. Under front illumination, the J_{SC} of the cell is significantly higher than the J_{SC} of the back-illuminated cell. The front-illuminated cell has a high J_{SC} (above 27 mA/cm²) even as the BSRV approaches 10^9 cm/s. The V_{OC} of the front-illuminated cell also increases as the back WF increases. The PCE of the front-illuminated cell is less impacted by BSRV and is able to stay higher than the back-illuminated cell's PCE as BSRV increases.

Band bending and the location of photogenerated carriers are likely the reason BSRV severely limits the J_{SC} and PCE of the back-illuminated cell and has an insignificant impact on the front-illuminated cell. Downward band bending at the back junction due to insufficiently high WF along with high BSRV of electrons are major problems limiting the performance of back-illuminated bifacial cells. Under back illumination, most of the carriers created by absorbed light are generated within

the back junction (BJ) depletion region [8]. When the back WF is below 5.5 eV, there is downward band bending at the BJ, and the electrons in the BJ depletion region get swept to the back surface, where they can easily recombine (especially when BSRV is high). Increasing the WF of the back contact decreases downward band bending at the BJ, which decreases the number of electrons that are pushed to the back contact where they recombine. Therefore, higher back WFs lead to better performance in both illumination directions.

Back WF Back Illumination

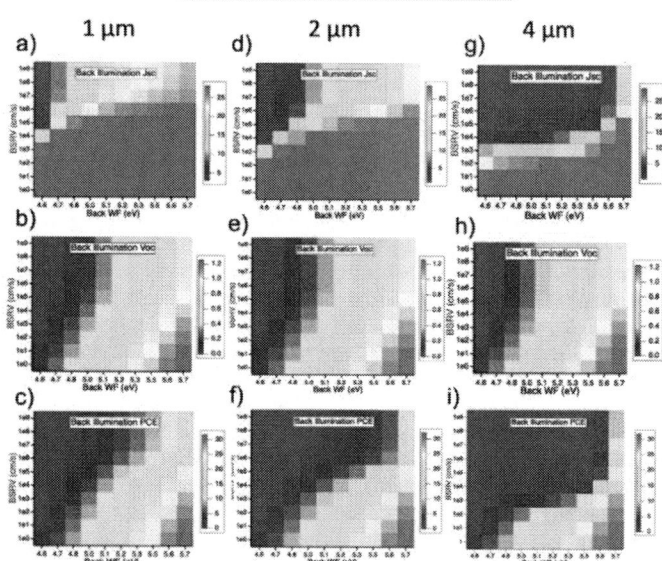

Fig. 2. PV parameters for (a-c) 1, (d-f) 2, and (g-i) 4 μm cells under back illumination when BWF and BSRV are varied, including (a, d, g) J_{SC}, (b, e, h) V_{OC}, and (c, f, i) PCE.

Back WF Front Illumination

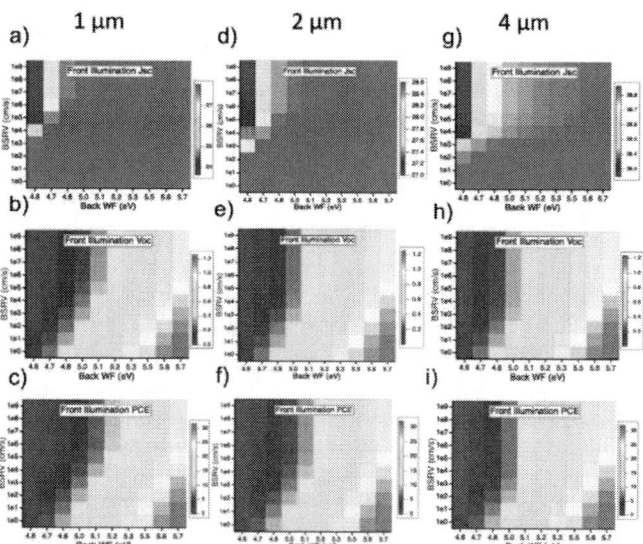

Fig. 3. PV parameters for (a-c) 1, (d-f) 2, and (g-i) 4 μm cells under front illumination when BWF and BSRV are varied, including (a, d, g) J_{SC}, (b, e, h) V_{OC}, and (c, f, i) PCE.

Next, we examined the impact of varying the BWF and BSRV of a cell on a CdTe cell with a thinner absorber layer thickness, i.e., 1 and 2 μm. Similar to the 4 μm absorber cell, a higher back WF for the thinner absorber cells led to less downward band bending at the BJ, which decreased the recombination of electrons at the back surface and improved the J_{SC}, V_{OC}, and PCE of the devices under back illumination. Increasing BSRV also decreased the J_{SC}, V_{OC}, and PCE of both the 1 and 2 μm absorber devices. However, BSRV impacted the thinner cells less than the 4 μm cell. The J_{SC} of the 1 μm cell stayed higher than that of the 2 and 4 μm cells as BSRV increased. The back-illuminated 4 μm cell had a J_{SC} lower than 15 mA/cm^2 once BSRV got higher than 10^4 cm/s (except for at very high BWFs), while the back-illuminated J_{SC} of the 1 μm cell was generally able to stay ~25 mA/cm^2 (as long as BWF was sufficiently high).

The values and trend of the V_{OC} of 1, 2, and 4 μm devices were essentially the same, so the higher J_{SC}'s of the 1 μm cell led to the 1 μm cell having higher efficiencies than the 2 and 4 μm cells. These results suggest that in the 1 μm cell, less of the excited electrons were pushed to the back contact by the downward band bending at the BJ. In devices with thinner absorber layers, more carriers are generated closer to the front junction (FJ) and within a diffusion length of the FJ region, allowing electrons to get to the front junction and get collected. This also meant fewer electrons were pushed back towards the back surface where they could recombine, so the J_{SC} and PCE of the thinner cells were less impacted by BSRV than the J_{SC} and PCE of the thicker cell.

B. Absorber Electron Mobility and BSRV

Next, we varied the μ_e of the absorber and BSRV under the front and back illumination. We first looked at a cell with a 4 μm absorber. Figs. 4 and 5 compare the J_{SC}, V_{OC}, and PCE of the back and front-illuminated cells, respectively.

When the 4 μm absorber cell is under back illumination, J_{SC} increases as μ_e increases. The back-illuminated cell's J_{SC} is also much more dependent on BSRV than the front-illuminated cell's J_{SC}. The front-illuminated cell has a high J_{SC} of ~28.5 mA/cm^2, even as BSRV increases and μ_e decreases.

For both the front and back-illuminated cells, V_{OC} decreases as μ_e increases. For the back-illuminated cell, the PCE of the cell is higher when μ_e is high and BSRV is low, while under front illumination, PCE decreases when μ_e increases. The PCE of the back-illuminated cell follows the trend of the back-illuminated J_{SC}, while the PCE of the front-illuminated cell follows the trend of the front-illuminated V_{OC}.

The increase in J_{SC} under back illumination when μ_e increases is likely due to the higher diffusion coefficient (D_e), which allows electrons to diffuse away from the BJ before they are pushed back to the back interface where they can easily recombine. This increases the J_{SC} of the back-illuminated cell significantly and slightly increases the J_{sc} of the front-illuminated cell. Yet, high μ_e decreases the V_{OC} of both front and back-illuminated cells.

After testing how μ_e impacted a cell with a 4 μm absorber, we changed the absorber thickness to 1 and 2 μm and repeated the same tests. Similar to the previous trial, the J_{SC} of the 1 μm

absorber cell was able to stay higher than the J_{SC} of the 4 μm cell when both were under back illumination.

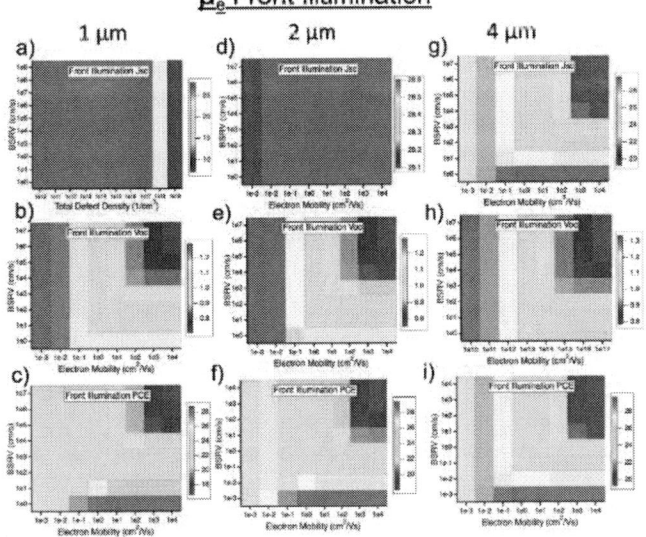

Fig. 4. PV parameters for (a-c) 1, (d-f) 2, and (g-i) 4 μm cells under back illumination when μ_e and BSRV are varied, including (a, d, g) Jsc, (b, e, h) Voc, and (c, f, i) PCE.

Fig. 5. PV parameters for (a-c) 1, (d-f) 2, and (g-i) 4 μm cells under front illumination when μ_e and BSRV are varied, including (a, d, g) Jsc, (b, e, h) Voc, and (c, f, i) PCE.

The J_{SC} of the 1, 2, and 4 μm cells increased as μ_e increased because electrons were able to travel away from the BJ and avoid recombination when electron mobility was high. The excited electrons in the 1 and 2 μm cells were generated closer to the FJ where they could be collected, so the thinner absorber cell's J_{SC} was able to stay high even as μ_e decreased and BSRV increased. The thinner absorber layer meant that fewer electrons were pushed to the back surface where they could easily recombine and electrons did not have to travel as far to reach the FJ, so low μ_e and high BSRV did not limit the 1 μm device's J_{SC} as severely as it limited the 4 μm device's Jsc.

The V_{OC} of all three devices generally had the same values and followed the same trend (lower μ_e's and lower BSRVs led to higher V_{OC}'s). The PCE of all three devices increased as μ_e increased and BSRV decreased, but the 1 μm cell had higher PCEs at lower μ_e's than the 2 and 4 μm cells since μ_e limited the J_{SC} of the thicker cells more than the J_{SC} of the thinner cells.

C. Absorber Total Defect Density

We also studied how the absorber's total defect density (N_t) impacted both front and back-illuminated cells with 1 - 4 μm absorber layers. Fig. 6 and 7 show the resulting PV parameters for cells under the back and front illumination, respectively.

Under front illumination, the 4 μm cell's J_{SC} stayed relatively high, except for very high N_t. The front-illuminated cell's J_{SC} decreases from about 25 to ~0 mA/cm² as N_t increases from about 10^{16} to 10^{19} cm⁻³. BSRV has an insignificant impact on the front-illuminated cell's J_{SC}, which is similar to the other trials. The back-illuminated cell's J_{SC} decreases as N_t increases, and increasing the BSRV also decreases the back-illuminated cell's J_{SC}.

Next, we studied how varying BSRV and N_t in the absorber impacted cells with 1 and 2 μm thick absorbers. Similar to the 4 μm cell, the J_{sc} and V_{oc} of both the 1 and 2 μm cells decreased as both N_t and BSRV increased. Once again, under back illumination when the device's absorber layer got thinner, the device had a higher J_{sc} and PCE; the 1 μm cell had the highest J_{sc} and PCE while the 4 μm absorber device had the lowest. The 1 μm cell's J_{sc} didn't decrease until N_t was above 10^{16} cm⁻³ and BSRV was above 10^4 cm/s, while the 4 μm cell's J_{sc} decreased once N_t got up to about 5.5×10^{14} and BSRV hit 5.5×10^2 cm/s. The PCE of the 1 μm cell was able to stay slightly higher than that of the 4 μm cell at higher values of N_t and BSRV, and the V_{oc} of the 1 μm cell was able to stay higher than the 4 μm cell's V_{oc} as BSRV increased. This suggests that because the electrons in the cells with thinner absorbers have a shorter path to travel to the FJ, there are fewer chances for them to recombine in the defects in the absorber layer. Additionally, fewer electrons end up at the BJ, so high N_t and high BSRV didn't increase recombination and weaken the performance of the cells with thinner absorbers as significantly as they did for the cells with thicker absorbers.

N_t Back Illumination

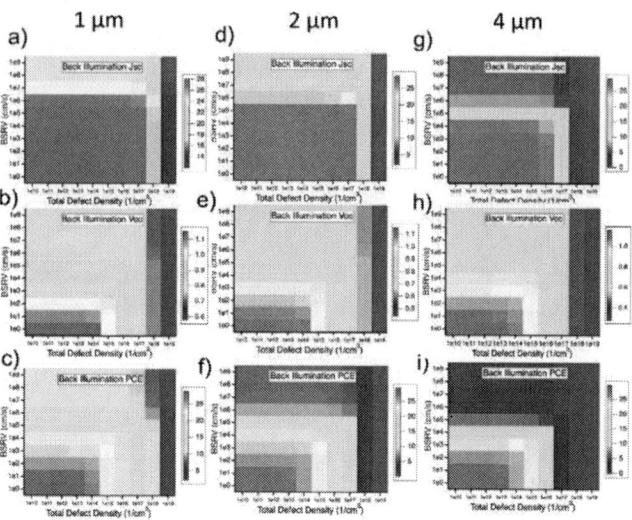

Fig. 6. PV parameters for (a-c) 1, (d-f) 2, and (g-i) 4 μm cells under back illumination when N_t and BSRV are varied, including (a, d, g) Jsc, (b, e, h) V_{OC}, and (c, f, i) PCE.

N_t Back Illumination

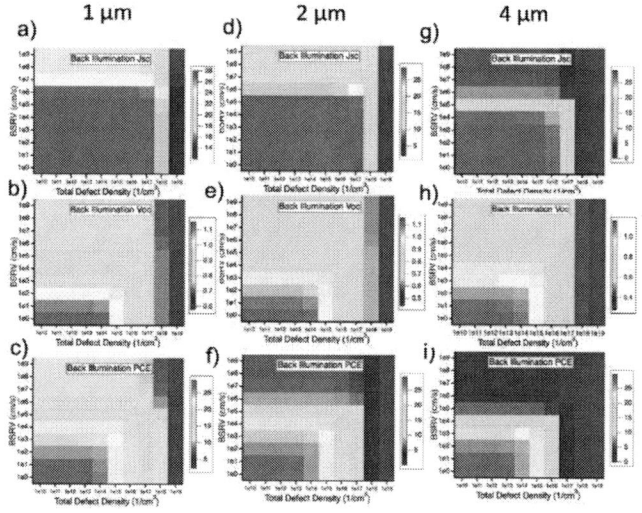

Fig. 7. PV parameters for (a-c) 1, (d-f) 2, and (g-i) 4 μm cells under back illumination when N_t and BSRV are varied, including (a, d, g) Jsc, (b, e, h) V_{OC}, and (c, f, i) PCE.

D. Shallow Acceptor Density

The final parameter we studied was the absorber's shallow acceptor density (N_a). We varied N_a from 10^{12} cm⁻³ − 10^{19} cm⁻³ and BSRV of electrons from 1 cm/s − 10^7 cm/s. The resulting J_{sc}, V_{OC}, and PCE graphs for 1, 2, and 4 μm absorber cells under the back and front illumination are shown in Figs. 8 and 9, respectively.

Under back illumination, the J_{sc} of the 1, 2, and 4 μm cells was higher when BSRV was lower (similar to the previous three trials) and when N_a was low. Increasing N_a of the absorber lowers the Fermi level of the absorber layer, which reduces band bending and the length of the depletion region at the FJ. A wider depletion region at the front junction increases the ability of the front contact to separate electrons and holes, so this decrease in downward band bending as N_a increases means the front contact is less carrier selective and the cell's J_{SC} decreases. The V_{OC} of all three cells increased as N_a increased and BSRV decreased, and the PCE of the three cells was highest when BSRV and N_a were low. The V_{OC} and PCE of the 1, 2, and 4 µm cells show similar trends and values, while the thicker absorber cells' J_{SC}'s were more impacted by varying N_a than the thinner absorber cells' J_{SC}'s. This seems to be a result of the fact that higher BSRVs hinder the J_{SC} of cells with thinner absorbers less than cells with thicker absorbers (similar to the previous three trials), which allows the J_{SC} of 1 µm cell to be higher than the Jsc of the 2 and 4 µm cells at higher BSRVs.

Fig. 9. PV parameters for (a-c) 1, (d-f) 2, and (g-i) 4 µm cells under front illumination when N_a and BSRV are varied, including (a, d, g) Jsc, (b, e, h) V_{OC}, and (c, f, i) PCE.

E. Bifaciality factor

Lastly, we summarize the bifaciality factor of r of 1, 2, and 4 µm cells with different BWF, μ_e, N_t, N_a, and BSRV in Fig. 10. In brief, lower BSRV leads to more efficient back-illuminated cells due to reduced back surface recombination current density. Higher BWF leads to less downward band bending at the back junction, which increases the Jsc and PCE of the back-illuminated cells. High μ_e and low N_a are beneficial for bifaciality but not front PCE. Lower N_t leads to high PCE under both the front and back illumination.

Back Illumination

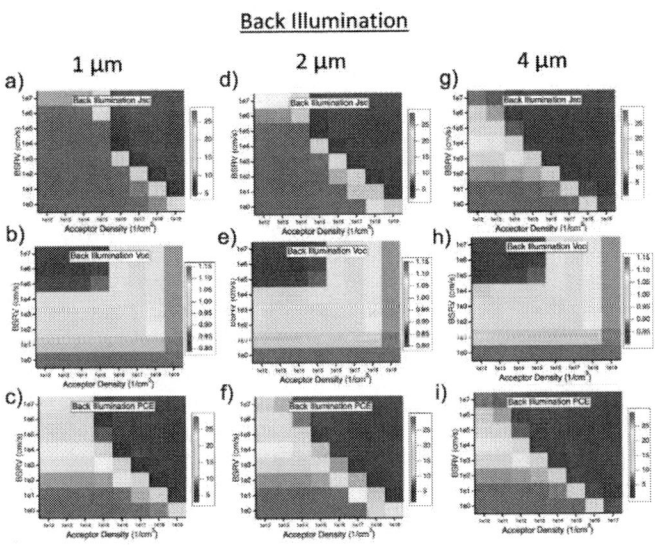

Fig. 8. PV parameters for (a-c) 1, (d-f) 2, and (g-i) 4 µm cells under back illumination when N_a and BSRV are varied, including (a, d, g) Jsc, (b, e, h) V_{OC}, and (c, f, i) PCE.

Front Illumination

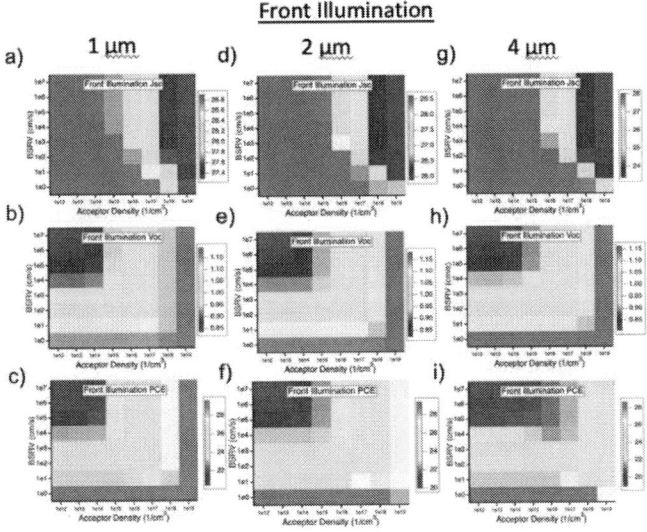

Bifaciality

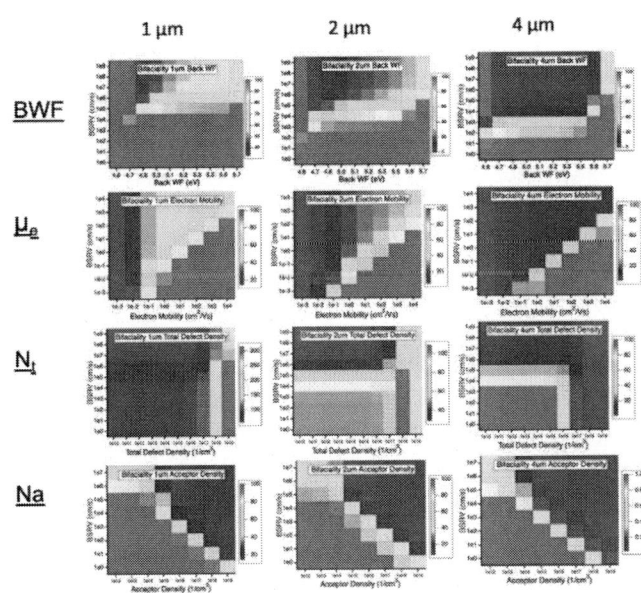

Fig. 10. Bifaciality factor of 1, 2, and 4 µm cells with different BWF, μ_e, N_t, N_a, and BSRV.

IV. CONCLUSION

Overall, by varying back WF, BSRV, μ_e, N_t, and N_a we found that BSRV has a major impact on the performance of back-illuminated bifacial solar cells. Increasing BSRV has an insignificant impact on front-illuminated cells, but it severely limits the J_{SC} and PCE of the back-illuminated cell. Higher BSRV also decreases the V_{OC} and PCE of both front and back-illuminated cells. Higher μ_e increases the J_{SC} but decreases the V_{OC} of the front and back-illuminated cells. Increasing N_t hinders the J_{SC}, V_{OC}, and PCE of both front and back-illuminated cells, while increasing N_a decreases the J_{sc} and PCE of the cells while increasing the V_{oc} of the cells. Optimizing these parameters can increase the performance of bifacial thin film solar cells.

Another potential way to improve the performance of back-illuminated cells specifically is to decrease the thickness of the absorber layer. Decreasing absorber thickness meant that excited electrons were generated closer to the FJ where they could be separated and collected, so high BSRV, high N_t, low μ_e, and low back WF limited the J_{SC} and PCE of the 1 μm cell less than the PCE of the 4 and 2 μm cells. This suggests that creating bifacial cells with absorber layers thick enough to collect a range of wavelengths of light while also being thin enough for excited electrons to avoid recombination and get collected can increase the efficiency of bifacial devices under back illumination.

ACKNOWLEDGMENT

This material is based upon work supported by the U.S. Department of Energy's Office of Energy Efficiency and Renewable Energy (EERE) under the Solar Energy Technology Office Award Number DE-EE0009832, by Air Force Research Laboratory under agreement FA9453-19-C-1002, and by The National Science Foundation for funding through the grant Research Experiences for Undergraduates #1950785. The views and conclusions contained herein are those of the authors and should not be interpreted as necessarily representing the official policies or endorsements, either expressed or implied, of the Air Force Research Laboratory, U.S. Department of Energy, or the U.S. Government.

REFERENCES

[1] R. Kopecek and J. Libal, "Bifacial Photovoltaics 2021: Status, Opportunities and Challenges," *Energies,* vol. 14, no. 8, p. 2076, 2021.

[2] Z. Song, C. Li, L. Chen, and Y. Yan, "Perovskite Solar Cells Go Bifacial—Mutual Benefits for Efficiency and Durability," *Adv. Mater.,* vol. 34, no. 4, p. 2106805, 2022.

[3] K. K. Subedi, E. Bastola, I. Subedi, Z. Song, K. P. Bhandari, A. B. Phillips, N. J. Podraza, M. J. Heben, and R. J. Ellingson, "Nanocomposite (CuS)x(ZnS)1-x thin film back contact for CdTe solar cells: Toward a bifacial device," *Sol. Energy Mater. Sol. Cells,* vol. 186, pp. 227-235, 2018.

[4] K. K. Subedi, A. B. Phillips, N. Shrestha, F. K. Alfadhili, A. Osella, I. Subedi, R. A. Awni, E. Bastola, Z. Song, D.-B. Li, R. W. Collins, Y. Yan, N. J. Podraza, M. J. Heben, and R. J. Ellingson, "Enabling bifacial thin film devices by developing a back surface field using CuxAlOy," *Nano Energy,* vol. 83, p. 105827, 2021.

[5] S.-C. Yang, T.-Y. Lin, M. Ochoa, H. Lai, R. Kothandaraman, F. Fu, A. N. Tiwari, and R. Carron, "Efficiency boost of bifacial Cu(In,Ga)Se2 thin-film solar cells for flexible and tandem applications with silver-assisted low-temperature process," *Nat. Energy,* vol. 8, no. 1, pp. 40-51, 2023.

[6] Z. Song, C. Chen, C. Li, S. Rijal, L. Chen, Y. Li, and Y. Yan, "Assessing the true power of bifacial perovskite solar cells under concurrent bifacial illumination," *Sustainable Energy Fuels,* 10.1039/D1SE00314C vol. 5, no. 11, pp. 2865-2870, 2021.

[7] C. Han, R. Santbergen, M. van Duffelen, P. Procel, Y. Zhao, G. Yang, X. Zhang, M. Zeman, L. Mazzarella, and O. Isabella, "Towards bifacial silicon heterojunction solar cells with reduced TCO use," *Prog. Photovoltaics,* vol. 30, no. 7, pp. 750-762, 2022.

[8] A. B. Phillips, K. K. Subedi, G. K. Liyanage, F. K. Alfadhili, R. J. Ellingson, and M. J. Heben, "Understanding and Advancing Bifacial Thin Film Solar Cells," *ACS Appl. Energy Mater.,* vol. 3, no. 7, pp. 6072-6078, 2020.

[9] M. Burgelman, P. Nollet, and S. Degrave, "Modelling polycrystalline semiconductor solar cells," *Thin Solid Films,* vol. 361–362, pp. 527-532, 2000.

[10] D.-B. Li, Z. Song, R. A. Awni, S. S. Bista, N. Shrestha, C. R. Grice, L. Chen, G. K. Liyanage, M. A. Razooqi, A. B. Phillips, M. J. Heben, R. J. Ellingson, and Y. Yan, "Eliminating S-Kink To Maximize the Performance of MgZnO/CdTe Solar Cells," *ACS Appl. Energy Mater.,* vol. 2, no. 4, pp. 2896-2903, 2019.

Passivating Contacts with Engineered Pinhole Enabled Transport

Harvey L Guthrey, Caroline Lima Salles, William Nemeth, Sumit Agarwal, David Young, Pauls Stradins

National Renewable Energy Laboratory, Golden, CO, United States

Colorado School of Mines, Golden, CO, United States

In this work we discuss a novel method for fabricating pinhole transport enabled passivating contacts in high performance silicon photovoltaic devices. The pinhole formation process occurs via metal assisted chemical etching (MACE) at room temperature and can potentially be integrated easily into existing fabrication lines. The contacts formed with the MACE process are known as Polysilicon on Locally Etched Oxide (PLEO) or PLENO when a SiNy/SiOx layer stack is used. Devices with these contacts have exceeded 22% conversion efficiency due to high open circuit voltages and low saturation current densities. In addition to electrical properties of devices with such contacts, we also present electron microscopy-based characterization (EBIC, TEM, STEM-EELS) analysis that reveal the variations in local charge carrier transport and directly connect these with the nanoscale structure of pinholes in these devices.

InP-Based Tunnel Junctions for Micro-Concentrator Photovoltaics

Kenneth J Schmieder, Thomas C Mood, Eric A Armour, Mitchell F Bennett, Margaret A Stevens, Martin Diaz, Ziggy Pulwin, Matthew P Lumb

US Naval Research Laboratory, Washington, DC, United States

Jacobs, Hanover, MD, United States

Veeco Instruments, Somerset, NJ, United States

Formerly with George Washington University, Washington, DC, United States

To further improve the performance of mechanically stacked micro-concentrator photovoltaic devices, we have studied tunnel junctions for inclusion in triple junction solar cells that are fully lattice-matched to the InP template. These tunnels are evaluated using both standalone tunnel diodes as well as full multijunction solar cells. Of particular focus herein is the p-type tunnel junction layer, which has proven particularly challenging to integrate in multijunctions with high electrical activity and an abrupt doping profile. Studies include the effect of polarity, tunnel diode dopant/composition, application of a hydride-free anneal, tunnel diode growth temperature, and cladding material. Resulting InP-based triple junction devices achieved up to 370 suns-equivalent tunneling capability, which is considered adequate for microconcentrator photovoltaic applications in the space environment.

978-1-6654-6060-6/23 $31.00 © 2023 IEEE

Residual Stress Limits Gridline Bridging in Cracked Solar Cells

Junki Joe, Timothy J Silverman, Michael Owen-Bellini and Nick Bosco

National Renewable Energy Laboratory, Golden, Colorado, 80401, USA

Abstract — Cracks in solar cells can be generated in various ways, but this does not mean immediate power loss. Previous studies showed that gridlines bridge cracked silicon cells, and the bridging behavior decreases during the contact and separation of gridlines within bare cells. In this study, we investigate bridging behavior in laminated monocrystalline cells. We characterize the behavior with Weibull analysis of critical crack opening distance (COD), at which the gridline is completely separated. The Weibull analysis of the laminated cell shows a good agreement with bare cells at the first cycle. However, we observe different behavior during cyclic bending. Bare cells show gradual decay of critical COD, while laminated cells show instant decay and plateau. We hypothesize that the difference is due to residual stress, squeezing gridlines, and causing plastic deformation. This is justified by the correlation between critical COD and gridline morphology. In the presentation, we shall present a Weibull analysis of the cyclic bending of a laminated cell with a reduced residual stress effect.

I. INTRODUCTION

Cracks in crystalline silicon solar cells are initiated due to various reasons from installation to weather conditions such as wind oscillation, hail impact, and temperature changes [1]. The crack generation does not mean immediate power loss because gridlines electrically bridge cracks [2]. However, the repetition of contact and separation between cracked gridlines during thermal [3] and mechanical [4] loading can eventually cause complete electrical isolation. Hence, it is important to study the relationship between the evolution of bridging behavior and dynamic mechanical loading to predict electrical failure. In [5], authors investigated this problem with bare cells. They defined critical crack opening displacement (COD), at which the gridline completely opens, and demonstrated decay of critical COD during the three-point bending test.

In this work, we extend the previous study [5] into the laminated monocrystalline cell. We employ Weibull analysis to compare characteristic critical COD with bare cells for monotonic and cyclic loading. The results agree well for monotonic loading, while we observe different behavior in cyclic loading. We hypothesize that the compressive residual stress after lamination may cause plastic deformation on the cracked gridline. At the time of abstract submission, this is justified by the correlation between the decay of critical COD with cyclic loading and the morphology evolution of the gridline fracture surfaces characterized by scanning electron microscopy (SEM). In the final presentation, we shall present critical COD results for the laminated cell by reducing the effect of residual stress.

II. MATERIALS AND METHODS

B. Laminated Cells

With laser micro-machining, six different types of crystalline silicon cells were scribed into ~25x156mm beams containing 21 gridlines. The cells were additionally laser-scored on the backside of the cell to increase the probability to generate only one crack along the scoring perpendicular to the gridlines. Two PV tabbing ribbons were soldered at the end-most busbars, then the cell was laminated within a 50x350mm sheet of tempered glass.

The laminated cells were loaded in four-point flexure with 338mm external span and 185mm internal span and a constant displacement rate of ~10μm/s while monitoring gridline electrical resistance. Once all the gridlines were disconnected, a light microscope was used to characterize crack openings with load-line displacement. For cyclic loading, the beams were loaded from zero external force to complete electrical failure for all gridlines with a constant displacement rate of ~10μm/s for 180 cycles. Since no interaction between gridlines is involved beyond the electrical failure, the upper bound of the cyclic test is less meaningful.

B. Bare Cells

Two types of cells were scribed into ~2.4x40mm beams with laser micro-machining, containing at most two gridlines. The bare cells were adhered to an acrylic beam with a notch at the center to confine the maximum strain to the center of the cell and promote the generation of only one crack. The beams were loaded in three-point flexure with a 60mm span and a constant rate of load line displacement ~10μm/s [5]. Again, a light microscope was employed to characterize the ratio between COD and load line displacement. For cyclic loading, the composite beams were similarly loaded from zero external force to the critical COD at 1st cycle with a constant rate of load line displacement ~3μm/s. Since the critical COD decays with proceeding the cyclic test, the value at 1st cycle guarantees electrical failure at subsequent cycles.

To investigate the fractography of the cracked gridline, we heated the strained acrylic beams with a heat-gun to relax their stress and allow for the silicon beams to be removed without the gridline fracture surfaces coming back into contact. This step was taken to ensure that plastic deformation between contacting gridlines does not occur and change the surfaces before fractography characterization in SEM.

978-1-6654-6060-6/23 $31.00 © 2023 IEEE

We note that all measurements were obtained from samples with only one crack.

III. RESULTS AND ANALYSIS

For the laminated cell, gridline electrical behavior must be discerned from a parallel resistance measurement. When parallel N gridlines are connected, the expected resistance R is approximately

$$R(N) = R_0 + R_{GRIDLINE}/N, \qquad (1)$$

where R_0 is the resistance from outside of the cell, and $R_{GRIDLINE}$ is the resistance for one gridline. These two parameters can be obtained from the monitored value of R(N=1) and R(N=21), which can be easily determined. Then, we obtain Nth critical COD, which is the onset of exceeding R(N). This method is justified by the fact that the transition through partial contact occurs within 1μm COD. For the bare cell containing two gridlines, one gridline disconnection exhibits obvious resistance chance, and hence no such equation (1) is needed.

We employ Weibull analysis to characterize critical COD for both bare cell and laminated cell,

$$f(COD) = \frac{b}{c}\left(\frac{COD}{c}\right)^{b-1} e^{-\left(\frac{COD}{c}\right)^b}, \quad (2)$$

where f is the failure probability, b is the Weibull modulus determining the shape of the distribution, and c is characteristic critical COD involving the scale.

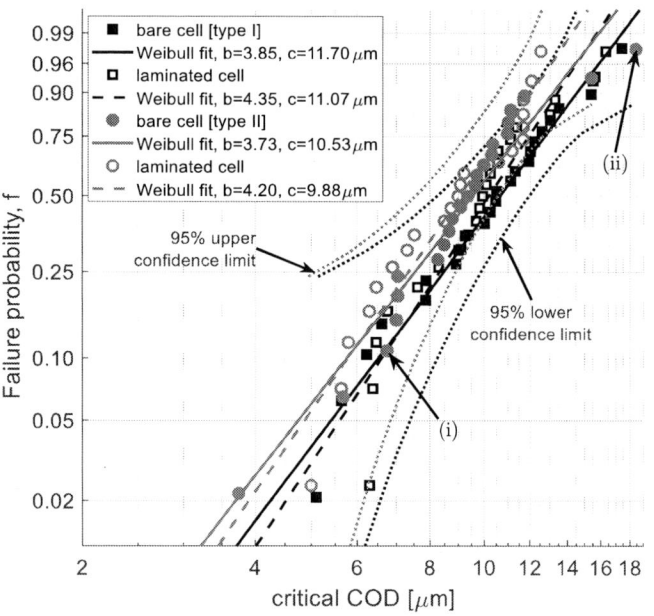

Fig. 1. Weibull distribution of critical COD for bare cell (empty marks) and laminated cell (solid marks) for two types (black, red color). Symbols denoted (i) and (ii) indicate a 6.7μm and 18μm critical COD, corresponding to fractography images presented in Fig. 2.

Dotted curves show the upper and lower 95% confidence limit of Weibull fit for the two laminated cells.

Figure 1 shows the results of the Weibull distribution curve of critical COD for bare cells (empty symbol) and laminated cells (solid symbol). The characteristic critical COD and Weibull modulus for bare cell and laminated cell agree well for two different types of cells, demonstrating the validity of the two techniques. The dotted curves are the upper and lower bounds of the 95% confidence limit of Weibull fit for the two laminated cells. Weibull analysis of bare cells is within the limits. In addition, we performed Weibull analysis for the other four types of laminated cells, observing characteristic critical COD as nearly 10.5μm with ~3 Weibull modulus. The previous study [5] refers to the low Weibull modulus as to the inhomogeneous nature of the fracture.

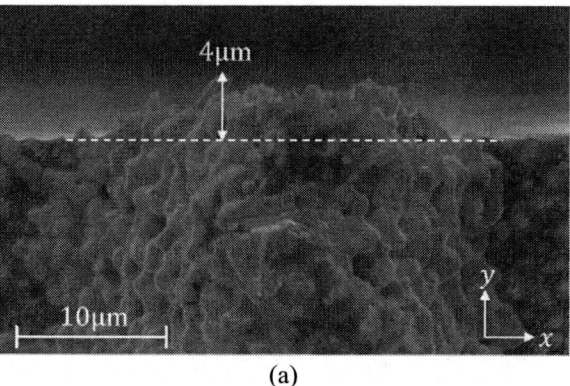

(a)

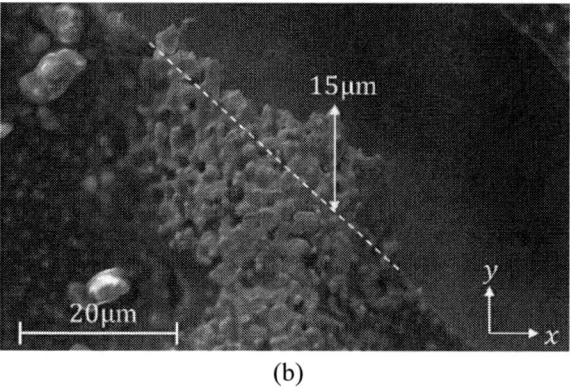

(b)

Fig. 2. SEM image of (i) and (ii) in Fig. 1 for (a) and (b) respectively. The crack opens in the y-direction. The dotted line is expected silicon crack. Notice that the magnification is different.

Figure 2 (a) and (b) show the SEM image of (i) and (ii) in Fig. 1 respectively. In the image, the crack opens in the y-direction, and we define the length of the bridge L_{bridge} as the maximum length of extra overhang material in the crack opening direction. In other words, it is the maximum length from the expected silicon crack (dotted line) in the -axis. L_{bridge} for Fig. 2 (a) and (b) are 4μm and 15μm, and the corresponding critical COD is 6.7μm and 18μm respectively.

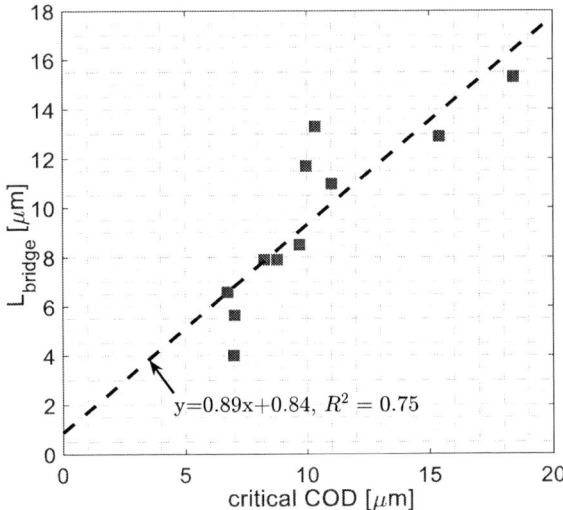

Fig. 3. Relationship between critical COD and L_{bridge}. Raw data [square marks] and linear fit [dotted line].

At the time of abstract submission, we further measured L_{bridge} of 12 cells from type II. The relationship between L_{bridge} and critical COD is presented in Fig. 3, showing linear dependency with 0.8 Kendall's τ (hence, strongly correlated). The proportionality is close to one. This may indicate that the decay of critical COD during the cyclic loading is caused by receding the extra overhang material presumably due to plastic deformation and then wear.

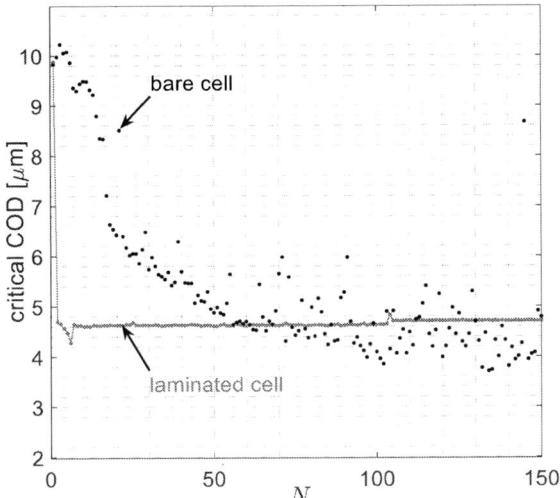

Fig. 4. Critical COD at Nth cycle for bare cell and laminated cell. For the laminated cell, the critical COD is the characteristic value of Weibull distribution.

Figure 4 shows the results of the cyclic bending test for both the laminated and bare cell samples. For the laminated cell, the critical COD is the characteristic value of Weibull distribution for all 21 gridlines. While the bare cell shows a gradual decay in critical COD, the laminated cell shows a dramatic decrease and then plateau at ~4.5μm. We hypothesize that this difference is due to the compressive residual stress in the laminate, which results in a sufficient contact force on the gridline fracture surfaces upon unloading the beams to plastically deform the mating fracture surfaces, effectively reducing their initial L_{bridge} and subsequent critical COD. In the final presentation, we will present the characterization of the critical COD for the laminated cell while reducing the effect of the residual stress.

IV. CONCLUSION

We demonstrate that the critical COD distribution of laminated cells is similar to the one previously presented using the bare cell evaluation method, thereby validating the two techniques. However, cyclic bending tests of the laminate results in a precipitous drop of critical COD, while the analogous bare cell test shows a gradual decay. We hypothesize that the difference is due to the compressive residual stress after lamination. In the final presentation, we shall present experimental results to support this hypothesis.

We also have shown the linear relationship between critical COD and the length of the extra overhang gridline for bare cells. This indicates that the receding of the overhang gridline may affect the decay of critical COD during cyclic loading.

REFERENCES

[1] Wohlgemuth, John H. "Photovoltaic module reliability." *John Wiley & Sons*, 2020.
[2] Köntges, M., Kunze, I., Kajari-Schröder, S., Breitenmoser, X., & Bjørneklett, B. "The risk of power loss in crystalline silicon-based photovoltaic modules due to micro-cracks." *Solar energy materials and solar cells* 95.4, 2011.
[3] Silverman, T. J., Bliss, M., Abbas, A., Betts, T., Walls, M., & Repins, I. "Movement of cracked silicon solar cells during module temperature changes." *2019 IEEE 46th Photovoltaic Specialists Conference (PVSC)*. IEEE, 2019.
[4] Silverman, T. J., Bosco, N., Owen-Bellini, M., Libby, C., & Deceglie, M. G. "Millions of Small Pressure Cycles Drive Damage in Cracked Solar Cells." *IEEE Journal of Photovoltaics*, 2022.
[5] Bosco, N., Chavez, A., Upadhyaya, V., & Han, S. "Fatigue-like behavior of silver metallization gridlines and proposed damage mechanics model." *2020 47th IEEE photovoltaic specialists conference (PVSC)*. IEEE, 2020.

Optimizing the Laser Scribing Process to Achieve a Certified Efficiency of 25.9% for Over 240 cm² Four-terminal Perovskite/Si Tandem Solar Cells

Yonglei Wang, Hongxu Zhang, Yang Liu, Yuan Qin, Jie Liu, Jiang Liu, Bo He*, Xixiang Xu*

Central R&D Institute, LONGi Green Energy Technology Co., Ltd., Xi'an, Shaanxi, 712000, China

Abstract — Due to the relatively independent fabrication of the semi-transparent top perovskite solar cells and the bottom HJT cells, the mechanically stacked 4-terminal perovskite/HJT tandem solar cells have been of great feasibility for the future industrialization. Optimized laser scribing technique is applied to achieve a series topology consisting of sub-cells in the large-scale top perovskite solar cells especially, which is the key to optimize the optical absorption for the bottom HJT cells and the carrier collection for the top perovskite cells. As a result, a semi-transparent perovskite mini-module with area over 240 cm² based on structure of FTO/NiOₓ/SAMs/CsFAMAPb(I,Br)₃/C₆₀/SnO₂/ITO and band gap of ~1.65 eV achieved a high power conversion efficiency over 16% with a fill factor of 78%. The encapsulated 4T tandem mini-module were certified by JET with a PCE up to 25.9%.

I. INTRODUCTION

As dominating technology in photovoltaic (PV) industry, c-Si based solar cells have achieved power conversion efficiency over 26.81% recently [1], getting closer to efficiency limit of c-Si solar cells (29.4%). Tandem solar cells with reduced thermallization loss by stacking wide-band-gap top cells and c-Si bottom cells have been considered as the most practical photovoltaic technologies to break the power-conversion-efficiency limit of silicon-based cells [2]. So far, electrically connected two-terminal (2T) and mechanically stacked four-terminal (4T) tandem solar cells with perovskite as wide band gap top cell and HJT as low band gap bottom cell have been intensively studied. The 2T cells with small size have achieved a world record power conversion efficiency (PCE) over 32.5% at Helmholtz-Zentrum Berlin für Materialien und Energie, showing the great efficiency potential and advantage of tandem PV technologies [3]. By contrast, the efficiency of 4T tandem solar cells has been lagging behind. A few reports presented 4T tandem cells with efficiency over 30% with cell area less than 1 cm² (typically 0.049 cm²) while efficiency of 4T tandem with cell area 4 cm² up to 23.7% has been achieved by Jaysankar et al.[4] The major challenges of 4T tandem solar cells include fabrication of high quality large-scale perovskite film, transparent back electrode with high conductivity for efficient charge collection and laser scribing technique to ensure good transportation through the sub-cells.

Laser scribing the top cells is a key technique to realize series configuration consisting of equal sub-cells for large-scale perovskite modules. Given no laser scribing processed

sub-cells, serious charge recombination due to poor lateral conductivity in perovskite film would cause big current loss. A large-scale single cell would also generate super high current density and low voltage, which is a big challenge for subsequent module assembling. In this work, we discuss the finely tunned laser scribing process along with the top cell fabrication and their influences on the resulting photovoltaic performances. With increased laser scribing widths, the current leakage is significantly avoided and the series resistance reduces, leading to improved FF as well as promising PCE over 16%. Additionally, the bottom cell receives more light incident through the widden laser scribing lines, increasing the short-circuit current of bottom cell. The resulting encapsulated stacks with identical aperture area of ~240 cm² exhibit a certified record PCE of 25.9% for large-scale perovskite/Si 4T tandem mini-modules.

II. RESULTS AND DISCUSSIONS

Figure 1 presents the schematic diagram of the top perovskite cell with a inverted device structure of FTO/NiOₓ/SAMs/CsFAMAPb(I,Br)₃/C₆₀/SnO₂/ITO, which is devided into a series topology consisting of equal sub-cells by using laser scribing technique. During the top cell fabrication, laser scribing is applied 3 times after the coating of NiOₓ/SAMs, low-power ITO, and high-power ITO, which are termed P1, P2, and P3 respectively.

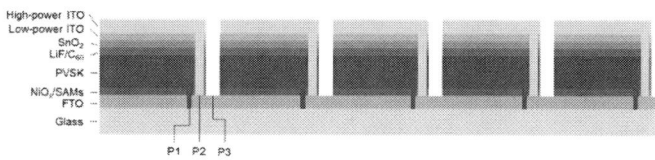

Figure 1. The schematic diagram of the top perovskite cell with a series topology by using laser scribing.

In a completed perovskite top cell, the collected carriers by the ITO back electrode of each sub-cell are directed to the FTO front electrode of the next sub-cell, forming a carrier transport path through the whole top cell series. To ensure good carrier collection, laser scribing should involve the reduction of undesired carrier loss due to current leakage between the neighbor sub-cells. In this regard, the electrical properties and deposition process of each layer in the perovskite cell should be carefully tuned to be compitable to the laser scribing processes.

P1 aims to scribe the FTO layer and form isolated substrates before the subsequent depositions. A conventional P1 process would scribe the FTO to a certain thickness and then deposite the following NiO_x/SAMs. However, NiO_x used for hole transport layer is often doped with Mg, Cu or Li to increase its conductivity. The deposited NiO_x after P1 may fill into the laser-scribing-formed slots, causing current leakage between neighbor sub-cells and resulting low FF for the whole top cell. To address this issue, P1 is applied right after deposition of the NiO_x layer in this work. Afterwards, the perovskite composition fills into the P1 processed slots during the perovskite deposition and forms low conductive region, which hinders lateral charge transportation between neighbor cells and improves the FF of the whole top cell.

Following functional layers including perovskite, $LiF/C60$, and SnO_2 are deposited by spin-coating, thermal evaporation, and ALD respectively. Presence of ITO in the perovskite module plays key role in the realization of charge transportation between the sub-cells. On the one hand, ITO is used as back electrode to collect the charges from the SnO_2 interface. On the other hand, ITO is also introduced into the P2 slots during deposition to direct the collected charges to the FTO substrate layer of the next cell, thus forming a carrier transport path through the whole top cell series. ITO is often deposited by using magnetron sputtering. High-power depositon of ITO (h-ITO) will damage the formed SnO_2 layer, which leads to the degraded device performances, while the low-power ITO (l-ITO) leads to less damage to the SnO_2 layer but often exhibits low carrier conductivity. To combine the advantages of both l-ITO and h-ITO layers, a bilayer ITO structure of l-ITO/h-ITO was introduced as shown in Figure 1. P2 is applied between l-ITO and h-ITO deposition (SnO_2/l-ITO/P2/h-ITO). Compared to the conventional SnO_2/P2/ITO process sequence, l-ITO in this new configuration avoids the damage of h-ITO deposition, while pure h-ITO filling in the P2-induced slots enables good carrier conductivity to reduce the series resistance, which is also beneficial for high FF.

The width of the P2 and P3 lines is found to influence the insulation effect and the light transmission to the bottom cell significantly. The broadened P2 slot loads more h-ITO, resulting in reduced series resistance between sub-cells and improved FF for the whole top cell. Broadening the width of P3 lines results in reduced current leakage with better insulation effect between the neighbor sub-cells. Broadening width of the P2 and P3 lines allows more light transmission to the bottom cell and thereby increases the I_{sc} of the bottom cell at cost of slightly decreasing the I_{sc} of the top cells. Additionally, reduction of the distance between P2 and P3 doesn't lead to extra current leakage.

The optimized top perovskite cell stacking with an in-house made HJT bottom cell was encapsulated into a monolithic 4T tandem mini-module (aperture area 241.4 cm^2) and certified at

JET (Japan Electrical safety & environment Technology Laboratory). Based on above mentioned optimizations of the laser scribing processes, the current leakage and the series resistance both were significantly reduced and the improved FF to 78% as well as PCE over 16.5% were achieved for the top perovskite cell (Figure 2). I-V measurement of the 4T tandem mini-module yielded stabilized power outputs of 4.022 W and 2.372 W for the top and bottom cells respectively (Figure 3), which translated into efficiencies of 16.7% and 9.8% for the top and bottom cells respectively. Together the 4T tandem mini-module delivered a certified efficiency of 25.9% from MPPT measurement. To our best knowledge, this is the first reported and the highest certified efficiency for large-scale 4T perovskite/c-Si tandem mini-modules.

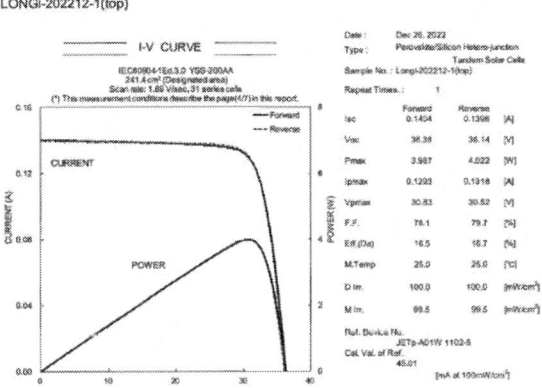

Figure 2. The certified I-V performance of the top perovskite solar cell of 241.4 cm^2.

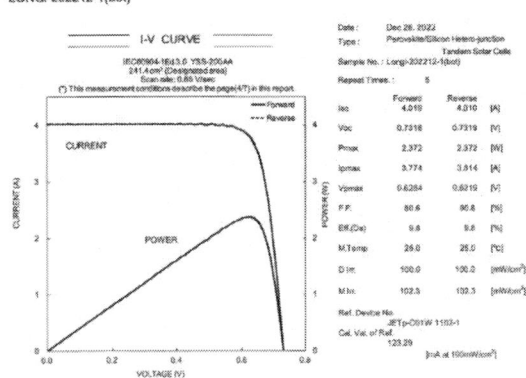

Figure 3. The certified I-V performance of the bottom HJT solar cell with aperture area of 241.4 cm^2.

In conclusion, laser scribing processes for wide band gap based semi-transparent perovskite modules were optimized, including: (1) Applying process sequence of FTO/NiO_x/SAMs/P1 avoids current leakage caused by high conductive NiO_x particles from lase scribing P1; (2) Applying process sequence of SnO_2/l-ITO/P2/h-ITO avoids the sputtering damage to SnO_2 while reduces the series resistance by increased h-ITO volume; (3) Broadening the P2 and P3 lines reduces current leakage and also allows more light transmission as well as higher Isc for the bottom HJT cell. As a result, the top perovskite solar cell exhibits an improved FF of 78% and a PCE over 16%. Finally, a record PCE of 25.9% was certified at JET with an encapsulated 4T tandem mini-module.

REFERENCES

[1] Martin A. Green, Ewan D. Dunlop, Gerald Siefer, Masahiro Yoshita, Nikos Kopidakis, Karsten Bothe, Xiaojing Hao, *Solar Cell Efficiency Tables* (Version 61), 2022.

[2] B. Chen, N. Ren, Y. Li, L. Yan, S. Mazumdar, Y. Zhao, X. Zhang, "Insights into the Development of Monolithic Perovskite/Silicon Tandem Solar Cells," *Advanced Energy Materials*, 12(4), p. 2003628, 2022.

[3] National Renewable Energy Laboratory, "Best Research-Cell Efficiencies: Emerging Photovoltaics," 2022, https://www.nrel.gov/pv/assets/pdfs/cell-pv-eff-emergingpv.pdf.

[4] M. Jaysankar, M.Filipic, B. Zielinski, R. Schmager, W. Song, W. Qiu, U. W. W. Paetzold, T. Aernouts, M. Debucquoy, R. Gehlhaar and J. Poortmans, "Perovskite–silicon tandem solar modules with optimised light harvesting" , *Energy Environ. Sci.*, 2018, 11, 1489-1498.

Predicting damp heat degradation in heterojunction PV modules using machine learning

Zubair Abdullah-Vetter, Felix O' Kearney, Priya Dwivedi, Robert Lee Chin, Brendan Wright, Thorsten Trupke, and Ziv Hameiri

University of New South Wales (UNSW), Sydney NSW 2052, Australia

Abstract— Due to the ongoing advancements in the efficiency of solar cells, photovoltaic-generated electricity is now the most affordable energy source globally. Nevertheless, to unlock the full potential of photovoltaic systems, their reliability needs to be improved. The capability to accurately predict the performance of photovoltaic modules over their years of operation using fast and cheap methods can be a game changer. In this study, the performance of photovoltaic modules is monitored during 1,500 hours of damp heat testing. Classical machine learning models were then developed to predict their performance at the end of the test, using ONLY 10% of the measurements. This research represents a crucial step toward predicting the long-term performance of photovoltaic modules in the field, a capability that will revolutionize the photovoltaic industry.

Keywords—extended damp heat testing, IEC 61215, end-of-life prediction

I. INTRODUCTION

The continued increase in conversion efficiency and decrease in cost has made photovoltaic (PV) energy the cheapest form of electricity in most countries [1]. When considering the cost of PV systems, their reliability has a critical impact as it is directly related to the expected lifetime of the systems (often 25-30 years of operation are assumed). The provided warranties, rebates, and paid purchase agreements all depend on the expected performance in the field across the anticipated lifetime duration. Developing a cost-effective capability to accurately predict long-term performance would drastically improve both the reliability and bankability of PV systems. Such capability will significantly advance the PV market, unlocking the full potential of this technology as the leading form of electricity of the future.

Accelerated degradation testing is a widely used method for evaluating the performance of PV modules under different outdoor conditions [2]. These tests are required by the International Electrotechnical Commission (IEC) and focus on various types of possible degradation modes. One common accelerated degradation test is damp heat (DH), which evaluates the performance under high temperature and humidity conditions [3]. To be approved for the market, IEC demands that a PV module must withstand conditions of 85°C and 85% relative humidity (RH) for 1,000 hours while maintaining at least 95% of its efficiency [3].

Common degradation modes found in PV modules due to DH include delamination and discoloration of the encapsulant, corrosion or breakages of the cell interconnections, as well as degradation of the surface passivation [4][5]. Current-voltage (I-V) measurements throughout a DH test can be modeled by analytical expressions that predict the overall performance or 'time to failure' [4][6]. However, complete DH tests are time-consuming and resource-intensive, usually running for multiples of the 1,000-hours standard [7] and therefore expensive. Furthermore, the fit parameters of these methods may not provide much insight to a wider audience.

This study proposes a machine learning (ML) approach for predicting the performance of PV modules undergoing DH testing. The ML models are trained to use only 10% of the measurements to predict the quality at the end of the test. The results demonstrate that the developed algorithms can predict degradation trends with high accuracy, even when utilizing a limited amount of data. The study is proposed as a step toward predicting the accurate long-term performance of PV modules in the field.

II. METHOD

A. Damp heat test

Two sets of measurements were conducted. For each set, 11 modules were placed inside an environmental chamber (ASLI TH-150C) to be measured at 85±0.5°C and 85±2% RH. The total 1,512 hours of DH testing were divided into 21 intervals of 72 hours. After each interval, the chamber was ramped down to room temperature and the modules were removed for measurements (Section *D*). Reference modules (Section *B*) were kept in a nitrogen cabinet for comparison.

B. Samples

The samples used in the DH tests are four-cell mini-modules (referred to as modules). Nine busbar hetero-junction (HJT) solar cells with measured efficiencies in the range of 24.1±0.02% were selected. The modules were fabricated using 3 mm thick soda lime glass (400×400 mm), an ethylene vinyl acetate (EVA; Lushan) encapsulant, and a polyethylene terephthalate (PET) based backsheet (Jolywood). A total of 24 modules were fabricated and 22 were subjected to DH. The remaining two modules were used as reference samples to monitor the repeatability of the various characterization systems. The data collected from the reference modules also

---

This work was supported by the Australian Government through the Australian Renewable Energy Agency (ARENA, Grant 2020/RND016). The views expressed herein are not necessarily the views of the Australian Government, and the Australian Government does not accept responsibility for any information or advice contained.

978-1-6654-6060-6/23 $31.00 © 2023 IEEE

helped to confirm that any degradation seen in the time series data was indeed caused by the DH conditions.

C. In-situ measurements

Throughout the test, *in-situ* dark I-V curves were measured using a customized characterization system. A Keithley 2561A source measurement unit (SMU) was connected to each of the modules via a series of relay channels. Each module had a 4-point probe connection, such that the collected dark I-V measurements excluded the series resistance due to the contacts and the cables. A LabVIEW code controlled the SMU such that the dark I-V measurement of each module could be collected every 20 minutes.

D. Ex-situ measurements

The environmental chamber was ramped down to room temperature every 72 hours for measurements under ambient conditions. Light I-V measurements were conducted using a flash module I-V tester (Eternal Sun; Spire) with an AM1.5G spectrum. All of the measurements were temperature corrected using the temperature coefficients of the cells. A BT Imaging M1 module imaging tool was used to collect electroluminescence (EL) and line-scan photoluminescence (PL) images. A line scanning speed of 40 mm/s with a forward bias of 8 A was applied to capture the EL images while an illumination intensity equivalent to 1-Sun photon flux was used to capture the PL images (at the same line scanning speed). The reference modules were measured together with the tested modules to assess the repeatability of the measurement systems.

III. MACHINE LEARNING MODELS

Typically, ML regression methods require a substantial amount of data while a single batch of the DH experiment only yields 11 sets of time series data. Therefore, the "sliding-window" method [8] was employed to augment the number of datasets available for ML training. The method involves extracting input-target pairs from time series data and treating them as separate training samples [8]. For example, from the efficiency dataset, the first three measurements (0th to 144th hour) are treated as the training input while the ninth last measurement (936th hour) is used as the *target value*. This is the first input-target pair. For the next pair, the "window" is shifted by 72 hours such that the measurements from the 72nd to 216th hours are the training input and the 1,008[th] efficiency measurement is the target value. This process is repeated, resulting in multiplying the time series dataset of one module by nine times. The timestamp difference between the input and target points for each pair is kept constant. The resulting complete dataset has 99 input-target pairs. Note that the input features can be any of the measured electrical parameters selected from any of the three available datasets (in-situ dark I-V, light I-V, and EL/PL images).

A support vector regressor (SVR) [9] model was used in this study. A single SVR model was trained on ten modules while one module was withheld for testing. This was done to avoid any data leakage between the training and testing sets. Once the nine individual target points were predicted, an exponential fit between the predictions was calculated to filter the measurement noise, which would otherwise propagate into the ML predictions. This curve of best fit was taken as the final

prediction of the module's efficiency at the end of the DH degradation. The performance of the model was evaluated using the mean absolute percentage error (MAPE) [10]:

$$\text{MAPE} = \frac{1}{N}\Sigma_{i=1}^{N}\frac{|y_i - \hat{y}_i|}{|y_i|}, \tag{1}$$

where N is the number of samples, i is the ith sample, y is the true value and $\hat{y}$ is the predicted value. The lower the MAPE value, the better the prediction performance.

IV. RESULTS

A small fraction of the time series data collected is displayed in Fig. 1. Clear trends can be seen in the normalized open circuit voltage (V_{oc}), efficiency (η), and mean pixel intensity of the module PL images ($\overline{PL}$). In contrast, the noise in the short circuit current (I_{sc}) and fill factor (FF) plots slightly obscures the trends. The reference module (black stars) is also included in each plot, indicating that (1) there is no degradation in that module, and (2) the changes within the measurement systems during that time were insignificant. Therefore, the trends seen in the other modules can be attributed to the degradation caused by the DH conditions. A computer vision algorithm was also used to extract the individual cells from each EL and PL image and their mean pixel intensity over time was also included in the training dataset.

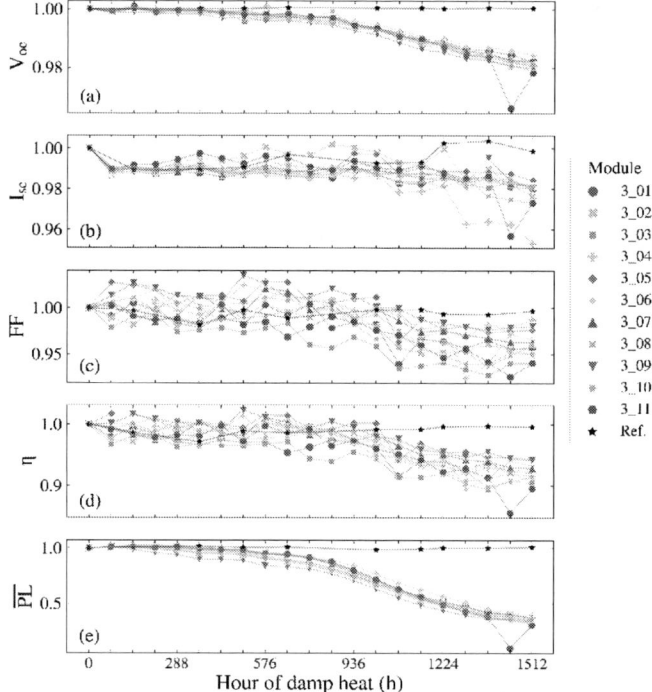

Fig. 1. Normalized key electrical parameters obtained from the time series data collected throughout the DH experiment: (a) V_{oc}, (b) I_{sc}, (c) FF, (d) efficiency, and (e) mean pixel intensity of the module PL images.

Results of withholding only one module as the test set is shown in Fig. 2 below. The normalized efficiency was used as the target for this model and each measured point represents a separate input-target pair from the sliding window method. The figure compares the predicted and the measured efficiency after more than 900 hours of DH test. Note that model was trained using only 10% of the input data at a time, hence, the presented measured points were not used for the training. The trained SVR model predicts the efficiencies with a MAPE of 0.0049. The trend of the predicted degradation closely mimics the measured trend of the test (withheld) module. Most importantly, there is only a small deviation of 0.004 between the predicted and normalized measured efficiency at the end of the DH test. This suggests that a trained ML model can predict the performance of PV modules using just 10% of the time series data, a novel capability that has not been reported previously. However, the accuracy of this approach and its ability to predict module degradation will have to be studied and demonstrated on larger sample sets before final conclusions can be drawn.

While the trained ML model has achieved great preliminary results, we note that the dataset used for training the ML models is just one portion of the dataset collected. Data such as the in-situ dark I-V parameters and EL images have not yet been used in the above ML method. Furthermore, convolutional neural networks (CNNs) are known to extract features from luminescence images that are more powerful than engineered features (e.g. $\overline{PL}$) [11]. The combination of a broader dataset and these CNN-extracted features is likely to produce enhanced results.

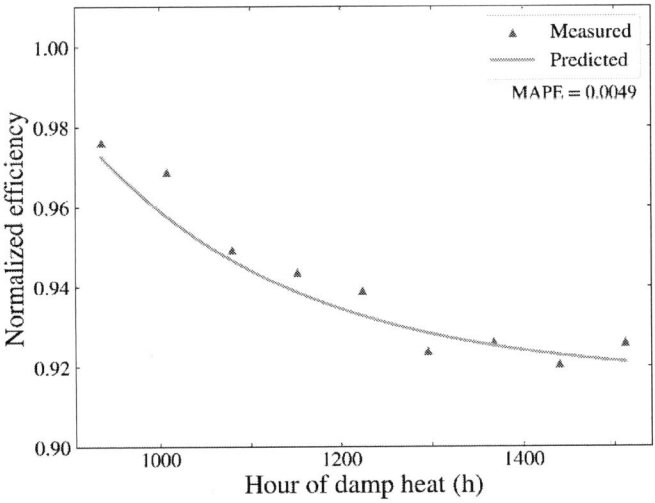

Fig. 2. The predicted and measured efficiency of the test module.

V. Conclusion

This study presents the use of ML to predict the performance of PV modules using just 10% of the measured data. The performance of modules was monitored across 1,500 hours of DH testing. Throughout the test, a variety of time series measurements were collected. The sliding window method was applied to increase the number of samples for the training. It was shown that classical ML models can predict the performance at the end of the DH test, achieving a great MAPE of ~0.005. If proven to work consistently, the developed ML model has the capability to predict the performance of PV modules after long-term field operation, which can contribute to improving the reliability and bankability of PV systems.

References

[1] P. Graham, J. Hayward, J. Foster, and L. Havas, "GenCost 2019-20," CSIRO, CSIRO publications repository, 2020.

[2] M. Kempe, D. C. Miller, S. V. Spataru, P. Hacke, and M. Owen-Bellini, "Combined-accelerated stress testing for advanced reliability assessment of photovoltaic modules," in *35th European Photovoltaic Solar Energy Conference and Exhibition*, 2018, pp. 1101–1105.

[3] International Electrotechnical Commission, "IEC 61215-1:2021." https://webstore.iec.ch/publication/61345 (accessed Dec. 25, 2021).

[4] M. Koehl, S. Hoffmann, and S. Wiesmeier, "Evaluation of damp-heat testing of photovoltaic modules," *Progress in Photovoltaics: Research and Applications*, vol. 25, no. 2, pp. 175–183, 2017.

[5] A. M. Karimi *et al.*, "Generalized and mechanistic PV module performance prediction from computer vision and machine learning on electroluminescence images," *IEEE Journal of Photovoltaics*, vol. 10, no. 3, pp. 878–887, 2020.

[6] N. Kyranaki *et al.*, "Damp-heat induced degradation in photovoltaic modules manufactured with passivated emitter and rear contact solar cells," *Progress in Photovoltaics: Research and Applications*, vol. 30, no. 9, pp. 1061–1071, 2022.

[7] N. Iqbal *et al.*, "Characterization of front contact degradation in monocrystalline and multicrystalline silicon photovoltaic modules following damp heat exposure," *Solar Energy Materials and Solar Cells*, vol. 235, p. 111468, 2022.

[8] L. Mozaffari, A. Mozaffari, and N. L. Azad, "Vehicle speed prediction via a sliding-window time series analysis and an evolutionary least learning machine: A case study on San Francisco urban roads," *Engineering Science and Technology, an International Journal*, vol. 18, no. 2, pp. 150–162, 2015.

[9] T. Hastie, R. Tibshirani, and J. Friedman, "Support vector machines and flexible discriminants," in *The Elements of Statistical Learning: Data Mining, Inference, and Prediction*, T. Hastie, R. Tibshirani, and J. Friedman, Eds. New York, NY: Springer, 2009, pp. 417–458.

[10] F. Pedregosa *et al.*, "Scikit-learn: machine learning in python," *Journal of Machine Learning Research*, vol. 12, pp. 2825–2830, 2011.

[11] J. Fioresi *et al.*, "Automated defect detection and localization in photovoltaic cells using semantic segmentation of electroluminescence images," *IEEE Journal of Photovoltaics*, vol. 12, no. 1, pp. 53–61, 2022.

Demand Following RE – A demand driven approach for rapid RE capacity addition in India

Prashant Kumar Upadhyay, Himanshu Gulati, and Yellasiri Bharath Kumar Reddy

Solar Energy Corporation of India Limited, New Delhi, India

Abstract— In the pursuit of its clean energy goals, India is on an ambitious path for transforming its energy mix. As part of commitments under the INDC, the country seeks to achieve the target of 50 percent cumulative electric power installed capacity from non-fossil fuel-based energy resources by 2030. The long-term goal is to become a carbon net-zero economy by 2070. Considering the current installed capacity of approx. 160 GW, this task entails rapid renewable generation capacity addition at the rate of more than 40 GW/year. Hitherto, RE capacity additions in the country have been largely driven by fulfilment of Renewable Purchase Obligations (RPOs) by utilities rather than a conscious appetite to add RE power to their power portfolios. These obligations are in the form of Solar and Wind RPOs. Accordingly, the power procurement demand has also been in the form of Solar and Wind power purchase agreement contracts. Demand - following RE Contracts seek to introduce a novel, demand-driven approach to RE power procurement by incorporating elements of firmness and flexibility in energy supply agreements, thereby encouraging utilities to actively seek addition of RE Power in their power purchase portfolios while putting the country on path to clean energy transition.

Keywords—RPO, demand-driven, demand-following, complementarity,

I. INTRODUCTION

Renewable Purchase Obligations are policy instruments that mandate power distribution companies, captive power plants and other large electricity consumers to meet a certain percentage of their requirements from renewable energy sources[1]. It is a top-down framework for promotion of renewable energy resources in the country which draws its force from the responsibility entrusted upon State Electricity Regulatory Commissions by The Electricity Act, 2003 enacted by the Indian Parliament. The Ministry of Power, Government of India, in June 2018 notified the long term trajectory of RPOs for solar as well as non-solar for three years from 2019 – 20 to 2021 – 22, reaching 21% of RPO by 2022, with solar being 10.50%. In July 2022, the Ministry notified Renewable Purchase Obligation (RPO) and Energy Storage Obligation Trajectory till 2029-30, requiring RPOs to be progressively increased from 24.61% in 2022-23 to 43.33% by 2029-30[2]. This includes wind RPO, hydropower purchase obligation (HPO) and other RPO. Distribution utilities strive to meet their obligations as per these targets which are notified for a medium to long-term period. This framework has served to create demand and thus, build RE generation capacities in the country as capacity additions have grown from a few MWs in 2010 to ~ 160 GW in 2022. However, this is proving increasingly difficult as the utilities strive to balancing of power to suit the demand with increased percentage of RE, along with conventional power capacities.

With new fossil-fuel based generation capacities not being taken up, the utilities are showing an increasing appetite for renewable energy in their energy portfolios but are worried on the quality (specifically, variability & dispatchability) of RE power. The Transmission Service Operators (CTU/STU) realize that the grid management is increasingly challenging due to higher penetration of Renewable Energy into their networks. Plain-vanilla Solar and Wind Power Sale Contracts lack two critical elements desired by the power utilities – firmness and dispatchability. This makes utilities to continue to depend on conventional sources for ensuring reliability of supply.

Starting in 2019, Solar Energy Corporation of India Limited, a Government of India Company under the Ministry of New and Renewable Energy - the nodal agency for renewable energy sector in India - seeking to address the specific concerns of the utilities invited bids for supply of RE Power with 'schedulable' characteristics. In 2019, SECI invited bids for Supply of Peak Power (6 hours of supply in non-solar hours as per the choice of the procuring utility) and Round-the-Clock Supply of RE Power (with minimum Plant Capacity Utilization Factor of 80%). These bids were significant milestones in India's RE capacity building trajectory as they have served to address specific needs of the utilities, paving the way for increasing the RE generation base in the country.

As next steps, bespoke RE supply contracts that accommodate Daily as well as seasonal variations as per the demand profile of utilities generate keen interest in procurers. At the same time, it would also encourage innovations in project configurations, utilizing the geographical diversity and resource diversity to a greater degree. Overall, to have more reliability & availability in the energy mix of the country, and to ease out the procurement challenges faced by the off-takers, it is important to formulate supply contracts that are conducive to the off-taker. Supply-Demand profile matching solutions & Complimentary Power Purchasing Agreements could pave the wave for new round of innovations in the RE sector as they address off-taker concerns. Project developers benefit through scaling up capacities under a single energy supply agreement With increasing volumes of power being traded on energy exchanges and the country set to add grid level storage capacities. The excess RE power, if any, from these projects are envisaged to replace the conventional power-based fossil fuels in the open market. To address this dynamic scenario in India, load-following RE Power generation concept is introduced so that

978-1-6654-6060-6/23 $31.00 © 2023 IEEE

Project Developer will supply the power to suit the load of the utility by deploying suitable combinations of RE with Energy Storage.

II. INTEGRATING RE INTO THE INDIAN GRID – ADDRESSING DISCOM CONCERNS

Promotion of RE RTC and other innovative projects requires involved understanding of the current position of the utility, its current power procurement portfolio, existing basket of simple RE capacities etc. Hitherto, the RPO mechanism has been a significant driver for Distribution utilities in India, when requisitioning for Renewable Energy. However, the utilities need to address their overall demand from various perspectives – Demand Growth, Demand Patterns, change in demand patterns (e.g., shifting crop patterns due to combination of climate and economic factors), new economic activities, changing consumer behavior, quantum, and nature of PPA as well as their tenures. Continued integration of simple RE capacities, however, is challenging as the DISCOMs need to plan for the Net-of RE demand curve in addition to the variability associated with RE sources. Consider below the seasonal Shortage/Excess Energy Supply Scenario in the north-Indian state of Punjab.

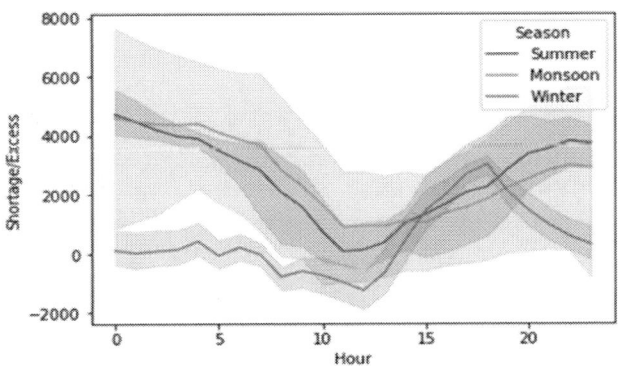

Figure 1: Energy Excess (-) /Shortage (+) Scenario - Punjab

While the state has excess power during the solar hours in Winter season (Oct-Mar), the demand peaks to nearly 4000 MW in the evening hours. In summer and monsoon seasons (Apr-Sep), there remains a subdued demand in the solar hours while it reaches levels of 4000 MW in non-solar hours. The demand profile is indicative of the state's current energy portfolio comprising of a strong solar component towards meeting solar RPOs. It also hints at the subsequent challenge of adding more RE power under a simple Solar or Wind supply contract given the significant seasonal variation of demand. This scenario is common with all states that have pursued a policy of Solar RPO fulfilment and ended with surplus power in the noon.

RE capacity building will be faced with addressing these challenges, especially because RE projects have been supported by the 'must run' status and other energy purchase agreements with 'built-in' variability. Utilities increasingly begin to express the need for specific reliability characteristics of RE Power like assured Peak Time Supply or specified hours of flexible supply. This trend is envisaged to gather more momentum as utilities seek to increase share of RE in their energy mix, while also

keeping the cost of procurement under check, where electricity distribution being a regulated subject.

III. DEMAND-SUPPLY PROFILE MATCHING - INNOVATING AND OPTIMIZING

Under the demand-supply profile matching framework, RE projects deliver an energy profile tailored to the Utility's demand profile. With stringent performance conditions (limited variability), the project developer must configure projects through a combination of generators and energy storage elements while limiting market exposure through excess power. To address large variations in the load pattern across the day & across the seasons strategic RE power generation combinations to be formulated with sufficient energy storage capacity. Key-Features of Demand-Following RE procurement contracts planned are:

Demand Driven Approach: Procurement documents are published with time-bock wise load curves of the off-taker/procurer, with min. and max. offtake guarantees.

Round the Clock Element: High fraction of demand met condition, e.g., more than 85-90% on monthly basis and > 90% on annual basis in energy terms, with reference to the demand curve specified in the procurement agreement.

Reliability and Schedulability: The procurement document to specify maximum variation (say +/-10%) from the Day Ahead Schedule in each time-block allowed to the supplier. Limited (and specified in the procurement agreement) Intra-Day Variability allowed to the Supplier and to be informed before two-time blocks of open market gate closure.

IV Project Configuration: Demand following energy agreements seek to enable innovative project configurations through a mix of geographical diversity, generation diversity (solar and wind in various fractions), incorporation of energy storage elements, simultaneously enabling a diversified generation base in the country. For e.g., the combined energy profile from different locations like Bhadla (Solar), Anantapur and Tuticorin could deliver an energy profile with longer hours of evening peak supply while also increasing reliability (illustration below hints at different peak generation hours in Anantapur and Tuticorin)

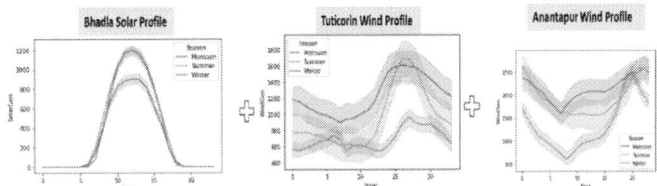

Figure 2: Combine Generation Profiles to mitigate variability and address specific demand patterns

As an illustrative example, the off-taker (Punjab), agrees to a month-wise min-max demand profile (500MW-1500 MW) which accounts for its seasonal variations. Therefore, the overall shortage-excess scenario (Fig. 1) is mapped to a 'demand-profile' that accommodates the seasonal variations. The figure below illustrates the season-wise demand-profile created for the state of Punjab.

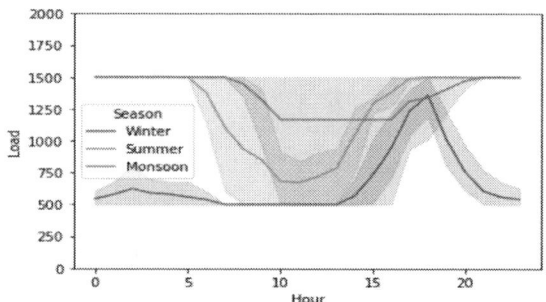

Figure 3: Punjab Demand Profile for RE Procurement

The total annual energy demand under the above demand profile is 8.4 BUs. Against the demand profile illustrated at Fig.3, a 1500 MW 'Flexible RE' contract with geographically distributed project configuration (In the present case Solar and Wind generation profiles of Anantapur, Andhra Pradesh was considered) comprising of 1000 MW Solar, 3500 MW Wind and 1200 MWh Energy Storage System delivers power as per below illustration, meeting nearly 99% of the total power demand.

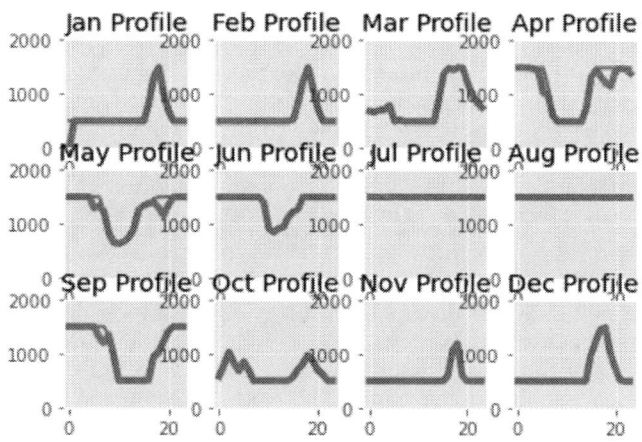

Figure 4: Month-wise Demand (Blue) Vs Supply (Red)

A typical day scenario is presented in the figure below:

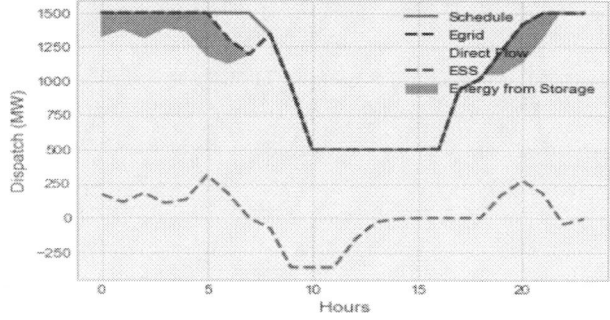

Figure 5: Demand Following Delivery Profile- Sample Day

From the figure illustration, the Project delivery profile (Egrid) is expected to match the Scheduled Injection profile (Schedule) with a combination of Direct Energy Flow (illustrated as filled area in yellow) from RE generator (Solar or Wind) and Storage element (illustrated as filled area in red). The off-taker actively stipulates a desired monthly time block-wise profile that the Project Developer must cater to. This is expected to address specific procurement planning issues of the Utilities' (e.g., reduced offtake during solar hours or seasonal variations in demand or peak hour supply requirement), thereby making RE more conducive to the demand.

The excess power from the project (~ 35%) can be supplied to another state with complementary demand. For e.g., Madhya Pradesh, another Indian state has a complementary demand.

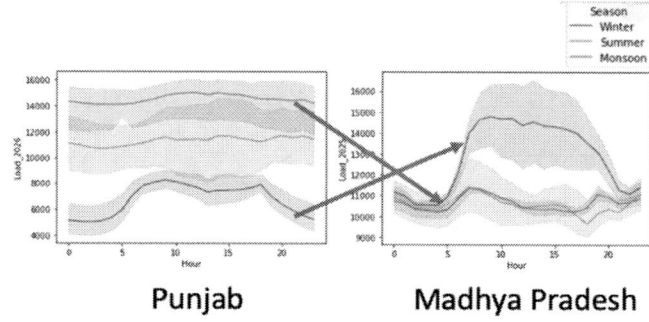

Punjab   Madhya Pradesh

Figure 6: States with complementarity of Demand

For future power procurement planning by utilities that find absorbing additional RE capacity to be challenging, incremental growth of demand and net of tied-up RE is analysed. In the manner illustrated above, the representative 'demand-profile' is constructed and Project Developers invited to supply as per the demand curve under long-term contracts. The treatment of excess power from the project could be mutually decided between the generator and procurer.

IV.   CONCLUSION

Demand-driven supply contracts of RE power can prove to be beneficial for the off-taker as it covers the cost of balancing power & thereby provides the off-taker a reliable solution completely with RE Power. In the case of federal structures like India, intermediary procurer could play the role of energy integrator to reduce overall project risks. For cases where large RE capacity additions are to be achieved, the model also helps achieve rapid capacity addition.

REFERENCES

[1]   https://rpo.gov.in/Home/Objective

[2]   https://powermin.gov.in/sites/default/files/webform/notices/Renewable_Purchase_Obligation_and_Energy_Storage_Obligation_Trajectory_till_2029_30.pdf

Perspectives on PV Adoption and Engaging Gen Z and Millennials in the Indian Scenario

Robins Anto[1] and Rhythm Singh[2]

[1,2] *Department of Hydro and Renewable Energy*

Indian Institute of Technology Roorkee, Roorkee, Uttarakhand, India. [1]r_anto@hre.iitr.ac.in, [2]rhythm@hre.iitr.ac.in

Abstract— **India is targeting a renewable energy capacity of 500GW to meet 50% of energy requirements from renewables by 2030. This will help in reducing the CO_2 emissions by a billion tonnes. The role of Solar PV in achieving the above target will be crucial. Being a country with 48% of the population under 40 years of age, successful implementation of the energy policies depends upon the perspectives of Millennials (aged 25-39 years) and Generation Z (Gen Z, aged 15-24 years). Hence, the timely engagement of the above segments of the population and their adoption of smart grid technologies and rooftop solar PV defines the success of the national smart grid and solar mission. This study engages Gen Z and Millennials to understand their views on energy saving, using electricity from renewables, and adopting PV and electric vehicles. The perspectives of EV proponents on adopting rooftop PV are also investigated. The study proposes a methodology for the early engagement of Gen Z and Millennials for knowledge dissemination at all levels and for workforce development. Responses from 1033 participants from 25 states in India were recorded and analysed. All of the participants were up-to-date with technology and were smartphone users.**

Keywords— *Consumer engagement, Gen Z, Global Solar PV adoption, Millennials, User behaviour*

I. INTRODUCTION

In November 2022, India reached the fourth position in the world solar PV deployment with an installed capacity of 62GW. In the same, 52GW is from grid-tied ground-mounted solar parks, another 7.8 GW from rooftop solar and the remaining 2.2 GW from off-grid solar projects [1]. Several new programs like canal bank and canal top solar PV schemes, solar-powered airports, grid-tied and off-grid rooftop solar PV are floated to increase the adoption of PV. The revised target through the national solar mission is 100GW. India is setting an example for all developing countries by following a least carbon-intensive economic development path. India has declared its net-zero emission targets by 2070. It is declared to meet 50% of electricity needs from renewable energy by 2030 [2].

A. Importance of engaging Gen Z and Millennials to achieve India's targets

The pace of technology adoption in the India greatly depends on the people's perspective towards technology. This perspective relies on the knowledge gained by individuals and the strong consumer engagement process undertaken by the agencies. In the case of grid modernization, it is quintessential to learn the consumers' perspective and engage them to be stakeholders of the process. India is a country with 48% of the people under the age of 40 years. This work focuses on understanding the perspective of Generation Z and Millennials regarding the adoption of solar PV. It is an integral part of a larger study that aims to investigate their attitudes towards various aspects of energy conservation, including the use of distributed energy sources (with a specific focus on solar PV), climate change, and the use of smart grid and smart home technologies. The knowledge and acceptance of the technology among the Millennials and Gen Z segments are important to increase the share of rooftop solar power in the grid-tied and stand-alone mode. For the success of all the future Government policies related to emission reduction, use of renewable energy and grid modernization, the above population segments have a critical role as future stakeholders. Hence systematic engagement of Gen Z and Millennials is essential. The primary step is learning their perspectives on all related areas.

II. THE PROPOSED WORK

This work tries to understand user behaviour and the perspectives on energy savings, use of PV and smart grid technologies concerning the Indian scenario. The work presented in this paper is unique, and no related published material is available in the public domain. The highlights are the following.

a) Presents a case study on the Millennial and Gen Z segments of the Indian population, its uniqueness and significance.

b) Consolidates the factors affecting PV deployment in the country.

c) Proposes a methodology for regularly engaging the millennials and Gen Z segments for successful workforce development in PV and grid modernization.

The layout of the work is as follows. Section III details the interactions and surveys conducted among the millennials and Gen Z segments of the Indian population. Section IV describes the result and discussions focusing on user behaviour and perception of PV. The methodology for early engagement of students and workforce development is presented in section V, and the conclusions are detailed in section VI.

III. ENGAGING MILLENNIALS AND GENERATION Z

Online and offline (direct) interactions with Millennials and Gen Z segments of the Indian population were carried out from April 2022 to December 2022. Interaction were made with more than 1500 people directly and more than 700 people in online manner, and finally received a total 1033 samples. The survey investigated the perspective of Millennial and Gen Z segments of populations on a broad range of aspects, including energy savings, climate change, perspective on the adoption of rooftop PV, electric vehicles, and smart home technologies. A brochure expressing interest in conducting an awareness program is sent to the institutions in the regions where the Government of India has proposed or implemented smart

978-1-6654-6060-6/23 $31.00 © 2023 IEEE

metering programs with the help of utilities. The student engagement programs were carried out based on invitations received from the institutions. Every one-to-one interaction involves dedicating 15 minutes to educate the individual followed by 15 minutes survey. The second method of engaging Millennials and Gen Z was by conducting awareness seminars at educational institutions. Two-hour interactive sessions and awareness programs on energy conservation, climate change, rooftop PV, electric vehicles, and smart home technologies were carried out followed by a fifteen-minute survey. Majority of the institutions, where the direct interaction program was conducted, belonged to the Uttar Pradesh, Uttarakhand and Delhi states of India. Samples from the 1033 people belonging to Millennials and Gen Z responses were recorded and processed. Similarly, online awareness programs and surveys were conducted in institutions from other states of India as well. Online sessions and surveys were done in small groups for 45 minutes, followed by 15-minute surveys.

A. Distinctiveness of the sampled population

The sample of population who participated in the study were well-educated, technologically up-to-date, and all of them were smartphone users. The most important fact was that 100% of the population participating in the program had a technical education background and were knowledgeable enough to understand and assess the questionnaire so as to provide a valid response. Fig.1(a) and (b) and fig. 2(a) shows the distribution of the sample population by age, educational qualification, and gender. Among the respondents, 40% of the samples belonged to Millennial population, and 60% belonged to Gen Z. Whereas, the number of females respondents in the survey is between 20-25% among both segments. Fig. 1(b) provides a clear depiction of the educational level of the survey respondents. 72% of the sampled population were about to complete their bachelor's degree, 24% were postgraduates and PhD pursuing people, and the remaining 4% hold a doctoral degree and are employed. The educational background of the participants points towards the quality of the samples. Respondents for the survey include millennials and Gen Z populations from 25 states all over the country.

IV. RESULTS AND DISCUSSION

The study emphasized engaging Gen Z and Millennials, constituting almost half of the Indian population. The perspectives of these population segments highly influence the country's sustainability plans towards 2030 through 2070. One such goal is to achieve a 30% penetration of Electric vehicles (EVs) in the country's market by 2030. This has an impact on PV adoption also as in fig. 2(b). All factors that influence PV deployment are briefed below.

A. Factors influencing the deployment of PV

Varied factors influence the adoption of PV among customers. The major factors influencing Millennials and Gen Z segments of population identified through this study, are summarized in fig. 3, which will affect the future of PV deployment in the country. The perspective of adopting rooftop PV by electric vehicle proponents is given in fig. 2 (b). It is evident from the fig. that EV proponents who are

willing to adopt rooftop solar PV are 49%. This points for high rooftop PV adoption with an increase in EV deployment in the country.

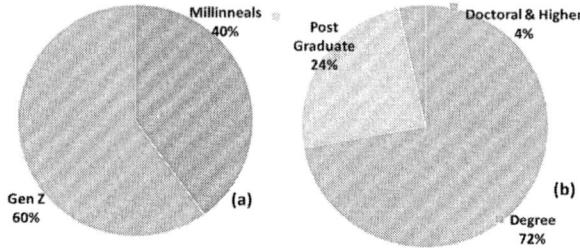

Fig. 1 (a). Millennials and Generation Z population segments participated. (b). Educational background of the sampled Indian population.

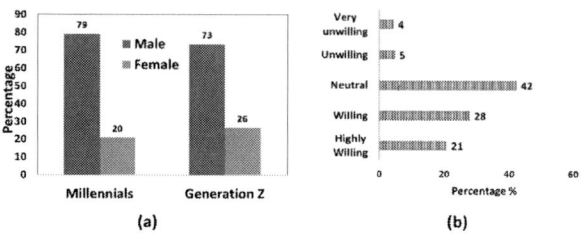

Fig. 2 (a). Proportion of males and females in the millennial and Generation Z who participated in the survey. (b). Perspective of adopting PV by EV proponents.

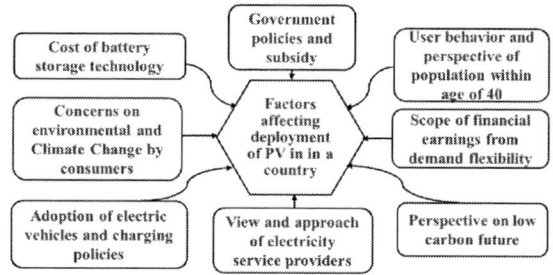

Fig. 3. Factors affecting the deployment of PV.

The promising fact is that only 9% of people are reluctant to adopt PV even if they own an EV in the future. Also, 42% of the population are ignorant about the technologies' benefits to energy, the environment and sustainability. Gen Z and Millennials should be well informed about the huge potential for demand flexibility with the rise of PV and EV adoption and the mutual benefits for the consumers and utility. The utilities can engage youngsters by conducting early awareness programs in educational institutions and hence influence their perspectives to adopt PV, EV and smart grid technologies.

Fig. 4(a) shows that 72% of the respondents are interested in using renewable-based electricity even at a slightly higher price. From Fig. 4(b), it is evident that 70% of the young generation are worried about India's growing energy demand. The above perspectives among Gen Z and Millennials certainly boost PV adoption. The interactions with the Millennials and Gen Z revealed that their attitude towards energy saving, use of renewables, and smart grid technologies vary widely. Many respondents believe they have to be technologically updated always and are keen to get access to the latest at the earliest.

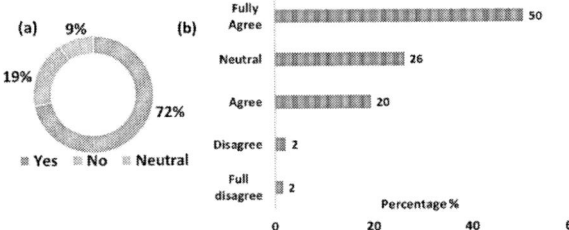

Fig. 4 (a). Percentage of respondents willing to use renewable-based electricity at a higher unit price. (b). Percentage of respondents believing that increasing energy demand is a serious future problem.

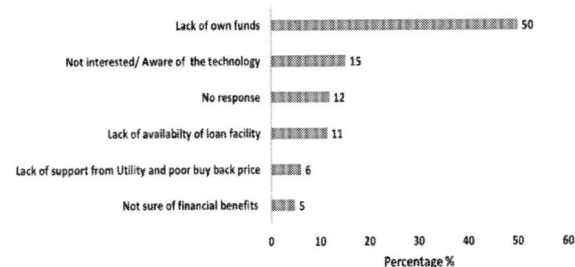

Fig. 5. Barriers to PV adoption in India.

At the same time, another group of respondents are aware of the technological developments but are hesitant to adopt them earlier. They act only based on feedback from their peers who have used the technology. The third category of respondents were aware of technological developments but want a 'status quo'. The fourth set of respondents are reluctant to change unless they are convinced about the financial benefits of adopting the technology. Another group is ready to adopt technological changes if their economic situation is conducive. In a context where Millennials and Gen Z are keenly interested in using renewable energy sources, it is essential to assess the factors that prevent them from adopting rooftop PV. Fig. 5 presents reasons to abstain from adopting PV and EV technologies, irrespective of all the positive perspectives and knowledge. 50% of the population believe that the lack of funds is the major hurdle, and 15% are not interested in the technology or are not aware of the benefits of the technology. Lack of appropriate loan facilities is considered a hurdle by 11%. Fear of not receiving a fair price for electricity exported to the network, poor utility support and awareness of the benefits are reasons for another 11% for not going for PV. The same was proved from an Australian case study that the fear of fair price will refrain consumers from adopting PV [4].

B. Proposed methodology for early engagement of students

The responsibility of engaging the Millennials and Gen Z to be a part of grid modernization is left with the regulators and electric utilities. To satisfy the objectives, respective utilities in the state and regions should establish 'institution-industry interaction' in every educational institution. This will help systematically engage the population segments and make them aware of renewables

and smart grid technologies. The government and regulatory bodies shall monitor the implementation. Fig. 6 depicts the proposed methodology adopted for this work.

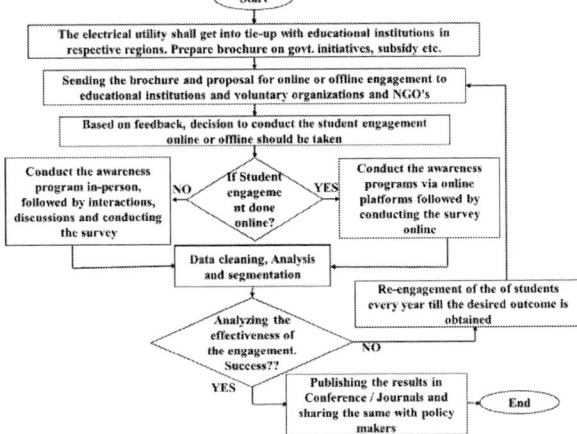

Fig. 6. Methodology for engaging the students from educational institutions for successful workforce development and knowledge dissemination.

Technical schools will be able to provide the needed workforce for the country as an outcome of the ongoing industry-institute engagement.

V. CONCLUSION

It was observed that the young population is inclined towards energy savings and renewables and that 49% of EV proponents are interested in PV adoption. However, there exists further scope for increased PV adoption by educating 42% neutral population in the segment to be prosumers. The study also identified the barriers to future PV adoption and found that the lack of funding is the major hurdle. The study observed that a greater emphasis on energy saving, climate change, emission reduction, electric vehicles, and a focus on smart grid technologies influences the perspective of the population to adopt PV. Early engagement of students will help grid modernization by motivating future generations to be prosumers and stakeholders of the utility. This in turn positively impacts solar PV deployment and future workforce development in the country. A methodology followed for early engagement is also presented. A detailed study by segmenting the population based on their perspectives is essential for the successful employment of the young population in the future.

REFERENCES

[1] MNRE," Government, 29 March 2023. [Online]. Available: https://mnre.gov.in/solar/solar-ongrid. [Accessed 25th March 2023].

[2] Dr Faith Briol, Amithab Kanth, "India's clean energy transition is rapidly underway, benefiting the entire world," IEA- International Energy Agency, 2022 January.

[3] L. B. K. M. D. V. Jeff Sommerfeld, "Influence of demographic variables on uptake of domestic solar photovoltaic technology," *Renewable and Sustainable Energy Reviews,* vol. 67, pp. 315-323, 2017.

[4] J. Clifton, "The emperor and the cowboys: The role of government policy and industry in the adoption of domestic solar microgeneration systems," *Energy Policy,* vol. 81, pp. 141-151, 2015.

An Analysis of the Current Status and Future Potential of Rooftop Solar Adoption in the United States

AC Lemay, BP Rand

Princeton University, Princeton, NJ, United States

We utilize a dataset that measures existing rooftop solar installations and classifies rooftops in terms of insolation, azimuth angle and pitch, shading, and size, from aerial imagery last updated in 2017. Analysis of these data reveals that rooftop solar adoption, defined as the number of buildings with existing photovoltaic (PV) installations divided by the total number of eligible buildings, is on average low (mean of 0.93% for 10,417 U.S. ZIP codes). Regarding potential electricity generation, fifteen states could meet their net residential electricity demand if panels were placed on all suitable buildings. We conduct a linear regression analysis to elucidate factors that positively (insolation, retail electricity price, Democratic voting fraction, net metering, fraction of science or engineering degree holders) and negatively (fraction of business or education degree holders) correlate with solar adoption. The results suggest anticipated electricity cost savings as a strong motivator for PV adoption, particularly in majority Republican areas. Installation cost and knowledge, however, remain barriers. Knowledge campaigns regarding technical aspects of installation and maintenance, as well as increased and stable financial incentives, may spur further PV deployment.

Investigation of Varying Se Vapor Pressure During Deposition of CdSeTe Thin Film PV Devices

Sushmakanth Myneni, Carey Reich, Daniel Shaw, Sampath Walajabad, Amit H. Munshi

Colorado State University, Fort Collins, Colorado, 80523, USA

Abstract — **Polycrystalline CdSeTe (CST) thin film material has shown External Radiative Efficiency (ERE) as high as 6% and carrier lifetimes as high as 4 microseconds measured using TRPL. However, resultant CST-only absorber devices exhibit poor cell performance. Due to the higher vapor pressure (V_p) of Se in comparison to Cd and Te, CST films are likely to have Se vacancies. Normalized emission reconstructed from PL exhibits low energy emission peak suggesting sub-bandgap features in CST. Device modeling using SCAPS 1-D suggests low hole mobility in CST. In this work, the role of vacancy Se (V_{Se}) that may be responsible for the low energy emission peak and poor carrier mobility in CST-only absorber devices is being investigated.**

I. Introduction

High efficiency polycrystalline CdTe-based thin film photovoltaic (PV) devices typically consist of CdSeTe(CST)/CdTe bilayer absorber. The highest efficiency device demonstrated by First Solar Inc. with 22.1% efficiency is believed to have a similar bilayer structure [1]. During the $CdCl_2$ passivation treatment, interdiffusion between CST and CdTe layers leads to Se gradient which also leads to grading of the bandgap within the absorber film [2]. Studies have shown that Se passivates grain boundaries as well as bulk defects [3]. CST has higher carrier lifetime in the bilayer CdTe-based solar cells. Gao et al [4] reported significantly higher minority carrier lifetime within the CST layer as compared to the CdTe layer within the same bilayer absorber film. Ablekim et al., [5] attributed 200ns of carrier lifetime to the presence of CST layer in the CdTe solar cells. Lifetimes up to several microseconds have been measured with CST-only devices prepared at Colorado State University (CSU) [7]. Fiducia et al., [6] reported higher CL intensity in the areas of higher Se concentration in comparison to areas with lower Se concentration suggesting alloying CdTe with Se reduces non-radiative recombination in the material. Overall, superior nature of CST material in terms of higher carrier lifetime, lower non-radiative recombination in comparison to CdTe material, encouraged researchers in the direction of investigating CST-only absorber devices. Higher ERE of 6% is reported in CST-only devices compared to ~0.1 % ERE in similar structured bilayer devices [7]. However, CST-only absorber results in poor devices.

A. Sub band gap defect in the CST bulk:

CST-only devices fabricated at (CSU) exhibit emission peaks at 1.37 eV (bandgap) and 1.1 eV (corresponding to sub band gap feature). Similarly, PL of bi-layer absorber exhibits low

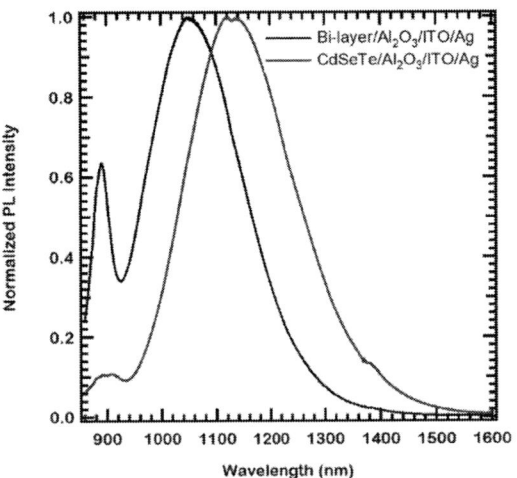

Fig. 1. Normalized PL emission of CST-only and Bi-layer device.

energy emission peak. However, CST-only shows very large low energy peak– indicating that alloying Se with CdTe is causing this low energy peak. In case of lower Se concentrations, Fiducia et al., [6] reported increment in the sub band gap peak with Se composition in the CST film suggesting that this sub band gap feature is Se related defect. Consequently, high Se vapor pressure among alloying elements in CST, composition of Se is lower than that in the composition of the sublimation source charge [8]. Particularly, possible presence of Se vacancies in the CST film have been reported [9]. Vacancy Te (V_{Te}) with activation energy $E_V+0.23$ eV is a hole trap in CdTe films [10]. Similarly, there is a possibility that V_{Se} (same group as Te) is a hole trap in CST films. Injection-dependent and wavelength dependent TRPL measurements suggested the presence of such traps [11]. Short initial decay in PL which specifies carrier separation and movement is absent in case of low energy peak indicating that the carriers are trapped before they can recombine radiatively [7]. Overall, Se vacancy defects may be associated to sub band gap feature in CST which may also be possible traps for carriers, resulting in poor open-circuit voltage (V_{OC}) and short-circuit current (J_{SC}) observed in polycrystalline CST-only thin film devices.

B. Similarity with CIGS technology:

In CIGS technology, several investigations of V_{Se} in the CIGS thin films have been reported [12][13]. Similar to CST, low energy peak corresponding to 1.05 eV is reported [14]. In fact, this low energy emission peak was reported extensively during early studies of $CuInSe_2$ (CIS). Multiple studies associated this peak to Vse[15][16]. Igalson et al., [17] stated that better performing devices can be made by reducing V_{Se}. Zhang et al., [14] reported deterioration of conversion efficiency in CIGS-based solar cells due to presence of V_{Se}. They attributed vacancy Se defect to the high vapor pressure of Se. Many strategies have been implemented in CIGS technology to eliminate V_{Se} and to correct the Se composition in the film. Krishna et al., [18] investigated change in Se fluxes during the deposition of CIGS films and concluded that hole concentration increased as resistivity decreased with increase in Se flux. Masse et al., [16] reported that V_{Se} were eliminated by annealing the CIS film in Se vapor. Consequently, improvement in the conductivity of the film was observed.

However, in case of CST, such efforts on eliminating Se vacancies in CST film by Se overpressure or annealing CST film over Se and their effect on the low energy emission peak and carrier mobility have not been reported. Addressing low carrier mobility by eliminating sub band gap features that are believed to be hole traps in CST-only devices with high ERE, carrier lifetime, implied voltage and implied efficiency, is essential step towards higher efficiency solar cells approaching the thermodynamic efficiency limit in CdTe-based PV technology [19].

II. EXPERIMENTAL

Two sets of CST-only thin film are being investigated:
- CST films deposited under excess vapor pressure of Se using CSU's proprietary co-sublimation process [20].
- Annealing CST under excess Se vapor.

More details on film deposition and device fabrication method are reported elsewhere [7].

A. Co-sublimation process

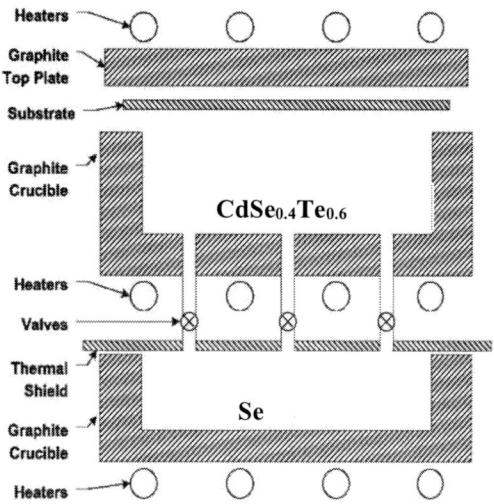

Fig. 2. Co-sublimation setup above is part of ARDs at Colorado State University [20][21][23].

Samples will be fabricated using co-sublimation hardware at CSU. Detail of the equipment is reported elsewhere [20]. Drew E. S et al., [21] deposited CST film with CdTe in the top crucible and Se in the lower crucible. In this study, we deposit CST film using $CdSe_{0.4}Te_{0.6}$ in the top crucible and Se in the bottom crucible as shown in the schematic in figure 2. CST films under excess Se vapor are being grown using co-sublimation method and hardware with an aim to eliminate Se vacancies. These experiments are currently underway and based on preliminary results it is promising.

B. Annealing over Se source

Samples are being fabricated by first depositing CST films and then annealing it over a Se vapor source.

Results and their analysis from these experiments will be reported at the conference.

C. Characterization

Photoluminescence (PL), Injection-dependent TRPL, External Radiative Efficiency (ERE), SEM/EDS, J-V characterization will be performed on the samples described above.

III. RESULTS AND DISCUSSION

A. SCAPS 1-D device modeling

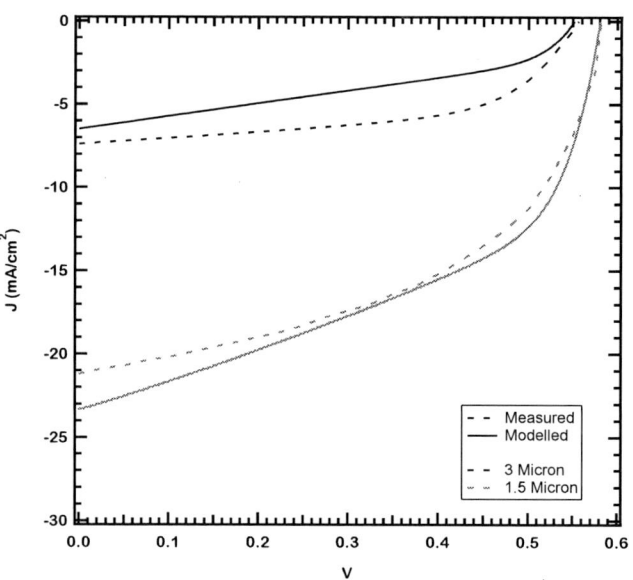

Fig. 3. Actual JV and JV from SCAPS modelling of CST only devices.

Reducing the hole mobility in the SCAPS model matches actual JV curves of devices made up of CST only absorber. In detail, modelled JV curves for different thickness of the CST layer (1.5 microns and 3 microns) is comparable to the actual JV curves by reducing the hole mobility to 0.05 cm²/Vs. The

978-1-6654-6060-6/23 $31.00 © 2023 IEEE

defects were modeled under the assumption that the hole mobility is limited due to V_{Se}. Further studies will be performed, and results reported at the conference to substantiate these models.

IV. SUMMARY

CST-only absorber devices fabricated at CSU have shown high ERE and carrier lifetimes but with efficiencies below 5%. Implied efficiency for such devices is around 25% measured by ERE at different intensities of light and correcting for sub band gap peak. PL showed low energy emission peak suggesting sub-band gap features. It is believed that Se vacancies are associated to these sub-band gap features. TRPL of this low energy peak suggests carrier traps in the CST film. SCAPS 1-D modelling, matching simulated JV to actual JV curves of the cells, attributed low performance of the devices to low hole mobility. Sebastian P. J et al., [9] have reported possible presence of V_{se} in CST films as a consequence of high Se vapor pressure. V_{Te} in CdTe reported as hole trap. Similarly, V_{se} believed to trap holes in CST films, resulting in low carrier mobility- a phenomenon which was already investigated and reported in CIGS films.

To eliminate V_{se} in CST films, we fabricate: a set of CST samples produced with Se over pressure and another set annealed over Se source. Their effects on low energy emission peak and the carrier mobility in the CST film are being investigated. Addressing charge mobility problem in CST-only devices is a crucial step in advancing to higher efficient CdTe-based solar cells.

ACKNOWLEDGMENTS

This conference abstract was developed based upon funding from the Alliance for Sustainable Energy, LLC, Managing and Operating Contractor for the National Renewable Energy Laboratory for the U.S. Department of Energy.

REFERENCES

[1] Green, MA, Dunlop, ED, Hohl-Ebinger, J, Yoshita, M, Kopidakis, N, Hao, X. Solar cell efficiency tables (version 59). *Prog Photovolt Res Appl.* 2022; 30(1): 3- 12. doi:10.1002/pip.3506

[2] Munshi, Amit H., et al. "Effect of CdCl2 passivation treatment on microstructure and performance of CdSeTe/CdTe thin-film photovoltaic devices." *Solar Energy Materials and Solar Cells* 186 (2018): 259-265

[3] Shah, Akash, et al. "Understanding the copassivation effect of Cl and Se for CdTe grain boundaries." *ACS Applied Materials & Interfaces* 13.29 (2021): 35086-35096

[4] Guo, Jinglong, et al. "Effect of selenium and chlorine co-passivation in polycrystalline CdSeTe devices." *Applied Physics Letters* 115.15 (2019): 153901.

[5] T. Ablekim et al., "Tailoring MgZnO/CdSeTe Interfaces for Photovoltaics," in IEEE Journal of Photovoltaics, vol. 9, no. 3, pp. 888-892, May 2019, doi: 10.1109/JPHOTOV.2018.2877982.

[6] Fiducia, Thomas AM, et al. "Understanding the role of selenium in defect passivation for highly efficient selenium-alloyed cadmium telluride solar cells." *Nature Energy* 4.6 (2019): 504-511.

[7] C. Reich, Ph.D. Dissertation, Colorado State University (2022).

[8] Munshi, Amit H., et al. "Advanced Co-sublimation of Low Bandgap CdSe x Te 1-x Alloy to Achieve Higher Short-Circuit Current." *2018 IEEE 7th World Conference on Photovoltaic Energy Conversion (WCPEC)(A Joint Conference of 45th IEEE PVSC, 28th PVSEC & 34th EU PVSEC)*. IEEE, 2018.

[9] Sebastian, P. J., and V. Sivaramakrishnan. "Oxygen adsorption on the surface of CdSe (x) Te (1− x) thin films." *Vacuum* 41.1-3 (1990): 647-649.

[10] Mathew, Xavier. "Photo-induced current transient spectroscopic study of the traps in CdTe." *Solar energy materials and solar cells* 76.3 (2003): 225-242.

[11] Darius et.al., in this conference.

[12] Lee, Seung-Kyu, et al. "Se interlayer in CIGS absorption layer for solar cell devices." *Journal of Alloys and Compounds* 633 (2015): 31-36.

[13] Shin, Young Min, et al. "Surface modification of CIGS film by annealing and its effect on the band structure and photovoltaic properties of CIGS solar cells." *Current Applied Physics* 15.1 (2015): 18-24.

[14] Zhang, Leng, et al. "The effects of annealing temperature on CIGS solar cells by sputtering from quaternary target with Se-free post annealing." *Applied Surface Science* 413 (2017): 175-180.

[15] Dagan, Geula, et al. "Defect level identification in copper indium selenide (CuInSe2) from photoluminescence studies." *Chemistry of Materials* 2.3 (1990): 286-293.

[16] Masse, G., and E. Redjai. "Radiative recombination and shallow centers in CuInSe2." *Journal of applied physics* 56.4 (1984): 1154-1159.

[17] Igalson, M., M. Cwil, and Marika Edoff. "Metastabilities in the electrical characteristics of CIGS devices: Experimental results vs theoretical predictions." *Thin Solid Films* 515.15 (2007): 6142-6146.

[18] Aryal, Krishna, et al. "Effect of selenium evaporation rate on ultrathin Cu (In, Ga) Se 2 films." *2014 IEEE 40th Photovoltaic Specialist Conference (PVSC)*. IEEE, 2014.

[19] Shockley, William, and Hans J. Queisser. "Detailed balance limit of efficiency of p-n junction solar cells." *Journal of applied physics* 32.3 (1961): 510-519.

[20] Munshi, Amit H., et al. "Advanced co-sublimation hardware for deposition of graded ternary alloys in thin-film applications." *2018 IEEE 7th World Conference on Photovoltaic Energy Conversion (WCPEC)(A Joint Conference of 45th IEEE PVSC, 28th PVSEC & 34th EU PVSEC)*. IEEE, 2018.

[21] Kobyakov, Pavel S., et al. "Deposition and characterization of Cd1− xMgxTe thin films grown by a novel cosublimation method." *Journal of Vacuum Science & Technology A: Vacuum, Surfaces, and Films* 32.2 (2014): 021511.

[22] Swanson, Drew E., James R. Sites, and Walajabad S. Sampath. "Co-sublimation of CdSexTe1− x layers for CdTe solar cells." *Solar Energy Materials and Solar Cells* 159 (2017): 389-394.

[23] Barricklow, Keegan Corey. *Advanced research deposition system (ARDS) for processing CdTe solar cells*. Diss. Colorado State University, 2014.

Performance Optimization of the CdSe$_X$Te$_{1-X}$/CdTe Solar Cell

Md Zahangir Alom, Sheikh Tawsif Elahi, Vasilios Palekis, Wei Wang, and Chris Ferekides

University of South Florida, Tampa, FL, 33620, USA

Abstract—The effect of CST (CdSe$_x$Te$_{1-x}$) Se (x) composition and thickness on the performance of CST/CdTe solar cells heat treated at different CdCl$_2$ temperatures has been studied. The superstrate cell structure was: ITO/MZO/CST/CdTe/Back Contact. CdTe and CST bi-layer absorbers were deposited by the CSS (close spaced sublimation) process. The CST Se composition and thickness varied from 10-40% and 0.25-1.0 μm respectively. With the increase of Se composition, V$_{OC}$ decreased and J$_{SC}$, fill factor increased due to the decrease in the band gap. A CST thickness of 0.50 μm resulted in the highest V$_{OC}$ and fill factor. The best performance was obtained for 27% Se with 0.50 μm CST. Performing the CdCl$_2$ heat treatment at 430^{0}C resulted in the highest minority carrier lifetime and improved performance.

Keywords—Thickness, Substrate Temperature, Heat Treatment, Bandgap, Se Composition, Lifetime

I. INTRODUCTION

The highest efficiency achieved for the CdTe solar cells is 22.1% by First Solar (recently a new record efficiency of 22.3% was announced) [1]. The maximum J$_{SC}$ (short circuit current density) for the CdS/CdTe solar cells is approx. 25.88 mA/cm^2, which is far below the theoretical limit [2]. The addition of the CST layer helps increase the J$_{SC}$ near the limit of CdTe solar cells (31.7 mA/cm^2). Moreover, Se improves the minority carrier lifetime to attain a higher V$_{OC}$ (open circuit voltage) [3]. Since the Se composition and CST thickness affect the bandgap of the absorber which also impact the V$_{OC}$. Therefore, an optimum CST Se composition and thickness should be sought to achieve higher efficiency.

The Se composition of the CST determines its bandgap which decreases with increasing Se composition till 40% due to the bandgap bowing effect and increases thereafter. Due to this bandgap change at different Se compositions, the conduction band offset (ΔE_C) between CST and the window/buffer layer also changes, thus controlling the carrier transport and ultimately the fill factor.

The CST thickness also affects performance. Because it controls the interdiffusion between CST and CdTe and can therefore change the final bandgap of the absorber. The interdiffusion can also lead to the formation of a graded absorber band structure.

The CdCl$_2$ heat treatment is an important process step for high-efficiency CST/CdTe solar cells. It can lead to increases in the grain size by recrystallization, and promotes the interdiffusion between CST and CdTe, thus controlling the final

bandgap of the absorber. Moreover, it passivates the grain boundaries to reduce recombination [4] and increase the minority carrier lifetime [5]. Therefore, the CdCl$_2$ heat treatment temperature should also be optimized.

This paper discusses the effect of Se composition, CST thickness, and CdCl$_2$ heat treatment on the CST/CdTe solar cells performance. Se composition of the CST film was varied from 10-40% and CST thickness was varied from 0.25-1.0 μm. CdCl$_2$ heat treatment temperature varied from 410-440^{0}C. Additionally, the light and monochromatic I-V and spectral response measurements were performed to examine the cell performance.

II. EXPERIMENTAL

The superstrate device structure used for this work was: ITO/MZO/CST(CdSe$_x$Te$_{1-x}$)/CdTe/Back Contact. The glass substrate used is the corning EagleXG. First, the glass was cleaned with dilute hydrofluoric acid (HF) and DI water. Then, ITO (Indium Tin Oxide) and MZO (Magnesium Zinc Oxide) were deposited by radio frequency (rf) sputtering. The CST was deposited by mixing CdTe (99.999%), and CdSe (99.999%) using CSS (close spaced sublimation) at source and substrate temperatures of 680 and 580^{0}C respectively. The Se (x) composition of the CST layer was varied by increasing the source deposition temperature, and the relative CdTe and CST amount. The CST thickness was controlled by varying the deposition time. The CdTe layer was deposited by CSS at source and substrate temperatures of 680 and 580^{0}C respectively. The CdCl$_2$ heat treatment was performed at 410-440^{0}C temperatures. The substrates were subsequently rinsed and etched in a dilute bromine solution, and then Cu-doped graphite back contact was applied. Finally, the back contact annealing was performed at 275^{0}C for the Cu diffusion into the absorber. The devices were characterized by light and monochromatic Current-Voltage (J-V), Spectral Response (SR) measurements. Hitachi SU800 and SU70 were used to see the SEM images, and Electron Dispersive Spectroscopy (EDS) detector was used to estimate the Se composition of the CST source. A PANalytical X'Pert MRT was used for obtaining the XRD spectra of the films.

III. RESULTS AND DISCUSSION

The different Se compositions of the CST films can be achieved by varying the source deposition temperature. Higher source temperature results in higher Se composition. The Se composition was confirmed by X-ray Diffraction (XRD) and EDS measurements.

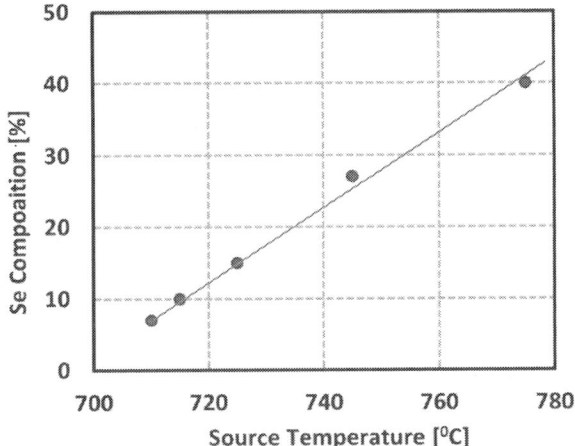

Fig. 1. Se composition at different source temperatures.

Figure 1 shows the relation between source temperature and Se composition. The maximum 40% composition was achieved for the 775^0C source temperature.

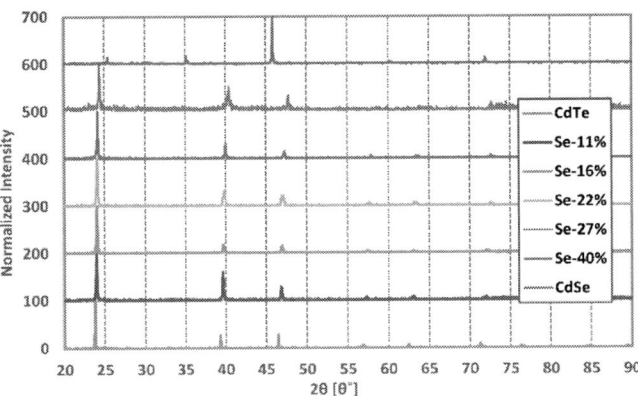

Fig. 2. XRD data of CdTe, CdSe and CST films of various Se compositions.

The XRD data of CdTe, CdSe, and CST (at different Se compositions) films are shown in Figure 2. From the XRD peaks, it was clear that CdTe, CdSe, and all CST films were polycrystalline. CdTe had a cubic structure, and CdSe had a hexagonal structure, and all CST films (Se composition at 10-40%) had a cubic structure with preferential orientation along the (111) direction. Similar behavior has been reported elsewhere [6].

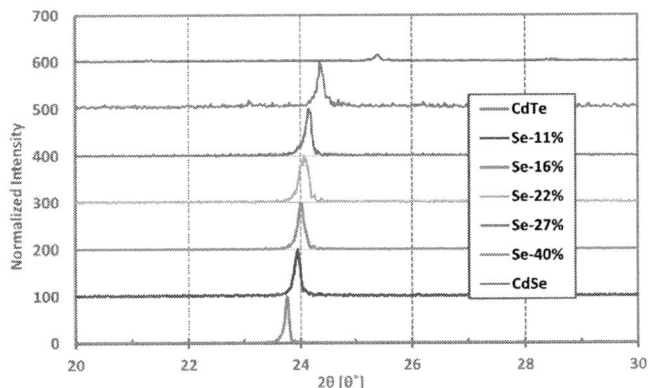

Fig. 3. (111) Peak of CdTe, CST and CdSe films.

Fig. 3 shows the (111) peak of CdTe, CdSe, and CST (Se composition- 10 to 40%) films. With the increase of Se composition, this peak shifted to the right. The Se composition was also confirmed with the EDS measurements.

The introduction of CST in the CdTe solar cells poses some great advantages for performance enhancement, such as it reduces the bandgap of the absorber which leads to higher J_{SC}. The addition of Se also results in improved lifetimes leading to higher V_{OC}. The bandgap of CST determines the conduction band offset (ΔE_C), thus controlling the interface recombination and fill factor. Moreover, the $CdCl_2$ heat treatment promotes interdiffusion between the CST and CdTe layers, thus changing the bandgap and can result in a graded absorber band structure [5]. That is why the Se composition, the thickness of the CST layer, and $CdCl_2$ heat treatment should be optimized for higher efficiency.

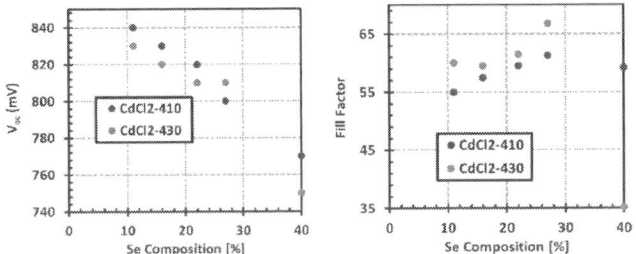

Fig. 4. V_{OC} (left) and fill factor (right) Vs Se composition (10-40%) for the CST/CdTe solar cells at 410 and 430^0C $CdCl_2$ heat treatment.

In Fig. 4, the effect of Se compositions on the V_{OC} and fill factor are shown for 410 and 430^0C $CdCl_2$ annealing temperatures. It is clear that the V_{OC} decreased gradually with the increase of Se composition. This was the consequence of the bandgap reduction in this Se range because of the bandgap bowing effect. The V_{OC} decreased significantly at 40% Se. This was not only because of bandgap, but also the conduction band offset (ΔE_C) between CST and MZO.

On the contrary, the fill factor increased with the increase of Se composition up to 27% Se. This was due to the higher minority carrier lifetime and efficient carrier collection in this Se range. Then, it decreased significantly at 40% Se due to lower carrier collection and higher ΔE_C.

Moreover, $CdCl_2$ heat treatment at 430^0C demonstrated an even higher fill factor because of the higher passivation effect at this temperature. Beyond 27% Se composition, the fill factor decreased significantly. This was because of the higher conduction band offset (ΔE_C) between CST and MZO. Higher ΔE_C reduced the carrier collection, which resulted in notable fill factor reduction as seen in Figure 4.

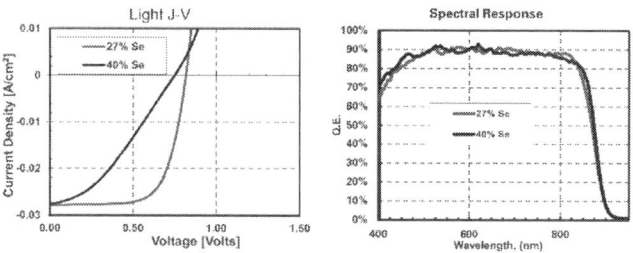

Fig. 5. Light J-V and Spectral Response at 27 and 40% Se composition.

Fig. 5 exhibits the light J-V and spectral response at 27% and 40% Se. From the figure, it was clear that at 40% Se, the value of ΔE_C limited carrier collection, which was responsible for the drastic reduction in the fill factor. The same results were also observed from the simulation. A 27% Se composition resulted in the optimum performance compared to all other Se compositions studied.

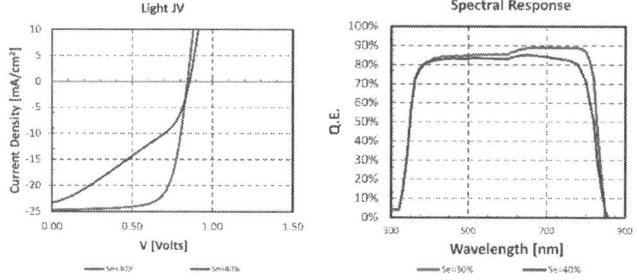

Fig. 6. Simulated Light J-V and Spectral Response (SR) at 27 and 40% Se composition.

Fig. 6 shows the simulated J-V and spectral response at 27% and 40% Se composition. The value of ΔE_C at 40% Se is higher compared to 27% Se because of the bandgap and electron affinity at these compositions. This higher ΔE_C at 40% Se caused a significant increase in the interface recombination, thus reducing the carrier collection. This was responsible for the decrease in J_{SC}, and the drastic reduction in fill factor at 40% Se.

Therefore, for the optimum performance device, the Se composition should be ~ 27% for effective band alignment of CST with MZO, and efficient carrier collection.

After optimizing the Se composition for optimum efficiency, the CST thickness was optimized. The thickness of the CST layer can be changed by varying the deposition time. The

thickness was varied from 0.25-1 μm (15-60 sec deposition time).

Table 1: V_{OC}, fill factor, and J_{SC} of 0.25-1 μm CST thickness at 410 and 430^0C heat treatment for 27% Se Composition.

Condition	V_{OC} [mV]	FF [%]	J_{SC} [mA/cm^2]
CST-0.25 micron, $CdCl_2$-410^0C	790	57.70	26.20
CST-0.25micron, $CdCl_2$-430^0C	810	62.70	26.75
CST-0.50 micron, $CdCl_2$-410^0C	790	63.00	27.73
CST-0.50 micron, $CdCl_2$-430^0C	820	68.30	28.11
CST-1 micron, $CdCl_2$-410^0C	780	63.30	27.86
CST-1 micron, $CdCl_2$-430^0C	790	65.90	28.50

From our previous work on CST thickness effect on the CST:As /CdTe solar cells, it was shown that 0.50 μm CST thickness produced higher V_{OC}, fill factor and higher efficiency [7]. Table 1 shows the CST thickness effect on the V_{OC}, fill factor, and J_{SC} at 27% Se for undoped cells (this work). A CST thickness of 0.50 μm produced higher V_{OC} because of the higher bandgap. At 1 μm thickness, the interdiffusion between the CST and CdTe is higher which caused the bandgap to decrease more than the 0.50 μm thickness. For this same reason, J_{SC} was a little bit higher for the 1 μm cell. However, 0.50 μm exhibited a higher fill factor. This may be because of producing a better-graded absorber layer at 0.50 μm thickness. Though 0.25 μm CST thickness gave the same V_{OC} compared to 0.50 μm thickness, it produced much lower J_{SC} and fill factor because of the very thin layer of CST. A 0.50 μm CST thickness produced an overall higher efficiency.

After the CST Se composition and thickness optimization, the $CdCl_2$ heat treatment temperature was also chosen carefully for optimum performance, because it controls not only the interdiffusion between CST and CdTe, but also impacts interface and grain boundary passivation for increased lifetime. The heat treatment was carried out in the temperature of 410 to 440^0C.

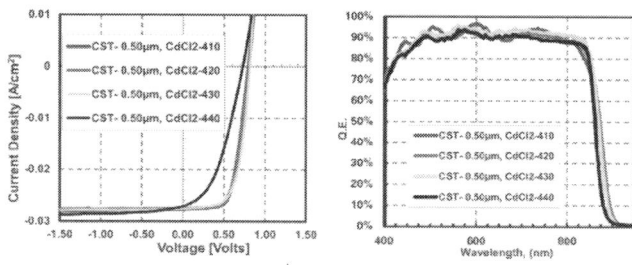

Fig. 7. Light J-V and Spectral Response (SR) at 410-440^0C $CdCl_2$ heat treatment.

From Fig. 7, it was observed that 430^0C resulted in higher V_{OC} and fill factor, presumably due to higher minority carrier lifetime than 410 or 420^0C. Temperature higher than 430^0C temperature treatment was presumed that excess chlorine in the bulk and front interface caused a significant drop in the performance, particularly on the V_{OC} and the fill factor.

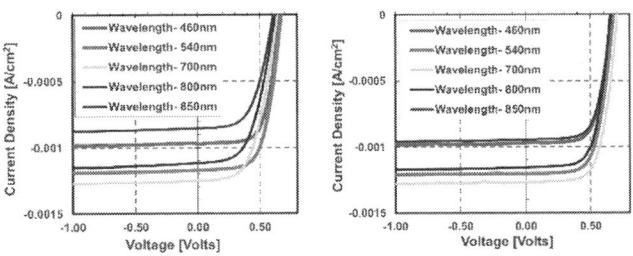

Fig. 8. Monochromatic J-V at 410 (left) and 430^{0}C (right) heat treatment.

Fig. 8 shows the monochromatic I-V measurements of CST/CdTe cells at 410 and 430^{0}C CdCl$_2$ heat treatment. Monochromatic I-V is a method to compare the collection efficiency and therefore show the minority carrier lifetime effects. This measurement was done with 20nm bandwidth filters of different wavelengths (460, 540, 700, 800, 850 nm).

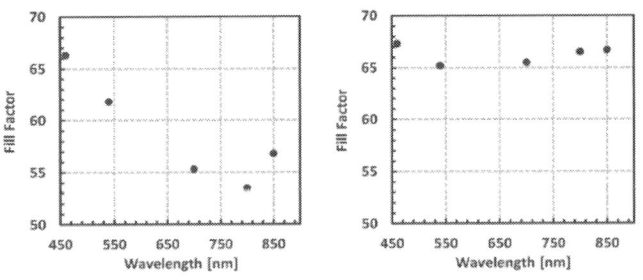

Fig. 9. Wavelength Vs fill factor (FF) at 410 (left) and 430^{0}C (right) heat treatment.

Fig. 9 demonstrates the relation between wavelength and fill factor. It is seen that at 410^{0}C, the fill factor decreased with the increase of wavelengths, and at 430^{0}C, the fill factor remained essentially constant. This can be interpreted as an indication that 430^{0}C CdCl$_2$ heat treatment leads to a higher lifetime compared to 410^{0}C.

IV. Conclusion

The optimization of the CST layer (Se composition, thickness) and CdCl$_2$ heat treatment was performed for CST/CdTe solar cells. The Se composition determined the bandgap of the absorber and the ΔE_C between the CST and MZO, thus controlling V_{OC}, J_{SC}, and fill factor. Since the addition of Se leads to an improved lifetime, the change in V_{OC} is smaller compared to the bandgap reduction. For Se compositions above 27%, a much lower V_{OC} and fill factor were measured due to higher ΔE_C between the CST and MZO. The CST thickness affects the interdiffusion between CST and CdTe thus controlling the bandgap and grading in the absorber. The

CdCl$_2$ heat treatment also impacts the interdiffusion between CST and CdTe to change the bandgap, in addition to interface passivation for a higher minority carrier lifetime. Optimum parameters established to-date are: 0.50 μm CST with Se composition 27% heat treated at 430^{0}C.

Acknowledgment

This research project was funded by the National Science Foundation (NSF) (grant no. EPMD-1711716).

References

[1] Green, Martin A., et al., "Solar cell efficiency tables (version 49)." *Progress in photovoltaics: research and applications* 25.1 (2017): 3-13.

[2] Morales-Acevedo, A., "Can we improve the record efficiency of CdS/CdTe solar cells?." *Solar energy materials and solar cells* 90.15 (2006): 2213-2220.

[3] T. Fiducia, A. Howkins, A. Abbas, B. Mendis, A. Munshi, K. Barth, W. Sampath, and J. Walls, "Selenium passivates grain boundaries in alloyed CdTe solar cells." *Solar Energy Materials and Solar Cells* 238 (2022): 111595.

[4] Amin, Nowshad, Mohammad Rezaul Karim, and Zeid Abdullah ALOthman., "Impact of CdCl2 Treatment in CdTe Thin Film Grown on Ultra-Thin Glass Substrate via Close Spaced Sublimation." *Crystals* 11.4 (2021): 390.

[5] Hsu, C. A., Palekis, V., Khan, I., Collins, S., Morel, D., & Ferekides, C., "The Effect of the CdCl2 Heat Treatment on CdSexTe1-x Solar Cells." *2017 IEEE 44th Photovoltaic Specialist Conference (PVSC)*. IEEE, 2017.

[6] Swanson, Drew E., James R. Sites, and Walajabad S. Sampath., "Co-sublimation of CdSexTe1− x layers for CdTe solar cells." *Solar Energy Materials and Solar Cells* 159 (2017): 389-394.

[7] Alom, M. Z., Elahi, S. T., Palekis, V., Wang, W., & Ferekides, C., "The Effect of CdSexTe1-x Thickness on CdSexTe1-x/CdTe Solar Cell Performance" *2022 IEEE 49th Photovoltaics Specialists Conference (PVSC)* (pp. 0976-0979). IEEE, 2022.

[8] Zheng, X., Kuciauskas, D., Moseley, J., Colegrove, E., Albin, D.S., Moutinho, H., Duenow, J.N., Ablekim, T., Harvey, S.P., Ferguson, A. and Metzger, W.K., "Recombination and bandgap engineering in CdSeTe/CdTe solar cells." *APL Materials* 7.7 (2019): 071112.

[9] Alom, M. Z., Elahi, S. T., Palekis, V., Wang, W., & Ferekides, C., "The Effect of Arsenic Doping on the Performance of CdSe x Te 1-x/CdTe Solar Cells." *2021 IEEE 48th Photovoltaic Specialists Conference (PVSC)*. IEEE, 2021.

[10] Hsu, C.A., Palekis, V., Levcenko, S., Morel, D. and Ferekides, C., "Cu-doping Effects in CdSe x Te 1-x/CdTe Solar Cells." *2019 IEEE 46th Photovoltaic Specialists Conference (PVSC)*. IEEE, 2019.

[11] Chen, Y., Tan, X., Peng, S., Xin, C., Delahoy, A. E., Chin, K. K., & Zhang, C., (2018). The influence of conduction band offset on CdTe solar cells. *Journal of Electronic Materials*, 47(2), 1201-1207.

Analysis of Optoelectronic Characterization Data via Bayesian Inference: a Desktop-scale MCMC Method

Calvin Fai, Gregory A. Manoukian, Jason B. Baxter, Anthony J. C. Ladd, Charles J. Hages

University of Florida, Gainesville, FL, United States

Drexel University, Philadelphia, PA, United States

The recovery of characteristic absorber parameters such as the carrier mobility, doping concentration, and rate constants of each carrier recombination mechanism from optical characterization measurements has historically been hindered by the complexity of the carrier dynamics. This necessitates the use of simplified, analytically solvable physics models that sacrifice some of the potential information content of the measurement data. In this work, we introduce a desktop-scale Markov Chain Monte Carlo (MCMC) sampler that utilizes simulation of the full carrier physics to recover material parameters with increased accuracy or which were previously inaccessible. From a "power scan" consisting of time-resolved photoluminescence (TRPL) data at varying excitation intensity, we recover the ambipolar carrier mobility, the doping concentration, and rate constants for Auger and bimolecular radiative recombination. We also obtain an effective lifetime for defect-assisted nonradiative recombination, which can be decomposed further into bulk and surface recombination components by introducing additional data from multiple material sample thicknesses. These results reaffirm the potential for simulation-driven statistical analyzers to greatly expand the utility of optical characterization measurements, though herein the need for expensive computational resources is not required.

Generalizability of Neural Network-based Identification of PV in Aerial Images

Joseph Ranalli[1] and Matthias Zech[2]

[1]Penn State Hazleton, Hazleton, PA, USA
[2]German Aerospace Center (DLR), Institute of Networked Energy Systems, Oldenburg, Germany

Abstract — **Identification of PV panels from aerial imagery is a potential strategy for building comprehensive behind-the-meter PV datasets. Several previous studies have utilized Convolutional Neural Networks with the goal of producing tools that can perform these identification tasks. Neural Network approaches rely on labelled data for training, with several aerial imagery datasets with labelled PV already available. This study aims to investigate generalizability of models trained on one set of labelled PV data to other datasets, to further understanding of how these models can be applied. Six different PV datasets were utilized, and test data results were compared. Overall, we find that generalizability suffers when models are presented with different data than they were trained on. We describe some dataset features that led to particularly poor generalization. This study highlights the need for further research to investigate strategies for improving generalizability of trained Neural Network models.**

I. INTRODUCTION

A multitude of academic and industrial contexts require knowledge of the location and configuration of distributed photovoltaic (PV) installations. Examples include evaluating renewable energy policy by tracking regional PV growth rates and providing technical data to support aggregate-level forecasts of distributed PV generation, necessary to simulate and operate climate-neutral energy systems. Generally, no large scale registry of distributed PV systems exists. Detailed data about behind-the-meter PV may be particularly difficult to obtain. Even in cases where data on PV installations exists, it may not be accessible to researchers or the general public. Instead, availability of data on these PV systems varies widely by jurisdiction and is subject to the policies of local stakeholders who directly collect and use this information.

Recent efforts to improve data availability for distributed PV have attempted to identify systems visually from satellite or aerial imagery. Progress has been made utilizing Deep Learning on these imagery sources, for instance for the United States [1], regions of Germany [2], [3], China [4] and even worldwide [5]. Global studies mostly rely on publicly available image sources with wide coverage, but low-detail spatial resolution. This typically limits detection to utility-scale PV systems with large capacity that are visible from satellites on rough spatial resolutions.

A suitable global PV inventory also needs to contain residential PV sites that are small, but numerous, making up

more than 40\% of the global PV capacity [6]. The rough spatial resolution of satellite imagery may limit its usefulness towards this task. Instead, aerial imagery may be a more reliable source for high resolution PV detection. Aerial imagery is becoming increasingly open-access, but due to its origin from airborne flights, is usually only available to a limited spatial extent. Furthermore, its localized nature may lead to greater diversity in the imagery, both in terms of acquisition equipment and processing, and resulting from local visual diversity in the area being imaged. To derive a global PV registry which includes small-scale and large-scale PV systems, it is therefore likely that PV detection systems need the ability to deal with a range of different image sources with variable spatial resolution and availability.

Regarding neural network models, many academic studies are based upon performance using individual data sources. But the need for a flexibility in generating a global PV registry raises the question how well models trained on single-site data generalize to other datasets, particularly for higher resolution aerial imagery datasets. Studies investigating the generalizability [7], [8] have already shown that neural networks generalize poorly when trained and tested on different cities, even when the images originate from the same data source. Different local characteristics, as geography and population density [8] may impact generalizability. A study testing the generalizability more broadly across different local sites and data sources is still lacking.

This study aims to fill this gap by systematically testing how well models trained on single datasets perform on other sites, using six different labeled aerial image datasets, including Northern Germany, Southern France, the United States (sets from California and New York City). These six datasets were labeled using different methods, by different research groups and have different original data sources. This provides a realistic testbed for verification metrics when a model is applied outside its trained dataset without any modifications.

II. METHODOLOGY

A. Model Architecture

978-1-6654-6060-6/23 $31.00 © 2023 IEEE

This study used a Fully Convolutional Neural Network architecture that had previously been applied for the purpose of identifying PV arrays in aerial images [3]. Specifically, the network architecture utilized was a u-net [9], which was designed for semantic segmentation tasks. A u-net architecture uses symmetric encoder and decoder paths, producing a structure that resembles the letter *u*. The implementation of u-net used in this study was based on the python library *segmentation_models* [10], which contained a version of u-net built using Tensorflow and Keras [11], along with pretrained weights from the *ImageNet* competition [12]. A backbone of ResNet-34 was used for the u-net model, as it was previously shown to produce similar quality results to more detailed backbones [3] for this task.

As the primary outcome of this study is the application of neural networks, rather than the neural network itself, a thorough optimization of the training process was not conducted. Rather, settings that produced reasonable and consistent results were used. Training was performed on a desktop computer with a single, consumer-grade GPU, and completed in a reasonable timeframe (2-3 hours per training dataset). During training, the encoder weights were frozen to their pre-trained values to reduce the number of weights being adjusted. Early stopping was employed to prevent overfitting, stopping the training after 10 consecutive epochs without a reduction in the loss. Data augmentation was used to simulate features of a larger dataset. The parameters used for augmentation were: rotation (up to 30°), zoom (factor of 0.2) and height and width shifts (factor of 0.1 each). The weights with the best value of validation loss observed during training were retained for evaluation.

B. Source Data

Data was utilized from six separate publicly available datasets consisting of labeled PV installations in aerial/satellite images: two cities in California, Fresno and Stockton (CA-F and CA-S) from [13], two datasets in France (FR-G and FR-I) from [14] using images from Google Earth and the French government (IGN) respectively, a dataset of Google Earth images from Germany (DE-G) used in a previous study by the authors [3] and a newly labelled dataset based on publicly available images for Queens, New York (NY-Q) [15].

In a few of the original datasets (CA-F, and CA-S and NY-Q) raw images were too large (e.g. 5000x5000 pixels) to be processed by the model as whole images. Thus, these large images were split into tiles of smaller size for the training and testing. Tiles that did not contain any labelled pixels were excluded from each dataset. The overall number of retained tiles and other details of the datasets are provided in Table I.

Due to the variable number of tiles in each dataset, a subset of 1000 tiles was randomly chosen from each to form the representative subset for that data source. This ensured that all models were trained on an equal quantity of data. Each set of 1000 images was randomly split into sets of 720 training, 80 validation and 200 test tiles. These split sizes were consistent across all datasets, and did not include any crossover. All images were resized to 576x576 pixels during processing to match the network architecture. Consequently, smaller images (e.g. FR-G and FR-I datasets) had an effective increase in scale of about 1.4, while the datasets with larger tiles were scaled down by a factor of around 0.9. The effective resolutions of each dataset after scaling are also given in Table I.

In addition to the 6 original datasets, a synthetic combination dataset was created by combining 133 of the processed tiles from each dataset (with two extras chosen from NY-Q to reach the correct total number of 1000) to compare training on the aggregate. After training, each trained model was evaluated on the test data for all the datasets and values for several performance metrics were recorded.

While fundamentally subjective in nature, an effort was made to describe contextual differences between the datasets. Five separate categories were created to describe the various types of characteristics found within the datasets: large structures and/or flat roofs (often commercial buildings), large open spaces (roughly more than 50\% of the image), agricultural (identified by visible patterns or rows), large bodies of water and utility scale PV. Images for each category from each dataset's 1000 tile subset were manually counted and rounded to the nearest ten to reflect the subjective nature of these judgments. A table summarizing these categories is in Table II. The remainder of images in each dataset primarily consisted of residential housing, for which the approximate number of structures per image is notated as well.

TABLE I
DATASETS

Dataset	Tot. Tiles	Tile Size	Resolution	Scaled Res.	Ref
CA-F	1,044	625x625	0.3 m/px	0.32 m/px	[13]
CA-S	4,192	625x625	0.3 m/px	0.32 m/px	[13]
FR-G	13,303	400x400	0.1 m/px	0.07 m/px	[14]
FR-I	7,865	400x400	0.2 m/px	0.14 m/px	[14]
DE-G	1,325	639x640	0.18 m/px	0.2 m/px	[3]
NY-Q[1]	1,007	625x625	0.15 m/px	0.16 m/px	[15]

[1] labeling of this dataset is still ongoing

TABLE II
CONTEXTUAL DIFFERENCES BY DATASET (APPROXIMATE)

Dataset	Large/Flat	Open Space	Agri.	Util. PV	# Bldg/Tile
CA-F	70	140	40	0	20-40
CA-S	70	80	10	0	20-40
FR-G	10	20	0	0	2-5
FR-I	20	90	20	0	5-10
DE-G	60	80	10	10	10-20
NY-Q	130	10	0	0	10-20

Despite the qualitative nature of these categorizations, a few generalizations can be made. All of the datasets primarily consisted of detached or semi-detached residential dwellings. The NY-Q dataset contained the lowest number of images that featured large open spaces and the highest number of large and flat-roofed structures (often commercial or industrial buildings in this case), which corresponds to its more urban character. Qualitatively, NY-Q also had less visible vegetation intermingled in the residential areas (i.e. trees) than the other datasets. CA-F had the highest frequency of images with large open areas, and many of these areas showed vegetation in rows/patterns that are presumed to indicate agricultural activity (and impacted the detection of PV as will be seen). The number of residential buildings per tile roughly corresponded to the scaled resolution of the images, with the two California datasets having the lowest zoom level (covering the widest area), FR-G having the highest zoom level, and FR-I, DE-G and NY-Q in the middle. It is also notable that the FR-G and FR-I datasets were both designed to have tiles centered on a positive PV system [14]. In part this explains some of the counts seen in Table II for FR-G; due to the high zoom level and smaller overall tile size, centering the tile on the PV array may have decreased the probability of observing some of the features present in some of the more randomly aligned datasets.

III. RESULTS AND DISCUSSION

As stated, each of the seven trained models was evaluated using 200 test images associated with each of the six datasets. Precision and recall are especially useful metrics in interpreting the performance of these models. Precision can be interpreted as the percentage of predicted positive pixels that are correct, defined as in Eq. 1. Recall can be interpreted as the percentage of the labelled positive pixels that are identified by the model, defined as in Eq. 2. The Jaccard index, also called the intersection over union, or IoU, is defined in Eq. 3. IoU represents the match between the the actual and predicted PV regions in the image and is useful in representing the overall performance of the model. Results of precision and recall for the models are presented in Table III and Table IV respectively. Results for the IoU are given in Table V.

$$p = \frac{True\ Positives}{Predicted\ Positives} = \frac{TP}{TP+FP} \qquad (1)$$

$$r = \frac{True\ Positives}{Labeled\ Positives} = \frac{TP}{TP+FN} \qquad (2)$$

$$IoU = \frac{Area\ of\ Intersection}{Area\ of\ Union} = \frac{TP}{TP+FP+FN} \qquad (3)$$

In conjunction, these metrics help us understand the model's performance. High values in both precision and recall indicate that the model identifies most ground truth pixels and does so correctly, while low values in both indicates that the model makes mostly incorrect positive identifications and fails to identify many ground truth positives. The crossed value cases on the two metrics provide useful information as well. A high precision with low recall indicates that the model is usually correct when it predicts positives, but that it fails to identify many of the ground truth positives (i.e. is very selective about predictions). For example, the FR-I trained model is correct 95\% of the time when predicting pixels to be PV in the FR-G test dataset (precision), but identifies only 36\% of the overall PV pixels (recall). Conversely, a low precision with a high recall indicates that a model identifies most of the positive pixels in the ground truth, but also incorrectly identifies many pixels as PV. An example is the CA-F trained model, which predicts 59\% of the PV pixels from the DE-G ground truth dataset (recall), but is only correct about its positive PV predictions 7\% of the time (precision).

A. Individual Model-to-Test Case Performance

Generally, models performed better on their own test data than on other test datasets, which is unsurprising, but does highlight the challenges in generalization of models that has been described by the literature. Additionally, each model performed best of the group on its corresponding test data, with a few minor exceptions in the precision metric. The combination dataset had relatively good performance for all test datasets. This is unsurprising as its more diverse training set seems to have helped it to produce more generalizable results. Figures 1 - 6 show results on one of the better performance images for each test dataset. The images in these figures are all examples of cases where all models performed reasonably well, and were selected from the top 5\% of average IoU scores for each test set.

The average IoU performance of models on their own test data was a value of 0.71. As measured by the IoU score, most generalization from trained models to different test sets was quite poor, though a few examples of moderate performance were observed. The best two examples of generalization when judging by IoU were FR-I model's performance on NY-Q data and CA-S predicting data from CA-F. In part, the connection between the two California cases may result from their shared data source and overall locale. However, the lack of the reciprocal boost to performance of CA-F on the CA-S test data suggests that other factors may be at play.

TABLE III
PRECISION RESULTS BY DATASET

	CA-F	CA-S	FR-G	FR-I	DE-G	NY-Q
CA-F	0.87	0.46	0.36	0.48	0.07	0.25
CA-S	0.82	0.79	0.51	0.31	0.22	0.24
FR-G	0.10	0.03	0.91	0.76	0.41	0.52
FR-I	0.63	0.64	0.95	0.79	0.67	0.77
DE-G	0.70	0.65	0.83	0.91	0.77	0.82
NY-Q	0.59	0.66	0.90	0.87	0.75	0.90

Train datasets in rows, test datasets in columns

978-1-6654-6060-6/23 $31.00 © 2023 IEEE

TABLE IV
RECALL RESULTS BY DATASET

	CA-F	CA-S	FR-G	FR-I	DE-G	NY-Q
CA-F	0.79	0.59	0.15	0.58	0.59	0.35
CA-S	0.62	0.72	0.13	0.47	0.59	0.37
FR-G	0.06	0.01	0.88	0.52	0.15	0.29
FR-I	0.15	0.23	0.36	0.84	0.37	0.67
DE-G	0.19	0.33	0.11	0.30	0.79	0.48
NY-Q	0.07	0.24	0.15	0.50	0.47	0.89

Train datasets in rows, test datasets in columns

TABLE V
IoU RESULTS BY DATASET

	CA-F	CA-S	FR-G	FR-I	DE-G	NY-Q
CA-F	0.71	0.35	0.11	0.36	0.06	0.16
CA-S	0.55	0.61	0.11	0.22	0.17	0.19
FR-G	0.03	0.00	0.81	0.45	0.13	0.26
FR-I	0.13	0.19	0.35	0.69	0.31	0.56
DE-G	0.18	0.29	0.11	0.29	0.63	0.44
NY-Q	0.07	0.22	0.15	0.47	0.40	0.81

Train datasets in rows, test datasets in columns

Investigating more deeply, the individual images with the best- and worst- IoU score performance were visualized for each of these combinations. In the case of FR-I predicting on NY-Q, the best performing images were dominated by large commercial rooftop systems (one shown in Fig. 6). The worst performing images were predominantly of residential housing, but no particular pattern was able to be discerned from qualitative inspection. In the case of CA-S predicting CA-F test data, the best performance was observed on large scale systems, as well as images with PV closely clustered, as opposed to scattered across the image. Several high performance images include instances of apparent agriculture, for which some models exhibited significant confusion (as shown in Fig. 7). No clear pattern was visible in the poorly performing images from CA-S to CA-F. Several ground mounted systems were missed, and many of the poor performing images did contain very small segments of PV. Though none of the models were a great success, the FR-I model was most generalizable overall, having the highest average IoU score across all other test sets (IoU = 0.31).

B. Overall Generalization

The worst individual example of generalization was the FR-G model, which had exceptionally poor performance on both of the CA test sets. Even FR-G's best performance on images within CA-S and CA-F showed essentially no skill at prediction. Since FR-G had the highest scaled detail level (0.07

m/pix) by a factor of two, this may suggest that the models are not adapting well to imagery at lower zoom levels. While FR-G generalized poorly overall, it showed its best test performance on the FR-I dataset, which in addition to having the closest scaled resolution, was also produced by the same labelers and for the same geographic region.

The average generalization performance on each metric was computed by excluding the results for each model's own test data and that for CMB-6. The highest IoU scores averaged across all models occurred for the FR-I test data (IoU = 0.36) and the NY-Q test data (IoU = 0.32), indicating that these test data were easiest for a general model to predict. It is important to note though that these numbers are well below the performance achieved by the specially-trained models. The easiest images for all models to predict in the FR-I test data were primarily relatively large systems, with very clear panel frames. The most predictable images in NY-Q were all large commercial-scale systems, also with clearly visible module frames.

The FR-G test data was the hardest for all other models to predict (average IoU = 0.17). However, most models had relatively good precision on this model, indicating that models were reluctant to predict positive values in FR-G, but were often correct when they did. One noticeable feature in the FR-G predictions is that most models tended to discretize the panels, rather than predicting overall areas of panels. This effect can be seen in Fig. 3. This may be a consequence of the higher zoom levels making the module frames appear to be gaps. FR-G's predictions on its own data did not exhibit this feature. This also points to how differences in labeling of module frames, and for larger systems, gaps between rows may impact the quality of predictions, especially when working with data sources at varying zoom levels. Despite the usage of different zooming levels in data augmentation, the models seem to be highly sensitive to the spatial resolution of the image source.

C. Combination Model Performance

As stated, the combination model (CMB-6) was trained on a random selection of 133 images from each other dataset. Besides representing a more generalized training source, this also provides the combination with a small sample of data from each test set it will be compared to. As seen in Tables III - V, this led to the combination model having relatively good precision across all test sets, meaning that it tended to avoid false positives. Except for the case of CA-F which showed exceptional generalization from CA-S, CMB-6 also had the best IoU score for each test dataset aside from specifically trained model for that dataset.

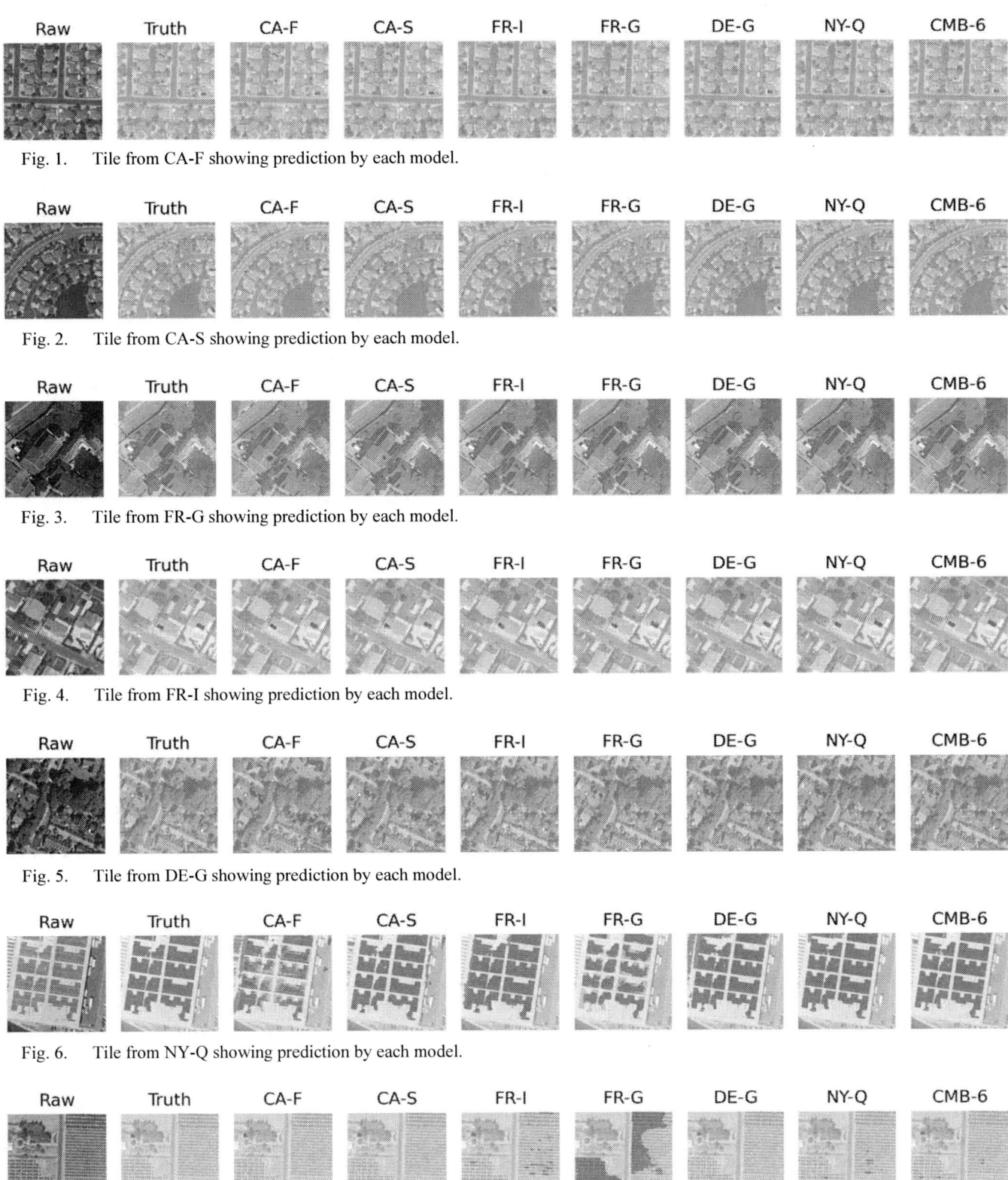

Fig. 1.   Tile from CA-F showing prediction by each model.

Fig. 2.   Tile from CA-S showing prediction by each model.

Fig. 3.   Tile from FR-G showing prediction by each model.

Fig. 4.   Tile from FR-I showing prediction by each model.

Fig. 5.   Tile from DE-G showing prediction by each model.

Fig. 6.   Tile from NY-Q showing prediction by each model.

Fig. 7.   Tile from CA-F showing incorrect identification of agricultural rows by some models.

Generally speaking, the images for which CMB-6 had its worst performance were comparable to those already described for other datasets. This is also true for the case of CA-F, the test data that was most challenging for CMB-6. As mentioned, the precision remained relatively high for CA-F indicating that inclusion of more general training data improved its selectivity in making predictions for that data. CMB-6 experienced the largest benefit on FR-G when compared to predictions of other datasets. Including a small quantity of the hardest-to-predict data improved the model substantially for that case. Looking at image-wise performance, CMB-6 avoided the discretized module predictions that other models tended to experience for FR-G. Its worst performing images were characterized by a high number of false positives.

III. Conclusion

Identification of solar PV from aerial or satellite imagery has potential to improve the quality of data on distributed solar installations and to broaden access to such data. A growing number of labeled datasets are available to support training of neural network models for this task. However, generalization of the models across locales and image sources is a necessary technical hurdle to clear to permit these datasets to find broad application.

This study found that generalization of models trained on a single dataset is relatively challenging when applied to other datasets. This result is consistent with conclusions made by other investigators [7], [8]. We observed some limited cases of strong generalization between models, but without an ability to draw strong conclusions on the basis of qualitative interpretation of the datasets alone, however some inferences were possible. The worst performing model was trained on a dataset that had a significant difference in zoom level as compared to others, which is likely to have had an impact on its performance. Overall, commercial rooftops and large scale systems tended to be best predicted across all models. A model trained on a combination of data from each of the test sets showed adequate (but not exceptional) performance across all sets of test data. Where most models showed modest precision, albeit with low recall, the combination model struck a middle ground to improve performance: raising the recall at the cost of some precision.

We have described some of the strengths and weaknesses of generalization across six separate aerial imagery datasets applied for identification of PV. However, no hard conclusions were able to be reached on the basis of this data alone. Further work is needed to test methodologies that may improve the generalizability of the trained models and address the differences found within the source data. These steps may lead to the ultimate goal of producing application-ready tools for computer-based aerial identification of PV.

Acknowledgement

J.R would like to acknowledge financial support from Penn State Hazleton and the Penn State School of Engineering Design and Innovation. The authors acknowledge Richard Ray and Brian Tylutke for help labeling the New York dataset.

References

[1] J. Yu, Z. Wang, A. Majumdar, and R. Rajagopal, "DeepSolar: A Machine Learning Framework to Efficiently Construct a Solar Deployment Database in the United States," *Joule*, vol. 2, no. 12, pp. 2605–2617, Dec. 2018, doi: 10.1016/j.joule.2018.11.021.

[2] K. Mayer *et al.*, "3D-PV-Locator: Large-scale detection of rooftop-mounted photovoltaic systems in 3D," *Appl. Energy*, vol. 310, p. 118469, Mar. 2022, doi: 10.1016/j.apenergy.2021.118469.

[3] M. Zech and J. Ranalli, "Predicting PV Areas in Aerial Images with Deep Learning," in *2020 47th IEEE Photovoltaic Specialists Conference (PVSC)*, Jun. 2020, pp. 0767–0774. doi: 10.1109/PVSC45281.2020.9300636.

[4] X. Hou, B. Wang, W. Hu, L. Yin, A. Huang, and H. Wu, "SolarNet: A Deep Learning Framework to Map Solar Plants In China From Satellite Imagery," in *Climate Change AI*, Climate Change AI, Apr. 2020. Accessed: Dec. 19, 2022. [Online]. Available: https://www.climatechange.ai/papers/iclr2020/6

[5] L. Kruitwagen, K. T. Story, J. Friedrich, L. Byers, S. Skillman, and C. Hepburn, "A global inventory of photovoltaic solar energy generating units," *Nature*, vol. 598, no. 7882, Art. no. 7882, Oct. 2021, doi: 10.1038/s41586-021-03957-7.

[6] S. Joshi, S. Mittal, P. Holloway, P. R. Shukla, B. Ó Gallachóir, and J. Glynn, "High resolution global spatiotemporal assessment of rooftop solar photovoltaics potential for renewable electricity generation," *Nat. Commun.*, vol. 12, no. 1, Art. no. 1, Oct. 2021, doi: 10.1038/s41467-021-25720-2.

[7] R. Wang, J. Camilo, L. M. Collins, K. Bradbury, and J. M. Malof, "The poor generalization of deep convolutional networks to aerial imagery from new geographic locations: an empirical study with solar array detection," in *2017 IEEE Applied Imagery Pattern Recognition Workshop (AIPR)*, Oct. 2017, pp. 1–8. doi: 10.1109/AIPR.2017.8457965.

[8] W. Hu *et al.*, "What you get is not always what you see—pitfalls in solar array assessment using overhead imagery," *Appl. Energy*, vol. 327, p. 120143, Dec. 2022, doi: 10.1016/j.apenergy.2022.120143.

[9] O. Ronneberger, P. Fischer, and T. Brox, "U-Net: Convolutional Networks for Biomedical Image Segmentation," in *Medical Image Computing and Computer-Assisted Intervention – MICCAI 2015*, N. Navab, J. Hornegger, W. M. Wells, and A. F. Frangi, Eds., in Lecture Notes in Computer Science. Cham: Springer International Publishing, 2015, pp. 234–241. doi: 10.1007/978-3-319-24574-4_28.

[10] P. Yakubovskiy, *Segmentation Models*. Github, 2019. [Online]. Available: https://github.com/qubvel/segmentation_models

[11] M. Abadi *et al.*, "TensorFlow: Large-Scale Machine Learning on Heterogeneous Distributed Systems," Google, 2015. [Online]. Available: https://tensorflow.org

[12] J. Deng, W. Dong, R. Socher, L.-J. Li, Kai Li, and Li Fei-Fei, "ImageNet: A large-scale hierarchical image database," in *2009 IEEE Conference on Computer Vision and Pattern Recognition*, Jun. 2009, pp. 248–255. doi: 10.1109/CVPR.2009.5206848.

978-1-6654-6060-6/23 $31.00 © 2023 IEEE

[13] K. Bradbury *et al.*, "Distributed Solar Photovoltaic Array Location and Extent Data Set for Remote Sensing Object Identification." figshare, Oct. 02, 2018. doi: https://dx.doi.org/10.6084/m9.figshare.3385780.

[14] G. Kasmi *et al.*, "A crowdsourced dataset of aerial images with annotated solar photovoltaic arrays and installation metadata." Zenodo, Jul. 20, 2022. doi: https://doi.org/10.5281/zenodo.7059985.

[15] "NYS Interactive Mapping Gateway." https://gis.ny.gov/gateway/mg/ (accessed Nov. 22, 2022).

Modal Analysis of GaAs Nanowire Solar Cells for Optimal Device Design

Venkata S. A. Chaluvadi, Eduardo Camarillo Abad, Hannah J. Joyce, Louise C. Hirst

University of Cambridge, Cambridge, United Kingdom

Nanowire solar cells have shown promise in next-generation applications, offering potential advantages in power-to-weight ratio, available form factors, cost and materials reduction, and radiation tolerance. However, nanowire solar cell performance remains limited by poor light absorption due to low dimensionality. Engineered light scattering through nanowire arrays could exploit the full phase-space, whereby scattered light, through constructive interference or coupling to waveguide modes, can result in stronger field localization within the active layer. To do so, the nanowire active layer demands optimization and careful design to fully take advantage of light-trapping in optical modes. This work presents potential trade-offs in device geometry guiding device design, particularly identifying fundamental modal structure for absorption enhancement through light management strategies. This work shows how nanowire parameters including nanowire diameter, array pitch, filling factor and substrate affect absorption in different spectral ranges due to changes in antireflection properties and escape cone losses. This analysis can further be applied to other III-V material systems and serve as a guide to more complex device fabrication in fully exploiting the available optical modes.

978-1-6654-6060-6/23 $31.00 © 2023 IEEE

Benchmark Tests for IV Fitting Algorithms

Clifford W. Hansen, Abigail R. Jones, Taos Transue, Marios Theristis

Sandia National Laboratories, Albuquerque, NM USA 87185-1033

Abstract — We propose a set of benchmark tests for current-voltage (IV) curve fitting algorithms. Benchmark tests enable transparent and repeatable comparisons among algorithms, allowing for measuring algorithm improvement over time. An absence of such tests contributes to the proliferation of fitting methods and inhibits achieving consensus on best practices. Benchmarks include simulated curves with known parameter solutions, with and without simulated measurement error. We implement the reference tests on an automated scoring platform and invite algorithm submissions in an open competition for accurate and performant algorithms.

I. INTRODUCTION

The current-voltage curve of a PV device is frequently modeled as an equivalent circuit comprising one or more diodes and resistors. The commonly-used model is that of a single-junction PV device (Figure 1). Applying Kirchoff's circuit laws, the current-voltage combinations at the device's output terminals are described by the single diode equation (Eq. 1) where V and I denote voltage (V) and current (A), respectively. The term N_S is the number of series-connected cells, V_{th} is the thermal voltage (V) given by Eq. 2 where k is the Boltzmann constant (J/K), q is the elementary charge (C) and T_C is the cell temperature in K.

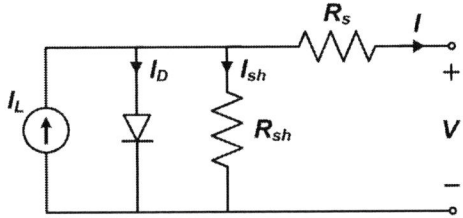

Figure 1. Singe diode equivalent circuit for a PV device.

$$I = I_L - I_O\left(\exp\left(\frac{V + IR_S}{nN_SV_{th}}\right) - 1\right) - \frac{V + IR_S}{R_{SH}} \quad (1)$$

$$V_{th} = \frac{k}{q}T_C \quad (2)$$

I-V curves are readily measured by a number of devices. The single-diode equation has five coefficients – photocurrent I_L, saturation current I_O, series resistance R_S, shunt resistance R_{SH} and diode ideality factor n – that can thus be determined by fitting Eq. 1 to measured data.

Fitting of Eq. 1 is a popular topic of interest as evidenced by hundreds of papers proposing new fitting techniques. New articles appear frequently across a large number of journals. To illustrate, 19 papers on this subject have appeared since 2020 in the Journal of Photovoltaics alone. A complete bibliography of published methods is far beyond the scope of this conference paper.

Comparison among proposed methods is practically impossible, due to the absence of common test cases, consistent metrics and validation practices. In our view, the lack of a common validation structure contributes substantially to the proliferation of papers on IV curve fitting, as authors, reviewers and editors cannot reasonably answer a basic question: "does this method improve upon the state of art or practice?"

Often, validation of fitting methods comprises fitting the single diode equation to a few measured I-V curves and computing metrics of the difference between the fitted curves and data. While indicative of a method's ability to yield reasonable results, this procedure overlooks two potential points of failure: misspecified models, and sensitivity to measurement error.

1. The single diode model represents an ideal single junction device with superimposed currents. The measured device's behavior may, or may not, be approximated well by this model. The fitted model may not be appropriate for the device being measured. A misspecified model may be detected by first applying the fitting procedure to synthetic curves calculated from the single diode equation; a successful fitting method should recover the known parameters used to generate the synthetic curves.

2. Measured IV curves always embody some degree of imprecise or inaccurate measurements. A fitting procedure that is overly sensitive to error in some, or all, of the measured values, may return parameters that vary significantly across measurements from the same device under the same conditions. Sensivity to measurement error may be detected by applying the fitting method to synthetic curves with simulated error and comparison of the fitted parameters with the known values used to generate the curves.

We propose a set of benchmark tests for fitting the single-diode equation to data and metrics to measure fitting accuracy. The benchmark tests explicity address the issues of model misspecification and sensitivity to error. We present an automated platform for scoring fitting methods against these benchmarks, with elements of competition including a leaderboard, to encourage progress toward consensus methods for accurate IV curve fitting. The competition platform is

978-1-6654-6060-6/23 $31.00 © 2023 IEEE

TABLE I
SUMMARY OF BENCHMARK TESTS

Test Set	Description	I_L (A)	I_O (nA)	n (-)	R_S (Ω)	R_{SH} (Ω)	Comments
1	cSi w/o noise	1.0, 8.0	0.5, 30	1.01, 1.3	0.1, 1.0	300, 3000	32 curves (all parameters combinations) 72 cells in series
2	Thin film w/o noise	0.5, 2.5	1, 10	1.3, 1.5	0.1, 1.0	300, 3000	32 curves (all parameter combinations) 140 cells in series
3A	cSi w/ noise	8.0	0.5	1.01	0.1	3000	50 realizations of base curve
3B	cSi w/ noise	1.0	30	1.3	1.0	300	50 realizations of base curve
3C	Thin film w/ noise	2.5	1	1.3	0.1	3000	50 realizations of base curve
3D	Thin film w/ noise	0.5	10	1.5	1.0	300	50 realizations of base curve

designed to accumulate a library of source code in order to enable independent verification and re-use of successful fitting methods.

II. BENCHMARK TESTS

Benchmark tests are used in a number of mathematical settings to provide both a consistent means of comparing algorithms and to ensure algorithms are tested against a wide range of problems. For example, in optimization, libraries are available containing numerous benchmarks (also termed "test functions") for constrained and unconstrained optimization problems for one- and higher-dimensional problems (e.g., [1]).

We propose three sets of benchmark tests, summarized in Table I:

1. Simulated I-V curves without noise (Test Set 1 and 2). These curves are formed by computing precise (abs. error $< 10^{-15}$) solutions to the single-diode equation for specified parameter sets. Parameters sets are chosen to represent both cSi and thin-film type modules with a wide range of variation in photocurrent, saturation current, resistances and diode ideality factor. These benchmarks measure an algorithm's capability to recover known parameters in the absence of any complicating factors, such as measurement error or model mis-specification, and is a frequently-omitted step in validation of published algorithms.

2. Simulated I-V curves with simulated measurement error. These curves are formed by adding simulated error to four curves selected from the first set of benchmarks; fifty realizations of each base curve are generated. For each set of fifty curves, the parameters for the underlying base curve are fixed and known. This set of benchmarks measures an algorithm's capability to recover known parameters from replicated measurements with reasonable measurement error, but without any complication from model mis-specification.

We intentionally do not include measured I-V curves for actual modules in the set of benchmark tests. Measured I-V curves are affected by both measurement error and the possibility of model mis-specification. Model mis-specification occurs when the equation being fit to the data (Eq. 1 in this case) does not accurately describe the physics of the device being measured. Eq. 1 relies on several assumptions (e.g., superposition of currents) and is itself an approximation (see [2]) of more refined descriptions of the relevant physics. For any actual device, it is difficult to know if the assumptions and approximation in Eq. 1 are appropriate. For these reasons, including measured I-V curves as benchmark tests serve only to confirm that the fitting algorithm conforms the equation to the data, in the presence of measurement error, and these aims are already accomplished by the simulated I-V curves.

Curves for Test Sets 1 and 2, and the base curves for Test Set 3, are computed to high precision using functions in pvlib-python [3] and python's mpmath package. The pvlib-python functions provide V, I pairs that solve Eq. 1 with relative error of approximately 10^{-12}; these solutions are then refined using the mpmath library to achieve V, I pairs with less than 10^{-15} absolute error. Code for these calculations is available at https://github.com/cwhanse/ivcurves.

III. IV CURVE FITTING COMPETITION

We have established an automated platform for scoring fitting methods at https://github.com/cwhanse/ivcurves. The platform envisions a competition where algorithms are scored for each category of benchmark tests and ranked on a Leaderboard by the summed scores, lowest (most accurate) score first (Figure 1). Instructions for participation are provided at https://cwhanse.github.io/ivcurves/participating.html.

Submissions must provide python code that reads the test sets, executes the fitting, and returns the fitted parameters as described by the user instructions. The fitting procedure may use code other than python, or may call external services.

978-1-6654-6060-6/23 $31.00 © 2023 IEEE

Scoreboard

Submissions are given a score for some or all test sets, and the sum of these scores is the submission's overall score. If a submission is not scored on a test set, that test set's score will be blank (–). Test sets **case1** and **case2** are scored by the distance between the known IV curve and the submission's fitted IV curve (see `ivcurves.compare_curves.score_curve()`). Test sets **case3a** through **case3d** are scored by the difference between the known and fitted single diode equation parameters (see `ivcurves.compare_curves.score_parameters()`).

Submission	Method Name	Overall Score	case1	case2	case3a	case3b	case3c	case3d	Links
cwhanse (#45)	sandia_simple	66.2551	30.0007	31.0003	4.25967	0.0568522	0.340993	0.596508	Code
cwhanse (#46)	sandia_simple	66.2551	30.0007	31.0003	4.25967	0.0568522	0.340993	0.596508	Code

Figure 1. Screenshot of scoreboard from IV curve fitting competition website.

However, it is are desireable that all source code be provided so that fitting procedures can be independently verified. Source code must be provided with the BSD 3-clause license. Submitters retain their copyright. Source code is accompanied by documentation that is rendered to html pages. Our vision is to accumulate a library of algorithms, with documentation, so that interested parties may select, download and apply suitable algorithms to their work.

IV. Conclusions

We propose a set of benchmark tests for fitting the single-diode equation to data and metrics to measure fitting accuracy. Benchmark tests provide a consistent, repeatable structure for comparing among fitting methods and for measuring improvement over time. We provide an open evaluation platform with elements of competition to facilitate use of these benchmarks and encourage public sharing of code for fitting algorithms. Our approach could be readily extended for other equivalent circuit models with, e.g., two diodes or infinite shunt resistance, or for evaluating fitting of full single diode models such as [4] or [5].

Acknowledgement

This work was supported by the U.S. Department of Energy's Office of Energy Efficiency and Renewable Energy (EERE) under the Solar Energy Technologies Office Award Number 38267. Sandia National Laboratories is a multimission laboratory managed and operated by National Technology and Engineering Solutions of Sandia, LLC., a wholly owned subsidiary of Honeywell International, Inc., for the U.S. Department of Energy's National Nuclear Security Administration under contract DE-NA-0003525.

References

[1] C. Floudas et al. Handbook of Test Problems in Local and Global Optimization. New York, New York. Springer-Verlag, 1999.

[2] J. L. Gray, The Physics of the Solar Cell, in Handbook of Photovoltaic Science and Engineering 2nd Ed., ed. A. Luque and S. Hegedus. John Wiley & Sons, Ltd. 2011.

[3] William F. Holmgren, Clifford W. Hansen, and Mark A. Mikofski. "pvlib python: a python package for modeling solar energy systems." Journal of Open Source Software, 3(29), 884, (2018). https://doi.org/10.21105/joss.00884

[4] K. J. Sauer, T. Roessler and C. W. Hansen, "Modeling the Irradiance and Temperature Dependence of Photovoltaic Modules in PVsyst," in IEEE Journal of Photovoltaics, vol. 5, no. 1, pp. 152-158, Jan. 2015, https://doi.org/10.1109/JPHOTOV.2014.2364133.

[5] Dobos, A. P. (March 6, 2012). "An Improved Coefficient Calculator for the California Energy Commission 6 Parameter Photovoltaic Module Model." ASME. J. Sol. Energy Eng. May 2012; 134(2): 021011. https://doi.org/10.1115/1.4005759

Validation Of Photovoltaic Plant Loss Estimation From Monitoring Data: String Faults, Shading and Degradation

Karel De Brabandere, Maitheli Nikam, Julien Deckx and Gofran Chowdhury

3E, Brussels, 1000, Belgium

Abstract — **Monitoring of a photovoltaic plant improves its performance by maximizing energy production as well as minimizing economic losses. Thus, identification and quantification of such performance losses is necessary to avoid them. 3E's monitoring platform provides a detailed loss breakdown analysis with performance losses due to soiling, shading, degradation, etc. The main focus of this study is the validation of string faults, shading and degradation losses against complementary results from thermal and visual drone imagery.**

I. INTRODUCTION

Solar is the fastest growing source of renewable energy where more than half of the renewable energy capacity installed in 2021 came from photovoltaics [1]. Globally solar hit the one Terawatt capacity in 2022 and is expected to more than double in the next few years [1]. On the other hand, utility-scale solar PV O&M cost came down a staggering 85% between 2005 and 2017 (numbers for Europe) [2]. This steep descent is an indication of the immense price pressure that O&M service providers are under. Since manpower is a large part of the cost, the managed capacity per operator is continuously increasing. The growth and consolidation of the market means that portfolios become larger and more geographically spread out. One of the ways that O&M service providers have been able to cope with this pressure is digitalization. Smart monitoring platforms allow to identify issues remotely, with less need for plant inspections. However, the evolution to larger, more diverse portfolios with less on-site presence makes it challenging to keep the operation of PV plants at an optimal level. With tight margins in a post-subsidy era, asset owners can no longer afford to lose revenue from avoidable losses.

There is still a significant potential to optimize O&M while improving the performance of solar plants through smart monitoring supported by advanced AI-driven analytics. While digital tools have helped to make the operation and maintenance of PV plants more efficient and effective, the reduced on-site presence also means that some production losses are left undiscovered for larger periods of time. Luckily, digital tools are further evolving. A new generation of AI-driven advanced analytics is capable of automatically and continuously providing a detailed breakdown of the root causes of production losses. Based on detailed monitoring data, such system can detect issues in a very early stage. This prevents PV plants from underperforming for large periods of time and increases overall profits.

The loss breakdown approach used by the 3E monitoring platform is unique in the sense that it combines advanced data-driven AI analytics with a detailed PV system model. Thus, it provides both expected and unexpected losses from plant to inverter input level and divides these losses into a long list of loss categories. Furthermore, validation of such a loss breakdown analysis by comparison with a complementary analysis from drone inspections has not been presented before.

II. AIM AND APPROACH

The performance of a solar plant is the result of many confounding factors. The key to identifying problems early on is to have a detailed simulation model, relying on an accurate digital twin, combined with equally detailed monitoring data. Measured current and voltage at MPPT level are compared with simulated MPPT current and voltage for the given environment conditions. Using pattern recognition methods, performance issues like shading, string faults, degradation and MPP mismatch can be detected, and corresponding losses estimated. Some examples of performance issues that are identified and quantified this way are:

- **String faults** are detected as a drop in DC current with respect to similar arrays of the plant, using the knowledge of the number of strings in the array [3]. Further intelligence is built in to avoid false positives, e.g., in the presence of shading. String faults can be detected reliably, as long as the number of strings per monitored string set is not too high.
- **Shading** is detected as a drop in DC current compared to the simulation model, with consistent patterns according to the position of the sun throughout the day and year [4].
- **Soiling** is detected as a day-to-day decrease of DC current compared to the simulation model, interleaved with sudden restoration of the DC current as a result of cleaning events [5].
- **Degradation** is identified as a consistently reduced DC current and voltage compared to the simulation model [6]. Having a split-up between current and voltage degradation is indispensable in order to reach detailed conclusions. For instance, normal annual degradation typically manifests itself mainly as current degradation. On the other hand, potential-induced degradation (PID) is characterized by MPP

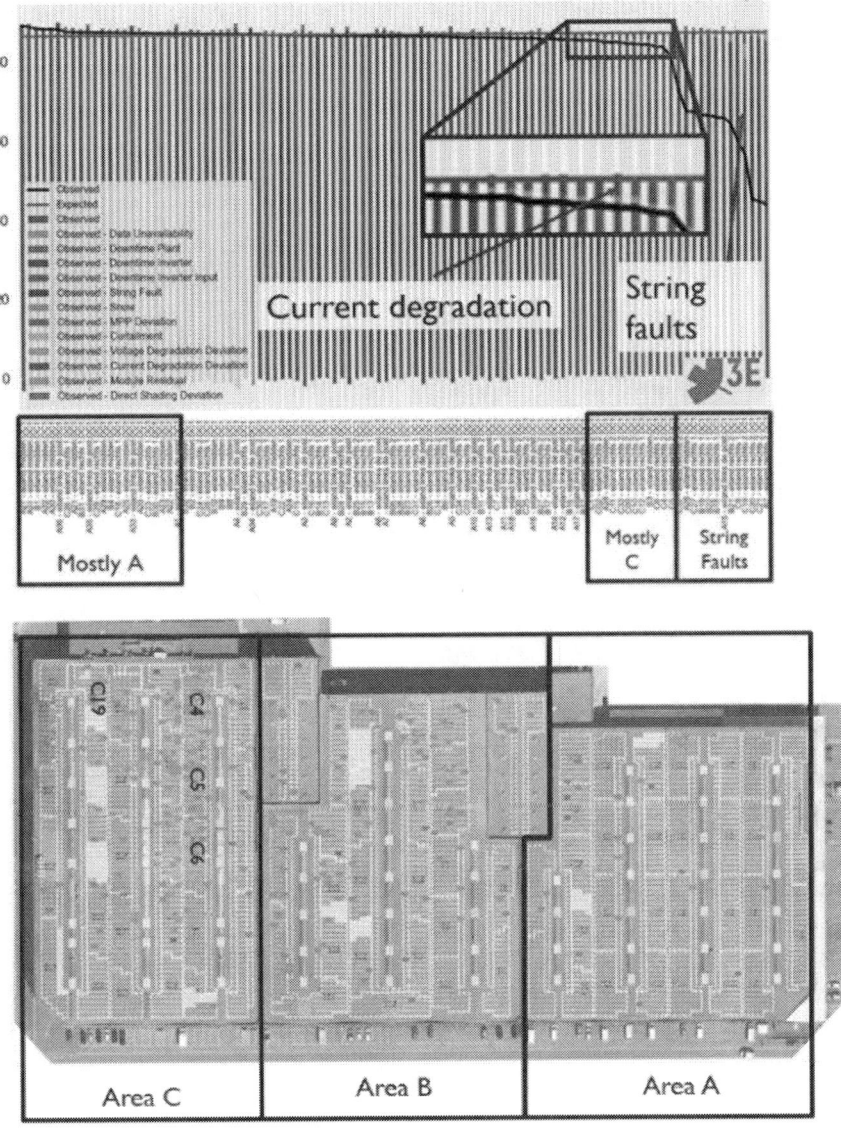

Figure 1: Automatic loss classification per inverter from 3E Solar Analytics (top); anomalies detected by drone inspection (courtesy of Sitemark) (down)

mismatch as a result of a non-uniform voltage degradation along each string [7].

- **MPP mismatch** refers to the power operating point not matching the Maximum Power Point under given conditions. MPP mismatch may be caused by the inverter MPP tracker deviating from the MPP, e.g., because of derating behaviour, but it also may be caused by PV module MPP power mismatch within a string/array, e.g., due to non-uniform degradation or irradiance conditions [8]. Using the available monitoring data, MPP mismatch is further classified in various subcategories, such as maximum AC power derating, inverter temperature derating, partial shading, etc.

This work reports on the validation of the AI-driven loss breakdown analysis of the 3E monitoring platform versus complementary results from thermal and visual drone imagery as part of the AnalystPV research project for three illustrative loss types: string faults, shading and degradation.

III. RESULTS AND CONCLUSIONS

Figure 1 shows the loss breakdown analysis for each inverter of a rooftop plant (top) and compares it with the anomalies detected by a drone inspection (down). A total of 16 string faults were identified by the loss analysis, which was 1 more than detected by the drone. Furthermore, current degradation was detected in those areas where the drone identified many hotspots, caused by bird droppings. The energy loss due to current degradation, as estimated by the loss analysis, shows a strong correlation with the number of hotspots detected by the drone.

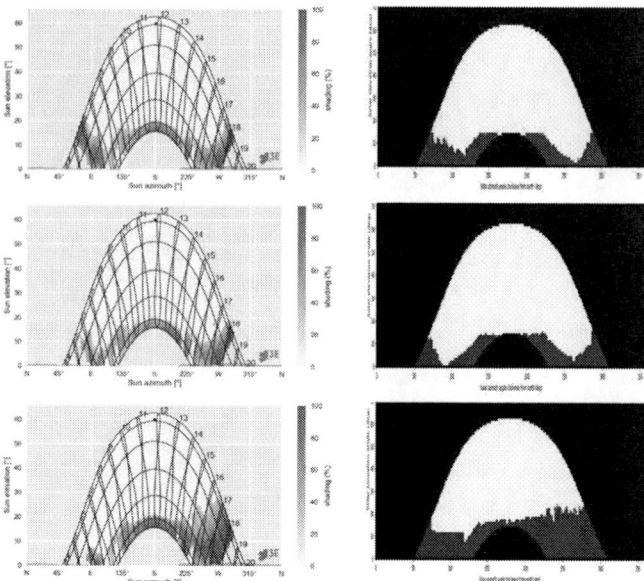

Figure 3: Shading maps of 3 sub-arrays in a plant as derived from monitoring data by 3E Solar Analytics (left); shading map of 3 corresponding points as obtained from a 3D simulation from drone images (right)

Figure 2 compares the shading maps, as derived from the monitoring data for three inverter inputs of a ground-mounted plant, with the shading maps, as obtained from a 3D simulation from drone images for three corresponding points in the plant. The results from both methods were found to be in good agreement (apart from a minor artefact in the 3D simulation due to an edge effect).

Figure 3 (top) shows the loss breakdown for each inverter of a ground-mounted plant for which a combination of degradation and MPP mismatch with a combined loss of ca. 10-15% was detected. Such a combination is characteristic for PID, which is indeed confirmed by the drone inspection indicating the presence of PID (down).

Overall, the loss breakdown analysis results correspond well with the issues detected by complementary drone inspections. More analytics and insights would be presented.

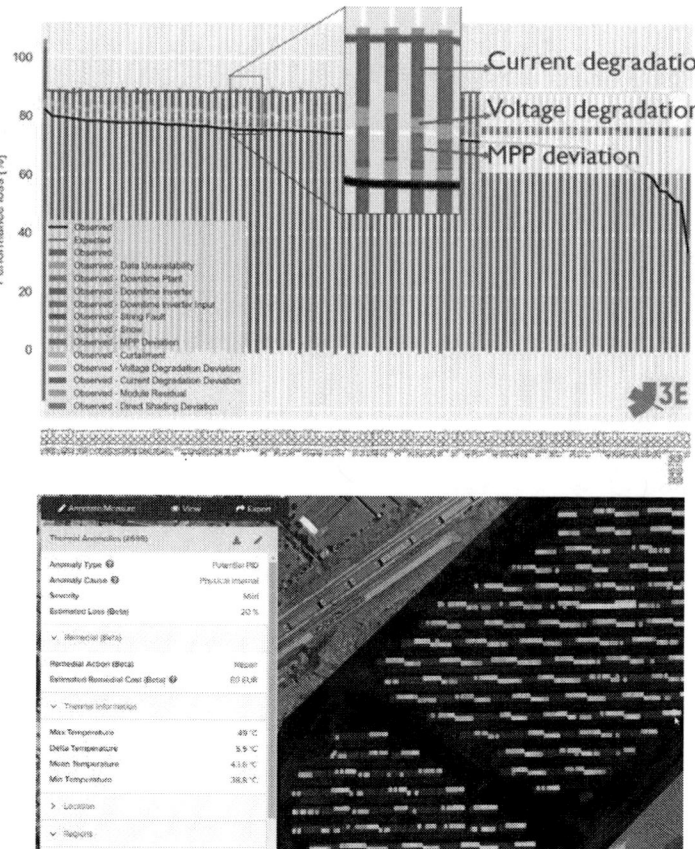

Figure 2: Automatic loss classification per inverter from 3E Solar Analytics (top); analysis from drone inspection (courtesy of Sitemark) (down)

REFERENCES

[1] Solar Power Europe, "Global market Outlook for solar power: 2022-2026", *Global Market Outlook* (2022), pp. 1-116

[2] IRENA (2020), *Renewable Power Generation Costs in 2019*, International Renewable Energy Agency, Abu Dhabi.

[3] A. Mellit, G.M. Tina, and S.A. Kalogirou. "Fault detection and diagnosis methods for photovoltaic systems: A review." *Renewable and Sustainable Energy Reviews* 91 (2018): 1-17.

[4] R. Hariharan, M. Chakkarapani, G. Saravana Ilango and C. Nagamani, "A Method to Detect Photovoltaic Array Faults and Partial Shading in PV Systems," in *IEEE Journal of Photovoltaics*, vol. 6, no. 5, pp. 1278-1285, Sept. 2016, doi: 10.1109/JPHOTOV.2016.2581478.

[5] M. Gostein, T. Düster and C. Thuman, "Accurately measuring PV soiling losses with soiling station employing module power measurements," *2015 IEEE 42nd Photovoltaic Specialist Conference (PVSC)*, New Orleans, LA, USA, 2015, pp. 1-4, doi: 10.1109/PVSC.2015.7355993.

[6] A. Ndiaye, C.M.F. Kébé, P.A. Ndiaye, A. Charki, A. Kobi, V. Sambou, "A Novel Method for Investigating Photovoltaic Module Degradation", *Energy Procedia*, 36, 2013, pp. 1222-1231

[7] S. Spataru, D. Sera, T. Kerekes, R. Teodorescu, "Diagnostic method for photovoltaic systems based on light I–V measurements", *Solar Energy*, 119, 2015, pp. 29-44

[8] T.S. Wurster, M.B. Schubert, Mismatch loss in photovoltaic systems, *Solar Energy*, 105, 2014, pp. 505-51

Single-junction bifacial and semitransparent $Sb_2(S,Se)_3$ solar cells

Chen Qian, Kaiwen Sun, Martin Green, Bram Hoex, Xiaojing Hao

University of New South Wales, Sydney, Australia

Abstract — $Sb_2(S,Se)_3$ has become a promising new photovoltaic (PV) material for not only high absorption coefficient, adjustable bandgap but non-toxic, eco-friendly and earth-abundant composition. However, high-efficiency $Sb_2(S,Se)_3$ solar cells reported up to now are opaque and only permit monofacial absorption, limiting the applicability in some circumstances where semitransparency and the capability of absorbing tilt incident light are demanded. Here, we demonstrate a bifacial and semitransparent $Sb_2(S,Se)_3$ solar cell enabled by an FTO substrate and ITO back contact. The ultra-thin inner n-i-p structure and transparent conductive oxides (TCOs) offer high transmittance at the long wavelength range. Although the non-ideal MnS/ITO interface brings slight band mismatch and an increased defect density, satisfying power conversion efficiencies (PCEs) of 7.41% and 6.36% are nevertheless attained for front and rear absorption respectively. Moreover, our further investigation reveals this innovative bifacial solar cell can well absorb the incident light with 2π tilt angles except the parallel one to the cell. These intriguing results suggest that bifacial and semitransparent $Sb_2(S,Se)_3$ solar cells have a significant advantage in practical applications.

I. INTRODUCTION

$Sb_2(S,Se)_3$ has emerged as a promising light-harvesting material due to its non-toxic, earth-abundant and stable properties. This quasi-1-D material exhibits an adjustable bandgap from 1.1 eV to 1.7 eV depending on the ratio of sulfide and selenium, and high absorption coefficient ($>10^5$ cm^{-1}). In 2020, $Sb_2(S,Se)_3$ solar cells achieved a power conversion efficiency (PCE) of 10% for the first time, realized by hydrothermal method[1]. In 2021, PCE has been further increased to 10.7% with alkaline metal fluoride assisted solution post-treatment[2]. The impressive and continuously increasing PCE make it feasible and necessary to start developing practical applications of $Sb_2(S,Se)_3$ solar cells, such as BIPV or indoor photovoltaic applications. BIPV normally requires certain transmittance of thin film solar cells to ensure the indoor brightness. However, the gold electrode commonly used in the high-efficiency $Sb_2(S,Se)_3$ solar cells block most of lights. Meanwhile, the indoor lights with various intensity usually come from all directions, partly including reflective lights from the albedo, which are difficult to be effectively absorbed for conventional opaque $Sb_2(S,Se)_3$ solar cells. Thus, an $Sb_2(S,Se)_3$ solar cell with semitransparent property and double-sided absorption is strongly in demand.

A bifacial solar cell is a good way to overcome the obstructions and support the practical applications. This concept was firstly put forward and investigated in 1960s[3]. Monocrystalline silicon material was used to create the first bifacial solar cell in 1980. Then, multicrystalline silicon was used as the source. In order to create flexible, light-weight, and semitransparent solar cells, this idea was later applied to thin film and dye-sensitized solar cells [4-6]. In previous literatures, bifacial thin film solar cells were usually realized through two ways: double-sided deposition, which involved symmetrically growing cells on each side of the substrate [7]; single-sided deposition, which used semitransparent front and back contacts [8, 9] to ensure the light absorption from both sides. Double-sided bifacial solar cells consume double amounts of materials, need complicated fabrication procedures and highly concern about the interaction from the alternating deposition of both-sided cells. Moreover, the potential voltage mismatch in the equivalent parallel circuit, resulting from uneven illumination from both sides in practical applications, could bring performance degradation and electrical risk, like hot spots. In contrast, single-junction bifacial solar cells exhibit cost-saving, easily acquiring and safely running properties.

In this study, we demonstrate a single-junction bifacial and semitransparent $Sb_2(S,Se)_3$ solar cell (abbreviated as "B-cell") with an ITO back electrode capping the MnS hole-transporting layer (HTL). Though the potential barrier and growing defect density at MnS/ITO interface cause detrimental effect on performance, particularly on open circuit voltage (V_{oc}), this semitransparent back contct enables the rear absorption. The ultrathin fully depleted absorber layer (350nm) made by the hydrothermal method allows carriers to drift rather than diffuse towards the functional layers, greatly increasing the bipolar transport property and bifaciality. Thus, the PCEs for front and rear absorption are thereby 7.41% and 6.36%, indicating a bifaciality of as high as 0.86. In contrast, the conventional gold-based monofacial and opaque solar cell (abbreviated as "M-cell") attained PCEs of 9.50% and 0.70% in respective conditions. Morevover, our further investigation reveals the B-cell can work well under incident light with a broad range of tilt angles. These unique advntages provide bifacial and semitransparent $Sb_2(S,Se)_3$ solar cells a great promise for flexible and widespread applications.

II. EXPERIMENT SECTION

CdS, $Sb_2(S,Se)_3$ and MnS layers were deposited as our previous article[10]. ITO layer was subsequently deposited by sputtering method. The specimens were stick on a round sample holder by double sticky tape. The sample holder was then inserted in the reserved position in the sputtering machine

(JGP-450A, Sky technology development Co., Ltd. Chinese Academy of Sciences) along the groove, with MnS side facing down towards the ITO target material (purchased from Beijing Zhongnuoxincai). Self-rotating was turned on to improve the uniformity of ITO thin film. The Ar gas flowed into the chamber once the pressure reached 5×10^{-4} Pa and the rate should be adjusted to stabilize the pressure at 0.7 Pa. Afterwards, we switched on the sputtering process and the power was 40W. The processing time for each batch was 15 minutes.

III. RESULTS AND DISCUSSION

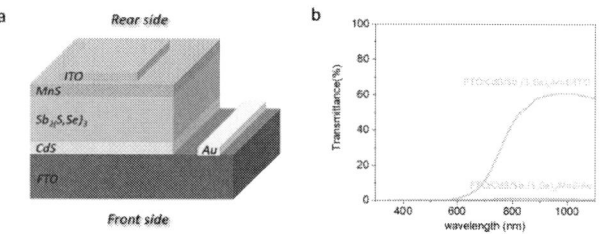

Fig. 1. a) The configuration of the $Sb_2(S,Se)_3$ B-cell. b) Transmittance of the B-cell (orange line) and the M-cell (blue line).

Figure 1a shows the eventual device architecture of B-cells, FTO/CdS/$Sb_2(S,Se)_3$/MnS/ITO, in contrast to FTO/CdS/$Sb_2(S,Se)_3$/MnS/Au in M-cells. Figure 1b displays transmittance of the B-cell starts to rise rapidly when the wavelength goes beyond the absorption edge of $Sb_2(S,Se)_3$ at around 850nm and finally reaches near 60%, implying its substantial superiority over M-cells at long-wavelength transmittance, owing to good transparency of the ITO back contact.

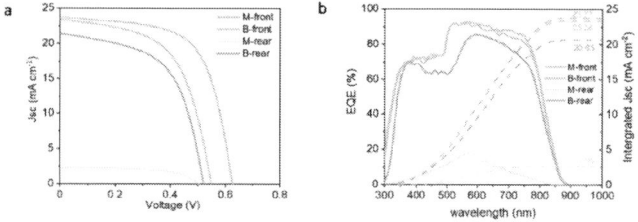

Fig. 2. a) J-V curve, b) EQE response of $Sb_2(S,Se)_3$ M-cells and B-cells with front and back illumination respectively.

To evaluate the bifaciality, we compared the PV performance (details in Table 1) and external quantum efficiency (EQE) response of B-cells and M-cells under front and back illumination respectively in Figure 2. J_{sc} of M-cell drops from 23.70 mA cm^{-2} under front illumination to only 2.29 mA cm^{-2} under back illumination because Au back contact blocks most of light, resulting in dramatically poor PV performance. In contrast, the rear absorption for B-cells remains at a satisfying

TABLE I
PV PERFORMANCE OF THE M-CELL AND THE B-CELL WITH FRONT AND BACK ILLUMINATION BY AM1.5G LIGHT RESPECTIVELY.

Device	V_{oc}/V	J_{sc}/mA cm^{-2}	FF/%	PCE/%
M-front	0.62	23.70	64.54	9.50
M-rear	0.50	2.29	61.47	0.70
B-front	0.54	23.36	58.29	7.41
B-rear	0.52	21.40	57.33	6.36

level. J_{sc} only drops from 23.36 to 21.40 mA cm^{-2}, resulting in a PCE of 6.36% and an exceptional bifaciality of 0.86.

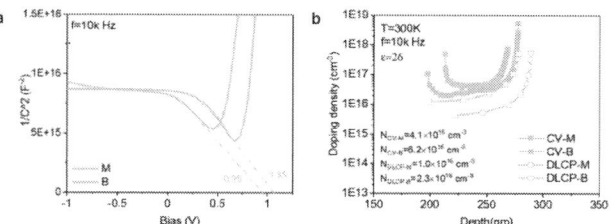

Fig. 3. a) M-S curves, b) defect concentration (N_d)-depth (W) of B-cells and M-cells respectively. (The capacitance values have been adapted in a unit area.)

Though a good bifaciality was obtained in our B-cell, the employment of semitransparent back contact did cause some detrimental effects on PV performance particularly on V_{oc} and FF compared to M-cells under identical front illumination.

Band alignment issue was thereby studied in response to the severe V_{oc} drop from 0.62 V to 0.54 V. The fermi level of ITO is 0.44eV higher than that of MnS[10]. The band inevitably bends downwards at the interface, suppressing the hole extraction to some extent. This Schottky contact at MnS/ITO interface builds a potential barrier and results in a diminished output potential, consistent with the fitted built-in potential from the Mott-Schottky (M-S) curves (0.95 eV and 1.15 eV for B-cells and M-cells respectively) in Figure 3a.

On the other hand, less efficient carrier extraction at MnS/ITO interface than at the MnS/Au interface is manifested by lower EQE response of the B-front between 600nm to 800nm wavelength compared with that of M-front in Figure 2b. To further figure out the quantity of interface defects, we implemented capacitance-voltage (C-V) and deep-level capacitance profiling (DLCP) to characterize interface quality through their deviation as DLCP is less sensitive to junction interface defects than C-V. The prior capacitance-frequency (C-f) measurement uncovers that B-cells have a lower critical frequency at around 10k Hz, compared with M-cells, which have a critical frequency of around 100k Hz. This difference may result from higher series resistance (Rs) for ITO. C-V measurements were thereby performed both at a frequency of 10k Hz. Figure 3b displays the profile of defect concentrations against depth and the concentration numbers at zero bias evaluated from CV and DLCP results. We calculated an interface defect areal density of 1.35×10^{12} cm^{-2} in B-cells, higher than 1.07×10^{12} cm^{-2} in M-cells, which is responsible for lower FF in B-front case.

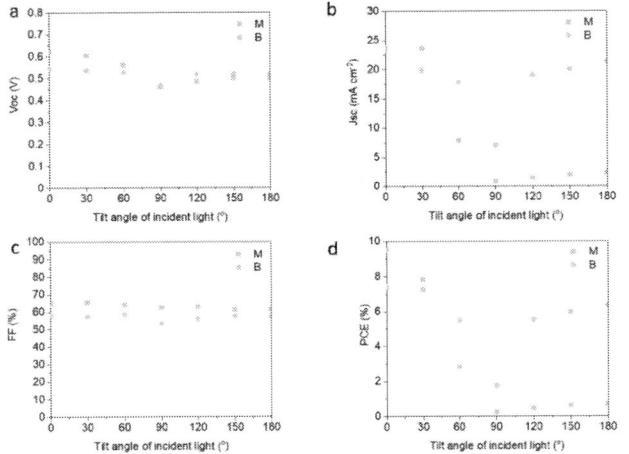

Fig. 4. a) V_{oc}, b) J_{sc}, c) FF and d) PCE of B-cells and M-cells against tilt angles of incident light.

The superiority of B-cells in practical application is reflected by its outstanding capability of absorbing incident light with tilt angles. We conducted a series of PV performance measurements under tilt incident light from 0 to 180 degrees to simulate the actual application scenario. Herein, zero degree denotes the front side faces the incident light. Figure 4 shows that PCEs at different tilt angles are strongly related to J_{sc}. Furthermore, the output power of the M-cell could become essentially zero once the tilt angle gradually goes beyond 90 degrees. In contrast, the B-cell exhibits a quasi-sin wave with a rationally expected period of 2π, benefiting from the great bifaciality. The tendency is in line with the results of the earlier study, which indicated a sin wave with a period of π for symmetrically growing double-sided bifacial solar cells[7]. Our B-cells can operate better than M-cells when the incident light illuminates with a tilt angle larger than 30 degrees. Particularly when the tilt angle exceeds 90 degrees, the advantage can be as much as nine times or greater.

IV. SUMMARY

In summary, we demonstrates a single-junction bifacial and semitransparent $Sb_2(S,Se)_3$ solar cell by employing a FTO layer as substrate and ITO layer as back contact. This sandwich structure provides a quasi-symmetric PV performance for incident light coming from front and rear side respectively. Specifically, the PCEs of 7.41% and 6.36% are obtained respectively. The excellent bifaciality of 0.86 contributes to well handling the incident light with various tilt angles, making $Sb_2(S,Se)_3$ applicable for practical applications with improved flexibility. Although V_{oc} is limited by the Schottky contact between MnS and ITO, tremendous potential of the single-junction bifacial and semitransparent $Sb_2(S,Se)_3$ solar cell has been uncovered owing to its non-toxic, stable, material-saving, and low cost characteristics.

REFERENCES

[1] R. Tang, X. Wang, W. Lian, J. Huang, Q. Wei, M. Huang, Y. Yin, C. Jiang, S. Yang, G. Xing, S. Chen, C. Zhu, X. Hao, M. A. Green, and T. Chen, "Hydrothermal deposition of antimony selenosulfide thin films enables solar cells with 10% efficiency," *Nature Energy,* vol. 5, no. 8, pp. 587-595, 2020.

[2] Y. Zhao, S. Wang, C. Jiang, C. Li, P. Xiao, R. Tang, J. Gong, G. Chen, T. Chen, J. Li, and X. Xiao, "Regulating Energy Band Alignment via Alkaline Metal Fluoride Assisted Solution Post-Treatment Enabling Sb2(S,Se)3 Solar Cells with 10.7% Efficiency," *Advanced Energy Materials,* vol. 12, no. 1, 2021.

[3] M. Hiroshi, "RADIATION ENERGY TRANSDUCING DEVICE," Patent Appl. US3278811, 1961.

[4] X. Wu, J. Zhou, A. Duda, J. C. Keane, T. A. Gessert, Y. Yan, and R. Noufi, "13·9%-efficient CdTe polycrystalline thin-film solar cells with an infrared transmission of ~50%," *Progress in Photovoltaics: Research and Applications,* vol. 14, no. 6, pp. 471-483, 2006.

[5] S. Marsillac, V. Y. Parikh, and A. D. Compaan, "Ultra thin bifacial CdTe solar cell," *Solar Energy Materials and Solar Cells,* vol. 91, no. 15-16, pp. 1398-1402, 2007.

[6] Z. Song, C. Li, L. Chen, and Y. Yan, "Perovskite Solar Cells Go Bifacial-Mutual Benefits for Efficiency and Durability," *Adv Mater,* vol. 34, no. 4, p. e2106805, Jan 2022.

[7] H. Deng, Q. Sun, Z. Yang, W. Li, Q. Yan, C. Zhang, Q. Zheng, X. Wang, Y. Lai, and S. Cheng, "Novel symmetrical bifacial flexible CZTSSe thin film solar cells for indoor photovoltaic applications," *Nat Commun,* vol. 12, no. 1, p. 3107, May 25 2021.

[8] S. H. Park, K.-H. Shin, J.-Y. Kim, S. J. Yoo, K. J. Lee, J. Shin, J. W. Choi, J. Jang, and Y.-E. Sung, "The application of camphorsulfonic acid doped polyaniline films prepared on TCO-free glass for counter electrode of bifacial dye-sensitized solar cells," *Journal of Photochemistry and Photobiology A: Chemistry,* vol. 245, pp. 1-8, 2012.

[9] W. Yang, X. Xu, Z. Tu, Z. Li, B. You, Y. Li, S. I. Raj, F. Yang, L. Zhang, S. Chen, and A. Wang, "Nitrogen plasma modified CVD grown graphene as counter electrodes for bifacial dye-sensitized solar cells," *Electrochimica Acta,* vol. 173, pp. 715-720, 2015.

[10] C. Qian, J. Li, K. Sun, C. Jiang, J. Huang, R. Tang, M. Green, B. Hoex, T. Chen, and X. Hao, "9.6%-Efficient all-inorganic Sb2(S,Se)3 solar cells with a MnS hole-transporting layer," *Journal of Materials Chemistry A,* vol. 10, no. 6, pp. 2835-2841, 2022.

The United States Renewable Energy Landscape: Siting, Management, and Potential Impacts

Jacob T. Stid, Anthony D. Kendall, Annick Anctil, Jeremy Rapp, David W. Hyndman

Michigan State University, East Lansing, MI, United States

The University of Texas at Dallas, Richardson, TX, United States

We are in the midst of a rapid shift in energy landscapes. As we make this transition, it will be imperative to develop an early understanding of how this change will impact society and all related ecosystem services. To develop this understanding, we first need a highly-detailed depiction of our current infrastructure and how it is distributed across the landscape. Here, we combine the best available United States spatiotemporal solar and wind installation datasets and assess the distribution of renewable energy, and the cohabitation of energy and landcover. We show that roughly 80% of all renewable energy infrastructure has been installed in privately owned land, predominantly agricultural in historical use with cropland (66%) and grassland (21%) dominating total converted area. While solar and wind can contribute collaboratively to utility demand profiles and permitting access when installed together, half of all solar arrays are 55 km from a wind turbine with 25% being further than 100 km from a wind power installation, a potential shortfall in current infrastructure. High-resolution image analysis of solar groundcover management suggests that 7% of solar direct area is actively managed with vegetative groundcover, 0.3% is managed with highly reflective/impervious surfaces, with the remainder is likely managed as barren (25%) or mixed landcover. We also show that local agricultural effects of wind turbines in agricultural fields diminish drastically around 90 to 120 m. This work enhances knowledge of our current energy landscape and can inform future placement and management strategies.

Interpreting Accelerated Tests on Perovskite Modules: Using Photooxidation of MAPbI3 as an Example

Ingrid L. Repins, Michael Owen-Bellini, Michael D. Kempe, Michael G. Deceglie, Joseph J. Berry, Nutifafa Y. Doumon, Timothy J. Silverman, Laura T. Schelhas

National Renewable Energy Laboratory, Golden, CO, United States

Solar panels based on metal halide perovskites are following a fast track to commercialization. However, unlike more established solar cell types, there are not yet decades-long field observations to increase consumer confidence. Thus, one must rely heavily on accelerated stress tests to demonstrate reliability. This work uses published experimental data for O2 diffusion and dry photooxidation of methylammonium-lead-iodide (MAPbI3) to estimate the acceleration factors (AF') of some common module accelerated tests. AF' are found to be low. These tests are thus only useful to screen for early failures. Commonly used dark damp heat will accelerate hydrothermal degradation but will not produce relevant photooxidative degradation. Module design choices affect degradation rate. Some degradation mechanisms may cause power loss at lower gas concentrations than dry photooxidation. For photooxidation of MAPbI3, diffusion of O2 into the package limits the degradation rate, allowing a simpler calculation. Steps for predicting decades of performance behavior without decades of field observation are presented.

Charge Carrier Diffusion and Recombination near Misfit Dislocations in GaAsP/GaInP Heterostructures

T.H. Gfroerer,[1] A. Edmondson,[1] Lilian Korir,[1] Fan Zhang,[2] Yong Zhang,[2] and M.W. Wanlass[3]

[1] Davidson College, Davidson, NC USA
[2] Univesity of North Carolina at Charlotte, Charlotte, NC USA
[3] Wanlass Consulting, Norwood, CO USA

Abstract—GaAsP alloys are good candidates for the upper junction in Si-based dual-junction tandem solar cells. However, monolithic growth of GaAsP on Si is limited by dislocation formation due to lattice mismatch. When metamorphic GaAsP is grown on GaAs substrates, dark crosshatch features can be observed in spatially-resolved electroluminescence and photoluminescence measurements. These features can be attributed to misfit dislocations in the epitaxial layers. We compare photoluminescence imaging with complementary confocal mapping measurements to test models of diffusion and recombination near misfit dislocations in a GaAsP/GaInP heterostructure. The crosshatch features become sharper with increased photoexcitation in accordance with reduced lifetimes and diffusion distances. However, the imaging and mapping experiments require different defect-related recombination rates in our computational models.

Keywords—electroluminescence, photoluminescence, imaging, confocal mapping, misfit dislocations, diffusion, recombination

I. INTRODUCTION

Silicon-based photovoltaic technology continues to dominate the solar cell market, but the efficiency of single-junction silicon devices is ultimately limited by poor conversion of high energy light. Future improvements in Silicon-based devices are likely to rely on the addition of one or more higher energy junctions to improve the conversion efficiency of light in the visible range of the solar spectrum. The top material candidates for these new junctions include III-V alloys and perovskites. [1] Within the family of III-Vs, GaAsP, [2] GaInP, and AlGaAs [3] are suitable alloys for achieving the 1.7 eV bandgap required for optimal Si-based dual-junction performance.

In this contribution, we report on photoluminescence (PL) imaging and PL confocal mapping of a 1.67 eV GaAsP/GaInP double heterostructure grown on a GaAs substrate. In both measurements, dark crosshatch features are observed in the plan-view luminescence emission pattern from the structure, which we attribute to dangling bonds and defect states associated with misfit dislocations. [4] We study how these features change with photoexcitation in both experiments and model the behavior with simple steady-state rate equation algorithms. While the preliminary simulations can reproduce the changing profiles that result from variation in carrier lifetime, different defect-related recombination parameters are required to fit the results of the two experiments.

II. EXPERIMENT

The GaAsP/GaInP double heterostructure (DH) incorporates 3 GaAsP step-grading layers between the GaAs substrate and the DH to accommodate approximately 0.5% lattice mismatch. The nominally lattice-matched DH is not intentionally doped and has a 2 μm $GaAs_{0.8}P_{0.2}$ active layer, with a room-temperature (RT) PL peak emission of approximately 1.67 eV.

For PL imaging, the sample is illuminated with an unfocused 532 nm laser having an approximate spot size of 5.2×10^{-3} cm^2. PL and EL images are obtained with a cooled Q-Imaging Rolera XR CCD camera, which has extended infrared sensitivity to 1μm. Two OG 570 long-pass filters are used to eliminate scattered laser light. The camera is fitted with an Edmund Scientific VZM 300 imaging lens set to 3X magnification, yielding image pixel dimensions of 4.4×4.4 μm^2.

PL mapping experiments are conducted at RT with a Horiba LabRAM HR confocal Raman microscope. In this system, a 532 nm laser is focused by a 50X long working distance lens to a spot with a diameter of approximately 2 μm. Neutral density filters in the laser path are used to control the incident optical power, which equals 114 μW unfiltered. The confocal geometry of the instrument ensures that the PL sample volume coincides with the photoexcitation volume. The PL spectrum is acquired via a CCD detector in a 24 nm window centered on the peak emission (726 – 750 nm). PL maps are obtained by laterally translating the sample under the objective with a precise, motorized stage.

III. RESULTS

Representative RT PL images and confocal maps of a prominent crosshatch feature are shown in Fig. 1, with the images on the left and the maps on the right. The upper results were obtained at low laser power and the lower results were obtained at high laser power. The dark spots and lines mark defective regions where non-radiative recombination dominates, carriers are depleted, and PL emission is reduced. In particular, we attribute the isolated spots to threading dislocations and the lines to more extended misfit dislocations. In both experiments, we see clear changes in the extent of defective regions (driven by changes in the diffusion of electrons and holes) when the laser power is increased. [5] Brighter illumination produces higher steady-state carrier densities and recombination rates, along with reduced lifetimes and diffusion lengths. Hence, defective features appear sharper at high excitation.

978-1-6654-6060-6/23 $31.00 © 2023 IEEE

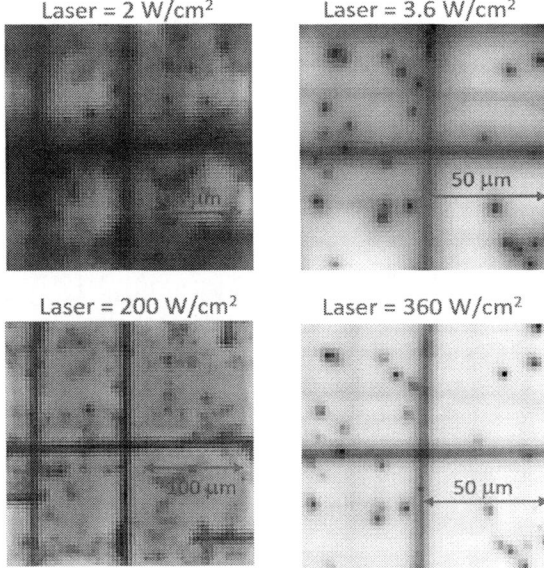

Fig. 1. Representative RT PL images and maps of a crosshatch feature in the emission pattern from the GaAsP/GaInP DH. PL images are on the left and confocal maps are on the right with lower excitation above and higher excitation below.

IV. ANALYSIS

We start by selecting a portion of the horizontal line having relatively few adjacent features above and below the line and then averaging the columns of pixels in this region. This procedure yields a line profile of the PL intensity across the horizontal line. We divide the PL count in each pixel by the integration time to obtain a relative measure of the radiative recombination rate in each pixel.

The local densities of electrons (n) and holes (p) in a given pixel are assumed to be equivalent ($n = p$) and independent of time in steady-state:

$$\frac{dn}{dt} = G + J - Bn^2 - A_P n = 0 \qquad (1)$$

where G is the electron-hole pair generation rate, J is the net diffusive flux into the pixel, Bn^2 is the radiative recombination rate, and $A_P n$ is the nonradiative recombination rate due to uniformly-distributed point defects. An additional nonradiative loss mechanism $A_L n$ is included in the rate equation for the pixel at the center of the line defect. For the imaging experiment, G is the same in every pixel, but in the mapping experiment, G is zero everywhere except the photoexcited pixel. We compute the diffusive flux using the carrier density in adjacent pixels:

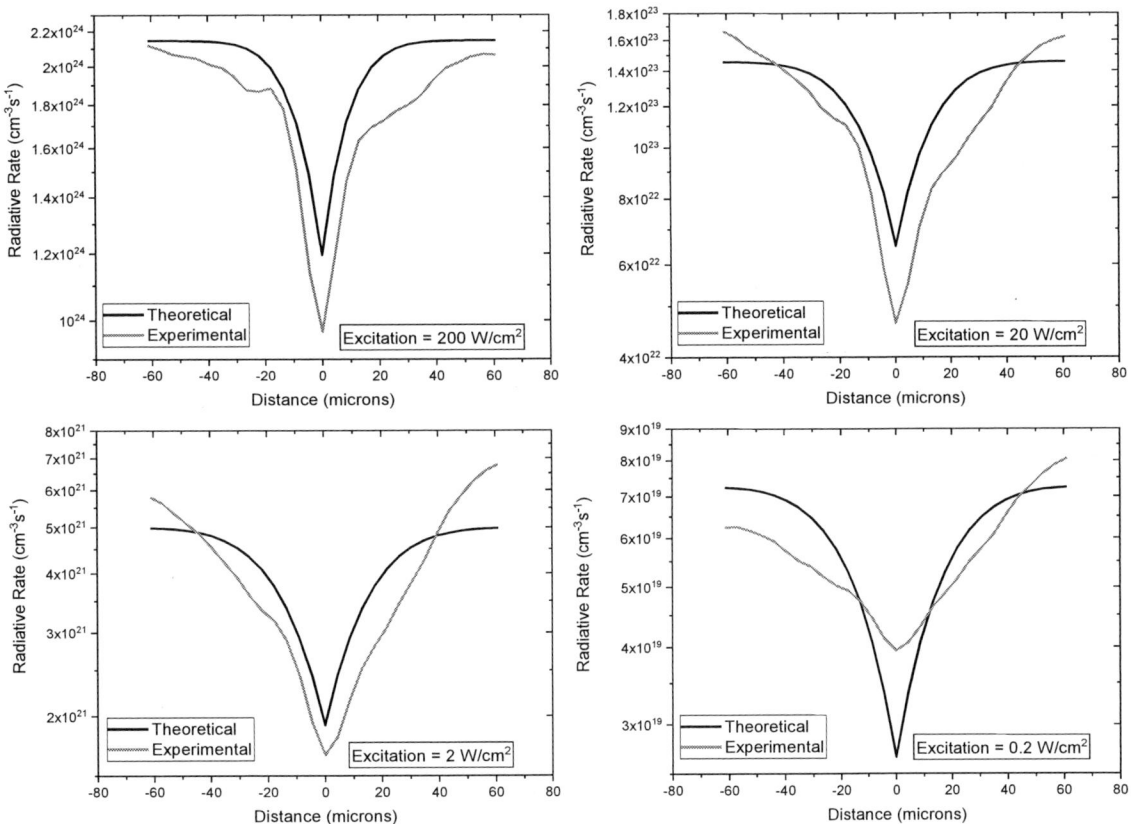

Fig. 2: PL imaging line profiles of the horizontal line defect under different photoexcitation conditions, showing how the radiative emission varies with distance from the misfit dislocation.

978-1-6654-6060-6/23 $31.00 © 2023 IEEE

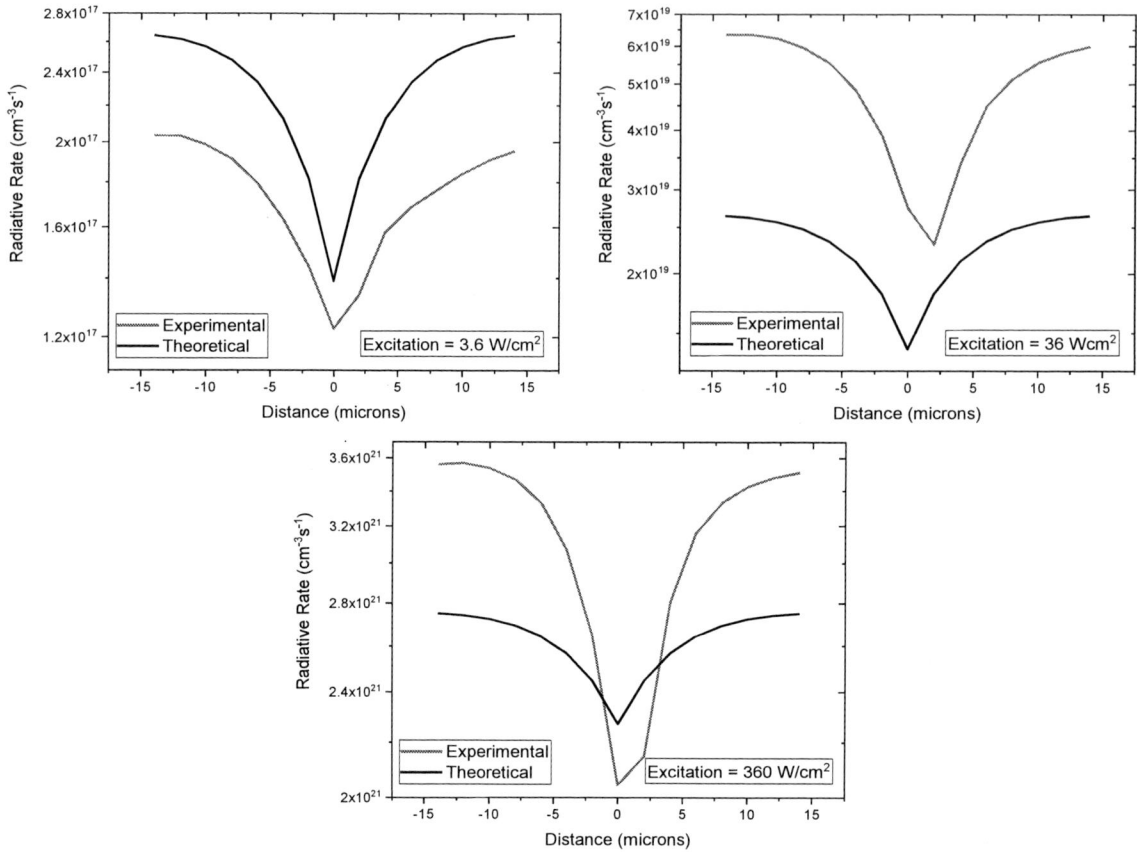

Fig. 3: PL confocal mapping line profiles of the horizontal line defect under different photoexcitation conditions.

$$J = D\left(n_{left} + n_{right} - 2n\right)/w^2 \qquad (2)$$

where D is the diffusivity and w is the width of the pixel. Using periodic boundary conditions, we iterate this calculation until a steady-state density distribution across the line defect is achieved. Then the theoretical radiative rate Bn^2 is divided by a loss factor LF, which accounts for light extraction efficiency and system response, and compared with the measured PL signal in each pixel. A basin-hopping algorithm [6] is used to identify the fitting parameters LF, A_P, and A_L that yield the minimum error between the experimental and simulated results. Fig. 2 shows the simulation results for the imaging experiment and Fig. 3 shows the results for the confocal mapping experiment. The optimized parameters are given in Table I.

TABLE I. COMPUTATIONAL FITTING PARAMETERS

Parameter	Experiment	
	Imaging	Mapping
B	2×10^{-10} cm^{3}s^{-1}	
D	1×10^2 cm^2 s^{-1}	
LF	2700	0.51
A_P	4.2×10^6 s^{-1}	1.2×10^8 s^{-1}
A_L	1.4×10^8 s^{-1}	4.5×10^8 s^{-1}

The density profiles for mapping are narrower as predicted by theory, [7] corroborating the diffusion algorithms. And the nonradiative recombination coefficients at the site of the line defect are generally consistent because the area of the defective pixel in the mapping experiment is approximately 4 times smaller and requires a proportionally higher rate. However, the discrepancy between the coefficients for nonradiative recombination at point defects points to limitations of the ABC recombination model [8] and suggests that different defect-related mechanisms are operating under the conditions probed by the two experiments. While the photoexcitation densities are comparable, diffusion out of the highly localized excitation volume in the mapping experiment yields steady-state carrier densities that are approximately two orders of magnitude smaller. The equivalent density assumption ($n = p$) is least accurate at low densities and prior work on GaAs/GaInP [5] has demonstrated that defect-related recombination can be strongly augmented in this extremely low-density regime.

Finally, we note that these plan-view measurements probe a fixed depth in the structure and cannot identify inclined misfits where the depth of the defective region may vary with lateral position. A multi-dimensional approach would provide a more complete picture of the misfit dislocation network in this system.

ACKNOWLEDGMENT

The work at the UNCC was supported by ARO/Electronics (Grant No. W911NF-16-1-0263).

REFERENCES

[1] Martin A. Green, Ewan D. Dunlop, Jochen Hohl‑Ebinger, Masahiro Yoshita, Nikos Kopidakis, and Anita W.Y. Ho‑Baillie, "Solar cell efficiency tables (Ver. 55)," *Prog. Photovoltaics* 28, pp. 3–15, 2020.

[2] Tyler J. Grassman *et al.*, "Toward >25% efficient monolithic epitaxial GaAsP/Si tandem solar cells," *Proc. 46th IEEE Photovoltaic Specialists Conference*, 2019.

[3] Ahmed Ben Slimane *et al.*, "AlGaAs/InGaP MBE-grown heterostructures for 1.73eV solar cells with 18.7% efficiency," *Proc. 46th IEEE Photovoltaic Specialists Conference*, 2019.

[4] T. H. Gfroerer, M. J. Romero, M. M. Al-Jassim, and M. W. Wanlass, "Band-to-band and sub-band gap cathodoluminescence from GaAsP/GaInP epistructures grown on GaAs substrates," *J. Lumin.* 122–123, pp. 348–351 (2007).

[5] T. H. Gfroerer, Yong Zhang, and M. W. Wanlass, "An extended defect as a sensor for free carrier diffusion in a semiconductor," *Appl. Phys. Lett.* 102 (1), 012114 (2013).

[6] https://docs.scipy.org/doc/scipy/reference/generated/scipy.optimize.basinhopping.html

[7] Fengxiang Chen, Yong Zhang, T.H. Gfroerer, A.N. Finger, and M.W. Wanlass, "Spatial resolution versus data acquisition efficiency in mapping an inhomogeneous system with species diffusion," *Sci. Rep.* 5, 10542 (2015).

[8] Fan Zhang, Jose F. Castaneda, Timothy H. Gfroerer, Daniel Friedman, Yong-Hang Zhang, Mark W. Wanlass and Yong Zhang, "An all optical approach for comprehensive in-operando analysis of radiative and nonradiative recombination processes in GaAs double heterostructures," *Light Sci. Appl.* 11, 137 (2022)

Improving Operational Stability of High-Efficiency Inverted Perovskite Solar Cells

Kai Zhu

National Renewable Energy Laboratory, Golden, CO, United States

We discuss three strategies to improve the efficiency and stability of the state-of-the-art inverted perovskite solar cells (PSCs), covering a wide range of bandgaps from 1.25 eV (narrow-bandgap) to 1.53 eV (mid-bandgap) to 1.75 eV (wide-bandgap). The three strategies include additive engineering, reactive surface treatment, and crystallization/growth control, with each one demonstrated for one bandgap. These approaches are effective at reducing defect densities, which is critical for improving device efficiency and operational stability. Finally, we discuss the importance of identifying a few selective stability testing protocols to enable rapid assessment of PSC technology advancement. Understanding operational stability at an elevated temperature is found most valuable to understand device operation under real-world conditions.

978-1-6654-6060-6/23 $31.00 © 2023 IEEE

Influence of Aluminum Co-Doping on Current-Induced Degradation and Regeneration Kinetics in Boron-Doped Cz PERC Solar Cells

August Weber[1], Mahsa Mohammadi[1], Matthias Trempa[2], Thomas Buck[3], Johannes Heitmann[1],

Matthias Müller[1]

[1]Institute of Applied Physics, TU Bergakademie Freiberg,
Leipziger Straße 23, D-09599 Freiberg, Saxony, Germany
[2]Fraunhofer Institute for Integrated Systems and Device Technology,
Schottkystraße 10, D-91058 Erlangen, Germany
[3]International Solar Energy Research Center Konstanz,
Rudolf-Diesel-Straße 15, D-78467 Konstanz, Germany

Abstract — The influence of aluminum co-doping in boron-doped Cz PERC solar cells on current-induced degradation and regeneration kinetics is investigated at different injection currents and varying elevated temperatures. A carrier-induced degradation (CID) setup for simultaneous degradation and in-situ measurement was utilized for long-term treatment experiments. Different treatment conditions resulted in varying CID behavior, thus showing a significant impact on the degradation and regeneration. Evaluating the degradation and regeneration curves, kinetic rates and activation energies are determined and compared to solely boron-doped PERC cells and aluminum co-doped carrier lifetime samples.

I. INTRODUCTION

Passivated emitter and rear cell (PERC) is the current PV mainstream solar cell technology. Several charge-carrier-induced degradation effects are known to it. To achieve a better understanding of these effects and thereby taking a step towards reducing or even completely negating aforementioned effects one investigates among other things the degradation kinetics, the chemical composition of possibly participating defects and the electrical properties of these. One prominent effect, which affects almost every type of silicon wafer [1], is the so-called "light- and elevated temperature-induced degradation" (LeTID). LeTID-sensitive solar cells show a carrier-induced degradation, which is followed by a distinctive regeneration, as shown by many researchers before, see e.g. references [2]-[4].

It has been shown, that the addition of Al can influence the behavior of degradation and regeneration effects in Cz Si wafers [5]. It is assumed, that the delayed degradation and regeneration is related to the different bonding strength between hydrogen and boron and hydrogen and aluminum. Building on these observations an experiment was designed to investigate the influence of aluminum co-doping on Cz PERC solar cells under various temperature and injection conditions.

The goal of the experiment is to make a statement about the involvement and influence of aluminum on defect related degradation and regeneration of PERC solar cells comparing kinetic rates and activation energies.

II. EXPERIMENTAL SECTION

In this study PERC solar cells made from boron-doped and aluminum co-doped Cz Si wafers with a bulk resistivity of about 1.3 Ωcm are investigated. The aluminum was added during the crystallization step, hereby bringing the concentration of Al in the material to approx. 3×10^{14} cm^{-3}. For comparison, boron doped reference samples with similar resistivity ($\approx$ 1.2 Ωcm) were also examined.

The influence of the added aluminum was investigated for a range of temperatures (75 °C … 115 °C) and injection currents (0.06 $\times I_{SC}$ … I_{SC}). The according design of experiment (DoE) is shown in Fig. 1.

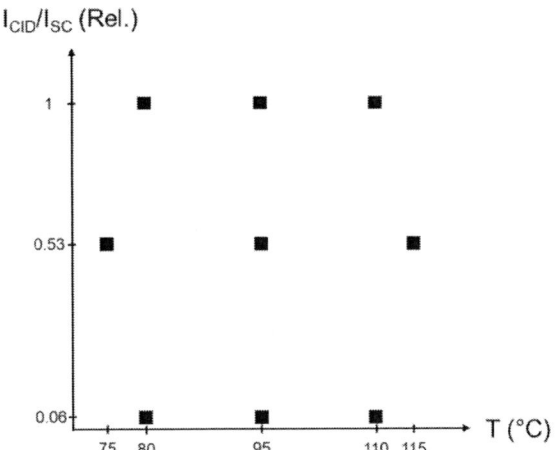

Fig. 1.   Visualization of the design of experiment. Treatment temperature and current injection, i.e. CID current, are varied.

For later evaluation of activation energies, it is of interest to examine and compare the behavior of cells treated at same injection conditions at different temperatures. A kinetic model considering injection conditions and temperature is going to be developed.

As seen in Fig. 2, a chuck was used to hold in place the sample to be treated which is placed on a hotplate. The chuck is positioned on top of a hotplate and designed to conduct heat.

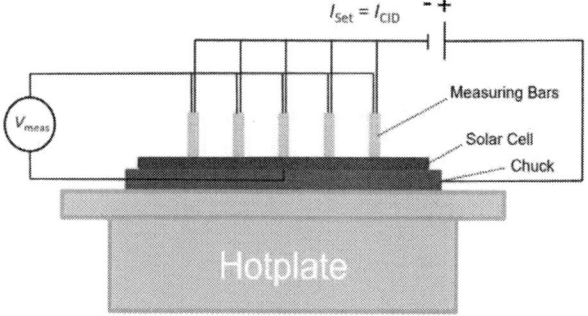

Fig. 2. Sketchy visualization of the CID setup.

Using measuring bars, the treatment current I_{CID} is injected through the busbars into the cell. Different kinds of pins on the measurement bars allow not only the injection of a current, but also the simultaneous sense voltage measurement of V_{CID}. The rear sens contact is a small pin inside of the chuck in contact with the back surface of the sample to enable a sense voltage measurement at the rear side of the sample. First cell degradations and regenerations under various treatment conditions on the Al co-doped samples have been performed. For fitting reasons, which will be explained later, the raw voltage data have to be calculated into a normalized defect density N^* using the example of Palmer et al. [6], as shown in eq. (1):

$$N^*(t) = \frac{I_{CID} \cdot N_A}{A_{Cell} \cdot q \cdot n_i^2 \cdot w_{Cell}} \cdot \left(\frac{1}{e^{\frac{qV_{meas}(t)}{nkT}}} - \frac{1}{e^{\frac{qV_{meas}(t=0)}{nkT}}} \right) (1)$$

In eq. (1) N_A (1.5E+16 cm^{-3}) stands for the doping concentration, which can be calculated via the given resistivity, A_{Cell} (240.82 cm²) describes the area of the sample, q is the electron charge, n_i is the intrinsic carrier concentration, w_{Cell} (160 µm) describes the cell thickness, n is the ideality factor, which was set to 1 for this work, and k is the Boltzmann constant.

Furthermore, an injection correction for the treatment time has to be introduced. This has been done by Tratnikov and Müller before [7, 8]. Eq. (2) shows the calculation of an injection correction rate and the injection corrected treatment time t^*.

$$R(t) \cdot t = R_{inf} \cdot \frac{\Delta n(t)}{\Delta n_{inf}} \cdot t = R_{inf} \cdot t^* (2)$$

III. RESULTS & DISCUSSION

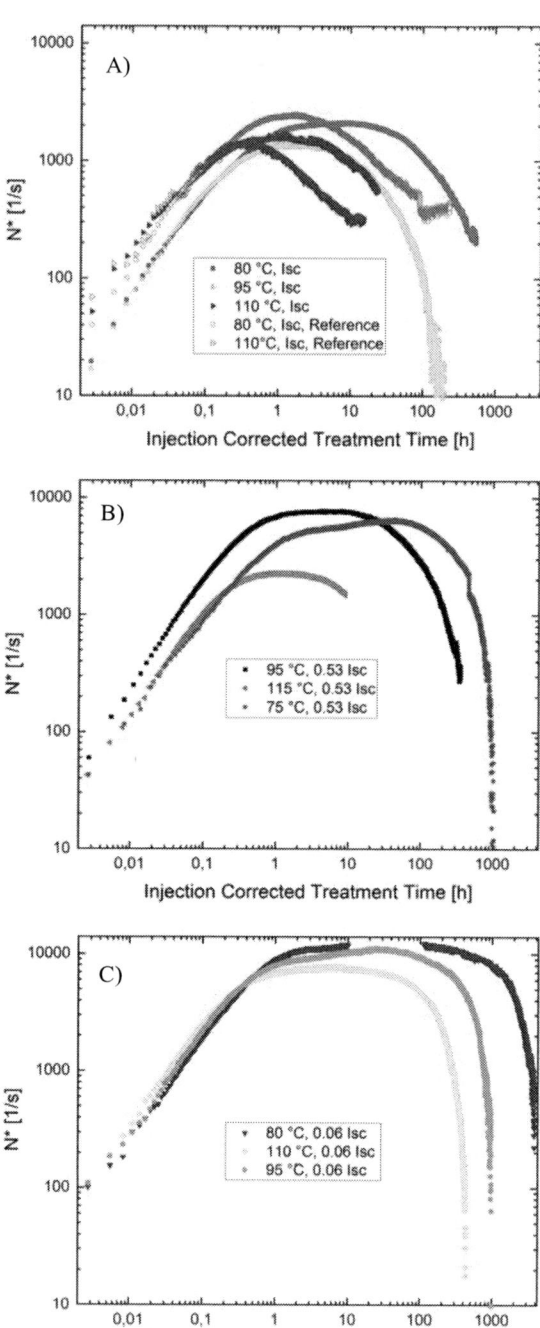

Fig. 3. Normalized defect density N^* of Al co-doped Cz PERC solar cells during CID under varying treatment conditions. A) $I_{CID} = I_{SC}$; B) $I_{CID} = 0.53\ I_{SC}$; C) $I_{CID} = 0.06\ I_{SC}$

In this section, CID results and the subsequent analysis of these data are shown and discussed. The analysis includes the fitting of the calculated normalized defect densities and the depiction of the obtained kinetic rates in an Arrhenius plot.

Also, the importance of an appropriate fitting function is mentioned.

Fig. 3 shows the behavior of the normalized defect densities N^* of the cells in dependence of the injection corrected treatment time. Under all circumstances a degradation and regeneration are observed. In Fig. 3 A) a delay in kinetics can be noticed for the with Al co-doped cells compared to the reference only boron-doped cells. A shift to faster kinetics is seen the higher the injection current and the treatment temperature are. The sample with the degradation conditions of 80 °C and 0.06 I_{SC} might show an interesting feature, as the behavior of the curve hints a second degradation mechanism at 3 to 10 hours of treatment. This sample is the one to degrade and regenerate the slowest, so certain features are easier to observe. The regeneration of the named sample is still ongoing.

Since the degradation and regeneration cannot be discussed separately, a three-state model is assumed which starts with a recombination-inactive initial state which transfers to a recombination-active state of the LeTID defect which can be stabilized into another recombination-inactive state. Thus, a fitting function needs to be applied which follows the model of a chain reaction. One appropriate approach is shown by Wehmeier et al. [9] in (3).

$$N^*(t) = a_{\mathrm{deg}} \cdot R_{\mathrm{gen,deg}} \cdot \frac{e^{-R_{\mathrm{gen,deg}} \cdot t} - e^{-R_{\mathrm{gen,reg}} \cdot t}}{R_{\mathrm{gen,reg}} - R_{\mathrm{gen,deg}}} \quad (3)$$

Using Eq. (3), kinetic rates for the degradation and for the regeneration can be obtained, which are independent of each other and thus more accurate than by simply adding them additively assuming our model is reasonable. One exemplary fit is shown in Fig. 4, to show the parameters and the goodness of the fit itself.

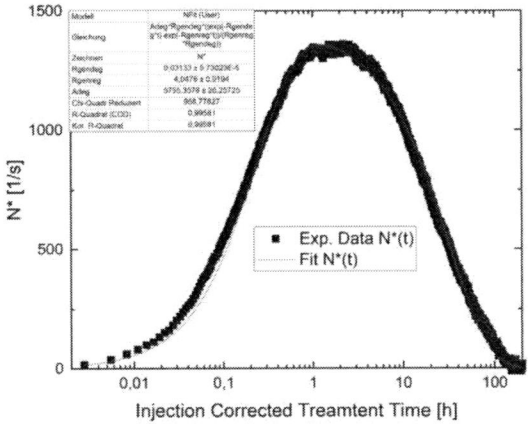

Fig. 4. Example of data fitting. Shown cell was degraded at 95 °C and $I_{\mathrm{CID}} = I_{\mathrm{SC}}$.

The described fitting model shows satisfying first conclusions. Both the degradation rate R_{gendeg} and the regeneration rate R_{genreg} show results with a reasonably small error in the range of 0.2% to 4.2%. Also, the $R^2_{\mathrm{adj.}}$ range of 89.9% for the worst fit to over 99.5% for the best one gives reason to believe, that within the given parameters the fit performed very well.

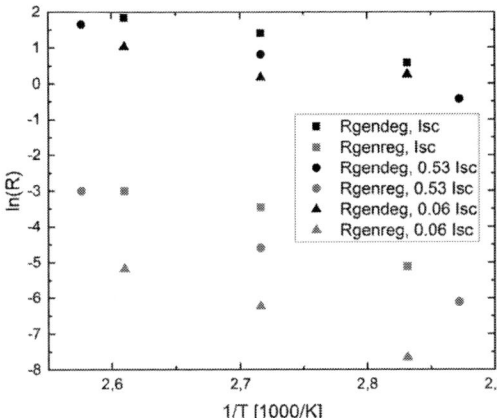

Fig. 5. Arrhenius plot of the calculated degradation and regeneration rates gained from the fitting results. Error bars are included in the graph, but are significantly smaller than the symbols themselves.

A summary of the fitting results can be found in Fig. 5, where an Arrhenius plot of the degradation and regeneration rates at different injections over the reverse temperature is shown.

Fig. 5 shows several things. On one hand, a clear distinction between degradation and regeneration can be observed, so that all the calculated degradation rates are significantly higher than the regeneration rates. On the other hand, two things can be confirmed with these results [10, 11] similar to only boron-doped PERC cells:

- With higher injection conditions the degradation and regeneration rates increase and
- With higher treatment temperature the kinetic rates also rise.

Using the Arrhenius plot, activation energies E_A for different injections can be calculated using the slope of linear fits for both the degradation and the regeneration rates. E_A are summarized in Tab. I:

978-1-6654-6060-6/23 $31.00 © 2023 IEEE

Tab. I.   Calculated activation energies for degradation and regeneration for different injection conditions using different CID currents.

I_{CID}	1.00 I_{SC}	0.53 I_{SC}	0.06 I_{SC}
$E_{A, DEG}$	0.52 ± 0.08 eV	0.59 ± 0.05 eV	0.28 ± 0.20 eV
$E_{A, REG}$	0.89 ± 0.30 eV	0.91 ± 0.06 eV	0.96 ± 0.05 eV

$E_{A,DEG}$ of the degradation process lies within a range of $0.28...0.59$ eV and $E_{A,REG}$ of the regeneration process within a range of $0.89...0.96$ eV for injections conditions at approximately maximum power point (MPP, corresponding to $0.06\,I_{SC}$) to approximately open-circuit voltage (V_{OC} , corresponding to $1.0\,I_{SC}$), respectively. Two of the calculated Activation energies show a relatively huge error and in general $E_{A,DEG}$ and $E_{A,REG}$ are very close considering the determined fit uncertainties. Thus, constant E_A for both processes are likely, especially as only three points per injection condition are available for fitting. As discussed in M. Mehler et al., [5], the delayed regeneration for Al co-doped boron-doped PERC cells compared to only boron-doped cells might be related to the higher binding energy [12] for Al–H pairs (1.44 ± 0.03 eV) compared to B–H pairs (1.28 ± 0.03 eV) assuming that hydrogen plays an important role for LeTID defect regeneration.

IV. SUMMARY

A setup for carrier-induced degradation is used to investigate Al co-doped and only boron-doped Cz PERC solar cells at different injection conditions. All curves show prominent degradation and regeneration features. The derived CID voltage curves versus treatment time have been translated into normalized defect densities, which, after applying an injection correction on the treatment time, have been evaluated through fitting. The fitting method has shown great promise. The obtained kinetic rates have been analyzed in an Arrhenius plot, from which activation energies for the different injection conditions were received.

Our experiment confirms a delayed regeneration of Al co-doped PERC cells compared to only boron-doped cells on solar cell level using current-induced degradation (CID). The Al co-doped cells show a similar temperature- and injection-behavior than only boron-doped cells. The activation energy of degradation lies within a range of $0.28...0.59$ eV and for the regeneration process within a range of $0.89...0.96$ eV for different injections conditions.

ACKNOWLEDGEMENTS

This work was funded by the German Federal Ministry for Economic Affairs and Climate Action BMWK (contract no. 03EE1051D).

REFERENCES

[1] D. Chen et al., "Progress in the understanding of light- and elevated temperature-induced degradation in silicon solar cells: A review" in *Progress in Photovoltaics: Research and Applications*, 29(11), 2021, Pages 1180—1201.

[2] Ramspeck et al., in *27th European Photovoltaic Solar Energy Conference and Exhibition (EUPVSEC)*, 2012, Pages 861—865.

[3] F. Kersten et al., in *2015 IEEE 42nd Photovoltaic Specialist Conference (PVSC)*, 2015, Pages 1—5.

[4] A. Zuschlag et al., "Degradation and regeneration in mc-Si after different gettering steps" in *Progress in Photovoltaics: Research and Applications*, 25(7), 2017, Pages 545—552.

[5] M. Mehler et al., "Delay of Regeneration by Adding Aluminum in Boron-Doped Crystalline Si" in *Phys. Status Solidi A*, 218: 2100603, 2021.

[6] D. W. Palmer, "Kinetics of the electronically stimulated formation of a boron-oxygen complex in crystalline silicon," Phsical Review, 2007.

[7] M. Tratnikov & M. Müller, 'Excess charge carrier injection densities in PERC solar cells at open-circuit voltage and maximum power point', AIP Conference Proceedings 2147, 020018, 2019.

[8] M. Tratnikov, 'Investigation on the kinetic of the boron-oxygen related defect in high-efficiency PERC solar cells', Master's Thesis, 2020.

[9] N. Wehmeier et al., "Kinetics of the Light and Elevated Temperature Induced Degradation and Regeneration of Quasi-Monocrystalline Silicon Solar Cells", in *IEEE Journal of Photovoltaics*, 11(4), July 2021, Pages 890—896.

[10] W. Kwapil, T. Niewelt, and M. C. Schubert, "Kinetics of carrier-induced degradation at elevated temperature in multicrystalline silicon solar cells," Sol. Energy Mater. Sol. Cells, vol. 173, pp. 80–84, 2017.

[11] C. Vargas, G. Coletti, C. Chan, D. Payne, and Z. Hameiri, "On the impact of dark annealing and room temperature illumination on p-type multicrystalline silicon wafers," Sol. Energy Mater. Sol. Cells, vol. 189, pp. 166–174, 2019.

[12] Zundel, T, and J Weber. "Dissociation energies of shallow-acceptor-hydrogen pairs in silicon." Physical review. B, Condensed matter vol. 39,18 (1989): 13549-13552

978-1-6654-6060-6/23 $31.00 © 2023 IEEE

Reducing the photovoltaic operation and maintenance costs through an autonomous control operation center

Andreas Livera[1], Álvaro Fernández-Solas[2], Joao G. Bessa[2], Jesús Montes-Romero[2], Eduardo F. Fernández[2], Vassilis Papaeconomou[3] and George E. Georghiou[1]

[1] PV Technology Laboratory, FOSS Research Centre for Sustainable Energy, Department of Electrical and Computer Engineering, University of Cyprus (UCY), 1678 Nicosia, Cyprus

[2] Advances in Photovoltaic Technology, CEACTEMA, University of Jaén (UJA), 23071 Jaén, Spain

[3] Alectris Hellas IKE, Industrial Area of Thessaloniki, 1344 Sindos, Greece

Abstract — An advanced control operation center to enable corrective, preventive and predictive maintenance, while also ensuring optimal photovoltaic (PV) plant performance was developed in this work. The developed software solution hosts innovative algorithms able to ensure data quality, while also allowing early failure and performance loss diagnosis without disrupting the normal operation of the PV plant. It is primarily based on real-time analysis of measurement data, machine learning and statistical analysis. The solution was validated experimentally against field measurements from an operating PV power plant of 1.8 MW$_p$ installed in Greece. The results showed technical availability and energy yield improvements of the test PV plant by handling intelligently the detected faults through the smart ticketing system. Optimal maintenance planning (e.g., optimum hardware replacement/maintenance, cleaning schedules, etc.) can thus lead to a reduction of operation and maintenance (O&M) costs and hence directly impacting positively the levelised cost of electricity (LCOE).

Keywords — *cleaning optimization, data analysis, diagnosis, fault detection, maintenance planning, monitoring, photovoltaic, ticketing system*

I. Introduction

The growing concerns about rising greenhouse gas emissions (GHG) affecting detrimentally the global climate, advocate the need for de-carbonization of the energy sector by reducing the GHG emissions, phasing out fossil fuels and accelerating the shift towards renewable technologies [1]. In fact, the future energy mix is expected to be heavily dependent on renewables, particularly solar photovoltaics (PV), which is set to become the "King of Renewables" [1]. For countries with high solar resource (such as Cyprus, Greece and Spain), PV is expected to play a central part in the future energy mix.

A vital factor that will enable the further growth of the PV technology is the reduction of PV electricity costs by increasing lifetime output, improving the operational efficiency and optimizing system operations [2]. This can be achieved by safeguarding the service lifetime performance through PV monitoring, supervision, maintenance and control of installed systems, hence directly impacting positively the investment cost, levelised cost of electricity (LCOE), and in general PV competitiveness [2].

To tackle the major challenges in increasing PV system performance (as it was recently reported that PV assets continue to underperform by up to 8%, thus highlighting the need for high-fidelity data and greater model transparency [3]) and technological competitiveness, machine learning algorithms and statistical techniques can be developed to enable corrective, preventive and predictive maintenance strategies [4]. The operation and maintenance (O&M) activities consist of two parts: (i) the operations that include remote monitoring, supervision, forecasting, communication and control of the PV power plant and (ii) the maintenance that includes the activities related to the health-state and optimum performance of PV plants. Therefore, ensuring cost-effective and online PV monitoring with automated data-driven operation functionalities is important for improving the LCOE through: (i) increased availability by the on-time triggering of losses/faults (hence increasing the energy yield) and (ii) reduced O&M costs by optimizing hardware replacement/maintenance, thus reducing the reaction and resolution times and hence the manual labor.

In this domain, the key battlegrounds of technical solutions that support high system performance are associated with the capabilities of operation centers that automatically analyze incoming data, provide real-time observability of PV assets, enable failure and health diagnostics, and handle intelligently the detected faults/errors through a smart ticketing system [5]. The scope of this paper is to address the fundamental challenge of automated PV plant operational-state management by developing an autonomous control operation center (i.e., an online software platform, powered by artificial intelligence algorithms and statistical analysis methods). The proposed software solution was validated using historical field data from a large-scale PV power plant installed in Greece. The results showed improvements in the availability and energy yield of the PV plant under study. This was achieved by utilizing the smart ticketing system, that provided the necessary information and steps to fix problems quickly and efficiently.

978-1-6654-6060-6/23 $31.00 © 2023 IEEE

II. METHODOLOGY

A. Experimental setup - Benchmarking

The developed software platform was benchmarked experimentally using historical field measurements from an operating PV power plant of 1.8 MW$_p$ installed in Larissa, Greece. It comprises of 7824 poly-crystalline-Silicon (poly c-Si) PV modules, each of nominal power of 230 W$_p$. The PV modules are south oriented, 25° tilted and connected in series to form 326 strings at the inputs of 4 grid-connected inverters (81 or 82 strings connected to the four inverters).

The performance of the PV system and the prevailing meteorological conditions are recorded according to the requirements set by the IEC 61724-1 [6]. The recorded data are stored using a measurement monitoring platform, that comprises of solar irradiance, wind, temperature and electrical operation sensors and stores data at a resolution of 1 second and accumulation steps of 15-minute averages. The meteorological measurements include the in-plane irradiance (G_I) measured with a pyranometer, ambient temperature (T_{amb}), wind speed (W_s) and direction (W_a). The PV system's operational measurements include the module back-surface temperature (T_{mod}), inverter temperature (T_{inv}), string DC current (I_{DC}), array DC current (I_A), voltage (V_A) and power (P_A) and AC output power (P_{out}). Additional yields and performance metrics such as the performance ratio (PR) and the temperature-corrected PR (PR_{TC}) were also calculated [7].

In this work, field data over different time periods were used. Over these time periods, different types of faults (e.g., communication errors, inverter shutdowns, low plant production and PR, equipment malfunctions, etc.) occurred during the operation of the system, which were resolved by technicians. Information about the outage's periods, fault types, O&M events/actions and technicians' feedback were kept in a maintenance log.

B. Autonomous control operation center

The autonomous control operation center integrates monitoring, supervision, maintenance, and fault diagnostic algorithms along with a smart ticketing system. The software platform leverages artificial intelligence algorithms and statistical analysis methods for the quality in operations and decision making. It is based on a modularized architecture to decouple the whole system and allows modules to interoperate autonomously. The architecture consists of four layers/modules as depicted in Fig. 1.

The control operation center continuously analyzes the incoming electrical and weather data for anomalies and outliers (Module 1). The data are initially pre-processed by the data quality assessment (DQA) stage to identify and treat invalid values, thus preparing the data for further performance analysis [7]. Subsequently, the data are aggregated into daily/monthly/yearly blocks. Then, machine learning and statistical algorithms are applied on the cleansed data to diagnose (detect and classify) the failure/performance loss and its type, triggering alarms in case of fault occurrences (Module 2) [8]. Afterwards, the smart ticketing system prioritizes the detected faults (based on the calculated energy and cost impact), derives an optimal maintenance planning, and suggests ways for resolving the detected incidents (Module 3). Finally, the detailed fault description, the criticality of the incidents and the list of suggestions for O&M field actions are visualized through the software platform (Module 4) and forwarded to the technicians.

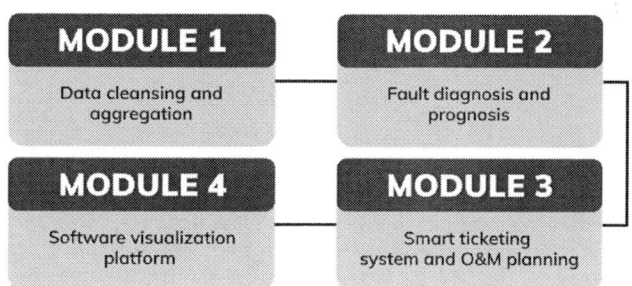

Fig. 1.   Module architecture of the control operation center.

C. Data quality assessment

DQA algorithms are initially applied to the available measurements to ensure high-fidelity time series data for further analysis. The DQA stage is used to detect/treat invalid measurements, which may indicate equipment malfunctions, sensors and/or PV faulty operation, thus reducing uncertainty and increasing confidence in energy estimates. The DQA process includes multiple algorithms for data consistency examination, data filtering and imputation/inference, outlier removal, etc. More details are provided by Livera et al. in [7].

The DQA stage also provides information about the technical availability (or uptime), energy-based availability along with insights about possible data errors, technical (performance) issues and fault root causes [9].

D. Fault diagnostics and alarms

Data-driven fault diagnostic algorithms (e.g., outlier detection, machine learning and comparative techniques) and open-source libraries (e.g., RdTools [10]) are applied to detect underperformance incidents (e.g., failures and performance losses), that cause power losses [4], [11]. Faults that can be detected by the fault diagnostic engine include inverter malfunctions, string disconnections, partial shading, soiling, performance degradation, etc. [4], [12]. Apart from the detection part, the diagnostic algorithms are also capable of categorizing the detected fault incidents into different root causes [8], providing also the energy loss breakdown list [4]. In case of underperformance incidents, alarms are generated by the fault diagnostic engine. These alarms are used along with the alarm signals generated by the inverter to determine the fault root cause [4]. The results of the fault diagnostic algorithms and the generated alarms are then forwarded to the ticketing system.

In this work, emphasis is given on the diagnosis of soiling, which was recently characterized as a multibillion-dollar issue in operating PV power plants [13]. Soiling losses caused at least a 3% to 4% loss to global annual PV energy production in 2018, accounting for €3-5 billion lost revenue [13]. And this is expected to further increase in the upcoming years due to the expanding number of deployed PV systems in regions highly prone to soiling.

E. Smart ticketing system

The smart ticketing system uses as inputs the detected data quality issues, underperformance incidents and the alarm signals to generate recommendations for field maintenance actions (to be performed by the technicians). The detected faults and errors are prioritized (i.e., incidents of low criticality indicated by green color, incidents of medium criticality indicated by yellow color and incidents of high criticality indicated by red color), based on the calculated energy and cost impact [4], deriving an optimal maintenance schedule in an attempt to optimize the O&M activities and reduce the associated costs.

The smart ticketing system finally alerts the technicians about the fault/loss root cause, while also providing a list of recommendations for field actions (accessible through the software platform) to resolve the problem [4], [5].

F. Visualization of autonomous operation center results through the software platform

ACTIS ERP is the user interface of the autonomous control operation center [14]. The ACTIS ERP is a comprehensive asset management solution, that integrates centralized real-time monitoring with alerting and ticketing, O&M activities, asset and project management in a single software platform. The software platform can be accessed remotely from any device, anytime and anywhere via internet.

It displays current and historical performance data, key performance indicators (KPIs) of PV assets and portfolios, financial and operational indications, alarms, detected incidents, breakdown list of energy losses, O&M events, and a list of recommendations for field actions. The health-state of PV components and the generated tickets are also displayed.

G. Economic impact of proposed O&M actions

To evaluate the economic impact of O&M actions, economic models based on metrics such as the LCOE and the Net Present Value (NPV) were used. The LCOE reflects the project's economic feasibility, while NPV evaluates the profitability of an investment (i.e., compares the revenues and costs over the project lifetime) [15], [16].

These economic metrics also allow the identification of the best time to conduct an artificial cleaning in a PV system (i.e., when the financial loss due to soiling surpasses the cleaning cost), thus optimizing the cleaning schedules by considering factors like the cleaning cost, the soiling rate, and the PV plant size [15].

The impact of the soiling on the LCOE and on the O&M costs was recently analyzed in [17]. Soiling and snow-related losses were found to be the second most severe fault category, accounting for approximately 25% of total lost energy. The study also showed that additional cleanings could reduce losses by up to 11%, but the economic viability depended on the cleaning costs and electricity prices.

Another recent study [18] demonstrated that actual PV cleaning can lead to an increase in the energy yield, having a positive impact on LCOE and NPV. Comparing the actual cleaning date with the optimal cleaning date, it was found that the actual cleaning was performed with a 7-day delay, resulting in lower improvement in NPV and LCOE.

III. RESULTS

A. Data quality assessment

DQA algorithms were initially applied to field measurements of the test PV plant. The data cleansing algorithms were used to restrict measurements within predefined physical limits [7] and to identify/treat invalid measurements. Over the period from January to December 2022, the DQA detected 5.28% invalid data points (e.g., erroneous and missing values). The application of DQA algorithms for inspecting and treating missing and erroneous data improved the PV plant availability.

The technical availability of the plant was then calculated (see Fig. 2). Over the yearly evaluation period, the uptime was higher than 98.30%. The whole plant (or part of it) was down for approximately 250 hours due to communication loss with the PV plant/inverters, grid problems and/or grid outage.

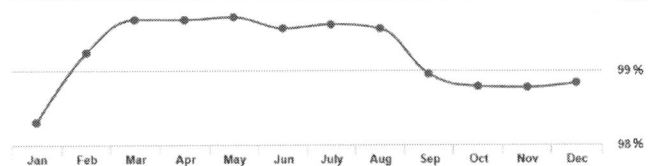

Fig. 2. PV plant availability over the period from January to December 2022.

B. Fault diagnostics and alarms

Over a 1-year period (January - December 2022), the test PV system produced 2,659 MWh. The fault diagnostic algorithms detected several fault incidents (e.g., plant was down, inverter shutdown failures, string problems, soiling, etc.), accounting for 24.38 MWh (0.92%) of lost energy. For the detected fault incidents, alarms were triggered by the engine and by the inverter itself (i.e., failed and warnings that appeared in the supervisory control and data acquisition system). The alarms were used to determine the fault root cause (e.g., inverter failures, string disconnections, soiling, shading, vegetation, etc.).

The classification of failures resulted to increased PV plant availability and optimized hardware replacement/maintenance.

The classification of the detected O&M incidents, 23 associated with corrective maintenance (i.e., faulty material, loose connections, malfunction of equipment, extreme weather conditions, defective PV modules, soiling, correction and maintenance works, grid problems/failure, grid undervoltage, inverter fault, ground fault, etc.) and 6 associated with preventative maintenance (e.g., vegetation), is shown in Fig. 3. Most of the detected incidents were due to PV plant related failures (41.38%). Other root causes included vegetation, monitoring system errors, power plant uninterruptible power supply (UPS) system problems, soiling, communication and electrical errors.

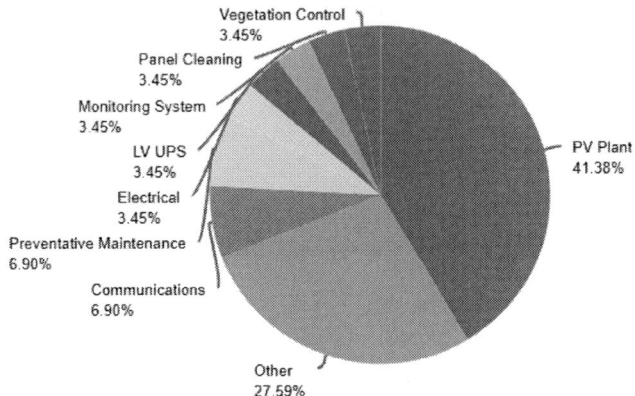

Fig. 3. Classification of detected O&M incidents for the test PV plant over the period from January to December 2022.

The RdTools python library [10] was then used to evaluate the PV production and to calculate rates of performance degradation and soiling loss. The Year-on-Year (YoY) method was used to estimate the performance loss rate (PLR), while the Stochastic Rate and Recovery (SRR) [19] method was used to identify soiling losses and cleaning events. The cleaning events were detected by observing positive shifts in the DC performance profile and using linear regression analysis to fit dry periods of at least 14 days. The SRR model generated potential soiling profiles through Monte Carlo simulation, and the median value of each day was extracted as the soiling profile per inverter.

Snow losses were also detected by analyzing PV performance parameters along with weather data.

For a reliable short-term performance evaluation, at least a 5-year time series data should be available to yield credible results [20]. To this end, field measurements over the period from February 2013 to January 2019 were used.

Over the investigated period (February 2013 to January 2019), an annual PLR of -0.90%/year was obtained. In parallel, the SRR model detected 34 cleaning events with the inverters experiencing low/limited soiling losses, with average(s) soiling rates of -0.26%/day to -0.0009%/day for the investigated period (see Fig. 4).

According to the maintenance log records, it was found that the O&M company regularly cleaned the PV modules twice a year. This practice can account for the significantly lower level of soiling losses observed, in contrast to the higher values commonly reported in the literature, which indicated soiling rate values up to -3%/day [21].

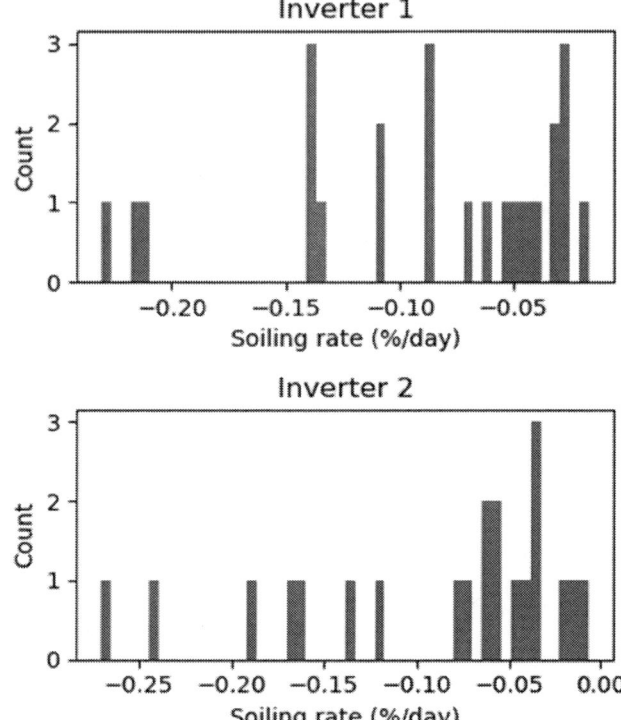

Fig. 4. Soiling losses experienced by inverters 1 and 2 of the test PV plant over the period from February 2013 to 2019.

C. Smart ticketing system and software visualization platform

The smart ticketing system processed the detected fault incidents and alarm signals to: (i) recommend specific field operations (e.g., corrective actions) to be performed by the technicians and (ii) provide insights and information regarding the fault root cause to the technicians to resolve the problem. It is worth noting here that the list of O&M recommendations is automatically generated. The field actions were also prioritized and scheduled based on the incident's severity (i.e., impact of the incidents on energy and cost) and finally forwarded to the responsible technicians. Eventually, the technicians executed the recommended field actions to improve operations (i.e., saving time, reducing O&M costs, and improving PV output production). The technicians were obliged to reach the PV plant within 4 daytime hours and 24 hours from the time of detecting the problem for medium and high criticality incidents, respectively.

An example of detected incidents and taken O&M actions is provided in Fig. 5 along with additional details (i.e., fault

Type	Title	Start Date	End Date	Action Taken	Status	Severity
Corrective Maintenance	String 2.2.20 is down	04/02/2022 08:00	04/02/2022 14:00	4/2/2022 10:00 πμ - The subcontractor replaced a faulty MC4 connector.	Resolved	●
Corrective Maintenance	Strings 1.3.1-1.3.8 have a defective communication card	04/02/2022 12:00	12/04/2022 15:00	12/4/2022 1:08 μμ - . Communication card has been replaced.	Resolved	●
Corrective Maintenance	Plant is down due to grid problems	22/02/2022 17:15	25/02/2022 17:30	25/2/2022 4:30 μμ The subcontractor rebooted the communication devices. Communication with the plant has been restored The technician found the plant in production when he restarted the communication equipment.	Resolved	●

Fig. 5. Screenshot depicting some of the detected O&M incidents for the test PV plant, their status, severity and action taken to resolve the incident.

description, severity of incident, date of occurrence, details of the dispatching technician to the field and action taken).

The results showed improvements in the availability and energy yield of the test PV plant by handling intelligently the detected faults through the smart ticketing system. Due to the smart management of incidents and optimal maintenance planning, only 0.92% (instead of 12.4% as simulated by Python Photovoltaic Reliability Performance Model [22]) of the produced energy was lost over the yearly period for the test PV system, while the plant's downtimes were minimized.

It is worth noting here that for longer evaluation periods (i.e., more than 1 year), the benefits would be greater (e.g., increased energy yield by ~6%, less downtime and reduced O&M costs by up to 10%) [5], [23].

D. Impact of cleaning optimization on economic metrics

In lack of information on the maximum extent of soiling (i.e., the losses in conditions of no mitigation), a cleaning optimization methodology was conducted on the available time series data to evaluate the impact of cleanings using two economic metrics (i.e., LCOE and NPV).

The results showed that after the first two cleanings, the inverters experienced low/limited soiling losses, with averages of 0.9% to 1.4% for the period between February 2013 to January 2019. This means that, for the given site, each cleaning can cost in between 0.6 and 1.3 €/kW, making regular soiling mitigation not profitable for this plant [5].

The actual cleaning activities could not only increase the energy output but also had a positive effect on LCOE and NPV, thereby reducing the cost of energy production from the PV system and increasing profits [18].

When considering LCOE, the optimization of cleaning is heavily influenced by the installation, and O&M costs. On the other hand, while the installation and O&M costs do impact profits, the NPV is not affected by them. Instead, the NPV is influenced by factors such as the cleaning cost, electricity prices, energy generation, degradation rates, and the recovery of losses through cleaning.

IV. SUMMARY OF THE WORK

A software platform was designed to optimise the O&M strategies and automate operations of PV systems. The proposed solution is predominantly progressing further the field of PV operational data quality, online fault diagnosis and automatic field operations. This is achieved through the development of an autonomous control operation center, that analyses the measurements collected from the constant monitoring of PV plants. The control operation center integrates monitoring, supervision, maintenance, and online fault diagnostic algorithms along with a smart ticketing system. The incorporated algorithms allow the early identification and classification of failures (through the fault detection engine) and ensure quality in operations and decision making (through the smart ticketing system). The smart ticketing system considers the alarms and the severity of the incident for maintenance planning and provides the necessary information and steps to fix problems quickly and efficiently. As such, improvements in the PV plant availability, energy yield, O&M costs and hence, LCOE are achieved.

To conclude, the development and operation of an autonomous control operation center helps to improve the PV plant availability, the intervention, response and resolution times. Therefore, O&M actions are taken effectively and timely by the corresponding asset owners or operators/contractors, thus safeguarding the PV performance and minimizing the investment risks, the associated costs and hence the LCOE.

ACKNOWLEDGMENT

This work was funded by the ROM-PV project (P2P/SOLAR/0818/0009). Project ROM-PV is supported under the umbrella of SOLAR-ERA.NET Cofund 2 by the General Secretariat for Research and Technology, the Ministry of Economy, Industry and Competitiveness-State Research Agency (MINECO-AEI) and the Research and Innovation Foundation (RIF) of Cyprus. SOLAR-ERA.NET is supported by the European Commission within the EU Framework Programme for Research and Innovation HORIZON 2020 (Cofund 2 ERA-NET Action, No. 786483).

Alectris IKE Hellas is kindly acknowledged for providing the field data of the test PV plant.

REFERENCES

[1] "International Energy Agency (IEA), Renewables 2021: Analysis and forecast to 2026," 2021.

[2] A. Livera, M. Theristis, G. Makrides, and G. E. Georghiou, "Recent advances in failure diagnosis techniques based on performance data analysis for grid-connected photovoltaic systems," *Renew. Energy*, vol. 133, pp. 126–143, 2019, doi:

10.1016/j.renene.2018.09.101.

[3] kWh analytics, "Solar Generation Index 2022," 2022.

[4] A. Livera, G. Tziolis, F. Jose G., and G. Ruben Gonzalez, Bernal George E., "Cloud-based platform for photovoltaic assets diagnosis and maintenance," *Energies*, vol. 15, no. 20, p. 7760, 2022, doi: https://doi.org/10.3390/en15207760.

[5] A. Livera, M. Theristis, L. Micheli, E. F. Fernández, J. S. Stein, and G. E. Georghiou, "Operation and maintenance decision support system for photovoltaic systems," *IEEE Access*, vol. 10, pp. 42481–42496, 2022, doi: 10.1109/ACCESS.2022.3168140.

[6] International Electrotechnical Commission (IEC), "Photovoltaic system performance - Part 1: Monitoring, IEC 61724-1:2017," 2017.

[7] A. Livera *et al.*, "Data processing and quality verification for improved photovoltaic performance and reliability analytics," *Prog. Photovoltaics Res. Appl.*, vol. 29, pp. 143– 158, 2021, doi: 10.1002/pip.3349.

[8] A. Livera, M. Theristis, L. Micheli, J. S. Stein, and G. E. Georghiou, "Failure diagnosis and trend-based performance losses routines for the detection and classification of incidents in large-scale photovoltaic systems," *Prog. Photovoltaics Res. Appl.*, vol. 30, no. 8, pp. 921–937, 2022, doi: 10.1002/pip.3578.

[9] A. Livera, M. Theristis, E. Koumpli, G. Makrides, J. S. Stein, and G. E. Georghiou, "Guidelines for ensuring data quality for photovoltaic system performance assessment and monitoring," in *37th European Photovoltaic Solar Energy Conference (EU PVSEC)*, 2020, pp. 1352–1356, doi: 10.4229/EUPVSEC20202020-5DO.2.4.

[10] M. G. Deceglie, D. Jordan, A. Shinn, and C. Deline, "RdTools : An Open Source Python Library for PV Degradation Analysis degradation rate," pp. 1–15, 2018.

[11] A. Livera, G. Paphitis, M. Theristis, J. Lopez-Lorente, G. Makrides, and E. George, "Photovoltaic system health-state architecture for data-driven failure detection," *Solar*, vol. 2, no. 1, pp. 81–89, 2022, doi: https://doi.org/10.3390/solar2010006.

[12] J. Lopez-lorente *et al.*, "Characterizing soiling losses for photovoltaic systems in dry climates : A case study in Cyprus," *Sol. Energy*, vol. 255, no. March, pp. 243–256, 2023, doi: 10.1016/j.solener.2023.03.034.

[13] C. Schill *et al.*, "Soiling Losses – Impact on the Performance of Photovoltaic Power Plants, Report IEA-PVPS T13-21:2022," 2022.

[14] Alectris, "ACTIS ERP - The ERP Platform for Renewables." [Online]. Available: https://actiserp.com/. [Accessed: 20-Dec-2022].

[15] L. Micheli *et al.*, "Improved PV Soiling Extraction through the Detection of Cleanings and Change Points," *IEEE J. Photovoltaics*, vol. 11, no. 2, pp. 519–526, 2021, doi: 10.1109/JPHOTOV.2020.3043104.

[16] M. Theristis, A. Livera, C. B. Jones, G. Makrides, G. E. Georghiou, and J. S. Stein, "Nonlinear Photovoltaic Degradation Rates: Modeling and Comparison Against Conventional Methods," *IEEE J. Photovoltaics*, vol. 10, no. 4, pp. 1112–1118, 2020, doi: 10.1109/JPHOTOV.2020.2992432.

[17] L. Micheli, E. F. Fernández, Á. Fernández-Solas, J. G. Bessa, and F. Almonacid, "Analysis and mitigation of nonuniform soiling distribution on utility-scale photovoltaic systems," *Prog. Photovoltaics Res. Appl.*, vol. 30, no. 3, pp. 211–228, 2022, doi: 10.1002/pip.3477.

[18] L. Micheli, E. F. Fernández, J. T. Aguilera, and F. Almonacid, "Economics of seasonal photovoltaic soiling and cleaning optimization scenarios," *Energy*, vol. 215, 2021, doi: 10.1016/j.energy.2020.119018.

[19] M. G. Deceglie, L. Micheli, and M. Muller, "Quantifying Soiling Loss Directly from PV Yield," *IEEE J. Photovoltaics*, vol. 8, no. 2, pp. 547–551, 2018, doi: 10.1109/JPHOTOV.2017.2784682.

[20] I. Romero-Fiances *et al.*, "Impact of duration and missing data on the long-term photovoltaic degradation rate estimation," *Renew. Energy*, vol. 181, pp. 738–748, 2022, doi: https://doi.org/10.1016/j.renene.2021.09.078.

[21] K. Ilse *et al.*, "Techno-Economic Assessment of Soiling Losses and Mitigation Strategies for Solar Power Generation," *Joule*, vol. 3, no. 10, pp. 2303–2321, 2019, doi: 10.1016/j.joule.2019.08.019.

[22] "PV-Reliability Performance Model – System Advisor Model." [Online]. Available: https://sam.nrel.gov/pvrpm.

[23] A. Livera, M. Theristis, A. Charalambous, J. S. Stein, and G. E. Georghiou, "Decision support system for corrective maintenance in large-scale photovoltaic systems," in *48th IEEE Photovoltaic Specialist Conference (PVSC)*, 2021, pp. 0306–0311, doi: 10.1109/PVSC43889.2021.9518796.

Large-Area Uniformity Mapping of High-Speed Flexography-Printed Perovskite Solar Cells via Scanning Photoluminescence

Julia E. Huddy and William J. Scheideler

Dartmouth College, Hanover, NH, 03755, USA

Abstract—**Perovskite solar cells have the potential to provide terawatt-scale energy capacity, however current scaling technology is limited by complex patterning requirements and a lack of reliable large-area fabrication methods. We present a method for scaling fabrication of $MA_{0.6}FA_{0.4}PbI_3$ absorbers linking device performance to photoluminescence for better areal monitoring of the fabrication process. Using high-speed (60 m/min) flexography, we achieve the fastest reported fabrication of perovskite thin films, engineering precursor rheology and optimizing drying parameters for optimal perovskite uniformity. This print method provides highly uniform films with low surface roughness, achieving high photovoltaic performance (PCE > 19%) outperforming those of spin-coated counterparts. These results provide opportunities to monitor and enhance large-area device performance for scalable photovoltaic device fabrication and guide patterning for module manufacture.**

Keywords—*perovskites, absorbers, patterning, flexography, scalable*

I. INTRODUCTION

Metal halide perovskite (PVSKs) provide fundamental material advantages over current solar technologies, including slow radiative recombination, high absorption coefficients, and amenability to low-cost solution-deposition, that could push them to the forefront of thin film solar technology and finally achieve terawatt-scale photovoltaic (PV) capacity. Improving global PV integration requires additional cost reductions and design flexibility that extend beyond the current capabilities of rigid Si solar panels [1]. PVSK materials could provide these advancements, however, to achieve deeper renewable integration with this technology, technological improvements in scalable fabrication of these devices must overcome the drop off in photovoltaic efficiency and reliability exhibited by large-area perovskite devices [2]. Recent advances in rapid fabrication of the PVSK absorber through slot die, inkjet, and gravure printing have exposed difficulties in the upscaling of devices through limitations with uniformity and drying of the perovskite, which are critically linked to perovskite crystallization—a key determinant of high efficiency and long-term operational stability in devices.

This work uses high-speed flexography to develop a new scalable method for fabrication of perovskite solar cells (PSCs) and integrates drying to control perovskite film quality in ambient air conditions. Flexography is advantageous for rapid fabrication of a range of absorber films and ultrathin charge transport layers without encountering the low-flow limit that exists when slot die coating [3]. This print method can also be used to pattern substrates over large areas, achieving low line edge roughness and high uniformity, which is essential to achieve integration with large-area modules and other flexible electronics. As a well-established printing technique, flexography requires lower capital expenditures (CapEx) than vacuum methods such as ALD, evaporation, or sputtering because of its exceptionally high speed (200 – 600 m/min) and capability to print on wide webs (> 2 meters) [4]. Similar roll-based methods such as offset and gravure printing can also provide methods of patterning at high-speeds, however only flexography is compatible with both rigid and flexible substrates. Pairing this with photoluminescence (PL) measurements allows us to monitor spatial uniformity and better understand how the drying process is coupled to both thickness and optoelectronic film quality. Variations in PL intensity are shown to reflect differences in electronic properties of the materials in a device stack as well as the quality of the architecture [5]. When applied in 2D, areal mapping of the PL intensity can be used to illustrate uniformity of printed films over large areas, directly linking film quality to device performance.

In this work, we present a method for flexographic printing of absorber materials for improving scalability of planar perovskite solar cells. This high-speed (60 m/min) method produces films with unmatched uniformity (< 2.5%) over large areas and low surface roughness. Engineering the precursor design and the drying process allows for optimization of film thickness while pairing this fabrication method with PL

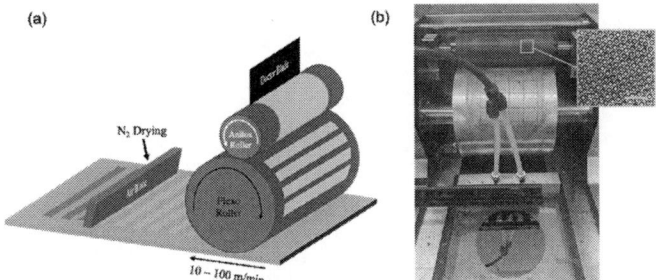

Fig. 1. a) Flexography schematic highlighting high speeds achievable with the printing method and showing the integrated airblade drying process. b) Image of flexographic printer set up used for printing. Image shows anilox roller, photopolymer stamp, and airblade used for drying, with inset displaying microscope image of engraved cells on the anilox roller (scale bar 250 μm).

978-1-6654-6060-6/23 $31.00 © 2023 IEEE

measurement allows for additional optimization of the film quality to achieve high photovoltaic efficiency (>19% PCE) while using the highest speed reported for any printing method applied to absorbers.

II. RESULTS AND DISCUSSION

In this work, we use high-speed flexography to reliably print and pattern perovskite thin films for use in PSCs. Pairing this with photoluminescence measurements, we observe how high-speed fabrication impacts device performance and uniformity. Using this process, we optimize fabrication and achieve devices with a champion efficiency of 19.3%. Flexography is a roll-based fabrication method that uses an anilox roller, doctored by a steel blade to meter ink volume, to transfer ink to a photopolymer stamp with raised features. This stamp then transfers the pattern to substrates, as shown in Fig. 1, allowing for more complex patterning of devices. Precise design of ink rheology allows for patterning features at the nanoscale while achieving films with tunable thickness. Perovskite films printed with this method achieve high thickness uniformity ($\pm$ 13.0 nm) over large areas (140 cm^2) with low areal pinhole densities of 0.40 pinholes/cm^2 as measured by large-area scanning microscopy (KeyenceVHX-7000). By integrating a 200 mm wide airblade into the printing setup, as illustrated in Fig. 1, we control the drying dynamics of the films *in-line* with the printing process, which directly impacts the crystallization dynamics of the material. Good perovskite crystallization is integral to production of high-performance PSCs and can be varied by changing the drying condition of the absorber film.

With our method we achieve highly uniform, large-area perovskite films that show low, micron-scale line edge roughness (< 40 μm) and gaps between features less than 200 μm. When integrated with module fabrication, this will facilitate device patterning, eliminating the need for complex laser scribing that can increase the CapEx. This process has the potential to be integrated with printing of other transport layers and metal bus bars to achieve a fully scalable process as well as with additional characterization methods to fully understand the differences in perovskite crystallization. Our high- speed (60 m/min) method deposits highly uniform films on both flexible and rigid substrates, allowing for integration of printed

MA$_{0.6}$FA$_{0.4}$PbI$_3$ absorber films with flexible single junction and tandem perovskite-silicon solar cell architectures.

A. Perovksite Ink Design

Scalable printing of perovskite absorber inks requires precise ink design to ensure highly uniform films with specific thickness. In this work, we observe changes in the morphology and thickness of the printed films as a function of the precursor ink viscosity, which strongly influences the amount of ink transferred during printing. Changing the concentration of the perovskite precursors in 2-ME from 1.0 M to 2.1 M, we observe that the viscosity of the ink increases from 4.7 to 17.4 cP due to additional coordination complexes formed between the precursor and the solvent. These variations in ink viscosity directly relate to changes in the resultant film thickness, allowing for precise tailoring of the perovskite within an acceptable range for PSC integration. Perovskite inks in this study reach thicknesses of 100 − 500 nm, with lower viscosity inks producing thinner films. Films thicker than 500 nm could be achieved by printing multiple layers or using an anilox roller with a higher ink volume than the 300 lpi anilox roller used for this work.

B. Drying of Perovksite Films

Perovskite precursor film drying is a key component of perovskite fabrication, with small variations to the process resulting in drastic differences in film quality. For perovskite fabrication, the drying and annealing processes aid in the crystallization of the perovskite, directly impacting the quality of the films. In this work, we explore the impact of various drying conditions on the quality of deposited perovskite films including the distance between the airblade and the substrate, the flow rate of the gas out of the airblade, and the timing of drying. By monitoring the corresponding photoluminescence (PL) response of the printed films, we optimize the drying process to produce smooth, uniform perovskite films.

Photoluminescence measurements indicate that perovskite films dried with nitrogen at close range are more uniform than those where the airblade was placed at a larger distance from the substrate. The measured PL spectra (Horiba LabRAM HR Evolution) for perovskite devices incorporating films dried at a distance of 1.5 cm shows higher peak intensity than those

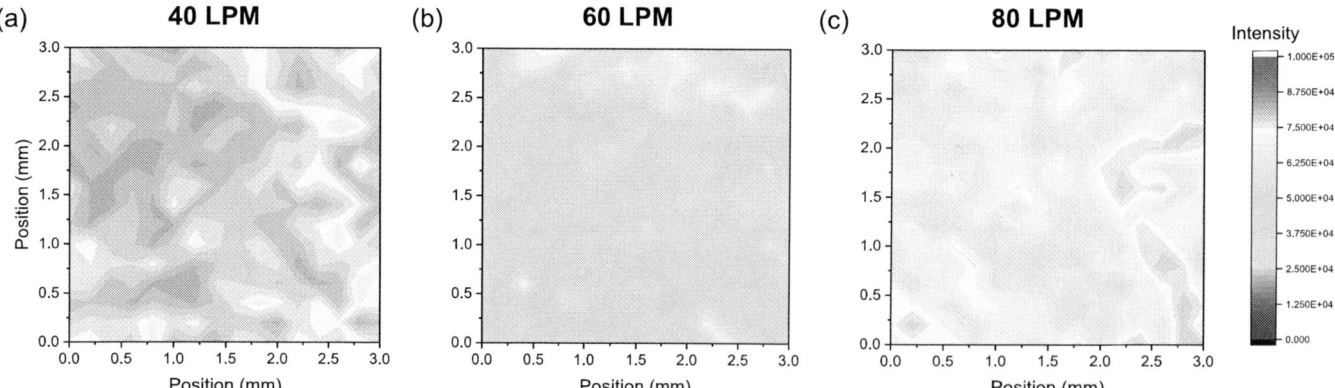

Fig. 2. 2D scanning photoluminescence maps of perovskite solar cells implementing flexographic printed *MA$_{0.6}$FA$_{0.4}$PbI$_3$* perovskite films dried at different nitrogen flow rates of (a) 40 LPM, (b) 60 LPM, or (c) 80 LPM over an area of 9 mm^2. Each data point in the maps of PL intensity represents a maximum intensity measured over the emission spectrum from 795 − 805 nm.

978-1-6654-6060-6/23 $31.00 © 2023 IEEE

incorporating films dried at a distance of 0.5 cm as measured with 533 nm excitation. This indicates that the films dried at a lower height are of higher optoelectronic quality since lower PL intensity corresponds to improved quenching of the PL signal in the device stack, indicating a higher charge collection efficiency and a better perovskite [5]. Additionally, comparing PL spectra for devices incorporating printed perovskite films with delayed drying shows increased PL spectrum peak intensity for longer delay times, indicating rougher films. This is consistent with the microscope images of the films that show decreased film quality. Delays up to three seconds have shown to produce films with better smoothness, many of which are best with a short delay between printing and drying.

The flow rate of the nitrogen over the surface of the substrate also impacts crystallization. Low flow rates do not provide enough nitrogen to quickly remove the solvent from the film, whereas high flow rates dry the film too quickly, resulting in visibly rougher and less specular perovskite films. In this work, we explore flow rates of 40, 60, and 80 LPM to understand how the flow rate of the nitrogen over the surface impacts the perovskite. Using photoluminescence, we observe changes to the PL spectrum peak intensity near the spectrum maximum consistent with other drying variations, where lower PL spectrum peak intensity corresponds to better film quality. Pairing this with 2D photoluminescence scanning, we spatially resolve the PL response of substrates with different drying conditions to understand the spatial uniformity of the drying method and compare the intensity of the PL response across various drying methods as well. Fig. 2 shows the PL response of films dried with different N_2 flow rates, indicating lower and more uniform PL emission for films dried at the optimal rate of 60 LPM.

C. Integration of Printed Perovskites into PSCs

These flexography-printed perovskite films were integrated as absorbers in double cation planar n-i-p PSCs in order to understand how the drying of the films affects device performance. When implemented in PSCs, these absorbers resulted in efficiencies meeting or exceeding those of devices fabricated with non-scalable deposition methods (spin-coating), indicating that flexography-printed perovskite can achieve the morphology and crystallization necessary for success as an absorber in PSCs. J-V curves for devices with printed and spun perovskite absorbers show these similarities for both PSCs, as shown in Fig. 3. Printed devices with optimal drying achieve an average power conversion efficiency (PCE) of 16.0 ± 2.5% with the champion device measuring 19.3%, whereas average spin-coated perovskite yields devices with a PCE of 16.8 ± 2.1%, with the champion device measuring 19.0%. Both devices show similar open circuit voltage and short circuit currents, with optimal printed devices outperforming spin-coated devices even though they are processed in an open-air environment at 50% relative humidity.

CONCLUSION

In summary, we present a method for rapid fabrication of perovskite absorbers for enhanced scalability and increased performance of planar perovskite solar cells. By employing high-speed flexography, we increase film deposition rate to a speed of 60 m/min, enhancing the potential for low CapEx. Patterning of films over areas exceeding 140 cm² with high uniformity shows that this is a promising method for large-area module fabrication. Pairing this fabrication method with 2D photoluminescence measurement allows for monitoring of film uniformity and location of defects to optimize the printing and drying processes and produce high efficiency (19.3% PCE) PSCs using the fastest reported absorber printing speed. This pair of processes has the potential to improve the understanding of perovskite crystallization and monitor in-line processing of perovskite solar cell devices for high-speed module fabrication.

ACKNOWLEDGMENT

This work was sponsored by a grant from the Arthur L. Irving Institute for Energy and Society at Dartmouth.

REFERENCES

[1] L. A. Zafoschnig, S. Nold, and J. C. Goldschmidt, "The Race for Lowest Costs of Electricity Production: Techno-Economic Analysis of Silicon, Perovskite and Tandem Solar Cells," *IEEE Journal of Photovoltaics*, vol. 10, no. 6, pp. 1632–1641, Nov. 2020.

[2] Z. Li *et al.*, "Scalable fabrication of perovskite solar cells," *Nat Rev Mater*, vol. 3, no. 4, pp. 1–20, Mar. 2018.

[3] J. E. Huddy, Y. Ye, and W. J. Scheideler, "Eliminating the Perovskite Solar Cell Manufacturing Bottleneck via High-Speed Flexography," *Advanced Materials Technologies*, vol. n/a, no. n/a, p. 2101282.

[4] P. Laden, *Chemistry and technology of water based inks.* London; New York: Blackie Academic & Professional, 1997.

[5] T. Kirchartz, J. A. Márquez, M. Stolterfoht, and T. Unold, "Photoluminescence-Based Characterization of Halide Perovskites for Photovoltaics," *Advanced Energy Materials*, vol. 10, no. 26, p. 1904134, 2020.

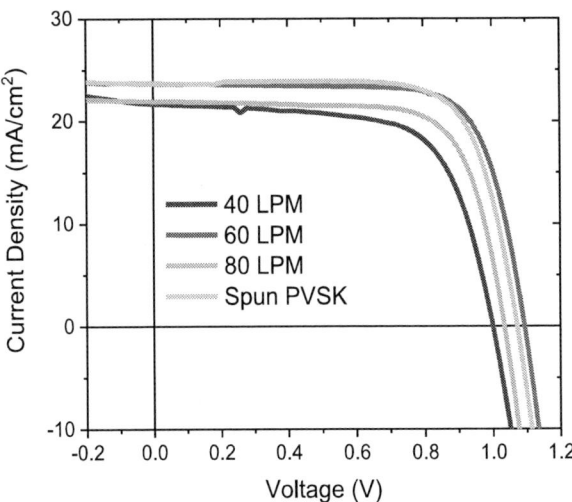

Fig. 3. Current density versus voltage measurements of PSCs with printed $MA_{0.6}FA_{0.4}PbI_3$ absorbers dried with nitrogen gas quenching at different flow rates compared with a spin-coated control cell (yellow).

Measuring the Doping Concentration of Si and CdTe Absorbers Using Lock-In Amplified Quantitative QSSPL

Mason P Mahaffey, Arthur Onno, Carey Reich, Adam Danielson, Walajabad Sampath, Zachary C Holman

Arizona State University, Tempe, AZ, United States

Colorado State University, Fort Collins, CO, United States

There are few doping concentration measurement techniques which are contactless and usable for all substitutionally doped semiconductors. In this work, we demonstrate the use of lock-in amplified quantitative quasi-steady state photoluminescence (QSSPL) to simultaneously measure the external radiative efficiency (ERE), lifetime, and activated dopant concentration of semiconductors. We first demonstrate that our doping concentration measurement agrees with the known doping concentration of a Si sample. Then, we demonstrate the use of our technique to measure the doping concentration of CdTe, which is much more difficult to accurately assess with other techniques.

Peer-to-Peer Energy Trading for PV Prosumers using Fuzzy Logic Inference Systems

Hector Lopez, Dr. Ali Zilouchian

Florida Atlantic University, Boca Raton, FL., 33431 , U.S.

Abstract — This paper presents a novel method using fuzzy inference systems (FIS) to help prosumers in Florida choose between a peer-to-peer (P2P) energy network and the traditional grid. With recent changes to net metering rules, it is less profitable for prosumers to always connect to the grid. The proposed method simplifies the FIS rules by breaking up the knowledge base into three separate systems: producer, consumer, and combined prosumer. This allows for more accurate modeling of the prosumer's trading profile and maximizing profit while considering cost, reliability, and environmental impact. The simulation results show the FIS solution's ability to maximize profit compared to solely connecting to the P2P market or the grid. This work provides a valuable contribution to the literature by considering recent policy changes and discussing potential solutions for recouping losses from roof-top solar owners in Florida.

I. INTRODUCTION

Prosumers, individuals and households that generate and consume electricity, are facing challenges in choosing between selling excess energy to the grid or a peer-to-peer (P2P) energy market. Recent changes to net energy metering (NEM) laws in Florida have increased the potential profitability of prosumers, but the decision is complex and involves a range of factors, including cost and reliability. In this paper, we propose a novel approach using fuzzy logic control systems to help prosumers make informed decisions about how to sell their excess energy and ensure they receive a fair price for it.

The prosumer problem of choosing between a P2P network or a traditional grid can be solved in alternate ways. Alternatives to the proposed fuzzy logic control system include decision tree analysis, linear programming, and multi-criteria decision analysis. These methods have been used in previous research and have limitations such as complexity and lack of personalization. Related works in P2P and Fuzzy Logic in energy markets, such as Liu et al. (2015)[1] and Rubio et al. (2016)[2], have highlighted the need for scalability and personalization in decision-making for prosumers. While Alam et al. (2017)[3] and Shahzad et al. (2021)[4] proposed methods with better scalability and additional trading factors, they lacked practical simulation of the system's impact on prosumer profitability. K.G.H. Kong et al. (2022)[5] used fuzzy optimization for multi-objective optimization in P2P systems, allowing for trade-offs between conflicting objectives. However, this paper presents a novel approach of breaking up the complexity of a prosumer fuzzy logic inference system into its fundamental components for simplification and

personalization of the decision-making process. Additionally, building on the direction of simpler control systems that can be robust and not rely on complex mathematical computations thtat may not scale. Our approach involves breaking down the FIS control system for prosumers into three separate systems: one for producers, one for consumers, and one for combined prosumers. This simplifies the rules and makes it easier for prosumers to understand and use the FIS control system, including personalized trading factors. Our main contributions are:

1. Using three different FIS systems to represent the various roles and preferences of prosumers as both producers and consumers of energy.
2. Simplifying the rules by breaking down the knowledge base into separate FIS systems, making it more user-friendly and adaptable.
3. Applying our method in a simulation for Florida prosumers, considering the NEM policy changes in 2023. The simulation allows prosumers to choose between P2P or grid payments and pricing for sales.

II. BACKGROUND

Florida's net metering rule was established in 2008 requiring IOUs to offer a standardized interconnection agreement. It promoted the adoption of customer-owned renewable generation by compensating the full amount of electricity returned to the grid. The credits flowed month to month until the 12th month. Any credits at that point were paid out at the full avoided costs of electricity [6]. In 2008 there where 577 customer-owned renewable generation interconnections. By 2020 that amount grew to roughly 150,000, but in comparison to the total number of electric utility customers in Florida, 10.5 million, it was still less than 1 percent [7,8]. Concerns of the cross-subsidization of net metered customers by non-net metered customers were raised to the PSC. Projections for cross subsidization among the general body of ratepayers for four of Florida's IOUs result in estimates of a cumulative cross-subsidy of over \$700 million by 2025 [9].

On January 1, 2023, a bill was passed that amends 366.91 and 163.04 of the Florida Statutes. The original net-metering credits where revised to only return the full-avoided cost to the prosumer for excess energy [6]. The decreasing of the credit amount from the retail rate to the *full avoided cost* may impact a prosumers decision to install a renewable generation system.

978-1-6654-6060-6/23 \$31.00 © 2023 IEEE

Additionally, if fewer customers purchase rooftop solar it can impact the solar installation and manufacturing industry.

A. Net-Metering

Net metering is a metering and billing methodology whereby customer-owned renewable generation is allowed to offset the customer's electricity consumption on site. Under net metering, customers are credited for excess energy produced which flows back to the grid. A meter is used to record both electricity drawn from the grid and excess electricity that flows to the grid from the customer-owned system. Florida's net metering rule was established in 2008 requiring IOUs to offer a standardized interconnection agreement for expedited interconnection and net metering of customer-owned renewable generation up to two megawatts [8 – NEM start].

The rule's purpose is to: Promote the development of small customer-owned renewable generation, particularly solar and wind energy systems; diversify the types of fuel used to generate electricity in Florida; lessen Florida's dependence on fossil fuels for the production of electricity; minimize the volatility of fuel costs; encourage investment in the state; improve environmental conditions; and, at the same time, minimize costs of power supply to investor-owned utilities and their customers [9 – Rule for NEM].

III. METHODOLOGY

A system that will leverage the P2P energy trading options in a micro-grid will need to make decisions between switching the load and excess generation between the grid interconnect and the micro-grid interconnect. The connected microgrid is an alternative to the grid connection. In Fig.1 an example of the setup is provided.

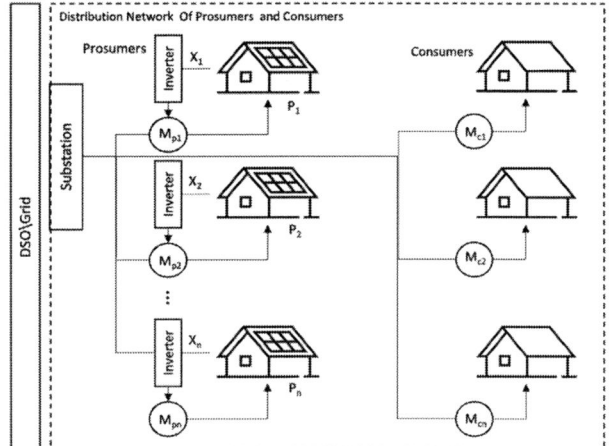

Fig. 1. Distribution network of prosumers with PV installed. The network facilitates an energy trading market that is local and can be an alternative to the connected grid connection. (P= Prosumers, C=Consumers, M=Meter, DSO=Distribution Service Oerator)

A. Prosumer Data Modeling

In this paper, we present a method to use fuzzy inference systems (FIS) to help prosumers choose between P2P or a grid connection. Our method involves breaking up the knowledge base of the prosumer FIS control system into three separate FIS systems: a producer FIS control system, a consumer FIS control system, and a combined prosumer FIS control system. By dividing the knowledge base in this way, we aim to simplify the rules and make it easier for prosumers to understand and use the FIS control system to include preferential trading factors. To model a prosumer, we use an aggregate dataset from the EIA.gov to determine the prosumer's load and generation curve. We then add statistical variance to the generation and load to create a unique prosumer data. In the simulation, we use the EIA.gov data for 14 months between 10/1/2019 and 7/1/2020. The sales and purchase price for P2P electricity is assumed to have more volatility and is calculated as a normal random variable with a mean at 75% of the retail price of electricity sold, and a normal distribution of 0.02 $/kWh of the retail price of electricity sold.

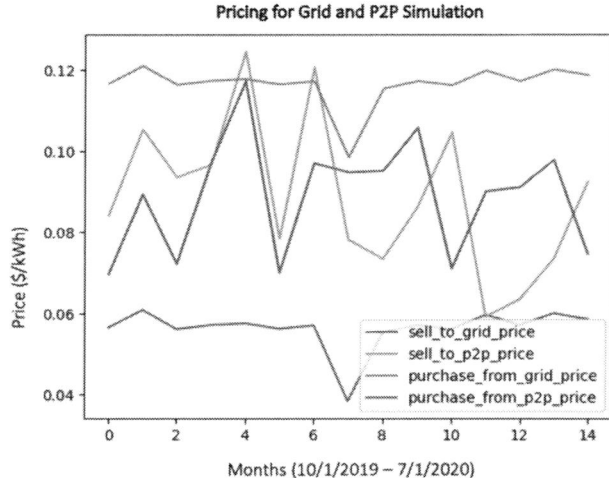

Fig. 2. Pricing for the Grid and P2P purchase and sales price for electricity. It is based on the retail electricity pricing sourced from EIA.gov. Pricing for P2P is a normal random variable sampling over the course of the same time window.

The advantages of a FIS are the low computational requirement, and the flexibility for changes in the control scheme based on unique user preferences. Fuzziness occurs in the uncertainty of linguistic ideas without clear boundaries. Most of the factors linked with energy trading can be optimally expressed using fuzzy logic theory. For example, factors, such as real-time price, times of the day, and buyers' interest, can be more naturally expressed in fuzzy logic than using the Boolean theory. The introduction of fuzzy logic will make the interaction clearer and more natural. Moreover, fuzzy logic is beneficial because there is no necessity for mathematical modeling compared to existing systems [12].

A test of all the FIS blocks joined a multi-block FIS system is provided in the following figure. The inputs chosen are arbitrary but are used to determine the membership function activation based on the rules. Finally, the output of the consumer and producer FIS are fed into the prosumer FIS where the final decision between P2P and Grid is made.

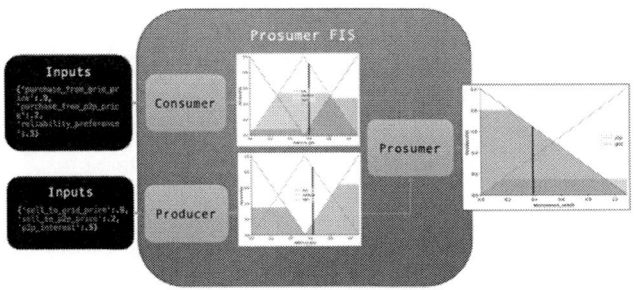

Fig. 3. Producer Block is a Fuzzy Inference Systems (FIS) that takes in the inputs specifically for the control to optimize the production of energy and the payment of that produced energy.

IV. RESULTS

The simulation shows that the total profit for a prosumer who is only connected to the grid is $3.36, while a prosumer who is only connected to P2P has a total loss of $63.81. However, using the FIS controller to switch between P2P and grid connections results in a total profit of $21.97, indicating that the controller is able to optimize the choice for the prosumer. The FIS controller is shown to be a more profitable choice than being exclusively connected to either the grid or P2P network.

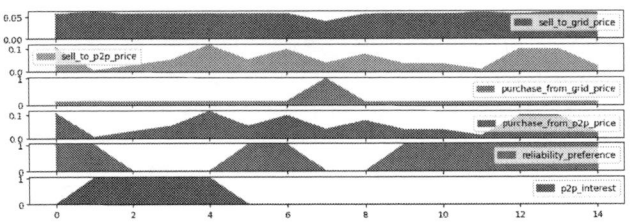

Fig. 4. Simulation of inputs to the FIS model over the course of 14 months. Shows the selected values each month for P2P interest, and Reliability preference by the user. The NEM and P2P pricing is provided month to month based on historical and synthesized data..

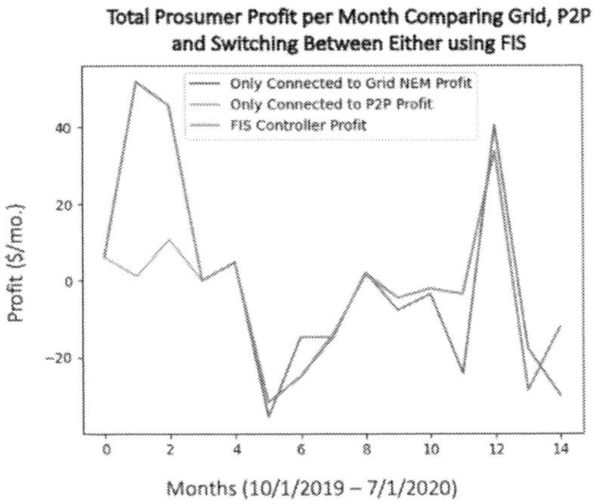

Total Prosumer Profit per Month Comparing Grid, P2P and Switching Between Either using FIS

Fig. 5. Prosumer Fuzzy Inference Systems (FIS) with a model that selects the FIS based on inputs and current month as grouped by seasonal FIS to improve accuracy of rules.

V. CONCLUSION

In conclusion, this paper presented a novel method for using fuzzy inference systems (FIS) to help prosumers choose between P2P or grid connections. By breaking up the knowledge base of the prosumer FIS control system into three separate FIS systems, we were able to simplify the rules and make it easier for prosumers to understand and use the system, while also incorporating preferential trading factors. The results of our simulation demonstrated the effectiveness of our method, showing that the FIS controller was able to optimize for profit by choosing the most suitable option between P2P and grid connections. Overall, our research highlights the potential for FIS to be a useful tool for prosumers in the face of changing NEM policies and the emergence of P2P energy markets. Future work can continue to explore the capabilities and limitations of using FIS in this context, and consider how it can be applied in real-world scenarios to help prosumers make informed decisions about energy trading.

REFERENCES

[1] Liu, T.; Tan, X.; Sun, B.; Wu, Y.; Guan, X.; Tsang, D.H.K. "Energy management of cooperative microgrids with P2P energy sharing in distribution networks." In Proceedings of the 2015 IEEE International

[2] Rúbio, Thiago RPM, Jonas Queiroz, Henrique Lopes Cardoso, Ana Paula Rocha, and Eugénio Oliveira. "TugaTAC broker: A fuzzy logic adaptive reasoning agent for energy trading." In European Conference on Multi-Agent Systems, International Conference on Agreement Technologies, pp. 188-202. Springer, Cham, 2016.

[3] Alam, Muhammad Raisul, Marc St-Hilaire, and Thomas Kunz. "An optimal P2P energy trading model for smart homes in the smart grid." Energy Efficiency 10, no. 6 (2017): 1475-1493.

[4] Shahzad, Khuram, Sohail Iqbal, and Hamid Mukhtar. "Optimal fuzzy energy trading system in a fog-enabled smart grid." Energies 14, no. 4 (2021): 881.

[5] Kong, Karen Gah Hie, Juin Yau Lim, Wei Dong Leong, Wendy Pei Qin Ng, Sin Yong Teng, Jaka Sunarso, and Bing Shen How. "Fuzzy optimization for peer-to-peer (P2P) multi-period renewable energy trading planning." Journal of Cleaner Production 368 (2022): 133122.

[6] Florida Department of Agriculture and Consumer Services, Electric Utilities, last visited Jan. 10, 2022, https://www.fdacs.gov/Energy/Florida-Energy-Clearinghouse/Electric-Utilities

[7] CS/SB 1024, 163.04, Section 350.001, F.S., The Florida Senate Bill analysis and Fiscal impact Statement: Regulated Industries Committee and Senator Bradley, Renewable Energy Generation, July 1, 2022.

[8] PSC, 2020 Interconnection and Net Metering Report, supra at n. 23.

[9] PSC, SB 1024 Analysis, supra at n. 10, p. 3 citing PSC, Review of 2021 Ten-year Site Plans of Florida's Electric Utilities, p.13, available at http://www.psc.state.fl.us/Files/PDF/Utilities/Electricgas/TenYearSitePlans/2021/Review.pdf (last visited Jan. 9, 2022).

[10] D. Arcos-Aviles, J. Pascual, L. Marroyo, P. Sanchis and F. Guinjoan, "Fuzzy Logic-Based Energy Management System Design for Residential Grid-Connected Microgrids," in IEEE Transactions on Smart Grid, vol. 9, no. 2, pp. 530-543, March 2018, doi: 10.1109/TSG.2016.2555245.

[11] Ma, Gang, Jie Lyu, Ying Wang, Jian Zhang, and Jie Xu. "The Prosumer Energy Management Method Based on Smart Load." IEEE Access 8 (2020): 117086-117095.

[12] Ross, Timothy J. Fuzzy logic with engineering applications. John Wiley & Sons, 2009.

Carbon Quantum Dots and their Possible Application in Perovskites Passivation

Maria Fernanda Villa-Bracamonte, Jose Raul Montes-Bojorquez, Alan Lopez-Becerra, Arturo Ayon

The University of Texas at San Antonio, San Antonio, TX, United States

High-quality perovskite films are the key factor in manufacturing high performance devices along with outstanding optoelectronic properties. However, methylammonium Lead Iodide (MAPbI3) perovskites are prone to defect formation during the iodide and methylamine ion defect migration. To achieve stable passivation in the perovskite film, additive engineering with nanoparticles can be used as an effective method to passivate the crystal defects. The carboxylic, hydroxyl and amino groups of Carbon Quantum Dots (CQDs) can bond the uncoordinated Pb atoms of the perovskite film creating strong and stable interactions inducing a low trap-state density improving the optoelectronic properties of the perovskite film. Therefore, we study the different functional groups in CQDs by Raman, FTIR, absorbance and photoluminice spectra and propose the addition of CQDs in the perovskite film to reduce the crystal defects and passivate the grain boundaries.

978-1-6654-6060-6/23 $31.00 © 2023 IEEE

A New Combined Accelerated Stress Test Sequence for Rapid Reliability Screening of Photovoltaic Materials

Yi Jiang*, Xuanji Yu*†, Ben Huang*, Ruirui Lv*, Yuanjie Yu*, Tao Xu*, Guangchun Zhang*

*Canadian Solar Inc. 1099 Xiangjiang Road, Suzhou, Jiangsu, China
†Case Western Reserve University, Cleveland, OH, USA

Abstract—As the global photovoltaic installed capacity continues to grow, reaching terawatt levels from gigawatt, the search for advancements in technology and cost reductions in the photovoltaic (PV) industry is ongoing. To keep up with these advancements, it is essential to develop accelerated testing methods for rapid reliability assessment of PV materials. To address this need, we propose a new combined accelerated stress test sequence that incorporates UV into stress steps of damp heat (DH), thermal cycling (TC), and humidity freeze (HF) based on the EC63209-1 sequence 3, as a rapid screening method for the reliability of PV materials. This approach has demonstrated its effectiveness by reproducing various failure modes including backsheet cracking, encapsulant delamination, yellowing and insulation failure of the junction box and connector.

Index Terms—Combined accelerated stress testing sequence, Photovoltaic materials, Failure modes, Rapid screening scheme

I. INTRODUCTION

The photovoltaic (PV) industry has experienced rapid development in the 21st century. Comprehensive methodologies for evaluating the reliability of PV materials have been gradually developed, including both module-level and material-level test standards or methods. [1]–[9] However, it is important to note that as the expectations for PV's role in the global energy transition increase, the well-established PV reliability tests need to be accelerated. For the purpose of efficiency increase or cost reduction, conducting conventional testing methods to evaluate the reliability of new materials or components with new designs can be very slow.

The IEC63209-1/2 and IEC62788-7-2 lab testing methods have been reported to successfully reproduce backsheet failures and several other critical module failures observed in the field, which the previous IEC 61215/61730 standards can not reproduce. However, they involve relatively long test cycles. To address this, we propose a new combined accelerated stress test sequence (CASTS) approach to more efficiently and quickly assess the risk of such module failures. As shown in Fig. 2, the CASTS method is similar to the IEC63209-1 stress test sequence but combines UV exposure with each stress steps, i.e., aging under UV and other stress simultaneously. It takes below 3 months to finish full cycles of CASTS, less than 1/3 of time needed for conventional methods.

Our studies show that CASTS is effective in reproducing currently known materials failures in field such as yellowing, delamination, and junction box insulation failure. PV modules subjected to CASTS or standard accelerated aging tests exhibited similar module failures in all cases, indicating that CASTS is a valid alternative test method for new module technologies and can be used to significantly shorten test time.

II. EXPERIMENTAL DESIGN

Fig. 3 shows the detailed stress test conditions of CASTS. We have incorporated UV exposure with a dose of 40 kWh/m2 into our aging sequences, with irradiation intensity of 150~220 W/m2. The UV lamp is kept on throughout the DH process, and it is only turned on when the temperature is above 0 °C during the TC and HF tests. Fig. 1 is a picture of our in-house CASTS equipment, compatible with both full-szie module level and material-level tests.

Fig. 1. Equipment for Combined accelerated stress test sequence (CASTS)

To replicate conventional failure modes of photovoltaic materials, we have conducted CASTS on full-size modules and mini-modules to study the consistency of this method at the module level. Additionally, we conducted CASTS on backsheet, adhesive film, junction box, and connector at the material level and analyzed the failure modes, compared to other conventional aging test methods.

978-1-6654-6060-6/23 $31.00 © 2023 IEEE

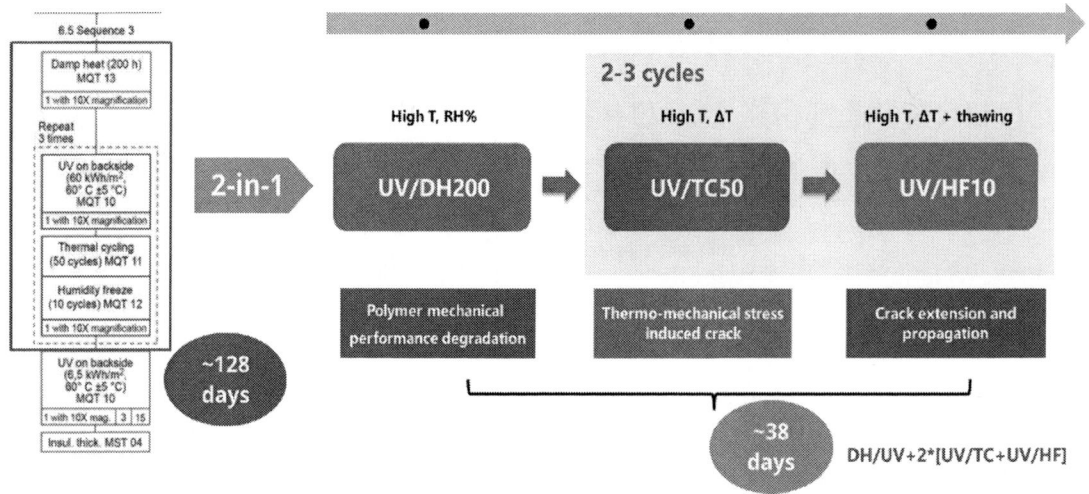

Fig. 2. Combined accelerated stress test sequence (CASTS) based on conventional stress test sequence from IEC63209-1

Item	Test conditions	Note
UV/DH	200 W/m², ~UV40 kWh/m² IEC61215-2, DH200h, 85℃/85%RH, 200h	UV irradiation of 200W/m² applied during 85℃
UV/TC	200 W/m², ~UV40 kWh/m² IEC61215-2, TC50 (no current*)	UV irradiation of 200W/m² applied above 0℃
UV/HF	200 W/m², ~UV40 kWh/m² IEC61215-2, HF10 (no current)	UV irradiation of 200W/m² applied above 0℃

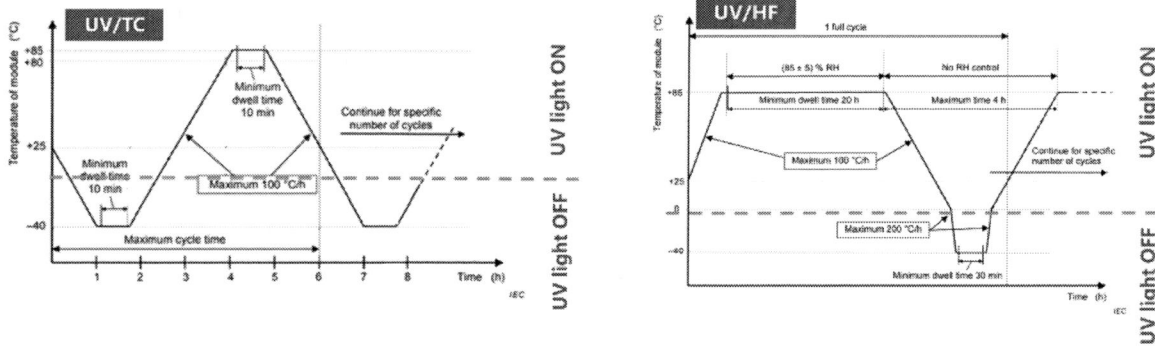

Fig. 3. Detailed stress test conditions of Combined accelerated stress test sequence (CASTS)

III. RESULTS AND DISCUSSION

A. Backsheet cracking

After undergoing combined stress testing, we observed backsheet cracking in both full-size (Fig. 4) and mini-modules, with a significant reduction in test duration, by a factor of 4. In the case of transparent backsheet, after the second cycle of TC/UV stress, we observed backsheet cracking as well as a loss of mechanical properties.

B. Other field failure modes

Furthermore, the CASTS scheme also reproduced failure modes such as delamination, yellowing, and junction box insulation failure. Additionally, the time for failure to occur under this scheme is notably shorter than the time required for failure under conventional aging methods (DH, UV, and TC).

IV. CONCLUSIONS

Our research shows that the new Combined accelerated stress testing sequence (CASTS) approach can quickly reproduce outdoor failure modes of materials and has a broader range of applications. Although the correlation to outdoor performance has not been fully verified yet, it can still meet the needs for rapid material screening in the rapidly developing photovoltaic industry.

REFERENCES

[1] IEC TS 63209-1:2021 | IEC webstore. [Online]. Available: https://webstore.iec.ch/publication/63120

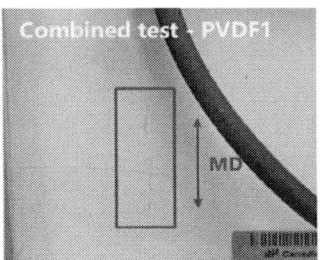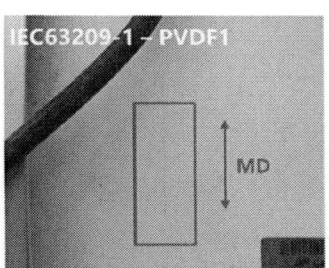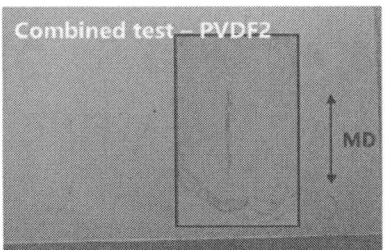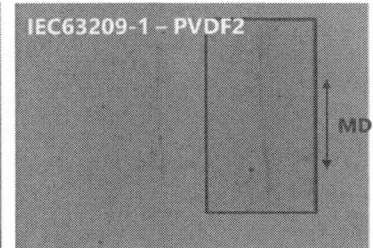

Fig. 4. Cracks observed on PVDF backsheets after Combined accelerated stress test sequence (CASTS) or conventional stress test sequence from IEC63209-1

[2] IEC TS 63209-2:2022 | IEC webstore. [Online]. Available: https://webstore.iec.ch/publication/65283

[3] IEC TS 62788-7-2:2017 | IEC webstore | LVDC. [Online]. Available: https://webstore.iec.ch/publication/33675

[4] M. Owen-Bellini, P. Hacke, D. C. Miller, M. D. Kempe, S. Spataru, T. Tanahashi, S. Mitterhofer, M. Jankovec, and M. Topič, "Advancing reliability assessments of photovoltaic modules and materials using combined-accelerated stress testing," vol. 29, no. 1, pp. 64–82, _eprint: https://onlinelibrary.wiley.com/doi/pdf/10.1002/pip.3342. [Online]. Available: https://onlinelibrary.wiley.com/doi/abs/10.1002/pip.3342

[5] P. Hacke, M. Owen-Bellini, M. Kempe, D. C. Miller, T. Tanahashi, K. Sakurai, W. J. Gambogi, J. T. Trout, T. C. Felder, K. R. Choudhury, N. H. Philips, M. Koehl, K.-A. Weiss, S. Spataru, C. Monokroussos, and G. Mathiak, "11 - combined and sequential accelerated stress testing for derisking photovoltaic modules," in *Advanced Micro- and Nanomaterials for Photovoltaics*, ser. Micro and Nano Technologies, D. Ginley and T. Fix, Eds. Elsevier, pp. 279–313. [Online]. Available: https://www.sciencedirect.com/science/article/pii/B9780128145012000116

[6] B. J. J. Liu and K. Hardikar, "A methodology for assessing field performance of flexible PV modules based on thermal cycling test results," pp. 1–1, ISBN: 9783936338416 Publisher: WIP. [Online]. Available: http://www.eupvsec-proceedings.com/proceedings?paper=37764

[7] J. Tsanakas, M. Karoglou, E. Delegou, P. Botsaris, A. Bakolas, and A. Moropoulou, "Assessment of the performance and defect investigation of PV modules after accelerated ageing tests," vol. 1, pp. 866–872. [Online]. Available: http://www.icrepq.com/icrepq'13/472-tsanakas.pdf

[8] IEC 61730-1:2016 | IEC webstore | solar panel, photovoltaic, PV, solar power, rural electrification, smart city, LVDC. [Online]. Available: https://webstore.iec.ch/publication/25674

[9] IEC 61215-1:2021 | IEC webstore. [Online]. Available: https://webstore.iec.ch/publication/61345

978-1-6654-6060-6/23 $31.00 © 2023 IEEE

Improved Soiling Rate Estimation by Calculating PV Module Temperature using a Distributed Thermal Model

Shoubhik De[1,2], Yogeswara Rao Golive[1,2], Narendra Shiradkar[1,2], and Anil Kottantharayil[1,2]

[1]Department of Electrical Engineering, Indian Institute of Technology Bombay, Mumbai, Maharashtra 400076, India
[2]National Centre for Photovoltaic Research and Education (NCPRE), Indian Institute of Technology Bombay, Mumbai, Maharashtra 400076, India

Abstract—**Soiling of PV panels results in energy production and revenue losses. Accurate measurements of soiling losses are essential for the optimised cleaning of solar panels. The Stochastic Rate and Recovery (SRR) model, used to measure soiling loss, uses King's model to estimate the PV cell/module temperature. King's model is a steady-state model and is likely to overestimate the changes in temperature under changing irradiance conditions. In this work, we evaluate a Distributed Model, which includes the thermal properties of a PV module to estimate the module temperature. The soiling rates obtained using the Distributed Model and King's model are compared. It is shown that the latter overestimates the soiling rate. King's model underestimates the temperature-corrected performance ratio compared to the Distributed Model. Further, the use of Distributed Model resulted in improved linear regression of the temperature-corrected performance ratio compared to the King's model.**

Keywords—*Kings Model (KM), Distributed Model (DM), Temperature-corrected Performance Ratio ($T_{cor}PR$), Soiling Rate (S_{Rate}).*

I. INTRODUCTION

Soiling of a Photovoltaic (PV) panel reduces its electrical power output, causing significant global economic loss [1]. In such situations, the panels need to be cleaned and deciding their cleaning frequency is a necessary step. To address this, soiling sensors can be installed in large-scale PV plants to monitor the soiling loss and provide useful information to the O&M team who in turn decides the cleaning frequency [2]. The soiling sensors uses a term called the Soiling Ratio (SR) to quantify the soiling loss of a PV plant. SR is a metric defined in the IEC 61724-1 standard [3] as the ratio of the power output of a soiled PV panel to that of a clean PV panel. The clean PV panel is cleaned daily. The SR has a value of 1 when the system is clean, and the value decreases when the system starts to soil. Monitoring the SR value over time can give insights about when to clean the PV system and the rate of soiling.

Deployment of the soiling sensors onto the PV plant indeed reduces the energy loss due to soiling, however, these are relatively new and require field validation. Moreover, they are costly, prone to human errors and equipment failure. An inexpensive alternative is to directly analyse the PV energy

generation data to gain insights about PV soiling and quantify the loss in terms of soiling rate (S_{Rate}). One such method which has become popular is the Stochastic Rate and Recovery (SRR) model [4]. The SRR model automatically detects Cleaning Events (CEs) in the daily time-series PV performance index (PI) data, thereby extracting the S_{Rate} between any two CEs. With time due to PV soiling, the PI reduces with time, and occurrence of CEs (manual/natural) improves the PI.

In the SRR model, the expected PV power output is calculated using the PVWatts model [5] which requires the input of PV cell or module temperature, and it is estimated using the King's model (KM) [6]. The KM is an empirical model which calculates the module temperature by ignoring the thermal inertia of the module. As a consequence, when one of the weather parameter (ambient temperature, plane of array (POA) irradiance, wind speed) changes, the module temperature also changes. Using the empirical based KM, Korab et al. [7] showed that the maximal temperature difference between the measured and estimated module temperature (estimated using KM) on a day when the POA irradiance showed high fluctuations was ~20°C. As a result, using these empirical models to evaluate the expected PV module power can result in uncertainty which in turn can make the daily PI time-series data noisy.

In this work, PV module temperature is estimated using the Distributed Thermal model (DM) described by Golive et al [8]. The DM takes into consideration the thermal inertia of the PV module, and thereby is able to estimate accurate module temperature on days when the POA irradiance shows high fluctuations. PV module temperatures evaluated using KM and DM is used to estimate PV soiling performance (comprising of parameters like SR and S_{Rate}). It is found that the accuracy of soiling performance improves when the module temperature is evaluated using the DM.

II. METHODOLOGY

A. Soiling Data

This work uses approximately four months of PV-SCADA data (6th February to 11th June 2020) of a 4.92 kW_p Evergreen solar system installed at the Desert Knowledge

TABLE I: PV Site details.

System Size	System Details	Parameters used for modeling
4.92 kW$_p$	• PV module: Evergreen solar (ES-A-205-fa3) [P_{max} = 205W_p] • No. of panels: 24 (1 String) • Inverter: 5kW (SMA SMC 5000A) • Array tilt: 20°, Azimuth: 0°	• Global Horizontal Irradiance (GHI) (W/m^2) • Ambient temperature (T_{amb}) (°C) • Wind Speed (WS) (m/s) • Accumulated Rainfall (mm)

Australia (DKA) Solar Centre, Alice Springs, Australia [9]. Additionally, the PV-SCADA data of the same installation on two days (2nd and 3rd) of February 2014 have also been used to highlight the fluctuations in the irradiance data.

Table 1 provides a detailed description of the site under study. POA irradiance for the site is estimated using Erbs model [10].

B. Evaluation of PV module temperature using the Distributed Thermal model

In this sub-section, the DM model is explained briefly. Thereafter, DM and KM are compared by estimating the module temperatures on two days when the POA irradiance show high variations. As described by Korab et al. [7], dynamic thermal models estimate accurate PV temperatures as these models are based on the energy balance between the PV module and the adjacent weather by taking into account the meteorological parameters, the module thermal capacity and the heat transfer mechanisms.

The DM described by Golive et al. [8] uses a 1-D Finite Difference Multi-physics method to evaluate PV module temperature by using the Optical, Electrical, Thermal and Physical properties of the solar panel. Ambient temperature, wind speed, and POA irradiance are used as the environmental inputs to the model. Sky temperature is calculated based on the environmental inputs and is used to evaluate the thermal radiation coefficient required to estimate PV module temperature. The model evaluates the temperature of all the layers of a PV module by dividing the whole structure into 107 nodes. Robustness of the model against changing irradiance conditions lies in the fact that it depends on the present and past input parameters. Moreover, the model uses dynamic sky temperature along with the appropriate physical properties of the module. For this study, the module's length, width and conversion efficiency have been taken from the module datasheet for the system studied in this work. All the remaining properties have been taken from [8].

Figures 1(a) and 1(b) show the variation of PV module temperatures evaluated using KM (KM-T) and DM (DM-T) on two days when the POA irradiance shows high fluctuations. While the module temperature estimated using the KM changes in lockstep with the irradiance, the temperature estimated using DM changes much slower, as anticipated. As

a result, the transients observed in the module temperature estimated using KM due to changes in POA irradiance, are higher than that observed in the module temperature estimated using DM. Therefore, the DM estimates a more realistic module temperature compared to the KM as it takes into account the thermal mass of the PV module.

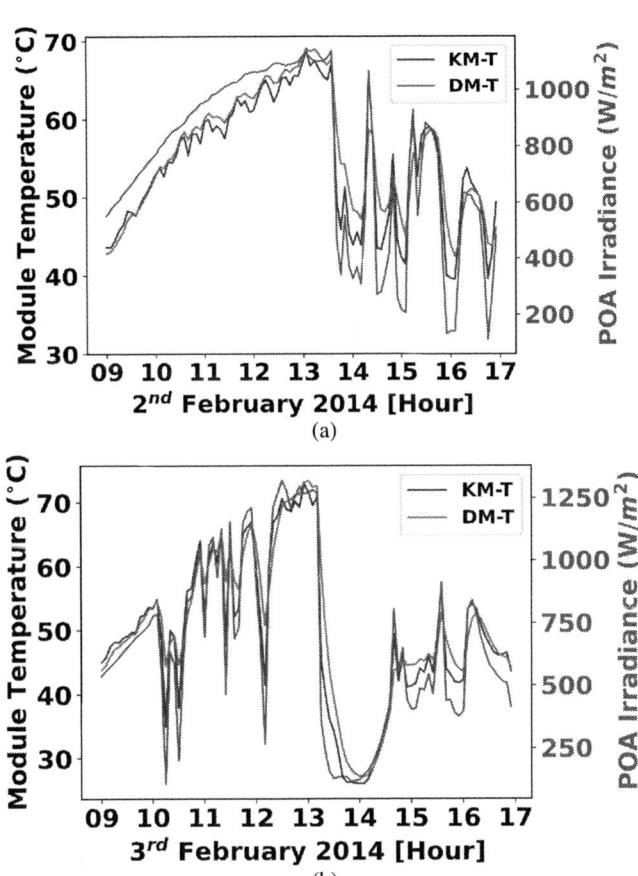

Fig. 1: Variation of PV module temperature evaluated using DM and KM on (a) 2nd February 2014 and (b) 3rd February 2014 when high fluctuations in Plane of Array (POA) irradiance were recorded. It is to be noted that the actual measurements of the PV module temperature isn't available for the site under study.

C. Generation of $T_{cor}PR$ profiles for S_{Rate} Estimation

In this work, we use Temperature-corrected Performance Ratio ($T_{cor}PR$) for our analysis. To remove shadows cast on the PV system or the irradiance sensor, Filter 2 described in [11] is used. Instead of using the insolation-weighted average [4, 12], median of the timestamps between 11 AM to 2 PM are taken to evaluate the daily $T_{cor}PR$ value (Median $T_{cor}PR$). To remove noise from the daily Median $T_{cor}PR$ time-series data, the noise-filtering algorithm described by De et al. [13] is used. Median T_{cor}PR is used to estimate the S_{Rate}s.

$$T_{cor}PR = \frac{Measured\ PV\ power}{Modeled\ PV\ power} \quad (1)$$

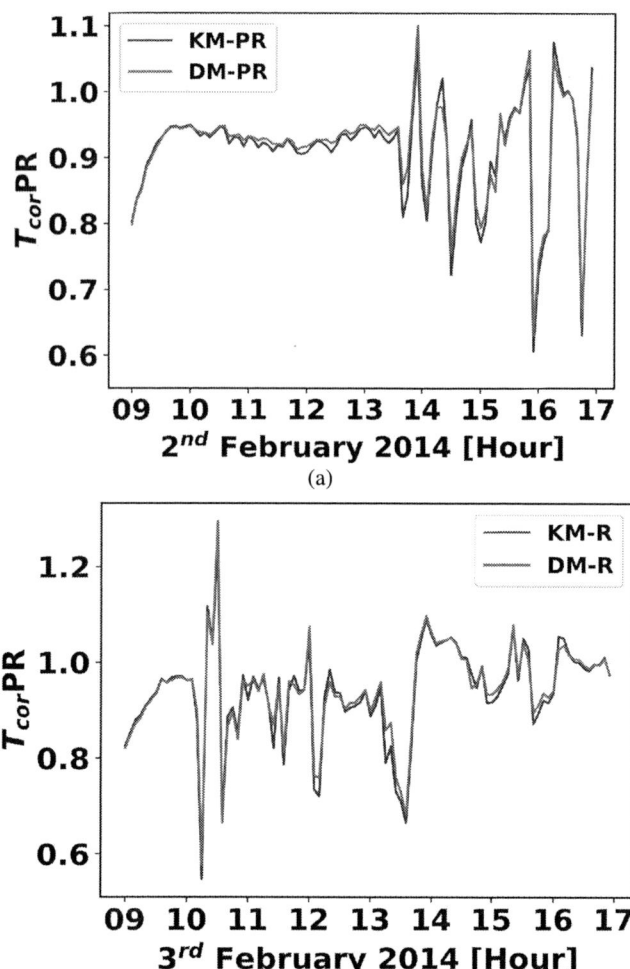

(a)

(b)

Fig. 2: Variation of $T_{cor}PR$ evaluated using DM and KM on (a) 2nd February 2014 and (b) 3rd February 2014. On both days, DM-PR showed lower fluctuations compared to KM-PR.

To study the effect of module temperature evaluated using DM and KM on the Median $T_{cor}PR$ time-series data, the plot of same for the site under study is shown. Difference between the daily Median $T_{cor}PR$ for the two models are plotted to point out that there is an improvement in the same, which signifies that the SR also improves. The results are shown in Section-III.

S_{Rate} of the soiling intervals is determined using scikit learn's linear regression model [14]. Coefficient of Regression (R^2) and Root Mean Squared Error (RMSE) are used to test the fit of linear regression model.

III. Results and Discussions

Plot of the $T_{cor}PR$ (described in equation 1) for 2nd and 3rd February 2014 are shown in figure 2. It can be seen that the $T_{cor}PR$ evaluated using DM (DM-PR) shows lower fluctuations as compared to the $T_{cor}PR$ evaluated using KM (KM-PR) during the transients. This is because the module temperature evaluated using the DM is more robust to fluctuations in POA irradiance compared to the temperature evaluated

using KM. This indicates that we can obtain a more accurate Median $T_{cor}PR$ estimate for both days when DM is used.

The time-series plot of Median $T_{cor}PR$s evaluated using KM and DM is shown in Fig. 3.The KM underestimates the $T_{cor}PR$ compared to DM. It can be observed that the time series plot of Median $T_{cor}PR$ obtained using DM is slightly above than the same obtained using KM. Therefore, it can be concluded KM underestimates the Median $T_{cor}PR$ compared to DM.

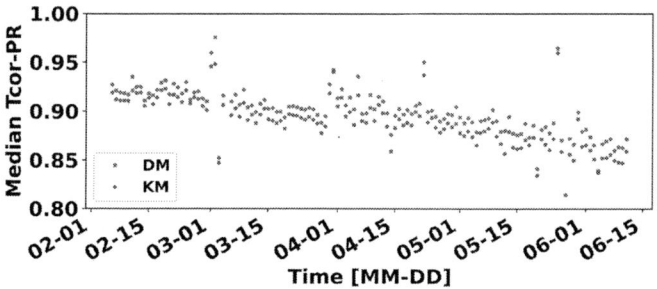

Fig. 3: Time-series plot of Median $T_{cor}PR$ for the system described in Table 1 using KM and DM.

To study the range of improvement of $T_{cor}PR$ when using DM, a box plot of the difference between Median $T_{cor}PR$ evaluated using DM and KM is shown in Fig. 4. It is seen that the difference varied between ~0.2% to 2% (a median absolute difference of 1.1% is obtained), which is significant for the determination of soiling rates, especially when the soiling rates are low.

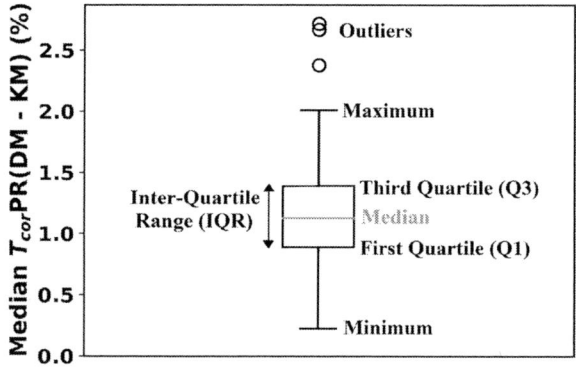

Fig. 4: A box plot highlighting the range of difference between Median $T_{cor}PR$ evaluated using DM and KM. The different parameters to read a box plot have also been shown.

To determine the S_{Rate}s of the soiling intervals, noise filtering algorithm described in [13] is applied. 'fil-DM' and 'fil-KM' represent the Median $T_{cor}PR$ data-points after filtering out noise for DM and KM respectively. The missing points are back-filled with the last known valid Median $T_{cor}PR$ point, and thereafter the CEs are detected. The detected CEs are shown separately for the soiling intervals extracted using DM and KM respectively. Fig. 5 shows the detected CEs for the period under study.

978-1-6654-6060-6/23 $31.00 © 2023 IEEE

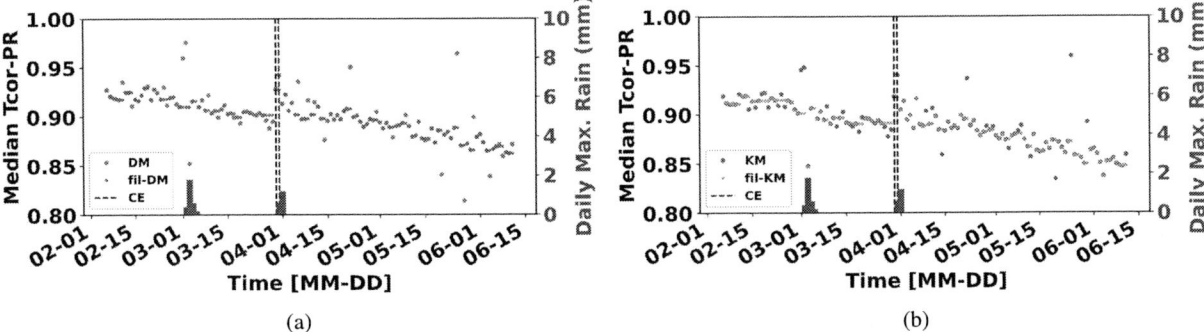

Fig. 5: Detected cleaning events for the Median $T_{cor}PR$ time-series plot evaluated (a) using DM, and (b) using KM for the site under study. The missing data points after filtering noise are filled by forward-filling with the last known Median $T_{cor}PR$ data-points.

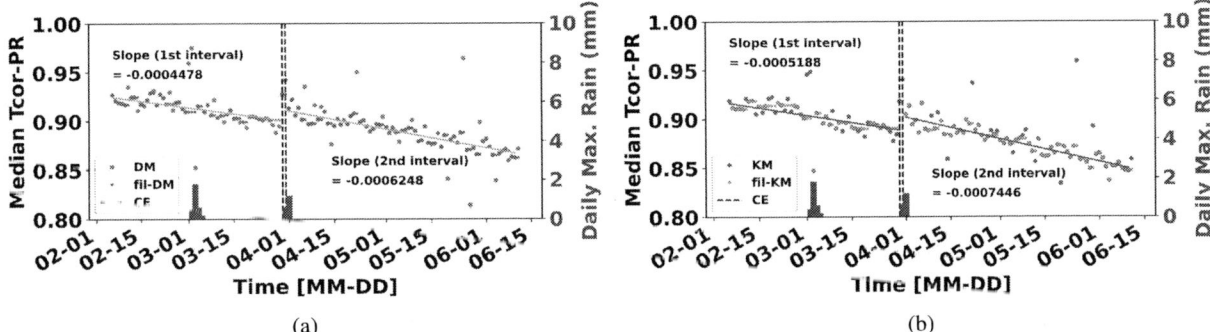

Fig. 6: Estimation of S_{Rate}s of the two soiling intervals for the Median $T_{cor}PR$ time-series plot evaluated (a) using DM, and (b) using KM.

From both figures 5(a) and 5(b), it can be seen that the period under study is a dry one, and the Median $T_{cor}PR$ decreases due to soiling. There are two precipitation events during this period. There is no improvement in the Median $T_{cor}PR$ due to the first precipitation event which occurred in the first week of March 2020. This signifies that the precipitation event weren't enough to clean the PV system. However, in the last week of March 2020, the second precipitation event were able to partially clean the PV system. A slight improvement in the Median $T_{cor}PR$ can be seen in figures 5(a) and 5(b), and this may be due to the reason that during the second precipitation event, the panels were more soiled than during the first event, and they were cleaned partially.

S_{Rate}s for the soiling intervals are determined for the daily Median $T_{cor}PR$ time-series data using linear regression algorithm. The results are shown in figure 6(a) and 6(b). The S_{Rate}s of the two soiling profiles evaluated using DM are seen to be lower than that estimated using KM. From this analysis, it can be concluded that using KM can result in higher S_{Rate}s which may not be accurate always, whereas, DM can give better estimation of S_{Rate}s. Reason to this lies in the fact that accurate module temperatures are evaluated using DM.

Table II shows the statistical comparison between the linear regression fit and the filtered Median $T_{cor}PR$ time-series data evaluated using DM and KM. The average R^2 value obtained for the linear regression fit using DM is found to be higher than KM. Again, lower RMSE values are obtained for the fit using

TABLE II: Statistical comparison of Median $T_{cor}PR$ time-series data evaluated using DM and KM

Module Temperature estimation model	Soiling interval	R^2	RMSE
DM	First	0.613	0.074
	Second	0.836	0.075
KM	First	0.414	0.098
	Second	0.846	0.081

DM. This indicates that the DM can reduce the noise in the Median $T_{cor}PR$ time-series data. As a result, the probability of accurate extraction of soiling loss occurring in a PV system is higher using DM than KM.

IV. CONCLUSION

Module temperature is an important variable for the estimation of soiling rates from PV power plant time series performance data. In this work, we evaluate the use of a distributed thermal model for the estimation of module temperature for extraction of soiling rates from temperature-corrected performance ratio. The King's model for estimating module temperature is used as the benchmark. King's model overestimates the temperature fluctuations of PV panels under varying irradiation conditions. Improved value of temperature-corrected performance ratio and lower soiling rate are obtained using the distributed model. Further, the confidence in the estimates is also improved with the distributed model.

V. ACKNOWLEDGMENT

This work was supported by the National Centre for Photovoltaic Research and Education (NCPRE) at IIT Bombay, funded by the Ministry of New and Renewable Energy of the Government of India through the Project No. 313-21/11/2022-Solar R&D. The authors would also like to thank the DKA Solar Centre, Australia for the data used in this work. SD acknowledges the PMRF scheme, Government of India for financial support.

REFERENCES

[1] K. Ilse *et al.*, "Techno-economic assessment of soiling losses and mitigation strategies for solar power generation," *Joule*, vol. 3, no. 10, pp. 2303–2321, 2019. doi: 10.1016/j.joule.2019.08.019.

[2] S. C. Costa, L. L. Kazmerski, and A. S. A. Diniz, "Estimate of soiling rates based on soiling monitoring station and pv system data: Case study for equatorial-climate brazil," *IEEE Journal of Photovoltaics*, vol. 11, no. 2, pp. 461–468, 2021. doi: 10.1109/JPHOTOV.2020.3047187.

[3] *Photovoltaic System Performance - Part 1: Monitoring*, IEC 61724-1:2017, International Electrotechnical Commission, Geneva, Switzerland, 2017.

[4] M. G. Deceglie, L. Micheli, and M. Muller, "Quantifying soiling loss directly from pv yield," *IEEE Journal of Photovoltaics*, vol. 8, no. 2, pp. 547–551, 2018. doi: 10.1109/JPHOTOV.2017.2784682.

[5] A. P. Dobos, "PVWatts version 5 manual," National Renewable Energy Lab. (NREL), Golden, CO (United States), Tech. Rep., 2014. doi: 10.2172/1158421.

[6] D. L. King, W. E. Boyson, and J. A. Kratochvil, "Photovoltaic array performance model, SANDIA Report SAND2004-3535," Tech. Rep. December, 2004. [Online]. Available: https://doi.org/10.2172/919131.

[7] R. Korab, M. Połomski, T. Naczyński, and T. Kandzia, "A dynamic thermal model for a photovoltaic module under varying atmospheric conditions," *Energy Convers. Manag.*, vol. 280, p. 116773, Mar. 2023. doi: 10.1016/J.ENCONMAN.2023.116773.

[8] Y. R. Golive, "Improving the Accuracy of PV Performance Estimation in the Field," Ph.D. dissertation, Indian Institute of Technology Bombay, 2022.

[9] "Desert Knowledge Australia Centre, Alice Springs (date accessed: 02/02/2014, 03/02/2014, 06/02/2020 to 11/06/2020)." [Online]. Available: https://dkasolarcentre.com.au/locations/alice-springs.

[10] D. Erbs, S. Klein, and J. Duffie, "Estimation of the diffuse radiation fraction for hourly, daily and monthly-average global radiation," *Solar Energy*, vol. 28, no. 4, pp. 293–302, 1982. doi: 10.1016/0038-092X(82)90302-4.

[11] S. De, P. Fuke, N. Shiradkar, and A. Kottantharayil, "Improved Shadow Filtering and Change-Point Detection Methods to Extract Soiling Loss from PV-Scada Data," in *8th World Conference on Photovoltaic Energy Conversion*, Milan, Italy, 2022, pp. 767–771. doi: 10.4229/WCPEC-82022-3BV.3.57.

[12] M. G. Deceglie, M. Muller, D. C. Jordan, and C. Deline, "Numerical validation of an algorithm for combined soiling and degradation analysis of photovoltaic systems," in *IEEE 46th Photovoltaic Specialists Conference (PVSC)*. IEEE, 2019, pp. 3111–3114. doi: 10.1109/PVSC40753.2019.8981183.

[13] S. De, P. Fuke, N. Shiradkar, and A. Kottantharayil, "Improved Cleaning Event Detection Methodology Including Partial Cleaning by Wind Applied to Different PV-SCADA Datasets for Soiling Loss Estimation," Manuscript under review.

[14] F. Pedregosa *et al.*, "Scikit-learn: Machine Learning in Python," *Journal of Machine Learning Research*, vol. 12, pp. 2825–2830, 2011.

978-1-6654-6060-6/23 $31.00 © 2023 IEEE

Electrical and Electroluminescence Evaluation of 17 Year Old Monocrystalline Silicon Building Integrated Photovoltaic Modules

Andrew M. Shore, Tali Schlenoff, Bakary Coulibaly, David Navon, Stephanie L. Moffitt, Brian Dougherty, and Behrang H. Hamadani

NATIONAL INSTITUTE OF STANDARDS AND TECHNOLOGY, GAITHERSBURG, MD, 20899, UNITED STATES

Abstract — Longterm reliable operation of photovoltaic modules is necessary to ensure the technology is key to reducing the cost of solar energy. However, degradation mechanisms in the field are not always consistent or predictable. We present a dataset of 72 building integrated photovoltaic (BIPV) monocrystalline silicon modules that were operated on the rooftop of a building at the National Institute of Standards and Technology (NIST) for 17 years. Electrical performance shows a majority of modules exhibit severe degradation beyond the typical ~1% loss per year. Xenon lamp flash solar simulator I-V curve measurements were compared to historical performance data to verify that the degradation seen in I-V measurements also existed during operation. While the open circuit voltage remained unchanged for most modules, a decrease in short circuit current and increase in series resistance led to deterioration of the short circuit current and maximum power. Hyperspectral electroluminescence (EL) imaging of three modules shows further evidence of significant series resistance increase for degraded modules. A module with a standard degradation rate has a clear EL image while a very degraded module has a very faint EL emittance at the same voltage due to series resistance losses. Further work will include calibrating the EL imaging system to quantify poor performing and current limiting regions within the module and determine the I-V curves of individual cells using the reciprocity relationship between EL and External Quantum Efficiency.

I. INTRODUCTION

As the installation of photovoltaic (PV) modules increases exponentially, understanding their longterm degradation mechanisms is critical for projecting the performance, power generation, and economic impacts. While manufacturing-related defects are inevitable, understanding how other, in-field defects form, grow and progress is crucial to understand the lifetime of the module. This is particularly relevant for silicon PV as it is the most pervasive solar technology. While the typical rate of degradation (~1% per year) is generally agreed upon [1], [2], some sets of modules will show a higher degree of variation despite typical operation and can be of particular interest for better understanding PV degradation mechanisms.

Frequently, just a small portion of the module, even just a couple of cells, will degrade and lead to a significant reduction in performance of the entire module. Therefore, techniques to analyze and determine the location of these defects are important. Current vs voltage (I-V) curve measurements, which describe the electrical performance of PV modules and cells,

are one technique. However, I-V measurements only show how the electrical performance of a whole module has been impacted by the degradation mechanisms and can only suggest what defects (e.g., increased series resistance) have occurred. Electroluminescence (EL) imaging is a commonly used technique to gain insight into where defects are located in the module. When current is sourced through a silicon module, the module emits a luminescence signal. Locations where the module does not emit signal indicates a defect region where current cannot flow well. When used together these techniques show where defects have developed and to what extent they impact the module performance.

We share a particular group of 72 modules with an atypically high degradation rate and, as a collection, have an electrical performance profile that is non-Gaussian. These 72 building integrated photovoltaic modules (BIPV) were operated and monitored as part of a 234 module array on the roof of a building in horizontal alignment at the National Institute of Standards and Technology in Gaithersburg, MD in the United States from 2001 through 2018 [3]. Along with performance data of the array, weather and irradiance data was also collected throughout the array's deployment. Unfortunately, the modules were incorrectly removed during a reroofing project. This means that many were lost and the original location of each module within the array is unknown. The spread of electrical performance was measured with a flash solar simulator as well as in-field measurements during operation to show that degradation was also seen in the field. We also share qualitative images from measuring module electroluminescence as a further means to show the degradation of the modules.

The BIPV modules are Siemens SP150 modules made up of 72 series connected monocrystalline cells. The modules are covered with tempered glass, have a multi-layered polymer backsheet, and an ethylene-vinyl acetate (EVA) encapsulant for environmental protection and electrical isolation. The modules have 4 branches, each with a bypass diode.

II. I-V MEASUREMENTS

I-V measurements under a solar simulator are an indicator for overall module electrical performance. Included in our data is a comparison between I-V measurements after removal and models of the on-site electrical performance data of the strings

978-1-6654-6060-6/23 $31.00 © 2023 IEEE

in the array fitted to standard reference conditions to show that the removal process had little impact on the module performance.

A. Flash Solar Simulator I-V Measurements

The I-V curves for all 72 modules were measured under a Xenon lamp flashed tower solar simulator with power adjusted to reach AM1.5G with the temperature held to $25°C \pm 1°C$. The system's spatial nonuniformity is less than 1.5% and temporal nonuniformity is less than 2%. IV curve measurements were conducted from short circuit (SC) current to open circuit (OC) voltage and from OC to SC to verify that hysteresis effects were negligible.

As seen in Fig. **1**., with a few representative modules from the batch included, there is a very wide range of performance across this subset from typical degradation seen in module BIPV-101-052, to more than normal in BIPV-101-011, to an amount that renders the module impractical to deploy like with BIPV-101-012.

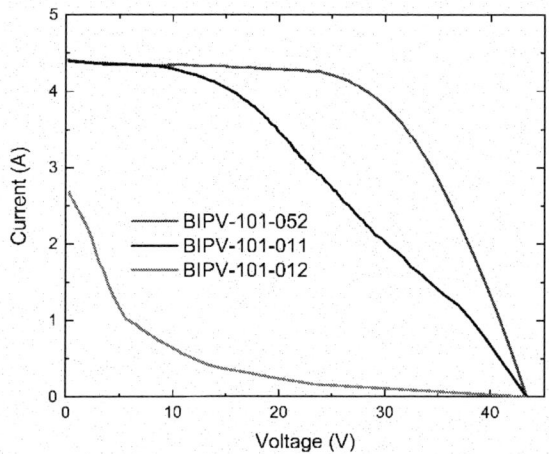

Fig. 1. I-V Curves of three modules in the dataset, showing large variability in performance.

The spread of the modules' performance, shown in Fig. **2**, is particularly interesting. The vast majority of the modules experienced very little degradation to the open circuit voltage (V_{oc}). There are a few modules that show the bypass diodes are used for one branch (indicated by the cluster of modules around 32 V). Otherwise, there are only a few modules that exhibit severe degradation from the rated 43.4 V at open circuit. The short circuit current (I_{sc}) is where we begin to see a more noticeable decrease in performance. The majority of modules still have I_{sc} close to the manufacturer's rated value. But, there are 20 modules with a short circuit current of less than 4.0 A. The degraded I_{sc} and rounding of the knee of the I-V curve (i.e., the Fill Factor) yields a wide range of maximum power (P_{max}) values for the 72 modules. Assuming a 1% loss every year for 17 years, one typical module would be expected to produce about 126 W at P_{max}. But the largest P_{max} of any of the 72 modules was 116 W and only 17 of the 72 modules have a P_{max} above 100 W. Meanwhile, another 30 modules have a P_{max}

between 50 W and 90 W and 25 produce 45 W or less showing very severe degradation. These I-V curves would suggest that, assuming the 17 best performing modules experienced typical degradation without any catastrophic failures, that the modules saw closer to 2.5% losses every year.

These results are very surprising given the consistent maintenance performed. Due to the removal of the array being unplanned and in an unexpected fashion, there are natural concerns that some of the measured degradation in output may be due to the removal of the modules and not from natural degradation experienced while deployed. We address these concerns below.

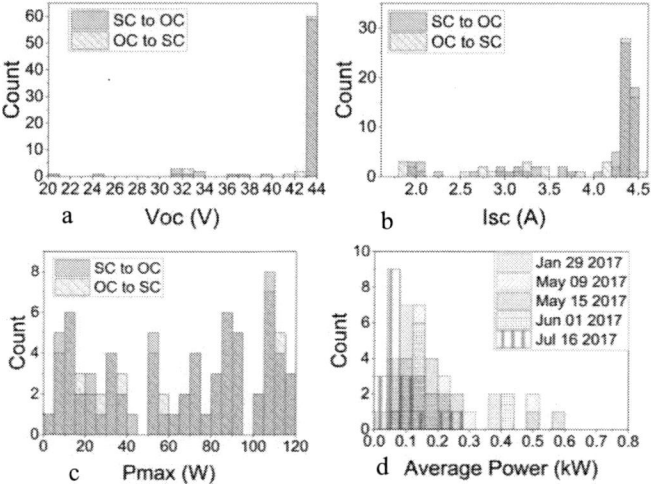

Fig. 2. Histogram summary of the 72 modules tested from Isc to Voc and from Voc to Isc. Open circuit voltage of the modules (a) did not degrade significantly. Short circuit current (b) shows some degradation and compounded with rounding of the IV curve, the maximum power shows (c) significant reduction in output with a notable spread. The module performance is constrasted with average power output from the 18 strings on 5 high irradiance days in (d).

B. Historical Performance Data

Before considering the granular data of each string, we did an immediate evaluation of the entire array's performance and found that the total degradation over the course of the 17 years was about 7% per year. To further verify that the removal of the array did not cause damage to the modules, we also evaluated the performance of the strings from the historical electrical and weather data before the array was taken down. This data was taken from the most recent data in 2017 on 5 sunny days when the irradiance was close to 1000 W/m². Since we are unsure of where each module was located within the array, we compared the spread of the data from the 18 strings to the spread among the 72 modules. Fig. 2(d) shows the average power from 11 am to 3 pm of the 18 strings. We see that the string performance is even more degraded than the I-V curve data in Fig. 2(c). Since each module has degraded at a different rate, there is an I-V curve mismatch between the modules in each string. So, the output of each string will be even worse with a decrease in

978-1-6654-6060-6/23 $31.00 © 2023 IEEE

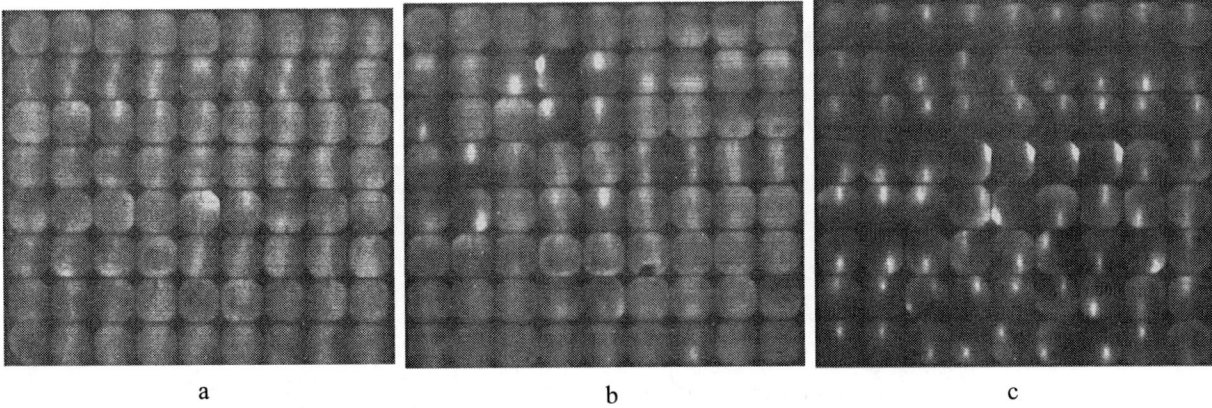

Fig. **3.** EL Images at 1140 nm of modules with 2.0 A applied with the same color scale. Yellow indicates a high luminescence, blue indicates low. BIPV-101-052 (a) operating at 44.4 A, BIPV-101-011 (b) at 51.2 V, and BIPV-101-012 (c) at 80.0.

overall maximum power. Our data backs this up as we mostly saw degradation rates of more than 10% per year for each string. This data gives us confidence that the module I-V data was not due to destruction upon removal, but from natural degradation mechanisms during deployment.

III. EL IMAGING AND FUTURE WORK

The defect mechanisms of these modules and reason for the variation is unknown at this time. So along with I-V measurements taken during operation in the field and under a flash solar simulator, we have also begun to utilize EL imaging to better understand failing regions of the modules in individual cells.

We used an in-house hyperspectral imaging system, with custom-designed optics, to capture electroluminescence images of an entire module. The module is placed into a rig at a distance of 4 meters from special optics that directs the luminsence signal into the hyperspectral imaging camera. A power supply was connected to the same modules as shown in Fig. **1** with the voltage adjusted until 2.0 A was run through the module. Under these conditions, EL images were acquired at 1140 nm (the peak emittance wavelength of silicon) with the HS imager. As seen in Fig. **3**, module BIPV-101-052, with the least degraded I-V curve, showed a very uniform luminescence across all cells suggesting the fewest defects. This module also required the lowest voltage (44.4 V) for 2.0 A to run through the module. BIPV-101-011 had more noticeable defect regions and dead areas and required 51.2 V to be applied for the same current. Finally, BIPV-101-012 had noticeable dark regions and defects on nearly every cell. It also required 80.0 V to be applied for a current of 2.0 A to flow through it. We can infer a much larger series resistance for module 012 than 011 or 052 given the higher required voltage to operate at 2.0 A. These results match the I-V curve measurements and further support the wide ranging degradation seen in these modules. These images are our first steps towards determining where the defects are within the module and working to quantify the electrical performance of each individual cell in the module in a nondestructive fashion.

IV. CONCLUSION

The 72 modules we were able to save from the building array at NIST exhibit severe degradation effects that are beyond typical degradation rates. We have further demonstrated with EL imaging that degradation of individual cells is significant in the modules with poor I-V curves. Plans include completing the EL imaging on all remaining modules and using the EL results to estimate series resistance losses or other degradation effects within each module. We plan to continue this work by modelling the I-V curves of all cells in each module using the reciprocity relationship between EL and EQE. Ultimately, module coring will be performed to verify the materials degradation that led to the performance degradation in these modules.

REFERENCES

[1] D. C. Jordan, S. R. Kurtz, K. VanSant, and J. Newmiller, "Compendium of photovoltaic degradation rates," *Progress in Photovoltaics: Research and Applications*, vol. 24, no. 7, pp. 978–989, Jul. 2016, doi: 10.1002/pip.2744.

[2] M. Dhimish and A. Alrashidi, "Photovoltaic degradation rate affected by different weather conditions: A case study based on pv systems in the uk and australia," *Electronics (Switzerland)*, vol. 9, no. 4, Apr. 2020, doi: 10.3390/electronics9040650.

[3] A. H. Fanney, E. R. Weise, and K. R. Henderson, "Measured impact of a rooftop photovoltaic system," *Journal of Solar Energy Engineering, Transactions of the ASME*, vol. 125, no. 3, pp. 245–250, Aug. 2003, doi: 10.1115/1.1591799.

Automated Photovoltaic Module Quality Assessment: Defect Identification and Classification from Luminescence Images using Machine Learning

Brendan Wright[1]*, James Petesic[2], Timothy Dawson[2], Ziv Hameiri[1]

[1]School of Photovoltaic and Renewable Energy Engineering, UNSW Sydney, Australia
[2]PV Industries Pty Ltd, Australia

Abstract — **To help address the important challenge of photovoltaic module end-of-life (EoL) management, an automated module quality assessment methodology to determine the preferred EoL paths of fielded photovoltaic modules using luminescence images has been developed. First, defects within the photovoltaic modules are identified and classified through feature extraction incorporating generative representation learning models. Second, the relative module power loss is estimated and the preferred EoL path is determined based on thresholds selected by the user. A method to calibrate module performance using the quality assessment results, and perform full module circuit simulations is then developed, and initial results using this novel approach are presented.**

I. INTRODUCTION

The challenge of photovoltaic (PV) module end-of-life (EoL) management is receiving increasing attention from PV plant owners, assets management and O&M (operations and maintenance) companies, module manufacturers, and government regulatory organisations. Fundamentally, in most cases, there is no definitive answer to the question "when do PV modules reach their EoL?", as often the modules still produce power when this question is asked. Furthermore, the decision regarding the future of these modules (reuse, resell, recycle) is not straightforward. To date, the methodology for making these decisions has not been developed for utility-scale PV plants, partly due to the lack of adequate large-scale characterisation of PV modules and associated analytical methods.

In this study, we develop a machine learning (ML)-based approach to determine the EoL of PV modules and their preferred future paths. The method is based on full-area PV module luminescence images that can be taken outdoors in the field or indoors. Based on these images, an automated analysis procedure has been developed to identify and classify defects within the PV module and to estimate the loss of power generation. The automated procedure includes: (1) a robust image pre-processing and segmentation procedure to extract individual cell images, (2) trained ML models to perform defect identification and classification, and (3) full module comparative simulation to estimate total power loss and remaining power generation potential. The extracted information is then used for making an informed decision regarding EoL. The final outcome of the developed

methodology is a report outlining options and risks for each module, including detailed supporting information. As discussed, the process is fully automated and is based only on luminescence images. It can be easily scaled to large batches and provides significant benefits to all stakeholders at almost no cost.

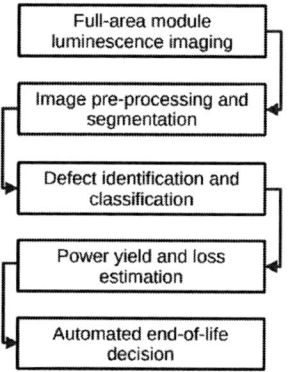

Fig. 1. A summary diagram of the developed automated EoL decision process, from module image pre-processing and segmentation to cell and module quality assessment, and full module performance simulation for power loss estimation.

II. LUMINESCENCE IMAGE PRE-PROCESSING

Outdoor module luminescence images often feature distortions and artefacts, as well as arbitrary perspective and rotation (shift relative to the centre-aligned module axis). Furthermore, as such images are typically obtained in the context of multi-module arrays, the existence of additional background features or adjacent modules must be accounted for, requiring feature (single module) identification and segmentation (background removal) prior to lens and perspective corrections. Hence, the first step in the developed methodology is a pre-processing procedure for full-area module luminescence images (see Figure 2). This procedure provides an effective and robust solution to prepare real-world outdoor images for further analysis. It is capable of overcoming the complexity and variability in image quality, providing a consistent undistorted and cropped full-area module image at native resolution, ready for segmentation and quality assessment.

978-1-6654-6060-6/23 $31.00 © 2023 IEEE

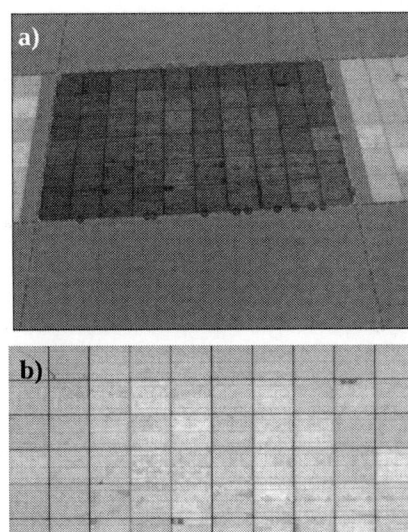

Fig. 2. (a) PV module luminescence image pre-processing (lens corrected, area isolation, edge detection, corner computation) ready for deskew (perspective transform) and crop; (b) isolated module area luminescence image (after pre-processing), cell dimensions and spacing computation, and generation of segmentation grid (blue lines).

III. DEFECT IDENTIFICATION AND CLASSIFICATION

After pre-processing, the first stage of module quality assessment involves the analysis of the luminescence images, to identify and classify defects. A database of individual cell images for each of the modules (obtained from the pre-processing and segmentation procedure) is prepared and normalised (per module or batch of equivalent modules), such that feature extraction can be performed. The statistics (area-mean, standard deviation, histogram, etc) of each image are then calculated while derivative images (area-gradient, Laplace, fast Fourier transform) are generated.

Generative representation learning (GRL) models are then used to output learned vectors that contain cell luminescence features. The GRL models employ deep convolutional neural network (CNN) architectures for the component encoder and decoder models. They are trained using a combination of variation autoencoder (VAE) [1] and generative adversarial network (GAN) frameworks to reproduce cell luminescence images through a compressed latent representation (feature vector). An example of the capability of the GRL models is presented in Figure 3, showing both input (a, c, e, g) cell luminescence images extracted from a module and the corresponding output (b, d, f, h) reproductions from a latent vector.

Clustering by similarity is then performed using the aggregate feature vector from the statistics and ML model results, with uniform manifold approximate projection (UMAP) embedding used for dimensionality reduction and hierarchical density clustering (HDBSCAN*) [2] for the

classification of defects and degradation. Hence, this step provides a detailed quality assessment of each cell based on its luminescence image statistic and identified features. The combination of the information about all the individual cells enables the assessment of the module's quality solely from luminescence images. In addition, the obtained information is used to estimate the electrical parameters of each of the cells to be used in full module simulations to determine the relative module power.

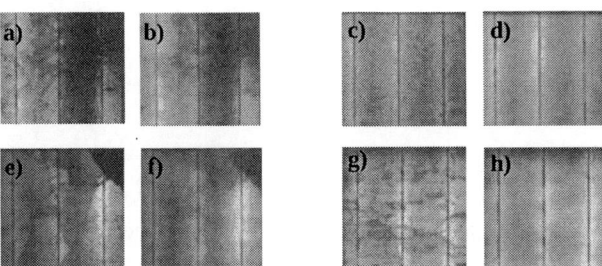

Fig. 3. Examples of generative representation learning capability used for feature extraction (individual cell luminescence images obtained via module pre-processing); four pairs of each: pre-processed cell input images [a,c,e,g] to the encoder model (outputs a compressed representation as a latent vector) and decoder model output images [b,d,f,h] to reproduce input image (reconstruction from the latent vector).

IV. RISK ASSESSMENT AND PERFORMANCE ESTIMATION

Automated module risk ratings are calculated using a 'risk matrix' (the number of cells within a module that are classified as 'risk' or 'warning'). Modules are assigned to a number of quality bins based on threshold ranges that can be adjusted by the user. Table 1 lists the threshold ranges used in this study. This risk assessment and quality bin assignment can be used directly to inform module EoL decisions (e.g. re-use: bins 1-2, re-sell: bins 3-4, recycle: bins 5-6).

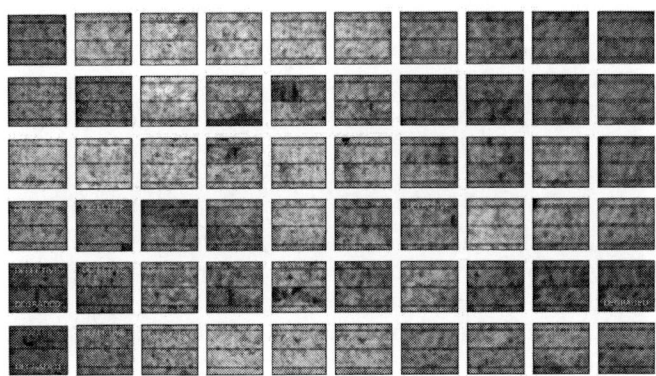

Fig. 4. Example module quality assessment from luminescence image after pre-processing and segmentation (analysis performed on individual cell images); identified defects are labelled by class (none, damaged or defective, dead or degraded, etc) and severity (coloured by increasing risk: yellow, orange, and red).

978-1-6654-6060-6/23 $31.00 © 2023 IEEE

An example of successful processing and assessment is displayed in Figure 4, with cell-level defect identification and quality assessment (no damage/defect, damaged or defective, dead or degraded), and risk-rating (coloured by increasing risk: yellow, orange, and red).

Automated power estimation can be used to further enhance the risk assessment, however, calibration is in progress to obtain the absolute degradation (to estimate the cell electrical parameters) using module circuit simulations developed with the PySpice framework. These results will be presented at the conference, demonstrating the estimation of module performance directly from luminescence images.

V. Validation of Module Quality Assessment Pipeline

The developed quality assessment methodology was validated using a highly varied real-world outdoor electroluminescence (EL) image dataset. A batch of 141 raw luminescence images of decade-old modules that were installed on a carpark in Sydney was passed to the automated pipeline, containing very challenging images (considering module-camera alignment, fore/background contrast, module quality/damage, etc). The summary of the module batch quality assessment (Table 1) illustrates the distribution of module qualities (identified defects, severity as risks/warnings). Only eight of the assessed modules were assigned risk-defect free (5% pass rate), containing zero identified risk cells. However, no modules were assessed as completely free of warnings, with a typical module having five risk and 20 warning cells. Adjusting the acceptable threshold for risk and warning cells can provide a method to filter for the highest quality modules.

Bin		1	2	3	4	5	6
Range							
	Risk	< 1	< 1	< 5	< 5	< 10	>= 10
	Warn	< 20	>= 20	< 20	>= 20	< 20	>= 20
Count							
	Auto	4	4	25	30	26	52
	Manual	22	3	28	3	74	11
Gap							
	Average	-2.5	-4.0	-0.7	-2.0	-0.2	+1.6

Table 1. Summary of threshold ranges for bin assignment by risk/warning cell count, module counts per quality bin for each automated and manual assessment, and magnitude of disagreement (average gap in bins of automatic assignment relative to manual).

A comparison between automated module assessment and manual assessment by a domain expert is presented in Table 1 and Figure 5. A breakdown of risk-ratings (quality bin assignment) calculated for each module is provided in Table 1. Both the manual and automated assignments show an increasing module count at higher bins (low quality). Interestingly, the magnitude of disagreement between automated and manual bin assignment (gap, as measured by average integer bin difference) shows a clear trend: a greater assignment bias to lower bins in manual assessment (-2.5) for the highest quality modules, which is reduced and eventually inverts at the lowest quality bin (+1.6) for the worst modules. These trends are visually illustrated in Figure 5.

While disagreement between manual and automated risk-rating assessment is typically minor, an analysis of representative examples is informative. Of particular note is the manual assessment of low-contrast defect features that, upon closer inspection, clearly represent risk defects, however, are not labelled as such where their distribution over the cells within the module is somewhat uniform and can be mistaken for multi-crystalline grain boundaries. Without image pre-processing and normalisation, visual assessment can be error-prone, and the task of maintaining consistent valid assessment throughout large batches of modules becomes increasingly difficult.

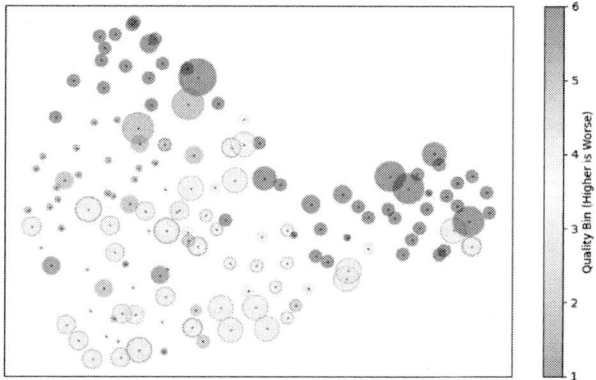

Fig. 5. Illustration of validation study (automated module batch quality assessment): module distribution by identified defects (each point is a module, point-to-point distance is high-dimensional module similarity based on the count of defect types); comparison between automated (colour) and manual (edge colour) quality bin assignment; point-size indicates the magnitude of disagreement between the bin assignment (automated vs manual).

VI. Conclusions

To summarise, the validation study results illustrate the accuracy of the developed automated quality assessment methodology, as well as the advantages provided (analysis speed and scale) while maintaining a consistent assessment. As such, the developed capability can enable the large-scale automated EoL assessment of PV module installations from in-field luminescence imaging at a low cost.

References

[1] I. Higgins *et al*, "β-vae: Learning basic visual concepts with a constrained variational framework", *ICLR 2017*, 2017.
[2] L. McInnes *et al*, "Accelerated Hierarchical Density Based Clustering" *2017 IEEE ICDMW*, pp 33-42. 2017

Influence of alkali iodide fluxes on Cu2ZnSnS4 monograin powder properties and performance of solar cells

Kristi Timmo, Katri Muska, Maris Pilvet, Mare Altosaar, Valdek Mikli, Mati Danilson, Reelika Kaupmees, Jüri Krustok, Maarja Grossberg-Kuusk, Marit Kauk-Kuusik

Tallinn University of Technology, Tallinn, Estonia

Molten salt synthesis-growth method is one possibility to produce very high-quality monocrystalline absorber materials in powder form for flexible solar cells. Results of the current study show the influence of different alkali salts on the CZTS monograin powder (MGP) properties and impact on the performance of monograin layer (MGL) solar cells. $Cu_{1.84}Zn_{1.09}Sn_{0.99}S_4$ powders were synthesized from CuS, ZnS and SnS by isothermal molten salt synthesis-growth method in the presence of molten LiI, NaI, KI, RbI and CsI salts as flux materials in sealed vacuum quartz ampoules at 740 oC. SEM and EDX studies showed that the morphology and composition of the formed crystals are influenced by the nature of the flux materials. Structural studies by XRD revealed a shift of all diffraction peaks towards lower angles for CZTS crystals grown in LiI and a larger lattice parameter values in comparison with powder crystals formed in CsI, RbI, NaI and KI. CZTS MGPs grown in LiI also showed the widest main Raman peak (FWHM=7.06 cm-1). In case of CsI, Raman peaks were sharper and narrower (FWHM=4.5 cm-1) compared to the other produced powders, showing a higher level of crystallinity. The estimated effective bandgap energy values from EQE measurements were ~1.57 eV and 1.66 eV for CZTS MGPs grown in NaI, KI, RbI, CsI and in LiI, respectively. In addition, detailed analysis of photoluminescence spectroscopy and temperature dependent solar cell parameters will be discussed. The highest efficiency of 10.88% was achieved with MGL solar cell based on CZTS grown in CsI.

978-1-6654-6060-6/23 $31.00 © 2023 IEEE

Pyrolyzer Assisted Vapor Transport Deposition of Antimony-doped Cadmium Telluride

Bin Du[1], Kevin Dobson[1], Brian McCandless[1], Aayush Nahar[2], Ujjwal Das[1], Shannon Fields[1], Aaron Arehart[2], William Shafarman[1]

[1]Institution of Energy Conversion, Newark, DE, 19716, USA

[2]The Ohio State University, Columbus, OH, 43210, USA

Abstract—A new method for in-situ Sb doping of CdTe that uses a modified vapor transport deposition system is described. This modification enables control of the Sb concentration with a pyrolysis stage to enhance the doping efficiency. CdTe:Sb films under different deposition conditions are characterized by SEM, XRD, and CV measurements for determining morphology, crystal structure, and hole concentration. Variations of the Sb dopant heater and pyrolyzer temperatures do not affect the CdTe morphology and crystal structure. However, CV measurements show that a higher dopant heater or pyrolyzer temperature leads to higher hole concentration. In this study, CdTe: Sb films achieve a hole concentration of 10^{16} cm^{-3} and 10% doping efficiency when the dopant heater is 600C and the pyrolyzer temperature is 1100C. This demonstrates a path to produce high hole concentration polycrystalline CdTe film with a low concentration of dopant-induced defect.

I. INTRODUCTION

In recent years, impressive cell efficiency improvements have been achieved by replacing the CdS front layer with the CdSe$_x$Te$_{1-x}$ (CST) alloy. This progress mostly comes from enhancing carrier lifetime and maximizing short circuit current (J_{sc}) [1]. State-of-the-art CdTe solar cells' J_{sc} can reach up to 30 mA cm^{-2} with a fill factor (FF) near 80%, which is close to its maximum value based on the Shockley-Queisser limit [2-3]. Despite this achievement, CdTe devices are still far from ideal performance, primarily due to low open-circuit voltage (V_{oc}). One pathway to overcoming CdTe's low V_{oc} problem is increasing the hole concentration while maintaining the carrier lifetime. Recent modeling and doped CdTe single crystal studies demonstrate that a high hole density and carrier lifetime can improve V_{oc} to greater than 1V with cell efficiency up to 25% [4-6].

Group V atoms have been explored as the substitutional p-type dopant in CdTe solar devices. From previous literature [7,8], As is commonly used as the dopant for CdTe solar cells since the earlier published values indicated that As had a much shallower acceptor transition energy compared to the Sb dopant [9]. However, new calculations show that As, P, and Sb have comparable transition levels (~100 meV) [10]. This is consistent with recent experimental work, which indicated that the accepter transition energy of Sb$_{Te}$ is shallower (100 meV) compared to the earlier value from the first principle calculation (230 meV) and that Sb doping is more stable than As and P

dopants [11,12]. Finally, the Sb atomic radius matches Te better than As (R_{Sb} = 140 pm, R_{As} = 120 pm, R_{Te} = 136 pm), which may reduce the dopant clustering and give higher doping efficiency [12].

The doping efficiency (absorber hole density divided by dopant concentration) is always an issue for group V-doped CdTe polycrystalline films. For ex-situ and in-situ doping studies, doping efficiency is generally 0.1 to 1 %, leaving a significant amount of dopant that is not incorporated at the Te site, which may form clusters, induce unknown defects, and limit the improvement of device performance from a higher hole density [7,8,11]. In this study, we demonstrate a new in-situ doping method that increases the hole concentration > 10^{16}cm^{-3} by Sb doping, and the doping efficiency is enhanced by an in-situ pyrolyzer doping source.

II. EXPERIMENTAL

A modified vapor transport deposition (VTD) system has been built for this study with a source configuration shown schematically in Fig. 1. This system incorporates a novel two-stage dopant source with a separate dopant heater and pyrolyzer.

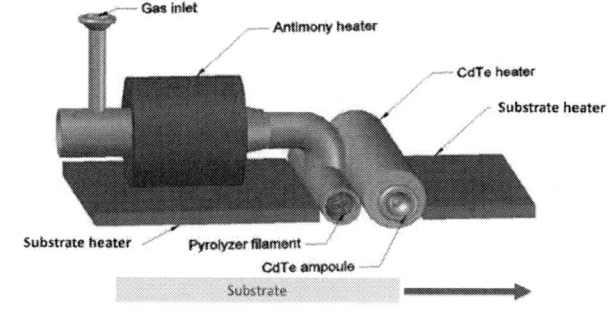

Figure 1. Schematic of dopant heater, pyrolyzer, and CdTe source in IEC's vapor transport deposition system.

The dopant heater allows independent control of the dopant flow rate during the VT deposition enabling precise control of the Sb concentration at any desired level. Inside the dopant heater is a graphite source container with two source beds that have different surface areas. The large source bed contains pure

978-1-6654-6060-6/23 $31.00 © 2023 IEEE

Sb and provides saturated Sb vapor during the deposition. The small source bed contains a pure Cd source and provides under-saturated Cd vapor to create a Cd-rich deposition condition. The pyrolyzer is designed to crack the Sb_x (Sb_4, Sb_2) species into monatomic Sb species. Compared to the Sb_x species, monoatomic Sb requires less energy for the formation of CdSb to facilitate Sb incorporation at the Te site and improve the doping efficiency.

An equilibrium calculation has been done to identify the pyrolyzer working temperature range at the target carrier concentration. The Van't Hoff equation is implemented for this calculation, using published values of $\Delta H°$ and $\Delta S°$ in this temperature range [13]. The equilibrium thermal dissociation equation and constant are given by:

$$Sb_4(g) \leftrightarrow 4Sb(g)$$

$$\ln(K_p) = \frac{-\Delta H°}{RT} + \frac{\Delta S°}{R}$$

$$K_p = \frac{P_{sb}{}^4}{P_{sb4}} = \frac{256\alpha^4 p^3}{1-\alpha}$$

$$\Delta H = 211.6 \text{Kcal mol}^{-1} \quad \Delta S = 85.5 \text{ kcal mol}^{-1} \text{ K}^{-1}$$

where K_p is the equilibrium constant, $\Delta H°$ is the change in enthalpy, $\Delta S°$ is the change in entropy, R is the gas constant, p is the initial Sb_4 vapor pressure and α is the conversion ratio.

This study focuses on the carrier concentration from 10^{15} to 10^{17} cm^{-3} corresponding to the Sb vapor pressure from 10^{-5} to 10^{-3} Torr. Based on the equilibrium calculation, the monoatomic Sb conversion ratio depends on the pyrolyzer temperature and target hole concentration as shown in Fig. 2.

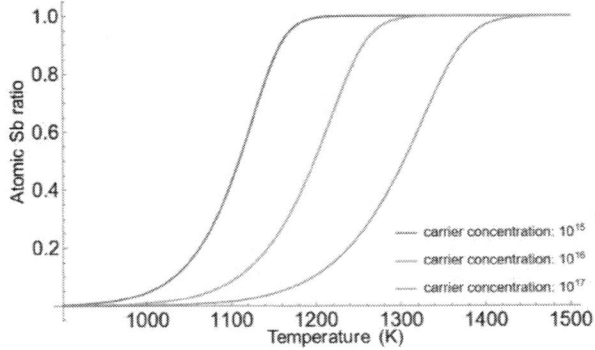

Figure 2. Monoatomic Sb ratio depends on the cracking temperature and target carrier concentration.

A glass/FTO/CdSe/CdTe:Sb structure is adopted, Using a 110 nm CdSe film and CdTe capping layer deposited onto a clean 4x4 inch Tec_12D substrate by thermal evaporation at 250C substrate temperature with a 10 nm/min deposition rate. The sample is transferred to the VTD system, and a 4 µm CdTe:Sb film is deposited at 500C substrate temperature. As-deposited CdTe:Sb samples are characterized by SEM, cross-section SEM, and X-ray diffraction measurements (XRD) to identify the effect of different Sb concentrations on the film morphology and crystal structure.

A CdCl$_2$ vapor treatment (435C, 12 min) followed by CuCl$_2$ annealing (250C, 30 min) and carbon paste contact are used to complete solar cell devices. The carrier concentrations of CdTe:Sb under different deposition conditions are determined from room temperature capacitance-voltage (CV) measurements. A measurement frequency of 100kHz was chosen based on capacitance-frequency analysis to minimize the effects of conductance, series resistance, and cable inductance.

III. Result

The dopant source is angled to provide overlap between Sb and CdTe flux during deposition. Before initiating the CdTe:Sb VT deposition, a 30 min stationary run was used to verify the Sb delivery and determine the location of Sb vapor reaching the substrate. A 400 nm Cu film on a 4x4 inch glass was used as the substrate. The Cu film will react with Sb vapor and form Cu$_2$Sb, preventing the Sb from re-evaporating. During the stationary run, the substrate stayed beneath the source heater at a temperature of 550C. The CdTe source heater was kept at 870C without any CdTe to mimic VT deposition conditions and prevent the Sb condensation. The Sb dopant was heated to 600C with a 3 sccm flow rate, generating enough Sb species to facilitate XRD and energy dispersive x-ray spectroscopy (EDS) measurement.

The stationary run sample's XRD pattern showed the predicted Cu$_2$Sb crystalline peaks, and EDS confirmed the Sb composition. Multiple EDS measurements were performed to determine the dopant distribution at the substrate. Fig.3 shows that the Sb was concentrated in a 1-inch-wide deposition zone underneath the CdTe source heater that enables the Sb and CdTe vapor to mix before reaching the substrate and on the surface during the Sb-doped CdTe VT deposition.

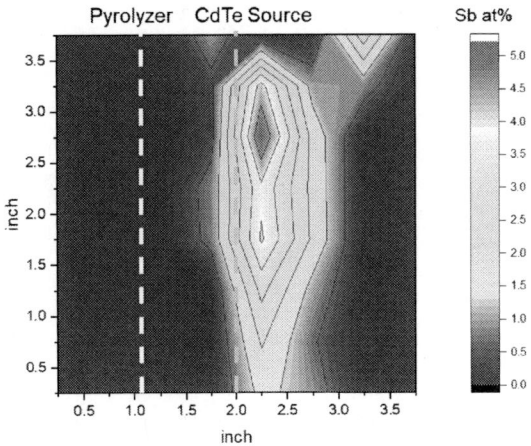

Figure 3. Sb atomic percentage (at%) distribution across 4X4 inch substrate. The white and purple dot lines indicate the position of the pyrolyzer and CdTe source heaters.

CdTe: Sb samples were deposited under the different dopant heater and pyrolyzer temperature combinations shown in Table I. The dopant heater temperature (T_D) varied from 510 to 600C, and the pyrolyzer temperature (T_P) ranged from 800 to 1100C. Table 1 also shows the Sb concentration ([Sb]^{calc}), and monoatomic Sb percentage calculate from a mass transfer model and equilibrium calculation on the different T_D and T_P combinations.

Table I. Dopant heater conditions and calculated Sb mole flow rate (Q_{Sb}), Sb concentration, and percentage of monoatomic Sb.

Combination #	1	2	3	4	5
T_D (C)	510	510	510	600	430
Q_{Sb} (mole/s)	1x10^{-9}	1x10^{-9}	1x10^{-9}	5x10^{-9}	2x10^{-10}
[Sb]^{calc} (cm^{-3})	1.5x10^{18}	1.5x10^{18}	1.5x10^{18}	7.5x10^{18}	3x10^{17}
T_P (C)	805	957	1100	1100	1100
Monoatomic Sb	10%	50%	100%	100%	100%

All the as-deposited CdTe:Sb films show a single-phase zincblende crystal structure with preferred (111) crystal orientation as in Fig. 4. SEM measurements show that the CdTe:Sb samples are densely packed, homogeneous, and free from pinholes or voids with an average grain size of 1.2 – 1.5 μm. No significant difference is observed by XRD and SEM measurements for the different T_D and T_P combinations.

Fig.5 shows the CV results for the CdTe: Sb devices with different T_D and T_P combinations. The net carrier concentration N_A-N_D is increased from 10^{15} cm^{-3} to 10^{16} cm^{-3} by increasing T_D from 430C to 600C with fixed T_P = 1100C. When keeping T_D = 510C and varying T_P from 800C to 1100C, the CV measurements show N_A-N_D increases from 1.5 x 10^{15} to 3.5 x 10^{15}cm^{-3}. The CV measurements demonstrate that the dopant heater can effectively control the carrier concentration of CdTe: Sb device. The pyrolyzer can enhance the carrier concentration by a factor of 2, indicating that a better doping efficiency can be achieved with a higher pyrolyzer temperature.

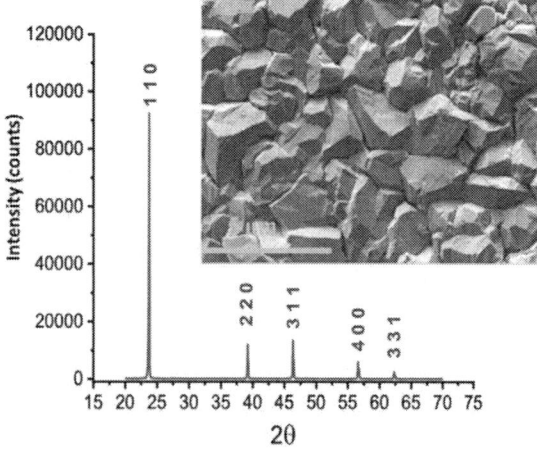

Figure 4. XRD and SEM data of as-deposited CdTe: Sb.

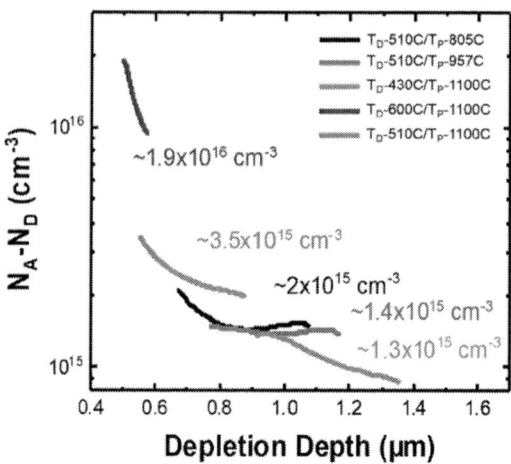

Figure 5. Room temperature CV measurement for CdTe: Sb samples under different dopant heater and pyrolyzer temperatures.

Fig.6 shows the SIMS results for CdTe:Sb device with different T_D and constant T_P = 1100C. The Te raw counts identify the interface between the CdTe absorber and the TCO layer. All three samples show relatively constant Sb concentrations through the bulk of the absorber. A one-order higher Sb concentration is measured at the first ~200 nm front and back interface, likely due to the matrix effects. Inside the absorber bulk, a uniform Sb concentration at 10^{17} cm^{-3} is measured for both samples, and the T_D setpoint does not affect the Sb concentration.

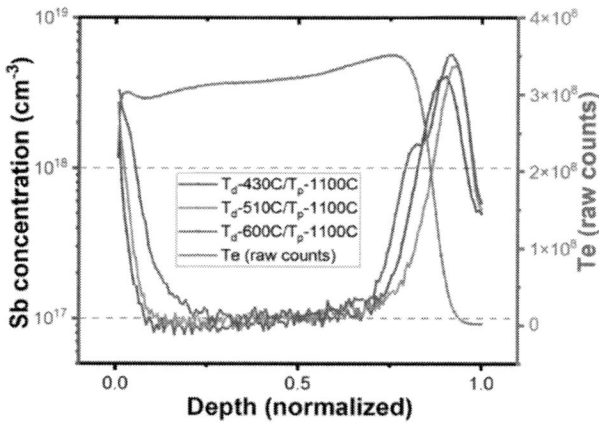

Figure 6. SIMS measurement for Sb concentration with different Dopant heater and consistent pyrolyzer temperature. The detection limit for Sb is 10^{14} cm^{-3}.

The SIMS result does not follow the same trend as the CV result for samples under different T_D, indicating that the dopant heater temperature does not control the Sb flow rate during the VT deposition even though the CV results show that a higher dopant heater temperature increases the carrier concentration. Our hypothesis to explain this is that during the VT deposition, the amount of Sb reaching the substrate is limited by condensation between the dopant heater and the pyrolyzer. From Fig.1, we can see that an L-shaped tube connects the dopant heater and pyrolyzer. Our current design has no additional heater to heat up this region, relying on radiant heat from other components to heat it up and prevent condensation. However, this does not stay true as we observe dopant vapor condensing on the tube indicating that the tube has a lower temperature than the dopant heater. During the CdTe:Sb VT deposition, the amount of Sb delivered to the substrate is controlled by the temperature of the L-shape tube instead of the dopant heater temperature setpoint.

IV. CONCLUSION & DISCUSSION

In this work, CdTe:Sb films were deposited by vapor transport deposition with in-situ Sb-doping under different dopant and pyrolyzer temperatures. All the samples under different conditions show an average 1.2 -1.5 um grain size with densely packed, pinhole-free morphology and a single-phase zincblende structure with random crystal orientation. No significant difference is shown in the films' morphology and crystal structure. CV measurements show a correlation between deposition conditions and net carrier concentration. A higher dopant or pyrolyzer temperature leads to a higher carrier concentration. The highest doped CdTe:Sb sample achieves a carrier concentration at 10^{16} cm^{-3} with 10% doping efficiency when T_D = 600C and T_P = 1100C.

In contrast, the SIMS measurements show a constant Sb concentration under different dopant heater temperatures. The carrier concentration enhancement by the higher dopant heater temperature may come from the higher excess Cd delivered to the substrate during the deposition since the Cd vapor from the dopant heater is unsaturated and will not be limited by the low-temperature L-shaped tube.

We successfully demonstrated a new in-situ doping method for VTD CdTe solar devices. By introducing a dopant heater with a pyrolyzer component, we control the carrier concentration during the CdTe VT deposition and demonstrate pyrolyzer enhanced doping efficiency. This method can simplify the process of doping CdTe solar cell fabrication since this doping method does not require a pre-made Sb-doped CdTe source or any additional step to activate the dopant.

We also note the limitation of our current dopant heater design, in which Sb dopant delivery is limited by the low-temperature L-shaped tube. An updated version of the dopant heater is under design in include a two-zone dopant heater with an additional heater for the full tube. The new dopant heater will give us the ability to decouple the Cd and Sb vapor pressure and study the effect of excess Cd on the CdTe: Sb device.

ACKNOWLEDGEMENT

This material is based upon work supported by the U.S. Department of Energy's Office of Energy Efficiency and Renewable Energy (EERE) under the Solar Energy Technologies Office Award Number DE-EE0009344.

REFERENCES

[1] X. Zheng, D. Kuciauskas, J. Moseley, E. Colegrove, D. S. Albin, Ferguson, and W. K. Metzger, "Recombination and bandgap engineering in CdSeTe/CdTe solar cells", *APL Materials* 7, 071112, 2019.

[2] R. M. Geisthardt, M. Topič, and J. R. Sites, "Status and Potential of CdTe Solar-Cell Efficiency", IEEE *Journal of Photovoltaics*, 2015.

[3] First Solar Press Release, First Solar Builds the Highest Efficiency Thin Film PV Cell on Record, 5 August 2014.

[4] A. Kanevce, M. O. Reese, T. M. Barnes, S. A. Jensen, and W. K. Metzger, "The roles of carrier concentration and interface, bulk, and grain-boundary recombination for 25% efficient CdTe solar cells", *Journal of Applied Physics 121*, 214506, 2017.

[5] J. Burst, J. Duenow, D. Albin, E. Colegrove, M. O. Reese, J. A. Aguiar, C. S. Jiang, M. K. Patel, M. M. Al-Jassim, D. Kuciauskas, S. Swain, T. Ablekim, K. G. Lynn, and W. K. Metzger, "CdTe solar cells with open-circuit voltage breaking the 1 V barrier", *Nature Energy* 1, 16015, 2016.

[6] Y. Zhao, M Boccard, S. Liu, J. Becker, X. Zhao, C. Campbell, E. Suarez, M. Lassise, Z. Holman, and Y. Zhang, "Monocrystalline CdTe solar cells with open-circuit voltage over 1 V and efficiency of 17%", *Nature Energy* 1. 16067, 2016.

[7] W. K. Metzger, S. Grover, D. Lu, E. Colegrove, J. Moseley, C. L. Perkins, X. Li, R. Mallick, W. Zhang, R. Malik, J. Kephart, C.-S. Jiang, D. Kuciauskas, D. S. Albin, M. M. Al-Jassim, G. Xiong, and M. Gloeckler, "Exceeding 20% efficiency with in situ group V doping in polycrystalline CdTe solar cells", *Nature Energy* 4, 837–845, 2019.

[8] D. B. Li, C. Yao, S. N. Vijayaraghavan, R. A. Awni, K. K. Subedi, R. J. Ellingson, L. Li, Y. Yan, and F. Yan, "Low-temperature and effective ex situ group V doping for efficient polycrystalline CdSeTe solar cells", *Nature Energy* 6, 715–722, 2021.

[9] S. H. Wei, and S. Zhang, "First-Principles Study of Doping Limits of CdTe", phys. stat. sol, (b), 229: 305-310. 2002.

[10] I. Chatratin, B. Dou, S. Wei, and A. Janotti, "Doping Limits of Phosphorus, Arsenic, and Antimony in CdTe", *The Journal of Physical Chemistry Letters*, 14, 273-278, 2023.

[11] A. Nagaoka, K. Nishioka, K. Yoshino, R. Katsube, Y. Nose, T. Masuda, and M. A. Scarpulla, "Comparison of Sb, As, and P doping in Cd-rich CdTe single crystals: Doping properties, persistent photoconductivity, and long-term stability", *Applied Physics Letters*, 116 (13): 132102, 2020.

[12] B. E. McCandless, W. A. Buchanan, C. P. Thompson, G. Sriramagiri, R. J. Lovelett, J. Duenow, D. Albin, S. Jensen, E. Colegrove, J. Moseley, H. Moutinho, S. Harvey, M. Al-Jassim, and W. K. Metzger, "Overcoming Carrier Concentration Limits in Polycrystalline CdTe Thin Films with In Situ Doping", *Sci Rep* 8, 14519, 2018.

[13] J. Kordis, K. A. Gingerich, "Mass spectroscopic investigation of the equilibrium dissociation of gaseous Sb_2, Sb_3, Sb_4, SbP, SbP_3, and P_2", *J. Chem. Phys.* 1973

Early degradation trend estimation of bifacial PV, investigating the seasonality effect

Gaetano Mannino[1], Giuseppe Marco Tina[1], Mario Cacciato[1], Lorenzo Todaro[2], Agnese Di Stefano[2], Fabrizio Bizzarri[2], Andrea Canino[2]

[1]DIEEI, University of Catania, 95124 Catania, Italy
[2]Enel Green Power SpA, Viale Regina Margherita, 125, 00198, Rome, Italy 2022

Abstract — It is well known that photovoltaic modules have a non-linear degradation trend which depends on the technology of the modules, on the manufacturer and on the place of installation. It is possible to use specific tools suitable for measuring the I-V curves and apply standards such as IEC 60891: 2021 to evaluate the degradation. However, in the absence of specific tools it is possible, even with greater uncertainty of the results, to estimate the degradation rate of the PV modules. One of the indices that can be used, is the Performance Loss Ratio (PLR). This index, through filtering and data processing procedures, allows to evaluate the loss of performance including both reversible and irreversible effects.

This study deals with the evaluation of the loss of performance considering the seasonality effect, that is here described using Fourier series. Filtered data from a string of PERC monocrystalline silicon bifacial modules are used for the work.

The usefulness of the method is to evaluate the seasonality effect on the early-stage degradation trend using different metrics, computing a dataset of electrical and environmental variables and without the aid of external tools.

I. INTRODUCTION

In the first period of operations, it is difficult to extrapolate the degradation trend of the PV modules, as stated in [1] the suggested minimum period of analysis should be two years, as in the case study analyzed.

Evaluating the degradation of photovoltaic modules starting from a dataset is useful if specific I-V measurement devices are not installed. In [2] an unsupervised machine learning approach for analyzing the time-series data applied to degradation estimation is presented, it requires just power signal as input.

Temperature and Rear irradiance are not always monitored but it is possible to estimate them through some models. In the case study, both module temperature and rear irradiance were monitored. In [3] the Seasonal ARIMA method is applied to develop a PR based degradation rate forecast method. A 5-year dataset is used. The results of the monthly PR forecasted for the next 3 years are compared with the values actually recorded.

In [4] the PLR is evaluated through the "5-setp method", where a linear fitting method is used for the PLR estimate, because of the brief time period analyzed.

In this study an attempt to identify seasonality effects in the trend of the metrics has been done.

Understanding the seasonality of the metric can help to:
- predict the approximate trend of the metrics such as PR during the months (therefore a greater awareness of the yield, for periods of the year characterized by similar climatic conditions, above all in terms of temperature).
- By combining a PLR estimation technique, once the trend has been extrapolated, by applying the seasonal variation of the identified metric, it is possible to know how the metric will vary over time. In the case study, the time unit on which the metrics are calculated are months.

In this study different models are tested in order to detect the combined effect of degradation trend and seasonality effect of three metrics for 2 years.

II. DATASET

Figure 1 Monitored Bifacial PV system

The monitored array of the PV system is made of 7 bifacial mono-PERC PV modules. The preprocessed data used in the study are: Front irradiance and Rear Irradiance, DC Current and Voltage and module temperature. Each quantity with 1-min of resolution is the metrics input and is the output of the filtering process applied in [4].

III. METHOD

To evaluate the degradation of PV modules it is possible to use different metrics: evaluation of I-V curves, empirical metrics or normalized and corrected metrics, in the present study three normalized metrics are adopted.

A. Metrics

Starting from a dataset, after the cleaning procedure and the irradiance range filter selection, the following metrics are applied, then the PLR is estimated.

$$PR = \frac{\frac{P_{dc}}{P_{stc}}}{\frac{G_{f,POA}}{G_{stc}}} \quad (1)$$

Where P_{dc} is the measured DC Power [W] and P_{stc} is the Power at STC [W], $G_{f,POA}$ is the front plane of array Irradiance [W/m^2] and G_{stc} is the Irradiance at STC (1000 W/m^2)

$$PR_T = \frac{\frac{P_{dc}}{P_{stc}[1+\gamma(T_c-T_{stc})]}}{\frac{G_{f,POA}}{G_{stc}}} \quad (2)$$

Where γ is the thermal coefficient for Power, T_c is the PV cell temperature [°C] and T_{stc} is Temperature at STC (25 °C)

$$PR_{Tb} = \frac{\frac{P_{dc}}{P_{stc}[1+\gamma(T_c-T_{stc})]}}{\frac{G_{f,POA}+\beta G_{r,POA}}{G_{stc}}} \quad (3)$$

where $G_{r,POA}$ is the rear plane of array Irradiance [W/m^2] and β is the bifaciality factor.

B. PLR statistical methods

As reported in [5], it is possible to apply different statistical methods to the calculated metrics in order to evaluate the PLR, among the best known are: Simple Linear Regression (SLR), Classical Seasonal Decomposition (CSD), HW Seasonal model, Autoregressive Integrated Moving Average (ARIMA) and Seasonal–Trend Decomposition Using LOESS.
In [4] The degradation rate values of bifacial modules are calculated using a linear fitting model applied to the metrics (1), (2) and (3). Subsequently, using the classic PR, the effect of seasonality is considered in the present study.
Classical Decomposition [6] separates the signal into three terms:

$$\text{Signal} = \text{Trend} + \text{Seasonality} + \text{Irregular}.$$

Using the PR metric there is a strong seasonality effect, it can be represented by a Fourier series:

ABOUT FOURIER SERIES MODELS

The Fourier series is a sum of sine and cosine functions that describes a periodic signal. It is represented in either the trigonometric form or the exponential form. [7] A generic trigonometric Fourier series form is:

$$y = a_0 + \sum_{i=1}^{n} a_i \cos(iwx) + \sum_{i=1}^{n} b_i \sin(iwx) \quad (4)$$

where a_0 models a constant (intercept) term in the data and is associated with the $i = 0$ cosine term, w is the fundamental frequency of the signal, n is the number of terms (harmonics) in the series, and $1 \leq n \leq 8$.

C. Equations

The following Fourier based models to interpret seasonality have been tested, finding the optimal coefficients for each one:

Simple Fourier:

$$PR(x) = a_0 + a_1 \cos((x \cdot w)) + b_1 \sin((x \cdot w)) \quad (5)$$

Fourier twice:

$$PR(x) = a_0 + a_1 \cos((x \cdot w)) + b_1 \sin((x \cdot w)) + a_2 \cos((2x \cdot w)) + b_2 \sin((2x \cdot w)) \quad (6)$$

Jordan/Fourier [6]

$$PR(x) = a_0 + a_1 \sin\left(\left(\frac{\pi}{6}\right) \cdot (x-w)\right) + b_1 \sin\left(\left(\frac{\pi}{3}\right) \cdot (x-w)\right) \quad (7)$$

Linear deg and Fourier:

$$PR(x) = a_0 + cx + a_1 \cos((x \cdot w)) + b_1 \sin((x \cdot w)) \quad (8)$$

Linear deg and Jordan/Fourier:

$$PR(x) = a_0 + cx + a_1 \sin\left(\left(\frac{\pi}{6}\right) \cdot (x-w)\right) + b_1 \sin\left(\left(\frac{\pi}{3}\right) \cdot (x-w)\right) \quad (9)$$

III. RESULTS AND DISCUSSION

TABLE I

Metric and fit	Data	R2	RMSE	Coef
Avg PR, linear	avgPR vs month	0.13706	0.023019	2
Avg PRt, linear	avgPRt vs month	0.2264	0.021974	2
Avg PRtb, linear	avgPRtb vs month	0.19944	0.019415	2
Avg PRtb, spline*	avgPRtb vs month	0.83342	0.013878	15.04
Avg PR, Eq (9)	avgPR vs month	0.83248	0.010913	5
Avg PR, Eq (8)	avgPR vs month	0.78086	0.012482	5
Avg PRt , Eq (9)	avgPRt vs month	0.32611	0.022069	5
Avg PRtb, Eq (9)	avgPRtb vs month	0.29734	0.019573	5
Avg PR, Eq (5)	avgPR vs month	0.63754	0.015647	4
Avg PR, Eq (6)	avgPR vs month	0.77855	0.012892	6
Avg PR, Eq (7)	avgPR vs month	0.6756	0.014802	4

*The spline was also included because it represents one of the best approximation functions, despite the extrapolated model is not usable due to the high "number of coefficients".

For the identification of the degradation trend, a linear fitting model was applied as in [4] because of the short monitoring period. For the identification of seasonality, the Fourier series in different forms have been applied. Among the methods that have been applied to identify the combined effect of degradation and seasonality, eq (9) was the most performing. The coefficients of the models that are presented in Table II have been obtained with the MATLAB curve fitting toolbox.

978-1-6654-6060-6/23 $31.00 © 2023 IEEE

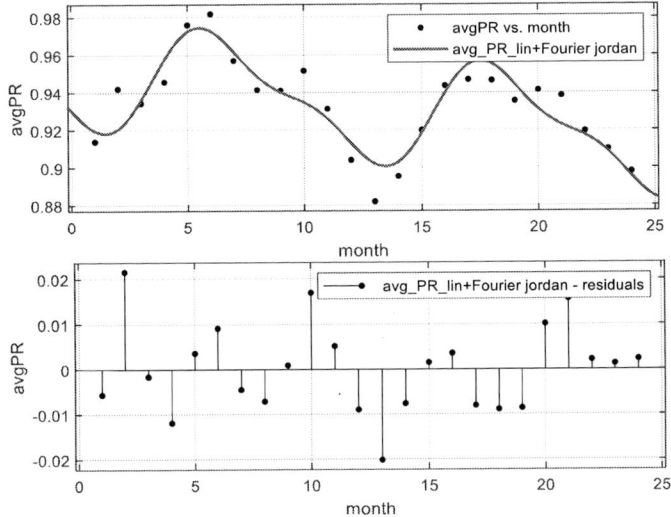

Fig. 1. Linear degradation rate combined with seasonality Jordan/Fouries (Eq 9)

coef	metric	Fourier based models coefficients				
		Eq (5)	Eq (6)	Eq (7)	Eq (8)	Eq (9)
a0	PR	0.9319	0.9313	0.9331	0.9495	0.9513
a1	PR	-0.02734	-0.0001884	0.02646	-0.02765	-0.02638
w	PR	0.5006	0.2458	3.198	0.498	-2.527
b1	PR	0.002686	0.01232	0.007814	-0.001988	0.009529
a2	PR	x	-0.02729	x	x	x
b2	PR	x	0.005654	x	x	x
c	PR	x	x	x	0.001421	-0.001457

IV. CONCLUSION

The present study starts from the application of the "5-STEP" method for the evaluation of degradation of bifacial modules.

The effect of seasonality on the value of metrics during two years of acquisition has been investigated. It has been verified that the effect of seasonality is marked on the PR metric, so it was chosen as the metric on which to investigate the effect of seasonality through Fourier series method, it was then combined with a linear term to obtain a linear decreasing and seasonality compound model.

Using the Fourier series the amplitudes of the harmonics are constant, working with more data it is possible to verify also if the amplitude of the harmonic decreases or increases over time, helping to quantify the change of seasonality over time, during the degradation process.

By multiplying the amplitude with decreasing or increasing exponential Fourier seasonality amplitude could be adjusted. With a longer acquisition time period it will be possible extract the degradation trend using methods more accurate model than linear fitting.

In addition to the reversible seasonality due to the effect of external factors causing reversible PLR, also a seasonality of the degradation is expected, even if less visible; since the environmental variables that stress the modules, such as the temperature and the amplitude of the thermal cycles undergone by the modules, are more marked in some months.

It was not possible to effectively identify the seasonality effect through Fourier series using the PRt and PRtb metrics, this is because the correction of the metric with respect to Temperature and Irradiance significantly reduces the seasonality effect. Using classical PR, Fourier series were able to describe the seasonality of the metric.

References

[1] S. Lindig, M. Theristis, and D. Moser, "Best practices for photovoltaic performance loss rate calculations," *Progress in Energy*, vol. 4, no. 2, 2022, doi: 10.1088/2516-1083/ac655f.

[2] B. Meyers, M. Deceglie, C. Deline, and D. Jordan, "Signal Processing on PV Time-Series Data: Robust Degradation Analysis without Physical Models," *IEEE J Photovolt*, vol. 10, no. 2, pp. 546–553, 2020, doi: 10.1109/JPHOTOV.2019.2957646.

[3] G. Mannino, G. M. Tina, M. Cacciato, L. Todaro, F. Bizzarri, and A. Canino, "A photovoltaic degradation evaluation method applied to bifacial modules," *Solar Energy*, vol. 251, no. December 2022, pp. 39–50, 2023, doi: 10.1016/j.solener.2022.12.048.

[4] S. Lindig, I. Kaaya, K. A. Weis, D. Moser, and M. Topic, "Review of statistical and analytical degradation models for photovoltaic modules and systems as well as related improvements," *IEEE J Photovolt*, vol. 8, no. 6, pp. 1773–1786, 2018, doi: 10.1109/JPHOTOV.2018.2870532.

[5] D. Jordan and C. Gotwalt, "JMP Applications in Photovoltaic Reliability (Presentation)," *JMP Discovery Summit*, 2011, [Online]. Available: http://www.osti.gov/energycitations/servlets/purl/1027684/%0Ahttp://www.nrel.gov/docs/fy12osti/51126.pdf

[6] E. Pieri, A. Kyprianou, A. Phinikarides, G. Makrides, and G. E. Georghiou, "Forecasting degradation rates of different photovoltaic systems using robust principal component analysis and ARIMA," *IET Renewable Power Generation*, vol. 11, no. 10, pp. 1245–1252, 2017, doi: 10.1049/iet-rpg.2017.0090.

[7] A. Kyprianou, A. Phinikarides, G. Makrides, and G. E. Georghiou, "Definition and Computation of the Degradation Rates of Photovoltaic Systems of Different Technologies with Robust Principal Component Analysis," *IEEE J Photovolt*, vol. 5, no. 6, pp. 1698–1705, 2015, doi: 10.1109/JPHOTOV.2015.2478065

978-1-6654-6060-6/23 $31.00 © 2023 IEEE

The rise and the decay of the photovoltage in perovskite solar cells

Uwe Rau[1,2], Lisa Krückemeier[1,2], Sandheep Ravishankar[1], Thomas Kirchartz[1,3]

[1]IEK-5 Photovoltaik, Forschungszentrum Jülich GmbH, Wilhelm-Johnen Straße, 52425 Jülich, Germany
[2]Jülich Aachen Research Alliance (JARA-Energy) and Faculty of Electrical Engineering and Information Technology, RWTH Aachen University, Schinkelstr. 2, 52062 Aachen, Germany
[3]Faculty of Engineering and CENIDE, University of Duisburg-Essen, Carl-Benz-Str. 199, 47057 Duisburg, Germany

Abstract— This contribution discusses experimental results obtained from a variety of transient optoelectronic techniques like transient photoluminescence (TRPL), transient photovoltage (TRPV), impedance spectroscopy (IS), and intensity modulated photovoltage (IMPV) applied to metal-halide materials and solar cells. While for the analysis of bare absorber films TRPL provides clear insight into the recombination kinetics of photogenerated charge carriers, this task gets more complex if the method is applied to layer stacks, including one or more contact layers. For finished devices it turns out that the amount of charge carriers extracted to the contact layers is, and should, be larger than the number of carriers that remains in the absorber. Therefore, a description of TRPL as well as of the other transient methods that can be applied to completed solar cells, needs to take into account a second independent variable, namely the number of charge carriers accumulated on the contacts of the solar cell. The resulting generic two-component model describes the experimental results, especially the fact that TRPV data exhibit an initial rise of the photovoltage followed by a decay.

Keywords—*perovskite solar cells, transient photoluminescence, transient photovoltage, charge carrier separation*

I. INTRODUCTION

Transient opto-electronic measurements are an important tool for the characterization of photovoltaic absorber materials and completed solar cells. Because of the high radiative efficiency of metal halide perovskites (MHP), these methods are especially convenient and popular for these materials. Transient photoluminescence (TRPL) has been successfully applied to perovskite absorbers to disentangle the different recombination mechanisms like Auger recombination, radiative recombination, and Shockley-Read-Hall recombination [1,2]. The interpretation of TRPL measurements for the combination of absorber films with contact layers or for completed solar cells is complicated by the fact that charge carriers accumulate at the interfaces and/or are extracted to those additional layers [3,4], as illustrated in Fig.1. Especially for completed cells the charging of the contact layers is an important, generic effect that is closely related to the overall functionality of the solar cell, namely the generation of the photovoltage. In addition, these contact effects are also present in the results from transient photovoltage (TRPV), impedance spectroscopy (IS), and intensity modulated photovoltage (IMPV) [5].

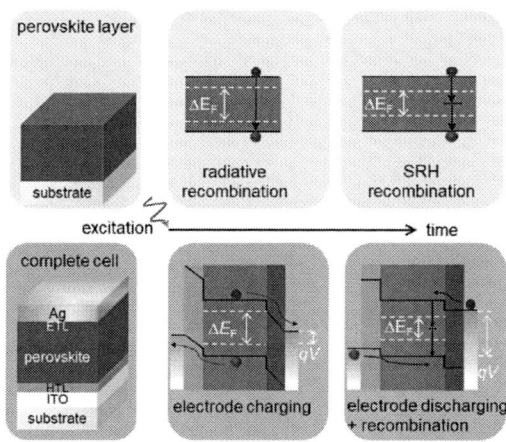

Fig. 1. *Charge carrier kinetics following a pulsed light excitation of a bare perovskite absorber film (upper row) and of a completed perovskite solar cell (lower row). For the absorber film, the recombination kinetics typically is described in dependence of only a single variable, namely the split ΔE_F of the quasi-Fermi levels. Different recombination mechanisms, like radiative recombination or Shockley-Read-Hall (SRH) recombination are dominant at different carrier concentrations. In a completed solar cell, charge carriers can be also extracted to the contact layers where they charge the electrodes and build up an external voltage V. As soon as the voltage exceeds $\Delta E_F/q$ these carriers are reinjected into the absorber where they subsequently recombine. This kinetics requires a description by a model that accounts for V and ΔE_F as two independent variables.*

The present contribution discusses experimental results for MHP absorbers and solar cells using a variety of time resolved (large signal and small signal) methods as well as frequency resolved methods. These results are discussed in terms of a novel, generic two-component model containing the external voltage V and the split of the quasi-Femi levels ΔE_F as two independent variables representing an electrostatic and a chemical potential respectively. With this, the model represents the fundamental photovoltaic action of a solar cell, namely the transformation of the initial excess carrier density in the absorber in the cell into an electrostatic potential build up by

978-1-6654-6060-6/23 $31.00 © 2023 IEEE

charges in the contacts of the cell. Two independent variables are also mathematically needed to describe the title action of the present paper: The rise and the decay of the photovoltage in solar cells upon excitation with a light pulse.

II. RESULTS AND DISCUSSION

Figure (2) shows large signal TRPL transients obtained from a bare MPH film compared to the transient from a completed MPH solar cell [3]. The interpretation of such data rises the immediate question: Should we take the faster TRPL decay of the cell as compared to that of the film as a bad sign as we possibly have introduced enhanced recombination at the interface between contact layers and the absorber? Or should we take it as a good sign because the free charge carriers are withdrawn swiftly from the absorber of the cell into the contacts? In fact, both answers are physically possible and a meaningful answer requires a model that is designed to describe recombination and carrier collection/injection quantitatively.

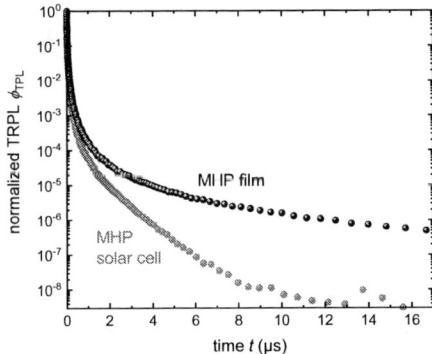

Fig. 2. Normalized TRPL signal of a bare MHP film (black) in comparison to a signal obtained from a completed MHP solar cell. The data are taken from Ref. [3].

The lower row in Fig. 1 sketches the physics that must be described by such a model: In excess to recombination free carriers can also be removed from the absorber by extraction to the contacts. Thus, we have for the chemical potential $\mu = \Delta E_F$ of the free carriers the equation

$$\frac{d}{dt}\frac{\mu}{q} = \frac{1}{C_\mu(\mu)}\left[J_{gen} - J_{rec}(\mu) - J_{exc}(\mu, V) \right] \quad (1)$$

that not only contains generation and recombination current densities J_{gen} and J_{rec}, but also an exchange current density J_{exc} between the absorber and the contacts, the capacitance $C_\mu(\mu)$ is the chemical capacitance of the free charge carriers in the absorber. The exchange current density independently depends on the external voltage V and the chemical potential μ via [6-8]

$$J_{exc}(\mu, V) = J_{exc,0}\left[\exp\left(\frac{\mu}{kT}\right) - \exp\left(\frac{qV}{kT}\right) \right]. \quad (2)$$

The charge carriers that are withdrawn from the absorber build up the electrostatic potential across the solar cell such that

$$\frac{d}{dt}V = \frac{1}{C_c(V)}\left[J_c + J_{exc}(\mu, V) \right] \quad (3)$$

where the capacitance $C_c(V)$ denotes the electrostatic capacitance of the cell and J_c the external current density. Solving the two combined differential equations [Eqs. (1) and (3)] describes the decay of the TRPL via the time dependence of $\mu(t)$ and simultaneously that of the TRPV via $V(t)$.

Whereas these non-linear differential equations account for the large signal decay of the photoluminescence and the photovoltage, the small signal analysis around a given light and voltage bias condition can is analyzed by a two-dimensional linear system according to

$$\begin{pmatrix} \delta\hat{J}_c \\ \delta\hat{J}_{gen} \end{pmatrix} = \begin{pmatrix} G_{exc} + i\omega C_c & -G_{exc}' \\ -G_{exc} & G_{exc}' + G_{rec} + i\omega C_\mu \end{pmatrix} \begin{pmatrix} \delta\hat{V} \\ \delta\hat{\mu}/q \end{pmatrix}$$
$$= \{ \mathbf{G} + i\omega\mathbf{C} \} \begin{pmatrix} \delta\hat{V} \\ \delta\hat{\mu}/q \end{pmatrix} = \mathbf{Y}\begin{pmatrix} \delta\hat{V} \\ \delta\hat{\mu}/q \end{pmatrix}. \quad (4)$$

Equation (4) describes the linear response of a solar cell under small signal excitation with angular frequency ω, while $G_{exc} = dJ_{exc}/d(\mu/q) = dJ_{exc}/dV$ and $G_{rec} = dJ_{rec}/d(\mu/q)$ are the conductivities related to carrier exchange and recombination, respectively. The current densities $\delta\hat{J}_c, \delta\hat{J}_{gen}$ and the potentials $\delta\hat{V}, \delta\hat{\mu}/q$ are complex valued variables, and the symbols $\mathbf{G}$, $\mathbf{C}$, and $\mathbf{Y}$ stand for the symmetric conductivity, capacitance, and admittance matrices, respectively. The matrix form of Eq. (4) allows us to calculate all variants of frequency domain small signal experiments. For instance, impedance spectroscopy measures the voltage following an electrical excitation $\delta\hat{J}_c$ of the solar cell with no optical excitation additionally to a possible bias illumination, i.e. $\delta\hat{J}_{gen} = 0$. Hence, we find from Eq. (4) by inversion of the admittance matrix

$$\begin{pmatrix} \delta\hat{V}/\delta\hat{J}_c \\ \delta\hat{\mu}/(q\delta\hat{J}_c) \end{pmatrix} = \mathbf{Y}^{-1}\begin{pmatrix} 1 \\ 0 \end{pmatrix} \quad (5)$$

where $\delta\hat{V}_c/\delta\hat{J}_c$ represents the complex admittance of the device. Analogously, IMPV measures the frequency dependent voltage response to a modulation $\delta\hat{J}_{gen}$ with $\delta\hat{J}_c = 0$. Thus, we have

$$\begin{pmatrix} \delta\hat{V}/\delta\hat{J}_g \\ \delta\hat{\mu}/(q\delta\hat{J}_g) \end{pmatrix} = \mathbf{Y}^{-1}\begin{pmatrix} 0 \\ 1 \end{pmatrix} \quad (6)$$

with the complex IMPV response $\delta\hat{V}/\delta\hat{J}_{gen}$.

It is noteworthy, that the solutions of Eqs. (5) and (6) also provide information on the response of the chemical potential, i.e. $\delta\hat{\mu}/q\delta\hat{J}_c$ in Eq. (5) represents the modulated electroluminescence response. Likewise, $\delta\hat{\mu}/q\delta\hat{J}_{gen}$ in Eq. (6) represents the frequency domain small-signal TRPL response, i.e. the Fourier transform of the small-signal TRPL in the time domain. The latter is also directly derived from Eq. (4) via

978-1-6654-6060-6/23 $31.00 © 2023 IEEE

$$\frac{d}{dt}\begin{pmatrix} \delta V \\ \delta\mu / q \end{pmatrix} = -\mathbf{C}^{-1}\mathbf{G}\begin{pmatrix} \delta V \\ \delta\mu / q \end{pmatrix}. \quad (7)$$

Equation (7) is solved by the diagonalization of the matrix $\mathbf{C}^{-1}\mathbf{G}$. The TRPL signal, the time dependence of the chemical potential of the free carriers after an initial excitation $\delta\mu_0$ in the absorber, reads

$$\delta\mu(t) = \delta\mu_0\left[a\exp\left(\frac{-t}{\tau_{decay}}\right) + (1-a)\exp\left(\frac{-t}{\tau_{rise}}\right)\right]. \quad (8)$$

While the time domain TRPV signal is given by

$$\delta V(t) = \delta V_x\left[\exp\left(\frac{-t}{\tau_{decay}}\right) - \exp\left(\frac{-t}{\tau_{rise}}\right)\right]. \quad (9)$$

The coefficients a, δV_x and the relaxation times τ_{rise} and τ_{decay} are defined by the eigenvalues and eigenvectors of the matrix $\mathbf{C}^{-1}\mathbf{G}$.

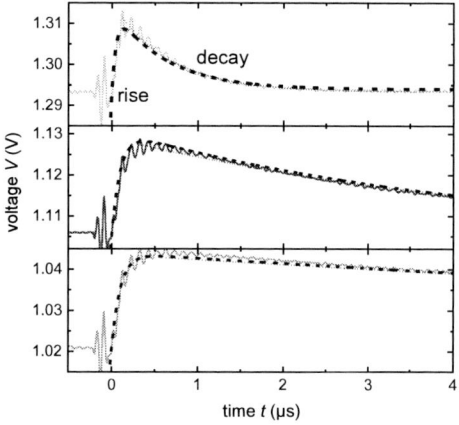

Fig. 3. *Experimental small-signal TRPV transients (taken from Ref. [3]) at different dc illumination (open-circuit) conditions in comparison to their fits to Eq. (9) (dashed lines).*

With Eq. (9) we have found a comprehensive, analytical solution for the title challenge of the present paper, as this equation describes the (small-signal) rise of the photovoltage from $V = 0$ at $t = 0$, followed by a decay towards zero for larger times. Note that we always have $\tau_{rise} < \tau_{decay}$, i.e., the rise is always faster than the decay. This fact is illustrated in Fig. 3 where we fit experimental small-signal TRPV data from a completed MHP solar cell to Eq. (9).

However, we emphasize that the knowledge of the relaxation times τ_{rise} and τ_{decay} alone is not enough to delineate the four independent quantities G_{exc}, G_{rec}, C_μ, and C_c, contained in Eq. (4). Instead, additional measurements, at different bias conditions and/or by different methods, have to be made in order to derive all unknown parameters.

III. Summary

The present paper has sketched a novel, generic model for the large-signal and small-signal optoelectronic response of solar cells. This model provides a comprehensive description of results from various experimental methods that investigate either the voltage or the luminescence response on either an electrical or an optical excitation. In that sense, all possible methods are covered. For the large-signal methods, we found a combination of two nonlinear differential equations that must be solved numerically. For the small-signal situations, in both the frequency domain as well as for the time domain, we found a linear matrix equation, that can be solved analytically in terms of eigenvalues and eigenvectors. Comparison to experimental TRPV data has demonstrated the viability of the method for perovskite solar cells.

References

[1] T. Kirchartz, J. A. Márquez, M. Stolterfoht, T. Unold, Photoluminescence - based characterization of halide perovskites for photovoltaics, Advanced Energy Materials 10, 1904134, 2020.

[2] F. Staub, H. Hempel, J. C. Hebig, J. Mock, U. W. Paetzold, U. Rau, T. Unold, T. Kirchartz, Beyond bulk lifetimes: insights into lead halide perovskite films from time-resolved photoluminescence, Physical Review Applied 6, 044017, 2016.

[3] L. Krückemeier, B. Krogmeier, Z. Liu, U. Rau, T. Kirchartz, Understanding transient photoluminescence in halide perovskite layer stacks and solar cells, Advanced Energy Materials 11, 2003489, 2021.

[4] L. Krückemeier, Z. Liu, B. Krogmeier, U. Rau, T. Kirchartz, Consistent Interpretation of Electrical and Optical Transients in Halide Perovskite Layers and Solar Cells, Advanced Energy Materials 11, 2102290, 2021.

[5] S. Ravishankar, Z. Liu, U. Rau, T. Kirchartz, Multilayer capacitances: How selective contacts affect capacitance measurements of perovskite solar cells, PRX Energy 1, 013003, 2022.

[6] U. Rau, Superposition and reciprocity in the electroluminescence and photoluminescence of solar cells, IEEE J. Photov. 2, 169, 2012.

[7] O. Breitenstein, An alternative one-diode model for illuminated solar cells, IEEE J. Photov. 4, 899, 2014.

[8] U. Rau, V. Huhn, and B. E. Pieters, Luminescence analysis of charge-carrier separation and internal series-resistance losses in Cu(In,Ga)Se$_2$ solar cells, Phys. Rev. Appl. 14, 014046, 2020.

Cadmium Zinc Telluride as an Electron Reflecting Back-Contact Layer for CdTe Solar Cells

Camden Kasik, Jennifer Drayton, and James Sites
Colorado State University, Fort Collins, Colorado, 80523, United States

Abstract— Cadmium Zinc Telluride (CdZnTe) with a 1.7 eV band gap was deposited at the back of thin CdTe-absorber devices to reduce back surface recombination and increase voltage. In these experiments we studied the effects CdZnTe thickness and substrate temperature during deposition have on device performance. In these experiments we show an efficiency increase of 2.4% above devices made without including CdZnTe. Furthermore, Photoluminescence data shows an order of magnitude increase after the deposition of CdZnTe, showing an increase in surface quality.

Index Terms—CdTe, Thin films, CdZnTe, Electron reflector

I. INTRODUCTION

Cadmium telluride (CdTe) solar devices have made significant advancements in recent years, with efficiencies reaching 22.1% [1]. These recent advancements can be attributed to work done at the front contact of devices, incorporating an MgZnO buffer layer and graded CdSeTe/CdTe absorber at the front of the cell. While the short circuit current density (J_{sc}) and fill factor are close to the Schokley-Queisser limits, device performance still falls well short of the theoretical maximum of 33%. The limitation on CdTe device performance is attributed to low open-circuit voltages (V_{oc}) compared to the Schokley-Queisser limit [2].

The back interface of CdTe devices remains the primary issue limiting the V_{oc}. A lack of passivation of the back interface in CdTe devices causes recombination, limiting the device performance [3]. The deep valance band of CdTe also presents problems. The CdTe valance band being lower than most contacting metals causes downward band bending at the back of the device, limiting the voltage carriers can be extracted at.

There are multiple theories on how to reduce recombination at the back of CdTe devices. One method, known as the electron reflection method, is adding a material with a higher conduction band minimum to repel electrons away from the back contact [4]. With a reduced number of electrons at the back there will be less opportunity for holes to recombine, thus reducing the recombination rate. Materials such as ZnTe are widely used for this purpose, but form a barrier to hole collection due to unfavorable positioning of the valance band maximum [5]. $Cd_{0.6}Zn_{0.4}Te$ potentially has both a conduction band barrier to electrons due to its 1.7-eV band gap, and favorable valance band alignment to transport holes to the back electrode making it an ideal back layer to CdTe devices.

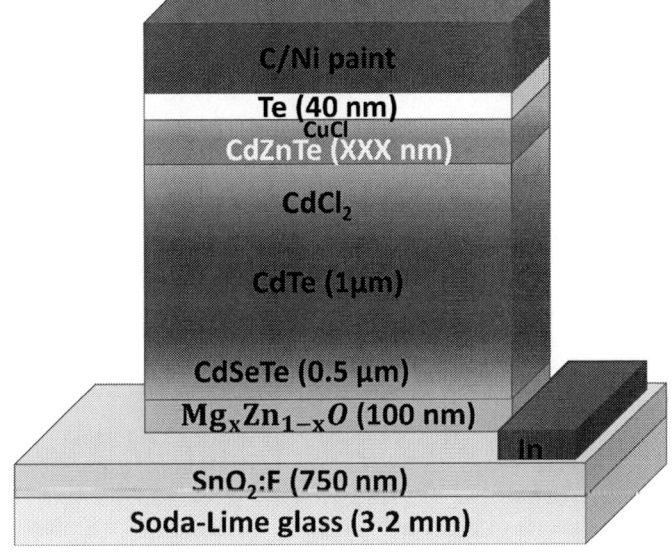

Fig. 1. Example structure of a CdZnTe device used in these experiments.

II. EXPERIMENT

A. Device Structure

The CdZnTe substrate temperature during deposition and the thickness were varied to explore the viability of CdZnTe as a back contact layer. Devices fabricated for these experiments have a 100-nm RF sputtered magnesium zinc oxide (MZO) layer deposited on fluorine-doped tin oxide. This was followed by an absorber layer of 0.5-μm cadmium selenium telluride (CST) and 1-μm of CdTe deposited using close-space sublimation (CSS). The deposition temperatures were 575°C and 555°C for CST and CdTe respectively. A cadmium-chloride ($CdCl_2$) passivation treatment followed without a vacuum break. A thin 1.5-μm absorber was chosen so the electron reflector benefits would be more apparent [4].

CdZnTe was then deposited using RF sputtering in layers of either 50 or 100-nm as shown in figure 1. An optimized, in another experiment not discussed, copper-chloride (CuCl) doping treatment, using CSS, followed. A thin 40-nm layer of Te was then deposited using thermal evaporation. Devices where then finished using a carbon paint polymer binding and nickle paint electrode. A second $CdCl_2$ treatment was not performed after the CdZnTe deposition due to zinc diffusion and loss at the high temperatures required for the treatment[6].

978-1-6654-6060-6/23 $31.00 © 2023 IEEE

To study the optimal fabrication procedure, the substrate temperature was varied during the CdZnTe depositions. Temperatures of 250°C, 200°C, 150°C, 100°C, and room temperature where used to determine the optimal fabrication parameters.

B. Characterization

The performance of these devices was measured with current voltage (J-V), under standard conditions. Photoluminescence (PL), with an excitation laser of 520-nm, was used to gain insight to the quality of the film and device. Transmission measurements were used to estimate the band gap of the deposited CdZnTe.

III. RESULTS AND DISCUSSION

A. Thickness of CdZnTe

Devices were made with both 50-nm and 100-nm of CdZnTe deposited to determine what thickness is ideal for device performance. Various copper doping treatments were performed on each thickness to account for the additional material and determine the optimal doping treatment. The J-V parameters of these cells were measured and on average the 50-nm layer outperformed the 100-nm layer. The best performing twenty five device plate with 50-nm had an average efficiency of 14.2% compared to 13.7% for the best 100-nm plate. The 50-nm layer also had greater uniformity across the plate with standard deviations of 0.3% and 1.4% for 50-nm and 100-nm respectively.

Based on this data, it was clear 50-nm of CdZnTe provided better consistent devices than the 100-nm layer. Future experiments use 50-nm of CdZnTe to improve consistency of results.

B. Substrate Temperature

The substrate temperature experiment utilized a 50-nm CdZnTe layer with the optimized CuCl treatment. This was based on the data presented in the previous section. Data from this experiment shows CdZnTe performed best when deposited onto a substrate heated to approximately 150°C as shown in figure 2. The average efficiency of the 150°C devices is 13.7% compared to 7% devices without any CdZnTe. The majority of the increase is due to an improved fill factor. This large increase in efficiency indicates the CdZnTe layer improves the back surface of our CdTe devices. The performance is notably more uniform, across the twenty five device plate, for plates that did include CdZnTe.

C. Photoluminescence

Photoluminescence, with front side illumination, increased by an order of magnitude after the deposition of CdZnTe, shown in figure 3. By using a thin 1.5-μm absorber, the impact of the back layers on photoluminescence is more readily apparent in front side illumination measurements. A similar increase, in data not shown, is seen for photoluminescence collected from the back of the device stack. This

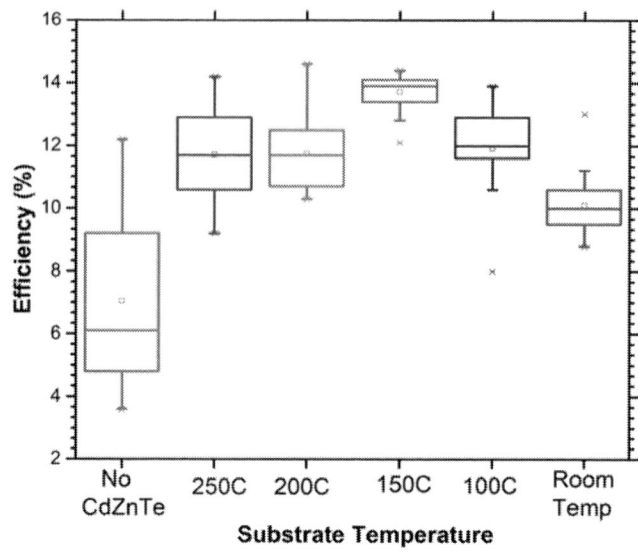

Fig. 2. Box plots of the efficiencies for different substrate deposition temperatures of CdZnTe. Each box represents twenty five cell devices made on a single plate of glass.

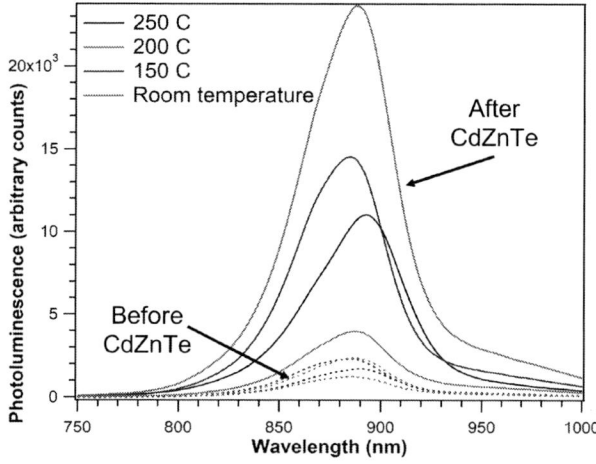

Fig. 3. Photoluminescence data taken before and after the deposition of CdZnTe at different substrate temperatures.Dotted lines indicate PL prior to CdZnTe deposition.

increase in photoluminescence indicates the CdZnTe layer reduces recombination at the back of the cell. The increase in photoluminescence is possibly due to electrons being repelled by the higher conduction band of the CdZnTe as theorized. If this is the case CdZnTe has tremendous potential to improve the back contact of CdTe solar devices.

IV. CONCLUSION AND CURRENT WORK

The back contact of CdTe solar cells has been a longstanding issue for the CdTe community. Using CdZnTe as a back layer we have demonstrated an increased efficiency of 14.6% compared to efficiencies of 12.2% from devices without

CdZnTe manufactured during the same day. The increased photoluminescence after the deposition of CdZnTe indicates a reduced recombination rate at the back surface of the device.

We believe this CdZnTe layer repels electrons away from the back surface, reducing the recombination rate, while providing a favorable valance band to allow holes to be collected at the back electrode. Photoelectron spectroscopy and temperature dependent J-V measurements are being worked on to provide information on the band positions compared to CdTe, as well as time resolved photoluminescence measurements to learn more about the impact of CdZnTe on the lifetime of carrier near the back surface. Future doping experiments utilizing hall measurements are also being worked on to explore possibilities of nitrogen doping of CdZnTe.

ACKNOWLEDGMENT

The authors would like to thank Dr. W.S. Sampath for use of his deposition systems and numerous helpful discussions, Dr. Tushar Shimpi for helpful discussions on CdZnTe depositions with the systems available, and 5N Plus for supplying the source material. This material is based on work supported by the U.S. Department of Energy's Office of Emerging and Renewable Energy under the Solar Energy Technologies Office Award Number DE-EE008974.

REFERENCES

[1] "Best research-cell efficiency chart," *NREL.gov*, 17-Nov-2021. [Online]. Available: https://www.nrel.gov/pv/cell-efficiency.html.

[2] J. M. Burst et al., "CdTe solar cells with open-circuit voltage breaking the 1v barrier," *Nature Energy, vol. 1, no. 3*, 2016.

[3] Alexandra M. Huss, Jennifer A. Drayton, and James R. Sites. "Front and Back Interface Recombination of MZO/CdTe/TeSolar Cells," *2018 IEEE 7thWorld Conference on Photovoltaic Energy Conversion(WCPEC) (A Joint Conference of 45th IEEE PVSC, 28th PVSEC 34th EU PVSEC)*, 2018.

[4] K.-J. Hsiao and J. R. Sites, "Electron reflector to enhance photovoltaic efficiency: application to thin-film cdte solar cells," *Progress in Photovoltaics: Research and Applications*, vol. 20, no. 4, pp. 486–489, 2012

[5] F. K. Alfadhili, A. B. Phillips, M. K. Jamarkattel, A. J. Snyder, J. M. Gibbs, G. K. Liyanage, and M. J. Heben, "Potential of cdznte thin film back buffer layer for cdte solar cells," in 2019 *IEEE 46th Photovoltaic Specialists Conference (PVSC)*, pp. 0140–0143, IEEE, 2019

[6] D. E. Swanson, C. Reich, A. Abbas, T. Shimpi, H. Liu, F. A. Ponce, J. M. Walls, Y.-H. Zhang, W. K. Metzger, W. S. Sampath, et al., "Cdcl2 passivation of polycrystalline cdmgte and cdznte absorbers for tandem photovoltaic cells," *Journal of Applied Physics*, vol. 123, no. 20, p. 203101, 2018.

Innovative Layouts for Utility-Scale PV Modules: Module Characteristics, Shading Tolerance, and Electricity Costs

Li C. Rendler, Christian Reichel, Matthew F. Berwind, Anna Heimsath, and Dirk H. Neuhaus

Fraunhofer Institute for Solar Energy Systems ISE, Freiburg, 79110, Germany

Abstract — Utility-scale photovoltaic (PV) modules with different electrical module layouts are investigated and compared in terms of module characteristics, shading tolerance, and electricity costs. Potential module layouts and their influence on PV module properties are analyzed in detail to better understand the effects of power-plant requirements and constraints. With the Cross String, the Cross String+, and the Cross Fox, three alternative module layouts to the most common Butterfly layout are introduced. For all module layouts, the dissipated power under hotspot conditions and the residual power under different shading conditions is analyzed through a modelling approach. With the Cross Fox layout a potential reduction of the dissipated power under hotspot conditions by up to -51% was found, substantially reducing the risk of severe damage due to high temperature. In addition, for solar cell row shading along the short edge of a PV module, all alternative layouts show a more than 20% higher residual power compared to the commonly used Butterfly module layout. A levelized cost of electricity (LCOE) analysis is performed with an example power plant configuration (2P static), calculating relative costs of each novel module type in field conditions, with the Cross String layout indicating maximum savings of approximately 1% over the Butterfly design.

I. INTRODUCTION

A trend toward increasing PV module formats in utility-scale power plants is currently observed due to new wafer formats and potentially reduced installation costs that result from them [1–3]. Module manufacturers accomplish this by increasing the size and number of solar cells in PV modules. The module layout defines the spatial arrangement and the interconnection of the solar cells and bypass diodes in a PV module, determining crucial properties, such as the dimensions, the electrical characteristics, the behavior under shading conditions, the mechanical stress in the solar cells under mechanical loads, etc.

Nowadays, most PV modules on the market feature half solar cells with the corresponding, so-called Butterfly layout (M1). This module layout was patented by REC Group and opposing procedures are currently running in Europe and in the U.S. [4, 5]. An exemplary high-power Butterfly PV module for a power plant includes 144 solar cells in twelve strings, each including twelve solar cells connected in series, and the typical main cross-connector line in the center between the two short edges of the rectangular PV module. In each module half there are six solar cell strings orientated parallel to the long edge with a parallel connection of opposing double-strings. Furthermore, there are three junction boxes and three bypass diodes, each one connected in parallel to 24 solar cells. When ignoring cell-to-

module (CTM) gains and losses [6], the assumed module voltage can be calculated by adding the voltage of 72 solar cells in one module half. As the solar cells are connected in series, the current is approximately the added current of two parallel solar cells, since there are always two strings connected in parallel. Fig. 1 shows a principle drawing of the module layout for the described PV module.

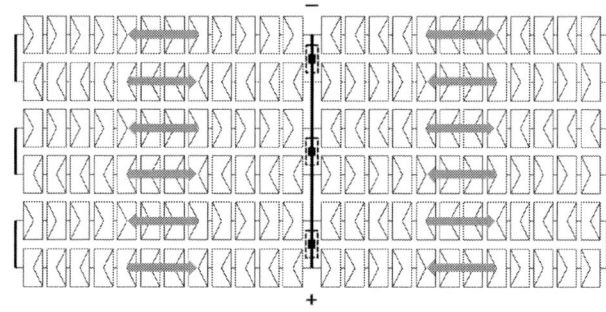

Fig. 1. Butterfly module layout (M1) with 144 half solar cells, including 12 strings, cross connectors, and three bypass diodes (black rectangles) in three junction boxes (dashes line boxes). The technical current flow direction in each of the strings is indicated by arrows.

In general, based on the electrical characteristics of the solar cells, the number of solar cells connected in series determines the expected voltage levels, and the number of electrically parallel solar cell strings define the maximum currents of a PV module. For a power plant the maximum voltage in a module string is limited and most utility-scale PV modules currently have a maximum system voltage of 1500 V. This limits the number of PV modules in a serial module string in a power plant. However, the maximum number of PV modules can be increased by decreasing the voltage, which at the same time increases the current of the module. This can be achieved by alternative interconnection layouts of the solar cells in a PV module. On the one hand, the used maximum power point (MPP) trackers and power inverters need to be able to work with the higher currents and the cable cross-section must be increased to reduce losses. On the other hand, the number of MPP trackers and power inverters per module for the same number of PV modules in a power plant is reduced. Hence, the costs in a power plant could be lower when using PV modules with a lower voltage.

Additional boundary conditions and limitations must be considered when designing a PV module. The trend to larger solar cells results in an increase of the dimensions of

978-1-6654-6060-6/23 $31.00 © 2023 IEEE

PV modules. The size of the solar cells, the number of solar cells per string, and the arrangement of strings define the geometric dimensions of the PV module, which is important for PV module manufacturers, installation technicians, and customers for several reasons. In some cases, countries have specific restrictions for the PV module size, for example a PV module with an area above 2 m² requires an additional type of approval to be installed on buildings in Germany [7]. Also, the efficient use of the space and the number of PV modules that can be transported in one shipping container is defined by their dimensions, which influences the module price.

In this study, three innovative module layouts for utility-scale high-power PV modules are introduced and compared with the commonly used PV modules with a Butterfly layout. The main goals are to identify promising alternatives for the Butterfly layout with reasonable dimensions and electrical properties, to analyze differences in shading resilience and hotspot risk, and to estimate potential cost reductions due to reductions of the voltage of a single module in a power plant although the current of such module is higher.

II. METHODS

A. Development and Analysis of Alternative Module Layouts

A PV module with a new layout must fulfil several requirements, such as manufacturability, low costs, high efficiency, and reliability. In the following, the most important requirements for this study are explained briefly:

1) Processing & Tools
Straightforward designs are preferred to enable manufacturing processes with commercially available manufacturing tools (stringers, cross connection tools, laminators, etc.), presupposing only minor adjustments on the machines and processes, as well as minimizing investments and enabling comparable specific module costs.

2) Dimensions
The dimensions of a PV module with a new layout should not differ much from the market standard products. First, this enables to use standard tools in the manufacturing process. Second, packaging and shipping can be performed as already established. Third, dimensioning of support-structures and installation does not require major adjustments.

3) Efficiency & Costs
An efficient area coverage with solar cells minimizing non-active areas (edge, string, and solar cell distances & cross connectors) leads to a high module efficiency. Furthermore, it results in low costs per watt for a PV module due to minimized material consumption (glass & polymer foils). In any case, module costs (processes and material consumption) should be comparable to industry standard modules.

4) Electrical Characteristics
With a typical system voltage limit of 1500 V, PV modules with a reduced voltage may result in lower cost per watt in a power plant, because more PV modules can be connected in series and less electronics can be used in a power plant. On the other hand, when the power stays constant, the current increases and larger cable cross sections and thicker isolation may be required. For this reason, this study compares the cost in a power plant for the different module layouts, all of them featuring identical numbers of solar cells, but with different output parameters for the voltage and current.

5) Mechanical Stress
Beinert et al. simulated mechanical stress in solar cells in PV modules under mechanical loads (e.g. snow or wind loads) and found significantly reduced mechanical stress for half-cell strings oriented parallel to the short edge of a rectangular PV module [8]. This potentially improves reliability of PV modules and increases yield in a power plant (less annual degradation).

6) Hotspots & Shading
According to Witteck et al. new wafer sizes and higher solar cell efficiencies could result in critical temperatures >170 °C under shading conditions in standard PV modules with Butterfly layout [9]. This could cause defects in PV modules, such as solar cell breakdown, polymer degradation, etc. For this reason, one layout with more than three bypass diodes and shorter solar cell strings protected by one bypass diode is considered.

B. Shading Simulations

1) Hotspots
Usually, the maximum power point is set as the operating point of the modules, but a shaded solar cell in a series-connected substring under short-circuit conditions represents the worst-case scenario according to the IEC 61215-2:2019 MQT 09 hotspot endurance test [10]. For this scenario, the operating point of the shaded solar cell can be constructed graphically by applying Kirchhoff's mesh rule and, as the voltage of the shaded solar cell is equal to the voltage of the remaining unshaded solar cells in the substring plus the forward voltage of the bypass diode as explained by Bishop [11–14].

2) Relative Power Loss
In a power plant, different shading scenarios are analyzed by simulation. First, the PV module power loss caused by shading of a single solar cell with area ratios between 0% and 100% is calculated. Second, shading of rows of solar cells along the PV module edges is evaluated. For both scenarios, the shading area is chosen that either one wafer length or two half M10 solar cells are affected by shading (no illumination) along the long edge or the short edge. Fig. 2 outlines the different shading scenarios.

In a power plant, the resulting power of a string of PV modules is most relevant for MPP tracking (MPPT). Even though MPPT algorithms are not part of this study the MPPT influences the operating point of a PV module in a power plant significantly [15]. For the analyzed shading scenarios, this is important because an affected PV module could be forced to be operated with electrical parameters that do not fit to its own MPP.

978-1-6654-6060-6/23 $31.00 © 2023 IEEE

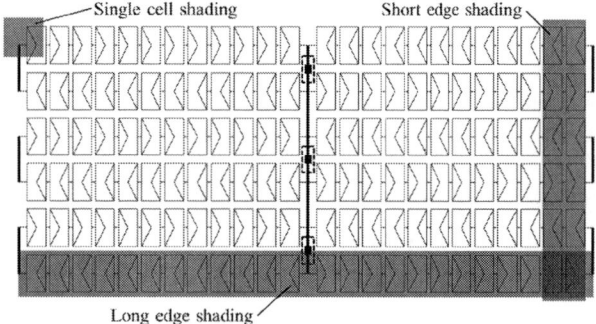

Fig. 2. Shading scenarios for the simulation of the relative power loss in a power plant for PV modules with different layouts: Single cell shading with variation of the area ratio, short edge shading, and long edge shading.

For this study, a single PV module with single cell shading (e.g. bird droppings in coastal areas) is assumed to be forced to remain at the I_{MPP} of the non-affected PV modules in the string due to current matching forced by the MPPT. Shading along the edges of PV modules can have different reasons. At low sun levels, PV module rows may shade the next module row behind them (self-shading). Soiling or snow could also cause shading along the edges of PV modules. For a portrait-oriented 2P-configuration two rows of PV modules with portrait-orientation are positioned on one table in a power plant, whereas for landscape-orientated 4L-configuration there are four PV modules with landscape orientation per table. For shading of rows of solar cells along the short or the long edges of the PV modules, it is assumed that all PV modules in a module string are affected equally and that all modules work at the same MPP with matched current. For this reason, the power at the individual MPP of one PV module is calculated for all four module layouts. For non-homogeneous shading of several PV modules in a power plant the situation could be different and is not analyzed in this study.

C. Cost Calculation for Power Plants

The LCOE calculation for a solar power plant involves the analysis of several key input parameters, as summarized in (1). In the numerator, investment, and operational costs as well as the residual value of the power plant are represented. In the denominator, results from the light and bifacial yield analyses are represented.

$$LCOE_{nom} = \frac{\left(CAPEX_{PV,total} + \sum\left[\frac{OPEX(t)}{(1+WACC_{nom})^t}\right] - \frac{ResValue}{(1+WACC_{nom})^N}\right)}{\sum\left[\frac{Yield(0)(1-Degradation)^t(Availability)}{(1+WACC_{nom})^t}\right]}$$ (1)

Firstly, the capital expenditures (CAPEX) are considered. This includes the upfront investment required for the construction and installation of the solar power plant, encompassing expenses for solar panels, inverters, mounting structures, electrical systems, and other infrastructure components. These costs can vary based on the scale of the project, the technology used, and any additional features or requirements. Secondly, the operational and maintenance (O&M, OPEX) costs are taken into account. These costs involve

the ongoing expenses associated with operating and maintaining the solar power plant throughout its lifetime. O&M costs include routine maintenance activities like cleaning solar panels, inspecting and repairing equipment, and ensuring optimal performance. It also encompasses monitoring systems, personnel salaries, training, and administrative overhead, among other common variables. Cost data are sourced from the IRENA report *"Renewable Power Generation Costs in 2021"* [16].

Another crucial factor in the LCOE calculation is the estimation of the lifetime energy production of the solar power plant. This involves considering factors such as solar irradiation levels, module efficiency, and the availability of sunlight at the project location. Energy production is estimated using historical meteorological data for the theoretical 5MW$_P$, two-in-portrait module arrangement (2P) installation site in Freiburg, Germany utilizing string inverters, as well as the time-series PV power plant simulation model Zenit™ [17]. A 25 year time span was considered for yield modelling, assuming a plant degradation rate of 0.5%/a. The longer the expected lifetime energy production, the lower the LCOE as the initial investment is spread over more energy units. Financing costs are also considered in the LCOE calculation. This includes the interest rates and loan terms, as they affect the overall cost of the solar power plant. Financial costs are considered within a weighted average cost of capital (WACC), and a fixed 6% was chosen as the comparison basis for this study.

By considering these factors, the LCOE calculation provides an assessment of the average cost of electricity per unit of energy produced by the solar power plant throughout its lifetime and is particularly useful when comparing *relative costs between different system technologies*, and less as an absolute cost prediction in the absence of company specific projecting data.

III. RESULTS AND DISCUSSION

A. Development and Analysis of Alternative Module Layouts

With the "Cross String" (M2), the "Cross String+" (M3), and the "Cross Fox" module layout (M4) three alternative, but straight-forward layouts for the common Butterfly layout are introduced. All module layouts feature an identical number of 144 solar cells and the dimensions of the PV modules are comparable to the market standard. However, in contrast to the Butterfly layout, for the alternative layouts the solar cell strings are parallel to the short edge of a PV module, which results in lower mechanical stress in the solar cells under mechanical loads [8] and a different shading behavior.

1) Cross String

The most straightforward module layout with solar cell strings along the short edge of a rectangular PV module is the Cross String layout as shown in Fig. 3 [18, 19]. First, there are only solar cells and no junction boxes or cross connectors in the center of the module potentially resulting in a comparable module efficiency as the common Butterfly layout. With three electrically parallel strings only two junction boxes each including one bypass diode is required, which reduces module costs. In addition, manufacturability is simple because the cross connectors are positioned only in one direction and only parallel to the long edges of the PV module.

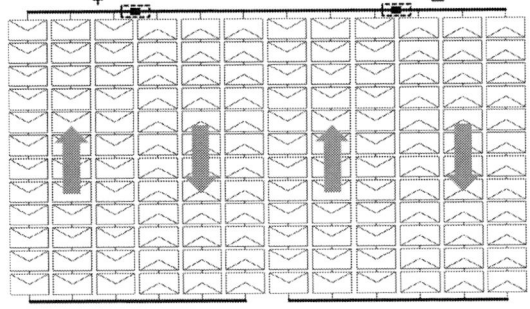

Fig. 3. Cross String module layout (M2) with 144 half solar cells, including 12 strings with 12 solar cells connected in series, cross connectors, and two bypass diodes (black rectangles) in two junction boxes (dashes line boxes). The technical current flow direction in each of the parallel string is indicated by arrows.

2) Cross String+

By shortening the solar cell strings to half the length and adding cross connectors in the center, a current path is added that could decrease the dissipated power in a solar cell under shading conditions. This could be one method to decrease the maximum temperature that is caused under shading conditions. Additionally, there are shorter strings to handle in the production process. However, these additional cross connectors cause an increase in size and material consumption and decrease in efficiency. This layout is called Cross String+ and is shown in Fig. 4.

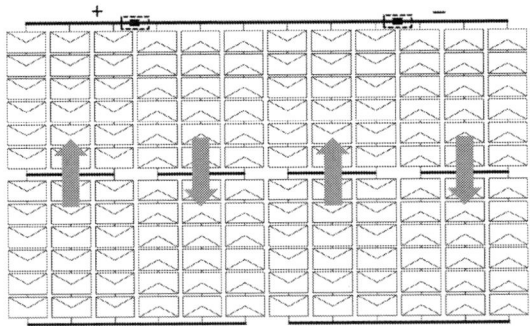

Fig. 4. Cross String+ module layout (M3) with 144 half solar cells, including 24 strings with six solar cells connected in series, cross connectors, and two bypass diodes (black rectangles) in two junction boxes (dashes line boxes). The technical current flow direction in each of the parallel string is indicated by arrows.

3) Cross Fox

The so-called Cross Fox layout was developed by Fraunhofer ISE (patent pending) and is shown in Fig. 5. It is characterized by two additional cross connectors between solar cell strings parallel to the short edges of a PV module. It enables to use four bypass diodes in two junction boxes, resulting in a maximum string length of 12 serial solar cells protected by one bypass diode. In comparison, for all other layouts the string length over one bypass diode is 24. A significant reduction of the dissipated

power in a solar cell under shading conditions is expected for this layout, due to half the maximum voltage of a solar cell string. This could be beneficial when shading is expected in a power plant, for example due to bird droppings in coastal areas, under inhomogeneous illumination conditions, or when single modules are affected by severe degradation or quality issues.

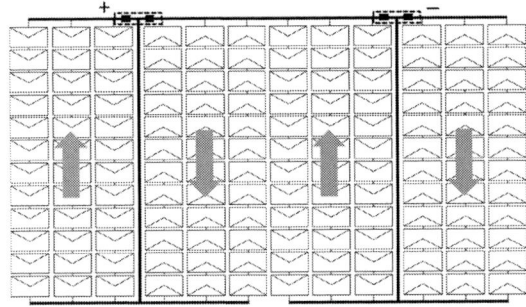

Fig. 5. Cross Fox module layout (M4) with 144 half solar cells, including 12 strings with 12 solar cells connected in series, cross connectors, and four bypass diodes (black rectangles) in two junction boxes (dashes line boxes). The technical current flow direction in each of the parallel string is indicated by arrows.

4) Dimensions

Compared to a PV module with Butterfly layout the three alternative layouts result in a shorter length of up to -43.5 mm and a maximum width increase of 34 mm. The relative area of a PV module is only changed by less than +1.2%. The dimensions of all four module layouts are shown in TABLE I.

TABLE I.   DIMENSIONS OF PV MODULES WITH DIFFERENT MODULE LAYOUTS

Module Layout	Dimensions		
	Length / mm	Width / mm	Area / m²
M1: Butterfly	2278.0	1134.0	2.58
M2: Cross String	2234.5	1158.5	2.59
M3: Cross String+	2234.5	1168.0	2.61
M4: Cross Fox	2253.5	1158.5	2.61

5) Electrical Characteristics

To calculate the approximate electrical properties for all modules commonly used PERC (Passivated Emitter Rear Cell) solar cells with an efficiency of 23% are assumed. When neglecting minor differences in CTM losses and gains, all modules deliver the same nominal power of 547 W at standard testing conditions (STC: 1000 W/m², temperature: 25 °C, air mass: 1.5). In addition, all alternative layouts include three instead of two electrically parallel solar cell strings, resulting in a reduced voltage (-33% relative) and an increase module current (+50% relative). The approximate electrical characteristics at the MPP (V_{MPP}, I_{MPP}, and P_{MPP}) were calculated and are shown in TABLE II. The Butterfly module shows a voltage of 43.0 V and a current of 12.7 A, whereas the other three module layouts feature a voltage of 28.7 V and a current of 19.1 A at their MPP under STC. However, the alternative layouts M2-M4 could be designed with two instead of three parallel strings resulting in comparable electrical properties to the typical layout M1.

TABLE II. IV PARAMETERS OF THE MODULE LAYOUTS (NEGLECTING OPTICAL AND ELECTRICAL GAINS AND LOSSES)

Module Layout	IV parameters		
	V_{MPP} / V	I_{MPP} / A	P_{MPP} / W
M1: Butterfly	43.0	12.7	547
M2: Cross String	28.7	19.1	547
M3: Cross String+	28.7	19.1	547
M4: Cross Fox	28.7	19.1	547

B. Shading Simulations

1) Hotspots

The IV curves of the PERC solar cell with an efficiency of 23.0% were measured and fitted to a two-diode-model and the split-cell model [20] to calculate the resulting IV curves for different shading ratios. The dissipated power in a solar cell with shading is analyzed for the PV modules and the results reveal large differences of the hotspot behavior.

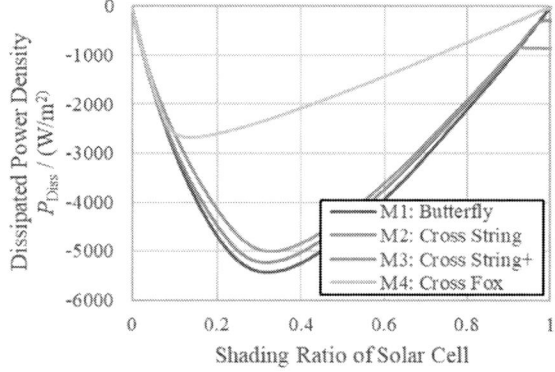

Fig. 6. Dissipated power density of a single shaded solar cell in dependence of the shading ratio in a hotspot endurance test for a Butterfly (M1), a Cross String (M2), a Cross String+ (M3), and a Cross Fox module (M4).

Fig. 6 shows that the worst-case shading scenario for the Butterfly module (M1) and the Cross String module (M2) exists when a solar cell in a series-connected string is shaded by 32%, dissipating 5430 W/m^2 for the Butterfly module, while for the Cross String module (M2) only 5160 W/m^2 is dissipated (-5% compared to M1) due to compensation currents which influence the maximum power dissipation in the shaded solar cell. The worst-case shading scenario for the Cross String+ module (M3) is obtained when a solar cell is shaded by 34%, but in this case only 5000 W/m^2 is dissipated (-8% compared to M1). In contrast, the worst-case shading scenario for the Cross Fox module (M4) exists when a solar cell is shaded by 14% with a much lower dissipated power density of 2670 W/m^2 (-51% compared to M1), showing the significant benefit of this module layout in terms of shading tolerance due to half the string length and the highest number of bypass diodes.

2) Relative Power Loss

First, the residual power of a PV module under shading of one single solar cell is simulated for all module layouts. It is assumed that only one module in a module string is affected so that the shaded module is forced by the MPPT to transmit the same current (I_{MPP}) of all unshaded PV modules in the module string. Fig. 7 shows the module power for shading ratios between 0% and 100%. Up to about 10% solar cell shading the residual power of all modules is not affected, but it decreases significantly above 10% shading area. Above 40% shading ratio of a single solar cell in one module all module layouts reach the maximum power reduction. The Butterfly module (M1) shows a residual module power of about 357 W (65%), whereas the Cross String and the Cross String+ module show a lower value of about 263 W (48%). Here, the Cross Fox layout shows an improved shading tolerance ending up at the highest power of 400 W (73%). This demonstrates the advantage of larger numbers of bypass diodes in a PV module for single cell shading in a power plant.

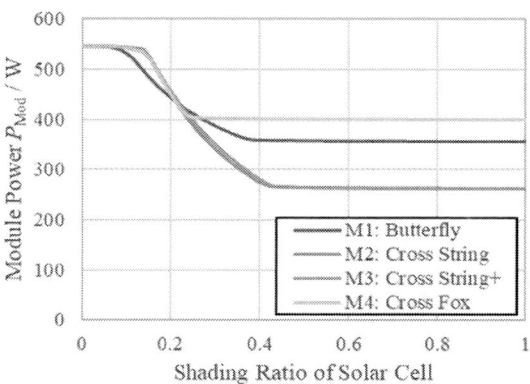

Fig. 7. Module power in dependence of the shading area ratio in a string of modules under the maximum power point current of the unshaded modules for a Butterfly (M1), a Cross String (M2), a Cross String+ (M3), and a Cross Fox module (M4).

The power of the different modules when not only one solar cell is shaded, but when parts of the module are shaded is shown in TABLE III. Here one equivalent full-size solar cells row is supposed to be completely shaded along the long side and short side of the module as indicated in Fig. 2.

TABLE III. RESIDUAL POWER FOR SHADING ALONG THE LONG AND THE SHORT EDGE OF THE DIFFERENT PV MODULES

Module Layout	Percentage of Initial Power	
	Long Side	Short Side
M1: Butterfly	65 %	50 %
M2: Cross String	0 %	71 %
M3: Cross String+	0 %	71 %
M4: Cross Fox	0 %	73 %

It is observed that the Butterfly module (M1) is more tolerant to shading along the long side of the module compared to the Cross String (M2), the Cross String+ (M3), and the Cross Fox module (M4). While the Butterfly module has about 65% of the initial power, the other modules have 0% of the initial power. The reason for this difference is that for the Butterfly layout the bypass diode in the shaded third is active and the module current stays constant (I_{MPP}) whereas for the alternative layouts, having all strings along the short module edge, all strings are shaded, reducing the residual module power to zero. This demonstrates that the Butterfly module layout could be favored for power plants with landscape-oriented PV modules (4L-configuration) with shading from the bottom.

In contrast, for shading along the short side of the module, the Cross String, the Cross String+, and the Cross Fox module have 71% and 73% of the initial power, which is more than 20% more than for the Butterfly module. For the Butterfly layout the bypass diodes are inactive, the voltage stays constant, and the current is reduced by about 50%, ending up in a residual power of 50%. In contrast, for the Cross String and the Cross String+ module layout the bypass diodes are not active, the current is reduced by about 32% and the module voltage is slightly increased resulting in a residual power of 71%. For the Cross Fox layout, the situation is different since the bypass diode in the shaded fourth is active, so that the voltage is reduced by about 25%. In this case, the current stays nearly constant and flows through the non-shaded strings and the active bypass diode. This results in a residual power of 73%. Therefore, power plants with portrait-oriented PV modules (2P-configuration) could have a significantly higher yield under shading conditions from the bottom when using alternative layouts with strings along the short edge, as the Cross String, the Cross String+, or the Cross Fox layout.

C. Cost Calculation for Power Plants

Initial cost estimates using SCost [21, 22] placed the Butterfly (M1) at 30.40 €ct/W_P, the Cross String (M2) at 30.30 €ct/W_P, the Cross String+ (M3) at 30.50 €ct/W_P, and the Cross Fox (M4) at 30.65 €ct/Wp respectively. Here, 41 low-voltage module variants could be safely strung together in a 1500 VDC system, whereas the Butterfly layout permitted only 27. LCOE cost differences due to different module string topologies, PV module costs, higher current cabling, etc. are listed in Fig. 8, showing a systematic cost benefit associated with low voltage module types for M2, M3, and M4.

Although the enhanced performance of different module string layouts under heterogeneous array irradiation conditions potentially has benefits for module and/or system lifetime, the effects on yield (and thus a pure LCOE comparison) are likely insufficient to justify new module layouts on their own. Enhanced shading performance would be of benefit (apart from soiling cases) when neighboring PV module rows shade each other in the morning and evening, times when incident solar irradiation and thus potential yield improvements are inherently low. This highlights the need for more enhanced parametric studies (not considered in this study) to determine an economic optimum for denser row packing (ground coverage ratio) which is enabled by such novel module layouts. Nevertheless, the cost benefits (0.5-1%) of such novel module layouts due to increased numbers of modules in the strings justify further examination.

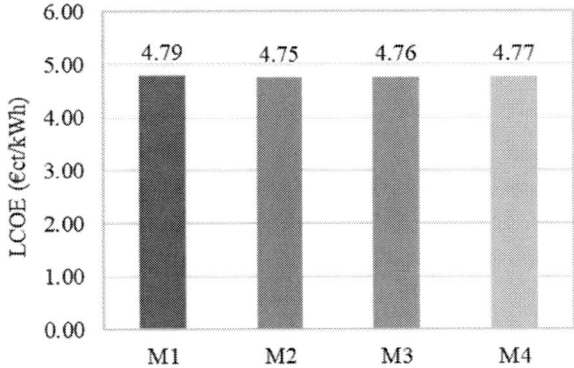

Fig. 8. Power plant nominal LCOE calculations based on (four) separate module layouts in an example static 2P power plant in Freiburg, Germany. The primary cost differences are resultant from the longer string lengths possible due to lower module voltages.

IV. Conclusion and Outlook

With the Cross String, the Cross String+, and the Cross Fox layout, three straightforward module layouts as alternatives to the most common Butterfly layout were introduces. All modules include 144 M10 half solar cells with a 23% efficiency and, neglecting expected minor deviations in CTM losses and gains, share the same P_{MPP} of 547 W. The dimensions show only small differences (1.2% maximum area increase). In contrast to the Butterfly layout, all alternative layouts feature three electrically parallel solar cell strings aligned in parallel to the short edge of a PV module. This results in lower voltages (-33% relative) and higher module current (+50% relative). However, these layouts could be easily modified to feature two parallel strings instead of three, resulting in comparable electrical properties as a PV module with Butterfly layout.

The dissipated power in a solar cell with shading is analyzed by simulations and the results reveal significant differences of the hotspot behavior. Whereas the Cross String and the Cross String+ layouts show only small reductions of the maximum dissipated power compared to the Butterfly, the Cross Fox module layout enables a reduction of more than 50%. This is due to its additional bypass diode and the resulting half string length enabling significant reductions of the maximum temperature under hotspot conditions.

The residual module power of a shaded PV module in a module string in a power plant is analyzed in detail. The simulation covers single cell shading with variation of the shading ratio, as well as shading along the short and the long edges of a PV module. The results reveal that single cell shading is most critical for the Cross String and the Cross String+ layout because both layouts only feature two bypass diodes. However, the Cross Fox layout enable an increase of the residual module power under single cell shading of up to 8% compared to the common Butterfly layout. For shading along the long edge, the alternative layouts are disadvantageous compared to the

Butterfly due to their string orientation parallel to the short module edge. However, for shading along the short module edge, all alternative layouts enable an increase of the residual module power of more than 20% compared to the Butterfly layout. For power plants with expected shading from the bottom of a module string (e.g. self-shading) this demonstrates that the alternative layouts are disadvantageous for PV modules in landscape orientation but could significantly increase energy yield for PV modules in portrait orientation.

The three novel module layouts display systematic cost benefits of approximately 0.5-1% compared to the Butterfly layout. However, analyzing all effects on cost performance is not possible without speculation, particularly when considering decreased degradation (effective yield increase over time) and a potential lifetime increase (OPEX benefit). When considering a single row-spacing, the benefits of increased shading performance are not highlighted, and further studies examining potential yield gains for more densely laid out systems is advisable to quantify true economic optima.

REFERENCES

[1] VDMA Photovoltaic Equipment, "International Technology Roadmap for Photovoltaic (ITRPV)," Results 2021, 13th edition.

[2] SolarPower Europe, "Global Market Outlook For Solar Power," 2021 - 2025.

[3] China Photovoltaic Industry Association, "China PV Industry Development Roadmap," (2021 Edition).

[4] S. G. Sridhara, Diesta, Noel G., Rostan, Philipp Johannes, and R. Wade, "Solar Cell Assembly," US10749060B2.

[5] Sandra Enkhardt, *Patentstreit: Hanwha Q-Cells erzielt positives Urteil in den USA gegen REC Solar*. [Online] Available: https://www.pv-magazine.de/2023/01/19/patentstreit-hanwha-q-cells-erzielt-positives-urteil-in-den-usa-gegen-rec-solar/. Accessed on: May 20 2023.

[6] I. Hädrich, U. Eitner, M. Wiese, and H. Wirth, "Unified methodology for determining CTM ratios: Systematic prediction of module power," *Sol Energ Mat Sol C*, vol. 131, pp. 14–23, 2014.

[7] Deutsches Institut für Bautechnik, *MVV TB: Technische Baubestimmungen*. [Online] Available: https://www.dibt.de/de/wir-bieten/technische-baubestimmungen. Accessed on: May 21 2023.

[8] A. J. Beinert, "Thermomechanic Design Rules for the Development of Photovoltaic Modules," Dissertation, Institut für Angewandte Mechanik - Werkstoff und Biomechanik, Karlsruher Institut für Technologie (KIT), Karlsruhe, 2021.

[9] R. Witteck, M. Siebert, S. Blankemeyer, H. Schulte-Huxel, and M. Kontges, "Three Bypass Diodes Architecture at the Limit," *IEEE J. Photovoltaics*, vol. 10, no. 6, pp. 1828–1838, 2020.

[10] IEC 61215-2:2016, "Terrestrial photovoltaic (PV) modules - Design qualification and type approval - Part 2: Test procedures,"

[11] C. Reichel, J. Forster, B. Artha, K. Inwersen, A. Tummalieh, J. Weber, E. Fokuhl, L. C. Rendler, D. H. Neuhaus, "Design Aspects Considering Hotspot Phenomena in Modern High-Performance Silicon Photovoltaic Modules," 13th International Conference on Crystalline Silicon Photovoltaics. Delft, Netherlands, 2023.

[12] J. W. Bishop, "Computer simulation of the effects of electrical mismatches in photovoltaic cell interconnection circuits," *Solar Cells*, vol. 25, no. 1, pp. 73–89, 1988.

[13] K. Mertens, *Photovoltaik: Lehrbuch zu Grundlagen, Technologie und Praxis*, 3rd ed. München: Hanser Verlag, 2015.

[14] Jacob Forster, "Design Aspects and Constraints Regarding Hotspot Resilient Interconnection Topologies for Integrated Photovoltaic Modules," Master Thesis, University of Freiburg, Freiburg, Germany, 2021.

[15] L. Liu, X. Meng, and C. Liu, "A review of maximum power point tracking methods of PV power system at uniform and partial shading," *Renewable and Sustainable Energy Reviews*, vol. 53, pp. 1500–1507, 2016.

[16] International Renewable Energy Agency, "Renewable Power Generation Costs in 2021," 2022.

[17] B. Müller, L. Hardt, A. Armbruster, K. Kiefer, and C. Reise, "Yield predictions for photovoltaic power plants: empirical validation, recent advances and remaining uncertainties," *Prog. Photovolt: Res. Appl.*, vol. 24, no. 4, pp. 570–583, 2016.

[18] S. Guo, J. P. Singh, I. M. Peters, A. G. Aberle, and T. M. Walsh, "A Quantitative Analysis of Photovoltaic Modules Using Halved Cells," *International Journal of Photoenergy*, vol. 2013, pp. 1–8, 2013.

[19] S. Guo, T. M. Walsh, A. G. Aberle, and M. Peters, "Analysing partial shading of PV modules by circuit modelling," in *38th IEEE Photovoltaic Specialists Conference*, 2012, pp. 2957–2960.

[20] C. E. Clement, J. P. Singh, E. Birgersson, Y. Wang, and Y. S. Khoo, "Illumination Dependence of Reverse Leakage Current in Silicon Solar Cells," *IEEE J. Photovoltaics*, vol. 11, no. 5, pp. 1285–1290, 2021.

[21] M. Mittag, A. Pfreundt, J. Shahid, N. Wöhrle, and D. Neuhaus, "Techno-Economic Analysis of Half Cell Modules - The Impact of Half Cells on Module Power and Costs," in *36th European Photovoltaic Solar Energy Conference and Exhibition (EU PVSEC)*, pp. 1032–1039.

[22] S. Nold, N. Voigt, L. Friedrich, D. Weber, I. Hädrich, M. Mittag, H. Wirth, B. Thaidigsmann, I. Brucker, M. Hofmann, J. Rentsch, and R. Preu, "Cost Modelling of Silicon Solar Cell Production Innovation Along the PV Value Chain," in *Proceedings of the 27th European Photovoltaic Solar Energy Conference and Exhibition*, Frankfurt, Germany, 2012.

978-1-6654-6060-6/23 $31.00 © 2023 IEEE

UV absorption utilizing a MoS₂/Ge Nano-Junction for solar applications

Ayman Rezk and Ammar Nayfeh

Khalifa University, Abu Dhabi, 127788, UAE

*Corresponding Author: *ammar.nayfeh@ku.ac.ae*

Abstract—We investigate the absorption capability of MoS₂/Ge nanoscale heterojunction using platinum (Pt) coated conductive atomic force microscope (c-AFM) tip. The MoS₂/Ge interface forms the proposed heterostructure of a nano diode-based ultraviolet (UV) absorber. The electrical characteristics observed at the nanoscale junction show the current response as well as the window between the illuminated and dark conditions, are larger after the insertion of the MoS₂ between the nanotip and Ge substrate. This research we believe will open a new path for the miniaturization of UV filters with high sensitivity for nano solar cells.

I. INTRODUCTION

Two-dimensional (2D) van der Waals heterostructures hold great promise as photodetectors due to their low cost and compatibility with the complementary metal oxide semiconductor (CMOS) technology [1-4]. This can be extended to use an UV optical filter for solar applications. MoS₂ captured a lot of attention in the past decade due to its intrinsic stability at nanoscale dimensions promising low power, and high density [5-10]. The claimed improvements in MoS₂-based devices can be related to their capacity to change the material interface properties and it's not limited by the absence of a band gap. Despite this, such devices are still in the early stages of development due to several issues, including growth scalability, passivation, doping, and metal junctions [11, 12]. Almost no work conducted on MoS₂/Ge-based systems which can be promising for various photonics applications, especially in the infrared regime.

Nanoscale resolution electrical characterizations utilizing conductive atomic force microscopy (c-AFM) provide a useful method to study electronic transport and conduction in MoS₂ structures. In this work, we demonstrate the photodetection capability of MoS₂/Ge nanoscale heterojunction using platinum (Pt) coated c-AFM tip.

II. METHODOLOGY

The cleaning and fabrication are conducted in a cleanroom of 100 level. We started by cutting square pieces with an area of 1×1 cm² from a 390 µm thick Ge bulk wafer. The native oxide from the top surface of the substrate is removed by ultrasonic cleaning in DI water. After removing the native oxide, pieces are cleaned by sonication for 1 minute in acetone, followed by rinsing in acetone, isopropanol then DI water. Immediately followed by spin coating 20µL of MoS₂ solution on the Ge substrates at 1000 RPM for 30 seconds. The (Sigma-Aldrich) MoS₂ solution concentration is 18 mg/L with flakes between 1-8 monolayers thick and 100-400 nm lateral size.

The photodetection effect in the fabricated samples is studied using a c-AFM tip on the sample surface. For the electrical measurements and topography, Pt-coated AFM tips with a 146 kHz resonance frequency and force constant of 1.2-29 N/m are used. The used Asylum Research MFP-3D c-AFM utilizes a probe holder (908.036) which is capable of measuring currents from ~1 pA to 20 nA, with a sensitivity of 2nA/V and a pre-amplifier output gain resistance R of 500M Ω. The bias was applied to the back of the substrate. To assure an ohmic contact, the scratched back of the Ge substrate is attached to gold-coated stainless steel using silver paint. During all electrical measurements, the tips were grounded, and a bias was applied to the substrate as depicted by the schematic in Fig. 1.

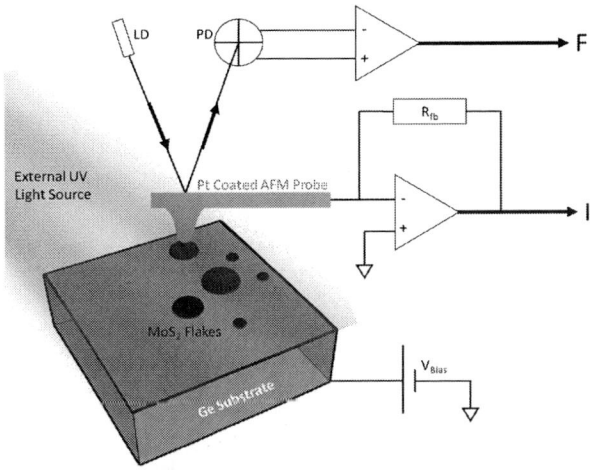

Figure 1: The Schematic illustration of the electrical measurement carried out by the Conductive atomic force microscope.

III. CHARACTERIZATION

The MoS₂ distribution on the surface was investigated using AFM and Scanning Electron Microscope (SEM). The topographic image of the substrate is carried out in the AC mode imaging of AFM, as shown in Fig. 2(a), with a scan length of 2.5 x 2.5 µm². The obtained SEM micrograph of the sample surface as shown in Fig. 2(b) shows the isolated MoS₂ flakes on the surface.

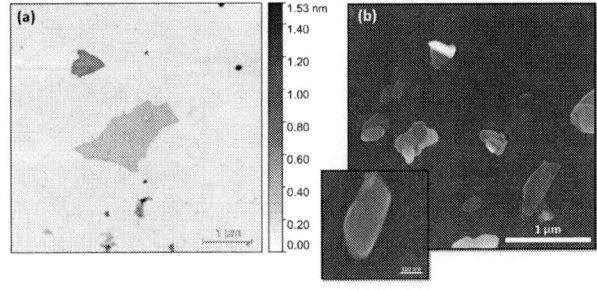

Figure 2: (a) AFM topography map and (b) SEM micrograph of MoS2 distribution on the Ge substrate

The total reflectance was measured by using a UV-vis-NIR spectrophotometer (LAMBDA 1050 UV/Vis/NIR Spectrometer, PerkinElmer) with an integrating sphere at a near-normal incidence angle of 0°. UV-Vis reflectance spectrum in Fig. 3 doesn't show a contribution from the

978-1-6654-6060-6/23 $31.00 © 2023 IEEE

MoS2 flakes due to their low coverage. However, a splitting of the 2 eV maximum of the Ge reflectance spectrum can be observed. The splitting can be linked to spin-orbit splitting of the $L_{3'}$ valence band extrema. The Λ_3 to Λ_1 transition has an energy close to the $L_{3'}$ to L_1 transition and has the most impact on the joint density of states function [13]. The observed structure in the reflectance spectra around 2 eV can be contributed to a combination of the L and Λ transitions or that the 2.3-eV peak corresponds to transitions at the A point and the peak at 2.1 eV corresponds to transitions at the L point. The inset in Fig. 3 depicts the absorption spectrum of the spin-coated MoS2 flakes.

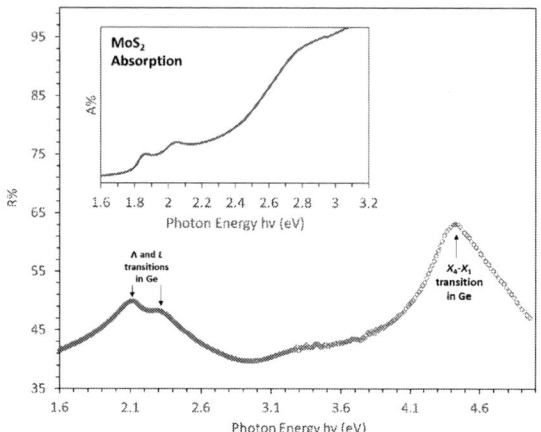

Figure 3: UV-Vis reflectance spectrum of the MoS₂/Ge structure. The inset shows the absorption spectrum of the spin-coated MoS₂ flakes.

Raman spectra of the investigated MoS₂ flakes (Fig. 4) is captured using a WITEC alpha300 R Confocal Raman system in air ambient conditions and obtained with an Ar-ion laser line of 488 nm. The measurements have a spectral resolution better than 1 cm⁻¹ and excitation laser power is kept around 1 mW.

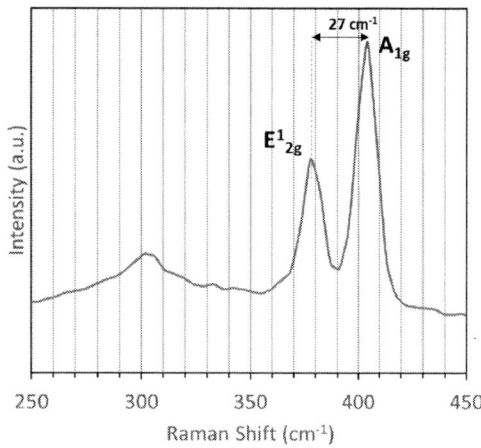

Figure 4: Raman spectra of the investigated MoS₂ flakes

IV. RESULTS AND DISCUSSION

To investigate the electrical characteristics of the van der Waals heterostructure in both dark and UV light conditions, all IV measurements are performed using an AFM system, placed in a vibration isolation black box. The voltage sweeps (-5V ~ 0 ~ +5V) are always applied on the substrate while the AFM tip is kept grounded. Figure 5 (a) and (b) show the

electrical characteristics of the Ge substrate and MoS₂ heterostructure respectively. We can readily see there is a prominent difference in the electrical characteristics under UV light and dark conditions. To ensure the reproducibility of the photodetection effect, we carried out the electrical characteristics three consecutive times for both structures, while alternating dark and UV light conditions every 1 minute. The photodetection phenomenon is reproducible and irrespective of the location of tip/structure contact.

The current response is significantly higher without the MoS₂ under dark conditions due to the much smaller tip area compared to the MoS₂ flake. A stronger electric field is generated at the tip-substrate interface, and the energy band bending becomes more prominent, which results in the thinning of the tunneling barrier. Thus, the smaller energy band width facilitates a higher tunneling probability for electrons from the tip into the Ge substrate, leading to a higher current response for the same voltage sweep on the substrate. However, under UV light, the current response is much higher with MoS₂ junction under the same voltage bias.

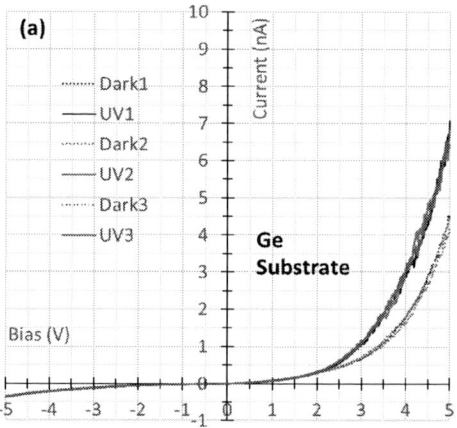

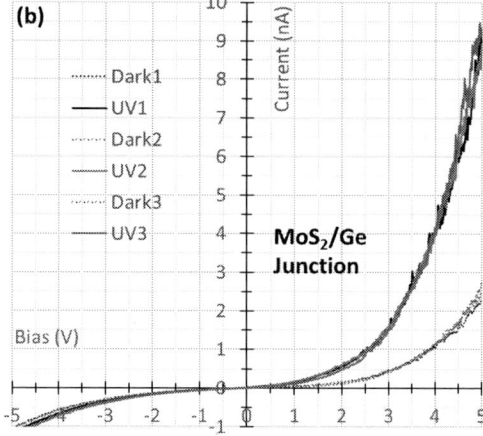

Figure 5: The electrical characteristics of the (a) Ge substrate and (b) MoS₂ heterostructure

Mott and Gurney's law [14] can justify the metal-semiconductor (MS) barrier height reduction and the low barrier height as compared to the metal work function, which causes space charge-limited current in semiconducting materials.

978-1-6654-6060-6/23 $31.00 © 2023 IEEE

Figure 6 demonstrates the Fowler-Nordheim (FN) emission in the Ge (red) and MoS$_2$/Ge junction (black) under dark (dotted) and UV light (solid). The linearity of the data confirms the conduction mechanism as being field-assisted tunneling.

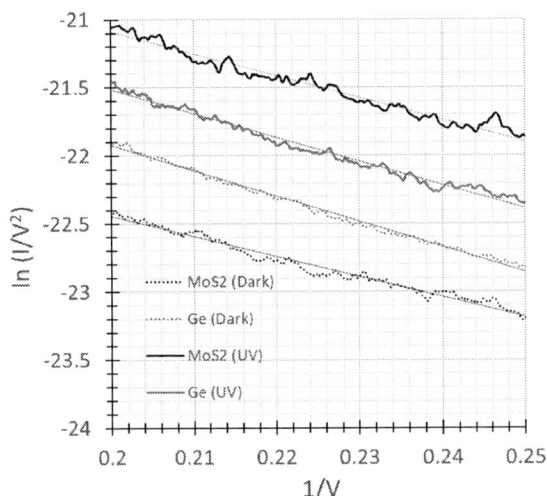

Figure 6: FN tunneling demonstrated by the ln(I/V^2) versus 1/V curves for Ge (red) and MoS$_2$/Ge junction (black) under dark (dotted) and UV light (solid).

IV. Conclusion

In conclusion, we have demonstrated UV absorption and filter effect in MoS$_2$ heterojunctions using a conductive AFM nanotip as well as the reproducibility of such an effect for solar applications.

References

[1] Wang Q H, Kalantar-Zadeh K, Kis A, Coleman J N and Strano M S 2012 Electronics and optoelectronics of two-dimensional transition metal dichalcogenides Nat. Nanotechnol. 7 699–712

[2] Si M, Liao P-Y, Qiu G, Duan Y and Ye P D 2018 Ferroelectric field-effect transistors based on MoS2 and cuinp2s6 two-dimensional van der waals heterostructure ACS Nano 12 6700–5

[3] Rezk A, Alhammadi A, Alnaqbi W and Nayfeh A 2022 Utilizing trapped charge at Bilayer 2d MoS2/SiO2 interface for memory applications Nanotechnology 33 275201

[4] Mukherjee B, Zulkefli A, Watanabe K, Taniguchi T, Wakayama Y and Nakaharai S 2020 Laser-assisted multilevel non-volatile memory device based on 2D van-der-waals few-layer-res 2 /h-bn/graphene heterostructures Adv. Funct. Mater. 30 2001688

[5] Lopez-Sanchez O, Lembke D, Kayci M, Radenovic A and Kis A 2013 Ultrasensitive photodetectors based on monolayer MoS2 Nat. Nanotechnol. 8 497–501

[6] Tsai M-L, Su S-H, Chang J-K, Tsai D-S, Chen C-H, Wu C-I, Li L-J, Chen L-J and He J-H 2014 Monolayer MoS2 heterojunction solar cells ACS Nano 8 8317–22

[7] ariwala D, Sangwan V K, Wu -C-C, Prabhumirashi P L, Geier M L, Marks T J, Lauhon L J and Hersam M C 2013 Gate-tunable carbon nanotube-MoS2 heterojunction p-n diode Proc. Natl Acad. Sci. 110 18076–80

[8] Widiapradja L J, Nam T, Jeong Y, Jin H J, Lee Y, Kim K, Lee S, Kim H, Bae H and Im S 2021 2D MoS2 charge injection memory transistors utilizing hetero-stack SIO2 /HFO2 dielectrics and oxide interface traps Adv. Electron. Mater. 7 2100074

[9] WooMH,JangBC,ChoiJ,LeeKJ,ShinGH,SeongH, Im S G and Choi S-Y 2017 Low-power nonvolatile charge storage memory based on MOS2 and an ultrathin polymer tunneling dielectric Adv. Funct. Mater. 27 1703545

[10] Radisavljevic B, Radenovic A, Brivio J, Giacometti V and Kis A 2011 Single-layer MOS2 transistors Nat. Nanotechnol. 6 147–50

[11] Das S, Chen H-Y, Penumatcha A V and Appenzeller J 2012 High performance multilayer MOS2 transistors with scandium contacts Nano Lett. 13 100–5

[12] Zhan Y, Liu Z, Najmaei S, Ajayan P M and Lou J 2012 Large-area vapor-phase growth and characterization of mos2atomic layers on a sio2 substrate Small 8 966–71

[13] Brust, D.; Phillips, J. C.; Bassani, F. Critical Points and Ultraviolet Reflectivity of Semiconductors. *Physical Review Letters* 1962, *9* (3), 94–97.

[14] Mott, N.F. and Gurney, R.W., 1940. Electronic Processes in Ionic Crystals. Oxford University Press, New York.

Defect Signatures in Admittance Spectroscopy of Perovskite Solar Cells

Rasha A. Awni[1,2], Zhaoning Song[1], Jared D. Friedl[1], Abasi Abudulimu[1], Adam B. Phillips[1], Randall J. Ellingson[1], Michael J. Heben[1], and Yanfa Yan[1]

[1] Department of Physics and Astronomy and The Wright Center for Photovoltaics Innovation and Commercialization (PVIC), University of Toledo, Toledo, OH 43606, USA

[2] Department of Physics, College of Education for Pure Sciences, Tikrit University, Tikrit, Salahedin, 34001, Iraq

Abstract — Understanding defect properties is key to improving the performance of solar cells. Thermal admittance spectroscopy (TAS) is widely used to analyze the properties of defect states in semiconductor devices. However, evaluating the defect capacitance signatures in admittance spectra is still problematic in emerging solar cells and needs to be clarified. The behavior of charge trapping and detrapping that introduces the capacitance step depends on the correlation between the defect properties with the applied temperature and frequency. Therefore, to understand the capacitance signatures of defects in the admittance spectra of perovskite solar cells, we perform detailed simulations using the one-dimensional (1D) solar cell capacitance simulator (SCAPS) software. The capacitance step can be introduced depending on the interplay between the defects and the measurement conditions. We found that the capacitance step at low frequencies is dominated by deep defects with large capture cross-sections, while the capacitance step of defects at high frequencies is dominated by shallow defects with small capture cross-sections. This indicates that not all deep or shallow defects are measurable by admittance spectroscopy due to the system's capability.

Keywords— admittance spectroscopy, perovskite, defects, capture cross-section, energy level, frequency, temperature

I. INTRODUCTION

Admittance spectroscopy (AS) is a powerful technique to understand the defect properties, such as the energy level, defect density of states (DOS), and their capture-cross section, from the AC response of semiconductor devices [1]. Admittance spectroscopy is commonly used to characterize the defects properties of inorganic semiconductor materials such as cadmium telluride (CdTe) [2], and, lately, in organic solar cells and perovskite solar cells (PSCs) [3]. However, determining the appropriate properties of the defects with the applied frequency and temperature of the experiment is one of the essential requirements for proper defect evaluation by admittance spectroscopy. The defect level (E_t) of a p-type semiconductor that locates inside the bandgap bends below the Fermi level (E_F) in the depletion region. The occupation status of the defect state can be changed as a response to the applied AC voltage (V_{AC}) at a certain temperature, i.e., the defect state can be filled and emptied from the charge, thus, contributing to the measured capacitance. Note that there is no charging or discharging and no contribution to the capacitance outside the depletion region where E_t is above E_F, because the defect state cannot be filled. Additionally, the defects that have enough thermal ionization energy can only keep up with the applied AC voltage and contribute to the capacitance if the emission rate of the defect state (e_h) is higher than the applied angular frequency (ω) of the AC voltage, $e_h \geq \omega$. However, if the emission rate of electrons is lower than the frequency of AC bias, they will not follow external bias and cannot contribute to the capacitance.

Most capacitance meters or impedance analyzers that are used to measure the AC signals of a semiconductor device can sweep AC voltage at a frequency ranging between 1 to 1×10^6 Hz. At frequencies higher and lower than the abovementioned range, the AC signals from the devices are noisy and cannot be analyzed. Usually, the samples can be cooled and heated in a cryo-chamber within a temperature range of 4 – 400 K. However, some materials may not introduce signals at very low temperatures, and others may degrade at high temperatures, resulting in system contamination and damage. This fact limits the capability of admittance spectroscopy to determine the defects in semiconductor devices. Thus, not all defects can be measured by admittance spectroscopy.

Here, we provide numerical simulations based on SCAPS software to show the correlation between the defect's response in PSCs and the applied AC signals. We focus on the influence of the applied frequency and temperature on the admittance spectroscopy measurements for the defect with various energy levels and capture cross-sections. Our simulations provide useful guidance in understanding the experimental limitations of AS measurements and defect characterization in PSCs.

II. SCAPS SIMULATIONS

SCAPS is simulation software that solves transport equations, continuity equations for electrons and holes, and the Poisson equation in 1D based on the drift-diffusion model [4]. We consider PSCs in the p-i-n architecture, ITO/PTAA/MAPbI$_3$/C$_{60}$, where ITO is indium tin oxide, a transparent conducting oxide that represents the front electrode, PTAA is a poly-triarylamine hole transport layer (HTL),

978-1-6654-6060-6/23 $31.00 © 2023 IEEE

MAPbI$_3$ is methylammonium led iodide, the absorber layer, and C$_{60}$ is a fullerene electron transport layer. The simulation parameters can be found in Ref [5]. The thickness of the perovskite layer is 650 nm. We considered an intrinsic perovskite layer such that the density of acceptors (N_a) equals to the density of donors (N_d). We introduced acceptors-like defect states with density (N_t) of 5×10^{16} cm^2 distributed uniformly. We varied the energy level location considering the valance band maximum as a reference. The capture cross-section of holes (σ_h) was also varied, whereas the capture cross-section of electrons ($\sigma_h=1\times10^{-19}$ cm^2) was fixed throughout the simulations. In our simulations, we assumed flat band contacts, which imply that there is no minority carrier injected due to an interface or Schottky barrier. By solving Poisson's equation in terms of the electrical potential over the entire device, the total charge (Q) is determined by the equation [6]:

$$Q = qA \int (p(x) - n(x) + N_d - N_a + N_{def}) \, dx \quad (1)$$

where $n(x)$ and $p(x)$ are the free electron and hole densities as a function of distance, and N_{def} is the total defect density. Under zero or moderate reverse bias conditions, the space charge region behaves like a capacitor, and its width changes with the applied bias voltage. The differentiation of the total charge (dQ) to the change in the applied bias voltage represents the obtained capacitance, $C(V_{bias}) = dQ/dV_{bias}$ originates from the shift of the depletion edge modulated by the small AC voltage. The simulation excludes capacitances associated with charges/ions migration and accumulation and interface polarization. Therefore, SCAPS simulation provides useful information about the AC response of charged defects, allowing focus on the correlation between the defect properties and the applied frequency and temperature.

III. ADMITTANCE SPECTROSCOPY ANALYSIS

The emission rate of a defect state depends on the capture cross-section (σ_h) and the energy level of the defect (E_t) and can be expressed by the following relationship [1]:

$$e_h = v_o \, exp(-\frac{E_t}{k_B T}) = \omega_{peak} \quad (2)$$

where v_o is the attempt-to-escape frequency, $v_o = v_{th} \, N_V \, \sigma_h$, which is a temperature-dependent parameter, v_{th} is the thermal energy, N_V is the density of states in the valance band, k_B is the Boltzmann constant, T is the absolute temperature, and ω_{peak} is the characteristic angular frequency. Note that $N_V \propto T^{3/2}$ and $v_{th} \propto T^{1/2}$, thus, $v_o = \xi T^2$, where ξ is a constant. The demarcation characteristic energy, E_t, represents the responded defect states to the AC voltage modulation at the transition frequency and temperature. Thus, temperature-dependent frequency sweeping resulted in different capacitance spectra. ω_{peak} can be extracted from the capacitance inflection point/ capacitance step that represents the maximum point of the peaks of the derivative of capacitance spectra, -ωdC/dω versus ω, see Fig. 1-2.

IV. RESULTS AND DISCUSSIONS

We present the AS simulation results for PSCs at room temperature. We introduced an acceptor-like defects state. The

energetic depth of this defect state was changed from a shallow level (0.1 eV) to a deep level near the midgap (0.7 eV), with a 0.1 eV energy step size. Here, the capture cross-section of holes was 1×10^{-15} cm^2. The resultant capacitance spectra and the corresponding derivative of capacitance spectra simulated at 300 K (as an example) are illustrated in Fig. 1. The capacitance step (Fig. 1a) due to the defect is shifted to the lower frequencies when the defect level becomes deeper. A similar trend is observed in the peaks (characteristic frequency, ω_{peak}) of the derivative of capacitance spectra in Fig. 1b. The characteristic frequency shifts to low frequencies when the defect level becomes deeper. This means the AC response at the low-frequency range is dominated by the deep defect level, while the AC response at the high-frequency range is dominated by shallow defect levels. For deep defect states, the emission rate decreases, which means the trap states have a slower emission rate. In contrast, shallow defects have a faster emission rate. Note that ω_{peak} of the shallow defects for $E_t < 0.4$ eV is shifted to frequencies > 10^7 Hz. Therefore, ω_{peak} cannot be determined.

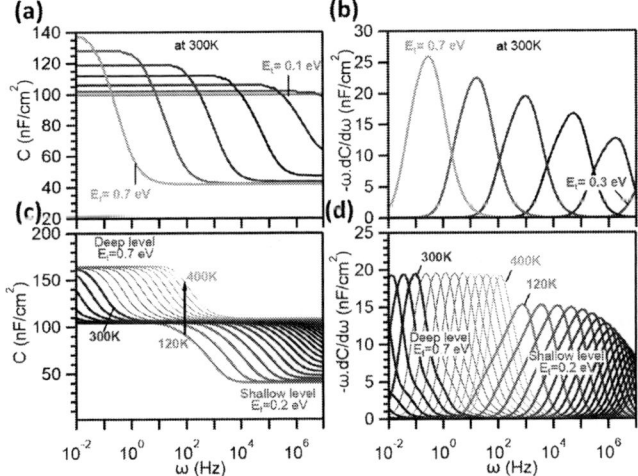

Fig. 1. Admittance spectroscopy results of PSC simulated under dark equilibrium. (a) The capacitance spectra and (b) the corresponding derivative of capacitance spectra simulated under 300 K with defect levels ranged between 0.2 – 0.7 eV with 0.1 eV energy level step. (c) The capacitance spectra, and the corresponding (d) derivative of capacitance spectra of PSC with two acceptor-like defect levels located at 0.2 and 0.7 eV, simulated with various temperatures, 120 – 400 K with 10 K temperature size. The defect capture cross section is 1×10^{-15} cm^2 and their density of states is 5×10^{16} cm^{-3}.

We further examined the temperature-dependent capacitance-frequency measurements with a case where there are two defect levels inside the bandgap: a shallow level located at 0.2 eV and a deep level located at 0.7 eV above the valance band edge. The results are shown in Fig. 1 c and d. The capacitance step and characteristic frequency in the derivative capacitance spectra shift from the low to the high frequencies with increasing temperatures from 120 to 400 K. Increasing the temperature leads to increasing the thermal ionization energy of the defect, and thus, leads to increasing the defect emission rate. In the experiment, we may not be able to calculate the defect activation energy because not all spectra that are shown in Fig. 1a and c can be measured for PSCs. A deep defect level similar to the one in Fig. 1c and d cannot be measured due to the shift of the capacitance step to the low frequencies, which is beyond many impedance analyzer system's frequency range limit.

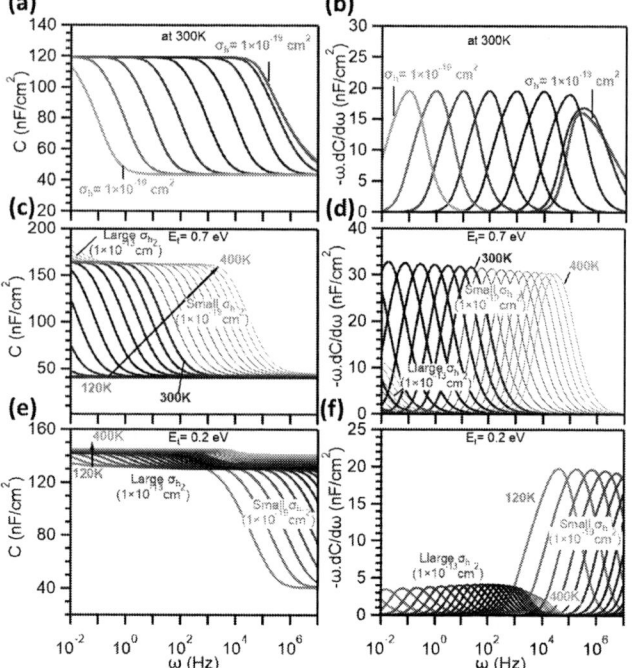

Fig. 2. (a, c, e) The capacitance spectra and (b, d, f) are the corresponding derivative of capacitance spectra for PSC. The AS measurements were simulated under 300K temperature with introducing defect level at 0.5eV inside the band gap of the perovskite layer and various σ_h ($1\times10^{-10} - 1\times10^{-19}$ cm^2 with 1×10^{-10} cm^2 log-stepped). Also, AS measurements were simulated under various temperatures ($120 - 400$K with 10K temperature step size) with two defect levels located at (c, d) 0.7 eV and (e, f) 0.2 eV above the valance band edge, and each defect level has different values of σ_h (1×10^{-13} and 1×10^{-19} m^2).

Since the defect emission rate is also influenced by the capture cross-section, we presented more simulations in Fig. 2 to illustrate the behavior of defects with various capture cross-sections. In Fig. 2a and b, the capacitance and the derivation of capacitance spectra for PSC with a defect energy level located at 0.5 eV above the valance band edge are illustrated. At 300 K, we notice that the capacitance step (Fig. 2a) and the transition frequency (Fig. 2b) are shifted to the low frequencies by increasing the defect capture cross-section. This is mainly due to the slow emission process of the defect state, and thus, the defects can only keep with low frequencies. The capacitance at high frequencies is dominated by defect signature with a small capture cross-section. Hence, the emission rate is fast, and the charges can only keep up with high frequencies.

We further examined the temperature-dependent capacitance-frequency measurements with a case where there are two defect levels inside the bandgap: a shallow level located at 0.2 eV and a deep level located at 0.7 eV above the valance band edge, with both small and large capture cross-sections. The results are shown in Fig. 1 c and d. The capacitance step and characteristic frequency in the derivative capacitance spectra shift from the low to the high frequencies with increasing temperatures from 120 to 400 K. In Fig. 2b and c, we can only see the capacitance signature of one defect level, which is the

one with small capture cross-section. The emission of deep defects with large capture cross section is much slower than any other type of defect, and their capacitance signature is shifted to low frequencies that are out of the frequency range limit, thus, it cannot be measured. While the capacitance signature of shallow defects with large and small capture cross sections (Fig. 2e and f) lies within the frequency range limits. Such defects can be measured experimentally by admittance spectroscopy.

Conclusions

In conclusion, we have shown by drift-diffusion simulations that the capacitance step in admittance spectroscopy can be introduced depending on the interplay between the defects and the measurement conditions, and thus, not all types of defects in thin PSCs are measurable. We found that the capacitance step at low frequencies is dominated by deep defects and defects with large capture cross-sections, while the capacitance step at high frequencies of defects is dominated by shallow defects with small capture cross-sections. If the emission rate of defects does not match the applied frequencies of the measurement system, then the defect properties are not measurable by AS measurement.

Acknowledgment

This material is based upon work supported by the National Science Foundation under contract number DMR-1807818, the U.S. Department of Energy's Office of Energy Efficiency and Renewable Energy (EERE) under the Solar Energy Technology Office Award Number DE-EE0008753 and by Air Force Research Laboratory under agreements FA9453-18-2-0037 and FA9453-19-C-1002. The views and conclusions contained herein are those of the authors and should not be interpreted as necessarily representing the official policies or endorsements, either expressed or implied, of the Air Force Research Laboratory, U.S. Department of Energy, or the U.S. Government.

References

[1] J. V. Li, and G. Ferrari, *Capacitance spectroscopy of semiconductors*, New York: Pan Stanford, 2018.

[2] R. A. Awni, D.-B. Li, C. R. Grice, Z. Song, M. A. Razooqi, A. B. Phillips, S. S. Bista, P. J. Roland, F. K. Alfadhili, R. J. Ellingson, M. J. Heben, J. V. Li, and Y. Yan, "The effects of hydrogen iodide back surface treatment on CdTe solar cells," *Solar RRL*, vol. 3, no. 3, pp. 1800304, 2019.

[3] R. A. Awni, Z. Song, C. Chen, C. Li, C. Wang, M. A. Razooqi, L. Chen, X. Wang, R. J. Ellingson, J. V. Li, and Y. Yan, "Influence of charge transport layers on capacitance measured in halide perovskite solar cells," *Joule*, vol. 4, no. 3, pp. 644-657, 2020/03/18/, 2020.

[4] M. Burgelman, P. Nollet, and S. Degrave, "Modelling polycrystalline semiconductor solar cells," *Thin Solid Films*, vol. 361-362, pp. 527-532, 2000/02/21/, 2000.

[5] Z. Ni, C. Bao, Y. Liu, Q. Jiang, W.-Q. Wu, S. Chen, X. Dai, B. Chen, B. Hartweg, Z. Yu, Z. Holman, and J. Huang, "Resolving spatial and energetic distributions of trap states in metal halide perovskite solar cells," *Science*, vol. 367, no. 6484, pp. 1352-1358, 2020.

[6] R. A. Awni, Z. Song, C. Li, L. Chen, S. Rijal, S. Bista, T. Zhu, X. Wang, and Y. Yan, "Numerical modeling of capacitance signatures of perovskite solar cells." 2022 IEEE 49th Photovoltaics Specialists Conference (PVSC).

978-1-6654-6060-6/23 $31.00 © 2023 IEEE

Modeling the Hardware Components of a Power Hardware-in-the-Loop Platform for Photovoltaic Applications

Edgardo Desarden-Carrero[1], Rachid Darbali-Zamora[2], Nelson E. Saavedra-Peña[1], Erick E. Aponte-Bezares[1]

[1]University of Puerto Rico-Mayagüez, Mayagüez, Puerto Rico 00682, USA
[2]Sandia National Laboratories, Albuquerque, New Mexico, 87185, USA

Abstract – Power Hardware-in-the-Loop (PHIL) enables the integration of a hardware device into a real-time (RT) simulation environment. This allows testing the performance of a physical device under real operating conditions before it is connected to an actual system. When developing the RT simulation models, there is no consideration of how the hardware device impacts the system because these are not typically modeled. To reduce the risks associated to instability from testing a physical device in a RT simulation environment using a PHIL platform, a high-fidelity simulation model that represents all the hardware components of the PHIL platform can be simulated and used before evaluating the hardware component. Herein, a simulation model of a PHIL platform is proposed, including power amplifier and photovoltaic (PV) inverter models. Simulation results show that the developed model of the PHIL platform is able to replicate the angle difference between the PV inverter current and the current measured on the RT simulation side. This causes a slight reduction in active power and a considerable difference in reactive power of 1.2 kVARs. A comparison between the simulation model and experimental results of a PHIL platform for a PV inverter operating with grid support functions is also presented.

Index Terms – *power hardware-in-the-loop (PHIL), voltage amplifier, system modeling, photovoltaic*

I. INTRODUCTION

The use of DC to AC conversion plays a critical role in power systems [1]. This has created a need for developing simulation alternatives that can accurately analyze the effects these devices could have on the power grid. Testing these devices without knowing if they are a potential safety hazard can be risky. The use of Power Hardware-in-the-Loop (PHIL) allows the integration of hardware components into a RT simulation environment, enabling a more accurate solution for testing the power behavior and performance of a physical device before they are connected to an actual system [2]. The hardware components of a typical PHIL platform are composed of a power amplifier, the equipment under test (EUT), and current and voltage measurement devices [3].

In most cases, when developing the real-time (RT) simulation models, there is no consideration for the impact from the hardware device because these are not typically modeled. In order to reduce the risks associated to instability from testing a physical hardware device in a RT simulation environment in a PHIL platform, a high-fidelity simulation model that represents all the hardware components of the PHIL platform can be simulated and used before evaluating the EUT. This would provide a glimpse into the interactions between the simulation model and the hardware components before performing the actual PHIL experiments. In order to achieve this, it is necessary to develop simulation models of each hardware

component of the PHIL platform. These models should replicate the hardware components with high-fidelity, capturing the delays attributed to the hardware interface.

Herein, a simulation model of a PHIL platform is presented. This article is organized in the following manner: Section II presents the power amplifier modeling, including the methodology for obtaining the power amplifiers transfer function. Section III summarizes the photovoltaic (PV) inverter modeling methodology, including the control architecture. Experimental and simulation results are illustrated in section IV and V respectively. Finally, section VI presents the conclusion.

II. POWER AMPLIFIER MODELING METHODOLOGY

High voltage amplifiers are one of the most important components used when implementing the ideal transformer method (ITM). The ITM is based on Thevenin's theorem. For this reason, the voltage measured in the software side must be re-constructed in the hardware side. The voltage amplifier acts as an actuator in a mechanical system [4]. Because the control signal generated in the software side is a voltage type, the amplifier is controlled by a voltage reference. The purpose of the voltage amplifier is to reproduce the system voltage at the point of common coupling on the software side. Fig. 1 shows a PHIL system with a physical hardware device connected. To ensure stability and accuracy, PHIL systems require power amplifiers with high accuracy and fast response. These characteristics are challenging because power, accuracy, and fast response are conflicting performance specifications [5].

Currently there are two power amplifier types commercially available for implementing a PHIL system: the switching power amplifier and the linear power amplifier. Switching power amplifiers are composed of a rectifier and an inverter, and they are cataloged as alternating-direct current converters. Switching power amplifiers typically offer high accuracy up to the medium power range and can reach high power ranges.

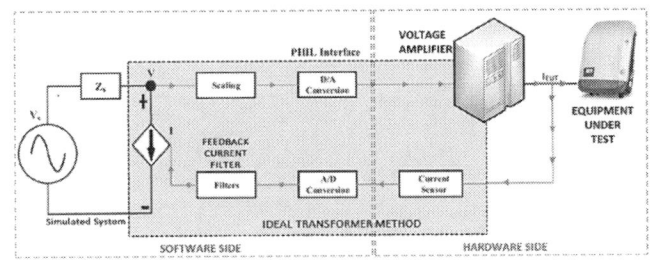

Fig. 1. Ideal transformer method with power amplifier and photovoltaic inverter connected in the hardware side.

The advantage of this power amplifier is its low cost with relatively high accuracy and power level. A disadvantage of this power amplifier is its susceptibility to harmonics and flickering caused by their internal switching components. Moreover, the switching reaction time of this power amplifier is lower compared to a linear power amplifier [4], [6].

Linear power amplifiers offer the same functionalities as switching power amplifiers and can operate in a linear range when required. This linear operation gives the power amplifier the capacity to handle low-level signals coming from the software side in a PHIL system [7]. Moreover, the linear power amplifier also provides a faster system response when compared with the switching power amplifier. Some disadvantages of these types of power amplifier include limited operational range, high cost and low efficiency (near 50%) [8].

In this work a MATLAB/Simulink model of an AMETEK RS 180 PWM three-phase power amplifier was developed. The power amplifiers supply a line-to-line voltage of 480V with a total power of 180 kVA. The power amplifier's harmonics frequency fundamental goes from 16 Hz to 820 Hz with an accuracy of 0.03%.

A. Power Amplifier Experimental Setup

In order to obtain the transfer function of a system, it is necessary to apply a signal at the input to see the system response at the output. The applied signal should be composed of a wide range of frequency components. Signals such as the impulse, the step, or the frequency sweep, offer a good frequency range to excite a system.

To avoid the Real-Time Simulator (RTS) processing delay in the model, an external two channel oscilloscope was used to collect the data. Channel one was connected to the RTS analog output and channel two to a voltage sensor connected to one phase of the power amplifier's output. The power amplifiers output is also connected to a parallel RLC load as illustrated in Fig. 2.

The Nyquist theorem, also known as the sampling theorem, stipulates that to accurately reproduce a signal the sampling frequency must be at least double the signal's frequency. The rule followed to guarantee optimal results is to use a sampling frequency at least ten times greater when possible. Based on this requirement, the oscilloscope sampling frequency was set to 8 kHz which is 10 times greater than the power amplifier maximum frequency of 820 Hz.

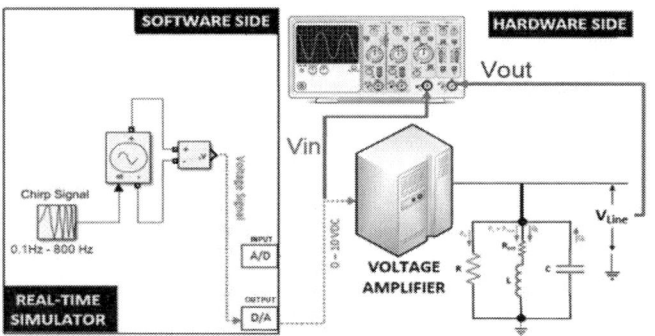

Fig. 2. Experiment setup to determine the power amplifier transfer function.

B. Input and Output Signal

Because the power amplifier has a maximum frequency of 820 Hz a combination of a step (ranging from 0 V to 1 V) and a frequency sweep (ranging from 0.1 Hz to 800 Hz) was used as the input signal. This input prevents the power amplifier from shutting down due to high frequency harmonics. The input signal is generated using the MATLAB/Simulink CHIRP function that controls the RTS analog output card. The analog output card can supply voltages between -10 V to 10 V. This analog voltage signal will be the power amplifiers control signal. When the power amplifiers control signal ranges from -10 V to 10 V, the power amplifier's output ranges between -430 V to 430 V. This means that for a 480/277 V_{rms} three phase system, the power amplifier voltage can reach 10% higher if necessary.

C. Tranfer Function Estimation

Transfer function identification based on frequency response is a widely used engineering tool. The algorithm (tfest) released in the MATLAB version 2016b was used in this work to estimate the power amplifier transfer function. The algorithm identification task is formulated as a nonlinear least squares problem. The algorithm implements a continuous time transfer function estimation from frequency response measurements using Sanathanan-Koerner and Instrumental Variable iterations combined with orthogonal rational basis functions. These functions lead to zero-pole-gain models and allow the transfer function estimation.

Before using the *tfest* MATLAB function, an identification data object is constructed with the *iddata* command. The *iddata* command has the structure *iddata(y, u, Ts)*, where *y*, *u*, and *Ts* are the output signal, the input signal and the sampling time respectively. Once the *iddata* object was created, the *tfest* function has the structure *tfest(data,np,nz)*, where *data*, *np*, and *nz* are the *iddata* object, the number of poles, and the number of zero, respectively. The *tfest* function will return the system transfer function together with the percent of curve fitting. Results with a positive curve fitting greater than 85% are acceptable.

After obtaining the estimated system transfer function, the power amplifier can be represented by a gain and a transfer function controlling three voltage sources. A very low resistance should be connected in series with each voltage source to prevent voltage sources connected in parallel and to guarantee equation system converge.

III. PHOTOVOLTAIC INVERTER MODELING METHODOLOGY

Besides the voltage amplifier, the other component in the MATLAB/Simulink model is the PV inverter. Grid tie PV inverters produce a voltage sine wave signal that closely matches the grid's voltage. The voltage is generated in the inverter using Pulse Width Modulation (PWM) and filtering techniques. Smart grid tied inverters are designed to control the power injected or absorbed to the grid and they also can help to maintain the system's voltage magnitude and frequency through the use of grid support functions (GSF) [9]–[11]. Based on a commercial PV inverter, a MATLAB/Simulink model was developed. This commercial PV inverter model doesn't consider the LCL output filter when regulating the specified power. The modeled PV inverter was represented using a three-phase voltage source controlled by a current control loop that regulates output power through an inductor. In series with the PV inverter model is the LCL filter designed to filter the PMW switching frequency and to provide reactive imbalance to improve the island condition detection [12].

A. Photovoltaic Inverter Mathematical Model

The PV inverter mathematical model was represented using a controlled voltage source connected to the grid through an $R+jL$ impedance. A current control loop regulates the PV inverter output power when modulating its voltage. The PV inverter model synchronizes with the grid using a phase-looked loop (PLL) control which follows the grid voltage, adjusting the phase angle until the quadrature component V_q becomes zero. The differential equation that represents the PV inverter dynamics is shown in equation (1).

$$\frac{dI}{dt} = -\frac{R}{L} \cdot I + \frac{V - V_g}{L} \tag{1}$$

In this equation, the variable L is the output filter inductance, and the variable R is the inductance internal resistance. Variable V is the inverter voltage before the at the terminals of the switching devices and variable V_g is the grid voltage. The system only has one state, which is provided by the inductor current [13].

The PV inverter model was develop using direct (d) and quadrature (q) reference frame theory as described in [14], [15]. The PV inverter equations expressed in matrix form are shown in equation (2) through equation (4) respectively.

$$L\begin{bmatrix} \dot{I}_d \\ \dot{I}_q \end{bmatrix} = \begin{bmatrix} -R & 0 \\ 0 & -R \end{bmatrix}\begin{bmatrix} I_d \\ I_q \end{bmatrix} + \begin{bmatrix} 0 & \omega \\ \omega & 0 \end{bmatrix}\begin{bmatrix} I_d \\ I_q \end{bmatrix}L + \begin{bmatrix} V_d - Vg_d \\ V_q - Vg_q \end{bmatrix} \tag{2}$$

$$\begin{bmatrix} V'_d \\ V'_q \end{bmatrix} = \begin{bmatrix} 0 & \omega \\ -\omega & 0 \end{bmatrix}\begin{bmatrix} I_d \\ I_q \end{bmatrix}L + \begin{bmatrix} V_d - Vg_d \\ V_q - Vg_q \end{bmatrix} \tag{3}$$

$$\begin{bmatrix} \dot{I}_d \\ \dot{I}_q \end{bmatrix} = \begin{bmatrix} \frac{-R}{L} & 0 \\ 0 & \frac{-R}{L} \end{bmatrix}\begin{bmatrix} I_d \\ I_q \end{bmatrix} + \begin{bmatrix} \frac{1}{L} & 0 \\ 0 & \frac{1}{L} \end{bmatrix}\begin{bmatrix} V'_d \\ V'_q \end{bmatrix} \tag{4}$$

The one-line diagram for the PV inverter model (including an LR impedances) is illustrated in Fig. 3.

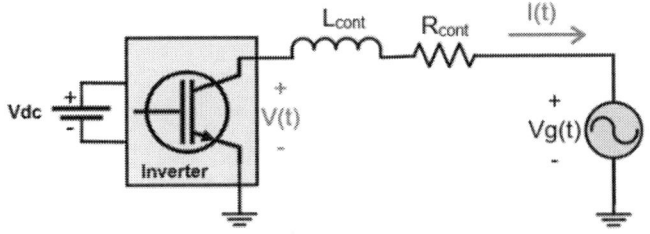

Fig. 3. Diagram of the photovoltaic inverter model.

B. Photovoltaic Inverter Control Strategy

The PV inverter output power is controlled by regulating the dq axis currents. Using the Park transformation, the total active power and reactive power injected to the grid are shown in equation (5) and equation (6), respectively [13].

$$P = \frac{3}{2}\left(V'_d I_d + V'_q I_q\right) \tag{5}$$

$$Q = \frac{3}{2}\left(V'_q I_d + V'_d I_q\right) \tag{6}$$

From the state-space system shown in equation (4), the matrix A and B, and vector C can be obtained. The values of the matrices A, B and vector C are shown in equation (7).

$$A = \begin{bmatrix} \frac{-R}{L} & 0 \\ 0 & \frac{-R}{L} \end{bmatrix}, B = \begin{bmatrix} \frac{1}{L} & 0 \\ 0 & \frac{1}{L} \end{bmatrix}, C = [1 \quad 1] \tag{7}$$

Replacing A, B and C into equation (8), it is possible to obtain the system transfer function, as shown in equation (9).

$$\frac{Y(s)}{U(s)} = C(sI - A)^{-1}B + D \tag{8}$$

$$\begin{bmatrix} V'_d \\ V'_q \end{bmatrix} = \begin{bmatrix} \frac{1}{Ls + R} & 0 \\ 0 & \frac{1}{Ls + R} \end{bmatrix}\begin{bmatrix} I_d \\ I_q \end{bmatrix} \tag{9}$$

Table I summarizes the Proportional-Integral (PI) controller gains and the RL filter values selected for the PV inverter model.

TABLE I:
PV INVERTER MODEL PARAMETER DESCRIPTION

Parameter	Description	Value	Units
K_i	Integral Gain	4.8	V/A·s
K_p	Proportional Gain	0.144	V/A
L_{cont}	Control Inductor	0.1	mH
R_{cont}	Inductor Resistance	10	μΩ
f_{nom}	Nominal Frequency	60	Hz
V_{LL}	Line to line voltage	480	V$_{rms}$
S_{tot}	Inverter Capacity	24	kVA

The control strategy for this system was achieved by using a PI current controller with a feedforward path of V_{set} as illustrated in Fig. 4.

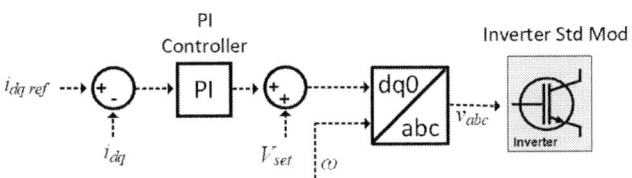

Fig. 4. Diagram of the photovoltaic inverter model control strategy.

978-1-6654-6060-6/23 $31.00 © 2023 IEEE

C. LCL Output Filter Design

The purpose of an LCL filter is to attenuate harmonics and to provide a sinusoidal output voltage waveform. It consists of two inductors (L) connected in a T configuration with a capacitor (C) in the center, forming the LCL configuration. The LCL filter is placed between the inverter output and the load. Although in most cases the LCL output current is considered in the PV inverter's control strategy, the behavior of commercial PV inverter used for this analysis proved to not consider the LCL filter in the controls. Hence, the LCL output filter is not considered as part of the PV inverter model. This means that the LCL filter should be designed in a very precise manner in order to verify that the filter dynamics are not affecting the PV inverter stability.

The parameters for the commercial PV inverter are necessary in order to properly estimate the LCL filter values. The PV inverter parameters are summarized in Table II.

TABLE II:
LABORATORY PV INVERTER PARAMETER DESCRIPTION

Parameter	Description	Value	Units
S_{tot}	Inverter Capacity	24	kVA
V_{LL}	Line to line voltage	480	V_{rms}
f_{sw}	Switching Frequency	20	kHz
f_{nom}	Nominal Frequency	60	Hz
η	Inverter Efficiency	99.5	%
V_{Bus}	DC Bus Voltage	680	V

As shown in Table II the values of the LCL filter were not available. Those values are estimated based on the power measurements, the switching frequency, the cutoff frequency, the PV inverter efficiency, and the DC bus voltage.

In an LCL filter the PV inverter side inductor (L_i) is responsible for filtering out the switching frequency's harmonics. The correct selection of L_i will depend on the ripple current ($\Delta i_{pp\ max}$), the switching frequency (F_{sw}) and the DC bus voltage (V_{Bus}) as shown in equation (10).

$$L_i = \frac{V_{Bus}}{4 \cdot F_{sw} \cdot \Delta i_{pp\ max}} \tag{10}$$

To compute L_i, we will assume 20% of ripple in the PV inverter full load output current. The calculated L_i value using equation (11) is shown in equation (11).

$$L_i = \frac{680\ V}{4 \cdot (20\ kHz) \cdot (16.29\ A)} = 520\ \mu H \tag{11}$$

The capacitor (C_f) and the grid side inductor (L_o) will form a lowpass filter that filters out the switching frequency. C_f should be kept small, usually equal or less than 5% of rated power because the capacitance determines de reactive power exchange when the PV inverter is offline. Five percent of the PV inverter capacity will represent a reactive power of 400 Vars per phase which gives a C_f value of 13.75 μF.

To calculate the grid side inductance L_o the resonance frequency in the LCL filter should be greater than one sixth of the switching frequency as shown in equation (12).

$$\frac{1}{2 \cdot \pi} \cdot \sqrt{\frac{L_i + L_o}{L_i \cdot L_o \cdot C_f}} > \frac{F_{sw}}{6} \tag{12}$$

A L_o value of 86 μH was obtained after setting the resonance frequency to 5 kHz. The inductors' internal resistance can be estimated from the PV inverter efficiency. At full load the PV inverter has an efficiency of 99.5%, which means 5% was lost due to the inductor's internal resistance. Based on efficiency and assuming the same internal resistance for both inductors in the filter, the real power dissipated on each inductor is 20 W. This means that each inductor will have an internal resistance of 0.025 Ω. Table III summarizes the LCL filter estimated values for the PV inverter model.

TABLE III:
LCL FILTER PARAMETERS DESCRIPTION

Parameter	Description	Value	Units
L_i	Inverter Side Inductor	520	μH
R_i	Parasitic resistance of L_i	0.025	Ω
L_o	Grid Side Inductor	86	μH
R_o	Parasitic resistance of L_o	0.025	Ω
C_f	Filter Capacitor	13.75	μF

The ability to regulate power is one of the most important capabilities an inverter can provide. When combined with GSFs, PV inverters are able to regulate output voltage and frequency within acceptable limits, by controlling its active and reactive power.

PV inverters should regulate the output voltage to provide a stable and consistent power supply. This is typically achieved by using feedback control mechanisms that monitor the output voltage and adjust the inverter's operation to maintain it within a specific range [16].

To guarantee accurate power management, the PV inverter model should be stable. Fig. 5 shows the inverter model with the RL control impedance connected in series with a LCL filter. Notice that the system transfer function's input is the PV inverter voltage while the output is the current flowing through the control impedance.

In order to validate the stability of the whole model presented in Fig. 5, a Root-Locus diagram was developed and presented in Fig. 6. The Root-Locus shows that the connection in series of the LCL filter to the PV inverter model is stable because there are no poles in the right-hand side of the imaginary plane. When considering the PI controller in the PV inverter models stability analysis, the pole added at the origin by the PI controller guarantees a zero steady state error and allows controlling the systems response.

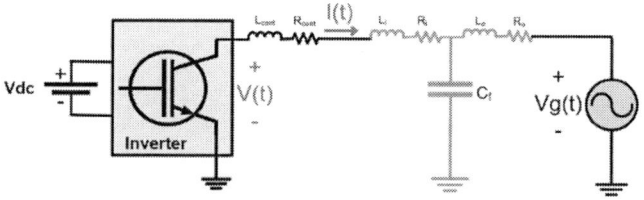

Fig. 5. Diagram of the photovoltaic inverter model connected in series with a LCL filter.

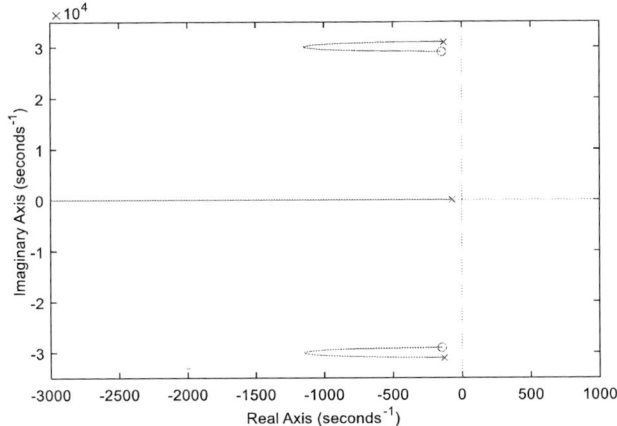

Fig. 6. Root-Locus of the photovoltaic inverter model connected to the LCL filter.

D. MATLAB/Simulink models validation through simulation

The IEEE Std 1547 UI Cat B. Test 5B was used to validate the power amplifier and the PV inverter models [17]. This test was simulated using MATLAB/Simulink, where the hardware component is the develop model of the power amplifier and the PV inverter including the output LCL filter.

To validate the amplifier model, test 5B was conducted with the PV inverters GSFs enabled and without any compensation in the PHIL. The objective is to observe and quantify the error introduced by the power amplifier model if it is not compensated with a lead controller.

To validate the inverter model, test 5B will be conducted with the inverter GSFs enabled and with a lead compensator to match the power quantities. The goal is to evaluate if the inverter tracks the real and reactive power imposed by the GSF algorithm to control the voltage and the frequency when the system goes to island. The simulation results in this test will be compared to the experimental results obtained in our previous work related to the evaluation of the Inverter GSF using a PHIL platform [18].

IV. EXPERIMENTAL RESULTS

A. Power Amplifier Model

The PHIL system one-line diagram used to perform test 5B is shown in Fig. 7. Prior to estimating the system transfer function using the *tfest* function, a sinusoidal excitation of the system is be performed. Fig. 8 (a) and Fig. 8 (b) illustrate the power amplifier's input and output signals respectively recorded by the oscilloscope. The output signal shows a gain of 43 when compared to the input signal. This gain is included in the transfer function.

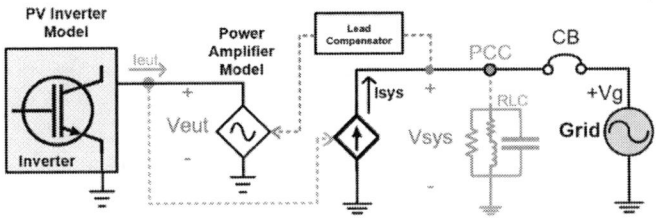

Fig. 7. Power Hardware-in-the-Loop one-line diagram.

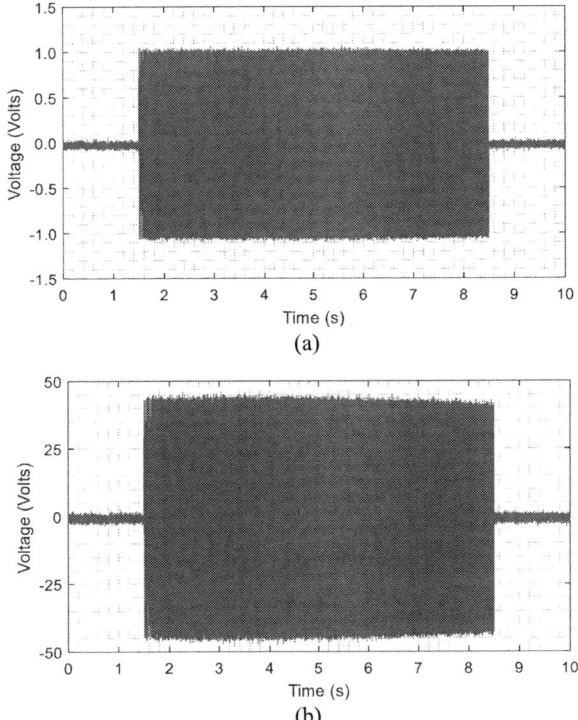

Fig. 8. Power amplifier frequency sweep signals. (a) Input signal. (b) Output signal.

To obtain the best approximation of the power amplifiers transfer function, the power amplifier is subjected to different load conditions, including resistive, inductive, and capacitive loads. Moreover, different combination of poles and zeros evaluated to obtain the transfer function with the best fit. After evaluating different configurations of pole-zero combinations during the estimation process, the most reasonable results were obtained with a single pole system that shows a fit of 95%. Systems with higher number of poles and zero complicated the system and only increased the fitting by values of less than 1%. Fig. 9 shows the bode plot magnitude and phase for the power amplifier under different load conditions.

After evaluating the bode plots results of Fig. 9, the transfer function that best distributes the error is obtained when the power amplifier is at full load (8 kW per phase). The bode plot results at full load show a magnitude gain of 32.7 db which is equal to a voltage gain of 43.24. The cutoff frequency indicated in the bode plot is 1.4 kHz which corresponds to 8,806 rads/sec. With the gain and cutoff frequency obtained from the bode plot results, the power amplifier transfer function can be obtained and is shown in equation (13).

$$tf_{amp} = 43.24 \cdot \frac{8,806}{S + 8,806} \qquad (13)$$

As previously mentioned in the methodology, the power amplifier can be represented by a gain and a transfer function controlling three Y connected voltage sources. A very low resistance should be connected in series with each source to prevent parallel connected sources. Fig. 10 shows the resulting power amplifier transfer function and how its implemented.

978-1-6654-6060-6/23 $31.00 © 2023 IEEE

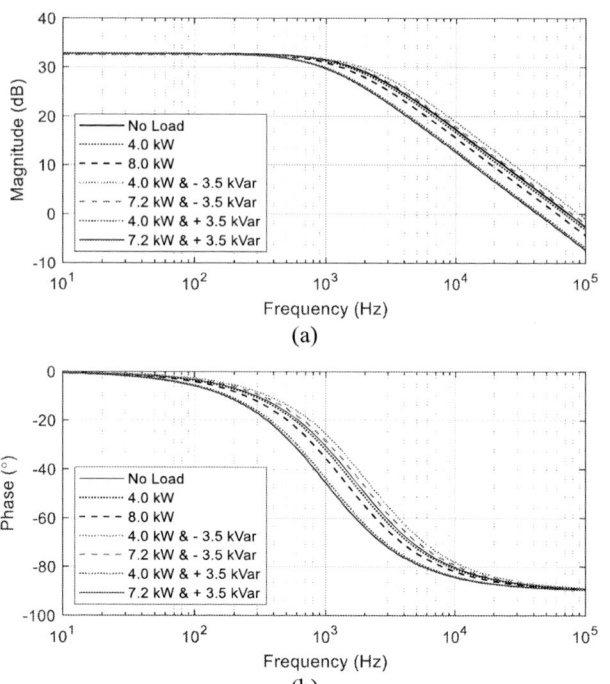

(a)

(b)

Fig. 9. The bode plots for different amplifier load conditions. (a) Magnitude. (b) Phase.

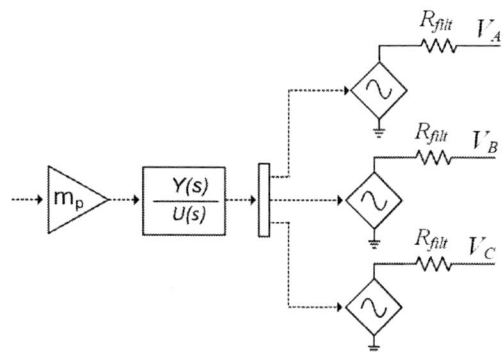

Fig. 10. Diagram of the power amplifier model.

V. SIMULATION RESULTS

A. Power Amplifier Model Simulation Results

Test 5B was conducted with the inverter GSFs disabled and without any compensation in the PHIL system. Fig. 11 (a) shows that the system voltage (V_{sys}) and the EUT voltage (V_{eut}) are in phase due to the PLL. The angle difference between the currents is approximately 2.89°, and it represents the reactive power mismatch. Fig. 11 (b) shows a slight reduction in active power and a reactive power difference of 1.2 kVAR. This difference in reactive power is the effect of the amplifier model without any compensation in the PHIL system. For this simulation, a resonant RLC load was connected in parallel with the PV inverter. This means that only active power should be demanded by the RLC load. The PV inverter LCL output filter will supply a small amount of 480 VARs to the system, but this amount should be reflected in the software side as well in the EUT side. A lead controller should be added to compensate for the system by reducing the lagging effect in the currents.

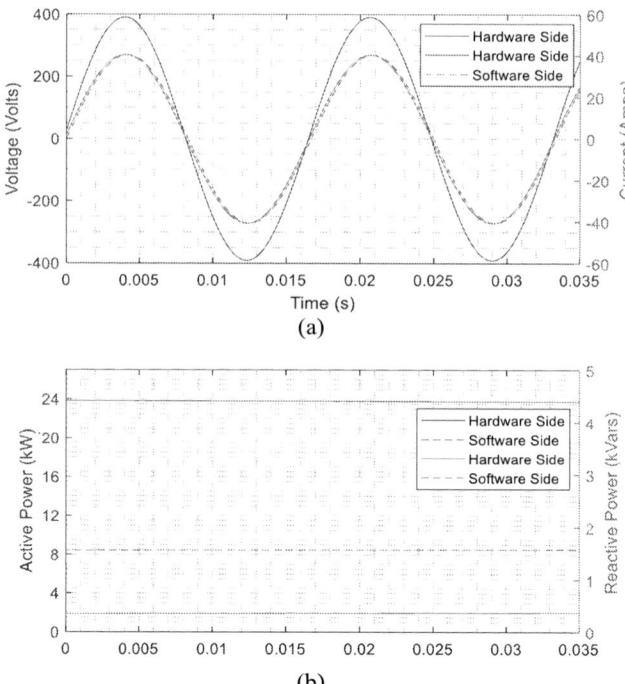

(a)

(b)

Fig. 11. Simulation results comparing the system side (software side model) with the equipment under test side (hardware side model). (a) The voltage and currents. (b) The active and reactive power.

B. Photovoltaic Inverter Model Simulation Results

Experimental results for the IEEE Std 1547.1 UI Test 5B are shown in Fig. 12 and were obtained from previous work [18]. This provides a comparison between the experimental and the simulated model under the same test conditions. The PV inverter is tested with GSFs (VVC, FWC) enabled at the default values. Unbalanced reactive power is added to the load, forcing the GSFs to start operating immediately after de islanded condition that is introduced at 1 second. Fig. 12 shows that after the island condition is introduced, the load voltage increases suddenly and goes outside of the upper deadband limit. The PV inverter activates the VVC, decreasing the reactive power and reducing the voltage. Once the PV inverter voltage is within the deadband, the VVC is suspended. The reduction in reactive power increases the frequency, forcing it outside of its deadband upper limit. The FWC activates in order to reduce the frequency, decreasing the active power and increasing the reactive power until frequency is within the deadband. The increase in reactive power will increase the system voltage, activating the VVC again which will create voltage and frequency oscillations close to its nominal values.

Simulation results for the IEEE Std 1547.1 UI Test 5B are shown in Fig. 13. Results show that after 1 second, the VVC activates but the system voltage decreases faster than in the experimental results shown in Fig. 12. This is due to the effect of the commercial PV inverters filter. After this point, the simulation behaves similar to the experimental results with the only difference that the VVC adjusts the voltage until it reaches the lower deadband instead the upper limit. Simulation results show that the PV inverter adjusts the active power and the reactive power according to the GSFs, ensuring the voltage and frequency are oscillating at nominal values.

978-1-6654-6060-6/23 $31.00 © 2023 IEEE

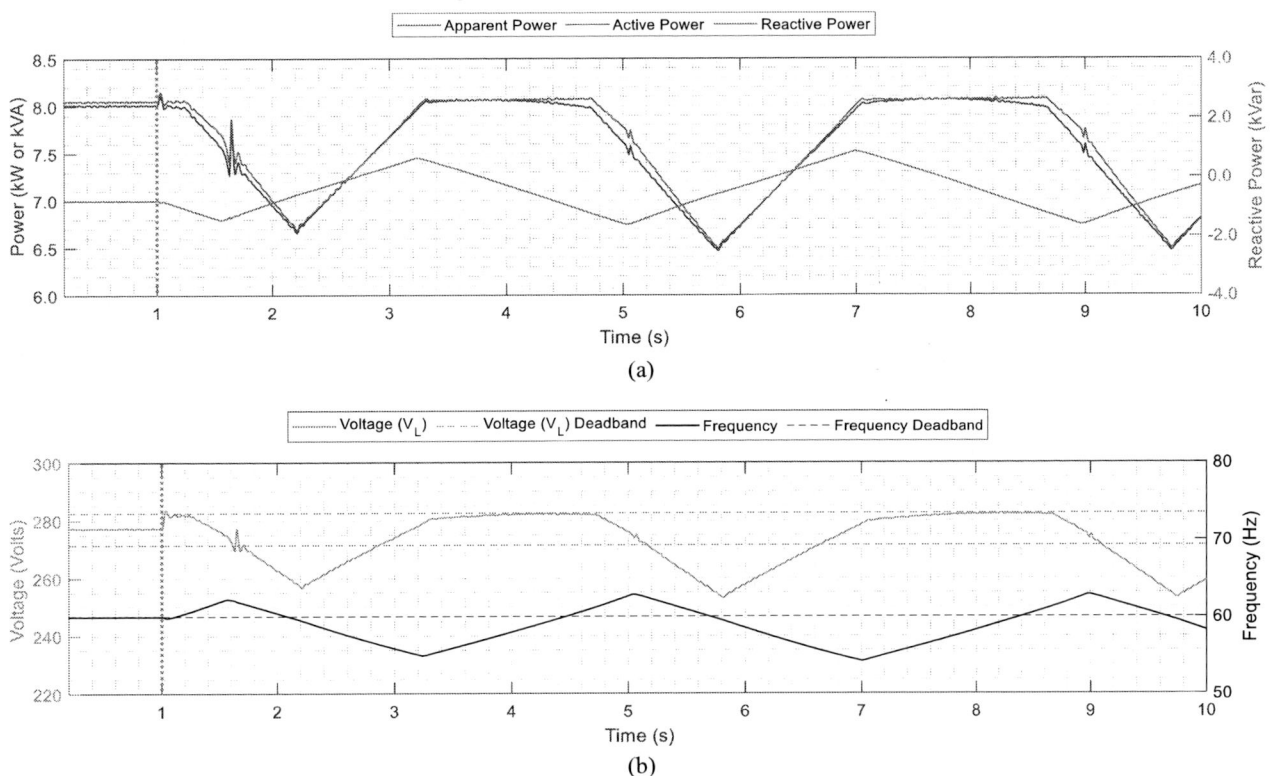

Fig. 12. IEEE Std 1547 UI Cat B. Test 5B implemented experimentally with Power Hardware-in-the-Loop (VVC and FWC). (a) Power at the load. (b) Load voltage and frequency.

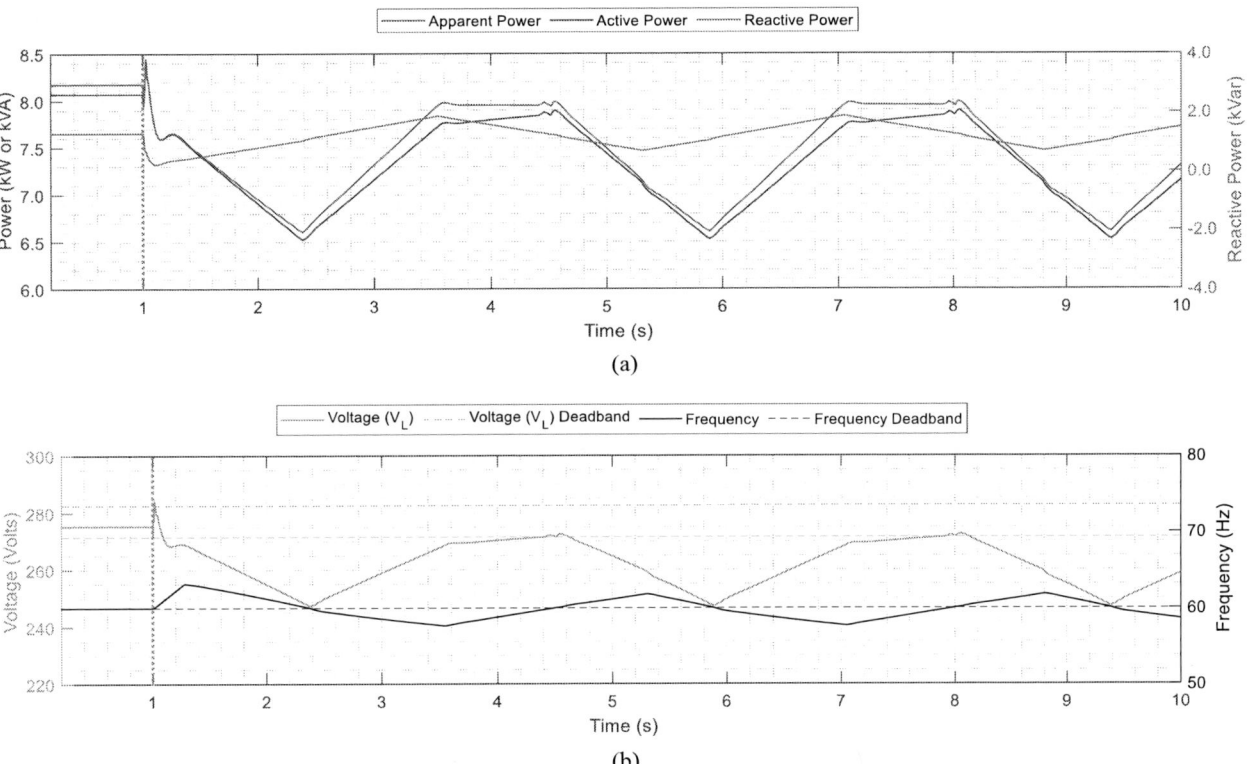

Fig. 13. Simulation of the IEEE Std 1547 UI Cat B. Test 5B implemented with Power Hardware-in-the-Loop (VVC and FWC). (a) Power at load. (b) Load voltage and frequency.

VI. Conclusion

Power amplifiers represent a considerable delay in a PHIL platform, altering the power measurement values. These measurement inaccuracies in the reactive power readings could lead the system to trip due to under or over-frequency as well as voltage. The pole that the power amplifier adds to the system will interact with other systems dynamics, including load, input filter, output filter, and RTS delay. If not considered, it could lead to system instability and low right-through performance from PV inverters. A simulation model of a PHIL platform that replicates the hardware components, capturing the delays attributed to the hardware interface, is proposed. This model includes both the power amplifier and the PV inverter. Simulation results show that the developed model of the PHIL platform is able to replicate the angle difference between the PV inverter current and the current measured on the RT simulation side. The developed PV inverter model behaves in a similar manner very well responding to the GSF commands. The fast and accurate tracking ability of the PV inverter model allows regulating the voltage and the frequency when subjected to islanded conditions.

Acknowledgment

This article has been authored by an employee of National Technology & Engineering Solutions of Sandia, LLC under Contract No. DE-NA0003525 with the U.S. DOE. The employee owns all right, title and interest in and to the article and is solely responsible for its contents. The U.S. Government retains and the publisher, by accepting the article for publication, acknowledges that the U.S. Government retains a non-exclusive, paid-up, irrevocable, world-wide license to publish or reproduce the published form of this article or allow others to do so, for U.S. Government purposes. The DOE will provide public access to these results of federally sponsored research in accordance with the DOE Public Access Plan.

This work was sponsored in part by the Consortium for Hybrid Resilient Energy Systems (CHRES) under grant number DE-NA0003982 from the National Nuclear Security Administration part of the U.S. Department of Energy.

References

[1] E. Desarden-Carrero, R. D. Zamora, E. Aponte-Bezares, and E. I. Ortiz-Rivera, "Seven-Level Cascaded H-Bridge Multilevel Single-Phase Inverter Implemented with an ATMEGA Microprocessor", *2022 IEEE 49th Photovoltaics Specialists Conference (PVSC)*, 2022, pp. 0916-0922.

[2] R. Darbali-Zamora, J. E. Quiroz, J. Hernández-Alvidrez, J. Johnson and E. I. Ortiz-Rivera, "Validation of a Real-Time Power Hardware-in-the-Loop Distribution Circuit Simulation with Renewable Energy Sources", *2018 IEEE 7th World Conference on Photovoltaic Energy Conversion (WCPEC) (A Joint Conference of 45th IEEE PVSC, 28th PVSEC & 34th EU PVSEC)*, 2018, pp. 1380-1385.

[3] E. Desarden-Carrero, R. Darbali-Zamora, N. S. Gurule, E. Aponte-Bezares and S. Gonzalez, "Evaluation of the IEEE Std 1547.1-2020 Unintentional Islanding Test Using Power Hardware-in-the-Loop", *2020 47th IEEE Photovoltaic Specialists Conference (PVSC)*, 2020, pp. 2262-2269.

[4] C. S. Edrington, M. Steurer, J. Langston, T. El-Mezyani and K. Schoder, "Role of Power Hardware in the Loop in Modeling and Simulation for Experimentation in Power and Energy Systems", *Proceedings of the IEEE*, vol. 103, no. 12, pp. 2401-2409, Dec. 2015.

[5] Z. Zhang, R. Schürhuber, L. Fickert, X. Liu, Q. Chen, and Y. Zhang, "Hardware-in-the-loop Based Grid Compatibility Test for Power Electronics Interface", *2019 20th International Scientific Conference on Electric Power Engineering (EPE)*, 2019, pp. 1-6.

[6] N. D. Marks, W. Y. Kong, and D. S. Birt, "Stability of a Switched Mode Power Amplifier Interface for Power Hardware-in-the-Loop", *IEEE Transactions on Industrial Electronics*, vol. 65, no. 11, pp. 8445-8454, Nov. 2018.

[7] N. D. Marks, W. Y. Kong, and D. S. Birt, "Interface Compensation for Power Hardware-in-the-Loop", *2018 IEEE 27th International Symposium on Industrial Electronics (ISIE)*, 2018, pp. 413-420.

[8] M. Orr, "Selecting a Linear or PWM Power Source", *Pacific Power Source, Technical Report*, Nov. 2008, pp. 1-4.

[9] R. Teodorescu, M. Liserre, and P. Rodríguez, "Grid Converters for Photovoltaic and Wind Power Systems", *Wiley-IEEE Press*, 2007.

[10] M. U. Rafique, S. Aslam and J. Anwer, "Implementation of a Novel Microcontroller-Based Voltage Source Sine-Wave Inverter", *2011 Frontiers of Information Technology*, 2011, pp. 167-172.

[11] A. S. K. Chowdhury, M. S. Shehab, M. A. Awal and M. A. Razzak, "Design and implementation of a highly efficient pure sine-wave inverter for photovoltaic applications", *2013 International Conference on Informatics, Electronics and Vision (ICIEV)*, 2013, pp. 1-6.

[12] N. E. Saavedra-Peña, R. Darbali-Zamora, E. Desarden-Carrero and E. Aponte-Bezares, "Towards an Islanding Detection Method Using a Digital Twin Concept", *2021 IEEE 48th Photovoltaic Specialists Conference (PVSC)*, 2021, pp. 0410-0415.

[13] N. E. Saavedra-Peña, R. Darbali-Zamora, E. Desarden-Carrero and E. Aponte-Bezares, "Development of Photovoltaic Inverter Model with Islanding Detection Using the Sandia Frequency Shift Method", *2022 IEEE 49th Photovoltaics Specialists Conference (PVSC)*, 2022, pp. 0398-0404.

[14] A. Yazdani and R. Iravani, "Voltage-Sourced Converters in Power Systems: Modeling, Control, and Applications", *Wiley-IEEE Press*, 2010.

[15] N. E. Saavedra-Peña, R. Darbali-Zamora, E. Desarden-Carrero and E. Aponte-Bezares, "Development of Photovoltaic Inverter Model with Islanding Detection Using the Sandia Frequency Shift Method", *2022 IEEE 49th Photovoltaics Specialists Conference (PVSC)*, 2022, pp. 0398-0404.

[16] E. W. H. Kahlane, L. Hassaine, and M. Kherchi, "LCL filter design for photovoltaic grid connected systems", *The 3rd international Semniar on New and Renewable Energies*, vol. 8, no. 2, pp. 227–232, 2014.

[17] E. Desardén-Carrero, R. Darbali-Zamora and E. E. Aponte-Bezares, "Analysis of Commonly Used Local Anti-Islanding Protection Methods in Photovoltaic Systems in Light of the New IEEE 1547-2018 Standard Requirements", *2019 IEEE 46th Photovoltaic Specialists Conference (PVSC)*, 2019, pp. 2962-2969.

[18] E. Desarden-Carrero, R. Darbali-Zamora and E. E. Aponte-Bezares, "Analysis of Grid Support Functionality Dynamics under Ride-Through Requirements Using Power-Hardware-in-the-Loop Implementation", *2021 IEEE 48th Photovoltaic Specialists Conference (PVSC)*, 2021, pp. 1795-1802.

978-1-6654-6060-6/23 $31.00 © 2023 IEEE

Effective and Equivalent Refractive Index Models for Patterned Solar Cell Films via a Robust Homogenization Method

Ekin Gunes Ozaktas, Sreyas Chintapalli, and Susanna M. Thon

Electrical and Computer Engineering, Johns Hopkins University, Baltimore, 21218, USA

Abstract — **Solar cells containing complex geometric structures such as texturing, photonic crystals, and plasmonics are becoming increasingly popular, but this complexity also creates increased computational demand when designing these devices through costly full-wave simulations. Treating these complex geometries by modeling them as homogeneous slabs can greatly speed up these computations. To this end, we introduce a simple and robust method to solve the branching problem in the homogenization of metamaterials. We start from the branch of the complex logarithm in the Nicolson-Ross-Weir method with the minimum absolute mean derivative in the low frequency range and enforce continuity. This is followed by comparing the reflectance, transmittance, and absorptance of the original and homogenized slabs. We use our method to demonstrate accurate and fast optical simulations of patterned PbS colloidal quantum dot solar cell films. We also compare patterned solar cells homogenized via equivalent models (wavelength-scale features) and effective models (sub-wavelength-scale features), finding that for the latter, agreement is almost exact, whereas the former contains small errors due to the unphysical nature of the homogeneity assumption for that size regime. This method can greatly reduce computational cost and thus facilitate the design of optical structures for solar cell applications.**

I. INTRODUCTION AND BACKGROUND

Complex geometries such as photonic crystals, plasmonic structures, and textured surfaces have gained popularity in solar cell design. However, the full-wave simulations required to study 2D/3D structures are computationally costly. In this work, we use effective and equivalent material models to homogenize patterned layers and reduce the problem to 1D, for which methods such as the Transfer Matrix Method (TMM) can be used. Once a homogenized model is produced, it can be reused in many simulations, reducing the computational cost drastically. Unlike equivalent models, effective models have an averaged electric field similar to that in the inhomogeneous medium, since the feature size (f.s.) of the inhomogeneities satisfies f.s. $\ll \lambda$, and homogenization is physically valid. In equivalent models f.s. $\approx \lambda$, and effects such as diffraction and scattering dominate [1].

Homogenizing complex materials simplifies their treatment. Many works use homogenization to model electrical properties of solar cells with multiple constituents [2]. Others use effective media to model complex geometries, such as textured [3], silicon nanowire [4], and plasmonic solar cells [5]. Many solar cell patterns satisfy f.s. $\approx \lambda$, necessitating a thorough investigation of equivalent models, which are insufficiently treated in the literature.

A popular homogenization method is the Nicolson-Ross-Weir (NRW) method, which starts from the S-parameters to obtain the refractive index $N_{eff}(\omega) = n_{eff}(\omega) + i\kappa_{eff}(\omega)$ of a corresponding homogeneous slab. $\kappa_{eff}(\omega)$ can be uniquely calculated but $n_{eff}(\omega)$ depends on a complex logarithm with multiple branches indexed by the integer branch number m [6]-[8]. The ambiguity of m results in a *branching problem*, which must be resolved. One approach takes m at each frequency such that $n_{eff}(\omega)$ is closest to the application of the Kramers-Kronig relations to $\kappa_{eff}(\omega)$. This can result in discontinuities, so a different approach is based on continuity [6], [9], but this leaves the starting branch ambiguous. It is common to start from the zeroth branch for a low enough frequency, since the slab is optically thin [1].

Some works consider an effective thickness d_{eff} different from the geometric thickness. One definition involves boundaries for which the incoming and outgoing waves are plane waves (which is not true near an inhomogeneous material) [6]. Another treatment rounds the branch number calculated from the Kramers-Kronig relations for permittivity and permeability and takes effective thickness as the value that minimizes the error associated with the rounding [10].

II. METHODS

We now introduce our solution to the branching problem. First, we obtain the S-parameters via finite-difference time-domain (FDTD) simulation [11] of the inhomogeneous material. Then we must choose the starting branch. Since the low-frequency behavior of $N_{eff}(\omega)$ of a wide range of materials will approach a constant, and since the branches differ by $2\pi m/k_0 d_{eff}$ (k_0 is the free space wavenumber), for low frequencies (small $k_0 d_{eff}$), the false branches become very steep while the correct branch is flat in comparison. Thus, we find the starting branch by minimizing the magnitude of the mean derivative across the low frequency portion of each branch (before the first discontinuity). Then, we use the continuity of $n_{eff}(\omega)$ for the remaining frequencies: for each point, the next branch index is recursively taken as that yielding the closest value to the present branch. This method is more robust than that commonly found in the literature, which starts from a low enough frequency to ensure that the slab is optically thin and thus takes the first branch index as 0. This is a "hard" requirement: that method fails if the starting branch is not 0 in reality. However, our method only has a "soft" condition that the correct branch be flat enough to be

978-1-6654-6060-6/23 $31.00 © 2023 IEEE

distinguishable, and thus the starting branch number can be nonzero.

Next, we analytically calculate the reflectance (R_{eff}), transmittance (T_{eff}), and absorptance (A_{eff}) of the homogenized slab. We compare these to R_{inh}, T_{inh}, A_{inh} from the FDTD simulation of the inhomogeneous material via mean squared error (MSE). We choose d_{eff} to minimize MSE. This improves matching of the optical spectra, especially for equivalent models, due to the greater region where plane wave behavior breaks down when effects such as diffraction dominate.

III. RESULTS AND DISCUSSION

We first demonstrate the steps of our method in Fig. 1 with a 790 nm thick homogeneous slab of PbS colloidal quantum dots (PbS CQDs) with exciton peak at 1200 nm. The agreement between the optical spectra from the FDTD calculation and our method is exact (MSE = 1.68×10^{-6}), verifying the self-consistency of our model. The refractive index of PbS CQDs was obtained via Variable Angle Spectroscopic Ellipsometry (VASE).

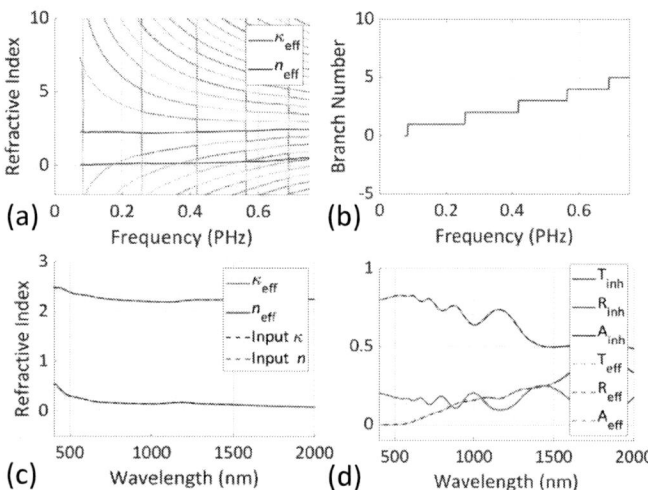

Fig. 1. The index model estimation method applied to a homogeneous slab of PbS CQDs. (a) Branches of the complex logarithm together with the real and imaginary parts of the refractive index; (b) branch number as a function of frequency; (c) extracted refractive index compared to the known refractive index of the PbS CQD (since it is homogeneous); (d) optical behavior from the FDTD simulation compared to the method of this work.

A more interesting example is an inhomogeneous material with f.s. $\approx \lambda$ so that assuming homogeneity is unphysical. Here, we want an equivalent slab model with the same R, T, and A as the inhomogeneous material. We investigate a PbS CQD film nanodisk array with ZnO filling the space in between. The nanodisks have radius 252.8 nm with periodicity 632 nm, and height 790 nm. The refractive index of ZnO was obtained from [12]. As visible in Fig. 2, the effective thickness of the homogenized slab is thicker than the geometric thickness due to the surrounding lack of plane wave behavior.

We also apply the Kramers-Kronig relations to $\kappa_{eff}(\omega)$, observing they are violated especially for shorter λ. However, this does not violate causality [13]. For an effective model, the averaged electric field in the inhomogeneous material corresponds to that in the homogenized one, and the Kramers-Kronig relations hold [13]. However, this is not so for equivalent models. Rather, it simply demonstrates that the homogenization process for f.s. $\approx \lambda$ is physically unjustified, resulting in an "unphysical" equivalent index model.

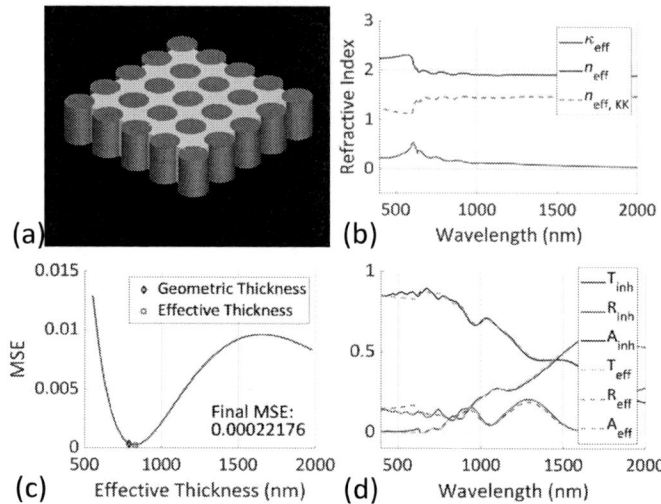

Fig. 2. Demonstration of the method for a patterned slab of PbS CQD nanodisks surrounded by ZnO. (a) The inhomogeneous structure with PbS CQDs (red) and ZnO (green); (b) the equivalent refractive index model with Kramers-Kronig relations; (c) MSE as a function of effective thickness; (d) comparison of the optical spectra.

We now simulate the behavior of a PbS CQD solar cell. Modifying the design in [14], the solar cell is composed of 225 nm of ITO as the front contact, 75 nm of ZnO, the above 790 nm slab of nanodisks, 100 nm of PbS CQDs, and 300 nm of Au as an electrode. The refractive indices of ITO and Au were obtained from [15], [16]. We replace the patterned layer by the equivalent slab calculated previously. Then we simulate the solar cell using a fast optical stack solver [11]. We compare the results to the FDTD simulation. We also compare (Fig. 3) with a solar cell that is identical except that the radius of nanodisks is 25.28 nm and the periodicity of the structure is 63.2 nm, for which f.s. $\ll \lambda$, and the homogenized medium is an effective model. As we might expect, for the solar cell simulated with an effective model, the comparison of FDTD to the stack solution is almost exact for the entire wavelength range of interest, due to the physically viable homogeneity assumption. However, for the solar cell simulated with an equivalent model, there is more error due to the unfounded assumption of homogeneity, and the dominance of effects such as diffraction and scattering that cannot be modeled with a homogeneous slab. Regardless, both simulations are substantially accurate and much faster than an FDTD simulation, running almost instantaneously.

978-1-6654-6060-6/23 $31.00 © 2023 IEEE

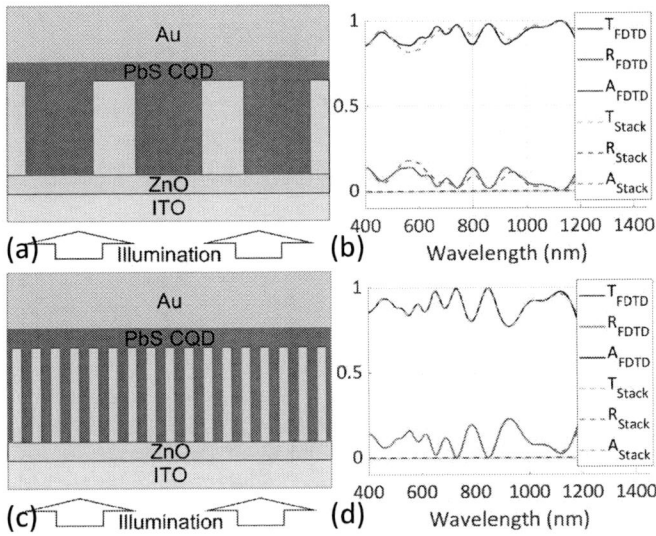

Fig. 3. Comparison of the optical behavior of the solar cell under consideration depending on simulation technique. (a) depicts a solar cell which was simulated with an equivalent model and (b) the corresponding FDTD results compared to the optical stack solver, while (c) contains the solar cell for which an effective model was used and (d) contains the optical behavior comparison.

IV. Conclusion

In this work, we have introduced a simple and robust solution to the branching problem in the NRW method that starts from the branch of the complex logarithm with the lowest absolute mean derivative in the low frequency region, and enforces continuity to recursively obtain the remaining branch numbers. We then compare the reflectance, transmittance, and absorptance of the homogenized and inhomogeneous slabs, and minimize the deviations between them to calculate the effective thickness of the homogenized slab. We investigate two regimes. When f.s. $\ll \lambda$, (i) the homogenization is physically valid, (ii) the averaged electric field in the inhomogeneous and effective models is the same [13], (iii) the optical behavior between the two agree, and (iv) the Kramers-Kronig relations hold. Meanwhile, in the regime with wavelength-scale features, these properties do not hold due to the unphysical homogeneity assumption. Regardless, an equivalent model giving the same optical behavior is still useful in simplifying complex models. We demonstrate this by reducing a patterned solar cell to a 1D problem and solving for its optical properties much more quickly than with FDTD simulations, thus improving computational efficiency during photovoltaic design. We expect that this model will be widely useful to the field in helping researchers design optical structures for a variety of solar cell applications.

Acknowledgements

We would like to acknowledge the National Science Foundation (ECCS-1846239) for funding this work.

References

[1] S. Arslanagic, T. V. Hansen, N. A. Mortensen, A. H. Gregersen, O. Sigmund, R. W. Ziolkonski, and O. Breinbjerg, "A review of the scattering-parameter extraction method with clarification of ambiguity issues in relation to metamaterial homogenization," *IEEE Antennas and Propagation Magazine*, vol. 55, pp. 91–106, 2013.

[2] G. Richardson, C. P. Please, and V. Styles, "Derivation and solution of effective medium equations for bulk heterojunction organic solar cells," *Euro. Jnl of Applied Mathematics*, vol. 28, pp. 973–1014, 2017.

[3] S. Rowlands, J. Livingstone, and C. Lund, "Optical modelling of thin film solar cells with textured interfaces using the effective medium approximation," *Solar Energy*, vol. 76, pp. 301–307, 2004.

[4] H. Wang, X. Liu, L. Wang, and Z. Zhang, "Anisotropic optical properties of silicon nanowire arrays based on the effective medium approximation," *International Journal of Thermal Sciences*, vol. 65, pp. 62–69, 2012.

[5] Y. A. Akimov, K. Ostrikov, and E. P. Li, "Surface plasmon enhancement of optical absorption in thin-film silicon solar cells," *Plasmonics*, vol. 4, pp. 107–113, 2009.

[6] X. Chen, T. M. Grzegorczyk, B.-I. Wu, J. Pacheco, and J. A. Kong, "Robust method to retrieve the constitutive effective parameters of metamaterials," *Physical Review E*, vol. 70, no. 016608, 2004.

[7] Z. Szabo, G.-H. Park, R. Hedge, and E.-P. Li, "A unique extraction of metamaterial parameters based on Kramers–Kronig relationship," *IEEE Transactions on Microwave Theory and Techniques*, vol. 58, pp. 2646–2653, 2010.

[8] G. Angiulli and M. Versaci, "Retrieving the effective parameters of an electromagnetic metamaterial using the Nicolson-Ross-Weir method: An analytic continuation problem along the path determined by scattering parameters," *IEEE Access*, vol. 9, pp. 77511–77525, 2021.

[9] Y. Shi, Z.-Y. Li, L. Li, and C.-H. Liang, "An electromagnetic parameters extraction method for metamaterials based on phase unwrapping technique," *Waves in Random and Complex Media*, vol. 26, pp. 417–433, 2016.

[10] S. Yoo, S. Lee, J.-H. Choe, and Q-H. Park, "Causal homogenization of metamaterials," *Nanophotonics*, vol. 8, pp. 1063–1069, 2019.

[11] Ansys, "Ansys Lumerical Photonics Simulation and Design Software," www.ansys.com/products/photonics.

[12] O. Aguilar, S. De Castro, M. P. F. Godoy, and M. R. Sousa Dias, "Optoelectronic characterization of Zn$_{1-x}$Cd$_x$O thin films as an alternative to photonic crystals in organic solar cells," *Optical Materials Express*, vol. 9, pp. 3638–3648, 2019.

[13] A. Alu, A. D. Yaghjian, R. A. Shore, and M. G. Silveirinha, "Causality relations in the homogenization of metamaterials," *Physical Review B*, vol. 84, no. 054305, 2011.

[14] T. Kim, X. Jin, J. H. Song, S. Jeong, and T. Park, "Efficiency limit of colloidal quantum dot solar cells: Effect of optical interference on active layer absorption," *ACS Energy Letters*, vol. 5, pp. 248–251, 2020.

[15] T. A. F. Konig, P. A. Ledin, J. Kerszulis, M. A. Mahmoud, M. A. El-Sayed, J. R. Reynolds, and V. V. Tsukruk, "Electrically tunable plasmonic behavior of nanocube–polymer nanomaterials induced by a redox-active electrochromic polymer," *ACS Nano*, vol. 8, pp. 6182–6192, 2014.

[16] P. B. Johnson and R. W. Christy, "Optical constants of the noble metals," *Phys. Rev. B*, vol. 6, pp. 4370–4379, 1972.

978-1-6654-6060-6/23 $31.00 © 2023 IEEE

Evaluation of Rear Contact Passivation Strategies via Surface Photovoltage Spectroscopy

Nathan Rock[1], Chien-Hsuan Chen[2], Kristopher Davis[2], Amit Munshi[3], Michael Scarpulla[1]

University of Utah, Salt Lake City, UT, 84104, USA

University of Central Florida, Orlando, FL 32816, USA

Colorado State University, Fort Collins, CO, 80523, USA

Abstract — The last several years have seen remarkable improvements to some of the core defects in CdTe photovoltaics. Improvements to carrier concentration, bulk lifetimes and emitter/absorber band alignment have led to increased short circuit current and fill factor. However, the most attractive feature of CdTe and $CdSe_xTe_{1-x}$ is its wide bandgap of 1.4-1.5 eV and correspondingly high open circuit voltage (V_{oc}). To date V_{oc} in polycrystalline films remains at only 76% of thermodynamic limit, despite these breakthroughs. With bulk lifetime and carrier concentration increased, interface passivation remains a likely source of loss, especially the rear interface, which has received less attention.

We present developments in surface photovoltaic spectroscopy (SPS), aimed at characterizing loss mechanisms and passivation strategies for CdTe cell interfaces – especially the rear contact. Effective rear contact passivation is predicted to increase device efficiency by 3% with flat bands or 3.5% in company with an electron reflector. We correlate and contrast the information gained from TRPL and SPS for known wet etchant strategies. We also explore passivating interface layers in CdTe and Si based devices. SPS signals are evaluated via SCAPS-1D modeling, as well as analytical modeling. We demonstrate sensitivity to surface recombination velocity, while also probing the potential barrier height, φ_s, and SCR recombination. This provides a more complete and dynamic picture of the effectiveness of passivation strategies..

Machine Learning, Unmanned Vehicles, and Energy: A Review

Brian L. Reyes Santiago
Department of Computer Science and Engineering
University of Puerto Rico
Mayagüez, PR
brian.reyes2@upr.edu

Eduardo Ortiz-Rivera, PhD
Department of Electrical and Computer Engineering
University of Puerto Rico
Mayagüez, PR
eduardo.ortiz7@upr.edu

Abstract—**Automation has recently increased with the rise of machine learning models. Tasks that are inherently dangerous or laborious can be automated to reduce costs and potential loss of life. Machine learning has also improved the effectiveness of unmanned vehicles for these tasks. However, these improvements necessitate increased energy to sustain them. This review aims to understand basic machine learning models and their effectiveness in unmanned vehicle and energy generation applications.**

Index Terms—**Machine Learning, Unmanned Vehicles, Drones, Energy**

I. Introduction

Machine Learning (ML) has been responsible for considerable technological development, cutting time and workloads. It allows automation to handle increasingly difficult tasks accurately. ML has a wide range of applications, including in the medical, business, science, and tech industries. However, it is also prone to errors and inconsistencies. Currently, applications require careful algorithm and data selection. Different learning models are more effective with discrete data sets. Because of this, the quality of data becomes just as crucial as the model selected for the application.

Unmanned Vehicles (UV) have benefited from these advancements, as ML allows them to have improved autonomy, thereby expanding the tasks that can be performed. These tasks, however, require more power from the system than previously estimated. Because of this need, predicting the energy requirements of a UV system has become a necessary part of the ML models being used today. The benefits of exploring the area of energy prediction expand to more than just UVs; they also include general energy generation and use.

II. Machine Learning Algorithms and Error Metrics

In this review, three machine learning algorithms and three error metrics are briefly described along with their use to predict energy consumption.

This research is supported by the University of Puerto Rico-Mayagüez's Innovative Wide Area Sensing/Mitigation (IWAS) Project and the Consortium of Hybrid Resilient Energy Systems (CHRES).

A. Support Vector Machines

Support Vector Machines (SVM) are supervised learning models that analyze data for classification and regression analysis. Their use is suitable for small to medium data sets [1]. Support vector machines are considered one of the most robust prediction methods because they are based on the statistical learning framework, Vapnik-Chervonenkis (VC) Theory. This learning model classifies information as belonging to a category or not, as a binary linear classifier.

B. Artificial Neural Networks

Artificial Neural Network (ANN) algorithms are a learning model that uses an artificial structure to simulate the structure and function of the human brain [1]. This complex simulation of the network structure of a large number of neurons is nonlinear and, therefore, can tackle problems differently than SVMs. ANNs consist of an input layer, several hidden layers, and an output layer [1]. Depending on the activation function, weight, and bias of the training model, the results can vary. In this review, the use of the Rectified Linear Units (ReLU) function commonly used in supervised learning is assumed.

C. Random Forests

Random Forests (RF) algorithms utilize a large number of decision trees that operate as a unit. Each decision tree provides a result, and the result with the most "votes" becomes the model's official prediction. Individually, each decision tree is not effective, but as a unit, performance increases considerably. Collaboratively, RF trees protect each other from individual errors [1].

D. Error Metrics

1) Mean Square Error (MSE): The mean square error provides the average of a set of errors and how close to the regression line a set of points is. It does this by squaring the distance between the points to the regression line. MSE can be described as follows.

$$\frac{1}{n} * \sum (actual - prediction)^2 \tag{1}$$

2) Mean Absolute Error (MAE): The mean absolute error provides the average of the absolute value of the difference between the predicted value and the actual value [1]. It measures the magnitude of the errors without consideration for their direction (+/-). MAE can be described as follows.

$$\frac{1}{n} * \sum |actual - prediction| \qquad (2)$$

3) Root Mean Squared Error (RMSE): The root mean squared error provides the standard deviation of the errors. RMSE measures how far from the regression line the errors are, and how concentrated the data is. RMSE can be described as follows.

$$\sqrt{\frac{\sum (actual - prediction)^2}{n}} \qquad (3)$$

III. UNMANNED VEHICLES AND MACHINE LEARNING

Unmanned Vehicles or drones play a leading role in today's technology developments due to their increasing coverage and mobility, as well as their decreasing costs compared to piloted aircrafts. They are used in information gathering, exploration, search and rescue, among others. By utilizing machine learning, the drone industry is developing further applications, like crop monitoring in agriculture and degradation surveying in photovoltaic panels [2]. These applications are making machine learning a key driver for improving autonomy.

Even though the potential of machine learning algorithms is great, it is not without its flaws. Out-of-the-box drones have constrained battery life and modest computing capabilities; these are significant inhibitors to ML-driven application performance [2]. When considering these issues, upgrading to a more powerful processor, or processing the computations remotely, may seem like viable fixes. However, these have their complications. The increased processing power would require more energy, as would the constant communication with a hub, affecting flight time. Similarly, remote computing incurs wait time, which can be dangerous in tasks like collision detection or search and rescue [2].

IV. ENERGY PREDICTION

Technological advancement and urban population growth have made effective resource management a bigger challenge [3]. However, this is not a new occurrence; humans have always adapted their dwellings to better suit their environment. Romans created aqueducts and water drainage, and more recently, the United States uses aerial photography, databases and data analysis to direct resources [3]. The need for this efficient use gives rise to the term "smart cities," which despite its current use, is not new.

Currently, smart cities include a wide range of new technology such as the Internet of Things (IoT), Artificial Intelligence (AI), Machine Learning (ML), among others. These technologies collect data in real time about infrastructures like power or water supply systems and store them so that they can be used for improvements [3]. UVs form a part of this system by functioning as mobile data centers or data collectors. Effective machine learning algorithms can enhance drone battery usage, thereby improving the efficiency of smart cities' energy usage.

To analyze this, [3] and [4] perform experiments on energy prediction using machine learning and deep learning models. The SVM, ANN, RF, and other models are used while MAE, MSE and RMSE are the error metrics utilized. Both studies are aiming for RMSE values closer to 0 to minimize the errors made by the models. The following tables show the results of both studies.

S/N	ALGORITHM	RMSE
1	CATBOOST	1.377
2	LIGHTGBM	1.119
3	ADABOOST	1.993

Figure 1. Results of the Analysis on the Test Set from [3].

Method	RMSE [kWh]	NRMSE [%]	MAPE [%]	PCC
Linear Regression	5434.89	25.81	7.39	0.90
Multiple linear regression	22566.99	94.15	29.86	0.57
SVR (RBF)	22452.64	87.08	30.63	0.78
SVR (polynomial)	10617.40	40.94	13.87	0.88
SVR (linear)	4443.11	23.63	6.34	0.91
Regression tree	14140.10	74.39	18.84	0.71
Ensemble regression tree	12791.76	67.53	17.42	0.75
Ensemble Tree LS Boost	12791.76	67.53	17.42	0.75
Ensemble Tree Bag	10285.76	52.50	14.60	0.81
Gaussian process (linear)	5529.98	27.13	7.54	0.87
Gaussian process	4326.72	18.85	5.35	0.93
Multilayer perceptron	2376.38	12.45	3.40	0.96

Figure 2. Average Results Over One Year Showing the Accuracy of each Prediction Method from [4].

In [3], Light Gradient Boosting Machine (LGBM) outperformed Category Boosting and Adaptive Boosting, while this model already outperforms SVM, ANN and RF. In [4], Multilayer Perception (MLP) a type of Deep Learning, is shown to outperform other forms of machine learning.

V. FUTURE WORK

Both [3] and [4] show that further investigation is necessary to properly estimate using either LGBM or MLP. ML offers the potential to study energy management on both a micro and macro level, while DL offers the opportunity to focus on additional methods, e.g., Convolutional Neural Networks to increase prediction accuracy. These studies could be expanded with the following.

- The models could be tested against each other under the same constraints. This test would provide a clearer image of accuracy and efficiency.
- Additional Deep Learning methods could be included to improve the representation of both models.

REFERENCES

[1] Z. Wu and W. Chu, "Sampling Strategy Analysis of Machine Learning Models for Energy Consumption Prediction," 2021 IEEE 9th International Conference on Smart Energy Grid Engineering (SEGE), 2021, pp. 77-81, doi: 10.1109/SEGE52446.2021.9534987.

[2] D. Trihinas, M. Agathocleous and K. Avogian, "Composable Energy Modeling for ML-Driven Drone Applications," 2021 IEEE International Conference on Cloud Engineering (IC2E), 2021, pp. 231-237, doi: 10.1109/IC2E52221.2021.00039.

[3] T. Ajagunsegun, J. Li, O. Bamisile and C. Ohakwe, "Machine Learning-Based System for Managing Energy Efficiency of Public Buildings: An Approach towards Smart Cities," 2022 4th Asia Energy and Electrical Engineering Symposium (AEEES), 2022, pp. 297-300, doi: 10.1109/AEEES54426.2022.9759759.

[4] N. G. Paterakis, E. Mocanu, M. Gibescu, B. Stappers and W. van Alst, "Deep learning versus traditional machine learning methods for aggregated energy demand prediction," 2017 IEEE PES Innovative Smart Grid Technologies Conference Europe (ISGT-Europe), 2017, pp. 1-6, doi: 10.1109/ISGTEurope.2017.8260289.

III-V Epitaxy on Detachable Porous Germanium 4" Substrates

Waldemar Schreiber[1], Jens Ohlmann[1], Patrick Schygulla[1], Stefan Janz[1], Jinyoun Cho[2] & Kristof Dessein[2]

[1]Fraunhofer Institute for Solar Energy Systems ISE, Freiburg, Germany; [2]Umicore Electro-Optic Materials, Olen, Belgium

Abstract — **Porous germanium multiple layer stacks were prepared using bipolar electrochemical etching. In dependance of the porosity within the individual porous layers they can be used after a high temperature process either for III-V epitaxy and/or detachment. Both characteristics are crucial to result in a thin detached multi-junction solar cell of which the germanium substrate occupies about 600 nm or less. We observe a strong correlation between the porosity of the detachment layer and the surface roughness after annealing. Finally an $Al_{0.5}Ga_{0.49}In_{0.01}As/Ga_{0.99}In_{0.01}As$ double heterostructure was grown in a MOVPE reactor on a reference CMP germanium sample and a comparable low roughness porous germanium substrate. An intensive material analysis was performed on both samples revealing a similar surface. Cathodoluminescence measurements expose a defect density of 5.2×10^4 cm^{-2} for the reference case and 2.4×10^5 cm^{-2} for the porous substrate. Time resolved photoluminescence measurements (TRPL) revealed a difference of approx. 2 ns in the lifetime of the minority carriers in the low injection regime for the two substrates (approx. 8 ns vs. 6 ns, respectively). In addition, XRD measurements confirmed the crystalline quality and in particular no change of the in-plane lattice parameter of the substrate due to the porosification or the annealing process. Finally, the results are very promising to achieve two goals at the same time. On the one hand, the weight reduction of a potential multi-junction solar cell especially for space applications. And on the other hand a cost reduction based on a lower consumption of the rare and valuable element germanium.**

Keywords—Germanium,bipolar etching, PorGe, III-V, Lift-off

I. INTRODUCTION

Due to its limited availability and at the same time high demand, germanium is a significant cost factor in the production of multi-junction solar cells [1]. Basically, there are a few promising methods to reduce the consumption of germanum for use as a substrate for III-V epitaxy. In addition to the reduction in manufacturing costs, would come a mass reduction. This would be a double advantage, especially for space applications. At this point, reactive ion etching (RIE), selective AlAs etching and epitaxial lift-off (ELO) should be mentioned. The ELO technique seems to be particularly suitable for this purpose [2]. The biggest advantage is the upscale possibility and thus less process costs per wafer. Combined with the freedom to change porosity over depth or the type of porous layer during the etcing process it becomes a powerful technique. Therefore the application for terrestrial purposes could become attractive, e.g. on pseudo square format as is the case for silicon. A similar procedure is not possible, for example, with the complex nanoimprint lithography (NIL) in combination with reactive ion etching (RIE) [3]. Here, only the etching depth can be controlled by the etching process. A change in the number of pores, their spacing and diameter must be determined by means of a photomask. In the case of ELO, the biporous layer etching also prevents unnecessary removal of germanium, since the diameter of the pores does not have to remain constant from top to bottom. Thus, only a part of the entire layer must have a higher porosity. First promising results on subsequent porosification, annealing and epitaxy have already been shown on Si [2] and on non-detachable porous Ge [4]. Targetly raising the porosity is more challenging with germanium, as due to hydroxide radical bonding, charge exchange is in principle not limited to the porous layer/unetched solid interface [5]-[8]. Rather, it must be ensured at all times that the existing layer remains protected and the etching process only takes place where necessary. Bipolar electrochemical etching is used for this purpose. Concerning this, an advanced understanding could be achieved. The envisioned process for using porous Ge for III-V epitaxy is shown in Fig. 1 c). In this case, the reuse of the substrate in a cyclic process is also ideally reproduced. Many approaches to successfully following this path revealed enormous difficulties [5]-[7]. Typically, either could be successfully grown but not detached, or vice versa. Which is due to an inappropriate etching process or resulting layer types. Furthermore, these experiments were limited to small areas of a few mm^2 to cm^2. However, in this work, almost fully processed 4" wafers will be used as a growth template. For this purpose, the quality of III-V heterostructures will be investigated using further developed and potentially removable layers. For this purpose, an overview of roughness is given using AFM measurements after etching, annealing and III-V epitaxial growth. Subsequently, CL, TRPL and XRD reciprocal space map measurements help to assess the opto-electronic and crystalline quality.

II. EXPERIMENTAL METHODS

A. Sample processing

Highly gallium p-doped Ge 4" (100)-orientated Cz-wafers ($\rho \approx 20$ mΩcm) with a thickness of ≈ 175 μm provided by Umicore were used for porosification and epitaxy. The porosifiction was performed in a double cell etching tank manufactured by AMMT. The used electrolyte is a solution of

978-1-6654-6060-6/23 $31.00 © 2023 IEEE

HF 50 wt.% and ethanol (5:1). Annealing after porosification was performed in a in-house made 4" RTCVD halogen oven under H₂ atmosphere. III-V epitaxy was performed in a G4-AIX2800-TM MOVPE reactor from Aixtron.

B. Experimnetal techniques

SEM cross sections were recorded with a Zeiss Auriga SEM. The surface roughness was measured in the non-contact mode with a Park XE-100 AFM. For defect density estimation CL with an acceleration voltage of 10 kV was performed in a Hitachi SU-70 SEM. For RSM a Panalytical X'Pert MRD in a triple-axis configuration using the copper K$_\alpha$ wavelength was used. Finally, minority carrier lifetimes were determined using TRPL setup with at an excitation wavelenght of 515 nm.

III. EXPERIMENTAL RESULTS

A. Bipolar electrochemical etched germanium multi-layers

As mentioned earlier, the main task is to get a smooth surface and at the same time a high porous layer for a subsequent detachment after annealing and epitaxy. For this purpose, a porosity gradient is inevitable. However, while increase the porosity there must be a suitable basic structure which won't be affected during this step. The vulnerability of the existing layer is based on the mentioned hydroxide passivation. But also the expansion of the depletion zones due to the doping during the different poling, is causal [8]-[9]. Taking these features into account, a suitable porous layer was etched. Fig. 1 a) & b) shows a SEM cross section micrograph and the corresponding schematical view with 2 significantly different regions.

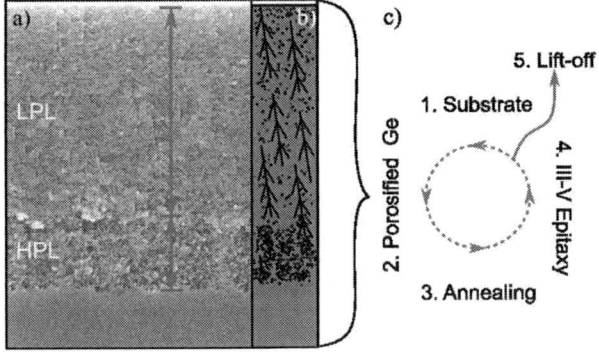

Fig. 1.   a) SEM cross section and b) schematic view of a porous germanium double layer stack: low porous layer (LPL) consisting of a fishbone-like type and underneath a high porous layer (HPL) mainly of a sponge-like type. c) Cyclic process of substrate reuse in combination with the ELO technique.

This example is intended to illustrate that the thickness of both layers can be varied. In particular, the LPL can be varied in thickness as mentioned before. Even though there was a significant increase in porosity at the interface to the unetched solid, the RMS roughness is still less than 500 pm (on 10×10 μm²). Consequently, this represents a suitable starting

position for achieving the two objectives mentioned above. If there is no significant diffusion from the LP-layer to the HP-layer, a suitable growth template would exist after annealing.

B. Annealing under H₂ athmosphere

To clarify whether this type of porous structure is suitable for III-V epitaxy, a high temperature step was performed. To match growing conditions in the MOVPE reactor while growing InGaP, the annealing temperature was chosen to be 720 °C. Fig. 2 a) shows a SEM crosssection of a porous multilayer structure after annealing for 5 minutes.

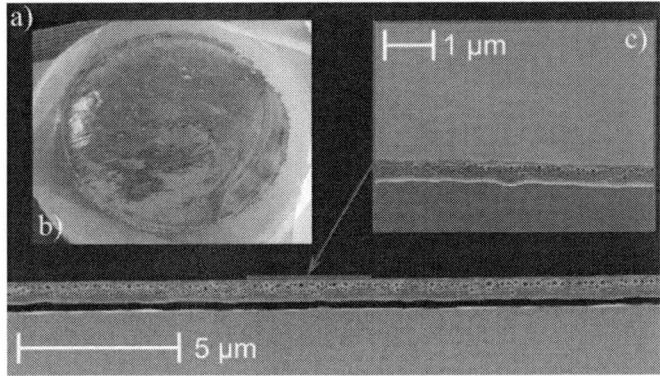

Fig. 2.   a) & c) SEM micrographs of a porous germanium double layer stack after annealing. b) Detached part of the top layer on 4" wafer connected to bluetape.

In the shown range there is now connection between the bulk structure and the top layer along the fracture. From Fig 2 c), it can be seen that the top layer is stable and closed. There are sufficient connections (not shown here) to the solid left on a larger area. The surface roughness on this type of structure is less than 700 pm on 1x1 μm² and therefore suitable as growth template. With the help of bluetape, a simple detachment experiment was performed. The result can be seen in Fig. 2 b). Almost the complete and formerly porosified portion of the total wafer could be detached. The only exceptions are areas where air was trapped between bluetape and wafer or insufficiently porosified areas due to hydrogen bubbles.

C. III-V epitaxy

Due to the duration of III-V epitaxy the final porosity within the high porous layer was decreased compared to the structure shown in Fig. 1. This was done in order to exclude the possibility of the top layer delamination during epitaxy. As shown in Fig. 3 a) the amount of connections between the top layer and the solid is obviously higher. At the same time the entire layer has not shown any early delamination. Furthermore the grown double hetero structure is clearly visible. The first direct comparison of III-V on porous germanium to a III-V layer on a CMP wafer is shown in Fig. 3 b) and c), respectively. The amount of halos for the III-V layer on porous germanium is obviously higher. However, large shiny areas can be found

978-1-6654-6060-6/23 $31.00 © 2023 IEEE   1342

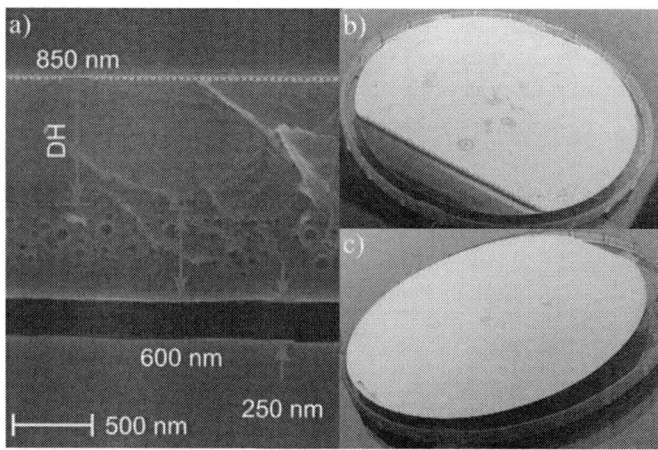

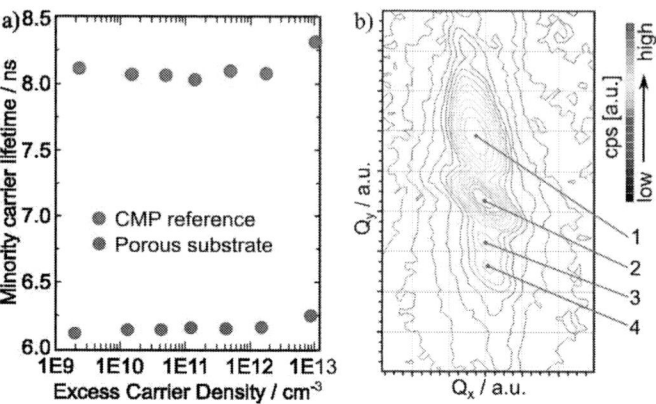

Fig. 3. a) SEM micrograph of a porous germanium double layer stack after annealing and III-V double hetero (DH) epitaxy. b) Top view photograph of the structure shown in a). c) Reference (CMP polished) wafer with same III-V layer.

Fig. 4. a) Minority carrier lifetime obtained from TRPL measurement. b) RSM of the (224) reflex measured in <110> direction parallel to the miscut. Peaks are indicated as follows: 1 porous peak, 2 substrate peak, 3 & 4 fringe peaks.

on which characterization was carried out. CL measurements revealed a higher dislocation density for III-V on porous substrate ($\approx 2.4 \times 10^5$ cm^{-2}) compared to III-V on the Ge reference ($\approx 5.5 \times 10^4$ cm^{-2}). The evaluation of the morphology of both surface types resulted in the same typical humped structure (not shown here). The roughness on porous substrate is ≈ 1.2 nm compared to the reference substrate with ≈ 0.6 nm on a 10×10 μm^2 area. The minority carrier lifetime in the GaInAs layer on porous substrate was 2 ns lower within the low injection regime, see Fig. 4 a). To establish a relationship of the electro-optic properties to the crystalline quality of the porous germanium, reciprocal space maps (RSMs) of the (004) and the (224) reflex were measured perpendicular and parallel to the <110> miscut. As an example, the RSM of the (224) reflection parallel to the miscut is shown in Fig 4. b). The porous peak 1, the substrate peak 2 and fringe peaks 3 & 4 are visible. Especially due to the same Q_x values of peak 1 & 2 no difference in the in-plane lattice constant is apparent. The slightly visible drift of the peaks is due tilted planes because of strain and the miscut.

V. CONCLUSION

We showed the suitability of the bipolar electrochemical etching process for porous multi-layer-stacks on germanium. In particular, maintaining a smooth surface while increasing porosity at the interface to the unetched solid. The possible adaption of thickness for every individual layer could also be shown. The demonstrated lift-off of the thin growth template (≈ 600 nm) on an almost completely porosified 4" wafer certainly takes on a special significance. In addition to the comparable dislocation density, the minority carrier lifetimes in particular suggest that their use as solar cell substrates is promising. Finally, the calculated in-plane lattice constant of the porous layer obtained by XRD could substantiate the high

crystal quality. Further efforts will aim at refining the etching process and reduce or increase the porosity as necessary. On the one hand, the focus will be on lowering the roughness in order to improve nucleation. On the other hand, it will be shown whether thinner substrate layers are also suitable for III-V epitaxy. This will most likely also lead to a further reduction in process time and thus in production costs.

REFERENCES

[1] K.A.W. Horowitz, T. Remo, B. Smith, A. Ptak, Techno-economic analysis and cost reduction roadmap for III-V solar cells, National Renewable Energy Laboratory, Golden, CO, 2018. NREL/TP-6A20-72103, https://www.nrel.gov/docs/fy19osti/ 72103.pdf.

[2] N. Milenkovic, M. Driessen, B. Steinhauser, et al, 20% efficient solar cells fabricated from epitaxially grown and freestanding n-type wafers. Sol Energy Mater Sol Cells, 2017; 159:570-575.

[3] V. Depauw, C. Porret, M. Moelants et al. Wafer-scale Ge epitaxial foils grown at high growth rates and released from porous substrates for triple-junction solar cells. Prog Photovolt Res Appl. 2022, 1 - 14.

[4] E. Winter, W. Schreiber, P. Schygulla, P.L. Souza, S. Janz, D. Lackner, J. Ohlmann, III-V material growth on electrochemically porosified Ge substartes. J. Cryst. Growth, 2023; 602:126980.

[5] E. Garralaga Rojas, B. Terheiden, H. Plagwitz, J. Hensen, C. Baur, G.F.X. Strobl, R. Brendel, Formation of mesoporous germanium double layers by electrochemical etching for layer transfer processes. Electrochem. Commun. 12 (2010) 231–233.

[6] S. Tutashkonko, A. Boucherif, T. Nychyporuk, A. Kaminski-Cachopo, R. Arès, M. Lemiti, V. Aimez, Mesoporous Germanium formed by bipolar electrochemical etching. Electrochim. Acta 88 (2013) 256–262.

[7] A. Boucherif, G. Beaudin, V. Aimez, R. Arès, Mesoporous germanium morphology transformation for lift-off process and substrate re-use. Appl. Phys. Lett. 102 (2013) 1, 11915

[8] S. Tutashkonko, S. Alekseev, T. Nychyporuk, Nanoscale morphology tuning of mesoporous Ge: electrochemical mechanisms. Electrochim. Acta 180 (2015) 545–554.

[9] W. Schreiber, T. Liu, S. Janz, The effect of passivation to etching duration ratio on bipolar electrochemical etching of porous layer stacks in Germanium. J. Phys. Chem. Solids (submitted 2022).

Effect of Soiling from Dust Particles on Solar Cell Efficiency in the United Arab Emirates (UAE)

Muntaser Abdelrahman Almansoori, Rawdah Almannaee, Mariam Aldhefairi, Meerh Alsuwaidi, Ayman Rezk , and Ammar Nayfeh

Khalifa University, Abu Dhabi, 127788, UAE *Corresponding Author: *Ammar.Nayfeh@ku.ac.ae*

Abstract— In this work, we investigate the effect of dust particle size on the performance of solar cells in the United Arab Emirates (UAE). Seven sand samples from different locations across the UAE are collected and analyzed. It's observed that each of the samples differs in grain size, color, and texture. After dispersing each of the samples on the surface of a solar panel, the cells' IV values are measured under one sun and dark conditions. The performance of the solar cells can be linked to the properties of the sand samples. The results show that dust buildup on the solar panel's surface lowers its efficiency by a percentage that goes up to 41.45%, as well as reducing the current. Furthermore, the size of the sand grains and agglomeration behavior has a direct effect on the cell's efficiency.

I. INTRODUCTION

Nations across the world already acknowledged the importance of combating climate change by signing the Paris Agreement at COP21[1]. Energy and power generation are responsible for the majority of CO_2 emissions globally [2]. The energy sector is one of the main sectors that can make a difference and aid in combating climate change by adopting cleaner, renewable, and sustainable energy sources. Not to mention, energy prices and demands are projected to be on the rise for the next two decades[2]. And, that's why we saw rapid adoption of renewable energy sources across the globe, with solar energy and photovoltaic (PV) cells being the top growing in capacity in 2021[3]. The United Arab Emirates (UAE) is focused on moving toward more sustainable and clean energy sources with a particular focus on solar energy[4]. When it comes to PV solar cells, the UAE is faced with harsh weather conditions and sandstorms which often can result in losses due to soiling at these PV power plants[5]. And while soiling is a well-studied problem, we are interested to investigate the effect of different types of soil and dust particles relevant to UAE's geographical location as they can influence the final optical transmittance into the solar cells[6].

In this paper, sand samples are collected from different locations around the UAE, and we study and analyze their grain size by utilizing of digital optical microscope and software image computations. Furthermore, we simulated soiling by dropping a constant volume of sand on the surface of a 6 by 6 centimeters commercial polycrystalline silicon solar cell, with a voltage of 2 Volts, a current of 150mA, and power of 0.3 Watts, and measured its I-V response using a sun simulator.

II. METHODOLOGY

Seven sand samples are taken from three primary cities in the United Arab Emirates: Abu Dhabi city, Al Ain city, and Dubai city as shown in table 1. The collected samples vary in shape, color, and size to determine how they impact the functionality of the solar cell. The samples were collected in glass containers since it has no synthetic chemicals and is hence inert.

Table 1: Summary of collected sand samples.

Sample number	City	location
1	Al Ain	Rimah
2	Abu Dhabi	Al Ruwais
3	Abu Dhabi	Al Zafaranah
4	Abu Dhabi	Al Shahama
5	Al Ain	Al Dhafer
6	Dubai	Al Barsha
7	Dubai	Jabal Ali

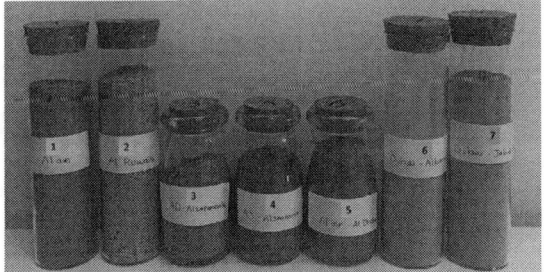

Figure 1: The seven sand samples.

To characterize the sand samples, each sample is pinch-dropped on a glass microscope slide as viewed in figure 2. Samples are then pressed against another glass slide to flatten it as much as possible and separate particles.

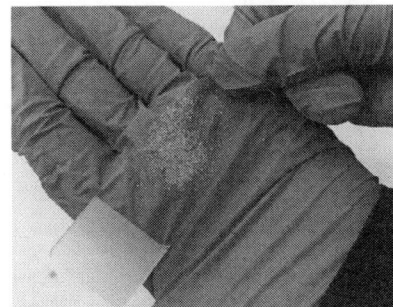

Figure 2: sand sample on a glass microscope slide.

The sand sample is analyzed using an OLYMPUS microscope SZX16, which provided us with a microscale view of the sand grains. Optical microscopy images for the seven samples of the samples are shown in figure 3. The figure displays the different appearances of the sand grains that exhibit different colors. Which will be further studied to find a relationship between grain color and light reflectivity.

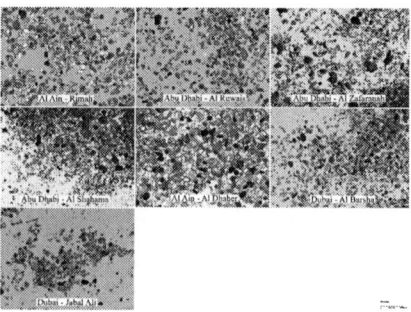

Figure 3. Optical microscopy images of the seven-sand sample.

Analysis of the sand grains images:

Images obtained through digital optical microscopy are then analyzed using the image processing toolkit ImageJ [7]. The first step is converting the image to an 8-bit image type. Then the image's scale is linked to the microscope magnification. Then the appropriate threshold value is adjusted to isolate the white particles from the black background. The watershed process is applied next. Finally, the Analyze Particle tool is used to extract and characterize the isolated sand grains. Figure 4 shows the output result of each step.

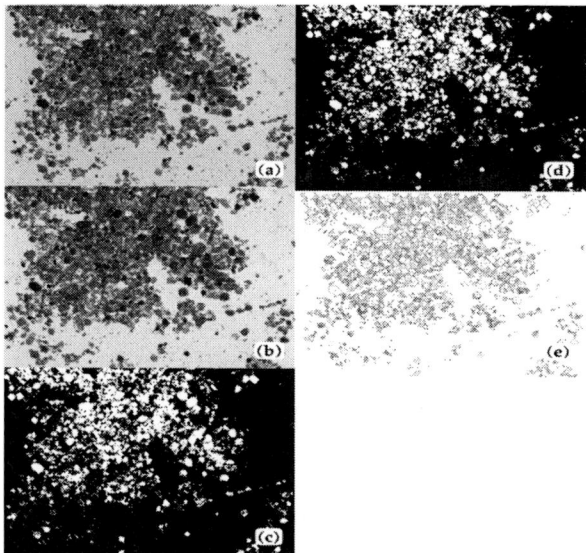

Figure 4. Image processing steps (a) optical microscopy digital image, (b) 8-bit type image, (c) after a threshold is set, (d) after the watershed, and (e) isolated grains after particle analysis.

Testing of the solar panel using a sun simulator:

In our experiment, a polysilicon solar panel of 60x60 mm is used to test the effect of different sand samples on the efficiency of the solar cell. Figure 5 depicted the tested solar panel. The polysilicon panel has a thickness of 3 mm, a working 2 V, and 150 mA for a total power of 0.3 W along with 2.4 V open circuit voltage (V_{oc}) and 167 mA short-circuit current (I_{sc}).

Figure 5. (a) the polysilicon solar panel and (b) the solar panel on the work surface of the sun simulator.

The seven samples collected across the emirates are dropped over the solar panel. And by using Sol3A 94123A solar simulator, we measure the I-V characteristics of the solar panel. Then we calculate and extract the solar panel's short circuit current, maximum power, and efficiency. Figure 6 demonstrates how the sand is dropped and distributed on the solar panel.

Figure 6: Solar panel under the sun simulator (a) 1 pinch of sand grains, (b) 5 pinches of sand grains, (c) 10 pinches of sand grains.

The I-V characteristics for each sample are measured by sweeping the voltage from -3 to 3 V after every sand dispersion for a total of ten. Open circuit voltage (V_{oc}), short circuit current density (J_{sc}), maximum power (P_{max}), fill factor (FF) and efficiency are extracted after each measurement.

III. RESULTS AND DISCUSSION

The solar IV curve under AM1.5G for Alin dust samples (multiple pinches) on solar cell is show in Figure 7.

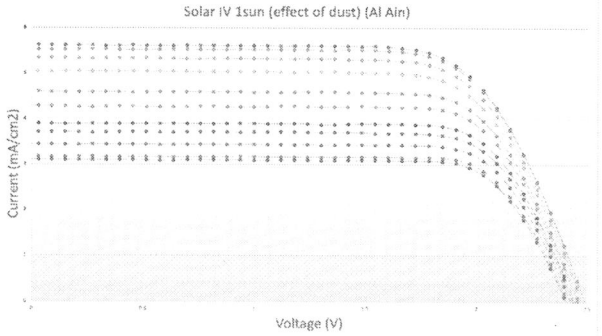

Figure 7 IV curve of solar cell with dust from al ain (adding dust)

The results found are summarized in table 2, including ImageJ particle analysis performed on each sample to determine its average size.

Table 2: Summary of Results

Sample Name (Average Grain Size in μm)		V_{oc} (V)	J_{sc} (mA/cm²)	Eff (%)
Al Ain – Rimah	Base	2.489	5.619	10.36

978-1-6654-6060-6/23 $31.00 © 2023 IEEE   1345

Location (grain size)				
(68.19 µm)	1-pinch	2.489	5.531	10.23
	5-pinches	2.455	4.274	8.229
	10-pinches	2.403	3.104	6.066
Abu Dhabi – Al Ruwais (61.36µm)	Base	2.492	5.515	10.38
	1-pinch	2.481	5.196	9.853
	5-pinches	2.436	4.255	8.245
	10-pinches	2.395	3.336	6.606
Abu Dhabi – Al Zafranah (56.23µm)	Base	2.473	5.523	10.28
	1-pinch	2.477	5.309	10.04
	5-pinches	2.445	4.637	8.887
	10-pinches	2.418	3.908	7.497
Abu Dhabi- Al Shahamah (60.37µm)	Base	2.466	5.420	10.06
	1-pinch	2.473	5.320	9.976
	5-pinches	2.443	4.643	8.916
	10-pinches	2.417	4.081	7.783
Al Ain- Al Dhaher (57.87µm)	Base	2.465	5.492	10.16
	1-pinch	2.461	5.388	10.01
	5-pinches	2.443	5.038	9.416
	10-pinches	2.414	4.458	8.465
Dubai – Al Barsha (62.17µm)	Base	2.459	5.439	10.04
	1-pinch	2.455	5.235	9.726
	5-pinches	2.414	4.494	8.462
	10-pinches	2.386	3.913	7.421
Dubai – Jabal Ali (59.34µm)	Base	2.452	5.479	10.05
	1-pinch	2.447	5.260	9.743
	5-pinches	2.416	4.635	8.626
	10-pinches	2.373	3.640	6.939

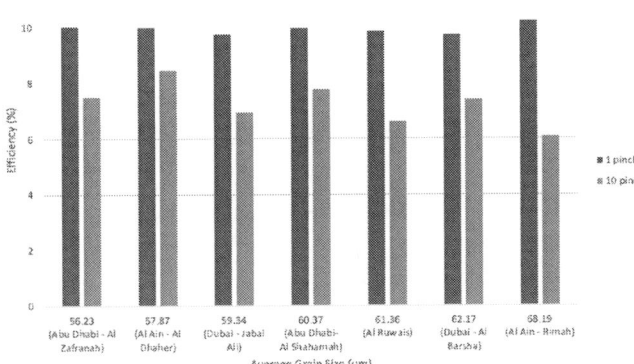

Figure 7: Efficiency vs. Average grain size bar plot

The analysis shows that the Al Ain, Rimah sample grain size is the largest among the samples, and Abu Dhabi, Al Zafranah sample is the smallest. Also, the sand sample from Abu Dhabi, Al Ruwais, and Dubai, Al Barsha, have a similar shape and average grain size. The size of the dust particles deposited on the surface of the solar panel affects the panel's efficiency. Many studies looked closely at the relationship between the solar panel's power losses and the quantity of dust that has formed on its surface. It is determined that there is a linear relationship between the two [5].

There is a significant change in efficiency and current. This is because the dustier the panels get; the less light penetrates the panel and the current flowing decreases. However, the changes in open circuit voltage are much more minor, with almost no change.

Results indicate that the Al Ain, Rimah sand sample with an average grain size of 68.19 µm has a huge percentage change in the efficiency equal to 41.45% after ten pinches. Contrarily, 16.68% is the minimum reduction in the efficiency found with the Al Ain, Al Dhaher sand sample with an average grain size of 57.87µm after ten pinches.

Figure 7 shows the difference in the solar cell efficiency for one pinch and ten pinches. Almost for all locations, the efficiency after one pinch is close to the base efficiency; therefore, for a very thin layer of dust, the efficiency is not much affected. However, we can witness a pattern where higher grain sizes of sand are having a bigger impact on efficiency reduction. These bigger particles might have a higher blocking chance initially considering they will occupy a larger area; however further soiling might result in further blockage caused by finer sand particles.

IV. CONCLUSION

Seven dust samples from various locations throughout the UAE are collected. A comparative study is completed in a laboratory with a solar simulator after the samples had grain size analyses done. The output power of solar panels is significantly reduced because of dust buildup on their surface. It is determined that even a thin layer of dust significantly lowers the effectiveness of a PV system. It is confirmed by ImageJ analysis and efficiency values that smaller particles block more sunlight, which lowers the effectiveness of solar panels and modules. Moreover, the solar panel under the natural sun and weather conditions confirm laboratory results.

REFERENCES

[1] "The Paris Agreement | UNFCCC." https://unfccc.int/process-and-meetings/the-paris-agreement/the-paris-agreement (accessed Dec. 22, 2022).

[2] "World Energy Outlook 2022," *IEA*. https://www.iea.org/reports/world-energy-outlook-2022 (accessed Dec. 22, 2022).

[3] "Renewable Energy and Jobs - Annual Review 2022." Accessed: Dec. 06, 2022. [Online]. Available: https://www.irena.org/Publications/2022/Sep/Renewable-Energy-and-Jobs-Annual-Review-2022

[4] "UAE Energy Strategy 2050." https://u.ae/en/about-the-uae/strategies-initiatives-and-awards/strategies-plans-and-visions/environment-and-energy/uae-energy-strategy-2050 (accessed Dec. 22, 2022).

[5] M. Gostein, J. R. Caron, and B. Littmann, "Measuring soiling losses at utility-scale PV power plants," in *2014 IEEE 40th Photovoltaic Specialist Conference (PVSC)*, Jun. 2014, pp. 0885–0890. doi: 10.1109/PVSC.2014.6925056.

[6] Z. Sun, J. Zhang, Z. Tong, and Y. Zhao, "Particle size effects on the reflectance and negative polarization of light backscattered from natural surface particulate medium: Soil and sand," *Journal of Quantitative Spectroscopy and Radiative Transfer*, vol. 133, pp. 1–12, Jan. 2014, doi: 10.1016/j.jqsrt.2013.03.013.

[7] "ImageJ." https://imagej.net/ (accessed Dec. 22, 2022).

Efficiency Limit of Transition Metal Dichalcogenide Solar Cells

Koosha Nassiri Nazif, Frederick U. Nitta, Alwin Daus, Krishna C. Saraswat, Eric Pop

Stanford University, Stanford, CA, United States

RWTH Aachen University, Aachen, Germany

Transition metal dichalcogenides (TMDs) exhibit promising optoelectronic properties for use as absorber material in high-specific-power photovoltaics, including their high optical absorption coefficients, desirable band gaps, and self-passivated surfaces. The ultimate performance limit of TMD solar cells is therefore of interest. In this study, we determine the power conversion efficiency limit of multilayer MoS_2, $MoSe_2$, WS_2, and WSe_2 solar cells under AM 1.5 G illumination as a function of thickness and material quality. We use an extended version of the detailed balance method which includes Auger and Shockley-Reed-Hall (SRH) recombinations in addition to the radiative losses, calculated from measured optical absorption spectra. We demonstrate that ultrathin TMD solar cells (as thin as 50 nm) can in practice achieve up to 25% power conversion efficiency with their current-stage material quality, already making them an excellent choice for high-specific-power photovoltaic applications.

Advanced health-state data analytic workflow for utility-scale photovoltaic power plants

Jesus Montes-Romero [1], Loucas Pikolos [1], Andreas Makrides [1], Nino Heinzle [2]
George Makrides [1], Juergen Sutterlueti [2], Steve Ransome [3] and George E. Georghiou [1]

[1] PV Technology Laboratory, FOSS Research Centre for Sustainable Energy,
Department of Electrical and Computer Engineering, University of Cyprus, Nicosia, 1678, Cyprus
[2] Gantner Instruments GmbH, Montafonerstraße 4, 6780 Schruns, Austria
[3] Steve Ransome Consulting Ltd, KT2 6AF #99 Kingston upon Thames, United Kingdom

Abstract—This work aims to present data analytic advances and next-generation workflows for utility-scale photovoltaic (PV) power plant monitoring. The proposed health-state architecture comprises of an integrated and scalable workflow that includes data enrichment, predictive modelling and fault detection modules applied to high-resolution data streams. The obtained results demonstrated high power output predictive accuracies of <1.2%, given by the average root mean square error (RMSE) relative to the nominal capacity of the test-bench PV system, across different weather patterns and time durations. Furthermore, the robustness and location independency of the architecture was verified at utility-scale PV power plants by exhibiting high predictive accuracies. Moreover, the architecture proved capable to identify power, voltage and current failures with a detection accuracy of over 90%, even for low loss magnitudes. Finally, useful information is provided for establishing effective workflows for the performance evaluation of utility-scale power plants.

Keywords—*data analytics, failures, machine learning, monitoring, performance, photovoltaic*

I. INTRODUCTION

Lifetime performance improvements and the optimization of photovoltaic (PV) system operations through next-generation condition monitoring workflows is a key enabling factor towards the future large-scale uptake of the technology. In this domain, the effectiveness and performance accuracy of solar analytic workflows to provide real-time performance observability for utility-scale PV power plants remains the main challenge. Advanced cloud-based monitoring solutions that integrate workflow analytics applied to real-time data streams present new pathways to ensure high-performance for PV power plants [1-2]. Workflow analytics of this type involve extensible pipelines of data-driven predictive models and failure detection algorithms to efficiently analyze large amounts of high-resolution monitored data.

Recent advancements in the field of machine learning (ML) and big data analytics have facilitated the evolution of PV monitoring systems by integrating data-driven decision functionalities. Large portfolio PV system monitoring companies have already identified this smart grid transitioning necessity towards digitally-enhanced interoperable systems, which facilitate real-time observability and on-line analysis of big data streams emanating from PV power plants. In particular, solar analytics based on ML principles are utilized in order to

optimize plant performances, ensure data validity and provision real-time operating state information (diagnosis of physical faults and under-performance through analysis of digital monitored data). Another important modelling technique that significantly enhanced the predictive accuracy of parametric performance models is the mechanistic approach applied on normalized performance data. Even though, leveraging ML principles the technical bottlenecks that emanate from the lack of PV system metadata are overcome since the performance is captured entirely from the data (black box approach), the application of mechanistic models presents the advantage of further providing meaningful insights of the deviations exhibited at coefficient level. Such deviations are useful to identify thermal, electrical and other PV operational faults.

Over the past years, several studies have presented the application of statistical outlier detection techniques that encompass comparisons between the actual and predicted performance to detect failures or the use of signal analysis methods and electrical circuit-based models [3-4]. Apart from signal processing techniques, ML classification techniques were also considered for analyzing detected failure conditions and patterns based on both supervised and unsupervised learning regimes [5-7]. However, most performance predictive approaches and failure diagnosis models are still site-specific and not fully demonstrated at different PV power plant scales.

In this context, data integrity and sanity are important pillars for the implementation of advanced health-state workflows. Missing data caused by outages or data acquisition failures is a common problem of monitoring systems. A common approach for handling outliers and missing data is to filter out and to impute using different techniques and models such as linear interpolation and Last Observation Carried Forward (LOCF) [8-12].

The scope of this work is to present an accurate, transferable and location-independent health-state analytical architecture for the real-time performance analysis of PV power plants. The advanced analytical workflow aims to fill-in the gap of knowledge by demonstrating high predictive performance accuracies and fault diagnostics at different system scales and locations. In particular, the application of the proposed architecture to high-resolution data streams enables the effective evaluation of PV power plants. Overall the major contributions from this paper are summarized as follows:

978-1-6654-6060-6/23 $31.00 © 2023 IEEE

- Innovative health-state architecture for PV power plants with predictive accuracy beyond the state-of-the-art.
- Novel data filtering and failure diagnosis algorithms (open- and short-circuit failures, inverter and clipping faults, shading and potential induced degradation).
- Verified failure diagnosis location- and technology-independency of the workflow at different scales.
- Low-latency demonstration of the workflow to analyze high-resolution utility-scale PV system data streams.

Ultimately, useful information is provided on the workflow's robustness when applied to different site topologies.

II. EXPERIMENTAL SETUP

The performance of the implemented data-driven models and integrated tools was validated at a test-bench PV system installed at the DER-Grid Smart Infrastructure of the UCY-FOSS nanogrid (nG). Specifically, the DER-Grid smart infrastructure of UCY-FOSS is a testing platform for smart grid technologies that includes amongst other, a test-bench grid-connected PV system of 1 kW$_p$ nominal capacity, as shown in Fig. 1.

Fig. 1. DER-Grid Smart Infrastructure of UCY-FOSS comprising of kW-scale test-bench PV systems.

The actual demonstration of the architecture was verified at data streams of utility-scale PV power plants administered by Gantner Instruments. At both sites the main PV operational parameters along with the prevailing meteorological conditions are recorded using high-performance edge controller for data acquisition (Gantner Instruments Q.station XT) and stored at high capacity, distributed streaming platforms.

III. METHODOLOGY

The methodology followed to develop the proposed health-state workflow consisted of a sequence of steps that involved the: (a) unique naming convention and normalization, (b) data enrichment, (c) development of predictive model and (d) performance evaluation. Data acquisition was performed using grid-edge programmable automation controllers that are equipped with cloud-connectivity and functionalities to allow the real-time acquisition and streaming of data, analysis and high-speed triggered data logging.

The developed data analytic predictive models and failure detection routines were integrated as a dynamic pipeline of modules integrated within a scheduled workflow. The deployment of the workflow enables the fast and effective analysis of high-resolution data streams within only a few seconds. In summary, the pipeline of modules of the proposed health-state workflow are illustrated in Fig. 2.

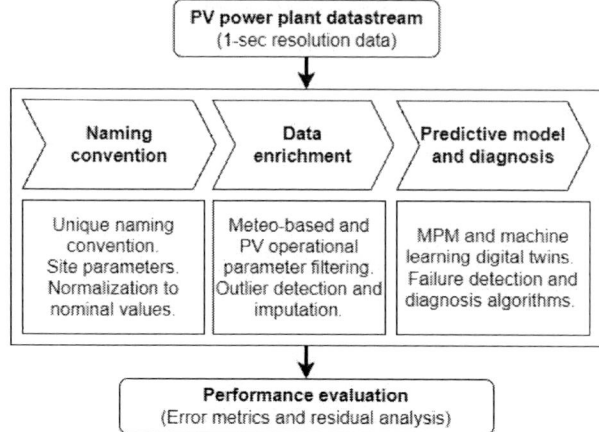

Fig. 2. Flowchart of the pipeline of modules of the health-state workflow.

A. Naming convention and normalization

An important initial stage of the workflow is to ensure a normalized state for all input variables which is used as the unique identifier. This is post-processed by an enrichment phase and all modelling steps are consuming this information as their input. Moreover, all PV system electrical parameters were normalized by nominal values in order to enable the scalability and transferability of the models irrespective of system size.

B. Data enrichment

All system data streams included the incident global irradiance (G_I) measured using pyranometers and referece cells, and the meteorological measurements of wind direction (W_a), wind speed (W_s) and ambient temperature (T_{amb}). The PV system's operational measurements include the module temperature (T_m), DC current (I_{DC}), voltage (V_{DC}) and power (P_{DC}) of the array at the maximum power point and AC power (P_{AC}). A sequence of filtering stages was applied to both meteo and PV operational parameters in order to ensure data fidelity by detecting invalid values.

C. Predictive and diagnostic model

Predictive models for each site were developed by applying the Mechanistic Performance Model (MPM) using a supervised learning regime of training and testing with a 70:30% data portion approach. The MPM is a mechanistic model that is applied to the normalized performance data in order to yield meaningful coefficients for the measured weather parameters of G_I, T_{mod} and W_s [13]. The MPM has been developed using the best features of existing models and it has 5 coefficients C_1 to C_5 defined in the equation below:

978-1-6654-6060-6/23 $31.00 © 2023 IEEE

$$PR_{DC} = C_1 + C_2 T_{mod} + C_3 \text{Log}_{10}(G_I) + C_4 G_I + C_5 W_s \quad (1)$$

where PR_{DC} is the performance ratio at the DC side.

Finally, the k-Nearest Neighbor (k-NN) supervised learning ML classifier was included in order to detect faults that were introduced to the data streams of each plant. The failures (100 imputed faults) were randomly introduced by declining the power, voltage and current at a relative magnitude of 5% during high irradiance conditions (G_I>600 W/m^2).

D. Performance Evaluation

The performance of the constructed models was evaluated using normalized RMSE relative to the maximum acquired power of each system. The nRMSE is calculated as follows:

$$\text{nRMSE} = \frac{1}{P_{nominal}} \sqrt{\frac{\sum_1^n (y_{predicted,i} - y_{measured,i})^2}{n}} \quad (2)$$

where n is the amount of data points, $y_{measured,i}$, $y_{predicted,i}$ is the measured and predicted power, respectively and $P_{nominal}$ is the nominal capacity of the PV system.

IV. RESULTS

The application of the MPM model to the test-bench system yielded P_{AC} predictive errors lower than 1.2% over the test set period (average RMSE normalized to the nominal capacity of the system), see Fig. 3a. In addition, the model was not affected by irradiance conditions since high accuracies were obtained for both clear and overcast days.

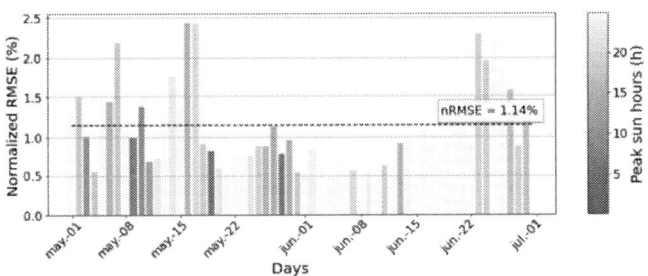

Fig. 3. Prdictive accuracy when applyng the MPM to the test-bench PV system given by the daily nRMSE.

The scalability of the developed method was further verified by applying the predictive model to a utility-scale PV power plant (3.5 MWp capacity). The comparative analysis between the investigated models (ANNs, XGBoost and MPM) showed that the highest performing was the XGBoost, presenting average P_{AC} predictive errors <2% over the test set period for the investigated utility-scale PV power plant (see Fig. 4). Moreover, all digital twin models demonstrated their capabilities to yield high accuracies when trained using only a monthly period and applying the filtered dataset.

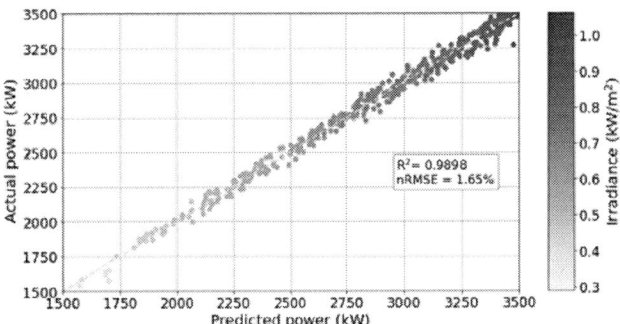

Fig. 4. Scatter plot of actual against predicted power for the investigated utility-scale PV power plant.

The plot of actual and predicted performance ratio (*PR*), presented in Fig. 5a, proved that the model predicted accurately the generated power at both low and high irradiance levels. Moreover, the residuals were plot and evaluated in order to diagnose fault occurrences, see Fig. 5b.

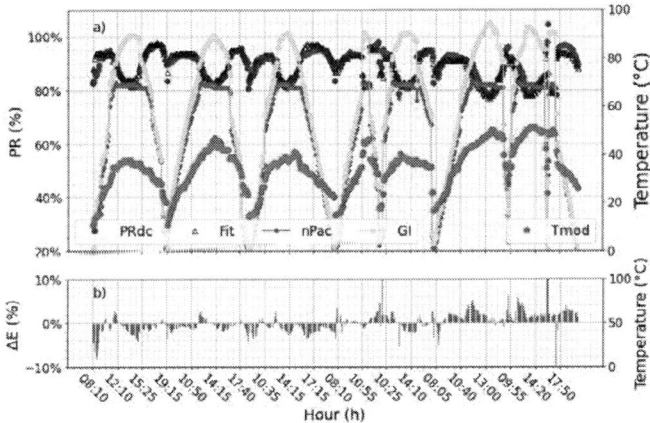

Fig. 5. (a) Plot of actual and predicted *PR* for the investigated utility-scale PV power plant. (b) Residuals depicted by the error deviation of measured from predicted value.

In addition, the predictive performance of the MPM model was evaluated against the calculated daily type categories. The results presented in Fig. 6 showed that the MPM model was capable to yield robust predictions in the range of 1% to 2% for almost all weather types. This further signifies the robustness of the model and its inherent ability to provide accurate predictions for all weather types and daily profiles.

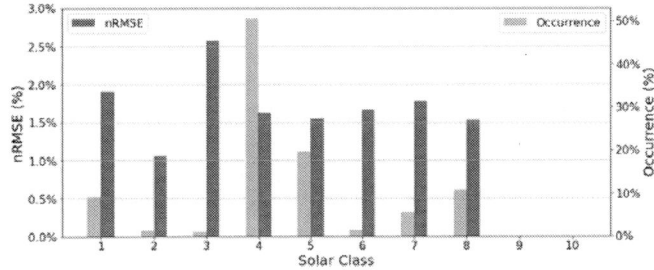

Fig. 6. Performance accuracy plots of MPM model for the test-bench PV system given by the daily nRMSE for all calculated weather types over a yearly period.

Similarly, the application of the MPM predictive model to utility-scale provided evidence that the implemented model trained only with less than 30 days of data (random data throughout the year) yielded robust predictions, since for all weather types and daily profiles error was less than 2% (see Fig. 7).

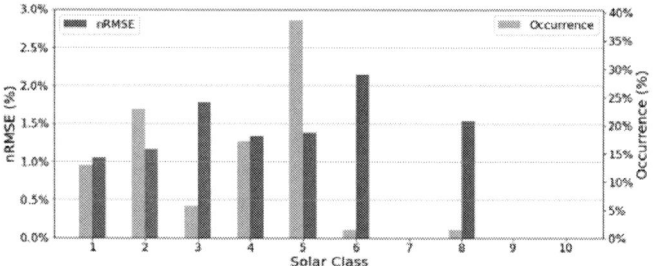

Fig. 7. Performance accuracy plots of MPM model for the utility-scale PV power plant given by the daily nRMSE for all calculated weather types over a yearly period.

Fig. 8 depicts the performance accuracy results obtained when applying the MPM and XGboost regressor for the detection of power faults at 5% level reduction for the kW-scale PV system at 1-, 15- and 30-minute resolutions. The results showed that at all resolutions the devised MPM model was able to detect the imputed faults at accuracies over 90%. Overall, for both models the detection accuracy was within the range of 90% to 92% and was not significantly affected by the increased data resolution.

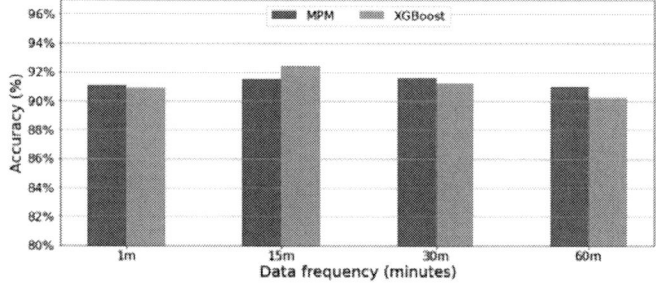

Fig. 8. Performance accuracy plots of MPM model for the utility-scale PV power plant given by the daily nRMSE for all calculated weather types over a yearly period.

The detection accuracy of both devised models (MPM and XGBoost) was further evaluated and presented in Fig. 9.

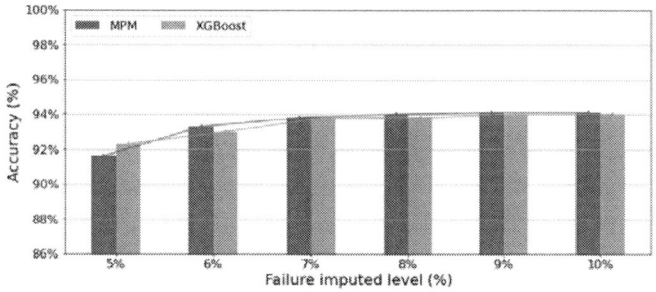

Fig. 9. Fault detection accuracy plots of MPM and XGBoost model for the utility-scale PV power plant at different failure imputed levels.

Specifically, by applying the models to datasets of the utility-scale plant that included Power faults at levels from 5% - 10%. The results exhibited in Fig. 9, demonstrated that high detection accuracies can be achieved by both models at 5% loss levels (detection accuracy >90%). Moreover, the detection accuracy of both models increased at higher fault levels.

The commenced residual analysis and application of the k-NN classifer proved capable of detecting imputed faults at an accuracy of 91.3%. The outliers are exhibited in black whereas inliers are confined within the red solid lines, see Fig. 10a. As an example, clipping was detected by correlating errors to high irradiance and constant power conditions. Fig. 10b further provides evidence that high predictive accuracies were obtained for the investigated utility-scale power plant since the fitted PR closely followed the perfromance. The model was capable to detect clipping by analyzing the outliers against the normalized voltage.

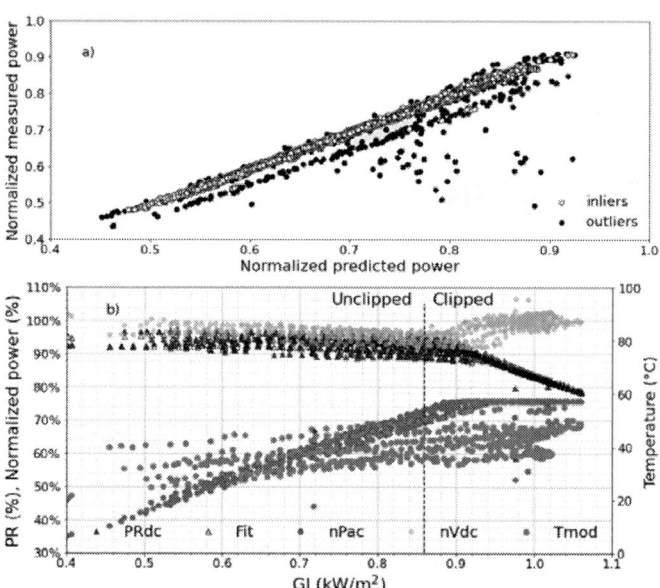

Fig. 10. Plots of: (a) Measured against predicted power for the utility-scale PV power plant and (b) PR and normalized power against irradiance levels. Normalized parameters of AC power (nP_{ac}) and DC voltage (nV_{dc}) are also presented.

Finally, Table I summarizes the classification results when applying the k-NN classifier to the constructed dataset of both the test-bench and utility-scale PV power plant including 100 imputed Voltage, Current and Power faults of 5% level. The results showed that the classifier is capable of diagnosing the nature of the faults at accuracies ranging from 95% to 98%. This demonstrates that the application of supervised learning regimes with fault labels improves the diagnostic process.

TABLE I
SUMMARY OF CLASSIFICATION PERFORMANCE OF k-NN CLASSIFIER

Fault Category	Performance Evaluation
	Accuracy (%)
Current	98
Voltage	95
Power	95

V. CONCLUSION

Active PV monitoring systems encompassing advanced health-state workflows can provide a clear indication whether the system is performing as expected and therefore safeguard optimal performance. This work presents the method followed to develop health-state architectures that comprise of advanced data analytic workflows applicable to utility-scale PV systems.

The results demonstrated that the workflow was capable to effectively analyze big data at latencies lower than 5 seconds. The application of MPM exhibited high maximum power predictive accuracies (average nRMSE < 1.2%) for both the test-bench PV system and utility-scale plants. Moreover, the application of the MPM predictive model to utility-scale levels provided evidence that with less than 30 days of data robust predictions can be obtained for all weather types and daily profiles.

Furthermore, the failure detection modules proved capable of detecting low magnitude imputed power faults with a detection accuracy of >90%. The detection accuracy of both devised the MPM and XGBoost models was within the range of 90% to 92% and was not significantly affected by the increased data resolution signifying that the models can be applied at any resolution (1-min to 60 min). An important note is that in terms of implementation, the MPM has a fairly low degree of complexity (simple mathematical equation) and presents meaningful coefficients for the system under study when fitting the data. On the other hand, the application of ML models has a higher complexity for implementation and there are no direct usable coefficients due to the "black box" modelling nature of these principles.

Finally, upon validation the analytics and workflow formed part of a Software as a Service (SaaS) platform that provides effective data sanity and integrity, health-state monitoring and real-time failure detection and classification.

ACKNOWLEDGMENT

This work was funded through the PV-ANALYTIC project. Project PV-ANALYTIC (P2P/SOLAR/0818/0012) is supported under the umbrella of SOLAR-ERA.NET Cofund by the Austrian Research Promotion Agency (FFG) and the Research & Innovation Foundation (RIF) of Cyprus. SOLAR-ERA.NET is supported by the European Commission within the EU Framework Programme for Research and Innovation HORIZON 2020 (Cofund ERA-NET Action, N° 691664). J. Montes-Romero collaboration was funded by the European Union -NextGenerationEU.

REFERENCES

[1] A. Chouder and S. Silvestre, "Automatic supervision and fault detection of PV systems based on power losses analysis," Energy Convers. Manag., vol. 51, no. 10, pp. 1929–1937, Oct. 2010, doi: 10.1016/j.enconman.2010.02.025.

[2] J. Sutterlueti et al., "Advanced system monitoring and artificial intelligent data-driven analytics to serve GW scale photovoltaic power plant requirements," in 2022 PV Performance Modelling Collaborative (PVPMC) Workshop, 2022, pp. 1-6.

[3] M. A. Zuniga-Reyes et al., "Photovoltaic failure detection based on string-Inverter voltage and current signals," IEEE Access, vol. 9, pp. 39939–39954, 2021, doi: 10.1109/ACCESS.2021.3061354.

[4] A. Livera et al., "Recent advances in failure diagnosis techniques based on performance data analysis for grid-connected photovoltaic systems," Renew. Energy, vol. 133, pp. 126–143, 2019, doi: 10.1016/j.renene.2018.09.101.

[5] S. Vergura and M. Carpentieri, "Statistics to detect low-intensity anomalies in PV systems," Energies, vol. 11, no. 1, 2018, doi: 10.3390/en11010030..

[6] A. Livera et al., "Data processing and quality verification for improved photovoltaic performance and reliability analytics," Prog. Photovoltaics Res. Appl., 2020, doi: 10.1002/pip.3349..

[7] H. Ding et al., "Local outlier factor-based fault detection and evaluation of photovoltaic system," Sol. Energy, vol. 164, no. January, pp. 139–148, 2018, doi: 10.1016/j.solener.2018.01.049.

[8] A. Karahalios, L. Baglietto, K. J Lee, D. R English, J. B Carlin and J. A Simpson, "The impact of missing data on analyses of a time-dependent exposure in a longitudinal cohort: a simulation study", in Emerging Themes in Epidemiology, Aug. 2013.

[9] T. McCandless, S. E. Haupt, G. Young, "Replacing missing data for ensemble systems - The Effects of Missing Data", in 7th Artificial Intelligence and Its Applications to the Environmental Sciences Conference, 2009.

[10] B. Wilmots, Y. Shen, E. Hermans, D. Ruan, "Missing data treatment: Overview of possible solutions", Diepenbeek: Policy Research Centre Mobility and Public Works, 2011.

[11] E.A.P. Gustavo, A. Batista and M. C. Monard, "An Analysis of Four Missing Data Treatment Methods for Supervised Learning", in Applied Artificial Intelligence, vol. 17, pp. 519-533, May 2003.

[12] A. Phinikarides, G. Makrides, and G. E. Georghiou, "Estimation of the degradation rate of fielded photovoltaic arrays in the presence of measurement outages," in 32nd EU-PVSEC, 2016, pp. 1754 - 1757.

[13] S. Ransome, J. Sutterlueti, " How to choose the best empirical model for optimum energy yield predictions," in IEEE 44th Photovoltaic Specialist Conference, 2017, pp. 652-657, doi: 10.1109/PVSC.2017.8366067.

Monolithic Bifacial Perovskite-CdSeTe Tandem Solar Cells

Zhaoning Song, Deng-bing Li, Sabin Neupane, Kamala Khanal Subedi, Samuel Seibert, Randy J. Ellingson, and Yanfa Yan

Wright Center for Photovoltaics Innovation and Commercialization, Department of Physics and Astronomy, The University of Toledo, 2801 W. Bancroft St., Toledo, OH 43606 USA

Abstract— Monolithic metal halide perovskite-cadmium selenide telluride (CdSeTe) tandem thin-film solar cells have not been demonstrated to date, primarily due to the challenge of the almost-fixed architecture of CdSeTe cells that is incompatible with conventional tandem designs. Here, we report a monolithic bifacial thin-film tandem design that not only enables monolithic integration of perovskite-CdSeTe tandems but also delivers a high equivalent power conversion efficiency by harvesting the albedo light. The bifacial design allows better current match and thus enables higher total power output density of the tandem devices than conventional tandem cells. Using our innovative bifacial tandem design strategy, we have fabricated proof-of-concept monolithic bifacial perovskite-CST tandem solar cells with a stabilized power output of 23 mW/cm² and a bifacial equivalent efficiency of 23% under bifacial illumination with 1-sun front illumination and an albedo of 0.4.

I. INTRODUCTION

Metal halide perovskite solar cells have been the focus of the photovoltaic (PV) research community in recent years, because of their excellent optoelectronic properties and projected low production costs. After a decade of rapid progress, the certified power conversion efficiency (PCE) of perovskite solar cells has improved to more than 25%, exceeding the record efficiencies of other established thin-film PV technologies, such as cadmium selenide telluride (CdSeTe) and copper indium gallium diselenide (CIGS). Moreover, the bandgap flexibility (1.25 to 3.1 eV) of perovskite by tuning the material composition and low-temperature fabrication process enables opportunities to produce monolithic tandem solar cells with theoretical PCEs exceeding 40%. Combining ultrahigh efficiencies and low production costs, perovskite-based thin-film tandem PV technologies may significantly contribute to the goal of lowering the costs of solar electricity and boosting the deployment of thin-film PV technologies.

Among all the possible perovskite-based tandem combinations, perovskite-CdSeTe thin-film tandems are particularly interesting because of the commercial readiness and success of CdTe-based thin-film solar modules and shared similarities in the production processes of these two dissimilar thin-film PV technologies. Since ~60% of the electric power produced by a tandem solar cell is from the top cell, pairing up an inexpensive wide-bandgap perovskite top cell with a low-cost lower-bandgap CdSeTe bottom cell into a hybrid thin-film tandem solar cell provides an unprecedented opportunity to achieve simultaneously high efficiency and low production costs, accelerating the introduction of perovskite PV technology to the market by incorporating into the existing capability of commercial CdTe production lines. Despite this attractive promise, thus far, no monolithic perovskite-CdSeTe tandem cells have been reported yet. Limited by the unfavorable downward band bending at the back interface of the absorber layer, the established CdSeTe cells are not able to achieve high efficiencies with conventional monolithic tandem designs.

Here, we report an innovative bifacial tandem thin-film PV cell design (Fig. 1(a)) that not only enables monolithic integration of perovskite-CdSeTe tandems but also overcomes the output power density limit of the conventional monolithic tandem designs by harvesting albedo light (reflection and deflection). The unique feature of bifacial tandem architecture allows the albedo light to be harvested at the front p-n heterojunction side of the CdSeTe bottom cell, which considerably boosts its photocurrent to match that of the top cell (Fig. 1(b)). Due to the high photovoltage generated by the monolithic tandem, the photocurrent contributed by the albedo light can have a large impact on the overall output power density. This unique working principle widens the optimal bandgap selection range for individual subcells and thus maximizes tandem efficiency.

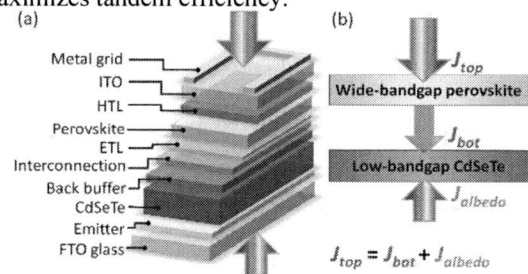

Fig. 1. (a) Device architecture and (b) working principle of bifacial perovskite-CdSeTe tandem thin-film solar cells.

II. EXPERIMENTAL DETAILS

A. Fabrication of bifacial perovskite-CdSeTe tandem cells

CdSeTe subcells were fabricated following our previous report [1]. In brief, a 60 nm oxygenated cadmium sulfide (CdS:O) and 80 nm cadmium selenide (CdSe) bilayer emitter was deposited on fluorine-doped tin oxide (FTO) coated glass substrates using radiofrequency (rf) magnetic sputtering. A ~3.5 μm CdTe absorber was then deposited by close-space sublimation (CSS), followed by a cadmium chloride (CdCl₂) treatment in dry air. A 50 nm copper thiocyanate (CuSCN) back buffer was spin-coated on CdTe, followed by rapid thermal annealing at 180 °C to facilitate Cu diffusion. A 120 nm indium tin oxide (ITO) interconnecting layer was deposited by rf-sputtering. Wide-bandgap perovskite subcells were constructed on top of the CdSeTe subcells according to our previously reported method [2]. The following layers were deposited in sequence, including a 20 nm tin oxide (SnO_2) electron transport layer (ETL) by atomic layer deposition, a 4000 nm $FA_{0.8}Cs_{0.2}Pb(I_{1-x}Br_x)_3$ (x = 0.1 to 0.4) perovskite absorber layer by spin-coating, a 150 nm 2,2',7,7'-Tetrakis[N,N-di(4-methoxyphenyl)amino]-9,9'-spirobifluorene (spiro-OMeTAD) hole transport layer (HTL) by spin-coating, a 10 nm molybdenum oxide (MoO_x) layer by thermal evaporation, a 150 nm ITO front contact by rf sputtering, and 200 nm silver (Ag) metal grids. The active area of the unit cell is 0.25 cm², defined by the patterns of the ITO front electrode. The detailed preparation procedures can be found in our previous work.

B. Device characterization

A home-built bifacial testing simulator was used for measuring tandem devices under concurrent bifacial illumination. The front illumination was simulated by a Xenon lamp (Oriel Instruments) equipped with an AM1.5G filter and calibrated to 100 mW/cm². The rear illumination was simulated by a multi-channel LED solar simulator (G2V Pico) calibrated to AM1.5G spectrum. A Keithley 2400 source meter was used for current density-voltage (J-V) measurements. External quantum efficiency (EQE) curves were measured using a modified commercial system (PV Measurements Inc.) with biasing light and voltage.

III. RESULTS AND DISCUSSION

A. Theoretical Efficiency Limits for bifacial perovskite-CdSeTe tandem solar cells

To evaluate the potential of bifacial perovskite-CdSeTe thin-film tandem solar cells, we calculate the detailed balanced equivalent efficiency limits of a double-junction tandem solar cell under bifacial illumination conditions. Fig. 2(a) shows the efficiency limits of a monofacial tandem cell as a function of bandgap combinations. For the 1.35-eV CdSeTe bottom cell, a top cell with a bandgap of 1.6 to 1.9 eV can reach theoretical efficiency limits of 20 to 40%. With different albedo conditions, the total output power density can boost to more than 50 mW/cm², corresponding to an equivalent bifacial power conversion efficiency of >50% with respect to 1-sun illumination (Fig. 2(b)). The maximum output power density of bifacial tandem cells can be more than 50% higher than the theoretical limit of a single-junction solar cell.

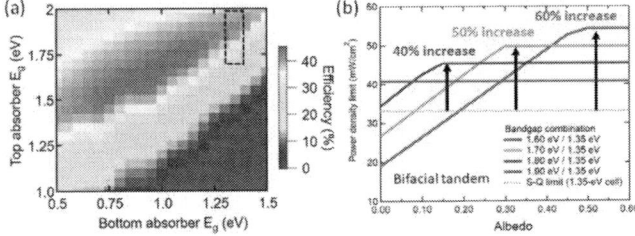

Fig. 2. (a) Detailed balanced efficiency limits of a double-junction tandem cell with different bandgap combinations under monofacial illumination. (b) Total output power density limits of bifacial tandem cells under different albedo conditions.

B. PV performance of bifacial perovskite-CdSeTe tandem solar cells

We fabricated proof-of-concept monolithic bifacial perovskite-CdSeTe thin-film tandem solar cells. Fig. 3(a) shows the photo of the front side of a prototype bifacial tandem cell. Fig. 3(b) shows the cross-sectional scanning electron microscopy (SEM) image of a bifacial perovskite-CdSeTe tandem solar cell. The layered structure of the tandem cell can be clearly seen in the image. The total thickness of the active layers of the thin-film tandem device is less than 5 μm.

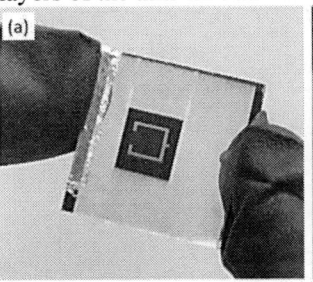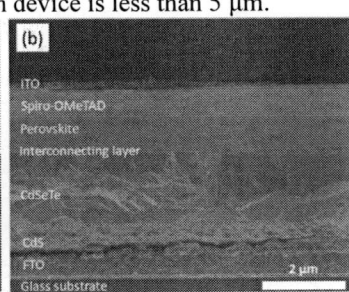

Fig. 3. (a) Photo and (b) cross-sectional SEM image of a bifacial perovskite–CdSeTe tandem solar cell.

Fig. 4 plots the J-V characteristics of a bifacial thin-film tandem cell constructed by a 1.62-eV perovskite ($FA_{0.8}Cs_{0.2}Pb(I_{0.9}Br_{0.1})_3$) top cell and a 1.35-eV CdSeTe bottom cell. The device shows substantially enhanced performance when concurrently illuminated from both sides of the device, with 1 sun intensity from the perovskite side as the primary light source and a 0.4 sun intensity light from the CdSeTe side to mimic the albedo light from the ground, compared with single-side illumination from either perovskite or CST side (Fig. 4(a)).

Fig. 4(b) compares the J-V characteristics of the bifacial perovskite-CdSeTe tandem cell under different albedo conditions. Table I lists the PV parameters of the tandem device under different illumination conditions. The measurements show that increasing the albedo light intensity increases the J_{SC}, open-circuit voltage (V_{OC}), and thus, equivalent PCE of a perovskite-CdSeTe tandem cell. This is ascribed to improved photocurrent matching by increasing the photocurrent density in the CdSeTe subcell.

TABLE I. J-V parameters of a bifacial perovskite-CdSeTe tandem solar cell under different bifacial illuminations.

Albedo	Equivalent PCE (%)	V_{OC} (V)	J_{SC} (mA/cm²)	FF (%)
0	3.83	1.601	3.68	65.0
0.1	7.10	1.605	6.82	64.9
0.2	10.58	1.671	9.85	64.2
0.3	15.33	1.704	12.81	70.2
0.4	20.18	1.758	15.85	72.4

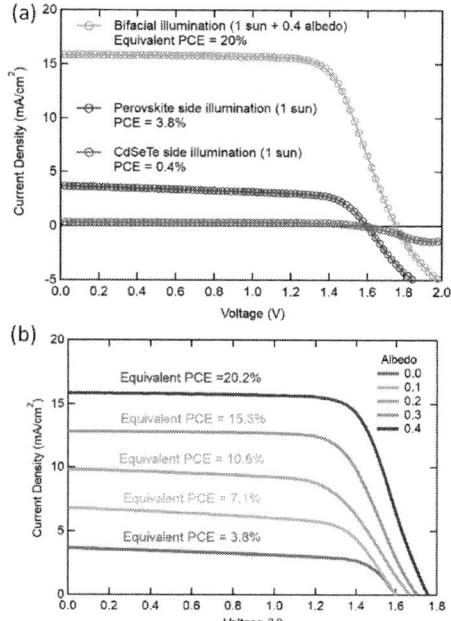

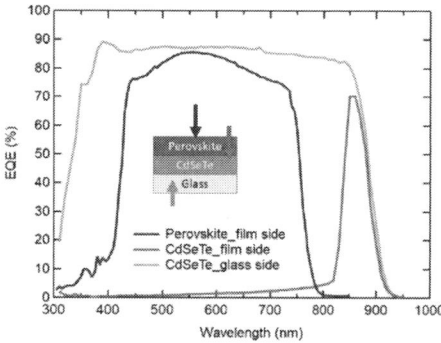

Fig. 4. J-V curves of (a) a bifacial perovskite-CdSeTe tandem cell under single-side and bifacial illumination and (b) under bifacial illumination with different albedo light conditions.

Fig. 5 shows the EQE curves of 1.62-eV perovskite and 1.35-eV CdSeTe subcells illuminated from both the front (film) and rear (glass) sides of the tandem device. Under the film side illumination, the integrated J_{SC} for perovskite and CdSeTe cells are 18 and 3 mA/cm², respectively. The high current mismatch is mainly caused by the poor back-illuminated efficiency of the CdSeTe subcell. When light is incident from the opposite side of the p-n heterojunction of CdSeTe cells, the EQE is typically very low because of the downward band bending caused by the deep valence band of CdSeTe and high surface recombination velocity at the interface [3]. Under the glass side illumination, CdSeTe can generate a high integrated J_{SC} of 28 mA/cm². The concurrent bifacial illumination on both the film and glass sides of the tandem device provides additional photocurrent in CdSeTe subcell to match the photocurrent generated in the perovskite top subcell, enabling higher total photocurrent density and thus total output power density or equivalent bifacial efficiency.

Motivated by the promising bifacial measurement results of the prototype devices, we further modified and optimized the metal oxides for the interconnecting layer. The champion tandem device delivers an equivalent bifacial PCE of 23.3%, with a V_{OC} of 1.875 V, a J_{SC} of 16.1 mA/cm², and a FF of 77.4% under bifacial illumination with 1-sun front illumination and an albedo of 0.4 (Fig. 6(a)). The device exhibits a stabilized power output of 22.8 mW/cm² under the same bifacial illumination condition (Fig. 6(b)). Details will be discussed during the presentation.

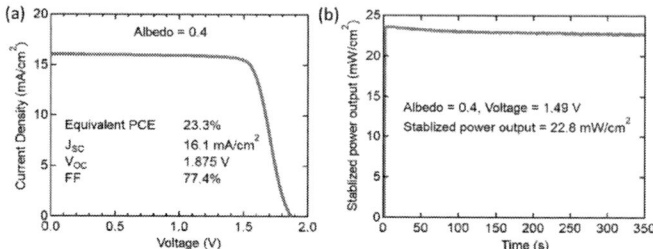

Fig. 6. (a) J-V curve and (b) stabilized power output of the champion bifacial perovskite-CdSeTe tandem cell under bifacial illumination with 1-sun front illumination and an albedo of 0.4.

IV. CONCLUSION

We developed a novel design of monolithic bifacial perovskite-CdSeTe tandem thin-film solar cells. Our prototype tandem devices exhibit stabilized power output of ~23 mW/cm², corresponding to a bifacial equivalent efficiency of 23% under concurrent bifacial illumination with 1-sun front illumination and an albedo of 0.4.

ACKNOWLEDGMENT

This material is based upon work supported by the U.S. Department of Energy's Office of Energy Efficiency and Renewable Energy (EERE) under the Solar Energy Technology Office Award Number DE-EE0009832 and by Air Force Research Laboratory under agreement FA9453-19-C-1002. The views and conclusions contained herein are those of the authors and should not be interpreted as necessarily representing the official policies or endorsements, either expressed or implied, of the Air Force Research Laboratory, U.S. Department of Energy, or the U.S. Government.

REFERENCES

[1] D.-B. Li, S. S. Bista, R. A. Awni, S. Neupane, A. Abudulimu, X. Wang, K. K. Subedi, M. K. Jamarkattel, A. B. Phillips, M. J. Heben, J. D. Poplawsky, D. A. Cullen, R. J. Ellingson, and Y. Yan, "20%-efficient polycrystalline Cd(Se,Te) thin-film solar cells with compositional gradient near the front junction," *Nat. Commun.,* vol. 13, no. 1, p. 7849, 2022.
[2] D. Zhao, C. Wang, Z. Song, Y. Yu, C. Chen, X. Zhao, K. Zhu, and Y. Yan, "Four-Terminal All-Perovskite Tandem Solar Cells Achieving Power Conversion Efficiencies Exceeding 23%," *ACS Energy Lett.,* vol. 3, no. 2, pp. 305-306, 2018.
[3] A. B. Phillips, K. K. Subedi, G. K. Liyanage, F. K. Alfadhili, R. J. Ellingson, and M. J. Heben, "Understanding and Advancing Bifacial Thin Film Solar Cells," *ACS Appl. Energy Mater.,* vol. 3, no. 7, pp. 6072-6078, 2020.

Fig. 5. EQE curves of perovskite and CdSeTe subcells with the light incident from film and glass sides.

Improving the Performance and Yield of Colloidal Quantum Dot Solar Cells through Electron Transport Layer Optimization

Dana Kachman,[1] Arlene Chiu,[1] Dhanvini Gudi,[1] Chengchangfeng Lu,[1] Eric Rong,[1] Sreyas Chintapalli,[1] Yida Lin,[1] Daniel Khurgin,[2] and Susanna M. Thon[1]

[1] Department of Electrical and Computer Engineering, Johns Hopkins University, Baltimore, Maryland, 21218, United States

[2] Department of Chemical and Biomolecular Engineering, University of California Los Angeles, Los Angeles, California, 90095, United States

Abstract — **Colloidal quantum dots (CQDs) are promising materials for photovoltaic applications due to their solution processibility and size-dependent band gap tunability. The electron transport layer (ETL) is an important component of PbS CQD solar cells, and the quality of the zinc oxide nanoparticle (ZnO NP) ETL film significantly impacts both the power conversion efficiency (PCE) and fabrication yield of CQD solar cells. We report on multiple methods to improve the quality of ZnO NP ETL films and demonstrate increased PCE and device yield in standard CQD solar cells employing optimized ZnO NP films. We also discuss the application of these methods in an inverted CQD solar cell architecture.**

I. INTRODUCTION

Lead sulfide colloidal quantum dots (PbS CQDs) are of interest for photovoltaic applications due to their low-temperature synthesis, solution processability, and band gap tunability. These properties allow for many applications beyond those of traditional solar cells, such as in hybrid multijunction solar cells, flexible solar cells, and building-integrated photovoltaics. The maximum efficiency achieved in a PbS CQD solar cell is 15.4% [1], as compared to 26.7% for market-leading silicon solar cells [2]. While significant progress has been made in improving CQD solar cell performance, their efficiency needs to be further improved to make them a commercially viable technology. Additionally, further study is required to analyze and improve the yield of current fabrication techniques, where yield is defined as the percentage of working devices per fabrication batch. We have identified that many popular zinc oxide nanoparticle (ZnO NP) synthesis methods yield inconsistent results for the electron transport layer (ETL), frequently including a large percentage of shorted and low PCE devices. We report on two methods to decrease agglomeration during ZnO NP synthesis and improve the PCE and yield of CQD solar cells.

We also report on the applications of these methods in the optimization of an inverted CQD solar cell architecture. Currently, material properties of the hole transport layer (HTL) are also limiting device PCE [3], so developing an inverted CQD solar cell to facilitate the incorporation of novel materials as the HTL is an important goal. However, this architecture presents new fabrication challenges, as the ZnO NPs must be deposited on top of the CQD layer. We demonstrate approaches to overcome the damage to the CQD film and resulting shunting and decreased performance when depositing ZnO NPs onto a PbS CQD film.

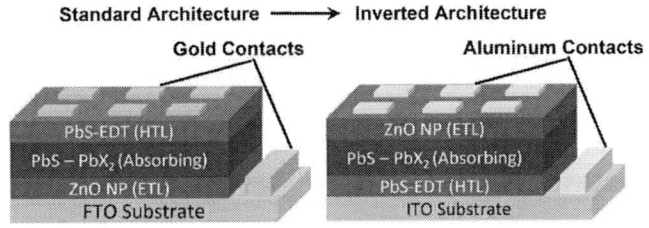

Fig. 1.   Schematics of a standard (left) and an inverted (right) PbS CQD solar cell.

II. ETL OPTIMIZATION IN A STANDARD CQD SOLAR CELL

We optimized synthesis time as well as the final solvent in the ZnO synthesis method popularized by Chuang et al. [4] to improve PbS CQD solar cell device performance and yield.

A. Synthesis Parameter Optimization

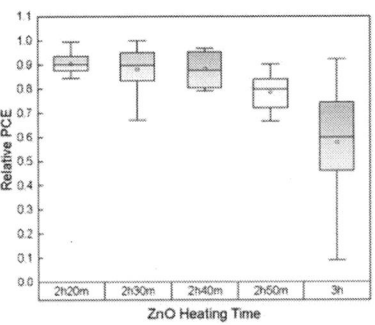

Fig. 2.   PbS CQD solar cell PCE relative to the best performing device for ETLs made from ZnO NPs heated for different amounts of time. 9 devices were measured at each heating time.

We first optimized the heating time for the ZnO synthesis. To explore all common heating times [4]-[5], we varied the ZnO synthesis time from 2 hours and 20 minutes to 3 hours. We observed that shorter heating times yielded higher average PCE in devices, but the ZnO NPs were hard to redisperse in methanol, which is used as the washing solvent. We found the best trade off at 2 hours and 40 minutes of heating time. This allowed for reliable washing in methanol without a significant reduction of device PCE (fig. 2).

We also performed an optimization of the final ZnO dispersion solvent. Chuang et al. used pure chloroform for their final solvent, while Lan et al. added methanol to the final solvent in 2016 [4],[6]. Lan et al. found that the optimal ratio was 1:1 chloroform to methanol. This ratio results in minimal agglomeration, leading to better film morphology and ultimately improved device performance. Our results agreed well with this finding, and we additionally found that while higher chloroform-to-methanol ratios can produce high-performance devices, device fabrication yield was significantly better at a 1:1 ratio than at other ratios (fig. 3).

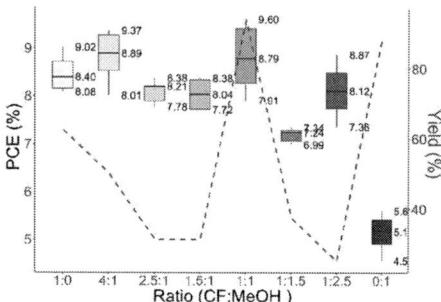

Fig. 3. Plots of the power conversion efficiency (PCE) of devices made with different ratios of chloroform (CF) and methanol (MeOH) for the final ZnO solution as well as the corresponding yield of working devices. 16 devices were measured at each ratio, except for the 0:1 ratio, for which 9 devices were measured.

III. ETL OPTIMIZATION IN AN INVERTED CQD SOLAR CELL

We next moved on to optimizing the performance of an inverted architecture PbS CQD solar cell. We chose to start with the method published by Ning et al. in 2018 [7], which uses the same ZnO film fabrication procedure discussed above, with minor modifications. For our devices, we used an amine mixture (10:3:2 butylamine: pentylamine: hexylamine) at a concentration of 300 mg/mL [8] for the PbS CQD absorbing layer.

A. SCAPS-1D Simulations

We studied the performance of inverted devices by conducting simulations using SCAPS-1D solar cell simulations [9]. We first directly inverted the solar cell structure, maintaining layer thicknesses and only changing the identity of the contacts. Our inverted devices use an ITO transparent contact and aluminum as the top contact, as seen in fig. 1. We found that directly inverting the CQD solar cell led to losses in both short circuit current density (Jsc) and open circuit voltage (Voc) (fig. 4). We attributed the drop in Jsc to non-optimized layer thicknesses. Since the *pn* junction is located at the absorbing layer/electron transport layer interface, it is further

from the illumination side of the device compared to the standard architecture. We thus expected that a thinner hole transport layer and absorbing layer would be needed to optimize charge extraction within the device. We confirmed this by conducting SCAPS simulations to optimize over a range of HTL and absorbing layer thicknesses (fig. 4).

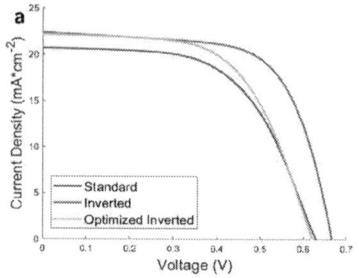

b	Standard Architecture	Optimized Inverted Architecture	Inverted Architecture
Absorbing Layer thickness (nm)	400	350	400
HTL thickness (nm)	60	25	60
$J_{sc}\left(\frac{mA}{cm^2}\right)$	22.4	22.2	20.7
V_{oc} (V)	0.67	0.62	0.63
FF	0.66	0.59	0.57
PCE(%)	9.78	8.13	7.49

Fig. 4. (a) JV curves for standard and inverted devices from SCAPS-1D simulations and (b) the corresponding device parameters. The inverted architecture was optimized by sweeping over a range of absorbing layer and hole transport layer thicknesses.

We saw significant improvement in the Jsc based on this optimization, but we still observed low Voc. Previous work on inverted devices has shown that the work function of the contacts has a significant impact on device performance [10], so we concluded that our current transparent contact did not have a sufficiently deep work function, which was limiting charge extraction from the hole transport layer. We confirmed this by conducting a SCAPS-1D simulation sweep of transparent contact work function, and we saw significant increases in the Voc as we increased the work function up to approximately 4.9 eV (fig. 5).

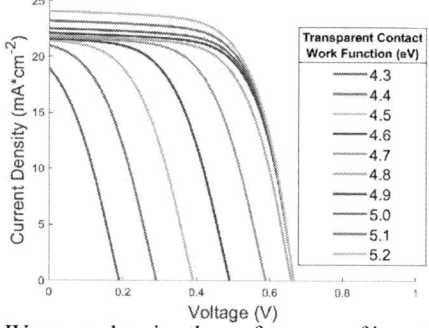

Fig. 5. JV curves showing the performance of inverted CQD solar cells with various work functions for the transparent contact, as indicated by the legend.

978-1-6654-6060-6/23 $31.00 © 2023 IEEE

B. Experimental Results

We also experimentally fabricated inverted CQD solar cell devices. We developed two approaches to improve the performance of the devices. The first approach was developed to address the low fill factor seen in our initial devices. We hypothesized that the low fill factor was a result of shunts in the device due to cracks in the absorbing layer film. We further hypothesized that the ZnO film deposition could be partially dissolving the absorbing layer if the absorbing layer was not fully dry before ZnO deposition. To address this issue, we introduced an annealing step after depositing the absorbing layer and let the film dry for at least 24 hours in a glovebox before depositing the ZnO film. Using this approach, we saw significant increases in both fill factor and Jsc (fig. 6).

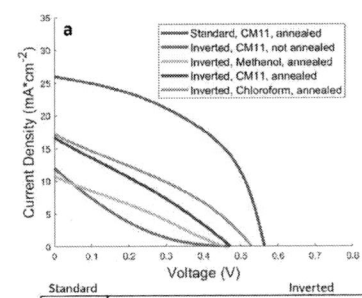

b ZnO Dispersion Solvent	CM1:1	CM1:1	CM1:1 (No anneal)	Chloroform only	Methanol only
$J_{sc}\left(\frac{mA}{cm^2}\right)$	22.7	13.3	9.5	14.6	9.7
V_{oc} (V)	0.49	0.40	0.43	0.52	0.36
FF	0.42	0.28	0.15	0.29	0.29
PCE(%)	5.1	1.5	0.6	2.2	1.0

Fig. 6. (a) JV curves and (b) solar cell parameters for experimentally fabricated inverted CQD solar cells. The data columns in (b) correspond to the blue, purple, orange, green, and yellow JV curves in (a), respectively.

We also conducted a solvent optimization study for the ZnO layer in an inverted device. We tested devices made with ZnO from just chloroform, just methanol, and a 1:1 chloroform and methanol solution. Our results show that using just chloroform significantly improved device performance, primarily through improved Voc. We are currently conducting further studies to determine the mechanism of Voc improvement and continue optimization of the inverted CQD solar cells to match performance in the standard architecture devices.

IV. CONCLUSION

We experimentally optimized both the heating time and final dispersion solvent for ZnO NPs used as the ETL for CQD solar cells. We showed that the fabrication yield of high-performance cells varies significantly as a function of both synthesis conditions. Device yield is an under-reported metric that will become increasingly important as CQD solar cells become commercially viable.

We additionally explored how these methods can be applied in an inverted CQD solar cell. We conducted SCAPS simulations to determine the optimal CQD layer thicknesses and determine the ideal work function for the transparent contact. Finally, we demonstrated two modifications to current fabrication techniques that improve the performance of inverted CQD solar cells, paving the way to further PCE improvements.

V. ACKNOWLEDGEMENTS

This work was funded by the National Science Foundation (DMR-1807342, ECCS- 1846239) and the US Department of Defense (W911NF2120213).

REFERENCES

[1] C. Ding, D. Wang, D. Liu, H. Li, Y. Li, S. Hayase, T. Sogabe, T. Masuda, Y. Zhou, Y. Yao, Z. Zou, R. Wang and Q. Shen, "Over 15% Efficiency PbS Quantum-Dot Solar Cells by Synergistic Effects of Three Interface Engineering: Reducing Nonradiative Recombination and Balancing Charge Carrier Extraction," Adv.Energy Mater., vol. 12, pp. 2201676, 2022.

[2] M. Green, E. Dunlop, J. Hohl-Ebinger, M. Yoshita, N. Kopidakis and X. Hao, "Solar cell efficiency tables (version 57)," Prog Photovolt Res Appl, vol. 29, pp. 3-15, 2021.

[3] E. Rong, A. Chiu, C. Bambini, Y. Lin, C. Lu and S. M. Thon, "New Chalcogenide-Based Hole Transport Materials for Colloidal Quantum Dot Photovoltaics," in - 2021 IEEE 48th Photovoltaic Specialists Conference (PVSC), pp. 750, 2021.

[4] C.M. Chuang, P.R. Brown, V. Bulović and M.G. Bawendi, "Improved performance and stability in quantum dot solar cells through band alignment engineering," Nature Materials, vol. 13, pp. 796-801, 2014.

[5] Y. Wang, K. Lu, L. Han, Z. Liu, G. Shi, H. Fang, S. Chen, T. Wu, F. Yang, M. Gu, S. Zhou, X. Ling, X. Tang, J. Zheng, M.A. Loi and W. Ma, "In Situ Passivation for Efficient PbS Quantum Dot Solar Cells by Precursor Engineering," Adv Mater, vol. 30, pp. 1704871, 2018.

[6] X. Lan, O. Voznyy, A. Kiani, F.P. García de Arquer, A.S. Abbas, G. Kim, M. Liu, Z. Yang, G. Walters, J. Xu, M. Yuan, Z. Ning, F. Fan, P. Kanjanaboos, I. Kramer, D. Zhitomirsky, P. Lee, A. Perelgut, S. Hoogland and E.H. Sargent, "Passivation Using Molecular Halides Increases Quantum Dot Solar Cell Performance," Adv Mater, vol. 28, pp. 299-304, 2016.

[7] R. Wang, X. Wu, K. Xu, W. Zhou, Y. Shang, H. Tang, H. Chen and Z. Ning, "Highly Efficient Inverted Structural Quantum Dot Solar Cells," Adv Mater, vol. 30, pp. 1704882, 2018.

[8] J. Xu, O. Voznyy, M. Liu, A.R. Kirmani, G. Walters, R. Munir, M. Abdelsamie, A.H. Proppe, A. Sarkar, F.P. García de Arquer, M. Wei, B. Sun, M. Liu, O. Ouellette, R. Quintero-Bermudez, J. Li, J. Fan, L. Quan, P. Todorovic, H. Tan, S. Hoogland, S.O. Kelley, M. Stefik, A. Amassian and E.H. Sargent, "2D matrix engineering for homogeneous quantum dot coupling in photovoltaic solids," Nature Nanotechnology, vol. 13, pp. 456-462, 2018.

[9] M. Burgelman, P. Nollet and S. Degrave, "Modelling polycrystalline semiconductor solar cells," Thin Solid Films, vol. 361, pp. 527-532, 2000.

[10] G. Kim, B. Walker, H. Kim, J.Y. Kim, E.H. Sargent, J. Park and J.Y. Kim, "Inverted Colloidal Quantum Dot Solar Cells," Adv Mater, vol. 26, pp. 3321-3327, 2014.

Uncertainty Considerations in Bifacial Photovoltaic Systems with High Albedo Seasonality

Javier Lopez-Lorente[1], Anja Neubert[2], Mike Hamer[3]

[1]DNV, Arnhem, 6812 AR, Netherlands
[2]DNV, Oldenburg, 26135, Germany
[3]DNV, Bristol, BS2 0PS, United Kingdom

Abstract—**Bifacial photovoltaic systems are experiencing an increased deployment with the promise of lower generation costs. The energy gain of these systems highly depends on the albedo, which is uncertain and changes over time. This study examines the deviations in the energy yield predictions as a function of albedo at different temporal resolution from local, nearby and satellite datasets. DNV's SolarFarmer is used to model a three-year public dataset of a bifacial horizontal single-axis tracking PV system and validate the plane-of-array irradiance and DC energy yield modelling. The results showed that sub-hourly calculation timesteps can provide more accurate predictions. Similarly, high temporal resolution of albedo can benefit intra-daily energy yield predictions, but daily and monthly albedo can be sufficient for accumulated energy yield over longer periods of time. Accuracy differences were found between albedo source data, where satellite-derived albedo reported up to 2.1% error increase compared to energy calculations using local site or nearby albedo data. The findings of the paper provide insights about the spatiotemporal resolution of albedo and its effects for energy assessment.**

Keywords—*albedo, bifacial, energy yield, rear-side irradiance, photovoltaics.*

I. INTRODUCTION

Bifacial photovoltaic (PV) technologies are set to have an increased adoption as a driver for solar PV competitiveness, particularly in utility-scale projects [1]. Bifacial PV technologies enable higher yield per installed capacity than monofacial technologies as electric power is converted from the light reaching both the front- and rear-sides of the module. The annual energy gain of bifacial PV technology compared to monofacial modules, a.k.a. bifacial gain, was found to be from 13.2-18.8% [2] in several test PV arrays in several climates of the USA. In addition, horizontal single-axis tracking (HSAT) can further enhance energy gains in latitudes over 20°, where the annual power gains were modeled in the range of 10-20% for latitudes over 40° compared to fixed-tilt bifacial PV plants [3]. The performance of bifacial PV systems depends on critical factors related to the module and the installation, such as the module's bifaciality, orientation, tilt, height above ground, ground coverage ratio, rear-side irradiance, and albedo effects [1], [4]. The understanding of the bifacial gain, its affecting factors and modeling is utterly important for the pre-construction energy assessment and design of PV plants.

This paper investigates the modeling of rear-side and reflected irradiance as affected by changing albedo conditions to evaluate the differences in PV power output at different temporal resolutions, from 1-minute to 1-hour time intervals. The study utilizes the 2-D view factor model as implemented in DNV's SolarFarmer tool [5] to model the rear-side plane-of-array (POA) irradiance, which uses a physical-based approach to estimate the DC power output of the array. The assessment utilizes multi-year data from a PV test site with HSAT systems in the USA. Moreover, the study assesses the impact of albedo's data source selection by evaluating in-situ measurements at the studied PV system, nearby measurements, and satellite-based albedo estimations at diverse temporal resolutions and aggregating techniques. Thus, this paper contributes to the understanding of bifacial energy gains and their modelling, while considering key factors, such as albedo and its spatiotemporal variability.

II. METHODS AND MATERIALS

A. Outdoor PV testing facility

The case study is NREL's Bifacial Experimental Single-axis Tracking (BEST) field in Golden, CO, USA (39.7°N, 105.2°W), warm summer continental climate "Dfb" according to the Köppen climate classification. The HSAT system has 75 kWp installed distributed in 10 rows of 20 modules, where 5 rows are bifacial technologies, and the remaining 5 rows are the equivalent monofacial technology for comparison. The ground-coverage ratio (GCR) of the site is 0.35, with a hub height of 1.3 m [6]. The available data are 1-minute resolution from June 2019 to September 2022. The variables include DC electrical parameters for some of the arrays including both monofacial and bifacial arrays, weather data and other relevant experimental variables (e.g., front and rear irradiances, albedometers and modules' temperatures). The data and a full description of NREL's BEST field testbed are available in the DuraMAT datahub [7].

The BEST test field provides a case study with a wide albedo range, where presence of recurrent snow episodes leading to high albedo (monthly average > 0.4) are observed particularly during fall and winter months. During spring and summer months, the albedo has low variation with values around 0.2 (soil and grass surface). The month-to-month albedo variation can be observed in Fig. 1, where the interquartile range (IQR, range between percentile 75th and percentile 25th) is illustrated to better indicate the variation within monthly values. The overall average albedo for the 3-year period was found of 0.262 and the median 0.195.

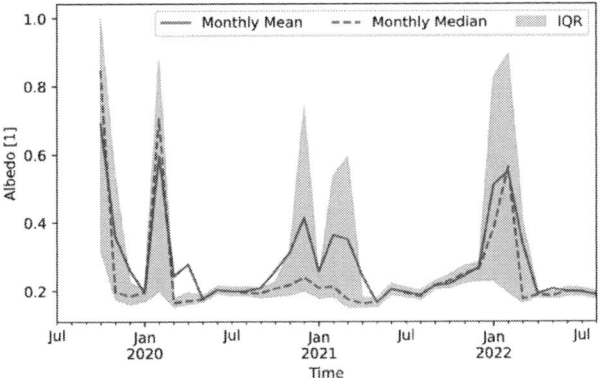

Fig. 1. Monthly mean and median albedo timeseries observed in NREL's BEST test site: October 2019 to August 2022. The interquartile range (IQR) illustrates the high albedo seasonality throughout the year. Higher albedo periods were given by snowfalls during fall/winter months.

Besides the data from the BEST test field, which is referred throughout the paper as site or in-situ data, other albedo data sources were used. The data referred as nearby albedo was based on NREL's Solar Radiation Research Laboratory (SRRL) Baseline Measurement System data [8], which are high-quality 1-minute measurements in the same facilities; the SRRL distances around 570 meters from the BEST site. In addition, satellite-derived albedo data with 5-minute temporal resolution at the coordinates of the BEST site was also used. The satellite data was retrieved from NREL's National Solar Radiation Data Base (NSRDB) [9]. NSRDB offers data over the United States and regions of the surrounding countries with a spatial gridded resolution of 2 km at hourly and sub-hourly time intervals up to 5 minutes until 2021.

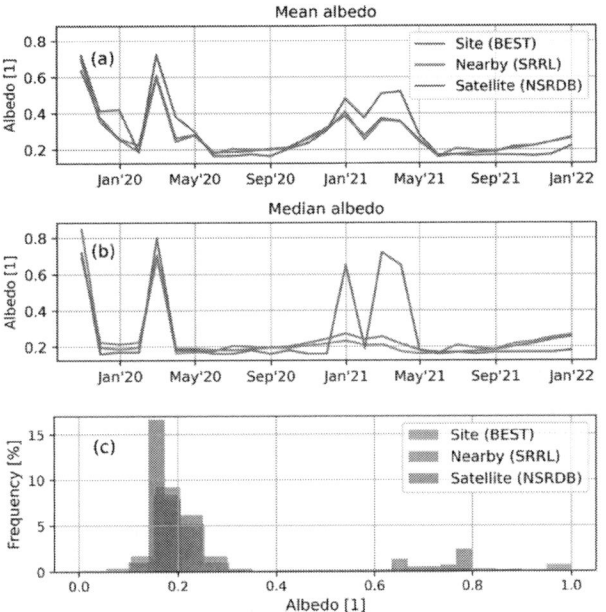

Fig. 2. Distribution of albedo by data source – site, nearby and satellite data: (a) monthly mean albedo; and (b) monthly median albedo; (c) histogram of 5-minute observations (20 bins, bin size of 0.05). Data from October 2019 to December 2021.

A timeseries of monthly albedo data and histograms for the three albedo datasets are presented in Fig. 2 for the period of simultaneous data availability, which ran from October 2019 to December 2021. The figure shows trend differences among the data sets, where site data and nearby data have slight differences either when looking at mean and median values. However, satellite-based albedo systematically reported higher values during winter months and lower values the rest of the year, where differences with site data were seen more substantial for median values (-0.12 to +0.5) than mean values (-0.07 to +0.16).

B. Methodology

The analysis of the energy yield variations in the PV HSAT system as a function of the temporal resolution and data source of albedo consisted of three differentiated stages: (i) data preparation and pre-processing of meteorological data for the specified resolution; (ii) the bifacial power modeling, including both the POA rear-side irradiance modeling and the array power output estimation for each temporal resolution and data source considered; and (iii) the classification of the months according to their albedo as a distinction for assessment. An overview of the methodological procedure is shown in Fig. 3, and each of the stages of the methodology are described below, as well as the performance metrics used.

1) Meteorological and field data pre-processing

The meteorological data used to model the power output of the BEST test site included global horizontal irradiance (GHI), diffuse horizontal irradiance (DHI), ambient temperature (T_{amb}), wind speed (WS), and albedo. Apart from albedo that changed based on the case evaluated, the meteorological variables used were from the SRRL, which are provided as part of the BEST dataset.

The albedo data from SRRL (nearby albedo) and the NSRDB (satellite albedo) already have pre-processing techniques that ensure that the end-user gets useable data. In the case of the raw data from the BEST site, the reported albedo from field measurements (site albedo) had to undertake a quality control routine. The applied quality control to aggregate at a given temporal resolution consisted of limiting the values to a 0-to-1 range and filtering the data for positive solar elevation angles. Given the high temporal resolution of the available data (1-minute averaged measurements for the BEST and SRRL data, and 5-minute intervals for the satellite albedo), multiple cases were evaluated in terms of the temporal resolution of the energy yield calculation (output power), the time resolution of albedo, the origin of albedo, and its scaling approach. For the albedo, the mean and median values were investigated as aggregating methods when scaling to a specified temporal resolution as for the differences observed between the variables in the historic timeseries (see Fig. 1 and Fig. 2). The averaged values were averaged without weighing for radiation income. This consideration does not affect the results of the study and conclusions thereof, as these are derived from the relative difference across simulations. The median albedo for a given period was determined as the P50 of the distribution.

The modeled cases are summarized in Table I. The field data was also pre-processed and only instances with a front

978-1-6654-6060-6/23 $31.00 © 2023 IEEE

irradiance in the HSAT system reported over 150 W/m² and with a DC power over 10% of the nominal power were kept. This filtering ensured removing low-irradiance instances with high contribution of diffuse sky irradiance only and any outages from the PV system.

TABLE I. Modeled Cases of Meteorological data for Energy Yield Estimations as a Function of Albedo Data Characteristics

Variable	Cases considered
Temporal resolution meteorological data[a]	1 min, 5 min, 10 min, 15 min, 30 min, 60 min
Albedo resolution[b]	1 min, 5 min, 10 min, 15 min, 30 min, 60 min (1H), 180 min (3H), daily (1D), and monthly (1M) value
Albedo scaling method	Mean and median aggregation
Albedo data source	Site (BEST), nearby (SRRL), satellite (NSRDB)

[a.] GHI, DHI, T_{amb}, and WS. The temporal resolution equals that of the resulting power timeseries.

[b.] The minimum resolution of albedo used was that of the meteorological data.

2) Bifacial power modeling approach

The modeling of the POA rear-side irradiance ($G_{POA,Rear}$) and array's DC power output to evaluate the bifacial gains were modeled using DNV's SolarFarmer version 1.0.240, which has been widely validated against operational data in the recent years [10]–[12]. The input meteorological variables for PV power modeling for a given case were GHI, DHI, ambient temperature, and albedo data. SolarFarmer can produce a timeseries of results, including irradiance components, multiple losses throughout the conversion from irradiance to electric power, and power output at a temporal resolution as high as the input meteorological data. Using this capability and SolarFarmer's API in Python, the multiple energy calculations (a total of 234 cases) presented in Table I were implemented.

Methodologically, SolarFarmer uses the 2-dimensional (2-D) view factor model infinite sheds [5] to estimate rear-side irradiance, whereas it uses SolarFarmer's 3-D hemicube model [13] for front-side shading of the sky diffuse irradiance and row-to-row shading. In the 2-D view factors models, the rear-side irradiance is quantified using geometric calculations and it is a common solution adopted in commercial PV simulation software as it is relatively simple and computationally fast [12]. The SolarFarmer model accounts for shade from PV racks on the ground, obstruction of the view factor of the sky from the ground, and blockage of the ground and sky from the PV panels due to adjacent rows. Additional effects like shading from support structures or transmission gains are considered as simple user-defined effects. It enables for additional parameters for the electrical mismatch due to non-uniform distribution of rear-side irradiance in the modules, which is often treated as a uniform (isotropic) distribution in 2-D view-factors-based models compared to 3-D ray-tracing models [14]. Finally, the power conversion is computed using the single-diode model [15] to predict performance at the module and submodule level depending on the PV module technology and the operating conditions.

For the modelling, the following simulation assumptions were used in SolarFarmer: bifacial transmission (gain) factor

was set to 5%, bifacial shade loss factor was 0.7%, model without soiling losses, HSAT with backtracking algorithm with tracker's angle limit [-50°, 50°]. The Perez diffuse radiation model was used. The cell temperature model used a constant heat transfer coefficient of 25.0 W/m²/K and a convective heat transfer coefficient of 1.2 W/m²/K/(m/s²). Since the module types are anonymized in the BEST testbed, a bifacial module of similar characteristics was chosen (72-cell silicon monocrystalline module with module bifaciality around 70%). Following the simulations in SolarFarmer, the output timeseries was produced for each case to assess the variables and metrics presented in the results given in Section III.

3) Classification of albedo months

Albedo is one of the key parameters used in the modeling of the rear-side irradiance. To examine the impact of albedo in the modeling, the 3-year dataset was split based on the distribution of monthly albedo (see Fig. 1). The classification was per month as albedo is commonly defined monthly in solar energy assessments. The months of the timeseries were clustered into high- and low-albedo months. It was noted that months with monthly median albedo around 0.2 (lower albedo months) have an IQR whose absolute range does not exceed 0.1. Therefore, a threshold of 0.1 of the IQR was used to classify each month of the dataset. This classification led to 10 high-albedo months (generally fall/winter months in the historical series) and 25 low-albedo months. Note that when evaluating satellite data, the period is considered until end of 2021 as that is the period of NSRDB data availability.

4) Evaluation metrics

The modeled rear-side irradiance and the DC power production were evaluated using the normalized versions of the mean bias deviation (MBD), mean absolute error (MAE) and root mean square error (RMSE), whose definition can be found in the literature [16]. In addition, we used a performance index (PI) to denote the modeling accuracy, which is the ratio of model over measured DC power as given by (1).

$$PI = \frac{Modeled\ Power\ [kWdc/kWp]}{Measured\ Power\ [kWdc/kWp]} \tag{1}$$

III. Results

A. Bifacial modeling of the PV HSAT system

The proposed binary classification of the months according to the albedo's IQR enabled the evaluation of the rear-side POA irradiance driven by intra-monthly variations. Fig. 4 shows a scatter plot of hourly averaged data for each group of albedo month type. Overall, rear-side POA irradiance resulted in an underestimation with an MBE of -24.2% and RMSE of 35.8% for the whole dataset. These results are lower than other view-factor-based simulation tools (i.e., SAM and PVsyst) previously evaluated in the BEST test field for 2019 [6], which reported RMSE of 53% for SAM and 61% for PVsyst.

Looking at the individual subplots, it is observed that higher dispersion in the data often coincides with albedo variations as illustrated by the color. It is worth mentioning that the higher scattering observed in high-albedo months (Fig. 4b) leads to lower MBD than in low-albedo months.

While the error in rear-side irradiance is high, its reduced contribution to the effective irradiance, which also receives the gain of the ground reflected irradiance to the front of the module, minimizes throughout the methodological modeling pipeline to estimate PV power. Fig. 5a demonstrates this by showing the normalized hourly averaged values of DC power for the 3-year period, where the errors dropped to MBD=0.2%, MAE=5.6% and RMSE=10.5%. Benchmarked to other simulation results in the BEST test field, the results for the 3-year period for SolarFarmer present a higher accuracy than those presented for SAM in [17] for the period July 2019 to March 2020, which showed MAE=7.5% and RMSE=19.7%. In Fig. 5a, it can be observed a positive bias in instances with high and low albedo values, which suggest that the other factors beyond albedo affect the modeling results. To investigate this further, Fig. 5b illustrates a performance index ratio for normalized DC power (modeled over measured) as a function of the tracker's rotation angle and colored by rear-side irradiance. The figure illustrates that such deviations often occur under conditions when rear-side irradiance is relatively low (darker shade color), more prone to happen when the tracker angle is negative (facing east), or at extreme tracking angles either east or west facing. This may suggest shading differences in the front-side of the module between the model and the test field (e.g., minor differences in the trackers' angles, hub height or projected row-to-row shading due to slightly different modules' sizes). The reader may note that tracker angles do not fall in the region around -5° to +5°, this was due to the averaging of the hourly intervals as trackers usually move in discrete steps every few minutes (e.g., every 5-10 minutes) depending on their control algorithm.

B. Albedo effects in bifacial PV performance

This section presents the results for the analysis of the energy yield variations as a function of changing characteristics in albedo, including variations in temporal resolution (from 1 minute to 1 monthly averaged values), scaling method (averaging or median aggregating) and origin of albedo data

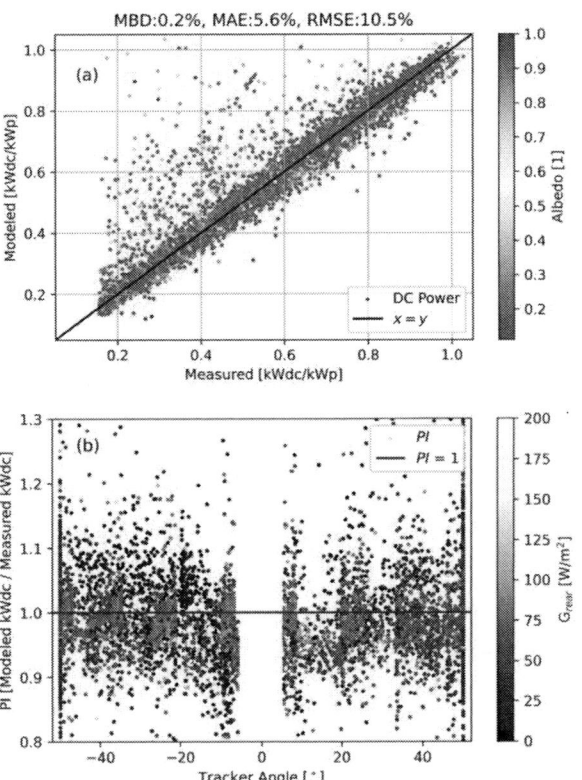

Fig. 5. Modeled hourly averaged normalized DC power for the PV HSAT system during the 3-year evaluation period: (a) scatter plot of modeled and measured DC power colored by albedo data; and (b) performance ratio (modeled over measured DC power) as a function of the tracker angle (positive orientation denotes west orientation) and colored by rear-side irradiance. The initial uncertainty in $G_{POA,Rear}$ is compensated throughout the modeling irradiance-to-power pipeline, leading to a reduced bias and error metrics (MBD: 0.2%; MAE 5.6%; RMSE 10.5%).

(site measurement, nearby measurement, or satellite-based estimation).

Fig. 6 presents six heatmaps presenting the deviation in DC energy yield production over the 3-year period, the metric shown corresponds to the percentage of error defined as *1-PI*, where *PI* is the performance index given by (1) using energy accumulated DC energy yield and corresponds to the mean absolute percentage error (MAPE). The results provide multiple insights in terms of the temporal resolution and nature of albedo data with respect of the energy calculation.

Focusing on a single heatmap with mean albedo data (e.g., Fig. 6a), it can be observed that the error increases progressively as the albedo time resolution increases from 1-minute to 3-hour data intervals, dropping again for monthly albedo. It is also observed that the error generally increases with higher calculation timesteps. In the case of Fig. 6a, the difference in error for hourly mean albedo data goes from MAPE 1.9% for 1-minute calculations to 2.8% for 1-hour calculations. Yet, the lowest MAPE is found in the 5-minute calculations either with mean 5-minute or monthly albedo, which may be related to the irradiance variability between the BEST site and the SRRL irradiance data used for the modeling. This difference around 1% error increase is found overall for all

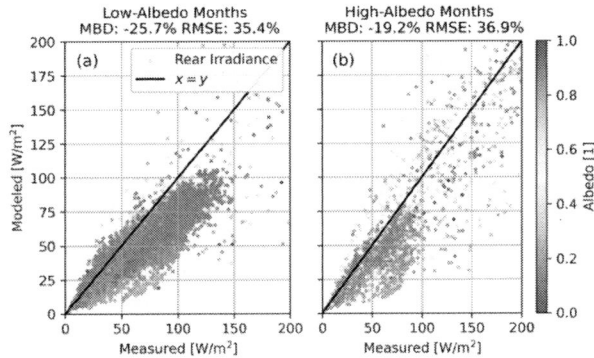

Fig. 4. Scatter plot of hourly average rear-side POA irradiance as a function of site albedo during the 3-year period evaluated for months classified as (a) overall low-albedo months, and (b) overall high-albedo months. The trend shows underestimation of the rear-side irradiance with an MBD around -24.2% and an RMSE 35.8% for the whole dataset. Higher albedo instances observe higher deviation from the *x=y* reference line, particularly for high-albedo months.

978-1-6654-6060-6/23 $31.00 © 2023 IEEE

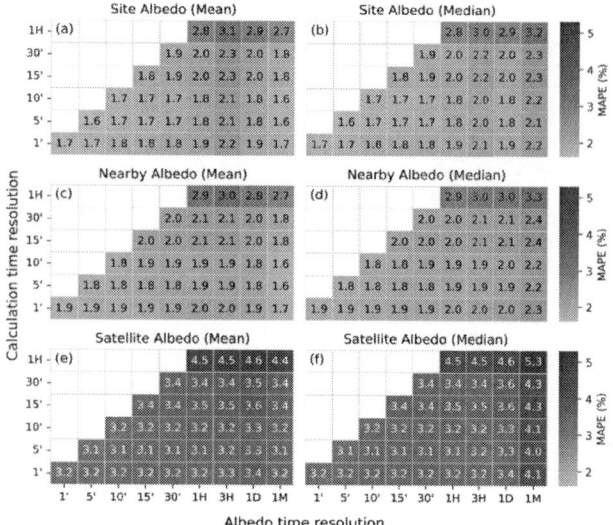

Fig. 6. Mean absolute percentage error ($1-|PI|$) in DC energy yield between the modeled results and the BEST testbed for the 3-year period evaluated as a function of time resolution of the calculation and albedo data. Each heatmap indicates a different albedo data source and aggregating technique: (a) site albedo – mean data; (b) site albedo – median data; (c) nearby albedo – mean data; (d) nearby albedo – median data; (e) satellite albedo – mean data; and (f) satellite albedo – median data.

subplots. It worth noting that an albedo resolution of 3-hour intervals (either mean or median aggregating) results in higher MAPE than hourly and daily intervals. This could be caused by the variability of albedo throughout the day, which seems not to have a significant impact at hourly intervals, and it can be captured in daily intervals (e.g., snowfall or snow melting during the day, where hourly and daily data would be more representative than 3-hourly intervals). Finally, it can be observed that monthly albedo calculation results reported lower error than hourly albedo ones. This suggests that while higher resolution albedo can be used to better estimate intra-daily PV plant output and bifacial contributions, it does not necessarily increase the accuracy of the energy yield calculation over a larger period. Thus, the results when focusing on a single heatmap suggest that it is more accurate the use of sub-hourly calculation timesteps, which showed improvements of up to 1.2% down to 5-minute calculation intervals. Regarding albedo, the best accuracies were achieved with monthly (or daily) mean albedo as these periods encompass most of the albedo variance, or with the same albedo and calculation time steps given that intra-day albedo variance can be fully captured.

Focusing on two heatmaps in the same row (e.g., Fig. 6a and Fig. 6b) provides insights of the scaling method of albedo data (mean or median aggregating), where a similar performance can be observed in the calculations except for monthly data. This can be understood from the differences identified in Fig. 1 and Fig. 2, where the monthly aggregation can lead to large variations in the albedo data. Thus, suggesting that aggregating albedo data for monthly intervals with the median can lead to energy yield inaccuracies, even though it could provide a better understanding of the statistical distribution in albedo. Looking at the comparison across all the subplots, the effect of albedo's

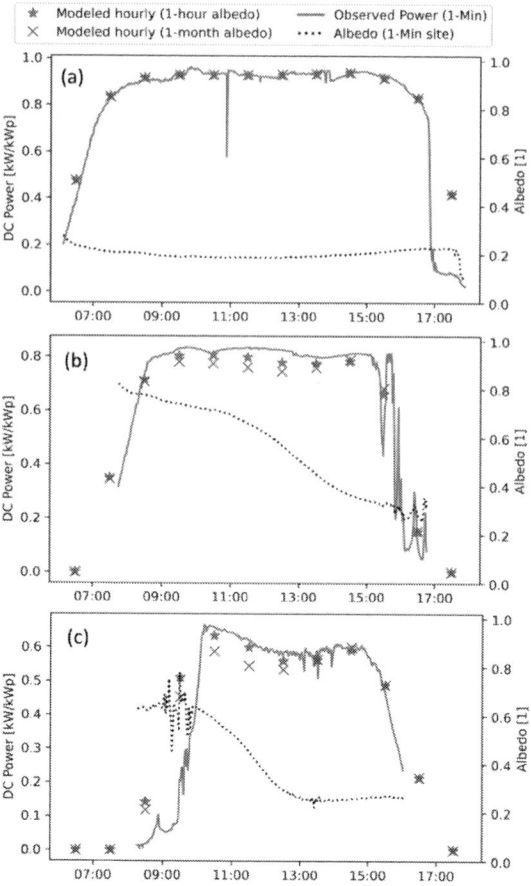

Fig. 7. Intra-daily profiles of simulated and observed hourly normalized DC power with their daily albedo profile for selected days of interest: (a) sunny day with low variation of albedo – April 17th 2022; (b) sunny day after high-albedo event – February 13th, 2022; and (c) partly sunny after high-albedo event – January 10th, 2021. The albedo for the simulations corresponded to site albedo with mean aggregation. Note that the simulated data are plotted at the mid point between timestamps (i.e., 30 min. past each hour).

data source can be analyzed. The results indicate that having local in-situ measurements of albedo is the best option for energy yield assessment. Deviations from non-local albedo data can be small (up to 0.2%) for nearby data or larger (up to 2.1%) for satellite-derived albedo. This highlights the importance of measurement campaigns to acquire meteorological datasets capturing the local characteristics of solar resource, which can be later complemented with satellite-derived datasets for longer periods. Comparing albedo measurements, it is worth mentioning that nearby albedo can lead to higher accuracy for 3-hour intraday observations than site albedo as observed when looking at Fig. 6a and Fig. 6c.

The results of Fig. 6 illustrate the sensitivity of different albedo products over different temporal resolutions of data used. Nevertheless, current industry practices on energy yield assessment are often based on hourly simulations, usually with typical meteorological years (TMYs), combined with additional corrections for sub-hourly effects (e.g., inverter clipping [18], [19]). Along that context, Fig. 7 illustrates the differences in intra-daily power estimations when modeling hourly power output using either hourly or monthly average albedo for several selected days of interest with changing daily albedo profiles. In days with low variations of albedo (Fig. 7a) using hourly and monthly albedo data would generally lead to similar estimates. That is not the case for periods when intra-monthly variations of albedo occur, particularly frequent within months classified as high-albedo months. Fig. 7b and Fig 7c illustrate such cases, where the use of hourly albedo can facilitate more accurate energy yield predictions than monthly albedo values, especially in the periods transitioning from higher albedo episodes (after a snowfall) to period with albedo closer to monthly averages. This can be notably important not only for prospecting of pre-construction PV plants, but operational monitoring based in historic data. The combined results from Fig. 6 and Fig. 7 demonstrate that intra-monthly power predictions can be highly affected by albedo's temporal variation, which is a finding consistent with those of [20] when looking at annual energy production.

IV. Summary of the Work

The aim of this paper was to evaluate the modeling implications in PV power output using albedo data of different temporal resolutions and origins for energy yield assessment. As part of the modeling methodology, the paper illustrated the calculation accuracies for the view factor bifacial model of the SolarFarmer tool from rear-side irradiance to DC energy yield. Regarding SolarFarmer's calculations, the modeling results (MBD 0.2%, MAE 5.6% and RMSE 10.5%) were lower compared to other view-factor modelling tools as reported in previous literature for the same case study in the USA. As part of the methodology, a simple albedo classification model was proposed; this can support the assessment of multi-year variability of albedo and seasonal performance of bifacial systems.

The results of the energy calculations as a function of the characteristics of albedo illustrated that sub-hourly resolution timesteps in solar energy assessments can improve the accuracy of the predictions compared to hourly simulations. Regarding albedo granularity, it was found that more frequently sampled albedo can improve intra-monthly yield estimates when calculation time step and albedo sampling are matched, but monthly albedo may be sufficient in areas without much albedo seasonality. Following current industry practices on hourly calculations, intra-daily profiles were investigated to evaluate the benefit of using hourly vs. monthly albedo for hourly simulations, where hourly albedo responded better in periods with changing albedo profiles at an intra-monthly or intra-daily basis. The aggregating technique for scaling albedo data seemed to have little impact for albedo resolutions other than monthly. Finally, energy predictions with local in-situ or nearby measured albedo were found to be more accurate than those satellite-derived albedo, which reported an increased error up to 2.1% compared to energy yield with site albedo.

The findings of this paper contribute to the understanding of bifacial PV energy systems and their modelling, illustrating the uncertainty (and added error) that albedo's spatiotemporal variability can have in the energy yield assessments.

Future work could be expanded to quantify the significance of intra-daily albedo variations with respect to cloudy and sunny days, including design parameters, such as ground coverage ratio to evaluate the contribution of ground reflected irradiance. In addition, site-adaption techniques of albedo based on local and satellite-derived datasets could be investigated.

Acknowledgment

The authors would like to thank NREL (USA) for making publicly available the dataset of the BEST test field, which was key for this study and can be accessed via the DuraMAT Data Hub [7].

References

[1] IRENA, "Renewable Power Generation Costs in 2021," Abu Dhabi, UAE, 2022. [Online]. Available: https://www.irena.org/publications/2022/Jul/Renewable-Power-Generation-Costs-in-2021

[2] J. E. Castillo-Aguilella and P. S. Hauser, "Multi-variable bifacial photovoltaic module test results and best-fit annual bifacial energy yield model," *IEEE Access*, vol. 4, pp. 498–506, 2016, doi: 10.1109/ACCESS.2016.2518399.

[3] M. T. Patel, H. Imran, Md. S. Ahmed, N. Z. Butt, M. A. Alam, and M. R. Khan, "When and where to track: A worldwide comparison of single-axis tracking vs. fixed tilt bifacial farms," in *2020 47th IEEE Photovoltaic Specialists Conference (PVSC)*, Calgary, AB, Canada: IEEE, Jun. 2020, pp. 1735–1737. doi: 10.1109/PVSC45281.2020.9300692.

[4] A. Neubert, M. Hamer, R. A. Kharait, and M. A. Mikofski, "Bifacial solar sensitivity to project capacity size," in *2020 47th IEEE Photovoltaic Specialists Conference (PVSC)*, Calgary, AB, Canada: IEEE, Jun. 2020, pp. 0703–0706. doi: 10.1109/PVSC45281.2020.9300712.

[5] M. A. Mikofski, R. Darawali, M. Hamer, A. Neubert, and J. Newmiller, "Bifacial performance modeling in large arrays," in *2019 IEEE 46th Photovoltaic Specialists Conference (PVSC)*, IEEE, Jun. 2019, pp. 1282–1287. doi: 10.1109/PVSC40753.2019.8980572.

[6] S. A. Pelaez *et al.*, "Field-array benchmark of commercial bifacial PV technologies with publicly available data," in *2020 47th IEEE Photovoltaic Specialists Conference (PVSC)*, Calgary, AB, Canada: IEEE, Jun. 2020, pp. 1757–1759. doi: 10.1109/PVSC45281.2020.9300379.

[7] S. Ovaitt and C. Deline, "NREL Bifacial Experimental Single-Axis Tracking (BEST) Field Dataset," *Duramat Datahub*, Sep. 2022. doi: 10.21948/1787805.

[8] A. Andreas and T. Stoffel, "NREL Solar Radiation Research Laboratory (SRRL): Baseline Measurement System (BMS)." NREL Report No. DA-5500-56488, Golden, Colorado (Data), 1981. doi: 10.5439/1052221.

[9] M. Sengupta, Y. Xie, A. Lopez, A. Habte, G. Maclaurin, and J. Shelby, "The National Solar Radiation Data Base (NSRDB)," *Renewable and Sustainable Energy Reviews*, vol. 89, pp. 51–60, 2018, doi: 10.1016/j.rser.2018.03.003.

[10] A. Neubert, M. A. Mikofski, M. Hamer, and P. Rainey, "Validation of the Solarfarmer software with operational data," in *37th European Photovoltaic Solar Energy Conference and Exhibition (EU PVSEC)*, Online, 2020, pp. 1621–1625. doi: 10.4229/EUPVSEC20202020-5CV.4.10.

978-1-6654-6060-6/23 $31.00 © 2023 IEEE

[11]   M. Mikofski and C. Chan, "Addendum to the 2021 Solar Energy Assessment Validation for Utility Scale Projects," DNV, Oslo, Norway, May 2023. [Online]. Available: https://www.dnv.com/publications/addendum-to-the-2021-solar-energy-assessment-validation-for-utility-scale-projects-243051

[12]   N. Riedel-Lyngskær *et al.*, "Validation of bifacial photovoltaic simulation software against monitoring data from large-scale single-axis trackers and fixed tilt systems in Denmark," *Applied Sciences*, vol. 10, no. 23, p. 8487, Nov. 2020, doi: 10.3390/app10238487.

[13]   A. Neubert, J. Lopez-Lorente, M. Hamer, and T. Mercer, "A model for efficient shading evaluation in large-scale PV plants based in hemicube geometry," in *2023 PV Performance Modeling Workshop*, Salt Lake City, USA: Sandia National Laboratories, May 2023.

[14]   K. Phetdee *et al.*, "Testbed validation of bifacial performance modelling methodology using ray-tracing methods," in *37th European Photovoltaic Solar Energy Conference and Exhibition*, 2020, pp. 1298–1304. doi: 10.4229/EUPVSEC20202020-5CO.9.4.

[15]   A. Mermoud and T. Lejeune, "Performance assessment of a simulation model for PV modules of any available technology," in *25th European Photovoltaic Solar Energy Conference and Exhibition / 5th World Conference on Photovoltaic Energy Conversion*, Valencia, Spain, Sep. 2010, pp. 4786–4791.

[16]   J. Lopez Lorente, X. A. Liu, and D. J. Morrow, "Worldwide evaluation and correction of irradiance measurements from personal weather stations under all-sky conditions," *Solar Energy*, vol. 207, pp. 925–936, 2020, doi: 10.1016/j.solener.2020.06.073.

[17]   S. Ayala Peláez *et al.*, "Ultimate bifacial showdown: 75 kW field results," in *2020 BifiPV Virtual Workshop*, Virtual, Jul. 2020. [Online]. Available: https://www.nrel.gov/docs/fy20osti/77486.pdf

[18]   R. Kharait, S. Raju, A. Parikh, M. A. Mikofski, and J. Newmiller, "Energy Yield and Clipping Loss Corrections for Hourly Inputs in Climates with Solar Variability," in *2020 47th IEEE Photovoltaic Specialists Conference (PVSC)*, IEEE, Jun. 2020, pp. 1330–1334. doi: 10.1109/PVSC45281.2020.9300911.

[19]   A. Parikh, K. Perry, K. Anderson, W. B. Hobbs, R. Kharait, and M. A. Mikofski, "Validation of Subhourly Clipping Loss Error Corrections," in *2021 IEEE 48th Photovoltaic Specialists Conference (PVSC)*, IEEE, Jun. 2021, pp. 1670–1675. doi: 10.1109/PVSC43889.2021.9518564.

[20]   V. Lara-Fanego, J. A. Ruiz-Arias, A. Skoczek, C. A. Gueymard, T. Cebecauer, and M. Suri, "Annual Energy Production Uncertainty of Bifacial PV Plants Caused by Inaccuracies in Albedo Data: Case Studies Using SAM," in *2022 IEEE 49th Photovoltaics Specialists Conference (PVSC)*, Philadelphia, PA, USA: IEEE, 2022, pp. 0923–0923. doi: 10.1109/PVSC48317.2022.9938944.

India as an Emerging Solar Manufacturing Country

Juzer Vasi[1], Mrunal Berad[1], Narendra Shiradkar[1], Anil Kottantharayil[1], Dinesh Kabra[1], Kedar Deshmukh[1], Aditi Chaubal[1], Rajeewa Arya[1,2], Probir Ghosh[3], Satyendra Kumar[3,4], and Lawrence L. Kazmerski[1,5]

[1]National Centre for Photovoltaic Research and Education, Indian Institute of Technology Bombay, Mumbai 400076 India, [2]Arya International, Beaverton, OR 97007, USA, [3]CASE-Bharat, Superior, CO 80027, USA, [4]Saurya EnerTech, Gurgaon, NCR 122016, India, [5]Renewable and Sustainable Energy Institute, University of Colorado, Boulder, CO 80309, USA

Abstract — **During the last two years, several policy initiatives by the Government of India have created significant industry interest to enhance solar manufacturing in India across the value chain of silicon from polysilicon to modules, as well as thin film solar modules. It is expected that India will be manufacturing about 100-110 GW of solar modules per year by 2027. This will not only be consistent with India's stated deployment goal of 500 GW of 'non-fossil fuel' electricity capacity by 2030, but will leave enough capacity for global export. This paper describes the present status of solar manufacturing in India.**

I. INTRODUCTION

During the last two years, COP-26 in 2021 and COP-27 in 2022 generated significant global interest in renewable sources of energy, especially in solar energy. Many countries have announced their net-zero deadlines. This energy transition, in addition to greenfield capacities which will be created in many countries like India as well as in Africa, South Asia and Latin America, means that the requirement of solar modules will grow tremendously till 2050 and beyond. Recent developments in Ukraine have propelled many countries to aim for further de-carbonization, enhancing the role of solar deployment.

The total global deployment of solar worldwide – mainly solar PV – has expanded exponentially during the last decade, crossing the much-anticipated 1 TW mark in 2022. The predictions for global deployment vary from 2.9 TW by 2027 [1] to 5.3 TW by 2030 [2] to 11 TW by 2050 [3]. It is predicted that by 2027, solar will be the largest installed electricity capacity globally, outstripping hydropower, coal and gas.

PV module production has grown rapidly between 2010 and 2021, as shown in Fig. 1 [4]. It was 240 GW/year in 2021, much of it from Asia (China, South-East Asia and India). Module production in 2022 was 310 GW [5], again largely from Asia (China itself accounted for about 260-270 GW). In order to reach 2.9 TW of installed capacity by 2027 (IEA's recent prediction [1]), there will need to be a global module production of about 1800 GW over the next 5 years, or an average of 360 GW per year. This may increase to about 600-700 GW/year by 2030. From current trends, it is quite possible for China and South-East Asia (ASEAN countries) to meet this global demand. Various geo-political and other

developments during 2020 to 2022 have, however, induced many other countries in the world, including India, the USA, Turkey and some in the EU to launch major initiatives for solar PV module manufacturing.

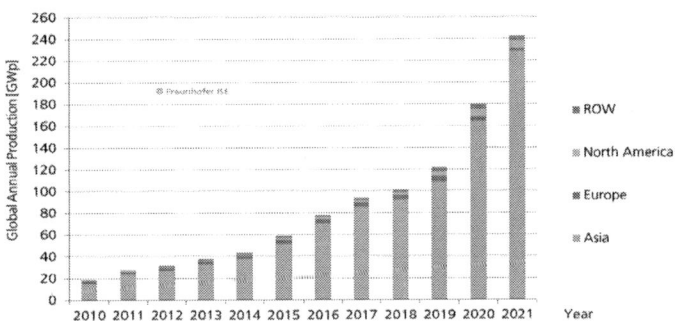

Fig. 1. Global production of modules 2010-2021 (From Fraunhofer ISE, "Photovoltaics Report" 2022 [4]). The 2020 production was 310 GW [5].

Energy security is doubtless an important reason for this diversification, but in addition, other issues like supply chain resilience, cost benefits arising from lower transportation costs, and lower carbon intensity of manufacturing, do contribute. Such a diversity will be welcome, as it would ensure continued and secure supply of PV at competitive prices in the coming decade and beyond.

India has announced several policy initiatives to encourage domestic solar manufacturing across the value chain. This paper describes these initiatives, and responses from Indian and global manufacturers to set up production in India. The paper also provides a prognosis for production up to 2027. It will be seen that the manufacturing capacity will cater not only to India's ambitious installation program, but can also generate sufficient capacity to export globally.

II. RECENT GLOBAL INITIATIVES FOR SOLAR MANUFACTURING

India, USA and Europe have all announced aspirations and plans for solar PV manufacturing. One obvious consideration is whether manufacturing in these countries, though desirable from the energy security perspective, will be cost competitive with manufacturing in China and South-East Asia (ASEAN countries like Vietnam, Malaysia and Thailand). IEA has

978-1-6654-6060-6/23 $31.00 © 2023 IEEE

estimated the cost of manufacturing mono-PERC modules as shown in Fig. 2 [1]. According to IEA, the higher cost of labor in the USA and Europe, the higher cost of capital in India, and the higher cost of energy in Europe account for the difference. In our reckoning, higher cost of electricity in India (especially important for polysilicon in an integrated solar production facility) also contributes, as does scale of manufacturing which may not initially match global plant capacities, mainly in China. The relative costs shown in Fig. 2 compare well with the estimates calculated by Shiradkar *et al.* in early 2022 [6].

However, the intrinsic costs of manufacturing are not always the prices at which modules from a particular country are available in another country, due to various factors such as tariffs (or customs duties) imposed, other non-tariff barriers, costs of transportation, etc. Specific cases of these are discussed later.

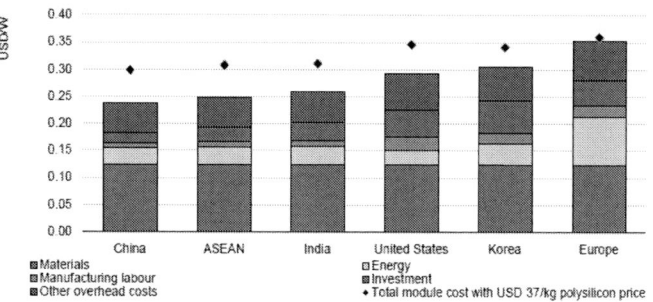

Fig. 2. Relative cost of manufacturing mono-PERC in different countries. (From IEA, "Renewables 2022" [1]

Some of the initiatives taken by various countries are briefly described below, before focusing specifically on India.

In USA, the Inflation Reduction Act (IRA) passed in August 2022 provides significant tax credits, and has created renewed interest to on-shore solar manufacturing. Several companies, including Meyer Burger, Hanwha Q-cells and others have indicated that they will start or enhance solar manufacturing in USA. Further, the Uyghur Forced Labor Prevention Act (UFLPA) of late 2021 has also curtailed import of modules from various countries. For Europe, Fraunhofer ISE's 2020 report [7] had already emphasized the benefits of manufacturing in EU, including a lower carbon footprint. With the 2022 energy crisis in Europe, the EU has launched the European Solar PV Industry Alliance [8], which will aim to boost solar manufacturing in EU to achieve the REPowerEU objectives of 320 GW of newly installed PV capacity by 2025. The European PV industry, in January 2023, issued a joint statement asking for political action to boost efforts for local PV manufacturing [9]. The target is to have 30 GW of solar manufacturing by 2025 [10].

In India, the Ministry of New and Renewable Energy (MNRE) announced, in 2021 and 2022, a slew of policy initiatives aimed to invigorate solar manufacturing in India.

These initiatives are in consonance with the target of achieving 500 GW of 'non-fossil fuel' electric power capacity by 2030, as announced at COP26 in Glasgow. This would entail installation of 280-300 GW of solar PV, compared to 63 GW at the end of 2022, and about 65 GW by March 2023. These policy initiatives attracted a great deal of attention from Indian as well as global companies. The policies and industry responses are described in the next two sections.

III. POLICY INITIATIVES IN INDIA FOR SOLAR MANUFACTURING

As described in detail elsewhere [11, 12], India is no stranger to solar PV manufacturing, which started with silicon solar cells and modules as far back as 1980. During the 2000's several Indian companies were manufacturing multicrystalline solar cells and modules in India, mainly for export to EU countries. During the 2010's, import of cells and modules from China and other countries decimated the domestic industry. Based on the expected demand in the coming decade, predicated on its 300 GW solar target, the Government felt it necessary to encourage solar PV production in India, as well as dis-incentivize import of solar PV components and modules. These policy initiatives are described below.

A. Production Linked Incentive (PLI)

MNRE announced the Production Linked Incentive (PLI) Scheme for solar PV in 2021, with a budget allocation of INR 45B (USD 0.55B). The winners of the PLI would be assessed on the basis of scale of production (minimum 1 and maximum 4 GW/year), and extent of integration (more 'marks' for greater backward integration). The winners would then receive an incentive amount depending on the efficiency of the modules actually manufactured, their temperature coefficient, whether greenfield or brownfield, and (like USA's IRA subsidy) the local-content proportion. The conditions are described in more detail by Shiradkar *et al.* [6]. The PLI tender received 18 applications (though 3 were later withdrawn), totaling 58 GW of production per year. The allocated budget could support only 3 companies, all of which had bid for the maximum capacity (4 GW) as well as full integration involving modules (M), cells (C), wafers (W) and polysilicon (P). This was subsequently called 'Tranche I' of the PLI.

The Government, noting the greater-than-anticipated interest in solar manufacturing, allocated up to a further USD 2.38B to the PLI Scheme. The structure of this 'Tranche II' of PLI, announced in September 2022, was somewhat different from the original Tranche I. Tranche II invited bids in one of 3 baskets: Basket 1 for full integration (P+W+C+M), Basket 2 for W+C+M, and Basket 3 for C+M. The distribution of funds was 62%, 23% and 15% for Baskets 1, 2 and 3, respectively. Fully integrated manufacturing was still prioritized, but smaller or more conservative companies had an opportunity to participate by bidding for Baskets 2 or 3. Doubtless, this aspect was introduced to obviate concerns expressed earlier that only large companies would be awarded the PLI. The maximum capacity which could be bid for was raised to 10 GW. The results of Tranche II were announced in March 2023, and a total of 11 companies bid for the PLI, of which 3 were for Basket 1, 5 for Basket 2 and 3 for Basket 3. The total amount of PLI awarded to these 11 companies would be USD 1.70B, less than the allocated budget of USD 2.38B.

The list of companies which bid successfully for PLI in both the Tranches is tabulated in Section IV.

B. Basic Customs Duty (BCD) on Imported Cells and Modules

In 2021, MNRE had announced that effective April 1, 2022 a basic customs duty (BCD) would be imposed on the import of solar PV components: 25% on solar cells and 40% on solar modules. The rate of BCD (and also the fact that there was to be no duty for import of wafers and polysilicon) reflected the existing capacities of manufacturing in India. These duties have now been in effect for over a year. It was expected that these duties would make solar manufacturing in India very competitive compared to China [7]. However, this situation has been complicated by the fact that import of solar PV components from countries with whom India has a Free Trade Agreement (FTA) would not be liable for BCD. Since India has an FTA with ASEAN countries, cells and modules from Vietnam, Thailand and Malaysia can be imported without duty. Indian solar manufacturers lobbied against this, but to no avail. India does not have an FTA with China, so imports directly from China do attract BCD, and are thus rendered expensive. Since the domestic production of modules and especially cells is currently not sufficient for installations in India, there is import from ASEAN countries and even China. A perverse result of the complex geopolitical dynamics has been that Indian module manufacturers currently prefer to export some of their PV production to the USA, where they do not fall afoul of the UFLPA and can get better margins, whereas some installations in India use modules imported from ASEAN countries which cannot prove that they do not violate UFLPA!

C. Approved List of Module Manufacturers (ALMM)

In 2019, MNRE had also notified the 'Approved List of Module Manufacturers' (ALMM). All government, net-metering and open-access installations must have modules from this list. In February 2023, there were 83 manufacturers in the list, all Indian, with a capacity of 21 GW. This meant that no imported modules could be used for the above types of installations. This was likely to result in a slow-down in the pace of installations (since the capacity of Indian manufacturers in the ALMM list producing advanced modules was not sufficient). Recognizing this, MNRE in March 2023 held in abeyance the requirement to use ALMM-compliant modules up to April 2024. This policy reversal caused much

consternation among domestic module manufacturers, who pleaded for policy consistency. For MNRE, it was a balancing act between encouraging solar PV production, and ensuring that the roadmap to the 2030 installation target of 280-300 GW remained viable.

IV. COMPANIES WHICH BID FOR PLI

Industry response to PLI Tranche I was overwhelming. Only 3 companies qualified for the limited funding available: Shirdi Sai Electricals Ltd., Reliance New Energy Solar Ltd., and Adani Infrastructure Ltd. All had bid for the maximum capacity (4 GW) fully integrated plants. For Tranche II, the response was good, but not 'over-subscribed'. The full list of bidders for both Tranche I and Tranche II is given in Table 1.

TABLE I
LIST OF COMPANIES BIDDING FOR PLI

Company	Capacity	Integration
Tranche I		
Shirdi Sai Electricals	4 GW	P+W+C+M
Reliance New Energy Solar	4 GW	P+W+C+M
Adani Infrastructure	4 GW	P+W+C+M
Total	**12 GW**	
Tranche II		
Shirdi Sai Electricals	6 GW	P+W+C+M
Reliance New Energy Solar	6 GW	P+W+C+M
First Solar (FS India)	3.4 GW	Thin Film
Waaree Energies Ltd.	6 GW	W+C+M
Avaada Ventures Pvt. Ltd.	3 GW	W+C+M
ReNew Solar	4.8 GW	W+C+M
JSW Renewable Tech. Ltd.	1 GW	W+C+M
Grew Energies Pvt. Ltd.	2 GW	W+C+M
Vikram Solar Ltd.	2.4 GW	C+M
AMPIN Solar One Pvt. Ltd.	1 GW	C+M
TP (Tata Power) Solar Ltd.	4 GW	C+M
Total	**39.6 GW**	

It may be noted that 2 companies, Shirdi Sai Electricals (and their Special Purpose Vehicle Indosol Solar Pvt. Ltd.) and Reliance New Energy Solar Ltd. were successful bidders in both Tranche I and Tranche II. In Tranche I they had bid for the maximum allowed (4 GW), and in Tranche II, they bid for an additional 6 GW to reach the maximum allowed in Tranche II (10 GW). First Solar (FS India) bid for Thin Film (CdTe), and it was deemed to fall into Basket 1 (P+W+C+M) since it would achieve full integration. The total capacity for Tranches I and II is 51.6 GW, and MNRE has estimated that the amount of incentive given may reach INR 184B (USD 2.24B).

The incentive to be given depends of several factors described in the MNRE notification of September 30, 2022

[13]: (a) extent of integration (b) greenfield or brownfield (c) manufacturing capacity (d) module performance in terms of efficiency and temperature coefficient (e) local value addition and (f) tapering factor going from 1.4 (first year) to 0.6 (fifth year).

As an example, we calculate that a greenfield plant with full integration (P+W+C+M), achieving highest performance (module efficiency > 23% and temperature coefficient better than −0.3%/°C), and LVA content > 90% would earn a PLI incentive of USD 0.019/W in the first year, tapering down linearly to USD 0.008/W in the fifth year (average USD 0.014/W). The PLI payable is for 5 years only, assuming that commissioning of the plant is on schedule (3 years for P, 2 years for W and 1.5 years for C+M). For a 10 GW plant, the PLI incentive works out to a total of USD 0.7B over 5 years. In the best possible case in the first year, the incentive is USD 0.19B.

As another example, a brownfield 4 GW plant having only C+M, achieving a modest performance (efficiency between 21.5% and 22.0% and temperature coefficient between −0.3%/°C and −0.4%/°C), and LVA of 80% was considered. It would earn a PLI of USD 0.037B (37M) over 5 years, the per watt inventive being about 0.2 cents/W.

We describe below salient points of some of the bidders.

Reliance New Energy Solar Ltd. is a newly created subsidiary of Reliance Industries Ltd, India's largest conglomerate, highly diversified in energy, natural gas, petrochemicals, telecommunication, retail and textiles. They are setting up the Dhirubhai Ambani Green Energy Giga Complex on 5000 acres in the state of Gujarat. Even prior to PLI, Reliance Industries Ltd. had already announced investment of at least USD 10B in green energy, including solar, green hydrogen and batteries. During late 2021 and 2022, they acquired several companies, including REC Solar Holdings, Faradion, and Lithium Werks, and made major investments in NexWafe GmbH, Ambri, Caelux, Sensehawk and Sterling & Wilson. They will take the HJT route, and are purchasing eight 600 MW HJT manufacturing lines from Maxwell. In August 2022, Reliance said that 10 GW of cell+module manufacturing will start in 2024, and scale up to 20 GW by 2026. They will install 100 GW of solar for captive use by 2030 and become a net zero company by 2035. This aggressive push into renewable energy is intended to decarbonise its oil- and chemical-dominated business. Their solar technology roadmap includes reaching 26% HJT module efficiency by 2026, and then 28% using tandems (probably silicon/perovskite). It should be noted that their eventual solar manufacturing capacity is expected to be double the capacity they have bid for in PLI.

Shirdi Sai Electricals is an Indian company in the area of electrical transmission and distribution, and a leading manufacturer of transformers. They have set up an SPV, Indosol Solar Private Limited to pursue their solar initiative. Though they do not have prior experience in solar, they have signed an MOU with Viridis-IQ, the German consultant company for silicon technology, and also have a tie-up with RCT Solutions of Germany. They have acquired land in the state of Andhra Pradesh. Their technology roadmap is to start with a mixture of PERC and TOPCon cell technologies [14].

Adani Infrastructure (Adani Solar) is part of another major Indian conglomerate with presence in infrastructure and energy. They have announced an investment of at least USD 20B in green energy. They are already India's largest solar cell and module manufacturer, using PERC cell technology. In their path to backward integration, in December 2022 they showcased India's first large-sized monocrystalline silicon ingots capable of producing M10 & G12 wafers. They report that they will have a 2 GW ingot+wafer facility up by the end of 2023, and will expand to 10 GW by 2025 [15]. Their bid for 4 GW integrated manufacturing in Tranche I was successful, but they did not bid for additional capacity in Tranche II. With these initiatives, they appear to be diversifying from their traditional coal-based power plants to renewables.

Vikram Solar, one of India's largest solar module manufacturers, is using the PLI route to increase its existing capacity from 2.5 GW to about 5 GW, and also undertake backward integration by setting up a 2 GW cell manufacturing line near Chennai. Similarly, Waaree Energies, also one of the largest module manufacturers in the country will expand their current capacity of 9 GW, and also backward-integrate to cells (mono-PERC) and wafers. They have entered into a multi-year agreement with CubicPV to supply 1 GW of wafers annually using CubicPV's wafer technology.

First Solar has progressed rapidly in setting up their 3.4 GW CdTe plant near Chennai. It is expected to start production in Q4 2023, and ramp up to full capacity within a year. The facility will produce the large format Series 7 modules (as in their Ohio plant), compared to the Series 6+ modules made by their facilities in Vietnam and Malaysia. A part of the production will be exported.

ReNew Power is the largest wind and solar installer in India, and decided to move into manufacturing to ensure predictable supply of high-quality modules. Their 2 GW cell+module plant with mono-PERC technology is coming up in Dholera in the state of Gujarat, and another 4 GW module-only plant is likely to be commissioned in early 2024 in the state of Rajasthan.

Tata Power (TP) Solar is one of India's oldest cell and module manufacturers (earlier named Tata BP Solar). Besides its existing plant in Bengaluru, it is setting up a greenfield 4 GW cell and module factory in the state of Tamil Nadu. The technology will be mono-PERC, transitioning to TOPCon.

V. YEAR-WISE SOLAR MANUFACTURING CAPACITY IN INDIA

India is expected to deploy between 25 and 35 GW of solar energy per year to meet its 280-300 GW target by 2030. In

978-1-6654-6060-6/23 $31.00 © 2023 IEEE 1369

order to have energy security and a dependable supply chain, it will be beneficial if this capacity can be fully met by domestic production. What is the year-wise solar manufacturing capacity which will come up? This is described below.

First, we calculate the build-up of manufacturing capacity for modules (M), cells (C), wafers (W) and polysilicon (P), based *only* on the PLI results, and assuming that all companies meet their capacity targets on time (3 years for P, 2 years for W and 1.5 years for C+M). The results are shown in Table II. The numbers for 2022 are the existing pre-PLI capacities (*operational* rather than name-plate).

It should be noted, however, that there are several other solar module manufacturers who have *not* bid for PLI. These include companies like Goldi, RenewSys, Premier, Emvee, Jupiter, Websol, Jakson, Premier, Navitas, and Gautam, among others. The capacity of each of these varies between 0.5 and 3 GW, and all have plans for capacity expansion. Some of these also manufacture cells. According to the report by Gulia *et. al.* [16], the total module capacity estimated by end-2026 (including PLI as well as non-PLI) is 110 GW. This compares reasonably well with the estimate by Mercom of 95 GW by the end of 2025 [17].

TABLE II
YEAR-WISE MANUFACTURING CAPACITIES IN INDIA

Year-end	Capacity based only on PLI (GW/year)				Total Capacity GW/year [16]	
	P	W	C	M	C	M
2022	0	0	4.3	18		
2023	0	0	4.3	18		
2024	0	12	16.3	30		
2025	12	40.8	52.5	69.6		
2026	24	40.8	52.5	69.6	59	110

A few caveats on the above Table are in order. The PLI numbers calculated by us assume that all the bidders achieve their capacities on time. This is not obvious, as delays cannot be ruled out. On the other hand, some companies (for example, First Solar) will achieve their capacity ahead of schedule. It is also possible that some companies may decide not to go for their bid capacity (even perhaps dropping out altogether, as the penalties are not excessive). On the other hand, some companies (for example, Reliance) are likely to go beyond the capacities in their bids.

Based on these projections, it is likely that India will have a module manufacturing base of around 100 GW/year by 2026 or 2027. It is important to mention that this does depend on consistency of policy being maintained over the coming decade. The cell technologies in production will be HJT, PERC/TOPCon and CdTe. Some companies are also exploring the possibility of silicon/perovskite tandems as a potential future technology. This 100 GW capacity leaves enough room for export of 60-70 GW/year. The export potential, which many companies will doubtless exploit, will ensure that the manufacturing in India is state-of-the-art, which also bodes well for the quality and performance of modules installed in India. The export capacity estimated here will make India one of the 3 largest exporters of PV modules, after China and probably USA. The cell capacity is expected to be about 50-60 GW by 2026/2027. This means that some of the made-in-India modules will use imported cells, probably from ASEAN countries, and perhaps from China (it is not clear yet how long the BCD regime will be in force). As mentioned earlier, these numbers will depend partly on how the complex geopolitical situation unfolds. It is already known that the YoY export of PV modules from India (mainly to the USA) in Q1 2023 has surged, while the YoY import of modules has declined [18].

A final point to touch upon is the carbon footprint of solar manufacturing in India. Since power supply in India will continue to rely on coal even as the transition to renewable energy takes place (it is estimated that coal will still provide 54% of electric energy in 2029-30), the carbon footprint is likely to be higher than in, say, Europe. An exception to this is Reliance, which has announced its intention to install 100 GW of solar for captive use by 2030 and become a net-zero company by 2035. This will embellish its sustainable credentials for solar PV manufacturing.

ACKNOWLEDGEMENTS

The authors thank the Ministry of New and Renewable Energy (MNRE) of the Government of India for funding provided to the National Centre for Photovoltaic Research and Education (NCPRE) at the Indian Institute of Technology Bombay since 2010. The views expressed in the paper, however, are the authors' views, and do not necessarily reflect the official views of the Ministry. The authors also thank many industry and research professionals in India and globally who have kindly provided their insights. Two of the authors (RA and LLK) thank Indian Institute of Technology Bombay for offering Visiting Professorships at NCPRE.

REFERENCES

[1] [1] International Energy Agency, "Renewables 2022" (revised version January 2023), https://www.iea.org/reports/renewables-2022

[2] J. Chase, "View from the solar industry: we don't need COP26 to shine, but what should we worry about?" *Joule* vol. 6 (2022) 495.

[3] International Energy Agency, "World Energy Outlook 2022" https://www.iea.org/reports/world-energy-outlook-2022

[4] Fraunhofer ISE, "Photovoltaics Report" (2022) https://www.ise.fraunhofer.de/en/publications/studies/photovoltaics-report.html

[5] F. Colville, "PV industry production hits 310GW of modules in 2022; what about 2023?" PVTech (2022) https://www.pv-

tech.org/pv-industry-production-hits-310gw-of-modules-in-2022-what-about-2023/

[6] N. Shiradkar *et al.*, "Recent Developments in Solar Manufacturing in India," *Solar Compass* vol. 1, 100009 (2022)

[7] Fraunhofer ISE, "Sustainable PV Manufacturing in Europe" (2020) https://www.ise.fraunhofer.de/content/dam/ise/de/documents/publications/studies/ISE-Sustainable-PV-Manufacturing-in-Europe.pdf

[8] European Commission, https://single-market-economy.ec.europa.eu/industry/strategy/industrial-alliances/european-solar-photovoltaic-industry-alliance_en

[9] Fraunhofer ISE, https://www.ise.fraunhofer.de/en/press-media/press-releases/2023/key-stakeholders-in-the-pv-Industry-call-for-measures-to-re-establish-upstream-solar-manufacturing-value-chain-in-europe.html

[10] McKinsey & Company, "Building a competitive solar-PV supply chain in Europe," (2022) https://www.mckinsey.com/industries/electric-power-and-natural-gas/our-insights/building-a-competitive-solar-pv-supply-chain-in-europe

[11] J. Vasi, "Solar Energy," in *Emerging Energy Resources in India* (K. Sai, S. Roy and H.K. Gupta, eds.), Geological Survey of India (2022).

[12] P. Ghosh *et al.*, "Scaling Sustainable Integrated PV Manufacturing Globally," 48th IEEE PVSC (2020).

[13] MNRE, Notification F.No. 282/63/2020-GRID SOLAR (2022) https://mnre.gov.in/img/documents/uploads/file_f-1619672166750.pdf

[14] S. Shetty, Mercom (2023) https://www.mercomindia.com/shirdi-sai-electricals-pli-ii-manufacturing-10-gw

[15] U. Gupta, PV Magazine (2022) https://www.pv-magazine.com/2022/12/22/adani-becomes-indias-sole-producer-of-large-monocrystalline-silicon-ingots/

[16] Gulia, J. et al., *India's Photovoltaic Manufacturing set to Surge,* JMK Research and Analytics (2023) https://jmkresearch.com/cpo_team/pv-manufacturing-in-india/

[17] Mercom, *State of Solar Manufacturing in India* (2023) https://www.mercomindia.com/product/state-solar-manufacturing-india

[18] A. Joshi, "Solar Exports surged 6,239% in Q1 2023, Imports fell 34%," Mercom (2023) https://www.mercomindia.com/solar-exports-surged-q1-2023-imports-fell-34

IBC Technology Targeting Fast and Effective Silver Reduction Applying Advanced Screen Printing

Radovan Kopecek, Florian Buchholz, Valentin D. Mihailetchi, Joris Libal, Ning Chen, Haifeng Chu, Christoph Peter, Dominik Rudolph, Thomas Buck, Tudor Timofte, Andreas Halm, Yonggang Guo, Xiaoyong Qu, Xiang Wu, Jiaqing Gao, Peng Dong

International Solar Energy Research Center Konstanz, Konstanz, Germany

SPIC Solar , Xining, China

We show that interdigitated back-contact technology, as the last evolutionary step in the field of single-juction crystalline silicon solar cells, can be produced cost-effectively and not only has the highest efficiency potential, but also offers several advantages over the emerging double-contacted n-type devices such as the tunnel oxide passivating contact and heterojunction technology. The main advantages are on the one hand the easier implementation of alternative metal contacts to silver such as copper and aluminium and on the other hand a simpler module interconnection for highest module efficiencies based on negative gap technology. Our low-cost ZEBRA-IBC cell and module technology is produced at SPIC Solar with average efficiencies of over 24%, with modules on the market reaching 22.3% efficiency. These are the highest efficiencies achieved with single junction crystalline silicon technology without charge carrier selective contacts. Open-circuit voltages of more than 700 mV are achieved with an advanced screen-printing process that also allows a simple and almost complete change from silver to copper or aluminium screen-printing metallization with a silver content of less than 5 mg/Wp. Such a silver reduction is necessary in order to enter a yearly 1 TW PV market from 2028 on. In this paper, we show the way and first resuls to interdigitated back-contact solar cells screen-printed with copper and aluminium for a necessary fast silver reduction.

978-1-6654-6060-6/23 $31.00 © 2023 IEEE

Development of Machine Vision System for Detection of Wrap-around in n-TOPCon Solar Cells

Junhee Kim, Han-Jung Kim, Yohan Ko, and Yoonkap Kim

Gumi Electronics & Information Technology Research Institute, Gumi, Gyeongbuk, 39171, Korea

Abstract — n-TOPCon solar cells have been the subject of extensive research by various groups in order to overcome the limitations on efficiency of p-PERC solar cells. The passivated contacts formed on the rear side of wafers for the production of n-TOPCon solar cells are deposited using LPCVD or PECVD, however, wrap-around of heavily doped polysilicon is inevitable and results in optical loss and increased shunt resistance. Therefore, post-etching process of wrap-around is necessary and detection of wrap-around after the etching process is also required. In this study, we report on the development of a machine vision system based on a digital microscope equipped with a high-magnification zoom lens and controlled by operation software to minimize human intervention as much as possible. Although further optimization of the machine vision system is needed, we expect to achieve meaningful results after optimization.

Index Terms — n-TOPCon, machine vision system, wrap-around.

I. INTRODUCTION

p-PERC (Passivated Emitter and Rear Contact) solar cells, which have been the most widely deployed in the world over the past decade, have encountered a constraint in their efficiency as a result of excessive carrier recombination at the interfaces where metal and silicon intersect. To mitigate this limitation in the efficiency of crystalline silicon solar cells, various strategies have been implemented, with TOPCon (Tunnel Oxide Passivated Contact) technology receiving extensive research attention since the publication of papers by F. Feldmann et al. in 2014 [1,2]. TOPCon, which comprises a tunnel oxide layer and heavily doped poly-Si layers, as depicted in Figure 1(a), augments charge carrier selectivity and mitigates carrier recombination at the metal contacts, thereby improving efficiency. According to the International Technology Roadmap for Photovoltaics (ITRPV) 2022 [3], the global market share of n-TOPCon is projected to continue to expand and comprise half of the world market by 2032.

In the conventional fabrication of n-TOPCon solar cells, low pressure chemical vapor deposition (LPCVD) is primarily utilized to create the tunnel oxide and n+-poly-Si layer, as noted in Ref. 4 and 5. During this process, wafers are loaded in a back-to-back configuration, leading to the formation of severe wrap-around on the front side, as depicted in Figure 1(b), which causes optical loss and shunt resistance. The use of plasma-enhanced chemical vapor deposition (PECVD) was initially considered as a promising strategy for mitigating the formation of unwanted wrap-around, as reported in Ref. 6. However, it was subsequently established that while the width

of the wrap-around was reduced, it still persisted, as illustrated in Figure 1(c). As a result, the elimination of the wrap-around has become an indispensable and ineluctable step in the fabrication process, as outlined in Ref. 7.

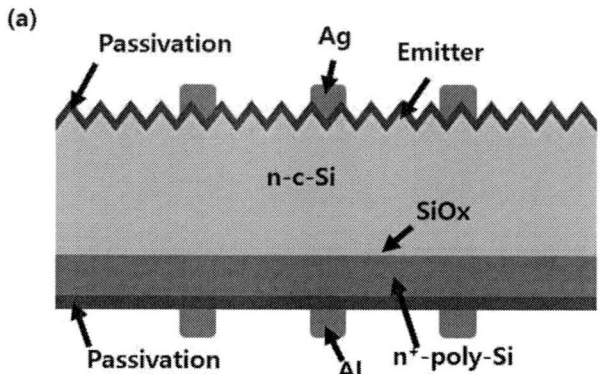

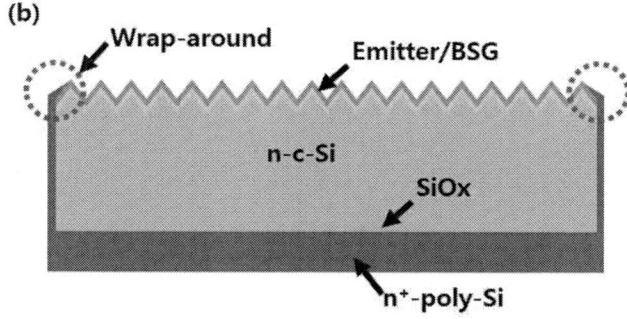

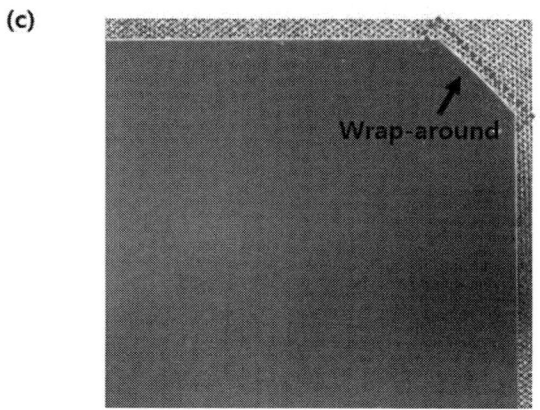

Fig. 1.   Schematic diagrams of the n-TOPCon solar cell structure (a), the wrap-around formed on the front side edges (b), and an image of the wrap-around formed through the utilization of PECVD (c).

Computer vision systems, which have been actively developed since the early 1980s, are now widely utilized in mass production lines. In mass production of crystalline silicon solar cells, a variety of inspection apparatus are implemented to identify and discard defective products, thereby reducing the consumption of unnecessary materials and costs. An example of such a system is the vision system employed post-metallization process, which is commonly implemented in solar cell production lines.

This paper presents a methodology for the visual inspection of the wrap-around formed on the front edges of solar cells fabricated with PECVD, and for making pass/fail determinations based on the observations. Initially, we describe the design and development of the machine vision system for the inspection of the wrap-around, and subsequently, we present the results of a comprehensive evaluation of its detection performance through both in-wafer and wafer-to-wafer assessment, in order to confirm its accuracy and reliability.

II. DEVELOPMENT OF MACHINE VISION SYSTEM

Machine vision is a cutting-edge technology that enables computers to process and comprehend visual information from images or videos, which can be used for a wide range of tasks such as object recognition, measurement, and feature extraction. As depicted in Fig. 2, we employed a digital microscope equipped with a high-magnification zoom lens to acquire high-resolution images of the wrap-around formation on the edges of the front side. Additionally, we utilized a ring-type LED lighting source to minimize the impact of external factors on the imaging process, ensuring the most optimal visual representation of the phenomenon. Finally, we employed linear actuators in x- and y-directions to automate the scanning of the entire area of the sample.

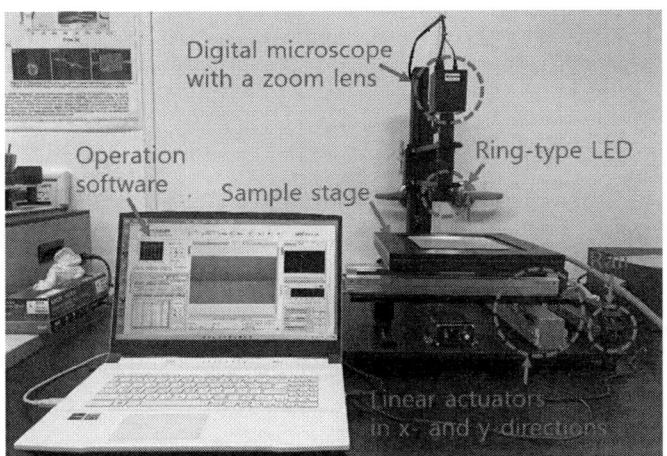

Fig. 2.   Machine vision system developed to investigate the wrap-around on the edges of the front side.

The machine vision system is controlled by the operation software, which was developed using LabVIEW to minimize human intervention as depicted in Fig. 3. We controlled the digital microscope settings, acquired images, and performed image processing tasks using the NI-IMAQdx driver (a software library developed by National Instruments that provides an Application Programming Interface (API) for communicationg with various types of machine vision cameras).

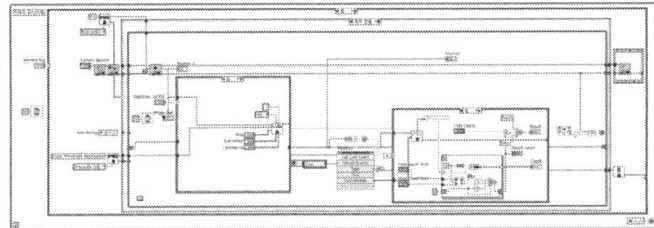

Fig. 3.   A graphical LabVIEW code for image acquision using NI-IMAQdx driver.

III. EXPERIMENTAL RESULTS

(a)

Wrap-around detection success

(b)

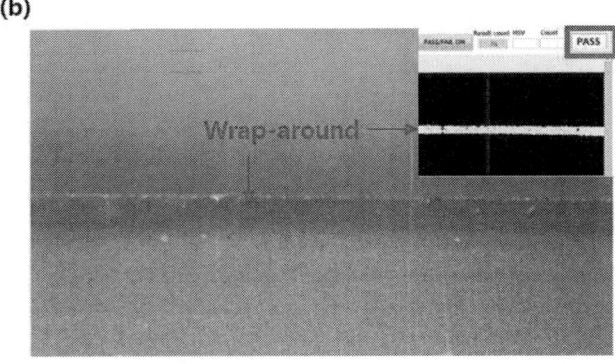

Wrap-around detection failure

Fig. 4.   The results of wrap-around detection as determined by the developed machine vision system, classified into (a) successful detection and (b) unsuccessful detection.

We investigated the wrap-around on the edges of the front side of the wafers during the production of n-TOPCon solar cells using a developed machine vision system. Specifically, we evaluated the in-wafer detection rate by conducting measurements on 1 wafer with wrap-around formation. We found that we were successful in detecting wrap-around in 11 out of 20 points, but failed in 9 points, resulting in a 55% in-wafer detection success rate. Additionally, we evaluated the wafer-to-wafer detection rate by measuring 40 points per wafer on 10 wafers with wrap-around formation. The results are presented in TABLE I.

TABLE I
INVESTIGATION RESULTS FOR WAFER-TO-WAFER
VARIABILITY IN 10 WAFER SAMPLES

No.	Success	Failure	Detection Success Rate [%]
01	26	14	65
02	23	17	57.5
03	25	15	62.5
04	25	15	62.5
05	22	18	55
06	26	14	65
07	25	15	62.5
08	23	17	57.5
09	24	16	60
10	25	15	62.5
			61

We developed an HSL(Hue, Saturation, Luminescence) image filter for precise detection of wrap-around and image-processing, however, it has not yet been optimized, thus resulting in a low detection rate for wrap-around. If further optimization of the HSL image filter is achieved, it is expected to yield higher detection rates above 61%, which would be considered meaningful results.

IV. SUMMARY AND CONCLUSION

In this paper, a method is introduced for detection and image processing of wrap-around formed on the edges of wafer front side in n-TOPCon solar cells. We developed a machine vision system based on a digital microscope equipped with a high-magnification zoom lens and controlled by an operation software developed using LabVIEW to minimize human intervention. We implemented in-wafer and wafer-to-wafer assessment for ensuring the ability of detection success rate, resulting in the 55% and average 61% in in-wafer and wafer-to-wafer, respectively. It is clearly evident that optimization of the machine vision system is necessary for enhancing the efficiency of n-TOPCOn solar cells. In particular, by enhancing the HSL image filter, it is expected that higher detection rates can be ontained compared to the current performance.

V. ACKNOWLEDGEMENT

This work was supported by the Technology Innovation Program (20016029) funded by the Ministry of Trade, Industry & Energy (MOTIE, Korea.). The authors would like to thank researchers at Korea Institute of Enery Research; especially H. Song, M. Kang, K. Jeong and S. Lee for producing the solar cells. The authors would like to also thank researchers at SEA(Korean wetstation manufacturer); especially I. Jang and J. Kim for wet etching experiment.

REFERENCES

[1] F. Feldmann, M. Bivour, C. Reichel, M. Hermle, and S. W. Glunz, "Passivated rear contacts for high-efficiency n-type Si solar cells providing high interface passivation quality and excellent transport characteristics," *Sol. Energy Mater. Sol. Cells*, vol. 120, pp. 270-274, 2014.

[2] F. Feldmann, M. Simon, M. Bivour, C. Reichel, M. Hermle, and S. W. Glunz, "Efficient carrier-selective p- and n-contacts for Si solar cells," *Sol. Energy Mater. Sol. Cells*, vol. 131, pp. 100-104, 2014.

[3] International Technology Roadmap for Photovoltaic (ITRPV) 2021 Results, 13th Edition, 2022.

[4] M. K. Stodolny, M. Lenes, Y. Wu, G. J. M. Janssen, I. G. Romijn, J. R. M. Luchies, and L. J. Geerligs, "n-Type polysilicon passivating contact for industrial bifacial n-type solar cells," *Sol. Energy Mater. Sol. Cells*, vol. 158, pp. 24-28, 2016.

[5] Q. Wang, W. Wu, D. Chen, L. Yuan, S. Yang, Y. Sun, S. Yang, Q. Zhang, Y. Cao, H. Qu, N. Yuan, and J. Ding, "Study on the cleaning process of n+-poly-Si wraparound removal of TOPCon solar cells," *Sol. Energy*, vol. 211, pp. 324-335, 2020.

[6] B. Grübel, H. Nagel, B. Steinhauser, F. Feldmann, S. Kluska, and M. Hermle, "Influence of plasma-enhanced chemical vapor deposition poly-Si layer thickness on the wrap-around and the quantum efficiency of bifacial n-TOPCon (tunnel oxide passivated contact) solar cells," *Phys. Status Solidi A*, vol. 218, pp. 2100156, 2021.

[7] B. Kafle, S. Mack, C. Teβmann, S. Bashardoust, L. Clochard, E. Duffy, A. Wolf, M. Hofmann, and J. Rentsch, "Atmospheric pressure dry etching of polysilicon layers for highly reverse bias-stable TOPCon solar cells," *Sol. RRL*, vol. 6, pp. 2100481, 2022.

Effect of CdS Annealing on the Performance of Antimony Selenosulfide Solar Cells

Alisha Adhikari, Suman Rijal, Sabin Neupane, Manoj K. Jamarkattel, Deng-Bing Li, Tamanna Mariam, Michael J. Heben, Zhaoning Song, and Yanfa Yan

Wright Center for Photovoltaic Innovation and Commercialization, Department of Physics and Astronomy, University of Toledo, Toledo, OH, 43606, USA

Abstract— Antimony Selenosulfide ($Sb_2(S, Se)_3$) has emerged as a promising light absorber material for thin-film photovoltaics due to its high stability, earth abundance, and low toxicity. Here, we investigate the effect of annealing the cadmium sulfide (CdS) buffer layer under different conditions on the performance of $Sb_2(S, Se)_3$ solar cells. Annealing CdS on a hot plate at 400 ºC in the air for 10 min improves the front heterojunction and enhance the power conversion efficiency of $Sb_2(S, Se)_3$ solar cells. Additionally, we study the impact of a $CdCl_2$ treatment on the CdS buffer layer under different atmospheres. The preliminary results indicate that combing $CdCl_2$ treatment and air annealing is promising to improve the performance of $Sb_2(S, Se)_3$ solar cells, but further device optimization is needed.

Keywords— $Sb_2(S, Se)_3$, CdS, hydrothermal growth

I. INTRODUCTION

Thin film chalcogenide solar cells have attracted substantial attention as an alternative to crystalline Si-based solar cells for terrestrial and space power applications [1]. Over the years, chalcogenide-based CdTe and $Cu(In, Ga)Se_2$ solar cells have already achieved power conversion efficiencies (PCEs) beyond 22% [2]. However, the shortcomings of these commercial thin-film photovoltaic (PV) technologies, such as the high cost of Ga and In, the toxicity of Cd and Te, and the relatively high capital cost of the equipment for production, drive the research effort towards exploring new chalcogenide-based materials for solar cells, which are low toxic, easy for synthesis, stable, and efficient [3].

Recently, antimony-based binary and ternary chalcogenide compounds like Sb_2Se_3, Sb_2S_3, and $Sb_2(S, Se)_3$ have emerged as promising materials for solar cell application. $Sb_2(S, Se)_3$ possesses a tunable band gap ($1.1-1.7$ eV) [4], high absorption coefficient $> 10^5$ cm^{-1}, and excellent material stability [5, 6]. In addition, low-cost constituents, low toxicity, and earth abundance properties make these materials suitable for terrestrial and space PV applications. Antimony chalcogenide solar cells can be fabricated via different approaches, including spin coating [7], chemical bath deposition (CBD) [8], hydrothermal deposition [4, 9, 10], injection vapor deposition (IVD) [1], sputtering [11], and close-spaced sublimation [12, 13]. Among various approaches, hydrothermal deposition is a simple approach to grow high-quality films at low fabrication costs. Tang et al. deposited $Sb_2(S, Se)_3$ thin films via a hydrothermal method and achieved an efficiency of 10% [4]. Rijal et al. studied the effect of post-annealing treatment on hydrothermally grown $Sb_2(S, Se)_3$ solar cells [10]. Here, we focus on optimizing the front heterojunction of hydrothermally grown $Sb_2(S, Se)_3$ devices. We study the effect of the annealing environment, i.e., ambient air and N_2 environment, on the CdS buffer layer for the improved performance of $Sb_2(S, Se)_3$ thin film solar cells.

II. EXPERIMENTAL DETAILS

We prepared devices in a superstrate configuration of glass/FTO/CdS/$Sb_2(S, Se)_3$/Spiro-OMeTAD/Au. Fluorine-doped tin oxide (FTO) coated glass substrates were cleaned with a diluted detergent solution (Micro-90) and deionized (DI) water in an ultrasonic bath and dried by air. Then, a CdS layer was deposited on an FTO substrate via a CBD approach [14]. For $CdCl_2$ treatment, a $CdCl_2$ in methanol solution (20 mg mL^{-1}) was spin-coated on the obtained CdS film at 4000 rpm for 30 s, followed by heat treatment at 400 ºC for 5 and 10 min in ambient air or N_2 environment. After that, an $Sb_2(S, Se)_3$ thin film was deposited on the CdS film via a hydrothermal approach. The Spiro-OMeTAD hole-transport layer (HTL) was spin-coated on the $Sb_2(S, Se)_3$ film at a speed of 3000 rpm for 30 s and left overnight in dry ambient air to get oxidized. 60 nm Au was deposited as a back contact on the top of spiro-OMeTAD using a thermal evaporator under the pressure of 5×10^{-6} Torr, and a shadow mask of 0.08 cm^2 was used to define the unit cell area.

For the $Sb_2(S, Se)_3$ thin film deposition, potassium antimony tartrate ($C_8H_4K_2O_{12}Sb_2.3H_2O$), sodium thiosulfate pentahydrate ($Na_2S_2O_3.5H_2O$), and selenourea (CH_4N_2Se) were used as Sb, S, and Se sources, respectively. 20 mM of ($C_8H_4K_2O_{12}Sb_2.3H_2O$), 40 mM of ($Na_2S_2O_3.5H_2O$), and 35 mg of ($CH_4N_2Se$) were mixed in 50 ml of DI water. These chemicals were stirred until the solution turned clear and yellowish. Then, the solution was transferred into the cleaned inner Teflon tank of an autoclave with the film side facing downward. Finally, it was kept in an oven heated at 135 ºC for 3 h. After 3 h, the autoclave was taken out of the oven and allowed to cool to room temperature. Then, the film was rinsed with DI water and dried with compressed air. After this, the films were annealed in a two-step process in a tube

978-1-6654-6060-6/23 $31.00 © 2023 IEEE

furnace. First, the graphite plate that holds the sample was heated at 100 °C for 1 min. Then, the tube was filled with 10 Torr of N_2, and the films were annealed at 350 °C for 10 min, then naturally cooled down to room temperature.

Current density-voltage (J-V) characteristics were measured using a Keithley 2440 digital source meter and a solar simulator with AM1.5G illumination. External quantum efficiencies (EQE) were measured using PV Instruments system (model IVQE8-C). Transmission data was measured using a PerkinElmer Lambda 1050 UV-VIS-NIR spectrometer.

III. RESULTS AND DISCUSSION

We first studied the impact of $CdCl_2$ treatment on the microstructure of the CdS film. Figure 1 compares the surface SEM images of the CdS films without and with $CdCl_2$ treatment. No significant differences were found in the grain size of the CdS layer annealed in ambient air without and with $CdCl_2$ treatment.

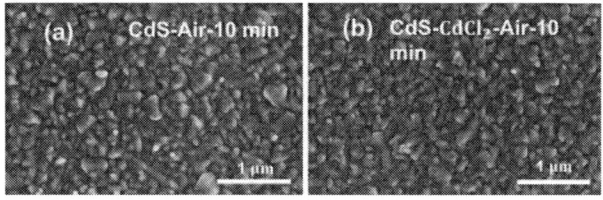

Fig 1. SEM images of the CdS layer annealed in the ambient air (a) without and (b) with $CdCl_2$.

Figure 2 shows the J-V measurements of $Sb_2(S, Se)_3$ solar cells with and without $CdCl_2$ treatment followed by thermal annealing of the CdS layer for 5 and 10 min in the ambient air or N_2 environment. Table I summarizes device performance. Among samples without the $CdCl_2$ treatment, the best PCE of 8.42% was obtained for the CdS-Air device annealed for 10 min, and the values of open-circuit voltage (V_{OC}), short-circuit current density (J_{SC}), fill factor (FF), series resistance (R_S), and shunt resistance (R_{SH}) are 659 mV, 20.70 mA/cm^2, 61.77%, 4.03 Ω cm^2, 841.1 Ω cm^2, respectively. Among samples with the $CdCl_2$ treatment, the best PCE of 8.13% was obtained for the CdS-$CdCl_2$-Air treated device annealed for 10 min, and the values of V_{OC}, J_{SC}, FF, R_S, and R_{SH} are 643 mV, 20.58 mA/cm^2, 61.46%, 4.01 Ω cm^2, and 554.4 Ω cm^2, respectively.

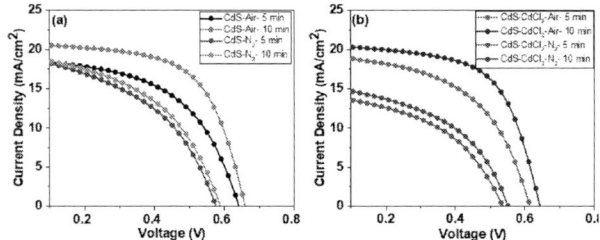

Fig 2. J-V measurements of $Sb_2(S, Se)_3$ devices (a) without and (b) with a $CdCl_2$ treatment on the CdS buffer layer annealed in ambient air and N_2 environment.

TABLE I. J-V parameters of the devices.

Annealing time	Device	V_{OC} (mV)	J_{SC} (mA/cm^2)	FF (%)	PCE (%)
5 min	CdS-Air	641	18.70	53.01	6.36
	CdS-N_2	576	19.12	44.99	4.96
	CdS-$CdCl_2$-Air	615	19.29	51.99	6.16
	CdS-$CdCl_2$-N_2	535	14.31	46.16	3.54
10 min	CdS-Air	659	20.70	61.77	8.42
	CdS-N_2	589	19.16	47.31	5.34
	CdS-$CdCl_2$-Air	643	20.58	61.46	8.13
	CdS-$CdCl_2$-N_2	551	15.53	46.69	3.99

Figure 3 shows the performance statistics of the devices. Clearly, CdS-Air and CdS-$CdCl_2$-Air devices annealed for 10 min show significantly better performances than other devices annealed for a shorter time and oxygen-deficient conditions. Figure 4 (a) shows the EQE spectra of $Sb_2(S, Se)_3$ devices. The EQE value of CdS-Air annealed for 10 min is higher than others at all wavelengths.

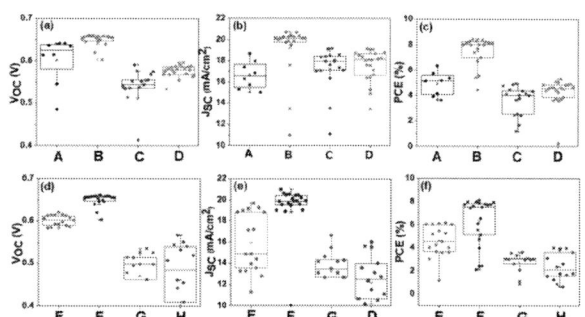

Fig 3. PV performance statistics of $Sb_2(S, Se)_3$ devices using CdS without $CdCl_2$ (a, b, c) and with $CdCl_2$ (d, e, f) annealed for 5 min and 10 min. Letters represent the following: A, CdS-Air-5 min; B, CdS-Air-10 min; C, CdS-N_2-5 min; D, CdS-N_2-10 min; E, CdS-$CdCl_2$-Air-5 min; F, CdS-$CdCl_2$-Air-10 min; G, CdS-$CdCl_2$-N_2-5 min; H, CdS-$CdCl_2$-N_2-10 min.

During the thermal annealing in the ambient atmosphere, CdS is partially oxidized to CdS:O, resulting from more oxygen incorporation in the film. Additionally, $CdCl_2$ treatment introduced chlorine (Cl) to the film. Thermal annealing increases the crystallinity of CdS. O and Cl incorporation passivates defects in the CdS buffer layer, thus improving the CdS charge transport properties and enhancing the overall device performances [15, 16]. Oxygen in the CdS layer may diffuse to the $Sb_2(S, Se)_3$ absorber layer during the hydrothermal growth, passivating defects in $Sb_2(S, Se)_3$. The improvements in EQE after annealing the CdS layer in the air. Figure 4 (a) confirms the positive effect of oxygen in $Sb_2(S, Se)_3$ solar cells. Additionally, O and Cl incorporation slightly increases the transmittance of the CdS buffer layer, as shown in Figure 4 (b). The increased transmittance leads to higher EQE at short wavelengths, in agreement with the EQE results in Figure 4 (a).

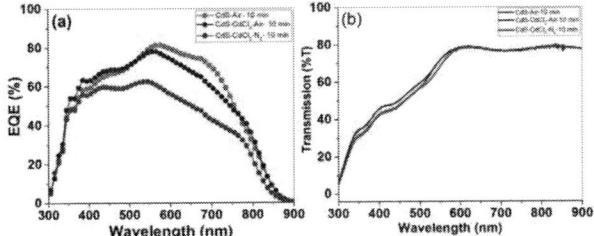

Fig 4. (a) EQE spectra of $Sb_2(S, Se)_3$ devices using CdS layer annealed at different conditions and (b) Transmission spectra of CdS layer annealed at different conditions.

IV. CONCLUSION

In this work, the effect of different annealing conditions of the CdS buffer layer on the performance of $Sb_2(S, Se)_3$ solar cells was studied. The performance of the air-annealed CdS layer without $CdCl_2$ treatment is better than other devices annealed in an oxygen-deficient condition. The $CdCl_2$ treatment on the CdS layer increases transmittance and short-wavelength spectral responses, but further device optimization is needed. Overall, ambient air annealing has a positive effect on the CdS layer for improving hydrothermally grown $Sb_2(S, Se)_3$ thin film solar cells.

ACKNOWLEDGEMENT

This work is based on research sponsored by Air Force Research Laboratory under agreement numbers FA9453-19-C-1002 and FA9453-21-C-0056. The U.S. Government is authorized to reproduce and distribute reprints for Governmental purposes notwithstanding any copyright notation thereon. The views expressed are those of the authors and do not reflect the official guidance or position of the United States Government, the Department of Defense or of the United States Air Force. The appearance of external hyperlinks does not constitute endorsement by the United States Department of Defense (DoD) of the linked websites, or the information, products, or services contained therein. The DoD does not exercise any editorial, security, or other control over the information you may find at these locations. Approved for public release; distribution is unlimited. Public Affairs release approval # AFRL-2023-2551.

REFERENCES

[1] Z. Duan *et al.*, "Sb2Se3 Thin‑Film Solar Cells Exceeding 10% Power Conversion Efficiency Enabled by Injection VaporDeposition Technology," *Advanced Materials,* vol. 34, no. 30, p. 2202969, 2022.

[2] M. Green, E. Dunlop, J. Hohl‑Ebinger, M. Yoshita, N. Kopidakis, and X. Hao, "Solar cell efficiency tables (version 57)," *Progress in photovoltaics: research and applications,* vol. 29, no. 1, pp. 3-15, 2021.

[3] K. Zeng, D.-J. Xue, and J. Tang, "Antimony selenide thin-film solar cells," *Semiconductor Science and Technology,* vol. 31, no. 6, p. 063001, 2016.

[4] R. Tang *et al.*, "Hydrothermal deposition of antimony selenosulfide thin films enables solar cells with 10% efficiency," *Nature Energy,* vol. 5, no. 8, pp. 587-595, 2020/08/01 2020, doi: 10.1038/s41560-020-0652-3.

[5] X. Wang, R. Tang, C. Wu, C. Zhu, and T. Chen, "Development of antimony sulfide–selenide Sb2 (S, Se) 3-based solar cells," *Journal of energy chemistry,* vol. 27, no. 3, pp. 713-721, 2018.

[6] X. Wang *et al.*, "Manipulating the electrical properties of Sb2 (S, Se) 3 film for high‑efficiency solar cell," *Advanced Energy Materials,* vol. 10, no. 40, p. 2002341, 2020.

[7] Y. C. Choi *et al.*, "Sb2Se3‑sensitized inorganic–organic heterojunction solar cells fabricated using a single‑source precursor," *Angewandte Chemie,* vol. 126, no. 5, pp. 1353-1357, 2014.

[8] Y. Zhao *et al.*, "Regulating Deposition Kinetics via A Novel Additive-assisted Chemical Bath Deposition Technology Enables 10.57%-efficient Sb2Se3 Solar Cells," *Energy & Environmental Science,* 2022.

[9] D. Liu *et al.*, "Direct hydrothermal deposition of antimony triselenide films for efficient planar heterojunction solar cells," *ACS Applied Materials & Interfaces,* vol. 13, no. 16, pp. 18856-18864, 2021.

[10] S. Rijal *et al.*, "Post‑annealing Treatment on Hydrothermally Grown Antimony Sulfoselenide Thin Films for Efficient Solar Cells," *Solar RRL,* 2022.

[11] G.-X. Liang *et al.*, "Sputtered and selenized Sb2Se3 thin-film solar cells with open-circuit voltage exceeding 500 mV," *Nano Energy,* vol. 73, p. 104806, 2020.

[12] S. Rijal, D.-B. Li, R. A. Awni, S. S. Bista, Z. Song, and Y. Yan, "Influence of post-selenization temperature on the performance of substrate-type Sb2Se3 solar cells," *ACS Applied Energy Materials,* vol. 4, no. 5, pp. 4313-4318, 2021.

[13] S. Rijal *et al.*, "Templated Growth and Passivation of Vertically Oriented Antimony Selenide Thin Films for High‑Efficiency Solar Substrate Configuration," *Advanced Functional Materials,* vol. 32, no. 10, p. 2110032, 2022.

[14] D.-B. Li *et al.*, "Stable and efficient CdS/Sb2Se3 solar cells prepared by scalable close space sublimation," *Nano Energy,* vol. 49, pp. 346-353, 2018.

[15] L. Wang *et al.*, "Ambient CdCl2 treatment on CdS buffer layer for improved performance of Sb2Se3 thin film photovoltaics," *Applied Physics Letters,* vol. 107, no. 14, p. 143902, 2015.

[16] J. Li *et al.*, "Hydrazine Hydrate‑Induced Surface Modification of CdS Electron Transport Layer Enables 10.30%‑Efficient Sb2 (S, Se) 3 Planar Solar Cells," *Advanced Science,* vol. 9, no. 25, p. 2202356, 2022.

20%-Efficient TOPCon Solar Cell with a Silicon Oxide Layer Deposited by Aerosol Impaction-Driven Assembly

Maria Angelica M Garcia, William Weigand, Zachary C Holman

Arizona State University, Tempe, AZ, United States

Tunnel oxide passivated contacts (TOPCon) have become an increasingly popular contact choice for use in silicon solar cells due to their carrier-selective capabilities and success in creating high-efficiency devices. TOPCon structures, fabricated using a thin silicon oxide (SiOx) passivation layer topped with a layer of doped polysilicon (poly-Si) have enabled cell efficiencies higher than 25%. In this work, we explore the use of a TOPCon rear contact-paired with a standard silicon heterojunction (SHJ) front contact to create hybrid cells-utilizing a novel oxide deposition technique, aerosol-impaction-driven assembly (AIDA). The best cells exceed 20% efficiency and are limited by the processing (in)compatibility of the TOPCon and SHJ contacts. Unusually for TOPCon, we find that AIDA oxide layers up to 6 nm thick provide both decent passivation and sufficient conduction pathways to produce cells with >75% fill factor, indicating a non-tunneling conduction mechanism.

978-1-6654-6060-6/23 $31.00 © 2023 IEEE

Understanding Practical Efficiency Limits for Tandem Solar Cells

Emily L. Warren, Sirazul Haque, Qi Jiang, William E. McMahon, Lorelle M. Mansfield

National Renewable Energy Laboratory, Golden, CO, United States

Many single-junction technologies are approaching their practical efficiency limits, motivating research into tandem solar cells for terrestrial energy generation. Tandem solar cells, where multiple single-junction cells are combined optically in series, provide a path to make cells with high areal efficiencies, with multiple material systems capable of achieving greater than 30% efficiency under 1-sun conditions. However, there are many different material combinations and configurations used to make a tandem, and it can be challenging to understand how advances in one material system will impact the performance of a tandem device. Spectral efficiency, a concept proposed by Yu et al. (Nature Energy, 2016), is a powerful tool to compare different tandem combinations without the need for actually fabricating tandem cells. The prior spectral efficiency analysis proposed a framework to calculate spectral efficiency (SE) of individual single-junction solar cells, either based on experimental data or simulations, and then combine different pairs assuming ideal optical coupling to predict an overall tandem efficiency. This provides a quick way to estimate the maximum tandem efficiency possible for the given subcell pair. In this work, we have developed open-source code (nrel.github.com/SEcalculator) to easily enable calculation of spectral efficiency for single junctions and predicted maximum efficiency of tandem pairs. We have expanded the analysis to calculate different optical coupling methods and include analysis of current-matched two-terminal (2T) tandems. We are interested in talking with researchers to understand what improvements or additional functionality would be beneficial to the community.

Effect of Angle and Direction of Solar Panels in the Desert Climate of Abu Dhabi, United Arab Emirates

Laith Nayfeh[1], Leia Nayfeh[1] and Ammar Nayfeh[2]*

[1]Dunecrest American School, Dubai UAE

[2]Khalifa University, Abu Dhabi, 127788, UAE

*Corresponding Author: ammar.nayfeh@ku.ac.ae

Abstract— In this work, we study the effect of angle and direction on the solar cell performance in Abu Dhabi, United Arab Emirates. Understanding the optimal angle and direction of solar panels in a desert climate can be key in mitigating some of the negative effects. These include very high temperature, humidity, and soiling. Our results show the optimal direction in Abu Dhabi, UAE is due south that provides the highest performance with an angle around 30 degrees. Moreover, a solar tracker can be used to keep the solar cell at the maximum efficiency.

I. INTRODUCTION

The negative effects of climate change and global warming are no longer a futuristic issue to deal with. This is the result of the continued and extensive use of fossil fuels and subsequent greenhouse gasses emissions such as CO_2. Earth has been warming at an alarmingly fast pace the last 40 years due to greenhouse gas emission and the trapping of heat in the atmosphere. To slow down and eliminate this unnatural warming rate a reduction in greenhouse gas emissions is essential. This is accomplished by an increase in the deployment and use of clean renewable energy. Solar energy continues to play a significant role in the clean energy shift due to the abundance and unlimited supply of energy provided by the sun. The use of solar energy is on an exponential rise thanks to recent government policies to reduce emissions. For example, solar energy production in the United Arab Emirates (UAE) from 2009-2019 shows an exponential rise starting in 2016 reaching almost 3500 GWh of production in 2019 [1]. The UAE is also home to the Mohammed Bin Rashid Al Maktoum solar park. The largest single site solar park the world. This project will help reduce more than 6.5 million tons of emissions and will have capacity of 5000 megawatts [2]. The United Arab Emirates is an ideal region for solar due the amount of sunshine and with very few cloudy/rainy days. However, desert climates have their own unique challenges, such as extreme heat and soling.

Temperatures can reach above 40 in summer months in the UAE with high humidity and effect of soiling from dust is not very well understood [3]. Also, what is important is finding the optimal angle and direction of roof top solar panels to maximize the power and limit these negative effects.

II. SETUP OF PANEL TEST

The first steps in the process of the solar cell test in Abu Dhabi was to get the materials we needed. We purchased a solar cell for local vendor in Dubai UAE. The panel is made up of 36 polysilicon solar cells with specs maximum power 3W, maximum power voltage 8.8V, maximum power current 0.35A, open circuit voltage 10.8V, short circuit current 0.38A. Figure 1 shows the backside of the solar cell from Dakodi.

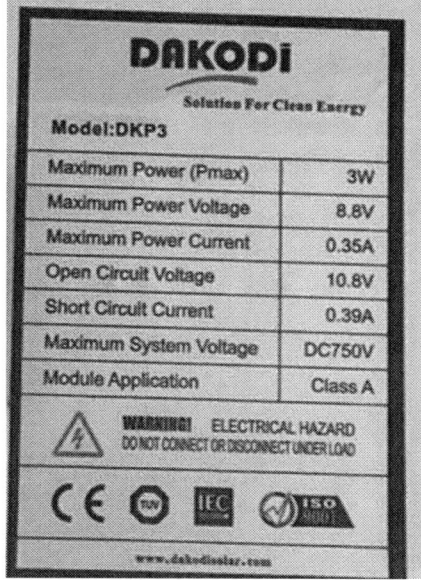

Figure 1 Backside of solar cell

978-1-6654-6060-6/23 $31.00 © 2023 IEEE

Figure 2: Image of Solar panel on rooftop of Khalifa University

We also purchased a solar panel multimeter from Elejoy to test the solar cells outside in real time.

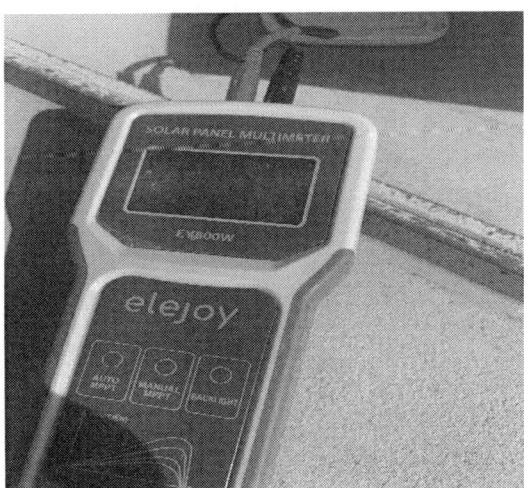

Figure 3. Outdoor multi meter to test solar cells

III. ROOFTOP SOLAR RESULTS

We placed the panel on a roof top in Abu Dhabi at Khalifa University SAN campus. To find the optimal angle and location we tested the panels at 12 noon in November. Figure 4 shows the testing rooftop setup.

Figure 4 solar panel setup at Khalifa University San campus

We first fixed the angle around 22° based on literature, and then modified the direction from north, south, west and east to find which direction gave the peak. Figure 5 shows the max power vs direction. We found that south gave the highest power. Figure 6 shows the max current vs direction also with peak at south.

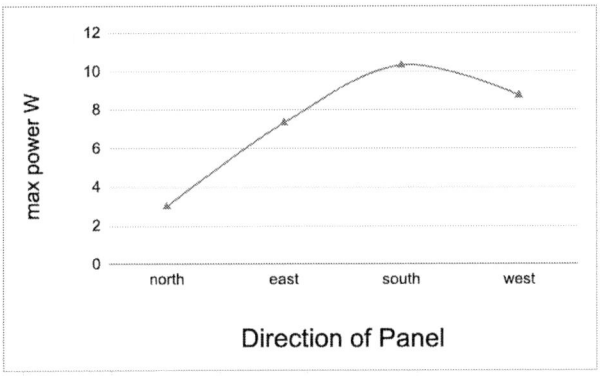

Figure 5. Maximum Power vs Direction of Panel

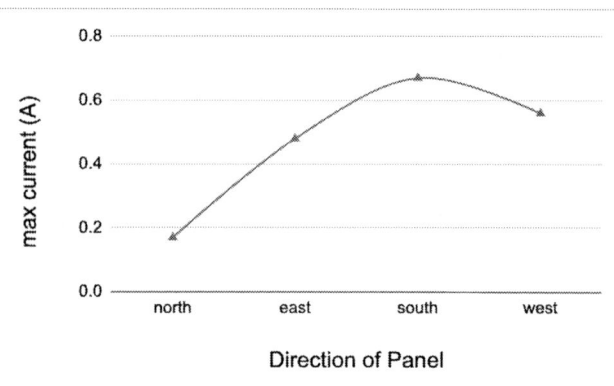

Figure 6 Maximum current vs Direction of Panel

After finding the optical direction, changed from angle from this data from 0°-90° in steps of 10°. Figure 7 shows the power max vs. angle of the panel.

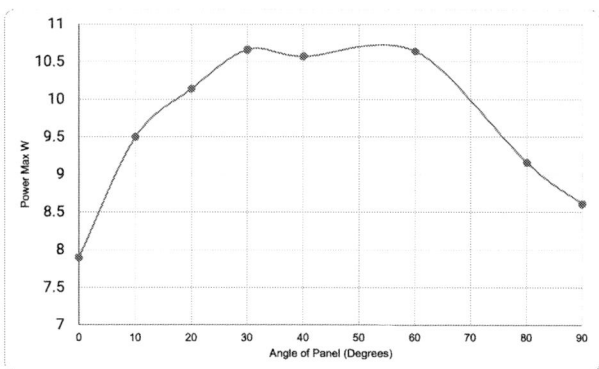

Figure 7: Power max vs angle of the panel

The maximum is around 30 degrees and remains somewhat stable between 30-60.

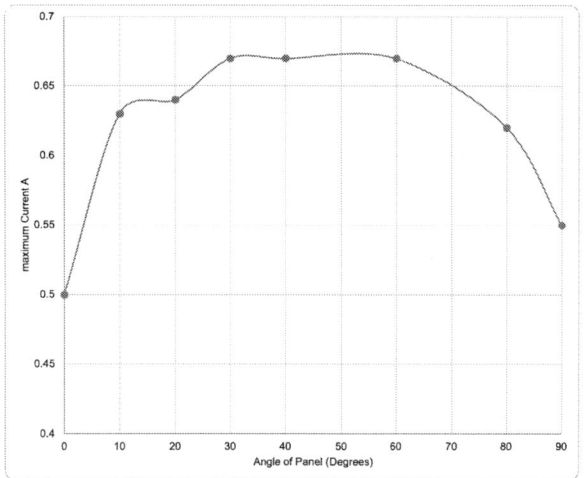

Figure 7 Maximum current vs angle of the panel.

The maximum current also peaks around 30 and stabilizes between 30-60 degrees around 0.7A. The drops between 60-90 degrees.

IV. CONCLUSION

In summary, the use for solar panels is key to help prevent the negative effect of climate change. In desert like conditions, while there is plenty of sunshine, the negative effects of soling and temperature is not fully studied yet. Finding the optimum, angle and direction is key it get the peak performance of the panel and helping to mitigate some of these negative effects. The results show here that the angle and direction can play a key role in the performance of the solar cell. Finally, knowing the negative effects can help researchers find new materials and solutions for solar cells for desert like environments.

REFERENCES

[1] E. Agency, "Renewables Information," July 2021. [Online]. Available: https://www.iea.org/data-and-statistics/data-product/renewables-information.

[2] T. National, "Road to Net Zero," 09 May 2019. [Online]. Available: https://www.thenationalnews.com/business/road-to-net-zero/2022/05/05/how-mohammed-bin-rashid-solar-park-plays-a-key-role-in-dubais-net-zero-strategy/.

[3] O. Albadwawi *et al.*, "Quantification of spectral losses of Natural soiling and detailed microstructural analysis of Dust collected from Different locations in Dubai, UAE," *2019 IEEE 46th Photovoltaic Specialists Conference (PVSC)*, 2019, pp. 3115-3118, doi: 10.1109/PVSC40753.2019.8980937.

Per-unit Dynamic Models for Grid-following Photovoltaic Inverters

Hyeonjung(Tari) Jung, D. Venkatramanan, Manish K. Singh, and Sairaj Dhople
Department of Electrical and Computer Engineering
University of Minnesota, Minneapolis, MN, 55455
{jungx367, dvenkat, msingh, sdhople}@umn.edu

Abstract—Per-unit models are frequently utilized in power-system analysis. The common approach that is followed is to outline base values for pertinent parameters and signals, and then normalize them to obtain dimensionless and per-unit models. While per-unit dynamic models for synchronous generators are widely documented, those for inverter-based resources are not as common. In this paper, we put forth a per-unit dynamic model for a standard grid-following photovoltaic inverter model. The topology we examine is a single-stage dc-ac voltage-source inverter interfaced to the grid via an *LCL* filter; the dc side includes the PV array interfaced to the inverter with a dc-link capacitor. The control architecture includes: a maximum power point tracking (MPPT) algorithm, a controller for regulating the dc-side voltage, a controller to regulate reactive power, an inner-loop current controller, and a standard synchronous-reference-frame phase locked loop for grid synchronization. We anticipate the proposed per-unit dynamic models being useful across scales (residential, commercial, and utility-level) and applications (modeling, time-domain simulations, control design, validation) to a wide range of stakeholders (researchers and practicing engineers).

Index Terms—Grid-following inverter, Maximum power point tracking, per-unit, photovoltaic inverter, three-phase inverter.

I. INTRODUCTION

Photovoltaic (PV) energy-conversion systems are deployed across end-use applications (residential, commercial, utility) and network types (distribution, sub-transmission, transmission) with the aid of different types of inverters (grid-following and grid-forming) [1]. The models of PV systems feature complex dynamics with interacting control- and physical-layer subsystems. The wide range of applications, network types, and inverter types induces combinatorial complexity to modeling the PV systems.

Modeling power networks at the system and component levels frequently involves the so-called per-unit analysis, whereby parameters and signals are rendered dimensionless and normalized via scaling with pertinent base values in SI units [2]. Invocations to per-unit analysis in power-systems analysis trace back significantly in time [3]. A particularly appealing aspect is that a principled choice of voltage base values picked in proportion to transformer turns ratios renders the transformers invisible in network-scale models. This reduces the need for bookkeeping different voltage and current levels that are typical across transmission and distribution networks. However, the conventional procedure, i.e., scaling all parameters and signals by base values, has several disadvantages: i) it requires an exhaustive base determination for all variables and parameters involved, ii)

there is no systematic procedure for determining control-loop-related bases, iii) it may involve redundant definitions for base values. Furthermore, the conventional approach appears to be more suited for network-level normalization in steady state. That said, per-unit models (including dynamics) for synchronous generators are widely utilized [4], [5]. With the proliferation of power-electronics circuits in power networks, there is a need to develop modeling frameworks to normalize them while acknowledging unique aspects that may challenge analysis. In this spirit, in [6] we offer an approach to per-unit analysis via state-space transformation. The basic idea is to transform the dynamic model of the system from SI units to one where states and inputs are normalized. A transformation matrix encodes base values for the states and inputs. Notably, parameters are not explicitly normalized; they emerge from the process involved to normalize the states and inputs. Specifically, the per-unit values of parameters can be inferred from the scalings that result from the state-space transformation. Counter to the limitations listed above for the conventional process, we note that the proposed approach: i) leads to direct normalization of system models (rather than parameters and variables individually) ii) is structure-preserving in that the model structure in code-based and block-diagram simulation environments is preserved, and iii) is rigorous, repeatable, constructive, and unified across timescales since it is grounded on system theory.

In this paper, we will apply an instance of the state-space transformation approach discussed above to obtain normalized dynamic models for grid-following (GFL) PV inverters. While a variety of inverter topologies and control strategies have been proposed in the literature, we focus our investigation on a GFL inverter with a synchronous reference frame phase-locked loop (PLL), PI-based current controllers, PV front end, and associated controllers. For the PV array, we consider a single-diode model. The inverter topology is assumed to be an H-bridge voltage-source converter with an inductive *LCL* filter on the AC side. The above cover salient features of a wide range of implementations; specific instances based on the application can be readily substituted. The baseline model is adopted from our previous effort [7] that looked into aggregation of parallel-connected PV inverters and [8] that dealt with model-aggregation of inverter dynamics in general.

Related prior work includes several isolated instances where per-unit dynamic models for power electronics are referenced, e.g., [9]–[11]. However, there is limited consen-

978-1-6654-6060-6/23 $31.00 © 2023 IEEE

sus on how to obtain per-unit models in a manner that is repeatable and standardized. While not disqualifying applications in which per-unit models are utilized, the lack of standardization—an aspect we attempt to remedy—does sow the seeds of confusion when different per-unit models are compared and parameterized; or when results from studies employing these models are compared.

The remainder of this paper is organized as follows. In Section II we overview the general procedure adopted to normalize the various dynamic and algebraic subsystems in the model. A set of rules that can be followed in general are listed to aid researchers and practitioners in normalizing other systems. In Section III we go through the individual subsystems of the PV GFL inverter and apply the procedure to normalize the system. In each subsystem, we briefly discuss the originating SI model and the per-unit counterpart. Section V concludes the paper and lists some avenues for future work.

Notation. Quantities in dq reference frame are denoted by superscripts d, q, and $X^{\mathrm{dq}} := [X^{\mathrm{d}} \, X^{\mathrm{q}}]^{\top}$. The rotation matrix is denoted by

$$\Gamma = \begin{bmatrix} 0 & 1 \\ -1 & 0 \end{bmatrix}.$$

SI-unit state variables and inputs are denoted by uppercase letters/symbols, say X, and their per-unit equivalents are denoted by lower-case letters/symbols, for instance x. While parameters are not explicitly normalized, upper-case symbols are generally utilized to represent SI-valued quantities in the originating models.

II. Summary of Technical Approach

In this section, we briefly outline our approach. The models we will contend with include both differential equations and algebraic equations. The governing differential equations for a wide suite of energy conversion devices including inverter-based resources abide by a control-affine form [6]. Consider a general nonlinear control-affine differential-equation model with all quantities represented in SI units:

$$\frac{\mathrm{d}X}{\mathrm{d}t} = F(X) + G(X)U, \tag{1}$$

where X and U are state and input, and $F(\cdot), G(\cdot)$ encode the functional dependence of the evolution of the state, and pertinent system parameters. Further, consider a general nonlinear algebraic dependence between the state and input:

$$H(X, U) = 0. \tag{2}$$

We follow a set of rules to normalize the differential and algebraic equations:

($\mathcal{R}1$) In the governing dynamic and algebraic equations, we substitute for each SI-valued state and/or input, X, by $X_{\mathrm{b}}x$, where X_{b} denotes the corresponding base value.

($\mathcal{R}2$) We do not normalize any constant and/or parameter explicitly. Reformulations of these are pursued as appropriate to preserve structure of the governing dynamic and algebraic equations.

($\mathcal{R}3$) Base values for controller-related state variables of the same type of controllers are—to the extent possible—maintained to be the same.

We next exemplify an abstract application of the aforementioned rules to normalize (1)–(2). Select the base value of states and inputs as X_{b} and U_{b}, respectively. Applying the transformation $X = X_{\mathrm{b}}x$ and $U = U_{\mathrm{b}}u$, we get the normalized dynamic model:

$$\frac{\mathrm{d}x}{\mathrm{d}t} = f(x) + g(x)u, \tag{3}$$

$$h(x, u) = 0 \tag{4}$$

where $f(x) = \frac{F(X_{\mathrm{b}}x)}{X_{\mathrm{b}}}$, $g(x) = \frac{G(X_{\mathrm{b}}x)U_{\mathrm{b}}}{X_{\mathrm{b}}}$, and $h(x, u) = H(X_{\mathrm{b}}x, U_{\mathrm{b}}u)$.

III. SI and Per-Unit Models

In this section, we overview all the constitutent elements of the GFL inverter model including physical- and control-layer subsystems. In each subsection, we first discuss the SI-unit model and present the per-unit counterpart following the rules outlined in ($\mathcal{R}1$)–($\mathcal{R}3$). A block diagram of the inverter model capturing both SI and per-unit representations is illustrated in Fig. 1.

A. Phase-locked Loop (PLL)

SI Model: Denote the PLL frequency and angle by Ω and Θ, such that

$$\Omega = \Omega_{\mathrm{s}} + \frac{\mathrm{d}\Theta}{\mathrm{d}t}, \tag{5}$$

where $\Omega_{\mathrm{s}} = 2\pi \times 60 \, \mathrm{rad/s}$ is the nominal synchronous frequency. In the adopted convention, the PLL observes the low-pass filtered d-axis grid voltage as

$$\frac{\mathrm{d}V_{\mathrm{pll}}}{\mathrm{d}t} = \Omega_{\mathrm{pll}}^{\mathrm{c}}(V_{\mathrm{g}}^{\mathrm{d}} - V_{\mathrm{pll}}), \tag{6}$$

where, V_{pll} is the PLL voltage, $V_{\mathrm{g}}^{\mathrm{d}}$ is the d-axis grid voltage, and $\Omega_{\mathrm{pll}}^{\mathrm{c}}$ is the filter cut-off frequency. The PLL dynamics ensure that $V_{\mathrm{g}}^{\mathrm{d}} \to 0$. The PLL internal state, denoted as Φ_{pll}, evolves as per a PI controller (with gains $K_{\mathrm{pll}}^{\mathrm{p}}$ and $K_{\mathrm{pll}}^{\mathrm{i}}$) as

$$\frac{\mathrm{d}\Phi_{\mathrm{pll}}}{\mathrm{d}t} = -V_{\mathrm{pll}}, \quad \frac{\mathrm{d}\Theta}{\mathrm{d}t} = K_{\mathrm{pll}}^{\mathrm{p}} \frac{\mathrm{d}\Phi_{\mathrm{pll}}}{\mathrm{d}t} + K_{\mathrm{pll}}^{\mathrm{i}} \Phi_{\mathrm{pll}}. \tag{7}$$

Per-unit Model: Denote the per-unit values of the PLL frequency, PLL angle, PLL voltage, d-axis grid voltage, and PLL internal state by ω, θ, v_{pll}, $v_{\mathrm{g}}^{\mathrm{d}}$, and φ_{pll}, respectively. Denote the base frequency by Ω_{b}, the base values for θ and φ_{pll} by Θ_{b} and Φ_{b}, respectively. Following ($\mathcal{R}1$), the PLL frequency in per unit is given by

$$\omega = \frac{\Omega_{\mathrm{s}}}{\Omega_{\mathrm{b}}} + \frac{\Theta_{\mathrm{b}}}{\Omega_{\mathrm{b}}} \frac{\mathrm{d}\theta}{\mathrm{d}t}. \tag{8}$$

Furthermore, the dynamics of v_{pll}, φ_{pll}, and θ are given by:

$$\frac{\mathrm{d}v_{\mathrm{pll}}}{\mathrm{d}t} = \Omega_{\mathrm{pll}}^{\mathrm{c}}(v_{\mathrm{g}}^{\mathrm{d}} - v_{\mathrm{pll}}), \tag{9}$$

$$\frac{\mathrm{d}\varphi_{\mathrm{pll}}}{\mathrm{d}t} = -\frac{V_{\mathrm{b}}}{\Phi_{\mathrm{b}}} v_{\mathrm{pll}}, \tag{10}$$

978-1-6654-6060-6/23 \$31.00 © 2023 IEEE

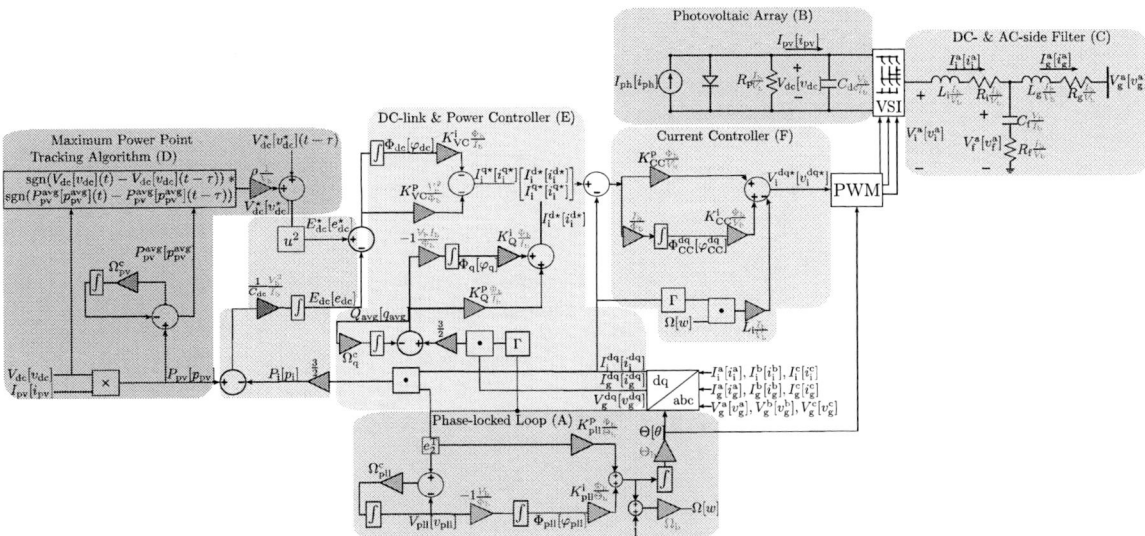

Fig. 1: Block diagram showing the overall PV system comprising of a grid-following (GFL) inverter with PV front end and output LCL filter. Both SI and per-unit models are depicted. Scaling factors of the parameters intended for the per-unit counterpart are highlighted in a shaded format; furthermore, states and inputs for the per-unit counterpart are enclosed in square brackets. A single-diode model is adopted for the PV array. For brevity, only a single leg of the output LCL filter is depicted. Variable superscripts denote the reference frames that they are represented in. Note that the process of normalization preserves the structure of the originating SI-unit model. Furthermore, parameters in the model are not explicitly parameterized.

$$\frac{d\theta}{dt} = \frac{\Phi_b}{\Theta_b} K^p_{pll} \frac{d\varphi_{pll}}{dt} + \frac{\Phi_b}{\Theta_b} K^i_{pll} \varphi_{pll}. \quad (11)$$

Note that, in line with ($\mathcal{R}2$), we do not normalize any of the parameters (i.e., Ω^c_{pll}, K^p_{pll}, and K^i_{pll}). We also take care to distinguish between the synchronous electrical frequency, Ω_s, which is tied to the prevailing electrical-grid operation and the base frequency, Ω_b, which is a modeling construct. While it is indeed prudent to set $\Omega_b = \Omega_s$ for the purpose of numerical simulations and analysis, there is no governing need to do so. In any case, substituting $\Omega_b = \Omega_s$ and the dynamics of θ from (11) in (8) yields the following expression for the PLL frequency:

$$\omega = 1 + \frac{\Phi_b}{\Omega_b} K^p_{pll} \frac{d\varphi_{pll}}{dt} + \frac{\Phi_b}{\Omega_b} K^i_{pll} \varphi_{pll}. \quad (12)$$

A pertinent question to pose is with regard to parameterization. Indeed, base values for voltage (and other physical-domain variables) follow from device and system ratings. On the other hand, base values of control-related states (e.g., Φ_b) can be set to unity without any loss of generality or compromising the overall modeling goal of normalizing system dynamics.

B. Photovoltaic Array

SI Model: The PV array is modelled with a single-diode model with parallel output resistance, R_p. The array constitutes of N_p parallel strings, each comprising of N_s series-connected modules. Each module comprises of N_m series-connected cells. Denote the irradiance-dependent photocurrent by I_{ph}, the reverse saturation current by I_{rs}, the dc-bus voltage by V_{dc}, and the module surface temperature by T.

Parameters in the model include the electron charge, q and the Boltzmann constant, K. The output current of the PV array, I_{pv}, is given by the algebraic equation: [12], [13]

$$I_{pv} = N_p \left(I_{ph} - I_{rs} \left(e^{\frac{q V_{dc}}{N_m K T N_s}} - 1 \right) - \frac{V_{dc}}{N_s R_p} \right). \quad (13)$$

Per-unit Model: Denote the per-unit values of the PV current, photocurrent, reverse saturation current, and dc-bus voltage by i_{ph}, i_{ph}, i_{rs}, and v_{dc}, respectively. The base values for current and voltage on the dc-side of the inverter are denoted by I_b and V_b, respectively. Following ($\mathcal{R}1$), we obtain from (13)

$$I_b i_{pv} = N_p \left(I_b i_{ph} - I_b i_{rs} \left(e^{\frac{q V_b v_{dc}}{N_m K T N_s}} - 1 \right) - \frac{V_b v_{dc}}{N_s R_p} \right). \quad (14)$$

Simplifying (14), we get the per-unit algebraic-equation model for the PV current:

$$i_{pv} = N_p \left(i_{ph} - i_{rs} \left(e^{\frac{q v_{dc}}{N_m k' T N_s}} - 1 \right) - \frac{v_{dc}}{N_s r_p} \right), \quad (15)$$

where $k' = \frac{K}{V_b}$ and $r_p = \frac{R_p I_b}{V_b}$ are parameters in the per-unit model.

C. DC- and AC-side Filters

SI Model: The DC-side filter is simply modeled as a capacitor C_{DC}. To model the dynamics of the filter, we focus on the (scaled) energy stored in this capacitor, $E_{dc} := V_{DC}^2$. This is governed by the power supplied by the PV source

$$P_{pv} = V_{dc} \cdot I_{pv} \quad (16)$$

and the inverter power output

$$P_{\mathrm{i}} = \frac{3}{2}(V_i^{\mathrm{d}} I_{\mathrm{i}}^{\mathrm{d}} + V_i^{\mathrm{q}} I_{\mathrm{i}}^{\mathrm{q}}). \tag{17}$$

Particularly, we have the dynamics:

$$\begin{aligned} \frac{\mathrm{d}E_{\mathrm{dc}}}{\mathrm{d}t} &= \frac{1}{C_{\mathrm{dc}}}(P_{\mathrm{pv}} - P_{\mathrm{i}}) \\ &= \frac{1}{C_{\mathrm{dc}}}\left(V_{\mathrm{dc}} I_{\mathrm{pv}} - \frac{3}{2}(V_i^{\mathrm{d}} I_{\mathrm{i}}^{\mathrm{d}} + V_i^{\mathrm{q}} I_{\mathrm{i}}^{\mathrm{q}})\right). \end{aligned} \tag{18}$$

On the AC-side, an *LCL* filter with the parameters shown in Fig. 1 is used. The states of this filter are the inverter terminal currents ($I_{\mathrm{i}}^{\mathrm{dq}}$), filter voltages ($V_{\mathrm{f}}^{\mathrm{dq}}$), and the grid-side currents ($I_{\mathrm{g}}^{\mathrm{dq}}$). The governing dynamics is given by

$$\frac{\mathrm{d}I_{\mathrm{i}}^{\mathrm{dq}}}{\mathrm{d}t} = \frac{1}{L_{\mathrm{i}}}(-R_{\mathrm{i}} I_{\mathrm{i}}^{\mathrm{dq}} + V_{\mathrm{i}}^{\mathrm{dq}} - V_{\mathrm{f}}^{\mathrm{dq}}) + \Omega\Gamma I_{\mathrm{i}}^{\mathrm{dq}}, \tag{19}$$

$$\frac{\mathrm{d}I_{\mathrm{g}}^{\mathrm{dq}}}{\mathrm{d}t} = \frac{1}{L_{\mathrm{g}}}(-R_{\mathrm{g}} I_{\mathrm{g}}^{\mathrm{dq}} + V_{\mathrm{f}}^{\mathrm{dq}} - V_{\mathrm{g}}^{\mathrm{dq}}) + \Omega\Gamma I_{\mathrm{g}}^{\mathrm{dq}}, \tag{20}$$

$$\begin{aligned} \frac{\mathrm{d}V_{\mathrm{f}}^{\mathrm{dq}}}{\mathrm{d}t} &= R_{\mathrm{f}}\Big(\frac{\mathrm{d}I_{\mathrm{i}}^{\mathrm{dq}}}{\mathrm{d}t} - \frac{\mathrm{d}I_{\mathrm{g}}^{\mathrm{dq}}}{\mathrm{d}t}\Big) - \Omega\Gamma R_{\mathrm{f}}(I_{\mathrm{i}}^{\mathrm{dq}} - I_{\mathrm{g}}^{\mathrm{dq}}) \\ &\quad + \frac{1}{C_{\mathrm{f}}}(I_{\mathrm{i}}^{\mathrm{d}} - I_{\mathrm{g}}^{\mathrm{dq}}) + \Omega\Gamma V_{\mathrm{f}}^{\mathrm{dq}}. \end{aligned} \tag{21}$$

Per-unit Model: Denote the per-unit (scaled) energy, inverter-side current, inverter-side voltage, capacitor voltage, and grid-side current by e_{DC}, $i_{\mathrm{i}}^{\mathrm{dq}}$, $v_{\mathrm{i}}^{\mathrm{dq}}$, $v_{\mathrm{f}}^{\mathrm{dq}}$, and $i_{\mathrm{g}}^{\mathrm{dq}}$, respectively. Elementary dimensional analysis reveals that the base value for E_{dc} ought to be V_{b}^2. Applying ($\mathcal{R}1$), we obtain the following per-unit dynamical equivalent to (18)–(21):

$$\frac{\mathrm{d}e_{\mathrm{dc}}}{\mathrm{d}t} = \frac{1}{c_{\mathrm{dc}}}\left(v_{\mathrm{dc}} i_{\mathrm{pv}} - \frac{3}{2}(v_i^{\mathrm{d}} i_{\mathrm{i}}^{\mathrm{d}} + v_i^{\mathrm{q}} i_{\mathrm{i}}^{\mathrm{q}})\right), \tag{22}$$

$$\frac{\mathrm{d}i_{\mathrm{i}}^{\mathrm{dq}}}{\mathrm{d}t} = \frac{1}{\ell_{\mathrm{i}}}(-r_{\mathrm{i}} i_{\mathrm{i}}^{\mathrm{dq}} + v_{\mathrm{i}}^{\mathrm{dq}} - v_{\mathrm{f}}^{\mathrm{dq}}) + \frac{I_{\mathrm{b}}\Omega_{\mathrm{b}}}{V_{\mathrm{b}}}\omega\Gamma i_{\mathrm{i}}^{\mathrm{dq}}, \tag{23}$$

$$\frac{\mathrm{d}i_{\mathrm{g}}^{\mathrm{dq}}}{\mathrm{d}t} = \frac{1}{\ell_{\mathrm{g}}}(-r_{\mathrm{g}} i_{\mathrm{g}}^{\mathrm{dq}} + v_{\mathrm{f}}^{\mathrm{dq}} - v_{\mathrm{g}}^{\mathrm{dq}}) + \frac{I_{\mathrm{b}}\Omega_{\mathrm{b}}}{V_{\mathrm{b}}}\omega\Gamma i_{\mathrm{g}}^{\mathrm{dq}}, \tag{24}$$

$$\begin{aligned} \frac{\mathrm{d}v_{\mathrm{f}}^{\mathrm{dq}}}{\mathrm{d}t} &= r_{\mathrm{f}}\left(\frac{\mathrm{d}i_{\mathrm{i}}^{\mathrm{dq}}}{\mathrm{d}t} - \frac{\mathrm{d}i_{\mathrm{g}}^{\mathrm{dq}}}{\mathrm{d}t}\right) - \Omega_{\mathrm{b}} r_{\mathrm{f}}\omega\Gamma(i_{\mathrm{i}}^{\mathrm{dq}} - i_{\mathrm{g}}^{\mathrm{dq}}) \\ &\quad + \frac{1}{c_{\mathrm{f}}}(i_{\mathrm{i}}^{\mathrm{dq}} - i_{\mathrm{g}}^{\mathrm{dq}}) + \Omega_{\mathrm{b}}\omega\Gamma v_{\mathrm{f}}^{\mathrm{dq}}, \end{aligned} \tag{25}$$

where $c_{\mathrm{dc}} = C_{\mathrm{dc}}\frac{V_{\mathrm{b}}}{I_{\mathrm{b}}}$, $\ell_{\mathrm{i}} = L_{\mathrm{i}}\frac{I_{\mathrm{b}}}{V_{\mathrm{b}}}$, $r_{\mathrm{i}} = R_{\mathrm{i}}\frac{I_{\mathrm{b}}}{V_{\mathrm{b}}}$, $\ell_{\mathrm{g}} = L_{\mathrm{g}}\frac{I_{\mathrm{b}}}{V_{\mathrm{b}}}$, $r_{\mathrm{g}} = R_{\mathrm{g}}\frac{I_{\mathrm{b}}}{V_{\mathrm{b}}}$, $c_{\mathrm{f}} = C_{\mathrm{f}}\frac{V_{\mathrm{b}}}{I_{\mathrm{b}}}$, $r_{\mathrm{f}} = R_{\mathrm{f}}\frac{I_{\mathrm{b}}}{V_{\mathrm{b}}}$. Following ($\mathcal{R}2$), we do not normalize parameters explicitly. A side effect of that is that there is no guarantee on parameters in the per-unit description being dimensionless.

D. Maximum Power Point Tracking (MPPT) Algorithm

SI Model: To maximally extract power from a PV source, MPPT algorithms determine reference values $V_{\mathrm{dc}}^{\star}$ for the DC-link voltage. Among various approaches for MPPT, in this work we adopt the widely used perturb and observe (P&O) algorithm, which incrementally adjusts $V_{\mathrm{dc}}^{\star}$ based on the increase/decrease in average values of P_{pv} observed from

previous actions. The average PV output power is computed via a low-pass filter as

$$\frac{\mathrm{d}P_{\mathrm{pv}}^{\mathrm{avg}}}{\mathrm{d}t} = \Omega_{\mathrm{pv}}^{\mathrm{c}}(P_{\mathrm{pv}} - P_{\mathrm{pv}}^{\mathrm{avg}}), \tag{26}$$

where $\Omega_{\mathrm{pv}}^{\mathrm{c}}$ is the filter cut-off frequency. The references $V_{\mathrm{dc}}^{\star}$ are updated at time interval, τ, as

$$V_{\mathrm{dc}}^{\star}(t+\tau) = V_{\mathrm{dc}}^{\star}(t) + \rho\,\mathrm{sgn}(\Delta V_{\mathrm{dc}}(t))\mathrm{sgn}(\Delta P_{\mathrm{pv}}^{\mathrm{avg}}(t)), \tag{27}$$

where ρ denotes the perturbation size, $\mathrm{sgn}(\cdot)$ denotes the signum function, and

$$\Delta V_{\mathrm{dc}}(t) = V_{\mathrm{dc}}(t) - V_{\mathrm{dc}}(t-\tau), \tag{28}$$

$$\Delta P_{\mathrm{pv}}^{\mathrm{avg}}(t) = P_{\mathrm{pv}}^{\mathrm{avg}}(t) - P_{\mathrm{pv}}^{\mathrm{avg}}(t-\tau). \tag{29}$$

Per-unit Model: Denote the per-unit values of the average PV power and the MPPT reference by p_{pv} and $v_{\mathrm{dc}}^{\star}$, respectively. The dynamics of p_{pv} and the update rule for $v_{\mathrm{dc}}^{\star}$ are given by:

$$\frac{\mathrm{d}p_{\mathrm{pv}}^{\mathrm{avg}}}{\mathrm{d}t} = \Omega_{\mathrm{pv}}^{\mathrm{c}}(p_{\mathrm{pv}} - p_{\mathrm{pv}}^{\mathrm{avg}}), \tag{30}$$

$$v_{\mathrm{dc}}^{\star}(t+\tau) = v_{\mathrm{dc}}^{\star}(t) + \rho'\,\mathrm{sgn}(\Delta v_{\mathrm{dc}}(t))\mathrm{sgn}(\Delta p_{\mathrm{pv}}^{\mathrm{avg}}(t)), \tag{31}$$

where $\rho' = \rho V_{\mathrm{b}}^{-1}$, and

$$\Delta v_{\mathrm{dc}}(t) = v_{\mathrm{dc}}(t) - v_{\mathrm{dc}}(t-\tau), \tag{32}$$

$$\Delta p_{\mathrm{pv}}^{\mathrm{avg}}(t) = p_{\mathrm{pv}}^{\mathrm{avg}}(t) - p_{\mathrm{pv}}^{\mathrm{avg}}(t-\tau). \tag{33}$$

E. DC-link & Power Controller

SI Model: The DC-link controller aims at steering the voltage V_{dc} to the reference $V_{\mathrm{dc}}^{\star}$ provided by the MPPT algorithm. It is composed of a PI controller, with PI gains ($K_{\mathrm{VC}}^{\mathrm{P}}$, $K_{\mathrm{VC}}^{\mathrm{i}}$), and controller state Φ_{dc}. The controller is implemented to treat the DC-link energy setpoint, $E_{\mathrm{dc}}^{\star} = (V_{\mathrm{dc}}^{\star})^2$, as the reference input. The output sets the reference for the quadrature component of inverter current, $I_{\mathrm{i}}^{\mathrm{q}}$. The controller dynamics is given by:

$$\frac{\mathrm{d}\Phi_{\mathrm{dc}}}{\mathrm{d}t} = E_{\mathrm{dc}}^{\star} - E_{\mathrm{dc}}, \tag{34}$$

$$I_{\mathrm{i}}^{\mathrm{q}\star} = -K_{\mathrm{VC}}^{\mathrm{p}}\frac{\mathrm{d}\Phi_{\mathrm{dc}}}{\mathrm{d}t} - K_{\mathrm{VC}}^{\mathrm{i}}\Phi_{\mathrm{dc}}. \tag{35}$$

The reference for $I_{\mathrm{i}}^{\mathrm{d}}$ needs to be computed as per the reactive power requirement from the inverter. For the considered setting of unity-power-factor operation, we aim at driving the (average) reactive power output to zero. The average is computed via a low pass filter with cut-off frequency $\Omega_{\mathrm{Q}}^{\mathrm{c}}$ as

$$\frac{\mathrm{d}Q_{\mathrm{avg}}}{\mathrm{d}t} = \Omega_{\mathrm{Q}}^{\mathrm{c}}(Q - Q_{\mathrm{avg}}), \tag{36}$$

where the instantaneous reactive power is measured as

$$Q = \frac{3}{2}(V_{\mathrm{g}}^{\mathrm{q}} I_{\mathrm{g}}^{\mathrm{d}} - V_{\mathrm{g}}^{\mathrm{d}} I_{\mathrm{g}}^{\mathrm{q}}). \tag{37}$$

A PI controller with gains ($K_{\mathrm{Q}}^{\mathrm{p}}$, $K_{\mathrm{Q}}^{\mathrm{i}}$) and state Φ_{Q} drives Q_{avg} to zero via the dynamics:

$$\frac{\mathrm{d}\Phi_{\mathrm{Q}}}{\mathrm{d}t} = -Q_{\mathrm{avg}} \tag{38}$$

$$I_{\mathrm{i}}^{\mathrm{d}\star} = K_{\mathrm{Q}}^{\mathrm{p}}\frac{\mathrm{d}\Phi_{\mathrm{Q}}}{\mathrm{d}t} + K_{\mathrm{Q}}^{\mathrm{i}}\Phi_{\mathrm{Q}}. \tag{39}$$

978-1-6654-6060-6/23 $31.00 © 2023 IEEE

Per-unit Model: Denote the per-unit values of the dc-link-controller state variable and the reactive-power-controller state variable by φ_{dc} and φ_{Q}, respectively. Furthermore, denote the per-unit value of $E_{\mathrm{dc}}^{\star}$ by $e_{\mathrm{dc}}^{\star}$. In line with ($\mathcal{R}3$), we adopt Φ_{b} as the base value for both controller states (recall that Φ_{b} was also chosen to be the base value for the PLL control state, Φ_{pll}). We can express the per-unit counterparts to (34)–(35), and (38)–(39) by:

$$\frac{d\varphi_{\mathrm{dc}}}{dt} = \frac{V_{\mathrm{b}}^2}{\Phi_{\mathrm{b}}}(e_{\mathrm{dc}}^{\star} - e_{\mathrm{dc}}), \tag{40}$$

$$i_{\mathrm{i}}^{\mathrm{q}\star} = -\frac{\Phi_{\mathrm{b}}}{I_{\mathrm{b}}}K_{\mathrm{VC}}^{\mathrm{p}}\frac{d\varphi_{\mathrm{dc}}}{dt} - \frac{\Phi_{\mathrm{b}}}{I_{\mathrm{b}}}K_{\mathrm{VC}}^{\mathrm{i}}\varphi_{\mathrm{dc}}, \tag{41}$$

$$\frac{d\varphi_{\mathrm{Q}}}{dt} = -\frac{V_{\mathrm{b}}I_{\mathrm{b}}}{\Phi_{\mathrm{b}}}q_{\mathrm{avg}}, \tag{42}$$

$$i_{\mathrm{i}}^{\mathrm{d}\star} = \frac{\Phi_{\mathrm{b}}}{I_{\mathrm{b}}}K_{\mathrm{Q}}^{\mathrm{p}}\frac{d\varphi_{\mathrm{Q}}}{dt} + \frac{\Phi_{\mathrm{b}}}{I_{\mathrm{b}}}K_{\mathrm{Q}}^{\mathrm{i}}\varphi_{\mathrm{Q}}. \tag{43}$$

Above, q_{avg} denotes the per-unit counterpart of Q_{avg}; translating (36)–(37), we obtain the following dynamics for it:

$$\frac{dq_{\mathrm{avg}}}{dt} = \Omega_{\mathrm{Q}}^{\mathrm{c}}\left(\frac{3}{2}(v_{\mathrm{g}}^{\mathrm{q}}i_{\mathrm{g}}^{\mathrm{d}} - v_{\mathrm{g}}^{\mathrm{d}}i_{\mathrm{g}}^{\mathrm{q}}) - q_{\mathrm{avg}}\right). \tag{44}$$

F. Current Controller

SI Model: The inverter current references ($I_{\mathrm{i}}^{\mathrm{d}\star}$, $I_{\mathrm{i}}^{\mathrm{q}\star}$) are inputs to current controllers that determine the references for inverter terminal voltages ($V_{\mathrm{i}}^{\mathrm{d}\star}$, $V_{\mathrm{i}}^{\mathrm{d}\star}$). The PWM modules and the inverter are assumed to instantaneously realize these voltage references ensuring $V_{\mathrm{i}}^{\mathrm{dq}} = V_{\mathrm{i}}^{\mathrm{dq}\star}$. The current controller consists of two PI controllers with gains ($K_{\mathrm{CC}}^{\mathrm{p}}$, $K_{\mathrm{CC}}^{\mathrm{i}}$) and states ($\Phi_{\mathrm{CC}}^{\mathrm{d}}$, $\Phi_{\mathrm{CC}}^{\mathrm{q}}$). The governing dynamics are:

$$\frac{d\Phi_{\mathrm{CC}}^{\mathrm{dq}}}{dt} = I_{\mathrm{i}}^{\mathrm{dq}\star} - I_{\mathrm{i}}^{\mathrm{dq}} \tag{45}$$

$$V_{\mathrm{i}}^{\mathrm{dq}\star} = K_{\mathrm{CC}}^{\mathrm{p}}\frac{d\Phi_{\mathrm{CC}}^{\mathrm{dq}}}{dt} + K_{\mathrm{CC}}^{\mathrm{i}}\Phi_{\mathrm{CC}}^{\mathrm{dq}} - \Omega L_{\mathrm{i}}\Gamma I_{\mathrm{i}}^{\mathrm{dq}}. \tag{46}$$

Per-unit Model: Denote the per-unit counterpart to the current-controller state variable by $\varphi_{\mathrm{CC}}^{\mathrm{dq}}$ and adopt base value Φ_{b} for it. Furthermore, denote the per-unit counterpart for the inverter reference voltage by $v_{\mathrm{i}}^{\mathrm{dq}\star}$. The per-unit dynamics corresponding to (45)–(46) are given by

$$\frac{d\varphi_{\mathrm{CC}}^{\mathrm{dq}}}{dt} = \frac{I_{\mathrm{b}}}{\Phi_{\mathrm{b}}}(i_{\mathrm{i}}^{\mathrm{dq}\star} - i_{\mathrm{i}}^{\mathrm{dq}}) \tag{47}$$

$$v_{\mathrm{i}}^{\mathrm{dq}\star} = \frac{\Phi_{\mathrm{b}}}{V_{\mathrm{b}}}K_{\mathrm{CC}}^{\mathrm{p}}\frac{d\varphi_{\mathrm{CC}}^{\mathrm{dq}}}{dt} + \frac{\Phi_{\mathrm{b}}}{V_{\mathrm{b}}}K_{\mathrm{CC}}^{\mathrm{i}}\varphi_{\mathrm{CC}}^{\mathrm{dq}} - \ell_{\mathrm{i}}\Omega_{\mathrm{b}}\omega\Gamma i_{\mathrm{i}}^{\mathrm{dq}}. \tag{48}$$

IV. SIMULATION RESULTS

In this section, we demonstrate our approach by simulating the SI and per-unit models of the PV system (Fig. 1) developed in Section III, and compare the results. Inverter parameters are based on the Schneider Electric Conext CL125 string inverter [14] (rated 125 kW, 600 V(rms)) while the parameters for PV modules are taken from the First Solar FS-6420 [15] PV module (rated 420 W with open-circuit voltage 218.5 V). The PV array comprised of $N_{\mathrm{p}} = 60$ parallel strings formed of $N_{\mathrm{s}} = 5$ series-connected PV modules. The number of cells per module was set to $N_{\mathrm{m}} = 72$. Filter and controller parameters and the base values are provided in Table I.

A 10 sec-long time-domain simulation was performed using ode23 solver in Matlab. Second-based solar irradiance data for this simulation was sourced from NREL's Oahu Solar Measurement Grid on July 24, 2011 (09:30:00–09:30:10 AM) [16]. The period and perturbation size of the MPPT algorithm were selected to be 0.005 seconds and 50mV, respectively. To obtain a smoother irradiance variation that is still tractable by the MPPT algorithm, the original second-based irradiance data was linearly interpolated to yield 0.1 sec granularity. Simulation time steps for SI- and per-unit models were identical.

Fig. 2 shows simulation results obtained for the SI and per-unit model. Both models yield nearly identical profiles for dc-link voltage, PV power, and grid current on the quadrature axis, with minor numerical deviations attributable to the solver's numerical tolerances.

V. CONCLUDING REMARKS & FUTURE WORK

This paper introduced a systematic approach to normalize the dynamics of grid-following photovoltaic inverters. A set of rules were outlined that can be followed to obtain per-unit dynamic models for systems described by differential-algebraic equations. As part of ongoing and future work, we are developing per-unit dynamical models for grid-forming inverters with a wide range of primary controllers.

ACKNOWLEDGEMENT

This work was supported in part by the U.S. Department of Energy (DOE) Office of Energy Efficiency and Renewable Energy through the award 38637 (UNIFI consortium).

REFERENCES

[1] D. Venkatramanan, M. K. Singh, O. Ajala, A. Domínguez-García, and S. Dhople, "Integrated system models for networks with generators & inverters," *arXiv preprint arXiv:2203.08253*, 2022.

[2] J. Glover, M. Sarma, and T. Overbye, *Power System Analysis and Design.* Boston, MA: Cengage Learning, 2011.

[3] I. Travis, "Per-unit quantities," *Trans. American Inst. Elect. Engineers*, vol. 56, no. 12, pp. 22–28, Nov 1937.

[4] P. Sauer and M. A. Pai, *Power System Dynamics and Stability.* Upper Saddle River, NJ: Prentice Hall, 1998.

TABLE I: CONTROLLER AND FILTER PARAMETERS

dc- and ac-side Filters		dc-link & Power Controllers		PLL	
L_{i}	23.1 μH	$K_{\mathrm{VC}}^{\mathrm{p}}$	0.0865 A/V^2	$K_{\mathrm{PLL}}^{\mathrm{p}}$	0.735 rad/V
R_{i}	0.016 Ω	$K_{\mathrm{VC}}^{\mathrm{i}}$	0.865 A/(V^{2}s)	$K_{\mathrm{PLL}}^{\mathrm{i}}$	5.88 rad/(V·s)
C_{f}	0.216 mF	$K_{\mathrm{Q}}^{\mathrm{p}}$	0.0059 V^{-1}	$\Omega_{\mathrm{PLL}}^{\mathrm{c}}$	7854 rad/s
R_{f}	0.462 mΩ	$K_{\mathrm{Q}}^{\mathrm{i}}$	0.059 (V·s)$^{-1}$	MPPT	
L_{g}	23.1 μH	$\Omega_{\mathrm{Q}}^{\mathrm{c}}$	50.26 rad/s	$\Omega_{\mathrm{PV}}^{\mathrm{c}}$	50.26 rad/s
R_{g}	2.77 mΩ				
C_{dc}	0.1 mF	Current Controller		Base Values	
		$K_{\mathrm{CC}}^{\mathrm{p}}$	0.6 V/A	I_{b}	80 A
		$K_{\mathrm{CC}}^{\mathrm{i}}$	35 V/(A·s)	V_{b}	490 V
				Φ_{b}	100
				Θ_{b}	1 rad
				Ω_{b}	377 rad/s

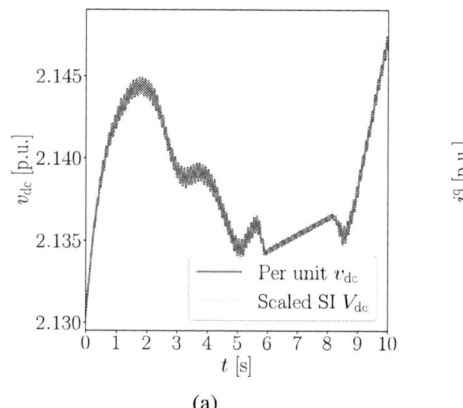 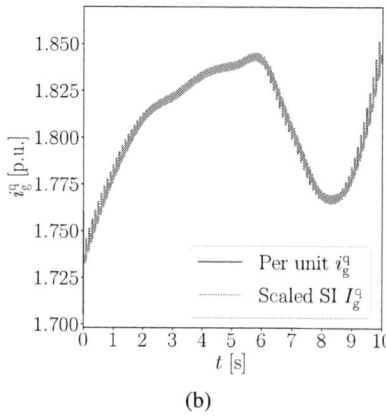 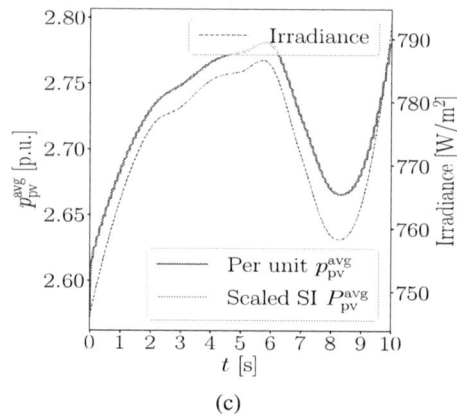

(a)	(b)	(c)

Fig. 2: Simulation results comparing SI and per-unit model of (a) dc-link voltage (b) grid current on the quadrature axis (c) PV power. SI values for all plots were scaled by its base values for comparison. The base values used for 2a, 2c, and 2b are $V_\mathrm{b} = 490$ V, $S_\mathrm{b} = 39200$ VA, and $I_\mathrm{b} = 80$ A, respectively.

[5] P. Kundur, *Power system stability and control*. McGraw-Hill, USA, 1994.

[6] D. Venkatramanan and S. Dhople, "Per-unit modeling via similarity transformation," *IEEE Trans. Energy Convers.*, pp. 1–13, Mar 2022, (early access).

[7] V. Purba, B. B. Johnson, and S. V. Dhople, "Reduced-order aggregate model for parallel-connected grid-tied three-phase photovoltaic inverters," in *IEEE Photovolt. Specialist Conf.*, Jun 2019, pp. 0724–0729.

[8] V. Purba, S. V. Dhople, S. Jafarpour, F. Bullo, and B. B. Johnson, "Reduced-order structure-preserving model for parallel-connected three-phase grid-tied inverters," in *2017 IEEE Workshop on Control and Modeling for Power Electron.*, Jul. 2017, pp. 1–7.

[9] S. Jafarpour, V. Purba, B. B. Johnson, S. V. Dhople, and F. Bullo, "Singular perturbation and small-signal stability for inverter networks," *IEEE Trans. Control of Network Systems*, vol. 9, no. 2, pp. 979–992, Jun 2022.

[10] O. Ajala, N. Baeckeland, B. Johnson, S. Dhople, and A. Domínguez-García, "Model reduction and dynamic aggregation of grid-forming inverter networks," *IEEE Trans. Power Syst.*, pp. 1–16, Dec 2022.

[11] O. Ajala, M. Lu, B. Johnson, S. V. Dhople, and A. Domínguez-García, "Model reduction for inverters with current limiting and dispatchable virtual oscillator control," *IEEE Trans. Energy Convers.*, vol. 37, no. 4, pp. 2250–2259, May 2022.

[12] M. G. Villalva, J. R. Gazoli, and E. R. Filho, "Comprehensive approach to modeling and simulation of photovoltaic arrays," *IEEE Trans. Power Electron.*, vol. 24, no. 5, pp. 1198–1208, Mar 2009.

[13] A. A. A. Radwan and Y. A. I. Mohamed, "Power synchronization control for grid-connected current-source inverter-based photovoltaic systems," *IEEE Trans. Energy Convers.*, vol. 31, no. 3, pp. 1023–1036, Mar 2016.

[14] "Conext cl125 string inverter," accessed: 2023-06-22. [Online]. Available: https://solar.se.com/us/en/product/conext-cl125-string-inverter/

[15] "First solar series 6™," accessed: 2023-06-22. [Online]. Available: https://www.firstsolar.com/-/media/First-Solar/Technical-Documents/Series-6-Datasheets/Series-6-Datasheet.ashx

[16] M. Sengupta and A. Andreas, "Oahu solar measurement grid (1-year archive): 1-second solar irradiance; Oahu, Hawaii (data)," National Renewable Energy Lab.(NREL), Golden, CO (United States), Tech. Rep., 2010.

978-1-6654-6060-6/23 $31.00 © 2023 IEEE

'There and Back again': Reusable Germanium Wafers with Ge-on-Nothing Structures for Triple-Junction Solar Cells

Valérie Depauw[1,2,3], Guillaume Courtois[4], Jinyoun Cho[4], Kristof Dessein[4], Clément Porret[1], Roger Loo[1]

[1] Imec, Leuven, Belgium | [2] University of Hasselt, imec imomec, Hasselt, Belgium,
[3] EnergyVille, Genk, Belgium | [4] Umicore, Electro Optic Materials, Olen, Belgium

Abstract — **Germanium-on-Nothing is proposed as a lift-off method to fabricate thin and lower-cost germanium templates for multijunction solar cells. This contribution presents how such foils, that are thickened epitaxially, can replace bulk wafers as III-V epitaxy seed, mechanical support, and bottom solar cell junction. Furthermore, by detaching large-area foils from reconditioned wafers, it demonstrates that the starting germanium wafer can be re-used and the recycling loop closed.**

I. INTRODUCTION

Germanium (Ge) is the major cost driver in lattice-matched multijunction solar cells [1]. It is also a scarce element, originating mostly from a single country, and is thus listed as a critical raw material [2]. Solutions to consume less, or none, are therefore of growing interest. Spalling and porous Ge are the two present solutions for thinner and kerfless Ge, both based on the lift-off of Ge films from a reusable wafer [1].

The present approach (Fig. 1) is a porous Ge approach where a regular array of macropores is formed by lithography and dry etching. With careful pattern engineering, these pores close and merge upon annealing, forming a suspended monocrystalline Ge membrane on top of one buried void, a so-called "Ge-on-nothing" (GeON) structure. This membrane can be thickened in-situ by homoepitaxy if required and, as the wafer miscut is preserved and no dislocations are induced, it can also be used as epitaxial template for III-V materials. This stack can then eventually be detached and transferred to another carrier for mechanical support. Afterwards, the Ge wafer is reconditioned and re-used to produce new foils. Contrarily to the electrochemical porous Ge route, this step results in a void- and hole-free membrane, on any wafer doping [3]. The morphology of the detachment layer can also be precisely controlled to tune the adhesion force between the wafer and the foil. All these properties have recently been demonstrated by detaching wafer-scale foils and small triple-junction solar cells [4]. Reaching a high detachment yield and a high device quality are in fact paramount to the added value of lift-off approaches. However, other factors are also essential but have not yet been addressed, namely the number of reuse cycles and the losses at wafer reconditioning.

In this contribution, we present updated technical highlights for this GeON approach and then focus on wafer reuse, demonstrating the completion of the recycling loop with a 19-cm wide foil detached from a reused substrate.

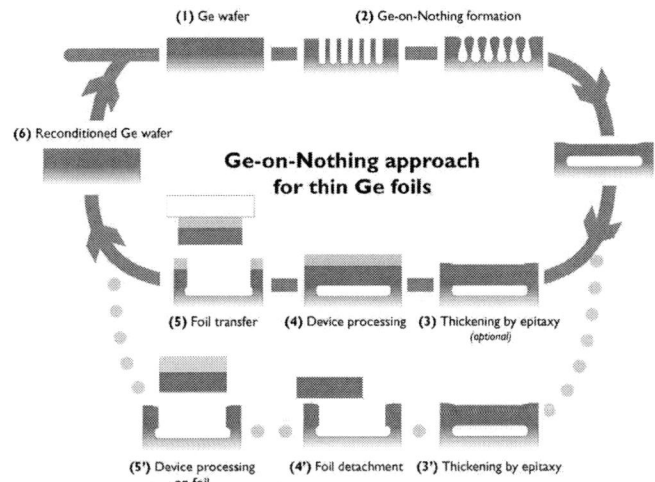

Fig. 1.   GeON approach, whereby a thin Ge foil is detached from a porosified wafer that is reused multiple times. Depending on the thickness of the epitaxial Ge layer, the detached foil may require mechanical support or be processed as a wafer equivalent.

II. FROM A BLANKET GE WAFER TO TRIPLE-JUNCTION SOLAR CELLS ON TUNABLE AND DETACHABLE FOILS

In triple-junction solar cells, Ge wafers fulfill three functions, namely mechanical support, bottom junction, and heteroepitaxy seed. As recently reported and as highlighted below, GeON foils can replace Ge wafers [4].

In this approach, mechanical support is provided by the bulk Ge wafer and, after detachment, by another carrier such as the glass cover (Fig. 1 steps 3-5). Alternatively, if the foil is thickened enough via Ge epitaxy, it may be processed free-standing (steps 3'-5'). In both cases, finding the suitable foil adhesion to withstand the detachment step without cracks, and the solar cell processes without lift-off, is essential. This can be achieved by introducing irregularities in the pore pattern inducing the creation of pillars that connect the foil to its parent (Fig. 2). Their density can be defined in function of the process requirements. Fig. 2 illustrates how foils with different pillar pitches (in a range 1-400 times denser) and strengths (full or hollow) are transferred, or not, to a Gel-Pak® tape.

978-1-6654-6060-6/23 $31.00 © 2023 IEEE

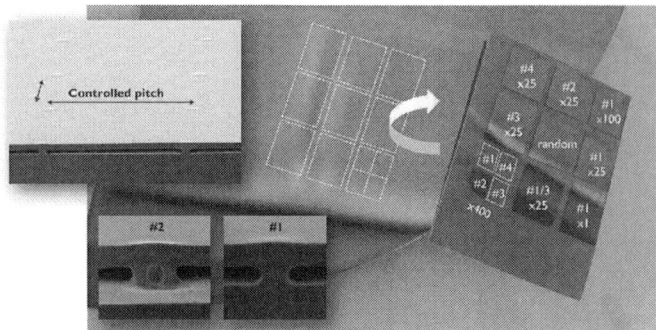

Fig. 2. Photograph of 12 GeON areas (9 mm × 9 mm) with 4 different pillar pitches and types (*e.g.*, SEM images of types #1 and #2 in lower insets), among which 9 were transferred to Gel-Pak®. The 3 areas with smallest pitch and largest pillar dimensions were too strongly attached.

The second function is that of bottom junction, which can be fulfilled by thickening and doping the membrane by homoepitaxy in-situ. By developing a dedicated atmospheric-pressure CVD process with $GeCl_4$ as precursor, growth rates up to ~190 nm/min have been achieved, with thicknesses from 1 up to 30 μm. Foils can currently be grown intrinsic (~ 1E14/cm³) or doped with boron up to 3E19/cm³.

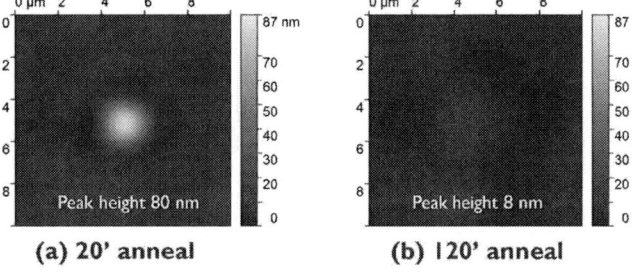

(a) 20' anneal   **(b) 120' anneal**

Fig. 3. 10 μm × 10 μm atomic force microscopy images plotted at the same scale in pairs, showing the reduction of (a, b) pillar bump height with increasing anneal time from 20 min to 2 h.

The last function of epitaxy seed for III-V materials is ensured by the smooth reflow of the porous surface, that preserves the crystalline quality of the parent wafer, including its miscut. This has been demonstrated by triple-junction InGaP/InGaAs/Ge solar cells grown by MOCVD on detachable 7-μm-thick GeON foils [4]. The GeON roughness is however about one order of magnitude larger than that of standard Cz Ge wafers, between 1 and 2 nm RMS. The foil surface presents a certain waviness and the additional pillars induce bumps or pits. By optimising the annealing conditions (temperature, time, atmosphere, ...), this topography may be reduced (Fig. 3). Owing to the etching properties of chlorine, epitaxial growth in $GeCl_4$ also results in significantly smoother layers (< 1 nm RMS beyond 2 μm epitaxy).

Under the surface, the foil quality was investigated by transmission electron microscopy and electron-channealing contrast imaging, by which no dislocation could be detected. Minority-carrier lifetimes confirmed the foil quality on a larger scale with 25 μs for intrinsic 16-μm-thick foils. Rutherford backscattering spectrometry channeling measurements are currently ongoing to further investigate the presence of crystal defects in the GeON seed and the epitaxial layer.

III. PARENT WAFER RECONDITIONING

If the foil is proven to be a proper replacement for bulk wafers, it is not yet clear whether its economical and environmental added values will be sufficient. The overall Ge consumption of this approach has to be kept much lower than that of the standard bulk-wafer approach, and this directly depends on the efficiency of the reconditioning process. The GeON process steps impact the parent wafer by in-diffusing foreign species, by creating a recess at the border between GeON and bulk areas, and by roughening the surface with pillar debris. Reconditioning (step 6 in Fig. 1) hence consists in removing a thickness of Ge from the front and/or rear sides which trades-off the wafer quality and the Ge losses.

A. Contamination of the wafer rear side

The step during which contamination can take place is essentially the epitaxial deposition of the III-V layers, the only high-temperature step involving foreign elements. The front-side is not the only side at risk, since the rear side - unless capped with a protective dielectric - could also get in contact with the precursors. The steps from metallization onwards, especially if involving Au, are also a contamination source. It is yet unsure what negative impact such contaminations could have on the successive foil quality. However, beyond foil quality, the front-end-of-line semiconductor equipment required to etch the porous structure have strict restrictions on metallic elements. An investigation of their concentration and penetration depth was thus initiated.

The in-diffusion of elements involved in heteroepitaxy was monitored by secondary ion mass spectrometry (SIMS) with two different stacks, namely a nucleation stack (GaInP:Si and GaInAs grown for 35 minutes at 640°C) and a stack for forming triple-junction cells (GaInAs, GaInP, grown for 2 hours at 640°C). Profiles at the wafer rear side for the latter case with the longest thermal budget are shown in Fig. 4. As expected from theory, column-V elements, and particularly As, diffused the deepest. If all contamination was to be removed, about 2 μm Ge would have to be etched-off. A comparison with the front-side and for shorter thermal budgets will further be reported.

B. Recess from the porosified wafer rim

Essential wafer properties that must be recovered are in-plane flatness and roughness, on which the lithography quality depends. Alterations originate from pillar debris, that leave protrusions or pits of a few hundred nm in width, and from the recess at the edge between porosified and non-porosified areas. The recess of this exclusion edge is in the micron scale and depends on the pore dimension and, mostly, the Ge epitaxial layer thickness.

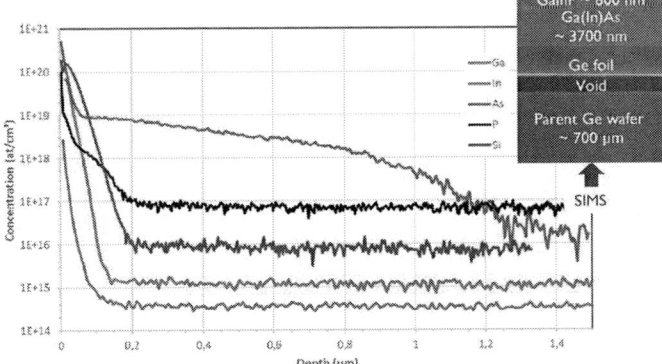

Fig. 4.   SIMS profiles of Ga, In, As, P and Si measured at the rear side of the GeON parent wafer after 2 hours of MOCVD, probed until the detection limit of As was reached.

Preliminary tests were performed to identify the reconditiong route. Four 100-mm wafers on which the recess was created by masking and dry etching were used. A depth of 2.5-3 μm was targeted to mimick the depth left by a 1-μm-thickened Ge foil. The first conditioning sequence was (1) front-side grinding, (2) chemical etching for stress relief, (3) front-side polishing and (4) cleaning. The first 2 wafers followed this procedure. After grinding, the recess was successfully removed (Fig. 5). In total, this sub-optimal sequence consumed ~ 40 μm.

To minimise the losses, polishing without grinding was tested on the other 2 wafers. Despite the multiplication of passes and a total removal of ~40-50 μm, the recess could not be levelled (Fig. 5), showing that a polishing-only approach is not suitable to recover from this exclusion edge and that a dedicated approach will be needed to reduce the losses below 10 μm.

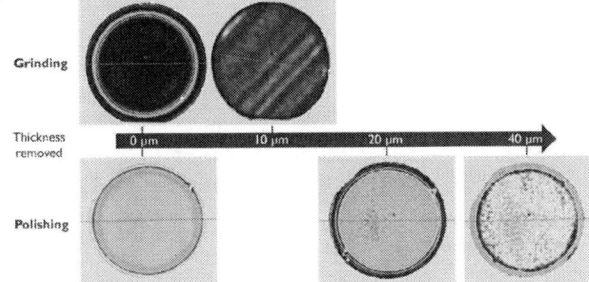

Fig. 5.   Waviness maps of 2 wafers with a 2.5-3 μm recess at the edge, after grinding or polishing for different process times, with the corresponding amount of Ge removed from the front side.

C. Demonstration of wafer re-use

After these preliminary tests, four 200-mm GeON-processed wafers were reconditioned and re-processed, resulting in the first "second-generation" foils. Foils of the first generation were formed without Ge thickening or III-V growth (a limitation from wafer size), and transferred to a Gelpak® (Fig. 6a). Grinding/polishing was then applied at the front side, removing a total of 38 μm. Their contamination and roughness/flatness specifications being recovered, they were reprocessed without issue. The GeON area was printed over the full wafer, which allowed for the transfer to glass with adhesive bonding of a 12 cm × 12 cm foil and a 19-cm round foil, with a few holes at the edges caused by pattern defects (Fig. 6b) [4].

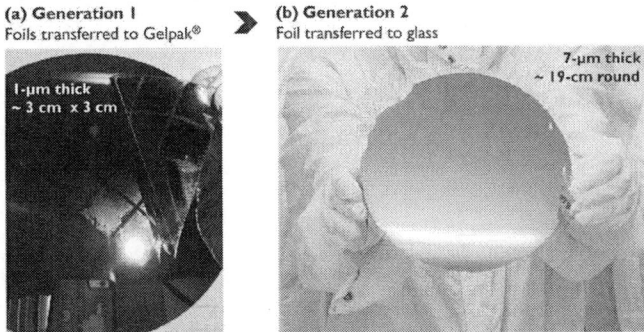

Fig. 6.   (a) Twelve GeON foils being peeled off from the parent wafer and (b) one GeON foil, 19-cm in diameter, transferred from the same wafer to glass after reconditioning and reprocessing.

V. Conclusion

Ge-on-Nothing structures are proven as an efficient lift-off layer to fabricate thin Ge wafers as templates for lattice-matched III-V layer growth. With the detachment of wafer-scale foils from re-used wafers, the process loop is now closed. If the suitability of this material is demonstrated for the formation of triple junction cells, its potential for cost and Ge consumption reductions is not yet clear. This potential will strongly depend on the number of re-use cycles that may eventually be achieved and the minimal losses of the reconditioning step.

Acknowlegments

I. García   from Universidad Politécnica de Madrid is acknowledged for the III-V layer growths. This project was carried out under a program of and funded by the European Space Agency (ESA) with contract no. 4000129924/20/NL/FE. The view expressed herein can in no way be taken to reflect the official opinion of the European Space Agency.

References

[1]   J. S. Ward et al., "Techno-economic analysis of three different substrate removal and reuse strategies for III-V solar cells," Progress in Photovoltaics: Research and Applications, vol. 24, no. 9, pp. 1284–1292, 2016, doi: 10.1002/pip.2776.

[2]   European Commission, "Critical Raw Materials Resilience: Charting a Path towards greater Security and Sustainability." Mar. 09, 2020. [Online].

[3]   N. Paupy et al., "Epitaxial growth of detachable GaAs/Ge heterostructure on mesoporous Ge substrate for layer separation and substrate reuse," in 49th IEEE Photovoltaic Specialists Conference, Philadelphia, Jun. 2022.

[4]   V. Depauw et al., "Wafer-scale Ge epitaxial foils grown at high growth rates and released from porous substrates for triple-junction solar cells," Progress in Photovoltaics: Research and Applications, vol. early print, no. n/a, 2022.

Characterizing Capacity Contribution of Renewable Resources Over Time in Renewable-Heavy Transmission System: MISO Case Study

Hyeonjung(Tari) Jung, Megan Pamperin, Elspeth McGarvey, Eduardo Ibanez, James Okullo, Armando Figueroa Acevedo, Jordan Bakke

{hjung, mpamperin, emcgarvey, eibanez, jokullo, afigueroa-acevedo}@misoenergy.org

Midcontinent Independent System Operator, Eagan, MN, 55404, USA

Abstract — The effects of solar photovoltaic (PV) and wind generation on the bulk electric system (BES) are varied and dependent on the grid network's conditions, including the resource mix and system uncertainties. As the grid evolves, characterizing the change in capacity contribution of each renewable resource is of vital importance in preparing for the reliability of the future grid. This research explores the capacity contribution of solar and wind units located in Midcontinent Independent System Operator (MISO)'s footprint for the resource mix between years 2031 and 2041. Several existing methods and the novel direct loss of load (DLOL) method were used to estimate the capacity contribution of solar PV and wind resources. The results were compared to capture the shifts in risk as the resource mix evolves. We also highlight the conditions of the BES that will influence the capacity contribution of renewable generation by examining the hours of high risk and probability of loss of load.

I. INTRODUCTION

Resource adequacy (RA) assesses sufficiency of generation, load modifying resources, and storage to meet the system demand across the year. Capacity contribution (CC) of a resource is determined based on the resource's ability to contribute during the system's most likely risk periods. These conditions normally occur when the electric demand is high and the availability of generation is scarce. The advent of variable generation (such as solar PV and wind) is shifting the timing of risk in the system and making it more difficult to understand the relation between resource's characteristics and their CC. For instance, in a portfolio with a high penetration of solar resources, as illustrated in Fig. 1, MISO's control center may see low reliability risk during the daytime when the solar generation is available. On the other hand, wind resources in the MISO footprint tend to perform stronger due to higher surface wind later in the day. Properly capturing the CC of individual resources is essential in system planning.

Risk hours and CC are both dependent on multiple system factors, such as its total resource mix, load demand, forced and maintenance outages, and weather conditions. Loss of load expectation (LOLE) is a reliability metric commonly used to forecast and characterize risk of a grid. LOLE represents the expected number of event-days in a year in which loss of load events may occur [15]. The LOLE metric allows the comparison of the risk level of systems with different conditions.

Effective load carrying capacity (ELCC) is a LOLE-based method to calculate the CC of the same resource under different grid conditions (usually with a system tuned to a LOLE of 0.1 days/year) [1, 11]. Average ELCC measures the CC of both existing and planned capacity of the resource has been a standard practice in the industry, but recent developments have raised questions as to whether average ELCC is an effective metric to capture the resource's projected impact on the grid.

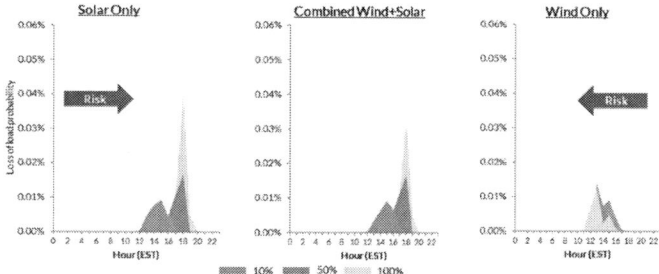

Fig. 1. Interdependencies between wind and solar resources and the risk distribution [16]

For instance, the ability of a new solar PV unit to contribute during highest load hours will likely vary if the system is already saturated with solar PV resources, and its highest net load hour shifted to later in the day, compared to a system with lower solar PV penetration. Marginal ELCC, the incremental contribution of a resource in the CC calculation, has been suggested as a suitable replacement by several industry entities [2,3].

In this paper, we conduct an RA analysis on the projected 2031 and 2041 models of the Midcontinent Independent System Operator (MISO). This work continues MISO's previous explorations of the resource accreditation methods with the increasing presence of renewable resources [17, 18]. The evolution of resource mixes in the fifteen-year horizon is summarized with emphasis on its impact on solar and wind CC, tested with both average and marginal ELCC methods. We share results from two approaches of marginal accreditation: marginal reliability impact (MRI) and direct loss of load (DLOL). Comparison of MRI and DLOL methods have not been studied in the existing literature to the best of the authors' knowledge.

The remainder of this manuscript is organized as follows: Four methods used to evaluate capacity contribution are presented with an illustrative example in Section II. Overview of simulation with the BES model assumptions is described in Section III. Section IV provides results of the RA study for solar and wind CCs. Section V includes our concluding remarks and our vision for the subsequent full-paper submission.

978-1-6654-6060-6/23 $31.00 © 2023 IEEE

II. CALCULATION OF CAPACITY CONTRIBUTION

In this section, an illustration of ELCC calculation will be presented. Definition of average, marginal, Marginal Reliability Impact (MRI), and DLOL methods are provided.

A. Steps for Average ELCC Calculation

The average ELCC method captures the reliability contribution of an entire resource class. This example illustrates the steps used to calculate average ELCC and is also a relevant reference for the calculation of marginal and DLOL methods. [4] shows the process used to estimate the ELCC in four steps. Step 1 records the LOLE of the Base System. Step 2 performs annual adjustments by adding or removing fixed load so that the average base system LOLE reaches 0.1 days/year. If the LOLE is less than the target, then a fixed load is added to the model. If the LOLE is greater than the target, then proxy combustion turbine generators with a typical size of 160 MW and class average equivalent forced outage rate (EFORd) are added to the model until the LOLE target is achieved [14]. Step 3 records the LOLE of the base system without the resource in question. Finally, step 4 adjusts the Base System without the resource to match the LOLE resulting from Step 2. The resulting ELCC is the difference between the fixed load adjustments from Step 2 to Step 4, divided by the resource's installed capacity.

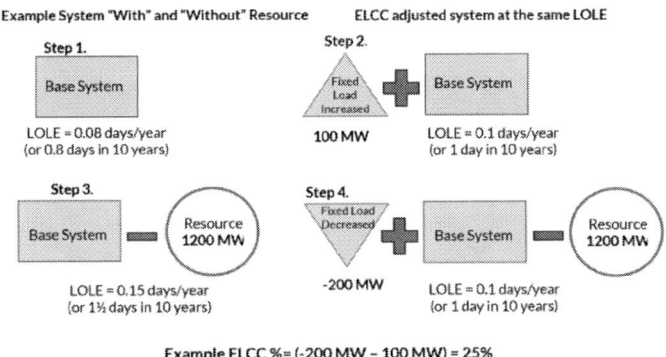

Fig. 2. Illustrative average ELCC calculation.

B. Marginal Methods: Marginal ELCC, MRI, and DLOL

While the average ELCC methodology captures the system's overall reliance on each resource class, the marginal method shows the capacity contribution of an incremental resource. There are multiple ways to calculate marginal accreditation of a resource type. Three of such methods, marginal ELCC, MRI, and DLOL, are examined in this paper.

All marginal methods assume that Steps 1 and 2 in the average ELCC calculation illustrated in Figure 2 has been completed. With the resulting base system of 0.1 LOLE, instead of subtracting the total resource capacity in Step 3, marginal ELCC method adds an incremental unit to the model, also

called a marginal unit. In this work, the marginal unit for all resource types was assumed to have an installed capacity of 1 GW, which represents between 5 and 10% of the existing wind and solar resources of MISO's system. The new system LOLE value is recorded after the incremental unit is added. Starting again from the adjusted system from Step 2 in Figure 2, a perfect unit is then added and its capacity is adjusted until the system LOLE reaches the same value as after the addition of the resource-specific marginal unit. The resource's marginal ELCC is defined as the amount of additional load needed to match the LOLE values. (1) shows the definition of $CC_{MarginalELCC}$, the resulting capacity credit of marginal ELCC.

$$CC_{MarginalELCC}$$
$$= \frac{Load\ to\ match\ LOLE_{Resource\ Marginal\ Unit}(MW)}{Capacity\ of\ Resource\ Marginal\ Unit(MW)} \quad (1)$$

$$where\ LOLE_{Resource\ Marginal\ Unit}$$
$$= LOLE_{Base}$$
$$+ LOLE_{Base+ResourceMarginalUnit}$$

The MRI method starts with Steps 1 and 2 in Figure 2, meaning the base system with all resources included is adjusted to an LOLE of 0.1 days/year. The next step is the same as the marginal calculation method, where a small marginal unit (1 GW for this study) of the resource type is added to the adjusted system and the LOLE is recorded. Starting again from the adjusted base system, a perfect unit of the same capacity as the marginal resource unit (1 GW) is added to the system and the new LOLE is recorded. This is the divergence from the marginal method, rather than adjusting the capacity of the perfect unit its capacity is set to the same as the marginal unit. The ratio of the difference between the adjusted LOLE and the marginal unit LOLE to the difference between the adjusted LOLE and the perfect unit LOLE is the capacity contribution of the resource. (2) shows the mathematical definition of capacity contribution calculated through the MRI method.

$$CC_{MRI} = \frac{LOLE_{Base} - LOLE_{MarginalCapacityChange}}{LOLE_{Base} - LOLE_{PerfectUnit}} \quad (2)$$

The MRI method approximates the same marginal accreditation that marginal ELCC is capturing with less computational resource. Instead of determining a difference in fixed load adjustment values, the MRI method compares the change in the system LOLE when the marginal unit and when the equivalent perfect unit are added, eliminating the iterative process of adjusting the marginal system's LOLE. Since the contribution of a resource can be deterministically compared as the change in LOLE between a perfect unit versus marginal capacity change, the process of MRI is easier to comprehend. The sensitive of the resulting accreditation to the size of marginal unit may drive uncertainty of the method.

The DLOL method, also LOLP Capacity Factor method, estimates the CC of a resource without explicitly modeling a marginal unit [10]. Instead of repeating Step 3 and 4, DLOL assumes that the profile of risk hours and total loss of load will remain relatively stable between the base model and the marginally adjusted model. This allows the DLOL capacity credit to be calculated as the average generation from that resource during scarcity events and it is based on the model achieved at the end of Step 2, when target LOLE is reached. (3) shows the mathematical definition of DLOL.

$$CC_{DLOL} = \frac{\sum_{hr=1}^{hr=8760} I(hour) * Hourly\ Generation(MW)}{Total\ Resource\ Capacity * \sum_{hr=1}^{hr=8760} I(hour)} \quad (3)$$

$$where\ I = \begin{cases} 1\ if\ LOLE\ hour \\ 0\ Otherwise \end{cases}$$

Because the accreditation can be directly calculated from the base system, DLOL method results in CC that is most closely aligned with the base model's assumptions, such as resource mix portfolio and base demand. It also aligns with the base LOLE model used to set the system requirementsA. These makes the interpretation of DLOL accrediation more straightforward and transparent when examining the system's driving factors of accreditation.

III. RESOURCE ADEQUACY SIMULATION

This section discusses the model input assumptions and outlines the RA simulation we implemented to test methods illustrated in Section II. While in this abstract we highlight results for 2031 and 2041 as a precursor, the full overview with additional study years will be provided in the full manuscript.

A. Model Assumptions

The future resource mix on the MISO grid in 2031 and 2041 was based on the capacity expansion modeling completed in MISO's 2022 Regional Resource Assessment study [5]. The projected growth of solar and hybrid generation is significant throughout the study years. Peak demand reaches a maximum of 135 GW in 2031 and 145 GW in 2041.

Hourly solar and wind generation profiles were created per load reserve zone (LRZ) based on the National Renewable Energy Laboratory (NREL) datasets for 2007 through 2012 and the Vibrant Clean Energy (VCE) datasets for the weather 2014 through 2018 [6,7,8]. Load profiles for the same years were based on MISO historical market data and were adjusted based on the annual peak forecasted for Future 1 of the 2021 MISO Futures Report [9]. The models used in this study assumed a perfect foresight of renewable generation and storage dispatch. The model did not include transmission constraints.

B. Model Assumptions

Planned and forced outage rates of the thermal resources were determined based on the five-year average of the plants, reported in Generating Availability Data System (GADS). When the individual plant data was not available due to missing mappings or because the plant has not existed for the past 5 years, a class average by resource type and capacity was assigned. Planned maintenance was scheduled optimally using a linear program method and kept constant per weather year for

each sample. Forced outages were the stochastic variable and were randomly generated for each sample of the Monte Carlo method. LOLE was calculated after 150 samples of Monte Carlo simulation per each weather year (11 total). This selection of sample size resulted in a standard error below 5%. PLEXOS 8.3R10 was used for all simulation [12].

IV. RESULTS

Table 1 and Table 2 present CC results from all four methodologies for wind and solar resources for the years 2031 and 2041, as well as the installed capacity (ICAP) of the resources in the future portfolios. Results of summer are provided as the largest contribution for all resources happened during the summer when the load peaks.

The marginal-type calculations result in much lower CC values for solar than the average calculation does, especially for 2041. The marginal ELCC methods best capture solar generation's contribution during actual risk hours in 2031 and 2041, which occur in the evening or overnight hours when solar energy is not available. Of the three marginal-type methods, DLOL shows lower CC value for 2031 compared to the other methods indicating that solar CC may decline most rapidly with this method. The average calculation method, on the other hand, provides a higher capacity value for both study years because it is based on the risk profile of the system before any solar generation was added and load (and risk) was high during daytime hours.

For wind, marginal-type calculations resulted in lower CC values compared to the average calculation for 2041. For 2031, the MRI method resulted in a higher CC for wind compared to the average method.

Generally, the accreditation resulting from MRI and Marginal ELCC aligned for all seasons in both portfolios. The mismatch of the exact results signals the capacity of the marginal unit used in the MRI method may need further calibration [13]. While 1GW marginal unit was selected since it was the smallest capacity to observe a meaningful change in the LOLE of the system, we observed CC_{MRI} is highly dependent on the capacity of perfect unit. Tuning it to match $CC_{MarginalELCC}$ is an arduous and arbitrary process of iteration.

DLOL and other marginal methods also share general trend of accreditation. The difference in result shows tuning a system to 1 day in 10 statistics— $LOLE = 0.1$— as the equivalent level of risk to measure the resource's contribution may obscure the comparison by changing the performance of resource and condition of the grid that are captured. In [10] and [17], using expected unserved energy (EUE) as the reliability metric, instead of LOLE, is noted as a potential improvement to align DLOL better with other marginal methods.

Table 1 PV Solar MRI-Marginal-DLOL Comparison

		Winter	Spring	Summer	Fall
2031	Average ELCC	1%	17%	23%	18%
	MRI	0%	0%	21%	0%

978-1-6654-6060-6/23 $31.00 © 2023 IEEE

		Winter	Spring	Summer	Fall
	Marginal ELCC	0%	0%	12%	0%
	DLOL	0%	2%	2%	2%
2041	Average ELCC	11%	11%	18%	20%
	MRI	0%	25%	0%	9%
	Marginal ELCC	0%	18%	0%	3%
	DLOL	2%	1%	0%	1%

Table 2 Wind MRI-Marginal-DLOL Comparison

		Winter	Spring	Summer	Fall
2031	Average ELCC	37%	12%	18%	21%
	MRI	14%	8%	21%	26%
	Marginal ELCC	10%	10%	12%	25%
	DLOL	14%	10%	12%	12%
2041	Average ELCC	12%	16%	21%	26%
	MRI	18%	25%	8%	31%
	Marginal ELCC	18%	18%	8%	20%
	DLOL	14%	9%	3%	21%

V. CONCLUSION

This work explored the potential capacity contribution of solar photovoltaic (PV) and wind generation using a wide range of existing and novel methods. The exercise demonstrated in this paper confirmed that a marginal accreditation method, compared to an average method, more meaningfully captures the incremental change of the system needs. Between the marginal accreditation methods, DLOL method was shown to be explainable and extendible as it reduces arbitrary complexity of adjusting marginal units.

Although the outcomes are subject on the grid network's conditions, including the resource mix and system uncertainties, the methods that reflect the risk profile of the actual system, such as the marginal ELCC, MRI and DLOL, result in a better alignment with the system-level risk distribution. The CC results shown in this work demonstrates the LOLE study-based accreditation is sensitive to stochastic variables, such as forced outage, and dispatch of energy capacity units, such as battery storage. Further work to determine the sufficient level of modeling these uncertainties for robust accreditation results will be pursued in the future.

VI. ACKNOWLEDGEMENT

The authors acknowledge with gratitude MISO Members for willingly providing information about their forward-looking resource plans and goals. MISO thanks the stakeholders who contributed their time and valuable perspectives to this work. The authors also thank Hilary Brown, Patrick Dalton and Laura Hannah from MISO for their contribution.

REFERENCE

[1] N. Schlag, Z. Ming, A. Olson, L. Alagappan, B. Carron, K. Steinberger, and H. Jiang, "Capacity and reliability planning in the era of decarbonization: practical application of effective load carrying capability in resource adequacy," Energy and Environmental Economics, Inc., 2020, https://www. ethree. com/elcc-resource-adequacy.

[2] Q. Duan, N. Wang, Q. Chen, P. Bie, H. Wang, Z. Xia, Y. Zhang, and X. Han, "Demand curve model based on system reliability in ISO-NE capacity market," *2021 IEEE Sustainable Power and Energy Conference (iSPEC)*, 2021.

[3] New York Independent System Operator, Inc., 179 FERC ¶ 61,102, 2022.

[4] M. Milligan, and K. Porter, "Determining the capacity value of wind: An updated survey of methods and implementation." No. NREL/CP-500-43433. National Renewable Energy Lab.(NREL), Golden, CO (United States), 2008.

[5] MISO. "2022 Regional resource assessment report - a reliability imperative report" Nov 2022, pp. 22-26, https://cdn.misoenergy.org/2022%20Regional%20Resource%20Assessment%20Report627163.pdf

[6] C. Draxl, A. Clifton, B. M. Hodge, and J. McCaa, "The wind integration national dataset (WIND) toolkit," *Applied Energy*, 151, pp. 355-366, 2015.

[7] M. Sengupta, Y. Xie, A. Lopez, A. Habte, G. Maclaurin, and J. Shelby, "The national solar radiation data base (NSRDB)," *Renewable and Sustainable Energy Reviews,* 89, pp. 51-60, 2018.

[8] C. Clack, A. Choukulkar, B. Coté, S. A. McKee, "Dataset overview: renewable generation, electric demand, transmission line ratings & losses, and climate change," Vibrant Clean Energy, LLC, 2020, https://vibrantcleanenergy.com/wp-content/uploads/2020/08/VCE-Weather-Dataset-Overview_August2020.pdf

[9] MISO. "MISO futures report," Apr 2021, MISO Futures Report538224.pdf (misoenergy.org).

[10] Energy Systems Integration Group. (2023). Ensuring Efficient Reliability NEW DESIGN PRINCIPLES FOR CAPACITY ACCREDITATION. *A Report of the Redefining Resource Adequacy Task Force.* https://www.esig.energy/new-design-principles-for-capacity-accreditation.

[11] Methods to model and calculate capacity contributions of variable generation for resource adequacy planning," North American Electric Reliability Corporation, Princeton, NJ, Rep., 2011. http://www.nerc.com/docs/pc/ivgtf/IVGTF1-2.pdf

[12] *PLEXOS Energy Modeling Software.* (n.d.). Retrieved May 28, 2023, from https://www.energyexemplar.com/plexos

[13] NYISO. (2022). Capacity Accreditation. *ICAPWG/MIWG.* Retrieved May 28, 2023, from https://www.nyiso.com/documents/20142/34087499/10-27-22%20ICAPWG%20Capacity%20Accreditation%20v2%20-%20repost.pdf/23474d78-642b-c476-8f4d-26953fe57bd5

[14] Madaeni, S. H., Sioshansi, R., & Denholm, P. (1998). *Comparison of Capacity Value Methods for Photovoltaics in the Western United States.* http://www.osti.gov/bridge

[15] Stephen, G., Tindemans, S. H., Fazio, J., Dent, C., Acevedo, A. F., Bagen, B., Crawford, A., Klaube, A.,

Logan, D., & Burke, D. (2022). Clarifying the Interpretation and Use of the LOLE Resource Adequacy Metric. *2022 17th International Conference on Probabilistic Methods Applied to Power Systems, PMAPS 2022.* https://doi.org/10.1109/PMAPS53380.2022.9810615

[16] MISO. (2021). *MISO's Renewable Integration Impact Assessment (RIIA).* https://cdn.misoenergy.org/RIIA%20Summary%20Report 520051.pdf

[17] Pickering, S., Figueroa-Acevedo, A., Rostkowski, I., Foley, S., Huebsch, M., Claes, Z., Ticknor, D., McCalley, J., Okullo, J., & Heath, B. (2019). Power System Resource Adequacy Evaluation under Increasing Renewables for the Midwestern US. 51st North American Power Symposium, NAPS 2019. https://doi.org/10.1109/NAPS46351.2019.9000346

[18] Heath, B., & Figueroa-Acevedo, A. L. (2018). Potential capacity contribution of renewables at higher penetration levels on MISO system. *2018 International Conference on Probabilistic Methods Applied to Power Systems, PMAPS 2018 - Proceedings.* https://doi.org/10.1109/PMAPS.2018.8440442

Device modeling of HTL/BaSi₂ heterojunction solar cells

Sho Aonuki[1], Carlos Mario Ruiz Tobon[2], Rudi Santbergen[2], Olindo Isabella[2], and Takashi Suemasu[1]

[1]University of Tsukuba, Tsukuba, Ibaraki, 305-8573, Japan

[2]Delft University of Technology, Mekelweg, Delft, 2628CD, the Netherlands

Abstract — We simulated the optical absorptance of BaSi₂-based heterojunction solar cells with transition metal oxides as hole transport layer (HTL) using GENPRO4 software and optimized the device structures. The complex refractive index of each layer was used as an input in the optical simulations. We adopted ITO (80 nm) / HTL / a-Si (3 nm) / n-BaSi₂ (500 nm) / TiN (250 nm) / glass substrates (200 μm) structures. First, the implied photocurrent density (J_{ph}) loss caused by parasitic absorption in 20-nm-thick p⁺-BaSi₂ layer was calculated to be 7.9 mA cm⁻². The J_{ph} increased to 29.1 mA cm⁻² by substituting p⁺-BaSi₂ with 2-nm-thick MoO₃. To figure out the optimal HTL materials and the structures for BaSi₂ solar cells, we simulated the absorption spectra as function of materials such as NiO, Cu₂O, MoO₃, V₂O₅, and WO₃, which have already demonstrated the HTL functionality, and their thicknesses. The highest J_{ph} was obtained with MoO₃, V₂O₅, or WO₃, meaning that these oxides are optically suitable HTL materials. By increasing the n-BaSi₂ absorber layer thickness to 2 μm and importing 3D random pyramidal texture structure with the height of 4 μm, the J_{ph} reached a maximum of 33.1 mA cm⁻². This is the largest value of all BaSi₂ solar cells ever reported.

I. INTRODUCTION

Safe, stable, and earth-abundant materials are of great importance to realize a decarbonized society fueled by photovoltaic (PV) technology. Under such circumstances, we have paid special attention to BaSi₂, a semiconducting silicide composed only of earth-abundant elements.[1] BaSi₂ has a suitable band gap for single-junction solar cell application (1.3 eV). In addition, both a high absorption coefficient of 3×10^4 cm⁻¹ at 1.5 eV and a prominent minority carrier diffusion length of 10 μm are realized. Solar cell materials possessing such properties are quite limited. According to the theoretical calculation, the conversion efficiency (η) of BaSi₂ homojunction solar cells exceeds 25%.[2] Moreover, the operation of BaSi₂ homo-junction solar cells was recently demonstrated[3] after the achievement of η = 9.9% for p-BaSi₂/n-Si solar cells architecture.[4] However, the η of BaSi₂ homojunction solar cells has been limited to 0.28% due to the parasitic absorption in BaSi₂ carrier collection layer containing a lot of defects.[5] The absorption coefficient of BaSi₂ is so high that such absorption loss is inevitable as long as BaSi₂ is employed as the surface layer. Therefore, it is crucial to explore the substitutional materials for the carrier collecting layer to fulfill high-η BaSi₂ solar cells. Since undoped BaSi₂ films

usually shows n-type conductivity, we focused on hole transport layers (HTLs) to form BaSi₂ heterojunction solar cells and suppress the parasitic absorption. Several kinds of HTL materials such as MoO₃ have been studied so far and the η over 23% have been achieved in crystalline silicon PV technology.[6] However, there has been limited reports on HTL/BaSi₂ solar cells. In this study, we thereby aim to model HTL/BaSi₂ heterojunction solar cells using several kinds of HTL materials from the theoretical viewpoint.

II. SIMULATION METHOD

In this work, we carried out the optical simulations using GENPRO4 developed at the Delft University of Technology.[7] The complex refractive index ($n + ik$) of each layer was used as an input. It is noted that the extinction coefficient k of BaSi₂ layer was fixed to zero at wavelength above 950 nm, corresponding to the fact that no photoresponsivity is obtained experimentally beyond such wavelength. The implied photocurrent density (J_{ph}) was calculated by integrating the simulated absorption spectra of n-BaSi₂ absorber layer over the AM1.5 spectrum range. We adopted an ITO (80 nm) / HTL / a-Si (3 nm) / n-BaSi₂ (500 nm) / TiN (250 nm) / Glass (200 μm) structure to simulate the optical properties of HTL/n-BaSi₂ heterojunction solar cells using several HTL materials and changing their thickness. We chose MoO₃, WO₃, V₂O₅, Cu₂O, and NiO as HTL materials, for which the HTL functionality has been demonstrated.[8] For comparison, an ITO (80 nm) / a-Si (3 nm) / p-BaSi₂ / n-BaSi₂ (500 nm) / TiN (250 nm) / Glass (200 μm) structure was also simulated. The a-Si capping layers were inserted at the top of the BaSi₂ layer to suppress the oxidation of BaSi₂ layer, which is experimentally implemented.[9] We adopted the TiN as a back contact layer.[10] According to the previous research, TiN layer blocked the diffusion of oxygen atoms from the glass substrate and suppressed the intermixing at the BaSi₂/TiN interface, leading to the growth of BaSi₂ films with high photoresponsivity on glass substrates. First, we simulated the solar cell device using flat surfaces; then, we introduced the typical 3D random pyramidal texture at the front surface of the glass substrates to clarify the effect of the anti-reflection and the light trapping by the 3D random pyramidal texture on the above structures. An atomic force microscopy image of a pyramid-type texture obtained by chemical-etching of a Si substrate was adopted for the simulations as shown in Fig. 1.

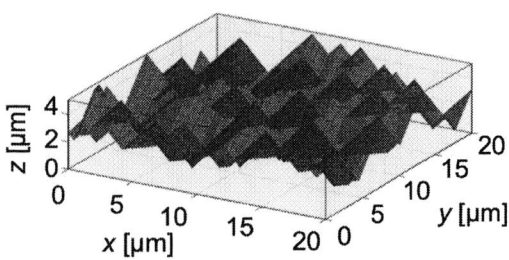

Fig. 1.   3D random pyramidal texture measured by atomic force microscopy.

III. RESULTS AND DISCUSSION

It was experimentally revealed that at least 20 nm is necessary to form flat BaSi$_2$ films; otherwise, the BaSi$_2$ islands grow.[11] This island layer decreases the shunt resistance, resulting in the deterioration of solar cell performance. Therefore, we first calculate the parasitic absorption of 20-nm-thick p$^+$-BaSi$_2$ layer. Fig. 2 (a) shows the absorption spectra of pn-BaSi$_2$ solar cell on TiN/glass substrate. Owing to the high absorption coefficient of BaSi$_2$, the absorption in p$^+$-BaSi$_2$ with the thickness of 20 nm is markedly observed in the wavelength range of 300 – 900 nm. This parasitic absorption causes loss in the J_{ph} of 7.9 mA cm^{-2}, leading to the limitation of J_{ph} in n-BaSi$_2$ absorber layer to 20.3 mA cm^{-2}. We substituted the p$^+$-BaSi$_2$ with MoO$_3$ as an HTL layer to increase the absorption in n-BaSi$_2$ absorber layer. Only 1.7-nm-thick MoO$_3$ worked as an HTL layer and over 23% of η have been achieved for MoO$_3$/c-Si heterojunction solar cells.[6] The absorption in n-BaSi$_2$ distinctly increased and the J_{ph} reached 29.1 mA cm^{-2} (Fig. 2 (b)). The absorption of MoO$_3$ was negligible thanks to the large band gap of MoO$_3$. To figure out the suitable HTL materials for BaSi$_2$ in terms of the absorption, we simulated the absorption spectra of HTL/BaSi$_2$ solar cells with various kinds of HTL materials. Fig. 2 (c) reports the J_{ph} as a function of the HTL thickness. The highest J_{ph} of all structures was obtained when MoO$_3$, V$_2$O$_5$, and WO$_3$ were used as an HTL layer. The structure with Cu$_2$O showed the slightly small J_{ph} compared to the above three materials. NiO and p$^+$-BaSi$_2$ greatly decreased the J_{ph} in n-BaSi$_2$ absorber layer with increasing its thickness. Therefore, MoO$_3$, V$_2$O$_5$, and WO$_3$ are promising HLT materials for maximizing absorption in BaSi$_2$.

Fig. 3 shows the J_{ph} caused in the n-BaSi$_2$ absorber layer with pn-BaSi$_2$ and MoO$_3$/n-BaSi$_2$ structure as a function of n-BaSi$_2$ thickness $t_{n\text{-BaSi2}}$. The J_{ph} saturated when $t_{n\text{-BaSi2}}$ reached 2 μm for each structure. The maximum J_{ph} of 22.0 mA cm^{-2} was obtained for p$^+$-BaSi$_2$(20 nm)/n-BaSi$_2$ solar cells with flat

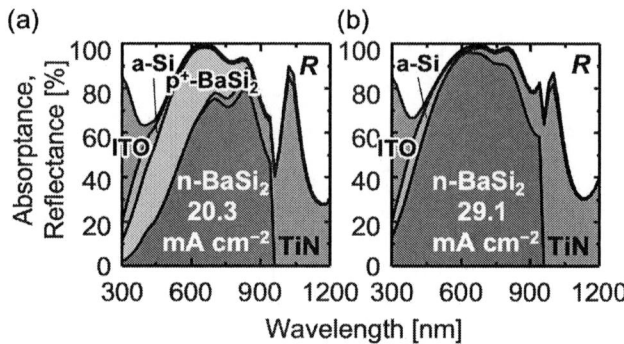

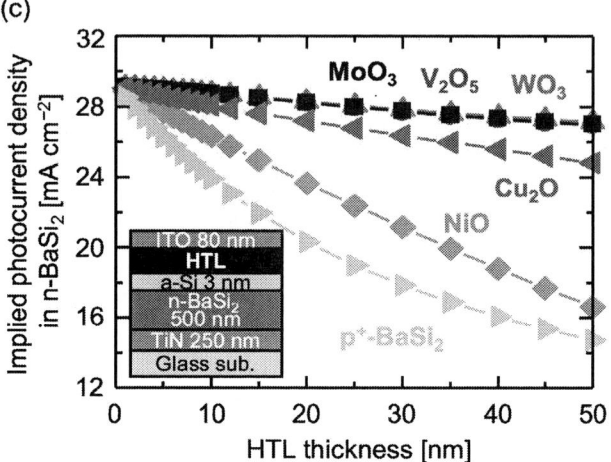

Fig. 2.   Absorptance spectra of BaSi$_2$ solar cells with (a) 20-nm-thick p$^+$-BaSi$_2$ and (b) 2-nm-thick MoO$_3$. (c) HTL thickness dependence of J_{ph} in n-BaSi$_2$ absorber layer contacted with several kinds of HTL materials.

structure. The J_{ph} markedly increased by replacing p$^+$-BaSi$_2$ with 2-nm-thick MoO$_3$ and reached 31.1 mA cm^{-2}. Moreover, 3D random pyramidal texture further increased the J_{ph} up to 33.1 mA cm^{-2}. The absorptance spectra of MoO$_3$ (2 nm) / n-BaSi$_2$ (2 μm) solar cell on a textured glass substrate were depicted in Fig. 4. Compared to Fig. 2 (b), the absorption in the n-BaSi$_2$ absorber layer increased both in the short (<600 nm) and long (>700 nm) wavelength regions. Both light trapping effect thanks to the 3D random pyramidal texture and increment of the n-BaSi$_2$ thickness are contributed to this improvement. The J_{ph} of 33.1 mA cm^{-2} is slightly higher than that of Al-doped n$^+$-ZnO/p-BaSi$_2$ structures on textured glass substrates simulated in the previous study.

IV. CONCLUSION

We modeled the HTL/n-BaSi$_2$ solar cells using optical simulation GENPRO4. The parasitic absorption of 20-nm-thick p$^+$-BaSi$_2$ layer caused the J_{ph} loss of 7.9 mA cm^{-2}. By replacing p$^+$-BaSi$_2$ with 2-nm-thick MoO$_3$, the J_{ph} increased up to 29.1 mA cm^{-2}. Among several kinds of HTL materials, MoO$_3$,

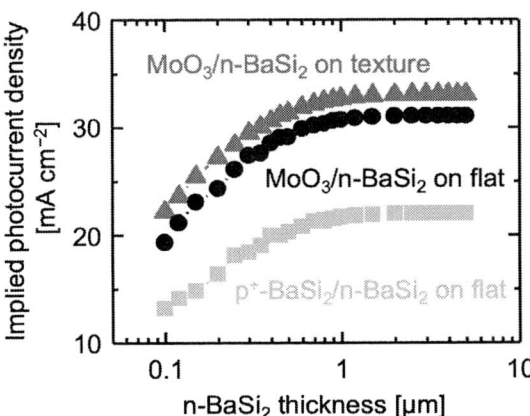

Fig. 3. n-BaSi₂ thickness dependence of the J_{ph} in n-BaSi₂ absorber layer contacted with 20-nm-thick p⁺-BaSi₂ and 1-nm-thick MoO₃ on flat or 3D random pyramidal texture glass substates.

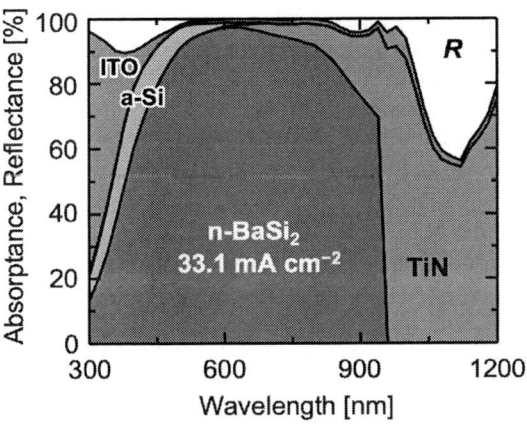

Fig. 4. Absorptance spectra of MoO₃(2 nm)/n-BaSi₂(2 μm) heterojunction solar cell on 3D random pyramidal texture structure.

V₂O₅, and WO₃ showed the lowest absorption, leading to the highest J_{ph}. Therefore, these oxides are promising HTL materials for BaSi₂ in terms of absorption. The J_{ph} increased with the thickness of n-BaSi₂ absorber layer and reached a maximum of 33.1 mA cm⁻² by introducing 3D random pyramidal texture structure. This value is the highest of all BaSi₂ solar cells ever simulated.

ACKNOWLEDGEMENTS

This work was financially supported in part by JSPS KAKENHI (Grants 19KK0104 and JP21H04548). A. S. are financially supported by Grant-in-Aids for JSPS Research Fellowships for Young Scientists (Grant 21J20404). We are grateful to the Scientific Grant Agency of the Ministry of Education, Science, Research and Sport of the Slovak Republic for financial support of projects VEGA 1/0532/19 and VEGA 1/0529/20.

REFERENCES

[1] T. Suemasu, and D.B. Migas, "Recent Progress Toward Realization of High-Efficiency BaSi₂ Solar Cells: Thin-Film Deposition Techniques and Passivation of Defects" *Physica Status Solidi A*, vol. 219, pp. 2100593, 2022.

[2] T. Suemasu, "Exploring the possibility of semiconducting BaSi₂ for thin-film solar cell applications" *Japanese Journal of Applied Physics*, vol. 54, pp. 07JA01, 2015.

[3] K. Kodama, Y. Yamashita, K. Toko, and T. Suemasu, "Operation of BaSi₂ homojunction solar cells on p⁺-Si(111) substrates and the effect of structure parameters on their performance" *Applied Physics Express*, vol. 12, pp. 041005, 2019.

[4] S. Yachi, R. Takabe, H. Takeuchi, K. Toko, and T. Suemasu, "Effect of amorphous Si capping layer on the hole transport properties of BaSi₂ and improved conversion efficiency approaching 10% in p-BaSi₂/n-Si solar cells" *Applied Physics Letters*, vol. 109, pp. 072103, 2016.

[5] Y. Yamashita, C.M. Ruiz Tobon, R. Santbergen, M. Zeman, O. Isabella, and T. Suemasu, "Solar cells based on n⁺-AZO/p-BaSi₂ heterojunction: Advanced opto-electrical modelling and experimental demonstration" *Solar Energy Materials and Solar Cells*, vol. 230, pp. 111181, 2021.

[6] L. Cao, P. Procel, A. Alcañiz, J. Yan, F. Tichelaar, E. Özkol, Y. Zhao, C. Han, G. Yang, Z. Yao, M. Zeman, R. Santbergen, L. Mazzarella, and O. Isabella, "Achieving 23.83% conversion efficiency in silicon heterojunction solar cell with ultra-thin MoOx hole collector layer via tailoring (i)a-Si:H/MoOx interface" *Progress in Photovoltaics: Research and Applications*, pp. 1–10, 2022.

[7] R. Santbergen, T. Meguro, T. Suezaki, G. Koizumi, K. Yamamoto, and M. Zeman, "GenPro4 Optical Model for Solar Cell Simulation and Its Application to Multijunction Solar Cells" *IEEE Journal of Photovoltaics*, vol. 7, pp. 919–926, 2017.

[8] Y. Liu, Y. Li, Y. Wu, G. Yang, L. Mazzarella, P. Procel-Moya, A.C. Tamboli, K. Weber, M. Boccard, O. Isabella, X. Yang, and B. Sun, "High-Efficiency Silicon Heterojunction Solar Cells: Materials, Devices and Applications" *Materials Science and Engineering: R: Reports*, vol. 142, pp. 100579, 2020.

[9] R. Takabe, H. Takeuchi, W. Du, K. Ito, K. Toko, S. Ueda, A. Kimura, and T. Suemasu, "Evaluation of band offset at amorphous-Si/BaSi₂ interfaces by hard x-ray photoelectron spectroscopy" *Journal of Applied Physics*, vol. 119, pp. 165304, 2016.

[10] R. Koitabashi, T. Nemoto, Y. Yamashita, M. Mesuda, K. Toko, and T. Suemasu, "Formation of high-photoresponsivity BaSi₂ films on glass substrate by radio-frequency sputtering for solar cell applications" *Journal of Physics D: Applied Physics*, vol. 54, pp. 135106, 2021.

[11] S. Yachi, R. Takabe, K. Toko, and T. Suemasu, "Effect of p-BaSi₂ layer thickness on the solar cell performance of p-BaSi₂/n-Si heterojunction solar cells" *Japanese Journal of Applied Physics*, vol. 56, pp. 05DB03, 2017.

Modulating Efficiency and Stability of Methylammonium/Br-Free Perovskite Solar Cells Using Fluoroarene Hydrazine

Dhruba B. Khadka[1], Yasuhiro Shirai[1], Masatoshi Yanagida[1] and Kenjiro Miyano[1]

[1]Photovoltaics Materials Group, Global Research Center for Environment and Energy based on Nanomaterials Science (GREEN), National Institute for Materials Science (NIMS), 1-1 Namiki, Tsukuba, Ibaraki 305-0044, Japan

Abstract — Halide perovskite solar cells (PSCs) with state-of-the-art efficiencies consist of thermally unstable methylammonium (MA). In this report, we have employed the surface passivation method with multifunctional fluoroarene molecule, which suppresses the formation of PbI_2 and δ-perovskite phase in MA/Br-free perovskite film. The penta fluoro-phenylhydrazine (5F-PHZ) passivation effectively mitigates the defects at surface or grain boundaries in perovskite film with fluoroarene embedded interfacial layer as a consequence of stronger halogen bonding with fluoroarene moieties or $NH-NH_2$ terminal. As a result, the PSC with a p-i-n configuration achieved superior operational thermal stability and a PCE exceeding 22 % with a large area of ~1 cm². This work underscores a universal strategy for defect passivation to further improvement of efficiency using a multifunctional passivator. This report gives insights into the film growth properties, device photo-physics, and defect analysis correlating with device performance and device stability.

I. INTRODUCTION

Lead perovskite solar cells (Pb-PSCs) have scaled up >25% benefiting from their exceptional optoelectronic properties. [1] However, this has imposed challenges for its practical application due to its lacking stability under heat and light stress as well as its susceptibility to a humid atmosphere. [2]–[6] The surface passivation approach has been widely employed in PSCs to improve the device parameters as well as stability. [7]

Several functional molecules have been used for passivating materials at the interfaces or additives in the perovskite precursor solution.[8]–[11] Gratzel and co-workers have used the fluoro in phenethyl chain as passivating materials for the improvement in device performance and its stability under a higher humid atmosphere.[12] Therefore, it is of great interest to explore the fluorinated functional materials in PSCs for modulating device performance and stability.

Here, we introduced a fluoroarene-anchored functional material; penta fluoro-phenylhydrazine (5F-PHZ) for interface treatment onto the MA-free Pb-HaP. This approach enhanced the device performance as high as 22.29 % (A~1 cm²) with superior operational stability. The 5F-PHZ treatment has shown a significant impact on the morphology, interface chemistry, and optoelectronic properties of HaP films. This report has discussed the synergetic effect in film growth and photo-physics of PSCs with interfacial passivation.

II. EXPERIMENTAL

A. Device fabrication

For the fabrication of MA-free RB-HaP; $FA_{0.84}Cs_{0.12}Rb_{0.04}PbI_3$: the precursor solution (1.05 M) was prepared by dissolving FAI (0.84 M), CsI (0.12 M), RbI (0.04 M), PbI_2 (1 M), and 5-AVAI (1 mM) in the mixture of dimethylformamide and dimethyl sulfoxide (4:1) solvent for 2 hours. The sputtered NiOx thin film was treated with MeO-2PACz by spin coating at 5000 rpm -50 s and subsequently dried at 100 ℃ for 10 min. For film deposition, the precursor was spin-coated at 1000 rpm-10 s and 5000 rpm-40 s followed by dripping 800 μl of CB at 34th s of 2nd step. Then, these as-grown films were simply placed on a hot plate at 60 ℃ for 1 min and at 100 ℃ for 45 min. For surface passivation, 5F-PHZ precursor solutions of different concentration (0.5 - 10 mol%/ml) was spin-coated onto the HaP film at 5000 rpm-40 s and annealed at 100 ℃-5 min. Then, we deposited C_{60} and BCP by thermal evaporation. Finally, Ag was thermally evaporated and get device. The detailed fabrication can be found in our earlier reports.[13], [14]

B. Materials and device characterizations

XRD patterns were measured using Rigaku Smart Lab, CuKα radiation, λ=1.5405Å. Scanning electron microscopy (SEM) images were obtained by a high-resolution scanning electron microscope (SEM) at 5 kV accelerating voltage (Hitachi, S-4800). The absorption and photoluminescence (PL) spectra were measured using UV-Vis-NIR spectrometer (UV-2600i, Shimadzu) and micro-PL spectrometer (HORIBA, LabRamHR-PL NF(UV-NIR). The current density–voltage (J-V) curves were measured under 1 sun with an AM1.5G spectral filter coupled with an MPPT system (Systemhouse Sunrise Corp.). Device certification was conducted in the National Institute of Advanced Industrial Science and Technology (AIST), Japan. It is registered as ISO / IEC 17025 accreditation laboratory (IA Japan ASNITE 0021 Calibration) according to international mutual recognition arrangements (MRA) for international laboratory (ILAC), and Asia pacific accreditation cooperation (APAC). Capacitance spectra (C–f) were collected using an LCR meter (IM3536, Hioki) under dark.

978-1-6654-6060-6/23 $31.00 © 2023 IEEE

III. RESULTS AND DISCUSSION

To examine the photovoltaic effect of 5F-PHZ treatment, the device structure is as depicted in Fig. 1a. The 5F-PHZ molecule is shown in the adjoining figure.

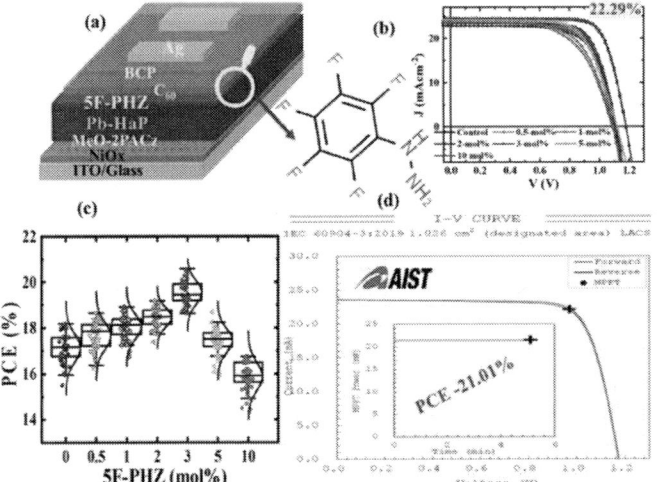

Figure 1. Device structure and molecular structure of 5F-PHZ (a). The J-V curves of Pb-PSCs with 5F-PHZ treatment concentartions (for x=0 – 10 mol%) (b). Device efficiency trend (c) and certified device efficiency (device of area 1.026 cm², at AIST).

Figure 1b presents the current density-voltage (J-V) curves of the control and 5F-PHZ passivation device with varying concentrations. The control device yields a PCE of 18.10%. The 5F-PHZ (3 mol%) treated device achieved PCE of 22.29% with an increase in $V_{OC} \sim$ 1.096 to 1.178 V, $J_{SC} \sim$ 22.88 to 24.51 mAcm⁻², and FF ~72.2 to 77.2%. The J-V curve with 5F-PHZ treated device has negligible hysteresis. It is reported that the fluorinated aromatic rings and NH-NH₂ terminals interact with perovskite and hence minimize the iodine vacancy, surface defect, and its migration with strong halogen bonding.[12] The PCE statistics as a function of the 5F-PHZ concentration is depicted in Figure 1c. The device efficiency was certified with PCE of ~21.01% (~1.026 cm²) in accredited independent photovoltaic test laboratory (AIST PV Lab, Japan) (Fig. 1d).

Figures 2a-c show SEM images of Pb-HaP with 5F-PHZ treatment. It indicates the film formation with a slight increase in grain size with a faint indication of the formation of an overlayer on the perovskite grain domain. A HaP Film with a higher concentration grows with overlayer surface features as a consequence of the adsorption of 5F-PHZ forming a 2D phase interacting with the lead iodide.

XRD patterns (Fig. 2d) were collected to investigate crystal growth. The control film grows with the dominant (110) plane of α-phase of HaP along with weak peaks of the δ- phase and residual PbI₂. Importantly, the characteristic diffraction peaks of δ- phase and PbI₂ disappeared on the 5F-PHZ treated (~3 mol%)

indicating the growth of better film quality. While an additional XRD peak appeared at <10° in the film 5F-PHZ (≥10 mol%) suggesting the formation of a 2D phase of (5F-PHZ)₂PbI₄.This observation underlines the importance of the 5F-PHZ treatment from the surface for improving perovskite film quality.

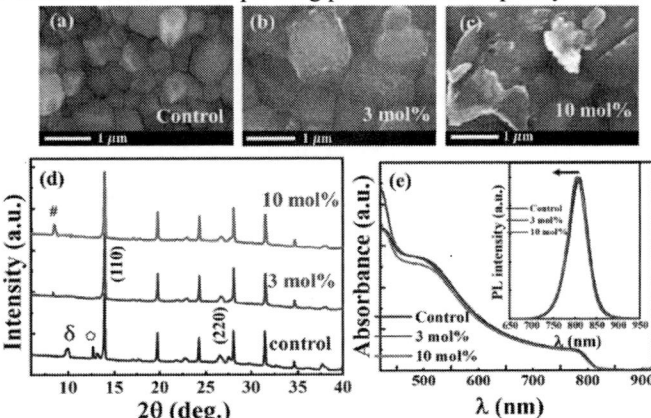

Figure 2. Effect of 5F-PHZ treatment: SEM image (a-c)), XRD patterns (#-2D phase, δ- non-photoactive perovskite phase, ⌂-PbI₂) (d), absorption spectra (e) (Inset- PL spectra).

The UV-vis spectra (Figure 2e) of respective films indicate no notable effect on absorption spectra of HaP films. The PL spectra (Fig. inset) also do not show any notable feature except a slight blue shift of PL characteristic peak (~819 to 817 nm).

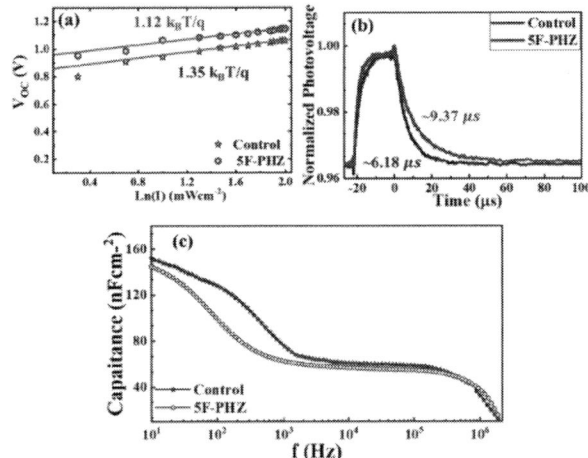

Fig. 3. Device characteristics; V_{OC}-I plot (a), TPV spectra (b), C-f spectra at room temperature (c).

To gain insight into photophysics, we investigated the light-intensity-dependent V_{OC} (Fig. 3a). The control device reveals a slope of 1.35 kBT/q which is higher than the 5F-PHZ treated device (1.12 kBT/q) indicating a reduction in trap-assisted recombination.[15], [16]

Figure 3b depicts the transient photovoltage (TPV) under transient illumination. The TPV decay signals demonstrates a carrier lifetime of 6.18 µs for the control device which is longer for the 5F-PHZ treated device (9.37 µs), indicating well

978-1-6654-6060-6/23 $31.00 © 2023 IEEE

consistency with device performance as a result of defect passivation.

Moreover, Figure 3c depicts the capacitance-frequency (C-f) spectra under the dark showing a slightly higher value in the plateau regime (1 to 100 kHz) that stems from the HaP layer accounting for defect dynamics. While the capacitance at a lower frequency reveals a much steeper feature for the control device. Thus, it indicates suppression of interfacial charge accumulation for the device with 5F-PHZ treatment.

IV. SUMMARY AND CONCLUSIONS

We demonstrated interfacial passivation on 3D-HaP to modulate the efficiency and stability of the inverted PSCs with sputtered NiO_x as HTL, enhanced PCE from ~18.10 to 22.29%. This surface treatment with 5F-PHz significantly modifies the surface chemistry and interfacial energy band due to strong halogen bonding induced by fluoroarene moieties coated on a 3D surface. This method is also applicable to PSCs with wide and narrow bandgap perovskite systems. The device analysis corroborates the suppression of defect densities for the 5F-PHZ treated device due to halogen bonding interaction with various fluoroarene derivatives.

ACKNOWLEDGMENT
This work was supported by JST-Mirai Program Grant Number JPMJMI21E6, Japan.

REFERENCES

[1] O. Almora *et al.*, "Device Performance of Emerging Photovoltaic Materials (Version 3)," *Adv. Energy Mater.*, vol. 13, no. 1, p. 2203313, Jan. 2023, doi: 10.1002/aenm.202203313.

[2] W. Tress *et al.*, "Performance of perovskite solar cells under simulated temperature-illumination real-world operating conditions," *Nat. Energy*, vol. 4, no. 7, pp. 568–574, 2019, doi: 10.1038/s41560-019-0400-8.

[3] D. B. Khadka, Y. Shirai, M. Yanagida, and K. Miyano, "Insights into Accelerated Degradation of Perovskite Solar Cells under Continuous Illumination Driven by Thermal Stress and Interfacial Junction," *ACS Appl. Energy Mater.*, vol. 4, no. 10, pp. 11121–11132, Oct. 2021, doi: 10.1021/acsaem.1c02037.

[4] D. B. Khadka, Y. Shirai, M. Yanagida, and K. Miyano, "Degradation of encapsulated perovskite solar cells driven by deep trap states and interfacial deterioration," *J. Mater. Chem. C*, vol. 6, no. 1, pp. 162–170, 2018, doi: 10.1039/C7TC03733C.

[5] I. Gueye *et al.*, "Chemical and Electronic Investigation of Buried NiO 1−δ , PCBM, and PTAA/MAPbI 3− x Cl x Interfaces Using Hard X-ray Photoelectron Spectroscopy and Transmission Electron Microscopy," *ACS Appl. Mater. Interfaces*, vol. 13, no. 42, pp. 50481–50490, 2021, doi: 10.1021/acsami.1c11215.

[6] D. B. Khadka, Y. Shirai, M. Yanagida, K. Uto, and K. Miyano,

"Analysis of degradation kinetics of halide perovskite solar cells induced by light and heat stress," *Sol. Energy Mater. Sol. Cells*, vol. 246, p. 111899, Oct. 2022, doi: 10.1016/j.solmat.2022.111899.

[7] Y. Liu *et al.*, "Ultrahydrophobic 3D/2D fluoroarene bilayer-based water-resistant perovskite solar cells with efficiencies exceeding 22%," *Sci. Adv.*, vol. 5, no. 6, p. eaaw2543, Jun. 2019, doi: 10.1126/sciadv.aaw2543.

[8] A. Q. Alanazi *et al.*, "Atomic-Level Microstructure of Efficient Formamidinium-Based Perovskite Solar Cells Stabilized by 5-Ammonium Valeric Acid Iodide Revealed by Multinuclear and Two-Dimensional Solid-State NMR," *J. Am. Chem. Soc.*, vol. 141, no. 44, pp. 17659–17669, 2019, doi: 10.1021/jacs.9b07381.

[9] D. B. Khadka, Y. Shirai, M. Yanagida, and K. Miyano, "Pseudohalide Functional Additives in Tin Halide Perovskite for Efficient and Stable Pb-Free Perovskite Solar Cells," *ACS Appl. Energy Mater.*, vol. 4, no. 11, pp. 12819–12826, Nov. 2021, doi: 10.1021/acsaem.1c02496.

[10] K. T. Cho *et al.*, "Water-Repellent Low-Dimensional Fluorous Perovskite as Interfacial Coating for 20% Efficient Solar Cells," *Nano Lett.*, vol. 18, no. 9, pp. 5467–5474, Aug. 2018, doi: 10.1021/acs.nanolett.8b01863.

[11] Z. Wang, Q. Lin, F. P. Chmiel, N. Sakai, L. M. Herz, and H. J. Snaith, "Efficient ambient-air-stable solar cells with 2D–3D heterostructured butylammonium-caesium-formamidinium lead halide perovskites," *Nat. Energy*, vol. 2, no. 9, p. 17135, Aug. 2017, doi: 10.1038/nenergy.2017.135.

[12] M. A. Ruiz-Preciado *et al.*, "Supramolecular Modulation of Hybrid Perovskite Solar Cells via Bifunctional Halogen Bonding Revealed by Two-Dimensional 19 F Solid-State NMR Spectroscopy," *J. Am. Chem. Soc.*, vol. 142, no. 3, pp. 1645–1654, Jan. 2020, doi: 10.1021/jacs.9b13701.

[13] D. B. Khadka, Y. Shirai, M. Yanagida, T. Tadano, and K. Miyano, "Interfacial Embedding for High-Efficiency and Stable Methylammonium-Free Perovskite Solar Cells with Fluoroarene Hydrazine," *Adv. Energy Mater.*, vol. 12, no. 38, p. 2202029, Oct. 2022, doi: 10.1002/aenm.202202029.

[14] D. B. Khadka, Y. Shirai, M. Yanagida, and K. Miyano, "Unraveling the Impacts Induced by Organic and Inorganic Hole Transport Layers in Inverted Halide Perovskite Solar Cells," *ACS Appl. Mater. Interfaces*, vol. 11, no. 7, pp. 7055–7065, Feb. 2019, doi: 10.1021/acsami.8b20924.

[15] D. B. Khadka, Y. Shirai, M. Yanagida, T. Masuda, and K. Miyano, "Enhancement in efficiency and optoelectronic quality of perovskite thin films annealed in MACl vapor," *Sustain. Energy Fuels*, vol. 1, no. 4, pp. 755–766, 2017, doi: 10.1039/C7SE00033B.

[16] D. B. Khadka, Y. Shirai, M. Yanagida, and K. Miyano, "Ammoniated aqueous precursor ink processed copper iodide as hole transport layer for inverted planar perovskite solar cells," *Sol. Energy Mater. Sol. Cells*, vol. 210, p. 110486, 2020, doi: https://doi.org/10.1016/j.solmat.2020.110486.

In-flight validation of End-Of-Life optimized Triple Junction solar cells onboard ASTROBIO cubesat

Luigi Schirone[1], Pierpaolo Granello[1], Matteo Ferrara[1], Matteo Avoli[1], Davide Imperatori[1], Nithin Maipan Davis[1], Lorenzo Iannascoli[1], Augusto Nascetti[1], Stefano Carletta[1], Claudio Paris[1], Erminio Greco[2], and Roberta Campesato[2]

[1]Sapienza University of Rome, Roma, Italy - [2]CESI, Milano, ITALY

Abstract — **Small satellites are known to provide low-cost access to space, enabling fast and cheap validation of new technologies for space missions. ASTROBIO is a 3U CubeSat (size 100x100x300 mm3), that was injected into an orbit crossing several times a day the cloud of radiation known as the Van Allen Belts. As a result, the solar panels are exposed to an enhanced dose rate and undergo in a few days the same degradation that would be observed after several years of space flight in Low Earth Orbit. We used this unique opportunity to validate the new End-Of-Life optimized Triple Junction solar cells (CTJ-EOL) produced in Italy by CESI.**

The spacecraft solar panel comprises two types of solar cells produced by the same manufacturer: the well-established CTJ-30 and the new CTJ-EOL. Their performance was separately monitored and the related telemetries were downloaded for off-line elaboration.

This manuscript reports and analyses the telemetries received by our Ground Station and by the networked stations distributed all over the globe. Data analysis clearly and unequivocally demonstrated the superior ruggedness of the End-Of-Life optimized devices: in the first week, the CTJ-30 performance dropped like it would be expected along a 10-years+ mission in the most-commonly-used orbits, whereas degradation of the CTJ-EOL was barely detectable.

The results also show that in-orbit testing can provide unique information about the operation of solar cells in the intended operational environment as well as an easy verification of the device's functionality, useful to complement standard characterization procedures for radiation hardness involving costly and time-consuming laboratory tests under particle beams.

I. INTRODUCTION

AstroBio CubeSat (ABCS) was a mission funded by the Italian Space Agency (ASI) aimed at validating the use of nanosatellites, developed by using the CubeSat standard, to perform autonomous research in astrobiology [1].

The spacecraft (Fig. 1) was built and managed by the School of Aerospace Engineering of Sapienza University of Rome. It was launched on 13 July 2022 from Guiana Space Centre as a secondary payload of the Vega-C launch vehicle, at his maiden flight and was deployed along a circular orbit with altitude of 5890 km, and inclination of 70°, with 224 minutes orbital period and eclipse duration of 28 minutes. Consequently, ABCS spent a significant fraction of its operative life within the the cloud of radiation known as the Van Allen Belts, a region of the magnetosphere characterized by high-energy protons and electrons [2].

Since from the beginning of mission analysis, it was clear that ABCS would collect a total ionizing dose (TID) significantly higher than that collected by CubeSats in low Earth orbit (LEO). Actually, the SPace ENVIronment Information System (SPENVIS) provided by the European Space Agency ESA estimated for the specific orbit of ASTROBIO, a TID at least three orders of magnitude larger than the average one for Low-altitude Earth Orbits (LEO) satellites. This lead to develop specific design techniques to provide an operative life longer than a few days.

Fig. 1. (a) The AstroBio CubeSat, ready for launch. (b) Detail of the solar panel.

II. SOLAR CELLS EMPLOYED IN THE ASTROBIO CUBESAT SOLAR ARRAY

The optimization of the solar cell structure as a function of the particular mission, is a key point for designing and sizing the solar array.

Considering the strong radiation environment of this mission, two different types of solar cells were used: the well-established CTJ-30 (fully qualified for GEO and LEO applications) and a new solar cell design, named CTJ-EOL, conceived to withstand high particle irradiation intensity, to provide improved performance at End Of Life (EOL). The CTJ-EOL solar cells exploit a novel design, expected to boost performance with respect to other previously proposed structures optimized for EOL performance [3-7].

Both CTJ-EOL and CTJ-30 solar cells are based on the well-known triple junction InGaP/InGaAs/Ge architecture. In this configuration, it is fully recognized that the middle InGaAs sub-cell is the most radiation sensitive, degrading faster under electrons and protons irradiation.

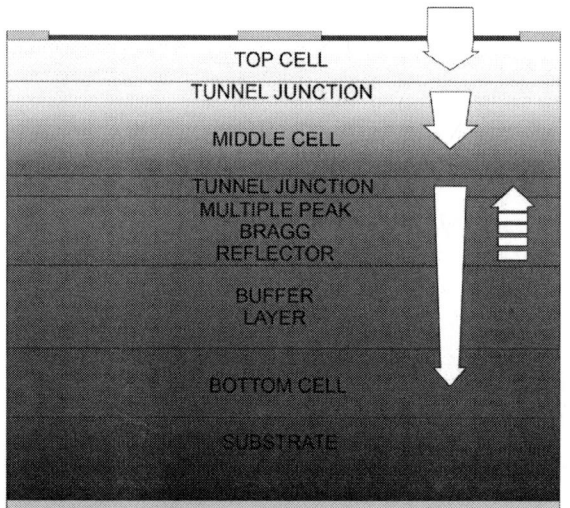

Fig. 2. Structure of the CTJ-EOL solar cell.

The CTJ-EOL design boosts the radiation tolerance of the middle InGaAs sub-cell. For this purpose, as represented in Fig. 2, the base layer of the middle InGaAs pn junction was reduced as much as possible, maximizing the optical absorption by means of light trapping structures placed underneath the base layer. More precisely, a multiple-peak Bragg reflector, having a reflectivity band that matches with the range of wavelengths transmitted by the thinned base layer, was designed and optimized for this solar cell [8].

The CTJ-EOL and CTJ-30 remaining factors (RF) of each electrical parameter are reported in Table I.

The figure of merit (RF$_{relative}$) used to compare the electrical performances of the new structures with the standard CTJ-30 is the ratio between their EOL efficiency and the BOL efficiency of the CTJ-30 devices:

$$RF_{relative} = \frac{Eff_{EOL}}{Eff(CTJ30)_{BOL}} \quad (1)$$

In the case of CTJ-30, the RF$_{relative}$ will correspond with the RPF (Remaining Power Factor) of the CTJ-30 solar cell.

TABLE I
REMAINING FACTORS VS. FLUENCE OF EQUIVALENT 1 MeV ELECTRONS

	FLUENCE (e⁻/cm2)	RF_Isc	RF_Voc	RF_Pmax
CTJ-30	1E+15	0.91	0.93	0.83
	3E+15	0.79	0.92	0.69
	1E+16	0.66	0.88	0.53
CTJ-EOL	1E+15	0.98	0.95	0.93
	3E+15	0.97	0.93	0.87
	1E+16	0.84	0.89	0.72

In Fig. 3, the RF$_{relative}$ of the CTJ-EOL design are reported and compared with those of the standard CTJ-30. It is evident that at BOL and up to 1E+15 e-/cm^2 the CTJ-30 standard structure is superior with respect to the CTJ-EOL design. On the other hand, at higher fluences (i.e. from 1E15 e-/cm^2 to 1e16 e-/cm^2) the CTJ-EOL design shows better performances.

Analysis of the telemetry data confirms the irradiation test results obtained for both the solar cell types.

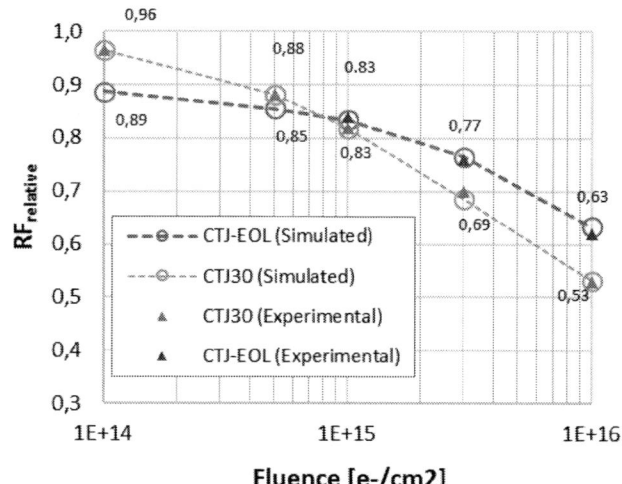

Fig. 3. RF$_{relative}$ of the solar cells in use versus fluence of equivalent 1 MeV electrons.

The different design also affects solar cells performance at Begin of Life (BOL), as it can be observed in the typical J-V characteristics reported in Fig. 4.

In particular, in CTJ-EOL solar cells the current density at BOL is appreciably reduced with respect to the CTJ-30 design (16.4 mA/cm^2 vs 17.7 mA/cm^2). Even if the current density, at BOL conditions, is lower in CTJ-EOL design, its value in EOL (1 MeV 3e15 e-/cm^2) is higher thanks to the improved radiation tolerance of the middle InGaAs sub-cell (15.9 mA/cm^2 vs 14.2 mA/cm^2).

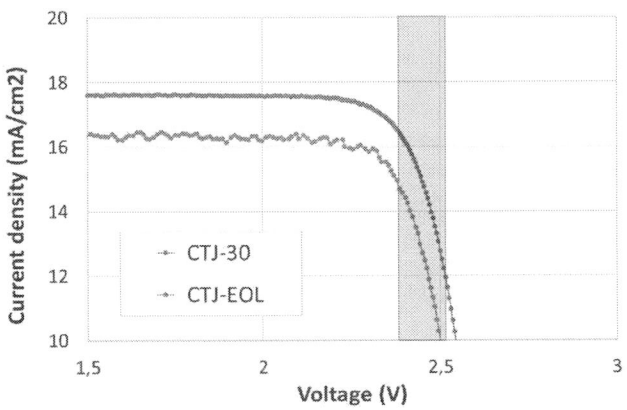

Fig. 4. Zoom on the Typical AM0 current-voltage characteristics at BOL for both kinds of solar cells (zoom around the operational voltage range). The shaded area represents the typical range of operating voltages during the ABCS mission.

III. The Solar Array of the ASTROBIO CubeSat

The solar array of ABCS consisted of four solar panels, integrated on the 100x300 mm² faces. As can be observed in Fig. 4, each of them comprised six CTJ-30 solar cells, with size 30.15 cm², and three CTJ-EOL solar cells [8], with size 8 cm².

The solar cells were arranged as independent strings consisting of a blocking diode and three cells in series, singularly protected by a bypass diode. Any solar panel comprised two strings with CTJ-30 and one with CTJ-EOL. In Fig. 5 also are schematically represented the current sensors, providing independent monitoring of the strings containing CTJ-EOL and CTJ-30 solar cells. At least two telemetries per minute were recorded and transmitted to Earth.

The bulk of primary power (max 7.5W per panel in AM0) was provided by the well-established CTJ-30, arranged as two

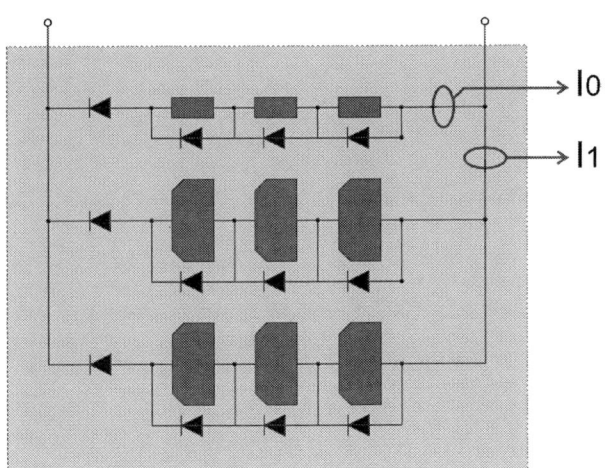

Fig. 5. Interconnections among solar cells in a solar panel.

independent strings comprising three cells connected in series. Otherwise. the CTJ-EOL strings provide a minor contribution to the power budget of the satellite (less than 1.3W in AM0 conditions), mainly for the small size of the solar cell. Moreover, the parallel connection of the strings accommodated in each solar panel forces the CTJ-EOL cells, having a smaller open circuit voltage, to operate at smaller current densities.

IV. The Observed Performance

The solar array currents for the CTJ-30 strings and for the CTJ_EOL strings were separately monitored, according to the setup highlighted in Fig. 5 (I0 and I1, respectively). The observed current densities are reported in the figures 6 and 7. The trend of the received telemetry data was smoothed in post processing by a low-pass filter, in order to remove any signature of short-term variations.

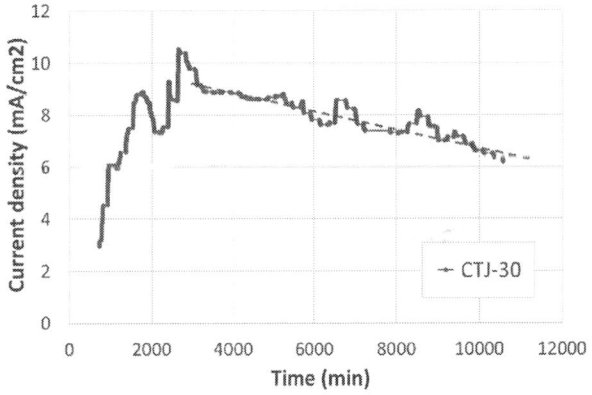

Fig. 6. Average current density in the CTJ-30 strings (I1) along the operation time.

Telemetry of current density in the CTJ-30 strings (I1) is reported in Fig. 6. The low average current observed at the begin of the operations is related to the initial battery voltage, close to its End-of-Charge value, that was causing the blocking diodes to be reverse biased most of the time.

A peak is reached after a couple of days. Later, the current starts decreasing with an average 15% drop from day two to day five (minutes from 3000 to 8500). According to the datasheet of the CTJ-30 solar cells, such loss of current could be related to an absorbed radiation fluence larger than 1E15 1MeV equivalent electrons per square cm. This is consistent with the particle fluxes estimated by the SPace ENVironment Information System (SPENVIS), that provides, for the specific orbit of AstroBio CubeSat a radiation fluence larger by three orders of magnitude than in common LEO missions.

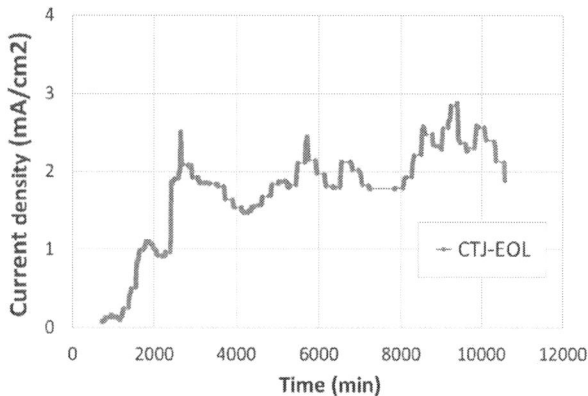

Fig. 7. Average current density in the CTJ_EOL string (I0) along the time of operation, given in minutes after the deployment in orbit.

The telemetries of the current in the CTJ-EOL string are reported in Figure 7. It is manifest that the decreasing trend observed in Fig. 6 no more is deceivable. On the other hand, these solar cells have been designed to feature a reduced sensitivity to radiations and the observed behavior seems to demonstrate that they met the expectations.

An accurate analysis of the telemetries reveals that the reported average current density is appreciably smaller than its AM0 values. This could be related to both the effect of eclipses and of spacecraft tumbling. Actually, during its flight in a circular orbit with altitude 5890 km, ABCS spends a total of 28 minutes out of 224 minutes in eclipse. Moreover, during the first days after the deployment, the passive magnetic attitude control system had not yet stabilized the spacecraft attitude with respect to the Earth's magnetic field, so that the angle of incidence of the solar light was chaotically varying [9]. In fact, the satellite's tumbling can be described by the combination of two rotations around orthogonal axes centered in the satellite center of mass. These oscillations of the tilt

angle shaped the amplitude of the generated current, as shown in Fig. 8. The fragmentary periodic sequences also provide information about the rotational state, useful to complement the outputs of the on-board Inertial Measuring Unit.

Indeed, the average current is also reduced by other effects: 1) the solar panels are only mounted on four faces; 2) current flow is stopped by the blocking diodes as soon as the string voltage drops below the battery voltage, and this happens before the solar panel goes in shadow; 3) increased reflectivity of glass surfaces at high tilt angles (a minor effect).

On the other hand, temperature of the solar panels remains in the range 0 13°C [10], as reported in Figure 9a. A detail of the trend observed for some orbits in a sequence, shown in fig. 9b, reveals that the maximum variation rate is approximately +5 deg/h in sunlight and 3 deg/h during eclipse. This is consistent with the typical behaviour of body-mounted solar panels.

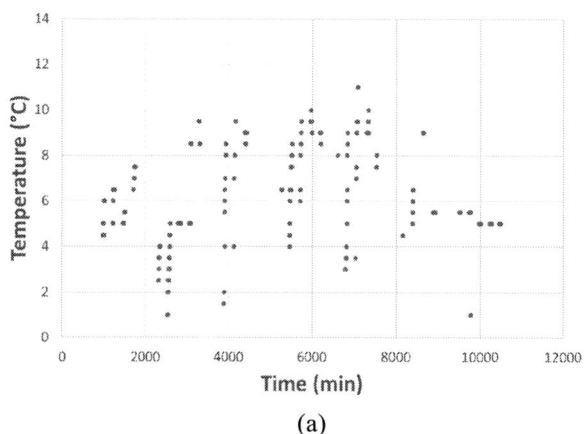

(a)

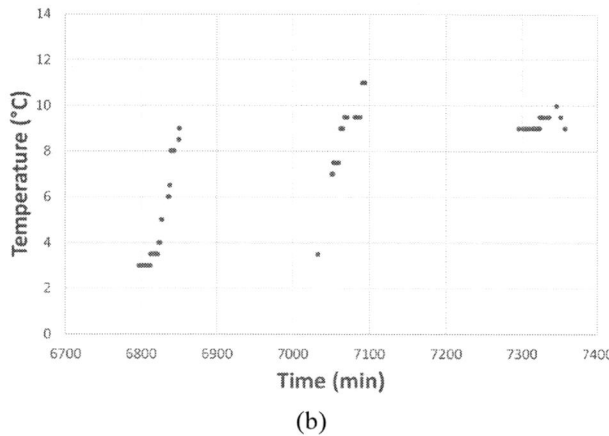

(b)

Fig. 9. Solar panels temperature: (a) general trend; (b) detail for three orbits.

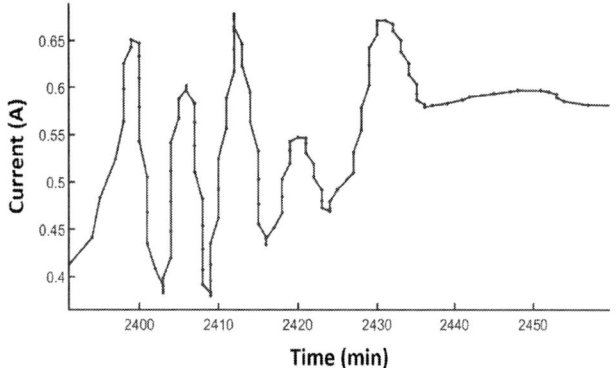

Fig. 8. Trend of the overall current generated by the solar array in a short time scale. The fluctuations are impressed by spacecraft tumbling.

V. SUMMARY

In-flight validation of the CTJ-EOL solar cells was one of the scientific goals of the ABCS mission. Analysis of the available data clearly demonstrated the superior ruggedness of the End-Of-Life optimized solar cells used for ABCS.

The obtained results also show that in-orbit testing campaigns can provide unique information about the operation of solar cells in the intended operational environment as well as an easy verification of the device's functionality, useful to complement standard characterization procedures for radiation hardness, involving costly and time-consuming laboratory tests under particle beams.

ACKNOWLEDGEMENTS

The authors acknowledge ASI, the Italian Space Agency, for supporting the ABCS mission and ESA (European Space Agency) for the launch opportunity.

REFERENCES

[1] L. Iannascoli, A. Nascetti, S. Carletta, L. Schirone, A. Meneghin, J.R. Brucato, D. Paglialunga, G. Poggiali, S. Pirrotta, G. Impresario, A. Sabatini, C. Pacelli, L. Anfossi, M. Mirasoli, M. Balsamo, L. Popova, A. Donati, A. Bardi, D. Caputo, G. de Cesare, "Astrobio cubesat: Enabling technologies for astrobiology research in space", Proceedings of the International Astronautical Congres (IAC), Virtual, On-line, 2020

[2] J. A. Van Allen, G. H. Ludwig, E. C. Ray and C. E. Mc Ilwain, "Observation of high intensity radiation by satellites 1958 Alpha and Gamma", Jet Propulsion 28, (1958) pp. 588-592

[3] B. C. Richards et al., "Performance and radiation resistance of quantum dot multi-junction solar cells," 2013 IEEE 39th Photovoltaic Specialists Conference (PVSC), Tampa, FL, USA, 2013, pp. 0158-0161, doi: 10.1109/PVSC.2013.6744119

[4] G. F. X. Strobl et al., "Development of lightweight space solar cells with 30% efficiency at end-of-life," 2014 IEEE 40th Photovoltaic Specialist Conference (PVSC), Denver, CO, USA, 2014, pp. 3595-3600, doi: 10.1109/PVSC.2014.6924884

[5] W. Guter, F. Dunzer, L. Ebel, K. Hillerich, W. Köstler, T. Kubera, M. Meusel, B. Postels, C. Wächter, "Space Solar Cells – 3G30 and Next Generation Radiation Hard Products ", E3S Web Conf. 16 03005 (2017), DOI: 10.1051/e3sconf/20171603005

[6] Jonas Schön, Gunther M. M. W. Bissels, Peter Mulder, Rosalinda H. van Leest, Natasha Gruginskie, Elias Vlieg, David Chojniak, David Lackner, "Improvements in ultra-light and flexible epitaxial lift-off GaInP/GaAs/GaInAs solar cells for space applications", Prog Photovolt Res Appl. 2022;30, p. 1003–1011, doi: 10.1002/pip.3542.

[7] Malte Klitzke, Jonas Schön, Rosalinda H. Van Leest, Gunther M.M.W. Bissels, Elias Vlieg, Michael Schachtner, Frank Dimroth, David Lackner, "Ultra-lightweight and flexible inverted metamorphic four junction solar cells for space applications", EPJ Photovoltaics 13, 25 (2022)

[8] R. Campesato, E. Greco, M. Casale and L. Schirone, "Advanced Triple Junction Solar Cells with Innovative Assembly for space application: development and flight test," 2021 IEEE 48th Photovoltaic Specialists Conference (PVSC), Fort Lauderdale, FL, USA, 2021, pp. 0316-0321, doi: 10.1109/PVSC43889.2021.9518590.

[9] S. Carletta, A. Nascetti, S. S. Gosikere Matadha, L. Iannascoli, T. Baratto de Albuquerque, N. Maipan Davis, L. Schirone, G. Impresario, S. Pirrotta, J. R. Brucato, "Characterization and Testing of the Passive Magnetic Attitude Control System for the 3U AstroBio CubeSat", Aerospace, 9 (11), (2022), 10.3390/aerospace9110723

[10] Grás, A. 1, Blanco, G. 1, 2, Baur, C., "Reverse characteristic of triple-junction solar cells under different insolation and temperature conditions", 28th European Photovoltaic Solar Energy Conference and Exhibition (28th EU PVSEC), Paris, France, 30th September to 4th October 2013, pp. 69-72

978-1-6654-6060-6/23 $31.00 © 2023 IEEE

Evaluation of Module Mismatch Losses and Generation Impact in Utility Scale PV Systems

Sara M. MacAlpine, David A. Bowersox

JUWI, Boulder, CO, United States

This work examines the impact of different photovoltaic (PV) module-to-module mismatch losses on the system level performance for single axis tracking PV systems with DC-AC ratios ranging from 1.0-1.5, typical of the utility scale PV market. Module-to-module mismatch losses arising from varied electrical characteristics as well as temperature and irradiance differences are simulated over a typical 35 year project lifespan. Results are presented evaluating the influence of site meteorological characteristics, DC-AC ratio, and module-to-module variation and binning when predicting utility scale PV system performance.

SCAPS-1D simulations of CdTe based Solar Cells with an Amorphous Silicon-based Back Buffer

Abdul Quader, Venkanna Kanneboina, Prabin Dulal, Madan Mainali, Bishal Shrestha, Ebin Bastola, Adam B. Phillips, Ambalanath Shan, Nicholas J. Podraza and Michael J. Heben

Wright Center for Photovoltaics Innovation and Commercialization (PVIC), Department of Physics and Astronomy, University of Toledo, Toledo, Ohio, 43606, USA

Abstract — **The Voc of CdTe solar cells is currently limited below 900 mV and several strategies to improve the Voc are being pursued. There are several strategies to achieve higher V_{OC} can be taken such as increasing the hole concentration to $2x10^{14}$ cm^{-3} with increasing minority carrier lifetime in the bulk by passivating active recombination centers, reducing recombination current densities at the interfaces with of window layer and back metal electrode with CdTe film. It is very hard to find a metal contact that can have an ohmic contact with CdTe. To overcome the problem a back buffer (BB) layer between CdTe and back metal contact can be used so that it can handle the initial Fermi energy level offsets and hence the band alignment both with the CdTe and the metal electrode. Therefore, it will be easier for the holes to reach the electrode. Here we deposited p-type hydrogenated amorphous Silicon based buffer layer between CdTe and the metal electrode. We used SCAPS-1D to identify the effect of majority charge carrier concentration in CdTe and a-Si:H(p) or a-SiC:H(p) and effect of interface recombination velocities at CdTe/ BB interface where s-kinks in the current density-voltage (J-V) curves were removed and overall performance of the devices were enhanced. The analysis also shows the effect of change in the bandgap that can be done both by hydrogenation and C incorporation. We successfully showed a pathway to achieve more that 900 mV and, hence, improve the efficiency of the device.**

I. INTRODUCTION

With a record efficiency of 22.1 %, CdTe has achieved only 67% of its detailed balance limit [1]. In contrast, record silicon and GaAs based solar cells have reached 87% of their detailed balanced limit [2]. CdCl$_2$ treatments improve the grain size and grain boundaries in the CdTe films, and CuCl$_2$ treatments can be used to dope the CdTe and reduce the back barrier height [3], but the problem of a low V_{OC} remains.

The charge carrier concentration in CdTe and increased majority carrier lifetime cannot help increasing the V_{OC} of the device [4]. Open circuit voltage can be increased when the recombination current densities at the interfaces can be reduced and the initial Fermi energy level (IFLO) [5] at the interface can also dictate the amount of V_{OC} extracted from the device. Theintial band alignment between the CdTe and hydrogenated amorphous Silicon at the back showed the hope to have higher V_{OC} and hence higher performance from the device.

Here, simulations were performed with SCAPS-1D software [6] to show the influence of hole concentrations of about 2x1013 cm-3 and 2x1014 cm-3 in CdTe and hole concentraions from 6x10-13 cm-3 to 6x10-18 cm-3 a-Si:H(p) or a-SiC:H(p) and CdTe/BB interface on the performance of the device.

II. DEVICE STRUCTURE

For the model, we chose a superstrate configuration with 300 nm of florine-doped tin oxide (FTO) on soda lime glass (SLG), 100 nm of Mg$_{1-x}$Zn$_x$O (x = 0.23) (MZO), 3.3 μm thick CdTe, and 10 nm thick a-Si:H(p) or a-SiC:H(p) back buffer (BB). In order to see the effect of CdTe/BB interface, the flat band condition was chosen for the BB/back metal contact. The band gap of MZO was set to 3.7 eV for the x = 0.23 composition and the conduction band offset (CBO) was +0.2 eV at the MZO/CdTe interface, following previous studies [7]. Fig 1

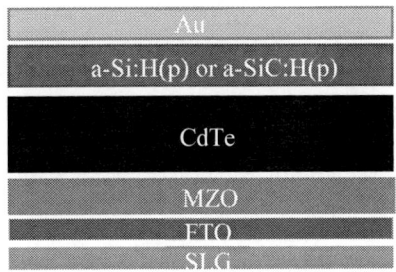

Fig. 1. Device configuration SLG/FTO/MZO/CdTe/a-Si:H(p)

III. RESULTS AND DISCUSSION

Plasma enhanced chemical vapor deposition (PECVD) can be used to deposit the a-Si:H(p) and a-SiC:H(p) films. PECVD technique, allows films to be desposited at low temperatures [8] which may be important for CdTe devices. The recipe used to deposit a-SiC:H(p) and a-SiC:H(p) employs gas mixture of silane (SiH$_4$), methane (CH$_4$), and diborane (B$_2$H$_6$) or trimethyldiborane (TMB – ((CH$_3$)$_2$BH)$_2$) as the sources for silicon, carbon, and boron, respectively, to prepare boron doped, p-type a-Si:H(p) and p-type a-SiC:H and additional source of H$_2$ [9].

In order to dive into details of the influence of the BB on the device, the effect of H2 should be discussed first. This is because in PECVD technique H2 plasma can etch the CdTe and this is evident from the previous experiments [10]. The rate of

978-1-6654-6060-6/23 $31.00 © 2023 IEEE

etching depends on the power used by the particular deposition

different BB-hole concentrations. For different BB-hole

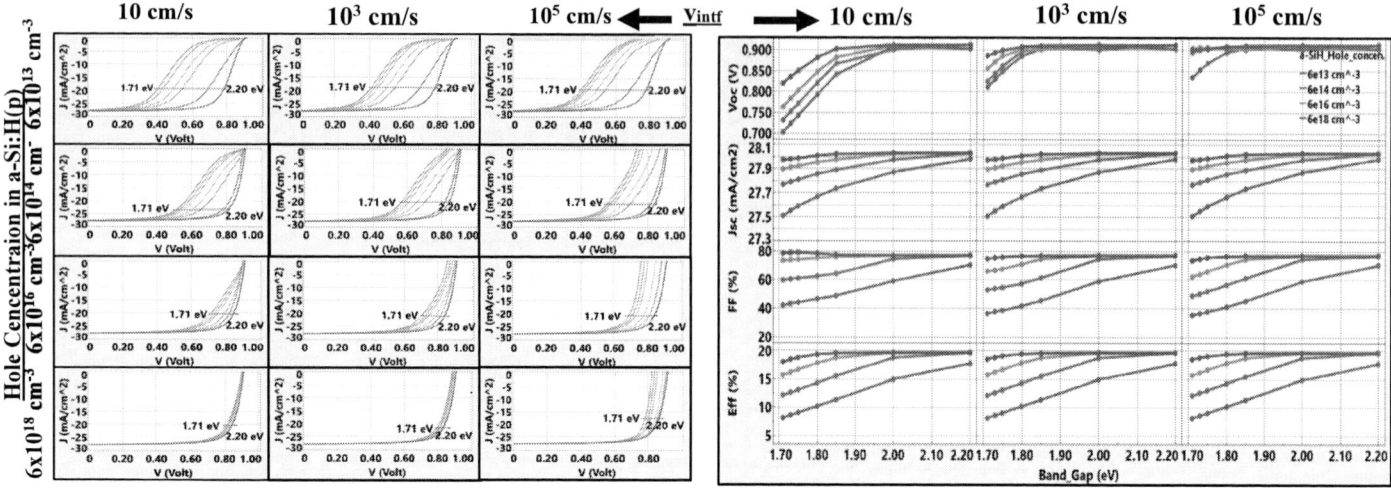

Fig. 2. For CdTe hole concentraionof $2x10^{13}$ cm^{-3} and effect of interface recombination velocities on : (a) J-V curves various hole concentration in p-type a-Si:H at different bandgap energies. (b) Variation in device parameters with bandgap at different BB-hole cocentraion

recipe and the etch rate increases if power increases. Additionally, the H2 may go into the CdTe film and passivate the active recombination centers in the film and obviously in the a-Si:H films as well. Besides etching the H$_2$ incorporation can increase the band gap of the material from 1.71 eV to 1.85 eV if fraction of H$_2$ goes from 16% to 30% [11].

The incorporation of C in a-Si:H has direct effect on the valence band location of the material. Previous experiments showed that with incorporation of C the valence band moves downward and that increases the band gap of the device. Both effects of incorporation of C and H$_2$ can be expressed as the increasing of band gap while location of the conduction band and electron affinity remain same [2].

Fig. 2(a) shows how the s-kink is removed as the hole concentration in the BB is changed when CdTe hole concentration is held constant at $2x10^{13}$ cm^{-3}. When the hole concentration in the BB is increased from $6x10^{13}$ cm^{-3} to $6x10^{18}$ cm^{-3}, the Fermi energy level in the p-type BB moves more towards the valence band and hence the back barrier height between CdTe and BB is increasesd. We also investigated the effect of varying Interface recombination velocities at the CdTe/BB interface that mimicks the quality of the interface.

Fig. 2(a) also shows the effect of interface recombination velocities, v$_{intf}$ on the J-V characteristics of the devices. It is clearly seen that the s-kinks in the curves remain for BB of $6x10^{13}$ cm^{-3} and for all v$_{intf}$s. We started seeing the removal of the s-kinks and higher BB-bandgap for $6x10^{14}$ cm^{-3} and vintf of 10^5 cm/s. After that the device performances keep getting improved for all other carrier concentrations and vintfs without any s-kinks.

Fig 2(b) shows the influence of the increase in the bandgap of the BB on the device parameters for different v$_{intf}$s and

concentration the trends in the change in the V$_{OC}$, J$_{SC}$, FF and Eff as a function of BB-bandgap energy are almost same. regardless of interface recombination velocities the V$_{OC}$ in the devices kept getting improved to more 900 mV where FF (Fill Factor) reaches maximum of 76. 80% and efficiency (Eff) reaches 19.64%. The JSC remains almost steady for the

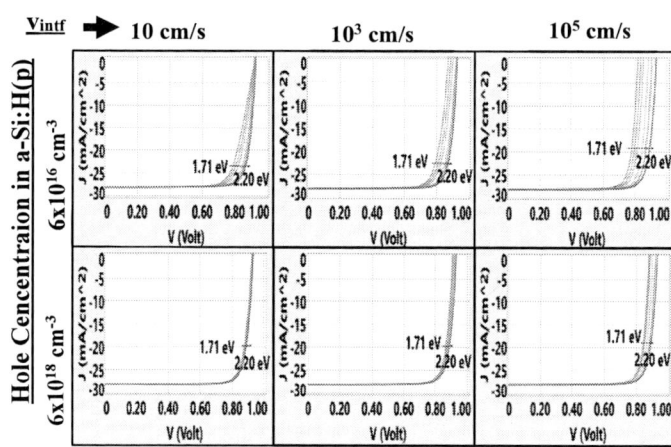

Fig 3. For CdTe hole concentraionof $2x10^{14}$ cm^{-3} and effect of interface recombination velocities on the J-V curves various hole concentration in p-type a-Si:H at different bandgap energies

configuraions. Fig 3 shows the J-V characteristics of the devices with BB that have hole concentrations of $6x10^{16}$ cm^{-3} and $6x10^{18}$ cm^{-3} and the hole concentration in the CdTe is $2x10^{14}$ cm^{-3}. Regardless of the carrierconcentrations in CdTe and BB and regardless of the interface recombination velocities at the CdTe/BB interface there is no evidence of the s-kinks in the J-V curves of the devices. The CdTe/BB carrier concentraions of $2x10^{14}$ cm^{-3}/$6x10^{18}$ cm^{-3} shows the best output behavior without s-kinks. The other charge configuration set up for the CdTe and p-type BB showed s-kinks and did not converge for all the back barrier heights but they showed s-kinks in the J-V curves. Fig 4 shows the effect of carrier concentrations in the BB and the vintfs. The V_{OC} in all configurations reach more than 900 mV mostly from 1.85 eV to 2.20 eV of BB-bandgap. The efficiency reaches app. 22% again for bandgap of BB from 1.85 eV to 2.20 eV. The FF reached around 84% for the same bandgap range, where JSC follows same treand as the V_{OC}.

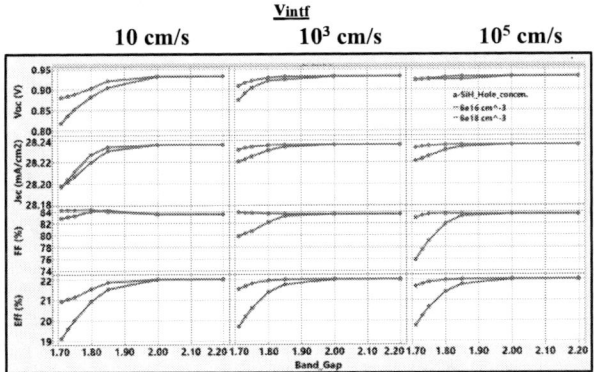

Fig 4. For CdTe hole concentraionof $2x10^{14}$ cm^{-3} and Variation in device parameters with bandgap at different BB-hole cocentraion

IV. CONCLUSION

The band gap of BB was varied from 1.71 eV to 2.2 eV corresponding to the effect of H$_2$ and C incorporation. It has been showed that the easily achievable hole concentration in CdTe can produce more that 900 mV open circuit voltage even though the J-V curves has S-kink in them where the FF increases from around 35% to 70%. If we increase the hole concentration in the BB to $6x10^{18}$ cm^{-3} the open circuit voltage increases to more than 900 mV but the efficiency of the device remained below 20%. For the the case of CdTe hole concentration $2x10^{14}$ cm^{-3} and hole contrations in BB to $6x10^{18}$ cm^{-3} the V_{OC} reaches more than 900 mV for all the band gaps of BB and efficiency of the devices remain around 22%.

ACKNOWLEDGEMENTS

This material is based on research sponsored by the U. S. DOE's office of Energy Efficiency and Renewable Energy (EERE) under Solar Energy Technologies Office (SETO) Agreement DE-EE0008974 and Air Force Research Laboratory under agreement number FA9453-18-2-0037 and FA9453-19-C-1002. The US government is authorized to reproduce and distribute reprints for governmental purposes notwithstanding any copyright notation thereon

REFERENCES

[1] W. Shockley and H. J. Queisser, "Detailed Balance Limit of Efficiency of p‑n Junction Solar Cells," *Journal of Applied Physics,* vol. 32, no. 3, pp. 510-519, 1961.Published.

[2] Y. Zhao *et al.*, "Monocrystalline CdTe solar cells with open-circuit voltage over 1 V and efficiency of 17%," *Nature Energy,* vol. 1, no. 6, p. 16067, 2016.Published.

[3] E. Bastola, A. V. Bordovalos, E. LeBlanc, N. Shrestha, M. O. Reese, and R. J. Ellingson, "Doping of CdTe using CuCl2 Solution for Highly Efficient Photovoltaic Devices," in *2019 IEEE 46th Photovoltaic Specialists Conference (PVSC)*, 16-21 June 2019 2019, pp. 1846-1850.

[4] J. Sites and J. Pan, "Strategies to increase CdTe solar-cell voltage," *Thin Solid Films,* vol. 515, no. 15, pp. 6099-6102, 2007.

[5] G. K. Liyanage, A. B. Phillips, F. K. Alfadhili, R. J. Ellingson, and M. J. Heben, "The Role of Back Buffer Layers and Absorber Properties for >25% Efficient CdTe Solar Cells," *ACS Applied Energy Materials,* vol. 2, no. 8, pp. 5419-5426, 2019.

[6] M. Burgelman, P. Nollet, and S. Degrave, "Modelling polycrystalline semiconductor solar cells," *Thin Solid Films,* vol. 361-362, pp. 527-532, 2000.

[7] T. Song, A. Kanevce, and J. R. Sites, "Emitter/absorber interface of CdTe solar cells," *Journal of Applied Physics,* vol. 119, no. 23, 2016.

[8] E. A. Filatova, D. Hausmann, and S. D. Elliott, "Understanding the Mechanism of SiC Plasma-Enhanced Chemical Vapor Deposition (PECVD) and Developing Routes toward SiC Atomic Layer eposition (ALD) with Density Functional Theory," *ACS Applied Materials & Interfaces,* vol. 10, no. 17, pp. 15216-15225, 2018.

[9] M. Boccard and Z. C. Holman, "Amorphous silicon carbide passivating layers for crystalline-silicon-based heterojunction solar cells," *Journal of Applied Physics,* vol. 118, no. 6, p. 065704, 2015.

[10] P. Koirala *et al.*, "Investigation of doped a-Si1−xCx:H as a novel back contact material for CdTe solar cells," in *2014 IEEE 40th Photovoltaic Specialist Conference (PVSC)*, 8-13 June 2014, pp. 2354-2359, 2014.

[11] T. F. Schulze, L. Korte, F. Ruske, and B. Rech, "Band lineup in amorphous/crystalline silicon heterojunctions and the impact of hydrogen microstructure and topological disorder," *Physical Review B,* vol. 83, no. 16, p. 165314, 2011.

Impact of current collecting grids on the scalability of 3-terminal perovskite/silicon tandems with bipolar transistor architecture

Gemma Giliberti, Federica Cappelluti

Department of Electronics and Telecommunications, Politecnico di Torino, Corso Duca degli Abruzzi 24, 10129 Torino, Italy

Abstract—The heterojunction bipolar transistor (HBT) structure is an attractive solution for developing three-terminal perovskite/silicon tandem solar cells compatible with dominant silicon photovoltaic devices, such as PERC and heterojunction. However, in contrast to three-terminal tandems based on interdigitated back contact silicon cells, the three-terminal HBT requires the implementation of the third contact at the base (middle) layer. To this aim, the simplest solution is to access the base layer from the cell front side by implementing a grid layout with top interdigitated contacts (TIC). In this work, we elaborate on the feasibility of the HBT structure for 3T perovskite/silicon tandem solar cells. We report, based on optical and drift-diffusion simulations, proof-of-concept designs with high efficiency potential, and we analyze, with the aid of circuit level simulations, the implications of a TIC grid layout in the perspective of scaling up to large areas. Our results show that the HBT architecture is a promising candidate for developing 3T perovskite/silicon tandem solar cells compatible with industry standard silicon photovoltaics.

Index Terms—perovskite-silicon tandem solar cell, three-terminal, power loss, physics-based simulation, circuit-level simulation.

I. INTRODUCTION

Perovskite/silicon (PVS) tandem solar cells are among the most promising candidates for next generation photovoltaics [1]. They recently achieved the efficiency record of 32.5% [2], approaching that one of costly III-V double-junction solar cells [2]. Besides two- and four-terminal approaches, three-terminal (3T) tandems based on interdigitated back contact (IBC) silicon cells are currently being investigated [3], in view of the advantages of the 3T approach in terms of energy yield in the field [4].

The 3T-HBT solar cell [5], [6] offers an attractive alternative for the development of 3T PVS tandem solar cells suitable for integration with industry standard photovoltaic technologies such as PERC and heterojunction (HTJ) silicon cells. The 3T-HBT solar cell has a theoretical (material-based) efficiency limit as high as that one of a perfectly matched 2-terminal tandem, but with a simpler structure because the top and bottom subcells are seamlessly connected, without the need of any additional interconnecting layer. However, in the perspective of large-area devices, the drawback might come from the realization of the third terminal at the base layer. In fact, implementing the 3T-HBT in the PERC or HTJ silicon plat-

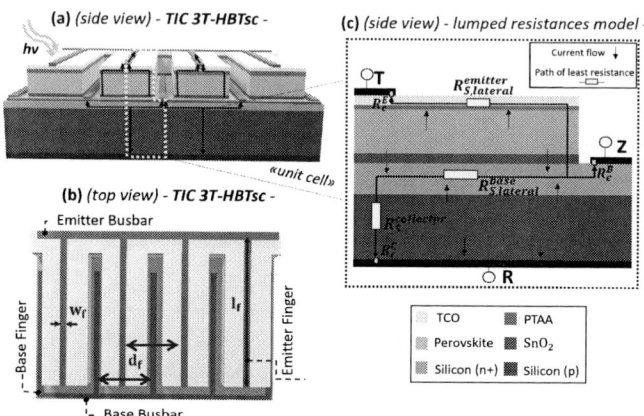

Fig. 1. TIC configuration of the 3T-HBT PVS tandem: side view (a), top view (b), side view with lumped resistance model (c).

forms requires, as simplest possible solution, adopting a top-interdigitated-contact (TIC) [7], whose implications in terms of shading and resistive losses must be carefully evaluated.

In this work, we report a simulation study of 3T-HBT PVS tandem solar cells aimed at investigating their efficiency potential and perspectives for scaling up to large areas. We present proof-of-concept 3T-HBT PVS devices designed by means of optical and transport simulations and analyze the power losses induced by the current collecting grid and how they scale with the cell area based on circuit-level simulations.

II. DEVICE & METHODS

Fig. 1(a) sketches a 3T-HBT PVS tandem made of a *n-i-p* PVK subcell on top of a planar homojunction n⁺/p Si bottom cell with design parameters in line with [1]. From the top, the TCO (34 nm)/ PTAA (11 nm) / PVK(480 nm) layer stack constitutes the p-emitter layer; the SnO_2 (25 nm) electron transport layer (ETL) and *n*-Si (150 nm) one form the base layer on top of the $150\mu m$ thick p-type c-Si collector. The cell uses a 92 nm thick MgF antireflection coating. As shown in Fig. 1(b,c), emitter and base are accessed from the top with a TIC layout, with the base contact (Z) placed on the *n*-Si layer and the emitter contact (T) on the TCO layer. A full-area contact (R) is used at the collector.

978-1-6654-6060-6/23 $31.00 © 2023 IEEE

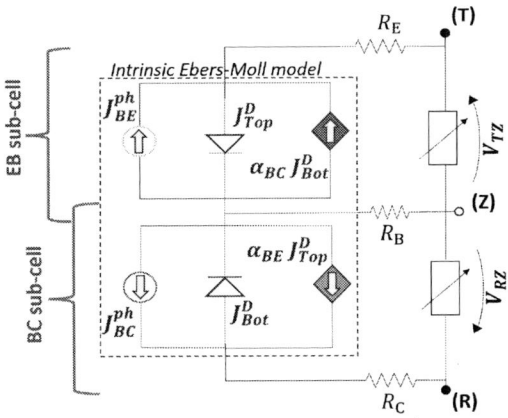

Fig. 2. Equivalent circuit of the 3T-HBT tandem.

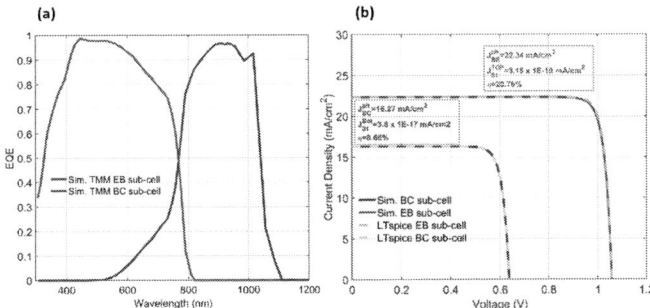

Fig. 3. (a) External Quantum Efficiency and (b) J-V characteristics obtained from TCAD simulations (solid lines) and Ebers-Moll circuit fit (dashed lines).

Under illumination, photogenerated electrons and holes induce current flows collected at the T, Z and R terminals. As depicted in Fig. 1, carrier transport through materials with finite conductivity and contacts cause power loss mechanisms that can be modeled as resistive effects. The resulting equivalent circuit model of the 3T-HBT is shown in Fig. 2. It consists of the Ebers-Moll circuit of the *intrinsic* device under illumination [5], [8] completed by parasitic series resistances at each terminal. In particular, as summarized in Table I, the equivalent base and emitter series resistance account for the contribution of lateral current flow ($R_{\text{S,lateral}}$) and of contact (R_c) and finger (R_{finger}) resistance of the current collecting grids. These are calculated based on the equivalent lumped circuit model of the unit cell of the metal grid (Fig. 1) according to the expressions given in Table I [7], [9]. At this stage, the analysis does not include the busbars.

The parameters of the Ebers-Moll model are extracted by fitting the current-voltage characteristics of the *intrinsic* 3T-HBT obtained by quasi-1D coupled optical-electrical simulations (which do not account for lateral transport neither for grid shading and resistive loss). Here, the photogeneration profile is computed by using the transfer matrix method and then given as input to the classical Poisson-drift-diffusion transport model. The approach was validated against experimental data in previous works [8], [10], where all the simulation material parameters assumed are available.

III. RESULTS & DISCUSSION

Fig. 3 shows the External Quantum Efficiency and J-V characteristics of the subcells of the 3T-HBT PVS tandem under study, and the J-V characteristics reproduced by the Ebers-Moll model. Transport simulations showed that band discontinuities and materials resistivity are such that the two subcells substantially work as independent, *i.e.* no appreciable transistor effect occurs through the SnO_2/n^+-Si base [8]. The *intrinsic* device has high efficiency of $\approx 29.4\%$.

In order to analyze the influence of the TIC grid on the device performance and scalability to large areas, we consider typical metal grid parameters [9], [11] as summarized in Table II. A shading loss factor of $2w_f/d_f$ is applied to the bottom sub-cell, and $3w_f/d_f$ to the top one, assuming for the etched region, needed to access the base layer, a width twice that of the metal finger.

First, we consider relatively small devices ($l_f = 1.5$ cm), in which resistive effects are dominated by lateral transport, and study the efficiency loss in terms of finger spacing and sheet resistance of the base layer, with fixed ITO/TCO emitter sheet resistance ($\approx 274\,\Omega/\square$). The HBT base region is composed of the bottom layer of the PVK subcell and the top layer of the Si one, therefore its sheet resistance is strongly dependent on the HBT configuration (p-n-p or n-p-n) and bottom cell technology. As seen in Fig. 4, for $R_{\text{sheet}}^{\text{base}}$ up to $100\,\Omega/\square$ (representative of PVS tandems on homojunction c-Si cells as in Fig.1) the efficiency loss can be minimized to less than 3%

TABLE I
EXPRESSIONS OF PARASITIC RESISTANCES

Resistive loss	Label	Expression [Ω cm^2]
Lateral conduction	$R_{\text{S,lateral}}$	$\frac{1}{12} R_{\text{sh}} d_f^2$
Contact	R_c	$\frac{1}{2} d_f \sqrt{\frac{\rho_c}{R_{\text{sh}}}} \coth\left(\frac{w_f}{2}\sqrt{\frac{R_{\text{sh}}}{\rho_c}}\right)$
Fingers	R_f	$\frac{1}{3}\frac{\rho_m}{t_f w_f} l_f^2\, d_f$
Emitter	R_E	$R_{\text{S,lateral}}^E + R_c^E + R_f^E$
Base	R_B	$R_{\text{S,lateral}}^B + R_c^B + R_f^B$
Collector	R_C	$\rho_{p\text{-Si}} t_{\text{collector}}$

TABLE II
GRID GEOMETRICAL AND MATERIAL PARAMETERS

Parameter	Label	Value
Finger width (μm)	w_f	40
Finger height (μm)	t_f	15
Contact resistivity (Ag/ITO) (mΩ cm^2)	ρ_c	1.27
Contact resistivity (Ag/Si) (mΩ cm^2)	ρ_c	1
Gridline resistivity (Ω cm)	ρ_m	2.65×10^{-6}
ITO resistivity (Ω cm)	ρ_{ITO}	9.31×10^{-4}
c-Si(n^+) resistivity (Ω cm)	ρ_{Si}	1.11×10^{-3}
SnO$_2$ resistivity (Ω cm)	ρ_{SnO_2}	5.2×10^{-3}

978-1-6654-6060-6/23 $31.00 © 2023 IEEE

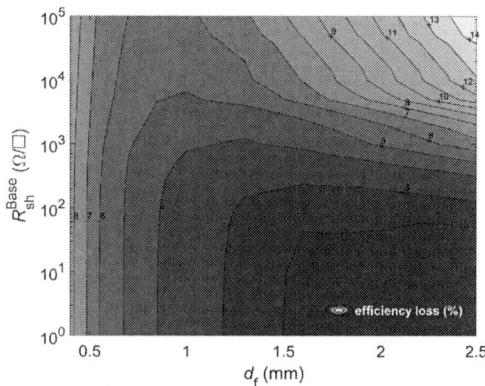

Fig. 4. Efficiency loss as a function of the finger distance (d_f) and base sheet resistance (R_{sh}^{Base}).

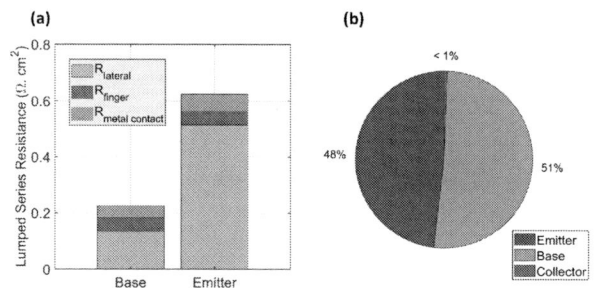

Fig. 5. (a) Emitter and base resistance components for a PVS 3T-HBT tandem with 1.5 cm x 1.5 cm area and (b) associated fractional resistive power loss.

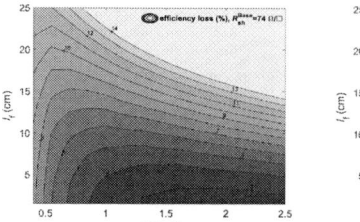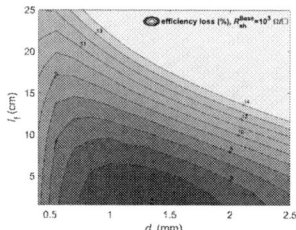

Fig. 6. Efficiency loss as a function of finger spacing (d_f) and finger length (l_f) for the case studyes of PVS 3T/HBT on (a) homojunction Si cell ($R_{sh}^{Base} = 74\ \Omega/\Box$) and (b) HTJ Si cell ($R_{sh}^{Base} = 1000\ \Omega/\Box$).

with finger distance of 1.5 mm or larger. On the other hand, for PVS 3T-HBT tandems on HTJ silicon cells, the high sheet resistance of the a-Si:H layers ($\approx 10^5\,\Omega/sq$) must be mitigated by exploiting a highly conductive transport layer for the PVK top cell, such as the SnO_2 layer used in the p-n-p HBT under study (sheet resistance $\approx 10^3\,\Omega/\Box$).

Fig. 5 shows the emitter and base resistance and the corresponding fractional power loss estimated for the 3T-HBT PVS in Fig.1 with $R_{sh}^{Base} = 74\,\Omega/\Box$, finger length 1.5 cm, finger distance 1.5 mm. Here, the dominant resistive path is associated to the lateral transport across base and emitter. The total base resistance is about 1/3 of the emitter one, but the fractional power loss ($\propto RJ^2$) associated to base and emitter grids is comparable since the current through the base, which corresponds to the sum of the top and bottom subcell currents, is about 70% higher than that one of the emitter. Therefore, as the cell area and finger length increase, the resistive effects of the base grid become the dominant cause of effiency loss.

In Fig. 6 we analyze the variation of efficiency loss with scaling up of the finger length for two values of the base sheet resistance, representative of 3T-HBT tandems on homojunction and heterojunction silicon cells. An optimized interdigitated front grid can keep power losses below 5% for finger lengths up to about 6-8 cm depending on the base sheet resistance. Such power loss could be reduced by material

and device optimization, and by developing more advanced grid designs, for example employing overlapped contact grids to reduce shading losses. Overall, these results show the potential, even in the perspective of scaling up to large areas, of using the HBT architecture to develop high efficiency three-terminal perovskite/silicon tandems that can be integrated with industry standard silicon bottom cells.

REFERENCES

[1] C. Messmer, B. S. Goraya, S. Nold, P. S. Schulze, V. Sittinger, J. Schön, J. C. Goldschmidt, M. Bivour, S. W. Glunz, and M. Hermle, "The race for the best silicon bottom cell: Efficiency and cost evaluation of perovskite–silicon tandem solar cells," *Progress in Photovoltaics: Research and Applications*, vol. 29, no. 7, pp. 744–759, 2021.

[2] NREL, "Best research-cell efficiencies: Emerging photovoltaics," https://www.nrel.gov/pv/assets/pdfs/cell-pv-eff-emergingpv.pdf, 2022.

[3] P. Tockhorn, P. Wagner, L. Kegelmann, J.-C. Stang, M. Mews, S. Albrecht, and L. Korte, "Three-terminal perovskite/silicon tandem solar cells with top and interdigitated rear contacts," *ACS Applied Energy Materials*, vol. 3, no. 2, pp. 1381–1392, 2020.

[4] F. Gota, M. Langenhorst, R. Schmager, J. Lehr, and U. W. Paetzold, "Energy yield advantages of three-terminal perovskite-silicon tandem photovoltaics," *Joule*, vol. 4, no. 11, pp. 2387–2403, 2020.

[5] A. Martí and A. Luque, "Three-terminal heterojunction bipolar transistor solar cell for high-efficiency photovoltaic conversion," *Nature communications*, vol. 6, no. 1, pp. 1–6, 2015.

[6] E. Antolín, M. H. Zehender, S. A. Svatek, M. A. Steiner, M. Martínez, I. García, P. García-Linares, E. L. Warren, A. C. Tamboli, and A. Martí, "Progress in three-terminal heterojunction bipolar transistor solar cells," *Progress in Photovoltaics: Research and Applications*, vol. 30, no. 8, pp. 843–850, 2022.

[7] M. H. Zehender, "Three-terminal heterojunction bipolar transistor solar cells = células solares de tres terminales tipo transistor bipolar de heterounión," PhD Thesis, Universidad Politécnica de Madrid, April 2022. [Online]. Available: https://oa.upm.es/70526/

[8] G. Giliberti, F. Di Giacomo, and F. Cappelluti, "Three terminal perovskite/silicon solar cell with bipolar transistor architecture," *Energies*, vol. 15, no. 21, 2022.

[9] L. Basset, "Contact electrodes for heterojunction silicon solar cells: Evaluations and optimizations of the electron contact," PhD Thesis, Université de Lille, Nov. 2020. [Online]. Available: https://theses.hal.science/tel-03159905

[10] G. Giliberti and F. Cappelluti, "Physical simulation of perovskite/silicon three-terminal tandems based on bipolar transistor structure," in *Physics, Simulation, and Photonic Engineering of Photovoltaic Devices XI*, A. Freundlich, S. Collin, and K. Hinzer, Eds., vol. 11996, International Society for Optics and Photonics. SPIE, 2022, p. 1199602.

[11] A. u. Rehman, E. P. Van Kerschaver, E. Aydin, W. Raja, T. G. Allen, and S. De Wolf, "Electrode metallization for scaled perovskite/silicon tandem solar cells: Challenges and opportunities," *Progress in Photovoltaics: Research and Applications*, 2021.

Siting Optimization of PV Recycling Plants for Supply Chain Security and Critical Material Recovery

Macarena Mendez Ribo, Silvana Ovaitt, Hope Wikoff, Heather Mirletz, Samantha Reese

National Renewable Energy Laboratory, Golden, CO, United States

Advanced Energy Systems Graduate Program, Colorado School of Mines, Golden, CO, United States

Climate change concerns are driving policymakers to commit to ambitious environmental goals. The US is aiming to reach net-zero emissions by 2050 with the deployment 1.6 TW of photovoltaics (PV). The projected amount of PV waste that will be generated due to this deployment could reach 8 million tonnes by 2050. Although some PV manufacturers are starting to develop their own initiatives, there are no widespread policies in the U.S. that encourage manufacturers or consumers to recycle solar panels and landfill disposal remains the cheapest option. Besides waste, the growing demand for renewable technologies will also increase the demand for virgin material, some of which are considered critical. In this study, we assess the economical and environmental feasibility of recovering glass, aluminum, silver, silicon, tellurium, cadmium from PV modules from a modeled recycling plant. Based on projected waste, estimated costs, and reverse logistics, we identify recycling facility candidates through different configuration scenarios that minimize capital investment risks and maximize profit. Since critical materials are prone to supply chain disruptions, a growing dependence on them will cause uncertainty in meeting future energy demands, which circular economy pathways can reduce.

Environmentally Controlled Electroluminescence/Photoluminescence Imaging System with Current Density-Voltage Capabilities for Quantitative Degradation Analysis of Perovskite Thin Film Solar Cells

Tamanna Mariam, Zahrah S. Almutawah, Adam B. Phillips, Sheng Fu, Jaehoon Chung, You Li, Manoj Rajakaruna, Kshitiz Dolia, Zhaoning Song, Randy J. Ellingson, Yanfa Yan and Michael J. Heben

Wright Center for Photovoltaics Innovation and Commercialization, Department of Physics and Astronomy, University of Toledo, 2801 W. Bancroft St., Toledo, OH, 43606

ABSTRACT — **Luminescence-based measurement techniques, such as electroluminescence (EL) and photoluminescence (PL), are great methods to evaluate the quality of solar cell materials and their electric contacts. Furthermore, imaging these responses can provide insights into the spatial character of the samples. In this work, we discuss and demonstrate the ability of our unique EL/PL system equipped with light/dark current density-voltage (JV) measurement capabilities built to observe the degradation mechanism of perovskite minimodules and the small area devices. Here, we report the fabrication of this system including the software capabilities and analysis methods. We briefly demonstrate the capabilities by presenting EL images of perovskite minimodules before and after stressing under constant current conditions. By converting the EL image into a histogram plot we provide a pathway for quantitative analysis of the EL images as a function of degradation time. In addition, we demonstrate EL, PL, and JV degradation as a function of time of four small area perovskite devices measured together at an elevated temperature. We show the acquired EL and PL images and how these and the device efficiency change as a function of time. These examples demonstrate the capability of the system and show that by repeating these measurements at multiple temperatures the degradation mechanisms and activation energies can be investigated.**

I. INTRODUCTION

Organic-inorganic hybrid halide perovskite materials are a promising candidate for next generation low-cost photovoltaics with a record small area perovskite solar cell (PSC) achieving efficiency 25.2% [1]. To advance this technology towards commercialization, development of large area deposition methods and monolithic integration has begun [2]. PSCs, though, are known to degrade under environmental stresses, such as humidity and temperature [3], and it is not known how these degradation mechanisms change when the PSCs are scaled from small areas to larger areas and monolithically integrated to form minimodules. Furthermore, it is essential to have a broader understanding of the degradation mechanism of the devices under various real-world conditions and how the deposition and fabrication techniques impact the device degradation.

Electroluminescence (EL) measurements provide a facile, rapid method to spatially identify defects in a device by forward biasing the device and taking a photograph of the emitted light. This method is widely used for silicon PV [4-5] to investigate, among other things, device degradation [6-8] and EL stress have been used to image perovskite devices at various points of degradation [9]. Likewise, photoluminescence (PL) imaging can provide similar information for samples under light excitation instead of voltage biasing. However, quantitative analysis to understand the degradation mechanism has not been reported, nor have the degradation been connected to the current density-voltage (JV) response devices.

In this work, we report the development of a unique characterization system to investigate the degradation of PSCs at several different temperatures. For this system the cells are held in an environmentally controlled box and can be interrogated using EL, PL, and light and dark JV as a function of time. To provide the quantitative analysis of the response, we capture EL images while recording the applied voltage required to maintain the constant current (or fix the voltage and record the current), measure the dark and light current density voltage (JV), and capture the PL image in situ during device stressing with temperature. By measuring the EL, PL, and JV curves, we can correlate the responses and determine if the degradation mechanism for each recombination mechanism is the same.

II. EXPERIMENTAL SYSTEM DETAILS

In our system a sample is mounted to a temperature-controlled stage inside an environmentally controlled box which allows the EL, PL, and dark and light JV measurements to be performed *in situ* in any combination. As the schematic diagram, Fig. 1, shows, the system consists of several modules, including the environmental chamber and sample stage, optical components, electrical components, control software, and data analysis software. In this section, we will discuss the role and detailed description of each section.

A. Environmental Chamber and Sample Stage

The environmental chamber consists of a gasket sealed box modified with a soda lime glass window for EL and PL measurements. In addition, the box includes electrical feedthroughs for four sample contacts and a single common

978-1-6654-6060-6/23 $31.00 © 2023 IEEE

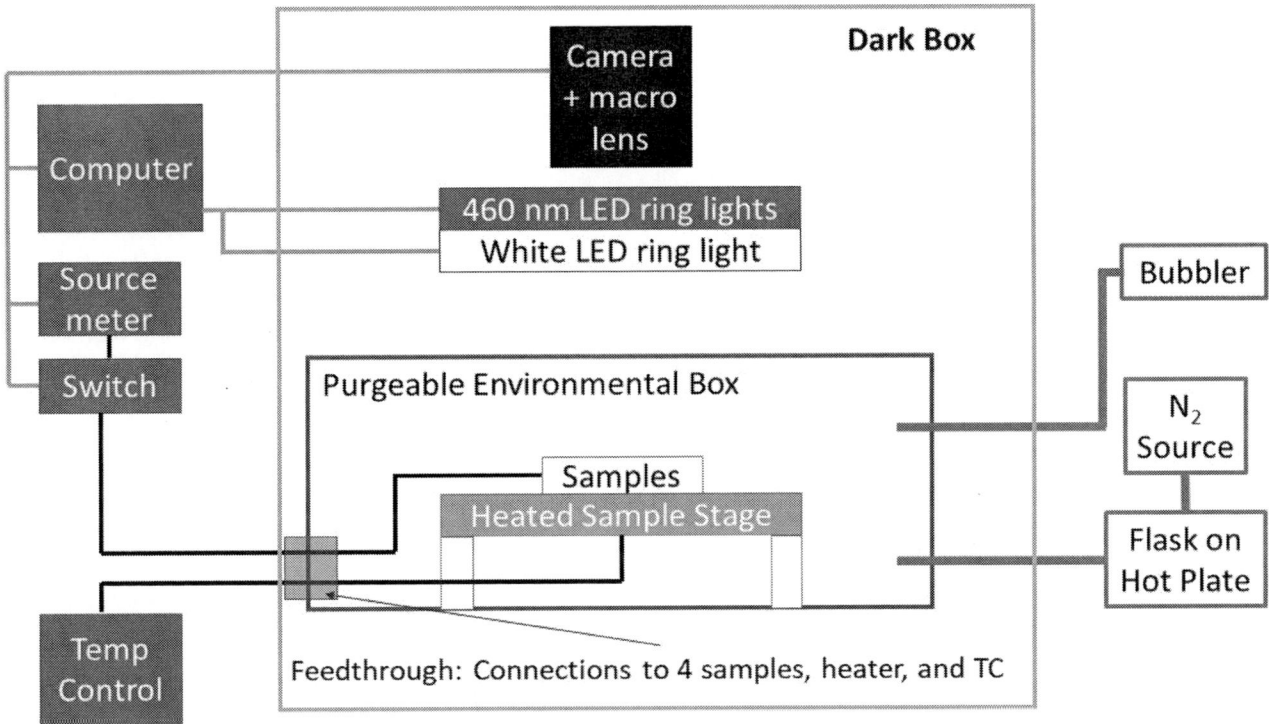

Fig. 1. Schematic diagram of the electroluminescence/photoluminescence measurement system. The system consists of an environmental chamber with heated sample stage and optical components consisting of the camera and light sources enclosed in a dark box to prevent other light from exciting the sample. The electrical components are outside of the dark box, and the connections to the system are shown with black lines. The control software connections are shown with red lines, and the purge gas connections are shown with the green lines.

contact for biasing samples for EL measurements and JV measurements, power to control the stage temperature, and a thermocouple. All the feedthroughs, as well as the glass top are sealed to the box with epoxy. In addition, the box is also connected to gas inlet and outlet.

To control the environment within the chamber, the gas inlet port is connected to a nitrogen source that flows through an Erlenmeyer flask filled with water on a hot plate. The relative humidity in the environmental chamber is controlled by the temperature of the hot plate. For humidity free experiments, the Erlenmeyer flask is bypassed, and the nitrogen flows directly into the chamber. A rotameter is used to control the gas flow through chamber. The gas outlet port is connected to an oil bubbler to prevent room air from entering the chamber.

The samples are mounted to one of several heated stage which is placed inside the chamber. The stages consist of a 3" x 4", 1/8" thick aluminum plate connected to a 3"x3" thin film heater on the underside. Alumina standoffs are used to lift the stage off the environmental box to thermally isolate the stage and allow for electrical connections to be made when necessary. The stage temperature is ready using a thermocouple mounted to the top of the stage.

Several stages have been developed for this system. One is for testing minimodules and is simply what is described above as the electrical leads are soldered to each cell during minimodule fabrication and are easily accessible with this

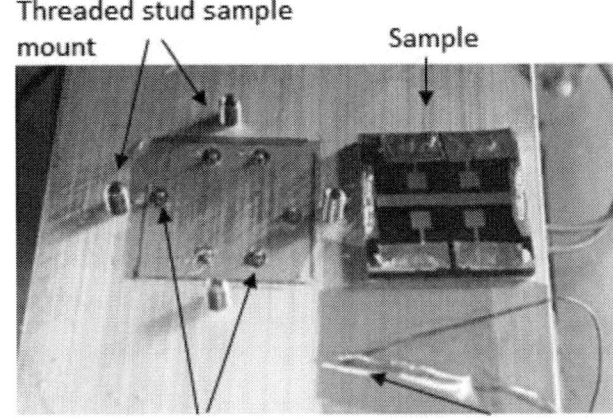

Fig. 2. Photograph of sample stage for small area device showing the electrical connections to the sample and thermocouple. The sample is also shown with the electrical connections facing upward. Kapton tape is used to prevent shorting the devices and mount the thermocouple.

system. A photograph of the stage designed to investigate small area cells is shown in Fig. 2. These devices are typically made in the superstrate configuration, so electrical connections must be made to the side of the device facing the sample stage. To accomplish this, pogo-pins were mounted through the thin film heater and stage and electrical connections between the stage

978-1-6654-6060-6/23 $31.00 © 2023 IEEE

and feedthroughs are made below the stage. Kapton tape is applied to the aluminum stage to provide electrical isolation between samples.

The stage temperature is monitored using a thermocouple, and the temperature is controlled using a variac transformer. With this configuration, the temperature can be increased from room temperature to ~120 °C in less than 2 minutes when the variac is set to 100 V. While the temperature is currently controlled by fixing the variac voltage, a PID temperature controller can easily be implemented if finer control is needed.

B. Optical Components

The heart of this EL/PL imaging system is the camera. Here, we use a 24-megapixel Nikon DS5300 DSLR camera modified by BrightSpot Automation LLC to allow scientific quality images in the infrared region. The detector of the camera is 4000 pixels by 6000 pixels, and the image can be saved in the raw 14-bit mode or compressed to 8-bit. We have two lenses for our system. A 40 mm lens is used for large area samples, such as our minimodules, while a 105 mm lens is used for small area modules. The 105 mm lens allows for one-to-one scaling, indicating the spatial resolution of the image will equal ~3.5 μm detector pixel size. Filters are added to the end of the lens to ensure no stray light is affecting the image.

The camera is controlled using Digicam Control software. This software allows the camera parameters such as the ISO, aperture setting, and shutter speed to be set. These settings are used to set the intensity scale of the image. In addition, Digicam Control is also used for the fine focus of the camera, which will be critical to measurements investigating spatial aspects of degradation.

The system also includes three light source to allow for PL and light JV measurements. A 5 W 60 mm diameter and 16.5 W 140 mm diameter microscope ring lights consisting of 60 and 267 blue light emitting diodes (LEDs) (BoliOptics), respectively, are used for the PL measurement. The 460 nm emission wavelength of these LEDs was specifically chosen to provide an excitation source wavelength that is short enough to be completely absorbed by the filters but not so short to degrade perovskite samples. The third light source is a 5 W 82 mm diameter microscope ring light with 96 white LEDs (BoliOptics) is used with the other two sources for JV measurements. All the lights are mounted approximately coplanar and sit a distance above the sample such that the intensity is approximately uniform over a 3" x 3" area. The three lights combine produce a photon flux of approximately 0.3 suns with the blue lights accounting for 75% of that total.

C. Electrical Components

While the camera is the heart of the optical components, the source meter is the heart of the electrical components. The source meter is used to provide the constant current or voltage during the EL measurement and to complete the JV measurements of each sample. In this system, the source meter is a Keithley 2401. The source meter is connected to the environmental chamber and each sample through a Keithley

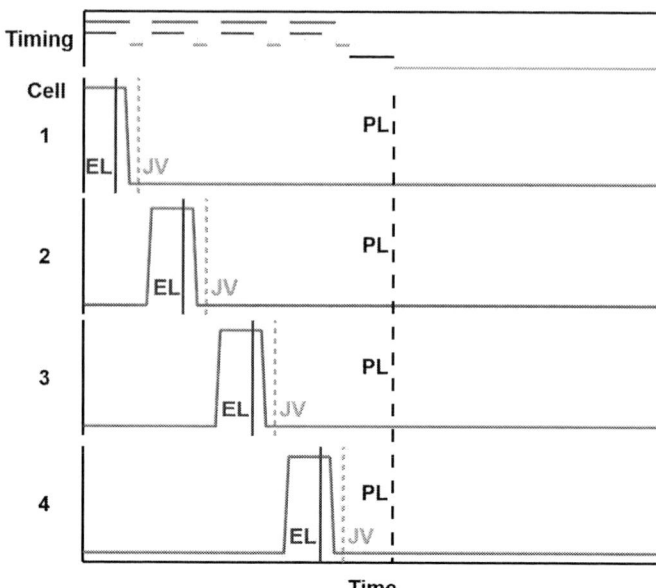

Fig. 3. Schematic diagram of the timing for each measurement of multiplexed devices for a single cycle. This process is repeated for the user input number of cycles. As shown by the bars at the top, the user inputs the total time the EL bias is applied to each cell, time the EL bias is applied before capturing the EL image (denoted with the blue vertical lines), time after EL bias is turned off until the JV measurements are made (green dashed vertical lines), the time between the JV measurement and acquisition of the PL image (black dashed line), and the time after the PL image until the end of the cycle. In the case for cells with only the JV and PL image, the EL bias is set to zero and no EL image is captured. The PL image is always captured after the last JV measurement regardless of number of cells interrogated.

7001 switch. The switch allows for the measurement of multiple samples during a single environmental condition. In addition, the switch is used to toggle the LEDs and to trigger image acquisition in the camera.

D. Control Software

The entire EL/PL system is controlled using LabVIEW software that allows for sequential and periodic acquisition of the EL, dark, light JV and PL measurements. This software must communication with some instruments directly and trigger third party software (Digicam Control) when necessary. Digicam Control, though, can be configured to download the image from the camera upon triggering the camera. Using this setting and remotely triggering the camera through the Keithley switch, the EL/PL software can control every aspect of the data acquisition.

For maximum flexibility of the system, many of the parameters are user inputs, such as the EL voltage/current bias, all timings between actions, the number of samples to measure, and the number of times each sample will be measured. Fig. 3 shows a schematic diagram of the timing and measurements with the input parameters noted. While this diagram shows EL, JV, and PL of all four samples being measured, this need not be the case. All these measurements can be made for fewer samples, and in this case, the PL image will be taken after the last sample is measured.

An additional feature is that not all measurement techniques need to be completed for each sample under interrogation. It is possible that one measurement affects subsequent measurements, such as the EL measurement potentially alter the JV response [10]. Consequently, the software is configured to allow different measurements for each sample. For example, the EL, JV, and PL can be measured for two samples, the JV and PL can be measured for a third sample, and only the PL can be measured for the fourth sample.

E. Data Analysis

The EL, JV, and PL are all measures as a function of time; therefore, all can be used to determine, for example, a degradation rate. However, not all the data is saved in the same format. For the JV measurements, two data sets are generated – the JV curves and a list of the photovoltaic (PV) parameters, open circuit voltage (V_{OC}), short circuit density (J_{SC}), fill factor (FF), and efficiency. The file containing the PV parameters can simply be graphed as a function of time in any software.

The EL and PL data, though, are in images that record the intensity of the luminescence from the sample as a function of position. There are two options for these data sets. The first is to consider the luminescence of the device as a whole, while the second is to consider the individual or a small subset of pixels individually. To convert the image data into intensity plots, we developed Igor Pro macros which will generate a histogram of the pixel intensity for a specified area of the figure, thereby cropping out unwanted or irrelevant pixels. This macro completes this calculation over a user defined series of images, thereby developing histogram plots as a function of time. Note that the area investigated in this analysis is user defined and can be as small as a single pixel.

In addition to comparing the luminescence intensity and PV parameters variation as a function of temperature, the acquisition software also records the applied current and subsequent voltage in the case of fixed current EL measurements or the converse for fixed voltage EL measurements during and after the bias is applied. This data may provide information on the transient response of the cell and is saved in an accessible format.

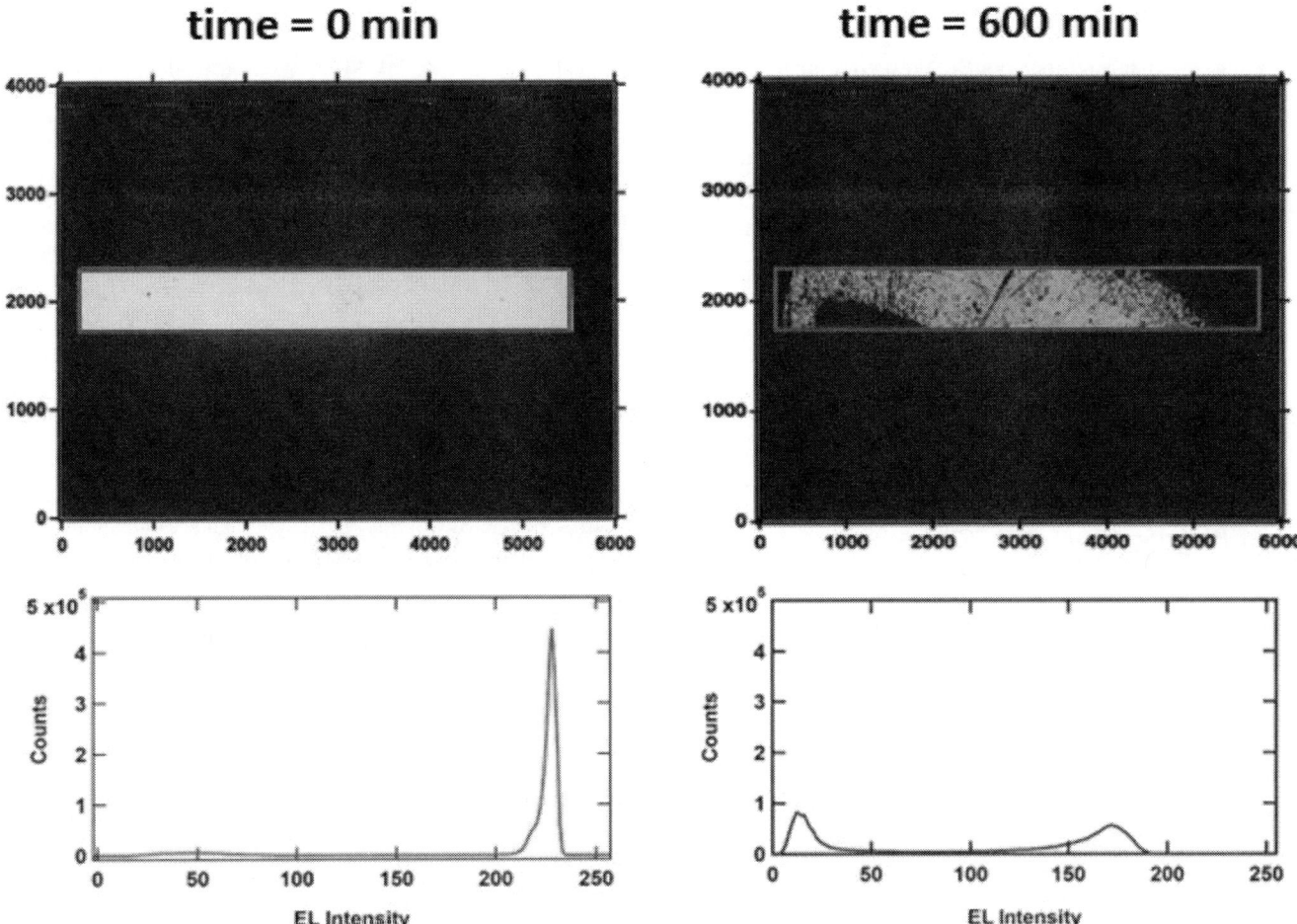

Fig. 4. EL images of a single cell in a monolithically integrated device as fabricated (left) and after 600 minutes (right) of EL current bias and the associated histogram curves used to quantify the EL intensity. The red boxes show the area used to calculate the histogram plot.

III. RESULTS AND DISCUSSION

To demonstrate the capability of EL/PL system, we provide two demonstrative examples.

Fig. 4 show the EL images of a single cell of an NIP perovskite minimodule as fabricated and after 600 minutes of maintaining a constant current flow through the device and the associated histogram generated using the analysis macros. The cropped area of the image is indicated in the EL image. From the images, it is clear that the device under investigation degrades after stressing and the histogram helps quantify the degradation. Prior to stressing the cell is highly luminescent, so the peak in counts occurs at a high bin number. After degradation the peak location decreases significantly. This change in peak location can be used to quantitate the degradation. Completing this analysis for smaller areas will allow to determine if specific areas of the sample, for example next to scribe lines in monolithically integrated minimodules, degrade faster than other parts of the device.

For the second demonstration, we look at the degradation of four small area PIN perovskite devices with an initial efficiency approximately 20%. Here we follow the timing diagram shown in Fig. 3 with the EL, JV, and PL measured for Cell A, JV and PL for Cell B, and PL for Cells C and D. Note that the PL of Cell B was quenched because the cell was held at zero bias during the PL measurement, which is an option during the measurements. For these measurements, the sample was mounted on the stage shown in Fig. 2 and placed in the environmental box, which was purged with nitrogen. The data was acquired for a sample temperature of 37 °C over a two-hour period.

The EL and PL images are shown in Fig. 5a. From the images, it is difficult to determine how the EL and PL changed with time, however, the histogram (not shown) shows a clear increase in luminescence, and the EL, PL, and efficiency (approximately scaled for the reduced light conditions used in this chamber) as a function of time are shown Fig. 5b. With this information, we can analyze the data to determine a rate constant for the change in response [11]. Completing these measurements at several temperatures will allow us to generate an Arrhenius plot and determine activation energies for EL, PL, and each PV parameters.

IV. CONCLUSIONS

Here, we describe the components of our unique multi-functional EL/PL system and provide some measurements to demonstrate some capabilities. We discuss how the environment of the sample, including the temperature, can be controlled allowing us to probe and compare several degradation mechanisms. We show that the spatial and temporal degradation of small area devices and minimodules can be investigated using EL and PL and provide a method to quantitatively analyze the luminescence data. The addition of the JV measurement capability allows us to connect EL and PL changes to device performance. Our system can provide more valuable information to better contribute to understand the

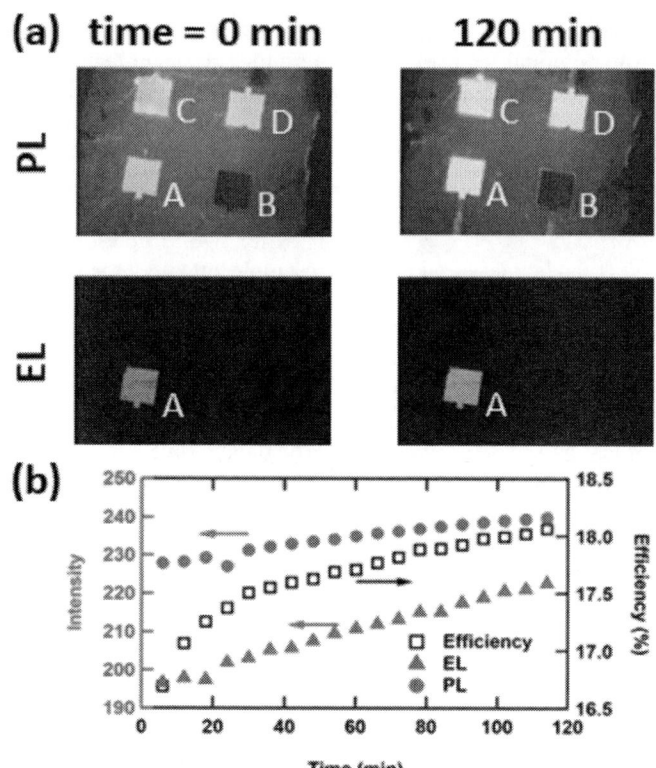

Fig. 5. (a) PL and EL images of a four cell sample held at 37 °C where only Cell A is held under current bias for the EL measurement and Cell B is held a short circuit during PL measurement. The left images show the initial measurements, and the right images show the response after 120 minutes at temperature. (b) Average PL and EL intensity and device efficiency for Cell A as a function of the heating time.

degradation mechanism and electrochemical changes on the perovskite solar cells to improve the stability of device and understand how degradation changes as perovskites are moved from small area devices to monolithically integrated modules.

ACKNOWLEDGEMENTS

This material is based on research sponsored by the U. S. DOE's office of Energy Efficiency and Renewable Energy (EERE) under Solar Energy Technologies Office (SETO) Award number 38050 and agreement number DE-EE0008970. The US government is authorized to reproduce and distribute reprints for governmental purposes notwithstanding any copyright notation thereon.

REFERENCES

[1] G.H. Kim. and D.S. Kim, "Development of Perovskite solar cell with >25% conversion efficiency", Joule, 5(5), 1033-1035, 2021.
[2] I. Zimmermann et al., "Industrially Compatible Fabrication Process of Perovskite- Based Mini-Modules Coupling Sequential Slot-Die Coating and Chemical Bath Deposition". ACS Appl. Mater. Interfaces, 14, 11636-11644, 2022.

[3] R. H. Ahangharnejhad et al., "Protecting perovskite material solar cells against moisture-induced degradation with sputtered inorganic barrier layers", ACS Appl. Energy Mater., 4,8, 7571-7578, 2021.

[4] J.A. Giesecke, M.C. Schubert, B. Michl, F. Schindler and W. Warta, "Minority carrier lifetime imaging of silicon wafers calibrated by quasi steady state photoluminescence". Sol. Energy Mater. Sol. Cells, 95, 1011, 2011.

[5] G. Kulesza, P. Panek and P. Ziezba, "Silicon solar cells efficiency improvement by the wet chemical texturization in the HF/HNO3/ diluent solution", Arch. Metall. Mater, 58, 291–295, 2013.

[6] M. Lipinski, G. Kulesza and Z. Starowicz, "Luminescence imaging for characterization of photovoltaic cells and modules" Elektronika (LV), 52,8, 2014.

[7] S. Spataru, P. Hacke, D. Sera, S. Glick, T. Kerekes and R. Teodorescu "Quantifying solar cell cracks in photovoltaic modules by electroluminescence imaging", Proc. IEEE 42nd Photovolt. Spec. Conf., 1–6, 2015.

[8] J.H. Jean, C.H. Chen, and H.L. Lin, "Application of an image processing software tool to crack inspection of crystalline silicon solar cells" Proc. Int. Conf. Mach. Learn. Cybern., 1666–1671, 2011.

[9] M.C. Schubert, L.E. Mundt, D. Walter, A. Fell and S.W. Glunz, "Spatially resolved Performance Analysis for perovskite solar cells", Adv. Energy mater., 1904001, 2020.

10] B. Xu, W. Wang, X. Zhang, H. Liu, Y. Zhang, G. Mei, S. Chen, K. Wang, L. Wang, and X.W. Sun, "Electric biased Induced Degradation in Organic-Inorganic Hybrid Perovskite Light-Emitting Diodes", Sci. Rep., 8, 15799, 2018.

[11] T.J. McDonald, C. Engtrakul, M. Jones, G. Rumbles and M.J. Heben, "Kinetics of PL quenching during Single-Walled Carbon Nanotube Rebundling and Diameter-Dependent Surfactant Interactions", J. Phys. Chem. B, 110, 25339-25346, 2006.

Method for Evaluating the Silicon Solar Cells Performances under AM0 thanks to AM1.5G Spectrum

Philippe Voarino*, Adem Dahi, Romain Cariou

Univ. Grenoble Alpes, CEA, Liten, Campus INES, 73375 Le Bourget du Lac, France
*philippe.voarino@cea.fr

Abstract — **Assessment of Solar Cells (SC) characteristics under indoor standard conditions has been a challenge since the beginning of the industrialization of photovoltaic technologies in 1954 by Bell. Nowadays, for spatial industry the use of silicon SCs becomes obvious for Low Earth Orbitrary (LE0) missions. Moreover, accurately knowing electrical performances is key to optimize the PhotoVoltaic Assembly (PVA) design (number of SCs and overall dimensions). This paper proposes a method to assess electrical performances under AM0 spectrum, using IV measurements carried out under a terrestrial solar simulator (AM1.5G) and the knowledge of External Quantum Efficiency for different technologies of silicon solar cells. This method is extended at SCs which have been irradiated with 1 MeV electrons at differents fluences. The robustness of this method is then evaluated.**

I. INTRODUCTION

The purpose of this work is to find a way to have access to the electrical performances of Silicon Solar cells under spatial spectrum ASTM G173. Nowadays, the AM0 solar simulators are complex and costly because they are designed for III-V Multi-Junctions SC (MJSC). A benchmark of sun simulator machines was conducted; lots of set-up are commercially available, from various companies such as (non-hexaustive list): Abet Technologies Inc., Alpha Omega, G2V Optics, Neonsee, Newport-Oriel, OAI, PASAN, Sciencetech Inc., Solar Light, Solaronix, Spectrolab, TS Space Systems, Wacom, etc.

From this list (91 sun simulators), 70 % of the set-up proposed on the market are AAA rated, and only 7 matched the specifications AAA ranked, with a large spectral range (300-1,800 nm), a spectrum close to AM0 even in the infrared region.

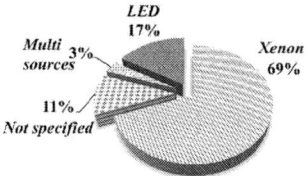

Fig. 1: Repartition of type of sources of 91 solar simulators.

The source of a solar simulator is very often base on Xe lamp (see fig.1), which the spectrum is not representative enough for a solar spectrum to test MJSCs under space conditions. However, for the LEO applications, we focus on Si SCs so the spectral range of interest is limited to ~ 300-1,200 nm.

To do so, solar simulators that can extract the electrical characteristics of a cell under an illumination close to the sun's spectrum are used. Because most solar simulators are designed for terrestrial applications, a given simulator does not necessarily have the optical and the calibration devices that allows the measurement under spatial conditions. The goal of this work is therefore to assess if there is a possible way to estimate the data (efficiency for instance) of cells under spatial conditions from the ones measured under terrestrial conditions and to experimentally evaluate the robustness of the corresponding method. In order to convert the characteristics of the SCs under AM1.5G spectrum to the data under AM0 spectrum, two models were studied: the "conversion" model, and the "translation" model.

II. SAMPLES

For this study, firstly a set of calibrated PERC (Passivated Emitter and Rear Cell) bifacial P-type half SC have been compared. Then, for the study, five silicon different types of SC were used: PERC bifacial p-type, Al-BSF p-type (Back Surface Field), PERT n-type (Passivated Emitter Rear Totally Diffused), Heterojunctions n-type SCs, and 1 MeV electron-irradiated Al-BSF p-type SCs (Fig. 2).

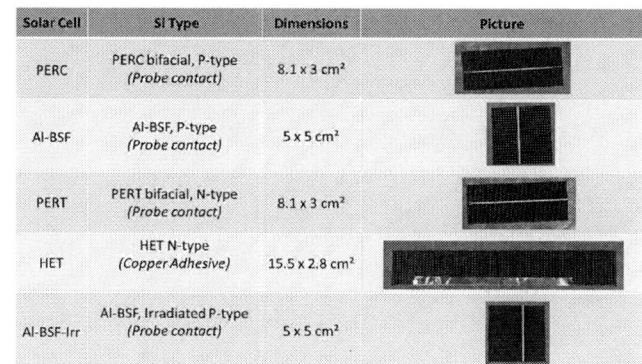

Solar Cell	Si Type	Dimensions	Picture
PERC	PERC bifacial, P-type *(Probe contact)*	8.1 x 3 cm²	
Al-BSF	Al-BSF, P-type *(Probe contact)*	5 x 5 cm²	
PERT	PERT bifacial, N-type *(Probe contact)*	8.1 x 3 cm²	
HET	HET N-type *(Copper Adhesive)*	15.5 x 2.8 cm²	
Al-BSF-Irr	Al-BSF, Irradiated P-type *(Probe contact)*	5 x 5 cm²	

Fig. 2: Main characteristics and pictures of the silicon solar cells used in this study.

III. METHODS

A. Applied Solar Cell Models

Two methods have been studied to extract electrical performances of a SC under AM0 spectrum with AM1.5G measurements: methods that we call "conversion" and "translation".

The "conversion" model is based on the derivation of the diode equation for an ideal SC (shunt and serial resistors equal to 0). The principle is to take some quantities of interest in the I-V curve (short circuit density current (J_{sc}), open circuit voltage (V_{oc}), fill factor (FF) and efficiency (η) and to convert each of them from their values under AM1.5G to their values under AM0 by using simple arithmetic relations and spectral response measurements [1]. The equations of the "conversion" model are presented below in the equations (1), (2), (3), and (4). P_{in}^0 is the incidence power density. The index, 0 or 1.5, presents the spectrum AM0 respectively AM1.5G. V_T is the ratio of the Boltzmann constant multiply by the temperature on the elementary charge. α is a spectral parameter defined by (1) and experimentally determined thanks to the QE measurements.

$$J_{sc}^0 = J_{sc}^{1.5}\alpha \tag{1}$$

$$V_{oc}^0 = V_{oc}^{1.5} + V_T \ln(\alpha) \tag{2}$$

$$FF^0 = FF^{1.5} \tag{3}$$

$$Eff^0 = \frac{J_{sc}^0 V_{oc}^0 FF^0}{P_{in}^0(\frac{mW}{cm^2})} \tag{4}$$

The "translation" model differs from the first in that it does not only convert few points of the curve but also translate it point by point [2]. It takes as an input an I-V curve measured under an illumination E_{in} and translates it to an estimation of the I-V curve measured under E_{out}, the concentration Ratio E_{out} / E_{in} is equal to the $I_L^0 / I_L^{1.5}$. This model is based on an equivalent circuit that takes into account R_s and sets $R_{sh} = +\infty$. R_s is calculated thanks to Suckow model [3], and R_{sh} may not be considered without influencing the model description as explained in the following papers [4] [5].

B. Equipments

The QE measurements are performed on an equipment manufactured by *Oriel* and optimised at the laboratory. For each wavelength, from 300 to 1,200 nm, intensity produced by the cell is measured, and QE is calculated. The sample is fixed on a thermally regulated support.

AAA class solar simulators used in this study are based both on continuous and pulsed technologies. The first one called *HELIOS 3030*, made by Solar Added Value, has a sample area of 25 cm² and uses a pulsed Xe lamp filtered with AM1.5G and AM0 filters. The wavelength range used is from 300 to 1,400 nm. The Isc is calibrated by using a Si SC as reference. The second one is made by *NEONSEE* and has the ability to measure only AM1.5G. For continous solar simulators, the measures are calibrated under AM1.5G thanks to a reference sample.

C. Manipulation plan

The experimental part has been carried out as follow:
- measurement of the quantum efficiency: in order to calculate the spectral parameter α for each SC,
- preliminary measurement of the references on the *NEONSEE* under AM1.5G and comparison with the calibrated data measured under AM0,
- measurement under AM1.5G and AM0 on the *Helios 3030*.
- Measurement with a variable resistance plugged in parallel, in order to study the sensibility of the models with the shape of the I-V curve, a potentiometer is plugged in parallel to the SC in order to induce a current leak that will affect the FF of the cell. The new equivalent circuit, based one-diode model is shown on fig 3.

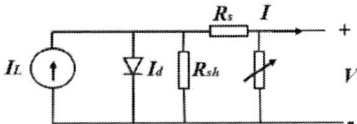

Fig. 3: Scheme representing a potentiometer plugged in parallel of a based one-diode model SC. I_L is the Illuminating current, I_d is t-he diode saturation current, R_{sh} is the shunt resistance, R_s is the serie resistance.

For the five cell types, each cell is measured five times to investigate measurement repetability. The five measurements are then averaged and the resulting I-V curve is processed by the models described above.

II. RESULTS

A. Results

Before starting the measurement series, the model was validated on data corresponding to calibrated cells presented in the table I.

TABLE I
Electrical parameters of the calibrated PERC SC measured by
CalLab PV Cells at Fraunhofer ISE.

Spectrum	Isc (mA)	Voc (mV)	FF(%)	Eff (%)
AM 1.5G	478.7±6.5	669.7±2.1	78.68±0.48	20.68±0.32
AM0	580.8±7.9	674.4±2.2	78.53±0.48	18.40±0.28

By applied both methods to this dataset, the folowing Fig. 4(a) presents IV curves of the references and the result when the translation method is applied on the calibrated AM1.5G measurements. The calculated AM0 curve overlap the measured AM0 curve. Results of some specific electrical parameters are presented for both method and a promising relative error for all electrical parameters under 0.3 % was get, as shown in fig. 4 (b).

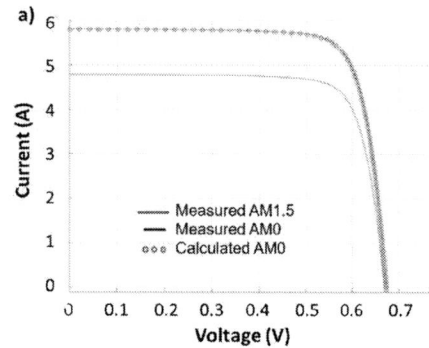

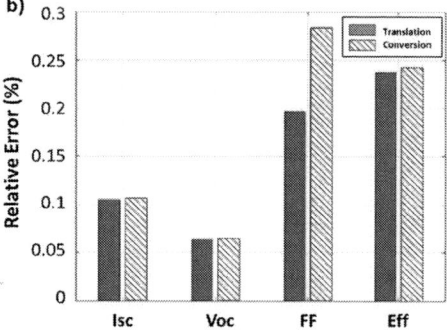

Fig. 4: (a) I.V. curves of half-SC calibrated measured and caulculated with translation AM0 model. (b) Comparaison of electrical parameters for conversion and translation models [6] based on AM1.5G measurements with the calibrated AM0 measurements

This value validate both methods for "perfect" datas. The next step was to evaluate the method on the SCs defined in fig. 2. So, thanks to the EQE curves, averaged α parameters have been determined (see table II). For the Al-BSF-Irr SCs, α cannot be averaged because irradiation is different for each SC.

TABLE II
α spectral parameters.

Type Sc	α	Cell E	α
PERC	1.219	**Al-BSF-Irr-1**	1.204
Al-BSF	1.214	**Al-BSF-Irr-2**	1.206
PERT	1.214	**Al-BSF-Irr-3**	1.216
HET	1.203	**Al-BSF-Irr-4**	1.228

B. Error Analysis

Figure 5 gives the relative error on efficiency at AM0 for the "conversion" model . For all SCs the median relative error is under 1.46 %. As for fig. 6 gives the relative error on efficiency at AM0 for the "translation" model with a value under 2.97 %. So it can be noticed that for the PERC and PERT SCs both models give a median efficiency error under 1 %. For Al-BSF and Al-BSF-Irr the error is under 1.87 %, and for HET is under 2.97 %. Analysis are in progress to understand and quantify if

errors are linked to the numerical methods, to the way to measuring (probe, references), or to the characteristics of the spectral response of the SCs. Then, to find a distribution that describes properly the experimental results, a kernel Density Estimation method had been used [7] and will be shown. Moreover, results with a variable resistance plugged in parallel wil be described and highlighted their added value to the method.

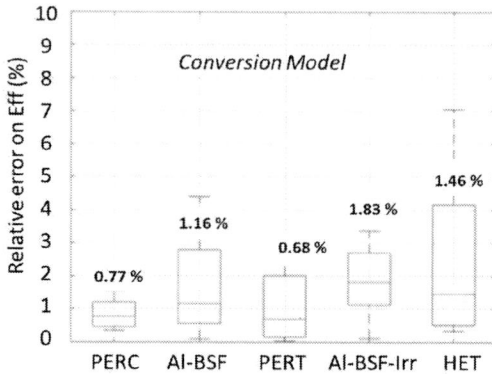

Fig. 5: Relative Error on efficiency (%) for the "conversion" model. The value in % corresponds to the median.

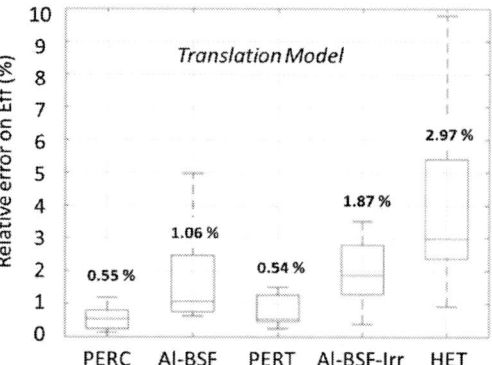

Fig. 6: Relative Error on efficiency (%) for the "translation" model. The value in % corresponds to the median.

C. Kernel Analysis

To go further into the statistical analysis, the errors induced by the two models on all kind of cell have been listed. (n= 24 samples). Only the HET cells have been eluded because of the adhesive copper which doesn't correspond to a real electrical connection for industrial photovoltaic. All these errors were represented on the histograms on figure 7 for the translation model and on figure 8 for the conversion model, along with boxplots and tables listing their main indicators (median, quartiles, minimum and maximum). The bin widths were settled to 0:1%.

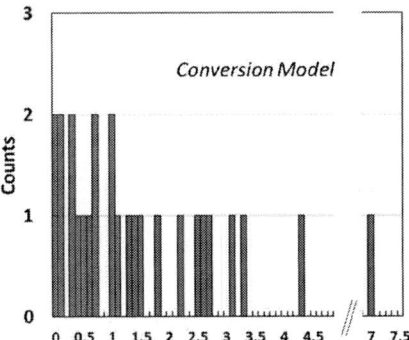

Fig. 7: Bowplots and histograms of the relative Error on efficiency for the "conversion" model.

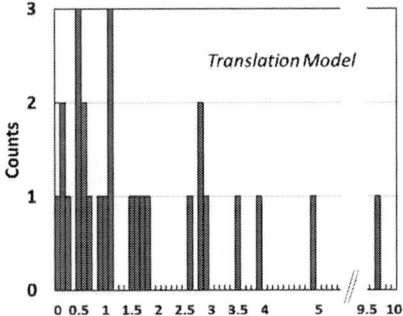

Fig. 8: Histograms of the relative Error on efficiency for all solar cells for the "Translation" model.

The next step is to find a distribution that describes properly the experimental . Because the data distribution is not normal and in the absence of a known parametric distribution that can fit them adequately, a nonparametric statistic approach is chosen. The goal is to come back to a probability density which can then be analyzed from the listed errors. The method used is the Kernel Density Estimation (KDE). A Kernel is a positive, symmetrical function and its area under the curve is equal to one. The principle is to center a kernel function on each point of the graph [7]. For this study a gaussian kernel is used from the equation 5 for each count. Then each K(x) (Kernel function) are summed to get the probability density function (PDF) shown in figure 9 and figure 10.

$$K(x) = \frac{1}{h\sqrt{2\pi}} e^{0.5(\frac{x-x_i}{h})^2} \qquad (5)$$

h is the bandwith, xi the position of the kernel function, x the value at which the kernel function is computed. The optimal bandwidth is computed automatically by software by a method that is detailed in [8]. The h value is 0.058 for the translation model and 0.58 for the conversion model.
Figure 9 and figure 10 presents the PDF and the cumulative distribution function (CDF) for the conversion and translation

models. The CDF has also been computed to determine the probability that the error (x) is less than, or equal to a certain value (Γ). The errors obtained for $\Gamma = 0.5$, $\Gamma = 0.9$, and $\Gamma = 0.95$ are also listed on the graphs. It can be noticed that for $\Gamma = 0.5$, the error value is slightly lower for the translation model than for the conversion (respectively 1.17 % and 1.24 %), but, for $\Gamma = 0.9$, and $\Gamma = 0.95$, the error value become signi_cantly higher for the translation than for the conversion model. It can be explained by the choice of the upper boundary condition: the conversion model gave better results for Heterojonction solar cells. The fact that this model has less extreme error values for very poor quality cells results in less error dispersion. It has been said in the analysis of figure 5 and 6 that the dispersion of the errors induced by the conversion seemed higher for all kind of cell but is lower when the cells are taken together. These two results are not incompatible since a trend that appears in several groups of data can disappears or reverses when the groups are combined as stated by the Simpson's paradox, a well known statistical bias [17].

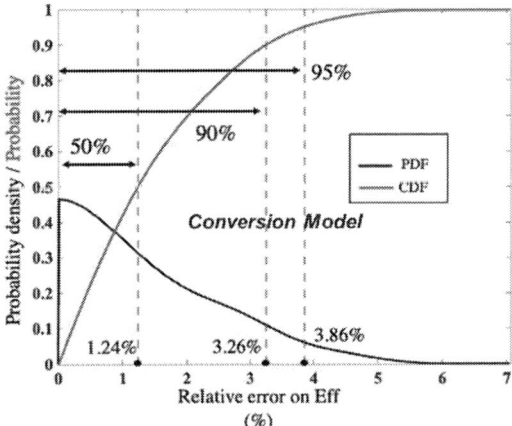

Fig. 9: Probability density function (PDF) and Cumulative distribution function (CDF) for the conversion model for the Relative error of efficiency for all solar cells measured in this study.

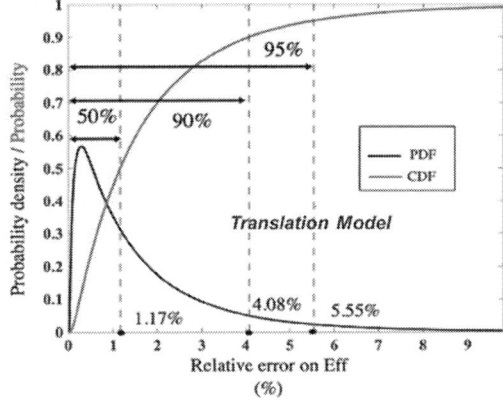

Fig. 10: Probability density function (PDF) and cumulative distribution function (CDF) for the translation model for the relative error of efficiency for all solar cells measured in this study.

D. Variable resistance

This experiment aims to analyse the effect of the FF variation for a given solar cell. For this measurement a Al-BSF solar cell has been used. I.V. curves were measured with four different resistance values: 10, 15, 20 and 40 Ω. The resistance value denoted ∞ corresponds to the open circuit of the Rshunt branch on the electrical diagram of the figure 3. For resistance values below 10 Ω, the solar simulator considers that there is a short circuit and prevents the measurement. It can be seen on the Table III that the Isc value is decreasing as the resistance decreases (from 827 mA to 0.872 mA). This increase is not intuitive, it can be explained by the increase of the shunt resistance and thus of the slope at V = 0 V, and by the fact that the output power remains almost constant for the four resistance values.

TABLE III

Electrical parameters of AL-BSF solar cell measured at AM1.5G at 25°C in function of the shunt resistance variations

Resistance (Ω)	10	15	20	40	∞
FF (%)	76.09	76.95	78.97	80.14	82.11
Isc (mA)	872	858	848	837	827
Power (mW)	383	382	387	388	393

By comparing both methods, the evolution of this error has been plotted against the FF (see figure 11). A clear decrease in error with increasing FF can be observed. This gives an element of validation as to the fact that the precision of the models is better on good quality cells.

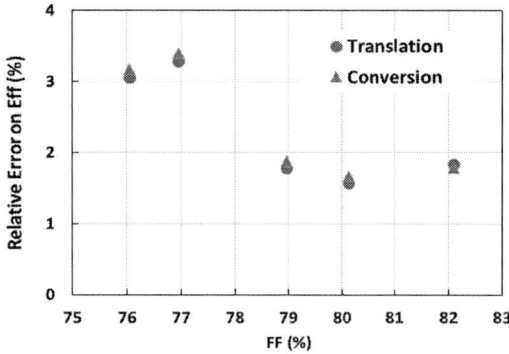

Fig. 11: Relative Error on efficiency in function of the FF for one Al-BSF solar cell tested by varying a shunt resistance fo ttconversion and translation models.

II. CONCLUSION

Two mathematical models were reviewed and evaluated experimentally to have access to the electrical performances under AM0 from data under AM1.5G. First results show that we can have access to a suitable preliminary results without performing calibrate measurements with two simulators. The sample used are silicon solar cells which are currently used in terrestrial applications, but also in space applications with some adjustments. A data analysis showed that conversion models can give better results on poor quality cells. However by adding resistance in parallel, the relative errors are comparable, and models give best results when the solar cell has a high FF. Results have to be more consolidated by increasing the sample statistics.

REFERENCES

[1] Alexis De Vos, « The fill factor of a solar cell from a mathematical point of view », *Solar Cells*, vol. 8, n° 3, p. 283-296, avr. 1983, doi: 10.1016/0379-6787(83)90067-4.

[2] C. Domínguez, I. Antón, et G. Sala, « Multijunction solar cell model for translating I-V characteristics as a function of irradiance, spectrum, and cell temperature », *Prog. Photovolt: Res. Appl.*, p. n/a-n/a, 2010, doi: 10.1002/pip.965.

[3] S. Suckow, T. Pletzer, et K. Heinrich, « Fast and reliable calculation of the two-diode model without simplifications », *Prog. Photovolt: Res. Appl.*, vol. 22, p. 494-501, 2014, doi: 10.1002/pip.2301.

[4] Y. Hishikawa, T. Takenouchi, M. Higa, K. Yamagoe, H. Ohshima, et M. Yoshita, « Translation of Solar Cell Performance for Irradiance and Temperature From a Single *I-V* Curve Without Advance Information of Translation Parameters », *IEEE J. Photovoltaics*, vol. 9, n° 5, p. 1195-1201, sept. 2019, doi: 10.1109/JPHOTOV.2019.2924388.

[5] K. Ishibashi, Y. Kimura, et M. Niwano, « An extensively valid and stable method for derivation of all parameters of a solar cell from a single current-voltage characteristic », *Journal of Applied Physics*, vol. 103, n° 9, p. 094507, mai 2008, doi: 10.1063/1.2895396.

[6] A. Dahi, P. Voarino, Y. Veschetti, et R. Cariou, « Measurement Methods Improvement for Solar Cells under Spatial Conditions », *JNPV*, 2021.

[7] S. Kamperis, « A Gentle Introduction to Kernel Density Estimation », 2023.

[8] Stathis, Kamperis. \A Gentle Introduction to Kernel Density Estimation | A Blog on Science." Accessed August 21, 2021. https://ekamperi.github.io/math/2020/12/08/kernel-density-estimation.html

[9] Clifford H. Wagner."Simpson's Paradox in Real Life". The American Statistician. 36 (1): 46{48. doi:10.2307/2684093. JSTOR 2684093.

Fatigue Debonding of EVA from Solar Glass at elevated PV Service Temperatures

Gernot Wallner, Gabriel Riedl, Philipp Haselsteiner, Robert Pugstaller

JKU-IPMT, CDL-AgePol, Linz, Austria

While the near-service conditions of photovoltaic (PV) modules consist of superimposed mechanical stresses and environmental factors such as elevated temperatures and humidity levels, established methods for characterization of debonding in PV modules are based on mechanical testing subsequent to environmental exposure. Hence, the service relevant loading conditions are not considered adequately. The main objective of this paper was to develop and apply a fracture mechanics test methodology allowing for the examination of the cyclic fatigue debonding of the ethylene vinylacetate copolymer (EVA) encapsulant from inorganic plies of PV modules under service-near superimposed loading conditions. An environmental fatigue test setup was implemented and used for characterization of double glass laminates with EVA encapsulation on specimen level. Double-cantilever beam specimen were prepared by vacuum lamination and characterized under displacement control at elevated temperatures ranging from 60 to 90°C and different humidity levels. Interestingly, the crack growth kinetics of the investigated glass/EVA-laminates revealed an anomalous behavior with a local minimum in debonding resistance under dry conditions at 70°C, which is in the upper range of the melting peak of the crosslinked EVA encapsulant. At 60 but also at 80°C, significantly higher threshold strain energy rate values were obtained. The local minimum was attributed to pronounced recrystallization effects within the melting peak of the slowly crystallizing, crosslinked EVA adhesive and viscosity instabilities within the multi-phase structure of the encapsulant. This anomaly was not observed under hot-humid conditions, which in general had an adverse effect on the crack growth behavior at the glass/EVA interface, especially at higher relative humidity levels. Under hot-humid or damp heat conditions, interface-near failure driven by pronounced local ageing effects within the fracture process zone were ascertained by X-ray photoelectron spectroscopy.

978-1-6654-6060-6/23 $31.00 © 2023 IEEE

Demonstration of Dual-Junction ELO Solar Cells with Strain-Balanced and Lattice-Matched Quantum Well Absorbers

Rao Tatavarti,[1] Andree Wibowo,[1] Mitsuru Imaizumi,[2]
Takeshi Ohshima,[3] David Wilt[4] and Roger Welser[5]

[1] MicroLink Devices, Inc. Niles, IL 60714 USA;
[2] Japan Aerospace Exploration Agency (JAXA), Tsukuba, Ibaraki, 305-8505 Japan;
[3] National Institutes for Quantum Beam Science and Technology (QST), Takasaki, Gunma, 370-1292 Japan;
[4] Consultant, Albuquerque, NM 02906 USA, and [5] Consultant, Providence, RI 02906 USA

Abstract — **Inverted InGaP/GaAs dual-junction (DJ) solar cells with quantum well absorbers were grown and fabricated via epitaxial liftoff (ELO) at MicroLink Devices, Inc.. ELO DJ cells have recently been identified as a game-changing, low-mass technology with reduced operating temperatures and a pathway to lower manufacturing costs. Collection from a strain-balanced quantum-well (SBQW) absorber in the GaAs subcell of ELO DJ cells is found to be remarkably insensitive to both electron and proton irradiation. ELO DJ cells with a lattice-matched quantum well (LMQW) absorber in the InGaP subcell are also demonstrated. The addition of SBQW and LMQW absorbers provide a promising pathway to improve both the beginning-of life (BOL) and end-of-life (EOL) performance of III-V multijunction solar cells.**

I. INTRODUCTION

Adding SBQW InGaAs/GaAsP absorbers to the depletion region of a GaAs subcell is a well-established approach for extending infrared absorption and increasing the short circuit current density (J_{sc}) of single- and multi-junction III-V solar cells [1-4]. Recent work at several laboratories has demonstrated GaAs-based SBQW solar cells with excellent V_{oc} and improved beginning-of-life (BOL) and end-of-life (EOL) performance [5-8]. In this work, we investigate the radiation tolerance of SBQW absorbers in a flexible, lightweight dual-junction (DJ) solar cell employing a radiation-tolerant n-i-p front-junction architecture in both the top and bottom subcells. Inverted InGaP/GaAs DJ cells with SBQW absorbers were grown and then fabricated via epitaxial liftoff (ELO) at MicroLink Devices. ELO DJ cells have recently been identified as a game-changing, low-mass technology with reduced operating temperatures and a pathway to lower manufacturing costs. Collection from a SBQW absorber in ELO DJ cells is found to be remarkably insensitive to both electron and proton irradiation.

Lattice-matched InGaAsP quaternary wells can also be add to the InGaP subcell [9]. Such LMQW absorbers in the top InGaP subcell can be used to optimize the subcell current ratios to further improve both BOL performance and EOL radiation tolerance. Here were present preliminary results from a first demonstration of a DJ solar cell incorporating quantum wells in both subcells: SBQW absorbers in the GaAs subcell and a

LMQW absorber in the top InGaP subcell. The results of this work strongly suggest that quantum well absorbers can be leveraged to improve the BOL and EOL performance of III-V DJ ELO solar cells.

II. DEVICE STRUCTURES AND EXPERIMENTAL DETAILS

Two different flavors of DJ ELO device structures are considered in this study: one with a LMQW absorber in the InGaP top subcell and one without. Both structures employ a radiation-tolerant n-i-p front-junction architecture in both the top InGaP and bottom GaAs subcells. The bottom GaAs subcell of each structure includes a SBQW absorber comprised of 9x pairs of 21-nm InGaAs wells and 38 nm GaAsP strain-compensating barrier layers. One of the structures also includes a LMQW absorber in the top InGaP subcell comprised of a single 9 nm InGaAsP quaternary well that is nominally lattice matched to the surrounding InGaP material.

All the of the device structures in this study were grown inverted via metalorganic chemical vapor deposition (MOCVD) on 6" GaAs substrates. Small area 0.25 cm^2 and 1.0 cm^2 cells were then built using MicroLink Devices' established ELO fabrication process [10]. The cells are tested before and after applying an antireflective (AR) coating.

After characterizing BOL AR coated performance, 1.0 cm^2 cells with a SBQW absorber (but without the LMQW) were sent to QST for proton and electron irradiation over a range of energies and fluence levels. A separate set of 0.25 cm^2 cells from the same wafer were also sent to NIST for 1-MeV electron irradiation. Irradiated cells were then recharacterized, testing both the external quantum efficiency (EQE) and the AM0 illuminated current-voltage (J-V) characteristics.

All the devices in this study utilize two back reflector structures. First, a phosphorus-based distributed Bragg reflector (DBR) provides a narrow band of high reflectance centered around 900 nm and covering both the GaAs and SBQW absorption edges. Second, broadband reflectance extending far into the infrared spectrum is provided by the reflective back metal electrode. As an example, Figure 1 shows the measured reflectance from a 1.0 cm^2 cell after 1-MeV electron irradiation with a fluence of 5 x 10^{14} e/cm^2. The high infrared reflectance

978-1-6654-6060-6/23 $31.00 © 2023 IEEE

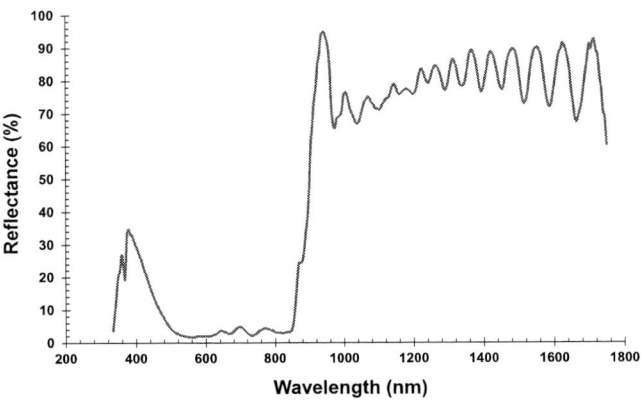

Fig. 1. Measured reflectance from a DJ SBQW ELO solar cell after 1-MeV electron irradiation with a fluence of 5×10^{14} e/cm^2.

of a DJ ELO minimizes absorption and heat generation. While a conventional triple-junction (TJ) solar cell on Ge absorbs virtually all infrared light, a DJ ELO cell can reject unabsorbed photons. First order estimates of the net heat transfer suggest this reduction in radiative heating should result in a > 30°C decrease in the operating temperature of a DJ cell relative to a TJ cell operating in space. Recent analysis of next generation solar arrays for LEO applications concludes that DJ ELO technology can enable operational EOL power density comparable to conventional TJ technology with half the specific mass [11].

III. RADIATION TOLERANCE OF SBQW ABSORBERS

The radiation tolerance of DJ SBQW ELO solar cells has been studied as a function of both electron and proton irradiation. As an illustration, Figure 2 compares the measured AM0 J-V and EQE characteristics of a 0.25 cm^2 DJ ELO solar cell with a SBQW absorber before and after irradiation with 1 $\times$ 10^{15} e/cm^2 of 1-MeV electrons. As seen in Figure 2(a), the degradation in the measured short circuit current density (J$_{sc}$) in this device is minimal. The impact of the SBQW absorber is evident in Figure 2(b), extending absorption into the infrared with a ~ 925 nm absorption edge. Moreover, while collection form the GaAs (and InGaP) base layers show clear signs of degradation, collection for the SBQW absorber is unchanged after irradiation. Similar behavior has been reported in single-junction GaAs solar cells incorporating SBQW absorbers [6].

To better quantify SBQW collection and the impact of irradiation, the integrated infrared (IR) J$_{sc}$ is computed and compared in Figure 3. By focusing only on collection at wavelengths longer than 870 nm, this integrated IR J$_{sc}$ metric captures the impact of the SBQW absorber on extending IR collection [7]. The integrated IR J$_{sc}$ is found to be unchanged after 1 MeV electron exposures up to 1 $\times$ 10^{15} e/cm^2 and 3 MeV proton exposures up to 1 $\times$ 10^{12} p/cm^2.

Excellent measured AM0 J$_{sc}$ remaining factors of 0.98 and 0.95 are also observed after 1 MeV electron (1 $\times$ 10^{15} e/cm^2)

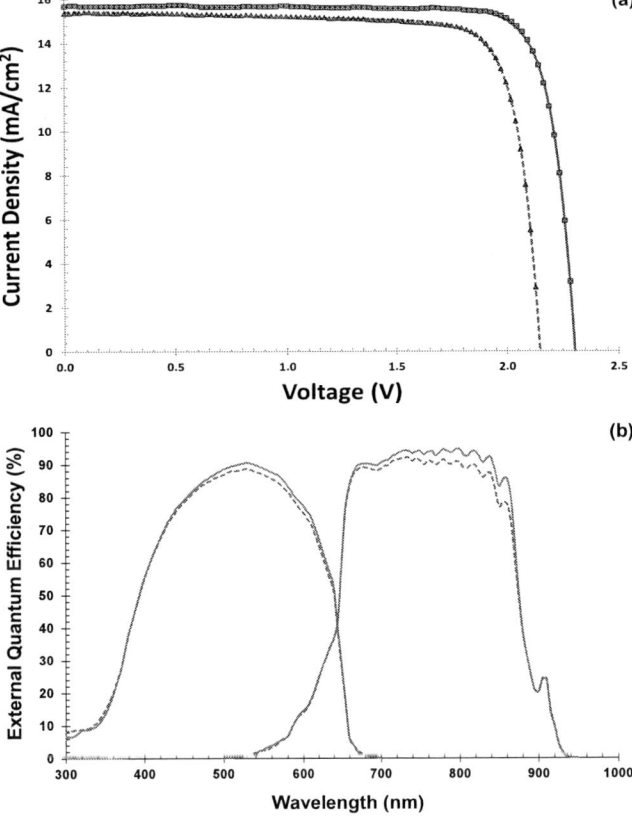

Fig. 2. Comparison of (a) AM0 J-V curves and (b) EQE spectrums from a DJ SBQW ELO cell before (solid red lines) and after (dashed blue lines) 1-MeV electron irradiation with a fluence of 1 x 10^{15} e/cm^2.

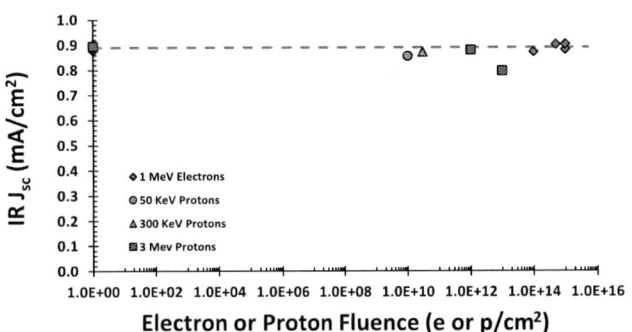

Fig. 3. Measured (solid shapes) variations in the integrated infrared J$_{sc}$ (>870 nm) as a function of electron and proton fluence. Also shown for reference the modeled (dashed line) variation assuming collection form the SBQW absorber is independent of irradiation.

and 3 MeV proton (1 $\times$ 10^{12} p/cm^2) exposures in thin DJ cells incorporating light trapping structures. The addition of the SBQW absorber does not appear to alter the degradation in open circuit voltage (V$_{oc}$) and fill factor (FF), as the observed changes in V$_{oc}$ and FF remaining factor are comparable or even higher than previously reported III-V solar cells without SBQW absorbers [12].

IV. ELO DJ DEVICE WITH SBQW AND LMQW ABSORBERS

DJ ELO solar cells with quantum wells in both subcells have also been demonstrated. The measured AM0 J-V and EQE characteristics of an initial uncoated SBQW + LMQW cell are shown in Figure 4. To the best of our knowledge, this is the first report of a DJ solar cell including both a SBQW absorber in the GaAs subcell and a LMQW absorber in the top subcell.

The addition of the LMQW did not degrade the V_{oc} relative to the SBQW-only structure, as both structures yield a V_{oc} ~ 2.34 V. At the same time, the EQE spectrum shows a clear extension of the InGaP subcell collection out to longer wavelengths (>720 nm). The GaAs subcell EQE of this uncoated cell exhibits strong Fabry–Pérot oscillations. Nevertheless, the extension of the infrared EQE out to ~ 925 nm is clearly visible.

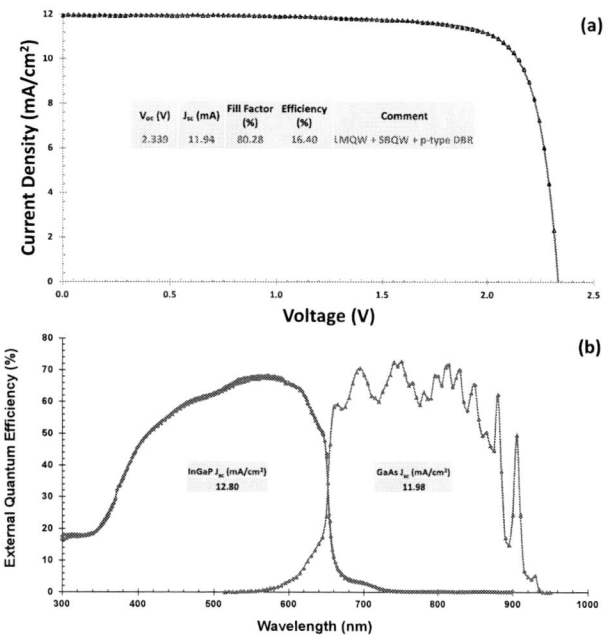

Fig. 4. Measured (a) AM0 J-V curve and (b) EQE spectrum from an uncoated DJ SBQW + LMQW ELO solar cell.

V. CONCLUSION

DJ ELO solar cells are a game-changing, low-mass technology with inherently lower operating temperatures in space environments and potentially reduced manufacturing costs. Adding SBQW and LMQW absorbers to the GaAs and InGaP subcells provides a pathway to improve both BOL and EOL performance. Radiation studies indicate that collection from optimized SBQW absorbers exhibits strong radiation

tolerance. A combined LMQW + SBQW DJ ELO solar cell has also been demonstrated exhibiting extended long wavelength collection in both subcells.

ACKNOWLEDGMENT

The authors acknowledge Air Force Research Laboratories (AFRL) for their support through SBIR contract FA9453-21-C-0717. Approved for public release; distribution is unlimited. Public Affairs release approval # _____.

REFERENCES

[1] D. B. Bushnell, T. N. D. Tibbits, K. W. J. Barnham, J. P. Connolly, M. Mazzer, N. J. Ekins-Daukes, J. S. Roberts, G. Hill, and R. Aiery. "Effect of Well Number on the Performance of Quantum-Well Solar Cells," *J. Appl. Phys.*, vol. 97, no. 124908 (June 2005).

[2] M. Sugiyama, Y. Wang, K. Watanabe, T. Morioka, Y. Okada, and Y. Nakano, "Photocurrent Generation by Two-Step Photon Absorption With Quantum-Well Superlattice Cell," *IEEE Journal of Photovoltaics*, vol. 2, pp. 298-302 (June 2012).

[3] I. Sayed and S. M. Bedair, "Quantum Well Solar Cells: Principles, Recent Progress, and Potential," *IEEE J. Photovotalics* 9, 402-422 (February 2019).

[4] R. E. Welser, S. J. Polly, M. Kacharia, A. Fedorenko, A. K. Sood, and S. M. Hubbard, "Design and Demonstration of High-Efficiency Quantum Well Solar Cells Employing Thin Strained Superlattices," *Scientific Reports*, vol. 9, no. 13955 (September 2019).

[5] M. A. Steiner, R. M. France, J. Buencuerpo, J. F. Geisz, M. P. Nielsen, Andreas Pusch, W. J. Olavarria, M. Young, and N. J. Ekins-Daukes, "High Efficiency Inverted GaAs and GaInP/GaAs Solar Cells With Strain-Balanced GaInAs/GaAsP Quantum Wells," *Adv. Energy Mater.*, vol. 2020, no. 2002874 (Dec. 2020).

[6] S. R. Tatavarti, K. Forghani, A. Wibowo, and R. E. Welser, "Radiation-Induced Degradation Mechanisms in Thin-Film Multiple-Quantum-Well Solar Cells with Wavelength-Selective Photonic Structures," *IEEE J. Photovoltaics*, vol. 12, no. 1192 (September 2022).

[7] R. E. Welser, S. J. Polly, B.M. Bogner, and S. M. Hubbard, "Impact of Well Number on High-Efficiency Strain-Balanced Quantum-Well Solar Cells," *IEEE J. Photovoltaics*, accepted for publication, August 2022.

[8] R. M. France, J. F. Geisz, T. Song, W. Olavarria, M. Young, A. Kibbler and M. A. Steiner, "Triple-Junction Solar Cells with 39.5% Terrestrial and 34.2% Space Efficiency Enabled by Thick Quantum Well Superlattices," *Joule*, vol. 6, pp. 1121-1135 (May 2022).

[9] I. E. H. Sayed, N. Jain, M. A. Steiner, J. F. Geisz, and S. M. Bedair, "100-Period InGaAsP/InGaP Superlattice Solar Cell with Sub-Bandgap Quantum Efficiency Approaching 80%," *Appl. Phys. Lett.*, vol. 111, no 082107 (August 2017).

[10] R. Tatavarti, G. Hillier, A. Dzankovic, G. Martin, F. Tuminello, R. Navaratnarajah, G. Du, D.P. Vu, and N. Pan, "Lightweight, Low-Cost GaAs Solar Cells on 4" Epitaxial Liftoff (ELO) Wafers," 33rd IEEE PVSC, pp. 1-4 (May 2008).

[11] M. Kroon, "Next Generation of Solar Arrays for LEO Applications," PVSEC-33 (November 2022).

[12] M. Imaizumi, T. Takamoto, N. Kaneko, Y. Nozaki, and T. Ohshima, "Qualification Test Results of IMM Triple-Junction Solar Cells, Space Solar Sheets, and Lightweight & Compact Solar Paddle," E3S Web Conf., vol. 16, no. 3012 (May 2017).

Cu2ZnSnS4 monograin layer solar cells for flexible photovoltaic applications

Marit Kauk-Kuusik, Kristi Timmo, Maris Pilvet, Katri Muska, Mati Danilson, Jüri Krustok, Raavo Josepson, Maarja Grossberg-Kuusk

Tallinn University of Technology, Tallinn, Estonia

Flexible photovoltaics have been and will be increasingly demanded in various applications in todays and future society. The search for an ideal flexible photovoltaic technology that can perfectly meet these expanding demands has long been an active branch of photovoltaic research. Monograin layer technology (MGL) is one possible path to develop the lightweight, flexible, and semi-transparent solar cells. The major innovations in the MGL solar cell technology are the light absorbing layer made of high quality micro-crystalline semiconductor powder enabling theoretically higher efficiencies and the low cost and easily up-scalable roll-to-roll PV module production process. In recent years, the main research focus of the monograin technology has been on the understanding of the synthesis and optoelectronic properties of Cu2ZnSnS4 absorber materials. The highest power conversion efficiency of this type of devices is 12.06% with output parameters as follows: V_{OC} =0.745 V, J_{SC}=28.36 mA/cm2 and FF = 57.10 %. In this study temperature dependence (T = 20- 320K) of current-voltage (J-V) characteristics of record efficiency Cu2ZnSnS4 MGL solar cell were investigated to clarify the main losses in CZTS, which are still not fully understood. The light J$-$V curve analysis was used to evaluate the quality of the p$-$n junction and losses related to resistive components of the device. In this study, the single exponential diode equation was employed to analyze the light J$-$V data. It turned out that at lower temperatures (T< 180 K) a second blocking diode appears and it is related to back contact barrier. We believe that this back contact barrier has an effect also at higher temperatures causing relative low values of FF. At T=300 K the diode ideality factor n has a value 2.58 and the series resistance R_s =2.6 Ωcm2.

Advances in Flexible and Lightweight III-V Multijunction Solar Cells for High Power Density Applications

Carlos Algora, Ivan Garcia, Clara Sanchez-Perez, Pablo Martin, Pablo Fernandez, Luis Cifuentes, Ivan Lombardero, Daniel Gomez, Mercedes Gabas, Ignacio Rey-Stolle

Instituto de Energía Solar-Universidad Politécnica de Madrid, Madrid, Spain

We are developing flexible and lightweight III-V multijuntion solar cells for high power density applications such as unmanned aerial vehicles (drones), High Altitude Pseudo Satellites (HAPS) and high altitude stratospheric balloons for satellite-like communications, cheap and widespread internet connectivity, etc. We focus our research on two different types of III-V solar cells covering the range from 1 kW/kg to 3 kW/kg (considering the weight of the solar cell itself and not that of the module). Our approach to achieve a goal of 3 kW/kg is based on 3J inverted metamorphic solar cells. We have recently developed GaInP/GaAs/GaInAs 3J IMM solar cells with a power density of 5.7 kW/kg under 1xAM1.5g for small areas. Demonstrating large area devices (tens of square cm) is the next challenge to tackle. In addition, we are developing lower-cost designs, targeting 1 kW/kg, based on GaInP/Ga(In)As/Ge 3J lattice matched solar cells. In this paper we present the advances in the modelling and manufacturing of this last type of solar cell. Modelling indicates that there are multiple efficiency-thickness combinations reaching the 1 kW/kg target. For example in the case of AM1.5g, a solar thickness of 46 microns requires an efficiency of 30% which is a demanding value. However, if the thickness is reduced to 40 microns, a more easily achievable efficiency of 27% would be required. We have demonstrated a process for thinning the Ge substrates reaching thicknesses as low as 14 microns with highly homogeneous thickness profiles. Preliminary results show that the 3J thinned solar cell exhibits an efficiency which is the 91% of the efficiency achieved by the non-thinned cell.

Improving the stability of polycrystalline silicon passivated contacts using titanium dioxide

Di Yan[1], Jesus Ibarra Michel[1], Yida Pan[1], Sieu Pheng Phang[2], Daniel Macdonald[2], Heping Shen[2], Leiping Duan[2], Kylie Catchpole[2], Jie Yang[3], Peiting Zheng[3], Xinyu Zhang[3], Hao Jin[3] and James Bullock[1]

1. Department of Electrical and Electronic Engineering, University of Melbourne, Victoria, 3010, Australia.

2. School of Engineering, The Australian National University, Canberra, ACT 2601, Australia

3. Jinko Solar, Haining, Zhejiang 314400, China

Abstract — Polycrystalline silicon (poly-Si) passivated junctions/contacts have been shown as a key technology to achieve high efficiency silicon solar cells. However, there remain some issues with this technology related to its thermal stability during/after metallization. In this work, we introduce the use of metal oxide based materials as an interlayer between the poly-Si and the metal electrode. We initially focus on TiO_2 interlayers, which we find can improve the thermal stability of n-type poly-Si contacts. The passivation quality of poly-Si contacts is maintained and even slightly improved for very thin (~8 nm) TiO_2 interlayers, when subjected to a post-metallization anneal at 400 °C. This improvement in passivation stability comes at the expense of higher contact resistivity, however, the obtained contact resistivity values are still acceptable for full area contacts. The thermal stability of the passivation can be further improved up to 500 °C by thickening the TiO_2 interlayer (~28 nm), resulting in increased contact resistivity. Nevertheless, we believe the high contact resistivities can be overcome by exploring different materials, for example in the families of the transparent conductive metal oxides and metal nitrides. The protective effects of this interlayer structure may allow the thinning of poly-Si layers, reducing their parasitic absorption, and permitting their usage on the sunward side of cells. This approach will also be impactful for silicon-based bottom cells in tandem solar cells as it can provide more flexibility in terms of top cell fabrication thermal budget.

I. INTRODUCTION

Polycrystalline silicon (Poly-Si) passivated junctions/contacts have been proven as one of the main technologies for advancing the efficiency of industrial silicon solar cells above 25%. Based on this technology, a record efficiency of 26.4% has been demonstrated on an industrial silicon wafer with a size of above 182 mm by 182 mm, and an average efficiency of 25% has also been achieved in mass production [1], [2]. However, after printing metal pastes and firing on poly-Si junctions/contacts, a significant passivation degradation is observed, particularly for thin poly-Si layers [3], [4]. The recombination current density factor, J_{oc}, increases by up to two orders of magnitudes, from 2 fA/cm^2 to 200 fA/cm^2, due to the penetration of metal into the poly-Si layers. Thus, in an industrial setting a thick poly-Si layer, above 100 nm, has to be implemented to improve the thermal stability of poly-Si junctions/contacts for the metallization process [3]–[5]. The thick poly-Si layer results in a low short circuit current density,

J_{sc}, due to its parasitic absorption [6], [7]. Even when placing the poly-Si junctions/contacts at the back side of a solar cell, a 100 nm poly-Si can cause J_{sc} losses of up to $0.3 - 0.5$ mA/cm^2 [8]. It has also been shown that the use of thermally evaporated metal electrodes can lead to passivation degradation for thin, < 50 nm, poly-Si contacts, which may be due to pinholes and grain boundary diffusion of metals and source contaminates [9], [10]. Thus, it is desirable to develop a contact structure which combines a thin poly-Si contact < 30 nm, with low parasitic absorption, with a protective interlayer to prevent/reduce interaction between the metal electrodes and poly-Si layers.

In this work, we explore the use of titanium oxide, TiO_2, interlayers between the n-type poly-Si passivated junctions/contacts and the aluminium electrode. TiO_2 is chosen as a starting candidate interlayer, since it has already been demonstrated as a good electron contact in silicon solar cells, with proven compatibility with n-type poly-Si junctions/contacts to collect electron [11]. Aluminium is used because of it has a low cost and easily react with poly-Si contacts to cause passivation degradation [12].

II. EXPERIMENTS

Two sets of samples were prepared; passivation samples and contact resistivity samples. Both passivation and contact resistivity samples use phosphorus doped CZ silicon wafers with a thickness of 170 μm and a resistivity of ~2 Ω·cm. Both have a symmetrical structure of LPCVD (low pressure chemical vapor deposition) n-type poly-Si (~50nm) /SiO$_2$ (1.5nm)/ c-Si/SiO$_2$ (1.5nm) / LPCVD n-type poly-Si (~50nm). The dopants in the n-type poly-Si are phosphorus. Samples were annealed at 920 °C for 30 mins in N$_2$ to activate dopants in the poly-Si film. After the high temperature activation process, a typical implied open circuit voltage of iV_{oc} of ~710 mV was obtained via QSSPC (quasi-steady state photoconductance) measurements. Prior to TiO_2 deposition, the oxides formed during the high temperature step were removed by a dilute HF solution. The TiO_2 interlayers were deposited by thermal ALD (atomic layer deposition) using titanium tetrakis isopropoxide (TTIP) and H$_2$O precursors at a temperature of 250 °C. Three TiO_2 thicknesses, 8 nm, 17 nm and 28 nm, were deposited over the

978-1-6654-6060-6/23 $31.00 © 2023 IEEE

full area of one side of the samples. From this point the fabrication process of passivation and contact resistivity samples differs. The contact resistivity samples were fabricated using a patterned shadow mask to form circular ~200 nm thick Al metal electrodes with different diameters on the TiO_2 deposited side, while the other side without TiO_2 was fully covered with a ~200 nm Al layer. The contact resistivity ρ_c was extracted from resistance measurements between the common rear electrode and the electrodes of different diameter dots using the Cox and Strack method [13]. The ρ_c was monitored for different interlayer thicknesses as a function of cumulative annealing steps from 250 °C to 500 °C in air.

For passivation samples the TiO_2 deposited side was partially coated with ~200 nm of Al to create adjacent regions with/without Al metallization. The impact of metal on the passivation quality of poly-Si junctions/contacts with and without TiO_2 was examined through photoluminescence (PL) imaging using a BT Imaging LIS-R1 tool. PL images were taken before and after cumulative high temperature annealing steps at 400 °C and 500 °C in air.

III. RESULTS AND DISCUSSION

A. Impact of TiO₂ interlayers on contact resistivity

The ρ_c of the contact samples, with 8 nm, 17 nm and 28 nm TiO_2 thicknesses, are shown as a function of cumulative annealing steps in Figure 1. A reference line for a sample without a TiO_2 interlayer as a function of the annealing steps is also included for comparison. The poly-Si contact without TiO_2 shows a low contact resistivity, in line with that typically found for such contacts [14]. This contact maintains a consistent low ρ_c of ~ 2 mΩ-cm², at the lower resolution limit of the Cox and Strack technique, with cumulative annealing steps. The samples with the 8 nm TiO_2 interlayer, have a ρ_c two orders of

magnitude higher than the samples without an interlayer, at ~200 mΩ-cm² in the as-deposited state. With increasing annealing temperature the ρ_c decreases, reaching a value of 50–60 mΩ-cm² at an annealing temperature of above 450 °C.

The samples with thicker TiO_2 have slightly higher ρ_c values compared to that of the samples with thin TiO_2. In the as-deposited state, samples with 17 nm and 28 nm TiO_2 both have a ρ_c ~ 240 mΩ-cm². Like the case of thin TiO_2, the ρ_c of the thicker TiO_2 interlayer contacts decreases with increasing annealing temperature, saturating at a ρ_c of ~110 mΩ-cm² above 400°C.

Thus, while the addition of TiO_2 interlayers leads to an increase in ρ_c, especially for thicker TiO_2 films, the obtained post-annealing ρ_c values are <150 mΩ-cm² which is still sufficiently low for full area contacts.

B. Impact of the TiO₂ interlayers on passivation

PL images of the passivation samples, with 8 nm, 17 nm and 28 nm TiO_2 thicknesses, are shown as a function of cumulative annealing steps in Figure 2. A reference passivation sample without a TiO_2 interlayer is included for comparison. The Al metallized regions are indicated by green rectangles. For the poly-Si samples without TiO_2, the Al metallized regions have significantly lower PL intensity than the non-metallized regions after annealing at 400 °C. This situation is improved for samples with just 8 nm TiO_2 interlayers, which show a PL

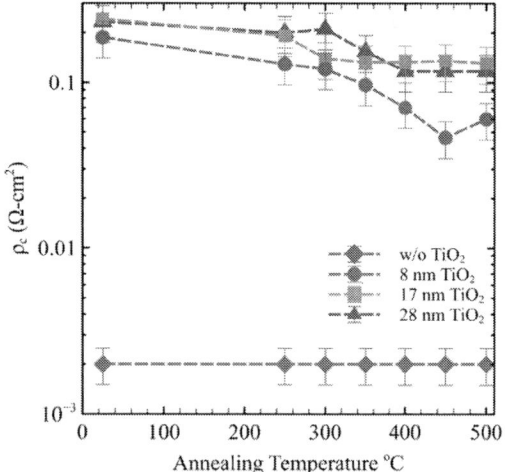

Figure 1. Contact resistivity ρ_c values of n-type poly-Si contacts with and without the TiO_2 interlayer shown as a function of cumulative annealing steps in air.

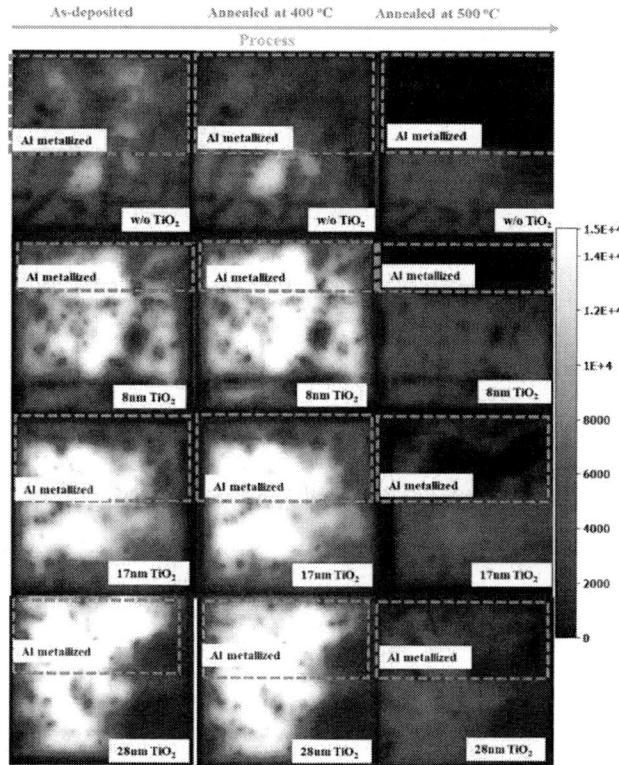

Figure 2. PL images of n-type poly-Si contacts without TiO_2 (1st row) and with 8 nm TiO_2 (2nd row), 17 nm (3rd row) and 28 nm TiO_2 (4th row) before and after cumulative annealing steps.

intensity in the metallized regions which is similar to, or even higher than, the non-metallized regions, after annealing at 400 °C. This demonstrates the effectiveness of the interlayer strategy. After an additional 500 °C annealing step, both above samples show low PL intensity, indicating poor passivation quality. This is possibly due to the penetration of Al through the poly-Si layers to the c-Si interface which degrades the surface passivation.

Thicker TiO_2 interlayers (i.e. 17 nm and 28 nm) exhibit similar protective behaviour to ~8 nm layers at 400°C. However, at 500°C the added thickness provides additional protection, and some PL intensity is maintained in the Al metallized region. In particular, the sample with the 28 nm TiO_2 interlayer maintains a similar magnitude of PL intensity in regions with/without Al metallization. Thus, TiO_2 interlayers can effectively reduce the interaction between the poly-Si layer and the Al overlayer.

IV. CONCLUSION AND FUTURE WORKS

As presented above, inserting an additional TiO_2 interlayer between the poly-Si layers and Al electrodes can improve thermal stability by limiting the interaction between poly-Si and Al. Based on PL intensity and ρ_c values, we found that thin ~8 nm TiO_2 can work as an effective protective layer for annealing steps at 400 °C. With the addition of the TiO_2 interlayer the passivation quality of metallized regions can be maintained, and even slightly improved, after the post-metallization annealing. This comes at the expense of increased ρ_c.

If stability up to 500°C is required, then thicker ~28 nm TiO_2 interlayers can be used. The use of thicker TiO_2 further increases the ρ_c to values of ~110 mΩ-cm^2 which is still suitable for full area contacts.

In this study, TiO_2 was utilised as a logical first proof-of-concept. In future work, we are trialling other materials which have already been utilised as selective contacts in silicon solar cells, such as conductive metal oxides and nitrides, as protective interlayers between poly-Si and Al electrodes. The optical properties of these materials will also be considered for sunward side applications. In addition, we will extend this TiO_2 interlayer study by using the standard industrial metallization processes, including firing steps and screen-printed aluminium pastes.

REFERENCES

[1] "JinkoSolar's High-efficiency N-Type Monocrystalline Silicon Solar Cell Sets Our New Record with Maximum Conversion Efficiency of 26.4%."

https://www.jinkosolar.com/en/site/newsdetail/1827 (accessed Jan. 16, 2023).

[2] "25% Efficiency For JinkoSolar's N-TOPCon Solar Cell | TaiyangNews." https://taiyangnews.info/technology/25-efficiency-for-jinkosolars-n-topcon-solar-cell/ (accessed Jan. 10, 2023).

[3] S. Mack, J. Schube, T. Fellmeth, F. Feldmann, M. Lenes, and J.-M. Luchies, "Metallisation of Boron-Doped Polysilicon Layers by Screen Printed Silver Pastes," *Phys. Status Solidi RRL – Rapid Res. Lett.*, vol. 11, no. 12, p. 1700334, 2017, doi: 10.1002/pssr.201700334.

[4] H. E. Çiftpınar *et al.*, "Study of screen printed metallization for polysilicon based passivating contacts," *Energy Procedia*, vol. 124, pp. 851–861, Sep. 2017, doi: 10.1016/j.egypro.2017.09.242.

[5] P. Padhamnath *et al.*, "Progress in screen-printed metallization of industrial solar cells with SiOx/poly-Si passivating contacts," *Sol. Energy Mater. Sol. Cells*, vol. 218, p. 110751, Dec. 2020, doi: 10.1016/j.solmat.2020.110751.

[6] S. Reiter *et al.*, "Parasitic Absorption in Polycrystalline Si-layers for Carrier-selective Front Junctions," *Energy Procedia*, vol. 92, pp. 199–204, Aug. 2016, doi: 10.1016/j.egypro.2016.07.057.

[7] Y. Larionova *et al.*, "Ultra-Thin Poly-Si Layers: Passivation Quality, Utilization of Charge Carriers Generated in the Poly-Si and Application on Screen-Printed Double-Side Contacted Polycrystalline Si on Oxide Cells," *Sol. RRL*, vol. 4, no. 10, p. 2000177, 2020, doi: 10.1002/solr.202000177.

[8] F. Feldmann, M. Nicolai, R. Müller, C. Reichel, and M. Hermle, "Optical and electrical characterization of poly-Si/SiOx contacts and their implications on solar cell design," *Energy Procedia*, vol. 124, pp. 31–37, Sep. 2017, doi: 10.1016/j.egypro.2017.09.336.

[9] B. Nemeth *et al.*, "Low temperature Si/SiOx/pc-Si passivated contacts to n-type Si solar cells," in *2014 IEEE 40th Photovoltaic Specialist Conference (PVSC)*, Jun. 2014, pp. 3448–3452. doi: 10.1109/PVSC.2014.6925675.

[10] W. Nemeth *et al.*, "Implementation of tunneling pasivated contacts into industrially relevant n-Cz Si solar cells," in *2015 IEEE 42nd Photovoltaic Specialist Conference (PVSC)*, Jun. 2015, pp. 1–3. doi: 10.1109/PVSC.2015.7356062.

[11] X. Yang, P. Zheng, Q. Bi, and K. Weber, "Silicon heterojunction solar cells with electron selective TiOx contact," *Sol. Energy Mater. Sol. Cells*, vol. 150, pp. 32–38, Jun. 2016, doi: 10.1016/j.solmat.2016.01.020.

[12] D. Yan, S. P. Phang, Y. Wan, C. Samundsett, D. Macdonald, and A. Cuevas, "High efficiency n-type silicon solar cells with passivating contacts based on PECVD silicon films doped by phosphorus diffusion," *Sol. Energy Mater. Sol. Cells*, vol. 193, pp. 80–84, May 2019, doi: 10.1016/j.solmat.2019.01.005.

[13] R. H. Cox and H. Strack, "Ohmic contacts for GaAs devices," *Solid-State Electron.*, vol. 10, no. 12, pp. 1213–1218, Dec. 1967, doi: 10.1016/0038-1101(67)90063-9.

[14] P. Zheng *et al.*, "Detailed loss analysis of 24.8% large-area screen-printed n-type solar cell with polysilicon passivating contact," *Cell Rep. Phys. Sci.*, vol. 2, no. 10, p. 100603, Oct. 2021, doi: 10.1016/j.xcrp.2021.100603.

Copper Oxide: A Potential Candidate for Hole Transport Material in Perovskite Solar Cells for Space

Daniel Muñoz-Pinzon[1], Rishabh Sahani[1], Mateo Ferreira[1], Neetesh Kumar[1], Cheng-Yu Lai[1], Daniela Radu[1,*], and Lyndsey McMillon-Brown[2,*]

[1]Florida International University, Miami, Florida, 33199, U.S.A

[2]NASA Glenn Research Center, Cleveland, Ohio, 44135, U.S.A

Abstract — With the upsurging interest in perovskite solar cells, the replacement of the organic hole transport material, Spiro-OMeTAD, in favor of robust inorganic material is crucial to improve the durability and stability of perovskite solar cells in space. Copper Oxide (Cu_2O) has exhibited significant potential, resulting in notable power conversion efficiencies when incorporated as inorganic hole transport layers (HTL). Herein, we present the performance of inorganic Cu_2O as a hole transport layer in laminated perovskite solar cells with initial characterization data for the Cu_2O material.

Keywords – Copper oxide, HTL, transport, durability, lamination.

I. INTRODUCTION

In recent years, perovskite solar cells (PSCs) have attracted wide attention from the scientific community. This can be attributed to their remarkable optoelectronic properties such as high absorption coefficients, charge carrier mobility, tunable bandgap, low exciton binding energy, and long electron and hole diffusion lengths.[1]-[3] Further, it has low fabrication cost and, solution-based processibility. PSCs have undergone rapid developments from their initial power conversion efficiency (PCE) of 3.8% [4] to above 25% [5] in a short timeframe. Due to the gradual advancement of the PCE in PSCs, investigations into stabilizing the metastable α-phase (photoactive phase) and substituting organic materials for more robust, inorganic materials, to further enhance the stability.

Under ambient conditions, PSCs experience a progressive regression from their black photoactive α-phase into their thermodynamically favorable yellow δ-phase. [1]-[4] Due to the wider bandgap and marginal photoactivity leads towards degradation forming charge trap-states eventually resulting in a greater hysteresis. [6] In addition to fragility demonstrated by the PSCs to humidity and heat, the solution-based fabrication is dependent on the device architecture. [7] Dunfield et al. introduced the use of lamination to fabricate PSCs which enables roll-to-roll processing at a PCE of 10.6%. In this methodology, the use of conductive material is no longer limited as the incorporation of two half-stacks to produce one cell opens up the possibility of implementing a material combination that was previously thought impossible [8]. Furthermore, it has been proved that perovskite films subjected to increased pressure improved the overall device efficiency by improving the interfaces between layers. [9] [10]

In an effort to move PSCs toward deployment in space, the incorporation of more durable inorganic materials in both ETL and HTL is essential. In this line, the use of inorganic copper oxide has the benefit of a more cost-effective material with low cost, no toxicity, and high p-type mobility which helped in achieving some of the notable PCE (19%) when using inorganic HTL. [11] Here we present the incorporation of inorganic copper oxide as a hole transport material in laminated perovskite solar cells via spray deposition and its impact on both the stability and performance of the devices.

II. EXPERIMENTAL

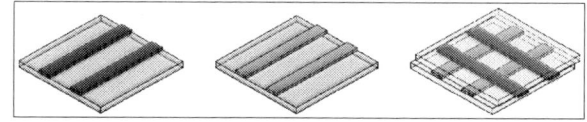

Fig. 1. Demonstrates the etching pattern used for laminated cells.

A. Fabrication Process

Substrate preparation: Fluorine-doped Tin Oxide glass substrates were etched using zinc powder and 2M HCl solution; the patterns are shown in Figure 1. Substrates were subsequentially washed via sonication through a multi-step cleaning process using nanopure water, acetone, and isopropanol, in this sequence. The substrates were further dried in a convection oven at 100°C.

Once substrates were prepared, UV-ozone treatment was conducted for 15 minutes to remove any other organic residues followed by the deposition of phenyl-C61-butyric acid methyl ester (PCBM) as an ETL on the first stack and copper oxide as HTL was sprayed on the second stack followed by annealing at 300 C respectively.

The addition of poly(3,4-ethylene dioxythiophene) polystyrene sulfonate (PEDOT: PSS) and D-Sorbitol as a "conductive glue" on the second stack [12] served as a bilayer between the two stacks was used to enhance the interface between both half-stacks once laminated. A triple cation perovskite $Cs_{0.04}(FA_{0.84}MA_{0.16})_{0.96}Pb(I_{0.84}Br_{0.16})_3$ was deposited via spin-coating on both half-stacks in order to create the perovskite-perovskite interface during lamination.

Cu$_2$O films were deposited via spray using a 0.05M Cu$_2$O precursor solution made of copper acetate monohydrate and D-glucose. Substrates were placed on a hotplate at 280°C and sprayed using an airbrush at varying cycles to determine the optimal thickness for HTL application. A phase conversion can be visually determined as the color of the substrate changes from its initial transparent color to black, brown, and lastly a slight yellowish hue. To ensure high crystallization, the substrates were annealed for an hour and then removed from the hotplate. Characterization was performed on uncoated substrates, as a control, and after spraying, to assess structural composition, thickness, morphology, and chemical composition.

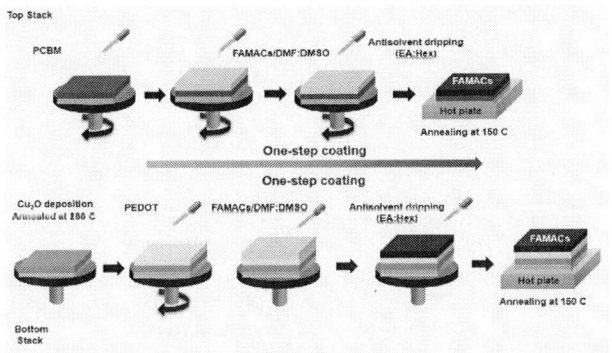

Fig. 2. Device fabrication route for PSCs.

B. Lamination Process

After the deposition of the perovskite layer on both half-stacks, substrates are prepared for lamination by placing polydimethylsiloxane (PDMS) membranes are placed above and below the two half-stacks to prevent any damage to the substrates from the heating or pressure element. The lamination is conducted for 10 min under ~ 6 MPa and 120°C. Once the process is finished, substrates are allowed to cool under the same pressure for another additional 15 min.

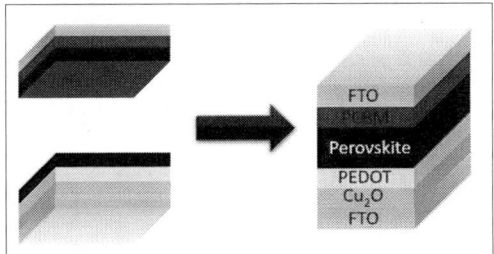

Fig. 3. Device architecture and materials used for PSCs.

Initial characterization performed on the Cu$_2$O HTL film showed pure Cu$_2$O formation. Using a Rigaku MiniFlex Benchtop X-Ray Diffraction Instrument (XRD), the scan showed high peak intensity in our Cu$_2$O film with the main peak located at around 18.5 degrees on the (111) plane as shown in Fig. 4. Absorbance measurements with a UV-Vis-NIR Spectrophotometer (UV-3600 Plus) showed a peak at around 580 nm as seen in Fig. 5. The bandgap of 2.13 eV extracted from the Tauc plot proves its potential to be used as an HTL indicated by Fig. 6. Scanning Electron Microscopy (SEM) images were taken using a Jeol JSM-IT700HR InTouchScope, and show a compact uniform morphology and an average particle size of 20-30 nm as demonstrated in Fig. 7.

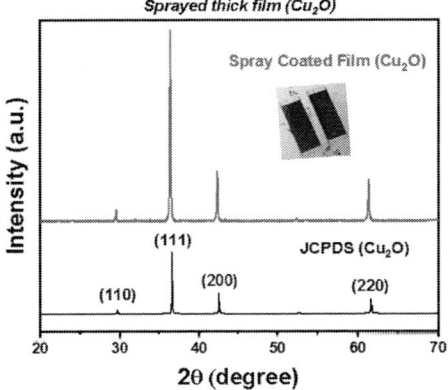

Fig. 4. XRD pattern for sprayed Cu$_2$O thin films.

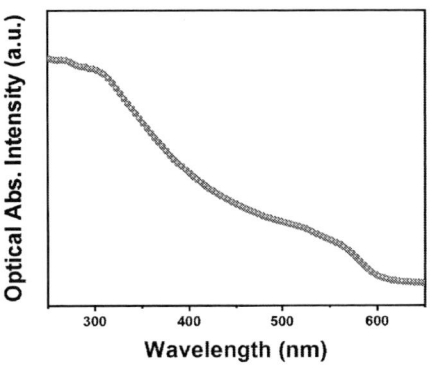

Fig. 5. UV-Vis Absorbance Curve and Tauc's Plot for Cu$_2$O.

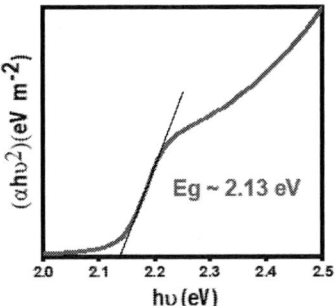

Fig. 6. Tauc's Plot for Cu$_2$O.

Fig. 7.   SEM images of Cu_2O at 90k magnification.

IV. SUMMARY AND CONCLUSION

We present a solution-processed methodology for Cu_2O film fabrication that enables its incorporation in laminated devices. Initial characterizations show that Cu_2O is a promising candidate as a hole transport material. The lamination method in perovskite photovoltaics has the potential to increase the stability of the devices with in-situ encapsulation. This methodology can reduce the cost of manufacturing as well as present a viable manufacturing methodology that can be employed in space.

REFERENCES

[1]   [1] H. B. Lee, N. Kumar, B. Tyagi, S. He, R. Sahani and J. W. Bulky organic cations engineered lead-halide perovskites: a review on dimensionality and optoelectronic applications Kang Materials Today Energy 2021 Vol. 21 Accession Number: WOS:000701812300002   DOI: ARTN   100759. 10.1016/j.mtener.2021.100759

[2] H. B. Lee, R. Sahani, V. Devaraj, N. Kumar, B. Tyagi, J. W. Oh, et al. Complex Additive ‐ Assisted Crystal Growth and Phase Stabilization of α ‐ FAPbI 3 Film for Highly Efficient, Air ‐ Stable Perovskite Photovoltaics Advanced Materials Interfaces 2022 Pages 2201658. DOI: 10.1002/admi.202201658

[3] A. Krishna, S. Gottis, M. K. Nazeeruddin, and F. Sauvage. Mixed Dimensional 2D/3D Hybrid Perovskite Absorbers: The Future of Perovskite Solar Cells? Advanced Functional Materials 2019 Vol. 29 Issue 8 Pages 1806482. DOI: 10.1002/adfm.201806482

[4] A. Kojima, K. Teshima, Y. Shirai and T. Miyasaka. Organometal Halide Perovskites as Visible-Light Sensitizers for Photovoltaic Cells Journal of the American Chemical Society 2009 Vol. 131 Issue 17 Pages 6050-6051. DOI: 10.1021/ja809598r

[5] H. Min, D. Y. Lee, J. Kim, G. Kim, K. S. Lee, J. Kim, et al. Perovskite solar cells with atomically coherent interlayers on $SnO2$ electrodes. Nature 2021 Vol. 598 Issue 7881 Pages 444-450. DOI: 10.1038/s41586-021-03964-8

[6] T. S. Sherkar, C. Momblona, L. Gil-Escrig, J. Ávila, M. Sessolo, H. J. Bolink, et al. Recombination in Perovskite Solar Cells: Significance of Grain Boundaries, Interface Traps, and Defect Ions. ACS Energy Letters 2017 Vol. 2 Issue 5 Pages 1214-1222. DOI:10.1021/acsenergylett.7b00236. https://www.ncbi.nlm.nih.gov/pmc/articles/PMC5438194

[7] N. K. Noel, B. Wenger, S. N. Habisreutinger and H. J. Snaith. Utilizing Nonpolar Organic Solvents for the Deposition of Metal-Halide Perovskite Films and the Realization of Organic Semiconductor/Perovskite Composite Photovoltaics. ACS Energy Letters 2022 Vol. 7 Issue 4 Pages 1246-1254 DOI: 10.1021/acsenergylett.2c00120

[8] S. P. Dunfield, D. T. Moore, T. R. Klein, D. M. Fabian, J. A. Christians, A. G. Dixon, et al. Curtailing Perovskite Processing Limitations via Lamination at the Perovskite/Perovskite Interface. ACS Energy Letters 2018 Vol. 3 Issue 5 Pages 1192-1197. DOI: 10.1021/acsenergylett.8b00548

[9] L. Shi, M. Zhang, Y. Cho, T. L. Young, D. Wang, H. Yi, et al. Effect of Pressing Pressure on the Performance of Perovskite Solar Cells. ACS Applied Energy Materials 2019 Vol. 2 Issue 4 Pages 2358-2363 DOI: 10.1021/acsaem.8b01608

[10] O. Y. Gong, M. K. Seo, J. H. Choi, S.-Y. Kim, D. H. Kim, I. S. Cho, et al. High-performing laminated perovskite solar cells by surface engineering of perovskite films. Applied Surface Science 2022 Vol. 591 DOI: 10.1016/j.apsusc.2022.153148

[11] H. Rao, S. Ye, W. Sun, W. Yan, Y. Li, H. Peng, et al. A 19.0% efficiency achieved in CuOx-based inverted CH3NH3PbI3−xClx solar cells by an effective Cl doping method. Nano Energy 2016 Vol. 27 Pages 51-57 DOI: 10.1016/j.nanoen.2016.06.044

[12] Bifacial Perovskite Solar Cells via a Rapid Lamination Process Tianyang Li, Wiley A. Dunlap-Shohl, and David B. Mitzi ACS Applied Energy Materials 2020 3 (10), 9493-9497 DOI: 10.1021/acsaem.0c00756

Innovative Methodology for an Advanced Characterization of Perovskite Systems to Reach Buried Interfaces: In-depth Profile by Coupling GD-OES and XPS

Mirella Al Katrib [1,2], Pia Dally [1,2], Armelle Yaiche [1,3], Jean Rousset [1,3], Muriel Bouttemy [1,2]

[1] Institut Photovoltaïque d'Île de France (IPVF), Palaiseau, 91120, France
[2] Institut Lavoisier de Versailles (ILV), Versailles, 78035, France
[3] EDF R&D, Institut Photovoltaïque d'Île de France (IPVF), Palaiseau, 91120, France

Abstract — **Interfaces properties in solar cells play a crucial role on the device's performance and stability, hence the importance of investigating the chemical behavior at the buried interfaces in solar devices. This work aims to develop an innovative methodology of coupling two in-depth profile characterizations, to accurately and reliably investigate the chemical composition from the surface to the interfaces. Hence, Glow Discharge Optical Emission Spectroscopy (GD-OES) and X-Ray Photoelectron Spectroscopy (XPS) were applied consecutively on half-cells. First, an optimization of the operating conditions was carried out to minimize the degradation of the perovskite layers. In the case of GD-OES not only the Radio Frequency power and the plasma gas pressure are changed, but also the nature of this gas (Ar, Ar/O). Secondly, the craters resulting from profiling by GD-OES were chemically studied by XPS in order to determine the chemical composition at different level of the layer as well as in the interface area. We observed that all the conditions employed for GD-OES profiling led to a systematic reduction or oxidation of lead as well as the degradation of the organic part and iodine loss, more or less pronounced depending on the plasma gas. A comparison of SEM (Scanning Electron Microscopy) images inside and outside the craters also showed a remarkable change in the surface morphology for a bombarded surface by Ar or Ar/O. Once this coupling optimized, it was applied on a complete solar device as a first step to sputter the Au contact electrode and reach the HTL/Perovskite interface with minimum degradation of the perovskite layer.**

I. INTRODUCTION

Perovskite-based solar cells showed an outstanding evolution in terms of their efficiency enhancement in the last decade. However, the main drawback of this technology resides in its intrinsic and extrinsic stability, more precisely at the interfaces. In addition to the solar cell's architecture, interfaces properties play a crucial role on the device's performance and stability [1]. This issue became nowadays the focus of many researchers; hence many characterizations has been optimized and performed on perovskite layers to access crystallographic, optical, electrical, and chemical information. Nevertheless, reaching the buried interfaces of perovskite-based devices without degrading the material remains a challenge. In this work, an innovative in-depth profiling methodology was developed by the coupling of two chemical characterization techniques, X-Ray Photoelectron Spectroscopy (XPS) and Glow Discharge Optical Emission Spectroscopy (GD-OES). Its aim is to get further insight into the critical role of perovskite / Electron Transport Layer (ETL) interface on device behavior

by probing the chemical composition from the surface to the interfaces. The GD-OES, semi-quantitative technique, assures a fast depth profiling to quickly reach specific areas of interest, which makes it possible to perform profiling on complete cells and quickly reach the perovskite layer. Different plasma gases can be used for the sputtering such as Ar, Ar/O, Cs, Ne, etc. On the other hand, the XPS allows to precisely probe the composition of the extreme surface and to have access to the atomic composition and the chemical environments. It is important to highlight that the sequences of GD-OES and XPS was already successfully applied on other photovoltaic absorbers such as CIGS or III-V materials [2][3].

II. RESULTS AND ANALYSIS

In-depth profiling was first applied on a half-cell constituted of glass/FTO/c-TiO_2/m-TiO_2/triple cation Perovskites ($Cs_{0.05}(MA_{0.14}FA_{0.86})_{0.95}Pb(I_{0.84}Br_{0.16})_3$). These in-depth profiles were realized by XPS using Ar^+ sputtering, or by GD-OES while optimizing the operating conditions and the nature of the plasma gas. To do so, qualitative GD-OES sputtering profiles were carried out on an identical perovskite stack using different pressure and power conditions, while maintaining the same gas in the plasma. It was noticed that working at lower pressure and power (250 Pa – 17 W instead of 650 Pa – 45 W) will generate coherent profiles where the TiO_2/perovskite interface is well defined. It was also detected that working with Ar/O as plasma gas will cause an artefact of insertion of oxygen in the qualitative profiles, while using Ar gas will lead to a profile in agreement with Ar^+ XPS profiling, as shown in Fig. 1.

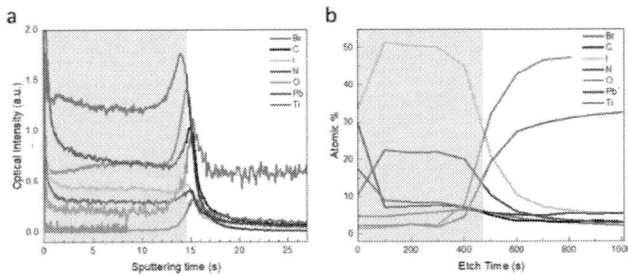

Fig. 1. In-depth profiles of the half-cell using a) GD-OES with Ar plasma and b) XPS with Ar^+ sputtering.

To verify the reliability of the GD-OES results and the effect of this sputtering on perovskite layers, XPS surface analysis was carried on inside three craters resulting from sputtering the perovskite layer for 5s, 10s and 15s using GD-OES. The results show that the sputtering leads to iodine loss and more importantly a degradation of organic compounds in the perovskite (MA and FA), whether Ar or A/O gas is used (Fig. 2.a). Variations in the Pb4f spectra were also detected: reduction of lead is confirmed by the formation of Pb^0 when Ar sputtering is used, whereas PbO_x is detected with Ar/O sputtering (Fig. 2.b).

Furthermore, modification of surface morphology was spotted using SEM, which can be correlated to thermal degradation due to the plasma etching (Fig. 2.c).

To sum up, it was confirmed that applying GD-OES sputtering directly on perovskite layers degrades the material. This does not prohibit however its application on a full stack device to help etching the top electrode before reaching the interface of interest. The in-depth profile could be then proceeded by using XPS sputtering inside the crater created by GD-OES [4].

As a proof of concept, the feasibility of this coupling methodology was confirmed by its application on a full solar cell with the following architecture: Au/PTAA/Perovskite/TiO2/FTO/Glass. The first step consisted of sputtering the 100 nm gold electrode by GD-OES which only took few seconds. This step was calibrated by controlling first the reproducibility of the GD profiling procedure to keep a small residual gold overlayer on the HTL interlayer. XPS in-depth profiling by Ar^+ sputtering was then carried on inside the GD-OES crater to reach the interface (Fig. 3.a). The HTL/perovskite interface was reached after 400s of etching and is identified by the decrease in the C content dominantly present in the PTAA layer, and the increase of the constitutive elements of the perovskite layer that are rapidly detected (I, Br, Pb...). With a closer look on the core level XPS spectra of Pb4f along the profile, a reduction of lead inside the perovskite was detected, operating in a progressive way, and attributed to an artefact of bombarding with Ar ions [5] (Fig. 3.b). Interestingly, a new contribution in the I3d core level spectra, associated to I_3^- signal, appears at 621 eV before reaching the interface (Fig. 3.c). This could result from a degradation behavior, where the iodine appears as I_2 which is non-volatile, so it accumulates on the surface as triiodine leading to the emergence of the feature centered at high binding energy in I3d core level spectrum [6]. A decrease in the atomic percentage of C1s was not only detected after reaching the HTL/perovskite interface, but also along the perovskite layer. This behavior was also noticed in the in-depth profile presented in Fig. 1.b., indicating a degradation of the organic compounds inside the perovskite due to the Ar^+ sputtering. This was also confirmed by checking the

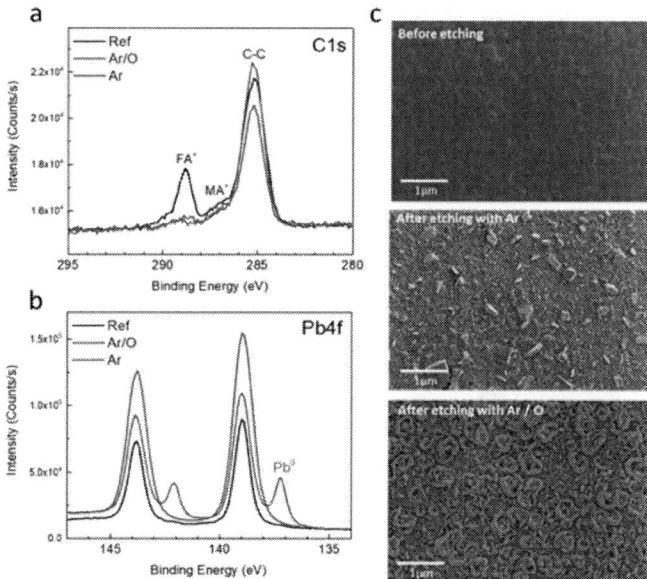

Fig. 2. a) C1s core level XPS spectra, b) Pb4f core level XPS spectra and c) Top-view SEM images of an intact perovskite, and after sputtering with Ar or Ar/O plasma gas.

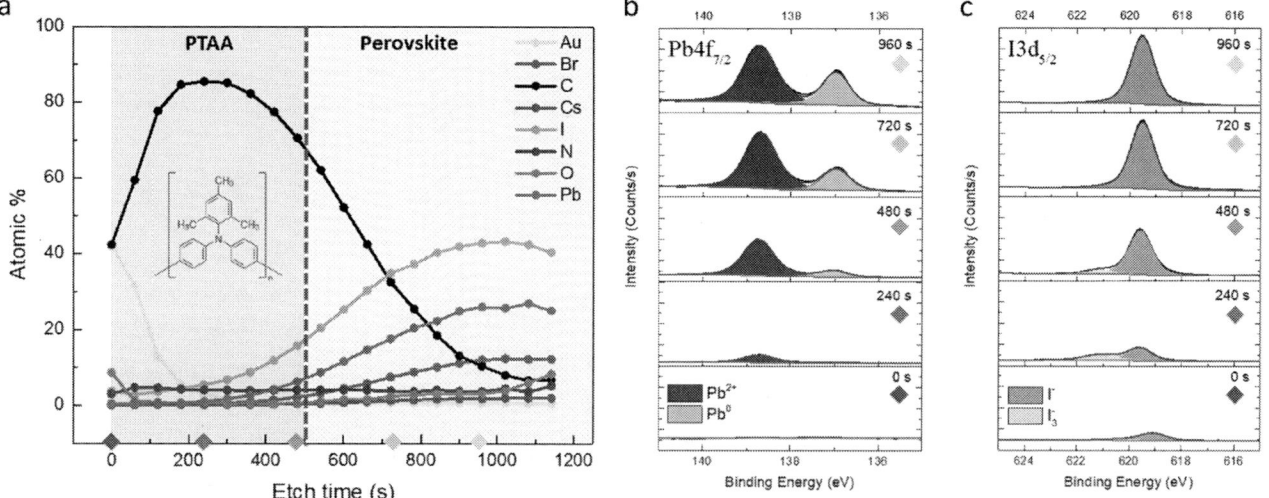

Fig. 3. a) XPS in-depth profile (Ar^+, 1000 eV) of the as-mentioned solar cell after GD-OES sputtering for 14s on the Au electrode, b) Pb4f$_{7/2}$ and c) I3d$_{5/2}$ core level XPS spectra at different sputtering times.

perovskite-related contributions in the C1s core level spectra along the perovskite layer (not presented here).

III. SUMMARY AND PERSPECTIVES

Perovskite materials emerged as a serious alternative for photovoltaic application in the last decade. However, enhancing its stability remains an important concern in the research field nowadays. Although perovskite layers might be more protected in an encapsulated solar device, the degradation due to charge extraction and internal heating is always detected, especially at the interfaces, which must be better understood to enhance the device's performance. These buried interfaces are hard to reach without further degrading the different layers in the device. Some characterizations might be applied on a cross section of the device, but cleaving the sample can deteriorate the materials and alter the interface definition and properties. Profiling techniques might be a key solution to reach buried interfaces, as long as the variations caused by sputtering the perovskite are limited and well identified.

This work presents a new methodology based on coupling GD-OES and XPS profiling to reach specific areas in-depth in a solar cell. XPS sputtering conditions were first optimized to limit the degradation of the perovskite material during analyses, by using monatomic ion gun (Ar^+) with different energies. Promising results were obtained by working with an energy of 1000 eV. Then, GD-OES pressure and power conditions were optimized, whether by using Ar or Ar/O plasma gas sputtering, leading to an equivalent profile with Ar^+ XPS one. The craters etched by GD-OES were also studied by XPS surface analysis to understand the effect of plasma sputtering on the perovskite. It was noticed that GD-OES profiling led to an iodine loss, a reduction or oxidation of lead as well as the degradation of the organic part in the perovskite layer. These modifications came along some morphological modifications detected using SEM.

The application of GD-OES and XPS coupling was then tested on a full stack device, where GD-OES was applied for etching the Au electrode and stopped before reaching the interface. Indeed, the live acquisition of the profile enables to precisely stop before the area of interest, making it possible to preserve the integrity of the chemical information registered through the (modified) residual overlayer. The in-depth profile was then proceeded by using XPS sputtering inside the crater created by GD-OES. Reduction of lead was detected all along the profile, and a new triiodide species emerged at the interface. The degradation of the organic compounds in perovskite after etching was also noticed, but this behavior was fully studied and understood in a previous work [4].

To understand the origin of the detected species and their behavior after a certain time, this coupling methodology must be first studied on a fresh reference device and then applied on aged solar devices for better comparison. It can also be carried out on tandem solar cells to reach different levels of the device. The fact that GD-OES profiling is applied in ambient conditions must also be taken into account during analysis to extract the best conclusions from this study.

REFERENCES

[1] P. Schulz, D. Cahen, A. Kahn, "Halide perovskites: is it all about the interfaces?". *Chemical reviews*, 119(5), pp.3349-3417, 2019.

[2] D. Mercier, M. Bouttemy, J. Vigneron, P. Chapon, A. Etcheberry, "GD-OES and XPS coupling: A new way for the chemical profiling of photovoltaic absorbers". *Applied Surface Science*, 347, pp.799-807, 2015.

[3] S. Béchu, C. Fypert, A. Loubat, J. Vigneron, S. Gaiaschi, P. Chapon, M. Bouttemy, A. Etcheberry, "Evaluation of the chemical and optical perturbations induced by Ar plasma on InP surface". *Journal of Vacuum Science & Technology B*, 37(6), pp.062902, 2019.

[4] S. Cacovich, P. Dally, G. Vidon, M. Legrand, S. Gbegnon, J. Rousset, J.P. Puel, J.F. Guillemoles, P. Schulz, M. Bouttemy A. and Etcheberry, "In-Depth Chemical and Optoelectronic Analysis of Triple-Cation Perovskite Thin Films by Combining XPS Profiling and PL Imaging". *ACS Applied Materials & Interfaces*, 14(30), pp.34228-34237, 2022.

[5] S. Béchu, M. Ralaiarisoa, A. Etcheberry and P. Schulz, "Photoemission spectroscopy characterization of halide perovskites". *Advanced Energy Materials*, 10(26), p.1904007, 2020.

[6] R.A. Kerner, P. Schulz, J.A. Christians, S.P. Dunfield, B. Dou, L. Zhao, G. Teeter, J.J. Berry, B.P. Rand, "Reactions at noble metal contacts with methylammonium lead triiodide perovskites: Role of underpotential deposition and electrochemistry". *APL Materials*, 7(4), pp. 04110, 2019.

Optimization of Optical and Electrical Properties of 2T Textured Perovskite/Silicon Tandem Solar Cell Structure

Chun-Hao Hsieh, Jun-Yu Huang, and Yuh-Renn Wu*

*Graduate Institute of Photonics and Optoelectronics and Department of Electrical Engineering,
National Taiwan University, Taipei 10617, Taiwan*
Author e-mail address: yrwu@ntu.edu.tw

Abstract—In this study, the optimization of referenced planar 2T perovskite/Si tandem cell with texture surface was done. This includes the optical optimization for the current matching condition using the RCWA solver and electrical optimization, including the nonradiative loss and tunneling junction issues with the 2D Poisson and drift-diffusion solver. With the texture structure optimization and considering current matching, the optimal textured structure improves the Jsc from 17.95mA/cm² to 20.42mA/cm² compared to the planar structure. It also improves the PCE from 21.54% to 27.23%. Furthermore, if the quality of the referenced perovskite thin film and tunneling junction can be improved, the efficiency can be further improved to 32.51%.

Keywords—*RCWA, perovskite/silicon tandem solar cell, textured structure, power conversion efficiency, fill factor*

I. INTRODUCTION

Crystalline silicon is the most widely used material for solar cell technology due to the advantage of stable material quality and low manufacturing cost. Although single-junction silicon solar cells are quite mature in the market, it has almost reached their limit. Today, the highest PCE of single-junction silicon solar cells is 26.7% [1], which is already close to the S-Q limit [2]. There is little room for future efficiency improvement, and cost increases are much higher than the improvement in efficiency. Hence, most commercially available Si single junction-based solar cells are around 20-22%. In these 20 years, some emerging photovoltaics began to appear; among these emerging PVs, perovskite/silicon tandem solar cells have the highest efficiency, and their efficiency has also grown from 20% to 29.15% [3] in the past decade. Thus Perovskite(PVK)/c-Si tandem solar cell (TSC) has great potential in the future.

For single-junction perovskite solar cells, the influence of adding a triangular texture surface has been mentioned in the previous study [4]. Therefore, this study will discuss the optimization of textured 2T PVK/c-Si TSC including the optical absorption, current matching, and electrical output balance of two junctions. The triangular structure applied on the textured surface will be mainly discussed. Furthermore, we will try to find the optimal structure with the highest power conversion efficiency(PCE) and better fill factor(FF) [5].

II. METHODOLOGY

A. Simulation of the optical field

For optical modeling, the 2D rigorous coupled-wave analysis (RCWA) method is applied to simulate the absorption of sunlight in the device. This method solves Maxwell's equations in the frequency domain and is widely used for optical stimulations of periodic structures [6]. This study applies the AM1.5G solar spectrum as the light source. After the electric field distribution is obtained from RCWA, we can then use Eq. (1) to calculate the generation rate of the electron-hole pair inside the device. The structure of the 2T PVK/Si solar cell is shown in Fig. 1(a), and the optical field distribution calculated from the electric field distribution is shown in Fig. 1(b).

$$G_{opt} = \frac{1}{\hbar\omega} Re(\nabla \cdot P) = \frac{nk}{\hbar}\epsilon_0 |\vec{E}|^2 \quad (1)$$

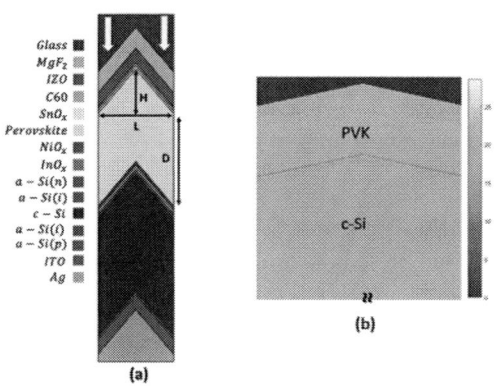

Fig .1 (a) PVK/c-Si TSCs structure of simulation, and H is the amplitude of triangular pattern, L is the period length (b) Optical field distribution (log scale)

B. Simulation of electrical properties

After obtaining the generation rate (G_{opt}), we applied 2D finite element Poisson and drift-diffusion solver (named 2D-DDCC software in our lab) to analyze the physical mechanism of PVK/Si TSCs. For the carrier exchange between the top and bottom cell, as shown in Fig. 2(b) tunneling junction is designed between perovskite and crystalline silicon two absorption layers.

Work was supported by MOST No. 109-2221- E -002-196-MY2.

978-1-6654-6060-6/23 $31.00 © 2023 IEEE

III. SIMULATION STRUCTURE

The materials selected for our simulated structure are based on the experiment results of PVK/c-Si TSCs published by Y. Hou [7] in 2020. These materials' tandem structures are shown in Fig. 1(a). The materials' band diagram for electrical simulation and carrier transport direction are shown in Fig. 2(a).

IV. RESULT AND DISCUSSION

A. The optical results

For the optical results, the TSCs can reduce the reflected photocurrent by adding a triangular pattern on the surface of the absorption layers. In this study, the period lengths L from 100 nm to 500 nm and amplitudes H from 100 nm to 300 nm are examed. In these different configurations, we know that when the period is fixed, the higher amplitude, the tip of the triangular pattern will be sharper. Note that due to the current matching requirement, when the H/L are changed, we need to run different thickness of PVK film to find the current matching condition. Hence, the thickness of PVK films cannot be the same, and the thickness of PVK is listed in Table I. From the reflected photocurrent results, as shown in Fig. 3, we can see the same trend in every period configuration. The sharper triangular pattern can reduce the reflected photocurrent.

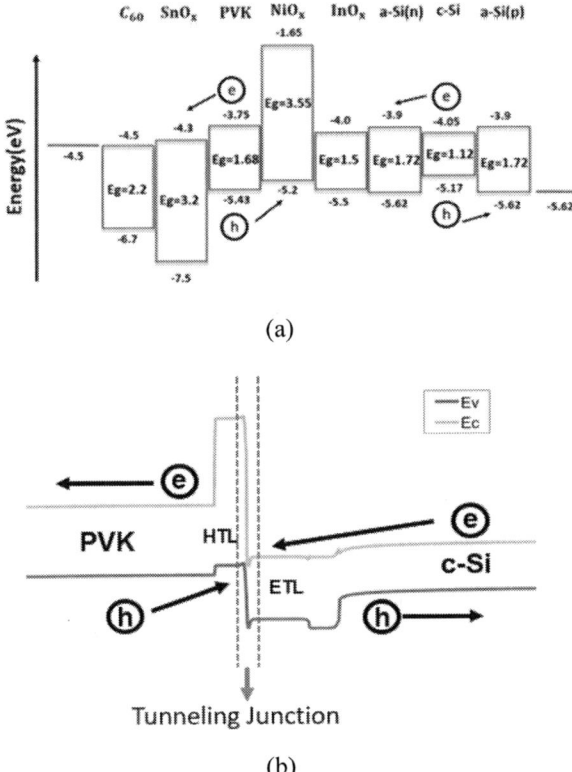

(a)

(b)

Fig. 2 (a) Simulation materials band alignment including bandgap and electron affinity. (a) The band diagram of the tunneling junction

When the amplitude is fixed at 300 nm, Fig. 4 compares the optical field distribution with periods from 100 nm to 500 nm. In the case of L= 100nm, the optical field of the overall PVK layer is very strong. As the period increases, some areas of the PVK layer begin to have a weaker optical field, affecting absorption. The result from optical absorption shows that the L from 100nm to 200nm has a smaller reflection loss. However, this may not be the case for the electrical results since we also need to consider the nonradiative recombination issues and tunneling layer issues.

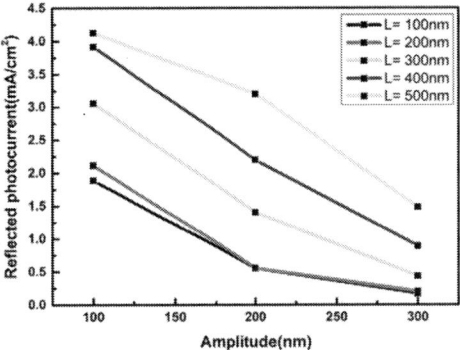

Fig. 3 Reflected equivalent photocurrent versus different amplitude H for different periods L.

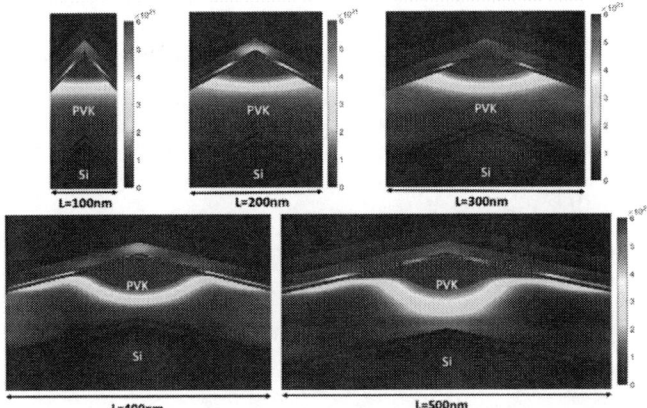

Fig. 4 Optical field distribution of amplitude H fixed at 300nm and the period L from 100 nm to 500 nm

B. The electrical results

The periods L from 100nm to 400nm were also evaluated for the electrical results. As discussed in Fig. 3, the L=100/200nm has a smaller reflection. However, the calculation of I-V does not have the same trend due to the nonradiative recombination loss affected by material geometry. Here we pick the configuration of the period, L, at 200 nm and 400 nm with different H for discussion. The characteristic parameters of these configurations are shown in Table I. And the J-V curves of each case are shown in Fig. 5(a) and Fig. 5(b). As the previous optical results mentioned, cases with less reflected photocurrent have higher short-circuit current density (Jsc). The highest Jsc 20.52 mA/cm^2 is at the case of L= 200 nm and H= 300 nm.

TABLE I CHARACTERISTIC PARAMETERS OF THE SIMULATION
FOR DIFFERENT CONFIGURATIONS

Structure	J_{SC} (mA/cm^2)	V_{OC} (V)	FF (%)	PCE (%)	PVK thickness(nm)
Planar	17.95	1.66	72.31	21.54	500
L= 200nm, H= 100nm	19.57	1.74	73.38	24.97	590
L= 200nm, H= 200nm	20.42	1.70	76.84	26.73	645
L= 200nm, H= 300nm	20.52	1.60	76.99	25.33	680
L= 400nm, H= 100nm	18.73	1.71	74.20	23.82	525
L= 400nm, H= 200nm	19.71	1.76	76.49	26.48	550
L= 400nm, H= 300nm	20.42	1.73	77.22	27.23	510

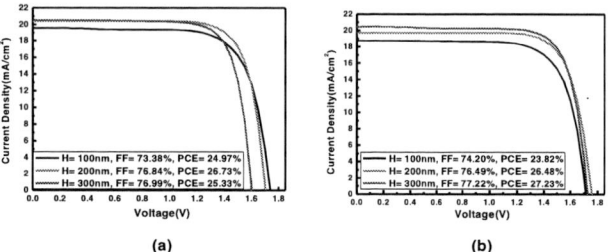

Fig. 5 J-V curves at (a) L= 200nm and (b) L= 400nm

However, from the electrical properties, we can also observe that the Voc decreases as the amplitude increases under the same period length. The decrease in V_{oc} has canceled the improvement of Jsc. For example, for L= 200 nm, the PCE maximum appears at H= 200 nm cases. This might be due to the strong nonradiative recombination in the PVK layer due to the geometrical issue (thicker PVK layer due to current matching issues), as shown in Fig. 6. For L= 400 nm cases, H= 300 nm case has a higher Jsc but also have a thinner PVK layer. The PCE of H= 300 nm is still higher even though it has a little bit smaller Voc.

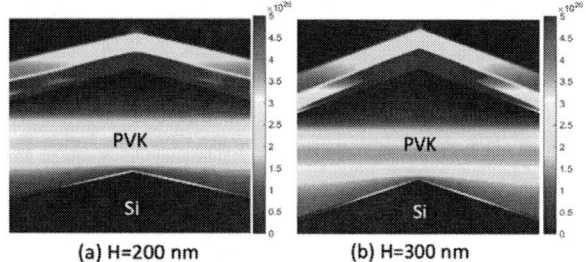

Fig. 6 L= 200nm, (a) H= 100nm and (b) H= 300nm cases nonradiative recombination distribution at MPP voltage

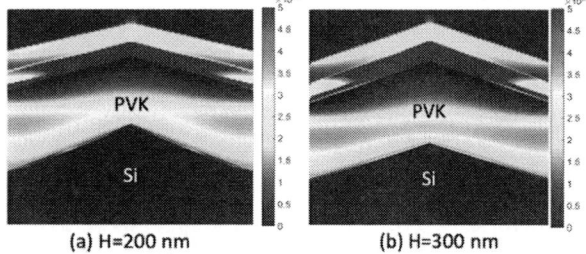

Fig. 7 L= 400nm, (a) H= 200nm, and (b) H= 300nm cases of nonradiative recombination distribution at MPP voltage

In addition, we found that the optimal texture structure which has the highest PCE of 27.23%, is L= 400 nm, H= 300 nm. Compared to the H=0 nm case (planar structure), the PCE has increased by 5.69% (21.54% to 27.23%). Furthermore, if the PVK thin film and tunneling junction(TJ) quality of the optimal structure mentioned earlier can be improved. As shown in Fig. 8, improving the two conditions at the same time can effectively improve the Voc from 1.73 V to 1.99 V, and increase the PCE from 27.23% to 32.51%. From the calculation result of the best PCE, it can be observed that in the future, PVK/silicon TSCs will have the opportunity to achieve an efficiency of 32.51% experimentally.

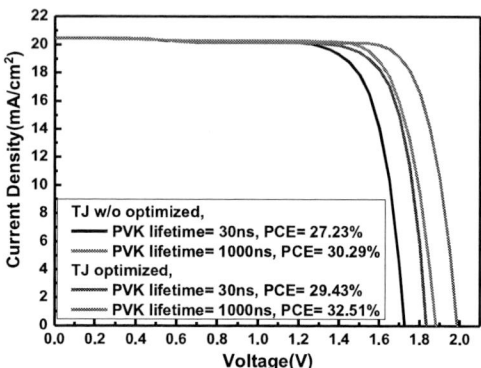

Fig. 8 J-V curves after improving PVK lifetime and tunneling junction quality.

ACKNOWLEDGMENT

This work was supported by the National Science and Technology Council under grant number NSTC 109-2221-E-002-196-MY2, 111-2221-E-002-075, and 112-2923-E-002-002

REFERENCES

[1] Yoshikawa, Kunta, et al. "Silicon heterojunction solar cell with interdigitated back contacts for a photoconversion efficiency over 26%." Nature Energy, 2(5), pp. 1-8, 2017.

[2] Sven Rühle, "Tabulated values of the Shockley–Queisser limit for single junction solar cells," Solar Energy, 130, pp.139-147, 2016.

[3] Al-Ashouri, Amran, et al. "Monolithic perovskite/silicon tandem solar cell with> 29% efficiency byenhancedhole extraction." Science 370.6522, pp. 1300-1309, 2020

[4] J. -Y. Huang, E. -W. Chang and Y. -R. Wu, "Optimization of MAPbI₃-Based Perovskite Solar Cell With Textured Surface," IEEE Journal of Photovoltaics, 9(6), pp. 1686-1692, 2019.

[5] F. Sahli, "Development of highly efficient perovskite-on-silicon tandem solar cells," p.240, 2020.

[6] M. Moharam and T. Gaylord, "Rigorous coupled-wave analysis of planar-grating diffraction," JOSA, 71(7), pp. 811–818, 1981.

[7] Y. Hou, E. Aydin, M. D. Bastiani, C. Xiao, F. H. Isikgor, D.-J. Xue, B. Chen, H. Chen, B. Bahrami, A. H. Chowdhury, A. Johnston, S.-W. Baek, Z. Huang, M. Wei, Y. Dong, J. Troughton, R. Jalmood, A. J. Mirabelli, T. G. Allen, E. V. Kerschaver, and E. H. Sargent, "Efficient tandem solar cells with solution-processed perovskite on textured crystalline silicon," Science, 367(6482), pp. 1135–1140, 2020

Development of an Ultra-Light Curvilinear Prismatic Window Which Mitigates Reflections and Glare for PV Modules and Other Surfaces

[1]Mark O'Neill, [2]Chris Youtsey, and [2]Robert McCarthy

[1]Mark O'Neill, LLC, Fort Worth, TX, 76244, USA

[2]MicroLink Devices, Inc., Niles, IL 60714, USA

Abstract—As presented at the last PVSC, a novel transparent window for PV modules and other surfaces has been developed to both eliminate glare and minimize front-surface reflections. The new window technology can be implemented with a thin, ultra-light, manufacturable curvilinear prismatic film which can be applied to both space and ground PV modules to increase power output, using appropriate materials and coatings. The new window can also be applied to non-PV-related surfaces to eliminate glare. This paper describes the key recent development of the new window under NASA funding.

Keywords—*PV module window, anti-reflection, anti-glare, performance enhancement*

I. INTRODUCTION

Conventional PV module windows typically comprise planar glass or polymer layers providing protection for the underlying PV cells from the environment. In space, the environment can include charged particles, solar ultraviolet radiation, atomic oxygen, etc. On the ground, the environment can include rain, snow, sleet, and hail, etc. Front-surface reflection losses from such conventional windows reduce the current and power output of the PV cells, especially for high solar ray incidence angles. The flat surfaces of conventional windows can also lead to glare due to the specular component of the reflected sunlight. Glare problems have led to cancellations of some terrestrial PV systems near airports, highways, and occupied buildings. Glare problems from constellations of spacecraft in low earth orbit (LEO) have also caused problems for ground-based telescopes. A new curvilinear prismatic window has recently been developed which overcomes the glare problem and minimizes the front-surface reflection power losses for PV modules both in space and on the ground [1 and 2].

The optical and thermal benefits of texturing window layers for PV modules have been recognized by many previous researchers [3 and 4]. Various textures have been analyzed and tested for transmittance improvements due to reduced front-surface reflections. Various textures have also been analyzed and tested for enhanced front-surface waste heat rejection due to the greater front-surface area of textured windows compared to flat windows. But these previous textured surface geometries have not incorporated curved surfaces which eliminate glare.

II. DESCRIPTION OF THE CURVILINEAR PRISMATIC WINDOW

The curvilinear prismatic window is a variation of the linear prismatic window achieved by changing the prismatic path from a straight line to a curved line. Fig. 1 shows the basic configuration with greatly exaggerated prism size. The curvilinear prisms follow a curvilinear path with orientation such that a liquid can run from top to bottom without encountering a barrier. For terrestrial applications, this liquid could be rain or cleaning water. For space applications, this liquid could be a cleaning fluid such as isopropyl alcohol used before launch. Other textured surface geometries such as inverted pyramids can trap liquids and dirt in regions where barriers to flow exist.

Fig. 1 also shows the triangular cross-sectional geometry of each prism. The sun's direction relative to the prismatic pattern is defined by the azimuth (az) and elevation (el) angles shown in Fig. 1. Compared to a linear prismatic pattern, the new curvilinear prismatic pattern offers major performance improvements for very small values of both az and el angles.

The curvilinear prismatic window can be mass-produced by embossing a polymer film or by casting a different polymer against an embossed polymer film. For terrestrial applications, the preferred material for the embossed polymer film is acrylic plastic. For space applications, the preferred material is space-grade silicone, which is easily cast and cured against embossed acrylic film. Thus, the same embossed film can be used directly for terrestrial applications or as a disposable molding tool for

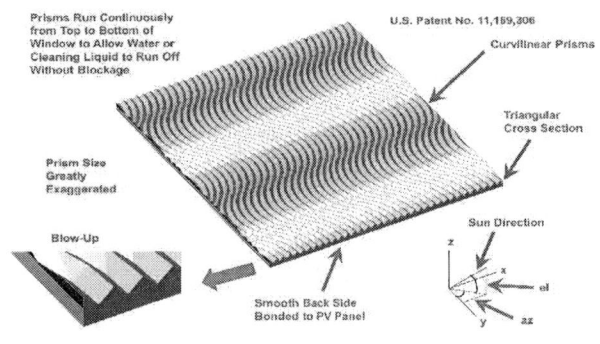

Figure 1

silicone prismatic windows for space applications. The preferred attachment method for a space PV module is still under evaluation. The preferred attachment method for a terrestrial PV module is to laminate an acrylic pressure sensitive adhesive (PSA) to the embossed acrylic curvilinear prismatic film to enable easy bonding to the PV module on top of the existing flat glass window. For terrestrial applications, the curvilinear prismatic film could be attached in the factory or in the field for already deployed arrays.

III. How the New Curvilinear Prismatic Film Works

The curvilinear prismatic film works in two ways:

- The triangular prisms minimize front-surface reflections for both large and small sun elevation angles for all sun azimuth angles.

- The curvilinear prismatic path eliminates glare by spreading the reflected light in all directions due to the curvature.

Fig. 2 shows how the prismatic film works for high lateral incidence angles. The prisms shown have 45° faces which work very well to minimize reflection losses for solar rays arriving at large lateral incidence angles. This minimization occurs because the tilted faces of each prism intercept the incoming rays at a more normal incidence angle than a flat window. The rays which enter each prism make their way to the PV cell below the window either directly or after total internal reflection (TIR) from the opposing prism face.

High Lateral Incidence Angle Light Is Efficiently Captured by 45° Prisms

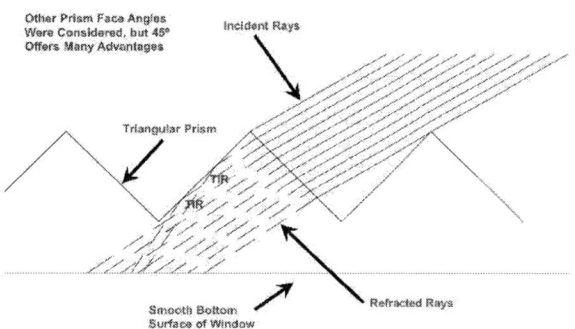

Figure 2

Fig. 3 shows how the prismatic film works for small lateral incidence angles. The prisms shown have 45° faces which work very well to minimize reflection losses for solar rays arriving at near normal incidence angles. This minimization occurs because the rays which are reflected by the outer surface of the tilted faces of each prism intercept the neighboring prisms which recover most of the reflected light and deliver it to the PV cell.

The curvilinear path of the prisms eliminates glare by spreading the reflected light into a wide range of departing angles, as shown in Fig. 4. Since there are no flat surfaces on the exposed face of the curvilinear prismatic window, there can be no glare from rays reflected by the exposed face of the window. Furthermore, rays which are reflected from the PV cells below the window are refracted by the curved surfaces of

Small Lateral Incidence Angle Light Is Efficiently Captured by 45° Prisms Including First Surface Reflections

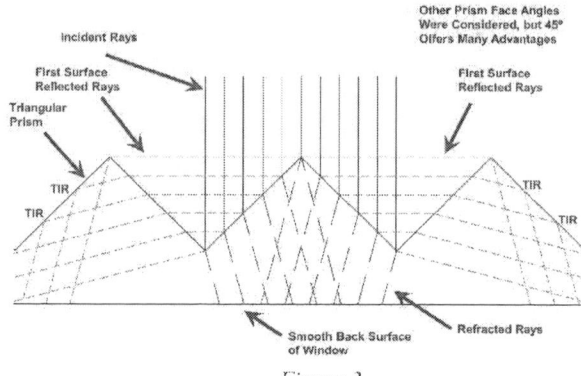

Figure 3

Glare Elimination Is Accomplished by Curved Surfaces Which Spread Light

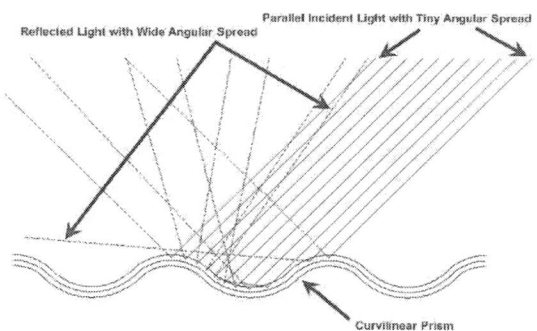

Figure 4

the new prismatic window and spread into a wide range of departing angles, eliminating glare from these reflections too.

A typical prototype multi-junction cell (about 3.1 cm wide x 6.5 cm long) with a curvilinear prismatic window is shown in Fig. 5. The appearance from a distance is similar to black velvet. The prismatic window for this prototype is relatively thick, but our team has developed processes for making very thin windows of any desired thickness.

Prototype Cell with New Curvilinear Prismatic Window

Figure 5

978-1-6654-6060-6/23 $31.00 © 2023 IEEE

IV. PERFORMANCE

The new curvilinear prismatic window provides excellent performance over the full range of possible sun azimuth and elevation angles of incidence. Fig. 6 shows results of parametric optical analysis of the preferred geometry of the new window made of silicone (1.4 refractive index). Note that the net transmittance into the window is much higher than for a flat ceria-doped glass (CMG) window over the full range of sun azimuth and elevation angles. Prototype testing has confirmed the accuracy of the optical analysis, as shown in Fig. 7.

Parametric Results for Curvilinear Prismatic Window Using Cosine Path with ±45° Max Slope

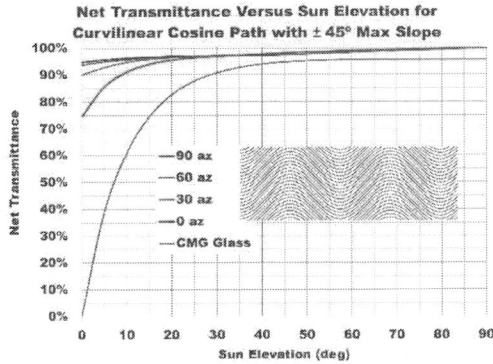

Figure 6

The standard window on the comparison cell without the prismatic window in Fig. 7 used FEP as the window material instead of CMG glass. FEP has a much lower refractive index than glass, resulting in a smaller reflection loss and therefore a smaller gain from the prismatic window. This difference was treated in the predicted results in Fig. 7. Since the two cells used to compare the prismatic window to the flat window were not current matched when bare, the results in Fig. 7 are normalized to the perpendicular incidence angle to provide a relative comparison instead of an absolute comparison, which would be larger.as shown in Fig. 6.

Fig. 8 shows the typical gain in current and power from application of the curvilinear prismatic window to a multi-

Measured and Predicted Performance Gain of Curvilinear Prismatic Window on MJ Cell

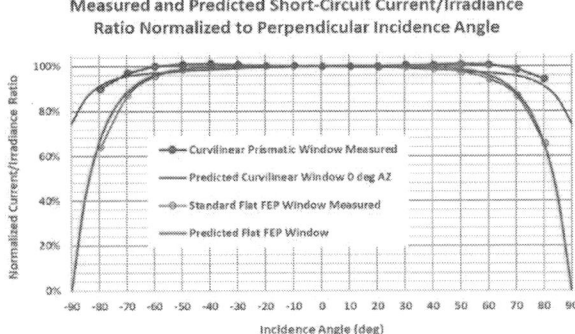

Figure 7

Curvilinear Prismatic Window Gain Compared to Bare Cell

Wafer	Cell	Voc (V)	Jsc (mA/cm2)	Fill Factor	Eff	Isc (mA)	Vmax (V)	Imax (mA)	Pmax (mW)	Comment
2-26810-3	I05	3.043	15.94	83.85	29.74	318.72	2.670	305.05	813.21	cell for curvilinear silicone prism test after lamination with S1001.5 + Sylgaard 184 curvilinear prisms from Mark O'Neill
2-26810-3	I05	3.038	15.26	84.46	28.65	305.25	2.690	290.86	783.29	cell for curvilinear silicone prism test before lamination

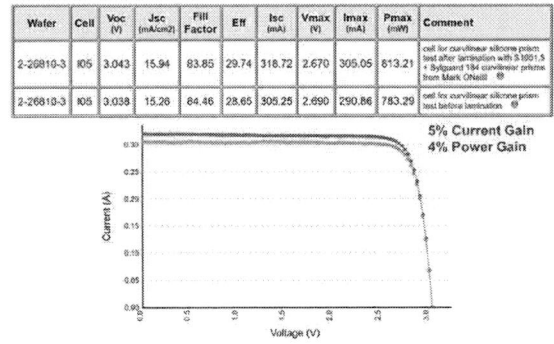

Figure 8

junction cell under normally incident space solar irradiance (AM0) in a simulator. The gains are close to expectations.

We have also successfully demonstrated glare elimination on multiple prototypes, with typical results shown in Fig. 9. This demonstration model used a 10 cm x 10 cm piece of glass with its back surface painted black to simulate a solar cell. A lamp illuminated the sample at an angle of incidence of about 60 degrees and the first surface reflection of the lamp light showed typical glare. But a smaller curvilinear prismatic window about 5 cm x 5 cm in size was bonded to the top surface of the glass and curtailed the reflected light from the lamp thereby eliminating the glare from the top surface as shown. Glare elimination is an important consideration for both space solar arrays and ground solar arrays.

Anti-Glare Demonstration

- 10 cm Square Glass Plate Painted Flat Black on Bottom Side to Simulate Solar Panel
- LED Lamp Light Shows Reflected Glare Off Top Surface of Glass
- 5 cm Square Silicone Curvilinear Prismatic Window Bonded to Top Surface of Glass Eliminates Glare for All Incidence Angles

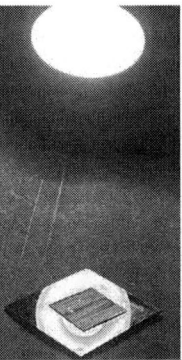
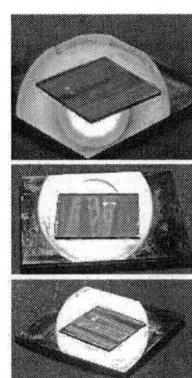

Figure 9

To quantify the glare properties of a MicroLink multi-junction (MJ) cell equipped with the new curvilinear prismatic window, the SPF Institute for Solar Technology in Switzerland tested a sample for Bidirectional Reflection Distribution Function (BRDF) under a variety of incidence angles and two azimuth angles. The test conditions simulated a clear day on the Earth's surface with bright sunshine at 100,000 lux. The reflected light distribution in the hemisphere above the test sample was measured for each set of incidence and azimuth angles to provide the reflected luminance distribution from all possible viewing directions. Fig. 10 shows the sample and describes the test.

Bidirectional Reflection Distribution Function (BRDF) Measurements for MJ Cell with Curvilinear Prismatic Window

- To Quantify the Glare Properties of a MicroLink Multijunction (MJ) Cell with the New Curvilinear Prismatic Window, BRDF Measurements Were Made by SPF Institute for Solar Technology in Switzerland

  ◆ The Test Sample Used a MicroLink Cell (about 3 cm x 6 cm) Shown Before (Top Photo) and After Window Application (Bottom Photo)

  ◆ Test Conditions Simulated a Clear Day on Earth with Bright Sunshine at 100,000 lux

  ◆ Incidence Angles Were Varied from 10 to 70 degrees and Tests Were Made at 0 and 90 degrees Azimuth Angles (Along the Cell and Across the Cell

- The Main Source of Reflections Is the Cell Itself, Including Its AR-Coated Surface and Gridlines

Figure 11

The results of the BRDF testing are summarized in Fig. 11. The two curves correspond to the two azimuth angles (0° and 90°) while the data points correspond to the various incidence angles (0° to 70°). Two example plots of the distribution of the reflected luminance for the brightest and dimmest data points show the small regions of relatively bright reflections and the much larger region of relatively dark reflections over the rest of the hemisphere above the sample. The brightest peak value corresponds to a very small spot shown by the red dot on the upper hemisphere in Fig. 11. The dimmest peak value likewise corresponds to a very small spot shown by the red dot on the lower hemisphere. To put the brightest peak value in perspective, Fig. 11 provides a comparison with a glass window over a multi-junction cell. The new prismatic window reduces the peak reflected brightness by more than 99.9% compared to the glass window.

Bidirectional Reflection Distribution Function (BRDF) Measurement Results

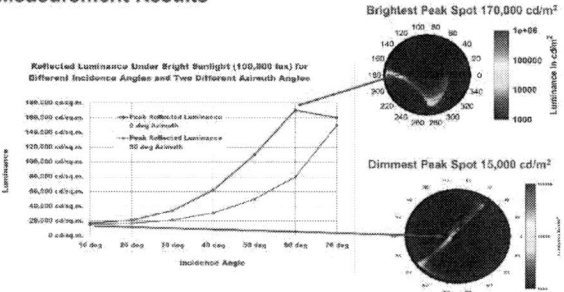

- The Solar Disk Brightness ≈ 100,000 lux / 6.8x10⁻⁵ steradians ≈ 1,500,000,000 cd/m²
- A Conventional PV Module with 60° Solar Incidence Would Reflect ≈ 8% at Its Outer Glass Surface. The Cell Surface Could Reflect ≈ 5% More Producing a Total Reflected Brightness ≈ 13% x 1,500,000,000 ≈ 195,000,000 cd/m²
- The Curvilinear Prismatic Window on a MJ Cell Therefore Reduces the Reflected Brightness by More than 99.9% Compared to a Glass Window on a MJ Cell at 60° Incidence

Figure 10

V. TOOLING AND PROCESS DEVELOPMENT

We are making progress on mass production of the new curvilinear prismatic window. We originally thought that diamond turning would be the best approach to make master tooling for the new curvilinear prismatic window, but we found another approach with many advantages, grayscale lithography (GSL), as implemented by Wavefront Technology (WFT). We then worked with WFT on the final design of the tooling to

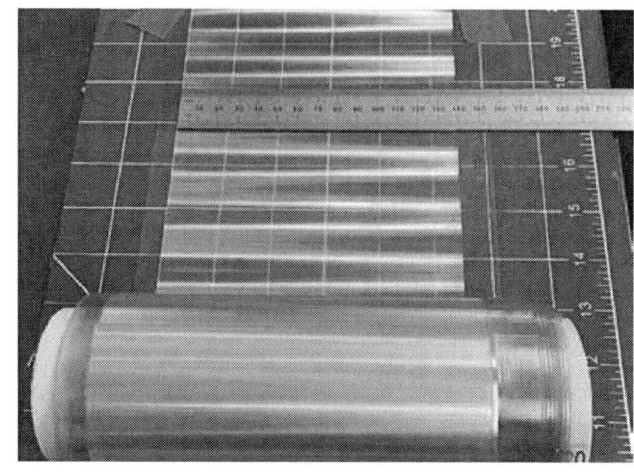

Figure 12

enable not only early prototypes like the one used in the testing shown in Fig. 7 above, but also later roll-to-roll production of acrylic prismatic film. The results were excellent. The first roll of acrylic lensfilm material is shown in Fig. 12. The blue tint is from a disposable film used to protect the prisms on the rolled-up acrylic "lensfilm" produced by the roll-to-roll process. The actual lensfilm is perfectly clear.

VI. ADVANTAGES FOR SPACE APPLICATIONS

The new curvilinear prismatic window offers significant advantages for space applications. One major advantage is the selectable thickness of the window for specific missions. Our team sees LEO missions as the first target for multi-cell modules using the new window. Different LEO missions (with differing altitude, orbital inclination, and lifetime) require different amounts of shielding to provide adequate protection against charged particle radiation, as shown in Fig. 13. Conventional windows use ceria-doped microsheet glass windows of specific thicknesses such as 75, 100, and 150 micrometers. The new silicone window can be cast to whatever thickness is desirable. Due to its lower specific gravity, the new window requires a larger thickness (about 2.47X) for the same equivalent radiation shielding. For some missions, the required thickness can be lower than the thinnest microsheet glass available, and the new window offers significant mass reductions.

Prismatic Window Thickness Can Be Optimized for Specific Missions

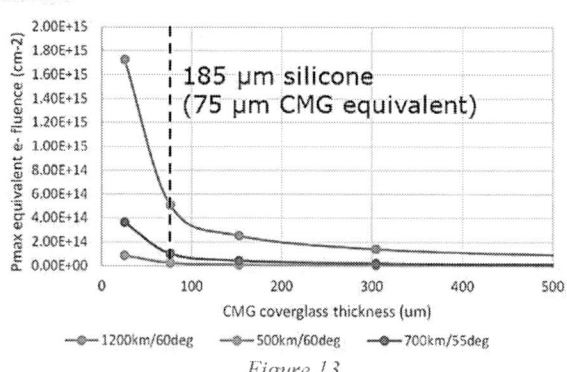

Figure 13

The new window also offers high robustness in the space environment as shown by the results of space environmental effects (SEE) testing. Atomic oxygen testing, 1 MeV election testing, 30 keV proton testing, and ultraviolet radiation testing results have all been very positive in demonstrating durability in the LEO environment. The new window uses a previously space-proven ultraviolet rejection (UVR) coating to block high-energy, low-wavelength vacuum ultraviolet (VUV) space solar radiation which is known to darken the space-qualified silicone material. The UVR coating is also useful in mitigating the effects of low-energy charged particle radiation.

Another major advantage of the new window technology is its ability to encapsulate the entire multi-cell module, including interconnects and gaps between cells, thereby maximizing protection of the complete cell circuit from arcing or other interactions with the space environment. Conventional modules use individual cell-level cover glass windows which cannot provide this module-level protection. Fig. 14 shows a prototype module with a single prismatic window over the entire module. The module is also flexible and offers substantial mass savings and specific power gains over conventional multi-junction cell modules.

37 x 12 cm Module with Ultra-Low Mass, Ultra-High Performance, Full Module Encapsulation, and Flexibility

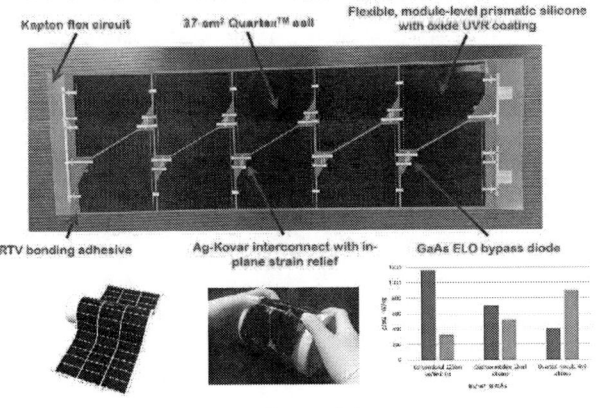

Figure 14

The new module also offers a small reduction in cell operating temperature on orbit due to its exposed structured surface, which slightly increases the effective emittance of this surface [1].

VII. ADVANTAGES FOR TERRESTRIAL APPLICATIONS

While our work to date has focused on space applications, we have identified a number of spin-off applications in the booming terrestrial solar energy area. A number of projects have drawn opposition due to glare, especially large solar farms near highways, airports, of occupied buildings. The new window in acrylic lensfilm form could be applied to terrestrial photovoltaic modules at modest additional cost in the factory to minimize glare for such applications. It would also provide performance gains due to reduced front-surface reflections and slightly reduced cell operating temperature. A conservative analysis predicts 5-10% additional power output for a solar farm equipped with the new curvilinear prismatic windows.

Another interesting potential terrestrial application is in the emerging technology area of solar-equipped electric vehicles. Several firms (e.g., Lightyear in Europe and Aptera in the U.S.) are introducing such vehicles. Since the solar cells are embedded on the roof and upward-facing surfaces of these vehicles, the angle of incidence of sunlight is generally very large, and the power output gains of the new window would be substantial. The minimization of glare is also an important consideration as such vehicles proliferate.

VIII. SUMMARY

The new curvilinear prismatic window offers power output advantages over conventional PV module windows for both space and ground applications for all sun incidence angles. It also eliminates glare, which is a significant and growing problem both on orbit [5] and on the ground [6].

The new window also provides small advantages in heat rejection, lowering cell operating temperature a few degrees compared to flat windows both in space and on the ground. Mass-production processes are being developed for both space and ground applications.

Space environmental effects testing is showing the new window to be durable in the space environment when made of space-qualified silicone with an ultraviolet rejection coating.

The new window can be optimized in thickness to meet the radiation shielding requirements of specific missions, thereby offering substantial mass savings and gains in specific power. Module-level specific power may exceed 1,000 W/kg.

The new window is flexible and provides module-level encapsulation and protection of the complete cell circuit.

Terrestrial spin-off applications include solar farms near highways, airports, and occupied buildings. Solar electric vehicles are another potential ground application.

IX. ACKNOWLEDGEMENT

Development of the new curvilinear prismatic window is being supported by NASA under a Phase II SBIR contract with MicroLink Devices.

X. REFERENCES

[1] M. O'Neill, "Curvilinear prismatic film which eliminates glare and reduces front-surface reflections for solar panels and other surfaces," U.S. Patent No. 11,169,306, November 9, 2021.

[2] M. O'Neill and C. Youtsey, "Curvilinear prismatic window which eliminates glare and reduces front-surface reflections for PV modules and other surfaces," 49th IEEE Photovoltaic Specialists Conference, Philadelphia, June 9, 2022.

[3] M. Duell, M. Ebert, M. Muller, B. Li, M. Koch, T. Christian, R.F. Perdichizzi, B. Marion, S. Kurtz, D.M.J. Doble, "Impact of structured glass on light transmission, temperature and power of PV modules," 25th European Photovoltaic Solar Energy Conference and Exhibition / 5th World Conference on Photovoltaic Energy Conversion, 6-10 September 2010, Valencia, Spain.

[4] Z. Zhou, Y. Jiang, N. Ekins-Daukes, M. Keevers and M. A. Green, "Optical and thermal emission benefits of differently textured glass for

photovoltaic modules," in IEEE Journal of Photovoltaics, vol. 11, no. 1, pp. 131-137, Jan. 2021.

[5] K. Whitt, "Satellites versus stars: which will dominate the sky?," https://earthsky.org/space/satellites-versus-stars-night-sky-kessler-syndrome/, September 28, 2021.

[6] K. Ulenhuth, "Glare study prompts utility Evergy to cancel solar array at Kansas City airport," https://energynews.us/2020/12/10/glare-study-prompts-utility-evergy-to-cancel-solar-array-at-kansas-city-airport/, December 10, 2020.

Evaluating Multi-Bias Modulation for Diagnostics of PV Modules in Daylight Electroluminescence Inspections

Rodrigo del Prado Santamaría, Gisele A. dos Reis Benatto, Thøger Kari, Peter B. Poulsen, Sergiu V. Spataru

Department of Electrical and Photonics Engineering, DTU Electro, Technical University of Denmark, Roskilde, Denmark

In this work we present a new method for modulating PV modules' current when performing Electroluminescence imaging in PV field inspections. This method allows recording images at different current bias levels in the same sequence. We have investigated the advantages for PV diagnostics of EL imaging PV modules affected by PID, at different current bias levels. Tests showed that PID affected modules have statistical differences between their cells compared to well-functioning modules when imaging them at different current bias levels. PID detection on the field could be enhanced by having multiple EL images at different biases and the proposed method allows acquiring these images in the same sequence, minimizing the variations in the images due to sunlight fluctuations.

Recomibiation Center Defects Induced by TCO Reactive Plasma Deposition in Carrire Selective Contact Solar Cells

Yoshio Ohshita, Tomohiko Hara, Taichi Tanaka, Keita Kimura, Yuto Ifuji

Toyota Technological Institute, Nagoya, Japan

Solar cells with a carrier selective contact (CSC) structure is one of the candidates to realize the theoretical limit conversion efficiency of crystalline Si cell, but some processes of cell fabrication deteriorate the cell performance and limit the conversion efficiency. In this paper, the authors discuss the properties of defects generated by the RPD process for TCO deposition. The defects are induced both at SiO2/Si interface and in the silicon bulk near the interface. The electrical characteristics, thermal stability, and formation mechanism of process-induced crystal defects are studied by DLTS using machine learning for spectrum analysis and high-low capacitance - voltage (C-V) method. The defects at the interface and three types of electron traps in the bulk are induced by the process. The interface states and the defects with deep energy around 0.5eV in the Si crystal near the interface act as recombination centers. Also, the fixed charges are induced in SiO2 and/or at the interface. They change the surface potential at the inteface layer of Si, which affect the recomination velocity at the interfafce between SiO2 and Si. These fixed charge disappeared by 10min annealing at 200 °C. There are two ahihilation processes of recombination centers, which have different activaion energies. At the initial stage of annealing, the density of defect swiftly decreases and then the concentration becomes lower with the reltively high activation energy. With regards to the degradation mechansim, it is suggested that the interaction between oxygen in the Si crystal and UV light during RPD process plays an important role in the defect formation.

978-1-6654-6060-6/23 $31.00 © 2023 IEEE

Predicting Site-Specific Adjustments to P50 Energy Production Estimates from Sub-Hourly Irradiance Data

Faisal Rashed

Leidos, Denver, CO, United States

Abstract—**Energy production models developed as part of the technical due diligence required for financing of photovoltaic solar projects are customarily based on hourly weather data, which—when compared to higher-frequency ("sub-hourly") data—have been shown by recent studies to overestimate energy production, particularly for systems with higher DC/AC ratios. Since long-term weather data at suitable sub-hourly resolutions are not widely available, techniques must be developed to adjust estimates based on long-term hourly data with insights derived from short-term sub-hourly data. This paper presents a multifaceted machine-learning approach to tailor such adjustments to a project site based on its corresponding hourly attributes. The method enables seamless integration of calculated factors into the energy production model and is suitable for use with both satellite-based and ground-measured sources of sub-hourly data.**

Index Terms—**photovoltaic, solar, modeling, performance, clipping loss, irradiance variability, sub-hourly, satellite, ground-measured, DC/AC ratio, classification, regression**

I. INTRODUCTION

The primary input to an energy production model for a photovoltaic ("PV") solar project is a typical meteorological year ("TMY") dataset, which describes the representative incident solar irradiance (among other weather characteristics) at its geographic location. The output of a TMY-based model is regarded as the median ("P50") production estimate (i.e., that which is expected to be exceeded in 50 percent of plausible scenarios). Because the underlying time series data from which a statistically acceptable TMY is constructed should have a period of record of at least 10 years [1], TMYs are usually based on satellite imagery from 1998 through present, which is available in the United States at a 30-minute temporal resolution. These datapoints are often aggregated to a 1-hour resolution for use in PV production modeling software. By contrast, higher-resolution ("sub-hourly") datasets with suitable periods of record for constructing TMYs are at present seldom available.

A. Sub-Hourly Adjustment Factor

In recent years, extensive literature has been published on the phenomenon that models utilizing hourly-resolution data tend to overestimate energy production [2]-[8]. The transient movement of clouds at finer temporal resolutions and the consequent fluctuations in incident irradiance are obscured by the hourly aggregation of the data (i.e., observed surges in irradiance are essentially smoothed away). As a result, in an hourly model, the portion of PV array output that exceeds the capacity limits imposed on the inverters (the "clipping loss") is underestimated, and thus the output ultimately realized at the inverter is overestimated. In systems with larger array capacities relative to their inverter capacities ("DC/AC ratios") [3] and at locations with higher variability in cloud coverage [5], there is a greater likelihood for an hourly interval to be subject to this phenomenon, and thus the magnitude of the associated net loss tends to be greater.

For any given hour, the difference between the clipping loss calculated using a sub-hourly model (C_{sh}) versus that of an hourly model (C_h) is termed the "sub-hourly loss" (Δ):

$$\Delta = C_{sh} - C_h \qquad (1)$$

The ratio between the inverter output from the sub-hourly (I_{sh}) and hourly (I_h) models is the "sub-hourly adjustment factor" (F):

$$F = \frac{I_{sh}}{I_h} \qquad (2)$$

These two terms are interchangeable with regard to discussion on the impact of sub-hourly data on modeled production.

B. Evolution from Prior Studies

Methods to calculate the sub-hourly adjustment factors for a given project and accordingly adjust its hourly production model have been the subject of several publications. Random forest regression approaches were presented in studies by Anderson & Perry [4] and Bradford et al. [5], the former of which was adapted by Parikh et al. [6]. In each of these, one universal regression model was trained on a network of existing locations for which high-resolution ground-measured data were available, which was used to identify trends to predict sub-hourly losses for new sites.

This paper utilizes a comparable approach involving a sequence of linear and ensemble machine learning ("ML") methods in conjunction with site-specific sub-hourly irradiance data to construct a customized model for any site based on the unique design and configuration of its PV system. As such, the detailed losses due to weather trends, optical and electrical characteristics of the array, inverter operation, AC equipment and wiring, and point-of-interconnection ("POI") limitations are all captured in the model, which ultimately computes a profile of sub-hourly adjustment factors tailored to the system's characteristics.

978-1-6654-6060-6/23 $31.00 © 2023 IEEE

II. DATA

The methods discussed in this paper are compatible with either satellite-based or ground-measured data sources for sub-hourly irradiance. Short-term satellite imagery at 5-minute temporal and 1-kilometer spatial resolutions throughout the United States has become increasingly available since the launch of the GOES-R Series of satellites in November 2016 [9]. Solar resource data providers are actively incorporating this higher-resolution imagery into their products, thus enabling wider access to sub-hourly irradiance data that capture intra-hour fluctuations due to cloud movement. In comparison, ground-measured pyranometer data is capable of capturing such trends with lower uncertainty at even higher spatial and temporal resolutions [10], though these typically span a shorter period of record (e.g., one year) and are unlikely to be available for a project site unless explicitly commissioned.

In this paper, we examine the impact of sub-hourly data to the modeled performance of several design variants of a utility-scale PV system (from DC/AC ratios of 1.0 through 1.5) at 506 locations throughout the contiguous United States and Hawaii. For each of these locations, the hourly TMY data is generated from the Meteonorm 8.0 database, and the sub-hourly data is generated from the National Solar Radiation Database ("NSRDB") at a 5-minute temporal and 2-kilometer spatial resolution.

III. METHODS

Various software packages are publicly available to translate irradiance data into PV array output through consideration of site geometry and equipment characteristics that impact the conversion of power. Regardless of whether the selected software supports computations at sub-hourly timesteps, recall that the P50 model is customarily based on TMY data, which is only available at hourly resolution. The sub-hourly analysis is therefore performed separately, and its resulting insights are incorporated back into the hourly TMY model.

This study utilizes the commercial PVsyst software for PV array modeling in conjunction with the Leidos-proprietary Solmoto application for postprocessing. Solmoto is built in the Python programming language and employs algorithms provided in the open-source pvlib [11] library for irradiance decomposition and transposition.

A. Calculating Sub-Hourly Factors from Time Series Dataset

The essence of quantifying the sub-hourly loss lies in identifying the relationship between the output of a sub-hourly model and that of an equivalent hourly model derived from the same irradiance dataset. This equivalent hourly model should be analogous in format to the TMY model, which itself will ultimately be adjusted through the identified relationship to simulate sub-hourly effects. Each hourly datapoint in the equivalent hourly time series is constructed by averaging the three sub-hourly datapoints that fall on the start-of-hour (:00),

half-hour (:30), and start-of-next-hour (:00) timestamps. Selecting these instantaneous snapshots at 30-minute intervals, rather than averaging all constituent sub-hourly datapoints, serves to emulate the fact that a TMY dataset is mainly derived from 30-minute imagery, and that no direct information from within the 30-minute intervals would have been available during its construction.

Next, the energy production model is run separately for the equivalent hourly and sub-hourly time series to determine power output at the inverter. Since PVsyst does not support sub-hourly computation, this is achieved in Solmoto by transposing time series global horizontal irradiance ("GHI") to incident plane-of-array ("POA") irradiance, and then transforming that to array output through a constant coefficient that is obtained from the linear relationship observed in the TMY model. Inverter capacity limits are determined by considering the operational characteristics of the equipment (e.g., temperature-based performance) and the impacts of downstream POI limitation.

Finally, with the time series modeling complete, the sub-hourly adjustment factors (2) are calculated for each time series hour, thereby forming a distribution that can be linked to hourly weather characteristics at the project site.

B. Predicting Sub-Hourly Adjustment Factors in TMY Dataset

To construe a relationship between the meteorological features of the hourly time series dataset and its corresponding sub-hourly factors, a two-step ML process is implemented. First, a binary classifier predicts whether a sub-hourly loss is realized at a given hour, and then if so a random forest regressor predicts the magnitude of that loss and, correspondingly, the sub-hourly adjustment factor, based on the historical occurrence of sub-hourly losses.

The model is trained on the time series dataset and used to predict the performance of the TMY dataset. Both ML processes utilized are driven by the same three predictor variables: incident POA irradiance, GHI, and solar elevation angle. This composite approach was found to perform better than if the first step was omitted and the zero-loss hours were handled directly by the regressor.

IV. RESULTS

The methodology outlined above is performed on a total of 3,036 scenarios, comprised of 506 locations for each of which six design variants are simulated. The six variants are distinguished by their DC/AC ratios, which range from 1.0 through 1.5 in increments of 0.1.

Fig. 1 shows a scatter plot of the distributions of sub-hourly adjustment factors from the training (i.e., time series) and predicted (i.e., TMY) datasets for one such simulation located in Northern Virginia with a DC/AC ratio of 1.3.

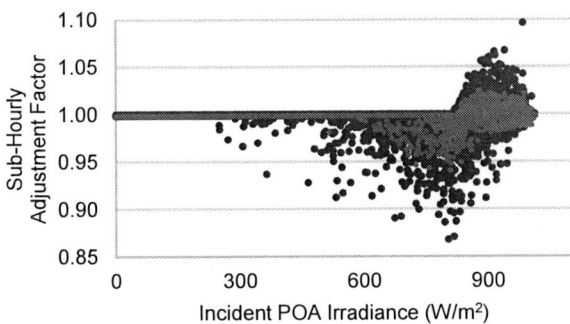

Fig. 1. Relationship between incident POA irradiance and sub-hourly adjustment factor for each hourly data point in training (i.e., time series; blue dots) and predicted (i.e., TMY; red dots) datasets.

TABLE I
SUMMARY OF ANALYSIS RESULTS

DC/AC Ratio	Median Classifier Accuracy	Median Regressor R^2 Value	Median Sub-Hourly Adj. Factor
1.0	1.0000	1.0000	1.0000
1.1	1.0000	1.0000	1.0000
1.2	0.9740	0.1773	0.9997
1.3	0.9963	0.3203	0.9963
1.4	0.9299	0.3171	0.9920
1.5	0.9256	0.2931	0.9888

Fig. 2 illustrates the variation of net annual (i.e., aggregated over the TMY year) sub-hourly adjustment factors throughout the locations and DC/AC ratios considered.

The median results of the analysis across the 506 locations are summarized in Table I for each DC/AC ratio. The classifier accuracy describes the proportion of datapoints correctly identified by the model as experiencing a nonzero sub-hourly loss. The R^2 value coefficient of determination (i.e., R^2 value) of the regressor describes the proportion of the variation in sub-hourly loss that is explained by the model's predictor variables, thus signifying the accuracy of the magnitudes of predicted sub-hourly losses.

V. DISCUSSION

A. Impacts of DC/AC Ratio and Geography

As described earlier (2), a sub-hourly loss is an adjustment to the clipping loss realized at the inverter in a PV system model. No sub-hourly loss occurs when the system is incapable of clipping due to the lack of overbuild in DC capacity, as is the case at a DC/AC ratio of 1.0. At higher DC/AC ratios of 1.1 and 1.2, the array is only able to produce at or beyond its DC

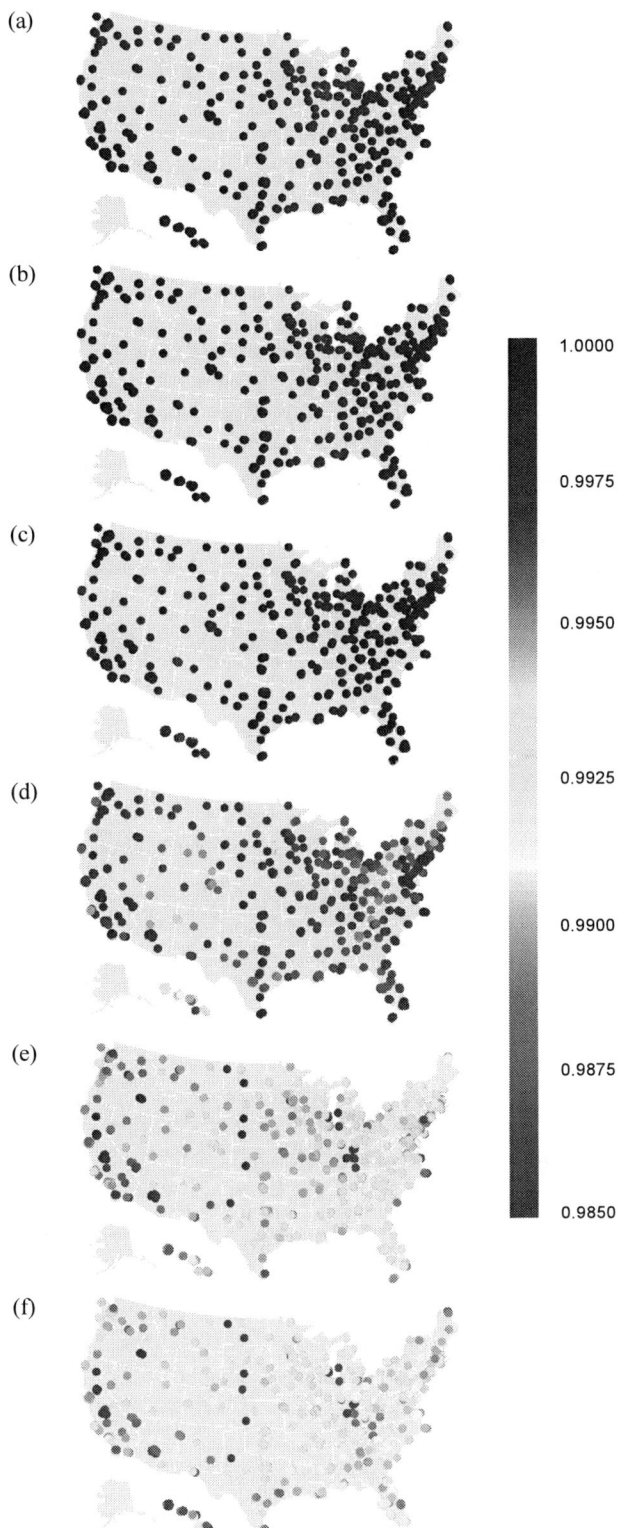

Fig. 2. Net annual sub-hourly adjustment factors by location for systems with DC/AC ratios of (a) 1.0, (b) 1.1, (c) 1.2, (d) 1.3, (e) 1.4, and (f) 1.5.

capacity for limited periods when incident irradiance is maximized. (Peak sun hours in North America occur during summer afternoons.) As such, the sub-hourly loss is again negligible.

The impact of considering sub-hourly data more prominently emerges at DC/AC ratios of 1.3 and greater in several geographic regions, as depicted in Fig. 2. In such systems, there are enough hourly datapoints throughout the year that experience a significant magnitude of clipping loss that the sub-hourly adjustment has a noteworthy influence on the modeled generation.

Regardless of DC/AC ratio, the relative sub-hourly adjustment is greater in Hawaii and the South Atlantic regions, which experience a high variability in cloud cover over a sub-hourly period. Conversely, much of the desert climate in the West is subject to far sparser cloud cover, and thus experiences far less variability. Climates near cities such as Seattle (Washington), Grand Rapids (Michigan), and Syracuse (New York) are often consistently overcast; these regions also experience minimal variability in cloud cover, and thus the corresponding impact of sub-hourly adjustment is lower. These insights are consistent with those observed in past literature.

In certain regions there can be considerable variation in sub-hourly loss even over a small geographic footprint. For example, for the locations considered within the city limits of San Diego (California), the sub-hourly adjustment factors in the 1.3 DC/AC ratio scenario ranges from 0.9975 to 0.9911. This discrepancy of 0.64 percent in the net power output from the inverters can result in significant implications for project financing, and thus reinforces the importance of site-specific considerations for sub-hourly data.

B. Model Performance

Per Table I, the classifier component of the ML model is shown to predict with reasonable accuracy which hours are subject to a sub-hourly loss. The R^2 values of the regressors are considerably lower, and while conventional wisdom would suggest that the model performs poorly when assessing the magnitudes of sub-hourly losses, the practical implication of this assessment is insignificant.

An R^2 value of 0.0 would suggest that none of the variability in sub-hourly loss could be explained by the ML model, thus the profile of predicted losses would be equivalent to the average loss observed in the training dataset. If this were the case, the predicted dataset would be subject to the identical net sub-hourly loss as that of the training. This result would still be useful, since the inherent randomness of cloud cover is indeed still being translated into a profile of site-specific adjustment factors, though they would fail to reflect any variability beyond that observed in the three-year sub-hourly dataset. As datasets with longer periods of record become available in the future, this associated uncertainty will be mitigated.

An improvement to this result is signified by a positive R^2 value, which indicates that a demonstrable relationship has been identified within the training dataset and that it can be translated to the predicted dataset. Certainly, the R^2 value can be strengthened with the use of predictor variables other than those considered in this paper that dictate atmospheric conditions (e.g., turbidity, relative humidity, terrain), though such variables are not commonly available in solar resource datasets and would thus be incompatible with most use cases.

C. Positive Sub-Hourly Adjustment Factors

Past literature has often dismissed the notion that sub-hourly data can cause a positive adjustment to estimated production [2], yet this is observed here for several datapoints, as illustrated in Fig. 1. This is due to the aforementioned averaging method, where the equivalent hourly dataset is constructed from instantaneous snapshots of sub-hourly data, and thus often either understates or overstates the actual irradiance during each hour, regardless of the occurrence of clipping. When the equivalent hourly datapoint understates the actual irradiance, a positive sub-hourly adjustment factor is introduced. However, the opposite behavior is just as often observed, so the net adjustment is zero, per this phenomenon, when aggregated across the entire dataset.

In all other cases, a sub-hourly adjustment is the result of clipping behavior, and occurs for any hourly timestamp in which its constituent sub-hourly readings include both values that are higher and values that are lower than the clipping limit. Conversely, when an hour's constituent sub-hourly readings are all higher (or all lower) than the clipping limit, then no sub-hourly adjustment is observed, as the sub-hourly data has not revealed any new insights about the clipping behavior during that hour.

An exception to this rule may arise if, at a given hour, the constituent sub-hourly clipping limits deviate from that of the equivalent hourly dataset. This can occur from temperature-based performance impacts to equipment (e.g., an inverter may derate beyond a certain ambient temperature) or dynamic software constraints on inverter capacity due to POI limits to optimize the useful power realized at the revenue meter. When such variation in intra-hour clipping limits occurs, the sub-hourly adjustment factor may occasionally be positive.

This phenomenon serves to counteract the downward adjustments that are more prevalent throughout the dataset, though is seldom dominant enough that the net annualized sub-hourly adjustment to a full TMY would be positive.

VI. CONCLUSION

This paper presents a method for utilizing sub-hourly irradiance datasets to correct hourly energy production models for clipping loss calculations on a site-specific basis, that considers its system design and local weather characteristics. With new PV systems being installed at increasingly larger DC/AC ratios, due to the continually decreasing cost of PV modules as well as the proliferation of PV systems that are integrated with battery energy storage systems, the need to accurately assess the impact of sub-hourly data is critical for project development stakeholders. The importance of using site-specific data is underscored by the observation that there is occasionally significant regional variation in the magnitudes of sub-hourly losses across relatively small areas (e.g., within a city).

The methodology presented demonstrates practical effectiveness in translating insights from short-term sub-hourly datasets to hourly TMY models based on long-term data, but its predictive performance is limited by the lack of availability in conventional solar resource datasets of meteorological parameters that are more closely correlated with variability of cloud cover.

The sub-hourly adjustment ratios presented in this paper were calculated from satellite-based sub-hourly irradiance datasets with 5-minute temporal resolutions. The paper did not explore the marginal impacts of using higher-resolution (e.g., 1-minute) datasets to the magnitudes of the sub-hourly losses assessed. Additionally, ground-measured data sources, rather than those that are satellite-based, are expected to more accurately reflect the sub-hourly characteristics of weather. Using the methods outlined herein, an investigation into the sensitivity of data source and granularity to the magnitudes of sub-hourly losses would make for a worthy sequel to this paper.

REFERENCES

[1] S. Wilcox and W. Marion, *Users Manual for TMY3 Data Sets (Revised)*, National Renewable Energy Laboratory, Golden, Colorado, 2008, p. 14.

[2] D. Cormode, N. Croft, R. Hamilton, and S. Kottmer, "A method for error compensation of modeled annual anergy production estimates introduced by intra-hour irradiance variability at PV power plants with a high DC to AC ratio," in *2019 IEEE 46th Photovoltaic Specialists Conference*, pp. 2293-2298.

[3] K. Anderson, W. Hobbs, W. Holmgren, K. Perry, M. Mikofski, and R. Kharait, "The effect of inverter loading ratio on energy estimate bias," in *2022 IEEE 49th Photovoltaic Specialists Conference*, pp. 0714-0720.

[4] K. Anderson and K. Perry, "Estimating subhourly inverter clipping loss from satellite-derived irradiance data," in *2020 IEEE 47th Photovoltaic Specialists Conference*, pp. 1433-1438.

[5] K. Bradford, R. Walker, D. Moon, and M. Ibanez, "A regression model to correct for intra-hourly irradiance variability at PV power plants with a high DC to AC ratio," in *2020 IEEE 47th Photovoltaic Specialists Conference*, pp. 2679-2682.

[6] A. Parikh, K. Perry, K. Anderson, W. Hobbs, R. Kharait, and M. Mikofski, "Validation of subhourly clipping loss error corrections," in *2021 IEEE 48th Photovoltaic Specialists Conference*, pp. 1670-1675.

[7] R. Kharait, S. Raju, A. Parikh, M. Mikofski, and J. Newmiller, "Energy yield and clipping loss corrections for hourly inputs in climates with solar variability," in *2020 IEEE 47th Photovoltaic Specialists Conference*, pp. 1330-1334.

[8] J. Allen and W. Hobbs, "The effect of short-term inverter saturation on modeled hourly PV output using minute DC power measurements," in *Journal of Renewable and Sustainable Energy*, vol. 14, 063503, 2022.

[9] National Aeronautics and Space Administration, *GOES-R Series Data Book*, Goddard Space Flight Center, Greenbelt, Maryland, 2019, p. 3-1.

[10] C. Hansen and A. Scheiner, "Uncertainty in annual energy resulting from uncertain irradiance measurements", in *2022 IEEE 49th Photovoltaic Specialists Conference*, pp. 0022-0026.

[11] W. Holmgren, C. Hansen, and M. Mikofski, "Pvlib python: a python package for modeling solar energy systems," *Journal of Open Source Software*, 3(29), 884, 2018.

Self-Assembled Monolayer Patterning for polySi/SiO₂ Passivated Contacts

B. Nemeth, D.L. Young, M.R. Page, V. LaSalvia, S. Theingi, and P. Stradins

National Renewable Energy Laboratory, Golden, Colorado, CO, 80401, USA

Abstract — **We utilize hexamethyldisilazane (HMDS) based self assembled monolayers to pattern polysilicon (polySi) passivated contacts. We find process conditions that allow for etching front side n/polySi between fingers; thereby increasing J_{sc}. Importantly, the V_{oc} does not degrade indicating the additional process steps do not introduce defects or impurities. HMDS layers remain on the surface for metallization without detriment to transport.**

| Texture, SC1-2 |
| LPCVD SiO₂ / polySi |
| PECVD n,p a-Si:H |
| 850C Anneal |
| ALD Al₂O₃ / FGA |
| HMDS / UV |
| TMAH |
| Metal |

Figure 1. Process flow

I. INTRODUCTION

Self-assembled monolayer (SAM) chemistries are used to controllably functionalize various material surfaces [1] to form an engineered monolayer. Once set, the SAM can: 1) enhance mechanical adhesion for subsequent layer application; 2) provide chemical resistance to etching; or 3) tailor electronic transport [2] when coupled with a metal or semiconductor.[3] Hexamethyldisilazane (HMDS) is one such SAM precursor that is used to functionalize surfaces with trimethyl-silyl groups, which can then be manipulated. In this contribution we successfully employ HMDS as a simple photoresist substitute for patterning the front side polysilicon of a both side polysilicon passivated contact solar cell. The patterned HMDS remains on the cell and is compatible with both subsequent hydrogenation dielectic layers and metallization.

II. EXPERIMENTAL

Silicon n-Cz wafer samples are KOH texture etched then cleaned in SC-1 and SC-2 solutions, and loaded into a low pressure chemical vapor deposition (LPCVD) tube to grow both the tunneling SiO₂ layer as well as intrinsic polySi layers on both sides of the wafer (Fig.1). Doped a-Si:H layers are deposited on the samples via 13.56 MHz RF plasma enhanced chemical vapor deposition (PECVD) using SiH_4, H_2, PH_3, and B_2H_6 at 1 Torr. The resulting structures are then annealed in a tube furnace at 850°C to crystallize and dope the polySi. Device samples are coated with atomic layer deposited Al_2O_3, annealed

at 400°C for hydrogenation, and HF dipped for subsequent metallization. HMDS (Gelest, Inc.) was then applied via vapor exposure in a closed vessel for various durations. UV exposure to transform the HMDS into SiO_2 was done with a Novascan-PSD unit through a silicon shadow mask creating patterns of SiO_x and HMDS on the wafer surface. PolySi etching was performed using 15% TMAH at various temperatures with SiO_2 acting as an etch resistant mask. Water contact angle (WCA) measurements were taken using a Kruss drop shape analyzer. Device J-V characteristics were measured under AM1.5G, and images were taken with a Keyence VK-X laser microscope.

III. RESULTS AND DISCUSSION

A Si surface can be functionalized with a SAM precursor via gas or liquid phase. The primary HMDS $[(CH_3)_3Si]_2NH$ functionalization mechanism occurs by the reaction with Si surface hydroxyl (OH-) groups, where the NH- leaving group from the HMDS reacts with the hydrogen from the OH- groups forming NH_3, and the $Si(CH_3)_3$ bonds to the remaining surface O-. (Fig. 2). The kinetics of this reaction are dependent on

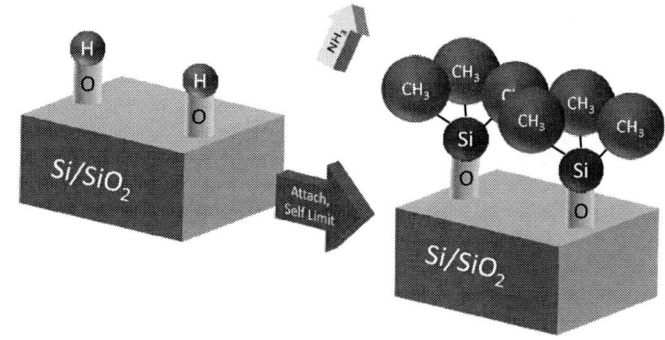

Figure 2. HMDS application of SAM to tunneling SiO₂.

typical experimental variables, so we choose ambient pressure and temperature to simplify the process utilizing vapor phase soaking. We determine, via water contact angle measurements (WCA), the time to achieve approximate conformal full area coverage on both SiO_2 and cSi to be over 10 hrs. This produces a hydrophobic surface, which we use as a simple visual feedback mechanism to develop our process further, where a hydrophilic surface implies either SAM removal or oxidation.

Next, we determine the duration that our functionalized surface can withstand HF and TMAH. We find that the SAM

978-1-6654-6060-6/23 $31.00 © 2023 IEEE

can withstand 2% HF for approximately 10 sec, whereas it can withstand 0.1% HF for at least 80 min, greatly widening the process window. We also determine that the temperature of the TMAH solution is critical by observing bubble formation on the Si surface which we assign to the breakdown of the SAM and initiation of polySi etching. At 50°C, etching begins within approximately 70 sec, while at 45°C this occurs around 150 sec.

To implement patterning, we expose the functionalized surface to UV light (through a Si shadow mask), which photocleaves [4] the methyl group bonds, promoting oxidation

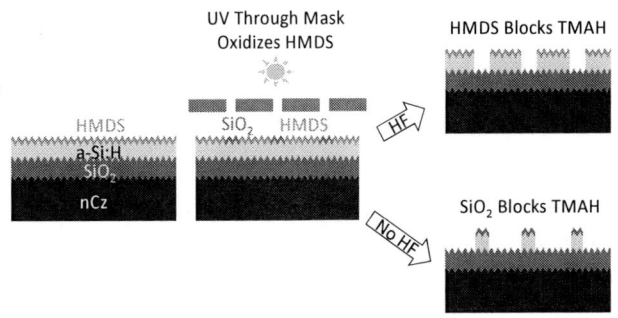

Figure 3. Patterning HMDS with UV-O₃ to achieve positive an negative subsequent etching.

of the underlying Si (Fig. 3). This SiOₓ is much more resistant to TMAH etching than a simple UV oxidized Si (or a-Si:H) surface, where the oxidized SAM withstands etching for approximately 5 minutes at 45°C, whereas the latter only resists for a few seconds. This is likely due to the density of each SiOₓ and is currently under investigation.

Since O₃ is naturally formed when O₂ is exposed to UV light (as in the case of ambient air), we control the reaction and number of O₃ molecules created by actively pumping the ambient within the UV source. Regardless, some O₃ will react in areas not directly exposed to UV (under the shadow mask), and we find 10 mins to be an optimum time of exposure. Less time does not sufficiently oxidize the SAM in the prescribed pattern, and more time results in oxidation outside of the

pattern, affecting the fidelity. Figure 4 shows the result of TMAH planarization of a textured surface between patterned fingers using the oxidized SAM as an etch mask. The shadow mask opening width in 4a) is around 80 μm and an etched sample in 4b) has a finger width of 106 μm. This gives approximately 15 μm of fidelity on each side. The thinner, brighter feature is the deposited metal through the accompanying shadow mask, which has purposely been shifted to discern true feature dimension for eventual incorporation into a cell.

Various approaches can be utilized for SAM implementation into a solar cell. Our initial focus was to first pattern intrinsic polySi fingers (preserve the thick polySi under the metal) then dope them with blanket a-Si:H prior to subsequent crystallization and hydrogenation, so that the area between fingers is capped with an eventual thin passivated polysilicon layer. Note that the nonoxidized HDMS may remain on the polySi during this a-Si deposition and annealing. The benefit of using TMAH to etch the intrinsic polySi prior to doping is that the tunneling SiO₂ is effectively an etch stop for around 1 to 2 minutes. Once we observe the cessation of bubble creation during etching, we remove the sample from the TMAH to preserve this tunnel SiO₂. We form doped polysilicon by depositing a thin (15-20 nm) doped a-Si:H layer as the dopant source onto intrinsic material and anneal to diffuse dopants.

Ideally, the area between metallized passivated contact fingers should have no polySi, thereby omitting parasitic absorption and maximizing J_{sc}. While this is the most desirable condition, a thin passivated polySi layer on tunnel SiO₂ buffers the process window for metallization, since any misalignment between the metal and polySi finger will result in direct metal contact on the cSi surface. Additionally, without polySi between fingers, a passivating dielectric layer will be required to be deposited and annealed to hydrogenate the cSi surface. If this is performed prior to metallization, the dielectric layer(s) will need to be removed only on the fingers, remaining on the desired nonmetallized surfaces. If dielectric passivation occurs after metallization, annealing a metallized polySi passivated contact oftentimes results in undesirable metal diffusion and even spiking, particularly for p-type polySi.

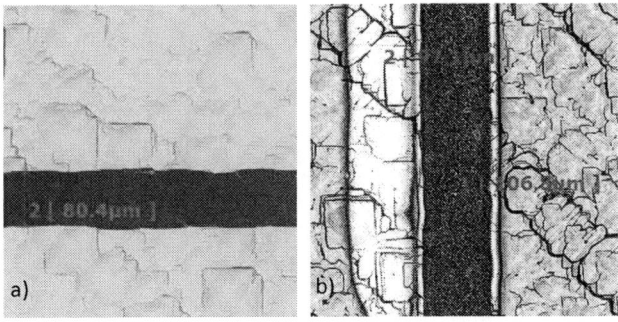

Figure 4. Image of a) Si shadow mask gridline width and b) resulting etched finger width (dark) next to masked metallization (light).

Our initial attempts to integrate finger patterning into an actual cell resulted in several practical hindrances such as mask inconsistency (where mask fingers were not completely open), mask misalignment, and under- or over- etching of the polySi. Regardless, we do observe increased J_{sc} (9-10 mA/cm^2) when comparing patterned (30 mA/cm^2) versus nonpatterned (20.6 mA/cm^2) devices; however, these are not acceptable values. Therefore, we employed rigorous mask design and fabrication to guarantee feature size consistency as well as alignment for subsequent experiments. This results in excellent alignment; however, it is apparent that the polySi is not etched completely to the tunneling SiO$_2$ as shown in the J-V curves (Fig. 5), where

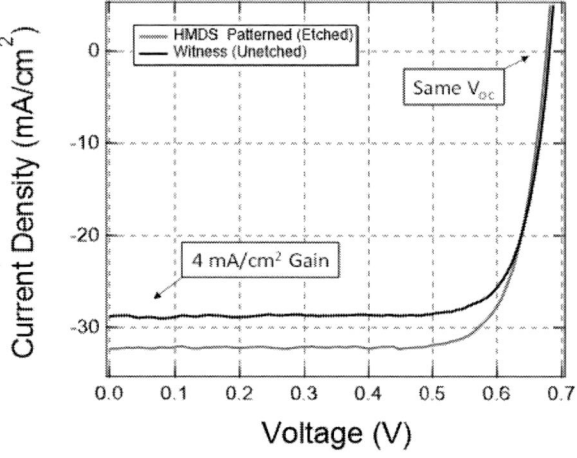

Figure 5. J-V curves of etched (blue) and unetched (black) gridlines.

a gain of only 4 mA/cm^2 (28 versus 32 mA/cm^2) is measured when comparing unetched to etched devices. As of this writing, the next batch of cells are currently in process, where we will pattern polySi fingers either before or after crystallization.

IV. CONCLUSIONS

We pattern the frontside n/polySi in a front-back passivated contact solar cell using UV exposed hexamethyldisilazane (HMDS), which remains on the surface for subsequent metallization. Experimental conditions such as etchant time and concentrations are identified and developed to increase the process window. Resulting cells show an increase in J_{sc} without significant V_{oc} degradation.

REFERENCES

[1] B. Arkles, "Tailoring Surfaces with Silanes," Chemtech, 7(12), 766 (1977).

[2] S.A. DiBenedetto, et al., "Molecular self-assembled monolayers and multilayers for organic and unconventional inorganic thin-film transistor applications," Advanced Materials, 21(14-15), pp. 1407-1433 (2009).

[3] S. Onclin et al., "Engineering silicon oxide surfaces using self-assembled monolayers," Angewandte Chemie International Edition, 44(39), pp. 6282-6304 (2005).

[4] J.M. Calvert, et al., "Deep UV photochemistry and patterning of self-assembled monolayer films," Thin Solid Films, 210, pp. 359-363, (1992).

Non-Ionizing Radiation Effects on the Room Temperature Surface Recombination Velocity of Unintentionally Doped AlGaAs/GaAs Heterostructures

Andrew Hudson[a)] and Daniele Monahan

The Aerospace Corporation, El Segundo, CA 90245, USA

a) **Author to whom correspondence should be addressed:** andrew.i.hudson@aero.org

ABSTRACT

We used time resolved photoluminescence to extract surface recombination velocities and bulk carrier recombination lifetimes for a series of AlGaAs/GaAs heterostructures exposed to neutron radiation. While the measured exposure effects were dominated by the bulk, a statistically significant increase in the surface recombination velocity was observed for neutron fluences of and exceeding 1×10^{13} n/cm^2. This result suggests that radiation effects on the surface recombination velocity become an important factor for modeling and qualification purposes when devices are exposed above a threshold fluence.

I. INTRODUCTION

An increase in carrier recombination upon exposure to radiation leads to performance degradation in detectors and other semiconductor based optoelectronic devices. Radiation damage can occur either in the bulk material or at the surfaces between bulk regions of different composition. The potential risk to device performance is expected to become more severe in complex structures such as in superlattice detectors or multi-junction solar cells.

The *surface recombination velocity* (SRV) describes the combined rate of carrier transport from a bulk absorber region toward defects at an interface and the rate of recombination at those interface defects. The susceptibility to forming electronic defects associated with interface displacement damage may differ from that in the bulk material, owing to the unique lattice constants and bond energies. Increases in SRV due to ionizing radiation has been well documented, especially in bipolar transistors. The importance of surface passivation for device performance is also well known. However, the effect of non-ionizing radiation exposure on surface recombination has been less well studied despite the importance of these effects on performance of optoelectronic devices in the space radiation environment. An improved understanding of the relationship between displacement damage induced by non-ionizing radiation and the formation of trapping/recombination defects on surfaces could improve our ability to evaluate the suitability of many-interface devices for on-orbit operation.

Here we examine the effects of non-ionizing neutron exposure on SRV in simple heterostructures. Using time-resolved photoluminescence spectroscopy (TRPL), we measure the carrier lifetime before and after radiation exposure. By performing this experiment on a series of heterostructures with different bulk thickness, but with the same number of interfaces, the component due to surface recombination can be isolated from that due to the bulk.

II. SURFACE STATE RECOMBINATION

Defects in semiconductors, including native defects and absorbed impurities[1,2] constitute deep, non-radiative surface state recombination sites. Defect identities and energies are often unknown, complicating the prediction and modeling of surface recombination properties for a given material[1]. Unlike the case for bulk Shockley-Reed-Hall (SRH) sites, which are non-radiative states with specific sub-bandgap energies, surface states can exhibit a wide range of energy level values.

When the free carrier density exceeds that at thermal equilibrium (via electrical or optical injection), "excess" carriers recombine in the bulk and at interfaces. Empirically derived models have successfully predicted the bulk[3] and surface[4] recombination processes. Recombination in the monolayers adjacent to the surface may be modeled as a combination of both surface and bulk effects, resulting in a carrier gradient with diffusion transport[4]. The diffusion transport model predicts the SRV, defined as the ratio of the rate of electron (or hole) flow into a unit surface area to the excess carrier density in the bulk just beneath the surface[5]. This depends on the bulk carrier densities, carrier thermal velocity, and the surface recombination state cross-section. The rate of carrier recombination at surface states depends on the

978-1-6654-6060-6/23 $31.00 © 2023 IEEE

capture process for both electrons and holes[6], with the slower of the two being the rate limiting process.

The total recombination rate (measured via TRPL) depends on bulk and surface recombination processes[1] as in Eq. (1),

$$k_{PL} = k_{bulk} + (S_1 + S_2)L^{-1}, \qquad (1)$$

for which k_{PL} is the observed carrier recombination rate for a sample with a bulk recombination rate k_{bulk}, L is the active region thickness, S_1 and S_2 are surface recombination velocities for the front and back interfaces (which may be replaced with $2S$. assuming the two are equal). Equation 1 applies to cases of uniform carrier excitation and materials with carrier diffusion lengths that exceed the active region thickness according to Eq. (2),

$$Lk_{bulk}^{1/2}\big/D^{1/2} \ll 1 \qquad (2)$$

for which D is the diffusion coefficient. Equation 1 may be rewritten as Eq. (3) assuming that the SRV has the same value at each interface.

$$k_{PL} = k_{rad} + k_{nr-SRH} + 2SL^{-1} \qquad (3)$$

III. RADIATION EFFECTS ON BULK AND SURFACE CARRIER RECOMBINATION

Radiation exposures common in the space environment can be ionizing and non–ionizing:

a) Ionizing radiation effects: electrons are liberated from host atoms via bond breakage. This can result in localized charge deposition and electronic device transients. Photons (gamma rays, x-rays) and charged sub - atomic particles (protons, electrons, nuclei) can act as ionizing radiation.

b) Non-ionizing radiation effects: atomic displacements from the original crystal lattice site — vacancy, interstitial, substitutional and complex point defect formation. Sub-atomic particles (electrons, protons, nuclei) can generate point defects.

The impact of ionizing radiation on device performance has been studied extensively in parts destined for space applications. To the best of our knowledge the effects of strictly non-ionizing radiation exposure on an SRV have not been examined. Such an investigation is necessary to better understand the importance of this potential source of performance degradation for specific materials and growth structures. Superlattices and other complex heterostructures with multiple epi-layers might be particularly prone to the radiation enhancement of interface recombination processes.

IV. TEST ARTICLES AND EXPERIMENTAL PROCEDURE

The test articles were grown via metal-organic chemical vapor deposition (MOCVD) at the NanoPower Research Labs of the Rochester Institute of Technology as part of a collaboration with The Aerospace Corporation on optoelectronic materials and device technology. Unintentionally doped (UID) AlGaAs/GaAs double heterostructures were grown for the primary purpose of investigating the dependence of radiation effects on base material properties. On-orbit optoelectronic devices (including solar cells) typically contain junctions formed by interfaces of differently doped layers. The use of passive heterostructures permits us to focus on material effects without the additional complexity of internal fields and their impact on carrier transport.

Prior to neutron exposure, each sample was characterizing using room temperature steady state photoluminescence (SSPL). The carriers were excited above the band edge using a 780 nm pump and monitored by sweeping the monochromator from 800 nm to 900 nm. The slit width was set to 300 µm to yield a wavelength resolution of ~1 nm.

The carrier lifetime data collected via the TRPL method was fit as a function of active region thickness, according to Eq. (3), to extract an estimate for the bulk carrier lifetime and the SRV. This process was repeated for samples exposed to a series of neutron fluences varying from 1×10^{10} to 5×10^{14} n/cm^2 to determine the effect of exposure on the SRV. The 1 MeV equivalent in Si neutron exposures were performed at McClellan Nuclear Research Center, University of California, Davis.

V. RESULTS

Linear fits of the extracted $1/e$ carrier recombination lifetime estimates as a function of the active region thickness according to Eq. (3) were used to extract the bulk carrier lifetime and the SRV for a given exposure fluence. These assume that $S_1 = S_2$, and that the surface recombination at each AlGaAs/GaAs interface is comparable. A comparison between the pre-exposed test articles and those exposed to fluences of 1×10^{13} n/cm^2 appears in Fig. 1. The change in slope is proportional to the SRV for each series. Similarly, a comparison of the data for test articles exposed to

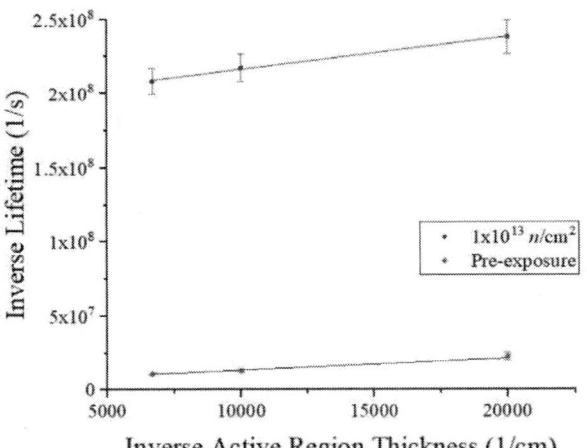

FIG. 1. SRV extraction. Pre-exposure: τ_{bulk} = 238 +/- 57 ns, SRV = 460 +/- 37 cm/s. 1×10^{13} n/cm²: τ_{bulk} = 5.1 +/- 0.1 ns, SRV = 1100 +/- 70 cm/s.

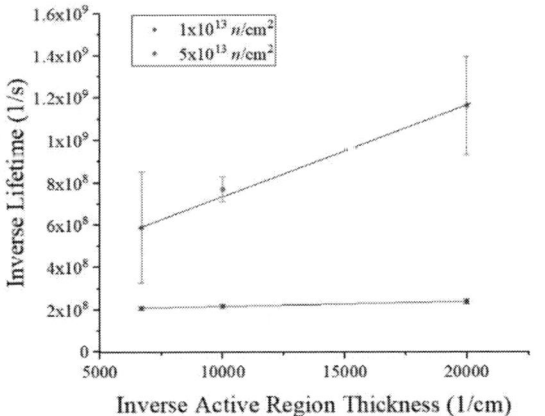

FIG. 2. SRV extraction. 1×10^{13} n/cm²: τ_{bulk} = 5.1 +/- 0.1 ns, SRV = 1100 +/- 70 cm /s. 5×10^{13} n/cm²: τ_{bulk} = 3.1 +/- 0.4 ns, SRV = 21000 +/- 1500 cm/s.

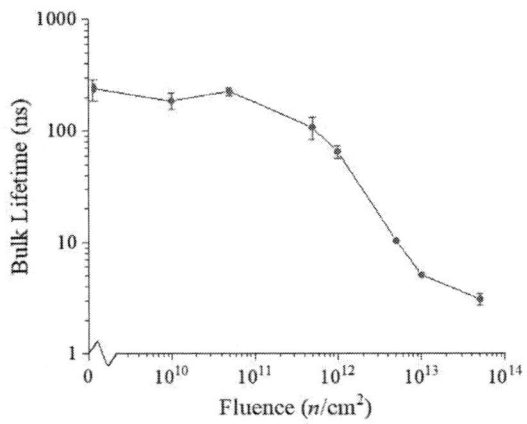

FIG. 3. Room temperature bulk carrier lifetime data extracted from linear data fits.

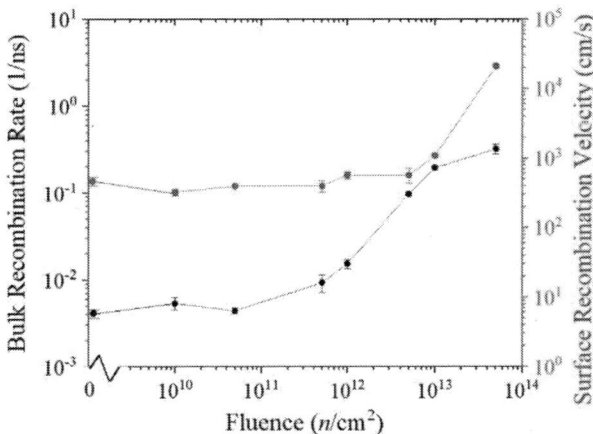

FIG. 4. Room temperature SRV estimates extracted from linear data fits.

fluences of 1×10^{13} n/cm² and 5×10^{13} n/cm² appears in Fig. 2. Note the considerable increase in the plot slope (the SRV). An examination of the extracted bulk carrier lifetime as a function of fluence appears in Fig. 3. A reduction in the carrier lifetime relative to the pre-exposure value significant to 2-σ becomes evident for fluences exceeding 5×10^{10} n/cm². An overall lifetime reduction by over two orders of magnitude occurs for the largest fluence of 5×10^{13} n/cm². The observed recombination rate for fluences of (and exceeding) 1×10^{14} n/cm² approach the detector IRF and cannot be deconvolved from it, so this data is not included in Fig. 3.

An examination of the SRV as a function of fluence appears in Fig. 4. An increase in the SRV significant to 1-σ becomes apparent for fluences of and exceeding 1×10^{13} n/cm². For comparison, the bulk carrier recombination rate (the inverse of the extracted bulk carrier lifetimes shown in Fig. 3) are plotted alongside the SRV. The results show that radiation-induced degradation begins to occur at a lower fluence in the bulk than at the AlGaAs/GaAs interfaces. At the highest fluence the estimated SRV increases by over one order of magnitude relative to the pre-exposure quantity.

VI. CONCLUSIONS

In this study it was determined that non-ionizing radiation has a measurable effect on the bulk recombination rate for the AlGaAs/GaAs test structures, likely through the generation of non-radiative recombination centers. Such exposure is also observed to increase the test article SRV for neutron fluences of and exceeding 1×10^{13} n/cm^2.

Our results suggest that optical detectors which incorporate complex growth structures (such as superlattices) may not experience an interface dependent performance degradation at lower exposure fluences. However, the sharp SRV threshold suggests that this loss mechanism might be substantial above an unknown critical exposure. Future studies are needed to examine how the surface-specific effect measured here scales for the case of increasingly complex growth structures.

REFERENCES

[1] J.M. Langer, W. Walukiewicz, "Surface recombination in semiconductors," in *18th ICDS Conference Proceedings*, Sendai, 1995.

[2] R. Ahrenkiel, "Minority-Carrier Lifetime in III-V Semiconductors," in *Minority Carriers in III-V Semiconductors: Physics and Applications, Semiconductors and Semimetals*, San Diego, USA: Academic Press, 1993, pp. 51–52.

[3] W. B. Shockley, W. T. Read, "Statistics of the recombination of holes and electrons," *Physical Review,* vol. 87, no. 5, pp. 835–842, Sept. 1952.

[4] W. H. Brattain, J. Bardeen, "Surface properties of germanium," *Bell Systems Technical Journal,* vol. 32, no. 1, pp. 1–41, Jan. 1953.

[5] A. Many, Y. Goldstein, N. B. Grover, Semiconductor Surfaces, 2nd ed., Amsterdam: North-Holland Publishing Company, 1965, pp. 171–173.

[6] R. F. Pierret, Advanced Semiconductor Fundamentals, 2nd ed., vol. 6, Upper Saddle River, NJ: Pearson Education Inc., 2003, p. 157.

978-1-6654-6060-6/23 $31.00 © 2023 IEEE

Flexible organic solar cells on Ti foil substrate

Huiying Jia, Lei Kerr, Benjamin Leever

Miami University, Oxford, OH, United States

Wright Patterson Air Force Base, Fairborn, OH, United States

The development of flexible solar cells is crucial in reducing costs and increasing practicality. In this work, titanium (Ti) foils was selected to be used as the conductive substrates for fabrication of inverted flexible polymer solar cells. The performance of the oxide layers on the metal foils as electron transporting layers was evaluated. We report that a power conversion efficiencies 2.5 % can be obtained by using properly polished Ti foil as solar cell substrates with a device structure of Ti/TiO2/P3HT:PCBM/thick layer PEDOT/Ag.

Novel approach to control environmental fatigue tests on glass/PV encapsulant laminates

Gabriel Riedl, Gernot M. Wallner, Robert Pugstaller

Christian Doppler Laboratory for Superimposed Mechanical-Environmental Ageing of Polymeric Hybrid Laminates, Institute of Polymeric Materials and Testing, University of Linz, Linz, Austria

Photovoltaic modules are exposed to superimposed mechanical loads and environmental influences. To mimic this service-relevant loading condition adequately, a novel fatigue, lab test methodology was developed. Double cantilever beam (DCB) specimen based on thermally toughened low iron float glass substrates and EVA or POE encapsulants were investigated under different environmental conditions. Displacement or force control of DCB fatigue delamination tests is commonly used. However, both approaches have specific disadvantages related to e.g., excessive forces at the beginning of a test and glass fracture at lower temperatures. Hence, in this paper a methodology for control of maximum strain energy release rate (SERR) was developed and implemented. User-defined SERR values were set with a control channel. The SERR was changed every 200,000 cycles or by exceeding crack length milestones. EVA laminates exhibited higher crack propagation rates than POE/glass laminates under constant SERR loading conditions. Especially in hot-humid (60°C, 80% rh) environment, POE was superior and showed no loss in delamination resistance. In contrast, EVA exhibited an order of magnitude higher crack propagation rate under hot-humid compared to hot-dry environment. Comparing SERR with displacement controlled test data revealed distinct deviations of crack propagation rates under hot-dry conditions. This was associated with high forces at the beginning of a displacement controlled test, which caused void formation and fibrillation of the encapsulant. This phenomenon was less pronounced under SERR control, which would also allow for characterization of delamination resistance of double glass specimen at temperatures well below the melting regime of the crosslinked encapsulants.

978-1-6654-6060-6/23 $31.00 © 2023 IEEE

Automated analysis of internal quantum efficiency measurements of GaAs solar cells using machine learning

Zubair Abdullah-Vetter, Priya Dwivedi, N.J. Ekins-Daukes, Thorsten Trupke, and Ziv Hameiri

University of New South Wales (UNSW), Sydney NSW 2052, Australia

Abstract— **Investigating the internal quantum efficiency (IQE) of solar cells is essential for identifying performance limitations and improving their efficiency. However, fitting IQE measurements of gallium arsenide solar cells using numerical simulation programs can be a laborious and tedious process, often limiting the depth of the analysis to only qualitative levels. In this study, we propose the use of machine learning to automate the fitting process and enable the extraction of key electrical quantities that represent the performance-limiting mechanisms of the cells. This novel method can help unlock the full potential of IQE measurements as a powerful characterization tool for further research and development of gallium arsenide solar cells.**

Keywords—GaAs cells, chain regression, spectral response, open-source

I. INTRODUCTION

The spectral response of a photovoltaic (PV) device reveals crucial information about its performance. This information is obtained by measuring the current of the device under illumination with variable monochromatic light [1]. The internal quantum efficiency (IQE) can then be determined as the ratio of the number of carriers collected to the number of photons absorbed [1]. To calculate IQE, the incident light intensity, and the wavelength-dependent current and reflection need to be measured [1]. IQE measurements are often analyzed to identify losses such as parasitic absorption in the short wavelength region or recombination losses in the long wavelength region [2]. Additionally, IQE measurements allow the determination of the device's expected short circuit current (I_{sc}) [2].

To fully understand the performance-limiting mechanisms of solar cells, IQE measurements are fit using different mathematical expressions [3]. To accomplish this, researchers can use simulation software such as PC1D [4], Griddler [5], SolCore [6], and others [7][8]. While these tools can provide valuable insights, the process can be time-consuming as it requires extensive knowledge and educated assumptions about a large number of input parameters [4]. Nevertheless, the ability to accurately analyze and understand these performance-limiting mechanisms is crucial for the continued development of efficient solar cell technologies.

This work was supported by the Australian Government through the Australian Renewable Energy Agency (ARENA, Grant 2020/RND016). The views expressed herein are not necessarily the views of the Australian Government, and the Australian Government does not accept responsibility for any information or advice contained.

There has been significant development of various methods for extracting loss parameters from IQE measurements of silicon (Si) cells [2][9]. However, the application of these methods to gallium arsenide (GaAs) cells has been limited. Therefore, researchers often utilize qualitative reporting such as the expected losses and relative comparisons of IQE measurements [10]. If quantitative data is provided alongside the IQE measurements, they are only related to the calculation of I_{sc} and current losses [11].

In this study, we propose a machine learning (ML) approach to obtain key electrical parameters that impact the performance of GaAs solar cells. The aim of this approach is to eliminate the need for manual manipulation of various parameters in simulation programs. This automated approach will enhance the understanding of the underlying mechanisms and unlock the full potential of IQE measurements of GaAs cells.

II. . METHODOLOGY

A. Dataset

A dataset of 20,000 IQE curves was generated using the Python program SolCore [6]. A default GaAs solar cell structure was simulated with a front $Al_{0.83}Ga_{0.17}As$ window (29.3 nm), p-type emitter (660 nm), n-type base (3,600 nm), and $Al_{0.3}Ga_{0.7}As$ rear buffer layer (1,000 nm) [11]. The reflectivity profile of this structure with a dual-layer ZnS/MgF_2 (50; 100 nm) antireflection coating (ARC) [11] was also applied in the simulation. A diagram of the structure is shown in Fig. 1.

Randomized combinations of the minority carrier diffusion lengths in the emitter (L_p) and base (L_n) as well as surface recombination velocities (S_p and S_n) were used to simulate a large variety of IQE measurements of GaAs cells. Table 1 summarizes the range of each of the simulated parameters. The effective diffusion lengths ($L_{p, eff}$ and $L_{n, eff}$) were also calculated [9]:

$$L_{eff} = L \cdot \frac{1 + S \cdot L/_D \cdot \tanh(\frac{W}{L})}{S \cdot L/_D + \tanh(\frac{W}{L})}, \qquad (1)$$

where L is the diffusion length and S is the recombination velocity of the minority carriers in a layer, W is the width of the layer (emitter or base) of interest, and D is the diffusion coefficient of the minority carriers. Combinations of L and S that resulted in cases where $L_{eff} > L$ were removed from the dataset.

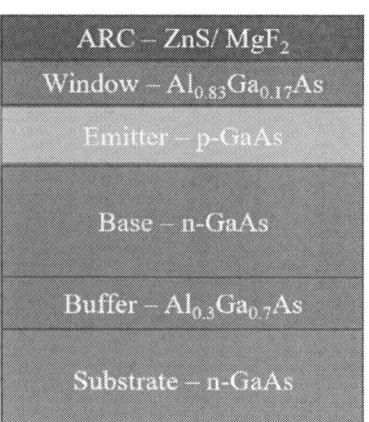

Fig. 1. A diagram of the default structure used in the simulations. The width of the layers in this diagram does not represent the actual thicknesses of the layers used.

TABLE 1. THE RANGE OF SIMULATION PARAMETERS

Parameter	Minimum	Maximum
L_p (um)	0.9	10
L_n (um)	3.6	10
S_p (cm/s)	29	1,000
S_n (cm/s)	2.2	1,000
$L_{p,\,eff}$ (um)	0.9	9
$L_{n,\,eff}$ (um)	3.6	9.3

B. Machine Learning

The developed ML method uses an array of random forest (RF) [12] models that have been trained to predict each of the key parameters. The RF models were trained in a specific sequential order such that the predictions of the first model were passed as extra training features for the training of the next model. This procedure, known as chain regression [13], allows the models to leverage any correlations between the parameters. As a result, the chain of models performed better than a single multi-output regression model. As the chain order affects the prediction performance, optimization is required. For the results shown here, the order was chosen based on the performance of the parameters from individual RF models, with the worst-performing parameter placed last, i.e. L_p, S_n, S_p, and L_n.

Prior to training, the dataset was split into training (80%) and testing (20%) sets. The discrete differences in the IQE values between each 10 nm wavelength step were added as extra features, which were found to improve prediction results. The training set was further split into four folds such that the four models of the chain regression were trained on separate randomized subsets of data. To evaluate the model capabilities, the test dataset was passed to the chain regressor, and the true and predicted parameters were compared. The performance was evaluated using the root mean square error (RMSE) formula [13]:

$$\text{RMSE}(y, \hat{y}) = \sqrt{\frac{1}{N} \sum_{i=1}^{N} (y_i - \hat{y}_i)^2}, \qquad (2)$$

where N is the number of samples, y_i is the true value of the ith sample, and $\hat{y}_i$ is the corresponding predicted value. The lower the RMSE value, the better the prediction, where zero indicates a perfect prediction across the entire test dataset.

III. RESULTS

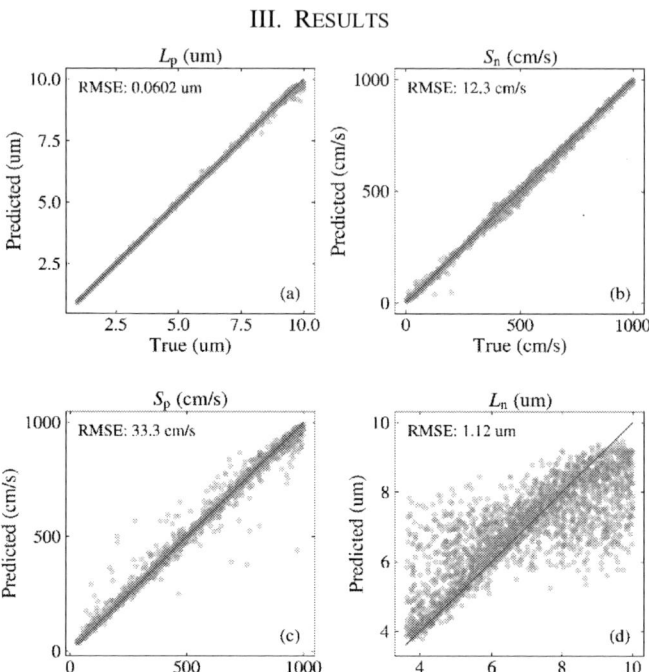

Fig. 2. The chain regression results are displayed as predicted vs true plots in the chain order of the regression model: (a) L_p, (b) S_n, (c) S_p, and (d) L_n.

The initial results of the four parameters are shown in Fig.2. As demonstrated, the chain regressor was able to predict L_p, and S_n with high accuracy (RMSE of 0.06 um and 12.3 cm/s, respectively). The predictions for S_p have higher errors (RMSE of 33.3 cm/s) while the predictions for L_n need to be improved (RMSE of 1.12 um). Nonetheless, the ML approach has been successful in accurately predicting three out of four key electrical parameters.

The errors in L_n were further investigated. It was found that in many cases, the provided fits of the IQE curve by the incorrect parameters are still very good. It seems that in these cases, a few combinations of parameters provide similar quality of fits – see Fig. 3 for an example (the true and predicted parameters are listed in Table. 2). Therefore, the ML model struggles to properly learn the differences in these cases. Identifying these cases and simulating a larger proportion of them within the training dataset may assist in improving the prediction. Other cases that have caused errors are when large changes in L_n do not significantly impact the IQE curve. These cases are more prevalent at values of $L_n \gg W_n$ (width of the base). In these cases, the ML model struggles again to differentiate the impact of L_n. The chain regression method will allow a user to provide expected parameter values, which could then help the other models in the array to predict more accurate parameters. Results using these approaches will be presented at the conference.

The errors in S_p can also be explained using Fig. 3. In this case, S_p seems to have a much smaller effect on the short wavelength region such that the IQE is almost the same. These cases were found to be occurring more when $L_p \gg W_p$ (thickness of the emitter). A larger dataset with more of these cases can be included to improve the ability of the ML models to predict S_p.

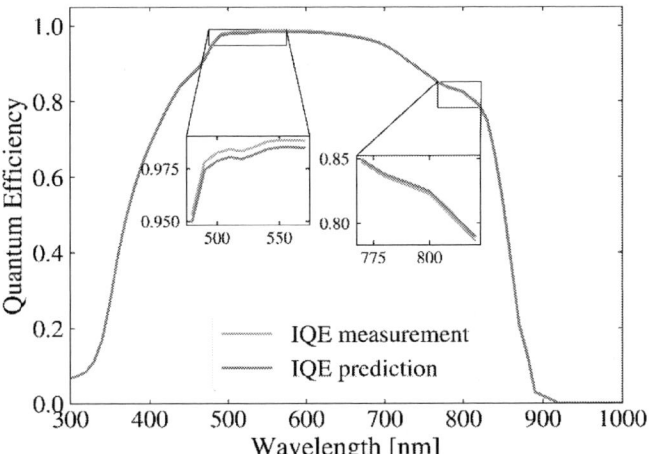

Fig. 3. An example where two sets of parameters produce very similar IQE curves. The insets highlight how similar the IQE curves are as a result of different combinations of $L_{p/n}$ and $S_{p/n}$.

TABLE 2. THE TRUE AND PREDICTED PARAMETERS OF THE EXAMPLE IN FIG. 3

Parameter	True	Prediction
L_p (um)	5.10	5.13
L_n (um)	3.93	7.63
S_p (cm/s)	54.85	229.3
S_n (cm/s)	662.06	631.02

IV. CONCLUSION

This study presents a novel use of ML to automatically analyze IQE measurements of GaAs solar cells. A chain of RF regression models was trained to extract four key electrical parameters from the measurements. Despite the early stage of development, the trained model achieved low RMSE scores for three of the four parameters, demonstrating that the model learned essential features of IQE measurements. The method can replace manual fitting when extracting performance-limiting mechanisms of GaAs cells, allowing for a more quantitative assessment of their development. By applying this ML method to automate the analysis of IQE measurements, their full potential can be unlocked.

REFERENCES

[1] J. S. Hartman and M. A. Lind, "Spectral response measurements for solar cells," *Solar Cells*, vol. 7, no. 1, pp. 147–157, 1982.

[2] B. Fischer, "Loss analysis of crystalline silicon solar cells using photoconductance and quantum efficiency measurements," Ph.D. Thesis, Konstanz University, 2003.

[3] W. J. Yang, Z. Q. Ma, X. Tang, C. B. Feng, W. G. Zhao, and P. P. Shi, "Internal quantum efficiency for solar cells," *Solar Energy*, vol. 82, no. 2, pp. 106–110, 2008.

[4] P. A. Basore, "Numerical modeling of textured silicon solar cells using PC-1D," *IEEE Transactions on Electron Devices*, vol. 37, no. 2, pp. 337–343, 1990.

[5] J. Wong, "Griddler: Intelligent computer-aided design of complex solar cell metallization patterns," in *39th IEEE Photovoltaic Specialists Conference*, 2013, pp. 0933–0938.

[6] D. Alonso-Álvarez, T. Wilson, P. Pearce, M. Führer, D. Farrell, and N. Ekins-Daukes, "Solcore: a multi-scale, Python-based library for modeling solar cells and semiconductor materials," *Journal of Computational Electronics*, vol. 17, no. 3, pp. 1099–1123, 2018.

[7] "SunSolve™." https://www.pvlighthouse.com.au/sunsolve (accessed Mar. 08, 2022).

[8] A. Fell, "A free and fast three-dimensional/two-dimensional solar cell simulator featuring conductive boundary and quasi-neutrality approximations," *IEEE Transactions on Electron Devices*, vol. 60, no. 2, pp. 733–738, 2013.

[9] P. A. Basore, "Extended spectral analysis of internal quantum efficiency," in *23rd IEEE Photovoltaic Specialists Conference*, 1993, pp. 147–152.

[10] V. Raj, T. Haggren, J. Tournet, H. H. Tan, and C. Jagadish, "Electron-selective contact for GaAs solar cells," *ACS Appl. Energy Mater.*, vol. 4, no. 2, pp. 1356–1364, 2021.

[11] S. P. Tobin *et al.*, "Assessment of MOCVD- and MBE-growth GaAs for high-efficiency solar cell applications," *IEEE Transactions on Electron Devices*, vol. 37, no. 2, pp. 469–477, 1990.

[12] L. Breiman, "Random Forests," *Machine Learning*, vol. 45, no. 1, pp. 5–32, 2001.

[13] F. Pedregosa *et al.*, "Scikit-learn: machine learning in python," *Journal of Machine Learning Research*, vol. 12, pp. 2825–2830, 2011.

978-1-6654-6060-6/23 $31.00 © 2023 IEEE

RF-Powered sputtering of iron pyrite for photovoltaic applications

*Awais Zaka [1], Ayman Rezk [1], Sabina Abdul Hadi [2], Saeed Alhassan [1], and Ammar Nayfeh [1, *]*

[1]Khalifa University, Abu Dhabi, 127788, UAE

[2]University of Dubai, Dubai UAE
[*]*Corresponding Author: ammar.nayfeh@ku.ac.ae*

Abstract—In this work, 2-Dimensional FeS$_2$ thin films were deposited on different substrates through a plasma-assisted, radio frequency (RF)-powered sputtering method intending to analyze the properties for photovoltaic applications. The FeS$_2$ films were characterized using several techniques. The as-prepared thin films were then tested for a solar device. Pyrite films showed excellent absorption in UV/Vis range, with the Tauc's plot calculations indicating a band gap of 1.2 eV. Moreover, solar device fabrication and analysis showed presence of photo-generated current, indicating a suitable material for photovoltaics.

I. INTRODUCTION

Iron pyrite, also known as fool's gold, is a naturally occurring iron disulfide (FeS$_2$) mineral with a metallic lustre and pale brass-yellow colour. It belongs to the family of transition metal dichalcogenides (TMDs), which have been a subject of a lot of research in various fields [1]. Though extensively used in commercial processes to produce iron and sulfur-based compounds, recent studies have shown that iron pyrite has the potential as a low-cost and efficient material for use in photovoltaic cells owing to its several suitable properties [2]. It is abundant and widely available, making it relatively inexpensive compared to other materials used in photovoltaic cells [3]. In addition, iron pyrite has a relatively high absorption coefficient, which means it can absorb a significant amount of sunlight. Finally, iron pyrite has a suitable bandgap, which can efficiently convert absorbed sunlight into electricity.

Despite these promising properties, there are also challenges to using iron pyrite in photovoltaic cells [4, 5]. The performance of pyrite in photovoltaics has remained poor, and though several theories have been presented in this aspect, the answer to why remains elusive. This allows researchers to explore this abundantly available resource.

Sputter deposition is a thin-film technique that involves bombarding a target material with high-energy ions. This process results in the sputtering or ejection of atoms from the target material, which are then deposited onto a substrate to form a thin film [6]. In this work, we have studied the properties and performance of pyrite thin films fabricated through RF-powered sputtering techniques.

II. METHODOLOGY

To deposit iron pyrite using sputter deposition, a target was custom-made and purchased from AJA Inc. U.S.A with a purity of 99.99% and placed in a sputter deposition system. The target is then bombarded with high-energy ions, which cause atoms of iron pyrite to be ejected from the surface. These atoms are then deposited onto a substrate, forming a thin film of iron pyrite. The target and the schematic diagram of the sputtering technique are shown in Fig. 1.

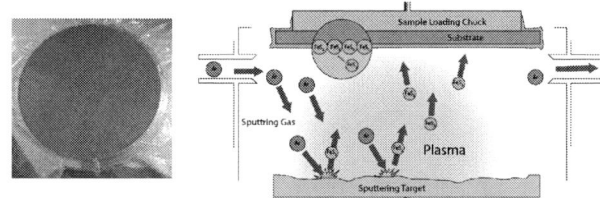

Fig 1. FeS$_2$ target used for sputter deposition

Three different kinds of substrates (p-type Si, Si with 300 nm thermally grown silicon oxide layer, and fused silica glass) were used in this study for characterization and analysis. The substrates were first cleaned thoroughly by sonicating them in acetone. The deposition rate and growth can be controlled through various parameters such as temperature, the pressure inside the chamber and the RF power. This work prepared and characterized several films with varying deposition times. A solar device was then fabricated using a FeS$_2$ layer deposited on a p-type silicon wafer (boron doped 1-10 Ω-cm) with gold contacts to measure the solar performance.

A solar device was fabricated by adding gold contacts at the top through physical vapour deposition (PVD). The step-by-step schematic of the solar device is shown in Fig 2.

978-1-6654-6060-6/23 $31.00 © 2023 IEEE

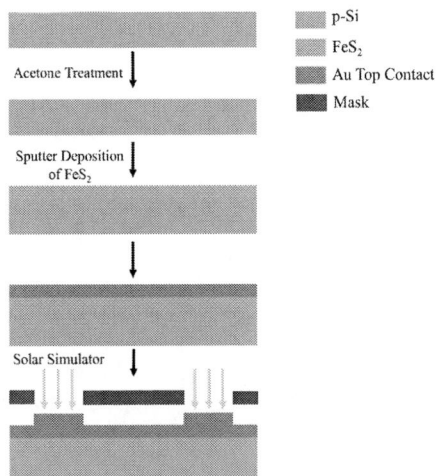

Fig 2. Schematic of FeS₂/Si solar cell preparation (*not to scale*). (*Si is boron doped 1-10 Ω-cm*)

III. CHARACTERIZATION

a) Variable Angle Spectroscopic Ellipsometry

Spectroscopic ellipsometry (SE) is widely used to characterize surfaces, interfaces, and multilayer structures non-destructively and non-invasively. It can offer precise measurements of the optical constants of the materials over a more comprehensive wavelength range from ultraviolet (UV) to infrared (IR). Essential parameters which can be obtained through variable angle spectroscopic ellipsometry (VASE) are film thickness, refractive index (n) and extinction coefficient (k). The first and foremost step in VASE is to develop an ellipsometry model for thin films. In this case, a b-spline model with activated Kramer's-Kronig mode, was used to fit the data and estimate layer thickness.

b) UV/Vis Spectroscopy

UV-vis Spectroscopy is another important technique that helps determine the absorbance for different film thicknesses over a range of wavelengths. The transmission and absorbance data for various film thicknesses (directly related to deposition time) are shown in Fig. 3. Measurements were obtained using LAMBDA 1050 UV/Vis/NIR spectrophotometer (Perkin Elmer Inc.).

c) Atomic Force Microscopy

Atomic Force microscopy was used in this work for two primary purposes. Firstly, it complements ellipsometry in terms of determining the thickness of the films and developing the

optical model. Secondly, the AFM is an excellent technique for calculating the mean roughness of the film and the topography. On deposited samples, during the FeS₂ film deposition, a step was created by masking some part of the substrate to develop a distinct topographical characteristic, which was then used during AFM measurements to estimate film thickness. Step on a sample and AFM measurements are shown in Fig. 4.

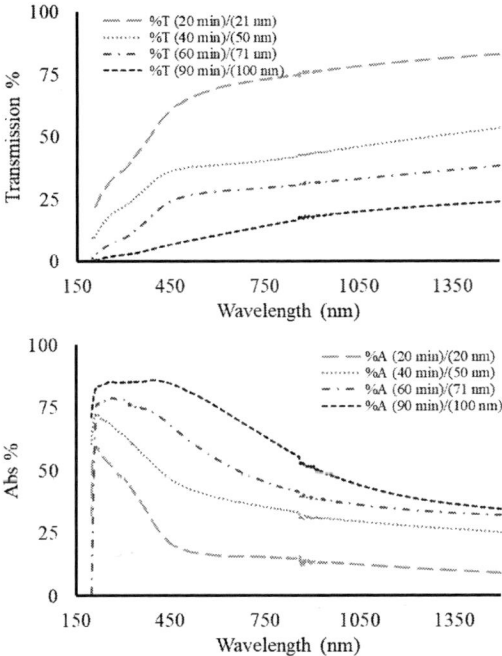

Fig 3. Effect of film thickness on the (a) transmission and (b) absorbance of light through pyrite thin films

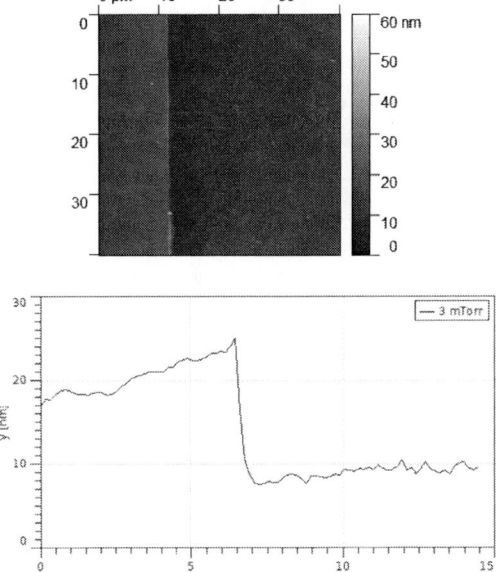

Fig 4. (Top) AFM image of the step formation on a sample (Bottom) Measurement of the step height of the sample prepared at three mTorr.

Other conditions: Room Temperature, 70 W RF, Deposition Time = 60 min.

IV. RESULTS AND DISCUSSION

The growth rate of pyrite film was obtained by controlling the deposition time. The growth showed a linear trend for deposition time, as shown in Fig. 6, where a ~100 nm film thickness was achieved after 90 minutes of deposition. The growth rate was found to be approximately 1.1 nm/min.

The absorbance data collected from UV/Vis spectrometer was used to establish Tauc's plot, assuming the estimated thickness from AFM and Ellipsometry measurements. Absorption coefficient was first determined using Beer Lambert's Law (Equation 1 and 2) and Max Planks Equation (Equation 3) [7].

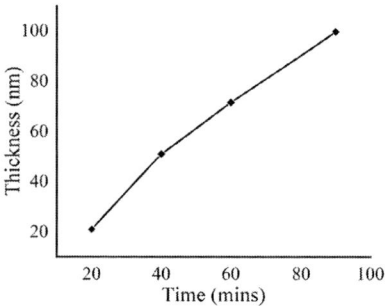

Fig. 5. Effect of film thickness w.r.t time. Other deposition conditions: Room Temperature, 70 W RF Power, Pressure = 3 mTorrs

$$\left(\frac{I}{I_o}\right) = e^{-\alpha d} \tag{1}$$

Where I is the intensity of the incident light, I_o is the intensity of transmitted light, d is the film thickness and α is the absorption coefficient. Simplifying eq. 1 we get

$$\alpha = 2.303 \frac{A}{d} \tag{2}$$

Where A is the measured absorbance for UV-Vis spectrometric measurements.

$$E_g = \frac{1240}{\lambda} \; (eV) \tag{3}$$

Where E_g is the bandgap and λ is the wavelength (in nm). Finally, Tauc's Equation is used for the measurement of direct band gap [2].

$$(\alpha h\nu)^m = K(h\nu - E_g) \tag{4}$$

where h is the Planck's constant, K is the energy independent constant and m is the nature of transition. The value of m used was 2 for direct band gap calculation [2]. The measured band gap showed a value of around 1.2 eV which was in good agreement with previously reported values [2].

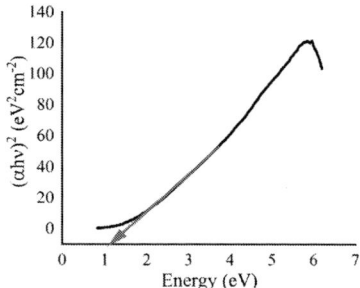

Fig. 6. Tauc's Plot for pyrite thin film

The final structure of the solar device is shown in Fig. 8. A solar simulator (Sol3a 94123A) was used to test the performance of the fabricated solar cell.

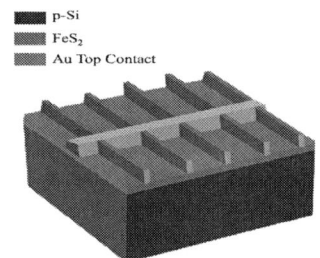

Fig 8. Configuration of solar cell

The photosensitivity was tested, and the current is measured with and without light, shown in Fig 9. Initial studies revealed polarity changes in while measurement, the work about that is in progress for further analysis. The increase is due to the significant absorption of light and the EHP generation in the FeS₂ layer. The solar performance and

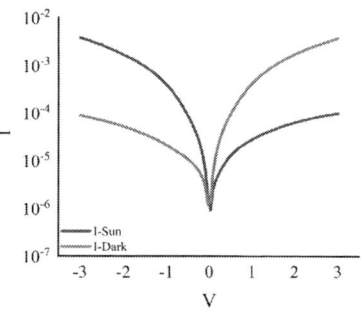

Fig 9. J-V characteristics of FeS₂ thin film solar cell under one sun and in the dark.

978-1-6654-6060-6/23 $31.00 © 2023 IEEE

V. CONCLUSION

Sputter deposition of iron pyrite can be used to produce thin films for various applications, such as photovoltaics or electronics. This work was desiccated to explore the growth of pyrite thin films and their photovoltaic applications. The sputtering growth rate was around 1.1 nm/min and the pyrite thin films showed a direct band gap of around 1.2 eV. The results demonstrated suitable photosensitivity and properties. As such, further research and development are needed to optimize the use of iron pyrite in sputter deposition and to overcome these challenges.

REFERENCES

[1] S. Khalid, E. Ahmed, Y. Khan, K. N. Riaz, and M. A. Malik, "Nanocrystalline pyrite for photovoltaic applications," *Chemistryselect,* vol. 3, no. 23, pp. 6488-6524, 2018.

[2] A. Zaka, S. M. Alhassan, and A. Nayfeh, "Iron Pyrite in Photovoltaics: A Review on Recent Trends and Challenges," *ACS Applied Electronic Materials,* vol. 4, no. 9, pp. 4173-4211, 2022/09/27 2022, doi: 10.1021/acsaelm.2c00489.

[3] C. Wadia, A. P. Alivisatos, and D. M. Kammen, "Materials availability expands the opportunity for large-scale photovoltaics deployment," *Environmental science & technology,* vol. 43, no. 6, pp. 2072-2077, 2009.

[4] M. Z. Rahman and T. Edvinsson, "What Is Limiting Pyrite Solar Cell Performance?," *Joule,* vol. 3, no. 10, pp. 2290-2293, 2019.

[5] G. Kaur, M. Kaur, A. Thakur, and A. Kumar, "Recent Progress on Pyrite FeS 2 Nanomaterials for Energy and Environment Applications: Synthesis, Properties and Future Prospects," *Journal of Cluster Science,* pp. 1-39, 2019.

[6] H. Liu and D. Chi, "Magnetron-sputter deposition of Fe3S4 thin films and their conversion into pyrite (FeS2) by thermal sulfurization for photovoltaic applications," *Journal of Vacuum Science & Technology A: Vacuum, Surfaces, and Films,* vol. 30, no. 4, p. 04D102, 2012.

[7] H. Qin, J. Jia, L. Lin, H. Ni, M. Wang, and L. Meng, "Pyrite FeS2 nanostructures: Synthesis, properties and applications," *Materials Science and Engineering: B,* vol. 236, pp. 104-124, 2018.

Analysis and identification of measurement uncertainty sources of a LED Sun Simulator with double-side illumination for bifacial PV Module Power Rating

Sebastian Dittmann[1,2], Giuliano L. Martins[1,3], Ralph Gottschalg[1,2]

[1]Hochschule Anhalt University of Applied Sciences, Bernburger Str. 55, 06366 Köthen, Germany
E-mail: sebastian.dittmann@hs-anhalt.de, Tel.: +49 3496 67 2354
[2]Fraunhofer-Center for Silicon Photovoltaics CSP, Halle, Germany
[3]Statkraft UK Ltd., London, United Kingdom

Abstract — **This paper gives an understanding of measurement uncertainty sources of LED sun simulators beyond uncertainty calculation for pulsed Xenon lamp systems. Technology-specific uncertainty sources will be identified, and an uncertainty budget calculation of a single-side and a double-side LED solar simulator will be presented. Results show significant differences between the two light sources. The expanded uncertainty (k=2) for power measurements at 1000 W/m² (front) and 200 W/m² (rear) is ±8.4%. Due to the 648 individual powered LED channels of each light source, the spectral distribution, the spatial non-uniformity, and the light stability during the I-V sweep are some of the main contributors to the expanded uncertainty.**

I. INTRODUCTION

Energy yield estimation and bankability of large Photovoltaik (PV) systems rely on precise input parameters such as the nominal power of the PV modules. PV modules should be measured with low measurement uncertainty to reduce financial risk. However, module technologies with a growing market share, e.g., bifacial PV modules, have new challenges in obtaining the nominal power. Due to the active rear-side of the module, new measurement methods and devices are required. The IEC 60904-1-2 [1] describes how the voltage-current (I-V) curves should be measured. The standard divides into two measurement possibilities. The first one is the Ge Method, where the measurement is performed using a sun simulator with one light source. This might not represent the physical behavior of a bifacial PV module due to the different properties of the module. The second method double-sided (DS) method, describes a measurement with a solar simulator with two independent light sources or a reflective surface to simulate the rear-side irradiance. Different spectra or different spatial non-uniformity could have a significant influence on the measurement uncertainty.

Novel devices, e.g., new sun simulator concepts with light sources based on light-emitting-diodes (LED), are introduced to the market. Due to the high number of individual LED channels, they offer high flexibility in spectral distribution. This opens new possibilities, such as measurement of the rear-side of the bifacial module beyond AM1.5g for different Albedo scenarios. However, it also provides new uncertainty sources due to the individual controlled LED channels which might

strongly influence the spectral distribution, the spatial uniformity across the plane of the PV module and the light stability during the I-V sweep.

Uncertainty calculation for pulsed Xenon based sun simulators has been continuously improved [1]-[5]. These systems are mainly tunnel-type systems with low or no diffuse light. Measurement uncertainties of accredited labs are for Power (Pm) 1.7-1.9% with spectral mismatch correction (MMF) and 2,5-3,6% without MMF correction [6]. LED-based sun simulator, which provides light with high diffuse components and multiple controlled LED channels, no proper uncertainty calculation for single light sources nor double-side light sources has been reported so far.

Our paper analyzes and identifies measurement uncertainties of a LED sun simulator with double-side illumination for bifacial PV modules. This will give an understanding of measurement uncertainty sources of LED sun simulators beyond uncertainty calculation for pulsed Xenon lamp systems (tunnel or table-top type) and the identification of technology-specific uncertainty sources for LED sun simulators.

II. METHODOLOGY

A. Measurement Equipment

This paper describes a pulsed-colored LED (light emitting diode) based sun simulator with two independent light sources used at the Anhalt Photovoltaic Performance and Lifetime Laboratory (APOLLO). The dimensions of each light source are 1150 m × 2100 mm. The main light source (further described as front-side light source) consists of 18 boards with 428 LEDs each. Each board has 36 individual powered LED channels. The total number of LEDs is 7704 for the front-side light sources. The secondary light source (further described as rear-side light source, rear-side light source) conciced of 18 boards with 296 LEDs each. Each board has 36 individual powered LED channels. The total number of LEDs is 5327 for the resr-side light sources. According to the simulator datasheet, both light sources have a Class AAA according to the

978-1-6654-6060-6/23 $31.00 © 2023 IEEE

TABLE I
MANUFACTURER'S SPECIFICATIONS

	Front-side	Rear-side
Dimensions of light source [mm]	1150 x 2100 mm	
Light source Individual LED Channels	18 board with 428 LEDs each, 7704 LEDs in total	18 board with 296 LEDs each, 5328 LEDs in total
No. of LED-channels	each board has 36 individual powered LED channels	
Cooling system	Active at 20°C	passive
Edge of light source	reflected mirrors	
Distance to PV module	25 cm	
Spectral range [nm]	350 – 1100 nm	400 – 1100 nm
Irradiance range [W/m²]	100 - 1300	50 - 1000
Maximum flash length [ms]	500 ms without cooling, continuously with cooling	100 ms*
Irradiance Sensor	Not available	
IEC Classification	A+AA+	
Temperature back of module sensor	one PT1000	
	Electronic load	
Measurement range current [A]	32A	
Measurement range voltage [V]	200 V	
Maximum Power [W]	500 W	

* Limitation in case of bifacial PV module measurement with both light sources

IEC 60904-9 classification at 1000 W/m² with pulse lengths of 500ms and 100ms for the front- and rear-side light source, respectively.

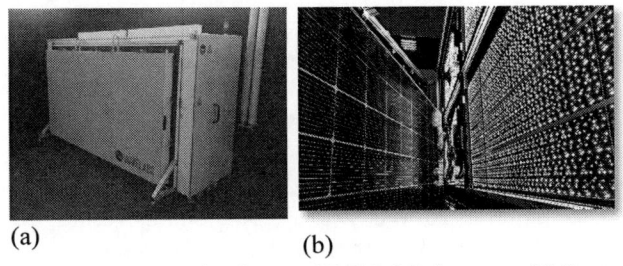

Fig. 1. (a) LED sun simulator at APOLLO Laboratory, (b) front side light source with LED boards

Furthermore, with an active cooling system, the font-side can be operated as a continuous light source. However, if the simulator is used for double-side measurements of bifacial PV modules, the flash duration is limited to the rear-side light source at 100 ms. The distance between the LED boards and the module plan is 25cm. Due to the high number of individual LED channels, the spectrum can be adjusted flexibly, and measurements beyond AM1.5g can be performed. Fig. 1 shows the simulator and the LED boards, and Table I shows the technical specification stated by the manufacturer.

The simulator has a passive electronic load (two quadrants) which measured the voltage-current (I-V) curve with four wires in a maximum range of 200V and 32A, respectively. Measurements are performed in a controlled environment with an ambient temperature of $25 \pm 2°C$. A PT1000 on the back side of the PV Module measures the module temperature. In the case of double-side measurement, the temperature sensor will be removed before the measurement. Since the simulator came without an irradiance measurement system, an irradiance measurement for both light sources has been implemented. It consists of calibrated WPVS (World Photovoltaic scale) reference cells connected via a trans-impedance amplifier. A correction routine has been implemented to correct I-V curves to the target irradiance.

With this measurement system, we perform STC measurements for monofacial PV modules according to IEC 60904-1 and performance measurements for bifacial PV modules according to IEC 60904-1-2 (Ge-method and double-side illumination). In an international round-robin (RR), the system and the measurement procedure have been compared with other well-known institutes. The RR shows comparable results within the measurement uncertainties [7].

III. EVALUATION OF SPECIFICATION

For this study, no further improvement has been made, e.g., improvement of the spectral and spatial uniformity or the temporal light stability. The simulator has been characterized with different state-of-the-art measurement technics to analyze and identify device-specific uncertainties and thus to understand the importance of measurement uncertainties.

Fig. 2. shows an overview of the identified uncertainty sources of the simulator as reported in the literature [1]-[5]. The main contributors are grouped into six categories with detailed contributors. In the first step, each light source must be analyzed, and the uncertainty budget has to be calculated individually. In the case of a double-side measurement, uncertainty sources might be influenced if two light sources are used simultaneously. These are, e.g., the spectrum and the spatial non-uniformity due to internal reflection or the introduction of a second reference cell. The yellow arrow represents the contribution of the rear-side light source in the case of double-side illumination.

On the other hand, some contributors may remain the same and not influence the overall uncertainty if two light sources are used simultaneously. Here indicated in green. Contributions include the data acquisition system, PV module temperature, or module properties, e.g., capacitive effects.

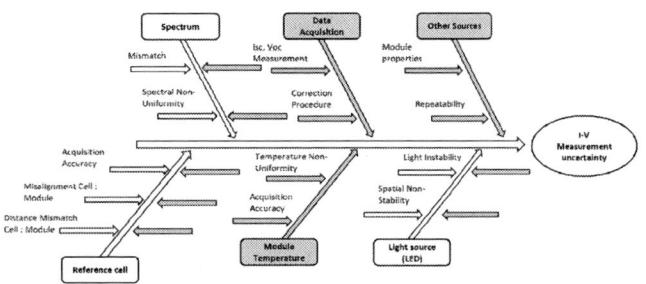

Fig. 2. Ishikawa diagram illustrating the identified uncertainty sources. The yellow arrow represents the contribution of the rear-side light sours in case of double-side illumination.

The following sub-sections show results for the spectrum's characterization and classification, the spatial non-uniformity, and the light temporal instability.

A. Spectrum

Fig. 3. Shows the measured spectrum of the front- and rear-side of the LED sun simulator compared to the spectrum of a Xenon base sun simulator and the standard spectrum AM1.5g, as well as the spectral response of the front- and rear-side of a bifacial PV module. In contrast, Table II shows the measured spectral match and corresponding IEC classification. It is visible that the measured values do not match the manufacturer's stated class A spectrum. Significant differences are in the higher wavelength range with up to class C. Measurements at different locations show variations in different wavelengths of the spectrum. It might be due to the individually powered LED channels, and this results in a spectral non-uniformity across the module plane.

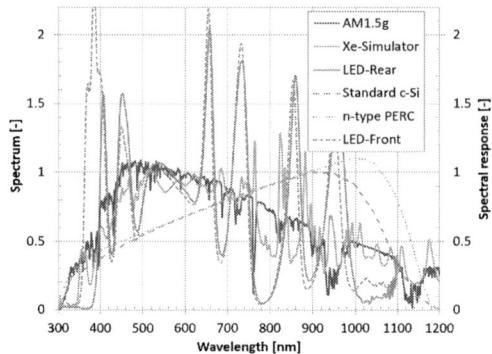

Fig. 3. Spectrum of Xenon and LED sun simulators, AM1.5 and spectral response of c-Si modules

TABLE II
MEASURED SPECTRAL MATCH AND CORRESPONDING IEC CLASSIFICATION

Spectral range [nm]	Front side		Rear side	
	Spectral match	IEC Class	Spectral match	IEC Class
400-500	0.70	B	0.88	A
500-600	0.79	A	0.82	A
600-700	0.83	A	0.84	A
700-800	0.74	B	0.79	A
800-900	0.56	C	0.72	B
900-1100	0.45	C	0.99	A

B. Spatial Non-uniformity

The spatial non-uniformity has been analyzed with two different measurement approaches. 1) with a uniformity module that consists of 72 cells with independent Isc measurements. Two different uniformity modules of the same type, both calibrated at the same institute. 2) a calibrated WPVS reference cell (smaller detector, 2x2cm) has been used. The reference cell was moved stepwise, and single flashes were performed. A second reference at a fixed position was used to guarantee stability and the level of irradiance.

Fig. 4. shows the results of the measured spatial uniformity with a uniformity module (a) measurement of the front-side light source by the manufacturer, (b and c) show the front-side and rear-side light source measurements be APOLLO, respectively. The measurement of the rear-side light source by the manufacturer is not available. Even if the measurement has been performed with the same type of uniformity module calibrated at the same institute, the results show a significant difference.

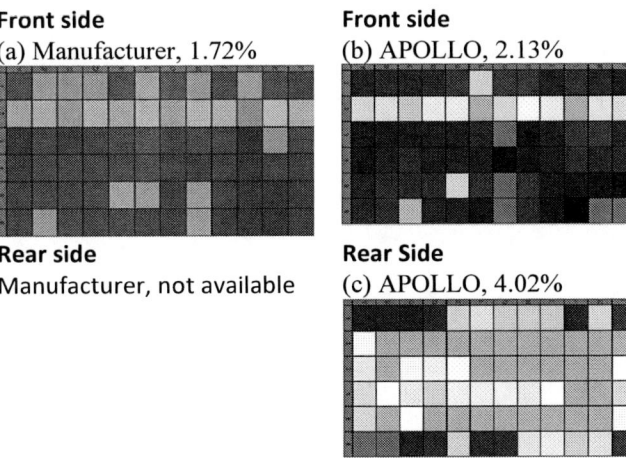

Front side
(a) Manufacturer, 1.72%

Front side
(b) APOLLO, 2.13%

Rear side
Manufacturer, not available

Rear Side
(c) APOLLO, 4.02%

Fig. 4. Visualization of the measured spatial uni-formity with a uniformity module, (a) measurements of the front-side light source by the manufacturer, (b and c) measurements of the front-side and rear-side light source be APOLLO, respectively.

Table III shows the spatial non-uniformity measured with a uniformity module. The measured values significantly differ from the manufacturer's, especially for low irradiance and the rear-side light source. The comparison to the measurement with the reference shows a significantly higher value for the front- and rear-side at 1000W/m² of up to 3.4% and 5.9%, respectively.

The measurement procedure to measure spatial non-uniformity must be well-defined and evaluated with a separate measurement uncertainty calculation. This uncertainty might have a higher impact on the overall measurement uncertainty if a module with smaller cells, e.g., a half-cells is used. However, the authors believe that the high values result from the high number of individual-powered LED channels. Future work is a detailed analysis of the spatial non-uniformity. Preliminary results show improvement by individual adjustment of single LED channels.

TABLE III
MEASURED SPATIAL UNIFORMITY WITH A UNIFORMITY
MODULE, VALUES IN %

Irradiance [W/m²]	Front side		Rear side
	Manufacture	APOLLO	APOLLO
1000	1.72	2.13	4.02
800	1.59	2.41	3.02
600	1.63	2.66	3.58
400	1.54	2.44	3.98
200	2.33	4.64	3.89
100	2.33	12.46	6.66

C. Light Temporal Instability

Fig. 5. Shows the stability of the two light sources. The rear-side has a much higher scattering than the front-side. However, for the rear-side, the flash possesses a linear pattern throughout the 100 ms compared to the stationary one, while the front-side

decreases until 80 ms. The longterm instability at 1000W/m² for the front- and rear-side are 1.15% and 0.45%, respectively. According to the IEC 60904-9, both simulators would be class B. As shown in TABLE IV the longterm instability increases with lower irradiance. For irradiances lower than 400 W/m², result in class C.

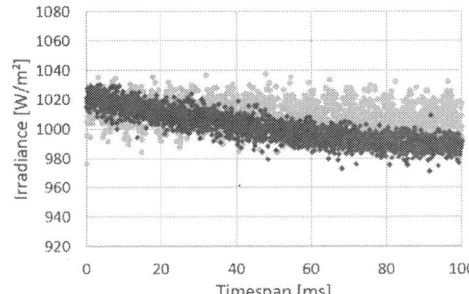

Fig. 5. Comparison of the front-side (blue) and rear-site (yellow) light sources.

TABLE IV
LONGTERM INSTABILITY (LTI)

Spectral range [nm]	Front side		Rear side	
	Average	Std.Dev	Average	Std.Dev
1100	0.73%	0.30%	-	-
1000	1.15%	0.50%	0.43%	0.03%
800	0.80%	0.19%	0.40%	0.08%
600	1.09%	0.10%	0.46%	0.02%
400	1.23%	0.05%	0.65%	0.12%
200	2.08%	0.47%	1.36%	0.12%
100	3.89%	0.71%	2.73%	0.29%

D. Expanded Uncertainty

Table V shows in detail the uncertainty sources and related uncertainty budgets for both the front- and the rear-side light source at 1000W/m² and for the double-side with 1000W/m² for the front-side and 200W/m² for the rear-side.

The spatial non-uniformity of both light sources has the highest contribution to the overall uncertainty budget. The spectral non-uniformity across the module plain might be another contributor, which needs future investigation. However, Table VI shows the expanded uncertainty (k=2) for both light sources at 1000W/m² and the double-side measurement. The expanded uncertainty for the front-site or monofacial measurement for Pm is ±3.5%, for the rear-side ±6.7%, and for the double-side at 1000/200W/m², it results in ±8.4%. The IEC classification for the front side is Class CBB, and for the rear-side it is BBB. For low irradiance, e.g., 200W/m², the classification is for front- and rear-side are CBC and BBC, respectively.

TABLE VI
UNCERTAINTY SOURCES RELATED TO SOLAR SIMULATOR'S UNCERTAINTY BUDGET

Standard uncertainty component	Source	Distribution	Front-side @1000 W/m²	Rear-side @1000 W/m²	Double-sided @1000/200 W/m²
Data Acquisition					
Data acquisition error - voltage	Manufacturer	Gaussian	± 0.03 %	± 0.03 %	± 0.03 %
Data acquisition error - current	Manufacturer	Gaussian	± 0.03 %	± 0.03 %	± 0.03 %
Light Source (LED)					
Front side - non-stability	Measurement	Rectangular	± 1.15 %	-	± 2.08 %
Front side - spatial non-uniformity	Measurement	Rectangular	± 2.42 %	-	± 3.64 %
Rear side - non-stability	Measurement	Rectangular	-	± 0.43 %	± 0.43 %
Rear side - spatial Non-Uniformity	Measurement	Rectangular	-	± 5.00 %	± 5.00 %
Module Temperature					
Acquisition accuracy	Manufacturer	Rectangular	± 0.34 %	± 0.34 %	± 0.34 %
Temperature non-uniformity	[4]	Rectangular	± 0.67 %	± 0.67 %	± 0.67 %
Measurement temperature	Manufacturer	Rectangular	± 0.67 %	± 0.67 %	± 0.67 %
Spectrum					
Spectral mismatch - reference cell front		Gaussian	± 1 %	-	± 1 %
Spectral mismatch – reference cell rear	Measurement	Gaussian	-	± 3%	± 3 %
Reference cells					
Acquisition Accuracy (USB 4704)	Manufacturer	Gaussian	± 0.10 %	± 0.10 %	± 0.10 %
Calibration - reference cell rear	Measurement	Gaussian	-	± 1.00 %	± 1.00 %
Calibration - reference cell front	Manufacturer	Gaussian	± 0.10 %	-	± 0.10 %
Angle misalignment - module and reference device	[3]	Rectangular	± 3.00 °	± 3.00 °	± 3.00 °
Distance Mismatch - module and reference device	Measurement	Rectangular	± 1.00 %	± 1.00 %	± 1.00 %

[1] No temperature control is currently in place. Smallest deviation from outdoor case is assumed.

TABLE VI
EXPANDED UNCERTAINTY (K=2) WITHOUT MISMATCH CORRECTION

Component	Front side @1000 W/m²	Rear side @1000 W/m²	Double-sided @1000/200W/m²
Current (I)	± 3.5 %	± 6.7 %	± 8.3 %
Voltage (V)	± 0.4 %	± 0.5 %	± 0.6 %
Power (P)	± 3.5 %	± 6.7 %	± 8.4 %

III. CONCLUSION

The paper identified and showed a stringent analysis of the measurement uncertainty of a LED sun simulator with double-side illumination for bifacial PV modules. This gives an understanding and the importance of measurement uncertainty sources of LED sun simulators beyond uncertainty calculation for pulsed Xenon lamp systems. Critical uncertainty sources for LED-based sun simulators are the spectral, the spatial non-uniformity, and the light temporal instability due to the high number of independently powered LED channels.

The measurement methods of the spatial non-unifmity with different uniformity modules of the same type or smaller detector show significant differences. This uncertainty might have a higher impact on the overall measurement uncertainty for LED-based sun simulators and the measurement of PV modules with smaller cells, e.g., half-cells. The uncertainty budget of this method must be calculated in detail and considered for the overall uncertainty budget. The spectral non-uniformity across the module plan needs further investigation and might have a more significant impact.

However, the results of a state-of-the-art uncertainty calculation show significant differences between the two light sources. The expanded uncertainty (k=2) for the front-site measurement for Pm is ±3.5%, for the rear-side ±6.7%, and for the double-side at 1000/200W/m², it results in ±8.4%. The IEC classification for the front-side is Class CBB, and for the rear-side it is BBB. For low irradiance, e.g., 200W/m², the classification is CBC and BBC, respectively.

As mentioned earlier, no further improvement has been made for this study, e.g., improvement of the spectral and spatial uniformity or the temporal light stability. The simulator has been characterized with different state-of-the-art measurement technics to analyze and identify device-specific uncertainties and thus to understand the importance of measurement uncertainties. Further work is a detailed investigation of the spectral and spatial non-uniformity and their improvement by adjusting the individually powered LED channels.

REFERENCES

[1] Photovoltaic Devices— Part 1-2: Measurement of current-voltage characteristics of bifacial photovoltaic (PV) devices, Int. Electrotech. Comm. (IEC), Geneva, Switzerland, IEC 60904-1 Ed. 1.0, 2019

[2] D. Dirnberger, U. Kräling, "Uncertainty in PV Modules Measurement – Part I: Calibration of Crystalline and Thin-Film Modules, IEEE Journal of Photovoltaics, Vol. 3, No. 3, 2013

[3] H. Müllejans, W. Zaaiman, and R. Galleano, "Analysis and mitigation of measurement uncertainties in the traceability chain for the calibration of photovoltaic devices," Meas. Sci. Technol., vol. 20, no. 7, p. 75101, 2009.

[4] N. Umachandran and G. TamizhMani, "Effect of spatial temperature uniformity on outdoor photovoltaic module performance characterization," in 2016 IEEE 43rd Photovoltaic Specialists Conference (PVSC), Portland, OR, USA, 2016, pp. 2731–2737.

[5] K. Emery, "Uncertainty analysis of certified photovoltaic measurements at the national renewable energy laboratory," Nat. Renewable Energy Lab. (NREL), Golden, CO, USA, NREL/TP-520-45299, Aug. 2009.

[6] C. Monokroussos, Harald Müllejans, S. Dittmann, et al., "Electrical characterization intercomparison of high-efficiency c-Si modules within Asian and European laboratories", Progress in Photovoltaics, 2019

[7] G. Koutsourakis, M. Rauer, T. R. Betts, J. C. Blakesley, S. Dittmann, W. Herrmann, S. Riechelmann, et al., "Results of the Bifacial PV Cells and PV Modules Power Measurement Round Robin Activity of the PV-Enerate Project,", 2020

The Effects of Global Damp Heat Ageing on Debonding of Polyolefin Glass Laminates

Martin Tiefenthaler, Gernot M. Wallner, Robert Pugstaller

Christian Doppler Laboratory for Superimposed Mechanical-Environmental Ageing of Polymeric Hybrid Laminates, Institute of Polymeric Materials and Testing, University of Linz, Austria

Abstract — The long-term reliability of photovoltaic (PV) modules depends significantly on the encapsulation material. Commonly used ethylene vinyl acetate copolymers (EVA) are prone to degradation such as environment-induced deacetylation. Furthermore, the polarity of EVA enables cation migration associated with potential induced degradation (PID) of PV modules. Less polar encapsulants allow for better water and ion barrier properties. Hence, polyolefin elastomers (POE) have been established as a promising alternative to EVA. However, so far only few studies have been dealing with the long-term reliability of POE encapsulants especially close to relevant interfaces of PV module laminates. Therefore, in this study global ageing experiments were performed on EVA and POE double glass laminates, which were exposed to damp heat (DH; 85°C, 85 %rh) for up to 10000h. The aged double glass specimens were characterized with microscopic methods, compressive shear testing and X-ray photoelectron spectroscopy (XPS). Both laminate types revealed pronounced haziness after 10000h of DH exposure. While the change in haziness was attributable to water uptake of the EVA/glass laminate, ageing at the POE/glass interface led to the formation of calcium and sodium carbonate crystals, which act as scattering domains. Interestingly, no significant change in compressive shear strength was ascertained for unaged (0h) and DH aged (10000h) EVA laminates. In contrast, POE specimens revealed a significantly higher compressive shear strength (+80%) after 10000h of DH exposure. XPS measurements confirmed excellent ion barrier properties of POE. No Na and Ca ions were detectable within the bulk, whereas a significant increase of these elements was found on the fractured surfaces of POE laminates failing in a cohesive manner within the glass substrates.

I. Introduction

Encapsulants significantly affect the long-term reliability of photovoltaic (PV) modules. Ethylene vinyl acetate copolymer (EVA) is the dominating encapsulation material due to its good prize to performance ratio. However, EVA is prone to environmentally induced degradation mechanisms as deacetylation, β-scission and hydrolytic depolymerization resulting in decreased optical and mechanical properties [1]. Furthermore, the polarity of EVA facilitates water ingress and ionic migration from the glass through the bulk polymer to the silicon solar cell, which can result in potential induced degradation (PID). Both, EVA degradation and PID highly affect the power output of PV modules. Therefore, current research is focusing on new materials based on polyolefin elastomers (POE). The less polar POE encapsulants exhibit good barrier properties hindering PID [2] and a high crack growth resistance under superimposed mechanical and environmental loading conditions [3]. Although, POEs are clearly an interesting replacement candidate for EVA encapsulants, only few durability data are available for this novel class of encapsulation material.

Hence, in this work a global ageing approach was applied to investigate the long-term reliability of EVA and POE double glass laminates. Therefore, specimens were exposed to damp heat (DH) conditions (85°C and 85 %rh) for up to 10000h. Microscopic, compressive shear testing and X-ray photoelectron spectroscopy (XPS) methods were implemented and applied to assess DH induced optical, mechanical and chemical changes within the laminates, respectively.

II. Experimental

Two peroxide crosslinking UV-transparent encapsulants based on ethylene vinyl-acetate copolymer (EVA) and polyolefin elastomer (POE) were investigated. Double glass laminates were manufactured (155°C, 800 mbar for 20 minutes) using the EVA and POE encapsulant films with thickness values of 0.49 (EVA) and 0.58 mm (POE). The thickness of the low iron solar glass substrates was 3.1 mm. Compressive shear specimens with dimensions of 25 x 25 x 7 mm³ were prepared by water jet cutting. Theses specimens were DH exposed (85°C, 85 %rh) in an environmental chamber WK3-340/40 (Weiss Umwelttechnik GmbH, Germany) for 1000, 2000, 5000 and 10000h (see Table 1). Due to a lack of specimens, only one specimen per material was tested after 10000h of global damp heat ageing.

TABLE I

SUMMARY OF THE GLOBAL AGEING STUDY (85°C, 85 %RH)

Global ageing duration (h)	0	1000	2000	5000	10000
No. of tested EVA/POE specimens	3	3	3	3	1

Direct light and polarized light microscopic (DLM and PLM) images were generated with the optical microscope VHX 7000 (Keyence, Japan). PLM measurements were performed in transmittance mode with the specimens being placed between crossed polarizers.

Compressive shear tests were carried out in tensile mode on a universal tensile testing machine Z020 (ZwickRoell,

Germany) equipped with a temperature chamber. The experiments were conducted at 1 mm/min. To prevent fracture of the glass substrate and to mimic critical service-relevant conditions, the test temperature was set to 60°C.

XPS survey scans were carried out using a Theta Probe system together with the Advantage data system (Thermo Fisher Scientific, USA) to assess the chemical composition of global aged surfaces and interfaces. The spot size diameter was set to 300 or 400 μm. The pass energy of the hemispherical analyzer was 50 eV for survey scans.

III. RESULTS

DH exposure of glass substrates is known to trigger a leaching process of the glass network modifier cations (e.g., Na, Ca) from the bulk to the water film on the glass surface. The pH value of the water film is raised through the cation migration induced change of the electrochemical potential between glass surface and bulk. At a pH value of 9, the migrated ions tend to form water-insoluble carbonate (e.g., Na_2CO_3, $CaCO_3$) clusters associated with a permanent whitening of the surface [4]. In Fig. 1 microscopic images of EVA laminates are illustrated after 0, 5000 and 10000h of damp heat exposure. Leaching induced surface whitening was clearly discernible after 5000h of DH exposure. The entire glass substrates showed a homogenous white appearance after 10000h. The surface whitening effect was not dependent on the encapsulant material. Quite similar effects were observed for POE double glass laminates.

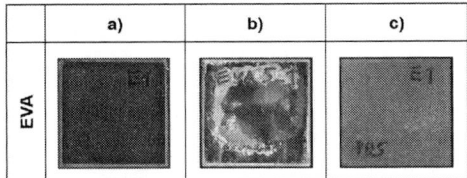

Fig. 1. EVA specimens exposed to DH for a) 0h, b) 5000h and c) 10000h.

To assess the global ageing induced changes at the glass-encapsulant interfaces and within the bulk, the double glass surface was polished. The light microscopic images revealed an increased haziness for both encapsulant types after 10000h of DH exposure. An edge clearing effect was discernible in EVA laminates exposed for 10000h. In contrast, the surface polished POE laminates appeared even more opaque at the edge regions (see Fig. 2a)). Polarized light microscopy characterization resulted in spherical microdomains (maltese-crosses) in both laminate types (see Fig. 2b)-c)). The microdomains differed in appearance, size and number. Compared to POE, EVA laminates exhibited less microdomains with bigger dimensions. The microdomains in EVA laminates were presumably related to water clusters. The edge clearing effect of EVA laminates

was dependent on the conditioning time at 23°C and 50% rh subsequently to DH exposure. The microdomains within the POE laminates were much smaller and more homogenous. In agreement with findings in [5], the microdomains in POE laminates were attributed to glass ageing induced formation of $CaCO_3$ and Na_2CO_3 crystals at the glass-POE interfaces. Interestingly, few black spots were detected in the POE laminates.

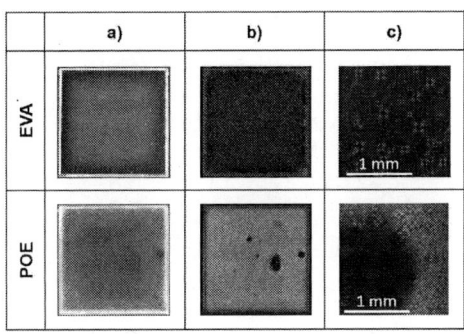

Fig. 2: Microscopic images of EVA and POE double glass laminates after 10000h of DH exposure: a) DLM: full specimen, b) PLM: full specimen, c) PLM: specimen detail.

In contrast to POE, which was reported to effectively barrier ion migration [2], the higher polarity of EVA favors ion migration into the bulk [1]- [3]. Additionally, the hydrolysis of EVA during DH exposure facilitates the formation of acetic acid, which effectively lowers the alkalinity of the pH value at the glass-EVA interface. These two mechanisms are considered to hinder the formation of sodium or calcium carbonates at glass-EVA interfaces by keeping the pH value well below 9.

Damp heat ageing up to 10000h did not affect the compressive shear failure mode of EVA specimens, which failed in an interfacial manner. The average maximum load difference of EVA specimens exposed for 0 and 10000h was about 8% (see Fig. 3c)), which was within the standard deviation of the tested 0h specimens (20%). On the contrary, POE laminates revealed two transitions in compressive shear failure mode, from purely interfacial (0h) to 1) combined interfacial and cohesive within the POE encapsulant (>1000h) and to 2) combined interfacial and cohesive within the polymer and the glass (>5000h). Distinct interfacial failure of the POE specimen DH aged for 10000h was observed at the black spots shown in Fig. 2b)-c). The transitions of failure mode correlated with the increased maximum shear loads after 1000h (+42%) and 5000h (+93%) of damp heat exposure. No further significant changes of the compressive shear behavior were detected between 5000 and 10000h (+80%) of DH exposure.

XPS survey scans of the reference surface and the white deposit on the aged glass substrate of POE laminates (illustrated in Fig. 4) revealed a significant damp heat induced increase of the Na (about 10 times higher) and the Ca content (about 2 times higher) together with a slightly reduced Si content (0.9 times lower). Interestingly, cohesively failed glass regions revealed higher Si, Ca and Na peaks at the polymer side (Fig. 4d)) than at the glass side (Fig. 4c)).

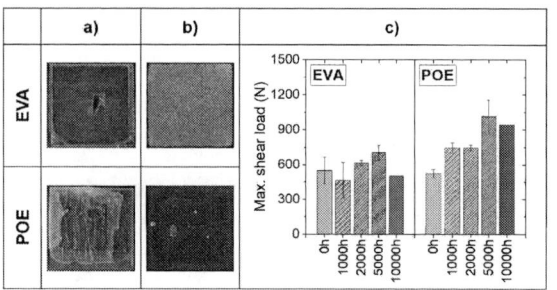

Fig. 3: DLM images of fractured surfaces: a) polymer side, b) glass side; c) max. compressive shear load as a function of damp heat ageing time.

Overall, the amount of Si, Na and Ca detected in cohesively failed glass regions was below the values obtained for the glass surface (Fig. 4b)). A cross-sectional cut of the 10000h aged POE layer with cohesive failure within the glass on both sides is illustrated in Fig. 4d)-e). XPS survey scans of the cross-sectional cut revealed high amount of Si, Na and Ca at the fractured surface (Fig. 4d)) and the absence of Na and Ca within the POE bulk (Fig. 4e)). The small amount of silicon (1.8 at%) within the bulk is presumably related to the silane adhesion promoter and/or contamination during compressive shear testing or cutting.

POE	a)	b)	c)	d)	e)
Si (at%)	25.2	23.3	7.8	13.1	1.8
Na (at%)	0.9	9.0	3.8	5.0	0.0
Ca (at%)	1.8	3.4	0.0	2.0	0.0

Fig. 4: XPS analysis of the Si, Na and Ca content from: a) unaged reference glass surface, b) white deposit on 10000h aged glass, c) cohesively failed in glass, d) polymer side of cohesively failed in glass, e) bulk POE.

IV. SUMMARY AND SIGNIFICANCE OF THE WORK

The damp heat ageing behavior of EVA and POE double glass laminates was investigated for exposure times up to 10000h. Distinct glass surface ageing was discernible for both laminate types. XPS analysis confirmed an enhanced amount of Na and Ca on the glass surface. By polarized light microscopy of surface polished specimens the formation of spherical

microdomains was ascertained within the laminates after 10000h of DH exposure. The larger microdomains in EVA were attributed to water uptake and clustering within the polar encapsulant. In contrast, the smaller microdomains in POE laminates were located close the interface and related to Na_2CO_3 and $CaCO_3$ aggregates. Compressive shear tests revealed no significant changes of the mechanical behavior of EVA laminates up to a damp heat exposure time of 10000h. On the contrary, the shear strength of POE laminates was increasing due to damp heat exposure up to 93% in a step-wise manner. The failure mode of POE laminates changed from purely interfacial (0h) to combined interfacial and cohesive within the glass and the POE layer after 10000h of DH exposure. XPS analysis of fractured surfaces of POE laminates confirmed the excellent ion barrier capability of POE along with an enrichment of Na or Ca carbonates close to the polymer-glass interface.

Finally, it should be emphasized that the damp heat ageing and degradation mechanisms are quite different for EVA and POE encapsulants. While the polar EVA allows for ion diffusion within the encapsulant and neutralization of Na and Ca ions by reaction with acetic acid residues, the less-polar POE encapsulant is an efficient water and Na or Ca ion barrier associated with an enhancement of the pH value at the polymer-glass interface and the formation of carbonates. A better compressive shear performance along with a damp heat induced improvement up to 10000h of exposure time was ascertained for POE double glass laminates, which are gaining importance in the PV industry.

REFERENCES

[1] P. Thornton, S. L. Moffit, L. T. Schelhas, R. H. Dauskardt, and S. L. Moffitt, "Dependence of adhesion on degradation mechanisms of ethylene co-vinyl acetate encapsulants over the lifetime of photovoltaic modules," *Solar Energy Materials and Solar Cells*, 2022.

[2] M. C. López-Escalante, L. J. Caballero, F. Martín, M. Gabás, A. Cuevas, and J. R. Ramos-Barrado, "Polyolefin as PID-resistant encapsulant material in PV modules," *Solar Energy Materials and Solar Cells*, vol. 144, pp. 691–699, 2016.

[3] G. Riedl, G. M. Wallner, R. Pugstaller, G. Säckl, and R. H. Dauskardt, "Methodology for local ageing and damage development characterization of solar glass/encapsulant interfaces under superimposed fatigue stresses and environmental influences," *Solar Energy Materials and Solar Cells*, vol. 248, pp. 1–9, 2022.

[4] V. Guiheneuf, F. Delaleux, O. Riou, P.-O. Logerais, and J.-F. Durastanti, "Investigation of damp heat effects on glass properties for photovoltaic applications," *Corrosion Engineering, Science and Technology*, vol. 52, no. 3, pp. 170–177, 2017.

[5] J. Harris, I. Mey, M. Hajir, M. Mondeshki, and S. E. Wolf, "Pseudomorphic transformation of amorphous calcium carbonate films follows spherulitic growth mechanisms and can give rise to crystal lattice tilting," *CrystEngComm*, vol. 17, no. 36, pp. 6831–6837, 2015.

Nanographene (NG)-based Hole Transporter with π- interface modifier for Thermally Stable Perovskite Solar Cells

Seul-Gi Kim, Thybault de Monfreid, Jeong-Hyeon Kim, Fabrice Goubard, Joseph J. Berry, Kai Zhu, Thanh‐Tuân Bui, Nam-Gyu Park

Chemistry and Nanoscience Center, National renewable energy laboratory, Golden, CO, United States

CY Cergy Paris Université, LPPI, Cergy, France

School of Chemical Engineering and Center for Antibonding Regulated Crystals, Sungkyunkwan University, Suwon, South Korea

SKKU Institute of Energy Science and Technology (SIEST), Sungkyunkwan University, Suwon, South Korea

Materials Science Center, National Renewable Energy Laboratory, Golden, CO, United States

Renewable and Sustainable Energy Institute, University of Colorado Boulder, Boulder, CO, United States

Department of Physics, University of Colorado Boulder,, Boulder, CO, United States

We report here thermally stable HTM for perovskite solar cell (PSC) which is based on Nanographene(NG) with functional substitution groups (coded NG-HTMs) to enhance its hole mobility and thermal property. Hole mobility of NG-HTM is optimized by changing functional groups and enhanced from 5.68×10^{-3} cm2V-1s-1 to 9.51×10^{-3} cm2V-1s-1 which shows much higher than spiro-MeOTAD (4.69×10^{-4} cm2V-1s-1). However, interface problem was detected when NG-HTM was applied to NIP perovskite solar cell due to interface property which originated from new chemical structure of NG-HTMs. NG-selective π-interface modifier (π-IM) which induces strong π-π interaction between perovskite and NG core significantly enhances charge extraction efficiency from perovskite layer. Therefore, power conversion efficiency (PCE) is significantly improved from 9.8% (stabilized PCE, w/o π-IM) to 23.06% (w/ π-IM) which shows higher than the spiro-MeOTAD-based device (20.56% (w/o π-IM) and 19.38% (w/ π-IM). In DSC measurement, our optimized NG-HTM shows 137.6 oC of Tg with very weak endothermic signal (6.00 μW/mg), while the spiro-MeOTAD shows clear endothermic signal (153.75 μW/mg) at 75.3 oC. Therefore, NG-HTM based device maintains 83.6% of initial PCE after 350 h at 75oC + 1000 h at 85 oC which is significant improvement compared to completely degraded spiro-MeOTAD device (which maintains 24.5% of initial PCE). These findings highlight the importance of core structure and proper substitution groups to design for new HTM and molecular-level design of interfacial modifier for highly efficient and stable PSCs.

Automated Workflows for Machine Learning on Photovoltaic Timeseries and UV Fluorescence Image Datasets Using FAIR Principles

William C. Oltjen*, Xuanji Yu*, Mengjie Li†, Dylan J. Colvin†‡, Yijia Sun*, Hubert Seigneur‡, Philip Knodle§,
Andrew M. Gabor§, Laura S. Bruckman*, Kristopher O. Davis†, Roger H. French*

*SDLE Research Center, Case Western Reserve University (CWRU), Cleveland, OH, 44106, USA
†University of Central Florida (UCF), Orlando, Florida, 32816, USA
‡Florida Solar Energy Center (FSEC), Cocoa, FL, 32922, USA
§BrightSpot Automation, Westford, MA, 01886, USA

Abstract—The long-term reliability and durability of photovoltaic (PV) modules have been critical to the recent growth of solar power. The analysis of timeseries data and ultraviolet fluorescent (UVF) images provide insights into critical degradation modes that affect the long-term performance of PV modules. From the timeseries data, we automate the calculation of performance loss rates for 67 different systems located throughout the state of Florida. From UVF imaging, we utilize a pre-trained Mask-RCNN deep learning network for the segmentation of PV modules out of noisy images. In order to perform both of these tasks, a considerable amount of time has to be put into cleaning and reorganizing the input data. This is important both for human understanding of the data and machine action-ability to perform analysis. FAIR principles can be used to assist in this cleaning process to produce datasets that are of high quality for use in machine learning tasks. Here, we apply the FAIR principles to timeseries and image analysis as a showcase for FAIRified workflows for the automation and standardization of datasets for use in machine learning.

Index Terms—FAIR, Timeseries, UV Fluorescence Imaging

I. INTRODUCTION

The impressive reliability and resilience of PV modules have been large contributors to the recent growth of photovoltaics. In order to understand long term trends in power output, timeseries analysis is often performed. A performance loss rate (PLR) is a common metric for assessing the long term stability of of PV systems. Because there is no IEC standard for the calculation of PLR, determining best practices for modeling PLR can be difficult [1]. There are many factors that can influence the performance of PV systems outside of module performance itself. Notably, the irradiance from the sun has the most effect, but other factors such as humidity, wind speed, and module temperature can influence the system. Issues with the timeseries dataset such as inverter cutoff, seasonality, and missingness can also have a significant impact

This material is based upon work supported by the U.S. Department of Energy's Office of Energy Efficiency and Renewable Energy (EERE) under Solar Energy Technologies Office (SETO) Agreement Number DE-EE0009347. The views expressed herein do not necessarily represent the views of the U.S. Department of Energy or the United States Government.

on the calculation of PLR. Therefore, calculating PLR is non-trivial given that small changes in power data must be isolated from years of timeseries power and weather data. There is a need for the automation of this process so that the calculation of PLR is simpler and more standardized.

UVF imaging has been used to identify defects in PV modules since 2010 [2]. It can be used to analyze common degradation mechanisms such as cracking, busbar corrosion, and hot spots. While electroluminescence (EL) imaging is the standard for the analysis of these kinds of degradation mechanisms in images, obtaining EL images in the field can be arduous and time-consuming. UVF images can be acquired much more easily in the field and at a much faster rate to provide information complimentary to that obtained by EL imaging [3]. While the use of UVF field imaging is relatively new, there is a demand to establish standards for the operation of novel UVF imaging systems. An important aspect of this process is determining analysis pipelines for the categorization of defects found in UVF images.

In the characterization of both timeseries data and UVF images, there is a need to adopt analysis pipelines for faster and more standardized results. In order to make this a possibility, it is necessary to FAIRify our data. FAIR data is data that is Findable, Accessible, Interoperable, and Reusable. These FAIR principles were introduced through the publication of Wilkinson et al.'s paper [4] which serves as the blueprint for our analysis pipeline. There have been efforts throughout the field of photovoltaics already to FAIRify data, especially through the OrangeButton Initiative which we build on here. FAIRifying our data allows us to circumvent common obstacles in data sharing, such as different data formats (csv, xlsx, etc), different databases (relational, non-relational), alternate parameter names, and more. There has been an extensive push throughout the science world to make data "FAIR" recently, as publishers, science funding agencies, and government agencies have begun to establish requirements for the proper management of metadata. Therefore, there exists a need to implement FAIR principles into the analysis of long term performance in PV modules.

978-1-6654-6060-6/23 $31.00 © 2023 IEEE

Fig. 1. Field Study site overview.

II. DATASET INTRODUCTIONS

The timeseries data was shared by the Florida Solar Energy Center (FSEC). This dataset includes information on 67 different PV systems throughout the state of Florida, either from FSEC test sites or the SunSmart schools program. This data includes 15-minute interval timeseries data with information about power output, ambient and reference temperature, irradiance, battery properties, and input and output current and voltage. The length of the timeseries varies between datasets, with the longest set including about 9 years of data, and most of the datasets including data on the order of about 2 years.

A field survey using a pole-mounted UVF imaging system designed by Brightspot Automation [3] was conducted at Case Western. The site consists of 4,000 modules made up of two brands, both of which account for half of the site (2,000 modules of each). We have imaged all of the modules from this site using two modules per image. Doing it in this way, we have insured high-quality images for use in machine learning analysis, where each image is 4,256 × 2,848 pixels large.

III. METHODS

A. Timeseries Analysis

We have designed a standardized pipeline for the analysis of timeseries data. We first go through and evaluate the quality of the data that we are working with using the data grading and heatmap plotting functions from the *PVPLR* R package developed by the SDLE lab at CWRU [5]. With the quality of the data in mind, we perform data cleaning and filtering. In this step, we remove any data below 1% of the max power and any data below an irradiance threshold of 200 W/m^2. We also convert the timestamp to a POSIXct and the other variables to numeric, and we add some helpful columns including day, week, and month. In the next step we use an X "by" X (XbX) + Universal Temperature Correction (UTC) model to perform power predictive modeling. Because power cannot be easily compared across systems with different environmental conditions, we need to use a power predictive model to make the data comparable. The XbX + UTC model batches data by

a user input (day by day, week by week, month by month, etc) using the following equations :

$$P_{cor} = \frac{P_{obs}}{1 + \gamma_T (T_{obs} - T_{rep})(\frac{G_{obs}}{G_{rep}})}$$

$$P_{cor} = \beta_0 + \beta_1 G + \epsilon$$

where the power is corrected based off of a universal temperature correction, and then that corrected power is used to model based off of irradiance. We use a month by month XbX + UTC model. Once the data has been batched by month, we perform a year-on-year regression, a technique that examines points exactly one year apart to determine PLR. This method avoids issues with outliers and seasonality, at the cost of needing long-term data in order to be meaningful. With our regression model, we run it 1,000 times on different subsets of our data in order to "bootstrap" our error measurement. We then report our performance loss rate along with its heatmap and data grades to provide context for the measurement.

B. UVF Image Analysis

We labeled 200 images for use in deep learning via the *Labelme* package in Python, where we have drawn bounding rectangles over the entire module area in an image. We use a Mask-RCNN for instance segmentation of module area out of the raw images. This model architecture has been used for this purpose in solar by previous groups, for example in the use of segmenting modules out of IR images [6]. The Mask-RCNN model has advantages over other image segmentation models in that it is fast and widely used as a benchmark (so we can utilize pre-trained models as a starting point for our Mask-RCNN model). This makes a Mask-RCNN ideal for segmenting modules out of noisy images.

C. FAIRification

In order to publish our data online for the use of the general public, there needs to be a strict structure imposed on that data. This way, the data is both intelligible to the humans looking to use the data and actionable to the algorithms researchers will apply to the data. In order to accomplish both of these things, we aim to implement data FAIRification (making our data Findable, Accessible, Interoperable, and Reusable). We do this by using an ontology to create a vocabulary and structure for our data which informs the creation of our metadata documents. An ontology is a formal dictionary of terms for a given industry or field that shows how properties are related. So not only does the ontology define terms, it defines the relationship between terms as well. Ontologies provide a shared understanding of the concepts and relationships in a particular domain, which helps to ensure that data is unambiguous and can be understood by humans and machines alike. This helps to ensure that data can be effectively integrated and reused across different applications and domains. Data FAIRification involves making data FAIR by implementing the standards created in an ontology. By applying data FAIRification principles, organizations can ensure that their data is accessible and usable by a wide range of stakeholders, including researchers,

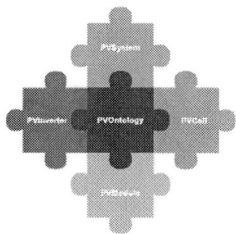

Fig. 2. The structure of PV ontology, where PVSystem, PVInverter, PVModule and PVCell are defined as their own sub-ontologies. The sub-ontologies are designed to be modular and interconnected.

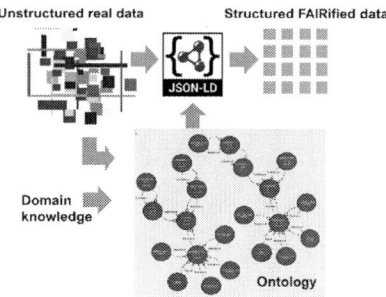

Fig. 3. FAIRification process of real materials science and energy science data using ontology based JSON-LD

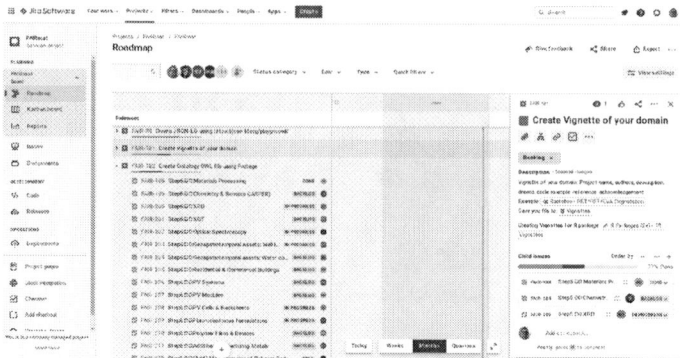

Fig. 4. Jira board of FAIRmaterials

Fig. 5. FAIRmaterials find-the-docs website

practitioners, and decision-makers. This helps to maximize the value of data and supports the development of new knowledge and insights. Data FAIRification is an essential component of effective knowledge management and representation, and it plays a critical role in enabling organizations to make better use of their data assets.

We combine domain knowledge in PV with data science expertise in order to FAIRify our data. The process of FAIRifying materials science and photovoltaic research data is outlined in Figure 3. We have chosen to use a JSON-LD (JavaScript Object Notation for Linked Data) text document to store our metadata, as it is a recommendation from the World Wide Web Consortium (W3C). JSON-LD has several benefits over other data formats, including being machine-readable, linked, easy to implement, interoperable, and optimized for search engines. To FAIRify data, we have created fillable JSON-LD templates in a number of domains in photovoltaics such as photovoltaic systems, inverters, modules, and cells. Each of these templates has an ontology associated with it that defines the structure and nomenclature used in the template document 2.

These templates convert unstructured real data into structured, FAIRified data. The detailed steps involved are outlined below.

- **Ontology Development**: Construct ontologies based on real datasets and domain knowledge, incorporating existing outside standards such as in the Orange Button Initiative and the ISO standards from the Basic Formal Ontology.
- **JSON-LD Template Creation**: Development of JSON-

LD templates with defined schemas for metadata structure and nomenclature, based on real datasets and ontologies.
- **R/Python Package Design**: Development of a "bilingual" FAIRmaterials code package for both Python and R to fill JSON-LD templates with user data in order to FAIRify it. These packages exist as tools with which users can FAIRify their own data based on our recommended standards.

The current team working on the FAIRification project consists of over 21 domain science researchers and 10 computer science researchers across many domains even outside of PV. To track progress and improve collaboration across such a large team, we use agile software development tools such as Jira and Bitbucket. Using these tools, we share tutorials, announcements, and create deadlines (Figure 4) to manage this collaborative effort.

We have been regularly updating our software packages on a near-monthly basis to include updates to the JSON-LD templates and new functionalities. All of our progress and documentation are updated on our find-the-docs website-https://cwrusdle.bitbucket.io/ as demonstrated in Figure 5.

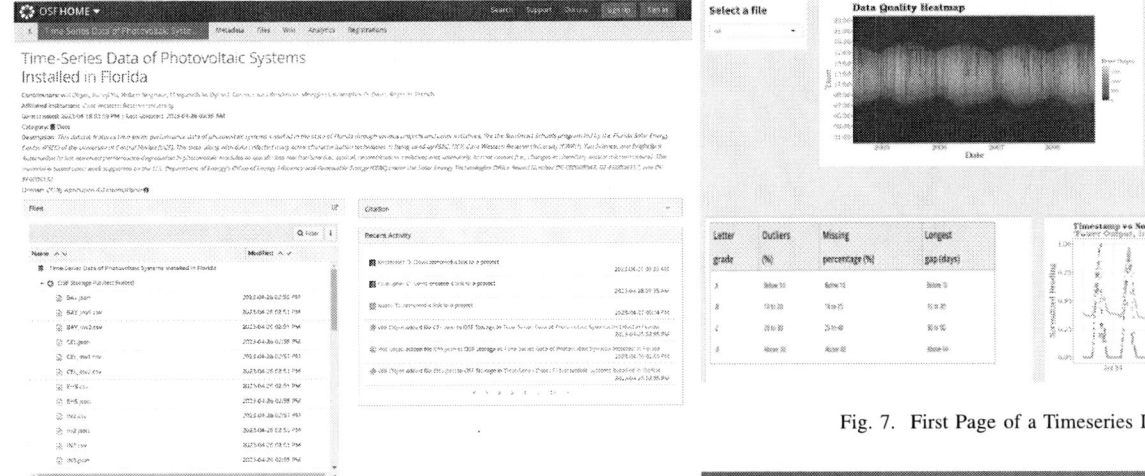

Fig. 6. Published Timeseries Data on Osf.io

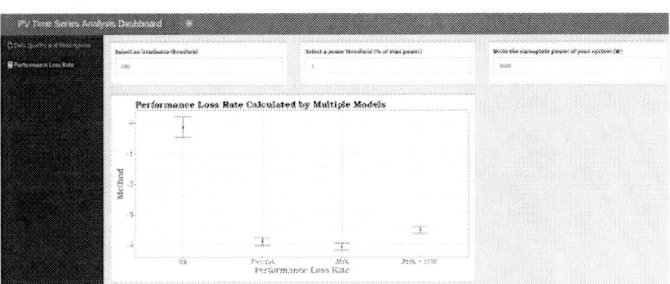

Fig. 7. First Page of a Timeseries Dashboard

Fig. 8. Second Page of a Timeseries Dashboard

1) Published Datasets on osf.io: Using our FAIRification process explained above, we fully FAIRified and published a series of timeseries datasets to osf.io. A screenshot of the page is shown in Figure 6. This data is openly accessible and provides a way for the general scientific community to examine a selection of the highest-quality timeseries datasets from our project. In total, there are 25 different timeseries datasets published from 15 different systems. These are systems from the Sunsmart Schools program, smaller rooftop sites scattered throughout Florida with 15-minute interval timeseries power data. The timeseries datasets include information on power, temperature, irradiance, battery properties, and more. The metadata includes information stored in JSON-LDs about the systems, inverters, and modules at these sites. Given our fully FAIRified dataset, users should be able to rapidly make their own insights about the data programmatically, without needing to reorganize the dataset themselves.

IV. DISCUSSION

A. Timeseries Analysis

After FAIRifying our data using the FAIRmaterials package, our analysis pipeline becomes much easier and more generalized. Using a list of common alternative variable names included as part of the PV power plant JSON-LD, column names from a number of different sources can be automatically renamed to those defined by our nomenclature. With a dataset including 67 different timeseries, manually renaming each column in each timeseries would normally be incredibly arduous. Because FAIRified metadata are stored in a defined structure, identifying where terms are stored is made simple. With minimal human intervention, a timeseries can then be shared from FSEC to Case Western, for example, and a dashboard such as in Figures 7 and 8 can be produced.

The method for the calculation of the performance loss rate is shown along with a heatmap and an analysis of the data grade. Using this automated process, performance loss rate can

be easily calculated and presented with the necessary context to understand the result.

B. UVF Image Analysis

Having FAIRified the UVF image dataset from the field study at Case Western, we can automate our image analysis. Certain information about images such as height, width, and sample ID are needed as inputs for deep learning algorithms. This information can be easily located from a FAIRified dataset. Furthermore, the manual segmentation of modules and cells out of field images is heavily resource intensive. We have trained an instance segmentation model on our labeled image data in order to pull individual modules out of our UVF images from field testing. Using the Detectron2 package in Python [7], we trained 156 Mask-RCNN models with different structures and hyper-parameters in order to find the best model for our application. We used 10 different neural network architectures pre-trained on 2 different datasets. Each of these different models was trained with varying hyper-parameters such as the number of iterations of the model and the learning rate. After training, the models were evaluated on the test set, and the average precision of the mask predictions was stored. Figure 9 shows a summary of the top 5 models from our analysis.

Each of the models produces incredibly high precision, with the R-50-DC5-1x model trending the most toward the highest accuracy. This model uses a ResNet conv5 backbone with dilations in the conv5 layers. The 1x indicates that the model was pre-trained using 12 epochs as opposed to 37. Taking the

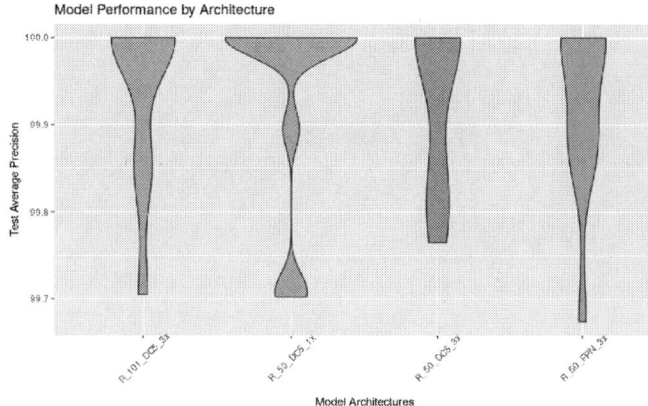

Fig. 9. Comparison of Top Model Architectures

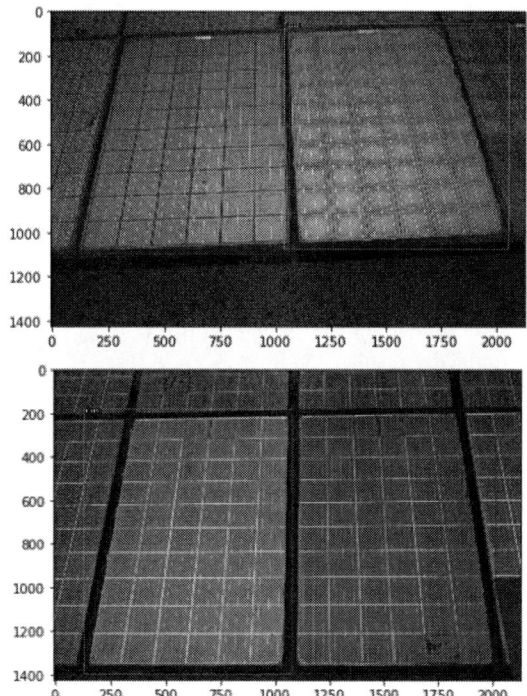

Fig. 10. Example Segmented Output from the Best Trained Model

V. CONCLUSIONS

After FAIRifying our data, automated workflows for the analysis of both timeseries and UVF image datasets were made possible. For timeseries analysis, the automated calculation of PLR directly from a shared dataset was demonstrated. For UVF image analysis, we were able to automate a normally resource intensive task in segmenting modules using a pre-trained Mask-RCNN to an average precision of 98.25 percent. FAIR data workflows enable automated analysis by giving data a standard structure and nomenclature that can be understood both by humans and computers. Through FAIRifying our data, we can unlock faster and more robust analysis for our research data.

REFERENCES

[1] S. Lindig, M. Theristis, and D. Moser, "Best practices for photovoltaic performance loss rate calculations," *Progress in Energy*, Apr. 2022.

[2] M. Köntges, A. Morlier, G. Eder, E. Fleiß, B. Kubicek, and J. Lin, "Review: Ultraviolet Fluorescence as Assessment Tool for Photovoltaic Modules," *IEEE Journal of Photovoltaics*, Mar. 2020.

[3] A. M. Gabor and P. Knodle, "UV Fluorescence for Defect Detection in Residential Solar Panel Systems," in *2021 IEEE 48th Photovoltaic Specialists Conference (PVSC)*, Jun. 2021.

[4] M. D. Wilkinson, M. Dumontier *et al.*, "The FAIR Guiding Principles for scientific data management and stewardship," *Scientific Data*, Mar. 2016.

[5] A. Curran, T. Burleyson, S. Lindig, D. Moser, R. French *et al.*, "PVplr: Performance Loss Rate Analysis Pipeline," Oct. 2020.

[6] L. Bommes *et al.*, "Computer vision tool for detection, mapping, and fault classification of photovoltaics modules in aerial IR videos," *Progress in Photovoltaics: Research and Applications*, 2021.

[7] Y. Wu, A. Kirillov, F. Massa, W.-Y. Lo, and R. Girshick, "Detectron2," https://github.com/facebookresearch/detectron2, 2019.

highest-performing model of the trained R-50-DC5-1x models, we can see example segmented results in Figure 10. While there is some "waviness" near the edges of the modules, the segmentation results from our Mask-RCNN model are very good.

In the future, we will use this segmentation algorithm to assist with cutting each individual cell out of a detected module in an image. This will enormously speed up our ability to perform machine learning for identifying defects in UVF images.

Post-Flight Analysis of Perovskite Solar Cells for NASA Materials International Space Station Experiment (MISSE)

Kaitlyn VanSant, Ahmad Kirmani, Severin Habisreutinger, Steve Johnston, Brian Wieliczka, Joseph Luther, Timothy Peshek, Lyndsey McMillon-Brown

NASA Glenn Research Center, Cleveland, OH, United States

National Renewable Energy Laboratory, Golden, CO, United States

Metal halide perovskite solar cells (PSCs) attract considerable attention as a photovoltaic technology that could provide high-efficiency, low-cost power for space missions. PSCs were passively flown on the exterior of the International Space Station (ISS) for eight months to evaluate cell durability with exposure to the ambient space conditions, then returned to Earth for characterization. This study will provide a post-flight assessment of those cells.

Cleaning optimization for photovoltaic powerplants: A novel approach combining techno-economic modelling with historic rain and soiling

Thore Müller, Kostiantyn Pogorelov, Franco Clandestino

Virtuous-Re Gmbh, Munich, Germany

Photovoltaic solar power has become the cheapest option for electricity generation, but especially in arid and desert areas, its profitability is threatened by high dust deposition on the surface of the modules. To mitigate this problem, project developers and operations and maintenance personnel need to define a cleaning strategy that maximizes the economic outcome of a power plant. This paper presents a novel approach to answering the question of which cleaning technology, in combination with which operating model, yields the highest net present value for a power plant, given its location and its physical and economic environment. The optimal cleaning strategy is defined as the strategy that will provide the best possible return if the past 20 years of rainfall and soiling would repeat in any order.

Characterization of Solar Cell Busbar Grid for Different Technologies by Time Domain Reflectometry Simulation: Transmission Line Approach.

A.M.C. Silveira[1] Student Member IEEE, M.R.M. Neves[2], R. Garcia[4], H. Alvarez[4],
M.G. Villalva[2] Member IEEE, F.C. Marques[3] and L.C. Kretly[1] Member IEEE

[1]*School of Electrical and Computer Engineering, Communications Department - DECOM, Campinas, Brazil*
[2]*School of Electrical and Computer Engineering, Dept. of Systems and Energy - DSE, Campinas, Brazil*
[3]*Institute of Physics Gleb Wataghin, Dept. of Applied Physics - DFA, Campinas, Brazil*
[4]*Build Your Dreams Company - BYD, Dept. Research and Development - R&D, Campinas, Brazil*

Abstract—**This work electromagnetically simulates the design of metallic grids for different Busbar technologies in standard solar cells existing in the photovoltaic market in order to characterize their grid impedance using a signal integrity technique known as Time Domain Reflectometry - TDR and addressing the design of metallized grid of solar cells within the foundations of transmission lines. The objective of the investigation is to parameterize the results obtained so that they can serve as a reference in studies of micro-resistance in series of solar cells and their effects on energy efficiency.**

Index Terms—**Solar cell, Wire Busbar, Ribbon Busbar, Serie Resistance, TDR Simulation and transmission line.**

I. INTRODUCTION

In photovoltaic (PV) context one of the ways to improve the energy efficiency of solar cells is to reduce the series resistances (R_s) [1]–[3]. Fig. 1 illustrates the solar cell equivalent circuit.

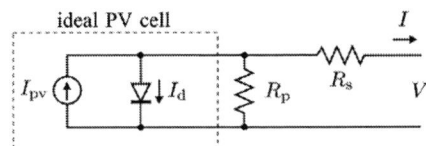

Fig. 1. Equivalent circuit for single-diode model. [adapted from 4]

Series resistance in a solar cell has three distinct origins, to know: current movement through the emitter and base of the solar cell; the contact resistance between the metal contact and the polycrystalline silicon; the resistance of the upper and rear metallic contacts [5], [6]. The third option involves finger resistors and busbar grids, the quest to improve energy efficiency in solar cells has driven the development of various busbar grid designs on top of solar cells that serve to interconnect the cells in assemblies of solar modules [5], [7], [8]. The resistance of the metal grid is determined by the resistivity of the material used along with the design of the metal grid technology. Fig 2 illustrates the metallic grid shapes of solar cell busbar and the micro resistances in solar cells.

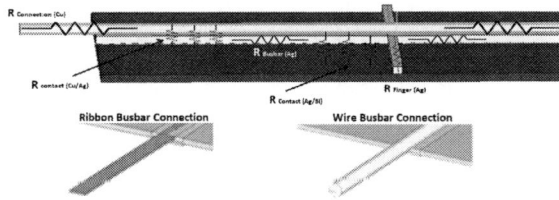

Fig. 2. Illustration of series resistance types and metallic connection shapes of solar cell busbar.[adapted from 9]

Low resistivity and high aspect ratio are desirable in solar cells metallization, but in practice are limited by the manufacturing technology used to make the solar cell, such as: 2BB, 3BB, 4BB, 5BB for ribbon shape and 12BB, 18BB, 22BB, 30BB for wire shape.

II. TDR MEASUREMENT TECHNIQUE

The TDR technique quickly reveals the characteristic impedance of the line and shows the position and nature (resistive, inductive or capacitive) of each discontinuity or load [10]. Since the reflection coefficient ρ is represented in the equation 1,

$$\rho = \frac{E_R}{E_i} = \frac{Z_L - Z_0}{Z_L + Z_0} \qquad (1)$$

Where E_R and E_i are reflected and incident energy, respectively Z_0 characteristic impedance of the line Z_L impedance of the load or equipment under test.

The distance D from the fault or anomaly can be defined by equation 2,

$$D = \frac{c}{\sqrt{\epsilon_r}} \frac{\Delta T}{2} = V_P \frac{\Delta T}{2} \qquad (2)$$

Where V_P is the propagation velocity of the pulse in the material, ΔT is the time interval to be monitored, c is the speed of light in vacuum and ϵ_r is the specific electrical permittivity of the material.

978-1-6654-6060-6/23 $31.00 © 2023 IEEE

The TDR technique does not analyze the DC resistance of the metallic busbar grid, but its AC impedance. The TDR pulse has a spectral component in relation to the rise time of the TDR pulse, which takes into account the skin effect of the transmission line [11], [12]. Due to this issue, the characterization of the solar cell is approached as an electronic signature derived from the impedance of the busbar grid as a function of time.

III. METAL GRID DESIGN OF SOLAR CELLS MODELED AS TRANSMISSION LINES

This work addresses the connection of solar cells in the form of microstrip, which is a type of transmission line and are commonly manufactured using printed circuit board (PCB) technology. The propagation of the reference signal and the busbar grid impedance is strongly linked to the substrate material and the radio aspect [13], [14], as shown in the equation 3, 4 and Fig. 4,

$$\epsilon_{eff} = \frac{\epsilon_r + 1}{2} + \left[\frac{\epsilon_r - 1}{2\sqrt{1 + 12\left(\frac{H}{W}\right)}} \right] \tag{3}$$

$$Z_0 = \frac{120\pi}{\sqrt{\epsilon_{eff}\left[\frac{W}{H} + 1.393 + \frac{2}{3}\ln\left(\frac{W}{H} + 1.444\right)\right]}} \tag{4}$$

Where ϵ_{eff} is effective electrical permittivity, ϵ_r is relative permittivity, H is substrate thickness, W is connection width and Z_0 is microstrip impedance.

The boundary conditions used in the simulations followed the standard configuration described below: the magnetic boundary in the Y direction, the electrical boundary in the Z direction and the wavefront in the X direction and microstrip excitation port methodology for TDR pulse coupling as shown in Fig. 4.

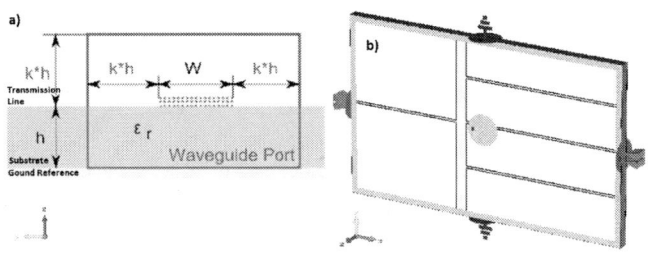

Fig. 3. a) Microstrip excitation port methodology for reference pulse TDR and b) Standard boundary conditions used for metallic grid simulation as transmission line.[adapted from 15]

Because the busbar grid is not a simple microstrip, but a transmission line with a complex arrangement, it becomes impracticable to estimate it in standard transmission line models. Thus, electromagnetic simulation software is used to determine the load impedance of the metal grid in solar cells.

The simulator applies a gaussian pulse to obtain the S-parameters and subsequently the TDR is calculated from the time integral of the pulse, which is known as the smooth step function [16], as shown in equation 5,

$$Z_{(t)} = \frac{\int_0^\infty i(\tau)d\tau + \int_0^\infty o(\tau)d\tau}{\int_0^\infty i(\tau)d\tau - \int_0^\infty o(\tau)d\tau} \tag{5}$$

Where $Z_{(t)}$ is the TDR impedance, Z_0 is the port line impedance, $i_{(t)}$ is input gaussian signal, $o_{(t)}$ is the reflection signal.

Using the TDR function from the S parameter, the electronic signature can be obtained for different solar cell busbar grid designs.

IV. SIMULATION RESULTS

This paper, first simulated the metallic grid designs under the standard FR-4 substrate $\epsilon_r = 4.3$ with $h = 1$ mm and copper reference ground from the transmission line universe and later on the polycrystalline silicon substrate $\epsilon_r = 11.9$ with $h = 0.5$ mm and aluminum reference ground. The simulated designs are shown in Fig. 4.

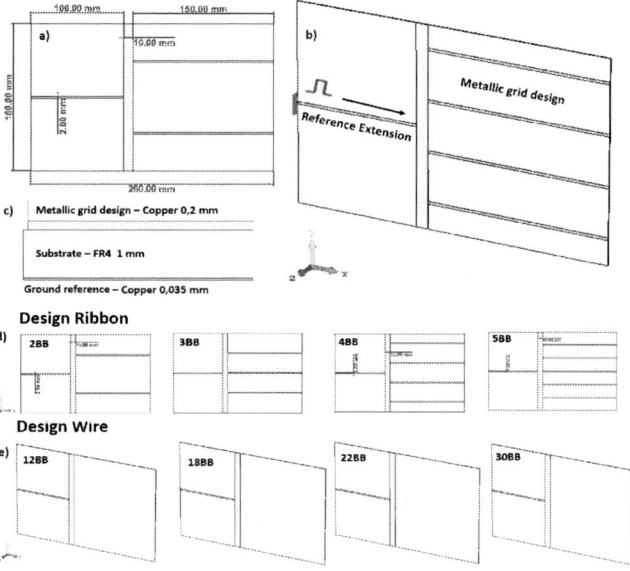

Fig. 4. Simulated design. a) Design dimensions, b) Design cross section, c) Isometric perspective and propagation direction of the TDR pulse, d) Ribbon busbar setup, e) Wire bus setup

The reference extension perfectly follows the microstrip transmission line model and can be sized to match the 50Ω impedance referenced to the RF connectors. The electronic signature of metal busbar technology starts after the reference extension. The parameterized results of simulations of metallic busbar designs typical of the PV market for solar cells are shown in Fig. 5 and 6.

978-1-6654-6060-6/23 $31.00 © 2023 IEEE

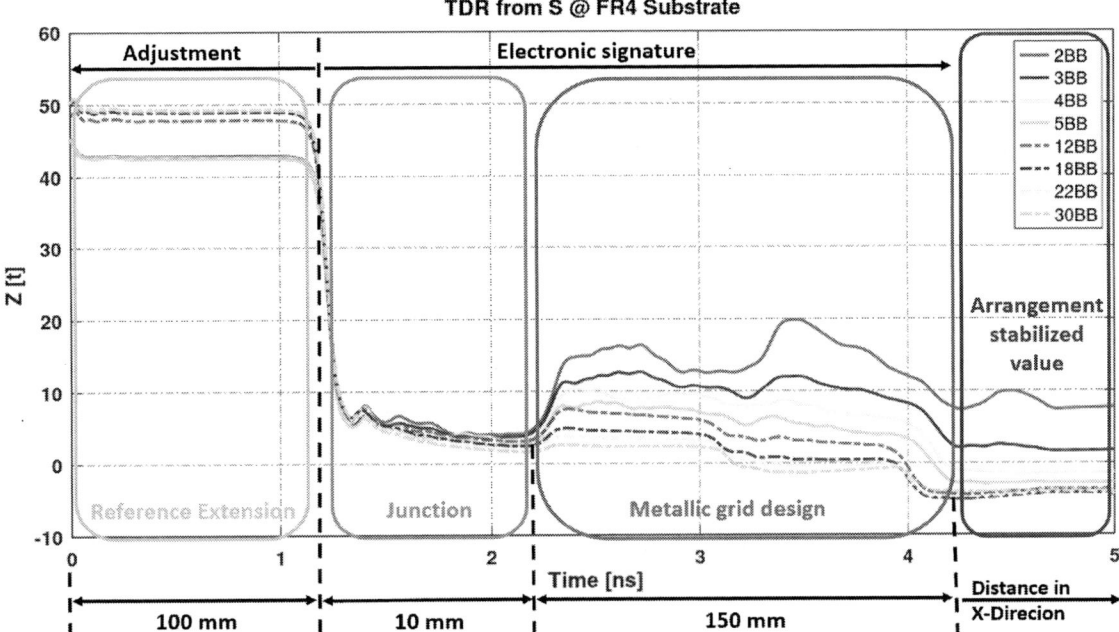

Fig. 5. Result of the parameterized simulations for the designs of 2BB, 3BB, 4BB, 5BB ribbon shape and 12BB, 18BB, 22BB, 30BB wire shape in FR4 substrate

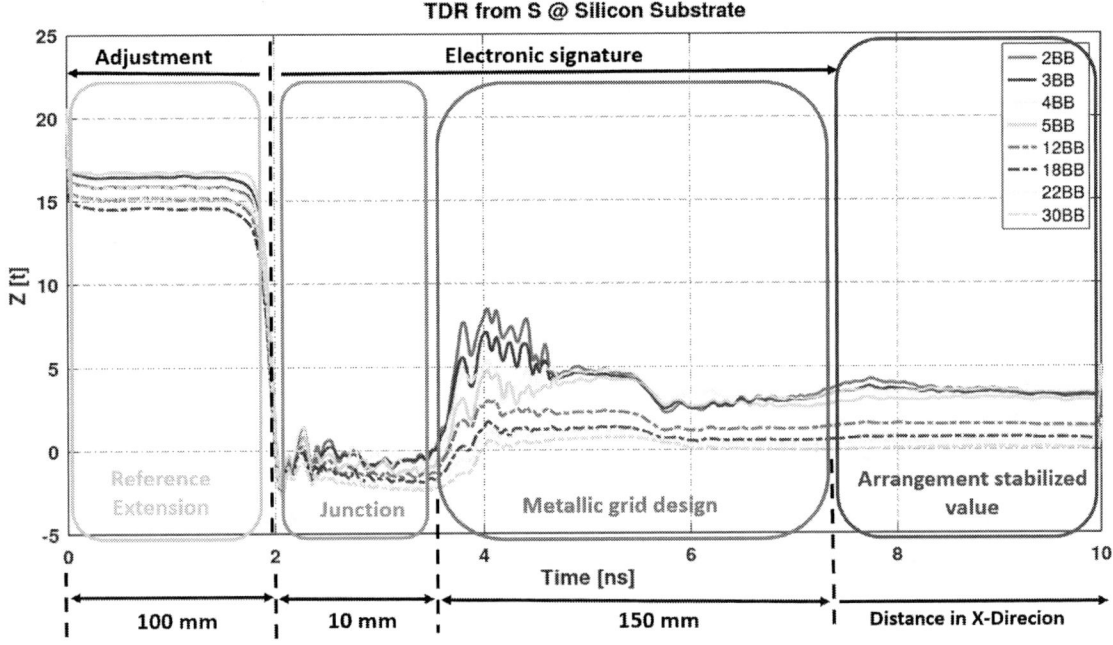

Fig. 6. Result of the parameterized simulations for the designs of 2BB, 3BB, 4BB, 5BB ribbon shape and 12BB, 18BB, 22BB, 30BB wire shape in polycrystalline silicon substrate

V. DISCUSSION OF RESULTS

The TDR pulse propagates throughout the design, which can be divided into 4 distinct structures, namely: reference extension, busbars junction, busbar grid and end of structure where the pulse reflection stabilizes. Fig 5 and 6 illustrate the propagation of the TDR pulse in the structure with polycrystalline silicon and FR4 substrate.

A. Energy reflected in the metal grid design.

Approaching the metallic grid of the solar cells as transmission lines, it can be seen in Fig 7 the path of the TDR pulse in the proposed design.

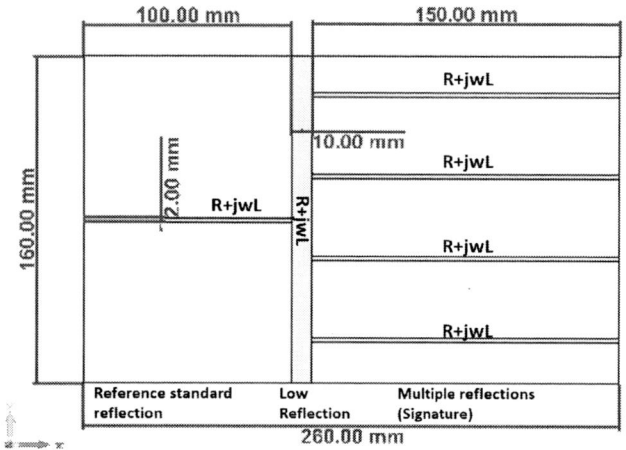

Fig. 7. Illustrative schematic of the transmission line approach to solar cell grids.

The busbar joint has a short and low reflection energy due to its geometric factor with W/h aspect ratio of 160 / 1 mm and a short length of 10 mm in the X direction. After the busbar joint, there are the busbars that produce multiple reflections in parallel, which accentuates their characteristic electronic signature.

B. Magnetic Field in the metal grid design.

The TDR pulse, because it is a pulse with a large spectral component and when traveling through the structure of the transmission lines, produces magnetic fields around the transmission line, which must be taken into account for a broad understanding of the electronic signature obtained by the TDR technique. Fig 8 illustrates the behavior of the magnetic fields in the metallic grid under test.

Due to the increase in the number of busbars in metallic grid design, electromagnetic fields can interact with each other intensely, as shown in Fig 8. From this particular electromagnetic overlapping of each busbar grid design, the grid impedance decreases as seen in figures 5 and 6 where tends to 0, thus eliminating the characteristic electronic signature for wire configurations. Note in Fig 8, the greater distance between parallel busbar in the structure present higher magnetic field strengths around them in relation to smaller distances, as well as greater impedance consequently.

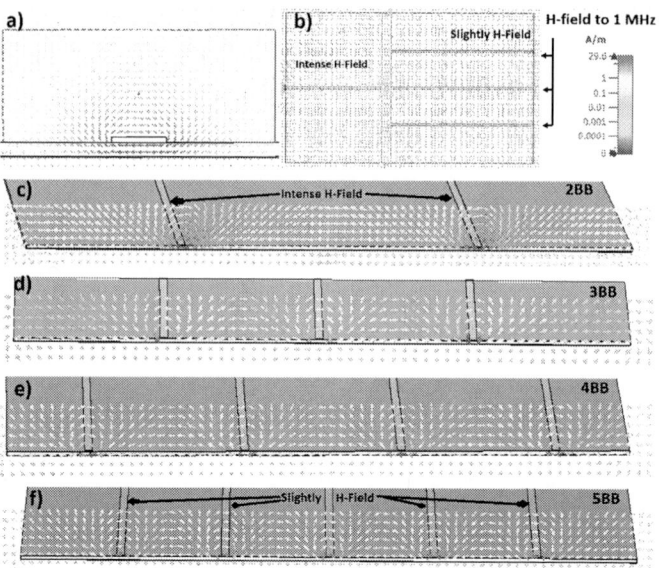

Fig. 8. Analysis of the behavior of magnetic fields on the metallic grid of the busbar. a) Reference extension cross section, b) plan view of metallic grid, c) 2BB cross section, d) 3BB cross section, e) 4BB cross section and f) 5BB cross section.

C. Substrate and Aspect ratio impact on Electronic Signature

In order to compare the behavior of the electronic signature on different substrates and aspect ratios, the results obtained in the 2BB configuration were parameterized, as shown in Fig 9.

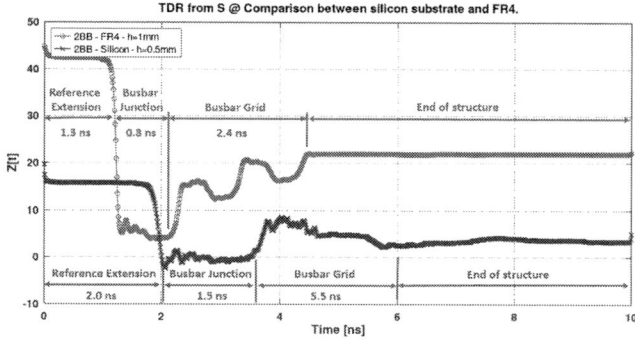

Fig. 9. Result parameterization on polycrystalline silicon substrate and FR4 in 2BB busbar technology.

It can be seen in Fig 9, the case of the metallic busbar grid under the FR4 substrate, where the reference extension with impedance close to 42Ω that propagates up to approximately $1.3ns$ in the graph, whereas the metallic busbar grid under the polycrystalline silicon substrate presents a variation in the signal propagation about $0.7ns$ and impedance amplitude about 36Ω, it is noted that the reference extension with impedance close to 16Ω that propagates up to approximately $2ns$ in the graph.

Pulse propagation speed V_p and impedance of microstrip are inversely proportional to the electrical permittivity of the substrate ϵ_r, as observed in equation 2 and 4, so the greater the electrical permittivity, greater the time used to travel the entire length of the transmission line in X direcion and lower the impedance of the transmission line. The ratio aspect W/h, also contributes to reducing the impedance of the transmission line as predicted in equation 4. The reference extension used in the design proposed in this work allows a clear visualization of this comparison.

VI. CONCLUSION

The transmission line approach proved to be suitable for electromagnetic simulation in the context of solar cells in relation to time domain reflectometry.

As can be seen in Fig. 5 and 6 the TDR simulations show clearly the ribbon eletronic signatures, however the wires eletronic signatures has little contour due to the low reflections of the parallel macrostrips.

The low impedance of the metallic grid arrangements in the wire shape, due to the large number of busbar in parallel, suggests lower electrical power losses and better energy efficiency within the simulated designs.

Thus, the TDR measurement can establish a pass-no-pass test mask for each type of solar cell busbar technology, in order to detect electrically by TDR inspection the non-visual defects, irregularities in the metallization of the solar cell grid or substrate degradation.

VII. ACKNOWLEDGMENT

This work was partially supported by the Research Project Unicamp-Funcamp-BYD Energy Brazil under contract nº 84406-22 and PADIS (Brazilian Program for the Technological Development of the Semiconductor Industry) of MCTI, Brazil. Also thanks to CNPq (National Council for Scientific and Technological Development) for granting of scholarship to the author with the process number 445074/2020-5.

REFERENCES

[1] B. Min, M. Müller, H. Wagner, G. Fischer, R. Brendel, P. P. Altermatt, and H. Neuhaus, "A roadmap toward 24% efficient perc solar cells in industrial mass production," *IEEE Journal of Photovoltaics*, vol. 7, no. 6, pp. 1541–1550, 2017.

[2] M. Mehos, C. Turchi, J. Vidal, M. Wagner, Z. Ma, C. Ho, W. Kolb, C. Andraka, and A. Kruizenga, "Concentrating solar power gen3 demonstration roadmap," National Renewable Energy Lab.(NREL), Golden, CO (United States), Tech. Rep., 2017.

[3] C. Messmer, A. Fell, F. Feldmann, N. Wöhrle, J. Schön, and M. Hermle, "Efficiency roadmap for evolutionary upgrades of perc solar cells by topcon: impact of parasitic absorption," *IEEE Journal of Photovoltaics*, vol. 10, no. 2, pp. 335–342, 2019.

[4] M. G. Villalva, J. R. Gazoli, and E. Ruppert Filho, "Comprehensive approach to modeling and simulation of photovoltaic arrays," *IEEE Transactions on power electronics*, vol. 24, no. 5, pp. 1198–1208, 2009.

[5] D. K. Schroder, *Semiconductor material and device characterization*. John Wiley & Sons, 2015.

[6] K. Mertens, *Photovoltaics: fundamentals, technology, and practice*. John Wiley & Sons, 2018.

[7] D. K. Schroder and D. L. Meier, "Solar cell contact resistance—a review," *IEEE Transactions on electron devices*, vol. 31, no. 5, pp. 637–647, 1984.

[8] T. Söderström, P. Papet, and J. Ufheil, "Smart wire connection technology," in *the 28th European Photovoltaic Solar Energy Conference*, 2013, pp. 495–499.

[9] A. Silveira, S. Barbin, and L. Kretly, "Using tdr-time domain reflectometry measurements to compare ribbon busbar versus wire busbar connections in polycrystalline solar cells: The signature approach," in *2018 IEEE-APS Topical Conference on Antennas and Propagation in Wireless Communications (APWC)*. IEEE, 2018, pp. 1–3.

[10] T. A. Note, "Tdr impedance measurements: A foundation for signal integrity," 2004.

[11] M. Buzuayene, "Rise time vs. bandwidth and applications," *Interference Technol. Available online: https://interferencetechnology. com/rise-time-vs-bandwidth-and-applications/(accessed on 30 October 2019)*, 2008.

[12] H. Casimir and J. Ubbink, "The skin effect," *Philips Technical Review*, vol. 28, no. 9, pp. P275–276, 1967.

[13] N. M. Din, C. K. Chakrabarty, A. B. Ismail, K. K. A. Devi, and W.-Y. Chen, "Design of rf energy harvesting system for energizing low power devices," *Progress In Electromagnetics Research*, vol. 132, pp. 49–69, 2012.

[14] R. Garg, I. Bahl, and M. Bozzi, *Microstrip lines and slotlines*. Artech house, 2013.

[15] C. M. Studio, "User manual," 2013.

[16] C. M. Studio and C. Tutorials, "Computer simulation technol," *Darmstadt, Germany*, 2003.

19.5% Efficient CdSeTe/CdTe Solar Cells Using ZnO Buffer Layers

Luksa Kujovic, Xiaolei Liu, Mustafa Togay, Luis C. Infante-Ortega, Kurt L. Barth, Jake W. Bowers and John M. Walls
CREST, Loughborough University, Loughborough, LE11 3TU, UK

Ochai Oklobia and Stuart J. C. Irvine
CSER, Swansea University, Swansea, SA2 8PP, UK

Wei Zhang, David W. Miller, Timothy Nagle, Rajni Mallick, Dingyuan Lu, Wyatt K. Metzger and Gang Xiong
First Solar, 1035 Walsh Ave, Santa Clara, CA 95050, USA

Abstract—Incorporating transparent n-type buffer layers in the CdTe photovoltaic device structure has led to significant efficiency improvements. In this study, we report on the use of ZnO buffer layers to achieve high device efficiencies. The 50 nm and 100 nm thick buffer layers were deposited on 3.8 mm thick NSG TEC™ 15 glass substrates and then fabricated into arsenic doped CdSeTe/CdTe devices using First Solar's absorber. The device incorporating the 50 nm ZnO buffer layer achieved an efficiency of 19.5% without the addition of an anti-reflective coating. Results show that the highly efficient ZnO based devices are stable and do not develop the anomalous J-V behavior frequently observed with MgZnO buffer layers. While intrinsic ZnO buffers can be used to fabricate high efficiency devices, the performance is limited by interface recombination.

Index Terms—CdTe solar cell, ZnO, buffer layer, As doped, CdSeTe/CdTe

I. INTRODUCTION

Since the development of photovoltaic (PV) technology, the global PV capacity has been increasing at an exponential rate [1]. This is due to the environmental and economic benefits that solar power provides. The greenhouse gas emissions from PV technologies are significantly lower than those caused by conventional fossil fuels, with thin film cadmium telluride (CdTe) modules having the lowest emission factors out of all PV technologies [2]. The cost of electricity generated from thin film PV is as low as 2.4 US cents per kWh at utility scale [3]. This is less expensive than power generated using conventional fossil fuels and other renewable energy technologies. Thin film CdTe is the most successful second-generation PV technology with a record cell efficiency of 22.1% [4]. To further reduce the cost of electricity and increase the level of PV deployment, the conversion efficiency needs to improve while maintaining low production costs. Cadmium sulphide (CdS) has been commonly used as the n-type buffer layer in CdTe devices. However, CdS causes parasitic absorption losses due to the low bandgap of 2.42 eV [5], [6]. The bandgap was widened by alloying the CdS with ZnS, resulting in performance improvements [7]. Further improvements were achieved by replacing the doped CdS layer with doped zinc

oxide (ZnO) [8], [9]. Furthermore, ZnO is a wide bandgap material which is environmentally friendly [10]. ZnO was identified as a suitable substitute due to the improved open-circuit voltage (V_{oc}) when compared to CdS/CdTe devices [11]. The approach of widening the bandgap has been reported in other research [12], this was achieved by replacing CdS with magnesium doped zinc oxide (MZO). Using alternative wide bandgap buffer layers, such as MZO, has led to efficiency improvements due to the decrease in optical losses and improved band alignment [12]. However, although it is known that MZO works well in PV devices, MZO has been observed to degrade when exposed to atmospheric conditions [13]. Magnesium oxide (MgO) absorbs the water from air to form magnesium hydroxide ($Mg(OH)_2$) [13]. This suggests that MgO doped films are not sufficiently environmentally stable. It is vital to explore alternative materials which can replace existing buffer layers to further improve device efficiencies.

This research focuses on developing ZnO based buffer layers for thin film CdTe devices. Before attempting to alloy with other materials, we have re-investigated the use of intrinsic ZnO buffer layers to establish a baseline. The ZnO buffer layers were fabricated into devices at First Solar, Inc. In this work, we demonstrate that intrinsic ZnO can be used as a stable buffer layer to achieve a conversion efficiency of 19.5% without the use of an anti-reflective coating.

II. EXPERIMENTAL DETAILS

A magnetron sputtering system (AJA International, Inc., ATC 2200-V) was used to deposit the ZnO films using radio frequency (RF) magnetron sputtering. The ZnO films were deposited on 5x5 cm substrates which include soda lime glass (SLG) and Eagle glass for optical, electrical, structural and chemical characterization. For device fabrication, the buffer layers were deposited on 5x5 cm F:SnO$_2$ coated glass (NSG TEC™ 15) substrates.

The ZnO (99.9%) target is 4 inches in diameter. The target was supplied by Plasmaterials, Inc. The substrates were set to rotate at 10 rpm. The tilt angle of the target was 9° and the distance between the target and substrates was 18.4 cm.

The base pressure in the chamber was between 2.0×10^{-7} and 4.0×10^{-7} Torr before each deposition. The working gas pressure was set to 1×10^{-3} Torr with 1% O_2 in Ar. The ZnO films were deposited at room temperature, using an RF power of 150 W. The 0613 GTC (AJA International, Inc.) power supply was used and a AIT-600-06R tuner (T&C Power Conversion, Inc.) was used for RF matching.

For device fabrication, a 50 nm ZnO layer was deposited on the 3.8 mm F:SnO$_2$ coated glass using RF sputtering. The CdSeTe/CdTe absorber was deposited on the ZnO buffer layer using First Solar's proprietary vapour-transport deposition (VTD) process [14]. Following the absorber deposition, the samples were activated with a CdCl$_2$ treatment. This was followed by a ZnTe and metal back-contact deposition [14]. The CdSeTe/CdTe absorber was doped with arsenic.

The devices were initially characterized using current density-voltage (J-V), capacitance-voltage (C-V) and external quantum efficiency (EQE) measurements. The illuminated J-V curves were measured using a 4-wire configuration under 1 Sun (AM 1.5G) with a class AAA Oriel Sol1A solar simulator. The Agilent E4980A LCR meter was used for the C-V measurements [14]. The EQE measurements were calibrated using a Si reference cell with a known (NIST-traceable) spectrum [14]. J-V characteristics were measured as a function of temperature and illumination intensity (JVTi). The measurements were taken over a temperature range of -70°C to 85°C. From the JVTi measurements, the V_{oc} was plotted as a function of temperature.

The unencapsulated ZnO/CdSeTe/CdTe device was exposed to damp heat for 1000 hours to perform accelerated lifetime testing. The damp heat conditions involved humidity of 85% at a temperature of 85°C.

III. RESULTS AND DISCUSSION

Devices incorporating MZO buffer layers often exhibit an 'S' shape behavior in the J-V characteristics [15]. Furthermore, the S-kink has been observed to form after exposing the MZO devices to atmospheric conditions for 30 days [13]. The 'S' shaped behavior can be removed after preconditioning. However, the recovery typically only remains for 3 days [16]. This instability is a drawback of using MZO in a device.

The J-V characteristics of the best performing cell in the ZnO/CdSeTe/CdTe device set are presented in Fig. 1. The J-V curve shows an improved stability when compared to MZO devices since no 'S' shape behavior is evident. Further, the J-V measurements taken before and after 1000 hours of damp heat exposure show that the device parameters did not degrade significantly and no 'S' shape behavior was observed. The ZnO/CdSeTe/CdTe device shows promising results, with a cell efficiency of 19.5%, fill factor (FF) of 82.0%, V_{oc} of 842 mV and a short-circuit current density (J_{sc}) of 28.2 mA cm^{-2}.

The EQE spectrum of the best performing cell in the ZnO/CdSeTe/CdTe device is presented in Fig. 2. Using the first derivative of the EQE as a function of wavelength [17], the bandgap of the absorber is estimated to be 1.425 eV.

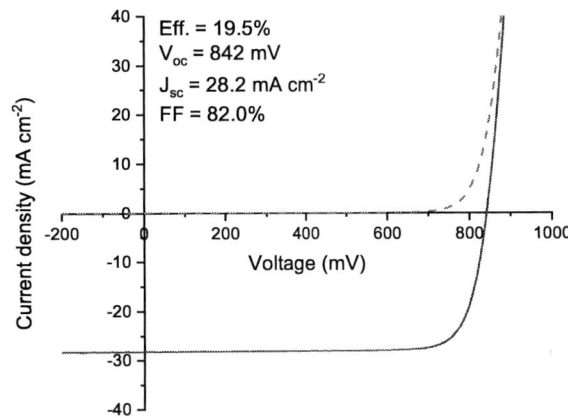

Fig. 1. Illuminated (solid) and dark (dashed) J-V characteristics for a ZnO/CdSeTe/CdTe solar cell.

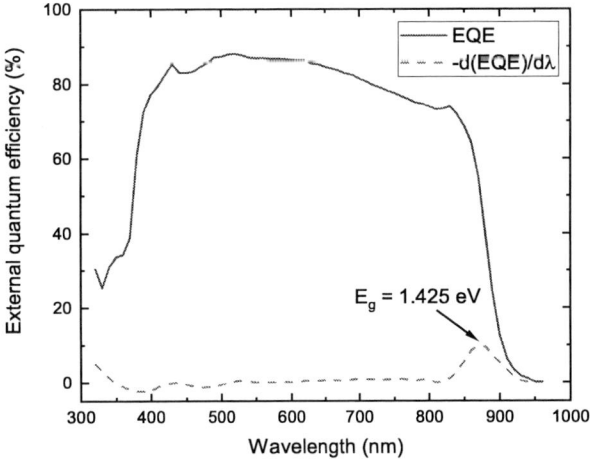

Fig. 2. EQE for a ZnO/CdSeTe/CdTe solar cell, the first derivative of the EQE as a function of wavelength is plotted to obtain the absorber bandgap.

The dominant recombination mechanisms in the device can be studied using JVTi measurements. The extrapolation of V_{oc} to 0 K gives the activation energy (E_a) of the recombination mechanism [18]. Fig. 3 shows the V_{oc} as a function of temperature, where the V_{oc} at 0 K is obtained by extrapolation. As shown in Fig. 3, the E_a (1.359 eV) is lower than the absorber bandgap (1.425 eV). This suggests that interface recombination is limiting the device performance [19].

The C-V measurements are used to plot the hole concentration as a function of depletion width. Fig. 4 shows the hole concentration versus the depletion width. The C-V profile shows the typical U-shape with a hole concentration of $\sim 1.35 \times 10^{16}$ cm^{-3}, extracted at zero-voltage bias.

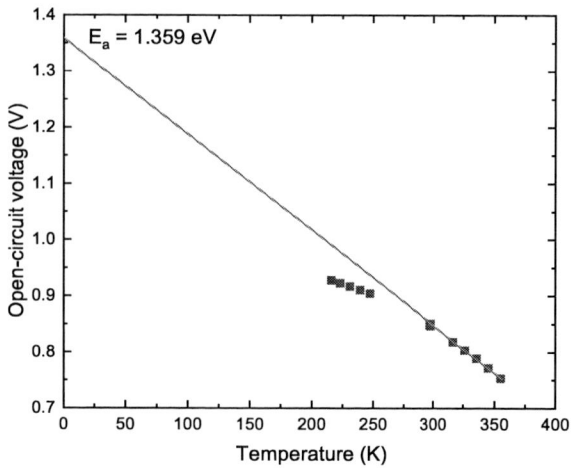

Fig. 3. V_{oc} versus temperature, a linear fit is applied to the high temperature data (355-315 K) to extrapolate V_{oc} at 0 K.

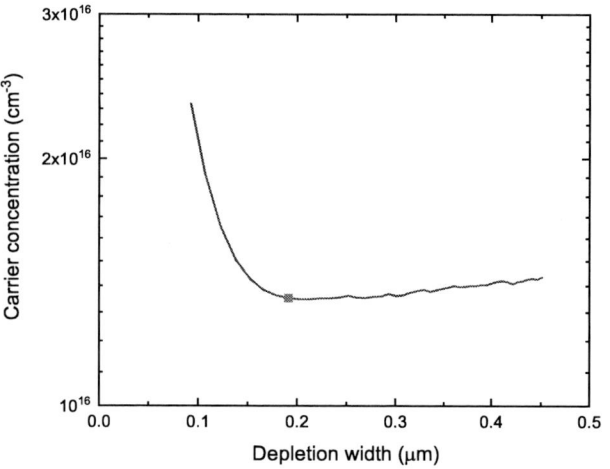

Fig. 4. Hole concentration as a function of depletion width, where the red square indicates the zero-voltage bias point.

IV. CONCLUSION

A cell efficiency of 19.5% has been achieved by incorporating a 50 nm thick ZnO buffer layer in a CdSeTe/CdTe device, without an anti-reflective coating. The J-V curve indicates an improved stability when compared to MZO-based devices. The 'S' shape behavior which is typically observed with MZO-based devices is not present. Furthermore, subjecting the ZnO device to damp heat did not cause S-kink formation. The full details of the accelerated lifetime testing will be provided elsewhere. The extrapolated V_{oc} at 0 K suggests that interface recombination is limiting the device performance. By re-investigating the intrinsic ZnO buffer layer, we find that

ZnO can be used with the As doped CdSeTe/CdTe absorber to achieve high device efficiency. In the future, alloying ZnO with suitable materials will be explored in an effort to reduce interface recombination and increase the efficiency further. The effect of varying buffer layer thickness, substrate temperature during buffer layer deposition and the use of alternative substrates, will also be studied in future research.

ACKNOWLEDGMENT

The authors acknowledge the use of facilities within First Solar's California Technology Center (CTC). The UK authors are grateful to UKRI and EPSRC for funding the project through EPW00092X/1.

REFERENCES

[1] IEA, "Renewables Data Explorer," [Online]. Available: https://www.iea.org/data-and-statistics/data-tools/renewables-data-explorer. [Accessed 4 January 2023].

[2] J. Nelson, A. Gambhir, and N. Ekins-Daukes, "Solar power for CO_2 mitigation," in Grantham Briefing Papers, no. 11, 2014.

[3] LAZARD, "Lazard's Levelized Cost of Energy Analysis - Version 16.0," 2023. [Online]. Available: https://www.lazard.com/research-insights/2023-levelized-cost-of-energyplus/. [Accessed 21 May 2023].

[4] NREL, "Best Research-Cell Efficiency Chart," 2021. [Online]. Available: https://www.nrel.gov/pv/cell-efficiency.html. [Accessed 4 January 2023].

[5] W. N. Shafarman and L. Stolt, "Cu(InGa)Se$_2$ Solar Cells," in Handbook of Photovoltaic Science and Engineering (eds A. Luque and S. Hegedus), John Wiley & Sons Ltd, pp. 567–616, 2003.

[6] B. E. McCandless and J. R. Sites, "Cadmium Telluride Solar Cells," in Handbook of Photovoltaic Science and Engineering (eds A. Luque and S. Hegedus), John Wiley & Sons Ltd, pp. 617–662, 2003.

[7] R. A. Mickelsen and W. S. Chen, Proc. 16th IEEE Photovoltaic Specialists Conf., 781–785, 1982.

[8] R. Potter, C. Eberspacher, and L. B. Fabick, Proc. 18th IEEE Photovoltaic Specialists Conf., 1659–1664, 1985.

[9] K. W. Mitchell and H. I. Liu, Proc. 20th IEEE Photovoltaic Specialists Conf., 1461–1468, 1988.

[10] J. Lang et al., "Fabrication and optical properties of Ce-doped ZnO nanorods", in Journal of Applied Physics, vol. 107, no. 7, 2010.

[11] C. Potamialis, "Process sensitivities and interface optimisation of CdTe solar cells deposited by close-space sublimation", PhD thesis, CREST, Loughborough University, 2020.

[12] A. H. Munshi et al., "Polycrystalline CdTe photovoltaics with efficiency over 18% through improved absorber passivation and current collection," in Solar Energy Materials and Solar Cells, vol. 176, March 2018, pp. 9–18, 2018.

[13] F. Bittau et al., "Degradation of Mg-doped zinc oxide buffer layers in thin film CdTe solar cells," in Thin Solid Films, vol. 691, 2019.

[14] W. K. Metzger et al., "Exceeding 20% efficiency with in situ group V doping in polycrystalline CdTe solar cells," in Nature Energy, vol. 4, pp. 837-845, 2019.

[15] D. -B. Li et al., "Eliminating s-kink to maximize the performance of MgZnO/CdTe solar cells," in ACS Applied Energy Materials, vol. 2, no. 4, pp. 2896-2903, 2019.

[16] M. Togay et al., "Transient Metastable Behavior Caused by Magnesium-Doped Zinc Oxide Emitters in CdSeTe/CdTe Solar Cells," in IEEE Journal of Photovoltaics, vol. 13, no.3, pp. 391-397, May 2023.

[17] R. Carron et al., "Bandgap of thin film solar cell absorbers: A comparison of various determination methods," in Thin Solid Films, vol. 669, pp. 482-486, 2019.

[18] R. E. Brandt, N. M. Mangan, J. V. Li, Y. S. Lee, and T. Buonassisi, "Determining interface properties limiting open-circuit voltage in heterojunction solar cells," in Journal of Applied Physics, vol. 121, no. 18, p. 185301, 2017.

[19] R. Scheer, "Activation energy of heterojunction diode currents in the limit of interface recombination," in Journal of Applied Physics, vol. 105, no. 10, p. 104505, 2009.

Numerical evaluation of optimal tilt angle for energy production and minimum shadowing for bifacial solar modules

Roberto Corso, Fabio Matera and Salvatore A. Lombardo

Institute for Microelectronics and Microsystems (IMM), National Research Council (CNR), Catania, 95121, Italy

Abstract — **Bifacial modules are becoming increasingly important in the photovoltaic market, especially in utility-scale applications. The installation of such modules, however, requires a determination of the optimal height and tilt angle in order to maximize the bifacial gain. Another important quantity to take into account is the extension and the shape of the shadow cast by the module, as these can strongly influence the structure of traditional solar farms, namely the array spacing, as well as of more innovative solutions, such as agrivoltaic solar farms. We performed numerical simulation with a 3D model of a bifacial module to determine the optimal tilt angle and the shadow cast in different configurations.**

I. INTRODUCTION

The bifacial technology is a new design for solar cell and modules which aims to harvest the albedo and diffuse light impinging on the back of the modules. This technology is gaining increasing momentum in the market, and is expected to become prevalent in the near future.

Of the many cell technologies currently present in the market, the Heterojunction Technology (HJT) is the one that presents the highest bifaciality factor [1], ranging between 90% and 95%, making it the most suited for this kind of applications.

However, much like in traditional monofacial systems, choosing the right module height and tilt angle is necessary to maximize the Energy Yield (EY) of the modules. A significant effort has been put towards the modeling of the EY of different installation configurations. Given that the EY increase of bifacial modules comes from the additional light collected by the back side, carefully estimating the albedo light impinging on the module becomes important in order to obtain significant insight from the modeling process. Therefore, 3D models are more suited than 2D models for bifacial arrays.

Another important aspect to be considered is the shadow projected by the module, which influences array spacing in conventional solar farms but also the productivity of agrivoltaic systems, in which the photovoltaic modules are installed in crop fields. The yield of these systems is heavily influenced by the interaction between the photovoltaic (PV) array and the field, and this interaction is mainly obtained by the shading effect of the modules. Accurately evaluating this effect is therefore crucial for the analysis of these systems [2].

In our previous works, we have presented a 3D numerical model which has been used to optimize the installation of bifacial minimodules [3] and to evaluate perimeter effect in large arrays [4]. In this work we use the model to evaluate the influence of the tilt angle on the EY of a bifacial module and to characterize the shadow cast by the module throughout the year.

II. MODEL DESCRIPTION

The PV system is defined by the number and disposition of the modules in the array and by the average height and tilt angle. A single module is defined by the number of cells and their size, whereas for a single cell the external quantum efficiency (EQE) spectra of both front and back side are needed in order to calculate its complete current-voltage (I-V) characteristics.

Moreover, the ground around the array is divided in finite elements from which the albedo light will be calculated. For this reason, the ground must also be characterized in terms of dispersive reflectivity.

At each day and time of the simulation, the position of the Sun is determined by its azimuth and zenith angles. From these angles the Air Mass (AM) coefficient is calculated. The total light impinging on the PV system is split in two components: the direct beam light and the diffuse light. The former is calculated from the AM value, whereas the latter is assumed to be isotropically diffused according to Liu and Jordan's model [5]. As the intensity of the direct beam light impinging on a surface depends on its tilt angle, the PV array and the ground will receive different irradiance at the same moment.

Given the Sun's azimuth and zenith angles, the shadow generated by the PV array is calculated as the projection of the array on the ground obtained from lines parallel to the Sun beams, as depicted in Fig. 1. Then, for each non-shadowed ground element, the albedo light is calculated as the reflected fraction of the incident horizontal irradiance. This albedo light is assumed to be isotropically emitted, so only a fraction of it will be collected by a PV module in the array. This fraction is equal to the solid angle under which the ground element sees the module. Since the solid angle depends on the relative position of the ground element and of the module, this calculation has to be performed for each ground element and for each module in the array.

Once the total light impinging on the module has been calculated, the total short-circuit current of each module is determined from the EQEs of the front and back side, obtaining the complete I-V characteristics for each module and for the whole array as the series connection of all modules.

978-1-6654-6060-6/23 $31.00 © 2023 IEEE

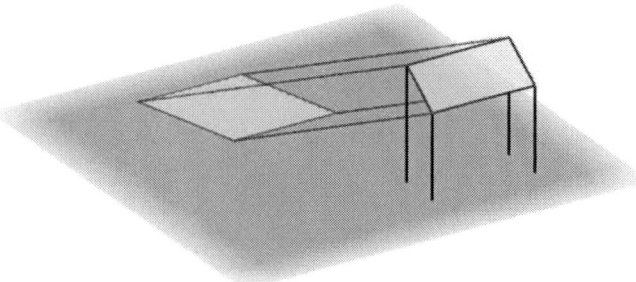

Fig. 1. Visual representation of a module (red) casting its shadow (blue) on the ground. The shadow is determined by intersecting lines parallel to the sun beams (dotted lines) with the ground.

III. SIMULATION RESULTS

A. Energy Yield optimization

The PV system considered in this work is a single module of 72 full cells in landscape configuration at an average height of 2 m over asphalt [6]. Different ground extensions have been considered, with surfaces ranging from 5 m x 5 m up to 40 m x 40 m. The EY of the module has been evaluated for a year for tilt angles ranging from 0° to 90°, with the module facing south, at a latitude of 35° N.

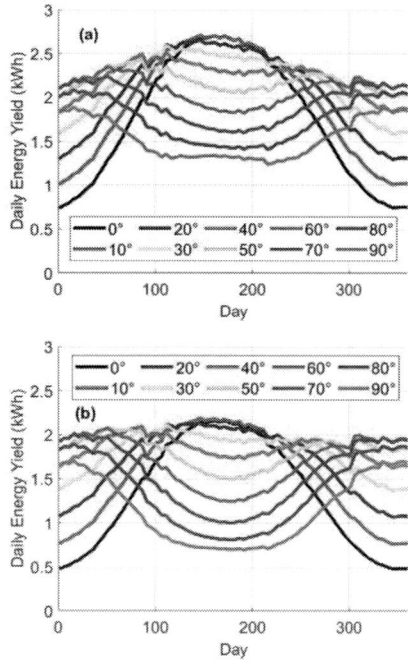

Fig. 2.   Daily EY over the year for tilt angles from 0° to 90° for a bifacial module (a) and a monofacial module (b) on a 20 m x 20 m asphalt surface.

The results of Fig. 2 show different EY trends throughout the year depending on the tilt angle for the case of the 20 m x 20 m asphalt surface: for low tilt angles the EY is higher in summer, whereas for high tilt angle the peak of EY is reached in winter. That is because the highest EY is obtained when the Sun is closest to the normal to the module front side: therefore, the higher azimuth angles reached in summer favor low tilt angle configurations, while the opposite happens in winter.

Fig. 3 shows the Bifacial Gain (BG), calculated as the percentage increase from the monofacial EY to the bifacial EY, during the year. The highest BG values are reached in summer for high tilt angles: that is because the direct beam component weights less on the total light collected by the module due to the high incidence angle, whereas the front and back sides collect similar amounts of diffuse light since the module is almost vertical to the ground.

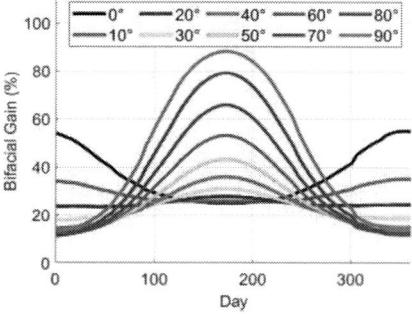

Fig. 3. Bifacial gain throughout the year for different tilt angles for a module on a 20 m x 20 m asphalt surface.

In Fig. 4 we compare the total EY of the bifacial module for different surface areas: the figure shows that the EY advantage of the wider surfaces increases as the tilt angle increases. At low tilt angles the area of ground that contributes to the albedo light with a sufficiently low incidence angle is limited, and therefore increasing the ground extension does not affect the EY of the module significantly; at high tilt angles the field of view of the modules extends both in front and behind the module, and the contribution of the albedo light to the EY increases.

Comparing a bifacial and a monofacial module on the same 20 m x 20 m surface shows a relative EY gain of the bifacial module at increasing tilt angles, as the higher tilt angles allows the bifacial module back side to see a greater ground surface, implying a higher amount of albedo light collected by the module.

The highest yearly EY (of about 172 kWh) is achieved by the bifacial module at a tilt angle between 40° and 50°, but the optimal angle decreases with the surface extension.

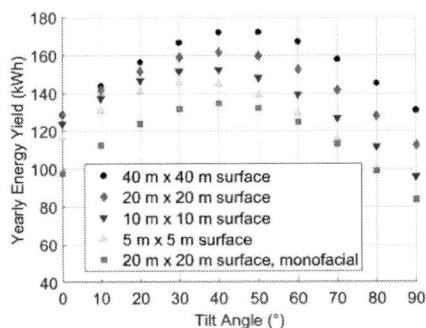

Fig. 4. Yearly EY of a bifacial module over different asphalt surfaces and of a monofacial module on a 20 m x 20 m asphalt surface for different tilt angles.

B. Ground shading

To evaluate the extension of the shadow cast by the module over the year, we consider 3 notable days: the spring/autumn equinox and the summer and winter solstices. For each day we report in Fig. 5 the fraction of the shadowed area compared to the 40 m x 40 m surface.

At the spring equinox, the shadowed area is between 0.3% and 0.5% of the total area (Fig. 4a), with the larger shadows being cast by the module at middle tilt angles. In summer, the fraction of shadowed area is less than 0.5%, but the largest shadow is cast by the module at low tilt angles (Fig. 4b). Conversely, in winter the shadow goes from 0.5% up to 2.2% of the total area, with longer shadows cast at high tilt angles (Fig. 4c).

IV. CONCLUSIONS

In this work we have applied a 3D numerical model to evaluate the Energy Yield of a bifacial module throughout the year at different tilt angles. The simulations showed that the optimal configuration corresponds, at a latitude of 35°, to a tilt angle between 30° and 50°, depending on the extension of the available ground surface. Moreover, we have calculated the extension of the shadow cast by the module, observing that while in spring and in summer the shadow is a very small fraction of the total ground surface, in winter it becomes much larger.

The present work reports the initial results of a more comprehensive study in which we evaluate the total area of the shadowed surface during the year to calculate the power of solar light that does not illuminate the ground. Assessing these quantities is of great interest for agrivoltaic applications, in order to determine the ground solar irradiation reduction due to the module shading and optimize the balance between electrical energy and crop production.

REFERENCES

[1] R. Kopecek and J. Libal, "Bifacial Photovoltaics 2021: Status, Opportunities and Challenges," *Energies*, vol. 14, no. 8, p. 2076, Apr. 2021.

[2] M. H. Riaz, H. Imran, R. Younas, M. A. Alam and N. Z. Butt, "Module Technology for Agrivoltaics: Vertical Bifacial Versus Tilted Monofacial Farms," in *IEEE Journal of Photovoltaics*, vol. 11, no. 2, pp. 469-477, March 2021.

[3] F. R. Galluzzo, A. Canino, C. Gerardi and S. A. Lombardo, "A new model for predicting bifacial PV modules performance: first validation results," *2019 IEEE 46th Photovoltaic Specialists Conference (PVSC)*, 2019, pp. 1293-1297.

[4] R. Corso, M. Leonardi, A. Scuto, S. M. S. Privitera and S. Lombardo, "Modeling of bifacial photovoltaic systems including temperature, albedo and perimeter effects: comparison with data," *2022 AEIT International Annual Conference (AEIT)*, 2022, pp. 1-5.

[5] S. Mousavi Maleki, H. Hizam, and C. Gomes, "Estimation of Hourly, Daily and Monthly Global Solar Radiation on Inclined Surfaces: Models Re-Visited," *Energies*, vol. 10, no. 1, p. 134, Jan. 2017.

[6] S. M. S. Privitera, M. Muller, W. Zwaygardt, M. Carmo, R. G. Milazzo, P. Zani, M. Leonardi, F. Maita, A. Canino, M. Foti, F. Bizzarri, C. Gerardi, and S. A. Lombardo, "Highly efficient solar hydrogen production through the use of bifacial photovoltaics and membrane electrolysis," *Journal of Power Sources*, vol. 473, p. 228619, 2020.

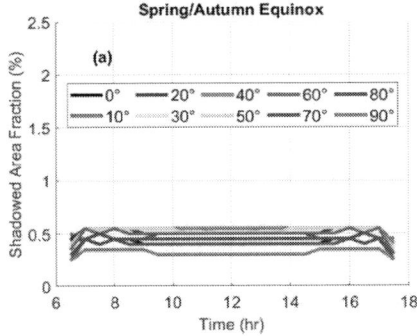
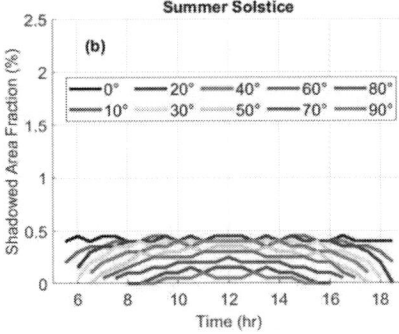
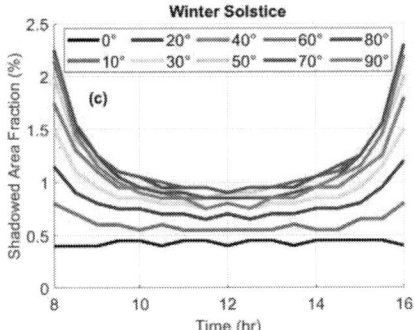

Figure 5 Fraction of shadowed area cast by the module on a 40 m x 40 m surface at the spring/autumn equinox (a), at the summer solstice (b) and at the winter solstice (e).

Optimization of 1-axis tracking with N-S rotating-axis orientation

Jiahui Shi, Xitao Liu, Teliang Mu, Xiaotong Feng, Vasilis Fthenakis

[1]Columbia University, 918 Mudd Bldg., 500 W, 120th street, New York, NY 10027,)

Abstract — This study shows that 1-axis E-W tracking installations with the axis of rotation inclined N-S (INS) towards the equator, can harvest significantly more solar energy than the same system with its rotating axis placed horizontally (HNS), and these systems can have a solar irradiation harvesting performance approaching that of 2-axes tracking systems at a lower cost than the later. The system performance is estimated using the Perez and the HDKR models for installations in locations of high solar irradiation. It is shown that a 15-degree INS tracker would increase the capacity factor of HNS from 29% to 36% in Phoenix and from 35% to 41% in Northern Chile (based on the percentage increase compared to latitude-tilt fixed system). The applicability of INS to long 1-axis tracker designs and the associated incremental cost are being investigated to assess the trade-off between increased performance and increased balance of system and installation costs.

I. INTRODUCTION

Solar tracking systems improves harvesting of solar irradiation as compared with fixed tilt systems and it is well known that 2-axes, altitude and azimuth orenting tracking systems offer a maximum solar energy capture but at much higher capital costs than fixed-plane systems. On the other hand, 1-axis altitude tracking systems have become the standard in large utility PV power plants because 1-axis tracking allows long-strings that can be installed with slightly higher cost (~10%) than the cost of fixed-tilt installations. Thus, a study of optimizing the performance of 1-axis traching systems is pertinent. The current study compares the performance of latitude fixed solar systems with 1-axis sun altitude tracking over a horizontal north-south (HNS) oriented axis of rotation, with the same 1-axis system with its axis of rotation inclined towards the equator (INS), and with 2-axis - azimuth and altitude- PV systems in two high insolation locations: Phoenix in Arizona and Atacama in Northern Chile. The HDKR and Perez anisotropic models were used to estimate the solar irradiation. The solar data used in these estimates were extracted from the National Solar Radiation Database (NSRDB) with a 30-minute frequency and a 4-km resolution. Estimations with the HDKR and Perez models of solar irradiation on the considered PV panels at these two locations were showed the same trends.

II. MODEL DESCRIPTIONS

This section sumarizes established calculations of solarlight incidence on fixed and sun-tracking planes; the calculations include diurnal sun trajectories and associated solar irradiation levels.

A. Sun position

Solar declination

$$\delta = 23.45 \, sin[\tfrac{360}{365}(n - 81)] \tag{1}$$

δ: solar declination
n: day of number, from 1 to 365

Hour angle

$$H = \left(\tfrac{15°}{hour}\right)(12 - LST) \tag{2}$$

H: hour angle
LST: Local solar time

Solar altitude

The position of the sun can be described in solar altitude angle α, and solar azimuth angle ϕ_s. In the northern hemisphere, azimuth angle is measured relative to south; whereas in the southern hemisphere, it is measured relative to north. It is calculated based on latitude L, solar declination δ, and hour angle H as shown by equations (3) and (4),

$$sin\,a = cos\,L\,cos\,\delta\,cos\,H + sin\,L\,sin\,\delta \tag{3}$$

$$sin\,\phi_s = cos\,\delta\,\tfrac{sin\,H}{cos\,\beta} \tag{4}$$

If $cos\,H \geq \tfrac{tan\,\delta}{tan\,L}$, then $|\phi_s| \leq 90°$, otherwise $|\phi_s| > 90°$ (5)

A solar zenith angle Z is defined as the complement angle of a as shown in equation (6),

$$Z = 90° - a \tag{6}$$

Angle of incidence on solar panel:

Two parameters are utilized to describe the orientation of solar panel. One is its tilt β_s, and s refers to the subarray of solar panels; the other is its azimuth ϕ_c, and c refers to collector. Same to the ϕ_s, in the north hemisphere, ϕ_c is relative to south direction. Angle of incidence of the fixed-tilt system can be estimated by equation (7),

$$cos\,\theta = cos\,\alpha\,cos(\phi_s - \phi_c)\,sin\,\beta_s + sin\,\alpha\,cos\,\beta_s \tag{7}$$

θ: Angle of incidence

However, this formulation can only be useful when β_s is specified, such as the fixed tilt. A discussion on equations for estimating sunlight incidence on sun-tracking planes follows.

Tracking tilt at different sun position

Three types of solar tracking systems are used to compare with a latitude fixed tilt solar energy system. The first is a 1-axis tracking system – Horizontal North and South (HNS). This tracking system set the axis in the North-South direction with angle of 0°, which means it rotates from east to west, exactly same as the daily solar pattern. Calculation of angle of

978-1-6654-6060-6/23 $31.00 © 2023 IEEE

Proceedings 50th IEEE PV Specialists Conference, San Juan, Puerto Rico, June 2023

incidence is based on the solar altitude angle α and the solar azimuth angle ϕ_s, as shown in equation (8),

$$\cos\theta = \sqrt{1-(\cos\alpha\cos\phi_s)^2} \qquad (8)$$

Subsequently, the surface tilt of the HNS tracking system is calculated from equation (9),

$$\cos\beta_s = \frac{\sin\alpha}{\cos\theta} \qquad (9)$$

The second tracking system is also 1-axis tracking but with an inclined north-south axis (INS). The calculation of incidence angle and tilt angle can be estimated by using equations (10) and (11).

$$\cos\theta = \sqrt{n_x^2 + n_y^2} \qquad (10)$$

$$\cos\beta_s = n_x \frac{\cos\beta_{axis}}{\sqrt{n_x^2+n_y^2}} \qquad (11)$$

β_{axis}: tilt of north-south axis
n_x, n_y: the unit vectors from the earth to the sun calculated as shown in equation (12) and (13),

$$n_x = \cos\delta\cos H\cos(L-\beta_{axis}) + \sin\delta\sin(L-\beta_{axis}) \qquad (12)$$

$$n_y = -\cos\delta\sin H \qquad (13)$$

The third tracking system is the 2-axis tracking system, in which the surface of the panels is perpendicular to the sun beam at all times; its surface tilt is calculated by equation

$$\beta_s = 90° - \alpha \qquad (14)$$

B. HDKR model

The beam irradiance on a horizontal surface is calculated by using equation (15),

$$I_{bh} = E_b \cos Z \qquad (15)$$

where Z is solar zenith angle which is the angle between the sun's rays and the vertical direction and E_b is beam irradiance (W/m2).

The ratio of Plane of Array Irradiance (POAI) to horizontal beam is equal to the ratio of the beam radiation on the pitched collector to that on a horizontal surface, as shown by equation (16),

$$R_b = \frac{\cos\theta}{\cos Z} \qquad (16)$$

where θ is angle of incidence which is the angle between the sun's rays and the normal on that surface.

The extraterrestrial irradiance can be calculated with a solar constant of 1367W/m2 from equations (17) and (18) [1],

$$G_{on} = 1367\left[1 + 0.033\cos\left(\frac{\pi}{180}\frac{360n}{365}\right)\right] \qquad (17)$$

$$H = \begin{cases} G_{on}\cos Z & \text{if } 0 < Z < \frac{\pi}{2} \text{ (sun is up)} \\ G_{on} & \text{if } Z = 0 \\ 0 & \text{if } Z < 0, \text{ or if } Z > \frac{\pi}{2} \end{cases} \qquad (18)$$

where G_{on} is the extraterrestrial radiation incident which occurred on the plane normal to the radiation, determined by day of number n.

The anisotropy index for forward scattering circumsolar diffuse irradiance depends on the beam horizontal irradiance

I_{bh} and extraterrestrial irradiance H as shown b equation (19),

$$A_i = \frac{I_{bh}}{H} \qquad (19)$$

A part of the horizontal diffuse irradiation classified as forward scattered is determined by the anisotropy index (A_i). The A_i will be high in clear weather, and the majority of the diffuse is forward distributed. The A_i will be zero if there is no beam present [1].

The total irradiance incident on a horizontal surface (I_{gh}) is equal to the sum of the beam irradiance on a horizontal surface (I_{bh}) and diffuse irradiance on a horizontal surface (E_d), as shown in equation (20),

$$I_{gh} = I_{bh} + E_d \qquad (20)$$

Applying a correction factor s to the isotropic diffuse allowed Temps and Coulson[2] to adjust for horizon brightening on clear days according to equation (21),

$$s = \sin^3\frac{\beta_s}{2} \qquad (21)$$

By manipulating factor f, Klucher [3] adjusted this correction factor as shown in equation (22),

$$f = \sqrt{\frac{I_{bh}}{I_{gh}}} \qquad (22)$$

The components of the diffuse irradiation can be calculated based on the equations (23),

$$cir = E_d A_i R_b$$
$$iso = E_d(1-A_i)\frac{1+\cos\beta_s}{2}$$
$$isohor = iso(1+fs) \qquad (23)$$

Finally, the result of the HDKR sky diffuse irradiance is shown in equation (24),

$$I_d = isohor + cir \qquad (24)$$

C. Perez model

As another anisotropic model, the overall solar irradiance with Perez diffusion model [4] is described by equation (25),

$$I = E_b\cos\theta + E_d(1-F_1)\frac{1+\cos\beta}{2} + E_d F_1\frac{a}{b} + E_d F_2\sin\beta + \rho(E_b\cos Z + E_d)\frac{1-\cos\beta}{2} \qquad (25)$$

where E_b and E_d is the direct normal irradiance and the diffused horizontal irradiance, respectively, $Z(Z(°))$ is the solar zenith angle, θ is the angle of incidence, β is the surface tilt angle, F_1 is the coefficient expressing the degree of circumsolar, F_2 is the coefficient expressing the degree of horizon/zenith anisotropy, a and b describe the view of the sky from the perspective of the surface, and ρ is surface albedo fraction.

The first term represents the direct beam irradiance. The second to the fourth term represent the isotropic, circumsolar, and horizon brighten component of the diffuse irradiance respectively. The fifth term represents the surface reflection irradiance (albedo).

a and b [4] are determined based on θ and Z as shown in equation (26),

$$a = \max(0, \cos\theta)$$
$$b = \max(\cos 85°, \cos Z(°)) \qquad (26)$$

The sky clearness ϵ [4] is calculated as shown in the

Proceedings 50th IEEE PV Specialists Conference, San Juan, Puerto Rico, June 2023

following equation (27),

$$\epsilon = \frac{\frac{E_d + E_b}{E_d} + \kappa Z(°)^3}{1 + \kappa Z(°)^3} \qquad (27)$$

which κ is a constant equaled to 5.534×10^{-6} [5].

ϵ is used to determine the empirical Perez model coefficients, f_{11}, f_{12}, f_{13}, f_{21}, f_{22}, and f_{23}.

The sky brightness Δ [5] is calculated by equation 28.

$$\Delta = E_d \frac{AM_0}{1367} \qquad (28)$$

AM_0 is the absolute optical air mass [5] and is calculated as shown in equation (29),

$$AM_0 = [\cos b + 0.15(93.9° - Z(°)^{-1.253})]^{-1} \qquad (29)$$

For each range of ϵ as shown above, there is a set of coefficients to calculate F_1 and F_2 as mentioned previously. F_1 and F_2 are calculated based on Δ, the specified Perez coefficients, and zenith angle [4] as show in equation (30),

$$F_1 = \max[0, f_{11}(\epsilon) + \Delta f_{12}(\epsilon) + Z f_{13}(\epsilon)]$$
$$F_2 = f_{21}(\epsilon) + \Delta f_{22}(\epsilon) + Z f_{23}(\epsilon) \qquad (30)$$

III. RESULTS

The amount of solar radiation incident on the surface of a PV collector is a major factor in determining the collectors' energy productiont. Using the HDKR and Perez models, the monthly average daily irradiation was calculated for two sunny sites to examine the improvement in Plane of Array (POA) solar irradiation achievable by incorporating 2-axis tracking and 1-axis tracking with horizontal rotation axis (HNS) and axis inclined 15° towards the equator (INS-15). Figure 1 shows predicted monthly averages of the POA irradiation for the three systems operating at the Atacama, Chile (27.36°S 70.33°W) and Phoenix, Arizona (33.45°N 112.06°W) and Table 1 shows annual irradiation for these locations and for Kansas City which represents average US irradiation. It is noted that the Perez model predicts POA irradiation levels that is 2-4% higher than those from the HDKR model.

Table 1 Annual irradiation levels at representative locations

HDKR model POAI Estimate:	Latitude	DNI	Diffuse	Fixed latitude	HNS	INS-15	2-Axis
Phoenix	33.45°N	2795	473	2408	3051	3245	3357
Atacama	27.36°S	2991	431	2480	3130	3312	3475
Kansas city	39.13°N	1875	516	1804	2191	2352	2458
Perez model POAI Estimates							
Phoenix	33.45°N	2795	473	2458	3174	3348	3568
Atacama	27.36°S	2991	431	2481	3379	3514	3688
Kansas	39.13°N	1875	516	1865	2277	2443	2652

As expected, for both locations, the 2-axis system receives the greatest POA irradiation over the course of the year. However, the 1-axis receives about the same POA irradiation as the 2-axis system during summer when declination is around its max value while the sunlight falls on the tropic of cancer around June 22nd. This position decreases the influence of the north-south tracking axis of the 2-axis tracking. Also, at some time, 1-axis may receive slightly more diffuse irradiation [6] and about the same beam irradiation. In summary, the performance of 1-axis-INS system approximates the 2-axis system and is superior to 1-axis-HNS system in winter season.

However, the implementation of 2-axis tracking carries extra costs that can increase the installed capital cost by 67% [7]. On the other hand, 1-axis tracking with 12% less POA irradiation than 2-axis, would carry only about 10% cost increase over a fixed-tilt [8].

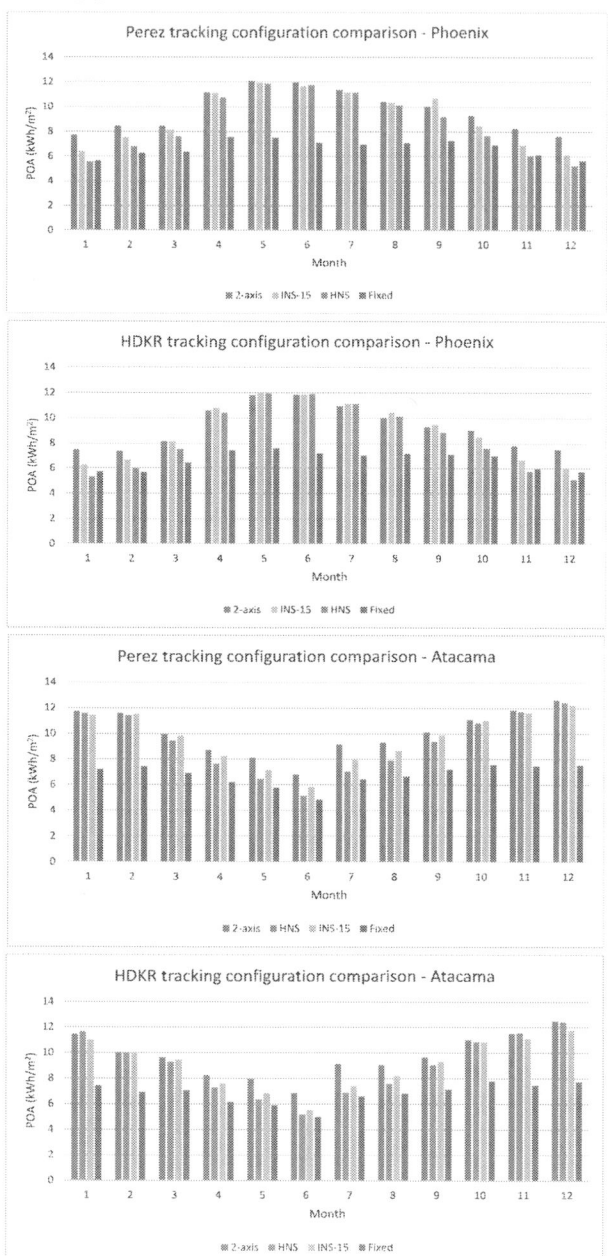

Figure 1 Monthly average daily irradiation of 2-axes, 1-axis HNS, 1-axis INS tracking and fixed tilt systems at at two locations from two models

Fig. 2. shows a detail of the POA irradiation differences on the autumnal equinox day. Overall, 2-axis tracker captures the largest irradiance and then 1-axis tracker. However, as the sun in the middle of the sky on the west-east axis at 12 p.m., that

978-1-6654-6060-6/23 $31.00 © 2023 IEEE

Proceedings 50th IEEE PV Specialists Conference, San Juan, Puerto Rico, June 2023

the POA incident irradiance is about identical for 2-axis and fixed tracking while the difference in irradiance between 1-axis and 2-axis achieves its maximum around noon. This is because tilt angle of 1-axis is 0 while the tilt angle of 2-axis and fixed is equal to the latitude of this place at noon. This highlights the advantages of inclining the rotating axis of 1-axis trackers towards the equator. As shown in Table 1, INS-15 systems receive 5.5-6.3% more irradiation than HNS systems in Phoenix, AZ.

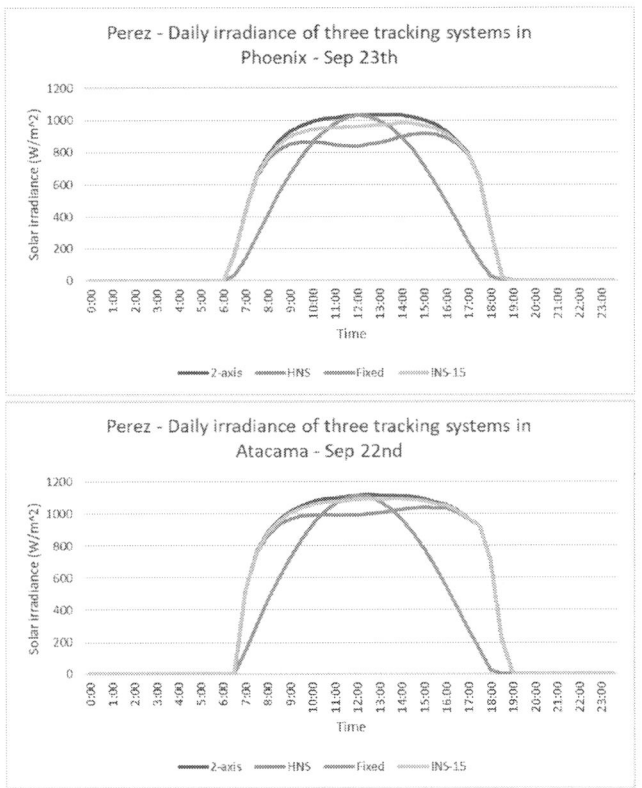

Figure 2 Daily irradiance using Perez model with 2-axis, 1-axis, and fixed tracking system, at fully sunny days; Phoenix, AZ on Sept. 23rd and Atacama, Chile on Sept. 22nd.

IV. SUMMARY

This study offers insights on implementing tracking systems When sunlight is strong, like in Phoenix and Atacama in Chile,

Figure .3 PV power plant with 80m long 1-axis HNS trackers.

solar irradiation of 2-axis tracker is about 40% higher comparing to the fixed system while 1-axis tracker is around 27% higher and its POA can be improved further by slightly inclining its rotating axis towards the equator. Considering the

high capital and operating and maintenance cost of 2-axis tracker, the 1-axis tracker is in general preferable for utility ground-mount applications, especially as long-axis designs (Fig. 3) decrease the cost-increment between 1-axis and fixed-tilt installations.

IV. LONG-TERM REVENUE AND COST ANALYSIS

Financial analysis was conducted to examine the tradeoff between the additional value creation (increased energy yield) vs additional cost on 1-axis INS versus HNS installations and area loss due to the need to increase spacing.

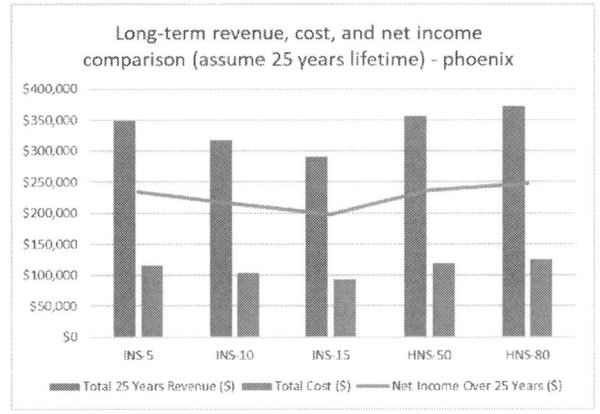

Figure 4 Long-term revenue, cost, and net income comparison for 50m and 80 m long HNS and INS tilted at 5, 10 and 15 degrees towards the equator. 25 years operation in Phoenix, AZ.

Using the annual solar irradation for Atacama, Chile, Phoenix, Arizona, and Kansas City, Kansas which were predicted by the Perez model for the considered HNS and INS systems, the total revenue of each system was calculated based on the following specifcations:

1. FS Series 6 modules with 18% efficiency
2. Performance ratio of 0.85
3. 25 years operating life with no degradation
4. Revenue from levelized cost of electricity (LCOE) of $40/MWh

Subssequently, the total revenue was calculated by multiplying the annual POA irradiation with panel efficiency, performance ratio, operating period and LCOE

For the capital cost, we considered the vertical support cost separately from the module and rest of the BOS costs. We assumed the system cost for HNS to be $900/kW and that the INS systems carry a higher cost due to the increased vertical support requirement. To quantify the additional length of supports we used the schematics shown in figures 5 and 6.

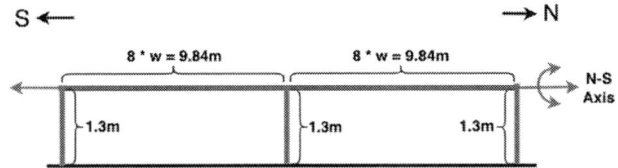

Figure 5. HNS tracking system side view schematics

Proceedings 50th IEEE PV Specialists Conference, San Juan, Puerto Rico, June 2023

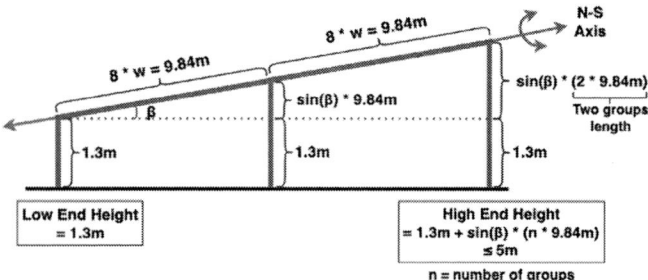

Figure 6. HNS tracking system side view schematics

We allow the INS systems to reach a maximum of 5 m off the ground at the high end and we estimated the spacing necessary to completely avoid shading between the hours of 9 am and 3 pm at the winter solstice (Dec 22). The system lengths and associated number of modules are shown in Table 2; the spacing shown in this Table corresponds to Phoenix, AZ; the spacing for Kansas is ~9.8 m and for Atacama ~6 m.

Table 2 Length & Spacing of Systems

System	INS-5	INS-10	INS-15	HNS 50m	HNS 80m
Length (m/system)	42.45	21.31	14.30	50	80
Spacing (m)	7.67	7.67	7.67	4	4
# of systems wihin 200 m	4.05	7.08	9.64	3.70	2.38
# of modules	276	241	212	296	310

INS-5	INS-10	INS-15	HNS
$26/kWdc	$27/kWdc	$29/kWdc	0.20/kWdc

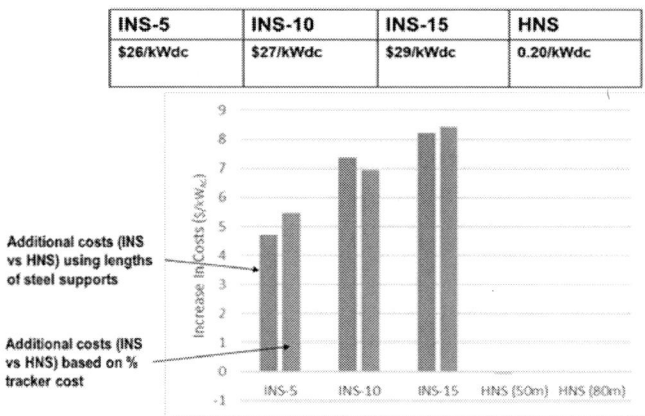

Figure 7 Estimation of incremental cost of vertical support in INS systems

The cost increment for additional supports in INS was estimated by aligning two approaches: a) Assuming the cost of

vertical supports to be 20% of the HNS tracker cost of $100/kW, increasing to $29/kW for INS-15 and b) assuming an installed cost of galvanized steel 4x4 inch posts of $11/m (Figure 7).

Based on these ssumptions, a 25-year revenue, cost, and income analysis of each system was estimated for Phoenix, AZ (figure 8). The revenue gained during 25 years of operation is decreasing with increasing tilt angle of the system due to increasing spacing and loss of active area. Comparing to the HNS-80, the gross income of INS-15 system over a 25 years

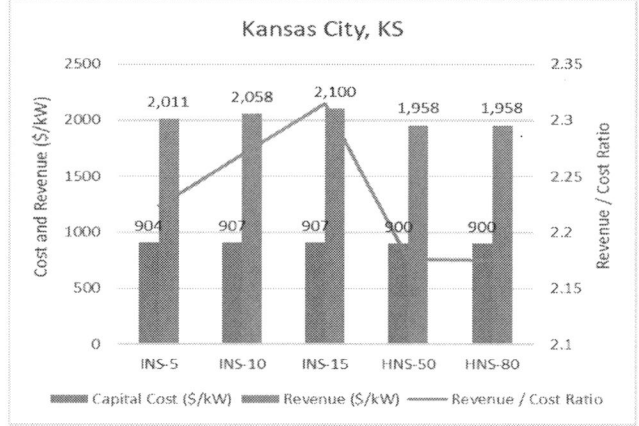

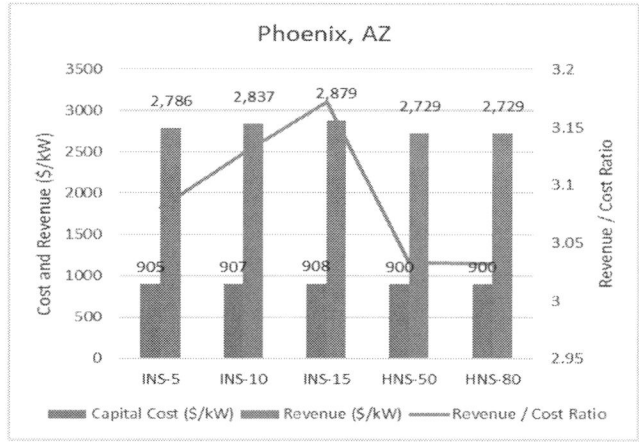

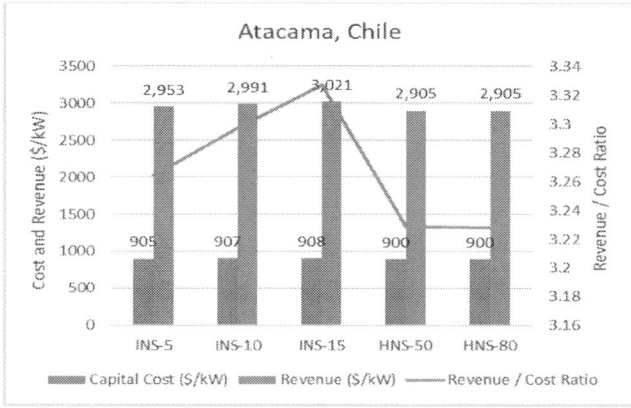

Figure 8 Normalized Revenue, capital cost, and revenue/cost ratio in Kansas City, Phoenix, and Atacama

Proceedings 50th IEEE PV Specialists Conference, San Juan, Puerto Rico, June 2023

lifetime has decreased by 25%. Even though the POA irradiation on the one-axis tracker system is improved by the inclination towards the equator, the total revenue is decreased by the reduction in the tracking length, additional supports and spacing requirements.

On the other hand, the INS systems include a smaller number of modules and BOS and correspondingly lower capital costs and it will be interesting to examine their profitability -ratio of revenue over cost- in comparison to HNS systems deployed on the same land area. Figure 7 shows the results of this analysis. The normalized cost per kWac was calculated by dividing the capital cost of each configuration by their power capacity. The power capacity was calculated by multiplying the module number in a 200 m long area with the power output of each module. In this study, we used the First Solar Series 6 module as a reference. The standardized power output is 440 Wdc/panel and the installed system cost is $900/kWdc.

The long-term revenue (assumed 25 years lifetime) was also normalized with the power capacity. The result is the revenue gained per kW power generated assuming LCOE=$40/MWhr which is typical for large utility-scale tracking systems in the US-SW. We divided the revenue per kWac by the cost per kWac to get the revenue/cost ratio. This cost analysis was performed for three locations: Kansas (latitude = 39.13), Phoenix (latitude – 33.45), and Atacama, Chile (latitude = -27.36). The ratios are listed in Table 3.

Table 3 Ratios of revenue over cost ($/$) per kWac of actual power production for Kansas, Phoenix, and Atacama Chile.

	INS-5	INS-10	INS-15	HNS-50	HNS-80
Kansas	2.22	2.27	**2.31**	2.18	2.18
Phoenix	3.08	3.13	**3.17**	3.03	3.03
Atacama	3.26	3.30	**3.33**	3.23	3.23

As shown in this table, among all tracking systems, INS-15 has the highest revenue/cost ratio in all three locations. This indicates that the increased irradiance received by the INS tracking system can cover the incremental cost associated with the longer vertical supports and also the loss of land area caused by spacing to prevent shading.

However, the difference in revenue return rate between INS and HNS systems is reduced as the latitude decreases and irradiation increases. At Kansas, the revenue/cost ratio is 2.18 for HNS-80 and increased 6% to 2.31 for INS-15. While at Atacama, a lower latitude location with greater irradiation, the percentage increase in revenue/cost ratio is only 3% from HNS-

80 to INS-15. Thus, the advantage of INS-15 in return on investment is reduced in lower latitude regions.

The POA irradiation on INS tracker systems tilted 5 to 15 degrees is 2%-7% higher than that of an HNS installation in the three considered locations.

Overall, this stsudy showed that based on the capacity (kW) of each system within a given land area, while accounting for area loss due to spacing, and increased costs of vertical support, the INS-15 can increase the revenue to cost ratio ($/$) by 3-6% over the HNS, depending on the location.

This assessment does not include potential installation labor increases and possible need for stronger (not only longer) supports at the high-end for pvercoming increased wind load. Nevertheles, our assessment shows that there is significant room for vertical support cost increases, which was assumed to be $11/m. Cost parity is achieved at cost of vertical post of $98/m for Kansas, $70/m for Phoenix and $50/m for Atacama.

REFERENCES

[1] J. A. Duffie and W. A. Beckman, *Solar Engineering of Thermal Processes*, vol. 3, no. 3. John Wiley & Sons, Inc., 2013.

[2] R. C. Temps and K. L. Coulson, "Solar radiation incident upon slopes of different orientations," *Solar Energy*, vol. 19(2), 179–184, 1977.

[3] T. M. Klucher, "Evaluation of models to predict insolation on tilted surfaces," *Solar Energy*, vol. 23(2), 111–114, 1979.

[4] R. Perez, P. Ineichen, R. Seals, J. Michalsky, and R. Stewart, "Modeling daylight availability and irradiance components from direct and global irradiance," *Solar Energy*, vol. 44 (5), 271–289, 1990.

[5] P. Gilman, A. Dobos, N. DiOrio, J. Freeman, S. Janzou, and D. Ryberg, "System Advisor Model (SAM) Photovoltaic Model Technical Reference Update," *National Renewable Energy Laboratory*,. March. p. 93, 2018.

[6] T. Huld, T. Cebecauer, M. Šúri, and E. D. Dunlop, "Analysis of one-axis tracking strategies for PV systems in Europe," *Prog in Photovoltaics: Resear and Applications*, vol. 18(3), 183–194, 2010.

[7] S. Ray and A. K. Tripathi, "Design and development of Tilted Single Axis and Azimuth-Altitude Dual Axis Solar Tracking systems," in *2016 IEEE 1st International Conference on Power Electronics, Intelligent Control and Energy Systems (ICPEICES)*, Jan. 2016, pp. 1–6.

[8] W. Nsengiyumva, S. G. Chen, L. Hu, and X. Chen, "Recent advancements and challenges in Solar Tracking Systems (STS): A review," *Renewable and Sustainable Energy Reviews*, vol. 81, pp. 250–279, Jan. 2018.

A Crucial Role of Spin-Dry Cleaning on the Surface Passivation Quality of Crystalline Silicon

Munan Gao, Vibhor Kumar, Ngwe Zin

School of Engineering, Rutgers University, New Brunswick, NJ, United States

Department of Electrical and Computer Engineering, Rutgers University, New Brunswick, NJ, United States

Present work investigates the impact of spin dry and N2 blow dry techniques on the quality and surface passivation performance of the silicon oxide grown in ozone dissolved deionized water. Though wafers were processed under the same batch for oxide growth, spin dry resulted in oxide thickness uniformity-averaged over 49-points-of ~1.39 nm ± 4.17% compared to 1.68 nm ± 21.67% in the case of N2 blow dry. The photoconductance decay measurements on spin-dried wafers provided poor lifetime and saturation current density as τ < 0.3ms and J0 (per side) with 26 to 45fA/cm2 . Above results are discussed in terms of basic phenomena involved in drying the wafers. Thereafter, an improved wafer spin drying process was established that yielded the lifetime and saturation current density as 1.4 ms and 5.6 fA/cm2 , respectively. Scanning electron microscopic imaging was also carried out that revealed the absence of pinholes in the oxide films dried using spin dry technique.

Holistic Assessment of Monocrystalline Silicon (mono-Si) Solar Panels with Recycled Content vs. Virgin-Grade Materials

Christopher C. Bondoc, Ross Lee, Mary E. McRae, and Pritpal Singh

Villanova University, Villanova, PA, 19085, United States

Abstract—With the rising demand for lower carbon energy technologies to combat global warming, the market for solar photovoltaics (PVs) has grown significantly. Inevitably, the amount of solar PV waste will increase as panels are reaching their performance end-of-life (EoL), posing major challenges in waste management, critical metal availability, and toxic material leakage. To implement the cradle-to-cradle (C2C) philosophy in this industry, several PV recycling solutions have emerged to prevent these issues from hindering the expansion of renewable energy technologies. A life cycle assessment (LCA) in this work seeks to compare the net environmental impacts (including carbon savings) of monocrystalline silicon panels (mono-Si) with virgin-grade materials compared to panels with a percentage of recycled material. A qualitative evaluation of recycling mono-Si solar panels will address the feasibility of implementation, regarding cost of material recovery, impact on human and environmental health, regulatory adjustments, and technical performance focusing on power conversion efficiency (PCE).

Keywords—cradle-to-cradle, crystalline silicon, end-of-life, life cycle assessment, material recovery, photovoltaics recycling, renewable energy, solar PV, waste management

1. Introduction

A. Background

As photovoltaic capacity expands to provide clean energy, there is a projected increase in PV waste as modules reach their end-of-life. Most panels have an expected operational lifetime of about 25 to 30 years without considering early replacement, and it is projected that 60 to 78 million tonnes of PV waste could be generated [1]. Different end-of-life scenarios apart from landfilling must be considered to prevent environmental leaching of toxic materials and the recovery of valuable components such as silver and semiconductor-grade silicon.

Fortunately, the EU has enacted legislation for PV material recovery in the form of the Waste Electrical and Electronic Equipment (WEEE) Directive, which calls upon manufacturers to their panels at the end of life. However, only the bulk materials such as aluminum frames and glass covers are recovered, which make up 80% of the panel's mass. The remaining 20% mass of the panel is incinerated, which includes the critical and expensive elements such as silver, copper, and silicon [2].

B. Recycling Methods

The PV recycling methods can be categorized into mechanical, chemical, and thermal methods. For mono-Si PV modules, the first step is frame removal, followed by delamination and material separation. The delamination stage is crucial in PV recycling since it is possible to recover an undamaged solar cell, which reduces the associated costs [3].

A full life cycle assessment of solar photovoltaic modules has not been conducted with recycling end-of-life scenarios considered. Much research in this space has only focused on cradle-to-gate or end-of-life disposal, but few studies have addressed the partial material recovery or full recycling of PV modules. The two assessed processes are Laminated Glass Recycling Facilities (LGRF) and Full Recovery End of Life Photovoltaic (FRELP). LGRF can only results in the recovery of glass, copper, and the aluminum frame, whereas FRELP recovers up to 97% of the PV module's total mass [4].

2. Goal and Scope definition

A. Goal of the Study

The goal of the study is to perform a comparative LCA between a mono-Si PV module containing virgin-grade materials with the same PV module containing recycled materials such as glass and aluminum.

B. Functional Unit

The functional unit is 1 kWh of electricity produced over the lifetime of the module. The production of the panel is normalized from area units (m^2) and the end-of-life is normalized from mass units (kg or tonne).

C. System Boundary

This LCA will involve a cradle-to-grave environmental assessment. For the PV module with recycled content, avoided carbon emissions will be accounted for. The system boundary as shown in Figure 1 includes raw material extraction (global market supply), manufacturing and assembly of the module (China), transportation, installation, use, and end-of-life (Germany). Key excluded components are the mounting system and additional electrical components.

978-1-6654-6060-6/23 $31.00 © 2023 IEEE

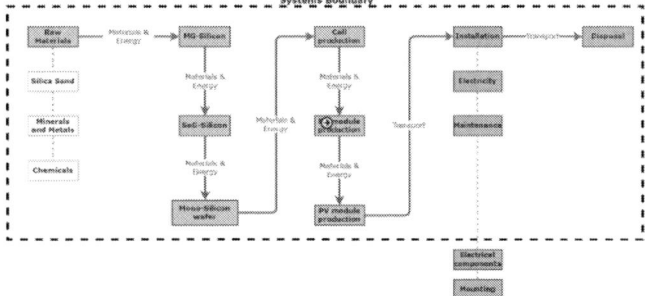

Figure 1. Systems boundary of mono-Si PV module supply chain

D. Cut-off Criteria

Any process whose total contribution to the final result, with respect to their mass and in relation to all considered impact categories, is less than 1% can be neglected. The sum of the neglected processes may not exceed 5% by mass and by 5% of the considered impact categories. For that a documented assumption is admissible.

E. Other Assumptions

The lifetime of the module is assumed to be 25 years with annual degradation rate of 0.5% per year. This lifetime is standard as PV manufacturers guarantee that modules will maintain 80-90% of their electricity power output rating at the end of 25-30 years [5].

3. LIFE CYCLE INVENTORY ANALYSIS

The LCI data was derived from the IEA PVPS Life Cycle Inventories and Life Cycle Assessments of Photovoltaic Systems 2020 report [6]. This LCI includes detailed inputs and outputs during the manufacture of the cell, wafer, and module, with additionally sections on recycling and avoided burden. The modeling software is the SimaPro Flow (version 2.22) online platform, which is a browser-based LCA tool currently in the beta phase [7]. The chosen library is ecoinvent 3.8 Allocation at the point of substitution, and the impact assessment method is ReCiPe 2016 Midpoint (H) version 1.07.1 [8].

4. LCA RESULTS

The hypothesized results for this LCA is that recycling will offer minimal environmental benefits when compared to the manufacturing of PVs with virgin-grade materials only. The impacts from recycling may be significantly underestimated as it is a novel technology at a very small scale. Additionally, a study in Australia suggests that if the construction impacts of dedicated PV recycling facilities were considered, the environmental impact would be much higher [9]. Another study of open and closed loop scenarios for PV supply chains stresses that wafer recovery should be the prioritized method of recycling PV panels, and that thermochemical methods during delamination should be avoided [3]. The silicon cells

should be recovered at end-of-life due to its high environmental impact in the production stage, as shown in Figure 2.

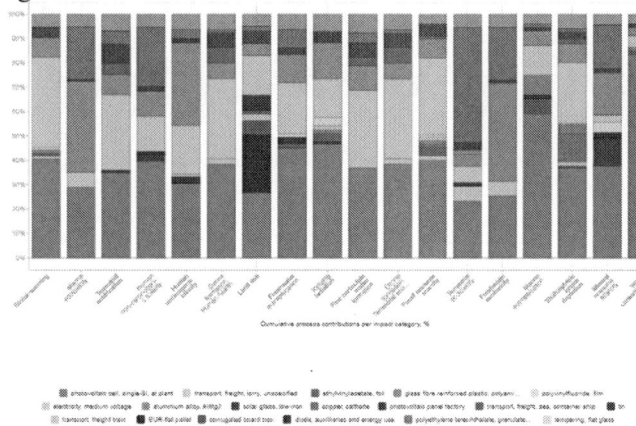

Figure 2. Cradle-to-gate environmental impacts of baseline mono-Si PV module using ReCiPe Midpoint 2016 (H) method

5. HOLISTIC ASSESSMENT

A qualitative assessment of recycling solar PVs will address the technical feasibility of implementation, socio-economic considerations, and regulations on handling PV waste. This will be done using the STEEP framework, which considers social, technological, environmental, economic, and political implications.

A. Social

From a social perspective, the PV recycling industry has the potential to create many new job opportunities. One study demonstrates that the exclusion of social factors underestimates the effect of decreased recycling prices on PV material circularity, and determines that reuse paired with recycling leads to the best social scenarios [10].

B. Technological

The technical performance of modules with recycled content has not been evaluated extensively as most modules have not reached their end of life. However, lab-scale studies are being conducted on related PV technologies. The Fraunhofer Institute along with the largest German recycling company for PV modules, Reiling GmbH & Co. KG, developed an industrial recycling solution for new PERC solar cells [11]. This groundbreaking process led to a 100% recycled silicon semiconductor without the need to add commercial ultrapure silicon. The resulting solar cell conversion efficiency was 19.7%, lower than the current 22.2% PCE found in new modules but much higher than the older generation modules.

C. Environmental

There are concerns that landfilled PVs at the end of life may leach toxic materials into the environment. For crystalline silicon PVs, there is generally little to no risk, with typical c-Si modules containing 99% non-hazardous materials [12]. Solder used to join cells in silicon modules contain lead in concentrations far below what the EPA considers hazardous, and there is very little risk. To ensure that there is no risk of these contaminants to enter waste streams, it is

D. Economic

From an economic perspective, current solar PV recycling is less cost effective than landfilling at the end of life. NREL estimates that it costs about $15-45 to recycle a PV module in the US, compared to $1-5 for landfilling [10]. A study from NREL proposes financially viable EoL recycling for PVs could be achieved with a subsidy of $10-18 per module. With an average of $28/module, this subsidy range could lead to profitability within 6 to 12 years [13].

It is also expected that the value of recycling PVs will increase over time with the growing demand for critical metals. The market value of recyclable materials in solar PVs is expected to exponentially increase from $170 million today to $2.7 billion in 2030 and $80 billion in 2050 [14].

E. Political

From a regulatory perspective, only regional policies have been put in place to deal with EoL of PVs. In 2012, the EU WEEE Directive expanded policies to include the collection and material recovery of PV panels. This mandates that PV producers (manufacturers, distributors/resellers, importers, internet vendors, etc.) comply with WEEE. Articles 1 and 4 of Directive 2008/98/EC lays down measures aimed at reducing overall impacts of raw material use and improving the efficiency of use. Generally, products with long life cycles such as PV modules should use existing collection and recovery systems similar to LGRF.

Currently, the US does not have equivalent legislation for PV producers to handle PV waste. The EPA Solar Energy Technologies Office has developed an End-of-Life Action Plan, with targets to accumulate 10 MW of PV EoL data by 2025 and cut module recycling costs to below $3 per module by 2030.

6. Conclusions

Based on the hypothesized LCA results and holistic assessment of recycling solar PVs, several key conclusions can be drawn. Firstly, recycling PV modules is a novel approach to end-of-life management. Landfilling is the currently the cheapest solution, but future incentives for recycling solar could make it a very profitable industry. The market for critical materials in out-of-warranty panels is projected to exponentially increase in value. This can lead to job creation and continual investment in solar PV manufacturing for decarbonization. Regulatory limitations need to be considered, however, as there are only policies in place in the EU currently. Environmentally, recycling extends the lifetime of carbon-intensive components such as silicon

cells. Both the LGRF and FRELP methods can recover necessary materials to manufacture new PV modules, aiming for continued decarbonization of the energy sector.

References

[1] L. Richardson, "How Long do Solar Panels Last? Solar Panel Lifespan 101 | EnergySage," *EnergySage Blog*, Oct. 08, 2021. [Online]. Available: https://news.energysage.com/how-long-do-solar-panels-last/. [Accessed: Dec. 08, 2022]

[2] "Solar panels face recycling challenge," *Chemical & Engineering News*. [Online]. Available: https://cen.acs.org/environment/recycling/Solar-panels-face-recycling-challenge-photovoltaic-waste/100/i18. [Accessed: Dec. 06, 2022]

[3] R. Contreras Lisperguer, E. Muñoz Cerón, J. de la Casa Higueras, and R. D. Martín, "Environmental Impact Assessment of crystalline solar photovoltaic panels' End-of-Life phase: Open and Closed-Loop Material Flow scenarios," *Sustainable Production and Consumption*, vol. 23, pp. 157–173, Jul. 2020, doi: 10.1016/j.spc.2020.05.008.

[4] "Research and development priorities for silicon photovoltaic module recycling to support a circular economy | Nature Energy." [Online]. Available: https://www.nature.com/articles/s41560-020-0645-2.epdf?sharing_token=gTbOszbDSFo3DPHaqB2oKtRgN0jAjWel9jnR3ZoTv0OtL9QEzcgFMYa45OUzo4NruhWqFy6wgWcztDdWim4XS3hCVOCBhp4d3QDIeKJkRd1qE3jMWfm99b10OTnspoqOjZXZdVAs0r5jaR6vhHiEl_Yr8DM9BCYUFb15u73ruAc%3D. [Accessed: Dec. 04, 2022]

[5] "Solar module lifetime predictions are getting better," *pv magazine International*. [Online]. Available: https://www.pv-magazine.com/2018/12/10/solar-module-lifetime-predictions-are-getting-better/. [Accessed: Jan. 22, 2023]

[6] "Life Cycle Inventories and Life Cycle Assessments of Photovoltaic Systems," *IEA-PVPS*. [Online]. Available: https://iea-pvps.org/key-topics/life-cycle-inventories-and-life-cycle-assessments-of-photovoltaic-systems/. [Accessed: Nov. 10, 2022]

[7] "SimaPro Flow," *SimaPro*. [Online]. Available: https://simapro.com/products/simapro-flow/. [Accessed: Jan. 22, 2023]

[8] "SimaPro | Methods manual," *SimaPro Help Center*. [Online]. Available: https://support.simapro.com/articles/Manual/SimaPro-Methods-manual. [Accessed: Dec. 06, 2022]

[9] J. K. Daljit Singh, G. Molinari, J. Bui, B. Soltani, G. P. Rajarathnam, and A. Abbas, "Life Cycle Assessment of Disposed and Recycled End-of-Life Photovoltaic Panels in Australia," *Sustainability*, vol. 13, no. 19, p. 11025, Jan. 2021, doi: 10.3390/su131911025.

[10] J. Walzberg, A. Carpenter, and G. A. Heath, "Role of the social factors in success of solar photovoltaic reuse and recycle programmes," *Nat Energy*, vol. 6, no. 9, pp. 913–924, Sep. 2021, doi: 10.1038/s41560-021-00888-5.

[11] "PERC Solar Cells from 100 Percent Recycled Silicon - Fraunhofer ISE," *Fraunhofer Institute for Solar Energy Systems ISE*. [Online]. Available: https://www.ise.fraunhofer.de/en/press-media/press-releases/2022/solar-cells-from-recycled-silicon.html. [Accessed: Jan. 22, 2023]

[12] Solar Energy Technologies Office, "Solar Energy Technologies Office Photovoltaics End-of -Life Action Plan," DOE/EE-2571, 1863490, 8846, Mar. 2022 [Online]. Available: https://www.osti.gov/servlets/purl/1863490/. [Accessed: Nov. 06, 2022]

[13] "Recycling solar panels: Making the numbers work," *pv magazine USA*, Sep. 21, 2021. [Online]. Available: https://pv-magazine-usa.com/2021/09/21/recycling-solar-panels-making-the-numbers-work/. [Accessed: Jul. 19, 2022]

[14] "There's big money in recycling materials from solar panels," *pv magazine International*. [Online]. Available: https://www.pv-magazine.com/2022/07/18/theres-big-money-in-recycling-materials-from-solar-panels/. [Accessed: Jan. 22, 2023]

Effect of Thickness of Electron Reflector Layer on the Efficiency of CdS/CdTe Heterojunction Thin-Film Solar Cell

Chaitanya Santosh Rampalli and Hamid Fardi

Department of Electrical Engineering, University of Colorado Denver, Campus Box 110, P.O. Box 173364, Denver, CO 80217-3364, USA

Abstract — ZnO/CdS /CdTe solar cell performance modeling has been done using AFORS-HET. Device simulation is used to investigate the effect of the thickness of ER layer on the efficiency of a CdS/CdTe Heterojunction Thin-Film Solar cell. Simulation analysis is used to optimize the experimental base device under AM1.5 solar spectrum. Efficiency of the device is observed by varying thickness of the extended electron reflector region from 100nm to 10,000nm with an Ohmic contact and a doping density of 7×10^{16} cm^{-3}. Efficiency obtained for a 500nm ER layer is 28.71% and FF is 79.37. The characteristic I-V results show that samples with higher CdTe thickness have higher efficiencies.

I. INTRODUCTION

The technology that transforms direct sunlight into electrical energy is known as a solar cell or photovoltaic gadget. Solar cells can function without releasing harmful substances into the atmosphere. Solar cells also have a lifespan of more than 30 years. Commonly used thin film solar cells are amorphous silicon, CuInGaSe2, and CdTe. Amorphous silicon solar cells have a great share in the market. CdTe has the advantage of being a direct bandgap material with high absorption (in comparison to silicon) and ease of device manufacturing. It can be formed on glass in thin films, often measuring only a few micrometers thick, at moderate temperatures. Experimental studies demonstrate that the Schottky barrier is formed by the CdTe cells at the back contact, lowering cell effectiveness and performance.

The simulation program AFORS-HET is the foundation for the modeling work on CdTe structures. Several solar cells have been developed and designed using the software in the past. The device modeling includes a number of position-dependent material and transport factors to make it resilient while solving a sizable number of nonlinear and tightly linked partial differential diffusion equations. Over a collection of mesh geometry, the device equations are solved. In order to accurately represent the properties of a solar cell, the model for solar cell architectures also incorporates optical absorption and a wide spectrum of light input that may be supplied. We will elaborate on how the software functions well for CdTe/CdS photovoltaic cells in the next sections.

By including a CdTe layer as an electron-reflector-extended region (ER), an offset to the CdTe absorption layer is taken into consideration for the device's optimization. Other materials, including as CdZnTe, CdMnTe, and CdMgTe, may be used in addition to CdTe to form an ER layer. The depletion region of the Schottky barrier contact is made narrow by a thin-layer electron reflector, allowing the majority of hole carriers to tunnel through and reducing loss.

II. DEVICE PARAMETERS

Fig.1 shows a schematic of a CdTe thin-film solar cell employed in the simulation with the TCO/CdS(n-type)/CdTe(p-type)/metal structure with ER layer.

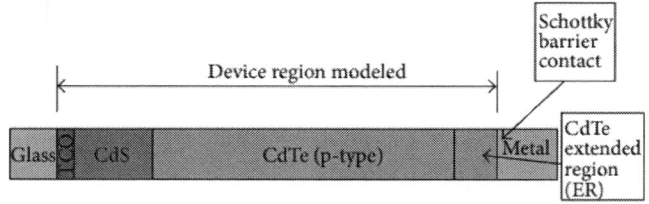

Fig. 1.   Structure of the device: CdTe thin-film solar cell. 0.5μm TCO(ZnO), 40nm CdS layer, 2 μm CdTe layer, varying thickness of ER layer.

CdTe being a direct band gap semiconductor having a large absorption coefficient, allows us to use thin films of micro-meter magnitudes. In this paper, this thickness will be varied from 100 nm to 10,000 nm with various parameters observed.

TCO (ZnO-TCO) layer's thickness is set to 500nm with a doping density of 1×10^{18} cm^{-3}. It has a bandgap of about 3.4 eV which reduces optical losses by preventing blue light to reaching the CdTe thick layer. Along with this CdS layer of thickness 40 nm is used with a doping density of 1×10^{18} cm^{-3} and a bandgap of 2.53eV.

It is assumed that the layer of CdTe has a bandgap of 1.42 eV and an electron affinity of 4.28 eV. A metal must have a work function (the sum of electron affinity and bandgap) larger than or equal to 5.7 eV in order to form an ohmic contact. Because of this, the majority of metal contacts used with p-type CdTe, such as copper (work function: 4.6 eV), nickel (work function: 5.1 eV), or titanium (work function: 4.3 eV), produce nonohmic Schottky type barrier connections that lower the CdTe cell's overall performance efficiency. CdTe layer and Extended CdTe layer(ER layer) both have a doping density of 7×10^{16} cm^{-3} of p-type. Also, lifetime of the carriers is assumed to be 1×10^{-9} s.

For maintaining optical properties of each layer, appropriate absorption files have been used.

III. EXPERIMENTAL RESULTS

The device structure shown in Figure 1 is considered for these experiments. The device structure and its parameters are similar to those of [1].

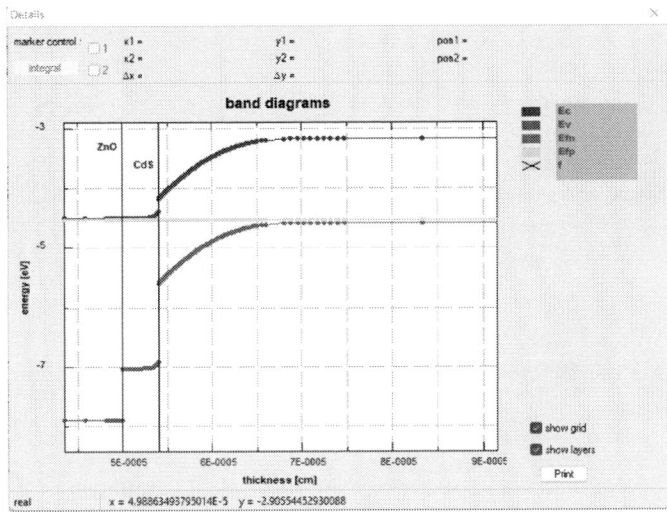

Fig 1: Band diagram of TCO(ZnO)/CdS/CdTe/Extended CdTe(ER) solar cell

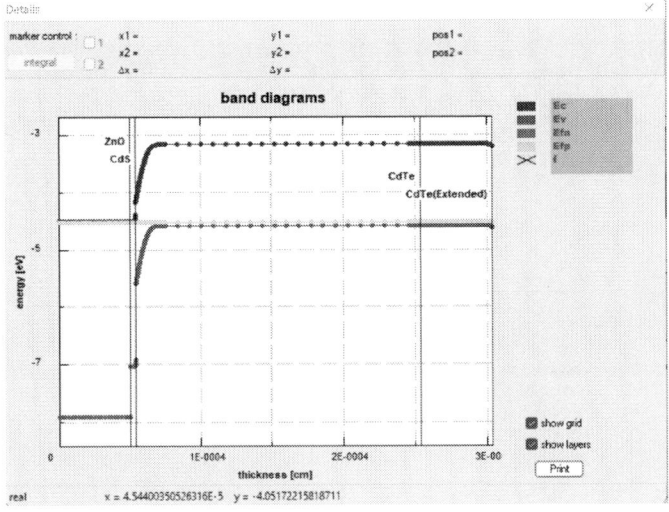

Fig 2: Full view of band diagram of TCO(ZnO)/CdS/CdTe/Extended CdTe(ER) solar cell

On Simulation in Afors-HET, with the values mentioned in the Device Parameters sections and reference values from [1], the band diagrams in Fig 1 and Fig 2 are obtained for the TCO/CdS/CdTe/Extended CdTe(ER) solar cell.

TABLE I
RESULTS OF THE DEVICE SIMULATION

ER THICKNESS (nm)	Voc(V)	Jsc(mA/cm^2)	FF%	EFF%
100	0.9189	39.26	79.36	28.63
500	0.9194	39.35	79.37	28.71
1000	0.9197	39.41	79.4	28.78
2000	0.9201	39.48	79.41	28.84
5000	0.9204	39.55	79.44	28.91
10000	0.9207	39.58	79.44	28.95

Table I tabluates the Voltage, Current density, FF and efficiency for thicknesses of CdTe Extended region i.e 100nm, 500nm, 1000nm, 2000nm, 5000nm, and 10000nm.

The voltage obtained is called the Open Ciruit voltage, Jsc is called the Short Circuit current. FF is called the "Fill Factor and shows the percentage of actual maximum power to the product of Voc and Jsc. And η represents the efficiency of the solar cell.

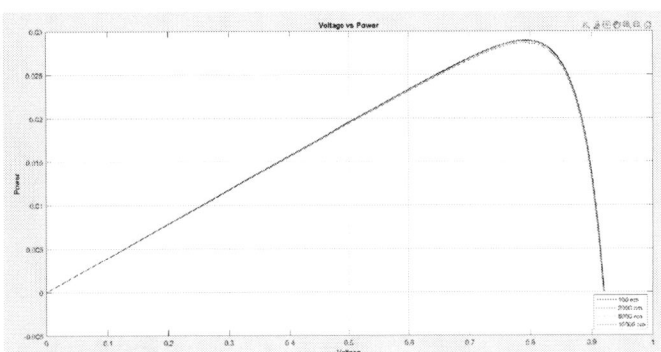

Fig. 3. Matlab plot of Voltage vs Power for 100nm, 2000nm,5000nm, and 10000nm

978-1-6654-6060-6/23 $31.00 © 2023 IEEE

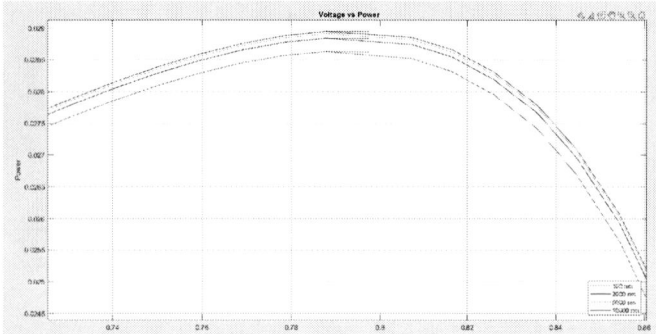

Fig. 4. Closer view of Matlab plot of Voltage vs Power for 100nm, 2000nm,5000nm, and 10000nm

The Fig.3 and Fig.4 represent the Voltage vs Power plot, the data points for which were obtained by running 100 iterations on Afors-HET and plotted using MATLAB. Fig.3 gives the total range of data from 0V to Voc V. Whereas Fig.4 gives a closer view of Fig.3, highlighting the increase in power with an increase in ER thickness.

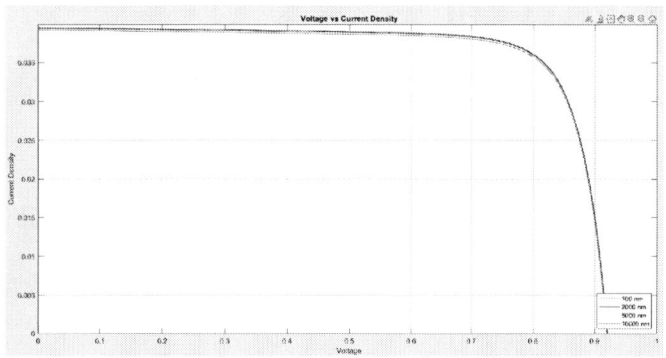

Fig. 5. MATLAB plot of Voltage vs Power for 100nm, 2000nm, 5000nm, and 10000nm.

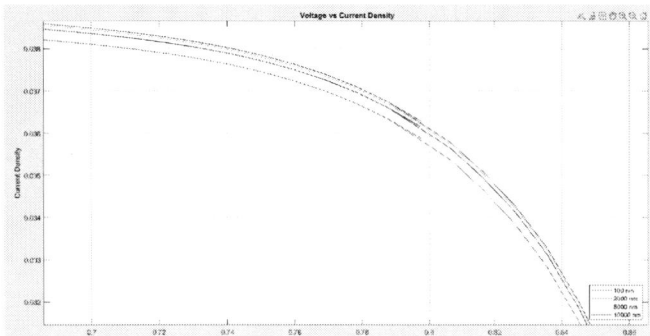

Fig. 6. Closer View of MATLAB plot of Voltage vs Power for 100nm, 2000nm, 5000nm, and 10000nm.

Fig.5 and Fig.6 represent the Voltage vs Current Density plot, the data points for which were obtained by running 100 iterations on Afors-HET and plotted using MATLAB. Fig.5 gives the total range of data from 0V to Voc V. Whereas Fig.6 gives a closer view of Fig.5, highlighting the increase in current density with an increase in ER thickness.

IV. CONCLUSION

CdTe layer and the ER layer have a doping of 7×10^{16} cm^{-3}. The analysis of 6 different thicknesses of ER layer which is shown in Table I shows that there is very less increment of efficiency when the device is illuminated with the AM1.5 spectrum. A device with 10000nm ER thickness has Voc 920.7 mV, Jsc 39.58 mA/cm^{2,} and an efficiency of 28.95%.

REFERENCES

[1] Hamid Fardi and Fatima Buny, "Characterization and Modeling of CdS/CdTe Heterojunction Thin-Film Solar Cell for High Efficiency Performance," *International Journal of Photoenergy Volume 2013, Article ID 576952.*

[2] N M D Putra, Sugianto, P Marwoto, R Murtafiatin and P D Rizaldi, "Performance profile analysis of ZnO/CdS/CdTe solar cells thin film: A review of absorber thickness and device temperature," Journal of Physics: Conference Series, Volume 1567, 6th International Conference on Mathematics, Science, and Education (ICMSE 2019) 9-10 October 2019, Semarang, Indonesia.

[3] G. C. Morris and S. K. Das, "Some fabrication procedures for electrodeposited CdTe solar cells," International Journal of Solar Energy, vol. 12, no. 1–4, pp. 95–108, 1992

Fabrication of Au/TiO$_x$ Nanoislands Systems by a Solid State Thermal Dewetting for Plasmonic Solar Cell applications

Brahim Aïssa[1,*], Mohammad I. Hossain[1], Adnan Ali[2]

[1]Qatar Environment and Energy Research Institute (QEERI), Hamad Bin Khalifa University (HBKU), Qatar Foundation, Doha, P.O. Box 34110, Qatar
[2]Department of Chemical Engineering, Jeju National University, Jeju 63243, Korea

Abstract — We report on the nucleation of Au nanoislands onto TiOx seed films surfaces which occurred in consecutive steps. Firstly, TiO$_x$ thin films were grown on quartz substrates reactively by e-beam evaporator and then thermally annealed at different temperatures, ranging from 300 ºC to 900 ºC. Subsequently, a nano-film of Au was deposited on the top of these annealed TiO$_x$ surfaces. The stacked Au/TiO$_2$ samples were then post-annealed at a temperature of 600 ºC for 1 hour, to study the thermal dewetting properties and the influence of the different TiO$_x$ morphologies on the formation of Au nanostructures and their plasmonic response. These Au/TiOx dual-systems were characterized accordingly to probe their topological, morphological, structural, and optical properties. The thermal dewetting effects on these systems were found to be more impactful at high temperatures (>500 ºC), where the Au nanoparticles size distribution was found to follow a Gaussian distribution centered around 30 nm. Finally, the absorption peak for Au nanoislands has shown a localized surface plasmon resonance close to 520 nm, along with a broad shoulder peak and a strong tail, thereby reflecting the wide distribution of the formed Au nanoparticles sizes. The integration of these Plasmonic systems into planar solar cell has demonstrated an increase up to 35.51 % in terms of the photocurrent generation and was found to be Au size and surface density' dependent.

I. INTRODUCTION

The mechanism of localized surface plasmon resonance (LSPR) occurs through an interaction between incident light and the electron cloud of a metal particle, at the surface. More specifically, it is attributed to the interaction between the fluctuating electro-magnetic waves of the light which oscillate the free electrons in the metal nanostructure. The driven away electrons from the nuclei results in the Coulomb force for a resonance condition [1,2]. Such mechanism occurs at the surface of the metal particles and significantly depends on the size and shape of the nanostructure, and also on the surrounding media [3,4]. Gold (Au) and silver (Ag) are the most competitive materials for such LSPR application though other metal can be also used [3].

In general, LSPR happens within the visible light spectrum, however it can be extended to ultra-violet or infrared spectrum, in the case, for instance, of Al, ITO, GaAs and InP [5]. Experimentally, only a few explicit parameters were found to control the resonance frequency, including the particle size, the overall size distribution (in case of network), the particle shape (i.e., geometry) and the density of the particle per unit area. As a matter of fact, the resonance frequency can be efficiently tuned towards shorter wavelength through the distance increment between the nanoparticles, and the plasmon energy can be shifted as well as split through the interactive coupling of the nanoparticles [6]. The intensity of the resonance response can be controlled through the diameter of the nanoparticles [7]. The shape of the particle also plays a significant role for the aforementioned effect, such as the frequency was demonstrated to shift to 480-560 nm for pentagon like-shape to 560-700 nm for triangle one [8]. Similar phenomenon was also observed for Ag nanoparticles (NPs) [9].

Various fabrication techniques have been used extensively to develop such plasmonic nanostructures, including e-beam lithography [10], nanoimprint lithography [11], and chemical synthesis [10]. Although the growth control is the strength of these synthesis methods, their associated fabrication cost if very high and their yield is rather low [4].

Physical vapor deposition growth processes are very efficient where hills of agglomerated films can be grown through introducing tensile stress [9]. The formation rate of the material agglomeration varies with respect to the annealing temperature, the initial thickness of the films, and strain of the grown NPs, and surface defect states [11]. The choice of the substrate is also an added factor which influences the growth dynamics, the size and distribution of the formed NPs, their structure, and their shape. Many works compared the size tunability of metal NPs as a function of different substrates [9-12]. Tanyeli et al. demonstrated that the dispersion of silver particles is lower on flat surface, whereas smaller Ag nanostructures were obtained in the case of thermally conductive substrates [12]. Hence, in this work, we have correlated the solid state thermal dewetting of e-beam evaporated Au thin films on the top of TiOx substrates, which were initially annealed at different temperatures to end up with different morphologies. The goal is to study the effect of the TiOx underneath layer on the morphology and plasmonic response of the formed Au nanostructures and to correlate both to the thermal dewetting temperature as a key parameter to control the LSPR. Once optimized, these structures have been implemented in planar solar cell and have demonstrated a clear improvement of the photocurrent generations (PV results will be detailed in the full version of this paper).

II. METHODOLOGY

978-1-6654-6060-6/23 $31.00 © 2023 IEEE

Metal oxide thin films (TiO$_x$) were grown using e-beam evaporation of Ti pellet at room temperature at a controlled deposition rate of 1 Å/s, under a constant oxygen flow rate to maintain a deposition pressure of 2×10^{-4} Torr. Metal layers (Au) were also grown on quartz substrates using the same e-beam evaporation, at room temperature, without any oxygen flow. The maximum flowrate was set to 20 sccm to restrict any arching due to free oxygen charge carriers. Samples were optically measured using ultraviolet–visible (UV–Vis) spectroscopy. Also, the surface wettability behavior was characterized using contact angle (CA) measurements. In addition, surface topology was assessed by atomic force microscopy (AFM), and the microstructure of the films was studied by field-emission scanning electron microscopy (FESEM). Structural characterization of the grown films was carried out using X-ray photoelectron spectroscopy (XPS) and X-ray diffraction (XRD). At the beginning, thickness optimization was performed for each layer to maximize the optical properties. Our tool allows us to deposit multi-layers in a stacked manner without breaking the vacuum. The substrates with a dimension of 1" × 1" were sonicated and cleaned using DI water, acetone, and isopropanol. Later, they were dried under inert nitrogen. The e-beam evaporation tool was a Denton Vacuum ExplorerTM evaporator. The deposition pressures were fixed at 2×10^{-4} Torr, while the oxygen flow was varied automatically by the evaporator to maintain the desired pressure. Ti and Au pellets were used to develop thin metal layers (Kurt J. Lesker, 99.9995% purity). Pristine TiOx films were annealed at different temperatures ranging from 300 °C to 900 °C, using muffle furnace, for 1 hour, whereas annealing temperature for Au/TiO$_2$ films was kept at 600 °C for 1 h.

II. RESULTS AND DISCUSSION

Figure 1(a) shows the grazing-incidence X-ray diffraction (GIXRD) pattern from 20 to 80° for Au-TiO$_x$ thin films (TiO$_x$ films were initially annealed at temperatures ranging from 300 to 900 °C, then Au onto TiO$_x$ samples were annealed all at the same temperature of 600 °C). In general, TiO$_x$ films confirm anatase phase without performing any annealing treatment. For all annealed samples, anatase phase has been confirmed at 25° with (101) orientation, anatase phase at 47° with (200) orientation, rutile phase at 65° with (310) orientation and rutile phase at 70.1° with (301) orientation. Furthermore, GIXRD analysis of gold thin films confirm the crystalline phase of the Au/TiO$_x$ as demonstrated by four peaks corresponding to standard Bragg reflections (111) at 38.1°, (200) at 44.3°, (220) at 64.5°, and (311) at 77.7° of face centers cubic lattice. The sharp peak at 38.1 with (111) orientation confirms the preferential growth of Au films. Higher crystallinity has been observed for both Au and TiOx samples at high dewetting temperature. Overall, the peak intensity is proportional to the level of crystallinity. As confirmed by the structural study,

samples annealed at high temperature resulted in larger crystallites.

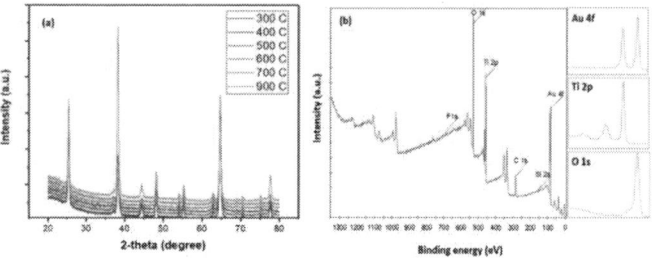

Fig. 1　(a) GIXRD patterns of the Au/TiOx thin films with 300°C–900 °C annealing temperature of TiOx films. (b) XPS survey of the Au/TiOx system (TiOx annealed at 900 °C, Au/TiOx annealed at 600 °C) along with Au 4f, Ti 2p and O 1s spectra.

Figure 1(b) shows the X-ray photoelectron spectroscopy (XPS) survey of a selected sample, namely the Au/TiO$_x$ system (TiO$_x$ annealed at 900 °C, Au/TiO$_x$ annealed at 600 °C) along with Au 4f, Ti 2p and O 1s spectra, after a monatomic etching, and with oxygen profiling from the XPS study of the thin films. Regarding the oxygen chemical state analysis, carbon spectra fitting has been used to identify the amount of oxygen related to carbon species (i.e., C − O and C═O), which can be deduced from the total oxygen to conclude the oxide-related oxygen (C-metal bond). All the samples show mainly the Au, Ti and O signals with clear TiO$_x$ chemical states, with Ti 2p located at 450 eV and Au located at 100 eV. The stoichiometry of the TiO$_x$ samples was found to be TiO$_{1.91}$ based on the annealing temperature. As confirmed, all samples show both Au and TiO$_x$ which indicates that Au is rather nanoislands like structure which were formed on the top of TiO$_x$.

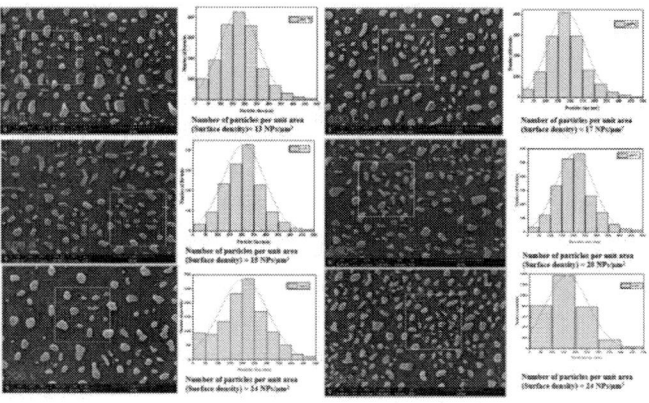

Fig. 2　SEM morphological results of Au/TiOx systems with TiOx films annealed at six different temperatures, from 300 °C to 900 °C. The size distribution and surface density of the Au NPs is associated with their respective histograms.

Figure 2 shows the SEM morphological results of Au/TiO$_x$ systems with TiOx films annealed at six different temperatures, i.e., ranging from 300°C to 900°C. First evaporated films covered the entire substrate and are uniform, which is essential to develop high-end optoelectronic devices. The size distribution and surface density of the Au NPs are associated with their respective histograms. As measured, TiO_x thin films resulted in larger grain with respect to the temperature, whereas Au NPs formed larger crystallites as a consequence of larger grain boundaries of the seed layer (i.e. TiO_x). In general, seed layers with smaller grains resulted in higher density of grain boundaries. It has been noticed that dewetting temperature of 900°C resulted in higher NPs density proportional to that of the density of the grain boundaries. It is mainly attributed to the diffusion of atoms from the edge to the pit.

The optical properties of the films were investigated through UV–Vis–NIR spectroscopy. The optical transmission was measured to be > 65%, while reflectance was below < 40% in the visible range, confirming thereby that the deposited films are rather transparent. All Au/TiO_x systems showed rather low absorptance (<10%) in the NIR regions of the spectra. Interestingly, the optical absorptance in the NIR was found to strongly decrease for lower annealing temperatures, which could be associated with an increase of the carrier concentration (Ne). The quality of the nanostructures resulting from the thermal treatment (in view of the presence of plasmon resonance) is reflected in the UV-Vis absorption spectra. The UV-Vis spectrum of Au thin film coated on TiO_x seed films which was dewetted at different temperatures is found to be complex, as it might be influenced by many factors, such as the size of nanostructures, their shape, the changes in their electronic structure and the dewetting temperature as well as the TiO_x phase, i.e., anatase or rutile.

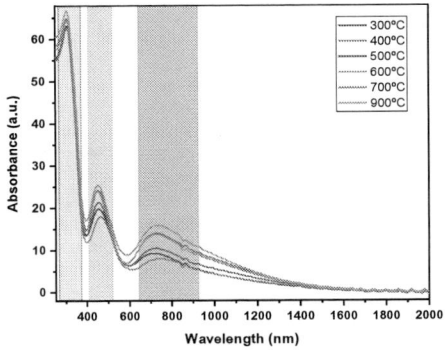

Fig. 3   Total reflectivity at the end of the test period by wavelength range of coupons. Values are the reflectivity of each (individual) coupon of that type and cleaning schedule.

Therefore, Au NPs absorption and scattering properties can be tuned by controlling the particle size, shape, and the local refractive index near the particle surface. The usual absorption peak for Au NPs has the LSPR absorption light band normally close to 520 nm and the broad peak with a strong tail is an indication of the wide distribution of nanoparticle size. For spherical shape where D ≪ λ, the plasmonic resonance reflects only the dipole mode of the collective oscillations of electrons. The absorbance of Au nanoparticles dispersion shown in Figure

3, where it shown that the LSPR activity of Au/TiOx systems drastically increases with increasing thermal dewetting temperature (except for 700 °C samples which do not follow such an observed trend) and with decreasing mean-Au nanoparticle size. Increasing Au surface area in spite showed that the LSPR significantly weakens. Finally, the integration of these plasmonic systems into planar solar cell has demonstrated an increase of 35.51 % of the photocurrent generation and was subject to the size and density of Au NPs. This section will be detailed in the full version of this paper.

III. CONCLUSIONS

The correlation of dewetting temperature and Au/TiO_x systems has been established in this work. In general, sharp XRD peak has been observed of all the films with the highest dewetting temperature related to the crystallinity of the films. Such results have been confirmed by the formation of nano-islands, surface wettability behavior, and average roughness. The crystallinity of Au depends significantly on the seed layer, which is in this case TiO_x. The optical measurements show that transmission increased to >65%, while reflectance stayed below <40% in the visible range as the dewetting temperature increased, confirming thereby that the deposited films become more transparent for increased dewetting temperature. Finally, the results confirm essential evidence of thermal dewetting efficiency for large scale and high-throughput fabrication. PV results will be discussed in the full version of this article.

REFERENCES

[1]   E. Petryayeva, U.J. Krull, Anal. Chim. Acta 706 (1) (2011) 8–24.

[2]   K.L. Kelly, E. Coronado, L.L. Zhao, et al., J. Phys. Chem. B 107 (3) (2003) 668–677.

[3]   J.N. Anker, W.P. Hall, O. Lyandres, N.C. Shah, J. Zhao, et al., Nat Mater 7 (6) (2008) 442–453.

[4]   T. Chung, Y. Lee, M.S. Ahn, W. Lee, S.I. Bae, C.S.H. Hwang, et al., Nanoscale 11 (18) (2019) 8651–8664.

[5]   G.V. Naik, V.M. Shalaev, A. Boltasseva, G.V. Naik, et al., Adv Mater. 25, 3264-3294 (2013).

[6]   M.M. Jiang, H.Y. Chen, B.H. Li, K.W. Liu, et al., J. Mater. Chem. C 2 (1) (2014) 56–63.

[7]   M.B. Ross, J.C. Ku, M.G. Blaber, C.A. Mirkin, G.C. Schatz, et al., Proc. Natl. Acad. Sci. U.S.A. 112 (33) (2015) 10292–10297.

[8]   H. Wei, S.M. Hossein Abtahi, P.J. Vikesland, Environ. Sci. Nano 2 (2) (2015) 120–135.

[9]   W. Yue, Y. Yang, Z. Wang, et al., J. Micromech. Microeng 22 (2012) 125007.

[10]   A.M. Lopatynskyi, V.K. Lytvyn, et al., Nanoscale Res Lett 10 (1) (2015) 1–9.

[11]   W.W. Mullins, J. Appl. Phys. 30 (1) (1959) 77–83.

[12]   I. Tanyeli, H. Nasser, F. Es, A. Bek, R. Turan, Opt Express 21 (S5) (2013) A798.

Detection and Analyze of Off-Maximum Power Points of PV Systems based on PV-Pro Modelling

Baojie Li, Xin Chen, Anubhav Jain

Energy Technologies Area, Lawrence Berkeley National Laboratory, Berkeley, CA, USA

Abstract — **Photovoltaic (PV) systems can operate off the maximum power point (MPP) for various reasons. Understanding when off-MPP behavior occurs is essential to the maintenance and operation (O&M) of PV systems. To detect off-MPP data, a reference power is usually needed, which can be obtained by system modeling that generally relies on physical model parameters. Traditional methods commonly obtain these parameters based on the initial condition of the PV system such as from the module datasheet. However, these parameters often do not reflect the current condition of the on-site PV system, which is likely to suffer from degradation and faults after years of operation with degraded parameters. Thus, we propose an off-MPP analysis algorithm based on the PV-Pro method, which can extract the model parameters (like series and shunt resistance) at the current operating condition only using the routine production data. In this way, the system power, current, and voltage can be accurately modeled. The off-MPP points are detected by comparing the measured power with the one modeled by PV-Pro. Points with large disagreement in power are further analyzed by deconvolving it into the error of the current and voltage at MPP, which allows tracing the error source of the off-MPP and provides valuable information for the O&M of PV systems. This off-MPP analysis is demonstrated on a 271kW PV field system, where it is shown that most of the off-MPP points are caused by the reduced DC current.**

I. Introduction

To detect anomalies in the output power of a PV system, a reference value of power is generally needed, which can be obtained via system modeling [1]. To perform PV system modeling, the equivalent model parameters are essential [2]. Traditional methods often estimate the model parameters based on the initial status of the PV module, such as from the manufacturer data sheets [3]. However, these model parameters do not reflect the actual status of an on-site PV system, which may suffer from degradation and various faults after years of field operation [4].

This paper presents a detection method for off-maximum power point (MPP) data based on accurate modeling of the PV system including model parameter estimation. The modeling is conducted by a methodology named PV-Pro [5] from our previous research. It can estimate the model parameters at the current operating condition using only the routine operation data. Thus, PV-Pro allows us to precisely model the output power, current, and voltage to perform the off-MPP analysis.

The remainder of the paper is organized as follows: Section II introduces the off-MPP algorithm. Section III demonstrates

the algorithm on a field PV system. Results are discussed in Section IV. Section V concludes the paper.

II. Methodology

The proposed off-MPP algorithm consists of three steps, as depicted in Fig. 1. The first step is to estimate the operation data (P_{mp}, V_{mp}, and I_{mp}) using the model parameters extracted by PV-Pro. The next step is to classify data points as off-MPP when the P_{mp} error exceeds a predetermined threshold. The final step is to deconvolve the P_{mp} error based on analyzing the V_{mp} and I_{mp} error to trace the root cause.

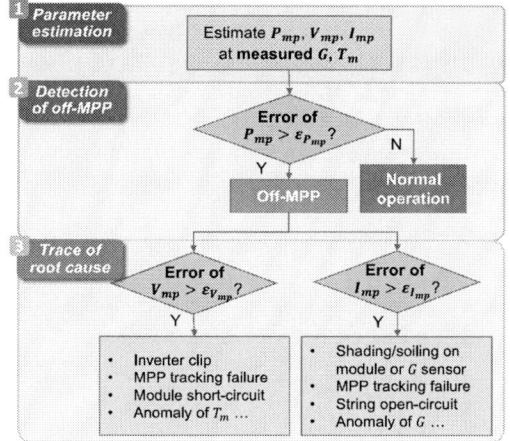

Fig. 1   Pipeline of three-step off-MPP analysis.

A. Estimation of operation data using PV-Pro

PV-Pro [5] is a methodology to extract the PV equivalent model parameters at the current operating condition from routine PV production data and environmental data (irradiance G and module temperature T_m). Extending the Suns-Vmp method [6], PV-Pro extracts the five single-diode model (SDM) parameters, allowing the PV system to be modeled under any environmental condition. It should be noted that the SDM parameters extracted by PV-Pro reflect the actual condition of the PV system, including degradation or faults. This makes the subsequent PV system performance modeling much more accurate compared to the methods that leverage the SDM parameters extracted from the module datasheet [3]. An example of the estimated P_{mp} of one field PV system [7] using the initial (via NREL PyPVRPM model [8] based on the

978-1-6654-6060-6/23 $31.00 © 2023 IEEE   1519

datasheet) and PV-Pro- extracted SDM parameters is plotted in Fig. 2. It is shown that after years of operation, the P_{mp} estimated using initial parameters can no longer fit the measured one. In contrast, the P_{mp} obtained by PV-Pro still well models the system performance. Thus, off-MPP analysis based on PV-Pro modeling will be more resistant to changes in system parameters over time.

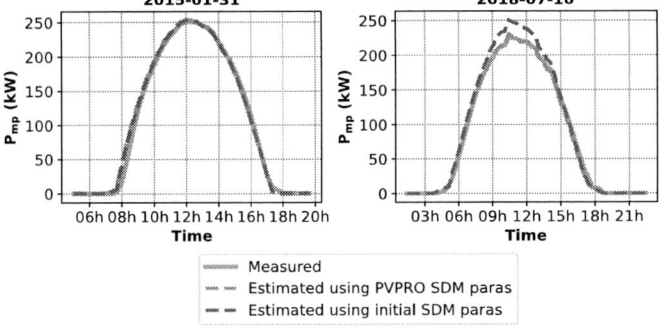

Fig. 2   Estimated P_{mp} of NIST ground array (began operation in 2015) using initial and PV-Pro-extracted SDM parameters at two different times. After three years of operation, the estimated P_{mp} using initial SDM parameters does not fit the measured one, while the P_{mp} modeled using PV-Pro still matches well.

In this study, based on the SDM parameters extracted by PV-Pro, P_{mp}, V_{mp}, and I_{mp} are estimated at the measured G and T_m for the off-MPP analysis.

B. Detection of off-MPP

To detect off-MPP points, the threshold method is adopted, i.e., when the error of power (ε_P) exceeds a pre-determined threshold, the data point will be identified as off-MPP. As the system power varies with the irradiance and temperature, we set the threshold as a ratio to the estimated power. The default value is 10%. It can be customized based on the specific requirements of the user on the off-MPP detection sensitivity.

C. Analysis of off-MPP

Because the DC power is the product of DC voltage and current ($P = VI$), after detecting the off-MPP points, it is also necessary to find out where the error of power comes from and quantify the contribution of each factor. First, the error of voltage (ε_V) and current (ε_I) are calculated based on the V_{mp} and I_{mp} estimated by PV-Pro. Then, the power error caused by voltage or current error is calculated independently, as presented in (2-3).

$$\varepsilon_{P_{\varepsilon_V}} = P - P_{\varepsilon_V} = V \cdot I - (V - \varepsilon_V) \cdot I = \varepsilon_V \cdot I \quad (1)$$

$$\varepsilon_{P_{\varepsilon_I}} = P - P_{\varepsilon_I} = V \cdot I - V \cdot (I - \varepsilon_I) = V \cdot \varepsilon_I \quad (2)$$

Next, the power error when both voltage and current error exist ($\varepsilon_{P_{\varepsilon_V \varepsilon_I}}$) is also calculated:

$$\varepsilon_{P_{\varepsilon_V \varepsilon_I}} = P - P_{\varepsilon_V \varepsilon_I} = V \cdot I - (V - \varepsilon_V)(I - \varepsilon_I)$$
$$= V \cdot \varepsilon_I + \varepsilon_V \cdot I - \varepsilon_V \cdot \varepsilon_I \quad (3)$$

Using $\varepsilon_{P_{\varepsilon_V}}$, $\varepsilon_{P_{\varepsilon_I}}$, and $\varepsilon_{P_{\varepsilon_V \varepsilon_I}}$, we define the contribution of the voltage and current error to the final power error (namely C_V, C_I) as expressed in (4-5).

$$C_V = \frac{\varepsilon_{P_{\varepsilon_V}}}{\varepsilon_{P_{\varepsilon_V \varepsilon_I}}} = \frac{\varepsilon_V \cdot I}{V \cdot \varepsilon_I + \varepsilon_V \cdot I - \varepsilon_V \cdot \varepsilon_I} \quad (4)$$

$$C_I = \frac{\varepsilon_{P_{\varepsilon_I}}}{\varepsilon_{P_{\varepsilon_V \varepsilon_I}}} = \frac{V \cdot \varepsilon_I}{V \cdot \varepsilon_I + \varepsilon_V \cdot I - \varepsilon_V \cdot \varepsilon_I} \quad (5)$$

It may be noted that the sum of C_V and C_I does not equal unity. This is because two factors (V and I) simultaneously impact the power. Thus, the ε_P when both factors act is not a simple sum of the ε_P when only one factor acts. Nevertheless, C_V and C_I reflect the relative contribution of the voltage and current error to the final power error.

III. CASE STUDY OF A 271KW PV FIELD SYSTEM

This section presents a demonstration of the proposed off-MPP analysis on a field PV system, i.e., the NIST ground array [7]. The array is located in Gaithersburg, Maryland, USA. It is ground mounted with a fixed tilt angle of 20°. 1152 modules (Sharp NU-U235F2, 235W, sc-Si) are installed, yielding 271 kW output. Data from 2015 to 2019 are available for analysis.

PV-Pro is applied to extract the SDM parameters based on every 2-week operation data over the 4-year data. Using these extracted parameters, the V_{DC} and I_{DC} are estimated and compared with measured values. The average relative error is less than 1%, which shows the overall good modeling capability of PV-Pro.

Next, the off-MPP points are detected. The P_{mp} error (i.e., difference in modeled and observed P_{mp}) of the NIST ground array is plotted in Fig. 3. It is shown that, when the threshold is 10% of the estimated power, the off-MPP ratio is 4.93% over the 4-year operation time.

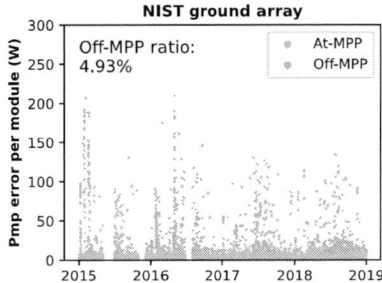

Fig. 3 P_{mp} error (per module) of the NIST ground array. PV-Pro identifies the presence of off-MPP in 4.93% of the operation time.

To further analyze these off-MPP data, the relative error (RE) of V_{mp} and I_{mp} is calculated and plotted in Fig. 4. We note that the relative error of V_{mp} varies in [0, 20%] while that of I_{mp} is much higher with the value in [0, 100%].

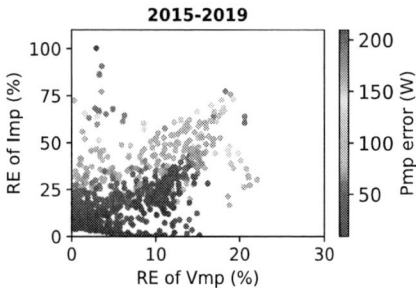

Fig. 4   Distribution of off-MPP points of the NIST ground array 2015-2019. The RE of I_{mp} varies more intensely than that of V_{mp}.

To quantify the contribution of the error of V_{mp} and I_{mp} to the P_{mp} error, Step 3 of the algorithm is performed. To illustrate, one day of the off-MPP data (Fig. 5 (a)) is selected as an example. Using (4) & (5), the contribution of the error of V_{mp} and I_{mp} is calculated and presented in Fig. 5 (b).

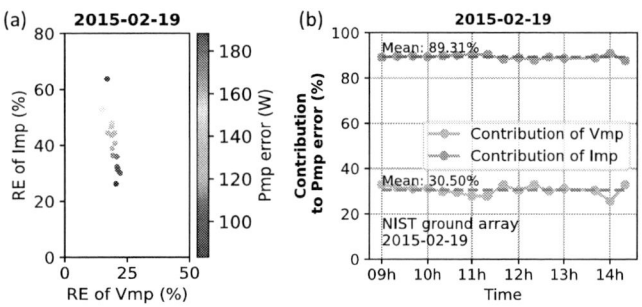

Fig. 5   (a) Relative error of V_{mp} and I_{mp} of one off-MPP cluster (b) Contribution of V_{mp} and I_{mp} error to P_{mp} error. The results show that I_{mp} error contributes more to the P_{mp} error on 2015-02-19.

It is revealed in Fig. 5 that, on 2015-02-19, although the relative error of I_{mp} varies from 25% to 65%, its contribution to the P_{mp} error (Fig. 5 (b)) is relatively stable and higher (89.3%) than that of V_{mp} (30.5%). This indicates that the I_{mp} error is the primary source of the off-MPP operation and proper O&M strategies to mitigate off-MP can focus on factors impacting the current generation, like shading, soiling, or module short-circuit.

IV. DISCUSSION

The core part of the off-MPP algorithm is the modeling using PV-Pro, which allows extracting the SDM parameters reflecting the current condition of the PV system based on the historical operation data. Therefore, this off-MPP algorithm is promising for application to real-time health monitoring of PV systems as accurate performance modeling is made possible by PV-Pro. This will also be the focus of future research.

V. CONCLUSION

This paper presents an off-MPP detection and analysis algorithm, which is based on accurate modeling of the current operating performance of the PV system using PV-Pro. The detected off-MPP are analyzed by quantifying the contribution of each error source (from current or voltage). This algorithm is demonstrated on a field PV system where the modeling error is below 1%. It is revealed that the primary source of the off-MPP is a decrease of current, which points out a clear direction for the planning of O&M for the PV system. The PV-Pro with the off-MPP algorithm is available on Github: https://github.com/DuraMAT/pvpro

REFERENCES

[1]   D. S. Pillai and N. Rajasekar, "Metaheuristic algorithms for PV parameter identification: A comprehensive review with an application to threshold setting for fault detection in PV systems," *Renewable and Sustainable Energy Reviews*, vol. 82, pp. 3503–3525, Feb. 2018, doi: 10.1016/J.RSER.2017.10.107.

[2]   I. de la Parra, M. Muñoz, E. Lorenzo, M. García, J. Marcos, and F. Martínez-Moreno, "PV performance modelling: A review in the light of quality assurance for large PV plants," *Renewable and Sustainable Energy Reviews*, vol. 78, pp. 780–797, 2017, doi: 10.1016/j.rser.2017.04.080.

[3]   A. R. Jordehi, "Parameter estimation of solar photovoltaic (PV) cells: A review," *Renewable and Sustainable Energy Reviews*, vol. 61, pp. 354–371, Aug. 2016, doi: 10.1016/J.RSER.2016.03.049.

[4]   A. Ameur, A. Berrada, A. Bouaichi, and K. Loudiyi, "Long-term performance and degradation analysis of different PV modules under temperate climate," *Renew Energy*, vol. 188, pp. 37–51, Apr. 2022, doi: 10.1016/J.RENENE.2022.02.025.

[5]   B. Li, X. Chen, T. Karin, and A. Jain, "Estimation and Degradation Analysis of Physics-based Circuit Parameters for PV Systems Using Only DC Operation and Weather Data," in *2022 IEEE 49th Photovoltaics Specialists Conference (PVSC)*, Nov. 2022, pp. 1236–1236. doi: 10.1109/PVSC48317.2022.9938484.

[6]   X. Sun, R. V. K. Chavali, and M. A. Alam, "Real-time monitoring and diagnosis of photovoltaic system degradation only using maximum power point—the Suns-Vmp method," *Progress in Photovoltaics: Research and Applications*, vol. 27, no. 1, pp. 55–66, Jan. 2019, doi: 10.1002/PIP.3043.

[7]   M. Boyd, T. Chen, and B. Dougherty, "NIST Campus Photovoltaic (PV) Arrays and Weather Station Data Sets," *National Institute of Standards and Technology. U.S. Department of Commerce, Washington, D.C. [Data set]*, 2017. https://doi.org/10.18434/M3S67G (accessed May 19, 2022).

[8]   B. Silva, P. Lunis, M. Theristis, and H. Seigneur, "PyPVRPM: Photovoltaic Reliability and Performance Model in Python," *J Open Source Softw*, vol. 7, no. 71, p. 4093, Mar. 2022, doi: 10.21105/JOSS.04093.

Laser Recycling of Silver from Waste Silicon Solar Cells

Mahantesh Khetri, Pawan K. Kanaujia, Mool C. Gupta

Department of Electrical and Computer Engineering, University of Virginia, Charlottesville, VA, United States

Recycling PV modules is highly important as total solar waste is expected to grow to tens of millions of tons. The economic management of solar waste is essential as they are currently discarded in landfills. Silver is one of the high-cost items in solar modules, and chemical methods for recovery have been investigated. This work demonstrates the recovery of silver using a laser ablation process in air and in a water medium. The laser ablation in a water environment allows for obtaining silver in nanoparticle form. The results of laser silver recovery in air and water are presented. The morphology and composition of the collected silver particles were analyzed using SEM and EDS techniques. The effect of laser power, scan speed, wavelength, and the pulse width was investigated in terms of the evaluation of throughput and purity of the laser-generated micro and nanoparticles. We have successfully demonstrated laser ablation as a viable method for the recovery of silver from waste solar cells.

Study on air gap effects on photovoltaic modules operating temperature on typical metal rooftop appliation

Quanzhi.Wang*, Yuanjie.Yu, Tao Xu

Canadian Solar, 199 Lushan Road, SND, Suzhou, Jiangsu, China, 215129

Abstract

The module's operating temperature is one of the major impact factors that dictate the system energy performance of photovoltaic (PV) modules. The operating temperature is impacted by various factors including array size, air gap to the rooftop, gaps between modules, roof features, air flow, mounting structure, and so on. In this paper, the effect of air gap on the performance of PV module has been predicted through a series of experiments with different mounting heights. The paper's goal is to quantitatively the influence of the air gap on the operating temperature of the metal rooftop modules and to provide a guide about minimum mounting height under different climatic conditions in different regions to meet relevant IEC63126 standards. A quantitative model is verified based on irradiance, ambient temperature, and wind speed conditions from six months of onsite testing data.

Keywords: operating temperature, air gap, PV modules, quantitative model

1. Introduction

Various previous studies[1-4] have been carried out to find out the effect of air gap which means the distance between roof top and module lowest side on the operating temperature.PVsyst does not provide enough information on the value of U0 and U1 for gaps between PV panels and roof.

In this paper, 3-parameter (ambient temperature, irradiance and wind speed) thermal model which from fitted the test data lasting for six months have been developed. This study aims to quantitatively expression of Tmod, Tamb, irradiation and wind speed taking into the influence of the air gap on the operating temperature of the metal rooftop modules. The quantitatively expression based on the Faiman[5] model given as:

Tmod=Tamb+G/(U0+U1*V)   (1)

The objectives of this paper is to identify the U0 and U1 expression by fitting the six month test data for different air gaps.

2. Experimental setup and procedure

Altogether there were five air gaps modules mounted on the metal rooftop structure designed and installed at Suzhou CSI factory roof, China. The five air gaps between the metal rooftop and the lowest surface of array modules were: 0mm, 50mm, 100mm, 150mm, 200mm. The arrays are equidistant from north to South. The data collected includes modules' temperatures, ambient temperature, irradiance and wind speed. The six month (Feb. to Aug. 2022) data collected was analyzed and report in the paper. Data analysis is carried out using python code, and data filtering code is compiled according to IEC 61853-2 standards. Data rules of different air gaps are obtained through summary analysis.

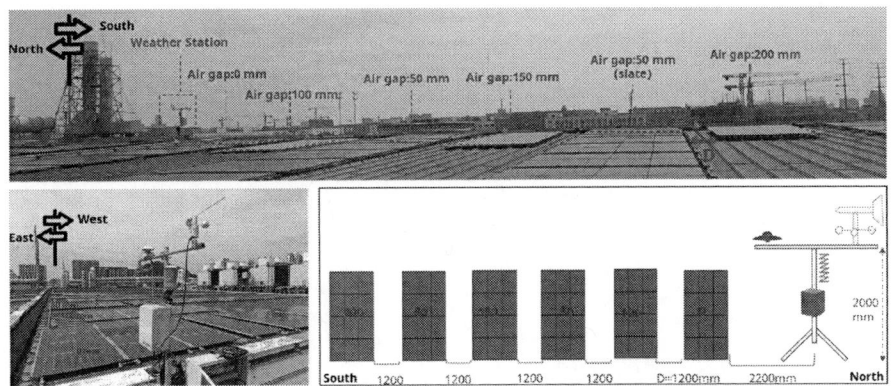

Figure 1. Photograph of air gaps location distribution

3. Results and thermal model

3.1 Data filtered and effect of air gap on module temperature

Data analysis is carried out using python code, and data filtering code is compiled according to IEC 61853-2 standards. There were 13943 data points (one minute data as a data point) filtered from the 6-month data and the filtered rate is 6.96% of total

3.2 U0&U1 for predict module temperature

The irradiance, ambient temperature and wind speed are included in the monitored data of different air gaps. The variation of the thermal model could be: $U0+U1*v=G/(Tm-Tamb)$. According to the variant of thermal model, with wind speed as the dependent variable and $G/(Tm-Tamb)$ as the independent variable, draw a fitting curve, and obtain U0 and U1 values according to the intercept and slope of the curve.U0 and U1 at different air gaps can be fitted from the scatter values at the five air gaps to obtain the relationship expression with air gap. Lower data scattering(higher R2 of about 0.985) have been got for U0 and U1. The fitting and the expression are shown in the Figure 2. So the predicted module temperature mounted at different air gaps could be done according to the relationship expression.

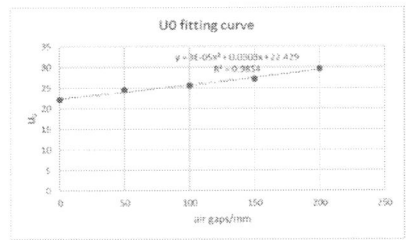 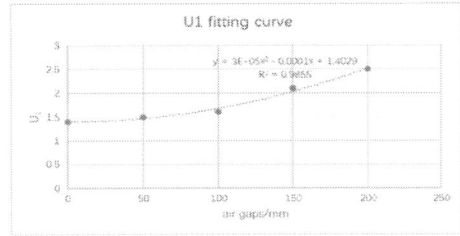

Figure 2. U0 and U1 fitting curve at different air gaps

So the relationship expression between U0&U1 and air gap are shown in the following two equations:

$$U0: \quad 3E\text{-}5x^2+0.0303x+22.429$$
$$U1: \quad 3E\text{-}5x^2-0.0001x+1.4029$$

3.3 Checking of U0&U1

The relationship expression between U0&U1 and air gap developed in this work was checked at the German field trial(48°11'26.99"N 12°26'18.15"E). The 210mm air gap of the test module mounted on the roof is shown in the Figure 3. So the U0 and U1 of 210mm are 30.1 and 2.7. The calculation module operating temperature can be get from the Faiman thermal model:

Tm=Tamb+G/(30.1+2.7*V). The ambient temperature(Tamb), irradiance(G) and test module temperature have been collected for a whole year in 2016. The data were filtered according to IEC 61853-2 standards using python code. In the filtered data, the irradiance should be greater than 400 W/m2, the variation of irradiation should be less than 10%. The comparation of the calulation and test actual operating module temperature is shown in the Figure 4 and Table 1. It is observed that the difference between actual and modeled temperatures which calculated from Tm=Tamb+G/(30.1+2.7*V) is found to 0.19%~5.9% for the German field trail.

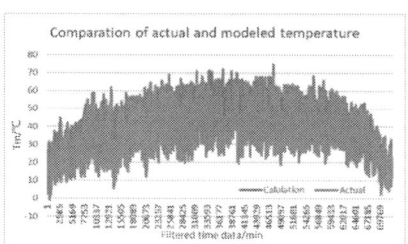

Figure 3. Air gap between the module and the metal rooftop Figure

Figure4. Comparison of the calulation and test actual operating module temperature

Air gap/mm	Percent	Actual temperature	Temperature calculated using U0&U1/℃	Temperature difference between actual and modeled
210	98th	61.5	59.91777	2.57%
	75th	47.1	49.89604	-5.94%
	50th	41.8	41.8811	-0.19%

Table 1. Temperature difference between actual and modeled at different percent

4. Conclusion

To predict the operating metal rooftop module temperature of different mounted air gaps, the relationship expression between U0&U1 in faiman thermal model and air gap have been developed in the work. The difference between actual and modeled temperatures is less than 5.9% in the German field trail checked. The analysis of rooftop material, array size, position in array, tilt angle, module type, and other factors affecting operating temperature will be carried out in the next work.

Reference

[1] Bijay L, Ernie G, G.TamizhMani, Temperature of rooftop photovoltaic modules: Air gap effects Proc. of SPIE Vol. 7412, 74120E-1-74120E-11(2009)

[2] Michael Kempe, Kent Whitfield, and Peter Seidel, TS 63126 Early Revision Project, Spring 2022, TC 82, WG2 Meeting (2022)

[3] Martin K. Fuentes, A Simplified Thermal Model for Flat-Plate Photovoltaic Arrays, SAND85-0330 Unlimited Release Printed May 1987

[4] M.S. Naghavi, A. Esmaeilzadeh, Singh, B.C. Ang, T.M. Yoon, K.S. Ong, Experimental and numerical assessments of underlying natural air movement on PV modules temperature, Solar Energy 216(2021)610-622

[5] Elena Barykina, Annette Hammer 2017, 'Modeling of photovoltaic module temperature using Faiman model: Sensitivity analysis for different climates', Solar Energy vol.146, pp. 401-416

Development of p-type Silicon Heterojunction Solar Cells with 26.6% Efficiency

Xiaoning Ru[1], Miao Yang[1], Yichun Wang[2], Jianqiang Wang[1], Chaowei Xue[1], Shi Yin[1], Chengjian Hong[1], Fuguo Peng[1], Minghao Qu[1], Junxiong Lu[1], Liang Fang[1], Tian Xie[2], Zhenguo Li[1], Xixiang Xu[1]

[1]Central R&D Institute, LONGi Green Energy Technology Co., Ltd., Xi'an, Shaanxi, 710000, China

[2]Wafer Business Unit, LONGi Green Energy Technology Co., Ltd., Xi'an, Shaanxi, 710000, China

Abstract — The development of high efficiency Si solar cells is seeing successful industrialization of carrier-selective and passivating contact technologies, including Tunnel Oxide Passivated Contact (TOPCon) and Silicon Heterojunction Technology (SHJ). Driven by cell technology innovation, the Si PV industry is making bold moves to see 26-27% efficiencies in mass production in the coming years. Undisputablly, Si wafer development provides strong support for the fast-upgrading cell technologies, and guarantees the technologies' deployment at affordable costs. To date, p-type Si wafer is dominating the market by 95% market share. However, a 5% to 50% uptaking of n-type Si wafer is predicted by ITRPV. One critical factor for this judgement is that the n-type Si solar cell efficiencies are leading ahead that of the p-type. It is well argued that the n-type Si wafer manufacturing can steadily ramp up given the demand, and the cost disadvantage can be alleviated to some extent with scaled production capacity. At the same time, the industry shall not ignore the inherent cost advantage of p-type wafers, and keep exploiting the possibilities to push up the p-type Si solar cell efficiency. In this work, we demonstrate that the p-type SHJ solar cell efficiency can reach 26.6%, which is just 0.2% behind of its n-type counterpart. This result establishes a solid foundation for further SHJ technology development on a cost-effective basis.

I. INTRODUCTION

Silicon heterojunction technology (SHJ) is one of the few candidates for next-generation high efficiency Si solar cells. First pioneered by Sanyo[1] and then developed by several R&D teams[2-4], the SHJ solar cell efficiency has increased steadily over the past decade and reached 26.8% in 2022[5]. A typical SHJ solar cell is fabricated by coating the passivating intrinsic silicon thin layers, carrier-selective amorphous or microcrystalline silicon layers, and transparent oxide conductive layers in sequence on both sides of the wafer. The fabrication process has advantages in simplicity and low thermal budget. Different from the diffused-junction solar cell technology, for example PERC (passivated emitter and rear contacts), the intrinsic amorphous silicon layers passivate the Si wafer surface excellently, which brings up the cells' open-circuit voltage to 750 mV. In other words, the minority carrier lifetime is more dependent on the Si bulk quality. This explains why the major efforts in SHJ technology development were based on n-type wafer, which is more defect tolerant and shows several times higher minority carrier lifetime than p-type wafer. Before 2020, the record efficiency of p-SHJ cell is almost 3% lower than n-SHJ (23.8% vs. 26.7%). Figure 1 displays the efficiency progression of p-SHJ in comparison to n-SHJ solar cells. From a techno-economical analysis by the UNSW team, to compete with n-SHJ in the market, the p-SHJ cell has to narrow the efficiency gap to below 0.3%[6]. In the past 2 years, the R&D team in LONGi introduced successfully the microcrystalline silicon (μc-Si) material to n-SHJ cell fabrication and obtained a series of efficiency records on M6 wafers. Meanwhile, the developed process is readily transferrable onto p-SHJ cell fabrication. As a result, the p-SHJ cell efficiency was levelled up to above 26.0% for the first time in 2022, and eventually reached 26.6% very recently, which is just 0.2% below that of the record 26.8% efficiency.

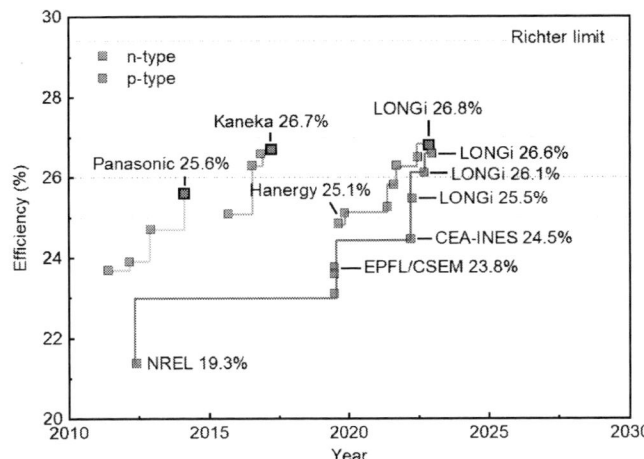

Fig. 1. Progression of n-SHJ and p-SHJ cell efficiency over the last 10 years.

In this work, we introduce the progress on state-of-the-art p-SHJ solar cell development. A roadmap for <0.1% efficiency gap between p-SHJ and n-SHJ is established based on specific process optimization for p-SHJ solar cell. A roadmap for p-type wafer improvement, specifically the gettering process development is introduced and discussed. Wafer cost analysis is conducted to predict the future p- and n-SHJ deployment in the market. Our study foresees a p- and n-SHJ parity in the future.

978-1-6654-6060-6/23 $31.00 © 2023 IEEE

II. Experiment

The cell structure is illustrated in Figure 2. The ultra-thin hydrogenated intrinsic amorphous Si (i:a-Si:H) passivation layers are grown on both sides of the crystalline silicon (c-Si) surface. The n-type microcrystalline silicon oxide layer μc-SiOx:H(n) is formed at front side for electron collection. The p-type microcrystalline Si layer μc-Si:H(p) is formed at rear side for hole collection. Transparent conductive oxide (TCO) and electrodes cover the carrier-selective contacts to transport light-generated currents. Radio-frequency plasma-enhanced chemical vapor deposition (RF-PECVD) tool was used to prepare intrinsic Si layers, while very high frequency PECVD (VHF-PECVD) for the nc-Si layers. The SHJ cell was fabricated using a standard M6 n-type Czochraski Si wafer from LONGi (150 μm, 1~2 Ωcm).

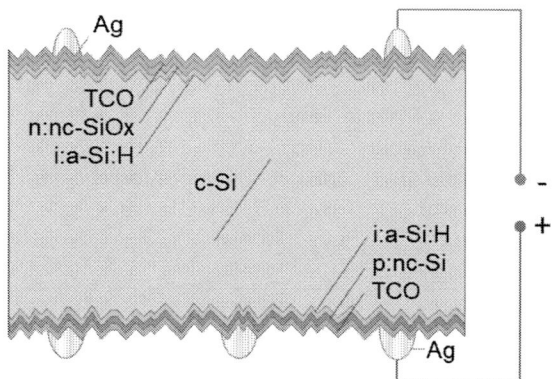

Fig.2. SHJ cell structure. The c-Si wafer is either p- or n- in this study.

A. Wafer gettering process

The p-type wafers were loaded in a tube furnace for the gettering process. The phosphorus gettering process was conducted at above 800 °C for a few hours. After high temperature gettering, the phosphorus glass layer was chemically etched off.

B. Characterization

Single layers of the solar cells were characterized by Raman spectroscopy and transmission electron microscopy (TEM) etc. The current–voltage (I–V) characteristics of the solar cells were taken by a Vision VS-6821S I-V tester under standard test condition (AM 1.5G, 1000 W/m², 25 °C). All cells were illuminated from the n-side and the conversion efficiency was calculated on the total area of 274.5 cm². External quantum efficiency (EQE) and reflectance spectra of SHJ solar cells were measured on the entire front surface including the grid-shaded area with Bentham PVE300-IVT system over the wavelengths ranging from 300 nm to 1170 nm. Sinton Suns-V_{oc} instrument was used for lifetime measurement. Quokka 3 was used to simulated the certified IV curve and perform power loss analysis.

III. Results and Discussion

Figure 3 shows the certified I-V and P-V curves of the record p-SHJ solar cell. The solar cell has a bifacial structure and reached efficiency of 26.56% (total area: 274.1 cm², J_{sc} 41.30 mA/cm², V_{oc} 751 mV, FF 85.59%).

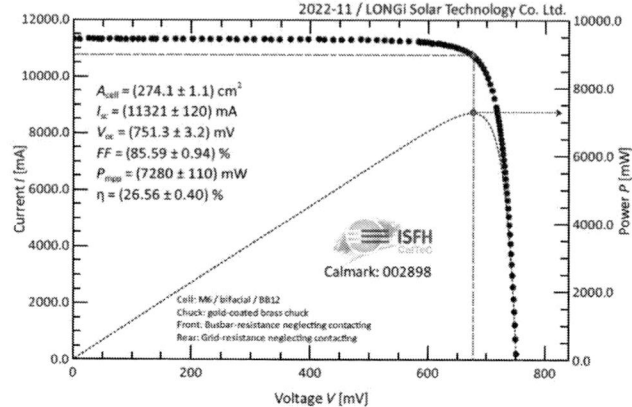

Fig.3. Certified IV and P-V curves of the record p-SHJ solar cell.

The primary efficiency improvement for p-SHJ solar cells comes from the improved wafer quality by the means of gettering process. The amount of metallic impurities in the p-type wafer can be significantly reduced, as indicated by improved minority carrier lifetime. There are several different methods to realize the gettering effect, for example, dielectric film coating and annealing, ion implantation, and P/B diffusion. Driven by the solubility and high temperature annealing, the metal impurities diffuse to the wafer surface and can be removed subsequently.

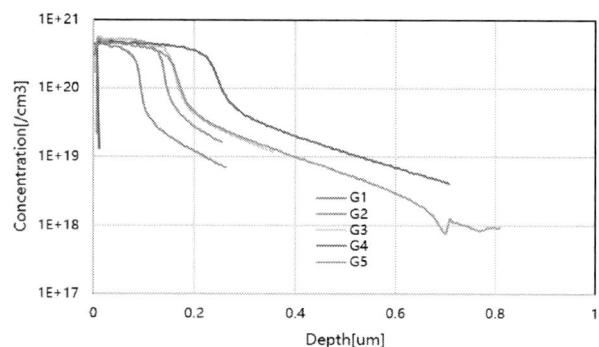

Fig. 4. Phosphorus concentration as measured from ECV.

After tuning of the gettering process, the p-type wafer lifetime can be improved from 1600 μs to 1900 μs. The p-SHJ cell efficiency was improved by a maximum of 0.53% at the optimized gettering conditions.

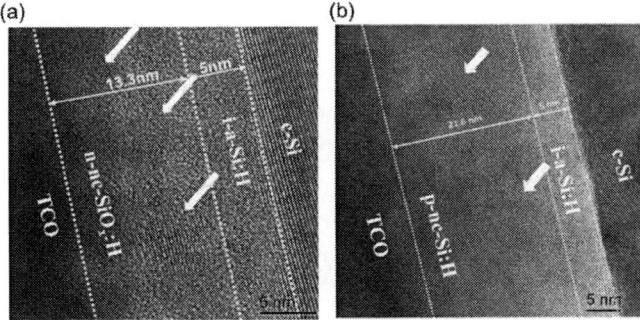

Fig.5. TEM images displaying the cross-section features of the component layers (a)front window structure; (b) rear emitter structure.

Based on the improved wafer quality, the optical and electrical properties of the subsequent component layers were optimized to improve cell performance. The major improvements comes from adoption of nc-Si layers, especially on the rear emitter side. Figure 5 dipicts the cross-section structures of the front window and rear emitter. As summarized in Table 1. The emitter side contact resistivity is reduced from 51 to 16 m$\Omega \cdot$cm^2 , contributing to FF% improvement. The interface recombination current is also significantly reduced by applying the μc-Si:H(p) layer.

Table 1. Recombination current density and contact resistivity of different passivation layers.

Passivation contact	J_0 (fA/cm^2)	ρ_c (m$\Omega \cdot$cm^2)
a-Si:H(i)/μc-SiO$_X$(n)	2.0	17
a-Si:H(i)/a-Si:H(p)	2.0	51
a-Si:H(i)/μc-Si:H(p)	0.5	16

IV. Conclusion and Outlook

As the p-SHJ efficiency is approaching that of the n-SHJ, its full potential on high efficiency needs to be exploited. At the moment p-type wafers have overall lower lifetimes than n-type wafers, due to the larger-than-unity electron/hole capture cross section ratios of most common metallic impurities. However, p-type wafers may have a higher potential for high-efficiency solar cells in the future, because 1) the impurity level is decreasing in the industrial manufacturing process, narrowing the gap between the p- and n-type lifetimes caused by such impurities; 2) higher voltages correspond to higher injections levels, where the Auger recombination, rather than the SRH recombination dominates; 3) p-type silicon is actually less prone to Auger recombination than n-type for the same doping concentration; and 4) p-type wafers are made into front-junction HJT solar cells, which may be less affected by bulk recombination comparing to n-type back-junction cells.

IV. Acknowledgement

The team in LONGi Central R&D Institute is acknowledged for dedications and continuous efforts on this work. Lei Feng and Genshun Wang are appreciated for device characterization. Chang Sun is acknowledged for fruitaful discussions.

References

[1] Maruyama, E., Taguchi, M., Sakata, H., Kanno, H., Tokuoka, N., Nakamura, Y., Matsuyama, K., Tohoda, S., Ogane, A., Yano, A., et al. (2011). The Approaches for High Efficiency HITTM Solar Cell with Very Thin (<100 µm) Silicon Wafer over 23%.

[2] Köhler, M., Pomaska, M., Procel, P., Santbergen, R , Zamchiy, A., Macco, B., Lambertz, A., Duan, W., Cao, P., Klingebiel, B., et al. (2021). A silicon carbide-based highly transparent passivating contact for crystalline silicon solar cells approaching efficiencies of 24%. Nature Energy 6, 529-537. 10.1038/s41560-021-00806-9.

[3] Haschke, J., Dupré, O., Boccard, M., and Ballif, C. (2018). Silicon heterojunction solar cells: Recent technological development and practical aspects - from lab to industry. Solar Energy Materials and Solar Cells 187, 140-153. https://doi.org/10.1016/j.solmat.2018.07.018.

[4] Yoshikawa, K., Kawasaki, H., Yoshida, W., Irie, T., Konishi, K., Nakano, K., Uto, T., Adachi, D., Kanematsu, M., Uzu, H., and Yamamoto, K. (2017). Silicon heterojunction solar cell with interdigitated back contacts for a photoconversion efficiency over 26%. Nature Energy 2, 17032. 10.1038/nenergy.2017.32.

[5] Green, M.A., Dunlop, E.D., Siefer, G., Yoshita, M., Kopidakis, N., Bothe, K., and Hao, X. (2023). Solar cell efficiency tables (Version 61). Progress in Photovoltaics: Research and Applications 31, 3-16. https://doi.org/10.1002/pip.3646.

[6] Chang, N.L., Wright, M., Egan, R., and Hallam, B. (2020). The Technical and Economic Viability of Replacing n-type with p-type Wafers for Silicon Heterojunction Solar Cells. Cell Reports Physical Science 1. 10.1016/j.xcrp.2020.10006

Oxy-fuel combustion: A threat or an opportunity for solar?

Mariela Colombo and Sarah Kurtz

University of California, Merced, Merced, California, 95343, United States

Abstract — **The rapid growth of solar, as well as other variable renewable energies, is key for achieving decarbonization targets. However, their integration to the grid and increased curtailment have become a major challenge for their further development in the most successful markets. In this study, the impacts of introducing an oxy-fuel combustion resource in California's energy grid are evaluated. To do so, we use RESOLVE, a capacity expansion model, to predict the energy grid mix for 2030, 2035, 2040, and 2045. Our results indicate that the model chooses to build the oxy-combustion resource when it is available at relatively low costs. While the introduction of oxy-fuel combustion, even at limited scale, reduces the selected operational capacity of solar PV and lithium-ion batteries, it substantially reduces the curtailment of the solar that *is* installed. Far from being a threat, this reduction could be an opportunity for solar to continue with robust growth, retaining its economic and environmental value even when the procurement of storage might otherwise limit its adoption rate.**

I. INTRODUCTION

To limit global warming to 1.5 °C, as outlined in the Paris Agreement, a comprehensive transformation of the global energy system is necessary. While the deployment of established clean energy technologies like solar and wind power is crucial in initiating a substantial decrease in emissions, additional technologies such as carbon capture and storage (CCS) are recognized as important contributors to achieving net-zero targets [1]-[2].

In California, ambitious targets had been set for decarbonizing its electricity sector. The state's Senate Bill 100 (SB100) sets targets of 60% of electricity generation with renewable energy by 2030, and 100% carbon-free retail electricity sales by 2045.

The case of California is of particular interest, as solar PV is the dominant technology leading the decarbonization. However, its variability and increased curtailment are creating some challenges for its expansion [3]-[4].

In this context, natural gas with CCS technologies could represent an opportunity for balancing the grid with low-carbon emission alternatives that can be used as dispatchable or baseload resources. However, today, CCS does not have a defined role in California's transition.

In this study, a particular CCS technology, oxy-fuel combustion, is introduced in a capacity expansion model for predicting 2030, 2035, 2040 and 2045 grid mix in California and evaluating the potential effect of this technology on solar PV growth.

II. METHODS

A. RESOLVE

For modeling California's grid in 2030, 2035, 2040 and 2045, RESOLVE, a capacity expansion model developed by Energy and Environmental Economics (E3) is used. The model is formulated as a linear optimization problem that co-optimizes investment and dispatch over a multi-year horizon to identify least-cost portfolios for meeting decarbonization targets and other system goals [5]. Fig. 1 shows a diagram with the main inputs and outputs of the model.

The inputs for the Baseline scenario are defined based on the Preferred System Plan (PSP) portfolio, considering the 38 million metric ton (MMT) 2030 electric sector greenhouse gas emissions target [6].

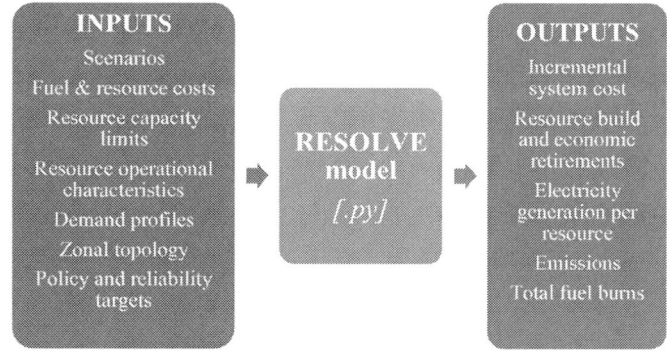

Fig. 1.   Schematic of RESOLVE modeling components.

To model the full year, capturing multiday and seasonal effects, while limiting computational requirements, a critical timesteps (CTS) approach was used, modeling every day of the year using two timepoints for each day: one hour after sunrise and one hour before sunset [7]. The profiles for the loads and electricity generation are integrated between those two timesteps for each day.

B. Oxy-fuel Combustion Resource Modelling

Oxy-fuel combustion or oxy-combustion involves the combustion of a fuel using nearly pure oxygen (not air); hence the flue gas is composed almost exclusively of CO_2 and water vapor [2]. This process has two main benefits:
•NO_X production is almost eliminated, due to the removal of N_2 from the air

• H_2O can be removed easily by means of dehydration to obtain a high-purity CO_2 stream which presents an opportunity to simplify carbon dioxide capture in power plant applications [8]-[9].

Of the technological alternatives of oxy-fuel combustion for power generation, the Allam Cycle is currently at the most advanced level of development [10].

In this process, a high-pressure supercritical CO_2 working fluid is used in a closed-loop cycle that retains all emissions by design. Apart from producing power, the by-products are liquid water and a stream of high-purity, pipeline-ready CO_2 [11]. The cycle can utilize a variety of carbon-based fuels, including natural gas and gasified solid fuels such as biomass and municipal solid waste. This design would allow efficiencies comparable to existing combined cycle gas turbines (CCGT), with no greenhouse gas emissions [11]-[12]-[13].

The Allam Cycle powered by natural gas is a proprietary power generation process. The company validated the technology in a testing facility in Texas and announced the first utility-scale project of 300 MWe planned to be operational in 2026, also in Texas [12]. A simplified diagram of the Allam Cycle with natural gas is shown in Fig. 2:

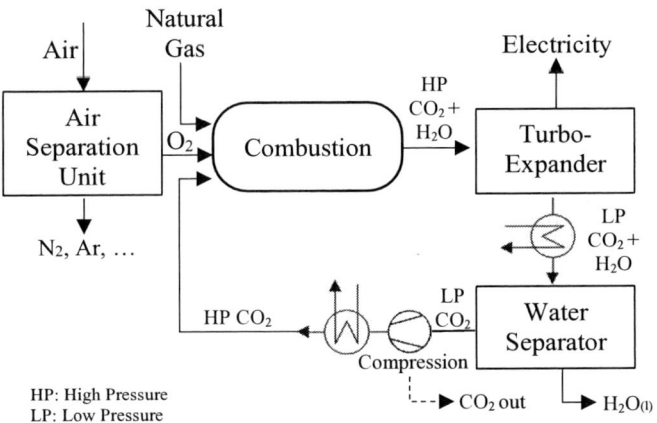

Fig. 2. Diagram of Allam Cycle with natural gas [12].

For evaluating the impact of this technology in California, an estimated projection of the maximum operational capacity was defined (Table I). We anticipate that the growth will be limited both by the implementation of the new Allam cycle technology as well as the infrastructure that will be needed to use or sequester the CO_2 that would be generated in large quantity.

TABLE I

MAXIMUM OPERATIONAL CAPACITY (GW) PROJECTED FOR OXY-COMBUSTION IN CALIFORNIA.

Year	Maximum Operational Capacity (GW)
2030	0.5
2035	1
2040	2
2045	4

Also, a cost sensitivity analysis was performed, considering a cost range from 1 to 2.5 times the cost of existing advanced CCGT.

III. RESULTS AND DISCUSSION

The operational capacity for California's grid in 2030, 2035, 2040 and 2045 has been modelled using RESOLVE.

For the Baseline scenario, no oxy-combustion resource is offered to the model. The Baseline results were compared to the case of having oxy-combustion available at different capital costs.

In Fig. 3, the operational capacity of oxy-fuel combustion selected by the model for the analyzed years is shown. For the lowest cost scenario, the maximum operational capacity is reached for all the years, as established in Table I. However, as the cost increases, the selected operational capacity is reduced reaching zero when the cost is more than 2 times the capital cost of existing CCGT facilities.

Of all the different electricity generation and storage resources, such as wind, geothermal, biomass, pumped-hydro and others, only solar PV and 4-hour lithium-ion batteries are affected by adoption of the oxy-combustion resource. In Figs. 4 and 5, the operational capacities of these resources are shown for the four years and cost scenarios.

The oxy-fuel combustion resource, even with limited capacity, reduces the need of 4-hour lithium-ion batteries as well as solar PV. The major differences are seen for the low-cost scenarios, 100% and 150% of CCGT cost. For example, in 2045, 4 GW of oxy-combustion reduces the need of solar PV by 19 GW and of lithium batteries by 14 GW. These reductions represent 12% of the overall capacity for both resources, compared to the baseline.

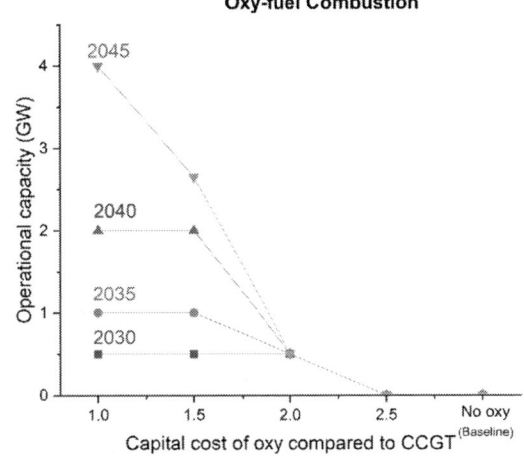

Fig. 3. Oxy-combustion operational capacity selected for 2030, 2035, 2040 and 2045 as a function of its capital cost.

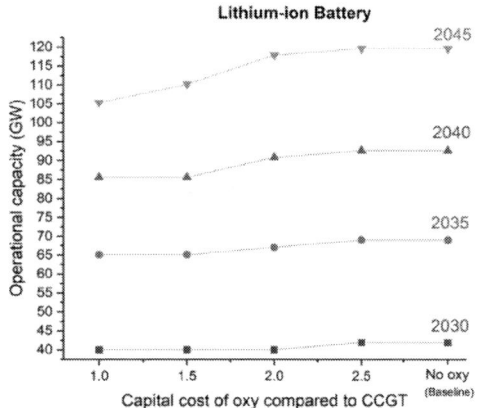

Fig. 4. Lithium-ion batteries operational capacity for 2030, 2035, 2040 and 2045 as a function of oxy-combustion capital cost.

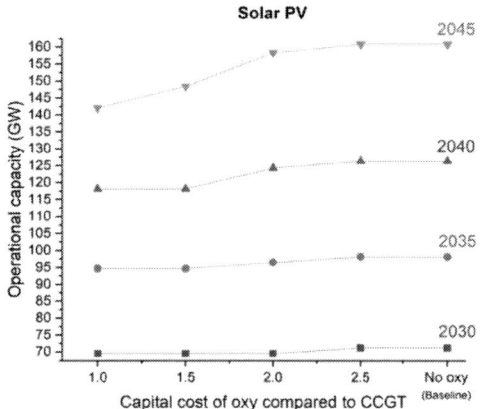

Fig. 5. Solar PV operational capacity for 2030, 2035, 2040 and 2045 as a function of oxy-combustion capital cost.

The capacity factor selected by the model for the oxy-combustion resource for each month and modeled year is shown in Fig. 6.

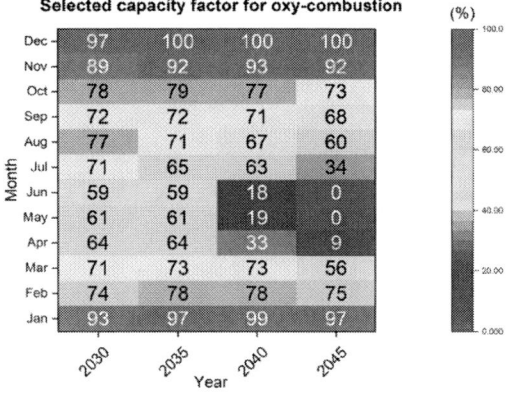

Fig. 6. Oxy-combustion capacity factor.

As expected in a grid with high solar PV penetration, the higher capacity factors for the oxy resources are found for fall and winter months (November, December, and January).

However, for spring and summer the capacity factors are smaller in general and experience a significant reduction from 2030 to 2045. As the deployment of solar PV and lithium-ion batteries increases, they can supply most of the electricity demand for these months. Moreover, the model chooses to shut down the oxy-combustion resources during these periods, even though there is an associated cost (which solar and batteries don't have).

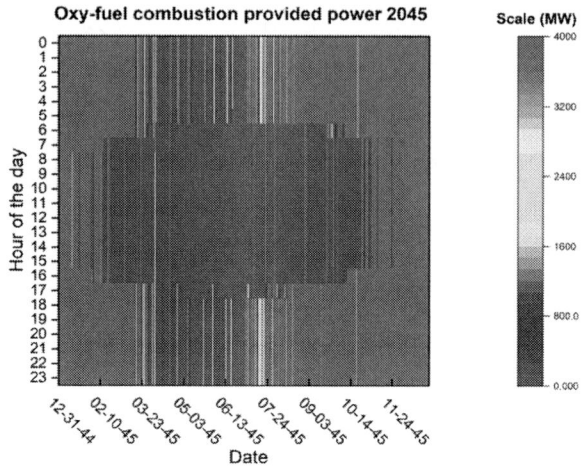

Fig. 7. Hourly dispatch of oxy-fuel combustion resource for 2045.

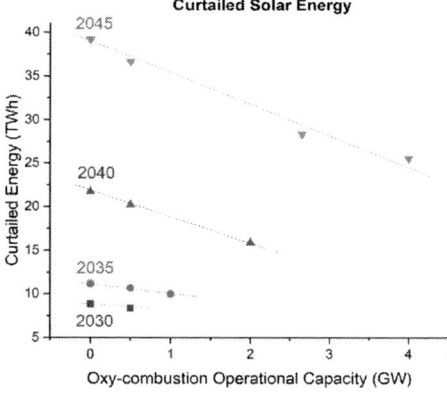

Fig. 8. Curtailed solar energy as a function of oxy-combustion operational capacity for 2030, 2035, 2040 and 2045.

Fig. 7 shows the hourly dispatch selected for oxy-combustion. Again, the role of this resource is different throughout the year. During winter it functions as a baseload type of resource, being ON and at maximum capacity most of the time. However, during spring and summer it is generating only during night-time or completely OFF, functioning more as a dispatchable resource.

As winter is the most challenging season in terms of renewable generation supply in California [13], having this flexible, low-emissions resource reduces the need of surplus capacities of solar and batteries. This helps to decrease the negative economic impact of overproduction during summer, in some cases leading to significant curtailment (Fig. 8). For

2045, 4 GW of oxy-combustion could reduce solar energy curtailment by 35%.

IV. CONCLUSIONS

In this study, the impact of oxy-fuel combustion on California's future grid was evaluated.

Even though there is high uncertainty regarding its ability to scale up in the state and regarding the speed of cost reduction in the short-term, the results indicated that even limited deployment of this technology will reduce selection of solar PV and storage operational capacities.

Even with these reductions, the opportunity for solar growth is huge. However, as solar penetration increases, so does its curtailment.

The use of oxy-fuel combustion reduces solar energy curtailment, which helps retain the economic and environmental benefits of solar electricity generation. Moreover, reducing the need in the short-term for batteries and solar, helps to ease the current pressure on the supply chain for raw materials allowing for a smoother transition. Effectively, assuming that the oxy-combustion can be used either as a baseload or as a dispatchable resource, reduces the need for storage, providing an excellent complement to solar.

In conclusion, oxy-fuel combustion technology is not in a position to grow to levels that could threaten solar development. In fact, the complement with solar electricity generation could be considered an opportunity for addressing the additional challenges for its increasing penetration.

ACKNOWLEDGMENTS

We thank R. Go for support with adapting the RESOLVE code. This paper was prepared as a result of work sponsored by the California Energy Commission. It does not necessarily represent the views of the Energy Commission, its employees, or the State of California. The Energy Commission, the State of California, its employees, contractors, and subcontractors make no warranty, express or implied, and assume no legal liability for the information in this document; nor does any party represent that the use of this information will not infringe upon privately owned rights. This paper has not been approved or disapproved by the Energy Commission nor has the Energy Commission passed upon the accuracy of the information in this report.

REFERENCES

[1] IPCC, "Synthesis Report of The IPCC Sixth Assessment Report (AR6)", 2023 [Online]. Available: https://www.ipcc.ch/report/ar6/syr/downloads/report/IPCC_AR6_SYR_LongerReport.pdf

[2] IEA, "About CCUS", 2021 [Online]. Available: https://www.iea.org/reports/about-ccus

[3] D. Millstein, R. Wiser, A. D. Mills, M. Bolinger, J. Seel, S. Jeong, "Solar and wind grid system value in the United States: The effect of transmission congestion, generation profiles, and curtailment," *Joule*, vol. 5, pp. 1749-1775, 2021.

[4] O. Ruhnau, "How flexible electricity demand stabilizes wind and solar market values: The case of hydrogen electrolyzers," *Applied Energy*, vol. 307, 118194, 2022.

[5] Energy and Environmental Economics (E3), "RESOLVE" [Online]. Available: https://www.ethree.com/tools/resolve/

[6] California Public Utilities Commission, "Decision adopting 2021 Preferred System Plan", 2021 [Online]. Available: https://www.cpuc.ca.gov/-/media/cpuc-website/divisions/energy-division/documents/integrated-resource-plan-and-long-term-procurement-plan-irp-ltpp/2019-2020-irp-events-and-materials/psp-decision-fact-sheet.pdf

[7] F. ZareAfifi, Z. Mahmud, S. Kurtz, "Diurnal, physics-based strategy for computationally efficient capacity-expansion optimization for solar-dominated grids," *Energy*, vol. 279, 128206, 2023.

[8] National Energy Technology Laboratory, "Oxy-combustion" [Online]. Available: https://netl.doe.gov/node/7477

[9] T. Evans, E. Grynia, "Carbon Capture – Purpose and Technologies", *Gas Liquid Engineering*, 2020 [Online]. Available: https://www.gasliquids.com/wp-content/uploads/2020_Carbon-Capture-Purpose-and-Technologies.pdf

[10] S. C. Gülen, M. Taher, L. G. Lyddon, "Oxy-fuel combustion supercritical CO2 power cycle – A critical look", *Global Power and Propulsion Society*, ISSN 2504-4400, 2022.

[11] R. Allam, S. Martin, B. Forrest, J. Fetvedt, X. Lu, D. Freed, G. William Brown Jr., T. Sasaki, M. Itoh, J. Manning, "Demonstration of the Allam Cycle: An Update on the Development Status of a High Efficiency Supercritical Carbon Dioxide Power Process Employing Full Carbon Capture," *Energy Procedia*, vol. 114, pp. 5948-5966, 2017.

[12] NET Power, 2023 [Online]. Available: https://netpower.com/

[13] M. Y. Abido, S. Kurtz, "Optimal Strategy for Using Biomass to enable California High Penetration Solar," *IEEE 49th Photovoltaics Specialists Conference (PVSC)*, pp. 0212-0212, 2022.

High-throughput in-line deposition of silicon oxide passivation layers in silicon TOPCon solar cells

Zachary B. Leuty, William J. Weigand, Jorge Ochoa, Joe V. Carpenter III, Mariana I. Bertoni, Zachary C. Holman

Arizona State University, Tempe, AZ, United States

Abstract - The ultra-thin silicon oxide layer in poly-Si passivated contacts plays a vital role in reducing recombination at the c-Si/oxide interface. We replaced this traditionally thermal or chemical oxide layer with a-SiOx:H deposited by a novel technique (AIDA) with a remarkably high wafer per hour throughput. This Poly-Si(n)/a-SiOx:H contact achieves excellent surface passivation with an iVoc = 726 mV and J0 = 8.8 fA/cm2. This technique can also uniformly cover textured pyramidal Si surfaces and has fine control over the deposition thickness. Due to several competing mechanisms, the oxide layer undergoes a small thickness change upon high-temperature annealing. The a-SiOx:H film desorbs hydroxyl (-OH) groups when annealed and increases the stoichiometric ratio towards SiO2 when annealed at high temperatures in nitrogen, compared to its as-deposited state.

978-1-6654-6060-6/23 $31.00 © 2023 IEEE

Measurement and Control of Mobile Ion Concentration in Halide Perovskites

Saivineeth Penukula, Nicholas Rolston

Arizona State University, Tempe, AZ, United States

The intrinsic ionic nature of metal halide perovskites (MHPs) is the main reason for ion migration which leads to severe degradation and stability challenges in perovskite solar cells (PSCs). Here, we provide a method to quantify the ionic nature of halide perovskites by measuring the ionic concentration of the devices. We also propose three mitigative strategies to reduce the ion migration in PSCs: the use of triple halide perovskites, small cation additives (K and Rb) to selectively bind to the halide species, and antioxidants to prevent adverse halide electrochemical reactions. We hypothesize and are demonstrating that these strategies increase operational stability and reliability.

2D-GaSe/In$_x$Se$_y$ Layer for Rapid ELO GaAs Technique

Nobuaki Kojima, Yoshio Ohshita, Masafumi Yamaguchi

Toyota Technological Institute, Nagoya, 468-8511, Japan

Abstract — In recent years, thin film III-V cells using epitaxial lift-off (ELO) technique and substrate re-use has attracted increasing attention as low-cost, high-efficiency, light-weight and flexible solar cells. Key problems for the cost reduction of this method are a long time process required and use of expensive GaAs substrate. We have developed a rapid ELO technique using mechanical cleavage via 2D-GaSe/In$_2$Se$_3$ double layer inserted between the III-V epitaxial layer and the substrate. In this paper, we discussed the role of In$_2$Se$_3$ in 2D-GaSe/In$_2$Se$_3$ double layer structure. We found the initial In$_2$Se$_3$ layer has an important role in the step-flow growth of GaSe. For the subsequent GaAs growth on 2D-material, the steps and terraces structure is impotant, since the crystal nucleation occurs at the step edges of 2D-material.

I. INTRODUCTION

In recent years, thin film III-V cells using epitaxial lift-off (ELO) technique and substrate re-use has attracted increasing attention as low-cost, high-efficiency, light-weight and flexible solar cells. In the common ELO technique, chemical etching of AlAs sacrificial layer is used for the epitaxial III-V layer separation from the substrate. Key problems for the cost reduction of this method are a long time process required and use of expensive GaAs substrate.

In our previous work, we have developed a rapid ELO technique using mechanical cleavage via 2D-layered metal selenide inserted between the III-V epitaxial layer and the substrate [1-4]. In this technique, processing time is only a few seconds and Si(111) can be used as a substrate, which has great potential for remarkable cost-reduction. Fig. 1 shows photographs of the transferred 1.5μm-thick GaAs layer onto polyimide sheet and the remained Si substrate. In the beginning of our study, we used In$_2$Se$_3$ as 2D-metal selenide layer, since the In$_2$Se$_3$ atomic structure of 2D unit layer surface is similar

with GaAs(111) surface. However, during GaAs growth, In$_2$Se$_3$ undergoes a structural change to non-2D phase due to high GaAs growth temperature. Then, we introduced 2D-GaSe/In$_2$Se$_3$ double layer instead of In$_2$Se$_3$ single layer for the heat resistant improvement. GaSe is more heat resistant than In$_2$Se$_3$. This structural modification led to the success of the GaAs layer transfer as shown in fig. 1. However, the need of In$_2$Se$_3$ in this structure remains unclear.

In this paper, we discuss the role of In$_2$Se$_3$ in 2D-GaSe/In$_2$Se$_3$ double layer structure for rapid ELO GaAs technique.

II. EXPERIMENTAL PROCEDURE

GaSe single layer or GaSe/In$_2$Se$_3$ double layer were grown on Si(111) 4° off substrate by molecular beam epitaxy (MBE). Typical growth temperature is 500 °C for GaSe and 450 °C for In$_2$Se$_3$, respectively. The In$_2$Se$_3$ layer thickness is around 8 nm. The surface morphology of these samples was compared by SEM images. The film crystal quality was characterized by X-ray diffraction (XRD).

After the growth of GaSe/In$_2$Se$_3$ double buffer layer, the sample was taken out into the air and loaded to the another MBE system for subsequent GaAs growth for limiting Se contaminations. GaAs layer was grown on the GaSe/In$_2$Se$_3$ buffer at 550°C. TEM image of near the interface between GaSe and Si substrate is analysed.

III. RESULTS AND DISCUSSION

Fig. 2 shows the SEM images of GaSe surface with or without In$_2$Se$_3$ layer. The steps and terraces structure is clearly seen in the GaSe/In$_2$Se$_3$ double layer sample. This indicates that

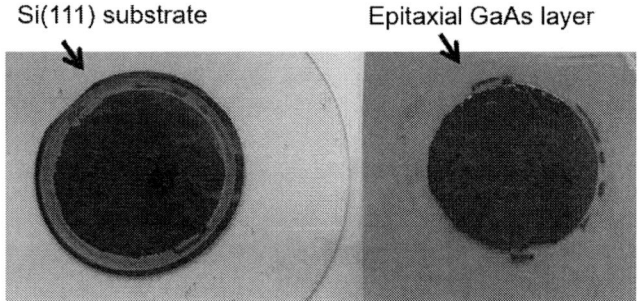

Fig. 1 Photographs of the transferred 1.5μm-thick GaAs layer onto polyimide sheet and the remained Si substrate using our developed rapid ELO technique.

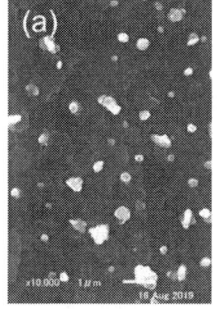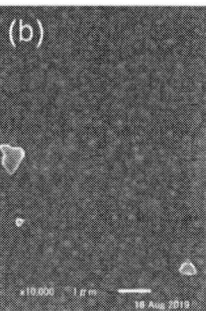

Fig. 2 SEM images of GaSe surface of the GaSe/In$_2$Se$_3$ double layer sample (a), and the GaSe single layer sample (b).

GaSe crystal growth is proceeding in a step-flow mode. For the subsequent GaAs growth on 2D-material, the steps and terraces structure is impotant, since the crystal nucleation occurs at the step edges of 2D-material. While the island growth of GaSe is seen in the GaSe single layer sample. Therefore, we can conclude the initial In_2Se_3 layer has an important role in the step-flow growth of GaSe.

Fig. 3 shows the HR-TEM image of $GaAs/GaSe/In_2Se_3$ structure near the interface between GaSe and Si substrate. The 2D-layered structures are seen from the Si interface. The calculated c-axis lattice constant of the film near the Si interface is closer to InSe's rather than In_2Se_3's. InSe also has the same layered crystal structure with GaSe. In fact, a weak diffraction peak was observed near the lattice constant of InSe in the XRD measurement. Therefore, we consider that the In_2Se_3 reconstruct into InSe during GaSe growth. The interface between GaSe and InSe is not clear in the TEM image, since these two material has the same crystal structure.

In conclusion, we consider the initial In_2Se_3 layer is grown epitaxially on the Si(111) substrate with a step-flow mode, then this layer reconstruct into InSe during the subsequent GaSe growth. GaSe can smoothly chemically bond with InSe at the step edges, and continues to grow in step-flow mode.

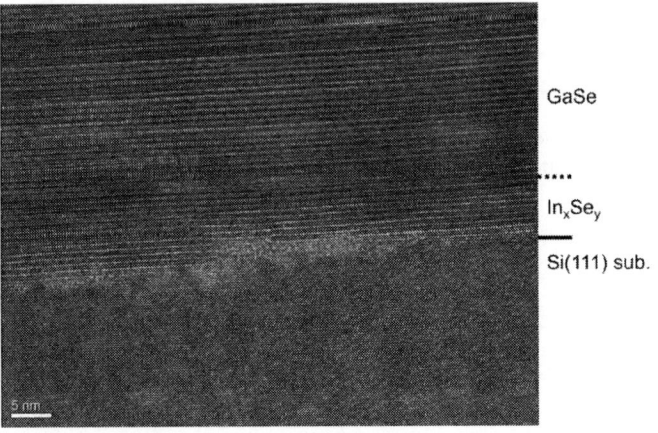

Fig. 3 HR-TEM image of $GaAs/GaSe/In_2Se_3$ structure near the interface between GaSe and Si substrate

IV. CONCLUSIONS

We have developed a rapid ELO technique using mechanical cleavage via $2D-GaSe/In_2Se_3$ double layer inserted between the III-V epitaxial layer and the substrate. In this paper, we discussed the role of In_2Se_3 in $2D-GaSe/In_2Se_3$ double layer structure. The initial In_2Se_3 layer has an important role in the step-flow growth of GaSe. For the subsequent GaAs growth on 2D-material, the steps and terraces structure is impotant, since the crystal nucleation occurs at the step edges of 2D-material. We consider the initial In_2Se_3 layer is grown epitaxially on the Si(111) substrate with a step-flow mode, then this layer reconstruct into InSe during the subsequent GaSe growth. GaSe can smoothly chemically bond with InSe at the step edges, and continues to grow in step-flow mode.

ACKNOWLEDGMENT

This work was supported in part by JSPS KAKENHI Grant Numbers JP22K04957.

REFERENCES

[1] N. Kojima, YC. Wang, Y. Ohshita, M. Yamaguchi, Proc. 37th European Photovoltaic Solar Energy Conference and Exhibition, pp. 615-617 (2020).
[2] N. Kojima, YC. Wang, K. Kawakatsu, A. Yamamoto, Y. Ohshita, M. Yamaguchi, Proc. 2019 IEEE 46th Photovoltaic Specialist Conference (PVSC), pp.1015-1017 (2019).
[3] N. Kojima, YC. Wang, K. Kawakatsu, Y. Ohshita, M. Yamaguchi, Proc. 2018 IEEE 7th World Conference on Photovoltaic Energy Conversion (WCPEC), pp. 0214-0215 (2018).
[4] N. Kojima, L. Wang, Y. Ohshita, M. Yamaguchi, Proc. 33rd European Photovoltaic Solar Energy Conference and Exhibition, pp.1295-1297 (2017).

NSF Industry-University Cooperative Research Center (IUCRC) for a Solar Powered Future 2050 (SPF2050)

Amit Harenkumar Munshi, Walajabad S. Sampath, Brian Korgel

Colorado State University, Fort Collins, CO, United States

University of Texas at Austin, Austin, TX, United States

We are a Phase III Industry/University Cooperative Research Center operating since 2010, previously under the name the Center for Next Generation Photovoltaics. With funding from the National Science Foundation and industry partners, our Center has supported over 200 students, published over 125 papers and developed several technologies that are now used by the solar power industry. Our researchers are committed to helping reach a carbon neutral future. Our goal is to enable the United States and the world to achieve a zero carbon footprint by 2050 with the use of solar energy as a resource. The NSF IUCRC program generates breakthrough research by enabling close and sustained engagement between industry innovators, world-class academic teams, and government agencies. The Center carries out research to enable higher performance and lower-cost solar cells; improved reliability of PV components and systems; develop effective ways to recycle end-of-life solar panels; inform the creation of a circular solar power economy; develop creative new approaches to land use in both rural and urban environments; integrate emerging storage technologies with solar power generation to alleviate problems of intermittency and variability; address techno-economic and socio-economic challenges; and identify new approaches to solar-powered electrification of transportation and heating/cooling sectors. SPF2050 catalyzes breakthrough pre-competitive research for the solar power industry by enabling close and sustained engagement between industry innovators, world-class academic teams and government agencies. Members have access to: (1) Major savings on institutional overhead charges, (2) Large pool of research funding, (3) Access to additional NSF supplemental funding, (4) Access to PhD students that could be future employees, (5) SPF2050 supports research across the entire solar power industry value chain, providing members with opportunities to gain insights that could benefit their business, (6) Access to world-class research teams and their extensive research networks across universities, national labs, and industry partners.

978-1-6654-6060-6/23 $31.00 © 2023 IEEE

Indoor and Outdoor Characterization of III-V/Ge Solar Cells Assembled on Glass Substrate for Concentrated Photovoltaic Applications

K. Kouame[1,2], J. Kinfack[1,2], D. Danovitch[1,2], P. Albert[1,2], T. Bidaud[1,2], A. Turala[1,2],
M. Volatier[1,2], V. Aimez[1,2], A. Jaouad[1,2], M. Darnon[1,2], G. Hamon[1,2]

[1] Laboratoire Nanotechnologies Nanosystèmes (LN2) - CNRS IRL-3463 Institut Interdisciplinaire d'Innovation Technologique (3IT), Université de Sherbrooke, 3000 Boulevard Université, Sherbrooke, J1K 0A5 Québec, Canada

[2] Institut Interdisciplinaire d'Innovation Technologique (3IT), Université de Sherbrooke, 3000 Boulevard Université, Sherbrooke, J1K 0A5, QC, Canada

Abstract — **We propose and evaluate a new assembly process to fabricate concentrated photovoltaic (CPV) modules. We replace wire bonding, alignment, and cell interconnexions with alternative microelectronic surface mount technologies (SMT). For this purpose, solar cells are first assembled on tempered glass printed circuit boards (PCB) and then characterized indoors. This PCB is then integrated into a CPV module with an efficiency of 38.07 % at CSOC measured outdoor.**

I. INTRODUCTION

Concentrated photovoltaic (CPV) consists in concentrating the incident sunlight on high efficiency multi-junction solar cells using concentration optics. The record for a solar cell under concentration is 47.6 ± 2.6% and was obtained for a 6-junction tandem solar cell under a concentration of 665× [1]. However, this technology remains less competitive in the photovoltaic market than flat crystalline silicon panels with only a maximum efficiency of 26.8% for a crystalline silicon solar cell [1]. This is partly due to the high material price, but also to the complexity of the technology. Indeed, module fabrication includes multiples wire bonding steps to connect the cells together which is limiting the throughput. The assembly costs represent 18% [2] of the manufacturing cost of CPV modules. In addition, current assembly methods require non-standard large surface and high precision alignment tools.

In this study, we propose a new assembly process which allows to: (1) Reduce assembly time by simplifying the process, (2) improves the performances of the modules (3) allows a better heat dissipation in the module.
We first present the characterization of solar cells fabricated with the assembly method, then we manufacture a 4 solar cells CPV module and characterize the module in Concentrated Standard Operating Conditions (CSOC [3]).

II. ASSEMBLY PROCESS OF SOLAR CELL

Our assembly process is based on SMT (Surface Mount Technologies). The solar cells are assembled on a tempered glass PCB. The assembly process can be summarized in 4 main steps as described Fig. 1 cross-sectional process flow: screen printing, pick and place, mass reflow, and underfill dispense.

1) <u>Screen printing</u> consists in using a stencil to locally deposit the solder paste on glass PCB as shown in Fig. 1-a and b. For prototyping, we used SAC305 solder paste alloy paste and 75 µm-thick Kapton as a stencil with a LPKF protoprint manual screen-printing equipment.

2) <u>Pick and place</u> is used for cell placement with coarse alignment accuracy (~200 µm). Cell is assembled face down (active surface in front of the glass plate). We used a Tresky 8800 placement tool.

3) <u>Mass reflow</u> is used for soldering. Simultaneously, fine alignment is obtained thanks to surface tension of the liquid solder joints. We use here a preheating of 170°C/170s, reflow to 280°C/125s and cooling to room air for 110s.

The steps 2) and 3) are described in Fig. 1-c.

4) <u>Underfill dispense</u> is used to fill the gap between the solar cell and the glass PCB with a transparent material as depicted in Fig. 1-d. This step was performed by capillary flow using a transparent underfill with viscosity between 225-425 cPs. The underfill was cured at 80°C for 3h.

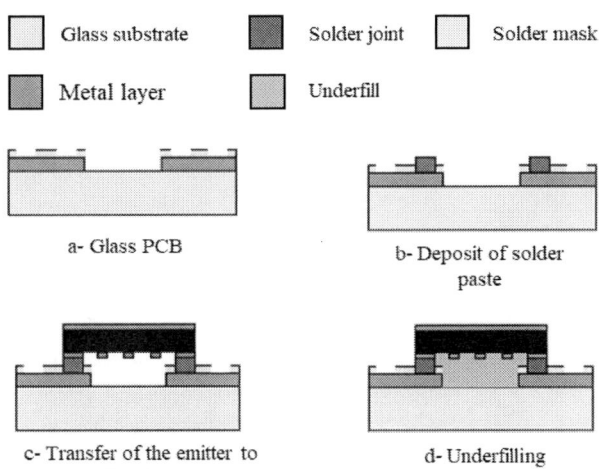

Fig. 1. Process flow of solar cell assembly on glass substrate.

978-1-6654-6060-6/23 $31.00 © 2023 IEEE

The solar cell assembled on the glass PCB is characterized by a 1-sun simulator, external quantum efficiency measurement (EQE) and under light concentration with a flash tester. As a reference, the same characterizations were performed on the bare cell, *i.e.*, before their assembly on the glass PCB.

For the fabrication of the CPV module, 4 solar cells were assembled on a dedicated glass PCB. Then, the following steps were performed:

5) A copper heat sink was glued on the back side of the solar cells with a conductive glue,

6) An EVA + Teldar lamination was made to protect the solar cells from the outside environment.

The 4 assembled solar cells (cell_A, cell_B, cell_C, cell_D) are aligned with each other on the PCB. These pre-aligned solar cells are then aligned with an array of four 160X concentration lenses.

III. EXPERIMENTAL RESULTS

A. Indoor Characterization

Fig. 2 represents the I-V characteristics of the bare and encapsulated solar cells. We obtained an open circuit voltage V_{oc} = 2.53V and a fill factor of 89.49% for both the bare and encapsulated solar cells. This confirms that the encapsulation process does not degrade the solar cells.

The quantum efficiency is the ratio of the number of carriers that are collected by the solar cell to the number of photons incident on the solar cell. J_{sc} can also be calculated from the EQE by the expression:

$$J_{sc} = -q \int_{\lambda_1}^{\lambda_2} EQE(\lambda) . \phi_{ph,\lambda}^{AM1.5D} . d\lambda \qquad (1)$$

With q the charge of an electron, $EQE(\lambda)$ = external quantum efficiency of each wavelenght, $\phi_{ph,\lambda}^{AM1.5D}$ the incident photon flow for an AM1.5D illumination.

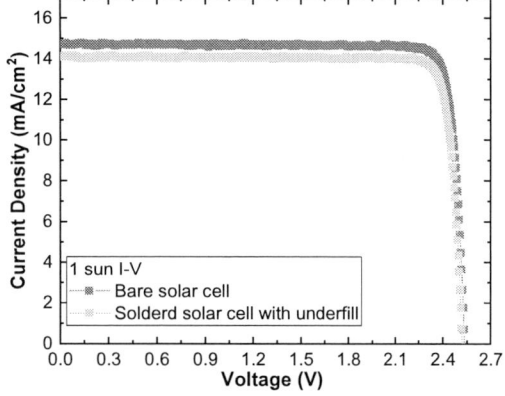

Fig. 2.   I-V characteristics of bare and encapsulated solar cells measured under 1 sun (AM1.5D) illumination.

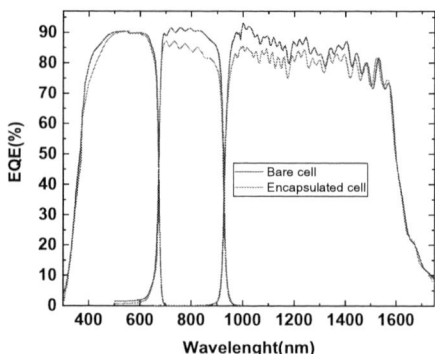

Fig. 3.   EQE measurements of bare and encapsulated solar cell.

The EQE has the advantage of a spectral shape that is independent of the light source and is also independent of the cell area. Fig. 3 shows the EQE measurements made as a function of wavelength of the bare cell and the encapsulated cell. We observe a decrease in EQE after assembly due to reflection and absorption. The calculations of J_{sc} made from the EQE measurements show that the limiting sub-cell of the bare cell is the Top Cell (TC) with J_{sc} = 13.72 mA/cm² and the limiting sub-cell of the encapsulated cell is the Middle Cell (MC) with J_{sc} = 12.98 mA/cm².

Since the assembly process developed here is for concentrated photovoltaic applications, measurements under concentration were performed using a flash tester. The results are shown in Fig. 4. As expected, J_{sc} increases linearly with increasing concentration and V_{OC} increases logarithmically with the concentration factor. The fill factor decreases from 89.23% to 85,16% for the bare cell and from 90.62% to 87.01% for the encapsulated cells, respectively, when the concentration increases for 100 to 900×. The efficiency is maximum at 500×. with 40.1% and 41% for the bare and encapsulated cells, respectively.

B. Outdoor Characterization Method

The characterization of the CPV module with 4 solar cells was carried out in the solar park of the Institut Interdisciplinaire d'Innovation Technologique (3IT) of the Université de Sherbrooke using a EKO Keithley 2-axis tracker. This tracker allows to follow the sun on two axes, in azimuth and in elevation with a precision of ±0.01°. The EKO tracker is driven by a Raspberry pi microcomputer on which a tracking program is implemented. The Raspberry also controls a Keithley multiplexer (3706A) and a Keithley source measurement unit (2601B) to measure electrical characteristics of each individual solar cells [4] The measurements were performed under COSC conditions, according to ASTM E2527 (American Standards of Technical Material) and the IEC 62670 (International Electrotechnical Commission) standard.

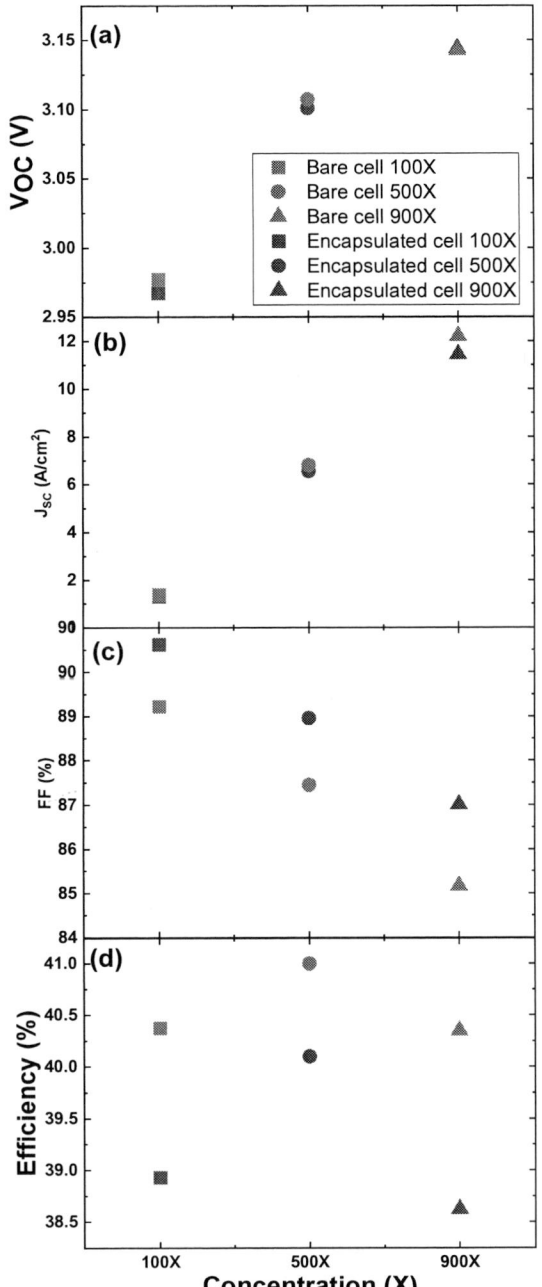

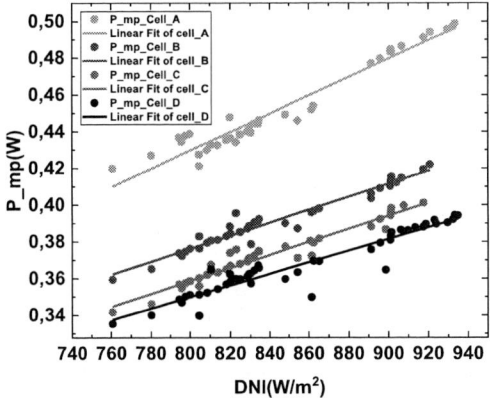

Fig.5.   MPPT measurements of the 4 cells of the CPV module.

Irradiance). The measurements are presented in Fig. 5. From these measurements, we deduce an efficiency of 38.07% for the best performing cell under a concentration of 160×, indicating state of the art performance of our module.

IV. CONCLUSION

We presented a new assembly method for concentrator photovoltaics modules fabrication, based on surface mount technology on a glass PCB. Characterization of the prototype under laboratory conditions showed an efficiency of 40.1% at 500× which decreases at high concentration due to the series resistance. The outdoor characterization in standard operating condition showed an efficiency of 38.07%, demonstrating the potential of this assembly method for high performance CPV modules fabrication.

Fig. 4.   (a) Open circuit voltage, (b) short circuit current density, (c) fill factor and (d) efficiency of bare and encapsulated solar cells estimated by flash test at 100, 500 and 900× concentration.

The I-V characteristics measurements allow to evaluate the power of the module according to the DNI (Direct Normal

REFERENCES

[1] M. A. Green, E. D. Dunlop, G. Siefer, M. Yoshita, N. Kopidakis, K. Bothe and X. Hao, "Solar cell efficiency tables (Version 61)", *Progress in photovoltaics,* 2022

[2] K. A. Horowitz, M. Woodhouse, H. Lee, and G. P. Smestad, "A bottom-up cost analysis of a high concentration PV module" in AIP Conference Proceedings, vol. 1679, p. 100001, AIP Publishing, 2015.

[3] IEC 62670-3. IEC 62670-3:2017, IEC Webstore energy efficiency, rural electrification, solar power, solar panel, photovoltaic, PV, smart city, LVDC. Int. Electrotech. Comm., 2017.

[4] A. J. Kinfack Leoga, A. Ritou, M. Blanchard, L. Dirand, Y. Prunier, P. St-Pierre, D. Chuet, P.-O. Provost, M. Volatier, V. Aimez, G. Hamon, A. Jaouad, C. Dubuc and M. Darnon, "Outdoor Characterization of Solar Cells with Micro-structured Anti-Reflective Coating in a Concentrator Photovoltaic Module", submitted to IEEE Journal of Photovoltaics.

Survey of Module and System Quality in Brazil PV Deployments

Lawrence L. Kazmerski[1], Denio Alves Cassini[2], Daniel Sena Braga[2], Suellen C.S. Costa[2], Vinícius Camatta Santana[2], and Antonia Sonia A.C. Diniz[2]

[1]Renewable and Sustainable Research Institiute (RASEI), University of Colorado Boulder, Boulder, CO 80309 USA; [2]Pontifícia Universidade Católica de Minas Gerais (PUC Minas), Belo Horizonte, Brazil

Abstract — **Brazil is a rapidly emerging solar-PV market and ranked fifth in added-PV power among world countries in 2021. As such, the population of power installations is growing rapidly with most PV-product is either imported or assembled in module production facilities in Brazil. The quality and reliability of these systems depends in part on the module integrity—and is a major basis for continued consumer confidence in this renewable-energy electricity source. This Fulbright-Scholar sponsored study surveyed installations in several climate locations in Brazil, and this paper focuses specifically on 11-sites in Minas Gerais State. Technologies include monofacial and bifacial Si (newly installed to 8-years operation) and thin-film a-Si:H (17-year operation). Tracking and non-tracking systems were evaluated. The methodology started with a site visual (spreadsheet) inspection, followed by a selection of individual modules and strings for performance evaluations based on the observations. The characterizations primarily included I-V and thermography (IR-mapping). Performance characteristics were compared to manufacturers' specifications. For a few cases, individual-module electroluminescence response was determined. Albedo at each site was measured because of the growing interest and deployment of bifacial technologies. In the cases of bifacial installations, the total and separate front- and rear-module I-V characteristics were measured. The relative soiling levels were documented (qualitatively), and the selected modules and strings were characterized in existing and cleaned conditions. The cleaning schedules and methods were recorded if available for the site. In general, the measurements confirmed that the site modules and strings were within the specifications. However, some issues (shading, module stresses, snail trails, contacts, wiring) were noted and discussed in this study that could lead to reliability issues.**

I. INTRODUCTION AND RATIONALE

Brazil has a diverse electricity-generation matrix. Through November 2022, the installed capacity in was ~190 GW, divided among 9000 power plants [1,2]. Brazil's electricity supply is based mainly upon renewable-energy resources. For decades, hydropower has been the main source of electrical power and through 2021 represented about 62% of total capacity [1,2]. Wind (12.5%) and bioenergy (7.2%) accounted for the next largest renewable energy contributions [1]. At the end of 2021, installed photovoltaics (PV) exceeded 13-GW [3]—and the growth in Brazil's cumulative PV is presented in Fig. 1. Though the 2021 PV capacity was only at 7.7% on the electrical power generation in the country (about 18-TWh in terms of the electricity generation), solar-PV electricity is the fastest growing of the renewable sources. However, Brazil has added more than 12 GW of PV in 2022, with a cumulative electrical capacity now exceeding 25 GW. At the end of 2022,

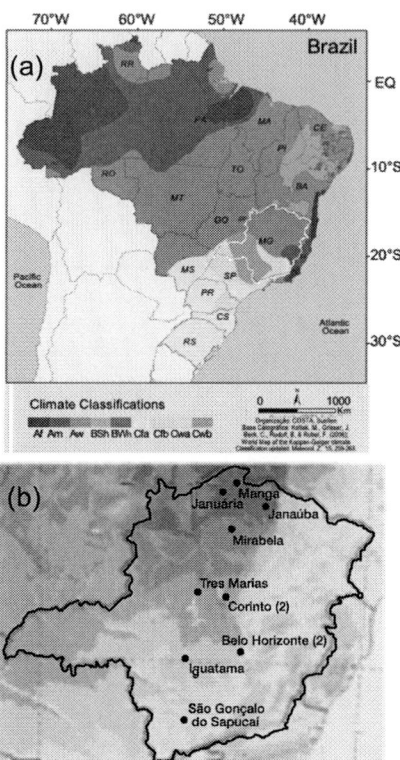

Fig. 1. Brazil: (a) Climate-zone map showing classifications with outline of Minas Gerais State; (b) PV sites surveyed in this study imposed on solar resource map of the state of Minas Gerais.

Brazil stood as the 11th-largest producer of solar electricity in the world [6]. Brazil ranked 5th in the world in installations that year and has been identified as among the top-10 emerging PV world markets [1-3]. The incredible growth of Brazil's solar PV is certainly indicated by the announcement that PV now surpassed the electric-power capacity of wind at the end of 2022 [1]. The emergence of PV is also enhanced by the issues surrounding the historically dominant hydroelectricity supplies. The onset of droughts and the concerns with climate and environmental effects has limited that sector's operations or expansions and had led to service interruptions and equipment issues. These issues have further brought focus on the quality and reliability of PV in meeting its electricity demands—and in continuing to build confidence of consumers and decision makers in renewable-energy electricity.

978-1-6654-6060-6/23 $31.00 © 2023 IEEE

The purpose of this Fulbright Scholar-sponsored project has been to examine the performance and quality of PV product being installed in the exapanding markets. As in most of the rest of the world, the PV modules being installed in systems in this country originate either from external country sources or from module manufacturing in Brazil that depends on import of the solar cells. This project has teamed with several Brazil installations to provide some confidence on the quality and the operation of the PV modules in system operations. The approach has been fashioned after the successful "All-India Surveys" [4] in order to provide an independent analysis of the technical aspects of the PV modules in operation. This paper is reporting on the *initial phases* of this mini-survey which was initiated in March 2022. Although the survey is being conducted over several Brazil States, this report focues on 10 installation-sites in the State of Minas Gerais (Fig. 1), which has been the initial and leading geographical location for this country's PV revolution.

The methodology for each site starts with a visual (spread-sheet) inspection to provide some overview of the condition of the installation. Based upon these observations, a group of PV modules and strings are selected for characterization (non-destructive electrical and themal). There is also some attention to the soiling of the modules which are measured in their "existing" and cleaned conditions for comparison to the manufacturers' specifications. (The maintenance aspects of cleaning periods and techniques are also recorded for information on severity of soiling and providing a basis for future examinations for effects of these procedures.)

The PV technologies included mainly crystalline-Si, both monofacial and bifacial modules in tracking and non-tracking installations. The systems ranged from newly installed to ones operating over the past 8-years. One early installation of thin-film a-Si:H modules was also evaluated, beginning operation in 2005.

II. FOCUS AND METHODOLODY

The locations covered in this paper are identified in the maps presented in Fig.1. These provide the climate zone map for Brazil [5] and the specific sites evaluated in this survey report on the solar-resource (GDI) map for the State of Minas Gerais [6]. Most of the surveyed sites were 5-MW, conforming to the distributed generation (DG) definition used in the Brazil regulations that provide specific financial incentives up to this power-generation capacity [7].

All test were performed and reported under standards: IEC 62446:2014, IEC-60891:2010, IEC 60904-1:2006, and Norm ABNT NBR 16274. I-V characteristics were measured with Daystar DS-1000, Seaward 210 and Solmetric PVA 1500V curve tracers. IR thermal mapping with FLIR M210 V2XT2 640 13 mm camera and IITB Canon IR camera. EL measurements used a modified Sony IITB design with control unit. All surveys were initiated with a visual inspection of the site using the NREL spreadsheet method [8]. In several cases, a Matrice V2 Drone Aircraft was used to overview the site—and gain IR images with the on-board FLIR M210 IR camera. Calibrated Eppley pyranometers

and Si reference cells were used to monitor the solar irradiance. Finally, a Hukseflux SRA30 albedometer was used to determining albedo.

Fig. 2. Three systems: (a) Belo Horizonte (a-Si:H); (b) Três Marias (monofacial-Si tracking); (c) Iguatama (bifacial-Si tracking).

III. RESULTS AND OBSERVATIONS

Table I presents a summary of the Brazil sites and other information on the systems and modules that are the focus of this survey report. The final paper will provide details on all these installations, including module and string characterization and other survey measurement. Three representative sites are : Belo Horizonte (thin film), Três Marias (crystalline-Si, non-tracking, monofacial), and Iguatama (crystalline-Si, tracking, bifacial)—with photos of these presented in Fig. 2. In general, the installations and modules at all Minas sites were found to meet manufacturers specifications with no serious issues on the module products. Some installation related issues are reported.

TABLE I. SUMMARY OF INSPECTION SITES AND OBSERVATIONS

Summary of Minas Gerais PV Sites

Site in Minas Gerais	Tech-nology	Module Structure (rating)	Track-ing	Capa-city (MWp)	Year*	Comments and Observations
Belo Horizonte (19.9181'S, 43.9367'W)	Multi-Si (Multi)	Monofacial (325-Wp)	No	1.3	2014	Degradation rate: ~1.5%-2%/year. Issues: High density of snail trails and shading by various permanent structures; corrosion of some contacts and minor delaminations. Issues with EVA discoloration.
Belo Horizonte	Thin-Film a-SiH	Monofacial (64-Wp)	No	0.003	2005	3-damaged modules did not meet specifications, although still functional: All others met or exceeded manufacturer's specs ("15% higher Pm initially" than after 8-10 weeks of operation).
Três Marias (13.2053'S, 45.2275'W)	Mono-Si	Monofacial (395-Wp)	Yes	2.5	2020	2 side-by-side sites each 2.5-MWp; Calibrated bifacial module measured for developer (reddish-dirt ground cover); Albedo: 0.2 (average 0.16). Proprietary "white cover" for future bifacial installation; Albedo: >0.5. Issues: Shading by ground vegetation in many power-plant areas.
Iguatama (16°2244'3'S, 45.7002'W)	Mono-Si	Bifacial (590-Wp)	Yes	5	2022	Awaiting commissioning; Albedo: low ground cover vegetation growth (about 75% coverage) on red dirt. Issues: Possible concern with bowing of modules. Minor shading in few locations by vegetation. Average albedo: 0.11-0.16.
Corinto 1 (18°2244'3'S, 44°2822'W)	Mono-Si	Monofacial (330-Wp)	No	5.28	2019	3200 modules examined in strings of 20 modules). Modules and strings all within specifications after corrections. System installation issues. Commissioned after correction.
Corinto 2	Mono-Si	Bifacial (355/360Wp)	Yes	6.44	2021	8550-355Wp and 9450-360Wp bifacial modules. (Total of 1800 modules). No module issues. [System analysis in progress]
Manga (14°48'11.32"S 43°67'34.15"W)	Multi-Si	Monofacial (330-Wp)	No	8.2	2019	Each UG connects 12 combiner boxes with 12 strings. 19200 modules; Vegetation & several system instal issues (corrected).
São Gonçalo do Sapucaí (21.8913"S 45.5977'W)	Mono-Si	Bifacial (590-Wp)	Yes	3.47	2021	Data analysis in progress and not yet released. 5882 modules. Issue noted with module overheating & module failure (damage).
Janaúba (18°41'48.85"S 43°28'0.04"W)	Multi-Si	Monofacial (325-Wp)	No	5.2	2019	Within specifications (99% of modules & strings for V and I-V test criteria). Issues with wiring, vegetation, cabling, cable clamps, bolts. One damaged backsheet module (replaced).
Januária (16°1253.13"S 44°0'44.26"W)	Multi-Si	Monofacial	No	5	2019	Modules within specifications. Some issues with vegetation, connectors and with J-boxes. Cold commissioning after soiling corrections. Labeling issues for components.
Mirabela (16°1253.13"S 44°0'44.26"W)	Mono-Si	Monofacial (385-Wp)	No	2.59	2020	6720 modules (system within specifications and commissioned). Separated into two GU units (1.294GWp each). Issues with shading by vegetation and growth; Labeling of components.

*Year of installation or commissioning

The ealiest of the installations surveyed was in **Belo Horizonte** at PUC Minas. This is a 1.9 kW system using a-Si:H modules manufactured in 2003 and installed in 2005. The system was reported to only have issues with the inverter. All modules on this system (and strings) were measured. Three of the modules had physical damage (cuts on the front surface)— and these were the only modules that did not meet or exceed the initial specification. It should be noted that the specification provided for a 15% light-generated degradation from the initial value—and most of these modules did not experience this degradation level. The 3-damaged modules had power levels about 20%-25% less than the stated specification—but were still functional.

The ***Três Marias*** site is about 250-km NW from Belo Horizone, situated near the San Francisco River (adjacent to a reservoir and hydro station). This plant consists of 2-2.5-MWp systems, with 360-Wp monofacial modules mounted on trackers. The system had a moderate dust layer (typical accumulations in this region are <0.1%/day). This site was commissioned in 2019—and the visual inspection reported no visual issues with the modules. The performaces of these relative new modules tested were all near or above the manufacurer's specification. Two issues were reported from this site: excessive vegetation growth that did lead to some hot spots on the modules; and some stress issues with the wiring. Also, the vegetation was observed to be growing through the inverters! However, the strings and total system was performing to the levels of the commissioning. The developer reported that the only issue that was encountered was with the tracking control—software issues that led to them developing their own control software. The developer was in the process of negotiating for two other systems. The first was a floating PV system planned for the reservoir. The second was for bifacial module system—possibly using a proprietary ground sheet that they developed. To test this, we brought two-bifacial reference modules to evaluate their operation with the existing ground cover (mainly a red earth) and with the "white" ground cover material. Parital results are included in Table 1, but the paper will have more extensive information on these measurements—including compleemntary laboratory-based results taken after this field survey.

Iguatama is 300 km WSW of Belo Horizonte. This system was just installed in 2022—and was awaiting commissioning. This gave complete access to the modules since they were not in an active mode. This is a tracking system (not yet in operation), with bifacial 440 Wp modules. The ground cover was mainly green vegiations covering 75%-80% of the ground area. The measured albedo was 0.18 ± 0.03. The modules tested all met the manfufactuer's specifications. Some interesting observations for these large glass-glass modules. As mounted on the tracker, all modules were noted to noticeably sag in the middle. The IR thermal mapping of modules indicated that all showed higher temperatures near the middle of the modules—though the modules still met specified outputs. This could eventually be a problem with the additional thermal and mechanical stress on the cells. The site ground area was not yet "groomed" for bifacial operation—and there were some minor issues with vegiation causing shading and hot spots. In these tropical regions, the vegetation grows rapidly—and maintenance attention is a priority. The developer was well aware of this. A return to this site is planned 6-9 months after the commissioning to evaluate the same strings and modules.

Of some interest is the other Belo Horizonte site that has had the Si-modules (non-tracking, monofacial) in operation for some 8+ years. These modules provide some contrast to the newer generation ones at the other sites. These modules and strings show significant degradation due to corrosion and some delaminations. Some surrounding structures caused shadowing and hot spots. Notably, the modules have a high density of snail trails—with the generation of the microcracks like due to the handling during installation in a confined site.

III. SUMMARY AND FINAL PAPER CONTENT

The major result of this survey is that module currently being delived into the Brazil markets meet manufacturers' specifications. There are no issues with the imported or Brazil assembled panes. Howerver, there are some concerns on installations and handling that have to be addressed. These obsevations are supported by module I-V measurements (all parameters), temperature characteristics, information on selected EL mapping, and descriptions of potential installation problems. Important the surveys to-date give strong indication that for both Brazil-assembled modules and those supplied from outside the country, there are no quality issues.

Acknowledgements: The authors gratefully acknowledge the Fulbright Foundation which supported this 2022 Fulbright Scholar project. The support of CNPq and CAPES-MEC is also acknowledged. We also thank the Graduate program in Mechanical Engineering, PUC Minas and GREEN PUC Minas for support and technical guidance. Finally, we acknowledge the cooperation and help of Márcio Eli Moreira de Sousa and Conerc (Mori), Brazil, for access to many of the solar plants.

REFERENCES

[1] ANEEL, Sistema de Informações da Generação da ANEEL (SIGA). https://dadosabertos.aneel.gov.br

[2] REN21 (2022). REN21 Renewables 2022 Global Status Report. (Paris, France: REN21 Secretariat). (ISBN 978-3-948393-04-5). https://www.ren21.net/wp

[3] Ernst and Young (2022). Renewable Energy Country Attractiveness Index (RECAI). https://www.ey.com/en_it/recai

[4] S. Chattopadhyay, R. Dubey, V. Kuthanazhi, S. Zachariah, S. Bhaduri, et al. ((2016). All-India Survey of Photovoltaic Module Reliability: 2016. National Centre for Photovoltaic Research and Education Indian Institute of Technology, National Institute of Solar Energy Gwalpahari, India. 2016.

[5] C.A. Alvares, J.L. Stape, P.C. Sentelhas, et al. (2013). "Köppen's climate classification map for Brazil." Meteorologische Zeitschrift 22, 711–728.

[6] E.B. Pereira, F.R. Martins, A.R. Gonçalves, R.S. Costa, et al. (2017). Atlas Brasilerio de energia solar. 2nd Edition. São Jose dos Campos: INPE, 2017. 80 pages.

[7] A.S.A.C. Diniz, L.V.B. Machado Neto, C.F. Camar, P.M.R. Morais, et al. (2011). "Review of the photovoltaic energy program in the state of Minas Gerais, Brazil." Renewable & Sustainable Energy Reviews 15, 2696-2706.

[8] J.H. Wohlgemuth, & S. Kurtz (2011). "Reliability testing beyond qualification as a key component in photovoltaics progress toward grid parity." In, Proceedings of the IEEE International Reliability Physics Symposium Monterey, 5E.3.1-5E.

690 Wp n-type i-TOPCon modules in mass production with >25% efficiency solar cells based on large-area 210 mm wafers

Yifeng Chen, Hong Chen, Shu Zhang, Le Wang, Chengfa Liu, Daming Chen, Jianmei Xu, Pietro Altermatt, Zhiqiang Feng, Pierre Verlinden

Trina Solar, Changzhou, China

Trina Solar, Changzhou, China

Trina Solar, Changzhou, China

Trina Solar, Changzhou, China

Trina Solar, Changzhou, China

Trina Solar, Changzhou, China

Trina Solar, Changzhou, China

Trina Solar, Changzhou, China

Trina Solar, Changzhou, China

AMROCK Pty Ltd, McLaren Vale, Australia

This paper presents the latest progress on the industrial mass production of the Industrial Tunnel Oxide Passivated Contacts (i-TOPCon) solar cells and modules, based on large-area 210 mm n-type silicon wafers. We have developed an industrial feasible fabrication technology for i-TOPCon cells. tunneling SiO_2 is thermally growth and covered in-situ by an intrinsic poly-Si layer, deposited by plasma enhanced chemical vapor deposition (PECVD) to solve the problem of quartz boat and tubes breakage in low-pressure LPCVD. We developed the industrial laser doped selective boron emitter with an efficiency improvement of 0.2%abs. We demonstrate that using thin wafers with a thickness of 140 μm is possible for large-area 210mm wafer. These enable us to achieve up to 25% efficiency in mass production. To improve the performance of i-TOPCon modules, we developed the 18-busbar (18bb) technology with low-damage laser cutting. To size the modules to fit perfectly into shipping containers, we developed the 210R technology (rectangular cell with dimension of 210 ×182 mm2). These enable us to develop modules with a power of 440, 590, and 680 Wp, which is the best modules design with lowest transportation cost. The 680 Wp module, to the best of our knowledge, is the modules with highest power output. And we demonstrate an aperture module efficiency of 24.24% with power over 680.5 Wp, which is independently confirmed by TÜV Nord, China. To the best of our knowledge, this is the highest efficiency TOPCon module on the market. For the fabrication of these 210/210R i-TOPCon cell/modules, a factory with a capacity of 8 GW has been built by Trina at the end of 2022.

Methylamine Post-Deposition Treatments of Vapor-Deposited Perovskite Thin Films

Chaiwarut Santiwipharat, Austin G. Kuba*, Kevin D. Dobson, Ujjwal K. Das, and William N. Shafarman

Institute of Energy Conversion, University of Delaware, Newark, Delaware, 19716, USA

Abstract — While the development of methylammonium lead iodide (MAPbI₃) perovskite for photovoltaics has grown rapidly, the growth of high-quality material from vapor processes continues to be difficult due to challenges including crystallinity control. Methylamine (MA) vapor post-deposition treatment is an approach to improve morphology and crystallinity of MAPbI₃ films, producing highly oriented, large grain-size perovskite films. Herein, experiments to characterize the liquefaction and recrystallization produced by treatment are described. Substrate temperature, MA partial pressure, and MA exhaust flow rate have substantial impacts on the film properties. Improved morphology and crystallinity after treatment indicates that substrate temperature plays a significant role in both liquefaction and recrystallization of the film. The result of this study is a useful pathway to fabrication of high-quality MAPbI₃ films from all-vapor process manufacturing.

I. INTRODUCTION

Due to their high efficiency and relatively simple fabrication processes, perovskite solar cells (PSCs) have attracted considerable attention. Vapor-processing of perovskite thin films potentially offers high throughput and versatility in the choice of material, and thus is a promising manufacturing approach. However, vapor-processed PSCs often show inferior performance to solution-based PSCs due to a complexity in sublimation control of organic precursor which is sensitive to deposition conditions [1].

To be competitive with solution processing, additional defect control and passivation can offer possible pathways to enhance vapor-based PSC performance. One novel technique, MA vapor-assisted post treatment, was reported to significantly enhance efficiency and stability of MAPbI₃ PSCs, with treated films exhibiting reduced trap densities and extended carrier lifetimes [2]. The treatment can be performed in a few seconds by exposing MAPbI₃ films to a MA-filled ambient and, therefore, could be integrated into all-vapor PSC fabrication.

During MA treatment, MAPbI₃ liquefaction and recrystallization occurs through MA adsorption and desorption, respectively. After an exposure to MA, the perovskite film turns to a transparent liquid phase of MAPbI₃ · x MA before crystallizing back to the solid MAPbI₃ film after removing MA gas from the ambient [3]. The effects of these liquefaction and recrystallization steps on the final film properties are dictated by treatment conditions, including temperature of substrate, concentration of MA in ambient, and removal rate of MA gas [4, 5]. This treatment has a significant potential to reduce the need of sublimation control during deposition. In this study, we characterize the influences of the MA treatment conditions on

final perovskite film properties including performance of solar cells from the treated films and evaluate the controlling factors of the treatment.

II. EXPERIMENTAL DETAILS

A. Vacuum deposition

300nm-ITO was deposited using radio frequency (RF) sputtering on soda lime glass (SLG) substrate at ambient temperature through a shadow mask. 20nm-SnO₂ electron transport layer (ETL) was deposited by RF sputtering.

MAPbI₃ perovskite films were deposited using a two-step close space vapor transport (CSVT) process [6]. In the first step, a PbI₂ source was sublimed and deposited on a 1"x1" glass/ITO/SnO₂ substrate placed above the source. The PbI₂ layer was then reacted with MAI vapor to form MAPbI₃. The thickness of the MAPbI₃ absorber is 350-450 nm. All perovskite samples were preserved in a N₂-filled glove box with <10 ppm O₂ and H₂O to prevent degradation.

To complete the MAPbI₃ cells, 60 nm of CuPC and 25 nm of MoOₓ were deposited by thermal evaporation under O₂ ambient as hole transport layer (HTL). 300 nm-Sb was evaporated through a mask as the back contact.

B. Methylamine Vapor Treatments

The MA vapor treatment system includes two main components, source container and treatment chamber, as shown in the schematic diagram in Fig. 1. This treatment enables control of the source and sample temperature, pressure of vapor in the source container and the reactor, and the gas flow rate in and out of the system. Both the source container and treatment reactor are placed on hot plates where they are heated to desired temperatures. Two flow meters with needle valves control gas flow rates in and out the system. Pressures of the vapor source container and the reactor are observed via pressure gauges. A connection valve between the source container and the reactor chamber is included.

To start the MA treatment, a MAPbI3 film is placed inside the treatment chamber which is then pumped down to remove air and filled with N2 gas to atmospheric pressure. To generate MA vapor, 33% methylamine in ethanol solution is added to the solution container. Pressure in the solution container is used to control MA gas concentration. After reaching the desired MA pressure, the connection between the container and the reactor is opened and MA gas transfers to the reactor chamber. A transparent MAPbI3· xMA liquid is formed in ≤ 5-7 sec due to

---

* Current address: Institute of Electrical and Micro Engineering, École Polytechnique Fédérale de Lausanne, Neuchâtel, Switzerland

978-1-6654-6060-6/23 $31.00 © 2023 IEEE

a complete collapse of MAPbI$_3$ crystal upon adsorption of MA [7]. To initiate recrystallization, the chamber's gas inlet and outlet is opened, and N$_2$ gas flows into the chamber, removing MA gas through the exhaust.

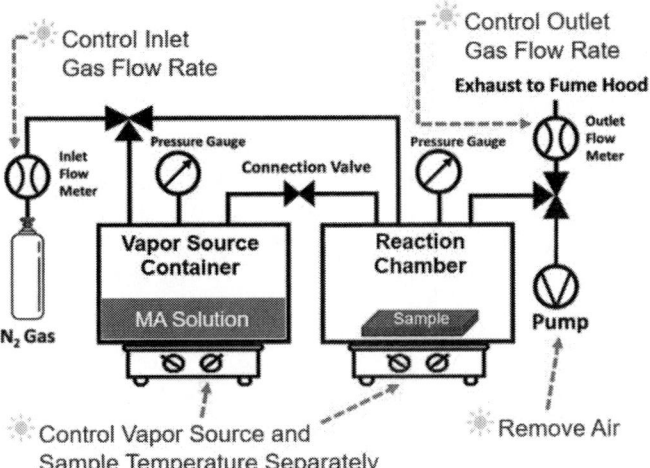

Fig. 1. Schematic of the vapor-assisted treatment setup

C. Characterization

Film morphology was observed using a JSM-7400F scanning electron microscopy (SEM) with 3kV accelerating voltage. A Rigaku D/Max 2200 system with Cu Kα radiation at 40 kV was used to obtain x-ray diffraction (XRD) measurements of films before and after treatments. J-V curves were obtained using a Keithley 2440 source metering unit with an AM 1.5 filtered light source under 1 sun condition.

III. RESULTS AND DISCUSSIONS

The substrate temperature, MA partial pressure in the solution container, and MA gas outlet flow rate were varied to characterize their effects on the perovskite film morphology after treatment and the ability to dissolve the film. To study the preliminary effect, the range of values for each factor was determined based on literature [3-5] and pre-experiment observations. The values are listed in Table I.

TABLE I
THE LOW-MIDDLE-HIGH VALUES FOR REACTION STUDY

Factor	Low	Mid	High
Substrate Temperature (T_{SS} [°C])	25	60	100
MA Partial Pressure (P_{MA} [kPa])	10	-	70
MA Exhaust Flow Rate (Q_{MA} [L/min])	2	-	10

The observation of MA treatment at each condition is presented in Table II. The table shows that at T_{SS} = 60°C, comparing to T_{SS} = 25°C condition, higher P_{MA} was needed to cause the film to liquefy, and a more rapid liquefaction was observed with increasing vapor concentration. More importantly, no liquefaction was observed in treatments at T_{SS} = 100 °C even at high MA vapor concentration. This finding corresponds to a report from Jacob et al. [4], that the MA gas cannot be absorbed due to the supersaturation of MAPbI$_3$ film above a critical temperature.

All the three factors in Table I have an impact on recrystallization. A low MA concentration in the chamber (P_{MA} = 10 kPa) caused haziness in some parts of recrystallized films. In contrast, P_{MA} = 70 kPA induced high MA concentration in the chamber and resulted in a relatively uniform and smooth morphology after recrystallization, potentially leading to lower surface defects. However, the mechanism for the haziness after the MA treatment is currently unknown.

Another important factor for recrystallization is Q_{MA}. As can be seen from SEM images in Fig. 2, low Q_{MA} of 2 L/min created large-grain morphology. Specifically, extra-large grains (>30 nm) were observed for T_{SS} = 60°C, Q_{MA} = 2 L/min treatment conditions at both pressure setpoints, some areas for P_{MA}=30 kPa and entire film for P_{MA} = 70 kPa. However, these large-grain films at these conditions exhibited numerous pinholes and deep grain boundaries throughout the treated-films.

A large-grain film can be obtained by a slow release of MA molecules from the intermediate MAPbI$_3\cdot x$MA liquid. In this case, the release rate is described by saturation of the intermediate liquid, which is controlled by T_{SS} and Q_{MA}. At T_{SS}

TABLE II
THE OBSERVATION OF LIQUEFACTION AND RECRYSTALLIZATION OF PEROVSKITE AT EACH SETPOINT

No.	T_{SS} [°C]	P_{MA} [kPa]	Q_{MA} [L/min]	Liquefaction	Recrystallized film
1	25	10	2	Dissolve in ~5-7 s	Haziness observed
2	25	10	10	Dissolve in ~5-7 s	Haziness observed
3	25	70	2	Dissolve in < 1 s	Smooth and packed grains
4	25	70	10	Dissolve in < 1 s	Smooth and packed grains
5	60	10	2	No liquefaction	
6	60	10	10	No liquefaction	
7	60	70	2	Dissolve in ~3-4 s	Large domain size observed
8	60	70	10	Dissolve in ~3-4 s	Large domain size observed
9	100	10	2	No Liquefaction	
10	100	10	10		
11	100	70	2		
12	100	70	10		

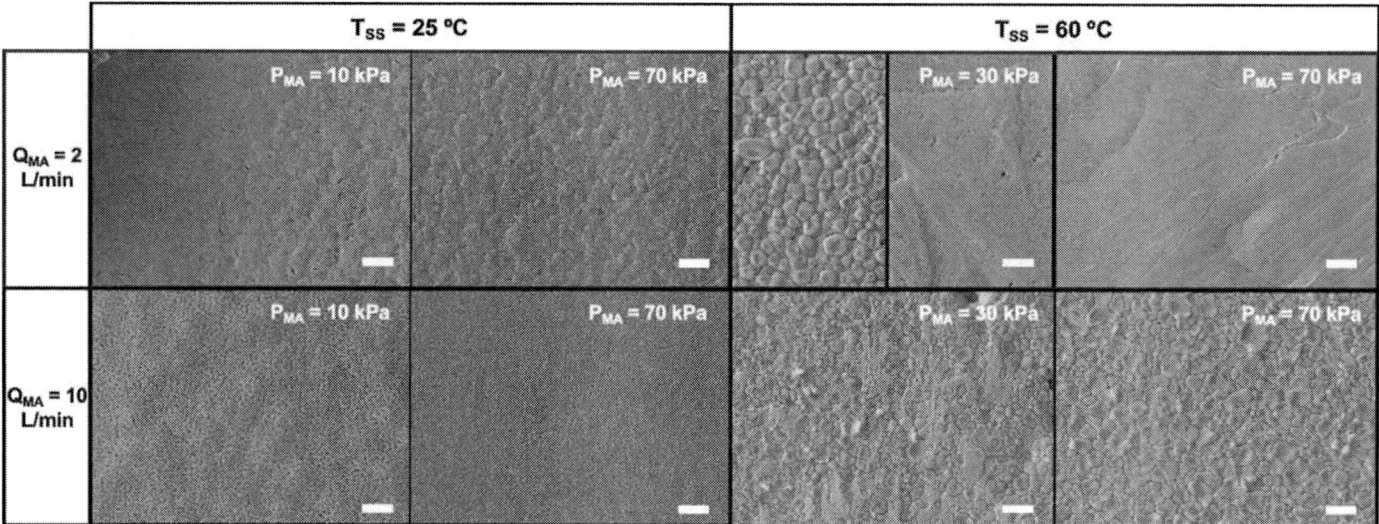

Fig. 2. SEM images of MA-treated MAPbI₃ films at different T_{SS}, P_{MA}, and Q_{MA} conditions. The reference bars have the length of 1 μm. In T_{SS} = 60 °C, Q_{MA} = 2 L/min, and P_{MA} = 30 kPa treatment condition, two different morphologies were observed on the same film.

= 60°C conditions, the saturation point is achieved faster than at T_{SS} = 25°C conditions, resulting in higher P_{MA} required for dissolving the MAPbI₃ film and slower liquefaction process. After MA gas is removed from ambient, the release of MA molecules from the liquid induces supersaturation, and recrystallization occurs as a result. Therefore, a slow reduction of MA partial pressure (low Q_{MA}) triggers formation of large MAPbI₃ nuclei due to a low nucleation density and long nuclei formation time. Fan et al. reported that MA-treated films with large grains were highly oriented, have low trap-state density, and long carrier lifetime [5]. The optimization is needed to fabricate large-grain pinhole-free perovskite films.

Fig. 3 shows semi-log plots of XRD patterns of MA-treated MAPbI₃ films at different T_{SS} compared with a non-treated film. Initially, both films exhibited residual PbI₂ from the MAI reaction process [8]. At T_{SS} = 20°C and 60°C where liquefaction and recrystallization occured, reduction of PbI₂ levels at these conditions indicated that the residual PbI₂ likely reacts with the adsorbed MA, and then either became amorphous or formed MAPbI₃ during the recrystallization process. In addition, the level of (112)/(200) and (330) planes of MAPbI₃ also decreased, showing that the formation of MAPbI₃ crystals were preferably oriented in (110)/(002) or (220) planes. On the other hand, at T_{SS} = 100°C where no liquefaction occurred, there was little to no change in crystallinity of MAPbI₃ and the amount of residual PbI₂.

J-V curves and parameters of solar cells from MAPbI₃ films treated at different conditions are presented in Fig. 4. Overall, there were boosts in V_{OC} for most treatment conditions due to reduction in recombination density after recrystallization. The V_{OC} boost can be as high as 138mV or 17% over the as-deposited case. Moreover, although the liquefaction did not happen at T_{SS} = 100°C, an increase in V_{OC} could result from healing of surface recombination sites of the MAPbI₃ interfacial layer since the exposure of MA gas was limited to surface. In contrast, the significant decrease in V_{OC} and FF at T_{SS} = 60°C and Q_{MA} = 2L/min could relate to the deep grain boundaries and

looser packing of the extra-large grains m, through inconsistent recrystallization and morphology, which may act as recombination sites.

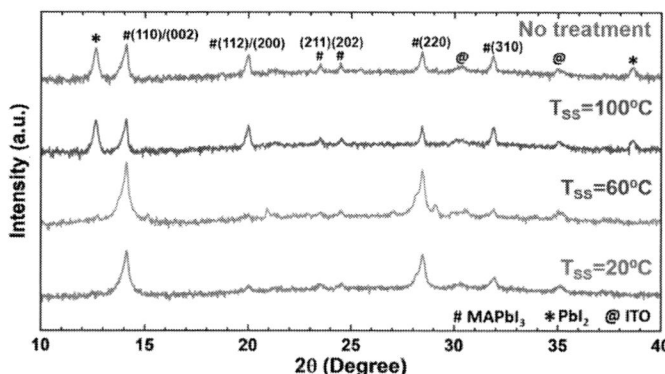

Fig. 3. Semi-log plots of XRD patterns of MA-treated films at different treatment temperatures.

Based on Fan et al., a large-grain and tightly-packed film leads to high solar cell performance and excellent device stability due to low trap-state density and long carrier lifetime [5]. As discussed earlier, the optimization of treatment condition is needed to construct large and packed grains which is possible in Q_{SS} ≤ 2L/min and 20°C < T_{SS} < 60°C condition.

IV. CONCLUSIONS

In this work, we have characterized the influences of three significant factors for MA treatment of MAPbI₃ thin films. T_{SS} contributes the prominent effect to both liquefaction and recrystallization processes as it dictates saturation point of the sample which affects abilities to absorb and desorb MA molecule, influencing the crystal grain shapes and sizes. Q_{MA} has important role in nucleation formation during the recrystallization process which impact the morphology and grain size. Lastly, P_{MA} or the MA concentration in chamber affects solely the liquefaction of the film. The change in crystallinity indicates that the MA-treated $MAPbI_3$ film with liquefaction and recrystallization altered the grain orientation, and potentially reduced defect concentrations. Solar cells made with treated films in most conditions result in high V_{OC} which could be a result from lower defect density after treatments. Optimization of treatment conditions is necessary to fabricate higher performance devices.

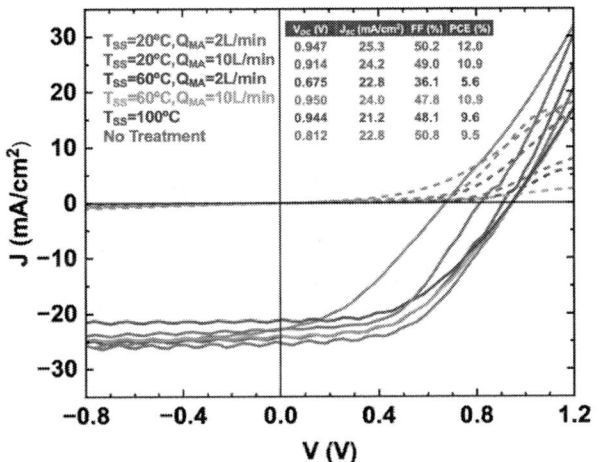

Fig. 4. Dark and light J-V curves and the breakdown of J-V parameters including V_{OC}, J_{SC}, FF, and PCE of MA treated solar cells.

ACKNOWLEDGEMENT

The authors acknowledge Shannon Fields for building the CSVT reactor and vapor-assisted treatment system, depositing ITO substrates, and other equipment maintenance.

REFERENCES

[1] Q. Guesnay, F. Sahli, C. Ballif, and Q. Jeangros, "Vapor deposition of metal halide perovskite thin films: Process control strategies to shape layer properties," Apl Materials, October 2021. DOI: 10.1063/5.0060642.

[2] Z. M. Zhou et al., "Methylamine-gas-induced defect-healing behavior of $CH_3NH_3PbI_3$ thin films for perovskite solar cells," Angewandte Chemie-International Edition, August 2015. DOI: 10.1002/anie.201504379.

[3] D. Bogachuk, L. Wagner, S. Mastroianni, M. Daub, H. Hillebrecht, and A. Hinsch, "The nature of the methylamine-$MAPbI_3$ complex: fundamentals of gas-induced perovskite liquefaction and crystallization," Journal of Materials Chemistry A, May 2020. DOI: 10.1039/d0ta02494e.

[4] D. L. Jacobs and L. Zang, "Thermally induced recrystallization of $MAPbI_3$ perovskite under methylamine atmosphere: an approach to fabricating large uniform crystalline grains," Chemical Communications, 2016. DOI: 10.1039/c6cc04521a.

[5] H. C. Fan et al., "Methylamine-assisted growth of uniaxial-oriented perovskite thin films with millimeter-sized grains," Nature Communications, November 2020. DOI: 10.1038/s41467-020-19199-6.

[6] A. J. Harding *et al.*, "The growth of methylammonium lead iodide perovskites by close space vapor transport," *Rsc Advances,* April 2020. DOI: 10.1039/d0ra01640c.

[7] M. Daub and H. Hillebrecht, "Understanding the "Molten Salt" synthesis of $MAPbI_3$ - characterization of new lead(II)-ammine complexes as intermediates," *European Journal of Inorganic Chemistry,* April 2021, DOI: 10.1002/ejic.202100077.

[8] A. G. Kuba et al., "Two-step close-space vapor transport of $MAPbI_3$ solar cells: effects of electron transport and residual PbI_2," Acs Applied Energy Materials, September 2022. DOI: 10.1021/acsaem.2c01468

Optimization of Sb2Se3 thin films prepared by selenization of Sb metallic precursors for photovoltaic application

Woo Kyoung Kim, Vasudeva Reddy Minnam Reddy, Sreedevi Gedi, Salh Alhammadi, Ignatius Andre Setiawan, Yujeong Ahn, Songhee Lee, Hyomin Kim

Yeungnam University, Gyeongsan, Korea

Sb2Se3 is an earth-abundant, non-toxic, and low-cost absorber material for solar cell applications. In this work, the physical properties of Sb2Se3 thin films were investigated for different tin metal layer thicknesses (300 nm - 1300 nm) deposited by DC sputtering followed by the selenization using rapid thermal processing (RTP). The X-ray diffractoin and Raman characterizations confirmed the formation of single-phase Sb2Se3 compound with orthorhombic structure. The selenized films of 1100nm-thick Sb (Sb-1100) had remarkable and homogeneous morphology with the optical band gap energy of 1.35 eV. The present results showed the larger grains with smooth morphology and better electrical and optical properties in the case of Sb2Se3 formed using Sb-1100. J-V characteristics of solar cell showed that the optimized Sb23 (deposition time of 23 min) device offered a JSC of 31.6 mA/cm2, a Voc of 317 mV, and an FF of 48%, thus achieving an efficiency of 4.85%. By varying the different metal tin layer thicknesses, this is the first report for Sb2Se3 thin films grown RTP.

978-1-6654-6060-6/23 $31.00 © 2023 IEEE

Narrow bandgap perovskite solar cell degradation monitoring by spectroscopic ellipsometry

Marie Solange Tumusange, Madan K. Mainali, Lei Chen, Zhaoning Song, Yanfa Yan, Nikolas J. Podraza

University of Toledo, Toledo, OH, United States

Mixed Pb+Sn narrow bandgap perovskites absorbers are used in high performance in single junction or the bottom junction in tandem solar cells. However, the instability of these materials with exposure to environmental conditions is still a major challenge, and it is necessary to monitor and understand the degradation mechanisms of these perovskites in devices. Spectroscopic ellipsometry measurements and analysis of complete narrow bandgap perovskite-based solar cells are used in combination with external quantum efficiency (EQE) measurements and simulations are used to track changes occurring from initial fabrication to 5 days aging in ambient air. The complex optical property spectra and thicknesses of all component layers of the solar cell are used as input for EQE simulations. No changes are observed in the ellipsometric spectra with aging, indicating that large structural or optical property variations are not occurring. Comparison of initial measured EQE to simulated EQE identifies reduced collection probability of photogenerated carriers in the ~10 nm and ~650 nm of perovskite material near the HTL and ETL interfaces, respectively. The carrier collection probability decreases with the aging of the device. SCAPS simulations indicate that the reduced collection probability and EQE can be attributed to the increase in trap state density with time.

Solution Processed n+ CdS/ n-CdTe/ Perovskite Heterojunction Thin-film Solar Cells

Isaiah Henry[1], Dakota Schwartz[1], Harry Larson[1], Shubhra Bansal[1,2*]

[1]University of Nevada Las Vegas, Las Vegas, NV 89052, U.S.A.

[2]Purdue University, West Lafayette, IN 47906, U.S.A.

*Email: bansal91@purdue.edu

Abstract— Here we demonstrate Tec 10/n+ CdS/n-CdTe/perovskite/Au device structure with two different perovskite hole transport layers (HTL) namely CsPbBr$_3$ and CsPtI$_3$. InCl$_3$-doped electrodeposited CdTe shows low carrier concentration of 3 x 10^{15} cm^{-3} measured using electrochemical capacitance profiling. CsPbBr$_3$ and CsPtI$_3$ exhibit a bandgap of 2.35 eV and 2.62 eV respectively and hole concentration on the order of 10^{16} cm^{-3}. A device structure of chemical bath deposited CdS on Tec 10, In-doped CdTe, CsPbBr$_3$ or CsPtI$_3$ HTLs with Au contacts results in device efficiency of and 5.39% and 6.85% respectively. Future research includes CdTe grain boundary and interface passivation, and tailoring perovskite HTL bandgap, thickness, and passivation to enable ultra-thin bifacial CdTe solar cells.

Keywords—n-CdTe, perovskite, hole transport layer

I. INTRODUCTION

CdTe exhibits a direct bandgap of ~ 1.4 eV with high absorption coefficient, however, efficiency of record devices is reported at 22.1% [1-3] significantly lower than the Shockley-Queisser detailed-balance limit. Record 16.5% devices in 2001 achieved open-circuit voltage (V$_{OC}$) of 0.848 V, which has marginally increased to 0.887 V for state-of-art devices, only 72% of 1.23 V calculated based on detailed-balance approach. Even though the V$_{OC}$ is limited in CdTe devices, recent advances in efficiency have resulted from improved photon transmission to CdTe layer, bandgap grading, novel contacts, and bifaciality using electron reflectors like CuAlO$_2$ or CuGaO$_2$. To improve p-type conductivity in CdTe, extrinsic doping with group-I (Li, Na, Cu, Ag) [4-6] and group-V elements (N, P, As, Sb, Bi) [7-9] has been explored in the literature. Group-I elements are expected to occupy Cd sites introducing shallow acceptor levels limited by self-compensation and formation of interstitial donors. Group-V elements introduce shallow acceptor levels by substitution in Te sub-lattice and can improve p-type conductivity when used in Cd-rich growth conditions, however, hole concentration is limited by formation of self-compensating AX centers. Yang et al.[10,11] have demonstrated above 10^{17}cm^{-3} hole density with P and As doping by shifting Fermi energy closer to acceptor levels through fast quenching, but these devices pose metastability challenge[12-14]. Cl and Cu treatments are beneficial for p-type CdTe devices enabling lower interface recombination and more optimum minority

carrier lifetime [15-17]. However, Cu adds to stability challenges due to fast diffusion of Cu_i and Cu_{Cd} type defects[18-20]. CdTe single crystals with P doping have shown ~ 400 ns carrier lifetime and V$_{OC}$ of 1.017 V, however device efficiency is limited to 13.6% [21]. Further, due to the high work-function of p-type CdTe, most metals form Schottky barrier, therefore Cu-containing contacts such as ZnTe:Cu are needed, which add to the device instability.

The most important factors limiting the efficiency of CdTe as an absorber material for photovoltaic devices is combination of small minority carrier lifetime, high surface recombination velocity and low dopability (~ 10^{14} cm^{-3}) resulting in V$_{OC}$ limited to less than 900 mV. Device modeling and experimental data has shown that increasing CdTe doping to > 10^{16} cm^{-3} while retaining carrier lifetimes near 2 ns can increase open-circuit voltage (V$_{OC}$) above 1 V, to push efficiencies > 25% [21-23]. Although significant progress has been made in development of high-efficiency CdTe solar cells with advances such as group V doping, replacement of CdS buffer with ZnMgO, bandgap grading with CdSeTe; no significant gains in V$_{OC}$ have been observed. Cohen et al. [24] have shown that CdTe can be doped n-type to 10^{16}-10^{19} cm^{-3} doping density, while maintaining carrier lifetime of 180 ns. Zhao et al. [25] have demonstrated n-type CdTe single crystal double-heterostructure with carrier lifetime of 3.6 μs and V$_{OC}$ of 1.096 V. The device structure with monocrystalline CdTe/MgCdTe double-heterostructure and a-SiC:H p-type contacts has shown device efficiency of 17%, highest reported for n-type CdTe. Palekis et al. [26] have demonstrated near 9% In-doped n-CdTe devices deposited by elemental vapor deposition (EVT) with p-type ZnTe contacts. Other p-type buffers that have demonstrated rectifying contacts with n-CdTe include NiO$_x$ [27] and MoO$_x$ [28]. Although doping CdTe n-type is easier, key challenge thus far has been discovery of p-type buffer materials or transparent conducting oxides (TCOs). CuAlO$_2$ or CuGaO$_2$ are promising options, however, sputtering of these materials can cause damage to already deposited CdTe in superstrate configuration.

Here, we demonstrate solution processed halide perovskite HTL using electrodeposited n-CdTe absorber. Devices with power conversion efficiency (PCE) of 6.85% and 5.39% and V$_{OC}$ > 500 mV have been demonstrated with CsPtI$_3$ and CsPbBr$_3$ HTL respectively.

978-1-6654-6060-6/23 $31.00 © 2023 IEEE

II. METHODOLOGY

Device structure used for this study is Tec10/ CdS/ n-CdTe/ p-perovskite/ Au as shown in **Figure 1**. For CdS chemical bath deposition, a mixture of 22 mL of 0.015 M $CdSO_4$ and 28 mL of NH_4OH in 150 mL of deionized water was heated at a temperature of 65 °C. Tec10 glass was submerged for 1 min and 22 mL of 0.75 M thiourea was added and additionally stirred for 12 min, with an expected CdS thickness of 50-80 nm. The sample was then dried with N_2 flow and annealed in ambient on a hot plate at 120 °C for 5 min to get rid of residual H_2O. CdTe was electrodeposited at a pH of 2 ± 0.02 and deposition potential of -0.65 V with respect to the reference Ag/AgCl electrode and current density of 0.1 mA/cm^2. Cd rod (Alfa Aesar 7440-43-9) was used as the counter electrode (anode) and the temperature of the electrolyte was maintained at 70 °C with continuous bi-directional stirring. Cd-precursor was prepared by dissolving 1M $Cd(NO_3)_2.4H_2O$ in 500 ml of deionized (DI) water and purifying for 100 hours. Te-precursor was prepared with TeO_2 in diluted HNO_3, which was then added to purified Cd-precursor. A two-step post-deposition treatment is used, first 10 second spray of 0.3 M $CdCl_2$ solution in DI water and additional 10 second spray of $InCl_3$ in DI water. The films were then annealed at 400 °C in air for 2 hours. P-type perovskite layers have been used, namely, $CsPtI_3$ and $CsPbBr_3$ deposited via spin coating on CdTe.

$CsPtI_3$ films were fabricated through precursor-based solution processing. 0.25 - 0.4 M solution of Cesium Iodide (Sigma Aldrich CAS 7789-17-5) and Platinum (II) Iodide (Sigma Aldrich CAS 7790-39-8) solutes with a molar ratio of

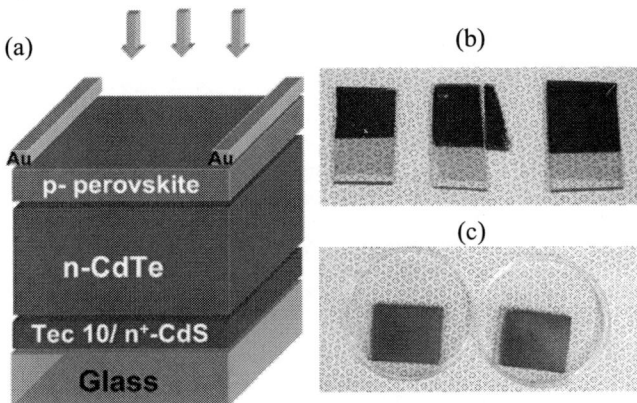

Figure 1: (a) Device schematic used in this study. (b) Pictures of electrodeposited CdTe samples on CdS on Tec 10 glass. (c) Picture of solution processed CsPbBr₃ on ITO glass.

1:1 was prepared in DMF and DMSO (Sigma-Aldrich CAS 68-12-2 and 67-68-5) and 50-50 volumetric mixtures thereof [29]. The precursor solution was mixed at 80 °C for 2.5 hours, and spin coated onto pre-heated glass substrate and electrodeposited CdTe at 70 °C. The substrates were then annealed in a vacuum oven at -15 in Hg and 100 °C for one hour. An ethylene diamine (EDA) spray treatment was applied on the films followed by a 100 °C anneal in vacuum for 30 min. Cesium bromide (CsBr, Sigma Aldrich CAS 7787-69-1) and lead-bromide (PbBr₂, Sigma Aldrich CAS 10031-22-8) solute mixture was used for fabrication of CsPbBr₃ films [30]. The device fabrication was

completed by depositing 70 nm thick Au electrode via thermal evaporation.

Electrochemical capacitance-voltage (CV) profiling was conducted by dipping the glass/Tec10/CdS/CdTe layer into a 0.5 M lactic acid electrolyte to form a solid/liquid junction to determine carrier concentration and conductivity type. Optical transmittance and reflectance measurements were performed using Shimadzu UV-2600 spectrophotometer followed by a Tauc analysis to determine optical bandgap of the thin-film samples on glass. X-ray diffraction (XRD) measurements were conducted on a Bruker diffractometer under ambient conditions using Cu-Kα radiation. Time resolved photoluminescence (TRPL) measurements were done using PicoQuant 532 nm laser with photodiode timing resolution of 40 ps. Quantum efficiency (QE) measurements were performed using a PV Measurements QEXL at 0V and JV sweeps were performed via a 4-probe setup, with typical DC bias voltage, V_{DC} = -0.5 to 0.8 V. Current-density vs. voltage (JV) measurements were taken with a Keithley 2400 source meter unit using an Abet Solar Simulator.

III. RESULTS AND DISCUSSION

A. Structure and Bandgap

X-ray diffraction pattern for as-deposited CdTe, $CsPtI_3$ and $CsPbBr_3$ are shown in **Figure 2a**. CdTe is identified as cubic F43m structure with Bragg reflections (111), (220), (311) and

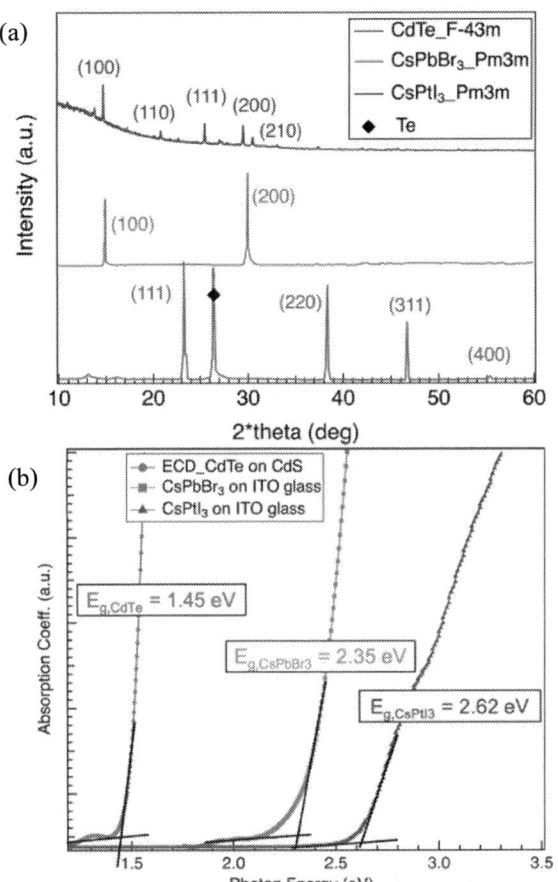

Figure 2. (a) XRD data for solution processed CdTe, CsPbBr₃ and CsPtI₃. (b) Tauc plot for CdTe, CsPbBr₃ and CsPtI₃.

(400) at around 23.8°, 38.4°, 46.7°, and 55.4° respectively. An additional peak at 26.4° corresponding to the (100) reflection from metallic Te is identified, which vanishes after post-deposition treatment. CsPbBr₃ exhibits crystalline films of cubic Pm3m phase with prominent (100) and (200) reflections. CsPtI₃ also shows Pm3m cubic structure with (100), (110), (111), (200), and (210) primary reflections, however, with poorer crystallinity compared to CsPbBr₃. Optical bandgaps have been measured for CdTe, CsPbBr₃ and CsPtI₃ of 1.45 eV, 2.35 eV and 2.62 eV respectively as shown in **Figure 2b**.

Figure 3 shows the cross-section SEM of electrodeposited and annealed CdTe on Tec 10/CdS and top-down image of spin coated CsPtI₃. CdTe is about 1 μm thick with ~ 90 nm thick CdS. CdTe layer shows pinholes and phase segregation indicated by dark and light regions in **Figure 3a**. The spin coated CsPtI₃ in **Figure 3b** also shows pinholes suggesting the need for further improvement in film quality. Carrier density and lifetime of the individual films was characterized using electrochemical capacitance-voltage and TRPL measurements, respectively.

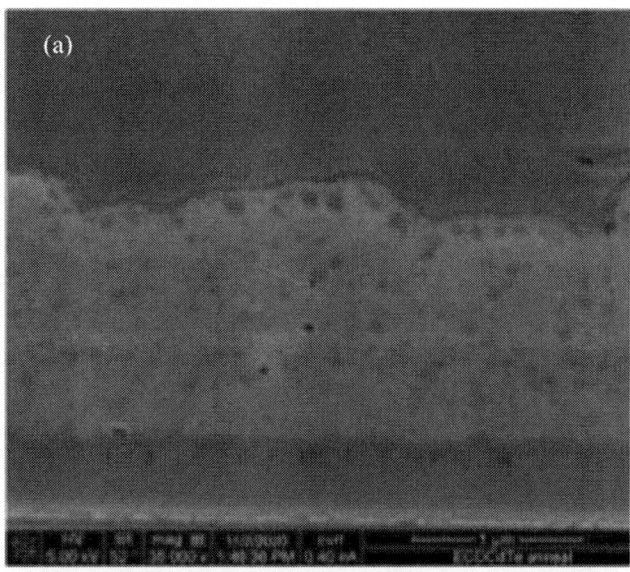

Figure 3. *(a) Scanning electron microscope (SEM) image of CdTe deposited on CdS/Tec10 superstrate annealed with CdCl₂ and InCl₃. (b) Top-down SEM image of spin-coated CsPtI₃.*

B. Carrier Density and Lifetime

Semiconductor-liquid electrolyte Schottky junction results in the formation of a space charge layer and the capacitance can be measured at a constant reverse bias. The semiconductor is electrolytically etched between capacitance measurements, leading to a depth profile, however, crystal quality can also affect the etch depth. 0.5M aqueous lactic acid solution with H₂O₂ and HI is used for CdTe electrochemical capacitance-voltage measurements, whereas 0.3M solution of poly(lactic acid) PLA was formed in chloroform for controlled etching of CsPbBr₃ and CsPtI₃. A small AC signal of 100 mV at a

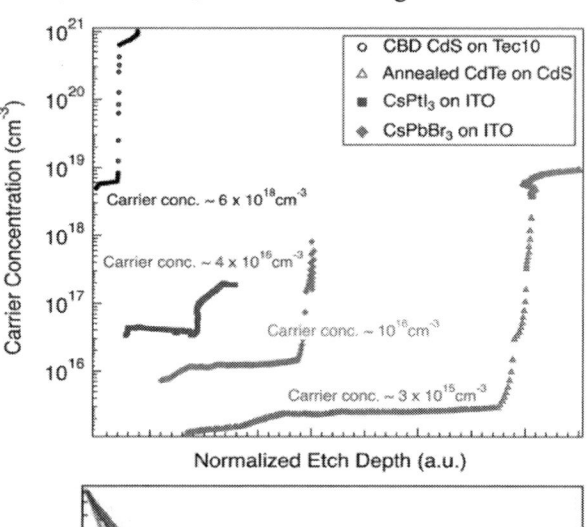

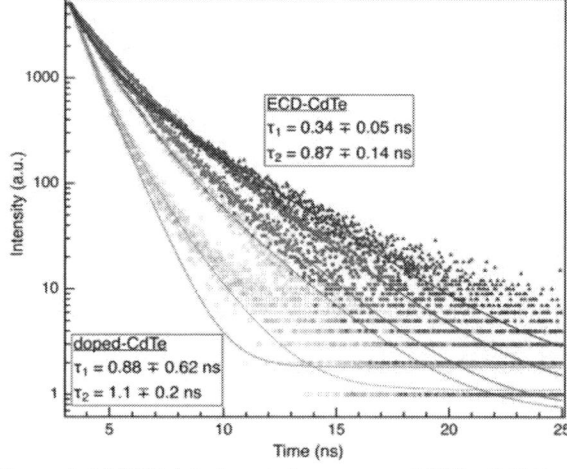

Figure 4. *(a) XRD data for solution processed CdTe, CsPbBr₃ and CsPtI₃. (b) Tauc plot for CdTe, CsPbBr₃ and CsPtI₃.*

frequency of 1 kHz is superposed on the DC bias during the capacitance measurement. The carrier concentration at the depletion width W_{dep} is extracted from the $1/C^2$ $vs.V$ curves according to Equation 1. The p-type CsPbBr₃ and CsPtI₃ perovskite films easily dissolve in the electrolyte with forward bias of the Schottky junction, however, to drive the process for n-type CdTe, holes are photogenerated by illuminating and reverse biasing the junction.

$$N(W_{dep}) = \frac{2}{qK_S\varepsilon_0 A^2} \frac{1}{\frac{d(1/C^2)}{dV}} \quad (1)$$

978-1-6654-6060-6/23 $31.00 © 2023 IEEE   1553

Where, q is the electron charge, K_s is the dielectric constant of the semiconductor (10 for CdTe, 25 for halide perovskites), ε_0 is the permittivity of free space and A is the area of the electrolyte-semiconductor junction. The sealing was about 500 X 500 μm^2, hence a contact area of 0.0025 cm^2 was used in the calculations for carrier concentration and etch depth.

Figures 4a shows the carrier concentration for CdS on Tec 10, CdTe on CdS and Tec 10, and perovskite on ITO. Electrodeposited and annealed CdTe shows low n-type In doping with carrier concentration of $3x10^{15}\ cm^{-3}$, indicating non-optimal film quality. Palekis et al. [26] have shown $2.1x10^{16}\ cm^{-3}$ net doping density with elemental vapor transport, whereas $10^{16} - 10^{19}\ cm^{-3}$ n-type doping has been demonstrated by molecular beam epitaxy methods [25, 31]. **Figure 4b** shows the TRPL data for as-deposited CdTe and after $CdCl_2$ and $InCl_3$ treatment. A bi-exponential fit is used over the decay part of the TRPL curves to estimate the slower decay time (τ_2). As-deposited CdTe shows average carrier lifetime of 0.87 ns which marginally increases to 1 ns with $CdCl_2$ and $InCl_3$ post-deposition treatment. Single crystal n-CdTe with $Mg_xCd_{1-x}Te$ double heterostructures have shown carrier lifetimes as high as 3.6 μs [25] indicating sub-optimal film quality through electro-deposition method.

C. Device Characteristics

In order to demonstrate the use of perovskite HTL, devices with the structure Tec10/CdS/CdTe/HTL/Au were completed with solution processed $CsPtI_3$ and $CsPbBr_3$ and evaporated Au. $CsPtI_3$ HTL resulted in best device efficiency of 6.85% with V_{OC} of 544 mV, J_{SC} of 16.8 mA/cm^2 and FF of 75%; whereas $CsPbBr^3$ HTL showed slightly lower PCE of 5.39% as shown in **Figure 5a**. Devices with $CsPtI_3$ HTL devices show higher FF and V_{OC} due to higher shunt resistance possibly due to more uniform coverage due to more stable intermediate phase formation with DMSO solvent [32]. External quantum efficiency (EQE) data in **Figure 5b** shows suppressed QE in 500-800 nm wavelength range for $CsPbBr_3$ HTL devices likely due higher recombination in these devices. J_{SC} for all devices is lower than expected likely due to the undesired interfacial layer between CdS and CdTe as shown in the SEM image. $CsPtI_3$ and $CsPbBr_3$ have an electron affinity close to 3.6 eV, with the measured bandgaps, the valence band maxima are expected at 6.22 eV and 5.95 eV respectively. CdTe is a high work-function semiconductor close to 5.9 eV, therefore $CsPtI_3$ should exhibit a positive valence band offset reducing the interface recombination. Electrodeposited CdTe layer is about 1 μm thick and the doping density is low ($\sim 10^{15}\ cm^{-3}$), which can allow for optimal band bending at the back contact to act as an electron reflector, and also enable a carrier selective back contact as shown in **Figure 5c**. If the buffer can be replaced by a material with wider bandgap than CdS, a carrier selective front contact can be formed enabling a drift dominated device structure with $\leq 1\ \mu$m thick CdTe. The absorber quality and thickness of the HTL is not optimized yet, nonetheless, all devices show > 5% efficiency with this preliminary device. Alternate halide perovskite compositions with $E_g \sim 3eV$ should be explored to enable bifacial devices.

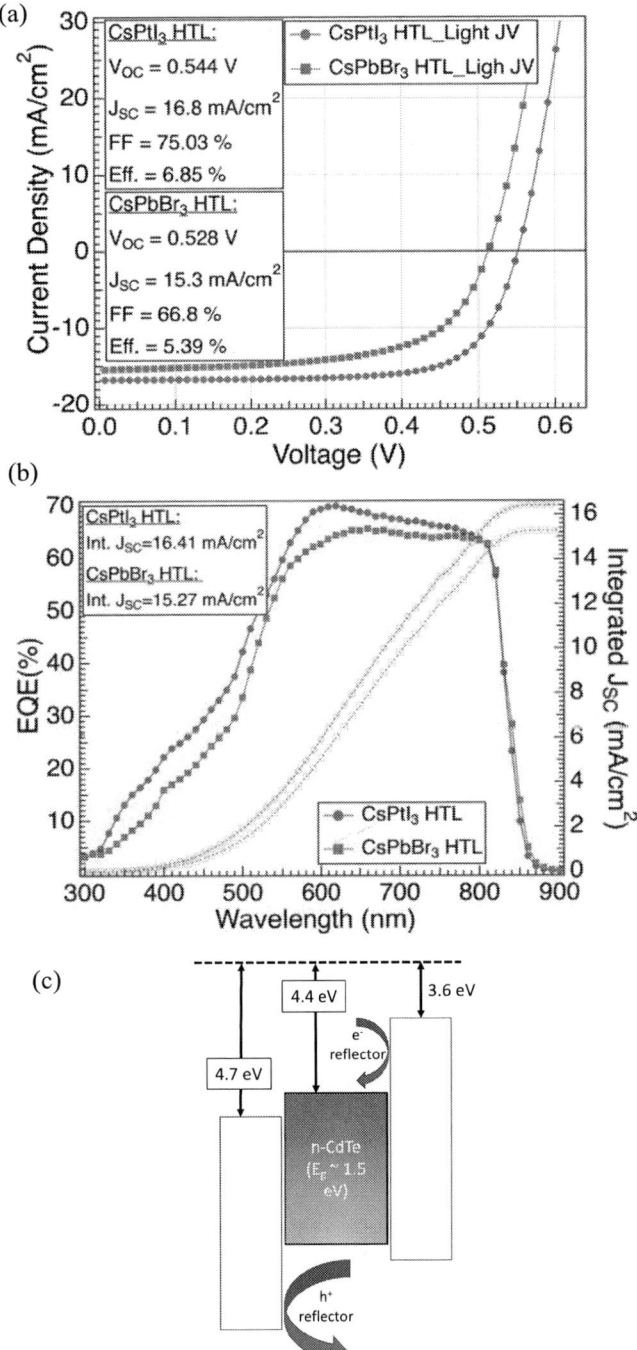

Figure 5. *(a) Light current density vs. voltage (J-V) plots CdS/CdTe devices with $CsPbBr_3$ and $CsPtI_3$ hole transport layers (HTLs) (b) Quantum efficiency plots and (c) Band diagram for CdS/CdTe/perovskite devices.*

IV. CONCLUSIONS AND FUTURE WORK

A proof-of-concept device with electrodeposited n-CdTe absorber and wide bandgap halide perovskite HTL is demonstrated here. Electrodeposited CdTe when annealed with $InCl_3$ and $CdCl_2$ show carrier density of 3 x 10^{15} cm^{-3} and carrier lifetime of 1 ns. Solution processed $CsPtI_3$ and $CsPbBr_3$ HTLs show hole density of ~ 10^{16} cm^{-3}. Our initial results have shown > 5% efficiency devices, however, improvements to CdTe film quality and optimized HTL are needed. Wide bandgap halide perovskites offer low temperature processable HTLs. Future work should include wide bandgap n-type buffer, improved CdTe quality and ways to increase bandgap and doping density of halide perovskite HTLs.

ACKNOWLEDGMENT

The project has been supported by UNLV Faculty Opportunity Award 2018 and UNLV NSF iCORP seed funding 2020 and NSF CAREER grant 2046944. The authors would like to thank Dr. Edward Barnard at Molecular Foundry for time resolved photoluminescence (TRPL) measurements.

REFERENCES

[1] M. Green, E. Dunlop, J. Hohl-Ebinger, M. Yoshita, N. Kopidakis, and X. Hao, "Solar cell efficiency tables (version 57)," *Prog. Photovoltaics Res. Appl.*, vol. 29, no. 1, pp. 3–15, Jan. **2021**, doi: https://doi.org/10.1002/pip.3371.

[2] National Renewable Energy Laboratory, "NREL Research Cell Efficiency Chart," *Webpage*, **2021**.

[3] W. K. Metzger *et al.*, "Exceeding 20% efficiency with in situ group V doping in polycrystalline CdTe solar cells," *Nat. Energy*, vol. 4, no. 10, **2019**, doi: 10.1038/s41560-019-0446-7.

[4] S. H. Wei and S. B. Zhang, "First-principles study of doping limits of CdTe," in *Physica Status Solidi (B) Basic Research*, **2002**, vol. 229, no. 1, 10.1002/1521-3951(200201)229:1<305::AID-PSSB305>3.0.CO;2-3.

[5] D. Krasikov and I. Sankin, "Defect interactions and the role of complexes in the CdTe solar cell absorber," *J. Mater. Chem. A*, vol. 5, no. 7, **2017**, doi: 10.1039/c6ta09155e.

[6] D. Krasikov, A. Knizhnik, B. Potapkin, S. Selezneva, and T. Sommerer, "First-principles-based analysis of the influence of Cu on CdTe electronic properties," in *Thin Solid Films*, **2013**, vol. 535, no. 1, doi: 10.1016/j.tsf.2012.10.027.

[7] M. A. Flores, W. Orellana, and E. Menéndez-Proupin, "Self-compensation in phosphorus-doped CdTe," *Phys. Rev. B*, vol. 96, no. 13, **2017**, doi: 10.1103/PhysRevB.96.134115.

[8] E. Colegrove *et al.*, "Experimental and theoretical comparison of Sb, As, and P diffusion mechanisms and doping in CdTe," *J. Phys. D. Appl. Phys.*, vol. 51, no. 7, **2018**, doi: 10.1088/1361-6463/aaa67e.

[9] R. B. Hall and H. H. Woodbury, "The diffusion and solubility of phosphorus in CdTe and CdSe," *J. Appl. Phys.*, vol. 39, no. 12, **1968**, doi: 10.1063/1.1655982.

[10] J. H. Yang *et al.*, "Enhanced p-type dopability of P and As in CdTe using non-equilibrium thermal processing," *J. Appl. Phys.*, vol. 118, no. 2, **2015**, doi: 10.1063/1.4926748.

[11] J. H. Yang, J. S. Park, J. Kang, W. Metzger, T. Barnes, and S. H. Wei, "Tuning the Fermi level beyond the equilibrium doping limit through quenching: The case of CdTe," *Phys. Rev. B - Condens. Matter Mater. Phys.*, vol. 90, no. 24, **2014**, doi: 10.1103/PhysRevB.90.245202.

[12] A. Nagaoka, D. Kuciauskas, and M. A. Scarpulla, "Doping properties of cadmium-rich arsenic-doped CdTe single crystals: Evidence of metastable AX behavior," *Appl. Phys. Lett.*, vol. 111, no. 23, **2017**, doi: 10.1063/1.4999011.

[13] A. Nagaoka, D. Kuciauskas, J. McCoy, and M. A. Scarpulla, "High p-type doping, mobility, and photocarrier lifetime in arsenic-doped CdTe single crystals," *Appl. Phys. Lett.*, vol. 112, no. 19, **2018**, doi: 10.1063/1.5029450.

[14] T. Ablekim *et al.*, "Self-compensation in arsenic doping of CdTe," *Sci. Rep.*, vol. 7, no. 1, **2017**, doi: 10.1038/s41598-017-04719-0.

[15] M. O. Reese *et al.*, "Intrinsic surface passivation of CdTe," *J. Appl. Phys.*, vol. 118, no. 15, **2015**, doi: 10.1063/1.4933186.

[16] M. Terheggen, H. Heinrich, G. Kostorz, D. Baetzner, A. Romeo, and A. N. Tiwari, "Analysis of bulk and interface phenomena in CdTe/CdS thin-film solar cells," in *Interface Science*, **2004**, vol. 12, no. 2–3, doi: 10.1023/B:INTS.0000028655.11608.c7.

[17] W. K. Metzger, D. Albin, M. J. Romero, P. Dippo, and M. Young, "CdCl 2 treatment, S diffusion, and recombination in polycrystalline CdTe," *J. Appl. Phys.*, vol. 99, no. 10, **2006**, doi: 10.1063/1.2196127.

[18] C. R. Corwine, A. O. Pudov, M. Gloeckler, S. H. Demtsu, and J. R. Sites, "Copper inclusion and migration from the back contact in CdTe solar cells," *Sol. Energy Mater. Sol. Cells*, vol. 82, no. 4, **2004**, doi: 10.1016/j.solmat.2004.02.005.

[19] S. S. Hegedus, B. E. McCandless, and R. W. Birkmire, "Analysis of stress-induced degradation in CdS/CdTe solar cells," in *Conference Record of the IEEE Photovoltaic Specialists Conference*, **2000**, vol. 2000-January, doi: 10.1109/PVSC.2000.915891.

[20] I. Visoly-Fisher, K. D. Dobson, J. Nair, E. Bezalel, G. Hodes, and D. Cahen, "Factors affecting the stability of CdTe/CdS solar cells deduced from stress tests at elevated temperature," *Adv. Funct. Mater.*, vol. 13, no. 4, **2003**, doi: 10.1002/adfm.200304259.

[21] J. M. Burst *et al.*, "CdTe solar cells with open-circuit voltage breaking the 1V barrier," *Nat. Energy*, vol. 1, no. 4, **2016**, doi: 10.1038/NENERGY.2016.15.

[22] A. Kanevce, M. O. Reese, T. M. Barnes, S. A. Jensen, and W. K. Metzger, "The roles of carrier concentration and interface, bulk, and grain-boundary recombination for 25% efficient CdTe solar cells," *J. Appl. Phys.*, vol. 121, no. 21, **2017**, doi: 10.1063/1.4984320.

[23] B. E. McCandless *et al.*, "Overcoming Carrier Concentration Limits in Polycrystalline CdTe Thin Films with In Situ Doping," *Sci. Rep.*, vol. 8, no. 1, **2018**, doi: 10.1038/s41598-018-32746-y.

[24] R. Cohen, V. Lyahovitskaya, E. Poles, A. Liu, and Y. Rosenwaks, "Unusually low surface recombination and long bulk lifetime in n-CdTe single crystals," *Appl. Phys. Lett.*, vol. 73, no. 10, **1998**, doi: 10.1063/1.122169.

[25] Y. Zhao *et al.*, "Monocrystalline CdTe solar cells with open-circuit voltage over 1 v and efficiency of 17%," *Nat. Energy*, vol. 1, no. 6, **2016**, doi: 10.1038/nenergy.2016.67.

[26] V. Palekis *et al.*, "Thin Film Solar Cells Based on n-type Polycrystalline CdTe Absorber," **2018**, doi: 10.1109/PVSC.2018.8548249.

[27] H. Parkhomenko, M. Solovan, V. Brus, E. Maystruk, and P. Maryanchuk, "Structural, electrical, and photoelectric properties of p-NiO/n-CdTe heterojunctions," *Opt. Eng.*, vol. 57, no. 01, **2018**, doi: 10.1117/1.oe.57.1.017116.

[28] M. M. Solovan, V. V. Brus, A. I. Mostovyi, P. D. Maryanchuk, E. Tresso, and N. M. Gavaleshko, "Molybdenum oxide thin films in CdTe-based electronic and optoelectronic devices," *Phys. Status Solidi - Rapid Res. Lett.*, vol. 10, no. 4, **2016**, doi: 10.1002/pssr.201600010.

[29] D. Schwartz *et al.*, "Air Stable, High-Efficiency, Pt-Based Halide Perovskite Solar Cells with Long Carrier Lifetimes," *Phys. Status Solidi - Rapid Res. Lett.*, vol. 14, no. 8, **2020**, doi: 10.1002/pssr.202000182.

[30] S. Bansal and M. Chiu, "Atmospherically Processed and Stable Cs-Pb Based Perovskite Solar Cells," in *MRS Advances*, **2017**, vol. 2, no. 53, doi: 10.1557/adv.2017.449.

[31] D. Hommel, A. Waag, S. Scholl, G. Landwehr, "Chlorine: A new efficient n - type dopant in CdTe layers grown by molecular beam epitaxy," *Appl. Phys. Lett.*, vol. 61, pp. 1546–1548, **1992**, doi:10.1063/1.107491.

[32] H. Chen *et al.* "Forming Intermediate Phase on the Surface of PbI_2 Precursor Films by Short-Time DMSO Treatment for High-Efficiency Planar Perovskite Solar Cells via Vapor-Assisted Solution Process", *ACS Appl. Mater. Interfaces*, vol. 10, no. 2, pp. 1781, **2018**. doi:10.1021/acsami.7b17781.

Effect of Double Cation Substitution on Nonradiative Recombination Losses in Cu2ZnSn(S,Se)4 Solar Cells

Vijay Karade, Kiwhan Kim, Jae Ho Yun, Jin Hyeok Kim

Department of Energy Engineering, Korea Institute of Energy Technology (KENTECH), Naju, South Korea

Optoelectronics Convergence Research Center and Department of Materials Science and Engineering Chonnam National University, Gwangju, South Korea

Photovoltaic Research Department, Renewable Energy Institute, Korea Institute of Energy Research, Deajon, South Korea

Recent developments In kesterItes encouraged the researcher to use Cu2ZnSn(S,Se)4 (CZTSSe) based photo absorber materials in diverse optoelectronic applications. However, the detrimental bulk and interface defects induced high carrier recombinations at corresponding regions stagnated further improvement in the device performance of kesterite solar cells. The present work demonstrates a facile Silver (Ag) and Germanium (Ge) incorporation approach to cure these defects. The Ag incorporation mainly reduced the interface defect states, whilst the Ge incorporation mainly cures the deep-level defect states, suppressing the carrier recombination within the space charge and quasi-neutral regions. As a result, an improved carrier separation process, minority carrier lifetime, and reduced nonradiative carrier recombination losses increased device performance by more than 20%. Finally, the champion device with double cation incorporation of Ag and Ge in the CZTSSe absorber layer delivers enhanced device performance from 9.11 to 11.32 %.

Mapping Spatial Variations of Wide Band Gap Perovskite Thin Films

Emily Miller, Kshitiz Dolia, Bailey Frye, Yanfa Yan, Zhaoning Song, Nikolas J. Podraza

The University of Toledo, Toledo, OH, United States

Wide band gap $FA_{0.8}Cs_{0.2}Pb(I_{0.65}Br_{0.35})_3$ perovskites are examined by mapping spectroscopic ellipsometry to determine spatial variations of structural and optoelectronic properties. Complex dielectric function ($\varepsilon = \varepsilon_1 + i\varepsilon_2$) spectra are highly sensitive to compositional variations, which may inadvertently cause nonuniformities in optical properties across the area of a device. Measurements of partial device-like structure consisting of glass substrate / indium tin oxide / NiOx / Me4-PACz monolayer / perovskite is sensitive to band gap and Urbach energy variations. Band gap values range from 1.77 to 1.80 eV from the sample edge to center. The Urbach energies range between 29 and 50 meV, from center to edge. This approach enables the detection of variations in perovskite material quality characteristics and layer thicknesses, both of are necessary during scale up for large area deposition.

Exploring Distributed PV Power Measurements for Real-Time Potential Power Estimation in Utility-Scale PV Plants

Michael Gostein[1], William B. Hobbs[2]

[1]Atonometrics, Austin, TX, USA; [2]Southern Company, Birmingham, AL, USA

Abstract — **We explore the use of distributed PV power measurements for real-time short-term forecasting of the maximum potential power output of a utility-scale PV power plant, to support future incorporation of PV plants in automatic generation control (AGC) systems. PV plants operating under AGC may run in a curtailed state but must be capable of accurately forecasting their maximum potential non-curtailed power output, or potential high limit (PHL), 10-20 seconds in advance – despite variable environmental conditions. One forecast approach estimates PHL from a subset of inverters, designated reference inverters, which are never curtailed. We propose an alternative approach using in-situ I-V tracers to measure maximum power point of reference modules distributed throughout a PV plant. This would allow for greater sampling of variations in irradiance, module temperature, soiling, and albedo, permitting better PHL prediction accuracy. To study potential benefits of this method, we analyzed a one-year data set from an approximately 100 MW PV power plant in the US Southeast, using non-curtailed string combiner outputs as a proxy for distributed module-level power measurements. We compare the reference inverter and distributed power measurement approaches. Results indicate that for highly flexible operation, substantial improvement in prediction accuracy can be achieved using distributed power measurements compared to using a low fraction of reference inverters. We review the study methodology, conclusions, and shortcomings and discuss prospective future work.**

Index Terms — **photovoltaic systems, grid integration**

I. INTRODUCTION

Flexible solar operation is a control strategy in which solar photovoltaic (PV) power plants operate at an output power below their maximum and then respond to signals from an automatic generation control (AGC) system to increase or decrease power output to the grid as requested [1]. As more solar energy is added to the grid, enabling flexible solar operation and participation in AGC systems can help improve grid stability and increase the value of solar energy [2]. However, effective operation within an AGC system requires accurate real-time determination of a PV plant's maximum potential output, or potential high limit (PHL). This is challenging because PHL depends on weather conditions and is highly variable.

In principle, real-time PHL could be estimated from meteorological data including irradiance and temperature measurements. However, this would require highly detailed modeling of a PV plant, which is impractical to perform with sufficient accuracy, especially in real time.

A PHL estimation method using a subset of a PV plant's inverters as reference inverters has been developed [3]. In this method, a subset of inverters is designated as reference inverters which are continuously operated at maximum power, and the potential output of the entire plant is estimated based on the output of the reference inverters. The method has been successfully demonstrated [4] and its potential accuracy was recently studied using data from several utility-scale PV plants in the US Southeast over the course of a year [5].

However, with the reference inverter method there is a tradeoff between minimum inverter utilization (maximum curtailment) and PHL prediction accuracy. Using a large fraction of reference inverters improves accuracy, especially during dynamic weather conditions, but when only a small fraction of reference inverters is used, prediction accuracy is degraded [3] [5]. Furthermore, the method has no prediction ability in a black start recovery from a zero-power state.

Here we propose an alternative method for PHL determination based on distributed PV power measurements using in-situ PV module current-voltage (I-V) tracers. This is made possible by recent advances in such equipment [6]. The method would involve taking I-V measurements at frequent intervals to determine maximum potential power output (P_{max}) of reference modules distributed throughout a PV plant and using these data to estimate the plant's PHL in real time. I-V measurements are performed only in brief time slices, so reference modules otherwise participate normally in the plant's power production.

The proposed method has several advantages. By measuring module P_{max} at many points throughout a PV plant, power output variations due to differences in irradiance, module

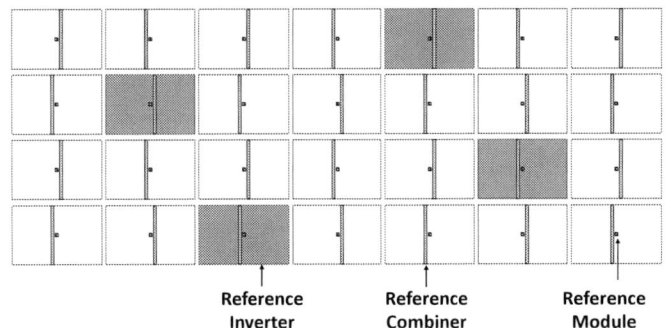

Fig. 1. Layout of reference inverters, reference combiners, and reference modules on a hypothetical PV array. Not to scale.

temperature, soiling, and albedo can be better sampled, leading to more accurate PHL prediction. There is no tradeoff between number of measurement points and inverter utilization (curtailment) range, since I-V measurements are performed regardless of the state of the inverters. This allows operators greater flexibility. Furthermore, this method inherently accounts for the effects of temperature on module output, including temperature differences between utilized and idle modules that could be different between reference and non-reference inverters.

The proposed method could be implemented using in-situ I-V measurements performed at different levels of aggregation, such as at the module, string, or even inverter level. However, measuring I-V at the module level offers a lower cost per physical measurement point, which allows for a greater distribution of measurement points throughout the plant, enabling greater sampling while remaining cost-effective. Furthermore, module-level I-V is easily integrated with all types of inverter systems and can also provide capability for additional applications including measurement of soiling losses and long-term tracking of module degradation.

II. STUDY DESIGN

To assess the potential of the proposed method, we have performed a retrospective study using one year of monitoring data from an approximately 100 MW monofacial utility-scale PV power plant in the US Southeast, which was previously studied in [5]. Data were from the one-year period beginning in February 2022. The plant includes 28 inverters. Because all inverters in the plant were operated at maximum power point for most of the study period, the data allow designating various subsets of the array as references and assessing how well those references can predict the total plant output. Our aim is to compare the prediction accuracy of scenarios using different numbers of reference inverters versus distributions of reference modules. However, the plant does not include module-level I-V tracing equipment. Therefore, we designated various string combiner outputs as references and used the power output of the reference combiners as a proxy for the reference module method. Fig. 1 illustrates the layout of reference inverters, reference combiners, and reference modules distributed across a hypothetical PV array similar to the subject array for the study. String combiners selected as references typically aggregated the output of 250 to 500 modules.

Monitoring data included the AC power output of each inverter, the DC current input to the inverter from each string combiner designated as a reference, and the DC voltage of the string combiner at the inverter input. In addition, meteorological data including plane-of-array (POA) irradiance were available. The AC power output of the plant equals the sum of the power output of all inverters. All data were sampled at an approximately one-second time interval and were stored and compressed in a PI data historian. For this study we exported data at six-second time intervals, corresponding to a typical six-second AGC communication and control cycle.

For the reference inverter method, we considered scenarios including varying numbers of reference inverters ranging from

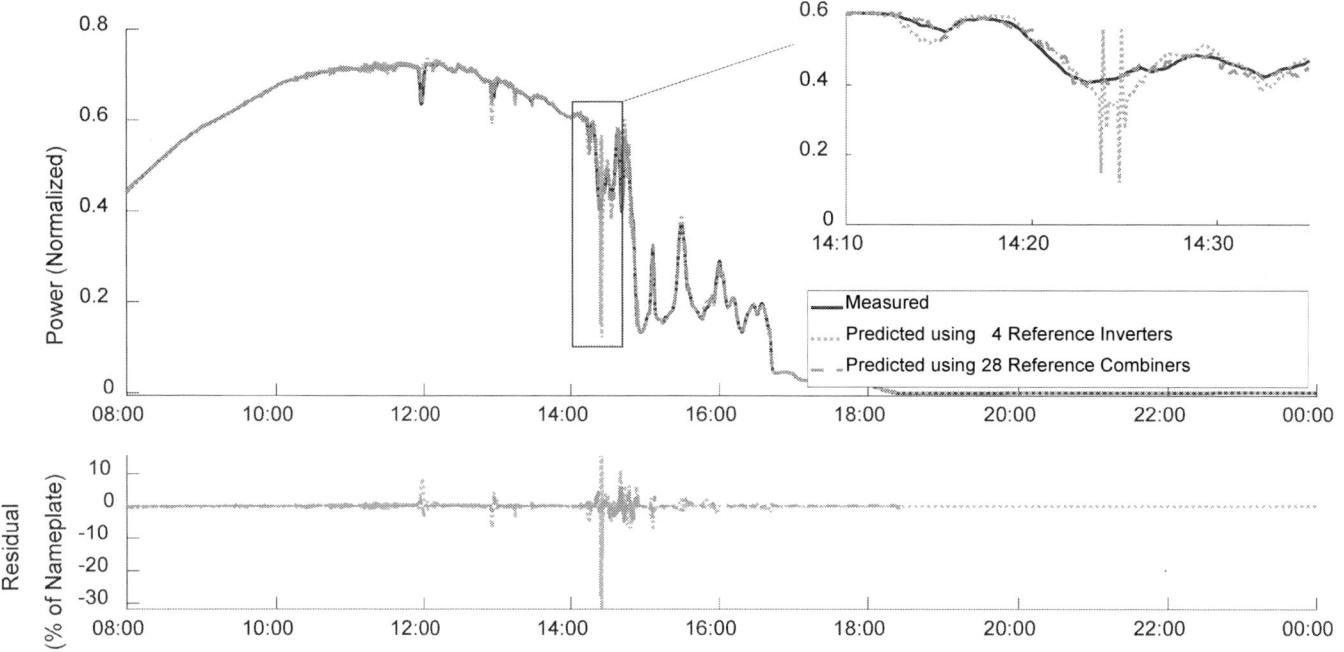

Fig. 2.   Normalized profiles from a sample day for measured plant power and predicted plant power using either 4 reference inverters or 28 reference combiners. The lower panel shows the residual difference between the predicted and measured power profiles.

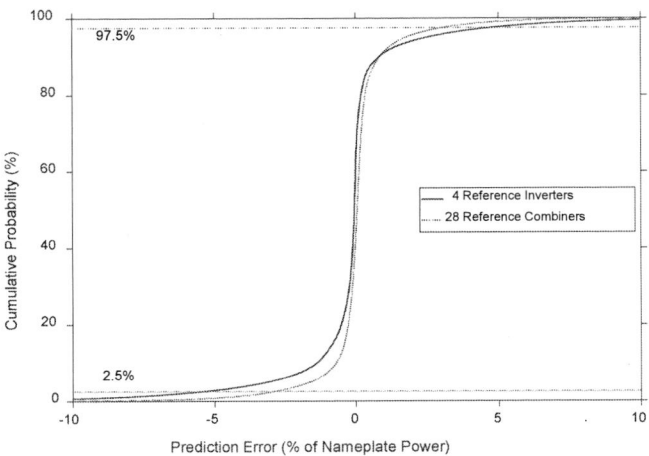

Fig. 3. Cumulative probability distribution of error between predicted and measured power, for two prediction scenarios. Red horizontal lines indicate bounds bracketing 95% probability.

15% to 100% of the total. Within each scenario, we chose reference inverters to be distributed across the plant for optimal spatial sampling, as in Fig. 1. We used one clear sunny day within the data set to establish a calibration factor relating each inverter's output power to the total output power of the plant.

For the reference combiner method, we selected one combiner near the center of each inverter block, as in Fig. 1. We calculated the DC power output of the combiner at each time point by taking the product of combiner current and inverter input voltage. We used one clear sunny day from the data set to determine a calibration factor relating each reference combiner DC output power to the plant's AC power output.

To perform predictions, we used either reference inverters or reference combiners as predictors. At each time point, we scale each of the predictors' power outputs up to the whole-plant power output, using the calibration factors described above, and average the output of predictions for all predictors in a scenario. We also determine the time rate of change of the prediction by comparing successive time points, and then calculate an estimated plant power 18 seconds in the future, corresponding to three AGC control cycles.

Night-time data was excluded by removing data points where more than ten inverters had zero power output. Also, at each time point, data for any inverter (or reference combiner feeding an inverter) was excluded from the analysis if the inverter was not operating at maximum power point or if the data comprised stuck or missing values. Approximately 8% of the daytime data points in the one-year data set were excluded for these reasons, but these excluded points were distributed substantially uniformly throughout the entire study period.

III. RESULTS

Fig. 2 illustrates key phenomena in the results. It shows comparisons of measured and predicted plant power over a

TABLE I
PREDICTION ACCURACY FOR VARIOUS SCENARIOS

Scenario	Ref. Inverters %	Error 95% Probability			Error 99% Probability		
		Min	Max	+/-	Min	Max	+/-
28 Ref. Inverters	100%	-0.9%	1.0%	0.9%	-2.8%	2.9%	2.9%
14 Ref. Inverters	50%	-2.3%	2.4%	2.4%	-4.8%	4.9%	4.9%
7 Ref. Inverters	25%	-4.6%	4.6%	4.6%	-9.2%	9.6%	9.4%
4 Ref. Inverters	15%	-5.4%	4.7%	5.0%	-10.7%	9.9%	10.3%
28 Ref. Combiners	n/a	-2.8%	3.0%	2.9%	-5.9%	6.2%	6.0%

representative day, using two different prediction scenarios including either 4 reference inverters (15% of the total) or 28 reference combiners (one combiner per inverter). The inset shows in greater detail a short period of time when clouds passed over the plant. The lower panel shows the residual difference between predicted and measured power. Note in the inset that rapid changes in the predicted power are slightly offset in time from corresponding changes in the measured power, either trailing or leading those changes slightly depending on how the rate of change is increasing or decreasing. During the morning period when the sky is clear and power output is changing slowly, both predicted results match the measured power closely. However, during the inset period we see that using the 4 reference inverters significantly under or overestimates plant power during the period of instability when a cloud passes, while using 28 reference combiners continues to track the plant power closely. This can be explained by the greater spatial sampling of the plant by the 28 reference combiners relative to the 4 reference inverters.

To quantify the accuracy of a prediction scenario, we use the cumulative probability distribution of errors between predicted

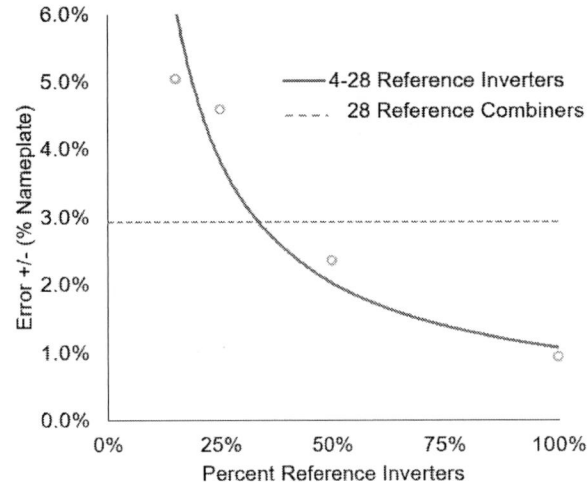

Fig. 4. Summary of 95% probability results from Table I illustrating trend in prediction error versus fraction of reference inverters. Results for reference combiners are independent of reference inverter fraction.

and measured power, as in [5]. Fig. 3 shows an example of results for two prediction scenarios, using all data over the one-year study period. Prediction error is normalized to the total plant nameplate power output. The red horizontal lines indicate bounds bracketing 95% probability.

Table I shows prediction error results for various scenarios, including scenarios with 4-28 reference inverters (15% to 100%) and a scenario with 28 reference combiners (one per inverter). The tabulated results are for the entire one-year data period. For each scenario, we found the lower and upper limits of error (min and max) for 95% probability using the cumulative probability of error distribution as illustrated in Fig. 3. Note that the error is not necessarily symmetric about zero. Note also that even using 28 reference inverters (100%), prediction error is not zero. This is primarily because the prediction is for a forecast of 18 seconds (three AGC cycles) in the future. The +/- error column is the average of the min and max columns, which we use for a simplified single metric. We also include the errors for 99% probability, calculated in a similar way only with different bounds on the distribution.

As a check on the analysis method, we also produced a version of Table I (not shown) for zero-second forecast time. This correctly showed that error for the 100% reference inverter scenario was negligible, with a residual error around 0.1-0.2% that we ascribe to uncertainty in determination of the calibration factors described above.

Fig. 4 illustrates the trends in the results tabulated in Table I. For the reference inverter method, as the fraction of reference inverters decreases, prediction error increases. When the fraction of reference inverters is less than approximately 30%, using the reference combiner method provided better accuracy.

Note that the results in Table I and Fig. 4 include predictions for all times and conditions in the dataset, and thus may be dominated by the low prediction error (as a percent of nameplate power) that is obtained when output power is low, such as during early morning or late afternoon. They also do not differentiate between accuracy on clear days with stable irradiance vs. cloudy days with unstable irradiance, as discussed in [3].

Therefore, we have segmented the data to determine the dependence of prediction error on plant output power level and irradiance variability. This dependence is shown by contour maps in Fig. 5, which were produced by calculating +/- prediction error as in Table I but filtering the data into bins representing different ranges of plant power output and irradiance variability. Irradiance variability was quantified as the difference between maximum and minimum plane-of-array (POA) irradiance in the minute preceding each time point, ratioed to the mean POA irradiance in this minute. Other metrics for irradiance variability could also be applied, as discussed further below. Results in Fig. 5 show that prediction error – as a percent of nameplate power – generally increases as plant output power increases, however it peaks at approximately 0.7 times nameplate power, likely because this corresponds to partly cloudy conditions when irradiance instability would be greatest. In these conditions the prediction error increases as POA variability increases. Considering the right-hand side of the contour maps, and in particular the lower

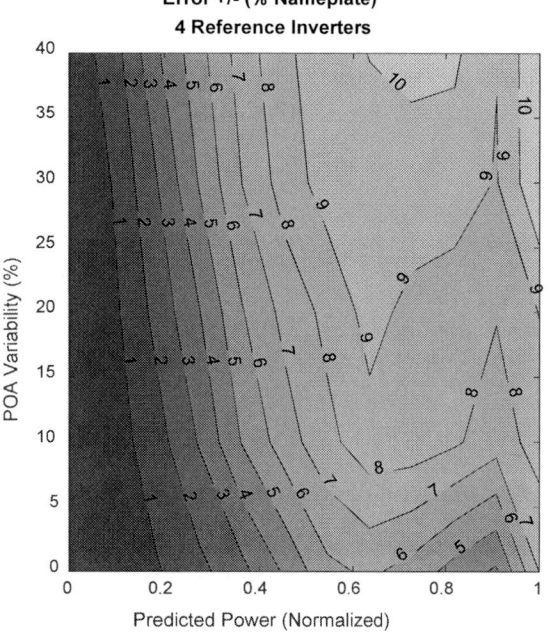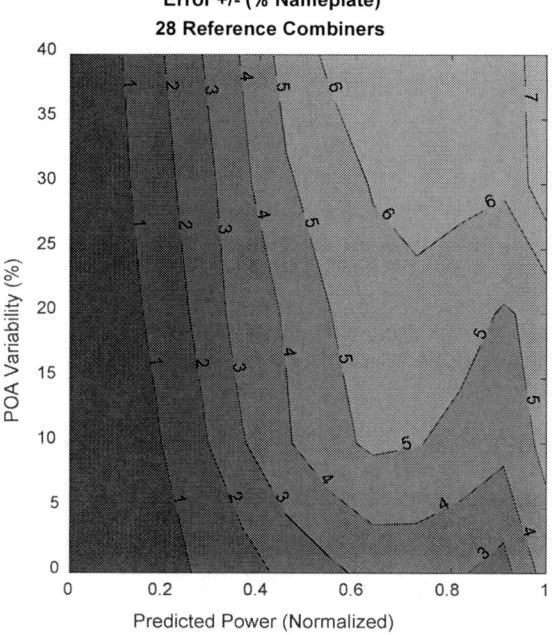

Fig. 5. Prediction error dependence on total power output and irradiance variability. Contour maps show prediction error as a percent of nameplate power versus predicted power normalized to the plant nameplate rating and plane-of-array (POA) irradiance variability, for two prediction scenarios. Left: 4 reference inverters. Right: 28 reference combiners.

right quadrant corresponding to high output power and moderate irradiance variability, to be of most importance, we see that the prediction scenario using 4 reference inverters (15% of the total) resulted in error up to 8% of nameplate, while the scenario using 28 reference combiners (one per inverter) was significantly better, with prediction error of 5% of nameplate. Even with minimal POA variability less than 5%, the reference combiner scenario was consistently more accurate.

IV. DISCUSSION

Our results demonstrate that using PV power measurements widely distributed throughout a plant can produce significantly more accurate estimates of plant output power as compared to estimates made using a small number of reference inverters. This is likely due to the greater spatial sampling provided by the distributed power measurements, which averages over varying irradiance conditions caused by cloud shadows which are smaller than the plant but potentially larger than an inverter block. When plant power output was significant (>0.5 nameplate) this result was maintained even at low levels of irradiance variability, as shown in Fig. 5.

However, while our aim is to identify the potential for using module-level in-situ I-V tracers to perform distributed power measurements, in this study our conclusions about distributed power measurements come from string combiner outputs which represent hundreds of modules. The hundreds of modules feeding the string combiner cover a much larger area than a single module and therefore average over at least some irradiance variability. Therefore, it is unclear whether using one reference module per inverter block would have produced the same results shown here for one reference combiner per inverter – by just as effectively averaging over the large-scale plant-wide differences in irradiance – or whether a larger number of reference modules would be required to obtain the same results. Nonetheless, since the scenario using one reference combiner per inverter corresponds to less than 3% of the modules in the plant yet produced better prediction results than the scenario using 15% of the inverters as references, we may assume that spatial distribution of the references is the primary contributor to improved prediction accuracy. This suggests that using even one reference module per inverter block could provide improved results versus a small number of reference inverters. The optimal fraction of reference modules is an open question.

In addition to the benefits of improved spatial distribution, using reference modules would lessen inaccuracies arising from temperature variations between modules used by reference inverters, which are held at maximum power point, and under-utilized modules of curtailed non-reference inverters, which will be at lower power output and therefore will operate at higher temperatures. We estimate that during typical mid-day periods of peak irradiance, completely idle modules could run approximately 4 °C hotter than modules at maximum power point, and therefore have approximately 1-2% lower potential power output. While in the reference inverter method modeling could be employed to account for this effect, if using reference modules the effect would be inherently included in the prediction of PHL.

For operation in AGC, it may be advantageous for a plant controller not only to predict PHL but also to determine an estimated uncertainty in its prediction, since prediction errors are not a constant but depend on instantaneous conditions, as shown in Fig. 5. Producing an uncertainty estimate would allow a plant controller, for example, to provide the AGC system with a lower bound on projected PHL within a specific confidence interval (e.g., 95% or 99%) thus avoiding over-prediction. Optimizing the tradeoff between maximizing the claimed PHL and minimizing the risk of over-prediction in any AGC cycle may depend on a techno-economic analysis involving contractual arrangements for the power plant. This could ascribe an economic value to estimates of uncertainty.

In Fig. 5 we used measured POA irradiance fluctuations in the minute preceding each time point to partition uncertainty from irradiance variability. This required using data from meteorological stations located at a small number of points within the plant. Using widely distributed reference modules could provide an improved method for estimating prediction uncertainty using a greater spatial sampling and using data closer in time to the AGC cycle. For example, at a single time point, the standard deviation of the maximum power potential of many reference modules distributed throughout the plant could provide an instantaneous metric of PHL uncertainty. In addition, using many distributed reference modules could allow for lower uncertainty that takes into account the motion of cloud shadows across the plant, for example by using a machine learning approach.

We aim to perform a field trial of the reference module approach for PHL prediction by deploying in-situ module-level I-V units at a utility-scale PV plant and collecting data over an extended period. This is the subject of ongoing work.

REFERENCES

[1] I. Chernyakhovskiy, S. Koebrich, and V. Gevorgian, "Grid-Friendly Renewable Energy: Solar and Wind Participation in Automatic Generation Control Systems," 2019. doi: 10.2172/1543130.

[2] Q. Wang, W. B. Hobbs, A. Tuohy, M. Bello, and D. J. Ault, "Evaluating Potential Benefits of Flexible Solar Power Generation in the Southern Company System," *IEEE J. Photovoltaics*, vol. 12, no. 1, pp. 152–160, Jan. 2022, doi: 10.1109/JPHOTOV.2021.3126118.

[3] V. Gevorgian, "Highly accurate method for real-time active power reserve estimation for utility-scale photovoltaic power plants," 2019. doi: 10.2172/1505550.

[4] C. Loutan, P. Klauer, S. Chowdhury, S. Hall, and M. Morjaria, "Demonstration of essential reliability services by a 300-MW

978-1-6654-6060-6/23 $31.00 © 2023 IEEE

solar photovoltaic power plant," 2017. doi: 10.2172/1349211.

[5] W. B. Hobbs, D. J. Ault, V. Gevorgian, and G. Saraswat, "Accuracy of Potential High Limit Estimation for Solar Plants in the Southeast US," in *2022 IEEE 49th Photovoltaics Specialists Conference (PVSC)*, Nov. 2022, pp. 0419–0423, doi: 10.1109/PVSC48317.2022.9938540.

[6] A. Marquis, M. Gostein, and B. H. King, "Validation of In-Situ I-V Measurement Unit for PV System Monitoring Applications," in *2022 IEEE 49th Photovoltaics Specialists Conference (PVSC)*, Nov. 2022, pp. 0291–0294, doi: 10.1109/PVSC48317.2022.9938898.

Controlling Photoexcited Carrier Relaxation through Phonon Management in GaAs/AlAs Superlattices

Muhammad Hanif, Milos Dubajic, Stephen P Bremner, Michael P Nielsen, Santosh Shrestha, Gavin J Conibeer

SPREE, UNSW Sydney, Kensington, Australia

Department of Physics, Cambridge University, Cambridge, United Kingdom

Superlattice structures have been subject of investigation to understand the hampered photoexcited carrier (PC) cooling since the first evidence of slow hot carrier cooling in AlGaAs /GaAs superlattices, compared to bulk GaAs superlattices by Rosenwack et al. It is believed that this slow carrier relaxation is somewhat caused by phonon-bottleneck effect and can be achieved by blocking one or more of energy relaxing mechanisms namely Klemens and/or Ridley. Both decays are highly material properties dependent and rely on the energy difference between lowest optical and highest acoustic phonon branches. Superlatticing provides a plausible way to tune phononic properties and provides a way to study phononic properties dependent carrier dynamics. Through spectral and time resolved photoluminescence measurements of phononically tailored GaAs/AlAs superlattices, we provide a novel way to achieve slower carrier cooling.

Controlling Residual Stresses for Scalable Open-Air Fabrication of Perovskite Solar Cells

Muneeza Ahmad, Carsen Cartledge, Nicholas Rolston

Arizona State University, Tempe, AZ, United States

Controlling Residual Stresses for Scalable Open-Air Fabrication of Perovskite Solar Cells Muneeza Ahmad, Carsen Cartledge, and Nicholas Rolston Arizona State University, Tempe, AZ, 85281, USA   Abstract - Perovskite solar cells have the potential to beat Si-based photovoltaic technologies because of their excellent optoelectronic properties and low-cost solution processing methods. However, challenges to upscaling include residual mechanical stresses which accelerate degradation, dominate at the module scale, and can lead to delamination or fracture. Studying the development and propagation of stresses during the crystallization process can help limit them to a desired value. Open-air blade coating of single-step coated perovskite eliminates the tensile stresses and introduces compression in the thin film. With the use of a polymer additive, gellan gum, we demonstrate how perovskite crystallinity, grain size, and orientation are tuned for a beneficial compressive residual stress that improves stability through an environment-friendly process.

Silver Recovery through a Fluoride Chemistry for Solar Module Recycling

Theresa K Chen, Meng Tao

Arizona State University, Tempe, AZ, United States

Recovery of pure metallic silver from end-of-life silicon solar cells through a fluoride chemistry is proposed. The process involves leaching and electrowinning. It is determined that silver can be completely leached into hydrofluoric acid with the addition of hydrogen peroxide in an hour. Cyclic voltammetry indicates that the reduction potential of silver is -0.41 V versus a silver/silver chloride reference electrode in a 0.05 M silver fluoride solution. The silver recovery rate by electrowinning depends on the potential and time. The highest recovery rate thus far is over 95%. However, the low coulombic efficiency suggests a parasitic reaction during electrowinning. A chemical analysis confirms that the deposit on the working electrode is 100% silver.

Thermal Models of Monofacial and Bifacial PV Modules: Machine Learning and Physical Estimation Models Comparison

Marco Grisanti[1], Gaetano Mannino[2], Giuseppe Marco Tina[2], Alessandro Ortis[1], Mario Cacciato[2], Sebastiano Battiato[1], Fabrizio Bizzarri[3], Andrea Canino[3]

[1] Department of Mathematics and Computer Science, University of Catania, 95125 Catania, Italy
[2] DIEEI, University of Catania, 95125 Catania, Italy
[3] Enel Green Power SpA, Viale Regina Margherita, 125, 00198, Rome, Italy

Abstract — The temperature of the photovoltaic modules influences the electrical power output, as the temperature increases, the power extracted will decrease, other environmental conditions being equal. The degradation of PV modules over time is also influenced by the operating temperature and thermal cycling experienced by the modules. The estimation of the module temperature is therefore of crucial importance if it is not measurable due to the absence of sensors or for preliminary studies. In the present work, conventional models with machine learning regressors are compared for the estimation of the temperature of monofacial and bifacial photovoltaic modules installed in two different locations.

I. INTRODUCTION

The influence of temperature on PV module power production and module degradation over time is well known. For the estimation of the temperature of the photovoltaic modules, physical models are conventionally used, which calculate the estimated temperature by inserting climatic variables as input. Different models are applied for temperature estimation, some of the best known are reported in the study [1]. However, it is possible to further increase the accuracy of temperature estimation models using Machine Learning (ML). In [2] the following physical models are applied: the Ross model, the Faiman model and the Sandia model; the Artificial Neural Network (ANN) and the Light Gradient Boosting Model (LightGBM) are instead applied as machine learning techniques. The Ross model requires as input only the ambient temperature, the NOCT from the module datasheet and the irradiance; the Faiman model does not use the NOCT but also considers the wind speed; the Sandia model has the same environmental input variables as the Faiman model.

Furthermore, using ML approaches, it was attempted to identify new relevant weather variables to improve thermal models accuracy. The physical and ML models were applied to three different PV plants in three sites characterized by different climatic conditions, the comparison showed that the results of the models vary, favoring the physical or ML models depending on the climatic characteristics of the sites.

The AI approach for PV module temperature prediction is also performed in [3], where a literature review on the application of AI to obtain models for estimating the temperature of photovoltaic modules is summarized in a table and a building integrated photovoltaic system is analyzed.

This study compares two ML regression techniques and some well-known physical models for estimating temperatures applied to a case study of real PV systems.

II. MODELS

Physical models

Monofacial module temperature

Sandia model:
$$T_{mod} = T_a + (G_f) \cdot (e^{a+bWs}) \qquad (1)$$

Where T_{mod} is the module temperature [°C], T_a is the ambient temperature [°C], G_f is the front plane of array irradiance [W/m^2], a and b are two empirical coefficients, Ws is the wind speed [m/s]

Faiman model:
$$T_{mod} = T_a + \frac{G_f}{U_0 + U_1 W_s} \qquad (2)$$

Where $U_0[W/(m^2K)]$ a $U_1[W/(m^3sK)]$ are empirical coefficients. Once the module temperature is calculated, it is possible to estimate the cell temperature using:
$$T_{cell} = T_{mod} + \Delta T \cdot \frac{G_f}{G_{stc}} \qquad (3)$$
ΔT is the temperature difference parameter, the value is typically between 2 and 3 °C for open-rack mounted PV modules.

Bifacial module temperature

Sandia bPV model [1]: $T_{mod} = T_a + (G_f + G_r) \cdot (e^{a+bWs})$ (4)

Where G_r is the rear irradiance [W/m^2].
Starting with the Sandia model, a coefficient c of G_r is introduced:
$$T_{mod} = T_a + (G_f + cG_r) \cdot (e^{a+bWs}) \qquad (5)$$

Further, the new term (1-nstc) is added to take into account the variation of T_mod due to the transformation of solar energy into electricity:
$$T_{mod} = T_a + (1 - \eta_{STC})(G_f + cG_r) \cdot (e^{a+bWs}) \qquad (6)$$

Machine learning regressors

Two widely used ML regressors for nonlinear problems are:

Random Forest Regressor: A Random Forest is a meta-estimator that fits a number of classification decision trees on various sub-samples of the dataset. The Random Forest Regressor is a machine learning algorithm based on the Random Forest meta-estimator that combines the output of multiple decision trees to reach a single result.

Gradient Booster Regressor: The Gradient Boosting Regressor is a machine learning algorithm, which allows to build an additive model in a progressive way allowing the optimization of the loss functions. Subdivided into several phases, each one fits a regression tree on the negative gradient of the loss function.

III. DATA TREATMENT AND RESULTS

The data used in this study refer to two PV plants analyzed respectively in [4], where PV system near Bolzano with polycrystalline silicon modules is monitored, and in [5], where a photovoltaic system mounting bifacial modules with PERC monocrystalline silicon cells is monitored.

The data processing, underwent the following simple steps:
1. Elimination of rows where there is at least one null value.
2. Conservation of values relating to a single year.
3. Preservation of values with a 15-minute sampling (in order to have the same sampling for EGP and EURAC).
4. Conservation of values with GF > 20 and GF < 1200.

Once the tables were obtained, only the input variables of the models were preserved. The empirical coefficients obtained through Python optimization algorithms are reported in TABLE 1. For each problem the chosen algorithm was one of BFGS[1], L-BFGS-B[2], SLSQP[3], depending on whether or not the problem has constraints or bounds [6].

TABLE 1: *Empirical coefficients of physical models*

MODEL	Bifacial modules			Monofacial modules	
	Coef. 1	Coef. 2	Coef. 3		
Sandia (1)	a= -3.47	b= -0.0594	—	-3.56	-0.075
Sandia (1) optimal a, b	a = -3.65	b = -0.0747	—	-3.39	-0.1125
Sandia bPV (4)	a = -3.47	b= -0.0594	—		
Sandia bPV (4) optimal a, b	a = -3.793	b = -0.0699	—		
Sandia bPV (5) Optimal, a,b,c	a= -3.579	b= -0.0768	c = -0.408		
Faiman (2)	U_0 =25	U_1=6.84	—	25	6.84
Faiman (2) Optimal U0, U1	U_0=36.83	U_1=3.81	—	26.65	6.0361

After finding the optimal coefficients of the empirical models applied to the two monitored PV plants, it was possible to compare the temperature estimates obtained through physical models with coefficients suggested in [7], physical models with optimal coefficients and the two ML regressors. Table 2 and Table 3-4 respectively show the statistical indices for the models applied to the modules of the two PV plants of Bolzano and Catania.

TABLE 2: *Statistical indices of physical and ML models, Bolzano*

EURAC Dataset: monofacial polycristalline						
Metric	Sandia M (1)	Sandia MFT (1)	Faimn M (2)	Faiman MFT (2)	RFR	GBR
MSE	13.43	10	10.3	10.11	7.85	7.44
RMSE	3.67	3.16	3.21	3.18	2.80	2.73
MAE	2.78	2.39	2.39	2.39	1.96	1.93
MAPE	0.26	0.29	0.30	0.30	0.18	0.21
R2	0.95	0.97	0.97	0.97	0.97	0.97

TABLE 3: *Statistical indices of physical and ML models, Catania*

Enel Green Power dataset: bifacial mono-PERC, neglecting Gr						
Metric	Sandia M (1)	Sandia MFT (1)	Faimn M (2)	Faiman MFT (2)	RFR	GBR
MSE	15.32	6.19	7.40	6.22	5.9	5.87
RMSE	3.91	2.49	2.72	2.49	2.43	2.42
MAE	3.05	1.81	1.95	1.82	1.76	1.76
MAPE	0.12	0.07	0.08	0.08	0.07	0.07
R2	0.89	0.96	0.95	0.96	0.96	0.96

TABLE 4: *Statistical indices of physical and ML models, Catania*

Enel Green Power dataset: bifacial mono-PERC, considering Gr					
Metric	Sandia B (3)	Sandia BFT (3)	Sandia BFT (4)	RFR	GBR
MSE	30.36	6.38	5.77	5.20	5.61
RMSE	5.51	2.53	2.40	2.28	2.37
MAE	4.52	1.85	1.73	1.61	1.70
MAPE	0.17	0.08	0.07	0.06	0.07
R2	0.79	0.96	0.96	0.96	0.96

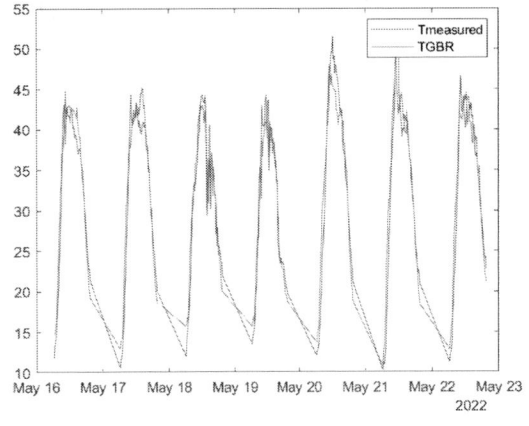

Fig. 1. Measured module Temperature vs Gradient Boosting estimated module Temperature.

---

[1] Broyden–Fletcher–Goldfarb–Shanno (BFGS),
[2] Limited-memory BFGS

[3] Sequential Least Squares Programming (SLSQP)

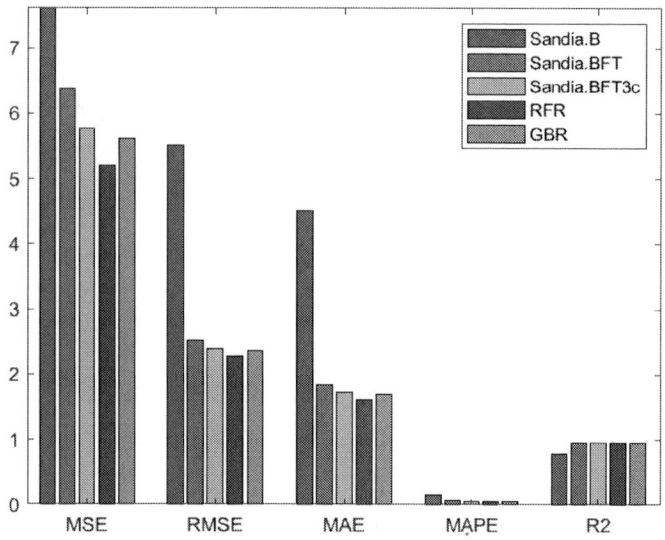

Fig. 1. Statistical indices of thermal models applied to bifacial modules considering the rear irradiance, Zoom

Discussion

The models with the prefixed empirical coefficients, as expected, gave less accurate results than both the same models in which the fine tuning was applied and the ML models.

Taking into account the irradiance hitting the back of the module has improved the accuracy of the results. However, it is important to note that the improvement deriving from considering the back irradiance in the models with optimal and ML coefficients, being small, could be compensated by a greater uncertainty when there is a small number of data or for climatic conditions different from those analyzed. Considering the different technology of the modules is important when the models of the present study are to be replicated.

IV. Conclusion

Physical models were compared with two ML regressors for estimating the temperature of photovoltaic modules, applied to the case of bifacial and monofacial photovoltaic modules and located in two different parts of Italy. The use of ML regressors improved the accuracy of the temperature estimation compared to physical models.

The physical models that use empirical coefficients obtained through fine tuning on the data has allowed to have an accuracy of the results which, although it is close to that of the estimation results from the ML regressors, is always exceeded. Further studies will allow a generalization of the use of ML regressors for temperature estimation for different technologies and climate zones.

References

[1] L. D. O. Santos, P. Cesar, M. De Carvalho, C. De Oliveira, and C. Filho, "Photovoltaic Cell Operating Temperature Models : A Review of Correlations and Parameters," vol. 12, no. 1, pp. 179–190, 2022.

[2] J. Ascencio-vasquez, K. Ismail, and M. Topic, "Application of Machine Learning to assess the thermal behaviour of PV modules in different climate zones," no. September, 2021.

[3] L. Serrano-Luján, C. Toledo, J. M. Colmenar, J. Abad, and A. Urbina, "Accurate thermal prediction model for building-integrated photovoltaics systems using guided artificial intelligence algorithms," *Appl Energy*, vol. 315, no. September 2021, p. 119015, 2022, doi: 10.1016/j.apenergy.2022.119015.

[4] S. Lindig *et al.*, "International collaboration framework for the calculation of performance loss rates: Data quality, benchmarks, and trends (towards a uniform methodology)," *Progress in Photovoltaics: Research and Applications*, vol. 29, no. 6, pp. 573–602, 2021, doi: 10.1002/pip.3397.

[5] G. Mannino, G. M. Tina, M. Cacciato, L. Todaro, and F. Bizzarri, "Experimental assessment of temperature estimation models of bifacial photovoltaic modules," IEEE, 2022, pp. 214–216. doi: 10.1109/PVSC48317.2022.9938792.

[6]

https://docs.scipy.org/doc/scipy/reference/generated/scipy.optimize.minimize.html.

[7] D. L. King, W. E. Boyson, and J. A. Kratochvil, "Photovoltaic array performance model," *Sandia Report No. 2004-3535*, vol. 8, pp. 1–19, 2004, doi: 10.2172/919131.

In-Situ Photostability Analysis of Perovskite Solar Cells by Time-Evolving Photoluminescence Imaging

Jackson W. Schall,[1,2] Amy Louks,[1,2] Goutam Paul,[1] Nikita S. Dutta,[1] Steve Johnston,[1] Chun-Shen Jiang,[1] Axel Palmstrom,[1] Mowafak Al-Jassim,[1] Dana B. Kern[1]

[1]National Renewable Energy Laboratory, Golden, CO, 80401, USA [2]Colorado School of Mines

Abstract— **In this work we investigate reversible metastabilities and irreversible degradation due to light soaking of triple-cation mixed-halide perovskite p-i-n solar cells. We use spatially resolved luminescence imaging to document the spatial inhomogeneities that exist even in high-performance devices with median efficiencies >20%. We monitor light-soaking degradation in-situ using time-evolving photoluminescence images, electro-luminescence images, and transient photovoltage. Post-stress measurements of external quantum efficiency and Kelvin-probe force microscopy suggest that long-term degradation primarily affects electrode interfaces rather than the absorber layer.**

Keywords—perovskite, reliability, electroluminescence, imaging

I. INTRODUCTION

Perovskite solar cells (PSCs) have quickly accelerated in power conversion efficiencies (PCE) from 3.8% in 2009 [1 [1] to efficiencies greater than 25.7% [2]. However, challenges in long-term operational stability remain a significant hurdle to commercialization. Although robust packaging to protect against oxygen and moisture can mitigate severe decomposition pathways, other intrinsic materials challenges exist in perovskite absorbers. In particular, perovskites differ from traditional photovoltaic (PV) materials due to linked ionic and electronic conduction. Under optical or electrical bias, this can create complex spatial variations both laterally and vertically across the PV device stack [3], e.g. in local chemical potential, trap density, material morphology or lattice strain, compositional variations, and/or interfacial accumulation of chemical species [3]. Although transient effects associated with ion migration are often reversible (e.g. causing current-voltage hysteresis [4]), it is still unclear if long-term degradation is largely independent of such metastabilities [5] or if the presumed reversible processes play a role in driving irreversible (photo)electrochemical degradation [6].

The task of assessing the possible relationship between metastabilities and long-term degradation is further complicated by macroscopic inhomogeneities across devices. As PSCs progress toward commercialization, active areas increase and there is a greater chance of including defects or spatial variation within a solar cell or PV module. For this reason, spatially resolved characterization is essential to understanding how imperfections impact device performance/ stability and to guide process optimization [7]. Spatially resolved characterization, such as luminescence imaging, is

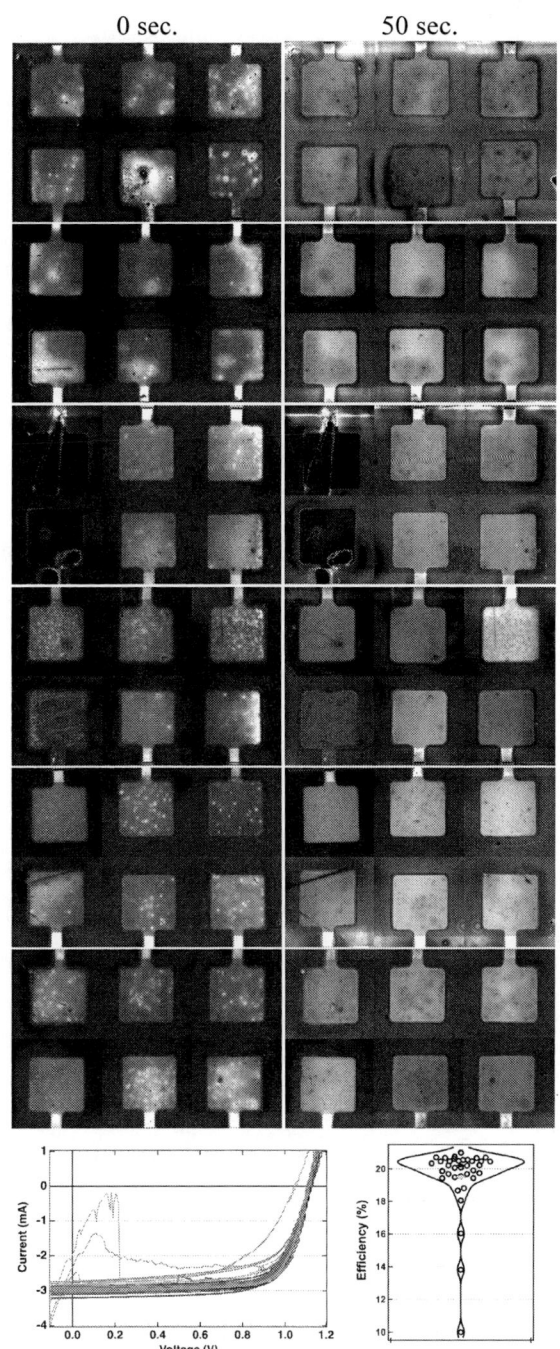

Fig. 1. Photoluminescence images and current-voltage characteristics of devices with area of 0.122 cm², 0 s (left) and 50 s (right) after light bias.

employed for commercial PV modules [8] with both internationally standardized methods [9] and equipment for in-line screening [10].

Here, we present methodology to evaluate perovskite photostability, considering both reversible and irreversible processes. We note that traditional steady-state luminescence methods do not account for perovskite metastability and thus are unreliable for assessing perovskite devices [11]. Therefore, we use time-evolving photoluminescence (PL) imaging under light/dark cycling to separate reversible degradation/recovery from longer-term irreversible losses. Our results document the spatial inhomogeneities that exist even in high-performing devices (median PCE > 20%), and our 24-hour in-situ stress studies assess severity of features upon light-induced degradation such as absorber defects or inhomogeneities, processing/handling defects, and electrical contact geometry.

II. EXPERIMENTAL

Devices underwent separate procedures for screening (non-destructive) and failure analysis (destructive). For screening, we capture both electroluminescence (EL) and PL images. During EL screening we applied forward bias at 10% of the short-circuit current (I_{SC}) for 50s while capturing 10 EL images and simultaneously recording voltage. The devices were then held at 0mA with voltage recovery monitored for 120s seconds. We repeated the image/rest cycle 2-3 times to ensure reproducibility and no degradation induced from screening. For PL screening, devices were illuminated at 1 Sun for 50s using a 450nm high power LED (Prismatix) while capturing 10 PL images and recording voltage at 0mA (open-circuit). The devices were then held in the dark at 0mA with voltage recovery monitored for 120s seconds. We similarly repeated the PL image/rest cycle repeated 2-3 times to ensure reproducibility and no degradation induced from the screening procedure.

During failure analysis, devices underwent 6 stress/rest cycles over 24 hours at room temperature under nitrogen atmosphere. Each stress/rest cycle was 2 hours at 5 Suns, followed by 2 hours in the dark. Devices were held at open circuit while continuously recording voltage using a Keithley 2401 sourcemeter. Images were collected using a Princeton Instruments PIXIS Silicon CCD camera with 715 nm long pass filter. The camera, LED, and sourcemeter were triggered by LabVIEW. Exposure times were 5ms for PL and 0.5s for EL.

III. RESULTS

A. Perovskite Device Screening

Fig. 1 shows PL images for a batch of triple-cation and mixed-halide perovskite solar cells all fabricated at the same time with the same process. Devices have the p-i-n construction of glass / indium tin oxide (ITO) / NiO$_x$ / N4,N4'-Di (naphthalen-1-y l)-N4,N4'-bis (4-viny lpheny l)bipheny l-4,4'-diamine (VPBN) / $FA_{0.77}Cs_{0.13}Pb(I_{0.95}Br_{0.05})_3$ / C$_{60}$ / bathocuproine (BCP) /Ag. The batch has median PCE of 20.2% (mean 19.5 +/- 2.1%), yet we observe significant variation in spatial PL distribution. We present images both immediately upon photoexcitation (left column) and after 50 seconds of illumination (right column), demonstrating spatial evolution

over time, consistent with effects of ion migration under optically induced fields [11], [12].

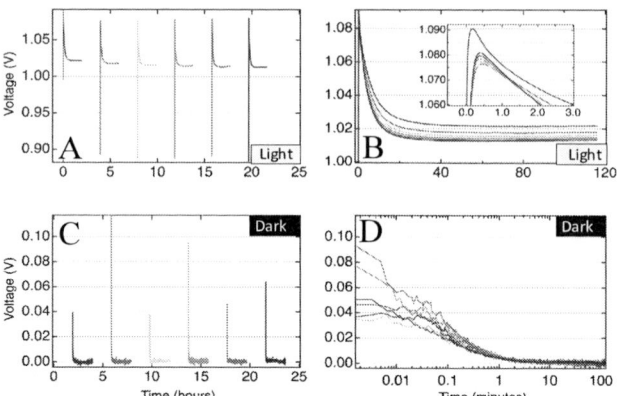

Fig. 2. Left: Photovoltage of a PSC during stress/rest cycling. Right: All voltage transients on a common time axis

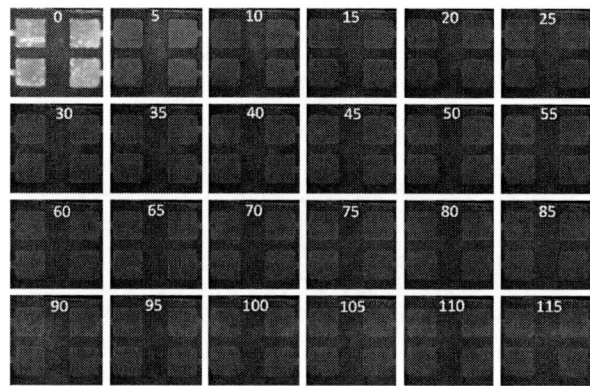

Fig 3. Photoluminescence images on a log intensity scale, showing evolution over the first two-hour stress. Labels represent time (minutes).

B. In-Situ Degradation Analysis with Stress/Rest Cycling

Next, we evaluate the importance of both the metastability and the spatial defects in these solar cells during a 24-hour light soaking test using 5 Suns at 450 nm alternated with dark rest periods. Fig. 2 shows voltage data during stress/rest cycling, and Fig. 3 shows the associated in-situ PL images during the first stress cycle.

Fig. 2a shows photovoltage evolution in 2-hour illumination cycles over 24 hours, and Fig. 2b overlays the same data to compare how the photovoltage evolves between cycles. Figs. 2b-c show the open circuit photovoltage decay during the 2-hour dark rest periods of our light soaking experiment. The high-level repeatability of voltage transients between cycles suggests reversible metastabilities are present. However, the slight change in shape of the voltage transients and decrease in final steady state photovoltage with subsequent cycles suggests light soaking has also irreversibly changed the devices.

Beyond the first cycle, a trend in the evolution of photovoltage transients is clear. For times greater than 100 s, the photovoltage settles to a lower value with continued stressing (Fig. 2b). This agrees with the lower open-circuit voltage (V_{OC}) in post-stress IV measurements (Fig. 4). At times less than 100 s (inset of Fig 2b) the voltage overshoot increases in magnitude with continued stressing. Similar to our previous studies [11], this change in photovoltage transient behavior at early times can be modeled with variations in ion diffusivity and effective mobile ion concentration. However, the long-timescale relaxation on the order of 1000 s cannot be explained by ionic motion alone [13]. Further work is underway on PSC drift diffusion modelling, using light-induced traps [12] to explain long-timescale transient phenomena.

C. Post-Stress Analysis

Fig. 4 and Table 1 summarize the changes in IV metrics before and after light soaking. The devices show significant degradation with most losses in short circuit current density (J_{sc}) and fill factor (FF). V_{OC} also suffers losses, and its gradual decay can be observed at the end of each light soaking cycle (Fig 2b), as discussed above. The significant loss of FF and J_{sc} with shallow slope in the IV curve approaching V_{oc} may suggest resistive issues and electrical contact problems, which we further investigate below.

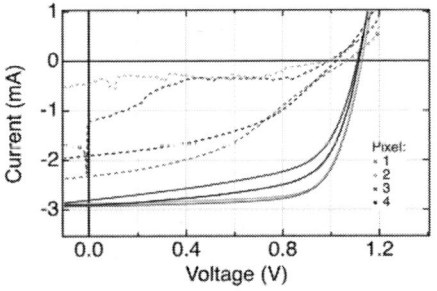

Fig. 4. Current-voltage curves before and after light soaking experiment.

Table I. IV parameters before and after light soaking experiment

Device	Voc (V)	Jsc (mA/cm²)	Pmax (mW)	Efficiency (%)	Fill Factor (%)
1	1.13	24.2	2.4	19.5	72.3
2	1.13	23.7	2.4	19.3	73.1
3	1.11	23.0	1.8	15.1	59.2
4	1.11	23.8	2.1	17.0	64.2
After LS					
1	1.07	19.0	1.0	8.0	39.6
2	1.05	2.6	0.2	1.2	43.2
3	0.92	12.5	0.4	3.3	28.7
4	1.02	15.6	0.9	7.2	45.0

We use the in-situ PL images to further understand the impact of spatial defects on the metastability and degradation. Fig. 5 shows the change in PL intensity from first to last cycle by dividing the images in last cycle by the first cycle. The change in luminescence is given by color scale of red (increase) white (no change) blue (decrease). We find a large increase in the early-time PL intensity for degraded samples, followed by overall decrease in PL intensity after the 2-hour light-soaking period. The results of image 6 (30s) indicates spatial changes at the time of maximum voltage overshoot from Fig. 2b.

Interestingly, the patterns in Fig. 5 at early times (image 1) and steady-state (image 38) imply mostly uniform changes in the devices' PL upon irreversible degradation. That is, the process of dividing the images largely removes contrast of defects. This may suggest that the spatially distributed defects

do not degrade faster than the rest of the device with respect to PL-relevant processes such as photon absorption or generation of photovoltage. However, the images at the time of maximum voltage overshoot (image 6) highlight some contrast from spot-like defects as well as from the electrode edges. We speculate that this variation the PL image can be explained by local changes in electric fields near electrode boundaries or electrode discontinuities as described in [3].

Fig. 5. Dividing PL images from last cycle by first cycle

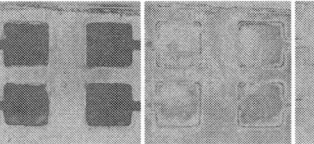

Image 1 (5 sec.)   **Image 6** (30 sec.)   **Image 38** (2 hours)

We next investigate the degradation in electrical contact quality using electroluminescence (EL) before and after the light soaking experiment (Fig. 6). While PL imaging is sensitive to the quality of the active semiconductor layer and the photoinduced charge generation process, EL imaging provides insight on contact quality as current is electrically injected/transported through the contact layers. The EL images show a severely degraded pattern post stress, where bright edges and cell defects contrast with lower-intensity mid-cell regions. This high-contrast pattern likely implies high series resistance, where current mainly flows through defects, and charge injection in the bulk areas has degraded. This result is consistent with the shallow slope in IV near V_{OC}, which also implies series resistance (Fig. 4).

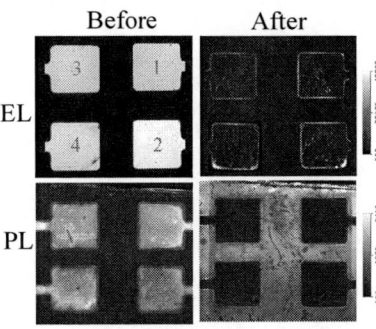

Fig. 6. EL and PL images of perovskite devices (0.122 cm²) before and after light soaking. PL images taken at the end of cycle 1 & 6, EL images taken after 50s of applied bias at 0.1x I_{sc}.

Further evidence of charge-transport interface degradation is shown with cross-sectional Kelvin-probe force microscopy (KPFM) and external quantum efficiency (EQE) measurements shown in Fig. 7. The shape of the EQE curve hardly changes after the light-soaking experiment, which implies minimal degradation of the absorber. In contrast, KPFM shows significant changes in interfacial electric field intensities. The result may be explained as either a leaky perovskite/hole-transport layer interface, or (more likely) as high equivalent resistance at the electron-transport layer side from degradations such as delamination or increased interface recombination. The KPFM results imply interfacial degradation and will be discussed in another PVSC submission [14].

Further evidence of charge-transport interface degradation is shown with cross-sectional Kelvin-probe force microscopy (KPFM) and external quantum efficiency (EQE) measurements shown in Fig. 7. The shape of the EQE curve hardly changes after the light-soaking experiment, which implies minimal degradation of the absorber. In contrast, KPFM shows changes in interfacial electric field intensities for multiple control and stressed devices. The result may be explained as either a leaky perovskite/hole-transport layer interface, or (more likely) as high equivalent resistance at the electron-transport layer side from degradations such as delamination or increased interface recombination. The KPFM results imply interfacial degradation and will be discussed in another PVSC submission [14].

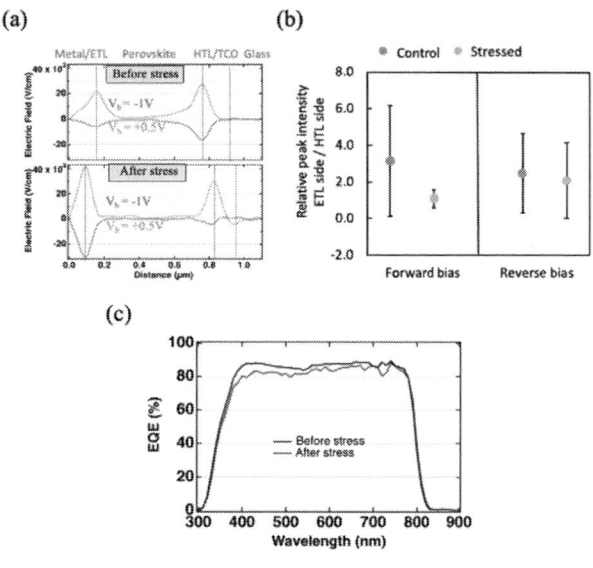

Fig. 7 (a) Cross-sectional KPFM of one control device and one device stressed under 450nm illumination at 5 Sun intensity. (b) Average value of peak ratio and standard deviation of several control and stressed devices. (c) EQE measurements.

IV. CONCLUSIONS

The investigation of luminescence images and photovoltage transient data presented here demonstrate the value of time-evolving PL imaging to understand reversible metastabilities and irreversible degradation of PSCs under stressors such as light soaking. Our results show a case study in which degradation appears to dominate at the interface between the perovskite and charge-transport layer.

REFERENCES

[1] A. Kojima, K. Teshima, Y. Shirai, and T. Miyasaka, "Organometal Halide Perovskites as Visible-Light Sensitizers for Photovoltaic Cells," *J. Am. Chem. Soc.*, vol. 131, no. 17, pp. 6050–6051, May 2009, doi: 10.1021/ja809598r.

[2] "Best Research-Cell Efficiency Chart." https://www.nrel.gov/pv/cell-efficiency.html (accessed May 31, 2023).

[3] D. A. Jacobs, C. M. Wolff, X.-Y. Chin, K. Artuk, C. Ballif, and Q. Jeangros, "Lateral ion migration accelerates degradation in halide perovskite devices," *Energy Environ. Sci.*, vol. 15, no. 12, pp. 5324–5339, Dec. 2022, doi: 10.1039/D2EE02330J.

[4] B. Chen, M. Yang, X. Zheng, C. Wu, W. Li, Y. Yan, J. Bisquert, G. Garcia-Belmonte, K. Zhu, and S. Priya, "Impact of Capacitive Effect and Ion Migration on the Hysteretic Behavior of Perovskite Solar Cells," *J. Phys. Chem. Lett.*, vol. 6, no. 23, pp. 4693–4700, Dec. 2015, doi: 10.1021/acs.jpclett.5b02229.

[5] B. Rivkin, P. Fassl, Q. Sun, A. D. Taylor, Z. Chen, and Y. Vaynzof, "Effect of Ion Migration-Induced Electrode Degradation on the Operational Stability of Perovskite Solar Cells," *ACS Omega*, vol. 3, no. 8, pp. 10042–10047, Aug. 2018, doi: 10.1021/acsomega.8b01626.

[6] S. Kim, S. Bae, S.-W. Lee, K. Cho, K. D. Lee, H. Kim, S. Park, G. Kwon, S.-W. Ahn, H.-M. Lee, Y. Kang, H.-S. Lee, and D. Kim, "Relationship between ion migration and interfacial degradation of CH3NH3PbI3 perovskite solar cells under thermal conditions," *Sci. Rep.*, vol. 7, no. 1, Art. no. 1, Apr. 2017, doi: 10.1038/s41598-017-00866-6.

[7] J. W. Schall, A. Glaws, N. Y. Doumon, T. J. Silverman, M. Owen-Bellini, K. Terwillinger, M. A. Uddin, P. Rana, J. J. Berry, J. Huang, L. T. Schelhas, and D. B. Kern, "Accelerated Stress Testing of Perovskite Photovoltaic Modules: Differentiating Degradation Modes with Electroluminescence Imaging," *Sol. RRL*, vol. n/a, no. n/a, doi: 10.1002/solr.202300229.

[8] S. Johnston, H. R. Moutinho, C. S. Jiang, H. L. Guthrey, A. Norman, S. B. Harvey, P. L. Hacke, C. Xiao, J. Moseley, D. Sulas, J. Liu, D. S. Albin, M. Nardone, and M. M. Al-Jassim, "From Modules to Atoms: Techniques and Characterization for Identifying and Understanding Device-Level Photovoltaic Degradation Mechanisms," NREL/TP--5K00-72541, 1572277, Oct. 2019. doi: 10.2172/1572277.

[9] D. B. Sulas-Kern, S. Johnston, and J. Meydbray, "Fill Factor Loss in Fielded Photovoltaic Modules Due to Metallization Failures, Characterized by Luminescence and Thermal Imaging," in *2019 IEEE 46th Photovoltaic Specialists Conference (PVSC)*, Jun. 2019, pp. 2008–2012. doi: 10.1109/PVSC40753.2019.8980840.

[10] "Products - English," *btimaging*. https://www.btimaging.com/products-c1g2d (accessed May 16, 2023).

[11] Y. Patikirige, D. B. Sulas-Kern, M. Nardone, S. Johnston, C. Xiao, H. Guthrey, K. Zhu, F. Zhang, and M. Al-Jassim, "Linking Transient Voltage to Spatially-Resolved Luminescence Imaging to Understand Reliability of Perovskite Photovoltaics," Piscataway, NJ: Institute of Electrical and Electronics Engineers (IEEE), NREL/CP-5K00-81134, Aug. 2021. doi: 10.1109/PVSC43889.2021.9519032.

[12] W. Nie, J.-C. Blancon, A. J. Neukirch, K. Appavoo, H. Tsai, M. Chhowalla, M. A. Alam, M. Y. Sfeir, C. Katan, J. Even, S. Tretiak, J. J. Crochet, G. Gupta, and A. D. Mohite, "Light-activated photocurrent degradation and self-healing in perovskite solar cells," *Nat. Commun.*, vol. 7, no. 1, Art. no. 1, May 2016, doi: 10.1038/ncomms11574.

[13] A. Pockett, G. E. Eperon, N. Sakai, H. J. Snaith, L. M. Peter, and P. J. Cameron, "Microseconds, milliseconds and seconds: deconvoluting the dynamic behaviour of planar perovskite solar cells," *Phys. Chem. Chem. Phys.*, vol. 19, no. 8, pp. 5959–5970, 2017, doi: 10.1039/C6CP08424A.

[14] G. Paul, J. W. Schall, C.-S. Jiang, A. E. Louks, A. F. Palmstrom, N. S. Dutta, S. Johnston, H. Guthrey, A. Norman, M. M. Al-Jassim, and D. B. Sulas-Kern, "Investigating Electric Field and Light Induced Degradation in Perovskite Solar Cells through Nanometer-Scale Potential Imaging," presented at the 2023 IEEE 50th Photovoltaic Specialists Conference (PVSC), Submitted.

Flexible Manufacturing of Colloidal Quantum Dot Solar Cells via Spray-Casting Techniques

Lulin Li[1], Botong Qiu[1], Yida Lin[1], Laura Shimabukuro[2], Alex Ozbolt[1], Keyi Kang Yao[3], Stephen Farias[4], Samuel Rosenthal[4] and Susanna M. Thon[1]

[1]Department of Electrical and Computer Engineering, Johns Hopkins University, 3400 N. Charles Street, Baltimore, Maryland 21218, USA

[2]Department of Electrical and Computer Engineering, University of California Davis, One Shields Avenue, Davis, California 95616, USA

[3]Department of Chemical and Biomolecular Engineering, Johns Hopkins University, 3400 N. Charles Street, Baltimore, Maryland 21218, USA

[4]NanoDirect LLC and Materic LLC, 1300 Bayard Street, Baltimore, Maryland 21230, USA

Abstract — **Colloidal quantum dots are a promising candidate material for solar energy generation because of their band gap tunability and solution-based processing flexibility. However, conventional colloidal quantum dot solar cell fabrication techniques are still limited by their lack of scalability, environment conditions, and difficult installation scenarios. Here, we develop spray-casting manufacturing methods for fabricating thin film solar cells, discuss the trade-off between conductivity and transmittance in transparent contact materials, and demonstrate the feasibility of spray-casting colloidal quantum dot layers. This work on flexible manufacturing methods paves the way for installing solar energy devices in a variety of novel scenarios.**

Keywords—colloidal quantum dots, spray casting, thin film solar cells

I. INTRODUCTION

Solar energy has been attracting the world's attention in recent decades due to its abundance, sustainability and environmental friendliness. It is now regarded as a significant replacement or supplement for traditional energy sources, especially in countries and areas with high solar irradiance. Expanding the feasibility of solar energy so that it can be used in more places and for more applications is a crucial challenge for society. Currently, the most popular and widely used solar energy conversion devices are silicon solar cells, which are highly commercialized and occupy the majority of the terrestrial solar market. Silicon solar cells, however, meet with limitations in a variety of new areas because of the heavy and rigid nature of the panels, even though their manufacturing costs have been continuously decreasing. For example, their application in building-integrated photovoltaics, major transportation methods, and mobile devices are quite limited. Additionally, the lack of mechanical flexibility makes it difficult to deploy silicon solar cells in wearable devices or on vehicle surfaces.

In order to deploy solar energy in these new applications, novel materials are needed. Colloidal quantum dots (CQDs) are thought to be one of the most promising candidate materials for thin film solar cells because of their solution-based processing capability. Critically, since the band gap of quantum dots can be tuned by varying their sizes, it is possible to fabricate multijunction solar cells using colloidal quantum dots with different sizes in both hybrid and single-material scenarios [1]. Moreover, colloidal quantum dots have high absorption in the infrared region [2], [3]. Thus, they can be important supplements to current conventional silicon solar cells which apply for the visible range better. However, in practice, the scalable fabrication of colloidal quantum dot solar cells must overcome certain limitations before they can be deployed commercially. Currently the best-performing cells use high-temperature-processed transparent conductive oxides (TCOs) which restrict the conditions of fabrication. In addition, non-portable and non-scalable spin-casting techniques are used in depositing the colloidal quantum dot absorbing layer and charge transport layers, while a vacuum environment is required for the physical evaporation of the top and bottom electrodes. These factors limit the fabrication process of conventional colloidal quantum dot solar cells to the laboratory environment.

The demands of flexible applications inspires the goal of building a colloidal quantum dot solar cell in which all layers can be fabricated via inexpensive manufacturing methods in ambient environmental conditions. Compared with traditional spin-casting processes, slot die coating processes or

978-1-6654-6060-6/23 $31.00 © 2023 IEEE

evaporation techniques, spray-casting shows the potential to fulfill this requirement because of several advantages for flexible thin film solar cell manufacturing. First, the starting liquid ink material has a higher utilization rate in a spray-casting process, while quite a large fraction is wasted during spin-casting [4]. Second, while spin-casting processes are limited in substrate size and require a rigid substrate, spray-casting processes have fewer limitations in terms of substrate sizes and types. Third, the use of room-temperature silver nanowire inks for the electrodes instead of bulk metal oxides and gold removes the requirements of time, temperature, pressure, and manufacturing costs for the electrodes and allows solar cells to be fabricated on almost any type of surface, expanding their feasibility in actual applications.

II. METHODS, RESULTS AND DISCUSSION

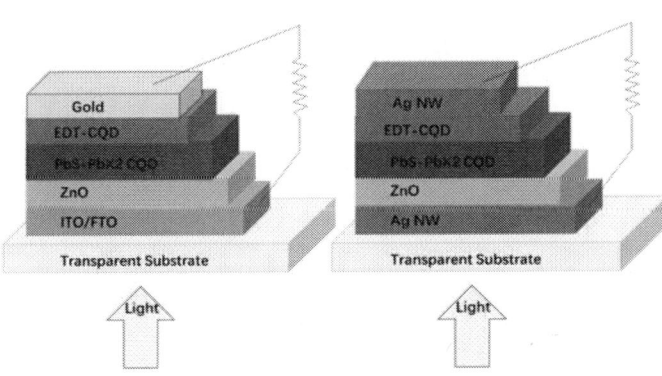

Fig. 1. (Left) Schematic of a conventional colloidal quantum dot solar cell structure and (right) a fully spray-cast colloidal quantum dot solar cell structure. EDT-CQD refers to the ethanedithiol-treated hole transport layer. PbS-PbX$_2$ CQD refers to the PbI$_2$/PbBr$_2$-treated light-absorbing layer. ZnO refers to the zinc oxide electron transport layer. ITO (indium tin oxide) and FTO (fluorine doped tin oxide) refer to the transparent conductive oxides. Ag NW refers to the silver nanowire-based transparent electrode.

In this work, we discuss the use of spray-cast manufacturing in building a fully spray-casted flexible CQD solar cell. Differences between the structures of a conventional CQD solar cell and a fully spray-cast CQD solar cell are shown in Figure 1. Generally, the crystalline TCO bottom electrode layer and the physically evaporated gold top electrode layer are both substituted by spray-cast silver nanowire electrode films. The zinc oxide (ZnO) electron transport layer and lead sulfide (PbS) CQD layers are all spray-cast in between the silver nanowire electrode layers. Hence, all layers of the CQD solar cell structure can be solution processed, and therefore, be spray-cast under ambient conditions.

Our work is based on the custom-built spray-casting setup shown in Figure 2. The use of a liquid siphon nozzle removes the requirement of an external liquid pressure supply. The air flow is able to trigger the liquid siphon effect, which pushes the solution into the nozzle, and the spray-casting process commences. The pilot air is set for controlling the on and off

function of the spray nozzle. In this setup, we can control the quality of the spray-casted film by modifying the air supply pressure level, the height difference between the solution holder and the nozzle, the spray duration, and the number of spray cycles.

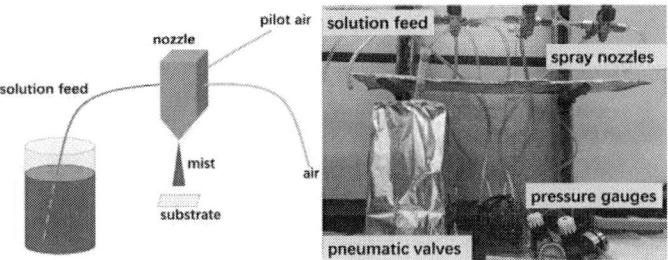

Fig. 2. Schematic (left) and photograph (right) of the spray-casting instrument. Air is supplied to trigger the siphon effect that pushes the liquid solution into the nozzle, and the spray-casting process commences. The pilot air switched by pneumatic valves which are used to turn on/off the nozzle.

A. Silver nanowire layer optimization

Silver nanowires are chosen as a substitute for conventional high-temperature-processed crystalline TCOs and evaporated metal contacts. Silver nanowires are solution-based and can be deposited under room temperature and ambient pressure conditions. The conductivity of a silver nanowire film is provided by the interconnection between single wires, while the transparency comes from the sparsity of the nanowire coverage. As in most transparent conductors, there is a trade-off between the conductivity and transparency of silver nanowire films. Our goal is to produce a film which provides sufficient conductivity for current collection in a solar cell while maintaining sufficient transparency for incident light collection.

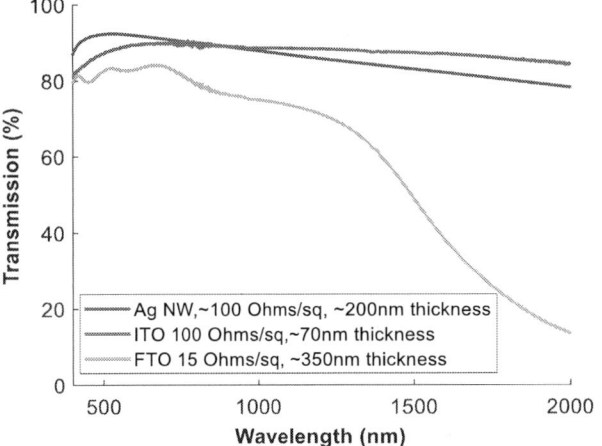

Fig. 3. Transmission curves for the best-performing spray-cast silver nanowire electrode and transparent conductive oxide control samples.

In this work, we use a silver nanowire solution provided by NanoDirect (5 mg/mL, ~20 nm diameter, ~30 μm length, dissolved in isopropanol) as our starting transparent conductor material. We find that diluting the solution to 0.5 mg/mL and annealing the film for 30 minutes after spray-casting leads to the best-performing films. The sheet resistance of the silver nanowire electrode is measured using a four-point-probe setup, and the transmittance is measured using a Cary 5000 UV-VIS-NIR spectrophotometer. Figure 3 illustrates that the best-performing spray-cast silver nanowire electrode provides transparency comparable to traditional transparent conductive oxides such as indium-doped tin oxide (ITO) and fluorine-doped tin oxide (FTO) in the visible range while maintaining an acceptable conductivity.

B. Absorbing layer optimization

After spray casting the ZnO electron transport layer [5], we moved on to developing spray-casting procedures for the PbS CQD absorbing layer. Oleic acid (OA) ligand capped PbS colloidal quantum dots are synthesized via previously published methods [6]. A solution-phase ligand exchange is performed using PbI_2, $PbBr_2$ and ammonium acetate in dimethylformamide (DMF) to get $[PbX_3]^-/[PbX]^+$-capped PbS colloidal quantum dots. This solution-phase ligand exchange increases the interconnection and decreases electronic trap state density in the CQD films. The ligand-exchanged quantum dots are then dispersed in butylamine for spray casting.

Fig. 4.   Effect of spray-casting pressure on the quantum dot absorbing layer film quality.

Figure 4 demonstrates the effect of carrier gas pressure in the spray casting process. As pressure decreases, droplets are less atomized leading to a very rough film surface. Therefore, high pressures (approximately 3 bar) are needed to ensure a uniform film and shorter drying time. We also find that a high solution concentration is needed to produce smooth, high-quality CQD films. However, islands begin to form at concentrations exceeding 60 mg/ml, likely due to quantum dot aggregation [7].

C. Discussion and ongoing work

We demonstrated spray-casting processes for the silver nanowire transparent conductor layer, the zinc oxide electron transport layer and the PbS CQD absorbing layer in a PbS CQD solar cell. The remaining tasks with regards to those materials include optimization of the film smoothness and integration into full solar cell devices. The next task has the goal of developing spray-casting processes for the CQD hole transport layer, in which deposition strategy development for the solid-state ligand exchange required for the hole transport layer will be the priority.

III. Summary

In this work, we demonstrate spray-casting processes for silver nanowire films to be used as transparent conductors in PbS CQD solar cells with an acceptable combination of electrical conductivity and optical transmittance, and the PbS CQD photovoltaic absorbing layer. This proof-of-principle demonstration should enable the flexible deployment of solar energy technologies on a variety of surfaces, especially in remote locations. A fully spray-cast CQD solar cell is under development and the processes developed for that will be scaled up for flexible solar cell manufacturing and other potential solution-processed optoelectronic technologies.

IV. Acknowledgements

We acknowledge the National Science Foundation (DMR-1807342, ECCS- 1846239) and the Maryland Energy Innovation Institute for funding.

References

[1] E. H. Sargent, "Colloidal quantum dot solar cells," *Nat. Photonics*, vol. 6, no. 3, Art. no. 3, Mar. 2012, doi: 10.1038/nphoton.2012.33.

[2] G. H. Carey, A. L. Abdelhady, Z. Ning, S. M. Thon, O. M. Bakr, and E. H. Sargent, "Colloidal Quantum Dot Solar Cells," *Chem. Rev.*, vol. 115, no. 23, pp. 12732–12763, Dec. 2015, doi: 10.1021/acs.chemrev.5b00063.

[3] E. H. Sargent, "Infrared photovoltaics made by solution processing," *Nat. Photonics*, vol. 3, no. 6, Art. no. 6, Jun. 2009, doi: 10.1038/nphoton.2009.89.

[4] I. J. Kramer *et al.*, "Efficient Spray-Coated Colloidal Quantum Dot Solar Cells," *Adv. Mater.*, vol. 27, no. 1, pp. 116–121, 2015, doi: 10.1002/adma.201403281.

[5] L. Li, B. Qiu, C. Lu, L. Shimabukuro, Y. Lin, and S. M. Thon, "Spray-Cast Electrodes in Colloidal Quantum Dot Solar Cells for Portable Solar Energy Manufacturing," in *2019 IEEE 46th Photovoltaic Specialists Conference (PVSC)*, Jun. 2019, pp. 0015–0018. doi: 10.1109/PVSC40753.2019.8980617.

[6] A. H. Ip *et al.*, "Hybrid passivated colloidal quantum dot solids," *Nat. Nanotechnol.*, vol. 7, no. 9, Art. no. 9, Sep. 2012, doi: 10.1038/nnano.2012.127.

[7] M.-J. Choi *et al.*, "Tuning Solute-Redistribution Dynamics for Scalable Fabrication of Colloidal Quantum-Dot Optoelectronics," *Adv. Mater.*, vol. 31, no. 32, p. 1805886, 2019.

Detection of PV Module Temperature Coefficient Using Machine Learning

John M. Obrecht, Julián Ascencio-Vásquez

Envision-Digital, Redwood City, CA, 94065, USA

Abstract — As the solar industry continues to mature, small improvements to performance modelling can provide increased visibility to production yields at solar plants, commercial & industrial (C&I) sites, and residential locations. Comparisons between performance models and production outputs are not only dependent upon sensor data, but also to system attributes, such as a PV module's temperature coefficient. The work presented here demonstrates a technique to detect individual temperature coefficients using a simple machine learning model. The results of technique are compared to a traditional surface-fitting technique and show consistency between the two methods, both in terms of accuracy and uncertainty quantification.

The results not only demonstrate the fact that a simple machine learning model can produce similar results to a well-accepted surface-fitting technique, but also show the power of the machine learning methodology for determining the sensitivity of a system to any given variable without limit to the complexity or to the number of degrees of freedom of the system – a powerful demonstration.

I. INTRODUCTION

Comparisons between performance models and production data is becoming ever more important as the solar industry continues to mature [1]. As models improve in accuracy, uncertainties become smaller, revealing previously undetected effects, which in turn can be modeled or studied, and eventually exploited or eliminated. This cycle of continuous improvement in solar performance modelling leads directly to a reduction in the levelized cost of energy to consumers – a win for the industry and the world in general.

One essential aspect of analyzing PV performance is the thermal behavior of PV modules [2], as temperature changes will trigger non-neglectable loss or gain, depending on the chosen technology and the specific geolocation [3]. This thermal behavior is characterized by the power-temperature coefficient, γ, expressed in %/°C: a parameter given by the PV module original equipment manufacturer (OEM) or one that can be inferred from on-site measurements or operational data [4, 5]. During the operational phase of a PV plant, the accurate quantification of γ will help to estimate the PV module temperature and physics-based degradation rates better [6], and at the same time, unleash new advanced analytics on PV module health by tracking the signature of specific degradation modes, like PID and hotspots [7].

The work presented here demonstrates a method of detecting a PV module's thermal sensitivity to power production using a simple machine learning (ML) model. This method is then compared to a traditional surface-fitting technique to benchmark the results in both the accuracy and uncertainty that are achievable with the technique. The approach presented hereby gives a new and effective angle on the usage of ML models on PV performance modelling.

Interestingly, one of the more powerful aspects of this method is not that it operates faster or more accurately than traditional surface fitting techniques (it doesn't), but rather that it easily scales to multi-variable functions, complex systems that would be nearly impossible to fit with analytic functions, or "black box" systems that may have unknown interdependencies. Scaling up by an additional degree of freedom simply requires the addition of one more feature (or target) column to the machine learning model.

The system that was chosen for study is a simple PV module whose AC power, P_{AC}, is recorded along with nearby weather station measurements of point-of-array (POA) irradiance and module temperature (T_{mod}).

The AC power can be described accurately with a simple function [8]:

$$P_{AC} = const \times POA \left(1 + \gamma(T_{CELL} - T_{REF})\right), \quad (1)$$

where the constant above is a mixture of the module's performance ratio, DC nameplate, and a reference irradiance value. The power is linearly dependent upon the POA irradiance, multiplied by a correction factor that depends upon the module's cell temperature.

II. APPROACH

For determining a PV module's temperature coefficient between 10-36 months of AC power production data is used, along with weather station data (POA irradiance and module temperature) from a weather station located in close proximity to the module. Data cleaning and filtering techniques (like the ones proposed in [9, 10]) are then applied to narrow down the data into that which will provide the clearest measurement.

The data is first cleared of stuck values, interpolated values, and non-physical data (by applying bounds to each data channel). A correction factor is applied to the POA irradiance that accounts for any measured differences in azimuthal orientation between the PV module and the weather station's POA irradiance sensor.

A clear-sky filter is then applied, which requires that rapidly changing data (in power, irradiance, and temperature) is removed, and that morning (evening) data is always increasing (decreasing) in value (for fixed-tilt systems only). The data is

978-1-6654-6060-6/23 $31.00 © 2023 IEEE

further filtered based on its operational status[1], allowing only 'normal operational' data to pass through. Finally, the data is filtered for irradiance (allowing values between 400-800 W/m²), and elevation angle (probably unnecessary).

A machine learning model is then trained to model the PV module's output AC power (*target* variable) as a function of its POA irradiance, and module temperature (*feature* variables), as shown in Fig. 1 below. A Random Forest Regressor (RFR) model [11] was chosen for its simplicity in its use, as well as its computational speed – a nice tradeoff for industrial analysis. The model's accuracy was not found to have a large sensitivity to the number of estimators in the RFR model, and a relatively small number of estimators (5) was chosen for speed purposes.

Once the ML model has been sufficiently trained, it is used to create two very similar sets of data, as shown in Fig. 1 below:

1. Modeled AC Power (variable temperature): created by applying the POA irradiance and module temperature training data to the model.
2. Modeled AC Power (fixed temperature): created by applying the POA irradiance training data and a fixed-reference temperature to the model.

The keen eye will notice that the only difference between these two sets of data is that the module temperature data is replaced by a fixed-reference temperature set of data (the reference temperature was fixed to 25°C here). This second set of data is, in essence, taking a slice through the data in a multi-dimensional parameter space. In this case, the slice can be viewed as a plane cutting a three-dimensional surface (shown as a color plot in Fig. 2).

Model Training

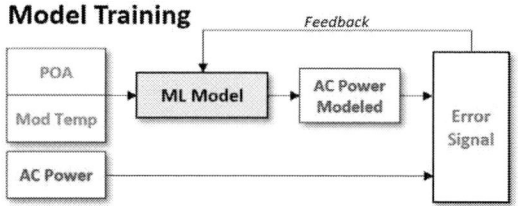

Model Evaluation

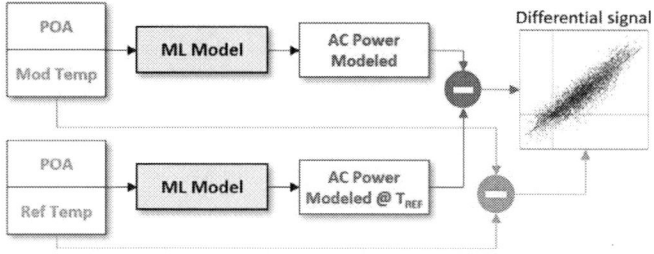

Fig. 1. Differential residual data is found by subtracting the outputs of the machine learning model from different sets of input: one that allows for variable module temperature and one that fixes the temperature to a reference value.

---

[1] As classified by algorithms in our proprietary software, EnOS Advanced Analytics for Solar.

The resulting set of data can then be analyzed to produce a single value for the module's power production sensitivity to temperature. A differential power, ΔP, can be calculated and scaled:

$$\Delta P = 100 * \left(\frac{P_{VAR}}{P_{FIX}} - 1\right) \tag{2}$$

and a differential temperature can be calculated as well:

$$\Delta T = T_{MOD} - T_{REF} \tag{3}$$

Figure 2 shows the measured AC power data as a function of irradiance (colored by module temperature). Overlaid on the data is the fixed-temperature modeled AC power (black line). The difference in power due to thermal effects is highlighted by a dashed arrow connecting to a single data point (orange point within the black circle). For a given irradiance, the temperature difference (in this case roughly 30°C) resulted in a lower power output by nearly 50 kW.

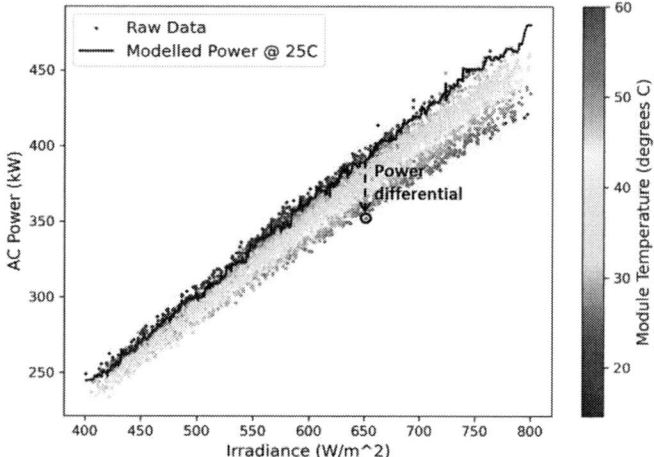

Fig. 2. AC Power data plotted as a function of POA irradiance, colored by module temperature. Overlaid on the data is the output of a machine learning model whose module temperature is evaluated at a reference temperature. The power differential is highlighted by an arrow for one exemplary point.

The comparison of the power difference data to the temperature difference data is shown in Fig. 3 below. A linear relationship is clearly visible, and standard fitting techniques can be applied to extract a slope (the PV module's thermal coefficient), along with its uncertainty (standard deviation or standard error).

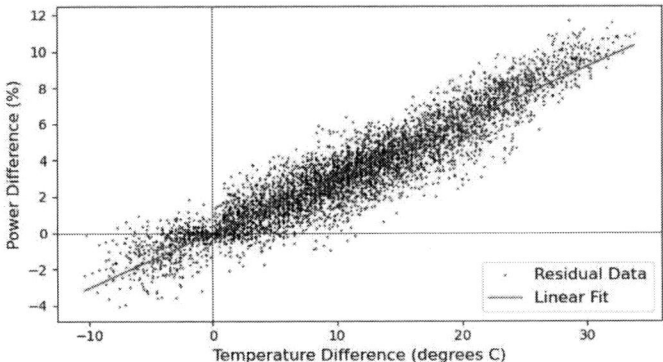

Fig. 3. Closer view of the differential data produced by analyzing the two sets of data outputs from the machine learning model. The slope of the differential data (shown as a solid red line) represents the PV module's temperature coefficient.

For comparison purposes, a simple first-order two-dimensional surface fit is also applied to the original dataset (AC power vs. irradiance and module temperature), according to Eq. 1. This surface fit produces a value for the PV module's thermal coefficient as well. Bootstrap analysis can also be employed to extract a standard deviation (or standard error) in the fit.

Both calculated values (and uncertainties) for the PV module's temperature coefficient can then be compared to known (OEM) values, as well as compared to one another across an entire farm or fleet.

Mathematically, this technique can be seen by looking at the analytic form of the power equation (Eq. 1), as it represents the trained machine learning model. The AC power with variable cell temperature is written,

$$P_{VAR} = const \times POA \left(1 + \gamma (T_{CELL} - T_{REF})\right), \quad (4)$$

While the AC power with fixed cell temperature is written,

$$P_{FIX} = const \times POA, \quad (5)$$

with the latter term in Eq. 1 cancelling to zero.

With some minor algebra, one can then express the relative power differential as

$$\frac{P_{VAR} - P_{FIX}}{P_{FIX}} = \frac{P_{VAR}}{P_{FIX}} - 1 = \gamma (T_{CELL} - T_{REF}), \quad (6)$$

The above equation[2] can then be compared to Eq. 2 & 3, to produce the following linear relationship found in Fig. 3:

$$\Delta P = \gamma \Delta T \quad (7)$$

---

[2] Note that the cell temperature and module temperature are very nearly identical, differing only by a small factor that is dependent upon the POA irradiance.

III. RESULTS

The results of the analysis are shown below in Fig. 4 for 91 inverters at six different sites (denoted by color). The ML model results (uncertainties) are shown with open circles (error bars), while the surface fit results (uncertainties) are shown with a dashed line (solid fill). In most, if not all, cases the results are indistinguishably similar between the two methods, except for the uncertainties at the fourth site (magenta color). In addition, it is easy to see that all the results are consistent with, and slightly lower in magnitude than the OEM values, as well as consistent with one another within a given site.

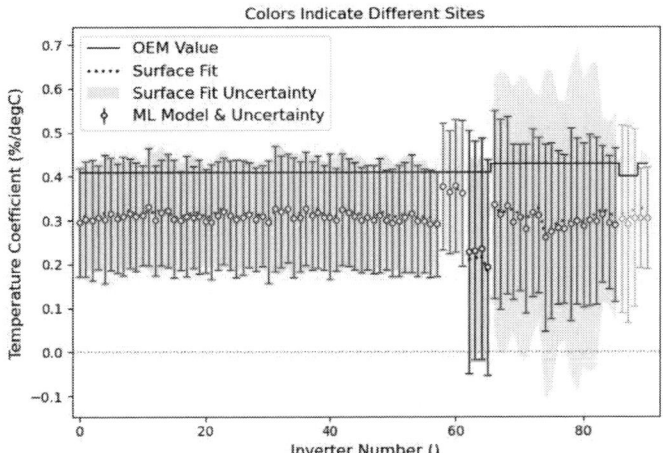

Fig. 4. Results for over 90 inverters from six different sites. The results show consistency between both fitting methods.

The computational run time for both methods was also investigated and showed that the surface-fit method was over 100 times faster than the machine learning method. This result was not unexpected, as the surface fitting technique is able to take advantage of the linear algebra construction of the problem, while the machine learning method requires both a training period before computing predicted values, as well as a first-order polynomial fit to the differential signal data.

Additionally, several challenges were observed in working with the data. First, the accuracy of the method was very sensitive to soiling behavior. Soiling losses are challenging to quantify, let alone accurately correct for. As a result, any sites that had significant soiling behavior required a deeper investigation.

Similarly, sites that had large temporal variations in performance ratio (PR) also required deeper investigation. The data could have been scaled by the site's recent PR value, but this step was not taken in this analysis.

Lastly, it is assumed that the irradiance and module temperature that is recorded by a nearby weather station is accurately representing these quantities for that module, when sometimes this assumption may not be entirely valid. Many systematic effects may cause discrepancies between what is measured and what would have been measured if each PV module had its own attached sensors. However, this type of investigation is beyond the scope of this work, and we assume here, with some confidence, that these quantities are being accurately recorded.

IV. CONCLUSION

We have shown a new and novel method for measuring the temperature sensitivity of a variety of PV modules across several different sites. The machine learning model was described in detail and showed results consistent with a standard surface-fitting technique in both measured value and uncertainties. These results demonstrated that the ML model method is a viable option for not only simple two-parameter sensitivity analysis, like that demonstrated here, but also for systems with higher degrees of freedom, or higher levels of complexity.

REFERENCES

[1] Uncertainty in Yield Assessments and PV LCOE, Report IEA-PVPS T13-18:2020, IEA PVPS Task 13 Performance, Operation and Reliability of Photovoltaic Systems.

[2] A. Driesse, M. Theristis and J. S. Stein, "PV Module Operating Temperature Model Equivalence and Parameter Translation," 2022 IEEE 49th Photovoltaics Specialists Conference (PVSC), Philadelphia, PA, USA, 2022, pp. 0172-0177, doi: 10.1109/PVSC48317.2022.9938895.

[3] H. Liu, C. D. Rodríguez-Gallegos, Z. Liu, T. Buonassisi, T. Reindl and I. M. Peters, "Worldwide theoretical comparison of outdoor potential for various silicon-based tandem module architectures," 2021 IEEE 48th Photovoltaic Specialists Conference (PVSC), Fort Lauderdale, FL, USA, 2021, pp. 0377-0381, doi: 10.1109/PVSC43889.2021.9518780.

[4] A.A. Belsky, D.Y. Glukhanich, M.J. Carrizosa, V.V. Starshaia, Analysis of specifications of solar photovoltaic panels, Renewable and Sustainable Energy Reviews, 159, 2022, https://doi.org/10.1016/j.rser.2022.112239.

[5] Mitterhofer, Stefan, Glažar, Boštjan, Jankovec, Marko And Topič, Marko, 2019, The development of thermal coefficients of photovoltaic devices. Informacije MIDEM. časopis za mikroelektroniko, elektronske sestavne dele in materiale. 2019. Vol. 49, no. 4, p. 219–227. DOI 10.33180/InfMIDEM2019.404

[6] I. Kaaya, J. Ascencio-Vásquez, K.A. Weiss, M. Topič, Assessment of uncertainties and variations in PV modules degradation rates and lifetime predictions using physical models, Solar Energy, Volume 218, 2021, Pages 354-367, ISSN 0038-092X, https://doi.org/10.1016/j.solener.2021.01.071.

[7] Segbefia, O.K.; Sætre, T.O. Investigation of the Temperature Sensitivity of 20-Years Old Field-Aged Photovoltaic Panels Affected by Potential Induced Degradation. Energies 2022, 15, 3865. https://doi.org/10.3390/en15113865

[8] A. P. Dobos, "PVWatts Version 5 Manual" http://pvwatts.nrel.gov/downloads/pvwattsv5.pdf (2014).

[9] Lindig, S, Moser, D, Curran, AJ, et al. International collaboration framework for the calculation of performance loss rates: Data quality, benchmarks, and trends (towards a uniform methodology). Prog Photovolt Res Appl. 2021; 29: 573– 602. https://doi.org/10.1002/pip.3397

[10] A. Livera, M. Theristis, E. Koumpli, G. Makrides, J. S. Stein, and G. E. Georghiou, "Guidelines for ensuring data quality for photovoltaic system performance assessment and monitoring," in 37th European Photovoltaic Solar Energy Conference (EU PVSEC), 2020, pp. 1352– 1356, doi: 10.4229/EUPVSEC20202020-5DO.2.4.

[11] Breiman, "Random Forests", Machine Learning, 45(1), 5-32, 2001.

Widegap CdSe solar cells with V_{OC} >750mV

Taylor Hill[1,2], Sachit Grover[2], James Sites[1]

1- Colorado State University, Fort Collins, Colorado, 80521, United Sates
2- First Solar, Santa Clara, California, 95050, United States

Abstract — With growing demand for solar energy and as single junction photovoltaic devices approach the SQ limit it is becoming increasingly important for the solar industry to enable multijunction tandems for terrestrial application. While there are numerous promising candidate technologies that can serve as the bottom cell with a bandgap of 1.1 eV, there are no at-scale options that can satisfactorily serve as the top-cell with a bandgap 1.7 eV. Cadmium Selenide (CdSe), with a direct bandgap of 1.72 eV, is a strong candidate material for top-cells in a tandem solar panel. We evaluate the applicability of CdSe thin-film on TCO coated glass as a photovoltaic absorber and find that it is inherently n-doped and wurtzite. The device architecture comprises a CdSe layer deposited onto SnO2:F that was annealed similar to CdTe for improving lifetime, followed by a hole contact material made with PTAA that works as a hole selective layer and a MoO3/Au bilayer that functions as a transparent, high work function contact layer. A bandgap of 1.72 eV was confirmed using PL and UV-Vis data, with Hall and C-V measurements revealing majority carrier concentrations of 10^{16} cm^{-3} and time resolved photoluminescence suggesting lifetimes close to 5ns. Based on the combination of $V_{OC,Ideal}$ extracted from sub-gap EQE and external radiative efficiency measurements, we estimate the absorber to be capable of achieving 1.2V V_{OC}. This suggests significant room for research in developing effective hole-selective contact for CdSe.

I. INTRODUCTION

Silicon (Si) photovoltaics has been the main driver of solar panel technology for the past century, with steady research allowing power conversion efficiencies (PCEs) to move from ~13% in the 70's to ~26% today, with modules up to ~25%. With such mature technology, Si dominates the global market, currently accounting for ~95% of the solar market [1]. However, detailed balance calculations indicate a limit of ~30% PCE for Si [2]. Meanwhile, demand for solar energy only continues to increase, with many countries aiming to have Net Zero Emission by 2050 [1]. One solution for overcoming this efficiency barrier is the development of tandem photovoltaics. Much like the detailed balance limit conducted by Shockley and Queisser, De Vos carried out a similar calculation for n-junction devices. For n=2, the pairing of 1.7 eV to 1.1 eV was found to be able to convert energy at an efficiency >40% [3], shown in fig. 1. This pairing of bandgaps (E_g) can take advantage of Si technology as a bottom cell, only requiring the development of a 1.7eV E_g top cell.

Of the known semiconductors with E_g = 1.7 eV, the binary chalcogenide CdSe stands out. Given the demonstrated success in industrializing binary chalcogenides for single junctions, it is quite reasonable to investigate this materials system for applications in a tandem setting. The closely related binary

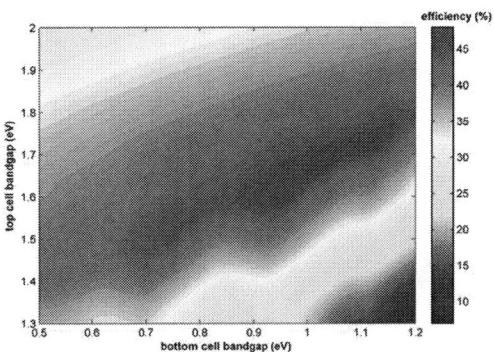

Fig. 1.   Efficiency contours for 2 cell tandem devices showing a peak >40% for the pairing of 1.1eV bottom cell to 1.7eV top cell. [4]

chalcogenide CdTe accounts for most of the solar market not taken by Si [5]. The highest efficiency CdTe cells already make use of CdSe as a tertiary alloy ($CdSe_xTe_{(1-x)}$) to grade the absorber bandgap and increase the amount of light which can be absorbed. When evaluating the bowing diagram of $CdSe_xTe_{(1-x)}$, we see that pure CdSe has the requisite 1.7eV bandgap [6]. Therefore, by removing the Te content from already commercialized $CdSe_xTe_{(1-x)}$, CdSe absorbers can easily be commercialized and integrated into tandem devices with Si, provided highly efficient CdSe single junctions can be realized. This gives merit to investigating CdSe as a potential top cell [7] – [10].

In this paper we describe the absorber qualities of CdSe films and then examine an example device with open-circuit voltage for CdSe greater than 750 mV. Absorber quality was examined by means of granularity through optical microscopy, photoluminescence (PL) before and after annealing process, and with general materials properties of carrier type, concentration, mobility, and lifetime. Device results are indicated by analysis of current density vs voltage (J-V) curves, capacitance vs voltage (C-V) curves, and external quantum efficiency (EQE) curves. Additional device metrology is performed through Kelvin Probe (KP) microscopy to determine certain layer work functions as well as the surface photovoltage (SPV) of devices through film side illumination up to 1 sun intensity. Following this evaluation of the absorber and devices based on it, we will consider the voltage impact on this absorber from sub-gap luminescence coupled with external radiative efficiency (ERE).

II. CdSe Absorber

CdSe films approximately 1.2 μm thick CdSe films were deposited using vapor transport deposition (VTD). These films were analyzed by photoluminescence (PL) with a 640nm laser excitation to determine their primary bandgap at 1.72 eV, Optical microscopy at 100x magnification shows the surface to have very low granularity, so a $CdCl_2$ heat treatment is performed much like is done in the typical CdTe technology [11]. Following this treatment, we see an increase in grain sizes up to 5μm, and an increase in peak PL by ~2 orders of magnitude, as shown in fig. 2.

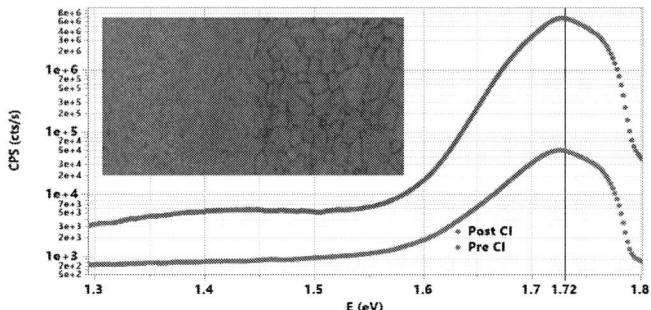

Fig. 2. PL curves before and after Cl treatment. Main peak shows a bandgap of 1.72eV, with an increase in peak height by two orders of magnitude following Cl treatment. Inlay shows an increase of granularity up to 5 μm following Cl treatment as well.

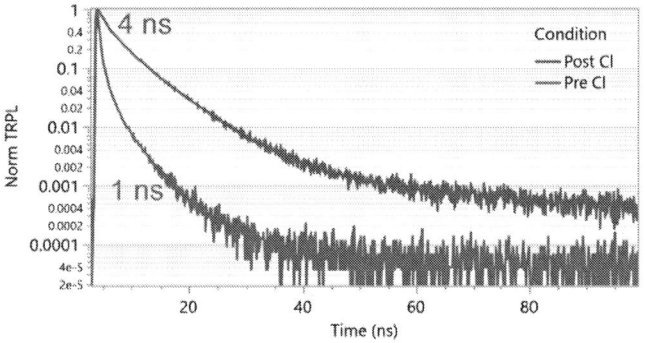

Fig. 3. TRPL curves before and after Cl treatment. Single exponential fit performed between 0 and 20ns demonstrates an increase in carrier lifetime from 1ns to 4ns for the shown device. Lifetimes >5ns have been measured on other devices.

The Cl-treated films were analyzed by Hall measurement to determine carrier type, concentration, and mobility. The CdSe films are n-type with an electron mobility 10-100 cm^2 / V*s at a concentration of 10^{16} cm^{-3}. This carrier concentration was also confirmed by C-V measurements taken in the dark at room temperature. Time resolved photoluminescence (TRPL) was used to determine the lifetime of carriers, which shows an increase from ~1ns to 4ns following Cl treatment, shown in fig. 3, with lifetimes greater than 5ns also having been measured.

UV-Vis spectroscopy was performed on CdSe absorbers, demonstrating absorption of 80% for photon energies above the bandgap and transmission of above 60% of photon energies below the bandgap out to 1.1 eV, which could be available for use by a Si bottom cell, as shown in fig. 4. This indicates CdSe is a strong candidate material for thin-film tandem top cells.

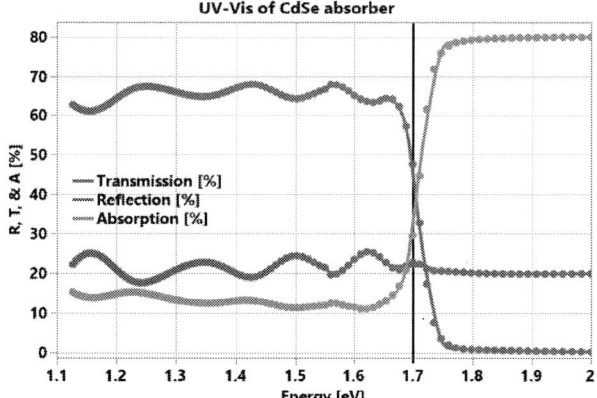

Fig. 4. UV-Vis spectroscopy of CdSe absorber shows a sharp increase in absorption at 1.7 eV, correlating with the main peak in PL emission. >60% of photons with E<E$_g$ are transmitted, which can then be available for absorption within a 1.1eV bottom cell in tandem structures.

III. Devices Based On N-type CdSe

In order to produce a high-efficiency tandem device, both the top and bottom cell themselves must be optimized. Therefore, we have explored devices based on single junction CdSe to determine the limits and possibilities for this absorber material for a tandem top cell.

As the CdSe films are n-type, a p-type widegap semiconductor with a minimized valance band offset is needed for making a successful junction. The ionization potential of wurtzite CdSe is 6.5 eV, and there is a general lack of materials that satisfy the valence band alignment criteria while remaining conductive, transparent, and p-type [12]. To circumvent this limitation, we design junctions with organic hole transport layers (HTLs) and high work function contact materials that serve as a proxy for deep valance band position. Organic semiconductors (OSCs) allow a large flexibility in experimentation due to their easy application and can be quickly modified through various solution additives. High workfunction transition metal oxides (TMOs) have been well studied as hole selective contacts in Si and various organic photovoltaic platforms [13] – [14].

Polytriaryl amine (PTAA) is a well-studied hole-transporting organic semiconductor (OSC) with reported hole mobilities of 10^{-5} cm^2/V*s and conductivity of 10^{-6} S/cm with HOMO/LUMO levels at 5.2/2.2 eV [15]. Although 5.2eV is a large HOMO level for OSCs, a valance band offset greater than 1 eV remains when contacting CdSe. To compensate for this, a bilayer contact comprising a high work function (WF) transition metal oxide (TMO) (referring to molybdenum oxide (MoO$_3$)) and a thin, semi-transparent gold (Au) film is used as a bilayer electrode on top of PTAA.

978-1-6654-6060-6/23 $31.00 © 2023 IEEE

IV. CdSe Devices Results

We now examine the outcome of variations in device architecture to determine if PTAA can operate as a sufficient p-type heterojunction with CdSe, and how to compensate for the large valance-band offset by use of high workfunction MoO_3. In general, devices were developed in a superstrate configuration, with SnO_2:F as the substrate which CdSe is grown on. Superstrate configuration was selected due to the difficulty in growing CdSe absorbers on top of the organic p-type heterojunction without degradation. This problem could be addressed by development of a proper inorganic p-type heterojunction, but that is outside the scope of this paper. PTAA, MoO_3, and Au films were deposited on CdSe through spin coating and thermal evaporation respectively. Given that the n-p junction exists at the front of the device, J-V measurements are performed through the film side, using a home-built J-V system with a xenon arc bulb in a Newport 67005 lamp housing and thin Accuprobe kelvin probes for contact. The system is controlled through LabView, which records the data and calculates the main J-V parameters.

PTAA is first evaluated without the presence of a high WF TMO to determine if it can form a proper heterojunction with CdSe. Concentrations between 1.25 and 10 mg/mL in chlorobenzene were examined, with the largest concentration of 10mg/mL showing the best performance with dark diode-like behavior. From this, we see that PTAA can act as a proper organic heterojunction partner with CdSe, likely since the HOMO level, while largely offset form the CdSe valance band, resides close to the CdSe fermi level. This allows PTAA to act as a hole selective buffer layer, conducting holes while reflecting electrons due to the barrier formed from the large (>2eV) conduction band offset (CBO). However, PTAA remains quite resistive, generating only 4.3 mA/cm^2 J$_{sc}$. The V$_{OC}$ of these devices is also low at only 110 mV, which is also reflected in Kelvin-Probe SPV of only 110 mV at 1 sun intensity. These results are shown in fig. 5 and 6 along with subsequent architectures.

Although PTAA shows proper n-p junction formation with CdSe, we want to evaluate the performance of a novel architecture which makes use of pure diffusion properties without the usual drift field associated with photocurrent. In this scheme, PTAA is completely removed and instead n-type, high work function MoO_3 is used along with a metallic contact of Au.

MoO_3 is thermally evaporated to a thickness of 5 nm, followed by evaporation of 10 nm of Au using a shadow mask, without breaking vacuum. Kelvin Probe microscopy was used to measure the work function of the portions of the MoO_3 which were covered by the shadow mask, and thus without any Au. MoO_3 films showed an average work function of 5.8 eV. These devices demonstrated a much larger J$_{sc}$ of 10.7 mV, demonstrating the benefit of a high workfunction TMO to extract charges in lieu of any p-type heterojunction. An increased V$_{OC}$ of 429 mV was recorded, also reflected in Kelvin-probe SPV of 430 mV at one sun intensity. These results

are shown in fig. 5 and 6. An increased fill factor from 22% for the PTAA/Au device to 49% for the MoO_3/Au device is likely due to the increased conductivity of the MoO_3/Au bilayer as measured by four-point probe (4PP). PTAA/Au films were at the limit of the 4PP system, saturating at 10^{10} $\Omega/\square$, while the MoO_3/Au films measured ~40 $\Omega/\square$. Although this novel device architecture was able to increase device performance, this system lacks any barrier to electrons at the CdSe/MoO_3 interface, likely allowing recombination at the surface.

In order to make the best of both worlds, PTAA as a hole selective heterojunction is first deposited with spin coating and a subsequent anneal, followed then by thermal evaporation of 5 nm MoO_3 and 10 nm of Au. In this structure, electrons are reflected by the large CBO between CdSe and PTAA, preferentially conducting holes, which can then be extracted through the high work-function, high conductivity bilayer of MoO_3 and Au. These devices showed an increased J$_{SC}$ of 11.5 mA/cm^2 and achieved a fill factor of 51%. Although KP SPV shows only 710 mV at 1 sun intensity, J-V revealed a V$_{OC}$ of 752 mV, as shown in fig. 5 and 6.

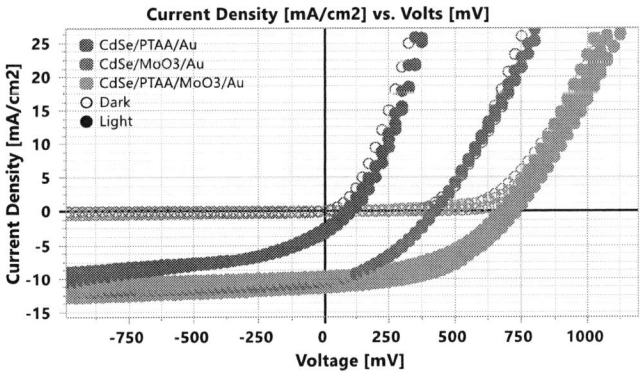

Fig. 5. JV curves of devices detailed within this manuscript. Champion devices of CdSe/PTAA/MoO_3/Au achieved J$_{SC}$ = 11.52 mA/cm^2, V$_{OC}$ = 752mV, and fill factor = 51%.

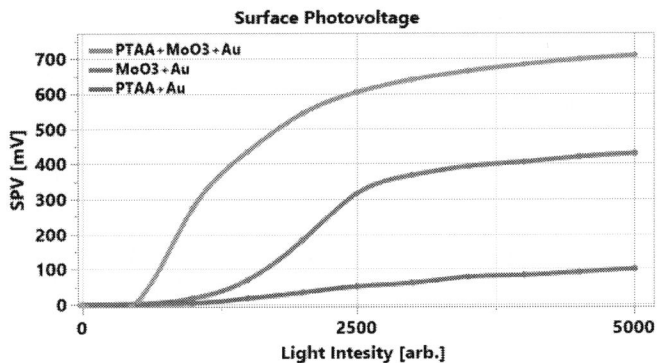

Fig. 6. Surface photovoltage for the three devices shown in fig. 4. Light intensity of 5000 above corresponds to ~1 sun intensity. The champion device shows a SPV value of 710 mV at 1 sun intensity, slightly lower than the actual V$_{OC}$ recorded.

IV. CELL VOLTAGE

While >750 mV V_{OC} has been achieved, which is slightly over half of the SQ limit for a 1.72 eV absorber, we must ask how to go beyond this level. We first note the presence of a sub-gap peak in PL, as seen in fig. 2 following Cl treatment. It is known that the presence of sub-gap peaks in PL can be indicative of trap states within the band gap which can be detrimental to V_{OC} [16]. To evaluate the impact this sub-gap peak could have on our V_{OC}, high sensitivity sub-gap EQE is done and coupled with filtered ERE data.

High sensitivity sub-gap EQE is performed by illuminating the device with light of energy 1.8 eV to 1.1 eV using a monochronometer, with current from the sample recorded through a preamplifier and a lock-in amplifier to reduce the noise level down to 1nA, which translates to an absorbance of 10^{-5} % for these samples. Sub-gap EQE measurements capture the band tail decay, with the slope of this band tail known as the Urbach edge energy (E_U). The Urbach energy is well known to be an indicator of the crystalline quality of semiconductor absorbers, with a smaller E_U, and thus more exponential decay of the band edge, to be indicative of the potential for high performance. E_U <20 mV has been identified as a critical parameter for high efficiency devices [17]. Along with this band edge, as light of $E<E_g$ is used to excite the sample, any recorded amperage can be assigned trap states which work to reduce the overall V_{OC} of a device [18].

Filtered ERE is performed by biasing the sample with a 640 nm pulsed laser to inject carriers, and a one sun light bias to reflect operating conditions. Emission from the sample is recorded by a detector above the sample. A 695-nm long-pass filter is used to capture the emission from the main peak and any sub gap features, and an 800-nm long-pass filter is used to remove the contribution from the main peak and only capture sub gap features.

Devices with the champion structure of CdSe/PTAA/MoO$_3$/Au were measured using sub-gap EQE with results shown in fig. 6. An exponential fit of the band edge gives $E_U = 14 \pm 2$ mV for a variety of cells and samples, indicating that CdSe has a high crystalline quality and $V_{OC} \sim$ 1.4V could be achieved, given the proper heterojunction and contact materials. However, trap states were not able to be properly recorded with the current instrumentation. The samples signal fell below the noise floor near 1.6 eV, and thus any trap signature which may be available at lower energies could not be deconvoluted from the background noise. This indicates that if we are limited by any trap emission from sub-gap states, their emission is not strong enough to be recorded at a signal level >1nA.

The ideal voltage for this absorber can be determined from the methods detailed by Onno et al. [18] by examining the ideal V_{OC}

$$V_{OC,Ideal} = \frac{k_B T}{q} ln\left(\frac{\int \alpha(\lambda)\phi_{exc}(\lambda)d\lambda}{\int \alpha(\lambda)\phi_{BB}(\lambda,T)d\lambda} + 1\right) \quad (1)$$

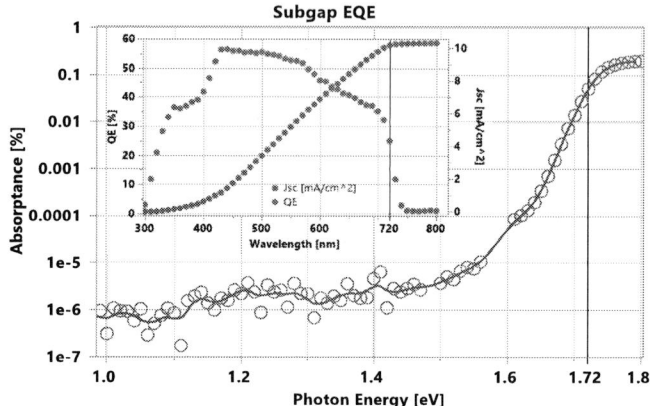

Fig. 7. Sub-gap EQE shows a sharp band edge with $E_U \sim$ 12mV, indicative of high absorber quality. Device signal fell below the noise floor, corresponding to $\sim 10^{-5}$ absorptance, near 1.6 eV. Inset graph shows full EQE spectrum with an integrated $J_{sc} \sim$ 11 mA/cm^2.

where k_B is the Boltzmann constant, T is the device temperature, q is the elementary charge, $\alpha(\lambda)$ is the samples absorptance, ϕ_{exc} is the excitation photon current, and ϕ_{BB} is the blackbody photon current. Given the low E_U of these samples, this calculation yields $\sim$1.4V using the sub-gap EQE absorptance shown in fig. 6.

Filtered ERE measurements done on this same device show that contribution from sub-gap emission accounts for less than 20% of the full spectrum ERE, as shown in table 1 where two long-pass filters are compared.

Table 1 – comparing filtered ERE		
ERE 695nm LP [%]	ERE 800nm LP [%]	ERE 800/695 [%]
0.28	0.05	18

Considering the full spectrum ERE at 0.28%, we can calculate the implied V_{OC} as follows

$$iV_{OC} = V_{OC,Ideal} - \frac{k_B T}{q}|ln(ERE)| \quad (2)$$

The second term in this equation, which is the reduction from $V_{OC,Ideal}$, is $\sim$150 mV. Combing this with (1), we calculate an implied V_{OC} of 1.25 V for this absorber. This results in a 60% selectivity based on a V_{OC} of 750 mV, indicating that the resistivity to one type of charge carrier is limited at either the front or back junction. This implied V_{OC} can also be calculated from carrier concentration as well, where

$$iV_{OC} = \frac{k_B T}{q} ln\left(\frac{n_e n_h}{n_i^2}\right) \quad (3)$$

Taking n_e equal to the carrier concentration measured in C-V, 10^{16}, with n_h equal to $5*10^{12}$ and n_i equal to 10^4, we find this value to be 1.2V, providing a confirmation of the iV_{OC} calculated through sub-gap EQE and ERE.

V. NEXT STEPS & CONCLUSIONS

While a $V_{OC} > 750$ mV has been measured, there is indication that $V_{OC} > 1$ V is achievable with this absorber. In order to achieve this, however, a heterojunction material is needed with CdSe which has a reduced valence-band offset with respect to CdSe and greater conductivity than the currently used PTAA. PTAA itself can be modified through dopants added to the solution prior to spin coating, which can enhance the conductivity of the deposited layer. Additionally, different PTAA chemistries, such as the fluorinated species 1F-PTAA & 2F-PTAA, have been shown to have decreased HOMO levels, which have led to increased V_{OC} in perovskite technologies [19]. Additional work can also be done to modify the contact layer between CdSe/PTAA such as with functionalization of the CdSe surface through self-assembled monolayers to modulate the workfunction at the surface [20].

While this paper is focused on the V_{OC}, there are obvious losses to J_{SC} due to reflection from the metallic contact and absorption losses due to the large bandgaps of both MoO_3 and PTAA, as seen in the dip in QE near 400nm in fig. 6. Switching from thin, semi-transparent Au to a proper transparent conductive oxide such as ITO or AZO will be an important next step. In addition to improving the J_{SC} and V_{OC} through these changes, the identification of an inorganic heterojunction material for CdSe could also enable the development of a substrate device, where illumination comes from the glass side.

CdSe remains an attractive candidate material for tandem top cells. Here we have demonstrated the fabrication of CdSe solar cells with open circuit voltages >750mV through use of PTAA as a hole transporting organic interface material, molybdenum oxide as a high work function p-type material, and thin gold as a semitransparent high work function contact. This work establishes a platform which needs an increased understanding of fundamental CdSe materials properties as well as identification of highly transparent and conductive p-type semiconductors to enable high efficiency CdSe.

ACKNOWLEDGEMENTS

Partial funding was provided by the National Science Foundation Industry/University Cooperative Research Center (I/UCRC) for Solar Powered Future and its Industrial Advisory Board under award number 2052735. Additional thanks to First Solar's California Technical Center for additional research support. Additional thanks to Dr. Angus Rockett of Colorado School of Mines for discussions on the project.

REFERENCES

[1] P. Bojek, "Solar PV – analysis," IEA, Sep-2022. Available: https://www.iea.org/reports/solar-pv.

[2] B. Ehrler, E. Alarcón-Lladó, S. W. Tabernig, T. Veeken, E. C. Garnett, and A. Polman, "Photovoltaics Reaching for the Shockley–Queisser Limit," ACS Energy Letters, vol. 5, no. 9. American Chemical Society (ACS), pp. 3029–3033, Sep. 01, 2020. doi: 10.1021/acsenergylett.0c01790.

[3] D. Vos, "Detailed balance limit of the efficiency of tandem solar cells," Journal of Physics D: Applied Physics, vol. 13, no. 5. IOP Publishing, pp. 839–846, May 14, 1980.

[4] S. P. Bremner, Levy, and Honsberg, "Tandem Cells," PVEducation. [Online]. Available: https://www.pveducation.org/pvcdrom/tandem-cells.

[5] "Global Cadmium Telluride Solar Cell (CdTe) market – industry trends and forecast to 2029," Cadmium Telluride Solar Cell (CDTE) Market Share, Future Analysis and Industry Size. Available: https://www.databridgemarketresearch.com/reports/global-cadmiumtelluride-solar-cell-cdte-market.

[6] J. Yang and S. H. Wei, "First-principles study of the band gap tuning and doping control in CdSexTe1-x alloy for high efficiency solar cell" Chinease Physics B, vol. 28, no 8. TOP publishing, p. 086106, Aug. 01, 2019

[7] J. D. Friedl, R. H. Ahangharnejhad, A. B. Phillips, and M. J. Heben, "Material Requirements for CdSe Wide Bandgap Solar Cells," 2021 IEEE 48th Photovoltaic Specialists Conference (PVSC). IEEE, Jun. 20, 2021

[8] K. Li et al., "Fabrication and Optimization of CdSe Solar Cells for Possible Top Cell of Silicon-Based Tandem Devices," Advanced Energy Materials, vol. 12, no. 26. Wiley, p. 2200725, May 30, 2022.

[9] B. Bagheri, "Research project to study cadmium selenide (CdSe) solar cells." Iowa State University. doi: 10.31274/etd-20200624-44.

[10] Jeedigunta, Sathyaharish, "Development Of Cadmium Selenide As An Absorber Layer For Tandem Solar Cells" (2004). USF Tampa Graduate Theses and Dissertations.

[11] Major, J., Al Turkestani, M., Bowen, L. et al. In-depth analysis of chloride treatments for thin-film CdTe solar cells. Nat Commun 7, 13231 (2016). https://doi.org/10.1038/ncomms13231

[12] N. Fioretti and M. Morales-Masis, "Bridging the p-type transparent conductive materials gap: synthesis approaches for disperse valence band materials," Journal of Photonics for Energy, vol. 10, no. 04. SPIE-Intl Soc Optical Eng, p. 1, Feb. 03, 2020.

[13] Greiner, M., Lu, ZH. Thin-film metal oxides in organic semiconductor devices: their electronic structures, work functions and interfaces. NPG Asia Mater 5, e55 (2013). https://doi.org/10.1038/am.2013.29

[14] M. Bivour, J. Temmler, H. Steinkemper, and M. Hermle, "Molybdenum and tungsten oxide: High work function wide band gap contact materials for hole selective contacts of silicon solar cells," Solar Energy Materials and Solar Cells, vol. 142. Elsevier BV, pp. 34–41, Nov. 2015. doi: 10.1016/j.solmat.2015.05.031.

[15] F. M. Rombach, S. A. Haque, and T. J. Macdonald, "Lessons learned from spiro-OMeTAD and PTAA in perovskite solar cells," Energy and Environmental Science, vol. 14, no. 10. Royal Society of Chemistry (RSC), pp. 5161–5190, 2021.

[16] J. Wong, S. T. Omelchenko, and H. A. Atwater, "Impact of Semiconductor Band Tails and Band Filling on Photovoltaic Efficiency Limits," ACS Energy Letters, vol. 6, no. 1. American Chemical Society (ACS), pp. 52–57, Dec. 02, 2020.

[17] J. Chantana, Y. Kawano, T. Nishimura, A. Mavlonov, and T. Minemoto, "Impact of Urbach energy on open-circuit voltage deficit of thin-film solar cells," Solar Energy

Materials and Solar Cells, vol. 210. Elsevier BV, p. 110502, Jun. 2020. doi: 10.1016/j.solmat.2020.110502.

[18] Onno, A., Reich, C., Li, S. et al. Understanding what limits the voltage of polycrystalline CdSeTe solar cells. Nat Energy 7, 400–408 (2022). https://doi.org/10.1038/s41560-022-00985-z

[19] Y. Kim, E. H. Jung, G. Kim, D. Kim, B. J. Kim, and J. Seo, "Sequentially Fluorinated PTAA Polymers for Enhancing V OC of High-Performance Perovskite Solar Cells," Advanced Energy Materials, vol. 8, no. 29. Wiley, p. 1801668, Sep. 14, 2018. doi: 10.1002/aenm.201801668.

[20] J. Hu, W. Fu, X. Yang, and H. Chen, "Self‐assembled monolayers for interface engineering in polymer solar cells," Journal of Polymer Science, vol. 60, no. 15. Wiley, pp. 2175‐2190, Jan. 31, 2022. doi: 10.1002/pol.20210938.

Eliminating the Need for Handling Individual Sub-Cells for Small Appliance PV Modules with Voltage Demands Above 12V

Jan Paschen, Andreas Brand, Matheus Melati Menegassi, Oliver John, Jan Nekarda

Fraunhofer Institute for Solar Energy Systems, Freiburg, Germany

Abstract — There is a high demand for small photovoltaic (PV) modules that can provide 12V or more to operate electronic devices with low power consumption, such as parking meters, street signs, and remote Internet of Things (IoT). However, state-of-the-art PV modules either require several step-up converters to achieve the necessary voltage or are put together from small sub-cells, requiring a significant handling and interconnection overhead during manufacturing.

The technology described in this work allows to interconnect small rear contact solar cells before their segregation, eliminating the need for the handling and interconnection of individual sub-cells. This is demonstrated by creating modules of thirty-six sub-cells from a single monolithic wafer. The resulting modules showed an open circuit voltage of over 24V and voltage at maximum power point of just under 20V while reaching a power of about 4.4W.

The novel Al-foil interconnection technique utilizes laser welding and Aluminum (Al) to substitute environmentally harmful materials such as solder including lead and solder-forming fluxes. Furthermore, it significantly reduces the use of silver, as large segments of the electrodes such as busbars or soldering pads are no longer necessary.

I. INTRODUCTION

Photovoltaics (PV) are gaining widespread adoption due to their cost-effectiveness in generating green-electricity. The improvements in solar cell technology resulting from this have made them a suitable option for a variety of niche applications, leading to new opportunities for their use.

One niche application is the use of PV to power small devices with low energy requirements such as in parking meters, street signs and other remote IoT objects. This can lead to significant savings, especially for remote devices that are not cost-effective to connect to the electricity grid. However, the required voltage makes it difficult to operate these devices directly with PV without further modification. Therefore, there is great demand for small modules that deliver voltages above 12V, to power these devices. Previous solutions for modules with a small number of large cells require step-up conversion [1], which is either expensive or inefficient. A practical alternative is to fabricate modules with long series circuits of small cells, but this involves a huge handling effort since each sub-cell must be connected individually.

Historically, the challenge of handling has been addressed by similar concepts that facilitate the interconnection of sub-cells on a monolithic wafer. An example of such a design is the "HighVo" concept [2]. In this concept, the sub-cells were isolated from each other using isolation cuts, but small ridges were left intact to ensure the integrity of the wafer. However,

there were persistent problems with this concept. The non-insulated ridges caused partial shunting of cells in the string, a problem that was particularly noticeable under low light conditions [3]. In addition, the wafers were mechanically weakened by the cuts, potentially complicating handling procedures. Although mechanical stability was somewhat restored by filling the cut planes with screen printing paste [4], the concept had inherent limitations. The concept proposed in this paper addresses these shortcomings by eliminating the need for ridges, thus completely isolating the wafers of each sub-cell from each other. It also removes the need for the wafer to maintain the integrity of the assembly, with aluminum (Al) foil performing this role instead.

The goal of the technology presented in this abstract is to connect rear contact solar cells using laser-welded and flexible Al-foil cell joints. The ability of Al-foil to conform to various shapes and the availability of both electrodes on the rear side of the cell enables the wiring process to be completed prior to the segregation of the sub-cells (See Fig. 1).

Not only does this eliminate the need for handling individual PV sub-cells, but it also substitutes environmentally detrimental materials, such as solder and cell connectors. Since large segments of the electrodes such as busbars or soldering pads are no longer necessary it also reduces the use of silver significantly.

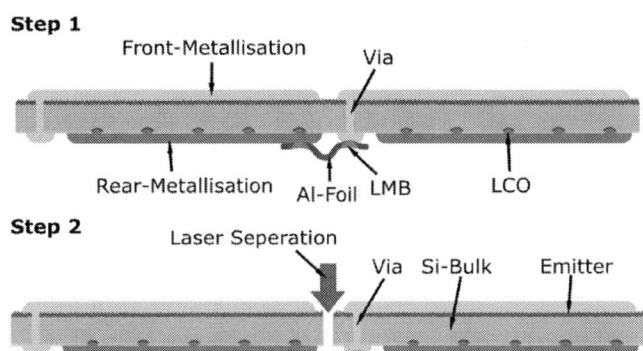

Fig. 1: Schematic illustration of the process steps for implementing the presented interconnection concept for MWT solar cells. Front and rear side metallization and arrangement of electrodes at the edges of adjacent sub-cells Step 1: The neighboring sub-cells are connected by an Al-foil. Step 2: Separation of the host cell using a laser. The sub-cells remain connected by the Al-foil and can be handled as an assembly.

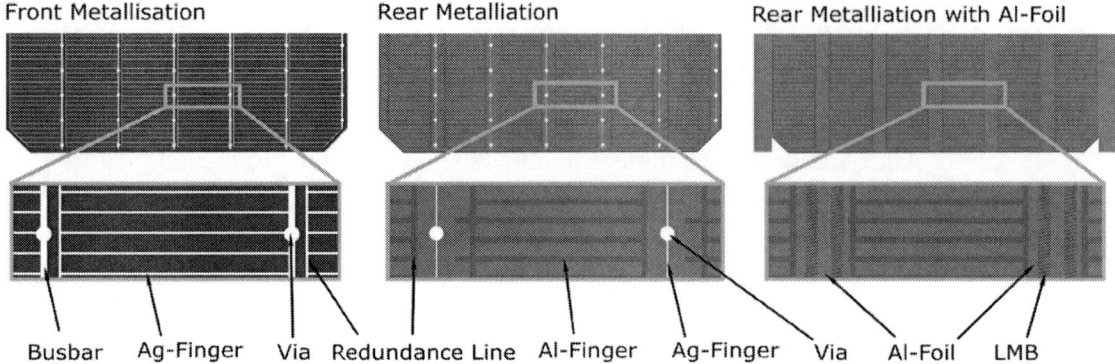

Front Metallisation   Rear Metalliation   Rear Metalliation with Al-Foil

Busbar   Ag-Finger   Via   Redundance Line   Al-Finger   Ag-Finger   Via   Al-Foil   LMB

Fig. 2: Schematic illustration of front and rear side layout and connection using Al-foil for MWT solar cells. (Left) Front side layout with 1.5mm finger spacing, a busbar connects the front side electrode to the rear side through vias, and a redundance line. (Middle) Rear side layout with screen printed Al fingers at 1mm pitch. The vias are connected by a continuous Ag finger. (Right) Illustration of the application of the Al-foil to the rear and its connection to the cell and electrodes by LMB.

II. TECHNOLOGY, DEVICE AND ASSEMBLY

The interconnection concept presented is applicable to most back-contact solar cell technologies. In this work, Metal Wrap Through (MWT) solar cells [5] are used because they are particularly suitable for rapid prototyping. However, it is unlikely that MWT cells will be used in industrial production. A more plausible candidate might be an Interdigitated Backcontact Cell (IBC) as described in [6, 7].

To implement this concept, both the front and rear side metallization must be chosen in such a way that the host cell can be divided into sub-cells. It is also necessary that the two electrodes are present at the edge of adjacent sub-cells (Fig. 1 Step 1). The adjacent electrodes of the sub-cells can then be joined together with an Al-foil. All these process steps can be performed on the non-separated host cell.

In the second step, the host cell is separated (Fig. 1 Step 2). As the sub-cells are still connected by the Al-foil, no handling of the individual sub-cells is necessary and they can always be moved as an assembly.

To enable series connection of the solar cells, it is essential that the MWT vias are placed on the edge following the current path. Fig. 2 shows three schematic drawings illustrating different aspects of the layout and interconnection. The first drawing shows the layout for the front side, where the fingers are spaced at 1.5 mm. All fingers have a redundancy line at one end and a busbar at the other, connecting the front side electrode to the rear side by vias. In the rear-side layout shown in the second drawing, the fingers of screen-printed Al have a spacing of 1mm. A continuous Ag finger is used to connect the vias, allowing effective electrical contact with the Al-foil using the previously published Laser Metal Bond (LMB) process [8].

The third drawing illustrates the application of the Al-foil to the rear of the solar cell and its connection to the cell and electrodes via LMB.

The interconnection concept presented not only enables the connection of cells in one dimension, but also more complex interconnections in two dimensions. To demonstrate this technology, we applied it in a proof-of-concept experiment on Passivated Emitter and Rear Cells (PERC) [9] with an edge length of 158.75 mm. Each host-cell is prepared to be divided into the number of sub-cells required for the voltage demand. The number of columns and rows should be chosen to create sub-cells that are as square as possible to minimize the edge-to-area ratio [10]. In this experiment, we chose a 6x6 design with

Fig. 3. This photograph shows the solar cell string composed of thirty-six sub-cells connected using laser-welded Al-foil. The Al-foil connectors forming a continuous circuit without the need for additional materials such as adhesive or solder. A multimeter attached across the string shows a V_{OC} of 16.69V. The photograph was taken indoor under light conditions much lower than sunlight and therefore the measurement of the multimeter differ from the calibrated measurements shown in Table 1.

978-1-6654-6060-6/23 $31.00 © 2023 IEEE

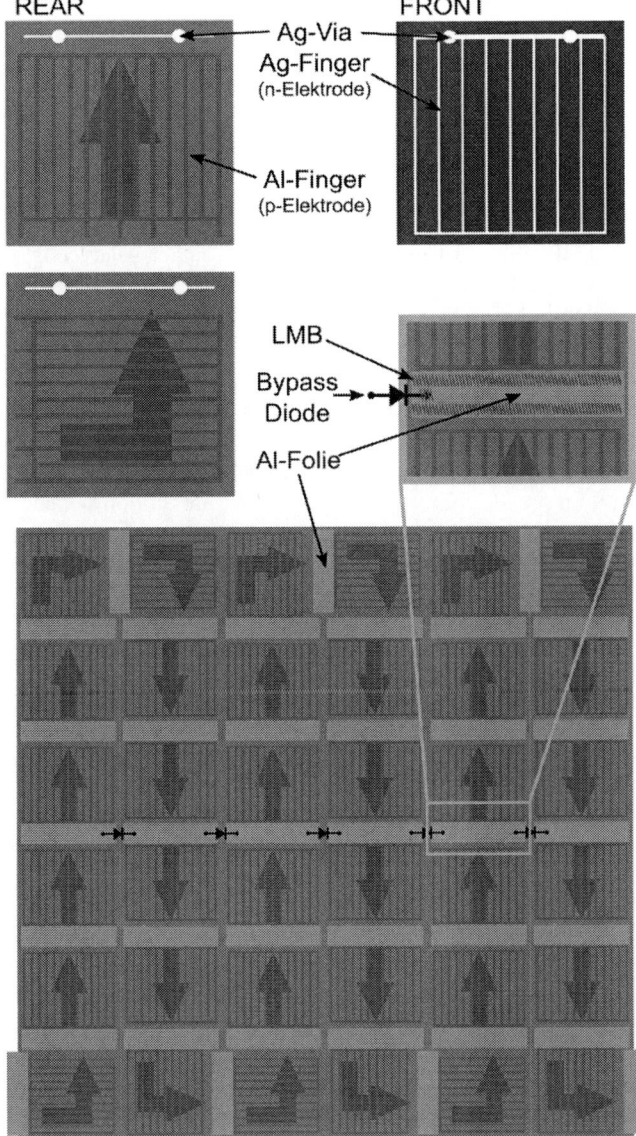

36 cells resulting in square cells with an edge length of 26.458 mm (See Fig. 3 and 4).

The front-side electrode is routed to the rear-side pads via the MWT process. To enable series interconnection, the MWT-vias are placed on the side following the current path. Since the current path runs in a serpentine pattern over the host cell, there are sub-cells in which the current path makes a 90° turn. This requires the rear-side metallization to be rotated 90° with respect to the front-side metallization. In this way, the n- and p-electrodes of the neighboring cells are always adjacent to each other in the series connection.

Bypass diodes are usually installed in parallel with the solar cells or cell strings within the solar module. The meandering course of the series connection makes it possible to place the bypass diodes in the middle of the straight sections, as shown in Fig. 4. If these diodes are placed on the undivided host cell before the Al-foil is applied, the connections can be wired together with the cell in one step without the need for an additional processing step.

Utilizing a nanosecond laser, a trench is scribed between the sub-cells to prepare for later breaking. However, the samples are not yet broken at this stage. The trench is located on the rear-side to prevent additional laser damage at the p-n junction.

In the following step, the sub-cells are interconnected by welding a 12μm thin Al-foil on the n-electrode of one cell and the p-electrode of the adjacent cell. The wafer is then divided along the established breaking points, though the sub-cells are still held in position by the Al-foil joints.

Separation of the sub-cells using Thermal Laser Separation (TLS) has not yet been successful since the Al-foil prevents the necessary movement [11].

The string is now fully manufactured and at no point do the individual sub-cells need to be manipulated, with handling being limited exclusively to the host cell.

Fig. 4. Schematic illustration of the rear side of a solar cell string consisting of 36 cells interconnected with Al-foil. Top right: Front layout of a sub-cell. Left: Rear layouts of the sub-cells with and without change of direction indicated by dark blue arrows. Bottom: Overall backside metallization layout for the whole cell. The gray Al-foil patches are welded on the interfaces between two adjacent cells, providing both a mechanical and electrical connection to connect all sub-cells in a meandering series connection of 36 cells. The arrows indicate the current path connecting all sub-cells. Right: Enlargement of an Al-foil connector showing the LMB layout. The joining process utilizes an IR laser, using only Al and laser light without the need for additional materials such as adhesive or solder.

III. RESULTS

The strings are now moved to an IV measurement station and measured. The results of the IV measurement are shown in Table 1. Both prototypes exceeded the required voltage of 12V with an open circuit voltage of over 24V and a voltage at Maximum Power Point (MPP) of just under 20V. Following the lamination of the strings, the modules display negligible difference in voltage. However, there is a slight reduction in current, which can be attributed to the introduction of the module glass. This glass causes a fraction of incident light to be reflected, reducing light absorption. Refer to Mittag [12] for a detailed cell-to-module loss analysis. This capability enables the modules to provide the necessary voltage even under low light conditions, making them an excellent choice for powering a variety of loads. When the MPP voltage is divided by the number of cells, the average voltage per sub-cell is approximately 547mV.

TABLE I
IV-Measurement of 36 Cell Sized Devices

Id	I_{sc} [mA]	V_{oc} [V]	I_{mpp} [mA]	V_{mpp} [V]	FF [%]	P_{mpp} [W]	Eta [%]
			String before lamination				
1	273.7	24.41	233.0	19.69	68,7	4.59	17,90
2	274.4	24.37	235.2	19.72	69,4	4.64	18,02
			Module after lamination				
1	262.8	24.36	225.3	19.70	69.3	4.44	17.31[*]
2	261.3	24.36	223.6	19.65	69.0	4.39	17.07[*]

[*] Calibrated on string size

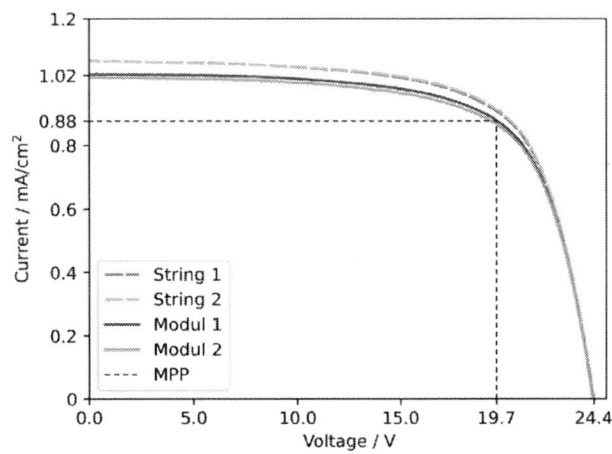

interconnection does not significantly elevate the series resistance.

Overall, the modules have a power output of around 4.4W, corresponding to a power density of around 18mW/cm² (See Fig 5).

VI. SUMMARY AND OUTLOOK

In this study, a technology is presented that utilizes laser-welded, flexible Al-foil to connect rear-contact solar cells in small appliance modules to achieve a voltage above 12V. The technology aims to provide a cost-effective and environmentally friendly alternative to traditional methods, such as step-up conversion and the use of solder in cell connection. The technology was demonstrated in a proof-of-concept experiment using PERCs, which were divided into thirty-six sub-cells resulting in strings with voltages at the maximum power point just under 20V on one Silicon wafer.

One notable aspect of this technology is that the separation of the smaller cells occurs only after they have been interconnected. This eliminates the need for handling individual cells, as they always remain connected as a unit. This represents a significant improvement over existing methods that require handling of individual cells, which can be time-consuming and labor-intensive. The use of laser-welded Al-foil to connect the cells has been proven to be a reliable and efficient method [13].

In the future, this technology will also be applied to Interdigitated Back Contact (IBC) cells [7] to further expand its capabilities and move this technology to a more modern vehicle than MWT cells. An additional improvement will be achieved by TLS as a separation process, which will reduce the recombination of charge carriers at the edges of the sub-cells. However, the Al-foil must be modified to allow for the necessary flexibility during separation.

This technology can help to increase the efficiency and reliability of solar-powered small appliances and might therefore contribute to the reduction of CO2 emissions and the fight against climate change.

The short-circuit current (I_{SC}) of around 262mA and the MPP current (I_{MPP}) of around 225mA fall within the expected range. When normalized to the area of a single sub-cell, the values for short-circuit current density (j_{SC}) and MPP current density (j_{MPP}) are approximately 37.4mA/cm² and 32.1mA/cm², respectively.

The fill factor (FF), which stands at approximately 69%, denotes the small size of the sub-cells and the absence of edge passivation. Nevertheless, it also demonstrates that the

References

[1] A. Kalirasu and S. Dash, "Simulation of closed loop controlled boost converter for solar installation," *Serb. J. Electr. Eng.*, vol. 7, no. 1, pp. 121–130, 2010.

[2] S. Keller, "Solarzellenanordnung," Germany.

[3] S. Keller, S. Scheibenstock, P. Fath, G. Willeke, and E. Bucher, "Theoretical and experimental behavior of monolithically integrated crystalline silicon solar cells," (en), *Journal of Applied Physics*, vol. 87, no. 3, pp. 1556–1563, 2000.

[4] S. Keller, M. Wagner, S. Scheibenstock, W. Jooss, A. Kress, G. Hahn, P. Fath, and E. Bucher, "Simple mini module fabrication schemes for high voltage silicon solar cells," in *16th European Photovoltaic Solar Energy Conference*, Glasgow, UK, 2000, pp. 1218–1221.

[5] E. Lohmuller, B. Thaidigsmann, M. Pospischil, U. Jager, S. Mack, J. Specht, J. Nekarda, M. Retzlaff, A. Krieg, F. Clement, A. Wolf, D. Biro, and R. Preu, "20% Efficient Passivated Large-Area Metal Wrap Through Solar Cells on Boron-Doped Cz Silicon," *IEEE Electron Device Lett.*, vol. 32, no. 12, pp. 1719–1721, 2011.

[6] R. Kopecek, F. Buchholz, V. D. Mihailetchi, J. Libal, J. Lossen, N. Chen, H. Chu, C. Peter, T. Timofte, A. Halm, Y. Guo, X. Qu, X. Wu, J. Gao, and P. Dong, "Interdigitated Back Contact Technology as Final Evolution for Industrial Crystalline Single-Junction Silicon Solar Cell," *Solar*, vol. 3, no. 1, pp. 1–14, 2023.

[7] R. Kopecek, J. Libal, J. Lossen, V. D. Mihailetchi, H. Chu, C. Peter, F. Buchholz, E. Wefringhaus, A. Halm, J. Ma, L. Jianda, G. Yonggang, Q. Xiaoyong, W. Xiang, and D. Peng, "ZEBRA technology: low cost bifacial IBC solar cells in mass production with efficiency exceeding 23.5%," in *2020 47th IEEE Photovoltaic Specialists Conference (PVSC)*, Calgary, AB, Canada, 2020, pp. 1008–1012.

[8] O. John, J. Paschen, A. de Rose, B. Steinhauser, G. Emanuel, A. A. Brand, and J. Nekarda, "Laser Metal Bonding (LMB) - low impact joining of thin aluminum foil to silicon and silicon nitride surfaces," *Procedia CIRP*, vol. 94, pp. 863–868, https://www.sciencedirect.com/science/article/pii/s2212827120312968, 2020.

[9] A. W. Blakers, A. Wang, A. M. Milne, J. Zhao, and M. A. Green, "22.8% efficient silicon solar cell," (en), *Appl. Phys. Lett.*, vol. 55, no. 13, pp. 1363–1365, 1989.

[10] M. Hermle, J. Dicker, W. Warta, S. W. Glunz, and G. Willeke, "Analysis of edge recombination for high-efficiency solar cells at low illumination densities," *3rd World Conference onPhotovoltaic Energy Conversion*, 2003.

[11] M. Oswald, M. Turek, J. Schneider, and S. Schoenfelder, "Evaluation of Silicon Solar Cell Separation Techniques for Advanced Module Concepts," *28th European Photovoltaic Solar Energy Conference and Exhibition, Paris, France*, pp. 1807–1812, 2013.

[12] M. Mittag, T. Zech, M. Wiese, D. Blasi, M. Ebert, and H. Wirth, "Cell-to-Module (CTM) Analysis for Photovoltaic Modules with Shingled Solar Cells," in *2017 IEEE 44th Photovoltaic Specialist Conference (PVSC)*, 2017.

[13] J. Paschen, P. Baliozian, O. John, E. Lohmüller, T. Rößler, and J. Nekarda, "FoilMet ® ‐ Interconnect: Busbarless, electrically conductive adhesive ‐ free, and solder ‐ free aluminum interconnection for modules with shingled solar cells," (en), *Progress in Photovoltaics*, vol. 30, no. 8, pp. 889–898, 2022.

Setting Priorities for Photovoltaic Reliability Research Using Criticality Analysis

Ingrid L. Repins, Michael G. Deceglie, Timothy J. Silverman, David C. Miller, Dirk C. Jordan, Michael Woodhouse, Teresa M. Barnes

National Renewable Energy Laboratory, Golden, CO, United States

A forward-looking research opportunity number (RON) is defined for photovoltaic (PV) reliability researchers. The RON enables researchers to prioritize their efforts toward the highest impact. For a given degradation mode, the RON is based on the worst-case impact on levelized cost of electricity (LCOE), the fraction of future module products that could be susceptible, and the status of accelerated tests that can detect and quantify the mode. Reporting bias is avoided, because the RON does not rely on polls. The RON is derived for three example cases: light and elevated temperature degradation (LETID), backsheet cracking, and anti-reflective (AR) coating abrasion. These examples demonstrate that targeted research has reduced the risk for these modes through the last several years.

Uncertainties in PV power simulation chain

Lubos Helienek, Jozef Rusnak, Branislav Schnierer, Martin Opatovsky, Lukas Dvonc, Vicente Lara Fanego,
Artur Skoczek, Tomas Cebecauer

Solargis s. r. o., Bratislava, Slovakia

Abstract — Considering the situation in the global renewable energy industry, photovoltaic (PV) power plants are becoming an increasingly significant player. In addition to the long-term PV system yield prediction, it is necessary to determine the uncertainty related to each step in the PV simulation chain, which directly impacts financial parameters of the project. This is an essential input in the feasibility studies during development, and performance evaluation during operational stage of the PV power plant. Currently, the industry deals with the analysis of uncertainty on a yearly basis. However, there is a lack of knowledge focused on uncertainty determination with monthly, daily, or hourly time resolution, needed in monitoring or fault-detection applications. This paper is concerned with the theoretical analysis of uncertainties determined in the actual operating conditions for each block of the PV simulation chain from the input solar irradiation resource to the electrical power output at grid connection point.

I. Introduction

The mathematical model of the conversion process from solar irradiance input (Global Tilted Irradiance/Irradiation – GTI) to photovoltaic electrical power output (PVOUT) consists of several modelling steps (the total number of the steps depends on the complexity and the purpose of the model), creating the simulation chain. The model used in the PV simulation presented in this article consists of the following parts, which further break down into simulation steps:

a) Optical simulation models – sky model, 3D calculation scene, ray-tracing, irradiance attenuation due to soiling and snow, angular reflections at the surface of the modules, and spectral correction model

b) Single diode model for conversion of irradiance to electricity

c) Electrical simulation models – calculation of DC circuit losses, inverter performance, auxiliary losses, transformer losses, AC circuit losses, and losses due to system unavailability.

Each of the simulation steps can be modelled as a block with input and output. Each block injects uncertainty into the overall model, regardless of how accurately its parameters are configured. Furthermore, the uncertainty associated with each block is dependent on its operating conditions. Therefore, when quantifying the uncertainties injected into the simulation solution by individual simulation blocks, it is necessary to analyze each simulation block separately. Knowing the uncertainties of the input values such as GTI in the optical part or current in the electrical part and the uncertainties injected

from the individual simulation blocks, it is possible to quantify the uncertainties associated with the losses in each block.

Conceptually, uncertainty can be categorized as either parameter uncertainty or model uncertainty [1]. Analysis of PV simulation uncertainties can be defined as a process consisting of:

1. Quantification for all uncertainties of all inputs, modelling steps and their respective parameters

2. Quantification of uncertainties of the simulation model itself; and

3. Combination of uncertainties to determine the overall uncertainty of the final result.

The calculation of the total uncertainty is a complex process which includes uncertainties of different time character. For example, the uncertainty of the GTI is different to the angular coefficient uncertainty, values of which depend on dust. Therefore, in determining the uncertainties it is necessary to make simplifications and assumptions that will allow us to arrive at a practicable result.

Two main assumptions have been made in the research presented in this article. The first assumption is that the uncertainty injected by a single simulation block has systematic character, it cannot be compensated over time like GTI, and its value depends on the value of input parameters. The second assumption is that all uncertainties have a normal distribution.

Considering the block-wise analysis of uncertainty and the assumptions stated above, the final uncertainty in the output power for the given time step will be composed of uncertainty of the input to the simulation chain (UNC_{GTI}) and uncertainties injected by each individual block (UNC_{BLOCKS})

$$UNC_{\text{PVOUT}} = \sqrt{UNC_{\text{GTI}}^2 + \sum UNC_{\text{BLOCKS}}^2} \qquad (1)$$

II. Analysis

Input irradiance data

Main input into the optical simulation part is GTI, calculated by numerical models, which are parameterized by a set of inputs characterizing the cloud transmittance, state of the atmosphere, and terrain conditions. An overview of the satellite model and related uncertainties are discussed in [2] and [3]. For the purpose of simplification in this article, input data uncertainty covers all factors and models included in satellite irradiance modelling, and it is represented by monthly estimates.

1. Sky irradiance model

The Perez all-weather sky model [4] also called Anisotropic model is used for determination of radiance distribution on the sky dome, using insolation conditions. Anisotropic model enables more detailed simulation of diffuse shadows and rear side irradiation cast on PV modules. This model is used for calculation of the diffuse contribution to tilted irradiance.

The core of the model is based on a large number of long-term high-quality experimental sky scans in various geographical and climatic conditions, and the time resolution of each scan is 15 minutes. Thus, the model accounts for most of the main predictable effects, but unpredictable changes such as random clouds pattern are not modelled yet. Mean deviation depends on sky conditions and PV module orientation, and can be as high as 30 % (see Tab. 2 in [4]). Sky irradiance model uncertainty is included in GTI uncertainty.

2. 3D calculation scene and ray-tracing

3D scene is used for mathematical representation of vertices, edges, and faces to simulate all objects involved in the simulation. Horizon shading is a part of the 3D scene and has a large influence on the uncertainty. It depends on the complexity of terrain, quality of data, and model resolution. Due to the complexity of this parameter, its uncertainty is represented by a conservatively estimated constant.

Ray-tracing models the behavior of rays taking into consideration the 3D scene and real physical principles. The algorithm has been tested against field measured data, and standards like the Radiance software [5]. The primary sources of uncertainty come from inputs and the ray-tracing itself does not generate significant uncertainty.

3. Irradiance attenuation due to soiling

The soiling model is based on that of Coello and Boyle [6]. However, it uses a parameterization configured in a particular way for the determination of the dry deposition velocities, which makes the obtained values of soiling ratio vary compared to other studies where this model has been evaluated [6], [7], [8]. Uncertainty of the model is the result of the contribution of input variables, natural cleaning effects (uncertainty and effectiveness), manual cleaning events, and uncertainty of observational soiling data used for validation. Furthermore, validation results also show that the level of uncertainty presents dependence on the geographical location, such that it would require regionalization.

4. Irradiance attenuation due to snow

The estimation of magnitude of snow losses and its uncertainty is not a trivial task. This is due to limited availability of precise on-site snow measurements and complex behavior of snow cover on the PV modules which is driven by the local environmental conditions (e.g. fresh snow fall, air temperature, humidity, wind speed, and solar irradiation), as well as the PV system configuration (such as the tilt and tracking type) and maintenance. Additionally, snow events exhibit high seasonal and inter-annual·variability.

The snow model is quite a complex part of simulation chain which is still in development, for that reason the uncertainty is defined as a constant so far.

5. Angular correction

To account for the losses due to angular reflectivity on the surface of PV modules, the model by Martin and Ruiz is used [9]. The magnitude of the resultant effect depends on the relative position of the sun, plane of the module, and the value of angular coefficient a_r (representing the technology of PV modules - cell material, module surface, spectral reflectance properties and surface cleanliness) ranging from 0.17 to 0.27 depending on the cleanliness of the panel surface. The factors with the highest influence on angular reflectivity losses are site position (latitude), sun position, and the module tilt angle [10].

Based on the values of these input parameters and the uncertainty of the angular coefficient a_r, the uncertainty of the angular correction coefficient can be quantified for each component of the irradiation (normal, diffuse, and reflected radiation) separately (Fig. 1). The uncertainty injected by angular block into the simulation can then be estimated by equation (2).

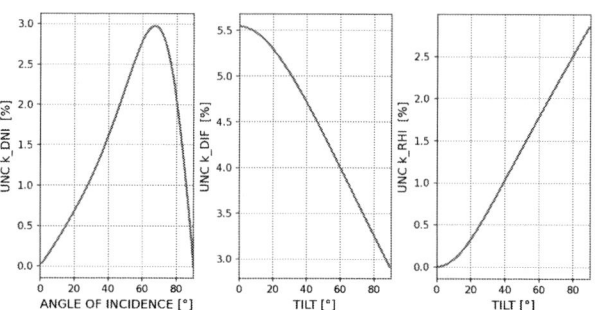

Fig. 1. Uncertainty of the angular correction coefficients for beam, diffuse and reflected parts of irradiation depending on angle of incidence and tilt (assuming that $UNC_{AR} = 0.05$ and $UNC_{TILT} = 5°$, $UNC_\alpha = 1°$).

$$UNC_{AOI} = \sqrt{\begin{array}{l}\left(DNI \cdot UNC_{k_{DNI}}\right)^2 + \left(DIF \cdot UNC_{k_{DIF}}\right)^2 + \\ + \left(RHI \cdot UNC_{k\_RHI}\right)^2 + (UNC_{Model})^2\end{array}} \quad (2)$$

6. Spectral correction

The model corrects for changes in the sunlight spectrum due to absolute air mass (AM), and precipitable water (PWAT) content. The air mass is determined based on the sun altitude and elevation of the site. The lowest uncertainty in the AM calculation is at the zenith and gradually increases towards the horizon. If we consider the AM calculation for 15-min data granularity with a moderately steep terrain and sun altitude close to the horizon, the uncertainty of the AM calculation is around 10 %. Considering PWAT, according to [11] the uncertainty of satellite model data with respect to ground measurements can be in the range of 2 % to 8 %.

The spectral correction uses the mathematical model defined in [12], determined based on a sensitivity analysis in the SMARTS model. The knowledge of the air mass, precipitable water uncertainty, and model uncertainty enables the estimation of the uncertainty of spectral losses for each combination of input parameters as given by equation (3) and plotted in Fig. 2.

$$UNC_{SPECTRAL} = \sqrt{\left(\frac{\partial f}{\partial PWAT}\right)^2 UNC_{PWAT}^2 + \left(\frac{\partial f}{\partial AM}\right)^2 UNC_{AM}^2 + UNC_{MODEL}^2} \quad (3)$$

It should be noted that the spectral correction model parameters were set only for cloud index > 0.7 and GTI > 200 W/m², and the model was validated only on 3 sites, where the highest uncertainty for monocrystalline silicon cells was 1.26 %. Thus, more validation data is needed to evaluate the model performance in different conditions.

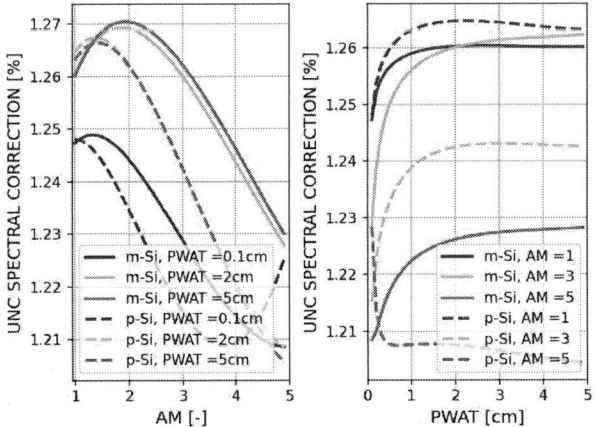

Fig. 2 Modelling of Spectral correction coefficient uncertainty depending on values of AM (left) and PWAT parameters assuming that UNC_{PWAT} = 8 %, UNC_{AM} = 10 %, UNC_{MODEL} = 1.20 %.

7. Conversion from solar energy to electrical energy

At this stage of the simulation chain irradiance is converted into electrical energy. Based on cell temperature and irradiance, the model reconstructs the I-V curve for each cell. The simulation uses the single diode model validated in [13]. The overall uncertainty is caused by the estimation of individual model parameters such as the light current, diode reverse saturation current, diode factor, and series and shunt resistances. Validation demonstrated that the uncertainty of the model can be as high as 5 % at lower values of irradiance when the PV module still produces useful power. With increasing irradiance the uncertainty decreases.

The second parameter which introduces uncertainty into the energy conversion calculation is the temperature coefficient, whose uncertainty can be around 10 % - 15 % of its value [14]. If the operating temperature of the panels deviates from the STC temperature at which the panels were validated, the uncertainty increases.

Another critical factor contributing to the uncertainty of this conversion step is the temperature of the silicon cell. A temperature difference of 5°C introduces an uncertainty of about 2.3% into the transferred power [15-17]. The final value of the conversion uncertainty is estimated as follows:

$$UNC_{CONVERSION} = \sqrt{UNC_{MODEL}^2 + UNC_{TEMP\_COEFF}^2 + UNC_{TEMP\_CELL}^2} \quad (4)$$

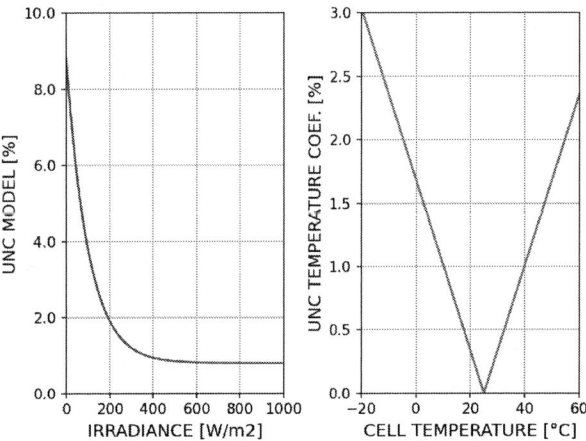

Fig. 3. Uncertainty of conversion depending on irradiance (left) and cell temperature (right).

8. DC circuit losses

A simple model based on Joule-Lenz law for heating of a conductor due to losses caused by current flow is used. However, for simplicity other parameters are neglected in the equations including temperature dependency, mechanical dimensions of the cables, material constants, and actual working conditions impacting the uncertainty of the simulation. Furthermore insufficient input information about DC wiring parameters, cabling layout, and manufacturing variation introduces uncertainty into the calculation itself. DC losses are defined by formula (4)

$$DC_{LOSSES} = 2 * R_{REF} * I^2 * [\alpha * (T_{CABLE} - T_{REF}) + 1] \quad (4)$$

Assuming that input variables in equation (4) can vary in ranges:

- temperature of cable T_{CABLE} = (20°C, 90°C)
- temperature coefficient $\alpha = \alpha_{CALC} \pm 5\%$
- reference resistance $R_{REF} = R_{REFCALC} \pm 10\%$

it is possible to calculate the maximum and minimum losses for any load. The difference between the maximum and minimum DC losses can reach 45 % of the reference value. (Fig. 4, left chart)

9. Inverter performance (DC to AC conversion)

Efficiency is a key parameter that defines the quality of an inverter. The manufacturers' datasheets usually state the efficiency as a single number, but for a PV simulation complete efficiency curves are required. For this reason the SANDIA inverter model [18] is used to derive the inverter efficiency curves.

For the DC to AC conversion model, two uncertainties are expected:

A. Model uncertainty based on [18], estimated at 0.1 %; and

B. Uncertainty of the component itself defined in IEC 61683 [19].

According to [19] the maximum tolerance for the efficiencies η guaranteed by manufactures (which is equivalent to component uncertainty) are defined by formula (5).

$$UNC_{\text{COMPONENT}} = 0.2 \cdot (1 - \eta) \cdot \eta \ [\%] \qquad (5)$$

Overall uncertainty injected to transferred power by the inverter only is estimated as per equation (6)

$$UNC_{\text{INVERTER}} = \sqrt{UNC_{\text{MODEL}}^2 + UNC_{\text{COMPONENT}}^2} \qquad (6)$$

Another effect that can occur in the inverter is clipping, when the inverter purposely curtails the energy delivered to the grid. The uncertainty of clipping losses is caused by the working conditions, mathematical model, and time granularity of solar data. We conservatively assume that it reaches up to 50 % of the loss value.

10. Auxiliary losses

Auxiliary losses represent consumption of active power by supporting equipment installed in the PV power plant (monitoring, lighting, security, PV module tracking, etc.). The difference between simulated and real auxiliary consumption during normal operation depends on the estimation accuracy of the expected auxiliary load, actual electrical voltage, but also on efficiency and degradation of equipment through ageing. Generally, auxiliary losses expressed in relative terms are inversely proportional to the capacity of the PV power plant. Systems with string inverter configuration will typically experience smaller auxiliary losses than central inverter configurations [20].

When setting up the simulation of a solar power plant there is often no information available that affects the auxiliary losses, so it is likely that the uncertainty of these losses will be high. In this study we consider uncertainties of up to 50 % of the loss value.

11. Transformer losses

From a mathematical point of view it is possible to model a transformer in several ways depending on the specific purpose. In PV systems calculation transformers are typically modelled by considering the so-called load losses and no-load losses.

If the uncertainties of measured losses are not given in the datasheet, it is necessary to rely on international standards. According to IEC 60076-1 [21], the variance of no-load loss and load loss should not exceed 15%. Further influences on the overall uncertainty of the transformer losses are the uncertainty of the mathematical model and the electrical voltage variability. Transformer losses uncertainty is defined by equation (7).

$$UNC_{\text{TR\_LOSSES}} = \sqrt{\begin{aligned} &UNC_{\text{MODEL}}^2 + \left(\frac{\partial f}{\partial TR_{\text{NOLOAD}}}\right)^2 UNC_{\text{TR}_{\text{NOLOAD}}}^2 + \\ &+ \left(\frac{\partial f}{\partial TR_{\text{LOAD}}}\right)^2 UNC_{\text{TR}_{\text{LOAD}}}^2 \end{aligned}} \qquad (7)$$

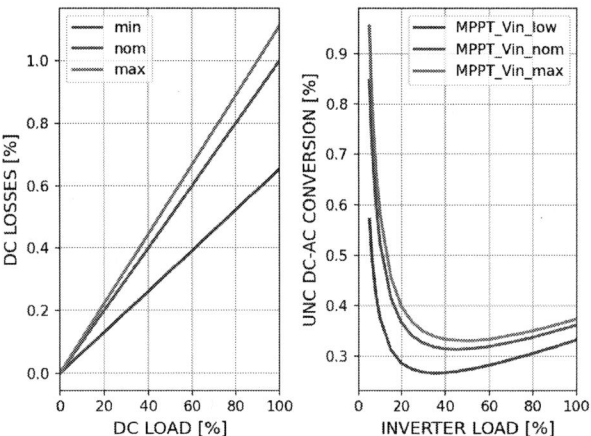

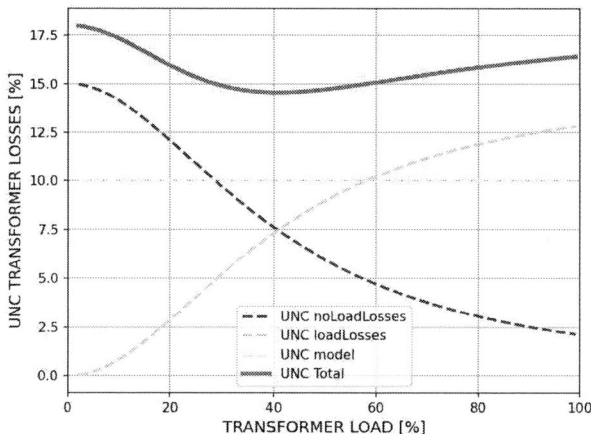

Fig. 4. Modelling of DC losses for min, nom, and max values of input parameters (left) and uncertainty of inverter output power depending on the value of the load (right) depending on the load.

Fig. 5. Modelling transformer uncertainty depending on the load

12. AC circuit losses

In addition to factors affecting uncertainty in the case of DC losses it is appropriate to take into consideration uncertainty of power factor (PF), harmonic current, and voltage variation. These parameters depend on network configuration, impedance of all parts creating the AC network and network operator requirements.

For a 3-phase system the AC losses are defined by formula (8).

$$AC_{LOSSES} = 3 * R_{REF} * I^2 * [\alpha * (T_{CABLE} - T_{REF}) + 1] \quad (8)$$

Assuming that input variables in equation (8) can vary in same range as in case of DC wiring. The difference between the maximum and minimum AC losses can reach 45 % of the reference value.

13. Losses due to system unavailability

System unavailability quantifies losses resulting from the PV system not being operational. In accordance with [22], the losses are considered due to internal and external reasons. The level of uncertainty depends on degradation of equipment through ageing, the electrical network configuration (complexity, back-up methods), maintenance regime, quality and reliability of equipment, time to repair of specific components [23], the number of interconnections between the internal and external electrical network, as well as the design of electrical protection system.

III. APPLICATION

For practical application of the described approach, a simulation of a 3,888 kWp PV power plant located in the East Devon region in the United Kingdom was performed. Plant parameters are detailed in Tab. 1. The simulation was performed using 15-min time series covering the full year 2022.

The uncertainties injected by the individual simulation blocks were calculated based on the 15-minute input data, and then aggregated to monthly values. The overall monthly uncertainty of the output power was calculated from the monthly uncertainties of the simulation blocks and the monthly uncertainty of the input GTI. This is presented in Tab. 2.

Tab. 2 shows the monthly values of uncertainties injected to transferred power from simulation blocks calculated based on the approaches described above. It can be observed that the largest source of uncertainty comes from GTI which has the highest uncertainty in the winter cloudy months (11 %) and the lowest in the summer sunny months (5 %).

Shading and soiling uncertainties are set as constant for simplicity, as the equations to calculate their precise values are very complex and hence impractical to calculate in the scope of a PV simulation. Uncertainty of these two parameters will be analyzed in detail later. Due to the location of the power plant, snow losses are not considered.

The calculated uncertainty of the spectral block after rounding is constant throughout the year, due to low sensitivity of the input parameters. The highest uncertainty injected by conversion block (3.4 %) is caused by lower values of GTI and bigger difference between simulated and STC cell temperatures in winter.

Tab. 1: Technical specification of the simulated PV plant

Parameter	Value
DC installed capacity	3,888 kWp
PV module count	7200
PV module tilt	37 °
String size	36 PV modules
String count	200
AC installed capacity	3,000 kW
DC/AC ratio	1.296
Grid power limit	No

Tab. 2: Monthly percentage uncertainties (relative to transferred power) injected into transferred power in simulation chain.

Month	JAN	FEB	MAR	APR	MAY	JUN	JUL	AUG	SEP	OCT	NOV	DEC
GTI	11.0	9.0	7.0	5.0	5.0	5.0	5.0	5.0	5.0	5.0	9.0	11.0
Shading	4.0	4.0	4.0	4.0	4.0	4.0	4.0	4.0	4.0	4.0	4.0	4.0
Soiling snd Snow	3.0	3.0	3.0	3.0	3.0	3.0	3.0	3.0	3.0	3.0	3.0	3.0
Angular	2.6	2.5	2.0	2.5	2.9	2.9	2.8	2.2	2.7	2.3	2.4	2.4
Spectral	1.2	1.2	1.2	1.2	1.2	1.2	1.2	1.2	1.2	1.2	1.2	1.2
Conversion	3.4	2.7	2.7	2.7	2.8	2.7	2.7	2.7	2.7	2.5	2.9	3.4
DC wiring	0.3	0.3	0.5	0.5	0.5	0.5	0.5	0.6	0.4	0.4	0.3	0.3
Clipping	0.0	0.0	0.4	0.7	0.7	0.4	0.6	0.5	0.0	0.0	0.0	0.0
DC/AC conversion	0.4	0.4	0.3	0.3	0.3	0.3	0.3	0.3	0.3	0.3	0.4	0.4
Auxiliary	0.6	0.4	0.3	0.3	0.3	0.3	0.3	0.3	0.3	0.4	0.5	0.6
AC (medium voltage)	0.0	0.0	0.1	0.1	0.1	0.1	0.1	0.1	0.1	0.1	0.0	0.0
Transformer	0.7	0.4	0.2	0.2	0.2	0.2	0.2	0.2	0.2	0.3	0.5	0.7
AC (high voltage)	0.1	0.1	0.1	0.1	0.1	0.1	0.1	0.1	0.1	0.1	0.1	0.0
Avaliability	0.0	0.0	0.0	1.1	0.0	0.0	0.0	0.0	0.0	0.0	0.0	0.0
UNC PVout	12.9	11.0	9.4	8.2	8.3	8.3	8.2	8.0	8.2	8.0	11.0	12.9

The uncertainties in the electrical part of simulations do not reach such high values as in the optical simulation blocks. The highest uncertainty in electrical part up to 0.7 % was calculated in the transformer block in winter months which was caused primarily by uncertainty of no-load losses transformer parameter. Uncertainty injected by DC to AC conversion also reaches higher values in the winter months due to lower load, and hence proportionally higher loss. Uncertainty of clipping losses affects the overall results only in the summer months when the value of uncertainty ranges from 0.4 % to 0.7 %.

IV. Conclusion

Thanks to the number of available software solutions, PV yield simulations can be performed easily nowadays. However, the uncertainty associated with these simulations has not yet been studied in great depth, and this should be the focus of future research.

In this study, the basic parts of a PV simulation chain are described, and the uncertainties associated with them analyzed and quantified. Among the publications on uncertainties in PV yield simulations, most are focused on error propagation. This study provides a contribution to the quantification and explanation of input uncertainties, which is a less-researched area.

Analysis in this study shows that the uncertainties injected to transferred power by individual simulation blocks are in general not constant and should be analyzed according to operational conditions. Based on the uncertainties of the input parameters of the simulation together with the uncertainty of the model, it is possible to calculate the uncertainties of the individual simulation blocks at their operational conditions. The simulation uncertainty of the resulting output power delivered to the grid is calculated from the uncertainties of the individual simulation parts.

Further research on this topic should focus on a more precise analysis of the optical part of the PV simulation. Moreover, detailed analysis of uncertainties of the simulation models should be performed based on precise laboratory or on-site measurements.

References

[1] Hansen Clifford W., Martin Curtis E., "Photovoltaic System Modeling: Uncertainty and Sensitivity Analyses", Sandia National Laboratories, Report SAND2015-6700, 2015.

[2] Perez R., Cebecauer T., Šúri M., "Semi-Empirical Satellite Models". *Kleissl J. (ed.) Solar Energy Forecasting and Resource Assessment.* Academic press, pp. 21 – 48, 2013.

[3] Cebecauer T., Suri M., Gueymard C., "Uncertainty sources in satellite-derived Direct Normal Irradiance: How can prediction accuracy be improved globally?". *Proc. Of the SolarPACES Conference 2011*, Granada, Spain, 20-23 Sept 2011.

[4] Perez, R., Seals, R., Michalsky, J., "All- weather model for sky luminance distribution—Preliminary configuration and validation", Solar *Energy*. vol. 50, Issue 3, pp 235-245. 1993.

[5] Cebecauer T et al. "Comparison of ray tracing rendering technique with ground measurements for improved solar radiation modeling", PVPMC workshop, Salt Lake City, 2022

[6] Coello, Merissa and Liza Boyle., "Simple Model for Predicting Time Series Soiling of Photovoltaic Panels." IEEE Journal of Photovoltaics 9 (2019), pp. 1382-1387.

[7] Polo, Jesús, Nuria Martín-Chivelet, Carlos Sanz-Saiz, Joaquín Alonso-Montesinos, Gabriel López, Miguel Alonso-Abella, Francisco J. Battles, Aitor Marzo and Natalie Hanrieder, "Modeling soiling losses for rooftop PV systems in suburban areas with nearby forest in Madrid." Renewable Energy, vol. 178, pp. 420-428, 2021.

[8] L. Micheli, G. P. Smestad, J. G. Bessa, M. Muller, E. F. Fernández and F. Almonacid, "Tracking Soiling Losses: Assessment, Uncertainty, and Challenges in Mapping," in IEEE Journal of Photovoltaics, vol. 12, no. 1, pp. 114-118, Jan. 2022, doi: 10.1109/JPHOTOV.2021.3113858.

[9] N. Martin, J.M. Ruiz, "Calculation of the PV modules angular losses under field conditions by means of an analytical model," *in Solar Energy Materials and Solar Cells*, pp. 25-38, 2001.

[10] IEC 61853-2:2016: Photovoltaic (PV) module performance testing and energy rating Part 2: Spectral responsivity, incidence angle and module operating temperature measurements

[11] Saunders, R. (2021). Assimilation of OLCI total column water vapour in the Met Office global numerical weather prediction system. Meteorological Applications,28(5), e2029

[12] M. Lee and A. Panchula, "Spectral Correction for Photovoltaic Module Performance Based on Air Mass and Precipitable Water", *in 43rd IEEE Photovoltaic Specialist Conference*, 2016

[13] Hansen Clifford W., "Parameter Estimation for Single Diode Models of Photovoltaics Modules", Report SAND2015-2065, p. 67, 2015

[14] Mihaylov, B., Betts, T. R., Pozza, A., Mullejans, H., & Gottschalg, "Uncertainty Estimation of Temperature Coefficient Measurements of PV Modules". IEEE Journal of Photovoltaics, 2016.

[15] Tuza A. Olukan, Mahieddine Emziane, A Comparative Analysis of PV Module Temperature Models, Energy Procedia, Volume 62, 2014, Pages 694-703, ISSN 1876-6102,

[16] García-López C, Álvarez-Tey G. Evaluation of the Uncertainty of Surface Temperature Measurements in Photovoltaic Modules in Outdoor Operation. Sensors. 2022

[17] D. Dirnberger and U. Kräling, Uncertainty in PV Module Measurement—Part I: Calibration of Crystalline and Thin-Film Modules. IEEE Journal of Photovoltaics, vol. 3, no. 3, pp. 1016-1026, July 2013, doi: 10.1109/JPHOTOV.2013.2260595.

[18] King, Gonzalez and Boyson, "Performance Model for Grid-Connected Photovoltaic Inverters", Report SAND2007-5036, p. 47, 2007

[19] EN 61683: 2000: "Photovoltaic systems – Power conditioners Procedure for measuring efficiency (IEC 61683: 1999)".

[20] International Finance Corporation: Utility-Scale Solar Photovoltaic Power Plants. A Project Developer`s Guide, 2015

[21] EN 60076-1: 2011: "Power transformers – Part 1: General (IEC 60076-1: 2011)".

[22] IEC TS 61724-3:2016: Photovoltaic system performance – Part 3: Energy evaluation method

[23] Geoffrey T. Klise, Olga Lavrova, Renee Gooding, "PV System Component Fault and Failure Compilation and Analysis", Report SAND2018-1743, p. 37, 2018

Characterizing TeO$_2$ Formation in CdTe Devices Using Transmission Electron Microscopy

John Farrell,[1] Ebin Bastola,[2] Manoj Jamarkattel,[2] Michael Heben,[2] Walajabad S. Sampath,[3] James Sites[3], Robert F. Klie[1]

[1]University of Illinois at Chicago, Chicago, IL, 60607, U.S.A., [2]University of Toledo, Toledo, OH, 43606, U.S.A., [3]Colorado State University, Fort Collins, CO, 80523, U.S.A.

Abstract— Poly-crystalline CdTe/Cd(Se)Te based thin film solar cells have shown to be competitive in terms of efficiency and cost of electricity production. Yet, the presence of hetero-interfaces in Cd(Se)Te structure and low minority carrier lifetime have limited the thin film devices from reaching their maximum theoretical efficiency of approximately 30 percent. The back-contact of CdSeTe devices has been identified as one significant limitation to increased device performance since no metal has been identified that has a sufficiently high work function to create an Ohmic contact with the CdTe absorber at the back-surface of the film stack. Here, we will explore the formation of native TeO$_2$ on the back contact of solar cells to draw inferences on device performance. Atomic-resolution imaging in a scanning transmission electron microscope (STEM) combined with electron energy-loss spectroscopy (EELS) are used to characterize these devices and to inform the production process. The goal is to identify the mechanisms of TeO$_2$ formation and its role in passivating the back contact.

Keywords—CdTe, TeO$_2$, Thin Film, HRSTEM, XEDS, Back Contact

I. INTRODUCTION

Polycrystalline CdTe thin film solar cells prove to be highly competitive technology because of the high absorption coefficient, nearly optimum direct band gap energy and simplicity of manufacturing. However, laboratory efficiencies of CdTe solar cells are still below the theoretical efficiency limit (30%)[1]. The back contact of the CdTe Photovoltaic (PV) cell has been a serious limit to performance due to a deep valence band at 5.8 eV, much higher than the work function of many metals (e.g. Ni at 5.2 eV)[2]. The semiconductor bands decrease near the interface and this creates a barrier for holes and accumulates electrons for higher recombination events. The objective is the development of major reductions in back-surface interfacial recombination. Partial solutions include Cu doping to move the Fermi level closer to the valence band edge and shrink the depletion region. However, Cu diffuses rapidly in the absorber bulk and its interstitials increase recombination events. The addition of an intermediate layer with a valence band maximum between CdTe and metal electrode is another band-engineering solution. A native TeO$_2$ thin film near the absorbed back-edge has been observed to form[3] and is investigated for its viability in this respect.

II. METHODOLOGY

The poly-crystalline CdTe thin film PV devices studied were deposited on glass coated with indium tin oxide (ITO). Either a CdS layer or a graded CdSeTe-MZO layer forms the emitter of the PV junction. A CdCl$_2$ heat treatment was applied to the CdTe absorber layer in an effort to passivate the grain boundary interfaces in the polycrystalline CdTe. In this study, the back contact of the device consists of a very thin (5nm) layer of Cu$_y$AlO$_x$ directly on the absorber followed by either another layer of ITO or by an Au contact.

The structural characterization of the CdTe devices and back-contact layers is performed using a JEOL ARM200CF aberration-corrected scanning transmission electron microscope (STEM) operated at acceleration voltage of 200 kV. The STEM images were acquired using a probe semi-convergence angle of 24 mrad and two annular detectors, a high-angle annular dark field (HAADF) detector and low-angle annular dark-field (LAADF) detector. X-ray energy dispersive spectroscopy (XEDS) maps and line-scans were obtained using a windowless XEDS silicon drift detector X-MaxN 100 TLE from Oxford Instruments. Electron energy-loss spectroscopy is performed using a Gatan post-column spectrometer, the GIF-Continuum. The combination of these techniques allows for spatially resolved characterization of interfacial atomic structures and chemical compositions with atomic resolution. Cross-sectional TEM samples were prepared using the focused ion-beam (FIB) lift-out technique in a ThermoFisher Helios 5CX dual-beam FIB/SEM system. The Helios 5CX is equipped with the STEM 3+ detector for high-precision lamella end-pointing.

III. RESULTS AND DISCUSSION

An Au-contact device is characterized below. This device is a CdTe absorber and CdS emitter. Figure 1 is a low-magnification HAADF image serving as an overview of the lamella. The Au electrode is the high signal area between the absorber and the Pt deposited as the shield for the FIB lift-out process.

This material is based upon work supported by the U.S. Department of Energy's Office of Energy Efficiency and Renewable Energy (EERE) under the Solar Energy Technologies Office Award Number DE-EE0008974.

978-1-6654-6060-6/23 $31.00 © 2023 IEEE

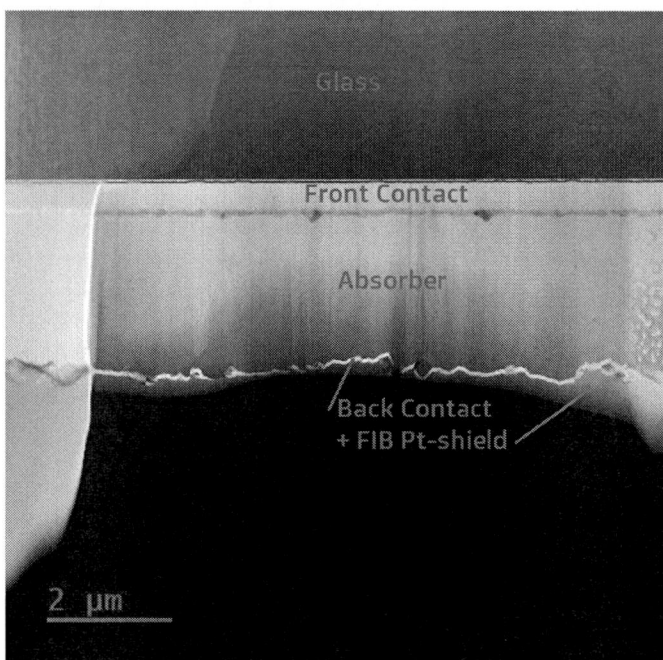

Figure 1: Low-magnification HAADF image of a CdTe/CdS device lamella

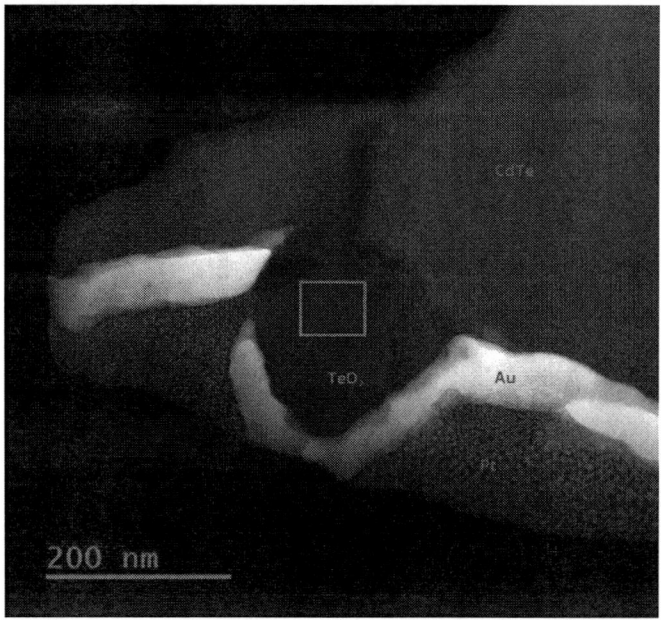

Figure 2: HAADF image of TeO₂ formation adjacent to the Au-electrode

Figure 2 shows a sub-micron scale region contrasted darkly in HAADF that is the site of native TeO_2 formation. The region in and surrounding the red square annotation is lower signal than the adjacent CdTe region due to the significant presence of Oxygen as well as thickness effects. The chemical species analysis is the result of EELS characterization in the proceeding figures.

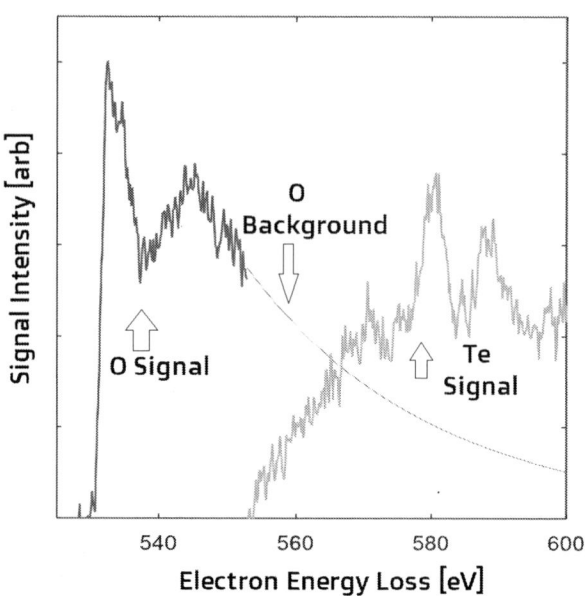

Figure 3: EELS signal showing the separated contributions from the O-*K* edge (blue) and Te-*M* edge (yellow)

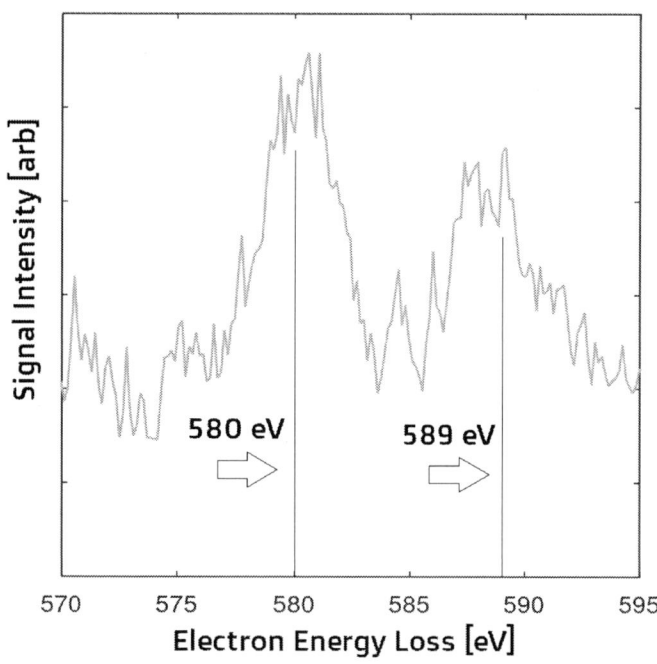

Figure 4: A magnified view of the spectral peaks from the Te-*M* edge associated with TeO_2

In Figure 3, the EELS signal of the O *K*- and Te *M*-edges taken from the area shown by the red rectangle in Figure 2. Figure 4 shows a higher resolution view of the Te *M*-edges, which can be seen as two sharp peaks, centered around 580 eV (M_4) and 589 eV (M_5). A comparison with previously published X-ray absorption spectroscopy work [4] taken from various Te and TeO_x materials shows that the Te M_4-edge for Te_0 should have its maximum at approximately 576, while the Te M_4-edge for Te^{4+} has its maximum at 580 eV. The values for the Te M_5-edge are 587 eV (Te^0) and 589 eV (Te^{4+}), respectively.[4] Therefore, we conclude that this area at the back-surface of the CdTe absorber is clearly made up of TeO_2.

Yet, the mechanism by which this native TeO2 layer is formed remains unknown, in particular the reason for the selective formation in the regions indicated in Figure 2. It needs to be pointed out here that we did not find a continuous native TeO_2 layer at the back-surface, but rather discrete areas where the oxide was detected. This could be either attributed to the fact that there is no continuous oxide layer formation, or that the native oxide layer is so thin that it can only be detected by STEM-EELS when the termination of the CdTe grains is precisely along the electron-beam direction.

We will explore this phenomenon further by deliberately oxidizing the back surface to create a thicker oxide layer or by depositing a TeO_2 layer. Moreover, the atomic-structure of the oxide layer and its interface with the CdTe grains is not resolved. Current studies are focused on identifying the local atomic and electronic structures of the hetero-interface and compare it to recent density functional theory studies [5].

IV. CONCLUSION

We utilized high-resolution STEM imaging and EELS to validate the formation of a TeO_2 layer at the back-surface of the CdTe absorber. Future work will attempt to identify possible mechanisms for this oxidation as well as attempt to correlate this process with device performance in regards to back-contact hole-selectivity and back-surface passivation. Comparative studies of different back-contact layer chemistries will be conducted to ascertain the prevalence of TeO_2 formation. The goal is to guide the fabrication process in the development of a back contact interface that erodes the valence band barrier to encourage charge carrier transport through the device and improve efficiency.

ACKNOWLEDGMENT

This material is based upon work supported by the U.S. Department of Energy's Office of Energy Efficiency and Renewable Energy (EERE) under the Solar Energy Technologies Office Award Number DE-EE0008974.

REFERENCES

[1] W. Shockley and H. J. Queisser, "Detailed balance limit of efficiency of p-n junction solar cells," J. Appl. Phys., vol. 32, no. 3, pp. 510–519, 1961.

[2] A. R. Davies and J. R. Sites. Effects of non-uniformity on rollover phenomena in CdS/CdTe solar cells. In the 33rd IEEE Photovoltaic Specialists Conference, 2008.

[3] F. A. Ponce, R. Sinclair, and R. H. Bube , "Native tellurium dioxide layer on cadmium telluride: A high-resolution electron microscopy study", Appl. Phys. Lett. 39, 951-953 (1981)

[4] Nie, Wells et al. "Impact of Valence States on Superconductivity of Oxygen Incorporated Iron Telluride and Iron Selenide Films" February 2011 Physical review. B, Condensed matter

[5] Anthony P. Nicholson, Akash Shah, Ramesh Pandey, Amit H. Munshi, James Sites, and Walajabad Sampath ACS Applied Materials & Interfaces 2022 14 (25), 29412-29421

Dense array TPV modules with alternating polarity InGaAs cells

Iván García, Aitana Cano, Víctor Orejuela, Pablo Martín and Ignacio Rey-Stolle

Instituto de Energía Solar, Universidad Politécnica de Madrid, 28040 Madrid, Spain

Abstract — Improving the cell coverage in TPV modules can increase the electrical power density and reduce the losses in thermal storage batteries. We propose a compact module design built using alternating polarity InGaAs solar cells connected in series. The integration of n/p and p/n cells with minimum current and series resistance losses is challenged by the differing carrier transport and recombination properties of n and p-type InGaAs. Since the photons reflected at the front metal of the cell are –in principle– reused in the TPV system, the front grid shadowing factor becomes an additional degree of freedom for the cell design with low impact on the TPV efficiency. We show that the areal and series resistance losses in the TPV system can be minimized in the proposed compact module by engineering n/p and p/n cell designs.

I. INTRODUCTION

Thermal energy storage batteries use electric energy to heat a material and take it to elevated temperatures. The energy stored can be converted back into electricity by using thermophotovoltaic (TPV) receiver modules [1], [2]. To maximize the performance of the battery, these modules must convert the infrared (IR) radiation into electricity with a high photovoltaic conversion efficiency. Moreover, the module must be able to reflect most of the unused low energy photons back to the emitter. Record-efficiency TPV InGaAs cells have been demonstrated, which use a back reflector that achieves an effective photon recycling in the receiver and enables returning the low energy radiation to the source [3], [4].

However, power losses in the TPV system are also affected by the receiver module configuration. TPV cells are characterized by the high current and low voltages produced. Hence, they are normally connected in series in a TPV module, to minimize the module current and the ohmic losses. For usual cells with metal contacts placed at the front and back, this

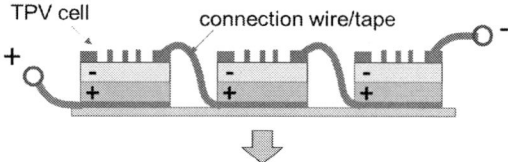

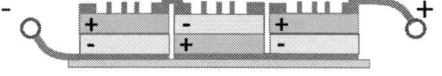

Fig. 1. Dense array TPV module concept based on alternating polarity InGaAs cells.

connection is achieved using bonded wires or tapes going from the rear side of a cell to the front side of the next one (see Figure 1). These connections require shifting the cells, creating inactive regions that lower the electrical power density and overall conversion efficiency of the module (Figure 2). These regions can be made highly reflective for the IR radiation to minimize their impact on the battery performance at the cost of a higher module manufacturing complexity. Densely-packed modules could also be achieved by using a shingled interconnection scheme or, alternatively, through-vias to place both contacts at the rear side of TPV cells, at the cost of a significantly higher cell manufacturing complexity.

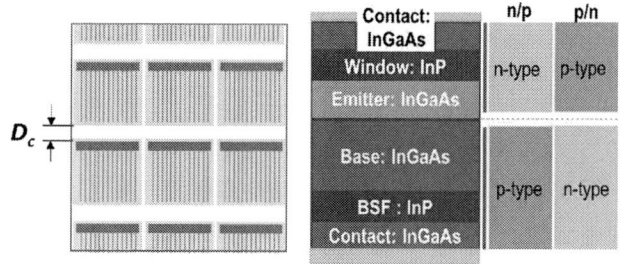

Fig. 2. Left: section of the TPV module with detail of the dimension that is reduced to achieve a higher compactness. Right: general structure of the InGaAs cells considered.

We propose an alternative dense array module design which significantly reduces the manufacturing complexity at both cell and module levels. This design relies on using series-connected TPV cells with alternating n/p – p/n polarity, enabling to minimize the spacing between cells in the module (see Figures 1 and 2) and use smaller cell area and lower currents. We examine the challenges and benefits of this approach at cell level and evaluate its potential regarding the performance of the TPV system.

II. LOSS ANALYSIS

The cell area size affects the conversion efficiency due to series resistance effects, while the separation between cells in the module affects the losses due to imperfect reflection of photons at inactive areas of the module. Moreover, an incomplete reflection of photons at the front grid metal causes power losses that scale with the shadowing factor, which increases with the cell area size.

978-1-6654-6060-6/23 $31.00 © 2023 IEEE

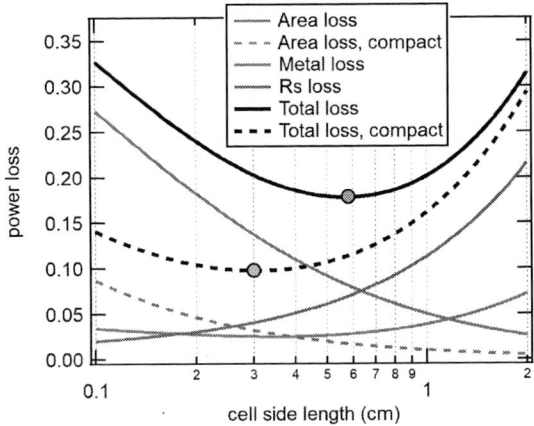

Fig. 3. Analysis of power loss in a TPV module.

TABLE I
LOSS ANALYSIS MAIN MODELING PARAMETERS

Parameter	Value
Short circuit current density (J_{SC})	6 A/cm^2
Inverse saturation current density (J_0)	$1.5 \cdot 10^{-8}$ A/cm^2
Emitter sheet resistance ($R_{s,E}$)	200 Ω/□
Metal sheet resistance ($R_{s,M}$)	$5 \cdot 10^{-3}$ Ω/□
Specific contact resistance (ρ_M)	$1 \cdot 10^{-6}$ Ω·cm^2
Averaged front metal reflectance	0.7
Cell spacing in standard module	1000 μm
Cell spacing in compact module	100 μm
Averaged reflectance around cells	0.5

A modeling study was carried out to understand the effect of InGaAs TPV cell size on the areal, series resistance and metal reflection losses. The analysis results depend on the current density, determined by the emitter temperature and cell view factor. The case example shown here (see relevant parameters in Table I) is in the range of those used in experimental systems [4]. Figure 3 shows a summary of the analysis, with the black solid line representing the global losses for usual separation between cells and the dashed black line the losses with narrower separation achieved if simpler connections between cells can be made (see Figure 1 and Figure 2).

Important conclusions can be extracted from this graph. An optimum area size that minimizes the system losses appears. The global losses can be reduced substantially by using a compact module. Furthermore, the advantage of the compact approach becomes larger if the cells are made smaller than the optimum for a standard module. The cell areas minimizing the losses, around 0.1 to 0.25 cm^2, appear appropriate for module fabrication. Note that the n/p – p/n cell interconnection scheme proposed allows using smaller cell areas with a lower impact in the module manufacture cost than for the traditional modules. Given the interconnection scheme proposed, low cost approaches such as screen printing can be used instead of costly wire or tape bonding.

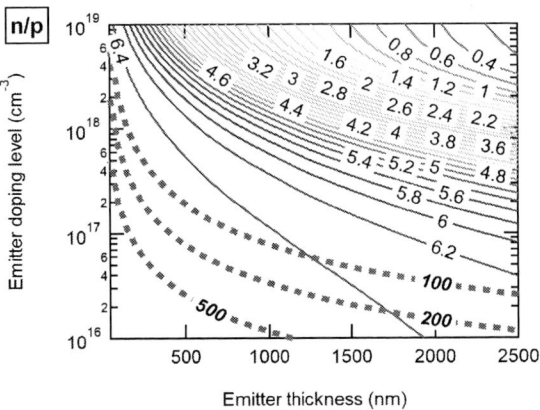

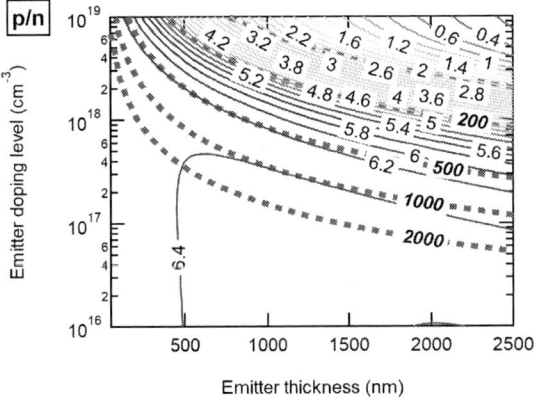

Fig. 4. Contour plots of J_{SC} of n/p and p/n InGaAs TPV cells structures (without front grid), plotted against the emitter thickness and doping level. The dashed lines represent the emitter sheet resistance. The total thickness of the absorber layers (emitter+base) is 2.5 μm.

II. COMPARISON OF N/P VS P/N INGAAS CELL PERFORMANCE

The proposed compact TPV module design requires integrating n/p and p/n InGaAs cells without compromising the module efficiency. Firstly, since the cells are connected in series, the short circuit current (I_{SC}) produced by both n/p and p/n cells must be equal to prevent losses. In addition, the series resistance must be minimized. The n/p cell is the preferred choice in applications handling high irradiances, since they achieve optimum carrier collection and I_{SC}, and low series resistance. The p/n configuration must match these characteristics to make our module design feasible.

The emitter sheet resistance (R_{sE}) contributes to a large fraction of the series resistance, together with the front grid characteristics. This sheet resistance is inversely proportional to the doping level, carrier mobility and thickness of the emitter and upper layers of the cell structure. An specific design for n/p and p/n structures is required to achieve similar I_{SC} and low series resistance in both cases, given the large difference in electron and hole mobilities in n and p-type InGaAs.

The emitter sheet resistance, short circuit current density (J_{SC}) and open circuit voltage (V_{OC}) of the InGaAs structures (i.e., without front grid) were computed using Hovel models and parameters from the literature and our own experimental results. Figure 4 shows the J_{SC} for the n/p and p/n cells plotted against the emitter doping level and emitter thickness of the structure shown in Figure 2. In all cases, the total absorber (emitter+base) thickness is constant at 2.5 µm, i.e. these plots show the performance for different junction positions. The J_{SC} contours for the n/p cell show that thin and highly doped emitters achieve the highest currents with low R_{sE}. Rear heterojunction designs, both in the solar cell and TPV cell fields [4], [5], have recently demonstrated also high currents and low R_{sE} with a boost in the V_{OC}, in agreement with out model results (V_{OC} plots not shown in this abstract). For the p/n design, a similar J_{SC} involves higher R_{sE}, with an expected impact on the cell design and R_S.

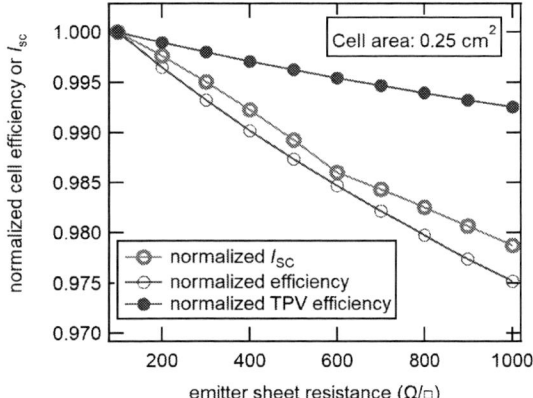

Fig. 5. Normalized I_{SC}, efficiency and TPV efficiency, against the emitter sheet resistance of the InGaAs cell. For each point in the curves the shadowing factor is optimized for maximum cell efficiency, which produces a constant fill factor.

IV. INTEGRATION OF N/P AND P/N CELLS

The effect of the higher R_{sE} on the series resistance of the p/n cells can be compensated for by using a denser front grid. This would cause a higher shadowing factor and lower I_{SC}, though. Using the full model of the cell, the impact on efficiency was studied (see Figure 5). The maximum efficiency attainable decreases with higher R_{sE}. Since the fill factor is constant, the main cause is the lower I_{SC}. However, the I_{SC} drop must not be computed fully as a loss regarding the cell TPV efficiency, since most of the light that is reflected on the front metal is returned to the emitter. Therefore, the TPV efficiency drop is reduced by a (1-R) factor, being R the averaged reflectivity of the front metal. For a conservative front metal reflectivity value of around 70%, the maximum loss caused by using a higher R_{sE} up to 1000 Ω/ □ is around 0.7%.

Therefore, a higher front grid shadowing factor can be used

in the p/n design to compensate for the higher R_{sE} with a limited impact on the TPV losses. However note that this higher shadowing factor actually produces a lower I_{SC} in the p/n than in the n/p cell, which would cause a loss in the series connection at module level. To fix this, the n/p cell has to use an appropriate means to lower the current without causing a TPV efficiency loss, i.e., by reflecting part of the light. This can be achieved by using a higher shadowing factor, despite not needed in terms of R_S. It can also be achieved by using an ad-hoc design of anti-reflection coating.

III. SUMMARY AND CONCLUSIONS

Dense array TPV module designs by using series connected, alternate polarity InGaAs cells are proposed aiming at higher TPV module efficiency and electrical power density, module assembly simplicity (and lower cost) and lower series resistance losses by enabling a cost-effective use of smaller area cells. Despite the large difference in carrier transport and lifetime parameters in n and p-type InGaAs, it is shown that engineered n/p and p/n cells can be integrated in a module with a minimum loss in TPV efficiency. An extended theoretical analysis and the experimental demonstration of this approach will be presented in an upcoming paper.

IV. ACKNOWLEDGEMENT

Project supported by a 2022 Leonardo Grant for Researchers and Cultural Creators from BBVA Foundation. The MOVPE reactor used is supported by the project EQC2019-005701-P, funded by AEI/ 10.13039/501100011033, MCIN and ERDF "A way to make Europe"

REFERENCES

[1] A. Datas, A. Ramos, A. Martí, C. del Cañizo, and A. Luque, "Ultra high temperature latent heat energy storage and thermophotovoltaic energy conversion," *Energy*, vol. 107, pp. 542–549, Jul. 2016, doi: 10.1016/j.energy.2016.04.048.

[2] C. Amy, H. R. Seyf, M. A. Steiner, D. J. Friedman, and A. Henry, "Thermal energy grid storage using multi-junction photovoltaics," *Energy Environ. Sci.*, vol. 12, no. 1, pp. 334–343, Jan. 2019, doi: 10.1039/C8EE02341G.

[3] D. Fan, T. Burger, S. McSherry, B. Lee, A. Lenert, and S. R. Forrest, "Near-perfect photon utilization in an air-bridge thermophotovoltaic cell," *Nature*, vol. 586, no. 7828, Art. no. 7828, Oct. 2020, doi: 10.1038/s41586-020-2717-7.

[4] E. J. Tervo, R. M. France, D. J. Friedman, M. K. Arulanandam, R. R. King, T. C. Narayan, C. Luciano, D. P. Nizamian, B. A. Johnson, A. R. Young, L. Y. Kuritzky, E. E. Perl, M. Limpinsel, B. M. Kayes, A. J. Ponec, D. M. Bierman, J. A. Briggs, and M. A. Steiner, "Efficient and scalable GaInAs thermophotovoltaic devices," *Joule*, vol. 6, no. 11, pp. 2566–2584, Nov. 2022, doi: 10.1016/j.joule.2022.10.002.

[5] J. F. Geisz, M. A. Steiner, I. García, S. R. Kurtz, and D. J. Friedman, "Enhanced external radiative efficiency for 20.8% efficient single-junction GaInP solar cells," *Appl. Phys. Lett.*, vol. 103, no. 4, p. 041118/2-041118/5, Jul. 2013, doi: 10.1063/1.4816837.

Influence of Spectral Albedo on the Performance of Lead-Free Perovskite Bifacial Tandem Solar Cell

Atanu Purkayastha*, and Arun Tej Mallajosyula

Department of Electronics and Electrical, Indian Institute of Technology Guwahati, Guwahati, Assam, 781039, India

Abstract—Bifacial tandem solar cell (BTSC) design, where the light can enter the device from both its front and back surfaces, is an effective way to boost the power conversion efficiency. Here, we report the performance of $MAGeI_3$ - 2D/3D $FASnI_3$ BTSC, concurrently illuminated with AM1.5G from the front surface and various albedo spectra from the back surface. The device structure was optimized using Silvaco TCAD software. At optimal $MAGeI_3$ and 2D/3D $FASnI_3$ thicknesses of 130 and 620 nm, respectively, the lead-free all-perovskite BTSC (LPBTSC) demonstrated a maximum efficiency of 31.21% under concurrent illumination of AM1.5G and dry grass albedo spectra. A maximum efficiency of 38.3% has been calculated for the LPBTSC with snow spectral albedo, which can be 22.34% more efficient than its monofacial counterpart. The efficiency of LPBTSC with other surfaces such as sandstone and white sand have been found to be 13.93% and 11.67% more than its monofacial counterpart. The energy payback time of the LPBTSC has been determined to be 0.45 years, which is 50% and 77.5% lower than the monofacial all-perovskite and perovskite-silicon TSCs. Furthermore, the LPBTSC has a green house gas emission factor of 0.065 Kg-CO_2.kWh^{-1}, which is 45% lower than that obtained for all-perovskite monofacial TSC.

Index Terms—Bifacial solar cell, Concurrent Illumination, Albedo, Tandem solar cell

I. INTRODUCTION

Bifacial design is an elegant technique for increasing the photoabsorption capability and, hence, the power conversion efficiency (PCE) of a solar cell. Bifacial solar cells (BSCs) have both their top and bottom contacts transparent. Thus, in contrast to their monofacial counterparts, these devices also absorb the light reflected off from the surface on which they are mounted (albedo) and, in the process, can generate a relatively higher photocurrent. The albedo, quantified by the ratio of the power of light reflected to the power of light coming in from all directions, is a measure of how well a surface reflects light. Silicon-based solar cells have also been reported to use the albedo, with a maximum PCE of 24.61% [1]. On the other hand, tandem solar cell (TSC) design, where two different semiconductors having complementary absorption spectra are integrated, can also increase the PCE effectively. Thus, it is imperative that a bifacial tandem solar cell (BTSC) has the potential to combine the advantages of both the BSC and TSC designs.

Monofacial Si-perovskite tandem solar cells with a PCE as high as 29.8 % have been demonstrated recently [2]. It is well-known that the use of halide perovskites for solar cells gives benefits over Si such as low cost, low temperature solution processability, and flexibility. Over the past five years, the

PCE of all-perovskite monofacial TSCs (MTSC) have reached 28% [2]. More recently, Chen *et al.* have reported an all-perovskite BTSC with a PCE of 29.3%, under a back-to-front irradiance ratio of 30 [3]. However, these devices contain lead (Pb), which is a carcinogen. Therefore, in this work, we have optimized and analyzed the performance of Ge- and Sn-based all-perovskite BTSCs under concurrent illumination with AM1.5G spectrum from the front surface and various types of albedos from the back surface. It may be noted that all albedos may not result in increased efficiency. In addition, we have done a life cycle analysis and calculated the environmental sustainability parameters such as energy payback time (EPBT) and green house gas emission factor (GEF) for solar modules using these cells.

A schematic of the lead-free all-perovskite BTSC (LPBTSC), illuminated concurrently with the AM 1.5G and dry grass albedo spectra at the Cu_2O and phenyl-C61-butyric acid methyl ester (PCBM) ends respectively, is shown in Fig. 1(a). Dry grass has a back-to-front irradiance ratio of 0.3, which has been used by various research groups to optimize the all-perovskite and perovskite-silicon BTSCs [3]. Transparent conducting indium tin oxide (ITO) has been used for both electrodes. The carrier concentration versus distance plot of the LPBTSC and lead-free all-perovskite MTSC (LPMTSC), calculated from the band diagrams of both devices under short circuit condition, is shown in Fig. 1(b). The integrated electron and hole concentrations of the LPBTSC were 6.32e19 and 6.66e16 cm^{-3} respectively, whereas the corresponding values for the LPMTSC were 6.30e19 and 6.49e16 cm^{-3} respectively.

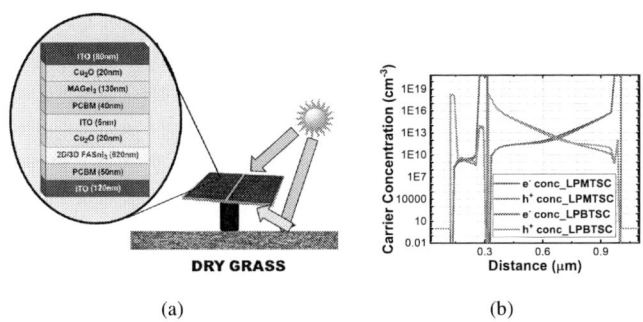

Fig. 1. (a) Schematic of the LPBTSC with dry grass albedo spectra from the PCBM end (inset shows the device structure), (b) Carrier concentration versus distance plot of the LPMTSC and LPBTSC with dry grass albedo spectra under short circuit condition.

Silvaco TCAD software has been used for device simulation. Parameters used for simulating this structure are listed in Table I [4]–[7]. The simulation has been carried out using the drift-diffusion model, the transfer matrix method (TMM), and the Shockley-Read-Hall (SRH) recombination model.

TABLE I
SIMULATION PARAMETERS OF LEAD-FREE ALL-PEROVSKITE BTSC

Parameters	Cu$_2$O	FASnI$_3$	MAGeI$_3$	PCBM
Thickness (nm)	20	100-500	100-1000	40-50
Shallow Acceptor Density (cm^{-3})	1e18	1e13	-	-
Shallow Donor Density (cm^{-3})	-	-	1e13	1e19
Band Gap (eV)	2.4	1.4	1.9	2.1
Electron affinity (eV)	3.2	3.5	3.98	3.9
Relative Dielectric Permitivity	7.1	32	10	3
CB Density of States (cm^{-3})	2.1e17	1e18	1e16	1e21
VB Density of States (cm^{-3})	1e19	1e18	1e15	1e22
Mobility of electron (cm^2/V-s)	200	22	162	0.01
Mobility of hole (cm^2/V-s)	80	22	102	0.01
Lifetime of electron (s)	1e-7	1.7e-9	1e-8	4e-6
Lifetime of hole (s)	1e-7	1.7e-9	1e-8	4e-6

II. RESULTS AND DISCUSSION

A. Active layer thickness optimization

The thicknesses of MAGeI$_3$ and 2D/3D FASnI$_3$ have been varied from 100 to 500 nm and 1000 nm respectively, and fill factor (FF), short-circuit current density (J_{SC}), open circuit voltage (V_{OC}), and PCE of the LPBTSC have been obtained. The dependence of PCE on the thicknesses of the two active layers under concurrent illumination of AM1.5G and dry grass albedo spectra is shown in Fig. 2.

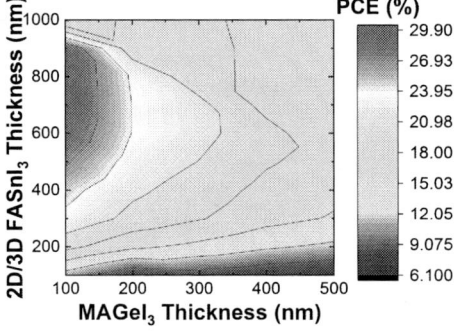

Fig. 2. PCE of the LPBTSC versus different thicknesses of MAGeI$_3$ and FASnI$_3$ layers

It has been observed that a maximum PCE (PCE$_{|MAX}$) of 31.21% is achieved by this cell when the thicknesses of MAGeI$_3$ and 2D/3D FASnI$_3$ are 130 and 620 nm, respectively. The corresponding efficiency parameters are V_{OC} = 2.57 V, J_{SC} = 19.80 mA.cm^{-2}, and FF = 77.92 %. The optimal thicknessess and efficiency parameters of LPMTSCs with Ag and ITO back contacts are listed in Table II for comparison.

TABLE II
PERFORMANCE PARAMETERS OF OPTIMIZED LPMTSCs WITH DIFFERENT
BACK CONTACTS AND ACTIVE LAYER THICKNESSES.

Back Contact	MAGeI$_3$ (nm)	FASnI$_3$ (nm)	V_{OC} (V)	J_{SC} (mA.cm^{-2})	FF (%)	PCE (%)
ITO	130	620	2.53	10.82	79.56	17.18
Ag	130	620	2.53	10.89	79.93	17.41
Ag	270	500	2.57	15.28	78.96	31.01
ITO	270	500	2.56	11.89	82.07	24.96

B. Effect of Spectral Albedo

Seven different mounting surfaces have been chosen in this work to study the effect of their spectral albedo on the performance of the LPBTSC. These are i) white sand, ii) soil, iii) snow, iv) sandstone, v) dry grass, vi) pond water, and vii) sea water. The spectral reflectivity and, thus, the albedo solar radiation of these surfaces have been taken from the literature and shown in Fig. 3 [8]. Albedo of snow has the highest spectral irradiance compared to the other surfaces in the wavelength range of 350 to 1000 nm. The current density (J) vs. voltage (V) characteristics of the LPBTSC illuminated on the front side by AM1.5G spectrum and on the rear side by different albedo spectra are shown in Fig. 4 and the efficiency parameters are listed in Table III.

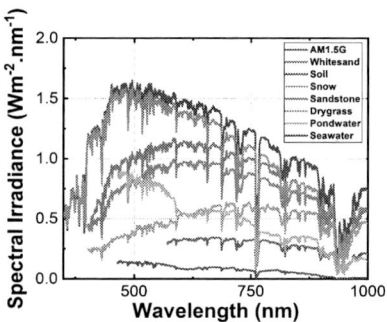

Fig. 3. Spectral albedo of the seven rear surface materials used to illuminate LPBTSC from the PCBM end.

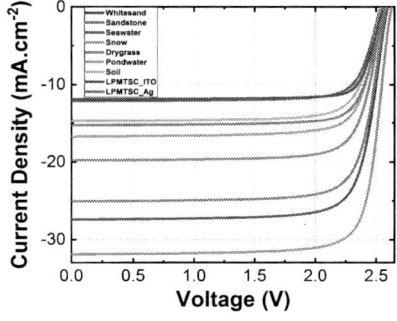

Fig. 4. J-V plots of the LPMTSC and LPBTSC illuminated concurrently with AM1.5G spectrum from the front surface and different spectral albedo from the back surface.

A PCE$_{|MAX}$ of 38.30% has been obtained with the snow rear surface which is 22.34% more than the optimized LPMTSC (31.01%) with Ag back contact. Also, the PCE$_{|MAX}$ of the

978-1-6654-6060-6/23 $31.00 © 2023 IEEE

device improved by 11.67% and 14.2% with the white sand and sandstone rear surfaces. On the other hand, it is decreased by 19.75%, 14.05%, and 24.68% with the soil, pondwater, and seawater respectively, and thus, such surfaces are not preferable for this device. Table III lists the efficiency parameters of the LPMTSCs and LPBTSCs with various albedo. From the table, we can see that the V_{OC} and FF remain nearly the same for all types of albedo. Consequently, the J_{SC} is the most important factor in determining the $PCE_{|MAX}$ of the device, which is dependent on the optical generation rate (G_{opt}). Fig. 5 depicts the G_{opt} for both the LPMTSC and LPBTSC with different albedo spectra.

The G_{opt} of the LPBTSC has been found to be maximum at the $Cu_2O|MAGeI_3$ and $FASnI_3|PCBM$ interfaces since the illumination is highest here. The best J_{SC} (31.91 mA.cm^{-2}) of the LPBTSC occurs with the snow rear surface owing to the maximum G_{opt} ($G_{opt|MAX}$). In this case, the $G_{opt|MAX}$ is 6.14e21 cm^{-3}.s^{-1}.

TABLE III
PERFORMANCE PARAMETERS OF THE LPMTSC AND LPBTSC ILLUMINATED CONCURRENTLY WITH AM1.5G SPECTRUM FROM THE FRONT SURFACE AND DIFFERENT SPECTRAL ALBEDO FROM THE BACK SURFACE.

Albedo	V_{OC} (V)	J_{SC} (mA.cm^{-2})	FF (%)	PCE (%)	$G_{opt|MAX}$ (cm^{-3} s^{-1})
Whitesand	2.59	27.43	78.13	34.63	3.92e21
Snow	2.63	31.91	79.97	38.30	6.14e21
Soil	2.54	14.68	77.43	25.17	2.0e21
Dry Grass	2.57	19.80	77.92	31.14	2.77e21
Sandstone	2.59	25.08	78.04	35.41	3.55e21
Pondwater	2.55	16.74	77.97	26.67	2.33e21
Seawater	2.54	12.13	78.97	23.37	1.78e21
LPMTSC_ITO	2.56	11.89	82.07	24.96	1.76e21
LPMTSC_Ag	2.57	15.28	78.96	31.01	2e21

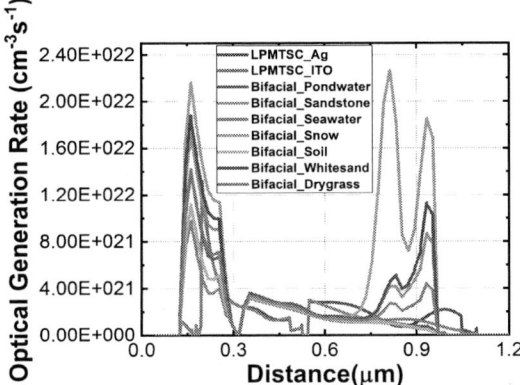

Fig. 5. Optical generation rates of LPMTSC and LPBTSC illuminated concurrently with the different albedo radiations and the AM1.5G spectrum from the rear and front ends, respectively.

On the other hand, $G_{opt|MAX}$ of the LPBTSC has been found to be minimum (1.78e21 cm^{-3}.s^{-1}) with the sea water albedo surface, and hence the minimum PCE of 23.37%. The $G_{opt|MAX}$ of the LPBTSC with other albedo surfaces is displayed in table III.

Using the parameters derived from the literature, the annual primary energy generation of the LPBTSC with the snow surface in India was calculated to be 152.20 kWh.m^{-2}.year^{-1}, which is 23.4% more than that of the optimized LPMTSC with the Ag back contact (116.59 kWh.m^{-2}) [9]. The primary energy demand to fabricate LPBTSC is 69.24 kWh.m^{-2}. Thus, the EPBT of the LPBTSC has been calculated to be 0.45 years. Similarly, the GEF of the LPBTSC has been measured to be 0.065 Kg-CO$_2$.kWh^{-1}, assuming the lifetime of the LPBTSC to be five years.

III. SUMMARY

Out of the seven different spectral albedo investigated here, snow, whitesand, and sandstone albedos have shown significant efficiency improvement for the lead-free all-perovskite BTSC over its MTSC counterpart with Ag back electrode. In contrast, soil and sea water albedos result in significantly reduced efficiencies. While the efficiency marginally increased with dry grass albedo, it marginally decreased with pond water albedo. Among all, snow albedo gave the best efficiency of 38.30% at an optimal $MAGeI_3$ and $FASnI_3$ thicknesses of 130 and 620 nm, respectively. The device optimization results under concurrent illumination show that the V_{OC} and FF are not affected much by the variations in spectral albedo and it is the J_{SC} that determines the overall improvement in PCE. The annual primary energy generation of the LPBTSC with the good reflecting surface has been found to be higher, thus capable of converting more solar energy into electricity than LPMTSC with an opaque rear contact. Also, the EPBT and GEF are significantly lower than the other existing TSCs.

REFERENCES

[1] Cruz, Alexandros, Darja Erfurt, Philipp Wagner, Anna B. MoralesVilches, Florian Ruske, Rutger Schlatmann, and Bernd Stannowski. "Optoelectrical analysis of TCO+ Silicon oxide double layers at the front and rear side of silicon heterojunction solar cells." Solar Energy Materials and Solar Cells 236 (2022): 111493.

[2] M. Green, E. Dunlop, J. Hohl-Ebinger, M. Yoshita, N. Kopidakis, X. Hao, Solar cell efficiency tables (version 60), Progress in photovoltaics: research and applications 30 (7) (2022) 687–701.

[3] Chen, Bo, Zhenhua Yu, Arthur Onno, Zhengshan Yu, Shangshang Chen, Jiantao Wang, Zachary C. Holman, and Jinsong Huang. "Bifacial all-perovskite tandem solar cells." Science Advances 8, no. 47 (2022): eadd0377.

[4] Karthick, S., Johann Bouclé, and S. Velumani. "Effect of bismuth iodide (BiI3) interfacial layer with different HTL's in FAPI based perovskite solar cell–SCAPS–1D study." Solar Energy 218 (2021): 157-168.

[5] Purkayastha, Atanu, Manoranjan Minz, Ramesh Kumar Sonkar, and Arun Tej Mallajosyula. "Investigation of lead-free 2D/3D mixed-dimensional tin perovskite solar cell embedded with plasmonic metal nanoparticles." In 2022 IEEE 49th Photovoltaics Specialists Conference (PVSC), pp. 0504-0506. IEEE, 2022.

[6] Pathak, Chetan, and Saurabh Kumar Pandey. "Design, Performance, and Defect Density Analysis of Efficient Eco-Friendly Perovskite Solar Cell." IEEE Transactions on Electron Devices 67, no. 7 (2020): 2837-2843.

[7] Zhao, Yu-Qing, Xuan Wang, Biao Liu, Zhuo-Liang Yu, Pe-Bing He, Qiang Wan, Meng-Qiu Cai, and Hai-Lin Yu. "Geometric structure and photovoltaic properties of mixed halide germanium perovskites from theoretical view." Organic Electronics 53 (2018): 50-56.

[8] Yao, Jizhong, Thomas Kirchartz, Michelle S. Vezie, Mark A. Faist, Wei Gong, Zhicai He, Hongbin Wu et al. "Quantifying losses in opencircuit voltage in solution-processable solar cells." Physical review applied 4, no. 1 (2015): 014020.

[9] K. Chakraborty, M. G. Choudhury, S. Paul, Life cycle assessment of a lead-free cesium titanium (iv) single and mixed halide perovskite solar cell based 1 m2 pv module, IEEE Transactions on Device and Materials Reliability 21 (4) (2021) 465–471.

Outdoor characterization of a bifacial four-terminal GaAs/Si mini-module under different albedo conditions

Roberto Corso[1], Fabio Matera[1], Andrea Scuto[2] and Salvatore A. Lombardo[1]

[1] Institute for Microelectronics and Microsystems (IMM), National Research Council (CNR), Catania, 95121, Italy

[2] 3SUN S.R.L. - Enel Green Power, Catania, 95121, Italy

Abstract — **Multijunction systems represent one of the most promising alternatives to overcome the theoretical efficiency limit of Si photovoltaics. In recent years, bifacial Si modules have been developed to increase the power output, but this approach can also be extended to multijunction modules. In this work we present the results of outdoor measurements on a four-terminal photovoltaic system based on combining a GaAs module with a bifacial silicon heterojunction module. Four grounds with different color have been placed under the four-terminal module, showing that the voltage match can be achieved in different conditions.**

I. Introduction

As the traditional Si-based photovoltaics (PV) is approaching its theoretical efficiency, with cells reaching efficiency values up to 26.7% [1], the research is focusing on finding new solutions in order to overcome this barrier and increase the power output of PV systems.

One of these solutions are the multijunction PV systems, in which two or more modules based on different bandgap semiconductors are joined together to collect solar light more efficiently and reduce thermalization losses. The two main module connection schemes are the two-terminal (2T) architecture, in which the modules are connected in series and that requires current matching among the modules, and the four-terminal (4T) architecture, in which the modules are connected in parallel and that requires voltage matching among the modules. Moreover, the modules can either be stacked in monolithic structures or be physically separated, in which case solar light has to be split by optic elements in order to send each portion of the solar spectrum to the semiconductor most suited to it.

Another possibility, which is already being implemented at utility scale, consists in employing bifacial cells. These cells are designed to collect albedo and diffuse light impinging on their back side. But this solution could also be extended to multijunction modules, and in particular to the bottom module. Moreover, between the two connection schemes described above, the 4T architecture is more suited to bifacial bottom modules, as the additional current produced by this module does not affect the voltage match required to maximize the power output of the multijunction system.

The feasibility of a 4T multijunction module with a bifacial bottom module has already been investigated in [2]. In this

work, we show that the bifacial 4T architecture can be implemented in several different conditions by monitoring the same 4T module in outdoor with four grounds, each characterized in terms of dispersive reflectivity.

II. Experimental Details

The 4T system is made up by a module of 2 series-connected GaAs cells connected in parallel with a module of 3 series-connected silicon heterojunction (SHJ) bifacial cells. SHJ cells are the most suited for bifacial applications, given their high bifaciality factor [3], ranging between 90% and 95%. Each GaAs cell is 2 cm x 2 cm, while each SHJ cell is 1.3 cm x 2 cm; therefore, each module as an area of 8 cm^2, and this will also be considered as the active area of the whole 4T device when calculating the power conversion efficiency (PCE). Direct solar light is split by two dichroic mirrors with a cut-on wavelength of 805 nm, reflecting lower wavelength light to the GaAs cells and transmitting higher wavelength light to the SHJ cells. We have measured the External Quantum Efficiency (EQE) of the SHJ module on both sides and of the GaAs module and the reflectivity and transmittance of the dichroic mirror from 300 nm to 1100 nm. These data are reported in Fig. 1.

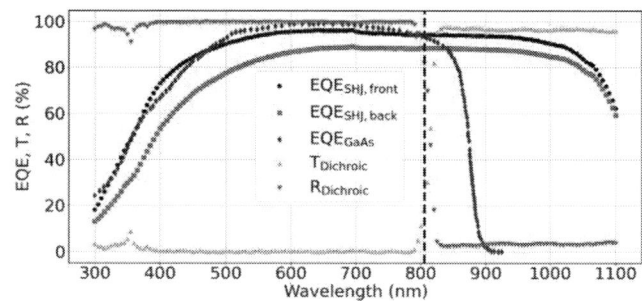

Figure 1 External Quantum Efficiency of the SHJ module front (black circles), of the SHJ module back (red square) and of the GaAs module (blue diamonds) and transmittance (cyan upward triangles) and reflectivity (magenta downward triangles) of the dichroic mirror. The dashed line corresponds to the cut-on wavelength of the dichroic mirrors.

To simulate different ground configurations, four paper sheets have been used, each with a different color: grey, white, green and orange. These paper sheets mimic asphalt, snow, grass and shingles respectively, four common albedo conditions for bifacial applications. The reflectivity of each sheet has been measured, obtaining the data reported in Fig. 2.

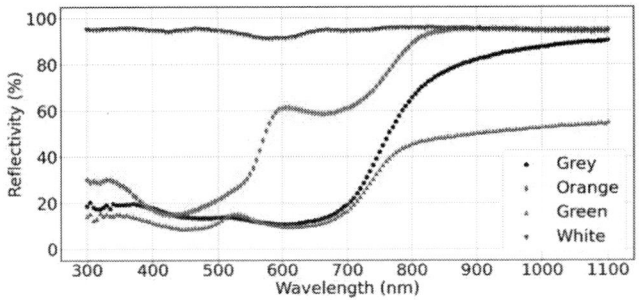

Figure 2 Reflectivity values of the four paper sheets in the 300 nm – 1100 nm range.

Of the four sheets, the white sheets exhibit the highest reflectivity, with values close to 100% over the 300 nm – 1100 nm wavelength range. The other three sheets showed a low reflectivity below 500 nm, with the orange sheet surpassing 50% reflectivity starting from 600 nm. Nevertheless, all these three samples achieved their highest reflectivity values in the IR range, with the green sheet being significantly less reflective than the other samples.

Fig. 3 reports pictures of the experimental setup with the different paper sheets. Due to the presence of the mirrors, the 4T device cannot redirect diffuse light efficiently, and it has been mounted on a bi-axial solar tracker, with the paper sheet being parallel to the back of the SHJ minimodule at a distance of 30 cm. Each paper sheet is 65 cm x 50 cm.

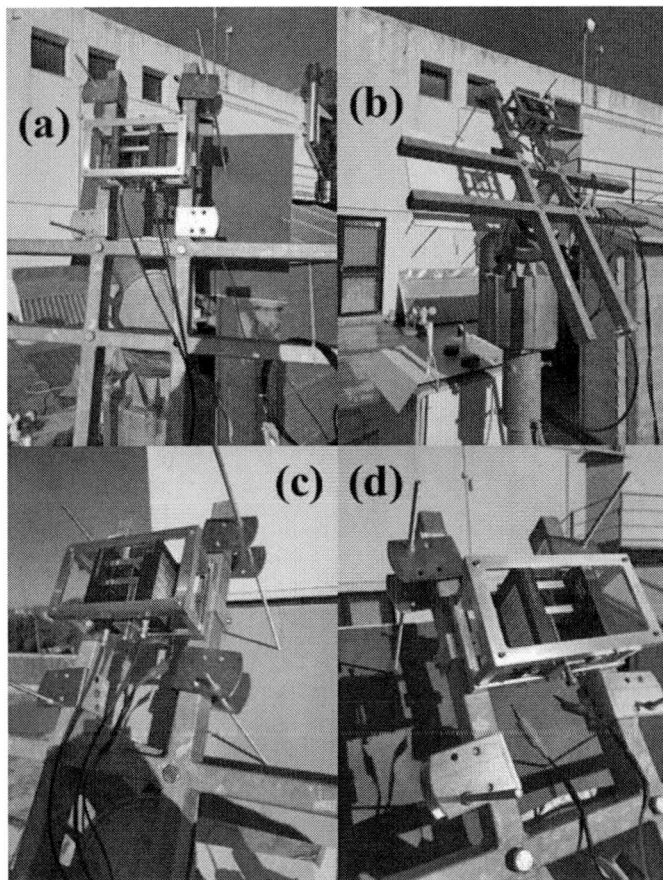

Figure 3 Pictures of the 4T device mounted on the bi-axial solar tracker with grey (a), white (b), orange (c) and green (d) paper sheet.

III. RESULTS

Fig. 4 reports the PCE values relative to the same 4T device operating in monofacial mode by covering the back of the SHJ module. The data show a PCE increase between 10% and 20% when the device is mounted above the grey sheet, an increase between 20% and 40% when mounted above the white sheet, an increase of about 10% when above the green sheet and an increase between 20% and 30% when above the orange sheet.

It has to be noted that the PCE increases between the bifacial mode and the monofacial mode are strongly correlated with the reflectivity ranking of the sheets reported in Fig. 2; moreover, as the EQE of the SHJ module back covers the entire wavelength range (Fig. 1), all the albedo light is collected by the SHJ module, so only this module produces additional current, whereas the GaAs module is not influenced by the albedo light. Nevertheless, the PCE increase in all cases considered, indicating that the voltage match is achieved in very different albedo conditions.

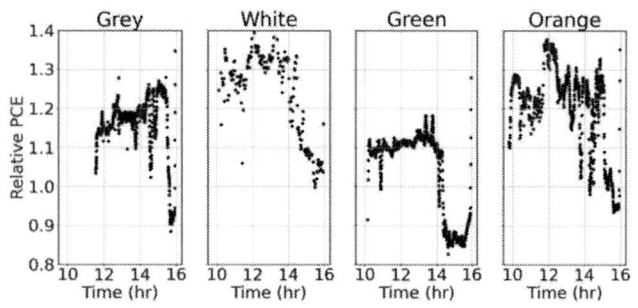

Figure 4 PCE values obtained with the four colored paper sheets in relation compared to the same 4T device operating in monofacial mode by covering the back of the SHJ module.

IV. CONCLUSIONS

In this work we have measured the power conversion efficiency of a 4T device in which a GaAs minimodule operates in conjunction with a bifacial silicon heterojunction minimodule. Direct solar light is split by dichroic mirrors, sending the visible portion to the GaAs minimodule and the infrared portion to the Si minimodule. Additionally, the latter also collects albedo light impinging on its back.

Four different albedo conditions have been investigated: these have been realized with colored paper sheets, each characterized in terms of dispersive reflectivity. The data show that, despite the higher current produced by the silicon heteojunction minimodule due to the additional albedo light, the consistency of the voltage match between the two minimodules allows the 4T device to achieve a PCE increase with respect to the same device in monofacial mode ranging from 10% to 40%.

REFERENCES

[1] M. A. Green, E. D. Dunlop, J. Hohl-Ebinger, M. Yoshita, N. Kopidakis, K. Bothe, D. Hinken, M. Rauer, and X. Hao, "Solar Cell Efficiency Tables (version 60)," *Progress in Photovoltaics: Research and Applications*, vol. 30, no. 7, pp. 687–701, 2022.

[2] A. Scuto, R. Corso, M. Leonardi, R. G. Milazzo, S. M. S. Privitera, C. Colletti, M. Foti, F. Bizzarri, C. Gerardi, and S. A. Lombardo, "Outdoor performance of GaAs/Bifacial Si heterojunction four-terminal system using optical spectrum splitting," *Solar Energy*, vol. 241, pp. 483–491, 2022.

[3] R. Kopecek and J. Libal, "Bifacial Photovoltaics 2021: Status, Opportunities and Challenges," *Energies*, vol. 14, no. 8, p. 2076, Apr. 2021.

Evaluation of PV Snow Loss Models in the East Coast of Canada using AI Computer Vision

Jessica Ma, Alexandre Khoury, and Marianne Rodgers

Wind Energy Institute of Canada (WEICan), North Cape, PE, C0B 2B0, Canada

Abstract — **Comprehensive understanding of the impact of snow cover on energy generation losses in photovoltaic arrays is needed to ensure realistic generation values are considered when developing, operating, or investing in solar systems in areas of colder climates. The current study uses an AI computer vision algorithm to process images capturing snow coverage on a 109 kW fixed tilt photovoltaic array in North Cape, PE, Canada. The algorithm calculates fractional snow coverage on the panels and is used in combination with site data to determine production losses due to snow cover. WEICan's results are compared to two predictive snow loss models prominent in the industry – the Townsend and the NREL models. It was observed that the absolute snow loss values on a monthly and annual basis were overestimated in both cases but could be significantly improved by using site specific model coefficients. The image analysis used in this study demonstrates how two prominent snow loss models perform using default and site specific model parameters and can also be used to indicate the snow losses that will be incurred for a windy coastal site in a northern latitude.**

I. INTRODUCTION

Solar capacity in North America has significantly increased over the last 10 years, with average annual growth rates of 21% and 31% in Canada and the US, respectively [1]. This growth has resulted in more sites being developed at mid to high latitudes in areas of colder climates. In winter, these areas typically see higher electricity consumption and have lower solar generation due to lower solar resources and snow, which can accumulate on panels and result in significant reduction in generation.

Previous studies have found that snow related losses are generally less than 10% annually [2]-[4] but can be as high as 34% in some areas [5]. Furthermore, the impact on daily energy production can be up to 100%, resulting in full days with no contribution from solar generation and a higher reliance on alternative energy generation [6]. Therefore, it is necessary to understand the impact of snow related losses to ensure realistic generation values are considered when developing, operating, or investing in solar photovoltaic (PV) systems.

Over the last decade predictive models have been developed to help provide realistic solar generation values for winter months. Many predictive snow loss models consider historic snow fall and meteorological data together with project design parameters (azimuth, tilt, racking, etc.) and often include a snow clearing mechanism (shedding or melting) to assess the potential future snow loss for a site [6]. Other snow loss models apply a curve fitting approach to estimate current or future snow loss but require on site data or data from nearby operational sites with similar characteristics to determine empirical constants [3]. This study considers two predictive snow loss models prominent in the industry: the Townsend model [7] which uses an empirical formula, and the National Renewable Energy Laboratory (NREL) model which uses a snow fall and clearing mechanism to assess potential snow loss. The NREL model is used in the System Advisor Model (SAM) [8] and is based on the Marion model [4]. Both models require snowfall and other meteorological input data.

Evaluations of the accuracy of these types of models has previously been conducted at a number of inland sites in North America, including Colorado[4], California[7], and Wisconsin[4]. As snow loss is dependent on several different variables, the magnitude and profile of the loss can vary significantly for different areas, regions, and climates. Thus, continued evaluation is needed to provide a better understanding of snow impact on winter PV generation across different regions and in coastal areas.

To evaluate operational snow-related losses, the aforementioned studies assessed the measured snow loss considering historic production and a baseline performance model for each site. In [7] snow-related losses were further isolated by conducting side by side panel measurements and manually clearing snow from one panel (baseline) while allowing the other to accumulate snow naturally. In [4] manual image inspection was conducted to assess snow cover amounts. Although effective, these methods are labor intensive and are dependent on the quality and consistency of the clearing and evaluation routine. More recently, studies have looked at automated image capture and processing to monitor and assess snow and soiling on PV Panels [3][9]. This is done by setting up a camera to capture images of the full or part of the array at regular intervals (e.g., 1 min, 5 min) and analyzing each image to calculate the amount of snow cover.

The current study uses images captured on a 2-minute basis to record snow coverage information for a 109 kW fixed tilt PV array with a mix of monofacial and bifacial panels at the WEICan site in North Cape, PE, Canada, which is characterized by high winds, average regional annual snowfall of 220 cm and temperatures below freezing from December to March. The images are processed using a machine learning algorithm in python to calculate the fraction of snow coverage across the array. The snow cover information is analyzed together with system Supervisory Control and Data Acquisition (SCADA)

978-1-6654-6060-6/23 $31.00 © 2023 IEEE

data to capture observed system losses and evaluate the Townsend and NREL predictive snow loss models for the site.

II. METHODOLOGY AND EXPERIMENTAL SETUP

Data were collected over a period of more than one year from January 2022 to April 2023, inclusive, from WEICan's Solar PV array and weather station. The following provides details on the available instrumentation, as well as the methodology for image capture, image analysis and data analysis.

A. WEICan 109 kW solar PV system and weather station

TABLE I
WEICAN PV SYSTEM

Location	47.033, -64.016
Altitude	15 m
Tilt angle	30 deg
Azimuth	30 deg
Orientation	landscape
PV modules	187 x Jinko JKM400M-72HL-TK 85 x Jinko JKM405M-72HL-V
Module capacity (DC)	109 kW
String size	17 modules
Inverters	2 x STP 62-US-41
Inverter capacity (AC)	124 kW
Power ratio	0.88
Racking	Fixed

WEICan has a 109 kW Solar PV system located at the northwestern most tip of Prince Edward Island, Canada. Details on the PV system are provided in Table I.

The array is sized so there is no inverter clipping; therefore, the impact of snow on the AC side is directly proportional to the impact of snow on the DC side.

Plane of Array (POA) solar irradiance and reflected irradiance are captured by reference cells mounted on the racking system in the middle of the array. A weather station is located approximately 5 m north of the PV system. The weather station captures wind speed and wind direction from an ultrasonic anemometer; temperature from a PT1000 sensor; and humidity from an HMP155A sensor. A ClimaVue all-in-one weather sensor is also mounted on the station; wind and temperature data captured by the ClimaVue are used as secondary sources for wind speed and direction. All data are recorded on a 1-minute basis with data available from November 2021. Snowfall intensity was also measured at the site from a Parsivel^2 OTT present weather sensor disdrometer from December 2022.

B. Image capture and AI Computer Vision

WEICan mounted a Campbell Scientific CCFC camera facing its PV system that captured images of the PV panels every two minutes from December 2021. The camera is mounted on a security pole next to the WEICan substation and looks down toward the PV system. The experimental setup of the camera is shown in Fig. 1.

Fig. 1   View of camera, mounting, and arrangement.

From the captured images, variations of snow coverage can be quantified and measured using AI computer vision. To do so, an algorithm was developed to determine the percentage of snow covering the PV array for any given time. This algorithm uses image segmentation with the unsupervised k-means algorithm from scikit-learn and a random walker segmentation model [10]. Using this algorithm, snow coverage information is recovered every two minutes from January 2022 to April 2023. For best performance, the algorithm was used to gather data between 11:00 and 18:00. This period correlates with peak power generation.

C. Identifying snow impact in terms of energy production loss.

To quantify snow losses, an operational performance baseline for the system was required. The performance model was calculated using a multilinear regression equation (1) as defined in ASTM E2848 − 13(2018) [11], which considers screened and filtered on-site meteorological measurements and power production for data periods not impacted by snow cover.

$$P = E(a_1 + a_2 * E + a_3 * T_a + a_4 * v) \qquad (1)$$

Where E is the POA irradiance term, T_a is ambient temperature, v is wind speed and a_1, a_2, a_3, a_4 are the linear regression coefficients.

Further, the NREL suggested modification for bifacial capacity testing [12] was applied according to equation (2), which accounts for reflected irradiance in the irradiance parameter, E, of (1).

$$E = E_{POA} + E_{Rear} * \varphi \qquad (2)$$

Where E_{POA} is the plane of array irradiance, E_{Rear} is the reflected irradiance and φ is the bifacial factor.

As the WEICan array is made up of a mix of monofacial and bifacial panels the contribution of E_{rear} in (2) was weighted by the number of bifacial panels relative to the overall array.

Given the seasonal variation in solar and weather patterns, the performance model was calculated on a rolling six-week basis to ensure data was seasonally representative of the assessed period. The resulting baseline performance model was compared to unfiltered SCADA power production. Outlier data records which were not impacted by snow but were identified as periods of underperformance by having a residual greater than two standard deviations were excluded from the regression calculations. The final performance model was used to assess the potential power production for snow impacted periods and to quantify the snow cover production loss for the site.

C. Predictive snow loss models

Model snow loss predictions were made with the NREL and Townsend models which are both prominent in the industry and available through the PVLIB python package [13]. The NREL model was implemented on an hourly basis using the default coefficients for a ground-based system [4]. A site-specific coefficient for whether snow can slide or not was also assessed based on measured POA irradiance and ambient temperature data for times where measured snow coverage was reduced relative to the previous record as evaluated from the computer vision results. The site-specific slide coefficient was used for a second implementation of the model.

The Townsend model derives snow loss through an equation considering the number of snow events, effective snow, ground interference, relative humidity, slant angle, temperature, and irradiance on a monthly basis. The model was initiated using no ground interference and a standard coefficient C1=5.7E04, which is based on a regression fit of the model to data from sites in Truckee, California [7]. Given the difference in geographic location and climate for WEICan compared to the Townsend reference site, a site-specific coefficient was also derived using a direct fit of the model to the operational snow loss for the site. As this is a direct fit to the data, the operational period was split into a 'training' period and a 'test' period for the evaluation of the model in this analysis.

Wind, temperature, humidity, and irradiance inputs for both models are from the site weather station. As site measured snow fall data were not available for the majority of 2022, modelled hourly snowfall values were calculated from the National Operational Hydrologic Remote Sensing Center (NOHRSC) modelled snow precipitation values for the nearby CWNE site location [14].

II. Results

A. Assessment of WEICan Snow Cover Production Loss

Fractional snow cover on the PV panels was calculated from January 2022 to April 2023 based on AI computer vision analysis. An example of the PV system setup for a period with partial snow cover and the resulting segmented image is shown

in Fig. *2* and results of the diurnal trend of snow cover for a clear sky day after overnight snowfall is shown in Fig. *3*.

(a)

(b)

Fig. 2   PV system setup at the WEICan site. (a) shows snow cover for 12:04 am February 9, 2022. (b) shows the segmented image for the same times, the yellow mask shows the part of the PV system that is not covered in snow as identified by the algorithm.

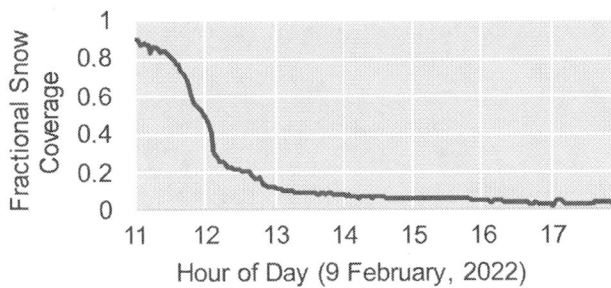

Fig. 3   Diurnal fractional snow cover on WEICan PV panels for February 9, 2022.

The machine learning algorithm developed for this study performed well for days with snow coverage but was sensitive to rain, fog, icing on the camera and sun glare off the solar panels which in some cases produced false positives. Most notably, false positives were observed for the months of October to November between 11:00 and 13:00, when there was significant glare on the solar panels. As these occurrences did not coincide with any snowfall periods, these data were manually set to have a snow flag of 0.

The periods where snow was identified on the panels align with periods of snowfall measured at the site and modelled at the NOHRSC CWNE location as shown in Fig. *4*. For the period of data shown, it is noted that small periods of very light snow occurred on the 6th of January resulting in a thin layer of snow across the panels, and that most of the snow fall on the 26th of January occurred in the early morning and quickly turned to rain so was not detected during the period considered for the image analysis.

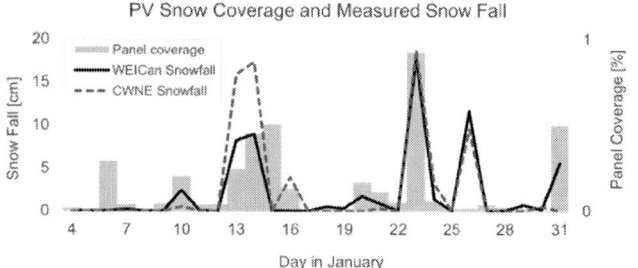

Fig. 4   Daily PV panel snow coverage detection compared to WEICan measured snow fall and NOHRSC snow fall for January 2023.

The resulting snow flags were used to identify periods of power production impacted by snow cover and assess the historic energy loss for the site using the site calculated performance model. Good data coverage for all variables was available for the performance model with a total availability of 95.6% for the evaluated hours.   Results of the baseline performance model compared to unfiltered SCADA power production is shown in Fig. 5. The resulting snow cover production loss for the site is shown in Table II and Fig. 6.

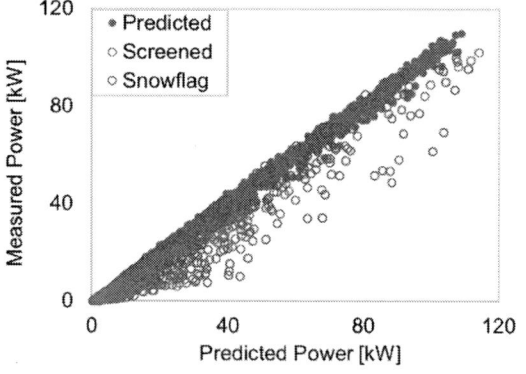

Fig. 5   Hourly Measured Power vs Predicted Baseline Power (solid blue) for curve fitting. Hollow data circles in red represent data records where snow was observed. Hollow data circles in blue represent data periods of underperformance with no visual indication of snow cover which were excluded for curve fitting only.

B. Evaluation of Snow Loss Models

The Townsend model [7] and the NREL model [8] were evaluated using WEICan's weather station data and modelled hourly snowfall values from NOHRSC [14]. To determine whether the data analyzed represented a typical snow year, the NOHRSC evaluated period was compared to historical data and was found to be on the lower end of the range but within the standard deviation with respect to mean annual snowfall going back to 2016. Table II presents the annual snow loss model results.

Two variations of the NREL model have been presented. The first considered default sliding parameters (default) and the second considered a site-specific sliding parameter of -50 W/m²/°C determined from site measured data (site specific). The NREL model overestimates snow loss in both cases; however, using the site-specific sliding parameter leads to significantly better energy loss estimates for the site.

TABLE I
ANNUAL SNOW LOSS

WEICan	1.3 %
NREL (default)	2.7 %
NREL (site specific)	1.7 %
Townsend	2.7 %

The Townsend model also overestimates snow loss for the site. A direct fit of the Townsend model to WEICan monthly PV snow loss was conducted for the 2022/2023 winter period (approximately 6 months with snowfall measurements), which resulted in a revised coefficient of 3.2E04 with a reasonable r^2 fit of 0.8.   The revised coefficient provides a potential alternative for the Townsend model for sites similar to WEICan and was used to assess the snow loss for a testing period from January to April 2022 as shown in Fig. 6; however, further data is required to better test the performance of the site specific coefficient and to see if it is applicable for different sites in the region.

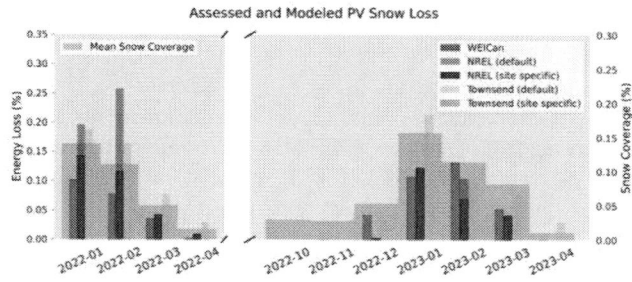

Fig. 6   NREL, Townsend and WEICan energy loss estimates due to snow. WEICan's algorithm for the monthly average of snow coverage is represented with shaded gray bars.

Fig. 6 presents the monthly snow loss model results for snow impacted months. It also presents the average snow coverage results from the machine learning algorithm on the secondary axis. As expected, due to the seasonal changes and rise in temperatures, there is a downwards trend for snow coverage and snow loss from January to April (for both 2022 and 2023) and an increasing trend from October to December. Although the two snow loss models follow the same seasonal trend as the detected site data, they result in significant overpredictions for all months when using default parameters. The monthly results of both models are considerably improved when using site specific parameters.

IV. SUMMARY

In this paper, imaging techniques using AI computer vision are used to analyze losses due to snow cover on the WEICan 109 kW fixed tilt PV system. Using a machine learning algorithm, the percentage of snow coverage on the PV system was determined from camera data and, in turn, used to assess snow cover production loss on a monthly and annual basis.

The NREL and Townsend models were evaluated for the WEICan site and compared to the site snow-loss values. The total energy loss for 2022 due to snow has been found to be 2.7% for both the NREL and Townsend models considering default parameters. These results overpredict the 2022 annual operational snow loss of 1.3% assessed for the site. The annual and monthly energy loss predictions were significantly improved with the use of site specific parameters based on site measured data for each model.

The image analysis used in this study is a repeatable process that can be applied to any site to give an indication of the snow losses for any operational PV system. If a site is not operational or the image capturing equipment is not available, this study demonstrates how two prominent snow loss models perform and can be used to indicate the snow losses that will be incurred on a coastal site in a northern latitude. It also provides an indication of the benefits of using site-specific data to improve model predictions compared to using default parameters, where data are available, and reports indicative alternative values that could be used for sites with similar characteristics to WEICan.

Next steps for this work are to improve the AI computer vision algorithm to include a longer time span, reduce sensitivities and consider individual monofacial and bifacial strings or panels separately from the full array; and conduct further analysis of additional site data or of data from nearby sites to further evaluate site specific parameters. In addition, WEICan has recently installed a pyrheliometer and a disdrometer which can be used to augment snowfall and irradiance data used in the study.

REFERENCES

[1] IRENA (2022), Renewable Energy Statistics 2022, The International Renewable Energy Agency, Abu Dhabi, ISBN: 978-92-9260-428-8.

[2] E. Janssen, D. Nixon, S. De Bruyn, G. Amdurski, and L. St. Hilaire, "Simple approach to estimating PV system snow losses applied to long-term PV generation datasets for different tilt angles and mounting styles," *Alternative Energy and Distributed Generation Journal*, vol. 2(2), pp. 59-76, 2021.

[3] R. W. Andrews, A. Pollard, and J. M. Pearce, "The effects of snowfall on solar photovoltaic performance," *Solar Energy*, vol. 92, pp. 84–97, 2013.

[4] B. Marion, R. Schaefer, H. Caine, and G. Sanchez, "Measured and modeled photovoltaic system energy losses from snow for Colorado and Wisconsin locations," *Solar Energy,* vol. 97, pp. 112–21, 2013.

[5] N. Heidari, J. Gwamuri, T. Townsend, and J. M. Pearce, "Impact of snow and ground interference on photovoltaic electric system performance," *Journal of Photovoltaics*, vol. 5, pp. 1680–1685, 2015.

[6] R. E. Pawluk, Y. Chen, and Y. She, "Photovoltaic electricity generation loss due to snow—A literature review on influence factors, estimation, and mitigation," *Renewable and Sustainable Energy Reviews*, vol. 107, pp. 171-182, 2019.

[7] Townsend and L. Powers, "Photovoltaics and snow: an update from two winters of measurements in the SIERRA," in *37th IEEE Photovoltaic Specialists Conference,* 2011, p. 25.

[8] D. Ryberg and J. Freeman, "Integration, validation, and application of a PV snow coverage model in SAM," *National Renewable Energy Laboratory*; 2017

[9] J. L. Braid, D. Riley, J. M. Pearce, and L. Burnham, "Image Analysis Method for Quantifying Snow Losses on PV Systems," in *47th IEEE Photovoltaic Specialists Conference*, 2020, pp. 1510.

[10] L. Grady, "Random walks for image segmentation*", IEEE Transactions on Pattern Analysis and Machine Intelligence.*, vol. 28, pp. 1768-1783, 2006.

[11] ASTM E2848-13, "Standard Test Method for Reporting Photovoltaic Non-Concentrator System Performance," 2013.

[12] M. Waters, C. Deline, J. Kemnitz and J. Webber, "Suggested modifications for bifacial capacity testing", in *46th IEEE Photovoltaic. Specialists Conference*, 2019, pp. 1-6.

[13] William F. Holmgren, Clifford W. Hansen, and Mark A. Mikofski. "pvlib python: a python package for modeling solar energy systems." Journal of Open Source Software, 3(29), 884, (2018). https://doi.org/10.21105/joss.00884

[14] A. P. Barrett, "National Operational Hydrologic Remote Sensing Center SNOw Data Assimilation System (SNODAS) Products at NSIDC." 2003

978-1-6654-6060-6/23 $31.00 © 2023 IEEE

Influence of NaF and KF post-deposition treatment on the sub-band gap absorption of Cu(In,Ga)Se₂ absorber layers

Sevan Gharabeiki, Michele Melchiorre and Susanne Siebentritt

Laboratory for Photovoltaics, Department of Physics and Materials Science,
University of Luxembourg, L-4422 Belvaux

Abstract — Urbach tails describe a density of states that decays exponentially into the band gap. Non-radiative and radiative recombinations in the Urbach tails contributes to voltage loss in the Cu(In,Ga)Se2 CIGSe solar cells. In the CIGSe semiconductor, the Urbach tails can be suppressed by alkali fluoride post-deposition treatment (PDT). We report the impact of NaF and KF-PDT on the quasi-Fermi level splitting (QFLS) and Urbach energy of the CIGSe absorbers, which are evaluated by absolute photoluminescence measurement. Our data reveal that both NaF and KF PDT have beneficial impacts on the QFLS. Samples with a low Na concentration show high Urbach energies, which are decreased by NaF-PDT. Sufficient KF PDT on the samples deposited on Mo-coated soda lime glass (i.e., with a high concentration of Na) further decreases the Urbach energy. The improvement of the QFLS is not solely due to the reduction of tail states, but it can also be due to surface passivation, grain boundary passivation or increase in the doping concentration.

I. INTRODUCTION

The efficiency of Cu(In,Ga)Se2 CIGSe solar cells has already surpassed 22% [1]. It has long been known that the addition of Na has beneficial effects on the efficiency of CIGSe solar cells. The main impact of Na is an increase in the open circuit voltage due to Na-induced higher carrier concentration [2]. The recent advancements in the efficiency of CIGSe solar cells are mainly due to heavy alkali post-deposition treatment (PDT). The first heavy alkali PDT was performed by KF [3]. Later, Rb and Cs were also introduced into the CIGSe absorbers by RbF and CsF PDT [1]. In all cases, the main effect of the heavy alkali PDT is increase in the open circuit voltage.

Tail states extend from the band edges into the forbidden gap. Part of these tail states is caused by phonon vibration i.e., thermal effect. However, an additional contribution can be made by static disorders such as alloy disorder, band gap fluctuation, electrostatic potential fluctuation, and grain boundaries [4, 5]. Tail states increase the radiative V_{OC} loss and act as non-radiative recombination centers, which have detrimental effects on Voc and, in turn on the efficiency of CIGSe solar cells.

In the CIGSe absorbers, the joint density of band tail states decays exponentially into the bandgap with a decay constant called Urbach energy. Higher Urbach energies mean that more density of states is available within the bandgap. In the ideal solar cell, the photons with energies lower than the band gap energy are not absorbed. However, in the presence of the Urbach tails, the absorption coefficient within the bandgap is not zero anymore. Since the Urbach tails contribute to the

photon absorption, the Urbach energy of the CIGSe absorbers can be extracted from the exponential decay of the absorption coefficient for energies lower than the band gap energy. Photoluminescence (PL) is a powerful tool that allows us to measure reliable values for Urbach energies deep into the bandgap [6].

In the first part of this contribution, we investigate the impact of NaF and KF PDT on the quasi-Fermi level splitting (QFLS) of CIGSe absorbers. Then, we extract the absorption coefficient and Urbach energy from the absolute PL spectrum. Finally, we discuss the effect of NaF and KF PDT on our samples' Urbach energy and QFLS.

II. EXPERIMENTAL SECTION

A. Sample Preparation

Overall, six different CIGSe samples have been investigated in this study. The sample properties are shown in Table I.

The CIGSe samples are prepared by 3-stage co-evaporation process. Mo-coated high temperature glass (HTG) and Mo-coated soda lime glass (SLG) were used as substrate. HTG contains very low concentration of Na compared to SLG. Additionally, 10 nm NaF was deposited as a precursor on Mo-coated SLG substrate. The maximum deposition (set) temperature is 580 and 560 °C for SLG and HTG, respectively. After the growth process, the samples were removed from the vacuum chamber. In each deposition run, four absorbers were prepared. Two absorbers were used as reference samples, and KF or NaF PDT was performed on the remaining pieces.

NaF PDT: 7.5 nm and 15 nm NaF was evaporated at room temperature on the absorbers in vacuum, followed by 20 minutes of annealing at 400°C in the Se atmosphere.

KF PDT: The KF was evaporated along with the Se on the surface of the absorbers at substrate temperature of 350 °C. The KF source temperature for the PDT was either 450°C or 460 °C.

B. Photoluminescence Characterization

Absolute photoluminescence measurements were carried out to extract the QFLS and Urbach energy of the absorbers. The samples were excited by a 660 nm diode laser. The incident photon flux density is equal to 1 sun photon flux density above the band gap. The QFLS is extracted from a fit of the high energy slope of the PL peak to Planck's generalized law. According to this method, we assume that absorptance for the

---

978-1-6654-6060-6/23 $31.00 © 2023 IEEE

photon energies higher than the band gap energy is equal to one. The temperature, which is related to the slope of this fit, is fixed at the measured temperature of 295 K. With the known QFLS the absorptance spectrum is extracted from Planck's generalised law and then the absorption coefficient from Lambert-Beer's law. More details about the QFLS and absorption coefficient calculation can be found elsewhere [7, 8].

TABLE I
SAMPLE INFORMATION

Name	Substrate	NaF PDT	KF PDT (KF source Temperature)
Ref NaF	HTG	No	No
NaF-7.5	HTG	7.5 nm	No
NaF-15	HTG	15 nm	No
Ref KF	SLG	No	No
KF-450	SLG	No	450 °C
KF-460	SLG	No	460 °C

III. RESULTS AND DISCUSSION

Fig. 1. depicts the PL spectra of the NaF and KF treated CIGSe samples. It can be seen that NaF and KF-PDT increases the PL quantum efficiency (PLQY) of these samples (see also Table II) . The dominant peak is attributed to the band-to-band transition. For the NaF series deposited on HTG, besides the main peak, another peak with lower intensities can be detected at lower energies. The origin of the second peak is likely a defect state. However, with the increase in the Na concentration, i.e., samples with 15nm NaF PDT, the second peak intensity decreases, and the main peak shifts to somewhat higher energies.

The PL spectra of the KF-series are displayed in Fig. 1. (b) The PL spectra of these samples only show the band-to-band transition. In contrast to the NaF-series on HTG, the low energy peak does not appear in these samples and the PL peak position does not change by KF-PDT.

Since the PL spectra of NaF series show a second peak at lower energies, the linear fit for the Urbach energy extraction was done at the energy range where the spectrum is not distorted by the low energy peak. However, it cannot be fully excluded that the low energy peak influences the slope in this region. 7.5 nm NaF-PDT has almost no effect on the Urbach energy. Since the Urbach tails are not passivated in this sample (i.e., the Urbach energy remained unchanged), the QFLS improvement of 80 meV is solely attributed to the Na-induced higher hole concentration. On the other hand, with 15 nm of NaF PDT, the Urbach energy of this sample decreases from 15.2 meV to 13.4 meV. The blue shift of the main PL peak also supports the decrease in the Urbach energy. It has been reported by many studies that sub-bandgap absorption can shift the PL

peak to lower energies [9, 10]. Thus, by passivating the Urbach tails, a blue shift is expected for the PL spectrum.

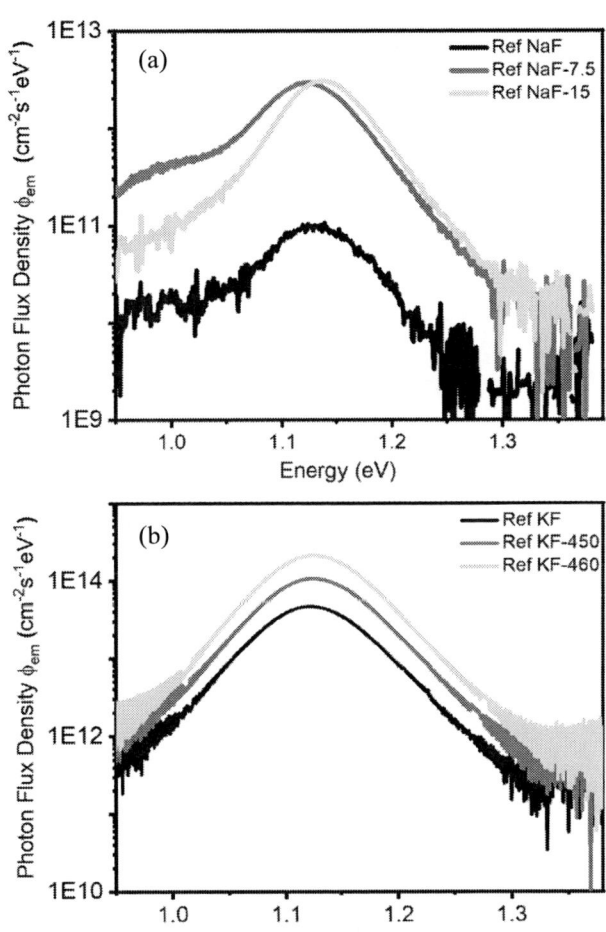

Fig. 1.   a) PL spectrum of NaF treated samples . b) PL spectrum of KF treated samples

Fig. 2. shows the absorption coefficient of the KF sample series. Since the absorption coefficient decays exponentially within the bandgap, the Urbach energy can be extracted from the linear fit of the absorption coefficient in the log scale.

For the KF series, the extracted Urbach energy show that Ref KF and KF-450 samples within the error have almost similar values (13.4 and 13.2 meV respectively). In these samples the PLQY and QFLS improvement should have another reason than urbach tails passivation. Surface and grain boundary passivation or increase in the doping level can be possible reasons for increase in the QFLS.

With the further increase in the KF source temperature (i.e., increase in the KF concentration), the Urbach energy decreases to 12.8 meV. Increase in the QFLS for this sample can be attributed to passivation of the Urbach tails. Additional time resolved PL shows that along with the decrease in the Urbach energy the life time of the samples increases.

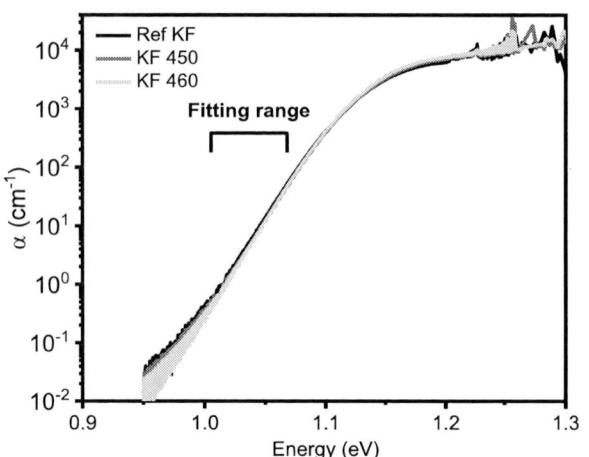

Fig. 2.   Absorption coefficient of the KF sample series

TABLE II
SUMMARY OF QFLS AND URBACH ENERGY

Sample	QFLS (meV)	PLQY	Urbach Energy (meV)
Ref NaF	447	4.1×10^{-8}	15.2
NaF 7.5	527	1.1×10^{-6}	15.3
NaF 15	535	9.7×10^{-7}	13.4
Ref KF	604	1.7×10^{-5}	13.4
KF-450	626	3.9×10^{-5}	13.2
KF-460	642	7.6×10^{-5}	12.8

IV. CONCLUSION

In summary, we have shown that KF and NaF post-deposition treatment increase the QFLS of CIGSe absorbers, independent of an additional change in Urbach energy. Low Na samples based on HTG have rather high Urbach energies that are reduced with sufficient NaF treatment. Samples on SLG with sufficient Na supply have already a lower Urbach energy before the treatment. KF treatment can additionally reduce the Urbach energy. Since an increase in QFLS was observed also if no decrease in Urbach energy was observed, other reasons such as increase in the doping or surface and grain boundary passivation have to be taken into consideration.

ACKNOWLEDGMENT

Financial support from the Luxembourgish Fonds National de la Recherche (FNR) in the framework of the projects TAILS.

REFERENCES

[1] P. Jackson, R. Wuerz, D. Hariskos, E. Lotter, W. Witte, and M. Powalla, "Effects of heavy alkali elements in Cu(In,Ga)Se 2 solar cells with efficiencies up to 22.6%," *physica status solidi (RRL) – Rapid Research Letters,* vol. 10, no. 8, pp. 583-586, 2016, doi: 10.1002/pssr.201600199.

[2] F. Pianezzi *et al.*, "Unveiling the effects of post-deposition treatment with different alkaline elements on the electronic properties of CIGS thin film solar cells," *Phys Chem Chem Phys,* vol. 16, no. 19, pp. 8843-51, May 21 2014, doi: 10.1039/c4cp00614c.

[3] A. Chirila *et al.*, "Potassium-induced surface modification of Cu(In,Ga)Se2 thin films for high-efficiency solar cells," *Nat Mater,* vol. 12, no. 12, pp. 1107-11, Dec 2013, doi: 10.1038/nmat3789.

[4] O. N. Ramírez, J.;Dingwell, F.; Wang, T.; Prot, Aubin Siebentritt, S.,, "On the origin of tail states and VOC losses in Cu(In,Ga)Se2.," *Submitted.*

[5] G. D. Cody, T. Tiedje, B. Abeles, T. D. Moustakas, B. Brooks, and Y. Goldstein, "Disorder and the Optical Absorption Edge of Hydrogenated Amorphous Silicon," *Le Journal de Physique Colloques,* vol. 42, no. C4, pp. C4-301-C4-304, 1981, doi: 10.1051/jphyscol:1981463.

[6] E. Daub and P. Wurfel, "Ultralow values of the absorption coefficient of Si obtained from luminescence," *Phys Rev Lett,* vol. 74, no. 6, pp. 1020-1023, Feb 6 1995, doi: 10.1103/PhysRevLett.74.1020.

[7] M. H. Wolter *et al.*, "How band tail recombination influences the open‐circuit voltage of solar cells," *Progress in Photovoltaics: Research and Applications,* 2021, doi: 10.1002/pip.3449.

[8] S. Siebentritt, T. P. Weiss, M. Sood, M. H. Wolter, A. Lomuscio, and O. Ramirez, "How photoluminescence can predict the efficiency of solar cells," *Journal of Physics: Materials,* vol. 4, no. 4, 2021, doi: 10.1088/2515-7639/ac266e.

[9] U. R. J. Mattheis, and J. Werner, "Light absorption and emission on semiconductors with band gap fluctuations - a study on Cu(In,Ga)Se2 thin films," *J. Appl. Phys,* vol. 101, no. 11, 2021, doi: doi.org/10.1063/1.2721768.

[10] S. Siebentritt *et al.*, "Photoluminescence assessment of materials for solar cell absorbers," *Faraday Discussions,* vol. 239, no. 0, pp. 112-129, Oct 28 2022, doi: 10.1039/d2fd00057a.

The Importance of Data Quality for Reducing the Uncertainty of Site-Adapted Solar Resource Datasets

Kristen Wagner[1], Alex Kubiniec[1], Tom McAlister[1], and Richard Perez[2]

[1] Clean Power Research, Kirkland, WA
[2] SUNY Albany, NY

Abstract — **In a 2016 study, Clean Power Research presented a statistically sound methodology for quantifying the uncertainty of satellite-derived solar resource datasets adapted to ground-measured data. Since that time, Clean Power Research has made numerous improvements to SolarAnywhere® data and reference data quality control (QC) procedures. Using the same methodology and site-adaptation procedure as the original study, we demonstrate a 40% reduction in the uncertainty of site-adapted solar resource data in North America. 83% of the reduction is attributable to the increased temporal consistency of SolarAnywhere V3.6 relative to prior versions. The remaining reduction in uncertainty (17%) is due to improved data QC. Together, the results demonstrate how using high-quality datasets reduces the uncertainty of site-adapted solar resource data.**

I. INTRODUCTION

Solar resource datasets are needed to model long-term solar PV energy production. For project financing, it is desirable to minimize risk because lower risk projects, all things being equal, have higher value. A driving factor of production risk is solar resource uncertainty. Therefore, accurate and bankable solar resource datasets are needed. SolarAnywhere® provides accurate satellite-derived solar resource data with known uncertainty for solar project financing [1]. Site-specific data from as early as 1998 through present is available on demand.

Most solar power project sites will deploy a high-quality solar meteorological (MET) station which provides the most accurate measurements of solar radiation. Obtaining site-specific data with ground-based sensors requires a significant upfront investment and the data from MET campaigns typically cannot be used directly for solar resource assessment due to the limited measurement period. These ground measurement stations must be well-calibrated and adequately maintained to ensure collection of valid data.

Combining these two sources of solar resource data using a site adaptation (also known as ground tuning) methodology can produce a solar resource dataset with lower uncertainty than either source on its own. Clean Power Research's solar resource data site-adaptation methodology was published in 2014 [2]. The non-linear approach derives separate satellite-to-ground site-adaptation input parameters (also known as tuning parameters) for periods of clear sky and cloudy conditions from the overlapping period of ground and satellite data. The parameters are then applied to the full historical SolarAnywhere dataset. Since the SolarAnywhere model operates independent of ground data input, the ground-to-satellite correlation studies are based on two independently derived measurement sources. In a 2016 study, we presented a statistically sound approach to estimating the uncertainty of site-adapted (also known as ground-tuned or tuned) data [3].

An International Energy Agency Photovoltaic Power Systems Programme task group performed a benchmarking study of site-adaptation techniques [4]. The study has several limitations for solar project developers. Mainly, the study does not show how the uncertainty of site-adapted data may be minimized. For example, the study treats the long-term datasets (the un adapted resource data) having various quality as a feature for evaluating the success of the site-adaptation techniques. In other words, how well does the site-adaptation correct for (sometimes large) errors in the data? For developers, the more relevant question is, what is the achievable uncertainty when leveraging the best available data?

Using the same methodology and site-adaptation procedure as the original study [2,3], this paper demonstrates the importance of solar resource data quality for minimizing the uncertainty of site-adapted data. In addition, we present the uncertainties achievable with current methods in North America as well as in Europe and South America. The results of this study represent improvements made to the SolarAnywhere satellite-to-irradiance model as well as data quality control methods.

II. METHODOLOGIES

A. Method for quantifying the uncertainty of site-adapted data.

The Clean Power Research method for quantifying the uncertainty of site-adapted data [3] demonstrates the non-linear relationship between the length of the ground campaign and the resulting uncertainty of the final site-adapted dataset. This relationship is shown in Fig. 1, where mean bias error (MBE) is used to quantify the uncertainty of the adapted solar resource data.

$$MBE = \frac{\sum_{i=1}^{N}(x_i{}^{SA} - x_i{}^{obs})}{N} \qquad (1)$$

978-1-6654-6060-6/23 $31.00 © 2023 IEEE

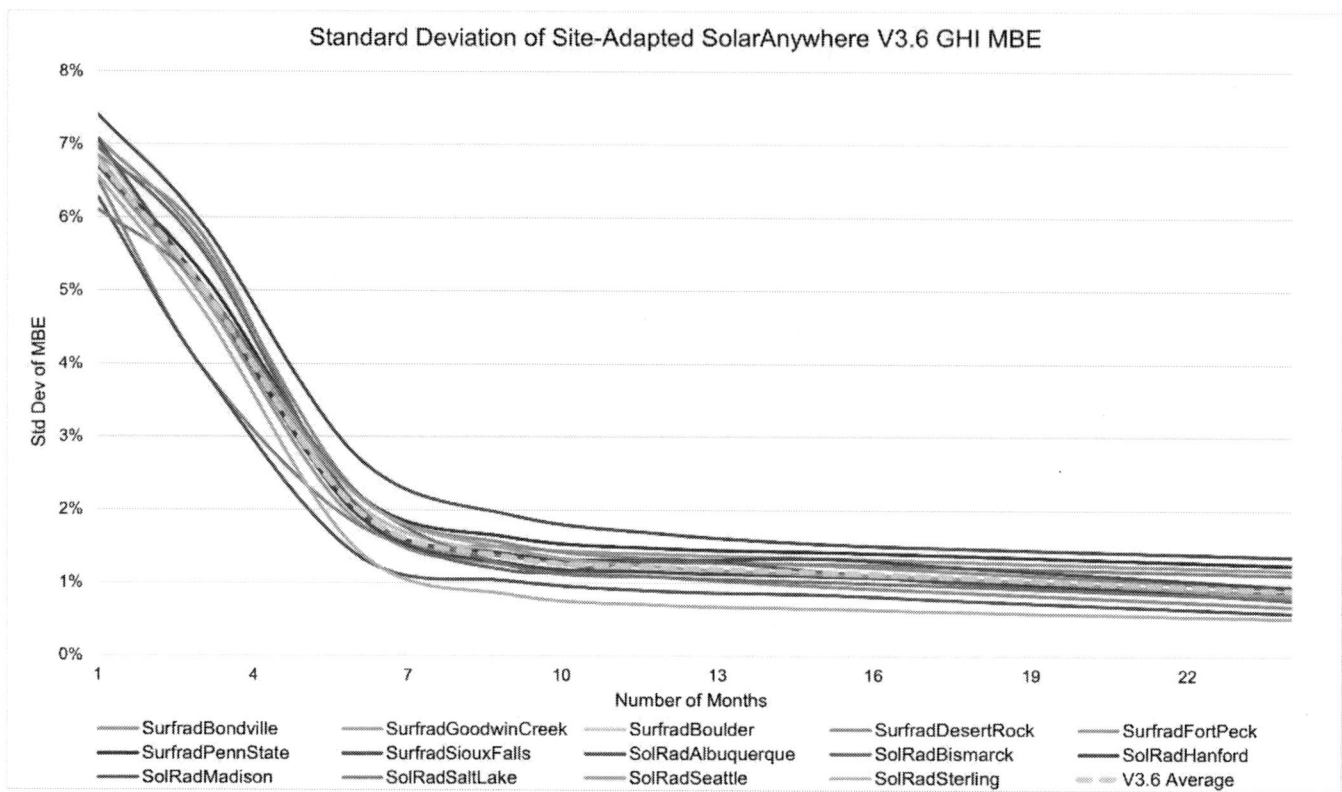

Fig 1. Standard deviation of adapted SolarAnywhere V3.6 GHI MBE at the United States SURFRAD and SOLRAD stations.

In (1), x represents the variable being considered (GHI in this case); N is the number of data points; and the superscripts *SA* and *obs* stand for SolarAnywhere site-adapted and ground observed data, respectively. The Clean Power Research method for quantifying uncertainty is summarized as follows:

1. Input Data Selection: High-quality, one-minute resolution, ground measured global horizontal irradiance (GHI) data were obtained from trusted ground stations. Corresponding hourly resolution SolarAnywhere data was collected for each location [1].
2. Ground Data Quality Control (QC): One-minute resolution ground measured data from each ground station were converted to hourly data using end-of-period averaging. The hourly data was inspected against hourly SolarAnywhere clear sky GHI data. Nighttime values and data deemed unreasonable during visual inspection or quality control were removed from the dataset. This step is described in more depth in section B.
3. Data Segmentation: The full length of ground data from each location was segmented into rolling, fixed-length periods of one to 24 months. For example, ground stations with 10 years of available data were broken up into 132 one-month segments, 131 two-month segments, 130 three-month segments, and so on.
4. Satellite Data Site Adaptation: For each ground data segment created in step three, site-adaptation input

parameters were first derived from the overlapping period of ground and satellite data, then applied to the full time series of SolarAnywhere data (1998-2021). For example, sites with 10 years of available ground data produced 132 site-adapted SolarAnywhere datasets based on 1-month overlapping periods, 131 site-adapted datasets based on 2-month overlapping periods, etc.
5. Standard Deviation Calculation: For every location, the standard deviation of mean bias errors (MBEs) is calculated for each segment length (1-24 months). The results are presented in Fig. 1.

By separating the long-term data at each location into rolling fixed-length segments, we can quantify the impact of using different times of the year and different years for site-adaptation. Additionally, the varying length of each segment accounts for the impact of the ground measurement period length. Generally, uncertainty decreases as the ground measurement period increases. However, as Fig. 1 shows, this relationship is non-linear and reductions in uncertainty begin to diminish after around nine months.

B. Input Data

To reproduce the site-adapted data uncertainty calculations for North America, this study uses ground measurement data from the SURFRAD and SOLRAD (formerly ISIS) sites used in the original study [3]. Results are generated using ground measurements collected between 1998-2021 for the seven

SURFRAD locations and 2010-2021 for the seven SOLRAD locations, as opposed to data collected from 2005-2015 for all locations. A total of 235 reference years are included from these sites in North America. More extensive quality control methods have allowed for the removal of data affected by soiling, shading, or calibration drift issues from these datasets. Periods of data with inconsistent sensor calibration were also identified by comparing measured GHI to GHI calculated from measured direct normal irradiance (DNI) and measured diffuse horizontal irradiance (DHI). When measured and calculated GHI values are equal, this indicates that the GHI, DNI and DIF irradiance measurements are clean. GHI measurements within 4% of calculated GHI, or within 20 W/m^2 of measured DNI are considered valid.

For the South America region, ground measurement data was collected at three BSRN stations and five stations maintained by the Energy Ministry of Chile. A combined total of 53 reference years are included from sites in South America. For the Europe region, ground measurement data was collected from seven BSRN stations with a total of 112 reference years. The South America and Europe ground measurement data were subject to the same QC process applied to the North America data. Table 1 of the Appendix details the number of years used per station and provides notes on data removed during the QC process.

This study uses SolarAnywhere V3.6 to present updated site-adapted data uncertainty values for North America, and to present uncertainty values for South America and Europe [1]. Since the release of SolarAnywhere version 3.2, several improvements have been made to the SolarAnywhere model to reduce the uncertainty of the native data and improve the temporal consistency of the dataset [1]. Key updates include updated satellite calibration parameters, a higher temporal resolution aerosol optical depth (AOD) input to the clear sky model, and a snow albedo input to the clear sky model to increase accuracy in locations with persistent snow cover.

III. RESULTS AND DISCUSSION

The improved consistency of the SolarAnywhere V3.6 model over the V3.2 model in combination with improved ground data quality control methodologies resulted in an overall reduction in site-adapted data uncertainty for North America. The original study [3] determined the average uncertainty (std. dev. of MBEs) of SolarAnywhere V3.2 data adapted to 12 months of ground measurements to be 2%. Fig. 1 presents a 12-month adapted data uncertainty of 1.2%. This represents a 40% reduction in the final uncertainty of a solar resource dataset adapted to 12 months of ground measured data as shown in Fig. 2. Most of this reduction (83%) is attributable to the increased temporal consistency of the SolarAnywhere model. The remaining reduction in uncertainty (17%) is due to improved data QC.

Fig. 2 presents the uncertainty of site-adapted SolarAnywhere P50 estimates averaged across the U.S.

SURFRAD and SOLRAD stations at various ground measurement campaign lengths. These uncertainties have been calculated for site-adapted SolarAnywhere V3.2 data (V3.2), SolarAnywhere V3.2 data adapted to ground measurements cleaned with improved QC methods (V3.2+QC), site-adapted SolarAnywhere V3.6 data (V3.6) and SolarAnywhere V3.6 data adapted to ground measurements cleaned with improved QC methods (V3.6+QC). Fig. 2 also presents the average uncertainty of site-adapted SolarAnywhere V3.6 data (applying the same updated data QC methodologies as used in the U.S.) across seven ground stations in Europe (V3.6+QC Europe) and eight in South America (V3.6+QC South America).

Although both versions of the SolarAnywhere data perform at similar uncertainties for short time periods, the updated model begins to result in significantly lower uncertainty at the six month ground measurement period. From the data presented in Fig 2, the average uncertainty of SolarAnywhere V3.6 data adapted to 6 months of ground measurements (w/ improved QC) is approximately equal to the uncertainty of SolarAnywhere V3.2 data adapted to 12 months of ground measurements (w/ improved QC). This demonstrates

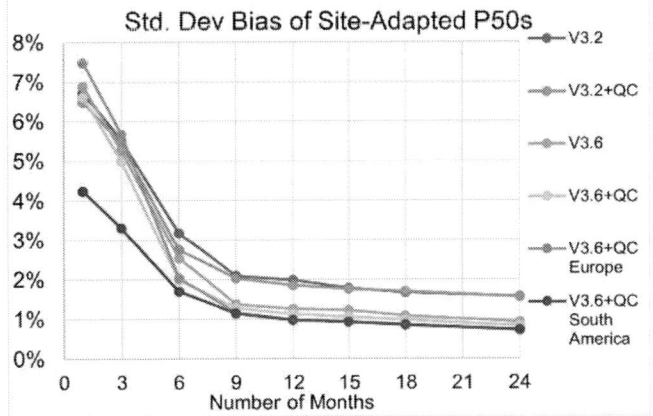

Fig 2. Reduction in site-adapted P50 bias due to improvements to the SolarAnywhere model and ground data QC.

the effect of improvements made to the SolarAnywhere clear sky model. Bias resulting from the use of ground measurement data periods shorter than 9 months is reduced in model version 3.6, indicating improved temporal consistency.

Temporal consistency in the satellite dataset is important because satellite-to-ground site adaptation processes commonly use a small fraction of the available satellite-based time series (9-12 months out of 20+ years) to derive input parameters. For example, if a satellite-to-solar irradiance model is shown to perform poorly during low irradiance years, input parameters derived from a 12-month low irradiance period could over or under correct the remainder of the time series.

Even with the use of a highly consistent satellite-based solar resource dataset, the site-adaptation process is sensitive to ground measurement campaign lengths shorter than nine months. Shorter campaign lengths do not capture the full seasonality of cloud conditions at a site, resulting in increased site-adapted data uncertainty.

The average standard deviation of MBEs at the 12-month segment is 1.0% across all South American sites and 1.0% across all European sites. These 12-month values are in good agreement with the North American 12-month value of 1.2%. This consistency across regions is seen at all segment lenths, as shown in Fig. 3. This demonstrates SolarAnywhere model consistency across varying geographic regions and the robustness of the site-adaptation methodology.

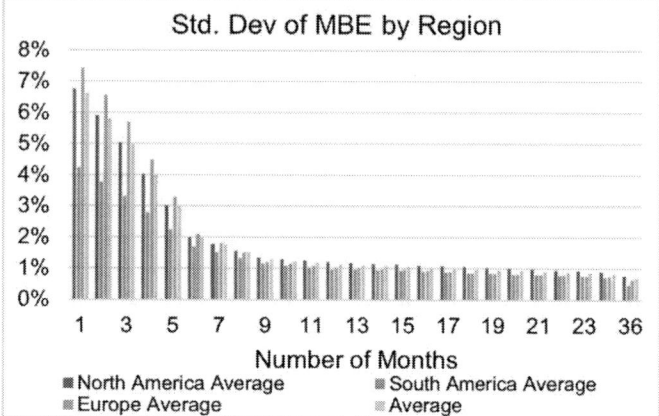

Fig 3. Comparison of all-site average standard deviations of MBEs between North America, South America, and Europe.

The results presented in this paper can be directly applied to assign uncertainty values to site-adapted SolarAnywhere datasets (V3.6 and on) according to the length of the ground measurement dataset used for the adaptation. However, when assigning a final uncertainty value to any site-adapted dataset, it is important to consider additional uncertainty that can be introduced by the ground measurements. The uncertainty values in this paper are generated using a trusted network of well-maintained ground measurement stations, meaning they do not capture additional uncertainty introduced if lower quality ground measurements are used for site-adaptation. The estimated uncertainty associated with the ground measurements can be combined with the site-adapted data uncertainty using the following root sum of squares equation.

$$combined\ uncertainty = \sqrt{a^2 + b^2} \qquad (2)$$

In (2), a is the site-adapted data uncertainty determined using the Clean Power Research methodology and b is the estimated uncertainty of the ground dataset. For example, 12 months of ground measurements from a Kipp & Zonen CMP22 pyranometer, which has a measurement uncertainty of 2.7% [5], combined with the site-adapted data uncertainty of 1.2%, results in an estimated final uncertainty of 3.0%.

IV. CONCLUSION

Reducing the risk associated with solar projects will remain important in obtaining financing as competition in solar project development continues to increase. Solar resource risk is one risk component that can be reduced by

obtaining a low-uncertainty solar resource dataset. Site-adaptation of satellite-based data with ground measurements is often used to obtain the lowest possible uncertainty in a solar irradiance dataset. The use of high-quality ground measurements over a short period of time in combination with the long-term data available with a satellite-to-irradiance model provides a lower uncertainty dataset than either of the two independent data sources.

Using the Clean Power Research site-adaptation methodology with SolarAnywhere satellite-based data and 12 months of ground measured reference data, it is possible to achieve a site-adapted data uncertainty as low as 1.2%. This represents a 40% reduction in the reported uncertainty of site-adapted solar resource data in North America. Most of this reduction can be attributed to updates to the satellite-based solar irradiance model, showing that the quality of the satellite-derived data used for site-adaptation is important. The remaining reduction in uncertainty can be attributed to the implementation of more extensive quality control methods. The low uncertainty demonstrated in this paper is only valid when using 1) the Clean Power Research site-adaptation methodology 2) high-quality ground data 3) SolarAnywhere V3.6 or later.

REFERENCES

[1] SolarAnywhere, (2022): SolarAnywhere Validation [White Paper].https://www.solaranywhere.com/ validation/leadership-bankability/data-validation/

[2] Perez, R., Kankiewicz, A., Dise, J., and Wu, E., (2014): Reducing Solar Project Uncertainty with and Optimized Resource Assessment Tuning Methodology. Proc. ASES Annual Conference, San Francisco, CA.

[3] Alfi, J., Kubiniec, A., Mani, G., Christopherson, J., He, Y., Bosch, J., (2016): Importance of Input Data and Uncertainty Associated with Tuning Satellite to Ground Solar Irradiation. Proc. IEEE PVSC 43, Portland, Oregon.

[4] Polo, J., Fernández-Peruchena, C., Salamalikis, V., Mazorra-Aguiar, L., Turpin, M., Martín-Pomares, L., Kazantzidis, A., Blanc, P., Remund, J., (2020): Benchmarking on improvement and site-adaptation techniques for modeled solar radiation datasets, Solar Energy, Vol. 201, Pg. 469-479, https://doi.org/10.1016/j.solener.2020.03.040.

[5] Habte, A., Sengupta, M., Andreas, A., Dooraghi, M., Reda, I., Kutchenreiter, M., (2017): Evaluating the Sources of Uncertainties in the Measurements from Multiple Pyranometers and Pyrheliometers. Atmospheric Radiation Measurement/Atmospheric System Research PI Meeting, Vienna, Virginia.

APPENDIX

Table 1. Years of data used from each ground measurement station used to produce the results in this study.

Region	Network	Name	Valid Reference Years	Comments
N. America	SURFRAD	Bondville	24	
		Boulder	24	
		Desert Rock	24	
		Fort Peck	23	2015 excluded due to measured GHI calibration issue.
		Goodwin Creek	24	
		Penn State	23	
		Sioux Falls	17	
	SOLRAD	Albuquerque	11	
		Bismarck	11	
		Hanford	11	
		Madison	11	
		Salt Lake	11	
		Seattle	11	
		Sterling	10	2011 excluded due to measured GHI calibration issue.
S. America	BSRN	Armazones	4	
		Brasilia	4	
		Florianopolis	9	
		São Martinho da Serra	6	
	Energy Ministry of Chile	Crucero 1	5	
		Crucero 2	6	
		Inca De Oro	9	
		Pozo Almonte	10	
Europe	BSRN	Cabauw	17	
		Camborne	13	
		Carpentras	21	
		Lindenberg	20	
		Paris	13	
		Payerne	16	
		Sarriguren	12	

Multifunctional Titanium Oxide Layers in Silicon Heterojunction Solar Cells by Selective Anodization

Leonie Jakob, Leonard Tutsch, Thibaud Hatt, Johan Westraadt, Markus Glatthaar, Martin Bivour, Jonas Bartsch

Fraunhofer ISE, Freiburg, Germany

PV2+ GmbH, Freiburg, Germany

Centre for HRTEM, Nelson Mandela University, Gqeberha, South Africa

In this contribution we introduce a new strategy to reduce the indium content of silicon heterojunction solar cells: an electrochemically oxidized TiOx layer is applied on top of the thinned transparent conductive oxide, working as anti-reflection coating and encapsulation layer. The TiOx layer is made by anodization of a thin Ti layer in an electrolyte. Two of the investigated electrolytes (aqueous and non-aqueous electrolyte 1) led to strong side reactions, namely oxygen evolution and metal dissolution, making them unsuitable for our application. The second applied non-aqueous electrolyte shows promising first results, exhibiting a JSC loss of 3 mA/cm² and a cell efficiency of only 1.5%abs lower than the reference cell.

Development of an Adaptive Droop Control Method for Interconnected Lunar DC Microgrids Using Power Hardware-in-the-Loop

Andrew R. R. Dow[1], Rachid Darbali-Zamora[1], Felipe Palacios II[1], Jack D. Flicker[1],
Marc A. Carbone[2], and Jeffrey T. Csank[2]

[1]Sandia National Laboratories, Albuquerque, New Mexico, 87185, USA
[2]NASA Glenn Research Center, Cleveland, Ohio, 44135, USA

Abstract – **NASA's Artemis Program outlines the need for a lunar habitat capable of sustaining human life as well as mining and producing raw materials on the lunar surface. This mission is a means towards deeper space exploration, with plans for reaching Mars and beyond. Human presence on the moon is not possible without the ability to generate and distribute energy, namely electricity, through a network of sources, loads, and power converters known as a microgrid. Multiple microgrids can be deployed on the moon based on location and need with interconnections to increase resiliency and reliability. A method for adaptive control through power converters connecting two DC microgrids is proposed.**

Keywords – *DC microgrid, voltage droop control, lunar habitat, DC/DC converters, photovoltaic arrays, energy storage, adaptive*

I. INTRODUCTION

Establishing an electrical grid for lunar and interplanetary habitation yields an engineering challenge unlike the traditional power systems found on Earth. The Artemis Program, a multi-mission plan led by NASA to further human presence into space, will approach this challenge of deploying power systems capable of sustaining human life for both the Earth's moon as well as Mars [1], [2]. The lack of familiarity to the terrestrial power grid allows for innovation and advancements as humans prepare for deep space exploration.

The development of a moon or Mars-based power grid does liken itself to recent innovations in terrestrial microgrids [3]. Connecting a series of distributed energy resources (DERs), power converters, and loads in a localized network allows for more adaptive power distribution methods common to remote and temporary deployments found on Earth. Leveraging the technology and experience gained through the development of these terrestrial microgrids, one can develop a resilient power system for other remote applications, such as space habitation.

The proposed lunar base provides accommodations for human habitation (LHAB) as well as in-situ resource utilization (ISRU), a means of mining and producing raw materials on the lunar surface [4]. Both LHAB and ISRU deployments will require independent microgrids capable of generating and storing energy for local power consumption. The two microgrids will be connected through a tie line to provide a means of power sharing in the event of increased loads, faults, or lack of generation.

Sandia National Laboratories has partnered with NASA to study the DC microgrid architecture intended for the Artemis Program's lunar base. Control methods are developed that allow for interconnection between multiple microgrids as well as utilizing hierarchical controls [5] over shared power resources between each microgrid. An adaptive droop control is developed that allows for the systems to operate in both islanded and configurable power share modes. This method is compared to a simpler autonomous droop control that lacks islanding capability, but still allows for power sharing between independent microgrids. Simulation results provide a basis for future experimental studies using DC Power Hardware-in-the-Loop capabilities.

In this paper, Section II outlines the methodology for both the power conversion models and control approaches. Section III covers preliminary simulation results for the model and Section IV offers a conclusion and path forward for additional studies to be conducted on this topic.

II. METHODOLOGY AND APPROACH

The following section details the design methodology for creating the lunar DC microgrid simulation model and control approaches taken for providing interconnected power flow.

A. Lunar DC Microgrid Model

The microgrid architecture includes photovoltaic (PV) arrays, battery energy storage systems (BESS), and loads distributed through a common bus using a series of DC/DC converters to regulate power and voltage. Each microgrid can service its internal loads and provide charge power to the energy storage based on generation and demand of the system. Fig. 1 illustrates the two independent microgrid connected via a tie line.

A major focus of this research is the design and modeling of the power converters that connect the two microgrids using an elevated voltage bus for bidirectional power flow. A state-space average model derived for a buck-boost converter [6] provided the ability for the microgrid's 120 V_{DC} bus to be connected through a 160 V_{DC} tie line to provide bidirectional current flow. Fig. 2 illustrates the tie line bidirectional converter circuit topology used for the microgrid interconnect.

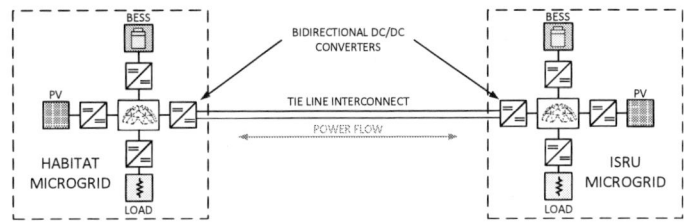

Fig. 1. Lunar DC microgrid architecture.

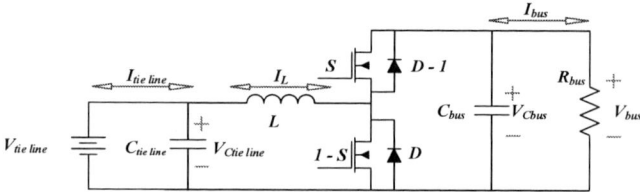

Fig. 2.　Bidirectional converter model for microgrid interconnect.

State-space averaging provides some advantage for power converter modeling by linearizing the internal switching states, while still providing accurate representation of the converter's continuous operation. Not only does this method provide a solution for efficient modeling of larger and more complex systems; state-space averaging provides a solution for future real-time simulation and power hardware-in-the-loop testing.

The two microgrids will ultimately be spaced up to several kilometers apart and the tie line interconnect will provide some impedance and losses due to the length of the conductors. A per-unit line impedance model consisting of series inductance (10 µH base), series resistance (10 mΩ base), and shunt capacitance (10 µF base) was employed in a 'T' configuration between the LHAB and ISRU tie line converters. A per-unit impedance of 16 p.u. provided an initial benchmark, however extended line lengths can be considered to model more significant power losses between the two microgrids.

B. Tie Line Converter - Autonomous Droop Control

An autonomous power-voltage droop curve was implemented to control power flow through the tie line converter. The droop curve utilizes both bus and tie line voltage nodes to create a target power setting used by the converter's PI controller for regulating converter output current. The droop curve consists of a linear power-to-voltage slope, a set point dead band at the nominal voltage level, and a limiter that saturates at minimum and maximum power set points [7]. The power set point is used to calculate a current set point. Error is calculated against the measured current and fed into a PI controller [8]. Fig. 3 details the controller, showing the inputs as the voltage nodes and the current measured in the converter output. The controller model is considered autonomous as it requires no higher-level intervention to change set point parameters. It relies solely on voltages measured in the converter to provide droop power control with fixed upper and lower power limits, eliminating the need for a hierarchical system-level controller.

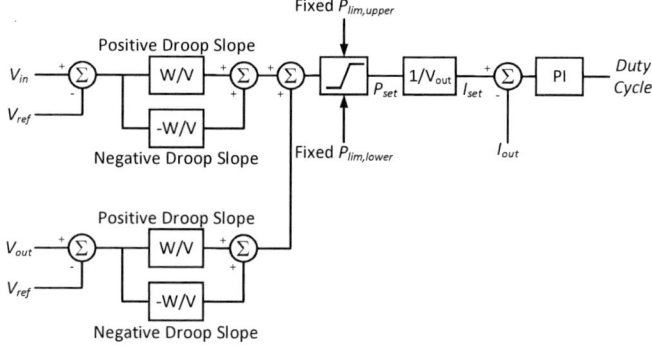

Fig. 3.　Autonomous droop control model for tie line converter.

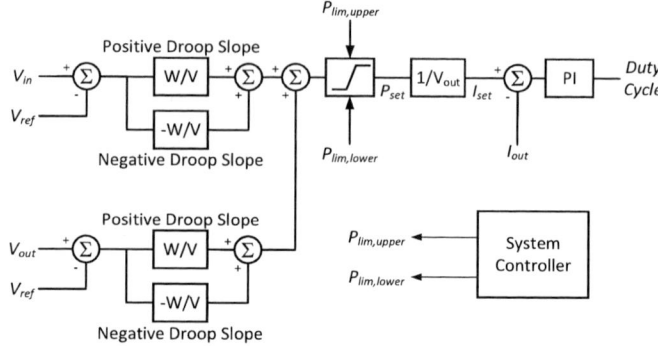

Fig. 4.　Adaptive droop control model for tie line converters.

C. Tie Line Converter – Adaptive Droop Control

A more coordinated control method was developed out of the original autonomous droop control model. The adaptive droop control model relies on a combination of primary and hierarchical system control to correctly provide power flow through the tie line converters [9]. Power set point limits are to be issued from a system-level controller to the primary controller of the tie line converter to adaptively control the bidirectional power flow. Fig. 4 details the controller design for the adaptive control model.

While the primary controller is still responsible for providing voltage and current regulation by means of average duty cycle control, a hierarchical controller can provide more guidance and insight based on the entire system. The saturation limits applied to the power-voltage droop control are modified during runtime to increase or decrease based on system events such as load scheduling or fault detection. Fig. 5 compares the droop control curve functionality between autonomous and adaptive droop controllers.

III.　PRELIMINARY SIMULATION RESULTS

A power flow test scenario for the two proposed control methods is implemented for the interconnecting tie line converters. Table I outlines the simulated event scenarios over time.

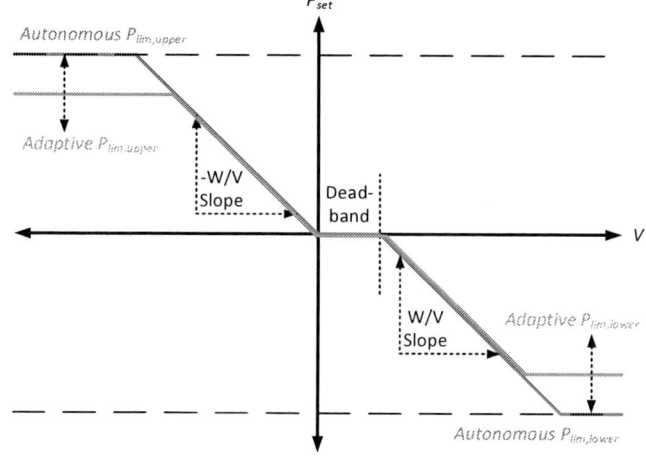

Fig. 5.　Comparison of autonomous and adaptive droop controllers.

TABLE I
SIMULATED TEST SCENARIOS

Schedule	LHAB Power (kW)			ISRU Power (kW)			LHAB-ISRU Power (kW)
	PV	Load	Surplus	PV	Load	Surplus	Power Share
0-230s	10	6	4	20	13	7	0
230-350s	10	6	4	20	21	-0.72	-0.72
350-420s	10	6	4	20	13	7	0
420-540s	5	6	-1	20	13	7	1
540-600s	10	6	4	20	13	7	0

The key difference between the two control methods is the ability for the adaptive control to accept power set point limits from a hierarchical control model. In the adaptive control simulation model, these power set point limits are provided as a response to the additional load and PV fault conditions that are introduced. During normal operation, the adaptive control model can be islanded to allow each microgrid to function independently. The autonomous droop control method does not have this capability, rather relying on fixed power limits regardless of system state. While both models are shown to effectively provide power sharing capability, the adaptive control method (shown in Fig. 6 in green) does provide a better means of islanding during periods of surplus. Coordination between the primary controller and a system-level controller allows for this functionality. The autonomous model (shown in Fig. 6 in red) can still share power across the tie line during surplus periods, which is not desirable when power sharing is not required.

IV. CONCLUSION AND FUTURE WORK

The DC microgrid architecture for the Artemis Program's lunar base is studied to develop control methods for power sharing between two interconnected microgrids. The interconnecting tie line is controlled through droop control as well as incorporating hierarchical control to initiate adaptive and islanded operation. This provides a novel solution for managed power sharing between two independent microgrids. Future work on this topic will continue the study of this power architecture and control methodology into the experimental test environment via power hardware-in-the-loop.

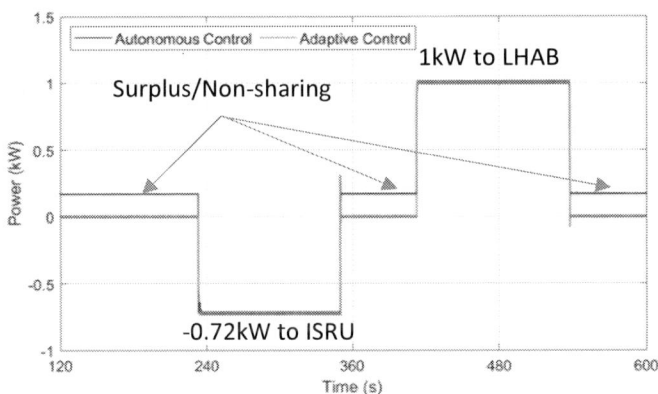

Fig. 6. Power flow comparison between autonomous and adaptive control methods for bidirectional tie line converters.

ACKNOWLEDGEMENT

This article has been authored by an employee of National Technology & Engineering Solutions of Sandia, LLC under Contract No. DE-NA0003525 with the U.S. Department of Energy (DOE). The employee owns all right, title and interest in and to the article and is solely responsible for its contents. The United States Government retains and the publisher, by accepting the article for publication, acknowledges that the United States Government retains a non-exclusive, paid-up, irrevocable, world-wide license to publish or reproduce the published form of this article or allow others to do so, for United States Government purposes. The DOE will provide public access to these results of federally sponsored research in accordance with the DOE Public Access Plan https://www.energy.gov/downloads/doe-public-access-plan.

The NASA Glenn Research Center, and NTESS of Sandia, LLC collaborate in accordance with the National Aeronautics and Space Act (51 U.S.C. § 20113(e)) under contract SAA3-1690.

REFERENCES

[1] J. Csank and J. H. Scott, "Power and Energy for the Lunar Surface," NASA, Cleveland, 2022.

[2] NASA, "Artemis," [Online]. Available: https://www.nasa.gov/specials/artemis/. [Accessed 28 November 2022].

[3] A. D. Bintoudi, C. Timplalexis, G. Mendes, J. M. Guerrero and C. Demoulias, "Design of Space Microgrid for Manned Lunar Base: Spinning-in Terrestrial Technologies," in *2019 European Space Power Conference (ESPC)*, Juan-les-Pins, France , 2019.

[4] NASA, "Overview: In-Situ Resource Utilization," 3 April 2020. [Online]. Available: https://www.nasa.gov/isru/overview. [Accessed 28 November 2022].

[5] J. Zhang, J. T. Csank and J. F. Soeder, "Hierarchical Control of Distributed Battery Energy Storage System in a DC Microgrid," in *2021 IEEE Fourth International Conference on DC Microgrids (ICDCM)*, Arlington, VA, 2021.

[6] A. R. R. Dow, R. Darbali-Zamora, J. D. Flicker and J. T. Csank, "Development of Hierarchical Control for a Lunar Habitat DC Microgrid Model Using Power Hardware-in-the-Loop," in *Photovoltaic Specialists Conference (PVSC)*, Philadephia, 2022.

[7] L. Ott and e. al., "An advanced voltage droop control concept for grid-tied and autonomous DC microgrids," in *IEEE International Telecommunications Energy Conference (INTELEC)*, Osaka, 2015.

[8] H. Peng, D. Wang and M. Su, "Comparative study of power droop control and current droop control in DC microgrid," in *The 11th IET International Conference on Advances in Power System Control, Operation and Management (APSCOM 2018)*, Hong Kong, 2018.

[9] M. Mokhtar, M. I. Marei and A. A. El-Sattar, "An Adaptive Droop Control Scheme for DC Microgrids Integrating Sliding Mode Voltage and Current Controlled Boost Converters," *IEEE Transactions on Smart Grid,* vol. 10, no. 2, pp. 1685-1693, 2019.

978-1-6654-6060-6/23 $31.00 © 2023 IEEE

Assessing Degradation in Bifacial Photovoltaic by Sequential Stress and Outdoor Aging

Dennice M. Roberts, Sona Ulicna, Michael Owen-Bellini, Paul Ndione, Helio Moutinho, Kent Terwilliger, Steve Johnston, Laura T. Schelhas, Dana B. Kern

National Renewable Energy Laboratory, Golden, CO, United States

Bifacial photovoltaic modules are increasing in deployment due to advantages in energy generation, and these advanced module packaging architectures present new reliability considerations. Here, we study bifacial packaging architectures and assess the effects of module processing quality, rear-side module coverings, and polymer encapsulant chemistry on aging and degradation. We determine that mini-modules aged by accelerated testing adapted from IEC 63209-2 show similar degradation trends to nominally identical modules aged outdoors. We find lamination processing conditions have a significant effect on early degradation. We investigate degradation modes non-destructively using luminescence imaging, current-voltage performance, and external quantum efficiency. Finally, we assess chemical and mechanical characteristics of various encapsulant formulations to compare adhesion changes that may result in cell delamination.

Damp Heat Exposure of Glass/Glass Coupons with Different Encapsulants

Chiara Barretta[1], Lisa Meinhart[1], Hannes Krebs[2], Andreas Brandstätter[3], Gernot Oreski[1]

[1] Polymer Competence Center Leoben GmbH (PCCL), Leoben, 8700, Austria, [2] CS Wismar GmbH Sonnenstromfabrik, Wismar, 23966, Wismar, Germany, [3] Lenzing Plastics GmbH & Co KG, Lenzing, 4860, Austria

Abstract — **Photovoltaic modules with glass/glass configuration can overcome some problems typical of glass/polymer backsheet modules by limiting, for instance, moisture ingress. However, it is necessary to adapt the bill of materials to extend the lifetime of glass/glass PV modules and to improve their reliability. The application of an edge seal and the use of encapsulants alternative to ethylene vinyl acetate might reduce the issues connected to humidity ingress and acetic acid production during exposure in the field. This study is focused on the analysis of the performance of test coupons laminated with four different encapsulants and exposed to damp heat test.**

I. INTRODUCTION

Photovoltaic (PV) technologies are fundamental to reduce the dependence from fossil fuels for electricity production. Additionally, to make photovoltaics even more sustainable, it is necessary to improve the reliability and therefore the lifetime of PV modules. Recently, glass/glass PV module configurations are becoming more popular because, especially thanks to the use of bifacial cells, they are able to provide higher power outputs compared to the traditional glass/non-transparent polymer backsheet configuration. However, there are also disadvantages in using glass/glass modules such as heavier weight, related to more difficulties in transportation and installation and chances of breakage. Furthermore, if on one hand the glass/glass configuration with additional edge seal has a positive effect because it prevents moisture ingress, on the other hand it does not allow degradation products such as acetic acid to permeate towards the outside. Ethylene vinyl acetate (EVA) is the most commonly used encapsulant and it might not be the best choice for glass/glass modules. The acetic acid that might be produced during exposure would be trapped inside the module, possibly causing corrosion of metallization and interconnections, potential induced degradation (PID), delamination and snail trails. Replacing EVA with alternative encapsulants that do not contain vinyl acetate moieties might be a solution to prevent the degradation modes related to acetic acid production and humidity ingress [1], but it is necessary to thoroughly test the new material combinations especially to analyze possible incompatibilities between the different module components. Additionally, to test the reliability of glass/glass PV modules that aim to operate in the field beyond 25 years and up to 40 years, it is necessary to perform artificial ageing tests or test sequences that can replicate the failures that occur during operation. The degradation modes that were mostly reported for glass/glass modules are delamination at different interfaces and electro-corrosion of interconnections and metallization mostly caused or favored by humidity ingress [2]. Damp heat (DH) test can be used to test the effectiveness of the edge seal in preventing moisture ingress and it is an important mean of simulating the humidity penetrating in the modules during 40 years of exposure in the field. [3, 4]

In this study, we produced test coupons using four different encapsulants and we exposed them to DH test to analyze the effects of humidity ingress on materials performances.

II. EXPERIMENTAL

The samples used in this study were produced in a custom-made vacuum laminator with membrane using different temperature protocols depending on the encapsulant used. Test coupons were laminated using double glass, encapsulant, fragment of bifacial cell and small piece of interconnection, as can be seen in Fig. 1. Some coupons were laminated with additional edge seal and some without.

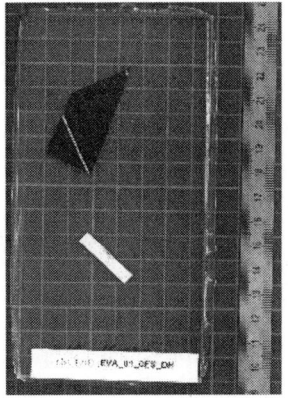 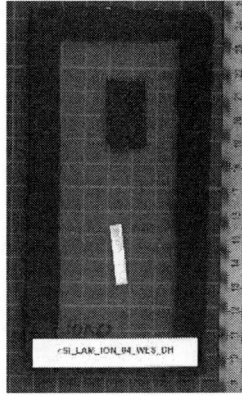

Fig. 1 Test coupon without edge seal (left) and with edge seal (right).

The coupons were laminated with different encapsulant materials to compare their degradation behavior. An EVA was selected as known-bad reference and three alternative encapsulants based on polyethylene copolymers were selected. The main encapsulants characteristics are described in Table 1.

978-1-6654-6060-6/23 $31.00 © 2023 IEEE

TABLE 1
SUMMARY OF ENCAPSULANTS USED AND THEIR
CHARACTERISTICS

Encapsulant	Main characteristics		
	Acetic acid	Crosslinking during lamination	Melting temperature [°C]
Ethylene vinyl acetate copolymer(EVA)	Yes	Yes (150 °C)	64
Ethylene copolymer (POE)	No	Yes (146 °C)	69
Ethylene copolymer (TPO)	No	No	62-121
Ionized ethylene acrylic acid copolymer (ION)	No	No	92

The samples were then stored in a climatic chamber, exposed to DH test at 85 °C and 85% of relative humidity for the first 2000 h, with sampling time of 500 h. It is planned to run the test for 5000 h in total to simulate the humidity ingress during 40 years of exposure. The coupons were characterized non-destructively by means of ultraviolet-visible-near infrared (UV-Vis-NIR) spectroscopy and UV fluorescence imaging to assess changes in optical properties and degradation pathways respectively. Additionally a coupon was taken out of the chamber every 500 h and destroyed to extract encapsulant materials for further analysis. A qualitative additive analysis was performed by means of thermo-desorption gas chromatography coupled to mass spectrometry (TD-GC/MS). Furthermore, investigation of changes in polymer's chemical structure and thermal behavior were carried out via Fourier Transform Infrared spectroscopy in Attenuated Total Reflectance mode (FTIR-ATR) and Differential Scanning Calorimetry (DSC), respectively.

III. RESULTS AND DISCUSSION

The coupons exposed to DH test containing EVA, POE and ionomer (ION) encapsulants did not show significant changes in optical properties. The coupons with TPO, with and without edge seal, showed a slight yellowing correlated with an increase in UV fluorescence.

The encapsulant analyzed showed a qualitatively different additive formulation. The EVA showed presence of a light stabilizer and a UV absorber, the POE only a UV absorber, the TPO an antioxidant and the ionomer did not show presence of additional stabilizers. Nevertheless, no significant changes in the qualitative additive composition could be detected before and after the exposure to the DH test. The chemical structure of the encapsulants laminated in the test coupons is shown in Fig. 2.

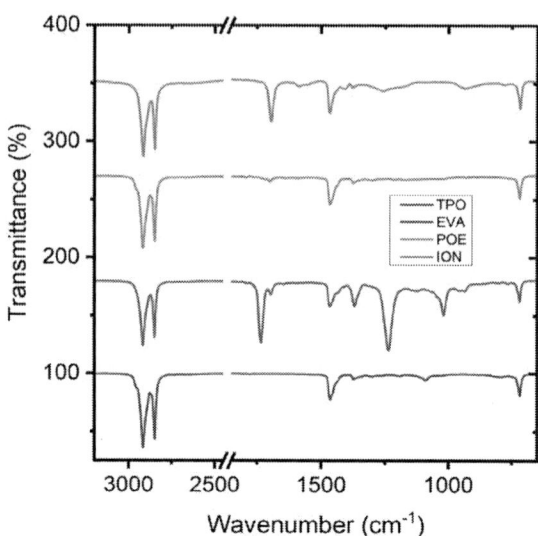

Fig. 2   FTIR ATR spectra of the encapsulant used.

The four encapsulant are characterized by a main polyethylene chain, as can be seen from the peaks at 2920 cm^{-1}, 2850 cm^{-1}, 1463 cm^{-1} and 720 cm^{-1}, corresponding to CH stretching, deformation and rocking vibrations [5]. Additional bands can be seen for EVA due to the vinyl acetate moieties at 1740 cm^{-1}, 1238 cm^{-1} and 1020 cm^{-1}. The ionomer encapsulant shows additionally presence of acrylic acid moieties, confirmed by the bands at about 1696 cm^{-1}, 1260 cm^{-1} and 1170 cm^{-1}. Despite the slight evidence of yellowing for the TPO, the measured FTIR-ATR spectra did not show significant differences between the reference encapsulant and the one extracted from the coupon exposed to 2000 h of DH test, Fig. 3.

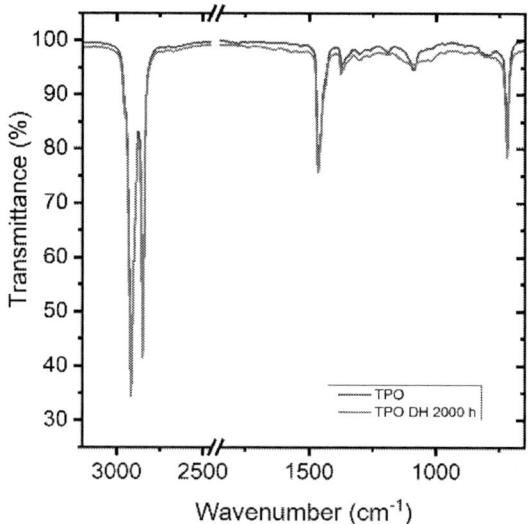

Fig. 3   FTIR ATR spectra of samples TPO encapsulant measured before and after the exposure to 2000 h of DH test.

The results of the thermal characterization did not show signs of chemical degradation or significant changes in crystallinity values. The exposure of the coupons to DH test is still ongoing and longer exposure times will allow to better understand the degradation mechanisms taking place in some of the materials tested. Additionally, it is planned to expose the coupons to UV irradiation to test the UV stability of the encapsulants.

IV. SUMMARY AND CONCLUSIONS

To summarize, test coupons were laminated using four different encapsulants with and without edge-seal. The objective was to investigate the effectiveness of the edge seal in preventing moisture ingress and to evaluate the performances of the different encapsulants under DH test. The samples will be exposed in total to 5000 h of DH to simulate the amount of humidity that is supposed to penetrate the PV over 40 years of operation in the field. The preliminary results after 2000 h of exposure showed good performances for all the encapsulants, but TPO. Slight yellowing and increase in UV fluorescence intensity were detected for the coupons laminated with TPO encapsulant. Nevertheless, no significant changes in chemical structure or qualitative additive composition could be detected. Furthermore, the coupons will be exposed to artificial ageing tests with UV to assess the stability of the encapsulants under UV irradiation.

ACKNOWLEDGEMENTS

This work was conducted as part of the Austrian "e!MISSION.at - Energy Mission Austria" project "PV40+" (FFG No. 881868) funded by the Austrian Climate and Energy Fund and the Austrian Research Promotion Agency (FFG).

REFERENCES

[1] D. B. Sulas‑Kern et al., "Electrochemical degradation modes in bifacial silicon photovoltaic modules," Progress in Photovoltaics, vol. 30, no. 8, pp. 948–958, 2022, doi: 10.1002/pip.3530.

[2] A. Sinha et al., "Glass/Glass Photovoltaic Module Reliability and Degradation: A Review," J. Phys. D: Appl. Phys., vol. 54, no. 41, 2021, doi: 10.1088/1361-6463/ac1462.

[3] M. Kempe, D. Panchagade, M. O. Reese, and A. A. Dameron, "Modeling moisture ingress through polyisobutylene-based edge-seals," Prog. Photovolt: Res. Appl., pp. n/a, 2014, doi: 10.1002/pip.2465.

[4] M. Koehl, S. Hoffmann, and S. Wiesmeier, "Evaluation of damp-heat testing of photovoltaic modules," Prog. Photovolt: Res. Appl., vol. 25, no. 2, pp. 175–183, 2017, doi: 10.1002/pip.2842.

[5] G. Socrates, Infrared and Raman Characteristic Group Frequencies: Tables and Charts, 3rd ed. Chichester, West Sussex, England: John Wiley & Sons, Ltd, 200

Trajectories to Reach 25% Efficiency CdTe Solar Cells with the Implementation of CdTe1-xSex Band Gradient in SCAPs 1-D

Joel Saucedo, Hasitha Mahabaduge

Georgia College & State University, Milledgeville, GA, United States

Thin film CdTe solar cells continue to be a promising candidate for solar energy conversion. Although CdTe photovoltaics is currently the only gigawatt scale thin-film technology, its 22.1% record efficiency did not achieve further improvements in the recent past. We used numerical simulations to show pathways to achieve conversion efficiencies over 25% by varying the band gap and acceptor densities of the absorber. The minority carrier diffusion equation was referenced for continuity, and reinforced by the results of the bulk simulations in selenium alloyed cadmium telluride. As the band gap was changed due to selenium incorporation, increase in both the Voc and Fill Factor contributed significantly to increase the overall efficiency of the device.

How do As-local structures in CdSexTe1-x respond to bias conditions under (X-ray) illumination?

Srisuda Rojsatien, Niranjana Mohan Kumar, Barry Lai, Dan Mao, Arun Mannodi-Kanakkithodi, Maria K. Y. Chan, Mariana Bertoni

Arizona State University, Tempe, AZ, United States

Argonne National Laboratory, Argonne, IL, United States

First Solar, Perrysburg, OH, United States

Purdue University, West Lafayette, IN, United States

X-ray microscopy is a powerful tool to study defect chemistry in solar cells since it is capable of probing structure, chemistry, and opto-electrical properties, correlatively, on a pixel-by-pixel basis. In this work, X-ray absorption near edge structure (XANES) was used to probe the As local structures across the depth of the CdSexTe1-x absorber under (x-ray) illumination and bias conditions. It was found that the As structures located near the front contact, which mostly have the spectral features of oxides, are susceptible to undergo structural change under different biasing conditions. While those located near the back contact largely remain unchanged - notably, Cd2AsCl2 and AsTe+Clx complex defect. Additionally, while we found no spectral evidence of the self-compensating AX center, we do observe a high fraction of AsTe+Clx complex around the middle of the absorber, in both low and high performing areas. This finding could explain why a high percentage of the incorporated As is inactive and could perhaps shed light on the self-compensating effect.

Trends in Solar PV Growth in Snowy Climates and Impact on Resource Adequacy

Shelbie Wickett and Ana Dyreson

Michigan Technological University, Houghton, Michigan, 49931-1295, USA

Abstract — Solar installations are increasing in cold-weather U.S. states. Snow cover decreases the amount of light received by photovoltaic (PV) panels, reducing their output. The widespread use of solar generation in a grid system may amplify this issue. However, researchers have yet to model the effect of snow cover on high-penetration PV grids. Although electricity grids with high levels of photovoltaic installations have been modeled, those models do not include the impact of snow. This research investigates the need for resource adequacy modeling that includes snow by utilizing current and projected utility PV capacity. High-penetration and mid-case utility PV scenarios were examined on a state level and compared to annual average snowfall. In both scenarios, Colorado, Michigan, New York, Wisconsin, and Minnesota stand out as high-snowfall states with utility PV compound annual growth rates of 5% or higher. Although the results show that there is a negative relationship between snowfall and projected PV growth per state, some states with high snowfall may still see a significant increase in utility PV capacity over the next 28 years. In addition to the state-level analysis, a utility-level analysis was performed using NSIDC snow data and utility IRP projections. The results of this step show that there are utilities in the northern Midwest with utility PV projections as high as 6500 MW by 2040. Even utilities with lower PV projections display elevated compound annual growth rates.

I. INTRODUCTION

Solar photovoltaics are typically associated with sunny environments and little shade. Therefore, it is no surprise that California, Texas, and Arizona lead the United States in solar power generation [1]. However, sunlight is not the only driver for solar installations. The falling cost of PV, government subsidies, and the high cost of electricity in some regions have made solar an attractive option. Due to these alternative market drivers, many colder states have seen an increase in solar power. Cold temperatures and the high albedo of snow-covered ground increase solar panel efficiency but come with a new complication – snow cover. With greater dependence on solar energy in these states, snowfall will reduce electricity generation and may increase the risk of inadequate power reaching customers. Energy resource managers will face a new challenge: how to ensure reliable electricity in snowy regions with growing solar installations. This research focuses on analyzing the projected utility-scale solar growth from 2022 to 2050 in the contiguous U.S. and contrasting that data with annual snowfall.

II. BACKGROUND

Grid modeling with high solar installation rates, such as that performed by Frew et al. [2], has yet to include the impact of snow. Solar snow research has been limited to the panel or individual system level. According to a recent 2020 paper, snow's effect on PV must be addressed in future grid modeling [3].

Snow cover reduces panel electricity output because the solar cells receive less direct sunlight. Researchers have studied the optical properties of snow for decades. Bergin performed one of the first studies by measuring light through undisturbed snow cover in 1971 [4]. Since then, other researchers have refined the calculations, including Warren [5] and Perovich [6]. Output reduction of solar panels due to snow cover has been researched extensively [7], [8], [9], [10], [11]. Many factors influence this metric, including ambient temperature and snow sliding off panels. In 2013, Marion et al. developed a mathematical model that projected solar panel output in snow environments. Their commercial PV projection over two years came within 0.5% of the generated output, and the residential PV model came within 1.5% of the generated output [8]. The Marion model will be a foundation for expanding solar-snow projection to a grid scale. However, before snow projection is modeled, the importance of future snow effects on the U.S. grid must be analyzed.

III. SUMMARY OF WORK

This research investigated the relationship between states with high annual snowfall and utility-level PV capacity growth. The analysis was completed in two stages – state level and utility level.

A. State-Level Analysis

In the state-level investigation, the capacity planning model, ReEDS, was used. ReEDS Standard Scenarios compare the current (2022) to the projected 2050 utility PV capacity per state [12]. This comparison used the ReEDS scenario that yielded the highest total utility PV capacity for the contiguous U.S. The scenario assumed significant electrification (high-demand growth), no nascent technologies, and 95% decarbonization by 2050. This projection serves as a worst-case scenario for snowy states in which high-PV penetration increases dependency on solar production year-round. In contrast, a mid-case, current policies, and no nascent technologies scenario was used as a "business as usual" projection.

978-1-6654-6060-6/23 $31.00 © 2023 IEEE

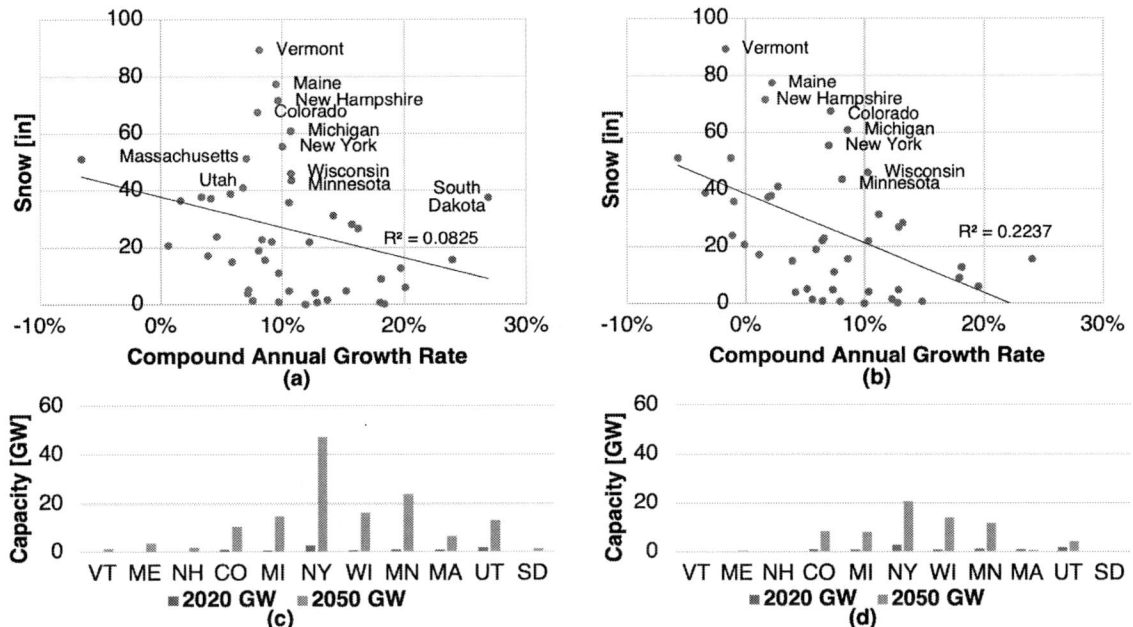

Fig. 1. The top two plots compare the CAGR of state utility solar with the state annual average snowfall. A line of best fit is overlaid. The red data points indicate states with a CAGR greater than 5% that receive more than 40 inches of snow on average annually. The bottom two plots show the increased utility PV capacity from 2022 to 2050 for selected states. Plots (a) and (c) represent the high PV scenario. Plots (b) and (d) represent the mid-case scenario.

PV utility compound annual growth rates (CAGRs) were calculated for each state and plotted against annual average snowfall [13]. States that showed significant snowfall and a high CAGR were further plotted to illustrate their total capacity growth from 2022 to 2050.

Current snowfall data was used since there is much debate over the future fluctuation of snowfall due to climate change. Some sources indicate there will be less average annual snowfall while the severity of snow events will increase [14]. Other studies have indicated that both average snowfall and snow event severity will decrease in the future [15].

B. Utility-Level Analysis

The five states with red markers from Fig. 1b were chosen for further analysis on a utility level. They are highlighted in both ReEds Scenarios, having a CAGR greater than 5% and receiving more than 40 inches of average annual snowfall.

Integrated resource plans (IRPs) were the chosen PV projection metric for the utility-level analysis. IRPs are either present or required in 30 states [16]. IRPs are documents that analyze projected peak load for a future timeframe and provide detail of how the utility will meet that load. They also address any state renewable generation requirements. Of the five states highlighted in Fig. 1b, three require IRPs to be filed by major utilities – Michigan, Minnesota, and Colorado. Wisconsin and New York do not require IRP filing, but some utilities that filed in Michigan and Minnesota cross state lines into Wisconsin, which provides some visibility into that state. One major utility in New York filed an IRP and was included in this analysis.

Although Colorado is an IRP state, it was not included, as the snow data was skewed by the Rocky Mountains. Further analysis is needed for this state.

Each utility was assigned a class based on its solar projection. The class definitions are as follows:

- Class A: Filed an IRP with high solar (>500 MW)
- Class B: Filed an IRP with medium solar (>100 MW)
- Class C: Filed an IRP with low solar (<100 MW)
- Class D: Filed an IRP with no solar

A full list of utilities and their classes can be found in Appendix B.

After the IRP projections were collected, snow data from the National Snow and Ice Data Center (NSIDC) was processed for the regions of interest [17]. This dataset provides a 4 km grid of daily snow depth for each year. The 2017 dataset was chosen for this analysis; the collection year spans from October 1st, 2016, to September 30th, 2017. The dataset was reduced to October 1st, 2016, to April 30th, 2017, for ease of processing. It is assumed that this reduction will capture adequate snow data for the winter of 2016-2017. This period was selected since it includes the 2017 North American Blizzard, also known as Winter Storm Stella [18]. Additionally, this winter provided one of Detroit's heaviest snowstorms. The overnight increase in snow depth for Detroit's snowstorm can be seen in Fig. 2. This figure was developed from the NSIDC dataset and shows that in a 24-hour period, some parts of Michigan accumulated 8 inches of snow. The National Weather Service verified this

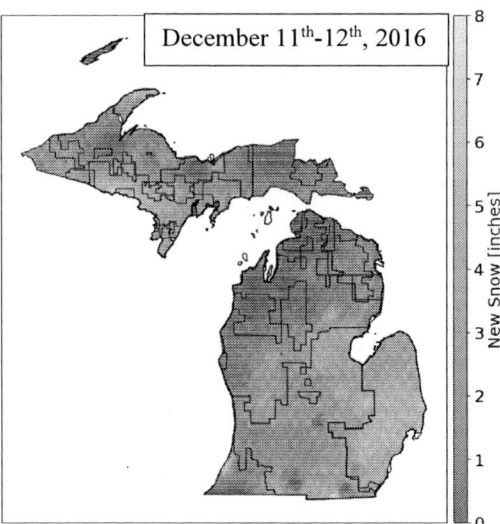

Fig. 2. Detroit snowstorm illustrating a widespread snow depth increase in Michigan.

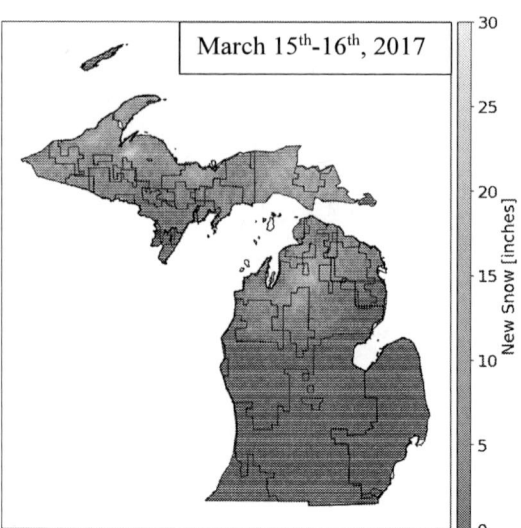

Fig. 3. Winter Storm Stella illustrating a concentrated snow depth increase in Michigan.

data, stating snowfall accumulations ranged from 8-11 inches in southeastern Michigan [19].

The NSIDC dataset was further verified by Winter Storm Stella. In Fig. 3, Michigan was plotted with NSIDC data and shows an incredible snow depth increase (29 inches) on March 16th, 2017, in a small area in Michigan's Upper Peninsula. This area is near Herman, MI which receives an average of 236 inches of snow per year [20]. Additionally, the National Weather Service radar shows a small concentration of clouds in the Herman area during Winter Storm Stella [19].

C. Utility-Level Python Calculations

The data was reduced to a dataframe from October 1st, 2016, to April 30th, 2017, that included latitude, longitude, snow depth, and time. Two methods were used to find the maximum increase in snow depth in a 24-hour period.

Method 1, non-coincident locational maximum changes: A Python script was developed to find the difference in snow depth between each day at each location within the utility. The code filtered the snow depth difference to only include positive changes. It then created a new dataframe including only the maximum positive change at each location within the utility. This method is shown in Fig. 4b and Fig. 6.

Method 2, maximum change in utility average: A new Python script was created to find the average snow depth in each utility area for each day. The code calculated the snow depth difference between each day and filtered the results to include only increased depth. It then identified the maximum snow depth increase and assigned the increase to each location in the utility. This method is shown in Fig. 4c and Fig. 7.

IV. RESULTS

The results of the state-level analysis show that there is a loose negative correlation between utility PV growth and annual snowfall. This is no surprise, as PV is an irradiance-driven energy resource. North Dakota and West Virginia were removed from both cases, as they were outliers with over 100% CAGR. Additionally, South Dakota and Idaho were removed from the mid case because they showed a CAGR of -100%. Currently, these states have few if any utility installations, which amplifies any change in capacity.

The states labeled in Fig. 1a and 1b show a relatively high annual snowfall and are also projected to have medium to high CAGRs. Many of these states are in New England, which already has a growing residential PV market [21]. The other states in this group are Midwest states where annual snowfall is high and more land may be available for utility PV installations.

Fig. 1c and 1d include many of the same states but show the absolute increase in utility capacity rather than the CAGR between 2022 and 2050. New York, Minnesota, Wisconsin, Michigan, and Colorado show significant increases in utility PV capacity by 2050. This is true for both the high-penetration PV case and the "business as usual" case.

The results of the utility-level analysis reveal two Class A utilities to be of interest. Consumers Energy and Northern States Power Company stand out among the utilities due to their significant PV projections and the occurrence of over 8 inches of snow accumulation in 24-hours using Method 2 and over 20 inches using Method 1. Consumers Energy plans to install 5,000 MW of solar by 2040, and Northern States Power

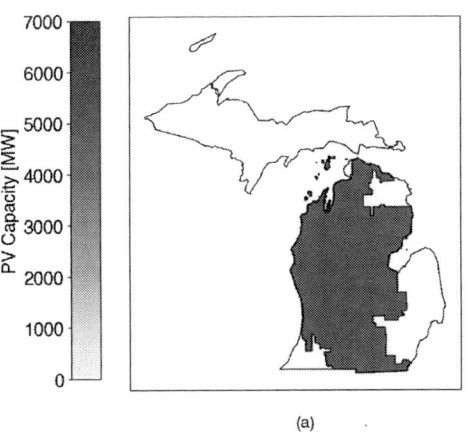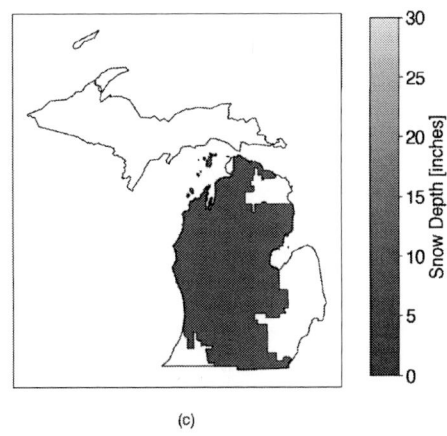

Fig. 4.   Plot (a) shows the IRP projected PV for Consumers Energy by approximately 2040. The utility plans to install 5000 MW of utility solar. Plot (b) shows method 1, the maximum change in snow in a 24-hour period for each location within the service territory in the winter of 2016-2017. Plot (c) shows method 2, the utility's maximum average snow depth change for a 24-hour period.

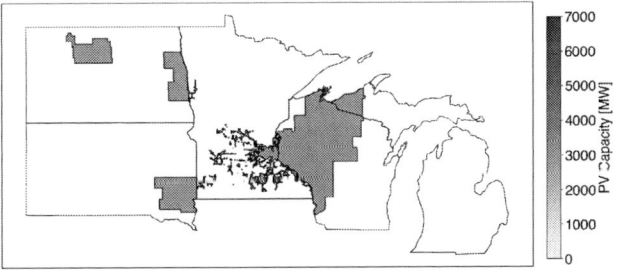

Fig. 5.   The Northern States Power Company IRP submitted to Michigan proposes 3150 MW of utility PV installations by 2034.

Company plans to install 3,150 MW of solar by 2034 [22], [23]. These increases can be seen in Fig. 4a and Fig. 5.

If we examine the worst-case condition, Consumers Energy could experience a 28-inch snow cover in one day. Similarly, the worst-case condition for Northern States Power Company is a 20-inch snow cover in one day. However, the intense snowfalls in the Consumers Energy territory occur primarily in the north, as seen in Fig. 4b. A more reasonable assessment is in Fig. 4c, where an average is taken using Method 2. This shows a maximum gain of approximately 8 inches in one day for Consumers Energy and 11 inches for Northern States Power Company shown in Fig. 7.

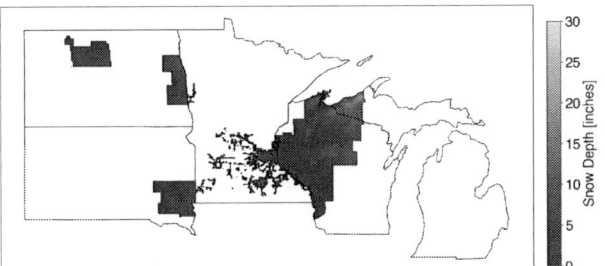

Fig. 6.   Method 1 suggests a maximum 24-hour snow depth increase of 20.157 inches in the Northern States Power Company service territory.

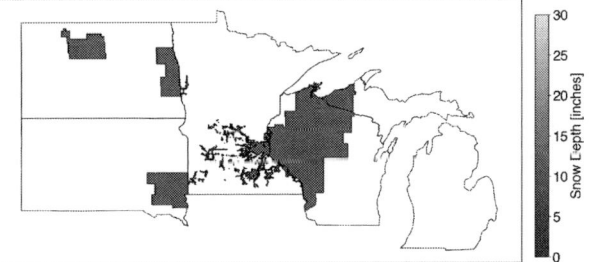

Fig. 7.   Method 2 suggests an average maximum 24-hour snow depth increase of 10.990 inches in the Northern States Power Company service territory.

Fig. 8 illustrates the relationship between snow depth increase and the IRP PV projections for Class A and B utilities. The snow depth increase in Fig. 8 is from Method 1. The same plot but with data from Method 2 can be found in Appendix A. It should be mentioned that both DTE and Con Edison (NY) have less snow compared to Consumers Energy and Northern States Power Company, but they still see approximately 7 and 10 inches of snow, respectively, for Method 1 and 7 and 9 inches of snow for Method 2. Therefore, all Class A utilities in this analysis experience at least 7 inches of accumulated snowfall in one day for the chosen analysis year.

V. DISCUSSION

With such capacity growth in states that experience high annual average snowfall, this data suggests that resource adequacy models will need to include snow effects to accurately plan grid adequacy. The results of this research indicate places where PV resource fluctuation should be investigated due to the shading from snow cover – Michigan, Wisconsin, Minnesota, and New York.

In Fig. 8, there are utilities that stand out on the right side of the plot – Northern States Power Company, Consumers

978-1-6654-6060-6/23 $31.00 © 2023 IEEE

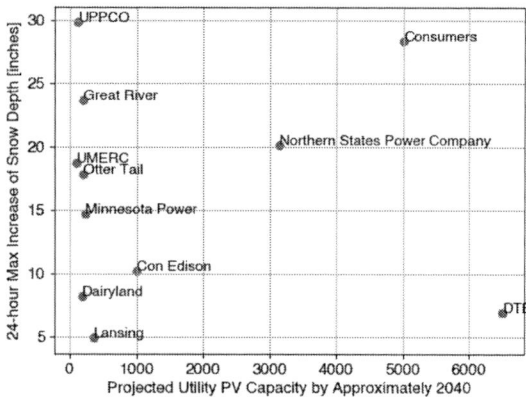

Fig. 8. Due to the significant snow increase in northern Michigan shown in Fig. 4b, Consumers Energy displays a high depth increase in Method 1. Additionally, the Class B utilities are grouped to the left due to low PV projections compared to Class A.

Energy, and DTE. All three are projected to have more than 3,000 MW installed by 2040. Although DTE receives less snow than Nothern States Power Company and Consumers Energy, it could still be affected by widespread snow accumulation. An example of this accumulation is the Detroit snowstorm in Fig. 2.

It is also important to note that the method used to calculate 24-hour snow accumulation changes the outcome for utilities with diversified snow behavior. For example, in Fig. 8, Consumers Energy has a maximum snow increase of 28.35 inches, but in Fig. A1 in Appendix A, it has a maximum snow increase of 8.45 inches. The difference is due to the high concentration of snow accumulation in northern Michigan which is shown in Fig. 4b. In future analyses, it will be important to choose a suitable calculation method for the application. A widespread average snow change (Method 2) may be more appropriate for a state-wide utility solar analysis, whereas a disaggregated snow depth increase (Method 1) may be appropriate for regional or individual system analysis.

In Fig. 8, a majority of utilities have a projected utility PV capacity of less than 500 MW. Their comparatively small projected PV capacity may seem inconsequential in this high-penetration PV discussion. However, Fig. 9 and Fig. A2 in Appendix A show that many of these same utilities have a high CAGR. This finding is significant because it indicates rapid occurrence of PV in utilities that traditionally have low or no utility PV installations. These new utility PV adopters may be faced with resource adequacy issues in the future due to snow cover if their solar resources continue to grow.

VI. Conclusion

The results of this research suggest that although projected PV growth is generally higher in states that do not receive much snow, some high snowfall states do have compound annual growth rates of over 5%, according to the state-level analysis. Further, some utility IRPs in snowy states show projected PV installations by 2040 to be as high as 6,500 MW,

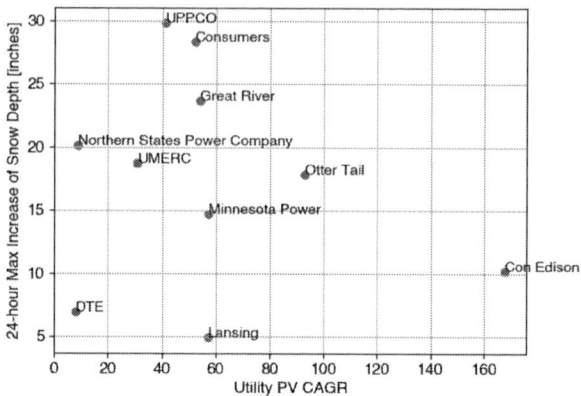

Fig. 9. CAGR compares current utility PV capacity to IRP projected utility PV capacity. The plot does not include Dairyland Power Cooperative, as the CAGR was 1093% and skewed the data.

while other utilities have a maximum utility solar CAGR of up to 1093% [24]. Additionally, we find that in some of these high-growth states, a utility-wide metric for snow events misses locationally extreme events. As solar capacity grows, the grid will depend on solar production to supply customers with adequate power. When snow covers the solar panels, their production is reduced, which may negatively affect grid managers' ability to meet load. Future research needs to include snow shed and snow shading metrics. Variables such as wind, sun, panel movement, and temperature need to be considered. Developing accurate snow-shedding models from these parameters will allow power system models to include snow events, ensuring resource adequacy.

References

[1] E. A. Soto, K. Arakawa, and L. B. Bosman, "Identification of target market transformation efforts for solar energy adoption," *Energy Rep.*, vol. 8, pp. 3306–3322, Nov. 2022, doi: 10.1016/j.egyr.2021.12.043.

[2] B. Frew, W. Cole, P. Denholm, A. W. Frazier, N. Vincent, and R. Margolis, "Sunny with a Chance of Curtailment: Operating the US Grid with Very High Levels of Solar Photovoltaics," *iScience*, vol. 21, pp. 436–447, Nov. 2019, doi: 10.1016/j.isci.2019.10.017.

[3] W. Cole, D. Greer, J. Ho, and R. Margolis, "Considerations for maintaining resource adequacy of electricity systems with high penetrations of PV and storage," *Appl. Energy*, vol. 279, p. 115795, Dec. 2020, doi: 10.1016/j.apenergy.2020.115795.

[4] J. D. Bergen, "The relation of snow transparency to density and air permeability in a natural snow cover," *J. Geophys. Res.*, vol. 76, no. 30, pp. 7385–7388, 1971.

[5] S. G. Warren, "Optical properties of snow," *Rev. Geophys. 1985*, vol. 20, no. 1, pp. 67–89, 1982.

[6] D. K. Perovich, "Light reflection and transmission by a temperate snow cover," *J. Glaciol.*, vol. 53, no. 181, pp. 201–210, 2007.

[7] L. Powers, J. Newmiller, and T. Townsend, "Measuring and modeling the effect of snow on photovoltaic system performance," in *2010 35th IEEE Photovoltaic Specialists Conference*, Jun. 2010, pp. 000973–000978. doi: 10.1109/PVSC.2010.5614572.

978-1-6654-6060-6/23 $31.00 © 2023 IEEE

[8] B. Marion, R. Schaefer, H. Caine, and G. Sanchez, "Measured and modeled photovoltaic system energy losses from snow for Colorado and Wisconsin locations," *Sol. Energy*, vol. 97, pp. 112–121, Nov. 2013, doi: 10.1016/j.solener.2013.07.029.

[9] L. B. Bosman and S. B. Darling, "Performance modeling and valuation of snow-covered PV systems: examination of a simplified approach to decrease forecasting error," *Environ. Sci. Pollut. Res.*, vol. 25, no. 16, pp. 15484–15491, Jun. 2018, doi: 10.1007/s11356-018-1748-1.

[10] E. Andenæs, B. P. Jelle, K. Ramlo, T. Kolås, J. Selj, and S. E. Foss, "The influence of snow and ice coverage on the energy generation from photovoltaic solar cells," *Sol. Energy*, vol. 159, pp. 318–328, Jan. 2018, doi: 10.1016/j.solener.2017.10.078.

[11] E. Mohammadi, J. Khodabakhsh, G. Moschopoulos, and R. Fadaeinedjad, "A Study on the Performance of PV Modules in Snowy Conditions Considering Orientation of Modules," in *2020 IEEE Canadian Conference on Electrical and Computer Engineering (CCECE)*, Aug. 2020, pp. 1–4. doi: 10.1109/CCECE47787.2020.9255738.

[12] P. Gagnon *et al.*, "2022 Standard Scenarios Report: A U.S. Electricity Sector Outlook," *Renew. Energy*, 2022.

[13] "U.S. Average Snow State Rank." http://www.usa.com/rank/us--average-snow--state-rank.htm (accessed Oct. 09, 2022).

[14] L. Quante, S. N. Willner, R. Middelanis, and A. Levermann, "Regions of intensification of extreme snowfall under future warming," *Sci. Rep.*, vol. 11, no. 1, Art. no. 1, Aug. 2021, doi: 10.1038/s41598-021-95979-4.

[15] W. S. Ashley, A. M. Haberlie, and V. A. Gensini, "Reduced frequency and size of late-twenty-first-century snowstorms over North America," *Nat. Clim. Change*, vol. 10, no. 6, Art. no. 6, Jun. 2020, doi: 10.1038/s41558-020-0774-4.

[16] "Energy Efficiency and Electric Infrastructure in the State of Michigan".

[17] Broxton P. ,. X. Zeng and N. Dawson, "Daily 4 km Gridded SWE and Snow Depth from Assimilated In-Situ and Modeled Data over the Conterminous US, Version 1." NASA National Snow and Ice Data Center Distributed Active Archive Center, 2019. doi: 10.5067/0GGPB220EX6A.

[18] "Winter Storm Stella was a Category 3 on Northeast Snowfall Impact Scale," *The Weather Channel*. https://weather.com/storms/winter/news/winter-storm-stella-northeast-blizzard-warning-noreaster-snow-forecast-march-2017 (accessed May 30, 2023).

[19] N. US Department of Commerce, "Winter Storm December 11-12, 2016." https://www.weather.gov/dtx/WinterStormDecember11-12 (accessed May 30, 2023).

[20] "Lake Effect Snow," *Snowtech Magazine*, Feb. 18, 2018. https://www.snowtechmagazine.com/lake-effect-snow/ (accessed May 30, 2023).

[21] "Solar Power in New England: Concentration and Impact." https://www.iso-ne.com/about/what-we-do/in-depth/solar-power-in-new-england-locations-and-impact (accessed Dec. 23, 2022).

[22] "Case No. U-20165 – In the Matter of the Application of Consumers Energy Company for Approval of an Integrated Resource Plan under MCL 460.6t and for other relief." Consumers Energy, Jun. 15, 2018. [Online]. Available: https://mi-psc.force.com/sfc/servlet.shepherd/version/download/068t000000231usAAA

[23] K. J. Sieben, V. Means, M. Schuerger, J. K. Sullivan, and J. A. Tuma, "Upper Midwest Integrated Resource Plan of Northern States Power Company d/b/a Xcel Energy."

[24] "2022 DTE Electric Integrated Resource Plan." DTE, 2022. [Online]. Available: https://dtecleanenergy.com/downloads/IRP_Executive_Summary.pdf

[25] "In the matter of Upper Peninsula Power Company for Approval of an Integrated Resource Plan under MCL 460.6t and for other relief." Upper Peninsula Power Company, Feb. 12, 2019. [Online]. Available: https://mi-psc.force.com/sfc/servlet.shepherd/version/download/068t0000003iRNDAA2

[26] "Alpena Power Company – Community Commitment Since 1881." https://www.alpenapower.com/ (accessed May 31, 2023).

[27] "2020 Integrated Resource plan." Lansing Board of Water & Light.

[28] "Keeping the Lights On," *Dairyland Power Cooperative*. https://www.dairylandpower.com/keeping-lights (accessed May 31, 2023).

[29] "2022 OPTIONAL-IRP COMPLIANCE REPORT OF DAIRYLAND POWER COOPERATIVE." Dairyland Power Cooperative, 2022. [Online]. Available: https://www.mncenter.org/sites/default/files/pdfs/20227-187147-01.pdf

[30] "2023-2037 Integrated Resource Plan." Great River Energy, Mar. 31, 2023. [Online]. Available: https://greatriverenergy.com/wp-content/uploads/2023/03/2023-IRP-FINAL.pdf

[31] "Notice of Changed Circumstances – 2017 Electric Integrated Resource Plan." Alliant Energy, Jan. 29, 2021. [Online]. Available: https://www.edockets.state.mn.us/edockets/searchDocuments.do?method=showPoup&documentId={90C44E77-0000-CF54-AE37-8C9EC89ED8CE}&documentTitle=20211-170416-03

[32] "APPLICATION FOR INTEGRATED RESOURCE PLAN APPROVAL 2019 - 2033." Minnesota Municipal Power Agency, Jul. 30, 2018. [Online]. Available: https://mmpa.org/wp-content/uploads/2019/08/2018-MMPA-IRP-Final-PUBLIC.pdf

[33] "2021 INTEGRATED RESOURCE PLAN." Minnesota Power, Feb. 01, 2021. [Online]. Available: https://www.edockets.state.mn.us/EFiling/edockets/searchDocuments.do?method=showPoup&documentId=%7b70795F77-0000-C41E-A71C-FD089119967C%7d&documentTitle=20212-170583-01

[34] "2022 INTEGRATED RESOURCE PLAN." Minnkota Power Cooperative, Inc. and Northern Municipal Power Agency, 2022. [Online]. Available: https://assets.website-files.com/5ef212e2cdca1e094063db4e/62d062e4a190aaacf6237297_2022%20Integrated%20Resource%20Plan.pdf

[35] "In the Matter of Southern Minnesota Municipal Power Agency's Submittal of its 2022–2036 Integrated Resource Plan." Southern Minnesota Municipal Power Agency, Nov. 02, 2022. [Online]. Available: https://www.edockets.state.mn.us/edockets/searchDocuments.do?method=showPoup&documentId=%7b40833984-0000-CC1A-98FD-E95CF9573022%7d&documentTitle=202211-190369-01

[36] "Long-Range Plan." Consolidated Edison Company, Jan. 2022. [Online]. Available: https://www.coned.com/-/media/files/coned/documents/our-energy-future/our-energy-projects/electric-long-range-plan.pdf

978-1-6654-6060-6/23 $31.00 © 2023 IEEE

APPENDIX A – METHOD 2 SCATTER PLOTS

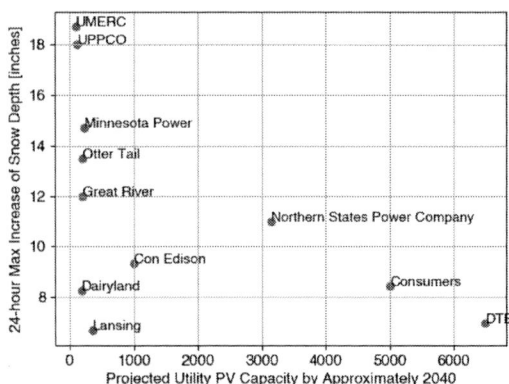

Fig. A1. The average snow depth for each utility is taken in Method 2 moving Consumers Energy down on the scatter plot compared to Fig. 8.

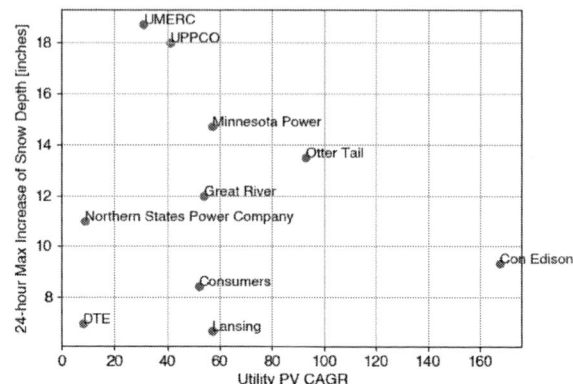

Fig. A2. CAGR compares current utility PV capacity to IRP projected utility PV capacity. The plot does not include Dairyland Power Cooperative, as the CAGR was 1093% and skewed the data.

APPENDIX B – UTILITY INFORMATION

IRP State	Utility	Full Utility Name	Class	Solar	Year	Notes	Citations
Michigan	UPPCO	Upper Peninsula Power Company	B	125 MW	2037		[25]
Michigan	Alpena	Alpena Power Company	D	No new capacity			[26]
Michigan	Consumers	Consumers Energy	A	5000 MW	2040		[22]
Michigan	DTE	DTE Energy	A	6500 MW	2042		[24]
Michigan	UMERC	Upper Michigan Energy Resources Company	B	100 MW	2037-2040		[25]
N/A	Lansing	Lansing Board of Water and Light	B	361 MW	2036	No IRP	[27]
Michigan	I&M	Indiana Michigan Power				Inconsistent coverage area	
Michigan Minnesota	NSPC	Northern States Power Company	A	3150 MW	2034		[23]
Minnesota	BEPC	Basin Electric Power Cooperative				Covers huge, multi-state region. Not suitable for this analysis.	
Minnesota	Dairyland	Dairyland Power Cooperative	B	194 MW	2025	Looking at 580 MW of combined wind and solar by 2035	[28], [29]
Minnesota	Great River	Great River Energy	B	200 MW	2037	Very wind heavy	[30]
Minnesota	Interstate Power and Light	Interstate Power and Light Company	B	400 MW	2023	Could not locate service territory in dataset	[31]
Minnesota	MMPA	Minnesota Municipal Power Agency	D	No capacity needed	2030		[32]
Minnesota	Minnesota Power	Minnesota Power	B	230 MW	2030		[33]
Minnesota	Minnkota	Minnkota Power Cooperative	D	No capacity needed	2035		[34]
Minnesota	Missouri River	Missouri River Energy Services				Covers huge, multi-state region. Not suitable for this analysis.	
Minnesota	Southern Minnesota	Southern Minnesota Municipal Power Agency	C	25 MW	2036		[35]
N/A	Con Edison	Consolidated Edison	A	1000 MW	2040	Located in New York	[36]

APPENDIX C – PV PROJECTIONS AND SNOW DATA

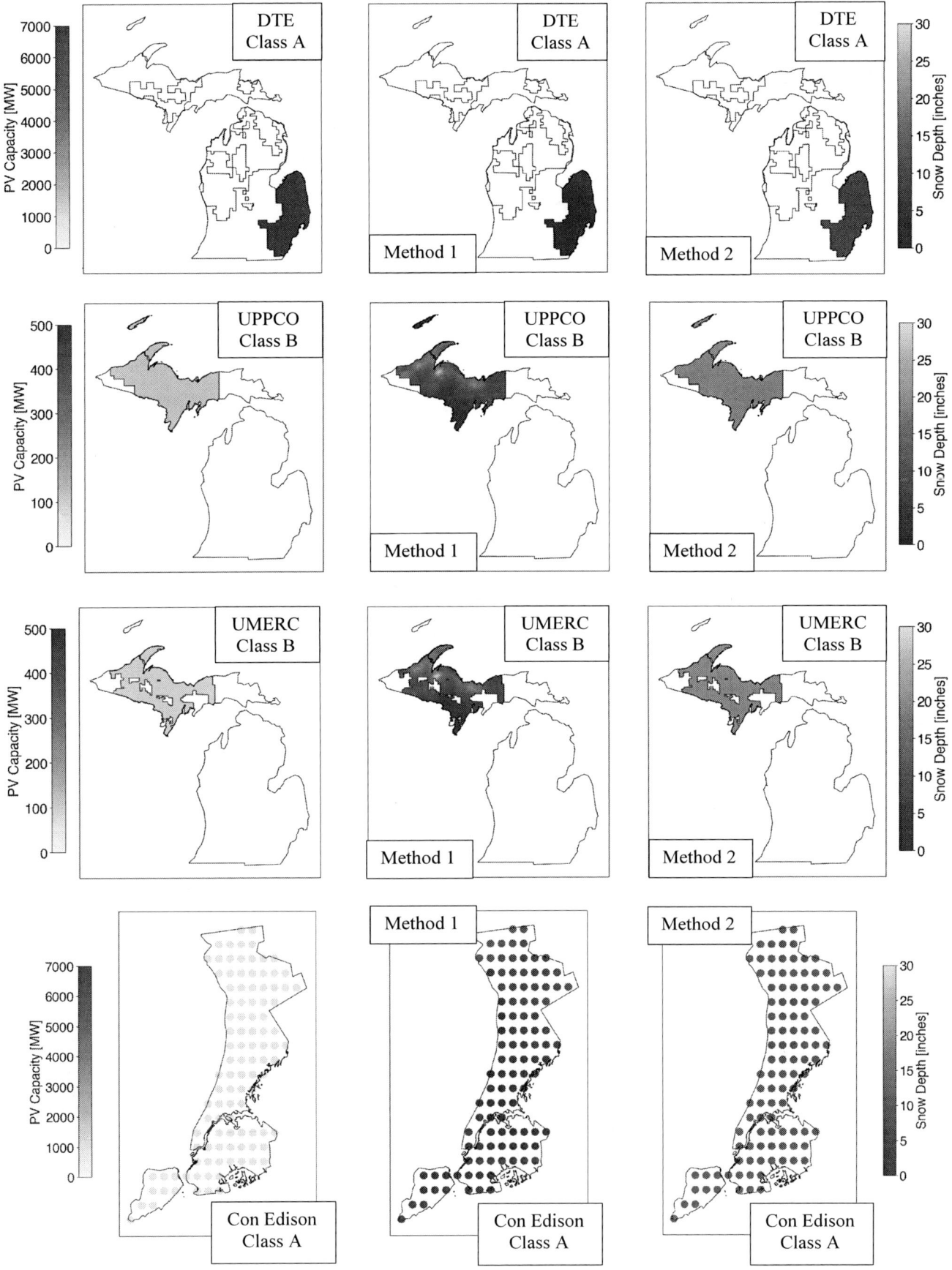

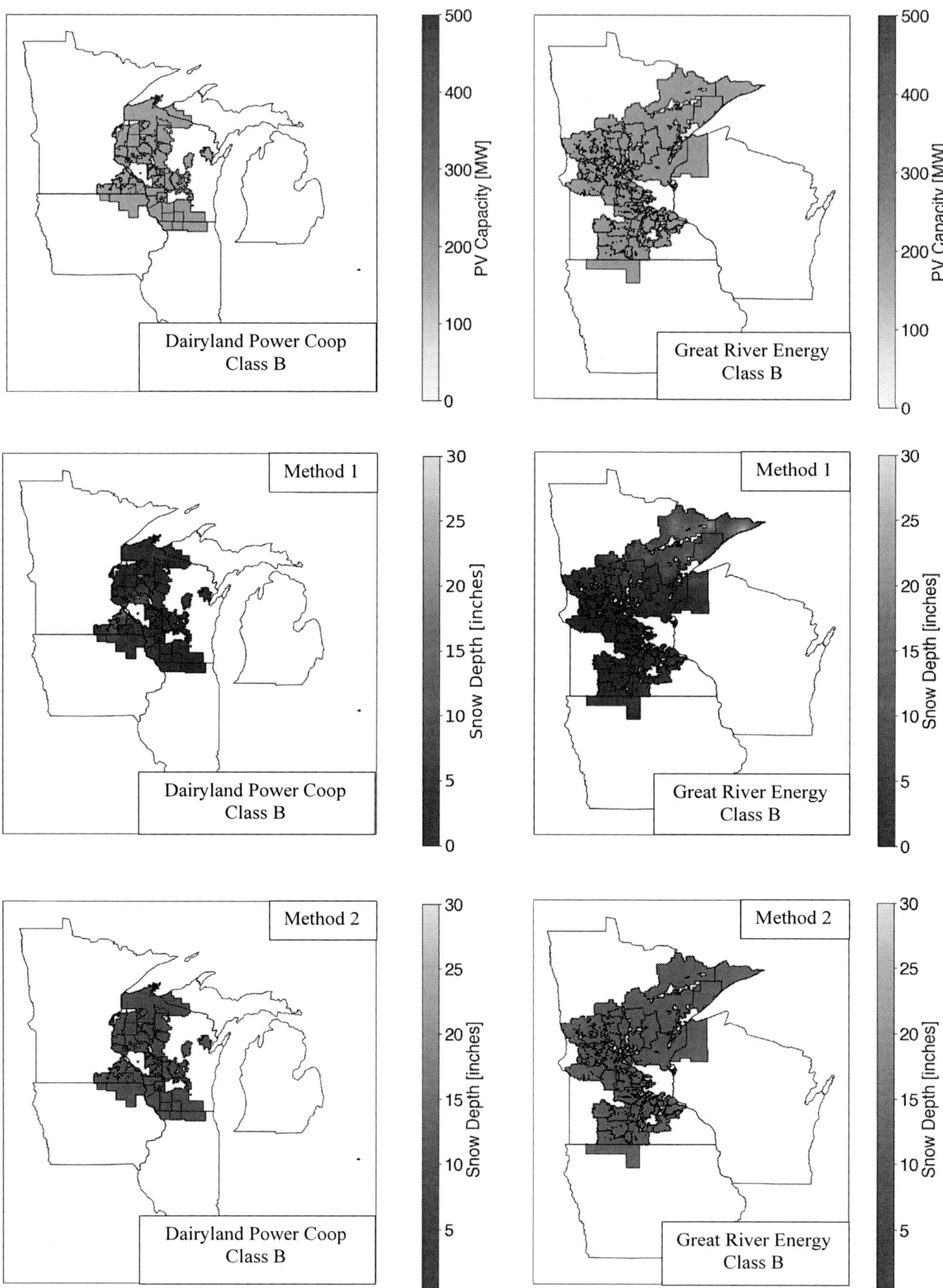

978-1-6654-6060-6/23 $31.00 © 2023 IEEE

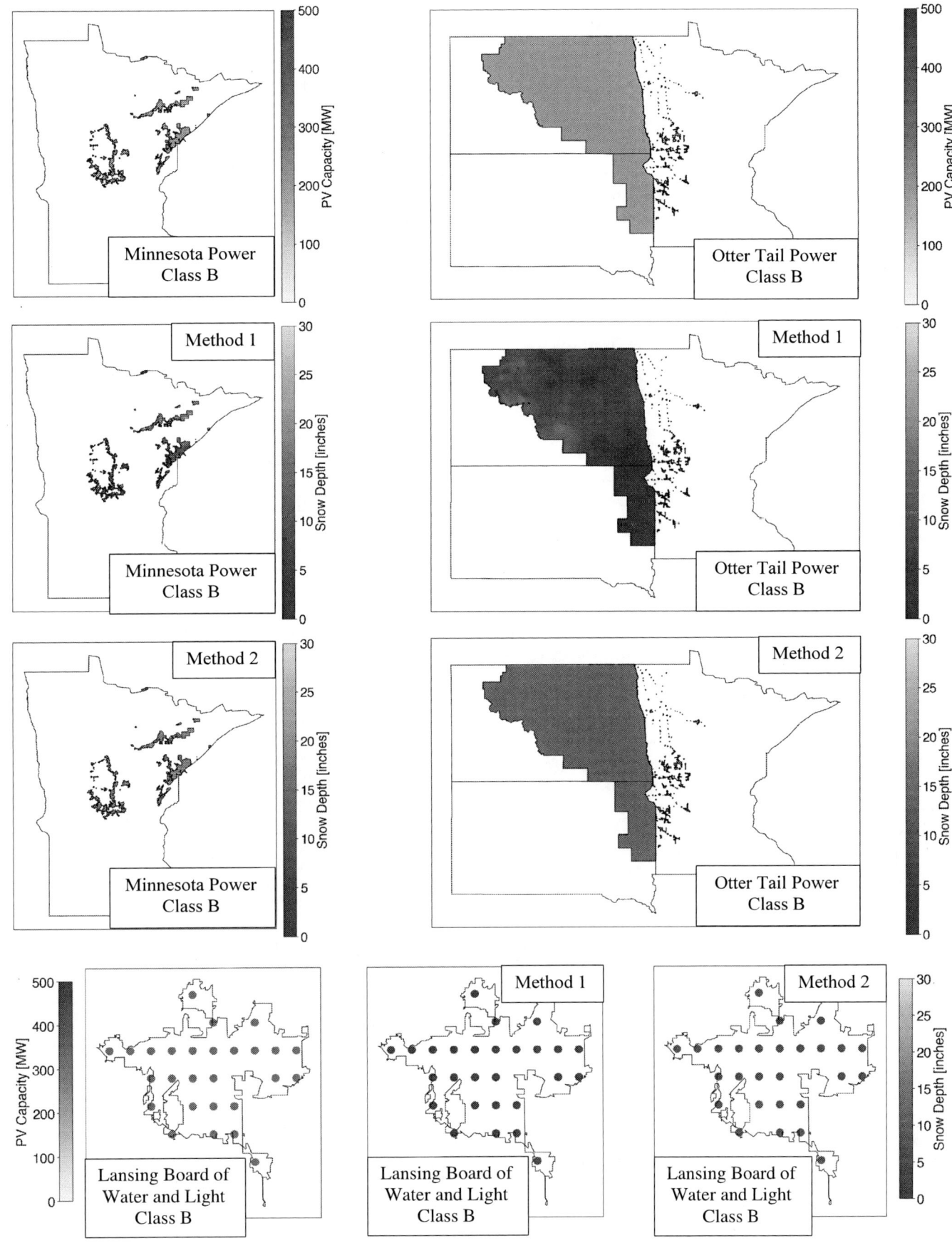

Modeling of Perovskite/Si Tandem Solar Cell

Yegao Xiao, Michel Lestrade, Zhiqiang Li, and Zhanming S. Li

Crosslight Software Inc, 230-3410 Lougheed Hwy, Vancouver, BC, V5M 2A4 Canada

Abstract — Based on a drift-diffusion simulator, the modeling of perovskite/Si tandem solar cell with tunnel junction is presented in this work. Current matching is explored between the two sub-cells. It is predicted that such a tandem cell can achieve conversion efficiency as high as 28.27% with open-circuit voltage and short-circuit current density as 2.04 V and 16.18 mA/cm², respectively. As approaches for cell design optimization, the results are also analyzed versus the thickness and the minority carrier recombination lifetime of the perovskite layer.

I. INTRODUCTION

Perovskite solar cells have attracted tremendous interest in recent years due to the involved low preparation cost and convenience of processing [1]. Although the power conversion efficiency for single-junction perovskite cell has been significantly enhanced within the last decade [2], it is actually restricted by the Shockley-Quieisser (S-Q) limit due to the absorption loss below the bandgap and the thermal relaxation loss of hot charge carriers [3]. To increase the efficiency further beyond the S-Q limit of single junction cell, tandem (or multiple-junction, MJ) solar cells have been proven to be promising strategies by integrating the wide bandgap top junction cell with the low bandgap bottom one.

The perovskite/Si tandem cell has been explored intensively worldwide because the metal halide perovskites are good light absorbers with tunable bandgaps (1.55-1.75 eV with top junction cell). The research efforts have been spent on perovskite absorber optimization [4], tunnel junction (TJ) [5], recombination layer studies [6], and bottom Si cells structure [7]. Record high efficiency for such cells has been reported [2].

However, the structure and material complexity as well as the tunneling and recombination mechanism involved makes the modeling of such tandem cells an indispensable but challenging task. Better modeling methods and software, especially compact packages with full breadth, are quite beneficial because of savings in R&D time and cost, and their capability to optimize cell design. Whereas one-dimensional calculation with simplified theoretical model can be a relatively easy work, modeling of TJ for tandem (or MJ) solar cells has only been done by a few of research groups (e.g. [8]) and commercial software providers (e.g. [9]). The Crosslight APSYS package [9] has been previously shown to be very effective in modeling MJ solar cells [10,11] with tunnel junction including tandem cells using Si bottom junction [12].

In this work, based on a drift-diffusion simulator, Crosslight APSYS package, modeling of perovskite/Si tandem solar cell is presented. The paper is organized with the following sub-sections.

II. CELL STRUCTURE AND MODELING DETAILS

The tandem cell structure is schematically shown in Fig. 1, where p^{++} a-Si/ n^{++} a-Si (with bandgap 1.7 eV) TJ is used. The top perovskite junction is assumed to be formed on the bottom Si wafer cell. The p^+ spiro-OMeTAD, the p^- perovskite and n^+ TiO$_2$ function as the hole transport, the absorber and the electron transport layers, which have bandgap/electron affinity combination as 2.95/2.2, 1.6/3.93, and 3.2/4.0 in eVs, respectively [13]. The relevant (n, k) index profiles is taken from Ref. [14]. The sun illumination is assumed to be solar.am15g. For optical simulation related to electron-hole generation, the software employs a transfer matrix method to calculate the light propagation in the multi-layers by taking into account the Fresnel reflections, transmissions, and standing wave effects.

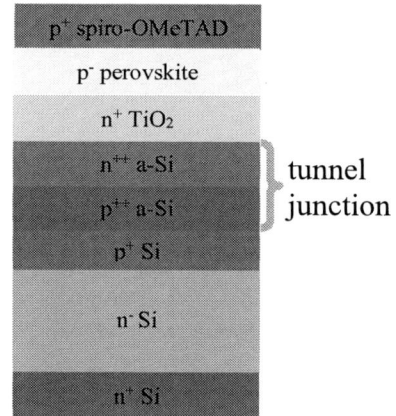

Fig. 1.   Schematic perovskite/Si tandem solar cell structure.

The band diagram (mostly for the TJ and top junction region) is shown in Fig. 2 at the the open-circuit bias poiont where the effective tunneling from conduction band to valence band across the TJ region is observed.

III. RESULTS AND ANALYSES

Once the tunnel junction works, the solar cell performance can be simulated and the relevant characteristic values can be extracted. Without optimization and current matching between the two junctions, the cell efficiency is usually low as shown

978-1-6654-6060-6/23 $31.00 © 2023 IEEE

by a current-mismatched case in Fig. 3, where the bottom junction is current limiting.

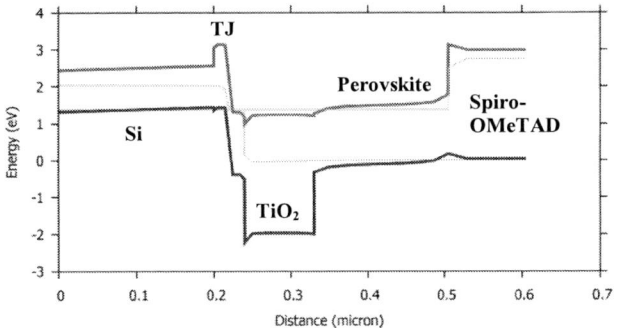

Fig. 2. Band diagram for the simulated tandem cell (most of the Si junction is truncated) at open-circuit bias.

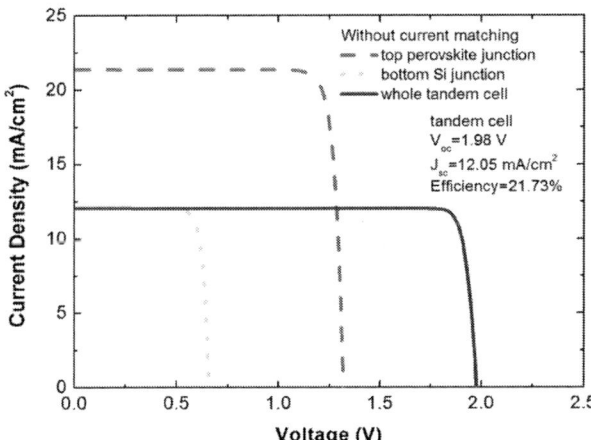

Fig. 3. J-V plots for the cell simulated without current-matching.

In order to find the current matching point, the perovskite layer thickness is varied to monitor the tandem cell performance. The tandem cell efficiency versus this thickness is shown in Fig. 4, and the highest point is aound 0.175 μm for the perovskite layer. This certainly correlates with the bottom Si junction cell which is assumed unchanged with a wafer-processed type (e.g., with thickness around 250 μm). In other words, if the bottom Si cell has texture and/or better growth quality with less defects (& therefore large minority carrier recombination time), it may allow a thicker top perovskite layer for current matching with even higher efficiency. Perhaps, one should optimize the bottom Si cell at first. From the modeling point of view, it should be also pointed out that the (n, k) index profiles may also affect the optimal perovskite layer thickness for the matching purpose. Some discrepancy for the index file is found in different references [14].

The J-V curves for the tandem cell with current matching is presented in Fig. 5 where the efficiency as high as 28.27% is obtained with open-circuit voltage and short-circuit current density as 2.04 V and 16.18 mA/cm², respectively. The corresponding external quantum efficiency (EQE) curves for the two sub-cells are also shown in Fig. 6. Both the top and the bottom junctions generate the same short-circuit current density close to 16.18 mA/cm². It should be noted that the EQE computation is peformed without any anti-reflection (AR) coating on the illumination surface.

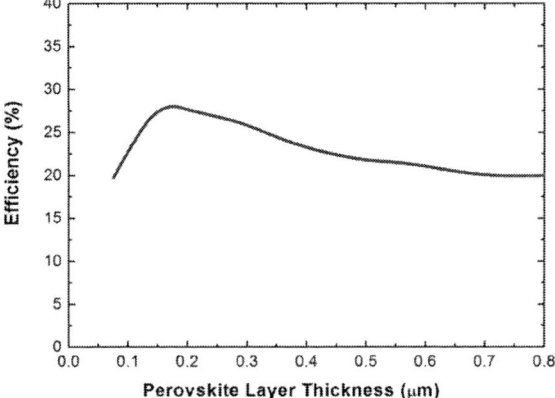

Fig. 4. Tandem cell efficiency vs. perovskite layer thickness.

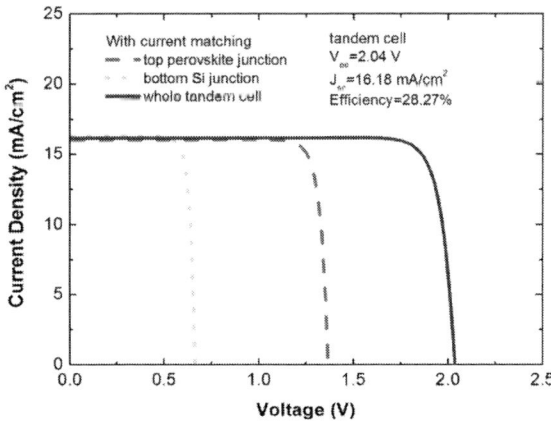

Fig. 5. J-V plots for the cell simulated with current-matching.

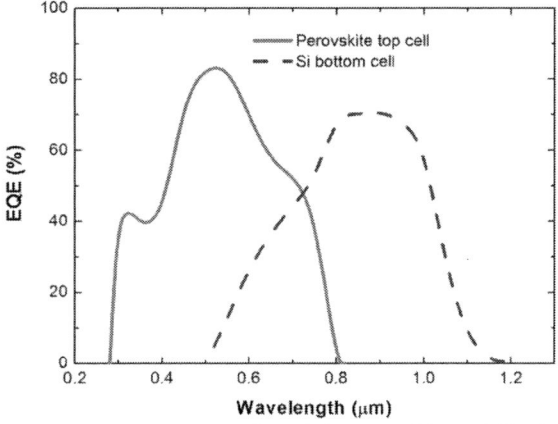

Fig. 6. EQE curves vs. wavelength for perovskite and Si sub-cells.

As a key material for the top junction absorber, it may be worthwhile to look at the effect of the minority carrier recombination lifetime for the perovskite material.

Particularly, the lifetime may change when its bandgap is tuned by composition engineering [15]. In Fig. 7, the fill fatcor and the open-circuit voltage are shown versus this change. The less the lifetime, the more recombination occurs which results in the reduced cell performance. This indicates that one needs to be cautions on the material growth and relevant quality when levying the benefits of bandgap engineering.

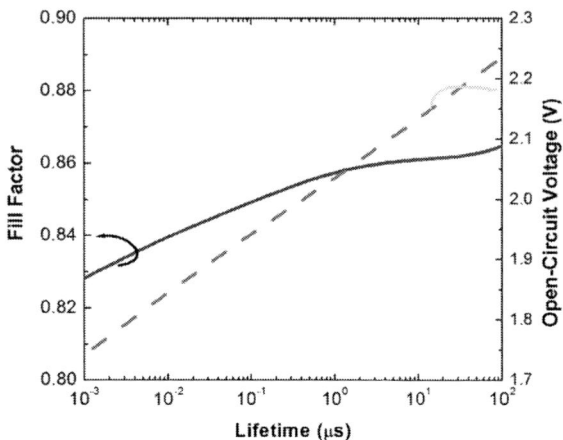

Fig. 7.   Fill factor and open-circuit voltage verus the minority carrier recombination lifetime of the perovskite layer.

Aiming to achieve higher cell efficiency than what is presented in this work, further modeling efforts to incorporate AR coating on the illumination front surace, to consider texture effect for the bottom Si cell as well as to look for alternative electron transport layer for the top junction are under way.

IV. Conclusions

In this work, modeling of two-terminal perovskite/Si tandem solar cell with TJ is presented based on a drift-diffusion simulator. When current matching is achieved between the top and the bottom junctions, the tandem cell shows a conversion efficiency as high as 28.27% with open-circuit voltagee and short-circuit current density as 2.04 V and 16.18 mA/cm2, respectively. The current matching is discussed versus the perovskite layer thickness and the tandem cell performance is further analyzed versus its minority carier lifetime. The modeling has also demonstrated the methodological capability of the involved software package.

References

[1] M. A. Green, A. Ho-Baillie, and H. J. Snaith, "The emergence of perovskite solar cells," *Nat. Photonics*, vol. 8, pp. 506-514, 2014.

[2] NREL, Best Research-Cell Efficiencies Chart 2023, https://www.nrel.gov/pv/assets/pdfs/cell-pv-eff-emergingpv.pdf.

[3] O. Malinkiewicz, A. Yella, Y. H. Lee, G. M. Espallargas, M. Graetzel, M. K. Nazeeruddin, and H. J. Bolink, "Perovskite solar cells employing organic charge-transport layers," *Nat. Photonics*, vol. 8, pp. 128-132, 2014.

[4] Z. Yang, Z. Yu, H. Wei, X. Xiao, Z. Ni, B. Chen, et al, "Enhancing electron diffusion length in narrow-bandgap perovskites for efficient monolithic perovskite tandem solar cells," *Nat. Commun.*, vol. 10:4498, 2019.

[5] J. P. Mailoa, C. D. Baillie, E. C. Johlin, E. T. Hoke, et al, "A 2-terminal perovskite/silicon multijunction solar cell enabled by silicon tunnel junction," *Appl. Phys. Lett.*, vol. 106:121105, 2015.

[6] J. Werner, A. Walter, E. Rucavado, S.-J. Moon, et al, "Zinc tin oxide as high-temperature stable recombination layer for mesoscopic perovskite/silicon monolithic tandem solar cells," *Appl. Phys. Lett.*, vol. 109:233902, 2016.

[7] Y. Wu, D. Yan, J. Peng, T. Duong, Y. Wan, S. P. Phang, et al, "Monolithic perovskite/silicon-homojunction tandem solar cell with over 22% efficiency," *Energy Environ. Sci.*, vol. 10, pp. 2472–2479, 2017.

[8] M. Zeman, and J. Krc, "Optic and electric modeling of thin-film silicon solar cells," *J. Mater. Res.*, vol. 23, pp. 889-898, 2008.

[9] APSYS 2022 and technical manuals, Copyright © Crosslight Software Inc., www.crosslight.com.

[10] Y. G. Xiao, Z. Q. Li, & Z. M. S. Li, "Modeling of GaInP/GaAs/Ge and the inverted-grown metamorphic GaInP/GaAs/GaInAs triple-junction solar cells," *Proc. SPIE*, vol. 7043, (8 pages) 2008.

[11] Y. G. Xiao, Z. Q. Li, M. Lestrade, and Z. M. S. Li, "Modeling of CdZnTe and CIGS and tandem solar cells," *35th IEEE PVSC*, pp. 001990-001994, 2010.

[12] Y.G. Xiao, K. Uehara, M. Lestrade, Z.Q. Li, and Z.M. S. Li, "Modeling of Si-based thin film triple-junction solar cells," *34th IEEE PVSC*, pp. 002154-002158, 2009.

[13] F. Liu, J. Zhu, J. Wei, Y. Li, M. Lv, S. Yang, B. Zhang, J. Yao, and S. Dai, "Numerical simulation: Toward the designof high-efficiency planar perovskite solar cells," *Appl. Phys. Lett.*, vol. 104:253508, 2014; See supplementary material at http://dx.doi.org/10.1063/1.4885367 for the three basic equations and solar cell structure diagram, etc.

[14] Y. Jiang, I. Almansouri, S. Huang, T. Young, Y. Li, Y. Peng, Q. Hou, L. Spiccia, U. Bach, Y.-B. Cheng, M. A. Green, and A. Ho-Baillie, "Optical analysis of perovskite/silicon tandem solar cells," *J. Mater. Chem. C*, vol. 4, pp. 5679-5689, 2016.

[15] T. J. Jacobsson, J.-P. Correa-Baena, M. Pazoki, et al, "Exploration of the compositional space for mixed lead halogen perovskites for high efficiency solar cells." *Energy Environ. Sci.*, vol. 9, pp. 1706–1724. 2016.

Assessment of a DER Inverter Model for IEEE 1547 Ride-Through Requirements Using a Model in the Loop Testbed

Nayeem Ninad and Eugene Desjardins Couture

CanmetENERGY – Natural Resources Canada (NRCan), Varennes, QC, J3X 1P7, Canada

Abstract — **In recent years, multiple interconnection standards have been published to resolve issues related to inverter-based resources (IBRs). (i.e., IEEE 1547-2018, CSA C22.3 No. 9-2020, IEEE 2800-2022, etc.). One of the required needs for IBRs is to integrate voltage, frequency, phase change and ROCOF ride-through (RT) capabilities. Electromagnetic transient (EMT) models of IBRs that are compliant with RT requirements are necessary. Validated EMT models of IBRs will allow system planners, researchers, and grid operators to assess the impact of these resources properly and quickly for different interconnection jurisdictions. In this regard, a real-time grid code testing architecture, which includes the modeling of a solar inverter with RT functions, is presented. The IEEE 1547-2018 RT functions of the overall model are validated using the RT test procedures from IEEE 1547.1 standard. The results were also compared with an actual commercial PV inverter unit.**

Index Terms-- **DER, IEEE 1547, inverter, ride-through, grid-support, modeling, PV, phase jump, solar, testing.**

I. INTRODUCTION

Worldwide, with the increasing number of inverter-based resources (IBRs), more synchronous machine-based fossil-fuel generators are retiring, leading to grids with less inertia and lower current contribution during faults [1]. In addition, the inverter-based distributed energy resources (DERs) cause additional challenges for managing the grid as these presents less visibility and controllability compared to conventional, centralized generation plants. Besides, recent tripping events of IBRs during power system faults have mandated ride-through requirements during voltage, frequency, and phase angle jump disturbances [2]. Therefore, interconnection standards or grid codes are being updated with new grid support functions (GSFs) requirement related to voltage, frequency, communications, and controls. For example, interconnection requirements per IEEE 1547-2018 std. [3] and Canadian CSA C22.3 No. 9-2020 std. [4] for distribution grid connected DERs have recently been updated and IEEE 2800-2022 [5] for bulk power grid connected IBRs was newly developed. However, it will still take a few years to have certified products that meet these new requirements.

Grid operators, researchers, system planners and associated stakeholders need appropriate models of IBRs in response to these new requirements so that they can conduct impact studies with different IBR penetration levels or determine appropriate parameters for GSF settings to meet grid code requirements. Realistic inverter models are essential for accurate, repeatable,

and comparable studies. With higher penetration of IBRs, the paradigm for these studies is also shifting to the use of electromagnetic transient (EMT) software [6]. Considering such need, inverter models were developed in [7] within the Matlab-Simulink environment which is suitable for both offline and real-time simulations. It includes different grid support functions from recent grid codes and interconnection standards.

In this work, considering the need for EMT models, a DER inverter model with IEEE 1547-2018 grid support requirements and RT capabilities is developed using the inverter toolbox from [7]. To validate the DER model RT capabilities, a simulated inverter laboratory (SIL) testbed was developed in a real-time model-in-the loop (MIL) setup. The SIL was inspired by the DER inverter test setup used in recent works where the test procedures from the IEEE 1547.1 std. have been used to analyze the voltage and frequency support functions of residential solar PV inverters [2], [8]–[10]. This study also includes the assessment of the DER inverter model following the test procedures of the IEEE 1547.1 std [11]. If the model meets the test procedures requirements, then it includes GSF solutions and adequately meets the GSFs requirement from the standard. In addition, the SIL architecture can provide a fast, accurate and efficient assessment tool for upcoming grid codes at different jurisdictions as it is difficult to have a hardware inverter to assess these new requirements.

The paper is organized as follows. Section II presents an overview of the System Validation Platform (SVP) and related activities with hardware inverter evaluation. Section III discusses the modeling of a grid-following PV inverter including the protection component of the inverter. Section IV presents the SVP-based simulated inverter laboratory testbed for grid code testing and model validation. Section V presents the results and analysis. Finally, Section VI presents the conclusion.

II. SYSTEM VALIDATION PLATFORM (SVP)

The system validation platform (SVP) is a versatile open-source DER testing and certification platform [12]. It uses a laboratory setup consisting of a grid simulator, a PV simulator, a data acquisition system (DAS), and equipment under test (EUT) as shown in Fig. 1. The SVP orchestrates automated testing of DERs by executing sequences of testing logic with AC and DC disturbances on the DER inverter [13] and evaluates the DER's responses. The user defines a test in SVP by selecting the appropriate test script and associated parameters.

Based on the test requirements and EUT rating, the SVP configures all test bench equipment. The SVP uses abstraction layers for all test bench components, thus enabling the use of the same test scripts at different laboratory testbeds by changing the equipment drivers [8]. The SVP software and drivers for different grid simulators, PV simulators, DASs, and commercial EUTs are available on the open-source GitHub repository [13] [14]. The SVP based IEEE 1547.1-2020 test scripts have been developed and were used to assess the GSFs of several commercial PV inverters. The test results and comments/recommendations have been shared with the IEEE 1547.1 working group for refinement. The Python test scripts are also available in the open-source GitHub repository [15].

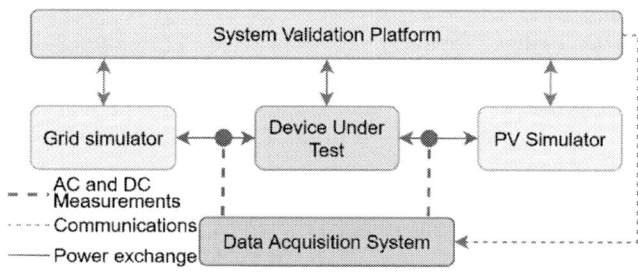

Fig. 1.   SVP generic laboratory configuration

III. PV INVERTER MODEL WITH IEEE 1547-2018 STD

A grid-connected solar PV inverter system was modeled in the Matlab-Simulink environment. The power components were modeled using the Simulink library components. Fig. 2 shows the block diagram of the 5 kW solar PV inverter system. It has a two-level three-phase voltage source converter that injects active and reactive powers to a 3-phase 60 Hz, 208 VLL AC grid. A 5.5 kW PV array acts as the primary energy source and is connected to the inverter system through an input DC-DC converter which is controlled by a PV controller block for operation either in maximum power point tracking (MPPT) mode or in power curtailment mode. The inverter is interfaced to the grid using an LCL output filter. Table I presents the inverter parameters.

The basic structure of the inverter control system is presented in [16]. It consists of the wave reference, the primary control, and the secondary control. Additionally, the model is enhanced in this work by including the voltage and frequency protection and associated signal conditioning blocks. In summary, the wave reference includes a current controller that generates the modulating signal for the power converter based on the real and reactive current provided by the primary

control. The primary control manages the active and reactive power injected by the inverter based on DC link voltage controller and AC side reactive power controller. Additionally, it can also limit active power based on a reference from the secondary control. The secondary control includes implementation of all the IEEE 1547-2018 Std's GSFs used for controlling the voltage (i.e., constant power factor, volt-var, etc.) and frequency (i.e., droop). [16] presented the assessment of six GSFs using the MIL. The GSF parameters are set based on the Category B definition of DER from IEEE 1547 std. [3].

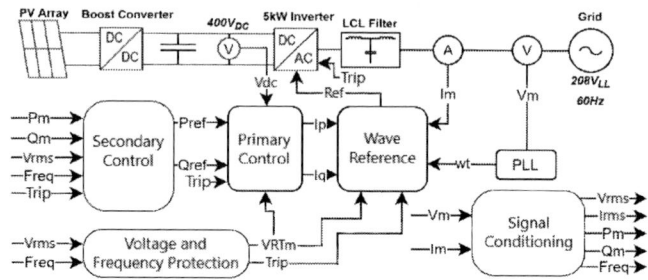

Fig. 2.   Graphical representation of the simulated inverter system

TABLE I
INVERTER SYSTEM PARAMETERS

Output LCL Filter	Inverter side inductor	L = 1.7 mH and R = 1 mΩ
	Capacitor	C = 0.15 µF and R = 0.7
	Grid side inductor	L = 71 µH and R = 10 µΩ
DC Link	Capacitor	12 mF
Current (PI) Controller	Proportional gain	1.13
	Integral gain	435
DC Link Voltage Controller	Proportional gain	100
	Integral gain	70
Reactive Power Controller	Proportional gain	1
	Integral gain	25

The voltage and frequency trip settings as well as the ride through (RT) settings with different modes are implemented using the protection components (voltage protection, frequency protection and latch) [7]. The protection block controls the inverter AC output current for cease to energize (CE), momentary cessation (MC), mandatory operation (MO), continuous operation (CO) or trip [7]. The parameters are chosen for Category III definition of DER from IEEE 1547-2018 and are presented in Table II.

TABLE II
VOLTAGE AND FREQUENCY SETTINGS

Voltage (V) RT Settings		Frequency (f) RT Settings	
Voltage (pu)	Operating mode	Frequency (Hz)	Operating mode
V > 1.20	CE	f > 62.0	CE
1.10 < V ≤ 1.20	MC	61.2 < f ≤ 61.8	MO
0.88 ≤ V ≤ 1.10	CO	58.8 ≤ f ≤ 61.2	CO
0.70 ≤ V < 0.88	MO	57.0 ≤ f < 58.8	MO
0.50 ≤ V < 0.70	MO	f < 57.0	CE
V < 0.50	MC		

IV. SVP BASED SIMULATED INVERTER TEST BENCH

The SVP based architecture is developed to be used with actual hardware testing equipment to test the DER unit in a laboratory environment. However, during the time when grid codes or standards are being developed with new grid support requirements, the existing DER inverter may not include all these required GSFs. One of the objectives of this work is to develop an SVP based real-time simulated inverter laboratory (SIL) testbed in a hardware in the loop (HIL) simulator that can serve as a good GSFs evaluation tool for the type testing of DER devices, thus helping with the development of grid code testing procedures. This approach is also known as model in the loop (MIL). The SIL can help refine those grid codes during the process of their development. The PV inverter system can easily be modeled with the required GSF parameter settings mandated by the particular grid code. Thus, it can accelerate the standard development process by providing feedback based on rapid evaluation of specific test cases. In addition, the PV inverter model discussed in Section III could be tested for the IEEE 1547.1 Std. If the model passes all the IEEE 1547-2018 evaluations/tests, then the PV inverter model would become compliant to the standard with the required GSFs and can be used for further power system studies.

measurements. Overall, the Simulink model interacts with the SVP through the RTLAB's application programming interface (API) and a dedicated HIL driver [10]. The SVP uses the same abstraction layers for the SIL as for any other test bed environment. The SIL drivers interact with the HIL driver instead of directly interacting with a physical equipment API. The developed Opal-RT HIL driver is available in the open-source GitHub repository [14].

The benefit of a real-time SIL test bench in comparison with a physical test bench is its simplicity and versatility. All the components of the model are designed in a Simulink environment. The components are easily accessible and can be modified to a specific grid code by changing the parameters of the component or even modifying the implementation of the GSF entirely. Moreover, those parameters can be modified automatically by the SVP as usually required by the type of testing procedures where the GSFs are evaluated with different settings (i.e., different category with different characteristic curves). The SIL can facilitate the real-time evaluation of the inverter system (including power components and control system or firmware) before the actual prototype is built, as testing and modifying the control system in the SIL is faster than doing the same using a physical/hardware inverter system.

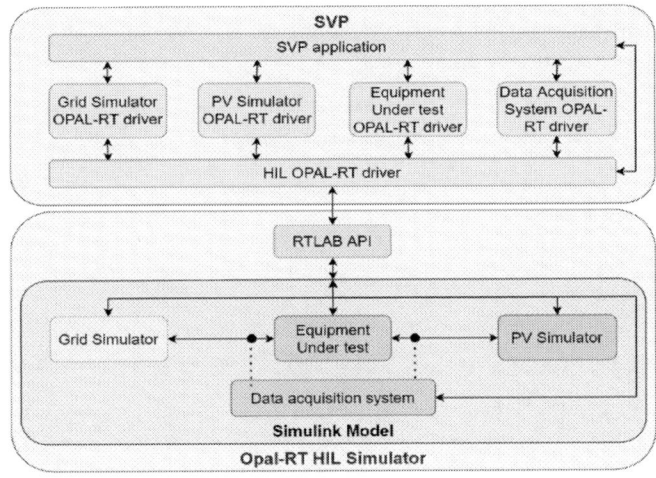

Fig. 3. Real-time Simulated Inverter Laboratory (SIL) testbed setup

The SIL consists of an Opal-RT's OP5700 real-time simulator hosting the SIL model and a computer for configuring and controlling the simulation with the SVP. Fig. 3 presents the block diagram for the SVP based SIL test bench. The solar PV inverter model acts as the EUT and the DAS acting as outputs to the SVP. The SVP applies AC side variations by controlling the AC grid of the SIL model and the PV production by changing the irradiance and temperature. It can also configure the inverter GSFs through the equipment under test (EUT) Opal-RT driver. The signal conditioning block acts as the DAS and provides the test bench

V. ASSESSMENT OF DER RIDE-THROUGH CAPABILITIES

In this section, the voltage, frequency and phase change RT capabilities of the PV inverter model are evaluated according to the IEEE 1547.1-2020 Std. The SVP based IEEE 1547.1 test scripts are available in open source [15] and were previously used to evaluate the RT of commercial PV inverters [2], [10]. The same test scripts are used for the inverter model in a MIL setup, and the results are compared against their counterparts from an actual commercial PV inverter. However, it should be mentioned that this commercial inverter is not certified for this IEEE 1547.1-2020 Std. Nevertheless, it still gives a good representation of how the GSFs are currently implemented with respect to the upcoming mandatory IEEE 1547 requirements. The hardware unit referred to in this section as the hardware under test or HUT is a 3 phase 10 kW solar PV inverter, with AC side ratings at 208 VLL, 60 Hz.

A. Voltage Ride-Through (VRT)

The VRT capability is the ability of the DER to RT voltage sags and swells without tripping. During the RT event, following the momentary cessation (MC), it shall restore total active current injection of at least 80% of pre-disturbance active current level within 0.4 s. All the VRT tests were conducted at 100% power output of the inverters. Fig. 4 presents applied voltage profile of the low voltage RT (LVRT) and the current response for the SIL and HUT inverters. It also includes three scenarios when the disturbance is applied at one phase, two phases and three phases. The inverter initially operates in normal voltage range and then a voltage of 0.03 pu

978-1-6654-6060-6/23 $31.00 © 2023 IEEE

is applied. When the disturbance is applied only on Phase A, the HUT inverter returns to pre-disturbance current level as soon as the voltage is beyond 0.46 pu. This ensures that it was in MC and returns to MO mode. However, the HUT trips for other disturbance cases (Phase AB and ABC). The SIL DER is programmed for MC for voltage less than 0.5 pu as shown in Table II. Therefore, it stays at MC for voltage value less than 0.5 pu and it returns close to the pre-disturbance current value within 5 ms after the voltage is higher than 0.5 pu. The MC region in the HUT is directly set by the manufacturer.

The high voltage RT (HVRT) test results for SIL inverter are presented in Fig. 5. The HUT could not be tested for HVRT due to testbench over voltage limitation. For the three different cases of applying HVRT test profiles on Phase A, AB and ABC, the SIL inverter goes into MC when the voltage magnitude is 1.18 pu. It comes out of MC and maintains output current close to the pre-disturbance value within 23 ms.

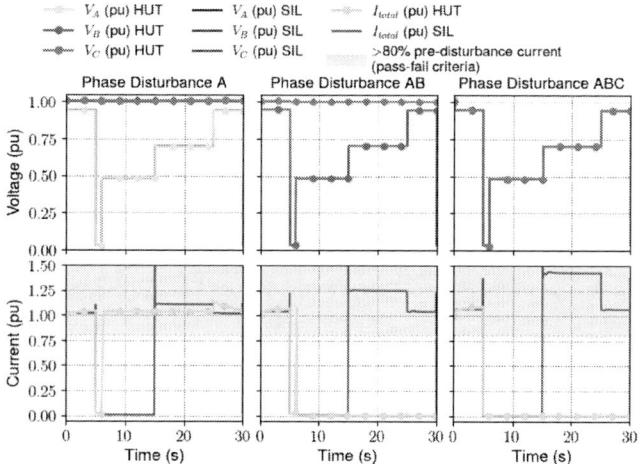

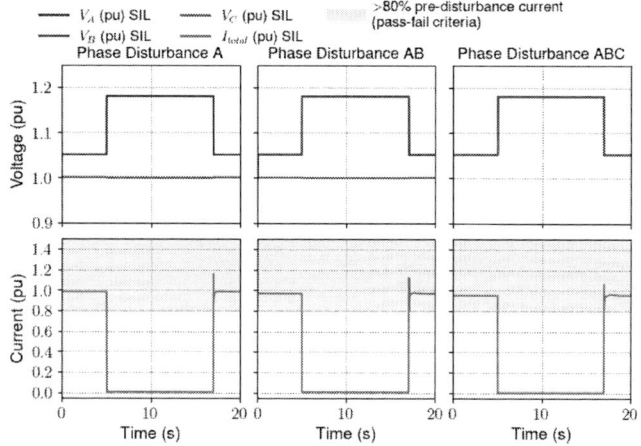

Fig. 4.   LVRT test results for SIL and HUT inverters.

Fig. 5.   HVRT test results for SIL inverter

B. Frequency Ride-Through (FRT).

The purpose of the frequency RT (FRT) test is to verify the capability of the DER to RT frequency excursions without tripping. The frequency droop function is programmed for the least aggressive (LA) setting to make the active power change with respect to frequency as small as possible [3]. The test also evaluates the rate of change of frequency (ROCOF) capability of DER by applying frequency variation at 3 Hz/s [3]. The DER is FRT-compliant if it restores/maintains output active power at the pre-disturbance active power output.

Fig. 6 presents the high FRT (HFRT) test results for both SIL and HUT inverters. To better evaluate the frequency support from DER, additional scenarios for droop with disable and default settings were also assessed. Both DERs ride through ROCOF of 3 Hz/s by continuing the current injections during the test. It should be mentioned that the HUT inverter employs a frequency-watt function, not a frequency droop function, as described in prior work [10]. With the least aggressive setting, when the frequency changes from 60 Hz to 61.8 Hz, following the frequency droop equation, the output power should be reduced by 0.26 pu. The SIL inverter follows the droop function, and the output power reduces to 0.67 pu, however, the HUT inverter output is 0.49 pu as it implements frequency-watt function. When the droop is disabled, both inverters maintain the same power output throughout the test. When the droop is set with default parameters, the output power of SIL inverter falls to 0.4 pu. When the frequency returns to 60 Hz, the output power returns to the pre-disturbance value after brief transient corresponding to the response time of droop function.

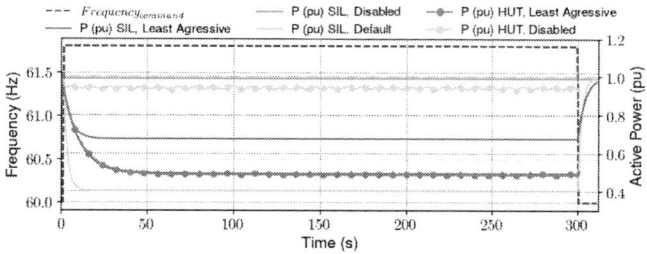

Fig. 6.   HFRT test results for HUT and SIL inverters

Both the HUT and MIL inverters do not respond to frequency excursions below the nominal value as solar inverters currently do not have this capability. Fig. 7 presents the HFRT test results for both inverters. The frequency variation dips to 57 Hz. The MIL and HUT inverters maintain the same output power value throughout the test, even though the droop was programmed for LA setting.

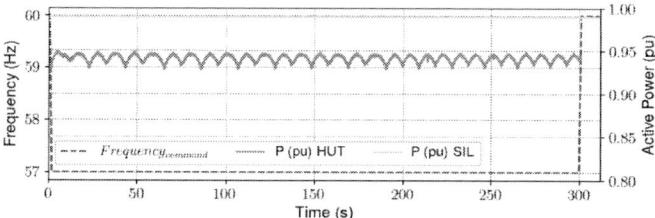

Fig. 7.   LFRT test results for HUT and SIL inverters

978-1-6654-6060-6/23 $31.00 © 2023 IEEE

C. Phase-Angle Change Ride-Through (PCRT)

The PCRT capability of DER is evaluated by applying balanced and unbalanced phase-angle jumps in the AC voltage. The DER is PCRT-compliant if it can restore its total rms current to a value higher than 80% of the pre-disturbance value in less than 0.5 s after the phase jump is removed.

Both the MIL and HUT inverters are assessed for an unbalanced PCRT test where a 60° phase jump is applied on Phase A for the duration of 0.5 s. They are also assessed for a balanced phase jump of 20° (applied for the duration of 50 s). Fig. 8 presents the voltage and current waveforms for the SIL and HUT inverters. Fig. 9 and Fig. 10 illustrates the RMS phase currents for both SIL and HUT inverters for 60° phase jump in Phase A and balanced phase jump of 20°, respectively. For both unbalanced and balanced phase jumps, the SIL inverter maintains balanced current throughout the test after brief transients. It also maintains the current above 80% of the pre-disturbance current throughout both tests. The rms current

of the MIL inverter increases during the unbalanced voltage (phase). However, the HUT inverter shows significantly different behaviour for unbalanced and balanced phase jumps. During the unbalanced phase jump, it injects unbalanced currents with observable oscillations and reduces the output current when the phase jump is removed, but barely meets the output current restoration requirements. The oscillations persist in post-fault behaviour. The standard does not provide any specific requirement about the DER current behaviour during the disturbance condition. The two commercial DERs tested in [2], show totally different current behaviours during the unbalanced PCRT tests. During the balanced phase jump test, the HUT three-phase currents are almost constant throughout the test, except for brief transients during the application and removal of the phase jump instants. Both inverters recover 80% of their total pre-disturbance currents within 0.5s and thus they are PCRT compliant.

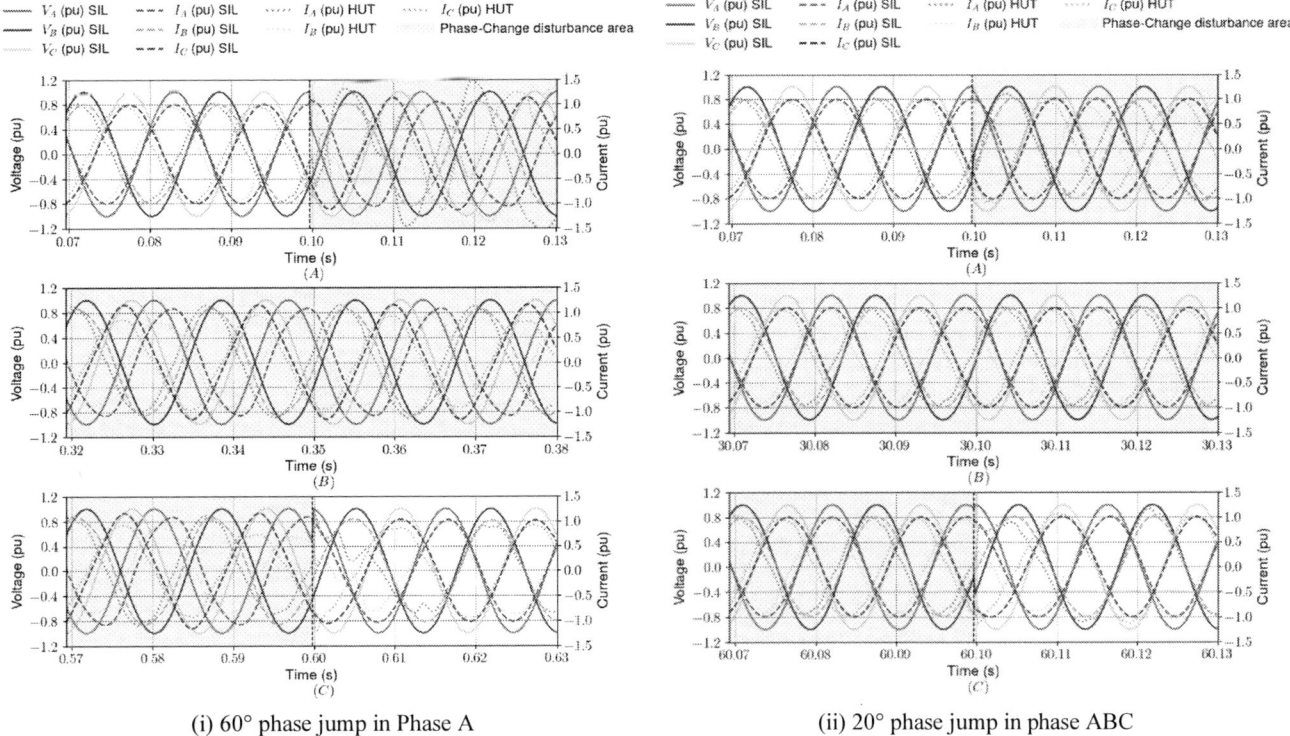

(i) 60° phase jump in Phase A

(ii) 20° phase jump in phase ABC

Fig. 8.   PCRT voltage and current waveforms for MIL and HUT inverters with unbalanced (on the left) and balanced (on the right) phase jump.

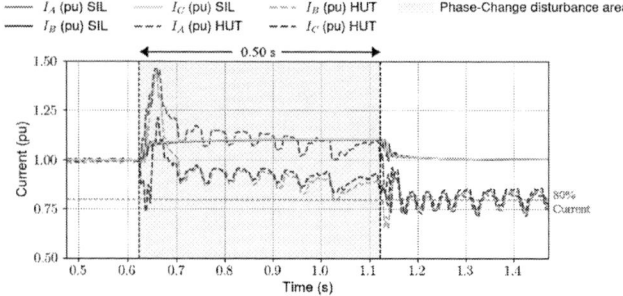

Fig. 9.   PCRT rms current values with 60° phase jump in Phase A.

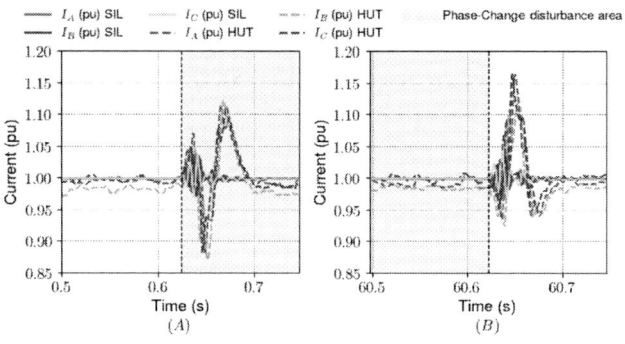

Fig. 10.   PCRT rms current values with 20° phase jump in Phase ABC.

VI. CONCLUSIONS

This paper presents the model of a solar PV inverter system with the grid support and ride through (RT) requirements from the IEEE 1547-2018 std. An SVP based real-time model in the loop (MIL) inverter laboratory test bench was also presented for assessing the IEEE 1547 compliance. The SVP-SIL architecture acts as a grid code test bench and has been used to evaluate the voltage, frequency and phase-jump RT capabilities of the PV inverter model following the test procedures from the IEEE 1547.1-2020 std. The results were compared with the performance of a commercial PV inverter, that was also tested using the SVP architecture in an actual laboratory setup. The results were analyzed in detail and the solar PV inverter model showed that it met the required RT capabilities per IEEE 1547. Thus, the model could be used as an IEEE 1547-compliant DER in different power system simulation studies. Moreover, the SIL was proven to be a useful tool to assess developing grid codes and standards (from different jurisdictions).

ACKNOWLEDGEMENT

Financial support for this research work was provided by the Government of Canada through the Program on Energy Research and Development (PERD) in the framework of the REN-2 Smart Grid and Microgrid Control for Resilient Power Systems Project.

REFERENCES

[1] B. Kroposki et al., "Achieving a 100% Renewable Grid: Operating Electric Power Systems with Extremely High Levels of Variable Renewable Energy," *IEEE Power Energy Mag.*, vol. 15, no. 2, pp. 61–73, Mar. 2017.

[2] R. Darbali-Zamora et al., "Evaluation of Photovoltaic Inverters Under Balanced and Unbalanced Voltage Phase Angle Jump Conditions," in *47th IEEE Photovoltaic Specialists Conference (PVSC)*, Jun. 2020, pp. 1562–1569.

[3] "IEEE Standard for Interconnection and Interoperability of Distributed Energy Resources with Associated Electric Power Systems Interfaces," *IEEE Std 1547-2018*, Apr. 2018.

[4] "CSA C22.3 No. 9 - 2020 : Interconnection of distributed energy resources and electricity supply systems." Canadian Standard Association (CSA), 2020.

[5] "IEEE Standard for Interconnection and Interoperability of Inverter-Based Resources (IBRs) Interconnecting with Associated Transmission Electric Power Systems," *IEEE Std 2800*, Apr. 2022.

[6] C. Shah et al., "Review of Dynamic and Transient Modelling of Power Electronic Converters for Converter Dominated Power Systems," *IEEE Access*, vol. 9, pp. 82094–82117, 2021.

[7] N. Ninad, J.-P. Bérard, and S. Q. Ali, "Smart Inverter Modelling Toolbox for EMT Simulation Studies of Power Systems," in *2021 IEEE Canadian Conference on Electrical and Computer Engineering (CCECE)*, Sep. 2021, pp. 1–6.

[8] N. Ninad, E. Apablaza-Arancibia, M. Bui, J. Johnsson, and et-al, "Development and Evaluation of Open-Source IEEE 1547.1 Test Scripts for Improved Solar Integration," Marseille, France, 2019.

[9] N. Ninad et al., "PV Inverter Grid Support Function Assessment using Open-Source IEEE P1547.1 Test Package," in *47th IEEE Photovoltaic Specialists Conference (PVSC)*, Jun. 2020, pp. 1138–1144.

[10] N. Ninad, E. Apablaza-Arancibia, M. Bui, and J. Johnson, "Commercial PV Inverter IEEE 1547.1 Ride-Through Assessments Using an Automated PHIL Test Platform," *Energies*, vol. 14, no. 21, Art. no. 21, Jan. 2021, doi: 10.3390/en14216936.

[11] "IEEE Standard Conformance Test Procedures for Equipment Interconnecting Distributed Energy Resources with Electric Power Systems and Associated Interfaces," *IEEE Std 15471-2020*, May 2020,

[12] "Sunspec System Validation Platform (SVP)," *Sunspec Alliance.* Aug. 2015. http://sunspec.org/sunspec-svp/.

[13] OpenSVP Software https://github.com/sunspec/svp.

[14] *SVP Drivers.* https://github.com/sunspec/svp_energy_lab/tree/dev37

[15] IEEE   1547.1-2020   std   Test   Script https://github.com/jayatsandia/svp_1547.1/tree/master

[16] N. Ninad, E. Desjardins-Couture, and E Apablaza-Arancibia, "Validating IEEE 1547 Capabilities of DER Inverter Model using a Real-time Simulated Inverter Laboratory Testbed," 10th International Conference on Smart Energy Grid Engineering, Oshawa, Canada, Aug. 10-12, 2022.

Methodology for the analysis of series arc fault algorithms

Paulo R. D. R. da Silva[1], Guilherme C. S. Prym[1], Geyciane P. de Lima[1], Andrei C. Ribeiro[1], Hugo da S. Alvarez[2], Rodrigo M. Garcia[2], Francisco C. Marques[1], João A. F. G. da Silva[1], Mauricio Taconelli[1], Tárcio A. dos S. Barros[1], and Marcelo G. Villalva[1]

[1] University of Campinas - FEEC, Campinas, São Paulo, Brazil, pauloricardo.drds@gmail.com

[2] Build Your Dreams (BYD), Campinas, São Paulo, Brazil, rodrigo.garcia@byd.com

Abstract — Solar energy has been growing in recent years, becoming increasingly common in family homes. However, improper installation, lack of preventive maintenance, and aging can accelerate the degradation of solar system components, increasing the risk of serious failures. Arc faults are among the most significant risks for residential solar systems due to the high risk of fire that can be initiated. The use of AI in fault detection systems is growing to mitigate these effects. Because of this, a typical scenario is needed for all algorithms to test their capabilities. Due to the existence of various arc fault models and the inherent components of the system aiding in the difficulty of detection, standardization in the evaluation of algorithms is necessary. This paper reviews the types of arc fault detection and proposes data collection to assemble a database to evaluate emerging detection algorithms.

I. INTRODUCTION

Renewable energies are becoming more and more common in residences, among which photovoltaic energy stands out, allowing applications in several power levels. Photovoltaics consists of parts of DC systems, and because of this generation stage, some challenges arise for the system to remain safe and efficient. An arc fault is the most destructive problem among the possible failures [1].

Arc faults can raise temperatures through the plasma generated by the arc. If dissipation does not occur, it can cause severe damage to system components and, worst cases, start fires [2]. There are two main types of arc faults, namely, series and parallel. A series arc fault is generated by separating a connection point of the conductors, while a parallel arc fault is the contact between two conductors of different potentials.

To maintain safety in DC systems, it is vital to have devices that can detect and interrupt circuits. These devices are called Arc Fault Circuit Detect (AFCI). There are several types of research on arc fault detection methods, the vast majority of which are series-type arcs. Mathematical models of arc faults have been developed using experimental data, as presented in [3]. However, these models do not describe the external features of the arc, limiting the use of an alternative methods for detection

With the advancement of technology and the development of more powerful computers, artificial intelligence (AI) methods are becoming popular, as they offer potentiator fault diagnosis in several areas, such as detecting faults in electrical machines [4].

Studies on detection methodologies for DC serial arcs using AI-based methods have shown great promise. At [5], the Wavelet packet decomposition methodology is adopted with a support vector machine (SVM) to detect arc faults. In [6]. The

weighted least squares SVM is trained using variables such as high-frequency current and energy to see series arcs effectively.

This research is based on the need for actual arc fault data. Due to cost, it is infeasible to use components from photovoltaic systems to generate the arcs. Physical models are also limiting. However, using a test bench to create arcs in the laboratory becomes a very attractive alternative. One can perform various tests and record the data for applying detection methodologies.

This work proposes a methodology of various tests with data collection, aiming to create a common database for performance testing of detection methods. Section II describes preliminary information on the occurrence of arc faults in photovoltaic systems. Section III proposes a methodology for the characterization of series arc faults. Finally, in section IV, the discussion of the work is presented.

II. DESCRIPTION PHOTOVOLTAIC SYSTEMS

A. Structure of a photovoltaic panel

Solar cells are arranged to build a solar module to generate solar energy. Figure 1 demonstrates the basic structure of a solar module. They are connected in series to form a string, thus allowing an increase in the DC voltage level of the system, and these strings can be connected in parallel, thus increasing the DC level. These procedures increase the power values generated for the inverter, resulting in more power injection into the AC system [7].

In a complete photovoltaic system, there are several ways to arrange the modules to meet the design criteria. Therefore, the system's number of cables and connectors can vary greatly. Each connection point is a potential failure creator for the system.

B. Faults in Photovoltaic Systems

The failures of photovoltaic systems can vary, such as power reduction due to shading and startup failure due to low voltage or current. However, more serious faults, such as arc faults, put the system and people nearby at risk. There are two types of arc faults for photovoltaic systems, series, and parallel arc faults, of

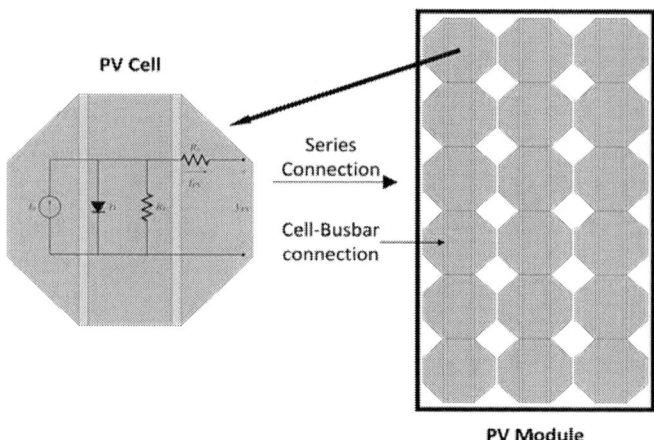

Fig 1. Typical structure of PV module.

which ground faults are included. Due to the way it occurs, the parallel arc fault usually consumes a large amount of fault current. When two conductors of different potentials touch each other, there is an intense current flow due to the low impedance [8]. However, because of the construction of the PV cell, the series arc fault current will not be sufficient to drive the protection fuse or activate combination box protection devices, i.e, the series arc fault will not consume a reverse current like the parallel arc fault [9].

Serial arc faults can be created from electrodes that simulate the connection point, one being mobile and the other fixed. The removal of the electrodes initiated the arc-creation process. At this time, the impedance injected by the arc fault begins to increase, already the current level will decrease, therefore, will not be able to reach 156% of current above the maximum nominal, which is set by the NEC so that the overcurrent protection fuse is triggered [10]. The report [11] shows that series arc faults and parallel-ground arc faults are the leading causes of fires in photovoltaic systems. But due to the probability of occurrence, series arc faults become the most significant hazard for residential PV systems.

Lack of preventive maintenance, natural aging of components, weathering, animal damage, poorly made connectors, and damage from mechanical loads can cause problems at connection points. These problems cause the cross-

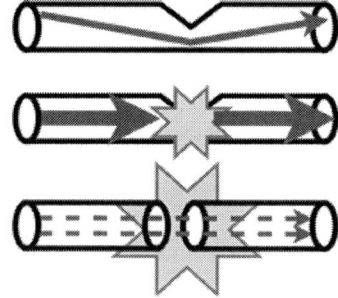

Fig 2. Illustration of series resistance types and metallic connection shapes of solar cell busbar.

sectional area of the connection to decrease, effectively increase the resistance of the connection, and cause increased heat loss. This process is the beginning of connection failure and the subsequent onset of an arc fault, as shown in Fig. 2.

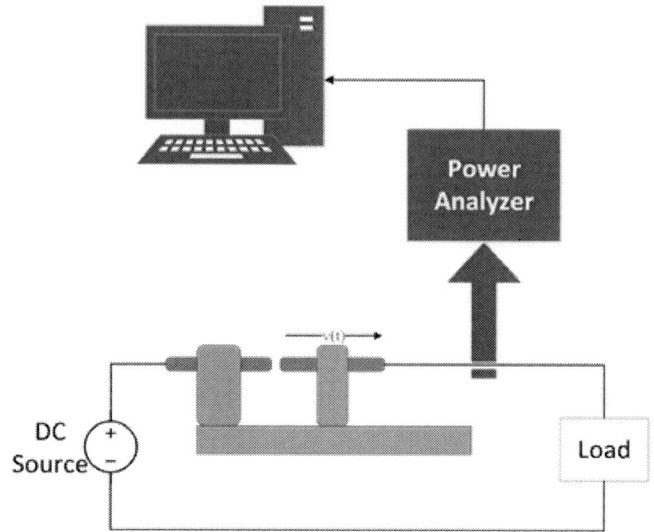

Fig 3. Arc fault generator circuit diagram.

III. DC SERIES ARC CHARACTERIZATION

To characterize series DC arcs, it is necessary to have an arc management circuit. Figure 3 shows the diagram of the course to be adopted. For data collection, an arc-generating circuit was designed by UL 1699B [12]. Moving the electrodes away from the arc generation circuit, a DC arc fault is caused and using a power analyzer, the arc occurrence data is collected and stored.

Table 1 presents the load specifications for generating different arc types. The values chosen aim at diversifying the possibilities of DC arc fault generation. Inverter-type loads aim to create natural disturbances in the frequency spectrum generated by the inverter due to the switching frequency.

The voltage parameters for arc generation vary depending on the desired power. Although the voltage values are far from the commercial importance of solar plant inverters, the initial objective of the method is to serve residential inverters. During the generation of arcs, some abnormal behaviors were noticed, such as sudden variations in voltage and current, growth of the harmonic components of the current, and distortion in the current waveform. It is also noted that the arc voltage stabilizes after a period while the current remains disturbed. The magnitude of the plasma generated depends on the type of load. The longer it lasts, the more heat is produced until it reaches the point of melting the shield and starting a fire. The initial

978-1-6654-6060-6/23 $31.00 © 2023 IEEE

TABLE I

SUMMARY OF THE PARAMETERS FOR PERFORMING THE TESTS

Load Types	Switching frequencies (kHz)	Current amplitude (A)	Power (W)
Three-phase Inverter	15, 20, 25, 30	4, 5, 6, 7, 8, 10, 12	50, 100, 200, 400
Single-phase Inverter	15, 20, 25, 30	2, 3, 4, 5, 8, 10, 12	50, 100, 200, 400
R	N/A	1, 2, 4, 8, 10, 12, 14	50, 100, 200, 400, 600
RL	N/A	1, 2, 4, 8, 10, 12, 14	50, 100, 200, 400, 600
RLC	N/A	1, 2, 4, 8, 10, 12, 14	50, 100, 200, 400, 600

behaviors described are already great allies in characterizing arc faults.

IV. CONCLUSION

Using AI in arc fault detection methodologies promises significant changes in the photovoltaic industry market and the regulatory landscape governing electrical installations. But there is still a need to develop methods to analyze devices containing AFCI robustly.

In this paper, a proposed methodology for testing arc faults for use in developing detection devices using AI is proposed. The proposed method aims to diversify the occurrence of arc faults so that the algorithms can characterize them efficiently. In addition, tests need to be defined considering the premature aging of components to obtain a more in-depth analysis of the reliability and longevity of the device in hostile environments. In this way, it is vital to ensure that the device undergoes a test that operates effectively in cases of arc fault and lasts longer than the lifetime of the photovoltaic system.

ACKNOWLEDGMENTS

This work was supported by BYD Energy Brazil (through the PADIS/MCTI program, Funcamp project N.5779), Total Energies (through ANP - Brazilian Oil & Gas Agency, Funcamp project N.6002), and CNPq (process N.407945/2022-9).

REFERENCES

[1] K. Jia, Z. Yang, Y. Fang, T. Bi, and M. Sumner, "Influence of Inverter-Interfaced Renewable Energy Generators on Directional Relay and an Improved Scheme," *IEEE Trans. Power Electron.*, vol. 34, no. 12, pp. 11843–11855, Dec. 2019, doi: 10.1109/TPEL.2019.2904715.

[2] L. Zhu, S. Ji, and Y. Liu, "Generation and developing process of low voltage series DC arc," *IEEE Trans. Plasma Sci.*, vol. 42, no. 10, pp. 2718–2719, 2014, DOI: 10.1109/TPS.2014.2330419.

[3] S. Lu, B. T. Phung, and D. Zhang, "A comprehensive review on DC arc faults and their diagnosis methods in photovoltaic systems," *Renew. Sustain. Energy Rev.*, vol. 89, pp. 88–98, June, 2018, DOI: 10.1016/j.rser.2018.03.010.

[4] R. Liu, G. Meng, B. Yang, C. Sun, and X. Chen, "Dislocated Time Series Convolutional Neural Architecture: An Intelligent Fault Diagnosis Approach for Electric Machine," *IEEE Trans. Ind. Inform.*, vol. 13, no. 3, pp. 1310–1320, June 2017, doi: 10.1109/TII.2016.2645238.

[5] K. Xia, S. He, Y. Tan, Q. Jiang, J. Xu, and W. Yu, "Wavelet packet and support vector machine analysis of series DC ARC fault detection in a photovoltaic system," *IEEJ Trans. Electr. Electron. Eng.*, vol. 14, no. 2, pp. 192–200, 2019, DOI: 10.1002/tee.22797.

[6] K. Yang, R. Zhang, J. Yang, C. Liu, S. Chen, and F. Zhang, "A Novel Arc Fault Detector for Early Detection of Electrical Fires," *Sensors*, vol. 16, no. 4, Art. no. 4, April. 2016, DOI: 10.3390/s16040500.

[7] G. Petrone, C. A. Ramos-Paja, and G. Spagnuolo, *Photovoltaic sources modeling.* Hoboken, NJ ; West Sussex, UK: John Wiley & Sons, Inc, 2017.

[8] J. Yuventi, "DC electric arc-flash hazard-risk evaluations for photovoltaic systems," *IEEE Trans. Power Deliv.*, vol. 29, no. 1, pp. 161–167, 2014, DOI: 10.1109/TPWRD.2013.2289921.

[9] C. Strobl and P. Meckler, "Arc Faults in Photovoltaic Systems," in *2010 Proceedings of the 56th IEEE Holm Conference on Electrical Contacts*, Oct. 2010, pp. 1–7. doi: 10.1109/HOLM.2010.5619538.

[10] S. Dhar, R. K. Patnaik, and P. K. Dash, "Fault detection and location of photovoltaic based DC microgrid using differential protection strategy," *IEEE Trans. Smart Grid*, no. 99, 2017.

[11] S. McCalmont, "Low-Cost Arc Fault Detection and Protection for PV Systems: January 30, 2012 - September 30, 2013," NREL/SR-5200-60660, 1110454, Oct. 2013. DOI: 10.2172/1110454.

[12] Underwriters Laboratories, "Photovoltaic (PV) DC Arc-Fault Circuit Protection." Standard, 2018.

Patterning the Front Polysilicon Contact for Silicon Solar Cells using Laser Oxidation

Sagnik Dasgupta[1], Pradeep Padhamnath[2], Vijaykumar Upadhyaya[1], Young-Woo Ok[1], Ruohan Zhong[1], Wookjin Choi[1], Kyu-Hyeon Im[3], and Ajeet Rohatgi[1]

[1]Georgia Institute of Technology, Atlanta, Georgia, 30332, USA
[2]Solar Energy Research Institute of Singapore, 7 Engineering Drive 1, 117574, Singapore
[3]Korea Institute of Energy Research, Daejeon 34129, Republic of Korea

Abstract — The use of poly-Si/SiO₂ carrier selective contacts has pushed up the efficiencies of commercial large-area silicon solar cells, primarily driven by their excellent passivation properties. However, the high absorption in the visible spectrum has restricted their use only to the rear side of the solar cell. The ability to pattern poly-Si on the front of the solar cells to restrict them only under the screen-printed metal fingers can significantly mitigate the detrimental effect of parasitic absorption while providing all the advantages of excellent passivation. In this paper, we demonstrate a novel laser oxidation process as a rapid and scalable for patterning poly-Si, achieving fingers as narrow as 35 μm. We also observe that the laser irradiation of in-situ doped poly-Si can increase the concentration of active dopants. The patterning process with laser modifies the morphology and the properties of the poly-Si layer, which is visible as a change in the solar cell parameters and SEM imaging. In selective area front contact cell design with <5% laser processed region, any negative impact will be limited relative to the benefit of rapid low-cost patterning and efficiency enhancement from selective TOPCon. We also used this method to fabricate a preliminary 19.8 % efficient rear junction solar cell with ex-situ poly-Si contacts on both sides.

Keywords — Selective area contact, TOPCon, Laser oxidation, Laser damage, passivating contacts, screen printed contacts, ECV profiles.

I. INTRODUCTION

Tunnel-oxide passivated contact (TOPCon) solar cells are quickly replacing conventional PERC-like structures in commercial production [1]. These contacts utilize a doped poly-Si/SiO₂ stack to physically isolate the metal contacts from the absorber while providing the appropriate band bending to enhance charge carrier selectivity and collection. This enables solar cells utilizing a TOPCon stack to achieve significantly higher open circuit voltages (V_{OC}) [2]. However, due to the high absorptance of poly-Si, the application of the TOPCon stack is restricted to the rear of the solar cells while utilizing a conventional diffused layer on the front.

Utilizing a poly-Si/SiO₂ contact on the front as well as the rear of the solar cell can further improve the passivation and mitigate metal-induced recombination on both sides. Patterning the front poly-Si such that it is present only under the metal grid can help minimize parasitic absorption while reaping the benefits of a front passivated contact under the metal grid [3]. While there are several traditional methods of patterning the poly-Si ([4], [5]), laser-oxidation is a unique, fast, and simple process to achieve this ([6], [7]). This method forms an ultra-thin SiO_x mask on the laser-processed regions that allows

etching the poly-Si selectively between the metal fingers using KOH, thereby achieving the desired patterned structure on the front side of the solar cell as shown in Fig. 1. The patterned poly-Si lines will hereafter be referred to as "poly-fingers".

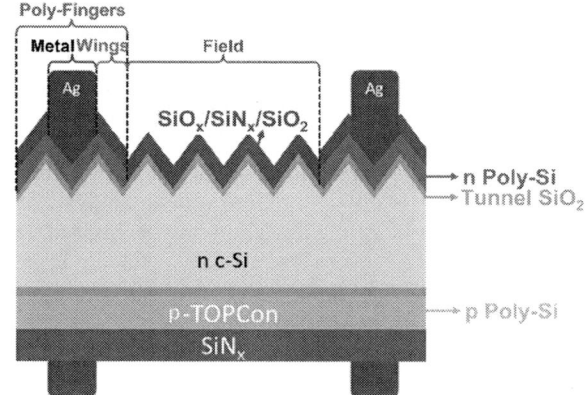

Fig. 1 Schematic of proposed ultimate cell structure.

In this work, we demonstrate the ability to form narrow poly-Si fingers using a laser-oxidation process. We present microscopic evidence that the laser-grown oxide can protect the underlying polysilicon from KOH etching. We also assess the impact of the laser-oxidation process on the electrical performance of the device through electrochemical capacitance-voltage (ECV) doping profile measurements. In addition, we fabricated some test aluminum rear junction PhosTop solar cells [8] with a full-area n+ poly-Si layer on the front surface, with and without laser irradiation (Fig. 2). Finally, we also used this method to fabricate preliminary selective front contact double-side TOPCon cells (Fig. 1).

II. FABRICATION AND CHARACTERISATION OF PATTERNED POLY-SI STRUCTURES

Both cell structures and microscopy samples were fabricated on random pyramid textured ~3 Ωcm n-type Cz-Si wafers. The wafers were then RCA cleaned and immersed in HNO₃ at 100 °C for 15 min, growing about 15 Å of SiO₂. Next, ~200 nm in-situ phosphorus-doped poly-Si was grown using low pressure chemical vapor deposition (LPCVD) in a Tystar system at 588 °C, followed by a crystallization and dopant activation anneal at 875 °C for 30 min in a Centrotherm tube furnace.

For the microscopy samples, after the poly-Si was grown, 150 μm wide poly-fingers were processed on the front side of the wafer using a Coherent Avia UV (355 nm) nanosecond-

978-1-6654-6060-6/23 $31.00 © 2023 IEEE

pulsed laser at a power of 4W (measured before the focusing optics) and a scan speed of 400 mm/s. Following this, the wafers were immersed in 9%wt KOH at 40 °C for two minutes. This was found to be enough to etch the 200 nm thick poly-Si in the field region between the patterned laser lines, probably without damaging the tunnel oxide. These patterned samples were then observed in a Thermo Helios 5 CX scanning electron microscope (SEM).

To assess the extent of laser damage without introducing complexities of metal alignment, the test cell structures were fabricated (Fig. 2) with the entire front surface of the wafer laser exposed using the process described earlier to study the cumulative impact of laser processing in the field and metallized regions on V_{OC} and FF, recognizing there will be significant light absorption in thick front poly-Si. Then, the front surface was masked with plasma-enhanced chemical vapor deposition (PECVD) SiN_x and immersed in 20 %wt KOH at 65°C for 10 min to remove the poly-Si wrap-around and planarize the rear surface of the wafer. Following that, the wafers were RCA cleaned, the SiN_x mask was removed in 10 %wt HF, and a double layer antireflection coating with 45 nm SiN_x capped with 90 nm SiO_x was grown through PECVD. For metallization, the cell was screen-printed with an H pattern (100 fingers, 35 μm) using Ag paste on the front and full area Al on the back.

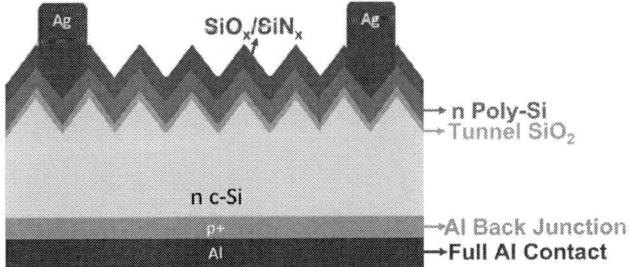

Fig. 2 Schematic of full-rear Al cell structure.

We also fabricated double-side TOPCon cells with patterned n-type poly-Si on the front. Starting with single side planarized wafers, following the nitric acid tunnel-oxide growth, LPCVD intrinsic poly-Si was grown at 588 °C resulting in a thickness of 200 nm on the textured front and 300 nm on the planar rear. After that, boro-silicate glass was deposited on the rear for doping using atmospheric pressure chemical vapor deposition (APCVD). The samples were then subjected to a two-step doping process to first drive in the boron in the rear at 900 °C followed by a $POCl_3$ pre-deposition and an 840 °C drive-in at the front. This co-anneal process has been shown to result in excellent ex-situ double-side TOPCon precursors [9]. Following the precursor fabrication, the wafers were laser-oxidized with 100 lines 150 μm wide each at the condition described above. The samples were then immersed in 9 %wt KOH at 40 °C for two minutes to remove the poly-Si from the field area while preserving the poly-Si in the lasered regions. This patterned precursor was then RCA cleaned and 8 nm of thermal SiO_2 was grown to aid passivation of the etched field region. The surfaces were then capped with the double layer anti-reflection coating described above on the front and 90 nm PECVD SiN_x on the back. After this, the front side was screen-

printed similarly as before, and the rear was metallized with 300 lines of AgAl paste.

Finally, all samples were fired in a belt furnace at a peak temperature of 760 °C and a belt speed of 240 inches per second. The completed cells were measured on a Sinton FCT-450 flash tester.

III. RESULTS AND DISCUSSION

A. Importance of Poly-Finger Geometry on Cell Performance

For brevity, we refer to the regions of the poly-fingers not covered by metal as the 'wing area.' The exposed poly-Si in the wing area incurs parasitic light absorption. Quokka 2 device simulations were performed to quantify the impact of poly-finger thickness and width on solar cell performance, using experimental details of samples and an optical model developed earlier [6]. The results of these simulations are summarized in Fig. 3. Note that the wing area fraction is calculated after subtracting the width of the metal lines (30 μm) from the width of the poly-finger, and then calculating the ratio of the exposed poly-Si area for 100 lines on a 156 mm wafer.

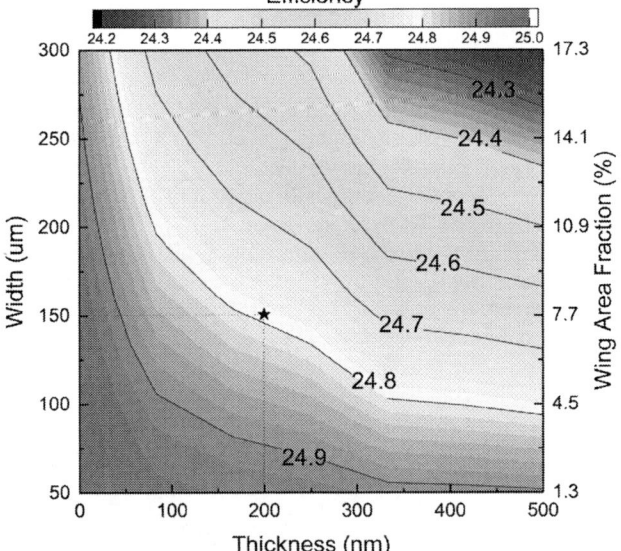

Fig. 3 Quokka 2 simulation of the effect of poly-finger width and thickness on cell efficiency.

As expected, we find that thicker and wider poly-fingers are detrimental to solar cell performance. While thicker poly-Si fingers cause increased parasitic absorption, they are beneficial for minimizing loss in V_{OC} with fire-through metallization. Additionally, having wider poly-fingers increases parasitic absorption while increasing its contribution to the J_0. On the other hand, we find that for narrower poly-Si fingers, the efficiency is less sensitive to poly-Si thickness, allowing the use of thicker poly-Si for lower metal-induced damage. This suggests that narrow, thick poly-Si fingers could help improve the performance of the double-side TOPCon solar cells. Poly-Si patterning with laser oxidation is ideally suited for this.

B. Micrographic Analysis of Laser Patterning

Fig. 4 shows that our laser can be used to pattern poly-Si fingers as narrow as 35 μm. The measured width obtained after patterning was found to exceed the target width of the pattern by approximately 10 - 15 µm. This is likely due to the output of the laser spot being a gaussian beam. Additionally, due to the spot-size limitations of our laser, the narrowest line that can be patterned is about 35 µm wide. This is close to the state of the art for the width of screen-printed grid metallization[1]. However, patterning of poly-Si using a screen-printed resist will require several additional steps.

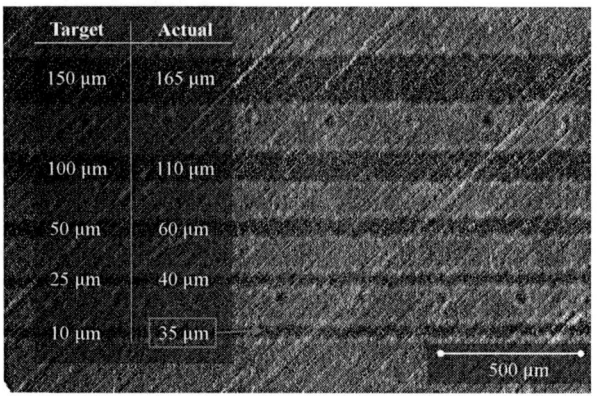

Fig. 4 Plan view scanning electron micrograph showing obtained widths of laser marked lines.

C. Protection of Poly-Si from KOH Etch due to Laser-Oxide

In the proposed device structure, after the laser patterning on the front, the poly-Si in the field region is etched in KOH (40°C, 9 %$_{wt}$). Subsequently, the anti-reflection dielectric is deposited rendering the wafer ready for metallization. Fig. 5 (b) shows that 2 min of etching in KOH is sufficient to remove the poly-Si without affecting the textured morphology of the wafer in the field region. Further, in Fig. 5 (c), we see that in the patterned poly-finger region, poly-Si is indeed present under the anti-reflection coating. However, we observe that the texturing in the laser-oxidized region is rounded, causing the poly-Si to redistribute, thereby resulting in non-uniform thickness. The rounding of the pyramids and resultant laser damage is consistent with prior observations [6].

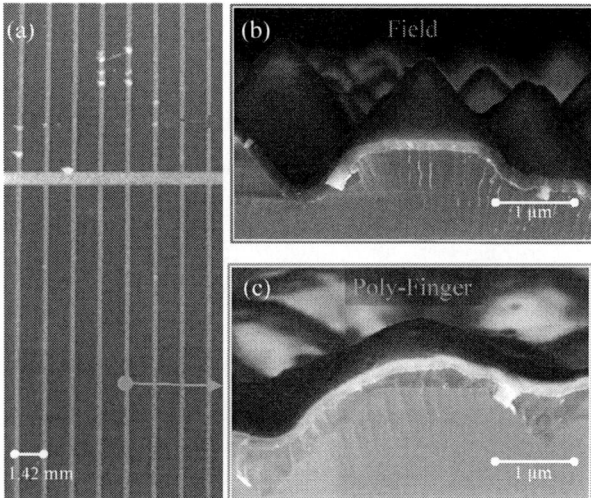

Fig. 5 (a) Plan view photograph of a laser patterned sample capped with silicon nitride showing multiple poly-fingers. (b) Cross-section image of a single texturing pyramid in the field region. (c) Rounded pyramid in the poly-finger region showing presence of poly-Si after KOH etching.

D. Effect of Laser Processing on Poly-Si Doping

Due to the morphological changes seen after laser processing, we investigated the change in the doping profile in the poly-Si and Si before and after laser irradiation. Electrochemical capacitance-voltage (ECV) measurements show the concentration of electrically active dopants in the sample as a function of depth from the surface. In Fig. 6 we see that for the reference sample (no laser irradiation) with in-situ poly-Si, the concentration of active phosphorus is constant until nearly 200 nm after which it falls rapidly. This shows that the activation of phosphorus in the poly-Si is uniform with depth ($\sim 4 \times 10^{20}$ cm^{-3}) and that there is minimal in-diffusion into the Si wafer through the tunnel oxide. For the sample irradiated with the UV laser, the surface concentration increased to 10^{21} cm^{-3}. This suggests the activation of inactive dopants that were trapped in regions of low crystallinity. This is consistent with the decrease in sheet resistance of the layer due to laser irradiation observed in previous work [6]. On the other hand, for ex-situ phosphorus doped poly-Si used in the fabricated double side TOPCon cell, the active dopant concentration increases from 2×10^{20} cm^{-3} to 4×10^{20} cm^{-3}. This change is a lot less than what was observed for in situ poly-Si due to fewer inactive dopants present in the ex-situ doped poly-Si. Furthermore, after laser irradiation, the phosphorus concentrations for both in-situ and ex-situ poly-Si begin to decrease at a shallower depth and the change in concentration is more gradual compared to the respective non-lasered reference samples. We hypothesize that this could be due to disruption or damage to the tunnel oxide layer causing the dopants to drive into the c-Si substrate.

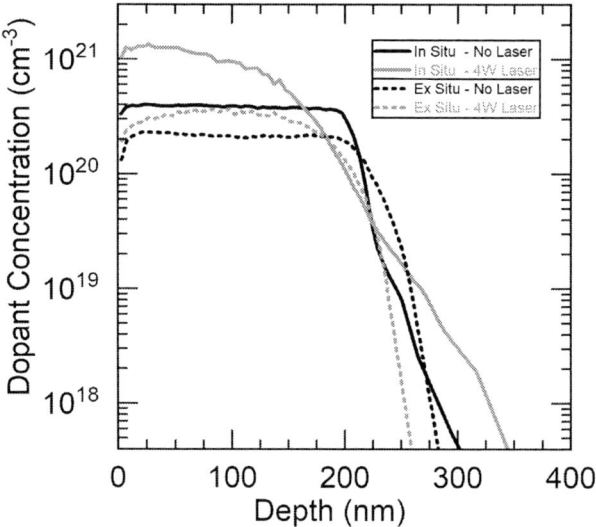

Fig. 6 ECV active dopant profile of phosphorus in the front n poly-Si.

E. Simplified PhosTop Solar Cells

The morphological and electrical changes in the poly-Si layer could impact solar cell performance. In our earlier work [6], we showed that laser irradiation caused a substantial deterioration in J$_0$ which was partially recovered after PECVD SiN$_x$ deposition.

978-1-6654-6060-6/23 $31.00 © 2023 IEEE

To further understand the impact of the laser patterning process on solar cell performance, we fabricated a full area laser irradiated solar cell (No patterning or KOH etching) with a full area Al contact on the rear side to compare with an otherwise identical non-laser processed solar cell (reference). Table I shows the lighted IV data under AM1.5g illumination.

Table 1 LIV Measurements of fabricated solar cells

Solar Cell Type	Laser Treatment	V_{oc} (mV)	J_{sc} (mA/cm^2)	FF (%)	Efficiency (%)
PhosTop (Full Rear Al)	No Laser	650.43	33.11	77.99	16.80
	4 W Laser	634.15	29.86	78.23	14.81
DS-TOPCon	4W Laser Patterned	670.30	38.55	76.52	19.77

We observed that both solar cells have very low short-circuit current densities. This is expected due to the significant parasitic absorption loss in the ~200 nm of full-area poly-Si on the front side. Additionally, the laser-irradiated solar cell has an even lower J_{SC} due to increased reflection and worse light trapping arising from the rounded pyramids. Furthermore, the open circuit voltage of the lasered cell is ~16 mV lower than the reference cell. This is likely due to the deterioration of passivation arising from laser damage. For the proposed device (Fig. 1), the area fraction of the laser-damaged region will be significantly lower (<5%) than it is in this device. As a result, the detrimental effect of the laser on V_{OC} would also be commensurately lower. Importantly, the fill factors for the two cells are comparable. This shows that the laser damage does not deteriorate metallization or contact formation.

F. Selective-Area Poly-Si Double-side TOPCon cells

Knowing that the lasered front poly-Si can be metallized with screen-printed silver, we fabricated our first double side TOPCon cells with a patterned front n type poly-Si based on the structure shown in Fig. 1. Fiducials were used at the top-left and bottom-right corners of the wafer to ensure alignment between the laser patterned poly-Si lines and screen-printed metal lines Fig. 7 shows the aligned 50 μm Ag line on the 150 μm poly-Si finger. Unlike the PhosTop cells, here we used 8 nm of thermally grown SiO$_2$ under the front double-layer antireflection coating for improved passivation in the field region.

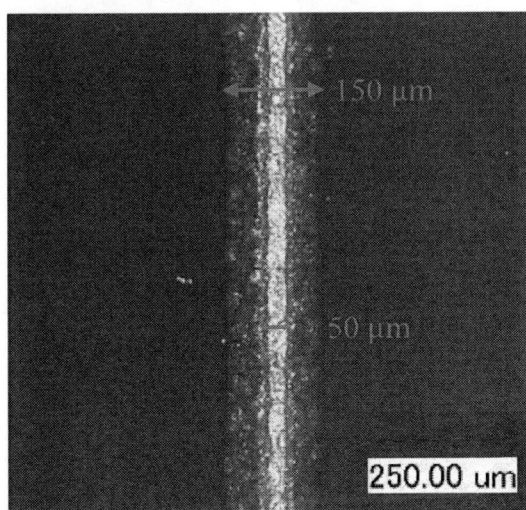

Fig. 7 Aligned metal line on poly-Si finger.

LIV measurements on these preliminary cells show a dramatic improvement in both open circuit voltage and short circuit current compared to the PhosTop cells discussed in the previous section resulting in a fabricated cell efficiency of 19.8%. While the vast majority of the V_{OC} improvement comes from the p TOPCon at the back of the solar cell, most of the J_{SC} benefit is due to significantly lower parasitic absorption at the front of the solar cell. The fill factor, however, was lower than expected due to high series resistance.

The implied open circuit voltage of the structure before metallization was 705 mV. The 30 mV drop in V_{OC} after metallization indicates the need for optimization of the screen-printing and firing processes, which is in progress.

IV. CONCLUSIONS

We have found that it is important to pattern narrow poly-fingers for minimizing parasitic absorption and the effects of laser damage on the front passivation of the rear junction cell. To this end, we have shown that using laser oxidation, we can pattern lines as narrow as 35 μm. Further, more precise laser systems can pattern significantly narrower lines allowing for minimal wing area exposed to light. Therefore, in practice, the line width of the poly-finger would be limited by the width of the metal fingers and the precision of alignment for metallization. We have also shown that while the laser-oxide can protect poly-Si from etching, it is at the expense of appreciable morphological and electrical changes, which manifests as a deterioration of passivation in the completed solar cell. Despite this, due to the small overall area fraction of the lasered region in the proposed selective area front contact, the effect of this damage could be minimized. Additionally, for narrower poly-fingers, it is possible to use thicker poly-Si to mitigate damage without incurring significant parasitic absorption losses. Furthermore, we fabricated preliminary double side TOPCon cells with patterned front n type poly-Si, resulting in a cell efficiency of 19.8%. This performance can be significantly improved through screen-printing and firing optimizations. These results advocate for the use of laser-oxidation for patterning poly-Si contacts, a fast, scalable, and tunable alternative to traditional patterning techniques.

V. ACKNOWLEDGMENTS

This material is based upon work supported by the U.S. Department of Energy's Office of Energy Efficiency and Renewable Energy (EERE) under Solar Energy Technologies Office (SETO) Agreement Number DE-EE0009350, DE-EE0008562, and DE-EE0008975. Additionally, we thank Mr. Richard Shafer for consistent assistance in the operation of the laser system used in this work and Ms. Pragna Bhaskar for the SEM imaging. SERIS is a research institute at the National University of Singapore (NUS). SERIS is supported by NUS, the National Research Foundation Singapore (NRF), the Energy Market Authority of Singapore (EMA), and the Economic Development Board of Singapore (EDB).

VI. REFERENCES

[1] ITRPV, "International Technology Roadmap for Photovoltaic 2022," 2022.
[2] M. A. Green *et al.*, "Solar cell efficiency tables (Version 60)," *Prog. Photovoltaics Res. Appl.*, vol. 30, no. 7, pp. 687–701, Jul. 2022, doi: 10.1002/PIP.3595.
[3] Y.-Y. Huang, A. Jain, W.-J. Choi, K. Madani, Y.-W. Ok, and A. Rohatgi,

"Modeling and Understanding of Rear Junction Double-Side Passivated Contact Solar Cells with Selective Area TOPCon on Front," in *2021 IEEE 48th Photovoltaic Specialists Conference (PVSC)*, Jun. 2021, pp. 1971–1976, doi: 10.1109/PVSC43889.2021.9518628.

[4] K. Chen *et al.*, "Self-Aligned Selective Area Front Contacts on Poly - Si/SiOx Passivating Contact c -Si Solar Cells," *IEEE J. Photovoltaics*, vol. 12, no. 3, pp. 678–689, May 2022, doi: 10.1109/JPHOTOV.2022.3148719.

[5] A. Ingenito *et al.*, "Silicon Solar Cell Architecture with Front Selective and Rear Full Area Ion-Implanted Passivating Contacts," *Sol. RRL*, vol. 1, no. 7, p. 1700040, 2017, doi: 10.1002/solr.201700040.

[6] S. Dasgupta *et al.*, "Novel Process for Screen-Printed Selective Area Front Polysilicon Contacts for TOPCon Cells Using Laser Oxidation," *IEEE J. Photovoltaics*, vol. 12, no. 6, pp. 1282–1288, Nov. 2022, doi: 10.1109/JPHOTOV.2022.3196822.

[7] S. Singh *et al.*, "Development of 2-sided polysilicon passivating contacts for co-plated bifacial n-PERT cells," *2020 IEEE 47th Photovolt. Spec. Conf.*, 2020, [Online]. Available: internal-pdf://91.219.111.60/1391-Singh-Development of 2-sided polysilicon.pdf.

[8] D. L. Meier *et al.*, "Aluminum alloy back p-n junction dendritic web silicon solar cell," *Sol. Energy Mater. Sol. Cells*, vol. 65, no. 1, pp. 621–627, 2001, doi: 10.1016/S0927-0248(00)00150-1.

[9] W.-J. Choi, Y.-W. Ok, K. Madani, S. Duttagupta, and A. Rohatgi, "Development of a co-anneal process for double-side TOPCon precursor fabricated by ex-situ POCl3 and APCVD boron diffusion," in *2022 IEEE 49th Photovoltaics Specialists Conference (PVSC)*, Jun. 2022, pp. 1068–1068, doi: 10.1109/PVSC48317.2022.9938560.

Evaluation of Motion-Induced Noise and Pixel-Bleeding in Electroluminescence Field Inspection of PV Modules

Thøger Kari[*1], Rodrigo Del Prado Santamaria[1], Gisele A. dos Reis Benatto[1], Pascal Koelblin[2], Liviu Stoicescu[3], Sergiu V. Spataru[1].

[1]Department of Electrical Photonics Engineering, DTU Electro, Technical University of Denmark, Frederiksborgvej 399, 4000 Roskilde, Denmark.
[2]Institut für Photovoltaik, Universität Stuttgart, Pfaffenwaldring 47, 70569 Stuttgart, Germany.
[3]Solarzentrum Stuttgart GmbH, Rotebühlstr. 145, 70197 Stuttgart, Germany.
*thkje@dtu.dk.

Abstract—**There is an increasing demand for accurate daylight diagnostics of larger photovoltaic (PV) plants with Electroluminescence (EL) imaging. Modulated contact-EL can remedy solar noise, but mobile platforms are required to increase inspection speed. However, any motion induces noise, even with perfect tracking of the PV modules in the images. This paper investigates the impact of motion and camera calibration on the quality of daylight contact-EL imaging since both introduce noise. Using the SNR_{50} and SNR_{Kari}' metrics and visual inspection, a total of 58 stationary and moving EL imaging series, 23 calibrated and 35 uncalibrated, were analyzed and investigated both with and without module tracking. SNR_{50} proved an unreliable predictor of image quality, but SNR_{Kari}' also revealed uncertainties when faced with camera motion. Motion severity was a superior metric due to an enhanced ability to predict non-tracked image quality, but more stable metrics must be developed. Without stabilization, image quality deteriorated rapidly at motion above 0.18 and 0.06 pixel/image with and without camera calibration, respectively. With stabilization, calibrated image series stayed at a quality level suitable for manual diagnostics, even at extreme motion. Still, the uncalibrated series did not; showing calibration vital for moving imaging inspection platforms. For reliable diagnostics and automated processing, better algorithms are needed.**

I. INTRODUCTION

Electroluminescence (EL) imaging is an important, emerging technique for fault detection in photovoltaic (PV) modules. EL imaging can provide much higher diagnostics detail than, e.g., infrared thermography [1, 2, 3]. Numerous papers have been written concerning modulated contact EL, where a modulated current is applied to a PV module or -string while recording an image series of the EL emission with a short-wave infrared (SWIR) camera. This modulation approach allows for both background subtraction techniques (lock-in) [4, 5, 6, 7] and Fast Fourier Transform (FFT) analysis [8, 9]. The latter can almost entirely eliminate the problem of solar noise.

Signal modulation allows daylight EL analysis in the field directly on a PV-string. This method significantly increases the prospect of inspecting and monitoring larger PV plants using EL. A primary obstacle to this is the speed between the imaging of each panel. Wheeled (e.g., rover) or aerial (drone) platforms can be utilized to improve acquisition speed. This approach, however, presents multiple noise problems: The first is the misalignment of the image subjects (PV modules) between

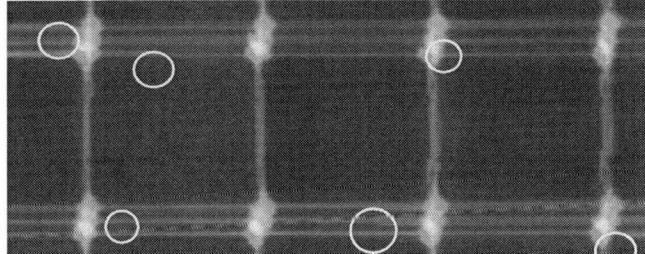

Fig. 1. Intensity differences between three consecutive frames with motion of 1.9pixel/image, enhanced in red, green and blue. The changes in effective line-width of the top and bottom lines (circles) are due to pixel-bleeding, as the underlying features have a fixed width.

frames. This must be solved by object tracking and realignment before any processing can be done. Second, any camera motion can induce pixel bleeding, an effect that is most noticeable near high-contrast edges but happens everywhere. As an object's alignment with the pixel-grid changes, information spills between neighboring pixels, causing local blurring. This problem is relevant primarily for low-resolution cameras like the SWIR cameras used for EL diagnostics.

Fig. 1 shows three consecutive frames of a fixed rectangular region, where a PV module (Fig. 2, left) moves between each frame. Intensity differences between frames are augmented by color. The thinning and thickening of the horizontal, colored lines are due to pixel-bleeding and cannot be remedied solely by re-aligning the module with tracking. This is exemplified in Fig. 3, which shows an image series with motion of 0.15 px/img before (left) and after (right) tracking (module stabilization).

Fig. 2. **Left**: A view of an image series with severe camera motion of 1.9 pixels/image. The red dashed square shows the cropping quadrilateral. **Right**: A cropping of 1/2 of the (failed) analysis result.

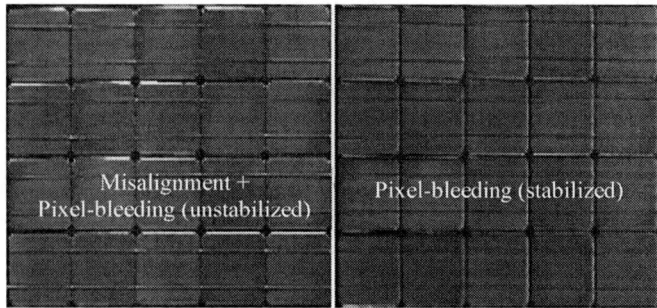

Fig. 3. **Left**: EL analysis of M2 on a non-stabilized PV module with motion of 0.15 pixel/image. $SNR_{Kari}' = 0.16$. **Right**: EL analysis of the same module after stabilization. $SNR_{Kari}' = 0.34$.

When considering that the busbars are actually bright in the original images, it becomes clear how this brightness spills into the edges of the cells when there is no stabilization. Yet, even after stabilization with zero visible motion, pixel bleeding causes spillover near the edges of the cells and blurs some of the busbars. In daylight, the EL signal is barely perceptible, and the relative intensity difference between pixel bleeding near module features and the EL signal can be orders of magnitude, making it challenging to reconstruct lab-quality EL images. The same is true for camera noise from improper calibration.

Significant harmonic noise can occur when pixel-bleeding fluctuations approach the modulation frequency. The severity depends on image-feature contrast and degree of frequency-aliasing, while the sign of the noise depends on the phase relative to the modulation. Negative noise can be removed simply by a zero threshold, while positive noise must be distinguished from the real signal. In addition, improper camera calibration can also create visually imperceivable noise, which is still very impactful on post-processing, given the low signal magnitude.

The relationship between motion noise on the SNR and image quality has not been properly investigated previously. In this paper, we will attempt to quantify it.

II. Camera Noise Sources

Motion noise is a problem in and of itself, even given perfect camera calibration. Imperfect camera calibration, however, exacerbates the problem. SWIR cameras are sensitive to temperature and settings, and a number of errors can occur if not adjusted and maintained appropriately.

- *Bias correction error*: The baseline noise of a SWIR sensor pixel is sensitive to temperature and should be adjusted according to the ambient environment. If not adjusted, the inherent offsets between neighboring pixel values can be much larger than the signal amplitude itself. Bias correction is adjusted while completely covering any light from reaching the sensor.
- *White correction error*: The individual pixel sensitivity of the camera's sensor depends on both temperature and integration time and must be set against a perfectly flat, white target after bias correction. It sets a linear correction factor to each pixel so there is a linear transition from black to white. The different factors will create potentially very large noise effects if adjusted in dark conditions but used in a lighter environment.

- *Bad scanlines and dead pixels*: Electronic circuitry errors and burnt-out pixels cause lasting defects that will turn into moving spots or lines when the panel is tracked and corrected. Fortunately, bad scanlines are rare, while dead pixels are somewhat more common. These can be accounted for at least to some degree using digital processing techniques.
- *Dust/scratches*: Dust and scratches on the lens or sensor cause artifacts that will impact image quality, similar to dead pixels and bad scanlines.
- *Software malfunctions*: Bad drivers or connectivity can cause image artifacts, making buffer transfers unstable. Artifacts can range from entirely black images to image-warping and lines akin to bad scanlines.

A clear demonstration of how camera noise impacts image quality under motion is shown in Fig. 4. The **left** image is four cells of a perspective-corrected module and contains an artifact caused by an improperly synchronized buffer that resembles a bad scanline. The **middle** image shows a complete, static analysis of the same area. Because the signal amplitude is relative to any background offset, the scanline makes no (noticeable) difference in the image quality. The image to the **right** shows how the panel's motion underneath the noise source causes dramatic artifacts in the analysis. This type of noise source is easy to spot, but camera calibration errors and dirt can be almost equally problematic. Multiple camera noise source indicators can be spotted in the rightmost image.

III. Method

The research is based on SWIR image recordings of PV modules subjected to a square-wave current oscillating at 1/8th of the camera framerate. A tripod imaging platform was used to enable static benchmarks with identical camera and panel positioning for all the motion series. Motion was caused simply by shaking the tripod at different intensities, which from experience, mimics wind and hovering drone movements quite well. The PV modules were stabilized across the image series for image processing using the tracking algorithm explained in [9], also meaning that all panels are cropped and perspective-corrected. Instead of the FFT algorithm described in the same paper, the lock-in method [4, 6] was used for EL image reconstruction, as it is more widespread than the FFT approach. A sub-dataset of 256 images (32 full periods) was used for each analysis, which should be sufficient to negate most solar noise through averaging.

For SNR calculation, we employ both the SNR_{50} [10] and the SNR_{Kari}' signal-to-noise metric shown in (1) and (2), which is a variant of the SNR_{Kari} metric explained in [8]. Subscripts 1 and 2 indicate data points corresponding to the upper and lower peaks of the EL modulation signal, respectively, while N_1 and N_2 are

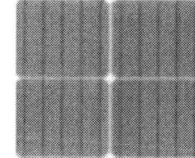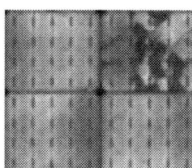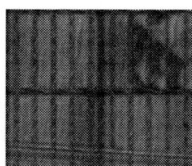

Fig. 4. **Left**: Bad scanline in the images at the bottom row. **Middle**: With zero-motion analysis, the line disappears. **Right**: With a movement-based analysis, the line moves around, causing excessive noise.

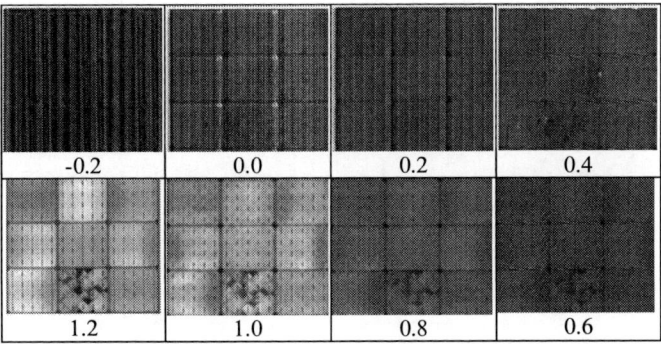

Fig. 5 Categories of image quality, q, clockwise from top left:

-0.2: Not distinguishable as PV cells.

0.0: Distinguishable as PV cells. High noise. Very low contrast.

0.2: Busbars and cell edges are visible, but noisy, contrast too low for diagnosis.

0.4: Busbars darkening. Contrast allows for low-confidence diagnosis.

0.6: Contrast allows for visual diagnosis of some features.

0.8: Busbars appearing dark, contrast good enough for visual diagnosis of most features. Noise too high for automated calculations.

1.0: Nearly lab quality contrast. Low noise. All essential faults visible. Suitable for automated calculations with little to no post-processing.

1.2: Lab quality contrast. No discernible noise. Very high contrast.

the sizes of these sets. For a given pixel, p, in an image series, μ_1 and μ_2 are the means of the upper and lower intensity sets, while $\overline{N}$ is the sum of the local average deviation from these means. The metric measures the average signal amplitude divided by the average noise amplitude across the panel image.

$$SNR_{Kari}' = \begin{cases} \frac{\sum_p |\overline{S}_p|\overline{S}_p}{\sum_p |\overline{S}_p|\overline{N}_p} & , \text{if } \sum_p \overline{S}_p > 0 \\ 0 & , \text{if } \sum_p \overline{S}_p \leq 0 \end{cases} \quad (1)$$

$$\overline{N} = \sum|x_{i,1} - \mu_1|/N_1 + \sum|x_{i,2} - \mu_2|/N_2, \quad \overline{S} = \mu_2 - \mu_1. \quad (2)$$

SNR_{Kari} estimates the general signal quality with low regard for the source area. It requires pre-confirming the presence of a substantial signal. Since all PV module images are cropped, and motion noise also impacts the *effective* signal-containing area, the SNR_{Kari}' is more appropriate, as it considers the relative emission area to a greater extent.

All image series were analyzed with and without PV module stabilization, even those without visible motion. The average magnitude of the quadrilateral corner adjustments required for perfectly tracking a module was chosen as a metric for motion severity. This procedure does not consider any total drift, which is a significant factor when the panel is not being tracked, but since all motion is oscillating, it should be sufficient.

The quality, q, of the reconstructed EL image was gauged visually and classified with a simple numerical scheme from [-0.2…1.2], as shown and listed in Fig. 5. Values of q between -0.2 and 0.4 can be considered useless. Between 0.6 and 0.8, it becomes possible to perform some diagnostics. At $q=1$, the quality should be sufficient for feeding into automated diagnostics algorithms, while $q=1.2$ is lab quality, with no noise.

IV. MEASUREMENTS

Uncalibrated image series were recorded with a C-RED 3 camera from First Light Imaging, adjusted to low light

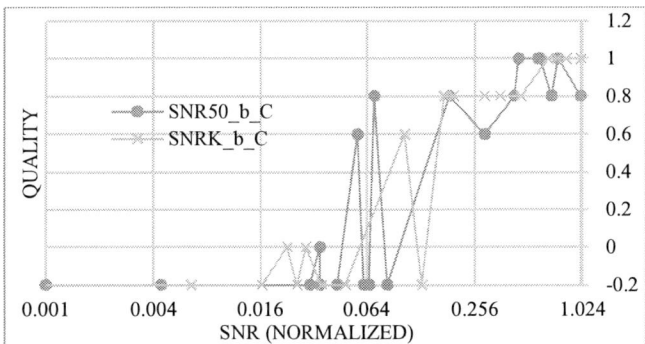

Fig. 6. Image quality as a function of (normalized) SNR_{50} and SNR_{Kari}', for the calibrated image series without module tracking.

conditions but applied in daylight. A Raptor Owl 640 Gen1 camera was used to acquire calibrated images with a built-in, automated calibration system. Both have a spatial resolution of 640x512 pixels.

Different PV module designs were investigated, each having different efficiency and EL signal intensity. Some had highly smooth features, which dramatically downscaled and postponed the relative changes in SNR as a function of movement. On the other end of the scale were multi-crystalline panels with prominent grain patterns, where the SNR dropped almost immediately. In general, all module types presented similar trends, but in highly different SNR domains, making easy visualization difficult. Therefore, three panels with somewhat similar emission intensity were selected for the results and conclusions of this paper: Trina TSM 305-DD050A (305 Wp), Atersa A-170M (170 Wp), Solon P220-6+ (220 Wp). The Solon panels are multi-crystalline, but with only moderately visible grain patterns that did not seem to skew the results.

In total, 35 uncalibrated and 23 calibrated image series were selected, with a broad variety of motion, ranging from stationary to severe. Some recordings were done at a high elevation of the tripod, where the wind would introduce low amounts of motion. The solar intensity remained relatively steady at around 730 W/m² − 800 W/m² throughout the capturing of the selected image series.

V. RESULTS

First, the SNR_{50} and SNR_{Kari}' metrics were tested against non-stabilized image series to potentially eliminate the most

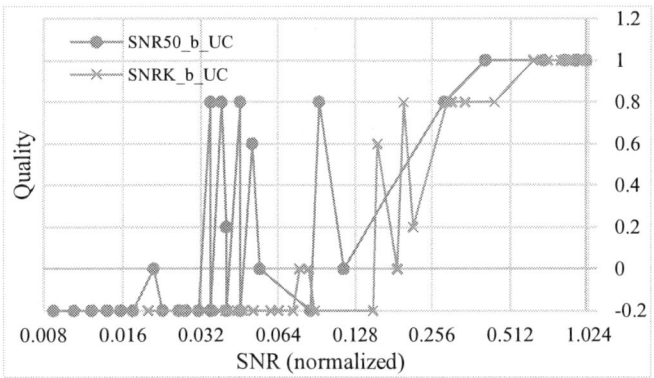

Fig. 7. Image quality as a function of (normalized) SNR_{50} and SNR_{Kari}', for the uncalibrated image series without module tracking.

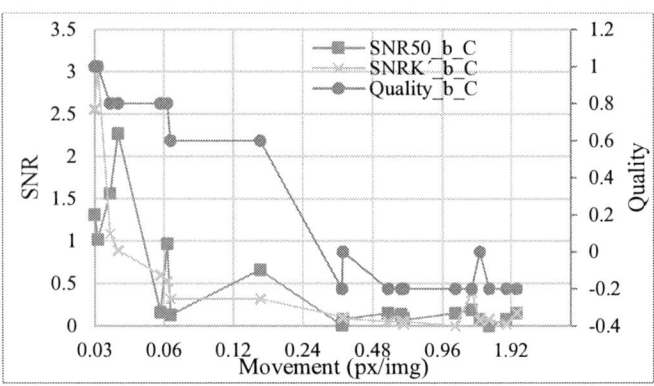

Fig. 9. SNR_{50}, SNR_{Kari}', and image quality as a function of motion severity, for the calibrated image series, without module stabilization.

uncertain metric from the analysis. Fig. 6 and Fig. 7 shows the relationship between the SNR metrics (predicted image quality) and visually confirmed image quality for the calibrated (_C) and uncalibrated (_U) image series respectively. Both graphs show the results without module stabilization (indicated by "_b"). Fig. 9 plots the SNR_{50}, SNR_{Kari}' and image quality as a function of motion severity for the calibrated image series.

Fig. 10 and Fig. 8 show the relationship between the SNR_{Kari}' and the image quality under increasing motion severity, for the uncalibrated and calibrated image series, respectively. Results after (_a) and before (_b) module tracking are shown for both, and trendlines are added for the tracked image series. Similarly, Fig. 11 shows the image quality of the analysis as a function of SNR_{Kari}' before panel image tracking, while Fig. 12 shows it post-tracking, including logarithmic trendlines.

The deviations from the trendlines were calculated, and the following error metrics are displayed in Table 1: The unbiased Root-Mean-Square-Error (RMSE), the Mean Absolute Error (MAE), and the Max Error (MaxE).

Table 1: The unbiased Root-Mean-Square-Error, Mean-Absolute-Error, and Max Error of the analysis-quality trendlines for figures Fig. 10, Fig. 8, and Fig. 12.

Predictor:	Motion (U)	SNRKari´(U)	Motion (C)	SNRKari´(C)
RMSE:	0.103	0.127	0.135	0.144
MAE	0.103	0.095	0.119	0.104
MaxE	0.331	0.339	0.322	0.300

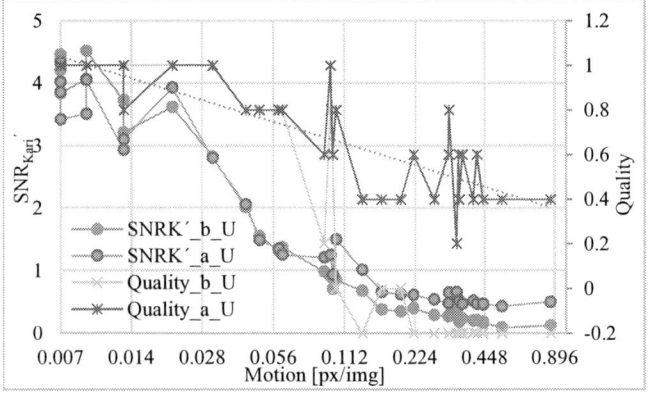

Fig. 10. SNR_{Kari}' and EL analysis quality plotted against motion severity. Trendline added for the post-tracking image quality (Quality_a_U).

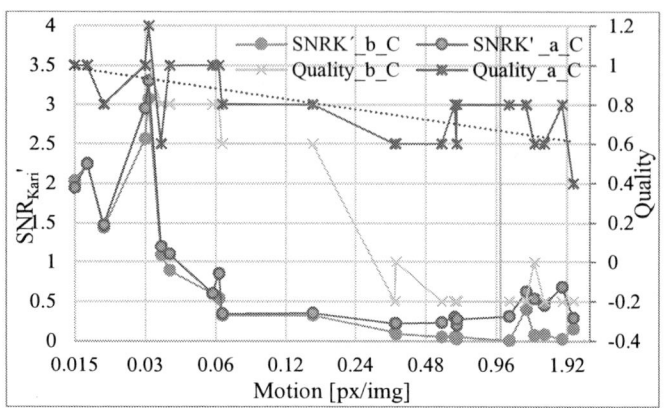

Fig. 8. SNR_{Kari}' and EL analysis quality as a function of motion severity (calibrated series). "_a" indicates module tracking "_b" indicates no tracking. Trendline added for post-tracking quality (Quality_a_C).

VI. DISCUSSION

SNR metrics: As Fig. 6 and Fig. 7 show, neither the SNR_{50} nor the SNR_{Kari}' can perfectly predict the resulting quality of an EL image analysis. However, the SNR_{50} metric is more unstable and prone to SNR predictions much lower than the image quality justifies. This limitation is particularly evident in Fig. 7, where the image SNR_{50} is almost incapable of predicting the image quality at low SNR values. Fig. 9 shows how the image quality drops much later as a function of motion than either SNR metrics, but the SNR_{Kari}' follows the trend more smoother. Due to this, the SNR_{50} was removed from further analyses.

In general, it became immediately clear from initial analyses on different panel types that the SNR metrics' ability to predict image quality is highly dependent on module type at their current state. To avoid operators having to return to the field for additional recording, more stable metrics are needed to verify a recorded image series.

Without tracking: Comparing Fig. 10 and Fig. 8, it is clear that the image quality and SNR_{Kari}' drop faster for the uncalibrated image series than the calibrated as a function of motion. For the uncalibrated series, the image quality takes a sharp dive at around 0.06 px/img, while the SNR seems to dive already at 0.022 px/img. For the calibrated series, the image quality drops to $q=0.6$ at around the same point but does not dive deeper until after 0.18 px/img.

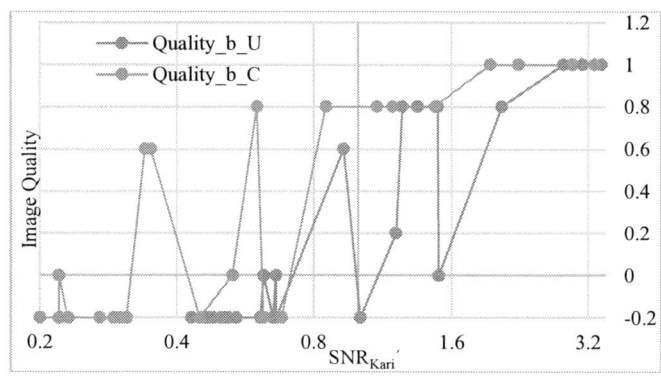

Fig. 11. Image quality as a function of SNR_{Kari}' without tracking the panel (_b), for both uncalibrated (_U) and calibrated (_C) image series.

978-1-6654-6060-6/23 $31.00 © 2023 IEEE

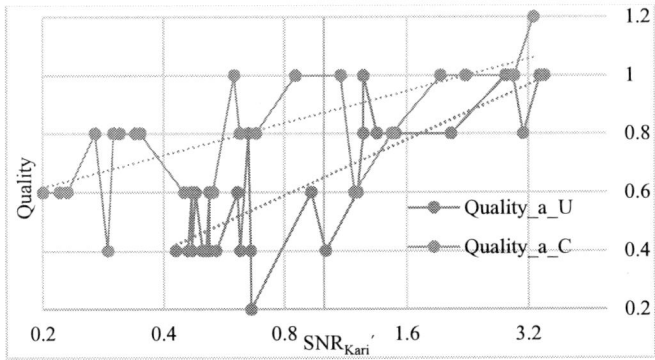

Fig. 12. Image quality for calibrated and uncalibrated data series as a function of SNR$_{Kari}$´, after PV module tracking. Log-trendlines are included.

With tracking: Comparing the two series after module image tracking, the image quality is improved for both, but for the calibrated series, the quality is kept much more stable. Except for the most extreme case of motion severity, the calibrated series stays within the usable range regardless of motion severity after 0.06 px/img, while only 1/3 of the uncalibrated image series are within the usable range starting from 0.133 px/img. This should be seen in the light of the facts that the motion severity of the calibrated series reaches a value twice as high, and the uncalibrated image series were done with the highest emission modules, which gives the uncalibrated series an edge in terms of SNR. One can assume that the lower power modules under similarly poor calibration conditions would fare even worse.

Tracking vs. no tracking: It is immediately apparent from both Fig. 10 and Fig. 8 how dramatically the image quality drops when the motion exceeds 0.06 px/img, which corresponds to very mild motion, like wind on a tall tripod. In addition, almost identical SNR values correspond to highly different image quality categories when comparing non-tracked (_b) and tracked (_a) panels.

When comparing the two untracked quality graphs in Fig. 10 and Fig. 8, as a function of motion, to the untracked quality as a function of SNR_{Kari}´ in Fig. 11, it is clear that the SNR_{Kari}´ is much worse at predicting the image quality, only reliably predicting the quality at very high or very low SNR numbers. As Fig. 12 shows, it fares much better when the modules are tracked, and the primary source of noise is pixel bleeding and any residual camera calibration issues. This behavior makes sense, as the metric was developed for stationary imaging.

SNR_{Kari}´ **vs. Motion Severity:** Superficially, the movement metric looks like a better predictor of image quality when observing the trendlines in Fig. 10 and Fig. 8 and comparing to Fig. 12, but Table 1 shows this to be incorrect. Motion severity and SNR_{Kari}´ score similar error metrics, and each come out best in 3 out of 6 lineups. If one were to assume the *MAE* to be a more appropriate measure than the *RMSE*, then indeed, the SNR_{Kari}´ scores slightly higher. On the other hand, it is immediately apparent from Fig. 11, that it performs much worse when the panel is not tracked. A combination metric could be a potential solution, as the SNR_{Kari} and SNR_{Kari}´ both have shown themselves excellent at predicting analysis quality based on solar noise analysis under static image capturing.

VII. Conclusion

In this paper, we have shown the substantial impact of motion noise on SNR characteristics and image quality. Moving the camera reduces the chosen SNR metrics and image quality significantly and rapidly. However, the image quality's rate and point of decline depend highly on the PV module type. Due to the lack of a stable, generalized SNR metric to provide a coherent analysis, it was decided to focus on select modules and image series with similar characteristics.

For the modules and image series used in this paper, the sensitivity to movement is very high without calibration and stabilization.

The image quality rapidly deteriorates to levels unsuitable for diagnostics starting from 0.06 px/img. A calibrated camera extends the range to around 0.18 px/img, at least for these panel types. Even after visually perfect tracking, an uncalibrated camera causes high noise-levels in the analysis due to relative pixel offsets, and the quality drops permanently to a mostly useless level already at 0.13 px/img.

For the calibrated camera series, tracking helps immensely to keep the quality at least in the manually inspectable domain from 0.06 px/img all the way up to 2 px/img. However, the quality is too low for reliable, automated processing and power loss prediction, and better and more robust analysis algorithms are required, possibly the FFT-based method.

Acknowledgments

This research has been carried out in the Eurostars-project "Automated Daylight Electroluminescence Inspection of Large Photovoltaic Systems" (E115687 ADELI.).

References

[1] A. Tsanakas et al., "Infrared and Electroluminescence Imaging for PV Field Applications: An Overview of the Latest Report by IEA PVPS Task 13," in *35th EUPVSEC*, 2018.

[2] S. Koch, T. Weber, T. Sobottka, A. Fladung, P. Clemens and J. Berghold, "Outdoor Electroluminescence Imaging of Crystalline Photovoltaic Modules," in *32nd EUPVSEC*, 2016.

[3] A. Fladung and J. Schlipf, *Large Scale Electroluminescence Inspection : Multi-Sensor Platforms and Automated Evaluation,* NREL PV Reliabilty Workshop 2019, 2019.

[4] L. Stoicescu, M. Reuter and J. H. Werner, "DaySy: Luminescence Imaging of PV Modules in Daylight," in *29th Eur. Photovolt. Sol. Energy Conf. Exhib.*, 2014.

[5] M. Guada et al., "Daylight luminescence system for silicon solar panels based on a bias switching method," *Energy Sci. Eng.,* p. 1–15, June 2020.

[6] J. Adams et al., "Non-Stationary Outdoor EL-Measurements with a Fast and Highly Sensitive InGaAs Camera," in *32nd Eur. Photovolt. Sol. Energy Conf. Exhib.*, 2016.

[7] G. A. dos Reis Benatto et al., "Drone-Based Daylight Electroluminescence Imaging of PV Modules," *IEEE J. Photovoltaics,* vol. 10, no. 3, p. 872–877, 2020.

[8] G. A. dos Reis Benatto et al., "Daylight Electroluminescence of PV Modules in Field Installations: When Electrical Signal Modulation is Required?," in *8th World Conference on Photovoltaic Energy Conversion*, 2022.

[9] T. Kari et al, "Computer Vision Method for Extracting an Induced Electroluminescence Signal from Photovoltaic Modules in Daylight Conditions Using Drone-Captured Images," in *37th EUPVSEC*, 2020.

[10] I. E. Commission, "IEC TS 60904-13:2018," 2018. [Online]. Available: https://webstore.iec.ch/publication/26703.

Interface Hydrogen and Passivation of Amorphous Silicon / Crystalline Silicon Heterojunction

Ujjwal K. Das[1,2,3], Tasnim K. Mouri[1,2], Marissa Pina[4], Tyler Parke[4], Dhamelyz R. S. Quinones[4], and Andrew V. Teplyakov[4]

[1]Institute of Energy Conversion, University of Delaware, Newark, DE 19716, USA
[2]Department of Materials Science & Engineering, University of Delaware, Newark, DE 19716, USA
[3]Department of Electrical and Computer Engineering, University of Delaware, Newark, DE 19716, USA
[4]Department of Chemistry and Biochemistry, University of Delaware, Newark, DE 19716, USA

Abstract — The Si surface passivation by amorphous Si (a-Si:H) layers is investigated, where the a-Si:H layers are grown with different plasma current and methods. A significant improvement in passivation quality (iV_{OC} = 738 mV, $\approx$ 40 mV increase from baseline) is achieved when the a-Si:H growth plasma was interrupted (OFF/ON) at the midway of deposition followed by a post deposition annealing. This improvement is attributed to the homogeneous uniform nucleation and growth of a-Si:H stack layer and layer densification (removal of some hydrogen) during annealing.

I. Introduction

Amorphous silicon (a-Si:H) / crystalline silicon (c-Si) heterojunction (HJ) solar cells have been one of the promising contenders for dominating the photovoltaics market in near future. This is due to the multifarious advantages of the technology, which includes record cell efficiency [1], low temperature fabrication process, high open circuit voltage (V_{OC}) > 740 mV and low temperature co-efficient that increases energy yield. High V_{OC}, the most critical aspect of the technology, stems from the superior interface passivation provided by the intrinsic a-Si:H over c-Si. Extensive studies have been afforded over the past decades on developing appropriate Si wafer surface preparation, and a-Si:H process-properties-device performance optimization that has led to the remarkable progress in cell performance. However, lately it has been reported that these high efficiency Si HJ solar cells and modules exhibit higher rate of performance degradation than the other established Si homojunction PV technologies [2, 3]. Faster degradation over long-term appears to be triggered by the loss of V_{OC}, unlike for other Si homojunction cells and thus necessitates further investigation of a-Si:H surface passivation layers and their interface kinetics with Si. As a first step, it is imperative to establish and understand the role, kinetics, and bonding types of interface hydrogen in Si HJ structure. In this work, we have studied surface passivation qualities of different a-Si:H layers, effect of post deposition annealing, and the interface hydrogen bonding microstructure.

II. Experimental Detail

The Si surface passivation in this study are performed on 140 μm thick, both side textured n-type CZ wafer having resistivity of 2 – 5 Ω.cm. The wafers were cleaned and textured by tetramethylammonium hydroxide using the process described elsewhere [4]. Additionally, clean pieces of 500 μm thick polished Si wafers were also included in each run. Different a-Si:H layers were deposited on both sides of the textured cleaned Si wafers using an in-vacuum flipper by a direct current (d.c.) plasma enhanced chemical vapor deposition (PECVD) reactor, details of which are given elsewhere [4],[5]. The polished Si wafers were loaded in the non-flipper position and thus received 2X deposition on the same surface in each run. The deposition temperature, pressure, silane, and hydrogen flow rates were fixed at 200 °C and 1250 mTorr, 10 sccm, and 25 sccm respectively, while the dc plasma current was varied from 30 – 150 mA, resulting in the power variation of $\approx$ 12 – 60 W. The a-Si:H deposition was performed in two different ways: a) continuous deposition (termed as "Single") for a desired duration to achieve $\approx$ 10 nm thick film and b) the deposition was interrupted by turning off the plasma power and turn on again after 60 s at the half way stage (2X 5 nm, termed as "Stack"). The surface passivation quality was assessed after a-Si:H passivation of textured wafers by effective minority carrier lifetime (τ_{eff}) and implied open circuit voltage (iV_{OC}) obtained from a Sinton WCT-100 tool using quasi-steady-state phtoconductance (QSSPC) decay in the "As-deposited" and also at "Annealed" state, after a thermal annealing at 300 °C for 30 mins in vacuum. The companion polished wafer samples were used to further characterize for film thickness by optical method, hydrogen bonding structures by fourier transform infrared spectroscopy (FTIR), and hydrogen depth profile by time of flight secondary ion mass spectroscopy (TOF SIMS).

III. RESULTS AND DISCUSSION

A. Surface passivation quality

The surface passivation quality was deteremined by τ_{eff} and iV_{OC} measurements in both "as deposited" and "Annealed" states for the "Single" and "Stack" a-Si:H layers deposited at different dc plasma current. Figure 1 shows the variation of τ_{eff} and iV_{OC} as a function of plasma current. The "Stack" layer passivation is found to be superior to the "Single" layer with ≈ 20 mV improvement in iV_{OC} at all plasma current conditions. Post deposition annealing ("Annealed") further improves τ_{eff} for the plasma current > 30 mA, leading to up to additional 20 mV improvement in iV_{OC} to reach iV_{OC} = 738 mV at 120 mA.

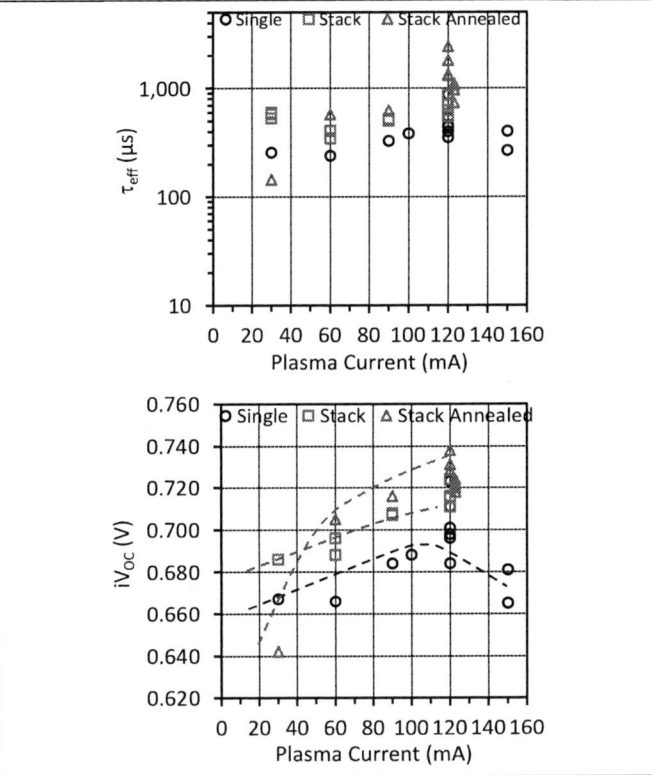

Fig. 1: (top) Effective minority carrier lifetime (τ_{eff}) at $\Delta n = 10^{15}$ cm^{-3} after the textured Si wafers passivated by different a-Si:H; Single (circle), Stack (square) and Annealed Stack (triangle) deposited with varying plasma current. (bottom) Implied V_{OC} (iV_{OC}) obtained from QSSPC from the same set of samples. The dashed lines are guide to the eye.

B. Hydrogen bonding structures

Figure 2 compares the FTIR spectra of Si–H stretching vibrational peaks in the wavenumber range of 1900 – 2200 cm^{-1} for different a-Si:H films deposited on polished Si wafers. The distinct peaks at 2000 cm^{-1} and 2080 cm^{-1} can be identified, which often referred as low stretching mode

(LSM) and high stretching mode (HSM), respectively. It is argued that the a-Si:H films with predominant LSM (Si-H monohydride in amorphous Si matrix) represents good electronic quality with compact films, while presence of higher HSM associates with more microvoids and deteriorates film quality. The a-Si:H films deposited with plasma current of 30 mA shows predominantly LSM and increase of plasma current increases HSM contribution with an associated decrease of LSM (Fig. 2(a)). Therefore, comparing these with the Fig. 1, it can be concluded that a-Si:H films with higher HSM provides better passivation quality with higher τ_{eff} and iV_{OC}. Similar results have been widely observed in the field [6, 7]. This is counterintuitive for the known trend of a-Si:H film's electronic quality, and indicates that the c-Si surface passivation critically depends on the type and amount of H-bonding or interface hydrogen and not necessarily on the thin a-Si:H layer propreties.

Comparison of single and stack a-Si:H layers deposited under same conditions with plasma current of 120 mA, shown in Fig. 2(b), interestingly show reduction of LSM, with no discernible change in HSM for stack layer. This is surprising due to the fact that the process conditions between them are exactly the same with the only difference of a brief interruption of plasma (i.e. deposition) after about ≈ 5 nm of film deposition for the stack layer. Two likely scenarios for this result could be; a) stack layer (interruption of plasma) promotes even and dense nucleation sites leading to better film coverage and/or b) H microstructural evolution from HSM to LSM during growth. The end result of the stack layer deposition is enhanced τ_{eff} and ≈ 20 mV increase in iV_{OC} compared to the single layer. The concept of stack with two different plasma conditions for the two layers for surface passivation has been quite common in literature [6, 7], but this work shows the two layers do not require separate conditions to achieve high quality surface passivation.

Fig.2(c) shows that the hydrogen microstructure does not change after post deposition annealing of the layers. But an uniform reduction of both LSM and HSM peak intensities is observed. Although only one set of data is presented, all samples studied here show the same trend. Thermal annealing likely causes removal of some hydrogen and densifies the films. This has a non-negligible improvement of τ_{eff} and iV_{OC} except for the films deposited at 30 mA (Fig.1). The sample deposited with 30 mA likely contains interface crystallinity, which cauases passivation loss after annealing, similar to the case of a-Si:H deposited at high hydrogen dilution, reported earlier [5].

C. Hydrogen depth profile

Figure 3 shows the TOF SIMS depth files for the as-deposited single (Fig. 3(a)) and stack (Fig. 3(b)) and stack annealed (Fig. 3(c)) states. The single a-Si:H layer deposited at 120 mA exhbits a noticeable slope of SiH$_2$ trace with a lower SiH$_2$ at the a-Si:H/c-Si interface (sputter time of ≈ 200

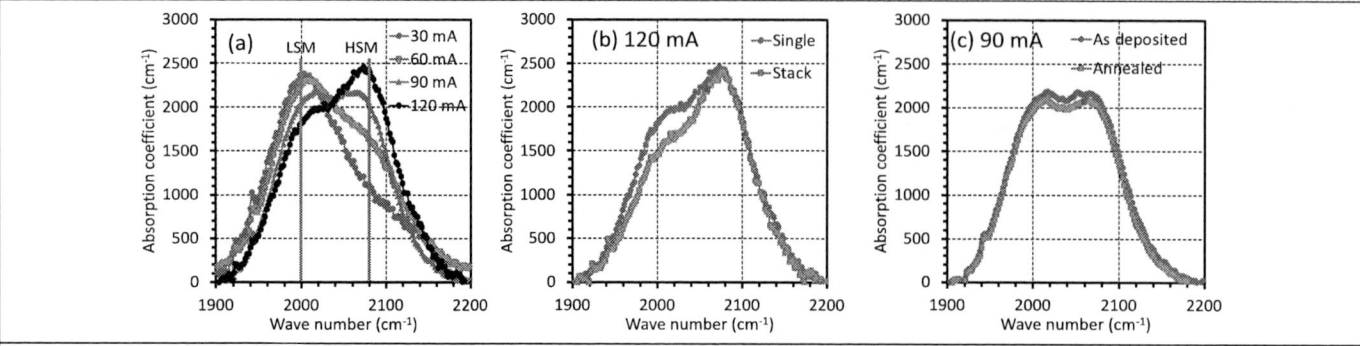

Fig. 2: (a) Si–H stretching vibrational peaks for thin a-Si:H films deposited at different PECVD plasma current. (b) comparison of FTIR peaks for "Single" and "Stack" a-Si:H films deposited with 120 mA plasma current. (c) comparison of FTIR peaks for "As-deposited" and "Annealed" a-Si:H film deposited with 90 mA plasma current.

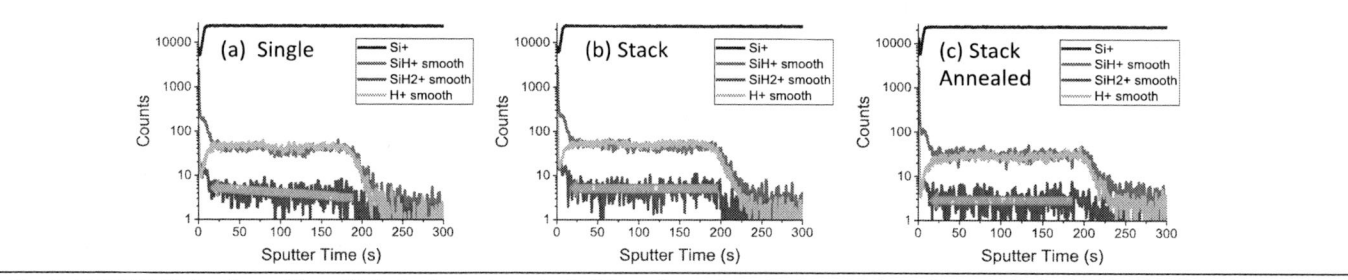

Fig. 3: TOF SIMS depth profiles of (a) single, as deposited, (b) stack, as deposited, and (c) stack, annealed a-Si:H thin films deposited at plasma current of 120 mA.

s). The stack a-Si:H layer, however, exhibits a flat and constant SiH_2 signal, representing a uniform and perhaps a more continuous film. This is likely the reason for better passivation with higher τ_{eff} and iV_{OC} for stack layer. The annealing of the stack layer results in an overall and uniform reduction of all hydrogen species (SiH and SiH_2). This TOF SIMS result agrees well with the observed reduction of both LSM and HSM H-realted peaks in FTIR. Further detail analysis of temperature dependent FTIR, TOF SIMS, and solar cell performances will be reported at the conference.

IV. Summary

A significant improvement of Si surface passivation quality is observed using a thin ($\approx$10nm) stack a-Si:H, where the plasma was interrupted OFF/ON in the middle of deposition. An iV_{OC} value of 738 mV is achieved after annealing the stack layer passivation structure. The stack a-Si:H layer deposition enables uniform conformal nucleation and growth with constant H microstructure at the interface and throughout the film as observed by FTIR and TOF SIMS depth analysis.

Acknowledgment

This work was supported by the US Department of Energy's Office of Energy Efficiency and Renewable Energy (EERE) under Solar Energy Technologies Office (SETO) Agreement Number DE-EE0010249.

References

[1] K. Yoshikawa et al., "Exceeding conversion efficiency of 26% by heterojunction interdigitated back contact solar cell with thin film Si technology," Sol. Energy Mater. Sol. Cells, vol. 173, pp. 37 – 42, 2017.

[2] D. C. Jordan et al., "Silicon heterojunction system field performance," IEEE J. Photovoltaics vol. 8, pp. 177 – 182, 2018.

[3] T. Ishii, and A. Masuda, "Annual degradation rates of recent crystalline silicon photovoltaic modules," Prog. Photovolt: Res. Appl., vol. 25, pp. 953 – 967, 2018.

[4] U. J. Nsofor et al., "Analysis of silicon wafer surface preparation for heterojunction solar cells using X-ray photoelectron spectroscopy and effective minority carrier lifetime," Sol. Energy Mater. Sol. Cells, vol. 183, pp. 205–210, 2018.

[5] U. K. Das et al., "Surface passivation and heterojunction cells on Si (100) and (111) wafers using dc and rf plasma deposited Si:H thin films," Appl. Phys. Letts, vol. 92, pp. 063504, 2008.

[6] W. Liu et al., "Underdense a-Si:H film capped by a dense film as the passivation layer of a silicon heterjunction solar cell," J. Appl. Phys., vol. 120, pp. 175301, 2016.

[7] H. Sai et al., "Impact of intrinsic amorphous silicon bilayers in silicon heterojunction solar cells," J. Appl. Phys., vol. 124, pp. 103102, 2018.

978-1-6654-6060-6/23 $31.00 © 2023 IEEE

AUTHOR INDEX

Abad, Eduardo Camarillo 1249
Abasi, Abudulimu 402
Abbas, Ali 245, 498, 656, 1193
Abdullah-Vetter, Zubair 1224, 1468
Aberle, Armin G. 1052
Abou-Ras, Daniel 432, 879, 939
Abrosimov, Nikolay V. 203
Abudulimu, Abasi 42, 46, 220, 275, 279, 683, 804,
.... 823, 909, 910, 1197, 1323
Acevedo, Armando Figueroa 1393
Adeleye, Damilola 1151
Adhikari, Alisha 1197, 1376
Adner, David 770
Afridi, Muhammad 237, 755
Afshari, Hadi 132, 161, 1156
Afzal, Syed Usama Bin 772
Agarwal, Shashank 1053
Agarwal, Sumit 30, 359, 475, 891, 1216
Ager, Joel W. 75
Aghaei, M. 737
Agrawal, Roshni 1053
Aguirre, Aranzazu 194
Ahlswede, Erik 1201
Ahmad, Muneeza 1109, 1565
Ahmad, S. 737
Ahmadi, Bahman 816
Ahn, Yujeong 1549
Aiello, Ashlee R. 282
Aimez, V. 1538
Aïssa, Brahim 27, 1516
Akiyama, Hidefumi 463
Akopian, Arkadi 641
Al Katrib, Mirella 1440
Alam, Fahad 173
Alam, M. A. 626
Alam, Muhammad A. 119
Alam, Muhammad Ashraful 772
Alam, Muhammad 1027
Alanis, Luis Eduardo 1060
Al-Baity, Shifaa M. 561
Albert, P. 1538
Alberts, Vivian 119
Albin, David S. 695
Albrecht, Steve 945
Aldalali, Bader S. 162
Aldhefairi, Mariam 1344
Alessi, Bruno 281
Algora, Carlos 1433

Alhammadi, Salh 1549
Alhassan, Saeed 1471
Ali, Adnan 1516
Al-Jassim, M. M. 315
Al-Jassim, Mowafak 1570
Allebé, Christophe 1129
Almannaee, Rawdah 1344
Almansoori, Muntaser Abdelrahman 321, 1344
Almonacid, Florencia 122, 437
Almutawah, Zahrah S. 1417
Alom, Md Zahangir 690, 1237
Alsuwaidi, Meerh 1344
Altenhöfer-Pflaum, Georg 749
Altermatt, Pietro 1544
Al-Thani, Hamda A. 561
Altosaar, Mare 1298
Alvarez, H. 1492
Alvarez, Hugo Da S. 1159, 1654
Alvarez, Hugo S. 1068
Alvarez, Hugo 1134
Aly, Shahzada Pamir 119
Amassian, Aram 332
Anctil, Annick 730, 743, 1095, 1259
Anderson, Caroline Lima 30
Anderson, Kevin S. 948
Anderson, Kevin 710
Andersons, Janis A. 348
Andreas, Afshin 600
Andreasen, Jens Wenzel 651
Anto, Robins 1230
Antón, Ignacio 746, 888
Aonuki, Sho 1398
Aponte, Erick 525
Aponte-Bezares, Erick E. 674, 1034, 1170, 1326
Aponte-Bezares, Erick 1082
Araki, Kenji 924
Aramaki, Ken 1164
Arehart, Aaron R. 107
Arehart, Aaron 1299
Arfaoui, Ines 426
Armour, Eric A. 1217
Arnold, Rachael 353
Arteaga, Jorge 415
Arya, Rajeewa 1366
Asadpour, Reza 626
Asahi, Shigeo 659, 895
Asami, Meita 374, 463, 935
Ascencio-Vásquez, J. 737

Ascencio-Vásquez, Julián 1578
Askins, Steve 746, 888
Astigarraga, Alexander 931
Athana, Dhawal 1009
Atwater, Harry A. 202, 1157
Augusto, André 5
Avoli, Matteo 1404
Awni, Rasha A. 1323
Ayala, Silvana 1141
Ayon, Arturo A. 769
Ayon, Arturo 1283
Azevedo, João Henrique Paulino 256
Azzolini, Joseph A. 13
Baan, Marzieh 917
Babayeva, Gülüsüm 1092
Babu, Balaashwin 671
Bach, Udo 298
Badosa, Jordi 268
Baer, Carsten 191
Baetzner, Derk 1
Baik, Sunhee 793
Bailie, Colin 740
Bainier, Camille 268
Baker, Wes 1096
Bakke, Jordan 1393
Balaji, Pradeep 5
Baldrias, Maristel 1024
Ball, James M. 298
Ballif, Christophe 1, 1129
Banerjee, Parag 629
Banks, Terry 740
Bansal, Shubhra 566, 1551
Barbose, Galen 793, 991
Barnes, Teresa M. 554, 1593
Barretta, Chiara 931, 1630
Barrit, Dounya 268
Barros, Tárcio A. Dos S. 1654
Barros, Tárcio A. S. 1159
Barros, Tarcio Andre Dos Santos 1134
Barth, Kurt L. 245, 981, 1497
Barth, Kurt 789
Barth, Vincent 352
Bartsch, Jonas 1625
Bastola, Ebin 220, 275, 279, 401, 402, 683, 804,
.... 823, 910, 1041, 1193, 1410, 1600
Batista-Alvarez, Natanael 370
Battiato, Sebastiano 1567
Baumann, Sara 733
Baxter, Jason B. 915, 1241
Beard, Matthew C 161, 1156
Beattie, Meghan N 941
Becker, Jan-Philipp 1201

Beltrán, Juan F. 761
Benatto, Gisele A. Dos Reis 16, 448, 1452, 1662
Benhaddou, Nada 395, 534
Bennett, Mitchell F. 1217
Benton, Brandon 587
Berad, Mrunal 1366
Bergin, Mike 437
Bernardini, Simone 768
Bernsen, Otto 749
Berriel, S. Novia 629
Berry, Joseph J. 298, 590, 1044, 1260, 1484
Bertoni, I. 291
Bertoni, Mariana I. 291, 400, 848, 970, 1533
Bertoni, Mariana 218, 333, 597, 689, 768, 782, 896, 1634
Berwind, Matthew F. 1313
Bessa, Joao G. 1270
Bessal, João Gabriel 437
Bevan, Geraint 591
Bhat, Akanksha 1147
Bhatia, Amandeep Singh 1027
Bidaud, T. 1538
Bidaud, Thomas 133, 781
Bila, Marine 295
Binyamin, Tal 879
Birgersson, Erik 1188
Bista, Sandip S 1197
Bista, Sandip Singh 46
Bittkau, Karsten 280
Bivour, Martin 770, 1625
Bizhanova, Gulzhan 39
Bizzarri, Fabrizio 1304, 1567
Black, Chloe L. 948
Blankemeyer, Susanne 652
Bloeck, Ulrike 939
Blom, Youri 1005
Blum, Adrienne 113
Boemer, Jens C. 1096
Bogner, Brandon M. 721
Bojorquez, Jose Raul Montes 769
Bolen, Michael 897
Bolink, Henk J. 938
Bondoc, Christopher C. 1510
Bonnassieux, Yvan 233
Borgers, Tom 48
Bosco, Nick 106, 1218
Bothe, Karsten 288
Bothwell, Alexandra M. 107
Bothwell, Alexandra 849
Boucherif, Abderaouf 133
Boucherif, Abderraouf 434, 1051
Bourarach, Fadi 426
Bourisli, Raed I. 162

Bournonville, Kenn Henrik 749
Bousselot, Jennifer 215
Bouttemy, Muriel 1440
Bowers, Jake W. 245, 395, 981, 1177, 1497
Bowers, Jake 534, 1193
Bowersox, David A. 1409
Boyce, Kenneth P. 1141
Boyd, Matthew 920
Boyd, Stephen P. 961
Boyer, Jacob T. 10, 291, 848, 994
Boz, Mesude Bayrakci 98
Brabec, Christoph J. 645, 1092
Bracamonte, Maria Fernanda Villa 769
Braga, Daniel Sena 1541
Braid, Jennifer L. 234, 367, 512, 572
Braid, Jennifer 861, 1088, 1131
Bramante, Rosemary C. 1158
Brand, Andreas 1588
Brandstätter, Andreas 1630
Brau, Tyler R. 909
Brau, Tyler 42, 220, 402, 804, 910
Braun, Anna K. 10, 71, 291, 848
Bredács, Marton 931
Bremner, Stephen P. 917, 1564
Bremner, Stephen 271
Brendel, Rolf 6, 288
Brenner, Tom 740
Brewer, Jeremy 178
Brewster, Charles 138
Breyer, Christian 708, 749
Brinck, Anna 1002
Brittman, Sarah 1045
Brockmann, Tim Lukas 652
Broderick, Robert 1170
Brooks, William 1009
Brown, Buck 240
Brown, Matthew 794, 1141
Bruckman, Laura S. 234, 597, 671, 861, 1088, 1141, 1485
Brueckner, Dennis 651
Bryan, Alex 1123
Buchholz, Florian 1372
Buck, Thomas 1266, 1372
Buerhop-Lutz, Claudia 540, 645
Bui, Thanh-Tuân 1484
Bullock, James 1434
Buño, Luis 1024
Burduhos, B. G. 737
Burnham, Laurie M. 572
Burnham, Laurie 391, 409, 1131
Burtone, Lorenzo 191
Bush, Meghan E. 698
Busko, Dmitry 996

Buster, Grant 474, 587
Butt, Nauman Zafar 772
Cabanillas, Juan 804
Cacciato, Mario 1304, 1567
Cai, Mengmeng 360
Calderon, Jose A. 914
Calili, Rodrigo Flora 256, 858
Calloquispe-Huallpa, Ricardo 674
Calvo-Barrio, Lorenzo 1059
Campbell, Robert 295
Campesato, Roberta 1404
Campo-Ossa, Daniel D. 916
Canino, Andrea 1005, 1304, 1567
Cano, Aitana 1603
Cao, Fangfang 822
Cappelluti, Federica 1413
Carbone, Marc A. 1626
Cariou, Romain 716, 934, 1423
Carletta, Stefano 1404
Carpenter, Joe V. 1533
Carron, Romain 651
Cartledge, Carsen 1109, 1565
Carvalho, Romullo R. M. 1159
Carvallo, Juan Pablo 793
Cassini, Denio Alves 1541
Castillo, Arnold 1024
Castillon, Jean 268
Castro-Sitiriche, Marcel 717
Catchpole, Kylie 1434
Cebecauer, Tomas 1594
Celi, Edoardo 871
Chacon, Sergio A. 132, 1156
Chadly, Assia 299
Chakar, Joseph 233
Chaluvadi, Venkata S. A. 1249
Chambers, Terrence L. 188
Chan, Maria K. Y. 1634
Chan, Maria 782
Chang, Atom 917
Chang, Nathan 289
Chapaneri, Kaushal 119
Chapon, Julien 426
Chapotot, Alexandre 434, 1051
Chard, Julie 1123
Chatzipanagi, Anatoli 550
Chaubal, Aditi 1366
Chaurasia, Harsh 182
Chen, Bin 518, 1019
Chen, Chien-Hsuan 597, 629, 1337
Chen, Christopher 179
Chen, Cong 822
Chen, Daming 1544

Chen, Gonglin 1021
Chen, Hong 1544
Chen, Kissenger 978
Chen, Lei 804, 909, 1197, 1550
Chen, Mike 740
Chen, Ning 1372
Chen, Stela 377
Chen, Theresa K 1566
Chen, Xin 1519
Chen, Yifeng 1544
Chen, Zeying 682
Chin, Robert Lee 227, 1224
Chintapalli, Sreyas 718, 1334, 1356
Chiu, Arlene 272, 718, 1356
Cho, Jinyoun 133, 434, 1051, 1341, 1390
Cho, Yunae 97
Choi, Wookjin 1657
Choi, Wook-Jin 179, 1030
Choudhury, Ashif 134
Choudhury, Kaushik R. 952
Choudhury, Kaushik Roy 377, 689
Chowdhury, Gofran 86, 1253
Chretien, Jeremie 133
Christiansen, Silke 194
Chu, Haifeng 1372
Chung, Jaehoon 844, 1417
Chutani, Ayush 391
Cieslak, Janko 191
Cifuentes, Luis 1433
Cimaroli, Alexander 412
Cioc, Sorin 402
Cira, Spencer 473
Clandestino, Franco 1491
Clausing, Roland 6, 12
Clemens, Daniel 240
Coco, Fabrizio 1005
Colegrove, Eric 245, 633, 849, 946
Coletti, Gianluca 917
Coll, Pablo G. 218, 291
Coll, Pablo Guimerá 848
Collavini, Silvia 289
Collin, Stéphane 610
Colomba-Colon, Luis 370
Colombo, Mariela 1529
Colón, Alanis M. 416
Colvin, Dylan J. 671, 758, 850, 1120, 1485
Conibeer, Gavin J. 271, 1564
Conibeer, Gavin 881
Coogan, Katrina 879
Cook, John P. D. 476
Coologeorgen, Alexander E. 918
Cooper, Emma C. 572

Cooper, Robert A. García 519
Correa-Baena, Juan-Pablo 969, 1183
Corso, Roberto 1500, 1609
Cosme, Damien 426
Costa, Carla 934
Costa, Suellen C. S. 1541
Coulibaly, Bakary 1292
Court, Philip 922
Courtois, Guillaume 434, 1390
Coutinho, Natália F. 290
Couture, Eugene Desjardins 1648
Crawford, Zachary 51
Crestani, Thais 290
Critchlow, Gary 174
Crowe, Iain F. 203
Crowe, Laura E. 919
Crowley, Kyle 415
Cruz, Edgar E. 420
Csank, Jeffrey T. 1626
Cudzinovic, Michael J. 1024
Cullen, David A. 46
Curran, Alan J. 903
Currie, Taylor M. 629
Curson, Kieran 245, 656
Da Silva, João A. F. G. 1654
Da Silva, Lucas Aló Rodrigues Araujo 858
Da Silva, Paulo R. D. R. 1159, 1654
Daenen, Michael 48
Dahi, Adem 1423
Dalal, Vikram 641
Dale, P. 995
Dale, Phillip J. 731, 975
Dalibor, Thomas 1067
Dally, Pia 1440
Danel, Adrien 716
Daniel, Valentin 133, 434
Danielson, Adam 1279
Danilson, Mati 1298, 1432
Danovitch, D. 1538
Dapprich, Karoline 113
Darawali, Renn 927, 1146
Darbali-Zamora, Rachid 334, 455, 674, 953, 984,
.... 1034, 1082, 1170, 1326, 1626
Darling, Halley C. 1146
Darling, Halley 927
Darnon, M. 1538
Darnon, Maxime 133, 781, 1051
Das, Ujjwal K. 91, 1545, 1667
Das, Ujjwal 1299
Dasgupta, Sagnik 1030, 1155, 1657
Daus, Alwin 1347
Davis, Kristopher O. 219, 597, 629, 671, 758, 850, 1485

Davis, Kristopher 1337
Davis, Melissa A 330
Davis, Nithin Maipan 1404
Dawson, Timothy 1295
De Albuquerque, Vanessa Cardoso 256, 858
De Brabandere, Karel 1253
De La Rosa, Angel 782
De Lafontaine, Mathieu 476
De Lima, Geyciane P. 1159, 1654
De Luna, Gabby 1030
De Monfreid, Thybault 1484
De Oliveira, Otávio J. 290
De, M. M. M. Modesto Ana Paula 290
De, Shoubhik 1287
Debije, Michael G. 809
Deceglie, Michael G. 921, 1260, 1593
Deckx, Julien 1253
Decristofaro, Eric R. 918
Degen, Ashley 467
Degenhart, Nick 113
Delazzer, Timothy 387
Deline, Chris 47, 920, 1145
Delmas, William 415
Demko, Michael 377
Demtsu, Samuel 914
Deng, Chenyang 743
Deng, Yuepeng 903
Depauw, Valérie 434, 1390
Dequilettes, Dane W 590
Desarden-Carrero, Edgardo 1082, 1326
Descoeudres, Antoine 1129
Deshmukh, Kedar 1366
Dessein, Kristof 133, 434, 1051, 1341, 1390
Deville, Lelia 188
Dhakal, Rabin 528, 897
Dhakal, Tara 682
Dharmadasa, Ruvini 38
Dhople, Sairaj 1384
Di Stefano, Agnese 1005, 1304
Diaz, Martin 1217
Dice, Paul W. 409
Dice, Paul 1131
Didier, Thevenard 861
Diederich, Marvin 6
Dierenbach, Jonas 1184
Dietsch, Tina 795
Diggs, Andrew 51
Digregorio, Steven J. 400
Digregorio, Steven 333
Ding, Kaining 177, 280, 419
Diniz, Antonia Sonia A. C. 94, 1541
Dippell, Torsten 1129

Dittmann, Sebastian 1475
Dobson, Kevin D. 1545
Dobson, Kevin 1299
Dobson, Wes 113
Dokken, Briana 1210
Dolia, Kshitiz 42, 1417, 1557
Doll, Bernd 645
Domínguez, C. 746
Domínguez, César 888
Don, Eric 981
Dong, Peng 1372
Dong, Shuan 240, 840
Donoso, Jose 749
Dougherty, Brian 1292
Doumon, Nutifafa Y. 1044, 1260
Dow, Andrew R. R. 1034, 1626
Drayton, Jennifer 1310
Drees, Martin 145
Drost, Christian 656
Druffel, Thad 38
Du, Bin 1299
Du, Liming 822
Du, Mohan 155
Duan, Leiping 1434
Duan, Weiyuan 177, 280, 419
Duan, Xiaomeng 1197
Dubajic, Milos 271, 1564
Dubois, Anne Migan 268
Dubois, Sébastien 716
Dubuc, Christian 133, 1051
Duenow, Joel N. 695, 946
Dulal, Prabin 1410
Duncan, Brent 492
Duncker, Klaus 191
Dunfield, Sean P. 644
Dunham, Scott 578
Duong, Calvin 508
Durant, Brandon K 1156
Dutta, N. S. 315
Dutta, Nikita S. 1570
Duzellier, Sophie 934
Dvonc, Lukas 1594
Dwivedi, Priya 472, 828, 1224, 1468
Dyreson, Ana 391, 1635
Eberspacher, Chris 740
Eberst, Alexander 419
Ebner, Rita 194
Ebong, Abafriseke 38
Echevarria, Angel 916
Edmondson, A. 1261
Eggink, Wouter 443
Einhaus, Lisanne M. 622

Ekins-Daukes, N. J. 1468
Ekins-Daukes, Nicholas J. 345, 976
Ekins-Daukes, Nicholas 90
Elahi, Sheikh Tawsif 1237
Elanzeery, Hossam 1067
El-Atab, Nazek 173, 327
Ellingson, Randall J. 1323
Ellingson, Randy J. 42, 46, 220, 275, 279, 402,
.... 683, 823, 844, 909, 1019, 1041, 1197, 1353, 1417
Ellingson, Randy 804, 910
Ellis, Brian 476
Elsehrawy, Farid 345
Engel, Bernd 64
Engelen, Tine 48
Engsig-Karup, Allan P. 16
Enjalbert, Nicolas 716
Eperon, Giles E. 132, 161, 919
Eperon, Giles 15
Erickson, Samuel 415
Escobar, D. Martinez 783
Espinet-Gonzalez, Pilar 11
Etgar, Lioz 879
Evans, Rhett 828
Facsko, Stefan 435
Fai, Calvin 915, 1241
Fairbrother, A. 737
Falkenberg, Gerald 651
Fan, Yangxin 861, 1088
Fanego, Vicente Lara 1594
Fang, Liang 195, 1526
Fang, Xin 840
Fardi, Hamid 1513
Farias, Stephen 1575
Farias-Basulto, Guillermo A. 1066
Farina, Angela 730
Farrell, Jack 771
Farrell, John 1600
Fassl, Paul 701
Fattah, Tarek O. Abdul 203
Faulwetter-Quandt, Björn 191
Fechner, Hubert 749
Fedoseyev, Alex 110
Fei, Chengbin 1044
Feng, Xiaotong 1503
Feng, Zhiqiang 1544
Fenning, David P. 644
Ferekides, Chris 690, 1237
Ferguson, Andrew J 341
Fernández, Eduardo F. 122, 437, 1270
Fernandez, Pablo 1433
Fernández-Solas, Álvaro 1270
Ferrara, Matteo 1404

Ferreira, Mateo 1437
Fertig, Fabian 191
Fevola, Giovanni 651
Fields, Shannon 1299
Filho, Neolmar De M. 94
Fischer, Benedikt 177
Fischer, Markus 191
Fisher, Kathryn 925
Flechas, Juan Pablo Medina 268
Fleming, Katelynn E 89
Flicker, Jack D. 984, 1034, 1626
Flicker, Jack David 634
Florakis, Antonios 426
Floren, Radovanovic-Peric 351
Fonoll-Rubio, Robert 1067
Fontanot, Tommasso 194
Forcade, Gavin P 941
Forcade, Gavin 476
Forchhammer, Søren 16
Forrest, Stephen R. 482
Forrest, Stephen 332
Forrester, Sydney 991
Fortmann, Charles M. 22
Foster, Michael 587
Foti, Marina 1005
Fox, Curtis 897
France, Ryan M. 71, 341, 394, 475, 994, 1081, 1150
Frasson, Nicola 352
Fregosi, Daniel 528, 555, 897
French, Roger H. 234, 671, 758, 861, 1088, 1485
Friedl, Jared D. 220, 275, 279, 1210, 1323
Friedl, Jared 683
Friedlmeier, Theresa M. 566
Friedlmeier, Theresa Magorian 199, 1201
Friedman, Daniel J. 394
Friedman, Daniel 485
Fritzsche, Helmut 476
Frye, Bailey 1557
Fthenakis, Vasilis 146, 1503
Fu, Oakland 377
Fu, Sheng 42, 1417
Fukaya, Shohei 775
Fürer, Sebastian O. 298
Furis, Madalina 161
Gabas, Mercedes 1433
Gabor, Andrew M. 671, 850, 1485
Gaddy, Edward 311
Galiazzo, Marco 352
Gamboa, Daniel H. 761
Gamel, M. 942
Ganguly, Subhankar 1074
Gao, David Wenzhong 840

Gao, Jiaqing 1372
Gao, Munan 1509
Gao, Ningchao 840
Gao, Tina 718
Gao, Zhiyu 822
García, I. 947
García, Iván 1433, 1603
Garcia, Maria Angelica M. 1379
Garcia, R. 1492
Garcia, Rodrigo M. 1068, 1159, 1654
Garcia, Rodrigo 1134
Garín, M. 942
Garrevoet, Jan 651
Gedi, Sreedevi 1549
Gehan, Tim 740
Gehrke, Aaron 578
Geissbühler, Jonas 1
Geistert, Kristina 144
Geisz, John F. 71, 291, 394, 709, 1158
Geisz, John 178, 485
Georghiou, George E. 194, 541, 1270, 1348
Gerardi, Cosimo 1005
Gerber, Andreas 166
Gerger, Andrew 311
Gerton, Jordan 134
Gevorgian, Vahan 360
Geyer-Klingeberg, Jerome 385
Gfroerer, T. H. 1261
Gharabeiki, Sevan 1617
Ghosh, Probir 1366
Ghosh, Sayantani 415
Giacchino, Evan S. 918
Gibbons, Daniel 861, 1088
Gil-Escrig, Lidon 938
Giliberti, Gemma 1413
Ginger, David S. 1020
Giraldo, Sergio 1059
Giridharagopal, Rajiv 1020
Giteau, Maxime 610
Giuri, Antonella 1109
Glatthaar, Markus 1625
Gloeckler, Markus 914
Goga, Adam 51
Gok, A. 737
Golive, Yogeswara Rao 1287
Golubev, Timofey 874
Gomez, Daniel 1433
Gonçalves, Felipe 256
Gong, Yuancai 1059
González, Alanis M. Colón 367
González, Emmanuel J. 416
Good, Brian 245

Goosay, Olivia 727
Gopal, Deepika 126
Gorman, John 922
Gorman, Will 793
Gostein, Michael 254, 295, 1558
Gotoh, Kazuhiro 292, 724, 775
Gottschalg, Ralph 486, 1475
Goubard, Fabrice 1484
Govaerts, Jonathan 48
Graeber, Dietmar 1184
Granello, Pierpaolo 1404
Grassman, Tyler J. 917
Greco, Erminio 1404
Green, Martin 302, 1256
Greenaway, Ann L. 75
Greenhalgh, R. C. 245
Greenhalgh, Rachael C. 981, 1193
Greenhalgh, Rachael 656
Gregory, Christopher 916
Grimm, Benjamin 6
Grisanti, Marco 1567
Grossberg-Kuusk, Maarja 1298, 1432
Grossklaus, Kevin A. 1152
Grover, Sachit 489, 1582
Grovogui, Jann A. 11, 198
Gruenhagen, Philip 662
Grundmann, Marius 12
Gu, Hangyu 366
Gu, Xiaohong 42, 282, 952, 1141
Guc, Maxim 1067
Gudi, Dhanvini 272, 1356
Guennou, M 995
Gueymard, Christian A. 54, 778
Guha, Mousumi 492
Guibin, Shen 1052
Guillemoles, Jean-François 233
Guillemot, Thomas 268
Gulati, Himanshu 259, 1227
Guo, Da 914
Guo, Yonggang 1372
Gupta, Apoorva 644
Gupta, Mool C. 1155, 1522
Gupta, Priya 830
Gurule, Nicholas S. 984
Gütay, Levent 1059
Guthrey, H. 315
Guthrey, Harvey L. 1216
Guthrey, Harvey 939
Gutzler, Rico 432, 1201
Hohn, Oliver 941
Haas, Benedikt 879
Haase, Felix 6, 733

Habisreutinger, Severin 1490
Habte, Aron 474, 587, 600, 778, 1106
Hacene, Benjamin 144
Hacke, Peter 648, 952
Hadi, Sabina Abdul 1471
Hadjipanayi, Maria 194
Hagemann, Johannes 651
Hagendorf, Christian 770
Hages, Charles J. 915, 1241
Hages, Charles 975
Hähnel, Angelika 770
Hajj, Adonis E. 81
Haley, Thomas 928, 1147
Halm, Andreas 1372
Halme, Janne 345
Halsall, Matthew P. 203
Hamadani, Behrang H. 695, 1292
Hameiri, Ziv 5, 227, 472, 828, 1224, 1295, 1468
Hamer, Mike 1359
Hamon, G. 1538
Hamon, Gwenaëlle 133, 781, 1051
Hanif, Muhammad 271, 1564
Hansen, Clifford W. 922, 1110, 1250
Hanuš, Tadeáš 434, 1051
Hao, Xiaojing 1256
Haque, Sirazul 1380
Hara, Tomohiko 724, 1453
Harada, Yukihiro 895
Harder, Nils-Peter 426, 1024
Harder, Ross 1183
Hare, Casey P 977
Hariskos, Dimitrios 432, 1201
Harper, Jim 14
Harrison, Jason 1002
Harrison, Samuel 352
Härtel, Marlene 945
Hartenstein, Matthew B 475
Hartenstein, Matthew 30
Hartweg, Barry 925
Harvey, Steve P. 946
Harvey, Steven P. 298
Hasan, Arif Yetkin 975
Haselsteiner, Philipp 1428
Hashad, Khaled 505
Hasoon, Falah S. 561
Hatt, Thibaud 1625
Hauch, Jens A. 645
Hauch, Jens 540, 1092
Hauschild, Dirk 91
Hawkins, Nicholas 298
Hayden, Steven 1044
He, Bo 1221

Heath, Garvin 594
Heben, Michael J. 35, 46, 220, 275, 279, 401, 402,
.... 683, 823, 844, 1041, 1193, 1210, 1323, 1376, 1410, 1417
Heben, Michael 804, 910, 1197, 1600
Hegedus, Steven 464
Heidrich, Robert 486
Heilscher, Gerd 1184
Heimsath, Anna 1313
Heinrich, Martin 1060
Heinzle, Nino 1348
Heitmann, Johannes 1266
Helder, Tim 199
Helfer, Eric 931
Helienek, Lubos 1594
Helmers, Henning 941
Helms, Clay 81
Henry, Isaiah 1551
Herasimenka, Stan 110
Heres, Geert C. 622
Hernández, Johann 761
Hernandez, Samuel I. 416
Hernandez-Alvidrez, Javier 984
Herrero, Rebeca 746
Heske, Clemens 91
Hettiaratchy, Elline C. 489
Hidalgo, Juanita 969
Hildreth, Owen J. 400
Hildreth, Owen 333
Hill, Blake 363
Hill, C. 995
Hill, Taylor D. 489
Hill, Taylor 1582
Hillhouse, Hugh W. 340
Hillhouse, Hugh 473
Hinken, David 288
Hinzer, Karin 47, 331, 476, 941
Hirst, Louise C. 1249
Ho, Kevin 1020
Hobbs, William B. 492, 921, 948, 1558
Hodges, Joseph 240
Hoerantner, Maximilian T. 919
Hoex, Bram 1256
Hoffman, Adam 861, 1088
Höger, Ingmar 191
Hoheisel, Raymond 894
Hoke, Andy 840, 1164
Hole, Jarand 749
Holman, Zachary C. 767, 1279, 1379, 1533
Holman, Zachary 925
Holmes-Smith, A Sheila 591
Holmgren, William F. 948
Holzhey, Philippe 289, 298

Hong, Chengjian 195, 1526
Hönig, René 191
Hool, Ryan D. 670, 997
Hörnlein, Stefan 191
Hoss, Jan 795
Hossain, Mohammad I. 27, 1516
Howard, Ian A. 172
Hsieh, Chun-Hao 1443
Hu, Hongjie 377
Hua, Amandee 91
Huaman-Rivera, Anny 477
Huang, Ben 1284
Huang, Jing 285
Huang, Jing-Shun 783
Huang, Jinsong 366, 1044
Huang, Jun-Yu 1443
Hubbard, Seth M. 63, 89, 721, 894
Hübner, Simon 1129
Huddy, Julia E. 1276
Hudry, Damien 996
Hudson, Andrew 1462
Hultqvist, Adam 1151
Huneycutt, Sandra 38
Hunter, Robert 941
Hunwick, Nicholas 981
Huque, Aminul 138, 835, 1096
Hussain, Zulkifl 683
Hwang, Jeong-Mo 179
Hyndman, David W. 1259
Iannascoli, Lorenzo 1404
Ibanez, Eduardo 1393
Ifuji, Yuto 1453
Ilahi, Bouraoui 133, 434, 1051
Im, Kyu-Hyeon 1657
Imaizumi, Mitsuru 76, 463, 721, 1429
Imenes, Anne G. 815
Imperatori, Davide 1404
Imran, Hassan 772
Infante-Ortega, L. C. 245
Infante-Ortega, Luis C. 981, 1497
Inoue, Kazuma 292
Ireton, Scott J 977
Irizarry-Rivera, Agustin 477
Irvine, Stuart J. C. 308, 1497
Irvine, Stuart 245
Irving, Richard 823
Isabella, Olindo 80, 655, 1005, 1398
Ishizuka, Shogo 939
Ital, Donald 38
Ito, Yuta 724
Jacob, David 1024
Jacobs, Janet 203

Jaeger-Waldau, Arnulf 924
Jäger-Waldau, Arnulf 550, 749
Jahandardoost, Mohsen 566
Jahangir, Jabir Bin 626, 1027
Jahelka, Phillip R. 202
Jahelka, Phillip 1157
Jain, Anubhav 1519
Jakob, Leonie 1625
Jamarkattel, Manoj K. 46, 220, 279, 401, 683, 823,
.... 1041, 1197, 1376
Jamarkattel, Manoj 910, 1600
Jannuzzi, Gilberto 256
Jansson, Peter Mark 380
Janz, Stefan 434, 1341
Jaouad, A. 1538
Jaouad, Abdelatif 133
Jarzembowski, Enrico 191
Jaubert, Jean-Nicolas 234, 794
Javier, Gaia Maria N. 472, 828
Jawinski, Tanja 12
Jay, Frédéric 716
Jayaraman, Sreenivas 855
Jeangros, Quentin 1
Jensen, Karissa 42
Jeon, Seokmin 134
Jhang, Song-Syun 282, 952
Ji, Liang 1141
Jia, Huiying 1466
Jiang, C.-S. 315
Jiang, Chun-Shen 1570
Jiang, Chun-Sheng 46, 946
Jiang, Fangyuan 1020
Jiang, Jessica Yajie 302, 345
Jiang, Qi 709, 812, 1158, 1380
Jiang, Yi 1284
Jimenez-Arguijo, Alex 1059
Jin, Hao 1434
Jin, Shuangshuang 240
Jin, Tan 240
Jiyun, Zhang 1092
Jo, Sangmin 436
Joe, Junki 1218
John, Jim Joseph 119
John, Oliver 1588
Johnson, Jay 334
Johnson, Samuel 938
Johnston, Michael B. 298
Johnston, S. 315
Johnston, Steve W. 10, 848
Johnston, Steve 219, 387, 702, 865, 1490, 1570, 1629
Jones, Abigail R. 1250
Jones, C. Birk 455

Jones, James	920
Jones, Luke O.	174, 245, 395, 498
Jones, Luke	656, 1177
Jones, Steve	308
Jordan, Dirk C.	219, 1110, 1593
Jordan, Dirk	794, 896, 1145
Jorgensen, Peter Stanley	651
Josepson, Raavo	1432
Jouanneau, Corentin	781
Joyce, Hannah J.	1249
Julien, Arthur	233
Jung, Hyeonjung Tari	1384, 1393
Jung, Kyung Taek	97
Junghänel, Matthias	191
Jurca, Titel	629
Kaaya, Ismail	86
Kabra, Dinesh	1366
Kachman, Dana	272, 1356
Kaewnukultorn, Thunchanok	464
Kaizuka, Izumi	749
Kakoulaki, Georgia	550
Kalizewski, Lauren M.	917
Kalpoe, Prashand	305
Kaltenbaugh, Jarod	758
Kaluarachchi, Prabodika N.	402, 683, 844
Kamal, Serene	272, 718
Kambley, Ankur	281
Kamikawa-Shimizu, Yukiko	939
Kanakkithodi, Arun K. M.	782
Kanaujia, Pawan K.	1522
Kaneko, Ryuji	442
Kanevce, Ana	107, 199, 1201
Kang, Min Gu	97
Kanneboina, Venkanna	1410
Karade, Vijay	1556
Karakaya, Sakir	634
Kari, Thøger	448, 1452, 1662
Kartopu, Giray	308
Kasher, Tal	917
Kashkimbayev, Ulan	39
Kasik, Camden	1310
Katakumbura, Nadeesha	402, 804
Kauert, Maximilian	191
Kauk-Kuusik, Marit	1298, 1432
Kaupmees, Reelika	1298
Kaur, Navdeep	606
Kawakami, Mizuto	895
Kaydanik, Katty	783
Kazim, S.	737
Kazmerski, Lawrence L.	94, 1366, 1541
Ke, Cangming	191
Kee, Jared	555
Keller, Jan	887
Kelzenberg, Michael D.	1157
Kempa, Heiko	12, 432
Kempe, Michael D.	794, 1141, 1260
Kendall, Anthony D.	1259
Kenyon, Jacques	395, 534
Kern, Dana B.	219, 702, 865, 1044, 1570, 1629
Kern, Dana	15, 387, 896
Kerr, Lei	1466
Kersten, Friederike	191
Kessler-Lewis, Emily	894
Kettle, J.	737
Khadka, Dhruba B.	324, 1401
Khan, M. Ryyan	626
Khenkin, Mark	1066
Khetri, Mahantesh	1522
Khoury, Alexandre	1612
Khulmann, Forrest	363
Khurgin, Daniel	1356
Kikelj, Miha	1
Kile, Kara B.	683
Kim, Bora	670
Kim, Boyoung	436
Kim, Dohyung	97
Kim, Han-Jung	1373
Kim, Hyomin	1549
Kim, Hyungoo	436
Kim, Jeong-Hyeon	1484
Kim, Jin Hyeok	1556
Kim, Jin Young	165
Kim, Junhee	1373
Kim, Kiwhan	1556
Kim, Mijung	670, 997
Kim, Munse	97
Kim, Sanggyun	1183
Kim, Seul-Gi	1484
Kim, Woo Kyoung	1549
Kim, Yoonkap	1373
Kimura, Keita	1453
Kinfack, J.	1538
King, Bruce H.	188, 254
King, Bruce	353, 409
King, Richard	916
Kinzer, Austin	920
Kipp, Tobias	651
Kirchartz, Thomas	199, 1307
Kirmani, Ahmad R	1156
Kirmani, Ahmad	1490
Kita, Takashi	659, 895
Klenk, Reiner	1066
Klenke, Christian	191
Klie, Robert F.	771, 1600

Kline, Michael	492
Klöter, Bernhard	569
Knebel, Kevin J.	769
Knodle, Philip J.	671, 850
Knodle, Philip	1485
Ko, Yohan	1373
Koch, Christoph	879
Kodalle, Tim	969
Koelblin, Pascal	166, 1662
Köhler, Matthias	191
Kojima, Haruki	724
Kojima, Nobuaki	924, 1535
Komoll, Felix	199
Kondzialka, Christoph	1184
Köntges, Marc	733
Kopecek, Radovan	1372
Kopidakis, Nikos	178, 485
Korgel, Brian	1537
Korir, Lilian	1261
Kornienko, Vlad	683
Kornienko, Vladislav	656, 1193
Körtgen, J.	546
Koschier, Linda	749
Koseki, Shuuichi	105
Koskey, Steven	555
Kottantharayil, Anil	1287, 1366
Kottokkaran, Ranjith	641
Kouame, K.	1538
Koumis, Anastasios	541
Krause, Timothy	415
Krebs, Hannes	1630
Kretly, L. C.	1492
Krich, Jacob J	941
Krishna, Anurag	194
Krishnani, Pramod N.	81
Kristensen, Sissel Tind	191
Kroeger, George F.	219
Krückemeier, Lisa	1307
Krustok, Jüri	1298, 1432
Kuba, Austin G.	1545
Kubiniec, Alex	928, 1620
Kuciauskas, Darius	107, 577, 633, 849
Kujovic, Luksa	245, 789, 981, 1497
Kumar, Akash	237, 755
Kumar, Akshay	622
Kumar, Dayanand	327
Kumar, Neetesh	276, 1206, 1437
Kumar, Niranjana Mohan	782, 1634
Kumar, Satyendra	1366
Kumar, Vibhor	1509
Kunkar, Alejandro	708
Kupets, Elaine	594

Kurokawa, Yasuyoshi	292, 775
Kurtz, Sarah R.	783
Kurtz, Sarah	209, 617, 1529
Kusch, G	995
Kwon, Ohjin	191
Lachenal, Damien	1
Lachowicz, Agata	352
Lackner, David	941
Ladd, Anthony J. C.	1241
Lahood, Catherine	22
Lahti, Gabriella D.	1158
Lai, Barry	782, 1183, 1634
Lai, Cheng-Yu	276, 606, 1206, 1437
Lakshmikanth, Balaji Bangolae	126
Lambertz, Andreas	177, 280, 419
Lampa, Nicole	191
Lan, Yucheng	272
Landis, Geoffrey A.	63
Lang, Tom	1002
Lange, Stefan	12, 770
Lao, Yao Y.	11
Lara-Fanego, Vicente	54
Larionova, Yevgeniya	701
Larson, Harry	1551
Lasalvia, V.	1459
Lasalvia, Vincenzo	359
Laufer, Felix	144
Lave, Matthew S.	674, 1170
Law, Adam M.	174, 498, 789, 981
Law, John M.	174
Laws, Nicholas D.	228
Le, Anh Huy Tuan	5
Leccisi, Enrica	146
Lee, Chungho	849, 998, 1008
Lee, Hyunjong	919
Lee, Hyunju	442, 724
Lee, Jehyun	436
Lee, Kyumin	391
Lee, Minjoo L.	670, 997
Lee, Ross	1510
Lee, Sang Hee	97
Lee, Sanghyun	1102
Lee, Songhee	1549
Leever, Benjamin	1466
Leijtens, Tomas	919
Leloux, J.	737
Lemay, AC	1233
Lemire, Amanda	1152
Lemos, Francisco V. E.	1159
Lenert, Andrej	482
Leonardi, Marco	1005
Lerat, Jean-Francois	133

Lestrade, Michel	1645
Leuty, Zachary B.	1533
Lewis, Mandy R.	47
Li, Baojie	1519
Li, Bo	794
Li, Bor	945
Li, Brian D.	670
Li, Can	822
Li, Chongwen	909, 1019
Li, Deng-Bing	275, 279, 823, 1041, 1197, 1353, 1376
Li, Dengbing	46, 804, 910
Li, Dinica	978
Li, Fang	648, 1120
Li, Gan	374, 935
Li, Lulin	1575
Li, Luxi	1183
Li, Mengjie	671, 758, 1485
Li, Minghui	822
Li, Muzhi	938
Li, Ning	1074
Li, Wayne	555, 897
Li, Xinjun	234
Li, Yongxi	332
Li, You	844, 1197, 1417
Li, Zelin	1141
Li, Zhanming S.	1645
Li, Zhen	822
Li, Zhenguo	195, 1526
Li, Zhiqiang	1645
Liao, Weilin	248
Libal, Joris	1372
Libby, Cara	921
Liggett, Max	671
Lightfoote, Stephen	607
Li-Kao, Zacharie Jehl	1059
Lim, Deokoh	436
Lim, Jihun	482
Limodio, Gianluca	305
Lin, Boris	903
Lin, Fen	1052
Lin, Yida	1356, 1575
Lindahl, Johan	749
Lindig, S.	737
Lindig, Sascha	931
Linke, Jonathan	795
Linss, Volker	795
Lipovšek, Benjamin	1
Liu, Baiqiang	248
Liu, Chengfa	1544
Liu, Jiang	1221
Liu, Jie	1221
Liu, Jiqi	234
Liu, Xiaolei	245, 789, 981, 1497
Liu, Xitao	1503
Liu, Yang	1221
Liu, Yangang	386
Livera, Andreas	1270
Lobato, K.	737
Loeding, Adam W.	918
Lombardero, Ivan	1433
Lombardo, Salvatore A.	1500, 1609
Lomuscio, Alberto	1067
Loo, Roger	434, 1390
López, G.	942
Lopez, Hector	1280
Lopez-Becerra, Alan	1283
Lopez-Cardalda, Guillermo	370
López-González, J. M.	942
Lopez-Lorente, Javier	1359
Louks, A.	315
Louks, Amy E	590
Louks, Amy	1570
Lu, Chengchangfeng	1356
Lu, Dingyuan	1197, 1497
Lu, Jianfeng	298
Lu, Junxiong	195, 1526
Lu, Meijun	248
Luderer, Christoph	770
Lüer, Larry	645
Lumb, Matthew P.	1217
Luo, Bin	48
Luo, Yanqi	1183
Luque, A.	947
Luque-Heredia, I.	947
Lustig, Zachary	123
Luther, Joseph M.	132, 1156
Luther, Joseph	415, 1490
Lv, Ruirui	1284
Ma, Depu	935
Ma, Jaliu	914
Ma, Jessica	1612
Ma, Yiwei	138
Macalpine, Sara M.	1409
Macdonald, Daniel	1434
Macías, Javier	746
Mack, Charles	178
Mack, Sebastian	795
Mack, Shawn	886
Madonna, Richard G.	1157
Magginetti, David	134
Mahabaduge, Hasitha	1633
Mahaffey, Mason P	1279
Mahamu, Hambalee	659
Mahesh, Suhas	298

Mahmood, Farrukh Ibne 237, 648, 755, 1120
Mahmoudi, Eslam.. 1134
Mahmud, Rasel.. 1074
Mahmud, Zabir 617
Mahoney, John... 1156
Maiberg, Matthias... 432
Mainali, Madan K. 401, 1550
Mainali, Madan.. 1410
Makita, Kikuo 105
Makrides, Andreas 1348
Makrides, George 541, 1348
Mallajosyula, Arun Tej.. 1606
Mallick, Rajni 1497
Mamun, Ashraful 134
Manceau, Matthieu 934
Mandic, Vilko 351
Manganiello, Patrizio.. 80, 655
Mangum, John S. 71, 994, 1150
Mannino, Gaetano.. 1304, 1567
Mannodi-Kanakkithodi, Arun 1634
Manoukian, Gregory A. 915, 1241
Mansfield, Lorelle M. 35, 1380
Mantel, Claire 16
Mao, Dan 782, 1634
Mapara, Varun N. 161
Maple, Larry 387
Marasini, Ganesh 1096
Marcos, Jesús... 888
Mariam, Tamanna 279, 683, 844, 1376, 1417
Mariotti, Davide... 281
Mariotti, Silvia 945
Markevich, Vladimir P. 203
Marques, F. C. 1492
Marques, Francisco C. 290, 1068, 1134, 1159, 1654
Marquis, Audrey 295
Marstell, Roderick J. 1024
Martel, Benoit... 352
Martín, Francisco 746
Martín, I. 942
Martin, Ina T. 597
Martín, Nazario 804
Martin, P. 947
Martin, Pablo 1433
Martinez, Daniel 15
Martinez-Szewczyk, Michael W. 333, 400
Martinez-Szewczyk, Michael 896
Martins, Giuliano L. 1475
Martír, Pablo 1603
Masi, Sofia 441
Masson, Gaëtan... 749
Masuda, Atsushi 442
Masuda, Taizo 924

Matam, Manjunath 671
Mate, Mayank 123, 363
Matera, Fabio 1500, 1609
Mathiak, Gerhard 119
Matsui, Takuya 281, 775
Matthews, Bryan 978
Matthews, David 767
Mayordomo, Alejandra A. 448
Mayyas, Ahmad 299
McAlister, Tom 1620
McCandless, Brian 1299
McCarthy, Robert F. 145
McCarthy, Robert 1446
McCulloch, Manuela.. 1184
McDanold, Byron... 47
McDonald, Calum 281
McGarvey, Elspeth 1393
McGlynn, Ruairi 281
McGott, Deborah L. 308, 633, 915
McKenna, Killian 1074
McKuin, Brandi 209
McMahon, William E. 291, 709, 848, 994, 1150, 1380
McMeekin, David P. 298
McMillon-Brown, Lyndsey 415, 1437, 1490
McNatt, Jeremiah 415
McRae, Mary E. 1510
Medjoubi, Karim 233, 268
Meeker, Rick 1002
Meidanshahi, Reza Vatan 1021
Meier, Rico 689, 970
Meila, Marina 340
Meinhart, Lisa 1630
Melchiorre, Michele.. 1151, 1617
Meléndez, Cristian R... 420
Mendez, Andres Felipe Castro 969
Méndez-Curbelo, Pablo... 717
Mendis, B G 995
Menegassi, Matheus Melati... 1588
Meng, Yuhuan 340, 473
Mercimek, Yavuzhan 655
Merkle, Arno P. 291
Mette, Ansgar 191
Metzger, Wyatt K. 1497
Meyer, Abigail... 359
Meyers, Bennet 710, 961
Michael, Sheri F 885
Michel, Jesus Ibarra 1434
Micheli, Leonardo...54, 122, 437, 1105
Mihailetchi, Valentin D. 1372
Mikeska, Kurt R. 248
Mikli, Valdek 1298
Mikofski, Mark 81

Miller, Chandler 793, 991
Miller, Clark 916
Miller, David C. 353, 1593
Miller, David W. 1497
Miller, Emily 1557
Miller, Jason 978
Miller, Michael F. 107
Mil'Shtein, Sam 1009
Min, Kwan Hong 179, 1030
Minuto, Alessandro 871
Mirletz, Brian T. 228
Mirletz, Heather M. 554
Mirletz, Heather 1416
Mitra, Suchismita 30
Mitterhofer, Stefan 282
Miyano, Kenjiro 324, 1401
Moffitt, Stephanie L. 42, 282, 952, 1141, 1292
Mogannam, Laura 467
Moghadamzadeh, Somayeh 701
Mohammadi, Mahsa 1266
Mohite, Aditya 938
Mohr, William 145
Mohsin, Muhammad Saeed 844
Molinero, R. 947
Molto, Cécile 597, 648, 1120
Monahan, Daniele 1462
Monnin, Ryan 914
Montes-Bojorquez, Jose Raul 1283
Montes-Romero, Jesús 1270, 1348
Mood, Thomas C 886, 1217
Mora-Seró, Iván 441
Mordvinkin, Anton 486
Morel, Don 690
Morlier, Arnaud 86
Morris, Kerrie M. 1177
Morris, Kerrie 981
Moscoso-Cabrera, Javier A. 420
Moser, D. 737
Moser-Mancewicz, Nicholas 896
Moses, Paul 665
Motes, Brandon T 590
Mouri, Tasnim K. 91, 1667
Mousumi, Jannatul Ferdous 629
Moutinho, Helio R. 219
Moutinho, Helio 1629
Mu, Teliang 1503
Mugnier, Daniel 749
Mule, Chirag 359
Müller, Jörg W. 191
Muller, Matthew 122, 356, 437, 710
Müller, Matthias 1266
Müller, Thore 1491

Mulloy, Eva M 577
Muñoz, Daniel 1206
Muñoz-Pinzon, Daniel 1437
Munshi, Amit H. 123, 363, 1234
Munshi, Amit Harenkumar 1537
Munshi, Amit 1337
Murakami, Takurou N. 253
Murphy, Alan 783
Muska, Katri 1298, 1432
Mussakhanuly, Nursultan 227
Muzzillo, Chris 1008
Myneni, Sushmakanth 1234
Nagarajan, Shreyas 81
Nagel, Henning 795
Nagle, Timothy 1497
Nahar, Aayush 1299
Nain, Preeti 743, 1095
Nakado, Takashi 924
Nakamura, Kyotaro 724, 924
Nakamura, Tetsuya 76, 463
Nakano, Yoshiaki 342, 374, 935
Nakarmi, Upama 662
Nakka, Laxmi 1052
Nambo, Apolo 38
Nardone, Marco 566, 577, 849
Nascetti, Augusto 1404
Nasser, Michael 498
Navon, David 1292
Nayfeh, Ammar 321, 327, 1320, 1344, 1381, 1471
Nayfeh, Laith 1381
Nayfeh, Leia 1381
Naylor, M 995
Nazeeruddin, Mohammad K. 804
Nazer, Afshin 80
Nazif, Koosha Nassiri 1347
Ndione, Paul F. 1158
Ndione, Paul 1629
Neal, Craig J. 671
Nekarda, Jan 1588
Nemeth, B. 1459
Nemeth, William 30, 359, 475, 891, 1216
Neto, Pedro O. C. M. 1159
Neubert, Anja 1359
Neuhaus, Dirk H. 1313
Neuhaus, Dirk Holger 1060
Neumann, Anica N. 218, 291, 848, 994, 1130, 1150
Neupane, Ganga R. 695
Neupane, Sabin 46, 220, 279, 1041, 1197, 1210,
.... 1353, 1376
Neves, M. R. M. 1492
Neves, Mendelsson R. M. 1068, 1159
Newberry, Milton G. 380

Ng, Annie	39, 502
Ni, Chaoying	248
Nicholson, Anthony P.	509
Niebergall, Larissa	191
Nielsen, Michael P.	271, 976, 1564
Nieves, Michael Y. Vazquez	367
Nigmetova, Gaukhar	39
Nihar, Arafath	758
Nikam, Maitheli	1253
Ninad, Nayeem	155, 1648
Nishihara, Tappei	442, 724
Nishinaga, Jiro	939
Nishioka, Kensuke	924
Nitta, Frederick U.	1347
Noack, Philipp	701
Nocerino, John C	198
Nolde, Jill A	886
Norman, A.	315
Norton, Matthew	194
Nowak, David	1059
Núñez, R.	746
Nuns, Thierry	934
Nuys, Maurice	177
Nyholm, Andrew W.	202
Obrecht, John M.	1578
O'Brien, Colleen	794, 1141
Ochoa, Jorge	768, 896, 1533
Ogura, Atsushi	442, 724
Ogut, Mehmet G.	961
Oh, Jaewon	794
Oh, Myeongchan	436
Ohlmann, Jens	434, 1341
Ohshima, Takeshi	76, 1429
Ohshita, Yoshio	442, 724, 924, 1453, 1535
Ok, Young-Woo	91, 179, 1030, 1657
Okada, Yoshitaka	463
O'Kearney, Felix	1224
Oklobia, Ochai	245, 308, 1497
Okoli, Fitzgerald C.	918
Okullo, James	1393
Oliver, R A	995
Oltjen, William C.	671, 758, 1485
Oltjen, William	234
Olzhabay, Yerassyl	502
O'Neill, Mark	1446
O'Neill-Carrillo, Efraín	57, 1170
Onno, Arthur	1279
Ooi, Tzy Wei	855
Opatovsky, Martin	1594
Orejuela, V.	947
Orejuela, Víctor	1603
Oreski, G.	737

Oreski, Gernot	931, 1630
O'Rourke, Michelena	330
Ortis, Alessandro	1567
Ortiz, Eduardo I.	420
Ortiz, Eduarto I.	416
Ortiz-Rivera, Eduardo I.	525
Ortiz-Rivera, Eduardo	370, 1338
Oshima, Ryuji	105
Oshima, Takeshi	463
Osowski, Mark	145
Ossig, Christina	435, 651
Ota, Yasuyuki	924
Ottoson, Larry	178
Ovaitt, Silvana	47, 554, 794, 920, 1416
Owen-Bellini, Michael	952, 1044, 1218, 1260, 1629
Oyewo, Ayobami S.	708
Ozaki, Ryo	924
Ozaktas, Ekin Gunes	1334
Ozbeytemur, Josh	240
Ozbolt, Alex	1575
Pacheco, Willian	717
Packard, Corinne E.	10, 71, 291
Padhamnath, Pradeep	1030, 1657
Padmakumar, Govind	305
Padmanaban, Dilli Babu	281
Paesa, Marta Casasola	48
Paetel, Stefan	107, 432, 1201
Paetzold, Ulrich Wilhelm	144, 701
Page, M. R.	1459
Page, Matthew	30, 359, 475
Pal, Shweta	622
Palacios II, Felipe	1626
Palacios, Felipe	1034
Palekis, Vasilios	690, 1237
Palmer, Jack R.	644
Palmiotti, Elizabeth	353
Palmstrom, A.	315
Palmstrom, Axel F.	590, 1158
Palmstrom, Axel	938, 1570
Pamperin, Megan	1393
Pan, Noren	145
Pan, Yida	1434
Panchalogaranjan, Vinushika	665
Pandey, Ramesh	123
Panzic, Ivana	351
Papaeconomou, Vassilis	1270
Parada, Gabor	981
Paraskeva, Vasiliki	194
Pareek, Devendrá	1059
Paris, Claudio	1404
Park, Chinho	749
Park, Nam-Gyu	1484

Park, Sungeun 97
Parke, Tyler 1667
Parra, Johan 268
Paschen, Jan 1588
Passarella, Bianca 1005
Patel, Aesha P. 402, 823
Patel, Jayeshkumar 476
Patel, M. Tahir 626
Paul, G. 315
Paul, Goutam 1570
Paul, Sritoma 482
Paupy, Nicolas 133, 434
Paviet-Salomon, Bertrand 1, 1129
Peaker, Anthony R. 203
Pearce, Phoebe M. 345, 976
Pearsall, N. 737
Pearson, Patrick 887
Pechmann, Sabrina 194
Peibst, Robby 6, 701, 733
Peña, Carlos 717
Peng, Fuguo 195, 1526
Peng, Hugh 727
Penukula, Saivineeth 1534
Penukula, Vineeth 1109
Perez, Marc 285, 662
Perez, Richard 662, 928, 1620
Perez-Rodriguez, Paula 305
Perini, Carlo A. R. 1183
Perini, Carlo Andrea Riccardo 969
Perkins, Craig L 577, 633
Perna, Allison N. 291
Perna, Allison. 71
Pernès, Nicolas 1129
Perret, Lionel 749
Perry, Kirsten 710, 1145
Perry, Lakesha N. 282
Perullo, Christopher 555
Peshek, Timothy J. 698
Peshek, Timothy 1490
Peter, Christoph 1372
Peters, Benjamin 634
Peters, Ian Marius 540, 645, 1092
Peters, Stefan 191
Peterson, Josh 600, 1123
Petesic, James 1295
Petter, Kai 191
Phang, Sieu Pheng 1434
Phillips, Adam B. 46, 220, 275, 279, 401, 402,
.....................683, 823, 1041, 1193, 1210, 1323, 1410, 1417
Phillips, Adam 804, 910
Pierce, Benjamin G. 512
Pierce, Benjamin 101

Pieters, B. E. 546
Pieters, Bart E. 166
Pikolos, Loucas 1348
Pilot, Nicholas 14
Pilvet, Maris 1298, 1432
Pina, Marissa 1667
Pineda, F. Brigham 783
Platzer-Björkman, Charlotte 887
Ploigt, Hans-Christoph 191
Podraza, Nicholas J. 1410
Podraza, Nikolas J. 401, 1550, 1557
Pogorelov, Kostiantyn 1491
Pohl-Hampel, Britta 191
Pokhrel, Dipendra 683
Polly, Stephen J. 63, 721, 894
Polly, Steve J 89
Polzin, Jana-Isabelle 795
Poortmans, Jef 48, 162
Pop, Eric 1347
Poplawsky, Jonathan D. 46
Porret, Clément 434, 1390
Posada, Jorge 233, 268
Pothoof, Justin 1020
Poulsen, Peter B. 448, 1452
Powell, Kaden 998
Prabakar, Kumaraguru 1164
Pradeep, Nisitaa Karen Clement 467
Prasanna, Rohit 919
Prathap, Nemalipuri Surya 182
Pravettoni, Mauro 1188
Prell, Henrik 939
Prettl, Michael 289
Price, Kent 1102
Prot, Aubin JC. M. 1067
Provost, Marion 268
Prym, Guilherme C. S. 1159, 1654
Ptak, Aaron J. 10, 71, 291
Puel, Jean-Baptiste 233, 268
Pugstaller, Robert 1428, 1467, 1481
Pulwin, Ziggy 1217
Purkayastha, Atanu 1606
Pusch, Andreas 976
Qazi, Suleman Sami 772
Qian, Chen 1256
Qian, Yang 134
Qin, Yuan 1221
Qiu, Botong 1575
Qiu, Feng 634
Qu, Minghao 195, 1526
Qu, Xiaoyong 1372
Quader, Abdul 1410
Queck, Martina 191

Quinones, Dhamelyz R. S. 1667
Quispe, David 767
Radhakrishnan, Hariharsudan 48
Radu, Daniela R. 276
Radu, Daniela 606, 1206, 1437
Raghoebarsing, A. 737
Ragonesi, Antonino..1005
Rahimi, Amirhossein 844
Rahman, Areefa 482
Rahman, Naveed 1183
Raikar, Subbarao 333, 400
Rajakaruna, Manoj....................... 402, 844, 1417
Raju, Sukhwant 855
Raker, David 402
Ramasubramanian, Deepak................................1096
Rametta, Francesco..1005
Ramirez-Iniguez, Roberto....................................591
Ramos, Wendy Reyes 682
Rampalli, Chaitanya Santosh 1513
Rana, Prem J. S. 1044
Ranalli, Joseph 263, 1242
Rand, BP 1233
Ransome, Steve..1348
Rapp, Jeremy 1259
Rashed, Faisal 356, 1454
Rashkin, Lee 1034
Rasmussen, Mirra 597
Rathgeber, Andreas 385
Rau, U. 546
Rau, Uwe 177, 199, 419, 1307
Raugewitz, Annika..6, 733
Ravello, Magdalena 969
Ravishankar, Sandheep 1307
Reagan, Jeremiah 209
Reddy, K. S. 182
Reddy, Vasudeva Reddy Minnam 1549
Reddy, Yellasiri Bharath Kumar 259, 1227
Reece, Peter J. 976
Reese, Matt 245
Reese, Matthew O.................... 35, 633, 695, 915, 946, 1008
Reese, Samantha 1416
Reeves, Adam 240
Reich, Carey 849, 1234, 1279
Reich, Gerly 1024
Reichel, Christian 1313
Reinders, A. H. M. E. 737
Reinders, Angèle H. M. E. 809
Reinders, Angele 443
Rendler, Li C. 1313
Rengifo, Fabio Andrade..519
Reno, Matthew J. 13, 334
Repins, Ingrid L. 702, 1260, 1593

Reyes-Colón, Ramón 57
Rey-Stolle, I. 947
Rey-Stolle, Ignacio.....................................1433, 1603
Rezk, Ayman 321, 1320, 1344, 1471
Rhee, Kurt 581
Ribeiro, Andrei C. 1159, 1654
Ribo, Macarena Mendez 554, 1416
Richards, Bryce S. 172, 996
Riedel, Maximillian.. 1066
Riedl, Gabriel 1428, 1467
Rienäcker, Michael.. 701
Rigby, O M 995
Rijal, Suman 1197, 1376
Rikhof, Anne 622
Riley, Daniel.. 101, 409, 1131
Ringel, Steven A. 917
Rippingale, Jan 922
Ritzer, David Benedikt 144
Rivera, Agustín Irizarry.. 519
Rivera, Eduardo I. Ortiz 630
Rivera-Matos, Yiamar 916
Rizk, Ayman 327
Rizzo, Aurora 1109
Ro, Jason 925
Roberts, Dennice M. 1629
Robles-Rivera, Emmanuel G............................... 1034
Rock, Nathan... 1337
Rockett, Angus A. 489
Rodgers, Marianne 1612
Rodriguez-Cabanas, Lissette 508
Rohatgi, Ajeet............ 91, 179, 1030, 1155, 1657
Rojas-Gatjens, Esteban 1020
Rojsatien, Srisuda................................... 782, 1634
Rolston, Nicholas.................. 938, 1109, 1534, 1565
Rome, Grace A. 75
Romer, Pascal 1060
Römer, Udo 917
Rong, Eric 272, 1356
Rosales, Bryan 740
Rosenthal, Samuel 1575
Roufberg, Lew 311
Rounsaville, Brian 91, 179
Rousset, Jean 268, 1440
Rout, Bibhudutta 132
Routhier, Alex 916
Rowell, David 145
Roy, Etee Kawna 998
Roy-Layinde, Bosun 482
Ru, Xiaoning 195, 1526
Rudolph, Dominik 1372
Ruhle, Ryan... 387
Runkana, Venkataramana 1053

Ruske, Florian...945
Rusnak, Jozef...1594
Russell, Annie C. J...331
Saavedra-Peña, Nelson E...........................1082, 1326
Sacchitella, Elijah...894
Saeed, Muhammad Mohsin.............................1210
Saenz, Theresa E....................... 994, 1130, 1150
Sahani, Rishabh........................ 276, 1206, 1437
Sai, Hitoshi...775
Saitta, Federica...305
Saive, Rebecca........................... 348, 622, 996
Sajja, Sunil...855
Saliba, Michael...289
Salles, Caroline Lima...................................1216
Sampath, Walajabad S................. 509, 1537, 1600
Sampath, Walajabad..................... 123, 387, 1279
San José, Luis J...746
Sanchez-Perez, C...947
Sanchez-Perez, Clara....................................1433
Sanci, Sal...978
Sankin, Igor...914
Santala, Annikki L...919
Santamaría, Rodrigo Del Prado.......... 448, 1452, 1662
Santana, Vinícius Camatta..............................1541
Santbergen, Rudi........................ 305, 1005, 1398
Santhanam, Lakshmi.......................................126
Santiago, Brian L. Reyes...............................1338
Santistevan, Kevin...409
Santiwipharat, Chaiwarut...............................1545
Santos, José...804
Saraswat, Govind...360
Saraswat, Krishna C......................................1347
Sargent, Edward H.......................................1019
Sargent, Ted...518
Sarkisov, Sergey...110
Sartor, Benjamin E.......................................1008
Sato, Shin-Ichiro..................................... 76, 463
Satymov, Rasul...708
Saucedo, Edgardo...1059
Saucedo, Joel...1633
Saw, Min Hsian...1188
Sazzad, Muhammad H.......................................976
Scarpulla, Michael A.......................................731
Scarpulla, Michael.......................................1337
Schall, J. W...315
Schall, Jackson W..1570
Schall, Jackson...1044
Schaper, Martin...191
Scharf, Jessica...191
Scheer, Roland...12
Scheibner, Michael...415
Scheideler, William J....................................1276

Scheidt, Rebecca A................................. 161, 1156
Scheiman, David...1045
Schelhas, Laura T............... 298, 353, 952, 1044, 1260, 1629
Schirone, Luigi...1404
Schlatmann, Rutger.......................................1066
Schlenoff, Tali...1292
Schley, Michael...191
Schmieder, Kenneth J.............................. 886, 1217
Schmitz, J...737
Schneble, Olivia D.......................................1150
Schneiderloechner, Eric...................................795
Schnierer, Branislav....................................1594
Schönmann, Antje...191
Schramm, Barbara...435
Schreiber, Waldemar............................... 434, 1341
Schropp, Andreas...651
Schüler, Marc Andre.......................................1060
Schulte, Kevin L............................... 10, 291, 848
Schulte-Huxel, Henning....................................652
Schulz, Susanne...191
Schurman, Matthew J.......................................311
Schütze, Matthias...191
Schwab, Andrew J...977
Schwabedissen, Axel.......................................191
Schwartz, Dakota...1551
Schwung, Julian...64
Schygulla, Patrick.......................................1341
Sciuto, Marcello...1005
Scuto, Andrea...1609
Seal, Sudipta...671
Sehirlioglu, Alp...758
Seibert, Samuel...1353
Seiboth, Frank...651
Seigneur, Hubert P...671
Seigneur, Hubert.................... 850, 1120, 1485
Sekulic, William...794
Sellers, Ian R....................... 132, 161, 1156
Selvidge, Jennifer...341
Senaud, Laurie-Lou...1
Sengupta, Manajit........... 152, 386, 474, 587, 600, 778, 1106
Seo, Seongrok...298
Sepúlveda-Mora, Sergio B..................................464
Sepúlveda-Vélez, Fredy A.................................1105
Serafini, Patricio...441
Sermarini, Anna Carolina De Paula.................. 256, 858
Serrano, Guillermo...525
Setiawan, Ignatius Andre.................................1549
Sevillano-Bendezú, Miguel Á..............................1066
Seymour, Kyle...................................... 928, 1147
Seyrich, Martin...651
Shafarman, William N.............................. 91, 1545
Shafarman, William.......................................1299

Shah, Akash 123
Shan, Ambalanath 1410
Shapiro, Finley R 915
Sharikadze, Saba 641
Shaton, Avishai 861, 1088
Shaw, Daniel 1234
Shen, Heping 1434
Sheppard, Scott 555
Sheyfer, Dina 1183
Shi, Jiahui 1503
Shi, Yangwei 1020
Shimabukuro, Laura 1575
Shimasaki, Takashi 342
Shimpi, Tushar 981
Shiradkar, Narendra 1287, 1366
Shirai, Yasuhiro 324, 1401
Shirazi, Eli 443, 816
Shojaei, D. 942
Shoji, Yasushi 105
Shore, Andrew M. 1292
Shrestha, Bishal 401, 1410
Shrestha, Santosh 881, 1564
Sidhik, Siraj 938
Siebentritt, Susanne 1067, 1151, 1617
Siegneur, Hubert 648
Siepchen, Bastian 656
Silhavy, Jake T. 918
Silva-Acuña, Carlos 1020
Silveira, A. M. C. 1492
Silveira, Allan 1068
Silverman, Timothy J. 106, 921, 1044, 1218, 1260, 1593
Simon, John 10
Sims, Jeremiah D 829
Singh, Luna 718
Singh, Manish K. 1384
Singh, Pritpal 1510
Singh, Rhythm 830, 1230
Sinha, Arpan 1155
Sinha, Parikhit 146, 855
Sinton, Ron 113
Sinton, Ronald A. 798
Sirkisoon, Sarah 22
Sites, James R. 509
Sites, James 1310, 1582, 1600
Sitiriche, Marcel Castro 519
Skoczek, Artur 1594
Slauch, Ian M. 970
Slauch, Ian 689
Slonopas, Andre 508
Smaine, Issam 426
Smets, Arno H. M. 305
Smith, David D. 1024

Smith, Emily 1045
Smith, Ryan 1120
Smith, Soshana 282
Snaith, Henry J. 298
Snell, Jeffrey 1009
Snuggs, Robert 823
Snyder, William 409
Sodabanlu, Hassanet 342, 374, 935
Soeriyadi, Anastasia H. 917
Song, Chang-Yun 432
Song, Hee-Eun 97
Song, Tao 178, 394, 485
Song, Zhaoning 42, 275, 683, 804, 812, 844, 909,
.... 1197, 1210, 1323, 1353, 1376, 1417, 1550, 1557
Sood, Mohit 1151
Sourabh, Shashi 132, 161
Sovetkin, E. 546
Sovetkin, Evgenii 166
Spaeth, Bettina 656
Spataru, Sergiu V. 448, 1452, 1662
Springer, Martin 794
Spurgeon, Ben 861, 1088
Sridhar, Seetharaman 554
Stall, Richard 311
Stanley, Bradley 508
St-Arnaud, Louis-Philippe 941
Stein, Joshua S. 188
Stein, Joshua 101
Steinebrunner, Udo 1060
Steiner, Myles A. 10, 75, 218, 291, 341, 394, 848,
.... 994, 1081, 1130
Stenzel, Florian 191
Stern, Jillian 467
Stevens, Margaret A 886, 1217
Stevens, Tristan 655
Steyn, Dirk 891
Stid, Jacob T. 743, 1259
Stoffel, Tom 1106
Stoicescu, Liviu 166, 1662
Stradins, P. 1459
Stradins, Paul 30, 475, 891
Stradins, Pauls 1216
Strandberg, Rune 815
Strandins, Pauls 359
Straub-Mueck, Michael 385
Strelow, Christian 651
Stroyuk, Oleksandr 540
Stuckelberger, Michael E. 435, 651
Sturm, Chris 12
Subedi, Kamala Khanal 46, 1019, 1353
Subramanian, Sivakumar 1053
Suemasu, Takashi 1398

Sugaya, Takeyoshi 105, 939
Sugimoto, Hiroki 76
Sugiyama, Masakazu 342, 374, 463, 935
Sulas-Kern, D. B. 315
Sun, Kaiwen 1256
Sun, Yijia 1485
Sun, Yukun 997
Sung, Li-Piin 282
Sunkari, Preetham P. 340
Sunkari, Preetham 473
Sunter, Deborah A. 467
Suppiah, Sam 476
Sutterlueti, Juergen 1348
Svrcek, Vladimir 281
Sweat, Rebekah 330
Syed, Faizan 606
Sytnyk, Mykhailo 1092
Szablewski, M 995
Szábo, Sandor 550
Szyszka, Bernd 945
Taconelli, Mauricio 1654
Taddei, Margherita 1020
Tafur, Lucila D. 927, 1146
Takahashi, Tadatoshi 611
Takamoto, Tatsuya 924
Talavera, Diego L. 1105
Tamizhmani, Govindasamy 237, 648, 755, 794, 1120
Tamuno-Ibuomi, Lewis Osikibo 591
Tan, Jin 840
Tan, Kelvin 85
Tanaka, Taichi 1453
Tanimoto, Tsutomu 924
Tank, Mehul 330
Tao, Meng 85, 1566
Tatavarti, Rao 1429
Taubmann, Rouven 1184
Tawsif, Sheikh Elahi 690
Tayagaki, Takeshi 253
Taylor, Andre D 1008
Taylor, P. Craig 359
Teasley, Corson 555
Teodor, Alexandra H. 11
Teplyakov, Andrew V. 1667
Ternes, Simon 144
Terry, Mason 783
Tervo, Eric J. 341
Terwilliger, Kent 702, 1629
Theelen, M. 737
Theingi, S. 1459
Theingi, San 475
Theocharides, Spyros 541
Theristis, Marios 122, 188, 1250

Thiagarajan, Ramanathan 1164
Thibodeau, Matthew R. 918
Thiel, Christian 924
Thiengi, San 30
Thind, Arashdeep S. 771
Thomas, Adam 682
Thomas, Sinju 432, 939
Thomsen, Vitus B. 16
Thon, Susanna M. 272, 718, 1334, 1356, 1575
Thon, Susanna 695
Tiefenthaler, Martin 1481
Tilli, Francesca 749
Timmo, Kristi 1298, 1432
Timò, Gianluca 871
Timofte, Tudor 1372
Tina, Giuseppe Marco 1304, 1567
Titus, Jochen 919
Tobail, Osama 191
Tobon, Carlos Mario Ruiz 1398
Todaro, Lorenzo 1304
Töfflinger, Jan A. 1066
Togay, Mustafa 245, 789, 981, 1177, 1193, 1497
Toh, Wei Wen 1188
Tokumasu, Takashi 292
Tomita, Yosuke 924
Tonita, Erin M. 331
Topic, Marko 1
Törndahl, Tobias 1151
Tovar, Michael 939
Tracy, Jared 952
Transue, Taos 922, 1250
Tremont-Brito, Rolando J. 1170
Trempa, Matthias 1266
Tresan, Jenner 492
Trimby, Pat 939
Troupe, Anthony T 590
Trupke, Thorsten 227, 472, 828, 1224, 1468
Tsakalids, Anastasios 924
Tse, Yau Yau 656
Tumusange, Marie Solange 1550
Tuomiranta, Arttu 162, 426
Turala, A. 1538
Turcotte, Dave 155
Tutsch, Leonard 1625
Ubukata, Akinori 105
Uddin, Md Aslam 1044
Uene, Naoya 292
Ukaegbu, Ikechi 502
Ulbrich, Carolin 1066
Ulicná, Sona 353, 952, 1044, 1629
Upadhyay, Prashant Kumar 259, 1227
Upadhyaya, Ajay D 1030

Upadhyaya, Ajay	91
Upadhyaya, Vijaykumar D	179, 1030
Upadhyaya, Vijaykumar	91, 1657
Ures, Sandra	888
Urs, Rahul R	299
Usami, Noritaka	292, 724, 775
Vaas, T. S.	546
Valdivia, Christopher E.	331, 476, 941
Valerino, Michael	437
Vallerotto, G.	746
Van De Voorde, Mathis	348, 996
Van Dyck, Rik	48
Van Nijen, David A.	655
Van Sark, W. J. G. H. M.	737
Van Swaaij, René A. C. M. M.	655
Van Velson, Nathan	1114
Van Vuure, Aart Willem	48
Vanderhaegen, Aline	1151
Vandervelde, Thomas E.	1152
Vansant, Kaitlyn	415, 1490
Vargas, Fernando J.	416
Vasconcelos, Cláudia K. B.	94
Vasi, Juzer	1366
Vazquez, Michael Y.	416
Venkat, Sameera Nalin	234
Venkatramanan, D.	1384
Verezhak, Mariana	651
Verkou, Maarten	1005
Verlinden, Pierre	1544
Vignola, Frank	600
Villa-Bracamonte, Maria Fernanda	1283
Villa-Ignacio, Armando	215
Villalva, M. G.	1492
Villalva, Marcelo G.	1068, 1159, 1654
Villalva, Marcelo Gradella	1134
Villalva, Marcelo	290
Voarino, Philippe	1423
Vogt, Malte	1005
Volatier, M.	1538
Von Gastrow, Guillaume	1024
Von Wenckstern, Holger	12
Voss, Stephen	607
Vuk, Dragana	351
Wadsworth, Matthew	330
Wagner, Kristen	1620
Wagner, Philipp	945
Wagner-Mohnsen, Hannes	569
Wahl, Tina	1201
Wakamiya, Atsushi	442
Walajabad, Sampath S.	981
Walajabad, Sampath	1234
Walker, Alexandre W	941

Walker, Don	198
Walkons, Curtis	566
Wallner, Gernot M.	1467, 1481
Wallner, Gernot	1428
Walls, J. Michael	981, 1193
Walls, John M.	245, 498, 789, 1177, 1497
Walls, Michael	656, 683
Wang, Jian	1020
Wang, Jianjian	1114
Wang, Jianming	248
Wang, Jianqiang	1526
Wang, Jing	1074
Wang, Le	1544
Wang, Liwei	240
Wang, Quanzhi	1523
Wang, Tonghui	332
Wang, Wei	690, 1237
Wang, Wenzong	835, 1096
Wang, Yichun	195, 1526
Wang, Yonglei	1221
Wang, Zhaoyu	634
Wanlass, M. W.	1261
Wargulski, Dan R.	879, 939
Warren, Emily L.	75, 218, 291, 709, 848, 994, 1130, 1150, 1158, 1380
Wasmer, Sven	569
Watanabe, Kentaroh	342, 374, 935
Watson, Stephanie S.	282
Wattenberg, Bianca	701
Weber, August	1266
Weber, Julian	1060
Weed, Emily	505
Weigand, William J.	1533
Weigand, William	1379
Weihrauch, Anika	191
Weinhardt, Lothar	91
Welch, Liam M.	395
Welch, Liam	534
Welser, Roger E.	145, 721
Welser, Roger	1429
Wenner, Scott L.	220
Wenner, Scott	910
Westerhof, Jelle	622
Westraadt, Johan	1625
Whalen, Devin C.	380
Wheeler, Aaron	783
Wheeler, Lance M.	1044
Whiteside, Vincent R	132, 161, 1156
Wibowo, Andree	145, 1429
Wickett, Shelbie	1635
Widrick, Devin	528
Wieghold, Sarah	1183

Wieliczka, Brian	1490
Wieser, Raymond J.	1141
Wieser, Raymond	861, 1088
Wietler, Tobias	6, 652, 733
Wikoff, Hope	1416
Wilcox, Stephen	1106
Williams, Henry J.	505, 727
Williams, Jennifer	415
Williams, Rafell	178
Wilson, Paige	476, 941
Wilson, Samantha S.	607
Wilson, Thomas	90
Wilt, David	1429
Wilterdink, Harrison	113
Winkelmann, Aimo	939
Winkler, Louisa	1060
Winter, Björn Oliver	64
Wirtz, L.	995
Witte, Wolfram	432
Witteck, Robert	709
Wittmann, Ernst	645
Wong, Johnson	978
Woodall, Mark	415
Woodhouse, Michael	1593
Wright, Brendan	1224, 1295
Wu, Xiang	1372
Wu, Yinghui	861, 1088
Wu, Yuh-Renn	1443
Wu, Zhenni	1092
Wyss, Patrick	1129
Xiang, Xiaofeng	578
Xiao, Chuanxiao	46, 822, 1019
Xiao, Yegao	1645
Xie, Tian	195, 1526
Xie, Yu	152, 386, 474, 587
Xiong, Gang	998, 1197, 1497
Xu, Binbin	419
Xu, Jianmei	1544
Xu, Tao	1284, 1523
Xu, Xixiang	195, 1221, 1526
Xu, Yawen	248
Xue, Chaowei	1526
Yagi, Shuhei	926
Yaguchi, Hiroyuki	926
Yaiche, Armelle	1440
Yamaguchi, Masafumi	924, 1535
Yamamoto, Kohei	253
Yan, Di	1434
Yan, Feng	1197
Yan, Yanfa	42, 46, 220, 275, 279, 683, 804, 812, 823, 844, 909, 910, 1019, 1041, 1197, 1210, 1323, 1353, 1376, 1417, 1550, 1557

Yanagida, Masatoshi	324, 1401
Yang, Guangtao	655
Yang, Jaemo	152, 386
Yang, Jie	1434
Yang, Miao	195, 1526
Yang, Ruiquan	975
Yang, Zhaoqing	1020
Yao, Dominique Akissi	758
Yao, Keyi Kang	1575
Ye, Jichun	822
Yelzhanova, Zhuldyz	39
Yermekov, Nurzhan	39
Yi, Chuqi	917
Yildirim, Murat	634
Yilmaz, P.	737
Yin, Shi	195, 1526
Yoon, Heayoung	134, 998
Yoon, Woojun	1045
Yoon, Yohan	134
Yoshita, Masahiro	253
Young, D. L.	1459
Young, David L	106
Young, David	30, 475, 891, 1216
Young, Ethan	921
Young, Matthew R.	1130
Young, Michelle	848
Youtsey, Chris	145, 1446
Yu, Li	1164
Yu, Xuanji	234, 597, 671, 758, 861, 1088, 1141, 1284, 1485
Yu, Yuanjie	794, 1284, 1523
Yu, Zhengshan J.	767
Yu, Zhibin	330
Yuan, Luyao	730
Yun, Changyeol	436
Yun, Jae Ho	1556
Yusuf, Jubair	13
Zabalza, Ruben	1141
Zaka, Awais	1471
Zawisza, Zachary W.	823
Zawisza, Zachary	910
Zech, Matthias	1242
Zehender, M.	947
Zelenina, Anastasia	1067
Zeman, Miro	655, 1005
Zeng, Yiyu	302
Zhang, Changgen	248
Zhang, Fan	1261
Zhang, Guangchun	1284
Zhang, Hongxu	1221
Zhang, K. Max	505, 727
Zhang, Kangping	248

Zhang, Shu 1544
Zhang, Wei 1497
Zhang, Xinyu 1434
Zhang, Yijun 718
Zhang, Yong 1261
Zhang, Zheyu 240
Zhao, Dewei 822
Zhao, Shijia 634
Zhao, Yong 248
Zhao, Zitong 1021
Zhaoning, Song 1019
Zheng, Jian-Yao 622
Zheng, Peiting 1434
Zhong, Ruohan 1030, 1657
Zhu, Kai 812, 1158, 1265, 1484
Zhu, Xitong 809
Zhu, Yan 227
Zilouchian, Ali 1280
Zimányi, Gergely T. 51, 1021
Zimmerman, Jeramy D. 994, 1130, 1150
Zimmermann, Gregor 191
Zimmermann, Iwan 804
Zin, Ngwe 1509
Zinßer, Mario 199
Zoppi, G 995
Zubieta, Diego 597

IEEE
445 Hoes Lane
Piscataway, NJ 08854-4141

ISBN 978-1-6654-6060-6